International Encyclopedia of

HEAT & MASS TRANSFER

International Encyclopedia of HEAT & MASS TRANSFER

Edited by
G.F. Hewitt
G.L. Shires
Y.V. Polezhaev

CRC Press
Boca Raton New York

Library of Congress Cataloging-in-Publication Data

International encyclopedia of heat and mass transfer / edited by G.F.
Hewitt, G.L. Shires, and Y.V. Polezhaev.
p. cm.
Includes bibliographical references.
ISBN 0-8493-9356-6 (hardcover : alk. paper)
1. Heat—Transmission—Encyclopedias. 2. Mass transmission—
Encyclopedias. I. Hewitt, G.F. (Geoffrey Frederick) II. Shires,
G.L. III. Polezhaev, IU. V.
TJ260.E467 1996
621.402′2′03—dc20 95-31097
CIP

International Standard Book Number 0-8493-9356-6
Library of Congress Card Number 95-31097
Printed in the United States of America 1 2 3 4 5 6 7 8 9 0
Printed on acid-free paper

DEDICATION

The Editors would like to dedicate this Volume to the memory of Mrs. Joy Shires whose untimely death on July 21st 1996 was a great shock and sadness to all of us. To organise an Encyclopedia with over a thousand articles and nearly three hundred authors is a formidable task and the credit for its successful completion must largely go to Joy Shires. As Contributors to this Volume will know well, Joy's excellent efforts in organising the production of the manuscript were exemplary and inspiring and we feel sure that the Contributors will join with us in recording our gratitude. In the future, whenever we pick up this Volume, we will remember Joy and particularly her zest for life, her sense of humour, and her determination and enthusiasm.

The Editors

The International Centre for Heat and Mass Transfer, ICHMT, is proud to be one of the sponsors of this Encyclopedia. We congratulate all the editors, the authors, CRC Press, and the others who are involved in the preparation and completion of this monumental work. We believe that the book will be an indispensable tool for all people in the field, and will soon find a place not only in the bookshelves but also on the tables of researchers and practitioners alike.

Frank Arinc
Secretary General of ICHMT

PREFACE

The idea for this book arose when Mr. William Begell (Head of Begell House, Inc. and a collaborator of CRC Press) asked one of the present editors and a colleague, Professor Brian Spalding, what would be the book on heat and mass transfer they would most like to see on their shelves. After some thought the idea for the encyclopedia was born; the concept was for a work which gave sufficient information on any given subject to answer a large fraction of queries but which, equally important, gave the reader an easy route into the most cogent review and tutorial literature sources.

The book has five main features:

1. It is broadly based geographically, containing the contributions of about three hundred authors from around the world, including many distinguished workers from Russia and East European countries whose achievements were previously largely inaccessible.

2. The contents, almost one thousand entries in all, represent a balance of viewpoints in that they have been written by authors drawn from both academic and industrial settings.

3. While some entries are necessarily highly specialized, where possible authors have provided a simple introduction, accessible to the general reader. It is our intention that the encyclopedia should be of use throughout the career of the engineer or scientist in this field, from undergraduate to seasoned practitioner.

4. The encyclopedia is uniquely structured. The reader may enter at any point and will be provided not only with a basic introduction but also with guidance to further information. Authors have added to their entry a short list of carefully chosen references, but also most major entries are shown as "following from" more general entries and "leading to" other entries concerned with more detailed information or associated topics.

5. An additional aid to the reader is provided by the use of cross-reference terms. Significant terms in each entry are listed alphabetically in the encyclopedia with references to all the entries in which they feature. In effect, the contents of the encyclopedia form a series of interlinked family trees, any of which may be accessed either by a topic title or alternatively by a single word or phrase. Some of these family trees are shown in the following pages as an illustration of the internal connections.

The editors would like to express their thanks to their publishing colleagues, notably Bill Begell, who initiated the project and the many staff of CRC Press who have helped to bring it to fruition (in particular, Gerry Jaffe, Jennifer Petralia, and Robert Stern). We would also like to thank the International Centre for Heat and Mass Transfer, whose members have taken an interest in the project and made a number of useful suggestions.

We have been most fortunate in persuading a most distinguished collection of authors to contribute to this book. The challenge of writing cogently and informatively, yet very concisely, is one of the most difficult in technical writing. We would like to thank our authors for rising to this challenge, despite the many calls on their time.

Finally, we would like to express our appreciation of the contribution of Joy Shires who acted as Editorial Co-ordinator. The task of bringing together the output of around 300 authors into a coherent volume was indeed a formidable one and the fact that it has been achieved is largely a result of Joy's commitment and tenacity.

G. F. Hewitt
G. L. Shires
Y. Polezhaev

INTERCONNECTIONS BETWEEN ARTICLES

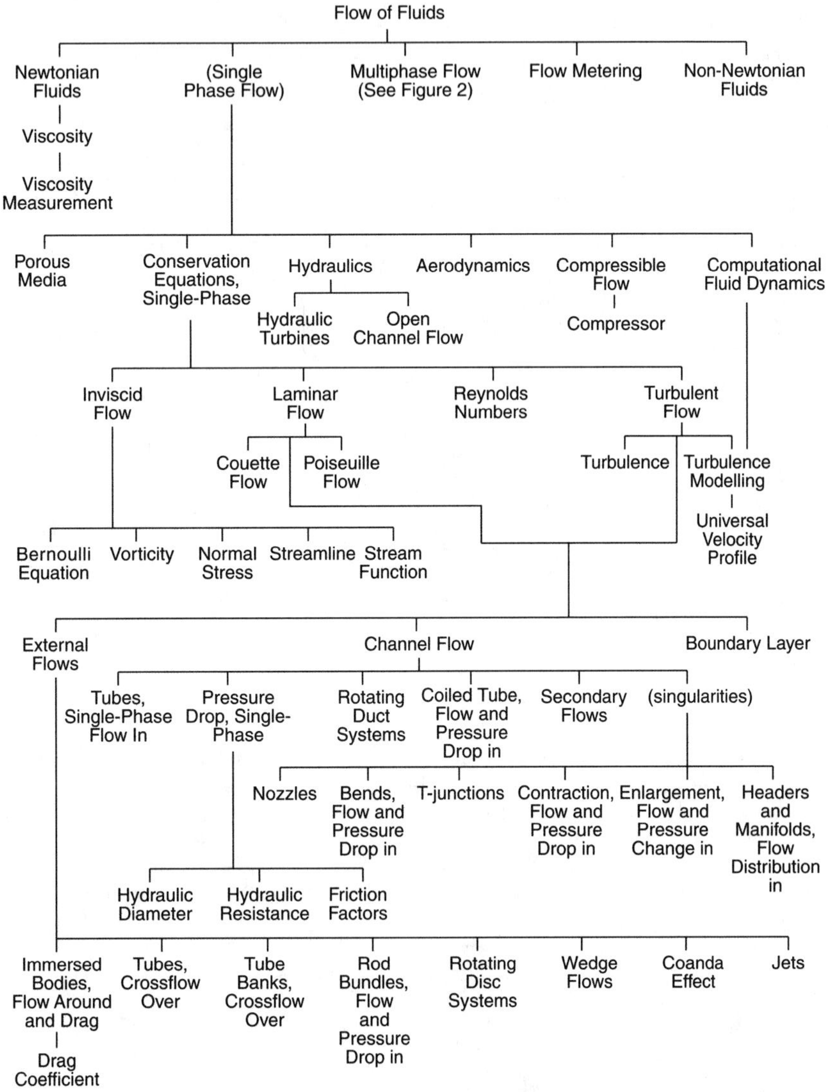

Figure 1. Flow of Fluids.

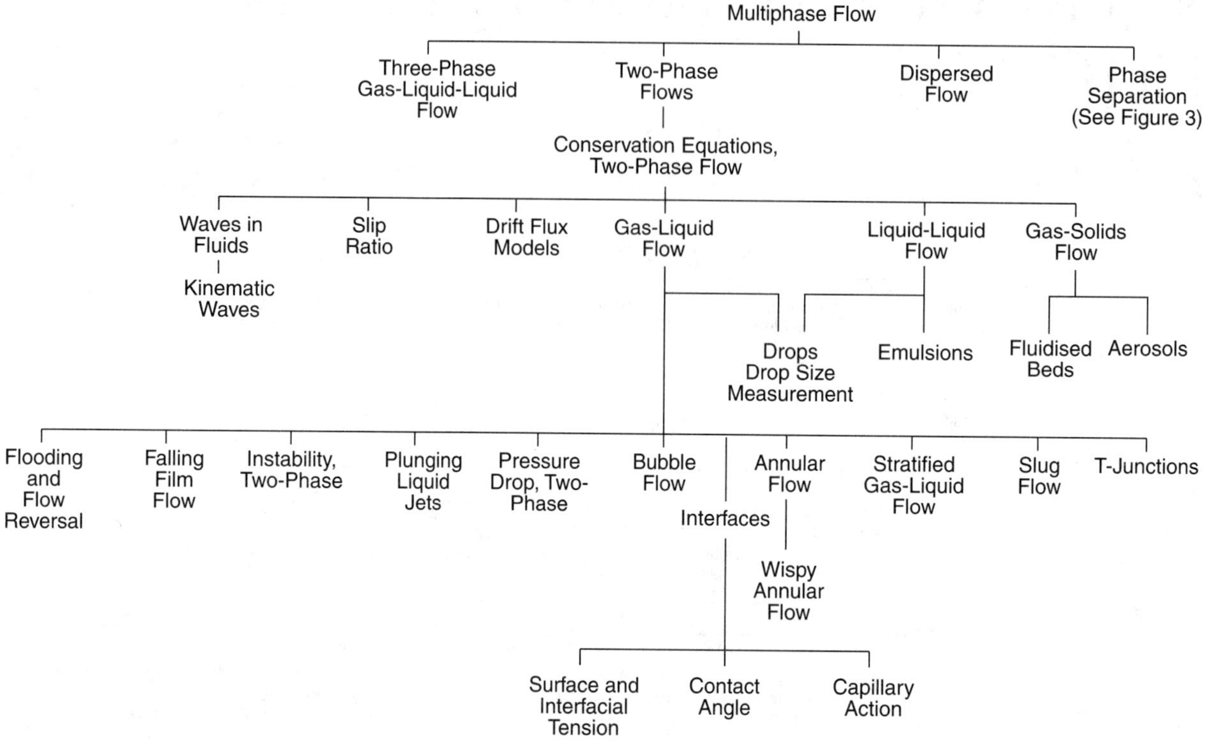

Figure 2. Multiphase Flow.

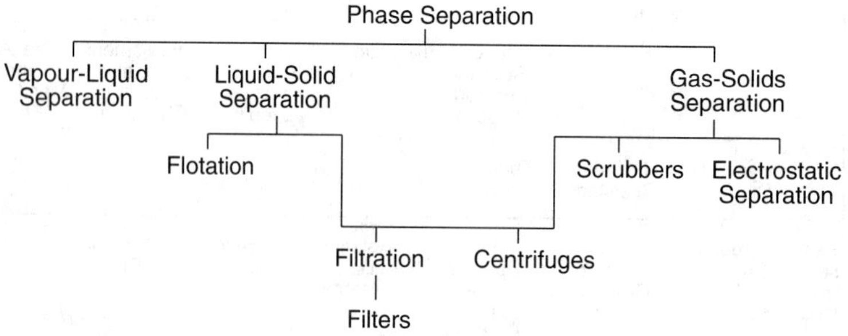

Figure 3. Phase Separation.

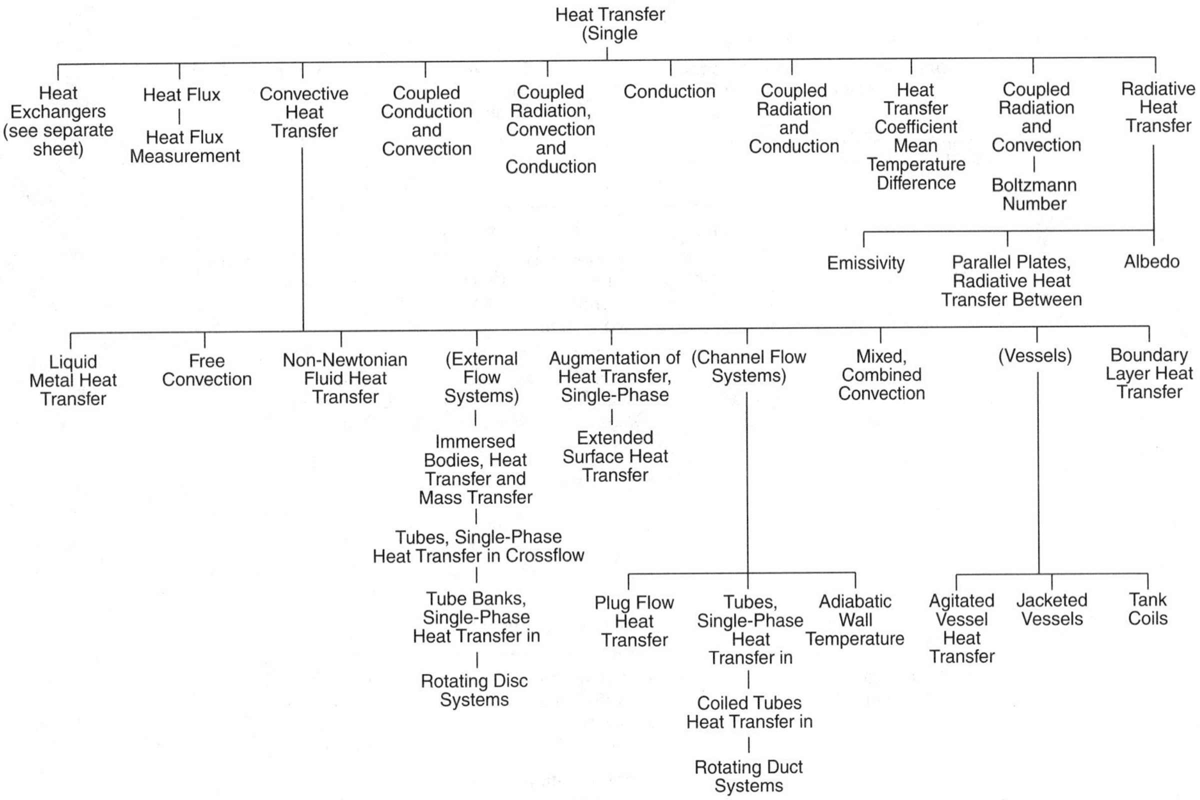

Figure 4. Heat Transfer.

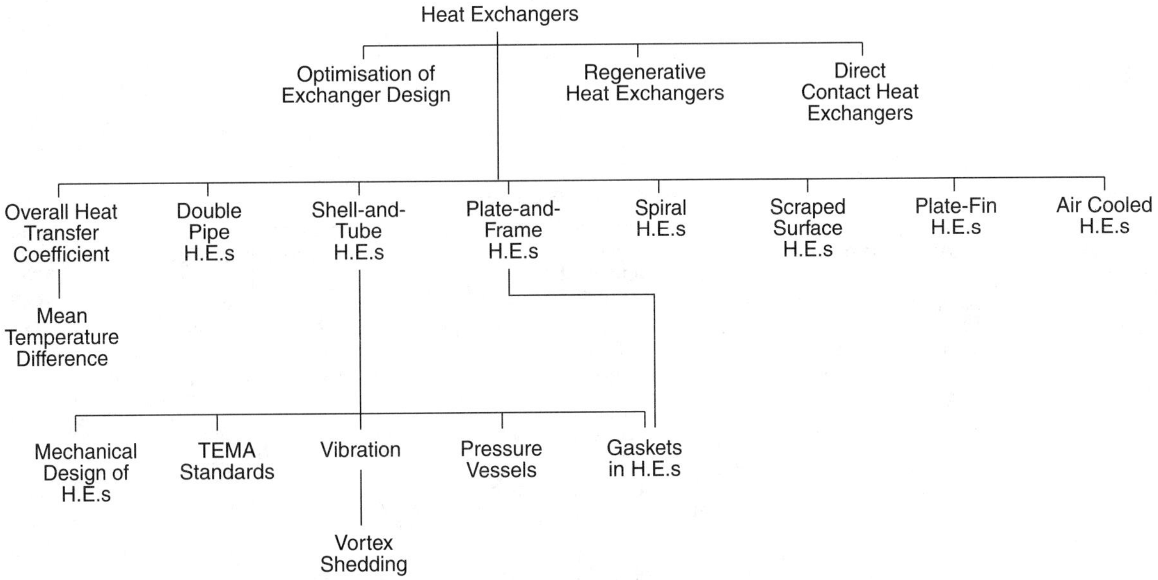

Figure 5. Heat Exchangers.

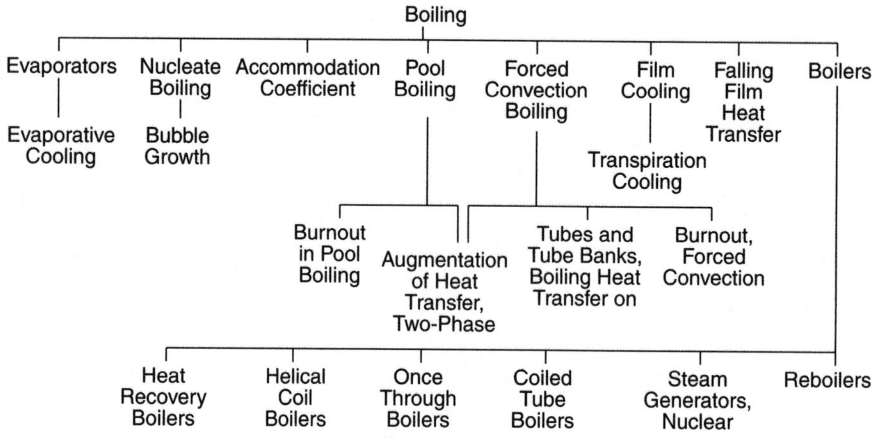

Figure 6. Boiling.

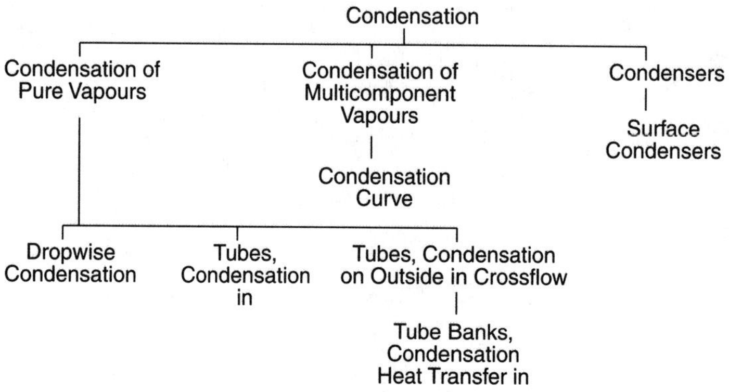

Figure 7. Condensation.

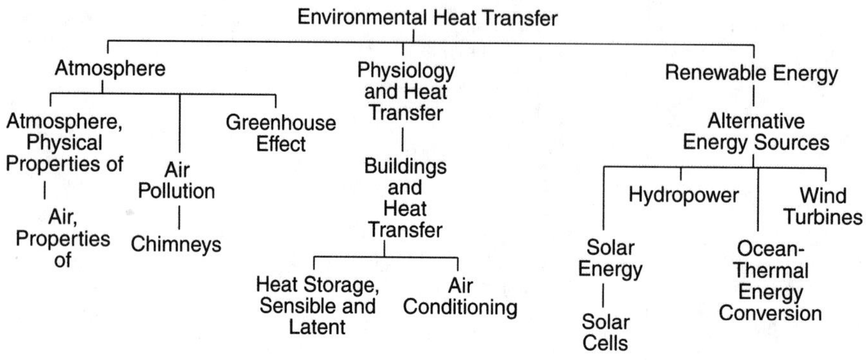

Figure 8. Exvironmental Heat Transfer.

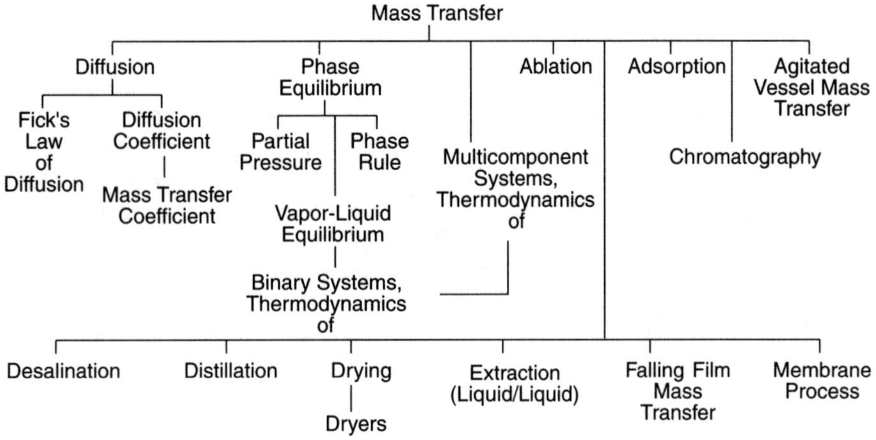

Figure 9. Mass Transfer.

CONTRIBUTORS

Professor A. A. Aleksandrov
MPEI
Moscow 111250
Russia

Dr. Dicos Arcoumanis
Department of Mechanical Engineering
Imperial College of Science, Technology and Medicine
London SW7 2BY
United Kingdom

Professor B. J. Azzopardi
Department of Chemical Engineering
The University of Nottingham
University Park
Nottingham NG7 2RD
United Kingdom

Professor A. I. Bailey
Department of Chemical Engineering and Chemical Technology
Imperial College of Science, Technology and Medicine
London SW7 2BY
United Kingdom

Mr. N. H. Balshaw
Scientific Research Division
Oxford Instruments Ltd.
Old Station Way
Eynsham
Witney
Oxon OX8 1TL
United Kingdom

Professor Howard Barnes
Liquid Processing Department
Unilever Research
Port Sunlight Laboratory
Quarry Road East
Bebington, Wirral
Merseyside L63 3JW
United Kingdom

Dr. C. J. Bates
School of Engineering
Division of Mechanical Engineering and Energy Studies
University of Wales College of Cardiff
Newport Road
Cardiff CF2 1TA
United Kingdom

Dr. Peter D. Baugh
Department of Chemistry and Applied Chemistry
University of Salford
Salford M5 4WT
United Kingdom

Professor T. V. Bazhenova
IVTAN
Moscow 127412
Russia

Dr. S. B. Beale
Thermal Technology Programme
National Research Council of Canada
Montreal Road
Ottawa, Ontario K1A OR6
Canada

Professor V. S. Beknev
MSTU
Moscow 107005
Russia

Professor K. J. Bell
College of Engineering, Architecture and Technology
Oklahoma State University
P.O. Box 1148
Stillwater, Oklahoma 74078-0533
U.S.A.

Dr. Hadj Benkreira
Department of Chemical Engineering
University of Bradford
Richmond Road
Bradford BD7 1DP
United Kingdom

Professor A. E. Bergles
Department of Mechanical Engineering, Aeronautical Engineering and Mechanics
Rensselaer Polytechnic Institute
110 Eighth Street
Troy, New York 12180-3590
U.S.A.

Professor J. T. Berry
Department of Mechanical Engineering
Mississippi State University
P.O. Drawer ME
210 Carpenter Engineering
Mississippi State, Mississippi 39762-5925
U.S.A.

Professor L. M. Biberman
IVTAN
Moscow 127412
Russia

Professor Z. Bilicki
Institute for Fluid Flow Machinery
Department of Gasdynamics
Polish Academy of Sciences
U1 Gen J Fiszera 14
PL-80-952 Gdansk
Poland

Dr. P. M. Birchenough
Process Studies Department
AEA Technology
Harwell Laboratory
Didcot
Oxon OX11 ORA
United Kingdom

Professor F. Bonetto
School of Engineering
Rensselaer Polytechnic Institute
Troy, New York 12180-3590
U.S.A.

Mr. A. W. Boothroyd
Alfa Laval Thermal Ltd.
Great West Road
Brentford
Middlesex TW8 9BT
United Kingdom

Dr. T. R. Bott
School of Chemical Engineering
University of Birmingham
Edgbaston
Birmingham B15 2TT
United Kingdom

Professor M. Yu. Boyarsky
MPEI
Moscow 111250
Russia

Mr. D. Bradley
CED Technology Ltd.
CED House
Taylors Close
Marlow
Bucks SL7 1PR
United Kingdom

Mr. I. Brittain
Hillyfield
2 Wareham Road
Owermoigne
Dorset DT2 8HL
United Kingdom

Mr. R. J. Brogan
Heat Transfer & Fluid Flow Service
Harwell Laboratory
Oxfordshire OX11 ORA
United Kingdom

Dr. A. Brown
ETSU
Building 154
Harwell Laboratory
Oxfordshire OX11 ORA
United Kingdom

Professor A. V. Bushman
IVTAN
Moscow 127412
Russia

Dr. Bruce Callander
Hadley Centre
Meteorological Office
London Road
Bracknell
Berkshire RG12 2SY
United Kingdom

Dr. A. W. Campbell
School of Science and Technology
University of Teeside
Middlesborough
Cleveland TS1 3BA
United Kingdom

Mr. M. Capobianchi
Department of Mechanical Engineering
State University of New York at Stony Brook
Stony Brook, New York 11794-2300
U.S.A.

Dr. K. J. Carpenter
Zeneca Fine Chemicals Manufacturing Organisation
North of England Works
P.O. Box A38
Leeds Road
Huddersfield
Yorkshire HD2 1FF
United Kingdom

Professor Maria da Graca Carvalho
Department of Mechanical Engineering
Instituto Superior Técnico
Avenue Rovisco Pais
1096 Lisboa cedex
Portugal

Professor J. C. Chen
Department of Chemical Engineering
Lehigh University
111 Research Drive
Bethlehem, Pennsylvania 18015-4791
U.S.A.

Mr. I. V. Chircov
IVTAN
Moscow 127412
Russia

Dr. J. Clark
Department of Physiology and Environmental Science
Sutton Bonnington Campus
University of Nottingham
Loughborough LE12 5RD
United Kingdom

Professor Roland Clift
Department of Chemical and Process Engineering
University of Surrey
Guildford
Surrey GU2 5XH
United Kingdom

Dr. Michael Cloke
Department of Chemical Engineering
University of Nottingham
Nottingham NG7 2RD
United Kingdom

Mr. John G. Collier
Nuclear Electric plc
Barnett Way
Barnwood
Gloucester GL4 7RS
United Kingdom

Mr. John Cooper
J. R. Cooper Associates
Cranston House
Mount Howe
Topsham
Exeter EX3 OBG
United Kingdom

Professor K. Cornwell
Department of Mechanical Engineering
James Nasmyth Building
Heriot-Watt University
Riccarton
Edinburgh EH14 4AS
United Kingdom

Professor Michael L. Corradini
Nuclear Engineering and Engineering Physics
University of Wisconsin-Madison
Engineering Research Building
1500 Johnson Drive
Madison, Wisconsin 53706
U.S.A.

M. Jean Costa
Département de Thermohydraulique et de Physique
Commissariat à l'Energie Atomique
17 rue des Martyrs
38054 Grenoble cedex 9
France

Mr. F. de Crécy
Commissariat a l'Energie Atomique
DRN/DTP/STR
17 rue des Martyrs
38054 Grenoble cedex 9
France

Professor Clayton Crowe
Department of Mechanical and Materials Engineering
Washington State University
Pullman, Washington 99164-2920
U.S.A.

Dr. T. W. Davies
School of Engineering
Exeter University
North Park Road
Exeter EX4 4QF
United Kingdom

Mr. Yu. A. Dedikov
IVTAN
Moscow 127412
Russia

Professor V. K. Dhir
Mechanical, Aerospace and Nuclear Engineering Department
School of Engineering and Applied Science
University of California, Los Angeles
405 Hilgard Avenue
RLos Angeles, California 90024-1597
U.S.A.

Dr. F. Dias
Institut Non-Lineaire de Nice
UMR 129 CNRS-UNSA
Université de Nice, Sophia Antipolis
1361 Route des Lucioles
06560 Valbonne
France

Dr. Gary E. Dix
General Electric Nuclear Energy
175 Curtner Avenue
San Jose, California 95125
U.S.A.

Mr. J. Dixon
Land Infra-Red
Wreakes Lane
Dronfield
Sheffield S18 6DJ
United Kingdom

Dr. P. Douglas
Chemistry Department
University of Wales, Swansea
Singleton Park
Swansea SA2 8PP
United Kingdom

Mr. V. N. Dvortsov
IVTAN
Moscow 127412
Russia

Dr. Alan Dyer
Department of Chemistry and Applied Chemistry
University of Salford
Cockcroft Building
The Crescent
Salford M5 4WT
United Kingdom

Dr. M. K. El-Daou
Applied Sciences Department
Faculty of Technological Studies
P.O. Box 42325 Shuwaikh
Kuwait

Professor Ezra Elias
Faculty of Mechanical Engineering
Technion—Israel Institute of Technology
Haifa
Israel 32000

Professor V. M. Epifanov
MSTU
Moscow 107005
Russia

Dr. P. R. Ereaut
AEA Technology
Harwell Laboratory
Oxfordshire OX11 ORA
United Kingdom

Mr. Robin Eycott
Houseman Ltd.
The Priory
Burnham
Slough SL1 7LS
United Kingdom

Professeur J. Fabre
I.N.P.—ENSEEIHT
Institut de Mecanique des Fluides de Toulouse
Av du Professeur Camille Soula 31400
Toulouse
France

Professor O. N. Favorsky
CIAM
Moscow 112250
Russia

Mrs. Rita Fehle
Lehrstuhl A für Thermodynamik
Technische Universität München
Arcistrasse 21
80290 München
Germany

Mr. A. J. Finn
Process Contracting Division
Costain Oil, Gas and Process Ltd.
Costain House
Styal Road
Wythenshawe
Manchester M22 5WN
United Kingdom

Professor Leroy S. Fletcher
Mechanical Engineering Department
Texas A & M University
College Station, Texas 77843-3123
U.S.A.

Professor Harvey M. Flower
Department of Materials
Imperial College of Science, Technology and Medicine
London SW7 2BP
United Kingdom

Professor V. E. Fortov
IVTAN
Moscow 127412
Russia

Mr. C. J. Fry
AEA Technology
Winfrith
Dorchester
Dorset DT2 8DH
United Kingdom

Professor Tohru Fukano
Department of Mechanical Engineering
Kyushu University 36
6-10-1 Hakozaki
Higashi-ku
Fukuoka 812
Japan

Professor O. A. Geratshenko
ITTF
Kiev 252057
Ukraina

Dr. Ghadiri
Department of Chemical and Process Engineering
University of Surrey
Guildford GU2 5XH
United Kingdom

Dr. Ian Gibson
3 Hardy's Close
Martinstown
Near Dorchester
Dorset DT2 9JS
United Kingdom

Dr. M. M. Gibson
Department of Mechanical Engineering
Imperial College of Science, Technology and Medicine
London SW7 2BX
United Kingdom

Mr. V. V. Gil
IVTAN
Moscow 127412
Russia

Mr. M. J. Gilkes
42 The Green
Southwick
Brighton
West Sussex BN42 4FR
United Kingdom

Dr. A. M. Godridge
Riverside House
Beechwood Road
Bartley
Southampton SO40 2LP
United Kingdom

Professor R. J. Goldstein
Department of Mechanical Engineering
University of Minnesota
111 Church Street SE
Minneapolis, Minnesota 55455-0111
U.S.A.

Mr. V. V. Golub
IVTAN
Moscow 127412
Russia

Professor W. Grassi
Dipartimento di Energetica
Universita di Pisa
Via Diotisalvi 2
56126 Pisa
Italy

Mr. Andrew J. Green
Oil and Gas Technology
BHR Group Ltd.
Cranfield
Bedfordshire MK 43 OAJ
United Kingdom

Dr. Andrew S. P. Green
CSMA
Rosemanowes
Herniss
Penryn
Cornwall TR10 9DU
United Kingdom

Mr. E. J. Gregory
Industrial Heat Exchanger Group
IMI Marston Ltd.
Wobaston Road
Fordhouses
Wolverhampton WV10 6QJ
United Kingdom

Professor Peter Griffith
Department of Mechanical Engineering
Massachusetts Institute of Technology
Cambridge, Massachusetts 02139
U.S.A.

Dr. U. Grigull
Lehrustuhl A für Thermodynamik
Technische Universität München
80290 München
Germany

Eur. Ing. J. H. Gummer
Mechanical Engineering Department
University of Southampton
Highfield
Southampton SO9 5NH
United Kingdom

Mr. A. Guy
Brown Fintube (U.K.)
P.O. Box 790
Wimborne
Dorset BH21 5YA
United Kingdom

Dr. A. R. W. Hall
National Engineering Laboratory (NEL)
East Kilbride
Glasgow
G75 006
United Kingdom

Mr. N. J. Hallas
Environmental and Process Engineering
AEA Technology
Building 404
Harwell Laboratory
Oxfordshire OX11 ORA
United Kingdom

Mr. B. A. Hands
Cryogenics Laboratory
Department of Engineering Science
University of Oxford
Parks Road
Oxford OX1 3PJ
United Kingdom

Mr. N. S. Harris
Edwards High Vacuum International
Manor Royal
Crawley
West Sussex RH10 2LW
United Kingdom

Mr. P. S. Harrison
Melverley Consultants Ltd.
Upper Bank
Melverley
Near Oswestry
Shropshire SY10 8PN
United Kingdom

Mr. Ian Harvey
EA (Electrical Association) Technology
Capenhurst
Chester CH1 6ES
United Kingdom

Professor R. I. Hawes
Rosecroft
Greenway Close
Weymouth
Dorset DT3 5BQ
United Kingdom

Mr. N. J. Hawkes
Department of Chemical Engineering
Imperial College of Science, Technology
and Medicine
London SW7 2BY
United Kingdom

Professor P. J. Heggs
Department of Chemical Engineering
UMIST
P.O. Box 88
Manchester M60 1QD
United Kingdom

Professor Morgan Heikel
Department of Mechanical and
Manufacturing Engineering
Cockcroft Building
University of Brighton
Moulsecoomb
Brighton BN2 4GJ
United Kingdom

Mr. Alan A. Herod
Department of Chemical Engineering
Imperial College of Science, Technology
and Medicine
London SW7 2BY
United Kingdom

Professor G. Hetsroni
Department of Mechanical Engineering
Technion—Israel Institute of Technology
Haifa 32000
Israel

Professor G. F. Hewitt
Department of Chemical Engineering and
Chemical Technology
Imperial College of Science, Technology
and Medicine
London SW7 2BY
United Kingdom

Dr. B. J. Holmes
AEA Technology Centre
Winfrith
Dorchester DT2 8DH
United Kingdom

Professor Hiroshi Honda
Institute of Advanced Material Study
Kyushu University
Kasuga
Fukuoka 816
Japan

Professor C. J. Hoogendoorn
Faculty of Applied Physics
Delft University of Technology
P.O. Box 5046
2600 GA Delft
The Netherlands

Dr. Campbell B. Hope
AEA Technology
CFDS Harwell Laboratory
Oxfordshire OX11 ORA
United Kingdom

Professor D. J. Howarth
Department of Computing
Imperial College of Science, Technology and Medicine
London SW7 2BZ
United Kingdom

Mr. J. S. Humphreys
National Engineering Laboratory
East Kilbride
Glasgow G75 OQU
United Kingdom

Professor M. H. Ibragimov
MSU
Moscow
Russia

Mr. V. A. Ibragimov
IVTAN
Moscow 127412
Russia

Professor T. F. Irvine
College of Mechanical Engineering and Applied Sciences
State University of New York at Stony Brook
Stony Brook, New York 11794-2300
U.S.A.

Mr. W. H. Isalski
Costain Oil, Gas and Process Ltd.
Costain House
Styal Road
Manchester M22 5WN
United Kingdom

Professor D. J. Jackson
Manchester School of Engineering
University of Manchester
Simon Building
Oxford Road
Manchester M13 9PL
United Kingdom

Professor H. R. Jacobs
College of Engineering
Colorado State University
Fort Collins, Colorado 80523
U.S.A.

Dr. Sreenivas Jayanti
Department of Chemical Engineering
Indian Institute of Technology
Madras 600 036
India

Mr. Peter Jenkins
"Steps"
The Highlands
Painswick
Glos GL6 6SL
United Kingdom

Dr. A. R. Jones
Department of Chemical Engineering
Imperial College of Science, Technology and Medicine
London SW7 2BY
United Kingdom

Dr. Peter L. Jones
EA Technology
Capenhurst
Chester CH1 6ES
United Kingdom

Professor T. V. Jones
Department of Engineering Science
University of Oxford
Parks Road
Oxford OX1 3PJ
United Kingdom

Professor D. N. Kagan
IVTAN
Moscow 127412
Russia

Mr. V. I. Kaljasin
IVTAN
Moscow 127412
Russia

Mr. I. V. Kalmikov
IVTAN
Moscow 127412
Russia

Dr. R. Kandiyoti
Department of Chemical Engineering
Imperial College of Science, Technology and Medicine
London SW7 2BY
United Kingdom

Professor Isao Kataoka
Department of Nuclear Engineering
Kyoto University
Yoshida, Sakyo
Kyoto 606-01
Japan

Mr. V. I. Katinas
LEI
Kaunas
Lithuania

Dr. R. B. Keey
Department of Chemical and Process Engineering
University of Canterbury
Private Bag 4800
Christchurch
New Zealand

Mr. R. L. Keightley
British Steel plc, Technical
Swinden Laboratories
Moorgate
Rotherham S60 3AR
United Kingdom

Dr. D. B. R. Kenning
Department of Engineering Science
University of Oxford
Parks Road
Oxford OX1 3PJ
United Kingdom

Professor L. S. Kershenbaum
Department of Chemical Engineering and Chemical Technology
Imperial College of Science, Technology and Medicine
London SW7 2BY
United Kingdom

Professor E. M. Khabakhpasheva
ITF
Novosibirsk -90
Russia

Dr. H. G. Khajah
Department of Applied Sciences
College of Technical Studies
P.O. Box 42325
Shuwaikh 70654
Kuwait

Mr. A. I. Kharitonov
IVTAN
Moscow 127412
Russia

Professor Hans-Joachim Kilger
George-Simon-Ohm Fachhochschule in Nuernberg
FB Maschinenbau
Kessler-Platz 12
D 90489 Nuernberg
Germany

Dr. N. F. Kirkby
Department of Chemical and Process Engineering
University of Surrey
Guildford GU2 5XH
United Kingdom

Mr. A. Knaani
Operating Analysis and Engineering
International Paper
11 Skyline Drive
Hawthorne, New York 10532
U.S.A.

Professor S. A. Kovalev
IVTAN
Moscow 127412
Russia

Mr. M. G. Kremlev
IVTAN
Moscow 127412
Russia

Mr. V. A. Kurganov
IVTAN
Moscow 127412
Russia

Professor Yu. A. Kuzma-Kichta
MPEI
Moscow 111250
Russia

Dr. J. A. Lacey
21 Edstone Close
Dorridge
Solihull B93 8DP
United Kingdom

Mr. J. Lafay
Commissariat a l'Energie Atomique
17 rue des Martyrs
38054 Grenoble cedex 9
France

Professor R. T. Lahey
School of Engineering
Rensselaer Polytechnic Institute
Troy, New York 12180-3590
U.S.A.

Mr. Brian Lamb
64 Muirfield Drive
Mickleover
Derby DE3 5YF
United Kingdom

Professor A. H. Lefebvre
Low Furrow
Pebworth
Stratford-upon-Avon CV37 8XW
United Kingdom

Professor R. Lemlich
Department of Chemical Engineering
697 Rhodes Hall
University of Cincinnati
Cincinnati, Ohio 45221-0171
U.S.A.

Mr. V. Ja. Leonidov
IVTAN
Moscow 127412
Russia

Professor P. E. Liley
School of Mechanical Engineering
Purdue University
West Lafayette, Indiana 47907-1288
U.S.A.

Professor A. Line
Institut de Mecanique des Fluides
Avenue du Professeur Camille Soula
31400 Toulouse
France

Dr. R. W. Lines
Coulter Electronics Ltd.
Northwell Drive
Luton
Beds LU3 3RH
United Kingdom

Mr. A. C. Lintern
Department of Chemical Engineering and Chemical Technology
Imperial College of Science, Technology and Medicine
London SW7 2BY
United Kingdom

Dr. T. J. Lockett
AEA Technology Petroleum Services
Harwell Laboratory
Oxfordshire OX11 ORA
United Kingdom

Dr. D. MacMahon
Imperial College Centre for Analytical Research in the Environment
Stillwood Park
Buckhurst Road
Ascot
Berks SL5 7TE
United Kingdom

Professor E. A. Manushin
MSTU
Moscow 107005
Russia

Mr. S. F. Marriott
ETSU
Harwell Laboratory
Oxfordshire OX11 ORA
United Kingdom

Professor O. G. Martynenko
ITMO
Minsk 220728
Belorussia

Professor F. Mayinger
Lehrstuhl A für Thermodynamik
Technische Universität München
Arcisstrasse 21
90290 München
Germany

Mr. J. McNaught
Rankine Building
National Engineering Laboratory
East Kilbride
Glasgow G75 OQU
United Kingdom

Professor S. A. Medin
IVTAN
Moscow 127412
Russia

Dr. M. A. Mendes-Tatsis
Department of Chemical Engineering and Chemical Technology
Imperial College of Science, Technology and Medicine
London SW7 2BY
United Kingdom

Professor W. Merzkirch
Lehrstuhl für Strömungslehre
Universität Essen
D-45117 Essen
Germany

Dr. J. P. Meyer
EPS Consulting Engineers
P.O. Box 9669
Hennopsmeer 0046
South Africa

Dr. H. J. Michels
Department of Chemical Engineering and Chemical Technology
Imperial College of Science, Technology and Medicine
London SW7 2BY
United Kingdom

Mr. D. S. Mikhatulin
IVTAN
Moscow 127412
Russia

Dr. J. Miles
SGS Redwood Technical Services
Rosscliffe Road
Ellesmere Port
South Wirral L65 3AS
United Kingdom

Dr. A. J. Miller
University of Brighton
Department of Construction, Geography and Surveying
Mithras House
Lewes Road
Brighton BN2 4AT
United Kingdom

Dr. B. C. Millington
Flow Centre
National Engineering Laboratory
East Kilbride
Glasgow G75 OQU
United Kingdom

Professor P. F. Monaghan
Computer Aided Thermal Engineering Research Unit
Manufacturing Research Centre
University College Galway
Galway
Ireland

Professor M. J. Moore
Driftwood
4 Rectory Close
Merrow
Guildford
Surrey GU4 7AR
United Kingdom

Mr. M. Morris
M. Morris & Associates
Bliss Cottage
20 Hine Town
Shillingstone
Dorset DT11 OSN
United Kingdom

Professor W. D. Morris
Department of Mechanical Engineering
University of Wales
Singleton Park
Swansea SA2 8PP
United Kingdom

Dr. Alfred Moser
Institut für Energietechnik
ETH-Zentrum Low E2
CH-8092 Zurich
Switzerland

Professor I. L. Mostinsky
IVTAN
Moscow 127412
Russia

Professor A. S. Mujumdar
Chemical Engineering Department
McGill University
3480 University Street
Montreal, Quebec
Canada H3A 2A7

Professor H. Müller-Steinhagen
Department of Chemical and Process Engineering
University of Surrey
Guildford
Surrey GU2 5XH
United Kingdom

Professor R. M. Nedderman
Department of Chemical Engineering
University of Cambridge
Pembroke Street
Cambridge CB2 3RA
United Kingdom

Professor G. Nicolis
Faculté des Sciences
Université Libre de Bruxelles
CP231 Boulevard du Triomphe
1050 Bruxelles
Belgium

Professor Brian Norton
Department of Building and Environmental Engineering
University of Ulster
Newtownabbey BT37 OQB
Northern Ireland
United Kingdom

Professor S. Oka
Belgrade
Yugoslavia

Dr. Smuel Olek
R & D Division
Ismael Electric Corporation Ltd.
P.O. Box 10
Haifa 31000
Israel

Ms. S. J. Oliver
c/o Mr. D. J. Archer
Secomak Ltd.
Honeypot Lane
Stanmore
Middlesex HA7 1BE
United Kingdom

Professor M. N. Orlova
IVTAN
Moscow 127412
Russia

Professor E. L. Ortiz
Department of Mathematics
Imperial College of Science, Technology and Medicine
London SW7 2BY
United Kingdom

Dr. E. S. Perez de Ortiz
Department of Chemical Engineering and Chemical Technology
Imperial College of Science, Technology and Medicine
Prince Consort Road
London SW7 2BY
United Kingdom

Professor J. M. Owen
School of Mechanical Engineering
University of Bath
Claverton Down
Bath BA2 7AY
United Kingdom

Dr. Lei Pan
Department of Chemical Engineering
Imperial College of Science, Technology and Medicine
London SW7 2BY
United Kingdom

Dr. Thanasis D. Papathanasiou
Department of Chemical Engineering and Chemical Technology
Imperial College of Science, Technology and Medicine
London SW7 2BY
United Kingdom

Mr. K. R. Parker
17 Somerville Road
Sutton Coldfield
West Midlands B73 6JD
United Kingdom

Mr. Roger Parker
CSM Associates Ltd.
Rosemanowes
Herniss
Penryn
Cornwall TR10 9DU
United Kingdom

Mr. Richard Paton
Flow Centre
National Engineering Laboratory
East Kilbride
Glasgow G75 OQU
United Kingdom

Dr. M. J. Pattison
Department of Chemical Engineering and Chemical Technology
Imperial College of Science, Technology and Medicine
London SW7 2BY
United Kingdom

Professor N. V. Pavljukevich
ITMO
Minsk 220728
Belorussia

Dr. R. S. Pease
Pease Partners Consultants
The Poplars
West Isley
Newbury
Berkshire RG16 OAW
United Kingdom

Professor V. A. Petrov
IVTAN
Moscow 127412
Russia

Dr. Paul Pickering
Baker Jardine
19 Heathmans road
Parsons Green
London SW6 4TJ
United Kingdom

Professor Yu. V. Polezhaev
IVTAN
Moscow 127412
Russia

Professor A. F. Polijakov
IVTAN
Moscow 127412
Russia

Dr. G. T. Polley
Department of Chemical Engineering
UMIST
P.O. Box 88
Manchester M60 1QD
United Kingdom

Professor V. S. Polonsky
IVTAN
Moscow 127412
Russia

Professor K. E. Porter
Department of Chemical Engineering and Applied Chemistry
Aston University
Aston Triangle
Birmingham B4 7ET
United Kingdom

Professor P. S. Poskas
LEI
Kaunas
Lithuania

Professor G. L. Quarini
Department of Mechanical Engineering
University of Bristol
Queen's Building
University Walk
Bristol BS8 1TR
United Kingdom

Dr. Michael J. Reader-Harris
Flow Centre
National Engineering Laboratory
East Kilbride
Glasgow G75 OQU
United Kingdom

Mr. Derek W. Reay
Oil Technology Centre
BP Oil International Ltd.
Chertsey Road
Sunbury-on-Thames
Middlesex TW16 7LN
United Kingdom

Dr. Michael Reeks
Nuclear Electric plc
Berkeley Technology Centre
Berkeley
Gloucestershire GL13 9PB
United Kingdom

Mr. A. Reeve
Sulzer (U. K.) Ltd.
West Mead
Farnborough
Hants GU14 7LP
United Kingdom

Dr. Graham Rice
Department of Engineering
University of Reading
Whiteknights
P.O. Box 225
Reading RG6 2AY
United Kingdom

Professor J. F. Richardson
Department of Chemical Engineering
University of Swansea
Singleton Park
Swansea SA2 8PP
United Kingdom

Professor S. M. Richardson
Department of Chemical Engineering and Chemical Technology
Imperial College of Science, Technology and Medicine
London SW7 2BY
United Kingdom

Mr. I. F. Roberts
Department of Chemical Engineering and Chemical Technology
Imperial College of Science, Technology and Medicine
London SW7 2BY
United Kingdom

Professor J. W. Rose
Department of Mechanical Engineering
University of London
Queen Mary and Westfield College
Mile End Road
London E1 4NS
United Kingdom

Professor Yu. N. Rudenko
RAN
Moscow 117334
Russia

Mr. A. E. Ruffel
Babcock Energy Ltd.
11 The Boulevard
Crawley
West Sussex RH10 1UX
United Kingdom

Professor C. Ruiz
Department of Engineering Science
University of Oxford
Parks Road
Oxford OX1 3PJ
United Kingdom

Professor Takeo S. Saitoh
Department of Aeronautics and Space Engineering
Tohoku University
Sendai 980
Japan

Mr. Massimo Salvatores
Nuclear Reactor Directorate
CEA—Cadarache
13108 Saint Paul lez Durance
France

Dr. H. Sandner
Lehrstuhl A für Thermodynamik
Technische Universität München
80290 München
Germany

Professor R. W. H. Sargent
Centre for Process Systems Engineering
Imperial College of Science, Technology
and Medicine
London SW7 2BY
United Kingdom

Dr. G. Saville
Department of Chemical Engineering and
Chemical Technology
Imperial College of Science, Technology
and Medicine
London SW7 2BY
United Kingdom

Dr. Keith Scott
Department of Chemical and Process
Engineering
University of Newcastle
Merz Court
Newcastle upon Tyne NE1 7RU
United Kingdom

Professor A. M. Semenov
MPEI
Moscow 111250
Russia

Professor A. Serizawa
Department of Nuclear Engineering
Kyoto University
Yoshida, Sakyo
Kyoto 606-01
Japan

Dr. J. P. K. Seville
School of Chemical Engineering
University of Birmingham
P.O. Box 363
Edgbaston
Birmingham B15 2TT
United Kingdom

Dr. Ramesh K. Shah
College of Engineering
University of Kentucky
520 Robotics Facility
Lexington, Kentucky 40506-0108
U.S.A.

Mr. Yu. L. Shekhter
IVTAN
Moscow 127412
Russia

Professor G. L. Shires
3 Silver Street
Sutton Poyntz
Weymouth
Dorset DT3 6LL
United Kingdom

Mrs. Joy Shires
3 Silver Street
Sutton Poyntz
Weymouth
Dorset DT3 6LL
United Kingdom

Professor Avraham Shitzer
Faculty of Mechanical Engineering
Technion—Israel Institute of Technology
Haifa 32000
Israel

Professor E. E. Shpilrain
IVTAN
Moscow 127412
Russia

Dr. R. Siegel
Mail Stop 5-9
NASA Lewis Research Center
21000 Brookpark Road
Cleveland, Ohio 44135
U.S.A.

Dr. J. R. Singham
54 Pensford Avenue
Kew Gardens
Richmond
Surrey TW9 4HP
United Kingdom

Professor J. M. Smith
Department of Chemical and Process
Engineering
University of Surrey
Guildford
Surrey GU2 5XH
United Kingdom

Professor T. Soendvedt
Department of Production Technology
Norsk Hydro
P.O. Box 200
N-1321 Stabekk
Norway

Professor P. L. Spedding
Department of Chemical Engineering
The Queens University of Belfast
David Keir Building
Stranmillis Road
Belfast BT9 5AG
United Kingdom

Dr. N. J. Spinks
AECL Research
Chalk River Laboratories
Chalk River, Ontario
Canada KOJ 1JO

Dr. S. Srichai
Department of Chemical Engineering
Faculty of Engineering
Chulalongkorn University
Phaya Thai Road
Bangkok 10330
Thailand

Dr. J. Stairmand
AEA Power Fluidics
Risley
Warrington
Cheshire WA3 6AT
United Kingdom

Professor K. Stephan
Institut für Technische Thermodynamik und
Thermische Verfarhrenstechnik
Universitat Stuttgart
70550 Stuttgart
Germany

Dr. C. D. Stewart
Flow Centre
National Engineering Laboratory
East Kilbride
Glasgow G75 OQU
United Kingdom

Mr. David J. Stockton
BT Laboratories
B67/117
Martlesham Heath
Ipswich IP5 7RE
United Kingdom

Dr. J. Straub
Lehrstuhl a für Thermodynamik
Technische Universität München
80290 München
Germany

Professor M. Streat
Department of Chemical Engineering
Loughborough University of Technology
Loughborough
Leicestershire LE11 3TU
United Kingdom

Dr. David Stuckey
Department of Chemical Engineering and
Chemical Technology
Imperial College of Science, Technology
and Medicine
London SW7 2BY
United Kingdom

Mr. C. Summers
61, Middle Lane
Epsom
Surrey KT17 1DP
United Kingdom

Professor B. Sunden
Division of Heat Transfer
Lund Institute of Technology
Box 118
22100 Lund
Sweden

Professor S. T. Surjikov
IPM
Moscow 117526
Russia

Professor Yu. S. Svirchuk
IVTAN
Moscow 127412
Russia

Dr. C. Swan
Department of Civil Engineering
Imperial College of Science, Technology and Medicine
London SW7 2BU
United Kingdom

Dr. J. Swithenbank
Department of Mechanical and Process Engineering
University of Sheffield
P.O. Box 600
Mappin Street
Sheffield S1 4DU
United Kingdom

Dr. Denis Tenchine
Department of Thermal-Hydraulics and Physics
CEA Nuclear Reactor Division
Grenoble Research Centre
17 rue des Martyrs
38054 Grenoble cedex 9
France

Dr. B. Terry
Materials Consultants (Surrey)
2 Juniper Close
Chessington
Surrey KT9 2AX
United Kingdom

Dr. M. Tezock
Courtalds plc
Research and Technology
P.O. Box 111
Lockhurst Lane
Coventry CV 5RS
United Kingdom

Professor J. D. Thornton
Battle Well
Greenhill
Evesham
Worcestershire WR11 4NA
United Kingdom

Mr. T. R. Tomlinson
Process Contracting Division
Costain Oil, Gas & Process Ltd.
Costain House
Styal Road
Manchester M22 5WN
United Kingdom

Dr. J. P. M. Trusler
Department of Chemical Engineering and Chemical Technology
Imperial College of Science, Technology and Medicine
London SW7 2BY
United Kingdom

Mr. K. W. Tupholme
British Steel plc
Swinden Laboratories
Moorgate
Rotherham S60 3AR
United Kingdom

Professor John Twidell
AMSET Centre
De Montfort University
Leicester LE1 9BH
United Kingdom

Professor G. de Vahl Davis
School of Mechanical and Manufacturing Engineering
University of New South Wales
P.O. Box 1
Kensington NSW 2033
Australia

Dr. Velisa Vesovic
Department of Mineral Resources Engineering
Imperial College of Science, Technology and Medicine
London SW7 2BP
United Kingdom

Dr. Jean-Michel Veteau
Commissariat a l'Energie Atomique
DRN/DTP/STR
17 rue des Martyrs
38054 Grenoble cedex 9
France

Professor W. A. Wakeham
Department of Chemical Engineering and Chemical Technology
Imperial College of Science, Technology and Medicine
London SW7 2BY
United Kingdom

Professor R. J. Wakeman
Department of Chemical Engineering
Loughborough University of Technology
Loughborough
Leicestershire LE11 3TU
United Kingdom

Professor G. B. Wallis
Thayer School of Engineering
Dartmouth College
Hanover, New Hampshire 03755-8000
U.S.A.

Dr. J. K. Walters
Department of Chemical Engineering
University of Nottingham
University Park
Nottingham NG7 2RD
United Kingdom

Mr. John A. Ward
Corporate Safety Directorate
AEA Technology
Harwell Laboratory
Oxfordshire OX11 ORA
United Kingdom

Dr. Muriel Watt
Centre for Photovoltaic Devices and Systems
School of Electrical Engineering
University of New South Wales
NSW 2052
Australia

Dr. David R. Webb
Department of Chemical Engineering
UMIST University of Manchester
Manchester M60 1QD
United Kingdom

Dr. D. M. Webber
Integral Science and Software Ltd.
484 Warrington Road
Culceth
Warrington
WA3 5RA
United Kingdom

Professor Manfred Weber
Institut für Fordertechnik abt Strömungsfördertechnik
Universität Karlsruhe
Hertzstrasse 16
76187 Karlsruhe
Germany

Professor J. H. Whitelaw
Department of Mechanical Engineering
Imperial College of Science, Technology and Medicine
London SW7 2BX
United Kingdom

Dr. David Wilkie
1 Randolph Place
Edinburgh EH3 7TQ
United Kingdom

Dr. Gordon Wilkinson
School of Chemical Technology
University of South Australia
The Levels Campus
Pooraka
South Australia 5095
Australia

Professor Alan Williams
Department of Fuel and Energy
Leeds University
Woodhouse Lane
Leeds LS2 9JT
United Kingdom

Dr. A. J. Willmott
Department of Computer Science
York University
Heslington
York YO1 5DD
United Kingdom

Mr. Winkler
Department of Chemical and Process Engineering
University of Surrey
Guildford
Surrey GU2 5XH
United Kingdom

Dr. Richard Winterton
School of Manufacturing and Mechanical Engineering
The University of Birmingham
Edgbaston
Birmingham B15 2TT
United Kingdom

Mr. David Woolnough
ETSU
Building 156
Harwell Laboratory
Oxfordshire OX11 ORA
United Kingdom

Mr. V. S. Yungman
IVTAN
Moscow 127412
Russia

Mr. I. G. Zaltsman
IVTAN
Moscow 127412
Russia

Mr. Yu. A. Zeigarnik
IVTAN
Moscow 127412
Russia

Professor M. F. Zhukov
ITF
Novosibirsk -90
Russia

Dr. Yuzhen Zhao
Department of Power Engineering
Box 513
Harbin Institute of Technology
Harbin 150006
P. R. China

Professor Y. Zvirin
Technion—Israel Institute of Technology
Faculty of Mechanical Engineering
Haifa 32000
Israel

A

AAAS (see American Association for the Advancement of Science)

AASE (see Association for Applied Solar Energy)

ABLATION

Following from: Mass transfer; Heat protection

Ablation is a means of thermal protection based on physico-chemical transformations of solid substances by convective or radiation heat flow. The heat-shield effect is the sum of the heat of phase and chemical transformations of the substance and the reduction of the heat flow when the ablation products are forced into the surround medium (see **Heat Protection**). Ablation can be referred to as a sacrificial method of heat protection, since in order to maintain acceptable heat conditions in a body, its surface layer is partially destroyed. Ablation can, as a rule, be allowed in objects of single application; for instance, the re-entry space vehicles, combustion chambers and the nozzle units of solid-propellant rocket engines. The use of ablative facing has a number of advantages over other methods of heat protection. The main advantage is the self-regulation process, i.e., the change in the ablation rate depending on the level of pressure and temperature of the gas flowing across the surface. Thanks to high values of heat of physico-chemical transformations and to the injection heat effect, the use of ablative facing materials exceeds substantially in efficiency that of systems functioning on the heat storage principle or on the principle of convective cooling (see **Heat Protection**). Together with penetrating cooling, ablative facings form the class of active heat protection, the basis for which is the direct effect on the process of heat transfer from the surrounding medium to the body.

The most commonly used ablative materials are the *composites*, i.e., materials consisting of a high-melting point matrix and an organic binder. The matrix can be glass, asbestos, carbon or polymer fibres braided in different ways. In some cases, a honeycomb construction can be used, filled with a mixture of organic and nonorganic substances and possessing high heat-insulating characteristics (as used, for instance, on the space vehicle "Apollo").

Shown in **Figure 1** is a schematic model of the destruction of a composite material from a high-melting point matrix and an organic binder. The characteristic property of such heat-shielding coverings is the presence of two fronts or zones, to be more exact, in which physico-chemical transformations take place. In convective heating, a viscous melt film can be formed on the surface of such composite materials. Despite its thinness, the film strongly affects the destruction process. In particular, the coalescence of particles of the surface layer prevents their erosion blow-off by the flow. The melt film also reduces the rate of oxidation of chemically-active components of the material by the incoming flow of gas.

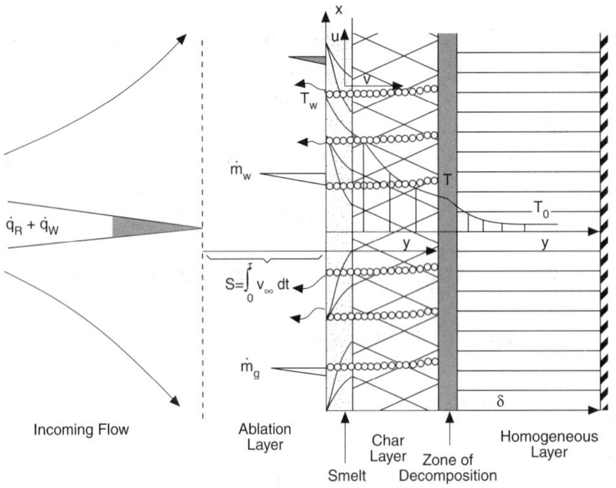

Figure 1. Schematic model for the destruction of an ablating composite material.

Further into the surface lies a comparatively thick layer of charred organic binder reinforced by high-melting fibres. Still deeper is the thermal decomposition zone, where a mixture of volatile and solid (coke) components is formed. The volatile components filtered through the porous matrix are injected into the boundary layer of the incoming gas flow. An intensive sublimation of glass or other oxides which form high-melting fibers occurs on the surface of the melt film. The fraction of gaseous ablation products in the total ablation mass can, therefore, be high. The particles of coke are practically pure carbon; thus, at the melting temperature of glass they remain solid. The spreading film of glass "breaks out" the porous structure of the charred layer and carries away the particles of coke. The latter, in turn, affects the flow of the melt, increasing its effective viscosity (see **Melting**).

At high temperatures, the coke particles in the melt film are not inert components—they interact actively both with glass and with any oxidant present in the gas flow. Tens of various strongly interacting components can exist in the boundary layer over the surface of the composite heat-shielding covering. The choice of a theoretical model for the destruction process of such materials, presents considerable difficulties. However, on the basis of extensive experimental and theoretical studies of thermophysical, thermodynamic and strength phenomena which attend the process of the incident flow effect, we have succeeded in creating a schematic model or a mechanism for the destruction of a heat-facing layer. Such a mechanism has been designed only for some classical representatives of the range of composites (see **Sublimation, Melting**). At the same time, advances in chemistry and materials technology extend the possibilities of selecting improved ablation materials. In this context, a demand arose for some unique parameter to compare various types of ablative materials convenient for both theoretical and experimental studies. One such parameter is the *effective enthalpy of destruction*, symbolized as h_{eff}.

The effective enthalpy defines the total thermal energy expenditure necessary to break down a unit mass of ablative material. The problem of comparing numerous ablative materials is most easily demonstrated for a *quasi*-stationary destruction (see **Heat Conduction**) when the velocity of all isotherms or destruction fronts inside the material coincides with the velocity of the outer surface displacement. In this case, the temperature profile inside the heat-shielding covering is described by a set of exponents, and the heat flux $\dot{q}_\lambda$ spent on heating inner layers does not depend on the material thermal conductivity λ_Σ.

Let us first consider a destruction process under conditions of exposure to convective heating. The thermal balance on a destructing surface (**Figure 2**) can be written as follows:

$$\dot{q} = (\alpha/c_p)_0(h_e - h_w) = \epsilon\sigma T_w^4 + \dot{m}_w \Delta Q_w + \dot{q}_b + \dot{q}_\lambda$$

$$= \epsilon\sigma T_w^4 + \dot{m}_w \Delta Q_w + \gamma\dot{m}_w(h_e - h_w)$$

$$+ \dot{m}_\Sigma[\bar{c}(T_w - T_0) + \Delta Q_\Sigma]$$

$$\{\text{re-emission}\} + \left\{\begin{matrix}\text{surface}\\ \text{destruction}\end{matrix}\right\} + \left\{\begin{matrix}\text{injection}\\ \text{effect}\end{matrix}\right\} + \left\{\begin{matrix}\text{bulk}\\ \text{destruction}\end{matrix}\right\}$$

Here, $(\alpha/c_p)_0$ is the heat transfer coefficient, and h_e and h_w are the enthalpies of the gases in the incoming flow and the wall, respectively. In contrast to a nondestructing ablative facing, the convective heat flux $\dot{q}_0$ supplied from without is expended not only for heating the material ($\dot{q}_\lambda$) and by radiant re-emission of the four heated surface ($\epsilon\sigma T_w^4$) but also for the surface (with mass loss rate $\dot{m}_w$) and bulk (with mass loss rate $\dot{m}_\Sigma$) physico-chemical transformations, whose thermal effects are evaluated as ΔQ_w and ΔQ_Σ. If a melt film is formed on the surface of a heat-shielding covering, then $\dot{m}_\Sigma = \dot{m}_w + \dot{m}_m$, where $\dot{m}_m$ is the mass loss rate of a substance in a molten form. The total thermal effect of the bulk failure ΔQ_Σ contains not only the heat of matrix melting, but also the thermal effect of the thermal decomposition of an organic binder, the heat of heterogeneous interaction between the glass and coke inside the charred layer, etc. In a similar manner, the thermal effect of surface destruction ΔQ_w must account for the thermal effect of evaporation

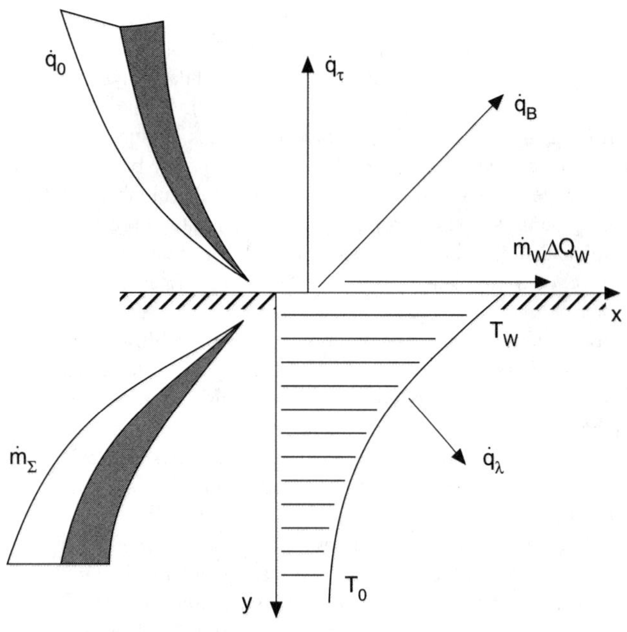

Figure 2. Destruction process with convective heating.

of a melted film and the burning of the coke particles in the incoming flow of gas.

Gaseous ablation products which penetrate into the boundary layer cause a reduction of a convective heat flow due to the so-called "injection effect." We can evaluate the blocking action of the injection effect by a linear approximation (see **Heat Protection**):

$$\dot{q}_b = \dot{q}_0 - \dot{q}_w = \dot{m}_w\gamma(h_e - h_w).$$

Here, γ is the dimensionless coefficient of injection ($\gamma < 1$), which in the general case depends on flow conditions in the boundary layer (laminar or turbulent) and the ratio of molecular masses of the gas injected and the incoming flow. Unlike other effects influencing the absorption of the heat energy supplied, the injection effect rises steeply with the increasing velocity or temperature of the incoming flow and finally becomes predominant.

If we denote the share of gaseous ablation products in the total mass loss of the substance by Γ ($\Gamma = \dot{m}_w/\dot{m}_\Sigma$), then we can obtain a generalized characteristic of destruction power, namely the effective enthalpy of destruction, h_{eff}:

$$h_{eff} = \frac{\dot{q}_0 - \epsilon\sigma T_w^4}{\dot{m}_\Sigma}$$

$$= c(T_w - T_0) + \Delta Q_\Sigma + \Gamma\{\Delta\dot{M}_w + \gamma(h_e - h_w)\}.$$

The effective enthalpy determines the amount of heat which can be "blocked" when breaking down a unit mass of covering material (whose surface temperature is T_w) through physico-chemical processes. The higher the effective enthalpy, the better the heat-shielding material. We place emphasis on the independence of the effective enthalpy from the geometrical dimensions or the shape of the body. Actually, as distinct from a heat flux whose value, with the given parameters of the incoming flow (p_e, h_e), is inversely proportional to $\sqrt{R_N}$ (where R_N is the typical dimension of the body; for instance, the radius of curvature in the vicinity of the critical point), the effective enthalpy is unaffected either by the shape or the dimension of the body. This qualifies it as a parameter for relating laboratory and real heat-loading situations.

We can see from the definition of effective enthalpy that in all cases when $\Gamma \neq 0$, it must increase substantially with the rise in the enthalpy of the stagnated flow h_e. The parameters of the incoming gas flow (pressure P_e and enthalpy h_e) can effect h_{eff} through changes in the temperature of the destructing surface T_w, the fraction of the ablation which in gaseous form Γ and the thermal effect of surface processes ΔQ_w. The effect of surface temperature T_w on h_{eff} can be considered to be rather limited. A typical dependence of T_w, Γ and h_{eff} on enthalpy h_e and pressure P_e in breaking down glass reinforced plastics in an air flow is shown in **Figures 3, 4** and **5**. The flow condition (laminar or turbulent) in the boundary layer determines the injection coefficient γ (see **Heat Protection**), which affects radically the dependence of h_{eff} on h_e (**Figure 6**). If the ablative material does not contain oxides, then, as a rule, the share of gasification Γ is close to unity. For graphite-like heat-shield covering in particular, $\Gamma = 1$. In this case, however, the thermal effect of surface processes ΔQ_w varies from a negative value on carbon burning $C + O_2 = CO_2$ to a positive value upon its sublimation. An extra liberation of heat upon burning brings about surface overheating

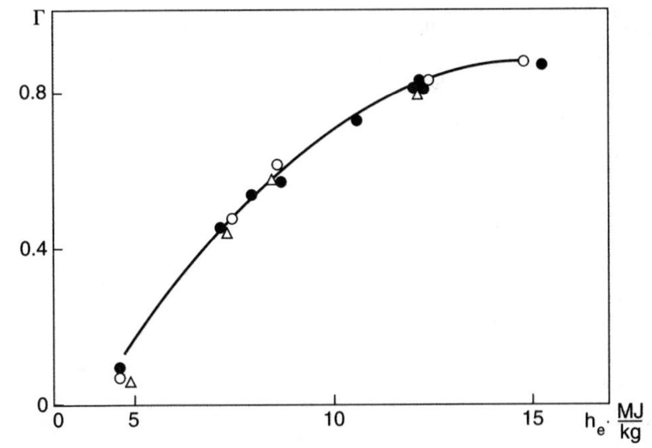

Figure 3. The share of gasification as a function of stagnation enthalpy of incoming gas h_e.

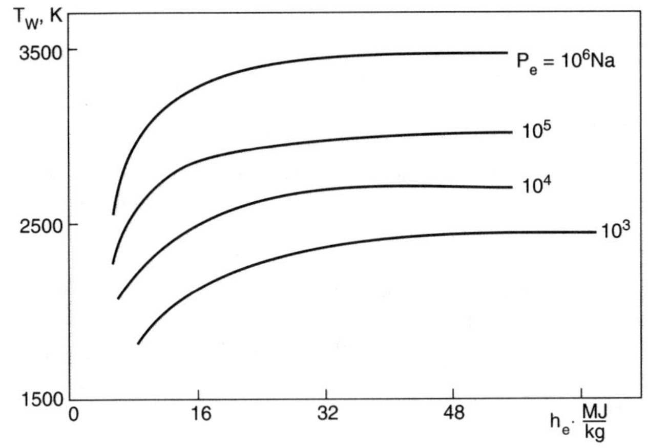

Figure 4.

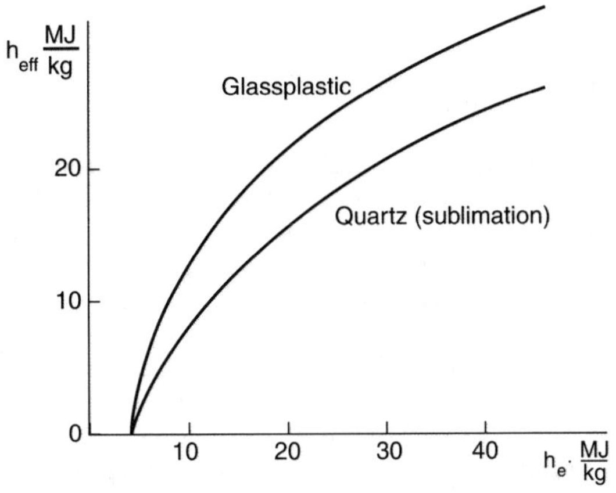

Figure 5.

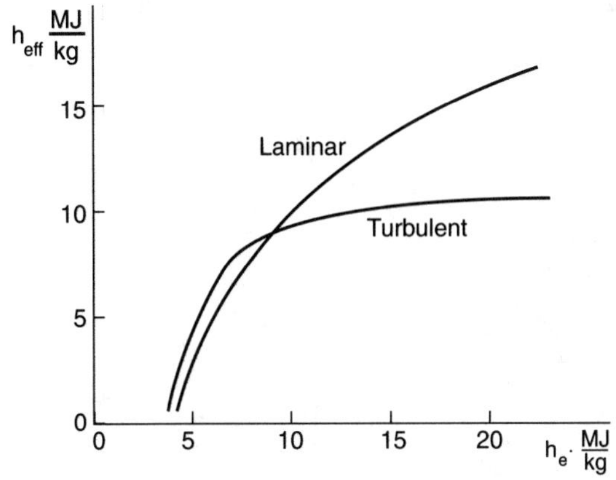

Figure 6.

relative to the equilibrium value of the temperature for a heat-insulated wall. In this case, the effective enthalpy becomes negative and the notion of h_{eff} loses practical sense. The dimensional rate of destruction is often used as an alternative parameter for generalizing the experimental and the design data:

$$\bar{\dot{m}}_\Sigma = \frac{\dot{m}_\Sigma}{(\alpha/c_p)_0} = \frac{(h_e - h_w)[1 - (\epsilon\sigma T_w^4)/\dot{q}_0]}{h_{eff}}.$$

Its advantage is that the function $\bar{\dot{m}}_\Sigma(h_e)$ is always positive and besides, the temperature of the destructing surface T_w and the degree of blackness ϵ is not warranted. Typical dependences of $\bar{\dot{m}}_\Sigma$ on the stagnation enthalpy h_e for Teflon, glass-reinforced plastic and graphite breaking down in air flow are shown in **Figure 7**.

Combined radiation-convective heating of the surface of an ablative material can considerably change the mechanism of its destruction. The injection of gaseous disintegration products in cases where they do not possess high absorption coefficients, slightly reduces the intensity of the radiation component $\dot{q}_R$ of the heat flow. As the $\dot{q}_r/\dot{q}_0$ ratio grows, the mechanism of destruction of the majority of ablative materials more closely resembles sublimation and thermal decomposition. This is due to a rapid decrease in the contribution of convective and diffusion transfer in the boundary layer while injecting gaseous products, to the ceasing of melt film flow and to the absence of burning on the destructing surface.

The heat balance on the surface of an ablative material in case of high levels of radiation of heat flows $\dot{q}_R$ is simplified as follows:

$$\dot{q}_r K_{\alpha,w} = \dot{m}_w \Delta Q_w + \dot{m}_\Sigma c(T_w - T_o) + \epsilon\sigma T_w^4.$$

Here, $K_{\alpha,w}$ is the absorption coefficient, which depends on the spectrum of incident radiation heat flow $\dot{q}_{R,\lambda}(\lambda)$ and on the spectral distribution of the destructing surface blackness $\epsilon_\lambda(\lambda)$:

$$K_{\alpha,w} = \int_0^\infty \epsilon_\lambda(\lambda)\dot{q}_{R,\lambda}(\lambda)d\lambda \Big/ \int_0^\infty \dot{q}_{R,\lambda}(\lambda)d\lambda$$

$$= \int_0^\infty \epsilon_\lambda(\lambda)\dot{q}_{R,\lambda}(\lambda)/\dot{q}_R d\lambda.$$

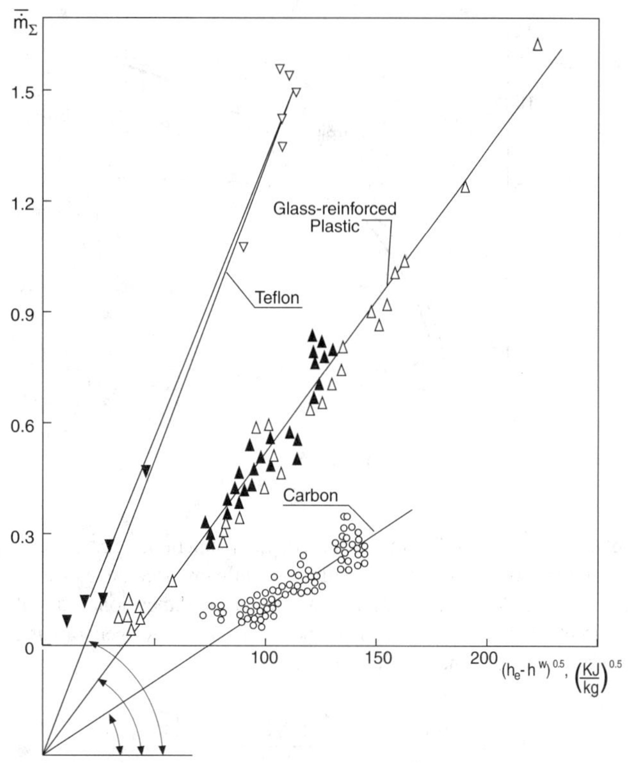

Figure 7. Dimensionless destruction rate ($\bar{\dot m}_\Sigma$) as a function of stagnation enthalpy (h_e) for various materials breaking down in an air flow.

When no mechanical cracking or melting of a heat-shielding material occurs, the total rate of ablation $\dot m_\Sigma$ coincides with $\dot m_w$ and the notion of effective enthalpy of the material under intensive radiation heat influence can be introduced as:

$$h_R = \frac{q_R - (\epsilon/K_{\alpha,w})\sigma T_w^4}{\dot m_\Sigma K\alpha,\, w} = \frac{\Delta Q_w + c(T_w - T_0)}{K_{\alpha,w}} = \frac{h}{K_{\alpha,w}}.$$

Table 1 shows the results of the evaluation of parameters h, $K_{\alpha,w}$ (in the $0.2 < \lambda < 1$ μm spectral range) and h_R for various substances.

An analysis of the data presented in Table 1 allows us to reach a paradoxical conclusion: under the influence of intensive radiation, the effective enthalpies of destruction of graphite and Teflon become about equal. We should note that the ablation rate of graphite, as compared to magnesium oxide, does not differ so strongly as the other values of the effective enthalpies given in the table. This is associated with the fact that the temperature of graphite destruction is almost half as great, and, therefore, the levels of the reemitted energy $\epsilon\sigma T_w^4$ differ by an order of magnitude. Nonetheless, the main conclusion that can be drawn in analysing **Table 1** is that by decreasing the absorption coefficient of the destructing surface ($K_{\alpha,w}$), we can obtain a greater efficiency of ablation than by increasing the heat of sublimation.

Table 1.

Material	h, kJ/kg	$K_{\alpha,w}$	h_R, kJ/kg
Graphite	30.000	0.85	35.000
Quartz	15.000	0.2	75.000
Magnesium oxide	15.000	0.13	115.000
Teflon	3.000	0.1	30.000

Yu.V. Polezhaev

ABSOLUTE PRESSURE (see Pressure)

ABSOLUTE TEMPERATURE (see Temperature)

ABSORPTION EFFICIENCY (see Optical particle characterisation)

ABSORPTIVITY (see Radiative heat transfer)

ACCELERATION PRESSURE GRADIENT (see Multiphase flow)

ACCIDENTS TO CHEMICAL PLANT (see Plumes)

ACCOMMODATION COEFFICIENT

Following from: *Boiling; Sublimation*

Accommodation coefficient is a physical quantity characterizing the behaviour of gas or vapour particles in their collisions with a solid or liquid body surface. The value of the accommodation coefficient depends on the surface nature and state as well as on the composition and pressure of the gas mixture in the environment and on other parameters.

One of the most commonly used models of the interaction of gas molecules with the surface is the Maxwell model. It implies that the fraction of the molecules, which after their impingement on the surface, are diffuse-reflected, is equal to α, and the fraction of the molecules which are specularly reflected is $1 - \alpha$. It appears here that α can also be interpreted as a loss coefficient of the tangential component of the momentum brought by the molecules' incident on the wall.

The *energy accommodation coefficient* is introduced analogously as:

$$\alpha_E = \frac{E_i - E_r}{E_i - E_w}$$

where E_i is the energy of the incident molecules, E_r is the energy carried away by reflected molecules, E_w is the energy which would be carried away by all reflected molecules if the gas has had time to come to thermal equilibrium with the wall. Assuming that the diffuse-reflected molecules reach the wall temperature, we have $\alpha = \alpha_E$ according to the *Maxwell model.* However, in reality the momentum and energy accommodation processes proceed differently. When multiatomic molecules collide with the surface, account must be taken of the possibility of a change in their internal energy.

If evaporation or vapour condensation occurs on the body surface, then evaporation and condensation coefficients are introduced to describe the transfer processes. The *condensation coefficient* is

Table 1. Accommodation coefficients α

Substance	α	Temperature range, K
Beryllium	1	1170–1550
Copper	1	1180–1460
Iron	1	1320–1870
Molybdenum	1	2070–2500
Nickel	1	1320–1600
Titanum	0.5–1.0	1650–1810
Tungsten	1	2520–3300
Carbon *C*	0.4	2670
Carbon C_2	0.3	2670
Carbon C_3	0.1	2670
Carbon C_5	10^{-3}	2670
Water (ice)	0.5–1.0	214–232
Water (ice)	0.94 ± 0.06	188–213
Phosphorus (red)	10^{-9}–10^{-7}	580–750
Iodine	0.055–0.208	310–340
Benzol	0.9	280
Chloroform	0.16	275
Camphor (synthetic)	0.139	260
Methyl alcohol	0.045	270
Naphtalene	0.135	310–340

defined as a share of the flow of molecules, which condensed on the surface, in the total flow of the particles which have impinged on it. The *evaporation coefficient* (which is smaller than unity) is a correction factor in the relation for maximum evaporation rate in the vacuum that accounts for various factors lowering the evaporation rate. In applied calculations, the values of these coefficients are assumed to be equal to each other, although this assumption holds strictly only under equilibrium conditions.

Accommodation coefficient values for some materials are presented in **Table 1**. It should be noted, however, that these data are related to pure substances. The presence of admixtures on the evaporation surface or of chemical reactions between the vapour and the components of the external gas flow, proceeding simultaneously with the evaporation, can significantly affect the accommodation coefficient.

It has been experimentally that in almost every case when α much less 1, the vapour molecules differ from the condensate molecules due to association, dissociation or polymerization. A temperature dependence of α has not been observed with experimental accuracy. The majority of experimental studies measure α in the vacuum, using the Knudsen-Langmiur equation (see **Sublimation**).

Yu.V. Polezhaev and N.V. Pavlyukevich

ACENTRIC FACTOR

The Acentric Factor was first proposed by Pitzer as a measure of the amount by which the thermodynamic properties of a particular substance differs from those predicted by the **Principle of Corresponding States**. This principle strictly applies only to a fluid (liquid or gas) comprised of spherical molecules. Fluids containing nonspherical molecules, or those with polar groups, show systematic deviations in their thermodynamic properties from their spherical counterparts. It is these deviations which are correlated with the acentric factor.

The acentric factor is defined as:

$$\omega = -1.000 - \log_{10}(P_\sigma/P^c)$$

where P^c is the critical pressure and P_σ is the vapour pressure at temperature T where $T/T^c = 0.7$ and T^c is the critical temperature.

For spherical molecules, ω is almost exactly zero. Nonspherical molecules have values above zero, but only the most severely nonspherical have values which approach unity.

If ω is small (say, less than 0.2), the departures from corresponding states are approximately linear in ω and, for example, the **Compressibility Factor** Z can be written as:

$$Z(P_r, T_r) = Z_0(P_r, T_r) + \omega Z_1(P_r, T_r)$$

where $P_r = P/P^c$, $T_r = T/T^c$, Z_0 is the compressibility factor of a substance comprising spherical molecules and Z_1 is the correction for nonsphericity. Graphs of the functions Z_0 and Z_1, and their equivalents for enthalpy and entropy, are given in many textbooks (see, for example, Bett, Rowlinson and Saville, 1975).

Reference

Bett, Rowlinson and Saville (1975) Thermodynamics for Chemical Engineers, Athlone Press.

G. Saville

ACETIC ACID (see Pyrolysis)

ACETONE (see Pyrolysis)

ACETYLENE COMBUSTION (see Flames)

ACHE'S (see Air cooled heat exchangers)

ACID RAIN (see Air pollution; Water)

ACID VIOLET 19, MONOMETHYL (see Photochromic dye tracing)

ACKERMANN CORRECTION FACTOR (see Condensation of multicomponent vapours; Wet bulb temperature)

ACOUSTIC FLOWMETERS (see Flow metering)

ACOUSTIC INSTABILITIES (see Instability, two-phase)

ACOUSTIC VIBRATION (see Vibration in heat exchangers)

ACOUSTIC WAVES (see Waves in fluids)

ACRYLIC CATION RESINS, ACRs (see Ion exchange)

ACTIVITY COEFFICIENT

The set of activity coefficients of the components in a fluid (gas or liquid) mixture is a measure of departure of the thermodynamic properties of that mixture from those of the ideal mixture.

An *ideal mixture* is defined as one for which:

$$\mu_i(P, T, \tilde{x}) = \mu_i^\circ(P, T) + \tilde{R}T \ln \tilde{x}_i \quad (1)$$

for each component i in the mixture. (μ_i is the **Chemical Potential** of component i in the mixture of mole fraction composition $\tilde{x}$, μ_i° is the chemical potential of pure component i and $\tilde{R}$ is the universal gas constant).

But since real mixtures differ from the ideal, the above equation is modified to give the real mixture equation:

$$\mu_i(P, T, \tilde{x}) = \mu_i^\circ(P, T) + \tilde{R}T \ln(\tilde{x}_i \gamma_i) \quad (2)$$

where γ_i is the **activity coefficient.** This equation can be regarded as the defining equation for activity coefficient. There will be an activity coefficient for each component in the mixture and, although it is not explicitly stated in the above equation, each of the activity coefficients will be a function of P, T and $\tilde{x}$. A necessary condition on γ_i is that

$$\gamma_i = 1 \quad \text{at} \quad \tilde{x}_i = 1$$

Although activity coefficients may be used for both liquid and gas, conventional use restricts activity coefficients to liquid phases and the alternative measure of deviations from ideality—**Fugacity**—for the gaseous phase.

Activity coefficients in liquid phases are usually measured by establishing a gaseous phase in thermodynamic and phase equilibrium with the liquid phase and then making use of the fact that under these conditions, the chemical potential of each individual component is the same in both the liquid and gaseous phases. The chemical potential of each component in the gaseous phase can be calculated from a knowledge of the equation of state of the gaseous mixture (or equivalently, from a knowledge of the virial *coefficients* if the pressure is not too high). For a binary mixture, this procedure leads to the final equation:

$$\tilde{R}T \ln(\tilde{x}_1\gamma_1 P_1^\circ/\tilde{y}_1 P) = (P - P_1^\circ)(B_{11} - \tilde{v}_1^\circ) + 2P\tilde{y}_2^2\delta B_{12} \quad (3)$$

and conversely for component 2, by interchanging indices. In this equation, $\tilde{x}$ and $\tilde{y}$ are the liquid and gas mole fractions, respectively; P° are the vapour pressures of the pure components at temperature T; $\tilde{v}_1^\circ$ is the molar volume of pure liquid 1 at temperature T; B are the second viral coefficients; and δB_{12} is defined as:

$$\delta B_{12} = B_{12} - \tfrac{1}{2}B_{11} - \tfrac{1}{2}B_{22} \quad (4)$$

Activity coefficients calculated from experimental measurements in this, or other, ways are usually correlated in the form of mathematical equations. Frequently encountered (expressed in the form appropriate for a binary mixture) are:

Margules':

$$\ln \gamma_1 = b\tilde{x}_2^2 + c\tilde{x}_2^3 \quad (5)$$

$$\ln \gamma_2 = (b + 3c/2)\tilde{x}_1^2 - c\tilde{x}_1^3 \quad (6)$$

and

van Laar's:

$$\ln \gamma_1 = A(1 + A\tilde{x}_1/B\tilde{x}_2)^{-2} \quad (7)$$

$$\ln \gamma_2 = B(1 + B\tilde{x}_2/A\tilde{x}_1)^{-2} \quad (8)$$

The adjustable parameters B and C in the first case, A and B in the second, are usually determined from a single set of measurements at either a fixed temperature or at a fixed pressure, but covering the whole binary composition range. However, their usage at other temperatures or pressures, or in situations in which more than the two components are present serious is problematical and may lead to serious errors. Various attempts have been made to overcome these difficulties, such as by making the parameters temperature-dependent, but making them composition dependent is essentially infeasible. Published values of activity coefficient correlations should, therefore, not be used in critical situations without first assessing their accuracy in a particular situation.

G. Saville

ADAPTIVE GRIDS (see Computational fluid dynamics)

ADDED MASS

Following from: *Flow of fluids*

Added mass is the additional mass that an object appears to have when it is accelerated relative to a surrounding fluid. For example, it adds to the effective inertia of boats, buoys, swimmers' limbs, airplanes and bubbles.

While added mass effects occur in all real fluids, they are most clearly defined in ideal irrotational incompressible inviscid flow that has a "potential", ϕ. For a nonrotating object of volume V moving at velocity **U** in such a fluid which is at rest at infinity, the impulse is:

$$\int \phi \mathbf{ds} = \underline{\underline{C}} \cdot \mathbf{U}V \quad (1)$$

and the kinetic energy of the surrounding fluid of density ρ is:

$$T = \frac{1}{2}\rho\mathbf{U} \cdot \underline{\underline{C}} \cdot \mathbf{U}V \quad (2)$$

where $\underline{\underline{C}}$ is a symmetric tensor, called the "coefficient of added mass," that depends on the shape of the object. If referred to principal axes, the tensor is diagonal with components C_{xx}, C_{yy}

and C_{zz}. Typical values for ellipsoids of revolution about the x-axis and with ratio of x- to y- axes of a/b are (Lamb, 1932):

a/b	C_{xx}	C_{yy}, C_{zz}
0.1	6.184	0.075
0.2	3.008	0.143
0.5	1.115	0.310
0.	0.651	0.434
1	0.5	0.5
1.5	0.304	0.622
2	0.210	0.704
4	0.082	0.860
8	0.029	0.945

In order to accelerate the object, both it and the surrounding fluid must be set in motion. If the volume and the added mass coefficient are constant, the required force is:

$$\mathbf{F} = V(\rho_0 \underline{\underline{I}} + \rho \underline{\underline{C}}) \cdot \frac{d\mathbf{U}}{dt} \quad \textbf{(3)}$$

where ρ_0 is the density of the object and $\underline{\underline{I}}$ is the unit tensor or Kronecker delta.

In a more general motion, the volume and added mass coefficient may change. For example, a cavitation bubble that is collapsing will accelerate.

If the object is moving steadily in a direction that is not parallel to a principal axis, it will experience a torque (Lamb, 1932; Wallis, 1994) from the fluid equal to:

$$M = \rho \mathbf{U} \times \underline{\underline{C}} \cdot \mathbf{U} \quad \textbf{(4)}$$

If it is moving steadily parallel to a principal axis, say the x-direction, and is not subject to internal (balancing) forces, the mean bulk stress (pressure) in the object is:

$$p = p_\infty - \frac{1}{2} \rho U^2 C_{xx} \quad \textbf{(5)}$$

where p_∞ is the pressure at infinity. For a more general steady motion, the mean bulk stress is:

$$p = p_\infty - \frac{1}{2} \rho \mathbf{U} \cdot \underline{\underline{C}} \cdot \mathbf{U} \quad \textbf{(6)}$$

as long as the restraining torque is applied in a way that does not contribute to the bulk stress (e.g., consisting of two equal and opposite forces acting at points joined by a line perpendicular to the forces).

The added mass coefficient is related to the *polarization*, **D** or net dipole moment of internal sources or dipoles that could represent the object moving at velocity **U** in a potential flow, by:

$$\mathbf{D}/V = \underline{\underline{D}} \cdot \mathbf{U} = (\underline{\underline{I}} + \underline{\underline{C}}) \cdot \mathbf{U} \quad \textbf{(7)}$$

where $\underline{\underline{D}}$ is a tensor "polarizability" (Wallis, 1993, 1994).

Let the object be at rest in a potential flow with velocity **U** that varies slightly so that changes in **U** are small over the scale of the object. Then the force on an internal source of strength m at location r is $-\rho m(\mathbf{U} + \mathbf{r} \cdot \nabla \mathbf{U})$ and the net force on the object is approximately, by summation, since $\Sigma\, m = 0$,

$$F = -\rho \mathbf{D} \cdot \nabla \mathbf{U} = \rho V \mathbf{U} \cdot \underline{\underline{D}} \cdot \nabla \mathbf{U} \quad \textbf{(8)}$$

$$= \rho(\mathbf{U} \cdot \nabla \mathbf{U} + \mathbf{U} \cdot \underline{\underline{C}} \cdot \nabla \mathbf{U})V \quad \textbf{(9)}$$

The first component is due to the overall pressure gradient while the second is the "added mass force" deduced by Taylor (1928). Effects due to the change in $\underline{\underline{D}}$ because of $\nabla \mathbf{U}$ are of second order.

Where there is circulation in the flow, the object cannot be represented by flux sources alone. This may introduce a "lift" force which Auton (1987) and Drew and Lahey (1987) show, for a sphere, to be:

$$\frac{1}{2} \rho V(\mathbf{U} - \mathbf{U}_s) \times \text{curl } \mathbf{U} \quad \textbf{(10)}$$

where $\mathbf{U}_s$ is the velocity of the sphere, which is small compared with lengths over which the velocity varies in the fluid.

Similar results may be derived for more complex objects, such as a porous sphere made up of an array of much smaller spheres (Wallis, 1991).

Added mass effects also occur in two-phase continua. For example, in one-dimensional uniform acceleration of a dispersed array in the x-direction, the equations of motion are (Wallis, 1994a; Zhang and Prosperetti, 1994):

$$\rho_1 \dot{U}_1 + \rho_1 C \alpha_2 (\dot{U}_1 - \dot{U}_2) = \frac{-dp}{dx} + \rho_1 g_1 \quad \textbf{(11)}$$

$$\rho_2 \dot{U}_2 - \rho_1 C \alpha_1 (\dot{U}_1 - \dot{U}_2) = \frac{-dp}{dx} + \rho_2 g_2 + f_2 \quad \textbf{(12)}$$

where α_1 and α_2 are volume fractions of the phases; p is the macroscopic pressure; g_1, and g_2 are body force fields; and f_2 is an external force per unit volume of the dispersed phase, 2, acting on phase 2. C is the component of an "inertial coupling" coefficient in the direction of motion that is closely approximated by 1/2 for isotropic arrays of spheres. The combination $C\alpha_2$ is Wallis' (1989) "exertia" while $C\alpha_1\alpha_2$ is Geurst's (1985) "added mass coefficient".

Similar equations are valid for three-dimensional motion in terms of vector velocities and forces, the pressure gradient, and a tensor form of C. More general forms have been proposed when there are gradients in velocities and volume fractions [e.g., Zhang and Prosperetti, (1994); Drew (1992)].

While C could be called an "added mass" coefficient, other definitions are possible. For example, if motion is due entirely to $f_2(g_1 = g_2 = dp/dx = 0)$, the first equation becomes:

$$\alpha_1 \dot{U}_1 = \alpha_2 \dot{U}_2 \left(\frac{C\alpha_1}{C\alpha_2 + 1} \right) \quad \textbf{(13)}$$

and the factor in parentheses is Wallis' (1989, 1994a) "added mass coefficient", C_W, that, in the same sense of the "drift" discussed by Darwin (1952), represents the ratio of the volumetric flux of fluid to the volumetric flux of particles induced by motion starting from rest. The second equation of motion becomes:

$$\dot{U}_2(\rho_2 + \rho_1 C_W) = f_2 \quad \textbf{(14)}$$

resembling the equivalent form for a single particle.

In an alternative frame of reference in which there is no net flux, or "bulk velocity":

$$\alpha_1 U_1 + \alpha_2 U_2 = 0 \quad \textbf{(15)}$$

the equations of motion, without body forces, may be combined to give:

$$\dot{U}_2\left[\rho_2 + \rho_1\left(\frac{C + \alpha_2}{\alpha_1}\right)\right] = f_2 \quad \textbf{(16)}$$

which resembles the single-phase case with an "added mass coefficient" equal to $(C + \alpha_2)/\alpha_1$, as used by Zuber (1964).

Geurst (1985, 1986) derived general equations of motion for two-phase inviscid flow based on the kinetic energy contributed by relative motion that was proportional to his "added mass coefficient", m. His results are valid if m is isotropic and depends only on the volume fraction of the dispersed phase. His equations are "ill-posed" and it is likely that a more realistic model should include a representation of changes in the structure of the dispersion in response to overall strain (Wallis, 1994b).

References

Auton, T. R. (1987) The Lift Force on a Spherical Body in a Rotational Flow, *J. Fluid Mech.*, **183**, 199–218.

Darwin, C. (1952) Note on Hydrodynamics, *Proc. Camb. Phil. Soc.*, **49**, 342.

Drew, D. A. (1992) Analytical Modeling of Multiphase Flows, in Boiling Heat Transfer, R. T. Lahey, Jr. (ed) Elsevier Science Publishers B.V.

Drew, D. A. and Lahey, R. T., Jr. (1987) The Virtual Mass and Lift Force on a Sphere in Rotating and Straining Inviscid Flow, *Int. J. Multiphase Flow*, **13**, 113–121.

Geurst, J. A. (1985) Virtual Mass in Two-Phase Bubbly Flow, *Physica*, **129A**, 233–261.

Geurst, J. A. (1986) Variational Principles and Two-Fluid Hydrodynamics of Bubbly Liquid/Gas Mixtures, *Physica*, **135A**, 455–486.

Lamb, H. (1932) *Hydrodynamics*, 6th ed., Cambridge University Press.

Taylor, G. I. (1928a) The Forces on a Body Placed in a Curved or Converging Stream of Fluid. *Proc. Roy. Soc.*, **A120**, 260–283.

Taylor, G. I. (1928b) The Energy of a Body Moving in an Infinite Fluid with an Application to Airships. *Proc. Roy. Soc.*, **A120**, 13.

Wallis, G. B. (1989) Inertial Coupling in Two-Phase Flow: Macroscopic Properties of Suspensions in an Inviscid Fluid, *Multiphase Science and Technology*, **5**, 239–361.

Wallis, G. B. (1991) The Averaged Bernoulli Equation and Macroscopic Equations of Motion for the Potential Flow of a Dispersion, *Int. J. Multiphase Flow*, **17**(6), 683–695.

Wallis, G. B. (1993) The Concept of Polarization in Dispersed Two-Phase Potential Flow, *Nuclear Engineering and Design*, **141**, 329–342.

Wallis, G. B. (1994a) Added Mass Coefficients for Uniform Arrays, *Int. J. Multiphase Flow*, **20**(4), 799–803.

Wallis, G. B. (1994b) The Particle Pressure of Arrays in a Potential Flow, *Nuclear Engineering and Design*, **151**, 1–14.

Zhang, D. Z. and Prosperetti, A. (1994) Averaged Equations for Inviscid Disperse Two-Phase Flow, *J. Fluid Mech*, **267**, 185–219.

Zuber, N. (1964) On the Dispersed Two-Phase Flow in the Laminar Flow Regime. *Chem. Eng. Sci.*, **19**, 897–917.

G.B. Wallis

ADDITIVES (see Augmentation of heat transfer, single phase)

ADENOSINE TRIPHOSPHATE, ATP (see Physiology and heat transfer)

ADHESION BETWEEN LIQUIDS (see Interfaces)

ADIABATIC CONDITIONS

Following from: *Thermodynamics*

Adiabatic conditions refer to conditions under which overall heat transfer across the boundary between the thermodynamic system and the surroundings is absent. Examples of processes proceeding under adiabatic conditions and applied in engineering are expansion and compression of gas in a piston-type machine, the flow of a fluid medium in heat-insulated pipes, channels and nozzles, throttling and setting of turbomachines and distribution of acoustic and shock waves. The flow of a viscous fluid or gas through a heat-insulated channel is often referred to as adiabatic.

A reversible adiabatic process in the thermodynamic system (see **Thermodynamics**) working only in expansion is described by the differential equation:

$$dq = du + pdv = T\,ds = 0, \quad \textbf{(1)}$$

i.e., the process is also isentropic ($ds = 0$, $s = const$). Equation 1 implies the following relations:

$$\gamma = \left(\frac{\partial h}{\partial u}\right)_s = \frac{v}{p}\left(\frac{\partial p}{\partial v}\right)_s = -\left(\frac{\partial \ln p}{\partial \ln v}\right)_s = \frac{c_p}{c_v}\left(\frac{\partial \ln p}{\partial \ln v}\right)_T, \quad \textbf{(2)}$$

where $h = u + pv$ is the enthalpy, c_p and c_v are the heat capacities at a constant pressure and volume, respectively. γ is referred to as an *isentropic exponent* (or *adiabatic exponent*, which is less strict). If $\gamma = const$ the system states are described by an adiabatic (Poisson) equation

$$pv^\gamma = const, \qquad v = 1/\rho. \quad \textbf{(3)}$$

Equation 3 is used for irreversible adiabatic processes too. In the case of gas expansion, an actual adiabatic exponent γ' is within $1 < \gamma' < \gamma$ and in the case of gas compression, $\gamma' > \gamma$. For solids and liquids, the values of γ are extremely high and change appreciably with the temperature. For instance, for water at 0°C $\gamma =$

3602000 and at 100°C $\gamma = 22300$. For perfect gases, $(\partial \ln p/\partial \ln v)_T = -1$, such that $\gamma = c_p/c_v$. For a monatomic gas, $\gamma \cong 1.67$; for diatomic gas, 1.40; for triatomic and polyatomic gases, 1.29. The variation of γ with temperature and pressure for gases is small except near the saturation curve and when the gas is in a dissociated and ionized state.

For constant γ, the work for system expansion in an adiabatic process is given by:

$$w_{1-2} = u_1 - u_2$$

$$= \frac{p_1 v_1}{\gamma - 1}\left[1 - \left(\frac{v_1}{v_2}\right)^{\gamma-1}\right] \tag{4}$$

$$= \frac{p_1 v_1}{\gamma - 1}\left[1 - \left(\frac{p_2}{p_1}\right)^{(\gamma-1)/\gamma}\right]$$

and

$$w_{1-2} = \frac{R(T_1 - T_2)}{\gamma - 1} = \frac{p_1 v_1}{\gamma - 1}\left(1 - \frac{T_2}{T_1}\right) = \frac{p_1 v_1 - p_2 v_2}{\gamma - 1} \tag{5}$$

for a perfect gas.

Useful work, that is the difference between the work of expansion and the work of pushing the working medium, is given by:

$$w'_{1-2} = -\int_{p1}^{p2} v\, dp - w_f = h_1 - h_2, \tag{6}$$

where w_f is the friction work ($w_f \geq 0$).

Adiabatic throttling is the adiabatic limit irreversible process for reducing the fluid pressure. Under adiabatic throttling, $h =$ *const*. The variation of the system thermal state in adiabatic throttling is described by the equation:

$$(\partial T/\partial p)_h = \mu_{jT} = \left[T\left(\frac{\partial v}{\partial T}\right)_p - v\right]/c_p, \tag{7}$$

where μ_{Jt} is the differential throttling (*Joule-Thomson*) coefficient. Variation of temperature in throttling is:

$$T_2 - T_1 = \int_{p1}^{p2} \mu_{jT}\, dp. \tag{8}$$

For a perfect gas, $\mu_{JT} = 0$ and $T_2 = T_1$. For imperfect fluids, μ_{JT} may be negative or positive, and may change from negative to positive depending on p and T. The locus of this change in sign of μ_{TP} forms, the $p - T$ diagram, an "inversion curve" of a given substance. The curves of adiabatic gas expansion and compression plotted in $p - v$ coordinates are shown in **Figure 1**.

An adiabatic flow of fluid in pipes and channels is described by the energy equation:

$$dh + u\, du + g\, dz + dw' = 0, \tag{9}$$

where u is the flow velocity, z the height above the ground level,

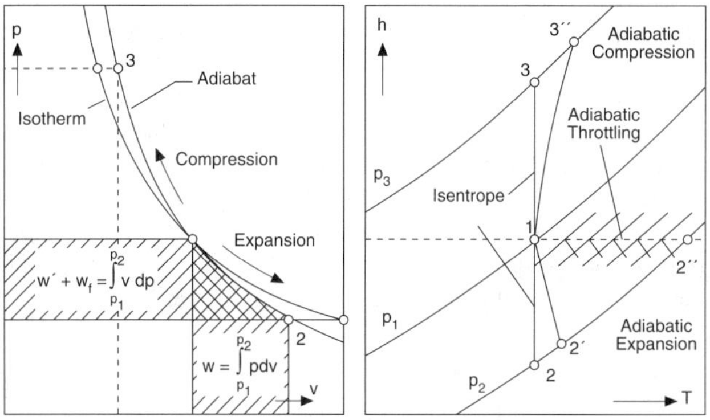

Figure 1.

and w' the useful work input or output. Equation 9 holds for both frictionless flows and flows with friction. In the particular case of a reversible adiabatic flow of incompressible fluid ($v =$ const, $dh = v\, dp$) and in the absence of useful work ($dw' = 0$), an integral of equation 3 has the form;

$$pv + \frac{u^2}{2} + gz = \text{const} \quad \text{or} \quad p + \frac{\rho u^2}{2} + \rho gz = \text{const} \tag{10}$$

and is known as **Bernoulli's Equation** or integral. $\rho u^2/2$ is the dynamic pressure and ρgz the hydrostatic pressure. The sum $p_0 = p + \rho u^2/2$ is called the *stagnation pressure*, p_0. The stagnation pressure can be measured by the **Pitot Tube** and the difference $p_0 - p$ by Pitot-Prandtl tube (which has a static reference tapping on the probe). For an adiabatic flow of compressible gas that performs no useful work (variation in gz is commonly neglected), from equation 9 we have:

$$dh + u\, du = 0, \qquad d\left(h + \frac{u^2}{2}\right) = 0,$$

$$h_0 = h + \frac{u^2}{2} = \text{const}, \tag{11}$$

or

$$u_2 = \sqrt{u_1^2 + 2(h_1 - h_2)}. \tag{12}$$

h_0 is called the stagnation enthalpy. Propagation of small disturbances in an elastic medium, i.e., acoustic waves, can be assumed adiabatic and isoentropic, which leads to the *Laplace equation for the velocity of sound*:

$$u_{sound} = \sqrt{(\partial p/\partial \rho)_s} = \sqrt{-v^2(\partial p/\partial v)_s} = \sqrt{\gamma p v}. \tag{13}$$

In the case of a perfect gas $u_{sound} = \sqrt{\gamma R_i T}$ where R_i in the gas constant for the specific substance. The finite **Velocity of Sound** is responsible for the critical regime of gas flow in the channel under which the local sound velocity and the maximum flow rate of gas (for given initial parameters) are achieved in the channel throat.

The increase in sound velocity with temperature causes large disturbances to propagate in gaseous media as shock waves travel, in relation to the undisturbed medium, with a supersonic velocity. This is because of the elevation in temperature in the compressed region. The shock wave is an irreversible adiabatic process of substance compression, accompanied by entropy growth. The variations of specific volume and pressure for perfect gas during the passage of the shock wave are related by the shock adiabatic equation (*Hugoniot adiabat*):

$$\frac{v_1}{v_2} = \frac{\dfrac{\gamma+1}{\gamma-1} + \dfrac{p_1}{p_2}}{1 + \dfrac{\gamma+1}{\gamma-1}\dfrac{p_1}{p_2}},$$

This equation makes it possible, using the Clapeyron equation, to determine the temperature variation.

There is much current interest in the adiabatic processes of high-velocity viscous fluid flow over an insulated wall. Friction brings about an elevation of wall temperature to a value at which there is a balance between heat generation and heat release to the external flow. This temperature is referred to as the **Adiabatic Wall Temperature** (see separate item in this topic) and characterizes an aerodynamic heating of the surface in the flow.

V.A. Kurganov

ADIABATIC DISC TEMPERATURE (see Rotating disc systems, basic phenomena)

ADIABATIC EXPONENT (see Adiabatic conditions)

ADIABATIC PROCESSES (see Carnot cycle)

ADIABATIC SATURATION TEMPERATURE (see Lewis relationship; Wet bulb temperature)

ADIABATIC THROTTLING (see Adiabatic conditions)

ADIABATIC WALL TEMPERATURE

***Following from:** Heat transfer*

Adiabatic wall temperature is the temperature acquired by a wall in liquid or gas flow if the condition of thermal insulation is observed on it: $(\partial T/\partial n)_w = 0$ or $\dot{q}_w = 0$. It is denoted as either T_r or T_{eq}, and is sometimes called an *equilibrium temperature* and, in aerodynamics, a *recovery temperature*. A distinction between the adiabatic wall temperature and a characteristic flow temperature may depend on the dissipative heat release in the boundary layer, on the existence of a different nature, in the flow of internal heat sources and on the thermal effect of other bodies (walls). In this case, if there is heat transfer between the wall and the flow, i.e., at $\dot{q}_w \neq 0$, the temperature field in the fluid can be represented as a superposition of the temperature field at $\dot{q}_w = 0$ on the natural field, which would be produced by the wall in the absence of disturbing factors, i.e., internal heat release or the effect of other walls. This superposition makes it possible to represent the law of heat transfer between the flow and the wall as:

$$\dot{q}_w = \alpha_0(T_w - T_r), \quad \textbf{(1)}$$

where T_r is the adiabatic wall temperature for given conditions of the problem and α_0 the heat transfer coefficient under undisturbed conditions.

The concept of adiabatic wall temperature is used in the field of high velocity aerodynamics. The temperature profile in the boundary layer of a high-velocity gas flow over an adiabatic surface is displayed in **Figure 1** (Curve 1).

Curve 2 in **Figure 1** shows a typical distribution of temperature for the case where heat is being added through the surface and Curves 3 and 4 are typical of cases where heat is being removed for the fluid via the surface. It is obvious that $T_r > T_\infty$, i.e., an adiabatic wall is heated in relation to the thermodynamic temperature of the flow or, in other words, aerodynamic heating of the surface immersed in the flow occurs. The adiabatic wall temperature is determined by the relation:

$$T_r = T_\infty + \tau\frac{u_\infty^2}{2c_p}, \quad \textbf{(2)}$$

where T_∞ is the thermodynamic temperature, u_∞ the velocity of the external flow and c_p the isobaric heat capacity of fluid. For a compressible gas, the equation is:

$$T_r = T_\infty\left(1 + r\frac{\gamma-1}{2}Ma_\infty^2\right), \quad \textbf{(3)}$$

where γ is the isentropic exponent, Ma_∞ the Mach number in the external flow and r is the *recovery coefficient*. r is not a constant but depends, in particular, on the character of the flow on the

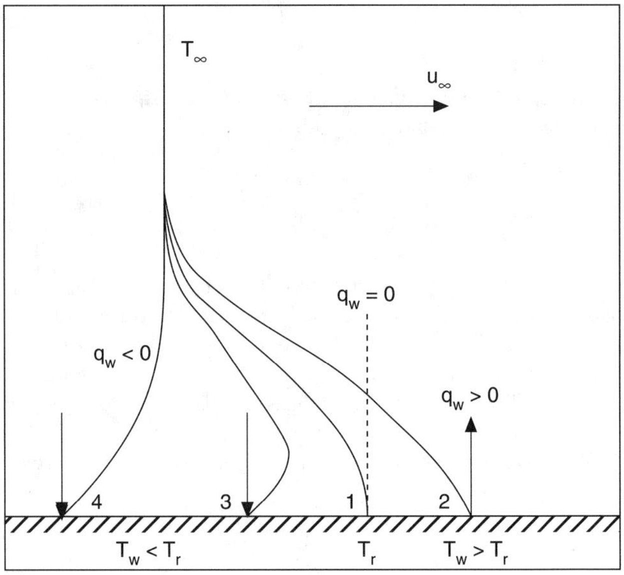

Figure 1.

surface, the flow regime, and the thermal properties of the medium. For some simple cases, its value can be estimated as follows: At the front stagnation point of bodies in the flow, r = 1; in a laminar boundary layer on a plane plate, $r \cong \sqrt{Pr}$ for Prandtl numbers 0.5 $< Pr < 10$; in a turbulent boundary layer on a plate, $r = \sqrt[3]{Pr}$ for Prandtl numbers close to 1. These estimates, with small corrections, also hold for gases flowing in pipes. For supersonic flows when variations of the thermal properties of gas become significant, the above relations hold for the enthalpy field

$$h_r = h_\infty + r\frac{u_\infty^2}{2}. \quad \textbf{(4)}$$

In estimating r, the Prandtl number is chosen at the reference *Eckert enthalpy* $h_* = h_\infty + 0.5(h_w - h_\infty) + 0.22\,(h_r - h_\infty)$.

References

Dorrance, W. H. (1962) Viscous Hypersonic Flow. Theory of Reacting and Hypersonic Boundary Layers, McGraw-Hill, New York.

Eckert, E. R. G. and Drake, R. M. (1972) Analysis of Heat and Mass Transfer, McGraw-Hill, New York.

V.A. Kurganov

ADSORBATE (see Adsorption)

ADSORBENT (see Adsorption)

ADSORBERS (see Adsorption)

ADSORPTION

Following from: *Mass transfer*

Adsorption is the process by which molecules of a substance from a gas mixture or liquid solution became attached to a solid or liquid surface. The substance being absorbed is an *adsorbate* and the absorbing substance, an *adsorbent*.

Adsorption is originated by the surface forces acting on the solid-gas, solid-liquid, gas-liquid or (in the case of immiscible liquids in contact) the liquid-liquid interface.

Adsorption of gases by solids

In adsorption of gas by a solid surface, one never distinguishes between *physical adsorption* and *chemisorption*. In physical adsorption, the main factors are the nature of interacting substances, the forces of interaction on the interface, and the interaction between adsorbate molecules. Of paramount importance is the degree of surface inhomogeneity and the capability of adsorbed molecules to be in translational, rotary, and oscillatory motion. The surface structure and, hence, adsorption properties may change during adsorption. This is most clearly pronounced when the so-called adsorption energy is of the same order as the surface energy of the adsorbent, e.g., ice, paraffin, and polymers.

Equilibrium in physical adsorption is rapidly established and is reversible. It should be born in mind, however, that the actual adsorption rate can be restricted by the mass transfer rate in the gaseous phase or inside the pores of a porous adsorbent, i.e., by the adsorbate supply to the surface. It is believed that physical adsorption is brought about by the same intermolecular forces as vapor condensation. Hence, a conclusion is drawn that the heat of physical adsorption is close to that of condensation and the amount of physical adsorption is particularly high when the gas has a temperature below critical i.e., in a vapor.

Gas adsorption by a solid surface is expressed quantitatively as a gas volume adsorbed by the adsorbent per unit mass or unit surface of the adsorbent.

Chemisorption, in contrast to physical adsorption, involves chemical reactions on the surface, which may result in its rearrangement. Cases are also known where physical adsorption proceeds over a chemisorbed layer.

The parameter describing the physical adsorption dynamics is adsorption time, which is the mean lifetime of the molecule on the adsorbent surface. It is $t = t_0 \exp Q/RT$, where t_0 is the time of molecular vibration ($t_0 = 10^{-13} - 10^{-12}$ s) and Q is the energy of the molecule-surface interaction, i.e., the heat of adsorption. If thermal equilibrium is established between the molecules and the surface, then the molecules being desorbed leave the surface in the direction independent of the incidence direction, i.e. the accommodation coefficient is assumed to be unity).

An important characteristic of adsorption is also a molecule concentration on an adsorbent surface. It is measured in mol/cm^2 and calculated as $\Gamma = zt$, where z is the number moles of gas colliding with unit surface in a second.

At a given temperature T, adsorption can be classified according to the parameters shown in **Table 1**.

An isotherm, i.e., the dependence graphed as a function $V = f(p)$ where V is the volume adsorbed and p the pressure, is the most common representation of adsorption. In 1918 *Langmuir* derived theoretically an *adsorption isotherm* equation that became known as the Langmuir equation. He assumed that the adsorbent surface possesses a certain number of adsorption sites s, part of which are occupied by adsorbed molecules s_1 and the other part are vacant s_2. He considered adsorption-desorption as a process of condensation and evaporation. The rate of condensation was taken

Table 1. Regions of adsorption

Q, kcal/mol	t, s	Γ, adsorbate concentration, mol/cm$_2$	
0.1	10^{-13}	0	No adsorption, specular reflection of molecules, the accommodation coefficient is zero
1.5	10^{-12}	0	The region of physical adsorption, the accommodation coefficient is one
3.5	410^{-11}	10^{-12}	
9.0	410^{-7}	10^{-8}	
20.0	100		The region of chemisorption
40.0	10^{17}		

to be proportional to the number of free surface sites s_2, i.e. k_2s_2 while, the rate of evaporation proportional to the number of occupied surface sites, i.e., k_1s_1. When the equilibrium was attained $k_1s_1 = k_2ps_2 = k_2p(s - s_1)$ or $\theta = bp/(1 + b)$, where p is the partial pressure of an adsorbate, $\theta = s_1/s$ is the fraction of the surface occupied by adsorbed molecules and b is the Langmuir constant. Taking into account the fact that the evaporation rate is proportional to the saturation vapor pressure p_G, the Langmuir adsorption isotherm equation can be reduced to $\theta = (cx)/(1 + cx)$, where $c = \exp Q/RT$ and $x = (p/p_G)$ is the ratio of the vapor pressures of a given substance in the mixture to its saturation vapor pressure at the same temperature T. The heat of adsorption Q can be assumed to be equal to the extent heat of vaporisation per mole. R is the universal gas constant related to a mole of vapour.

The next step in the description of adsorption isotherm is the allowance for formation of multimolecular layer, which was realized in the *BET (Brunauer-Emmet-Teller)* equation.

$$\frac{V}{V_m} = \frac{cx}{(1 - x)[1 + (c - 1)x]}$$

where V and V_m are respectively the actually absorbed gas volume and its limit value in a monolayer. This equation has gained recognition as a most reliable adsorption isotherm equation and has been widely employed for determining the surface per unit volume from the experimental data. It can be transformed to the form

$$\frac{x}{V(1 - x)} = \frac{1}{cV_m} - \frac{(c - 1)x}{cV_m}$$

Plotting $x/V(x - 1)$ against x gives a straight line the intercept and slope of which is used to find the constants V_m and c.

The diversity of experimental adsorption isotherms is depicted in **Figure 1**. Type 1 isotherms correspond to the Langmuir equation and are characterized by a monotonic approach to a limit value conforming to a complete monolayer surface coverage by a monolayer of adsorbed molecules. Type 2 corresponds to the case when first, a monolayer is formed, and then (from the point B) a multimolecular layer predominantly develops. Isotherms of Type 3 are relatively uncommon and correspond to the formation of multimolecular layer when heat of adsorption is less than or equal to adsorbate heat of condensation. Types 4 and 5, as a rule, are characteristic of capillary adsorbate condensation in porous solids.

Adsorption in porous solids with pores substantially exceeding the free path length of molecules is of the same nature as it is on an open surface of a nonporous solid. In particular, at temperatures higher than the critical temperature, no multimolecular layers are formed. An adsorbent with molecular size pores behaves like a **Molecular Sieve**, and its apparent surface per unit volume depends on the size of adsorbed molecules: small molecules pass into pores and adsorb on their surfaces, while large ones are trapped and no adsorption occurs in pores.

Of great interest are *zeolites* composed of tetrahedral groups $(Al, Si)O_4$. They have large voids into which the molecules of adsorbed gas have to pass through much smaller holes. Thus, in chabazite $CaAl_2Si_4O_{12}$, there are six inlet holes about 4Å in diameter in the void 10Å in diameter. These holes accommodate only monatomic or diatomic gases, H_2O, and H-alkanes, but trap larger molecules. Changing the hole size makes it possible to separate gases.

At temperatures below critical, for multimolecular adsorption in small size pores not only is the number of layers restricted but also so-called capillary condensation may be observed. For capillary condensation to occur, the effective curvature of the meniscus in the micropores must be sufficient to entirely fill the pores with condensing liquid; however, since the curvature is variable even with a narrow range of pore size, the desorption and adsorption curves do not coincide leading to a hysteresis in adsorption (**Figure 2**).

Chemisorption gives rise to bonds between adsorbate and adsorbent approaching chemical bonds in strength. In this case, the chemical properties of adsorbate and adsorbent may differ from their initial ones. Chemisorption conspicuously shows up in formation of the first molecular layer because in this case, an enhanced heat of adsorption characteristic of chemisorption comes into play. Considerable variation may occur because of inhomogeneities in the surface or in the energy distribution of adsorption centers. The inhomogeneity may be attributed to surface defects and impurity inclusions, i.e., it depends on the material treatment and quality.

As the second and subsequent molecular layers are formed, chemisorption is reduced to physical adsorption. The distinguishing feature of chemisorption, in contrast to physical adsorption, is the fact that it can proceed with gases, whose temperature is above critical.

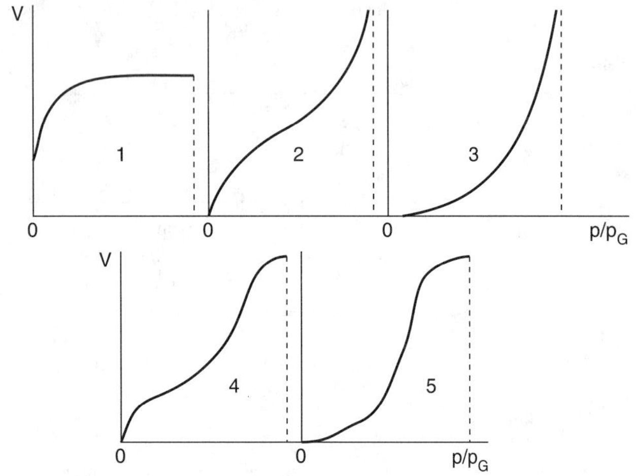

Figure 1. Forms of experimental adsorption isotherms.

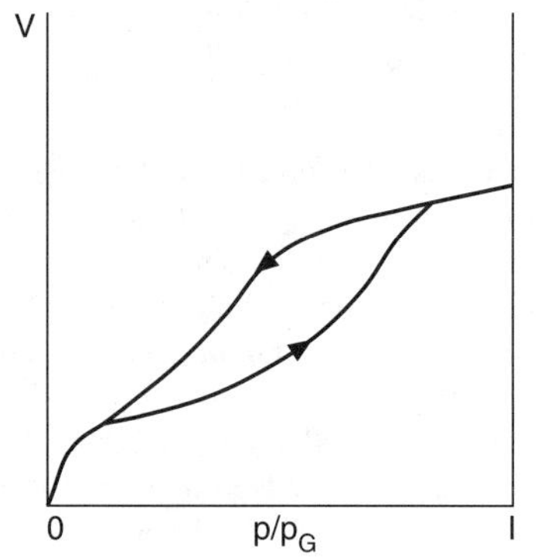

Figure 2. Adsorption isotherm hysteresis due to capillary condensation.

The rate of chemisorption is described by the expression

$$R_\alpha = d\theta/d\tau = k_2 f(\theta)p$$

where $f(\theta)$ is a function depending on the nature of adsorption; k_2 the constant of adsorption rate; $k_2 = k \exp(-E_a^*/RT)$; and E_a^* is the activation energy which, taking into account its dependence on the surface coverage θ, is described by the binomial $E_a^* = E_a^* + \alpha\theta$. Hence, at a given temperature, the rate of adsorption varies with surface coverage by molecules in proportion to $f(\theta) \exp(-\alpha\theta)$.

The activation energies of desorption and adsorption differ by the heat of adsorption Q: $E_d^* = E_a^* + Q$. The rate of desorption is described as $R_d = d\theta/d\tau = kf'(\theta)\exp(-E_a^* + Q/RT)$, where $f'(\theta)$ is analogous to $f(\theta)$.

Chemisorption is an important underlying phenomenon in heterogeneous catalysis.

Adsorption from solutions

In adsorption from solutions, the behaviour depends greatly on whether the solutions are of nonelectrolytes or electrolytes. Adsorption from nonelectrolyte solutions depends on adsorbate concentration and, in the case of dilute solutions, is similar to gas adsorption. The solvent properties manifest themselves at high concentrations.

Two basic models of nonelectrolyte adsorption exist. Adsorption in terms of the first model is restricted to a monolayer directly on the surface, whereas the consecutive layers are merely an ordinary solution. This resembles gas chemisorption, but the heat of adsorption from a solution is low in relation to the energy of chemical reaction. In the second model, adsorption occurs in a fairly thick (up to 100Å) multimolecular interphase layer in a decreasing potential field of the solid surface.

Both models are consistent with experiment and yield the same mathematical expressions, e.g., an adsorption isotherm: $\theta/(1 - \theta) = ka_2/a_1$, where a_1 and a_2 are activities of the solvent and the solute in the solution. This is obviously an expression similar to the Langmuir isotherm equation.

Adsorption in *electrolyte solutions* is considered using several approaches. If an electrolyte is assumed to be adsorbed as an entity bound by opposite charged ions, then this resembles adsorption in nonelectrolyte solutions. However, more frequently, the ions of one sign are more strongly attracted by the surface than the ions of the other sign. The latter produce near the surface a diffuse layer through which the adsorbed ions are forced to diffuse.

Absorbers

A wide range of *adsorbents* are used for adsorption from gases and liquids. Special adsorbents are often developed for specific purposes. Among the most well-known adsorbents are activated carbon used for adsorbing gases, including hydrocarbons, and silica gel for adsorbing water vapor.

A technological apparatus in which adsorption is implemented is called an *adsorber*. Adsorbers can have an immovable adsorbent bed and operate either with periodic regeneration under an adsorption-desorption regime or without regeneration but with periodic change of the adsorbent.

In fixed-bed adsorbers, the adsorption process is represented by the breakthrough curve (**Figure 3**). The concentration of pollutant (C) in the exit gas stream is plotted versus time. Initially, the adsorbent adsorbs the pollutant gas from the stream readily and efficiently, so that the pollutant concentration of the outlet stream is close to zero (period o − a on the abscissa in **Figure 3**). In this period, the bulk of the adsorption is taking place near the inlet with the rest of the bed removing the traces. Eventually, the inlet region becomes saturated and the main region of adsorption moves towards the outlet. Traces of the adsorbate begin to appear at the outlet. When the outlet concentration begins to rise rapidly (B), the so-called "break through point" has been reached. Though, at this time, the exit pollutant concentration may be below the required emission standard C_{std} after the point of breakthrough, the exit concentration rises rapidly towards the inlet concentration C_0. Thus, at time B, regeneration is required. A steep breakthrough curve is more desirable than a flat one. For a steep curve, the bed saturation may reach 80 per cent, but with a flat curve only 15–20 per cent may be typical before breakthrough. The calculation methods for adsorption, particularly in porous adsorbents, must allow for heat and mass transfer to the adsorbed substance in pores.

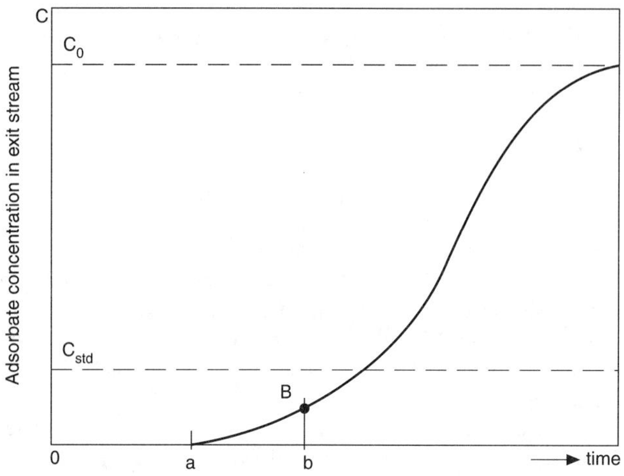

Figure 3. The breakthrough curve.

References

Adamson, A. (1976) Physical Chemistry of Surfaces, 3rd ed., Wiley, New York.

Perry, R. H., Green, D. W., and Maloney, J. O. (eds). Perry's Chemical Engineers Handbook (1984), McGraw-Hill, New York.

I.L. Mostinsky

ADSORPTION OF GASES (see Adsorption; Molecular sieves)

ADSORPTIVE BUBBLE TECHNIQUES (see Flotation)

ADSUBBLE TECHNIQUES (see Foam fractionation)

ADVANCED BOILING WATER REACTOR, ABWR (see Pressure suppression)

ADVANCED GAS-COOLED REACTOR

Following from: *Nuclear reactors*

Introduction

The advanced gas-cooled reactor (AGR) was developed in the United Kingdom (UK) as a successor to the Magnox gas-cooled reactor. (See **Magnox Power Station**) As the aim was to operate at higher fuel, cladding and gas temperatures, it was necessary to change both materials and design. In the initial design stage uranium was to be replaced by uranium oxide ceramic fuel and magnox cladding by beryllium. Later, because of developmental problems and the reduction in the price of enriched UO_2, the cladding selected was stainless steel, its additional neutron absorption being offset by higher enrichment. The lower thermal conductivity of UO_2 necessitated smaller diameter fuel in order to keep centre fuel temperature down. The introduction of hollow fuel pellets also helped to limit temperature and had the additional advantage of accommodating released fission gases without unduly increasing internal gas pressure. The small diameter fuel was arranged in a cluster of fuel pins in each coolant channel in order to generate the required power.

Windscale AGR (WAGR)

The first AGR to be built was the pilot reactor, WAGR, designed by the UK Atomic Energy Authority and built on their site at Windscale, Sellafield, Cumbria. Research and development progressed in parallel with the design work on many aspects, including heat transfer. Heat transfer enhancement was achieved by means of small height discrete roughening of the cladding, as a means of disturbing the boundary layer. This led to the development of an isolated square-rib form of roughness, which offered advantages over other forms such as screw threads, sine waves and studs, in terms of economy in steel, relative ease of manufacture and thermal performance. Thus, the original idea of 36 smooth beryllium cans came to be replaced by 21 roughened stainless steel cans of larger diameter. (See **Augmentation of Heat Transfer, Single-Phase Systems**)

By the time the commercial AGR (CAGR) was being designed (in the early stages of operation of WAGR), research work had shown that the optimum rib pitch to height ratio was about 7 with another optimum at about 2.5. The former was preferred because of its much lower steel content. The value of 7.2 specified in manufacture was arrived at by averaging manufacturing tolerances on pitch and height.

When this optimum roughness, which had a rib height of 0.011 inches (and hence a ratio of rib height to hydraulic equivalent diameter of only 0.007), was chosen it meant that WAGR—a test bed for CAGR—had to opt for a new pin design. The first arrangement, with 12 of the larger pins on an approximately square array with the 8 outer pins located very close to the channel wall, was made without involving heat transfer expertise and resulted in overheating and pin failure. It was soon replaced by a single-ring, nine-pin design in an annular passage formed between the graphite channel and an inner graphite cylinder. It is this design, based like the earlier 21-pin design on computer prediction using the HOTSPOT code developed at the Windscale heat transfer laboratories, that provided the bulk of irradiation experience for CAGR. In-pile measurement and out-of-pile fuel element examination, backed up by research, provided information on pin bowing and its causes; material properties; fission gas release; graphite deposition (arising from the chemical reactions between the graphite and the carbon dioxide coolant, enhanced by radiation); oxidation; fuel pin failure; and heat transfer and fluid flow. The reactor was instrumented to provide measurements of channel flow rate, inlet and outlet gas temperatures and can temperatures.

Ribbed Can Manufacture

The roughening on the can was first produced by machining, later by grinding. The metre-long cans were ground down from thick tube stock in two stages, using a half metre-long grinding wheel containing slots at the specified rib pitch and width. Hundreds of pins could be ground before the wheel wore down, ensuring good repeatability between pins with regard to rib pitch. A quicker and cheaper method, called thread-whirling, was later developed. It produced a single-start spiral rib at an angle of 2.5° to the transverse direction. Tests showed that the angle had a negligible effect on performance. A further development of the rib design was the multi-start rib (see Wilkie, 1966) manufactured by a process akin to thread whirling.

The commercial AGR

The UK CAGRs, namely Dungeness B, Hinkley Point B, Hunterston B, Heysham, Hartlepool and Torness, all use 36-pin clusters formed by three rings of six, twelve and eighteen roughened pins with a central smooth tie tube all contained within a machined graphite sleeve forming the coolant flow passage. Eight of these metre-long elements are stacked vertically in a reactor channel. Although the fuel element assemblies can be aligned before loading so that like pins in neighbouring assemblies are in line, the elements are free to turn and indeed do turn, especially during on-power loading. Less than 5° rotation is sufficient to cause increases in pressure drop and in temperature variations around the pins, as a result of the flow from a lower assembly meeting the blunt pin ends of the next assembly. These effects are taken into account when assessing operating conditions.

The initial fuel employed single-start ribs. For some replacement fuel, 12 start multi-start ribs 0.163 in high, with a pitch to height ratio of 6.5 and a lead angle of 34°, have been adopted to promote coolant mixing. The Reynolds number is in the region of 10^6. (Wilkie and Mantle, 1979)

Refuelling can take place off-load or on-load; the latter became possible after the machining of ribs onto the outer surface of the graphite sleeve (Wilkie et al, 1989).

Research on roughened surfaces

Prior to the adoption of roughened surfaces in AGRs, it had been maintained that roughening introduced a penalty in performance in that **Friction Factor**, f, was increased much more than **Nusselt Number**, Nu, or **Stanton Number**, St, and therefore was only of value in specific situations where pressure drop was unimportant or small size essential. What had not been realised was that while keeping coolant mass flow rate and temperature rise fixed, the pumping power could be reduced for the same thermal performance by increasing the flow area—provided that St^3/f is improved by roughening (Walker and Wilkie, 1967). Also, the thermal performance could be improved for the same pumping power by increasing the flow area if $St/f^{1/3}$ is improved by roughening (Wilkie, 1971). Variations of these criteria, when parasitic pressure losses caused by support features were taken into account, are also given. The ribbed surfaces chosen for CAGR give a Stanton number about 2.5 times higher than a smooth surface and a friction factor about 7 times higher at the reactor Reynolds number. This is equivalent to an increase in thermal performance $St/f^{1/3}$ of some 40%. The data for a wide variety of ribbed surfaces is contained in Wilkie, 1966a.

The multi-start ribbed design, first conceived in 1965 (see Wilkie, 1966a, for friction factor data), produced a marked reduction in friction factor compared with a transverse ribbed design, with a concomitant reduction in Stanton number. Several designs gave improved thermal performance (White and Wilkie 1970).

Non-uniform boundary conditions of roughness and heat flux

In all AGRs, there are flow passages with a mixture of boundary conditions. For example, the flow between the outer facing surfaces of the outer ring of pins is bounded by a roughened surface on the pins and a smooth surface, or rather a slightly roughened machined surface, on the inner surface of the graphite sleeve. A similar situation applies between the inner facing surfaces of the inner pins and the tie tube, and between the ribbed outer surface of the graphite sleeve and the "smooth" graphite moderator block. Similarly, because of decreasing neutron absorption from the outer pins to the inner pins and the slowing down of neutrons by the graphite, all pin surfaces and all graphite surfaces are subject to different heat fluxes. These vary laterally, circumferentially and longitudinally throughout the reactor core. Little information has been available on these effects and a solution had to be found.

The seminal contribution was made by Hall in 1958, who developed the idea of dividing the annular flow zone into two parts, separated by the surface of zero shear stress taken to be located at the peak of the velocity profile in the annulus. The coolant temperature distribution, which had zero gradient at the adiabatic insulated unheated channel wall, was transformed to provide a profile with the same temperature at the heated roughened wall, but with zero gradient at the proposed zero shear stress surface. The Stanton number, friction factor and Reynolds number—all based on a hydraulic equivalent diameter defined by the perimeter of the roughened cylinder and the flow area between the roughened cylinder and the zero shear stress surface—were the basic data defining the performance of the roughened surfaces tested (Wilkie, 1966a).

The labour of measuring velocity and temperature profiles for all future tests was reduced by correlating the data to provide the position of the zero shear stress surface in terms of more easily measured quantities. A calculation procedure was then laid down (Wilkie, 1966b). This procedure was more accurate than the Hall transformation since it was free of the random experimental error attached to the measurements.

However, it became clear that the large apparent increase in the smooth outer wall friction factor, as the inner wall friction factor increased (K_3 factor), presented a gap in understanding. Further study (Wilkie et al, 1963) of friction factor with a large aspect ratio rectangular channel, the walls of which could be made identical or non-identical in degree of roughening, suggested that one explanation for the large increase in smooth friction factor was the non-coincidence of the maximum velocity and the zero shear stress surfaces. This was later confirmed by direct measurement (Kjellstrom, 1965). Warburton and Pirie, 1973, later amended the K_3 factor to a lower but non-zero value so that the transformation calculation implicitly used the correct zero shear stress surface.

The transformation procedure, in effect, permits data to be applied to a fully roughened and fully heated passage, e.g., a circular pipe, although there are effects of passage shape, defined by the ratio of roughened cylinder radius to zero-shear surface radius for convex surfaces (Rapier, 1963–4).

Instead of correlating Stanton number and friction factor in terms of the several dimensionless groups required to define a roughened surface (e.g. pitch to height), a more general correlation involving roughness and passage shape parameters has been devised (Firth, 1979, and Dawson et al, 1983).

In applying the data to passages with nonuniform boundary conditions, a reverse of the transformation procedure is required (Wilkie and White, 1965, and Rapier, 1963–4). This, together with the roughness correlations for the Stanton number and friction factor, the treatment of turbulent mixing of heat, conduction in the fuel and can and thermal radiation between surfaces is embodied in the computer code HOTSPOT devised by Rapier (for initial version see Cowking, 1970).

References

Cowking, C. B. (1970). HOTSPOT—An IBM computer programme for calculation of systematic can and fuel temperatures in gas-cooled rod-cluster fuel channels. UK Atomic Energy Authority TRG Report 1961(R)

Dawson, J. T., Firth, R. J., Langley, M. J., Jackson, G. F., Rapier, A. C., and Wilkie, D. (1983) A comparison of measured and predicted can temperatures in simulated CAGR fuel elements. Paper 144, *Br. Nucl. Energy Soc. Conf.* 'Gas-cooled Reactors Today', London.

Firth, R. J. (1979) A method for analysing heat transfer and pressure drop data from partially roughened annular channels. UK Atomic Energy Authority Report ND-R-301(W).

Hall, W. B. (1958) Heat transfer in channels having rough and smooth surfaces. UK AEA Report ICR-TN/W832. Also *J. Mech. Eng. Sci.* (1962) Vol 4, 287.

Kjellstrom, M. B. and Hedberg, S. (1965) On shear stress distributions for flow in smooth or partially rough annuli. AE-RTL-796.

Rapier, A. C. (1963–4) Forced convection heat transfer in passages with varying roughness and heat flux around the perimeter. Thermodyn. and Fluid Mech. Convention, Cambridge, Proc. Instn. Mech. Engrs. Vol 178, Pt 31, 12–20.

Walker, V. and Wilkie, D. (1967) The wider application of roughened heat transfer surfaces as developed for advanced gas-cooled reactors. Symposium on High Pressure Gas as a Heat Transfer Medium. London 9–10 March I. Mech. E.

Warburton, C. and Pirie, M. A. (1973) An improved method for analysing heat transfer and pressure drop tests on roughened rods in smooth channels. Central Electricity Generating Board, Berkeley. Report RD/B/N2621.

White, L. and Wilkie, D. (1967) The heat transfer and pressure loss characteristics of some multi-start ribbed surfaces. UKAEA TRG Report 1504 (W) Also 'Augmentation of convective heat and mass transfer'. American Society of Mechanical Engineers, New York, 1970, Dec. 55–62.

Wilkie, D. (1966a) Forced convection heat transfer from surfaces roughened by transverse ribs. Third Int. Heat Transfer Conf. Chicago. Amer. Inst.

Chem. E. New York Aug. Paper No 1. pp 1–19. Also see UKAEA TRG Report 781 (w), 1966.

Wilkie, D. (1966b) Calculation of heat transfer and flow resistance of rough and smooth surfaces contained in a single passage. Third Int. Heat Transfer Conf. Chicago. Amer. Inst. Chem. E. New York Aug. Paper No 2 pp 20–31.

Wilkie, D. (1971) Criteria for choice of surface for for gas-cooled reactors. Nuc. Eng. Int. Vol 16, March, No. 177, 215–217.

Wilkie, D., Parkin, M. W., and Goldthorp, R. H. (1989) UK Patent No. GB 2168192 B

Wilkie, D., Cowin, M., Burnett, P. and Burgoyne, T. (1963) Friction factor measurements in a rectangular channel with walls of identical and non-identical roughness. UKAEA TRG Report 519(w) Also Int. J. Heat Mass Transfer 1967 Vol 10, May, 611–621.

Wilkie, D. and Mantle, P. L. (1979) Multi-start helically-ribbed fuel pins for CAGR. Nucl. Energy Vol 18, Aug., No. 4, 277–282.

Wilkie, D. and White, L. (1965) Calculation of flow resistance of passages bounded by a combination of rough and smooth surfaces. UKAEA TRG Report 1113(w) Also *J. BNES* (January 1967), 48–62.

D. Wilkie

AEA TECHNOLOGY

Harwell Laboratory
Didcot
Oxon OX11 ORA
UK
Tel: 01235 821111

AELOPILE OF HERO (see Steam engines)

AERATION

Aeration refers to the use of atmospheric air to supply an oxygen demand, almost always relating to the biological *treatment of waste water* based on aerobic cultures. Most, though not all, organic materials can serve as nutrients for bacterial colonies that oxidise contaminants producing biomass, water and carbon dioxide. The naturally occurring biological systems of streams and rivers utilise residues as nutrient sources, with the oxygen needs supplied from the air by mass transfer through the free liquid surface. Enhancement of the natural aeration rate is needed if there are significant levels of *pollution*.

Waterfalls and cascades increase the purification capacity of a stream. Drops or jets that impact a free water surface with velocities above about 2 m s^{-1} entrain air and carry bubbles below the liquid surface. There is a bi-phasic region of intense turbulence, though of short residence times. Gas that accumulates and coalesces below this escapes in a rising bubble cloud surrounding the plunging point of a vertical jet. In terms of energy use, plunging jets are efficient (Bin, 1994). The entrainment and subsequent mass transfer depend on the velocity at impact, as well as the length and degree of disturbance to the falling jet. Installations of practical size (with jets of say 2 cm diameter) can achieve atmospheric oxygen take up rates in the order of 3 kg per kWh of power consumed. (see **Plunging Liquid Jets**)

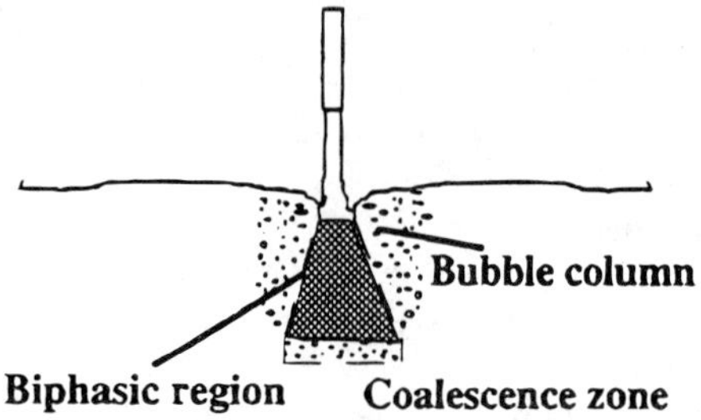

Figure 1. Plunging jet principle.

Most waste water treatment systems use these aerobic biological processes. Oxidation ditches are often used if there is space. The water is repeatedly aerated as it circulates in an oval or labyrinthine channel while a biologically active sludge is maintained. The circulation imposes a cycle of repeated oxygenation and rest that can be exploited to assist the removal of nitrates and other oxidising materials.

Various mechanical devices that directly disturb the liquid surface have been used to enhance the contact between liquid and gas. The most common designs use rotating paddles mounted at or just below the free surface. These are shaped so that they scoop water from the surface and project it upwards and outwards. The sprayed water takes up oxygen as it passes through the air and entrains more as it impacts with the surface. Another design uses horizontally-mounted cylindrical beaters to thrash the water surface. Like plunging jets, these aerators can generate large-scale circulation of the water in the channel, mixing the bulk liquid and suspending the active sludge. Unfortunately, the energy efficiency falls in large scale installations: values fall below around 1 kg O_2 per kWh.

A different approach is used when air is injected beneath the surface of the water to form a *bubble plume*. In large effluent ponds, the plume is unconfined and induces large scale circulation. The volume of water Q_W brought to the surface by injection of a volume of gas Q_G at a depth H is approximately given by the dimensionless relationship $Q_w^3/(gH^5Q_g) = 3.4 \times 10^{-3}$.

The circulation is driven by buoyancy forces and at first approximation is independent of the bubble size distribution. However, the oxygen transfer does depend on the bubble size, so some installations use *venturi ejectors* or *static mixers* mounted above the air injection nozzles to ensure that the gas stream is well broken up.

When the *bubble column* is confined in a pipe, the water can be subjected to a more intense level of oxidation. Bubble columns are widely used for fermentation and in biotechnology, in general,

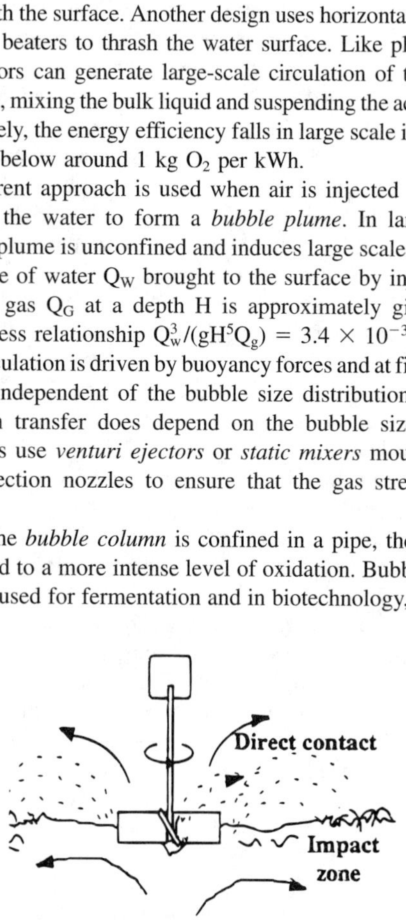

Figure 2. Rotary surface aerator.

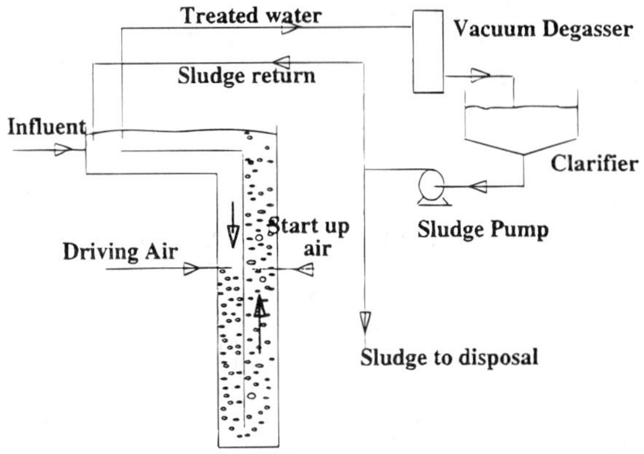

Figure 3. ICI-Zeneca deep shaft.

since they are suitable for large systems with relatively slow kinetics or large volumes of process fluids—important factors in waste water purification. One such type is the *Deep Shaft*. Top-to-bottom circulation is induced in divided deep (or tall) columns, typically 1–5 m diameter and 100 m deep. Once an initial circulation has been established with start-up air, the main air supply is introduced into the downcomer. The air is carried to the bottom of the shaft by the momentum of the circulating fluid before it enters the riser. The pressure increases the driving force for mass transfer, though at the cost of some loss of surface area. Water velocities in a typical installation are of the order of 2 m s^{-1}, giving liquid circulation times of about two minutes.

Respired CO_2 is removed at the top of the shaft. A vacuum degasser ensures that the sludge separates satisfactorily in the classifier. Circulation instability is avoided by retaining a small injection of air directly into the riser.

References

Bin (1993) "Gas Entrainment by Plunging Liquid Jets," Chem. Engrg. Sc., 48, 3585–3630, Pergamon.

Deckwer, W. D. (1992) "Bubble Column Reactors," John Wiley.

Goossens, L. H. J. and Smith, J. M. (1982) "The Mixing of Ponds with Bubble Columns," Proc. 4th European Mixing Conference, Cranfield, U.K.

Leading to: *Plunging liquid jets*

J. M. Smith

AERODYNAMIC COEFFICIENTS (see Aerodynamics)

AERODYNAMIC EFFICIENCY (see Aerodynamics)

AERODYNAMIC FLOW SPECTRUM (see Streamline)

AERODYNAMIC RESISTANCE OF ATMOSPHERE (see Environmental heat transfer)

AERODYNAMICS

Following from: *Flow of fluids*

Aerodynamics is the branch of hydrodynamics that deals with the laws of air motion and with the forces acting on the surfaces of streamlined bodies. Aerodynamics generally studies motion with velocities which are far short of the sound velocity (340 m/s, 1200 km/h). This is in contrast to *gas dynamics*, which deals with the motion of compressible gaseous media and with their interaction with solids.

One of the major problems facing aerodynamics is to determine forces acting on bodies when they move through an atmosphere. It is, therefore, a part of the science concerned with external flows. Nevertheless, the methods and results of classical "force" aerodynamics also enjoy wide use in investigating heat transfer between a body and a fluid flowing around the body and in analyzing the structure of single- and multi-phase flows.

Aerodynamics developed in response to the demands of practical applications. Aerodynamics as a branch of hydrodynamics appeared early in the 20th-century with the advent of aircraft construction. The *aerodynamics of aircraft* and, in general, of a *flying vehicle* deals with the definition of aerodynamic forces and moments acting on the entire vehicle and on its parts, wing, fuselage and tail unit.

Depending on the methods of solving aerodynamics problems, these are classified as experimental or theoretical. The basis for the experimental methods is the *reversibility principle* of motion, which states that "the force acting on a body when the fluid moves around it with constant velocity is equivalent to the force which a body experiences in uniform and translatory motion in a stationary atmosphere." In this case, it is obvious that the velocity V of flight must correspond in magnitude and direction relative to the body axis, with the velocity of an incoming (converted) flow V_∞ (**Figure 1**). The effect of the gaseous flow around the body manifests itself as a pressure p and a friction (shear force) τ (**Figure 2**). The pressure p is directed along the normal to an elementary area on the surface of the body, and the friction (shear) τ lies in the plane tangent to the body surface. The friction results from the viscosity of the air flow and from the adhesion of particles of gas to the body surface.

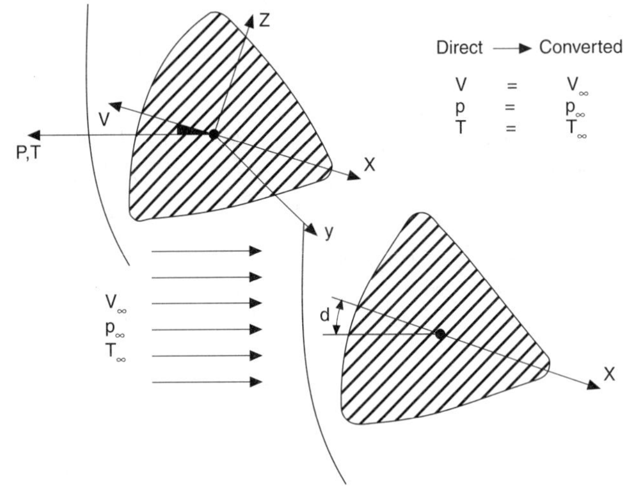

Figure 1. The reversibility principle.

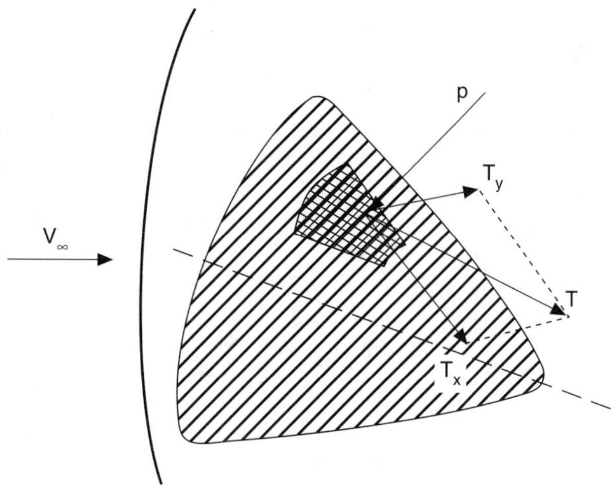

Figure 2. Pressure and friction forces.

The pressure forces and the frictional forces acting on the surface of a body can be reduced to the resultant $\overline{R}$ of these forces, called the aerodynamic force, and to a couple with a moment $\overline{M}$, called the aerodynamic moment. The aerodynamic moment plays a leading role in designing air vehicles, defining their stability and controllability. Both the aerodynamic force $\overline{R}$ and the aerodynamic moment $\overline{M}$ can be resolved into components in a rectangular Cartesian coordinate system, associated either with the vector of flight velocity $\overline{v}$ (wind coordinate system) or with the body itself (body axis system).

In the wind coordinate system, the force directed along the flow opposite to the direction of motion is called the *drag force* X; the force perpendicular to it and lying in the vertical plane, the *lift force* Y; and the force perpendicular to both, the lateral force Z. The projections of the moment $\overline{M}$ onto the axes of the body axis coordinate system are called the *roll moment* M_x, the *yawing moment* M_y and the pitching moment M_z.

The shape of a body which when it moves in air results in a lift exceeding the force which impedes the flight (for instance, gravity) is called the aerodynamic profile (or lifting surface). The lift-drag ratio is called the *aerodynamic efficiency* K. Aerodynamic efficiency depends on the profile of the air vehicle. Thus, the aerodynamic efficiency of the spherical spacecraft Vostok, piloted by Yu. A. Gagarin, was close to zero. The spacecrafts *Soyuz* and *Apollo* were symmetrical vehicles of low efficiency (of the order of 0.3). The aerodynamic efficiency of such space vehicles as the *Space Shuttle* and *Buran* is considerably higher. Aerodynamic efficiency also depends on the angle of attack α, formed by the vehicle axis or the chord of its wing and the direction of velocity of the undisturbed flow of air.

To simulate the motion of a body in a stationary air, special experimental set-ups called wind tunnels are used. The simulation method is based on the similarity theory, according to which the aerodynamic forces and moments acting on a flying vehicle can be determined from the test results of a small-scale model of this object. In this case, conditions which will ensure the possibility of transferring the results obtained on a laboratory set-up to a full-scale object should be met.

In addition to the geometrical similarity of the natural object and the model, the Mach number and the Reynolds number found from the parameters of the incoming flow must be the same. The results of measurement of forces and moments are represented in the from of dimensionless ratios:

$$C_R = |\overline{R}|/\left(\frac{1}{2}\rho V^2 S\right), \qquad C_X = |\overline{X}|/\left(\frac{1}{2}\rho V^2 S\right),$$

$$C_Y = |\overline{Y}|/\left(\frac{1}{2}\rho V^2 S\right), \text{ etc.}$$

The dimensionless quantities C_R, C_X and C_Y, . . . , which characterize aerodynamic forces and moments acting on the body moving through a gaseous atmosphere are called the *aerodynamic coefficients*, C_x being the drag coefficient (see also **Drag Coefficient**).

A typical dependence of aerodynamic coefficients on the Mach and Reynolds numbers is given in **Figures 3** and **4**. The use of the velocity head of the air flow ($\rho v^2/2$) and of typical cross-sectional area S of the vehicle is of prime importance for making and planning aerodynamic investigations. The aerodynamic force can vary by several orders of magnitude, while the aerodynamic coefficient may remain constant. In this case, the possibility exists of generalizing as single parameters all varieties of vehicles and their models and of obtaining convenient relationships for predicting coefficients.

For subsonic flight velocities of large-sized vehicles, the aerodynamic coefficients depend mainly on the shape of the vehicle and on the angle of attack. In general, the achievement of two similarity criteria, the Mach and the Reynolds number, in wind tunnels is next to impossible. Because of a reduction of the model dimensions, in order to satisfy the Reynolds number condition, there is a need either to increase the velocity V (but then the Mach number condition is not met) or to increase the density of gas or to decrease the viscosity of air. In practice the problem can be overcome if the aerodynamic coefficient is independent of Re or Ma over a sufficient range. Thus, the drag coefficient of a sphere is practically independent of the Reynolds number in the range of $10^3 < \text{Re} < 10^6$ (see **Figures 3** and **4**).

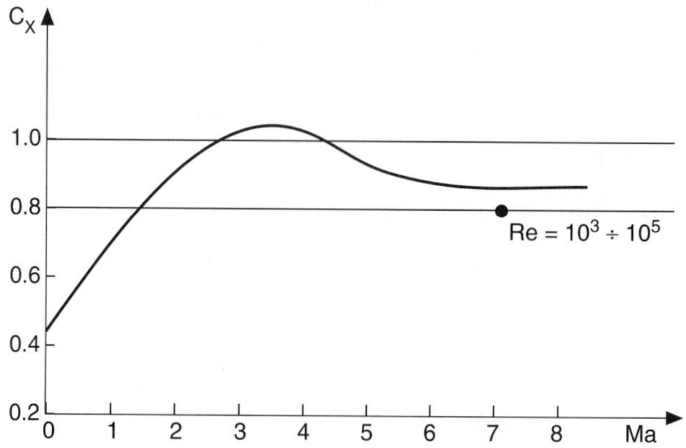

Figure 3. Variation of the drag coefficient of a sphere with Mach member.

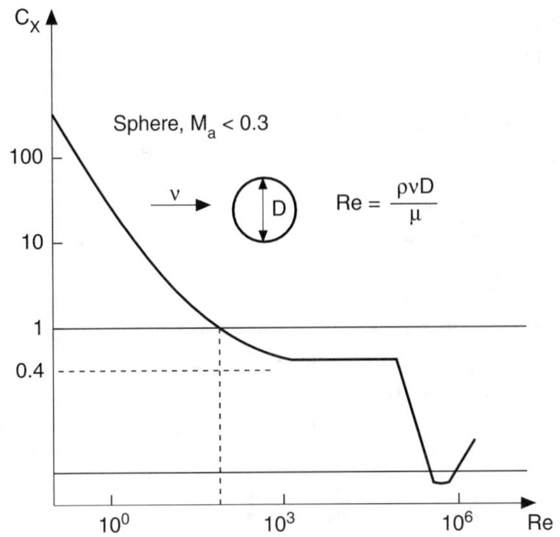

Figure 4. Aerodynamic drag coefficient for a sphere as a function of Reynolds number.

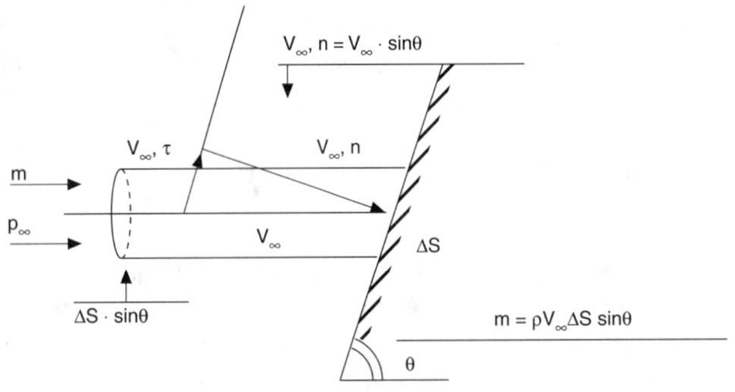

Figure 5. Forces on a body in a fluid stream.

The presentation of test results as functional equations between dimensionless similarity criteria, on the one hand, reduces the number of independent variables, and on the other, allows us to find the range of self-similarity and thus to increase the reliability of transferring the laboratory test data to natural conditions.

There has been a tremendous growth in modern theoretical aerodynamics. Based on the hypothesis suggested by Prandtl in 1904, the entire space around the streamlined body is divided into two areas: the boundary layer (see **Boundary Layer**), where the effect of air viscosity and thermal conductivity and also of thermal condition of the body is significant; and the area outside the boundary layer, where the air can be considered as an ideal gas (see **Inviscid Flow**).

Analysing the basic laws of hydrodynamics (see **Conservation Laws**) formulated in the form of equations by Euler, Lagrange, Stokes, and Prandtl, one can obtain the solution for a large number of problems on the motion of a body through the atmosphere of the Earth and of the other planets. In many cases, however, the intricate design of the vehicle and the proximity of the Earth surface presents difficulties in the theoretical analysis even with the use of modern computers. Therefore, approximate methods of analysis are extensively employed.

The first theoretical definition of the law of the atmosphere resistance through which a body moves was formulated by Isaac Newton, who suggested that it is associated with the impact of particles against the frontal surface of the body. He showed that resistance is proportional to the square of the body velocity and the angle of inclination θ of its surface to the direction of motion (**Figure 5**). In this case, for instance, for a wedge with an apex angle 2θ we have $C_X = 2 \sin^2 \theta$. For bodies of irregular shape, the quantity C_X can be estimated by approximating the real contour of the body to a certain cone or wedge. Therefore, for a cylinder with a plane end face, $C_X = 2$ and for a sphere, $C_X = 1$. The experimental data, however, are in good agreement with the Newton model only for supersonic velocities. Nevertheless, Newton laid the foundation for explaining the problems of aerodynamics on the basis of the laws of mechanics.

Modern aerodynamics has gained experience in defining the aerodynamic characteristics of a large set of simple casings and wings, and also of their various combinations. This substantiated the proposed principle of separating a flying vehicle into components or constituents for which the definition of aerodynamic coefficients presents no problems, and integral parameters are calculated by the additivity principle. In some cases, a second approximation is used in which interference corrections are defined.

This principle is also used as the basis for the calculation of nonstationary aerodynamic characteristics using additional terms of the series defining the aerodynamic characteristics reflecting the nonstationary flows.

Aerodynamic design has assumed an independent significance in connection with the production of vehicles of the "flying-wing" type. Of special interest is the study of casings of body-of-revolution type, because among these are the majority of rocket-type units. As to the velocities of irregular shape, experimentation shows that as the flight velocity increases (especially in the supersonic range), the effects of the interference interaction become less significant. This is because the zones of mutual effect become narrower. It should be remembered, however, that this conclusion cannot be extended to heat transfer problems.

Modern controlled flying vehicles have numerous aerodynamic, gasdynamic (jet) and combined controls. To find the control force is the key element in the aerodynamic calculation; because in this case, the arrangement and flying characteristics of the vehicle are defined. Nevertheless, this problem is rather intricate and can often be solved only by experimentation.

On increasing the velocity of flight and on its approach to sound velocity, consideration must be given to the compressibility of the atmosphere. The supersonic flight of the body is characterized by a number of peculiarities: *shock waves* arise (see **Shock Tubes**) which increase the drag, the flying body warms up through air friction and due to radiation from the highly compressed gas behind the shock wave. In flights with high supersonic (or hypersonic) velocity, dissociation and ionization of air molecules occur. All these problems are usually reffered to a branch of hydrodynamics called "gas dynamics."

A broad field of non-aviation applications of aerodynamics represents a science called *industrial aerodynamics*. It deals with the problems associated with the design of blowers, wind motors, fans and air conditioners and also ejectors.

A *blower* is designed to compress and to feed air at moderate pressures. The air produced by blowers is mainly consumed in ferrous metallurgy. To produce 1 ton of pig iron, 2 tons of compressed air at a pressure of up to 0.5 MN/m^2 is required. Modern blowers have compressors driven by gas or steam turbines, with a power of 20 MW and higher. Their capacity is 5000–10,000 m^3/min of air at a temperature of 500°C.

A *fan* is the machine designed to deliver air or other gases at an excess pressure not higher than 12–15 kN/m^2 (0.12–0.15 atm). Fans serve to ventilate buildings and mines, to deliver air to boiler rooms and furnace apparatuses, to remove flue gases, to dry materials, to cool machine parts and mechanisms, to produce air screens, to transport pneumatically loose and fibrous materials, etc.

The centrifugal or *radial fan* has a wheel located in a volute casing; on rotating the wheel, the gas delivered through an inlet enters into the channels between the plates and, acted upon by the arising centrifugal force, moves into the volute casing and is directed into the outlet along the tangent to the casing (**Figure 6**).

The *axial-flow fan* has a blade wheel located in a cylindrical casing (**Figure 7**). When the wheel rotates, the supplied gas moves in the axial direction. The axial-flow fan, as compared to the centrifugal one, is more simple, has a higher efficiency, but cannot achieve high pressures. To increase the fan capacity and pressure, the fans connected in either series or in parallel.

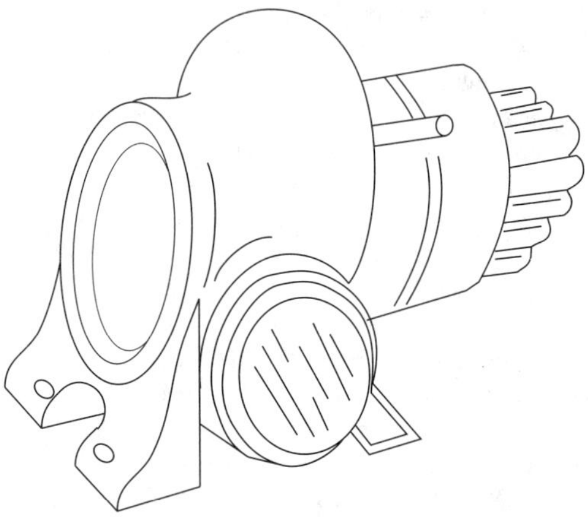

Figure 6. Centrifugal or radial fan.

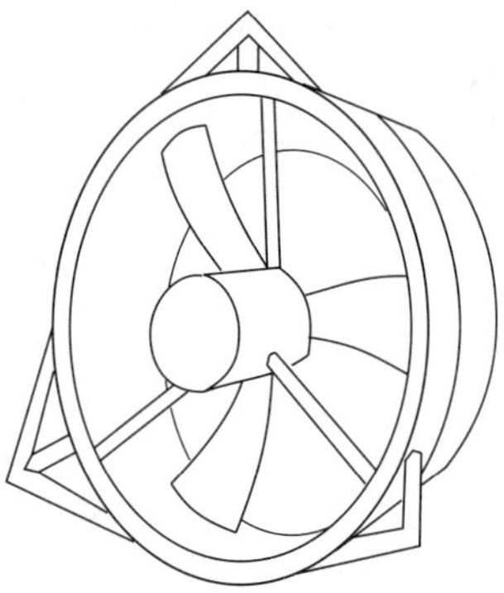

Figure 7. Axial flow fan.

As a preliminary estimate of the optimum conditions for fan operation, we can use the generalized plot given in **Figure 8**. Presented in the pressure p[N/m^2] – capacity $\dot{V}$ [m^3/s] coordinate system are the angular velocities ω[rad/s] and the efficiency η. From the given $\dot{V}$ and p, we can find the point on the plot which defines the required velocity ω and the efficiency η, from which the required power is calculated.

The air screw or propeller can also be used as the motor for subsonic aircraft, helicopters or transport *air cushion* vehicles (*ground-effect* machines, hovercraft). Two methods of air cushion formation are recognized the static and dynamic. In the static method, the pressure in the air cushion is built up by a fan or a compressor; in the dynamic method, it is created by the relative velocity of the air flow. The schemes of air cushion formation that have enjoyed the widest application are the: (a) chamber; (b) nozzle; (c) slit; and (d) wing (**Figure 9**).

In a chamber scheme, the lift is produced by the static pressure of the air delivered by the fan under the casing base. The chamber is raised by the lift and the air flows out through a clearance formed between the chamber edges and the bearing surface. Since the efflux erea is rather large, a considerable body of air is required even with considerably small clearances.

In a nozzle scheme, the air cushion is formed because of the air efflux from the nozzle located along the periphery of the nozzle unit. The lift is the sum of the static pressure at the cut of the nozzle unit and the reactive force of the air efflux from the nozzle. This scheme allows larger clearances between the apparatus and the bearing surface with smaller air expenditure.

In a slit scheme, the air cushion is formed in a thin clearance through which the air flows out in all directions. In order to form the slit clearance, both the lifting and the bearing surfaces are shaped. An increased pressure in the air cushion is maintained due to the viscosity of the air passing through the slit.

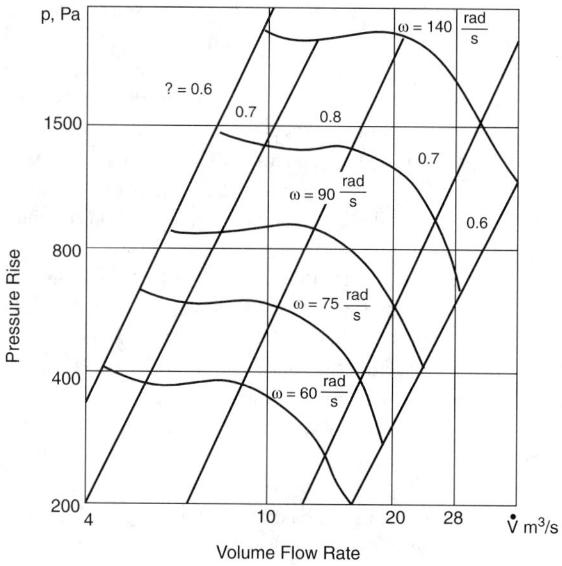

Figure 8. Angular velocity (ω) and efficiency (η) for fans providing a given volume flow rate $\dot{V}$ at a given pressure rise (*p*).

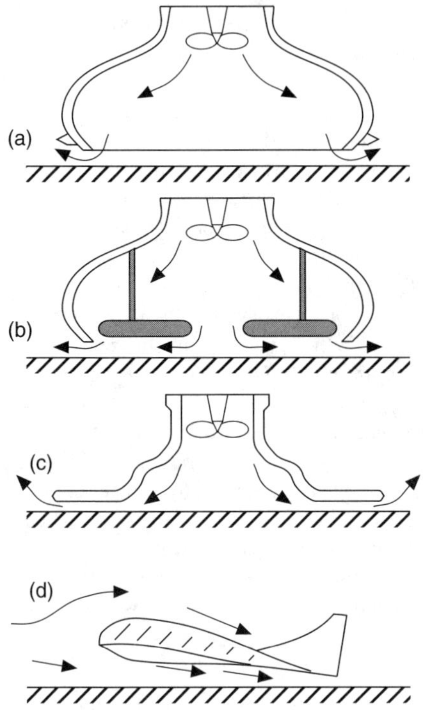

Figure 9. Schemes for air cushion formation.

In a wing scheme, the air cushion is formed under the wing of the air vehicle when it moves with an angle of attack near the bearing surface.

The *air ejector* is the device in which the total pressure of air flow is increased under the action of the jet of another high-pressure flow. The energy from one flow is transferred to another flow through turbulent mixing. The ejector is simple in design, can operate in a wide range of change of gas parameters, permits easy control of the operating process and a change from one mode of operation to the other. Depending on the purpose, ejectors are made by different ways.

In some wind tunnels, a number of ejectors functioning as pumps may be set in series as illustrated in **Figure 10**. Bottles (1) contains air at high pressure. The ejector (2) sucks in atmospheric

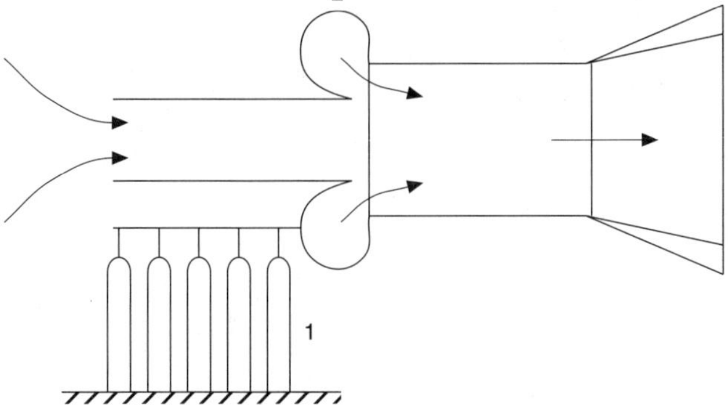

Figure 10. Air ejection system for wind tunnel operation.

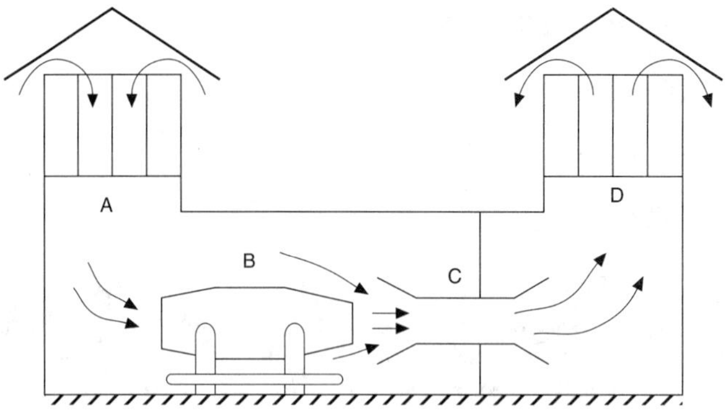

Figure 11. Air movement using a jet engine as an ejector.

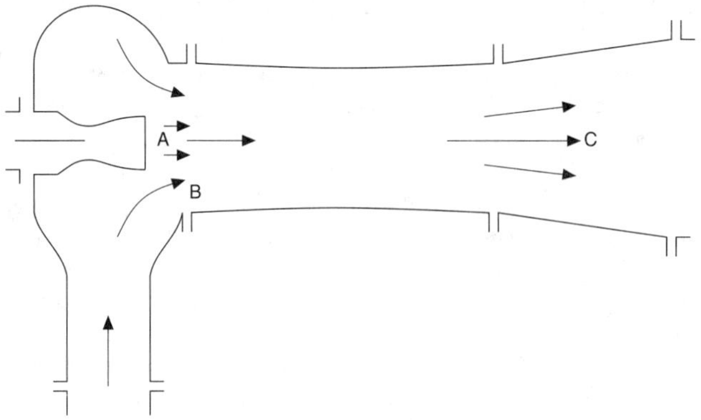

Figure 12. Vacuum ejector system.

air as shown. This leads to a volumetric flow rate $\overline{V}$ which in many times the flow from the bottles, thus allowing larger models to be tested.

Figure 11 shows a schematic diagram of a jet engine used to feed an ejector to create airflow in the test stand for the engine. The exhaust gas jet flowing out of the propulsive nozzle draws in the air from the pit A into the ejector C, thus ensuring ventilation of the stand location and cooling of the engine B. Hot gases intermix with the atmospheric air, which results in the drop of the gas temperature in the pit D and improves the conditions of the noise suppressor operation.

Finally, in vacuum systems, the *vacuum ejector* functions as an exhauster (**Figure 12**). Especially high vacuum (down to one millionth of an atmosphere) can be obtained mercury vapor is used as the ejecting gas. The ejecting A and the ejected B gas enter into a mixing chamber as two separate flows: in general, they can differ in chemical composition, velocity, velocity, temperature and pressure. Mixing of flows results in the equalizing of gas parameters across the entire section C of the jet outlet section. For other versions of ejector application, in condensation systems of steam power plants, in particular, see **Ejectors**.

Yu.V. Polezhaev

AEROGENERATORS (see Wind turbines)

AEROSOL FILTRATION (see Filtration)

AEROSOLS

Following from: *Gas-solids flow*

Aerosols are stable systems consisting of a gaseous (air) medium and miniscule suspended solid and liquid particles. Aerosols are conventionally classified into *dusts*, *mists*, and *smokes*, although a number of systems can refer to both types at once, e.g. dusts and smokes.

Aerosols are formed in two main ways: 1) by dispersion, when fine particles are formed as a result of crushing a solid or atomization of a liquid; and 2) by condensation, when the aerosol particles originate as a result of molecule aggregation.

The first method is predominant in dust generation. In nature, dusts are the products of mineral and soil erosion, of volcanic eruptions and dust storms. *Dusts* commonly consist of particles of irregular, sometimes crystalline, shape and constitute polydisperse systems with particles from fractions of a micron (µm) to 100 microns in size. However, coarse particles (over 10 µm) are liable to sedimentation and rapidly settle. Thus the size of aerosol particles proper is restricted to a maximum of about 10 microns with a minimum size of tenths or even hundredths of a micron.

Mists are mainly formed by condensation. Air containing water vapor is cooled below the saturation temperature and the vapor becomes supersaturated. If the air contains fine dust particles, they serve as the sites of condensation and minute liquid droplets arrear on them, which afterward may grow due to either persisting condensation or as a result of coagulation. The absence (or a low number density) of solid particles causes inhibition of vapor condensation and, as the gas is cooled, the *supersaturation* value rises. Finally, at a certain, critical, supersaturation condensation centers nucleate homogeneously in the vapor itself, and further condensation, known as *spontaneous condensation*, proceeds on them. In this case, the finest particles are produced abundantly, which facilitate their coagulation and, consequently, growth. The range of droplet size in natural mists lies within 0.01 to 10 µm. As the droplets become larger, they show a tendency to fall, i.e., the mist as an aerosol system decays. Mist droplets are initially spherical and remain so after coagulation. Only very large droplets may be slightly flattened or extended during falling.

A further type of mist is formed as a result of chemical reactions. In nature, minute droplets of sulfuric acid are generated in this way. Sulfur oxides, chiefly SO_2, ejected into the atmosphere, are re-oxidized to form SO_3 which, reacting with atmospheric water vapor, produces H_2SO_4, whose vapors are at once in a supersaturated state and condense as droplets. These, in turn, may react with other substances in the atmosphere; for instance, ammonia to produce salts, e.g., ammonium sulfate, settling as submicron solid particles.

Smokes are finely dispersed, extremely stable aerosol systems produced in chemical reactions between gaseous substances that lead to generation of solid particles. In nature, the main source of smokes is combustion. Thus, one way of generating submicron soot particles is the reaction $2CO \rightarrow CO_2 + C$ proceeding in the flame, where other solid particles, including benzapyrene, are generated. The particle size characteristic of smokes is from 0.001 to 5 µm; the shape may be either crystalline or some other.

Human activities, that is, mining, industry, mistreatment of soil, explosions for military and peaceful purposes, have substantially increased the amount of particles ejected in the atmosphere, forming aerosols in the environment. In some industrial regions, dust loading is so high that the solar radiation attenuates, and at a high air humidity (e.g., in north-western Europe) a persistent mist known as *smog* is formed.

The most widespread mechanism for aerosol generation is vapor condensation within the gas continuum. In a pure vapor-gas mixture in which there are no particles, molecular aggregates (nuclei) of the vapor itself, arising as a result of a high supersaturation, become the sites of condensation. The nucleus, which may give rise to a droplet growth at a given supersaturation, is said to be of a critical size. This critical size is the smaller, the higher the supersaturation. **Figure 1** demonstrates the critical radius of nucleating droplets versus the supersaturation value for water (1) and dioctyl phthalate (2) Critical supersaturation depends on the properties of the substance being condensed:

$$\ln\left(\frac{p}{p_\infty}\right) = K\left(\frac{\sigma}{T}\right)^{3/2}\frac{\tilde{M}}{\rho}, \qquad (1)$$

where σ and ρ are the surface tension and the condensate density, respectively; $\tilde{M}$, the molecular weight of the substance; p and p_∞, the vapor pressure over the nucleus surface and over a plain liquid surface at the ambient temperature T.

If the vapor contains particles whose surface is wetted by the condensate, the process proceeds at a lower supersaturation. A still lower supersaturation is needed if particles consist of a substance soluble in the condensate.

The presence of charged particles in a vapor-gas medium substantially affects the value of $(p/p_\infty)_c$ (**Figure 2**). The charge sign has been shown experimentally to play a significant role: negative ions are more efficient.

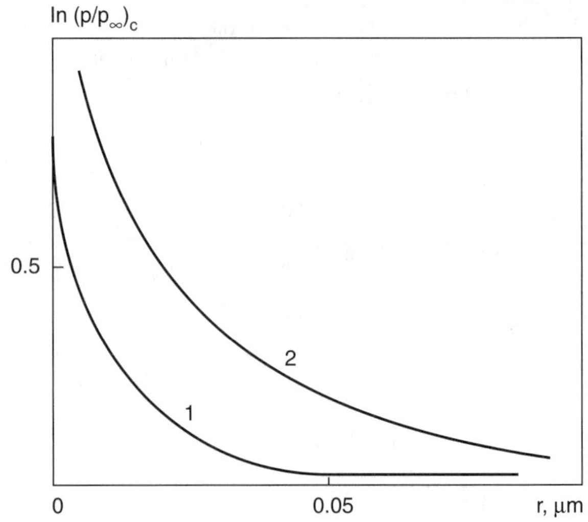

Figure 1. Critical supersaturations ratio for water (1) and dioctyl phthalate (2).

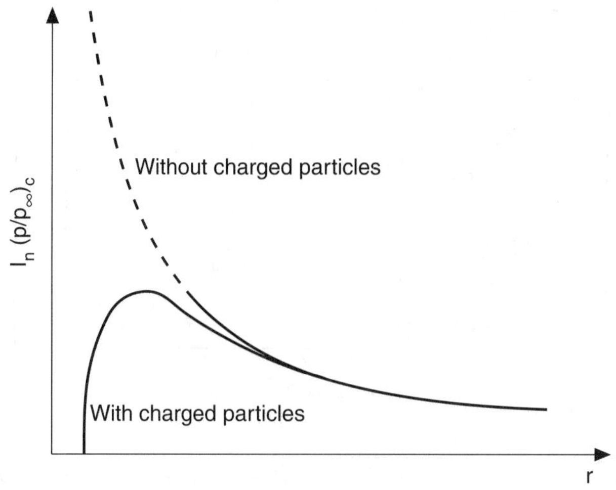

Figure 2. Effect of charged particles on critical supersaturation ratio.

The above effects are all used for artificial generation of aerosols. The simplest technique is to inject a vapor into a cool air stream, either pure or containing particles; this produces a polydisperse aerosol. Generation of *monodisperse aerosols* needs a more intricate technology. An example is the Sinclair-La Mez generator (**Figure 3**), which facilitates the production of aerosols with particles of about 1 μm, differing from one another by no more than 10%, by thoroughly controlling vapor condensation on appropriate nuclei.

A specific feature of aerosols is the spontaneous *coagulation* of particles, resulting in sticking together, growing in size, and, finally, sedimentation of the contacting particles. The main factor underlying coagulation is the **Brownian motion** of particles. For

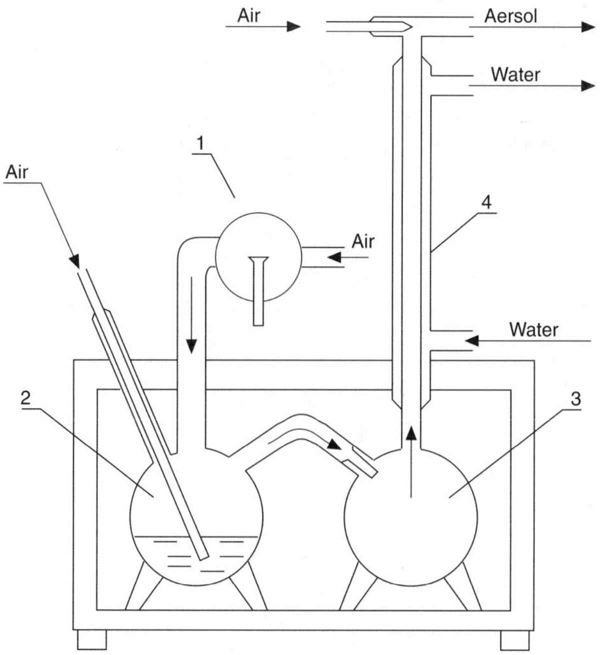

Figure 3. Sinclair-La Mez aerosol generator.

coagulation of monodisperse aerosols, the following equation holds:

$$\frac{1}{n} - \frac{1}{n_0} = \frac{2\tilde{R}Tst}{3\eta N}, \qquad \textbf{(2)}$$

where n and n_0 are, respectively, the number of particles in a unit volume at time t and those present at t = 0, $\tilde{R}$, the gas constant; T, the temperature; N, the Avogadro number; η, the medium viscosity; and, s, the ratio of the particle effective range to the radius of the particle itself (if the particles coagulate only at s = 2r). If the molecule mean free path λ is commensurable with the particle radius r, the right-hand side of the equation is multiplied by the Cunningham correction $(1 + A(\tilde{\lambda}/r))$ where A is a constant equal to 0.9 for smoke.

For polydisperse aerosols, coagulation occurs at a higher rate. In addition to temperature and pressure, it is also affected by active mixing, including the superposition of acoustic fields, and by electric charges on the particles.

Studying aerosol generation, coagulation, and other processes, as well as the practical utilization of aerosols, make it necessary to measure their basic parameters which, first and foremost, include particle size and shape, particle number density and mass fraction in a gas (see **Drop Size Measurement**).

The simplest and most frequently used devices for drop size measurement are impactors and thermal and electric precipitators. Impactors employ the inertia effect: the finer the particles, the higher the velocity to be imparted to an aerosol flow for a particle to be captured by the surface. Precipitators employ the effect of thermophoresis and electrophoresis, which have different effects on particles of different sizes and make possible that sizing of the particles.

Currently the term "aerosol" also covers gas-liquid mixtures with large droplets, used in the household, in industry and in agriculture. The mixtures are produced by liquid atomization and injection of this dispersed liquid in the gas flow. These mixtures are unstable since large droplets fall rapidly. The generators of such aerosols are simple in design, easy to handle and possess a high capacity which militates in favor of their use in agriculture for pesticide distribution over widespread areas such as fields and meadows.

References

Green, H. and Lane, W. (1964) Particulate Clouds: Dusts, Smokes, and Mists, 2nd ed., London.

I.L. Mostinsky

AEROSOLS, CLIMATIC EFFECTS (see Greenhouse effect)

AFTER BURNING (see Flames)

AGGLOMERATES (see Crystallization)

AGGLOMERATION OF PARTICLES (see Dispersed flow)

AGITATED VESSEL HEAT TRANSFER

Following from: *Agitation devices; Heat transfer*

Introduction

Heat transfer in agitated vessels can be carried out either through an *external jacket* on the vessel or by *internal coils.* Where a jacket or coils cannot provide the surface area required, a recirculation loop with an external heat exchanger may be used. In this case the heat exchanger would be designed by the normal methods and will not be covered further in this chapter. (See **Heat Exchangers**)

A jacket may be either a full conventional jacket, a dimpled jacket, or a half-pipe jacket, often called a limpet coil, as illustrated in **Figure 1a**, **1b** and **1c**. The designs are compared by Markovitz (1971).

A conventional jacket has the advantage that it covers the full wall and base surface and is very simple to construct. A dimpled jacket allows construction from light gauge metals while maintaining strength. A half-pipe jacket may be cheaper for a high pressure on the service side and has the advantage that more than one service can be supplied to different sections of the wall. However, a limited amount of the surface will be covered by a half-pipe jacket, the large amount of weld can cause mechanical concerns where thermal cycling occurs and the jacket welding must be spaced from the dished end main welds to maintain mechanical integrity of the vessel wall.

Internal coils may be full helical coils, or a number of smaller, ringlet coils, **Figure 2a** and **2b**.

A full helical coil is the more usual design, allowing the maximum surface to be installed, but requires a two-piece vessel with a relatively expensive main flange. Smaller ringlet coils can be designed to be inserted through large branches on the upper vessel dished end, but can leave quiescent, unmixed regions within their circumference.

The choice between a jacket and coils is based on a number of considerations. For highly corrosive or highly reactive materials, a jacket has the advantage that there are no extra materials of construction and no extra metal surface in contact with the process other than the normal vessel wall. There is also less risk of cooling fluid coming into contact with the reaction mass. For the manufacture of pharmaceuticals, fine chemicals and performance products, a jacket minimises contamination as there are no extra surfaces to clean. For materials with difficult rheology the full range of agitator designs can be used with a jacket without difficulty. However, a jacket has a lower heat transfer performance than a coil as there will be a lower process side coefficient, usually a greater wall thickness, and a smaller surface area. A jacket may also require a higher service side flow. For exothermic reactions, a jacketed vessel has the disadvantage that the area/volume ratio decreases with increasing scale. The use of a greater height/diameter ratio at larger scale can help to reduce this problem, but only to a limited extent. A coil has the advantage that a large surface area can be provided, for example in one particular highly exothermic reaction $18 m^2 m^{-3}$ has been installed in a 5 m^3 reactor. However, it is important not to pack the coil so tightly as to form a false wall.

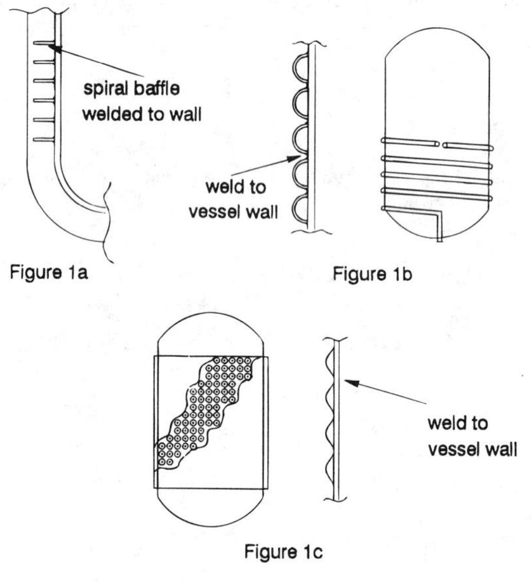

Figure 1.

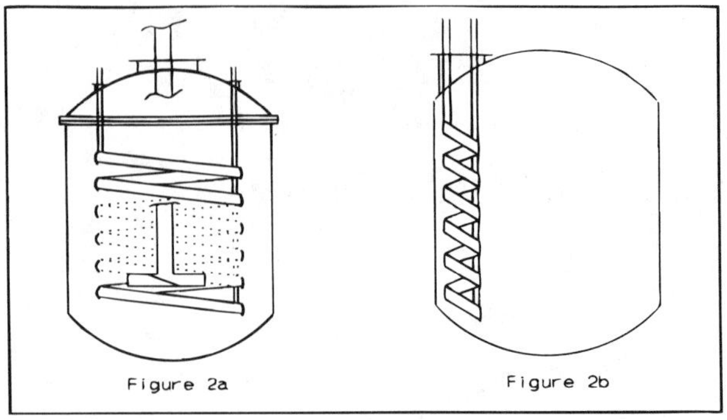

Figure 2.

Agitated vessel heat transfer is commonly used in batch manufacture where it is frequently necessary to calculate the time to heat or cool a batch or the cooling capacity required to hold an exothermic or endothermic reaction at constant temperature. It may also be necessary to define the stable operating region or acceptable reagent addition rate for an addition controlled highly exothermic semi-batch reaction. The heat removal rate is defined by:

$$\dot{Q} = UA\Delta T_m \tag{1}$$

For the simple case of the time to cool or heat a batch of mass, M:

$$\dot{Q} = Mc_p \frac{\delta T}{\delta t} = UA\Delta T_m \tag{2}$$

For a constant service side temperature, T_s for example steam heating:

$$\frac{\delta T}{\delta t} = \frac{UA}{Mc_p}(T_s - T) \tag{3}$$

The time to reach a temperature, T from a starting temperature $T_{t=0}$ is:

$$t = \frac{Mc_p}{UA} \ln\left(\frac{T_s - T_{t=0}}{T_s - T}\right) \quad \textbf{(4)}$$

For more complex situations numerical integration may be required, but there are many suitable dynamic simulation languages available. For coils and jackets, the **Overall Heat Transfer Coefficient** can be calculated in the usual way:

$$\frac{1}{U} = \frac{1}{\alpha} + \frac{\delta}{\lambda} + \frac{1}{\alpha_s} + \frac{1}{\alpha_f} \quad \textbf{(5)}$$

Where α and α_s are the process and service side heat transfer coefficients respectively. The service side fouling resistance, $1/\alpha_f$, will be available from local experience or from Kern (1950) for example. As a general guide, approximate overall coefficients typical of agitated jacketed vessels are given in **Tables 1** and **2**.

Table 1. Typical overall coefficients for jacketed glass lined steel vessels

Duty	U ($W\ m^{-2}K^{-1}$)
Distillation/Evaporation	350
Heating	310
Cooling	200
Cooling (chilled service)	100

Table 2. Typical overall coefficients for jacketed carbon and stainless steel vessels

Duty	U ($W\ m^{-2}K^{-1}$)
Heating	400
Cooling	350
Cooling (chilled service)	150

A typical overall coefficient for a well designed coil would be 400 to 600 $W\ m^{-2}K^{-1}$.

Service heat transfer coefficient

Under normal circumstances the overall coefficient should be dominated by the process side. However, for the service side the following guidelines should be observed to ensure good performance:

For jackets:

1. A conventional jacket should be fitted with baffles (see **Figure 1a**)
2. Service injection nozzles should be used to direct the service flow, especially for a glass-lined steel vessel (see **Figure 3**)
3. A vent should be fitted and maintained to prevent gas blanketing
4. For a plain jacket with liquid service the target circumferential velocity should be 1–1.5 ms^{-1}
5. For a half-pipe jacket with liquid service the minimum target velocity should be 2.3 ms^{-1}
6. For a dimpled jacket with liquid service, pressure drop may limit the velocity to 0.6 ms^{-1}

For an internal coil:

1. For a liquid service, the minimum target velocity should be 1.5 ms^{-1}

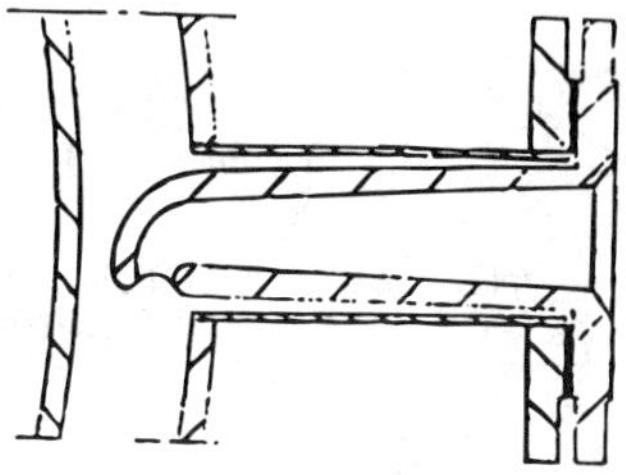

Figure 3. Service injection nozzle.

There are several correlations available for the service side coefficient, depending on the jacket or coil design and the flow regime of the service fluid. Fletcher (1987) provides some useful guidance and the research club, HTFS has also produced a design report for members. The following correlations are recommended:

1. **Conventional unbaffled jacket, liquid service with high flow (Lehrer 1970)**

$$Nu = \frac{0.03Re^{3/4}Pr}{1 + 1.74Re^{-1/8}(Pr - 1)}\left(\frac{\eta}{\eta_w}\right)^{0.14} \quad \textbf{(6)}$$

where:

$$Nu = \frac{\alpha_s d_e}{\lambda},$$

$$Re = \frac{d_e\rho(\sqrt{v_i v_A} + v_B)}{\eta}, \quad \textbf{(7)}$$

$$d_e = 0.816(D_j - D_T)$$

v_i is the velocity at the inlet nozzle or branch and v_B is the velocity component due to buoyancy, D_j is the jacket diameter and D_T is the vessel diameter.

$$v_i = \frac{4\dot{V}}{\pi d_i^2}, \qquad v_B = 0.5\sqrt{2z\beta g\Delta T_s} \quad \textbf{(8)}$$

Where $\dot{V}$ is the volumetric flow rate, d_i the diameter of the inlet, z the wetted height of the jacket, β the fluid thermal expansion coefficient and ΔT_s the temperature rise of the service fluid. v_A is the rise velocity in the jacket annulus and depends on the inlet orientation:

$$\text{Radial; } v_A = \frac{4\dot{V}}{\pi(D_j^2 - D_T^2)}, \qquad \text{Tangential; } v_A = \frac{2\dot{V}}{(D_j - D_T)z} \quad \textbf{(9)}$$

2. **Conventional unbaffled jacket, liquid service with low flow dominated by convection (Barton and Williams 1950)** (see also **Free Convection**)

$$\mathrm{Nu} = \mathrm{K}\left(\frac{z^3\rho^2\beta g\Delta T_m}{\eta^2}\right)^{1/3} Pr^{1/3} \quad \textbf{(10)}$$

Here,
K = 0.15 for upward flow, heating; downward flow cooling
K = 0.128 for downward flow heating; upward flow cooling;

ΔT_m is the mean temperature difference between the service and the vessel wall

$$\mathrm{Nu} = \frac{\alpha_s z}{\lambda} \quad \textbf{(11)}$$

3. **Baffled or dimpled jacket, liquid service.**The service side coefficient for a baffled or dimpled jacket will be greater than for an unbaffled jacket with high flow, therefore the above can be used as a conservative estimate. Using the correlation for a half-pipe coil with the flow area equivalent to the baffle channel is not recommended as it may give an overestimate.
4. **Half-pipe coil, liquid service**

$$\mathrm{Nu} = 0.023\mathrm{Re}^{0.8}\mathrm{Pr}^{1/3}\left(\frac{\eta}{\eta_w}\right)^{0.14} \mathrm{E} \quad \textbf{(12)}$$

This is the normal Sieder-Tate equation, (see **Forced Convective Heat Transfer**) applied to the whole jacketed area with the effectiveness factor (E) from Kneale (1969), typically 0.8 − 1. Nu and Re are based on the hydraulic mean diameter, d_e:

$$d_e = \frac{\pi d_i}{2} \quad \textbf{(13)}$$

5. **Condensing service**
A condensing coefficient in a jacket should be extremely high compared to the process side and normally $\alpha_s^{-1} \sim 0$. A conservative estimate can be obtained from the Nusselt analysis (see Kern 1950):

$$\mathrm{Nu} = 0.943\left[\frac{z^3\rho^2 h_{LG}\, g}{\eta\lambda\Delta T}\right]^{1/4} \quad \textbf{(14)}$$

The physical properties refer to the liquid phase and $\Delta T = T_{sat} - T_{wall}$ (See **Condensation**)

Process side heat transfer coefficient

The process side coefficient will be determined by the agitator type and speed. (See also **Agitation Devices**) For low viscosity fluids, most turbine-type high-speed agitators can give good performance. For high viscosity and non-Newtonian fluids, larger diameter agitators will be required. Harnby et al. (1985) and Oldshue (1983) provide guidance on agitator selection. For **Non-Newtonian Fluids**, there are two important considerations:

Table 3. Metzner-Otto constants for impellers

Impeller type	K_{mo}
propeller	10
disc or flat blade turbine	11.5
angled turbine	13
anchor	25*
helical ribbon	30
EKATO Intermig ($0.8D_T$)	40

*The value for an anchor depends on the rheology of the material, for more detail see Nienow (1988).

1. A mean apparent viscosity is required to calculate Nu from Re. The normal practise is to estimate it using the Mezner-Otto approach, where:

$$\eta_a = F(\gamma);\ \text{and}\ :\gamma = K_{mo}N \quad \textbf{(15)}$$

K_{mo} depends on the agitator type, N is the rotational speed (s^{-1}):

The viscosity correction only allows for the effect of temperature near the wall and NOT to the distribution of shear rate. It will normally be sufficient to assume an homogeneous distribution of shear rate. Calculation of a wall shear rate to predict a local η_a is beyond the scope of this article.

2. It is essential to ensure that a material with a significant yield stress is maintained fluidised right up to the vessel wall. Nienow (1988) reviews the design techniques. Use the vessel diameter (D_T) as the diameter of the region to be fluidised:

$$\left(\frac{D_T}{D}\right)^3 = \frac{\mathrm{PoRe}_y}{\left(0.55 + \frac{1}{3}\right)\pi^2} \quad \textbf{(16)}$$

Po is the impeller Power number (Harnby et al(1985)), and:

$$\mathrm{Re}_y = \frac{N^2D^2\rho}{\tau_y} \quad \textbf{(17)}$$

For the vessel wall surface, the process side coefficient can be calculated from:

Table 4. Heat transfer constants for impellers

Impeller	A	B
propeller	0.46	1.4
45° turbine	0.61	1.4
disc turbine	0.87	1.4
retreat curve	0.33	0.87
anchor	0.33	
Intermig	0.54	

$$Nu = ARe^{2/3}Pr^{1/3}\left(\frac{\eta}{\eta_w}\right)^{0.14} \quad \textbf{(18)}$$

$$Re = \frac{ND^2\rho}{\eta}; \qquad Nu = \frac{\alpha D_T}{\lambda} \quad \textbf{(19)}$$

For transfer to an internal coil:

$$Nu = BRe^{0.62}Pr^{1/3}\left(\frac{\eta}{\eta_w}\right)^{0.14} \quad \textbf{(20)}$$

Wall resistance

The conductivity of the wall material can be found in standard texts (Kern 1950). The resistance may be significant for some vessels linings, for example glass lined steel, where the manufacturer's data should be consulted. There will also be some limitations on the ability of glass lining to withstand thermal shock.

References

Barton, E. and Williams, E. V. (1950) Experimental Determination of Film Heat Transfer Coefficients, *Trans. I. Chem. E.*, 17:3.

Fletcher, P. (1987) Heat Transfer Coefficients for Stirred Batch Reactor Design, *The Chemical Engineer*, April.

Harnby, N., Edwards, M. F., and Nienow A. W. (1985) *Mixing in the Process Industries*, Butterworths.

Kern, D. Q. (1950) *Process Heat Transfer*, New York: McGraw-Hill.

Kneale, M. (1969) Design of Vessels with Half Coils, *Trans. I. Chem. E.*, 47.

Lehrer, I. H. (1970) Jacket-side Nusselt Number, *Ind. Eng. Chem. PDD*, 9:4.

Markovitz, R. E. (1971) Picking the Best Vessel Jacket, *Chemical Engineering*, Nov 15th.

Nienow, A. W. (1988) Aspects of Mixing in Rheologically Complex Fluids, *Chem. Eng. Res. Des.* 66:1.

Oldshue, J. Y. (1983) *Fluid Mixing Technology*, New York: McGraw-Hill.

K.J. Carpenter

AGITATED VESSEL MASS TRANSFER

***Following from:** Agitation devices; Mass transfer*

Mass transfer in agitated vessels usually involves dispersed gases absorbing into and often reacting with a continuous liquid phase, e.g. in oxidation or chlorination. Interfacial area and contact time depend on the operating conditions. Even when the relevant mass transfer is to or from suspended solids, the rate may be affected by segregation. The rate of mass transfer N_B of a component B between phases 1 and 2 can be expressed simply as an overall mass transfer coefficient K multiplied by the total interfacial contact area A and the overall equivalent concentration driving force.

$$N_B = K.A.(c^*_{B1} - c_{B2}) = K.a.V.(c^*_{B1} - c_{B2})$$

In the latter case a is the specific interfacial area, i.e., the total contact area per unit volume of the dispersion and V is the volume of the reactor. The notation c_{B2} refers to the actual concentration of component B in phase 2 while c^*_{B1} refers to the concentration of B that would be present in phase 2 if it were in equilibrium with the bulk concentration in phase 1.

The K term embodies the resistances to mass transfer on each side of the interface which are controlled by the local turbulence and the local rates of diffusion of the transferred material, $1/K = 1/k_1 + 1/k_2$. In processes involving contaminated interfaces there may be a third term to allow for a diffusional resistance in a surface film.

Depending on the nature of the system, particularly the ease of saturating one of the phases, or the presence of relatively fast reactions, it is usual to simplify the model and consider the controlling resistance as being on one side or the other. For most physical absorption systems (like oxygen dissolving in clean water) liquid phase conditions are most important so k_L is the "film coefficient" of interest. With reacting systems (e.g., atmospheric oxygen into a waste water with a significant concentration of a strong reducing agent like sulphite) k_G may be more relevant. In general, liquid phase agitation has little influence on the gas side coefficient, though it does affect the contact area term of course.

The area A results from a balance between the dispersing actions of the fluid forces in the system and complex coalescence processes that depend on the frequency and duration of bubble or droplet collisions. The physical and chemical nature of the system is very relevant, as are the agitation conditions, usually characterised in terms of the local turbulence or energy dissipation rate. Unfortunately, the critical influence of contamination on dispersion size distribution and the cleanliness of interfaces has not been quantified and most of the reliable data available in the literature refer to clean air-water systems.

Since it is difficult to separate the effects of agitation on K and A, it has become usual to group the terms and express values of k_La or k_Ga as functions of the process variables. The concentration driving force is also difficult to define accurately, especially in large vessels which may be very far from homogeneous and for which it is usually necessary to select some appropriate model that allows for possible segregation in either or both phases.

Hydrodynamic considerations

Mass transfer processes in agitated vessels are almost invariably turbulent, with *mixing Reynolds numbers* above 10^4 in the uniform ungassed *Power number* regime ($Re = ND^2\rho/\eta$ where D and N are the impeller diameter and speed, and ρ and η are the liquid density and viscosity respectively, see **Agitation Devices**). Energy dissipation controls both surface area and turbulent mass transfer coefficients, so it is necessary to know the power demand of gassed agitated systems, a parameter that is considerably influenced by operation in a two-phase regime.

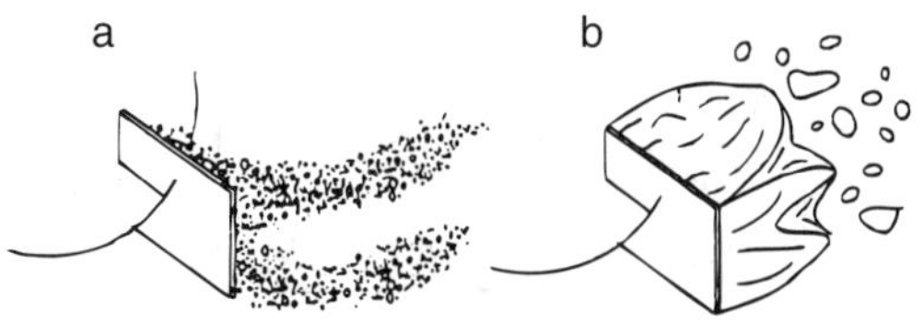

Figure 1. The principal cavity forms.

Gas dispersion

Radial flow turbines (see **Agitation Devices**) are usually used for gas dispersion. Efficient dispersion is not sensitive to the details of sparger design providing that the supplied gas reaches and is distributed by the impeller. Down pumping axial flow impellers are less suitable because rising gas tends to disrupt the liquid outflow unless the sparger ring is larger than the agitator so that only gas that has circulated around the tank reaches the impeller.

Blade cavities and flow regimes

Gas-liquid dispersion can be characterised in terms of the *gas flow number* Fl_G ($= \dot{Q}_G/ND^3$) and the **Froude Number** Fr ($= N^2D/g$) in which $\dot{Q}_G$ is the gas supply rate, g gravitational acceleration. It is probable that the Froude Number ought also to include a density ratio term for the two phases though since $\Delta\rho/\rho \sim 1$ in gas/liquid systems this is usually ignored. At low values of Fl_G, common in small scale laboratory investigations, the trailing vortices behind the blades of radial flow impellers accumulate and redisperse gas, **Figure 1a**. Beyond a well defined value of Fl_G gas accumulates in large cavities behind the blades, **Figure 1b**. This regime is the usual at industrial scale.

Moderately high impeller speeds are needed to recirculate rising gas to the impeller or to carry bubbles down to the bottom of the vessel. **Figure 2a** and **Figure 2b**.

Very high gas rates, or insufficient impeller speeds, can lead to the pumping action of the agitator being overwhelmed by buoyancy forces, a condition known as flooding, **Figure 2c**. The flow transitions are characterised by the following equations; cavities are of the vortex form if $Fl_G < 0.18(D/T)^2$, flooding will occur if $Fl_G > 30Fr(D/T)^{3.5}$. Recirculation requires that $Fl_G < 13Fr^2(D/T)^{-5}$. If conditions are too gentle, $Fr < \sim 0.045$, buoyancy forces allow the gas to escape and no stable cavities are retained. These equations are used as the basis for **Figure 3**, which shows a flow regime map for Rushton turbines with $D/T = 0.4$.

Flow maps can ensure that hydrodynamic conditions met in a full-scale plant match those used in laboratory investigations.

Power demand in turbulent gassed systems

For almost all designs of impeller the power required to maintain a given rate of rotation is less when loaded with a dispersion of gas or vapour in liquid than when operating in liquid alone. It has been shown that this is not merely a density effect. The *relative power demand*, (RPD), generally falls smoothly until the impeller is finally swamped by gas **Figure 4**. shows typical results for a Rushton turbine loaded with different gas rates while maintaining constant speeds.

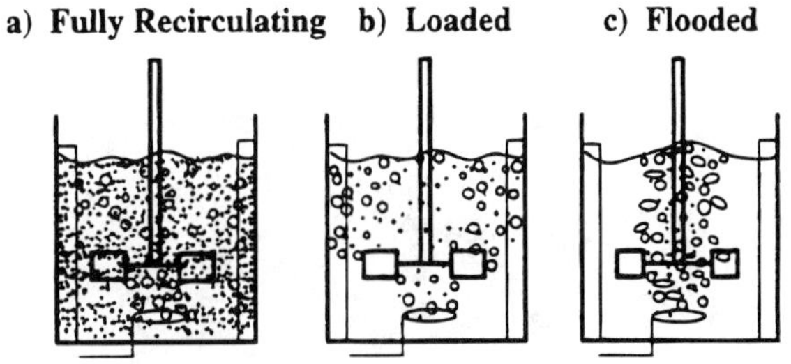

Figure 2. Vessel flow regimes as $\dot{Q}_G$ increases or N is reduced.

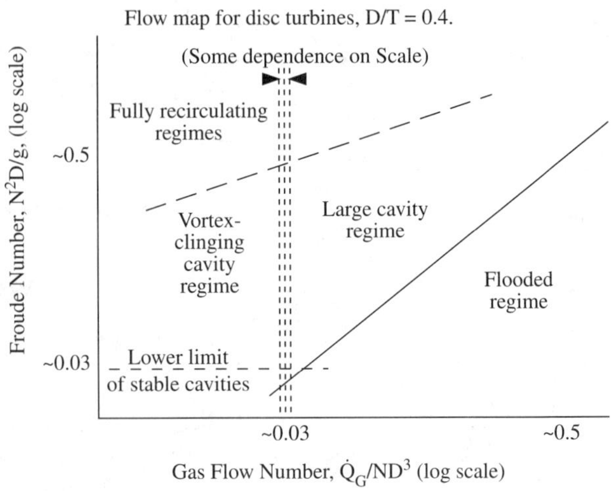

Figure 3. A flow map for radial flow turbines.

The RPD levels off at about $0.25 \times Fr^{-0.3}$ as the flooding transition is approached. The moment of large cavity development corresponds to a point of inflection in the curve, about halfway towards the final level of RPD. On this basis accurate estimates can be made of the power input of a Rushton impeller for any operating conditions.

Gas supplied beneath a downward pumping axial impeller opposes the agitator action as it rises against the downflow. The impellers have less intense tip vortices than a Rushton so that vortex cavities are less effective in their dispersing action. The development of large cavities, which severely affect the efficiency of pitched blade turbines and narrow blade hydrofoils, is hastened by direct loading if gas is supplied from a small sparger mounted close to the impeller.

Figure 5 shows a set of RPD curves for a 30° pitched Blade Turbine. Operating near the catastrophic fall in power level when large cavities develop would be highly undesirable from the mechanical viewpoint. Beyond this point the competing actions of pumping and buoyancy make the impeller function as a rather inefficient radial impeller.

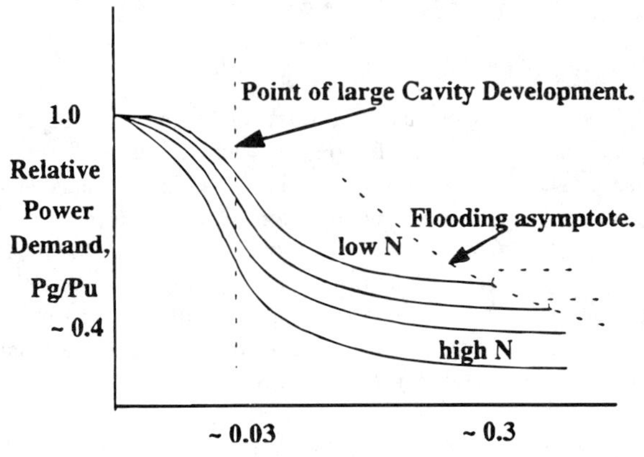

Figure 4. Relative power demand for a Rushton turbine at various constant speeds.

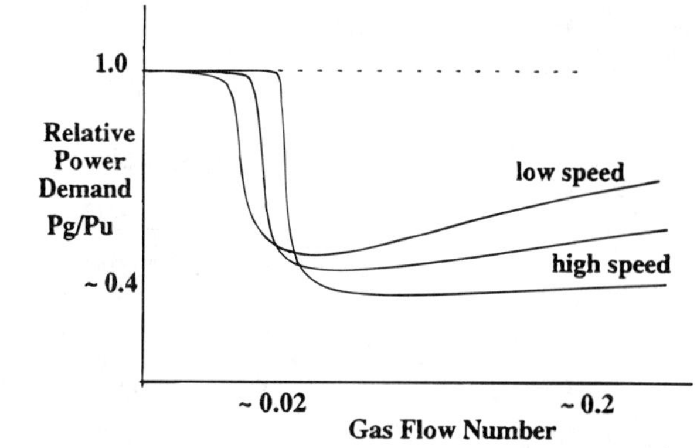

Figure 5. Relative power demand for a 30° pitched blade turbine.

Multiple impeller systems

Many reactors use more than one impeller on a common shaft, usually in an attempt to improve the quality of mixing throughout the vessel (see Figure 6). The favoured geometry is that of one or more downward pumping axial impellers above a single radial flow impeller which is mounted just above a sparger. In modern practice the upper impellers are usually wide blade hydrofoils. RPD in multiple impeller equipment can be estimated by treating the lowest impeller as if it were alone in a "standard" tank. The upper impellers are only loaded with about 40% of the sparged gas throughput because of by-passing (Warmoeskerken and Smith, 1988).

Mass transfer in agitated systems

Power dissipation is the key to mass transfer and gas holdup. In terms of overall performance a widely used correlation for clean ("coalescing") systems (Harnby et al. 1992) is :

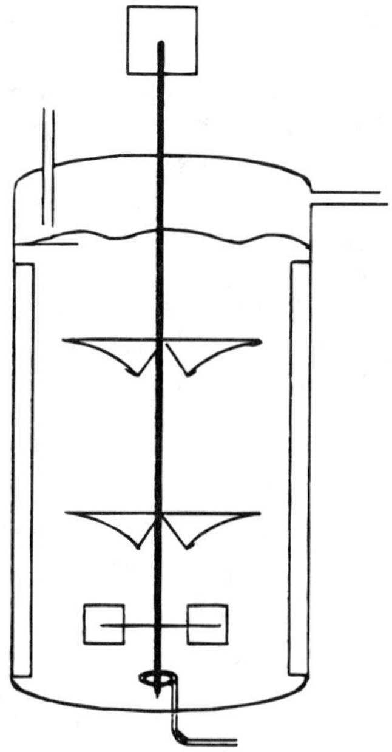

Figure 6. Multiple impellers.

$$k_La = 1.2(P_G/V_L)^{0.7}v_s^{0.6}$$

In some solutions with reduced coalescence, e.g. aqueous ionic solutions above 2 mole% concentration, the smaller bubble sizes lead to mass transfer rates perhaps twice that represented by this equation. Other solutions that generate stable small bubble dispersions, notably those of proteins and surfactants, do not produce the opposite effect if transfer through the interface is hindered.

Mass transfer rates in multiple impeller installations are much more difficult to characterise because of the uncertainties in the local concentration driving force. One basis for design is to consider the gas either to be well mixed (implicitly then everywhere at its outlet concentration) or in plug flow, with the liquid compartmentalised into regions centred on each impeller with appropriate estimates of the local energy dissipation rate.

Boiling and hot sparged systems

Many stirred tank reactors involve boiling systems. In the absence of sparged gas the RPD can be estimated on the basis of an *Agitation Cavitation number* defined as $CAgN = (2Sg/v_t^2)$ in which S is the submergence to the impeller mid plane and v_t is the tip velocity. Providing this Agitation Cavitation number is less than some critical value, characteristic for each design of impeller, the RPD can be expressed to a sufficient accuracy by $RPD = A \times CAgN^{0.4}$ in which the constant A has the value of 1.2 for a Rushton Turbine and 0.8 for a 6 blade PBT, (Smith and Katsanevakis, 1993). The power draw in sparged conditions can be estimated by assuming that the total loading provided by vapour and gas in a sparged hot system is that corresponding to complete saturation of the sparged gas at the bulk liquid temperature. This will always be lower than the boiling point at the reactor pressure. It seems likely that in boiling systems the mass transfer will be gas film controlled so the conventional equations for predicting mass transfer rates, which are based on k_La rather than k_Ga, need to be used with caution.

References

Harnby, Nienow, and Edwards, (1992) *Mixing in the Process Industries*, Massachusetts: Butterworth Heinemann.

Smith, J. M. and Katsanevakis, A. (1992) *Trans. I. Chem. E.*

Warmoeskerken, M. M. C. G. and Smith, J. M. (1988) Proc. 2nd. Conf. *Bioreactor Fluid Mechanics*, Elsevier, 79–197.

J.M. Smith

AGITATED VESSELS (see Crystallizers)

AGITATION CAVITATION NUMBER (see Agitated vessel mass transfer)

AGITATION DEVICES

Agitation is the key to many heat and mass transfer operations that rely on mixing. Process requirements vary widely, some applications requiring *homogenisation* at near molecular level while

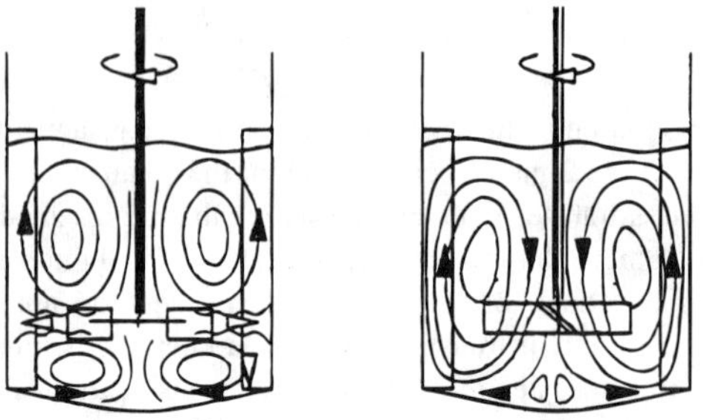

Figure 1. Alternative flow fields.

other objectives can be met as long as large scale convective flows sweep through the whole vessel volume. Performance is crucially affected both by the nature of the fluids concerned and on how quickly the mixing or dispersion operation must be completed. For these reasons a wide variety of agitation devices have been developed.

Conventional, mechanically agitated, *stirred tank reactors* may be used for either batch or continuous processes, though the design and operating constraints are different in the two cases.

Low viscosity fluids can usually be mixed effectively in baffled tanks with relatively small high speed impellers generating turbulent flows, while high viscosity (typically above about 10 Pa s) and non-Newtonian materials require larger, slow moving agitators that work in the laminar or transitional flow regimes. It is convenient to classify *impellers* as radial or axial pumping depending on the flow they generate in baffled tanks, **Figure 1**.

Figure 2 shows: a) a radial flow, *"Rushton", turbine* which produces considerable turbulence near the impeller, b) a "pitched blade" impeller with flat, angled blades that generates a diverging but generally axial flow, c) a hydrofoil impeller with carefully profiled blades that develop a strong, more truly, axial flow of low turbulence. Impellers suitable for viscous fluids are: d) a helical ribbon with a blade that travels close to the wall of the tank to force good overall circulation and e) an anchor that produces strong swirl with poor vertical exchange, even when baffled with stationary breaker bars or "beaver tail" baffles.

Energy transfer

The power input P to an impeller of diameter D driven at a rotational speed N in a fluid of density ρ and viscosity η can be

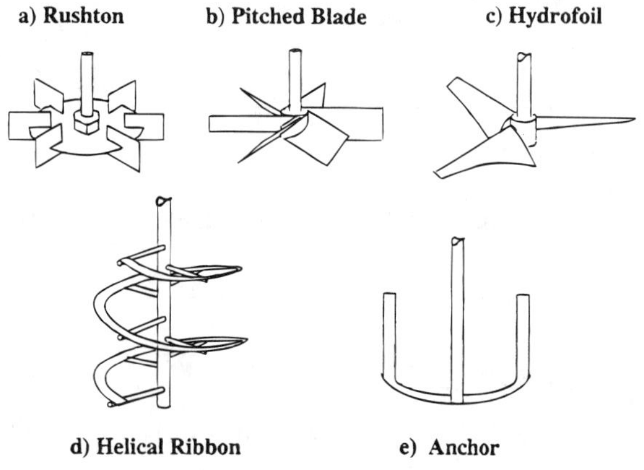

Figure 2. Various impellers.

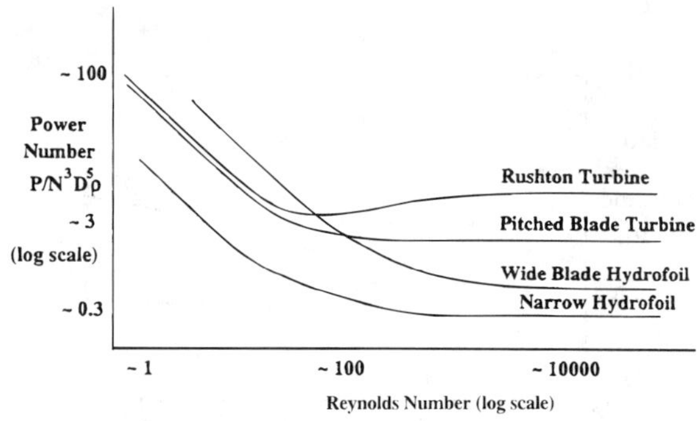

Figure 3. Log-log plot of power number as a function of Reynolds number for Rushton, pitched blade and hydrofoil turbines.

expressed in terms of a dimensionless *Power number*, $P/(N^3D^5\rho)$. This is a form of drag coefficient and is a function of the *mixing Reynolds number* $(ND^2\rho/\eta)$. For a given pattern of impeller the Po vs. Re function is always of the same form.

Figure 3 shows typical graphs of the relationship for four impellers, [see e.g. Harnby et al. (1992)]. Each shows the transition from laminar flow, where the drag coefficient is dominated by viscosity to turbulent conditions. At high Reynolds Numbers, $>\sim 10^5$ the Power Numbers become relatively constant at values that reflect the local turbulence generation. Although these curves are typical, that for a particular design of impeller is affected by the fine detail of construction.

Paddle impellers with six flat blades are ancient (Agricola 1553). The popular, simple, robust and effective design with six blades mounted on a disc was adopted by Rushton as a standard for comparative tests. Providing the tank is fitted with wall baffles to eliminate gross swirl—there are usually four baffles, each with a width of about one tenth of the tank diameter—the overall flow fields sketched in **Figure 1** develop.

The detail of the flow behind the blades is interesting, **Figure 4a**. Flow around the blade of a Rushton turbine separates with a pair of trailing vortices from the tips of each blade. These vortices, which are rather less developed in a turbine without a disc, are the locations of the greatest shear rates and most intense turbulence in the vessel (van't Riet and Smith). In the fully developed turbulence region ($Re > \sim 10^5$) the power number of a radial flow turbine is about 5. Much of the energy is small scale turbulence associated with the vortices; this decays rapidly in the outflowing stream so that about 80% of the energy transmitted from the shaft is dissipated in about 20% of the vessel volume.

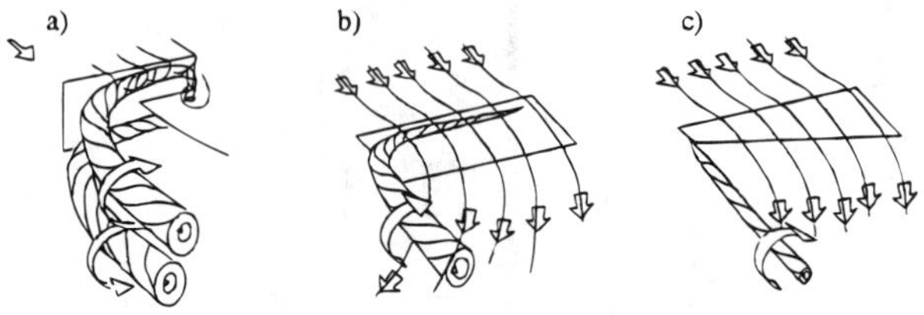

Figure 4. Details of the flow around turbine blades a) a Rushton turbine, b) a pitched blade impeller. c) a hydrofoil.

Axial flow impellers

Pitched Blade Turbines have lower turbulent power numbers than Rushton turbines, about 1 in the fully turbulent regime, in part at least because of the lower intensity of the single vortex from each blade end, **Figure 4b**. Providing the tank is baffled (without this the bulk of the liquid will swirl around with little axial motion) the axial flow will develop. However, the blades are not streamlined and there is some separation of the boundary layer giving a discharge that diverges as it leaves the impeller plane. This divergence becomes more marked the higher the fluid viscosity.

Hydrofoils

In recent years improved, more streamlined, axial flow impellers, usually with three or four blades, have been developed. These generate good axial flow with very little turbulence and are widely used when effective bulk motion if the liquid is required. The tip vortices are weaker than those of a pitched blade impeller and the energy is dissipated very uniformly throughout the vessel volume, **Figure 4c**. Narrow blade hydrofoils have been used for heat transfer, solid suspension and dissolution while wider blade versions are more successful in gas-liquid systems.

Helical Ribbons and other proximity agitators

The near impossibility of generating turbulence in viscous and non-Newtonian materials means that effective mixing depends on ensuring that all the fluid is moving. This can only be achieved with impellers that are large and which sweep out the whole vessel volume. Several patterns have been developed, amongst them helical ribbons and anchors (**Figures 2d, e**). Driving these large diameter, slow moving impellers requires gearboxes that deliver a large torque; mixing equipment for these fluids is therefore usually heavy and expensive.

Batch mixing time

One measure of mixing performance is the batch mixing time. This has to convey some assessment of the asymptotic approach to homogenisation, usually achieved in terms of the addition of a tracer and the subsequent approach to homogenisation. The 95% mixing time is often used for this criterion.

The moment of tracer addition, is taken as zero, t_d is the dead time before the addition is first detected, t_c the circulation time and t_m the 95% mixing time, defined by the last measurement that lies outside the 5% band of the total concentration change.

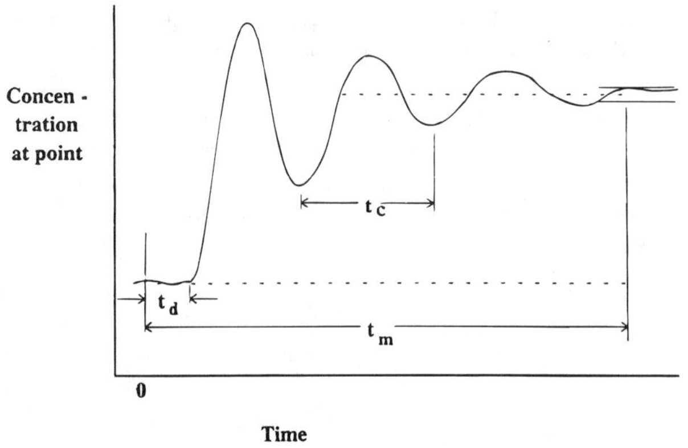

Figure 5. An idealised mixing time experiment.

Suitable tracers are pH changes, dyes and salts; even hot liquid can be used. The major difficulties include avoiding spurious affects due to density differences and establishing a reproducible protocol that can be related reliably to the desired process application.

References

Agricola, De Re Metallica, (1553) (rep. Dover, 1950).

Harnby, N., Nienow A. W., and Edwards, M. F. (1992) *Mixing in the Process Industries,* Butterworth-Heinemann.

van't Riet, K. and Smith, J. M. Real and pseudo turbulence in the discharge stream from a Rushton turbine, *Chem. Eng. Sci.* 31:407–412.

Leading to: *Agitated vessel heat transfer; Agitated vessel mass transfer*

J.M. Smith

AGR (see Advanced gas cooled reactor)

AICHE (see American Institute for Chemical Engineers)

AIR CARRYUNDER (see Plunging liquid jets)

AIR CONDITIONING

Following from: *Buildings and heat transfer*

Introduction

Heating, Ventilating and Air Conditioning (HVAC), generally means the provision of an acceptable thermal environment within buildings. It includes heating, cooling, humidifying, dehumidifying, filtering, and distribution of air at suitable conditions for the maintenance of human comfort or for the undertaking of a particular process.

In order to design an air conditioning system the appropriate heating, cooling and other environmental loads must first be calculated and a variety of other factors such as initial and running costs, plant room and distribution space, control requirements etc. must be assessed. In practice, there is a wide variety of air conditioning systems available and selection is often dictated by factors other than the air conditioning loads.

Need for air conditioning

The need for air conditioning is dictated by the required conditions within the space and the incident loads. In the case of process conditioning the requirements are often quite critical where even a small deviation in the temperature or humidity might result in catastrophic damage to the process. In fact, many products of today could not be produced at all were the environment not controlled within narrow limits.

On the other hand, conditioning for human comfort can be less critical and acceptance of diurnal swings in temperature and humidity can result in considerable savings in plant costs and in energy during operation. Careful design of thermal mass of the building fabric and of natural ventilation can often reduce or even obviate the need for air conditioning altogether.

Until recently, air conditioning was generally considered to be a prestigious component of a building, with "fully air-conditioned" buildings commanding far higher rents than their "heated and ventilated" counterparts. However, in climates such as the UK there has been a swing in demand towards low energy naturally ventilated buildings. This together with the changes in work practices away from the use of large, centralised, prestigious headquarter buildings has lead to less significance being attached to air conditioning in the eyes of the building owners and occupiers. However, there will always be a need for air conditioning in harsher climates, but even in the UK when building on restricted, noisy, polluted centre city sites that require deep plan, high rise buildings. Under these circumstances daylight does not penetrate the centre of the building and the continuous use of artificial lighting produces high cooling loads. In addition, natural ventilation is restricted as traffic noise and external pollution often preclude the use of openable windows. Air conditioning applications and their general criteria and design considerations are discussed in detail in the ASHRAE Handbook, HVAC applications volume (1991).

Thermal comfort and indoor air quality

The provision of thermal comfort to a building occupant depends on the correct balance of air temperature, mean radiant temperature, air velocity and vapour pressure to suit the level of activity and clothing worn by the occupants. Much research has been undertaken to identify the ranges and combinations of these six factors that result in comfortable environments. Fanger (1972) described an extensive study of thermal comfort and presented a method of calculating the Predicted Mean Vote (PMV) of building occupants with respect to their thermal comfort on a scale of -3 to $+3$ with zero being the comfortable equilibrium. From this figure the Predicted Percentage Dissatisfied (PPD) can be determined. An instrument has been developed to measure the appropriate factors, integrate their impact on comfort of the occupants and produce readings of PMV and PPD.

Other factors that are controlled by the HVAC system involve the maintenance of clean, healthy, and odour-free indoor environments. These factors are often what is intended by the term *indoor air quality* or IAQ. Maintaining good indoor air quality involves keeping gaseous and particulate contaminants below some acceptable level. A description of common contaminants and methods of controlling their levels is given in McQuinston (1994).

The Chartered Institution of Building Services Engineers (CIBSE) in the United Kingdom and the American Society of Heating, Refrigeration and Air conditioning Engineers (ASHRAE) in the United States of America each publish recommendations for achieving comfort in buildings. CIBSE (1979) and ASHRAE (1993) Chapter 8.

Building loads

Air conditioning systems are required to overcome the heat loads on the building to maintain acceptable comfort conditions within the space. The systems should provide heating to overcome the heat losses and cooling to overcome the heat gains.

Heat losses

Heat losses from the building are usually calculated on a steady state basis using an internal design temperature defined by the comfort requirements of the occupants and an external design temperature appropriate for the geographical location.

The total steady state heat loss of a building is the sum of the Fabric and Ventilation heat losses (see **Buildings and Heat Transfer**). The fabric loss is the heat loss through the walls, floor and roof of the building and is dependent upon the overall thermal transmittance or 'U' value of each component and their respective surface areas. The ventilation heat loss is dependent upon the fresh air ventilation rate into the space and the need to raise it to room temperature. (See **Overall Heat Transfer Coefficient**)

These calculations are documented in CIBSE (1980), ASHRAE (1993) Chapter 8, and McQuinston (1994). They produce a calculated heat load in kW which can be used to select the appropriate boiler plant and heat exchangers for the heating cycle.

Heat gains

The heat gains to the building are rather more complex and must be calculated on a dynamic basis considering variations in external conditions and time lags of heat flow through the building fabric. The instantaneous heat gain into a conditioned space, is quite variable with time, because of the strong transient effect created by the hourly variation in solar radiation. There may be an appreciable difference between the heat gain of the structure and the heat removed by the cooling equipment at a particular time. This difference is caused by the storage and subsequent transfer of energy from the structure and contents to the circulated air. Ignoring these factors can lead to gross oversizing of equipment.

In addition to the external heat gains from solar radiation and ventilation, internal gains caused by occupants, lighting and machinery must be included. (See **Physiology and Heat Transfer**) The internal gains can be determined from the density of occupation and activity levels together with the expected lighting and equipment levels for the conditioned space. It is, however, necessary to predict future demands especially for equipment which has increased rapidly in modern office buildings.

Solar radiation is dependent upon the intensity of incident solar radiation, its angle of incidence and the properties of the building envelope. (See **Solar Energy**) Solar gains can be broadly subdivided into direct and diffuse components. The direct component is directly transmitted through the earth's atmosphere and as such has a defined angle of incidence on the surface of the building. The diffuse component is scattered by the earth's atmosphere and has no discernible angle of incidence. Nevertheless, it is absorbed at the building surface and can be conducted through to the inside. In addition, it serves to warm the external temperature which adds to the heat gain through the intake of ventilation air.

Direct solar radiation received at the outside surface of the building will be reflected, absorbed and directly transmitted in proportions dependent upon the surface materials and the angle of incidence. The component directly transmitted through glass is of greatest importance to the air conditioning cooling load.

The cooling loads for different configurations of single and double glazing with different types of glass, clear, absorbing, reflecting etc. are tabulated in the CIBSE guide CIBSE (1979a).

In addition to the sensible heat gains described above there is a need to evaluate the latent heat gains based on the moisture generated from the occupants, ventilation and any process moisture that is generated within the space.

CIBSE (1986), CIBSE (1986a) and Chapter 26 of ASHRAE (1993) provide detailed calculation procedures for estimating heat gains. These procedures and others have been computerised so that most heat gain analysis is carried out by computer calculation or modelling. Simpler methods suitable for manual calculations are described in McQuinston (1994).

Estimation of energy consumption

The selection of air conditioning plant is carried out on the basis of design conditions, ensuring that the plant is of adequate capacity to maintain the required comfort conditions within the space under the normal range of climatic conditions for the region. The estimation of energy consumption on the other hand requires detailed information of the diurnal and seasonal variations in external temperature and incident solar radiation as well as the pattern of occupation of the building.

Energy consumption will depend upon the efficiency of the air conditioning plant and of the controls employed to maintain the comfort conditions.

Many computer codes are available that model building systems quite well, and they may be used with all types of structures. The building simulation is usually carried out for a whole year considering heating, cooling and other energy requirements. There are cases, however, where computer simulation cannot be justified. In these cases, reasonable results can be obtained using the simpler degree day or bin method described in McQuinston (1994). Energy estimating is discussed in detail in Chapter 28 of ASHRAE Handbook of Fundamentals ASHRAE (1993).

Air conditioning psychrometrics

Comfort conditions are maintained within a building by supplying conditioned air at an appropriate state to overcome the incident heating and cooling loads. The required state will have specified conditions of temperature and moisture content and the processes required to achieve that supply state can be evaluated by air conditioning psychrometrics. Tabulated data defining the psychrometric

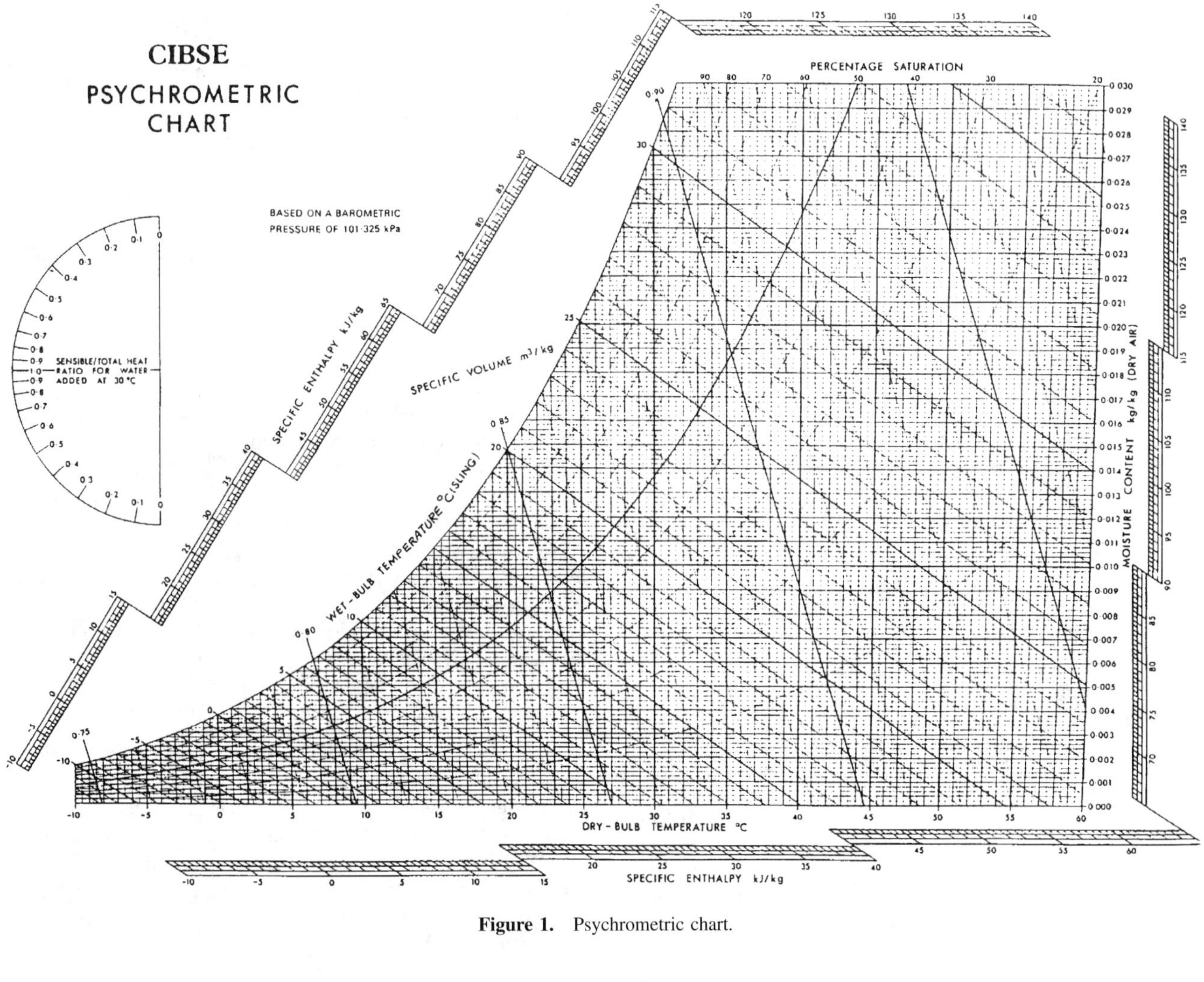

Figure 1. Psychrometric chart.

properties of moist air can be found in the data section of the CIBSE Guide (CIBSE 1975).

The changes in psychrometric properties of a sample of air as it passes through an air conditioning system can best be visualised by use of a psychrometric chart which, for a given barometric pressure, presents the psychrometric properties of moist air over a range of conditions. The standard psychrometric chart used in the UK is that produced by the Chartered Institution of Building Services Engineers (CIBSE) and reproduced here as **Figure 1**. Any point on the chart is defined by two psychrometric properties and from the chart all other properties of the state point can be determined. A schematic representation of the various air conditioning processes is given in **Figure 2**.

A full description of the psychrometrics of various air conditioning plants is given in Jones (1985), Look and Sauer (1986), Eastop and McConkey (1993) and McQuinston (1994).

Supply state

The required supply state of air conditioned air to maintain thermal comfort within a room will depend upon the sensible and latent energy gains.

There is an infinite number of combinations of supply temperature and moisture content and volume flow that will satisfy the heat load requirements. These conditions can be identified on the psychrometric chart by determining the Room Sensible Heat Factor (RSHF).

$$\mathrm{RSHF} = \frac{\dot{Q}_s}{\dot{Q}_t} = \frac{\dot{Q}_s}{\dot{Q}_s + \dot{Q}_l}$$

where
$\dot{Q}_s$ = sensible heat gain (kW)
$\dot{Q}_l$ = latent heat gain (kW)
$\dot{Q}_t$ = Total head gain (kW)

The cosine of this factor gives the slope of a line on the psychrometric chart which, when drawn through the required room conditions is known as the Room Line. Supply of air at any point on that linewill provide the correct combination of temperature and moisture content to overcome the loads.

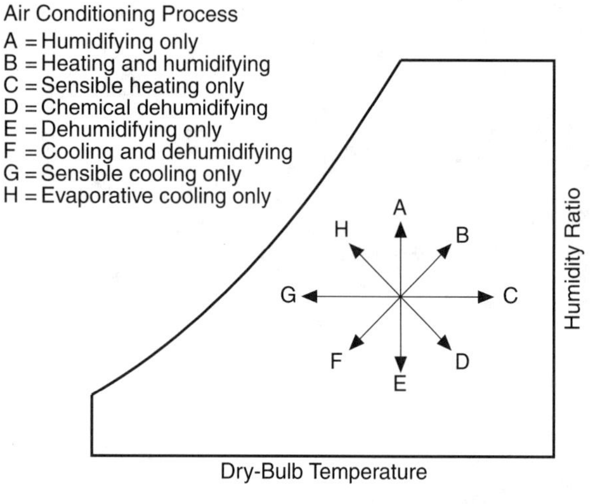

Figure 2. Schematic representation of air conditioning processes.

The volume flow rate can then be determined from the following equations (Jones 1985)

$$\dot{V} = \frac{\dot{Q}_s}{t_r - t_s} \times \frac{T}{358}$$

or

$$\dot{V} = \frac{\dot{Q}_l}{g_r - g_s} \times \frac{T}{856}$$

where
$\dot{V}$ = volume flow rate at temperature t (m^3/s)
t_r = room temperature (°C)
t_s = supply temperature (°C)
T = Temperature at which the flow rate volume flow rate is calculated (k)
g_r = room moisture content (g/kg)
g_s = supply moisture content (g/kg)

Air Conditioning Systems

Air conditioning systems are generally characterised by their thermofluid distribution medium, air or water, and by their means of controlling heating and cooling. They may also be classified as central or local depending on the plant employed.

Four major categories of air conditioning systems may therefore be identified

1. Centralised all air
2. Centralised air and water
3. Centralise all water
4. Localised

Centralised all air systems

These systems consist of a central plant room where incoming air is conditioned to a controlled supply state and then distributed throughout the building. They are classified (ASHRAE 1992) into single duct, dual duct and multizone systems which are then subdivided into constant and variable air volume systems.

Constant volume systems meet changing air conditioning loads by varying the supply state of the air. Thus, the heat gain to the space rises so the supply temperature is reduced. On the other hand, variable air volume systems meet changing loads by keeping the same supply state but varying the volume supplied.

Centralised air and water

These systems again utilise a central plant room to generate chilled or hot water and conditioned air to distribute throughout the building. However, the volumes of air distributed are usually much less than for the all air systems because it is only the intake of fresh air for ventilation that is conditioned centrally. Additional room air is treated within the conditioned room by passing it through a terminal plant served by the chilled or heated water distributed from the central plant room.

The systems can be further categorised by the mode of operation of the terminal plant. There are two main types of terminal plants; fan coil units or induction units. Full description of these systems can be found in ASHRAE (1992).

Centralised all water systems

These systems operate on the distribution of chilled and heated water from a centralised plant room to terminal plant such as fan coil units where heating and cooling of the air takes place locally. The requirement for fresh air ventilation must be provided locally either by natural infiltration or by an opening through the wall.

It is not possible with these systems to control the humidity within the space as the terminal plant includes only sensible heat exchangers.

Localised air conditioning systems

As the name suggests these systems are stand alone units that are located in or near the space to be conditioned. They are based upon the vapour compression cycle (see **Refrigeration**). In this case the evaporator is housed within the conditioned space, cooling the room air that is drawn across it. The condenser on the other hand is positioned outside the conditioned space dissipating the heat to atmosphere.

If the roles of the two heat exchangers, evaporator and condenser, are reversed the unit can be used as a heat pump heating the inside air from the external source.

Heat recovery

Wherever air conditioning systems employ return air ductwork there is a potential for heat recovery from the exhaust air before disposal to atmosphere.

This heat recovery is accomplished through the use of air-to-air heat exchangers (see **Heat Exchangers**) Under certain climate conditions, when the return air is nearer the supply conditions than the external air, maximum recirculation will reduce energy consumption. However, there will always be a need to exhaust some air as there is always a need for fresh air intake for ventilation and the well being of the occupants.

The viability of the introduction of heat recovery depends upon the expense of routing the ductwork, the capital and installation costs of the heat recovery plant and the usefulness of the recovered heat.

Control and instrumentation

Since the loads in various zones of a building vary with time, control systems are used to match the output of the HVAC system to the loads. A HVAC system is designed to meet the extremes in the demand, but most of the time it operates at part load conditions. A properly designed control system maintains good indoor air quality and comfort under all anticipated conditions with the lowest possible cost of operation.

Controls may be pneumatic, electric, electronic or they may even be self contained, where no external power is required. Developments in both analogue and digital electronics and in computers, have allowed control systems to become much more sophisticated. These systems also offer additional monitoring capability allowing efficient energy management. Control systems are described in detail in ASHRAE (1991) and Haines (1983).

References

ASHRAE (1993) *ASHRAE Handbook of Fundamentals.* American Society of Refrigeration and Air-Conditioning Engineers Inc. Atlanta, GA.

ASHRAE (1992) *ASHRAE Handbook of HVAC Systems and Equipment.* American Society of Refrigeration and Air-Conditioning Engineers Inc. Atlanta, GA.

ASHRAE (1991) *ASHRAE Handbook HVAC Applications.* American Society of Refrigeration and Air-Conditioning Engineers Inc. Atlanta, GA.

CIBSE (1986) *CIBSE Guide Section A7 Internal Heat Gains.* Chartered Institute of Building Services Engineers, London.

CIBSE (1986a) *CIBSE Guide Section A8 Summertime Temperatures in Buildings.* Chartered Institute of Building Services Engineers, London.

CIBSE (1982) *CIBSE Guide Section A2 Weather and Solar Data.* Chartered Institute of Building Services Engineers, London.

CIBSE (1980) *CIBSE Guide Section A3 Thermal Properties of Building Structures.* Chartered Institute of Building Services Engineers, London.

CIBSE (1979) *CIBSE Guide Section A1 Environmental Criteria for Design.* Chartered Institute of Building Services Engineers, London.

CIBSE (1979a) *CIBSE Guide Section A9 Estimation of Plant Capacity.* Chartered Institute of Building Services Engineers, London.

CIBSE (1975) *CIBSE GUIDE Section C1 Properties of Humid Air,* Chartered Institute of Building Services Engineers, London.

Eastop, T. D. and McConkey, A. (1993) *Applied Thermodynamics,* Longman Scientific and Technical, Harlow.

Fanger, P. O. (1972) *Thermal Comfort.* McGraw Hill, New York.

Jones, W. P. (1985) *Air Conditioning Engineering.* Edward Arnold.

Look, D. L. and Sauer H. J. (1988) *Engineering Thermodynamics.* Van Nostrand Reinhold (International), Wokingham.

McQuinston, F. C. and Parker, J. D. (1994) *Heating, Ventilating and Air conditioning.* Fourth Edition, New York: John Wiley & Sons, Inc.

Roger, W. H. (1983) *Control Systems for Heating, Ventilating and Air Conditioning.* 3rd Edition, New York: Van Nostrand Reinhold.

Leading to: *Space heating*

M.R. Heikal and A. Miller

AIR COOLED CONDENSERS (see Air cooled heat exchangers; Condensers)

AIR COOLED HEAT EXCHANGERS

Following from: *Heat exchangers*

Introduction

Water shortage and increasing costs, together with more recent concerns about water pollution and cooling tower plumes, have greatly reduced industry's use of water cooled heat exchangers. Consequently, when further heat integration within the plant is not possible, it is now usual to reject heat directly to the atmosphere, and a large proportion of the process cooling in refineries and chemical plants takes place in *Air Cooled Heat Exchangers* (ACHEs).

There is also increasing use of *Air Cooled Condensers* for power stations. The basic principles are the same but these are specialised items and are normally configured as an A-frame or "roof type". These condensers may be very large—the condensers for a 4000MW power station in South Africa have over 2300 tube bundles, 288 fans each 9.1m in diameter and a total plot area 500m × 70m.

ACHEs for process plants are normally just called *Aircoolers,* but should not be confused with devices for cooling air (best described as Air Chillers).

The design of an ACHE is more complex than for a **Shell** and **Tube Heat Exchanger,** as there are many more components and variables.

Construction

The principle component of an ACHE is the tube bundle, of which there may be many, normally comprising *finned tubes* terminating in header boxes. The fins are most commonly spirally wound aluminium strips 12.7×10^{-3} m or 15.9×10^{-3} m high and with 275 to 433 fins/m. There are two main types of wound fin which are usually known as *L-fin* and *G-fin.* There are several variations of the former type—single, overlapped and knurled, but all suffer a high contact resistance, which increases with temperature due to differential expansion between the fin and the core tube. Embedded fins (G-fins) are wound into a groove in the core tube which is then peened back providing a mechanical bond. This gives better heat transfer but requires a thicker core tube. Integral fins extruded from an aluminium sheath are often used for more severe environments, and instead of embedded fins with expensive core tubes. When an exceptionally long life is required in aggressive environments galvanised steel fins can be the best choice, and these frequently use elliptical tubes which also have improved airflow characteristics. Core tubes may be carbon steel, stainless steel or various alloys and are usually of 25.4×10^{-3} m outside diameter. For low pressure or highly viscous applications the tubes can be up to 50.8×10^{-3} m diameter. Tube lengths vary to suit the installation, which will often be over a piperack, but generally do not exceed 15m. (See also **Extended Surfaces Heat Transfer**).

Unlike most other pressure vessels an ACHE header box is normally rectangular in cross-section, and the most widely used type has threaded plugs opposite each tube for access. Various coverplate types may be used for low pressures, and for high pressures (up to 500 bar) manifold headers made from thick walled pipe or forged billets are needed. When they could be a large temperature drop across a multipass tube bundle, split headers may be necessary to accommodate differential expansion between passes.

The air is moved over the tubes in a single crossflow pass by axial flow **fans**, which may be arranged for forced or induced draught. Forced draught is suitable for most applications, has easier maintenance and is by far the more common. Induced draught gives a more even air distribution across the tubes, but requires more power as the fans are in the hot air stream. This latter point also means that induced draught is not suitable for high process temperatures, but is advisable for a close temperature approach as exit velocities are higher and hot air recirculation less likely. For induced draught installations with fan diameters greater than 2.4m the motor and speed reducer will normally be mounted below the tube bundles, with an extended drive shaft, as shown in **Figure 2**. There are normally at least two fans in each exchanger bay so that significant cooling is maintained in the event of a partial failure, and it is preferable for fans to cover at least 40% of the total bundle face area.

Installation

An ACHE is a large piece of equipment compared to other types of heat exchangers, and requires free space around it for the cooling air flow. In refineries and chemical plants ACHEs are usually mounted over a piperack, saving plot space at grade and ensuring free airflow. A further advantage of this elevated mounting is shorter pipe runs for column overheads, saving both cost and pressure drop. In some cases an ACHE may be mounted on top of a column to keep pressure loss to an absolute minimum, but this can make maintenance more difficult. Rooftop mounting is sometimes used, particularly for turbine steam condensers. When no suitable supporting structure is available, or where there is ample space available, the cooler may of course be ground mounted.

Design features

A typical face velocity for the air flowing across the tube bundle is 3 m/s. Higher air flows increase both the heat transfer coefficient and the mean temperature difference, thereby reducing the surface area required, but at a higher power consumption. Increased airflow and power also mean greater fan noise, which is an increasingly important factor.

The choice of design ambient temperature is the most critical factor affecting the size of an ACHE. A dry bulb temperature that is not exceeded for 95% of the year is the usual choice, accepting that there may be a cooling shortfall on the hottest days. In some cases the plant loading may be reduced in the summer, so that a

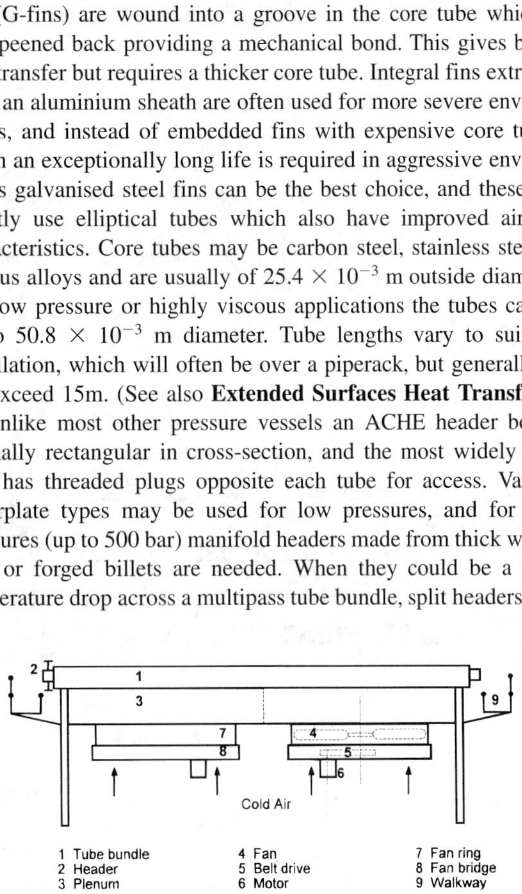

Figure 1. Typical forced draught air cooled heat exchanger.

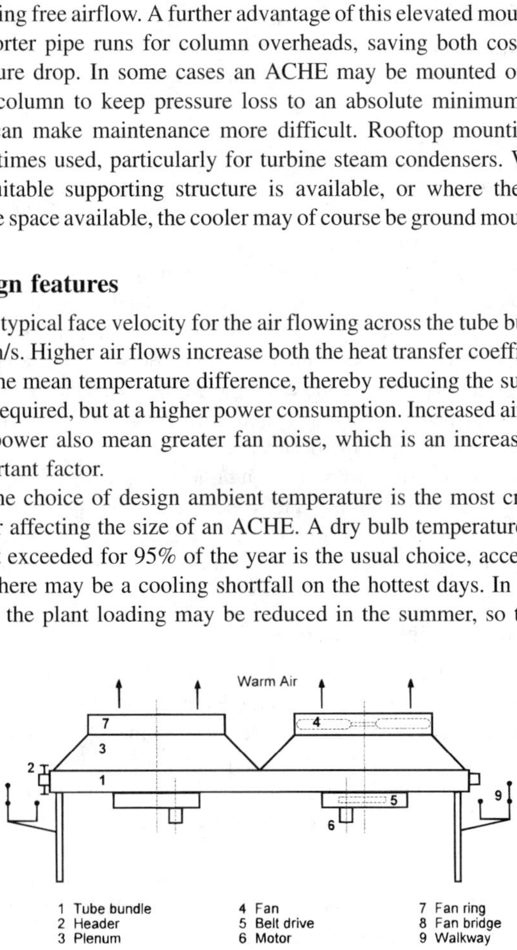

Figure 2. Typical induced draught air cooled heat exchanger.

lower design air temperature is appropriate. The majority of ACHE designs have between 4 and 6 rows of tubes (in the airflow direction). This may rise to 8 rows or more if there are plot restrictions, but successive rows become less and less effective for heat transfer and costs increase. If the core tubes are of high value material, fewer rows and increased plot area will certainly be cheaper.

Small independent ACHEs can be quite expensive, and it is therefore normal practice to install two or more small units in a shared fan bay. This is particularly useful when several exchangers are to be mounted in a bank with a common tubelength.

Noise

Sound pressure level limits in work areas within a plant are usually about 85 dB(A), but community noise levels need to be much lower and frequently necessitate an analysis of overall sound power levels. In Europe the sound power limits now tend to be more severe than the local sound pressure limits, and in some cases control the ACHE design.

The principal source of noise in ACHEs is the fans. Moderate reductions in noise levels can be achieved by reducing the fan speed and using more blades or wider chord blades. Very low noise designs necessitate low face velocities, with a consequent increase in surface area, so that the fans can run very slowly and still generate sufficient pressure.

The extremely low noise restrictions now being applied on some sites has led to the development of special fan designs, which are much quieter than conventional fans while maintaining a reasonable airflow.

Thermal design

The tubeside heat transfer and pressure drop are calculated in the same way as for **Shell & Tube Heat Exchangers**. For the airside heat transfer rate a number of calculation methods are available, including correlations by Briggs and Young (1963), PFR (1976) and ESDU (1986). As there is a temperature gradient along the fin the calculated heat transfer is adjusted using the concept of fin efficiency, which is the ratio of the actual heat transfer from a given surface, to the heat which would be transferred from the same surface at a uniform temperature equal to the fin root temperature—for details see **Extended Surfaces Heat Transfer**. The fin efficiency is in the range 0.8 to 0.9 for fin types and dimensions generally used in ACHEs.

Several correlations exist for predicting the airside pressure loss across the finned tube bank—those most commonly used are by Robinson and Briggs (1966), PFR (1976) and ESDU (1986).

Typical values of overall heat transfer coefficient for various fluids are given in ESDU (1993) and these may be used to obtain approximate sizes. This item also describes the *C-value method* of comparing costs for various heat exchanger types.

Control

Several options are available for controlling ACHEs. Simply switching fans on and off is adequate in many cases, and can give quite close control if the item has a large number of fans. The addition of louvre shutters, which can be manually or pneumatically operated, will provide further improvement, and two-speed motors are sometimes used.

The best control is obtained by the use of auto-variable pitch fans or variable speed motors, both of which provide gradual airflow adjustment. Improved electronics have made variable speed much more popular in recent years, with the additional benefits of power consumption and noise always being minimised.

The method of flooding condensers as frequently employed in shell and tube heat exchangers is not practical for ACHEs, and reductions in the effective surface area can only be achieved by valving off bundles.

Large variations in ambient temperature throughout the year will have a considerable effect on the available range of control, especially if there is a close approach at the design condition. Process engineers should be aware of this and avoid building in large design margins when a high degree of turndown is required, since for most of the year the ACHE will be massively oversurfaced and a control problem created.

Controlled recirculation

If there is a possibility of freezing, waxing or hydrate formation, it will be necessary to maintain a sufficiently high tube wall temperature to avoid this under all conditions. In many cases this will not be a problem, or may be easily solved by using reduced finning and/or co-current flow. However, in extreme cases hot air recirculation will be required. This is achieved by enclosing the ACHE in a cabin with inlet and outlet louvres, and a duct to redirect some of the exhaust air to mix with the cold inlet air. The normal arrangement is shown in **Figure 3**, although the recirculation duct may occasionally be at a header end (external over end).

Process side enhancement

In the majority of ACHE designs the airside heat transfer coefficient is controlling (i.e. much lower than the tubeside coefficient), and enhancement of the inside coefficient would give very little overall improvement, such that the additional cost of the enhancement device cannot be justified. However, for viscous fluids where the flow in plain tubes would be laminar, wire wound turbulator inserts are frequently used. The improved heat transfer coefficient these inserts provide can also help to avoid pour point problems, since the tube wall temperature will be closer to the bulk fluid temperature.

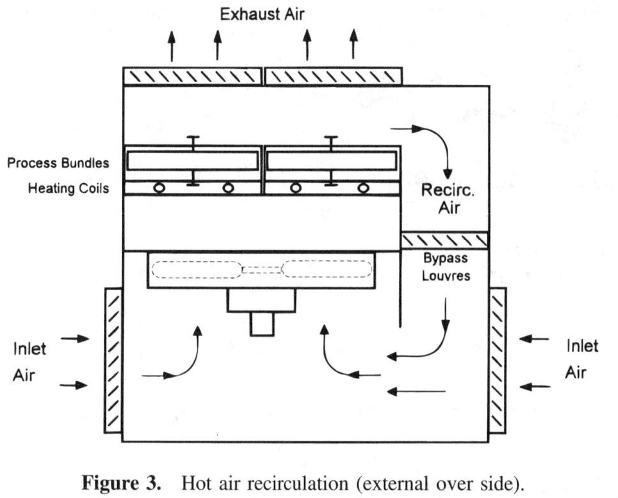

Figure 3. Hot air recirculation (external over side).

Fouling

Tubeside fouling factors normally follow shell and tube standard practice. Airside fouling factors are sometimes specified but have little effect on the already low airside heat transfer coefficient. The restriction to airflow of fouling on the finned tubes is of greater significance, and occasional cleaning is advisable to maintain cooling efficiency. In order to avoid fin damage, particularly with wound aluminium fins, this cleaning should be carried out by specialists.

Standards

The internationally accepted standard specification for refinery ACHEs is API 661. Many user companies now base their own specifications on this standard, with their preferences given as amendments/supplements to the API 661 clauses.

References

API Standard 661, (1992) *Air-Cooled Heat Exchangers for General Refinery Service,* 3rd Ed., Washington D.C: American Petroleum Institute.

Briggs, D. E. and Young, E. H. (1963) Convection heat transfer and pressure drop of air flowing across triangular pitch banks of finned tubes, *Chem. Engng. Progr., Symp. Ser.,* 59 (41): 1–10.

ESDU (1986) High-fin staggered tube banks: Heat transfer and pressure drop for turbulent single phase gas flow, Item No. 86022, London: Engineering Sciences Data Unit.

ESDU (1993) Selection and costing of heat exchangers, Item No. 92013, London: Engineering Sciences Data Unit.

PFR Engineering Systems Inc. (1976) Heat transfer and pressure drop characteristics of dry tower extended surfaces, Part II: Data analysis and correlation, Report BNWL-PFR-7-102, Marina del Rey, California.

Robinson, K. K. and Briggs, D. E. (1966) Pressure drop of air flowing across triangular pitch banks of finned tubes, *Chem. Engng. Progr., Symp. Ser.,* 62 (64): 177–184.

Leading to: *Extended surfaces; Fans*

C.J. Summers

AIR COOLERS (see Air cooled heat exchangers, ACHEs)

AIRCRAFT, AERODYNAMICS OF (see Aerodynamics)

AIRCRAFT, PARABOLIC FLIGHTS (see Microgravity conditions)

AIR CUSHIONS (see Aerodynamics)

AIR CYCLE HEAT PUMPS (see Heat pumps)

AIR CYCLE REFRIGERATION (see Refrigeration)

AIR EJECTOR (see Aerodynamics)

AIR JET ENGINES (see Jet engine)

AIRLESS DRYING (see Drying)

AIR POLLUTANTS (see Carbon monoxide)

AIR POLLUTION

Following from: *Atmosphere*

The air in the environment is never absolutely pure; it always contains traces of pollutants the force of of suspended solid particles (dust and smoke), droplets of liquid (mist), vapors, and contaminant gases. The sources of *pollution* are grouped into those of natural and anthropogenic origin, i.e. the products of human activities. The former include volcanic eruptions, dust storms, wood fires and rotting of plants. They have been known for millions of years, and the atmosphere spontaneously cleans in some finite period of time as a result of sedimentation, chemical oxidation or, conversely, reduction, and by absorption by the ocean and soil. Ultimately an equilibrium concentration of atmospheric pollution of all kinds is established for ongoing natural pollution. Essential disturbances of the equilibrium in the earth's atmosphere were brought about in the past only by large-scale volcanic eruptions such as Krakatau eruption in 1883.

An intensified level of human activity, i.e. a vigorous growth of production, the mining industry, metallurgy, power generation, chemistry, and transport, has given rise to ejection into the atmosphere of vast amounts of dust, gases, and condensed vapors including those containing toxic and radioactive substances. As a rule, these ejections are concentrated over confined geographic regions (viz. industrial centers) and carried over the earth's surface by atmospheric air flows. Therefore, their concentration in various regions is not identical; if it substantially exeeds the natural background, the atmosphere, in the process of self-cleaning, fails to dispose of these substances. This results in the global pollution of the atmosphere.

Figure 1 shows as an example the global cycle of *sulfur*, which typifies the main sources of *pollution*, and shows how various substances find the way from the earth's surface to the atmosphere and back.

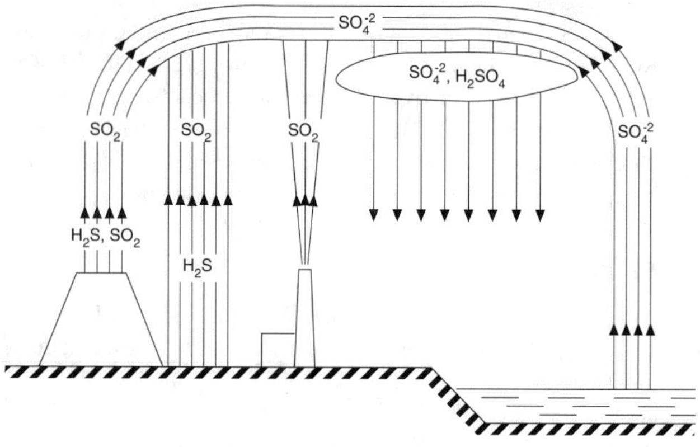

Figure 1. Global cycle of sulfur pollution.

At present, the term "pollution of the atmosphere" implies the presence in the surrounding atmosphere of one or more substances and their combinations in such amounts and during such span of time that they can have an injurious effect on the environment and human health.

Until the mid-70s neither water vapor nor CO_2 were thought to be atmospheric contaminants, but currently *carbon dioxide* attracts attention not only as a substance whose high local ejections lead to a reduction of oxygen concentration but also as one producing, on the global scale, a "**Greenhouse Effect**" due to a reduction of the transmission of thermal radiation from the earth's surface in the infrared region. Water vapor, when ejected in large quantities, produces the same effect.

Concentration of a gaseous contaminants in air is coventionally expressed as the number of parts per million (ppm) and that of condensed contaminator (dust, droplets), in $\mu g/m^3$. The main contaminants are suspended particles of various chemical composition, gaseous sulfur and nitrogen compounds, vapor of organic substances, as well as halogen compounds and radioactive substances. The degree of natural and anthropogenic pollution of the atmosphere can be judged from the data of **Table 1**, where the annual discharge of basic contaminants is presented.

Suspended particles are classified into coarse *dust* (over 100 μm in size), and so-called aerosols, i.e. fine dust (under 100 μm), mists (from 0.01 to 10 μm), and *smokes* (from 0.001 to 1 μm). These particles can be in suspension from a few seconds to an infinitely long time. Thus, 0.1 μm and smaller particles are subject to Brownian diffusion and virtually do not settle. The particles from 0.1 to 1 μm have sedimentation rates much lower than even the weak wind velocity, and their sedimentation by gravity is negligible. Slow sedimentation is also observed for particles from 1 to 5 μm, and only particles tens of microns in size and larger actively settle from the atmosphere back to the earth. Atmospheric precipitation such as rain, snow and drizzle play an important role in purifying air from suspended particles, particularly, if account is taken of the fact that particles are the main condensation sites of water vapor. This can be observed in the atmosphere over industrial centers with large dust ejection. Here, rainfall is lower on days when the industry is not operational and when the particle concentration in the atmosphere falls.

The negative impact of suspended particles on life manifests itself not only in reduced visual range and solar radiation due to light absorption and scattering, but also in a direct influence on animal and human health. Solid particles are inhaled and retained in the lungs. They may be toxic owing to their chemical and physical properties; for instance, particles of heavy metals, may hinder clearing of respiratory tracts. Solid particles may be the carriers of toxic substances adsorbed on them. The latter is most dangerous because the statistical analysis of diseases has shown that suspended particles combined with other contaminants (e.g. SO_2, NO_x) disturb human health more gravely than each contaminant taken separately.

Carbon monoxide, if its concentration in the air is higher than 750 ppm, causes pathological changes in human organs and is ultimately lethal. The reason is that CO, combining with haemoglobin, reduces the capacity of the blood in transporting oxygen.

The injurious effect of sulfur is still more diverse. Sulfur dioxide ejected into the atmosphere is re-oxidized to form SO_3 and, in humid air, sulfuric acid forms, which condenses on either suspended particles or, at high supersaturation, on its own nuclei, and generates mist. Production of particles is also attributed to atmospheric photochemical reactions between SO_2, NO_x, hydrocarbons, and suspended particles. As a result, aerosols of sulfuric acid and its salts constitute up to 20% of the total number of suspended particles in the air of cities. They are the source of the so called *smog* that is characteristic of many cities of north-western Europe, where sulfur-bearing coal is used as fuel. Smog is formed, as a rule, in winter in windless weather. Smoke particles that are the sites of condensation of moisture and sulfuric acid become heavy and hang over the city gradually descending on buildings and streets. They are responsible for metal corrosion, fracture of marble, limestone, and other construction materials as well as fabrics primarily, nylon. The destructive effect of sulfuric acid aerosol, though weaker, is observed even in the absence of smog.

Table 1. The main sources of atmospheric pollution

	Sources of pollution		Annual production, mill.ton/y	
Gases	Anthropogenic	Natural	Anthropogenic	Natural
Sulfur dioxide	Burning of coal and oil, recovery of metals from sulfide ores	Volcanoes	146	6–12*
Hydrogen sulfide	Chemical technologies, processing of waste waters	Volcanoes, biological processes in swamps	3	100–300
Carbon monoxide	Burning, car exhausts	Wood fires, terpene reactions	300	3000
Nitrogen oxide	Burning	Bacterium activity in soil, thunderstorms	50	60–270
Ammonia	Processing of waste products	Biological decomposition	4	100–200
Nitrous oxide	Use of nitride fertilizers	Biological processes in soil	17	100–450
Hydrocarbons	Burning, car exhausts, chemical reactions	Biological processes	88	*CH* 300–1600 terpenes 200
Carbon dioxide	Burning	Biological decomposition	1.5×10^4	15×10^4

*To this nearly all the hydrogen sulfide must be added since it oxidizes in air, forming an intermediate product SO_2 (Fig. 1).
Source: W.Strauss, S.J. Mainwaring. Air Pollution. Edward Arnold Ltd., London, 1984.

In windy weather a substantial portion of sulfur-containing aerosols is carried at long distances. Their precipitation on the earth is facilitated by rains which are of a clearly pronounced acid nature and, therefore, are known as *acid rains*. Their effect on soils with $pH > 7$ is reduced to a considerable extent by the soil alkalinity, while for soils with, $pH < 7$ precipitation of acid rains is extremely unfavorable since an increase in soil acidity reduces fertility and an increased acidity of water basins leads to the mortaility of fish, primarily of the most valuable varieties. Precipitation of acid rains is injurious to plant leaves.

Sulfure dioxide and mixtures of sulfur-containing components, in particular, heavily irritate the rispiratory tract in humans, stimulating diseases of respiratory tract and order diseases, and lead to a higher mortality. Among nitrogen oxides the most harmful for men is nitrogen dioxide that, together with atmospheric moisture, forms a strong nitric acid the effect of which on men is similar to that of sulfuric acid.

Methane, halogen compounds, and NO_2 are regarded currently to be destructive of the earth's ozone protective layer.The main sources of atmospheric air pollution are power generation, metallurgy, chemical industry, and transport. The former three deliver to the atmosphere, first and foremost, dust, sulfur and nitrogen oxides, whereas vehicles chiefly deliver CO.

Monitoring of the steadily increasing pollution of the atmosphere required the establishment of permissible limiting values. In addition to a local pollution (maximum allowable concentration determined in the near-earth layer or the "breathing layer"), a limit also has to be established for the total discharge.

Scattering of polluting substances in the atmosphere to a great extent depends on convective and turbulent mixing. The height of the air layer in which active mixing occurs depends on the season, the weather, the time of day, and the topography of the ground. The greater the height on which the pollutant is entitled, the greater the space into which the pollution is diluted.

The jet configuration of stacks under different weather conditions is demonstrated in **Figure 2**. **Figure 2a** depicts the most characteristic, "ordinary" conditions, when dispersion in the horizontal and vertical directions is approximately the same and the jet cross section approaches a circle. Under inversion when a stack mouth rises above the *inversion layer* (IL) "hanging" over the earth (**Figure 2b**) the plume may extend only horizontally and upward, therefore, its section is of a triangular shape. **Figure 2c** illustrates the plume shape when the pollutant is emitted below the inversion layer.

The mathematical description of dispersion is often represented in a simplified form, e.g. the problem is considered one of the propagation of an impurity from a point source in a final direction with the wind blowing with a constant velocity and direction (**Figure 3**). The jet is round in cross section and distribution of impurities along the radius is assumed to be Gaussian. Since the gases discharged from the stack mouth have some velocity, of the order of 10 m/s, and a temperature higher than the ambient air temperature, the jet rises upward by Δh, and this is the distance by which the effective point source for the emission (A in Fig. 3) is raised over the stack. An equation allowing to calculate the impurity concentration at any point of the jet C(x, y, z) has the form

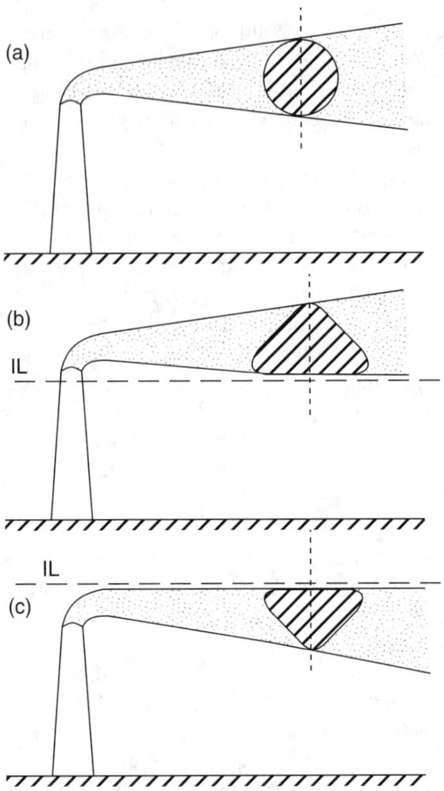

Figure 2. Effect of emission position relative to inversion layers.

$$C = \frac{\dot{M}}{2\pi u S_y S_x} \exp\left\{ -\frac{y^2}{2S_y^2} - \frac{(z-H)^2}{2S_x^2} - \frac{(z+H)^2}{2S_z^2} \right\},$$

where x, y, z are the distances along the appropriate axes (m) S_y and S_z the diffusion in the y and z directions (m) $\dot{M}$ the intensity of the source (g/s) C the concentration at a given point (g/m^3) u the wind velocity (m/s) and H the effective height of discharge (m). H is equal to the sum of the stack height h and an additional jet rise height Δh as discussed above.

Of particular interest is the calculation of impurity concentration in the near-earth layer when the jet touches the earth's surface. For this parameter empirical formulas, based on a vast body of experimental evidence, of the following form are used:

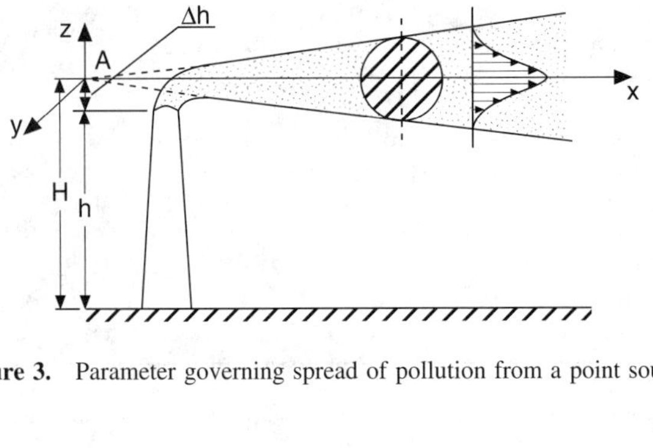

Figure 3. Parameter governing spread of pollution from a point source.

$$C = \frac{AF_m \dot{M}_T}{h^2} \sqrt{\frac{N}{\dot{V}\Delta T}}, \qquad mg/m^3$$

are used, where A is a coefficient allowing for vertical and horizontal dispersion conditions, F_m a coefficient relating to physical state of a given harmful substance m, $\dot{M}_T$ the maximum overall discharge of harmful impurity from all the stacks (g/s) h the geometric height of the chimney (m) N the number of stacks of identical height, $\dot{V}$ the overall volumetric flow of the flue gases (m^3/s) and ΔT the difference between the temperature of discharged gases and that of the ambient air (K).

High buildings not far from the pollution source may essentially alter the aerodynamic situation, and the jet may descend to the earth.

Unfavorable environmental conditions in a city may result from a large accumulation of heat by bulky buildings in the daytime. In this way a stable self-maintaining system of polluted air recirculation, a so called "*heat island*", upset only by a strong wind, is established in and over the city.

For monitoring atmospheric pollution in the near-earth layer and for collecting and analysing data there is a large network of observation stations equipped with analytical automatic instruments regularly sending information to a central processing system.

References

Wark, K. and Warner, C. F. (1976) Air Pollution: Its Origin and Control, IEP-Dun-Donnelley, New York, London.

Brimblecombe, P. (1986) Air Composition and Chemistry, Carbridge Unit Press, London, New York.

Strauss, W. and Mainwaring, S. J. (1984) Air Pollution. Edward Arnold Ltd., London.

Leading to: *Chimneys*

I.L. Mostinsky

AIR (PROPERTIES OF)

Following from: *Atmosphere, physical properties of*

Atmospheric air is a mixture of nitrogen and oxygen being the earth atmosphere. Main components of air which are practically the same throughout the globe are nitrogen (78.08 volume per cent) and oxygen (20.95 v.%). Along with them air contains 0.94 v.% of inert gases and 0.03 v.% of carbon dioxide. The air of such a composition is named dry. Its molecular mass is regarded to be M = 28.96 g/mole.

In the lower atmosphere strata the air contains also water vapor, its concentration is substantially variable depending on the partial water vapor pressure at the appropriate temperature and relative humidity. For instance at 20°C and relative humidity 80% air contains about 0.02 v.% of water vapor. In the air layers adjacent to the earth surface other components may be present being in most cases of antropogenic origin.

At ambient pressure and temperature air can be regarded as a perfect gas, its properties may be described by equations:

Table 1. Thermodynamic properties of air along the saturation curve

T,K	p_b	p_c	v'	v''	h'	h''	s'	s''
1	2	3	4	5	6	7	8	9
65	0.1468	0.08613	1.065	2154	−149.6	+64.6	.669	6.040
70	0.3234	0.2052	1.090	968.4	−140.6	+69.2	.797	5.867
80	1.146	0.8245	1.148	269.6	−122.6	+77.8	3.034	5.585
90	3.036	2.397	1.216	100.2	−103.5	+84.8	3.251	5.376
100	6.621	5.599	1.302	44.67	−83.3	+89.3	3.456	5.204
110	12.59	11.22	1.418	22.15	−61.9	+90.1	3.649	.045
120	21.61	20.14	1.569	11.45	−37.5	+84.8	3.850	4.877
130	34.16	33.20	2.075	5.425	+0.4	+66.1	4.136	4.644
132	—	37.69	—	3.196	—	+37.4	—	4.410

p_b, p_c, bar,—pressure respectively at the boiling and condensing curves;
v', v'', $10-^3$ m^3/kg—specific volumes respectively of the liquid and vapor;
h', h'', kJ/kg—specific enthalpies of the liquid and vapor;
s', s'', kJ/kg · K—specific entropies of the liquid and vapor.

$$v = RT \qquad \text{and} \qquad \left(\frac{\partial u}{\partial p}\right)_T = 0,$$

where v denotes specific volume; u is specific internal energy; R is the gas constant for air.

At low temperatures the air is liquified. The normal (at 0.1013 MPa) boiling (condensation) temperature of the oxygen is equal—183°C, that of the nitrogen −195.8°C. Liquid air at atmospheric pressure behaves practically as an ideal solution following the **Raoult's Law**. The normal condensation temperature of air is −191.4°C, the normal boiling temperature −194°C.

At elevated temperatures air undergoes some physico-chemical transformations. The nitrogen reacts with oxygen producing various oxides: N_2O, NO, NO_2, NO_3. Their equilibrium concentration can be derived from the isotherm equations of the respective reactions.

At temperatures higher than 2000K and moderate pressures the nitrogen and oxygen start to dissociate, and at temperatures exceeding 4000K and atmospheric pressure the ionisation of oxygen, nitrogen and other components becomes evident. This implies the transition of air into the plasma state. The equilibrium dissociation degree can be calculated according to the Saha equation.

The thermodynamic properties of air along the saturation curve are given in **Table 1**; these properties for the liquid and gaseous —in **Table 2**.

The enthalpy is taken as zero at an arbitrary point. The entropy is taken zero for the solid air at 0K.

Air is a mixture mainly consisting of diatomic gases. Therefore its heat capacity at close to normal temperatures and pressures may be with good accuracy taken equal to

$$c_p = \frac{7}{2} R,$$

where

$$R = \frac{8.314}{28.96} = 0.287 \text{ kJ/kg} \cdot \text{K}.$$

Table 2. Thermodynamic properties of liquid and gaseous air

T,K	v	h	s	v	h	s	v	h	s
	p=1 bar			p=10 bar			p=100 bar		
75	1.119	−131.7	2.918	1.1171	−131.1	2.913	.0976	−124.5	2.867
100	281.2	+98.3	5.759	1.2995	−83.2	3.452	1.2448	−79.4	3.376
200	572.8	199.7	6.463	56.07	+195.2	5.786	4.674	148.8	4.949
300	861.0	300.3	6.871	85.89	298.3	6.204	8.551	79.9	5.486
400	1148	401.2	7.161	115.1	400.2	6.497	11.84	91.3	5.807
500	1436	503.4	7.389	144.0	502.9	6.727	14.94	99.0	6.048
600	1723	607.5	7.579	172.9	607.3	6.917	17.97	06.0	6.244
700	2010	713.8	7.743	252.0	713.9	7.145	20.95	15.3	6.412
800	2297	822.5	7.888	287.9	822.7	7.291	23.90	25.8	6.559
900	2585	933.5	8.019	323.9	933.8	7.421	26.83	38.2	6.691
1000	2872	1046.8	8.138	359.8	1047.1	7.541	29.75	052.4	6.812
1100	3159	1161.7	8.248	395.8	1162.1	7.650	32.66	168.2	6.922
1200	3446	1287.4	8.349	431.7	1278.9	7.752	35.56	285.6	7.024
1300	3733	1396.6	8.444	467.6	1397.1	7.847	38.46	404.4	7.120

The units for *v, h, s* are the same as in Table 1

With increasing temperature the heat capacity slightly increases due to exciting of the vibrational degrees of freedom in the oxygen and nitrogen molecules. **Table 3** gives air heat capacity values for a wide range of temperatures and pressures.

As for all pure substances in the supercritical region the isobars and isotherms of the heat capacity c_p have maximums the steeper the closer to the critical point.

The temperature dependence of the *viscosity* of air is qualitatively the same as for pure substances: in the liquid phase the viscosity decreases with temperature following an approximately exponential function; in the gas phase at low pressures the viscosity increases according to equation:

$$\eta(t) = \eta(T_0)(T/T_0)^{0.85}$$

with increasing pressure at constant temperature the viscosity of gaseous viscosity increases. This dependence is most strong in the vicinity of the critical point. Air viscosity values at various temperatures and pressures are given in **Table 4**.

The behavior of the thermal conductivity of air is similar to the viscosity: in the liquid phase with growing temperature the heat conductivity decreases whereas in the gas phase-increases. At low pressures the temperature dependence is described by the equation:

$$\kappa(T) = \kappa(T_0)(T/T_0)^{0.9}$$

Along the isotherm with increasing pressure the thermal conductivity increases. In **Table 5** the air thermal conductivity is given at various temperatures and pressures.

At low pressures and high temperatures the thermal conductivity sharply increases due to dissociation. With growing temperature the

Table 3. Air heat capacity c_p, kJ/kg · K

T,K	p = 1, bar	p = 10, bar	p = 100, bar	p = 1000, bar
100	1.032	2.041	1.852	—
200	1.007	1.049	1.650	—
300	1.007	1.021	1.158	1.303
400	1.014	1.021	1.087	1.217
500	1.030	1.034	1.073	1.175
600	1.051	1.055	1.080	1.164
700	1.075	1.077	1.096	1.168
800	1.099	1.100	1.114	1.179
900	1.121	1.122	1.133	1.192
1000	1.141	1.142	1.151	1.204
1100	1.159	1.160	1.167	1.215
1200	1.175	1.175	1.181	1.225
1300	1.189	1.189	1.194	1.234

Table 4. Air viscosity η, · 10^7 N · s/m²

T,K	p = 1, bar	p = 10, bar	p = 100, bar	p = 1000, bar
100	71.1	837.8	1019	—
200	132.5	134.6	181.2	—
300	184.6	185.9	205.0	545.5
400	230.1	231.2	243.4	463.4
500	270.1	270.9	279.9	437.5
600	305.8	306.4	313.6	434.7
700	338.8	339.4	345.3	442.7
800	369.8	370.3	375.4	456.4
900	398.1	398.5	403.0	472.0
1000	424.4	424.8	428.7	488.7
1100	449.0	449.3	452.8	505.8
1200	473.0	473.2	476.5	523.7
1300	496.0	496.3	499.2	541.8

Table 5. Air thermal conductivity $\lambda \cdot 10^3$, W/m · K

T,K	p = 1, bar	p = 10, bar	p = 100, bar	p = 1000, bar
100	9.34	110.3	122.0	—
200	18.1	18.9	29.9	—
300	26.3	26.8	31.4	86.2
400	33.8	—	37.3	75.6
500	40.7	—	43.3	71.8
600	46.9	—	49.0	72.5
700	52.4	—	54.1	74.1
800	57.3	—	58.8	75.8

thermal conductivity goes through maximums which are connected with maximum heat transfer by the heats of respective reactions. Thermal conductivities of air at dissociation conditions are given in **Table 6**.

Table 6. Air thermal conductivity at high temperatures $\lambda \cdot 10^3$ W/m · K

T,K	p = 0.01 bar	p = 0.1 bar	p = 1.0 bar	p = 10 bar	p = 100 bar
1500	100	100	100	100	100
2000	165	144	137	135	134
3000	812	813	486	304	238
4000	520	417	620	701	489
5000	3281	1424	718	604	663
6000	3191	4287	2272	1087	749

References

Additional information about air properties can be found in the following literature:

Handbook, edited by V. P. Glushko (1978) "Nauka" Publishing House, Moskow, (in Russian).

Wassermann, A. A. and Rabinovitch, V. A. (1968) Thermophysical properties of liquid air and its components. Standarts Publishing House, Moscow, (in Russian).

Handbook Thermophysical Properties of Gases and Liquids. (1972) Edited by N. B. Vargaftic. "Nauka" Publishing house, Moscow, (in Russian).

E.E. Shpilrain

AIR-TO-AIR HEAT PUMPS (see Heat pumps)

AIR-TO-WATER HEAT PUMP (see Heat pumps)

AIR-WATER SYSTEM (see Lewis relationship)

AISI (see American Iron and Steel Institute)

ALBEDO

***Following from:** Radiative heat transfer*

Electromagnetic radiation (for instance thermal radiative in the infra-red range) may be attenuated in passing. The *albedo* ω is the dimensionless value that is equal the ratio of scattering coefficient (β) to the sum of absorption (k) and *scattering* (β) coefficients:

$$\omega = \frac{\beta}{(k + \beta)}.$$

The single scattering albedo defines the fraction of energy scattering in total energy attenuation. The dependencies of ω on frequency and temperature are determined by the corresponding dependencies of k and β. Albedo is widely used in radiative heat transfer calculations.

References

Seigel, R. and Howell, J. R. (1992) *Thermal Radiation Heat Transfer,* third edition, Washington D.C., Hemisphere.

Ozisik, M. N. (1985) Radiative transfer and interaction with conduction and convection. New York: Werbel & Peck.

V.A. Petrov

ALBEDO, OF EARTH (see Atmosphere)

ALCOHOL (see Ethanol)

ALDEHYDES (see Pyrolysis)

D'ALEMBERT PARADOX (see Flow of fluids; Tube, crossflow over)

ALFVEN NUMBER (see Magneto hydrodynamics)

ALFVEN WAVES (see Magneto hydrodynamics)

ALGEBRA, FUNDAMENTAL THEOREM OF (see Polynomials)

ALIPHATIC HYDROCARBONS (see Hydrocarbons; Paraffin)

ALKANES (see n-Butane)

ALLEN FLOW (see Immersed bodies, flow around and drag)

ALLOY STEELS (see Steels)

ALPHA PARTICLES (see Ionizing radiation)

ALTERNATIVE ENERGY SOURCES

Interest in sources of energy that provide alternatives to conventional fossil and nuclear fuel has been growing steadily over the past twenty years. In particular emphasis has been placed on the development of renewable sources of energy. **Renewable Energy** is the term used to cover those energy flows that occur repeatedly in the environment and can be harnessed for human benefit. The ultimate sources of most of this energy are the sun, gravity and the earth's rotation.

Today few would perceive a future without the Renewables contributing to our energy provision and many believe that renewable energy will make a substantial contribution to our energy supplies in the longer term. This interest has been stimulated by concerns over the use of conventional energy technologies and their environmental impacts. With little or no net emissions of polluting gases the Renewables are seen as part of a solution to these problems.

The technologies to harness these renewable energy flows are many and various. Brief descriptions of some of the renewable energy technologies that have been considered candidates for development and deployment are presented below.

Wind Power

Wind power is an intermittent resource which is strongly influenced by geographical effects such as the local terrain. The amount of wind energy is a strong function of the wind speed, the instantaneous power in the wind increasing as the cube of the wind speed.

Wind power has been harnessed by Man for over 2,000 years and is one of the most promising renewable energy sources for electricity generation. There are two basic design configurations—horizontal axis machines and vertical axis machines. Horizontal axis designs are at a more advanced stage of development and the evidence is increasing that they are also more cost-effective. Apart from the need to demonstrate adequate lifetimes, there is no doubt about the technical feasibility of harnessing wind power.

The existing technology offers a range of power ratings from a few kilowatts up to several megawatts. The technology is well established, with over 20,000 grid connected machines in operation world-wide. Current development work is concentrating on reliability, the further reduction of cost and noise levels, aspects of the electrical connection into the grid and overall performance.

There are limitations on the availability of land for wind turbine sites due both to physical constraints—such as the presence of towns, villages, lakes, rivers, woods, roads and railways—and institutional constraints such as the protection of land areas designated as being of national importance. Offshore there is potentially a very large wind resource but it will require additional technology development before it can be effectively exploited and it will cost more than onshore wind. (See also **Wind Turbines**)

Water Power

Hydro power

Hydro power comes from the energy available from water flowing in a river or in a pipe from a reservoir. Evidence of the use of hydro power as a source of energy has been found in primitive devices from the first century BC. During the Industrial Revolution, small-scale hydro power was commonly used to drive mills and various types of machinery. The first large-scale hydro scheme in the UK was built in Scotland in 1896.

Hydroelectric technology can be regarded as being fully commercialised. Turbine plant, engineering services and turnkey systems are sold by many organisations world-wide.

Tidal power

Tides are caused by the gravitational attraction of the moon and sun acting on the oceans of the rotating earth. The relative motions of these bodies cause the surface of the oceans to be raised and lowered periodically. In the open ocean, the maximum amplitude of the tides is about 1 m. Towards the coast, the tidal amplitudes are increased by the shelving of the sea bed and funnelling of estuaries. For example, in the Severn Estuary the maximum amplitude is about 11 m.

The energy obtainable from a tidal scheme varies with location and time. The available energy is approximately proportional to the square of the tidal range, and the output changes not only as the tide ebbs and floods each day but can vary by a factor of four over a spring-neap cycle. However, this output is exactly predictable in advance. Extraction of energy from tides is considered to be practical only at those sites where the energy is concentrated in the form of large tides and in estuaries where the geography provides suitable sites for tidal plant construction. Such sites are not commonplace, but a considerable number have been identified in the UK, which probably has the most favourable conditions in Europe for generating electricity from the tides. This is the result of an unusually high tidal range along the west coast of England and Wales, where there are many estuaries and inlets which could be exploited.

A tidal barrage is a major construction project built across an estuary, consisting of a series of gated sluices and low-head turbine generators. Several locations around the world have been studied as potential barrage sites, but relatively few tidal power plants have been constructed. The first and largest (240 MW) tidal plant was built in the 1960s at La Rance in France, and has now completed more than 25 years of successful commercial operation.

Wave energy

Ocean waves are caused by the transport of energy from winds as they blow across the surface of the sea. The amount of energy transferred depends upon the speed of the wind and the distance over which it acts. As deep ocean waves suffer little energy loss, they can travel long distances if there is no intervening land mass. Therefore the western coastline of Europe has one of the largest wave energy resources in the world, being able to receive waves generated by storms throughout the Atlantic.

Wave energy is still in the RD&D phase. Currently there are two types of device known to be operating in Europe. The Norwegians have developed a tapered channel device (Tapchan) but the concept is limited to use in areas where there is a small tidal rise and fall and having suitable shoreline topography. In the UK the Government has funded work on a 75 kW oscillating water column device incorporating a *Wells air turbine* developed by the Queens University, Belfast. This device is now connected to the grid on Islay in the Inner Hebrides.

World-wide, installed devices are limited to experimental plants of less than 100 kW, including oscillating water column devices incorporated in sea defence breakwaters.

Solar Energy

Photovoltaics

Photovoltaic (PV) materials generate direct current electrical power when exposed to light. Power generation systems using these materials have the advantage of no moving parts and can be formed from thin layers (1 to 250 microns) deposited on readily available substrates such as glass. To date, the photovoltaic effect has been widely exploited where the low power requirements, good solar resource and simplicity of operation outweigh the high cost of PV systems. Current applications include consumer goods, such as calculators and watches and, on a larger scale, power systems for lighting and water pumping in developing countries and in remote areas with no grid supply. Other applications include powering of "professional" systems such as remote telecommunications facilities and cathodic protection of pipelines.

There is world-wide interest in developing PV systems for future power generation because of the huge potential renewable resource available and the environmental benefits offered by a technology which avoids the emissions and pollution associated with fossil-fuelled plant. However, PV is still a relatively young technology. Much research and development will be necessary if world-wide system costs (modules and associated components) are to be reduced to acceptable levels and significant new markets are to be established.

PV could contribute to electricity supply in two ways—through the use of central PV generating plant (PV power stations) or through building-integrated systems where PV units would be located in the facades of domestic and commercial buildings. Building integrated systems could supply power for use inside the buildings for applications such as appliances, air conditioning and lighting, with any excess available for export to the grid. (See also **Solar Cells**)

Active solar and thermal solar power

Active solar thermal systems consist of solar collectors, which transform solar radiation into heat, connected to a heat distribution system. Due to the nature of the UK climate, such heating systems are best suited to applications at temperatures below 100°C. High temperature applications, such as thermal solar power for electricity generation, are not practical in the UK.

There is a developed technology and an existing small market for systems to supplement the heating energy demands of buildings. This market is served by a small number of manufacturers and installers, but many of the installers see solar heating as a secondary activity associated with another business, such as central heating installation.

Passive solar design

Passive solar design (PSD) aims to maximise free solar gains to buildings so as to reduce their energy requirements for heating or cooling and lighting. It is most effective when used with energy efficiency measures as an integral part of energy-conscious design of new buildings. However, some PSD features, such as conservatories and roof space collectors, can be retrofitted.

The concept of PSD is not new. However, its potential energy benefits, as distinct from its use for aesthetic or health reasons, have only recently become a focus of attention. To maximise these benefits in terms of the heating requirements of a building, PSD seeks to orientate and arrange glazed surfaces so as to make full use of shortwave solar radiation for heating interior spaces and to avoid heat loss resulting from siting windows on shaded walls. To cool buildings it uses solar heated air to assist natural convection, thus providing natural ventilation and cooling. For lighting, it uses glazing to reduce the need for artificial lighting whilst still maintaining a comfortable environment. More complex approaches such as mass walls, atria or conservatories are basically extensions of these simple design principles. Effective use of PSD depends on sympathetic interior design and on grouping buildings to minimise shading and gain protection from prevailing winds. (See also **Solar Energy**)

Geothermal Energy

Geothermal hot dry rock

There is a large amount of heat just below the earth's surface—much of it stored in low permeability rocks such as granite. This source of geothermal heat is called "*hot dry rock*" (HDR). Attempts to extract the heat have been based on drilling two holes from the surface. Water is pumped down one of the boreholes, circulated through the naturally occurring, but artificially dilated, fissures present in the hot rock, and returned to the surface via the second borehole. The superheated water or steam reaching the surface can be used to generate electricity or for combined heat and power systems. The two boreholes are separated by several hundred metres in order to extract the heat over a sizeable underground volume. A typical HDR power station would produce about 5 MW of electricity and be expected to operate for at least 20 years.

The engineering of the underground "heat exchanger" has turned out to be a formidable technical problem which has not yet been satisfactorily solved after more than ten years of intensive research in the UK, the USA and elsewhere. Because of the technical difficulties, there are no commercial HDR schemes in existence anywhere in the world.

Geothermal aquifers

Geothermal **Aquifers** extract heat from the earth's crust through naturally occurring ground waters in porous rocks at depth. A borehole is drilled to access the hot water or steam, which is then passed through a heat exchanger located on the surface. If the temperature of the hot fluid exceeds about 150°C it can be used for generating electricity; otherwise it can be tapped as a source of warm water. In the UK, there are very few sources with temperatures above 60°C and the resource would be exploitable mainly in district heating systems or industrial processes.

The use of aquifers is well established in certain geologically favoured parts of the world, such as Iceland, Hungary, Italy, the USA and the Paris Basin of France. Some 6 GW of electrical generating capacity is currently installed overseas in several regions where both steam and water are produced at temperatures over

200°C. In addition, geothermal resources are used in many district and process heating schemes.

Bioenergy

There are a number of ways in which biological systems can be used to produce energy. Agricultural wastes can be used to produce energy by combustion or by biological processing to produce liquid or gaseous fuels. It is also possible to grow crops specifically for energy purposes.

Agricultural and forestry wastes fall into two main groups, dry combustible wastes such as forestry wastes and straw, and wet wastes such as 'green' agricultural crop wastes (i.e. root vegetable tops) and farm slurry. The former group of biofuels are utilised using thermal processes to give heat directly (via **Combustion**), or converted into a second fuel either gaseous (via gasification) or liquid (via **Pyrolyis**). The latter group of biofuels are best utilised via anaerobic digestion to produce methane ('biogas').

Currently, very little use is made of these materials as sources of energy, despite the fact that frequently there is a cost associated with their clean disposal. For example, surplus straw now has to be ploughed into the soil and animal slurries must be contained to prevent water course adulteration. Conversion of these *wastes* into fuels can generally be accommodated within existing agricultural and forestry practice. Thus in future these wastes may be considered as additional income earners.

The use of fuels derived from agricultural and forestry residues could create markets which energy crops might then supply at a later date. In Denmark today, there is 50 MWe of straw burning plants and in the USA wood fuelled power stations total approximately 6,000 MWe.

Crops which may be grown to produce energy range from food crops grown for energy purposes to woody *biomass.* From these sources solid, liquid or gaseous biofuels may be derived. Many methods for the conversion of biomass are available, reflecting the diversity of the resource. The drier, lignin-rich materials (e.g. wood) are best suited to combustion, gasification or pyrolysis conversion processes. Wetter biomass can be converted through anaerobic digestion to a methane-rich biogas fuel. Other fermentation techniques produce liquid fuels such as **Ethanol**.

Well-developed systems of varying scale are available to provide direct heat or electricity from the crop. Conventionally, electricity is produced by burning wood in a boiler to generate steam that is fed to a turbine. More advanced technologies involving **Gasification** are ready for demonstration. They should allow electricity production at higher efficiency and lead to significant reductions in costs.

There are now significant opportunities for the production of energy crops. In Europe one important factor is of the reform of the Common Agricultural Policy, a central element of which is the reduction of food overproduction. New measures will result in some farm land becoming potentially available for non-food crops, including energy crops. The benefit of this approach is that such land will remain productive and not fall derelict. This will go some way to maintaining farm incomes and rural economies.

World interest in biomass is growing rapidly with major programmes underway in Scandinavia, the USA and Europe. Significant energy crop enterprises already exist in Brazil and Sweden with development programmes underway in Scandinavia, Europe and North America. Over 6,000 ha of short rotation coppice have been established in Sweden alone.

There are a number of so called advanced conversion technologies that are being considered as alternatives to conventional steam raising plant for converting biofuels. Principal amongst these are pyrolysis, gasification and liquefaction. These thermochemical processes produce solid, liquid and gaseous intermediates from biofuels, which can be used to produce electricity and/or heat, or be upgraded to directly substitute for fossil fuels. The intermediates can be used in an engine to produce power, avoiding the use of a steam cycle. This gives a very significant increase in conversion efficiency with comparable capital and operating costs at the scales relevant to biofuels (less than 50 MWe). In this way, the economic viability of electricity production from energy crops, forestry wastes and straw is significantly improved.

These processes promise to be inherently less polluting than conventional incineration. By reducing the costs of pollution abatement and offering a more secure disposal route in the light of ever more stringent pollution legislation, the incorporation of these technologies in energy from waste schemes is likely to increase. For sewage sludge, gasification may prove to be the best option for both disposal and power generation as conventional incineration pushed to the limits is barely autothermic.

In the Scandinavian paper industry over 100 MW_{th} of biomass gasification plant has been in operation for many years. Major advanced biomass gasification projects for power generation are now underway in Sweden, Finland and Hawaii.

References

An Assessment of Renewable Energy for the UK, ETSU R82—HMSO, ISBN 0-11-515348-9.

Renewable Energy Resources: Opportunities and Constraints, 1900–2020, World Energy Council.

Renewable Energy: Sources for Fuels and Electricity, (1993) T. B. Johansson et al., Island Press.

Leading to: *Hydro power; Renewable energy; Wind turbines*

A. Brown

ALUMINA (see Aluminum oxide)

ALUMINUM

(L. *alumen, alum*), Al; atomic weight 26.98154; atomic number 13; melting point 660.37°C; boiling point 2467°C; specific gravity 2.6989 (20°C); valence 3. The ancient Greeks and Romans used *alum* in medicine as an astringent, and as a mordant in dyeing. In 1761 de Morveau proposed the name *alumine* for the base in alum, and Lavoisier, in 1787, thought this to be the oxide of a still undiscovered metal. Wohler is generally credited with having isolated the metal in 1827, although an impure form was prepared by Oersted two years earlier. In 1807, Davy proposed the name *alum-*

ium for the metal, undiscovered at that time, and later agreed to change it to *aluminum.* Shortly thereafter, the name *aluminium* was adopted to conform with the "ium" ending of most elements, and this spelling is now in use elsewhere in the world. *Aluminium* was also the accepted spelling in the U.S. until 1925, at which time the American Chemical Society officially decided to use the name *aluminum* thereafter in their publications.

The method of obtaining aluminum metal by the electrolysis of alumina dissolved in *cryolite* was discovered in 1886 by Hall in the U.S. at about the same time by Heroult in France. Cryolite, a natural ore found in Greenland, is no longer widely used in commercial production, but has been replaced by an artificial mixture of sodium, aluminum, and calcium fluorides. *Bauxite,* an impure hydrated oxide ore, is found in large deposits in Jamaica, Australia, Surinam, Guyana, Arkansas, and elsewhere. The Bayer process is most commonly used today to refine bauxite so it can be accommodated in the Hall-Heroult refining process, used to produce most aluminum. Aluminum can now be produced from clay, but the process is not economically feasible at present.

Aluminum is the most abundant metal to be found in the earth's crust (8.1%), but is never found free in nature. In addition to the minerals mentioned above, it is found in feldspars, granite, and in many other common minerals. Pure aluminum, a silvery-white metal, possesses many desirable characteristics. It is light, nontoxic, has a pleasing appearance, can easily be formed, machined, or cast, has a high thermal conductivity, and has excellent corrosion resistance. It is nonmagnetic and nonsparking, stands second among metals in the scale of malleability, and sixth in ductility. It is extensively used for kitchen utensils, outside building decoration, and in thousands of industrial applications where a strong, light, easily constructed material is needed. Although its electrical conductivity is only about 60% that of copper per area of cross section, it is used in electrical transmission lines because of its light weight.

Pure aluminum is soft and lacks strength, but it can be alloyed with small amounts of copper, magnesium, silicon, manganese, and other elements to impart a variety of useful properties. These alloys are of vital importance in the construction of modern aircraft and rockets. Aluminum, evaporated in a vacuum, forms a highly reflective coating for both visible light and radiant heat. These coatings soon form a thin layer of the protective oxide and do not deteriorate as do silver coatings. They have found application in coatings for telescope mirrors, in making decorative paper, packages, toys, and in many other uses.

The compounds of greatest importance are aluminum oxide, the sulfate, and the soluble sulfate with potassium (alum). The oxide, alumina, occurs naturally as ruby, sapphire, corundum, and emery, and is used in glassmaking and refractories. Synthetic ruby and sapphire have found application in the construction of lasers for producing coherent light. In 1852, the price of aluminum was about \$545/lb, and just before Hall's discovery in 1886, about \$11.00. The price rapidly dropped to 30c and has been as low as 15c/lb.

Handbook of Chemistry and Physics, CRC Press

ALUMINIUM COMBUSTION (see Flames)

ALUMINUM OXIDE

Aluminum Oxide (Al_2O_3) or *alumina* is a naturally occurring compound that is a constituent of a variety of minerals (corundum, rubys bauxite, etc.). It is a white, non-toxic powder of high melting point (2303K) and its molecular weight is 101.96. It is insoluble in water and only sparingly soluble in strong acids or alkalis.

V. Vesovic

ALUMINOSILICATE ZEOLITES (see Molecular sieves)

AMAGAT'S LAW

Following from: *Perfect gas*

When two or more pure gases, held at a temperature T and a pressure p, are mixed to form a homogeneous gas mixture at the same temperature and total pressure, the relationship between the volume of the mixture and the volume of the pure components is, generally, complicated. However, if the pure components and the mixture conform to the perfect gas equation of state then the volume of the mixture is simply the sum of the volumes of the pure components (Bett et al., 1975) so that

$$V(T, p, n_1, n_2 \ldots n_N) = \sum_i^N V_i(T, P, n_i)$$

This result is known as Amagat's Law of Additive Volumes. While it is strictly valid only for the mixing of perfect gases, the law is often a useful approximation for real gases.

References

Bett, K. E., Rowlinson, J. S., and Saville, G. (1975) Thermodynamics for Chemical Engineers, Athlone Press, London.

W. A. Wakeham

AMERICAN ASSOCIATION FOR THE ADVANCEMENT OF SCIENCE, AAAS

1333 H Street NW
Washington DC 20005
USA
Tel: 202 326 6400

AMERICAN INSTITUTE FOR CHEMICAL ENGINEERS, AIChe

345 East 47th Street
New York 10017
USA
Tel: 212 705 7338

AMERICAN IRON AND STEEL INSTITUTE, AISI

11001 17th Street NW
Washington DC 20036
USA
Tel: 202 452 7100

AMERICAN PETROLEUM INSTITUTE, API

1220 L Street NW
Washington DC 20005
UK
Tel: 202 682 8000

AMERICAN SOCIETY OF HEATING, REFRIGERATION AND AIR-CONDITIONING ENGINEERS, ASHRAE, INC

1791 Tullie Circle NE
Atlanta GA 30329
USA
Tel: 404 636 8400

AMERICAN SOCIETY OF MECHANICAL ENGINEERS, ASME

345 East 47th Street
New York NY 10017
USA
Tel: 212 705 7722

AMMONIA COMBUSTION (see Flames)

AMMONIUM NITRATE FERTILIZER (see Prilling)

AMMONIUM PERCHLORATE (see Flames)

AMPLIFIED SPECTRUM (see Spectral analysis)

ANABOLISM (see Physiology and heat transfer)

ANALOGY BETWEEN HEAT AND MASS TRANSFER (see Condensation, overview)

ANEMOMETERS (LASER DOPPLER)

Following from: *Velocity measurement*

Yeh and Cummins (1964) first demonstrated that gas lasers could be used to measure the velocity of fluids by observing the Doppler shift in the frequency of *light scattered from small particles* moving with the fluid. Since then the technique, which became known as *laser Doppler anemometry (LDA)* has been greatly developed and today specialised optical and signal processing systems are routinely used in almost every fluid mechanics laboratory in the world.

Gas lasers provide the high energy monochromatic light source for most LDA systems, different optical techniques being used depending upon whether or not forward scatter or back scatter light is collected by the photodetector. In order to scatter light, particles are required in the flow. These may occur naturally in some situations; in other instances small seed particles have to be introduced into the flow. Small seed particles are usually larger than the laser wavelength and for conventional LDA, it is assumed that the small seed particles are capable of following the small scale fluctuations in the flow. The velocity of the particles is therefore taken to be the velocity of the flow at the point of measurement. A typical scattering diagram associated with a single particle is shown in **Figure 1**.

The intensity of scattered light received varies with angular position of the photodetector. An angle of 0° denotes forward scatter and 180° is known as backscatter. For forward scatter the flow has to be accessible through a minimum of two viewing directions, whereas with backscatter at 180° a single window is sufficient. The figure shows that the intensity of scattered light decreases with angular location of the detector, being a maximum at 0° and a minimum at 90°. The ratio of forward to backscattered light is of the order of 10^2 to 10^3; a high powered laser is usually required for backscatter measurements.

Optical techniques based on forward scatter (reference beam, dual beam and dual scattered beam) and backscatter geometries have been fully developed, see Durst et al (1976). All transmission optical components are corrected for spherical aberration, surfaces are coated to provide optimal transmission characteristics for the selected wavelengths. Today the dual beam optical geometry is most widely used for both forward and backscatter optical configurations.

The concept of the *fringe model* was introduced by Rudd (1969) to explain the operation of the dual beam LDA, in which each of the two, equal intensity, Gaussian beams can be considered to contain a series of wavefronts. When correctly aligned the two beams come to a focus or waist at their intersection, so that in this region the wavefronts are very nearly parallel. The wavefronts from each beam form a series of parallel interference fringes in the plane normal to the plane of the beams. The fringe spacing (Δx) is given by

Figure 1. Scattering diagram for a particle greater than the laser wavelength.

$$\Delta x = \frac{\lambda}{2\sin\theta/2} \tag{1}$$

where λ is the laser wavelength and θ is the intersection angle of the two focussed beams. **Figure 2** shows the interference fringes formed in the intersection region, as **Equation 1** shows Δx decreases with increasing θ.

The characteristic dimensions of the length ($2\sigma_y$), width ($2\sigma_x$) and height ($2\sigma_z$) of the scattering control volume, are given by

$$2\sigma_x = \frac{4\lambda}{\pi}\frac{F}{B\cos\theta/2}$$

$$2\sigma_y = \frac{4\lambda}{\pi}\frac{F}{B\sin\theta/2}$$

$$2\sigma_z = \frac{4\lambda}{\pi}\frac{F}{B},$$ and the number of fringes

$$N_f = \frac{8}{\pi}\frac{F}{B}\tan\theta/2.$$

F is the focal length of the focussing lens and B is the diameter of the laser beam at the e^{-1} point of the Gaussian intensity distribution. The introduction of beam expander units into the transmission optical geometry is essential for applications where the size of the measuring volume should be small. The beam expander expands both the laser beam diameter (B) and the beam spacing by the expansion ratio (E). θ is increased by the expansion ratio so that the measurement volume length decreases by a factor equal to the square of the expansion ratio. The light intensity in the control volume will be increased by the square of the expansion ratio. Particles in the flow cross the beam intersection region, which is known as the measurement volume, and scatter light with an intensity which is modulated by the Doppler frequency (f_D). A typical *Doppler burst* or signal from a particle is shown in **Figure 3**. If the optical system is correctly aligned the Doppler frequency f_D associated with each individual fringe (Δx), within the measurement volume is a constant. This means that the velocity of each individual particle may be represented as

$$V = f_D \Delta x, \tag{2}$$

V is the instantaneous velocity normal to the fringes and f_D is known as the instantaneous Doppler frequency. This leads to the relationship that

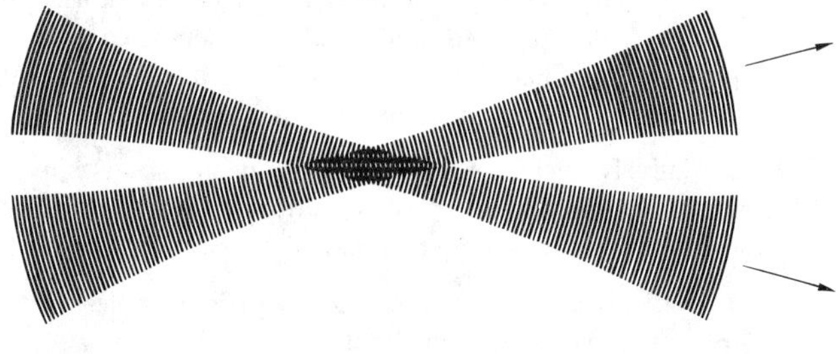

Figure 2. Interference fringes formed in the intersection region.

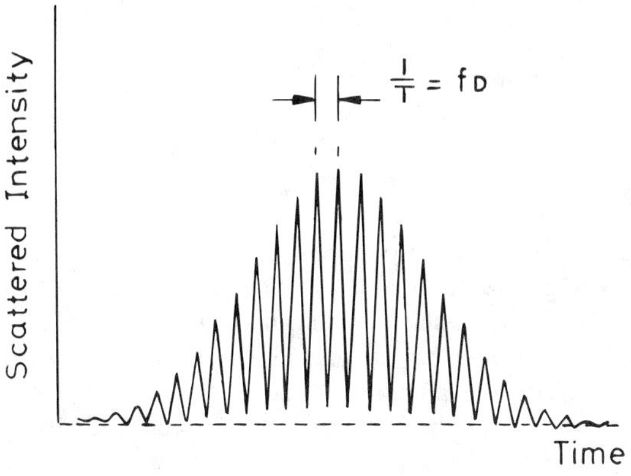

Figure 3. Typical Doppler burst.

$$V = f_D\Delta x = \frac{f_D\lambda}{2\sin\theta/2} \tag{3}$$

from which it is clear that the instantaneous particle velocity V varies linearly with f_D, since for any optical geometry and laser $\lambda/2\sin\theta/2$ is a constant. This relationship also shows that no system calibration is necessary.

Representing the instantaneous quantities as the sum of a mean (V_m) plus a fluctuation (v'), then

$$V = V_m + v' \quad \text{and} \quad f_D = f_{D_m} + f'_D,$$

and

$$(V_m + v') = (f_{D_m} + f'_D)\frac{\lambda}{2\sin\theta/2}$$

or for mean quantities

$$V_m = f_{D_m}\frac{\lambda}{2\sin\theta/2}$$

and for fluctuating quantities

$$v'^2 = f'^2_D\left(\frac{\lambda}{2\sin\theta/2}\right)^2.$$

These relationships show that a one dimensional LDA system is capable of providing the mean velocity as well as the corresponding Reynolds normal stress. In addition other turbulence parameters, such as the skewness and flatness factors, together with the spectrum of turbulence and the auto or cross-correlations, are available via sophisticated software packages.

For many measurements made by LDA the velocity statistics are compiled from N discrete instantaneous velocity measurements made at random time intervals, each of which corresponds to the passage of a particle through the measurement volume. Mean quantities are calculated by performing ensemble averages, that is

$$V_m = \frac{1}{N}\sum_{i=1}^{N} V_i, \quad \text{and}$$

$$v'^2 = \frac{1}{N}\sum_{i=1}^{N} (V_i - V_m)^2$$

suffix i denotes the velocity from the ith Doppler burst in the sample.

The signal to be analysed is provided by the photodetector, usually as a current plulse, which contains the required frequency information relating to the velocity. Photodetector tubes have a band width up to approximately 120 MHz. Interference filters are required to restrict the light reaching the photo sensitive surface to that from the required wavelength or colour.

Using the blue and green lines simultaneously from an Argon-ion laser enables a two-dimensional LDA system to be assembled, the blue and green fringes being orthogonal to one another. A particle crossing the orthogonal fringes will scatter blue and green light, providing two orthogonal velocity components. Again using two photodetectors and ensemble averaging this optical configuration provides the capability to measure two mean velocity components (U_m, V_m), two normal stresses (u'^2, v'^2) and a Reynolds shear stress ($u'v'$).

A third colour and photodetector enables a fully three-dimensional system to be constructed which will measure, at a point in the flow, ensemble averages of the three mean velocity components (U_m, V_m, W_m) and all the components which represent Reynolds normal and shear stresses in the general form of the Navier Stokes equations.

Modern laser Doppler anemometer systems incorporate a *Bragg cell* for frequency shifting of a laser beam, at a frequency of 40 MHz, in order to remove any directional ambiguity. This facility enables regions of positive, zero and negative flow velocities to be clearly identified, the importance of this capability is highlighted in **Figure 4**. An exceptionally small control volume $30 \times 17 \times 15\ \mu m$ was used to provide detailed measurements within the cavity between two 5 mm square ribs on a roughened surface, the pitch to height ratio (x/e) being 7.2. These near wall measurements to within 100 μm of the surface enable the various flow regimes to be clearly identified and confirms that flow reattachment does not occur along the base of the cavity [see Martin and Bates, (1992)].

With recent advances in fibre optic technology a new generation of LDA probes have been introduced with the fibre used as a link between the laser and the optical head, which incorporates the focussing lens. The fibres which transmit the laser beams are polarisation-preserving single mode fibres. A multimode fibre is then used to collect the scattered light for transfer to the photodetector. The use of fibre optics has improved the inherent safety of LDA systems, since the high powered laser beams are confined within

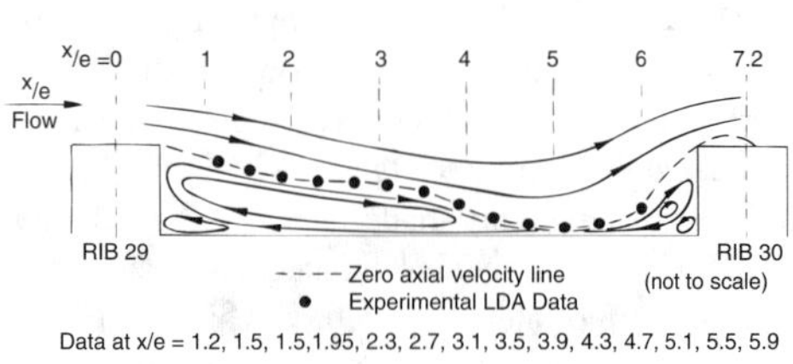

Figure 4. Near wall measurements identifying various flow regimes.

the fibre which may be run through a laboratory from a centrally located high power laser.

Non-intrusive LDA measurements have been undertaken on a large number of carefully controlled experiments of laminar and turbulent flows which may be isothermal or combusting. Good quality experimental data has been used as a basis for comparison and validation of modern computational fluid dynamic codes based upon the numerical solution of the three-dimensional Navier Stokes equations.

After a decade of research conventional LDA techniques have been successfully extended to measure simultaneously the size and velocity of spherical droplets and particles suspended in two-phase flows. *Phase Doppler anemometry (PDA)* is today the most widely used technique for two-phase flow investigations.

Modern PDA systems comprise two or more photodetectors which are located at different points in space, but each of the detectors sees the light scattered from the same particle as it travels across the measurement probe volume. Each detector will therefore generate an identical Doppler burst. As with conventional LDA the Doppler frequency provides the velocity of the particle or droplet. Each Doppler signal carries a phase difference with respect to one another. Saffman et al. (1984), Bachalo (1980) and Durst and Zareé (1975) showed, either theoretically or experimentally, that the phase difference varied linearly with particle or droplet diameter. The linear relationship between phase and diameter depends upon both the transmission and collection optical geometries, the angle at which the scattered light is collected as well as, for refraction only, the ratio of the refractive index of the droplet to that of the transporting medium. Two matched photodetectors enable phase differences up to 360° to be considered, whilst the inclusion of a third matched photodetector eliminates phase ambiguity and doubles the sizing range. Modern PDA receivers are integrated units, which accommodate up to four photodetectors. A range of micrometer screws control the size of the aperture, over which scattered light is collected, as well as the polarisation of the system and the focussing of the unit. Drain (1985) showed how the scattered light intensity varied with the scattering angle at which the light was collected, taking into account refraction and reflection due to the droplet as well as the polarisation of the laser. In general 30°, 70° and 150° scattering angles, from the forward direction are the most appropriate for scattering from water droplets. Scattering at both 30° and 70° takes into account refraction and reflection associated with each droplet. The collection of scattered light at 70° minimises the influence of reflection on the experimental measurements.

The most obvious advantages associated with the use of PDA techniques for multiphase flows are:

a. A small single-colour control volume, such as that used in a one dimensional conventional LDA system is capable of providing simultaneously the velocity and diameter of individual droplets suspended in a two-phase flow.

b. a relatively cheap low-power helium-neon laser is appropriate. To improve the PDA's sensitivity to small droplets a high powered laser may be required.

c. a relatively large dynamic range of 40:1 is possible, with a minimum and maximum droplet size of approximately 1 and 10000 μm respectively.

d. new high speed co-variance signal processors provide the ability to measure in real time at data rates of up to 200 kHz. The autocorrelation of a Doppler burst provides the Doppler frequency, whereas the cross-correlation of two Doppler bursts from a phase Doppler system provides the phase difference. The data processing software converts the phase difference into a droplet or particle diameter based on the linear calibration curve.

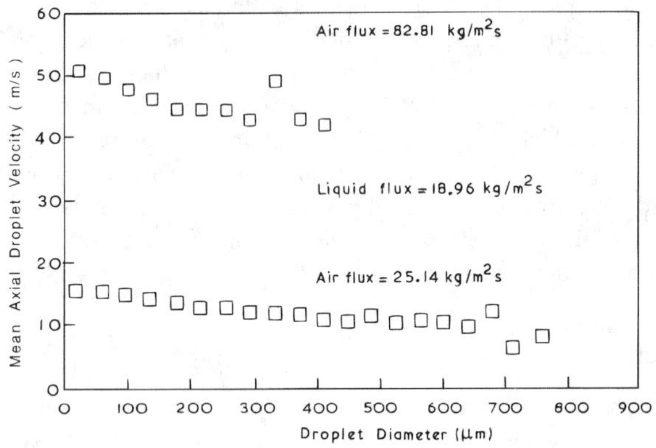

Figure 5. Correlations of droplet velocity versus droplet diameter for an annular two-phase flow.

Tayali and Bates (1990) have reviewed the development of optical techniques for particle sizing in multiphase flows. Detailed PDA experimental measurements, of the diameter and velocity of individual droplets or particles found in a wide range of two-phase and multi-phase flows, have been published in the research literature. This data provides a new insight into the complex flow phenomena whereby droplets:

a. may break-up
b. may be entrained into a flow
c. may coalesce
d. may deposit on to a surface.

New data collected using PDA systems will enable the performance of existing empirical models to be investigated. **Figure 5** shows measurements of the correlation of droplet velocity versus droplet diameter for an air/water annular two phase flow, the pipe diameter being 32 mm and the hydrodynamic development length is 228 diameters. The figure confirms that the larger droplets lag the smaller ones and clearly shows the effect of the variation of the gas mass flux, on the correlation, at a constant water flux. The size range of the droplets observed in the flow decreases with increasing gas flowrate (Bates and Sheriff (1992), Bates and Ayob (1994)).

References

Bachalo, W. D. (1980) Method for Measuring the Size and Velocity of Spheres by Dual Beam Light Scattering Interferometry, *Appl. Opt.* 19 3:p363.

Bates, C. J. and Ayob, R. (1995) Annular Two-Phase Flow Measurements using Phase Doppler Anemometry with Scattering Angles of 30° and 70°, Flow Meas. Instrum. (6):21.

Bates, C. J. and Sheriff, J. M. (1992) High Data Rate Measurements of Droplet Dynamics in a Vertical Gas-Liquid Annular Flow, *Flow Meas Instrum* (3):247.

Drain, I. E. (1985) Laser Anemometry and Particle Sizing, *Int Conf on Laser Anemometry—Advances and Applications,* 7.

Durst, F., Melling, A., and Whitelaw, J. H. (1976) Principles and Practice of Laser Doppler Anemometry, Academic Press.

Durst, F. and Zaré, M. (1975) Laser-Doppler Measurements in Two-Phase Flows, *Proc LDA Symp,* Copenhagen, 403.

Martin, S. R. and Bates, C. J. (1992) Small Probe Volume Laser Doppler Anemometry Measurements of Turbulent Flow near the Wall of a Rib Roughened Channel, Flow Meas Instrum, 2, 81.

Rudd, M. J. (1969) A New Theoretical Model for the Laser Doppler Meter, *J Phys E: Sci Instrum,* 2, 55.

Saffman, M., Buchhave, P., and Tanger, H. (1984) Simultaneous Measurement of Size, Concentration and Velocity of Spherical Particles by a Laser-Doppler Method, 2nd ISALAFM, Lisbon, Paper 8.1.

Tayali, N. E. and Bates, C. J. (1990) Particle Sizing Techniques in Multiphase Flows: A Review, *Flow Meas Instrum,* 1, 77.

Yeh, Y. and Cummins, H. Z. (1964) Localised Fluid Flow Measurements with a He-Ne Laser Spectrometer, *Appl Phys Lett,* 4, 176.

C.J. Bates

ANEMOMETERS (PULSED THERMAL)

***Following from:** Velocity measurement*

The method of measuring velocity with the help of the *pulsed thermal anemometer* is based on measuring the transit time of particles heated by the wire to which electric pulses are sent up to the wire transducers operating in the resistance thermometer mode. The advantage of the method is its utilization for the flows with a high degree of turbulence including flows with sign-variable averaged velocity, i.e., in conditions when the traditional method of diagnosis of turbulent flows, a hot-wire anemometer, cannot be used. Furthermore, the velocity is measured directly and is absolutely independent of the properties of a fluid.

A sensing element of a transducer of a *pulsed thermal anemometer* consists of three wires (see **Figure 1**). The central wire is the source of thermal pulses, and on each side of it, the receiving wires are arranged at distance l apart, their axes being perpendicular to the pulse source axis. For the central wire, platinum or nickel wires 10 μm in diameter are used, and 5 μm wires $L/l = 5–6$ long and spaced $1.5 \leq l \leq 2.5$ mm apart are used for the peripheral wires. The temperature of the heated wire must not exceed 500–600°C in order to maintain its strength. The pulse length of the voltage supplied to the central wire for receiving warmed gas particles is 5–30 μs, the front rise time being not less than 0.5 μs and with a current of 4–9A. The receiving wires are supplied with a 0.5–3 mA constant current and are connected to the bridge arms. The polarity of a signal depends on the transducer conductors which receive a thermal pulse and determine the flow direction.

At currents lower than 0.5 mA the influence of electron noise becomes significant and at currents higher than 5 mA the transducer starts to respond to velocity fluctuations as a hot-wire anemometer.

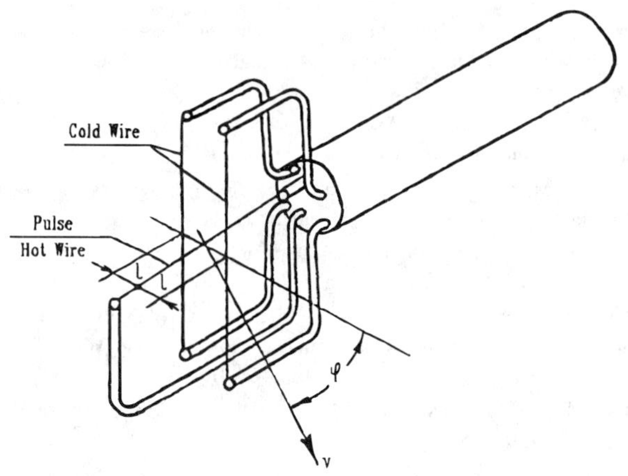

Figure 1.

In an ideal case the temperature of a pulse wire rises instantaneously when the temperature of the surrounding medium increases to a higher value. The particles of the air warmed by it are carried away with an instantaneous velocity of the flow streamlining the transducer. With the effects of diffusion and thermodiffusion being neglected, the time t required for the front of warmed particles to reach the receiving wire is $t = l/u \cos \varphi$, where φ is the angle formed by the instantaneous velocity vector u and the normal to the transducer plane. The use of two receiving wires on each side of the pulse wire ensures an unambiguous definition of the flow direction at the time moment being considered. If a sufficient number of measurements of the transit time for a given position of a transducer are made, then the averaged set of realizations will allow us to determine the time average and the fluctuation components of the velocity vector. The disadvantage of the method is that the transducer does not respond to turbulent pulsations which have typical dimensions smaller than the dimension of the transducer. The effect of thermal diffusion causes an additional barrier in applying the method; the limit is $Pe = yl/\kappa = 50$.

Reference

Bradbury, L. J. S., and Castro, I. P. A. (1971) A Pulsed-Wire Technique for Velocity Measurement in Highly Turbulent Flows, *J. of Fluid Mechanics*, 4.

Yu.L. Shekhter

ANEMOMETERS (VANE)

Following from: *Velocity measurement*

The vane anemometer is an instrument designed to measure velocity utilizing the kinetic energy of a flow. Its main element is a vane with blades of different shape. To measure velocity in the atmosphere, the blades are made in the form of hollow hemispheres with their concavity facing away from the flow; the plane of rotation is parallel to the direction of motion. In measuring the velocity in pipes and channels, vanes with flat blades are used, arranged at an angle of attack of 40–45° to the direction of flow in the plane perpendicular to it. In hydraulic laboratories microvanes with blades diameters of 6–8 mm are used for local measurements. The speed of vane rotation is proportional to the flow velocity. The characteristic is linear within the range of from a minimum defined by friction in the mechanism to a maximum. The range of mean velocity $\bar{u}$ for application in air is $\bar{u} = 2.0 - 50$ m/s. The accuracy of velocity measurement is $\pm 0.5 + 0.035\bar{u}$. For exact measurements to be performed, the anemometers is calibrated against a standard velocity value. The deviation of the plane of rotation from a nominal value by $\pm 10°$ causes an error of not higher than 1%. The vane rotation is transmitted to a special mechanism or an electronic counter, which counts the number of revolutions per unit time, proportional to flow velocity. If the density of gas on measuring ρ_1 differs from the density of gas in calibration ρ_0, the actual value of velocity u_1 is defined from the measured u_0 (read off from the calibration curve) $u_1 = u_0(\rho_0/\rho_1)^{0.5}$.

Reference

Fluid Meters. Their Theory and Application. Report of ASME, Research Committee on Fluid Meters, New York, Published by ASME, 1971.

Yu.L. Shekhter

ANEROID BAROMETER (see Pressure measurement)

ANL (see Argonne National Laboratory)

ANNEALING (see Steels)

ANNULAR DISPERSED FLOW (see Annular flow; Gas-liquid flow)

ANNULAR FIXED BEDS (see Fixed beds)

ANNULAR FLOW

Following from: *Gas-liquid flow*

Annular flow is a flow regime of two-phase gas-liquid flow (see gas-liquid flow). It is characterized by the presence of a liquid film flowing on the channel wall (in a round channel this film is annulus-shaped which gives the name to this type of flow) and with the gas flowing in the gas core. The flow core can contain entrained liquid droplets. In this case, the region is often referred to as *annular-dispersed flow,* where the entrained fraction may vary from zero (a pure annular flow) to a value close to unity (a dispersed flow). Often both types of flow, pure annular and annular-dispersed, are known under the general term of annular flow.

In vertical and inclined channels, the near-wall liquid film and the gas core may be both concurrent and countercurrent. The true volumetric gas concentration (void fraction) of the annular flow ϵ_G, determined as a fraction of the cross section occupied by the gas phase, is high and, as a rule, exceeds 75–80%. In horizontal channels the film thickness is non-uniform around the channel perimeter due to gravity. Both adiabatic annular flows and diabatic annular flows occur in industrial applications; eg. in steam generators, evaporators, condensers and boiling water reactors. In the latter case a phase transition (evaporation, condensation) is observed at the liquid film-gas core interface. The phase change may also occur in the liquid film when vapor bubbles appear on the heated channel wall (nucleate boiling).

In the case of high wall superheats a regime of film boiling sets in when a vapor film flows along the wall and a liquid core flows in the middle of the channel. In this case we are concerned with an *inverse annular flow.*

A general diagram of annular flow is presented in **Figure 1** in which $\dot{M}_G$, $\dot{M}_{LF}$, and $\dot{M}_{LE}$ are the flow rates of the gas (vapor) phase, the liquid in the wall film, and liquid entrained in the core respectively. The velocity profile in a gas core, as in the case in the flow of single-phase fluid, obeys the logarithmic law. However, as $\dot{M}_{LE}$ grows, it becomes more peaked which is related to both the increasing number of droplets in the core and a rougher film-core interface.

The thickness δ of the liquid film is determined by the ratio of $\dot{M}_{LF}$ to $\dot{M}_G$ (the appropriate Reynolds numbers Re_L and Re_G) and by the shear stress τ_i on the interface (related to the pressure gradient in a two-phase flow dp_f/dx). The relation between δ, $\dot{M}_{LF}$ and dp_f/dx is known as a *triangular relationship*. Given any two of the parameters, we can determine the unknown one making use of a model of shear stress distribution in the film (for fairly thin films the assumption $\tau = const$ is justified). Various forms of the triangular relationship are discussed by Hewitt and Hall Taylor (1970) and by Hewitt (1982).

In order to determine ϵ_G in a pipe of diameter D for low ratios of working pressure to critical pressure the empirical *correlation of Armand* may be used

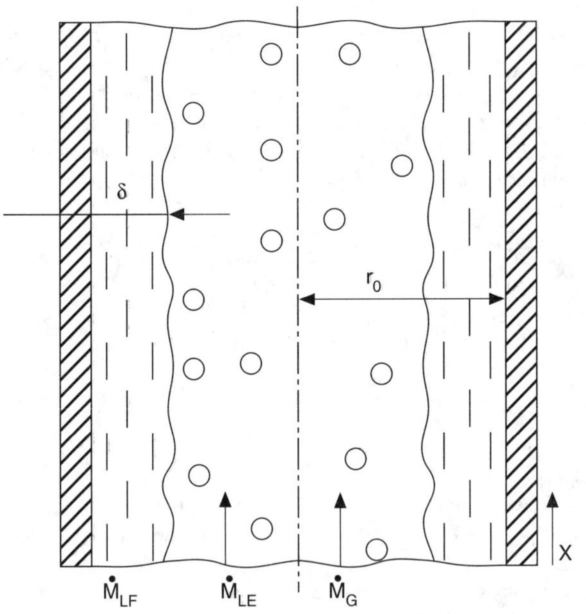

Figure 1. Parameters of annular flow.

$$1 - \epsilon_G = \frac{4 + \frac{8}{7}b}{5 + b\left[\frac{\beta}{1-\beta} + \frac{8}{7}\right]}, \tag{1}$$

where

$$\beta = \frac{\dfrac{\dot{M}_G}{\rho_G}}{\dfrac{\dot{M}_G}{\rho_G} + \dfrac{\dot{M}_{LF} + \dot{M}_{LE}}{\rho_L}},$$

$$b = 4a(Re_L)^{0.125}\left(\frac{\rho_G}{\rho_L}\right)^{0.5}, \qquad a = 0.69 + (1 - \beta)(4 + 21.9\sqrt{Fr_{0L}})$$

$$Fr = \frac{\dot{m}_L^2}{\rho_L^2 gD}; \qquad \dot{m}_L = \frac{\dot{M}_{LF} + \dot{M}_{LE}}{A}$$

$\dot{m}_L$ is the mass flux of the liquid, A the channel cross section, ρ_L and ρ_G are the densities respectively of the liquid and the gaseous phase, and η_L the dynamic viscosity of the liquid phase. The mean thickness δ of the film is related to ϵ_G by an obvious relation

$$1 - \epsilon_G = \frac{4\delta}{D}. \tag{2}$$

The above formulas pertain to the case of concurrent phase motion. If we deal with a liquid film flowing down the channel wall and with the gas counterflow, then the moment comes, as the gas velocity increases, when on the film surface there are generated large waves which are carried upward by the gas stream, i.e., liquid transfer is observed upward of the place of its injection. This is known as flooding. As the gas velocity increases, the fluid flow reverses and the annular flow becomes concurrent, i.e., upward.

The film surface may be covered with an intricate *wave* system. The waves arise on a fairly thick film and make interface rough. Two types of waves are distinguished, viz. small-scale waves or ripples and large-scale or *disturbance waves* whose amplitude is five- or sixfold higher than the mean thickness of the film. The spacing between the disturbance wave crests is a function of a film thickness alone and does not depend on flow velocity. Experimental findings show that the disturbance waves are extremely nonuniform both axially and and circumferentially. It is the disturbance waves, from the crests of which liquid breaks away, which are responsible for droplet entrainment to the core. *Droplet entrainment* may also occur as a result of bubbles bursbury from the film, giving rise a range of drop sizes.

The range of droplet sizes in the flow core is fairly broad. Their maximum size is determined by the critical value of the Weber number $We = \rho u_r^2 D_d/\zeta 2\sigma$, where u_r is the relative droplet velocity in the carrier vapor flow and D_d the droplet diameter. During acceleration of a droplet broken away from the surface it may be broken up. The final size distribution of the droplets depends on their initial size and the path traveled along the channel.

The entrained liquid mass flux and velocity are distributed over the channel cross section non-uniformly. According to measurements, the highest liquid flow rate is observed near the channel axis, but in approaching the liquid film-vapor core interface the flow rate of entrained liquid phase is reduced. Nonuniformity of the mean mass velocity profile of droplets grows with increasing fraction of liquid entrained.

The droplets in the flow core deposit on the film surface. The deposition rate $\dot{m}_D$ per unit peripheral as depends on a combined action of buoyancy, inertial, and Magnus forces. The behavior of the finest droplets depends on turbulent diffusion.

For estimating $\dot{m}_D$ Hewitt and Hall Taylor (1970) suggested the formula

$$\dot{m}_D = k_D c, \tag{3}$$

where

$$c = \frac{\dot{M}_{LE}}{\dfrac{\dot{M}_{LE}}{\rho_L} + \dfrac{\dot{M}_G}{\rho_G}}$$

is the mass concentration of droplets in the flow core. The deposition coefficient k_D has the dimensions of a velocity and is a function of fluid physical properties and flow rates. It also appears to depend on concentration of droplets at high concentrations. Correlations for k_D and for the rate of liquid entrainment in E are discussed by Hewitt and Govan (1990a and 1990b).

The relationship between $\dot{M}_{LE}$ and quality, x, is illustrated in **Figure 2**. Straight line 1 shows the total mass of liquid which decreases linearly with increasing vapor quality x. Curve 2 describes entrainment in a long, adiabatic channel in which deposition and entrainment reach equality ($\dot{M}_D = \dot{M}_E$). Thus, at a quality denoted by C, the entrained droplet flow (M_{LE}) is represented by CB and the liquid film flow (M_{LF}) is represented by BA. Curve 3 describes entrainment in a channel with heating. In this case the hydrodynamically equilibrium state is not attained. Here, the wall film is additionally depleted as a result of evaporation. With an allowance for the factors indicated above an overall mass balance is described by

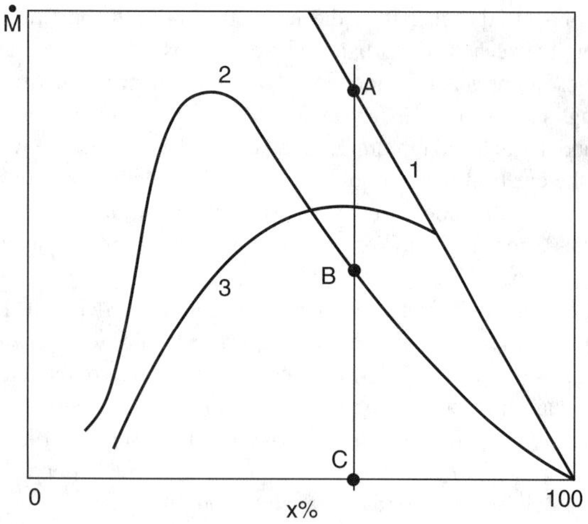

Figure 2. Entrained liquid flow rate in adiabatic equilibrium (Curve 3) and diabatic (heated) flow (Curve 1).

$$\frac{d\dot{M}_{LE}}{dx} = \pi D\left(\dot{m}_D - \dot{M}_E - \frac{\dot{q}}{h_{LG}}\right), \tag{4}$$

where h_{LG} is the latent heat of vaporization and $\dot{q}$ the heat flux on the wall. It should be pointed out that, due to the presence of radial vapor flow from the film surface $\dot{m}_D$ may be somewhat lower than under adiabatic conditions. Conversely, because vapor bubbles burst from the liquid film $\dot{m}_E$ may increase compared to its value for adiabatic flow.

A complete analytical description of annular flow is beyond current capabilities. Most often, annular flow is calculated using one-dimensional approximation. In this case, the continuity equation becomes.

$$\dot{m} = \rho_G u_G \epsilon_G + \rho_L u_{LE} \epsilon_{LE} + \rho_L u_{LF} \epsilon_{LF} = const \tag{5}$$

The momentum equation is:

$$\frac{d}{dz}(\rho_G \epsilon_G u_G^2) + \frac{d}{dz}(\rho_L \epsilon_{LE} u_{LE}^2) + \frac{d}{dz}(\rho_L \epsilon_{LF} u_{LF}^2)$$

$$= -\frac{dp}{dz} - dp_F/dz - g \sin\theta[\rho_G \epsilon_G + \rho_L(1 - \epsilon_G)], \tag{6}$$

and the energy equation is:

$$\frac{d}{dz}(\rho_G u_G \epsilon_G h_{0G}) + \frac{d}{dz}(\rho_L u_{LE} \epsilon_{LE} h_{0LE})$$

$$+ \frac{d}{dz}(\rho_L u_{LF} \epsilon_{LF} h_{0LF}) = \frac{4\dot{q}}{D}. \tag{7}$$

In the above equations u_G, u_{LE}, and u_{LF} are the mean velocities of gas phase, droplets, and liquid film respectively; ϵ_G, ϵ_{LE}, and ϵ_{LF} are the fractions of the channel cross section occupied by an appropriate phase ($\epsilon_G + \epsilon_{LE} + \epsilon_{LF} = 1$). $h_0 = h + (u^2/2)$ is the stagnation enthalpy, z the axial distance, and θ the angle of inclination of the channel from the horizontal. dp/dz is the total pressure gradient and dp_F/dz is the frictional pressure gradient. These parameters are related to vapor quality of the flow

$$x = \frac{\dot{M}_G}{\dot{M}_{LE} + \dot{M}_{LF} + \dot{M}_G}$$

and the fraction of liquid phase entrained as droplets is given by

$$\psi = \frac{\dot{M}_{LE}}{\dot{M}_{LE} + \dot{M}_{LF}}$$

Other obvious relations are $\rho_G u_G \epsilon_G = \dot{m}x$, $\rho_L u_{LE} \epsilon_{LE} = m(1 - x)\psi$, $\rho_L u_{LF} \epsilon_{LF} = m(1 - x)(1 - \psi)$ where in is the lot. It should be borne in mind that the velocity of droplets u_{LE} differs slightly from u_G and substantially exceeds u_{LF}.

Equations (6) and **(7)** take into account mass and energy transfer between phases, external input of heat, change of phase fraction, and change of physical properties with pressure. The equations need to be supplemented by the relevant relations for physical properties and flow parameters.

It is obvious that in order to use **Eqs. (5)–(7)** we must know the values ϵ_G, ϵ_{LF}, ψ_L and dp_F/dz. Their authentic description can be obtained mainly from experimental data. In view of extremely diverse experimental conditions (geometrical, regime, kind of fluid) universal relations are not available to date while those available should be used with caution. The value ϵ_G can be estimated using Armand's correlation (1). Alternatively, the liquid film thickness (and hence ϵ_G) can be determined from the liquid film flow rate and from the pressure gradient (the "triangular relationship—see Hewitt and Hall Taylor, 1970 and Hewitt, 1982). Values of ψ may be estimated using the relationships of Hewitt and Govan (1990a, 1990b). The frictional pressure gradient dp_F/dz can be calculated from a relationship suggested by Wallis (1969) which takes account of the additional interfacial shear exerted by the gas in the liquid film as a result of interfacial waves:

$$\frac{dp_F}{dz} = \frac{2f_G[1 + 80(1 - \epsilon_G)]\dot{m}_G^2}{D\rho_G} \quad \textbf{(8)}$$

where f_a is the friction factor for gas flow alone in the tube and in the gas mass flux (kg/m²s calculated to the fuel correction. **Heat transfer** in annular flows may occur both with and without phase change.

Annular flow heat transfer with phase change (evaporation or condensation), is encountered in a wide range of industrial plant. Heat transfer in evaporation in the annular flow regime is noteworthy for high heat transfer coefficients. The mechanism of heat transfer may be nucleate boiling dominated, forced convection boiling dominated (ie. without nucleation) or be a combination.

At given combinations of $\dot{m}$, p, x, and $\dot{q}$ the wall liquid film is exhausted, the wall is dried out, and "dryout" or "burnout" sets in (see **Burnout-forced convection**). This is accompanied with a rapid rise in wall temperature if the wall heat flux is fixed.

References

Hewitt, G. F. and Hall-Taylor, N. S. (1972) Annular Two-Phase Flow, Pergamon Press, Oxford.

Hewitt, G. F. (1982) Chapter 2 of Handbook of Multiphase System. McGray Hill, New York.

Wallis, G. B. (1969) One-Dimensional Two-Phase Flow, McGraw-Hill Book Company, New York.

Hewitt, G. F. and Govan, A. H. (1990a) Pnenomenological modelling of non-equilibrium flows with phase change. Int. J. Heat Mass Transfer *133,* 229–249

Hewitt, G. F. and Govan, A. H. (1990b) Phenomena and prediction in annular two-phase flow. Invited Lecture, Symposium are Advances in gas-Liquid Flow, ASME Winter Annual Meeting Dallas, November 1990 (Published ASME FED Vol 99/HTD Vol 157 pp 41–46).

Leading to: *Wispy annular flow; Film flowrate measurement; Film thickness measurement*

Yu.A. Zeigarnik

ANNULAR FLOW, IN LIQUID-METAL BOILING (see Liquid-metal heat transfer)

ANODE (see Electrochemical cells)

ANTIDERIVATE FUNCTION (see Integrals)

ANTI-FREEZE (see Ethylene glycol)

ANTI-NEUTRINO (see Neutrons)

ANZAAS (see Australian & New Zealand Association for the Advancement of Science)

API (see American Petroleum Institute)

API GRAVITY (see Petroleum)

APPLICATORS (see Atomization)

AQUIFER

Aquifer is the name given to a geological formation that contains water in sufficient quantity to supply wells. The water may be contained in unconsolidated gravel, in porous or cracked rock or in underground caves. The amount of water that can be drawn from an aquifer varies from less than a gallon per minute for watering cattle in the desert to thousands of gallons per minute for supplying the domestic and industrial needs of a town.

In some parts of the world aquifers contribute significantly to the natural water reserves. There are large aquifers in several regions of the USA, for example the sand and gravel formations in the South Eastern coastal plains stretch for hundreds of miles and are several hundred feet thick.

Aquifers are a valuable natural reservoir for the storage of large amounts of water and do not suffer the evaporation losses of surface waters. Care has to be taken not to draw too much from the reserve. Dangers of overpumping are seepage into the aquifer of saline water, and subsidence causing damage to wells and to surface structures.

G.L. Shires

ARCHIMEDES FORCE

Following from: *Hydrostatics*

Archimedes force is a particular case of mass or *volume forces*. By mass (volume) forces we mean forces acting upon each element of a volume or mass of a body. In mechanics of continua the mass is introduced proceeding from the volumetric density $\rho(x_i, t)$, determined for each time instant t and for any space point x_i. The mass in a volume v at time t is given has the integral $\int_V \rho(x_i, t)dV$.

Similarly, the mechanics of continua is concerned not with forces themselves, but with densities of forces, and their distributions in space and in time.

Thus, we may interpret the distribution density of volume forces $\vec{F}(x_i, t)$ at the given point x_i of a medium as the ratio of the basic vector of forces applied to a small volume including point x_i, to the mass of the volume, when the latter tends to zero.

As an example of volume forces, we may consider *gravity force* $\rho\vec{g}$ dV, where ρ is the average value of density in volume dV, $\vec{g}$ is the vector of the acceleration due to gravity. Other examples include centrifugal forces or electromagnetic forces on a fluid carrying an electrical charge.

The pressure gradient is the analog of the volume force (taken with a reverse sign) affecting the liquid element if the pressure itself varies from point to point. If we separate a certain volume in the fluid, then the force, acting upon this volume, is equal to the integral $-\int p dA$, where dA is the surface element, where the integral is taken over the surface surrounding this volume. Transforming the surface integral into a volume one, we find.

$$-\int_A p dA = -\int_V \text{grad } p dV.$$

The last integral is the volume force acting on the whole volume.

The fact that the equation of motion includes not the pressure itself, but only its gradient, shows that the pressure value in the liquid is determined only with respect to an arbitrary constant.

If the external volume force is $\vec{F} = \int_V \rho\vec{f}\, dV$, where $\vec{f}$ is force referred to unit mass, and F_A is the force affecting volume V from the surrounding medium through the boundary surface A, then the sum of the forces acting upon the separated volume will be as follows

$$\vec{F} + \vec{F}_A = \int_V (-\text{grad } p + \rho\vec{f})dV.$$

Force $\vec{f}$ is assumed to be a known function of time and space. If field $\vec{f}$ is related to a potential Φ which is independent of time, then $\vec{f} = -\text{grad } \Phi$.

In hydromechanics, in many practically interesting cases the main driving force is caused by the presence of a temperature or concentration field. Variation of temperature or concentration leads to a change in density; this leads to the development of a buoyant force, which is formed because of the presence of the volume force field. The buoyant force is also called the *Archimedes* or the *hydrostatic lift force.* At small temperature drops the dependence of the fluid density on temperature may be taken to be linear, ie. $\rho = \rho_0[1 + \beta(T_0 - T)]$, where ρ_0 is the fluid density at temperature T_0; β is the coefficient of cubic expansion. The product βg is called also the buoyancy parameter.

In the case of the gravitation field a hot substance moves under the action of the Archimedes force $F = (\rho - \rho_0)gV$ with reference to a cooler substance in the direction opposite to the gravity force direction. The convection intensity depends on the temperature difference between layers, thermal conductivity and on the medium viscosity.

O.G. Martynenko

ARCHIMEDES NUMBER

Following from: *Free convection*

Archimedes number Ar is the similarity criterion of two thermal or hydrodynamic phenomena, in which determining factors are the buoyant (Archimedes) force and the viscosity force. Thus

$$\text{Ar} = g\frac{l^3}{\nu^2}\frac{\rho - \rho_0}{\rho_0},$$

where l is the characteristic linear dimension, ν is the kinematic viscosity ρ and ρ_0 are the medium densities at two points and g is the acceleration due to gravity.

If the variation of the density is caused by a change of temperature T, then at small temperature drops $(\rho - \rho_0)/\rho_0 = \beta\Delta T$ (where β is the volumetric expansion coefficient), the Archimedes number becomes the **Grashof Number** Gr.

By comparing the volume (mass) forces and the kinetic energy of a moving free-convective flow, we may obtain an estimate of the characteristic velocity of motion

$$U_0 = \sqrt{\frac{g(\rho - \rho_0)l}{\rho_0}}.$$

The Archimedes number may then also be represented in the form

$$\text{Ar} = \left[\frac{(\rho - \rho_0)gl}{\eta U_0/l}\right]^2.$$

It is possible to consider this expression as ratio of the square of the volume (buoyant) forces to the viscous friction forces.

The molecular friction forces prevent the development of disturbances. The volumetric forces, on the other hand, intensify the disorder of the motion. Therefore, viscous friction and volume forces exert the opposite effect on the liquid flow. For small values of Ar the motion may be laminar but with increasing Ar the stability of this motion decreases. At a certain critical value of Ar_c the laminar motion turns into the turbulent one.

Ar is sometimes also characterised as a ratio of the lift force to internal friction force, however, its numerical value in this case is different at different points of the flow.

O.G. Martynenko

ARCHIMEDES PRINCIPLE (see Hydrostatics)

ARGON

(Gr. *argos,* inactive), Ar; atomic weight 39.948; atomic number 18; freezing point −189.2°C; boiling point −185.7°C; density 1.7837 g/l. Its presence in air was suspected by Cavendish in 1785, discovered by Lord Rayleigh and Sir William Ramsay in 1894. The gas is prepared by fractionation of liquid air, the atmosphere containing 0.94% argon. The atmosphere of Mars contains 1.6% of Ar^{40} and 5 p.p.m. of Ar^{36}.

Argon is two and one half times as soluble in water as nitrogen, having about the same solubility as oxygen. It is recognized by the characteristic lines in the red end of the spectrum. It is used in electric light bulbs and in fluorescent tubes at a pressure of about 3 mm, and in filling photo tubes, glow tubes, etc. Argon is also used as an inert gas shield for are welding and cutting, as a blanket for the production of titanium and other reactive elements, and as a protective atmosphere for growing silicon and germanium crystals.

Argon is colorless and odorless, both as a gas and liquid. It is available in high-purity form. Commercial argon is available at a cost of about 3c per cubic foot.

Argon is considered to be a very inert gas and is not known to form true chemical compounds, as do krypton, xenon, and radon, However, it does form a hydrate having a dissociation pressure of 105 arm at 0°C. Ion molecules such as (ArKr)*, (ArXe)*, (NeAr)* have been observed spectroscopically. Argon also forms a clathrate with β hydroquinone. This clathrate is stable and can be stored for a considerable time, but a true chemical bond does not exist. Van der Waals' forces act to hold the argon. Naturally occurring argon is a mixture of three isotopes. Five other radioactive isotopes are now known to exist.

Handbook of Chemistry and Physics, CRC Press

ARGON-ION LASER (see Lasers)

ARGONNE NATIONAL LABORATORY, ANL

9700 South Cass Avenue
Argonne IL 60439
USA

ARMAND CORRELATIONS, FOR VOID FRACTION IN ANNULAR FLOW (see Annular flow)

AROMATIC CHEMICALS, OR AROMATICS (see Pyrolysis)

AROMATIC HYDROCARBONS (see Hydrocarbons)

ARRHENIUS EQUATION

Following from: *Chemical reaction*

Chemical reaction rate expressions are usually given in terms of temperature dependent—concentration independent reaction rate constant(s) and terms expressing the concentration dependence of the rate of reaction. An empirical observation is that the temperature dependence of reaction rate constants can be written in terms of:

$$k = k_0 \exp(-E_a/RT),$$

the **Arrhenius Equation**, where k is the reaction rate constant; k_o the pre-exponential 'frequency factor' (units similar to k); E_a the energy of activation (kJoule/mol; 50–250 kJoule/mole for most reactions); R, the gas constant, kJoule/(mol)(degree Kelvin); T, degree Kelvin (Atkins, 1986). Units of k are normally defined by the formulation of the reaction rate expression. Collision Theory and Transition State Theory require the frequency factor to contain a temperature dependent component (e.g., $k_o = k_o' T^m$; $4 > m > 0$). However, in experimental determinations of k, rapidly changing values of the exponential term tend to mask the temperature dependence of the frequency factor.

Experimental data on ln(k) vs. 1/T can usually be plotted as a straight line. k_o and E_a are calculated from the intercept and the slope, respectively:

$$\ln k = \ln k_o - E/RT.$$

The constants can also be calculated from adjacent data points; where these calculations are performed using experimental data, however, this method would lead to severe inaccuracies unless some form of averaging or data smoothing is introduced, since the differentiation would tend to amplify the effects of experimental scatter.

Departures from linearity may be encountered (i) if the temperature dependence of k_o cannot be ignored, e.g., over large ranges of temperature, (ii) if the rate expression does not adequately represent the actual reaction scheme, e.g., the reaction scheme is more complex, or, (iii) if the reaction rate is not entirely kinetically controlled, e.g., mass transfer resistances should not have been ignored. (Perry and Chilton, 1984; Richardson and Peacock 1994)

References

Atkin, P. W. (1986) Physical Chemistry, OUP, Oxford, UK.

Perry, R. H. and Chilton, C. H. (Eds.). (1984) Chemical Engineers' Handbook, McGraw-Hill, New York, USA.

Richardson, J. F. and Peacock, D. G. Eds., (1994) Coulson and Richardson's Chemical Engineering, Volume 3 (3rd edition), Pergamon, Oxford UK.

A.A. Herod and R. Kandiyoti

ASH FORMATION (see Pulverised coal furnaces)

ASH LAYER MODEL (see Ion exchange)

ASHRAE (see American Society of Heating, Refrigeration and Air Conditioning Engineers)

ASME (see American Society of Mechanical Engineers)

ASPECT RATIO

The general definition is the ratio of the longer to the shorter dimension of a rectangle, i.e. L/B

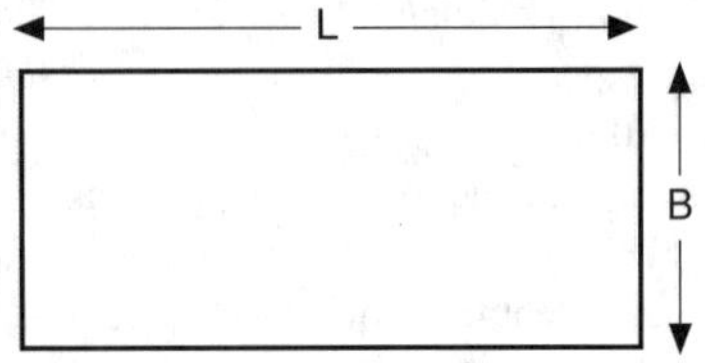

Figure 1.

The term has specific usages. In aerodynamics it is the ratio of the span of a wing to its width or the ratio of the square of the span to total area. In relation to television screens it is the ratio of the frame width to height. In the motor industry it is used for the ratio of height of a tyre to its width, otherwise called the tyre profile.

G.L. Shires

ASSOCIATIVE CATALYSIS (see Catalysis)

ASYMPTOTE

An asymptote is the limiting position of a tangent to a curve where the point of contact is at an infinite distance from the origin.

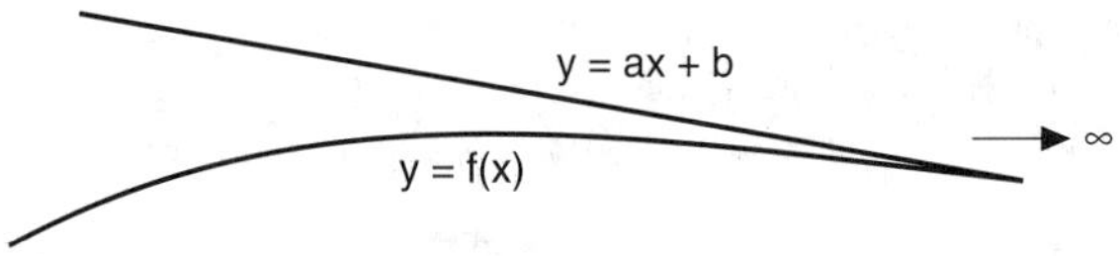

The asymptote has the equation

$$y = ax + b$$

where $a = \lim_{x \to \infty} f'(x)$

$$b = \lim_{x \to \infty} [f(x) - xf'(x)]$$

G.L. Shires

ASYMPTOTIC EXPANSION

Asymptotic expansion is one of the methods of approximating functions by their asymptotic expressions. The function g(x) is an asymptotic expression of the function f(x) for $x \to a$, if

$$\lim_{x \to a} \frac{g(x)}{f(x)} = 1.$$

Hence it follows that

$$\lim_{x \to a} \frac{f(x) - g(x)}{f(x)} = 0,$$

i.e., a relative error of replacing the function by its asymptotic expression for $x \to a$ tends to zero.

An asymptotic expansion is a special kind of an asymptotic expression, in which the function f(x) is approximated by partial sums of some convergent or divergent series

$$\sum_{k=0}^{\infty} \varphi_k(x) = \varphi_0(x) + \varphi_1(x) + \ldots,$$

so that partial sums $S_k = \phi_0(x) + \phi_1(x) + \cdots + \phi_k(x)$ are the asymptotic expressions of the function f(x) and

$$\lim_{x \to a} \frac{f(x) - S_{k+1}(x)}{f(x) - S_k(x)} = 0,$$

i.e., each successive partial sum is the best asymptotic expression for the function f(x).

If the form of the function $\varphi_i(x)$ is unknown beforhand there exists a broad range of possibilities for choosing particular asymptotic expansions.

In the theory of fluid dynamics and heat mass transfer, and also in other branches of mechanics of continuous media the method of perturbation has found extensive application, in which approximate solutions of the problem, which is defined by a single or a system of (integro) differential equations and by the corresponding boundary conditions, are found through asymptotic expansions of dependent variables when one or several parameters of the problem (for instance, numbers M, Re, Pr, Sc etc., or their complexes) are small or large (in this latter case reciprocal values of the parameters are considered). Usually, if suffices to only define the first, or more rarely the first and the second approximations; the rest of the approximations serve for adjusting the first one. As a rule, the application of perturbation theory brings about a solution of more simple differential equations, of a smaller dimention or order, and also for differential equations in partial derivatives, of more simple equations of another type (for instance, parabolic equations of a boundary layer approximation instead of Navier-Stokes elliptic equations in problems of dynamics of viscous liquid for large Reynolds numbers).

If the asymptotic expansion in terms of a small parameter is uniformly exact in the entire domain of definition of independent variables, a problem of regular perturbations holds; if it is non-uniformly exact the problem is of singular perturbations. Among the methods of solution of the last problem the methods of deformed coordinates, of joint asymptotic expansions and of many scales are widely employed.

References

Nayfeh, (1973) Perturbation Methods. J. Wiley & Sons, New York. London, Sydney, Toronto.

Van Dyke, M. (1964) Perturbation Methods in Fluid Mechanics. Acad. Press, New York, London.

Cole, J. D. (1968) Perturbation Methods in Applied Mathematics. Blaisdell Pull. Comp., Toronto, London.

I.G. Zaltsman

ATMOSPHERE

Following from: *Environmental heat transfer*

The atmosphere is a layer of gas surrounding the Earth which extends to a height of several hundred kilometres. On the basis of composition, it can be divided into three main regions, the *homosphere,* the *heterosphere* and the *exosphere.* **Figure 1** shows these regions and the temperature profile through the atmosphere. The vertical structure of the atmosphere is discussed in more detail by Wallace and Hobbs (1977).

The homosphere is the lowest layer and accounts for virtually all (i.e. 99.999%) the mass of the atmosphere. This layer comprises approximately 78% Nitrogen, 21% Oxygen and 1% Argon, with a variable amount of water vapour. Many other gases are found in small quantities, the most important of which is carbon dioxide (~0.03%). In the heterosphere, the photo-decomposition leads to a high concentration of atomic oxygen, which is the dominant constituent over much of this layer. The exosphere is composed mainly of the light elements, helium and hydrogen.

The homosphere can be subdivided into three strata, the *troposphere*, the *stratosphere*, and the *mesosphere*. The troposphere is separated from the stratosphere by the tropopause, which is a transition from an unstable to a stably stratified layer. Very little mixing between the two strata takes place. Most of the atmospheric ozone is found in the stratosphere; this allotrope is important in that it shields the Earth against UV radiation. However the ozone is currently being depleted through the effects of freons and other pollutants which have been released into the atmosphere.

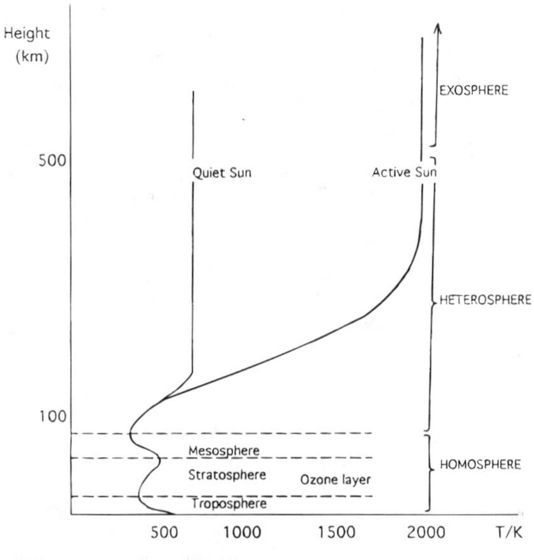

Figure 1. Structure of the atmosphere.

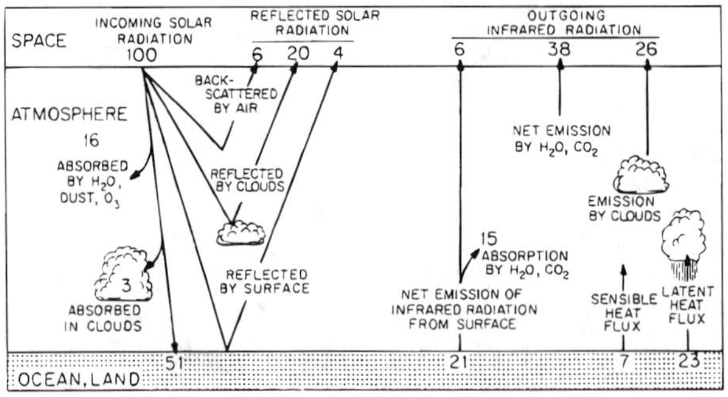

Figure 2. Heat balance of the earth.

The *synoptic-scale* (i.e. ~1000km) *circulation* of the atmosphere is governed largely by the Earth's rotation (*Coriolis Effect*) and this leads to cyclonic and anticyclonic weather systems where the fluid velocity is perpendicular to the pressure gradient. Thermal circulations and centrifugal forces are important in the small-scale features of the atmospheric motions. The dynamics of the atmosphere are discussed at length by Holton (1979). It is worth mentioning that the popular belief that the direction of rotation of water flowing down a plug hole is determined by the rotation of the Earth is a myth.

A discussion of the heat balance of the atmosphere can be found in Wallace and Hobbs (1979). The **Albedo** of the earth is heavily dependent on the amount of water vapour and droplets in the atmosphere, but on a global average, 30% of the incident solar radiation is reflected (**Figure 2**). The outgoing solar radiation is mainly in the infra-red part of the spectrum, and some of this is absorbed by carbon dioxide and other gases. Over the last few decades, there has been great concern that increases in atmospheric carbon dioxide, methane and other gases which absorb infra-red could upset the radiation balance and so bring about an overall warming.

Further information about the atmosphere is given in the article **Atmosphere (Physical Properties of).**

References

Holton, J. R. (1979) An Introduction to Dynamic Meteorology. Academic Press, New York.

Wallace, J. M. and Hobbs, P. V. (1977) Atmospheric Science—An Introductory Survey, Academic Press Inc, Orlando, Florida.

Leading to: *Atmosphere (physical properties of); Air pollution; Environmental heat transfer; Greenhouse effect*

M.J. Pattison

ATMOSPHERE (PHYSICAL PROPERTIES OF)

Following from: *Atmosphere*

The various regions of the atmosphere are described in the article on **Atmosphere**. The mass of the atmosphere is about 5.15

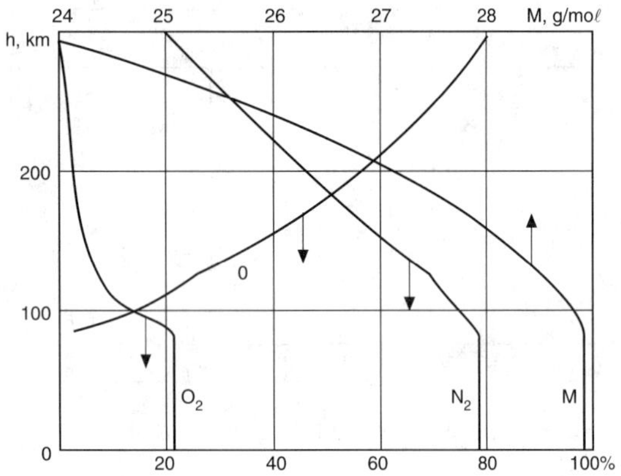

Figure 1. Variation of molecular weight (M) and % of various gases as a function of distance above the ground (h).

× 10^{15} tonne. The Earth's atmosphere is a mixture of gases and vapour (air), and also of some amount of aerosols (dust, smoke, condensation products of vapour). The percentage ratio of main gases of a dry atmosphere (see **Figure 1**) changes slightly up to an altitude of about 100 km (in homosphere). At an altitude of 20 − 25 km an ozonal layer is situated which prevents living beings on the Earth from a harmful short-wave radiation. A share of light gases rises above 100 km (in heterosphere) and at very high altitudes helium and hydrogen prevail; a part of molecules decay into atoms and ions thus forming an ionosphere.

The vapour content is subject to variations to a greater extent then others. One of the most vital characteristics of the climate is the air humidity, the content of vapour in it. "Absolute" humidity can either be defined as mass of water vapour per unit volume or as mass of water vapour per unit mass of dry air. "Relative" humidity is the ratio of the absolute humidity to that for saturation. The highest mean value of humidity at the Earth's surface is 30 g/m^3 (absolute) (or 3% relative) and is characteristic of the equator area, the lowest

Table 1.

Altitude cloud h, km	Temperature T, K	Pressure p, Pa	Density, kg/m^3	Altitude cloud h, km	Temperature T, K	Pressure p, Pa	Density, kg/m^3
−2	301.15	1.277×10^5	1.478				
−1	294.65	1.139×10^5	1.347				
0	288.15	1.013×10^5	1.225				
1	281.65	9.545×10^4	1.112	55	270.56	4.57×10^1	5.893×10^{-4}
2	275.14	7.949×10^4	1.007	60	253.40	2.41×10^1	3.316×10^{-4}
3	268.64	7.012×10^4	9.094×10^{-1}	65	236.26	1.21×10^1	1.794×10^{-4}
4	262.13	6.165×10^4	8.194×10^{-1}	70	219.15	5.83×10^1	9.275×10^{-5}
5	255.63	5.404×10^4	7.365×10^{-1}	75	202.06	2.64×10^1	4.649×10^{-5}
6	249.13	4.721×10^4	6.602×10^{-1}	80	195.00	1.11×10^1	2.098×10^{-5}
7	242.63	4.109×10^4	5.901×10^{-1}	85	185.00	0.45×10^1	8.530×10^{-6}
8	236.14	3.564×10^4	5.259×10^{-1}	90	185.00	1.84×10^{-1}	3.473×10^{-6}
9	229.64	3.079×10^4	4.671×10^{-1}	95	185.00	7.52×10^{-2}	1.417×10^{-6}
10	223.15	2.649×10^4	4.136×10^{-1}	100	209.22	3.24×10^{-2}	5.399×10^{-7}
12	216.66	1.939×10^4	3.118×10^{-1}	110	257.36	7.82×10^{-2}	1.058×10^{-7}
14	216.66	1.416×10^4	2.278×10^{-1}	120	332.24	2.56×10^{-2}	2.659×10^{-8}
16	216.66	1.034×10^4	1.664×10^{-1}	130	552.04	1.21×10^{-2}	7.505×10^{-9}
18	216.66	7.562×10^3	1.216×10^{-1}	140	968.00	7.39×10^{-4}	3.277×10^{-9}
20	216.66	5.527×10^3	8.887×10^{-2}	150	980.05	5.12×10^{-4}	1.768×10^{-9}
22	216.66	4.040×10^3	6.497×10^{-2}	160	1155.3	3.81×10^{-4}	1.108×10^{-9}
24	216.66	2.954×10^3	4.750×10^{-2}	170	1175.0	2.92×10^{-4}	8.279×10^{-10}
26	219.40	2.162×10^3	3.434×10^{-2}	180	1193.2	2.25×10^{-4}	6.233×10^{-10}
28	224.87	1.594×10^3	2.470×10^{-2}	190	1210.6	1.75×10^{-4}	4.728×10^{-10}
30	230.35	1.184×10^3	1.790×10^{-2}	200	1226.8	1.36×10^{-4}	3.611×10^{-10}
32	235.82	8.853×10^2	1.308×10^{-2}	210	1245.0	1.07×10^{-4}	2.771×10^{-10}
34	241.28	6.669×10^2	9.630×10^{-3}	220	1262.0	8.48×10^{-5}	2.141×10^{-10}
36	246.74	5.056×10^2	7.139×10^{-3}	230	1277.4	6.75×10^{-5}	1.666×10^{-10}
38	252.20	3.855×10^2	5.324×10^{-3}	240	1290.9	5.40×10^{-5}	1.304×10^{-10}
40	257.66	2.959×10^2	4.000×10^{-3}	250	1302.8	4.35×10^{-5}	1.027×10^{-10}
42	263.11	2.285×10^2	3.024×10^{-3}	260	1316.2	3.52×10^{-5}	8.111×10^{-10}
44	268.56	1.772×10^2	2.298×10^{-3}	270	1328.8	2.86×10^{-5}	6.447×10^{-10}
46	274.00	1.383×10^2	1.758×10^{-3}	280	1340.0	2.34×10^{-5}	5.155×10^{-10}
48	274.00	1.082×10^2	1.375×10^{-3}	290	1349.5	1.93×10^{-5}	4.146×10^{-10}
50	274.00	8.458×10^1	1.075×10^{-3}	300	1358.0	1.59×10^{-5}	3.352×10^{-10}

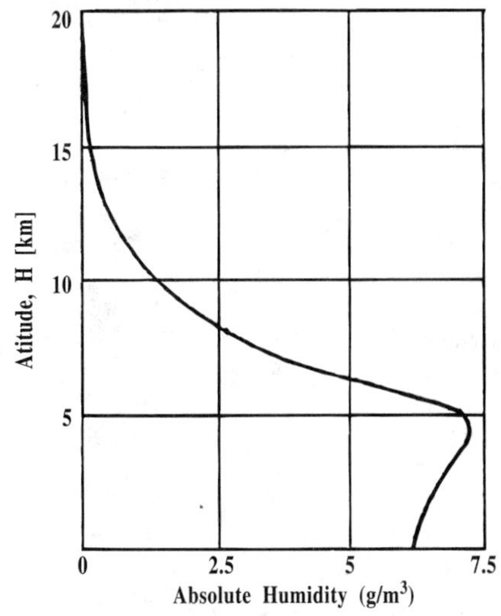

Figure 2. Variation of absolute humidity with altitude in a rain cloud.

(2×10^{-5}% relative) is in Antarctica. The atmospheric formations (clouds) are the aggregation of water drops and ice crystals suspended in the atmosphere. The diameters of cloud drops are of the order of several μm. The enlarged drops fall out as (rain, snow, hail). The size of rain drops varies from 0.5 to 6 − 7 mm; with a smaller size of drops the rainfall is called drizzle. The cooling of air below 0°C brings about snow fall-out.

A typical distribution of absolute humidity with height in a rain cloud is represented in **Figure 2**.

The atmospheric pressure is the pressure of air on the objects in it and on the earth surface. The pressure at each point of the atmosphere is equal to the weight of the air column lying above it; it is measured in Pascals (1 Pa = 1 N/m^2).

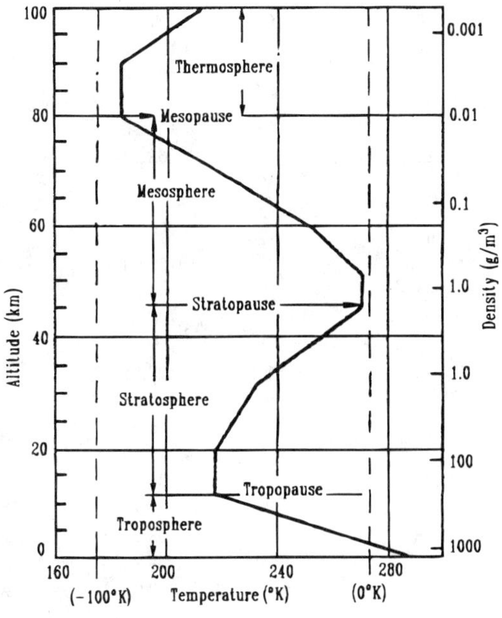

Figure 3. Variation of atmospheric temperature with altitude.

The international standard atmosphere to be used in calculation and designing the flying vehicles, in reducing the results of flight testings to the same conditions, in graduating the altimeters and in solving other technical and thermo-physical problems is accepted, in which the distribution of parameters with height is calculated from the mean sea level (see **Table 1**) with certain suppositions of temperature distribution along the vertical.

The variation of temperature of the atmosphere with distance from the earth is shown in **Figure 3**.

Leading to: *Air, properties of*

D.S. Mikhatulin

ATOM

The atom is the basic building block of chemistry. In the simplest terms, it consists of a single positively-charged nucleus, containing Z *protons* and zero or more *neutrons,* surrounded by a cloud of *electrons*. The number Z, known as the *atomic number,* characterises the atom uniquely as being that of a particular element. In the neutral atom, the number of electrons is equal to the number of protons; however, electrons may be removed or acquired to form a positively- or negatively-charged ion of the same element.

The chemical activity of atoms is determined almost entirely by electronic structure. The energy required to add or remove electrons from the neutral atom determines the ability to form ions and to take part in the simplest kind of chemical bonding: *ionic bonding.* Many compounds have molecules in which the atoms are bound together by *covalent bonding* in which electrons are shared.

The chemical physics of the atom is determined theoretically by quantum mechanics and may be investigated experimentally by **Spectroscopy**. The electronic structure of atoms so determined is explained well by a simple model in which the electrons are located within a series of shells. Each electron is then identified by a set of four quantum numbers which determine its energy, orbital angular momentum, and spin angular momentum (Atkins, 1990).

The mass of the atom is determined by the total number of *nucleons* (protons and neutrons) and electrons. Many exist in several forms, known as *isotopes,* which differ in the number of neutrons present. Unstable isotopes may decay through radioactive nuclear reactions in which elementary particles may be emitted from the nucleus thereby transforming the atoms of one element into those of another.

References

Atkins, P. W. (1990) *Physical Chemistry,* 4th ed., Chapter 13, Oxford University Press, Oxford.

Leading to: *Atomic weight; Molecule*

J.P.M. Trusler

ATOMIC ENERGY AUTHORITY (see AEA Technology)

ATOMIC NUMBER (see Atom; Atomic weight)

ATOMIC SPECTROSCOPY (see Spectroscopy)

ATOMIC WEIGHT

Following from: Atom

The atomic weight $A_r(E)$ of element E is defined by the International Union of Pure and Applied Chemistry (IUPAC) as the *relative molar mass* $12 \cdot M(E)/M(^{12}C)$, where $M(E)$ is the molar mass of the element and $M(^{12}C)$ is the molar mass of pure carbon-12 (IUPAC, 1994). The molar mass is defined as the mass per unit amount of substance of the specified entity; thus, $M(E) = A_r(E) \times 10^{-3}$ kg· mol^{-1}.

Whereas the atomic weights of the nuclides are invariant, those of the elements depend, through variations in the *isotopic composition*, on the origin and treatment of the material. Values of $A_r(E)$ recommended by IUPAC are given in **Table 1** for the first 103 elements in order of increasing *atomic number Z*. The values given generally relate to the elements as they exist naturally on earth and the uncertainty in the last digit is then given as the first note; other notes elaborate the kinds of variation to be expected for each element. In the case of unstable elements, the atomic weight of the longest-lived known isotope is given.

Table 1. Atomic weights $A_r(E)$ of the elements E (IUPAC, 1994)

Z	Name	E	$A_r(E)$	Notes
1	Hydrogen	H	1.00794	7 g m r
2	Helium	He	4.002602	2 g r
3	Lithium	Li	6.941	2 g m r
4	Beryllium	Be	9.012182	3
5	Boron	B	10.811	5 g m r
6	Carbon	C	12.011	1 g r
7	Nitrogen	N	14.00674	7 g r
8	Oxygen	O	15.9994	3 g r
9	Fluorine	F	18.9984032	9
10	Neon	Ne	20.1797	6 g m
11	Sodium	Na	22.989768	6
12	Magnesium	Mg	24.3050	6
13	Aluminium	Al	26.981539	5
14	Silicon	Si	28.0855	3 r
15	Phoshorous	P	30.973762	4
16	Sulfur	S	32.066	6 r
17	Chlorine	Cl	35.4527	9
18	Argon	Ar	39.948	1 g r
19	Potassium	K	39.0983	1
20	Calcium	Ca	40.078	4 g
21	Scandium	Sc	44.955910	9
22	Titanium	Ti	47.867	1
23	Vanadium	V	50.9415	1
24	Chromium	Cr	51.9961	6
25	Manganese	Mn	54.93805	1
26	Iron	Fe	55.845	2
27	Cobalt	Co	58.93320	1
28	Nickel	Ni	58.6934	2

Table 1. Atomic weights $A_r(E)$ of the elements E (IUPAC, 1994)—*continued*

Z	Name	E	$A_r(E)$	Notes
29	Copper	Cu	63.546	3 r
30	Zinc	Zn	65.39	2
31	Gallium	Ga	69.723	1
32	Germanium	Ge	72.61	2
33	Arsenic	As	74.92159	2
34	Selenium	Se	78.96	3
35	Bromine	Br	79.904	1
36	Krypton	Kr	83.80	1 g m
37	Rubidium	Rb	85.4678	3 g
38	Strontium	Sr	87.62	1 g r
39	Yttrium	Y	88.90585	2
40	Zirconium	Zr	91.224	2 g
41	Niobium	Nb	92.90638	2
42	Molybdenum	Mo	95.94	1
43	Technetium	Tc	97.9072	A
44	Ruthenium	Ru	101.07	2 g
45	Rhodium	Rh	102.90550	3
46	Palladium	Pd	106.42	1 g
47	Silver	Ag	107.8682	2 g
48	Cadmium	Cd	112.411	8 g
49	Indium	In	114.818	3
50	Tin	Sn	118.710	7 g
51	Antinomy	Sb	121.760	1
52	Tellurium	Te	127.60	3 g
53	Iodine	I	126.90447	3
54	Xenon	Xe	131.29	2 g m
55	Caesium	Cs	132.90543	5
56	Barium	Ba	137.327	7
57	Lanthanum	La	138.9055	2 g
58	Cerium	Ce	140.115	4 g
59	Praseodymium	Pr	140.90765	3
60	Neodymium	Nd	144.24	3 g
61	Promethium	Pm	144.9127	A
62	Samarium	Sm	150.36	3 g
63	Europium	Eu	151.965	9 g
64	Gadolinium	Gd	157.25	3 g
65	Terbium	Tb	158.92534	3
66	Dysprosium	Dy	162.50	3 g
67	Holmium	Ho	164.93032	3
68	Erbium	Er	167.26	3 g
69	Thulium	Tm	168.93421	3
70	Ytterbium	Yb	173.04	3 g
71	Lutetium	Lu	174.967	1 g
72	Hafnium	Hf	178.49	2
73	Tantalum	Ta	180.9479	1
74	Tungsten	W	183.84	1
75	Rhenium	Re	186.207	1
76	Osmium	Os	190.23	3 g
77	Iridium	Ir	192.217	3
78	Platinum	Pt	195.08	3
79	Gold	Au	196.96654	3
80	Mercury	Hg	200.59	2
81	Thallium	Tl	204.3833	2
82	Lead	Pb	207.2	1 g r
83	Bismuth	Bi	208.98037	3
84	Polonium	Po	208.9824	A
85	Astatine	At	209.9871	A
86	Radon	Rn	222.0176	A
87	Francium	Fr	223.0197	A

Table 1. Atomic weights A_r(E) of the elements E (IUPAC, 1994)—*continued*

Z	Name	E	A_r(E)	Notes
88	Radium	Ra	226.0254	A
89	Actinium	Ac	227.0278	A
90	Thorium	Th	232.0381	1 g r X
91	Protactinium	Pa	231.03588	2 X
92	Uranium	U	238.0289	1 g m X
93	Neptunium	Np	237.0482	A
94	Plutonium	Pu	239.0522	A
95	Americium	Am	243.0614	A
96	Curium	Cm	247.0703	A
97	Berkelium	Bk	249.0703	A
98	Californium	Cf	251.0796	A
99	Einsteinium	Es	252.083	A
100	Fermium	Fm	257.0951	A
101	Mendelevium	Md	258.10	A
102	Nobelium	No	259.1009	A
103	Lawrencium	Lr	262.11	A

g Geologically exceptional specimens are known in which the element has an isotopic composition outside the limits for normal material. The difference between values of *M* in such specimens and that given might considerably exceed the stated uncertainty.

m Modified isotopic compositions can be found in commercially available material because it has been submitted to an undisclosed or inadvertent isotopic separation. Substantial deviation from the given *Ar* can occur.

r Range in isotopic composition of normal terrestrial material prevents a more precise value of *Ar* being given; the tabulated value of *Ar* should be applicable to any normal material.

A Radioactive element, having no stable nuclide, that lacks a characteristic terrestrial isotopic composition. The atomic weight of the longest-lived nuclide is given. See IUPAC (1994) for the atomic weights of other known nuclides.

X An element, having no stable nuclide, that exhibits a range of characteristic terrestrial compositions of long-lived radionuclide(s) such that a meaningful value of *Ar* can be given.

Reference

IUPAC: Pure and Applied Chemistry (1994) 66, 2433–2444.

J.P.M. Trusler

ATOMIZATION

Introduction

The conversion of bulk liquid into a dispersion of small droplets ranging in size from submicron to several hundred microns (micrometres) in diameter is of importance in many industrial processes such as spray combustion, spray drying, evaporative cooling, spray coating, and crop spraying; and has many other applications in medicine, meteorology, and printing. Numerous spray devices have been developed which are generally designated as *atomizers, applicators, sprayers,* or *nozzles.*

A *spray* is generally considered as a system of droplets immersed in a gaseous continuous phase. Sprays may be produced in various ways. Most practical devices achieve atomization by creating a high relative velocity between the liquid and the surrounding gas (usually air). All forms of *pressure nozzles* accomplish this by discharging the liquid at high velocity into quiescent or relatively slow-moving air. *Rotary atomizers* employ a similar principle, the liquid being ejected at high velocity from the rim of a rotating cup or disc. An alternative method of achieving a high relative velocity between liquid and air is to expose slow-moving liquid to a high-velocity stream of air. Devices based on this approach are usually termed air-assist, airblast or, more generally, *twin-fluid atomizers.*

Most practical atomizers are of the pressure, rotary, or twin-fluid type. However, many other forms of atomizers have been developed that are useful in special applications. These include "*electrostatic*" devices in which the driving force for atomization is intense electrical pressure, and '*ultrasonic*' types in which the liquid to be atomized is fed through or over a transducer which vibrates at ultrasonic frequencies to produce the short wavelengths required for the production of small droplets. Both electrical and ultrasonic atomizers are capable of achieving fine atomization, but the low liquid flow rates normally associated with these devices have tended to curtail their range of practical application.

Basic Processes

There are several basic processes associated with all methods of atomization, such as the conversion of bulk liquid into a jet or sheet and the growth of disturbances which ultimately lead to disintegration of the jet or sheet into ligaments and then drops. These processes determine the shape, structure, and penetration of the resulting spray as well as its detailed characteristics of droplet velocity and drop size distribution. All these characteristics are strongly affected by atomizer size and geometry, the physical properties of the liquid, and the properties of the gaseous medium into which the liquid stream is discharged. The liquid properties of importance in atomization are surface tension, viscosity, and density. Basically, atomization occurs as a result of the competition between the stabilizing influences of surface tension and viscosity and the disruptive actions of various internal and external forces. In most cases, turbulence in the liquid, cavitation in the nozzle, and aerodynamic interaction with the surrounding gas (referred to henceforth as air), all contribute to atomization. In all cases, atomization occurs when the magnitude of the disruptive force just exceeds the consolidating surface tension force. Many of the larger drops produced in the initial breakup of the liquid jet or sheet are unstable and undergo further disintegration into smaller droplets. Thus, the drop size characteristics of a spray are governed not only by the drop sizes produced in primary atomization but also by the extent to which the largest of these drops are further disintegrated during secondary atomization.

Breakup of liquid jets

Rayleigh (1878) was among the first to study theoretically the breakup of liquid jets. He considered the simple situation of a laminar jet issuing from a circular orifice and postulated the growth of small disturbances that produce breakup when the fastest growing disturbance attains a wavelength λ_{opt} of 4.51 *d,* where *d* is the initial jet diameter. After breakup, the cylinder of length 4.51 *d* becomes a spherical drop, so that

$$4.51d \times (\pi/4)d^2 = (\pi/6)D^3 \quad \textbf{(1)}$$

and hence *D,* the drop diameter, is obtained as

$$D = 1.89d \tag{2}$$

Figure 1a shows an idealization of Rayleigh breakup for a liquid jet. Observations of actual jets show good agreement with this theory, but also reveal the presence of "satellite" drops which are created as the individual cylinders neck down and separate. Thus, the end result is a pattern of large drops with much smaller single drops between them.

Rayleigh's analysis takes into account surface tension and inertial forces but neglects viscosity and the effect of the surrounding air. *Weber* (1931) later extended Rayleigh's work to include the effect of air resistance on the disintegration of jets into drops. He found that air friction shortens the optimum wavelength for drop formation. For zero relative velocity he showed that the value of λ_{opt} is 4.44d, which is close to the value of 4.51d predicted by Rayleigh for this case. For a relative velocity of 15m/s, Weber showed that λ_{opt} becomes 2.8d and the drop diameter becomes 1.6d. Thus the effect of relative velocity between the liquid jet and the surrounding air is to reduce the optimum wavelength for jet breakup which results in a smaller drop size.

Weber also examined the effect of liquid viscosity on jet disintegration. He showed that the effect of an increase in viscosity is to increase the optimum wavelength for jet breakup. We have

$$\lambda_{opt} = 4.44d(1 + 3Oh)^{0.5} \tag{3}$$

where

$$Oh = \mu_L/(\rho_L \sigma d)^{0.5} \tag{4}$$

This group is sometimes referred to as the *Z* number, the stability number, or the *Ohnesorge number* (Oh).

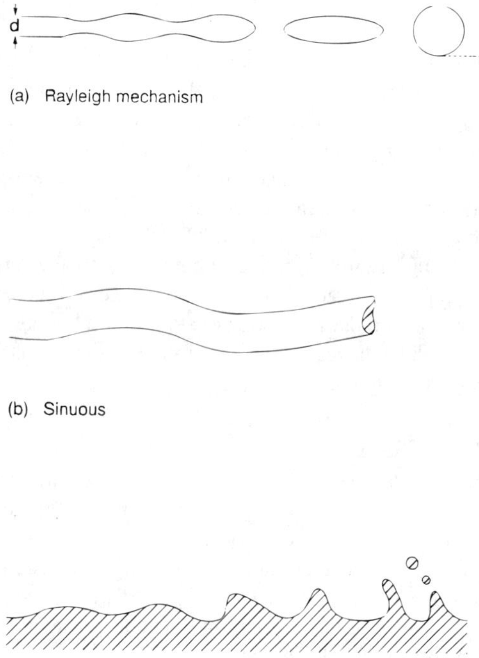

(a) Rayleigh mechanism

(b) Sinuous

(c) Surface breakup

Figure 1.

At higher jet velocities, breakup is caused by waviness of the jet (**Figure 1b**). This mode of drop formation is associated with a reduction in the influence of surface tension and increased effectiveness of aerodynamic forces. The term "*sinuous*" is often used to describe the jet in this regime. At even higher velocities, the atomization process is enhanced by the effect of relative motion between the surface of the jet and the surrounding air. This aerodynamic interaction causes irregularities in the previously smooth liquid surface. These irregularities or ruffles in the jet surface become amplified and eventually detach themselves from the liquid surface, as illustrated in **Figure 1c**. Ligaments are formed which subsequently disintegrate into drops. As the jet velocity increases, the diameter of the ligaments decreases. When they collapse, smaller droplets are formed, in accordance with Rayleigh's theory.

Thus the various modes of atomization may be classified into four groups according to the relative velocity between the jet and the surrounding air, as follows:

1. At low velocities, the growth of axisymmetric oscillations on the jet surface cause the jet to disintegrate into drops of fairly uniform size. This is the Rayleigh mechanism of breakup. Drop diameters are roughly twice the initial jet diameter. Drop sizes are increased by increases in liquid viscosity and are reduced by increases in jet velocity.
2. At higher velocities, breakup is caused by oscillations of the jet as a whole with respect to the jet axis. The jet has a twisted or sinuous appearance This mode occurs over only a fairly narrow range of velocities.
3. Droplets are produced by the unstable growth of small waves on the jet surface caused by interaction between the jet and the surrounding air. These waves become detached from the jet surface to form ligaments which disintegrate into drops. Mean drop diameters are much smaller than the initial jet diameter.
4. Atomization. At very high relative velocities atomization is complete within a short distance from the discharge orifice. A wide range of drop sizes is produced, the mean drop diameter being considerably less than the initial jet diameter.

Another factor influencing jet breakup is the turbulence of the jet as it emerges from the nozzle. When the liquid particles flow in streams parallel to the main flow direction, the flow is described as laminar. Laminar flow is promoted by low flow velocity, high liquid viscosity and the absence of any flow disturbances. With laminar flow, the velocity profile varies across the jet radius in a parabolic manner, rising from zero at the outer surface to a maximum at the jet axis. If a laminar jet is injected into quiescent or slow-moving air, there is no appreciable velocity difference between the outer surface of the jet and the surrounding air. Consequently the necessary conditions for atomization by air friction do not exist. Eventually, surface irregularities develop that cause the jet to disintegrate into relatively large drops.

If the liquid particles do not follow the flow streamlines but cross each other at various velocities in a random manner, the flow is described as turbulent. Turbulence is promoted by high flow velocities, low liquid viscosity, surface roughness, and cavitation. If the flow emerging from the atomizer is fully turbulent, the strong radial velocity components quickly disrupt the jet surface, thereby promoting air friction and rapid disintegration of the jet. It is of interest to note that when the issuing jet is fully turbulent, air friction is not essential for breakup. Even when injected into a vacuum, the jet will disintegrate solely under the influence of its own turbulence.

Breakup of liquid sheets

Many atomizers do not form jets of liquid, but rather form flat or conical sheets. Flat sheets can be produced by the impingement of two liquid streams or by feeding the liquid to the centre of a rotating disc or cup. Conical sheets can be generated by imparting a tangential velocity component to the flow as it issues from the discharge orifice.

The mechanisms of sheet integration are broadly the same as those responsible for jet breakup, as discussed above. If the liquid sheet is flowing at high velocity, the turbulence forces generated within the liquid may be strong enough to cause the sheet to disintegrate into drops without any aid or intervention from the surrounding air. However, the principal cause of sheet breakup stems from interaction of the sheet with the surrounding air, whereby rapidly growing waves are superimposed on the sheet. Disintegration occurs when the wave amplitude reaches a critical value and fragments of sheet are torn off. Surface tension forces cause these fragments to contract into irregular ligaments which then collapse into droplets according to the Rayleigh mechanism. The dependence of the drop sizes produced in this mode of atomization on air and liquid properties can be expressed as

$$\frac{D}{\delta} \propto We^{-0.5} \quad \textbf{(5)}$$

where δ is the sheet thickness and We, the Weber number, is $U_A{}^2\rho_A\delta/\sigma$.

Drop Size Distribution

Owing to the random and chaotic nature of the atomization process, the threads and ligaments formed by the various mechanisms of jet and sheet disintegration vary widely in diameter, and their subsequent breakup yields a correspondingly wide range of drop sizes. Most practical atomizers produce droplets in the size range from a few microns up to several hundred microns. A simple method of illustrating the distribution of drop sizes in a spray is to plot a histogram in which each ordinate represents the number of droplets whose dimensions fall between the limits $D - \Delta D/2$ and $D + \Delta D/2$. As ΔD is made smaller, the histogram assumes the form of a frequency distribution curve, provided it is based on sufficiently large samples.

Because the graphical representation of drop size distribution is laborious, many attempts have been made to replace it with mathematical expressions that provide a satisfactory fit to the drop size data. All of these distribution parameters, which include normal, lognormal, and upper limit distributions, have drawbacks of one kind or another, and no single parameter has yet been found which has clear advantages over the others. At present the most widely used expression for drop size distribution is one proposed by *Rosin* and *Rammler* (1933). It may be expressed in the form

$$1 - Q = \exp - (D/X)^q \quad \textbf{(6)}$$

where Q is the fraction of the total spray volume contained in drops of diameter less than D, and X and q are constants. This expression allows the drop size distribution to be described in terms of the two parameters X and q. For most sprays the value of q lies between 1.5 and 4. The higher the value of q, the more narrow is the distribution of drop sizes in the spray.

For most engineering purposes the distribution of drop sizes in a spray may be described satisfactorily in terms of two parameters (as in the Rosin-Rammler expression, for example), one of which is a representative diameter and the other a measure of the range of drop sizes. There are many possible choices of representative diameter, of which the most widely used is the mass median diameter (MMD) or volume mean diameter (VMD). These terms denote the drop diameter such that 50 percent of the total mass (or volume) of the spray is in drops of smaller diameter.

In many calculations of mass transfer it is convenient to work in terms of mean diameters instead of the complete drop size distribution. The most common of these is the **Sauter Mean Diameter** (SMD) which is the diameter of the droplet whose surface to volume ratio is the same as that of the entire spray.

In summary, it is important to recognize that no single parameter can completely define a drop size distribution. Thus, two sprays are not similar just because they have the same VMD or SMD. However, if a Rosin-Rammler distribution is assumed, the distribution of drop sizes in a spray may be expressed by two parameters, a representative or mean diameter and a measure of drop size distribution.

Atomizers

The following discussion is confined to atomizers of the pressure, rotary, and twin-fluid types, as shown schematically in **Figure 2**. Information on other spray devices, such as electrostatic and ultrasonic atomizers is contained in Lefebvre (1989).

Pressure

When a liquid is discharged under pressure through a small orifice, pressure energy is converted into kinetic energy. If the pressure drop across the discharge orifice is sufficiently high, the issuing liquid jet or sheet will disintegrate into droplets. Combustion applications for plain-orifice atomizers, as shown in **Figure 2a**, include diesel, rocket and turbojet engines.

The narrow spray cone angles of around 10° produced by discharging the liquid through a simple circular orifice are disadvantageous for many spraying applications. Much wider cone angles of between 30° to 150° an be achieved with *pressure-swirl nozzles* in which a swirling motion is imparted to the liquid so that as it emerges from the discharge orifice it spreads radially outward to form a hollow conical spray. The simplest form of hollow-cone atomizer is the so-called *simplex atomizer*, as illustrated in **Figure 2b**.

A drawback to all types of pressure nozzles is that doubling the liquid flow rate requires a fourfold increase in injection pressure. Due to practical limits on injection pressures, this seriously restricts the range of liquid flow rates that any given atomizer can handle. This basic drawback has led to the development of various *"wide-range" atomizers* which are capable of providing good atomization over ratios of maximum to minimum flow rate in excess of 20 without having to resort to impractical levels of injection pressure. The most common form of wide-range atomizer is the

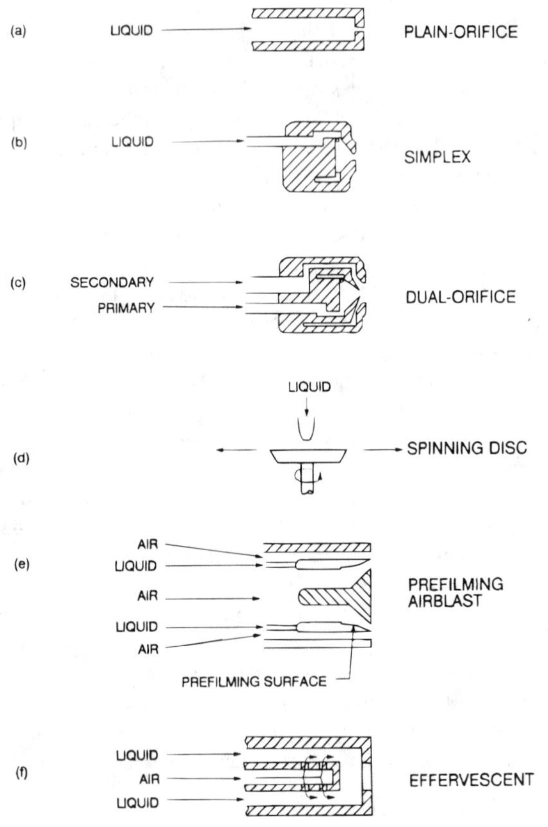

Figure 2.

dual-orifice nozzle, shown schematically in **Figure 2c**, which has been widely used on many types of aircraft and industrial gas turbines. Essentially, a dual-orifice comprises two simplex nozzles that are fitted concentrically one inside the other. When the liquid flow rate is low it all flows through the inner, primary nozzle, and atomization quality is high because the small flow passages dictate a high injection pressure. As the liquid flow rate is increased, an injection pressure is eventually reached at which the pressurizing valve opens and admits liquid to the outer, secondary nozzle. This nozzle has large flow passages which allow high flow rates to be achieved without resorting to excessively high injection pressures.

Rotary

Rotary atomizers utilize centrifugal energy to achieve the high relative velocity between air and liquid that is needed for good atomization. A rotating surface is employed which may take the form of a flat disc, vaned disc, cup, bell, or slotted wheel. A simple form of rotary atomizer, comprising a spinning disc with means for introducing liquid at its centre, is shown in **Figure 2d**. The liquid flows radially outward across the disc and is discharged at high velocity from its periphery. Several mechanisms of atomization are observed with a rotating flat disc, depending on the liquid flow rate and the rotational speed of the disc. At low flow rates the liquid is discharged from the edge of the disc in the form of droplets of fairly uniform size. At higher flow rates, ligaments are formed along the entire periphery which subsequently disintegrate into droplets according to the Rayleigh mechanism. With further increase in flow rate the condition is eventually reached where the ligaments can no longer accommodate the flow of liquid, and a thin continuous sheet is formed that extends beyond the rim of the disc. This sheet eventually breaks down into ligaments and drops, but because the ligaments are formed from a ragged edge, the resulting spray is characterized by a wide range of drop sizes. Serrating the edge of the disc delays the transition from ligament formation to sheet formation.

Twin-fluid

Most twin-fluid atomizers employ the kinetic energy of a flowing airstream to shatter a liquid jet or sheet into ligaments and then drops. Atomizers of this type are usually called "airblast" or "air-assist", the main difference between them being the amount of air employed and its flow velocity. Air-assist atomizers are characterized by the use of a relatively small quantity of high-velocity air. This air does not flow continuously but is used only as and when needed to supplement some other mode of atomization. Airblast atomizers on the other hand employ large quantities of atomizing air flowing continuously at relatively low velocities (20 − 120 m/s).

Airblast atomizers have many advantages over pressure atomizers in combustion applications. They require lower injection pressures and produce a finer spray. Moreover, because the atomization process ensures thorough mixing of fuel drops and air, the ensuing combustion process is characterized by low soot formation, low flame radiation, and clean combustion products.

In most airblast atomizers the liquid is first spread out on a "prefilming" surface to form a thin continuous sheet and then subjected to the atomizing action of high-velocity air, as shown in **Figure 2e**. Two separate airflows are provided to allow the atomizing air to impact on both sides of the liquid sheet. Swirling airflows are often used, not to improve atomization, but to deflect the droplets formed in atomization radially outward to create a conical spray.

With air-assist and airblast atomizers, high velocity air is used either to augment atomization or as the sole driving force for atomization. An alternative approach is to introduce low-velocity air directly into the bulk liquid at some point upstream of the nozzle discharge orifice, as illustrated in **Figure 2f**. The injected air forms bubbles which produce a two-phase bubbly flow at the nozzle exit. As the air bubbles flow through the discharge orifice they assist atomization by squeezing the liquid into thin shreds and ligaments. When the air bubbles emerge from the nozzle they "explode", thereby shattering the liquid shreds and ligaments into small droplets.

Influence of Liquid and Air Properties on Atomization

The three liquid properties of relevance to atomization are density, surface tension, and viscosity. In practice, the significance of density for atomization performance is diminished by the fact that most liquids exhibit only minor differences in this property. Surface tension is important to atomization because it resists the formation of new surface area which is fundamental to the atomization process. Whenever atomization occurs under conditions where surface tension is important, the **Weber Number** is a dimensionless parameter for correlating drop size data. From a practical standpoint viscosity is an important liquid property. The main role of viscosity is to inhibit the development of instabilities in the liquid jet or sheet emerging from the nozzle, and generally to delay the onset of atomization. This delay causes atomization to occur further downstream from the nozzle where conditions are less conducive to the production of small drops. Another important practical consideration is that whereas the variations normally encountered in surface tension are only about three to one, the corresponding variations in viscosity could be as high as three orders of magnitude

The most important air property influencing atomization is density. With air-assist and airblast atomizers, an increase in air density improves atomization by increasing the Weber number. With pressure-swirl atomizers it is found that drop sizes increase with ambient air density up to a maximum value and then decline with any further increase in density.

Mean Drop Size

The physical processes involved in atomization are not yet sufficiently well understood for mean diameters to be expressed in terms of equations derived from first principles. In consequence, the majority of investigations into the drop size distributions produced in atomization have bean empirical in nature and have resulted in empirical equations for mean drop size. The most authentic of these equations are those in which mean drop size (usually SMD, MMD, or VMD) is expressed in terms of dimensionless groups such as Reynolds number, Weber number, or Ohnesorge Number ($Oh = We^{0.5}/Re$). Most of the mean drop size equations published before the 1970's should be regarded as suspect due to deficiencies in the methods available for drop size measurements. Even equations based on accurate experimental data should only be used within the ranges of air properties, liquid properties, and atomizer operating conditions employed in their derivation.

The equations for mean drop size (SMD) presented below are considered as being among the best available in the literature. More detailed information on drop size equations for all types of atomizers may be found in Lefebvre (1989).

Pressure atomizers

Due to the formidable problems involved in making drop size measurements in the dense sprays produced by plain-orifice nozzles, few equations for mean drop size have been published. According to Elkotb (1982)

$$SMD = 3.08\nu_L^{0.385}(\sigma\rho_L)^{0.737}\rho_A^{0.06}\Delta P_L^{-0.54} \quad \textbf{(6)}$$

For pressure-swirl nozzles, mean drop sizes are usually correlated using empirical equations of the form

$$SMD \propto \sigma^a \nu_L^b \dot{m}_L^c \Delta P^d \quad \textbf{(7)}$$

A typical example, which has an advantage over most other equations in that it is dimensionally correct, is the following:

$$SMD = 2.25\sigma^{0.25}\eta_L^{0.25}\dot{m}_L^{0.25}\Delta P_L^{-0.5}\rho_A^{-0.25} \quad \textbf{(8)}$$

Rotary atomizers

For this type of atomizer, any equation for mean drop size should take into account the effects of variations in disc or cup diameter and rotational speed, in addtion to liquid properties and liquid flow rate.

For atomization by direct drop formation, Tanasawa et al,(1978) obtained a good correlation between their experimental data and the following expression for mean drop size

$$SMD = \frac{0.45}{N}\left(\frac{\sigma}{d\rho_L}\right)^{0.5}\left(1 + 0.003\frac{\dot{m}_L}{d\eta_L}\right) \quad \textbf{(9)}$$

For atomization by ligament formation, these same workers propose the following equation

$$SMD = \frac{0.50}{N}\left(\frac{\sigma}{d\rho_L}\right)^{0.5}\left(\frac{\dot{m}_L}{\eta_L}\right)^{0.1} \quad \textbf{(10)}$$

An interesting feature of this equation is that it predicts the mean droplet size to increase slightly with increase in liquid viscosity

Twin-fluid atomizers

The mean drop sizes produced by twin-fluid atomizers are usually correlated in terms of the Weber and Ohnesorge numbers and the air/liquid mass ratio (ALR), as illustrated below

$$\frac{SMD}{L_c} = (A.We^{-0.5} + B.Oh)\left(1 + \frac{1}{ALR}\right)^c \quad \textbf{(11)}$$

where A, B, and c are constants whose values depend on atomizer design and must be determined experimentally. L_c is a characteristic dimension of the atomizer. For prefilming types of airblast atomizer we have (El-Shanawany and Lefebvre, 1980)

$$\frac{SMD}{D_h} = \left[0.33\left(\frac{\sigma}{\rho_A U_A^2 D_p}\right)^{0.6}\left(\frac{\rho_L}{\rho_A}\right)^{0.1} + 0.68\left(\frac{\eta_L}{\rho_L \sigma D_p}\right)^{0.5}\right]\left[1 + \frac{1}{ALR}\right] \quad \textbf{(12)}$$

where D_h is the hydraulic mean diameter of the atomizer air duct at its exit plane, and D_p is the prefilmer diameter.

References

Elkotb, M. M. (1982) Fuel Atomization for Spray Modelling, Progress in Energy and Combustion Science, 8, 61–91.

El-Shanawany, M. S. M. R. and Lefebvre, A. H. (1980) Airblast Atomization: The Effect of Linear Scale on Mean Drop Size, *J. Energy,* 4, 184–189.

Lefebvre, A. H. (1989) Atomization and Sprays, Hemisphere Publishing Corporation, New York.

Rayleigh, Lord. (1878) On the Instability of Jets, *Proc. London Math. Soc.* 10, 4–13.

Rosin, P. and Rammler, E. (1933) The Laws Governing the Fineness of Powdered Coal, J. Inst. Fuel, 7, 29–36.

Tanasawa, Y., Miyasaka, Y. and Umehara, M. (1978) Effect of Shape of Rotating Discs and Cups on Liquid Atomization, Proceedings of First International Conference on Liquid Atomization and Spray Systems, Tokyo, 165–172.

Weber, C. (1931) Disintegration of Liquid Jets, Z. Angew. *Math. Mech.*, 11, (2):136–159.

Nomenclature

ALR	air/liquid ratio by mass
D	drop diameter, m
d	initial jet diameter, m
MMD	mass median diameter or disc diameter, m.
m	flow rate, kg/s
N	rotational speed, rps
ΔP_L	pressure differential across nozzle, Pa.
Oh	Ohnesorge number
Re	Reynolds number
SMD	Sauter mean diameter, m
u	velocity, m/s
VMD	volume median diameter, m
We	Weber number
δ	sheet thickness, m
λ	wavelength, m
η	dynamic viscosity, kg/ms
υ	kinematic viscosity, m^2/s
ρ	density, kg/m^3
σ	surface tension, kg/s^2

Subscripts

A	air
L	liquid

A.H. Lefebvre

ATOMIZERS (see Atomization)

ATTENUATION COEFFICIENT, PHOTON TRANSMISSION (see Ionizing radiation)

ATTENUATION, OPTICAL (see Fibre optics)

ATTRACTORS (see Non-linear systems)

AUGMENTATION OF HEAT TRANSFER, SINGLE PHASE

Following from: *Coupled conduction and convection; Heat transfer*

Energy and materials saving considerations, as well as economic incentives, have led to efforts to produce more efficient heat exchange equipment. Common thermo-hydraulic goals are to reduce the size of a heat exchanger required for a specified heat duty, to upgrade the capacity of an existing heat exchanger, to reduce the approach temperature difference for the process streams, or to reduce the pumping power.

The study of improved heat transfer performance is referred to as heat transfer *augmentation, enhancement,* or *intensification.* In general, this means an increase in heat transfer coefficient. Attempts to increase "normal" heat transfer coefficients have been recorded for more than a century, and there is a large store of information. A survey (Bergles et al., 1991) cites 4345 technical publications, excluding patents and manufacturers' literature. The recent growth of activity in this area is clearly evident from the yearly distribution of the publications presented in **Figure 1**.

Augmentation techniques can be classified either as passive methods, which require no direct application of external power (**Figure 2**), or as active methods, which require external power. The effectiveness of both types of techniques is strongly dependent on the mode of heat transfer, which may range from single-phase free convection to dispersed-flow film boiling. Brief descriptions of these methods follow.

Treated surfaces involve fine-scale alternation of the surface finish or coating (continuous or discontinuous). They are used for boiling and condensing; the roughness height is below that which affects single-phase heat transfer.

Rough surfaces are produced in many configurations ranging from random sand-grain type roughness to discrete protuberances. See **Figure 2a**. The configuration is generally chosen to disturb the viscous sublayer rather than to increase the heat transfer surface area. Application of rough surfaces is directed primarily toward single-phase flow.

Extended surfaces are routinely employed in many heat exchangers. See **Figures 2b and 2c**. Work of interest to augmentation is directed toward improvement of heat transfer coefficients on extended surfaces by shaping or perforating the surfaces. (See also **Extended Surface Heat Transfer**)

Displaced enhancement devices are inserted into the flow channel so as indirectly to improve energy transport, at the heated surface. They are used with forced flow. See **Figures 2e** and **2f**.

Swirl-flow devices include a number of geometric arrangements or tube inserts for forced flow that create rotating and/or secondary flow: coiled tubes, inlet vortex generators, twisted-tape inserts, and axial core inserts with a screw-type winding.

Surface-tension devices consists of wicking or grooved surfaces to direct the flow of liquid in boiling and condensing.

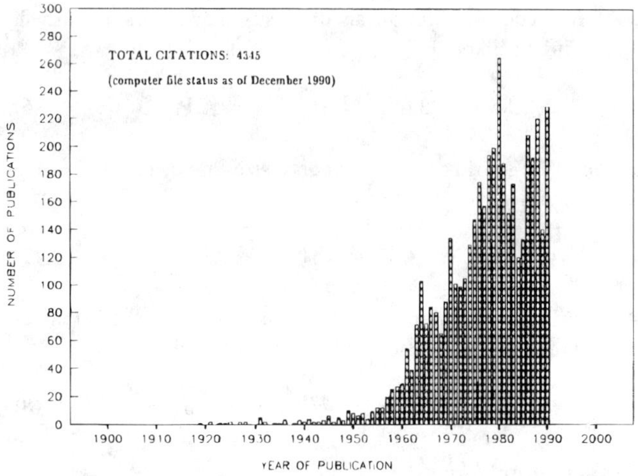

Figure 1. References on heat transfer augmentation versus year of publication (to late 1990) (Bergles et al., 1991).

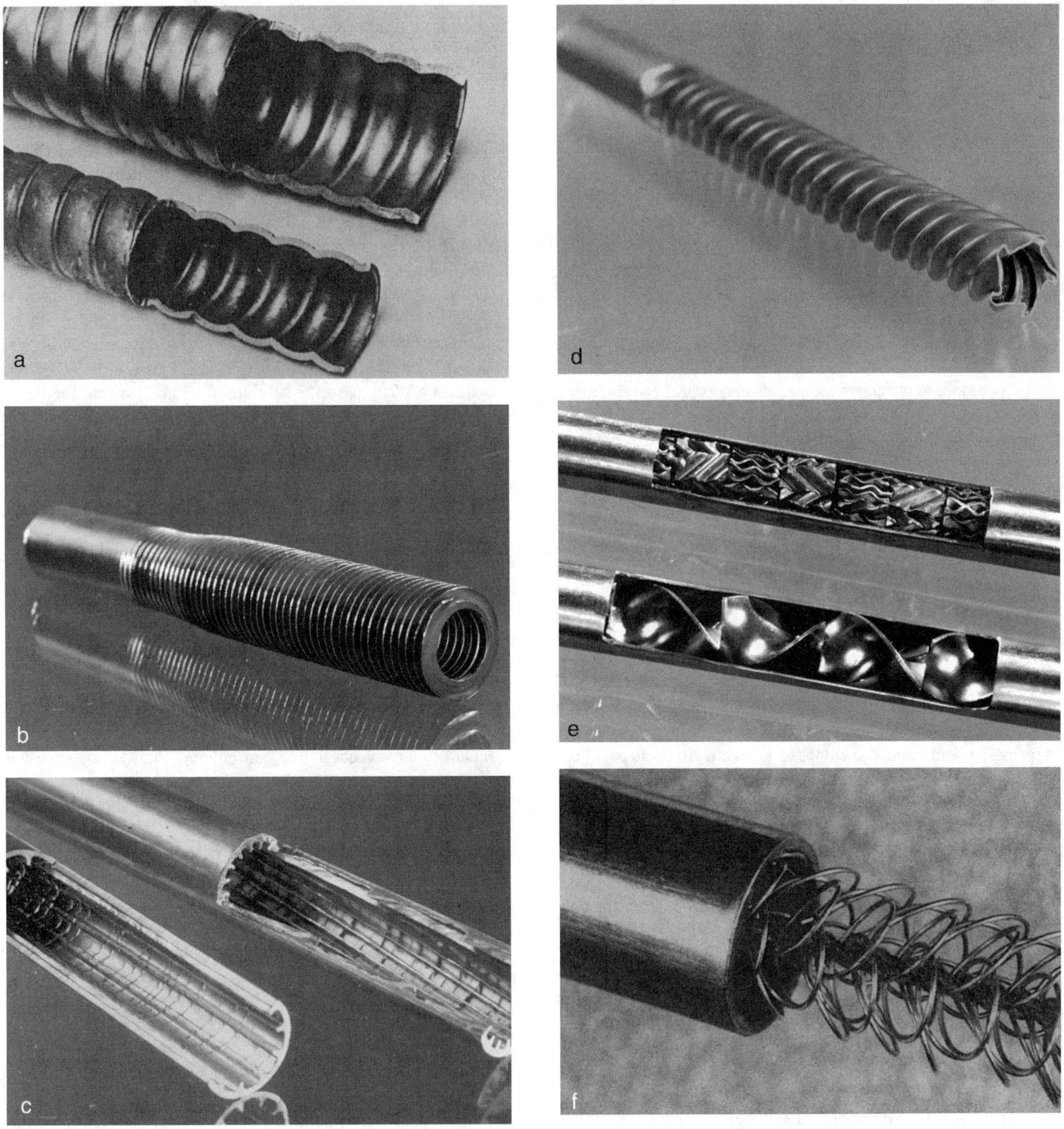

Figure 2. Enhanced tubes for augmentation of single-phase heat transfer. (a) Corrugated or spirally indented tube with internal protuberances (b) Integral external fins. (c) Integral internal fins. (d) Deep spirally fluted tube. (e) Static mixer insert. (f) Wire-wound insert.

Additives for liquids include solid particles and gas bubbles in single-phase flows and liquid trace additives for boiling systems.

Additives for gases are liquid droplets or solid particles, either dilute-phase (gas-solid suspensions) or dense-phase (fluidized beds).

Mechanical aids involve stirring the fluid by mechanical means or by rotating the surface. Surface "scraping," widely used for batch processing of viscous liquids in the chemical process industry, is applied to the flow of such diverse fluids as high-viscosity plastics and air. (See **Scraped Surface Heat Exchanger**). Equipment with rotating heat exchanger ducts is found in commercial practice.

Surface vibration at either low or high frequency has been used primarily to improve single-phase heat transfer.

Fluid vibration is the practical type of vibration augmentation because of the mass of most heat exchangers. The vibrations range from pulsations of about 1 Hz to ultrasound. Single-phase fluids are of primary concern.

Electrostatic fields (DC or AC) are applied in many different ways to dielectric fluids. Generally speaking, electrostatic fields can be directed to cause greater bulk mixing of fluid or disruption of fluid flow in the vicinity of the heat transfer surface, which augments heat transfer.

Injection is utilized by supplying gas to a stagnant or flowing liquid through a porous heat transfer surface or by injecting similar fluid upstream of the heat transfer section. Surface degassing of liquids can produce augmentation similar to gas injection. Only single-phase flow is of interest.

Suction involves vapor removal, in nucleate or film boiling, or fluid withdrawal, in single-phase flow, through a porous heated surface.

Two or more of the above techniques may be utilized simultaneously to produce an augmentation that is larger than either of the techniques operating separately. This is termed *compound augmentation.*

It should be emphasized that one of the motivations for studying augmented heat transfer is to assess the effect of an inherent condition on heat transfer. Some practical examples include roughness produced by standard manufacturing, degassing of liquids with high gas content, surface vibration resulting from rotating machinery or flow oscillations, fluid vibration resulting from pumping pulsation, and electrical fields present in electrical equipment.

The preceding is a general introduction to the augmentation of single-phase heat transfer as well as the subsequent entry on augmentation of **Heat Transfer, Two-Phase**. The surfaces in **Figure 2** have been used for both single-phase and two-phase heat transfer augmentation. The emphasis is on effective and cost-competitive (proved or potential) techniques that have made the transition from the laboratory to commercial heat exchangers. Broad reviews of developments in augmented heat transfer are available (Bergles, 1985; Bergles, 1990; Webb, 1994).

Free Convection

With the exception of the familiar technique of providing extended surfaces, the passive techniques have little to offer in the way of augmented heat transfer for free convection. This is because the velocities are usually too low to cause flow separation or secondary flow. The limited data for free convection from machined or formed rough surfaces, with water, and oil, indicate that increases in heat transfer coefficient up to 100% can be obtained with air, but that the increases with liquids are very small.

Design procedures for single fins and fin arrays are well established (see **Extended Surface Heat Transfer**); however, little testing or analysis has been directed at interrupted extended surfaces. The restarting of thermal boundary layers is expected to increase coefficients, so as to more than compensate for the lost area. The effectiveness of this concept has been demonstrated in the wire-loop fins used for base-board hot water heaters or "convectors". This problem is also of interest in electronic cooling, where the "heat sinks" are often discontinuous fins, and in natural draft cooling of finned tube banks, necessitated by loss of fan power.

Mechanically aided heat transfer is a standard technique in the chemical and food industries when viscous liquids are involved. (See also **Agitated Vessel Heat Transfer**)

Surface vibration has been extensively studied in the laboratory. The predominant geometry has been the horizontal cylinder vibrated either horizontally or vertically. Heat transfer coefficients can be increased 10-fold for both low-frequency/high-amplitude and high-frequency/low-amplitude situations. Although the improvements can be dramatic, it must be recognized that natural convection is inherently an inefficient mode of heat transfer. Since at maximum enhancement, the average velocity of the surface over a cycle is less than 1 m/s, it is more practical to provide steady forced flow. The mechanical designer is also concerned that such intense vibrations could result in equipment failures.

Rotating surfaces may also be used near the prime heat transfer surface to aid the free convection flow, thereby augmenting the heat transfer. Since it is usually difficult to apply surface vibrations to practical equipment, an alternative technique is utilized whereby vibrations are applied to the fluid and focused toward the heated surface. Generators employed range from flow interrupters to piezoelectric transducers, thus covering the range from pulsation of 1 Hz to ultrasound of 106 Hz. Much research has been reported on the effects of acoustic vibrations on heat transfer to gases from horizontal cylinders. Increases in average coefficients are observed only above an intensity of about 140 db, which is far in excess of human ear tolerance. The maximum increases reported are usually 100 to 200%. With proper ultrasonic transducer design, it is also possible to improve heat transfer to simple heaters immersed in liquids by several hundred percent. Cavitation is generally the dominant enhancement mechanism. In general, with liquids, there is considerable difficulty in designing systems to transmit vibrational energy to large surfaces.

Electric fields can be utilized to increase heat transfer coefficients in free convection. The configuration may be a heated wire in a concentric tube maintained at a high voltage relative to the wire, or a fine wire electrode may be utilized with a horizontal plate. *Dielectrophoretic* or *electrophoretic* (especially with ionization of gases) *forces* cause greater bulk mixing in the vicinity of the heat transfer surface. Heat transfer coefficients may be improved by as much as a factor of 40.

Recent activity has centered on the application of corona discharge cooling to practical problems. Cooling of cutting tools by point electrodes has been proposed. Also, parallel-wire electrodes have been used to improve the heat dissipation of a standard, horizontal finned tube; heat transfer coefficients can be increased by several hundred percent when sufficient electrical power is supplied. It would appear, however, that the equivalent effect could be produced at lower capital cost and without the hazards of 10,000–100,000 V by simply providing forced convection with a blower or fan.

Gas injection into a liquid through a porous heated plate, which has been used to similate nucleate boiling, can be regarded as an augmentation technique. Although coefficients can be increased several hundred percent, the practical applications of injection would appear to be rather limited, because of the difficulty of supplying and removing the gas.

Forced Convection

The present discussion emphasizes augmentation of heat transfer *inside* ducts that are primarily of circular cross section. The book by Zukauskas and Kalinin (1990) can be consulted for information on heat transfer in external flow, including augmented tube banks.

Surface roughness has been used extensively to augment forced convection heat transfer. Integral roughness may be produced by the traditional manufacturing processes of machining, forming, casting, or welding. Various inserts can also provide surface protuberances. In view of the infinite number of possible geometric variations, it is not surprising that, even after more than 300 studies, no completely satisfactory unified treatment is available.

In general, the maximum augmentation of laminar flow with many of the techniques is the same order of magnitude, and seems to be independent of the wall boundary condition (Joshi and Bergles, 1980). The augmentation with some rough tubes, corrugated tubes, inner-fin tubes, various static mixers, and twisted-type inserts is about 200 percent. Spikes and ripples have been used to enhance nominally laminar flow of air in parallel-plate channels of large aspect ratios (~ plate heat exchangers). Most plate heat exchangers utilize corrugated surfaces, for structural reasons as well as augmentation. It is generally agreed that the heat transfer and pressure drop characteristics of commercial corrugated surfaces used in plate exchangers are quite similar. The improvements in heat transfer coefficient with turbulent flow in rough tubes (based on nominal surface area) are as much as 250%. Analogy solutions for sand-grain type roughness and for square repeated-rib roughness have been proposed. A statistical correlation is also available for heat transfer coefficient and friction factor.

Much work has been done to obtain the augmented heat transfer of parallel angled ribs in short rectangular channels, simulating the interior of gas turbine blades.

Jets are frequently used for heating, cooling, and drying in a variety of industrial applications. A number of studies have reported that roughness elements of the transverse-repeated rib type mitigate the deterioration in heat transfer downstream of stagnation.

Extended surfaces can be considered "old technology" as far as most applications are concerned. The real interest is in increasing heat transfer coefficients on the extended surface. Compact **Heat Exchangers** of the plate-fin or tube-and-center variety use several augmentation techniques: offset strip fins, louvered fins, perforated fins, or corrugated fins. Coefficients are several hundred percent above the smooth-tube values; however, the pressure drop is also substantially increased, and there may be vibration and noise problems. For more details on the current status of air-side (external) heat transfer in finned-tube heat exchangers, the review of Webb (1994) should be consulted.

Internally finned circular tubes are available in aluminum and copper (or copper alloys). Correlations (for heat transfer coefficient and friction factor) are available for laminar and turbulent flow, for both straight and spiral continuous fins.

A numerical analysis of turbulent flow in tubes with idealized straight fins was reported. The necessary constant for the turbulence model was obtained from experimental data for air. Further improvements in numerical techniques are expected, so that a wider range of geometries and fluids can be handled without resort to extensive experimental programs.

Many proprietary surface configurations have been produced by deforming the basic tube. The "convoluted," "corrugated," "spiral," or "spirally fluted" tubes have *multiple-start spiral corrugations,* which add areas, along the tube length. A systematic survey of the single-tube performance of *condenser tubes* indicates up to 400% increase in the nominal inside heat transfer coefficient (based on diameter of a smooth tube of the same maximum inside diameter); however, pressure drops on the water side are about 20 times higher.

Displaced enhancement devices are typically in the form of inserts, within elements arranged to promote transverse mixing (static mixers, (**Figure 2e**). They are used primarily for viscous liquids, to promote either heat transfer or mass transfer. There are no broad-based correlations available, because of the many geometric arrangements and the strong influence of fluid properties and heating conditions. In general, the higher the heat transfer coefficient, the higher the pressure drop. Similar inserts or packings have been used for turbulent flow; however, this application is recommended only for short sections with high heat fluxes, since the pressure drop is so high.

Displaced promoters are also used to enhance the radiant heat transfer in high-temperature applications. In the flue-tube of a hot gas-fired hot water heater, there is a trade-off between radiation and convection.

Another type of displaced insert *generates vortices,* which enhance the downstream flow. Delta-wing and rectangular wing promoters, both co-rotating and counter-rotating, have been studied

Wire-loop inserts (**Figure 2f**) have also been used for augmentation of laminar and turbulent flow.

Heat transfer coefficients can be substantially higher in coiled tubes than in straight tubes because of the secondary flow generated by the curvature.

Twisted-tape inserts have been widely used to improve heat transfer in both laminar and turbulent flow. Correlations are available for laminar flow, for both uniform heat flux and uniform wall temperature conditions. Turbulent flow in tubes with twisted-tape inserts has also been correlated. Several studies have considered the heat transfer augmentation of a decaying swirl flow, generated, say, by a short twisted-tape insert.

Modest improvements in heat transfer are observed when gas bubbles or solid particles are added to liquids.

Gas-side heat transfer can also be augmented by adding a small volumetric fraction of solid particles. The particles are carried along with the stream and separated for reuse, in the case of a once-through system, or circulated, in the case of a closed system. The enhancements of up to four times the pure gas heat transfer coefficients are attributed to thinning of the viscous sublayer and higher thermal conductivity in that layer. It does not appear that any practical applications of dilute-phase gas-solid heat transfer are currently being considered.

Fluidized Beds are used in many industrial applications. Heat transfer coefficients to tubes within a bed can be enhanced by a factor of 20, compared to pure gas flow at the same flow rate.

When liquid droplets are added to a flowing gas stream, heat transfer is enhanced by sensible heating of the two-phase mixture, evaporation of the liquid, and disturbance of the boundary layer. When spray cooling was applied to a compact heat exchanger core, the maximum improvement of 40% was attributed to formation of a partial liquid film and sensible heating of that film. In general, the large flow rate of liquid required tends to limit practical application of this technique.

Under active techniques, mechanically aided heat transfer in the form of surface-scraping can increase forced convection heat transfer. Unfortunately, the necessary hardware is not particularly compatible with most heat exchangers. Surface-scraping has been used with air flow on flat plates.

The other aspect of this technique is *rotating surfaces*. Moderate increases in heat transfer coefficients have been reported for laminar flow in a straight tube rotating about its own axis, a straight tube rotating around a parallel axis, a rotating circular tube, and the rotating, curved, circular tube. Maximum improvements of 350% were recorded for laminar flow, but for turbulent flow the maximum increase was only 25%. In general, these are examples of naturally occurring phenomena that result in enhancement: cooling windings of rotating electrical machinery, cooling of gas turbine rotor blades, and so on. (See **Rotating Duct Systems, Heat Transfer In**)

The passive augmentation techniques may be compromised by use in fouling situations, because augmented heat transfer often means augmented mass transfer. However, some surfaces display "antifouling" behavior. See Bergles and Somerscales (1995).

Surface vibration has been demonstrated to improve heat transfer to both laminar and turbulent duct flow of liquids. The largest improvements (up to 200%) are observed with laminar or transitional flow utilizing a concentric-tube heat exchanger with the inner tube vibrated transversely or a rectangular channel with a flexible, vibrating side. The complexity of the vibrational equipment and the relatively large power expenditure would seem to rule out this technique for practical application.

Fluid vibration has been extensively studied for both air (loudspeakers and sirens) and liquids (flow interruptors, pulsators, and ultrasonic transducers). The gas results are not encouraging, as intensities above 120 db are required, and the effect is largely one of triggering fury turbulent flow at transitional Reynolds numbers. On the other hand, pulsating jets have significant augmentation as compared to steady impinging jets.

Pulsations are relatively simple to apply to low-velocity liquid flows, and improvements of several hundred percent can be realized. Turbulence triggering and cavitation appear to be important enhancement mechanisms. Application of high-frequency vibrations is difficult, and only modest improvements in heat transfer coefficient are recorded.

Of course, heat transfer rates in pulse combustion tailpipes and in other oscillating turbulent flows have been found to be significantly higher than those of steady, turbulent flow.

Some very impressive enhancements have been recorded with electrical fields, particularly in the laminar flow region. Improvements of at least 100% were obtained when voltages in the 10-kV range were applied to transformer oil. Although it is desirable to take advantage of any naturally occurring electrical fields in electrical equipment, it would appear to be quite difficult to introduce this augmentation in practical equipment.

It is found that even with intense electrostatic fields, the heat transfer augmentation disappears as turbulent flow is approached in a circular tube with a concentric inner electrode. There is little effect of corona wind, even at low air velocities, with the exception of tests with three electrodes under a finned tube. In this case, a 60% increase in heat transfer coefficient was noted. A comprehensive survey of the subject is given by Ohadi (1991).

Single-phase heat transfer can be enhanced by injecting gas into a liquid through an porous heated surface. Up to five-fold increases in local heat transfer coefficients have been observed by injecting similar fluid into a turbulent tube flow. In pipe flow, this is done by having injectors spaced along the pipe.

Large increases in heat transfer coefficient are predicted for laminar and turbulent flow with surface suction. The general characteristics of the latter predictions were confirmed by experiments. However, suction is difficult to incorporate into practical heat exchange equipment.

Compound techniques are slowly emerging area of enhancement that hold promise for practical applications, since heat transfer coefficients can usually be increased above each of the several techniques acting alone. Some examples that have been studied are as follows:

1. Rough tube wall with twisted-tape inserts
2. Rough cylinder with acoustic vibrations
3. Internally finned tube with twisted-tape insert
4. Finned tubes in fluidized beds
5. Externally finned tubes subjected to vibrations
6. Rib-roughened passage being rotated
7. Gas-solid suspension with an electrical field
8. Fluidized bed with pulsations of air
9. Rib-roughened channel with longitudinal vortex generation.

One may consider the use of augmentation techniques to satisfy any of the following thermal-hydraulic objectives: (1) to reduce prime surface area, (2) to increase heat transfer capacity, (3) to reduce the approach temperature difference for the process streams, or (4) to reduce pumping power. Having defined a basic objective, the designer will establish the parameters that are fixed and the basic constraints that must be satisfied. Through manipulation of the data or correlations for heat transfer coefficients and friction factors, performance ratios can be calculated, for example, the ratio of the prime surface area of the enhanced heat exchanger to that of the normal or reference exchanger at constant pumping power. Design objectives and performance ratios are discussed in detail in Nelson and Bergles (1986).

Augmentation usually will not be attractive unless the augmented heat exchanger offers a cost advantage relative to the use of conventional heat transfer configurations. Additional factors entering into the ultimate decision to use an augmentation technique are materials limitations, fouling potential, safety, and reliability. The status and prospects for commercial development of augmentation techniques are discussed by Bergles (1988).

References

Bergles, A. E. (1985) Techniques to Augment Heat Transfer, *Handbook of Heat Transfer Applications* (Ed. W. M. Rohsenow, J. P. Hartnett, and E. N. Ganic) McGraw-Hill, New York, NY, 3-1–3-80.

Bergles, A. E. (1988) Some Perspectives on Enhanced Heat Transfer-Second-Generation Heat Transfer Technology, *Journal of Heat Transfer,* 110, 1082–1096.

Bergles, A. E. (1990) Augmentation of Heat Transfer, *Heat Exchanger Design* (Ed. G. F. Hewitt) Hemisphere, New York, NY 2.5.11-1–12.

Bergles, A. E. and Somerscales, E. F. C. (1995) The Effect of Fouling on Enhanced Heat Transfer Equipment, To be published in *Enhanced Heat Transfer.*

Bergles, A. E., Jensen, M. K., Somerscales, E. F. C., and Manglik, R. M. (1991) Literature Review of Heat Transfer Enhancement Technology for

Heat Exchangers in Gas-Fired Applications, Gas Research Institute Report, GR191-0146.

Joshi, S. D. and Bergles, A. E. (1980) Survey and Evaluation of Passive Heat Transfer Augmentation Techniques for Laminar Flow, *The Journal of Thermal Engineering,* 1, 105–124.

Nelson, R. M. and Bergles, A. E. (1986) Performance Evaluation for Tube-side Enhancement of a Flooded Evaporator Water Chiller, *ASHRAE Transactions,* 92, Part 1B, 739–755.

Ohadi, M. M. (1991) Heat Transfer Enhancement in Heat Exchangers, *ASHRAE Journal* (December) 42–50.

Webb, R. L. (1994) *Principles of Enhanced Heat Transfer,* Wiley, New York, NY.

Zukauskas, A. A. and Kalinin, E. K. Eds. (1990) *Heat Transfer: Soviet Reviews, Vol. 2. Enhancement of Heat Transfer,* Hemisphere, New York, NY.

A.E. Bergles

AUGMENTATION OF HEAT TRANSFER, TWO-PHASE

Following from: *Heat transfer; Forced convective boiling*

Introduction

Selected passive and active augmentation techniques have been shown to be effective for pool boiling and flow boiling/evaporation. Most techniques apply to nucleate boiling; however, some techniques are applicable to transition and film boiling.

It should be noted that phase-change heat transfer coefficients are relatively high. The main thermal resistance in a two-fluid heat exchanger often lies on the non-phase-change side. Fouling of either side can, of course, represent the dominant thermal resistance. For this reason, the emphasis is often on augmentation of single-phase flow. On the other hand, the overall thermal resistance may then be reduced to the point whereas significant improvement in the overall performance can be achieved by augmenting the two-phase flow. Two-phase augmentation would also be important in double phase-change (boiling/condensing) heat exchangers.

As discussed elsewhere, surface material and finish have a strong effect on nucleate and transition pool boiling. However, reliable control of nucleation on plain surfaces is not easily accomplished. Accordingly, since the earliest days of boiling research, there have been attempts to relocate the boiling curve through use of relatively gross modification of the surface. For many years, this was accomplished simply by area increase in the form of low helical fins. The subsequent tendency was to structure surfaces to improve the nucleate boiling characteristics by a fundamental change in the boiling process (Webb, 1981). Many of these advanced surfaces are being used in commercial shell-and-tube boilers. A book on the subject was prepared by Thome (1990). The application to high-heat-flux systems is discussed by Bergles (1992).

Pool Boiling

Numerous types of *surface treatment* and *structuring* have been utilized to intentionally reduce the wall-minus-saturation temperature difference ΔT_s. For water, the placement of small nonwetting spots (teflon or epoxy), either splattered on the surface of placed in pits, reduces ΔT_s at constant q by a factor of 3–4. This comparison, as well as others in this section, is based on heat flux evaluated for the surface area of the plain tube.

The usual surface modifications are of a rather fine scale, in keeping with the nature of the nucleate boiling process. In the present classification, these surfaces mostly qualify as treated surfaces; however, some of them could be considered rough. Several manufacturing processes have been employed: machining, forming, layering, and coating. **Figure 1** indicates seven of the presently commercialized surfaces.

In **Figure 1a**, standard low-fin tubing is shown. **Figure 1c** depicts a tunnel-and-pore arrangement produced by rolling, upsetting, and brushing. An alternative modification of the low fins is shown in Fig. 1d, where the rolled fins have been split and rolled to a T shape. Knurling and rolling are involved in producing the surface shown in Fig. 1g. The earliest example of a commercial structured surface, shown in Fig. 1b is the porous metallic matrix produced by sintering or brazing small particles.

The relative performance of three of these surfaces, tested as single tubes, is shown in **Figure 2**. Wall superheat reductions of up to a factor often are common with these surfaces. The advantage is seen to be not only a high nucleate boiling heat transfer coefficient, but the fact that boiling can take place at very low temperature differences.

In all cases, a complex liquid-vapor exchange is involved. The liquid flows at random locations around the helical aperture or through selected pores to the interior of the structure, where thin-film evaporation occurs over a large surface area. The vapor is then ejected through other paths by bubbling. The latent heat transport is complemented by agitated free convection from the exposed surfaces. Data indicate that these surfaces are effective for pure liquids and mixtures with a wide range of boiling points.

The behavior of structured surfaces is not yet understood to the point where correlations are available to guide custom production of the surfaces for a particular fluid and pressure level. For example, a model for nucleate boiling from Thermoexcel-E surfaces requires

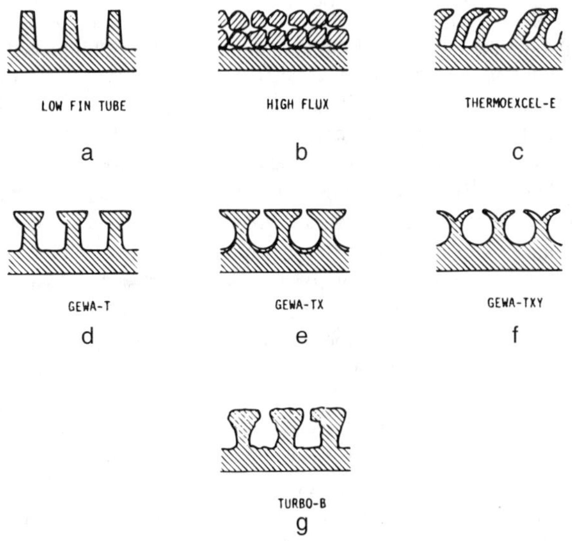

Figure 1. Examples of commercial structured boiling surfaces.

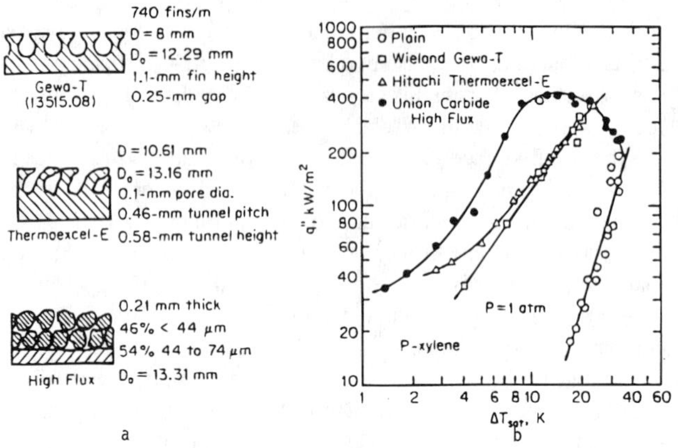

Figure 2. Pool boiling from smooth and structured surfaces on the same apparatus (Yilmaz et al., 1980). (a) Sketch of cross-sections of the three enhanced heat transfer surfaces tested. (b) Boiling curves for the enhanced tubes and a smooth tube.

eight empirically determined constants. Such models are useful in confirming the probable physics of the boiling process, but are of no use in engineering design. Some manufacturers have accumulated sufficient experience to provide optimized surfaces for some important applications, such as flooded evaporators for refrigerant dry-expansion chillers and thermosiphon reboilers for hydrocarbon distillation columns. The structured boiling surfaces developed for refrigeration and process applications have been used as "heat sinks" for microelectronic chips (Bergles, 1990).

The behavior of tube bundles is often different with structured-suface tubes. The nucleate boiling augmentation dominates, and the convective boiling enhancement, found in plain tube bundles, does not occur.

Structured surfaces are not exempt from temperature overshoots and resultant boiling curve hysteresis. However, the superheats required for incipient boiling with the highly wetting liquids are generally lower than for plain surfaces, due to the greater probability of finding active sites over the larger surface area. The practical problem is to provide, at least on a transient basis, the necessary superheat to initiate boiling. In some cases, the stimulus can be provided by injected vapor, either, naturally, as with a dry-expansion chiller or, intentionally, through sparging.

Structured surfaces have recently been applied to *thin film evaporation.* Here, in contrast to the pool experiments noted above, the liquid to be vaporized is sprayed on or dripped on heated horizontal tubes to form a thin film. Structured surfaces promote nucleate boiling in the film at modest temperature differences.

Little attention has been given the use of surfaces with discrete roughness elements in pool boiling. Improvement is not expected due to the microscale of the nucleate boiling. On the other hand, *knurled surfaces* are available commercially, and they apparently improve nucleate boiling. This would be expected, however, from the area increase. More commonly, rough surfaces have been applied to horizontal tube spray film evaporators. Longitudinal ribs or grooves may promote turbulence in the film; however, there is the possibility that film drainage is impeded. Increases of 100 percent in heat transfer coefficient have been referred with knurling. This is apparently a convective enhancement due to turbulence promotion within the film, as well as a substantial increase in surface area.

Low fin tubes (**Figure 1**) have been the standard enhancement technique for pool boiling of refrigerants and organics. Heat transfer coefficients (based on total area) are typically greater than those for the reference plain tube.

Circumferential or helical fins with various profiles have been considered for augmentation of horizontal-tube, spray film evaporation. Area increases are typically 100 percent. Surface tension causes redistribution of the liquid so that thin films are present at the fin peaks; accordingly, heat transfer coefficients (based on projected area) can increase by more than the area increase. For example, V-shaped grooves or threads increase heat transfer coefficients up to 200 percent. It was found that heat transfer coefficients for falling film evaporation inside vertical tubes could be increased by more than a factor often with loosely attached internal fins.

Active augmentation techniques include heated surface rotation, surface wiping, surface vibration, fluid vibration, electrostatic fields, and suction at the heated surface. Although active techniques are effective in reducing ΔT_s and/or increasing $\dot{q}$, the practical applications are very limited, largely because of the difficulty of reliably providing the mechanical or electrical effect. Perhaps the major contribution of the many studies in this area is the information that is available to predict changes in pool boiling when these effects occur naturally in heat transfer equipment.

Compound augmentation, which involves two or more techniques applied simultaneously, has also been studied. The addition of surface roughness to the evaporator side of a rotating evaporator-condenser increased the overall coefficient by 10 percent. *Electrohydrodynamic augmentation* was applied to a finned tube bundle, resulting in nearly a 200 percent increase in the average boiling heat transfer coefficient of the bundle (Cheung et al., 1995). The working fluid was an alternate refrigerant R-134a, and the maximum EHD power consumption was 1.2 percent of the bundle heat transfer rate.

Convective Boiling/Evaporation

The situation of main interest is boiling or vaporization inside tubes. Forced flow normal or parallel to enhanced tubes has received little attention, in spite of the fact that it might be desirable to augment heat transfer on the shell side of a shell-and-tube heat exchanger. Both process and power applications benefit directly from single-tube testing. In other words, there is no difference between single and multiple channels, assuming that the flow is known and steady.

The structured surfaces described in the previous section are generally not used for intube vaporization, due to the difficulty of manufacture. One notable exception is the High flux surface in a vertical thermosiphon reboiler. The considerable increase in the low quality, nucleate boiling coefficient is desirable, but it is also important that more vapor is generated to promote circulation.

Helical repeated rib tubes have been found to increase local heat transfer coefficients for vaporization of R-12 by up to 100 percent. The dryout heat flux was increased by 200 percent.

The use of helically *coiled wire inserts* to increase critical heat flux was reported. Increases in critical heat flux up to 50 percent were reported, with a pressure drop increase of only 25 percent.

In an important power application, widely spaced helical ribs with a gentle twist considerably increase the dryout heat flux in a once-through boiler. The probably mechanism is stabilization of the annular liquid film through the partially confined rotation. Pseudo-film boiling of supercritical water is a similar process; hence, it is not surprising that the ribbed tubes suppress this condition. Heat transfer coefficients in the post-dryout or mist-flow film boiling region are increased with roughness.

Numerous tubes with internal *fins,* either integral or attached, are available for refrigerant evaporators. Original configurations were tightly packed, copper, offset strip fin inserts soldered to the copper tube or aluminum, star-shaped inserts secured by drawing the tube over the insert. Examples are shown in **Figure 3**. Average heat transfer coefficients (based on surface area of smooth tube of the same diameter) for typical evaporator conditions are increased by as much as 200 percent. Although the integral fins represented in this figure are small, the trend has been toward more numerous and even smaller fins.

A cross-sectional view of a typical *"micro-fin" tube* is included in **Figure 3**. These tubes are 9.5 mm outside diameter and have 60 to 70 spiral fins ranging from 0.10 to 0.19 mm in height. The average evaporation boiling coefficient is increased 30–80 percent. The pressure drop penalties are less; that is, lower percentage increases in pressure drop are frequently observed. The outstanding thermal-hydraulic performance is related to increased surface area, increase in the film turbulence level, delay in film dryout, and alteration of the flow pattern. These tubes also have promise for micro-gravity situations, where the capillary forces associated with the small fins promote a favorable flow pattern, even at reduced levels of local acceleration.

Twisted-tape inserts are generally used to increase the burnout heat flux for subcooled boiling at high imposed heat fluxes (10^7–10^8 W/m^2), as might be encountered in the cooling of fusion reactor components. Increases in burnout heat flux of up to 200 percent were obtained. Unfortunately, the augmentation of the twisted tape drops off rapidly as pressure is increased above 2 MPa. A further problem is promotion of the burnout condition by virtual insulation of the heated wall at the wall-tape interface.

Local heat transfer coefficients for the evaporation of R-113 are increased by as much as 90 percent with twisted-tape inserts. An increase in the dryout heat flux is generally observed, because the rotating flow centrifuges the core droplets to the tube wall.

The effect of the centrifugal force field on the droplets is dramatic in the case of mist flow (or dispersed flow) film boiling. There are reductions in wall temperature for both the traditional dryout, at some point along the tube, and for dryout at the tube inlet.

Coiled-tube vapor generators have advantages of higher heat transfer performance as well as compactness. (See also **Coiled Tubes, Heat Transfer in**) Modest improvements in axially local,

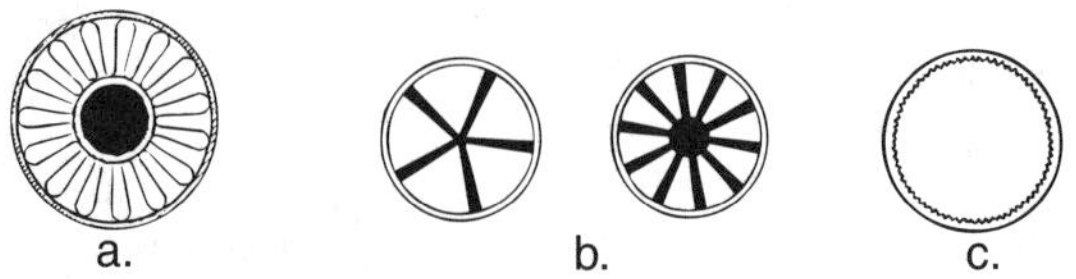

Figure 3. Inner-fin tubes for refrigerant evaporators. (a) strip-fin inserts, (b) star-shaped inserts, (c) micro-fin.

but circumferential average, heat transfer coefficients are obtained, especially when the coil diameter is small. The dryout heat fluxes are substantially higher than the straight-tube values at interrupted at the leading edge of a strip, and the film is thinner when it reforms past the trailing edge.

Rough surfaces augment condensation primarily by introducing turbulence in the film. The average heat transfer coefficients for steam condensing on vertical tubes can be nearly doubled by knurling the surface. Condensing coefficients for R-11 on horizontal tubes have been found to be four to five times the smooth-tube values. Part of the improvement, of course, is due to the increase in surface area.

Surface extensions are widely employed for augmentation of condensation. The integral low fin tubing (**Figure 1**), used for kettle boilers, is also used for horizontal tube condensers. With proper spacing of the fins to provide adequate condensate drainage, the average coefficients can be several times those of a plain tube with the same base diameter. These fins are normally used with refrigerants and other organic fluids that have low condensing coefficients, but which drain effectively because of low surface tension.

The fin profile can be altered according to mathematical analysis to take full advantage of the *Gregorig effect,* whereby condensation occurs mainly at the tops of convex ridges. Surface-tension forces then pull the condensate into concave grooves where it runs off. The average heat transfer coefficient is greater than that for an axially uniform film thickness. The initial application was for condensation of steam on vertical tubes used for reboilers and in desalination. The analysis of Mori et al. (1981) is typical of those that outlet qualifies above 0.2. Post-dryout heat transfer coefficients are also higher with helical coils.

Vapor Space Condensation

The horizontal and vertical condensers found in both the power and process industries involve condensation on the outside of tubes. Of course, due to the high vapor velocities normally encountered, the condensation process in an actual heat exchanger is different from the classical laboratory experiment using a single tube in a large vapor space.

As discussed elsewhere, condensation can be either filmwise or dropwise. In a sense, dropwise condensation is augmentation of the normally occurring film condensation by surface treatment. The only real application is for steam condensers because nonwetting coatings are not available for most other working fluids. Even after much study, little progress has been made in developing permanent hydrophobic coatings for practical steam condensers. The augmentation of dropwise condensation is pointless, because the heat transfer coefficients are already so high.

It has been found that average coefficients for film condensation of steam on horizontal tubes can be improved up to 20 percent by strategically placing strips of teflon or other nonwetting material around the tube circumference. The condensate film is interrupted at the leading edge of a strip, and the film is thinner when it reforms past the trailing edge.

Rough surfaces augment condensation primarily by introducing turbulence in the film. The average heat transfer coefficients for steam condensing on vertical tubes can be nearly doubled by knurling the surface. Condensing coefficients for R-11 on horizontal tubes have been found to be four to five times the smooth-tube values. Part of the improvement, of course, is due to the increase in surface area.

Surface extensions are widely employed for augmentation of condensation. The integral low fin tubing (**Figure 1**), used for kettle boilers, is also used for horizontal tube condensers. With proper spacing of the fins to provide adequate condensate drainage, the average coefficients can be several times those of a plain tube with the same base diameter. These fins are normally used with refrigerants and other organic fluids that have low condensing coefficients, but which drain effectively because of low surface tension.

The fin profile can be altered according to mathematical analysis to take full advantage of the *Gregorig effect*, whereby condensation occurs mainly at the tops of convex ridges. Surface-tension forces then pull the condensate into concave grooves where it runs off. The average heat transfer coefficient is greater than that for an axially uniform film thickness. The initial application was for condensation of steam on vertical tubes used for reboilers and in desalination. The analysis of Mori et al. (1981) is typical of those that can be performed to obtain the optimum geometry. According to their numerical solutions, the optimum geometry is characterized by a sharp fin tip, gradually changing curvature of the fin surface from tip to root, wide grooves between fins to collect condensate, and periodic condensate strippers. **Figure 4** schematically presents the configuration. Optimum fin pitches and stripper spacing were calculated for water and R-113.

Experience has shown that vertical tube condenser performance can be improved by a factor of five to ten by a variety of axial flutes, not just those that have the mathematically correct shape. For example, the effectiveness of simply tying small, vertically oriented wires around the circumference has been demonstrated.

The same concepts for essentially two-dimensional-shaped surfaces can be applied to horizontal tubes; however, the improvements are usually less than a factor of five. This is important in steam surface condenser technology.

Recent interest has centered on three-dimensional surfaces for horizontal-tube condensers. A rolling and upsetting process is used to produce the surface shown in **Figure 5**. (In fact, the *Thermoexcel-C* condensing surface is the *Thermoexcel-E* boiling surface without the final high-speed brushing operation.) It is seen that the coefficients for R-113 are as much as seven times the smooth-tube values.

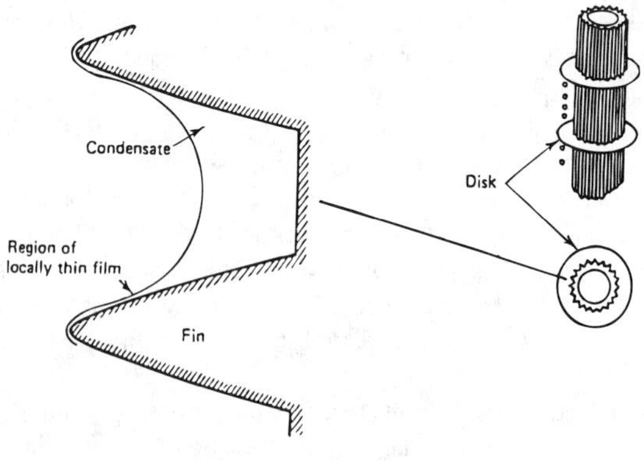

Figure 4. Recommended flute profile and schematic of condensate strippers, according to Mori et al. (1981).

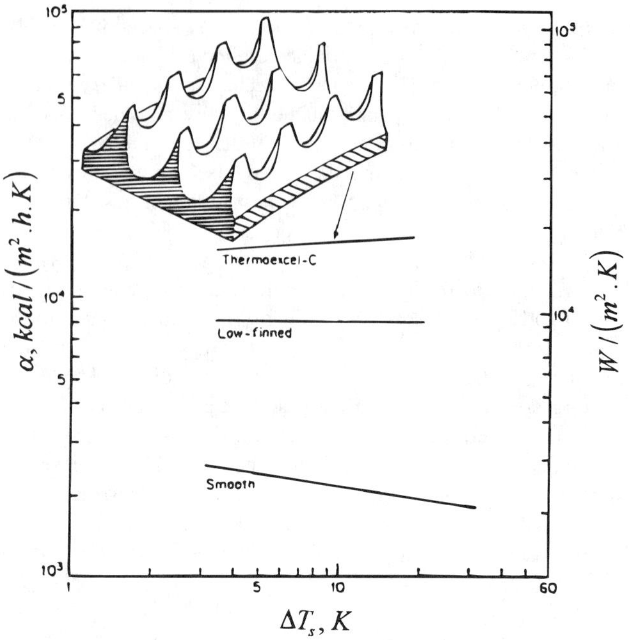

Figure 5. Performance of enhanced tubes compared to a smooth tube for condensation of R-113.

The considerable improvement relative to low fins or other two-dimensional profiles is apparently due to multidimensional drainage at the fin tips. Other three-dimensional strips include circular pin fins, square pins, and small metal particles that are bonded randomly to the surface.

Convective Condensation

This final section on augmentation of the various modes of heat transfer focuses on in-tube condensation. The applications include horizontal kettle-type reboilers, moisture separator reheaters for nuclear power plants, and air-conditioner condensers. In the latter application, the tubes must also perform well during evaporation, if a heat pump system is involved.

Internally grooved and knurled tubes have been studied, and average heat transfer coefficients were increased for several of the configurations; however, it appears that further improvements can be realized by optimizing the geomerty.

It has been found that average coefficients for complete condensation are increased 80 percent above smooth-tube values when helical repeated ribs, were used. For an extreme case of roughness, deep spirally fluted tubes were found to have coefficients 50 percent above those of smooth tubes of the same maximum inside diameter.

Random roughness has been applied by means of metallic particles with about 50 percent area density. With R-12, the condensing coefficient was increased 300 percent for average qualities greater than 0.6, and 140 percent for lower qualities.

Conventional inner-fin tubes have been used for condensation of steam and other fluids. The fins are relatively long but few in number, and may be straight or spiralled. Up to 150 percent increases in average condensing coefficients were observed for complete condensation. Design correlations for both heat transfer and pressure drop were developed. With similar tubes with R-113, coefficient increases up to 120 percent were observed.

The micro-fin tubes mentioned earlier have also been applied successfully to in-tube condensing. As in the case of evaporation, the substantial heat transfer improvement is achieved at the expense of a lesser percentage increase in pressure drop. By testing a wide variety of tubes, it has been possible to suggest some guidelines for the geometry, e.g., more fins, longer fins, and sharper tips; however, general correlations are not yet available. Fortunately for heat-pump operation, the tube that performs best for evaporation also performs best for condensation.

Data have been reported for condensation of R-113 in tubs with Kenics static mixer inserts. The average heat transfer coefficients were increased considerably; however, the increases in pressure drop were very large. Surface renewal/penetration theory (with one empirical constant) is in good agreement with the experimental data.

Twisted-tape inserts result in rather modest increases in heat transfer coefficient for complete condensation of either steam or R-113. The pressure drop increases are large due to the large wetted surface. Coiled tubular condensers provide a modest improvement in average heat transfer coefficient.

Concluding Remarks

These two entries give some indication as to why heat transfer augmentation is one of the fastest growing areas of heat transfer. Many techniques are available for improvement of the various modes of heat transfer. Fundamental understanding of the transport mechanism is growing, but more importantly, design correlations are being established. Heat transfer augmentation has indeed become a second-generation heat transfer technology that is becoming widely used in industrial heat exchangers, particularly those that involve boiling. New journals, e.g., *Enhanced Heat Transfer and International Journal of Heating, Ventilating, Air-Conditioning and Refrigerating Research* feature this technology.

References

Bergles, A. E. (1988) Heat Transfer Augmentation, *Two-Phase Heat Exchangers. Thermal-Hydraulic Fundamentals and Design,* (Ed. S. Kakac, A. E. Bergles, and E. O. Fernandes) Kluwer, Dordrecht, The Netherlands, pp. 343–373.

Bergles, A. E., Ed. (1990) *Heat Transfer in Electronic and Microelectronic Equipment,* Hemisphere, New York, NY.

Bergles, A. E. (1992) Enhanced Heat Transfer Techniques for High-Heat-Flux Boiling, *High Heat Flux Engineering* (Ed. A. M. Khounsary), Vol. 1739, SPIE, Bellingham, WA, pp. 2–16.

Cheung, K. H., Ohadi, M. M., and Dessiatoun, S. (1995) Compound Enhancement of Boiling Heat Transfer of R-134a in a Tube Bundle, to be published in *ASHRAE Transactions,* Part 1.

Pate, M. B., Ayub, Z. H., and Kohler, J. (1990) "Heat Exchangers for the Air-Conditioning and Refrigeration Industry": State-of-the-Art Design and Technology, *Compact Heat Exchangers,* (Ed. R. K. Shah, A. D. Kraus, and D. Metzger) Hemisphere, New York, NY, pp. 567–590.

Thome, J. R. (1990) *Enhanced Boiling Heat Transfer,* Hemisphere, New York.

Webb, R. L. (1981) The Evolution of Enhanced Surface Geometries for Nucleate Boiling, *Heat Transfer Engineering,* Vol. 2, (Nos. 3–4), pp. 46–69.

Yilmaz, S., Hwalck, J. J., and Westwater, J. W. (1980) Pool Boiling Heat Transfer Performance for Commercial Enhanced Tube Surfaces, ASME Paper 90-HT-41.

A.E. Bergles

AUSTRALIAN & NEW ZEALAND ASSOCIATION FOR THE ADVANCEMENT OF SCIENCE, ANZAAS

GPO Box 2816
Canberra ACT
2601
Australia
Tel: 6 248 5846

AUTO CORRELATION (see Correlation analysis)

AUTO-IGNITION (see Explosion phenomena; Internal combustion engines)

AUTOMATIC SAMPLING (see Sampling)

AUTOMOTIVE GAS TURBINES (see Gas turbine)

AVERAGE FILM FLOW RATE MEASUREMENT (see Film flow rate measurement)

AVERAGE PHASE VELOCITY (see Multiphase flow)

AVERAGE VOID FRACTION MEASUREMENT (see Void fraction measurement)

AVOGADRO NUMBER

Avogadro's hypothesis, first stated in 1811, was that equal volumes of gases under fixed conditions of temperature and pressure contain equal numbers of molecules. Half a century passed before the significance of this statement was fully appreciated, and Avogadro did not live to see his hypothesis confirmed. Today his name is given to the number of molecules in a gram-mole (or atoms in a gram-atom), or more precisely by SI standards the number of entities in one mole of carbon 12. This is known as Avagadro number or Avagadro constant, N_A.

In the past N_A has been measured experimentally by various methods, each producing different values of about 6×10^{23} per mol (Moelwyn-Hughes, 1957). Today the accepted value is

$$N_A = 6.022 \times 10^{23} \text{ mol}^{-1}$$

Note that N_A can also be derived from the relationship

$$N_A = \frac{\text{universal gas constant, } \tilde{R}}{\text{Boltzmann constant, k}} = \frac{8.3143\ JK^{-1}\ mol^{-1}}{1.3806 \times 10^{-23}\ JK^{-1}}$$

$$= 6.022 \times 10^{23}\ mol^{-1}$$

Reference

Moelwyn-Hughes, E. A. (1957) Physical Chemistry Pergamon Press.

G.L. Shires

AVOGADRO'S LAW (see Gay Lussac's law)

AXIAL FLOW COMPRESSOR (see Compressors)

AXIAL FLOW FANS (see Aerodynamics)

AXIAL TURBINE (see Turbine)

AZEOTROPES (see Distillation; Vapour-liquid equilibrium)

B

BAAS (see British Association for the Advancement of Science)

BACKWARD FACING STEP (see Contraction, flow and pressure loss in; Convective heat transfer)

BACKWARD RETURN CONDENSER (see Dephlegmator)

BAFFLE RING CENTRIFUGE (see Centrifuges)

BAFFLE TRAY COLUMNS (see Direct contact heat exchangers)

BAFFLES (see Mechanical design of heat exchangers; Shell-and-tube heat exchangers)

BAG FILTERS (see Filters)

BALANCING HEADER (see Benson boilers)

BALL FLOWMETERS (see Flow metering)

BALLOONING (see Rod bundles, heat transfer in)

BAND DRYER (see Dryers)

BANKER TYPE EVAPORATOR (see Evaporators)

BANKS OF OBJECTS, FLOW PAST (see Crossflow)

BARN (see Nuclear reactors)

BAROMETER (see Pressure measurement)

BARRA-CONSTANTINI SYSTEM (see Solar energy)

BASSET HISTORY TERM (see Particle transport in turbulent fluids; Stokes law for solid spheres and spherical bubbles)

BATCH CRYSTALLIZERS (see Crystallizers)

BATCHELOR NUMBER (see Liquid metals)

BATCHELOR TYPE FLOW (see Rotating disc systems, applications)

BATCH SEDIMENTATION (see Kinematic waves)

BATCH TYPE FILTERS (see Filters; Filtration)

BATH SMELTING (see Smelting)

BATTELLE MEMORIAL INSTITUTE, BMI

505 Kings Avenue
Columbus OH 43201
USA
Tel: 614 424 6424

BED VOIDAGE (see Fixed beds)

BEER-LAMBERT LAW (see Optical particle characterisation; Radiative heat transfer)

BELL-DELAWARE METHOD (see Shell-and-tube heat exchangers)

BENARD CELLS (see Free convection)

BENDS (FLOW AND PRESSURE DROP IN)

Following from: *Channel flow*

Single-Phase Flow

The main feature of flow through a bend is the presence of a radial pressure gradient created by the centrifugal force acting on the fluid. Because of this, the fluid at the centre of the pipe moves towards the outer side and comes back along the wall towards the inner side. This creates a double spiral flow field shown schematically in **Figure 1**. If the bend curvature is strong enough, the adverse pressure gradient near the outer wall in the bend and near the inner wall just after the bend may lead to flow separation at these points, giving rise to a large increase in pressure losses. Even for fairly large-radius bends, the flow field in the bend will be severely distorted as illustrated by the data of Rowe (1970) shown in **Figure 2**.

The pressure losses suffered in a bend are caused both by friction and momentum exchanges resulting from a change in the direction

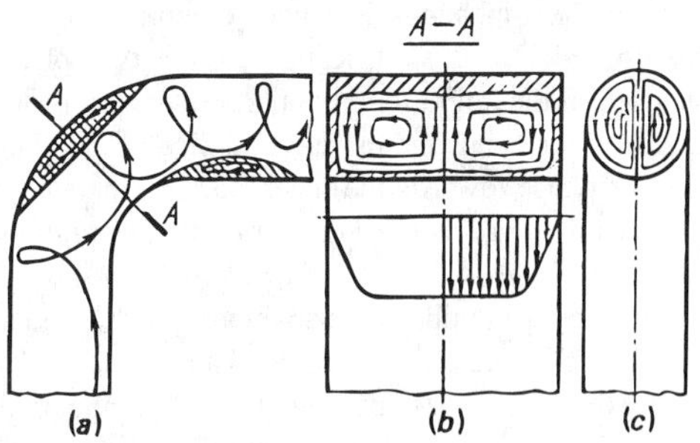

Figure 1. Schematic diagram of a double spiral flow in a bend: a) longitudinal section; b) cross-section (rectangular section); c) cross-section (circular cross-section) (Idelchik, 1986).

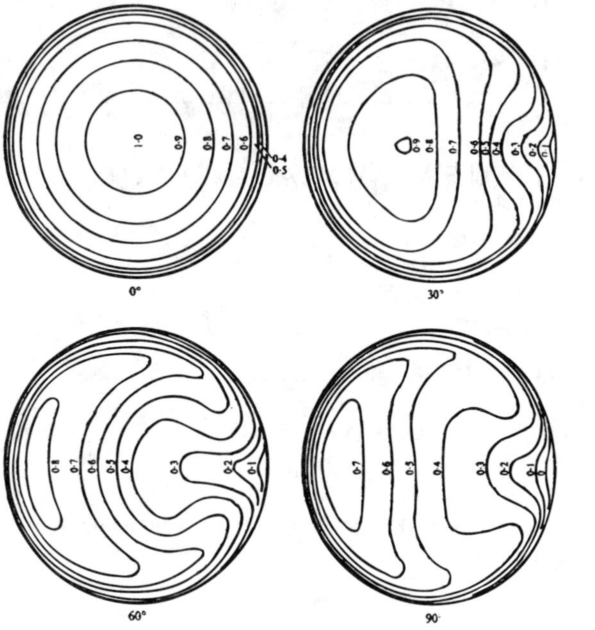

Figure 2. Total pressure contours in a U-bend of a bend-to-pipe diameter ratio of 24; Reynolds number = 236000 (Rowe, 1970).

of flow. Both these factors depend on the bend angle, the curvature ratio and the **Reynolds Number**. The overall *pressure drop* can be expressed as the sum of two components: 1) that resulting from friction in a straight pipe of equivalent length which depends mainly on the Reynolds number (and the pipe roughness); and 2) that resulting from losses due to change of direction, normally expressed in terms of a bend-loss coefficient, which depends mainly on the curvature ratio and the bend angle. The pressure loss in a bend can thus be calculated as:

$$\Delta P = \frac{1}{2} f_s \rho u^2 \frac{\pi R_b}{D} \frac{\theta}{180°} + \frac{1}{2} k_b \rho u^2 \quad (1)$$

where f_s is the Moody friction factor in a straight pipe; ρ, the density; u, the mean flow velocity; R_b the bend radius; D, the tube diameter; θ, the bend angle; and k_b, the bend *loss coefficient* obtained from **Figure 3**. Extensive data on loss coefficient for bends are given by Idelchik (1986).

Two-Phase Flow

Two-phase flow in bends is rendered more complex by the centrifugal-force-induced stratification of the two phases. This manifests itself in the migration of bubbles in bubbly flow towards the inner side of the bend, and more paradoxically, in the migration of the heavier phase towards the inner side in separated flows (a process known as *film inversion*) under certain conditions.

The pressure losses suffered in two-phase flow through bends are influenced by a number of parameters, and no generalised method is available to calculate them accurately. The usual practice is to multiply the single phase pressure losses by a factor known as the *two-phase multiplier*, which is empirically correlated. Chisholm (1980) has examined a number of data sets and proposed the following method:

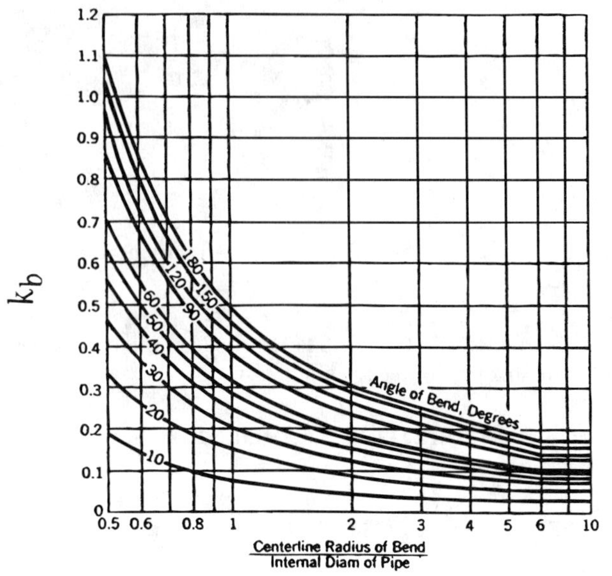

Figure 3. Bend loss coefficients for a pipe (Babcock & Wilcox Co., 1978).

$$\frac{\Delta p_{TP}}{\Delta p_{LO}} = 1 + \left(\frac{\rho_L}{\rho_G} - 1\right) \times \left[\left\{1 + \frac{2.2}{k_{LO}\left(2 + \frac{R_b}{D}\right)}\right\} x(1 - x) + x^2\right] \quad (2)$$

where Δp_{TP} is the pressure drop in two-phase flow, Δp_{LO} is that in a single phase flow of the total mass flux and liquid properties, k_{LO} is the bend loss coefficient for single phase flow, and x is the quality.

References

Babcock and Wilcox Company (1978) *Steam: Its Generation and Use.*

Chisholm, D. (1980) Two-phase flow in bends, *Int. J. Multiphase Flow,* vol 6:363–367.

Idelchik, I. E. (1986) *Handbook of Hydraulic Resistance,* Hemisphere.

Rowe, M. (1970) Measurement and computation of flow in pipe bends, *J. Fluid Mech.,* 43:771–783.

S. Jayanti

BENEDICT-WEBB-RUBIN EQUATION (see PVT relationships)

BENJAMIN BUBBLE (see Slug flow)

BENSON BOILER

A Benson boiler is a type of **Once-Through Boiler** patented by Marc Benson in Germany in 1923 and incorporates recirculation characteristics at part-load operation below about 60% MCR.

The Benson boiler usually consists of small diameter tubes (ca. 25mm bore) spirally wound to form the furnace envelope. Feed water enters the bottom of the furnace at high sub-critical or super-critical pressure and is evaporated to high quality in the spiral section. A balancing header is commonly provided near the top of the furnace to alleviate any differences in steam quality resulting from variations in heat absorption in different parallel spiral circuits before the steam/water mixture is introduced to the open boiler pass. Here, it is superheated in the upper parts of the furnace envelope and subsequently in pendant tube banks. The balancing header also serves as a means of separating excess liquid when, at low loads, the furnace flow rate exceeds the steam demand from the boiler as a whole. As in other forms of **Fossil Fuel-Fired Boilers**, the flue gases are used for reheat, economiser and air-heating duties.

Along with other types of **Once-Through Boiler**, the convective heating surfaces may be mounted either in a vertical up-pass above the furnace (a tower boiler) or in a horizontal and vertical down-pass behind the furnace (a two-pass boiler). Illustrations of both types appear in the article on **Once-Through Boilers**.

A.E. Ruffell

BENT TUBES, HEAT TRANSFER IN (see Tubes, single phase heat transfer in)

BERNOULLI EFFECT (see Magnus force)

BERNOULLI EQUATION

Following from: *Conservation equations, single phase*

If the force-momentum equation is applied to an inviscid, incompressible fluid in steady flow, it may be shown that along any one streamtube:

$$\frac{u^2}{2g} + \frac{P}{\rho g} + z = H = \text{constant} \qquad (1)$$

where u is the velocity, P is the pressure and z is the height above a pre-determined datum. This equation expresses the conservation of mechanical work-energy and is often referred to as the incompressible steady flow energy equation or, more commonly, the Bernoulli equation, or Bernoulli's theorem. All the quantities appearing within this equation have the physical dimensions of length and may be regarded as the energy per unit weight of fluid.

In hydraulic engineering, such quantities are referred to as "heads", and the sum of all such terms as the *total head* H. Bernoulli's theorem expresses the conservation of total head along a given streamtube, and defines the balance between the kinetic energy represented by $u^2/2g$, the potential energy, z, and the flow-work, $P/\rho g$, associated with the pressure forces.

The energy transfers described by Bernoulli's theorem are all reversible (or conservative), and no account is taken of the non-reversible (or non-conservative) changes through which mechanical work-energy is transferred out of the system or converted to internal energy. In particular, the work done against the viscous shear stresses or external forces (other than gravity) are neglected. These omissions provide the greatest restriction to the applicability of Bernoulli's theorem. All real fluids are to some extent viscous, and consequently, there will be some reduction in the total head due to the shear stress. This may be taken into account within Bernoulli's theorem by incorporating a head loss (ΔE), which is either determined experimentally or assessed from previous engineering practice. Tabulated values of typical head losses in a wide variety of pipe flows and open channel flows are given by Idelchik (1986) and Zipparro and Hasen (1993).

In many practical cases, the energy loss (ΔE) will be small in comparison with other terms in Bernoulli's equation. In particular, if the cross-sectional area of the flow reduces (i.e., the flow is convergent and therefore accelerates), the turbulence levels will tend to reduce. In this case, the energy losses are small and the changes in both the pressure and velocity are well predicted by Bernoulli's theorem. If the flow diverges, however, the turbulence levels will increase and significant energy losses may result. In this case, Bernoulli's theorem is inappropriate, and should not be applied.

The conservation of mechanical work-energy may also be applied to unsteady flows. Integration of the *Euler equations* produces the so-called unsteady Bernoulli equation. The most commonly used form of this equation arises when the flow is also irrotational (zero vorticity):

$$-\rho\frac{\partial \Phi}{\partial t} + \frac{1}{2}\rho u^2 + \rho g z + P = \text{constant} \qquad (2)$$

where ϕ represents the velocity potential. In this case, the equation is expressed in terms of the component pressures rather than the energy heads. However, the limitations discussed previously still apply, and the equation should not be applied if there are significant non-conservative energy losses. A full derivation of the Bernoulli's theorem, in both its steady and unsteady form, is given by Duncan et al. (1970).

References

Duncan, W. J., Thom, A. S., and Young, A. D. (1970) *Mechanics of Fluids.* Edward Arnold, London.

Idelchik, I. E. (1986) *Handbook of Hydraulic Resistance.* Hemisphere Publishing Corp., Washington.

Zipparro, V. J. and Hasen, H. (1993) *Davis' Handbook of Applied Hydraulics.* McGraw-Hill Inc., New York.

Leading to: *Dynamic pressure; Velocity head*

C. Swan

BERNOULLI THEOREM (see Bernoulli equation)

BERNOULLI'S INTEGRAL (see Inviscid flow)

BESSEL FUNCTIONS

Following from: *Differential equations*

The Bessel functions of order λ (Cylindrical functions of the first kind) are defined by the following relationships:

$$J_\lambda(x) = \sum_{k=0}^{\infty} \frac{(-1)^k}{\Gamma(k+1)\Gamma(\lambda+k+1)} \left(\frac{x}{2}\right)^{\lambda+2k}. \quad (1)$$

$J_\lambda(x)$ is an analytic function of a complex variable for all values of x (except maybe for the point $x = 0$) and an analytic function of λ for all values of λ. It is represented in the form $x^\lambda f_\lambda(x^2)$, where $f_\lambda(x^2)$ is an integer function.

Bessel functions are the partial solution of the Bessel differential equation:

$$x^2y'' + xy' + (x^2 - \lambda^2)y = 0$$

which appears in problems of heat transfer when solving the **Laplace Equation**, written in cylindrical coordinates by the method of separation of variables.

For Bessel functions, the following recurrent relationships are valid:

$$\frac{2\lambda}{x} J_\lambda(x) = J_{\lambda-1}(x) + J_{\lambda+1}(x)$$

$$2J'_\lambda(x) = J_{\lambda-1}(x) - J_{\lambda+1}(x). \quad (2)$$

The equation $J_\lambda(x) = 0$, for $\lambda > -1$ has infinitely many real roots. All these roots are simple (except for 0, possibly) and are limited by the roots of the function $J_{\lambda+1}(x)$. They are arranged symmetrically about the point 0 and have no finite limit points.

The following asymptotic expression is valid for a Bessel function with large x:

$$J_\lambda(x) = \sqrt{\frac{2}{\pi x}} \cos\left(x - \frac{\pi\lambda}{2} - \frac{\pi}{4}\right) + O(x^{-3/2}). \quad (3)$$

Hence, an approximate formula for the roots $J_\lambda(x)$ is:

$$\mu_k^{(\lambda)} \approx \frac{3\pi}{4} + \frac{\pi\lambda}{2} + k\pi \quad (4)$$

For real positive values of x and λ, a Bessel function is real, with its curve in the form of decaying oscillations.

For an integer $\lambda = n$, the $J_n(x)$ is the coefficient for t^n in expanding its generating function to:

$$e^{x12(t-1/t)} = \sum_{-\infty}^{\infty} J_n(x)t^n. \quad (5)$$

In particular,

$$J_{-n}(x) = (-1)^n J_n(x). \quad (6)$$

For a Bessel function $J_n(x)$, an integral presentation is valid:

$$J_n(x) = \frac{1}{n}\int_0^\pi \cos(n\theta - x\sin\theta)\, d\theta. \quad (7)$$

For even n, Bessel functions are even; for odd n, they are odd.

Bessel functions of order obtained ($\lambda = m + \frac{1}{2}$) are expressed in terms of elementary functions:

$$J_{m+1/2}(x) = (-1)^m \sqrt{\frac{2}{\pi}}\, x^{m+1/2}\left(\dot{x}\frac{d}{dx}\right)^m \frac{\sin x}{x}$$

$$J_{-m-1/2}(x) = \sqrt{\frac{2}{\pi}}\, x^{m+1/2}\left(x\frac{d}{dx}\right)^m \frac{\cos x}{x}. \quad (8)$$

In terms of Bessel functions are expressed.

Other types of cylindrical functions, which are of great importance for solving problems of heat transfer include *Neumann functions* (cylindrical functions of the second kind) and *Hankel functions* (cylindrical functions of the third kind).

V.N. Dvortsov

BESSEL'S SERIES

Following from: *Series expansion*

Bessel's series is the expansion of a function f(x) square integrable and weight integrable in the interval (0, a) into a series in terms of **Bessel Functions** of order λ, $J_\lambda(\mu k_a^x)$, where $k = 1 \div \infty$

$$f(x) = \sum_{k=1}^{\infty} C_k J_k\left(\frac{\mu_k x}{a}\right). \quad (1)$$

Here, μ_k are the positive roots of Bessel's function J_λ arranged in an increasing order ($\lambda > -\frac{1}{2}$). The coefficients of a series have the form:

$$C_k = \frac{1}{[aJ'_k(\mu_k)]^2}\int_0^a xf(x)J_\lambda\left(\frac{\mu_k x}{a}\right) dx. \quad (2)$$

A system of Bessel's functions is orthogonal to weight x and is complete on the interval (0, a). This means that a sequence of partial sums of the series converge on the interval (0, a) in a root-mean-square (with weight x) to the function f(x).

The orthogonality relationships for the Bessel function have the following form:

$$\int_0^a xJ_\lambda\left(\frac{\mu_k x}{a}\right) j_\lambda\left(\frac{\mu_m x}{a}\right) dx = 0 \qquad k \neq m$$

$$\int_0^a xJ_\lambda^2\left(\frac{\mu_k x}{a}\right) dx = \frac{a^2}{2}[J'_k(\mu_k)]^2 \qquad k = m. \quad (3)$$

A Bessel's series for a function f(x) can be obtained by expanding a function $\sqrt{x}\ f(x)$ into a **Fourier series** with respect to an orthogonal system $\sqrt{x}\ J_\lambda(\mu_k x/a)$; $k = 1 \div \infty$, which makes it possible to transfer onto the *Fourier-Bessel series* the results known for the Fourier series from orthogonal functions.

V.N. Dvortsov

BET, BRUNAUER-EMMET-TELLER, EQUATION (see Adsorption)

BETA PARTICLES (see Ionizing radiation)

BEYERLOO CORRELATION (see Granular materials, discharge through orifices)

BIFURCATION (see Non-linear systems)

BILLETS (see Casting of metals)

BIMETALLIC THERMOMETER (see Temperature measurement, practice)

BINARY DIFFUSIVITY (see Maxwell-Stefan equations)

BINARY MASS TRANSFER (see Condensation of multicomponent vapours)

BINARY SYSTEMS, THERMODYNAMICS OF

Following from: *Multicomponent systems, thermodynamics of; Vapor-liquid equilibrium*

A binary system is a particular case of the more general **Multicomponent System** in which only two components are present. Such systems are sometimes called *mixtures,* with the implication that both substances present are to be treated on an equal footing, and sometimes as *solutions,* in which the excess component is called the *solvent* and the other, the *solute.* The thermodynamics, however, contain much in common and for the most part, one need not distinguish between them.

The general characteristic of binary systems is that when, say, two liquids are mixed together at the same pressure and temperature, the extensive properties (volume, enthalpy, entropy, etc.) of the mixture are not the sum of those of the unmixed components. Nevertheless, it is convenient to ascribe part of the volume, say, to component 1 and the rest to component 2. There is no unique way in which this can be done, but one way which is useful to thermodynamics is via partial molar quantities. For example, the partial molar volume of component i is defined as:

$$\tilde{v}_i = (\partial V/\partial n_i)_{P,T,N_j} \tag{1}$$

where n_i is the amount of component i present in the mixture, and the n_j constraint to the partial derivative means that all amounts of components other than i, are to remain fixed. On this basis, the total volume of the (binary) mixture is:

$$V = n_1\tilde{v}_1 + n_2\tilde{v}_2. \tag{2}$$

A similar set of equations exists for all other extensive thermodynamic quantities.

Expressed in molar terms, this last equation becomes:

$$\tilde{v} = \tilde{x}_1\tilde{v}_1 + \tilde{x}_2\tilde{v}_2 \tag{3}$$

where $\tilde{x}$ are the mole fractions. If we differentiate this equation, partially with respect to each of the mole fractions and invoke the **Gibbs-Duhem Equation**, it becomes:

$$\tilde{v}_1 = \tilde{v} - \tilde{x}_2(\partial\tilde{v}/\partial\tilde{x}_2)_{P,T} \tag{4}$$

$$\tilde{v}_2 = \tilde{v} - \tilde{x}_1(\partial\tilde{v}/\partial\tilde{x}_1)_{P,T} \tag{5}$$

providing a means of determining the partial molar quantity from experimentally observable results. A similar set of equations can be obtained for the other extensive thermodynamic properties, but the quantities involved are usually less experimentally accessible.

One partial molar quantity of particular importance is the **Chemical Potential**:

$$\mu_i = (\partial G/\partial n_i)_{P,T,n_j} \tag{6}$$

where G is *Gibbs free energy.*

The thermodynamics of binary systems is intimately linked with the desire to predict the thermodynamic properties of such systems. There exists a hierarchy of methods for making these predictions, ranging from the extremely simple, but limited in scope, to the highly complex, but of wide applicability.

The simplest case is that of the *ideal mixture.* This is defined as one for which:

$$\mu_i(P, T, \tilde{x}) = \mu_i^\circ(P, T) + \tilde{R}T \ln \tilde{x}_i \tag{7}$$

for each component i in the mixture. (μ_i is the chemical potential of component i in the mixture of mole fraction composition $\tilde{x}$, μ_j° is the chemical potential of pure component i). $\tilde{R}$ is the universal gas constant.

It follows from this definition that: the equation for Gibbs free energy, entropy, enthalpy and volume, respectively are:

$$G = \Sigma n_i\mu_i^\circ + \tilde{R}T\ \Sigma n_i \ln \tilde{x}_i \tag{8}$$

$$S = \Sigma n_i\tilde{s}_i^\circ - R\ \Sigma n_i \ln \tilde{x}_i \tag{9}$$

$$H = \Sigma n_i\tilde{h}_i^\circ \tag{10}$$

$$V = \Sigma n_i\tilde{v}_i^\circ \tag{11}$$

where $\tilde{s}$, $\tilde{h}$ and $\tilde{v}$ are the molar entropies, enthalpies and volumes of the pure components i and n_i are the number of moles of each component present in the mixture. Note, that whereas enthalpies and volumes are additive, this is not true for G and S. It can be shown from the above that C_p and C_v, the heat capacity at constant pressure and at constant volume, respectively, are also additive.

In the gas phase, it is often sufficient to work at a low enough pressure, say ambient, in order for the approximation of an ideal mixture to be adequate. However, more stringent conditions apply for the liquid phase. In general, the ideal mixture approximation only applies when the molecular species present are very similar.

When the ideal mixture approximation is inadequate, more precise expressions must be used for the partial molar quantities. See, for example, **Activity Coefficient, Fugacity**.

G. Saville

BINDING ENERGY, ELECTRON (see Ionization)

BINGHAM FLUIDS (see Non-Newtonian fluids)

BINGHAMIAN SLURRIES (see Liquid-solid flow)

BINGHAM PLASTIC (see Non-Newtonian fluids)

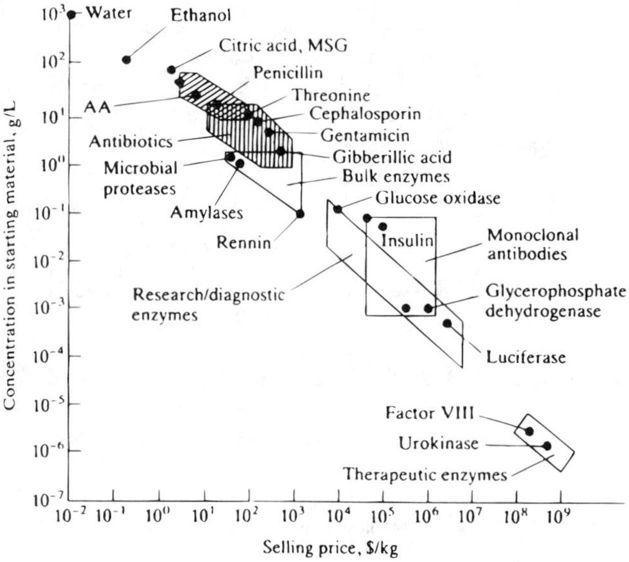

Figure 1. Concentration in the broth versus final selling price. (Reprinted by permission from J. L. Dwyer, "Scaling Up Bioproduct Separation with High Performance Liquid Chromatography," Bio/Technology, vol. 2, p. 957, 1984).

BIOCHEMICAL ENGINEERING

In a new field such as Biochemical Engineering, considerable confusion often arises over the definition of a number of terms. *Biotechnology* is usually taken to mean the use of sophisticated genetic techniques outside the cell to result in the production of useful products. However, many people broaden the term to include applied biology, or even engineering. Strictly speaking, *biochemical engineering* is usually defined as the extension of chemical engineering principles to systems using a biological catalyst to bring about desired chemical transformations. It is often subdivided into reactor design and downstream separation.

Despite the appearance that it is a new discipline, biochemical engineering as such dates back to the turn of the century, when certain rudimentary principles were used in the biological treatment of wastewater and in the production of acetone and butanol for explosives. Furthermore, fermentation ethanol was one of the basic raw materials for the chemical industry until the Second World War. The modern genesis of the discipline dates back to the early 1940s when penicillin was required in vast quantities.

The recent development of biochemical engineering is the direct result of advances in molecular biology (e.g., recombinant DNA, tissue culture, protoplast fusion, monoclonal antibodies) and protein engineering. Biochemical engineers work in a wide cross-section of industry including; pharmaceuticals, food, fine chemicals, wastewater treatment, mining and energy. Hence, biochemical products vary from high volume, low cost to very low volume and extremely high cost (e.g., Factor 8, the blood clotting protein costs around $\$10^8$/kg!). **Figure 1** details this wide variation in cost.

Biochemical engineering involves the use of a catalyst, either an enzyme or whole cell, which is either freely suspended in an aqueous medium or "immobilised" in a gel or attached to a solid surface. In contrast to an enzyme, whose activity decreases over time, cells are autocatalytic ($dx/dt = \mu x$, where x = cell concentration and μ = growth rate, t^{-1}) and can grow exponentially with excess nutrients. Reactor conditions are typically: a pH of 6–8; a temperature of 15–60° C; atmospheric pressure; and a cell concentration of 2–40 g/l. Reactors can be operated in either batch mode (most common in the pharmaceutical industry), typically over 24–48 hours, or continuously (e.g., wastewater treatment). Reactors commonly used vary from CSTRs (continually-stirred tank reactors), through plug flow (whose efficiency is similar to CSTRs due to back mixing), to specialised reactors which immobilise the cells or enzymes to prevent them from being washed out of the reactor due to their small size and density (SG = 1.04). These types of reactors essentially "de-link" the hydraulic retention time from the cell retention time, thereby enabling process intensification to occur.

The downstream separation of biological products is one of the major challenges in biochemical engineering. These products are often present in very low concentrations (100 mg/l), and are extremely labile to mechanical shear and extremes in pH and temperature. The difficulty of the separation process depends very strongly on the product itself and whether it is the biomass, an extracellular product excreted by the cell, or an intracellular protein or inclusion body. Since most fermentation broths are extremely complex mixtures of a variety of organic and inorganic constituents and most products have to meet extremely stringent standards in terms of purity—especially the therapeutics—recovery and concentration can often involve 10–15 unit operations. This can lead to quite low overall recoveries (10–20%); hence, separation can constitute as much as 70% of the overall product cost. The separation techniques used include those based on size, diffusivity, charge, surface activity, density and polarity, and often include techniques rarely seen in the chemical process industry (see **Figure 2**). (See also **Liquid-Solid Separation.**)

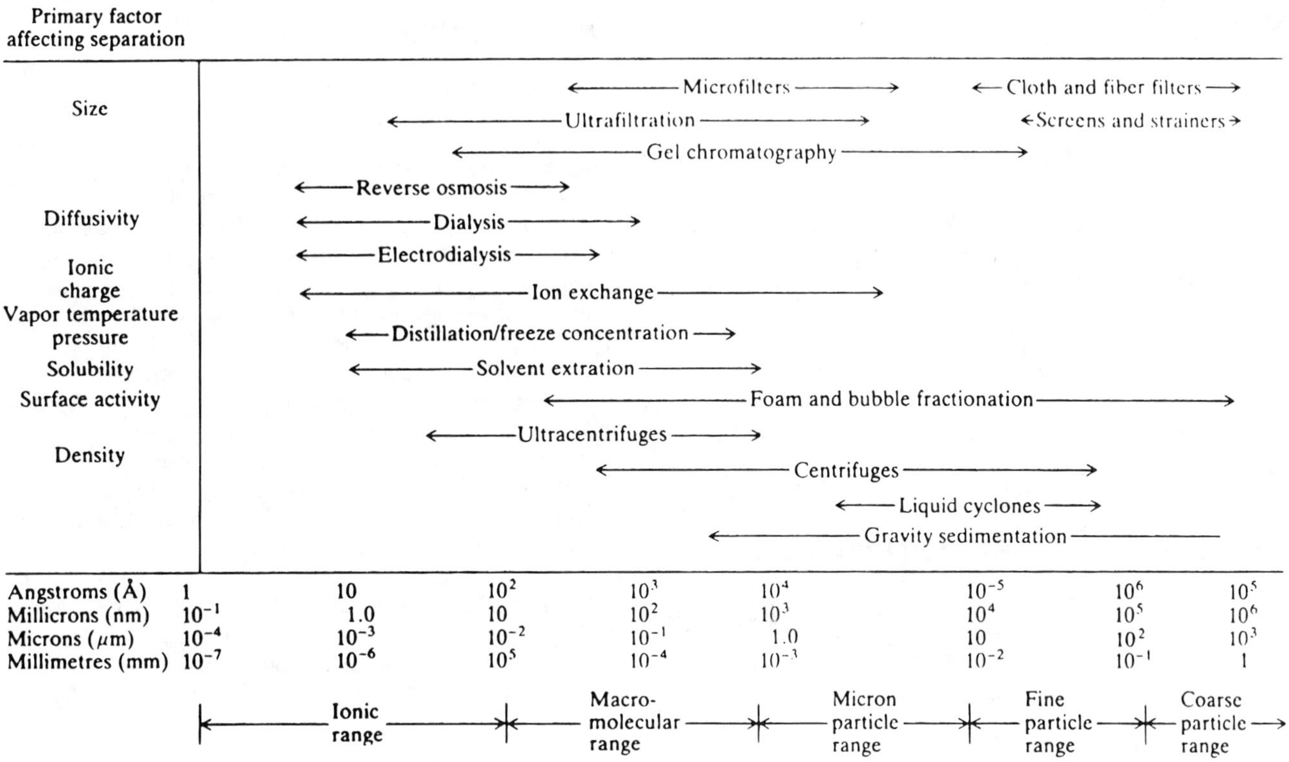

Figure 2. Ranges of applications of various unit operations. (After B. Atkinson and F. Mavituna, Biochemical and Biotechnology Handbook, Macmillan Publ. Ltd., Surrey. England, 1983, p. 935.)

The *modeling* of most biochemical engineering processes is still fraught with difficulties, although simple *enzymatic reactions* can be represented by the theoretically derived *Michaelis-Menten relationship.*

$$\nu = \frac{\nu_{max} c}{K_s + c}$$

where

ν = rate of reaction, t^{-1};
ν_{max} = maximum rate of reaction, t^{-1};
K_m = Michaelis-Menten, or half rate constant, ML^{-3};
c = concentration of substrate, ML^{-3}.

However, the modelling of whole cell reactions is far more complex since cell metabolism consists of many sequential reactions (10^3–10^4), any one of which may be rate limiting at any point in time. In addition, the release of product may be kinetically associated with cell growth ("growth associated"), or occur after growth has virtually ceased ("secondary metabolite"), or be a complex mixture of both. In modelling growth kinetics, it is possible to take two approaches:using an "unstructured" model, which takes an uncritical "black box" stance; or developing a "structured" model, which is based more on the basics of cell metabolism. Due to it's simplicity and relative success, the most commonly-used cell growth model is the unstructured one developed empirically by Monod:

$$\mu = \frac{\mu_{max} c}{K_s + c}$$

where

μ = rate of cell growth, t^{-1};
μ_{max} = maximum rate of cell growth, t^{-1};
K_s = half rate constant, ML^{-3};
c = concentration of rate limiting substrate, ML^{-3}.

While empirically derived, the equation has obvious similarities to the Michaelis-Menten equation for enzyme kinetics.

References

Bailey, J. E. and Ollis, D. F. (1986) *Biochemical Engineering Fundamentals,* 2nd edition. McGraw-Hill.

Atkinson, B. and Mavituna, F. (1991) *Biochemical Engineering and Biotechnology Handbook,* 2nd Edition. Macmillan Publishers.

Kennedy, J. F. and Cabral, J. M. S. (1993) *Recovery Processes for Biological Materials.* John Wiley and Sons.

Moo-Young, M. (1985) *Comprehensive Biotechnology.* Pergamon Press.

D. Stuckey

BIO-ENERGY (see Alternative energy sources)

BIOHEAT EQUATION (see Body human, heat transfer in)

BIOLOGICAL FOULING (see Fouling)

BIOFUELS (see Renewable energy)

BIOMASS (see Alternative energy sources; Pyrolysis; Renewable energy)

BIOT AND SAVART'S LAW (see Biot, J.B.)

BIOTECHNOLOGY (see Biochemical engineering)

BIOT, JEAN BAPTISTE V. (1774–1862)

French physicist J. B. Biot best known for his work in polarization of light, was born in Paris on April 21, 1774. In 1800, he became Professor of Physics at the College de France through the influence of Laplace, from whom he had sought and obtained the favour of reading the proof sheets of the *"Mecanique Celeste"*.

JEAN BAPTISTE BIOT 1774–1862

J. B. Biot, although younger than Fourier, worked on the analysis of heat conduction even earlier—in 1802 or 1803. He attempted, unsuccessfully, to deal with the problem of incorporating external convection effects in heat conduction analysis in 1804. Fourier read Biot's work and by 1807 had determined how to solve the problem.

In 1804, he accompanied Gay Lussac on the first balloon ascent undertaken for scientific purposes. In 1820, with Felix Savart, he discovered the law known as *"Biot and Savart's Law"*. He was especially interested in questions relating to the polarization of light and for his achievements in this field, he was awarded the Rumford Medal of the Royal Society in 1840. He died in Paris on Feb. 3, 1862.

U. Grigull, H. Sandner, and J. Straub

BIOT NUMBER

Following from: *Conduction*

The Biot Number is a dimensionless group named after J.B. Biot who, in 1804, analysed the interaction between conduction in a solid and convection at its surface. The numerical value of the Biot Number (Bi) is a criterion which gives a direct indication of the relative importance of conduction and convection in determining the temperature history of a body being heated or cooled by convection at its surface. Bi should always be enumerated at the outset to identify transient conduction problems which may be treated simply as lumped parameter problems, for which $Bi < 0.1$ and for which it is seldom necessary to solve the conduction equation, i.e., convection is the rate controlling process.

Figure 1 may be used to impute physical meaning to Bi, and shows the limiting steady-state temperature distributions in two plates [(a) high λ and (b) low λ] which have cooled from an initial uniform temperature T_1 as a result of exposure to a cooling flow at T_2 along one face.

For both plates, a heat balance at the cooled surface can be constructed by using **Fourier's Law** for the conduction flux in the solid at the surface and **Newton's Law of Cooling** for the convective loss at the surface:

$$\frac{\lambda}{L}(T_1 - T_s) = \alpha(T_s - T_2). \qquad \textbf{(1)}$$

This heat balance can be rearranged to produce the dimensionless Bi:

$$\frac{(T_1 - T_s)}{(T_s - T_2)} = \frac{L/\lambda}{1/\alpha} = \frac{R_{cond}}{R_{conv}} = \frac{\alpha L}{\lambda} = Bi, \qquad \textbf{(2)}$$

from which it can be seen that Bi may be considered to be the ratio of the resistance to heat transfer presented by the conduction R_{cond} and the convection R_{conv} processes.

Figure 1a exemplifies the temperature distribution in a low Bi system (e.g., a steel plate (λ = 35 W/mK) of 5 cm thickness cooling in air [(α = 10 W/m^{2}K) giving Bi = 0.0023)] where, because the resistance to heat flow within the solid is small relative to the resistance presented by the convection processes at the surface, the internal temperature distribution in the solid is relatively uniform. Generally, in bodies of simple geometry, e.g., plates,

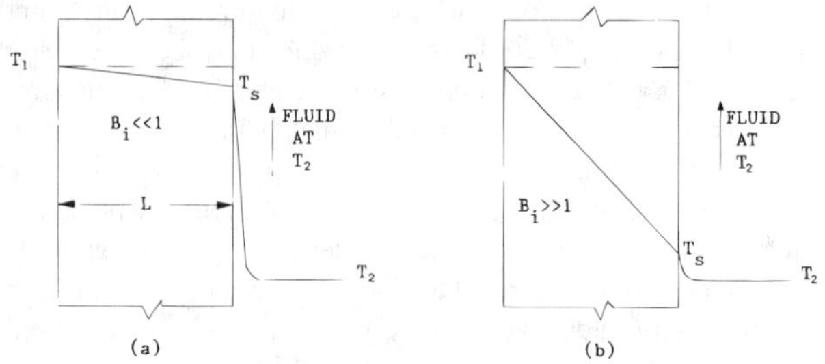

Figure 1. Comparison of fully developed temperature profiles in two plates cooled by the same fluid.

cylinders, spheres, the error introduced by the assumption of uniform body temperature will be less than 5% when the internal resistance is less than 10% of the external resistance, i.e., when the Bi < 0.1.

Leading to: *Transient conduction*

T.W. Davies

BITUMEN (see Oils)

BLACK BODY RADIATION (see Stefan, J.; Radiative heat transfer)

BLACK BODY SPECTRAL INTENSITY (see Emissivity)

BLADE COOLING (see Transpiration cooling)

BLADE COOLING, IN GAS TURBINES (see Gas turbine)

BLASIUS EQUATION (see Flow of fluids; Friction factors for single phase flow; Turbulence)

BLASIUS EQUATION, FOR FRICTION FACTOR (see Tubes, single phase flow in)

BLASIUS EQUATION, MODIFIED (see Triangular ducts, flow and heat transfer)

BLASIUS SOLUTION, FOR BOUNDARY LAYERS (see Boundary layer; Boundary layer heat transfer)

BLENDING (see Mixing)

BLOOD FLOW MEASUREMENT (see Electromagnetic flowmeters)

BLOWDOWN

Following from: *Nuclear reactors*

Nuclear reactor coolant circuits normally operate at pressures well above atmospheric. Blowdown is the depressurisation following the opening of a breach in a circuit, either by accident or by design.

Blowdown in Pressurised Water Reactor (PWR)

This arises as part of an important class of safety problems known as *LOCAs (Loss-Of-Coolant Accidents).* The PWR primary circuit, shown diagramatically in **Figure 1**, operates in single phase at a pressure of about 15.5 MN/m^2 and a maximum temperature of $\simeq 320°C$. Upon a rupture of the circuit, the important thermal hydraulic considerations are the timescale of the depressurisation, the amount of coolant lost from the circuit and the magnitude of the heat transfer in the reactor core. The effect of the forces generated on the reactor structures is also important, but this is not discussed further here.

For safety purposes, the maximum breach size deemed to be credible is the double-ended offset shear of a hot or cold leg (DECLB). In a modern 4-loop design, this gives a discharge area of about $2 \times 0.75\ m^2$ in a circuit containing some 250,000 Kg of coolant. Following such a breach, the pressure falls very quickly (<1 sec) to about 11 MN/m^2, the saturation pressure corresponding to the highest temperature points in the circuit (core outlet and hot legs). After this point, two-phase conditions appear and the rate of depressurisation is determined by two-phase critical flow at the breach. After about 25 seconds, the pressure has fallen to the containment equalisation value of 0.2–0.3 MN/m^2 and the blowdown phase is over. Smaller breach sizes can occur almost anywhere in the circuit and are much more probable than the DECLB. In safety studies, breaches leading to blowdown taking hundreds to tens of thousands of seconds are considered. For a given size of smaller breach, the location of the breach becomes important. Breaches at the bottom of a component tend to discharge liquid and give rise to relatively slow rates of depressurisation. Breaches at the top tend to discharge steam and give rise to relatively higher rates of depressurisation, but less loss of coolant.

Depressurisation of the hot coolant to near-atmospheric conditions leaves some 70% of the coolant in the liquid phase, sufficient to keep the reactor core completely covered if all the liquid remains in the circuit. In the worst case, however, depressurisation occurs so rapidly that flashing occurs throughout the circuit, and the high steam flows thus generated carry much of the liquid around the circuit and out of the breach. At the end of blowdown, very little liquid remains in the circuit and the reactor core is completely uncovered. It is the function of the *Emergency Core Cooling System (ECCS)* to refill the circuit and re-cover the core. One component of the ECCS consists of large tanks, known as accumulators, which contain water at a pressure of about 4MN/m^2, and which automatically inject into the cold legs when the circuit pressure falls below this value. This injection thus begins before the blowdown phase is over, and some of the injected water is carried out of the breach by the blowdown flows in the phenomenon known as Downcomer Bypass.

In the DECLB, a large reverse flow is very quickly established at the vessel inlet nozzle, causing the whole core flow to reverse within a fraction of a second. Within a second, the core is voided and the fuel dries out and heats up rapidly. The initial rise in clad temperature, see **Figure 2,** is due to radial temperature equalization across the fuel and clad, which has a time-constant of a few seconds. In the absence of any significant heat transfer, the clad temperature would continue to rise by a few tens of degrees per second due to the decay heat even though the fission reactions in the core stop almost instantaneously in this accident. The temperatures would thus reach the 1200°C limit, beyond which a runaway oxidation would occur, before the end of the blowdown period. In fact, analysis shows that the core flows—two-phase and single-phase steam—during blowdown give rise to heat transfer which is sufficient to halt the rise in clad temperature well before the 1200°C

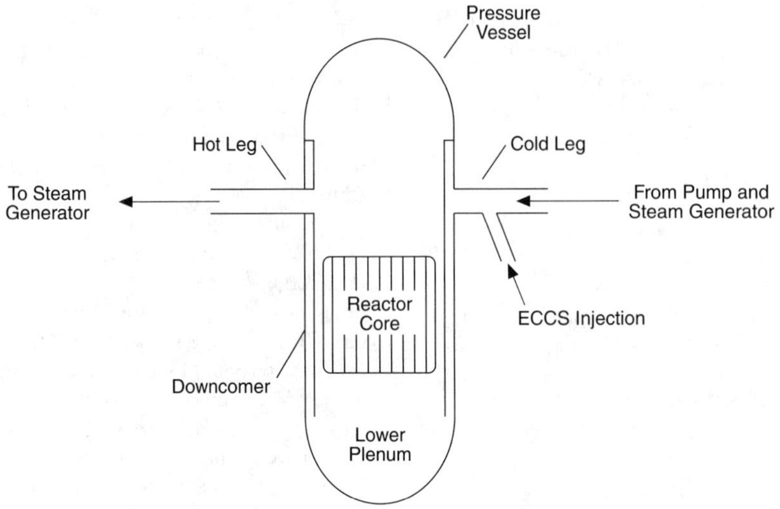

Figure 1. Simplified circuit diagram for PWR.

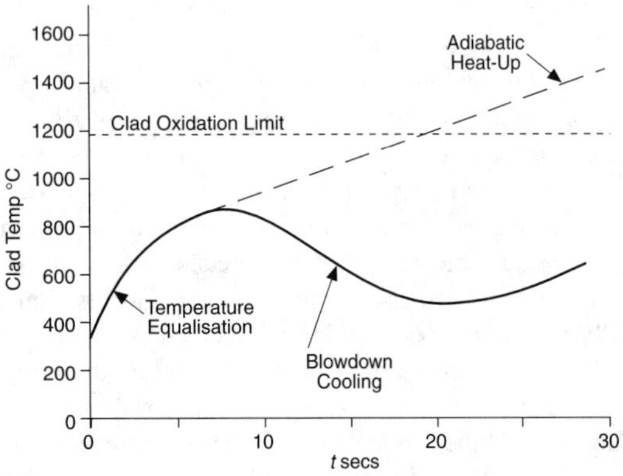

Figure 2. Clad temperature variation following a DECLB.

limit is reached. Over the first few seconds of the transient, the breach flow rapidly reduces from its initial peak as the pressure falls and conditions become two-phase. However, the circulating pumps in the intact legs continue to rotate under inertia even if the electrical supplies have been lost. The result is that a net forward flow into the vessel is re-established, and some liquid flows back into the bottom of the core. This leads to partial rewetting of that region and some heat removal throughout the core. Later in the transient, the core flow again reverses, and some residual liquid may drain back from the steam generators and hot legs to cause partial quenching at the top of the core. Even in the high power regions which stay in dryout, some 50% of the heat is removed by blowdown cooling, and temperatures at the end of blowdown are well below danger levels.

At the end of blowdown in this worst case senario, the core is in dryout and heating up due to decay heat. The pressure vessel may contain very little water, but the ECCS will be injecting liquid into the intact cold legs. This will take the transient into the Refill and **Reflood** (q.v.) phases.

Reference

Collier, J. G. and Hewitt, G. F. (1989) *Introduction to nuclear power.* Hemisphere Publishing Corporation.

Leading to: *Reflood*

Ian Brittain

BLOWERS (see Aerodynamics)

BLUFF BODIES, FLOW AROUND AND DRAG (see Immersed bodies, flow around and drag)

BLUFF BODIES, FLOW OVER (see Crossflow)

BLUFF BODIES, CONVECTIVE HEAT AND MASS TRANSFER (see Immersed bodies, heat transfer and mass transfer)

BMFT (see Bundesministerium für Forschung und Technologie)

BMI (see Battelle Memorial Institute, USA)

BNL (see Brookhaven National Laboratory)

BODY FORCES (see Conservation equations, single phase)

BODY (HUMAN) HEAT TRANSFER

Following from: *Physiology and heat transfer*

Heat is continuously generated in the human body by metabolic processes and exchanged with the environment and among internal organs by conduction, convection, evaporation and radiation. Transport of heat by the circulatory system makes heat transfer in the body—or bioheat transfer—a specific branch of this general science. As in all entities, the principle of conservation of energy yields:

$$\dot{Q}_{generated} = \dot{Q}_{stored} + \dot{Q}_{lost} + \dot{W} \quad \textbf{(1)}$$

where the terms denote, from left to right, the rate of heat generation due to *metabolic processes;* rate of heat stored in body tissues and fluids; heat lost to the environment and adjacent tissues; and the rate of work performed by the tissue. This latter quantity is usually negligible at the tissue level.

In the tissue element shown in **Figure 1**, heat due to metabolic processes (5–10,000 W/m^3) is generated at a variable rate, $\dot{q}_m(\bar{x},t)$ which, when integrated over the entire control volume obtains:

$$\dot{Q}_{generated} = \int_V \dot{q}_m(\bar{x}, t)dv \quad \textbf{(2)}$$

where $\bar{x}$ denotes the spatial coordinate and t is time.

Under unsteady conditions, part of the heat flow will be stored in the control volume:

$$\dot{Q}_{stored} = \int_V \rho c(\bar{x}) \frac{\partial T(\bar{x}, t)}{\partial t} dv \quad \textbf{(3)}$$

where ρ is tissue density (900–1060 kg/m^3); c is tissue specific heat (2.1–3.8 kJ/kg K); and T is tissue temperature.

The term representing heat lost to adjacent tissues and to the environment contains a number of components, one of which is heat exchanged by diffusion (**Fourier's Law** of conduction):

$$\dot{Q}_{conducted} = -\int_A k\nabla T(\bar{x}, t) \cdot \hat{n} dA \quad \textbf{(4)}$$

where k is the thermal conductivity of the tissue (0.29–1.06 W/m K); $\nabla T(\bar{x},t)$ is tissue temperature gradient; $\hat{n}$ is outward-pointing unit vector; and A is control volume surface area.

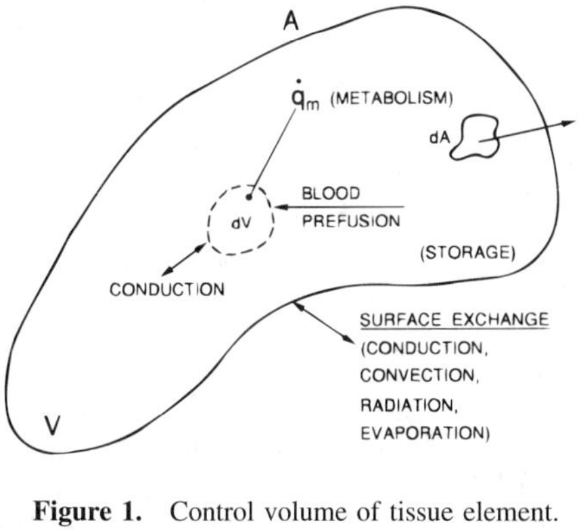

Figure 1. Control volume of tissue element.

A second component of the heat lost to adjacent tissues is due to blood perfusion. Blood circulates in a variety of vessels ranging in lumen diameter from the 2.5 cm aorta to the 6–10 μm capillaries. Due to this four-fold size distribution, heat transport effects of blood are coupled to the specific group of vessels under consideration. A common approach to modeling this effect is to assume that the rate of heat taken up by the circulating blood at the capillary level equals the difference between the venous and arterial temperatures times the flow rate (**Fick's Law**):

$$\dot{q}_{blood} = \rho_b c_b \dot{w}_b (T_a - T_v) \quad \textbf{(5)}$$

where $\rho_b c_b$ is blood heat capacity (~4000 kJ/m^3 K) and $\dot{w}_b$ is volumetric blood perfusion rate (0.17–50 kg/m^3 s). At the capillary level, blood flow velocity is very slow and thermal equilibration with surrounding tissue occurs. Thus, Equation 5 may be modified by setting $T_v = T$ (Pennes, 1948). When all the terms are substituted into Equation 1 and integrated over the entire volume and surface area, the well known *bioheat equation* is obtained:

$$\rho c \frac{\partial T}{\partial t} = \nabla \cdot k\nabla T + \rho_b c_b \dot{w}_b (T_a - T) + \dot{q}_m. \quad \textbf{(6)}$$

Equation 6 has been very useful in the analysis of heat transfer in various body organs and tissues characterized by a dense capillary bed. Other thermal effects due to blood flow are not adequately accounted for by Equation 6, including: 1) counter-current heat transfer between adjacent vessels; 2) directionality effects due to the presence of larger blood vessels; and 3) heat exchange with larger vessels in which complete thermal equilibrium may not be assumed. These issues have been analyzed by Chen and Holmes (1980) and by Weinbaum, et al. (1984).

Heat is exchanged with the environment through a complex combination of conduction, convection, radiation and evaporation. Clothing worn by humans and natural integuments also play a role. As a good approximation, these effects may be calculated by an equation of the form:

$$\dot{Q}_i = h_i A_D (\bar{T}_s - T_o) \quad \textbf{(7)}$$

where $\dot{Q}_i$ is the amount of heat exchanged and h_i is the heat exchange coefficient (2.3–2.7 W/m^2 K for free convection, $7.4V^{0.67}$ for forced convection, where V is wind velocity, m/s, and 3.8–5.1 for radiation); $\bar{T}_s$ is average body surface temperature; T_o is environmental temperature; and A_D is *Dubois' body surface* given by:

$$A_D = 0.202m^{0.425}h^{0.725} \quad \textbf{(8)}$$

where m is body mass in kg and h is height in m.

References

Chen, M. M. and Holmes, K. R. (1980) Microvasculature Contributions in Tissue Heat Transfer, *In: Thermal Characteristics of Tumors: Applica-*

tions in Detection and Treatment, R. K. Jain and P. M. Gullino Eds., Ann. N.Y. Acad. Sci., 335:137–150.

Pennes, H. H. (1948) Analysis of Tissue and Arterial Blood Temperatures in the Resting Human Forearm, *J. Appl. Physiol.,* 1:93–122.

Shitzer, A. and Eberhart, R. C., Eds. (1985) *Heat transfer in medicine and biology—analysis and applications,* Plenum Press, New York.

Weinbaum, S., Jiji, L. M., and Lemons, D. (1984) Theory and Experiment for the Effect of Vascular Microstructure on Surface Heat Transfer, *ASME J. Biomech. Eng.,* 106:321–330 (Pt. 1); 331–341 (Pt. 2).

***Leading to:** Physiology and heat transfer*

A. Shitzer

BOILERS

Boilers are equipment in which a fluid, normally water, is heated and usually evaporated and superheated to produce steam for power generation and/or heating purposes. The energy required is transferred from a heating fluid by combinations of radiation, convection and conduction by one of the following means: either the heating fluid is passed through tubes mounted within a drum which contains the water (a *shell boiler* or *fire-tube boiler*) or the water is contained in arrangements of tubes over which the heating fluid is constrained to flow (a **Water-Tube Boiler**). The heating fluid may be a product of combustion (as in **Fossil Fuel-Fired Boilers**), a chemical reaction or have been used to cool plant like a nuclear reactor or equivalent device, such as a chemical reaction vessel (as in **Heat Recovery Boilers**). **Heat Recovery Boilers** are also commonly found in combination with gas turbines where they are used to extract valuable energy from the exhaust gas.

An important class of equipment in which the heated fluid is not usually water is the **Reboiler**, which is generally found in chemical plants in association with distillation columns.

The classification of boilers is often based on their application, viz. Industrial (for steam generation, heat and/or power where the upper limit is usually considered to be about 150 Te/hr or 50 MW(e)), or utility (for electricity generation where the size is usally in excess of 200 MW(e)).

Boilers may also be sub-divided according to the type of fuel being burned or the type of firing system used. In furnace-fired boilers, fuel—either coal, oil or gas—is fired through a number of burners mounted in the walls of the furnace, whereas solid fuels can also be fired on grates or stokers at the base of the furnace, or they may be burned in fluidised beds. Similarly, waste materials may be burned on grates, in rotating drums or fluidised beds.

Boilers are also classified as *recirculation boilers* or **Once-Through Boilers**, depending on whether the heated fluid is partially evaporated or completely evaporated and superheated during a single pass through the unit.

Applications of boilers range from domestic heating units with capacities of a few kilowatts to utility boilers for power generation, which have capacities of up to 3000 megawatts of heat.

A.E. Ruffell

BOILING

Boiling is the process of turning a liquid at its saturation temperature into vapour by applying heat. A similar process occurs when the pressure applied to a hot liquid is reduced: this would normally be called flashing. Production of cavities in a cold liquid by the application of negative pressures (tensions) is called cavitation.

In the popular imagination, boiling is associated with the rapid formation of vapour bubbles in a rather chaotic manner. This is normally correct, though boiling can also occur by evaporation at large liquid-vapour interfaces without the appearance of vapour bubbles: in particular, in high velocity annular flows where there is only a thin liquid film on the heated surface. In both cases, there is liquid in contact with the surface and the heat transfer is good. When the temperature of the heated surface is very high, the characteristics of the boiling process change. It is no longer possible for the liquid to come into good contact with the surface. There is a film of vapour separating solid and liquid, preventing good heat transfer. If there is plenty of liquid present, this vapour film may be quite smooth and orderly.

Boiling is a widespread phenomenon in industry. The reasons for using the process vary. In a power station using steam turbines, the vapour itself is the desired product (for a description of boilers as equipment see **Boilers**). When cooling electronic components, it is the good heat transfer characteristics of boiling that are important. On the other hand, in the analysis of hypothetical loss of coolant accidents in water-cooled nuclear power stations, it is the rather poorer heat transfer that can occur at high surface temperatures that is of interest.

It is usual to base more detailed discussion of the types of boiling on a *boiling curve* such as that shown in **Figure 1**. The boiling curve is a graph of heat flux $\dot{q}$ versus wall superheat ΔT_{SAT}, the difference between the wall temperature and the saturation temperature (or boiling point). The curve is often drawn with log scales to accommodate the rather large range of variables. The general shape of the curve remains much the same for a variety of boiling situations. The actual values of $\dot{q}$ and ΔT_{SAT} will depend on other details, in particular, what fluid is being boiled and what the flow rate is. For the moment, Pool Boiling is assumed, i.e., there is no pump causing large-scale convection of the liquid.

Before discussing the boiling curve in any detail, two observations must be made. Firstly, it is not easy to do an experiment to

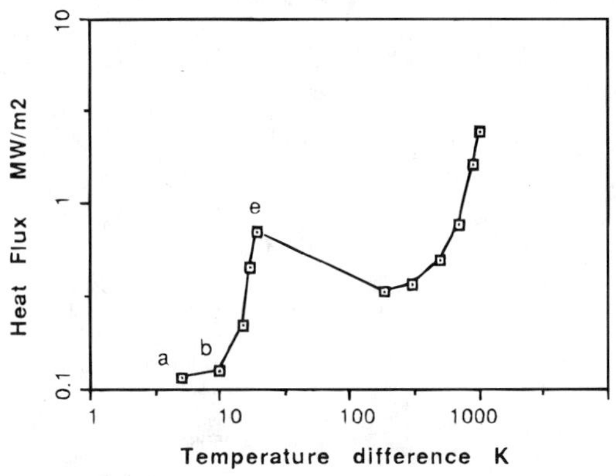

Figure 1. Boiling curve for a heated, 0.14 mm diameter, platinum wire in water at atmospheric pressure (Nukiyama 1934).

get results resembling **Figure 1**; if one attempts a simple experiment with an electrical heater immersed in a pool of liquid and varies the power supplied to the heater, one might obtain a curve that looks markedly different. Nukiyama did not obtain any results in the region of negative slope in **Figure 1**, but he argued that such a region must exist. Prior to his work, there was disagreement as to whether there was a maximum on the curve—very high heater powers are needed to reach it. The second observation is that, for many practical purposes, the entire curve to the right of Point **e** in the figure can be ignored. The reason for this is that the designer is normally interested in good heat transfer, i.e., high heat removal rates at low temperature differences. This can be achieved using the portion of the curve to the left of Point **e**.

Consider now the boiling curve in more detail. Starting on the left at a low temperature difference, in the region a–b, there is no boiling taking place; heat transfer is by convection. It might seem odd that there is no boiling when the wall superheat ΔT_{SAT} is positive, i.e., the temperature of the wall and of the liquid in contact with the wall are both above the saturation temperature. However, it turns out that there is a problem of nucleating the phase change, connected with the existence of surface tension. Suppose that somehow a small spherical vapour bubble had appeared in the liquid.

It is a standard result that mechanical equilibrium of the *vapour bubble* requires:

$$p_G - p = 2\sigma/r \qquad \textbf{(1)}$$

where p_G is the vapour pressure, p the external pressure, σ the surface tension and r the radius of the bubble. But the definition of saturation temperature is that $p_G = p$ (i.e., the requirement for *thermal* equilibrium is $p_G = p$). Even with p_G slightly greater than p, the vapour region cannot grow. In the initial stages of bubble formation r will be small and the $2\sigma/r$ term very large.

In practice, two conditions have to be met for nucleation. Firstly, p_G must be greater than p, and secondly, a pre-existing vapour region is needed to provide a *nucleation site,* such that Equation (1) can be satisfied. It is generally considered that these nucleation sites are *cavities* or cracks in the solid surface which have not been completely filled with liquid. They are too small to be seen by the naked eye and anyway, in boiling, things happen too quickly to be easy to follow. (However, the process of dissolved gas coming out of solution is in principle similar, but occurs more slowly. The stream of bubbles of gas, that seem to all come from the same point on the inside of a glass of beer or lemonade, have been nucleated at a surface cavity.)

The reason that some surface cavities always contain a vapour embryo is also connected with surface tension. The analogy between the two cases in **Figure 3** is not particularly close. The window sill design is clearly strongly influenced by gravity—one

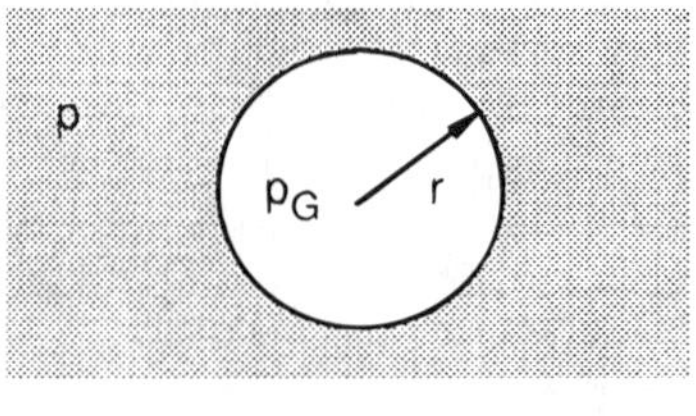

Figure 2. Bubble in a liquid.

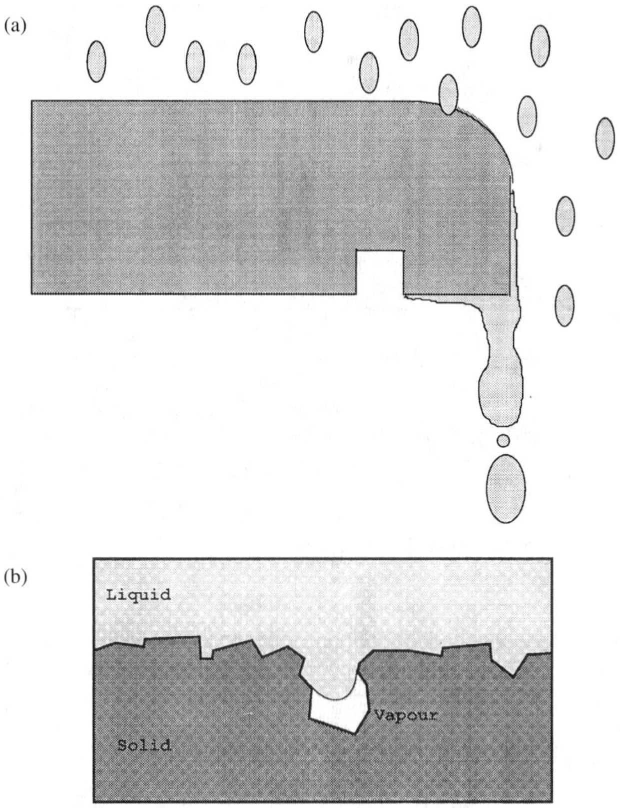

Figure 3. a) Macroscopic. The cavity prevents rain running under the windowsill. b) Microscopic. Surface tension opposes liquid entering cavity.

would not install it upside down. Active surface cavities have dimensions of around 1 μm; gravity is now unimportant compared with surface tension. Surface tension will oppose liquid entry, in the intervals between bubble nucleation, provided the curvature of the meniscus is as shown, i.e., concave on the liquid side. For further information the article on **Nucleation** should be consulted.

Returning to the boiling curve, e.g., **Figure 4**, the significance of Point **b** can now be understood. The liquid superheat, ΔT_{SAT} is high enough to initiate boiling at the first surface cavities, i.e., the ones with the largest radius **r.** As the temperature or power are raised further, the number of active sites increases until at Point **c** boiling is in evidence all over the surface. The region c–d is called the **nucleate boiling** region. Initially, the role of the individual nucleation sites is fairly clear. At higher power levels, the rate of generation of vapour bubbles is such that they coalesce to form vapour columns streaming away from the surface. Nucleate boiling is the type of boiling familiar in the kitchen; for example, on the element of an electric kettle.

By Point **d (Figure 4)**, the rate of vapour generation is such that it is starting to impede the flow of fresh liquid onto the surface. At Point **e** a crisis is reached. In pool boiling, this crisis is interpreted as a maximum possible flow rate in the counter-flow of vapour away from the surface and liquid towards the surface. This is the *critical heat flux.* If the power to the heater is increased further then, in the steady-state, the only point on the boiling curve corresponding to this higher power level is at g and the wall temperature will jump to a much higher value. In pool boiling, and with normal

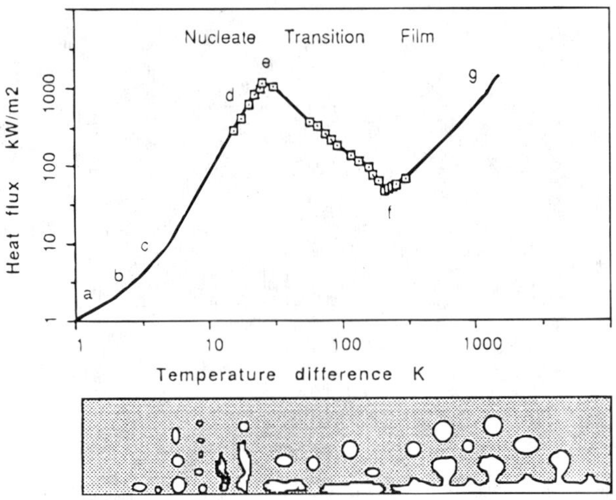

Figure 4. Pool boiling of water on a horizontal plate at 1 atm. (data points from Abbassi et al. 1989).

heater construction, this jump in temperature may well be several hundred degrees and sufficient to destroy the heater. For this reason, the critical heat flux is often referred to as **Burnout**. Alternative names are *departure from nucleate boiling, DNB* and *dryout.* An ability to predict the critical heat flux is clearly very important.

Rather than considering region e–f, it seems appropriate to consider region f–g next. As already mentioned, the full curve of **Figure 4** is not easy to obtain experimentally and in a power (i.e., heat flux) controlled experiment, an increase in heat flux above the critical value lead to Point g. This is in the *film boiling* region. The heated surface is now covered with a film of vapour, which impedes the heat transfer. Once film boiling is established, the entire film boiling region can be covered by increasing or reducing the power supplied to the heater. Point **f** is known as the *minimum film boiling point.*

The region e–f is known as *transition boiling* (sometimes called unstable film boiling). It has the very unusual characteristic of a negative slope, i.e., as the driving temperature difference increases, the heat transfer rate worsens, contrary to all experience of single-phase heat transfer. The physical explanation is that as the surface temperature increases, the situation changes rapidly from near-complete liquid contact with the surface in nucleate boiling to near-complete vapour contact in film boiling. Liquid contact leads to good heat transfer and vapour contact to poor heat transfer. Measurements of the extent of liquid-solid contact have been made by, amongst others, Alem Rajabi et al. (1988). They have found, in pool boiling of methanol, that the degree of liquid-solid contact decreases rapidly from 65% at the critical heat flux to less than 1% at minimum film boiling point.

The instability of transition boiling in a heat flux-controlled experiment follows directly from the negative slope. The system cannot recover from small, random, excursions in temperature. Suppose that a fluctuation caused the surface temperature to increase slightly. The heat flux removed by the boiling process reduces. Since the heat supplied continues at the same level, there is a net input of heat to the surface and the surface temperature rises, i.e., the initial small fluctuation is reinforced. The process continues until the point on the film boiling curve corresponding to the given heat flux is reached. Equally, a small negative fluctuation in temperature would have taken the boiling process all the way to the nucleate boiling curve.

How then can be transition boiling part of the curve be studied? Either by controlling the temperature of the surface instead of the heat flux, or by doing transient experiments. If the temperature is controlled—either by using a fluid at a given temperature as the heat source or by using an electrical heater with a temperature controller—then any temperature can be set and the boiling curve is single-valued—there is only one heat flux corresponding to any set temperature. Alternatively, in a transient experiment, the thermal capacity of the heater means that there cannot be instantaneous jumps in the wall temperature. All wall temperatures and all points of the boiling curve must be passed through during the transient. Transient cooling assumes some importance in *quenching,* when a hot solid object is immersed in a bath of cold liquid, and in the related phenomenon of **Rewetting** i.e., the point during cooling when transition boiling is reached, liquid starts to contact the surface again, and heat transfer starts to improve.

Mention has already been made of the difference between a surface with a constant heat flux and one with a constant temperature. It is worth stressing the difference this makes when considering the critical heat flux. With the constant heat flux boundary condition, arising, for example, from electrical heating or from nuclear fission in a fuel rod in a nuclear reactor, exceeding the critical heat flux will lead to a large and possibly dangerous increase in temperature, i.e., the possibility of burnout. Consider now the steam generator in a sodium-cooled fast nuclear reactor (where the boundary condition on the boiler tube wall approximates the constant temperature case). The steam generator is supplied with liquid sodium at, say, 500 °C. This temperature might be sufficient to place the operating point on the appropriate boiling curve for water slightly to the right of Point **e**; in other words, the temperature corresponding to the critical heat flux has been exceeded. Operation in this mode, in transition boiling, might be considered perfectly acceptable. There is no possibility of the wall temperature exceeding that of the heat source, 500 °C. (The above examples relate to flow boiling but the principle is unchanged).

The main features of the pool boiling curve are now considered in a more quantitative way. Large numbers of correlations and prediction methods have been suggested over the years. Here, just a few simple equations are given. In the case of nucleate boiling, all the heat transfer mechanisms increase in effect as the number of bubbles increases, i.e, as the number of active nucleation sites increases. This leads to important practical problem in predicting the heat transfer, since predicting the number of nucleating cavities is difficult. One approach is to experimentally determine a surface constant appropriate to different heater materials. Another is to relate this effect to surface roughness. In particular, because of the steepness of the nucleate boiling curve, any prediction of heat flux based on known temperatures is liable to show large errors.

Many studies of nucleate pool boiling are consistent with the heat flux $\dot{q}$ being proportional to ΔT_{SAT}^3, although in individual measurements the exponent can vary considerably from 3. Cooper (1984) analysed data from over a hundred experiments and concluded that the accuracy of the data was insufficient to justify the complexity of many of the equations that had previously been proposed for predicting the heat transfer. Arguing that the physical properties of the boiling liquid that appear in the earlier correlations are themselves functions of parameters like the molecular weight M and the reduced pressure p_r, he succeeded in producing a correlation that does not require values of surface tension, enthalpy, density, etc., as follows:

$$\alpha = 55\ \dot{q}^{0.67} p_r^{(0.12-0.2\log R)} (-\log p_r)^{-0.55} M^{-0.5} \qquad Wm^{-2}K^{-1} \qquad (2)$$

where R is the roughness in μm. Rearranging the equation, with regard to the dependence of $\dot{q}$ on ΔT_{SAT}, gives $\dot{q} = \alpha \Delta T_{SAT}$ or $\dot{q}$ proportional to $\dot{q}^{0.67}\ \Delta T_{SAT}$ or $\dot{q}$ varying as ΔT_{SAT}^3. If the roughness of the surface is unknown, one could assume R = 1 μm, i.e., log R = 0.

Commercial heat transfer surfaces exist on which the number of nucleation sites has been increased, e.g., by the deposition of a thin porous coating. These give better heat transfer. A fuller discussion of this topic is given in the article on **Nucleate Boiling**.

Often more important than the level of nucleate boiling heat transfer is the value of the critical (or burnout) heat flux. Zuber (1958) used the idea of instability in the counterflow of liquid and vapour to derive the following expression for the pool boiling critical heat flux on a flat plate:

$$\dot{q}_{crit} = 0.131\ h_{LG} \rho_G^{0.5} [\sigma g (\rho_L - \rho_G)]^{0.25} \qquad Wm^{-2}. \qquad (3)$$

For water at atmospheric pressure, this gives $\dot{q}_{crit} = 1.1 \times 10^6$ Wm^{-2}.

In general, the Zuber theory works well, although there is evidence for much lower values of $\dot{q}_{crit}$ on nonwetted surfaces, e.g., for water on surfaces coated with silicone grease or PTFE (Gaertner (1965)). Also for normal, well-wetted surfaces, some authors recommend values that are 10 to 20% higher. A further discussion of the area is given in the article on **Burnout (Pool Boiling)**.

The next region of the boiling curve is transition boiling. Methods of predicting the heat transfer here are not well developed. A common approach is to locate the critical heat flux and minimum film boiling points and say that transition boiling follows a straight line linking the two (on a log $\dot{q}$ versus log ΔT_{SAT} plot).

One method of predicting the heat flux at the minimum film boiling point was developed by Zuber (1958). Previous work had shown that the liquid-vapour interface in film boiling can become unstable, with waves appearing. These waves develop and, in practice, lead to vapour mushrooms which grow and detach from the film at regular intervals. The theory leads to a prediction for the growth rate of the vapour region. Zuber had argued that a minimum heat flux is required to evaporate this amount of liquid and provide this amount of vapour, otherwise the film would collapse. This gives a minimum film boiling heat flux of:

$$\dot{q}_{mfb} = 0.13\ \rho_G h_{LG} \{\sigma g (\rho_L - \rho_G)/(\rho_L + \rho_G)^2\}^{0.25} \qquad Wm^{-2} \qquad (4)$$

This works reasonably for a number of liquids, but measurements on hydrocarbons support Berenson's (1960) suggestion that the constant should be 0.09.

A completely different approach is to say that at the minimum film boiling point, the liquid is at the homogeneous nucleation temperature. In general, as explained earlier, vapour cannot spontaneously nucleate in pure bulk liquid. However, as the temperature approaches the critical temperature, the distinction between liquid and vapour is becoming less pronounced (at the thermodynamic critical temperature, there is no difference between the phases). A complete and quite complex theory of homogeneous nucleation exists (see article on *Nucleation*). It is sufficient here to point out that one would expect the homogeneous nucleation temperature to be close to the critical temperature, T_c, and that detailed calculations give results close to:

$$T_{mfb} = 0.90\ T_c \qquad (5)$$

The physical argument is that once the liquid temperature reaches T_{mfb}, the boiling process is no longer constrained by the availability of surface cavity nucleation sites and liquid in contact with the surface will spontaneously flash into vapour, i.e., film boiling has started. There is evidence that Equation 5 gives a reasonable estimate of the minimum film boiling point at high pressures, but Equation 4 works better at low pressures. Of course, neither equation is enough to completely define the minimum film boiling point by itself: additional information, such as the film boiling heat transfer coefficient, is needed.

Analysis to find the film boiling heat transfer coefficient is possible, in simple geometries at least. A laminar layer of vapour builds up on the surface, with the evaporation rate at the liquid-vapour interface controlled by heat conduction and radiation though the vapour layer. The inefficiency of these processes accounts for the low heat transfer coefficients. Film boiling on a horizontal cylinder, diameter D as shown in **Figure 5**, has been analysed by Bromley in 1950. The theory is the same as Nusselt's earlier analysis for laminar film condensation on a cylinder.

The film boiling heat transfer coefficient is:

$$\alpha = 0.62\{k_G^3 \rho_G h_{LG} g (\rho_L - \rho_G)/(\mu_G \Delta T_{sat} D)\}^{0.25} \qquad Wm^{-2}. \qquad (6)$$

This equation assumes purely conduction heat transfer and is likely to be valid near the minimum film boiling point. At higher temperatures, the effect of radiation must be included. For a short vertical plate, of height L, D in Equation 6 is replaced by L and the constant becomes 0.94 instead of 0.62 (for a tall vertical plate, the film tends to break up).

A number of refinements to the theory behind Equations 3, 4 and 6 have been made over the years. Lienhard is particularly associated with this work, and a summary of many of the results is given in the chapter on boiling in his book (1981).

Thus far, the discussion has centred on pool boiling. **Figure 6** for **Forced Convective Boiling** shows that in most cases, the basic shape of the boiling curve is retained. All the different regions still

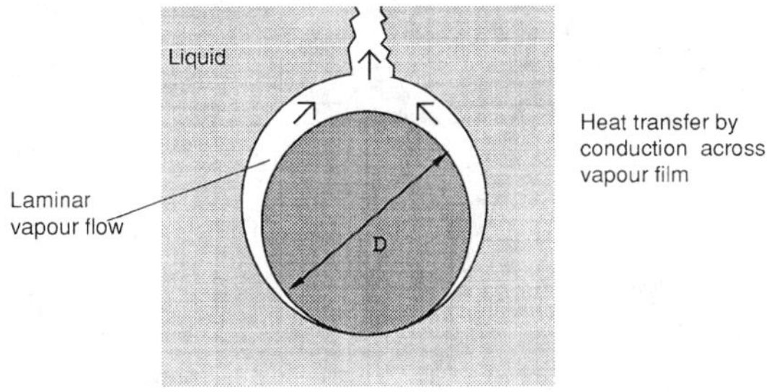

Figure 5. Model for film boiling on horizontal cylinder.

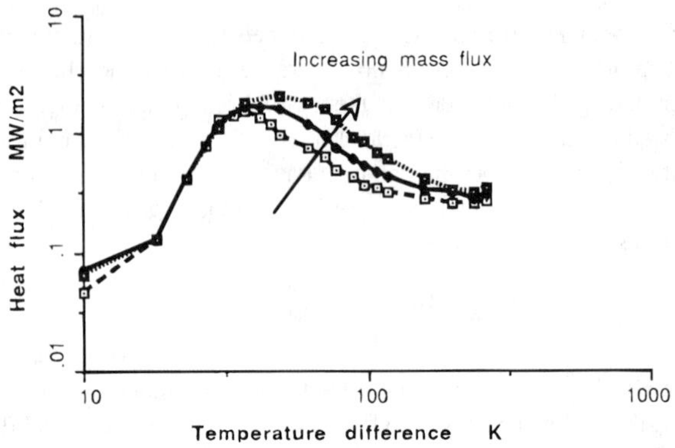

Figure 6. Effect of mass flux (68, 136 and 203 $kgm^{-2}s^{-1}$) on the flow boiling of saturated water at 1 atm. (Cheng et al. 1978).

exist although the terminology may not be exactly the same. An improvement in heat transfer due to the forced convection is seen in all parts of the boiling curve; though it is often scarcely noticeable in nucleate boiling. A parameter that assumes a greater importance in flow boiling is the *subcooling,* i.e, the difference between the saturation temperature and the liquid bulk temperature. Increasing the subcooling has a similar effect to increasing the flow rate. For saturated boiling, the amount of vapour in the flow, i.e., the quality, becomes important (quality = x = vapour mass flow rate/total mass flow rate).

Although it is simplest, and usual, to continue to refer to the nucleate boiling region, a better name might be the wet wall region, i.e., where the liquid is in good contact with the heat transfer surface. In subcooled boiling, and also in saturated boiling at low flow rates and qualities, bubble formation close to the wall still dominates the heat transfer, and macroscopic parameters such as subcooling and flow velocity have little effect. At higher qualities, when an annular flow pattern exists, the layer of liquid on the wall can become very thin. Also, the flow velocities are now much higher than in the single-phase liquid case (because of the presence of low density vapour). In this situation, *suppression of nucleate boiling* can occur. With high velocities in both vapour and liquid, the convective heat transfer in the thin liquid layer becomes very effective due to the intense turbulent mixing. Heat is transferred through the liquid layer to cause evaporation at the bulk liquid-vapour interface. This cooling process can be sufficiently effective to reduce the wall temperature below the minimum needed to activate the largest nucleation sites, i.e., nucleate boiling is suppressed, as shown in **Figure 7** below.

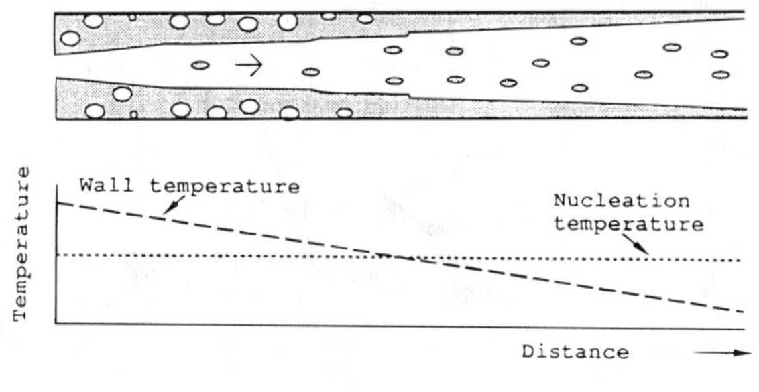

Figure 7. Suppression of nucleation in high quality annular flow.

The heat transfer mechanism when nucleate boiling is fully suppressed is called *forced convection vaporisation.* The existence of two very different mechanisms of heat transfer makes prediction a little involved. The most successful approach appears to be the superposition approach, i.e., contributions from the two mechanisms are combined. An early correlation of the type has been published by Chen in 1966. Although this division into two mechanisms—nucleate boiling and forced convection vaporisation—is widely accepted, it must be recognised that in the vast majority of cases where heat transfer measurements have been made there have been no visual observations of the boiling process.

The equations required for some of these predictions methods are quite lengthy. In addition, there is often the need for a wide range of fluid property values. A simpler method, in reasonable agreement with a large amount of data, has been suggested by Liu et al. in 1991. The two-phase heat transfer coefficient is given by:

$$\alpha_{TP}^2 = (S\alpha_{pool})^2 + (F\alpha_L)^2 \quad \textbf{(7)}$$

The nucleate boiling term, α_{pool}, is given by equation (2) with R = 1. The forced convection term is given by the Dittus-Boelter equataion:

$$\alpha_L = 0.023(\lambda/D)Re^{0.8}Pr^{0.4} \quad \textbf{(8)}$$

based entirely on liquid properties and with the total mass flow considered to be flowing as liquid, S and F, the empirical suppression and enhancements factors, as given by:

$$F = [1 + xPr(\rho_L/\rho_G - 1)]^{0.35} \quad \text{and}$$

$$S = [1 + 0.055F^{0.1}Re^{0.16}]^{-1} \quad \textbf{(9)}$$

The correlation is valid for saturated boiling in vertical tubes and also for horizontal tubes if the **Froude Number** is over 0.05. Simple extensions to lower flow rates in horizontal tubes, to flow in annuli and to subcooled boiling are explained in the original paper. A further discussion of heat transfer in the region is given in the article **Forced Convection Boiling**.

Prediction of the burnout point is covered in a separate article [**Burnout (Forced Convection)**]. Again, it may be sufficient in many cases to know the value of the critical or burnout heat flux, ensure that there is no danger of exceeding it, and then the higher temperature part of the boiling curve can be ignored. The negative slope portion of the curve is again called transition boiling.

The last part of the boiling curve is normally calaled the *post-dryout* or **Post Burnout** region. Certainly, the term film boiling is no longer universally appropriate. The heat transfer situataion may not be that of a thin vapour film between the heat transfer surface an the bulk liquid; indeed, there may well not be any continuous bulk liquid present, only liquid drops. This situation is shown in **Figure 8**. The critical heat flux, or burnout, corresponds physically to the drying out of the thin liquid film on the wall in annular flow (hence the term *dryout*). Direct confirmation of this effect was achieved by Hewitt et al. in 1963. The flow rate in the liquid film was measured by drawing it off through a porous section of wll. The critical heat flux was shown to correspond to the liquid film

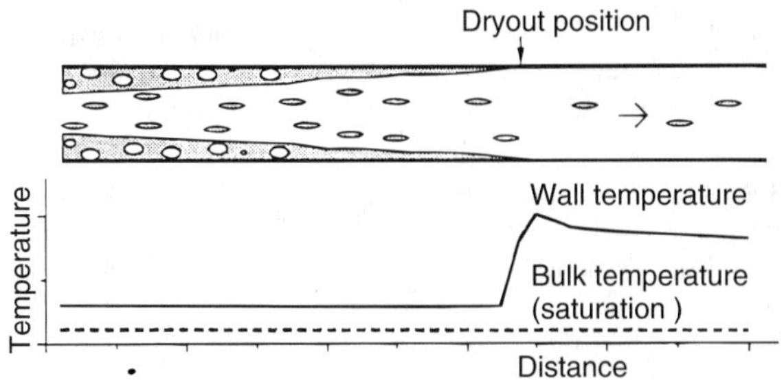

Figure 8. In high quality flows the critical heat flux is equivalent to dryout of the annular liquid film.

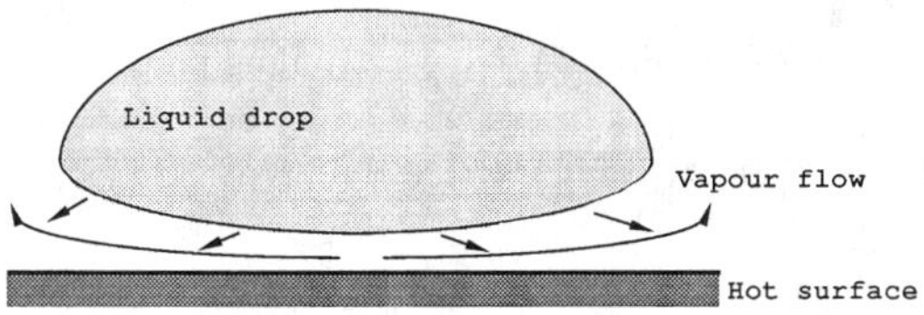

Figure 9. Leidenfrost drop.

flow rate falling to zero. Beyond the dryout position, there is a tendency for the vapour, being directly in contact with the wall, to superheat.

There is a sudden increase in the wall to bulk temperature difference once the critical heat flux (burnout or dryout) point is passed. The size of the temperature jump is likely to be larger if heat flux is the controlling parameter, rather than surface temperature. Although there are still liquid drops present in the flow, there is evidence that they do not contribute much to the heat transfer. At these high wall temperatures, vapour formation, as the drop approaches the wall, may be sufficient to prevent contact (**Figure 9**). The flow of vapour out of the narrow gap creates a pressure drop. The raised pressure in the gap prevents contact. For the simple case of a drop resting on a hot horizontal surface, this phenomenon is known as the Leidenfrost effect (observed by Leidenfrost in 1756) (see article on **Leidenfrost Phenomena**). In fact, unless the surface is slightly dish-shaped, the drop will not stay there. Drops of water spilt on a very hot surface, such as the heating element of an electric cooker, shoot off in all directions. They are not in contact with the solid surface; they are floating on a film of vapour. These topics are further discussed in the article in **Post-Burnout Heat Transfer**.

Although the mechanism at the critical heat flux in flow boiling may often be dryout of the annular liquid film, there are other possibilities. At sufficiently high heat fluxes, the critical heat flux may occur in low quality or even subcooled flows. Bulk liquid continues past the critical heat flux point, separated from the wall by a vapour film, i.e., a type of film boiling.

The above discussion of forced convection boiling implicitly assumed vertical flow. Provided the flow rate is reasonably high, all parts of the tube are still well wetted in horizontal flow, and the heat transfer behaviour is very similar. At sufficiently low flow rates in horizontal tubes, the liquid will tend to flow along the bottom of the tube and, roughly speaking, only the bottom wetted portion of the circumference will be cooled. Burnout is likely to occur earlier and will appear on the top of the tube. Even at higher flow rates, premature burnout may occur in annular flow if the annular film becomes thinner on the top of the tube.

References

Abbassi, A., Alem Rajabi, A. A., and Winterton, R. H. S. (1989) Effect of confined geometry on pool boiling at high temperature. *Experimental Thermal and Fluid Science* 2:127–133.

Alem Rajabi, A. A. and Winterton, R. H. S. (1988) Liquid-solid contact in steady-state transition pool boiling *Int. J. Heat and Fluid Flow* 9:215–219.

Berenson, P. J. (1961) Film boiling heat transfer from a horizontal surface, Trans. *ASME J. Heat Transfer* 83:351–358.

Bromley, L. A. (1950) Heat transfer in stable film boiling *Chem. Engng. Progress* 46:221–227.

Chen, J. C. (1966) *Ind. Eng. Chem. Process Design and Development,* 5:322–329.

Cheng, S. C. Ng, W. W. L., and Heng, K. T. (1978) Measurements of boiling curves of subcooled water under forced convective conditions, *Int. J. Heat and Mass Transfer* 21:1385–1392.

Cooper, M. G. (1984) *Saturation Nucleate Pool Boiling—A Simple Correlation.* 1st U.K. National Conf. on Heat Transfer, Inst. Chemical Engineers 785–793.

Gaertner, R. F. (1965) Photographic study of nucleate pool boiling on a horizontal surface, *J. Heat Transfer* 87:17–29.

Hewitt, G. F., Kearsey, H. A. P., Lacey, M. C., and Pulling, D. J. (1963) Burnout and nucleation in climbing film flow, *Int. J. Heat and Mass Transfer* 8:793–814.

Lienhard, J. L. (1981) *A Heat Transfer Textbook,* Prentice-Hall, New York.

Z. Liu and R. H. S. Winterton (1991) A general correlation for saturated and subcooled flow boiling in tubes and annuli based on a nucleate pool boiling equation. *Int. J. Heat and Mass Transfer* 34:2759–2766.

Nukiyama, S. (1966) The maximum and minimum values of the heat Q transmitted from metal to boiling water under atmospheric pressure, *Int. J. Heat and Mass Transfer* 9:1419–1433.

Zuber, N. (1958) On the stability of boiling heat transfer, *Trans. ASME,* 80:711–720.

Additional References

The following are three recently published books on the subject:

J. G. Collier and J. R. Thome (1994) *Convective Boiling and Condensation,* 3rd edition, Oxford.

K. Stephan (1992) *Heat Transfer in Condensation and Boiling,* Springer-Verlag.

P. B. Whalley (1987) *Boiling, Condensation and Gas-Liquid Flow,* Oxford.

***Leading to:** Accommodation coefficient; Forced convective boiling; Pool boiling*

R.H.S. Winterton

BOILING COOLANTS, FOR NUCLEAR REACTORS (see Coolants, reactor)

BOILING CURVE (see Boiling)

BOILING, FORCED CONVECTIVE (see Forced convective boiling)

BOILING HEAT TRANSFER ON TUBES AND TUBE BUNDLES (see Tubes and tube bundles, boiling heat transfer on)

BOILING LOOPS, INSTABILITIES (see Instabilities, two-phase)

BOILING MIXTURES, HEAT TRANSFER (see Multi-component mixtures, Boiling in)

BOILING NUMBER

The boiling number is a dimensionless group involving the heat flux $\dot{q}$, defined by $Bg = \dot{q}/(h_{LG}\dot{m})$. It can be thought of as a ratio mass flow rates per unit area, $(\dot{q}/h_{LG}) / (\dot{M}/A_C)$, i.e.,

$$Bg = \frac{\text{mass of vapour generated per unit area of heat transfer surface}}{\text{mass flow rate per unit flow cross-sectional area}}.$$

This dimensionless group was used by Davidson, et al in 1943. They argued that it represented the stirring effect of the bubbles upon the flow. Later, the group was named the boiling number.

The boiling number is used in convective boiling. Shah (1982) uses an empirical boiling number correction to the single-phase liquid heat transfer coefficient in order to predict the two-phase heat transfer. This approach avoids the introduction of a nucleate pool boiling term into the correlation.

Another application is in modeling the critical heat flux. Measurements of critical heat flux can be expressed in dimensionless form, which, to a first approximation is valid for all fluids, if they are put in the form of a boiling number, i.e., $Bg_{crit} = \dot{q}_{crit}/(h_{LG}\dot{m})$. Tables can then be constructed of Bg_{crit} in terms of other dimensionless groups (e.g., ESDU, 1986).

References

Davidson, W. F. et al. (1943) Studies of heat transmission through boiler tubing at pressures from 500 to 3300 pounds, *ASME Trans.* 65:553–579.

ESDU Data Item 86032. (1986) *Boiling Inside Tubes: Critical Heat Flux for Upward Flow in Uniformly-Heated Vertical Tubes.*

Shah, M. M. (1982) Chart correlation for saturated boiling heat transfer: equations and further study, *ASHRAE Trans.* 88:185–196.

R.H.S. Winterton

BOILING OF LIQUID METALS (see Liquid metal heat transfer; Liquid metals)

BOILING POINT

Following from: *Boiling*

A liquid at its boiling point is in thermodynamic and phase equilibrium with its coexisting vapour. For a liquid of fixed composition, only one of either pressure or temperature are independent. Thus, one can refer to the *boiling temperature* or the *boiling pressure,* in each case implying that the other variable has been specified.

Conventionally, the term boiling point is most frequently used to mean the boiling temperature at a specified pressure. The normal boiling point is the boiling temperature at 1 atmosphere pressure. (See also **Boiling**.)

G. Saville

BOILING PRESSURE (see Boiling point)

BOILING TEMPERATURE (see Boiling point)

BOILING TRANSITION (see Boiling water reactor, BWR)

BOILING WATER REACTOR, BWR

Following from: *Nuclear reactors*

A Boiling Water Reactor (BWR) circulates water past nuclear fuel in a large pressure vessel at 6.9 MPa. The *nuclear fuel* is configured as cylindrical pellets contained within long metallic tubes, referred to as *fuel rods.* The fuel rods are combined in various arrangements within flow channels to create *fuel assemblies.* Several hundred such fuel assemblies are connected in parallel between inlet and outlet plena. These parallel fuel assemblies are referred to as the *core* of a BWR. During passage of water along the nuclear fuel, significant boiling and vapor formation occurs—typically producing 10% to 20% vapor flow that exits into the outlet plenum. The resultant vapor is separated and directed to steam turbines for electrical generation. The remaining liquid is circulated back, along with condensate liquid returning from the turbines, to the inlet plenum. Key components of a jet-pump BWR steam supply system are illustrated in **Figure 1**.

Normal Operation Analyses

Normal operation of a BWR depends upon accurate prediction of several key thermal hydraulic parameters in the fuel assemblies. In particular vapor void fractions, transition boiling limits and fuel assembly pressure drops must be accurately predicted to support BWR design and operation. Prediction methods must address the range of two-phase flow regimes from subcooled liquid at the fuel assembly inlet to annular flow at the outlet. (See **Forced Convective Boiling**.)

The surrounding channels on BWR fuel assemblies effectively isolate flows within each fuel assembly. This allows the necessary two-phase flow characteristics to be evaluated by reproducing realistic thermal hydraulic conditions in isolated fuel assembly simulation tests. Such simulations require large experimental facilities that can reproduce the powers and flows of full-scale BWR fuel assemblies, using electrically heated fuel rod simulators. Data from such full scale experiments are used to provide very accurate empir-

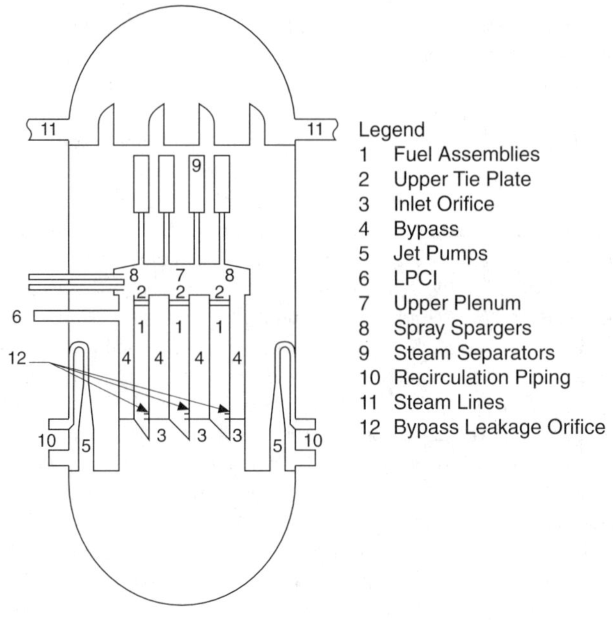

Figure 1. Typical jet pump BWR steam supply system.

ical relationships for the two-phase flow parameters required for BWR design analyses.

The vapor **Void Fraction** is the fraction of the flow area that is occupied by vapor. This is directly related to the average velocity of the vapor. In a BWR, the circulating water serves both as cooling medium and neutronic moderator for the nuclear fuel. This produces close coupling between power generation and two-phase flow conditions in the fuel assemblies. The primary coupling parameter is the density of the steam-water mixture surrounding the nuclear fuel. Predicting the two-phase density requires accurate prediction of the void fraction.

Fuel assembly **Pressure Drop** is one of the most important parameters determining conditions in the BWR core. Fuel assembly powers and flows can differ significantly due to mechanical and nuclear design differences, as well as locations within the core. To assure appropriate flow distribution to all of the fuel assemblies, it is important to have accurate predictions of pressure drops over the entire range of fuel assembly operating conditions. (See **Pressure Drop, Two-Phase**).

Boiling transition refers to conditions where typical boiling heat transfer starts to deteriorate. This has also been referred to as *critical heat flux* (see **Burnout**). The boiling transition limit for BWR operation is usually associated with deterioration of the liquid film flowing on a fuel rod under annular flow conditions. The fuel rods of a BWR fuel assembly utilize *zirconium alloy,* which minimizes parasitic capture of neutrons, for the outer metallic tubes. This material has good corrosion resistance for temperatures existing under typical boiling conditions. However, if boiling transition conditions are exceeded, the deteriorated heat transfer causes higher temperatures in the fuel. If such conditions persist for an extended period of time, the increased corrosion rate of the metallic tubes can cause fuel failures. Therefore, accurate predictions of boiling transition limits are very important for reliable BWR operation.

The elimination of cross-flows between BWR fuel assemblies allows for both realistic experimental simulations of fuel assembly thermal hydraulics and simplified analyses methods. BWR analyses methods typically utilize either one dimensional representations of fuel assemblies or subchannel representations. One-dimensional analyses utilize correlations based on cross-sectional averaged flow quantities, with empirical parameters to account for radial variations where necessary. Nuclear coupling based upon one-dimensional values is usually sufficiently accurate for typical BWR fuel assemblies. Representative solution techniques and correlations are discussed by Collier (1980) and Lahey (1993). Subchannel analyses incorporate more detailed mechanistic descriptions of two-phase flow phenomena to predict cross flows and local parameters within fuel assemblies. Subchannel methods are summarized by Shiralkar (1992).

One-dimensional void fraction predictions are often based on the **Drift Flux** concept of Zuber (1965). Average vapor velocity is characterized by the average of local vapor drift velocities and the volumetric flux adjusted by a distribution parameter. The distribution parameter incorporates radial phase and velocity distribution effects. These parameters are determined from experimental void fraction data. While the original drift flux work anticipated discrete values of these parameters for various flow regimes, considerable success has been achieved by Dix (1971) and Chexal (1991) with continuous correlation relations which implicitly reflect the effects of flow regimes. These correlations are usually assumed to be insensitive to power distribution changes across the fuel assemblies. And has been confirmed by recent data from Yagi (1992) which demonstrated average void fraction results were unchanged even when large variations were imposed in local void fractions within a typical BWR fuel assembly.

One-dimensional pressure drop predictions typically use average values for void fraction and flow quality in combination with standard two-phase flow formulations and correlations. Single-phase pressure drop data associated with wall friction and local losses are necessary to provide the reference bases for each specific fuel assembly design.

Boiling transition limits reflect local liquid film disruptions on individual fuel rods. Those limits are dependent upon both local film flow characteristics and local power distributions across fuel assemblies. Mechanical *spacers,* which maintain fuel rod spacing along their axial lengths, have important effects on boiling transition limits. Detrimental thinning of the fuel rod liquid films can be caused just upstream of the spacers, while favorable deposition of droplets into the liquid films can be caused just downstream of the spacers. The detrimental upstream effects are usually sufficient to cause initial boiling transitions to occur just upstream of spacers. However, the favorable downstream effects usually dominate, such that boiling transition improves with closer axial pitch of the spacers. Boiling transition limits can vary by 10% due to mechanical features of the spacers.

Predictions of boiling transition are often based on one dimensional averaged parameters, such as critical quality and boiling length, with empirical treatments of local effects. Such correlations can provide very accurate predictions but require extensive calibration with full scale test results for each specific fuel assembly design and radial power distribution range. This extensive testing requirement is one of the major motivations for more general *subchannel analyses* methods.

Subchannel analyses methods are based upon radial solution meshes within the fuel assemblies. Subchannel methods divide fuel assemblies into a number of interacting flow regions for analyses. The models then track liquid films, core vapor and core entrained droplets as separate fields for each of the subchannel regions. The goals for subchannel methods are to provide predictions which are mechanistically-based and less dependent upon full-scale calibration experiments, particularly for the development of new fuel designs.

Since BWR boiling transition limits are related to the deterioration of annular liquid films flowing on fuel rod surfaces, the subchannel methods incorporate mechanisms to predict and track the net liquid flows in those films, as well as criteria for the minimum film flows corresponding to film disruption and boiling transition. Key aspects for these film predictions are liquid evaporation, entrainment of liquid from the film to the vapor/droplet core and deposition of droplets from the core onto the surface films. Flow distributions among the subchannel regions usually require empirical mechanisms such as 'void drift' to achieve the overall flow distributions observed in BWR fuel assembly experiments.

The effects of fuel rod spacers are very important in determining boiling transition limits of BWR fuel assemblies, as previously discussed. Current subchannel methods incorporate various modeling assumptions to describe these spacer effects, including empirical parameters which must be calibrated with full scale boiling transition data for the specific spacer to be analyzed. General modeling of these spacer effects in subchannel analyses will require a further level of detail and mechanistic understanding which is not yet available.

Current subchannel modeling methods provide excellent predictions of boiling transition for limited variations from their calibration bases. Void fraction distributions are also predicted quite well by these methods, except for highly heterogeneous conditions across fuel assemblies. In general, subchannel methods are appropriate tools for BWR analyses, but further developments are necessary to significantly reduce dependence on full-scale testing for final design applications.

Stability analyses

Parallel flow channels and close coupling between neutronic power and two-phase flow conditions can cause instability conditions in a BWR. Accurate predictions and avoidance of potentially unstable conditions are important for reliable BWR operation. Two types of instability can occur. Density wave oscillations can be driven by the two-phase hydraulics in individual fuel assemblies, with only minor effects from neutronic coupling. Alternatively, core-wide neutronic coupling can cause instabilities in many or all of the fuel assemblies. The fuel assemblies can all have oscillations in-phase, or local regions can oscillate. Accurate predictions of these oscillation conditions require coupled solutions of neutronics and two-phase flow equations. The same basic one-dimensional formulations for two-phase flow parameters are also applied in stability predictions. Both frequency domain and time domain solution techniques are used successfully for these analyses, as discussed by Lahey (1993). (See **Instability, Two-phase**.)

Loss of coolant analyses

One of the most challenging and well-researched areas of BWR analyses is the prediction of safety system and fuel cooling responses to a wide range of accidents that might result in the loss of coolant water from the core. Postulated breakage of pipes connected to the pressure vessel are the primary bases for these accident evaluations. BWR designs include emergency systems to distribute coolant above the core, as well as systems to refill the pressure vessel and reflood the core if it does become uncovered during an accident. Modern BWR designs also include recirculation pumps within the pressure vessel (rotary or jet pumps) to minimize the size of potential breaks to pressure vessel connections.

Predictions of fuel cooling under these postulated *loss of coolant accident (LOCA)* conditions require detailed system simulation methods to analyze the transient two-phase fluid conditions adjacent to the fuel as the postulated accident proceeds. Accurate predictions require modeling of flow and heat transfer with surface temperatures well above **Leidenfrost** conditions. Elements include falling liquid films, reflooding from below and counter current flows of the two phases. Modeling requirements for accurate BWR/LOCA predictions are summarized by Andersen (1985).

Large scale experiments have demonstrated that condensation of vapor by subcooled liquid and three-dimensional distribution effects are important aspects determining the progression of emergency coolant to the fuel. A significant effect for BWR/LOCA is the accumulation and drainage of coolant water in the plenum region above the core. *Counter-current flow limitation (CCFL)* conditions at the top of fuel assemblies cause accumulation of emergency coolant water in a pool above the core. Some fuel assemblies then experience downward flow of subcooled liquid, while others experience counter-current flows. The results of this complex pattern are rapid reflooding and cooling of the BWR core. These effects and other LOCA studies are summarized by Dix (1983). (See also **Flooding and Flow Reversal**.)

Passive safety features

The *Simplified Boiling Water Reactor (SBWR)* is a new reactor design which incorporates passive safety features rather than active safety systems used previously. This design includes a gravity drain pool inside the containment to provide liquid to the core in case of a LOCA. Heat exchangers submerged in large water pools outside the containment absorb energy from within the reactor containment and transport it to the atmosphere. The latter energy transport occurs without any activation in case of a LOCA. The water pools have sufficient capacity to transport decay heat for several days and can be replenished from outside the containment.

Since the SBWR is designed to maintain the core covered during a LOCA, the primary design limit is containment structure pressure capability. Prediction of containment pressure responses for such events requires computer programs that address the entire coupled system. Condensation on the containment concrete walls and structures, as well as interaction effects between steam and other noncondensible gases, are particularly important for accurate predictions of these slow transient responses.

References

Andersen, J. G. M., Chu, K. H., Cheung, Y. K., and Shaug, J. C. (1985) *BWR Full Integral Simulation Test (FIST) Program,* TRAC-BWR Model Development Volume 2—Models, GEAP-30875-2,NUREG/CR-4127-2, EPRI NP-3987-2.

Chexal, B., Lellouche, G., Horowitz, J., Healzer, J., and Oh, S. (1991) *The Chexal-Lellouche Void Fraction Correlation for Generalized Applications,* NSAC Report-139.

Collier, J. G. (1980) *Convective Boiling and Condensation,* McGraw Hill.

Dix, G. E. (1971) *Vapor Void Fractions for Forced Convection with Subcooled Boiling at Low Flow Rates,* General Electric NEDO-10491 (PhD Thesis, University of California Berkeley).

Dix, G. E. (1983) *BWR Loss of Coolant Technology Review,* Proceedings ANS Symposium Thermal Hydraulics of Nuclear Reactors, Volume 1.

Lahey, R. T. and Moody, F. J. (1993) *The Thermal Hydraulics of a Boiling Water Reactor,* ANS Monograph.

Shiralkar, B. S. (1992) *Recent Trends in Subchannel* Analysis, Proceedings of the International Seminar on Subchannel Analysis, Tokyo.

Yagi, M., Mitsutake, T., Morooka, S., and Inoue, A. (1992) *Void Fraction in BWR Fuel Assembly and Evaluation of Subchannel Code,* Proceedings of the International Seminar on Subchannel Analysis, Tokyo.

Zuber, N. and Findlay, J. (1965) *Average Volumetric Concentration in Two-Phase Flow Systems,* Transactions of ASME Volume 87 Series C.

Gary E. Dix

BOILING WATER REACTOR INSTABILITIES (see Instability, two-phase)

BOLTZMANN CONSTANT (see Entropy)

BOLTZMANN DISTRIBUTION

Following from: *Thermodynamics*

The distributions laws of *statistical mechanics,* of which Boltzmann's is one, are concerned with the distribution of energy within a system of molecules. Knowledge of the true distribution function is of fundamental importance and permits evaluation of the thermodynamic properties of the system from statistical mechanics. Boltzmann's distribution law refers specifically to a system of noninteracting molecules in a state of *thermodynamic equilibrium.* It was postulated before the discovery of quantum mechanics and, in its original form, relied upon a classical description of molecular energies. In this description, the distribution function for a system of structureless molecules is specified by the probability P that a molecule will, at any instant, be located within the element of volume dxdydz and have velocity components in the ranges u to u + du, v to v + dv, and w to w + dw. According to Boltzmann's distribution law, this probability is given by:

$$P = \frac{\exp(-\varepsilon/kT)dxdydzdudvdw}{\int \exp(-\varepsilon/kT)dxdydzdudvdw}, \quad (1)$$

where ε is the total (kinetic + potential) energy of the molecule, k is a positive constant known as *Boltzmann's constant,* and the integral is performed over all possible positions and velocities of the molecule. This law states that the probability of a molecule acquiring an energy ε declines with increasing energy in proportion to the Boltzmann factor $\exp(-\varepsilon/kT)$. For molecules with rotational and vibrational modes of motion, the classical distribution law may be extended, but a quantum-mechanical generalisation is to be preferred.

In *quantum mechanics,* molecular energies are restricted to discreet levels ε_i and the counterpart of Equation (1) is:

$$P_i = \frac{g_i \exp(-\varepsilon_i/kT)}{q}. \quad (2)$$

Here, P_i is the probability that, at any instant, a given molecule will be found in a quantum state having energy ε_i, g_i is the *degeneracy* of that energy level (i.e., the number of quantum states with the same energy ε_i), and q is the *molecular partition function* given by:

$$q = \Sigma_i\, g_i \exp(-\varepsilon_i/kT). \quad (3)$$

Since the energy levels of the molecule depend on the volume of the system, q is a function of both T and V. It turns out that the *Helmholtz free energy* (**Free Energy**) A of N noninteracting molecules is related to q by the simple formula:

$$A = -kT \ln(q^N/N!). \quad (4)$$

Thus, all of the thermodynamic properties of the perfect gas may be determined from q(T,V) (see Hill, 1960).

An important simplification arises when we assume that the energy of each molecule may be written as the sum of independent terms for translations (t), rotations (r), vibrations (v) and other internal *degrees of freedom.* In this case, q factorises into the product $q_t q_r q_v \cdots g_0 \mathrm{ext}(-\varepsilon_0/kT)$. Here, g_0 and ε_0 are the degeneracy and energy of the lowest energy level and all other energies are measures relative to ε_0. Each of the factors q_t, q_r, q_v, etc, in q is given by an equation similar to (3), but with the summation running only over energy levels of the specified kind. The separation of molecular energies and consequent factorisation of q is exact if rotations are treated as those of a rigid body and molecular vibrations are assumed simple harmonic. These are fair approximations for many molecules. For a diatomic molecule of mass m, moment of inertia I, vibration frequency ν, and symmetry number s, the factors of q are:

$$\left.\begin{aligned} q_t &= (2\pi mkT/h^2)^{3/2}V \\ q_r &= (8\pi^2 IkT/sh^2) \\ q_v &= \{1-\exp(-h\nu/kT)\}^{-1} \end{aligned}\right\}. \quad (5)$$

Here, s = 2 for homonuclear molecules, s = 1 otherwise and the zero-point energy ε_0 is taken to be the energy of the molecule in its lowest vibrational state. A similar but more complicated treatment is applicable to polyatomic molecules (Hill, 1960). It should be noted that the formula for q_r is based on a semi-classical treatment, but that a full quantum treatment may be needed at low temperatures for molecules with small moments of inertia (e.g. H_2).

Although Equation (2) is based on quantised energy levels, it is not strictly consistent with quantum-mechanical restrictions. The correct distribution function depends upon whether the molecules in question contain an even or an odd number of elementary particles (protons, neutrons and electrons): the first case leads to the *Bose-Einstein distribution,* while the second gives the *Fermi-Dirac distribution.* However, both of these reduce to the Boltzmann distribution at the limit where the number of available quantum states greatly exceed the number of molecules. This limit is approached with increasing molecular mass and increasing temperature; in practice, Boltzmann statistics apply with great accuracy, except for the isotopes of hydrogen and helium at temperatures below 10 K.

It is possible to make a generalisation of Boltzmann's distribution law applicable to a system of N; interacting molecules. We now define a new probability function Π_i as the probability that, at a given instant, the entire system will be in a quantum state with energy E_i. This probability too is proportional to a Boltzmann factor, $\exp(-E_i/kT)$, and we have:

$$\Pi_i = \frac{\Omega_i \exp(-E_i/kT)}{Q}, \quad (6)$$

where

$$Q = \sum_i \Omega_i \exp(-E_i/kT). \quad (7)$$

Here, Ω_i is the degeneracy of E_i, Q is called the *canonical partition function* and the summation in Equation (7) is carried out over all possible energy levels. According to statistical thermodynamics, the Helmholtz free energy and all of the other thermodynamic properties of the system may be obtained from Q(N,T,V) (see Hill, 1960).

Reference

Hill, T. L. (1960) *An Introduction to Statistical Thermodynamics,* Addison-Wesley Publishing, Reading (Mass.).

J.P.M. Trusler

BOLTZMANN EQUATION (see Kinetic theory of gases)

BOLTZMANN NUMBER

Following from: *Coupled radiation and convection*

Boltzmann number Bo, is a dimensionless parameter used in the problem of heat transfer by radiation and convection formally showing the radiation contribution to the overall heat transfer. It appears when dimensionless values are introduced into the equation of energy transfer in the radiating and absorbing gas. For instance, for a stationary subsonic flow in one-dimensional channel, the energy equation written for simplicity in the "narrow channel" approximation is of the form:

$$\rho c_p(uTx + vTy) = y(\lambda Ty) - (\dot{q}_{Rx}x + \dot{q}_{Ry}y).$$

If we introduce the dimensionless values $x' = x/D$; $y' = y/D$; $u' = u/u_0$, $p' = p/\rho_0 u_0^2$; $c_p' = c_p/c_{p0}$ $\lambda' = \lambda/\lambda_0$; $\rho' = \rho/\rho_0$; $\mu' = \mu/\mu_0$; $T' = T/T_0$; $\dot{q}'_{Rx} = \dot{q}_{Rx}/\sigma T_0^4$; $\dot{q}T_{Ry} = \dot{q}_{Ry}/\sigma T_0^4$; where D is the characteristic dimension (the height of the channel) and the zero subscript denotes characteristic conductions, e.g., at the inlet to the channel, then the equation takes the form (the primes are omitted):

$$\rho c_p(uTx + vTy) = \frac{1}{Pe} y(\lambda Ty) - \frac{1}{Bo}(\dot{q}_{Rx}x + \dot{q}_{Ry}y)$$

where Pe = RePr is the **Peclet number**, $Bo = \rho_0 c_{p0} y_0/\sigma T_0^3$ is the Boltzmann number.

A high value of Boltzmann number shows a weak effect of the radiation on gas temperature.

To evaluate the role heat transfer by radiation plays in the flow of optically "grey" medium (this is a rarely applicable approximation in which the optical properties are taken to be independent of frequency), it is necessary, apart from Boltzmann number, to consider an additional parameter—optical thickness $\tau = kD$ (where k is the volume coefficient of absorption of the medium in the channel), which is the ratio of the characteristic dimension D to the mean path length of radiation k^{-1}. In the case of optically-thin gas layer ($\tau << 1$), the density of radiation heat flux from the medium to the channel walls can be estimated to be $2\tau_0\sigma T_0^4$, and in the case of optically-thick layer ($\tau_0 >> 1$) to be $16\sigma T_0^4/3\tau_0$. The density of convective heat flux is proportional to ρ_0, u_0, c_{p0}, T_0. Then the effect of heat transfer by radiation is characterized by the parameters:

$$\frac{2\tau_0\sigma T_0^4}{\rho_0 u_0 c_{p0} T_0} = 2Bo^{-1}\tau_0 \qquad \text{for } \tau_0 << 1,$$

$$\frac{16\sigma T_0^4/3\tau_0}{\rho_0 u_0 c_{p0} T_0} = \frac{16}{3}(Bo\ \tau_0)^{-1} \qquad \text{for } \tau_0 >> 1.$$

Therefore, even for low Bo the impact of heat transfer by radiation may not be strong if for $\tau << 1$, $\tau_0/Bo << 1$ and for $\tau_0 >> 1$, $(\tau_0\ Bo)^{-1} << 1$.

In the general case of selective (nongrey) gas when $k = k(\lambda)$, where λ is the radiation wave length, the degree of radiation effect on heat transfer generally cannot be described by only two parameters, Bo and τ_0. It also depends on optical thickness $\tau_0\lambda$ of the medium in energy-carrying wavelength bands, the values of the latter and their positioning with respect to the maximum point of the intensity function of the black body radiation.

Sometimes, in problems of heat transfer by radiation and convection or by radiation and conduction, a dimensionless parameter

$$N_R = \lambda k/\sigma T_0^3 = Bo\ \tau_0/Pe$$

is used, known as a conduction-radiation parameter or the *Stark number.*

References

Özisik, M. N. (1973) *Radiative Transfer and Interactions with Conduction and Convection,* J. Wiley & Sons, New York, London, Sydney, Toronto.

Siegel, R. and Howell, J. R. (1972) *Thermal Radiation Heat Transfer,* McGraw Hill Book Comp., New York et al.

I.G. Zaltsman

BOND NUMBER

Bond number, Bo, is a dimensionless group which arises from the analysis of the behaviour of *Bubbles* and **Drops**. It is represented as:

$$Bo = gL^2(\rho_L - \rho_G)/\sigma$$

where g is the acceleration due to gravity; ρ_L, the liquid density; ρ_G, the gas density; σ, the interfacial surface tension; and L. is the appropriate linear dimension, e.g., bubble diameter.

It represents the ratio of gravitational force to surface tension force.

This same group is sometimes referred to as the *Iötvös number.*

G.L. Shires

BOOLEAN ALGEBRA (see Risk analysis techniques)

BOSE-EINSTEIN DISTRIBUTION (see Boltzmann distribution)

BOTTOM FLOODING (see Rewetting of hot surfaces)

BOURGUER-HAWKES LAW (see Radiative heat transfer)

BOUNDARY LAYER

***Following from:** Flow of fluids*

A boundary layer is a thin layer of viscous fluid close to the solid surface of a wall in contact with a moving stream in which (within its thickness δ) the flow velocity varies from zero at the wall (where the flow "sticks" to the wall because of its viscosity) up to U_e at the boundary, which approximately (within 1% error) corresponds to the free stream velocity (see **Figure 1**). Strictly speaking, the value of δ is an arbitrary value because the friction force, depending on the molecular interaction between fluid and the solid body, decreases with the distance from the wall and becomes equal to zero at infinity.

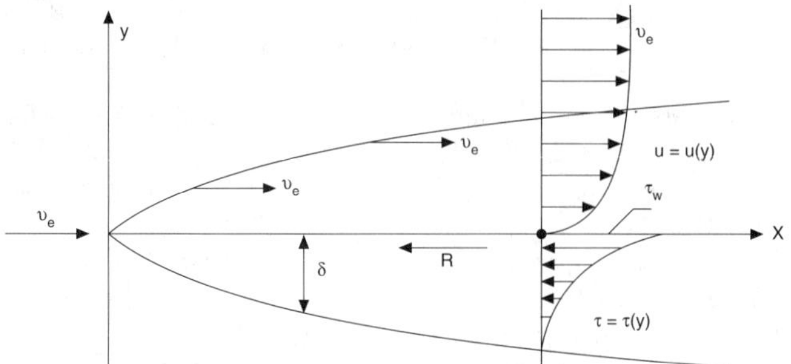

Figure 1. Growth of a boundary layer on a flat plate.

The fundamental concept of the boundary layer was suggested by L. Prandtl (1904) defines the boundary layer as a layer of fluid developing in flows with very high **Reynolds Numbers** Re, that is with relatively low viscosity as compared with inertia forces. This is observed when bodies are exposed to high velocity air stream or when bodies are very large and the air stream velocity is moderate. In this case, in a relatively thin boundary layer, friction **Shear Stress** (viscous shearing force): $\tau = \eta[\partial u/\partial y]$ (where η is the dynamic viscosity; u = u(y) − "profile" of the boundary layer longitudinal velocity component, see **Figure 1**) may be very large; in particular, at the wall where u = 0 and $\tau_w = \eta[\partial u/\partial y]_w$ although the viscosity itself may be rather small.

It is possible to ignore friction forces outside the boundary layer (as compared with inertia forces), and on the basis of Prandtl's concept, to consider two flow regions: the boundary layer where friction effects are large and the almost **Inviscid Flow** core. On the premises that the boundary layer is a very thin layer (δ << L, where L is the characteristic linear dimension of the body over which the flow occurs or the channel containing the flow, its thickness decreasing with the growth of Re, **Figure 1**), one can estimate the order of magnitude of the boundary layer thickness from the following relationship:

$$\delta/L \simeq Re^{-0.5} \quad \textbf{(1)}$$

For example, when an airplane flies at U_e = 400 km/hr, the boundary layer thickness at the wing trailing edge with 1 metre chord (profile length) is δ ≃ 0.015 m. As was experimentally established, a laminar boundary layer develops at the inlet section of the body. Gradually, under the influence of some destabilizing factors, the boundary layer becomes unstable and transition of boundary layer to a **Turbulent Flow** regime takes place. Special experimental investigations have established the existence of a transition region between the turbulent and laminar regions. In some cases (for example, at high turbulence level of the external flow), the boundary layer becomes turbulent immediately downstream of the stagnation point of the flow. Under some conditions, such as a severe pressure drop, an inverse phenomenon takes place in accelerating turbulent flows, namely flow relaminarization.

In spite of its relative thinness, the boundary layer is very important for initiating processes of dynamic interaction between the flow and the body. The boundary layer determines the aerodynamic drag and lift of the flying vehicle, or the energy loss for fluid flow in channels (in this case, a hydrodynamic boundary layer because there is also a thermal boundary layer which determines the thermodynamic interaction of **Heat Transfer**).

Computation of the boundary layer parameters is based on the solution of equations obtained from the *Navier-Stokes equations* for viscous fluid motion, which are first considerably simplified taking into account the thinness of the boundary layer.

The solution suggested by L. Prandl is essentially the first term of power series expansion of the Navier-Stokes equation, the series expansion being performed for powers of dimensionless parameter (δ/L). The smaller parameter in this term is in zero power so that the boundary layer equation is the zero approximation in an **Asymptotic Expansion** (at large Re) of the boundary layer equation (asymptotic solution).

A transformation of the Navier-Stokes equation into the boundary layer equations can be demonstrated by deriving the Prandtl equation for laminar boundary layer in a two-dimensional incompressible flow without body forces.

In this case, the system of Navier-Stokes equations will be:

$$\begin{cases} \dfrac{\partial u}{\partial t} + u\dfrac{\partial u}{\partial x} + v\dfrac{\partial u}{\partial y} = -\dfrac{1}{\rho}\dfrac{\partial p}{\partial x} + \nu\left(\dfrac{\partial^2 u}{\partial x^2} + \dfrac{\partial^2 u}{\partial y^2}\right) \\ \dfrac{\partial v}{\partial t} + u\dfrac{\partial v}{\partial x} + v\dfrac{\partial v}{\partial y} = -\dfrac{1}{\rho}\dfrac{\partial p}{\partial y} + \nu\left(\dfrac{\partial^2 v}{\partial x^2} + \dfrac{\partial^2 v}{\partial y^2}\right) \\ \dfrac{\partial u}{\partial x} + \dfrac{\partial v}{\partial y} = 0. \end{cases} \quad (2)$$

After evaluating the order of magnitude of some terms of Equation (2) and ignoring small terms the system of Prandtl equations for laminar boundary layer becomes:

$$\begin{cases} \dfrac{\partial u}{\partial t} + u\dfrac{\partial u}{\partial x} + v\dfrac{\partial u}{\partial y} = -\dfrac{1}{\rho}\dfrac{\partial p}{\partial x} + \nu\left(\dfrac{\partial^2 u}{\partial x^2} + \dfrac{\partial^2 u}{\partial y^2}\right) \\ \dfrac{\partial u}{\partial x} + \dfrac{\partial v}{\partial y} = 0 \end{cases} \quad (3)$$

in which x, y are longitudinal and lateral coordinates (**Figure 1**); v is the velocity component along "y" axis; p, pressure; t, time; and ν the kinematic viscosity.

The boundary layer is thin and the velocity at its external edge U_e can be sufficiently and accurately determined as the velocity of an ideal (inviscid) fluid flow along the wall calculated up to the first approximation, without taking into account the reverse action of the boundary layer on the external flow. The longitudinal pressure gradient $[\partial p/\partial x] = [dp/dx]$ (at $p(y) =$ *const*) in Equation 3 can be depicted from the *Euler equation of motion of an ideal fluid.* From the above, Prandtl equations in their finite form will be written as:

$$\begin{cases} \dfrac{\partial u}{\partial t} + u\dfrac{\partial u}{\partial x} + v\dfrac{\partial u}{\partial y} = \dfrac{\partial U_e}{\partial t} + U_e\dfrac{\partial U_e}{\partial x} + \nu\dfrac{\partial^2 u}{\partial y^2} \\ \dfrac{\partial u}{\partial x} + \dfrac{\partial v}{\partial y} = 0 \end{cases} \quad (4)$$

This is a system of parabolic, nonlinear partial differential equations of the second order which are solved with initial and boundary conditions

$$\text{at } t = 0,\ u = u(0, x, y); \qquad y = 0,\ u = 0,\ v = 0;$$

$$y = \delta,\ u(t, x, y) = U_e(t, x); \qquad x = x_0,\ u = u_0(t, y).$$

The system of Equation 4 is written for actual values of velocity components u and v. To generalize the equations obtained for turbulent flow, the well-known relationship between actual, averaged and pulsating components of turbulent flows parameters should be used. For example, for velocity components there are relationships connecting actual u and v, averaged $\bar{u}$ and $\bar{v}$ and pulsating u' and v' components:

$$u = \bar{u} + u' \quad \text{and} \quad v = \bar{v} + v'.$$

After some rearrangements, it is possible to obtain another system of equations [Eq. (6)] from system (3), in particular for steady flow:

$$\begin{cases} \bar{u}\dfrac{\partial \bar{u}}{\partial x} + \bar{v}\dfrac{\partial \bar{u}}{\partial y} = -\dfrac{1}{\rho}\dfrac{\partial \bar{p}}{\partial x} + \nu\dfrac{\partial^2 \bar{u}}{\partial y^2} - \dfrac{1}{\rho}\dfrac{\partial(\rho\overline{u'v'})}{\partial y} \\ \dfrac{\partial \bar{u}}{\partial x} + \dfrac{\partial \bar{v}}{\partial y} = 0. \end{cases} \quad (6)$$

Using the following relation for friction shear stress in the boundary layer:

$$\tau = \eta\dfrac{\partial \bar{u}}{\partial y} + \rho(-\overline{u'v'}) \quad (7)$$

and taking into account that in the laminar boundary layer $u = u'$ and $\rho(u', v') = 0$, it is possible to rewrite the Prandtl equations in a form valid for both laminar and turbulent flows:

$$\begin{cases} \bar{u}\dfrac{\partial \bar{u}}{\partial x} + \bar{v}\dfrac{\partial \bar{u}}{\partial y} = -\dfrac{1}{\rho}\dfrac{\partial \bar{p}}{\partial x} + \dfrac{1}{\rho}\dfrac{\partial \tau}{\partial y} \\ \dfrac{\partial \bar{u}}{\partial x} + \dfrac{\partial \bar{v}}{\partial y} = 0 \end{cases} \quad (8)$$

The simplest solutions have been obtained for a laminar boundary layer on a thin flat plate in a two-dimensional, parallel flow of incompressible fluid (**Figure 1**). In this case, the estimation of the order of magnitude of the equations terms: $x \sim L$, $y \sim \delta$, $\delta \sim \sqrt{\nu L/U_e}$, allows combining variables x and y in one relation

$$\xi = y\sqrt{u_e/(4\nu x)} \quad (9)$$

and to reduce the solution of Eq. (8) (at $dp/dx = 0$) to determining the dependencies of u and v upon the new parameter ξ. On the other hand, using well-known relations between velocity components u, v and stream function ψ

$$u = \partial\psi/\partial y, \qquad v = -\partial\psi/\partial x$$

it is possible to obtain one ordinary nonlinear differential equation of the third order, instead of the system of partial differential equations (8)

$$2f'''(\xi) + f(\xi)f''(\xi) = 0. \quad (10)$$

Here, $f(\xi)$ is the unknown function of ξ variable: $f = f = \psi/\sqrt{u_e \nu x}$.

The first numerical solution of Eq. (10) was obtained by Blasius (1908) under boundary conditions corresponding to physical conditions of the boundary layer at $y = 0$: $u = 0$, $v = 0$; at $y \to \infty$: $u \to U_e$ (Blasius boundary layer).

Figure 2 compares the results of *Blasius solution* (solid line) with experimental data. Using these data, it is possible to evaluate the viscous boundary layer thickness. At $\xi \simeq 2.5$, $(u/U_e) \simeq 0.99$ (**Figure 2**); consequently from Eq. 9 we obtain: $\delta \simeq 5\sqrt{\nu x/U_e}$.

From the Blasius numerical calculations of the value of the second derivative of $f(\xi)$ function at the wall friction shear stress, the relationship in this case is:

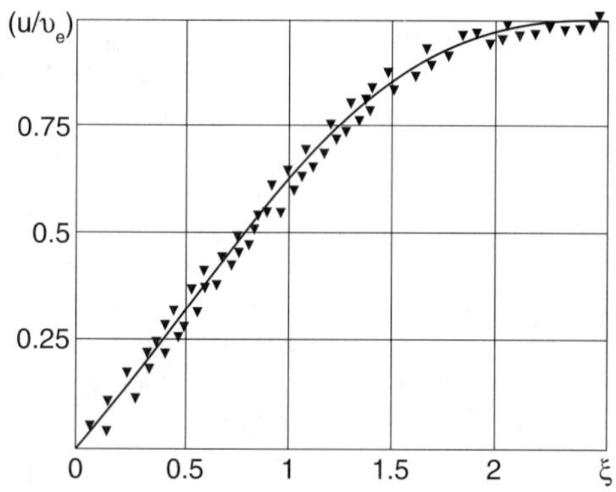

Figure 2.

$$\tau_w = \eta[\partial u/\partial y]_{y=0} = \eta \frac{U_e}{4} \sqrt{U_e/(\eta x)} f''(\xi)_{y=0} \quad \text{or}$$

$$\tau_w = 0.664\rho(U_e^2/2)\sqrt{\nu/(U_e x)}. \qquad \textbf{(11)}$$

Friction force R, acting on both sides of the plate of L length (**Figure 1**), is also determined from Eq. (11):

$$R = 2\int_0^L \tau_w \, dz = 0.664\sqrt{\rho\eta U_e^3 L},$$

as in the friction coefficient for flat plates:

$$\zeta = R[(\rho U_e^2/2)L]^{-1} = 1.328\sqrt{\nu/(U_e L)}.$$

Despite the fact that Prandtl equations are much simpler than Navier-Stokes equations, their solutions were obtained for a limited number of problems. For many practical problems, it is not necessary to determine velocity profiles in the boundary layer, only thickness and shear stress. This kind of information may be obtained by solving the integral momentum equation

$$\frac{d}{dx}\left[\int_0^\delta \rho u^2 \, dy\right] - U_e \frac{d}{dx}\left[\int_0^\delta \rho u \, dy\right] = -\tau_w - \delta\frac{dp}{dx}. \qquad \textbf{(12)}$$

The integral relationship (12) is valid both for the laminar and turbulent boundary layer.

Functions which were not known *a priori* but which characterize distribution of fluid parameters across the layer thickness δ are under the integral in Equation 12. And the error of calculating the integral is less than the error in the approximately assumed integrand function $\rho u = \rho u(y)$. These create conditions for developing approximate methods of calculating boundary layer parameters which are less time-consuming than the exact methods of integrating Prandtl equations. The fundamental concept was first suggested by T. von Karman, who introduced such arbitrary layer thicknesses as: displacement thickness δ^*

$$\delta^* = \int_0^\delta \left[1 - \frac{u}{U_e}\right] dy \qquad \textbf{(13)}$$

and momentum displacement thickness δ^{**}

$$\delta^{**} = \int_0^\delta \frac{u}{U_e}\left[1 - \frac{u}{U_e}\right] dy. \qquad \textbf{(14)}$$

thus, we can transform Equation (12) for two-dimensional boundary layer of incompressible fluid to:

$$\frac{d}{dx}(\delta^{**}) + \frac{U_e'}{U_e}(\delta^* + 2\delta^{**}) = \frac{\tau_w}{\rho U_e^2}. \qquad \textbf{(15)}$$

There are three unknown functions in Equation (15), namely, $\delta^* = \delta^*(x)$, $\delta^{**} = \delta^{**}(x)$ and $\tau_w = \tau_w(x)$ [functions of $U_e(x)$ and correspondingly $U_e'(x)$, which are known from computations of flow in the inviscid flow core].

The solution of an ordinary differential equation like Equation (15) usually requires assumption (or representation) of velocity distribution (velocity profile) across the boundary layer thickness as the function of some characteristic parameters (form-parameters), and it also requires the use of empirical data about the relationship between *friction coefficient* $C_f = 2\tau_w/(\rho U_e^2)$ and the arbitrary thickness of the boundary layer (friction law).

Some definite physical explanations can be given as far as the values of δ^* and δ^{**} are concerned. The integrand function in Equation 13 contains after rearrangement, a term $(U_e - u)$ which characterizes the velocity decrease. The integral in Equation 14 can thus be considered as a measure of decreasing the flow rate across the boundary layer, as compared with the perfect fluid flow at the velocity U_e. On the other hand, the value of δ^* can be considered as the measure of deviation along a normal to the wall (along "y" axis) of the external flow stream line under the influence of friction forces. From this consideration of the integral structure Equation 14, it is possible to conclude that δ^{**} characterizes momentum decrease in the boundary layer under the influence of friction.

The following relations are valid:

$$\delta > \delta^* > \delta^{**} \quad \text{and} \quad H = (\delta^*/\delta^{**}) > 1,$$

where H is the form-parameter of the boundary layer velocity profile. For example, for linear distribution $u = ky$,

$$\delta^* = (1/2)\delta, \qquad \delta^{**} = (1/6)\delta, \qquad H = 3.0.$$

At present, so-called semi-empirical theories are widely used for predicting turbulent boundary layer parameters. In this case, it is assumed that total friction stress τ in a turbulent boundary layer is a sum

$$\tau = \eta\frac{\partial \bar{u}}{\partial y} + \tau_T. \qquad \textbf{(16)}$$

Here τ_T is additional (turbulent or Reynolds) friction stress, in particular, in an incompressible flow $\tau_T = -\rho\overline{(u'v')}$, see Equation (7).

This representation is directly connected with the system of equations of motion in the boundary layer (6). In the compressible boundary layer, density pulsations can be considered to be the result of temperature pulsations

$$\tau_T = -\rho(\overline{u'v'})(1 - \beta), \tag{17}$$

where $\beta = (1/T)$ is the volumetric expansion coefficient.

Additional semi-empirical hypotheses about turbulent momentum transfer are used for determining τ_T. For example,

$$\tau_T = \eta_T \frac{\partial \bar{u}}{\partial y},$$

where η_T is the dynamic coefficient of turbulent viscosity introduced by J. Boussinesq in 1877.

On the basis of the concept of similarity of molecular and turbulent exchange (*similarity theory*) Prandtl introduced the *mixing length (die Mischungsweg) hypothesis.* The mixing length l is the path a finite fluid volume ("mole") passes from one layer of averaged motion to another without changing its momentum. In accordance with this condition, he derived an equation which proved to be fundamental for the boundary layer theory:

$$\tau_t = \eta_T \frac{\partial \bar{u}}{\partial y} = \rho l^2 \left|\frac{\partial \bar{u}}{\partial y}\right| \frac{\partial \bar{u}}{\partial y}. \tag{18}$$

For turbulent region of the near wall flow boundary layer, L. Prandtl considered the length l proportional to y

$$l = \kappa y, \tag{19}$$

where κ is an empirical constant.

Close to the wall, where $\eta_T << \eta$, viscous molecular friction [the first term in Equation (15)] is a determining factor. The thickness of this part of the boundary layer δ_l, which is known as laminar or viscous sub-layer, is $\simeq 0.01\delta$. Outside the sub-layer, the value of η_T increases, reaching several orders of magnitude larger than η. Correspondingly, in this zone of the boundary layer known as the turbulent core $\tau_T > \tau_0 = \eta[\partial\bar{u}/\partial y]$. Sometimes the turbulent core is subdivided into the buffer zone, where the laminar and turbulent friction are the comparable value, and the developed zone, where $\tau_T >> \tau_0$. For this region, after integrating Equation 18 and taking into account Equation 19, it is possible to derive an expression for *logarithmic velocity profile*:

$$u = \frac{1}{\kappa}\sqrt{\frac{\tau_w}{\rho}}\ln y + C. \tag{20}$$

If dimensionless (or universal) coordinates are used,

$$u^+ = u/v^* \quad \text{and} \quad y^+ = v^* y/\nu,$$

where $v^* = \sqrt{(\tau_w/\rho}$ is the so-called dynamic velocity (or **Friction Velocity**), Equation 20 can be rewritten in the following form:

$$u^+ = \frac{1}{\kappa}\ln y^+ + B. \tag{21}$$

Velocity distribution representation in universal coordinates and mathematical models for turbulent viscosity coefficient are dealt with in greater detail in the section on **Turbulent Flow**.

One of the current versions of the semi-empirical theory of turbulent boundary layer developed by S.S. Kutateladze and A.I. Leontiev is based on the so-called asymptotic theory of turbulent boundary layers at $Re \to \infty$, where the thickness of laminar (viscous) sub-layer δ_l decreases at a higher rate than δ as a result of which $(\delta_l/\delta) \to 0$.

Under these conditions, a turbulent boundary layer with "vanishing viscosity" is developing. In this layer, $\eta \to 0$ but is not equal to zero and in this respect, the layer differs from perfect fluid flow. The concept of relative friction law, introduced by S.S. Kutateladze and A.I. Leontiev (1990), indicates

$$\Psi = (C_f/C_{f0}) \text{ at } Re^{**} = \text{idem}. \tag{22}$$

The law is defined as the ratio of friction coefficient C_f for the condition under consideration to the value of C_{f0} for "standard" conditions on a flat, impermeable plate flown around by incompressible, isothermal flow. Both coefficients being obtained for $Re^{**} = U_e\delta^{**}/\nu$. It is shown that at $Re \to \infty$; $\eta \to 0$; and $C_f \to 0$, the relative variation of the friction coefficient under the influence of such disturbing factors as *pressure gradient, compressibility,* nonisothermicity, *injection (suction)* through a porous wall etc., has a finite value.

The equations derived for calculating the value of Ψ have one important characteristic which makes Ψ independent of empirical constants of turbulence. In accordance with the fundamental concept of the integral "approach", the integral momentum equation is transformed into:

$$\frac{d}{d\bar{x}}(Re^{**}) + (1 + H)\frac{Re^{**}}{U_e}U_e' = Re_L\frac{C_{f0}}{2}[\psi + b]. \tag{23}$$

Here, $\bar{x} = x/L$, $Re_L = U_eL/\nu$, $b = (2/C_{f0})(\rho_w v_w)/(\rho_e U_e)$ are the permeability parameters for the case of injecting a gas at density ρ_w through a permeable wall at the velocity of v_w. For determining the function $Re^{**} = Re^{**}(\bar{x})$, it is necessary to calculate the distribution $\Psi = \Psi(\bar{x})$. For this purpose, the principle of superposition of disturbing factors applies

$$\Psi = \Psi_M \cdot \Psi_T \cdot \Psi_B \cdot \Psi_P \cdot \cdots \tag{24}$$

In Equation 24, each multiplier represents the relative friction law, taking into account the effect of one of the factors, among them *compressibility* Ψ_M, temperature (or **Enthalpy**) head Ψ_T, injection Ψ_B, pressure gradient Ψ_P and others.

The fundamental concepts of the boundary layer create conditions for explaining such phenomena as *flow separation* from the surface under the influence of the flow inertia, deceleration of viscous flow by the wall and adverse pressure gradient acting in the upstream direction $[\partial p/\partial x] = [dp/dx] > 0$ or $[\partial u/\partial x] < 0$.

If pressure gradient is adverse on the surface location between sections '1–4' (see **Figure 3**), the velocity distribution $u = u(x,y)$ in the boundary layer changes gradually; becoming "less full," decreasing the inclination in the fluid jets which are closer to the wall and possessing less amount of kinetic energy (see velocity profile shapes in **Figure 3**) which penetrate far downstream into the region of increased pressure. In some sections, for example section '4', fluid particles which are on the 'a-a' stream line (dotted

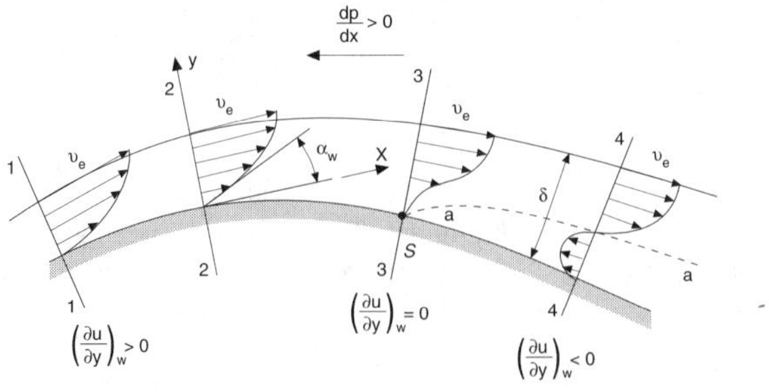

Figure 3. Boundary layer in flow over a covered surface.

line in **Figure 3**)—having completely exhausted their supply of kinetic energy become decelerated ($u_a = 0$).

Static pressure and pressure gradient value do not vary across boundary layer thickness. Therefore, fluid particles which are closer to the wall than line 'a-a' and possessing still less amount of energy begin to move in the opposite direction under the influence of the pressure gradient in '4-4' section (see **Figure 3**). Thus, the relationship:

$$\tan \alpha_w = [\partial u/\partial y]_w < 0.$$

In this way, at some locations of the surface, the velocity profile changes. This change is characterized by the alteration of the sign of the derivative $[\partial u/\partial y]_w$ from positive (section 2, **Figure 3**) to negative (section 4). Of course, it is also possible to define the section where $[\partial u/\partial y]_w = 0$ (section 3, **Figure 3**). This is referred to as the *boundary layer separation* section (correspondingly point 'S' on the surface of this section is the separation point). It is characterized by the development of a reverse flow zone—the flow around the body is no longer smooth, the boundary layer becomes considerably thicker and the external flow stream lines deviate from the surface of the body flown around. Downstream of the separation point, the static pressure distribution across the thickness of the layer is not steady and the static pressure distribution along the surface does not correspond to the pressure distribution in the external, inviscid flow.

The separation is followed by the development of reverse flow zones and swirls, in which the kinetic energy supplied from the external flow transforms into heat under the influence of friction forces. The flow separation, accompanied by energy dissipation in the reverse flow swirl zones, results in such undesirable effects as increases in the flying vehicles' drag or hydraulic losses in channels.

On the other hand, separated flows are used in different devices for intensive mixing of fluid (for example, to improve mixing of

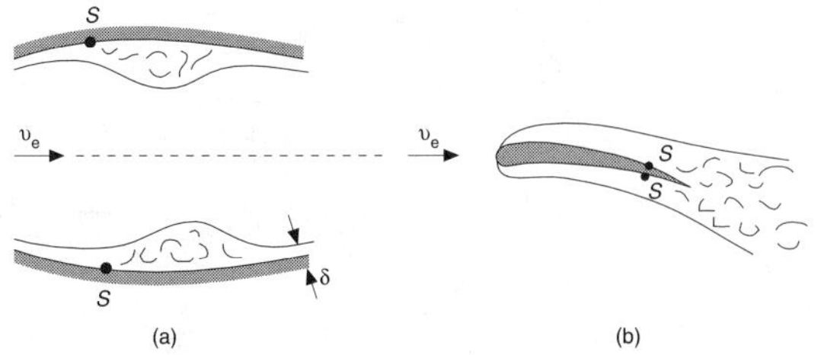

Figure 4. Boundary layer separation phenomena.

fuel and air in combustion chambers of engines). When viscous fluids flow in channels with a variable cross-section (alternating pressure gradient), the separation zone may be local if the diffusor section is followed by the confusor section, where the separated flow will again re-attach to the surface (see **Figure 4a**). When the flow separates from the trailing edge of the body (for example, from the wing trailing edge), the so-called *wake* is formed by "linking" boundary layers (see **Figure 4b**).

References

Prandtl, L. (1904) *Über Flüssingkeitsbewegungbeisehr Kleiner Reibung*: Verhandl. III Int. Math. Kongr.—Heidelberg.

Blasius, H. (1908) Grenzschichten in Flüssigkeiten mit Kleiner Reibung: *Z. Math. Phys.*, 56:1–37.

Kutateladze, S. S. and Leontiev, A. I. (1990) *Heat Transfer, Mass Transfer and Turbulent Boundary Layers,* Hemisphere Publishing Corporation, New York, Washington, Philadelphia, London.

Leading to: *Boundary layer heat transfer; Suction effect*

V.M. Epifanov

BOUNDARY LAYER HEAT TRANSFER

Following from: *Convective heat transfer; Boundary layer*

The time-averaged differential equation for energy in a given flow field is linear in the temperature if fluid properties are considered to be independent of temperature. Thus, the concept of a **Heat Transfer Coefficient** arises such that the heat transfer rate from a wall is given by:

$$\dot{q} = \alpha(T_w - T_r) \qquad (1)$$

where the heat transfer coefficient, α, is only a function of the flow field. T_w is the wall temperature and T_r the recovery or adiabatic wall temperature. The above is also true of the **Boundary Layer** energy equation, which is a particular case of the general energy equation. When fluids encounter solid boundaries, the fluid in contact with the wall is at rest and viscous effects thus retard a layer in the vicinity of the wall. For large **Reynolds Numbers** based on distance from the leading edge, these viscous layers are thin compared to this length.

When the wall is at a different temperature to the fluid, there is similarly a small region where the temperature varies. These regions are the *velocity* and *thermal boundary layers.* In 1905 **Prandtl** showed that this thin region could be analysed separately from the bulk fluid flow in that pressure variation normal to the wall may be neglected and the pressure is given by that impressed by the free-stream. Velocity normal to the wall is also, of order, the thickness of the boundary layer, the characteristic velocity being that of the free-stream and the length being the distance from the leading edge. Thus, the boundary layer equations for steady incompressible laminar flow in two dimensions may be approximated to be:

$$\frac{\partial u}{\partial x} + \frac{\partial u}{\partial y} = 0 \text{ continuity} \tag{2}$$

$$u\frac{\partial u}{\partial x} + v\frac{\partial u}{\partial y} = -\frac{1}{\rho}\frac{dp}{dx} + \mu\frac{\partial^2 u}{\partial y^2} \text{ momentum} \tag{3}$$

$$u\frac{\partial T}{\partial x} + v\frac{\partial T}{\partial y} = \frac{\lambda}{\rho c_p}\frac{\partial^2 T}{\partial y^2} + \mu\left[\frac{\partial u}{\partial y}\right]^2 \text{ energy} \tag{4}$$

p, T, u and v are the flow pressure, temperature and velocities along and perpendicular to the surface, respectively. λ, μ, c_p and ρ are similarly the thermal conductivity, viscosity, specific heat and density. x and y are *Cartesian coordinates* along and perpendicular to the surface.

The classical laminar solution to the momentum equation was provided by Blasius for the case of a semi-infinite flat plate aligned with uniform flow. The velocities normalised by the free-stream value u_o are plotted in **Figure 1** vs. the nondimensional quantity $\eta = y/xRe_x^{-1/2}$; Re_x is the Reynolds number based on distance from the leading edge of the plate.

The velocity gradient at the wall gives the *skin friction,* τ, which can be expressed as the skin friction coefficient:

$$c_f = \frac{\tau}{\frac{1}{2}\rho u_o^2} = 0.664\ Re_x^{-1/2}. \tag{5}$$

The suffix o refers to the free-stream value.

The energy equation may be solved using the *Blasius solution* to give the heat transfer in terms of the **Nusselt Number**, Nu_x, when the dissipation term, $\mu(\partial u/\partial y)^2$ is neglected.

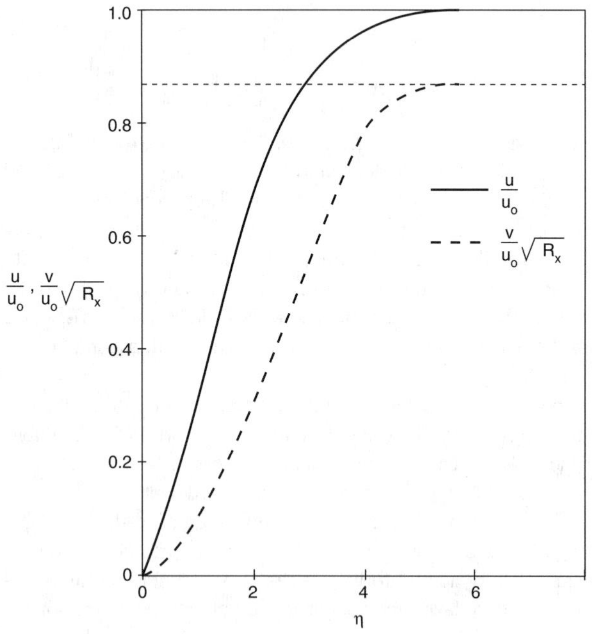

Figure 1. Laminar boundary layer normalised velocities along and perpendicular to a flat plate from Young (1989).

$$Nu_x = \frac{\alpha x}{\lambda} = 0.332\ Pr^{1/3}Re_x^{1/2} \qquad (0.5 < P_r < 15). \tag{6}$$

This may also be expressed in terms of a **Stanton Number**, $St = \alpha/\rho u c_p$, as, Re St Pr = Nu. **Figure 2** shows temperature profiles for different **Prandtl Numbers**. There is thus an analogy between heat transfer and skin friction (Reynolds analogy) which may be expressed as:

$$St = \frac{1}{2}c_f Pr^{1/3} \tag{7}$$

Expressions for the boundary layer deficit thicknesses of mass, momentum and temperature are, respectively:

$$\frac{\delta_1}{x} = 1.72\ Re_x^{-1/2}; \qquad \frac{\delta_2}{x} = 0.664\ Re_x^{-1/2}; \tag{8}$$

$$\frac{\delta_3}{x} = 0.664\ Pr^{-2/3}Re_x^{-1/2}. \tag{9}$$

The deficit thickness represents the height of free-stream fluid which carries the boundary layer deficit in the relevant quantity

$$\rho u_o \delta_1 = \int_o^\infty \rho(u_o - u)dy; \qquad \rho u_o^2 \delta_2 = \int_o^\infty \rho u(u_o - u)dy; \tag{10}$$

$$\rho u_o c_p(T_o - T_w)\delta_3 = c_p\int_o^\infty \rho u(T_o - T)dy. \tag{11}$$

By integrating the boundary layer equations with respect to y, the integral boundary layer equations are generated in terms of the boundary layer thicknesses. These form the basis for many approximate solutions [Young (1989), Kays and Crawford (1993)].

The analysis of boundary layers on wedges, where the external velocity varies as a power law from the leading edge, has been performed by Falkner and Skan. Approximate methods are avail-

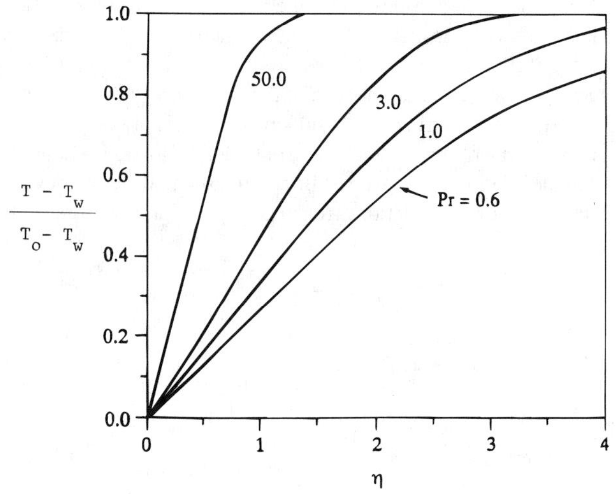

Figure 2. Normalised temperature profiles in a laminar boundary layer for different Prandtl numbers from Kays and Crawford (1993).

able for arbitrary free-stream velocity variations [see Young (1989), Kays and Crawford (1993), and Schlichting (1987)]. (See also **Wedge Flows**.) The two-dimensional stagnation point may be treated as the special case of a wedge of 90° half angle and gives a constant boundary layer thickness. This result is also given by the Hiemenz solution [Schlichlting (1987)]. The stagnation point Nusselt number, based on R for a cylinder of radius R, may thus be evaluated as:

$$\mathrm{Nu_R} = 0.81\ \mathrm{P_r^{0.4} Re_R^{1/2}}. \qquad \textbf{(12)}$$

For an axisymmetric stagnation point boundary layer, the result becomes:

$$\mathrm{Nu_R} = 0.93\ \mathrm{P_r^{0.4} Re_R^{1/2}}. \qquad \textbf{(13)}$$

The boundary layer analysis described above may be considered as a particular case of general *perturbation methods* in solving the differential equations of fluid mechanics Van Dyke (1964). Perturbation methods add a mathematical rigour to the boundary layer concept and also allow solutions to be determined for variable property flows. Gersten (1982) made a review of perturbation methods in heat transfer, showing that the effects of variable properties may be taken into account with property ratios raised to a relevant power. i.e.,

$$\mathrm{Nu} = \mathrm{Nu_o}\left(\frac{\mu_o}{\mu_w}\right)^a\left(\frac{\lambda_o}{\lambda_w}\right)^b\left(\frac{c_{po}}{c_{pw}}\right)^c. \qquad \textbf{(14)}$$

Suffix w refers to the wall value.

Herwig et al. (1989) have developed this analysis further. Ziugzda and Zukauskas (1989) used experimental data to derive empirical relationships that take temperature-dependent properties into account for laminar and turbulent boundary layers.

The boundary becomes turbulent in response to external disturbances. This *transition* to a *fully turbulent boundary* may take place over a significant length of the surface, and is an important factor in the heat transfer to turbine blades for example (Mayle (1991)). The transition takes place in a characteristic manner by the formation of *turbulent spots.* These spots convect along the surface and grow, so as to coalesce, to finally form the turbulent boundary layer as shown in **Figure 3**.

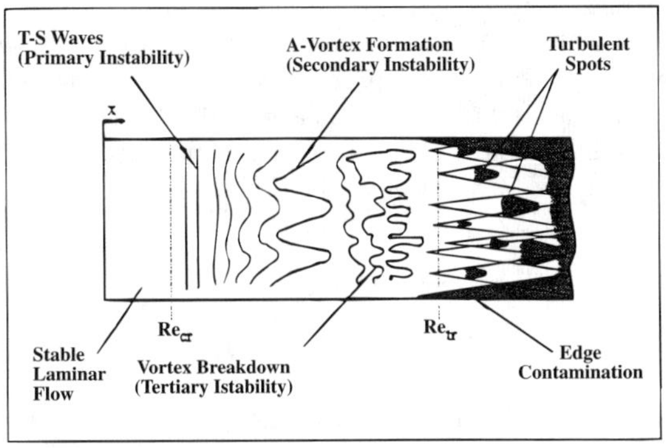

Figure 3. Idealised sketch of the transition process from White (1991).

The dynamics of the turbulent spot are presently the subject of much experimental and theoretical study [Smith (1994), Kachanov (1994)]. The spreading of spots is inhibited at high free-stream *Mach number* and also by accelerating flow (favourable pressure gradients). Adverse pressure gradients, on the other hand, cause very rapid formation of the turbulent boundary layer. An interesting property of turbulent spots is that there appears to be a *wake* region in which generation of spots is inhibited.

It is an experimental observation that after a short inception stage, the heat transfer to the surface under the spot is closely given by that under a continuous turbulent boundary layer, which has grown from the point where spots are first initiated. A typical variation of heat transfer in the transition region is given in **Figure 4**. Turbulent spots can be seen in this figure passing over consecutive gauges and growing to coalesce to a turbulent boundary layer.

The fraction of time for which the boundary layer is fully turbulent at a point is called the *intermittency,* γ, and this follows a universal law on a flat plate.

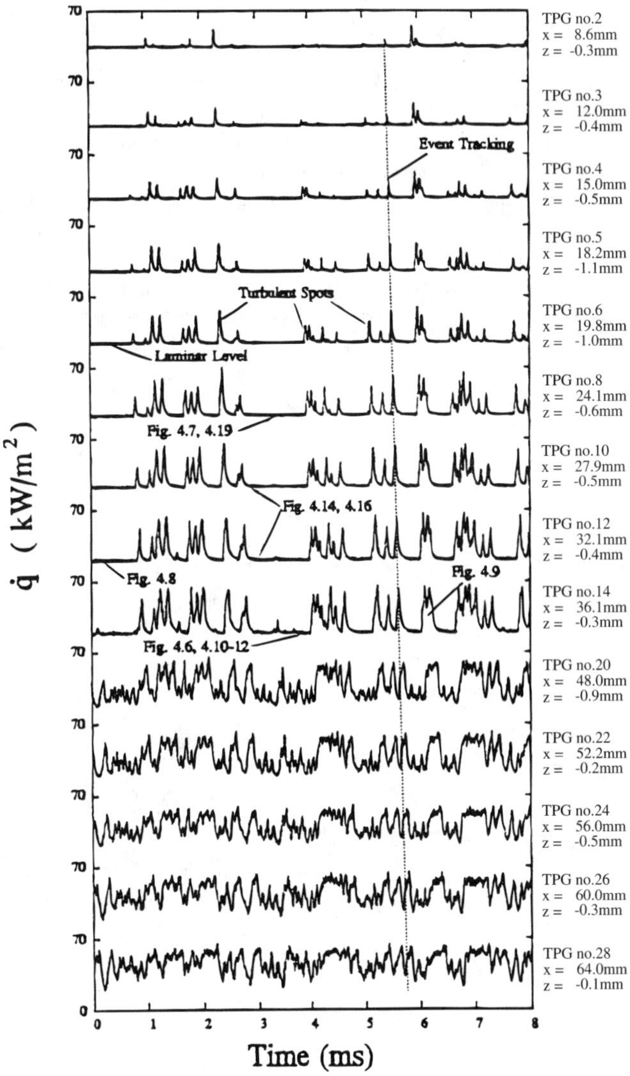

Figure 4. Unsteady simultaneous heat flux signals due to turbulent spots passing over gauges mounted downstream of one another from Clark (1993).

$$\gamma = 1 - e^{-0.41\eta^2} \quad \mathbf{(15)}$$

where $\eta = (x - x_t)/\lambda$, x_t being the point of transition and λ the distance between the 25% and 75% intermittency points. The heat transfer may be estimated by assuming that the local Stanton number is given by that time average of the laminar and turbulent values, St_L and St_T indicated by the intermittency.

$$\text{i.e.,}\quad St = St_L(1 - \gamma) + \gamma St_T. \quad \mathbf{(16)}$$

A review of such engineering formulae is given in Fraser et al. (1994).

Figure 5 shows a typical intermittency plot and **Figure 6** gives the momentum thickness Reynolds number at the start of transition as a function of free-stream turbulence.

The start of transition and the length of the transition region are notoriously difficult to predict accurately. Classical instability analysis may be used to find those disturbances which first form in the boundary layer as *Tollmien-Schlichting waves* (Schlichting (1987)). However, determining when these grow to become non linear and form turbulent spots has not been perfected and empirical methods are based on the predicted growth of these waves to a critical point. At free-stream turbulence levels above 1%, the Tollmein-Schlichting wave instability may possibly not have a controlling effect and, so-called, "bypass" transition takes place. In this case, the transition point is a function of free-stream turbulence as indicated in **Figure 6**. Predictions may be made in this case using low Reynolds number models of turbulent boundary layers, which can also reproduce a laminar boundary layer. Thus free-stream turbulence is entrained, generating a turbulent boundary layer without recourse to a stability analysis, and predictions are given in Figure 6. Other instabilities occur on concave surfaces and Saric (1994) has reviewed these vortex structures, usually termed Goertler vortices. (See **Goertler-Taylor Vortex Flow**.)

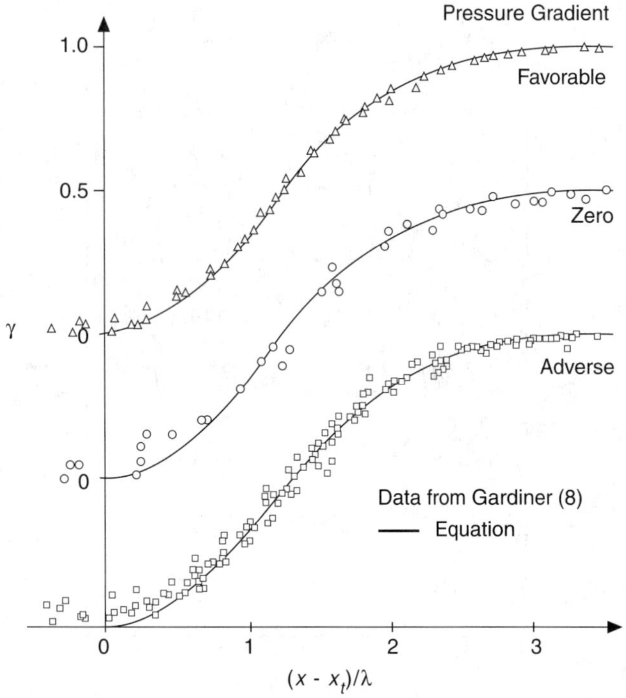

Figure 5. The universal intermittancy curves and experimental data from Fraser et al. (1994).

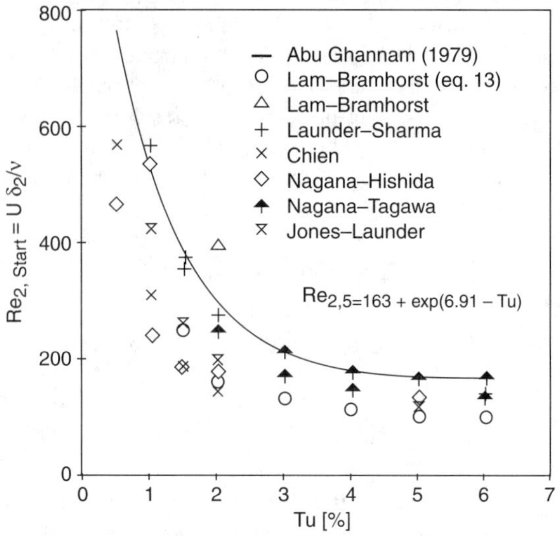

Figure 6. Momentum thickness Reynolds number at the start. The solid line is experimental and the points are Low Reynolds Number Models, from Seiger (1992).

When the boundary layer becomes fully turbulent, the heat transfer through the boundary layer is dominated by the transport associated with the turbulent eddies. Very close to the wall, however, molecular conduction still prevails as the eddies are inhibited by the wall. The time-averaged boundary layer equations become the incompressible constant property case without dissipation.

$$\frac{\partial \bar{u}}{\partial x} + \frac{\partial \bar{u}}{\partial y} = 0 \quad \mathbf{(17)}$$

$$\bar{u}\frac{\partial \bar{u}}{\partial x} + \bar{v}\frac{\partial \bar{u}}{\partial y,} = -\frac{\partial}{\partial x}\overline{u'v'} + \mu\frac{\partial^2 \bar{u}}{\partial y^2} - \frac{1}{\rho}\frac{\partial p}{\partial x} \quad \mathbf{(18)}$$

$$\bar{u}\frac{\partial \bar{T}}{\partial x} + \bar{v}\frac{\partial \bar{T}}{\partial y} = -\frac{\partial}{\partial y}\overline{(T'u')} + \frac{\lambda}{\rho c_p}\frac{\partial^2 \bar{T}}{\partial y^2} \quad \mathbf{(19)}$$

The $^-$ denotes the time-averaged quantity whereas the $'$ denotes the the fluctuating value. The physical modelling of the terms $\overline{u'v'}$ and $\overline{T'v'}$, and additional terms in the fully compressible equations are the subject of much study [Cebeci and Bradshaw (1984), Wilcox (1993), Huang Bradshaw and Croakley (1994)]. The traditional treatment is to express the turbulent transport in terms of an *eddy viscosity,* μ_i and *mixing length,* l, such that:

$$\overline{u'v'} = \frac{\mu_t}{\rho}\frac{\partial \bar{u}}{\partial y} = \epsilon_M \frac{\partial \bar{u}}{\partial y} = Kl\,\frac{\partial \bar{u}}{\partial y}\cdot\frac{\partial \bar{u}}{\partial y}. \quad \mathbf{(20)}$$

ϵ_M is the diffusivity and K is a constant.

In the region close to the wall, it can be assumed that the **Shear Stress** is constant and equal to the wall value, τ_o. Also the mixing length is taken to be proportional to the distance from the wall. This leads to the *law of the wall* which may be written in non dimensional terms as:

$$u^+ = 2.44 \ln y^+ + 5.0 \quad \textbf{(21)}$$

$$u^+ = \frac{u}{u_\tau}, \qquad y^+ = \frac{yu_\tau}{v} \quad \text{where} \quad u_\tau = \sqrt{\frac{\tau_0}{\rho}}. \quad \textbf{(22)}$$

Similar arguments apply to the temperature field. The region close to the wall is represented by a laminar relationship and that far from the wall, the outer region is characterised by essentially a constant mixing length. Pressure gradients have an influence on the outer region. The mixing length concept may be extended through the viscous sublayer using the *Van Driest damping* formula for the mixing length, which reduces this to zero exponentially as the wall is approached. **Figure 7** shows experimental results confirming the law of the wall. A similar relationship may also be determined for the temperature profiles through the boundary layer, although pressure gradients influence the law of the wall region significantly.

From the above analysis of the turbulent boundary layer, the heat transfer to a flat plate in uniform gas flows may be derived.

$$\text{St}\,\text{Pr}^{0.4} = 0.0287\,\text{Re}_x^{-0.2}. \quad \textbf{(23)}$$

Boundary layer development is now largely predicted by computing solutions to the boundary layer equations with the relevant boundary conditions [Cebeci and Bradshaw (1984), Wilcox (1993)]. More complex turbulence modeling is now routinely used in such codes and subsidiary differential equations for turbulence quantities are solved. For example, the $k - \epsilon$ model takes ϵ_M = constant. k^2/ϵ, where k is the local turbulence kinetic energy and ϵ is the turbulence dissipation rate. Thus, free-stream acceleration and freestream turbulence may be taken into account. This hierarchy of turbulence models, which allow closure of the governing differential equations, are classified by the number of subsidiary differential equations

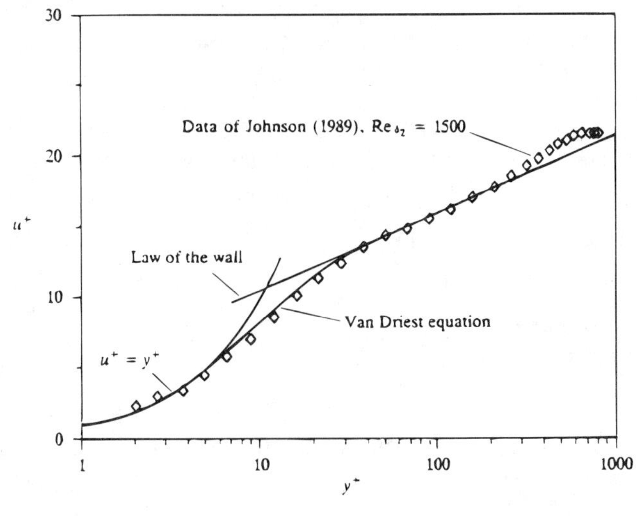

Figure 7. Experimental results compared with the Law of the Wall and the Van Driest sublayer modification, from Kays and Crawford (1993).

employed in this closure. Thus the $k - \epsilon$ model is a two-equation model employing separate differential equations for both k and ϵ. Boundary conditions for the turbulence quantities are necessary as the viscous sublayer is approached, and those arising from the law of the wall (i.e., a wall function) may be employed. Alternatively, modifications to the differential equations give rise to low Reynolds number formulations which apply down to the wall and may also represent a laminar boundary layer (Schmidt and Patankar (1991)).

It is possible to model to the term $\overline{u'v'}$ directly without the concept of an eddy viscosity. *Reynolds stress transport models* make use of the transport equation for the eddy stress and model terms within this equation so as to give the eddy stress term in the boundary layer. A brief survey of these methods and modeling of the transition region is given by Singer (1993).

Compressibility effects occurring in high-speed gas flows may be taken into account in a straightforward manner in computational predictions of the boundary layer through the **Equation of State**. The dissipation term now plays a dominant role in the governing energy equation. Analytical procedures usually allow transformation of the boundary layer equations to facilitate solution. The *Crocco transformation* is an example of this for a laminar boundary layer and leads to the conclusion for Prandtl numbers close to unity—which is typical of gases—that the incompressible relationship between heat transfer and skin friction may be employed for a flat plate in a uniform stream. The *recovery temperature* in this case is found from.

$$T_r = T_\infty\left(1 + \frac{1}{2}(\gamma - 1)M^2P_r^{1/2}\right). \quad \textbf{(24)}$$

The suffix ∞ refers to the free-stream static value. M is the free-stream Mach number.

The relationship between Stanton number, St, and skin friction coefficient, c_f, still applies.

$$\text{St} = \frac{1}{2}c_fP_r^{1/3}. \quad \textbf{(25)}$$

This is independent of Mach number when properties are determined at free-stream conditions. The value of skin friction coefficient is found from the incompressible expression, except that properties are evaluated at a reference temperature.

$$T_{ref} = .45T_\infty + .55T_w + 0.09(\gamma - 1)M^2P_r^{1/2}T_\infty. \quad \textbf{(26)}$$

In turbulent compressible flow, the law of the wall is recovered by a transformation of coordinates proposed by Van Driest. The performance of two-equation turbulence models in the compressible case has been reported by Huang, Bradshaw and Croakley (1994). Simple corrections to the incompressible heat transfer equations using reference temperatures and temperature ratios may also be employed (Kays and Crawford (1993)).

The boundary layer displacement thickness gives the aerodynamic influence of the boundary layer on the external flow. Usually, this change is small and has little bearing on the growth of the boundary layer itself. At hypersonic speeds, there can be strong effects of displacement as the heating of the boundary layer gas decreases the density and increases this thickness. A parameter on which laminar hypersonic phenonema depend is the hypersonic viscous interaction parameter, $\bar{x} = (M^3/\sqrt{R_e})\sqrt{\rho_w\mu_w/\rho_\infty\mu_\infty}$. Weak and strong interactions are recognised. In the former case, there is an effect on the external flow but little on the boundary layer. In the latter, there is a significant influence on both. Anderson (1989) gives an account of this phenonema.

Free-stream turbulence influences the transition from laminar to turbulent flow in the boundary layer, but it also has a very significant effect on the levels of heat transfer within a laminar boundary layer prior to transition. The effect on the stagnation point heat transfer is pronounced and is taken into account using semi-empirical correlations of the form below, where Tu is the free-stream turbulence intensity $u'/\bar{u}$,

$$\frac{Nu_R}{\sqrt{Re_R}} = f(Tu\sqrt{Re_R}). \qquad \textbf{(27)}$$

Mayle (1994) has generalised this stagnation point enhancement by relating this to the local free-stream acceleration. The effects of fluctuations in free-stream velocity along the surface of a flat plate have been examined by Lighthill (1954) and do not produce such a large increase in surface heat transfer. The influence of free-stream turbulence on turbulent boundary layer heat transfer is not as significant as for a laminar boundary layer. In this case, turbulent intensity and length scale are relevant. Moss and Oldfield (1992) have discussed previous works and produced an empirical correction for the enhancement of heat transfer on a flat plate.

Buoyant flow producing *free convection boundary layers* is considered in both Cebeci and Bradshaw (1984) and Kays and Crawford (1993). Time-dependent problems are also covered in the texts of Schlichting (1987) and Young (1989). Note that the time to establish a steady laminar boundary layer—when the free-stream is suddenly started—is, of order, the transit time of the free-stream from the leading edge to the point under consideration [Jones et al. (1993)].

References

Anderson, J. D. Jr. (1989) *Hypersonic and High Temperature Gas Dynamics,* McGraw-Hill.

Cebeci, T. and Bradshaw, P. (1984) *Physical and Computational Aspects of Convective Heat Transfer,* Springer-Verlag, New York.

Clark, J. (1993) *A Study of Turbulent—Spot Propagation in Turbine Representative Flows,* D. Phil Thesis, Univ. of Oxford, U.K.

Dullenkopf, K., and Mayle, R. J. (1994) *An Account of Free Stream Turbulence Length Scale on Laminar Heat Transfer,* ASME, Paper 94-GT-174.

Fraser, C. J., Higazy, M. G., and Milne, J. S. (1994) End-Stage Boundary Layer Transition Models for Engineering Calculations, Proc. Instn. Mech. Engrs. Part C: *Journ. Mech Eng. Science,* 208:47–58.

Gerston, K. (1982) *Advanced Boundary Layer Theory in Heat Transfer,* 7th Int. Heat Transfer Conf. Vol 1, Hemisphere pp. 159–179.

Herwig, H., Voigt, M., and Bauhaus, F. J. (1989) *The Effect of Variable Properties on Momentum and Heat Transfer in a Tube with Constant Wall Temperature, Int. J. Heat Mass Transfer,* 32/10:1907–1915.

Huang, P. G., Bradshaw, P., and Croakley, J. J. (1994) *Turbulence Models for Compressible Boundary Layers, AIAA Journal,* Vol.32, No.4, April.

Jones, T. V., Oldfield, M. L. G., Ainsworth, R. W., and Arts, T. (1993) *Transient Cascade Testing, Ch5, Advanced Methods for Cascade Testing,* AGARDOGRAPH 328, Ed. Ch. Hirsch, AGARD.

Kachanov, Y. S. (1994) Physical Mechanics of Laminar Boundary Layer Transition, *Ann. Rev. Fluid Mech.* 26:471–482.

Kays, W. M. and Crawford, M. E. (1993) *Convective Heat and Mass Transfer,* 3rd Edition, McGraw Hill.

Lighthill, M. J. (1954) The Response of Laminar Skin Friction and Heat Transfer to Fluctuations in the Freestream Velocity, *Proc. Roy. Soc.,* 224A:1–23.

Mayle, R. E. (1991) The Role of Laminar-Turbulent Transition in Gas Turbine Engines, *Journal of Turbomachinery, ASME,* 113:509–537.

Moss, R. W. and Oldfield, M. L. G. (1992) *Measurements of the Effects of Free-Stream Turbulence Length Scales on Heat Transfer.* ASME, Paper 92-GT-244.

Saric, W. S. (1994) Görtler Vortices, *Ann. Rev. Fluid Mech,* 26:379–409.

Schlichting, H. (1987) *Boundary Layer Theory,* McGraw-Hill.

Schmidt, R. C. and Patankav, S. V. (1991) Simulating Boundary Layer Transition with Low Reynolds Number $k - \epsilon$ Turbulence Models, Part I—An Evaluation of Prediction Characteristics, *J. Turbomachinery, ASME,* 113:10–17, Part II—An Approach to Improving the Prediction Characteristics, 18–26.

Seiger, K., Schulz, A., Crawford, M. E., and Wittig, S. (1992) *Comparative Study of Low-Reynolds Number $k - \epsilon$* Turbulence Models for Predicting Heat Transfer along Turbine Blades with Transition, 1992 Symposium on Heat Transfer in Turbomachinery, Int. Centre for Heat and Mass Transfer (Dubrovnik), Marathon.

Singer, B. A. (1993) *Modelling the Transition Region,* NASA CR4492.

Smith, F. T. (1994) Special Issue on Transitional—Turbulent Spots, *Journ. Engineering Mathematics,* Vol. 28, 1994.

Van Dyke, M. (1964) *Perturbation Methods in Fluid Mechanics,* Academic, New York.

White, F. M. (1991) *Viscous Fluid Flow,* 2nd Ed., McGraw Hill, New York.

Wilcox, D. C. (1993) *Turbulence Modelling for CFD,* DCW Industries Inc., La Canada, Ca.

Young, A. D. (1989) *Boundary Layers,* B.S.P. Professional Books, Oxford.

Ziugzda, J. J. and Zukauskas, A. A. (1989) *The Influence of Fluid Physical Parameters and of their Variation on Heat Transfer at Forced Convection,* in "Heat Transfer—Soviet Reviews—1 Convective Heat Transfer" Eds. OG. Martymenko and A. A. Zukauskas, Hemisphere.

T.V. Jones

BOUNDARY LAYER MASS TRANSFER (see Mass transfer)

BOUNDARY LAYER SEPARATION (see Cross flow; Cross flow heat transfer; Tube banks, crossflow over; Tubes, single-phase heat transfer to in crossflow)

BOUNDARY LAYERS, IN FREE CONVECTION (see Free convection)

BOUNDARY LAYERS, IN SUPERSONIC AND HYPERSONIC FLOW (see Boundary layer heat transfer)

BOUNDARY LAYERS, SUCTION EFFECTS IN (see Suction effects)

BOUNDARY LAYER THICKNESS (see Boundary layer heat transfer)

BOURDON GAUGES (see Pressure measurement)

BOUSSINESQ (see Turbulence models)

BOUSSINESQ APPROXIMATION (see Solidification; Turbulence)

BOUSSINESQ ASSUMPTION (see Particle transport in turbulent fluids)

BOUSSINESQ HYPOTHESIS (see Computational fluid dynamics; Turbulent flow)

BOUSSINESQ NUMBER

The Boussinesq number, B, is a dimensionless group representing the square root of the ratio of inertia force to gravitational force. It is represented as

$$B = u/(gL)^{1/2} \quad \text{or} \quad u/(2gR_H)^{1/2}$$

where u is the fluid velocity, g is the acceleration due to gravity, L is a representative length or hydraulic diameter and R_H is the hydraulic radius. The group occurs frequently in the analysis of **Fluid Flow** with an interface between liquid and gas (see for example, **Channel Flow** and **Waves in Fluids**).

Boussinesq number is the same as **Froude Number**.

G.L. Shires

BOWEN RATIO METHOD OF SURFACE HEAT BALANCE (see Environmental heat transfer)

BOYLE'S (BOYLE-MARIOTTE) LAW

Following from: *PVT relationships; Perfect gas*

One of the main empirical ideal-gas laws was established in 1662 by R. Boyle, and independently in 1676 by E. Mariotte. According to Boyle's Law, at constant temperature (and low pressure) the volume V of a gas mass M is inversally proportional to its pressure p, i.e., pV = const. This law represents a description of the isothermal process of ideal gas.

The joint analysis of Boyle's law and the empirical *Charles'*, **Gay-Lussac's** and *Avogadro's laws* can lead to the universal **Equation of State** of ideal gas—*the Clapeyron (Clapeyron-Mendeleyev) equation.* $pV = (M/\tilde{M}) RT$, where $\tilde{M}$ is the molecular mass of gas, R is the universal gas constant (R = 8.314 J/mole K), and T is the absolute temperature.

This equation was subsequently rigorously substantiated by the molecular kinetic theory according to which p = nkT where n is the number of molecules in unit volume, k is the *Boltzmann constant* ($k = R/N_A$, $N_A = 6.022\ 10^{23}$ mole^{-1} is the Avogadro number, $k = 1.381\ 10^{-23}$ J/K), from which follows the ideal gas equation of state and, accordingly, the expression for Boyle's law.

As with the other ideal-gas laws, Boyle's law is applicable at pressures far from critical, i.e., in that region of state, where one may neglect: a) the proper size of particles as compared to the interparticle distance; and b) the forces of interparticle interaction, where the energy of this interaction is considerably lower than the kinetic energy of particles.

D.N. Kagan

BRAGG CELL (see Anemometers, laser Doppler; Lasers)

BRAYTON CYCLE (see Jet engine)

BRAZED HEAT EXCHANGERS (see Plate fin heat exchangers)

BREEDER REACTORS (see Liquid metal cooled fast reactor)

BREEDING (see Nuclear reactors)

BREMSSTRAHLUNG

Following from: *Electromagnetic waves*

Bremsstrahlung is *electromagnetic radiation* (as are X-rays and gamma rays) produced by the acceleration or deceleration of moving charged particles, such as electrons or positive ions. The word '*bremsstrahlung*' is German and literally means 'braking radiation'. Bremsstrahlung is produced when, for example, a high energy electron (or beta particle) is slowed down in matter by the interaction between the electric field of the electron and that of the atomic electrons and nuclei in the matter. The resulting bremsstrahlung has a continuous energy spectrum between zero and the electron's initial kinetic energy.

The existence of bremsstrahlung leads to problems in shielding sources of high energy beta particles. It is possible to determine the thickness of the shield necessary to stop all beta particles of a particular energy. The bremsstrahlung produced however, being electromagnetic radiation, can travel much further than the initial electron which produced it. It will be attenuated in an exponential manner and cannot be stopped completely by extra shielding; but it can be reduced to an acceptable level by an appropriate thickness.

Bremsstrahlung is also known as '*synchrotron radiation*' when it is produced by the acceleration of charged particles in a particle accelerator. Some particle accelerators are constructed specifically as sources of synchrotron radiation.

T.D. MacMahon

BREWSTER WINDOW (see Lasers)

BRIDGEMAN TABLES (see Thermodynamics)

BRINKMAN NUMBER (see Flow of fluids)

BRITISH ASSOCIATION FOR THE ADVANCEMENT OF SCIENCE, BAAS

Fortress House
23 Saville Row
London W1X 1AB
UK
Tel: 0171 494 3326

BRITISH STANDARDS INSTITUTION, BSI

2 Park Street
London W1A 2BS
UK
Tel: 0171 629 9000

BRITTLE FRACTURE (see Fracture of solid materials)

BROMINE

Bromine—(Gr. *bromos,* stench), Br; atomic weight 79.904; atomic number 35; melting point −7.2°C; boiling point 58.78°C; density of gas 7.59 g/l, liquid 3.12 kg/l (20°C); valence 1, 3, 5, or 7. Bromine was discovered by Balard in 1826, but not prepared in quantity until 1860. A member of the halogen group of elements, it is obtained from natural brines from wells in Michigan and Arkansas. Little bromine is extracted today from seawater, which contains only about 85 ppm.

Bromine is the only liquid nonmetallic element. It is a heavy, mobile, reddish-brown liquid, volatilizing readily at room temperature to a red vapor with a strong disagreeable odor, resembling chlorine, and having a very irritating effect on the eyes and throat; it is readily soluble in water or carbon disulfide, forming a red solution; it is less active than chlorine but more so than iodine; it unites readily with many elements and has a bleaching action; when spilled on the skin, it produces painful sores. It presents a serious health hazard, and maximum safety precautions should be taken when handling it.

Much of the bromine output in the U.S. is used in the production of ethylene dibromide, a lead scavenger used in making gasoline antiknock compounds. Lead in gasoline, however, is presently being drastically reduced, due to environmental considerations. This will greatly affect future production of bromine.

Bromine is also used in making fumigants, flameproofing agents, water purification compounds, dyes, medicinals, sanitizers, inorganic bromides for photography, etc. Organic bromides are also important.

Handbook of Chemistry and Physics, CRC Press

BROMLEY EQUATION FOR FILM BOILING (see Boiling)

BROOKHAVEN NATIONAL LABORATORY, BNL

Upton, Long Island
NY11973
USA

BROWNIAN DIFFUSION (see Diffusion)

BROWNIAN DIFFUSIITY (see Gas-solids separation, overview)

BROWNIAN MOTION

Discovered in 1828 by the botanist Robert Brown, **Brownian motion** is a continuous but random motion exhibited by sufficiently small particles when suspended in a liquid. The phenomenon is due to agitation of the particles by the molecules of the liquid, and the first adequate theory was put forward by Einstein in 1905. This theory establishes a connection between the mean square distance $\langle x^2 \rangle$ that a Brownian particle travels in time t and the viscosity of the liquid, η, through which it travels. For a spherical particle of radius a, this relation is:

$$\langle x^2 \rangle = \frac{kT}{3\pi a \eta} t \quad (1)$$

where k is *Boltzmann's constant.* The quantity $\langle x^2 \rangle/t$ is thus a kind of diffusivity for the suspended particles. As a numerical example, consider particles of radius 1 μm in liquid water at 300 K (η = 1 mPas); in this case, $\langle x^2 \rangle/t = 4.4 \times 10^{-13}$ m²/s and the RMS distance travelled in 10 s is about 2 μm.

Reference

Kennard, E. H. (1938) *Kinetic Theory of Gases,* McGraw-Hill, New York.

J.P.M. Trusler

BRUNAUER-EMMET-TELLER, BET, EQUATION (see Adsorption)

BRUNT-VAISALA FREQUENCY (see Turbulence)

BSI (see British Standards Institution)

BUBBLE COALESCENCE (see Gas-liquid flow)

BUBBLE COLUMN (see Aeration; Direct contact heat exchangers)

BUBBLE DISPERSION (see Plunging liquid jets)

BUBBLE FLOW

Following from: *Gas-liquid flow*

General Description

Bubble flow is defined as a **Two-Phase Flow** where small bubbles are dispersed or suspended as discrete substances in a liquid continuum. Typical features of this flow are moving and deformable interfaces of bubbles in time and space domains and complex interactions between the interfaces, and also between the bubbles and the liquid flow. According to the magnitude of these interactions, bubble flow is classified into four different flow regimes: i.e., *ideally-separated bubble flow; interacting bubble flow; churn turbulent bubble flow;* and *clustered bubble flow*, as shown in **Figure 1**.

In ideally-separated bubble flow, the bubbles do not interact with each other directly or indirectly. The bubbles thus behave like single bubbles. In interacting bubble flow, bubble number density becomes so large that the bubbles begin to interact with each other directly or indirectly due to collisions or the effects of wakes caused by other bubbles. With a further increase in bubble number density, the bubbles tend to coalesce to form so-called *cap bubbles*, and the flow changes to churn turbulent bubble flow. The flow contains cap bubbles formed in this way and also smaller bubbles; it is highly agitated due to the interactions between bubble motions and turbulent flow. The large bubbles ocassionally form clustering of bubbles, as shown in **Figure 1**, and they behave like a single gas slug. After a certain travel, they sometimes coalesce to form a gas slug and sometimes, they separate into individual bubbles. This flow regime is thus a transition from bubble flow to slug or churn flow. Quantitative criteria for the transitions from ideally-separated to interacting bubble flow, and from interacting bubble flow to churn turbulent bubble flow are roughly 0.01 and 0.06 in void fraction, respectively. Detailed discussions of single bubble behavior are given by Wallis (1969).

Bubble flow is characterzed by phase distribution phenomena, which exhibit different lateral void fraction profiles, depending on the volumetric flow rate of gas and liquid phases. Typical lateral void distribution patterns are given in **Figure 2**, representing wall void peaking, core void peaking and intermediate void peaking. For given flows of the two-phases in a given channel, the lateral void profile can take more or less any of these forms.

Figure 3 summarizes the trends of these void distribution patterns observed in vertically upward air-water flow in pipes using conventional types of bubble generators.

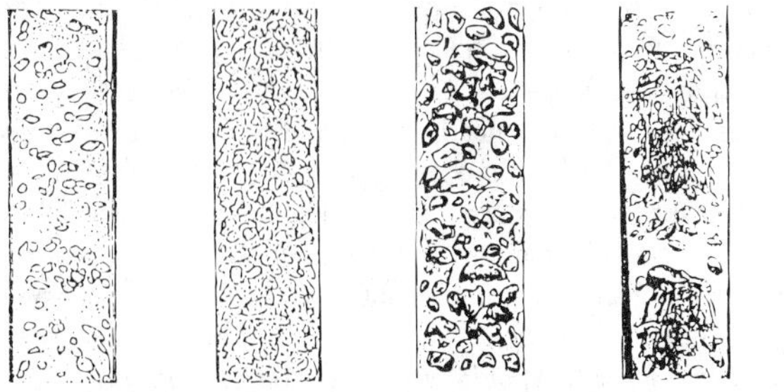

Figure 1. Bubble flow regimes in vertical pipe flow.

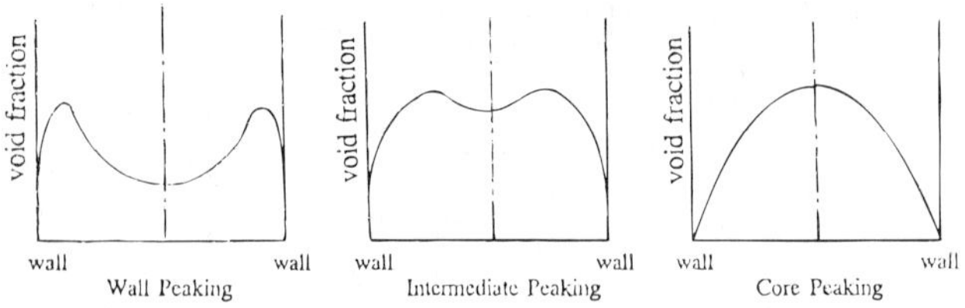

Figure 2. Typical lateral void fraction distribution patterns.

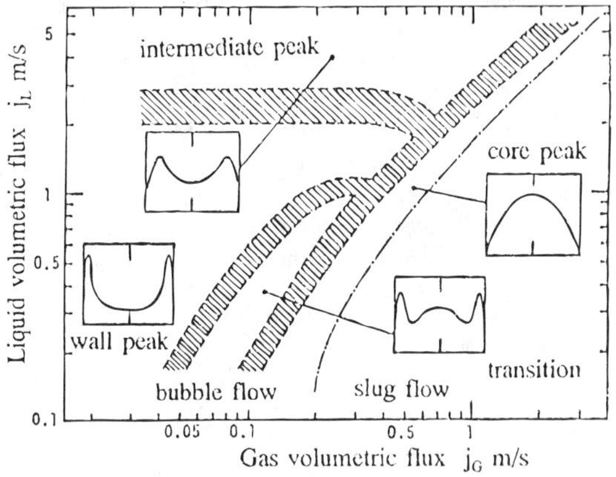

Figure 3. Lateral void distribution patterns map for air-water flow in vertical pipes, obtained by Serizawa and Kataoka, (1988).

Types of void peaking are due to a bubble segregation phenomenon which depends significantly on the size and shape of bubbles, as shown in **Figure 4**, obtained by Sekoguchi et al. (1974). The bubbles ranging from 2 to 5 mm in diameter tend to collect towards the wall. If the bubble size is carefully controlled under fixed flow conditions, the transition boundaries in **Figure 3** shift towards lower or higher gas or liquid flow rate and therefore, existing flow pattern maps are no longer applicable.

The most important aspect of phase distribution phenomena is a close connection with turbulence and interfacial structures. This triangular link, as illustrated in **Figure 5**, strongly depends on the bubbles' configurations; therefore, the bubble size and shapes are key parameters which determine the bubble flow characteristics such as turbulence (including wall shear stress) and interfacial structures (including interfacial shear stress) through different phase distribution profiles. Detailed discussions are given by Serizawa and Kataoka (1988).

Phase distribution

Figure 6 represents some typical lateral phase distributions in air-water flows in channels. In horizontal flows, the buoyancy effect is predominant at low liquid flow rates. At higher liquid flows, the inertia effect overcomes the buoyancy effects and the profile thus becomes similar to vertical flows. The effects of physical properties of the fluids and turbulence fields are discussed by Serizawa and Kataoka (1988). The mechanisms of the phase distribution phenomena are supposed to be: 1) bubble segregation due to a lateral lift force acting on the bubbles; 2) nonuniform pressure distribution induced by nonuniform turbulence field (Lahey, 1988); 3) a force similar to diffusion force related to void

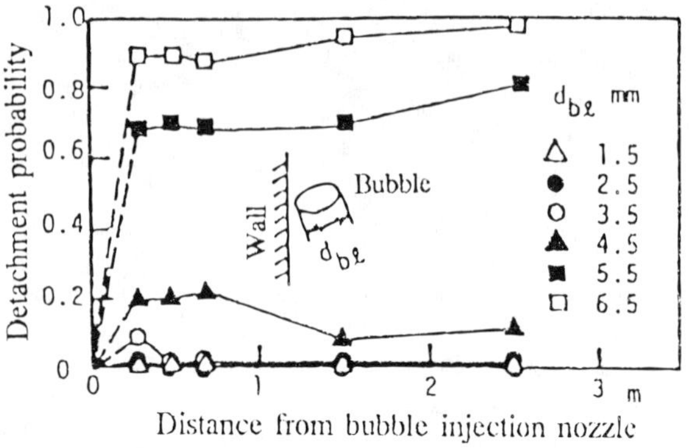

Figure 4. Bubble detachment probability (Sekoguchi et al. 1974).

fraction gradient; 4) bubble coalescence; and 5) effect of bubble trapment by eddies (Zun, 1986). The transverse lift force is usually expressed by:

$$F_L = -C_T\rho_L(u_G - u_L)(\partial u_L/\partial y) \quad \textbf{(1)}$$

where C_τ is the transverse lift coefficient. For spherical bubbles immersed in an invisid liquid, $C_\tau = 1/2$. This coefficient is generally an experimentally determined constant. The lift force given by the above equation yields the force pushing the bubbles towards the wall for vertically-upward flow, and towards the tube center for vertically-downward flow. However, it does not always explain the trends shown in **Figure 6**. A vortex shedding model is discussed by Serizawa and Kataoka (1992).

A bubble diffusion model generally assumes under steady-state condition a balance of bubble flux due to diffusion ($= D_B\ \partial\ \epsilon/\partial\ y$) with bubble migration due to lateral lift force. Based on a simple assumption of bubble-eddy collisions, the bubble dispersion coefficient D_B is given by:

$$Pe\ (= d_B u_L'/D_B) = 2.9 \quad \textbf{(2)}$$

where d_B and u_L' are **Sauter Mean Diameter** of bubbles for bubble size ranging $2 \sim 5$ mm in diameter and turbulence, respectively. [see also **Figure 7** cited by Serizawa and Kataoka (1992) from Hinata (1980)]. Other approaches are reviewed also by Serizawa and Kataoka (1992).

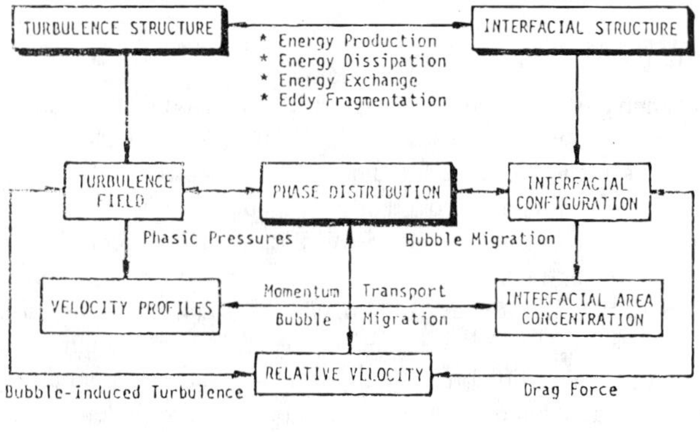

Figure 5. Triangular linkage in bubble flow structures.

Bubble coalescence is another important phenomenon which we should consider in predicting flow development along a channel. Chesters and Hofman (1982) presented a criterion for this by considering the formation and thinning of a liquid film formed between the two bubbles with diameters d_{B1} and d_{B2}. Here, u_{rel} is the relative velocity between the two deformed bubbles.

$$We\ (= \rho_L u_{rel} R_{eq}/\sigma) \leq 10^{-2} \quad \textbf{(3)}$$

$$R_{eq}^{-1} = 1/d_{B1} + 1/d_{B2}. \quad \textbf{(4)}$$

The numerical prediction of phase distribution in bubble flow has progressed on the basis of the two-fluid model, in collaboration with turbulence models. Details should be referred to Lahey (1988) for the application of $k - \varepsilon$ *model* and Lance and Lopez de Bertodano (1992) for *Reynolds stress model.* Several bubble diffusion models have been also proposed (reviewed by Serizawa and Kataoka, 1992).

Turbulence

Typical 2D turbulence measurements in upward air-water bubble flows in a vertical tube are demonstrated in **Figure 8** (Serizawa and Kataoka, 1988). The numerical predictions of multidimensional turbulence can be made by a two-fluid model, coupled with an appropriate turbulence model. Typically, either two equation model like a $k - \varepsilon$ model and a $\tau - \varepsilon$ model or one equation model based on Reynolds stress conservation equation are used for this purpose. (See **Turbulence** and **Turbulence Modelling**). Detailed discussions are given by Lahey (1988), Lance and Lopez de Bertodano (1992), and Serizawa and Kataoka (1992). These models predict also the time-averaged liquid velocity and phase distributions. A linear superposition is often assumed for two-phase **shear stress** as follows (Sato et al. (1981):

$$\tau = \tau_{B1} + \tau_{S1} \quad \textbf{(5)}$$

where τ_{B1}, τ_{S1} are the bubble-induced turbulence and the shear-induced turbulence, respectively. Other approaches are reviewed by Lance and Lopez de Bertodano (1992).

The turbulence modification in bubble flow is either to increase or to decrease local turbulence depending on the flow conditions, as shown in **Figure 8**. A criterion for turbulence reduction or promotion is constructed for air-water flows in **Figure 9** in terms of the liquid flux j_L versus gas flux j_G.

Figure 10 is a generalized expression for turbulence modification proposed by Gore and Crowe (1989) for dispersed flows. The coordinate is a nondimensional particle or bubble size where the integral scale of turbulence l_e is chosen as the characteristic scale. The turbulence reduction occurs for d_p/l_e (or d_B/l_e) ≤ 0.1.

Local turbulence energy production, dissipation and diffusion rates are important quantities in turbulence reduction/enhancement. Relative motions between the bubbles and the liquid flow produce both additional turbulence and dissipation due to viscous effect, depending on the scale of interfaces and turbulence eddies. Discussions are given and reviewed by Serizawa and Kataoka (1992). However, generally accepted physical explanations for the mechanisms have not yet been developed, although turbulence modifica-

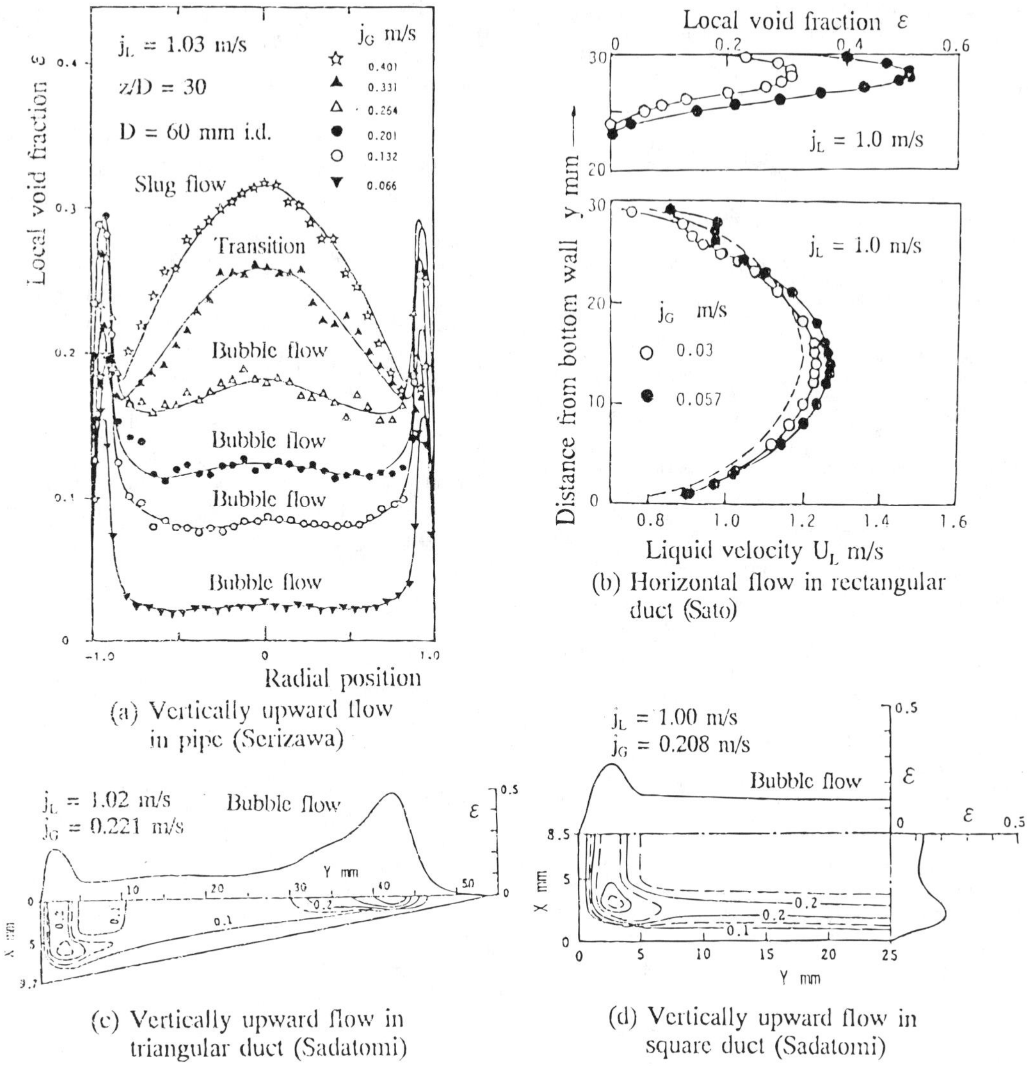

Figure 6. Typical phase distributions in channels.

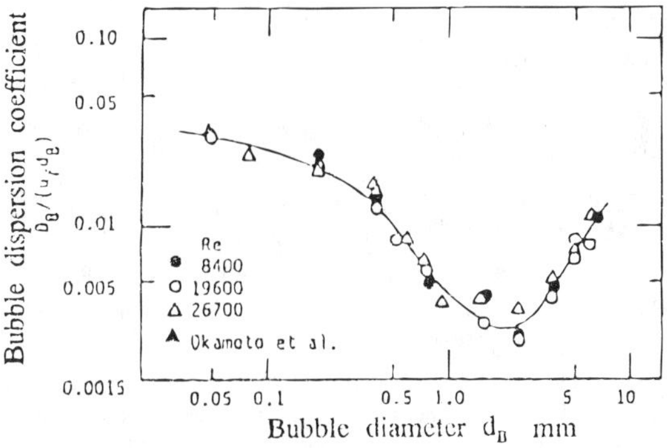

Figure 7. Bubble dispersion coefficient (Hinata 1980).

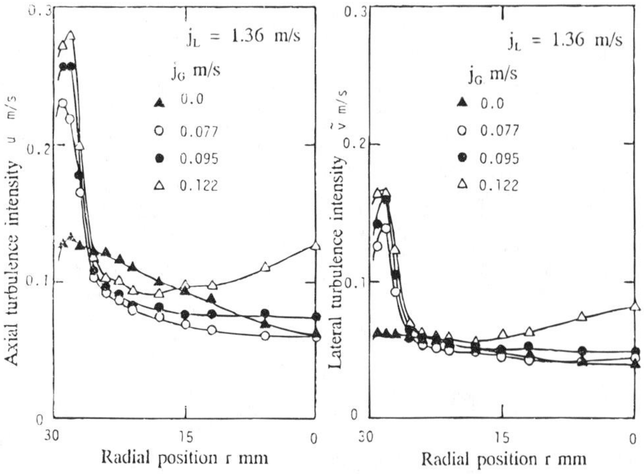

Figure 8. 2-D turbulence measurements in vertical upward air-water flow (Serizawa and Kataoka 1988).

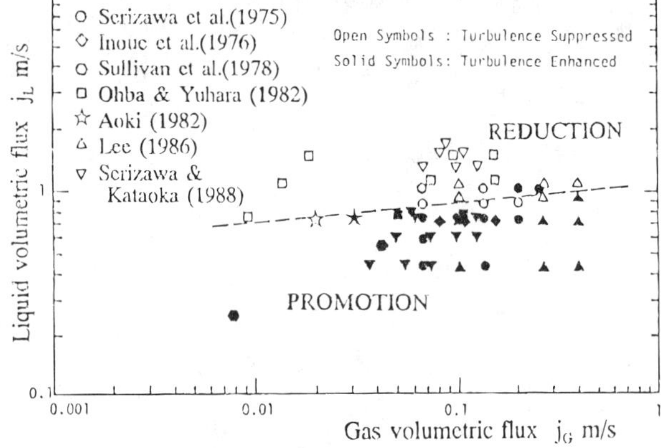

Figure 9. A map of turbulence modification (Serizawa and Kataoka 1988).

tion may be predicted to some extent by numerical methods (Lahey, 1988; Lance and Lopez de Bertodano, 1992).

Turbulence scales, mixing length and turbulence energy spectra are reported by Michiyoshi and Serizawa (1986) and are reviewed by Serizawa and Kataoka (1988, 1992).

Interfacial Area Density and Transfer Terms

The *interfacial area density* (or concentration) is defined as the sum of the interfaces per unit volume of two-phase mixture. This term appears in basic conservation equations in two-fluid model formulations. In order to close the equation system, this term is usually given as a closure law. With an assumption of spherical bubbles, the local interfacial area concentration a_i is given as:

$$a_i = 6\varepsilon/d_B \tag{6}$$

Experimental results modify the above equation. Detailed discussions are given by Serizawa and Kataoka (1992). Interfacial transfer terms appearing in two-fluid model, are discussed by Lahey and Moody (1975).

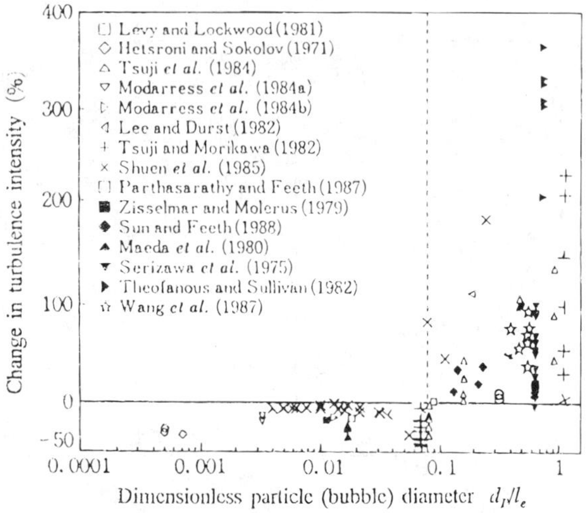

Figure 10. Change in turbulence intensity obtained by Gore and Crowe (1989).

Bubble Flow with Phase Change

Multidimensional bubble flow characteristics with phase change are less known than adiabatic bubble flows. Particularly, the structure and temperature field in near heated wall region is poorly understood. In one-dimensional calculation of void fraction distribution along flow direction, the Saha and Zuber correlation is useful to predict the point of net vapor generation.

Thermally-controlled region Pe $(= GD_e c_{pL}/\lambda_L) \leq 70{,}000$:

$$Nu\ (= \dot{q}_w D_e/\lambda_L \Delta T_B) = 455 \tag{7}$$

Hydrodynamically-controlled region Pe $> 70{,}000$:

$$St\ (= Nu/Pe) = 0.0065 \tag{8}$$

The calculation methods to predict axial void distribution are given by Lahey and Moody (1975). Several interfacial transfer terms are referred also to this.

Properties of Bubbly Mixtures

According to homogeneous flow model, the *viscosity* of bubbly mixture is given by either of the following equations:

$$1/\eta = x/\eta_G + (1 - x)/\eta_L, \qquad \eta = \eta_G\varepsilon + \eta_L(1 - \varepsilon). \tag{9}$$

For small bubbles with diameters less than 1 mm suspended in a liquid, the mixture viscosity is given by the *Einstein equation:*

$$\eta = \eta_L(1 + 2.5\varepsilon). \tag{10}$$

For emulsion, the *Taylor equation* can be used:

$$\eta = \eta_L\{1 + 2.5\varepsilon(\eta_p + 0.4\,\eta_L)/(\eta_p + \eta_L)\}. \tag{11}$$

Electrical conductivity is estimated by the following equations:

$$\text{Hall-Taylor: } \sigma = \sigma_L/(1 + 1300\,x) \tag{12}$$

$$\text{Maxwell: } \sigma = 2\sigma_L(1 - \varepsilon)/(2 + \varepsilon) \tag{13}$$

References

Chesters, A. K. and Hofman, G. (1982) Bubble Coalescence in Pure Liquids: Mechanics and Physics of Bubbles in Liquids (Ed.L.van Wijingaarden). Matinus Nijhoff Publishers, pp. 353–361.

Gore, R. A. and Crowe, C. T. (1989) Effect of Particle Size on Modulating Turbulent Intensity, *Int. J. Multiphase Flow,* 15;279–285.

Lahey, R. T. Jr. and Moody, F. J. (1975) The Thermal-Hydraulics of a Boiling Water Nuclear Reactor. ANS Monograph.

Lahey, R. T. Jr. (1988) Turbulence and Phase Distribution Phenomena in Two-Phase Flow: Transient Phenomena in Multiphase Flow (Ed. N. H. Afgan). Hemisphere Publishing Corporation, pp.139–177.

Lance, M. and Lopez de Bertodano, M. (1992) Phase Distribution Phenomena and Wall Effects in Bubbly Two-Phase Flows, Proc. of the 3rd Int. Workshop on Two-Phase Flow Fundamentals, June 15–19, 1992, London, UK.

Michiyoshi, I. and Serizawa, A. (1986) Turbulence in Two-Phase Bubbly Flow, Nucl. Eng. and Design, 95:253–267.

Sato, M., Sadatomi, M., and Sekoguchi, K. (1981) Momentum and Heat Transfer in Two-Phase Bubbly Flow-I, *Int. J. Multiphase Flow,* Vol.7.

Sekoguchi, K., Sato, Y., and Honda, T. (1974) Two-Phase Bubble Flow (First Report), *Trans Japan Soc. Mech. Engrs.* 40:1395–1403.

Serizawa, A. and Kataoka, I. (1988) Phase Distribution in Two-Phase Flow: Transient Phenomena in Multiphase Flow (Ed. N. H. Afgan). Hemisphere Publishing Corporation, pp. 179–224.

Serizawa, A. and Kataoka, I. (1992) *Dispersed Flow,* Proc. of the 3rd Int. Workshop on Two-Phase Flow Fundamentals, June 15–19, 1992, London, UK.

Wallis, G. B. (1969) One-Dimensional Two-Phase Flow. McGraw Hill, N.Y.

Zun I. (1986) Influence of Void Fraction on The Pressure Drop Prediction in Bubbly Flow, Proc. of the 5th Chemical Engineering Congress, Toronto, pp. 155–162.

A. Serizawa and I. Kataoka

BUBBLE GROWTH

Vapour bubble growth in a superheated liquid is part of the process of nucleate boiling. It is also related to depressurization (flashing) processes. The early stages of bubble growth—when the bubble is still microscopic—are complicated by the uncertainties surrounding the nucleation process itself, and by the fact that the excess pressure due to surface tension across the liquid-vapour interface, $2\sigma/R$, is comparable with the pressure difference driving bubble growth. p_G-p_∞. Once this initial period is over the process of bubble growth, in a uniformly superheated liquid at least, is straightforward.

Obviously, a vapour bubble in a superheated liquid is unstable. It will grow while superheated liquid remains. There are two main constraints to the rate of growth. There is the inertia of the surrounding liquid, which has to be pushed out of the way, and there is the need for heat to diffuse from the surrounding liquid to the interface to cause evaporation.

Inertia-controlled growth is more likely to exist in the early stages, particularly at low pressures when only a small amount of heat is needed to create a large volume of vapour. If the bubble radius is R then, as shown in **Figure 1**, the velocity of the bubble boundary is $\dot{R}$.

Continuity requires that the outward flow rate of liquid at every radius r is the same, i.e.,

$$4\pi r^2 u = \text{constant} = 4\pi R^2 \dot{R}$$

where u is the velocity at r.

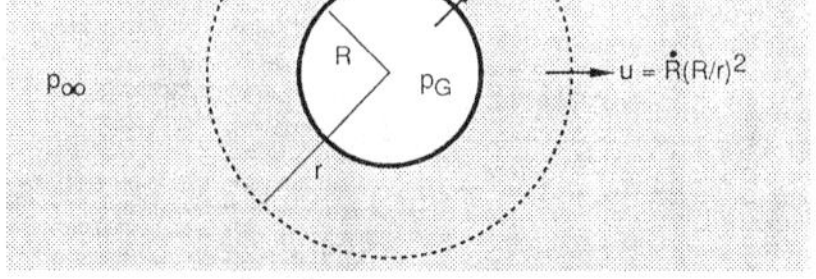

Figure 1. Growing bubble.

The kinetic energy of this moving liquid is:

$$\int_R^\infty 4\pi r^2 \rho_L u^2/2 \, dr = 2\pi R^3 \rho_L \dot{R}^2. \tag{1}$$

This energy is supplied by the expanding vapour. Oversimplifying the derivation, with vapour pressure p_G expanding against the external pressure p_∞, it yields

$$\int p dv = (p_G - p_\infty) 4\pi R^3/3. \tag{2}$$

Equating expressions 1 and 2 gives

$$\dot{R}^2 = 2(p_G - p_\infty)/3\rho_L = \text{constant} = A^2 \tag{3}$$

and the bubble radius is given by

$$R = At \tag{4}$$

which is essentially the solution originally derived, much more rigorously, by Rayleigh in 1917. The pressures can be replaced by temperatures using the *Clausius-Clapeyron equation*, i.e.,

$$A^2 = 2h_{LG}\rho_G(T_\infty - T_{sat})/(3\rho_L T_{sat}). \tag{5}$$

In practice the diffusion, heat transfer controlled, regime is more important. The solution has been obtained by Plesset and Zwick (1954). One may observe that to supply the latent heat required for a bubble of radius R, the heat has to diffuse over a distance proportional to R. The time taken to diffuse such a distance is itself proportional to R^2, so the expected solution is:

$$R = Bt^{0.5} \tag{6}$$

The *Plesset and Zwick solution* gives B as

$$B = \left[\frac{12}{\pi} \frac{(Ja)^2 \lambda_L}{\rho_L c_{pL}}\right]^{1/2}$$

where Ja is the Jacob number given by:

$$Ja = \frac{(T_W - T_{sat})\rho_L c_{pL}}{h_{LG}\rho_G}$$

A particularly convenient solution, that avoids the problem of deciding whether Equation 4 or 6 is the more appropriate has been given by Mikic et al. in 1970. Defining the dimensionless variables

$$R^+ = AR/B^2 \quad \text{and} \quad t^+ = A^2 t/B^2,$$

the solution is:

$$R^+ = 2[(t^+ + 1)^{3/2} - (t^+)^{3/2} - 1]/3 \tag{7}$$

which in the limit of $t^+ << 1$ gives Equation 4 and for $t^+ >> 1$ gives Equation 6.

Most practical situations do not involve isolated bubbles surrounded by infinite seas of uniformly superheated liquid. One complication, with bubble growth on a solid surface, is that the spread of vapour over the surface often leaves a microlayer of liquid behind. The effect of this microlayer on bubble growth is discussed in the article on nucleate boiling.

Recent papers that discuss more complex situations include one that analyses a bubble on a wall in a linear temperature gradient, taking into account the effect of the wall thermal conductivity (Klimenko 1989), and another that assumes a falling liquid pressure, i.e., depressurization (Wang and Bankoff 1991).

References

Fyodorov, M. V. and Klimenko, V. V. (1989) Vapour bubble growth in boiling under quasi-stationary heat transfer conditions in a heating wall, *Int. J. Heat Mass Transfer*, 32:227–242.

Mikic, B. B., Rohsenow, W. M., and Griffith, P. (1970) On bubble growth rates, *Int. J. Heat Mass Transfer*, 13:657–666.

Plesset, M. S. and Zwick, J. A. (1954) Growth of vapour bubbles in superheated liquids, *J. Applied Phys.* 25:493.

Rayleigh, L. (1917) *Pressure due to Collapse of Bubbles*, Phil. Mag. 34:94.

Wang, Z. and Bankoff, S. G. (1991) Bubble growth on a solid wall in a rapidly depressurizing liquid pool, *Int. J. Multiphase Flow*, 17:425–437.

R.H.S. Winterton

BUBBLE NUCLEATION (see Interfaces)

BUBBLE PLUME (see Aeration)

BUBBLES, DRAG ON (see Stokes law for solid spheres and spherical bubbles)

BUBBLING FLUIDISED BED (see Fluidised bed)

BUBBLES, IN FLUIDISED BEDS (see Fluidised bed)

BUCKINGHAM EQUATION (see Liquid-solid flow)

BUGER'S LAW, FOR LIGHT ATTENUATION (see Dropsize measurement)

BUILDING HEATING (see Space heating)

BUILDINGS AND HEAT TRANSFER

Following from: Environmental heat transfer; Heat transfer and physiology; Heat transfer

Introduction

The original purpose of a building is to provide shelter and to maintain a comfortable or at least liveable internal temperature. Other purposes include security, privacy and protection from wind and weather. To feel comfortable in a thermal sense, a human has to be able to release a well-defined amount of **Heat**. If this gets difficult, a person will either feel cold or hot. The human body operates as a chemical reactor that converts chemical energy of food and respiratory oxygen into mechanical work and heat. Heat output can vary from about 100 W for a sedentary person to 1000 W for an exercising person (ASHRAE Fundamentals, 1993). (See also **Physiology and Heat Transfer**.)

To maintain body temperature within a narrow band, the heat produced by an occupant must be released to the indoor environment. If too much heat is lost, room temperature should be increased or warmer clothes be worn. The heat transfer on the human skin, the indoor temperature and the heat transfer through the building envelope are factors that influence thermal comfort (Mayer 1991).

Figure 1 shows schematically the ranges of temperature variations of the human body, of the room air and outdoor air. The adjustment of heat transfer around the human alone (by variation of clothing or sweating) is not normally sufficient to control body heat release at large outdoor temperature variations without the thermal protection of the building envelope and heating or cooling. The dynamic storage of heat in building components is important to control indoor temperature variations.

Heat, the Ultimate Form of Energy

Heat is one form of energy. It is a "lower" form of energy because no system can convert heat fed to it by heat transfer into mechanical work completely and continuously. Heat is the ultimate form of energy because systems tend toward a state where all energy is transformed into heat. The burning of fuel in a boiler turns chemical energy into heat. Heat is generated by the process of combustion, but the total amount of energy is unchanged.

In a closed system, energy is conserved. The energy contained in a building is increased, for instance, by sunshine, by the supply of electricity and fuel for heating and other purposes, and also by food brought in by occupants for preparing meals. The total amount of enclosed energy is reduced by heat losses and other transport mechanisms listed below. In the long-term, the average energy content of a building is almost constant. Energy flows "in transit" through the control volume of a building. The balance of heat transfer and energy flow determines the temperature level at which the interior settles.

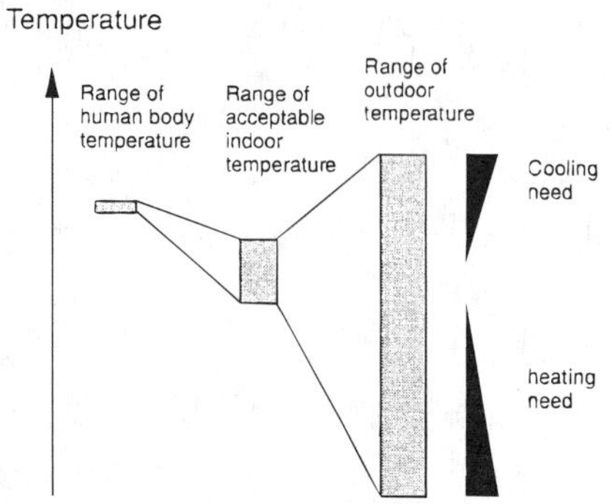

Figure 1. Temperature ranges in a building. Heat transfer at the building envelope and on the human body determine thermal comfort. High thermal resistance of insulating layers reduces temperature amplitudes felt on the human skin.

Heat Loss of a Building in Simple Terms

According to the *Second Law of Thermodynamics*, heat transfer is only possible in the direction from a higher temperature to a lower one. It becomes zero if temperatures are equal. The heat loss through an envelope should therefore be proportional to the difference $T_{inside} - T_{outside}$, or to a positive power of it for small differences. For a simple formula, a linear dependence on temperature difference is sufficient. Accepting further that heat loss grows linearly with surface area A, one finds:

$$\dot{Q} = AU(T_{inside} - T_{outside}). \quad (1)$$

The constant of proportionality, U, is the **Overall Heat Transfer Coefficient** in $W/(m^2K)$. In the example of **Figure 2**, a building is represented by a cube of 5m × 5m × 5m. If no heat is lost into the soil and with $U = 0.4\ W/(m^2K)$, the total heat loss is:

$$\dot{Q} = 5 \times 5^2 \times 0.4 \times 20\ W = 1000\ W \quad (2)$$

Equation 1 suggests three ways to reduce heat loss: 1) As the heat loss is proportional to the inside-outside temperature difference, the set-point for the indoor temperature can be reduced during the heating season; 2) The insulation of the envelope can be improved to reduce the overall heat transfer coefficient U; And 3) If possible, the surface area should be reduced without changing the enclosed volume. A spherical igloo would be optimal, but a cubical shape is still better than an elongated building with many wings. The opposite is true for the design of heat exchanger surfaces or *fin-tubes*, where the effective surface should be maximised.

The cumulated amount of lost heat is the time integral of the instantaneous heat flow,

$$Q = \int \dot{Q}\ dt. \quad (3)$$

The quantity of heat, Q, is measured in J (Joule). In the construction sector, it is often converted into kWh (kW-hours). The fuel consumed for heating is roughly proportional to the difference between Q and the sum of internal heat gains from sun, occupants, lights, equipment, and so forth. Therefore, the time-average of the temperature difference $\Delta T_{bal} = T_{balance} - T_{outside}$ during the heating season is of importance for estimating heating cost. Here, $T_{balance}$ is that outside temperature at which no heating is required to maintain a prescribed inside temperature at given internal heat gains (Chapter 28, ASHRAE 1993).

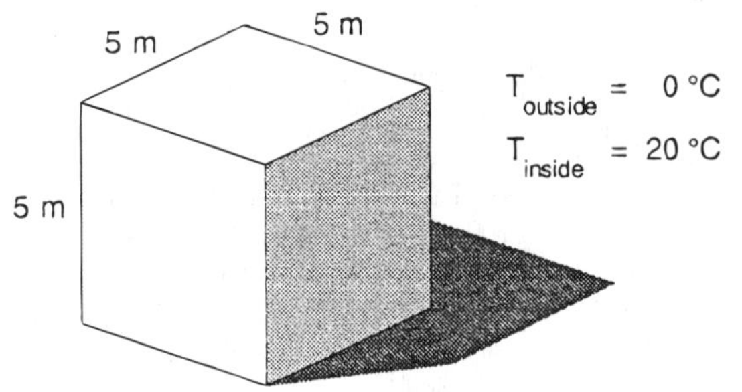

Figure 2. Estimation of the heat loss of a two-storey building. At 20 K temperature difference, this cube loses 1000 W to the atmosphere at an assumed overall heat transfer coefficient of 0.4 $W/(m^2K)$, and if heat loss to the ground is neglected.

The thermal performance of buildings can be compared on the basis of *degree-days* (ASHRAE 1993). Heating degree-days, DD_h, for a geographical site between given dates correspond to a time integral of ΔT_{bal} in which only positive values of the difference are counted.

Different Modes to Transport Heat

In building **Heat Transfer**, many different types of energy transport are effective. Often, heat is transported by different modes to or from the same place. Energy that reaches a point via different paths and modes may be added up for the heat balance. For instance, the heat loss of a human body is the sum of convection, radiation and latent heat released by sweating and so forth.

Primary heat transport modes are:

- **Conduction** (heat flow on a molecular scale. Medium at rest or moving);
- **Convection** (heat conveyed as internal thermal energy of mass that is displaced by mean or turbulent motion);
- **Radiation** (heat transfer by **electromagnetic waves** such as infra-red or visible light).

(See entries on **Conduction, Convective Heat Transfer** and **Radiative Heat Transfer**.)

In buildings, heat is also transported by the following mechanisms, which basically belong to the convective mode:

- Transfer of latent heat by transport of water or water vapour.
- Thermal energy associated with the air replaced in a building by ventilation or by air leakage (infiltration).
- Thermal energy associated with fresh and used domestic water and combustion air (including flue gases), and fluids feeding **Heat Pumps**.

The transport of energy in the above list is limited to energy in the form of *sensible* or *latent* heat. A change of *sensible* heat is characterised by a change of temperature while a change of *latent* heat is associated with some mass altering its phase. Phases are *gaseous, liquid, solid.* Transport of energy in forms other than heat are not considered.

Heat transfer in buildings may involve the listed types of transport. For an energy balance, other forms of energy—often referred to as energy sources or heat loads—and dynamic (time-dependent) storage of heat in solid, liquid, or gaseous media have to be taken into account.

Two Models for Buildings: Network and Continuum

Equation (1) above may be useful for a rough overall energy balance, but not for a detailed description of energy flows in a real building. Energy transport in a building is often analysed by a *network* model, **Figure 3**. To each flow path, a mode of energy transport may be assigned.

In reality, the air in a building is a *continuum* and the building structure can be divided into regions that may be considered contin-

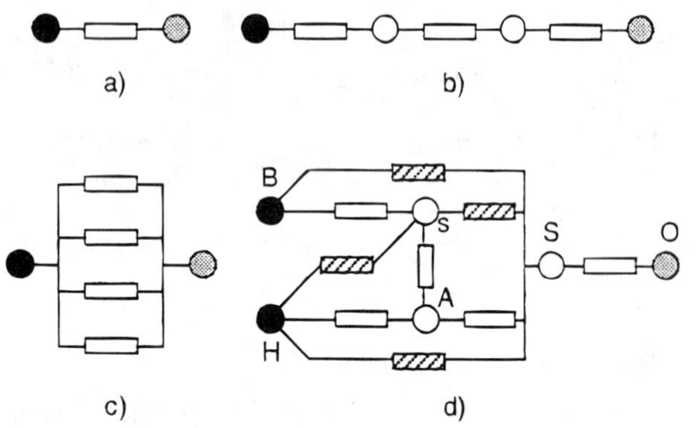

Figure 3. Flow paths of energy transport in a building are represented as networks: a) Single path, heat flows from a warmer place (black symbol) to a colder place (shaded circle) along a flow path (rectangle); b) Heat flow across a roof may pass through several layers in series; c) Parallel paths as through several windows in the same wall; d) Example with a human body (B), surface of clothes (s), a heater (H), the room air (A), and the inside (S) and outside (O) wall surfaces. The hatched flow links in d) represent radiation.

uous in themselves. So the network model is far too simple for a true description of heat transfer in a building. However, the network approach is quite successful in simulation of thermal building dynamics. For an introduction, see, e.g., Clarke (1985).

In a *network*, as sketched in **Figure 3**, the energy balance must be satisfied at each node. This leads to a system of algebraic equations. Their solution is a set of node temperatures, which may vary with time or be constant in a state of equilibrium. In a continuum, the temperature field and the associated energy transport are defined by partial differential equations. These equations, which contain derivatives with respect to all spatial co-ordinates and the time, must be satisfied at each point within the continuum.

To calculate the temperature distribution in a continuum and velocity field in air, the domain is subdivided into small cells. Discretization produces large systems of difference equations that are solved by computers (finite-difference, finite-volume and finite-element methods).

Local Heat Flux to Quantify Heat Transfer

Heat transfer is best described by the local **Heat Flux**, $\dot{q}$, which amounts to a heat flow density. It is the heat flow per unit cross section. In **Figure 4** the heat flow rate, $d\dot{Q}$, that passes through an infinitesimal surface element dA is illustrated in two different situations: In (**Figure 4a**) the heat is transferred across an interface between different materials of the same or different phase as, e.g., from air to a wall; in (**Figure 4b**) heat flows within one medium, as within a solid body.

The local heat flux, measured in W/m^2, is:

$$\dot{q} = \frac{d\dot{Q}}{dA}. \quad (4)$$

The local heat flux, $\dot{q}$, is the basic quantity for the analysis of heat transfer. It is associated with a point on a surface, as in **Figure 4a**, or with a point and a direction, as the arrow dn in **4b**. In

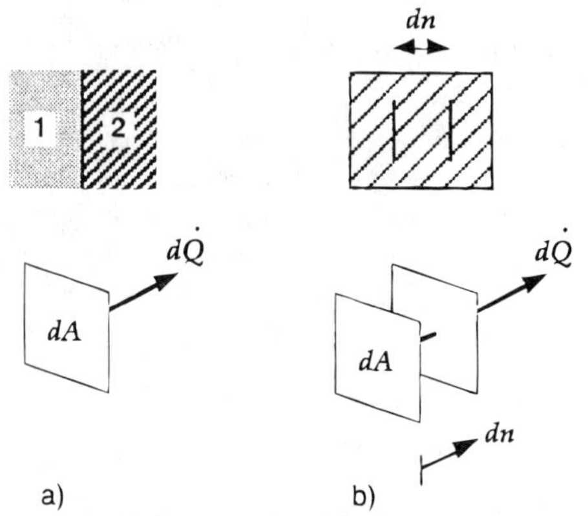

Figure 4. Heat flux across an interface, a), and within a homogeneous medium, b). *dA* is the surface element; $d\dot{Q}$, the heat flow rate.

mathematical terms, heat flux in a continuum is a vector with components in x, y, z directions:

$$\vec{\dot{q}} = (\dot{q}_x, \dot{q}_y, \dot{q}_z). \quad (5)$$

The $\dot{q}$ at an interface is the vector component normal to the surface. Heat flow rate, $\dot{Q}$, across a finite area may be obtained from $\dot{q}$ by integration or by multiplying the area with the mean value of heat flux over that area.

For the **Conduction** transfer mode, the heat flux in a continuum (**Figure 4b**) is:

$$\dot{q} = -\lambda \partial T/\partial n, \text{ or in vector form } \vec{\dot{q}} = -\lambda \nabla T \quad (6)$$

with the thermal conductivity, λ, in W/(mK) and the gradient of the temperature field, ∇T.

The heat transfer from room air to the wall surface is an example for the interface heat flux illustrated in **Figure 4a**. In this particular case, the interface is thought to comprise the surface and the boundary layer. This is a thin layer of air flow retarded by wall friction. Depending on room size, the boundary layers may have a thickness of several centimetres.

The local heat flux, $\dot{q}$, in this example is a function of temperatures T_1 and T_2 on either side of the interface and of properties of air and of the boundary layer. The air temperature is measured outside of the boundary layer, for instance, 0.1 m from the surface point where $\dot{q}$ is determined.

$$\dot{q} = f(T_1, T_2, \text{flow parameters}) \quad \text{or} \quad \dot{q} = -\alpha(T_2 - T_1) \quad (7)$$

The **Heat Transfer Coefficient**, α, is itself a function of T_1, T_2 and of flow parameters. The relation at right defines α such that $\dot{q}$ is positive when heat flows from 1 to 2, and vanishes if the two temperatures are equal (**Figure 4a**). On the inside surface of a wall, α is of the order of 3 to 5 $W/(m^2K)$ if it does not include radiation. Heat transfer coefficients for different situations have been measured and correlated by non-dimensional parameters, as in Chapter 3 of ASHRAE (1993).

The transmission of heat through a building wall, Equation 1, may now be considered as a network with three resistances in series (**Figure 3b**). The overall heat transfer coefficient, U, becomes:

$$U = 1/R \quad \text{and} \quad R = R_1 + R_2 + R_3 \tag{8}$$

where

$$R_1 = 1/\alpha_{inside} \qquad R_2 = d/\lambda_{wall} \qquad R_3 = 1/\alpha_{outside}. \tag{9}$$

These equations are obtained by combining Equations 6 and 7 and eliminating intermediate temperatures.

Heat Transport within a Building

To reduce heat loss of a building, different modes of transport have to be considered. The example of **Figure 2** does not account for thermal energy associated with the air replaced by ventilation and air leakage. This additional heat loss is now estimated for the example of **Figure 2**. At an air change rate of $n = 1\ h^{-1}$ for ventilation and leakage together (i.e., all the air in the building is, on the average, renewed once every hour), the ventilation heat loss is:

$$\dot{Q} = \frac{n}{3600} V\rho c_p(T_{inside} - T_{outside}). \tag{10}$$

At a volume, $V = 125\ m^3$; air density, $\rho = 1.2\ kg/m^3$; specific heat capacity of air at constant pressure, $c_p = 1000$ J/(kgK); and a temperature difference of 20 K, the ventilation heat loss becomes:

$$\dot{Q} = 830\ W.$$

This is of the same order as the transmission heat loss estimated at 1000 W. If only a small portion of the air change is caused by leakage, exhaust air heat recovery may reduce this loss.

The transport of latent heat by humid air can amount to a substantial heat loss (Hens 1991, 1995). The heat required to evaporate 1 kg water is 2500 kJ. Saturated air of 20°C holds about 0.016 kg water vapour per kg air. Condensation of this water releases about $0.016 \times 2500 = 40$ kJ latent heat (per kg air).

The ventilation heat loss was estimated above for *dry air.* For the same example, latent heat transport is computed for a relative humidity of 50% indoors and outdoors. The warm indoor air holds more water vapour than cold air. To maintain indoor relative humidity, water must be evaporated. This evaporation energy has to balance the loss of latent heat in the extract air. This additional heat loss is $\dot{Q} = 550$ W. In this example, latent heat loss accounts for about 40% of the ventilation loss.

Figure 5 shows an example of a faulty insulation layer between warm and cold surfaces. A small-scale but steady circulation of humid air may develop in cavities of building components, or between heated and unheated spaces of a house. These streams transport latent heat, that is released when vapour condenses on cool surfaces. (See also **Humidity**.)

Heat Transport within a Room

In the heat transfer mode of **Convection**, moving packets of air transport energy as their internal thermal energy. The *"vector"* or carrier of energy is the air. In turbulent convection, the individual packets are small and their size is of the order of the length scale of the **Turbulence**. These are the turbulent eddies. Heat is also

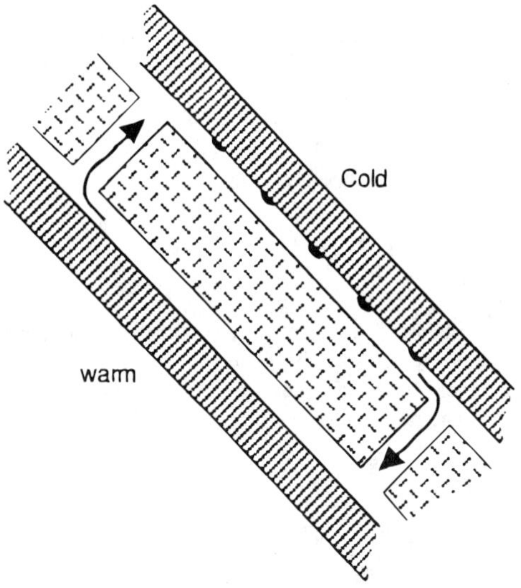

Figure 5. Roof insulation with a faulty vapour barrier: Air circulation develops through cracks in the insulation layer. Moisture evaporates on the warm surface and condenses on the cold surface. This heat leak operates like a heat pipe if water is allowed to drip back onto the warm surface.

convected by the mean motion of turbulent or laminar flow. Depending on the driving forces for the motion of the packets, two types of convection are defined:

- Heat transfer by forced convection (air motion is independent of heat transfer);
- Heat transfer by natural or **Free Convection** (driven by buoyancy forces acting on heated or cooled air).

Heat transfer in a hair-dryer or in a room with mechanical ventilation is an example of forced convection; the upward flow that develops on a vertical radiator induces natural convection.

A third heat transfer mode active in a room is *radiation.* The heat exchanged by radiation between two black surfaces 1 and 2 is:

$$\dot{Q}_{1-2} = A_1 F_{1-2}\sigma(T_1^4 - T_2^4). \tag{11}$$

The geometric configuration factor F_{1-2} accounts for the distance between and for the relative orientation of the two surfaces and for the size of surface 2 (see Siegel and Howell 1992). The *Stefan-Boltzmann constant* is $\sigma = 5.7 \times 10^{-8}\ W/(m^2K^4)$. Each surface has a uniform surface temperature, T_1 or T_2, respectively, that is measured in degree K. (See **Radiative Heat Transfer.**)

A simplified example shows the contributions to the heat transfer by radiation and convection between walls of a room (**Figure 6**). In the illustrated cavity, heat is exchanged between the opposing walls by radiation and natural convection. The other bounding surfaces of the cube are assumed to be perfect adiabatic mirrors. The radiative transfer may therefore be evaluated by Equation 11 with a geometric configuration factor $F_{1-2} = 1$. According to a correlation by Henkes (Henkes 1990), the turbulent free-convection heat flow rate in air is approximately proportional to the 4/3-power

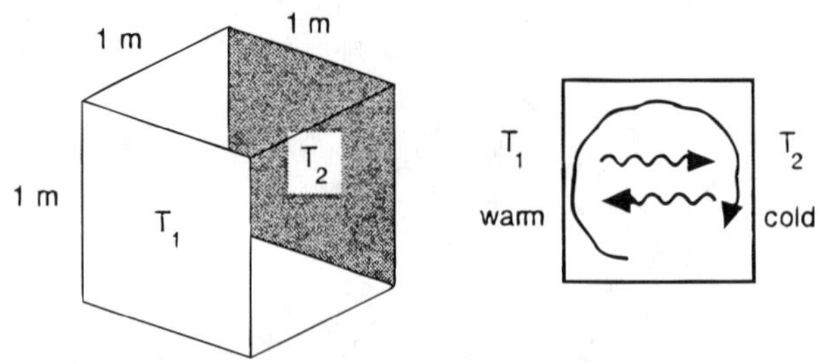

Figure 6. A cube with edges of 1 m has hot and cold opposing black walls. The other sides are reflective for thermal radiation and well insulated. The air in the cavity moves up near the warm surface and down at the cold surface. The heat transfer by natural convection and by radiation are compared.

of the temperature difference. The heat flow rates in the example of **Figure 6** are:

T_1	T_2	$\dot{Q}_{convection}$	$\dot{Q}_{radiation}$ (black body)	$\dot{Q}_{radiation}$ $1 \rightarrow 2$	$\dot{Q}_{radiation}$ $2 \rightarrow 1$
25°C	20°C	4.7 W	29 W	447 W	418 W
40°C	20°C	30 W	126 W	544 W	418 W

The radiation rate is given for perfectly black surfaces. For grey surfaces, the actual rate may be as low as 50% of the listed value. Even then, radiation accounts for two to three times the free-convection heat transfer. And the radiation in *one direction* is even larger and demonstrates that radiation—even at room temperatures—acts as a "short-cut" transfer mechanism between walls. The radiation from far surfaces to the human skin is of significance for thermal comfort.

Long-wave radiation may also be absorbed and emitted by a gas. A small portion of the radiation that crosses a room is absorbed by the room air and by water vapour, or CO_2. Depending on its temperature, the so-called participating gas emits radiation diffusely into all directions. Gas radiation may have an effect in air with high moisture content or in large rooms, and it may affect the transient reaction of air temperature to a sudden temperature change of a bounding surface. Moisture participates effectively at wave lengths of its absorption spectrum. For water vapour, each absorption line must be considered.

References

ASHRAE (1993) ASHRAE Handbook, *1993 Fundamentals, SI Edition*; one of four volumes published periodically by the American Society of Heating, Refrigeration and Air-Conditioning Engineers, Inc., Atlanta, GA 30329, USA.

Clarke, J. A. (1985) *Energy Simulation in Building Design.* Adam Hilger Ltd., Bristol and Boston.

Henkes, R. A. W. M. (1990) *Natural-Convection Boundary Layer*, Ph.D. Thesis, Delft University of Technology, The Netherlands.

Hens, H. editor (1991) *Condensation and Energy, Volume 1: Sourcebook.* International Energy Agency, IEA, Energy Conservation in Buildings and Community Systems (BCS), Annex 14.

Hens, H. editor (1995) *Heat, Air and Moisture Transfer in Insulated Envelope Parts*, several volumes. International Energy Agency, IEA, Energy Conservation in Buildings and Community Systems (BCS), Annex 24.

Mayer, E. (1991) *Ueberprüfung einer neuen Bewertungsgrösse für Luftbewegungen in Räumen*, Bericht 3/1/34/91 der Forschungsvereinigung für Luft- und Trocknungstechnik, Frankfurt.

Siegel, R. and Howell, J. R. (1992) *Thermal Radiation Heat Transfer*, 3rd edition, Hemisphere Publishing Corp., Washington DC 20005, USA.

Leading to: *Air-conditioning; Space heating*

A. Moser

BULK POROSITY (see Porous medium)

BULK VISCOSITY (see Conservation equations, single phase)

BUNDESMINISTERIUM FÜR FORSCHUNG UND TECHNOLOGIE, BMFT

Postfach 200706
5300 Bonn 2
Germany
Tel: 49 228 590

BUNKERS, GRANULAR FLOW FROM (see Granular materials, discharge through orifices)

BUNSEN BURNER FLAME (see Flames)

BUOYANCY PARAMETER, BU (see Archimedes force; Rotating duct systems, orthoganal, heat transfer in)

BURNERS

Following from: *Combustion; Flames; Radiative heat transfer; Turbulence; Mixing*

Introduction

Burners are used to fix the location of the **Combustion** region within a **Furnace, Boiler, Gas Turbine Combustion Chamber**, or other device requiring heat from **Flames**.

Ignition

The most important **Fuels** used to produce heat are coal, oil and gas. These fuels and air react very slowly at atmospheric temperatures, but will react very rapidly producing a flame when raised to high temperatures. A burner is a device designed to ensure that the flame is stabilised by establishing a suitable flow field to produce the initial temperature rise. The flame is used as the source of heat to preheat the fuel/air mixture to the ignition temperature. In the case of a simple laminar flame propagating through mixture, the heating is caused by the thermal conductivity of the mixture transmitting heat upstream of the flame front. Most burners operate with a turbulent flame, and in this case, hot combustion products are recirculated by a reverse flow region in the burner flow field; mixing of these hot gases with unburned mixture raises its temperature to the ignition point. The presence of highly reactive radicals in the combustion products also contributes to the ignition process. A reliable ignition process generally leads to a stable flame.

Combustion Efficiency and Mixing

In addition to ensuring a stable flame, the burner is also required to achieve a high completeness of combustion, which is often referred to as high 'combustion efficiency.' This depends on thorough mixing of the fuel and air, and the retention of high enough temperatures until the reaction is complete. A fundamental feature of the mixing process is that it requires power. The source of this power is usually the fan, which supplies pressurised air to the upstream region of the burner. The burner is designed in such a way that this pressure is converted into a high speed jet (or jets) in the downstream region. There is a steep velocity gradient in the edge of the jet (known as the **Shear Layer**) and provided that the size of the shear layer is significantly larger than the size of the *Kolmogorov eddies,* whose Reynolds Number is unity, turbulent eddies will be formed. Most of the energy of the jet is transferred to the **Turbulence** within about twenty jet widths from the burner head.

Turbulence decays very rapidly through the small scale Kolmogorov dissipation eddies, so there is little turbulent kinetic energy or dissipation beyond this twenty jet width region. At the molecular level, the dissipation of turbulence is caused by molecules carrying their momentum from one eddy to the next. In a qualitative sense, it is apparent that they will also carry their species to the adjacent eddy and thus turbulent dissipation results in mixing at the molecular level.

Chemical reactions depend on mixing at the molecular level rather than on the average of large-scale rich and lean regions, hence the turbulent dissipation region coincides with the region at which molecular mixing and reaction take place. Of course, it is possible that the mixing process consists of 'like' mixing with 'like.' In order to minimise this possibility, it is important that adjacent Kolmogorov eddies should consist of the different materials to be mixed. Since small-scale eddies are derived from the large-scale eddies by a stretching process, it must be ensured that adjacent energy containing eddies consist of the materials to be mixed. This can be accomplished by designing the fuel injection system so that the maximum separation of the fuel elements is smaller than the size of the energy containing eddies. These energy containing eddies are comparable in size to the thickness of the shear layers; hence, they can also be decided by the burner designer. Thus, these fundamental principles can be used to determine the geometry of the burner, including such features as the number of gas jets required to distribute the gas in a gas burner. (See also **Mixing**.)

The aerodynamic design of the burner must, therefore, accomplish two major tasks. The first is to ensure recirculation of hot combustion products to stabilise the flame by some feature, such as a bluff body. The second is to thoroughly mix the fuel and the air in the shear layers downstream of the burner to ensure a high combustion efficiency.

The burner designer must take other factors into account, such as the required shape of the flame or the required heat distribution on the load to be heated. Thus, high velocity jet flames (*tunnel burners*) are used for intensive convective heating, whilst a 'wall flame' may be used when we do not want flame impingement to occur.

One important class of industrial flames rely on strong swirl to stabilise the flame by virtue of the fact that swirl causes a low static pressure on the axis compared to the local ambient static pressure. The radial differential pressure may be calculated by integrating the radial pressure gradient expression:

$$\frac{dp}{dr} = \frac{\rho v_t^2}{r}$$

where v_t denotes the tangential velocity which is a function of radius. Since the axial static pressure quickly rises to the ambient pressure, it is apparent that there is an adverse pressure gradient along the axis which leads to reverse flow for high levels of swirl. A recirculation region is thus formed on the axis of the burner extending downstream from the burner throat. The *swirl burner* has the advantage that the hot recirculation region is formed away from walls, whereas a bluff body flame stabiliser, such as a 'V' gutter, has flame in contact with the metal walls. (See also **Vortices**).

Other features which must be taken into account in the design of the burner is the minimisation of *pollutants* such as **Carbon Monoxide**, soot and oxides of nitrogen (NO_x).

Carbon monoxide and *soot* are formed when the oxygen available for combustion is inadequate to oxidise all the fuel to carbon dioxide. The solution to this problem is to operate the burner on the lean side of the stoichiometric mixture point, whilst ensuring good mixing so that there are no locally rich regions. The provision of excess air to minimise carbon monoxide and soot formation brings with it the disadvantage that the quantity of flue gases is increased. At the **Chimney**, these gases have to be maintained at reasonably high temperatures to avoid condensation and to provide a buoyant plume that will disperse high in the atmosphere. The gases carry sensible heat away from the region where it could be used, and thus decrease the efficiency of the boiler or furnace. An optimum quantity of excess air is, therefore, employed at which the efficiency is as high as possible consistent with acceptable levels of pollutant formation.

The minimisation of the oxides of nitrogen (NO_x) is slightly more complicated and involves special techniques, such as staged combustion, to avoid high local temperatures, since oxides of nitrogen are formed from the nitrogen in the air at high temperatures. (See **Nitric Oxide, Nitrogen Dioxide** and **Nitrous Oxide**).

Burners designed for oil firing are generally similar to gas fired burners, except that the fuel must be sprayed into the shear layers where the drops can be mixed thoroughly with the air. It is also important that an appropriate proportion of the fuel is carried into the recirculation region to maintain flame stability. The fuel atomiser is thus a critical component, and its design is still somewhat empirical. The energy to atomise the fuel is either derived from a pump, which raises the fuel to a high pressure, or a second fluid, such as high pressure steam, may be used. In either case, shear forces at the surface of a thin film of the liquid lead to instability of the surface, which breaks up into small drops. The drop sizes required for rapid combustion is usually less than 50 μm and are difficult to produce with heavy oil fuel. Heating heavy fuel oil dramatically reduces its viscosity and aids the atomisation process; thus, heavy oil fuels are heated before they are burned. (See also **Atomization**.)

Burners designed for pulverised coal firing are based on the same principles as described above; however, the coal particles are of the order of 70 μm in diameter and require about one second to burn. The furnace chamber must therefore allow sufficient residence time for burnout to occur, and sufficient oxygen must be present. The residual ash from the particle will be either liquid or solid, depending on the local temperature and the ash fusion temperature. It is important that liquid ash does not impinge on boiler components, such as the superheater tubes. This is achieved by removing heat from the hot gases by radiation in the early stages of the boiler. Once the gases have cooled to an acceptable level, then convective heat transfer can be employed. (See also **Boilers**.)

Burners for gas turbines are required to produce hot gases with a near uniform temperature profile at the turbine entry. Since the acceptable turbine entry temperatures are less than the stoichiometric flame temperature, the hot gases from the first stage of combustion must be diluted by mixing with additional air. As explained above, the dilution mixing process again requires mixing power derived from the pressure difference across the combustion chamber wall. Furthermore, the chamber wall must be kept cool by some mechanism, such as the use of a film of cool air injected through slots parallel to the wall, or through the use of some form of effusion cooling.

One of the problems which may be encountered with burners is the phenomenon of oscillatory combustion. The oscillations may be either random or periodic.

Random fluctuations result from the turbulent combustion process itself and noise is produced due to the rate of change of the rate of heat release. Although the efficiency of this process is very small, typically about 10^{-8}, large burner installations may have capacities of hundreds of MW and several watts of combustion noise is very noisy indeed.

Periodic oscillations result from positive feedback from the acoustic field in the furnace. The mechanism of the feedback typically consists of three stages:

a. The acoustic sound field consists of standing waves in which the velocity fluctuations are 90° out of phase with the pressure fluctuations.
b. The velocity fluctuations cause changes in the fuel/air mixing pattern and the shape or location of the flame.
c. The changes in the flame result in fluctuations in heat release which have a component in phase with the pressure fluctuations. Energy is then transferred into the pressure fluctuations, thus sustaining the acoustic field in the furnace. All the harmonics of the furnace chamber, for which the acoustic energy input overcomes the damping, will be driven.

The solution to this problem, when it occurs, involves identifying the specific coupling mechanism and modifications to the burner or to the chamber acoustic damping.

J. Swithenbank

BURNING VELOCITY (see Flames)

BURNOUT (See Heat pipes)

BURNOUT (FORCED CONVECTION)

Following from: *Forced convective boiling*

The term "burnout" indicates a change in the boiling regime, the disturbance of contact between the liquid and the heated wall resulting in an abrupt drop of the heat transfer coefficient. If the surface heat flux is controlled (e.g., electrically or nuclear heated systems), burnout leads to an inordinate rise in the wall temperature. If there is a controlled temperature heat source (e.g., the tube is surrounded by hot fluid on the outside), the burnout is accompanied by a considerable reduction of heat flux carried away by the boiling fluid.

Burnout is the most important factor in designing steam generators for steam power and atomic power plants, systems for engine cooling, radio-electronic equipment, cryogenic systems and many other thermally stressed devices.

The mechanism for changing boiling regimes is rather diversified. Besides "burnout," the terms "boiling crisis," "*departure from nucleate boiling*" (*DNB*), "*critical heat flux (CHF)*," "maximum heat flux," and "peak heat flux" are also used. The further term *dryout* suggests that the film of a liquid on the heated surfaces dries. The term burnout has received the widest acceptance.

The behaviour of the wall temperature and the temperature of a liquid along the heated pipe ($\dot{q}$ = const) is shown in **Figure 1**. A cool liquid ($T_L(z = 0) < T_{SAT}$ where T_{SAT} is the saturation temperature) is supplied into the pipe, and a superheated steam leaves the pipe. Nucleate boiling begins in the pipe cross-section, where $T_W \cong T_{SAT}$. After burnout occurs there exists a transition zone where an annular dispersed flow is replaced by a dispersed one. In the transition zone, drops of liquid fall out periodically onto the wall of the pipe. This is the zone of considerable pulsations of the wall temperature. After the transition zone, the liquid is concentrated in the flow core and does not come in contact with the wall.

In short pipes, not all the types of flow patterns shown in **Figure 1** can be realized. **Figure 2** illustrates how the burnout process differs with the rise in the vapour content of the flow with the motion along the vertical pipe.

(a) The mean-mass temperature of the flow is below the saturation temperature (**Figure 2a**). At large heat flows, the wall temperature can exeed the saturation temperature and in the boundary layer, bubbles are formed slipping over the heating surface. With the rise in heat flux, the number of bubbles increases and at $\dot{q} = \dot{q}_{c1}$ the bubbles collapse together and form a continuous vaporous film. The burnout mechanism is approximately similar to the burnout in conditions of pool boiling (DNB). It usually occurs at the tube outlet, where the subcooling of the liquid relative to the saturation temperature is lowest. In the zone of film boiling, the flow presents a jet of cold liquid separated from the wall by a continuous vaporous film. If x is close to zero, then a dispersed flow sets in.

(b) The slug or the plug flow (**Figure 2b**). The bubbles are formed on the wall and are entrained in the flow core. The bubbles

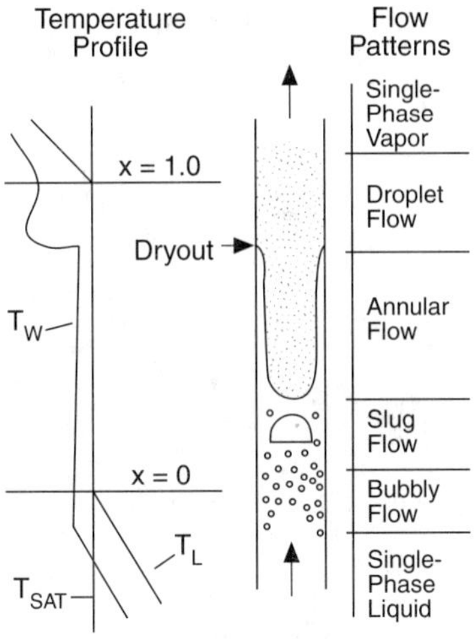

Figure 1. Regimes of heat transfer and two phase flow in a heated channel.

are unevenly distributed along the flow section. At the boundary layer edge, the bubbles are accumulated. The burnout can occur as a result of evaporation of moisture in the boundary layer and of the formation of a vaporous film.

Dry spots on the wall formed as a result of film breakdown can cause burnout.

(c) Annular-dispersed flow. Steam generation occurs both as a result of evaporation on the surface of the film and vapour bubble formation on the wall. Moisture exchange between the film and the flow core takes place. The liquid is lost by the film due to boiling, evaporation from the film surface and separation of drops from the wave crests by action of the vapour. The film is filled up with moisture due to the fall-out of drops from the flow core. It thins along the heated pipe as the vapour content increases. Burnout occurs at the flow section in which the liquid flow rate in the film approaches zero.

(d) At large x, practically the entire liquid is concentrated in the flow core in the form of droplets. The wall is covered by a very thin film of liquid. Bubbles are not formed in the film; heat is removed by evaporation from the film surface. In this flow, $\dot{q}_c$ are low and are determined by the intensity of droplets falling on the heated wall.

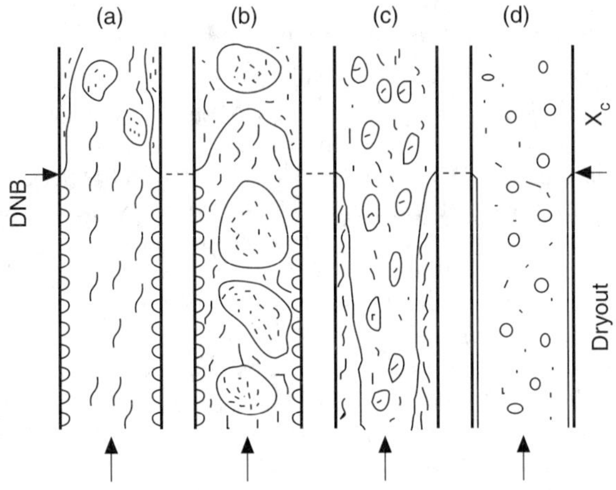

Figure 2.

Burnout has been most thoroughly studied in water boiling. Here, $\dot{q}_c$ depends on the flow's vapour content (x), mass flux ($\dot{m}$), pressure (p) and the geometry of the channel. Typical dependencies $\dot{q}_c(x)$ for fixed values of p and $\dot{m}$ are given in **Figure 3a,b**.

As can be seen from the figures, the critical heat flux drops with the rise in x. The lines $\dot{q}_c(x)$ showing different flow rates have a point of intersection, the point of inversion B (sometimes, this is a certain region rather than a point). To the left of the point, $\dot{q}_c$ increases with the rise in $\dot{m}$; to the right, $\dot{q}_c$ decreases. Changes in the dependence of $\dot{q}_c$ on $\dot{m}$ can be explained by different mechanisms of burnout.

In case of departure from nucleate boiling (DNB), the increase of $\dot{m}$ is to be followed by the increase of $\dot{q}_c$ because bubble departure from the heated surface is easier. If the burnout occurs as a result of liquid film drying out (Dryout), the increase in $\dot{m}$ corresponds to liquid entrainment from the wall. Therefore, the film flow rate becomes lower and $\dot{q}_c$ is lower too.

Lines $\dot{q}_c(x_c)$ may be divided into different parts, with each of them corresponding to a certain burnout mechanism or a certain flow pattern. Thus, within pressure range 3 − 15 MPa and 500 $< \dot{m} <$ 3000 kg/m^{2}s, five different parts can be distinguished (**Figure 3a**). Part 1 (AB), DNB takes place. Part 3 line CD, the burnout is connected to the liquid film drying out (Dryout); the

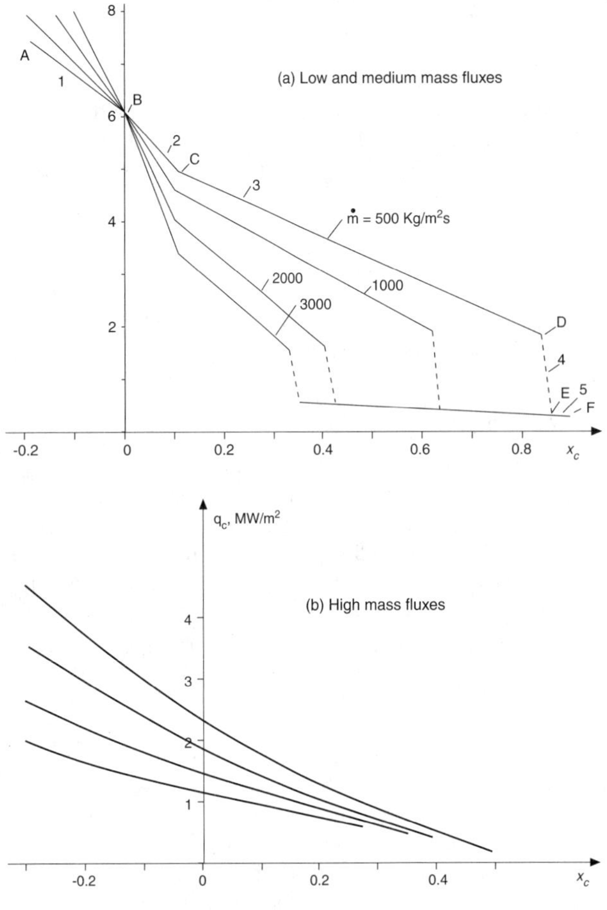

Figure 3. Variation of critical heat flux with quality.

annular dispersed flow is replaced by a dispersed one. There is some transition part, BC between 1 and 3, which corresponds to slug or churn flows.

In Part 5 (EF), tube surface is covered with a thin liquid film. The value of $\dot{q}_c$ is determined by droplet deposition onto the surface from the flow core.

The transition zone Part 4 (DE) has a particularity. In this region, the moisture exchange between the film and the flow core becomes negligible and ceases. The burnout occurs because the liquid in the film dries out. According to experimental data under conditions of no moisture exchange, the dessication occurs either at one and the same value of x, or independently of $\dot{q}$, or in the narrow range of Δx. Zone 4 is plotted in coordinates $\dot{q}(x)$ either by a vertical line or by a strongly inclined line. The vapour content at which the moisture exchange ceases is called the boundary quality vapour, x_b.

At high pressures ($15 < P < 20$ MPa) and high mass fluxes ($3000 < \dot{m} < 6000$ kg/m^{2}s), zone 4 practically disappears and Parts 1 and 5 fill each other (**Figure 3b**).

Influence of Pressure

The character of the dependence $\dot{q}_c(x)$ varies with changes in pressure. The point of inversion moves to the zone of higher x with increase in pressure. At pressure over 18 MPa, it disappears.

The $\dot{q}_c$ gradually drops in absolute value, beginning with $p = 1 - 2$MPa, with the rise in pressure.

Influence of Tube Diameter

Data for the influence of the tube diameter (D) on the $\dot{q}_c$ value are contradictory. Most of the investigators believe that $\dot{q}_c$ decreases with the rise in D. The influence of the diameter can be accounted for under the formula:

$$\dot{q}_c(D)/\dot{q}_c(D_0) = (D/D_0)^{0.5}.$$

Here $D_0 = 8$ mm, $4 < D < 20$ mm, $\dot{q}_c(D)$, $\dot{q}_c(D_0)$ is the critical heat flux in the pipe with diameter D and 8 mm.

Influence of Pipe Length

Pipe length only slightly affects q_c. In long pipes ($L/D < 20$), q_w does not depend on length. In short pipes ($L/D < 20$), $\dot{q}_c$ increases with the decrease in length. The method of treatment of the wall surface, and the thermo-physical properties of materials used in heat and power engineering slightly effect the $\dot{q}_c$ value. The oxidation of the wall material and a slight deposition of scale leave $\dot{q}_c$ unaffected.

Nonuniform Heating Along the Channel Length

The distribution of vapour and liquid along the channel cross-section depends on the heat flux density. The complicated and ambiguous effect of the above parameters on $\dot{q}_c$ can be inferred from the entrainment diagram shown in **Figure 4** obtained by Benett et al. Here, the entrained liquid flux ($\dot{m}_{2E}$) is treated as a function of quality x. The distribution of liquid between the film and the flow core in a fully developed adiabatic flow is shown by

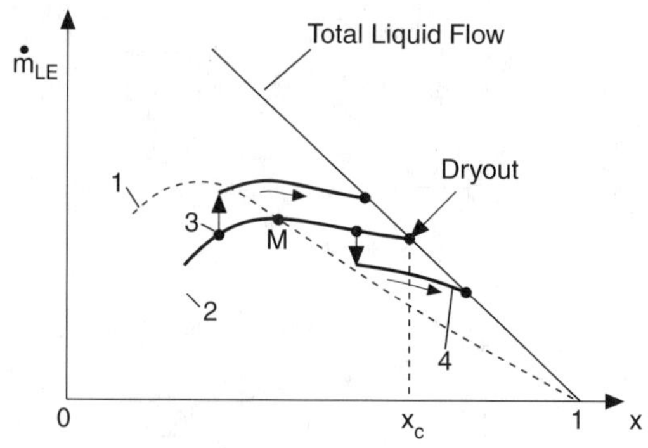

Figure 4. Variation of entrained liquid mass flux ($\dot{m}_{LE}$) with quality (x).

line 1. A redistribution of liquid between the film and the flow core (line 2) occurs in a heated channel. The burnout is observed in the flow section in which the liquid flow rate in the film equals zero (dryout point). If there is a nonheated section in the central part of uniformly heated pipe, then in this section the phase equilibrium will tend to restore. If a nonheated section (line 3) lies to the left from point M (line 3), then the amount of moisture in the film will decrease and the burnout will occur at smaller x. If a nonheated section is to the right of the point M (line 4), then the amount of liquid in the film in the non-heated section will increase and x_c increases (see Hewitt and Whalley 1989).

There exist several ways for accounting for the effect of nonuniformity of heating along the channel length on burnout. For instance, we can take into account the effect of the previously discussed parameters with the help of the influence function:

$$\frac{\dot{q}_c}{\dot{q}_{c0}} = \frac{1}{\dot{q}(z)} \int_0^z \dot{q}(\xi) w(\xi, z) d\xi,$$

where $\dot{q}_{c0}$ is the critical density of heat flux with uniform heating; $\dot{q}(z)$ is the local density of heat flux in the z section; $w(\xi, z) = \exp[(\xi - z)/L_r]/L_r$ is the influence function; L_r is the length of influence relaxation of those parameters. In case of a cosinusoidal profile, $L \geq 100$ D and $x \geq 0.2$, $L_r = 40$ D.

Correlation for Forced Convective Burnout

There are a great many experimental data on critical heat fluxes on water boiling in pipes. Tables for $\dot{q}_c$ and x_c, which embrace practically the entire range of parameters p and ρ_w used in power engineering (see IVTAN (1980), Groeneveld, et al. (1986)), have been compiled from experimental observations. Numerical relationships have also been suggested, but since these relationships are valid only for a small range of operating conditions and their accuracy is less than that of the tabulated data, the data given in the tables are preferred.

In case of nonuniform heating along the channel length, the critical heat output of the channel and the place of burnout can be determined as follows. The dependence $\dot{q}_c = f(x_c)$ for the given type of channel with the given rate and pressure (line ABC, **Figure 5**) is plotted on the basis of experimental data and design recommendations. Given the vapour content at the inlet, the pro-

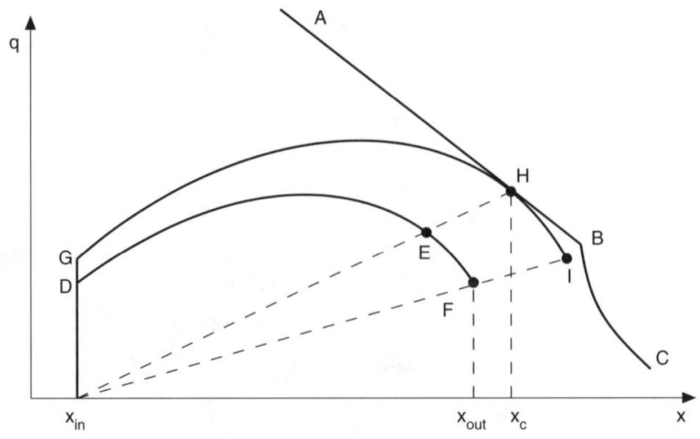

Figure 5. Burnout with non-uniform heating.

file of heat flux density is plotted versus vapour content (DEF). By gradually increasing the power delivered to the channel, the GHI profile can be plotted, which will be in contact with the line ABC. As can be seen from the figure, the burnout will occur at the point of contact H. The critical power of the channel will correspond to the GHI profile. The burnout margin can be determined as the $\dot{q}_H/\dot{q}_E$ ratio. However, in some cases, this procedure can lead to large errors. It may be better to use an analytical prediction (Hewitt and Whalley, 1989).

Modelling Burnout using Refrigerants

Large values of $\dot{q}_c$ and a high saturation pressure for water present certain difficulties for investigation. Therefore, in order to simplify the experiment in designing new steam-generating devices, it is advisable to perform simulations using refrigerants. Such refrigerants have lower latent heats and thus require less power. In this simulation, geometric and hydrodynamic similarity must be observed. If the same vapour content $x_W = x_F$ is maintained at the outlet of the channel with water and of the channel with the refrigerant fluid, and the subcooling at the channel inlet is taken to be proportional to the heat of vaporization $K_{\Delta h} = \Delta h_w/\Delta h_F = h_{lg,w}/h_{lg,F}$, then the lengths of boiling sections for both heat transfer agents will be the same. The pressure of water and freon is chosen such that the condition $p_{L,w}/p_{G,W} = p_{L,Ref}/p_{G,Ref}$ is fulfilled. These conditions make it possible to fulfill the conditions of hydrodynamic similarity, to obtain the equality of volume rate and true vapour contents in channels with water and refrigerant. Experimentally, the mass velocity ratio $K_m = \dot{m}_u/\dot{m}_F$ is been selected so that the condition $\dot{q}_{cw}/\dot{q}_{cf} = K_m K_{\Delta h}$ is met.

Burnout in Annular Channels

An annular channel can be heated either from one side (inner or outer surface) or from both sides. Usually, the critical densities of heat flux for unilateral and bilateral heatings are about the same, all other factors being equal. The dependence $\dot{q}_c(x_c)$ is similar to that for circular pipes. The effect of annular gap width within the range of 1–4 mm on $\dot{q}_c$ is small. At smaller gaps, $\dot{q}_c$ is reduced. An influence of the channel length is observed for $L/D_H < 100$. The occurrence of eccentricity reduces $\dot{q}_c$.

Burnout in Rod Clusters

In nuclear power reactors, the fuel pins are combined into clusters. Inside the cluster, the position of the rods is fixed by means of spacer grids. A two-phase flow moves along parallel "subchannels" formed by the adjacent pins. In this case, boiling differs from that in a single-pipe because of constant exchange between the coolant paths. The quantity $\dot{q}_c$ depends on such factors as the heat flux distribution, the relative position of the fuel pins, the conditions of coolant input, etc. It is natural that the critical conditions will be achieved first in the subchannels with higher local $\dot{q}$ and with high local vapour content of the flow.

There are a number of computer programs available which allow the estimation of the local parameters of the flow and the critical power of the cluster. In order to define local parameters of the flow the entire space between the fuel pins and the shell surrounding them is divided into parallel subchannels. The boundaries of the channels are drawn through the centres of the adjacent pins or around separate pins. The cluster is divided axially into several sections. The axial and radial changes of enthalpy and of mass velocity are determined by solving the conservation equations of mass, energy and momentum for each subchannel. The conservation equations take into account the transverse transfer of the coolant, and also the exchange of heat and momentum between subchannels due to turbulent mixing. The local conditions are compared with available empirical relationships for $\dot{q}_c$. The solutions are made for a series of regimes, with power supplied to the cluster increasing in succession. The power at which the local critical heat flux is reached is determined.

Full-scale tests are carried out in order to develop and validate the design procedures. The cluster of heat-releasing fuel elements is simulated by a set of hollow cylinders, through which electric current passes. By adjusting the distribution of the electric resistance of the cylinders, nonuniform heat release along the length and the radius of the reactor core, and also the different stages of fuel burnup can be replicated. The wall temperature is measured by a sensing transducer. By gradually increasing power supply the initial temperature rise in the wall due to burnout is detected and the burnout location, the local parameters of the flow and the local heat flux at the burnout location are determined as well as the total power of the assembly.

Critical Heat Fluxes for Forced Convection of Other Fluids

A wide range of coolants, from helium ($T_{SAT} = 4$ K) to sodium ($T_{SAT} = 1150$ K), has been studied. The boiling of nitrogen, hydrogen, freon, ammonia, nitrogen tetraoxide, potassium, cesium and other fluids have been included. It was found that for any fluid, as with water, burnout of the DNB or dryout types can be observed. By virtue of a considerable difference in the thermal properties of the fluids, only limited relationships can be obtained, which only describe burnout for a finite number of fluids and for a narrow range of parameters.

In boiling, the dependence $\dot{q}_c(x_c)$ for helium is about the same as of water. There exists an inversion area in which the different lines obtained for $\dot{q}_c(x_c)$ intersect. With an increase in pressure, x_{inv} increases. The existence of a vertical section x_b of the dependence $\dot{q}_c(x_c)$ is noted as in Figure 3b. The size of the vertical section decreases with increases in pressure and mass velocity. At pressures above 0.16 MPa and mass velocities exceeding 160 kg/cm^{2}s, the vertical section on the graph $\dot{q}_c(x_c)$ degenerates and the dependence is smooth.

Burnout with boiling liquid metals has a number of characteristic properties. Metals (sodium, potassium, caesium) differ from other heat transfer agents in their high boiling temperature. The pressures used are usually not high ($P < 1$ MPa). For boiling, considerable superheating of fluids is usually required. After boiling, an annular dispersed fluid flow sets in practically immediately. A thin fluid film moves along the wall of the channel. Due to the high thermal conductivity of liquid metal, the superheat of the liquid is negligible and gas bubbles do not form in the film. The processes of evaporation and droplet deposition from the flow core take place on the surface of the film. The burnout occurs due to dryout of the film.

Increasing $\dot{q}_c$

In order to increase $\dot{q}_c$ and to avoid strong overheating of the wall, an additional inflow to the wall must be ensured. Use can be made of flow swirl, which enhances the deposition of drops from the flow onto the wall because of centrifugal forces. Spiral twisted metal tapes placed at the end of the zone of developed boiling are used for this purpose. The twisted tapes operate well at high quality where the high vapour velocity gives rise to larger centrifugal forces. An alternative method for increasing the dryout quality is to use coiled tubes. Here, the droplets are deposited by centrifugal action and the liquid film is spread around the wall under the influence of circumferential shear stress and pressure gradient. By optimising the ratio of coil diameter to tube diameter, burnout qualities near 100% can be obtained.

Flow turbulisation by transverse or spiral corrugations, fins and inverts may also be used. For instance, in an annular channel with the inner surface heated, turbulizing projections may be attached to the channel. The projections release the liquid from the outer wall into the flow core, thus ensuring a strong inflow to the heated wall. In nuclear fuel pin assemblies, the spacer grids exert a turbulizing effect. They enhance the mixing of the two-phase flow and the levelling-off of the enthalpy between adjacent subchannels.

In order to increase both $\dot{q}_c$ and the heat transfer coefficient, porous coatings can be applied to the wall. The coatings are made by sintering metal particles (copper, bronze, steel powders are used) to the wall. Porous coatings are most efficient in boiling a subcooled liquid. A more than twofold increase in $\dot{q}_w$ has been observed. Porous coatings can be used in cases when, in boiling, no scale is formed on the wall or no contaminants clog up the pores.

References

Hewitt, G. F. and Whalley, P. B. (1989) *Vertical Annular Two Phase Flow,* Chapter 2 of Multiphase Science and Technology, Vol. 4 (Ed. G. F. Hewitt, J. M. Delhaye and N. Zuber), Hemisphere Publissing Corporation.

IVTAN-I-57, (1980) *Recomendation for Critical Heat Flux Transfer Calculation for Water Boiling in Tubes,* Moscow.

Groeneveld, D. C., Cheeng, S. C., and Doan, T. (1986) AECI-40 *Critical Heat Flux Transfer Lookup Table,* Heat Transfer Eng., Val. 7, p. 46.

Groeneveld, D. C., and Leung, L. K. H. (1989) *Tabular Approach for Predicting CHF and Post-Dayout Heat Transfer* Proc. NURETH-4, Val. 1, pp. 109–114, Karlsruhe, FRG.

S.A. Kovalev

BURNOUT HEAT FLUX (see Burnout in pool boiling)

BURNOUT, IN COILED TUBE (see Coiled tube, heat transfer in)

BURNOUT IN POOL BOILING

Following from: *Pool boiling*

Introduction

The *peak heat flux, maximum heat flux,* or the *burnout heat flux* represents the upper limit of fully developed **Nucleate Boiling**. In an engineering system, this heat flux also defines the limit for the safe operation of a component. After the occurrence of the maximum heat flux condition, heat removal from the heater surface degrades substantially. This, in turn, leads to a large increase in wall superheat, or melting, or burnout of the heater. Fully developed nucleate boiling of saturated liquids is generally associated with the formation of vapor columns and mushroom type bubbles on the heater surface. Vapor columns form as a result of the merger of bubbles normal to the heater surface, whereas mushroom type bubbles form as a result of merger of bubbles at neighboring sites. The mushroom type bubbles are believed to be attached to the heat surface through several stems.

According to Zuber (1958), the maximum heat flux condition occurs when vapor velocity in the large jets leaving the heater surface reaches a critical value. At this value of the vapor velocity, the jets in a counter-current flow situation become unstable and cannot support additional outflow of vapor. This leads to the accumulation of vapor at the heater surface and a drastic reduction in the rate of heat transfer from the surface. The instability of the vapor jets occurs away from the heater surface and surface conditions do not affect the maximum rate of heat removal. By assuming that 1) all of the energy dissipated at the heater surface is used in phase change; 2) the vapor jets are located on a square grid with a spacing equal to a Taylor wavelength (critical or "most susceptible", see **Taylor Instability**); and 3) the jet diameter is equal to half of the Taylor wavelength, Zuber obtained an expression for the maximum heat flux on an infinite horizontal plate as:

$$\dot{q}_{max_Z} = \frac{\pi}{24}\rho_G h_{LG}\sqrt[4]{\frac{\sigma g(\rho_L - \rho_G)}{\rho_G^2}} \times \left\{\frac{\rho_L(16-\pi)}{\rho_L(16-\pi)+\rho_G\pi}\left(\frac{\rho_L+\rho_G}{\rho_L}\right)^{1/2}\right\}. \quad (1)$$

where ρ_G is the vapour density, ρ_L the liquid density, σ the surface tension and h_{LG} the latent heat of evaporization.

Equation 1) has been found to be quite successful in predicting maximum heat flux on well-wetted, large horizontal surfaces. The numerical constant in Equation 1) resulted from averaging of the results for critical and most susceptible wavelengths. The theory supporting Equation 1) is now known as the *hydrodynamic theory of boiling*. Lienhard and Dhir (1973) have shown that Equation 1) predicts the flat plate data better if a lead constant of 0.15, instead of $\pi/24$, is used. Also, Equation 1) is applicable as long as the heater is at least 3 Taylor wavelengths wide. For smaller heaters observed, maximum heat fluxes can vary significantly from those given by Equation 1).

The hydrodynamic theory has been developed further by Lienhard and co-workers (1973) to account for the geometry and finite size of the heaters. According to this extended hydrodynamic theory, maximum heat flux on finite heaters (cylinders, spheres, ribbons, etc.), can be written as:

$$\frac{\dot{q}_{max}}{\dot{q}_{max_Z}} = F(L'), \qquad \textbf{(2)}$$

where L' is the characteristic dimensionless length of the heater (e.g., radius of a cylinder or sphere, height of a ribbon, etc.) and is defined as $L' = L\sqrt{g(\rho_L - \rho_G)/\sigma}$. **Figure 1** shows the predictions of maximum heat flux for several geometries.

Hydrodynamic theory does not account for the effect of surface conditions on maximum heat flux. There is substantial evidence in the literature [e.g., Costello and Frea (1965) and Hasegawa, et al. (1973)] that maximum heat flux is influenced by the degree of wettability of the heater. The recent work of Liaw and Dhir (1986), in which a static contact angle was used as an indicator of surface wettability, shows that maximum heat flux on a vertical surface decreases as the surface becomes less-wetting. Similar observations on horizontal surfaces have been reported by Maracy and Winterton (1988), and on horizontal cylinders by Hahne and Diesselhorst (1978). **Figure 2** shows the maximum heat flux data as a function of contact angle. It is seen that maximum heat flux approaches that given by the hydrodynamic theory when the contact angle is less than 20°. Dhir and Liaw (1989) have inferred that for partially wetted surfaces, the rate of evaporation near the heater surface sets the upper limit of nucleate boiling heat flux, whereas for well wetted surfaces, this limit is probably set by the rate of vapor overflow.

The soundness of the assumption of instability of large vapor jets in hydrodynamic theory has been questioned in recent years by Haramura and Katto (1983) and several others, on the grounds that the vapor jets are too blunt to allow the development of the critical wave on the vapor-liquid interface. Haramura and Katto have suggested that instability of the vapor stems underneath the mushroom type bubbles determines the maximum heat flux. No verification of the latter assertion exists yet.

Several other variables, aside from those described above, influence the burnout heat flux and are discussed below.

Surface contamination

Data from Bui and Dhir (1985) and Berenson (1962) show that maximum heat flux is generally higher on dirty surfaces. During boiling on a vertical surface placed in a pool of saturated water exposed to laboratory environment for a long period of time, Bui and Dhir have noted that maximum heat flux on a surface contaminated with dust particles in water was about 25% higher than that on a clean surface. The cause of such an enhancement in maximum heat flux is not yet understood.

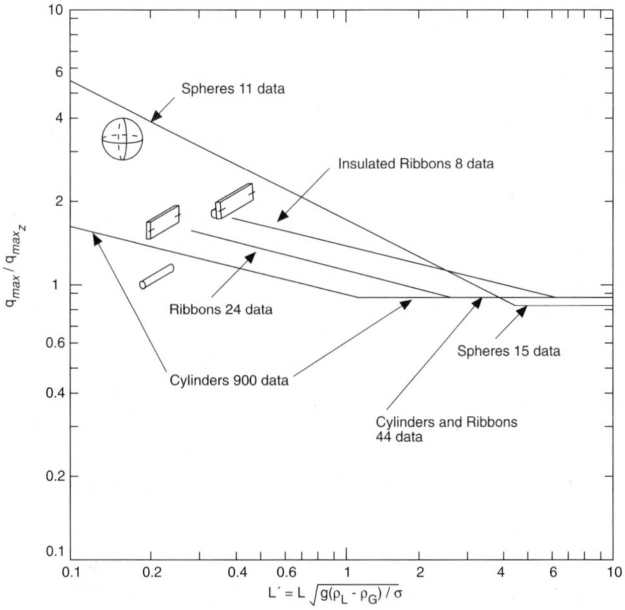

Figure 1.

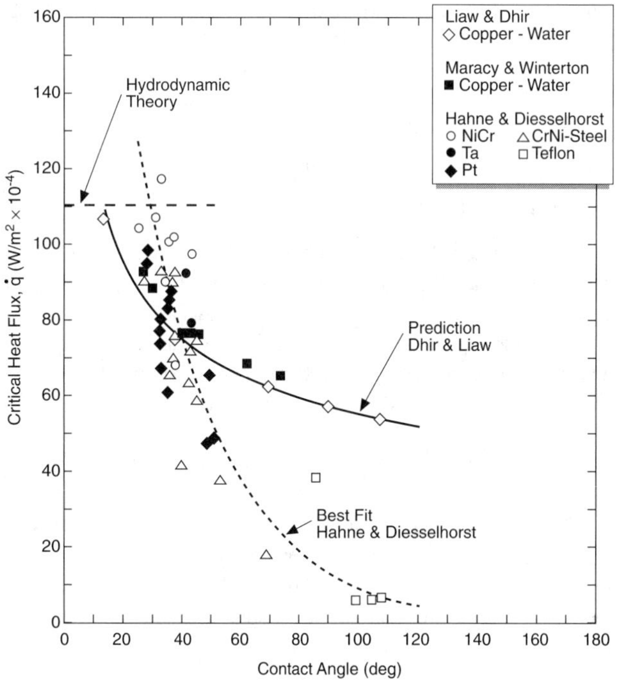

Figure 2.

Heater thickness and material

Several investigators [e.g., Houchin and Lienhard (1966) and Tachibana et al. (1969)] have found that on thin heaters, maximum heat flux occurs prematurely. It is postulated that hot spots under vapor bubbles or jets are unstable at high heat fluxes, and advance to increase the heater fraction that is dry. Maximum heat fluxes obtained under steady-state conditions on well-wetted surfaces show little effect of the heater material. However, Lin and Westwater (1982) have found that maximum heat flux obtained during quenching is influenced by the heater material properties. They correlated the quasi-steady maximum heat flux data with the product of density, specific heat and thermal conductivity of the material.

System pressure

The maximum heat flux initially increases with system pressure, attains its highest value between reduced pressures of 0.3 and 0.4, and thereafter, decreases with further increase in pressure. For well-wetted surfaces, Equation 1, based on the hydrodynamic theory, yields predictions for the maximum heat flux that are in good agreement with the data obtained over a wide range of pressures.

Gravity

According to equation (1), the maximum heat flux should vary with gravity as $\sqrt{4g}$. The high gravity centrifuge data confirm this dependence. However, the drop tower data of Siegel and Usiskin (1959) shows that the value of the exponent of gravity decreases as the magnitude of gravitational acceleration is reduced; at very low gravity, the maximum heat flux is independent of gravity. Recent low gravity data of Straub et al. (1992) shows that observed heat fluxes are significantly higher than those obtained from Equation 1. At present, no rational basis exists to describe the observed behavior. For short duration microgravity conditions, many internal and external factors (e.g., thermocapillary forces, vibrations, etc.) can influence the process. (See also *Microgravity Conditions*).

Mode of heating the surface

The maximum heat fluxes obtained during rapid heating of a surface are generally higher than those obtained under steady-state conditions. Studies of Sakurai and Shiotsu (1977) on platinum wires submerged in a pool of water show that for exponential heating periods of less than 100 ms, the maximum heat fluxes increase with a decrease in the exponential time period.

Liquid subcooling

The maximum heat flux is found to increase with liquid subcooling. Zuber et al. (1961) have extended equation (1) to subcooled liquids by accounting for heat lost to the liquid in a transient manner. Their expression for maximum heat flux in subcooled liquid is:

$$\frac{\dot{q}_{max,sub}}{\dot{q}_Z} = [1 + C_1(g(\rho_L - \rho_G)\rho_G^2\kappa_L^4/\sigma^3)^{1/8}Ja], \quad (3)$$

where κ_L is the liquid thermal diffusivity and,

$$Ja = \frac{\rho_L c_{PL} \Delta T_{sub}}{\rho_G h_{LG}}$$

and C_{PL} is the specific heat of the liquid and ΔT_{sub} the degree of subcooling. From analysis, the constant, C_1, was found to have a value of 5.33.

According to Equation 3 maximum heat flux increases linearly with subcooling. Elkassabgi and Lienhard (1988), from their experiments on horizontal cylinders, have found that only at low subcoolings and maximum heat flux increased linearly with subcooling. At very high subcoolings, the maximum heat flux attains an asymptotic value. The asymptotic value is related to molecular effusion limit. Separate correlations for different ranges of subcoolings have been developed.

Flow velocity

Liquid velocity over heaters in the direction of gravity enhances maximum heat fluxes. However, for low velocities in the direction opposite to the gravity, the maximum heat flux can be lower than that obtained under pool boiling conditions. Correlations for maximum heat flux on discs cooled by impinging jets have been developed by Katto and Yokoya (1988), and on cylinders subjected to cross flow by Lienhard and co-workers (1988). (see also **Boiling**).

Finally, it should be stated that Equations 1 and 2 have generally been accepted for engineering applications involving pool boiling. However, their basis is still under debate.

References

Berenson, P. J. (1962) Experiments on pool boiling heat transfer, *Int'l. J. Heat and Mass Transfer,* 5:985–999.

Bui, T. D. and Dhir, V. K. (1985) Transition boiling heat transfer on a vertical surface, *ASME J. Heat Transfer,* 107:756–763.

Costello, C. P. and Frea, W. J. (1965) A salient non-hydrodynamic effect in pool boiling burnout of small semi-cylinder heaters, *Chem. Eng. Prog. Symp. Series,* 61:258–268.

Dhir, V. K. and Liaw, S. P. (1989) Framework for a unified model for nucleate and transition pool boiling, *ASME J. Heat Transfer,* 111;739–746.

Elkassabgi, Y. and Lienhard, J. H. (1988) The peak pool boiling heat flux from horizontal cylinders in subcooled liquids, *ASME J. Heat Transfer,* 110:479–486.

Hahne, E. and Diesselhorst, T. (1978) Hydrodynamic and surface effects on the peak heat flux in pool boiling, *Proc. 6th Int'l. Heat Transfer Conf.,* Toronto, Ontario, Canada, 1:209–219.

Haramura, Y. and Katto, Y. (1983) "A new hydrodynamic model of critical heat flux, applicable widely to both pool and forced convection boiling on submerged bodies in saturated liquids," *Int'l. J. Heat and mass Transfer,* 26:389–399.

Hasegawa, S., Echigo, R., and Takegawa, T. (1973) Maximum heat fluxes for pool boiling on partly ill-wettable heating surfaces, *Bull. JSME,* 96:1076–1084.

Houchin, W. R. and Lienhard, J. H. (1966) Boiling burnout in low thermal capacity heaters, ASME Paper No. 66-WA/HT-40.

Katto, Y. and Yokoya, S. (1988) Critical heat flux on a disk heater cooled by a circular jet impinging at the center, *Int'l. J. Heat and Mass Transfer,* 31:219–227.

Liaw, S. P. and Dhir, V. K. (1969) Effect of surface wettability on transition boiling heat transfer from a vertical surface, *Proc 8th Int'l. Heat Transfer Conf.,* San Francisco, CA, 4:2031–2036.

Lienhard, J. H. and Dhir, V. K. (1973) Extended hydrodynamic theory of the peak and minimum pool boiling heat fluxes, NASA CR-2270.

Lienhard, J. H. (1988) Burnout on Cylinders, *ASME J. Heat Transfer,* 110:1271–1286.

Lin, D. Y. T. and Westwater, J. W. (1982) Effect of metal thermal properties on boiling curves obtained by the quenching method, *Proc. 7th Int'l. Heat Transfer Conf.,* Munich, Germany, 4:155–160.

Maracy, M. and Winterton, R. H. S. (1988) Hysterisis and contact angle effects on the peak heat flux in pool boiling of water, *Int'l J. Heat and Mass Transfer,* 31:1443–1449.

Sakurai, A. and Shiotsu, M. (1977) Transient pool boiling heat transfer, Part 2. Boiling Heat Transfer and Burnout, *ASME J. Heat Transfer,* 99:554–560.

Siegel, R. and Usiskin, C. (1959) A photographic study of boiling in the absence of gravity, *ASME J. Heat Transfer,* 81:230–238.

Straub, J., Zell, M., and Vogel, B. (1992) Boiling under microgravity conditions, Proc. 1st European Symposium for Fluids in Space, Ajaccio, France, ESA SP-353, pp. 269–297.

Tachibana, F., Akiyama, M., and Kawamura, H. (1967) Non-hydrodynamic aspects of pool boiling burnout, *J. Nucl. Sci. and Tech.,* 4:121–130.

Zuber, N., Hydrodynamic aspects of boiling heat transfer, Ph.D. diss., University of California, Los Angeles, (also published as USAEC Report No. AECU-4439).

Zuber, M., Tribus, M., and Westwater, J. W. (1961) The hydrodynamic crisis in pool boiling of saturated liquids, *Proc. 2nd Int'l. Heat Transfer Conf.,* Denver, CO, Vol. 27.

Vijay K. Dhin

BURNUP, FAST REACTOR (see Liquid metal cooled fast reactor)

BUTANE (see Oil refining)

n-BUTANE

n-Butane (C_4H_{10}) is a colourless gas that, unlike the first three *alkanes,* is very soluble in water. The principal raw materials for its production are petroleum and liquefied natural gas. It forms an explosive and flammable mixture with air at low concentrations. Its main uses in industry are as a raw material in the production of butadiene and acetic acid. It is also used as a domestic fuel, as a gasoline blending component, as a solvent and as a refrigerant.

Butane is a major constituent of *liquefied petroleum gas* (LPG) and as such, it is increasingly stored in liquefied form at low temperatures in specially adapted natural underground caverns. Its physical characteristics are listed below:

Molecular weight: 58.12
Melting point: 134.82 K
Normal boiling point: 272.66 K
Critical temperature: 425.16 K
Critical pressure: 3.796 MPa
Critical density: 225.3 kg/m^3
Normal vapor density: 2.59 kg/m^3 (@ 0°C, 101.3 kPa)

Reference

Beaton, C. F. and Hewitt, G. F. (1989) *Physical Property Data for the Design Engineer,* Hemisphere Publishing Corp., New York.

V. Vesovic

Table 1. n-Butane, Values of thermophysical properties of the saturated liquid and vapor

T_{sat}, K	273.15	289	305	321	337	353	369	385	405	425.16
P_{sat}, kPa	103	184	304	469	706	1 023	1 526	1 925	2 739	3 797
ρ_l, kg/m^3	603	587	571	551	529	504	475	441	388	225.3
ρ_g, kg/m^3	2.81	4.81	7.53	11.6	17.4	25.1	35.6	51.3	80.7	225.3
h_l, kJ/kg	−1 194	−1 158	−1 121	−1 081	−1 040	−997	−945	−896	−821	−665
h_g, kJ/kg	−809	−789	−769	−747	−725	−706	−681	−663	−648	−665
$\Delta h_{g,l}$, kJ/kg	385	369	352	334	315	291	264	233	173	
$c_{p,l}$, kJ/(kg K)	2.34	2.47	2.59	2.68	2.80	2.95	3.11	3.36	3.80	
$c_{p,g}$, kJ/(kg K)	1.67	1.76	1.88	2.00	2.15	2.33	2.62	3.03	4.76	
η_l, μNs/m^2	206	179	154	131	112	95	80	65	51	
η_g, μNs/m^2	7.35	7.81	8.32	8.87	9.44	10.20	11.25	12.77	16.30	
λ_l, (mW/m^2)/(K/m)	114.6	109.8	104.9	100.1	95.1	90.4	85.5	80.7	74.6	
λ_g, (mW/m^2)/(K/m)	13.69	15.19	16.82	18.57	20.47	22.49	24.69	27.24	31.2	
Pr_l	4.20	4.02	3.80	3.51	3.30	3.11	2.89	2.72	2.59	
Pr_g	0.90	0.90	0.93	0.96	1.00	1.06	1.19	1.42	2.48	
σ, mN/m	14.8	12.8	11.0	9.10	7.29	5.54	4.03	2.75	1.34	
$\beta_{e,l}$, kK^{-1}	1.73	2.01	2.37	2.80	3.45	4.31	7.31	9.87	10.0	

The values of thermophysical properties have been obtained from reference [1].

BUTANOL

1-Butanol (C_4H_9OH) is a colourless liquid, slightly miscible with water. It is obtained either from butylaldehyde or by the fermentation of corn products. It has low toxicity, but it can act as an irritant to the eyes and the skin. Its main uses are as a raw material in the production of butyl acrylate, butyl glycol ethers and butyl acetate. Butanol is also used as a solvent for resins and paints and as a hydraulic fluid.

The physical characteristics of butanol are as follows:

Molecular weight: 74.12
Melting point: 183.2 K
Normal boiling point: 390.65 K
Critical temperature: 561.15 K
Critical pressure: 4.960 MPa
Critical density: 270.5 kg/m^3

Reference

Beaton, C. F. and Hewitt., G. F. (1989) *Physical Property Data for the Design Engineer,* Hemisphere Publishing Corp., New York.

V. Vesovic

BY-PASS ENGINES (see Jet engine)

BY-PASSING (see Tube banks, crossflow over)

Table 1. Butanol, Values of thermophysical properties of the saturated liquid and vapor

T_{sat}, K	390.65	410.2	429.2	446.5	469.5	485.2	508.3	530.2	545.5	558.9
P_{sat}, kPa	101.3	182	327	482	759	1 190	1 830	2 530	3 210	4 030
ρ_l, kg/m^3	712	688	664	640	606	581	538	487	440	364
ρ_g, kg/m^3	2.30	4.10	7.9	12.5	23.8	27.8	48.2	74.0	102.3	240.2
h_l, kJ/kg	0.0	64.8	135.0	206.8	315.3	399.6	541.9	700.2		
h_g, kJ/kg	591.3	629.8	672.3	716.5	784.1	836.8	924.4	1 015.3		
$\Delta h_{g,l}$, kJ/kg	591.3	565.0	537.3	509.7	468.8	437.2	382.5	315.1	248.4	143.0
$c_{p,l}$, kJ/(kg K)	3.20	3.54	3.95	4.42	5.15	5.74	6.71	7.76		
$c_{p,g}$, kJ/(kg K)	1.87	1.95	2.03	2.14	2.24	2.37	2.69	3.05	3.97	
η_l, μNs/m^2	403.8	346.1	278.8	230.8	188.5	144.2	130.8	115.4	111.5	105.8
η_g, μNs/m^2	9.29	10.3	10.7	11.4	12.1	12.7	13.9	15.4	17.1	28.3
λ_l, (mW/m^2)/(K/m)	127.1	122.3	117.5	112.6	105.4	101.4	91.7	82.9	74.0	62.8
λ_g, (mW/m^2)/(K/m)	21.7	24.2	26.7	28.2	31.3	33.1	36.9	40.2	43.6	51.5
Pr_l	10.3	9.86	9.17	8.64	10.2	8.10	8.67	9.08		
Pr_g	0.81	0.83	0.81	0.86	0.87	0.91	1.01	1.17	1.56	
σ, mN/m	17.1	15.6	13.9	12.3	10.2	7.50	6.44	4.23	2.11	0.96
$\beta_{e,l}$, kK^{-1}	1.69	1.92	2.19	2.46	2.98	3.48	4.90	8.45	14.7	

C

CAD, PLASTICATING SCREWS (see Extrusion, plastics)

CAF, COMPRESSED ASBESTOS FIBRE JOINTING (see Gaskets in heat exchangers)

CALANDRIA (see CANDU nuclear power reactors)

CALCIUM

Calcium—(L. *calx,* lime), Ca; atomic weight 40.08; atomic number 20; melting point 839 ± 2°C; boiling point 1484°C; specific gravity 1.55 (20°C); valence 2. Though lime was prepared by the Romans in the first century under the name calx, the metal was not discovered until 1808. After learning that Berzelius and Pontin prepared calcium amalgam by electrolyzing lime in mercury, Davy was able to isolate the impure metal.

Calcium is a metallic element, fifth in abundance in the earth's crust, of which it forms more than 3%. It is an essential constituent of leaves, bones, teeth, and shells. Never found in nature uncombined, it occurs abundantly as *limestone* ($CaCO_3$), *gypsum* ($CaSO_4 \cdot 2H_2O$), and *fluorite* (CaF_2); *apatite* is the fluophosphate or chlorophosphate of calcium. The metal has a silvery color, is rather hard, and is prepared by electrolysis of the fused chloride to which calcium fluoride is added to lower the melting point. Chemically it is one of the alkaline earth elements; it readily forms a white coating of nitride in air, reacts with water, burns with a yellow-red flame, forming largely the nitride.

The metal is used as a reducing agent in preparing other metals such as thorium, uranium, zirconium, etc., and is used as a deoxidizer, desulfurizer, or decarburizer for various ferrous and nonferrous alloys. It is also used as an alloying agent for aluminum, beryllium, copper, lead, and magnesium alloys, and serves as a "getter" for residual gases in vacuum tubes, etc. Its natural and prepared compounds are widely used.

Quicklime (CaO), made by heating limestone and changed into slaked lime by the careful addition of water, is the great cheap base of chemical industry with countless uses. Mixed with sand, it hardens as mortar and plaster by taking up carbon dioxide from the air. Calcium from limestone is an important element in Portland cement.

The solubility of the carbonate in water containing carbon dioxide causes the formation of caves with stalactites and stalagmites and hardness in water.

Other important compounds are carbide (CaC_2), chloride ($CaCl_2$), cyanamide ($CaCN_2$), hypochlorite [$Ca(OCl_2)_2$], nitrate [$Ca(NO_3)_2$], and sulfide (CaS).

Handbook of Chemistry and Physics, CRC Press

CALDER HALL (see Gas-graphite reactors)

CALORIE

CALORIE

The calorie is a unit measure of the quantity of heat; 1 cal = 4.1868 **Joule** (J).

Before the universal adoption of the **SI systems**, the calorie was included in the dimensions of some functions which characterize thermodynamic and translational properties of substances and heat transfer processes. These functions are: **internal energy, enthalpy**, the *Gibbs energy* (isobaric-isothermal potential), the *Helmholtz energy* (isochoric-isothermal potential), the partial molar Gibbs energy (chemical potential), **entropy**, heat capacity, heat conduction and **heat transfer coefficients**.

Historically, the calorie has been defined as the quantity of heat required for heating up 1 g of distilled water at atmospheric pressure by 1°C. The reason for the discrepancy in the value of calorie—which existed in literature (in particular, the so-called *thermochemical calorie,* 1 cal = 4.1840 J)—is due to the temperature dependence of the heat capacity of water. At present, the Joule, which represents the energy measurement unit in the SI system, is most frequently used in measurements and tabulations of the functions mentioned above.

D.N. Kagan

CALORIFIC VALUE OF FUEL (see Flames)

CALORIMETRY (see Specific heat capacity)

CANDU NUCLEAR POWER REACTORS

Following from: *Nuclear reactors*

Introduction

Two basic features of the CANDU (CANada Deuterium Uranium. Registered trademark.) nuclear power reactor are the use of heavy water as neutron moderator and the use of pressure tubes to contain the reactor fuel and coolant. Contemporary CANDU reactors also use heavy water as primary coolant. As **Figure 1** shows, heat is transferred from the fuel to the primary coolant, which transports the heat to steam generators. The **Steam Generator** secondary side forms part of a conventional steam power cycle.

The low-pressure heavy-water moderator, contained in a vessel called the *calandria,* provides efficient fuel utilization and permits the use of natural rather than enriched uranium. The moderator is separate from the high-pressure coolant.

A CANDU fuel channel contains 12 or 13 fuel bundles end-to-end. Each fuel bundle comprises an array of fuel elements held together by Zircaloy end-plates. As **Figure 2** shows, a fuel element is made from UO_2 fuel pellets contained in a thin cylindrical Zircaloy sheath. The coolant, at a pressure of about 10 MPa, is contained in zirconium alloy pressure tubes, each of which is insulated from an outer calandria tube by CO_2 gas.

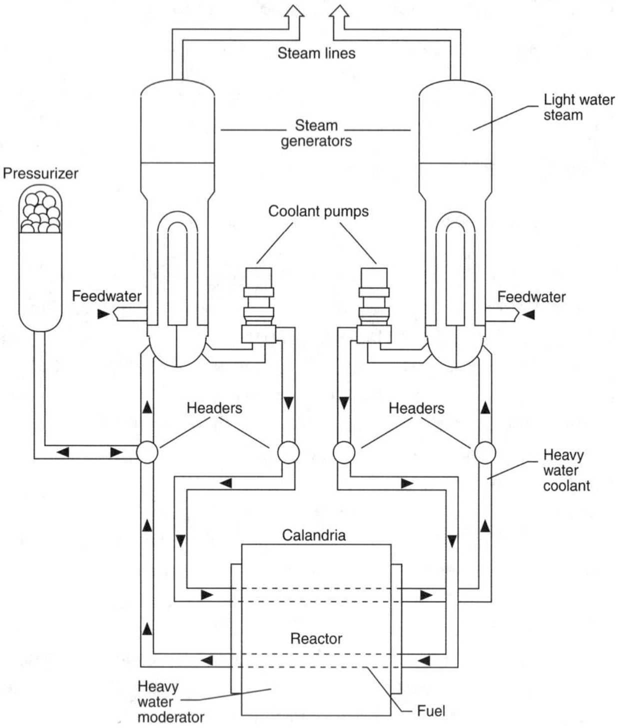

Figure 1. CANDU simplified flow diagram.

A CANDU reactor employs several hundred horizontal fuel channels, only two being shown in **Figure 1**: the horizontal orientation facilitates on-power refueling.

Heat Transfer from Fuel to Coolant

A fuel channel in a CANDU reactor is normally supplied with up to 25 kg/s of coolant and produces up to 7 MW of power. The coolant can boil slightly, which increases steam generator temperatures, thereby increasing the plant's thermodynamic efficiency.

Under normal operating conditions, the heat transfer from fuel to coolant is high and the sheath temperature is only a few tens of degrees higher than coolant temperature. However, at abnormally high-channel power or abnormally low-channel flow, the rate of heat transfer to the coolant can deteriorate due to the development of a vapour film at the fuel surface. This condition, called *dryout,*

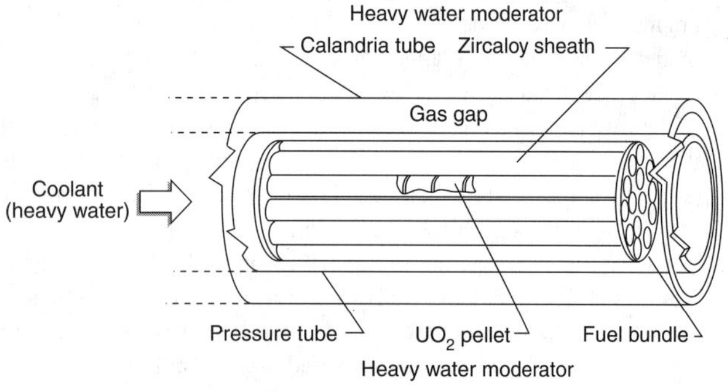

Figure 2. Fuel bundle in a fuel channel.

is avoided with margins in power and flow. The reactor is tripped (power reduced) if measurements show that the dryout condition is being approached. (see **Burnout**.)

The channel condition at dryout is measured in CANDU-specific full-scale tests using electrically-heated fuel-string simulators (Leung et al., 1985). The tests are done over a range of mass flows and pressures, and for several channel axial power distributions.

Light water, instead of heavy water, is used as coolant in the tests. This requires a conversion of the dryout data, which is done using the fluid-to-fluid *modeling techniques* of Ahmed (1971). The boiling length/critical quality approach of Bertoletti (1965) is used to correlate the data: dryout is assumed to have occurred if the steam quality in the fuel channel exceeds the dryout steam quality (flow-cross-section averaged). The dryout steam **Quality** X_c is correlated as a function of the boiling length L (the distance along the channel from the onset of boiling to the location of dryout). This approach suggests the existence of an upstream effect—instead of dryout being a purely local phenomenon—and is necessary in accurately correlating data over a range of axial power shapes.

The correlation has the functional form:

$$X_c = aL/(b + L)$$

where a and b depend on pressure and mass flux. In the tests, the location of dryout is identified using sliding thermocouples. The location of onset of boiling and X_c are calculated from measurements of power, pressure, coolant inlet temperature and flow.

The above dryout correlation is used in a reactor-system thermohydraulic code. The code is used to simulate a CANDU reactor under a large number of operating conditions, with different three-dimensional neutron flux shapes and power distributions. It generates the channel power at dryout (called the *critical channel power*) for all channels and all operating conditions. This information, together with detailed information on the three-dimensional neutron flux shapes, is used to establish the trip setpoints for neutron flux as measured by detectors at a number of in-core locations. The detectors are part of a protection system that provides a 99% probability that the reactor will be tripped before dryout in any channel, given an accident leading to overpower.

In applying the thermohydraulic code, an assumption generally used is that the condition in each reactor header stays constant as power is varied in a particular fuel channel. (**Figure 1** shows fuel channels connected to headers via feeder pipes.) In particular, the header-to-header pressure drop, ΔP, is conservatively assumed to be constant. However, if the power were to increase, the extra boiling would cause a reduction in flow and an increase in the head from the pumps. The ΔP would increase.

When setting overpower trips in CANDU reactors, dryout is defined and measured as the first small additional rise in fuel-sheath temperature as power is being increased. However, dryout occurs at a level of flow and steam quality such that any further increase of temperature with power is "slow," rather than "fast," as described by Groeneveld and Borodin (1979). Some wetting of the surface still occurs while the temperature is less than 382°C, the minimum film-boiling temperature (Groeneveld and Stewart (1982)). From the full-scale tests, sheath temperatures are typically less than 500°C, even for powers 10% beyond dryout. (See also **Postdryout Heat Transfer**.)

References

Ahmed, S. Y. (1971) *Fluid-to-fluid Modeling of Critical Heat Flux: A Compensated Distortion Model,* AECL Report, AECL-3663.

Bertoletti, S. et al. (1965) *Heat Transfer Crisis with Steam-Water Mixtures,* Energia Nucleare, 12, 3.

Groeneveld, D. C. and Borodin, A. S. (1979) *The Occurrence of Slow Dryout in Forced Convective Flow,* Second Multi-Phase Flow and Heat Transfer Symposium Workshop, Miami.

Groeneveld, D. C. and Stewart, J. C. (1982) *The Minimum Film Boiling Temperature for Water During Film Boiling Collapse,* Proceedings of the Seventh Int. Heat Transfer Conference, Munich.

Leung, A., Merlo, E. E., Gacesa, M., and Groeneveld, D. C. (1985) *Critical Channel Power Evaluation Methodologies at Atomic Energy of Canada Limited,* Proceedings of the 6th Annual CNS Conference, Ottawa, June.

N.J. Spinks

CANONICAL PARTITION FUNCTION (see Boltzmann distribution)

CAPACATIVE HEAT EXCHANGERS (see Heat exchangers)

CAP BUBBLES (see Bubble flow)

CAPILLARITY (see Surface and interfacial tension)

CAPILLARY ACTION

Following from: *Interfaces*

Capillary action is the physical phenomenon arising due to surface tension on the interface of immiscible media. Commonly, capillary phenomena occur in liquid media and are brought about by the curvature of their surface that is adjacent to another liquid, gas, or its own vapor.

Surface curvature in a fluid gives rise to an additional so-called capillary pressure p_σ whose value is related to an average surface curvature H

$$H = \frac{1}{2}\left(\frac{1}{R_1} + \frac{1}{R_2}\right)$$

(R_1 and R_2 are the radii of curvature of principal normal sections) via the *Laplace equation*

$$p_\sigma = p_1 - p_2 = 2\sigma_{12}H, \quad (1)$$

where σ_{12} is the surface or interfacial tension (see **Surface and Interfacial Tension**) on the boundary between phases 1 and 2, p_1 and p_2 are the pressures in the phases. The pressure is higher in the phase to which the concavity of the interface is presented. For a plane interface, $H = 0$ and $p_\sigma = 0$.

Capillary phenomena include various cases of equilibrium and flow of fluid surface under the action of surface tension forces and of external forces, primarily gravity. In the simplest case when the external forces are absent, e.g., under weightlessness conditions, a limited fluid volume takes the shape of a sphere due to surface tension forces. This state corresponds to a stable equilibrium of the fluid since a sphere has the minimum surface and, consequently, the minimum surface energy.

As shown in **Figure 1**, when the liquid (1) comes in contact with the gas vapor (2) and the solid (3), the shape of the free liquid surface depends on wetting. Interaction of surface tension forces at the liquid-gas σ_{12}, liquid-solid σ_{13}, and gas-solid σ_{23} interfaces is responsible for a curved area of the liquid surface (meniscus) near the solid. An angle θ made by the tangent to the liquid surface and the solid surface, known as a wetting angle, is determined by the balance of forces within the surface tension. According to the *Young equation,*

$$\sigma_{23} = \sigma_{13} + \sigma_{12}\cos\theta.$$

If $\sigma_{23} > \sigma_{13}$, i.e., if the surface tension between gas and a solid is greater than between a solid and a liquid, then $\theta < \pi/2$ and the solid surface is said to be wetted by the liquid. If $\sigma_{23} < \sigma_{13}$, then $\theta > \pi/2$, and this indicates that the solid is not wetted by the liquid. The cases $\theta = 0$ and $\theta = \pi$ correspond to absolute wetting and nonwetting of the surface by the liquid.

The wetting angle depends on the liquid and the surface it is in contact with. For instance, liquid alkaline metals and cryogenic liquids wet metal surfaces nearly absolutely and the angle θ approaches zero. Teflon and paraffin are virtually nonwettable by water and some other liquids. It should be borne in mind that the wetting angle strongly depends on the surface state, i.e., surface contamination—the presence of surface active agents—and surface roughness, which enhances wettability.

As a result of displacement of the interface of the three phases, the wetting angle displays *hysteresis;* that is, on the surface wetted earlier, the wetting angle appears to be smaller than in the case of displacement of the interphase boundary on an initially dry surface.

A most commonly encountered and visual example of capillary phenomena are the suction of liquid in narrow tubes with wettable walls (**Figure 2a**) and the expulsion of the liquid out of them if the walls are nonwettable (**Figure 2b**). At each point of the curved free liquid surface, *capillary pressure* is built up that leads to a liquid rise in the tube. The height of rise h is determined by the relation between external (in this case, gravitational) and surface tension forces. This relation is expressed by the **Bond Number**

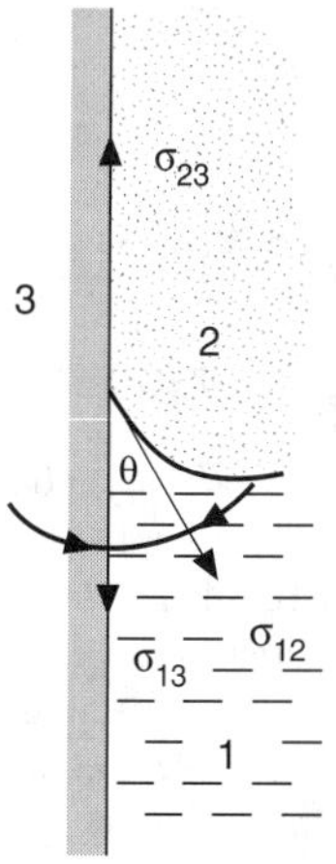

Figure 1. Solid surface in contact with gas and liquid gases.

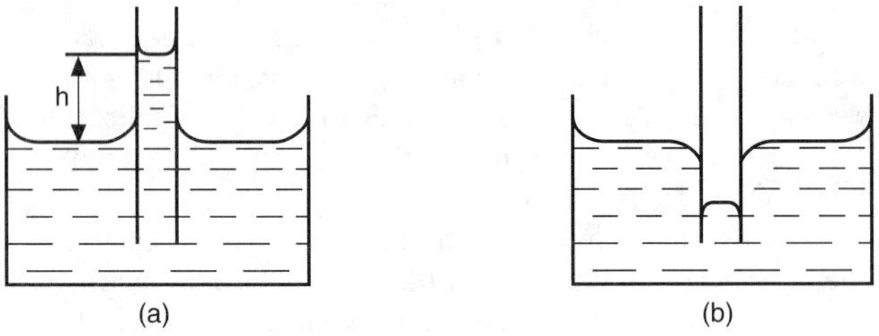

Figure 2. Level rise and depression due to capillary action.

$$Bo = g(\rho_l - \rho_g)L^2/\sigma_{12},$$

where ρ_l and ρ_g are the densities of the liquid and gaseous (vapor) phases, respectively, and L, the characteristic linear dimension of interface. The condition Bo = 1 governs the linear dimension of a at which the indicted forces are equal

$$a^2 = \frac{2\sigma_{12}}{g(\rho_l - \rho_g)}.$$

a^2, known as the capillary constant, is a constant of the liquid-vapor or liquid 1-liquid 2 interface which is independent of the tube diameter and material its wall is made of. Under terrestrial conditions, at pressures substantially lower than critical a = 1 − 3 mm for most liquids.

Tubes with diameter $d_c < a$ are said to be capillary. If the capillary radius $r_c < 0.05$ a, then the meniscus has the shape close to a sphere with the curvature radius

$$R = r_c/\cos\theta \quad \textbf{(2)}$$

and the height h of the liquid is determined by *Jurin's formula*

$$h = \frac{4\sigma_{12}\cos\theta}{g(\rho_l - \rho_g)d_c}. \quad \textbf{(3)}$$

As r_c increases, the meniscus shape steadily departs from the spherical and the capillary rise also steadily deviates from the value calculated via Equation (3). **Rayleigh** suggested a formula in the form of a power series which allows for this effect, and is valid for $r_c < 0.46$ a:

$$a^2 = r_c(h + r_c/3 - 0.1288\, r_c^2/h + 0.1312\, r_c^3/h^2).$$

The terms on the right-hand side, beginning with the second, are the correction for deviation of the meniscus shape from the sphere.

In very narrow ($d_c \ll a$) capillary tubes, the vapors of an absolutely wetting liquid absorbing on the walls form a multimolecular liquid film. The presence of the film decreases the curvature radius of the meniscus as compared to that calculated through Eq. (2) at $\theta = 0$. In this case, the capillary pressure on the meniscus exceeds the value calculated by the Laplace equation (1) at $R_1 = R_2 = r_c$.

In a plane capillary tube, i.e., the gap between two plates at distance d_p apart, the meniscus surface is of cylindrical shape with the radius

$$R = d_p/\cos\theta,$$

therefore, the height of the capillary rise is half that in a cylindrical capillary at $d_c = d_p$.

If the capillary walls are not wetted, the liquid meniscus is convex and the level of liquid in it deviates below the level of a free liquid by a value also determined by Eq. 3.

In vessels with the characteristic dimension $L \gg a$, the liquid surface is plane except the small area near the vessel walls, where the liquid rises or lowers by about the quantity a.

The difference in capillary pressures due to different curvature of liquid menisci may give rise to a liquid flow in the capillary (**Figure 3**), where the flow of wetting liquid is directed toward the meniscus with the smaller curvature radius. This is used, for instance, when entraining the working liquid from the condensation to the evaporation zone in heat pipes (see **Heat Pipes**).

The *capillary pressure* can be calculated theoretically only for capillaries of the simplest form: a plane slot which has been considered above; a circular cylindrical capillary; and capillaries of triangular, square, and some other shapes of cross-section. For capillaries with cross-sections of intricate irregular shape (such as the case, in particular, with capillary paths of heat pipe wicks), an accurate value of capillary pressure can be found only experimentally.

Capillary absorption plays an essential role in liquid movement in the soil, ground and porous coatings of heat-releasing surface with boiling liquids on them. Capillary impregnation of materials is widely applied in chemical technology.

Many properties of disperse systems, such as permeability and strength, depend to a large extent on capillary phenomena because in the fine pores of these bodies, a high capillary pressure is realized.

At a given temperature T, the intermolecular cohesive forces in a concave surface layer are stronger than on the plane surface. Therefore, the number of molecules leaving the surface and, as a result, the saturation pressure over the concave surface, is lower than those over the plane surface. For a convex surface, this pressure is higher. The relation between pressures p and p_0 over curved and plane surfaces, respectively, is described by the Kelvin equation

$$\frac{p}{p_0} = \exp\left[\frac{\sigma_{12}}{\rho_l\tilde{R}T}\left(\frac{1}{R_1} + \frac{1}{R_2}\right)\right],$$

where $\tilde{R}$ is the individual gas constant and ρ_l, the liquid density.

In narrow pores, the vapors of wetting liquid are condensed at the temperature of vapor saturation over the plane surface (capillary condensation). This underlies vapor collecting by fine-pored sorbents.

Curvature of a free liquid surface exposed to external forces gives rise to so-called *capillary waves,* that is the rippled surface of liquid.

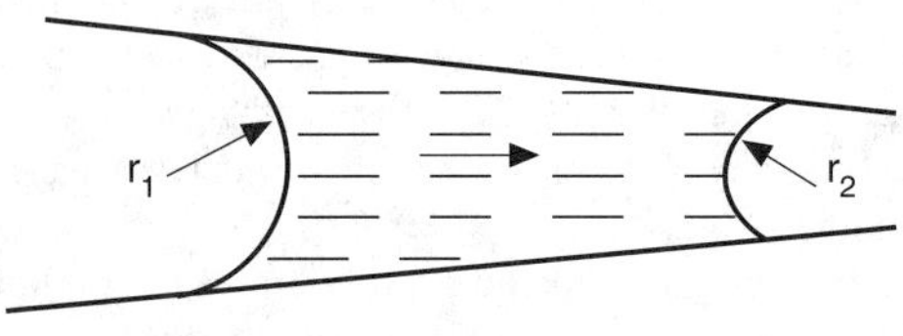

Figure 3. Difference in radius of curvature inducing flow in a capillary.

The joint action of surface tension, **viscosity,** inertia, and gravitational forces and their correlation account for the diverse flow regimes, i.e., phase distribution throughout the volume in **Two-Phase Flows** in channels (see **Multiphase Flows**).

The measurement of some physical values characterizing capillary phenomena forms the basis for experimental methods determining **surface tension** coefficients of liquids. For instance, the measurement of the height h of liquid rise in the capillary Eq. (3) is used to find the surface tension coefficient σ_{12}.

Leonardo da Vinci (16th century), B. Pascal (17th century) and Jurin (18th century) pioneered investigations into capillary phenomena in experiments with capillary tubes. The theory of capillary phenomena was further developed by P. Laplace (1806), T. Young (1879) and I. S. Gromeka (1876).

References

Adamson, A. W. (1976) *Physical Chemistry of Surfaces,* Wiley and Sons, New York.

I.V. Kalmykov

CAPILLARY CONDENSATION (see Interfaces)

CAPILLARY COLUMN CHROMATOGRAPHY (see Gas chromatography)

CAPILLARY EFFECTS (see Marangoni effect)

CAPILLARY FLOWMETERS (see Differential pressure measurement)

CAPILLARY FORCES (see Gas-solids separation, overview)

CAPILLARY-POROUS MATERIAL (see Drying)

CAPILLARY PRESSURES (see Capillary action)

CAPILLARY WAVES (see Capillary action; Stratified gas-liquid flow)

CARBOLIC ACID (see Phenol)

CARBON

Carbon—(L. *carbo,* charcoal), C; atomic weight 12 exactly (C^{12}); atomic weight (natural carbon) 12.011; atomic number 6; melting point ~ 3550°C, graphite sublimes at 3367 ± 25°C; triple point: (graphite-liquid-gas), 3627 ± 50°C at a pressure of 10.1 MPa and (graphite-diamond-liquid), 3830–3930°C at a pressure of 12–13 GPa; specific gravity amorphous 1.8 to 2.1, graphite 1.9 to 2.3, diamond 3.15 to 3.53 (depending on variety); gem diamond 3.513 (25°C); valence 2, 3, or 4.

Carbon, an element of prehistoric discovery, is very widely distributed in nature. It is found in abundance in the sun, stars, comets and atmospheres of most planets. Carbon in the form of microscopic diamonds is found in some meteorites. Natural diamonds are found in *kimberlite* of ancient volcanic "pipes," such as found in South Africa, Arkansas and elsewhere. Diamonds are now also being recovered from the ocean floor off the Cape of Good Hope. About 30% of all industrial diamonds used in the U.S. are now made synthetically. The energy of the sun and stars can be attributed at least in part to the well-known carbon-nitrogen cycle.

Carbon is found free in nature in three allotropic forms: amorphous, graphite and diamond. A fourth form, known as "white" carbon, is now thought to exist. Graphite is one of the softest known materials while diamond is one of the hardest. Graphite exists in two forms: alpha and beta. These have identical physical properties, except for their crystal structure. Naturally occurring graphites are reported to contain as much as 30% of the rhombohedral (beta) form, whereas synthetic materials contain only the alpha form. The hexagonal alpha type can be converted to the beta by mechanical treatment, and the beta form reverts to the alpha on heating it above 1000°C.

In 1969 a new allotropic form of carbon was produced during the sublimation of pyrolytic graphite at low pressures. Under free-vaporization conditions above ~2550 K, "white" carbon forms as small transparent crystals on the edges of the basal planes of graphite. The interplanar spacings of "white" carbon are identical to those of a carbon form noted in the graphitic gneiss from the Ries (meteoritic) Crater of Germany. "White" carbon is a transparent birefringent material. Little information is presently available about this allotrope.

In combination, carbon is found as carbon dioxide in the atmosphere of the earth and dissolved in all natural waters. It is a component of great rock masses in the form of carbonates of calcium (limestone), magnesium, and iron. Coal, petroleum, and natural gas are chiefly hydrocarbons.

Carbon is unique among the elements in the vast number of variety of compounds it can form. With hydrogen, oxygen, and nitrogen, and other elements, it forms a very large number of compounds, carbon atom often being linked to carbon atom. There are upwards of a million or more known carbon compounds, many thousands of which are vital to organic and life processes. Without carbon, the basis for life would be impossible. While it has been thought that silicon might take the place of carbon in forming a host of similar compounds, it is now not possible to form stable compounds with very long chains of silicon atoms. The atmosphere of Mars contains 96.2% CO_2.

Some of the most important compounds of carbon are carbon dioxide (CO_2), carbon monoxide (CO), carbon disulfide (CS_2), chloroform ($CHCl_3$), carbon tetrachloride (CCl_4), methane (CH_4), ethylene (C_2H_4), acetylene (C_2H_2), benzene (C_6H_6), ethyl alcohol (C_2H_5OH), acetic acid (CH_3COOH), and their derivatives.

Carbon has seven isotopes. In 1961 the International Union of Pure and Applied Chemistry adopted the isotope carbon-12 as the basis for atomic weights. Carbon-14, an isotope with a half-life of 5730 years, has been widely used to date such materials as wood, archeological specimens, etc. Carbon-13 is now commercially available.

Handbook of Chemistry and Physics, CRC Press

CARBONACEOUS FUELS (see Fuels)

CARBON DIOXIDE

Carbon Dioxide (CO_2) is a colourless, odourless and nonflammable gas heavier than air. It is classified as one of the major air pollutants when it exceeds its naturally-occurring concentration. There are a number of processes for its production including combustion, fermentation and heating of organic matter. Its main use is in the food and beverage industry, although it is also used as a refrigerant, in fire extinguishing equipment, as a pressure source, aerosol propellant, and also in welding and some metallurgical processes.

Below are some of its major physical characteristics:

Molecular weight: 44.0098 **Critical temperature:** 304.21K
Melting point: 216.5K **Critical pressure:** 7.3825MPa
Normal boiling point: 194.67K **Critical density:** 466 kg/m^3
Normal vapour density: 1.98 kg/m^3 (@ 273.15K; 1.0135MPa)

References

Angus, S., deReuck, K. M., and Armstrong, B. (1972) *International Thermodynamic Tables of the Fluid-State - 3,* Carbon Dioxide, Pergamon Press, Oxford.

Beaton, C. F. and Hewitt, G. F. (1989) *Physical Property Data for the Design Engineer,* Hemisphere Publishing Corporation, New York.

Vesovic, V. et al. (1990) The Transport Properties of Carbon Dioxide, *J. Phys. Chem. Ref. Data, 19:763.*

V. Vesovic

CARBON DIOXIDE, AS A POLLUTANT (see Air pollution)

CARBON DIOXIDE POLLUTION (see Air pollution)

CARBON DISULPHIDE COMBUSTION (see Flames)

CARBON MONOXIDE

Carbon Monoxide (CO) is a colourless and odourless gas slightly heavier than air. It is highly toxic and at low to medium concentration in air, it forms a flammable and explosive mixture. It is classified as one of the major *air pollutants*. There are a number of processes for its production from carbon dioxide, oxygen, coke and steam. It is also produced by combustion of organic compounds in an oxygen deficient atmosphere. It is used in metallurgy and in the organic synthesis of methanol, acetic acid, aldehydes and a number of other compounds.

Listed below are some of its physical characteristics:

Molecular weight: 28.011 **Critical temperature:** 133.16 K
Melting point: 68.16 K **Critical pressure:** 3.498 MPa
Normal boiling point: 81.66 K **Critical density:** 301 kg/m^3

Reference

Beaton, C. F. and Hewitt, G. F. (1989) *Physical Property Data for the Design Engineer,* Hemisphere Publishing Corporation, New York.

V. Vesovic

Table 1. Carbon dioxide; values of thermophysical properties of the saturated liquid and vapour

T_{sat}, K	216.6	230	240	250	260	270	280	290	300
P_{sat}, kPa	520	892	1279	1779	2410	3191	4146	5301	6696
ρ_l, kg/m^3	1179	1130	1090	1046	998.1	944.7	882.4	803.5	678.8
ρ_g, kg/m^3	13.83	23.25	33.18	46.42	64.04	87.80	120.9	170.8	266.7
h_l, kJ/kg	−422.0	−397.3	−378.1	−357.8	−336.3	−313.5	−288.7	−260.3	−222.6
h_g, kJ/kg	−75.9	−71.4	−69.3	−68.4	−69.3	−72.4	−78.9	−91.0	−117.5
$\Delta h_{g,l}$, kJ/kg	346.1	325.9	308.8	289.3	267.0	241.0	209.8	169.3	105.0
$c_{p,l}$, kJ/(kg K)	1.84	1.87	1.97	2.10	2.24	2.44	2.81	3.68	8.50
$c_{p,g}$, kJ/(kg K)	0.95	1.04	1.13	1.25	1.42	1.70	2.22	3.48	10.60
η_l, μNs/m^2	258.7	208.7	178.2	152.3	130.0	110.1	91.8	73.8	52.9
η_g, μNs/m^2	10.95	11.7	12.3	12.9	13.6	14.5	15.6	17.3	21.2
λ_l, mW/(K m)	180.5	163.5	150.7	138.1	125.7	113.3	100.9	87.9	77.6
λ_g, mW/(K m)	11.0	12.1	13.1	14.3	15.8	17.9	21.1	27.5	48.1
Pr_l	2.64	2.38	2.33	2.31	2.31	2.37	2.56	3.09	5.79
Pr_g	0.95	1.00	1.05	1.12	1.22	1.38	1.64	2.19	4.67
σ, mN/m	17.1	13.8	11.4	9.16	7.02	5.01	3.19	1.61	0.33
$\beta_{e,l}$, kK^{-1}	2.94	3.49	3.98	4.59	5.44	6.81	9.38	15.8	54.8

NOTE: *The values of thermodynamic properties have been obtained from reference [1], the values of transport properties have been obtained from reference [3], while the values of the surface tension have been obtain from reference [2].*

Table 1. Carbon monoxide; values of thermophysical properties of the saturated liquid and vapor

T_{sat}, K	81.66	90	95	100	105	110	115	120	125	133.16
p_{sat}, kPa	101.3	245	437	548	776	1070	1433	1875	2423	3498
ρ_l, kg/m^3	789	751	728	702	675	646	613	574	526	301
ρ_g, kg/m^3	4.40	9.99	15.4	22.4	31.0	42.6	57.2	77.5	153	301
h_l, kJ/kg	150.4	168.6	179.8	192.3	204.7	216.1	227.1	239.3	258.1	314.33
h_g, kJ/kg	366.0	369.7	371.2	372.9	374.3	374.6	374.0	370.9	363.7	314.33
$\Delta h_{g,l}$, kJ/kg	215.6	201.1	191.4	180.6	169.6	158.5	146.9	131.6	105.6	
$c_{p,l}$, kJ/(kg K)	2.15	2.17	2.20	2.26	2.34	2.46	2.61	2.79	3.02	
$c_{p,g}$, kJ/(kg K)	1.22	1.35	1.52	1.60	1.79	2.03	2.42	3.18	5.25	
η_l, μNs/m^2	154	120	105	93.5	83.9	75.9	69.1	63.3	58.3	
η_g, μNs/m^2	7.08	7.51	7.80	8.11	8.44	8.80	9.23	9.78	11.7	
λ_l, mW/(Km)	141	125	116	107	98.2	89.3	80.3	71.4	62.5	
λ_g, mW/(Km)	6.89	8.22	8.93	9.71	10.5	11.5	12.6	14.0	18.4	
Pr_l	2.35	2.08	1.99	1.97	2.00	2.09	2.25	2.47	2.82	
Pr_g	1.25	1.23	1.33	1.34	1.44	1.55	1.77	2.22	3.34	
σ, mN/m	9.47	7.73	6.71	5.71	4.73	3.78	2.86	1.97	1.13	
$\beta_{e,l}$, kK^{-1}	5.46	6.65	7.52	8.44	9.63	11.6	14.9	20.7	31	

From reference [1] with permission

CARBON STEELS (see Steels)

CARBON SUBNITRIDE COMBUSTION (see Flames)

CARBONTETRACHLORIDE

Carbontetrachloride (CCl_4) is a colourless liquid insoluble in water. It was one of the first organic chemicals to be produced on a large scale. Nowadays, it is usually produced by the chlorination of methane at high temperature. It is very toxic and narcotic. Its usage has dramatically decreased in recent years due to its toxicity, but it is still used on a small scale as a solvent.

Its physical characteristics are as follows:

Molecular weight: 153.8
Melting point: 250.25 K
Normal boiling point: 349.85 K
Critical temperature: 556.35 K
Critical pressure: 4.560 MPa
Critical density: 588 kg/m^3

Reference

Beaton, C. F. and Hewitt, G. F. (1989) *Physical Property Data for the Design Engineer,* Hemisphere Publishing Corporation, New York.

V. Vesovic

CARBON THERMOMETERS (see Resistance thermometry)

Table 1. Carbontetrachloride; values of thermophysical properties of the saturated liquid and vapor

T_{sat}, K	349.9	370	390	410	430	450	470	495	525	556.35
p_{sat}, kPa	101.3	184	307	473	701	1020	1390	2020	3160	4560
ρ_l, kg/m^3	1484	1442	1397	1351	1303	1250	1199	1107	989	588
ρ_g, kg/m^3	5.44	9.40	15.2	23.4	34.8	50.3	71.2	108.5	184.5	588
h_l, kJ/kg	−36	−17	−1	16	38	53	72	92	123	177
h_g, kJ/kg	159	169	179	188	197	205	212	218	221	177
$\Delta h_{g,l}$, kJ/kg	195	188	180	172	159	152	140	126	98	
$c_{p,l}$, kJ/(kg K)	0.92	0.94	0.97	1.01	1.06	1.14	1.24	1.36	1.57	
$c_{p,g}$, kJ/(kg K)	0.58	0.60	0.62	0.65	0.68	0.73	0.80	0.91	1.30	
η_l, μNs/m^2	494	407	352	309	274	241	205	154	98	
η_g, μNs/m^2	11.9	12.5	13.3	14.1	14.9	15.7	16.7	18.9	21.0	
λ_l, (mW/m(K m)	92	87	83	78	74	70	65	57	45	
λ_g, (mW/(K m)	8.6	9.3	10.0	10.7	11.5	12.3	13.2	14.3	16.3	
Pr_l	4.94	4.40	4.16	4.08	3.93	3.92	3.91	3.67	3.42	
Pr_g	0.80	0.81	0.82	0.85	0.88	0.93	1.01	1.20	1.67	
σ, mN/m	20.2	17.6	15.4	13.1	10.9	8.8	6.9	4.4	2.0	
$\beta_{e,l}$, kK^{-1}	1.36	1.50	1.68	1.90	2.19	2.58	3.14	4.30	7.83	

From reference [1] with permission.

CARNOT CYCLE

Following from: *Thermodynamics*

The aim of all clockwise operating cycle processes is to produce work by transferring heat from a high-temperature energy reservoir to a low-temperature energy reservoir.

According to the *First Law of Thermodynamics:*

$$\Delta U = 0 = \Sigma Q - \Sigma W = Q_H - Q_L - W. \qquad \textbf{(1)}$$

The maximum amount of heat converted into work is determined by the *Second Law of Thermodynamics.* For a reversible process,

$$\Delta S_{tot} = 0 = \Delta S_{syst} + \Delta S_{sur}. \qquad \textbf{(2)}$$

For a system operating in a cyclic process,

$$\Delta S_{syst} = 0 \qquad \textbf{(3)}$$

and hence:

$$\Delta S_{sur} = 0 = \frac{Q_L}{T_L} - \frac{Q_H}{T_H}. \qquad \textbf{(4)}$$

Combining Equations (1) and (4) provides the maximum amount of work obtainable for any quantity of heat supplied Q_H.

$$W = Q_H - Q_L = Q_H\left(1 - \frac{T_L}{T_H}\right) \qquad \textbf{(5)}$$

A cycle process to fulfil the above criteria was suggested by the French military engineer **Sadi Carnot**. It is depicted in **Figure 2**.

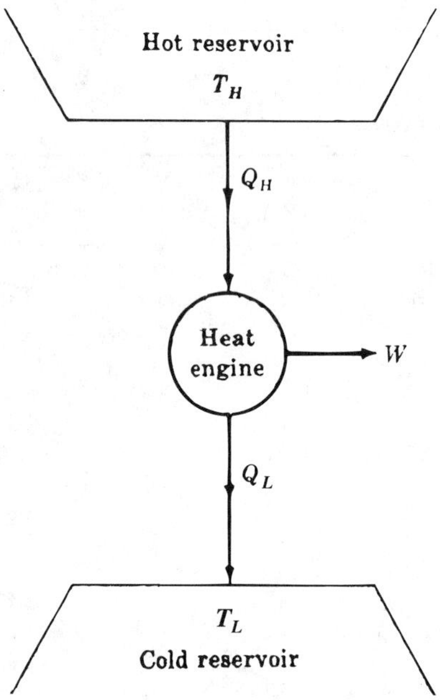

Figure 1. Power cycle.

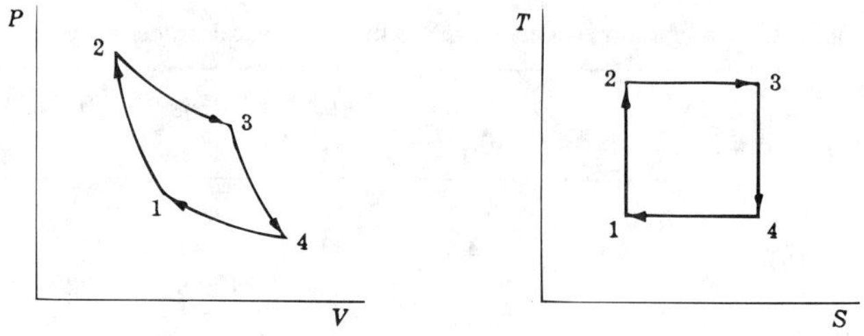

Figure 2. p-V and T-S diagram of Carnot cycle.

1-2. Gas is compressed *reversibly* and *adiabatically* from the initial state. Work is done on the system and the temperature increases from T_L to T_H.

2-3. The gas at temperature T_H expands *reversibly* and *isothermally* from state 2 to state 3. Work is done by the gas and heat is transferred from a high temperature energy reservoir to the system. Since the process is reversible, the temperature of the energy reservoir can only be infinitesimally higher than T_H.

3-4. The gas expands *reversibly* and *adiabatically* from state 3 to state 4, doing work. During this process the temperature drops from T_H to T_L.

4-1. The gas is compressed *isothermally,* rejecting heat at T_L to the low temperature energy reservoir. The temperature T_L is only infinitesimally higher than the temperature of the reservoir, because the process is reversible.

The Carnot cycle is not limited to gas cycles; it can be executed in many different types of systems, such as gas, liquid, electric cell, soap film, steel wire and rubber band, as long as there are two energy reservoirs, a part of the surrounding that can do and absorb work and some means to periodically insulate the system. While the Carnot cycle discussed above produces work, a reversed Carnot cycle (operating counterclockwise) acts as a heat pump, requiring work to transfer heat from a low-temperature energy reservoir to a high-temperature energy reservoir. The *efficiency of the Carnot cycle* η_c is obtained from Eq. (5) as:

$$\eta_o = \frac{W}{Q_H} = \frac{T_H - T_L}{T_H} \qquad \textbf{(6)}$$

Note that the conditions of reversible **Heat Transfer** and engine operation imply infinitely large heat transfer surfaces and/or infinitely slow operation/heat transfer. Since these conditions are never met in practical applications, the efficiency of practical power cycles is always lower than the Carnot efficiency.

H. Müller-Steinhagen

CARNOT, NLS (see Steam engines)

CARRYUNDER (see Steam generators, nuclear; Vapour-liquid separation)

CARTESIAN CO-ORDINATES (see Conservation equations, single phase; Coordinate system)

CASTING OF METALS

Following from: Solidification

Casting of metals and alloys is concerned with the production of shaped castings (For example, an automotive cylinder block) or of simply shaped *billets* intended for subsequent metalworking into mill products (For example, into bars, plates, sheet, tubing, etc.).

In the production of shaped castings, a molten metal or alloy of desired composition is poured into a *mould* or *die,* which contains a cavity in the form of the component to be cast. The molten material is conveyed into the cavity via a gating system (**Figure 1**). Adjacent to the casting—often at the point or points (gates) where the gating system enters the casting cavity—are placed additional cavities called risers or feeders, the object of which is to provide a reservoir of molten metal. The function of these reservoirs is to compensate for shrinkage. (The majority of alloys shrink during solidification, grey cast-iron and type-metals being notable exceptions)

The mould is most often of silica sand bonded with a bentonite type clay, which has been activated with water. However, many shaped castings are made in metallic molds by processes called

1. *Gravity die-casting,* known as *permanent molding* in the USA.
2. *Pressure die-casting,* known as *die-casting* in the USA.

A hybrid of the above methods, *low pressure die-casting,* is widely used for the production of automobile wheels. As the nomenclature implies in the above metallic-mould processes, the delivery of the molten alloy may be by the action of gravity or pressure, the latter generated by a piston or by a gas.

Metal casting is most often chosen as a production method when the component concerned is of complex shape, although many mill products are initially cast into simple rectangular-sectioned billets by *continuous casting* machines (**Figure 2**). Such billets are subsequently worked plastically into the mill product concerned by rolling, drawing, extrusion, etc. In shaped-casting production, the component usually receives no further shaping operation other than finish machining.

Although the beginnings of metal casting stretch back to the Chalcolithic period (5000–3000 B.C.), which preceded the Bronze Age, attempts to develop science-based models of pouring and solidification only began in the present century. Early researchers were limited by the sheer complexity of the multidimensional coupled problems of fluid flow, heat transfer and solidification encountered. In spite of such limitations, a number of basic rules evolved, which permit practitioners to pour a wide variety of complex shapes of good reliability. For example, the US metal casting industry currently produces annually castings worth $23 000 million, which range from humble valves and fittings to sophisticated gas turbine and prosthetic hardware.

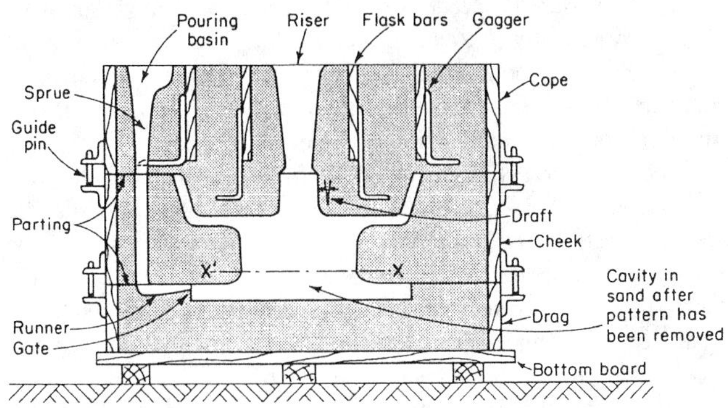

Figure 1. A sand mould for a wheel which is to be cast in steel. The line shown indicates where the two halves of the pattern are parted to facilitate their withdrawal from the mould formed around them.

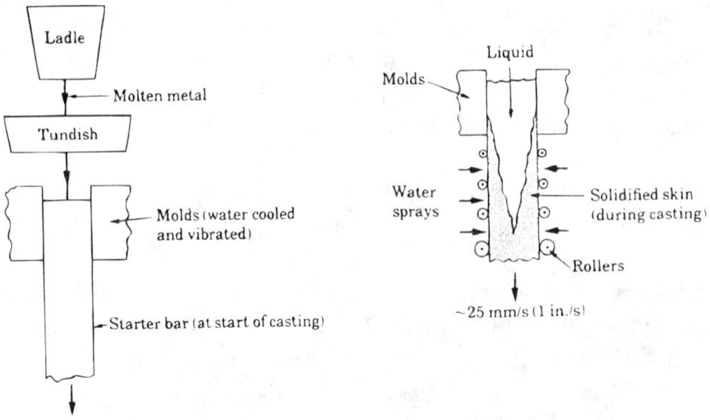

Figure 2. A continous caster for producing steel billets for subsequent working into mill products. The mould enclosure is an internally-cooled hollow copper box type structure.

The production of defect-free shaped castings depends on the design of the gating and feeding arrangements. Prior to the middle of the present century, such arrangements were left almost entirely to skilled craftsmen. Ruddle (1956) reported that the earliest published scientific work appeared in the 1930s. He quoted Harding (1949) as stating that 50% of all rejected castings directly resulted from poor gating. Gating design should provide systems that deliver molten metal of good quality at the appropriate rate and velocity to the mold cavity. The metal should be free from entrained gases, slag and dross, etc., and should not undergo gross oxidation in pouring.

The propensity of most molten metals to form oxide films almost instantaneously complicates this design process and can lead to inconsistent mechanical properties if not taken into account. Campbell (1991) suggests that careful control of surface turbulence in the molten metal stream must be practiced. He suggests this may be accomplished by ensuring that the **Weber Number** is confined to the range of 0.2 to 0.8. This number relates the action of inertial pressure in the stream to the effects of surface tension, to unit:

$$W_e = u^2 \rho_1 L/\sigma_1. \quad (1)$$

The design of the feeding system owes much to Chvorinov (1940), who demonstrated that, assuming planar freezing, the solidification time for casting could be related to its volume to surface area ratio:

$$t_s = B\left(\frac{V}{A}\right)^2 \quad (2)$$

The *mould constant,* B, is a function of the thermophysical properties of the mould and the metal concerned, and the average temperature of the mould-metal interface during freezing, etc. Good thermal contact is assumed between the casting and the mould.

$$B = \left\{ \frac{\rho_1 \sqrt{\pi\kappa_2}[h_{1s} + c_{p1}(T_1 - T_0)]}{2\lambda_2(T_i - T_o)} \right\} \tag{3}$$

where h_{ls} is the latent heat of fusion, c_{pl} is specific heat of the molten metal, κ is the thermal diffusivity of the mould, and T_o denotes the initial temperature of the mould.

Conductivity, diffusivity and other thermophysical properties are assumed to remain constant during freezing. It is also assumed that no temperature loss occurs during pouring. The subscripts one and two denote, respectively, the metal and the mold while the letter i refers to the interface. Equation (2) has found extensive use in feeder design, where the freezing time of the feeder must always be greater than that of the casting. Chvorinov validated Equation 2 by measuring the freezing times of a variety of low-carbon steel castings. It was later shown that the divergent heat flow, which occurs with curved surfaces and also at corners and edges of rectangular castings, should be taken into account. (See Berry, et al., 1959)

Chvorinov also showed that the progress of freezing in planar solidification of a pure metal or other congruent freezing material is governed by:

$$\delta_s = K\sqrt{t}. \tag{4}$$

The thickness of the frozen layer is thus related to the square root of the elapsed time. The so-called *solidification constant* is given by:

$$K = \frac{2\lambda_2(T_i - T_o)}{h_{ls}\rho_1\sqrt{\pi\kappa_2}}. \tag{5}$$

It is noted that the equation for skin thickening holds true only for minimal superheat in the molten metal.

At present, casting technology is poised at an important juncture in its history. This has been brought about by the impact of computer modeling. The earliest examples of computer modeling have employed the analogue computer. The contributions of Pehlke and his co-workers (See Pehlke et al. 1976) have provided a major insight on the possibilities of model experiments. However, the application of digital computer to the solution of metal casting problems has opened a new epoch for casting producers. Progress over the last decade using various computational techniques now renders possible a variety of options. Among them are the prediction of solidification profiles, the design of gating and feeding systems, the prediction of microstructure and finally the linkage of microstructure to mechanical properties.

Many of these developments are described in reports of conferences held periodically under the title The Modeling of Casting, Welding and Advanced Solidification Processes (1981 onwards).

References

Berry, J. T., Kondic, V., and Martin, G. (1959) Solidification Times of Simple Shaped Sand Castings in Sand Moulds, *Trans. Amer. Foundrymen's Soc.*, 67:449–476.

Brody, H. D. and Apelian, D. (1981) *Modeling of Casting and Welding Processes I,* The Metallurgical Society of AIME, Warrendale, Pa., 1981 and subsequent volumes in series.

Campbell, J. (1991) *Castings,* Butterworth-Heinemann, Oxford, 1991.

Chvorinov, N. (1940) *Theory of Casting Solidification,* Giesserei, 27:(10)177–186, 11:201–208, No. 222–225.

Harding, E. W. (1949) *Foundry Trade Jnl.,* 69:343.

Pehlke, R. D., Marrone, R. E., and Wilkes, J. O. (1976) *Computer Simulation of Solidification, Amer. Foundrymen's Soc.*

Ruddle, R. W. (1956) *The Running and Gating of Sand Castings,* Inst. of Metals, Monograph and Report Series No. 19, London.

Leading to: *Computational fluid dynamics; Coupled fluid flow and heat transfer; Numerical analysis of heat transfer*

J.T. Berry

CATALYSIS

Following from: *Chemical reaction*

Catalysis is the alteration of the rate of a chemical reaction using substances known as *catalysts,* which can repeatedly participate in the elementary steps of a chemical process changing the number and character of such steps; the catalyst is regenerated after the transformation is completed. A *positive catalysis* promotes the chemical reaction and a *negative catalysis* retards it. The term "catalysis" is more frequently applied to the former.

Catalysis is classified in terms of phase as *heterogeneous* and *homogeneous.* In homogeneous catalysis, reactants and catalyst are in the same phase while in heterogeneous catalysis, reactants and catalyst are in different phases and the reaction proceeds at the interface. Diffusion is an indispensable stage of catalysis in the heterogeneous system. Catalysis with colloid particles in the liquid phase, that is, a *micellar catalysis,* is intermediate between the two kinds mentioned above. Here, the reaction begins on the surface of solid catalyst, subsequently proceeding to the bulk of the reaction medium.

The important properties of catalysts include catalytic activity, selectivity, sensitivity to small amounts of extraneous substances (promoters and modifiers), resistance to external factors, (e.g., reaction medium, temperature, pressure, mechanical action) and regeneration after the reaction terminates.

The *catalytic activity* is determined by the change of the rate of the chemical reaction under the action of a catalyst. A specific catalytic activity is an activity per unit volume or mass of the catalyst. Heterogeneous catalytic reactions are characterized by the number of reaction recycles, that is, a specific activity related to a single active center. The reaction rate in the homogeneous catalysis is proportional to the catalyst concentration.

Under catalysis, the activation energy E_A is normally lower than in the noncatalytic reaction. A reduction in E_A in the catalytic reaction in relation to the noncatalytic reaction is schematically shown in **Figure 1**.

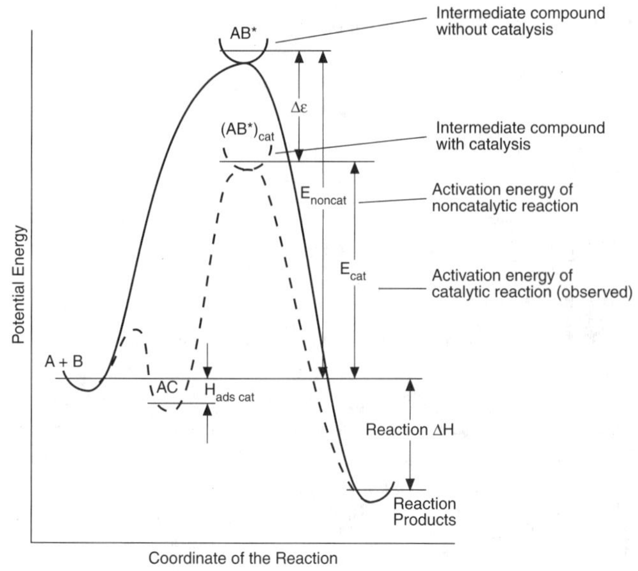

Figure 1. Energy changes in catalyzed and uncatalyzed reactions.

The catalyst does not change the chemical equilibrium, but it alters the rates of the direct and reverse reactions, thereby facilitating faster attainment of an equilibrium state.

Each catalyst alters the rates of specific reactions, i.e., catalysts are *selective* in action.

One of the main points in various catalytic theories is an insight into the formation of intermediate compounds of a catalyst and a reagent, and their decomposition (with catalyst regeneration) in the course of a reaction.

Two mechanisms of catalytic action via formation of an intermediate complex are suggested for the $A + B \rightarrow C + D$ reaction types. One is accomplished in stages

$$A + K \rightarrow \quad [AK]^* \quad \rightarrow AK$$

$$\rightarrow \begin{bmatrix} \text{activated} \\ \text{complex} \end{bmatrix}^*$$

$$AK + B \rightarrow [AKB]^* \rightarrow D + C + K,$$

and the other is *associative* (synchronous)

$$A + B + K \rightarrow [ABK]^* \rightarrow D + C + K.$$

Heterogeneous catalytic reactions are characterized by an intricate macrokinetics. Macrokinetic stages commonly considered are transport of reagents from the flow core to the outer surface of the catalyst grain, diffusion in the grain pores, adsorption of initial reagents on the catalyst grain surface, the chemical interaction proper that may proceed in a few stages, desorption of reaction products from the catalyst surface, diffusion of products from the grain inner surface and transport of products from the outer surface of the catalyst particle to the flow core of the reaction mixture. The stage with the lowest rate determines the mechanism of the entire process. The kinetic characteristics of heterogeneous catalysis depend not only on the nature and the state of the catalyst surface, but also on the laws of heat and mass transfer.

Depending on the nature of the rate controlling stage, outer and inner kinetic, outer and inner diffusion, and adsorption limiting macrokinetic regions of heterogeneous catalysis are distinguished.

There exists no unified theory of heterogeneous catalysis that can account for all the properties of all the catalysts and which can help select an appropriate catalyst for a given reaction. The well-known theories fall into three groups.

The first includes the *geometrical theories* of Taylor, Balandin, and Kobozev that are based on the geometrical correspondence of catalytically active centers; the role of such centers can be played by several atoms or ions on a nonhomogeneous catalyst surface, and depends on the arrangement of atoms in the reactant molecules.

The second group covers the *electronic theories* of Schwab, Dowden, Turner, and Volkenshtein that proceed from the fact that reactant molecules adsorbed on the catalyst surface, exhibit free valences due to redistribution of electron density and gives rise to a new bond. Reagent molecules dissociate and atoms are added to the surface. An atomic migration on the surface and the interaction of molecules that are weakly bound to the surface may result in formation of the reaction products that are later desorbed.

These two groups of theories rest on physical approaches. Recently, an ever-increasing preference is given to *chemical theories* that consider a catalyst as a reagent forming, with the reactants, unstable surface intermediate complexes which, in the course of the reaction, are decomposed into the reaction products and the starting catalyst.

All the heterogeneous catalysis theories assign an essential role to the stage of reactant adsorption on the catalyst surface. Adsorption is assumed to result in the weakening of the bonds in the molecules of the reactants and the production of surface intermediate compounds that decompose into the products and are desorbed.

For heterogeneous catalytic reactions of the $A + B \rightarrow C$ type, two adsorption mechanisms are considered. One is described by the *Rideal-Eley model,* involving formation on the catalyst surface of an intermediate complex with the atoms of one of the reagents ($A + Z \rightleftharpoons [ZA]^*$), with subsequent interaction of an adsorbed atom and the molecule of the other reagent incident from the main gas flow ($[AZ]^* + B \rightarrow C + Z$). The other mechanism is described by the *Langmuir-Hinshelwood model,* based on the assumption made for deriving the *Langmuir adsorption isotherm.* In terms of this model, starting reagents are adsorbed on the catalyst surface to produce ($A + Z \rightleftharpoons [AZ]$ and $B + Z \rightleftharpoons [BZ]$) surface compounds. Migration of compounds on the catalyst surface causes the $[AZ] + [BZ] \rightarrow C + 2Z$ transformations. In this case, it is assumed that the rate of the chemical reaction is lower than the rates of reagent and product adsorption and desorption.

Currently, the study of complex compound catalysis is proceeding vigorously. Interaction of atoms and ions of a transition metal with those of other substances leads to the formation of complex metal compounds with one or several ligands, the orbitals of which are overlapped by the orbitals of the central atom or a metal ion. The catalytic effect of complex catalysts is accounted for by addition in the coordination sphere of reagent molecules with substitution of less strongly bound ligands. The catalytic reaction is promoted by a mutual orientation of reagents, their polarization in the central atomic field, an easier electron transition due to participation of the central atom, reduction of E_A and compensation of the energy of bond rupture by a simultaneous formation of other bonds. An advantage of complex catalysts is the alteration of catalytic properties via the addition of other ligands. This makes it possible to achieve high specificity and selectivity of the catalysts. Complex compounds can serve as catalysts in both homogeneous and heterogeneous catalysis.

Catalysis underlies many industrial processes. Catalysts are employed for industrial synthesis of ammonia from N_2 and H_2, production of HNO_3, and for production of H_2SO_4 by contact technology. The conversion processes $CH_4 + H_2O \rightarrow CO + 3H_2$, $C + H_2O \rightarrow CO + H_2$, $CH_4 + CO_2 \rightarrow 2CO + 2H_2$, and $C + CO_2 \rightarrow 2CO$ also proceed in the presence of catalysts. Catalytic techniques are used in organic and petrochemical synthesis, production of petroleum derivatives (catalytic cracking and reforming), polymers, methanol, acetic acid, etc. In recent years, catalysts are widely employed in developing environmentally-important processes. Thus, they are used for cleaning waste gases in power generation and metallurgy (elimination of detrimental impurities such as NO_x, SO_x, H_2S) and for eliminating CO and NO_x from exhaust gases of internal combustion engines.

References

Rideal, E. (1968) *Concepts in Catalysis,* New York.

Satterfield, C. (1981) *Heterogeneous Catalysis in Practice,* New York.

V.A. Ibragimov

CATALYSTS (see Molecular sieves)

CATALYTIC ACTIVITY (see Catalysis)

CATALYTIC CONVERSION (see Oil refining)

CATALYTIC CONVERTERS (see Catalytic cracker; Internal combustion engine; Oil refining)

CATALYTIC RICH GAS PROCESS, CRG (see Substitute natural gas)

CATHODE (see Electrochemical cells)

CAUCHY'S CONVEYENCE PRINCIPLE (see Series expansion)

CAUCHY'S THEOREM (see Conservation equations, single phase)

CAUCHY SURFACE (see Differential equations)

CAUSTIC SODA (see Sodium hydroxide)

CAVITIES, FOR NUCLEATION (see Boiling)

CEA (see Commissariat a l'Energie Atomique)

CEC (see Commission of the European Communities)

CELL GROWTH (see Biochemical engineering)

CELL POTENTIAL (see Electrochemical cells)

CELLULOSIC FIRES (see Fires)

CELSIUS TEMPERATURE SCALE

Following from: *Temperature*

The Celsius temperature scale is one of three measures for determining the temperature of materials. Named after Swedish astronomer and physicist Andres Celsius (1701–1744). This scale has 100 intervals, or Celsius degrees, between the temperature of ice melting (0°C) and of water boiling (100°C). It is also known as the *Centigrade temperature scale.*

In the International practical temperature scales of 1968 and 1990, Celsius degree is defined as $t_C = T_K - 273.15$. The dimension of the unit of the Celsius and *Kelvin scales* is the same.

M.P. Orlov

CENTIGRADE TEMPERATURE SCALE (see Celsius temperature scale)

CENTRIFUGAL FILTERS (see Filters)

CENTRIFUGAL FLOWMETERS (see Differential pressure flowmeters)

CENTRIFUGAL SEPARATORS (see Vapour-liquid separators)

CENTRIFUGAL SCRUBBER (see Scrubbers)

CENTRIFUGES

Following from: *Liquid-solid separation*

There are two groups of centrifuges for liquid-solid separation: filtering centrifuges and sedimenting centrifuges. Purchas (1981) has given full account of the designs and applications of each type. The major difference between the two is that the former utilises a perforated bowl through which the fluid (centrate) can pass while the solids are retained inside the bowl. The latter is equipped with a solid (impermeable) bowl, and separation of the fluid is done by forcing it to overflow from the bowl while the solids are retained on its walls. The method of transport of solids and the control of liquid flows can vary widely in both types of centrifuge, and these factors are the ones generally used to subclassify the different machines.

Sedimenting centrifuges

Sedimenting centrifuges remove solids from liquids by causing the particles to migrate radially towards the walls of the centrifuge bowl. The basic types of centrifuges in this category are a) the high speed, tubular bowl type with manual discharge of solids; b) the skimmer pipe/knife discharge types; c) the disc-type centrifuge; and d) the continuous scroll discharge machines.

A) The *tubular bowl centrifuge* generally has a bowl with a diameter of between 15 and 50 cm which rotates at high-speed to generate a settling acceleration of up to about 18,000 g (where g is the acceleration due to gravity) for industrial models and 65,000 g for laboratory models. (This compares to accelerations of $<$1,000 g for cylindrical solid bowl machines). The feed slurry jets into the bottom of the bowl through the bowl neck, and a distributor disperses the feed to prevent travel too far along the bowl length. The centrate discharges from the top of the bowl by overflowing into a collecting cover.

B) Some designs use a *skimmer pipe and knife* to discharge the solids. Bowl diameters can be up to about 1.5 m for larger industrial units, and rotate about the vertical axis to generate up to 1,600 g. The feed enters at the bottom of the bowl and discharges over a lip ring at the top. The solids sediment to the walls of the bowl, where they are allowed to accumulate until the centrate clarity is adversely affected, then the feed is stopped and the solids discharged. When the feed is stopped, the skimmer pipe engages the layer of liquid above the solids and the rotational energy in the liquid causes it to flow up the skimmer pipe and be discharged. The skimmer is lowered into the rotating liquid; but before it contacts the solids layer, the discharging stream is diverted and those solids which are sufficiently fluid are then discharged. If the solids are compacted and will not flow, a knife is used to discharge the solids after the skimmer pipe has been retracted and the bowl slowed to an idling speed of 1 to 2 Hz. After solids discharge the knife is retracted, the bowl accelerated, and the cycle repeated. The cycle is usually fully automated.

C) The *disc type centrifuge* thickens the suspension to form a thick slurry by generating an acceleration of up to about 12,000 g. The bowl contains between 50 and 150 conical discs, spaced approximately 2 mm apart. The suspension flows towards the axis of the machine through the spaces between the discs; solids settle to the underside of each disc, slide along the disc in the outward direction and from the edge of the disc, are thrown to the wall of the bowl by the centrifugal field. The clarified liquid passes to the centre of the bowl and discharges over a weir at the top or bottom. The solids (thickened slurry) are usually discharged through ports at the periphery of the bowl, which open at a timed cycle (opening bowl centrifuge), or continuously through nozzles (nozzle discharge centrifuge) (**Figure 1**).

D) The *continuous scroll discharge* (or decanter) centrifuge provides a settling acceleration of up to 4,000 g. Bowl diameters are generally in the range of 10 to 100 cm and are cylindrical with a conical end (a few designs have wholly conical bowls). The feed is introduced through a concentric pipe to an appropriate point along the bowl; the centrifugal force causes the solids to sediment to the wall to leave a clear liquid in the bowl. Inside the bowl, a helical screw conveyor rotates at a speed of up to 2 Hz slower than the bowl and scrolls the sedimented solids to one end of the bowl, up the *beach* (the conical section), out of the *pond* (the liquid) and discharges them from the bowl at the end of the conical section. The clarified liquid spills over a weir at the opposite end of the bowl (although in some designs the liquid flows cocurrently with the solids).

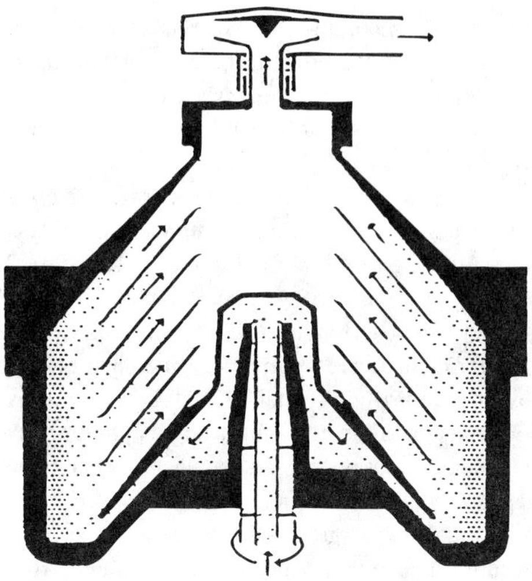

Figure 1. A disc stack centrifuge.

The Σ theory concept is used for the scale-up of sedimenting centrifuges. For 50% capture of particles of diameter D, density ρ_s, suspended in a fluid of density ρ_l and viscosity η, the volumetric flow rate Q is given by:

$$Q = 2\Sigma u$$

where u is the settling velocity of the particle in a gravitational field as given by **Stokes' Law**

$$u = \frac{D^2 g}{18\eta}(\rho_s - \rho_l),$$

Σ has the units of m^2. For a given feed and for centrifuges of the same geometry, the ratio Q/Σ is considered constant and permits scale-up calculations.

For the tubular bowl, skimmer pipe and scroll discharge machines, Records (1986) gives Σ as:

$$\Sigma = 2\pi L \frac{\omega^2}{g}\left(\frac{3}{4} r_2^2 + \frac{1}{4} r_1^2\right)$$

where L is the clarifying length; r_1 and r_2 are the radii of the liquid surface and the bowl, respectively; and ω is the angular velocity of the bowl. For the disc type machine Alt (1986) gives

$$\Sigma = \frac{2\pi n}{3g}\omega^2(r_o^3 - r_i^3)\tan\phi$$

where n is the number of discs at an angle of inclination ϕ, and r_i and r_o are the inner and outer radii of the discs, respectively. It is important that one uses Σ or modified Σ values as recommended by the equipment manufacturer.

Filtering centrifuges

Centrifugal filters have developed from simple batch types of machines to sophisticated continuous plant; both batch and continuous equipment are in common usage. Basket, peeler and pusher-peeler centrifuges are examples of batch-operated machines, and conical screen and *pusher centrifuges* are examples of continuous units. There are many variants within some of these categories of centrifuges.

A) Batch-type centrifugal filters are comparatively simple in design and versatile in application, and represent the first generation of machines (although they are still widely available). The main problem of this type of centrifuge is stabilization of the imbalances resulting from unequal loading—hence the names which have evolved: buffer centrifuges (foundations mounted on rubber buffers), gyroscopic centrifuges (rotating shaft which permits nutation movement), and pendulum centrifuges (rigid mounting in a housing from which the basket is hung). Basket machines can be subdivided into solid and open basket designs; solid basket centrifuges tend to require manual discharge of cake whereas open baskets allow knife discharge, and either type may be operated from beneath through a base bearing or from overhead via a link suspension or pendulum.

The main advantages of batch centrifuges are high separation efficiency and high purity of the separated products. They are generally operated at variable speeds and are most suitable for washing. A typical cycle is:

1. acceleration to loading speed;
2. screen rinsing;
3. loading;
4. acceleration to filtration speed (higher than loading speed);
5. cake formation;
6. cake washing;
7. cake dewatering;
8. progressive deceleration to low speed and thence to rest; and
9. unloading.

Peeler centrifuges mostly operate at a constant speed to avoid time losses and higher power consumption associated with accelerating and decelerating. The solids are discharged by being peeled from the basket by a knife edge and then dropped into a discharge chute which carries them out of the basket.

The theory of batch centrifuges is not well-developed and is limited to providing an estimate of the instantaneous flow rate of centrate discharging from the machine. The expression for the volume flow rate of centrate is:

$$Q = \frac{\rho_l \omega^2 \pi l (r_2^2 - r_1^2)}{\eta \left(\alpha \ln \frac{r_2}{r_1} + \frac{R}{r_2} \right)}$$

where ρ_l is the density of the centrate, ω is the angular velocity of the basket, ℓ is the thickness of the cake which has a specific resistance α, r_2 is the inner radius of the basket, r_1 is the inside radius of the liquid phase and R is the resistance of the filter medium.

B) Continuous centrifuges constitute the second- and third-generations. Second-generation centrifuges, from which there is continuous discharge of both liquid and solids from the basket, are the pusher and conical screen types. The third-generation centrifuges include the screen decanter, baffle ring, screen baffle and siphon centrifuges.

Conical screen centrifuges are available in a wide variety of designs. A first classification of these is the angle of the screen——wide-angle (slip discharge and guide channel centrifuges) and small-angle screens. Small-angle screen centrifuges are then subdivided according to the method of solids discharge used: vibration (vibration centrifuge); oscillation (oscillating and tumbler centrifuges); metering (worm or scroll screen centrifuges); and pushing (pusher conical screen centrifuge). Conical screen centrifuges may rotate about either the vertical or the horizontal axis.

The pusher centrifuge consists of a rotating perforated drum lined with a slot screen and a push plate reciprocating with a frequency of about 1 Hz, with a variable advance of between 30 and 60 mm. At first sight, the mechanical design of a tumbler appears complicated, but the construction is simple in comparison with other types of continuous centrifuge. This is reflected in low capital and running costs and its applications, which have covered freely filtering materials such as iron ore, coal fines and coarser crystalline materials. Although the shape of the drum is conical like that of a conical basket centrifuge, the tumbler typically enables longer residence times.

The geometry and complex motions of many continuous centrifuges make them extremely difficult to formulate; hence few attempts have been made to do so and adequate design equations do not exist. The motion of particles through pusher centrifuges has been analysed in detail by Deshun et al. (1991), and through cone centrifuges by Wakeman et al. (1991).

The third-generation represents improvements in process technique being built-in into the continuous centrifuges. In the screen decanter centrifuge, sedimentation and filtration are combined to reduce the total time for separation (and limiting their application to easily filterable products). In the baffle ring centrifuge, the particles bounce against rotating rings to effect a further release of liquid. In the screen baffle centrifuge, the particles bounce against screens. Applications of both the ring and screen baffle machines are limited to granular particles. The siphon centrifuge is an adaptation of peeler and pendulum machines. The siphoning action downstream of the filter medium increases the pressure difference across the basket, leading to an increase in capacity.

References

Deshun, F., Wakeman, R. J., and Zhong, H. (1991) *An Analysis of Gyratory Forces and Wobble Angles in Tumbler Centrifuges, Trans IChemE,* 69, Part A, 409–416.

Purchas, D. B. (1981) *Solid/Liquid Separation Technology,* Uplands Press, London.

Records, F. A. Chapter 6, Hultsch, G. and Wilkesmann, H. (1986) Chapter 12, *Solid/Liquid Separation Equipment Scale-up,* (Eds. D. B. Purchas and R. J. Wakeman), Uplands Press & Filtration Specialists, London.

Wakeman, R. J. and Deshun, F. (1991) *The Control Ring Centrifuge—A New Type of Conical Basket Centrifuge, Trans IChemE,* 69, Part A, 403–408.

R.J. Wakeman

CENTRIPETAL BUOYANCY (see Rotating duct systems, orthoganal heat transfer in; Rotating duct systems, parallel heat transfer in)

CENTRIPETAL FORCE (see Rotating duct systems, parallel heat transfer in)

CERAMIC CRUCIBLE PLASMA FURNACE (see Plasma arc furnaces)

CERENKOV RADIATION

Following from: *Electromagnetic waves*

Cerenkov radiation is emitted when a charged particle passes through a medium with a velocity greater than the velocity of light in that medium. In contrast to **Bremsstrahlung** emission—which is mainly due to interactions between charged particles and nuclei—the emission of Cerenkov radiation is a property of the gross structure of the absorber material. As a charged particle passes through the bound electrons of a material, it induces a polarisation of the medium along its path. The time variation of this polarisation can lead to radiation. If the charge is moving slowly, the phase relations will be random for radiation from different points along the path and no coherent wave front will be formed. However, if the particle velocity (v) is greater than the *velocity of light* in the medium (c/n), where n is the **Refractive Index**, then a coherent wave front can be formed and radiation emitted. The radiation will be in phase and will be emitted in a forward cone of semi-angle θ where:-

$$\cos\theta = \frac{c}{vn}.$$

As the velocity increases, cosθ decreases and θ increases; thus, the cone opens out in contrast to bremsstrahlung emission. The radiation is mainly in the blue end of the visible spectrum and in the ultraviolet.

Cerenkov radiation may be observed around the core of a water-cooled nuclear research reactor of the swimming pool type. In this case, it is caused by high energy beta particles from the beta decay of fission products in the fuel rods.

Cerenkov detectors, operating in a similar way to scintillation detectors, can be used to detect high energy beta particles.

T.D. MacMahon

CERMETS (see Electroplating)

CFCS, CHLOROFLUOROCARBON (see Methylchloride)

CFD (see Computational fluid dynamics)

CFD MODELS (see Two-phase flows)

CHAIN REACTION (see Nuclear reactors)

CHANNEL CONTROL (see Weirs)

CHANNEL FLOW

Following From: *Flow of fluids; Laminar flow; Turbulent flow*

Channel flow is an internal flow in which the confining walls change the hydrodynamic structure of the flow from an arbitrary state at the channel inlet to a certain state at the outlet. The simplest illustration of internal flow is a laminar flow in a circular tube (see **Poisseuille Flow**), while a turbulent flow in the rotor of a centrifugal compressor is an example of the most intricately-shaped internal flow (see **Rotating Ducts**). Engineering devices employ channels with cross-sections of various geometrical shapes. Circular tubes [see **Tubes (Single-Phase Flow in)**; **(Single-Phase Heat Transfer in)**] are the most extensively-used. The variation in diameters of concentrically-arranged tubes makes it possible to form annular channels of various wall curvature. Curved circular tubes coiled as a spiral are used as coils in heat exchangers (see **Coiled Tubes**). Placing twisted strips in the channels produces rotational flows. Various combinations of plane surfaces make it possible to form rectangular (see **Parallel Plates**, **Flow between**) and triangular ducts (see **Triangular Ducts**) with different short and long sides ratio of the perimeter. An intricate geometrical shape of ducts appears in bundles of fuel rods in a core of a nuclear reactor or in bundles of circular tubes (heat exchangers and vapor-generating equipment) (**Figure 1**). A characteristic geometrical parameter of a symmetrical bundle is a pitch ratio of the lattice s/D (s is the space between rod centers and D the rod diameter). Depending on the s/D value, the core lattices are classed into tight ones ($s/D < 1.2$) and extended ones ($s/D > 1.2$). A limiting case of tight lattices is dense packing of rods ($s/D = 1$). The rods, arranged in a triangular (**Figure 1a**) or a square (**Figure 1b**) lattice with different combination of shells (1) and inserts (2), form channels of various geometrical shape. A special case is the flow in ducts with porous walls, in which there is a mass transfer through the surface (see **Porous Wall Ducts**).

With flows in ducts, there always arise a hydraulic resistance which hinders fluid motion. Overcoming these losses requires expenditure of mechanical energy on the flow, which is responsible for the pressure drop over the duct length. Flows in ducts, characterized by a total pressure gradient normal to streamlines, fall into the category of cross shear flows. A relation between the pressure drop Δp and a **shear stress** on the wall $\tau_w = \Delta p D_H/4l$ (D_H and l are the hydraulic diameter and the channel length) can be found from the balance of forces acting on the flow in the duct. The comparison of this relation with Darcy's formula yields $\tau_w = (\bar{f}\rho \bar{u}^2)/8 = \rho u_*^2$, where $\bar{f}$ is the *friction factor* (see **Friction**

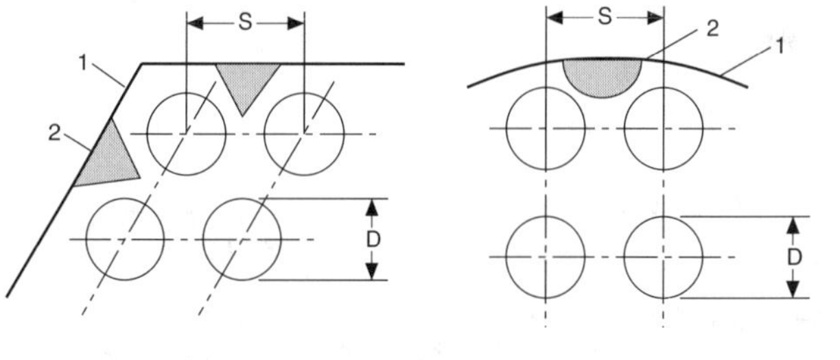

Figure 1. Channels formed in spaces between tubes.

Factor), $u_* = \sqrt{\tau_w/\rho}$ the dynamic or **Friction Velocity.** When considering channels of different geometrical shapes, the hydraulic diameter D_H (see **Hydraulic Diameter**) is commonly used as an independent dimension. In the general case, however, D_H is not a parameter which can set a unified relation between the hydrodynamic characteristics of channels of arbitrary shape and of those a circular tube. The universal character of hydraulic diameter does not hold for channels with narrow cross-sections, in which stagnation zones are originated (e.g., close packings of rods), and for channels with a high curvature of the perimeter (e.g., for extended rod bundles), in which the velocity profile normal to the wall substantially differs from that in a circular tube.

Mathematical models of flows in ducts are based on differential equations of motion, energy balance and mass conservation, which observes two general properties of the medium, viz. continuity and fluidity (see **Conservation Laws**).

For a laminar flow regime, the equations of motion can be solved analytically and enable the calculation of different hydrodynamic characteristics of the flow: velocity and pressure profiles, the friction factor, flow rate, shear stresses, etc. [see **Tubes (Single-Phase Flow in)** and **Parallel Plates, Flow between**]. In an annular duct, the flow rate and the velocity distribution depend on the relation of coaxial cylinder radii r_1 and $r_2 > r_1$

$$\dot{m} = \pi\Delta p[r_2^4 - r_1^4 - (r_2^2 - r_1^2)^2/\ln(r_2/r_1)]/8\eta l,$$

$$u = \Delta p[(r_2^2 - r_1^2)\ln(r/r_1)/\ln(r_2/r_1) - (r^2 - r_1^2)]/4\eta l.$$

The friction factor in a circular tube with diameter D for laminar flow is determined by the relation $\bar{f}_0 = 64/\text{Re}$ ($\text{Re} = \bar{u}D/\nu$). In order to calculate $\bar{f}$ for channels with a noncircular cross-section, a correction factor k_F is introduced such that $\bar{f} = k_F\bar{f}_0$. k_F for a planar slit equals 1.5. For a symmetrical annular duct (radii ratio $\theta = r_1/r_2$), $k_F = (1 - \theta^2)/[1 + \theta^2 + (1 - \theta^2)/\ln\theta]$; for an eccentric annular duct,

$$k_F = [1 - 0.25e^2(1 - \theta)]/[1 + e^2(4 + \theta)/2(1 + \theta)]$$

where e is the relative eccentricity. For relatively narrow eccentric annular ducts ($\theta > 0.25$), an approximate formula $\bar{f}$ Re x$(1 + 1.5e^2) \cong 96$ can be used. In the channels formed by rod bundles, the coefficients of a cross-section shape substantially depend on the pitch ratio s/D of the lattice (**Table 1**).

Calculated values of the hydrodynamic characteristics for laminar fluid flow in channels of various shape, e.g., a rectangle, a triangle, an ellipse or a sector of a circle, can be found in the article **Hydraulic Resistance.**

Experimental investigations have demonstrated the existence of a universal velocity profile in the near-wall region $(y/r_0) < 0.2$ under stabilized turbulent flow in a circular pipe [see **Tubes (Single-Phase Flow in)**].

Table 1. k_F values for channels formed by rod bundles

s/D	1.0	1.05	1.10	1.20	1.40	2.00
Triangular lattice	0.407	0.966	1.27	1.56	1.83	2.46
Square lattice	0.406	0.678	0.913	1.26	1.70	2.52

For turbulent flow in an annular duct, the velocity profile remains the same for both parts of the flow adjacent to the external wall and to the circular tube. At the same time, the velocity profile for that part of the flow in contact with the inner wall depends to a great extent on the perimeter curvature, i.e., $\theta = r_1/r_2$. The difference between the friction factors in the annular duct and in the circular pipe ($\bar{f}$) depends on the parameter θ

$$\bar{f}/\bar{f}_0 = \left[(1 - \theta)/\left(1 + \frac{1 - \theta^2}{\ln \theta^2}\right)\right](1 + 0.04\theta).$$

Turbulent fluid flow in straight channels of intricate cross-sections displays some specific features distinct from the flow in circular tubes and concentric annular ducts. One of them is that the shear stress varies along the perimeter of intricately-shaped channels. In narrow sections of the channel, the interference of *boundary layers* is responsible for the creation of stagnation zones which have, in some cases, even the laminar flow regime. Another specific feature is the convective fluid transfer across the flow. This is not related directly to the local velocity gradient in the tangential direction, but is brought about by the motion of large-scale vortices and by secondary currents. These secondary flows arise due to the turbulent transport of fluid particles along isotaches; therefore, there are no secondary flows the laminar flow. In case the isotach curvature changes across the flow section—in particular, in angular zones of the channel—secondary flows appear that cause the characteristic curvature of isotaches (**Figure 2**, a solid undulated line).

The velocities of secondary currents are low (1–3%) relative to the mean flow velocity along the channel, but they, together with large-scale vortices, facilitate a pronounced mixing of fluid across the channel cross-section, equalize the velocity and shear stress distribution, and strongly affect the turbulent characteristics of the flow. The techniques of *fluid dynamic* calculation in intricately-shaped channels are most divergent. Conventionally, they can be divided into two groups: those using turbulent transfer models and those based on analysis and generalization of experimental data.

Experiments have shown that the velocity profile normal to the wall surface of intricate geometry channels is governed by a universal law, as is the case with a circular tube, if use is made of the local over the channel perimeter dimensionless velocity and length scales (u^+ and y^+). The channel of any arbitrary shape is

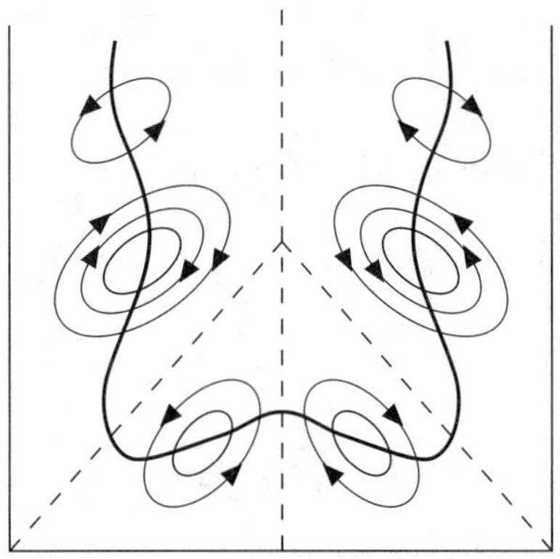

Figure 2. Secondary flows inducing curvature of isotaches.

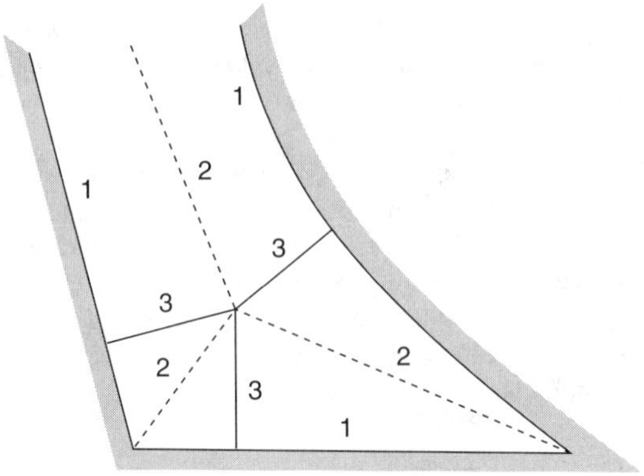

Figure 3. Cells in channel of arbitrary shape.

considered as a combination of unit cells (**Figure 3**) bounded by the wetted perimeter (1), the maximum velocity line (2), and normals (3) from the perimeter points at which the sign of the derivative dy/dL reverses (y_0 is the distance from the wall to line 2; L, the coordinate along the perimeter; and S, the cross-sectional area of the cell). Distribution of the shear stress over the channel perimeter takes into account the specific features of fluid flow in different regions of the cell: $d\tau_w/dy_0 \cong$ *const* in narrow regions and $\tau_w \cong$ *const* in broad regions. These conditions correspond to the relation $\tau/\bar{\tau}_w = c[1 - \exp(-by_0/\bar{y}_0)]$, where b is the empirical value characterizing the geometrical shape of the channel, $b = 7.7/(S/\bar{y}_0^2)^{0.8}$. The constant c is determined from the condition of normalizing τ_w to the channel perimeter P: $(1/P)\int_P(\tau_w/\bar{\tau}_w)dP = 1$.

In specific cases when $y_0/\bar{y}_0 = 1$ and this relation does not vary along the channel perimeter, distribution of τ_w is also constant over the perimeter (the tube, plane slit, concentric annular duct). The calculated distribution of the shear stress over the perimeter enables the estimation of u* and the velocity profile in the channel of any geometrical shape. Proceeding from the generalization of experimental data and the results of calculation, the formulas suggested to determine the friction factor in a unit cell $\bar{f}_i$ and in intricately-shaped channel $\bar{f}_k$ consisting of i cells are:

$$\bar{f}_i/\bar{f}_0 = [0.58 + 0.42\exp(-0.021K_i^3)][1 + 0.01(\beta_i + 1)^{4/3}],$$

$$(\bar{f}_k/\bar{f}_0)^{-4/7} = \sum_i (S_i/S_k)(D_{H,i}/D_{H,k})^{6/7}(\bar{f}_i/\bar{f}_0)^{-4/7},$$

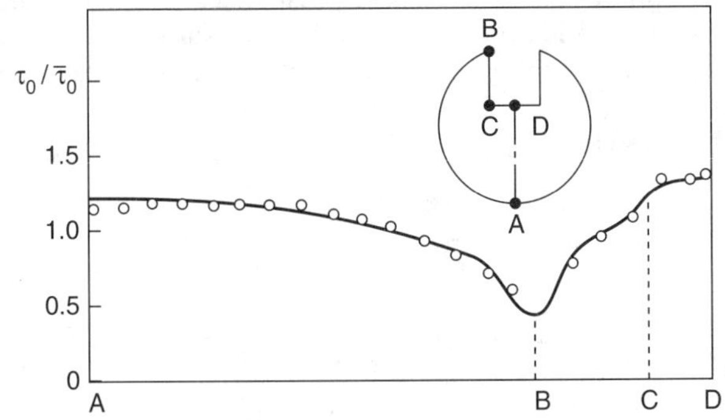

Figure 4.

Table 2. Friction factors in complex channels

The shape of the channel cross-section	$\left(\frac{\bar{f}_k}{f_0}\right)_{exp}$	$\left(\frac{\bar{f}_k}{f_0}\right)_{cal}$
1. Triangular packing of N rods.		
1.1. Hexahedral shell, N = 37, s/D = 1.4	1.15	1.16
1.2. Irregularly-shaped shell, N = 19, s/D = 1.13	1.18	1.08
1.3. Hexahedral shell, N = 19, s/D = 1.12	0.97	1.02
1.4. Irregularly-shaped shell, N = 37, s/D = 1.05	1.00	0.98
1.5. Hexahedral shell, N = 37, s/D = 1.05	0.84	0.87
2. Triangular shell with a bundle of rods.		
2.1. N = 3, s/D = 1.41	0.57	0.55
2.2. N = 3, s/D = 1.30	0.73	0.69
2.3. N = 3, s/D = 1.15	0.99	1.00
3. Square shell with a bundle of tubes.		
3.1. N = 4, s/D = 1.45	0.97	0.94
3.2. N = 4, the tube touches the wall	0.71	0.68
3.3. N = 4, tube-to-tube touch	0.68	0.66
4. Rectangle with an aspect ratio a/b.		
4.1. a/b = 0.2	1.00	0.94
4.2. a/b = 0.1	1.00	1.02
5. Isosceles triangle with an angle at vertex α.		
5.1. α = 12°	0.73–0.85	0.77
5.2. α = 20°	0.89	0.84
5.3. α = 40°	0.95	0.95

where k_i, β_i are the geometrical parameter of the cell: $k_i = (Y^{0i}_{max} - Y^{0i}_{min})(S/\bar{y}_0^2)^{0.25}/\bar{y}_{0i}$, $\beta_i = \pm\bar{y}_{0i}/R_i$ (R_i is the curvature radius of the wetted perimeter; the "+" sign is taken for the cell with a convex perimeter; the "−" sign for one with a concave perimeter).

Calculation results using these formulas have shown consistency with experiment ±10% for channels of various shapes (**Table 2**).

The structure of the turbulent flow is essentially complicated if spacers are present in the channels, e.g., in the case of helical finning of fuel elements in the assembly (**Figure 5**). In bundles of finned rods (the fin-on-fin touch), a linear growth of the friction factor can be observed when increasing the lattice pitch $\bar{f}_l/\bar{f}_{l,0} = 1 + 600[(s/D) - 1]/(T/D)^2$, where T is the pitch ratio of winding. $\bar{f}_{l,0}$ values for the corresponding smooth (nonfinned) bundle of rods

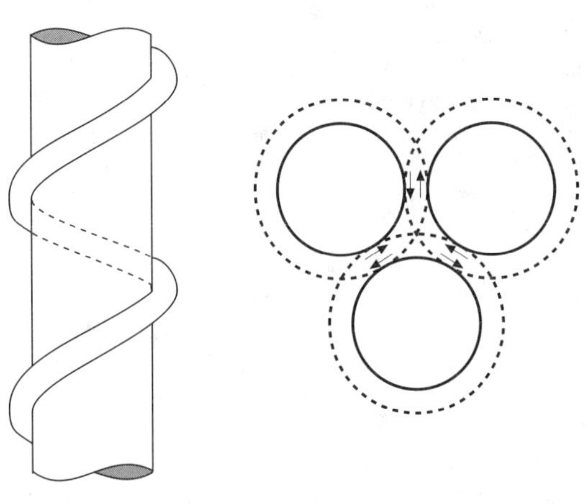

Figure 5.

are calculated by empirical formulae, e.g., for a triangular lattice (m/D > 1.02):

$$\bar{f}_{1,0}/\bar{f}_0 = 0.57 + 0.18[(s/D) - 1] + 0.53[1 - \exp(-a)],$$

where $a = 0.58 + 9.2[(s/D) - 1]$.

At a sufficient length of the channel, the hydrodynamic characteristics of the flow change from arbitrary inlet to stabilized values, which are determined by the flow regime and the geometrical shape of the channel. The length on which hydrodynamic stabilization occurs is referred to as an *entrance length* l_e. Stabilization of hydrodynamic flow parameters due to the effect of the channel walls occurs continuously according to an exponential law; therefore, an entrance section has no clear end boundary. As a rule in calculations, the distance from the channel inlet, such that the difference between the parameter in question and the stabilized parameter is 1%, is taken as the entrance length. For a laminar fluid flow in channels, specifically in a circular tube, the entrance length is fairly large: $l_e/D = 0.065$ Re. In the case of turbulent flow, hydrodynamic stabilization is of a more involved nature and depends on many factors, viz. the shape of the channel cross-section, the state of the flow at the inlet to the channel, entrance effects, etc. Experiments show that the length of the stabilization section for different hydrodynamic characteristics of the flow substantially differs. The longitudinal pressure gradient, or the shear stress on the wall, is the fastest to stabilize. According to different authors, the stabilization length of this parameter in a circular tube with a sharp entrance edge is $(5 - 15)D_H$. Under the same entrance conditions, the velocity profile is stabilized at the length $(40 - 60)D_H$, the dependence of the entrance section length on the Re number being weak. The greatest stabilization length is exhibited by fluctuating characteristics of the turbulent flow (l_e/D_H is equal to 100 and higher). Different lengths of entrance sections for different flow characteristics represent the physics of turbulent processes proceeding in the channels. The velocity field is a result of integral action of the turbulent properties of the flow while the shear stress is an integral result of velocity field in the wall region. In this connection, an appreciable variation with length of fluctuating flow characteristics, particularly in the center of the channel, causes a weak variation of the velocity profile, and more so of the longitudinal pressure gradient (of the shear stress on the wall). Flow in the section of hydrodynamic stabilization of intricately-shaped channels possesses specific features of its own, and they are due to the fact that the flow properties are formed both normal to the wall (radial stabilization) and over the channel perimeter (tangential stabilization). This leads, in certain cases, to an increased entrance length in relation to a circular tube. For instance, for the channel formed by a dense packing of rods, the stabilization length for the velocity profile is $(80 - 90)D_H$.

In the general case of flows in noncircular cross-section channels, heat transfer depends not only on the flow characteristics but also on the properties of the heat-transmitting wall since there is a **Heat Flux** normal to the wall surface and in the tangential direction. Therefore, the temperature field in the flow and the heat-transmitting wall results from the thermal interaction of the flow and the wall (a conjugate problem). Under such conditions, when the wall temperature varies along the channel perimeter, the heat transfer coefficient cannot adequately characterize the temperature regime of the heat transfer surface. In some cases, e.g., in channels with dense packing of rods, the local *heat transfer coefficients* may assume zero or even negative values, losing their physical significance. Therefore, the concept of a dimensionless wall temperature $\Delta T_w = (\bar{T}_w - T_w)/(q_w D_H)$ is introduced.

Of primary importance in the conjugate problem is consideration of the specific hydrodynamic features of intricately-shaped channels, such as stagnation and laminar zones, secondary currents, and large-scale vortices responsible for anisotropy of turbulent transfer. Therefore, heat transfer under conditions of turbulent flow in such channels is a complex phenomenon and cannot be described by simple formulae.

References

Reynolds, A. J. (1974) *Turbulent Flows in Engineering,* Wiley, London.

Heat Exchanger Design Handbook 1983. Vol. 1 and 2, Hemisphere Publishing Corporation.

Leading to: *Bends; Coiled tubes, flow and pressure drop in; Contraction; Enlargement; Headers and manifolds; Nozzles; Plug flow heat transfer; Pressure drop, single phase; Rotating duct systems; Secondary flow; T-Junctions; Tubes, single phase flow in*

M.Kh. Ibragimov

CHAOS (see Non-linear systems)

CHAR (see Pyrolysis)

CHARACTERISTIC DRYING CURVE (see Drying)

CHARACTERISTIC EQUATIONS, FOR SUPERSONIC FLOW (see Inviscid flow)

CHARACTERISTICS, METHOD OF

Following from: *Differential equations*

The method of characteristics is a classical technique for solving problems involving the dynamics of a supersonic two-dimensional steady **Inviscid Flow** of a compressible fluid, which in the general case of a plane irrotational flow of an ideal gas, can be described by a nonlinear system of two differential equations with partial derivatives of the first order

$$(u^2 - u_{sound}^2)\frac{\partial u}{\partial x} + uv\left(\frac{\partial u}{\partial y} + \frac{\partial v}{\partial x}\right) + (v^2 - u_{sound}^2)\frac{\partial v}{\partial y} = 0$$

$$\frac{\partial u}{\partial x} - \frac{\partial u}{\partial y} = 0 \quad (1)$$

where u, v are the components of velocity vector in the x y plane, and u_{sound} is the sound velocity.

In the case of a supersonic flow, when $(u^2 + v^2)/u^2_{sound} = v^2/u^2_{sound} = Ma^2 > 1$ (Ma is the **Mach number**) the system of Equations [1] is of a hyperbolic type; in the case of a subsonic flow when $Ma < 1$, it is elliptic.

A characteristic feature of *hyperbolic equations* is the presence of characteristic surfaces (lines). For a system of equations (1) in the case of supersonic flow ($Ma > 1$), the characteristic lines (characteristics) used in the x, y plane are the *Mach lines,* i.e., the lines which at a given point form with the velocity vector angles equal to the Mach angles for perturbation propagation

$$\alpha = \pm \arcsin Ma^{-1},$$

and the velocity vector itself is directed along the bisector of the angle between the characteristics. Thus, the projections of the velocity vector on a normal to the characteristics at a given point are equal in absolute value to a local sound velocity.

The dependent variable satisfies key reciprocity relationships for the calculation method

$$-\theta + \sigma(Ma) = R, \qquad \theta + \sigma(Ma) = Q$$

where

$$\sigma(Ma) = \sqrt{\frac{\gamma+1}{\gamma-1}} \arctan \sqrt{\frac{\gamma+1}{\gamma-1}(Ma^2-1)} - \arctan \sqrt{Ma^2-1}$$

is the Prandtl-Mayer function, $\gamma = C_p/C_v$, θ is the angle between the velocity fector $\vec{v}$ and the axes x, R and Q are the values (Rieman's invariant) constant along the characteristics of the first and second families, respectively,

$$\begin{cases} \alpha_1 = \arcsin Ma^{21} \\ \alpha_2 = -\arcsin Ma^{-1} \end{cases} \begin{cases} \dfrac{du}{dx_1} = \dfrac{uv + u_{sound}\sqrt{v^2 - u^2_{sound}}}{u^2 - u^2_{sound}} \\ \dfrac{du}{dx_2} = \dfrac{uv - u_{sound}\sqrt{v^2 - u^2_{sound}}}{u^2 - u^2_{sound}} \end{cases}.$$

For axisymmetric flows and other cases, the reciprocity relationships can be written in differential form which cannot be solved by quadratures, as in the plane case—but can easily be represented in finite-difference form in a numerical solution. A characteristics method of solution of the Coushi problem, in which the components of the velocity vector at some point P must be found from the known boundary values of a certain noncharacteristic line, is shown schematically in **Figure 1a**.

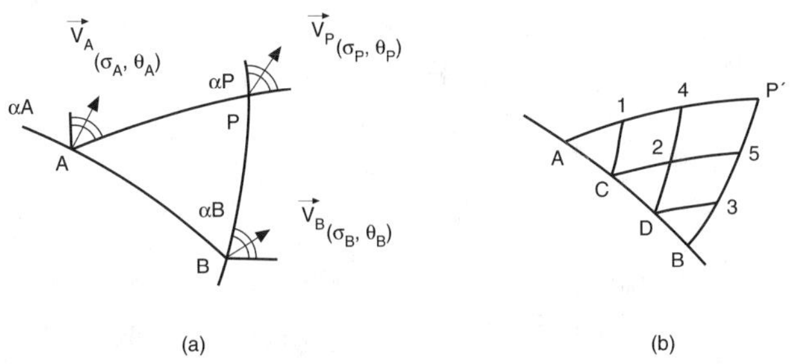

Figure 1. Solution of Coushi problem using method of characteristics.

The solution of the problem in variables (σ, θ), which can be transformed into physical variables (u, v), is done with the help of two characteristics of different families—AP and BP, according to which the value of the invariants Q_A and R_B are translated into point P. Hence,

$$\sigma_p = \frac{1}{2}(Q_A + R_B), \qquad \theta_p = \frac{1}{2}(Q_A - R_B).$$

However, because of the absence of a solution to the problem inside the region APB—as a consequence of which the characteristics AP and BP cannot be given beforehand in the numerical solution—this region is divided into small triangular elements (actually, this region is constructed from such triangulars) and the solution at point P′ (Fig. 1b) is found when the problem for points (1–5) is solved.

The accuracy of the solution obtained depends on the accuracy of construction of a net from the elements of characteristics, and can be increased by various means, for instance by decreasing the size of the cells.

The region bounded by the characteristics of different families, drawn from the end points of the curve on the boundary (in **Figure 1**, the region APB and the curve AB, respectively), is called the *effect region.*

Through the use of various modifications, the method of characteristic allows the solution of problems with fixed and free boundaries, with shock waves, with nonisentropic flows, etc.

The advantage of the method is that it makes it possible to follow the breaks (shock waves) and carefully calculate them. However, the procedure of break separation differs from the procedure of solving the problem in the remaining region and if numerous breaks occur, the efficiency of the method drops.

At present, a number of finite-difference methods successfully compete with the characteristics methods. These include the methods of Lax-Vendroff and Godunov; TVD-methods, flow charts with flux splitting, etc., which are applied to more difficult problems, three-dimensional flows included.

Characteristic methods are also used for solving problems of one-dimensional, nonstationary isentropic motion of a compressed gas.

Reference

Liepmann, H. W. and Roshko, A. (1957) *Elements of Gas Dynamics,* New York, London.

I.G. Zaltsman

CHARCOAL (see Pyrolysis)

CHARGE CARRIERS (see Electrical conductivity)

CHARGE COUPLED DEVICES, CCD (see Photographic techniques)

CHARLES LAW (see Gay Lussac's law)

CHARACTERISTICS, OF DIFFERENTIAL EQUATIONS (see Differential equations)

CHEBYSHEV EQUATION

Following from: *Differential equations*

The Chebyshev equation is a linear homogeneous differential equation of the second order

$$(1 - x^2)y'' - xy' + \alpha y = 0 \quad (1)$$

and is a particular case of a hypergeometric equation.

The fundamental system of solving the Chebyshev equation in the interval $-1 < x < 1$ for $\alpha = n^2$, where n is a natural number, consists of **Chebyshev polynomials** of the first kind of the degree n $T_n(x) = \cos(n \arccos x)$ and the functions $U_n(x) = \sin(n \arccos x)$ related to Chebyshev polynomials of the second kind $1/n + 1\ T'_{n+1}(x)$. Chebyshev polynomials of the first kind $T_n(x)$ serve as an effective solution to Chebyshev equation with $\alpha = n^2$ and on the entire real axis.

Chebyshev polynomials are a sequence of *eigen functions* for a certain Sturm-Liouville's problem, from whence their orthogonality follows. Chebyshev polynomials have the completeness property on the interval $[-1, 1]$. In this case, any continuous function which satisfies a Lipschitz condition can be expanded into a Fourier-Chebyshev series $f(x) = \sum_{n=0}^{\infty} a_n \hat{T}_n(x)$, uniformly converging on this interval. $\hat{T}_n(x)$ are the orthonormal Chebyshev polynomials $\hat{T}_0(x) = \sqrt{\frac{1}{\pi}}\,T_0(x)$; $\hat{T}_n(x) = \sqrt{\frac{2}{\pi}}\,T_n(x)$. The coefficient of this series is defined as:

$$a_n = \int_{-1}^{1} f(t)\hat{T}_n(t)\frac{dt}{\sqrt{1 - t^2}}. \quad (2)$$

V.N. Dvortsov

CHEBYSHEV POLYNOMIAL EXPANSION (See Tau method)

CHEBYSHEV POLYNOMIALS

Chebyshev polinomials of the first kind are the trigonometric polynomials defined by:

$$T_n(x) = \cos(n \arccos x)$$

$$= \frac{1}{2}\left[(x + i\sqrt{1 - x^2})^n + (x - i\sqrt{1 - x^2})^n\right] \quad (1)$$

whence

$$T_n(x) = 2^{n-1}\left[x^n - \frac{n}{1!2^2}x^{n-2} + \frac{n(n-1)}{2!2^4}x^{n-4} - \frac{n(n-4)(n-5)}{3!2^6}x^{n-6} + \cdots\right]. \quad (2)$$

For Chebyshev polynomials, a generalized *Rodrigues formula* is valid

$$T_n(x) = (-1)^n 2^n \frac{n!}{(2n)!}\sqrt{1 - x^2}\frac{d^n}{dx^n}(1 - x^2)^{n-1/2}. \quad (3)$$

A recurrent relationship holds for Chebyshev polynomials

$$T_{n+1}(x) = 2xT_n(x) - T_{n-1}(x). \quad (4)$$

Chebyshev polynomials for a negative value of n are defined by the relationship:

$$T_{-n}(x) = T_n(x) \quad (5)$$

Chebyshev polynomials of the first kind are orthogonal with respect to a weight function $1/\sqrt{1 - x^2}$ on the interval $[-1, 1]$. The orthogonality relationship is:

$$\int_{-1}^{1} T_n(x)T_m(x)\frac{dx}{\sqrt{1 - x^2}} = \begin{cases} 0, & m \neq n \\ \frac{\pi}{2}, & m = n \neq 0 \\ \pi, & m = n = 0. \end{cases} \quad (6)$$

The roots of the polynomial T(x), defined by the equality $x_k^{(n)} = \cos(\frac{2k - 1}{2n})$, $k = 1,2, \ldots, n$ are often used as cusps of quadrature and interpolation formulas.

Chebyshev polynomials of the first kind with a unit coefficient of the higher term, i.e., $\bar{T}_n(x) = \frac{1}{2^{n-1}}T_n(x)$ are the polynomials least deviated from zero on the interval $[-1, 1]$, i.e., for any other polynomial $F_n(x)$ of degree n with unit heading coefficient the following relationship holds:

$$\max|F_n(x)| > \max|\bar{T}_n(x)| = \frac{1}{2^{n-1}} \quad (7)$$

$$x \in [-1, 1]$$

This property of Chebyshev polynomials is used for constructing optimal iteration algorithms in solving problems of **heat transfer** with the help of numerical methods.

V.N. Dvortsov

CHELATION (see Foam fractionation)

CHEMICAL EQUILIBRIUM

Following from: *Thermodynamics; Chemical reaction*

Chemical equilibrium is the thermodynamic equilibrium in a system where direct and reverse chemical reactions are possible. If chemical equilibrium takes place in the system, the rates of all reactions proceeding in two opposite directions are equal. Therefore, the macroscopic parameters of the system do not change and the relationship between concentrations of reacting substances remains constant at a given temperature. Equilibrium for any chemical reaction is expressed by an equality $\Sigma \nu_i \mu_i = 0$, where μ_i is the chemical potential of each reagent ($i = 1,2, \ldots$) and ν_i is the stoichiometric coefficient of each substance in an equation of chemical reaction (it is positive for initial substances and negative for products of a reaction).

The dependence of chemical equilibrium on external conditions is expressed by the *Le Chatelier-Braun principle* (1885–1886). It consists of the following correlation: Let equilibrium take place and then influence the system, changing some external conditions (temperature, pressure, concentrations of reacting substances). The equilibrium of a reaction tends to follow such direction that allows the reduction of an external influence. A temperature increase will cause a displacement of the equilibrium to the direction of such reaction that proceeds with heat absorption. A pressure increase will cause equilibrium displacement to follow the direction of such reaction that leads to a volume decrease. The introduction of any additional reagent in the system will propel equilibrium displacement to a direction where this reagent is consumed.

The total *Gibbs energy* change of a chemical reaction aA + bB = cC + dD (when temperature and pressure are constant) is expressed by the equation

$$\Delta_r G_{p,T} = \Delta_r G^0_{p,T} + RT \ln(\alpha_C^c \alpha_D^d / \alpha_A^a \alpha_B^b),$$

where R is the gas constant, p is the pressure, T is the absolute temperature, α_i refers to the activities of the reacting substances and $\Delta_r G^0_{p,T}$ is the standard Gibbs energy's change of that reaction ($\alpha_i = 1$). The value of $\Delta_r G^0_{p,T}$ can be calculated on the basis of standard values of the Gibbs energies of formation ($\Delta_f G^0$) of the reagents at 298.15 K and of known thermodynamic relationships that determine the temperature and pressure dependencies of Gibbs energy change.

If equilibrium is attained, then

$$\Delta_r G^0_{p,T} = -RT \ln \frac{(\alpha_{C\ \text{equil.}})^c (\alpha_{D\ \text{equil.}})^d}{(\alpha_{A\ \text{equil.}})^a (\alpha_{B\ \text{equil.}})^b} = -RT \ln K_\alpha.$$

Here, $\alpha_{i\text{equil.}}$ are the activities corresponding to the equilibrium state and K_α is the equilibrium constant expressed in terms of activities. Hence, it follows that $\Delta_r G_{P,T} = RT[\ln(\alpha_C^c \alpha_D^d/\alpha_A^a \alpha_B^b) - \ln K_\alpha]$. The last relationship is the van't Hoff isotherm equation (or *van't Hoff equation*). It permits the determination of a probable direction of the reaction under given conditions. The process will take place when $\Delta_r G_{P,T} < 0$, i.e., when $K_\alpha > (\alpha_C^c \alpha_D^d/\alpha_A^a \alpha_B^b)$. Analogous relationships can be obtained when the equilibrium constant (K_p) is expressed in terms of partial pressures (P_i) of the reagents:

$$\Delta_r G_{P,T} = RT\left[\ln \frac{P_C^c P_D^d}{P_A^a P_B^b} - \ln K_p\right]; \qquad \Delta_r G^0_{P,T} = -RT \ln K_p.$$

The "equilibrium constant of reaction" is the result of the **mass action law**, which determines a correlation between the masses of reacting substances under equilibrium. According this law, the reaction's rate depends on the concentrations of reacting substances. The rate constant of a given reaction at fixed temperature is a constant value; therefore, the relationship of the rate constants of direct and reverse reactions is a constant value too. This relationship is a function of temperature only.

The equations that express a relationship between the $\Delta_r G^0_{P,T}$-value and equilibrium constant of reaction allow the calculation of the equilibrium of chemical reactions, avoiding expensive and prolonged experiments. For such calculations, it is necessary to have reliable values of thermodynamic functions for all reacting substances.

Various experimental methods are used to determine equilibrium constants of chemical reactions. There are static and dynamic methods as well as the circulation method. The last is a specific combination of the static and dynamic methods. When static methods are used, the reaction mixture stays at a given temperature until an adjustment of the equilibrium takes place. Then "tempering" and chemical analysis of the reaction mixture are carried out. "Equilibrium tempering" is the fast-cooling of the reaction mixture to a low temperature where the rate of reaction is very small.

The more common dynamic method of defining equilibrium constants has often been called the transportation method. A steady stream of inert gas is passed over the mixture of substances that is maintained at a constant temperature. This "carrier" gas removes the volatile components of the reaction at a rate that depends on the rate of gas flow. The vapours of the reagents are condensed or collected by absorption or chemical combination at the colder portion of the apparatus. The experiments are carried out at different rates of gas flow. The equilibrium pressures of volatile reagents are determined by extrapolation of the results up to zero rate of the carrier gas.

A modification of the dynamic method used for investigating heterogeneous equilibria is the circulation method. The gas mixture is circulated in a closed space; circulation is carried out by means of electromagnetic pump. Equilibrium is attained where passing this mixture many times over the solid phase into the furnace. Tempering of the gas mixture is done when it is taken out from the hot zone and passed through a capillary. In view of the large linear rate of gas flow, this mixture becomes cold rapidly and its' composition is not changed.

The most direct way of measuring equilibrium constants of chemical reactions is through the measurement of electromotive forces (the e.m.f. method). For example, the reaction

$$Zn(\text{cryst.}) + CuSO_4(\text{solution}) = ZnSO_4(\text{solution}) + Cu(\text{cryst.})$$

is a process of potential generation for the Daniel galvanic element: $Zn^0/Zn^{2+}//Cu^0/Cu^{2+}$. A zinc plate (one electrode) is immersed into a solution of zinc sulfate and a copper plate (the other electrode) is immersed into a solution of copper sulfate. A galvanic element (source of electromotive force) can be created if both electrodes are connected by a tube that contains a solution-conductor. The dissolution of zinc (process: $Zn^0 = Zn^{2+} + 2e$) takes place at one electrode; the precipitation of copper (process: $Cu^{2+} + 2e = Cu^0$) takes place at the second electrode. Therefore, the common potential forming reaction is: $Zn^0 + Cu^{2+} = Zn^{2+} + Cu^0$. The Gibbs energy change for such reaction is given by the formula $\Delta_r G^0_T = -nFE_T$, where n is the number of gramme-equivalents of reagent; F is Faraday's constant (nF is the number of coulombs of electricity passed); and E_T is the electromotive force of the galvanic element at a given temperature. The value of the Gibbs energy of reaction can be used for calculating its equilibrium constant (K): $\ln K = -\Delta_r G_T^0/RT = nFE_T/RT$.

The equilibrium state is a thermodynamic state of a system that is a permanent in time. This invariability is not connected with some external process taking place. There are different kinds of equilibria. If the equilibrium is "steady," then any adjacent states of the system are less steady. It would be necessary to spend external work for transition from the equilibrium state to these adjacent states. It is also typical that steady equilibrium can be approached from two opposite directions. However, this discussion is concerned with steady equilibria only or "chemical equilibria." From the physicist's point of view, steady equilibrium is dynamic. It is attained when the rates of direct and reverse reactions are equal, but not under conditions when the process is stopped in general. The equality $dG = 0$ is a general condition for "steady" and "unsteady" equilibria, but the value of the second differential of Gibbs energy is positive under steady equilibrium ($d^2G > 0$) and negative under unsteady equilibrium ($d^2G < 0$). The conditions of stability of the equilibrium can be deduced using the *second law of thermodynamics.* These are: 1) the pressure increases at a constant temperature if volume decreases [$(\partial P/\partial V)_T < 0$]; and 2) the value of heat capacity is positive ($C_p > 0$).

The degree of stability of the different states of chemical systems can vary. States which possess some relative stability are called "metastable" states. Such states have often arisen due to kinetic factors, which create difficulties for the transition of a system from the metastable (unsteady) state to a steady equilibrium state.

The development of thermodynamic theory of equilibria—in particular, equilibria of chemical reactions—owed much to J. W. Gibbs (1873–1878) and Le Chatelier (1885), who discovered the principle of displacement of equilibria under conditions of external change. The theory of chemical equilibria was developed further by F. H. van't Hoff (1884–1886).

References

Gibbs, J. W. (1950) *Thermodynamic Works.* Translation from English, Gostekhteorisdat, M.-L.

Munster, A. (1971) *Chemical Thermodynamics.* Translation from German.

Kubaschewski, O. and Alcock, C. B. (1979) *Metallurgical Thermochemistry,* Pergamon Press Ltd., Oxford. New York. Toronto. Sydney. Paris. Frankfurt.

Karapetyans, M. Kh. (1981) *Introduction to Theory of Chemical Processes,* Vysshaya Shkola, M. (in Russian).

Vasiliev, V. P. (1982) *Thermodynamic Properties of Solutions of Electrolytes,* Vysshaya Shkola, M. (in Russian).

V.Ya. Leonidov

CHEMICAL LASERS (see Lasers)

CHEMICAL POTENTIAL

Following from: *Thermodynamics*

The chemical potential of the i-component of a thermodynamic system in a given phase is a thermodynamic state function. It defines changes of the *Gibbs energy* and other thermodynamic potentials when the number of particles of a corresponding component is changed. Chemical potential of the ith-component of the system is the derivative of any thermodynamic potential divided by the quantity (or number of molecules) of this component when the values of the other thermodynamic variables, given a thermodynamic potential, are constant, e.g., $\mu_i = (G/n_i)_{T,p,n_j'}$, where G is the Gibbs energy of the phase; n_i, the number of moles of the i-component of the phase; T, the absolute temperature, p, pressure, and n_j, the number of moles of all other components ($j = i$). It is obvious that chemical potential (μ_i) is a partial molar Gibbs energy. The sum of $\Sigma\mu_i dn_i$, which enters into the expression for the total differential of all thermodynamic potentials, has been called the fundamental Gibbs equation, e.g.,:

$$dG = -S\,dT + V\,dp + \Sigma\,\mu_i dn_i$$

where S is the entropy and V the volume.

The conditions of thermodynamic equilibrium can be determined in the simplest way by using chemical potentials. If the sum of the products of chemical potentials and stoichiometric coefficients for all reagents is equal to zero, then this equality is a condition of chemical equilibrium. It is assumed that the values of stoichiometric coefficients of the reaction products are negative. The condition of phase equilibrium is the equality of chemical potentials for each component in all phases. If an equilibrium system is homogeneous, the chemical potential of each component in all points of the system is identical.

Chemical potential is expressed in energy units per unit of the substance mass (Joule/kg) or per one mole of the substance (Joule/mole) or per molecule of the substance. The term "chemical potential" was coined by J. Gibbs (1875), a prominent American physicist.

V.Ya. Leonidov

CHEMICAL REACTION

Chemical reaction is a process of conversion of some chemical compounds into others. In general, chemical reaction can be represented by the equation

$$n_1A_1 + N_2A_2 + \cdots = n_1'A_1' + n_2'A_2' + \cdots + Q$$

where A_i are the reactants; A_i' are the products; n_i and n_i' are the corresponding stoichiometric coefficients; and Q is the thermal effect of the reaction. Both reactants and products can be atoms, molecules or charged particles (ions or electrons); in photochemical reactions, photons are included as well. The sign of Q can be positive (which means that heat is evolving through the reaction; such reactions are called *exothermic*) or negative (when heat is absorbed through the reaction; such reactions are called *endothermic*). The thermal effect of the reaction does not depend on the method of transformation from reactants to products (Hess' law, 1836).

Any chemical reaction proceeds simultaneously in two directions: forward and reverse. At equilibrium, both reactions have the same rates, thus the rate of the total process is equal to zero. The relationship between concentrations (or partial pressures for reactions in the gaseous state) at equilibrium is called the **mass action law** (Huldberg and Waage, 1867):

$$\prod_i p_i^{Nni'}(A_i')/\prod_i p_i^{ni}(A_i) = K_p(T)$$

or

$$\prod_i c_i^{ni'}(A_i')/\prod_i c_i^{ni}(A_i) = K_c(T),$$

where $p_{i_i}^{n}(A_i)$[or $c_{i_i}^{n}(A_i)$] are the partial pressures (concentrations) of A_i and $K_p(T)$[or $K_c(T)$] is the equilibrium constant which depends on the temperature T, but does not depend on partial pressures (concentrations) of the components A_i. Strictly, the mass action law can be applied only to the ideal case where no interaction occurs between the molecular particles of the gas or the solute. This assumption is usually justified for gaseous reactions if the total pressure is not considerably higher than 0.1 MPa. Generally, it is not applied to reactions in the solution phase, except for very dilute solutions. The more correct form of this law applied to real systems requires the inclusion of **fugacity** f_i instead of the partial pressure p_i, or activity a_i instead of concentrations c_i.

When doing mass action law calculations, a distinction must be drawn between homogeneous and heterogeneous reactions. A homogeneous reaction is one which takes place completely in one phase, such as $H_2 + I_2 = 2HI$, or $CuSO_4(aq) = Cu^{2+}(aq) + SO_4^{2-}(aq)$. For them, the mass action law can be written as:

$$K_p = p^2(HI)/p(H_2)p(I_2)$$

and

$$K_c = c(Cu^{2+}, aq)c(SO_4^{2-}, aq)/c(CuSO_4, aq),$$

where K_p and K_c are the mass action constants at a certain temperature, p denotes partial pressure and c is the molar concentration. In heterogeneous reactions, one or more condensed phases appear in addition to the gaseous or solution phase. The mass action relationship is then expressed as in the following example for the reaction $Fe_2O_3(cr) + 3C(cr) = 2Fe(l) + 3CO(g)$:

$$K_p = p^3(CO, g).$$

In practice, equilibrium can be shifted when only the products of the reaction are present and the concentration of the reactants is so small that it can be neglected. In this case, only the forward reaction takes place. These reactions are called kinetic one-sided or kinetic unreversible. In such reactions, one of the products formed is constantly removed, thereby continuously displacing the system from equilibrium.

Written in its usual form, the equation of a chemical reaction shows only the initial and final states of the process. This equation can be considered as a symbolic expression of the *mass action law*. Actually, many reactions proceed through a series of intermediate stages. These individual stages may be referred to as elementary reactions. In most cases, the exact mechanism of the reaction is unknown because of the difficulty in correctly revealing all the intermediate products which the reaction is going through.

The rate of reaction v is proportional to the concentrations of each substance $[A_1]$, $[A_2]$, . . . participating in the reaction raised to a power which is equal to a corresponding stoichiometric coefficient n_1, n_2, . . . :

$$v = -k[A_1]^{n1}[A_2]^{n2}\cdots$$

This equation is often called the mass action law for the rate of the chemical process. The factor k is the rate constant of the chemical reaction. For simple reactions, the degrees n_1, n_2, . . . are integer. Each of these values is called the order of the reaction for a given substance. The sum $N = n_1 + n_2 + \ldots$ is called the overall order of the reaction. For more complex reactions, this expression can also be applied, but the order of such a reaction may not be an integer. For example, consider a reaction with reactants A_1 and A_2. If the reaction is of the first order in A_1 and the first order in A_2, the overall order of the reaction is equal to two.

There are reactions of zero order, in which the rate is unaffected by changes in the concentrations of one or more reactants because it is determined by some limiting factor other than concentration, such as the amount of light absorbed through a photochemical reaction or the amount of catalyst in a catalytic reaction.

To determine the change of the concentrations of the components with time for the case of a complex chemical reaction, the system of differential equations must be solved. This system is defined through the reaction mechanism and the equations of material balance, and usually can be very complicated. If a multistage process proceeds through the formation of short-lived molecules, the initial differential equation system can be simplified assuming that $dc/dt = 0$ for these molecules, and thus facilitate the process of solving the system.

The dimensions of the rate constants depends on the order of the reaction N and on the dimensions of the concentration. Concentrations are usually expressed either in a number of molecules per unit of volume (m^{-3}) or in a number of moles of the substance per m^3 (mol m^{-3}). The units of the rate constant k for chemical reactions of the overall order N can be expressed by the general formula $(m^3mol^{-1})^{N-1}s^{-1}$.

The dependence of the rate constant on temperature is given by the **Arrhenius equation** (1889)

$$K = A\exp(-E_a/kT),$$

where A is a factor which may also vary with temperature, E_c is the energy of activation, k is the *Boltzmann constant* and T is the absolute temperature. The exponential dependence of the rate constant on temperature had been detected experimentally before Arrhenius, but he was the first who proposed its theoretical interpretation. According to his hypothesis, molecules whose energy exceed E_a are the only ones undergoing chemical conversion. The fraction of such molecules in a **Boltzmann distribution** is directly proportional to $\exp(-E_a/kT)$. The dependence of the constant A with temperature can be described by the relation $A = A_0T^n$, where n is defined by the characteristics of a process.

Calculation of rate constants is the typical task of the theory of elementary chemical processes. The process of chemical reaction depends on the electronic, vibrational, and rotational states of the colliding particles participating in the reaction. If the density of gas is not too high, the time of the collision ($10^{-12} - 10^{-19}$s) is sufficiently smaller than the time between the collisions of the molecules (10^{-10}s under normal conditions). Therefore, the problem can be separated in two parts: determination of the reaction cross-sections σ and microscopical rate constants (i.e., the value which characterizes the rates of reaction for a particle in a given state) on the one hand, and evaluation of the partition function F of molecules, on the other. Such a problem is usually very difficult to solve. In the method of transitional states, the following basic assumptions are used:

1. The *phase space* of the molecular system is divided into parts which correspond to the reactants, products and intermediate species;
2. These parts of phase space are separated by critical surfaces. The region of phase space which corresponds to the location of the system on the critical surface is called an activated complex;
3. Concentration of the activated complex is assumed to be at equilibrium and can be calculated;
4. The rate of the elementary process coincides with the flux from the region of reactants into the region of products.

Using all these assumptions, the rate constant of the chemical reaction can take the following form:

$$k = \chi \frac{kT}{h} \frac{Z^{\neq}}{Z} - \exp(-E_a/kT).$$

This is the basic formula of the transitional states method; Z and $Z^{\neq}$ are the partition functions for the initial molecular system and transitional state, respectively. The transmission factor χ in a classical case cannot be greater than 1. Since the calculation of the value of χ is a rather difficult problem, the critical surface is usually chosen so that χ is equal to 1.

Chemical reactions are usually classified in correspondence with the number of molecules participating in each elementary chemical process. The reaction in which only one molecule undergoes chemical transformation is called monomolecular. The most typical transformation of this kind is a reaction of thermal dissociation. If the pressure is high enough, the reaction will be of the first order; if the pressure is low, the order of the reaction is close to two. Bimolecular reactions form the most common class of chemical reactions. In these reactions, both colliding particles undergo chemical transformation. Usually the reactions are of the second order. The simplest process of this class is a reaction of A + BC = AB + C. In trimolecular reactions, all three particles in collision undergo chemical transformation. Under normal conditions, the rate of such reactions is approximately two orders less than for bimolecular ones. Therefore, trimolecular reactions occur very rare. A classical example of this reaction is $2NO_2 + X_2 = 2NOX$ where X is the halogen or oxygen atom.

In open systems, the rate of concentration change of a substance is not equal to the rate of the chemical reaction, and **mass transfer** should been taken into account.

References

Benson, S. W. (1976) *Thermochemical Kinetics.* 2nd ed. John Wiley and Sons, Inc.

Daniels, F. and Alberty, R. A. (1975) *Physical Chemistry.* 4th ed. John Wiley and Sons, Inc.

***Leading to:** Chemical equilibrium; Catalysis*

V. Yungman

CHEMICAL REACTION FOULING (see Fouling)

CHEMICAL THEORIES, FOR CATALYSIS (see Catalysis)

CHEMICAL THERMODYNAMICS (see Thermodynamics)

CHEMISORPTION (see Adsorption)

CHEN CORRELATION (see Forced convective boiling)

CHEVRON SEPARATORS (see Vapour-liquid separation)

CHEZY FORMULA

***Following from:** Open channel flow*

Experimental measurements suggest that the frictional resistance experienced by a uniform flow is approximately proportional to the square of the volumetric flow rate, provided the velocity of the flow is not too small. If u defines the mean velocity and m the mean hydraulic depth, the **Chezy formula** gives:

$$u = C\sqrt{mi}$$

where i is the gradient of the total head line and C is the Chezy coefficient. In the context of open channel flow, the Chezy formula may be written as:

$$u = C\sqrt{ms}$$

where the bed slope (s) is equivalent to i for steady uniform flows.

Although the Chezy formula was originally defined empirically, it is directly related to the *friction factor* (f). The Chezy coefficient is thus dependent upon surface roughness, the **Reynolds number**, and the hydraulic mean depth. Tabulated values of C and the related coefficient n (usually referred to as *Mannings coefficient*) are given by Chow (1959).

References

Chow, V. T. (1959) *Open-Channel Hydraulics.* McGraw-Hill Inc., New York.

Leading to: *Friction factors for single phase flow in smooth and rough tubes*

C. Swan

CHF, CRITICAL HEAT FLUX (see Burnout, forced convection; Post dryout heat transfer)

CHF CORRELATIONS (see Rod bundles, heat transfer in)

CHILTON-COLBURN ANALOGY (see Condensation of multi-component vapours; Reynolds analogy)

CHIMNEY PLUMES (see Plumes)

CHIMNEYS

Following from: *Combustion; Air pollution*

Environmental Considerations

The potential for damage from flue gas emissions can be avoided by dispersion and dilution from a tall chimney.

Pollutants emitted from a chimney must first clear the area of turbulent air created by the wind around the chimney top and any turbulent areas caused by the wind over adjacent structures.

The function of the chimney is to discharge *flue gases* to the atmosphere at such a height and velocity that the concentration of *pollutants,* such as sulphur dioxide, is kept within acceptable limits at ground level.

After leaving the top of the chimney, the gases are carried higher by their own buoyancy compared to the surrounding air and the momentum of the flue gases emitted.

For large power station plants, flue gas is typically at about 120°C for a coal-fired plant and 150°C for oil-fired applications, and contains appreciable quantities of SO_2 for oil and SO_2 and HCl for coal. Although these efflux temperatures are set to avoid dew point acid condensation within the duct work, downwash within the flue and heat loss from the stack cause acid deposition within the flues themselves.

Determination of Chimney Heights

For a new plant, chimney heights are determined from a mathematical model calculating ground level concentration of pertinent contaminants, usually SO_2 or NO_x. The function of the chimney is to reduce the resulting ground level concentration of each constituent of the emission so that it is below the threshold which gives rise to a health hazard or nuisance.

The calculation of this concentration is technically complex and takes into account the topography surrounding the power station, the local meteorology, the presence of tall buildings and other emission sources. In some cases, wind tunnel investigations are required.

In many applications, the height may be estimated with reference to empirical formulae dependant on the height of surrounding buildings and levels of emissions.

Under ideal conditions, and so far as gases are concerned, the ground level concentrations from a chimney discharge vary directly with the mass rate of emission, and inversely with wind speed and the square of the effective chimney height. Maximum ground level concentrations occur at about 10 to 15 times the effective chimneys' height distance down wind. For large power station chimneys in the UK, the design efflux velocity should not be less than 15 m/s at maximum continuous rating.

Design and Construction

Brickwork has, in the past, made a suitable structure for free-standing chimneys up to about 60 m high. For taller chimneys, the overturning moment due to increased wind load can be more economically resisted by a reinforced concrete shaft. Due to the need to deal with acid condensates within the flue, it is necessary to provide a lining to protect the concrete shaft internally from heat and acid attack. This lining is most often constructed from free-standing, acid-resisting brickwork about 100 mm thick. In some cases a separate concrete lining is provided, itself protected with a suitable chemical liner, for example, a synthetic resin. An acid brick lining is self-supporting up to a maximum height of 10 m. Consequently, the lining is built as a series of truncated cones carried on corbels inside the concrete windshield at 10 m intervals.

There is typically a cavity 50 mm wide between the concrete shaft and the brickwork lining, which may be filled with an insulating material or left as an air gap.

The lining is usually specified as dense acid-resisting brick laid in potassium silicate mortar.

The basic parameters for flue design are the height of the flue, the temperature, the efflux velocity and the rate of emission of the gases. The diameter at the top of the flue will be determined from the rate of emission and the efflux velocity; the latter being kept as high as practicable to minimise downwash and to enhance dispersion. In order to maintain velocities as high as possible, it is common to provide multiflues within a common windshield on larger installations.

The pressure head which causes the flow of gases up the flue is the result of the difference in density between the flue gases and external atmosphere. It is good practice to maintain a slight negative pressure inside the flue to reduce gas leakage; therefore, a balance must be maintained between the head available through density difference and the losses at entry and exit and by friction in the flue.

In the UK, multiflue chimneys for large power stations have been constructed up to 260 m in height. For a 2000 MW station of 4 × 500 MW units, flues about 200 m high, 6 m in diameter and efflux velocities of 23 m/s have been constructed.

References

National Society for Clean Air and Environmental Protection, 1994 Pollution Handbooks.

Modern Power Station Practice. Third Edition.

British Electricity International, London, Pergamon Press.

J.R. Cooper

CHLOR-ALKALI ELECTROLYSIS (see Electrolysis)

CHLORINE

Chlorine—(Gr. chloros, greenish yellow), Cl; atomic weight 35.453; atomic number 17; metric point −100.98°C; boiling point −34.6°C; density 3.214 g/l; specific gravity 1.56 (−33.6°C); valence 1, 3, 5, or 7. Discovered in 1774 by Scheele, who thought it contained oxygen; named in 1810 by Davy, who insisted it was an element. In nature, it is found in the combined state only, chiefly with sodium as common salt (NaCl), *carnallite* ($KMgCl_3 \cdot 6H_2O$) and *sylvite* (KCl). It is a member of the halogen (salt-forming) group of elements and is obtained from chlorides by the action of oxidizing agents and more often by electrolysis; it is a greenish-yellow gas, combining directly with nearly all elements. At 10°C one volume of water dissolves 3.10 volumes of chlorine, at 30°C only 1.77 volumes.

Chlorine is widely-used in making many everyday products. It is used for producing safe drinking water the world over. Even the smallest water supplies are now usually chlorinated. It is also extensively used in the production of paper products, dyestuffs, textiles, petroleum products, medicines, antiseptics, insecticides, foodstuffs, solvents, paints, plastics and many other consumer products. Most of the chlorine produced is used in the manufacture of chlorinated compounds of sanitation, pulp bleaching, disinfectants, and textile processing. Further use is in the manufacture of chlorates, chloroform, carbon tetrachloride, and in the extraction of bromine.

Organic chemistry demands much from chlorine, both as an oxidizing agent and in substitution, since it often brings desired properties in an organic compound when substituted for hydrogen, as in one form of synthetic rubber.

Chlorine is a respiratory irritant. The gas irritates the mucous membranes and the liquid burns the skin. As little as 3.5 ppm can be detected as an odor, and 1000 ppm is likely to be fatal after a few deep breaths. It was used as a war gas in 1915. Exposure to chlorine should not exceed 1 ppm (8-hr time-weighted average—40 hr week.)

Handbook of Chemistry and Physics, CRC Press

CHLOROFLUOROCARBON, CFC (see Heat pumps; Methylchloride; Refrigeration)

CHLOROFORM

Chloroform ($CHCl_3$) is a colourless, sweet-smelling, moderately volatile liquid that is slightly soluble in water. It is produced by the chlorination of methyl chloride. It is moderately toxic and narcotic. Prolonged inhalation can be fatal. Its usage has been dramatically decreased in the last 50 years and it is nowadays mainly used as a raw material in the production of HCFC-22.

Some of its physical characteristics are:

Molecular weight: 119.4
Melting point: 209.9 K
Normal boiling point: 334.5 K
Critical temperature: 536.4 K
Critical pressure: 5.470 MPa
Critical density: 498 kg/m^3

Reference

Beaton, C. F. and Hewitt, G. F. (1989) *Physical Property Data for the Design Engineer,* Hemisphere Publishing Corporation, New York.

V. Vesovic

CHOKED FLOW (see Compressible flow; Critical flow)

CHROMATIC DISPERSION (see Fibre optics)

Table 1. Chloroform; values of thermophysical properties of the saturated liquid and vapor

T_{sat}, K	334.5	360	380	400	420	440	460	480	505	536.4
p_{sat}, kPa	101.3	221	374	599	914	1321	1893	2598	3725	5470
ρ_l, kg/m^3	1415	1361	1333	1282	1248	1184	1114	1050	969	498
ρ_g, kg/m^3	4.50	9.33	15.3	23.4	36.1	53.0	77.3	109	158	498
h_l, kJ/kg	−108	−79	−61	−41	−17	4	27	45	67	136
h_g, kJ/kg	141	154	164	173	181	188	194	197	197	136
$\Delta h_{g,l}$, kJ/kg	249	233	225	214	198	184	167	152	130	
$c_{p,l}$, kJ/(kg K)	1.00	1.03	1.07	1.11	1.15	1.21	1.32	1.43	1.59	
$c_{p,g}$, kJ/(kg K)	0.60	0.63	0.66	0.69	0.73	0.79	0.87	1.00	1.28	
η_l, μNs/m^2	400	342	299	267	220	197	181	165	150	
η_g, μNs/m^2	11.2	12.2	13.0	13.7	14.6	15.5	16.5	17.9	19.8	
λ_l mW/(K m)	111	107	103	98.4	94.2	91.3	81.6	73.3	69.1	
λ_g mW/(K m)	8.71	9.74	10.6	11.5	12.5	13.5	13.6	15.9	17.5	
Pr_l	3.60	3.29	3.11	3.01	2.69	2.61	2.93	3.22	3.45	
Pr_g	0.77	0.79	0.81	0.81	0.85	0.91	1.06	1.13	1.45	
σ, mN/m	22.5	18.6	16.0	13.6	11.2	9.5	6.8	5.9	3.1	
$\beta_{e,l}$, kK^{-1}	1.38	1.58	1.78	2.03	2.36	2.82	3.54	4.63	6.86	

From reference [1] with permission.

CHROMATOGRAPHY

Following from: *Mass transfer*

Chromatography is a physico-chemical differential-migration method of separation, analysis and investigation of substances. Credit for discovering the chromatographic method belongs to Russian botanist M.S. Tsvet (1903). According to him, the prerequisite for separating components of a mixture is a different kind of adsorption. Modern chromatography employs separation of substances, apart from molecular adsorption, and other physico-chemical phenomena.

The term "chromatography" refers to all kinds of separation based on the division of components of the analyzed mixture between an eluent phase that can be a gas (vapor) or a liquid, and an eluate phase that can be a solid or a liquid. Different equilibrium and kinetic distribution of substances between them are prerequisites of chromatographic separation.

Depending on the phases of the substances involved, a distinction is made between gas and liquid chromatography. **Gas chromatography** includes gas-adsorption (gas-solid) and gas-liquid chromatography. Liquid chromatography is subdivided into liquid-liquid, liquid-solid and liquid-gel chromatography. The former characterizes the phase state of the eluent; the second, that of the eluate.

Chromatography is classified by the separation mechanism, that is the nature of interaction between a sorbent (an absorbing substance) and a sorbate (a substance being absorbed) as follows

1. Adsorption chromatography, in which separation is based on the different adsorptive properties of mixture components separated by a solid adsorbent;
2. Partition chromatography, separation based on the different solubility of substances separated in eluates (gas chromatography) as well as in liquid eluents and eluates;
3. **Ion exchange** chromatography, in which the different capabilities of mixture components in ion exchange are used as basis for separation;
4. Penetration chromatography, in which separation occurs due to the different molecular size and shape of separated substances, e.g., in molecular sieves.

Other types include precipitation chromatography (separation based on precipitation of substances with different solubility with a sorbent), adsorption and chelation chromatography (separation based on formation of coordination compounds of different strength on an adsorbent surface). Thus, as a rule, separation of substances proceeds by several mechanisms.

Depending on the geometry of an eluate's sorption layer and correspondingly, the equipment used, chromatography is subdivided into column and one-dimensional chromatography. The latter includes paper chromatography, in which separation of substances takes place on a special paper and *thin-layer chromatography*, in which separation is by virtue of a thin layer of sorbent. *Column chromatography* is used to separate a mixture in special columns, as in *capillary chromatography.* Column and thin-layer chromatography employ any separation mechanism, while paper chromatography uses partition and ion exchange mechanisms.

Chromatography is performed using special devices known as chromatographs which have, as main components, a chromatographic column and a detector (**Figure 1**). **Thermal conductivity**, electron capture, spectral (flame-photometric), density and other types of detectors are usually used. When the sample is fed, the mixture of substances is at the inlet to the chromatograph column. Eluent flow makes the mixture components migrate at different velocities due to the different adsorbing capabilities of the adsorbent surface. A detector at the column outlet continuously determines the concentration of separated compounds in the eluent. The detector signal, as a rule, is self-recorded and a chromatogram is obtained (**Figures 2 and 3**).

The three techniques for obtaining chromatograms are: frontal, elution and displacement. When the frontal technique is used, the

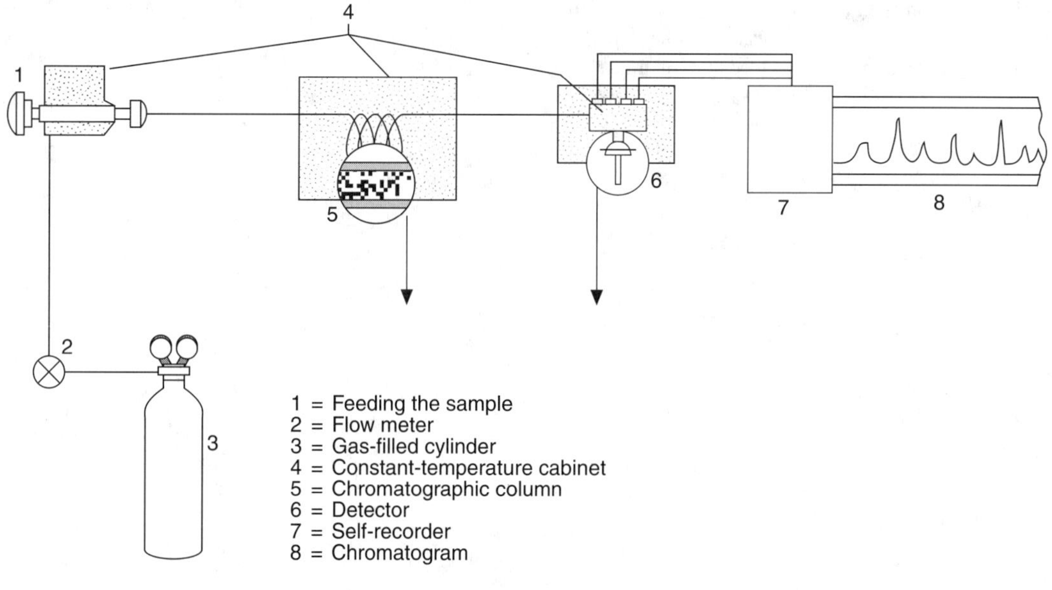

Figure 1. Schematic diagram of chromatography system.

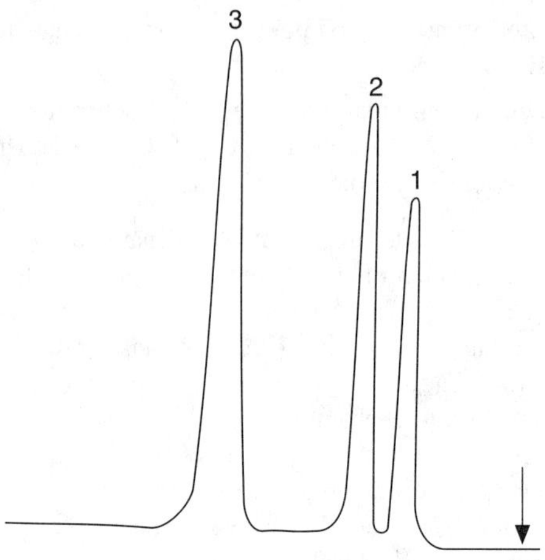

Figure 2. Typical chromatogram.

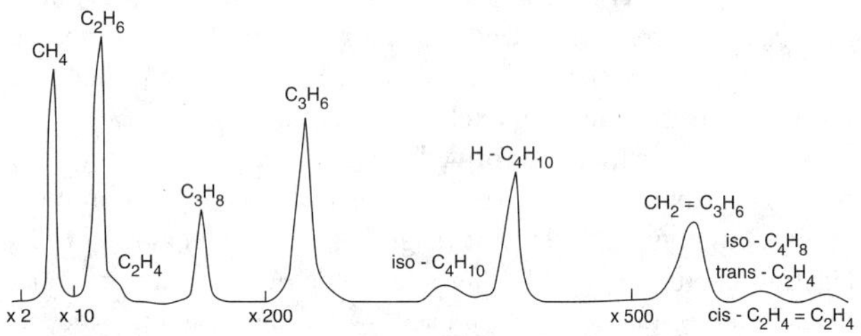

Figure 3. Typical chromatography.

eluent containing several components is continuously fed to the column. The substance derived at the column outlet is known as an effluent. For instance, if the eluent contains three substances B, C and D with the sorption ability order B < C < D, then separation in the column and on the chromatogram can be represented schematically (**Figure 4**).

Elution chromatography is the most commonly-encountered. The chromatograph column is first washed by the eluent flow the eluent having lower sorption power than any of the separated substances. Then the analyzed mixture is added at regular intervals

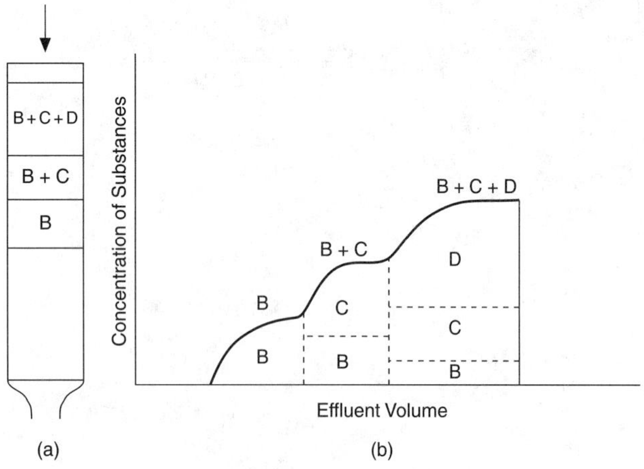

Figure 4. Schematic representation of frontal chromatography.

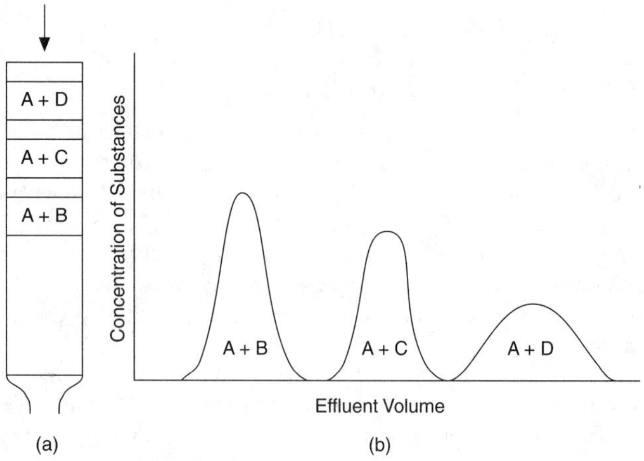

Figure 5. Schematic diagram of elution chromatography.

into the eluent flow, the mixture in the column being separated into components with eluent zones between them. For instance, a mixture contains substances B, C and D with the sorption ability order B < C < D and the substance A is an eluent, its sorption ability being lower than that of B, i.e., A < B < C < D. Separation of substances in the column and on the chromatogram is shown in **Figure 5**.

When *displacement chromatography* is used, a small amount of mixture is fed to the column after which a displacer that is adsorbed better than any other mixture components is continuously passed through it. This results in the appearance of adjoining zones of separated substances.

Figure 6 schematically shows *displacement chromatography* where B, C and D are the separated substances; E is the displacer; and B < C < D < E, the sorption ability order. The frontal and displacement techniques require regeneration of the column before the next experiment is started.

The main parameters governing chromatography are retention, efficiency and separation degree. They can be determined from a chromatogram. One of the basic characteristics of substance

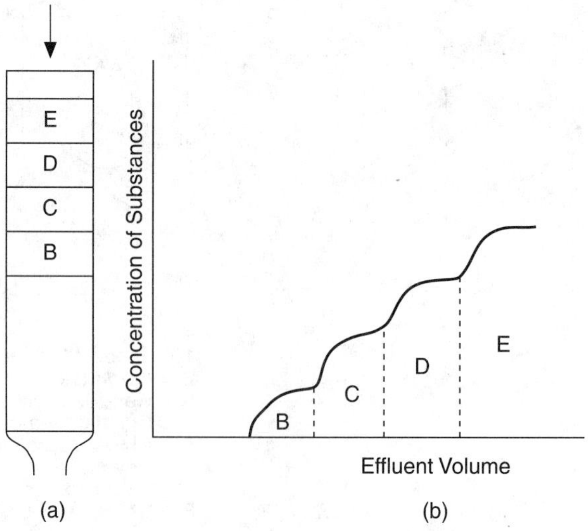

Figure 6. Schematic diagram of displacement chromatography.

separation is the retention time of a given component t_R (**Figure 7**), that is, the time elapsed from the moment the sample is fed to the column up to the moment the chromatographic zone of the substance output reaches its maximum.

An important factor in chromatography for a given substance is retained volume V_R, which is determined by the equation $V_R = \dot{V}t_R$, where $\dot{V}$ is the volume velocity of the flow. Peak sharpness depends on column efficiency, and the distance between the maxima is determined by its selectivity. Column efficiency is understood as a restricted washout of the chromatographic zone in the column. Efficiency can be expressed quantitatively by the number of "thermal trays" N into which the chromatographic column is divided

$$N = \left(\frac{t_R}{\sigma}\right)^2,$$

where σ is the standard deflection of the peak ($\sigma = L/4$). The length H of column in which an equilibrium between the substance concentrations in the eluent and the eluate is established is called the height of the equivalent theoretical tray $H = L/N$, where L is the column length. The overall height H of the tray is made up of components due to the nonhomogeneous flow of the eluent H_p, molecular diffusion H_d, mass exchange in the eluent H_m and in the eluate H_s. Thus, $H = H_p + H_d + H_m + H_s$.

Selectivity is a common measure of substance separation during chromatography and a measure of relative retention. Selectivity is determined by:

$$a = \frac{t_{R_2} - t_0}{t_{R_1} - t_0} = \frac{D_2}{D_1},$$

where D_2 and D_1 are the separation coefficients of the second and the first mixture components, respectively. $D = C_s/C_m$, where C_s and C_m are the substance concentrations in eluates and eluents, respectively. Thus, the separation of mixture components is a function of D, V_s, H, and L.

A high rate and a high degree of separation make it possible to apply chromatographic techniques for qualitative and quantitative analyses of organic and inorganic compounds in scientific research and industrial processes. The main disadvantage of chromatographic techniques is the cyclic character of the analysis.

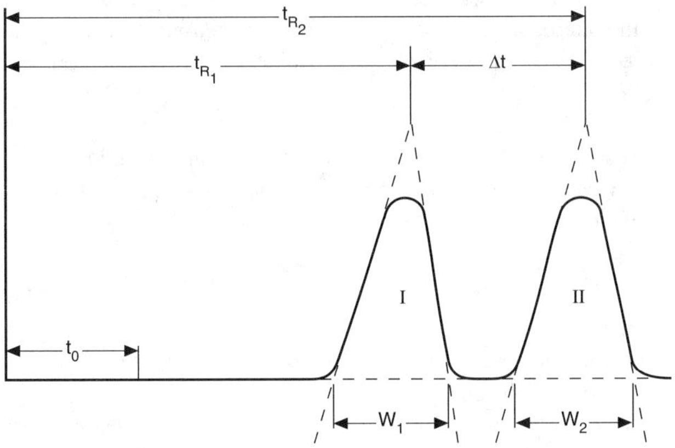

Figure 7. Retention time in chromatography

Chromatography is most frequently used for determining the concentrations and physico-chemical properties of substances and for deriving high-purity chemical elements and complex compounds. Chromatographic techniques are widely-used in controlling and automating production processes in the fields of environmental monitoring, agriculture, medicine, geology, etc.

References

Perry, S., Amos, R., and Brewer, P. (1972) *Practical Liquid Chromatography,* New York, London.

Belyaevskaya, T. A., Bol'shova, T. A., and Brykina, G. D. (1975) *Chromatography of Inorganic Substances,* Van Nostrand Reinhold, New York.

Heftmann, E. (ed.) (1975) *Chromatography, a Laboratory Handbook of Chromatographic and Electrophoretic Methods,* Van Nostrand Reinhold, New York.

V.A. Ibragimov

CHUGGING INSTABILITIES (see Instability, two-phase)

CHURN FLOW

Churn flow, also referred to as *froth flow* and *semiannular flow* is a highly disturbed flow of gas and liquid. It is characterised by the presence of a very thick and unstable liquid film, with the liquid often oscillating up and down. As an established flow regime, it appears only in vertical and near-vertical tubes and is usually bounded by the slug and the annular flow regimes. In contrast to these two regimes, both of which have a well-defined structure, churn flow appears chaotic and is one of the least understood of gas-liquid flow regimes. Due to its complexity, it is often thought merely as a transitional regime (Dukler & Taitel, 1986), but recent experimental evidence shows **Flooding** of the film (and the consequent entrainment of large chunks of liquid and their subsequent deposition) as one of its main features. Conditions in which churn flow occurs and some correlations for determining the pressure drop and the holdup in this flow regime are discussed below.

There are three bounds to the churn flow regime, these are, (1) the slug-churn transition; (2) the churn-annular transition; and (3) the maximum liquid flow rate limit above which churn flow does not occur. Of these, the slug-churn transition has received the most attention and is subject to some controversy. The major schools of thought on this have been reviewed recently by Jayanti & Hewitt (1992). Recent evidence suggests that **Slug Flow** is destabilised when the (downcoming) liquid film surrounding the Taylor bubble is flooded by the gas flowing upwards. Jayanti & Hewitt (1992) have proposed a model based on this mechanism, which may used to determine the condition for the slug-churn transition. The conditions for the flooding of the film in the Taylor bubble in slug flow and the subsequent transition to churn flow are described by Equation (Eq.) [1]:

$$\sqrt{U_{bs}^*} + m\sqrt{U_{fs}^*} = 1 \qquad (1)$$

where U_{bs}^* and U_{fs}^* are the nondimensionalised velocities of the

Taylor bubble and the film surrounding it, respectively, and m is a constant which depends on the length and the diameter of the test section. Further details of the model can be found in Jayanti & Hewitt (1992).

The churn-annular transition is often taken as the point at which the film ceases to flow downwards even intermittently, and is therefore associated with the *flow reversal* point in counter-current flow (Wallis, 1969). This criterion is expressed in the form of an upper limit of gas velocity above which churn flow does not occur:

$$U^*_{gs} = U_{gs}\sqrt{\frac{\rho_g gD}{\rho_l - \rho_g}} = 1 \qquad (2)$$

where U_{gs} is the gas superficial velocity; ρ, the density; g, the acceleration due to gravity; and D, the pipe diameter.

There appears to be an upper limit on the liquid velocity also above which churn flow does not occur (Govier & Aziz, 1972). This can be expressed roughly as:

$$U^*_{ls} = U_{ls}\sqrt{\frac{\rho_l gD}{\rho_l - \rho_g}} = 1. \qquad (3)$$

Understandably, the frictional *pressure gradient* is very high in churn flow, reflecting the violent interaction between the gas and the liquid phases. The nature of this interaction is not well-understood and can only be represented in the form of empirical correlations. There are, however, few data for pressure drop and holdup in churn flow. The correlations and methods suggested here are based on a small number of data, notably, of Bharathan (1978) and Govan et al. (1991). The interfacial friction factor, f_i, can be correlated as a function of the void fraction, ε_g, as follows:

$$f_i = \frac{1}{2}(f_{iB} + f_{iW}) \quad \text{where} \quad f_i = \frac{2\tau_i\varepsilon_g^2}{\rho_g U_{gs}^2} \qquad (4)$$

and

$$f_{iB} = 0.005 + 24(1 - \varepsilon_g)^2$$

$$f_{iW} = 0.005 + 0.375(1 - \varepsilon_g).$$

The pressure gradient, dp/dz, and the liquid holdup, $\varepsilon = (1 - \varepsilon_g)$, can then be calculated by solving the following force balance equations on the inner core and on the pipe:

$$-\frac{dp}{dz} = \frac{4\tau_i}{D\sqrt{\varepsilon_g}} + \rho_g\varepsilon_g g \qquad (5)$$

$$-\frac{dp}{dz} = \frac{4\tau_w}{D} + g[\rho_l\varepsilon_l + \rho_g\varepsilon_g] \qquad (6)$$

where the wall shear stress, τ_w, is given by:

$$\tau_w = \frac{1}{2} f_l \rho_l \left(\frac{U_{ls}}{\varepsilon_l}\right)^2 \qquad (7)$$

and

$$f_l = 16Re_{ls}^{-1} \qquad \text{if } Re_{ls} < 2100$$

$$f_l = 0.079Re_{ls}^{-0.25} \qquad \text{if } Re_{ls} > 2100$$

where Re_{ls}, the liquid **Reynolds number** is based on the pipe diameter and the superficial liquid velocity.

References

Bharathan, D. (1978) *Air-Water Counter-Current Annular Flow in Vertical Tubes,* EPRI Report no. EPRI NP-786.

Dukler, A. E. and Taitel, Y. (1986) *Flow Pattern Transition in Gas-Liquid Systems: Measurement and Modelling,* in Multiphase Science and Technology, 2, 1–94, 1986, ed. G. F. Hewitt, J. M. Delhaye & N. Zuber, Hemisphere Publishing Corp.

Govan, A. H., Hewitt, G. F., Richter, H. J. and Scott, A. (1991) *Flooding and Churn Flow in Vertical Pipes,* Int. J. Multiphase Flow, 17, 27–44.

Govier G. W. & Aziz K. (1972) *The Flow of Complex Mixtures in Pipes,* Van Nostrand Reinhold.

Jayanti, S. and Hewitt, G. F. (1992) *Prediction of the Slug-to-Churn Flow Transition in Vertical Two-Phase Flow,* Int. J. Multiphase Flow, 18, 847–860.

Wallis, G. B. (1969) One-Dimensional Two-Phase Flow, McGraw-Hill.

S. Jayanti

CIRCULATION RATIO (see Steam generators, nuclear)

CISE CORRELATIONS (see Void fraction)

CLADDING (see Advanced gas cooled reactor; Magnox power station)

CLAPEYRON-CLAUSIUS EQUATION

Following from: *Thermodynamics; Vapour pressure*

The Clapeyron-Clausius equation is a differential equation giving the interdependence of the pressure and temperature along the phase equilibrium curve of a pure substance. This equation was suggested by B. Clapeyron in 1834 and improved by R. Clausius in 1850.

According to the general conditions of thermodynamic equilibrium, when two phases of a pure substance are in equilibrium, the following equations are valid:

$$T_1 = T_2; \qquad p_1 = p_2; \qquad g_1 = g_2 \qquad (1)$$

where the subscripts 1 and 2 refer to the respective phases; g is the specific Gibbs energy.

Since the specific Gibbs energy is a function of temperature and pressure, it follows that there is an additional interdependence between temperature and pressure:

$$g_1(p, T) = g_2(p, T). \tag{2}$$

This interdependence can not be expressed in an explicit form because the specific Gibbs energy is defined in thermodynamics only in terms of arbitrary constants included in the enthalpy and entropy expressions. Therefore, the interdependence between T and p is considered in a differential form.

Equation [2] has to be obeyed in every state of phase equilibrium, which means that the following equation has to be valid:

$$dg_1 = dg_2 \tag{3}$$

where dg is the change of g along the phase equilibrium curve when both temperature and pressure are changed. Therefore:

$$dg = \left(\frac{\partial g}{\partial T}\right)_p dT + \left(\frac{\partial g}{\partial P}\right)_T dp. \tag{4}$$

From thermodynamics, it is known that:

$$\left(\frac{\partial g}{\partial T}\right)_p = -s; \qquad \left(\frac{\partial g}{\partial P}\right)_T = v$$

where s and v are, respectively, the specific entropy and specific volume of the phase.

As a result, the equilibrium condition can be written as follows:

$$-s_1 dT + v_1 dp = -s_2 dT + v_2 dp$$

or

$$\frac{dp}{dT} = \frac{s_2 - s_1}{v_2 - v_1}. \tag{5}$$

Equation (5) is the initial form of the Clapeyron-Clausius equation.

Proceeding from the fact that an equilibrium phase transition in a pure substance is an isothermal and isobaric one, it can be concluded:

$$s_2 - s_1 = \frac{h_{12}}{T} = \frac{h_2 - h_1}{T} \tag{6}$$

where h_{12} is the specific enthalpy of phase transition from Phase 1 to Phase 2 and h_2 and h_1 are the respective specific enthalpies of the phases.

According to Eq. (6), the Clapeyron-Clausus equation for different phase equilibria can be written:

- **solid-vapor equilibrium**

$$\frac{dp}{dT} = \frac{h_{SG}}{T(v_G - v_S)}, \tag{7}$$

- **solid-liquid equilibrium**

$$\frac{dp}{dT} = \frac{h_{LS}}{T(v_L - v_S)}; \tag{8}$$

- **liquid-vapor equilibrium**

$$\frac{dp}{dT} = \frac{h_{LG}}{T(v_G - v_L)}. \tag{9}$$

Here h_{SG}, h_{LS} and h_{LG} are respectively the heats of sublimation, melting and evaporation. The subscripts G, S and L—refer respectively to the vapor, solid and liquid phases.

To integrate the Clapeyron-Clausius equation, it is in general necessary to know the explicit temperature and pressure relations for the enthalpy of phase transition and specific volume. As a rule, the respective functions are complicated and unknown; but in some particular cases, the integration can be carried out.

At moderate pressures, the vapor specific volume is several orders of magnitude greater than the liquid or solid specific volume. It is therefore possible to neglect the values v_S and v_L in Eqs. (7) and (9). At the same time, the vapor's specific volume can, with reasonable accuracy, be derived from the perfect gas equation of state:

$$v_G = \frac{RT}{p}.$$

Using these assumptions, Eqs. (7) and (9) can be written as:

$$\frac{d \ln p}{dT} = \frac{h_{SG}}{RT^2}; \tag{10}$$

$$\frac{d \ln p}{dT} = \frac{h_{LG}}{RT^2}. \tag{11}$$

As a first approximation, the heats of phase transition h_{SG} and h_{LG} at moderate pressures can be regarded as constants. With this assumption, Eqs. (10) and (11) are easily integrated:

$$\ln p = -\frac{h_{SG}}{RT} + C_1; \tag{12}$$

$$\ln p = -\frac{h_{LG}}{RT} + C_2. \tag{13}$$

The integration constants C_1 and C_2 can be found when the temperature and pressure in the triple point T_{tr}, P_{tr} and the normal boiling temperature $T_{n \cdot b}$ are known. Equations (12) and (13) are then expressed as:

$$\ln \frac{p}{p_{tr}} = -\frac{h_{SG}}{R}\left(\frac{1}{T} - \frac{1}{T_{tr}}\right); \tag{14}$$

$$\ln p = -\frac{h_{LG}}{R}\left(\frac{1}{T_{n.b}} - \frac{1}{T}\right). \tag{15}$$

In a semi-logarithmic plot, both equations are presented as

segments of straight lines with slopes h_{SG}/R and h_{LG}/R having an intersection in the triple point.

Since the specific volumes v_L and v_S are of the same order, the above approach is invalid for the solid-liquid equilibrium. Moreover, there exist substances (for instance, water) having $v_S > v_L$. In this case, $dp/dT < 0$ i.e., with growing pressure the melting temperature for such substances decreases.

E.E. Shpilrain

CLAPEYRON EQUATION (see Latent heat of vaporisation)

CLARIFICATION (see Liquid-solid separation; Sedimentation)

CLARIFIERS (see Decanters)

CLASSIFICATION OF HEAT EXCHANGERS (see Heat exchangers)

CLASSIFIERS

***Following from:** Gas-solids separation*

Classification is a process of dividing a particle-laden gas stream into two, ideally at a particular particle size, known as the cut size. An important industrial application of classifiers is to reduce overgrinding in a mill by separating the grinding zone output into fine and coarse fractions. Many types of classifier are available, which can be categorised according to their operating principles. A distinction must be made between gas cleaning equipment, in which the aim is the removal of all solids from the gas stream, and classifiers in which a partition of the particle size distribution is sought. Prasher (1987) identifies the following categories: a) *screens,* b) cross-flow systems, c) *elutriation,* d) inertia systems, e) centrifugal systems without moving parts, f) centrifugal systems with rotating walls, and g) mechanical rotor systems. A classification process may combine these alternative principles, sometimes within a single separator, to achieve a desired result.

Screens

These contain apertures which are uniformly-sized and spaced, and which may have circular, square or rectangular shapes. Particles which are smaller than the aperture in at least two dimensions pass through, and larger ones are retained on the surface. The screen is shaken or vibrated to assist motion of particles to the surface, and continuous screens are often tilted to further aid particle bed motion along the screen surface. Static (or low-frequency) screens or grizzlies have a different construction. They are comprised of parallel bars or rods with uniformly clear openings, often tapered from feed to discharge ends. The bars may lie horizontally above a bin, or be inclined to provide the feed to a crusher.

Cross-flow systems

It is in principle possible to winnow out fines from a falling curtain of material of constant density by a cross-current of air. In practice, humidity of the air (and moisture on the particles) leads to blockage of the narrow ducts necessary to give a thin enough falling curtain for winnowing. It is possible to winnow thin flakes; Etkin et al. (1980) has successfully classified mica particles with an aspect ratio greater than 30.

Elutriation

Gravity counter-current classifiers (elutriators) have been reviewed by Wessel (1962). A simple example, the Gonnell (1928) classifier, consists of a long vertical cylindrical tube with a conical transition zone located at the bottom end. Air flows up the tube, carrying with it the finer particles. The disadvantage of this and many other gravity counter-current classifiers is the presence of a laminar velocity profile in the gas, a large cone angle leading to *flow separation* and eddy formation, settling out of fines due to the retarded velocities near the walls, and the noise of vibrators necessary to prevent particle adhesion to the walls. Their advantage lies in the good dispersion of powders achieved in the cylindrical section. In the zig-zag classifier, vortex formation leads to the acceleration of the main flow owing to a reduction of the effective tube cross-section. Fines follow the main gas stream and coarse particles travel to the wall, and fall back against the main gas flow. In this design, the sharpness of cut is low at each stage (zig-zag), but a required cut size is generally achievable even at high velocities.

Inertia systems

In an inertial classifier, the particle-laden gas stream is turned through 180° by appropriate internal baffling. In order to reach the exit port, the gas passes through a further 180° to continue in the same direction it was travelling before it was diverted. The fines are able to follow, more or less, the same route as the gas. However, the momentum of coarser or denser particles prevents them from following the same trajectory and they fall into a collection zone after the first turn.

Centrifugal systems with no moving parts

The capacities of these types of classifiers cover a wide range. Generally, higher-capacity machines have a poorer sharpness of cut. Typical high-capacity industrial units are the *cone classifier* (often built into some types of mills) and the cyclone. The feed is given a high tangential velocity and is introduced near to the top of the unit. The gas flows in a spiralling fashion towards the bottom end where it experiences a flow reversal and passes up as a central core. In the cone classifier, the central core of gas actually flows in a reverse spiral up the wall of a central feed. Under the influence of centrifugal force, coarse particles are thrown to the inner wall of the cone or cyclone. Particles less than the cut size are carried up the central vortex and are carried out of the unit by the bulk of the gas flow. The diameter and position of the vortex finder at the top of the unit are critical in the determination of a specified cut size. Further information on cyclones is given in the overview article on **Gas-Solid Separation**.

The Larox classifier is another high capacity system, shown in **Figure 1**. The particles are dispersed by the feed falling across an

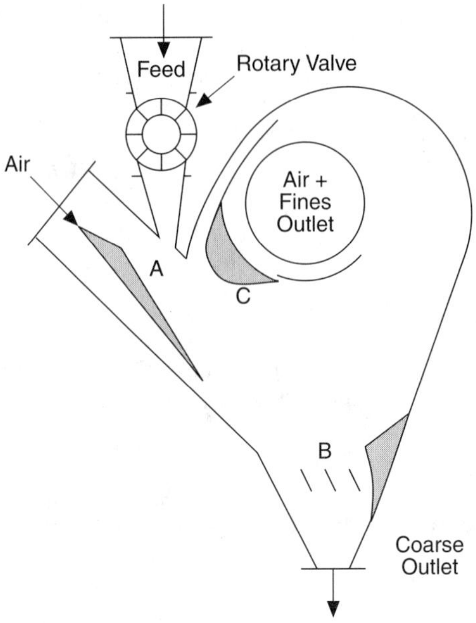

Figure 1. Larox classifier (Hukki, 1976).

inlet gas; the coarsest particles fall through the gas stream and into an outlet chute, and are thereby separated. Classification of the remainder occurs in a horizontal cyclone. There are three adjustable flights (A, B and C) to be positioned to give the best cut.

Centrifugal systems with rotating walls

Spiral classifiers, such as the Alpine Mikroplex design for separation in the superfine region, were developed to partially overcome undesirable boundary layer effects associated with spinning fluids at stationary walls (Rumpf and Leschonski (1967)). Air is introduced tangentially at the periphery into a flat cylindrical space and moves along spiral flow lines into the centre, from where it is drawn off. The fines follow the flow while the coarse particles spin round at the circumference; in some designs, this recirculating coarse stream is reclassified by passage of the incoming air through it. The coarse fraction leaves through a slit at the periphery (as in the Walther Classifier) or is removed using a screw extractor (as in the Alpine Mikroplex Classifier). The cut size theoretically has a stable circular trajectory in the classifying zone, but (in common with most other classifiers) separation is poorer with higher solids loadings.

Mechanical rotor systems

To extend effective separation over a wider range of operating parameters, many classifiers are designed with a mechanical rotor built into them. The rotor has several effects: 1) large particles are deflected back into the classifier, thereby reducing the proportion of coarse particles in the fine product, 2) it aids recirculation of the air stream in some classifier types, and 3) the generation of a *forced vortex* keeps large particles at the periphery, but fines follow a helical trajectory to the centre where they pass out with the exiting air.

References

Etkin, B., Haasz, A. A., Raimondo, S., and D'Eleuterio, G. M. T. (1980) *Air Classification of Thin Flakes,* Inst. Chem. Eng. Symp. Series 59, Dublin, 5:3/1–5:3/23.

Gonnell, H. W. (1928) *Ein Windsichtverfahren zur Bestimmung der Kornzusammensetzung staubförmiger Stoffe,* Z. VDI, 72, 945–950.

Prasher, C. L. (1987) *Crushing and Grinding Process Handbook,* Wiley, Chichester.

Rumpf, H. and Leschonski, K. (1967) *Prinzipien und neuere Verfahren der Windsichtung,* Chem. Ing. Techn., 39, 1231–1241.

Wessel, J. (1962) *Schwerkraft-Windsichter,* Aufbereitungstechnik, 3, 222–230.

R.J. Wakeman

CLAUSIUS (see Entropy)

CLAUSIUS-CLAPEYRON EQUATION (See Clapeyron-Clausius equation; Condensation of pure vapour; Heat pipes)

CLAUSIUS-MOSOTTI EQUATION (see Visualization of flow)

CLAUSIUS NUMBER

Clausius number, C1, is a dimensionless group which appears in the analysis of heat conduction in fluid flow. It is expressed as:

$$Cl = u^3 L\rho/\lambda\Delta T$$

where μ is the fluid velocity; ρ fluid density; λ, fluid thermal conductivity; Δ_T, temperature difference; and L is the appropriate linear dimension.

Clausius number represents the ratio of energy transport associated with fluid momentum to energy transfer by thermal conduction, i.e.,

$$Cl = (\rho u^2 \times u)/(\lambda\Delta T/L).$$

G.L. Shires

CLEANING TECHNIQUES, HEAT EXCHANGERS (see Fouling)

CLIMATISATION (see Air conditioning)

CLIMBING FILM EVAPORATOR (see Evaporators)

CLOSED CYCLE GAS TURBINE (see Gas turbine)

CLOSED CYCLE MHD GENERATORS (see MHD electrical power generators)

CLOSED SYSTEM (see Thermodynamics)

CLOSURE LAWS (see Conservation equations, two-phase)

CLOUD POINT SPECIFICATION (see Oils)

CNEN (see Comitato Nazionale per l'Energie Nucleare)

COAGULATION (see Liquid-solid separation; Sedimentation)

COAGULATION, OF AEROSOLS (see Aerosols)

COAGULATION, OF DROPS (see Drops)

COAL (see Fuels)

COAL BURNERS (see Pulverised coal furnaces)

COAL CARBONISATION (see Pyrolysis)

COAL COMBUSTION (see Fossil fuel fired boilers)

COAL GAS (see Pyrolysis)

COAL GASIFICATION (see Gasification)

COAL RESEARCH ESTABLISHMENT, CRE

Stoke Orchard
Cheltenham GL52 4RZ
UK
Tel: 01242 673361

COANDA EFFECT

***Following from:** External flows (overview)*

The Coanda effect describes the tendency of a jet to follow the contours of an adjacent boundary even when this boundary curves away from the initial jet axis. This effect either arises due to a) the pressure gradient perpendicular to a curved streamline, or b) differential entrainment and the development of a partial vacuum.

In the first case it may be shown using the *Euler equation* (or the **Bernoulli Equation** in an incompressible fluid) that a curved streamtube experiences a net force towards the centre of curvature (**Figure 1a**). Since no component of the viscous force acts perpendicular to a streamtube, this implies that $P_1 > P_2$. It is this pressure gradient that causes the jet to be deflected from its initial axis, and accounts for the so-called Coanda effect.

Alternatively, if a jet is discharged in the vicinity of a solid boundary (**Figure 1b**), the entrainment of fluid into the jet will be restricted on one side (A). This creates a partial vacuum so that $P_a < P_b$, and consequently the jet attaches to the flow boundary.

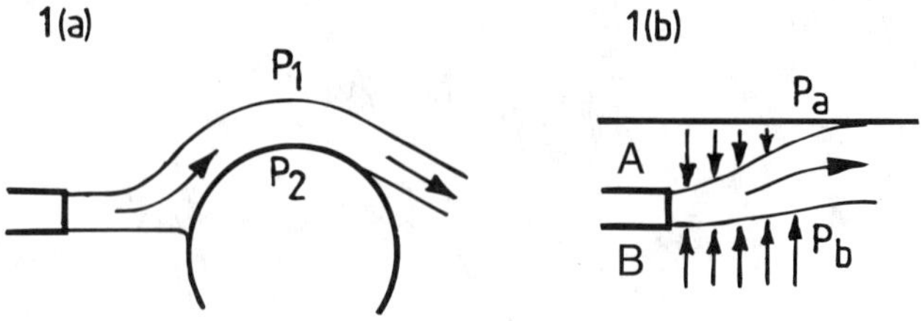

Figure 1. Coanda effect; (a) Streamline curvature, (b) restricted entrainment.

This effect is particularly apparent in 2-D jets. Schlichting (1968) provides further details of flows near solid boundaries.

Reference

Schlichting, H. (1968) *Boundary Layer Theory.* McGraw-Hill Inc., New York.

C. Swan

COEFFICIENT OF PERFORMANCE, COP (see Heat pumps; Refrigeration)

CO-GENERATION SYSTEMS (see Rankine cycle)

COHERENCE FUNCTION (see Spectral analysis)

COHERENCE, OF RADIATION (see Optics)

COHERENCE STRICTURES, IN TURBULENT FLOW (see Turbulent flow)

COHERENT SYSTEM OF UNITS (see Dimensional analysis)

COIL IN TANK (see Tank coils)

COILED TUBE BOILERS

The coiled tube boiler is a type of **Water Tube Boiler** in which a cylindrical furnace envelope is formed by multistart, helically-coiled tubes in which water is boiled. Firing is by oil or gas and the burner is provided usually in the base, although sometimes in the roof of the furnace. The output from the boiler is usually saturated steam, but superheat may be provided by returning the saturated steam to a separate coiled section concentrically-mounted within the furnace envelope. Typical arrangements of both types are shown in **Figure 1** and **Figure 2**.

Coiled tube boilers have also been used in nuclear applications, the most significant of which are the pod-boilers in the Hartlepool and Heysham **Advanced Gas-Cooled Reactor** (AGR) power stations in the UK. In these power stations, the boilers are of the **Once-Through Boiler** type and consist of concentric multistart helical coils, alternately handed around a central spine which houses the feed water tubes. Above the boiler bank, a coiled reheater is

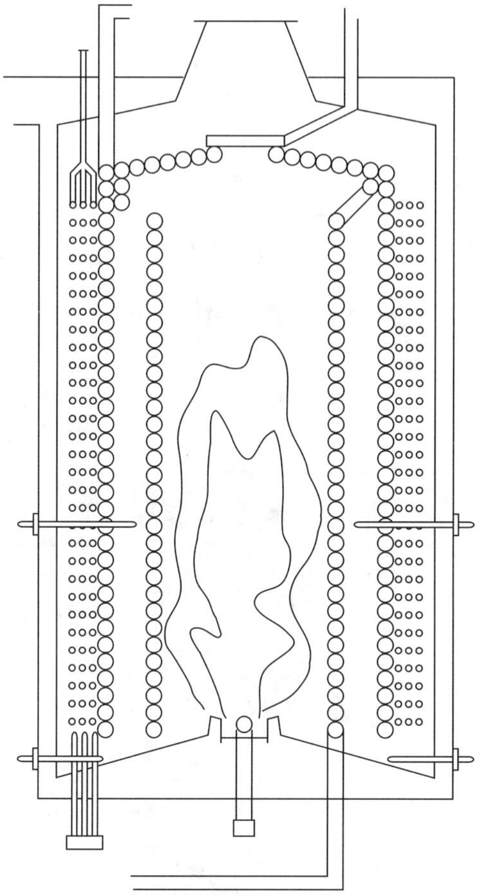

Figure 1. Typical arrangement of coiled tube boiler.

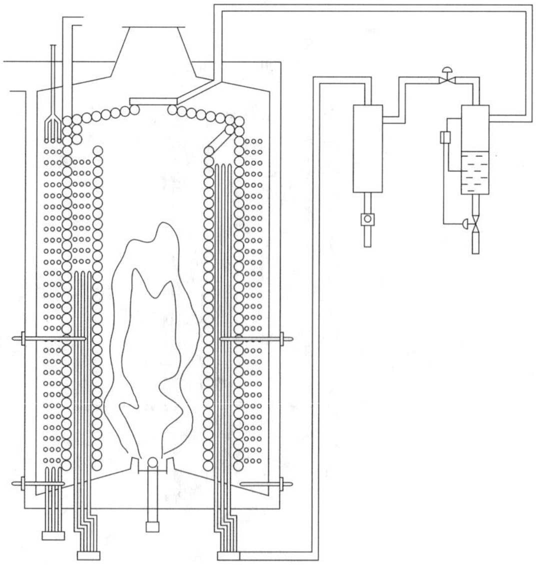

Figure 2. Typical arrangement of coiled tube boiler.

mounted. Hot gas from the reactor outlet is circulated down over the heating surfaces by high-powered blowers situated below the boilers. By using the pod-boiler concept, it has been possible to incorporate the units within the reactor's concrete pressure vessel. The general arrangement of the boilers is shown in **Figure 3**. (See Figure 3 on page 168).

A.E. Ruffell

COILED TUBE (FLOW AND PRESSURE DROP IN)

Following from: *Channel flow*

Vapor-generating surfaces in the form of a coiled tube have advantages in compactness, in compensation for thermal expansion and in giving enhanced heat transfer.

Inertial forces perpendicular to the axis of liquid flow act on the liquid in *curvilinear channels*, in particular in a coiled tube. A larger centrifugal force acts on the faster-moving fluid near the tube center than on the slower moving fluid near the wall. As a result, the fluid in the central part of the tube moves towards the outer generatrix while that near the wall moves towards the inner generatrix. *Secondary flow* arises in the form of a pair of symmetrical vortices in the cross-section (**Figure 1**); along the tube axis, the fluid trajectory is in the form of a double coil. The maximum axial flow velocity in a coiled tube takes place near the outer generatrix (**Figures 2 and 3**) the secondary flow velocity $V(r/r_o)$ is constant in the core, but changes near the wall.

In a coiled tube, the distribution of turbulent fluctuations is nonuniform. Near the outer generatrix, turbulence intensity is lowest.

The *friction factor* for a coiled tube with a relative diameter $(D_{coil}/D) > 4$ (D_{coil} is the coil diameter, D is the tube diameter) is calculated as:

$$f = f_0 K_\phi,$$

where f_0 is the *friction factor* in a straight tube with the same diameter, as determined from the **Reynolds Number** and K_ϕ is the coefficient of a coiled tube. The value of coefficient K_ϕ depends on the flow regime. In laminar flow, $K_\phi = 1$.

In curvilinear channels, turbulent flow begins at a higher Reynolds number Re given by

$$Re_c = \frac{2300}{1 - \left[1 - \left(\frac{D_0}{2000D}\right)^{0.4}\right]^{2.2}}.$$

and in turbulent flow, the coiled tube coefficient K_ϕ is given by the equation

$$K_\phi = 1 + 1.68\left(\frac{D}{D_{coil}}\right)^{0.65}$$

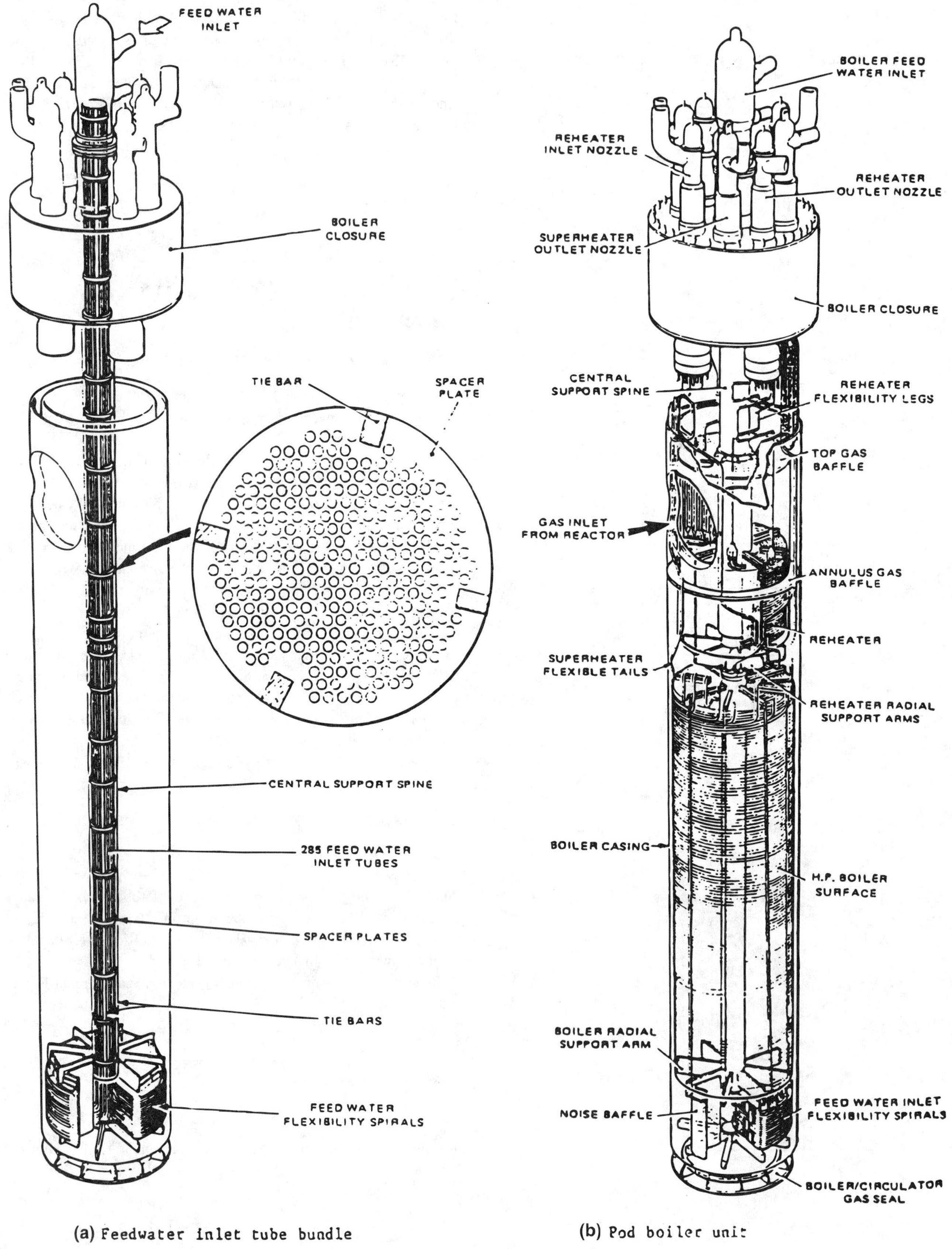

Figure 3. Coiled tube pod boilers used in advanced gas cooled reactor (Acme).

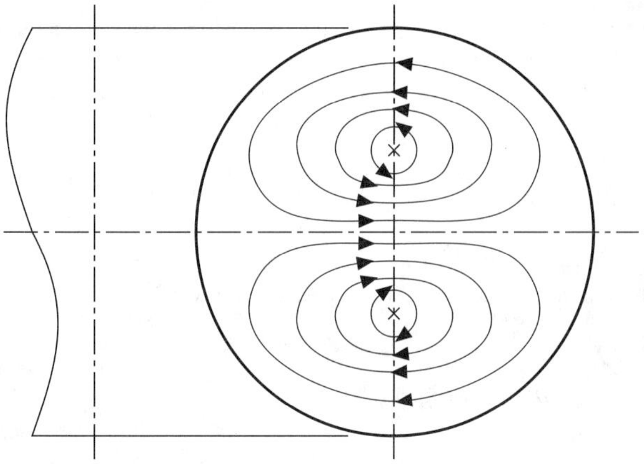

Figure 1. Secondary flow in a coiled tube.

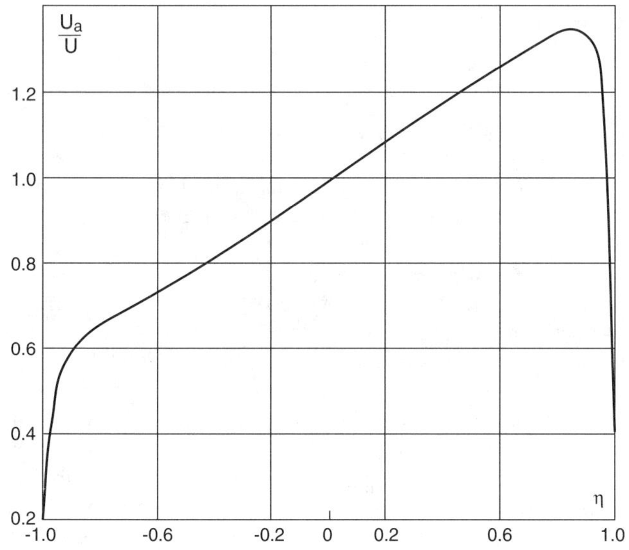

Figure 2. Typical variation of axial velocity Ua with radial position ($\zeta = r/r_0$ where r_0 is the radius of the tube and U is the velocity for $z = 0$).

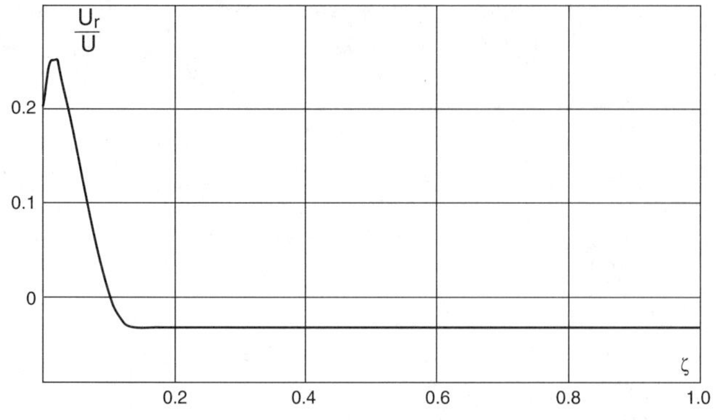

Figure 3. Typical variation of secondary velocity Ur with radial position.

In two-phase flow, the flow pattern in a coiled tube has peculiarities determined by the inertial force in the cross-section and by the secondary circulation. The centrifugal force promotes phase stratification and liquid transfer to the outer generatrix; secondary circulation in the gas phase leads to phase mixing and liquid shifts to the inner generatrix. These processes depend on mass velocity, pressure, ratio D_{coil}/D, and vapor content. A more detailed discussion is given by Hewitt and Jayanti (1992).

Pressure losses of a two-phase flow in a coiled tube are determined by the equation

$$\Delta p = \frac{2f_{LO}\, L\, \dot{m}^2}{D\rho_L}\left[1 + x\left(\frac{\rho_L}{\rho_G} - 1\right)\right]\psi$$

where f_{LO} is the Fanning friction factor (corrected as above for the coiled tube flow) for the total mass flux flowing with liquid properties, L the length of the tube, $\dot{m}$ the mass flux of the fluid in the tube, D the tube diameter, ρ_L and ρ_G the liquid and gas densities and x the quality. ψ is a factor which connects the equation for departures from homogeneous flow ($x = 1$ for homogeneous flow) (See **Pressure Drop, Two-Phase Flow**).

References

Hewitt, G. F. and Jayanti, S. (1992) Prediction of film inversion in two-phase flow in coiled tubes. J. Fluid Mech vol. 236, pp 497–511.

Schlichting, M. (1973) *Grenzchicht-Theorie.* Verlag G. Brawn. Karlsrube.

Heat Exchange Design Hand Book. v.1,2. Hemisphere Publishing Corporation.

Yu.A. Kuzma-Kichta

COILED TUBES (HEAT TRANSFER IN)

***Following from:** Tubes, single phase transfer in*

The heat transfer characteristics in coiled tubes are determined by the peculiarities of the axial velocity distribution and of secondary flow (see **Coiled Tube, Flow and Pressure Drop In**).

The mean heat transfer of coefficient for laminar flow in coiled tubes is given by the equation:

$$\overline{Nu} = 0.06\, Re^{0.7} Pr^{0.43} (Pr_f/Pr_w)^{0.25} \left(\frac{D}{D_{coil}}\right)^{0.18},$$

where $Re = \bar{u}D/\nu$, $\overline{Nu} = \bar{\alpha}D/\lambda$, D is the tube diameter, D_{coil} is the coil diameter, and $\bar{\alpha}$ is the mean heat transfer around the tube perimeter.

The **Nusselt Number** in turbulent flow of water and superheated vapor ($Re > Re_c$) in coiled tube, is

$$\overline{Nu} = Nu_0 F(D/D_{coil})$$

where Nu_0 is the Nusselt number in straight pipe which for $Pr = 0.7 - 2$, may be determined from the equation

$$Nu_0 = 0.023\, Re^{0.8} Pr^{0.4}.$$

The multiplier $F(D/D_{coil})$ can be approximated as:

$$F(D/D_{coil}) = 1 + 3.5D/D_{coil}$$

For turbulent flow in coiled tubes, the *heat transfer coefficient* distribution around the tube perimeter is essentially nonuniform. The nonuniformity is caused by the nonhomogeneity of the flow velocity and temperature distributions which can cause heat transfer from the inner to the outer generatrix. It is necessary to consider this when determining the heat transfer coefficient from surface temperature measurements. The Nusselt number distribution along the coiled tube perimeter for a vertical axis coil with $D_{coil}/D = 16$ and for water flow ($Re = 2 \times 10^4$) is presented in **Figure 1**. This shows that the heat transfer coefficient in the vicinity of the outer generatrix is approximately constant. Upon approaching the inner generatrix, the heat transfer coefficient decreases. The ratio of heat transfer coefficient for the outer and inner generatrices is approximately equal to 4. The heat transfer coefficient in the vicinity of the inner generatrix of a coiled tube coincides nearly with value α for a straight tube. The Nusselt number $\overline{Nu}$ for straight tube is shown by Line 2 on **Figure 1**. The mean Nusselt number $\overline{Nu}$ for a coiled tube is greater and the heat transfer coefficient increases with a decrease in the ratio D_{coil}/D.

The heat transfer coefficient α_{LG} for forced convective boiling of water in a coil for $p = (0.1 - 0.2)$ MPa, $\dot{m} = (80 - 3000)$ kg/(m^{2}s), $\dot{q} = (60 - 800)\ 10^3$ W/m^2, $D_{coil}/D = 7 - 50$ can be calculated from the equation

$$\frac{\alpha_{LG}}{\alpha_c} = \sqrt{1 + 7\,10^{-9}\left(\frac{0.7\alpha_0}{\alpha_c}\right)^2\left(\frac{u_m\rho_L h_{LG}}{q}\right)^{3/2}}$$

where $\alpha_c = \sqrt{\alpha_b^2 + (0.7\alpha_0)^2}$; α_b is the heat transfer coefficient for pool boiling of water at the same pressure and wall temperature and α_0 is given by

$$\alpha_0 = 4.34\dot{q}^{0.7}(p^{0.14} + 1.37\ 10^{-2}p^2),$$

$u_m = (\dot{m}(1 - x)/\rho_L + \dot{m}x/\rho_G)$ is the mean velocity of vapor-water mixture, h_{LG} is the latent heat of evaporation and x is the quality. Wall temperature fluctuations appear when *burnout* in coiled tube takes place, and they are essentially greater in the vicinity of the inner generatrix. The maximum intensity of wall temperature pulsations in a coiled tube is lower than in a straight tube. The length of the transition region where the burnout occurs develops quickly after burnout is first detected.

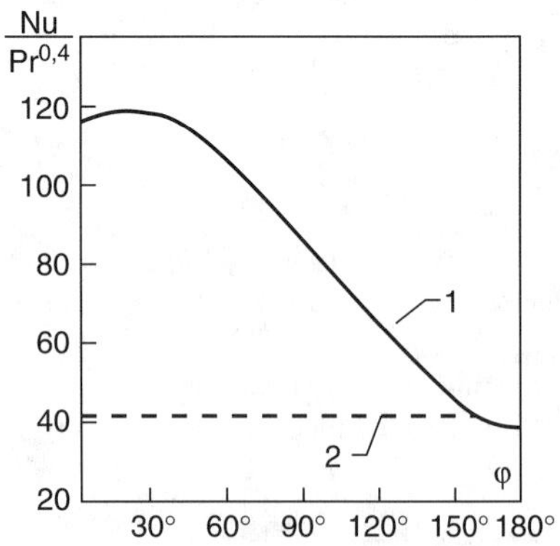

Figure 1. Variation in average heat transfer coefficient from the outer generatrix (x = 0°) to the inner generatrix (x = 180°) for water flow in a coiled tube (dotted line (2) shows value for the straight tube).

References

Schlichting, M. (1973) *Grenzchicht-Theorie.* Verlag G. Brawn. Karlsrube.

Heat Exchange Design Hand Book. v.1,2. Hemisphere Publishing Corporation, 1983.

Leading to: *Rotating ducts, heat transfer in*

Yu.A. Kuzma-Kichta

COILED WIRE INSERTS (see Augmentation of heat transfer, two-phase)

COKE (see Fuels; Pyrolisis)

COKE-OVEN GAS

Coke-oven gas is a fuel gas having a medium calorific value that is produced during the manufacture of metallurgical coke by heating bituminous coal to temperatures of 900°C to 1000°C in a chamber from which air is excluded. The main constituents are, by volume, about 50% hydrogen, 30% methane and 3% higher hydrocarbons, 7% carbon monoxide, 3% carbon dioxide and 7% nitrogen. The gas has a heating value of about 20,000kJ/m^3.

Typically, coke-oven gas is obtained from a battery comprising a number of narrow, vertical chambers, or ovens (0.5m wide, 5m high and 12m long) built of silica brick that are separated by heating ducts, such that heat is transmitted to the coal through both sides of the chamber walls. The ovens are slightly tapered so that one end is wider than the other to facilitate the horizontal discharge of the coke. Crushed coal is charged from overhead bunkers into the ovens, which are sealed at each end by refractory-lined sheet doors and heated for about 24 hours. The hot coke is then discharged. About 12%, by weight, of the coal is converted into gas. The hot gases evolved from the coal pass through a gas space at the top of the oven and into a collecting main prior to quenching and treatment to remove dust, tar and oil, and gaseous impurities such as ammonia and hydrogen sulphide.

References

Himus, G. W. (1972) *The Elements of Fuel Technology* Leonard Hill London.
Gas Making and Natural Gas BP Trading Ltd.

J.A. Lacey

COKE OVENS (see Pyrolysis)

Figure 1.

COLBURN, ALLAN PHILIP (1904–1955)

Allan Philip Colburn was an American engineer, born in Madison, Wisconsin, on June 8, 1904. He graduated from the University of Wisconsin in chemical engineering with high honors in 1926 and was awarded an engineering fellowship for graduate studies. He received his Master of Science in 1927 and his Ph.D. in 1929.

His research was on condensation of water vapor from saturated air streams, a topic that in its broader aspects interested him to the end of his life. He brought together for the first time in American engineering work the fundamentals of momentum, heat and mass transfer along with thermodynamic principles to deal with this complex problem.

Although not known formally, perhaps, as a dimensionless parameter, the empirical *Colburn J-Factor* is indeed an operational one.

Colburn joined the chemical engineering department at the University of Delaware in 1938. He was appointed as assistant to the president of the university in 1947, acting president in 1950 and provost and coordinator of scientific research until his death in 1955.

U. Grigull, H. Sandner, and J. Straub

COLBURN-CHILTON ANALOGY (see Cross flow heat transfer)

COLBURN CORRELATION (see Tubes, condensation in)

COLBURN FACTOR (see Extended surface heat transfer)

COLBURN HEAT TRANSFER FACTOR (see Tube banks, single phase, heat transfer in)

COLBURN J-FACTOR (see Dimensionless groups)

COLD ROD EFFECTS (see Rod bundles, heat transfer in)

COLEBROOK-WHITE EQUATION, FOR FRICTION FACTOR (see Tubes, single phase flow in)

COLEBROOK-WHITE FORMULA (see Corrosion; Friction factors for single phase flow)

COLLECTION EFFICIENCY (see Gas-solids separation, overview)

COLLIGATIVE PROPERTIES (see Solubility)

COLLIGEND (see Foam fractionation)

COLLOCATION (see Tau method)

COLLOIDAL DISPERSIONS (see Liquid-solid flow)

COLUMN CHROMATOGRAPHY (see Chromatography)

COLUMNS (see Distillation)

COMBINED CYCLES (see Rankine cycle)

COMBINED HEAT AND MASS TRANSFER (see Condensation of multicomponent vapours)

COMBUSTION

Introduction

Combustion is the burning of materials (**Fuels**), usually by reaction with the oxygen in air. During the reaction heat is released, and some of the common chemical products of burning are **Carbon Dioxide**, water vapour and *soot.*

Combustion is probably the oldest science, since soot deposited about 800,000 years ago on the roof of a cave near Beijing in China could only have been due to fires deliberately created by man for warmth and to cook his food. The primitive fuel technologists who lit these fires must have been aware of the role of draughts of air and the buoyancy of the products. They would also quickly discover the role of radiant **Heat Transfer** since a lone piece of charcoal will not remain alight, whereas a group of pieces of charcoal would radiate heat to each other and thus maintain a viable fire.

Combustion may be classified as:

1. Intentional, where the purpose is to release the chemical energy.
2. Unintentional, as in property destroying fires.
3. Naturally-occurring, as in forest fires caused by lightning.

Intentional combustion is one of the most important technologies in modern society as illustrated by the fact that developed countries spend more on fuel than on food. The main applications are: electrical power production, warmth for buildings and industrial process heat. In these applications, the fuel and air for the reaction are fed to burners fitted in **Boilers** or **Furnaces**. After the heat is removed, the products of combustion pass into the flue and are normally dispersed high into the atmosphere from the **Chimney**.

Combustion products may include a number of pollutants such as: **Carbon Monoxide**, oxides of nitrogen, **Sulphur Dioxide**, particulates, unburned hydrocarbons and dioxins (if **Chlorine** is present). The permissible levels of the various *pollutants* emitted to the atmosphere are governed by legislation and for the foreseeable future, these levels will be progressively reduced. It is now strongly suspected that the increasing level of carbon dioxide in the atmosphere is leading to a significant increase in atmospheric temperature by the so-called **Greenhouse Effect**. For this reason, carbon dioxide is now regarded as a pollutant even though it occurs naturally in the atmosphere. It is the task of the combustion engineer to maximise the efficiency of the combustion process and simultaneously minimise the production of pollutants. (See also **Air Pollution**.)

The science of combustion involves complex interactions between many constituent disciplines, including: chemical kinetics; fluid dynamics—including **Turbulent Flow**; **Thermodynamics**; heat and **Mass Transfer**; and **Two-Phase Flow** (i.e., the gas phase, and a liquid or solid fuel phase).

Each of these topics may be represented mathematically by equations, however since all the processes are interdependent, these governing equations must be solved simultaneously to model the overall combustion process. Furthermore, since most of the equations are nonlinear partial differential equations, it is necessary to evaluate the equations on a digital computer at a number of grid nodes distributed throughout the geometrical domain of interest.

Nowadays it is generally considered to be more satisfactory to use a numerical model to study combustion systems since other techniques, such as physical models, can only represent a part of the problem. For example, there is no simple method to scale a combustion system from small- to large-scale since simple correlation parameters such as geometry, residence time, velocity or radiant heat transfer follow different *scaling* laws and hence do not result in satisfactory scaling criteria. Nevertheless, at the present time mathematical modeling must always be checked experimentally to verify that the equations used correctly represent all the relevant physical and chemical phenomena.

Governing Equations

Many of the equations governing the behaviour of combustion systems are well established and they can be used with reasonable confidence. These include the equations of conservation of mass, momentum and energy. (See **Conservation Equations**.) There is more doubt about the equations of **Turbulence** and **Mixing** since these are models which attempt to use tractable equations to represent the fluctuating phenomena in the combustor.

In the case of two-phase flow when using liquid or solid fuels, the governing equations are well established for certain well-defined processes. Thus the evaporation rate of a drop of pure fluid can be computed with some confidence, but the drag and devolatilisation rate of a complex coal particle are not so well known and some discrepancies can be expected between computed and experimental results. It is important to recognise that such discrepancies do not invalidate mathematics! Rather, they indicate areas in which further research is required to improve the particular governing equation which gives rise to the discrepancy.

The governing equations for the mean motion of a fluid are based on the following laws of conservation:

1. Conservation of Mass;
2. Conservation of Momentum;
3. Conservation of Energy.

The Continuity Equation. Applying the Conservation of Mass to a fluid passing through an infinitesimal, fixed-control volume, the following equation is obtained:

$$\frac{\partial \rho}{\partial t} + \frac{\partial}{\partial x_i}(\rho u_i) = 0 \tag{1}$$

where *e* is fluid density, *u* velocity and t time.

The Momentum Equation. Conservation of momentum states that the total momentum within a system remains constant during the exchange of momentum between two or more masses in the system. For a fluid passing through an infinitesimal, fixed-control volume, the following momentum equation applies:

$$\frac{\partial}{\partial t}(\rho u_j) + \frac{\partial}{\partial c x_i}(\rho u_i u_j) = \rho g_j + \frac{\partial \delta_{ij}}{\partial x_i}. \tag{2}$$

The stress tensor δ_{ij} represents the external stresses, consisting of normal stresses and shearing stresses. For a Newtonian fluid, the *stress tensor* is related to **pressure** and velocity in the following tensor form:

$$\delta_{ij} = \eta\left(\frac{\partial u_i}{\partial x_j} + \frac{\partial u_j}{\partial x_i}\right) + \gamma_{ij}\eta'\left(\frac{\partial u_k}{\partial x_k}\right) - p\gamma_{ij} \tag{3}$$

where η is viscosity and γ_{ij} is the Kronecker delta function (γ_{ij} = 1 if i = j and γ_{ij} = 0 if i $\neq$ j). The equations of conservation of mass and conservation of energy are known as the *Navier-Stokes equations*

The Energy Equation. The *First Law of Thermodynamics* states that the increase of energy in the system is equal to heat added to the system plus the work done on the system. If the flow in a *Cartesian coordinate system* is assumed to be incompressible with a constant coefficient of thermal conductivity, then the energy equation is:

$$\rho\frac{De}{Dt} = \lambda\nabla^2 T + \Phi + S_h \tag{4}$$

where

$$\Phi = \eta\Bigg[2\left(\frac{\partial u}{\partial x}\right)^2 + 2\left(\frac{\partial v}{\partial y}\right)^2 + 2\left(\frac{\partial w}{\partial z}\right)^2 + \left(\frac{\partial v}{\partial x} + \frac{\partial u}{\partial y}\right)^2 + \left(\frac{\partial w}{\partial y} + \frac{\partial v}{\partial z}\right)^2 + \left(\frac{\partial u}{\partial z} + \frac{\partial w}{\partial x}\right)^2 - \frac{2}{3}\left(\frac{\partial u}{\partial x} + \frac{\partial v}{\partial y} + \frac{\partial w}{\partial z}\right)^2\Bigg] \tag{5}$$

and the last term, S_h, the source term, includes the heat of chemical reaction, radiation, and any interphase exchange of heat.

The Equation of State. As there are many unknowns in the conservation equations, additional equations are required for their solutions. The relationships between the thermodynamic variables pressure, density, temperature, internal energy and enthalpy (p, ρ, T, e, h), and the relationships between transport properties viscosity and thermal conductivity (η, λ) are utilised to generate variables. The equations utilised are known as equations of state, examples of which for a PERFECT gas are as follows:

$$p = \rho\tilde{R}T \qquad e = c_vT \qquad h = c_pT \tag{6}$$

Where $\tilde{R}$ is the universal gas constant and c_v and c_p the specific heat capacity at constant volume and constant pressure, respectively.

Turbulent Motion. Turbulent fluid motion is an irregular condition of flow in which various quantities show a random variation with time and space. In practical systems, turbulent flows are inevitable even in the absence of chemical reaction, causing considerable difficulties in practical flow simulation. When the unsteady Navier-Stokes equations are applied to turbulent flows, the time and space scales of the turbulent motion are so small that the number of grid points and the small size of the time steps required in the calculation are outside the realm of current computer technology. Present-day computational fluid mechanics is thus achieved through the use of time-averaged Navier-Stokes equations; The gross effects of turbulence on time-mean flow is considered, while the detailed structure of the turbulence is disregarded. (See also **Turbulence**.)

The time-averaged Navier-Stokes equations are referred to as the *Reynolds equations of motion.* They are of the same form as the fundamental Navier-Stokes equations, but incorporate the terms of apparent stress gradients and heat-flux quantities associated with turbulent motion. As these quantities must be related to the mean flow variables through turbulence models, the Reynolds equations are derived by decomposing the dependent variables in the conservation equations into time mean and fluctuating components. The entire equation is then time averaged. Thus,

$$\Theta = \overline{\Theta} + \Phi' \tag{7}$$

$\overline{\Theta}$ and Θ' are the time mean and fluctuating components of dependent variables u, v, w, T, p, ρ. Fluctuations in other fluid properties such as viscosity, specific heat and thermal conductivity are usually negligible.

Taking the continuity equation, for example, the instantaneous fluid velocity u_i is first decomposed into its time mean and fluctuating components. The velocity components are substituted into the equation. By definition, the time-average of a fluctuating quantity is zero, thus the continuity equation after time-averaging is transformed to:

$$\frac{\partial\bar{\rho}}{\partial t} + \frac{\partial}{\partial x_i}(\bar{\rho}\bar{u}_i + \overline{\rho' u_i'}) = 0 \tag{8}$$

For steady incompressible flows, $\rho' = 0$ and the above equation can be reduced to:

$$\frac{\partial\bar{u}_i}{\partial x_i} = 0 \tag{9}$$

The time-averaged momentum equation can also be written as follows:

$$\frac{\partial}{\partial t}(\rho\bar{u}_j) + \frac{\partial}{\partial x_i}(\rho\bar{u}_i\bar{u}_j) = -\frac{\partial\bar{p}}{\partial x_j} + \frac{\partial}{\partial x_i}\left[\eta\left(\frac{\partial\bar{u}_i}{\partial xj} + \frac{\partial\bar{u}_j}{\partial x_i}\right)\right] + \frac{\partial}{\partial x_i}(\rho\,\overline{u_i'u_j'}) \tag{10}$$

where the effect of turbulence is incorporated into the fundamental momentum equation through the *Reynolds stresses* $(\rho\,\overline{u_i'u_j'})$

Turbulence Modeling

In turbulent flows, the velocity at a point is considered as the sum of the mean and fluctuating components. In the time-averaged momentum equation, the effect of the fluctuating components is to introduce the effect of turbulence through a term involving the Reynolds stresses, $\rho\overline{u_i'u_j'}$. Turbulence modeling relates the Reynolds stresses to mean flow quantities so that the turbulent flow field can be calculated without calculating the full detail of the fluctuating flow. (See also **Turbulence Models**.)

In the $k - \varepsilon$ model, Reynolds stresses are related to the mean flow via the *Boussinesq hypothesis:*

$$k = \frac{1}{2}\overline{u_i'u_i'} \quad \text{and} \quad \varepsilon = \nu\,\overline{\frac{\partial u_i'\partial u_i'}{\partial x_j\partial x_j}} \tag{11}$$

$$\rho\overline{u_i'u_j'} = \rho\,\frac{2}{3}k\delta_{ij} + \eta_i\left(\frac{\partial'''u_i}{\partial x_j} + \frac{\partial u_j}{\partial x_i}\right). \tag{12}$$

The effective or "turbulent" viscosity (η_t) is computed from a velocity scale ($k^{1/2}$), and a length scale ($k^{3/2}/\varepsilon$) which are predicted by the solution of transport equations for k and ε:

$$\frac{\partial k}{\partial t} + \bar{u}_i\frac{\partial k}{\partial x_i} = P - \varepsilon + \frac{\partial}{\partial x_i}\left(\frac{\nu_T}{\sigma_k}\frac{\partial k}{\partial x_i}\right) \tag{13}$$

$$\frac{\partial\varepsilon}{\partial t} + \bar{u}_i\frac{\partial\varepsilon}{\partial x_i} = C_i\frac{\varepsilon}{k}P - C_2\frac{\varepsilon^2}{k} + \frac{\partial}{\partial x_i}\left(\frac{\nu_T}{\sigma_\varepsilon}\frac{\partial\varepsilon}{\partial x_i}\right) \tag{14}$$

where,

$P = 2\nu_T\overline{S_{ij}S_{ij}}$ is the turbulence production, and

$\bar{S}_{ij} = \frac{1}{2}\left(\frac{\partial\bar{u}_i}{\partial x_j} + \frac{\partial\bar{u}_j}{\partial x_i}\right)$ is the mean rate of strain tensor.

These well known equations involve empirical constants which have the following values:

$C_1 = 1.44, \quad C_2 = 1.92, C_\mu = 0.09, \sigma_k = 1.0, \quad \sigma_e = 1.3.$

In highly-swirling flows such as are used in the near-field region of a burner, μ_t is strongly directional and the isotropic $k - \varepsilon$

model may be inadequate. For such cases, we may solve additional differential equations for the Reynolds stresses.

Chemical Reaction Modeling

The conservation equation for a chemical species (see **Chemical Reaction**) can be written as:

$$\frac{\partial}{\partial t}(\rho X_i) + \frac{\partial}{\partial x_i}(\rho u_i X_i) = \frac{\partial}{\partial x_i} J_i + R_i + S_i \quad (15)$$

where X_i is the mass fraction of chemical species i; R_i is the mass rate of creation or depletion by chemical reaction; and S_i is the rate of creation by addition from a dispersed phase. J_j is the diffusion flux of species i, and is given by:

$$J_i = \rho D_{im}\frac{\partial X_i}{\partial x_i} - \frac{D_i^T}{T}\frac{\partial T}{\partial x_i} \quad (16)$$

where D_i^T is the thermal diffusion coefficient and D_{im} is the diffusion coefficient of species i in the mixture. In multistep and multispecies reactions, the source of chemical species i due to reaction, R_i is computed as the sum of the reaction sources over the k reactions that the species may participate in. The overall rate of production of species i is:

$$R_i = \sum_k R_{ik}. \quad (17)$$

Reaction may occur in the gas phase between gaseous species, or at surfaces resulting in the surface deposition of a chemical species. The reaction rate, R_{ik}, is controlled either by kinetics or mixing. The Arrhenius reaction rate (see **Arrhenius Equation**) is calculated as:

$$R_{ik} = \nu'_{ik} M_i T^{\beta_k} A_k \prod_{j\ \mathrm{reactants}} [C_j]^{\nu_{ik}} \exp(-E_k/\tilde{R}T). \quad (18)$$

When the reaction is mixing-controlled, the influence of turbulence on the reaction rate can be taken into account by employing the *Magnussen and Hjertager, 1976, model.* In this model, the rate of reaction R_{ik} is computed as:

$$R_{ik} = A\rho\frac{\varepsilon}{k}\frac{X_i}{\nu'_{ik}} \quad (19)$$

where A is an empirical constant.

Laminar flame structures are reasonably well-understood for both premixed and diffusion flames. (See **Flames.**) On the other hand, turbulent flame structures are much more complicated. The basic problem in reaction rate models is due to the nonlinearity of the chemical equations, thus:

$$W = AT^{\beta} \prod_{j\ \mathrm{REACTANTS}} [C_i]^{\nu_1} \exp(-E/\tilde{R}T); \quad (20)$$

$$\overline{W} \neq A\overline{T}^{\beta} \prod_{j\ \mathrm{REACTANTS}} [\overline{C}_i]^{\nu_i} \exp(-E/\tilde{R}\overline{T}). \quad (21)$$

Solutions to this problem can be based on stirred reactor models, Reynolds decomposition, eddy break-up models and statistical descriptions. Over the last few years, attention has turned to the relationship between the turbulent flame and details of the turbulence structure, such as length scales and the frequency spectrum. These aspects have been surveyed by Bradley (1992)

- The effect of turbulence is to increase the area of laminar flamelets due to wrinkling of the flame. In an un-stretched flame this gives an increase in the flame speed.
- At sufficiently high rates of stretch, the flame is quenched; however, the effects of stretching in turbulent flames is difficult to generalise.
- The laminar flamelet model, with assumed probability density functions (PDFs) which are functions of the reaction progress variable and the stretch rate, appears to be valid over a wide range. It has been used successfully to predict lift-off heights of turbulent diffusion flames, the combustion field in premixed swirling flow, and flame blow-off.

Heat Transfer

The energy equation is solved in the form of a transport equation for static enthalpy, h:

$$\frac{\partial}{\partial t}(\rho h) + \frac{\partial}{\partial x_i}(\rho u_i h) = \frac{\partial}{\partial x_i}(k + k_t)\frac{\partial T}{\partial x_i} + \frac{Dp}{Dt} + \tau_{ik}\frac{\partial u_i}{\partial x_k} + S_h \quad (22)$$

where k is the molecular conductivity, k_t is the effective conductivity due to turbulent transport ($k_t = \eta_t/Pr_t$) and S_h consists of source terms including heat of chemical reaction, any interphase exchange of heat, and other user-defined heat sources.

Enthalpy is defined as:

$$h = \sum_j m_{j'} h_{j'} \quad (23)$$

where $m_{j'}$ is the mass fraction of species j′,

$$h_{j'} = \int_{T_{ref}}^{T} c_{p,j'} dT. \quad (24)$$

Enthalpy sources due to reaction are defined as:

$$S_{h,\mathrm{reaction}} = \sum_j h_{j'}^{o} R_{j'} \quad (25)$$

where h_j^o is the enthalpy of formation of species j′ and $R_{j'}$ is the volumetric rate of creation of species j′.

The laminar flow heat transfer to the fluid from the wall is calculated by a first order approximation of the heat flux:

$$\dot{q}'' = \lambda\frac{\partial T}{\partial n}\Big|_{wall} \approx \lambda\frac{\Delta T}{\Delta n}. \quad (26)$$

In turbulent flows, the above equation is replaced by a log-law formulation based on the **Analogy Between Heat and Momentum** transfer:

$$\frac{\lambda(\Delta T/\Delta n)}{\dot{q}''} = \frac{1}{\kappa y^+}\frac{Pr_t}{Pr}\ln(Ey^+)$$

$$+ \frac{1}{y^+}\left(\frac{Pr_t}{Pr}\right)^{5/4}\frac{\pi/4}{\sin \pi/4}\left(\frac{A}{\kappa}\right)^{1/2}\left(\frac{Pr}{Pr_t} - 1\right) \quad \textbf{(27)}$$

where Pr is **Prandtl Number** and Pr_t turbulent Prandtl number. Within solid conducting regions, a simple conduction equation is used that includes the heat flux due to conduction and volumetric heat sources within the solid:

$$\frac{\partial}{\partial t}\rho_w c_w T = \frac{\partial}{\partial x_i}\lambda_w\frac{\partial T}{\partial x_i} + \dot{q}'' \quad \textbf{(28)}$$

where $\dot{q}$ is the rate of heat production per unit volume. This equation is solved simultaneously with the static enthalpy equation to yield a fully coupled conduction/convection heat transfer prediction.

Radiation Heat Transfer

Surface-to-surface radiation with participating media may be calculated by use of a Discrete Transfer Radiation Model (DTRM). The DTRM usually assumes no wavelength dependency (grey radiation), with the radiant intensity representing an integrated intensity across all wavelengths. (See also **Radiative Heat Transfer.**) The change in radiant intensity, dI, along a path ds, is given by:

$$\frac{dI}{ds} = -(\alpha_{abs} + \alpha_s)I + \frac{\alpha_{abs}\sigma T^4}{\pi}. \quad \textbf{(29)}$$

where α_{abs} and α_s are coefficients which account for absorption and scattering and σ is the Stefan Boltzmann constant.

This equation is integrated along a series of rays emanating from a single point in each discrete control volume on a surface. This series of rays defines a hemispherical solid angle about that point. The integrated intensity along each ray is:

$$I(s) = \frac{\sigma T^4}{\pi}\left(\frac{\alpha_{abs}}{\alpha_{abs} + \alpha_s}\right)(1 - \exp[-(\alpha_{abs} + \alpha_s)s])$$

$$+ I_o \exp[-(\alpha_{abs} + \alpha_s)s] \quad \textbf{(30)}$$

where I_o is the radiant intensity at the start of the incremental path ds.

Enthalpy produced by radiation in the fluid is computed by summing the change in intensity along the path of each ray through each fluid control volume.

The intensity of radiation approaching a point on a wall is integrated to yield the incident radiant heat flux,

$$\dot{q}^- = \int_\Omega I D\Omega \quad \textbf{(31)}$$

where Ω is the hemispherical solid angle and I is the intensity of the ray approaching the wall. The net radiation heat flux leaving the wall surface, $\dot{q}^+$, is computed as the sum of the reflected portion of $\dot{q}^-$ and the emissive power of the surface:

$$\dot{q}^+ = (1 - E)\dot{q}^- + E\sigma T_p^4 \quad \textbf{(32)}$$

where T_p is the temperature of the surface at point P, and E is the wall emissivity. The radiation heat flux calculated in this equation may be incorporated in the prediction of the wall surface temperature, T_p. From this equation, the radiation intensity I of a ray emanating from point P is:

$$I = \dot{q}^+/\pi. \quad \textbf{(33)}$$

The radiative absorptivity, α, of a gas is related to the absorption coefficient by:

$$\alpha = 1 - \exp(-\alpha_{abs}L) \quad \textbf{(34)}$$

where L is the radiation path length.

Solution Procedure

This set of equations together with any additional equations required to describe the behaviour of the second-phase (such as the evaporation of liquid drops) are generally evaluated using a computer package. The particular geometry and flow boundary conditions are input to define the particular problem to be modeled. The output from the calculation is available in both graphical and numerical form and can be used for experimental investigations, or to guide the design of burners installed in combustion systems.

References

Magnussen, B. F. and Hjertager, B. H. (1976) "*On Mathematical Models of Turbulent Combustion with Special Emphasis on Soot Formation and Combustion*" 16th International Symposium on Combustion. Cambridge, MA.

Bradley, D. (1992) "*How Fast Can We Burn*" 24th International Symposium on Combustion. Sydney.

Leading to: *Combustion chamber; Burners; Fires; Flames; Furnaces; Incineration*

J. Swithenbank

COMBUSTION CHAMBER

Following from: *Combustion; Jet engines*

Combustion chambers are one of the main units of air jet and rocket engines or gas-turbine plants that heat up the original components (working medium) from an initial temperature T_o to a preset T_g temperature through the calorific power of the burnt fuel H_u. In an *air jet engine,* the heat delivered to 1 kg of air in a typical combustion chamber at a constant pressure—and with an allowance for combustion efficiency and heat losses ξ through the walls—is determined by the equation

$$q = \left(1 + \frac{1}{\alpha L_0}\right) c_{p_g} T_g - C_{p0} T_o = H_u \frac{1}{\alpha L_0},$$

where C_{po} and C_{pg} are the specific heat capacities of the original working medium and the combustion products respectively; the product αL_0 is the ratio of working medium to fuel flow rate and depends on the oxidizing medium, e.g., air. The theoretical quantity of oxidising medium needed for complete burning of 1 kg of fuel is L_0. α is the excess coefficient (is the factor by which the stoichiometric air requirement is multiplied to take account of excess air). Thus, burning hydrocarbon (petroleum) fuel in air requires $L_0 = 0.115C + 0.345H - 0.043O$, where C, H, and O are, respectively, the mass fractions of carbon, hydrogen, and oxygen in the fuel. For instance, $L_0 = 14.9$ for aviation kerosene (84–86% C, 14–16% H). For CH_4 and H_2, $L_0 = 17.2$ and 34.5, respectively.

Calorific power, or the lowest heat of fuel combustion, is defined as the quantity of heat in Joules that is released as a result of the complete combustion of 1 kg of fuel in air at $t_O = 15°C$ and $p = 0.1$ MPa during cooling of combustion products to 15°C. This does not consider the heat of condensation and water vapor content. It is roughly estimated by:

$$H_u = 339\ C + 1030\ H - 109\ O.$$

For instance, $H_u = 42900$ to 43100 kJ/kg for aviation kerosene and 49500 and 116700 for CH_4 and H_2, respectively.

The combustion products of hydrocarbon fuels are CO_2 and CO, NO and NO_2, water, hydrocarbons C_xH_y, etc. Their composition affects the combustion chamber from the environmental standpoint. A deterioration of combustion efficiency, $\xi < 1$, raises the quantity of CO, C_xH_y and gives rise to *soot* and *smoke.* Ejection of nitrogen oxides NO_x increases as combustion temperature rises and the length of time combustion products are in the combustion zone grows. The allowable levels of NO_x, CO, C_xH_y and smoke for most types of engines are thus subject to state control.

The oxidizer (air) excess coefficient $\alpha = G_a/G_f$ represents the combustion regime. The mixture produced as a result of combustion is *stoichiometric* at $\alpha = 1$; rich at $\alpha < 1$; and lean at $\alpha > 1$. With an excess or a deficient oxidizer, the temperature of combustion products T_g is lower than the maximum closest to stoichiometric due to the heat consumption of the surplus fuel and oxidizer. With a significant change in α, the steady-state combustion in the chamber stops. These are called "rich" and "poor" *flame-out,* respectively.

Combustion chambers of power units should be able to provide a high combustion efficiency (in up-to-date gas-turbine engines, $\xi = 0.995$ and higher), low pressure losses of the working medium flow across the chamber ($\sigma = p_{out}/p_{in}$ in gas-turbine engines, is 0.94 – 0.96), high reliability and longer service life (in gas-turbine engines, up to 10,000 hours). These can be assured by the absence of overheat, carbon deposit, etc. Variation of ξ and σ coefficients with the air flow rate (or $\dot{M}_{comb.ch}$) and the value of preheating T_g/T_0 are said to be the characteristics of the combustion chamber. As $\dot{M}_{comb.ch}$ and T_g/T_0 grow, σ drops. Combustion efficiency ξ is enhanced with increasing T_g/T_0 and attains a flat optimum if plotted versus $\dot{M}_{comb.ch.}$

Of particular significance in gas-turbine engines is a high uniformity of the fields of circumferential gas temperatures at the combustion chamber outlet (for a reliable operation of a nozzle device) and of the temperature versus radius profile (for reliability of blades), with temperature diminishing toward the upper and lower ends of the blade. The fields are produced by the development of oxidizer (air) and fuel flows in the combustion and mixing zones.

Flammability of homogeneous hydrocarbon fuel-air mixtures ranges between $0.5 < \alpha < 1.7$. The velocities of flame front propagation are not high: 0.5 to 2.0 m/s for kerosene and 210 m/s for hydrogen. Therefore, to ensure stable combustion at mean flow velocities much higher than the velocity of the flame front propagation, a combustion stabilizer with a reverse current zone can be devised that assures a reliable *mixture* inflammation in the combustion zone for all operating regimes of the combustion chamber. **Figure 1** shows the structure of such a flow in the combustion zone behind the combustion stabilizer. Addition of air to combustion products in the mixing zone reduces average temperature values and raises α values. For example, the characteristic values of α for a poor flame-out in the combustion chamber of an air jet engine normally vary from 20 to 50. In rocket engines, the flame front stabilization is generally affected by the system of vortices near the oxidizer and fuel jets.

Combustion chambers are classified according to engine type (air jet, rocket, and other engines), the purpose it is designed for (the main combustion chamber or afterburner in an air jet engine), combustion character (subsonic or supersonic), fuel pressure (high- and low-pressure), the type of atomizers and fuel atomization (cen-

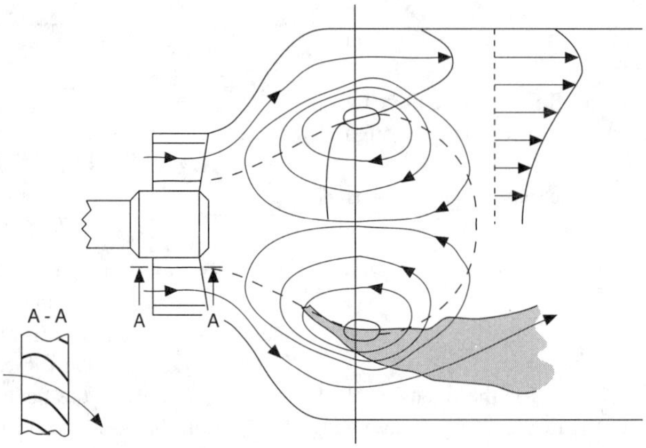

Figure 1. Combustion in the presence of a flame stabilizer.

trifugal, high-turbulence, evaporation), the number of combustion zones, and design (axial, radial, reverse-flow, tubular-type, annular, etc.).

Reference

Lefevre, A. H. (1983) *Gas Turbine Combustion,* McGraw Hill, 1983.

O.N. Favorskii

COMBUSTION PRODUCTS (see Combustion)

COMFORT CONDITIONS (see Space heating)

COMITATO NAZIONALE PER LA RICERCA E PER LO SVILUPPO DELL'ENERGIA NUCLEARE E DELLE ENERGIE ALTERNATIVE, ENEA

via Regina Margherita 125
00198 Rome
Italy

COMMISSARIAT A L'ENERGIE ATOMIQUE, CEA

31–33 Rue de la Federation, BP510
75752 Paris cedex 15
France
Tel: 1 40 56 10 00

COMMISSION OF THE EUROPEAN COMMUNITY, CEC

200 Rue de la Loi
1049 Bruxelles
Belgium
Tel: 295 1111

COMMON MODE FAILURE (see Risk analysis techniques)

COMMONWEALTH SCIENTIFIC AND INDUSTRIAL RESEARCH ORGANISATION, CSIRO

PO Box 225
Dickson ACT 26009
Australia

COMPACT HEAT EXCHANGERS (see Energy efficiency; Helical coil boilers)

COMPILER (see Computer programmes; Computers)

COMPLEX COMPOUND CATALYSIS (see Catalysis)

COMPLEXING IONS (see Electroplating)

COMPLEXITY (see Non-linear systems)

COMPOSITE MATERIALS, ABLATION OF (see Ablation)

COMPOUND AUGMENTATION (see Augmentation of heat transfer, single phase; Augmentation of heat transfer, two phase)

COMPRESSED ASBESTOS FIBER JOINTING, CAF (see Gaskets in heat exchangers)

COMPRESSIBILITY EFFECTS (see Boundary layer heat transfer)

COMPRESSIBILITY FACTOR

Compressibility factor, usually defined as $Z = pV/RT$, is unity for an ideal gas. It should not be confused with the isothermal compressibility coefficient. In most engineering work, the compressibility factor is used as a correction factor to ideal behavior. Thus, $v_{real} = Z\ v_{id}$ is used to calculate the actual volume, v_{real}, as the product of the compressibility factor and the ideal gas volume, all at the same pressure and temperature. Z is most commonly found from a generalized compressibility factor chart as a function of the reduced pressure, $p_r = p/p_c$, and the reduced temperature, $T_r = T/T_c$ where p_r and T_r are the reduced variables and the subscript 'c' refers to the critical point. (See **Corresponding States, Principle of.**)

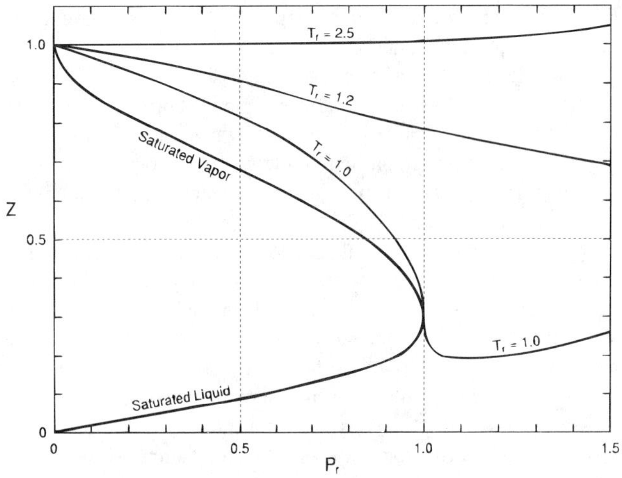

Figure 1. Generalized compressibility factor chart.

Figure 1 shows the essential features of a generalized compressibility factor chart. The most widely-used compressibility factor charts are apparently those of Nelson and Obert (1954, 1955). These have been extended [see, e.g., Liley (1987)] to include the saturated liquid. A three-parameter correlation $Z = f(P_r, T_r, \omega)$, where ω = acentric factor = $-\log_{10}p_r(T_r = 0.7) - 1$, involves the use of two compressibility factor charts so that $Z = Z_0(p_r, T_r) + w Z_1(p_r, T_r)$. [See, e.g., Sonntag, R. E. and van Wylen, G. J. (1991).]

References

Nelson, L. C. and Obert, E. F. (1955) Trans. A.S.M.E., *76* (10), 1057–1066 (1954); A.I. Ch. E. J. *1* (1), 74–77.

Liley, P. E. (1987) *Chemical Engineering* (NY), *94*(10), 123–126.

Sonntag, R. E. and van Wylen, G. J. (1991) *Introduction to Thermodynamics, Classical and Statistical,* Wiley, NY, 800 pp.

P.E. Liley

COMPRESSIBLE FLOW

Following from: *Flow of fluids*

All fluids are compressible and when subjected to a pressure field causing them to flow, the fluid will expand or be compressed to some degree. The acceleration of fluid elements in a given pressure gradient is a function of the fluid density, ρ, whereas the degree of compression is determined by the isentropic bulk modulus of compression, κ. The speed of sound in a medium is given by, $a = (\kappa/\rho)^{1/2}$ and compressibility effects are apparent when the flow velocity, u, becomes significant compared to the local speed of sound. The local **Mach number** $M = u/a$ is the primary parameter which characterises the effects of compressibility. Under normal atmospheric conditions, the speed of sound in water is $1500 ms^{-1}$ and that in air is $345 ms^{-1}$. Thus, it can be expected that compressibility manifests itself in gas flows more readily than in liquid flows and the discussion below deals predominantly with gas flows. Transients in hydraulic systems are an example of compressible liquid flow which is of some importance. The case of liquid-gas mixtures is of interest and is discussed below.

The role of Mach number in compressible gas flow may be derived from the governing equations of motion and state. However, the physics of these processes are clear when gas flow from one chamber to another is considered. Flow from a constant pressure reservoir, a, is produced by reducing the pressure in chamber b below that in a (**Figure 1**).

An element of gas, **c**, will accelerate from **a** to **b**, and while doing so increases its volume and decreases temperature. The local velocity of sound reduces as a result of this fall in temperature as $a = (\gamma RT)^{1/2}$, where γ is the ratio of specific heats, T is the absolute temperature and R is the specific gas constant. Initially, at low-pressure differentials, the flow is essentially incompressible and pressure falls as the gas passes through the throat, **d**. But it recovers somewhat upon diffusion in the nozzle into chamber **b**. When the pressure in **b** is reduced further, this fact is conveyed by sound waves which travel back through the nozzle into **a**. The flowfield then responds by passing more gas through the nozzle. This process will continue as the pressure in **b** is reduced up to a point when the local speed of sound has *fallen* to a level equal to the local flow velocity. This will first occur at the throat, **d**, and henceforth, the flowfield upstream of the throat will remain frozen as sound waves with the information about conditions in **b** cannot travel through the throat against the flow. From then on, mass flow through the nozzle remains constant and the nozzle is said to be *choked.* Reducing the pressure in **b** further does not increase the mass flow. The flow at the throat has a unity Mach number, i.e., $a = u$; upstream the flow is *subsonic*, $M < 1$, and downstream the flow becomes *supersonic*, $M > 1$. (See also **Nozzles**.)

The same conclusion may be drawn from a one-dimensional isentropic analysis of the steady nozzle flow in **Figure 1**. This leads to the expression below for du, the change in velocity, with u, resulting from a change, dA, in the nozzle cross-sectional area, A.

$$\frac{du}{u} = \frac{-dA/A}{1 - M^2}. \tag{1}$$

At low Mach numbers, $M << 1$, u varies inversely with A as demanded by incompressible continuity. For subsonic conditions, $M < 1$, u increases with decreasing A and vice versa. However, the reverse is true for the supersonic case when $M > 1$. Also, it is only possible for U to keep increasing through the throat, $dA = 0$, if $M = 1$; otherwise du must be zero. The conclusion is therefore the same as that arrived at above; the nozzle is *choked* and the Mach number is unity at the minimum area if the pressure ratio exceeds a particular value.

The isentropic nature of the flow is based on the assumption that no heat transfer between elements of the gas occurs and that the expansion is reversible, i.e., the normal conditions for an isentropic change. Thus, no heat conduction or viscous effects occur. This is true in the free-stream remote from the walls and boundary layer. However, when shock waves or low-density effects are present, this will not be the case.

In any flowfield, a narrow stream tube may be taken such that conditions may be considered one-dimensional (see **Figure 1**). The Steady Flow Energy Equation (i.e., a control volume for energy) may be applied from the *stagnation conditions* at the inlet (suffix 0) to any point along the stream tube to give the result:

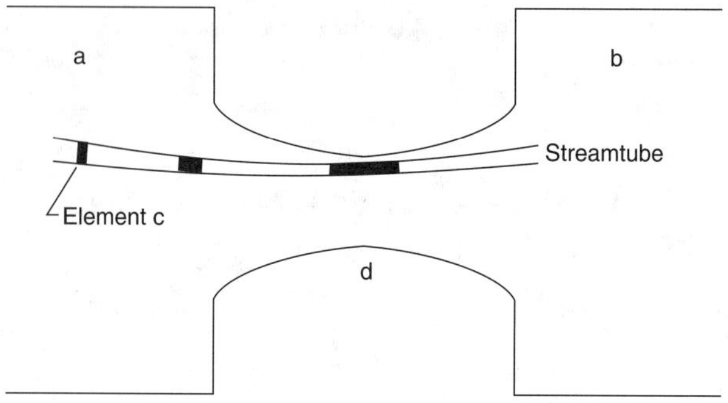

Figure 1. Flow of element C from chamber A to chamber B.

$$h_0 = h + \frac{1}{2}u^2 \qquad (2)$$

where h is the specific enthalpy of the fluid. For a perfect gas, $h = \gamma RT/(\gamma - 1)$. Hence, the variation of local static temperature, T, throughout the flowfield—in terms of the local *stagnation temperature* and local Mach number—immediately follows.

$$\frac{T_0}{T} = 1 + \frac{\gamma - 1}{2}M^2. \qquad (3)$$

The assumption of isentropic flow then gives the local pressure, P, and density, ρ, in terms of the stagnation values of these quantities. Thus, the ratio of *stagnation pressure* to local pressures.

$$\frac{p_o}{p} = \left(\frac{T_o}{T}\right)^{\gamma/\gamma-1} = \left(1 + \frac{\gamma - 1}{2}M^2\right)^{\gamma/\gamma-1} \qquad (4)$$

Values calculated from these relationships are tabulated in many texts. Anderson (1990), Shapiro (1954), Liepmann and Roshko (1960) have provided such tables. As mentioned in the physical description of nozzle flow, information about boundary conditions is transmitted by sound waves travelling through the flow to adapt the flowfield. In supersonic flow, these waves will always be swept downstream. If a body is introduced into supersonic flow, there has to be a mechanism whereby the upstream flow becomes aware of the presence of the body. *Shock waves,* being discontinuities in flow parameters, provide this mechanism. In the case of slender and pointed bodies, the shock wave may attach to the leading edge whereas in blunt bodies, the shock wave detaches and stands upstream as a normal shock (**Figure 2**).

The flow passing through a normal shock is subject to large gradients in temperature and the assumption of isentropic flow is not tenable. The energy conservation equation, together with the mass continuity and momentum equations, and the equation of state for an ideal gas lead to relationships between like properties on either side of the shock; u_2/u_1, T_2/T_1, P_2/P_1 (suffices 1 and 2 refer to static conditions upstream and downstream). These may be given in terms of $M_s = u_1/a_1$, the flow Mach number relative to the stationary shock wave, as the independent variable. Tables of these quantities are given in all standard texts. For a normal shock, the downstream flow is always subsonic in shock relative coordinates.

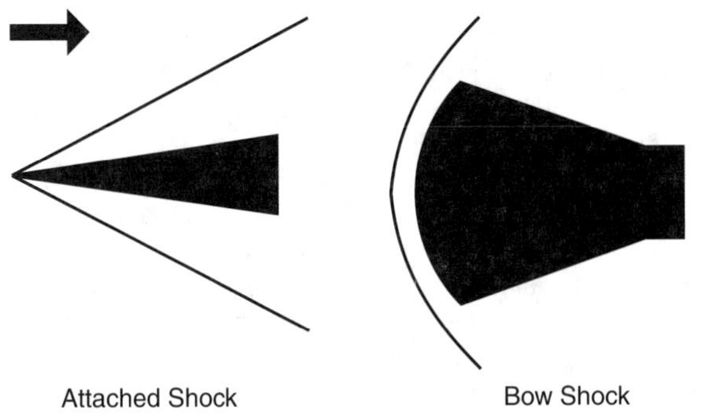

Figure 2. Shock waves around sharp and blunt bodies.

When supersonic flow expands around a surface convex to the flow (**Figure 3**), the information is again transmitted along the sound waves, called *Mach waves,* which travel into the flow.

Conditions along a streamline obey the isentropic equation mentioned previously. However, there is now a direct relationship between the angle the flow has turned through and the local flow Mach number. This is the *Prandtl-Meyer relationship,* which gives the Prandtl-Meyer angle, ν (M), as a function of the Mach number. The angle ν is the angle the flow has turned through from sonic conditions (M = 1) to the local Mach number. Thus, if flow at M_1 turns through a further angle, θ, the change in Prandtl-Meyer angle equals θ and this enables the new Mach number M_2 to be determined.

$$\theta = \nu(M_2) - \nu(M_1). \qquad (5)$$

The Mach waves travel into the flow causing the turning as shown in **Figure 3** and have an inclination to the local flow, μ, called the Mach angle, given by $\mu = \sin^{-1}(1/M)$.

When two such convex surfaces are arranged to constitute a nozzle, the Mach waves can be seen to interact. This is shown in **Figure 4** where it is clear that the streamline on the centreline has been subjected to Mach waves from both surfaces, which give no net turning. Nevertheless, the local Mach number will increase according to the sum of the modulus of the turning angles from

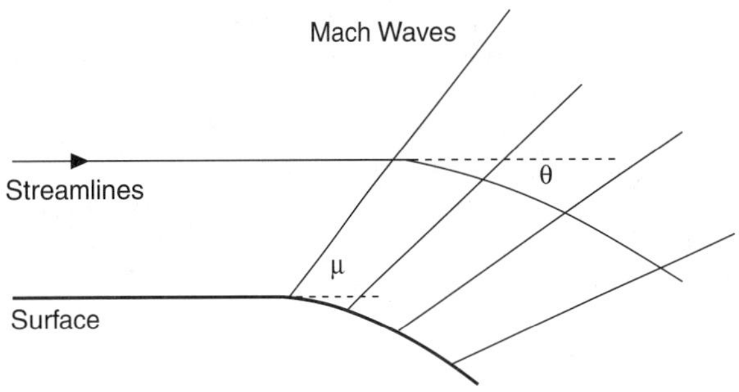

Figure 3. Expansion around a corner.

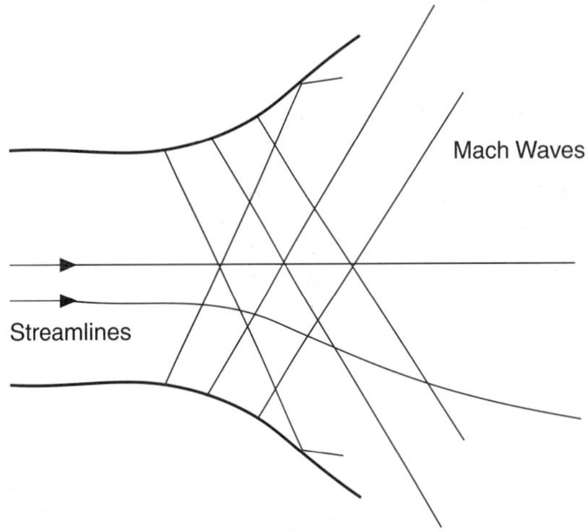

Figure 4. Interaction of Mach waves in a nozzle.

both walls for there is no physical difference in the flow being turned and expanded by the upper or lower wall.

The process described can be used to design nozzle shapes by calculating the trajectory of the Mach waves and the resulting turning of the flow and is a simple example of the *Method of Characteristics* for solving supersonic flowfields.

In two dimensions, the shock waves may be inclined and determination of the conditions behind the shock wave is achieved by considering those components of velocity along and perpendicular to the shock separately. The momentum along the shock is unaltered whereas the normal component may be considered as in a normal shock. The results are often given graphically as β, the angle of the shock to the flow as a function of the flow deflection angle, θ. The upstream Mach number, M_1, is the independent variable. Such a plot is given in **Figure 5**.

It can be seen that there are two solutions of β for each value of θ at a given free-stream Mach number, M_1. For attached shocks, it is usually the lower value of β—the weak shock—which is relevant. Flow is predominantly supersonic behind the weak shock. When the flow deflection angle is increased above a certain angle, there is no solution. Thus, for sharp bodies of large angle, the shock wave cannot be attached and a normal shock is formed (**Figure 6**).

Inclined shocks may reflect off solid boundaries in a "regular" manner, which satisfies the condition that the flow remains in contact with the wall after reflection. As there is a restriction on the maximum value of θ, as shown in **Figure 5**, some regular reflections are not possible and a normal shock forms which is normal to the wall. This automatically satisfies the requirement that the flow remains in contact with the wall. The normal shock then leaves the wall and curves to join the incident shock as shown in **Figure 6**. These shock formations have their counterparts in supersonic jet plumes. Here, there may be regular reflections from the centerline or a normal shock, called a Mach disk in the axisymmetric situation, could be present. These are also shown in **Figure 6**.

Isentropic flow which has constant stagnation enthalpy can be shown to be irrotational. This results from *Crocco's theorem* and

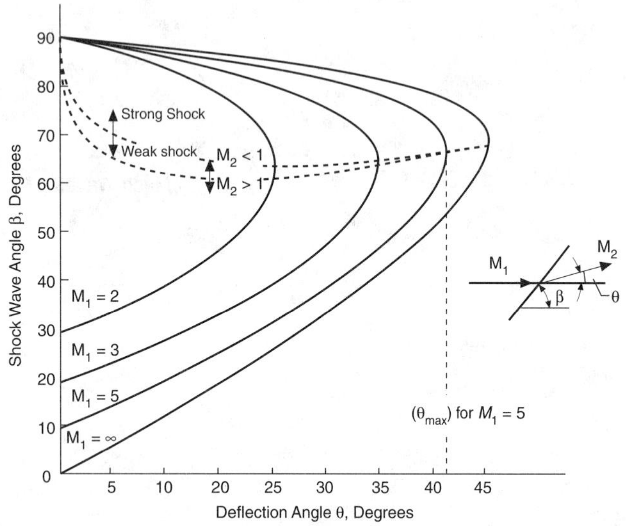

Figure 5. Inclined shock properties from Anderson (1990).

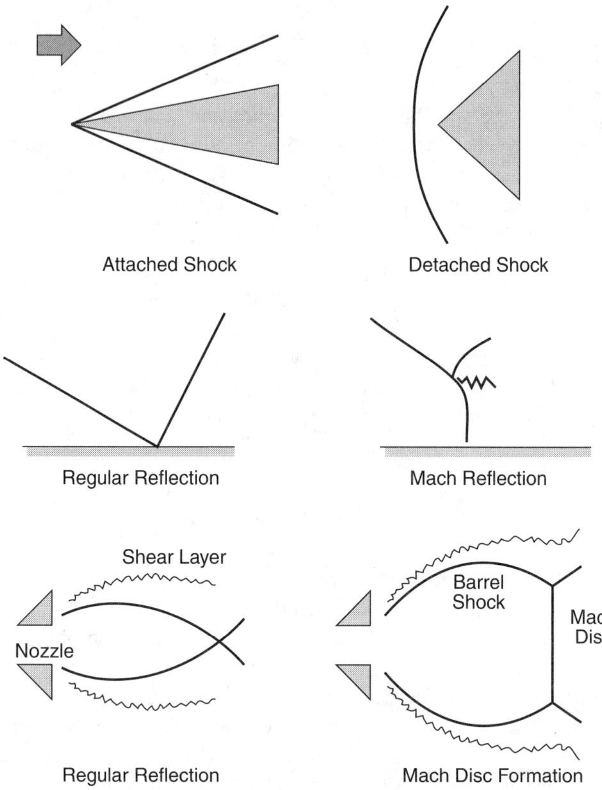

Figure 6. Examples of regular and Mach reflections.

can apply to a wide range of compressible flows, such as nozzle flow. The irrotational nature of the flow means that the velocity may be derived by taking the gradient of a potential function (i.e., a scalar function of position). Thus, the governing equations of motion—the Euler equations—may be represented by a differential equation for the potential function, commonly called the velocity potential equation. In subsonic flow, the equation is elliptical whereas in supersonic flow, the equation is hyperbolic. In the latter case, it can be shown that along Mach lines certain quantities are invariant. In **Figure 4**, two Mach lines or characteristics arise from any point in the flowfield. These will be at angles $\theta + \mu$ and $\theta - \mu$, where θ is the flow angle and μ, the Mach angle. The quantities

$$\theta + \nu(M) = K_+ \quad \textbf{(6)}$$

$$\theta - \nu(M) = K_- \quad \textbf{(7)}$$

are constant along the −ve and +ve characteristics, respectively. Given an initial upstream boundary condition for the flow, it is possible to march downstream, taking into account constraining walls, to solve the flowfield. In the past, this method of characteristics has been solved graphically but is solved numerically at present. The method of characteristics may also be employed in axisymmetric and three-dimensional flowfields. The characteristics are the mathematical counterparts of the sound waves discussed earlier.

The velocity potential equation may be linearised to solve problems where deviations in a uniform stream are small. This applies to thin airfoils in a uniform stream. In this case, it can be shown that with a simple transformation of geometry the linearised equation is the same as that for incompressible flow, i.e., the Laplace equation.

As a result, the pressure coefficient, Cp, for Mach number, M, is related to the incompressible value, Cp_O, by

$$Cp = Cp_o/\sqrt{1 - M^2} \qquad M < 0.8. \qquad \textbf{(8)}$$

This is known as the *Prandtl-Glauert Rule*. In supersonic flow, the pressure coefficient from linearised theory gives the airfoil surface pressure coefficient as a function of deflection angle, α, of the airfoil surface

$$Cp = \frac{2\alpha}{\sqrt{M^2 - 1}} \qquad M > 1.2. \qquad \textbf{(9)}$$

Transonic flow occurs beyond the point when the airfoil, for example, becomes critical—i.e., when sonic conditions appear on the airfoil. Shock waves can then exist in the supersonic patch which occurs. This usually results in separation of the airfoil boundary layer, with a consequent increase in drag (**Figure 7**). In the transonic region, the Mach number is extremely sensitive to small changes in flow area.

In hypersonic flow, the free-stream velocity is much greater than the local velocity of sound. This may roughly occur at Mach numbers greater than 5. When expanding from fixed stagnation conditions, the flow velocity tends to the constant value $a_O\sqrt{2/\lambda - 1}$ where a_O is the stagnation velocity of sound. In the hypersonic limit, the density ratio across a normal shock approaches a constant and the angle of an oblique shock is linearly related to the deflection angle. Compressible flows are described in standard texts. Anderson (1990), Anderson (1989), Liepmann and Roshko (1957) and Shapiro (1954).

Compressible effects are also very important in long ducts subjected to large pressure ratios. Choking of the ducts may occur if this is of sufficient length, as is often the case. Here, viscous effects are important and in the analysis it is usual to assume that fully-developed viscous flow is present. If the flow is adiabatic and the viscous effects are characterised by a constant friction factor, f, this leads to the *Fanno flow* solution. The salient features of this flow are.

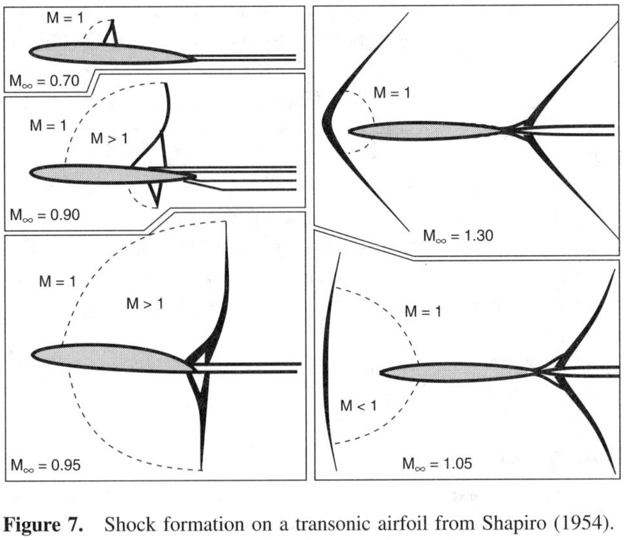

Figure 7. Shock formation on a transonic airfoil from Shapiro (1954).

1. For a given subsonic inlet Mach number to the duct, there is a maximum length of the duct, L_{max}, in nondimensional terms as given by f L/D max where D is the duct diameter, at which the flow becomes sonic, i.e., it is choked.
2. Similarly, if supersonic flow enters the duct, the flow decelerates and there is also a nondimensional length at which the flow chokes.

Conditions in the duct relative to the choked conditions for different inlet Mach numbers for Fanno flow are tabulated in most texts. These are widely-used for the design of flow in pipework. Even if choked exit conditions do not prevail, the compressible effects may be deduced from these tables. Another extreme is *Rayleigh flow,* which is frictionless but with heat addition. Shapiro (1954) has covered a wide range of such flows.

Two-phase bubbly flow can have interesting properties since it is highly compressible, due to the gaseous component, and a high density, due to the liquid present. Thus, the velocity of sound may be much lower than the gas or the liquid in isolation. The subject is covered by Van Wijngaarden (1972) and Drew (1983). The flow does act as a compressible fluid of low speed of sound and exhibits shock structure. This velocity may be as low as 20 ms^{-1} in bubbly water; however, the bubbles may slip with respect to the liquid.

A wide range of unsteady compressible flow phenomena exist and these are covered in the works of Glass and Sislian (1994) and Kentfield (1993). (See also **Shock Tubes**.)

References

Anderson, J. D. Jr. (1989) *Hypersonic and High Temperature Gas Dynamics, McGraw Hill* New York.

Anderson, J. D. Jr. (1990) *Modern Compressible Flow, McGraw Hill* New York.

Drew, D. A. (1983) *Mathematical Modeling of Two-Phase Flow,* Ann. Rev. Fluid Mech. 15, 261–91.

Glass, I. I. and Sistian, J. (1994) *Non Stationary Flows and Shock Waves,* Oxford Engineering Series 39, Oxford University Press.

Kentfield, J. A. C. (1993) *Nonsteady, One-Dimensional Internal, Compressible Flows*, Oxford Engineering Science Series, 31, Oxford University Press. New York.

Liepmann, H. W. and Roshko. A. (1956) *Elements of Gasdynamics,* Galcit Aeronautical Series, Wiley, New York.

Shapiro, A. H. (1954) *Compressible Fluid Flow,* Volumes I and II, Ronald Press, New York.

Wijngaarden, L. V. (1972) *One-Dimensinal Flow of Liquids Containing Small Gas Bubbles,* Ann. Rev Fluid Mech 4,369.

Leading to: *Compressors; Critical flow; Shock tubes*

T.V. Jones

COMPRESSION-IGNITION ENGINES (see Internal combustion engines)

COMPRESSION POINT (see Sedimentation)

COMPRESSION ZONE (see Thickeners)

COMPRESSORS

***Following from:** Compressible flow; Aerodynamics*

Compressors are devices for compression and delivery of gases. They are widely-used as separate units and as important parts of different types of heat engines. Compressors are driven by different types of prime movers (electric motors, steam and gas turbines, diesel engines, etc). There are two main principles of compressor action—the displacement principle with a cyclic process and the dynamic principle with a continuous process of gas compression. The pressure ratio (PR) for the displacement machine is the result of decreasing the volume that the trapped gas occupies. The dynamic machine develops pressure when power supplied to the rotor increases the angular momentum of the gas flowing through the rotor cascade. The stator cascade transforms this angular momentum of the gas into static pressure. There are also heat (thermosorptive) compressors in which pressure is increased after heating the matrix which absorbed the gas at lower temperature.

According to the developed pressure difference (PD), there are three types of compressors: low PD, medium PD and high PD units. If absolute pressure at the compressor inlet is significantly lower than atmospheric pressure, this unit is referred to a vacuum pump or turbomolecular pump. The inlet pressure may vary from 10^{-10} Pa to several Pa. Low PD compressors are termed blowers; the PR for them is equal to 1.05 to 1.15. For medium PD units, the PR is usually equal to 1.3 to 4.0. For high PD units, the PR is equal to 6 to 10 and more. Multicasing designs are used to achieve higher values of PR.

Displacement machines may be divided into reciprocating and rotary compressors and adsorptive units while dynamic machines are divided into axial and radial flow compressors and jet units. Shear force radial compressors and magnetohydrodynamic pumps for plasma transportation by a moving electromagnetic field can also be classified as dynamic machines.

Reciprocating compressors

The *reciprocating compressor,* (**Figure 1a**), consists of a working cylinder (4) with two valves—a suction valve (3) and a discharge valve (2) in corresponding cavities in the cylinder head (1). The piston (5) moves reciprocatingly inside the cylinder with the help of the piston rod (6), crank (9), connecting rod (8) and crosshead (7). The discharge cavity is hermetically separated from the suction cavity. When the piston moves from the top, the cylinder is being filled with the gas at the suction pressure through valve 3. When the piston reaches its extreme position, valve 3 is closed. When the piston moves back, the cylinder volume is decreased, the gas pressure is increased, and the gas is compressed. When gas pressure exceeds the discharge pressure, valve 2 is opened and gas flows from the cylinder into the discharge line until the piston has reached the end of its stroke. This process takes place during one revolution of the shaft or double stroke of the piston. The extreme positions of the piston are called the top and bottom of the stroke. The volume between the top and the piston at its nearest point to the top position is a waste space. Reciprocating compressors can be one-stage and multistage. There can be single- and double-acting designs (in the latter, both sides of the piston are acting). Different positions of the cylinders are possible (vertical, horizontal, angular, V- or W-shaped, etc.).

The working process of the ideal reciprocating compressor may be shown on a p − V diagram (**Figure 1b**). The suction process takes place at a constant pressure (line 4-1), then the gas is compressed inside the cylinder (line 1-2) and later is discharged (line 2-3). After valve 2 is closed, the gas pressure in the cylinder immediately drops from $P_3 = P_2$ to $P_4 = P_1$ and this process is repeated. Lines 4-1 and 2-3 are for constant gas parameters; only the volume and mass of the gas are changed. These lines are not the process lines. Line 1-2 is the compressing process line with a constant mass. The compressing process may be presented depending on the heat-exchange conditions. Line I is for the isothermic process ($pV = $ const); line III is for the isoentropic process ($pV^\gamma = $ const); and line II is for the polytropic process ($pV^m = $ const). Polytropic compression work for the ideal compressor is calculated using the following equation:

$$W = \int_1^2 V\,dp = \frac{m}{m-1} R(T_2 - T_1).$$

For isothermal compression ($m = 1$), the work

$$W_T = RT_1 \ln \pi_c,$$

where $R = \tilde{R}_0/\tilde{M}_0$ is the gas constant; $\tilde{R}_0$, the universal gas constant; $\tilde{M}_0$, the molar mass of the gas; and $\pi_c = p_2/p_1$, the pressure ratio. For isoentropic compression,

$$W_S = \frac{\gamma}{\gamma - 1} RT_1(\pi_c^{(\gamma-1)/\gamma} - 1)$$

where γ is the specific ratio. In an actual reciprocating compressor, there are always some losses leading to the signification difference between the real diagram and the ideal one (**Figure 1b**). The existence of a heat flux between the cylinder wall and the gas should also be taken into account. The heat flux varies as far as its amount and direction are concerned. At the beginning of the compression process, heat is supplied to the gas ($m > \gamma$) while at the end of this process, the heat flux changes its direction ($m < \gamma$). This results in the increase of the indicator diagram area in p − V coordinates. The indicator diagram area A_1 is proportional to the indicator compressor power $P_i = A_i n$, where n is a rotative speed (RPM). The power absorbed by the compressor is equal to $P = P_1/\eta_M$, where the mechanical efficiency $\eta_M = 0.82$ to 0.95. The indicator power is related to the isoentropic and isothermic power by an expression $P_{S,T} = P_i \cdot \eta_{S-i,T-i}$, where $\eta_{S-i} = 0.88$ to 0.95 and $\eta_{T-i} = 0.7$ to 0.8.

The volume flow rate $\dot{V}$ of the compressor is proportional to the cylinder volume V_c and n. It can be calculated as $\dot{V} = \lambda_v V_c n$, where λ_v is the coefficient of admission, ($\lambda_v = 0.88$ to 0.92). For atmospheric suction pressure, the reciprocating compressor can increase gas pressure to 10 MPa and more with a volume flow of from 10 to 200 m^3/min. Reciprocating compressors are required for many gas storage, transmission, process, energy conversion and refrigeration systems.

Rotary compressors

There are many different types of rotary compressors. The principal types are the lobe (Roots type), screw (Lysholm type),

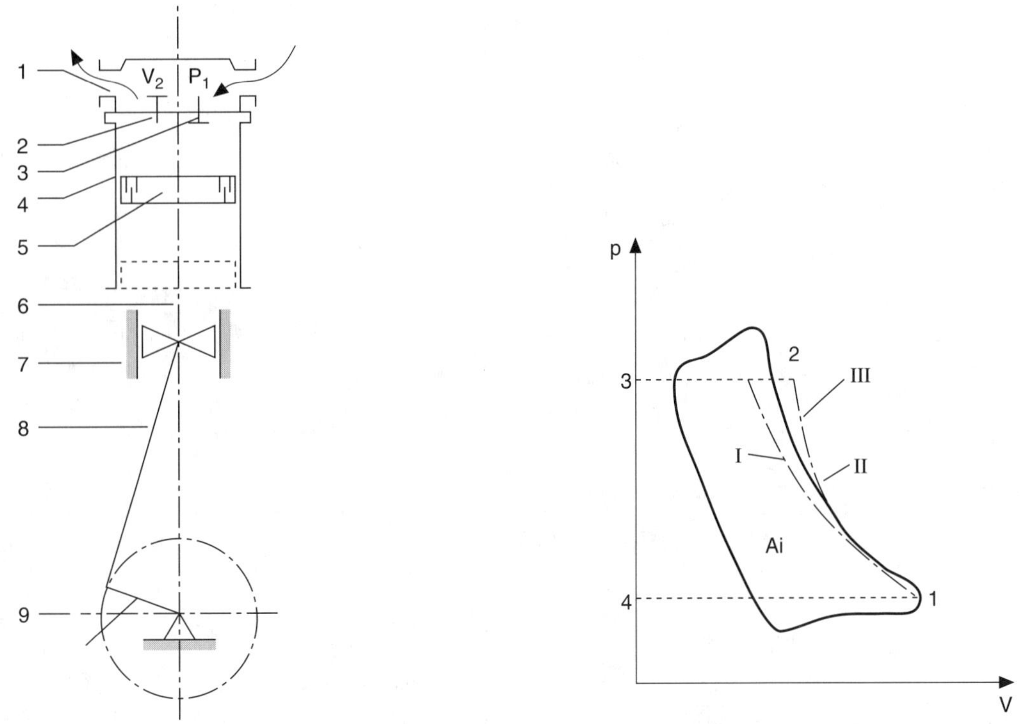

Figure 1. Reciprocating compressor.

sliding vane and gear (external and internal). Compared with reciprocating compressors, rotary compressors have better dynamic balancing, smaller sizes and mass, need no valves and have more uniform gas discharge.

The *Roots type compressor* has two rotors with parallel axes, (**Figure 2a**). Each rotor has two or three rounded lobes and revolve in opposite directions. A maximum clearance of 0.001D, where D is a rotor diameter, exists between the rotors and the casing. During rotation, the rotors transport the increments of gas from the suction side to the discharge side under an isochoric process. The volume flow rate for two lobe rotary compressor under suction condition is $\dot{V} = 2\lambda_v(0.785D^2 - A_r \cdot n \cdot l$, where A_r is the rotor cross-section area and l is the rotor length. The isochoric compression power is $P_V = \rho_1 \dot{V}RT_1(\pi_c - 1)$. The absorbed power $P = P_v/\eta_M$, where $\eta_M = 0.85$ to 0.95. The presence of radial clearances limits the allowable value of $\pi_c \leq 1.8$ for the rotary compressors without lubricant supply to the working cavity. Lubricant supply permits the increase of π_c, but the lubricant has to be separated from the compressed gas.

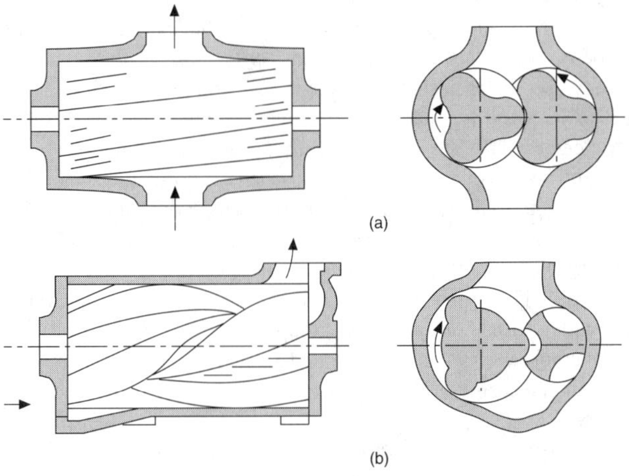

Figure 2. Roots type compressor.

The *screw rotary compressor,* **Figure 2b** has different shape of rotor lobe and positions of the inlet and discharge holes. The compression process of the screw compressor is close to an isoentropic one and takes place inside the compressor simultaneously with gas transport towards the discharge hole. The mechanical efficiency of the screw compressor is in the range $\eta_M = 0.92$ to 0.98 and the pressure ratio $\pi_c = 3$ to 4. The tip velocity of the lobe on the leading rotor is as much as 120 m/s and more. Water injection inside the screw rotor reduces power consumption and leakage.

The process of compression in the sliding vane compressor is close to that of the screw compressor. This compressor has a rotor which revolves eccentrically inside a cylinder. The vanes (plates) can freely move inside radial or oblique slots. When the rotor revolves, these vanes come out of the slots and press on the internal surface of the cylindrical casing, forming cavities with changing volumes. The tip velocity of the vane is equal to 12 to 15 m/s. Gas compression gas discharge and gas filling take place on successive arc lengths in the rotation.

The *water-ring compressor,* **Figure 3**, is similar in its process to the vane compressor but has some differences in design. A rotor

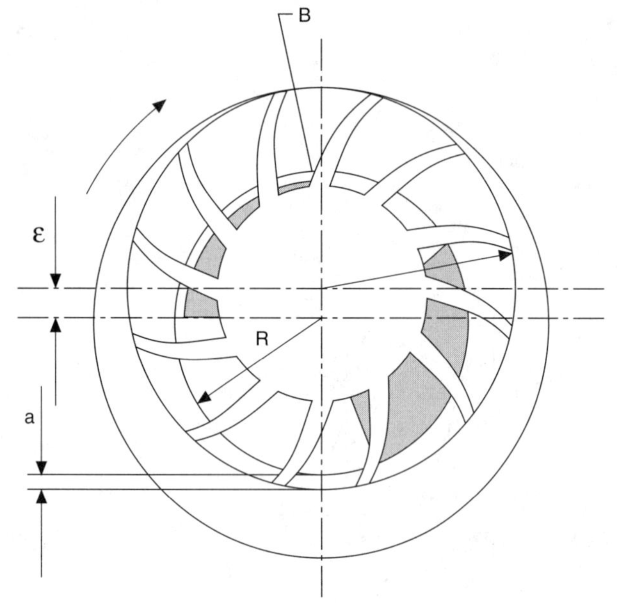

Figure 3. Water ring compressor.

with solid blades of the R_t tip radius revolves in the cylindrical casing with some eccentrity ε. Water or any liquid with low viscosity is in the space between the rotor and the casing. During the revolution of the rotor, the water assumes a ring shape with the internal radius R. The minimum depth of the blade immersion is equal to a. The point B indicates a minimum distance between the hub of the rotor and the water-ring. Gas is sucked through the large port in the cylinder flange (shadowed) and is discharged through the smaller port. Working cavities are located between the rotor hub and the water-ring. The compression process is close to the isothermic one. These machines have long service life and are suitable for any dusty gas compression. They are widely-used as vacuum-pumps and create up to 98% vacuum. Their pressure ratio may be up to 2.5 to 2.8 when they are used as blowers; but the efficiency is only 40 to 45% due to high losses because of the interaction between the blades and the water-ring and due to friction in the casing.

The *thermosorptive compressor* uses the ability of some substances or metal alloys to absorb some gases at room temperature and pressure, with subsequent discharge of the gases at much higher pressure after some slight heating of the matrix. For instance, the $LaNi_5$ alloy is capable of absorbing hydrogen-forming $LaNi_{5x}$ hydride. After the slight heating of the powdery matrix, the hydrogen leaves at an elevated pressure. A similar absorptive process can take place when methane is dissolved in water to form a hydrate. A slight heating of the hydrate releases the methane at higher pressure.

The *axial flow compressor,* **Figure 4**, consists of a casing (1) with the inlet (8) and outlet (10) ducts, supplied with the axisymmetrical inlet confuser channel (7) to obtain uniform flow (2) and an axisymmetrical outlet diffuser channel (9) for partial transformation of the kinetic energy of the gas flow into static pressure. The rotor has two bearings (12) and sealings (11) to reduce the leakage of the gas between the rotor and the casing. Compressor blading usually consists of several stages placed one after another. A stage

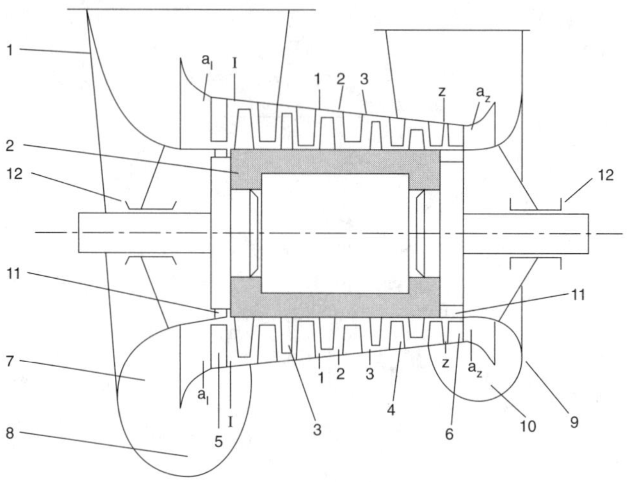

Figure 4. Axial flow compressor.

has two rows of the wing-shaped blades, which form aerodynamic cascades for the rotor (R) (3) and the stator (S)(4). Both cascades have curvelinear diffuser channels. In front of the first rotor blade row, there is often an inlet guide vane (IGV) row with the turbine-type cascade. At the entrance to the outlet axisymmetrical channel, there is an additional stator (S′) blade row (6) to get the axial flow in front of the outlet duct.

Figure 5a shows the distribution of the static pressure p, the absolute velocities v and the relative velocities w of the gas for several cascades along the compressor axis. The axial velocity u is related to w and v via the vector triangle (see Fig. 5b). From the continuity equation, one can see that the increase of the gas density ρ leads to a decrease of the blade height towards the compressor exit. The main parameters of the cascade are solidity $\sigma = c/s$ [ratio of the blade chord c to the blade spacing (pitch) s], the stagger angle ξ and the camber angle θ for the profile mean line. Compression work of the stage can be calculated either by the expression

$$W_{st} = C_p T_1(\pi_{st}^{(\gamma-1)/\gamma} - 1)/\eta_{st},$$

where C_p is the specific heat at constant pressure π_{st}, the stage pressure ratio and π_{st}, the stage efficiency. The approach of relative velocity to the speed of sound leads to decrease of the stage efficiency η_{st}. The pressure ratio for the subsonic stages is less than 1.2 to 1.25 and for the transonic stages, about 1.75 to 2.0 for the efficiencies of 0.88 to 0.92. Generally-speaking, the stage efficiency is influenced by the flow coefficient (the ratio of the mean flow rate velocity to the tip velocity U_t of the blade); the head coefficient (the ratio of the work of the stage to U_t^2); and the degree of reaction (the ratio of the rotor static pressure increase to the static pressure increase). The vortex and the constant degree of reaction designs are widely-used for axial flow compressors. The multistage axial flow compressors can be designed to achieve the total pressure ratio in one casing of the order of 10 to 15 and efficiencies of 0.86 to 0.88.

The *radial compressor,* consists of a casing of a complicated shape, in which the rotor induces a radial flow towards an outlet stator ring.

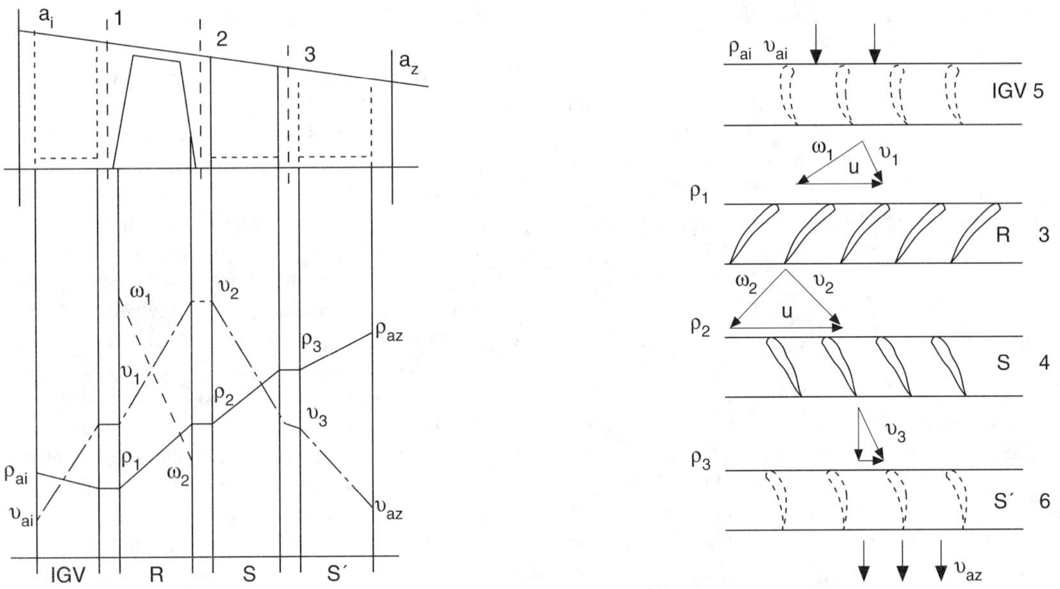

Figure 5.

Due to the action of centrifugal forces a radial stage produces a higher pressure ratio than the axial one, in which the compression is mainly connected with the diffusion effect at the cascade. The efficiency of the radial compressor stage is usually equal to 0.79 to 0.83 for a pressure ratio of 3 to 5.

Static pressure in the flow can be increased with the help of a jet device which acts as the injector.

V.S. Beknev

COMPTON SCATTERING (see Gamma rays)

COMPUTATIONAL FLUID DYNAMIC MODELS (see Two-phase flows)

COMPUTATIONAL FLUID DYNAMICS

Following from: *Flow of fluids; Computer programmes*

Computational Fluid Dynamics (CFD) is the term given to the task of representing and solving the fluid flow and associated equations on a computer. Although the equations controlling fluid flow have been known for over 150 years significant advances in CFD were delayed until the 1960s when digital computers became available to the scientific comunity. Since then CFD has attracted an ever-increasing level of resources and has generated real benefits for industry sectors that have invested in it. The power and relatively low price of modern work stations, together with the high quality of commercial CFD codes now available, make CFD a very attractive tool for designers and engineers in the process industries, and an effective vehicle for many research workers in the heat and mass transfer fields. Although CFD is about solving complex equations, the real challenges revolve around understanding the physics and how the essential elements of the problem can be represented in terms of equations and boundary conditions. The nonlinearities present in the flow equations and the complexity of the physics are such that CFD is not likely to replace all physical experiments in the foreseeable future. CFD is, however, likely to reduce the volume of expensive experimental work and help design better experiments, as well as increase our understanding and predictive abilities.

The two essential components of CFD are *mathematical modeling* and *numerical analysis*, although it is sometimes difficult to separate them fully. Mathematical modeling is about expressing the problem in a mathematical form with reasonably correct differential equations and adequate boundary conditions. Although this appears straightforward, it is in fact the most difficult and demanding task most engineers face when using CFD. Decisions have to be made about how detailed the CFD calculation is going to be, and indeed how detailed it needs to be to represent the significant processes involved in a problem. Some of these decisions are easy to make: is the problem two- or three-dimensional?, even if it is three-dimensional, will a two-dimensional representation suffice? Others are very difficult and may lead to lengthy subsidiary work: is the standard turbulence model adequate? are the boundary conditions imposed for heat transfer in the reattachment zone reasonable for my application? The continuity, momentum and scalar transport equations are nonlinear and coupled and take the following form:

$$\frac{\partial \rho}{\partial t} + \frac{\partial}{\partial x_j}(\rho U_j) = 0$$

$$\frac{\partial(\rho U_i)}{\partial t} + \frac{\partial}{\partial x_j}(\rho U_j U_i) - \frac{\partial}{\partial x_j}\left(\eta \frac{\partial U_i}{\partial x_j} - \overline{\rho u_i u_j}\right) - \frac{\partial P}{\partial x_i} + S_i = 0$$

$$\frac{\partial(\rho \Phi)}{\partial t} + \frac{\partial}{\partial x_j}(\rho U_j \Phi) - \frac{\partial}{\partial x_j}\left(\frac{\mu}{Pr}\frac{\partial \Phi}{\partial x_j} - \overline{\rho \phi u_j}\right) + S_\phi = 0$$

where U_i and u_i, are the mean and fluctuating components of velocity in the x_i direction; Φ and ϕ are the mean and fluctuating components of a passive scalar, such as temperature; P is the pressure; ρ is the density, η_t is the viscosity; Pr is the Prandtl number, and the S terms represent sources for the momentum or scalar equations. The overbar indicates that an averaging procedure has been applied to the cross-correlation of the fluctuating components. Constitutive relationships are required for the correlation terms; for example, the *Boussinesq hypothesis* leads to:

$$-\rho\overline{u_i u_j} = \eta_t\left(\frac{\partial U_i}{\partial x_j} + \frac{\partial U_j}{\partial x_i}\right)$$

and η_t is the so-called turbulence viscosity which must, in turn, be modeled (see also **Conservation Equations**).

Turbulence Modeling is a field in its own right, and the complexity of turbulence models adopted reflects the complexity of the physics and the computing resources available. For a typical single-phase problem with heat transfer, there are three momentum, one conservation and one energy or temperature equations. The complexity of the turbulence model adopted determines the number of equations used to determine η_t, the well known k-ε turbulence adds two further equations to the problem——one for the turbulent kinetic energy k and one for the turbulent dissipation rate ε. If the problem also involves mass transfer, then more transport equations similar in form to the energy equation are required to represent the various chemical species transported. Near-wall treatment and boundary condition interpretation may need special attention when heat and mass transfer are taking place. (See also **Turbulence** and **Turbulence Modeling**.)

The second component of CFD is related to the numerical aspects of representing the above equations on a computer and solving them. The first task is to choose a coordinate system and mesh which will be able to give adequate resolution of the geometry and physics of the problem. The second task is to discretize the differential equations into their difference form, in a manner which will result in an accurate and robust (stable) set of algebraic equations suitable for numerical manipulation. The final numerical task is to use a solution procedure which will solve the discretized equations quickly and 'accurately' without making undue demands on the hardware (memory, disk space and central processor speed). There are essentially two solution methods: one is classified as uncoupled and the other as a coupled method. In the uncoupled method, the discretized equations for each variable are solved separately for the whole field so that each velocity component is found separately. Pressure is then obtained through a procedure which uses the mass conservation equation. In the coupled method, velocities and pressures are solved simultaneously. (See also **Numerical Methods**.)

CFD has matured to a point where most CFD calculations are undertaken on commercial packages. Most CFD software vendors offer body-fitted, multiblock structured mesh or totally unstructed mesh capabilities, which provide excellent geometric resolution; most codes are able to use geometries and meshes set up on the large commercial solid body modeling software packages. Vendors also offer a choice of discretisation schemes; in general, the more accurate the scheme the greater the demands it will make on computing resources. Unless there is a research need to modify or use alternative discretisation algorithmic, it is usual to use the vendors' offerings. The same is true of the solution procedures. Vendors have coded up a number of algorithms and allow the user to choose between them. (See also **Computer Programmes**.)

The CFD user is faced with a three-component task: setting up the problem; using the CFD software to solve the equations; and examining the CFD solutions. For most engineers, the first component is the the most time-consuming; but this is changing. The laborious task of setting up geometries and grids has been mechanised and it is now possible to use numerical geometric information from other software packages to quickly generate appropriate CFD grids. The technology has reached a point where commercial vendors are offering *adaptive grids*. These are grids which move their positions during the calculation so as to optimise the resolution of the physical phenomena being modeled. There are a number of well-tried and proven numerical schemes available which have been coded by commercial vendors. Thus, this is generally no longer a problem area for most CFD users.

Analysis and assessment of CFD predictions is surprisingly difficult. Given the inherent three-dimensional nature of CFD, and the large number of variables normally computed, good interactive graphical capabilities become essential. Even with these, it is difficult to display the vast quantities of information in a manner which facilitates clear understanding of the problems.

G.L. Quarini

COMPUTER AIDED DESIGN, CAD (see Extrusion, plastics)

COMPUTER PROGRAMMES

***Following from:** Computers*

The advent of the digital *computer* has had a profound influence on the complexity and the way in which problems are tackled in the heat and mass transfer field. It is now common to solve most groups of equations by numerical means, even though analytical solutions may exist, and to regard the solution of sets of hundreds of linear simultaneous equations as an everyday event. The heart of the computer is the central processor, which is responsible for carrying out arithmetic operations in correct sequence. The central processor also has the responsibility to organise the transfer of information between various parts of the computer; for example, it will transfer numbers from memory stores to the arithmetic operations unit and vice versa. The central processor takes its instructions from the computer programme held in store.

A computer programme consists of a set of instructions which enables the computer to perform the tasks the programmer requires. Typically, a programme will comprise input instructions, instructions to control the processing of the data and output instructions. All of these instructions have to be precise and unambiguous. In their simplest forms, computer programmes instruct the computer in a way in which there is a direct correspondence between instruction and control circuits in the computer. Such programmes are written in what is referred to as *machine language.* A machine language instruction includes the address of the location in the store containing the piece of information to be acted upon. Many of the programmes for early computers were written in machine language. Creation of these programmes was time-consuming and required a great deal of skill. The programmes themselves, once written, tended to be efficient in both the usage of store space and the time required for the machine to undertake a given operation.

As computers became more powerful (faster, with larger memories and relatively cheaper), the need for *higher level languages,* which would enable the programmer to specify sequences of operations in concise terms, became apparent. Examples of such high-level languages include FORTRAN, ALGOL, BASIC and C. These languages allow the programmer to create a programme in the same way a numerical analyst would write his algorithm based on the original equations. Instructions encapsulated within the high-level languages can only be used by the computer if these are translated into machine language statements. This process of translation is achieved by a *compiler.* A compiler is, itself, a programme written in machine language which, after translating from high-level language to machine language, organises the allocation of storage for the translated programme instructions. The compiler is also responsible for detecting 'obvious' errors. High-level languages reduce the task of organising storage allocation and save the programmer enormous amounts of time; however, it is generally-accepted that machine language programmes produced by compilers are not as efficient as those crafted by experienced programmers. A typical programme will consist of a main programme segment and a number of *subprogrammes* or *subroutines.* The role of the main programme segment is to steer or control the course of the computations and to call the subprogrammes as required. This is quite an elegant way of organising the work since it is likely that the subroutines will be called many times to do repeat operations during the execution of the complete programme.

There has been an impressive development in the number and quality of commercial subroutine libraries, which contain very efficient programmes for specific tasks. These tasks vary from specific numerical manipulation of arithmetic data, such as inverting a matrix, to the plotting of complex three-dimensional functions. These subroutines are likely to be efficient and robust, and offer the further advantages of saving programmer time and effort. It is therefore common practice to adopt appropriate commercial subroutines and integrate them into customised computer programmes.

The continual increase in the performance of computers has made it possible to sacrifice some efficiency for improved utilisation of programmer effort. Further, increases in central processor speed and faster memory have enabled the development of programmes which fit users' needs and expectations. Thus, many computer users these days expect programmes to be built into a *Windows* environment, to be able to use click-on-icons, and to have access to pull down menus. Computer programmers are now faced with the challenge of writing the specific parts of their code in one of the common high-level languages, and then creating an attractive and useful package by exploiting the *graphical user interface (GUI)* of the Windows operating system. There are sophisticated software packages available which provide the programmer with appropriate tools to create graphical user interfaces for specific applications.

A modern large commercial computer programme is likely to consist of a core programme, written in possibly more than one high-level language, with a number of subroutines taken from a specialist subroutine library. This will be interfaced with a GUI, which will give it a 'user friendly' appearance. The programme may also be interfaced with a number of data banks, where it has access to thermal properties data (or other appropriate data).

Leading to: *Computational fluid dynamics*

G.L. Quarini

COMPUTERS

It was just 50 years ago that the first electronic computer was under development at the Moore School of Electronics in Philadelphia, as a tool to assist in the accurate targeting of long range artillery. Developments in computer technology and use since then have been a sensational demonstration of the pace of advance in science and engineering. Computers now impact on every aspect of life—as tools in scientific research, as dominant components of increasingly-complex business and commercial activities, as tools for personal use and as intelligent components of machines (ranging from automated factories to washing machines). It has been estimated that an equivalent rate of progress in the automobile industry would have provided luxury cars priced at £25 and with a fuel consumption of 400 miles per gallon!

This growth has been made possible by developments in hardware technology, which enable the construction of machines of appropriate performance, scale, reliability and cost, and by developments in software technology which provide cost-effective interfaces to users for a wide variety of requirements and expertise. Some of these advances are outlined briefly below and linked to the development of the discipline of computing.

Early computing systems which developed following the demonstration of the ENIAC system in 1946 (for example, EDSAC in the UK, EDVAC in the US) were simple systems of limited scale; their simplicity arose from limited knowledge of the exploitation of computers; their limited scale was due to the components used in their construction (electronic valves with cathode ray tubes or mercury delay lines as storage) and the associated problems of size, power consumption and reliability. They were prototypes which provided experience for gaining new methods of exploiting their capabilities. The advent of the transistor and the use of ferrite magnetic core memory devices made possible significant increases in scale and the incorporation of more sophisticated facilities, such as floating point arithmetic. These also made possible the creation of scaled-down versions of large *mainframe* machines, which became known as *minicomputers.* The DEC PDP-11 was a good example of such a system—small in scale but quite sophisticated in facilities.

The advent of microelectronics, whereby circuits involving a large number of transistors could be constructed by suitable doping of a single chip of a semiconductor, introduced dramatic changes to the scale of hardware systems as well as to their cost and reliability. Integrated circuits replaced discrete components. These could be mass-produced in sufficient quantities for use not only within arithmetic units but also as storage devices, with the result that stores became much larger (by a factor of 10^5 compared with early systems) and arithmetic units became more complex to improve their capability. Thus, the mainframe computer became more powerful without an increase in cost; the minicomputer became a *maxicomputer* such as VAX-11.

Perhaps more significantly, it became econimically-feasible to produce *microcomputers* which were sufficiently simple that an entire processor—complete with fast storage in the form of registers—could be implemented on a single chip. Microprocessors, incorporated in microcomputer systems, also had a profound effect on the market and in shaping future system development. Their scale allowed computers to be used economically to replace mechanical control of simple analogue processes, such as clocks, cameras and washing machines. They extended the range of computing systems by supporting desk-top computers (to the "mini" PDP-11 was added the micro version LSI-11) heralding the way for an explosive increase in computer usage. The impact of the microprocessor on computer system development, however, was a mixed blessing. The limited scale of early microprocessors led to the introduction of very simple architectures, such as the Intel 8080 and the Z-80, with short word lengths and primitive register structures and arithmetic units. In many ways, these were comparable with features of the early computing systems of the 1950s, and the tools used to exploit them were equally primitive due to limitations of scale.

The rapid growth in microelectronics capability made possible an expansion in the capabilities of microprocessors. The Intel 8080 formed the starting point of a range of processors used to support personal computers and work stations, with the same capabilities as earlier minicomputers and "main frames" (the top-of-the-range Intel 80486 was a 32-bit system with a very wide range of complex features; the Micro-VAX 11 and Motorola 68000 became the equivalent of mini-systems). In an attempt to exploit microelectronic capability, the resulting systems became more complex. A typical example was the case of Intel 80x86 range. Its complexity was in part due to compatibility with early microprocessors whose features had never been designed for future expansion. In the case of the Motorola 68000, its complexity arose from attempts to combine sophisticated and primitive facilities with conservative use of resources, such as a store to hold instructions.

The increased complexity of processor architecture, using microcode to enable the decoding and execution of a wide range of basic instructions, was enhanced further by the introduction of the *Reduced Instruction Set Computer* or RISC architecture. It was suggested that complex operations such as multiplication or string handling could be decomposed by a compiler at the time of compilation instead of by microcode or complex circuitry at execution time. The executing system could incorporate only simple primitives which could be executed at high speed and the space saved on the processor chip could be used for fast storage in the form of registers and slave stores. Provided that the hardware-executed primitives are suitable as a target for compilers (not for mythical assembler writers), the net effect could be a more cost-effective computing system. The widespread growth of RISC architectures had shown that this was indeed a powerful argument. But the increasing complexity of RISC architectures to achieve higher performance—as seen in the DEC ALPHA processor—also suggested that the cycle of computer development may well be repetitive. Useful studies on computer development can be found in the work of Siewiorek, Bell and Newell (1982) while a thorough analysis of current trends in computer design has been undertaken by Hennesey and Patterson (1990).

Although the core of any computing system is the hardware of the processor and stores, it became clear that a cost-effective system with a wide range of expertise must incorporate software components. Initially, these took the form of library subroutines which could be copied into a user's deck of cards; the user could then book time to run the task on the computer system. Hardware development of faster processors and of bulk storage capacity using magnetic tapes and discs made the manual procedures of library maintenance and of organising a queue of users increasingly ineffective; they were therefore replaced by automatic procedures, implemented by a resident system software in the form of an operating system, which could organise the maintenance and use of library procedures and the processing of a sequence of user tasks.

Early *operating systems,* such as the Fortran Monitor System, were designed to process a sequence of user tasks comprising decks of cards separated by "End-of-Job" cards. The mismatch between processor speeds and input/output operation using slow mechanical devices was be partially overcome by copying input material to a magnetic tape which has a faster interface with the main processor. The advent of random access fast storage devices, such as discs, enabled a more flexible job assembly system to be supported, in the form of "Spooling" systems. The Atlas system was an early example of this.

Such systems were inspired by the replacement of manual procedures by automatic procedures. But the resulting "batch systems" suffered a severe disadvantage in that the user was denied access to the computing system during execution of a job. User interaction is clearly useful in many applications and essential in others, but its handicap was that the processor was synchronised with the reaction time of a user. Thus, processing time became totally dominated by the time taken by the user to analyse output, and in the light of that analysis to direct future processing. To address this problem, an alternative organisation of job processing was introduced. This was paralleled by the development of batch systems, in which a single processor-store system could be switched to service many users, each interacting via their own terminal and enabling the system to overlap "think time" with processing of other active tasks. Time-sharing systems supporting multiple users were pioneered in the Multics System, which directly influenced the design of the UNIX and OS-2 systems, both of which are currently in widespread use in a variety of computer systems.

With the creation of microcomputers, which made available low-cost, high-performance processing, further development of operating procedures has become possible. One approach is to provide multiple "stand-alone" systems in the form of desk-top personal computers, each operated by users using manual control techniques in much the same way as early computer systems were used. An alternative strategy can be regarded as a development of time-sharing: the terminal is replaced by a work station physically similar to a personal computer, but providing only part of a user requirements; the rest are supplied by one or more servers which operate on a time-shared basis. Typically, an X-Terminal will support the graphical user interface to one or more tasks, with functions such as file store access and maintenance, compilation etc. being carried out by by servers (which again may be physically identical to a personal computer or may be mini or mainframe computing systems). The resulting network of computing resources—under the control of an operating system software—presents the appearance of a single overall system with multiple "intelligent" access points. Supported operating systems, such as UNIX, MS-DOS and OS-2, are now commonly available.

It might be supposed that the implementation of distributed processing has made irrelevant the capabilities implemented in batch and time-sharing systems on large mainframe computers. But this has not been the case for several reasons. Firstly, workstations are commonly used to support multiple tasks for a single user, as in the *Windows* system, and system support is similar to that developed in multiuser systems. More important, such systems are only effective if they conceal some properties of the real machine from access by user tasks, thus replacing the real machine by a *Virtual Machine* which reflects the properties of the underlying hardware and software mechanisms. For example, if several tasks share a single physical memory, it is clearly inappropriate to access the instructions and data by the real address of the appropriate memory cells, for these will vary depending upon the mix of tasks under execution. It becomes more appropriate to support a *Virtual Memory* with mechanisms to convert a virtual address into a real address at time of use. The virtual memory can reflect the logical property of the information, whilst mapping to a real store hierarchy is a system function; for example, there is benefit if the elements of an array occupy contiguous cells of virtual memory, whereas the array can be located in real store as blocks or pages which can be independently located.

Support by system software of a "virtual machine" is one of the most important provisions of a system software. It is commonplace in all modern systems, although the nature of the virtual machine and its relationships to the characteristics of the real machine differ in different systems. UNIX supports virtual machines which are almost independent of the hardware of the real machine; the VM-370 system supports virtual machines which are almost identical qualitatively to the real IBM 370 range of machines. Virtual machines contribute significantly to the ease the programming of tasks for execution by any computing system—whether a stand-alone personal computer, a network of work stations or a multiuser mainframe system. A useful exposition of the structure of various operating systems is provided by Deitel (1990).

The creation of sequences of instructions which can be stored and subsequently executed by a processor can be achieved in a number of ways. A sequence could be manually created by means of a set of switches; more realistically, the sequence could itself be produced by the computing system, taking as input an expression of the required sequence supplied by a user. A *compiler* is a program converting a source *language* expression of a problem into a machine-executable sequence of instructions.

Early source languages were a mnemonic form of machine code. This was easily extended to form an Assembly Language, which had a close relationship to machine code but included some higher-level facilities such as naming and macroinstructions or subsequences of machine instructions. This level of problem expression was unsatisfactory in that the task of converting a problem to this form is exceedingly complex and error-prone and it is specific to the architecture of a particular system. The growth of computing was therefore accompanied by the development of higher-level languages providing a more natural description of the problem in a largely machine independent form, but allowing translation to machine code by a compiler executed by the computing system. Fortran and Cobol were two early languages suitable for scientific and commercial problems, respectively. Basic was a simpler language, popular in small-scale systems such as personal computers because of the simplicity of the required translation. These languages were static languages, in which work space and store for variables were allocated at the start of the job (by the compiler).

A preferable approach is to allocate space dynamically as required during execution, and procedural languages such as Algol, Pascal, and Modula 2 exhibit this characteristic. The C language is also a procedural language which enjoys current popularity. It is modeled on the Unix virtual machine—a largely machine-independent system—and can be used at a variety of levels of abstraction.

These languages are all "imperative" languages in that they consist of a sequence of statements which, when translated, correspond to sequences of machine instructions. An alternative form of problem description is that found in "declarative" languages. Here, the rules of problem solution are described and the compiler uses these rules to construct appropriate instruction sequences. For example, consider the problem of evaluation of the factorial N!. This could be expressed in imperative form as a sequence of multiplications by 2, 3, 4. . . . . .N. In declarative form, the functional definition

$$N! = 1 \text{ if } N = 0 \text{ else } = N*(N - 1)!$$

allows the compiler to create an instruction sequence appropriate to any particular machine configuration. One such declarative language enjoying considerable popularity among knowledged-based systems is PROLOG; the declarations consist of Facts and Rules stored in a Knowledge Base and operated on by an Inference Engine, which applies the rules to a particular problem.

The ever-increasing scale of computer application has the inevitable consequence of increasing the scale and complexity of problem description. The discipline of *Software Engineering* addresses this requirement of producing complex problem descriptions which can be generated, tested, maintained and modified in an effective manner. Languages can help in this process; currently, Object-Orientated Languages such as SMALLTALK and C++ are becoming widely-used. Their principle of operation is to move away from the customary description of data structures and procedures which operate on the data to a description of Objects which incorporate data and the procedures that manipulate that data, and which communicate with each other by passing messages. The resulting modularisation of complex problems is one tool that can be used in the quest for manageable production of reliable and maintainable programs. Further discussions of the various types of programming languages are given by Horowitz (1987) and Wilson and Clark (1988).

Computing systems, whose original objective was to provide a calculation mechanism for use in scientific and engineering research, now pervade the whole of society. Their partially unforeseen expansion in scope has been the result of the increased cost-effectiveness of computing systems themselves over the past 50 years; in turn, this expansion in scope has influenced the development of computing systems.

Leading to: *Computer programmes*

D.J. Howarth

CONCENTRATING COLLECTOR (see Solar energy)

CONCAVE SURFACE, FLOW OVER (see Görtler-Taylor vortex flows)

CONDENSATE INUNDATION (see Tube banks, condensation heat transfer in)

CONDENSATION COEFFICIENT (see Accommodation coefficient; Condensation of pure vapour)

CONDENSATION CURVE

Following from: *Condensation of pure vapour; Condensation of multicomponent vapours; Condensation, overview*

A Condensation Curve, **Figure 1**, is a plot of temperature against specific enthalpy, h, or cumulative heat removal rate, $\dot{Q} = \dot{M}(h_{in} - h)$, for a pure vapour or a mixture. A superheated vapour at A is cooled, becoming saturated at B, the dew point T_{dew}, by removal of heat at rate $\dot{Q}_B$. Condensation takes place along the curve BC until the bubble temperature, T_{bub}, is reached at C. Thereafter, a subcooled liquid is produced at D, corresponding to an overall heat removal rate, $\dot{Q}_T$. EF shows a cooling curve for the coolant.

Figure 1 is the *Integral Condensation Curve.* It describes condensation in a device where vapour and condensate follow a parallel path with complete mixing so that equilibrium is maintained. For a binary, the curve ABCD of **Figure 1** is derived from the temperature–composition diagram (**Figure 2**), which gives the state, (vapour, liquid or two-phase), the dew and bubble temperatures and the vapour fraction, θ. Thus at E, one mole of feed separates into θ moles of vapour and (1–θ) moles of liquid, of molar composition $\tilde{y}$ and $\tilde{x}$, respectively. θ is given by the division of the tie line.

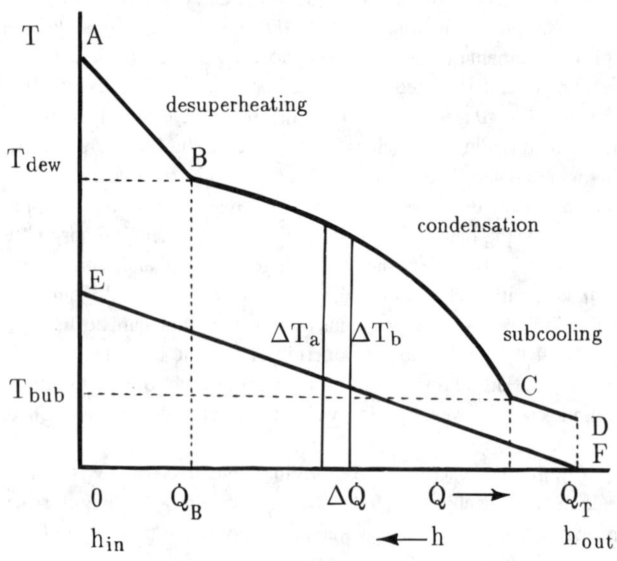

Figure 1. Integral condensation curve.

Other condensation curves may be defined. A *Differential Condensation Curve* arises when vapour and condensate are separated within the condenser, and depart from overall equilibrium. Such a curve is also derived from **Figure 2**. Tie lines are constructed to give a number of equilibrium stages. Each stage divides the mixture into a condensate and vapour, which are then considered separately. The next separation is applied to the remaining vapour. **Figure 2** shows a process giving 10 equal condensates with saturated vapour. An auxiliary curve might be constructed to show how the condensates are mixed and cooled. The curve C_oG shows all prior condensate fully-mixed and cooled to the vapour temperature. A differential curve is the limit of an infinite number of stages.

Such cooling curves are convenient because they do not depend on condenser geometry. They may sometimes approximate the true cooling curve of a real condenser, **Fig. 3**. Vertical condensers with turbulent condensate films are likely to approximate the integral curve while the differential curve, C_oG, might be a good description of a horizontal condenser, where condensate forms a layer in the bottom of the unit, or a vertical condenser with condensate in laminar flow. It is clear that differential condensation must be avoided, because of the rectification involved. The vapour becomes richer in the more volatile component with a fall in saturation temperature and driving force so that a larger area is required for a given condensation.

The condensation curve defines the mean temperature driving force (**Figure 1**). Consider the interval $\Delta\dot{Q}$, small enough that the slope of the cooling curve is constant. The appropriate mean temperature driving force is the log mean of the terminal temperature differences between hot and cold streams (See **Mean Temperature Difference**):

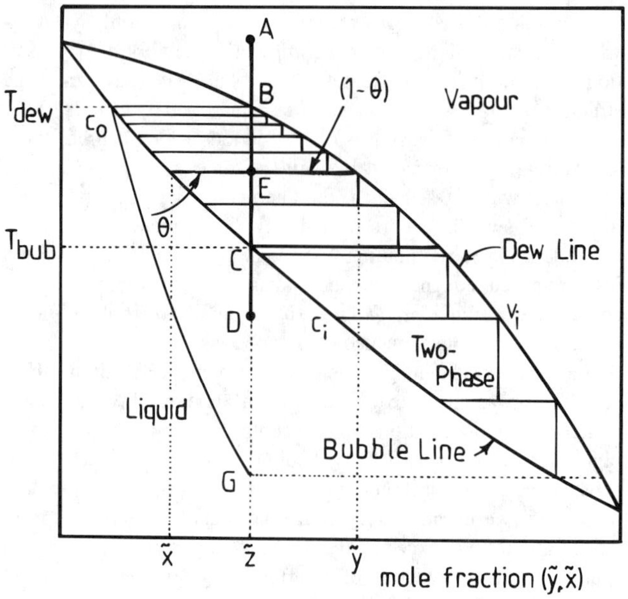

Figure 2. Integral and differential condensation.

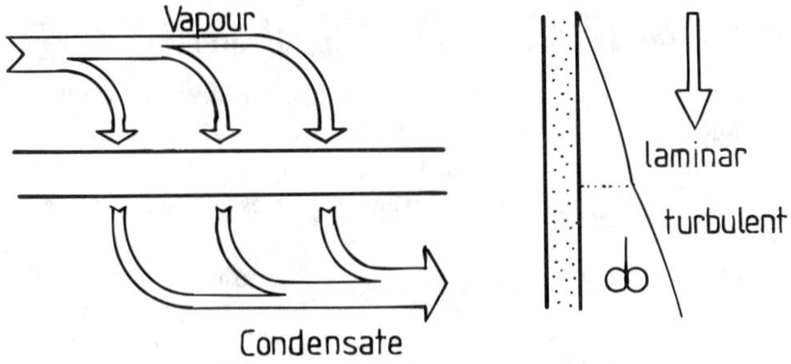

Figure 3. Condensation processes.

$$\frac{1}{\Delta T_m} = \frac{1}{\Delta \dot{Q}} \int_{\Delta \dot{Q}} \frac{d\dot{Q}}{\Delta T}$$

$$\Delta T_m = \frac{\Delta T_a - \Delta T_b}{\ln(\Delta T_a / \Delta T_b)} = \Delta T_{ln}. \quad (1)$$

Further, the slope of the cooling curve determines the ratio of the sensible to total heat transfer rates in an interval, Z, which fixes the gas side heat transfer resistance, Z / α_g,

$$Z = \left\{ \frac{d\dot{Q}_g}{d\dot{Q}} \right\}_{sat} = \dot{M}_g c_{pg} \frac{dT_g}{d\dot{Q}} \quad (2)$$

$$\frac{1}{U_o} = \frac{1}{\alpha_o} + \frac{Z}{\alpha_g}. \quad (3)$$

General Determination of the Integral Cooling Curve

The overall and component mass balances show the split of total flow, $\dot{N}$, composition $\tilde{z}_i$, into condensate and vapour flows, $\dot{N}_c$, and $\dot{N}_g$, compositions $\tilde{x}_i$ and $\tilde{y}_i$, respectively,

$$\dot{N} = \dot{N}_c + \dot{N}_g = \Sigma\,(\dot{N}_{ic} + \dot{N}_{ig}) \quad (4)$$

$$\tilde{z}_i = (1 - \theta)\,\tilde{x}_i + \theta \tilde{y}_i = \dot{N}_i / \dot{N} \quad (5)$$

with definition of K values, $K_i = K_i(T,P,\tilde{x}_i,\tilde{y}_i)$,

$$\tilde{y}_i = K_i \tilde{x}_i \quad (6)$$

then

$$\tilde{x}_i = \tilde{z}_i / (1 + \theta\,[K_i - 1]) \quad (7)$$

A quantity, $S = \Sigma\,(\tilde{y}_i - \tilde{x}_i)$, which is a decreasing function of θ, is determined. When $S = 0$, the mass balance, Equation (5), summed over all components, guarantees that the individual $\tilde{x}_i$ and $\tilde{y}_i$ sum to unity. For various θ at a given temperature, S is calculated from $\tilde{z}_i$, by Eqs. (6) and (7). **Figure 4** then allows diagnosis of the state of the mixture—all liquid, all vapour or at a two-phase equilibrium. In the latter case, the particular θ which makes $S = 0$ has to be found.

The mixture enthalpy and hence, the cooling curve may now be determined. At any specified temperature, T,

$$\tilde{h} = \theta\,\Sigma\,\tilde{y}_i \tilde{h}_{ig} + (1 - \theta)\,\Sigma\,\tilde{x}_i \tilde{h}_{il}. \quad (8)$$

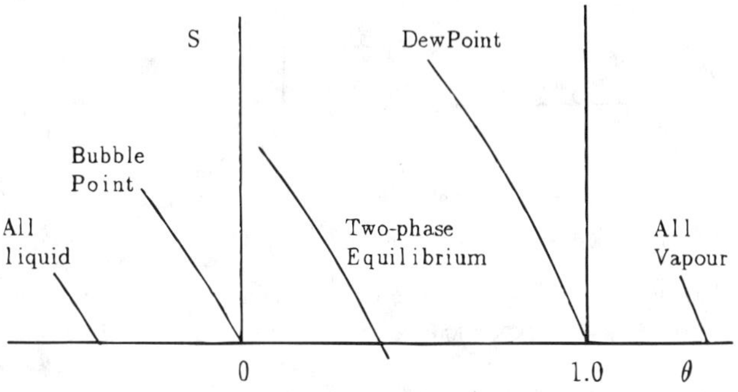

Figure 4. Diagnosis of mixture state.

Determination of the Mean Temperature Driving Force

Figure 5 shows how the mean temperature difference, ΔT_m, may be calculated for an 'E' shell with one-coolant pass.

The energy balance maps the cooling curve, $T = T(h)$, of the cold fluid to $T = T(h')$. The area required follows when the cooling curve is split into intervals where ΔT is linear with h',

$$(h - h_{in}) = \dot{M}'(h' - h'_{out})/\dot{M}$$

$$A_T = \sum_j \dot{M}' \Delta h'_j / [U_{m,j} \Delta T_{ln,j}]. \quad (9)$$

With two passes, **Figure 6** shows the essentials of a procedure used to calculate the appropriate ΔT_m.

The coolant curve, T(h) is mapped to T(h′) by energy balance with the assumption that the heat transfer rates to each pass, δh^I and δh^{II}, are in the ratio of the temperature differences. The arithmetic

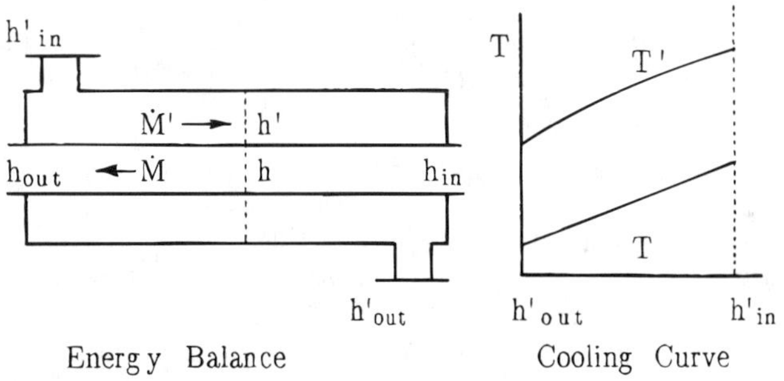

Figure 5. TEMA 'E' shell with one coolant pass.

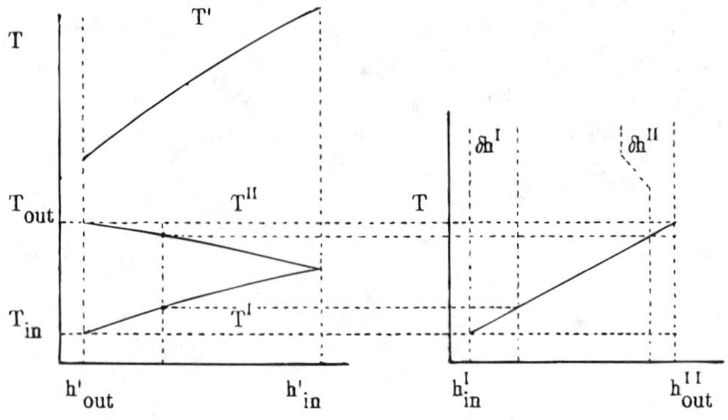

Figure 6. TEMA 'E' shell with two coolant pass.

average, $(T^I + T^{II}) / 2$, is the effective coolant temperature. These relationships are expressed as:

$$\delta h^{II}/\delta h^{I} = (T' - T^{II})/(T' - T^{I})$$
$$h' = h'_{in} - \dot{M}(h^{II} - h^{I})/\dot{M}'$$
$$\Delta T = [T' - (T^{I} + T^{II})/2]. \quad (10)$$

With more than two passes, the temperature change in a pass is so small that the mean temperature is independent of h'. With $\bar{T}'$ and $\bar{T}$ the effective temperatures of the two streams, the mean temperature driving force can be found by solving

$$\Delta T_m = \frac{(T_{out} - T_{in})}{\ln\left(\dfrac{\bar{T}' - T_{in}}{\bar{T}' - T_{out}}\right)} = (\bar{T}' - \bar{T}) = \dot{Q}_T/\Sigma \frac{\Delta \dot{Q}_j}{\Delta T_{ln,j}}. \quad (11)$$

D.R. Webb

CONDENSATION IN TUBE BANKS (see Tube banks, condensation heat transfer in)

CONDENSATION, OF DROPS (see Drops)

CONDENSATION OF MULTICOMPONENT VAPOURS

Following from: *Condensation, overview; Diffusion; Maxwell Stefan equations*

Introduction

Two methods of design of condensers for multicomponent mixtures are used in industrial practice: the equilibrium method of Silver (1947) and the rate model of Colburn et al. (1934) and (1937). They are both film-based models in which the resistance of a series of layers are added, but the former is essentially a heat transfer model in which the mass transfer rate is approximated, whereas the latter provides an accurate description of heat and mass transfer processes.

Fundamental Mass Transfer Considerations

Mass transfer is important in many condensation processes and the fundamentals are given in the entries on **Diffusion** and the **Maxwell Stefan Equations**. In a mixture of n components, the *Diffusion Law* gives the n independent fluxes of condensation

$$\dot{n}_i = J_{ig} + \tilde{y}_{ig}\dot{n}_t, \qquad i = 1 \text{ to } n$$

$$\begin{bmatrix}\text{molar}\\ \text{flux}\end{bmatrix} = \begin{bmatrix}\text{diffusive}\\ \text{flux}\end{bmatrix} + \begin{bmatrix}\text{convective}\\ \text{molar flux}\end{bmatrix}. \quad (1)$$

It is the $\dot{n}$ which are needed in design and the above equation shows a diffusive and a convective contribution to these fluxes. The theories of diffusion allow the evaluation of J_g, the diffusive fluxes, and therefore the prediction of mass transfer effects in condensation. In most cases of condensation,

$$J_{ig} < \tilde{y}_{ig}\dot{n}_t.$$

Convection, the condensation flux towards the interface, induced by removal of energy, dominates the transfer process and the diffusion is a small effect riding on top. However, there are many condensation processes, e.g., reflux condensers, where the diffusional transfer rate becomes important. It is worth emphasizing that there are n independent fluxes, $\dot{n}_i$, but only (n − 1) independent diffusive fluxes, J_i. Condensation is not fully defined by diffusional processes and the additional determinacy relation required is the total condensation rate.

Prediction of the Diffusive Flux

There are three starting points for the description of multicomponent mass transfer. The following form of **Fick's Law** describes multicomponent mass transfer in a dilute mixture of vapours in a noncondensing gas, where it is a form of *Effective Diffusivity Method*. It is exact for binary mixtures.

$$J_i = -\tilde{c}\mathcal{D}_{in}\frac{d\tilde{y}_i}{ds} \qquad \text{for } i = 1 \text{ to } (n-1). \quad (2a)$$

The second is the Generalised Fick's Law equation. The *Linearised Theory* uses this equation as starting point.

$$J_i = -\tilde{c}\sum_{k=1}^{n-1} D_{ik}\frac{d\tilde{y}_k}{ds} \qquad i = 1 \text{ to } (n-1). \quad (2b)$$

Thirdly, the **Maxwell-Stefan Equations** may be used. Their advantage is that there is an *Exact Matrix Solution* of the M-S equations for a film model.

$$\frac{d\tilde{y}_i}{ds} = \sum_{\substack{k=1\\k\neq i}}^{n} \frac{\tilde{y}_i\dot{n}_k - \tilde{y}_k\dot{n}_i}{\tilde{c}\mathcal{D}_{ik}}. \quad (2c)$$

Equation [2a] differs from the other two equations in that independent diffusion is implied. Each component transfers under the action of its own composition driving force alone. This is acceptable in most cases, but sometimes the latter two interactive methods give better predictions. No significant difference between these two arises in film methods.

Film Method Solution for Mass Transfer

A one-dimensional, steady film of thickness, s_f, is considered adjacent to the condensate surface. The transfer of mass follows the diffusion law and all changes in concentration occur across this film; from a vapour which is completely mixed to a condition of equilibrium at the condensate surface. The thickness of the hypothetical film is not required, but is incorporated within the binary mass transfer coefficient, β_{ik}. See **Figure 1**.

The solution of Eq. (1) for the J_i has been achieved with each of the three above methods of determining the diffusive fluxes to give a solution of the same form,

$$(J_g) = [B][\Phi]\{\exp[\Phi] - \ulcorner I \lrcorner\}^{-1}(\tilde{y}_g - \tilde{y}_s). \qquad \textbf{(3)}$$

Here, braces and square brackets show (n − 1) column and square matrices, and the usual rules of matrix multiplication apply. In this equation, [B] is a matrix of mass transfer coefficients. In general, it will have off-diagonal elements of finite size so that it will show an interaction between species, where the transfer of any component is influenced by all the independent driving forces in the mixture.

The matrix [Φ] is present in the equation to show the effect of a finite rate of transfer within the film. It is akin to the *Ackermann correction factor*, which arises when mass transfer takes place in the direction of a temperature gradient. With [M(m)] defined as a function of $\beta_{ik} = \tilde{c}\,\mathcal{D}_{ik}/s_f$ by,

$$M_{ii} = \frac{m_i}{\beta_{in}} + \sum_{\substack{k=1\\k\neq i}}^{n} \frac{m_k}{\beta_{ik}} \qquad M_{ik} = -m_i\left[\frac{1}{\beta_{ik}} - \frac{1}{\beta_{in}}\right],$$

Table 1 shows how to evaluate matrices $[B_g]$ and [Φ] in Eq. (3) by the methods of Eqs. (2a), (2b), and (2c).

In the effective diffusivity method, the matrices are diagonal, shown by $\ulcorner\ \lrcorner$, and each species transfers independently. More detailed descriptions of the developments leading to the above equation are given in Webb (1980) and Taylor and Krishna, (1993).

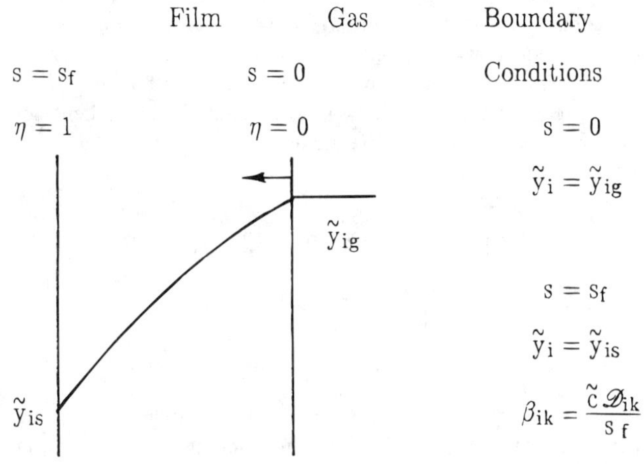

Figure 1. Simple film model.

Table 1. Evaluation of mass transfer coefficients

Method	Eq.	Matrix[B]	Matrix[Φ]
Eff. Diff.	(2a)	$\ulcorner\beta_{in}\lrcorner$	$\ulcorner\dot{n}_t/\beta_{in}\lrcorner$
Linearised	(2b)	$[M(\bar{y})]^{-1}$	$[M(\bar{y}\ \dot{n}_t)]$
Exact matrix	(2c)	$[M(\tilde{y}_g)]^{-1}$	$[M(\dot{n})]$
where	$\bar{y} = (\tilde{y}_g + \tilde{y}_s)/2$		

Film Method of Combined Heat and Mass Transfer

Figure 2 shows the region near the interface and defines nomenclature when condensation is treated by a film model.

The energy equation, with terms which describe steady-state conduction and convection, is quoted and integrated across the film to give the conductive flux across the interface, $\dot{q}_s$.

$$\dot{q}_s = -\lambda_g \frac{dT}{ds} + \Sigma\, \dot{n}_i \tilde{c}_{pfi}(T - T_s)$$

$$\begin{bmatrix}\text{heat}\\\text{flux}\end{bmatrix} = \begin{bmatrix}\text{conductive}\\\text{flux}\end{bmatrix} + \begin{bmatrix}\text{convective}\\\text{heat flux}\end{bmatrix}. \qquad \textbf{(4)}$$

$$\dot{q}_s = \alpha_g \frac{\varepsilon e^{\varepsilon}}{(e^{\varepsilon} - 1)}(T_g - T_s). \qquad \textbf{(5)}$$

where

$$\alpha_g = \frac{\lambda_g}{s_t} \quad \text{and} \quad \varepsilon = \frac{\Sigma\, \dot{n}_i \tilde{c}_{pfi}}{\alpha_g} = \frac{\dot{n}_t \tilde{c}_{pf}}{\alpha_g}.$$

The heat transfer coefficient, α_g, is that for conductive heat transfer alone because in the limit of zero mass transfer rate,

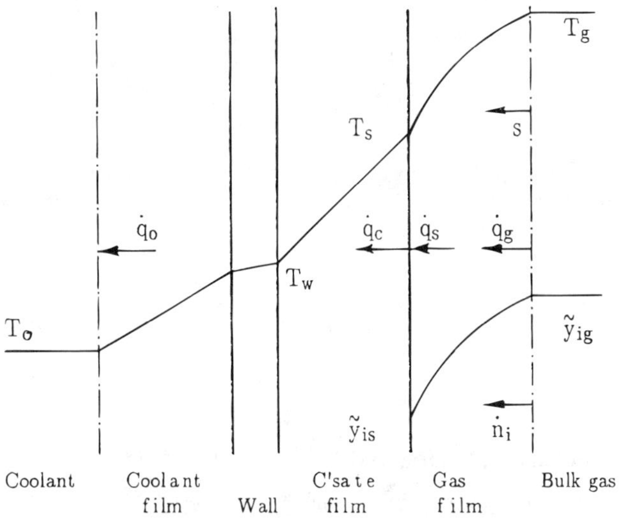

Figure 2. Film model of condensation.

$$\underset{\varepsilon \to 0}{\mathrm{Lt}} = \alpha_g \frac{\varepsilon e^{\varepsilon}}{(\varepsilon^{\varepsilon} - 1)} = \alpha_g.$$

For simplicity, liquid subcooling effects are not included here, but are treated later. It is assumed that continuity of energy across the interface may be written as $\dot{q}_o = \dot{q}_c = \dot{q}_s + \dot{n}_t\,\Delta\tilde{h}_v$. An overall liquid side heat transfer coefficient, α_o, is defined including condensate, wall, coolant film and dirt,

$$\frac{1}{\alpha_0 A_{wo}} = \frac{1}{\alpha_c A_{wo}} + \frac{s_w}{\lambda_w A_{mw}} + \frac{1}{\alpha_w A_{wi}} + r_f$$

$$\alpha_o(T_s - T_o) = \alpha_g \frac{\varepsilon e^{\varepsilon}}{(e^{\varepsilon} - 1)}(T_g - T_s) + \dot{n}_t\,\Delta\tilde{h}_v. \qquad \textbf{(6)}$$

The rate of cooling of the gas phase, $\dot{q}_g$ may be obtained. The two heat fluxes, $\dot{q}_g$ and $\dot{q}_s$, differ by the sensible heat change over the gas film, so that the rate of gas cooling is less than the heat flux at the condensate surface.

$$\dot{q}_g = \alpha_g \frac{\varepsilon}{(e^{\varepsilon} - 1)}(T_g - T_s) = \alpha_{\dot{g}}(T_g - T_s).$$

The term $\varepsilon/(e^{\varepsilon} - 1)$ is the *Ackermann Correction Factor.* It is the factor by which the heat transfer is modified to account for mass transfer. A similar factor, $\phi/(e^{\phi} - 1)$ occurs in Equation (3). Comparison of $\dot{q}_g$ and $\dot{q}_s = e^{\varepsilon}\,\dot{q}_g$ shows that the higher the rate of condensation, the slower the rate of gas cooling.

Determination of Transfer Coefficients

Standard correlations may be used for the geometry in question. The correlations define the unknown thicknesses, s_t and s_f, of the thermal and mass transfer films.

$$\mathrm{Nu} = \frac{\alpha d}{\lambda} = k_1\,\mathrm{Re}^{k2}\mathrm{Pr}^{k3} = \frac{d}{s_t}$$

$$\mathrm{Sh} = \frac{\beta_{12} d}{\tilde{c}\mathcal{D}_{12}} = k_1 \mathrm{Re}^{k2}\mathrm{Sc}_{12}^{k_3} = \frac{d}{s_f}. \qquad \textbf{(7a)}$$

With an analogy, only a single correlation is needed. The analogy effectively links s_f and s_t. The *Chilton-Colburn analogy* (1934) may be written as follows:

$$J_h = \frac{\alpha_g \mathrm{Pr}^{2/3}}{\dot{N}_g \tilde{c}_{pg}/S} = J_d = \frac{\beta_{12}\mathrm{Sc}^{2/3}}{\dot{N}_g/S} = f(\mathrm{Re}). \qquad \textbf{(7b)}$$

Condensate Mixing and Interfacial Conditions

Vapour liquid equilibrium must be satisfied at the interface. The behaviour of the condensate lies between two limiting conditions of liquid mixed and unmixed. These are compared in Figure 3.

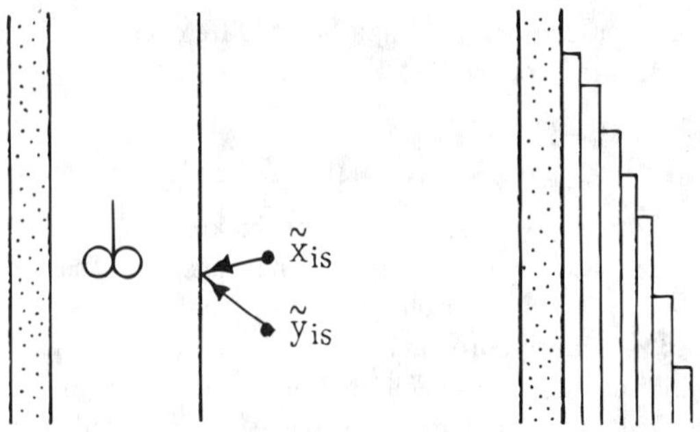

Vertical condensers with laminar condensate	Vertical condensers with turbulent condensate
	Horizontal condensers
Integral Condensation	Differential Condensation
Well mixed liquid	Unmixed liquid

Figure 3. Mixed and unmixed condensates.

$$\tilde{x}_{is} = \tilde{x}_{ic} = \dot{n}_{ic}/n_c \qquad \tilde{x}_{is} = \dot{n}_i/\dot{n}_t \qquad \textbf{(8)}$$

Local Equations of Mass and Energy Transfer

At any location in a condenser, gas temperature, T_g, gas composition, $\tilde{y}_{ig}$, and cooling fluid temperature, T_o, are known. The calculation of the local rates of mass transfer, $\dot{n}_i$, and energy transfer, $\dot{q}$, involves the estimation of interfacial conditions, T_s, $\tilde{y}_{is}$ and $\tilde{x}_{is}$, a total of $(2n + 1)$ unknowns. Solving the $(2n + 1)$ equations allows the solution of the local problem. The equation of continuity of energy, Eq. (6), may be rewritten by defining a dimensionless interfacial temperature, $\chi = (T_s - T_o)/(T_g - T_o)$, between 0 and 1.

$$\dot{n}_t = \alpha_o \chi \Big/ \left\{\frac{\Delta\tilde{h}_{vf}}{(T_g - T_o)} + \frac{\tilde{c}_{pf}(1 - \chi)}{\{1 - \exp(-\varepsilon)\}}\right\}. \qquad \textbf{(9)}$$

The physical properties $\Delta\tilde{h}_{vf}$, $\tilde{c}_{pf}$ and ε are flux dependent because the components contribute through their condensation rate. Moreover, the equation is implicit in that ε depends on $\dot{n}_t$. Neither is a problem in that the equation is usually solved within an iterative sequence to determine the fluxes, and the implicit dependence on $\dot{n}_t$ in ε is weak. With diffusive fluxes,

$$(J_g) = [B][\Phi]\{\exp[\Phi] - \lceil I \rfloor\}^{-1}(\tilde{y}_g - \tilde{y}_s) \qquad \textbf{(3)}$$

$$(\dot{n}) = (J_g) + \dot{n}_t(\tilde{y}_g) \text{ for } i = 1, n - 1, \qquad \textbf{(1)}$$

it is seen that Eq. (9) is a determinacy condition allowing the determination of the n independent $\dot{n}_i$ from the J_i. The interfacial vapour composition is defined by the assumption of equilibrium with the liquid.

$$(\tilde{y}_s) = {}^{\ulcorner}K_{\lrcorner}(\tilde{x}_s) \text{ for } i = 1, n \quad \textbf{(10)}$$

$$K_i = K_i(T, p, \tilde{x}_s, \tilde{y}_s) = \frac{p_i^o(T_s)\gamma_i}{\phi_i\, p} \quad \textbf{(11)}$$

$\gamma_i = \gamma_i(\tilde{x}_s)$ Liquid activity coefficient

$\phi_i = \phi_i(\tilde{y}_s)$ Vapour fugacity coefficient

The $(2n + 1)$ equations required are:

- Continuity of energy (1);
- Mixed or unmixed liquid conditions (n); and
- Vapour-liquid equilibrium at the interface (n).

Computer algorithms for determining the equilibrium state of an n component mixture usually involve two iteration loops and this is the structure of methods to solve these equations.

Solution of Local Equations

Factors that affect the solution of the local problem are:

- The liquid mixing condition, Mixed or Unmixed;
- Presence of non-condensing gas;
- Miscible or immiscible condensate.

Liquid mixing has a strong effect and changes the structure of the algorithm required to calculate the transfer rates. A non-condensing gas is usually a component above its critical temperature. However, in the case of vapours of immiscible liquids with one condensate phase present, but above the azeotropic temperature, all components that will form the second immiscible phase behave as non-condensing species.

Mixed Condensates

The case of mixed condensate is easier because the interfacial liquid condition is defined by prior condensation. In the absence of non-condensing gas, irrespective of whether one- or two-liquid phases are present, a bubble point calculation defines the interfacial state. The above equations are then explicit in transfer rates, iteration being required only to determine the flux dependent quantities in Equations (1), (3) and (9), [B] [Φ], and properties $\tilde{c}_{pf}$ and $\Delta\tilde{h}_{vf}$.

When noncondensing gases are present, and these may be vapours having the potential to form a second condensate phase, the calculation of interfacial temperature becomes iterative. For each non-condensing gas, the liquid composition is $\tilde{x}_i = 0$, and the vapour composition can be obtained directly from the Maxwell Stefan equations as:

$$\tilde{y}_{is} = \tilde{y}_{ig} \exp\left\{\sum_{k=1}^{\ell} \frac{\dot{n}_k}{\beta_{ik}}\right\} \qquad i = \ell + 1 \text{ to } n$$

where ℓ is the number of species present in the liquid.

Unmixed Condensates

Condenser design for the case of an unmixed condensate is similar to the vapour-liquid phase evaluation calculation and can be done in the same way, despite being rate-controlled. It involves finding an interfacial temperature χ (between 0 and 1), and compositions $\tilde{x}_{is}$ and $\tilde{y}_{is}$ for specified vapour conditions, T_g and $\tilde{y}_{ig}$, and coolant temperature, T_o.

In comparison, the phase evaluation seeks the vapour fraction, θ (between 0 and 1), and compositions $\tilde{x}_i$ and $\tilde{y}_i$ for specified overall composition, $\tilde{z}_i$, temperature and pressure. A phase evaluation uses two nested iterations (See **Cooling Curve**):

1. Outer loop to converge θ
2. Inner loop on composition.

Condenser design algorithms may be devised with similar structure. The unmixed liquid condition, Eq. (8), is combined with Equations (1) and (3) to give

$$(\tilde{y}_s) = \{{}^{\ulcorner}1/K_{\lrcorner} + [f(\chi)]\}^{-1}\{{}^{\ulcorner}I_{\lrcorner} + [f(\chi)]\}(\tilde{y}_g)$$

$$[f(\chi)] = [B][\Phi]\{\exp[\Phi] - {}^{\ulcorner}I_{\lrcorner}\}^{-1}/\dot{n}_t. \quad \textbf{(12)}$$

The term $f(\chi)$ is a function only of the interfacial temperature.

The following shows an algorithm for the case of an unmixed, single condensate with or without a noncondensing gas:

Step			
1	Guess	χ, $\tilde{c}_{pf}$, $\Delta\tilde{h}_{vf}$	{Comp av.}
2	Calculate	$\dot{n}_t$	{Eq 9}
3	Calculate	$K_i(T,p)$	
4	Calculate	β_{ij}, [B], [Φ]	{Table 1}
5	Calculate	$\tilde{y}_{is}$, $i = 1,n - 1$	{Eq 12}
		$\tilde{y}_{ns} = 1 - \sum^{n-1} \tilde{y}_{is}$	
6	Calculate	$\tilde{x}_{is} = \tilde{y}_{is}/K_i$	
7	Calculate	(Jg),($\dot{n}$)	{Eq 1 & 3}
8	Calculate	$K_i(T,p,\tilde{x}_{is}/\Sigma\tilde{x}_{is},\tilde{y}_{is})$	
9	Calculate	$\tilde{c}_{pf}$, $\Delta\tilde{h}_v$,	{Flux av.}
NO	Is $\Sigma\, \tilde{x}_i$ constant?	YES	
NO	Is $\Sigma\, \tilde{x}_i = 1$	SOLN	

In such an algorithm, the following must be taken into account:

- The guess of χ in the outer loop is made as follows, If $\Sigma\, \tilde{x}_{is} > 1.0$ increase χ and vice versa.
- Properties $\tilde{c}_{pf}$ and $\Delta\tilde{h}_{vf}$ are flux-averaged. Their values are updated in the inner loop as fluxes are defined.
- There are only $(n - 1)$ independent, $\tilde{y}_{is}$, in Step 5. The last composition $\tilde{y}_{ns}$ is obtained by difference from 1.0,
- If the value of $\Sigma\, \tilde{x}_{is} \neq 1.0$ in the inner loop, K_i is obtained using the normalised estimate, $\tilde{x}_{is}/\Sigma\, \tilde{x}_{is}$.
- In the first iteration, evaluate the mass transfer coefficients and rate factors using known quantities,

$$[B] = [M(\tilde{y}_g)]^{-1} \qquad [\Phi] = [M(\dot{n}_t\tilde{y}_g)].$$

The correct dependence, as shown in **Table 1**, can be introduced as the dependent values become available.

Design and Rating Calculations

Design or rating a condenser involves integration of the ordinary differential equations expressing the downstream development of the independent variables. Here, we consider T_g - Gas Temperature, T_o - Coolant Temperature, $\dot{N}_{ig}$ - Gas Flowrates.

A full treatment should also include pressure, if pressure drop causes a significant loss of saturation temperature, and condensate temperature to allow for subcooling, (See later). Flowrate, $\dot{N}_{ig}$, and temperatures, T_g and T_o, must be known at the vapour inlet to define an initial value problem, **Figure 4**.

Design involves the calculation of the area to achieve a given degree of condensation while rating involves the calculation of the degree of condensation produced in a given area. Design is usually achieved as a series of rating calculations of guessed condensers until a satisfactory performance is achieved. Rating calculations are presented below.

Mass and energy balances are carried out over an element of differential area, δA, as shown in **Figure 5**.

Mass balances over the area δA give for any component, for the mixture as a whole and for a non-condensing gas,

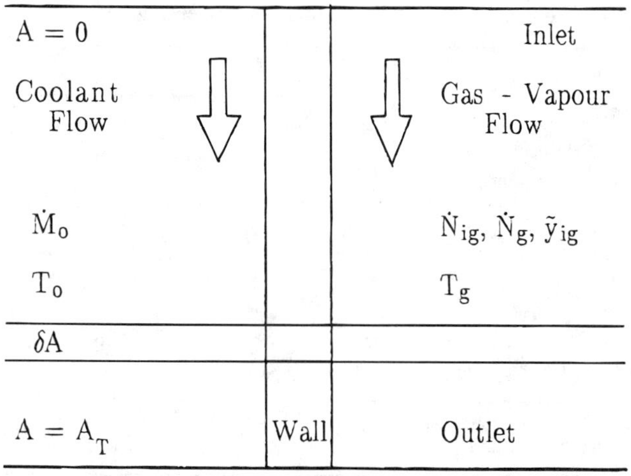

Figure 4. Development of downstream conditions.

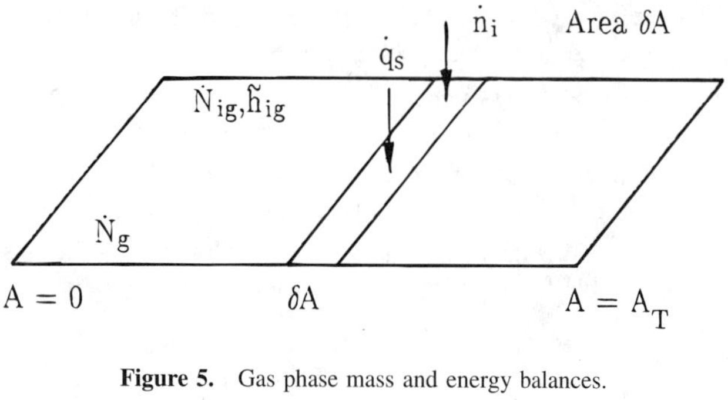

Figure 5. Gas phase mass and energy balances.

$$\frac{d\dot{N}_{ig}}{dA} = -\dot{n}_i \qquad \text{for } i = 1, n$$

$$\frac{d\dot{N}_g}{dA} = -\dot{n}_t \qquad \Sigma\, \dot{N}_{ig} = \dot{N}_g$$

$$\frac{d\dot{N}_{ng}}{dA} = 0 \tag{13}$$

Energy balance over the same increment of area give,

$$\frac{dT_g}{dA} = -\frac{\alpha_g^{\bullet}(T_g - T_s)}{\dot{N}_g \tilde{c}_{pg}}$$

where

$$\alpha_g^{\bullet} = \alpha_g \frac{\varepsilon}{(e^{\varepsilon} - 1)}. \tag{14}$$

The rate of gas cooling in the presence of mass transfer is suppressed by the above factor, relative to heat transfer in the absence of mass transfer. For the coolant in co(+) or countercurrent (−) flow,

$$\frac{dT_o}{dA} = \pm \frac{\alpha_o(T_s - T_o)}{\dot{M}_o c_{po}}. \tag{15}$$

The integration of the above equations as a means of condenser design has been described in many works, e.g., Schrodt (1973), Price and Bell (1974) and Webb and McNaught (1980).

Prediction of Fogging in Condensers

Colburn and Edison (1941) showed that, based on Eqs. (13) and (14), it is possible to calculate the condensation path as the rate of change of partial pressure with gas temperature,

$$\frac{dp_{1g}}{dT_g} = \frac{(p_{1g} - p_{1s})}{(T_g - T_s)} \left(\frac{Pr}{Sc_{12}}\right)^{2/3} \frac{\varepsilon}{\phi} \frac{(e^{\phi} - 1)}{(e^{\varepsilon} - 1)}. \tag{16}$$

This allows verifying the likelihood of fogging by comparing the vapour pressure of the component with the condensation path predicted by the above, **Figure 6**.

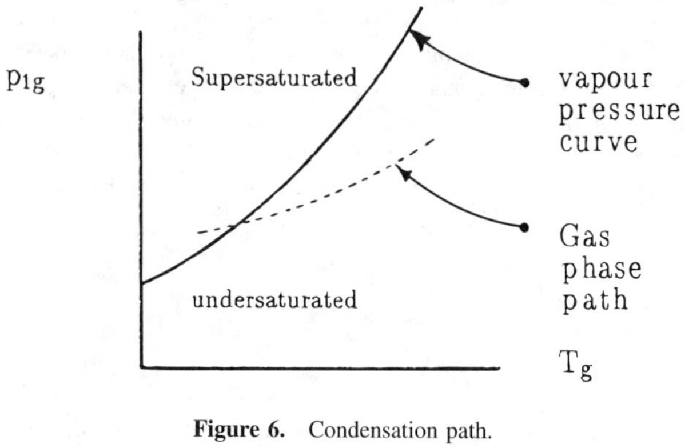

Figure 6. Condensation path.

An excursion into the supersaturation zone does not, of itself, ensure fogging. Steinmeyer (1972) defines Supersaturation Ratio, S, for clean systems that must be exceeded for fogging to occur; but dust or fine spray may hasten the onset,

$$S = \frac{\text{Partial pressure of vapour}}{\text{Vapour pressure at } T_g}$$

$$S > \exp\left(\frac{C\tilde{M}}{\rho_c}\left(\frac{\sigma}{T_g}\right)^{2/3}\right) \tag{17}$$

where
$C = 1.76 \times 10^7$ for water, alcohols etc.,
$C = 1.38 \times 10^7$ for nonpolars.

The Colburn-Edison equation also shows the conditions that must be satisfied by a vapour if it is to remain saturated during a condensation process. These are small driving forces and $Le = Sc/Pr = 1$, rarely satisfied in real processes. This implies that the equilibrium approach can only approximate condenser behaviour.

Colburn-Hougen and Colburn-Drew Methods

The methods initiated by these authors are special cases of the above. For binary mixtures, the condensation rate can be expressed, following Colburn and Drew (1937) as

$$\dot{n}_t = \beta_{12} \ln\left(\frac{z_1 - \tilde{y}_{1s}}{z_1 - \tilde{y}_{1g}}\right) \text{ where } z_1 = \dot{n}_1/\dot{n}_t. \tag{18}$$

When considering the special case of an unmixed condensate, z_i becomes the liquid composition. Alternatively, if component 2 is a non-condensing gas, where $z_1 = 1$ and $\dot{n}_1 = \dot{n}_t$, the Colburn-Hougen, (1934) equation is obtained. In each of these special cases, a determinacy condition has been added and as above, this is the specification of total flux. It is this binary form of the equations that has found the widest application in industrial design practice.

Silver-Bell and Ghaly Methods

In this approach, the detailed mass transfer equations are not used. However, a gas side resistance is included by assuming that the sensible heat change of the saturated gas mixture must be conducted across the gas film. It has been stated by Bell that this assumption is usually conservative. The overall heat transfer coefficient is then calculated by:

$$\frac{1}{U_o} = \frac{1}{\alpha_o} + \frac{Z}{\alpha_{g,eff}}, \tag{19}$$

where Z is the ratio of the heat flux of the gas side to the total heat flux in an interval. Consider **Figure 7**; by definition,

$$\dot{q}_g = \frac{\Delta \dot{Q}_g}{\Delta A} \qquad \dot{q} = \frac{\Delta \dot{Q}}{\Delta A};$$

hence

$$Z = \frac{\Delta \dot{Q}_g}{\Delta \dot{Q}} = \frac{\dot{q}_g}{\dot{q}} = \left\{\frac{\dot{M}_g \tilde{c}_{pg} \Delta T_g}{\dot{M} dh'}\right\}_{sat} \tag{20}$$

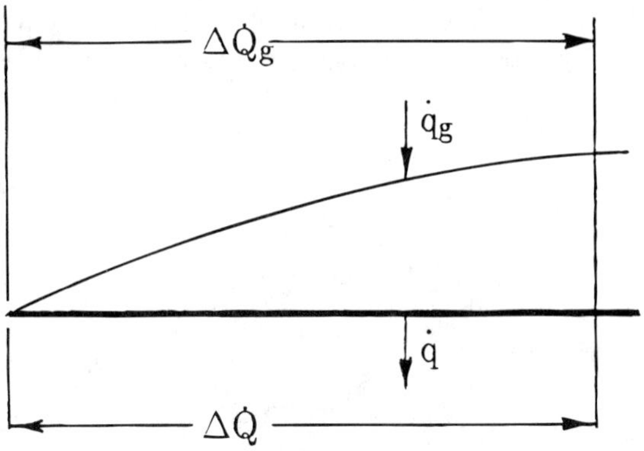

Figure 7. Heat fluxes in an interval.

The appropriate heat transfer coefficient, $\alpha_{g,eff,}$ in Equation (19) remains unspecified. In early work, it has been taken as the gas film heat transfer coefficient, α_g, for the geometry in question. However, McNaught (1979) has shown that better results are given when it is corrected by the Ackermann factor.

$$\alpha_{g,eff} = \alpha_g \frac{\varepsilon}{(e^{\varepsilon} - 1)} < \alpha_g. \tag{21}$$

Care should be taken with superheated vapours. The early practice was to provide a desuperheating zone which was designed as a gas cooler, but this was too conservative because it ignored wet wall desuperheating. McNaught (1981) has suggested the following equation to allow the use of the equilibrium model in cases of wet wall desuperheating.

$$\frac{1}{U_o} = \frac{1}{\alpha_o} + \frac{Z}{\alpha_{g,eff}}\left\{\frac{(T_g^* - T_s)}{(T_g^* - T_s) + Z(T_g - T_g^*)}\right\}. \tag{22}$$

Caution is also needed to distinguish the gas and saturated gas temperatures, T_g and T_g*, respectively. The gas temperature T_g can only be followed using the equations of the film model. The application of the equilibrium method requires the calculation of the appropriate mean temperature driving force and this is considered in the section on **Cooling Curve**.

The equilibrium model is applied very widely in condenser design practice. Examination of the Colburn-Edison equation shows that it is only exact in the limit of low driving forces and $Le = 1$. In fact, it is only conservative if the Lewis Number is less than 1 and its accuracy for $Le > 1$ is not published.

A second case that has not been treated satisfactorily in the literature is the condensation of one liquid alone from vapours that form immiscible liquids. This is similar to the wet wall desuperheating case and the method may be very conservative.

Heat Transfer Coefficients

The presence of multicomponent vapours does not cause special problems unless immiscible condensates are formed. In this case, the following methods have been applied.

Bernhardt et al. (1972), considered the two condensates to form separate layers around the surface, and recommended the following combination of the Nusselt coefficients of the two condensates, α_{c1} and α_{c2}:

$$\alpha_c = v_1\alpha_{c1} + (1 - v_1)\alpha_{c2}, \tag{22a}$$

with v_i = Volume fraction of the phases.

Akers and Turner (1962) proposed a model which essentially considers a homogeneous condensate layer with the viscosity of the surface wetting layer,

$$\alpha_c = 0.94\ \lambda_{av}\left(\frac{\rho_{av}{}^2 g}{\eta_c \Gamma}\right)^{1/3} \tag{22b}$$

where ρ_{av} is the mass fraction average density, η_c the viscosity of the surface wetting fluid and λ_{av} is the volume average thermal conductivity. These simple equations seem to fit most data within about ± 30%. The influence of turbulence, vapour shear and inundation on immiscible condensates are not known.

Condensate Subcooling

Condensate subcooling has not been included in the above film model treatment. It is implicit in the Silver method, when the integral cooling curve is applied, because the entire mixture is assumed to be cooled to the equilibrium temperature.

With the assumption of laminar flow and a linear temperature profile in the condensate, the heat flux, $\dot{q}_c$, at the condensate surface is related to $\dot{q}_o$, passing to the coolant (See **Figure 2**) by,

$$\dot{q}_o = \dot{q}_c + \frac{3}{8}\dot{n}_t\tilde{c}_{pc}(T_s - T_w) - \dot{N}_c\tilde{c}_{pc}\frac{dT_c}{dA}. \tag{23}$$

Two terms are added to the previous treatment where it was assumed that $\dot{q}_o = \dot{q}_c$. They describe subcooling of new condensate to the mean temperature, $T_c = T_s - 3(T_s - T_w)/8$, and subcooling of all prior condensate, respectively.

A new method is presented. It is assumed that all heat of subcooling is conducted across the full condensate layer, which is conservative. Both terms above are included, with the second estimated by Equations (14) and (15) to predict downstream change, assuming constant relative film heat transfer resistances. Equation (9) holds and all previous algorithms apply if modified quantities, $\alpha_o^{\bullet}$, $\Delta\tilde{h}_v^{\bullet}$ and $\tilde{c}_p^{\bullet}$ are defined,

$$\dot{n}_t = \alpha_{\dot{o}}\chi\Big/\left\{\frac{\Delta\tilde{h}_{vf}^{\bullet}}{(T_g - T_o)} + \frac{\tilde{c}_{pf}^{\bullet}(1 - \chi)}{\{1 - \exp(-\varepsilon)\}}\right\} \tag{24}$$

$$\alpha_o^{\bullet} = \alpha_o\left\{1 \pm \frac{\dot{N}\tilde{c}_{pc(1-W)}}{\dot{M}c_{po}}\right\} \quad \begin{matrix}+\text{Co-current}\\ -\text{Counter}\end{matrix}$$

$$\Delta\tilde{h}_{vf}^{\bullet} = \Delta\tilde{h}_{vf}\left\{1 + \frac{3\tilde{c}_{pc}\alpha_o(T_s - T_o)}{8\alpha_c\Delta\tilde{h}_{vf}}\right\}$$

$$\tilde{c}_{pf}^{\bullet} = \tilde{c}_{pf}\left\{1 + \frac{\dot{N}_c\tilde{c}_{pc}}{\dot{N}_g\tilde{c}_{pg}}\frac{W}{e^{\varepsilon}}\right\} \qquad W = \chi\left\{1 - \frac{3\alpha_o}{8\alpha_c}\right\}.$$

Choice of Design Method

The Silver approach is preferred for industrial design of multicomponent condensers. With unspecified mixtures, where only a cooling curve is provided, it must be used. Its advantage is that no further vapour-liquid equilibrium calculations are needed, irrespective of geometry. The more physically realistic film model has been used to correct the equilibrium method; but the equilibrium method remains less reliable. These are: Where condensate composition must be predicted in a partial condenser (reflux condensers); where pressure drop is a large fraction of pressure; where vapours of immiscible liquids condense above the azeotropic temperature; and where the Lewis Number exceeds unity.

References

Ackermann, G. (1937) *Combined Heat and Mass Transfer at High Temperature and Partial Pressure Differences,* Forsch.Ing.Wes., VDI, 8, 382, 1.

Akers, W. W. and Turner, M. M. (1962) *Condensation of Vapours of Immiscible Liquids,* AIChEJ., 5, 587–589.

Bell, K. J. and Ghaly, M. A. (1972) *An Approximate Generalised Design Method for Multicomponent—Partial Condensers,* AIChE Symp. Ser., 69, 72.

Bernhardt, S. H., Sheridan, J. J., and Westwater, J. W. (1972) *Condensation of Immiscible Mixtures,* A.I.Ch.E.J. Symp. Ser., 68, 118, 21–37.

Chilton, T. H. and Colburn, A. P. (1934) *Mass Transfer Coefficients: Prediction from Data on Heat Transfer and Fluid Friction,* Ind. Eng. Chem., 26, 1183.

Colburn, A. P. and Hougen, O. A., (1934) *Design of Cooler Condensers for Mixtures of Vapours with Non-Condensing Gases,* Ind.Eng.Chem., 26, 1178.

Colburn, A. P. and Drew, T. N. (1937) *The Condensation of Mixed Vapours,* Trans.Am.Inst.Chem.Engnrs., 33, 197–215.

Colburn, A. P. and Edison, A. G. (1941) *Prevention of Fog in Cooler—Condensers,* Ind.Eng.Chem., 33, 457.

Mc Naught, J. (1979) *Mass Transfer Correction Terms in Design Methods for Multicomponent-Partial Condensers, in Condensation Heat Transfer,* ASME-AIChE, 111.

Mc Naught, J. (1981) *Assessment of Design Methods for Condensation of Vapours from Non-condensing Gas,* ICHMT Seminar, Dubrovnik.

Price, B. C. and Bell, K. J. (1974) *Design of Binary Vapour Condensers Using the Colburn-Drew Equations,* AIChEJ Symp. Series, 70, 163.

Schrodt, J. T. (1973) *Simultaneous Heat and Mass Transfer from Multi-Component Condensing Vapour-Gas Systems,* AIChEJ, 19(4), 753.

Silver, L. (1947) *Gas Cooling with Aqueous Condensation,* Trans. Inst. Chem. Eng., 25, pp 30–42.

Steinmeyer, D. E. (1972) *Fog Formation in Partial Condensers,* Chem. Engng. Prog., 68, 7, 64–68.

Taylor, R. and Krishna, R. (1993) *Multicomponent Mass Transfer,* Wiley Interscience.

Webb, D. R. and McNaught, J. M. (1980) *Condensers, Developments in Heat Exchanger Technology,* App. Sci. Pub. Ltd., Barking, England.

Webb, D. R. (1990) *Multicomponent Condensation,* Plenary Lecture, 9th Int. Heat Trns. Conf., Jerusalem, August.

Leading to: *Condensation curve*

D.R. Webb

CONDENSATION OF A PURE VAPOUR

Following from: *Condensation, overview*

Many factors are involved in determining the vapour-to-condenser surface heat transfer coefficient during condensation. The problem is even more complex when the vapour contains different molecular species which may or may not condense (see **Condensation of Multi-Component Vapours**). The term "pure vapour" implies that only one species is present.

In industrial **Condensers**, vapour and condensate flows are, in general, three-dimensional and involve effects of gravity, shear stress at the condensate surface due to vapour velocity, interface temperature difference due to nonequilibrium and *inundation,* i.e., condensate from higher or upstream surfaces impinging on lower or downstream surfaces. For profiled surfaces (e.g., finned tubes), surface tension effects are also important. Condensate and vapour flows may be either laminar or turbulent. The condensate may form a continuous film on the surface or, when the surface is not wetted by the condensate, form discrete droplets (see **Dropwise Condensation**). Condensation may occur on external surfaces, e.g., on the outside of the tubes in a shell-and-tube condenser or on internal surfaces, e.g., in-tube condensation. (See **Tubes, Condensation on Outside in Cross Flow, Tube Banks**; **Condensation Heat-Transfer in Tubes**; **Condensation in.**) In "direct contact" condensation, liquid at a temperature below the saturation temperature, is brought into contact with the vapour. The extent to which the condensation process is understood and the accuracy with which heat transfer coefficients can be calculated depend on the circumstances. After almost a hundred years of research, accurate predictions can now be made for relatively simple geometry (e.g., flat-plate or single horizontal tube) and for well-defined flow conditions.

The fact that, during condensation, equilibrium conditions cannot prevail at the vapour-condensate interface, means that a temperature difference must occur in the vicinity of the interface. Extrapolations to the interface of the temperature distributions in the vapour and liquid would indicate a discontinuity, the "*interface temperature drop.*" As discussed by Niknejad and Rose (1981), the problem has been studied for many years and it is now established that the interface temperature drop is essentially confined to a region in the vapour having a thickness of a few mean free paths of the vapour molecules. Various related theoretical approaches lead to the expression

$$\dot{m} = \xi\{P_v - P_{sat}(T_s)\}/(RT_s)^{1/2}, \quad (1)$$

where $\dot{m}$ is the net condensation mass flux, P_v is the vapour pressure, $P_{sat}(T_s)$ is the saturation pressure at the liquid surface temperature, T_s and R is the specific gas constant of the vapour. ξ is a dimensionless quantity of order unity, and can be considered as an *interface mass transfer coefficient.* Experiment and theory show that ξ depends on $\dot{m}$, ξ *increasing* with *increasing* $\dot{m}$, the dependence being most marked at low pressure. However, as $\dot{m}$ tends to zero (equilibrium), ξ approaches a constant value which is independent of pressure and condensing fluid.

A difficulty which arises in the theory of interphase matter transfer is that it is not certain that all vapour molecules striking the liquid surface remain in the liquid phase, i.e., some molecules may be reflected. Some theoretical models incorporate the parameter σ, the "*condensation coefficient*" defined as the fraction of those vapour molecules impinging on the liquid surface which remain in the liquid phase, σ being assumed to have a constant value. In these models, ξ_o (the value of ξ in Equation (1) for $\dot{m} \to 0$) is a function of σ. Other approaches assume that no vapour molecules are reflected at the liquid surface when $\sigma = 1$ and ξ_o is a constant. Various theoretical models lead, with $\sigma = 1$, to values of ξ_o between about 0.6 and about 0.8. While earlier experimental investigations gave much lower values—which were interpreted as indicating low values of σ (down to around 0.01)—more recent studies indicate a value near 0.7, confirming the validity of the theory and suggesting that σ is equal to or near unity. The most probable explanation of the earlier low values is the presence in the vapour of noncondensing gas, leading to a significant temperature drop in the vapour which was erroneously attributed to the interface.

For a saturated vapour and for moderate departure from equilibrium at the interface when the vapour temperature $T_v \approx T_s$, Equation (1) with the **Clapeyron-Clausius Equation,** leads to the following expression for the interface heat transfer coefficient α_i:

$$\alpha_i = \xi_o h_{lg}^2/(v_{lg}T_v\sqrt{RT_v}). \quad (2)$$

where h_{lg} is the specific latent heat of evaporation, v_g is the specific volume of saturated vapour, v_l is the specific volume of saturated liquid and $v_{lg} = v_g - v_l$, and R is the specific ideal-gas constant. When the vapour is treated as an ideal gas and with $v_g \gg v_l$, Equation (2) gives:

$$\alpha_i = \xi_o P_v h_{lg}^2/[T_v(RT_v)^{3/2}]. \quad (3)$$

With $\xi_o = 0.7$, Equation (3) gives values of α_i much higher than the heat transfer coefficient for the condensate in film condensation, except in the case of liquid metals. For example, for steam at 30 °C, $\alpha_i \approx 1.1$ MW/m^2 K, which may be compared with typical values of the heat transfer coefficient in the range 10–50 kW/m^2 K for the condensate film during film condensation. The interface resistance is therefore, apart from the case of liquid metals, generally negligible for practical purposes.

The problem of laminar film condensation was first treated by **Nusselt** (1916) for condensation on a vertical plane surface and on a horizontal tube: For a saturated vapour with uniform condenser wall temperature and assuming that the motion of the condensate is controlled solely by gravity and viscosity; that acceleration of the condensate film is negligible; that shear stress on the condensate surface due to the vapour is negligible; that heat transfer across the condensate film is by pure conduction; and that the condensate properties do not vary across the film, the following expressions for local film thickness and for the local and average heat transfer coefficient and Nusselt number are obtained:

For a vertical plane surface:

local condensate thickness,

$$\delta_x = \left(\frac{4\eta\lambda x\Delta T}{\rho\Delta\rho g h_{lg}}\right)^{1/4} \quad (4)$$

local heat transfer coefficient

$$\alpha_x = \left(\frac{\rho\Delta\rho g h_{lg}\lambda^3}{4\eta x\Delta T}\right)^{1/4} \quad \textbf{(5)}$$

mean heat transfer coefficient

$$\overline{\alpha}_L = \frac{4}{3}\left(\frac{\rho\Delta\rho g h_{lg}\lambda^3}{4\eta L\Delta T}\right)^{1/4} \quad \textbf{(6)}$$

mean Nusselt number

$$\overline{Nu}_L = \frac{4}{3}\left(\frac{\rho\Delta\rho g h_{lg}L^3}{4\eta\lambda\Delta T}\right)^{1/4} \quad \textbf{(7)}$$

where η is liquid viscosity; λ is the liquid thermal conductivity; ρ, the liquid density; $\Delta\rho$ is the difference between liquid and vapour density; ΔT is the vapour-to-condensing surface temperature difference; x is the distance from top of vertical plate; L is the plate height and g is the specific force of gravity.

For a horizontal tube:

local condensate thickness

$$\delta_\theta = \left(\frac{2\eta\lambda d\Delta T}{\rho\Delta\rho g h_{lg}}\right)^{1/4} z_\theta^{1/4} \quad \textbf{(8)}$$

where

$$z_\theta = \frac{1}{(\sin\theta)^{4/3}}\int_0^\theta (\sin\phi)^{1/3}\,d\phi \quad \textbf{(9)}$$

and θ is the angle measured from the top of the tube,

local heat transfer coefficient

$$\alpha_\theta = \left(\frac{\rho\Delta\rho g h_{lg}\lambda^3}{\eta d\Delta T}\right)^{1/4} z_\theta^{-1/4} \quad \textbf{(10)}$$

mean heat transfer coefficient

$$\overline{\alpha} = 0.728\left(\frac{\rho\Delta\rho g h_{lg}\lambda^3}{\eta d\Delta T}\right)^{1/4} \quad \textbf{(11)}$$

mean Nusselt number.

$$\overline{Nu} = 0.728\left(\frac{\rho\Delta\rho g h_{lg}d^3}{\eta\lambda\Delta T}\right)^{1/4} \quad \textbf{(12)}$$

where d is the tube diameter.

In the treatment of horizontal tube, the additional approximation that the condensate film thickness is much less than the tube radius is required. Since Equation (9) indicates that $\delta_\theta \to \infty$ for $\theta \to \pi$, this approximation is evidently invalid near the bottom of the tube. However, as the film becomes thicker, the local heat flux becomes small and the contribution of the erroneous values near the bottom of the tube are insignificant in the calculation of the total heat transfer rate or mean heat flux for the tube. For quiescent vapour, Equations (6), (7), (11) and (12) have been very well verified by experiment.

The advent of the computer has made possible more complete solutions to the problem of laminar film condensation. These have shown that the Nusselt assumptions lead to very accurate results for the practical range of operating conditions. A generally small correction has been proposed—i.e., that h_{lg} in Equations (4–7) should be replaced by $h_{lg} + (3/8)c_p\Delta T$ and in Equations (8–12), by $h_{lg} + 0.68c_p\Delta T$, where c_p is the isobaric specific heat capacity of the condensate. A satisfactory reference temperature for the evaluation of properties of condensate film is:

$$T^* = (1/3)T_v + (2/3)T_w \quad \textbf{(13)}$$

where T_v is the temperature of the bulk vapour and T_w that of the condensing surface, and h_{lg} is evaluated at T_v.

The Nusselt assumptions have been used for the case where the surface heat flux, rather than surface temperature, is uniform. This leads, for the vertical plate, to the same expressions as Equations (6) and (7), with the uniform heat flux and mean temperature difference replacing the mean heat flux and uniform temperature difference. For the horizontal tube case, the coefficient in Equations (11) and (12) becomes 0.695. However, as indicated by Rose (1988), the validity of the solution is questionable since the mean value of the temperature difference is significantly affected by the erroneously large values near the bottom of the tube where the condensate film becomes thick. In practice, the surface temperature variation around a horizontal condenser tube approximates to a cosine distribution. It is interesting to note that in this case, the coefficient in Equations (11) and (12) (with mean heat flux and mean surface temperature) is negligibly affected [see Memory and Rose (1991)].

When the vapour velocity over the condensing surface is significant, account must be taken of the shear stress at the condensate surface. This problem is discussed and reviewed in some detail by Rose (1988).

For the case of condensation with vapour flow along a *horizontal plane surface*, the surface shear stress, τ_δ, is given by

$$\frac{\tau_\delta}{\rho_v U_\infty^2}\cdot Re_x^{1/2} = 0.332 \text{ where } Re_x = \frac{U_\infty\rho_v x}{\eta_v} \quad \textbf{(14)}$$

for very low condensation rates, and by

$$\frac{\tau_\delta}{\rho_v U_\infty^2}\cdot \tilde{R}e_x^{1/2} = \frac{1}{2}\left(\frac{\lambda\Delta T}{\eta h_{lg}}\right) \text{ where } \tilde{R}e_x = \frac{U_\infty\rho x}{\eta} \quad (15)$$

for high condensation rates, where U_∞ is the free stream velocity of the vapour, ρ_v is the vapour density and η_v, the vapour viscosity and x is distance from the leading edge. In the general case, the problem requires simultaneous solution of the conservation equations for the vapour and condensate with appropriate conditions at the interface [see Rose (1988)]. When the Nusselt assumptions are used for the condensate film, Equation (14) leads to

$$Nu_x\tilde{R}e_x^{-1/2} = 0.436\ G^{-1/3} \quad \textbf{(16)}$$

where

$$G = \left(\frac{\lambda\Delta T}{\eta h_{lg}}\right)\cdot\left(\frac{\rho\eta}{\rho_v\eta_v}\right)^{1/2}, \quad \textbf{(17)}$$

and Equation (15) gives

$$Nu_x\tilde{R}e_x^{-1/2} = 0.5. \quad \textbf{(18)}$$

Equations (16) and (18) are both conservative, i.e., they underestimate the heat transfer coefficient; Equation (16) is more accurate when $G <$ about 1 and Equation (18) is more accurate when $G >$ about 1. Solutions have also been obtained for vapour downflow on a *vertical* plate. The results are summarised by the interpolation formula:

$$Nu_x\tilde{R}e_x^{-1/2} = K(1 + 0.25K^{-4}F_x)^{1/4}, \quad \textbf{(19)}$$

where

$$K = 0.45(1.2 + G^{-1})^{1/3}. \quad \textbf{(20)}$$

and $Fx = \eta h_{fg}\ gx/\lambda\Delta T\ U_\infty^2$

For condensation on a *horizontal tube,* the problem is complicated by separation of the vapour boundary layer. An approximate solution, given by Shekriladze and Gomelauri (1966), uses the high condensation rate limiting value of the surface shear stress [analogous to Eq. (15) for the flat plate]. This underestimates the surface shear stress, and hence the heat transfer coefficient, and should lead to conservative results. The mean heat transfer coefficient for vertical vapour downflow on a horizontal tube, using this approach, is given to within 0.4% by the equation

$$Nu_d\tilde{R}e_d^{-1/2} = \frac{0.9 + 0.728F_d^{1/2}}{(1 + 3.44F_d^{1/2} + F_d)^{1/4}} \quad \textbf{(21)}$$

where $F_d = \eta h_{lg}\ gd/\lambda\Delta TU_\infty^2$.

Equation (21) is in fair agreement with experimental data for various fluids. A more detailed discussion of this problem is given in **Tubes, Condensation on the Outside in Crossflow**.

An additional factor affecting the performance of multitube condensers is inundation i.e., condensate from higher or upstream tubes falling under gravity or being carried by the vapour flow onto lower or downstream tubes. For the case of negligible vapour velocity, Nusselt has extended his analysis by assuming that condensate drains from a tube as a continuous sheet onto the tube below and obtained, for the mean heat transfer coefficient for a column of n tubes,

$$\overline{\alpha}_n = \alpha_1 n^{-1/4} \quad \textbf{(22)}$$

where α_1 is the heat transfer coefficient for the top tube. Experimental data are very scattered but indicate that Equation (22) is very conservative. An index of $-1/6$ is frequently used in design, but this is probably also conservative.

Turbulent film condensation is much less well-understood than the laminar case, particularly in the presence of significant vapour shear stress. Transition from laminar to turbulent film condensation occurs at a film Reynolds number ($4\Gamma/\eta$) between 1 000 and 2 000. Γ is the condensate mass flow rate per width of surface. Various theoretical models have been proposed which treat the turbulence and the effect of surface shear stress in somewhat different ways. Experimental data are generally for condensation inside tubes. In circumstances where the condensate forms a thin film on the tube wall (stratified flow), and particularly for the case of a vertical tube with vapour downflow, the data are suitable for validation of turbulent condensation models. Experiment and theory both indicate that the heat transfer coefficients for turbulent film condensation are appreciably higher than for the laminar case. Various models and calculation methods are described by Butterworth (1983), Marto (1991) and Stephan (1992).

Except in the case of steam and liquid metals, for condensation on the outside surface of tube, the vapour-side resistance is usually controlling owing to the low thermal conductivity of the condensate. For a water-cooled condenser tube, the coolant-side, heat transfer coefficient might typically be around 5 000 W/m^2 K. This is of similar magnitude to that found with film condensation of steam (5 000–15 000 W/m^2 K). For organic vapours, condensing-side heat transfer coefficients are lower typically by a factor of 5 or more. There is therefore considerable incentive to enhance the condensing-side heat transfer, particularly for organic vapours such as refrigerants.

A commonly-used enhancement method, particularly in refrigeration and air-conditioning plant, is the use of horizontal low-finned tubes. Here, the tube diameter is typically around 20 mm while the height of the radial integral fins is normally around 1 mm. Increase in the condensing-side heat transfer coefficient by factors of 7 or more are achieved for fluids with low surface tension. The enhancement arises from the increase in surface area, the low effective height of the fins [see Eq. (6)] and surface tension-enhanced condensate drainage due to the streamwise pressure gradient set up in the film by surface curvature changes. Surface tension also has a detrimental effect due to capillary retention of condensate between fins on the lower part of the surface. For this reason enhancement factors for condensation of steam are lower (typically 2 to 3) owing to the high surface tension of water. (See also **Augmentation of Heat Transfer, Two-Phase Systems**.)

To date, techniques for promoting dropwise condensation have only been successful on the laboratory scale for fluids with high surface tension, notably steam. The fact that heat transfer coefficients for dropwise condensation of steam are 10 to 20 times those for film condensation continues to stimulate research in this field. More information on dropwise condensation is given in the article (**Dropwise Condensation**.)

Direct contact condensation, where vapour and liquid are brought into contact without an intervening wall, is sometimes used for special applications. Liquid may be sprayed into an atmosphere of vapour or vapour may be injected into a pool of liquid. In both cases, the limiting heat transfer mechanism is conduction in the liquid. Direct contact condensation has been discussed by Jacobs (1985).

References

Butterworth, D. (1983) Film Condensation of a Pure Vapour. *Heat Exchanger Design Handbook*, Ed. E. U. Schlunder, Hemisphere, New York, 1989.

Jacobs, H. R. (1985) Direct Contact Condensers. *Heat Exchanger Design Handbook,* Ed. E. U. Schlunder, Hemisphere, New York, 1989.

Marto, P. (1991) Heat Transfer in Condensation, in Boilers, Evaporators and Condensers. Ed. S. Kakac, Wiley, 525–570.

Memory, S. B. and Rose, J. W. (1991) Free Convection Laminar Film Condensation on a Horizontal Tube with Variable Wall Temperature. *Int. J. Heat Mass Transfer*, 34, 2775–2778.

Niknejad, J. and Rose, J. W. (1981) Interphase Matter Transfer—an Experimental Study of Condensation of Mercury. *Proc. R. Soc. Lond.* A 378, 305–327.

Nusselt, W. (1916) Die Oberflachenkondensation des Wasserdampfes. *Z. Vereines Deutch. Ing.* 60, 541–546, 569–575.

Rose, J. W. (1988) Fundamentals of Condensation Heat Transfer: Laminar Film Condensation. *JSME Int. Journal, Series II,* 31 (3), 357–375.

Shekriladze, I. G. and Gomelauri, V. I. (1966) Theoretical Study of Laminar Film Condensation of Flowing Vapour. *Int. J. Heat Mass Transfer,* 9, 581–591.

Stephan, K. (1992) Heat Transfer in Condensation and Boiling. Springer-Verlag.

Leading to: *Condensation curve; Dropwise condensation; Tube banks, condensation heat transfer in; Tubes, condensation in; Tubes, condensation on outside in cross flow*

J.W. Rose

CONDENSATION ON OUTSIDE OF TUBES IN CROSSFLOW (see Tubes, condensation on outside in cross flow)

CONDENSATION, OVERVIEW

Condensation heat transfer plays an important role in many engineering applications, notably electric power generation, process industries, refrigeration and air-conditioning. Many different physical phenomena are involved in the condensation process, their relative importance depending on the circumstances and application. (For examples, see **Air Conditioning**, **Condensers**, **Desalination**, **and Refrigeration**.)

When a liquid and its vapour are in contact, molecules pass from liquid to vapour and from vapour to liquid. **Condensation** occurs when the number of molecules entering the liquid phase exceeds that of leaving molecules. Under these circumstances, the temperature of the vapour in the immediate vicinity (a few mean free paths) of the vapour-liquid interface is higher than that of the liquid. The *interface temperature drop* increases with increasing condensation rate and with decreasing pressure but in most circumstances (an exception being the case of liquid metals), this is very small and equilibrium conditions at the interface can be assumed. A brief summary of interface matter transfer during condensation is given by Niknejad and Rose (1981).

The most common and best understood case of condensation heat transfer is that of film *condensation* of a pure quiescent vapour on a solid surface. The problem of calculating the heat transfer rate for a plane vertical surface and for a horizontal cylinder with uniform surface temperatures, and where the condensate flow is laminar and governed only by gravity and viscous forces, has been solved by **Nusselt** (1916). Nusselt's main assumptions are that heat transfer across the condensate film is by pure conduction, the effect of vapour drag in supporting the falling condensate film is negligible, and the properties of the condensate may be taken to be uniform across the film, i.e., are essentially independent of temperature.

The well-known *Nusselt equations* are:

For the vertical plane surface, the mean value of Nusselt number is

$$\mathrm{Nu} = 0.943\left\{\frac{\rho\Delta\rho g h_{lg} L^3}{\eta\lambda\Delta T}\right\}^{1/4}. \qquad \textbf{(1)}$$

and for the horizontal tube,

$$\mathrm{Nu} = 0.728\left\{\frac{\rho\Delta\rho g h_{lg} d^3}{\eta\lambda\Delta T}\right\}^{1/4} \qquad \textbf{(2)}$$

where ρ is the liquid density, g is acceleration due to gravity, h_{lg} is the latent heat of evaporation, η is the liquid viscosity, λ is the liquid thermal conductivity, $\Delta\rho$ is the density difference between vapour and condensate, ΔT is the vapour-to-surface temperature difference; L is the plate height; and d, the tube diameter. For the case of the tube, the additional assumption that film thickness is small compared with the tube radius is needed. Since the theory predicts that the radial film thickness tends to infinity at the bottom of the tube, this assumption is evidently invalid for the lower part of the tube. The fact that the heat transfer rate is inaccurate where the film becomes thick is relatively unimportant because it is small and makes a minor contribution to the total heat transfer rate for the tube. Equations (1) and (2) have been well verified experimentally for condensation of pure (only one molecular constituent) vapour. In order to obtain the constant in Equation (2), numerical integration is required twice. Nusselt's slightly inaccurate value $(0.8024 \times (2/3)^{1/4} = 0.725)$ is due to his use of planimetry. (See **Condensation of Pure Vapor**.)

As discussed by Rose (1988), more recent theoretical studies in which the effects of inertia and convection in the condensate film and vapour drag on the condensate surface are included have shown that these effects are unimportant. More recently, it has been shown by Memory and Rose (1991) that the effect of variable wall temperature, which occurs in practice during condensation on a horizontal tube, also has a negligible effect on the *mean* heat transfer rate. Thus, Equations (1) and (2) can be used with confidence for condensation of pure "stationary" vapours when the condensate flow is laminar.

In the case of significant vapour flow rate over the condensing surface, the effect of drag on the condensate becomes significant. In some circumstances, the effect of vapour drag overwhelms that of gravity. Since 1960, *vapour shear stress effects* have been studied extensively. Some of the more important contributions are described by Rose (1988).

The relative importance of vapour shear stress and gravity on the motion of the condensate film is measured by the dimensionless parameter $F = \eta h_{lg} g x / \lambda \Delta T u_\infty^2$, where x is the relevant linear dimension (plate height or tube diameter) and u_∞ is the free-stream vapour velocity. For downward vapour flow over a horizontal tube, an approximate analysis gives

$$Nu_d \tilde{Re}_d^{-1/2} = \frac{0.9 + 0.728F^{1/2}}{(1 + 3.44F^{1/2} + F)^{1/4}}. \quad \textbf{(3)}$$

Nu_d is the mean Nusselt number for the condensate film and $\tilde{Re}_d$ is a Reynolds number using the vapour approach velocity and condensate properties. Equation (3)—which indicates that for $F > 10$, gravity dominates while for $F < 0.1$, vapour shear stress is controlling—agrees quite well with experimental data from several investigations using various condensing fluids. (See **Tubes, Condensation on Outside in Crossflow.**)

When the vapour contains more than one molecular species, the problem is complicated by diffusion of species in the vapour. For example, for a two-constitutent vapour where only one constituent condenses (e.g., steam-air), the mixture is rich in the *noncondensing gas* near the interface where vapour molecules are removed. The tendency is for noncondensing gas molecules to diffuse away from the interface so that in the steady-state, the rate at which gas molecules arrive at the surface with the condensing vapour is equal to their diffusion rate away from the surface. Even in the absence of forced convection of the vapour-gas mixture, the density difference, which results from the composition difference between that of the bulk vapour and that of the vapour adjacent to the interface, leads to natural convection. The process by which the steady-state is reached is therefore one of diffusion in the presence of convection. The fact that the vapour-gas mixture adjacent to the condensate surface is rich in noncondensing gas causes the temperature at the interface to be lower than in the bulk. Assuming equilibrium at the interface, the temperature is equal to the saturation temperature corresponding to the partial pressure of the vapour, which may be significantly lower than in the more remote vapour. Composition (or partial pressure) and temperature boundary layer are set up in the vapour adjacent to the interface. This gives an effective heat transfer resistance since the temperature drop across the condensate film, and hence the heat transfer rate, is significantly reduced. Detailed boundary layer solutions of this problem for free and forced convection, notably by Koh, Sparrow, Fujii et al., have been given. Earlier works are discussed and approximate equations are given by Rose (1969) for the free convection problem, and Rose (1980) for the forced convection case. When two or more constitutents of the vapour condense together, the situation is similar to that described above since the more volatile constituent is more concentrated at the condensate surface. Extensive treatments of these problems for the case of the plane vertical condensing surface and laminar flow of vapour and condensate have been given bu Fujii (1991). (See **Condensation of Multicomponent Vapours.**)

Approximate methods, based on the *analogy* between heat and diffusive mass transfer in the vapour, are widely used for multiconstituent problems. The essence of the method is that the differential equations expressing conservation of energy and molecular species can be arranged in identical form by appropriate nondimensionalization. Known results (theoretical or experimental) for heat transfer problems are used to infer results for the corresponding mass transfer diffusion problems. The method has wide utility but is approximate since the boundary condition on the normal vapour velocity at the surface is not the same for the heat and mass transfer problems. For the heat transfer problem, the normal velocity at the (solid) heat-exchanger wall is zero. In the case of condensation, where molecules pass through the vapour-liquid interface, the normal velocity is not zero. The validity of the analogy depends on the smallness of the "suction parameter" $(-v_o/u)Re^{1/2}$, where v_o is the normal outward (i.e., negative for condensation) vapour velocity at the condensate surface, u is the free-stream velocity parallel to the surface and Re is the free-stream vapour **Reynolds number**. The results are strictly correct only in the limiting case of zero condensation rate. A widely-used approximate *stagnant film model* extends the range of validity of the analogy. The heat-mass transfer analogy and the stagnant film model are discussed by Lee and Rose (1983), Butterworth (1983) and Webb (1990).

The foregoing refers exclusively to laminar flow conditions. For tall condensing surfaces, or under conditions of high vapour shear stress, transition to turbulent flow in the condensate film may occur. This brings to the problem unresolved difficulties associated with the general problem of **Turbulence**. For moderate or low vapour velocities, the "effective height" of the surface for a horizontal tube ($\sim$d) is small and laminar flow of the condensate is expected. Moreover, since turbulent mixing enhances heat transfer across the condensate film, the Nusselt solution [Equation (2)] is conservative and is widely used in design calculations. Various models for turbulent film condensation from an essentially stationary vapour exist in the literature and predict somewhat different results. For gravity-dominated flow, transition to turbulence has been found to occur at film Reynolds numbers, $4\Gamma/\eta$, rather lower than 2 000, where Γ is the condensate flow rate per width of surface. In the presence of high-vapour shear stress, the problem is more complicated. Turbulent film condensation on a vertical surface under free-convection conditions can, in principle, be analysed by an approach similar to that used for single-phase pipe flow. However, this problem is relatively unimportant since in practice, condensing surfaces are usually not sufficiently tall for turbulence to occur under purely free-convection conditions. Significant shear stress, due to vapour flow along the condensate surface, promotes the onset of turbulence. The analysis is then more complicated, particularly when the vapour flow is not in the same direction of gravity.

Turbulence is more often encountered for condensation inside a tube. In this case, the problem is generally complicated by the presence of significant vapour shear stress on the condensate film since even when all of the vapour is condensed in the tube, the shear stress for a portion of the tube towards the inlet end is generally significant owing to the high vapour velocity resulting from the small tube cross-section. Condensation inside tubes is beset with all the problems and uncertainties of **Two-Phase Flow**. The only case which can be analysed wholly satisfactorily is that of downward vapour flow in a vertical tube with a laminar condensate film on the wall (stratified flow). In this case, the problem is the same as that for external flow, except that account must be taken of the progressive reduction of vapour flow rate, and hence shear stress, due to condensation. Approaches used in other cases are outlined by Butterworth (1983). (See **Tubes, Condensation in.**)

In many practical applications, condensation occurs in bundles or banks of horizontal tubes (shell-side condensation). In these cases, there is the additional complication of *inundation* (condensate from higher or upstream tubes falling or impinging on lower or downstream tubes). This leads to thicker condensate films on the inundated tubes. At the same time, the condensate film on inundated tubes is disturbed and the heat transfer coefficient may be enhanced. Nusselt's (1916) approach for a vertical, in-line column of horizontal tubes assumes that condensate drains to lower tubes in the form of a continuous laminar film. This leads to a simple expression for the average heat transfer coefficient for a column of N tubes:

$$\overline{\alpha}_N = \alpha_1 N^{-1/4} \tag{4}$$

where α_1, for the uppermost tube, is given by Equation (2). In view of the more probable mode of *drainage* or inundation with condensate film disturbance due to splashing from droplets, columns or unstable broken films of liquid, it is not surprising that Equation (4) has been found to be overconservative. Many experimental studies of condensation on tube banks have been made. The data are widely-scattered owing primarily to the effects of noncondensing gases, turbulence and vapour velocity. Various correlations have been proposed and approximate methods used in practiced are discussed by Butterworth (1983). (See **Tube Banks, Condensation Heat Transfer in.**)

Numerous techniques for *enhancement of film condensation heat transfer* have been proposed [see, for instance, Webb (1981)]. Notable amongst these, for shell-side condensation, are low (fin height small in relation to tube diameter) integral-fin tubes. In this case, it is found that for horizontal tubes, the enhancement of the heat transfer coefficient can significantly exceed the increase in surface area due to the presence of the fins. The reasons for this are: 1) the vertical or near-vertical fin flanks have small heights so that the heat transfer coefficients are large (see Equations 1 and 2) surface tension effects give rise to an additional mechanism for draining condensate from parts of the surface. The latter arises from the pressure gradient set up in the presence of a condensate surface of varying curvature, e.g., for the two-dimensional case:

$$\frac{dP}{ds} = \sigma \frac{d(r^{-1})}{ds} \tag{5}$$

where P is pressure, s is distance measured along the surface, σ is surface tension and r is the local radius of curvature of the condensate film. At the same time, surface tension has a detrimental effect on the heat transfer coefficient due to capillary retention of condensate between the fins and the consequent "blanketing" of heat transfer surface on the lower part of the surface. The extent of condensate retention can be calculated from:

$$\phi = \cos^{-1}\{(4\sigma\cos\beta/\rho g b d_o) - 1\} \tag{6}$$

as formulated by Honda et al. (1983). In Equation (6), ϕ is the angle measured from the top of the tube to the position where the interfin tube space is filled with retained condensate; β is the angle between the fin flank and radial plane; b is the distance between adjacent fins measured at the fin tip; and d_o is the tube diameter over the fins.

In an early theoretical solution of the problem of condensation on low-finned tubes by Beatty and Katz (1948), the vertical fin-flanks and horizontal interfin tube spaces were treated on the basis of the Nusselt theory and effects of surface tension were ignored. This simple approach proved quite successful in practice for relatively low-surface tension fluids. This is partly because condensate retention in this case is small, and partly because the beneficial and detrimental effects of surface tension tend, to some extent, to nullify each other. More detailed and complicated models have been proposed, notably by Honda and Nozu (1987). These require numerical solution and are less readily applied than the simple analytical result of Beatty and Katz. The various models are discussed by Marto (1988). A recent semi-empirical approach, which includes surface tension effects, has been given by Rose (1994). The result, in the form of an equation for the "enhancement ratio," is in good agreement with experimental data from seven investigations using four condensing fluids and 41 tube/fin geometries. (See also **Augmentation of Heat Transfer**, **Two-Phase Systems**.)

The foregoing relates to the case when the condensate wets the condensing surface and forms a continuous film. When the surface is not wetted, a quite different mode, namely **Dropwise Condensation**, may occur. In this case, minute droplets form at nucleation sites on the surface and growth takes place by condensation and coalescence with neighbours until drops reach a size at which they are removed from the surface by gravity or vapour shear stress. Moving drops sweep up stationary drops in their path, making available new area for condensation. The maximum-to-minimum drop size is around 10^6. To date, dropwise condensation has only been obtained with a few high-surface tension fluids (notably water). A nonwetting agent or "promoter" is required to promote dropwise condensation on metal surfaces. Heat transfer coefficients for dropwise condensation are much higher than those for film condensation under the same conditions. For steam at atmospheric pressure, the factor is around 20. Not surprisingly, this has stimulated research work over the past 60 years with the aim of finding an effective promoter. Although good promoters (e.g., dioctadecyl disulphide) are available, which form stable monomolecular layers on copper or copper-containing surfaces, and give dropwise condensation for hundreds or thousands of hours under clean laboratory conditions, effective promoters for use under industrial conditions have yet to be found. Recent surveys of dropwise condensation heat transfer have been given by Tanasawa (1991) and Rose (1994).

In some applications, such as desalination and geothermal power plant, use is made of direct contact condensation. This is a term applied to processes wherein vapour condenses on subcooled liquid drops, sprays, jets, films or in a liquid pool. Various types of equipment and the relevant mechanisms for direct contact condensation are given by Jacobs (1985). (See also **Direct Contact Heat Transfer**.)

References

Beatty, K. O. and Katz, D. L. (1948) Condensation of Vapors on the Outside of Finned Tubes. *Chem. Eng. Prog.* 44, 55–70.

Butterworth, D. (1983) *Film Condensation of Pure Vapour.* Heat Exchanger Design Handbook, Ed. E. U. Schlunder, Hemisphere Publishing Corporation, New York.

Fujii, T. (1991) *Theory of Laminar Film Condensation.* Springer-Verlag, New York.

Honda, H., Nozu, S. and Mitsumori, K. (1983) Augmentation of Condensation on Horizontal Finned Tubes by attaching a Porous Drainage Plate. Proc. ASME-JSME Thermal Engng. Joint Conf., 3, 289–296.

Honda, H. and Nozu, S. (1987) A Prediction Method for Heat Transfer during Film Condensation on Horizontal Integral Fin Tubes. *J. Heat Transfer,* 109, 218–225.

Jacobs, H. R. (1985) *Direct Contact Condensers.* Heat Exchanger Design Handbook, Ed. E. U. Schlunder, Hemisphere Publishing Corporation, New York.

Lee, W. C. and Rose, J. W. (1983) Comparison of Calculation Methods for Non-Condensing Gas Effects in Condensation on a Horizontal Tube. *IChemE Symp.* Ser. No. 75, 342–355.

Marto, P. J. (1988). An Evaluation of Film Condensation on Horizontal Integral-Fin Tubes. *J. Heat Transfer,* 110, 1287–1305.

Memory, S. B. and Rose, J. W. (1991) Free Convection Laminar Film Condensation on a Horizontal Tube with Variable Wall Temperature. *Int. J. Heat Mass Transfer,* 34, 2775–2778.

Niknejad, J. and Rose, J. W. (1981) Interphase Matter Transfer—an Experimental Study of Condensation of Mercury. *Proc. R. Soc. Lond.* A 378, 305–327.

Nusselt, W. (1916) Die Oberflachenkondensation des Wasserdampfes. *Z. Vereines Deutch. Ing.* 60, 541–546, 569–575.

Rose, J. W. (1969) Condensation of a Vapour in the Presence of a Non-Condensing Gas. *Int. J. Heat Mass Transfer,* 12, 233–237.

Rose, J. W. (1980) Approximate Equations for Forced-Convection Condensation in the Presence of a Non-Condensing Gas on a Flat Plate and Horizontal Tube, *Int. J. Heat Mass Transfer,* 23, 539–546.

Rose, J. W. (1988) Fundamentals of Condensation Heat Transfer: Laminar Film Condensation. *JSME Int. Journal,* Series II, 31 (3), 357–375.

Rose, J. W. (1994) An Approximate Equation for the Vapour-Side Heat Transfer Coefficient for Condensation on Low-finned Tubes. *Int. J. Heat Mass Transfer,* 37, 865–875.

Rose, J. W. (1994) Dropwise Condensation. *Heat Exchanger Design Update,* 2.6.5, 1–11, Begell House.

Tanasawa, I. (1991) Advances in Heat Transfer, *Advances in Condensation Heat Transfer.* Vol. 21. Eds. J. P. Hartnett and T. F. Irvine. Academic Pres Inc., New York.

Webb, D. (1990) Multicomponent Condensation, *Proc. Ninth Int. Heat Transfer Conf.* Jerusalem, Vol. 1, 287–304.

Webb, R. L. (1981) *The Use of Enhanced Surface Geometries in Condensers: An Overview.* Power Condenser Heat Transfer Technology, Eds. P. J. Marto and R. H. Nunn, Hemisphere Publishing Corporation, New York.

Leading to: *Condensation curve; Condensation of pure vapour; Condensation of multicomponent vapours; Condensers; Dropwise condensation; Tube banks, condensation heat transfer in; Tubes, condensation in; Tubes, condensation on outside in crossflow*

J.W. Rose

CONDENSATION IN TUBES (see Tubes, condensation in)

CONDENSATION SHOCKS (see Critical flow)

CONDENSERS

Following from: *Condensation, overview*

Introduction

Tubular condensers used in power plant to condense exhaust steam are known as **Surface Condensers**. However, there are many other applications in which condensers are used, and a wide variety of condenser types has been developed.

Direct-Contact Condensers

The *direct-contact condenser* is one in which the coolant is brought into contact with the vapour. It has the advantage of low cost and simplicity of mechanical design, but its use is restricted to those applications in which mixing of the vapour and coolant is permissible.

The various types of direct contact condensers are:

1. The Spray Condenser. The coolant is sprayed, using nozzles, into a vessel to which the vapour is supplied. This is shown schematically in **Figure 1**. It is important that the spray nozzles and vessel are designed to produce a fine spray of liquid (to give a large interfacial area for heat transfer), and a long enough residence time of liquid droplets in the vessel.
2. The Baffled Column. This is similar to the spray condenser, except that the coolant is directed to flow over a series of trays in a column (see **Figure 2**). The vapour is supplied to the bottom of the column. It has the advantage of counter-current flow of vapour and coolant, though care must be taken to avoid flooding. (Flooding is an unstable condition when the vapour flow is such that the downward flow of condensate is interrupted and held up).
3. The Packed Column. A packed column may consist of tightly-packed metal rings to increase the interfacial area

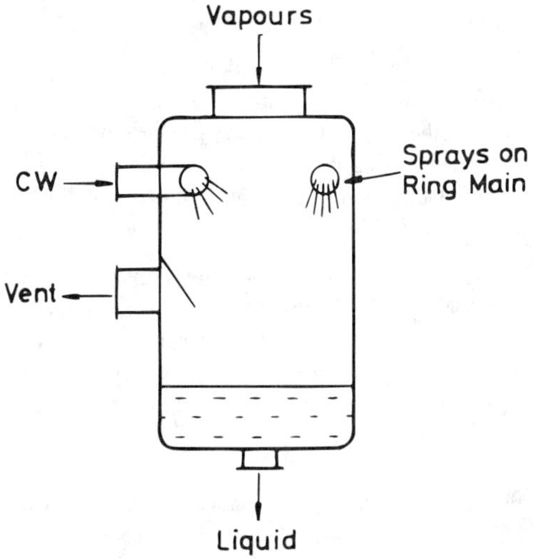

Figure 1. Spray condense.

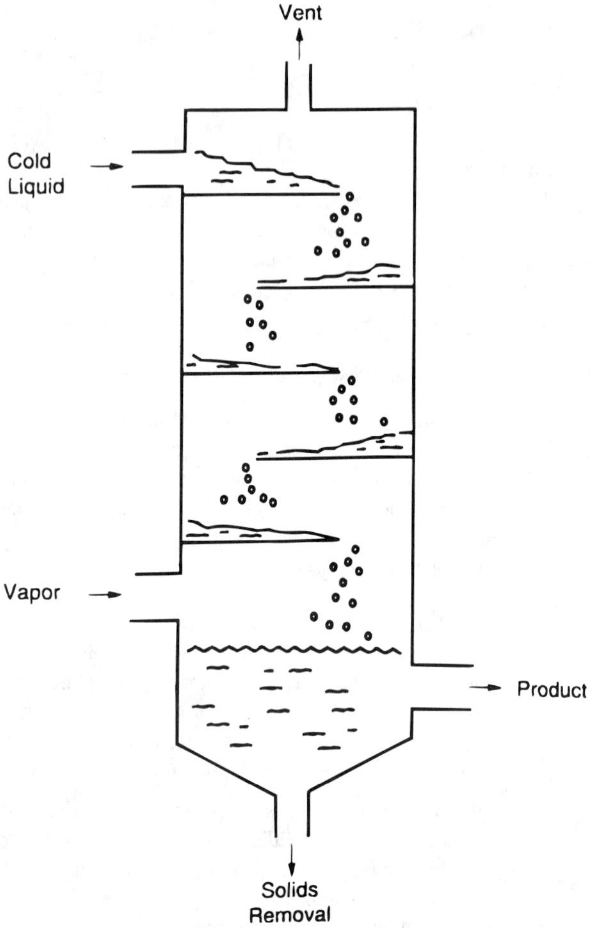

Figure 2. Tray type condenser. *Source:* G. F. Hewitt, G. L. Shires, and T. Bott *Process Heat Transfer* (1994).

for heat transfer. Liquid is supplied to the top of the column and vapour is supplied to the bottom. The disadvantage of this type of condenser is that the pressure drop is higher than in other types of direct-contact condenser.

4. The Jet Condenser. This is a device in which a jet of liquid is directed into a vapour stream, usually with the objective of desuperheating the vapour. A jet of liquid is injected into a pipeline carrying vapour via a small bore pipe and a nozzle located at the centre line. The liquid is usually injected in counterflow to the vapour.
5. The Sparge Pipe. The sparge pipe consists of a pipe with holes for injecting bubbles of vapour into a pool of liquid. This is a simple method of condensing a vapour, but there are practical problems associated with generating a good distribution of bubbles of small size, which are required for efficient heat transfer.

The design of direct-contact condensers is well described by Fair (1972). Most equipment of this type is designed on the basis of empirical information from experimental and operating data.

Shell-and-Tube Condensers

Shell-and-tube condensers are used extensively in the process industries, typically to condense the overhead vapour from a distillation column. There are three main types.

Crossflow shell-side condenser

The crossflow condenser is similar to the surface condenser. It consists of a shell containing tubes through which the coolant flows. The shell-side flow path is designed such that the vapour flows mainly in crossflow direction to the tubes. The crossflow condenser is typically used for low-pressure applications, in which there is a large volume flow of vapour and a low-pressure drop is required.

The tubes are supported at intervals by plates to prevent sagging of the tubes and to avoid vibration.

The vapour enters at the top of the shell. Often, more than one nozzle is used to minimise pressure loss and promote good distribution.

It is particularly important to ensure that the crossflow condenser is properly vented.

Baffled shell-and-tube condenser

The baffled shell-and-tube heat exchanger, with shell-side condensation, is the most common type of condenser used in the process industries. It is most often mounted horizontally. A typical shell-side condenser is shown in **Figure 3**. This is a shell of the TEMA E-type in which the vapour enters at one end of the shell and flows to the outlet end, where the condensate and any uncondensed vapour and noncondensing gases are removed.

The baffles are normally plates with a single segmental cut. The cut is usually vertical to allow the condensate to flow along the bottom of the shell to the outlet. Double segmental baffles may be specified to achieve reduced pressure drop. The space between the baffles is determined by considerations such as:

1. The tubes must be supported to avoid tube vibration;
2. Pressure drop depends on vapour velocity, and hence on baffle spacing;

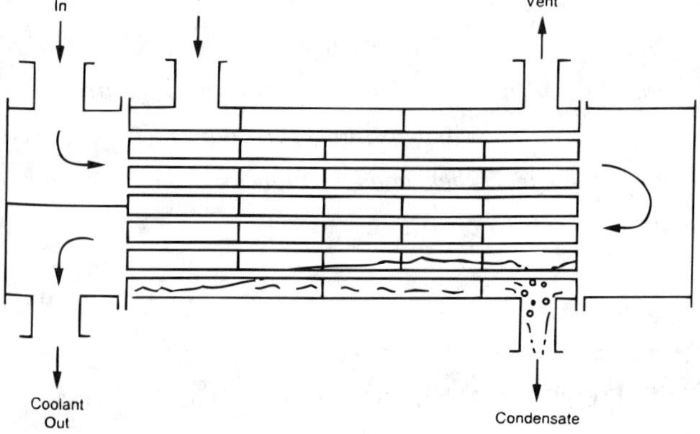

Figure 3. Horizontal shell-side condenser. *Source:* G. F. Hewitt, G. L. Shires, and T. Bott *Process Heat Transfer* (1994).

3. The resistance to heat transfer due to the presence of noncondensing gases is inversely proportional to the vapour velocity.
4. High vapour velocities can reduce resistance to heat transfer through the shear effect on the condensate film.

Thus, if adequate pressure drop is available, it may be possible to reduce baffle spacing to obtain increased heat transfer rates.

The size of the baffle cut is generally chosen such that the vapour velocity in the baffle window is roughly equal to that in the overlap zone between the baffle plates.

The number of tube passes is determined by the required coolant velocity. Plain tubes are generally used, though low-finned tubes can be used to obtain increased heat transfer rates when the condensate resistance to heat transfer with plain tubes is significantly greater than that of the coolant.

The vapour vent should be situated at the end of the vapour flow path. A typical vent location is shown in **Figure 3**.

It is possible to use other TEMA shell types, such as the J-type, to minimise pressure drop. A possible configuration for low pressure drop is the J-shell with double segmental baffles.

Tube-side condensers

Condensation on the tube-side is preferred when the coolant is a gas, such as air. It may also be preferred if the condensing fluid is at a higher pressure than the coolant, since it is usually less expensive to contain a higher pressure inside tubes than inside a shell. An air-cooled condenser is typical of a tube-side condenser. It consists of a tube bundle, normally with finned tubes, over which air flows in crossflow. The air flow is driven by fans, either in forced- or induced-draft mode. A typical forced-draft, air-cooled condenser is depicted in **Figure 4**.

If the tubes are vertical, the two configurations in common use are:

1. Co-current downward vapour and liquid flow;
2. Counter-current flow with the vapour flowing upward and the liquid flowing downward. This is commonly known as the reflux condenser.

It is not advisable to attempt co-current upward vapour and liquid flow unless there is a high vapour velocity at the outlet of the tubes.

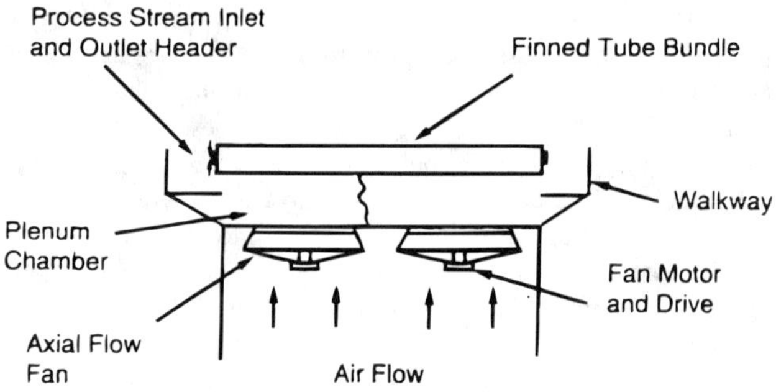

Figure 4. Schematic of air-cooled condenser operating in forced-draft mode. *Source:* G. F. Hewitt, G. L. Shires, and T. Bott *Process Heat Transfer* (1994).

The reflux condenser must be designed such that vapour velocity is less than the flooding velocity. This is a limiting velocity above which drainage of condensate is irregular. (See **Flooding and Flow Reversal**.)

Thermal Design of Shell-and-Tube Condensers

Area calculation

In condensation of a single pure vapour, provided that the pressure drop is small compared to the absolute pressure, the temperature of the condensing stream is a constant value determined by the saturation pressure.

If the coolant is single-phase, and if the overall heat transfer coefficient is reasonably constant, then the assumptions underlying the "logarithmic mean temperature difference (LMTD)" are valid. (See **Mean Temperature Difference**.) This means that the surface area requirement, A, of the condenser can be determined from:

$$A = \dot{Q}_T/(U[LMTD]) \qquad \textbf{(1)}$$

where Q_T is the total heat load and U is the mean overall heat transfer coefficient.

In condensation from mixtures, with or without a noncondensing gas, the variation of the equilibrium temperature with enthalpy can be highly nonlinear. Also, the heat transfer coefficient of the condensing stream can vary by an order of magnitude over the condensing path. This means that it is not possible to assign a single representative temperature difference and overall heat transfer coefficient to the exchanger, and that a zonal or stepwise calculation of the surface area is required.

The thermal design of condenser is therefore considerably more complicated than that of a single-phase heat exchanger.

Figure 5 shows a typical temperature/enthalpy relationship for a mixture which is superheated at entry. This relationship, and the corresponding physical properties, are normally obtained from specialist computer programs which perform vapour-liquid equilibrium calculations and determine the compositions of the vapour and liquid phases along the condensing path. The corresponding relationship for a single-phase coolant flowing in a single pass in counterflow to the condensing stream is also shown.

An outline of the procedure to determine the surface area requirement for such a condensing duty is as follows:

1. Divide the temperature/enthalpy diagram into a number of zones such that the curves of both the condensing stream and the coolant can be regarded as being reasonably linear. An example of how this is done is shown in **Figure 5**.
2. Specify the principal geometric parameters of the design, such as the number of tubes, tube outside diameter and wall thickness, tube pitch, baffle cut, baffle pitch and shell diameter.
3. Calculate the local overall heat transfer coefficients at the zone boundaries.
4. For each zone, calculate the zonal overall heat transfer coefficient, U_z, from the arithmetic average of the local overall coefficients at the zone boundaries.

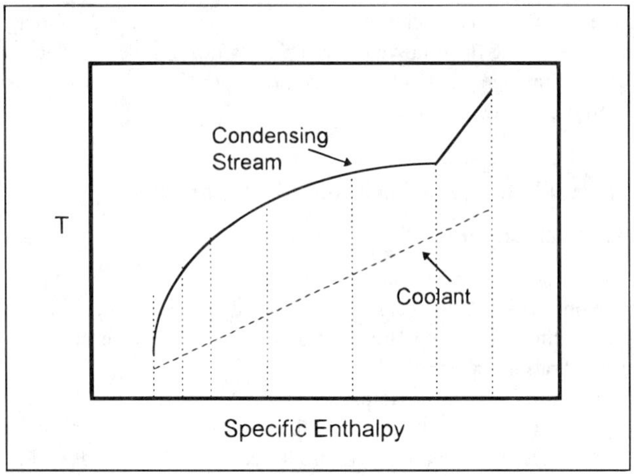

Figure 5. Example of temperature/enthalpy curves for a condensing stream and coolant, showing subdivision into zones.

5. Calculate the LMTD for each zone from the temperatures at the zone boundaries.
6. Calculate the surface area requirement for each zone by applying Equation (1) above.
7. Calculate the total surface requirement by summing the zonal surface areas.

The design process consists of repeating the above process, varying the main geometrical parameters to both meet the pressure drop constraints and minimise either area or cost. The procedure is generally carried out using a computer program.

If, as is often the case, condensation occurs on the shell-side with multiple tube-side passes, the calculation of temperature profiles and zonal surface areas is more complicated. A suitable methodology is described by *Bell and Ghaly* (1972).

Local heat transfer coefficients

The above procedure for calculating surface area requires evaluation of the local heat transfer coefficient at the zone boundaries. This, in turn, requires calculation of:

1. The coolant heat transfer coefficient;
2. The resistance to heat transfer due to the tube wall;
3. The condensate heat transfer coefficient associated with heat transfer through a film of condensate on the tube wall;
4. The resistance to heat transfer associated with the presence of noncondensing gases, or with a mixture of more than one vapour.

In addition, it is necessary to estimate the resistance to heat transfer due to fouling on both sides of the tube wall. (See **Fouling** and **Fouling Factors**.)

Calculation of the condensate heat transfer coefficient depends strongly on condenser geometry. Appropriate methods are described by Hewitt, Shires and Bott (1994), who also prescribed the calculation of the gas-phase resistance. Use of a computer program is required for all but the simplest calculations.

Pressure gradient

The pressure gradient in condensers is due to both frictional and accelerational effects. The frictional effect, which gives rise to a pressure loss, can be calculated by applying a two-phase multiplier to the pressure gradient for single-phase flow. The accelerational effect gives rise to a pressure increase due to the deceleration of the flow. It is usually only significant at low pressures, and can be calculated on the basis of the vapour flow rate alone.

Plate Type Condensers

Some types of plate heat exchangers, which have been traditionally used in other applications, are increasingly used as condensers. For example, the plate-and-frame and brazed-plate heat exchangers, have advantages of lower cost and lower fluid inventory. The use of plate heat exchangers as condensers is discussed by Kumar (1983). Normally, a special plate passage and inlet port configuration will be required to handle the high-vapour volume flow rate at a condenser inlet.

The plate-fin heat exchanger, traditionally used in cryogenic applications, can also be applied to some general condensing applications in which compactness is required.

Problems in the Design of Condensers

Removal of gases

It is important to provide a condenser with a vent for removal of gases, either during start-up and/or for continuous operation. The vent must be located near the end of the vapour flow path. It is particularly important in crossflow shell-and-tube condensers to avoid regions of very low vapour velocity, where stagnant zones of noncondensing gas can form and consequently render some of the heat exchanger surface ineffective.

Condensate drainage

The condensate drain from a condenser must be carefully designed to ensure that it is adequately-sized for the condensate flow rate, and to avoid entrainment of uncondensed vapour or gas into the condensate pipework.

Fog formation

Fogging in a condenser is due to the formation of tiny droplets of liquid in the vapour. Fog formation can occur when the temperature of the vapour-gas mixture falls significantly below the local saturation temperature during the condensation process. This tends to happen when the transport properties of a vapour-gas mixture and the process conditions are such that more heat than mass is removed from the mixture. Fog formation represents undesirable loss of product and may, in some circumstances, represent a pollution problem. The actual onset of fog formation will depend on the presence or otherwise of nucleation sites. Removal of fog may require special separation methods. Steinmeyer (1972) gives a good practical account of how fog forms and how its effects can be minimised.

References

Bell, K. J. and Ghaly, M. A. (1972) *An Approximate Generalised Design Method for Multicomponent/Partial Condensers,* AIChE Symp. Series, 131 (69), 72.

Fair, J. R. (1972) *Designing Direct-Contact Coolers/Condensers, Chem. Engng.*, 12 June, 91–100.

Hewitt, G. F., Shires, G. L., and Bott, T. R. (1994) *Process Heat Transfer,* CRC Press, Boca Raton, Florida.

Kumar, H. (1983) *Condensation Duties in Plate Heat Exchangers,* Instn. *Chem Engnrs. Symposium Series,* 75, 2, 1275.

Steinmeyer, D. E. (1972) *Fog Formation in Partial Condensers, Chem. Engng. Progress,* 68, 7, 64–68.

Leading to: *Surface condensers*

J.M. McNaught

CONDUCTANCE, ELECTRICAL (see Ohm's Law)

CONDUCTANCE PROBES, FOR LOCAL VOID FRACTION (see Void fraction measurement)

CONDUCTION

Following from: *Thermal conductivity mechanisms; Heat transfer*

Conduction is a diffusion process by which thermal energy spreads from hotter regions to cooler regions of a solid or stationary fluid. A range of microscopic diffusive mechanisms may be involved in heat conduction (Gebhart (1993)) and the observed overall effect may be the sum of several individual effects, such as molecular diffusion, electron diffusion and lattice vibration.

A simple model of the mechanism of heat conduction is provided by the Kinetic Theory of Gases. Consider the system consisting of a stationary mass of an ideal monatomic gas confined between two parallel isothermal walls in a zero gravity field (i.e., no natural convection possible) and initially at a uniform temperature T_i throughout, as illustrated in **Figure 1**.

Suppose that at time $t = 0$, the left hand wall is raised to and kept at a higher temperature T_1 so that at some subsequent time,

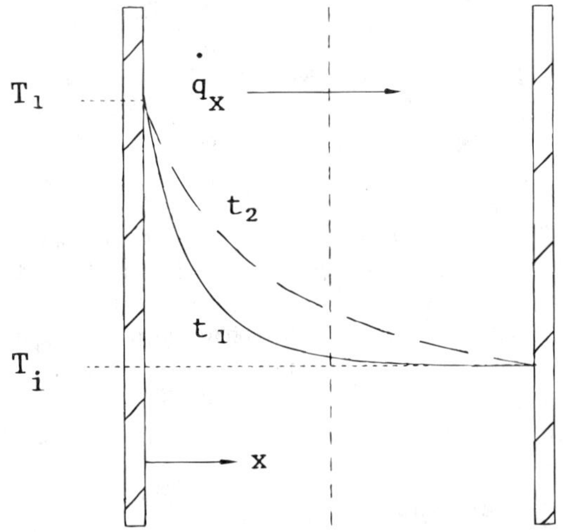

Figure 1. Heating of a stationary gas mass by diffusion.

t_1, a typical temperature distribution such as that represented by the full curve develops, indicating a spread of heat from the hot wall in the direction of the cold wall. This energy transfer process in a simple gas is the result of the diffusive effects of random molecular motion. Consider the molecular traffic across a vertical plane, such as that represented by the vertical dashed line in **Figure 1**. Because of the random nature of molecular motion and because there can be no bulk flow of gas, molecules must cross the imaginary plane at an equal rate in both the positive and negative x-directions. Those molecules crossing the plane in the positive x-direction possess more translational/rotational/vibrational energy than those crossing in the opposite direction. Molecular collisions lead to energy transfer from more energetic molecules to less energetic ones. By this diffusive action, the thermal wave strengthens and advances so that at some later time, t_2, the temperature distribution has evolved as indicated by the broken curve in **Figure 1**.

Energy diffusion in more complex fluids and solids is not so mechanistically simple, but whatever the actual mechanism of energy transport at a microscopic level, a macroscopic formulation of the rate process of diffusion, known as **Fourier's Law**, provides a simple mathematical description and is crucial for the analysis of conduction problems. The differential form of Fourier's Law for one-dimensional conduction in an isotropic medium with constant thermal conductivity, such as the process represented in **Figure 1** is:

$$\dot{q}_x = -\lambda \frac{\partial T}{\partial x} \quad (1)$$

where it is clear that for the most part $\dot{q}_x$ varies with x and t until steady-state is approached ($t \rightarrow \infty$), whereupon $\dot{q}_x$ becomes constant with both x and t and the temperature distribution becomes linear.

Many practical conduction problems are of the undeveloped or unsteady-state type, with the fully-developed solution forming a simple limiting case. Typically, a conduction problem will be posed in the form of a specified geometry, an initial temperature distribution throughout the region of interest, and information on external and internal heating effects. The normal requirement will be for information on the temperature distribution at some elapsed time from the commencement of the heating or cooling process. The starting point for the analysis of such a problem is the differential form of the general conduction equation, which is derived by applying the principle of conservation of energy to a representative differential control volume, such as the one illustrated in **Figure 2**.

The control volume (dxdydz) is drawn in the rectangular Cartesian coordinate system, but could equally well be drawn in the cylindrical or polar coordinate systems [see Bird, Stewart and Lightfoot (1960)]. The control volume is embedded in a system which exemplifies the most general case of conduction i.e., unsteady three-dimensional conduction in a material which may be consuming or producing thermal energy internally by virtue of some process, such as endothermic chemical reaction or ohmic heating. The volumetric rate of any such internal energy conversion will be denoted by $\dot{q}^*$.

The only way that energy can pass into or out of the control volume across the six imaginary faces is by the process of conduction. An energy balance on the control mass in the control volume can be written as:

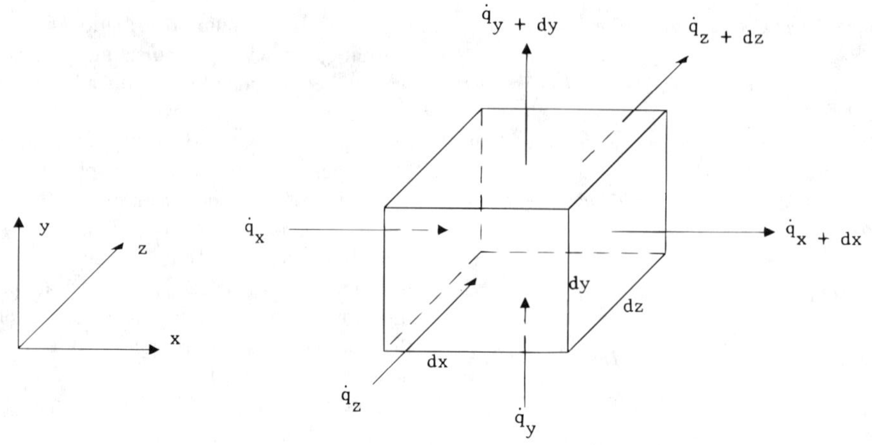

Figure 2. Cartesian differential control volume used to formulate an energy balance for a general conduction process.

Net rate of energy gain by conduction + Rate of energy production/loss by internal processes = Rate of energy increase/decrease of control mass

Each of these terms can be quantified to produce a differential energy balance, as follows:

(1) The net rate of energy gain (loss) by conduction in the x-direction is $(\dot{q}_x - \dot{q}_{x+dx}) \times (dydz)$. The typical parabolic temperature profile associated with the transient conduction process illustrated in **Figure 1** indicates that the local heat flux $\dot{q}_x$ is a function of x (see Fourier's Law) and so a Taylor Series expansion can be used to approximate $\dot{q}_{x+dx}$:

$$\dot{q}_{x+dx} = \dot{q}_x + \frac{\partial \dot{q}_x}{\partial x} dx + \text{higher order terms.} \tag{2}$$

Accounting for the energy gains due to conduction in the y and z directions gives the total net rate of energy gain by the control mass to be:

$$-\left\{\frac{\partial \dot{q}_x}{\partial x} + \frac{\partial \dot{q}_y}{\partial y} + \frac{\partial \dot{q}_z}{\partial z}\right\} dxdydz. \tag{3}$$

This expression can be rewritten with T as the dependent variable by using Fourier's Law to substitute for $\dot{q}$:

$$\left\{\frac{\partial}{\partial x}\left[\lambda \frac{\partial T}{\partial x}\right] + \frac{\partial}{\partial y}\left[\lambda \frac{\partial T}{\partial y}\right] + \frac{\partial}{\partial z}\left[\lambda \frac{\partial T}{\partial z}\right]\right\} dxdydz. \tag{4}$$

If it is assumed that λ is invariant with T (a reasonable assumption in many cases where the temperature changes are not large, see **Thermal Conductivity**), then the conduction term in the energy balance simplifies to:

$$\lambda\left\{\frac{\partial^2 T}{\partial x^2} + \frac{\partial^2 T}{\partial y^2} + \frac{\partial^2 T}{\partial z^2}\right\} dxdydz. \tag{5}$$

(2) The second term in the energy balance which accounts for the rate of energy gain or loss in the control volume due to possible internal energy conversion processes is simply written as $\dot{q}^*(dxdydz)$.

(3) The third term in the energy balance is the rate at which the control mass stores energy as a net result of conduction and internal energy conversion, which is quantified as: mass $\times$ specific heat $\times$ rate of rise of temperature, or,

$$\rho(dxdydz)c_p \frac{\partial T}{\partial t}.$$

The complete energy balance can now be written as:

$$\lambda\left\{\frac{\partial^2 T}{\partial x^2} + \frac{\partial^2 T}{\partial y^2} + \frac{\partial^2 T}{\partial z^2}\right\}(dxdydz) + \dot{q}^*(dxdydz) = \rho(dxdydz)c_p \frac{\partial T}{\partial t} \tag{6}$$

or, simplifying:

$$\left\{\frac{\partial^2 T}{\partial x^2} + \frac{\partial^2 T}{\partial y^2} + \frac{\partial^2 T}{\partial z^2}\right\} + \frac{\dot{q}^*}{\lambda} = \frac{1}{\kappa}\frac{\partial T}{\partial t}. \tag{7}$$

This energy balance often appears in vector form:

$$\nabla^2 T + \frac{\dot{q}^*}{\lambda} = \frac{1}{\kappa}\frac{\partial T}{\partial t} \tag{8}$$

where the operator ∇^2, known as the Laplacian, may also be expressed in cylindrical or spherical coordinates. This is the *conduction equation* and is one of the most important in applied mathematics. It is the simplest approximation to bulk processes governed at the microscopic level by random spatial variations. .

Special Cases of the Conduction Equation

1. Fourier Equation

For conduction problems in which there is no internal energy conversion, the energy balance simplifies to:

$$\left\{\frac{\partial^2 T}{\partial x^2}+\frac{\partial^2 T}{\partial y^2}+\frac{\partial^2 T}{\partial z^2}\right\}=\frac{1}{\kappa}\frac{\partial T}{\partial t}. \quad (9)$$

2. Poisson Equation

For steady state conduction with internal energy conversion, the energy balance may be written as:

$$\left\{\frac{\partial^2 T}{\partial x^2}+\frac{\partial^2 T}{\partial y^2}+\frac{\partial^2 T}{\partial z^2}\right\}+\frac{\dot{q}^*}{\lambda}=0. \quad (10)$$

3. Laplace Equation

For steady state conduction with no internal energy conversion, the energy balance reduces to its simplest form:

$$\left\{\frac{\partial^2 T}{\partial x^2}+\frac{\partial^2 T}{\partial y^2}+\frac{\partial^2 T}{\partial z^2}\right\}=0. \quad (11)$$

This is a potential field equation and the analogy with other potential field systems can be utilised to obtain solutions to conduction problems.

Boundary Conditions

The analysis of a conduction problem involves the solution of the appropriate form of the differential energy balance, subject to the boundary conditions applicable to the problem at hand. A well-posed, soluble and general heat conduction problem will have the following characteristics:

1. It is required to find T(x,y,z,t) such that:

$$\nabla^2 T+\frac{\dot{q}^*}{\lambda}=\frac{1}{\kappa}\frac{\partial T}{\partial t} \quad (12)$$

with $0 < (x,y,z,t) < \infty$.

2. The initial temperature distribution throughout the region of interest is known i.e., $T = T_i(x,y,z)$ at $t = 0$, except for steady-state problems and problems where $\dot{q}^*$ or the boundary conditions vary periodically with time.
3. T must also satisfy two boundary conditions for each coordinate, which are often of the following three common types (see **Figure 3**):
(1) Boundary conditions of the first kind (or *Dirichlet conditions*) where T is specified around the edge of the conduction domain for $t > 0$, e.g. $T = T_1$ at $x = 0$, as indicated in **Figure 3a**.

(2) Boundary conditions of the second kind (or *Neumann conditions*) where the gradient of T normal to the boundary is specified around the boundary of the solution domain for $t > 0$. A common example is a surface which is subjected to a constant heat flux i.e., $\dot{q}_{x=0} = -\lambda\left\{\frac{\partial T}{\partial x}\right\}_{x=0} = \text{constant}$, this situation being represented in **Figure 3b**.

(3) Boundary conditions of the third kind where the gradient of T normal to the boundary is directly proportional to the temperature on the boundary, as is the case when convection occurs at the surface. A heat balance at a conduction boundary where the heat flux due to conduction at the edge of the conduction domain (at T_s), which varies with time, is equated to the convective heat flux from the surrounding fluid at T_a, gives $\dot{q}_{x=0} = -\lambda\left\{\frac{\partial T}{\partial x}\right\}_{x=0} = \alpha(T_a - T_s)$. This boundary condition is represented in **Figure 3c**.

Although many problems will have boundary conditions which fall into the above three categories other, more complex boundary conditions can arise, [see for example Carslaw and Jaeger (1980)].

Solution Procedures

Good introductions to the main features of the most widely-used methods for solving the conduction equation can be found in textbooks, such as that by Kreith and Bohn (1993). Analytical solutions of conduction problems in simple geometric shapes (plates, cylinders and spheres) have been compiled by Carslaw and Jaeger (1980), and many solutions are presented in the form of charts by Schneider (1963), which have considerable utility for estimation and design. A wide range of techniques have been employed to obtain solutions to the conduction equation; the most common ones are clearly described in Hill and Dewynne (1987) and Myers (1987). Numerical techniques have been developed for the solution of heat conduction problems and these numerical methods are capable of dealing with complex geometries and non-linearities such as temperature dependent physical properties, anisotropy and boundary conditions which are nonlinear in T, as when radiative heating/cooling is involved. A list of commercially-available software packages for the numerical solution of a wide range of conduction (and convection) problems is given by Kreith and Bohn (1993). Solution methods suitable for dealing with an important class of transient conduction problems, known as *Stefan problems,* and involving the additional feature of a moving phase-change boundary (melting or freezing fronts) are described by Crank (1984) and Ockenden and Hodgkins (1975).

References

Carslaw, H. S. and Jaeger, J. C. (1980) *Conduction of Heat in Solids,* 2nd ed., Clarendon Press, Oxford.

Crank, J. (1984) *Free and Moving Boundary Problems,* Clarendon Press, Oxford.

Hill, J. M. and Dewynne, J. N. (1987) *Heat Conduction,* Blackwell Scientific Publications, Oxford.

Kreith, F. and Bohn, M. S. (1993) *Principles of Heat Transfer,* 5th ed., West Publishing Company, St. Paul.

Myers, G. E. (1987) *Analytical Methods in Conduction Heat Transfer,* Genium Publishing Co., Schenectady, New York.

Ockenden, J. R. and Hodgkins, W. R. (1975) *Moving Boundary Problems in Heat Flow and Diffusion,* Clarendon Press, Oxford.

Schneider, P. J. (1963) *Temperature Response Charts,* John Wiley, New York.

Leading to: *Biot number; Coupled conduction and convection; Coupled radiation and conduction; Coupled radiation,*

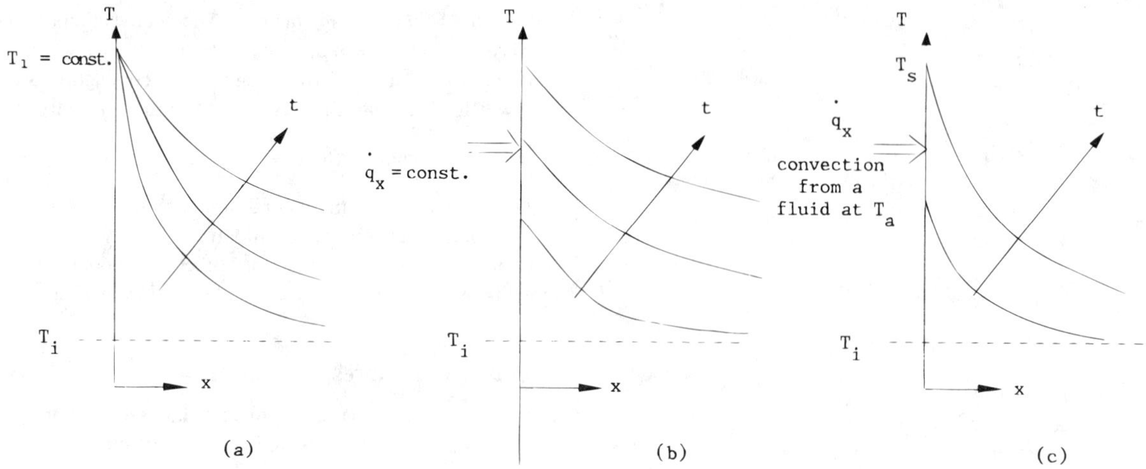

Figure 3. Schematic representation of common surface boundary conditions.

convection and conduction; Fourier's law; Fourier Number; Thermal diffusivity; Numerical heat transfer

T.W. Davies

CONDUCTION AND CONVECTION COMBINED (see Coupled conduction and convection)

CONDUCTION COMBINED WITH RADIATION (see Coupled radiation and conduction heat transfer)

CONDUCTION DRYING (see Dryers)

CONDUCTION EQUATION (see Conduction; Thermal diffusivity)

CONDUCTION IN HEAT EXCHANGER WALLS (see Coupled conduction and convection)

CONDUCTIVE HEAT FLUX (see Heat flux)

CONDUCTIVITY (see Electromagnetic flowmeters)

CONDUCTIVITY, ELECTRICAL (see Ohm's Law; Semiconductors)

CONDUCTIVITY, OF PLASMA (see Plasma)

CONDUCTIVITY RATIO (see Coupled conduction and convection)

CONE CLASSIFIER (see Classifiers)

CONFORMAL MAPPING

Conformal mapping or conformal transformation describes a mapping on a complex plane that preserves the angles between the oriented curves in magnitude and in sense. That is, the images of any two intersecting curves, taken with their corresponding orientation, make the same angle of intersection as the curves, both in magnitude and direction. A mapping that preserves the magnitude of each angle but not necessarily the sense is described as *isogonal.*

The engineering usefulness of conformal mapping is both in grid generation and in solving certain boundary value problems in two-dimensional potential theory, heat conduction and electrostatic potential by mapping a given complicated region (problem plane) onto a much simpler one (solution plane). An important role in the mapping task is performed by linear transformations, bilinear transformations, the Schwarz Christoffel transformation and other special functions [see Churchill, Brown & Verhey (1974)].

For instance, the problem of calculating the free surfaces of a two-dimensional inviscid fluid jet is extremely complicated analytically because the dynamic and nonlinear boundary conditions depend on the free surface, whose shape must be determined as part of the solution. However, by conformally mapping the physical flow region onto a unit circle, a solution can be found numerically with relative ease (Dias, Elcrat & Trefethen, 1987). Analytic conformal mapping functions between the problem plane and the solution plane only exist for a limited range of problems. For most practical problems, the conformal map is represented by an integral equation that has to be solved numerically. Some numerical solution schemes and the resolution of practical boundary conditions are given in a monograph by Trefethen (1986).

References

Churchill, R. V., Brown, J. W., and Verhey, R. F. (1974) *Complex Variables and Applications*, McGraw Hill, New York.

Dias, F., Elcrat, A. R., and Trefethen, L. (1987) Ideal Jet in Two Dimensions *Journal of Fluid Mechanics,* 185, 275.

Trefethen, L. (1986) *Numerical Conformal Mapping,* North-Holland, Amsterdam.

M.C. Tezock

CONFORMAL POTENTIALS (see Reduced properties)

CONICAL SHOCK WAVE (see Inviscid flow)

CONJUGATE HEAT TRANSFER (see Coupled conduction and convection)

CONSERVATION EQUATIONS (see Couette flow; Poiseuille flow)

CONSERVATION EQUATIONS, SINGLE-PHASE

Background

Continuum Hypothesis

In *continuum mechanics,* the focus of study is the behaviour of a body on a length scale large compared with the size of molecules or the distances between them. The assumption therefore is that the body has a continuous structure. This is the *continuum hypothesis.*

All variables are then continuous functions of position in the fluid and a definite meaning for the notion of properties at a point.

Reference frames and coordinate systems

Deformations are defined only relative to a *reference frame.* A reference frame is a body whose component parts are fixed relative to each other.

There is an infinite set of reference frames, of which a subset are *inertial reference frames.* An inertial reference frame is a reference frame in which linear and angular momentum are conserved. Alternatively, they can be defined as reference frames in which a particle moving without action from any forces travels in a straight line with a constant velocity.

For some purposes, the Earth can be regarded as an inertial reference frame whereas a rotating object cannot.

Given a reference frame, a three-dimensional *Euclidean space* can be fixed to it and a three-dimensional **Coordinate System** selected to span the space. A three-dimensional coordinate system is defined by an origin O and three noncoplanar directions, ε_1, ε_2 and ε_3. There exists an infinite set of coordinate systems. Always, an *orthogonal right-handed* system is chosen. An orthogonal system is one for which the directions associated with ε_1, ε_2 and ε_3 are mutually perpendicular or normal. A right handed system is one such that, if ε_1 and ε_2 are orientated in the plane of the page, ε_3 is orientated out of the page (i.e., toward the viewer). The three quantities $\mathbf{i}_{\varepsilon 1}$, $\mathbf{i}_{\varepsilon 2}$ and $\mathbf{i}_{\varepsilon 3}$ denote *unit vectors* (i.e., vectors of unit magnitude) aligned in the ε_1, ε_2 and ε_3 directions, respectively.

The two most common coordinate systems are:

1) *Rectangular (or Cartesian) coordinates* (x,y,z), see **Figure 1**.
2) *Cylindrical polar coordinates* (r, θ, z), see **Figure 2**.

The directions of unit vectors $\mathbf{i}_r$ and $\mathbf{i}\theta$ in cylindrical polar coordinates change with position in space. Cylindrical polar coordinates are related to rectangular ones as follows:

$$x = r\cos\theta \qquad y = r\sin\theta \qquad z = z \tag{1}$$

$$r = \sqrt{(x^2 + y^2)} \qquad \theta = \tan^{-1}\frac{y}{x} \qquad z = z.$$

Since the choice of coordinate systems is motivated by convenience, it follows, for example, that rectangular coordinates would reasonably be used in problems involving flows past flat plates and cylindrical polar coordinates in problems involving flows through pipes of circular cross-section.

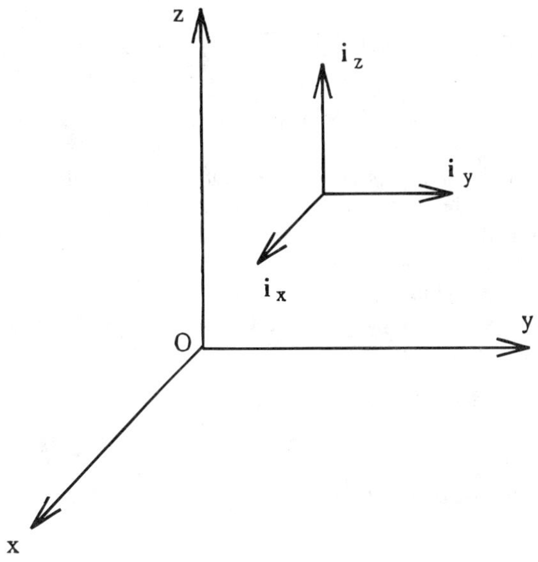

Figure 1. Rectangular coordinates (x,y,z).

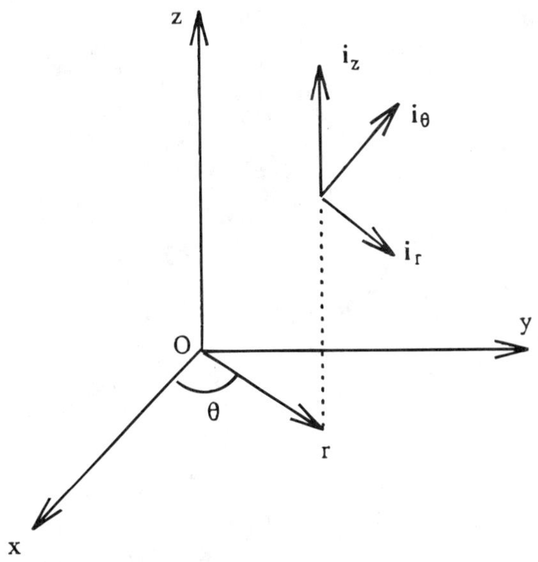

Figure 2. Cylindrical coordinates (r, θ, z).

Scalar, vector and tensor fields

A scalar (or 0th order tensor) field is a quantity with which only magnitude can be associated. An example is the density field, ρ. Scalar fields have one component, s.

A vector (or 1st order tensor) field is a quantity with which magnitude and direction is associated. An example is the velocity field, **u**. Vector fields have three components.

A tensor (or 2nd order tensor) is a quantity with which magnitude and two directions are associated. An example is the stress field, τ, since this is a force (a vector with one associated direction) per unit area (another vector with another associated direction). Tensor fields have nine components corresponding to the nine possible combinations of the two base vectors, i.e.,

$$\begin{aligned}\mathbf{T} = {} & T_{11}\mathbf{i}_{\varepsilon_1}\mathbf{i}_{\varepsilon_1} + T_{12}\mathbf{i}_{\varepsilon_1}\mathbf{i}_{\varepsilon_2} + T_{13}\mathbf{i}_{\varepsilon_1}\mathbf{i}_{\varepsilon_3} \\ & + T_{21}\mathbf{i}_{\varepsilon_2}\mathbf{i}_{\varepsilon_1} + T_{22}\mathbf{i}_{\varepsilon_2}\mathbf{i}_{\varepsilon_2} + T_{23}\mathbf{i}_{\varepsilon_2}\mathbf{i}_{\varepsilon_3} \\ & + T_{31}\mathbf{i}_{\varepsilon_3}\mathbf{i}_{\varepsilon_1} + T_{32}\mathbf{i}_{\varepsilon_3}\mathbf{i}_{\varepsilon_2} + T_{33}\mathbf{i}_{\varepsilon_3}\mathbf{i}_{\varepsilon_3}.\end{aligned} \tag{2}$$

There is no simple physical interpretation of the entity called a tensor; it is a mathematical entity that represents a physical entity and is defined such that its components transform in a certain way during a change of reference frame.

Vectors and tensors are written in boldface.

Operations with scalars, vectors and tensors. A tensor, **T** may be written as:

$$\mathbf{T} = \sum_i \sum_j T_{ij}\mathbf{i}_{\varepsilon_i}\mathbf{i}_{\varepsilon_j}, \tag{3}$$

the transpose of **T** is then

$$\mathbf{T}^T = \sum_i \sum_j T_{ji}\mathbf{i}_{\varepsilon_i}\mathbf{i}_{\varepsilon_j}. \tag{4}$$

For all scalar fields, s, vector fields, **v**, **a** and **b** and tensor fields, **T**, **A** and **B:**

Scalar products:

$$\mathbf{a}\cdot\mathbf{b} = a_1b_1 + a_2b_2 + a_3b_3 \tag{5}$$

$$\mathbf{v}\cdot\mathbf{T} = \mathbf{T}^T\cdot\mathbf{v} = \sum_i \sum_j v_j T_{ji}\mathbf{i}_{\varepsilon_i} \tag{6}$$

Trace:

$$\text{trace}(\mathbf{T}) = T_{11} + T_{22} + T_{33} \tag{7}$$

$$\mathbf{A}:\mathbf{B} = \mathbf{B}:\mathbf{A} = \text{trace}(\mathbf{A}\cdot\mathbf{B}) \equiv \sum_i \sum_j A_{ij}B_{ji} \tag{8}$$

Vector product:

$$\begin{aligned}\mathbf{a}\wedge\mathbf{b} = -\mathbf{b}\wedge\mathbf{a} = {} & (a_2b_3 - a_3b_2)\mathbf{i}_{\varepsilon_1} \\ & + (a_3b_1 - a_1b_3)\mathbf{i}_{\varepsilon_2} + (a_1b_2 - a_2b_1)\mathbf{i}_{\varepsilon_3}\end{aligned} \tag{9}$$

Unit tensor:

$$\mathbf{I} = \mathbf{i}_{\varepsilon_1}\mathbf{i}_{\varepsilon_1} + \mathbf{i}_{\varepsilon_2}\mathbf{i}_{\varepsilon_2} + \mathbf{i}_{\varepsilon_3}\mathbf{i}_{\varepsilon_3} \tag{10}$$

$$\mathbf{I}\cdot\mathbf{v} = \mathbf{v} = \mathbf{v}\cdot\mathbf{I} \qquad \mathbf{I}\cdot\mathbf{T} = \mathbf{T} = \mathbf{T}\cdot\mathbf{I} \tag{11}$$

Tensor symmetry:

$$\sum_i \sum_j T_{ij} = +\sum_i \sum_j T_{ji} \text{ tensor symmetric} \tag{12}$$

$$\sum_i \sum_j T_{ij} = -\sum_i \sum_j T_{ji} \text{ tensor antisymmetric} \tag{13}$$

Dyadic product:

$$\begin{aligned}\mathbf{v}\cdot(\mathbf{ab}) &= (\mathbf{v}\cdot\mathbf{a})\mathbf{b} \\ (\mathbf{ab})^T\cdot\mathbf{v} &= \mathbf{b}(\mathbf{a}\cdot\mathbf{v})\end{aligned} \tag{14}$$

Vector Laplacian:

$$\Delta\mathbf{v} = \nabla(\nabla\cdot\mathbf{v}) - \nabla\wedge\nabla\wedge\mathbf{v} \tag{15}$$

Identities:

$$\nabla\wedge\nabla s = 0 \tag{16}$$

$$\nabla\cdot\nabla\wedge\mathbf{v} = 0 \tag{17}$$

$$\begin{aligned}\nabla(\mathbf{a}\cdot\mathbf{b}) = {} & (\mathbf{b}\cdot\nabla)\mathbf{a} + (\mathbf{a}\cdot\nabla)\mathbf{b} + \mathbf{b}\wedge(\nabla\wedge\mathbf{a}) \\ & + \mathbf{a}\wedge(\nabla\wedge\mathbf{b})\end{aligned} \tag{18}$$

$$\nabla\cdot(\mathbf{ab}) = \mathbf{a}\cdot\nabla\mathbf{b} + \mathbf{b}(\nabla\cdot\mathbf{a}) \tag{19}$$

$$\nabla\cdot(s\mathbf{v}) = \nabla s\cdot\mathbf{v} + s(\nabla\cdot\mathbf{v}) \tag{20}$$

$$\nabla\cdot(\mathbf{T}\cdot\mathbf{v}) = \mathbf{T}:\nabla\mathbf{v} + \mathbf{v}\cdot(\nabla\cdot\mathbf{T}) \tag{21}$$

$$\nabla\cdot(s\mathbf{T}) = \nabla s\cdot\mathbf{T} + s(\nabla\cdot\mathbf{T}) \tag{22}$$

Divergence theorem:

For a volume V enclosed by an orientable (two-sided) surface, A,

$$\int_V \nabla\cdot\mathbf{v}\,dV = \int_A \mathbf{n}\cdot\mathbf{v}\,dA \tag{23}$$

$$\int_V \nabla\cdot\mathbf{T}\,dV = \int_A \mathbf{n}\cdot\mathbf{T}\,dA \tag{24}$$

where **n** is the unit outer normal to an element dA of the surface.

Kinematics

Kinematics is the study of motion without considering how it is caused.

Langrangian and Eulerian descriptions. There are two possible descriptions of fluid motion:

1) To identify a material point (fluid element), X such that X is at $\mathbf{x}_0$ at time t_0, then at time t, X is at $\mathbf{x}_0 + \Delta\mathbf{x}$ (see **Figure 3**). The velocity of X could be written

$$\hat{\mathbf{u}}(x, t) \equiv \hat{\mathbf{u}}(\mathbf{x}_0(X, t_0), t - t_0), \tag{25}$$

this is the *Lagrangian* description and gives the velocity of material points as a function of time.

2) The velocity of the fluid at a particular point in space and at a particular time is specified by $\mathbf{u}(\mathbf{x},t)$, which, if specified at all points $\mathbf{x}$ in a fluid, would give a complete picture of fluid motion. This is the *Eulerian* description.

In practise, the Lagrangian description is usually cumbersome and the Eulerian description is almost invariably adopted.

Differentiation following the motion. In an Eulerian description, if the velocity at x over a period of time is observed velocity changes can be noted, but $(\partial\mathbf{u}/\partial t)_x$ is not the acceleration of the fluid because particles at x change with time. However, in a Lagrangian description for a particular fluid particle X, $(\partial u/\partial t)_x$ is the acceleration of fluid.

In order to use the Eulerian description, the acceleration of the fluid must be known. For small Δt, and hence small $\Delta\mathbf{x}$, a Taylor series expansion in space and time leads to

$$\mathbf{u}(\mathbf{x} + \Delta\mathbf{x}, t + \Delta t) = \mathbf{u}(\mathbf{x}, t) + \Delta t\left(\frac{\partial\mathbf{u}}{\partial t}\right)_{\mathbf{x}} + \Delta x_1\left(\frac{\partial\mathbf{u}}{\partial x_1}\right)_t + \Delta x_2\left(\frac{\partial\mathbf{u}}{\partial x_2}\right)_t + \Delta x_3\left(\frac{\partial\mathbf{u}}{\partial x_3}\right)_t, \tag{26}$$

and so

$$\frac{1}{\Delta t}[\mathbf{u}(\mathbf{x} + \Delta\mathbf{x}, t + \Delta t) - \mathbf{u}(\mathbf{x}, t)] = \left(\frac{\partial\mathbf{u}}{\partial t}\right)_{\mathbf{x}} + \frac{1}{\Delta t}\Delta\mathbf{x}\cdot\Delta\mathbf{u}. \tag{27}$$

Taking the limit as $\Delta t \to 0$,

$$\left(\frac{\partial\mathbf{u}}{\partial t}\right)_{\mathbf{x}} = \frac{\partial\mathbf{u}}{\partial t} + \mathbf{u}\cdot\nabla\mathbf{u} = \frac{D\mathbf{u}}{Dt} \tag{28}$$

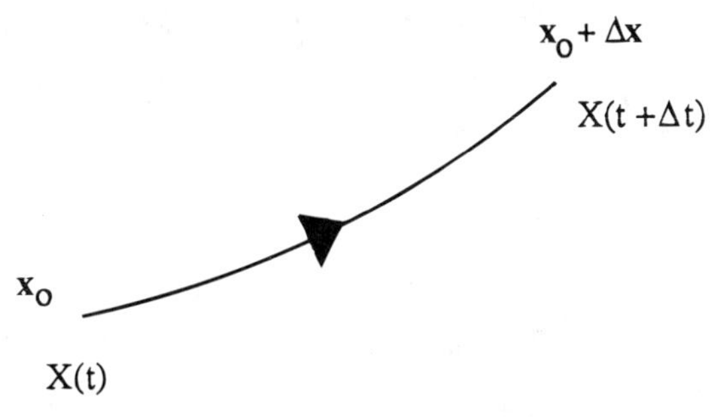

Figure 3. Motion of a material point.

where D/Dt is the substantial (material) derivative written as:

$$\frac{Df}{Dt} = \frac{\partial f}{\partial t} + \mathbf{u}\cdot\nabla f \tag{29}$$

for any scalar, vector or tensor quantity, f.

Rate of strain and vorticity. The *rate of strain* (or strain rate, or rate of deformation) tensor field $\mathbf{e}$ is defined as:

$$\mathbf{e} = \nabla\mathbf{u} + (\nabla\mathbf{u})^T \tag{30}$$

and the *vorticity* (or spin) tensor field, $\mathbf{w}$, as:

$$\mathbf{w} = \nabla\mathbf{u} - (\nabla\mathbf{u})^T. \tag{31}$$

Evidently $\mathbf{e}$ is symmetric ($\mathbf{e} = \mathbf{e}^T$) and $\mathbf{w}$ is antisymmetric ($\mathbf{w} = -\mathbf{w}^T$), also

$$\nabla\mathbf{u} = \frac{1}{2}(\mathbf{e} + \mathbf{w}) \tag{32}$$

and so, to within a factor of two, $\mathbf{e}$ is the symmetric part and $\mathbf{w}$, the antisymmetric part of $\nabla\mathbf{u}$.

It can be shown that there exists a vector field, ω, also called the vorticity such that, for an arbitrary vector field, $\mathbf{v}$

$$\omega \wedge \mathbf{v} = -\mathbf{v} \wedge \omega = \mathbf{v}\cdot\omega = \mathbf{v}\cdot(\nabla\mathbf{u} - (\nabla\mathbf{u})^T) \tag{33}$$

and equivalently,

$$\omega = \nabla \wedge \mathbf{u}. \tag{34}$$

In rectangular coordinates, with velocity components u_x, u_y and u_z in the x, y and z directions, respectively,

$$\mathbf{e} = e_{xx}\mathbf{i}_x\mathbf{i}_x + e_{xy}\mathbf{i}_x\mathbf{i}_y + e_{xz}\mathbf{i}_x\mathbf{i}_z + e_{yx}\mathbf{i}_y\mathbf{i}_x + e_{yy}\mathbf{i}_y\mathbf{i}_y + e_{yz}\mathbf{i}_y\mathbf{i}_z + e_{zx}\mathbf{i}_z\mathbf{i}_x + e_{zy}\mathbf{i}_z\mathbf{i}_y + e_{zz}\mathbf{i}_z\mathbf{i}_z \tag{35}$$

$$e_{xx} = 2\frac{\partial u_x}{\partial x} \quad e_{yy} = 2\frac{\partial u_y}{\partial y} \quad e_{zz} = 2\frac{\partial u_z}{\partial z}$$

$$e_{xy} = e_{yx} = \frac{\partial u_x}{\partial y} + \frac{\partial u_y}{\partial x} \quad e_{xz} = e_{zx} = \frac{\partial u_x}{\partial z} + \frac{\partial u_z}{\partial x} \tag{36}$$

$$e_{yz} = e_{zy} = \frac{\partial u_z}{\partial y} + \frac{\partial u_y}{\partial z}$$

$$\omega = \omega_x\mathbf{i}_x + \omega_y\mathbf{i}_y + \omega_z\mathbf{i}_z \tag{37}$$

$$\omega_x = \frac{\partial u_z}{\partial y} - \frac{\partial u_y}{\partial z} \quad \omega_y = \frac{\partial u_x}{\partial z} - \frac{\partial u_z}{\partial x} \quad \omega_z = \frac{\partial u_y}{\partial x} - \frac{\partial u_x}{\partial y}. \tag{38}$$

In cylindrical polar coordinates with velocity components u_r, u_θ and u_z in the r, θ and z directions, respectively,

$$\begin{aligned}\mathbf{e} &= e_{rr}\mathbf{i}_r\mathbf{i}_r + e_{r\theta}\mathbf{i}_r\mathbf{i}_\theta + e_{rz}\mathbf{i}_r\mathbf{i}_z \\ &+ e_{\theta r}\mathbf{i}_\theta\mathbf{i}_r + e_{\theta\theta}\mathbf{i}_\theta\mathbf{i}_\theta + e_{\theta z}\mathbf{i}_\theta\mathbf{i}_z \\ &+ e_{zr}\mathbf{i}_z\mathbf{i}_r + e_{z\theta}\mathbf{i}_z\mathbf{i}_\theta + e_{zz}\mathbf{i}_z\mathbf{i}_z\end{aligned} \tag{39}$$

$$e_{rr} = 2\frac{\partial u_r}{\partial r} \qquad e_{\theta\theta} = \frac{2}{r}\frac{\partial u_\theta}{\partial \theta} + \frac{2}{r}u_r \qquad e_{zz} = 2\frac{\partial u_z}{\partial z}$$

$$e_{r\theta} = e_{\theta r} = r\frac{\partial}{\partial r}\left(\frac{1}{r}u_\theta\right) + \frac{1}{r}\frac{\partial u_r}{\partial \theta} \qquad e_{rz} = e_{zr} = \frac{\partial u_r}{\partial z} + \frac{\partial u_z}{\partial r}$$

$$e_{z\theta} = e_{\theta z} = \frac{1}{r}\frac{\partial u_z}{\partial \theta} + \frac{\partial u_\theta}{\partial z} \tag{40}$$

$$\omega = \omega_r\mathbf{i}_r + \omega_\theta\mathbf{i}_\theta + \omega_z\mathbf{i}_z \tag{41}$$

$$\omega_r = \frac{1}{r}\frac{\partial u_z}{\partial \theta} - \frac{\partial u_\theta}{\partial z}$$

$$\omega_\theta = \frac{\partial u_r}{\partial z} - \frac{\partial u_z}{\partial r} \tag{42}$$

$$\omega_z = \frac{1}{r}\frac{\partial}{\partial r}(ru_\theta) - \frac{1}{r}\frac{\partial u_r}{\partial \theta}.$$

Irrotational and solenoidal fields. An *irrotational* flow is one in which the angular velocity and hence the vorticity vector, ω (which is a measure of the local rotation in a flow) vanishes everywhere, i.e.,

$$\omega = \nabla \wedge \mathbf{u} = 0. \tag{43}$$

For any scalar field, s, $\nabla \wedge \nabla s = 0$ so a function ϕ is defined such that

$$\mathbf{u} = \nabla\phi, \tag{44}$$

which satisfies

$$\nabla \wedge \mathbf{u} = \omega = \nabla \wedge \nabla\phi = 0 \tag{45}$$

so that **u** is irrotational.

Conversely, it can be shown that if **u** is irrotational, a scalar field exists such that Equation (44) is true. The scalar field, ϕ, is called a *scalar velocity potential.*

A *solenoidal flow* is one for which

$$\nabla\cdot\mathbf{u} = 0. \tag{46}$$

It will be shown later (in conservation equations) that any incompressible flow is solenoidal.

Since $\nabla \cdot \nabla \wedge \mathbf{v} = 0$ for any vector field, **v**, defining

$$\mathbf{u} = \nabla \wedge \mathbf{\Psi} \tag{47}$$

will have

$$\nabla\cdot\mathbf{u} = \nabla\cdot\nabla \wedge \mathbf{\Psi} = 0. \tag{48}$$

So if the flow is solenoidal, a *vector velocity potential,* Ψ, exists such that Equation (47) is true. Note that the definition is not unique, but this does not cause any problems in practice.

Stress tensor

There are two main types of force which may act on a body: *long range,* e.g., gravitational, and *short range,* e.g., intermolecular forces.

Long range forces, often called *body* or *volume* forces, vary slowly over the deforming body and are relatively easy to deal with.

Short range forces of molecular origin are generally negligible unless there is direct contact between interacting parts of a body, since molecular forces only extend over very small distances. Short range forces are viewed as *surface forces* and are the focus here.

Consider an elementary tetrahedron of the body, as shown in **Figure 4**, in which **t** is the stress tensor and **n** is the *unit outer normal* to face ABC. Stress is the force per unit area on the surface. The convention is that **t** is exerted by material on the side from which **n** points on the material to which **n** points and is positive when in the direction of **n**.

Thus $\mathbf{t}_1$, $\mathbf{t}_2$ and $\mathbf{t}_3$ can be written in component form as:

$$\begin{aligned}\mathbf{t}_1 &= \tau_{11}\mathbf{i}_{\varepsilon_1} + \tau_{12}\mathbf{i}_{\varepsilon_2} + \tau_{13}\mathbf{i}_{\varepsilon_3} \\ \mathbf{t}_2 &= \tau_{21}\mathbf{i}_{\varepsilon_1} + \tau_{22}\mathbf{i}_{\varepsilon_2} + \tau_{23}\mathbf{i}_{\varepsilon_3} \\ \mathbf{t}_3 &= \tau_{31}\mathbf{i}_{\varepsilon_1} + \tau_{32}\mathbf{i}_{\varepsilon_2} + \tau_{33}\mathbf{i}_{\varepsilon_3}.\end{aligned} \tag{49}$$

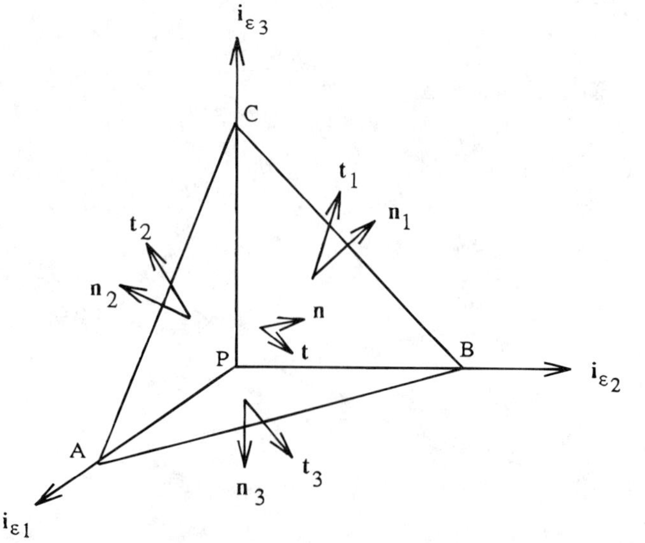

Figure 4. Elementary tetrahedron in a deforming body.

Denoting the surface area of ABC by A, then the areas of PCB, PAC and PAB are $A(\mathbf{n}\cdot\mathbf{i}_{\varepsilon 1})$, $A(\mathbf{n}\cdot\mathbf{i}_{\varepsilon 2})$, and $A(\mathbf{n}\cdot\mathbf{i}_{\varepsilon 3})$, respectively. If the tetrahedron is small enough, then the force acting on a given face is the product of the component of the stress vector and the area of the surface.

A force balance on the material inside the tetrahedron leads to

$$\int_A \mathbf{t}dA = \mathbf{t}A - \mathbf{n}\cdot\mathbf{i}_{\varepsilon_1}A\mathbf{t}_1 - \mathbf{n}\cdot\mathbf{i}_{\varepsilon_2}A\mathbf{t}_2 - \mathbf{n}\cdot\mathbf{i}_{\varepsilon_3}A\mathbf{t}_3 \to 0 \quad \textbf{(50)}$$

as surface area tends to zero. There are no inertial or body force terms in this expression since these vary as volume (i.e., l^3, where l is a typical linear dimension of the tetrahedron) whereas the surface forces vary as area (l^2), and so the former becomes negligible when l is very small. Thus,

$$\mathbf{t} = (A\cdot\mathbf{i}_{\varepsilon_1})\mathbf{t}_1 + (A\cdot\mathbf{i}_{\varepsilon_2})\mathbf{t}_2 + (A\cdot\mathbf{i}_{\varepsilon_3})\mathbf{t}_3 \quad \textbf{(51)}$$

$$= \mathbf{n}\cdot(\mathbf{i}_{\varepsilon_1}\mathbf{t}_1 + \mathbf{i}_{\varepsilon_2}\mathbf{t}_2 + \mathbf{i}_{\varepsilon_3}\mathbf{t}_3) \quad \textbf{(52)}$$

$$= \mathbf{n}\cdot\tau \quad \textbf{(53)}$$

where τ is a *second order tensor* known as the *stress tensor.* This is *Cauchy's fundamental theorem for stress.*

The diagonal elements of τ (τ_{11}, τ_{22} and τ_{33}) are referred to as the normal stress components and the remainder, as the shear stress components.

Conservation equations

Having assumed the continuum hypothesis, it can now be supposed that mass, momentum and energy are conserved in any motion or flow. To mathematically represent the conservation of these quantities, the notion of *control volumes* must be introduced.

A control volume is a volume located in the body which is fixed with respect to a frame of reference (see **Figure 5**). Any convenient frame of reference may be chosen for the conservation of mass and energy, but an inertial frame of reference is required for momentum conservation. Note that control volumes should not be confused with *material volumes;* the latter contains a fixed amount of material, but are not fixed with respect to a reference frame and moves with the material.

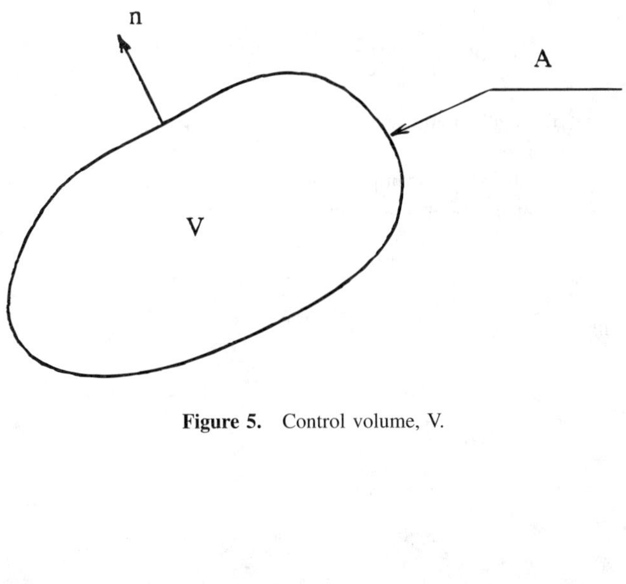

Figure 5. Control volume, V.

The volume, V, is enclosed by the surface, A, where A is orientable (i.e., two-sided).

Conservation of mass

Considering the flow of mass through the control volume, V;

$$\begin{matrix}\text{rate of accumulation} \\ \text{of mass in V}\end{matrix} + \begin{matrix}\text{net flux of mass} \\ \text{out through A}\end{matrix} = 0$$

or

$$\frac{\partial}{\partial t}\int_V \rho dV + \int_A \rho\mathbf{u}\cdot\mathbf{n}dA = 0 \quad \textbf{(54)}$$

and since V is independent of time,

$$\int_V \frac{\partial\rho}{\partial t}dV + \int_A \rho\mathbf{u}\cdot\mathbf{n}dA = 0. \quad \textbf{(55)}$$

Applying the divergence theorem, Equation (23), to the second term leads to

$$\int_V \left(\frac{\partial\rho}{\partial t} + \nabla\cdot(\rho\mathbf{u})\right)dV = 0. \quad \textbf{(56)}$$

Since V is arbitrary, the only way for this expression to be always true is if the integrand itself is zero, i.e.,

$$\frac{\partial\rho}{\partial t} + \nabla\cdot(\rho\mathbf{u}) = 0. \quad \textbf{(57)}$$

This expression of the conservation of mass is often referred to as the *continuity equation.* Another form of this expression may be derived by applying the mathematical identity Equation (20) to $\nabla\cdot(\rho\mathbf{u})$ giving

$$\nabla\cdot(\rho\mathbf{u}) = \mathbf{u}\cdot\nabla\rho + \rho\nabla\cdot\mathbf{u} \quad \textbf{(58)}$$

which, upon substitution into Equation (57), results in

$$\frac{\partial\rho}{\partial t} + \mathbf{u}\cdot\nabla\rho + \rho\nabla\cdot\mathbf{u} = 0. \quad \textbf{(59)}$$

Recalling the definition of the substantial derivative, Equation (29), this may be written as:

$$\frac{D\rho}{Dt} + \rho\nabla\cdot\mathbf{u} = 0. \quad \textbf{(60)}$$

Equations (57) and (60) are alternative forms of the continuity equation.

The continuity equation can be simplified if the material is incompressible (i.e., if the density of each material point is constant with respect to time), to give

$$\nabla \cdot \mathbf{u} = 0. \quad (61)$$

Recalling Equation (46), it can be seen that the velocity field of an incompressible material is solenoidal. Note that when reference is made to an incompressible flow it really means an isochoric (volume preserving) flow.

The mass conservation equation for a fluid of constant density, ρ, in rectangular coordinates with velocity components u_x, u_y and u_z in the x, y and z directions, respectively, is

$$\frac{\partial u_x}{\partial x} + \frac{\partial u_y}{\partial y} + \frac{\partial u_y}{\partial y} = 0 \quad (62)$$

and in cylindrical polar coordinates with velocity components u_r, u_θ and u_z in the r, θ and z directions, respectively,

$$\frac{1}{r}\frac{\partial}{\partial r}(ru_r) + \frac{1}{r}\frac{\partial u_\theta}{\partial \theta} + \frac{\partial u_z}{\partial z} = 0. \quad (63)$$

Linear momentum conservation

Considering the flow of momentum (linear as opposed to angular) through control volume V;

$$\begin{matrix}\text{rate of accumulation}\\ \text{of momentum in V}\end{matrix} + \begin{matrix}\text{flux of momentum}\\ \text{out through A}\end{matrix} =$$

$$\begin{matrix}\text{rate of gain of momentum}\\ \text{due to body forces}\end{matrix} + \begin{matrix}\text{rate of gain of momentum}\\ \text{due to surface stresses}\end{matrix}$$

which is written mathematically as:

$$\frac{\partial}{\partial t}\int_V \rho \mathbf{u} dV + \int_A \rho \mathbf{u}\mathbf{u}\cdot\mathbf{n} dA = \int_V \rho \mathbf{g} dV + \int_A \mathbf{t} dA, \quad (64)$$

where it has been assumed that all body forces are gravitational in origin. Since V is independent of time, the time derivative inside the integral sign in the first term can again be taken. Also, by using Cauchy's fundamental theorem for stress, Equation (53), the expression becomes

$$\int_A \mathbf{t} dA = \int_A \mathbf{n}\cdot\tau dA; \quad (65)$$

using the divergence theorem, Equation (24):

$$\int_A \mathbf{n}\cdot\tau dA = \int_V \nabla\cdot\tau dV \quad (66)$$

and

$$\int_S \rho\mathbf{u}\mathbf{u}\cdot\mathbf{n} dA = \int_V \nabla\cdot(\rho\mathbf{u}\mathbf{u}) dV. \quad (67)$$

Thus,

$$\int_V \left(\frac{\partial}{\partial t}(\rho\mathbf{u}) + \nabla\cdot(\rho\mathbf{u}\mathbf{u}) - \rho\mathbf{g} - \nabla\cdot\tau\right) dV = 0 \quad (68)$$

and since V is arbitrary

$$\frac{\partial}{\partial t}(\rho\mathbf{u}) + \nabla\cdot(\rho\mathbf{u}\mathbf{u}) = \rho\mathbf{g} + \nabla\cdot\tau. \quad (69)$$

Applying the mathematical identity, Equation (19), to the $\nabla \cdot \rho\mathbf{u}\mathbf{u}$ term results in

$$\nabla\cdot(\rho\mathbf{u}\mathbf{u}) = \rho\mathbf{u}\cdot\nabla\mathbf{u} + \mathbf{u}(\nabla\cdot\rho\mathbf{u}) \quad (70)$$

and substituting this expression into Equation (69) leads to

$$\mathbf{u}\frac{\partial \rho}{\partial t} + \rho\frac{\partial \mathbf{u}}{\partial t} + \rho\mathbf{u}\cdot\nabla\mathbf{u} + \mathbf{u}\nabla\cdot(\rho\mathbf{u}) = \rho\mathbf{g} + \nabla\cdot\tau. \quad (71)$$

Substitution of the conservation of mass, Equation (57), into this expression gives

$$\rho\frac{\partial \mathbf{u}}{\partial t} + \rho\mathbf{u}\cdot\nabla\mathbf{u} = \rho\mathbf{g} + \nabla\cdot\tau, \quad (72)$$

which is often referred to as the *equation of motion* and is an expression of *Cauchy's first law of motion.*

A scalar field, p, can now be defined which is conceptually similar to the static pressure and which reduces to the static pressure when the fluid is at rest:

$$p = \frac{1}{3}(\tau_{11} + \tau_{22} + \tau_{33}) = \frac{1}{3}\text{trace}(\tau); \quad (73)$$

it is thus the mean normal stress in a fluid. It is now possible to decompose the total stress, τ, thus;

$$\tau = -p\mathbf{I} + \tau_D \qquad \text{trace}(\tau_D) = 0 \quad (74)$$

where **I** is the unit tensor (see above section on **scalar, vector and tensor fields**) and τ_D is the *deviatoric* stress tensor field. Note that the isotropic components from the total stress tensor have been stripped out and these are identified with the pressure (the negative sign accounts for the fact that pressure is positive). Henceforth only deviatoric stress is referred to unless otherwise stated, so the subscript D is dropped.

Substituting this decomposition into Equation (72) gives

$$\rho\frac{\partial \mathbf{u}}{\partial t} + \rho\mathbf{u}\cdot\nabla\mathbf{u} = \rho\mathbf{g} - \nabla p + \nabla\cdot\tau, \quad (75)$$

where the mathematical identity given by Equation (22) has been applied to the $-p\mathbf{I}$ term.

If a material is incompressible, gravitational acceleration can be eliminated by defining a modified pressure, $\tilde{p}$, such that

$$\tilde{p} = p - \rho\mathbf{g}\cdot\mathbf{x} \quad (76)$$

thus

$$\rho\frac{\partial \mathbf{u}}{\partial t} + \rho\mathbf{u}\cdot\nabla\mathbf{u} = -\nabla\tilde{p} + \nabla\cdot\tau. \quad (77)$$

Conservation of angular momentum

In an inertial frame of reference, angular momentum is conserved. By following the same procedure for mass and momentum conservation, the expression for a nonpolar material becomes

$$\tau = \tau^T, \quad (78)$$

i.e., the stress tensor is symmetric (see section on **Scalar, vector and tensor fields**). A nonpolar material is one in which the torques within it arise only as the moments of direct forces. Most materials, and certainly most fluids, are nonpolar; so it is generally assumed that conservation of angular momentum implies symmetry of the stress tensor.

Conservation of energy

Taking the dot product of each term in the linear momentum conservation, Equation (75) results in

$$\mathbf{u}\cdot\left(\rho\frac{\partial \mathbf{u}}{\partial t}\right) + \mathbf{u}\cdot(\rho\mathbf{u}\cdot\nabla\mathbf{u}) = -\mathbf{u}\cdot\nabla p + \mathbf{u}\cdot(\rho\mathbf{g}) + \mathbf{u}\cdot(\nabla\cdot\tau) \quad (79)$$

or

$$\frac{1}{2}\rho\frac{\partial}{\partial t}(\mathbf{u}\cdot\mathbf{u}) + \frac{1}{2}\rho\mathbf{u}\cdot\nabla(\mathbf{u}\cdot\mathbf{u}) = -\mathbf{u}\cdot\nabla p + \mathbf{u}\cdot(\rho\mathbf{g}) + \mathbf{u}\cdot(\nabla\cdot\tau) \quad (80)$$

or

$$\rho\frac{D}{Dt}\left(\frac{1}{2}|\mathbf{u}|^2\right) = -\mathbf{u}\cdot\nabla p + \mathbf{u}\cdot(\rho\mathbf{g}) + \mathbf{u}\cdot(\nabla\cdot\tau) \quad (81)$$

where $(1/2)|\mathbf{u}|^2$ is the kinetic energy per unit mass. These equations express the conservation of mechanical energy.

The energy of a material comprises internal, kinetic and potential energy. Defining u as the internal energy per unit mass and $-\mathbf{g}\cdot\mathbf{x}$ as the potential energy per unit mass, in the absence of heat generation by chemical or nuclear reaction, conservation of energy requires that:

rate of accumulation of energy in V	+	flux of energy out through A	=	flux of heat in through A	+	rate of gain of energy due to surface forces

Mathematically,

$$\frac{\partial}{\partial t}\int_V\left(\rho u + \frac{1}{2}\rho\mathbf{u}^2 - \rho\mathbf{g}\cdot\mathbf{x}\right)dV + \int_A\left(\rho u + \frac{1}{2}\rho\mathbf{u}^2 - \rho\mathbf{g}\cdot\mathbf{x}\right)\mathbf{u}\cdot\mathbf{n}dA = -\int_A \dot{\mathbf{q}}\cdot\mathbf{n}dA + \int_A \mathbf{n}\cdot(\tau\cdot\mathbf{u})dA, \quad (82)$$

where **q** is the heat flux vector and τ is the total (not deviatoric) stress. The last term is the rate at which surface stresses do work on the material inside A, i.e., force × velocity ($\int_A \mathbf{t}\cdot\mathbf{u}dA$) which is $\int_A \mathbf{n}\cdot(\tau\cdot\mathbf{u})dA$ by Cauchy's fundamental theorem for stress [Equation (53)].

Applying the divergence theorem and noting that V is independent of time leads to

$$\frac{\partial}{\partial t}\left[\rho\left(u + \frac{1}{2}\mathbf{u}^2 - \mathbf{g}\cdot\mathbf{x}\right)\right] + \nabla\cdot\left[\rho\mathbf{u}\left(u + \frac{1}{2}\mathbf{u}^2 - \mathbf{g}\cdot\right)\right] = -\nabla\cdot\mathbf{q} + \nabla\cdot(\tau\cdot\mathbf{u}). \quad (83)$$

Considering the left-hand side of this equation; expanding the internal and kinetic energy terms gives

$$\frac{\partial}{\partial t}\left(\rho u + \frac{1}{2}\rho|\mathbf{u}|^2\right) + \nabla\cdot\left[\left(\rho u + \frac{1}{2}\rho|\mathbf{u}|^2\right)\mathbf{u}\right] = \rho\frac{\partial}{\partial t}\left(u + \frac{1}{2}|\mathbf{u}|^2\right) + \rho\nabla\cdot\left[\left(u + \frac{1}{2}|\mathbf{u}|^2\right)\mathbf{u}\right] + \left(u + \frac{1}{2}|\mathbf{u}|^2\right)\left(\frac{\partial\rho}{\partial t} + \nabla\cdot(\rho\mathbf{u})\right) \quad (84)$$

which, upon substitution of mass conservation [Equation (57)] becomes

$$\frac{\partial}{\partial t}\left(\rho u + \frac{1}{2}\rho|\mathbf{u}|^2\right) + \nabla\cdot\left[\left(\rho u + \frac{1}{2}\rho|\mathbf{u}|^2\right)\mathbf{u}\right] = \rho\frac{D}{Dt}\left(u + \frac{1}{2}|\mathbf{u}|^2\right). \quad (85)$$

Similarly for the potential energy terms,

$$\frac{\partial}{\partial t}(\rho\mathbf{g}\cdot\mathbf{x}) + \nabla\cdot(\rho\mathbf{u}\mathbf{g}\cdot\mathbf{x}) = \mathbf{g}\cdot\mathbf{x}\frac{\partial\rho}{\partial t} + \rho\frac{\partial}{\partial t}(\mathbf{g}\cdot\mathbf{x}) + \mathbf{g}\cdot\mathbf{x}\nabla\cdot(\rho\mathbf{u}) + \rho\mathbf{u}\cdot\nabla(\mathbf{g}\cdot\mathbf{x}) \quad (86)$$

which, upon substitution of mass conservation and noting that **g** is independent of time and position, results in

$$\frac{\partial}{\partial t}(\rho\mathbf{g}\cdot\mathbf{x}) + \nabla\cdot(\rho\mathbf{u}\mathbf{g}\cdot\mathbf{x}) = \rho\mathbf{u}\cdot\mathbf{g}. \quad (87)$$

Substituting Equations (85) and (87) into Equation (83) and decomposing the stress tensor (as described in **linear momentum conservation**) gives

$$\rho \frac{D}{Dt}\left(u + \frac{1}{2}|\mathbf{vu}|^2\right) =$$

$$-\nabla \cdot \dot{\mathbf{q}} + \nabla \cdot (\tau \cdot \mathbf{u}) - \nabla \cdot (p\mathbf{u}) + \rho \mathbf{u} \cdot \mathbf{g} \quad \textbf{(88)}$$

where τ is now the deviatoric stress.

Substitution of the mechanical energy [Equation (81)] into this result gives

$$\underset{(a)}{\rho \frac{Du}{Dt}} = \underset{(b)}{-\nabla \cdot \dot{\mathbf{q}}} \underset{(c)}{- \rho \nabla \cdot \mathbf{u}} + \underset{(d)}{\tau : \nabla \mathbf{u}} \quad \textbf{(89)}$$

where Equation (21) has been used and the double dot product of two tensors is defined in the section on **scalar, vector and tensor fields.** This is the equation of conservation of energy, it is an expression of the first law of thermodynamics. The physical meanings of the terms in this equation are as follows:

- Term (a) is the rate of gain of internal energy.
- Term (b) is the rate of internal energy input by conduction.
- Term (c) is the reversible rate of internal energy conversion.
- Term (d) is the irreversible rate of internal energy increase by surface forces.

Developments

Constitutive equations

The conservation equations themselves are not sufficient to solve problems in fluid mechanics. This is because they describe the behaviour of a completely general material which may be a fluid, but could equally well be a solid. What are required then, are additional relationships that represent the intrinsic response of a particular material. Such relationships are known as *constitutive equations* or, equivalently, **Equations of State**, because they describe the constitution or state of a material. Specifically, relationships expressing p, $\dot{\mathbf{q}}$ and τ in terms of ρ, $\mathbf{u}$ and u or T (and, perhaps, each other) are required. Then, there will be sufficient equations to determine all the unknowns and the equation set is said to be *closed.* Constitutive equations cannot be obtained from continuum mechanics alone, instead they are generally based on phenomenological arguments and experiment.

Pressure. For compressible materials, the following constitutive equation for pressure applies:

$$p = p(\rho, T). \quad \textbf{(90)}$$

An example is that for a perfect gas:

$$p = \frac{\rho RT}{M}, \quad \textbf{(91)}$$

where R is the universal gas constant and M, the molecular weight or relative molar mass of the gas.

For incompressible material,

$$p = \text{constant} \quad \textbf{(92)}$$

and p, which is called the 'pressure' but is not necessarily the thermodynamic pressure, is determined as part of the solution of the equations of motion.

Heat flux. An example of a constitutive equation for **heat flux** is **Fourier's law:**

$$\dot{\mathbf{q}} = -\lambda \nabla T \quad \textbf{(93)}$$

where λ denotes the thermal conductivity of the fluid. This experimental law states that heat flux is proportional to the temperature gradient, with the minus sign indicating that heat flows down a temperature gradient from a region of higher temperature to one of lower temperature. Most materials, and certainly most fluids, obey this law under normal conditions. Note however that $\dot{\mathbf{q}}$ in the energy balance equation includes heat transmission due to radiation as well as conduction, whereas Fourier's law incorporates only conduction. The justification for using Fourier's law is that, since radiation is proportional to T^4, where T is the absolute temperature, heat flux due to radiation can be neglected where modest temperature differences are involved.

The energy equation [Equation (89)] for a Fourier fluid becomes

$$\rho \frac{Du}{Dt} = -\lambda \nabla^2 T - \rho \nabla \cdot \mathbf{u} + \tau : \nabla \mathbf{u}. \quad \textbf{(94)}$$

It is easy to show that

$$\tau : \nabla \mathbf{u} = \frac{1}{2} \eta(\mathbf{e} : \mathbf{e}), \quad \textbf{(95)}$$

and furthermore, heat capacity can be defined as:

$$c = \frac{1}{\rho}\left(\frac{\partial u}{\partial T}\right), \quad \textbf{(96)}$$

so

$$u = u_0 + \rho c(T - T_0) \quad \textbf{(97)}$$

where c is assumed constant. Generally, $c \equiv c_p$, the heat capacity at constant pressure, is used. Equation 94 thus becomes

$$\rho c_p \frac{\partial T}{\partial t} + \rho c_p \mathbf{u} \cdot \nabla T = \lambda \nabla^2 T + \frac{1}{2} \eta \mathbf{e} : \mathbf{e} \quad \textbf{(98)}$$

which expresses conservation of energy for a Fourier fluid of constant density, ρ, constant viscosity, η and constant thermal conductivity, λ. This equation is, in fact, unchanged if the fluid is of variable viscosity. In rectangular coordinates with velocity components u_x, u_y and u_z in the x, y and z directions, respectively, it becomes

$$\rho c_p\left(\frac{\partial T}{\partial t} + u_x\frac{\partial T}{\partial x} + u_y\frac{\partial T}{\partial y} + u_z\frac{\partial T}{\partial z}\right)$$

$$= \lambda\left(\frac{\partial^2 T}{\partial x^2} + \frac{\partial^2 T}{\partial y^2} + \frac{\partial^2 T}{\partial z^2}\right)$$

$$+ 2\eta\left[\left(\frac{\partial u_x}{\partial x}\right)^2 + \left(\frac{\partial u_y}{\partial y}\right)^2 + \left(\frac{\partial u_z}{\partial z}\right)^2\right]$$

$$+ \eta\left[\left(\frac{\partial u_x}{\partial y} + \frac{\partial u_y}{\partial x}\right)^2 + \left(\frac{\partial u_x}{\partial z} + \frac{\partial u_z}{\partial x}\right)^2 + \left(\frac{\partial u_y}{\partial z} + \frac{\partial u_z}{\partial y}\right)^2\right] \quad \textbf{(99)}$$

and in cylindrical polar coordinates with velocity components u_r, u_θ and u_z in the r, θ and z directions, respectively,

$$\rho c_p\left(\frac{\partial T}{\partial t} + u_r\frac{\partial T}{\partial r} + \frac{1}{r}u_\theta\frac{\partial T}{\partial \theta} + u_z\frac{\partial T}{\partial z}\right)$$

$$= \lambda\left[\frac{1}{r}\frac{\partial}{\partial r}\left(r\frac{\partial T}{\partial r}\right) + \frac{1}{r^2}\frac{\partial^2 T}{\partial \theta^2} + \frac{\partial^2 T}{\partial z^2}\right]$$

$$+ 2\eta\left[\left(\frac{\partial u_r}{\partial r}\right)^2 + \left(\frac{1}{r}\frac{\partial u_\theta}{\partial \theta} + \frac{1}{r}u_r\right)^2 + \left(\frac{\partial u_z}{\partial z}\right)^2\right]$$

$$+ \eta\left[\left(\frac{\partial u_r}{\partial z} + \frac{\partial u_z}{\partial r}\right)^2 + \left(\frac{\partial u_\theta}{\partial z} + \frac{1}{r}\frac{\partial u_z}{\partial \theta}\right)^2\right.$$

$$\left.+ \left(\frac{1}{r}\frac{\partial u_r}{\partial \theta} + r\frac{\partial}{\partial r}\left(\frac{1}{r}u_\theta\right)\right)^2\right]. \quad \textbf{(100)}$$

Stress. The simplest constitutive equation for stress is that of the **inviscid fluid**, in which there is no stress so that

$$\tau = 0. \quad \textbf{(101)}$$

Obviously, no such fluid exists in practice; however this idealisation is very useful due to its simplicity and is often utilised, most notably in the theory of gas dynamics (where viscous effects are generally small). It is also used for theories of rapid liquid flows. (See **Flow of Fluids**.)

The next simplest constitutive equation for stress is where the stress is linearly related to the rate of strain, thus

$$\tau = \eta\mathbf{e} \quad \textbf{(102)}$$

where η is the (dynamic) viscosity of the fluid. A fluid obeying this constitutive equation is said to be **Newtonian**; many common fluids are Newtonian. It is apparent that a Newtonian fluid is analogous to a Fourier material, for which it will be recalled that heat flux, $\dot{\mathbf{q}}$ is linearly related to temperature gradient, ∇T.

If this result is used in the linear momentum conservation equation [Equation (75)], the following expression is obtained:

$$\rho\frac{\partial \mathbf{u}}{\partial t} + \rho\mathbf{u}\cdot\nabla\mathbf{u} = \rho\mathbf{g} - \nabla p + \eta\nabla\cdot\mathbf{e}. \quad \textbf{(103)}$$

Using mass conservation for an incompressible fluid [Equation (61)] along with Equations (15), (30), (14) and (18), it can be shown that

$$\nabla\cdot\mathbf{e} = \Delta\mathbf{u} \quad \textbf{(104)}$$

and so

$$\rho\frac{\partial \mathbf{u}}{\partial t} + \rho\mathbf{u}\cdot\nabla\mathbf{u} = -\nabla p + \eta\Delta\mathbf{u} \quad \textbf{(105)}$$

which are the *Navier-Stokes equations.* They cannot be solved exactly for the general case owing to the nonlinearity in **u** and thus, recourse is usually made to the following two methods:

1. Obtain approximate solutions to the exact equations by numerical simulation (computer methods).
2. Obtain exact solutions to simplified forms of the equations.

A basis for the latter approach is discussed briefly below in the section on *Euler Equation.*

The Navier-Stokes equations for a fluid of constant density, ρ and constant viscosity, η may be written in rectangular coordinates as:

$$\rho\left(\frac{\partial u_i}{\partial t} + u_x\frac{\partial u_i}{\partial x} + u_y\frac{\partial u_i}{\partial y} + u_z\frac{\partial u_i}{\partial y}\right)$$

$$= \rho g_i - \frac{\partial p}{\partial i} + \eta\left(\frac{\partial^2 u_i}{\partial x^2} + \frac{\partial^2 u_i}{\partial y^2} + \frac{\partial^2 u_i}{\partial z^2}\right) \quad \textbf{(106)}$$

for each direction, i (i = x, y, z) where u_x, u_y and u_z are the velocity components in the x, y and z directions, respectively.

In cylindrical polar coordinates with velocity components u_r, u_θ and u_z in the r, θ and z directions, respectively,

$$\rho g_r - \frac{\partial p}{\partial r} + \eta\left(\frac{\partial}{\partial r}\left[\frac{1}{r}\frac{\partial}{\partial r}(ru_r)\right] + \frac{1}{r^2}\frac{\partial^2 u_r}{\partial \theta^2} - \frac{2}{r^2}\frac{\partial u_\theta}{\partial \theta} + \frac{\partial^2 u_r}{\partial z^2}\right)$$

$$= \rho\left(\frac{\partial u_r}{\partial t} + u_r\frac{\partial u_r}{\partial r} + \frac{1}{r}u_\theta\frac{\partial u_r}{\partial \theta} - \frac{1}{r}u_\theta^2 + u_z\frac{\partial u_r}{\partial z}\right) \quad \textbf{(107)}$$

$$\rho g_\theta - \frac{1}{r}\frac{\partial p}{\partial \theta} + \eta\left(\frac{\partial}{\partial r}\left[\frac{1}{r}\frac{\partial}{\partial r}(ru_\theta)\right] + \frac{1}{r^2}\frac{\partial^2 u_\theta}{\partial \theta^2} - \frac{2}{r^2}\frac{\partial u_r}{\partial \theta} + \frac{\partial^2 u_\theta}{\partial z^2}\right)$$

$$= \rho\left(\frac{\partial u_\theta}{\partial t} + u_r\frac{\partial u_\theta}{\partial r} + \frac{1}{r}u_\theta\frac{\partial u_\theta}{\partial \theta} - \frac{1}{r}u_r u_\theta + u_z\frac{\partial u_\theta}{\partial z}\right) \quad \textbf{(108)}$$

$$\rho g_z - \frac{\partial p}{\partial z} + \eta\left(\frac{1}{r}\frac{\partial}{\partial r}\left[r\frac{\partial u_z}{\partial u_r}\right] + \frac{1}{r^2}\frac{\partial^2 u_z}{\partial \theta^2} + \frac{\partial^2 u_z}{\partial z^2}\right)$$

$$= \rho\left(\frac{\partial u_z}{\partial t} + u_r\frac{\partial u_z}{\partial r} + \frac{1}{r}u_\theta\frac{\partial u_z}{\partial \theta} + u_z\frac{\partial u_z}{\partial z}\right). \quad (109)$$

Euler equation

In this section, order of magnitude arguments based on a dimensionless formulation are used to obtain approximate forms of the Navier-Stokes equations.

In order to make the Navier-Stokes equations dimensionless, a characteristic length, l_c, and a characteristic velocity, u_c need to be defined. These should be chosen such that the dimensionless variables are all of order one. The choice is usually straightforward; for flow in a pipe for example, l_c might be the pipe radius and u_c, the mean axial velocity. The following dimensionless variables are thus obtained:

$$\tilde{\mathbf{u}} = \frac{1}{u_c}\mathbf{u} \qquad \tilde{p} = \frac{1}{\rho u_c^2}p \qquad \tilde{\mathbf{g}} = \frac{1}{g}\mathbf{g} \qquad \tilde{t} = \frac{u_c}{l_c}t \quad (110)$$

$$\tilde{\nabla} = l_c\nabla \qquad \tilde{\Delta} = l_c^2\Delta \qquad \tilde{e} = \frac{l_c}{u_c}\mathbf{e}$$

where g is the magnitude of the gravitational acceleration.

Substitution into Equation (105) yields

$$\rho\frac{\tilde{u}_c^2}{l_c}\frac{\partial\tilde{u}^2}{\partial\tilde{t}} + \rho\frac{u_c}{l_c}\tilde{\mathbf{u}}\cdot\tilde{\nabla}\tilde{\mathbf{u}} = \rho g\tilde{\mathbf{g}} - \rho\frac{u_c^2}{l_c}\tilde{\nabla}\tilde{p} + \eta\frac{u_c}{l_c^2}\tilde{\Delta}\tilde{\mathbf{u}}. \quad (111)$$

Defining the following dimensionless numbers (see **Froude Number** and **Reynolds Number**):

$$Fr = \frac{u_c^2}{gl_c} \qquad Re = \frac{\rho u_c l_c}{\eta}, \quad (112)$$

Equation (105) can be written

$$\frac{\partial\tilde{\mathbf{u}}}{\partial\tilde{t}} + \tilde{\mathbf{u}}\cdot\tilde{\nabla}\tilde{\mathbf{u}} = \frac{1}{Fr}\tilde{g} - \tilde{\nabla}\tilde{p} + \frac{1}{Re}\tilde{\Delta}\tilde{\mathbf{u}}. \quad (113)$$

Since Re may be interpreted as the ratio of inertial to viscous forces, the viscous terms in the Navier-Stokes equation could be neglected when Re is very high (and conversely, the inertial terms neglected when Re is very low). This can be seen from Equation (113) since the inertial terms (first two) will be of order one and the viscous terms (last one) of order 1/Re, i.e., much less than one when Re is large. The Navier-Stokes equations become

$$\rho\frac{\partial\mathbf{u}}{\partial t} + \rho\mathbf{u}\cdot\nabla\mathbf{u} = \rho\mathbf{g} - \nabla p, \quad (114)$$

which is the *Euler equation.* It describes the motion of an inviscid fluid.

Bernoulli equation

From Equation (18) comes

$$\nabla(\mathbf{u}\cdot\mathbf{u}) = 2\mathbf{u}\cdot\nabla\mathbf{u} + 2\mathbf{u}\wedge\omega \quad (115)$$

where ω is the vorticity (see section on **Rate of Strain and vorticity**) defined by

$$\omega = \nabla\wedge\mathbf{u}. \quad (116)$$

Substitution of Equation (115) into (114) gives

$$\rho\frac{\partial\mathbf{u}}{\partial t} + \frac{1}{2}\rho\nabla|\mathbf{u}|=+^2 - \rho\mathbf{u}\wedge\omega = \rho\mathbf{g} - \nabla p, \quad (117)$$

where $|\mathbf{u}|$ is the speed, $|\mathbf{u}| = \sqrt{(\mathbf{u}\cdot\mathbf{u})}$.

If the flow is irrotational such that the vorticity vanishes, then from the above section on **Irrotational and Solenoidal forms, u** can be expressed in the form

$$\mathbf{u} = \nabla\phi \quad (118)$$

where ϕ is a *scalar potential* field. Noting that $\mathbf{g} = \nabla(\mathbf{g}\cdot\mathbf{x})$, the equation becomes

$$\rho\frac{\partial}{\partial t}(\nabla\phi) + \frac{1}{2}\rho\nabla(|\nabla\phi|^2) = \rho\nabla(\mathbf{g}\cdot\mathbf{x}) - \nabla p \quad (119)$$

which, since the operators $\partial/\partial t$ and ∇ are commutative

$$\frac{\partial\nabla}{\partial t} \equiv \nabla\frac{\partial}{\partial t}, \quad (120)$$

yields

$$\nabla\left(\frac{\partial\phi}{\partial t} + \frac{1}{2}|\nabla\phi|^2 - \mathbf{g}\cdot\mathbf{x} + \frac{1}{\rho}p\right) = 0. \quad (121)$$

This can be integrated to give

$$\frac{\partial\phi}{\partial t} + \frac{1}{2}|\nabla\phi|^2 - \mathbf{g}\cdot\mathbf{x} + \frac{1}{\rho}p = \zeta(t) \quad (122)$$

where ζ is, at most, a function of time. This is the *Bernoulli equation.* It can often be simplified; for example, if the flow is steady (independent of time) then

$$\frac{1}{2}|\mathbf{u}|^2 - \mathbf{g}\cdot\mathbf{x} + \frac{1}{\rho}p = \zeta \quad (123)$$

where ζ is a constant and $\mathbf{u} = \nabla\phi$ has been used again.

The terms in this equation may be identified with kinetic, potential and pressure energy, respectively. The equation admits to the following physical interpretation: for an incompressible, inviscid fluid in steady flow, the sum of the kinetic, potential and pressure energies is constant.

Nomenclature

a	vector field
A	area
A	tensor field
b	vector field
B	tensor field
c	specific heat
e	rate of strain
f	arbitrary scalar, vector or tensor quantity
Fr	Froude number
g	gravitational acceleration
i	unit vector
I	unit tensor
l	length
M	molecular weight or relative molecular mass
n	unit normal
p	pressure
$\tilde{p}$	modified pressure
$\dot{\mathbf{q}}$	heat flux
r,θ,z	cylindrical polar coordinates
R	universal gas constant
Re	Reynolds number
s	scalar field
t	time
t	stress (vector)
T	temperature
T	tensor field
u	internal energy per unit mass
u	velocity
v	vector field
V	volume
w	vorticity (tensor)
x,y,z	rectangular coordinates
x	position
X	material point
ζ	constant
λ	thermal conductivity
η	viscosity
$\epsilon_1,\epsilon_2,\epsilon_3$	right-handed orthogonal coordinates
ρ	density
τ	stress (tensor)
φ	scalar potential
Ψ	vector potential
ω	vorticity (vector)

Subscripts

c characteristic value

Superscripts

^ Lagrangian specification

~ dimensionless variable

References

Aris, R. (1962) *Vectors, Tensors and the Basic Equations of Fluid Mechanics.* Prentice-Hall, Englewood Cliffs, N.J.

Bird, R. B., Stewart, W. E., and Lightfoot, E. N. (1960) *Transport Phenomena.* Wiley.

Kay, J. M. and Nedderman, R. M. (1985) *Fluid Mechanics and Transfer Processes.* Cambridge University Press.

Richardson, S. M. (1989) *Fluid Mechanics.* Hemisphere, New York.

Slattery, J. C. (1972) *Momentum, Energy, and Mass Transfer in Continua.* McGraw-Hill.

Leading to: *Conservation equations, two-phase; Inviscid flow; Turbulent flow; Laminar flow*

I.F. Roberts

CONSERVATION EQUATIONS, TWO-PHASE

Following from: *Conservation equations, single-phase*

Local Instantaneous Equations

Local instantaneous equations form the foundation for almost all two-phase modeling procedures. They may be used directly, in the study of bubble dynamics or film flows for example. More commonly however, they are used in averaged form as in the study of flow in pipes and conduits. Averaged forms of the local instantaneous equations will be considered in the section below on *averaged equations*.

The formulation of local instantaneous equations involves deriving the appropriate conservation equations and then closing the set, the latter problem discussed in the section below on closure and applications. Conservation equations for a two-phase flow can be derived by writing integral balances in much the same way as for the single-phase case (see **Conservation Equations, Single-Phase**). The problem is how to take into account behaviour at the interface.

Treatment of Phase Interface

A phase interface may be considered as a three-dimensional region which separates the bulk portions of two phases and in which the constitutive equations may differ from those applicable in the bulk portions of each phase.

A surface is said to be singular with respect to a quantity at a point if the limiting value of this quantity at a point on the surface—obtained by approaching the point along a path restricted to one side of the surface—differs from that obtained by approaching the point from the other side of the surface. The phase interface may be envisaged as a singular surface.

The above allows integral balances to be written encompassing both phases and the interface. These result in *phase equations* which are identical to their single-phase counterparts and *jump conditions* which are unique to multiphase flow analysis. The derivation leads to the local instantaneous equation set.

Integral balances

The formulation of integral balances for single-phase flow under the continuum hypothesis has well-proven validity. Here, these balances must be generalized to the case of two-phase flow. This is done by considering two-phase flow as a field which is subdivided into single-phase regions with moving boundaries between the phases. The standard single-phase balances hold for each subregion and are matched by the interfacial jump conditions.

First, a *general balance* must be written for any quantity, Ψ, over a material volume V

$$\frac{d}{dt}\int_V \rho\Psi dV = -\int_A \mathbf{n}\cdot\mathbf{J}dA + \int_V \rho\phi dV \tag{1}$$

where $\mathbf{J}$ is the *influx* of Ψ through A while ϕ is the *supply* of Ψ within V.

Consider now a material volume V within which there occurs a surface $A_I(t)$ splitting V into V_1 and V_2, and splitting A into A_1 and A_2, (**Figure 1**). The surface is a persistent singular surface with respect to a quantity Ψ, and possibly also with respect to $\mathbf{u}$. It is assumed to be smooth and may be in motion with any speed of displacement, $\mathbf{u}_I$.

The next step is to generalise the well-known *transport theorem* [Slattery (1972), Truesdell and Toupin (1960)] to regions containing such a singular surface. The transport theorem in its usual, single-phase formulation may be written as:

$$\frac{d}{dt}\int_V \Psi dV = \int_V \frac{\partial\Psi}{\partial t}dV + \int_V \Psi\mathbf{u}\cdot\mathbf{n}dA \tag{2}$$

for any Ψ. In other words, this states that: the rate of change of the total Ψ over a material volume V equals the rate of change of Ψ over V, plus the flux of $\mathbf{u}\Psi$ out of the bounding surface.

Turning back to the region divided by a singular surface, consider the case where the region is arbitrarily small and arbitrarily smooth. Thus it can be assumed that Ψ and $\mathbf{u}$ are continuously differentiable in the two regions, V_1 and V_2. In general, the regions and surfaces V_1, V_2, A_1, A_2 are not material. The fields $\mathbf{u}_1$ and $\mathbf{u}_2$ can now be defined as follows:

$$\mathbf{u}_1 \equiv \begin{cases}\mathbf{u} \text{ on } A_1\\ \mathbf{u}_I \text{ on }_I\end{cases} \qquad \mathbf{u}_2 \equiv \begin{cases}\mathbf{u} \text{ on } A_2\\ \mathbf{u}_I \text{ on } A_I\end{cases}. \tag{3}$$

Since A is a persistent common boundary of V_1 and V_2, the expression may be written as:

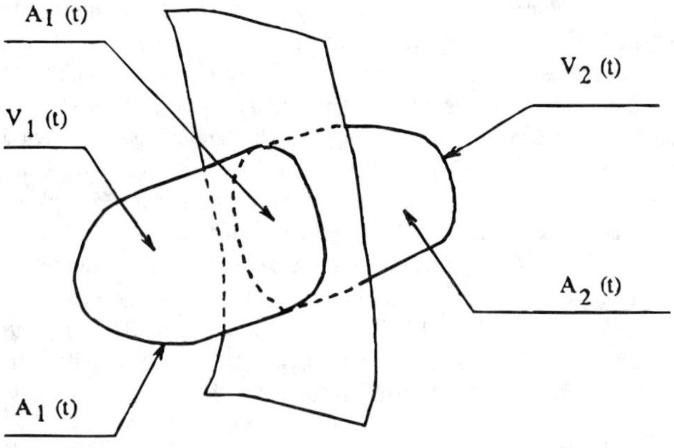

Figure 1. Diagram for proof of the transport theorem for a region containing a singular surface.

$$\frac{d}{dt}\int_V \Psi dV = \frac{d_{\mathbf{u}_1}}{dt}\int_{V_1}\Psi dV + \frac{d_{\mathbf{u}_2}}{dt}\int_{V_2}\Psi dV, \tag{4}$$

where the operator $d\mathbf{u}_1/dt$ indicates that the time derivative of the integral over a region that instantaneously coincides with V_1 and is material with respect to the field $\mathbf{u}_1$, must be taken while $d\mathbf{u}_2/dt$ is analogously defined. To each of the integrals on the right—since Ψ and $\mathbf{u}$ are continuously differentiable in V_1 and V_2 and approach continuous limits on the entire boundaries $A_1 + A_I$ and $A_2 + A_I$—the basic transport theorem, Equation (2), may be applied to obtain

$$\frac{d_{\mathbf{u}_1}}{dt}\int_{V_1}\Psi dV = \int_{V_1}\frac{\partial\Psi}{\partial t}dV + \int_{A_1}\Psi\mathbf{u}\cdot\mathbf{n}dA - \int_{A_I}\Psi^+\mathbf{u}_I\cdot n dA, \tag{5}$$

$$\frac{d_{\mathbf{u}_2}}{dt}\int_{V_2}\Psi dV = \int_{V_2}\frac{\partial\Psi}{\partial t}dV + \int_{A_2}\Psi\mathbf{u}\cdot\mathbf{n}dA + \int_{A_I}\Psi^-\mathbf{u}_I\cdot\mathbf{n}dA \tag{6}$$

These results can then be substituted into Equation (4) to give

$$\frac{d}{dt}\int_V \Psi dV = \int_V \frac{\partial\Psi}{\partial t}dV + \int_A \Psi\mathbf{u}\cdot\mathbf{n}dA - \int_{A_I}[\Psi]\mathbf{u}_I\cdot\mathbf{n}dA \tag{7}$$

where $[\Psi]$ is the *jump* of quantity Ψ across the singular surface denoted by

$$[\Psi] \equiv \Psi^+ - \Psi^-, \tag{8}$$

where Ψ^+ is the limiting value of Ψ as the interface is approached from one side and Ψ^-, the limiting value from the other side. The sign of the jump is a matter of convention. Equation (7) is the transport theorem written over a region containing a singular surface.

A general balance at a surface of discontinuity must be written next. It is assumed that a general balance equation of the form Equation (1) holds for an arbitrary material volume, irrespective of whether it contains a singular surface or not. Then, combining Equation (1) and Equation (7) for a sufficiently small material volume V containing a singular surface A_I, will yield

$$\int_V \frac{\partial\rho\Psi}{\partial t}dV + \int_A \rho\Psi\mathbf{u}\cdot\mathbf{n}dA - \int_{A_I}[\rho\Psi]\mathbf{u}_I\cdot\mathbf{n}dA$$

$$= -\int_A \mathbf{n}\cdot\mathbf{J}dA + \int_V \rho\phi dV. \tag{9}$$

It can now be assumed that in the neighbourhood of A_I, the quantities $\partial\rho\Psi/\partial t$ and $\rho\phi$ are bounded, while on each side of A_I the quantities $\rho\Psi$, $\mathbf{u}\cdot\mathbf{n}$, and $\mathbf{n}\cdot\mathbf{J}$ approach limits that are continuous functions of position. Under these conditions, let A_1 and A_2 shrink down to A_I (**Figure 2**) so that the volume of $V_1 + V_2$ vanishes while the area of A_I remains finite in the limit. The volume integrals then vanish in the limit, and Equation (9) becomes

$$\int_{A_I} ([\rho\Psi \mathbf{u}\cdot\mathbf{n}] - [\rho\Psi]\mathbf{u}_I\cdot n + [\mathbf{J}\cdot\mathbf{n}])dA = 0. \quad \mathbf{(10)}$$

This equation holds for an arbitrary area on A_I, thus the integrand must be zero. The general balance for a singular surface may then be written as:

$$\sum_{k=1,2} (\dot{m}_k\Psi_k + \mathbf{n}_k\cdot J_k) = 0 \quad \mathbf{(11)}$$

with

$$\dot{m}_k = \rho_k(\mathbf{u}_k - \mathbf{u}_I)\cdot\mathbf{n}_k \quad \mathbf{(12)}$$

where $\dot{m}$ is the mass transfer per unit area of interface and per unit time. By choosing appropriate values for the quantities Ψ and $\mathbf{J}$, Equation (11) may be used to express the balances of mass, momentum and energy across the phase interface for a two-phase flow, i.e., the *interfacial jump conditions.*

Phase equations

The balances of mass, momentum and energy for a single-phase flow (see **Conservation Equations, Single-Phase**) may be expressed by a single *general balance equation,* thus

$$\frac{\partial(\rho_k\Psi_k)}{\partial t} + \nabla\cdot(\rho_k\Psi_k\mathbf{u}_k) - \rho_k\phi_k + \nabla\cdot\mathbf{J}_k = 0 \quad \mathbf{(13)}$$

for a single-phase, k. As previously noted, phase equations of two-phase flow are identical to their single-phase flow counterparts; the former may be extracted from the general balance equation by using the values of Ψ_k, $\mathbf{J}_k$ and ϕ_k given in **Table 1**. In this table, $\mathbf{u}$ is the internal energy per unit mass; $\dot{\mathbf{q}}$ is the heat flux vector; τ is the total stress tensor; and $\mathbf{g}$ is the gravity vector (all body forces have been assumed to be of gravitational origin).

Jump conditions

The interfacial jump conditions follow from the general balance for a singular surface, Equation (11), by using the appropriate values for the quantities Ψ_k and $\mathbf{J}_k$ given in **Table 1**.

Closure and applications

Closure laws may be considered as those necessary and sufficient relations which must be added to conservation equations to allow calculation.

Table 1. Parameters for general balance equations

Balance	Ψ_K	$\mathbf{J}_K$	ϕ_K
Mass	1	0	0
Momentum	$\mathbf{u}_k$	$-\tau k$	$\mathbf{g}$
Energy	$u_k + \frac{1}{2}\mathbf{u}_k^2$	$\dot{\mathbf{q}}_k - \tau_k\cdot\mathbf{u}_k$	$\mathbf{g}\cdot\mathbf{u}_k$

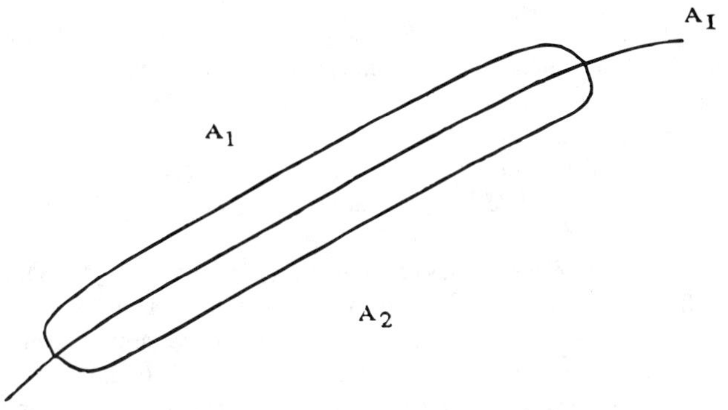

Figure 2. Diagram for derivation of the general balance at a singular surface.

For each of the phases, three constitutive equations are sufficient to effect closure. These are the same as the closure laws for single-phase flow, and one set might be: One scalar equation of state for ρ, one vectorial constitutive law for $\dot{\mathbf{q}}$ and one tensorial constitutive law for τ. However, the problem of achieving a correct and consistent description of the interfaces is much less simple.

The practical applications of the local instantaneous equation set are rather limited. This is because although the set is closed, the resulting initial moving boundary value problem is intractable except in the simplest of cases. The difficulties stem from the existence of deformable moving interfaces, with their motions unknown, and fluctuations of variables due to turbulence and to the motion of the interfaces. The former leads to complicated coupling between the field equations of each phase and the interfacial conditions, whilst the latter inevitably introduces statistical characteristics. In addition, the resulting set of partial differential equations is generally singular when the void fraction is identically one or zero since many coefficients vanish in the equations of the corresponding phase. They are also numerically ill-conditioned for small and large values of the void fraction. They are of use, however, in theoretical studies since they do not involve any further algebra. Typical examples are found in situations where the interfacial geometry is particularly simple, such as the study of a single bubble or of a laminar separated flow.

Averaged equations

For most practical purposes, for example modeling flow in a pipe, the local instantaneous formulation is not very useful largely due to the intractability of the set as discussed in the preceding section. Moreover, the local instantaneous behaviour of the various flow variables is often not required; the prediction of averaged quantities appears sufficient. Thus, the almost universal approach is to average the local instantaneous equations using one or more of the following averaging operators: ensemble, time and distance. The resulting averaged equation set is much simplified, but information is inevitably lost during the averaging procedure. This information must be resupplied in the form of auxiliary relationships, i.e., additional closure laws. The problem is exactly analogous to that encountered in single-phase turbulent flow. The closure problem is a considerable one and is discussed in a separate section below.

Here, focus is only on *Eulerian averaging* since it is closely related to human observations and instrumentations. Eulerian averaging is based on time-space description of physical phenomena. Since in the Eulerian description time and space coordinates are taken as independent variables, it is natural to consider averaging with respect to these independent variables, i.e., time and spatial averaging.

The local instantaneous equation set is spatially averaged in the following section. Then in the succeeding section, the spaced-averaged equations are averaged over time to give a more useful *composite-averaged equation set.* It is also possible (although it is not given here) to time average the local instantaneous equation set and then volume average the resultant time-averaged, leading to the same result.

Instantaneous, space-averaged equations

The local instantaneous equations may be averaged over either area or volume. In practice, volume-averaged equations are generally more useful than their area-averaged counterparts. One reason for this is that discontinuities may arise in area averages. Discontinuities do not arise if averaging is performed over a thin slice with a finite volume, rather than over an area. Consider a fixed tube with axis O_z (unit vector $\mathbf{n}_z$) in which a volume V_k^* is cut by two cross-sectional planes located a distance Z apart over area A_{k1} and A_{k2} (see **Figure 3**). Let V_k be the volume limited by A_{k1}, A_{k2}, and the portions A_I and A_{kW} of interface and wall enclosed between the two cross-sectional planes. The unit vector normal to the interface and directed away from phase k is denoted by $\mathbf{n}_k$. The cross-sectional planes limiting the volume V_k are not necessarily fixed and their speeds of displacement are denoted by $-\mathbf{u}_{A_{k1}} \cdot \mathbf{n}_z$ and $\mathbf{u}_{A_{k2}} \cdot \mathbf{n}_z$.

The transport theorem [Equation (2)], may be generalised to give

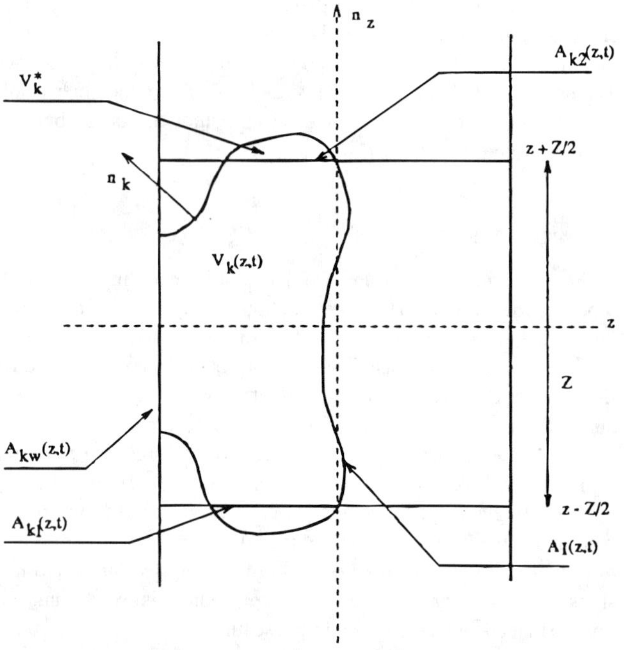

Figure 3. Volume cut by two cross-sectional planes.

$$\frac{\partial}{\partial t}\int_{V_{kv}(z,t)} f(x, y, z, t)dV = \int_{V_k(z,t)} \frac{\partial f}{\partial t} dV + \int_{A_k(z,t)} f\mathbf{u}_k \cdot \mathbf{n}dA. \quad \textbf{(14)}$$

Applying to volume V_k leads to

$$\frac{\partial}{\partial t}\int_{V_k(z,t)} f(x, y, z, t)dV = \int_{V_k(z,t)} \frac{\partial f}{\partial t} dV + \int_{A_I(x,t)} f\mathbf{u}_I \cdot \mathbf{n}_k dA$$
$$- \int_{A_{k1}(x,t)} f\mathbf{u}_{A_{k1}} \cdot \mathbf{n}_z dA + \int_{A_{k2}(z,t)} f\mathbf{u}_{A_{k2}} \cdot \mathbf{n}_z dA \quad \textbf{(15)}$$

where $\mathbf{u}_I \cdot \mathbf{n}_k$ is the speed of displacement of the interface A_I and the integral over the wall vanishes since the velocity of the wall is assumed to be zero.

The Gauss divergence theorem applied to volume V_k for any vector $\mathbf{v}$ gives

$$\int_{V_k(z,t)} \nabla \cdot \mathbf{v}dV = \int_{A_I(z,t)} \mathbf{n}_k \cdot \mathbf{v}dA + \int_{A_{kW}(z,t)} \mathbf{n}_k \cdot \mathbf{v}dA$$
$$- \int_{A_{k1}(z,t)} \mathbf{n}_z \cdot \mathbf{v}dA + \int_{A_{k2}(x,t)} \mathbf{n}_z \cdot \mathbf{v}dA. \quad \textbf{(16)}$$

The volume average operator is defined as follows:

$$\langle f_k \rangle_3 = \frac{1}{V_k}\int_{V_k} f_k dV. \quad \textbf{(17)}$$

Integrating the general balance, Equation (13), over the volume V_k leads to

$$\int_{V_k(z,t)} \frac{\partial}{\partial t}(\rho_k \Psi_k)dV = -\int_{V_k(z,t)} \nabla \cdot (\rho_k \Psi_k \mathbf{u}_k)dV$$
$$- \int_{V_k(z,t)} \nabla \cdot \mathbf{J}_k dV + \int_{V_k(z,t)} \rho_k \phi_k dV. \quad \textbf{(18)}$$

Applying Equations (15) and (16) leads to

$$\frac{\partial}{\partial t}\int_{V_k(z,t)} \rho_k \Psi_k dV = \int_{A_I(z,t)} \rho_k \Psi_k \mathbf{u}_I \cdot \mathbf{n}_k dA$$
$$- \int_{A_{k1}(z,t)} \rho_k \Psi_k \mathbf{u}_{A_{k1}} \cdot \mathbf{n}_z dA + \int_{A_{k2}(x,t)} \rho_k \Psi_k \mathbf{u}_{A_{k2}} \cdot \mathbf{n}_z dA$$
$$- \int_{A_I(z,t)} (\mathbf{n}_k \cdot \rho_k \Psi_k \mathbf{u}_k + \mathbf{J}_k)dA - \int_{A_{kW}(z,t)} \mathbf{n} \cdot \mathbf{J}_k dA$$
$$+ \int_{A_{k1}(z,t)} \mathbf{n}_z \cdot (\rho_k \Psi_k \mathbf{u}_k + \mathbf{J}_k)dA$$
$$- \int_{A_{k2}(x,t)} \mathbf{n}_z \cdot (\rho_k \Psi_k \mathbf{u}_k + \mathbf{J}_k)dA \quad \textbf{(19)}$$
$$+ \int_{V_k(z,t)} \rho_k \phi_k dV.$$

By using the definition of the volume averaging operator, Equation (17) and also Equation (12), this result can be written as

$$\frac{\partial}{\partial t} V_k\langle \rho_k \Psi_k \rangle_3 - V_k\langle \rho_k \phi_k \rangle_3$$

$$= \int_{A_{k1}(x,t)} \mathbf{n}_z \cdot [\rho_k \Psi_k (\mathbf{u}_k - \mathbf{u}_{A_{k1}}) + \mathbf{J}_k] dA$$

$$- \int_{A_{k2}(x,t)} \mathbf{n}_z \cdot [\rho_k \Psi_k (\mathbf{u}_k - \mathbf{u}_{A_{k2}}) + \mathbf{J}_k] dA$$

$$- \int_{A_I(x,t)} (\dot{m}_k \Psi_k + \mathbf{n}_k \cdot \mathbf{J}_k) dA - \int_{A_{kW}(z,t)} \mathbf{n}_k \cdot \mathbf{J}_k dA. \quad \textbf{(20)}$$

Use of the appropriate values for Ψ_k, ρ_k and ϕ_k (see **Table 1**) give the instantaneous volume-averaged equations for the balances of mass, momentum and total energy.

Composite-averaged equations

Composite-averaged equations are those which, for example, have been averaged over both space and time. The vast majority of practical situations are analysed with a composite-averaged equation set.

Here, the volume-averaged equations will be time-averaged to give a composite-averaged equation set. A *single* time-averaging operator will be used, which is defined as follows:

$$\overline{f_k} = \frac{1}{T} \int_{[T]} f_k dt, \quad \textbf{(21)}$$

where [T] is the time interval over which averaging occurs. If f_k has no jump discontinuities, as is the case here since volume averaging has been performed, then

$$\frac{\partial \overline{f_k}}{\partial t} = \overline{\frac{\partial f_k}{\partial t}}, \quad \textbf{(22)}$$

i.e., the partial differentiation of a time-averaged function and the time-average of a partial derivative commute if the function is smooth.

Consider the volume-averaged general balance, Equation (20), for the case when the cross-sectional planes are fixed such that the velocities $\mathbf{u}_{A_{k1}}$ and $\mathbf{u}_{A_{k2}}$ vanish. Integrating the resultant expression over the time interval [T] gives

$$\int_{[T]} \frac{\partial}{\partial t} V_k\langle \rho_k \Psi_k \rangle_3 dt - \int_{[T]} V_k\langle \rho_k \phi_k \rangle_3 dt$$

$$= \int_{[T]} \int_{A_{k1}(z,t)} \mathbf{n}_z \cdot (\rho_k \Psi_k \mathbf{u}_k + \mathbf{J}_k) dA dt$$

$$- \int_{[T]} \int_{A_{k2}(z,t)} \mathbf{n}_z \cdot (\rho_k \Psi_k \mathbf{u}_k + \mathbf{J}_k) dA dt$$

$$- \int_{[T]} \int_{A_I(z,t)} (\dot{m}_k \Psi_k + \mathbf{n}_k \cdot \mathbf{J}_k) dA dt \quad \textbf{(23)}$$

$$- \int_{[T]} \int_{A_{kW}(z,t)} \mathbf{n}_k \cdot \mathbf{J}_k dA dt$$

which, using the above definition of the time averaging operator along with Equation (22) and also the commutativity properties of integrals, gives

$$\frac{\partial}{\partial t} \overline{V_k\langle \rho_k \Psi_k \rangle_3} - \overline{V_k\langle \rho_k \phi_k \rangle_3}$$

$$= \int_{A_{k1}(z,t)} \overline{\mathbf{n}_z \cdot (\rho_k \Psi_k \mathbf{u}_k + \mathbf{J}_k)} dA - \int_{A_{k2}(z,t)} \overline{\mathbf{n}_z \cdot (\rho_k \Psi_k \mathbf{u}_k + \mathbf{J}_k)} dA$$

$$- \overline{\int_{A_I(z,t)} (\dot{m}_k \Psi_k + \mathbf{n}_k \cdot \mathbf{J}_k) dA} - \overline{\int_{A_{kW}(z,t)} \mathbf{n}_k \cdot \mathbf{J}_k dA}. \quad \textbf{(24)}$$

The two terms in the above expression involving integration over the cross-sectional planes A_{k1} and A_{k2} may be combined into a differential term since the distance Δz is arbitrarily small. This results in

$$\frac{\partial}{\partial t} \overline{V_k\langle \rho_k \Psi_k \rangle_3} - \overline{V_k\langle \rho_k \phi_k \rangle_3} = -\frac{\partial}{\partial z} V_k\langle \overline{\mathbf{n}_z \cdot (\rho_k \Psi_k \mathbf{u}_k + \mathbf{J}_k)} \rangle$$

$$- \overline{\int_{A_I(z,t)} (\dot{m}_k \Psi_k + \mathbf{n}_k \cdot \mathbf{J}_k) dA} - \overline{\int_{A_{kW}(z,t)} \mathbf{n}_k \cdot \mathbf{J}_k dA}. \quad \textbf{(25)}$$

In practice, it is generally more useful to have a composite-averaged equation set expressed in terms of average over the control volume, V, rather than the phasic volume, V_k. Fortunately, the transformation is a simple one which, defining

$$\langle\langle f_k \rangle\rangle_3 = \frac{1}{V} \int_V f_k dV \quad \textbf{(26)}$$

results in

$$\frac{\partial}{\partial t} A\langle\langle \varepsilon_k (\overline{\rho_k \Psi_k})^X \rangle\rangle_3 - A\langle\langle \varepsilon_k (\overline{\rho_k \phi_k})^X \rangle\rangle_3$$

$$+ \frac{\partial}{\partial t} A\langle\langle \mathbf{n}_z \cdot \varepsilon_k (\overline{\rho_k \Psi_k \mathbf{u}_k})^X \rangle\rangle_3 + \frac{\partial}{\partial z} A\langle\langle \mathbf{n}_z \cdot \varepsilon_k \overline{\mathbf{J}_k}^X \rangle\rangle_3 \quad \textbf{(27)}$$

$$= -\frac{1}{\Delta z} \overline{\int_{A_I(z,t)} (\dot{m}_k \Psi_k + \mathbf{n}_k \cdot \mathbf{J}_k) dA} - \frac{1}{\Delta z} \overline{\int_{A_{kW}(z,t)} \mathbf{n}_k \cdot \mathbf{J}_k dA}$$

where ε_k is the time fraction of phase k and $\overline{f_k}^X = \overline{f_k}/\varepsilon_k$. This is the space/time-averaged equation expressed in a more useful form.

Averaging of the Interfacial Jump Conditions

Integrating Equation (11) over space and time results in

$$\sum_{k=1,2} \left(\overline{\int_{A_I} \dot{m}_k \Psi_k dA} + \overline{\int_{A_I} \mathbf{n}_k \cdot \mathbf{J}_k dA} \right) = 0. \quad \textbf{(28)}$$

This is the composite-averaged general interfacial jump condition which may be used in conjunction with the composite-averaged general balance equation [Equation (27)].

Closure

A brief indication of how averaged equation sets may be *closed* in order to allow calculation is provided below, but restricted to the composite (volume/time)-averaged equation set derived above.

Classification of closure laws

The closure laws of the composite-averaged equation set may be subdivided into three distinct types:

1. *Constitutive laws,* these are so called due to their similarity with the constitutive laws of single-phase flow. They describe the constitution or state of a material.
2. *Transfer laws* which express the boundary terms in the balance equations.
3. *Topological laws* which restore necessary information on flow structure lost during the averaging process.

The term *topological law* is not a standard one and the motivation for its introduction merits further discussion. It has been popularised by Bouré (1986) who used the term to highlight the difference in the closure problem between local instantaneous and averaged equation sets. The local instantaneous equation set is closed very classically by true constitutive laws, as discussed in the section on **closure and applications.** Averaging the local instantaneous equation set to obtain equations of practical usefulness inevitably entails a loss of information. In particular, details of flow structure are smoothed out, i.e., fluctuations whose characteristic times are shorter than the averaging operator characteristic time and distributions whose characteristic lengths are smaller than the averaging operator characteristic length. This information must be resupplied in the form of auxiliary relationships since some of it plays an essential role in the thermohydraulic behaviour of the flow (e.g., geometry of the interfaces, velocity and temperature gradients within each phase). Closure laws that are primarily responsible for resupplying interfacial structure information are the topological laws.

Practical equations and the closure problem

The Phase equations and jump conditions can be made dimensionless, using for example characteristic values l^c, t^c, ρ^c, u^c_{kz} and Δh^c as units for lengths, times, densities, velocities and enthalpies, where u_{kz} is the projection of the velocity vector along the z axis ($u_{kz} = \mathbf{u}_k \cdot \mathbf{n}_z$). The only difference between dimensional and dimensionless balance equations lies in the appearance of the ratio

$$\eta = \frac{(u^c_{kz})^2}{\Delta h^c} \tag{29}$$

in energy equations. The equations presented here may be regarded as dimensionless, although setting $\eta = 1$ in energy equations results in the dimensional form again.

The first difficulty concerns the averages of products of the dependent variables in the composite-averaged general balance equation [Equation (27)]. What is needed is the products of the averages of the dependent variables since it is the averaged variables that are generally measured experimentally and that are related to physically. In fact, it is very difficult to relate the averages of products to the product of averages and the almost universal approach is to assume that they are equal. This is equivalent to making a statement something like the following:

$$\langle\langle \varepsilon_k \overline{(\rho_k \Psi_k)}^X \rangle\rangle_3 = \langle\langle \varepsilon_k \rangle\rangle_3 \langle\langle \overline{\rho_k}^X \rangle\rangle \langle\langle \overline{\Psi_k}^X \rangle\rangle_3. \tag{30}$$

Whenever fluctuations and/or transverse distributions are present, this practice is obviously questionable. In this discussion, the above convention is followed and henceforth, the averaging symbols are omitted. Thus, the classical phase fraction, $\lll \varepsilon_k \ggg_3$ is denoted by ε_k.

Composite (space/time)-averaged phase equations are obtained by substituting the appropriate values of Ψ_k, $\mathbf{J}_k$ and ϕ_k from **Table 1** into Equation (27). The composite (space/time)-averaged interfacial jump conditions are obtained by substituting the appropriate values of Ψ_k, $\mathbf{J}_k$ and ϕ_k from **Table 1** into Equation (28). No mass transfer is assumed to take place between the fluid and the walls of the pipe, although this could easily be included if required.

Balance of mass, phase equation

$$\frac{\partial}{\partial t}(A\varepsilon_k\rho_k) + \frac{\partial}{\partial z}(A\varepsilon_k\rho_k u_{kz}) - \lim_{\Delta z \to 0}\left(-\frac{1}{\Delta z}\overline{\int_{A_I} \rho_k(\mathbf{u}_k - \mathbf{u}_I)\cdot \mathbf{n}_k dA}\right) = 0. \tag{31}$$

(A_I is the instantaneous interfacial area present in volume V.)

Balance of mass, jump condition

$$\sum_k \lim_{\Delta z \to 0}\left(-\frac{1}{\Delta z}\overline{\int_{A_I} \rho_k(\mathbf{u}_k - \mathbf{u}_I)\cdot \mathbf{n}_k dA}\right) = 0. \tag{32}$$

Balance of momentum, phase equation

Here, the phasic total stress tensor τ_k is decomposed in the classical manner in the bulk phase (i.e., not at the walls or interface), thus

$$\tau_k = -p_k\mathbf{I} + \tau^D_k \tag{33}$$

where $\mathbf{I}$ is the unit tensor. This decomposition holds for any scalar field, p_k and any tensor field τ_k; p_k is identified with the pressure and τ^D_k, with the deviatoric stress tensor field. The momentum balance is thus:

$$\frac{\partial}{\partial t}(A\varepsilon_k\rho_k\mathbf{u}_k) + \frac{\partial}{\partial z}(A\varepsilon_k\rho_k u_{kz}\mathbf{u}_k) + \frac{\partial}{\partial z}(A\varepsilon_k p_k \mathbf{n}_z) - \frac{\partial}{\partial z}(A\varepsilon_k\tau^D_k\cdot\mathbf{n}_z) - \lim_{\Delta z \to 0}\left(\frac{1}{\Delta z}\overline{\int_{A_{kW}} \tau_k\cdot\mathbf{n}_k dA}\right) - A\varepsilon_k\rho_k\mathbf{g} - \lim_{\Delta z \to 0}\left(-\frac{1}{\Delta z}\overline{\int_{A_I} [\rho_k(\mathbf{u}_k - \mathbf{u}_I)\cdot\mathbf{n}_k]\mathbf{u}_k + \tau_k\cdot\mathbf{n}_k dA}\right) = 0 \tag{34}$$

Balance of momentum, jump condition

Neglecting surface tension effects;

$$\sum_k \lim_{\Delta z \to 0}\left(-\frac{1}{\Delta z}\int_{A_I} \overline{[\rho_k(\mathbf{u}_k - \mathbf{u}_i)\cdot\mathbf{n}_k]\mathbf{u}_k + \tau_k\cdot\mathbf{n}_k dA}\right) = 0. \tag{35}$$

Balance of energy, phase equation

Again, the stress tensor is decomposed. Thus,

$$\frac{\partial}{\partial t}\left[A\varepsilon_k\rho_k\left(h_k + \frac{\eta}{2}\mathbf{u}_k^2\right)\right] + \frac{\partial}{\partial z}\left[A\varepsilon_k\rho_k u_{kz}\left(h_k + \frac{\eta}{2}\mathbf{u}_k^2\right)\right]$$

$$-\eta\frac{\partial}{\partial t}(A\varepsilon_k p_k) - \eta\frac{\partial}{\partial z}[A\varepsilon_k(\tau_k^D\cdot\mathbf{n}_z)\cdot\mathbf{u}_k] + \frac{\partial}{\partial z}(A\varepsilon_k\dot{\mathbf{q}}_k\cdot\mathbf{n}_z)$$

$$+\lim_{\Delta z\to 0}\left(\frac{1}{\Delta z}\int_{A_I}\overline{[\rho(\mathbf{u}_k + \mathbf{u}_I)\cdot\mathbf{n}_k]h_k - \frac{1}{2}\eta[\rho_k(\mathbf{u}_k - \mathbf{u}_I)\cdot\mathbf{n}_k]\mathbf{u}_k^2 dA}\right)$$

$$-\eta\lim_{\Delta z\to 0}\left(\frac{1}{\Delta z}\int_{A_I}\overline{(\tau_k\cdot\mathbf{n}_k)\cdot\mathbf{u}_k - \dot{\mathbf{q}}_k\cdot\mathbf{n}_k dA}\right)$$

$$-\lim_{\Delta z\to 0}\left(-\frac{1}{\Delta z}\int_{A_{kW}}\overline{\dot{\mathbf{q}}_k\cdot\mathbf{n}_k dA}\right) - \eta A\varepsilon_k\rho_k\mathbf{u}_k\cdot\mathbf{g} = 0 \tag{36}$$

where velocity of the wall has been assumed to be zero.

Balance of energy, jump condition

$$\sum_k\lim_{\Delta z\to 0}\left(-\frac{1}{\Delta z}\int_{A_I}\overline{[\rho(\mathbf{u}_k - \mathbf{u}_I)\cdot\mathbf{n}_k]h_k - \frac{1}{2}\eta\rho_k(\mathbf{u}_k - \mathbf{u}_I)\cdot\mathbf{n}_k]\mathbf{u}_k^2 dA}\right)$$

$$+\lim_{\Delta z\to 0}\left(\frac{1}{\Delta z}\int_{A_I}\overline{\eta(\tau_k\cdot\mathbf{n}_k)\cdot{=}{+}\mathbf{u}_k - \dot{\mathbf{q}}_k\cdot\mathbf{n}_k dA}\right) = 0 \tag{37}$$

where, once again, the stress tensor has been decomposed and surface tension effects have been neglected.

Further simplifications

Possibly the most common, simplified two-fluid model for pipe flows is the *single-pressure* model. In single-pressure models, it is recognised that the averaged phasic pressures are not very different in practical pipe flows (except for surface tension effects). Thus it has become customary to assume

$$\langle\langle\overline{p_1}^X\rangle\rangle_3 = \langle\langle\overline{p_2}^X\rangle\rangle_3, \tag{38}$$

which enables elimination of one dependent variable and serves as a substitute for the void fraction (topological) closure law.

This is particularly convenient since a satisfactory void fraction closure law has yet to be found, as discussed by Bouré (1986). Unfortunately, assuming that the phasic pressures are equal dictates that their partial derivatives must also be equal. Thus, under this assumption, pressure disturbances have the same averaged effect on the two phases; in particular, they propagate at the same velocity within the two phases. This is clearly unrealistic in many cases and very restrictive where propagation phenomena are important.

Another assumption commonly made is that the field and viscous conduction terms

$$\frac{\partial}{\partial z}(A\varepsilon_k\tau_k\cdot\mathbf{n}_z),\ \frac{\partial}{\partial z}[A\varepsilon_k(\tau_k\cdot\mathbf{n}_k)\cdot\mathbf{u}_k],\ \frac{\partial}{\partial z}(A\varepsilon_k\dot{\mathbf{q}}_k\cdot n_k) \tag{39}$$

may be neglected. This assumption is motivated by the fact that these terms are generally very small compared to the other terms of momentum and energy balances. Care must be exercised, however, since in some cases they are not negligible; the sizeable conduction term for liquid metals is one such example.

With these simplifications, the following closure laws are required: two equations of state (one per phase), one interfacial mass transfer law, three interfacial momentum transfer laws, seven interfacial energy transfer laws, two wall momentum transfer laws and four wall energy transfer laws. They are usually obtained in a rather *ad hoc* manner, and are often empirical. Further simplifications may be possible for a given situation.

Nomenclature

Roman Letters

A	surface area
A_I	interfacial area
f	arbitrary quantity
f^+	limiting value of f as the interface is approached from one side
f^-	limiting value of f as the interface is approached from other side
$\mathbf{g}$	gravity vector
h	specific enthalpy
$\mathbf{I}$	unit tensor
$\mathbf{J}_k$	flux term
l	length
$\dot{m}_k$	mass transfer per unit of interface area and per unit time
$\mathbf{n}$	unit normal to a surface, principle normal of a curve, etc.
p	pressure
$\dot{\mathbf{q}}$	heat flux vector
T	time interval
t	time
$\mathbf{u}$	velocity vector
$\mathbf{u}_I$	velocity vector of surface point
$\mathbf{u}_k$	velocity vector of phase k
u_x	component of u in x direction
u_y	component of u in y direction
u_z	component of u in z direction
u	internal energy per unit mass
V	volume
$\mathbf{v}$	arbitrary vector
$\mathbf{x}$	spatial displacement vector
x,y,z	spatial coordinates
z	spatial displacement in the z direction

Greek

ε	time fraction or classical phase fraction
η	ratio appearing in dimensionless form of energy equation

ρ density
Ψ arbitrary scalar, vector or tensor
τ total stress tensor
τ^D deviatoric stress tensor
ϕ source term
$[\Psi]$ jump of Ψ across a surface

Operators

$\bar{f}$ time average operator defined by equation 21
$\langle f_k \rangle_3$ phasic volume average of f_k defined by equation 17
$\lll f_k \ggg_3$ phasic volume average of f_k defined by equation 26

Subscripts

I interface
κ phase index
ω wall

Superscripts

c characteristic value

References

Bouré, J. A. (1986) *Two-phase flow models: The closure issue.* In European Two-Phase Flow Group Meeting, Munchen, Germany, June.

Bouré, J. A. and Delhaye, J. M. (1982) General equations and two-phase flow modeling. In G. Hetsroni, editor, *Handbook of Multiphase Systems.* Hemisphere.

Ishii, M. (1975) *Thermo-Fluid Dynamic Theory of Two-Phase Flow.* Eyrolles, Paris.

Slattery, J. C. (1972) *Momentum, Energy and Mass Transfer in Continua.* McGraw-Hill.

Truesdell, C. and Toupin, R. A. (1960) The classical field theories. In S. Flugge, editor, *Handbuch der Physik.* Springer-Verlag.

Leading to: *Waves in fluids*

I.F. Roberts

CONSERVATIVE SYSTEMS (see Non-linear systems)

CONSTANT RATE PERIOD, DRYING CURVE (see Drying)

CONSTITUTIVE EQUATIONS (see Conservation equations, single phase; Flow of fluids; Non-Newtonian fluids)

CONSTITUTIVE RELATION, THERMODYNAMICS (see Non-linear systems)

CONTACT ANGLE

Following from: *Surface and interfacial tension*

When a drop of liquid is placed on the flat surface of a solid, it may: 1) spread out to form a thin film; or 2) remain as a drop forming a finite angle with the solid surface. The angle, measured in the liquid is defined as the contact angle, Θ.

The concept also applies to a system consisting of two immiscible liquids in contact with a solid. In this case, Θ may apply to either liquid.

Methods of Measuring Contact Angle

Direct measurement

Fox et al. (1950) has designed a system in which a small drop of the liquid is placed on the surface of a flat solid in a small chamber which has optically worked faces or windows. The chamber is placed on a table which forms part of an optical bench. The drop is illuminated by a collimated beam of monochromatic light. The silhouette of the drop is then examined using a telescope fitted with a goniometer eyepiece or it may be photographed and mea-

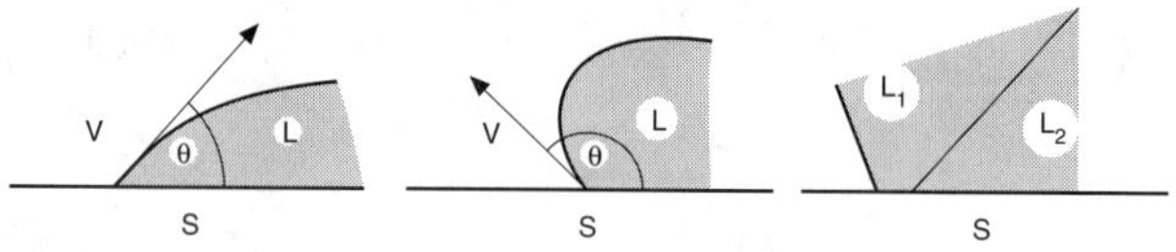

Figure 1.

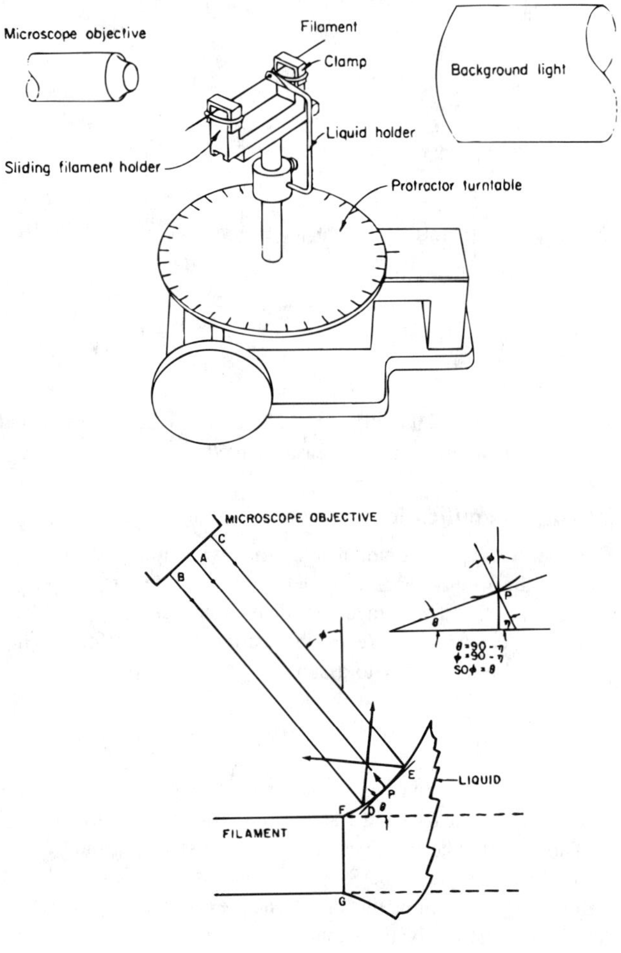

Figure 2.

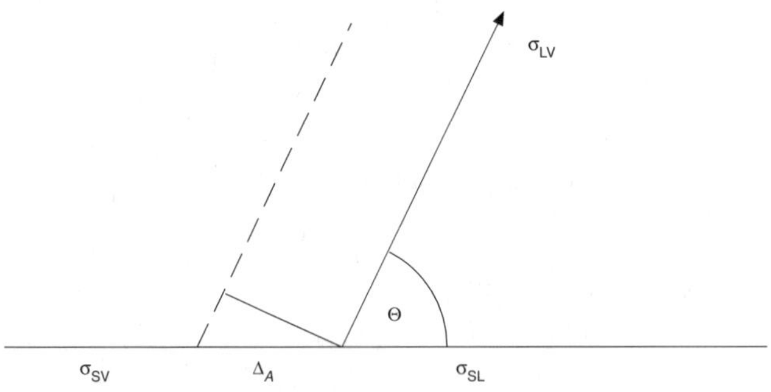

Figure 3.

sured later. In Fox and Zisman's device, the chamber could be evacuated or filled with an inert atmosphere. Hysteresis of the contact angle can be determined by adding or removing fluid from the drop.

The tilting plate method

When the amount of liquid available is not important and the solid is available as a transparent plate, the plate may be partially immersed in the liquid and rotated about an axis parallel to the surface of the liquid until the liquid remains flat right up to the surface of the plate. This position may be ascertained by viewing a grid immersed in the liquid, which should show no distortion.

Interference method

For very small angles of contact, collimated monochromatic light may be used to illuminate the edge of the film. Light rays reflected from the liquid-vapour interface and from the solid-liquid interface will interfere and produce Fizeau fringes, from the spacing of which the angle made by the film on the solid may be determined.

Method for solids in the form of fibres

Jones and Porter (1967) have devised a method suitable for routine measurements on fibres. The fibre is stretched between two holders fixed to a sliding table (see **Figure 2**). An eyelet is fitted round the fibre and filled with the liquid. The meniscus is viewed with a microscope having vertical illumination. Only rays incident perpendicular to the surface are reflected back to the microscope and a pin-point of light is observed at P.

The system is rotated until the point P just reaches F, where the liquid touches the solid surface. At this point, the spot vanishes. The method is suitable for angles up to about 75°. Hysteresis of the contact angle may be measured by sliding the fibre forwards and backwards.

Methods suitable for powders or packed beds

If the surface of a packed bed or compressed powder touches the surface of a liquid which wets or partially wets the solid, the liquid will flow into the system into the action of the Laplace pressure. If the system has an effective pore radius of r, then $\Delta P = \dfrac{2\sigma\cos\Theta}{r}$. The volume rate of flow w is given by the *Poiseuille equation*

$$w = \frac{\pi \Delta \rho r^4}{8\eta h}, \tag{1}$$

where η is the viscosity of the liquid and h, the height to which the liquid has risen. The rate of advance of the liquid front, dh/dt, is:

$$\frac{dh}{dt} = \frac{w}{\pi r^2} = \frac{\Delta \rho r^2}{8\eta h}. \tag{2}$$

This is the *Washburn equation* [Washburn (1921)]. By making measurements using a liquid, which completely wets the solid, one may obtain a value for the average pore radius. The system may then be used to measure Θ for any other liquid. In a modification of Washburn's method made by Bartell (1932), the pressure needed to prevent the liquid from penetrating the capillary system is measured.

Hysteresis of contact angle

Measured values of the contact angle show that they are dependent on the direction of liquid motion before coming to rest, i.e., hysteresis occurs. When the liquid is advancing over the surface, the advancing angle, Θ_A, is measured and as liquid withdraws from the surface the receding angle, Θ_R, is obtained. There are five main causes of hysteresis:

1. Heterogenity in surface energy of the solid surface;
2. The liquid may remove adsorbed molecules from the solid surface;
3. Molecules of the liquid may form an oriented adsorbed layer at the solid-liquid interface;
4. Molecules of the liquid may form an oriented monomolecular film on the solid surface, which is not wetted by the liquid;
5. The surface may be rough.

Spreading of liquids on solids

The work of adhesion between a liquid and a solid may be defined in a similar way to that between two liquids, viz.:

$$W_{SL}^{adh} = \sigma_{SV} + \sigma_{LV} - \sigma_{SL}. \tag{3}$$

The angle of contact is an excellent measure of the wettability of a solid surface. If the liquid spreads over the surface without limit, it is zero; otherwise, a finite angle is formed. The interfacial free energies, σ_{SV} and σ_{SL}, are not in general easy to measure. Young has derived an expression which relates these quantities to the contact angle.

The contact line between three phases in equilibrium—solid, liquid and vapours—is shown in **Figure 3**. Consider a virtual displacement which causes $\Delta\,\mathcal{A}$ m^2 of solid surface to be covered by the liquid. Then the solid-liquid and liquid-vapour interfacial energies increase by $\sigma_{SL}\Delta\mathcal{A}$ and $\sigma_{LV}\cos\Theta\Delta\mathcal{A}$, respectively, while the solid-vapour interfacial energy decreases by $\sigma_{SV}\Delta\mathcal{A}$.

By the principle of virtual work,

$$\sigma_{SL} + \sigma_{LV} \cos \Theta = \sigma_{SV} \quad (4)$$

$$= \sigma_{SO} - \pi \quad (5)$$

where π is the reduction in the surface free energy of the solid due to the adsorption of vapour. This is the *Young-Dupré equation.*

Hence,

$$W_{SL}^{adh} = \sigma_{LV}(1 + \cos \Theta) \quad (6)$$

and the spreading coefficient S is then

$$S = W_{SL}^{adh} - 2\sigma_{LV} \geq 0. \quad (7)$$

Zisman and his coworkers (1964), made a systematic study of the variation of contact angle of a series of liquids, in particular solid surfaces. They observed a linear relationship between the surface tension of the liquids and the cosine of the angle of contact formed by them on any particular solid [see **Figure 4**].

They extrapolated the lines to cos Θ = 1. Zisman defined this corresponding value of σ_{LV} as the *critical surface tension* of the solid. It is the surface tension of a liquid which just spreads on the surface and is characteristic of the solid. It is **not** the surface free energy of the solid since if the interaction between the solid and the liquid is high, the solid has a high surface energy and a large value of σ_{SL} could exist. If the critical surface tension is to equal solid surface tension, σ_{SL} would be zero.

References

Bartell, F. E. and Miller, F. L. (1932) *Industr. Engin. Chem.*, 24, 335.
Fox, H. W. and Zisman, W. A. (1950) *J. Coll. Sci.*, 5, 514.
Jones, W. C. and Porter, M. C. (1967) *J. Coll. Int. Sci.*, ***24,*** 1–3.
Washburn, E. W. (1921) *Phys. Rev. Ser.*, 2, 17, 273.
Zisman, W. A. (1964) *Advances in Chemistry,* No. 43, American Chemical Society, Washington D. C.

A.I. Bailey

CONTACT CONDUCTANCE (see Thermal contact resistance)

CONTACT DISCONTINUITIES (see Magneto hydrodynamics)

CONTACT RESISTANCE (see Thermal contact resistance)

CONTRACTORS (see Extraction, liquid/liquid)

CONTAINMENT

Following from: *Nuclear reactors*

Introduction

Nuclear reactors are designed to respond in a completely safe manner to a wide range of possible transients arising from all reason-

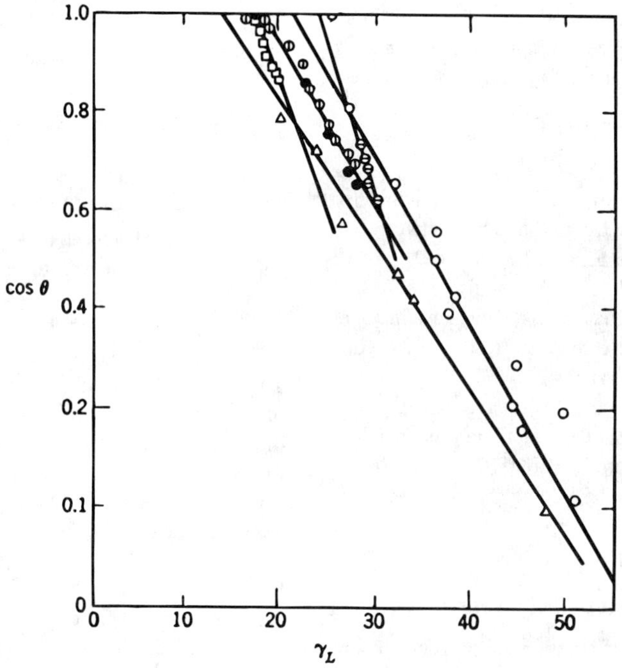

Figure 4. Zisman plots of the contact angles of various homologous series on Teflon: ○, RX: ⊖, alkylbenzenes; ∅, n-alkanes; ●, dialkyl ethers; □, siloxanes; △, miscellaneous polar liquids.

ably conceivable initiators. These initiators will include failure of important valves or pumps and rupture of any part of the piping. In a PWR, main coolant pipes are some 1,000 mm in diameter and in safety analysis, the unlikely event of a guillotine rupture of one of these pipes is considered. The consequence could be the rapid ejection of most of the water from the primary circuit, which would flash into steam and may also be radioactive. The **containment** structure is designed to capture this steam and to limit containment pressure, which may be a value consistent with the design strength of the containment. Most water reactors have this type of containment. The issues in gas-cooled reactors (which have a single-phase coolant) and sodium-cooled reactors (which are not pressurised) are rather different, and are not included in this discussion.

As reactor safety rationale developed, it has become necessary to consider so-called 'Severe Accidents,' which result from the extremely unlikely failure of safety systems to cool the reactor fuel adequately following some accident initiator. Containment has been shown to have a strong influence on the subsequent course of events and could well limit the consequences of such an accident. Designs have progressed by considering whether there are some extra features which could be added to further reduce the consequences of a severe accident, without having a large additional effect on reactor cost. Examples of such additions are provisions for cooling and solidification of any molten core material which might penetrate the reactor vessel, or for the release of excessive internal pressure through filters. (See also **Blowdown**)

Containment Systems and their Function

According to the International Atomic Energy Agency (IAEA) Safety Guide, the containment system consists of the containment structure and associated subsystems including:

- the containment structure and extensions, such as external passive fluid retaining boundaries which, together, form an envelope around the reactor coolant system;
- the active features of the containment isolation system, which in general, provide closure of openings in the containment envelope upon demand,
- energy management features (collectively referring to the functions of pressure suppression, containment atmosphere pressure and temperature reduction and containment heat removal after a postulated accident);
- radionuclide confinement features, which are provided to reduce the release of radionuclides to the external environment, after their release to the containment volume;
- combustible gas control features, which limit the build up of combustible gases, such as hydrogen, in the containment envelope and prevent uncontrolled combustion or detonation of these gases;
- other prevention or mitigation features for impulsive loads produced by severe accidents (such as ex-vessel steam explosions).

The main functions of the containment system are:

- to prevent uncontrolled large releases to the environment in all those plant conditions which need to be taken into account, and to minimise controlled releases;
- to maintain the system's structural integrity to assure the required leak-tightness and the necessary support to systems and components;
- to allow the removal of decay heat and the cool-down of the reactor to safe shutdown conditions;
- to prevent radioactive releases to the environment as a consequence of external events of natural and man-induced origin;
- to provide a biological shield for operating personnel and the public.

Additionally, the containment system must allow access for operating and maintenance staff and equipment.

Types of containment

There are a large number of different designs of containment, but they can be divided into a few broad headings. The first major category is Dry Containments and Pressure Suppression Containments. In addition, containment systems may be either single or dual; in the latter case, a second structure is provided to catch any gases which may leak from the primary structure and also, in some cases, to provide extra protection against aircraft crashes.

Containment systems need to have provisions for removal of radioactive decay heat in the event of more extreme transients and severe accidents. In most existing reactors, these systems require the operation of pumps and heat exchangers and hence, require guaranteed electrical supplies. In many advanced designs being considered, there is a move away from dependence on such supplies towards reliance on natural processes, such as natural convection in which gravity is the motive force. The specifications for such systems allow for operation without operator intervention for at least 72 hours. Most reactors with these containment designs claim a cost reduction due to simplification of emergency systems.

The following brief descriptions indicate the most important types of containment in existence or being designed (1994).

Dual dry containment

The principle of dry containment is to provide a structure which is large enough and strong enough to contain without rupture all the contents of the primary coolant system which have flashed to an equilibrium pressure. Design pressure is in the range of 0.2 MN/m^2 and the vessel is either spherical or cylindrical in steel or concrete, with a steel liner for leak tightness. An outer structure permits an interspace between the two structures, which is maintained at subatmospheric pressure by pumps which discharge through filters. Any activity escaping from the primary vessel is trapped on the filters. The In-containment Refuelling Water Storage Tank (IRWST) provides water for residual heat removal and for cooling of molten core material in the remote event of such an occurrence.

BWR Pressure suppression containment

In a pressure suppression system, much of the vapour emitted from the ruptured primary system is condensed by passing these through a water tank before reaching the main volume of the containment. The primary reactor vessel is situated in the so-called dry-well. The water is in the wet-well and there are ducting arrange-

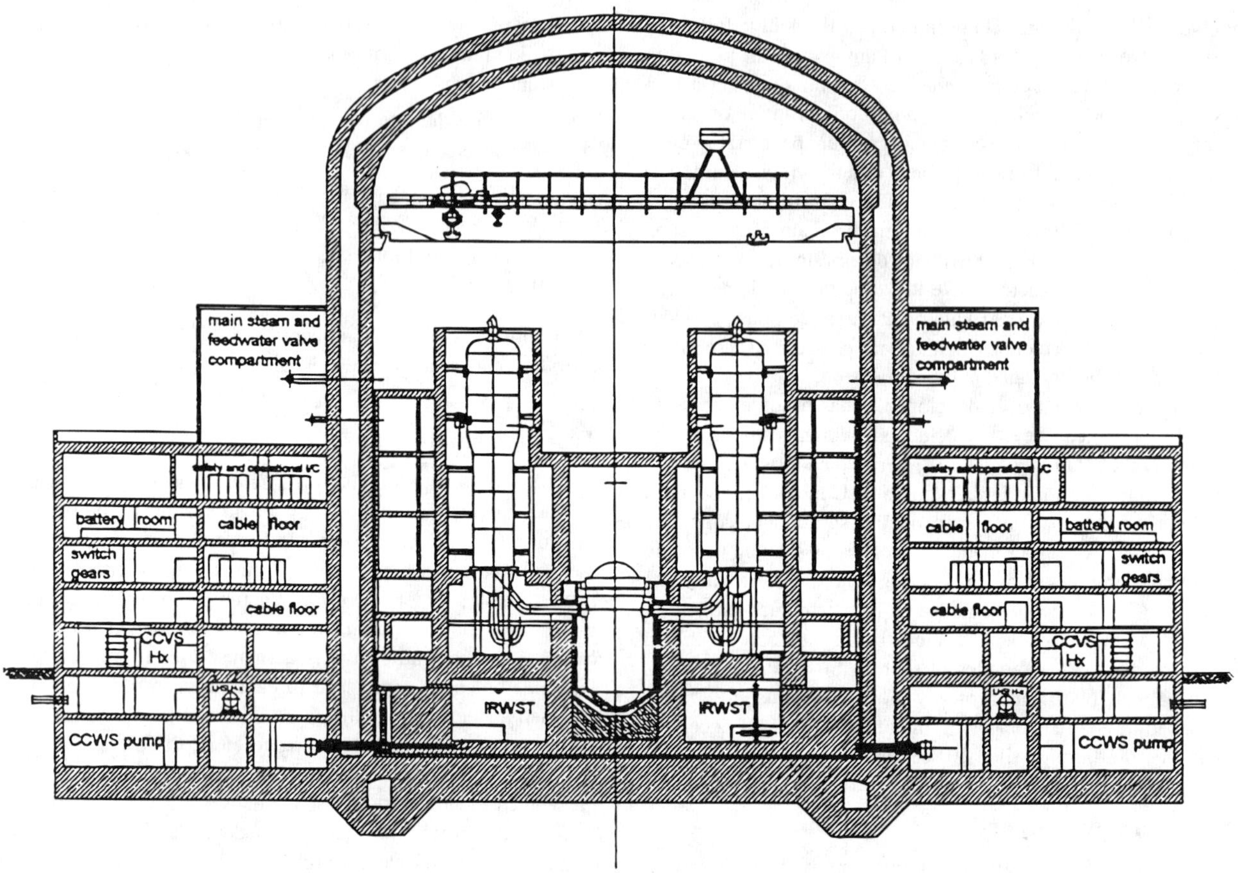

Figure 1 Dual dry containment design. (With permission of Nuclear Power International, Paris.)

ments to channel the vapour between the two parts of the containment. The advantage is a smaller and less expensive containment than a dry containment. It is particularly well-suited to BWRs since the maximum pipe sizes are smaller than in PWRs and depressurisation through a ruptured pipe takes place more slowly, giving time for condensation in the pressure suppression system to occur without an unacceptable pressure transient. There have been several different designs embodying the same principle. Designs for PWRs also exist; for example, the PWR Bubbling Condenser Containment has tanks of water half way up the containment building through which steam from the reactor space has to pass.

Ice condenser containment

This is a form of pressure suppression containment used in some of the early PWRs. Banks of ice are maintained at a high level inside the main containment building. Steam from a ruptured pipe is constrained to pass through these banks of ice to condense as much as possible and so reduce the ultimate pressure. A variant is to replace the ice by banks of gravel which form a completely passive, but limited, heat sink.

Negative pressure containment

This has been developed for the Canadian design of Pressure Tube, Heavy Water Reactors. See **CANDU Nuclear Power Reactor**. The main reactor building is maintained at subatmospheric pressure. Attached to this building by a duct is another building maintained as a pressure less than atmospheric. In the event of a pressure rise in the main building, valves open to connect the vacuum building to the reactor building which, combined with a spray system in the reactor building, ensures that the pressure does not exceed the design pressure.

In multiunit stations, several reactors are connected to a single vacuum building.

Passive containment

Many concepts of advanced reactor design have adopted the principle that reactors should safely contain the consequences of accidents considered in the design without need of electrical supplies for at least 72 hours. This implies that the provisions for removing residual heat during this period must rely solely on natural processes, such as heat conduction and natural convection. The containment atmosphere can be cooled by conduction through a metal containment structure. The amount of heat lost in this way, and hence the power of the reactor which can be operated, depends on the area of the containment structure and the temperature which can be tolerated within it. Heat transfer can be enhanced by spraying the outside of the structure with water from a high-level tank and by inducing an enhanced flow of air through a secondary enclosing structure and a chimney effect. Many variants of this principle are under development (1994).

The above cooling system has a single steel containment only, and some safety authorities regard it as inadequate based on integrity grounds. A dual containment is not possible. An alternative is to install a heat exchanger system operating between the containment atmosphere and an external heat sink, such as a cooling tower. The components have to be configured so as to allow it to operate by natural convection.

Severe Accidents

Containment systems have been designed to alleviate the consequences of so-called design basis accidents, such as the rupture of any pipework. The Three Mile Island accident, coupled with the development of Probabilistic Safety Assessment (PSA), have led to the consideration of a greater range of extremely-improbable accident sequences, such as overheating, failure and possibly melting of the fuel. These would require consideration of a set of physical processes concerned with the emission and dispersion of radioactive material from the fuel and with the progression of molten fuel through the reactor core structure, the reactor pressure vessel and its interaction with concrete and other structural materials.

Much of the radioactive material is emitted in the form of aerosols, in which the different volatile fission products interact with each other and with their environment. They get carried along in the flow of steam and gases and are deposited in the reactor pipes and within the containment building. It is important to know how much of the material remains in suspension to allow for the possibility of its discharge to the external environment through any breaches in the integrity of the containment.

If there were to be a complete failure of the emergency core cooling systems, there is the possibility that some of the core material may melt due to the continued generation of decay heat from the radioactive decay of radioactive fission products and actinides. All volatile species will be released and other materials may be caught up in the aerosols. The molten material itself may penetrate through the core structure and may reach and melt through the reactor vessel. Some cooling processes do still occur, such as natural convection cooling by steam and other gases generated. There is also significant conduction through the core materials. It is not necessarily a foregone conclusion that the vessel will be penetrated, but if it is, the molten core material will fall onto the concrete of the reactor cavity and will interact with it, producing further aerosols and hydrogen.

Considering these processes, the objective has been to assess the degree of protection afforded by existing containment designs. A large number of computer codes have been developed to consider these complex and linked processes. They are of two types: Mechanistic codes attempt to treat all the processes using the basic laws of physics in an exact a way as possible. These codes are expensive to run; Much faster codes are needed for assessment purposes, where a large number of accident sequences have to be developed. These codes make use of empirical data from experiments which have been carried out to study all aspects of the phenomena. These studies have shown that existing containment designs offer a large degree of protection against severe accidents.

In some new containment designs, the objective is to include additional systems in the containment to further mitigate the consequences of these severe accidents. One example is the provision of containment venting through a filter. This avoids the possibility of rupture of the containment through excessive build-up of pressure in an accident, but retains radioactive species in the filter. Such filters—using gravel beds in the filter—have been back-fitted to existing reactors in some countries. Another example is the provision of a core catcher in the reactor cavity underneath the reactor vessel. This causes the molten core to take up a predetermined geometry if it penetrates the reactor vessel and provides a cooling system to stabilise it.

I. Gibson

CONTINUITY EQUATION (see Conservation equations, single phase; Flow of fluids)

CONTINUITY SHOCKS (see Kinematic waves)

CONTINUITY WAVES (see Kinematic waves)

CONTINUOUS CASTING (See Casting of metals)

CONTINUOUS CRYSTALLIZERS (see Crystallizers)

CONTINUOUS FILTERS (see Filters)

CONTINUOUS WAVE LASERS (see Lasers)

CONTINUUM MECHANICS (see Conservation equations, single phase)

CONTINUUM MODELS (see Buildings and heat transfer)

CONTRACTION (FLOW AND PRESSURE LOSS IN)

Following from: *Channel flow; Pressure drop, single-phase; Pressure drop, two-phase flow*

The flow of fluid through a contraction (decrease in pipe diameter) results in an increase in the velocity and consequently, a pressure drop greater than the value for the equivalent straight pipe. If the contraction is sharp or sudden, the behaviour of single-phase flow is as shown in **Figure 1** and involves two *recirculation* regions. The first starts about 1.5 inlet pipe diameters upstream whilst the second starts at the contraction and extends up to 15 outlet pipe diameters downstream. The dissipation of energy caused by these recirculation regions means that not all the pressure drop is converted to kinetic energy (and thence recoverable at a subsequent enlargement) and reversible, and irreversible components of pressure drop must be considered. If the contraction is being used to create kinetic energy from pressure, it is necessary to employ a more gradual change in diameter so as to eliminate or minimise recirculations and thus losses.

Lighthill (1986), in discussing the calculation of pressure drop through a generalised contraction, points out that it is not possible

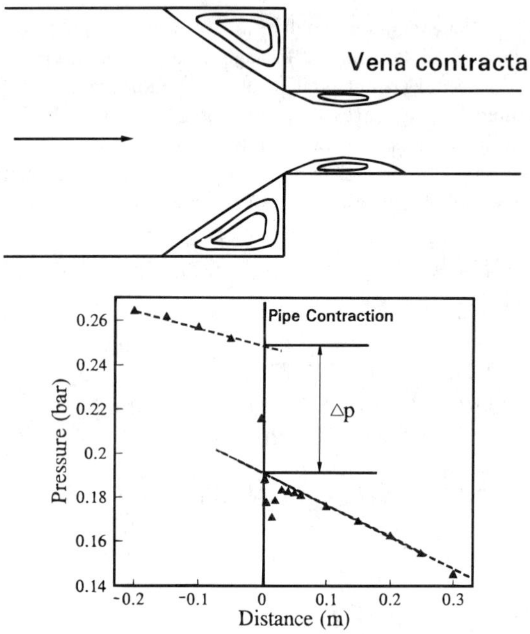

Figure 1. Flow structure and pressure profile for single-phase flow through a sudden contraction.

to use a momentum balance as there is an unknown reaction force from the walls to be accounted for. An energy balance gives a computable expression for the (reversible) pressure drop. To account for both the reversible and irreversible components of pressure drop, a balance is carried out from upstream to the minimum flow area point at the *vena contracta* (with no irreversibility) and a second balance downstream where most of the dissipation occurs. The combination of these results in

$$\Delta p = -\frac{\dot{m}^2}{2\rho}\left[S^2\left(\frac{1}{C_c} - 1\right)^2 + (S^2 - 1)\right] \qquad (1)$$

where $\dot{m}$ is the mass flux; ρ, the density; S, the area ratio between upstream and downstream pipes; and C_c, the contraction coefficient (the ratio of areas of the *vena contracta* and the outlet pipe. Equations for the contraction coefficient in terms of S are given by, for example, Benedict (1980) and Chisholm (1986).

In gas/liquid flow, the sudden contraction can act as a homogeniser mixing the phases and making their velocities more equal. For annular flows, the contraction can cause an increase in the proportion of liquid travelling as drops.

For two-phase pressure drop, Chisholm (1983) provides an equation equivalent to (1), derived using a separated flow approach. Comparison with experimental data shows that the homogeneous version of this equation gives the best results. However, there is recent evidence, Schmidt (1993), that the vena contracta does not always occur. The pressure profile shown in **Figure 2** provides confirmation of this as it lacks the characteristic minimum seen in **Figure 1**, which is characteristic of the vena contracta.

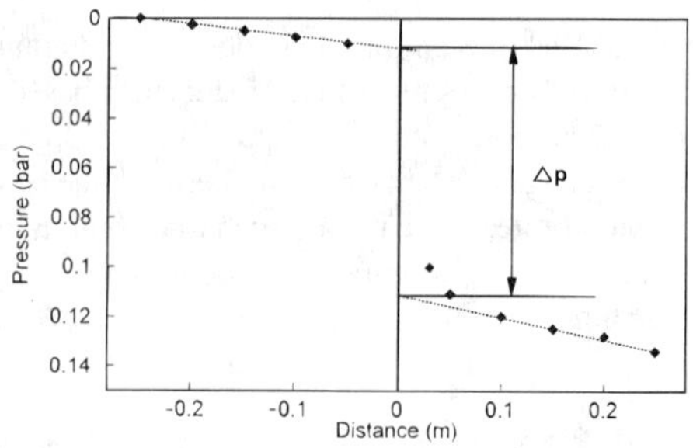

Figure 2. Pressure profile for gas/liquid flow through a sudden contraction. Quality = 0.5.

References

Benedict, R. P. (1980) *Fundamentals of Pipeflow,* Wiley-Interscience, New York.

Chisholm, D. (1983) *Two-Phase Flow in Pipelines and Heat Exchangers,* Pitman Press Ltd, Bath, England.

Lighthill, J. (1986) *An Informal Introduction to Theoretical Fluid Mechanics,* Oxford University Press, Oxford.

Schmidt, J. (1993) *Berechnung und Messung des Druckabfalls uber plötzliche scharfkantige Rohrerweiterungen und -verengungen bei Gas/*Dampf-Flüssigkeitsströmung, VDI-Forschungsheft.

B.J. Azzopardi

CONTROL THEORY (see Nyquist stability criterion; Nyqust diagrams)

CONVECTION DRYING (see Dryers)

CONVECTIVE HEAT FLUX (see Heat flux)

CONVECTIVE HEAT TRANSFER

Following from: *Heat transfer*

Introduction

This article is concerned with the transfer of thermal energy by the movement of fluid and, as a consequence, such transfer is dependent on the nature of the flow. Heat transfer by convection may occur in a moving fluid from one region to another or to a solid surface, which can be in the form of a duct, in which the fluid flows or over which the fluid flows. Convective heat transfer may take place in boundary layers, that is, to or from the flow over a surface in the form of a boundary layer, and within ducts where the flow may be boundary-layer-like or fully-developed. It may also occur in flows which are more complicated, such as those which are separated; for example, in the aft region of a cylinder in cross-flow or in the vicinity of a *backward-facing step.* The flow may give rise to convective heat transfer where it is driven by a pump and is referred to as forced convection, or arise as a consequence of temperature gradients and buoyancy, referred to as natural or free convection. Examples are given later in this Section and are shown in **Figure 1** to facilitate introduction to terminology and concepts.

The boundary layer on the flat surface of **Figure 1** has the usual variation of velocity from zero on the surface to a maximum in the free-stream. In this case, the surface is assumed to be at a higher temperature than the free-stream and the finite gradient at the wall confirms the heat transfer from the surface to the flow. It is also possible to have zero temperature gradient at the wall so there is no heat transfer to or from the surface but heat transfer within the flow. If the flow is laminar, heat transfer from the surface is given by the Fourier flux law, that is:

$$\dot{q} = -\lambda \, dT/dy$$

where $\dot{q}$ represents the rate of heat transfer per unit surface area, λ is the thermal conductivity, T is the temperature and y is distance measured from the surface. The same expression applies to any region of the flow and also in the case of the adiabatic wall where zero temperature gradient implies zero heat transfer. It should be noted that the surface can be horizontal as shown, with air flow driven by a fan or a liquid flow by a pump, and that it can equally be vertical, with buoyancy providing the driving force for the flow. In the latter case, the free-stream velocity would be zero so that the corresponding profile would have zero values at the wall and far from the wall.

The backward-facing step of **Figure 1** results in a more complicated flow and several boundary layers can be identified within the flow as a consequence of separation and reattachment. The details of flows of this type are not well-understood so it is difficult to identify the characteristics of the boundary layers and it can be imagined that the shapes of the velocity and temperature profiles—and therefore of the local heat transfer within the fluid and to the wall—will vary considerably from one location to another. It is known, for example, that the rate of heat transfer can become high at the location of reattachment of the upstream flow on to the surface of the step, as is also the case at the leading edge of a cylinder in cross-flow, but the detailed mechanisms remain incompletely understood and research continues.

It is well known that even comparatively simple geometrical configurations, such as those of **Figure 1,** can give rise to heat transfer rates which vary considerably depending on the nature of the flow and of the surface. With laminar flows, heat transfer to or from the wall varies with distance from the leading edge of a

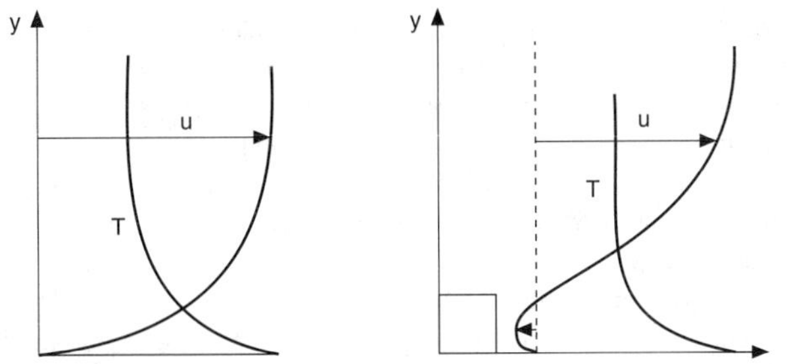

Figure 1. Velocity and temperature profiles in boundary-layer and separated flows.

Table 1. Typical values of heat transfer coefficient

Flow type	α (W/m^{2}K)
Forced convection; low speed flow of air over a surface	10
Forced convection; moderate speed flow of air over a surface	100
Forced convection; moderate speed cross- flow of air over a cylinder	200
Forced convection; moderate flow of water of water in a pipe	3,000
Forced convection; boiling water in a pipe	50,000
Free convection; vertical plate in air with 30°C temperature difference	5

boundary layer. Turbulent flows can give rise to heat transfer rates which are much larger than those of laminar flows, and are caused by the manner in which the turbulent fluctuations increase mixing; they also affect the heat transfer to and from the surface, especially where the free-stream fluid is able to penetrate to the wall even for short periods of time. The nature of the surface, for example the degree or type of roughness, usually affects heat transfer to or from it, and in some circumstances to a large extent. It is convenient, therefore, to represent the heat transfer at the wall by the expression

$$\dot{q} = \alpha(T_w - T_\infty)$$

where $\dot{q}$ again represents the rate of heat transfer from the wall, this time over unit area of surface; the temperature difference refers to that between the wall and the free stream; and α is the 'heat transfer coefficient' which is a characteristic of the flow and of the surface. The two temperatures can vary with x-distance and it can be difficult to identify a free-stream temperature in some complex flows. Typical values of α are shown in **Table 1**, from which it can be seen that increases in velocity generally result in increases in heat transfer coefficient, so that α is smallest in natural convection and increases to 100 and more on flat surfaces with air velocities greater than around 50 m/s. The heat transfer coefficient is considerably greater with liquid flows and greater again with two-phase flows.

It should be noted that the above equations are expressed in terms of dimensional parameters. And it is easy to see that a combination of the two will lead to a nondimensional parameter $\alpha x/\lambda$, where α is a wall heat transfer coefficient, x is a characteristic distance and λ is the conductivity of the fluid; this is known as the Nusselt number and can readily be devised from dimensional analysis as well as from nondimensional forms of conservation equations as suggested in the following section. The heat transfer coefficients of **Table 1** can be expressed in terms of this nondimensional number, the Nusselt number, and analytical and correlation equations are usually expressed in this way as will be shown below.

It is also useful to note that the heat transfer coefficient and the Nusselt number can be used to refer to local values at a location x on a surface, or to an integrated value up to the location x.

The concept of dimensional analysis gives rise to several nondimensional groups, to which reference will be made in this section, and it is convenient to introduce them here. In addition to the Nusselt number, reference will be made to the following:

Prandtl number $Pr = \eta c_p/\lambda$

Reynolds number $Re = \rho ux/\eta$

Nusselt number $Nu = \alpha x/\lambda$

Stanton number $St = \alpha/\rho c_p u = Nu/Pr\ Re$

Grashof number $Gr = g\beta(T_w - T_\infty)y^3/\nu^2$

The Prandtl number is dependent only on fluid properties; the Reynolds number is a ratio of inertial to viscous forces and is relevant throughout the subject of fluid mechanics and convection; the Stanton number is a combination of Nu, Pr and Re; and the Grashof number characterises natural convection with the gravitational acceleration, g, and β, the coefficient of volumetric thermal expansion, and is a combination of inertial, u^2/y, frictional, vu/y^2, and buoyancy, $g\beta\Delta T$, scales. These nondimensional groups may be obtained from conservation equations and are convenient in the representation of results and correlations of experimental data.

Conservation Equations

It is useful to examine the equations which represent conservation of mass, momentum and energy and these are written below for rectangular Cartesian coordinates with simplification of uniform properties.

$$\partial u/\partial x + \partial v/\partial y + \partial w/\partial z = 0$$
$$Du/Dt = \eta/\rho[\partial^2u/\partial x^2 + \partial^2u/\partial y^2 + \partial^2u/\partial z^2] - 1/\rho\ \partial p/\partial x$$
$$Dv/Dt = \eta/\rho[\partial^2v/\partial x^2 + \partial^2v/\partial y^2 + \partial^2v/\partial z^2] - 1/\rho\ \partial p/\partial y$$
$$Dw/Dt = \eta/\rho[\partial^2w/\partial x^2 + \partial^2w/\partial y^2 + \partial^2w/\partial z^2] - 1/\rho\ \partial p/\partial z$$
$$DT/Dt = \lambda/\rho c_p[\partial^2T/\partial x^2 + \partial^2T/\partial y^2 + \partial^2T/\partial z^2] - S$$

where

$$D\phi/Dt = \partial\phi/\partial t + u\partial\phi/\partial x + v\partial\phi/\partial y + w\partial\phi/\partial z.$$

The three equations representing conservation of momentum and the equation representing conservation of energy have the same form, with the terms on the left-hand sides representing convection of momentum and energy. It should be noted that these convective terms are nonlinear, thereby presenting difficulties for any solution and that there are four individual parts of convection corresponding to variations in time and in the three directions. The terms on the right-hand side are slightly simplified forms of those representing transport by diffusion together with pressure forces and sources or sinks of thermal energy. Terms for buoyancy may be added as shown in a following section. It is easy to see that nondimensional velocities and distances in the momentum equations will lead to the inverse of Reynolds number and of the temperatures, velocities and distances in the energy equation to a nondimensional group which comprises (1/PrRe). In later sections, these equations will be simplified to deal with convective heat transfer in steady, laminar flows of forced and free convection.

It is evident from the above that there is some similarity between the equations for conservation of momentum and thermal energy so that the solutions of the two equations will have similar forms when the source terms are zero, the Prandtl number is unity and the solutions are presented in nondimensional form. The presence of buoyancy is often limited to the second momentum equation into which an additional term of the form $\rho\beta g\ (T_w - T_\infty)$ must be added. Where the surface which gives rise to the temperature difference—and therefore to the buoyant force—is not vertical, the angle of the surface to the direction of the gravitational force must be considered. This will lead to the resolution of forces so that part of the buoyancy term will appear in the first momentum equation with that in the second equation, multiplied by the sine of the angle to the vertical. This will give rise to an additional nondimensional group, the Grashof number.

In the absence of convection terms, the energy equation reduces to that for heat conduction and the momentum equations are no longer relevant where the conduction takes place in a stationary material. Many other simplifications of the above equations are possible, including those for two-dimensional flows and for boundary-layer flows, as will be seen below. Also, it is possible to integrate the equations and, in their simpler forms, this can have some merit; for example, in the integral momentum and energy equations where the dependent variable is devised so as to be represented in terms of one independent variable, and therefore solvable by simple numerical methods. More complicated forms may also exist as discussed in the following section.

Laminar and Turbulent Flows

Most flows in nature and in engineering equipment occur at moderately high Reynolds numbers, so they are described as turbulent. Thus, the properties of the flow at any point are time dependent with scales which vary from very small, the Kolmogorov scale, to that corresponding to the largest possible dimension of the flow. In a room, for example, the Kolmogorov scale may be of the order of a fraction of a mm or less than a 1 ms time scale if the velocity is of the order of 1 m/s, and the largest, of the order of several metres or more than 10^3 larger. There are two important implications for this: the first, that the rate of heat transfer from a surface to a flow will be considerably greater than if the flow were laminar at the same Reynolds number; and secondly, that the conservation equations are even more difficult to solve than for the laminar flow since any numerical solution must now consider physical and time scales which encompass three orders of magnitude. The first means that turbulent convection is important, much more important than laminar convection; and the second, that the conservation equations cannot be solved in their general form except where the boundary conditions allow them to be reduced to simpler forms and even then, with additional problems. This conclusion has led to the widespread use of correlation formulae based on measurements and these, of necessity, encompass limited ranges of flow. Some examples are presented and discussed in the following section. It has also led to widespread attempts to solve complicated forms of the conservation equations with assumptions which represent the turbulent aspects of the flow. The following paragraphs provide an introduction to this approach.

The introduction of Reynolds averaging, that is, to rewrite the time-dependent variables as sums of mean and fluctuating components, to introduce the new dependent variable into the conservation equations and to average overall time results in equations of the form:

$$\partial U_i/\partial x_i = 0$$

$$U_j\partial U_i/\partial x_k = \partial/\partial x_k[\nu\partial U_i/\partial x_k - \overline{u_iu_k}] - \partial P/\partial x_k$$

$$U_j\partial T/\partial x_k = \partial/\partial x_k[\kappa\partial T/\partial x_k - \overline{qu_k}] + S$$

where the upper case symbols refer to time-averaged quantities; the lower case, to fluctuating quantities with q, the temperature fluctuations; κ is $\lambda/\rho c_p$; and the overbars, to averages of multiplications of two time-dependent quantities. The equations have been written in tensor notation to render them more compact, but the similarity between the conservation of the time-averaged momentum and energy equations is still evident. The terms representing convection are still on the left-hand side, with diffusion on the right-hand side. There are now two diffusion terms in each equation: one representing laminar diffusion; and the second, the correlations between fluctuating components. There are still five equations, but now there are more than five unknowns since the correlations imply six terms in the momentum equations and three in the energy equation. Thus, it is evident that these equations do not represent a soluble set without assumptions which reduce the number of unknowns to the number of equations. These require models for the Reynolds stresses, $\rho\overline{u_iu_k}$, and the turbulent heat fluxes, $\rho c_p\,\overline{qu_k}$, and, as shown elsewhere, it is possible to derive equations for these correlation terms. Each gives rise to higher order correlations so that a decision must be made about closure as well as the introduction of model assumptions.

By analogy with laminar flow, it is possible to write the turbulent momentum flux and turbulent heat flux in the forms

$$\rho\overline{u_iu_k} = \mu_{turb}\partial U_i/\partial x_k \quad \text{and} \quad -\rho c_p\overline{qu_k} = \lambda_{turb}\partial T/\partial x_k$$

or

$$\overline{u_iu_k} = \nu_{turb}\partial U_i/\partial x_k \quad \text{and} \quad \overline{qu_k} = -\alpha_{turb}\partial T/\partial x_k$$

and nondimensional forms of these expressions with turbulent viscosity and turbulent conductivity will lead to Reynolds and Prandtl numbers, where the latter is frequently referred to as a turbulent Prandtl number.

The turbulent Prandtl number has found considerable use in engineering calculations of convective heat transfer since it can be assigned a value of unity. With the laminar Prandtl number also near unity for air—and often of secondary importance since laminar diffusion is less important than turbulent diffusion—the momentum and energy equations can be solved once for flows where there is no pressure gradient and no sources or sinks of energy, with similar results if presented in nondimensional variables. This approach applies to complex flows with difficult numerical solutions and to simple boundary-layer flows as will be shown.

With assumptions of high Reynolds numbers and local equilibrium, so that the influence of one region of flow on another is small, it is possible to simplify the time-averaged conservation equations. Assuming two-dimensional boundary layers yields:

$$-\overline{uv} = C_\mu \ell_m^2(\partial U/\partial y)^2$$

and

$$-\overline{qu} = C_t\ell_m\ell_t(\partial U/\partial y)(\partial T/\partial y)$$

where C_μ and C_t are constants, ℓ_m is the mixing length for the transfer of momentum and ℓ_t is a corresponding mixing length for the transfer of thermal energy. These equations reduce to the effective viscosity and Prandtl number equations referred to above when the length scales and constants are equal and the Prandtl number is unity. Thus, the concept of a turbulent Prandtl number is limited in its applicability, as is that of a turbulent viscosity. But the range of acceptance for engineering calculations remains large.

Correlations

As will be shown below, the exact solution of the equation appropriate to the laminar flow over a flat plate, where the free-stream and plate temperatures are constant and different, may be written as:

$$Nu_x = 0.331\ Pr^{0.33}Re_x^{0.5},$$

which recognises the importance of the Reynolds and Prandtl numbers and expresses the heat transfer coefficient in terms of the Nusselt number. The corresponding result for laminar natural convection over a vertical plate with similar boundary conditions is:

$$Nu_x = 0.508\ Pr^{0.5}(0.95 + Pr)^{-0.25}Gr_x^{0.25}$$

In turbulent flows, approximations appropriate to a flat plate with forced convection have led to expressions of similar form; for example,

$$Nu_x = 0.0292\ Pr\ Re^{0.8}/[1 + 2.12\ Re_x^{-0.1}(Pr - 1)].$$

As a consequence, equations used to represent measurements of complex flows—where analytic and numerical solutions are either impossible or subject to large inaccuracy—tend to have this form. Several examples are provided in the following sections.

Forced Convective Heat Transfer

Forced convection is associated with flows which are driven by pumps and fans or by the movement of a body through stationary fluids, as in an aeroplane or ship where each has substantial means at its disposal to cause it to move. It is in contrast to natural convection where gravity provides the driving force, although it is possible to have mixed convection in a limited number of flows where the pressure and gravitational forces are of the same order of magnitude, that is Gr/Re^2 is approximately unity. All exact analytical solutions are simplified forms of conservation equations and for laminar flows. Some other cases are discussed below.

Boundary Layer Heat Transfer is discussed in the relevant article.

Heat transfer between parallel plates

The flow between flat plates is portrayed in **Figure 2**. It comprises boundary layers which begin at the leading edges, grow on each of the two surfaces until the potential core narrows to zero and then continues towards a fully-developed laminar flow, after which all gradients in the x-direction become zero.

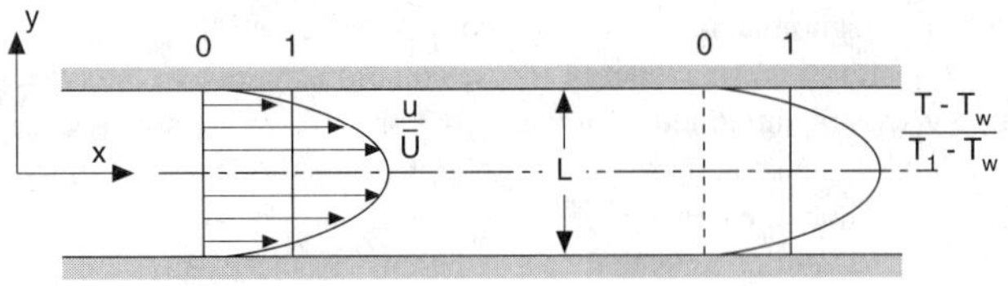

Figure 2. Laminar flow between flat plates.

The boundary layers are represented by the boundary layer equations

$$\partial u/\partial x + \partial v/\partial y = 0$$

$$u\partial u/\partial x + v\partial u/\partial y = \eta/\rho\partial^2 u/\partial y^2 - 1/\rho\ dp/dx$$

$$u\partial T/\partial x + v\partial T/\partial y = \lambda/\rho c_p\partial^2 T/\partial y^2$$

with boundary conditions

$x = 0$, $u = u_1$, $v = o$, $T = T_1$
$y = 0$ and L, $u = 0$, $v = 0$, $T = T_{w1}$ and T_{w2} = constants

and

$y = L/2$

corresponds to a symmetry boundary condition.

In the initial region where the boundary layers are separated by a region of potential flow, the analysis is similar to that for a boundary layer, with the free-stream condition represented by the potential core velocity and temperature. Further downstream, the flow becomes fully-developed so that the velocity and temperature profiles will not change if expressed in terms of appropriate dimensionless quantities. This will be demonstrated below. It is useful to note, however, that there is an intermediate region where there is no potential core and where the flow is not fully-developed. In this region, it is necessary to solve the equations representing conservation of mass and momentum so that each is satisfied; this may require an interactive approach.

In the case of fully-developed laminar flow, the convective terms become zero since

$$\partial u/\partial x = -\partial v/\partial y = 0$$

and the momentum equation becomes

$$0 = \eta/\rho\ \partial^2 u/\partial y^2 - 1/\rho\ dp/dx,$$

with the pressure gradient constant so that integration with the boundary conditions at the wall and on the symmetry line leads to:

$$u = -(L^2/2\eta)dp/dx[y/L(1 - y/L)],$$

and, if one plate moves parallel to the other with a constant velocity U, the solution becomes

$$u = (yU/L) - (L^2/2\xi)dp/dx[y/L(1 - y/L)].$$

In the former case, the temperature profile has the simple form

$$(T - T_{w1})/(T_{w2} - T_{w1}) = y/L.$$

This, too, may be complicated by considering the effect of viscous heating, which requires the addition of a term of the form $\eta(\partial u/\partial y)^2$ to the conservation equation for energy, and—for zero pressure gradient and constant values of U—leads to

$$u/U = yU/L$$

and

$$(T - T_{w1})/(T_{w2} - T_{w1}) = qy/L$$
$$+ [\eta U^2/\{2\lambda(T_w - T_1)\}][y/L(1 - y/L)].$$

This last result must be regarded as an approximation since no account has been taken of variations likely to occur in transport and thermodynamic properties.

The velocity profile for fully-developed laminar flow is a parabola when the walls are stationary, provided that the fluid properties are constant and the velocities are low; it is linear when the pressure gradient is absent and the wall moves with constant speed with respect to the other. The effect of a moving surface is to provide a force which can act against or with that exerted by pressure. This is reflected in the velocities which can be in positive and negative directions. The temperature profile is expressed in terms of surface temperatures and it is clear that the bulk temperature will increase if one or both of the walls is hotter than the initial temperature, T_1. Thus, the temperature profile is often expressed in terms of the initial temperature and the bulk-mean temperature, defined as:

$$T_b = (1/LU)\int uT\ dx$$

where U is the bulk velocity as discussed further below.

Heat transfer in a pipe

Flow and heat transfer in a pipe are of rather more importance than those between parallel plates since they are found more frequently in engineering practice. The flow may again begin at the leading edge so that laminar flow solutions can be obtained as for parallel plates, but this time to equations in cylindrical coordinates and without the prospect of one surface moving with respect to another. At small values of Reynolds number, $\rho ud/\eta$, the length required to achieve fully-developed laminar flow may be given by the expression

$$l_{fd}/d = 0.06\ Re_d$$

and originates from asymptotic solutions of the boundary layer equations. The flow in small-diameter pipes required to achieve these small Reynolds numbers comes from larger diameter pipes or from plenum chambers, so it is likely that boundary layers do not have their origins at the beginning of the small diameter pipe. Rather, there is a sudden contraction for which the flow is properly represented by more complete forms of the conservation equations than their boundary-layer forms. Indeed, the flow may separate inside the pipe with a more rapid movement towards fully-developed conditions than would be the case with attached boundary layers.

The region of developing flow can be small in many cases and fully-developed flow is usually more important than the developing flow. The conservation equations in cylindrical coordinates may be reduced for fully-developed flow in the same way as between two plates, with the result

$$0 = -dp/dx + \eta/r\ d/dr\ (r du/dr)$$

and this, with boundary conditions

$$r = 0, \qquad du/dr = 0,$$

and

$$r = R, \qquad u = 0$$

may be integrated to yield

$$u/U = 2[1 - (r/R)^2],$$

where

$$U = 1/\pi R^2 \int 2\pi r u\ dr = \text{constant}.$$

The Moody friction factor, defined as $f = -(dp/dx)/0.5\ \rho U^2/D$, commonly represents the relationship between pressure drop, geometry and fluid properties, and may be deduced for fully-developed laminar pipe flow as:

$$f = 64/Re_D,$$

which is sometimes referred to as the Hagen-Poiseuille friction law.

The energy equation in cylindrical coordinates has the form

$$u\partial T/\partial x + v\partial T/\partial r = \lambda/\rho c_p(\partial^2 T/\partial x^2 + \partial^2 T/\partial r^2 + 1/r\ \partial T/\partial r)$$

and this reduces for fully-developed flow to

$$u\partial T_b/\partial x = (\lambda/\rho c_p r)\ \partial/\partial r(r\partial T/\partial r),$$

where T_b is the bulk temperature, defined as:

$$T_b = (1/\pi r^2 U) \int 2\pi r u T\ dr.$$

Integration of the differential equation with boundary conditions corresponding to symmetry at the centre line, and for the particular condition that

$$\partial T/\partial x = \partial T_w/\partial x = \partial T_b/\partial x = 0$$

leads to

$$T/T_w = (2U/\alpha)\partial T/\partial x[3R^2/16 + r^2/4 + r^4/16R^2]$$

and to

$$Nu_d = 4.364,$$

which is independent of the Reynolds and Prandtl numbers, provided the flow remains laminar. An iterative solution is required to solve the equations for the boundary condition

$$T_w = \text{constant}$$

and leads to the result

$$Nu_d = 3.658,$$

which shows that the solution depends on the thermal boundary condition.

Of course, the flow will remain laminar only if the Reynolds number is less than around 2 300 or to larger values if it is so free from disturbances that none are available to propagate and cause turbulent flow, as is usually the case. Where turbulent flow occurs because a disturbance has propagated and led to fluctuations in all regions of flow except in the viscous sublayer, the nature of the flow and of the problem has changed. It is possible to return to considering the consequences of the onset of transition and of the transitional region in the context of the boundary layer in the entrance region of the pipe. But the overall effect will be to induce turbulent flow rapidly so that the emphasis is again, and even more so, on the region of fully-developed flow—which now corresponds to turbulent and not laminar flow. It is possible to retain a boundary-layer flow, possibly with transitional regions for some distance, but the common shape of slightly-rounded entrance geometry usually leads to a fully-developed turbulent flow in distances not much more than 50 diameters, and in shorter distances for engineering calculations.

Correlation of measurements of pressure drop against bulk velocity and diameter has led Blasius to propose the expression

$$f = 0.316\ Re_D^{-0.25}$$

which, together with the laminar flow result of

$$f = 64\ Re_D^{-1}$$

allow **Figure 3** to be drawn and in which the laminar-flow result can be extended to Reynolds numbers well in excess of 10^5, provided care is taken with the nature of the initial conditions, with the smooth surface of the pipe and with the absence of disturbances of any kind. More usually, laminar flow does not exist at Reynolds numbers larger than 2,300, above which transition to the turbulent-flow curve takes place with a transitional region which can be short or long depending on the nature of the disturbances. The friction factor thus varies with the Reynolds number, based on the diameter of the pipe, and with laminar, transitional and turbulent regions as shown on **Figure 3**.

The skin-friction coefficient (or Fanning friction factor) is related to the friction factor by

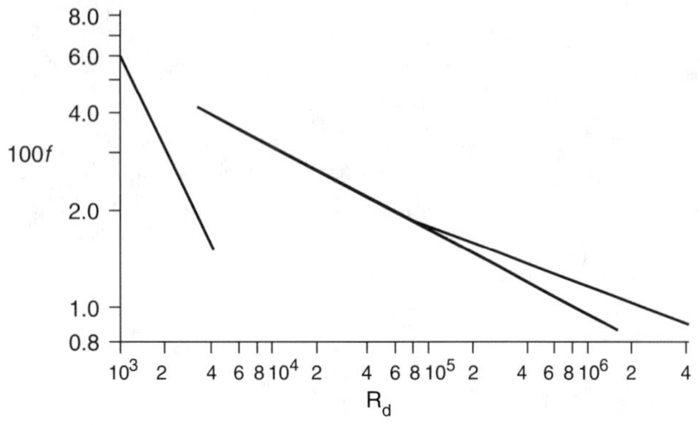

Figure 3. Variation of friction factor with Reynolds number for flow in a pipe.

$$C_f = \tau_w/0.5\rho U^2 = 0.25\ f$$

so that the coefficient for turbulent flow may be expressed as

$$C_f = 0.079\ Re_D^{-0.25}$$

with the constants stemming from consideration of experimental results and are, therefore, of limited applicability. **Figure 3** shows the variation of friction factor with Reynolds number based on the pipe diameter, and the distinction between those for laminar and turbulent flow. At high Reynolds number, the results become less certain as indicated by the two lines, but the graph is adequate for many design purposes.

Consideration of the similar nature of the equations representing conservation of momentum and energy implies that the variation of Nusselt number will also be dependent upon Reynolds number, together with the Prandtl number where it is different from unity. An example of an expression describing the variation of Nusselt number with turbulent flow in a pipe is:

$$Nu_s = 0.04\ Re_D^{0.75} Pr^{0.33}.$$

As with **Figure 3** and the friction factor and skin-friction coefficient, uncertainty increases at high Reynolds numbers and also in the transitional region where the difference between the results for laminar and turbulent flows are widely divergent. This may occur over a range of Reynolds numbers depending on the initial and boundary conditions. It should be noted that rough surfaces increase the values of skin-friction coefficient and Nusselt number. Related calculations can be made for noncircular ducts with an hydraulic diameter replacing the geometric diameter.

Leading to: *Heat transfer coefficient; Tubes, single phase heat transfer in; Coupled radiation and convection; Coupled radiation, convection and conduction; Free convection; Liquid metal heat transfer; Mixed (combined) convection*

J.H. Whitelaw

CONVECTIVE BOILING (see Forced convective boiling)

CONVERGENCE OF SERIES (see Power series; Series expansion)

CONVERSION FACTORS

Following from: *SI units*

The foot, pound, second (fps) system of units is widely-used, particularly in the UK and the USA, in many industries and in most public transactions. When SI units are used, fps units are often offered as an alternative. In this section, fps units are listed and conversion factors relevant to heat and mass transfer are tabulated.

The **foot,** ft, is one third of the **Imperial Standard yard,** yd, which is defined as 0.9144 metre exactly. The foot is therefore 0.3048 metre.

The **pound,** lb, is defined as 0.453 592 37 kilogram.

The **gallon,** gal, is a unit of volume. The British gallon is defined as the volume of 10 lbs of water of density 988.859 kg/m^3 weighed in air with a density of 1.217 kg/m^3 against weight of density 8 136 kg/m^3, i.e., 1 gal = 4.546 09 × 10^{-3} m^3. The US gallon is smaller than the British; 1 USA gallon = 3.785 412 × 10^{-3} m^3 = 0.8327 gal.

Table 1 gives a list of the most common secondary units in the British System, and **Table 2** lists factors for converting fps units to SI units and vice-versa. (See Table 2 on pages 244 and 245).

Table 1. Secondary units in the British FPS system

Units of Length	
12 inches	= 1 foot (ft)
3 feet	= 1 yard (yd)
22 yards	= 1 chain
10 chains	= 1 furlong
8 furlongs or 1760 yards	= 1 mile (mi)
6080 feet	= 1 UK nautical mile*
6 feet	= 1 fathom
Units of Mass	
16 ounces (oz)	= 1 pound (lb)
14 pounds (lb)	= 1 stone
28 pounds	= 1 quarter
4 quarters or 112 pounds	= 1 hundredweight
20 hundredweight (cwt) or 2240 lb	= 1 ton
Units of Area	
4840 square yards	= 1 acre
640 acres	= 1 square mile
Units of Volume	
20 fluid ounces (fl. oz)	= 1 pint
2 pints (pt)	= 1 quart
4 quarts (qt)	= 1 gallon

*The Nautical mile is the average distance on the earth's surface subtended by one minute of latitude. The UK nautical mile is 6080 ft but the International nautical mile, which is used by the Admiralty and most other nations, measures 1852 m.

Table 2. Conversion factors, FPS/SI

Physical quantity	Given in Gives	Multiplied by Divided by	Gives Given in	Approximate or useful relationship
Length	ft	0.3048	m	$3\frac{1}{4}$ ft ≃ 1 m
	in	25.4 (exact)	mm	1 in ≃ 25 mm
	mil	0.0254	mm	
	yard	0.9144	m	
	mile (mi)	1609.3	m	1 mi ≃ 1.6 km
	km	0.621388	mi	
Area	ft^2	0.092903	m^2	100 ft^2 ≃ 9 m^2
	in^2	645.16	mm^2	1 in^2 ≃ 650 mm^2
	acre	4 047.0	m^2	
Volume	ft^3	0.028317	m^3	35 ft^3 ≃ 1 m^3
	U.S. gal	0.003785	m^3	260 gal ≃ 1 m^3
	U.S. gal	3.785	liter (L)	1 gal ≃ $3\frac{3}{4}$L
	L (liter)	0.2642	U.S. gal	1 L ≃ 0.26 gal
	Brit. gal	0.004546	m^3	
	U.S. gal	0.13368	ft^3	
	barrel (U.S. pet.)	0.15898	m^3	
	barrel (U.S. pet.)	42	U.S. gal	
Velocity	ft/s[a]	0.3048	m/s	10 ft/s ≃ 3 m/s
	m/s	3.2808	ft/s	
	ft/min	0.00508	m/s	100 ft/min ≃ 0.5 m/s
	mi/h	1.6093	km/h	30 mi/h ≃ 48 km/h
	km/h	0.6214	mi/h	50 km/h ≃ 31 mi/h
	knots	1.852	km/h	
Mass	lb_m	0.45359	kg	1 lb_m ≃ .45 kg
	kg	2.2046	lb_m	1 kg ≃ 2.2 lb_m
	metric ton	2 204.6	lb_m	metric ton = 10^3 kg
	ton (2 000 lb_m)	907.18	kg	
Force	lb_f	4.44822	N = kg m/s^2	
	lb_f	0.45359	kg_f	1 N ≈ 0.1 kg_f
	kg_f	2.2046	lb_f	≈ 0.22 lb_f
	kg_f	9.80665	N	
	dyne	0.00001 (exact)	N	
Amount of substance	lb_m-mol	453.6	kmol	
	g-mol	1.000	mol	
	kg-mol	1.000	kmol	
	mol	1 000	kmol	
Mass flow rate	lb_m/h	0.0001260	kg/s	10^3 lb/h ≃ .13 kg/s
	kg/s	7 936.51	lb_m/h	
	lb_m/s	0.4536	kg/s	
	lb_m/min	0.00756	kg/s	
Volume flow rate	U.S. gal/min	6.309×10^{-5}	m^3/s	
	U.S. bbl/day	0.15899	m^3/day	
	U.S. bbl/day	1.84×10^{-6}	m^3/s	
	ft^3/s	0.02832	m^3/s	
	ft^3/min	0.000472	m^3/s	
Mass velocity	lb_m/h ft^2	1.356×10^{-3}	kg/s m^2	
(mass flux)	kg/s m^2	737.5	lb_m/h ft^2	
Energy (work)	Btu[b]	1 055.056	J = N m = W s	1 Btu ≃ 1 000 J
(heat)	Btu	0.2520	kcal	1 kcal ≃ 4 Btu
	Btu	778.28	ft lb_f	
	kcal	4 186.8	J	1 kcal ≃ 4 000 J
	ft lb_f	1.3558	J	
	W h	3 600	J	
Power	Btu/h	0.2931	W = J/s	10^6 Btu/h ≃ 300 kW
	W	3.4118	Btu/h	
	kcal/h	1.163	W	
	ft lb_f/s	1.3558	W	1 000 kW ≃ 3.5×10^6 Btu/h
	hp (metric)	735.5	W	
	Btu/h	0.2520	kcal/h	
	tons refrig.	3 516.9	W	

Table 2. Conversion factors, FPS/SI—*continued*

Physical quantity	Given in Gives	Multiplied by Divided by	Gives Given in	Approximate or useful relationship
Heat flux	Btu/h ft^2	3.1546	W/m^2	1 000 Btu/h ft^2 ≃ 3.2
	W/m^2	0.317	Btu/h ft^2	kW/m^2
	kcal/cm^2 s	41.868	W/m^2	
Heat transfer coefficient	Btu/h ft^2 °F	5.6784	W/m^2 K	1 000 Btu/h ft^2 °F ≃
	W/m^2 K	0.1761	Btu/h ft^2 °F	5 600 W/m^2 K
	kcal/cm^2 s °C	41.868	W/m^2 K	
Heat transfer resistance	(Btu/h ft^2 °F)$^{-1}$	0.1761	(W/m^2 K)$^{-1}$	0.001 (Btu/h ft^2 °F)$^{-1}$ ≃
	(W/m^2 K)$^{-1}$	5.6784	(Btu/h ft^2 °F)$^{-1}$	0.000 18 (W/m^2 K)$^{-1}$
Pressure	lb$_f$/in^2 (psi)	6.8948	kN/m^2 = kPa	1 psi ≃ 7 kPa
	kPa	0.1450	psi	14.5 psi ≃ 100 kPa
	bar	100	kPa	
	lb$_f$/ft^2	0.0479	kPa	
	mm Hg (torr)	0.1333	kPa	1 000 kPa = 1 MPa ≃
	in Hg	3.3866	kPa	150 psi
	mm H_2O	9.8067	Pa	
	in H_2O	249.09	Pa	1 in H_2O ≃ .25 kPa
	at (kg$_f$/cm^2)	98.0665	kPa	
	atm (normal)	101.325	kPa	atm = 760 mmHg
Mass flux	lb$_m$/ft^2 s	4.8824	kg/m^2 s	
	lb$_m$/ft^2 h	0.001356	kg/m^2 s	
		Physical and Transport Properties		
Thermal conductivity	Btu/ft h °F	1.7308	W/m K	steel ≃ 50 W/m K
	W/m K	0.5778	Btu/ft h °F	water (20°C) ≈ 0.6 W/m K
	kcal/m h °C	1.163	W/m K	air (STP) ≃ 24 mW/m K
Density	lb$_m$/ft^3	16.0185	kg/m^3	
	kg/m^3	0.06243	lb$_m$/ft^3	62.4 lb$_m$/ft^3 ≃ 1 000
	lb$_m$/U.S. gal	119.7	kg/m^3	kg/m^3
Specific heat capacity	Btu/lb$_m$ °F	4 186.8	J/kg K	1 Btu/lb$_m$ °F ≃ 4.2
	kcal/kg °C	4 186.8	J/kg K	kJ/kg K
Enthalpy	Btu/lb$_m$	2 326	J/kg	
	kcal/kg$_m$	4 186.8	J/kg	
Dynamic (absolute) viscosity	centipoise (cP)	0.001	kg/m s	kg/m s = N s/m^2 = Pa s
	poise (P)	0.1	Pa s	
	cP	1.000	mPa s	
	cP	1 000	μPa s	water (100°C), 0.31 cP
	lb$_m$/ft h	0.0004134	Pa s	
	lb$_m$/ft h	0.4134	cP	
	cP	2.4189	lb$_m$/ft h	air (100°C), 0.021 cP
	lb$_m$/ft s	1.4482	Pa s	
Kinematic viscosity	stoke (St), cm^2 s	0.0001	m^2/s	
	centistoke (cSt)	10^{-6}	m^2/s	
	ft^2/s	0.092903	m^2/s	
Diffusivity	ft^2/s	0.092903	m^2/s	
Thermal diffusivity	m^2/h	0.0002778	m^2/s	
	ft^2/s	0.092903	m^2/s	
	ft^2/h	25.81×10^{-6}	m^2/s	
Surface tension	dyne/cm	0.001	N/m	
	dyne/cm	6.852×10^{-5}	lb$_f$/ft	
	lb$_f$/ft	14.954	N/m	

Temperature relations: $°C = \frac{5}{9}[°F - 32]$ $\quad °C = (°F + 40)\frac{5}{9} - 40$ $\quad \Delta T(°C) = \frac{9}{5}\Delta T(°F)$ $\quad K = °C + 273.15$

$°F = \frac{9}{5}(°C) + 32$ $\quad °F = (°C + 40)\frac{9}{5} - 40$ $\quad \Delta T(°F) = \frac{5}{9}\Delta T(°C)$ $\quad R = °F + 459.67$

Miscellaneous:

Acceleration of gravity (standard):	g = 9.806 65 m/s^2
Gas constant:	R = 8 314.3 m N/K kmol
Stefan-Boltzmann constant:	$5.669\,7 \times 10^{-8}$ W/m^2 K^4
	1.714×10^{-9} Btu/ft^2 h R^4

[a]Even though the abbreviations s and h were introduced only with the SI, they are used here throughout for consistency.

[b]Note: the calorie and Btu are based on the International Standard Table values. The thermochemical calorie equals 4.184 J (exact) and is used in some older texts.

Source: J. Taborek, HEDH, 1984.

Reference

Taborek, J. (1984) *Heat Exchange Design Handbook*, Hemisphere Publishing Corp.

G.L. Shires

COOLANTS, REACTOR

Following from: *Nuclear reactors*

A reactor coolant picks up the heat from fuel elements in the reactor core and carries it to the boilers where it gives up that heat to water carried in pipes to turn it into steam. The efficiency of this process depends upon the flow rate of the coolant, the flow cross-section of the fuel element, the difference in temperature between the cladding and the coolant and the choice of the material cladding the fuel. The fuel and its cladding are designed so they operate at the highest satisfactory and safe temperature. Similarly, the coolant is heated to the highest acceptable temperature to increase the plant's efficiency by providing steam to the turbines at high temperature and pressure. A special case is that in which the coolant is in the form of a boiling liquid. The vapour of the *boiling fluid coolant* is often used as the working fluid in the turbine [e.g., steam generated within a boiling water-cooled reactor is used directly in a steam turbine].

The general features that make a particular gas or liquid attractive as a reactor coolant are as follows:

- Chemically compatible and noncorrosive to circuit materials;
- Cheap and readily-available in pure state;
- Good nuclear properties and stable under radiation;
- Well-defined phase state and high boiling point;
- High rates of heat transfer;
- High specific heat and ease of pumping.

No practical fluids meet all these requirements. All known coolants have one or more disadvantages. The thermodynamic and heat transfer characteristics of a coolant can be compared conveniently by using a parameter called the *figure of merit* [F], which derives from the heat transfer process and the associated pumping power required

$$F = \frac{C_p^{2.8}\rho^2}{\mu^{0.2}}$$

Table 1 lists possible reactor coolants and their figure of merit. Of those listed, water is the most common. It has three main disadvantages: its low boiling point [which requires operation at high pressure]; its neutron absorption [which requires enrichment of the fuel]; and its corrosive nature [which requires specific steels and cladding]. Of the gases, carbon dioxide and helium have been widely-used. Liquid metals, whilst being excellent coolants, present a whole range of novel handling problems because of their chemical reactivity. Their use has been mainly restricted to fast reactors.

Table 1. Figure of merit for physical properties of coolants

Material	Figure of merit (Relative to Na)
Liquids	
Water, H_2O or D_2O (300 °C)	60
Sodium hydroxide, NaOH (350 °C)	13
Dowtherm oil (300 °C)	7
Potassium and lithium chlorides, KCl, LiCl (400 °C)	2.5
Sodium, Na (300 °C)	1
Tin, Sn (300 °C)	0.45
Mercury, Hg (300 °C)	0.31
Lead, Pb (500 °C)	0.27
Bismuth, Bi (300 °C)	0.21
Potassium, K (300 °C)	0.21
Gases	
Carbon Dioxide, CO_2(300 °C, 5 MPa)†	2.8×10^{-3}
Helium, He (300 °C, 5 MPa)†	1.9×10^{-3}
Hydrogen, H_2 (300 °C, 1 MPa)†	4.0×10^{-4}
Carbon dioxide, CO_2 (300 °C, 1 MPa)†	1.1×10^{-4}
Helium, He (300 °C, 1 MPa)†	7.6×10^{-5}
Air (300 °C, 1 MPa)†	4.5×10^{-5}

† 1 MPa is approximately equal to 10 times atmospheric pressure.

References

Collier, J. G. and Hewitt, G. F. (1987) *Introduction to Nuclear Power,* Hemisphere Publishing Corp.

Nuclear Power Technology, Volume 1, Ed. by W. Marshall (1983) Clarendon Press. Oxford.

J.G. Collier

COOLING TOWERS

It is essential for the operation of a wide variety of industrial plants that heat be transferred in some way from the plants to their surroundings. Sometimes, this is done by direct convection to the atmosphere, as in an air-cooled internal combustion engine. More frequently, a stream of water, "cooling water", is used. If a plant is near a river, a lake or the sea, an abundant supply of such cooling water can be supplied to it at a low temperature and returned to where it came from at a somewhat higher temperature. But often the choice of site for a plant is dominated by other factors, such as the location of raw materials, coal for example. There may be a river or a lake nearby but it may be unacceptable to raise the temperature of its waters any further. In such cases, the cooling water itself has to be cooled after use by heat transfer to the atmosphere and returned to the plant for reuse. Cooling towers are designed for precisely this purpose: to transfer heat from a stream of cooling water to the atmosphere.

In most large cooling towers, the cooling water and air are in *direct contact* (**Figure 1**). The cooling water is pumped into a system of sprayer pipes and nozzles within the tower and is drawn by gravity into a pond below. Air from the atmosphere enters the base of the tower and flows through the falling water. A "packing"

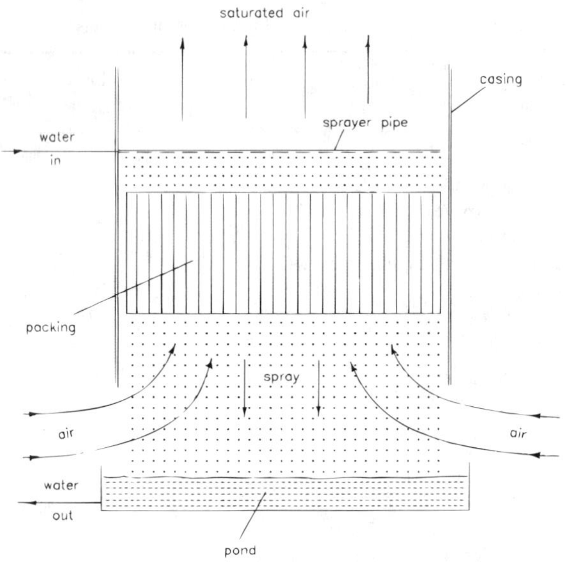

Figure 1. Essential features of a cooling tower.

is provided across the whole cross-section of the tower to ensure that the water, in its descent, presents a large surface area to the air. Water flows as a film or splashes and trickles over the surfaces of the packing while air flows between the surfaces. Depending on the particular design, air may flow predominantly upwards (counter-flow) or horizontally (cross-flow). As water temperature decreases, air temperature increases. When, as in most towers, the water and air are in direct contact, the "driving force" for heat transfer is not the local temperature difference but a local enthalpy difference; namely, the difference ($h_S - h_G$) where h_S is the specific enthalpy of saturated air at the local temperature of the water/air interface and h_G is the local bulk specific enthalpy of the moist air. The local heat flux from the water to the air is the product of this driving force and an empirical coefficient having the units of a mass flux. It is several times greater than it would be if the local temperature difference were the driving force, as is the case with "dry" heat transfer where the fluids are separated by a solid wall and no evaporation occurs.

Flow of air through the tower can be created in two ways: by natural draft or by mechanical draft. In natural draft towers, the packing region is located inside the base of what is, in effect, a large chimney (**Figure 2**). After its contact with warm cooling water, the density of the air above the packing is 5% or so less than that of the atmosphere. The difference in weight between the air in the "chimney" and the air outside provides the driving force to overcome pressure losses that resist the flow of air through the tower. The typical "hyperbolic" profile of such towers (as sketched in **Figure 2**) is chosen mainly for structural and economic reasons—it is much more resistant to wind-induced stress and vibration than a plain cylindrical shell and requires less material.

Mechanical draft towers use fans, driven by electric motors, to produce the flow of air (**Figure 3**). The tower is called "forced draft" when the fan is located in the air entry at the base of the tower; "induced draft" when the fan is located in the air exit at the top of the tower. Both centrifugal and axial flow fans are used, but axial fans are usual with induced draft.

The relative merits of natural and mechanical draft towers are summarised in **Table 1**. In some situations, there may be an economic compromise in the form of an "assisted draft" tower—a natural draft tower with fans added around the base in the air entry, the size of the tower being much less than would be necessary without the fans.

In the majority of towers, i.e., direct contact or "wet" towers of the type discussed above, about 1% of the water flow rate is lost to the atmosphere by evaporation and by entrainment of fine droplets of water by the air. (The larger droplets are intercepted by "eliminators," typically an array of slats covering the whole cross-section of the tower just above the water spray system). This loss has to be supplemented by an external supply and, to keep the concentration of salts in the water acceptably low, the supply of make-up water may have to be around 3%. In some locations, such a quantity may be too difficult or too costly to provide and a "dry" tower may be economic.

In dry towers, the water and air are not in direct contact so there is no loss of water by evaporation. In effect, the packing is replaced by a heat exchanger in which metal walls separate the two streams. Dry towers may be natural, mechanical or assisted draft. But there is a big penalty for eliminating water loss by evaporation. The evaporative cooling effect is also eliminated and it is estimated that the air flow rate required to achieve the same cooling capacity will have to be three or more times greater than in a wet tower; so the tower will have to be much larger and much more expensive. Moreover, the set of heat exchangers for the tower will be more costly to produce than a corresponding packing. "Wet-dry" towers have been proposed as a compromise between wet and dry towers. [Singham (1990)]

A wide variety of materials and geometric design have been used for packing: corrugated roofing sheets made of cement-based or plastic material; timber laths of triangular or rectangular cross-section; plastic-impregnated paper "honeycomb"; complex cellular geometries made of thin plastic material. [Singham (1990); Hill, Pring & Osborne (1990) and Johnson (1990)]

The required performance of a proposed cooling tower can be specified by listing the values of the following five quantities:

1. the mass-flow rate of the water;
2. the inlet temperature of the water;
3. the exit temperature of the water;
4. the atmospheric wet-bulb temperature;
5. the atmospheric dry-bulb temperature.

Except in special circumstances, atmospheric pressure is assumed to be the standard atmospheric pressure. The mass flow rate of the air is not specified and it is the first task of the designer to determine its value. For prediction of the performance of an existing tower, the mass flow rate of the air replaces the exit water temperature in the above list and it is the latter that has to be determined. *Merkel's Equation* enables both these calculations—for proposed and for existing towers—to be carried out. It relates the above variables to variables associated with the heat transfer performance of the proposed or existing packing. It is based on the assumption of uniform, one-dimensional counter-flow, but can

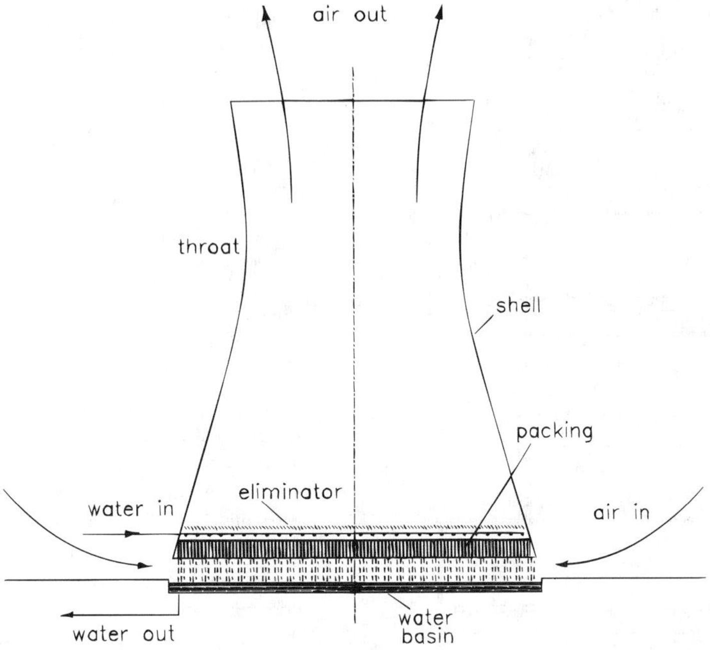

Figure 2. Natural draft cooling tower.

be adapted to other conditions. The equation can be stated as [Singham (1990)]:

$$I_M = I_P \qquad \textbf{(1)}$$

where

$$I_M \equiv \int_{h_{w,out}}^{h_{w,in}} \frac{dh_w}{h_S - h_G} \qquad \textbf{(2)}$$

and

$$I_P \equiv \frac{\dot{m}_G}{\dot{m}_W} \text{NTU} \qquad \textbf{(3)}$$

In Equations (2) and (3),

h_W = specific enthalpy of the water at any level;
h_G = specific enthalpy of bulk moist air at the same level;
h_S = specific enthalpy of saturated air at temperature of water at same level;
$\dot{m}_W$ = mass flow rate of water per unit area;
$\dot{m}_G$ = mass flow rate of moist air ("gas") per unit area;
NTU = "*number of transfer units*" for the packing.

The quantity I_M, defined by Equation (2), is a measure of the cooling requirement. The quantity I_P, defined by Equation (3), is a measure of the performance of the packing. Merkel's equation, Equation (1), requires them to be equal.

NTU is an empirical quantity and may be thought of as the product of a **Stanton Number** and a geometric feature of the packing, namely, the surface area of the water-air interface per unit plan area of packing. Often packing performance is expressed instead as another empirical group for which the conventional symbols are "KaV/L". The relation between NTU and KaV/L is a simple one [Singham (1990)]. (See also NTU.)

With the above five design quantities specified and with enthalpy property data for water and moist air, I_M can be evaluated for any value of the water/air mass flow ratio. The performance of a chosen packing, expressed as I_p, is also an empirical function of flow ratio. Thus the equality of these two quantities, required by Merkel's equation, allows flow ratio to be found. Since the water mass flow rate is specified, air mass flow rate is now known, without need for other details concerning tower design. This is true for both natural and mechanical draft towers. The rest of the design procedure, however, depends on which type of tower is under consideration. For a natural draft tower, the height and cross-sectional area of the tower have to be such as to make the sum of the estimated pressure losses that occur with this now-known air flow rate equal to the driving pressure difference caused by the buoyancy of warm moist air. For a mechanical draft tower, a fan has to be selected to supply the now-known air flow rate with a pressure rise equal to the estimated pressure loss.

Any design which satisfies these heat transfer and pressure loss requirements is technically valid. An infinite number of designs is possible, differing from one another in shape, height, cross-sectional areas, type of packing, depth of packing, etc. The final choice is made after taking into account economic, environmental and operational considerations, often in conjunction with some cost-optimisation procedure.

With an existing tower a slightly different calculation procedure is required. The aim is to determine the value of the third variable in the above list—water exit temperature—for any set of the other four that may be of interest. The first step is to determine air mass flow rate by matching buoyancy and pressure loss for a natural draft tower; or fan pressure rise and pressure loss for mechanical

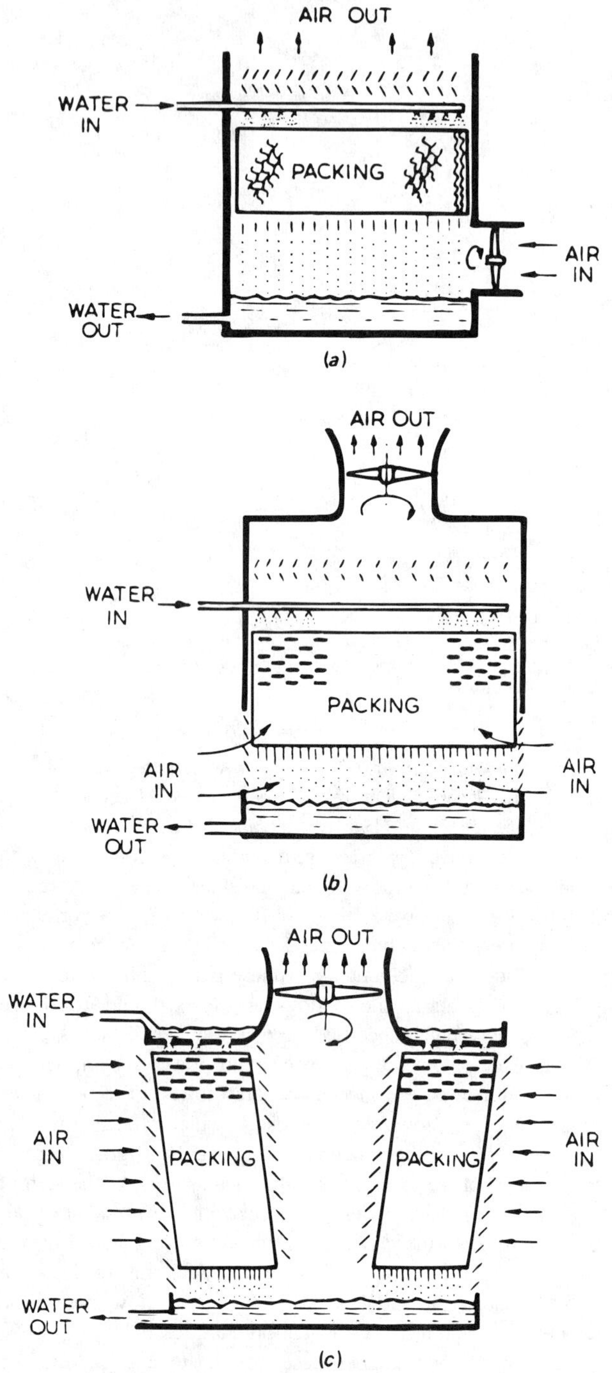

Figure 3. Mechanical draft cooling tower layouts. (a) Forced draft, counter flow; (b) Induced draft, mixed flow; (c) Induced draft, cross flow. From Singham, J. R. (1990) in Heat Exchanger Design Handbook, Hemisphere.

draft. Since the water mass flow rate is already known (specified), the water/air mass flow ratio can be calculated and used to find I_p from the known characteristics of the packing; and hence I_M, from Merkel's equation. The final step is similar to that outlined above: find by trial the value of exit water temperature that results in the required value of I_M.

Table 1. Comparison of natural and mechanical draft

	Natural Draft	Mechanical Draft
Height	Tall	Short
Construction	Much on-site work	Prefabrication possible
Noise	Quiet	Noisy
Air Recirculation	No problem	Can be serious
Range of Size	Small size uneconomic	Large size range
Initial Cost	Higher	Lower
Running cost	Lower	Higher

It is well-established that air and water flows in real towers are far from the uniform one-dimensional flows often assumed in design and analysis. The discrepancy can be taken into account by the introduction of correction coefficients whose values are estimated from full-scale test data on towers. Such data are in short supply and even when available, there remains some doubt as to the validity of the correction; some "over-design" may be prudent depending on contractual terms. Another approach is to return to the fundamental equations of fluid mechanics and heat and mass transfer and to arrive at numerical solutions with the aid of **Computational Fluid Dynamics** (CFD) techniques. These solutions can, in principle, be used as the sole basis of design; or they can be used to modify and improve existing simpler methods.

References

Singham, J. R. (1990) Hemisphere Handbook of Heat Exchanger Design, Section 3.12 (Cooling Towers), Hemisphere Publishing Corporation, New York.

Hill, G. B., Pring, E. J., and Osborn, P. D. (1990) Cooling Towers: Principles and Practice, Third Edition, Butterworth-Heinemann, London.

Johnson, T. (1990) *Plastic Packings for Large Cooling Towers,* Chem. Eng. (London), pp. 18–24.

British Standard 4485, (1988) Water Cooling Towers, Part 2, *Methods for Performance Testing*; Part 3, *Code of Practice for Functional and Thermal Design.*

Bibliography of Technical Papers (1993) Cooling Tower Institute, Houston, Texas 77273, USA.

J.R. Singham

COOPER CORRELATION, FOR NUCLEATE BOILING (see Boiling)

COORDINATE SYSTEM

A coordinate system is a system of surfaces in n-dimensional space which allows the association of every point in that space point with an ordered set of numbers ($\xi_1, \xi_2, \ldots, \xi_n$) (in customary three-dimensional space—three numbers (ξ_1, ξ_2, ξ_3). The proper choice of a coordinate system in a mathematical description of a process in some volume can significantly simplify its' mathematical formulation and problem solution.

Cartesian (rectangular) coordinates are formed by three families of self-perpendicular planes: x = *const;* y = *const;* z = *const.*

If one puts three other families of surfaces, which are not parallel with each other, on this system, then the location of every points (x,y,z) can be defined by a crossing of three surfaces from these families, which form a new system of curvilinear coordinates. If the surfaces of curvilinear coordinates are described by the equations $\xi_1(x,y,z) = const$, $\xi_2(x,y,z) = const$, $\xi_3(x,y,z) = const$, the location of some point is fixed by coordinates (ξ_1, ξ_2, ξ_3), as well as by coordinates (x,y,z). The Jacobian transformation $(x,y,z) \rightarrow (\xi_1, \xi_2, \xi_3)$, which is a coefficient of an elementary volume change through the transformation, isn't equal to zero.

The distance between two close points is predicted as:

$$dr = \sqrt{dx^2 + dy^2 + dz^2} = \sqrt{\sum_{i,j=1}^{3} H_{ij} d\xi_i d\xi_j}$$

where H_{ij} are the Lamé coefficients, describing the metrics of the coordinate system.

For orthogonal coordinate systems with self-perpendicular coordinate surfaces, $H_{ij} = 0$ for $i \neq j$ and

$$dr^2 = \sum_{i=1}^{3} H_i^2 \, d\xi_i^2,$$

where $H_i = H_{ii}$. Then, the values of surface and volume elements are:

$$d\sigma_{ij} = dS_i dS_j = H_i H_j \, d\xi_i \, d\xi_j,$$

$$dv = dS_1 \, dS_2 \, dS_3 = H_1 H_2 H_3 \, d\xi_1 \, d\xi_2 \, d\xi_3.$$

In orthogonal coordinate systems, a function gradient is written as:

$$\vec{\nabla}\varphi(\xi_1, \xi_2, \xi_3) = \vec{e}_1 \frac{1}{H_1} \frac{\partial \varphi}{\partial \xi_1} + \vec{e}_2 \frac{1}{H_2} \frac{\partial \varphi}{\partial \xi_2} + \vec{e}_3 \frac{1}{H_3} \frac{\partial \varphi}{\partial \xi_3},$$

where $\vec{e}_i$ are unit vectors normal to surfaces $\xi_i = const.$

A vector divergence is written as:

$$\vec{\nabla}\vec{v}(\theta_1, \xi_2, \xi_3) = \frac{1}{H_1 H_2 H_3} \times \left[\frac{\partial}{\partial \xi_1}(H_2 H_3 v_1) + \frac{\partial}{\partial \xi_2}(H_1 H_3 v_2) + \frac{\partial}{\partial \xi_3}(H_1 H_2 v_3)\right],$$

a Laplacian function

$$\Delta\varphi(\xi_1, \xi_2, \xi_3) = \frac{1}{H_1 H_2 H_2} \times \left[\frac{\partial}{\partial \xi_1}\left(\frac{H_2 H_3 \partial\varphi}{H_1 \partial \xi_1}\right) + \frac{\partial}{\partial \xi_2}\left(\frac{H_1 H_3 \partial\varphi}{H_2 \partial \xi_2}\right) + \frac{\partial}{\partial \xi_3}\left(\frac{H_1 H_2 \partial\varphi}{H_3 \partial \xi_3}\right)\right].$$

Components of a vector rotor are:

$$\vec{\nabla} \times \vec{v}\,|_1 = \frac{1}{H_2 H_2}\left[\frac{\partial}{\partial \xi_2}(H_3 v_3) - \frac{\partial}{\partial \xi_3}(H_2 v_2)\right],$$

$$\vec{\nabla} \times \vec{v}\,|_2 = \frac{1}{H_1 H_3}\left[\frac{\partial}{\partial \xi_3}(H_1 v_1) - \frac{\partial}{\partial \xi_1}(H_3 v_3)\right],$$

$$\vec{\nabla} \times \vec{v}\,|_3 = \frac{1}{H_1 H_2}\left[\frac{\partial}{\partial \xi_1}(H_2 v_2) - \frac{\partial}{\partial \xi_2}(H_1 v_1)\right].$$

In the Cartesian coordinate system, x,y,z $H_1 = H_2 = H_3 = 1$.

In a cylindrical coordinate system, ρ, φ, z ($x = \rho \cos \varphi$, $y = \rho \sin \varphi$, $z = z$) $H_1 = 1$, $H_2 = \rho$, $H_3 = 1$ (an equivalent coordinate system on a plane ρ, φ is called the polar coordinate system).

In a spherical coordinate system, r, θ, φ ($x = r \sin \theta \cos \varphi$, $y = r \sin \theta \sin \varphi$, $z = r \cos \theta$) $H_1 = 1$, $H_2 = r$, $H_3 = r \sin \theta$.

For some problems characterized by particular volume geometry, some other coordinate systems can be employed, such as ellipsoidal (elliptic), paraboloidal, bicylindrical (bipolar), conical, etc.

I.G. Zaltsman

CO-ORDINATE TRANSFORMATION METHODS (see Solidification)

COPPER

Copper—(L. *cuprum,* from the island of Cyprus), Cu; atomic weight 63.546; atomic number 29; melting point 1083.4 ± 0.2°C; b.p. 2567 °C; specific gravity 8.96 (20°C); valence 1 or 2. The discovery of copper dates from prehistoric times; it is said to have been mined for more than 5,000 years. It is one of man's most important metals.

Copper is reddish colored, takes on a bright metallic luster and is malleable, ductile and a good conductor of heat and electricity (second only to silver in electrical conductivity). The electrical industry is one of the greatest users of copper.

Copper occasionally occurs native, and is found in many minerals such as *cuprite, malachite, azurite, chalcopyrite,* and *bornite.* Large copper ore deposits are found in the U.S., Chile, Zambia, Zaire, Peru, and Canada.

The most important copper ores are the sulfides, oxides, and carbonates. From these, copper is obtained by smelting, leaching, and by electrolysis. Its alloys, brass and bronze, long used, are still very important; all American coins are now copper alloys; monel and gun metals also contain copper. The most important compounds are oxide and sulfate, blue vitriol; the latter has wide use as an agricultural poison and as an algicide in water purification. Copper compounds such as Fehling's solution are widely used in analytical chemistry in tests for sugar. High-purity copper (99.999 + %) is available commercially.

Handbook of Chemistry and Physics, CRC Press

CORE, NUCLEAR REACTOR (see Boiling water reactor, BWR)

CORED BRICK HEAT EXCHANGERS (see Regenerative heat exchangers)

CORIOLIS EFFECT, IN ATMOSPHERIC CIRCULATION (see Atmosphere)

CORIOLIS MASS FLOWMETER (see Mass flowmeters)

CORLISS VALVE (see Steam engines)

CORONA DISCHARGE, ELECTROSTATIC PRECIPITATION (see Gas-solids separation, overview)

CORRELATION (see Correlation analysis)

CORRELATION ANALYSIS

The term "correlation" is used to indicate the degree of interrelation between two or more variables. The procedure of calculating quantitatively the degree of the interrelation is called correlation analysis. Correlation analysis can be carried out for both continuous variables and discrete data, and the analysis for discrete data most often found in engineering practice and digital calculations is described below.

Auto-correlation function. Assuming a discrete time-series with finite number of samples of N and an average value of $\bar{x}$,

$$x_n = \{x_0, x_1, \ldots, x_{N-1}\}, \quad (1)$$

the auto-correlation function of which is defined as:

$$R_{xx}(k) = \sum_{n=0}^{N-1-k} (x_n - \bar{x})(x_{n+k} - \bar{x}) \quad (k = 0, 1, \ldots, N-1) \quad (2)$$

which effectively averages all possible products of the time-series and its time-shifted version separated by a time lag k. In practice, formula (2) is preferred in its normalized form

$$\rho_{xx}(k) = \frac{R_{xx}(k)}{R_{xx}(0)} = \frac{\sum_{n=1}^{N-1-k} (x_n - \bar{x})(x_{n+k} - \bar{x})}{\sum_{n=1}^{N-1} (x_n - \bar{x})^2} \quad (3)$$

$$(k = 0, 1, \ldots, N-1).$$

The value of ρ_{xx} is such that $-1 \le \rho_{xx} \le 1$. The auto-correlation function is an average measure of the time-domain properties of the time-series, and is related to the power spectral density function in the frequency domain by the Fourier transform (see **Spectral Analysis**). If the magnitude of the auto-correlation function ρ_{xx} decreases with increasing time lag k, there is some degree of randomness in the time series. If the ρ_{xx} changes sign at regular time intervals, then the time-series is periodic, and a combination of the two may imply that the time-series is quasi-periodic, which is often the case in real engineering problems.

Cross-Correlation Function

The cross-correlation function for two sets of time-series data

$$x_n = \{x_0, x_1, \ldots, x_{N-1}\} \quad \text{and} \quad y_n = \{y_0, y_1, \ldots, y_{N-1}\}$$

is defined as

$$R_{xy}(k) = \sum_{n=0}^{N-1-k} (x_n - \bar{x})(y_{n+k} - \bar{y}) \quad (k = 0, 1, \ldots, N-1) \quad (4)$$

The correlation function and cross-spectral function are equivalent measures in time and frequency domains which are related to each other by the Fourier transform (see SPECTRAL ANALYSIS).

Correlation Coefficient

The correlation coefficient is defined as the normalized version of formula (4) and is given by

$$\rho_{xy}(k) = \frac{R_{xy}(k)}{\sqrt{R_{xx}(0)R_{yy}(0)}} = \frac{\sum_{n=1}^{N-1} (x_n - \bar{x})(x_{n+k} - \bar{x})}{\sqrt{\sum_{n=1}^{N-1} (x_n - \bar{x})^2 \sum_{n=1}^{N-1} (y_n - \bar{y})^2}} \quad (5)$$

$$(k = 0, 1, \ldots, N-1),$$

the value of which at a particular time corresponding to k is a measure of similarity of the strength of components in x_n and y_n at that time. The value of ρ_{xy} is such that $-1 \le \rho_{xy} \le 1$, and the larger the ρ_{xy}, the more strongly correlated are the x_n and y_n at a given time.

It should be emphasized that the concept of correlation is different from that of regression. The procedure of finding a best fit curve is called regression, whereas the accuracy of the regression curve is measured by correlation.

References

Gardner, W. A. (1988) *Statistical Spectral Analysis, a Non-probabilistic Theory,* Prentic-Hall, Inc., New Jersey.

Linn, P. A. (1989) *An Introduction to the Analysis and Processing of Signals, 3rd Edition,* Macmillan Press Ltd., London.

Schwartz, M. and Shaw, L. (1975) *Signal Processing: Discrete Spectral Analysis, Detection and Estimation,* McGraw-Hill, Inc., USA.

Lei Pan

CORRELATION COEFFICIENT (see Correlation analysis)

CORRELATION, FOR CONVECTIVE HEAT TRANSFER (see Convective heat transfer)

CORRESPONDING STATES, PRINCIPLE OF

If the pressure, volume and temperature (absolute) of a fluid are expressed as fractions of the critical values, giving the so-called reduced pressure, $\pi = p/p_c$, volume, $\phi = v/v_c$ and temperature, $\theta = T/T_c$, then fluids having the same reduced pressure, volume and temperature are said to be in *corresponding states*. If the equation of state of the fluid is based upon two characteristic properties of it (apart from the mass), then there is a relation of the form $p_r = f(v_r, T_r)$ and it is sufficient to say that if any two independent reduced quantities are the same, the fluids are in corresponding states.

Other reductions are possible. In terms of intermolecular force parameters, $p^* = p(\sigma^3/\varepsilon)$, $v^* = v/(N\sigma^3)$, $T^* = kT/\varepsilon$, and if $p^* = f(v^*, T^*)$ then equality in v^*, T^* implies corresponding states [Hirschfelder, Curtiss & Bird (1964)]. For quantum fluids, $p^* = f(v^*, T^*, \Lambda^*)$ where $\Lambda^* = h/\sigma\sqrt{m\varepsilon}$ while for polar fluids, $\psi^* = f(v^*, T^*, \mu^*)$ where Ψ^* is any property at $\mu^* = \mu/\sqrt{\varepsilon\sigma^3}$) = reduced dipole moment. In these equations N, is **Avogadro's Number**, σ is the collision diameter, ε is the well depth in the inter-molecular potential function and μ is the dipole moment. Close to the critical point, "scaling" [see, e.g., Bejan, (1988)] can be used. Extensive charts and tables for several thermodynamic properties in terms of p_r, v_r and Z_c are available. For **Compressibility Factor**, see the separate section on that property.

References

Hirschfelder, J. O., Curtiss, C. F., and Bird, R. B. (1964) *Molecular Theory of Gases and Liquids,* Wiley, NY.

Bejan, A. (1988) *Advanced Engineering Thermodynamics,* Wiley Interscience, NY.

P.E. Liley

CORROSION, PREDICTION METHODS FOR

Introduction

This article deals with the prediction of corrosion inside carbon steel pipes carrying unprocessed hydrocarbons and in pipes used for injection of water into reservoirs (the injected water will maintain the pressure in the reservoir).

Several forms of corrosion may occur in such systems. Depending on the form of corrosion, the predictions may result in: a) "yes" or "no" rating of a material; or b) a predicted corrosion depth after a certain length of service. An example for a) is if the partial pressure of H_2S is larger than 0.003 bar and the pH and temperature are smaller than certain magnitudes, stress corrosion cracking of carbon steel may occur. A carbon steel with a specified heat treatment and hardness will then obtain a "yes" rating.

For b), an example is general CO_2 corrosion. This form *may* allow operation in an environment which induces significant corrosion of the carbon steel. The expected lifetime is now compared with a predicted cumulative corrosion depth. In order to prevent reduction of the operating pressure in the pipe, the predicted corrosion depth must be equal to or smaller than the corrosion allowance, i.e., the extra thickness of pipe which may be consumed by corrosion.

Prediction of corrosion depth in carbon steel relies largely on experiments. Based on experience, present-day mechanistic models do not allow estimates of corrosion depth outside the parameter range covered in experiments. One reason is the inability to predict how protective a corrosion product will be—i.e., its porosity and strength—without experiments. No mechanistic model can simulate how a film-forming inhibitor will influence the corrosion process. If a film-forming inhibitor is added to water, the inhibitor may, for example, modify the characteristics of the corrosion product in a manner specific for the given inhibitor. It is even possible that certain inhibitors may prevent formation of this product. Further partial filming may induce pitting. Thus, mechanistic models are limited to the estimation of how, for example, certain additives to the water can modify the pH and hence the corrosivity of a fluid.

Below, two important forms of corrosion which can be predicted by mechanistic models will be demonstrated. These are:

- **Production systems** - CO_2 corrosion of carbon steel
- **Injection systems** - O_2/Cl_2 corrosion of carbon steel

Corrosion of carbon steel in a CO_2 environment (production of hydrocarbons) and of such steel in an O_2/Cl_2 environment may serve as examples of corrosion processes when mass transfer of the corrosion species through electrolytes may partly or completely control the corrosion rates.

Production Systems - Carbon Dioxide Corrosion of Carbon Steel

In practice, a model is used for prediction if the corrosivity of the environment allows the use of carbon-steel when an effective inhibitor is applied. An effective inhibitor should reduce the general corrosion rates by 80–90% and should not induce pitting in the steel.

The next step is to verify the corrosivity and select the inhibitor by realistic experiments, i.e., in high-pressure hydrocarbon gas/liquid flow loop [Gramme (1994)]. Later, inhibitor effectiveness should be verified during production for actual field application.

Here, however, only the description of a mechanistic model will be given. The model described by Nordgaard and Søntvedt (1994) will serve as a basis:

Mechanistic corrosion model

Kinetics of surface reactions

The corrosion rate of steel in CO_2 environment appears to be mainly controlled by cathodic reaction, which most probably is the hydrogen evolution reaction [Smith and Rothmann (1979)].

There are discrepancies in the literature as to the species reacting at the electrode to form hydrogen atoms. The applied cathodic reaction mechanism is based on the work by de Waard and Milliams (1975) who suggested that the undissociated acid plays a catalytic role in the cathodic process

$$2H_2CO_3 + 2e^- \rightarrow H_2 + 2HCO_3^- \quad \textbf{(1)}$$

$$H_2CO_3 \leftrightarrow H^+ + CO_3^- \quad \textbf{(2)}$$

with Equation (1) as the rate determining step. The anodic reaction is as follows:

$$Fe \rightarrow Fe^{2+} + 2e^-. \quad (3)$$

From these reactions, it can be seen that H_2CO_3 is consumed and Fe^{2+} and HCO_3^- are generated by the reaction.

According to the *electrochemical theory of corrosion,* the corrosion process takes place at a mixed potential E_{corr}. It is assumed that the Bulter-Volmer expression can be applied to the kinetics of system reactions [Newman (1973)]. Based on experimental data from de Waard and Milliams, the reactions flux can then be given by

$$j_{r,H_2CO_3} = 1187 \cdot 10^{-A \cdot pHs} \cdot e^{(5385/T)} \cdot c_{s,H_2CO_3} \quad (4)$$

$$j_{r,Fe^{2+}} = -0.5 \cdot j_{r,H_2CO} \quad (5)$$

$$j_{r,HCO_3^-} = j_{r,H_2CO} \quad (6)$$

where

j_r = reaction flux at the surface;
A = constant describing the pH dependence of surface reaction;
pH_s = pH at surface;
T = bulk fluid temperature; and
$c_{s,i}$ = concentration of species i at the surface.

Thermodynamic calculations

The thermodynamic model is based on a work by Liu and High (1992). The bulk concentrations are calculated, assuming that the solution is saturated with CO_2 and chemical equilibrium is established for the system, by

$$H_2CO_3 \Leftrightarrow H^+ + HCO_3^- \quad (7)$$

$$HCO_3^- \Leftrightarrow H^+ + CO_3^{2-} \quad (8)$$

$$H_2O \Leftrightarrow H^+ + OH^-. \quad (9)$$

If the system satisfies the equation involving the solubility product criterion, an additional equation is required,

$$Fe^{2+} + CO_3^{2-} \Leftrightarrow FeCO_3. \quad (10)$$

At present, the species involved are H_2CO_3, H^+, Fe^{2+}, Na^+, OH^-, HCO_3^- and CO_3^{2-}, and electroneutrality can be expressed as:

$$H^+ + 2Fe^{+2} + Na^+ = HCO_3^- + 2CO_3^{2-} + OH^-. \quad (11)$$

The equilibrium constants are taken from the work by Edwards et al. (1978). The concentration of H_2CO_3 can be obtained from Henry's law

$$[H_2CO_3] = \frac{P_{CO_2}}{H_{CO_2}} \quad (12)$$

where P_{CO2} is the partial CO_2 pressure and H_{CO2} is the Henry constant.

In order to calculate the bulk concentrations, the activity coefficients need to be evaluated. The calculations are the same as in SOLMINEQ88 [Kharaka et al. (1988)].

Convective diffusion in liquids, including passage of current through electrolytic solutions

General formulation. The corrosive gas, CO_2, will dissolve into the liquid phase, H_2CO_3. H_2CO_3 will disassociate into its corresponding ions, HCO_3^- and CO_3^{2-}. Ions and H_2CO_3 are transported to the pipe wall where corrosion will occur. In the corrosion process, surface electrochemical reactions will set up an electric field. A feature of ions transfer in a moving solution that differentiates it from the transfer of dissolved neutral particles is the fact that the motion of ions is affected by the electric field in the solution. Ion transfer in the solution is produced by convective diffusion and by the migration of ions in the electric field. Convective diffusion in the solution is governed by two quite different mechanisms. Molecular diffusion is the mechanism of transport of species to the electrode as a result of concentration differences causing a diffusional flux

$$\mathbf{j}_{md,i} = -D_{md,i} \cdot \nabla c \quad (13)$$

where D_{mdi} is the molecular diffusion coefficient for species i and c_i is the concentration of species i. ∇c_i is given by

$$\nabla c_i = \left(\frac{\partial C_i}{\partial x}, \frac{\partial C_i}{\partial y}, \frac{\partial C_i}{\partial z} \right). \quad (14)$$

In a moving liquid, the solute is entrained and transported by the flowing stream, causing a convective flux

$$\mathbf{j}_{conv,i} = c_i \cdot \mathbf{u} \quad (15)$$

where the velocity field, **u,** is defined as:

$$\mathbf{u} = (u, v, w). \quad (16)$$

The migration of i-type ions can be written in the form

$$\mathbf{j}_{migr,i} = \frac{D_{md,i}}{R \cdot T} \cdot z_i \cdot F \cdot \mathbf{E} \cdot c_i \quad (17)$$

where R is the universal gas constant, F is the Faraday constant and z_i is the valency of ion i. The electric field, **E,** is given by

$$\mathbf{E} = \left(-\frac{\partial \varphi}{\partial x}, -\frac{\partial \varphi}{\partial y}, \frac{\partial \varphi}{\partial z} \right) \quad (18)$$

where φ is the electric potential. The total flux of particles of the species i in the moving medium, $\mathbf{j}_i$, is equal to

$$\mathbf{j}_i = c_i \cdot \mathbf{u} - D_{md,i} \cdot \nabla c_i + \frac{D_{md,i}}{R \cdot T} \cdot z_i \cdot F \cdot \mathbf{E} \cdot c. \quad (19)$$

Diffusion in turbulent flow in a pipe. Levich (1962) has shown that turbulent flow has a four-layered structure. Far from the surface, there is a zone of developed turbulence in which the concentration remains constant. Closer to the surface, in the turbulent boundary layer, both average velocity and average concentration decreases slowly according to a logarithmic law. In this zone, neither molecular viscosity nor diffusion play a noticeable part. Both momentum and matter are transferred by means of turbulence eddies. Still closer to the wall, in the viscous sublayer, turbulence eddies become so small that the momemtum transferred by molecular viscosity exceeds that transferred by turbulence eddies. However, the molecular diffusion coefficient is 1,000 times smaller than kinematic viscosity, Sc >> 1, and the remaining turbulence eddies still transfer substantially more solute than molecular diffusion. The coefficient of turbulent diffusion (eddy diffusivity) in the viscous sublayer is proportional to y^4 and decreases rapidly as the wall is approached. At a certain distance from the wall, δ_{dl}, eddy diffusivity must equal the coefficient of molecular diffusion.

$$\varepsilon_D = \gamma \cdot u^* \cdot \frac{\delta_{dl}^4}{\delta_{vl}^3} = D_{md,Fe^{2+}} \quad \mathbf{(20)}$$

where

$\varepsilon_{D,i}$ = eddy diffusivity;
γ = constant;
u^* = friction velocity;
δ_{dl} = diffusion layer thickness; and
δ_{vi} = viscous sublayer thickness.

The viscous sublayer thickness can be expressed as:

$$\delta_{vl} = a \cdot \frac{\nu}{u^*} \quad \mathbf{(21)}$$

where ν is the kinematic viscosity. Solving for the diffusion layer thickness gives

$$\delta_{dl} = \frac{a^{3/4}}{\gamma^{1/4}} \cdot \frac{1}{u^*} \cdot D_{md,Fe^{2+}}^{1/4} \cdot \nu^{3/4} \quad \mathbf{(22)}$$

where $\gamma = 1$ and $a = \frac{4}{3} \cdot 10^{3/4}$. The friction velocity is given by the following relation

$$u^* = \sqrt{\frac{\tau_w}{\rho}} = U_\infty \cdot \sqrt{\frac{C_f}{2}} \quad \mathbf{(23)}$$

where

τ_w = Wall shear stress
ρ = Liquid density; and
U_∞ = Pipe velocity.

The skin-friction coefficient, C_f, ($f = 4 \cdot C_f$) is a function of the Reynolds number and pipe roughness, k. An equation by *Colebrook and White* encompasses all flow regimes encountered in a Moody chart. This equation is:

$$\frac{1}{\sqrt{f}} = 1.74 - 2 \cdot \log\left(\frac{2 \cdot k}{D} + \frac{18.7}{Re_D \cdot \sqrt{f}}\right). \quad \mathbf{(24)}$$

The molecular diffusion coefficients are obtained from dilute electrolyte theory

$$D_{md,i} = \frac{R \cdot T \cdot \lambda_i}{F^2 \cdot |z_i|} \quad \mathbf{(25)}$$

where λ_i is the limiting ionic conductance of species i in water at 298 K. In the innermost portion of the viscous sublayer at $y < \delta_{dl}$, the molecular mechanism predominate over the turbulence mechanism. **(See also Boundary Layer; Fraction Factors for Single Phase Flow; Turbulent Flow)**

Mass Transfer and Ion Migration in Diffusion Layer

Flux and mass balance equations. At present, it is assumed that the diffusional layer provides all the resistance to mass transfer. Therefore, outside this layer, the concentration is uniform for all species and equal to the bulk concentration. A one-dimensional problem is considered. Equation 19 can then be simplified to yield

$$j_i = D_{md,i} \cdot \frac{dc_i}{dy} + \frac{D_{md,i} \cdot z_i \cdot F}{R \cdot T} \cdot c_i \cdot \frac{d\varphi}{dy} \quad \mathbf{(26)}$$

and the mass balance is reduced to

$$D_{md,i} \cdot \frac{d^2c_i}{dy^2} + \frac{D_{dm,i}}{R \cdot T} \cdot z_i \cdot F \cdot \frac{d}{dy}\left(c_i \cdot \frac{\partial\varphi}{\partial y}\right) = 0. \quad \mathbf{(27)}$$

These equations involve an electrical potential term. Then, one more equation is necessary to completely specify the system. For a corrosion system, the assumption that the net current flow is zero (open-circuit condition) is always true [Levich (1962)]

$$\sum_j z_j \cdot j_j = -\sum_j z_j \cdot D_{md,j} \cdot \frac{dc_j}{dy} - \sum_j \frac{z_j^2 \cdot D_{dm,j} \cdot F \cdot c_j}{R \cdot T} \cdot \frac{d\varphi}{\partial y} = 0. \quad \mathbf{(28)}$$

From the latter, the distribution of the field in the solution can be derived

$$\frac{d\varphi}{dy} = \frac{R \cdot T}{F} \cdot \frac{\sum_j z_j \cdot D_{dm,j} \cdot \frac{dc_j}{dy}}{\sum_j z_j^2 D_{md,j} \cdot c_j}. \quad \mathbf{(29)}$$

Boundary conditions. *At pipe wall surface (y = 0):*

The total reaction flux of species i in the moving medium in the diffusion layer is equal to

$$j_{r,i} = D_{md,i} \cdot \frac{dc_{s,i}}{dy} + \frac{D_{dm,i} \cdot z_i \cdot F}{R \cdot T} \cdot c_{s,i} \cdot \frac{d\varphi}{dy} \quad \mathbf{(30)}$$

If the species are nonreactive at the wall, $j_{r,i}$ is zero.

The diffusion layer boundary ($y = \delta$):

At the interface of the diffusional layer and the turbulent layer (outer part of the viscous sublayer), $c_i = c_{\delta,i} = c_b$,
where $c_{\delta,i}$ is the concentration of species i at the diffusion layer boundary and $c_{b,i}$ is the concentration of species i in the bulk.

Calculation procedure

The model needs bulk temperature, CO_2 partial pressure, bulk velocity, bulk iron concentration and added sodium bicarbonate as input. Initial concentration values for all species in the bulk is calculated by assuming the activity coefficients to be unity. Based on these initial values, the actual concentrations are calculated. After calculating the physical properties (viscosity, density and diffusion coefficients), the diffusion layer and flow condition are calculated.

The diffusion layer is divided into grids, and initial values for all the grid points are calculated. The surface reaction fluxes are calculated for the given initial values. The grid point concentrations throughout the diffusion layer are calculated by a finite difference technique Patankar (1980).

Because ordinary differential equations are highly nonlinear and coupled, they are solved one by one. The specific equation generally converges in two or three iterations. When all the equations are solved, the concentration of the species are updated. As a convergence criteria, the total number of iterations are used. The program checks if convergence really occurred. If not, more iterations need to be performed.

Comparison with experimental data

In order to test out the model, experimental results from Kjeller Sweet Corrosion have been used. These experiments are documented elsewhere [Dugstad and Videm (1990)]. Simulation results of the pH and the resulting corrosion rate along with the measured values are given in **Table 1**.

Experiments with temperatures up to 60°C have been applied. Above this value, the formation of protective layers is seen to lower the corrosion rate. The pH calculations agree very well with the values given for the experiments. The corrosion rate simulations are within 20% of the experimental results, which is regarded as a very good agreement.

Corrosion in Seawater Injection Systems

Model formulation

Seawater used for injection into the reservoir is treated to avoid microbiological activity, corrosion and reservoir formation blockage. The need for both microbiological and corrosion control can be in conflict because chlorine, which is corrosive, is one of the chemicals used for microbiological control. Typical chlorine concentrations are 0.3–0.5 ppm [Hudgins (1971) and Mitchell (1978)].

Chlorine can be produced by electrochlorinators or added as hypochlorite. When chlorine is added to seawater, an equilibrium between chlorine (Cl_2), hypochlorous acid (HOCl) and hypochlorite (OCl^-) is established. At $pH > 7.5$, hypochlorite predominates whereas at pH 6.5, OCl^- is present with considerable quantities of HOCl [Mitchell (1978)]. Another complicating factor is the presence of approximately 70 mg/l bromide in seawater. The bromide will have the effect of reducing the added hypochlorite to Cl^- while Br^- is oxidised to hypobromite (OBr^-) [Gundersen et al. (1989)].

Several cathodic reactions can take place in chlorinated seawater.

Reduction of hypochlorite:

$$OCl^- + H_2O + 2e^- \ 2OH^- + Cl^- \quad \textbf{(31)}$$

Or hypobromite:

$$OBr^- + H_2O + 2e^- \ 2OH^- + Br^- \quad \textbf{(32)}$$

Reduction of oxygen:

$$O_2 + 2H_2O + 4e^- \ 4OH^- \quad \textbf{(33)}$$

Reactions (31), (32) and (33) are flow-dependent. Flow dependence will not only include the velocity, but also the geometry (i.e., obstacles with local enhanced mass transfer and wall shear stress). Corrosion rates have been measured in a simulated water-injection system where seawater has been chlorinated and deoxidised [Andersen and Søntvedt (1993)]. The influence of chlorine, oxygen, flow velocity and geometry have been investigated.

The following conclusions were made from the experiments:

- Oxygen alone is more corrosive than in combination with chlorine.
- In injection systems with chlorinated seawater, two types of layers are formed on the steel surface. In seawater with low oxygen concentrations, magnetite will be the corrosion product. It was also observed that the pipes were covered with a layer of organic matter when chlorine was added to seawater.

Table 1. Comparison between IFE data and mechanistic corrosion model

Input parameters				IFE data		Mechanistic model	
T [°C]	P_{CO2} [bar]	U [m/s]	Fe^{2+} [ppm]	pH [−]	CR [mm/y]	pH [−]	CR [mm/y]
30	2.9	3.1	10	4.5	9.2	4.5	6.2
40	1.2	3.1	2	4.0	12.2	4.1	8.7
60	3.0	3.1	~0	3.9	29.7	3.9	31.0
60	1.5	3.1	~0	4.1	12.8	4.1	14.5
60	0.7	3.1	~0	4.2	9.9	4.2	7.7
60	1.5	3.1	8	4.3	11.7	4.3	13.5
60	3.9	8.5	180	5.4	19	5.3	16.1

- The corrosion rate seems to be controlled by mass transfer of oxygen and chlorine to a protective layer, and through this layer to the metal surface.
- The corrosion rate is mainly controlled by the concentration of oxygen and chlorine and local mass transfer of corrosive species. The corrosion rates at flow disturbances are much larger than on smooth sections.

It is assumed that the corrosion rate is limited by transport of chlorine and oxygen through a protective layer. The flux density of each corrosive species through the diffusion boundary layer and through pores in the protective layer with thickness H and permeability K_p is given as follows:

$$J_i = \frac{\tilde{c}_b}{(1/\beta_m) + H/(D_{md,i} \cdot K_p)} \quad (34)$$

where
β_m = local mass transfer coefficient to the protective layer;
H = height of layer made by both oxygen and chlorine;
$\tilde{c}_b$ = bulk concentration of species;
$D_{md,i}$ = molecular diffusivity; and
K_p = porosity of layer.

The total corrosion rate:

$$CR = K_2 \cdot F \cdot [(z \cdot J \cdot \tilde{c}_b)_{O_2} + (z \cdot J \cdot \tilde{c}_b)_{Cl_2}] \quad (35)$$

where
K_2 = corrosion rate corresponding with unit cathodic diffusion limited current;
z = valency change; and
F = Faraday constant.

It can now be assumed that the thickness of the protective layer (H) is increased by mass transfer of each species to the layer (β_m) and is decreased by wall shear (τ_w) from the liquid. In steady-state, this gives H equal to

$$H = K \cdot \beta_m/\tau_w \quad (36)$$

where K is a constant.

Apparently the porosity of magnetite can be constant as described by Surman and Castle (1969) and Sanchez-Caldera (1984). The porosity is set equal to 0.3. For turbulent flow with a moderate flow normal to the wall, the wall shear is related to the mass transfer coefficient by:

$$\tau_w = K_3 \cdot (\beta_m)^{2.3}. \quad (37)$$

When two pipes with different diameters have the same friction velocity, the wall shear on the pipe walls are identical. From the above relations, it follows that the height of the protective layer is given as

$$H = K_4/(u^*)^{1.1} \quad (38)$$

where K_4 is a constant for a given liquid density;

$$u^* = (\tau_w/\rho)^{0.5} \quad (39)$$

where ρ is the density of the fluid.

A main variable, if the corrosion process is diffusion-controlled, is the local mass transfer distribution on a specific section of the piping. This local mass transfer can be very different from smooth-section magnitudes. This, in turn, governs the velocity dependence by enhanced transport to the protective layer and the equilibrium height (or porosity) of the protective layer.

Figures 1–2 contain estimated magnitudes of the mass transfer expressed as Sherwood number (Sh) for geometries used in the corrosion tests [Søntvedt (1991)].

The main unknown in this model is the height of the protective layer. This must be determined by correlation with experiments. Derivation of apparent height characteristics for different geometries follows below.

Correlation with Experiments

Smooth Pipe

The model presented above has been correlated with several experimental points.

If no chlorine is present, the calculated limiting corrosion rates fit with the recorded corrosion rates, i.e., no protective layer is found on the smooth sections.

To correlate with the experiments when chlorine is present, a protective layer must be introduced in the simulation of the experiments. The derived layer thickness is large and given as follows:

$$H = 1.6/(u^*)^{1.1}. \quad (40)$$

Sudden reduction in diameter:

The geometry considered is illustrated in **Figures 3** and **4**. The results are included in **Table 3**. The corrosion rates are significantly smaller than the limiting value.

Analysis of the experiments reveals that the effective protective layer height varies with friction velocity as follows:

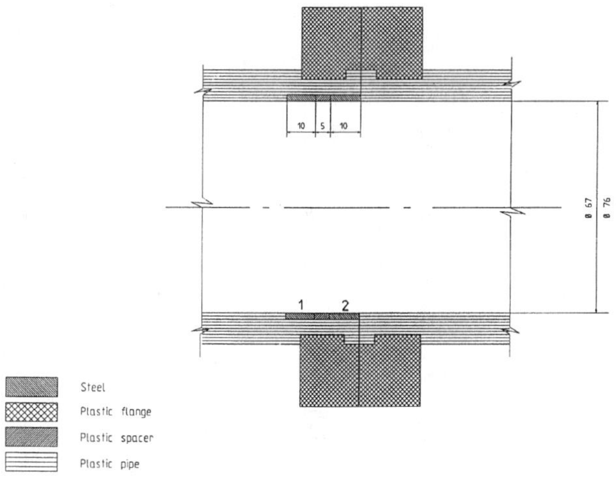

Figure 1. Geometry A – flush mounted specimens.

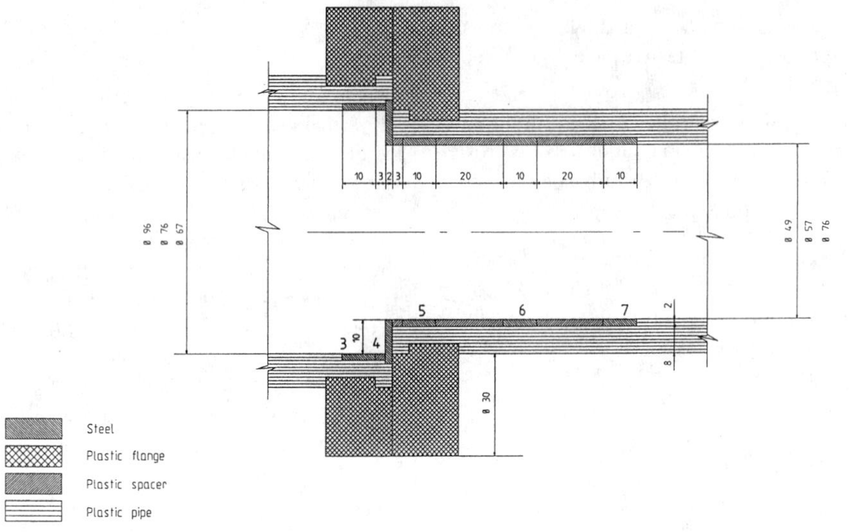

Figure 2. Geometry B – cross-over (diameter reduction).

$$H = 1.758/(u^*)^{1.1}. \quad \textbf{(41)}$$

It has been observe that H is reduced as u* increases with an exponential of similar value to that given by Equation (38) (i.e., u* to a power of 1.1).

Bends

The results are presented in **Table 4**.

The correlated protective layer height is described with the same type of formula as for sudden reduction in diameter:

$$H = 2.48/(u^*)^{1.1}. \quad \textbf{(42)}$$

H is zero when no chlorine is present. Again, the corrosion rates

Table 2. Smooth pipe—Test conditions and results from measurements and calculations

Specimen no.	C_{Cl2} [ppb]	C_{O2} [ppb]	Pipe diameter [mm]	Velocity [m/s]	Friction velocity [m/s]	Limiting corr. rate [mm/y]	Measured corr. rate [mm/y]	Calculated corr. rate [mm/y]
1	1500	20	96	2.70	0.118	0.40	0.30	0.14
2	1500	20	96	2.70	0.118	0.40	0.27	0.14
1	1500	20	76	4.30	0.183	0.61	0.30	0.23
2	1500	20	76	4.30	0.183	0.61	0.25	0.23
1	1500	20	67	5.50	0.232	0.76	0.24	0.29
2	1500	20	67	5.50	0.232	0.76	0.27	0.29
13	500	20	50	1.00	0.050	0.08	0.04	0.02
13	500	20	50	2.00	0.096	0.13	0.04	0.04
13	500	20	50	3.00	0.138	0.19	0.03	0.06
14	500	20	50	1.00	0.050	0.08	0.04	0.02
14	500	20	50	2.00	0.096	0.13	0.04	0.07
14	500	20	50	3.00	0.138	0.19	0.04	0.06
15	500	20	50	1.00	0.050	0.08	0.04	0.02
15	500	20	50	2.00	0.096	0.13	0.04	0.04
15	500	20	50	3.00	0.138	0.19	0.04	0.06
21	500	20	25	3.97	0.190	0.27	0.05	0.09
21	500	20	25	7.93	0.355	0.46	0.10	0.17
21	500	20	25	7.93	0.355	0.46	0.12	0.17
21	500	20	25	11.93	0.512	0.64	0.13	0.24
21	500	6	25	7.93	0.355	0.41	0.08	0.15
21	250	100	25	7.93	0.355	0.56	0.09	0.19
21	0	100	25	7.93	0.355	0.36	0.25	0.36

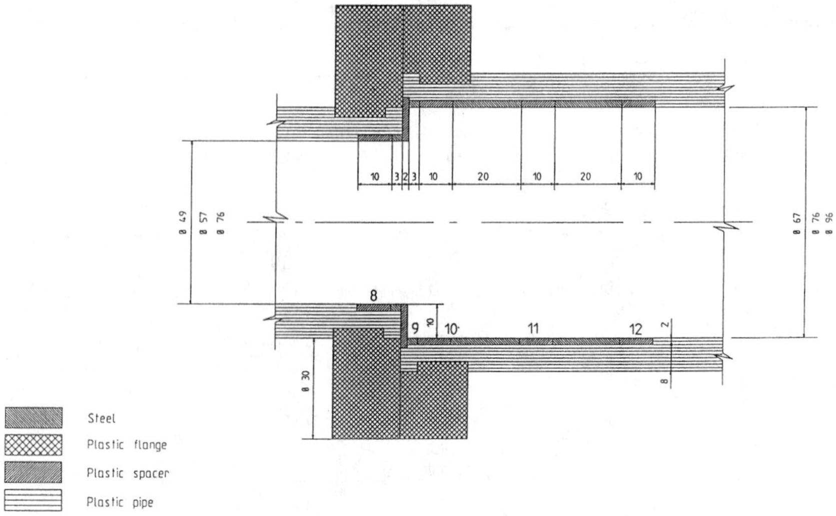

Figure 3. Geometry C – cross-over (diameter increase).

with Cl_2 present are significantly smaller than the limiting values.

Welds

The geometry considered is sketched in **Figure 5**. The correlation for thickness of the protective layer is given by:

$$H = 1.15/(u^*)^{1.1}. \quad \textbf{(43)}$$

This leads to corrosion rates much smaller than magnitudes corresponding with no/inefficient protective layer (see **Table 5**).

Sudden Increase in Diameter

The geometry in question is sketched in **Figures 4** and **6**.

In the reversed flow zone, the increase of Sh/Sh_0 is given as follows:

$$Sh/Sh_0 = 24.9(Re)^{-0.225} \quad \textbf{(44)}$$

The high wall shear stress fluctuations in this region seem to give a thinner protective layer, H,

$$H = 0.42(u^*)^{1.1}. \quad \textbf{(45)}$$

H is zero when no chlorine is present.

For the edge before the enlargement, the significant corrosion rates indicate a local high increase of mass transfer on the edge while the wall shear stress fluctuations should be small. To fit the data, Sh/Sh_o is given as follows:

$$Sh/Sh_0 = 24.9(Re)^{-0.7}. \quad \textbf{(46)}$$

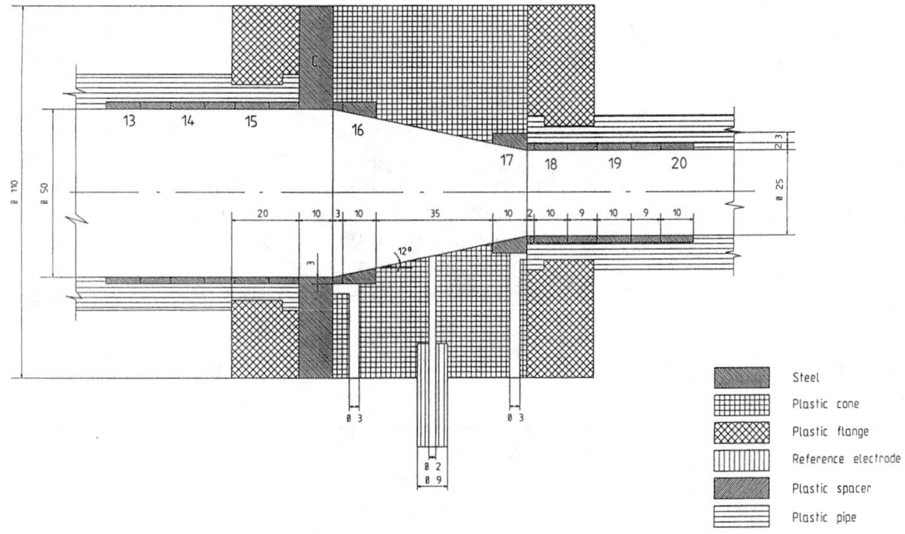

Figure 4. Geometry D – concentric reducer.

Table 3. Cross-over (reduction in diameter)—test conditions and results from measurements and calculations

Specimen no.	C_{Cl2} [ppb]	C_{O2} [ppb]	Pipe diameter [mm]	Velocity [m/s]	Friction velocity [m/s]	Sh/Sh_o	Measured corr. rate [mm/y]	Calculated corr. rate [mm/y]
4	1500	20	96	2.70	0.120	3.50	0.76	0.46
4	1500	20	76	4.30	0.180	3.50	0.97	0.98
4	1500	20	67	5.50	0.230	3.50	1.55	1.42
5	1500	20	96	2.70	0.120	3.50	0.68	0.46
5	1500	20	76	4.30	0.180	3.50	0.94	0.98
5	1500	20	67	5.50	0.230	3.50	1.46	1.42
24	500	20	25	3.97	0.190	3.50	0.44	0.41
24	500	20	25	7.93	0.355	3.50	1.02	1.03
24	500	6	25	7.93	0.355	3.50	0.66	0.92
24	250	100	25	7.93	0.355	3.50	0.46	0.31
24	0	100	25	7.93	0.355	3.50	0.85	0.75
25	500	20	25	3.97	0.190	3.50	0.48	0.41
25	500	20	25	7.93	0.355	3.50	0.93	1.03
25	500	6	25	7.93	0.355	3.50	0.60	0.92
25	250	100	25	7.93	0.355	3.50	0.14	0.31
25	0	100	25	7.93	0.355	3.50	0.57	.75

Table 4. Bend—Test conditions and results from measurements and calculations

Specimen no.	C_{Cl2}[ppb]	C_{O2}[ppb]	Pipe diameter [mm]	Velocity [m/s]	Friction velocity [m/s]	Sh/Sh_o	Measured corr. rate [mm/y]	Calculated corr. rate [mm/y]
30	500	20	25	3.97	0.190	1.36	0.11	0.04
30	500	20	25	7.93	0.355	1.36	0.17	0.14
30	500	20	25	11.90	0.512	1.36	0.19	0.28
33	500	20	25	3.97	0.190	1.20	0.09	0.03
33	500	20	25	7.93	0.355	1.20	0.16	0.10
33	500	20	25	11.90	0.512	1.20	0.21	0.19
33	500	6	25	7.93	0.355	1.20	0.06	0.09
33	250	100	25	7.93	0.355	1.20	0.22	0.11
33	0	100	25	7.93	0.355	1.20	0.40	0.43
35	500	20	25	3.97	0.190	1.37	0.07	0.04
35	500	20	25	7.93	0.355	1.37	0.24	0.30
35	500	20	25	11.90	0.512	1.37	0.25	0.53
35	500	6	25	7.93	0.355	1.37	0.16	0.13
35	250	100	25	7.93	0.355	1.37	0.12	0.16
36	500	20	25	3.97	0.190	1.29	0.06	0.03
36	500	20	25	7.93	0.355	1.29	0.23	0.12
36	500	20	25	11.90	0.512	1.29	0.18	0.24
36	500	6	25	7.93	0.355	1.29	0.15	0.11
36	250	100	25	7.93	0.355	1.29	0.10	0.13
37	500	20	25	3.97	0.190	1.25	0.06	0.03
37	500	20	25	7.93	0.355	1.25	0.15	0.11
37	500	20	25	11.90	0.512	1.25	0.15	0.22
37	500	6	25	7.93	0.355	1.25	0.11	0.10
37	250	100	25	7.93	0.355	1.25	0.09	0.12

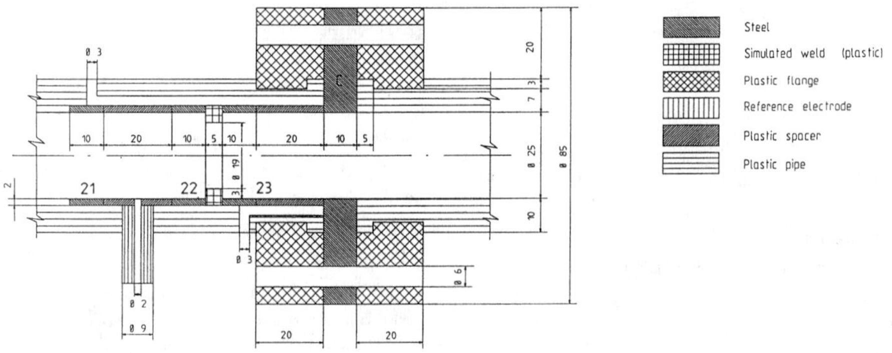

Figure 5. Geometry E − smooth pipe + weld.

Table 5. Weld—Test conditions and results from measurements and calculations

Specimen no.	C_{Cl2}[ppb]	C_{O2}[ppb]	Pipe diameter [mm]	Velocity [m/s]	Friction velocity [m/s]	Sh/Sh_o	Measured corr. rate [mm/y]	Calculated corr. rate [mm/y]
22	500	20	25	3.97	0.190	1.07	0.05	0.05
22	500	20	25	7.93	0.355	1.07	0.15	0.17
22	500	20	25	11.90	0.512	1.07	0.22	0.06
22	500	6	25	7.93	0.355	1.07	0.13	0.15
22	250	100	25	7.93	0.355	1.07	0.15	0.19
22	0	100	25	7.93	0.355	1.07	0.34	0.38
23	500	20	25	3.97	0.190	1.09	0.08	0.06
23	500	20	25	7.93	0.355	1.09	0.09	0.06
23	500	20	25	11.90	0.512	1.09	0.16	0.33
23	500	6	25	7.93	0.355	1.09	0.13	0.15
23	250	100	25	7.93	0.355	1.09	0.12	0.20
23	0	100	25	7.93	0.355	1.09	0.34	0.39

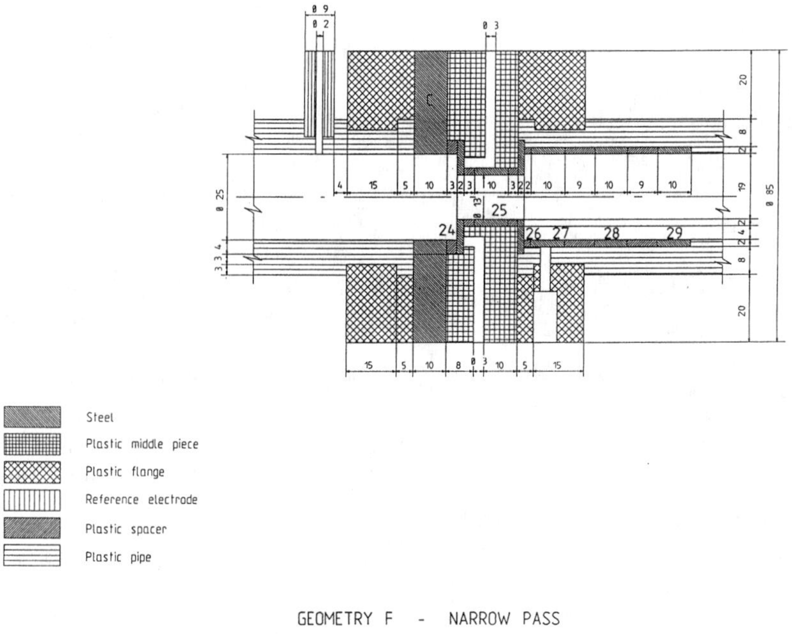

Figure 6. Geometry F − narrow pass.

Nomenclature

- A Constant describing the pH dependence of surface reaction (−)
- $\tilde{c}_b$ Bulk concentration of species (mol/m^3)
- $c_{b,i}$ Concentration of species i in the bulk (−)
- $c_{\sigma,i}$ Concentration of species i at the diffusion layer boundary (−)
- C_f Skin-friction coefficient
- $c_{s,i}$ Concentration of species i at the surface (−)
- CR Corrosion rate (mm/year)
- D Pipe diameter (m)
- $D_{md,i}$ Molecular diffusivity (m^2/s)
- **E** Electric field (V/m)
- F Faraday constant (C/mol)
- f Friction factor (−)
- H Protective layer (μm)
- H_{CO2} Henry constant
- J_i Flux density of corrosive species i (mol/m^{3}s)
- $j_{conv,i}$ Flux due to convective diffusion of species i (m/s)
- j_i Total flux of species i (m/s)
- jmd,i Flux due to molecular diffusion of species i (m/s)
- $j_{migr,i}$ Flux due to migration of species i (m/s)
- j_r Reaction flux at surface (m/s)
- K Constant (−)
- K_p Permeability of protective layer (−)
- K_2 Corrosion rate corresponding with unit cathodic diffusion limited current (A/m^2)
- P_{CO2} Partial CO_2 pressure (N/m^2)
- pH_s pH at surface (−)
- R Universal gas constant (J/K mol)
- Re Reynolds number (−)
- Sh Sherwood number (−)
- Sh_o Sherwood number for smooth pipe (−)
- T Bulk fluid temperature (°C)
- U_∞ Pipe velocity (m/s)
- **u** Velocity field (m/s)
- u Velocity component in x-direction (m/s)
- u* Friction velocity (m/s)
- v Velocity component in y-direction (m/s)
- w Velocity component in z-direction (m/s)
- x xcoordinate (m)
- y y coordinate (m)
- z z coordinate (m)
- z_i Valency of ion i (−)

Greek Symbols

- β_m Mass transfer coefficient (m/s)
- $\varepsilon_{D,i}$ Eddy diffusivity (m^2/s)
- δ_{dl} Diffusion layer thickness (m)
- δ_{vl} Viscous sub layer thickness (m)
- γ Constant
- λ_i Limiting ionic conductance of species i in water at 298 K (m^{2}C^2/J mol s)
- φ Electric potential (V)
- ρ Liquid density (kg/m^3)
- ν Kinematic viscosity (m^2/s)
- τ_w Wall shear on the layer (N/m^2)

References

Andersen, T. R. and Søntvedt, T. (1991) The influence of chlorine, oxygen and flow on corrosion in seawater injection systems, UK Corrosion Manchester Conference 22–24 Oct.

deWaard, C. and Milliams, D. E. (1975) "Carbonic Acid Corrosion of Steel". Corrosion, 31, 177–181.

Dugstad, A. and Videm, K. (1990) Kjeller Sweet Corrosion-II Final Report IFE/KR/F-90/008.

Edwards, T. J., Maurer, G., Newman, J., and Prausnitz, J. M. (1978) Vapor-liquid Equilibrium in Multicomponent Aqueous Solutions of Volatile Weak Electrolytes AlChE Journal, 24, 996–976.

Eriksrud, E. (1980) Internal Corrosion of Offshore Pipelines, Veritas Report No 80-0517.

Fischer, W. and Siedlarek, W. (1978) "Werkstoffe und Korrosion" 28, 822.

Gundersen, R., et al. (1989) The effect of sodium hypochlorite on the electrochemical properties of stainless steel and on the bacterial activity in seawater. NACE Corrosion '89, Paper no 108, New Orleans, Louisiana, April 17–21.

Henriksen, N. et al. (1987) Efficient new processing system for water deoxidation. Off-shore Technology Conference, Paper no. OTC 5591, Houston, Texas, April 27–30.

Horner, R. A. et al. (1994) The Forties Water Injection System—A Review of Corrosion Control (Late Paper), UK Corrosion Wemberly Conference Centre 12–14 Nov. 1984.

Hudgins, C. M. (1971) The Oil and Gas Journal, February 15, 71.

Kharaka, Y. K., Gunter, W. D., Aggarwal, P. K., Perins, E. H., and DeBraall, F. D. I. (1988) "SOLMINEQ88: A Computer Program for Geochemical Modelling of Water-Rock Interactions". U.S. Geological Survey, Water-Resources Investigations Report 88-4227, Menlo Park-Ca.

Levich, V. G. (1962) Physicochemical Hydrodynamics. Prentice Hall, Inc.

Liu, G. and High, M. S. (1992) Final Report: Down hole Corrosion Modelling. A report to Norsk Hydro a.s. School of Chemical Engineering, Oklahoma State University.

Manner, R. and Heitz, E. (1978) Werkstoffe und Korrosion 29, 783.

Mitchell, R. W. (1978) Journal of Petroleum Technology, June, 887.

Newman, J. S. (1973) Electrochemical Systems. Prentice Hall, Engellewood Cliffs, N.J.

Nordgaard, A. and Søntvedt, T. (1994) Mechanistic CO_2 corrosion modelling. Model formulation and comparisons with experiments up to 60 °C. Norsk Hydro Publication, Oil and Gas Division, Stabekk, Norway.

Patankar, S. V. (1980) Numerical Heat Transfer and Fluid Flow, Hemisphere publishing corporation, McGraw-Hill book company.

Sanchez-Caldera, L. E. (1984) The mechanism of corrosion-erosion in steam extraction lines of power plants, Massachusetts Institute of Technology.

Schmitt, G. and Rothmann, B. (1978) Werkstoffe und Korrosion 29, 98 and 237.

Schwenk, W. (1974) Werkstoffe und Korrosion, 25. 643.

Surman, P. L. and Castle, E. J. (1969) Corrosion Science, 9.

Søntvedt, T. (1991) Prediction of corrosion rate in deoxidized and chlorinated sea water. Development of model. Prod. Technology, U&P Division, Norsk Hydro.

T. Søntvedt

CORROSION FOULING (see Fouling)

CORRUGATED CONDENSED TUBES (see Augmentation of heat transfer, two phase)

CORRUGATIONS, PLAIN, PERFORATED, AND SERATED (see Plate fin heat exchangers)

COUETTE FLOW

Following from: *Flow of fluids*

Couette Flow is drag-induced flow either between parallel flat plates or between concentric rotating cylinders. Drag-induced flow is thus distinguished from pressure-induced flow, such as **Poiseuille Flow**. Flow between parallel flat plates is easier to analyse than flow between concentric cylinders. Thus, it is assumed that there is **Laminar Flow** of an incompressible **Newtonian Fluid** of density ρ and viscosity η between two parallel flat plates a constant distance H apart. The lower plate is of length L >> H and infinite width; it is stationary. The upper plate is of infinite length and infinite width; it moves at constant speed U in its own plane in the direction of the length of the lower plate. There is no variation in pressure p anywhere in the flow.

Because of the geometry, Couette flow is analysed using rectangular or Cartesian coordinates (x,y). The x-direction is aligned with the direction of motion of the upper plate and the y-direction is aligned vertically upwards (see **Figure 1**). Because L >> H, Couette flow is fully-developed, that is the velocity **u** is independent of axial position x everywhere except near the ends of the stationary plate (at x = 0 and x = L). Solution of the mass and linear momentum **Conservation Equations,** specifically the *Navier-Stokes equations,* with boundary conditions of no-slip at both plates (y = 0 and y = H) yields [see Richardson (1989)]:

$$u_x = U\, y/H. \tag{1}$$

Thus, the axial velocity profile is linear (see **Figure 2**) and depends only on flow geometry and speed of the moving plate: it is independent of fluid properties. The mean axial velocity $\bar{u}_x$ is given by:

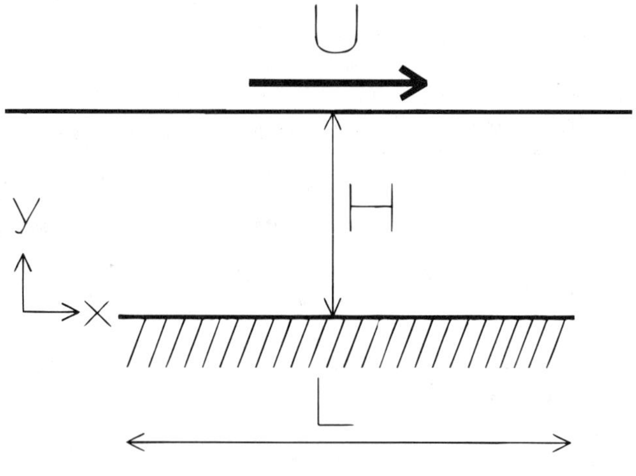

Figure 1.

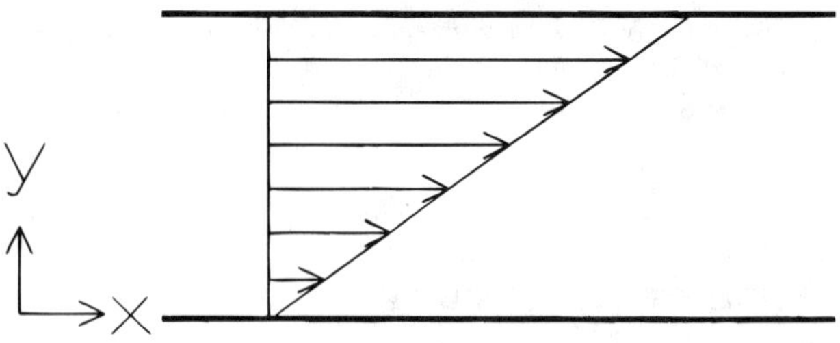

Figure 2.

$$\bar{u}_x = \frac{1}{2} U. \tag{2}$$

The volumetric flow rate per unit width $\dot{V}$ between the plates is given by:

$$\dot{V} = \frac{1}{2} UH. \tag{3}$$

Practical realisation of Couette flow between parallel flat plates is difficult. A good approximation can, however, be obtained in the annular gap between two rotating concentric cylinders of slightly differing radii. Thus, if the inner cylinder has an outer radius R_1 and the outer cylinder has an inner radius R_2, the flow between them is very similar to that given by Equation (1) provided that $R_2 \simeq R_1$. Flow between rotating concentric cylinders is susceptible to a centrifugal instability called the **Taylor Instability** [see Chandrasekhar (1961) and Drazin & Reid (1981)], which occurs when the inner cylinder is rotated and the outer cylinder is kept stationary and the **Taylor Number** Ta defined by:

$$Ta = 4\omega^2 R_1^2 \rho^2 (R_2 - R_1)^4 / \eta^2 (R_2^2 - R_1^2) \tag{4}$$

exceeds a critical value $Ta_c \simeq 1700$, where ω denotes the angular velocity of the inner cylinder. Flow between rotating concentric cylinders forms the basis of the *Couette viscometer.* In order to avoid the Taylor instability, it is usual to rotate the outer cylinder and keep the inner cylinder stationary. Then, if there is a torque or moment F × r on the inner cylinder, it can be shown that [see Fredrickson (1964)]:

$$F \times r = 2\pi R_2^3 \omega L \eta / (R_2 - R_1), \tag{5}$$

where ω denotes the angular velocity of the outer cylinder and L the length of the inner cylinder. Measurement of F × r, ω, L, R_1 and R_2 permits evaluation of η. Note, incidentally, that Equation (5) makes no allowance for end effects; these can, however, be rendered small by making L >> R_2 > R_1.

References

Chandrasekhar, S. (1961) *Hydrodynamic and Hydromagnetic Stability,* Clarendon Press, Oxford.

Drazin, P. G. and Reid, W. H. (1981) *Hydrodynamic Stability,* Cambridge University Press, Cambridge.

Fredrickson, A. G. (1964) *Principles and Applications of Rheology,* Prentice-Hall, Englewood Cliffs.

Richardson, S. M. (1989) *Fluid Mechanics,* Hemisphere, New York.

S.M. Richardson

COUETTE VISCOMETER (see Couette flow)

COULTER COUNTER

Following from: *Particle technology*

Coulter Counter is a registered trademark of Coulter Corporation for a device which provides particle count and size measurement using the so-called electrical sensing zone (or Coulter) principle. Several manufacturers produce particle size measurement instruments based on this original method, due to Coulter (1953, 1956).

A simplified representation of an apparatus design is shown in **Figure 1**. The particles to be measured must be suspended at low concentration in an electrically-conductive fluid, usually a solution of electrolyte in water or in an organic liquid. They are then made to pass, essentially one at a time, through a small aperture (or orifice) in the wall of an electrical insulator which is typically called the aperture (or orifice) tube. Across this aperture is also passed an electrical current, creating in and around the aperture an electrical sensing zone with a certain base impedance. A direct current is often used. As each particle enters the aperture, it has essentially displaced a volume of electrically-conductive fluid equal to its own volume, and the base impedance is therefore modulated by an amount proportional to the volume of the particle. This results in an electrical pulse of short duration being created by each particle; the height of the pulse being essentially proportional to the volume of the particle. The pulse may be measured as a change in resistance, current or voltage across the electrodes.

The passage of a number of particles produces a train of pulses which can be observed on an oscilloscope and analyzed by counter and pulse height analyzer circuits to provide the particle volume (or equivalent diameter) distribution. An integrated volume or mass percentage against size distribution can also be obtained.

For accurate concentration measurements, e.g., as particle counts per unit volume of fluid in a specific particle size range, the device is fitted with a volumetric metering system. This may typically take the form of a simple U-tube filled with mercury, which is connected to the aperture tube, or a piston displacement mechanism.

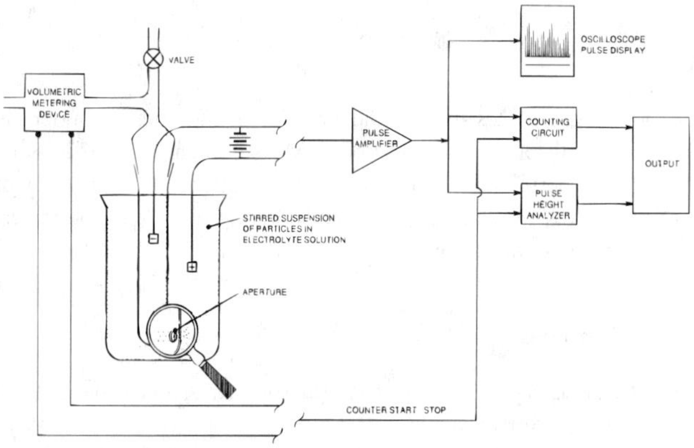

Figure 1. Simplifed diagram of one form of analyzer.

Particle resistivity has very little effect on electrical response, unless it is very close to the resistivity of the fluid. [See Lines, 1992].

Electrical response is virtually linear with particle volume up to some 80% of the aperture diameter, at least for spherical test particles used in the study by Harfield et al. (1984).

The lower size-limit of detection of particles is some 1.5–2.0% of the aperture diameter. Thus, a 70 μm aperture may be used with some instrument designs to count and size particles in the size range of some 1.2–56 μm equivalent spherical diameter. Performing particle size distribution measurements on a very wide size-range of particulate material requires the use of two (or more) apertures, with the results from each blended together.

Concerning calibration of such instruments, it is important to note that for any particle size distribution measurements there are two potential variables: particle quantity and particle size. Usually, volumetric metering and counting systems of electrical sensing zone instruments can be validated separately, so the particle counting measurement does not require calibration. However, although particle size measurement is in theory absolute (Kachel 1979), in practice it is always calibrated. There are two preferred methods; the more common is to use particles such as latex spheres (which are themselves, or can often be related to, Certified Reference Materials). The other is to allow the device to be self-calibrating by using a size fraction of the sample material under test and then to relate the results to data obtained by analytical balance and volumetric pipettes. Both methods are described in, for instance, British Standard BS 3406 : Part 5 (1983).

The electrical sensing zone method has become a reference method for particle size distribution analysis and for blood-cell counting and sizing. It is widely used in the industrial, biological and medical fields [Lines (1992), Richardson-Jones (1982), Coulter Electronics Ltd. (1991, 1992)].

References

British Standard BS 3406:Part 5 (1983) British Standards Institution, London, 32 pp.

Coulter, W. H. (1953) U.S. Patent 2,656,508.

Coulter, W. H. (1956) *Proc. Natl. Acad. Sci., 12,* 1034.

Coulter Electronics Ltd., Luton, England. (1991) Medical and Biological Bibliography, 4815 references.

Coulter Electronics Ltd., Luton, England. (1994) Industrial and Scientific Bibliography, 2000 references.

Harfield, J. G., Wharton, R. A., and Lines, R. W. (1984) Part. Charact., *1,* 32.

Kachel, V. (1979) *Flow Cytometry and Cell Sorting,* (eds. Melamed, M. R. et al.), Wiley, New York, 64.

Lines, R. W. (1992) *Particle Size Analysis,* (eds. Stanley-Wood, N. G. and Lines, R. W.), Royal Society of Chemistry, London, 350.

Richardson-Jones, A. (1982) *Advances in Hematological Methods: The Blood Count,* (eds. Van Assendelft, O. W. and England, J. M.), CRC Press, Inc., Boca Raton, Fl., 49.

R.W. Lines

COUNTER CURRENT FLOW LIMITATION, CCFL (see Boiling water reactor, BWR)

COUNTER CURRENT TWO-PHASE FLOW (see Flooding and flow reversal)

COUNTERIONIC ATTRACTION (see Foam fractionation)

COUPLED CONDUCTION AND CONVECTION

Following from: *Conduction convection*

In heat exchanger applications, it is often necessary to analyse the way in which heat conduction in solid components interacts with the convection processes in the fluid medium, with the objective of optimisation of design or performance. The context of this analysis may be the study of extended surface design (fins) or the design of laminar flow heat exchangers, where heat conduction along separating walls may have a significant effect on heat exchanger performance.

Coupled Conduction and Convection

A simple method of dealing with coupled conduction/convection heat transfer may be exemplified by the analysis of heat dissipation from a plain fin, such as the one illustrated in **Figure 1**.

In a surface extension such as a fin, there exists an internal distribution of energy transfer by conduction which is dependent on convective dissipation around the fin boundary. The function of the fin is to enhance heat transfer beyond that possible with a plain surface, and the most effective fin would be one in which the entire surface is at the same temperature as the base T_b. (See

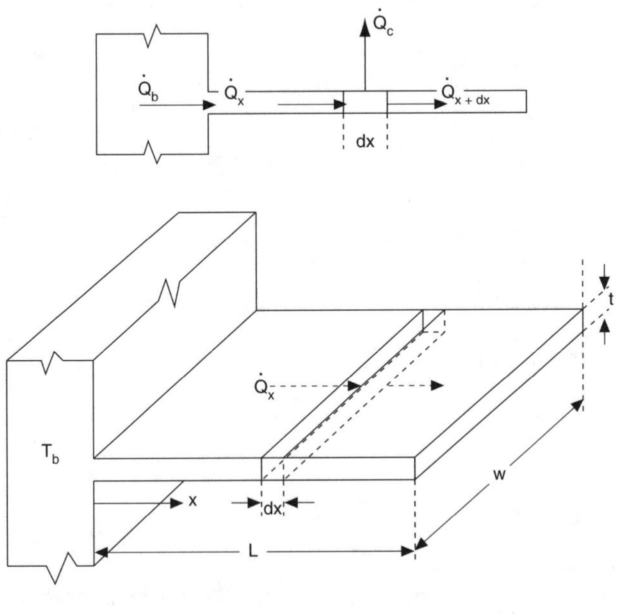

Figure 1. Convention from a thin conducting fin.

Augmentation of Heat Transfer.) To determine heat dissipation rate from a hot fin, it is necessary to know the temperature distribution along the fin. This may be predicted from the principle of conservation of energy applied to a representative differential section of the fin with width dx. The analysis is easier if simplifying assumptions are made: that conduction is quasi-one-dimensional in the x-direction (reasonable for thin fins with high conductivity); that the system has reached a steady-state; that λ and α are constant; and that there is no internal energy conversion and radiation effects are negligible. The energy balance on the differential element is thus:

$$\dot{Q}_x = \dot{Q}_{x+dx} + \dot{Q}_c \tag{1}$$

where the convective losses from the perimeter of the element [surface area P.dx, where $P = 2(w + t)$] to the ambient fluid at T_a are denoted by $\dot{Q}_c$. These losses may be calculated, using Newton's Law of Cooling, as $\dot{Q}_c = \alpha\,(P.dx)\,(T - T_a)$ where T is the temperature of the fin material at position x. A Taylor Series expansion may be used to approximate $\dot{Q}_{x+dx}$:

$$\dot{Q}_{x+dx} = \dot{Q}_x + \frac{d\dot{Q}_x}{dx}dx + \text{higher order terms} \tag{2}$$

and **Fourier's Law** may be used to give $\dot{Q}_x = -\lambda A\dfrac{dT}{dx}$, where A ($=$w.t) is the cross-sectional area of the fin, thus enabling the energy balance on the element to be written as:

$$\frac{d^2T}{dx^2} - \frac{\alpha P}{\lambda A}(T - T_a) = 0. \tag{3}$$

Defining $m^2 = \dfrac{\alpha P}{\lambda A}$ and introducing a new dependent variable $\theta = (T - T_a)$ this energy balance can be written as:

$$\frac{d^2\theta}{dx^2} - m^2\theta = 0 \tag{4}$$

which is a linear, homogeneous, second-order differential equation with constant coefficients having a general solution of the form:

$$\theta = C_1e^{mx} + C_2e^{-mx}. \tag{5}$$

The constants C_1 and C_2 are evaluated using boundary conditions which proscribe the fin temperature distribution. Four possible tip boundary conditions are illustrated schematically in **Figure 2**.

1) The temperature at the base of the fin ($x = 0$) is T_b, i.e.,

$$\theta(0) = T_b - T_a = \theta_b = C_1 + C_2. \tag{6}$$

2) There are a number of possible conditions which may exist at the tip of the fin. The simplest condition is that which would occur at the end of a very long ($L \to \infty$) fin when the tip would approach the ambient fluid temperature, i.e., $\theta(L) = 0$ as indicated in **Figure 2a**. The tip may be insulated, in which case, the tip temperature gradient is zero, **Figure 2b**. Alternatively, the tip could be maintained at some prescribed temperature T_L, **Figure 2c**. How-

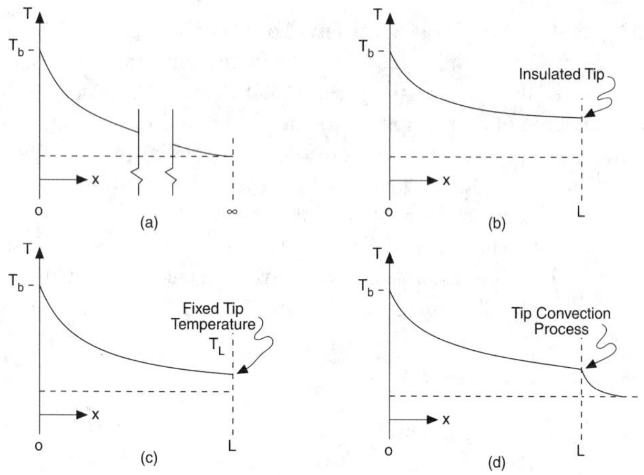

Figure 2. Schematic representation of four different heat transfer situations at the fin tip.

ever, a more likely condition is that the tip temperature is unknown but it can be assumed that energy arriving at the tip by conduction is lost by convection through the tip end area, where a heat balance would give

$$\alpha(\theta_L - \theta_a) = -\lambda\left[\frac{d\theta}{dx}\right]_{x=L}. \tag{7}$$

Using this latter tip condition gives:

$$\alpha(C_1e^{mL} + C_2e^{-mL}) = \lambda m(C_2e^{-mL} - C_1e^{mL}). \tag{8}$$

The particular solution for fin temperature for these boundary conditions can then be obtained as:

$$\frac{\theta}{\theta_b} = \frac{\cosh[m(L-x)] + (\alpha/m\lambda)\sinh[m(L-x)]}{\cosh[mL] + (\alpha/m\lambda)\sinh[mL]}. \tag{9}$$

The form of this temperature distribution is shown in **Figure 2d**, which illustrates the way in which convective heat loss from the fin, including the tip, affects conduction rate along the fin. The amount of heat dissipated by the fin may be calculated by noting that the conduction rate into the fin through the base must equate to convective loss from the fin surfaces in contact with the fluid, i.e., heat dissipation rate $= \dot{Q}_b = -\lambda A \{\frac{d\theta}{dx}\}_{x=0}$. Differentiating the expression for the fin temperature distribution with respect to x, and then setting $x = 0$ gives:

$$\dot{Q}_b = \alpha P\lambda A\frac{\sinh[mL] + (\alpha/m\lambda)\cosh[mL]}{\cosh[mL] + (\alpha/m\lambda)\sinh[mL]}. \tag{10}$$

The effectiveness of finning is usually assessed by a *fin efficiency factor,* defined as the ratio of the fin assembly heat transfer rate (as exemplified by the above expression) to the heat transfer rate of the finless system, in this case $\alpha A\ (T_b - T_a)$, which should generally be greater than 2. A comprehensive review of the coupling effects occurring in extended surface systems is given by Kern and Krauss (1972).

Conjugated Heat Transfer

In the above example of heat conduction in a fin interacting with convection from the fin surfaces, it has been assumed that the heat transfer coefficient was both constant and known *a priori* so that temperature distribution in the solid was easily obtained with this single piece of information about thermal interaction with the fluid. Problems in which the thermal boundary condition at a fluid boundary is not known *a priori* are called conjugate problems and require simultaneous solution of temperature fields in both the solid and adjacent convecting fluid. Such problems are of particular interest in the design of compact heat exchangers where the flow is *laminar* since it is then possible that *conduction along the walls of the exchanger* can compromise the exchanger performance.

In the design of heat exchangers, it is usual to assume that the film coefficients have constant values estimated from the appropriate correlations and sizing is then usually satisfactory for turbulent flow conditions. For laminar flow conditions, thermal inlet lengths can be of the same order of magnitude as flow path length, and there can therefore exist continuous variation in film coefficient within the exchanger. Since the two streams are thermally-coupled by the wall, the boundary conditions for both streams cannot be defined *a priori* and axial conduction in either or both the wall and the fluid may also be significant. The body of knowledge on conjugated heat transfer problems has grown since the early discussion of Shah and London (1978), and Pagliarini and Barozzi (1991) have given a recent literature survey. **Figure 3** illustrates conjugation in a simple and idealised counterflow heat exchanger system in which two flow passages are separated by an insulated wall containing a conducting window through which hot and cold fluids thermally interact, and along which energy may be transferred by conduction.

A simplfied hydrodynamic system serves to illustrate the effects of conjugation. Both flows are uniform in velocity and temperature as they approach the conducting window, which is sharply defined by the adjacent insulated walls. The heat capacity flow rates ($C = \dot{m}.c_p$) of the hot and cold streams are identical. The plate may be one of many making up an idealised plate heat exchanger,

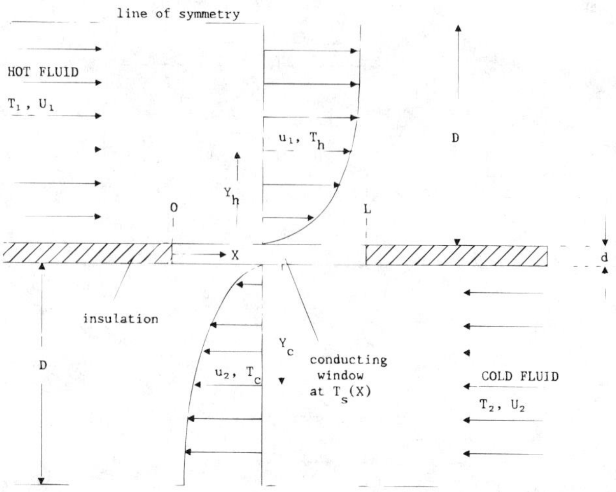

Figure 3. Model of an idealised counterflow heat exchanger. From Demko, J. A. and Chow, L. C., *AIAA J.*, 22(5), 705, (1984). Copyright © 1984 AIAA. Reprinted with permission.)

and the flow centerlines may be considered as axes of symmetry across which no energy is transferred. The inlet fluid temperatures are known and uniform, but the convective heat transfer coefficients at the solid/fluid interfaces are not known, nor is the plate temperature distribution here assumed to be one-dimensional, i.e., the window is long and thin. The temperature fields in both fluids and the solid must be obtained simultaneously with no prior knowledge of the temperature or heat flux at the two solid/fluid interfaces.

Three energy balances need to be solved simultaneously; one for each of the flow streams and one for the intervening wall. The dimensionless energy balances for each convective flow are both of the form:

$$u\frac{\partial\theta}{\partial x} + v\frac{\partial\theta}{\partial y} = \frac{1}{Pe}\left[\frac{\partial^2\theta}{\partial x^2} + \frac{\partial^2\theta}{\partial y^2}\right]. \tag{11}$$

Assuming that the wall is relatively thin and that the conductivities of the hot and cold fluids are equal, the energy balance for the solid is:

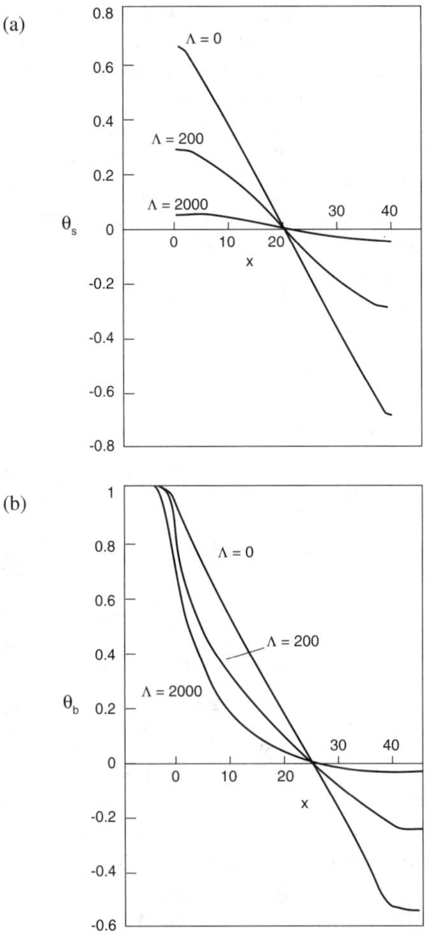

Figure 4. Effect of conjugation on temperature distributions in a simple counterflow heat transfer device. From Demko, J. A. and Chow, L. C., *AIAA J.*, 22(5), 705, (1984). Copyright © 1984 AIAA. Reprinted with permission.

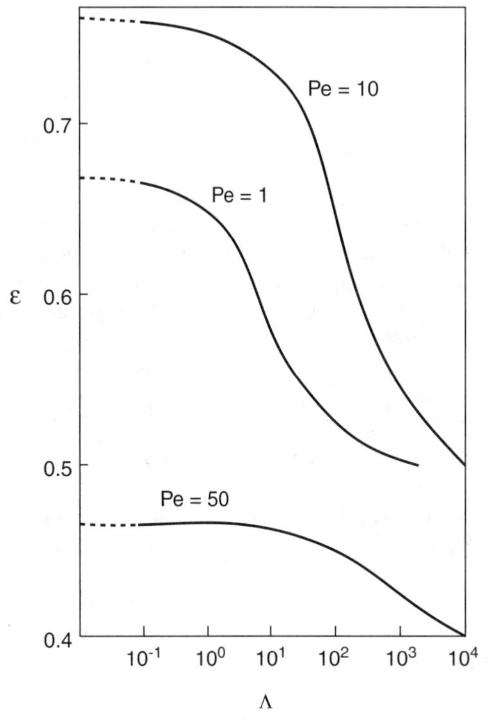

Figure 5. Influence of conjugation on heat exchanger effectiveness. From Demko, J. A. and Chow, L. C., *AIAA J.*, 22(5), 705, (1984). Copyright © 1984 AIAA. Reprinted with permission.

$$\frac{\partial\theta_c}{\partial y} - (d/D)(\lambda_s/\lambda_f)\frac{\partial^2\theta_s}{\partial x^2} - \frac{\partial\theta_h}{\partial y} = 0 \tag{12}$$

where λ_s is the thermal conductivity of the window material and λ_f is the thermal conductivity of both fluids. Referring to **Figure 3**, the following dimensionless variables are employed in the above energy balances: streamwise distance, $x = X/D$; cross-stream distance, $y = Y_h/D$ or Y_c/D; streamwise velocity component, $u = u_1/U_1$ or u_2/U_2; cross-stream velocity component, $v = v_1/U_1$ or v_2/U_2; hot stream temperature at (x,y); $\theta_h = 2(T_h - T_{av})/(T_1 - T_2)$; cold stream temperature at (x,y), $\theta_c = 2(T_c - T_{av})/(T_1 - T_2)$; window temperature at (x,y), $\theta_s = 2(T_s - T_{av})/(T_1 - T_2)$ where $T_{av} = (T_1 + T_2)/2$. The **Peclet Number,** Pe(=RePr), and the conductivity ratio, $\Lambda = (d/D)(\lambda_s/\lambda_f)$, emerge as important parameters. **Figures 4a** and **4b** show the effect of Λ on the plate and hot bulk temperature distributions for Pe = 10.

For Peclet Numbers less than 50, axial fluid conduction can be significant compared with axial bulk convective transport so that hot fluid may lose energy and cold fluid may gain energy prior to their entry into the conducting window area and heat transfer across the flow to and from the wall will be higher than at high Pe values. The wall conduction parameter determines, in part, how much of the energy from the hot fluid will be conducted axially along the wall. The effects of wall conduction are more significant at low Pe. For laminar flow, heat conducted along the plate may be of the same magnitude as the energy transferred to the cold fluid.

Numerical solution of the above system of equations is possible [Demko and Chow (1984)] and the influence of Pe and Λ on heat exchanger effectiveness ε is shown in **Figure 5**, using the usual definition of effectiveness (see **Heat Exchangers**).

$$\varepsilon = \frac{C_h}{C_{min}}\left[\frac{T_{h_{in}} - T_{h_{out}}}{T_{h_{in}} - T_{c_{in}}}\right]. \qquad (13)$$

where C represents the heat capacity of the stream, $\dot{M}c_p$, and the subscripts min and h signify the stream with lower and higher capacities, respectively.

The effectiveness ε is the ratio of the total amount of heat actually transferred in a heat exchanger to the maximum possible amount of heat exchange between the flow streams at their respective inlet temperatures. For $\Lambda = 0$ (the nonconjugated case), the extrapolations to the left ordinate (at $-\infty$) show that ε is not monotonic with Pe; but for all Pe, ε decreases as Λ (or wall conduction) increases, the rate of decrease depending on Pe. As $Pe \to 0$, fluid conduction dominates the entire process and the temperature of both the hot and cold fluids approach each other and that of the wall; consequently $\varepsilon \to 0.5$. As $Pe \to \infty$, axial bulk convection dominates and the hot and cold fluid temperatures will not change much so that $\varepsilon \to 0$ and the effect of wall conduction is small.

In general, axial wall conduction lowers the bulk temperature of the hot fluid near the exchanger entrance but results in higher exit temperatures, with the net result that effectiveness is decreased.

References

Demko, J. A. and Chow, L. C. (1984) *Heat Transfer Between Counterflowing Fluids Separated by a Heat Conducting Wall,* AIAA Jnl., 22(5).

Kern, D. Q. and Kraus, A. D. (1972) *Extended Surface Heat Transfer,* McGraw-Hill, New York.

Pagliarini, G. and Barozzi, G. S. (1991) Thermal Coupling in Laminar Flow Double-Pipe Heat Exchangers, *J Heat Transfer,* 113,526–534.

Shah, R. K. and London, A. L. (1978) *Laminar Flow Forced Convection in Ducts,* Academic Press, New York.

Leading to: *Augmentation of heat transfer, single phase; Heat exchangers*

T.W. Davies

COUPLED (COMBINED) RADIATION AND CONDUCTION

Following from: *Heat transfer; Radiative heat transfer*

Coupled or combined radiation and conduction heat transfer takes place in heated semitransparent media that have a spectral range of partial transparency (frequency range where the value of absorption coefficient k is approximately in the interval $0.01 < k < 100\ cm^{-1}$) and where radiation is an appreciable part of the total energy flux.

The presence of a semitransparency range in the spectrum of thermal radiation is typical for dielectrics and semiconductors in condensed phases and for multiatomic gases with asymmetrical molecules.

Combined radiation and conduction heat transfer is of great practical importance for some semitransparent material manufacturing and heat treatment processes carried out at high temperatures: melting, moulding and fritting of glasses, growth of dielectric and semiconductor single crystals, sintering of ceramics, drawing out of fibers and light guides.

Also, combined radiation and conduction heat transfer is important where semitransparent materials are used at high-temperature conditions (ceramics and fiber thermal protection systems of reusable space vehicles entering Earth atmosphere; thermal insulation of high-temperature furnaces and other industrial high-temperature equipment; intense radiation sources; solar volumetric receivers; radiative converters, packed beds of powders of oxides and other materials during remelting in solar and arc image furnaces; ceramic materials subjected to cutting and other processing by intense laser radiation).

The purpose of solving problems of combined radiation and conduction heat transfer is the calculation of temperature distribution and energy fluxes in the volume of semitransparent medium and on its boundaries. Side by side with this, inverse problems of combined radiation and conduction heat transfer are important, especially for determining true thermal conductivity of materials.

The theory of combined radiation and conduction heat transfer in media capable of absorbing radiation, as well as scattering radiation on various heterogeneities (pores, insertions of other phases), is based usually on simultaneous solution of the radiation transfer equation and of the energy conditions.

The formulation of radiative transfer equation for combined radiation and conduction heat transfer does not differ from the formulation for radiation transfer only.

The integro-differential energy conservation equation describing an energy balance on an elemental volume of medium has the form:

$$C_p(M)\rho(M)\frac{\partial T(M, t)}{\partial t} = -\text{div}\,\vec{\mathbf{q}}(M, t), \qquad (1)$$

where $\vec{q}$ (M,t) is the total heat flux vector at point M:

$$\vec{\mathbf{q}}(M, t) = -\lambda(T(M))\text{grad}\,T(M, t) + \vec{\mathbf{F}}_i(M, t) \qquad (2)$$

and $\vec{F}_i$ (M,t) is total radiation flux vector:

$$\vec{\mathbf{F}}_i(M, t) = \int_Q \int_{\Omega=4\pi} I(M, \vec{l}, t)\vec{l}\, d\Omega\, d\nu, \qquad (3)$$

where Ω is the solid angle and ν the frequency. In Equation (3), the integration on frequency ν for calculation of vector $\vec{F}_i$ is fulfilled over the frequency range Q, corresponding to the region of media semitransparency.

It follows from Equation (1) that for calculating temperature distribution over the volume of the semitransparent medium, it is necessary to know the radiation intensity distribution. Also, from the radiation transfer equation and its boundary conditions, it follows that for calculating radiation intensity distribution the temperature distribution must be known for all points inside the medium and on its boundaries. This is how the combination of heat transfer by radiation and conduction is manifested mathematically.

The simultaneous analytical solution of the radiation transfer and energy conservation equations is impossible and numerical solutions present a very difficult problem. The solution of the radiation transfer equation demands a large volume of calculations as radiation intensity is a function of point coordinates, direction and frequency of radiation. During combined radiation and conduction transfer calculations, the radiation transfer equation must be solved iteratively. A general solution of combined radiation problems for the general three dimensional case considering the frequency, direction and temperature dependence of optical properties and temperature dependence of thermophysical properties has not so far been published. To decrease the computational effort, combined radiation and conduction heat transfer problems are usually simplified by considering one-dimensional cases, as a rule a plane layer. Combined radiation and conduction heat transfer problems in a plane layer have been solved successfully for nonscattering media. The computer codes developed give the possibility of calculating transient temperature distributions, considering the frequency and temperature dependence of the optical properties of the medium and its boundaries' optical properties.

For plane layers of scattering media, the problem is more difficult since the radiation transfer equation includes not only the absorption coefficient k and refractive index n, but also the scattering coefficient β and phase function of scattering $P(\vec{l}, \vec{l}')$—the latter depending on the frequency ν and the directions of incident $\vec{l}$ and scattered $\vec{l}'$ radiation. Therefore, some simplifications and approximations have to be used for solving the radiation transfer equation.

In addition to the mathematical difficulties, the lack of data on frequency dependence of k, β and $P(\vec{l}, \vec{l}')$ is a significant obstacle for analysis of combined radiation and conduction heat transfer. Obtaining these characteristics by calculations using Mie theory is possible only for media with perfect spherical form and for cylindrical scatterers when the clearance between scatterers is bigger than the wavelength and the properties n and k of scatterer substance and surrounding medium are known. In practice this is very seldom possible. In most cases, the values k, β and $P(\vec{l}, \vec{l}')$ must be measured experimentally, but this is very difficult. Therefore, most calculations of combined radiation and conduction heat transfer use model "grey" media, where the scattering is assumed to be wavelength independent and isotropic.

Instead of using the strict radiation transfer equation, approximations may be applied for decreasing the many parameters describing the optical properties. The two-flux *Schuster-Schwarzchild* approximation uses four parameters: the averaged absorption coefficient, the scattering coefficient, the back-scattered fraction factor and the refractive index. The two-flux *Kubelka-Munk* approximation is used for describing radiation transfer in a plane layer of highly-scattering medium. The parameters are the refractive index n and two constants, K and S, defining absorption and scattering. The *radiation diffusion approximation* may be used not only for one-dimensional transfer problems but also for two- and three-dimensional cases. It uses three parameters: the refractive index n, the absorption coefficient k and the radiation diffusion coefficient D. The optical properties K and S or k and D may be considered as the new phenomenological parameters evaluated from experiments. Here, they may be used for describing combined radiation and conduction heat transfer even if the radiation transfer equation is not applicable (eg., for media with dense packing of scatterers).

The radiation thermal conductivity approximation is often used for describing combined radiation and conduction heat transfer in scattering media and is sometimes employed for two- and three-dimensional problems in nonscattering media. This approximation is applicable in the interior of an optically-thick medium when the influence of radiation at the boundaries disappears beyond the region adjacent to the surface. In this case, the combined radiation and conduction heat transfer calculation reduces to solving the Fourier thermal conductivity equation, where thermal conductivity is a temperature-dependent "effective" thermal conductivity λ_{eff} measured experimentally. It is a sum of "true" (conductive) thermal conductivity λ_c and "radiation" thermal conductivity λ_R:

$$\lambda_{eff} = \lambda_c + \lambda_R. \quad \textbf{(4)}$$

If the radiation transfer equation is applicable and optical properties of medium are known, the radiation thermal conductivity may be calculated using the formula:

$$\lambda_R = \frac{4\pi}{3} \int_Q \frac{n^2}{k + \beta(1 - \bar{\mu})} \frac{\partial I^p(T)}{\partial T} d\nu, \quad \textbf{(5)}$$

where μ is the mean cosine of the scattering angle and $I^p(T)$ is the intensity of equilibrium (Planckian) radiation at temperature T; the integration is performed over the frequency range Q corresponding to the region of medium semitransparency.

To use experimentally-measured values of λ_{eff}, local radiation equilibrium must apply for every point inside the medium. When the medium is highly scattering but its absorption coefficient is small (for example, the absorption coefficient of silica glass fiber thermal insulation is equal to $k \simeq 10^{-3}\ cm^{-1}$ or less), local radiation equilibrium may be disturbed even though the mean free-path of radiation is within the limits, from a few to hundreds of microns, and thickness of the layer is in centimeters. Here, combined radiation and conduction heat transfer calculations using the radiation thermal conductivity approximation may lead to errors and it is preferable to use the diffusion approximation.

References

Ozisik, M. N. (1985) *Radiative Transfer and Integration with Conduction and Convection.* New York: Werbel and Peck.

Seigel, R. and Howell, Y. R. (1992) *Thermal Radiation Heat Transfer,* third edition, Washington D.C.: Hemisphere.

Sparrow, E. M. and Cess, R. D. (1978) *Radiation Heat Transfer,* Belmont: Wadsworth.

Viskanta, R. and Anderson, E. E. (1975) *Heat Transfer in Semitransparent Solids: Advances in Heat Transfer,* New York: Academic Press, 317–441.

V.A. Petrov

COUPLED RADIATION AND CONVECTION

Following from: *Heat transfer; Radiative heat transfer; Coupled radiation, convection and conduction*

Coupled or combined radiative and convective heat transfer is a particular case of simultaneous radiative, convective and conductive heat transfer which occurs when heat transfer by conduction is negligibly small compared with that by radiation and convection. Combined radiation and convection occurs between the boundary surfaces and the surrounding medium, between the surfaces separated by the moving medium and inside the moving medium. Depending on conditions in the medium, geometrical factors and surface state, regimes of strong and weak interaction between radiative and convective heat transfer are possible. For weak interaction, problems of heat transfer by radiation and convection can be solved successively and independently while for strong interaction, these processes are essentially interdependent. Thus, energy transfer by one mechanism can influence heat exchange by the other mechanism and vice versa.

Simultaneous radiative and convective heat transfer is taken into account when solving many problems of heat transfer for hypersonic flows around aircrafts, for steam boilers, for aircraft and rocket engines for plasma generators, as well as of heat transfer in shocks, deflagration and detonation fronts and laser deflagration waves for various discharges.

Mathematical formulation of combined radiation and convection problems includes the continuity equation, the momentum (in the Euler or Navier-Stokes forms) and energy conservation equations, the state equation and relations for thermodynamical, thermophysical transport and optical properties.

The energy conservation equation has no terms corresponding to conductive heat transfer and is of the dimensionless form

$$\rho c_p \frac{dT}{dt} = -\frac{1}{Bo}\frac{\partial}{\partial x_i} q_{R,i} + Ec\frac{dp}{dt} + \frac{1}{Bo^*} S + \frac{Ec}{Re}\mu\Phi. \quad (1)$$

All the letters and symbols in the above equation are defined in the article on **Coupled Radiation, Convection and Conduction**. That fact that there are no conductive terms in (1) means that heat conductivity is of no importance in the problem, both for the temperature field formation in the moving medium and for its heat transfer with the surface. Thus, heat conductivity can be ignored for small temperature gradients and when the sharp temperature front is replaced by a contact discontinuity. Obviously, in the latter case, the information on temperature distribution at the front is lost and the medium thermal state on both sides of the contact discontinuity should be determined without heat conductivity.

When analysing combined radiation and convection, the Boltzmann criterion can be represented in the form of the convective to radiative heat flux component ratio

$$Bo = \rho_0 u_0 c_{p0} T_0 / \tilde{\sigma} T_0^4. \quad (2)$$

The classical problem of combined radiation and convection is the problem of steady-state gas flow across a flat plate at a constant temperature [Sparrow and Cess, (1970)].

The flow scheme is shown in **Figure 1** and the mathematical problem is to solve the energy conservation equation

$$\rho u_\infty c_p \frac{\partial T}{\partial x} = -\frac{\partial q_{R,y}}{\partial y} \quad (3)$$

subject to the following boundary conditions: at $x = 0$, $T = T_\infty$, $u = u_\infty$; at $y = 0$, $u = u_\infty$, $T = T_w$, $v = 0$, and the condition for radiation energy intensity (for example, the black-body assumption for the surface). Even such simplified and idealized problem have been addressed by a wide spectrum of methods with a range of results obtained; this is caused by the very complicated radiation problem. When formulating the complete problem for a radiation-scattering medium (for the optical thickness $\tau_0 \sim 1$), it is simplier to use numerical methods. The assumptions of weak light scattering or implementation of limited optical thickness regimes ($\tau_0 \gg 1$ and $\tau_0 \ll 1$) make it possible to obtain analytical solutions. For example, in a "grey" gas without scattering and using the exponential approximation of the kernel of the integro-differential equation obtained from (3) by substituting $q_{R,y}$ for the plane layer, the following solution has been obtained in by [Sparrow and Cess, R. D. (1970)]:

$$q_{R,y=0}/\tilde{\sigma}(T_w^4 - T_\infty^4) = \exp(-4\xi)[I_0(4\xi) + I_1(4\xi)]. \quad (4)$$

Here I_0 and I_1 are the modified Bessel functions, $\xi = 2\tilde{\sigma}\kappa T_\infty^3 x/\rho c_p u_\infty$, and κ is the volumetric gas absorption coefficient. When gas heat conductivity is excluded, any solution of (3) has a temperature discontinuity.

When solving the heat protection problem for flight vehicles entering the dense atmosphere at a superorbital velocities the combined radiation and convection approximation appears to be very useful and fruitful. When a vehicle with bluntness radius $\sim$ 1 m enters the Earth is atmosphere at superorbital velocities of $v_\infty = 10 - 16$ km/s at an altitude of 75 km, radiation heating of its heating surface exceeds convection heating. Therefore, the thin boundary layer where the convective flow forms can be neglected. The problem where heat protecting material failure is taken into account is the more realistic one. In this case, in a stagnation streamline the shock layer structure can be represented as two oppositely-directed mass flows separated by a thin transient layer in which the viscosity is essential. Replacing the thin transient layer by a contact surface allows the exclusion of viscosity and thermal conductivity. Considering the flow to be steady-state, the governing equation system can be formulated [Olstad (1971)] as:

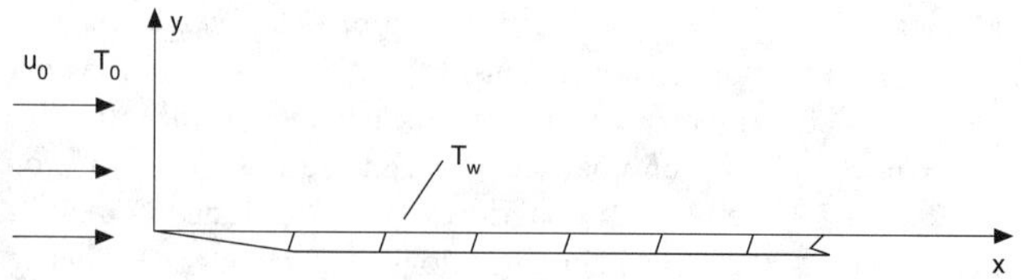

Figure 1. Combined radiation and convection in flow across a flat plate.

$$\nabla(\rho\mathbf{V}) = 0,$$

$$\rho(\mathbf{V}\nabla)\mathbf{V} = -\nabla p, \tag{5}$$

$$\rho c_p \mathbf{V}\nabla T + \rho\mathbf{V}\nabla \frac{1}{2}\mathbf{V}^2 = \nabla \mathbf{q}_R$$

$$\rho = \rho(p, T)$$

assuming Hugoniot conditions on the shock and the condition of flow nonpenetration or a given injection rate trough the surface. In problems of radiating shock layers, the radiation transfer is considered, as a rule, to be as in a plane-parallel layer, this is justified a small thickness of the layer. Solving equations (5) makes it possible to obtain flow variable distributions inside the shock layer and the radiation fluxes to the surface and the shock front. In this situation, radiation transfer strongly influences the shock layer parameters, hence, the problem can be attributed to the class of strong interaction of radiation and convection.

Problems on laser wave deflagration applied in laser rocket engines and in the optical plasmotron are also problems involving the strong interaction regime. The plasma cloud, localized in space and with a size of $\sim$ 1 cm and a temperature of $\sim$ 15,000 K, moves to meet a laser beam (e.g., from a CO_2-laser) due to heating of the surrounding cold gas by the heat radiation of the plasma itself. The plasma exists due to laser energy absorption. The heating rate of the cold gas by heat radiation essentially exceeds that by heat conductivity, even for extremaly high temperature gradients in the laser deflagration wave. The process runs usually at atmospheric or higher pressures, and this makes it possible to use the equilibrium approximation for the thermodynamic medium state and an equation system similar to (5).

The main difference is that the integral radiation flux vector is represented as a sum

$$\mathbf{q}_R = \mathbf{q}_{R,T} + \mathbf{q}_{R,L}$$

where $\mathbf{q}_{R,T}$ and $\mathbf{q}_{R,L}$ are the vectors of integral heat flux density and laser radiation flux density, respectively. To obtain the function $\mathbf{q}_{R,T}$ the two-dimensional equation for selective radiation transfer must be solved while for $\mathbf{q}_{R,L}$ the geometrical optics approximation may be employed.

Combined radiation and convection problems concerned with calculations of heat transfer in devices such as steam boiler furnaces are widely encountered in engineering practice [Siegel and Howell (1972); Ozisik, (1973)]. The temperature of the furnace medium (multicomponent disperse system of gas and solid phases) is maintained by combustion and the general configuration of the temperature field in the furnaces is determined mainly by convective processes formed when the gas-dust component flows in a working space. The surfaces are heated principally by radiation heat transfer from the furnace medium; therefore, the heat conductivity contribution is neglected when computing real furnace steam boilers. Analyzing the combined radiation and convection processes in furnaces, the transfer equation for this complicated geometry should take into account the radiation scattering and absorptive properties of the furnace medium with various additives (the properties of which are hardly known) as well as the optical-physical characteristics of the heat receptive surfaces.

References

Olstad, W. B. (1971) Nongrey Radiating Flow about Smooth Symmetric Bodies, *AIAA Journal*, 1, 122–130.

Ozisik, M. N. (1973) *Radiative Transfer and Interaction with Conduction and Convection*, A Wiley-Interscience Publication.

Sparrow, E. M. and Cess, R. D. (1970) *Radiation Heat Transfer*, Brooks/ Cole Publishing Company.

Siegel, R. and Howell, J. R. (1972) *Thermal Radiation Heat Transfer,* McGraw-Hill Book Company.

S.T. Surzhikov

COUPLED RADIATION, CONVECTION AND CONDUCTION

Following from: *Radiative heat transfer*

Coupled radiation, convection and conduction (CRCC) is a self-consistent heat transfer process by the mechanisms of thermal radiation, convection and thermal conductivity. CRCC occurs between a surface in contact with a moving medium and between various components of dust-loaded flows, the moving medium being considered not only as gas and plasma, but the condensed state as well.

Here, self-consistent heat transfer means, in essence, that each of the above-mentioned mechanisms influences to the same extent the energy balance inside and (or) on the boundary of the domain considered and thereby changes the energy exchange intensity by the other mechanisms.

The combined action of radiative, convective and conductive heat transfer must be considered when solving a wide class of heat and mass transfer problems in such fields as power, aerospace and process engineering. These include:

1. Heat transfer in hypersonic flows around bodies moving in planetary atmospheres;
2. Heat transfer in arcs, microwaves and optical plasma generators, in combustion chambers and nozzles of rocket engines (gaseous, liquid and solid propelled rocket engines; plasma, nuclear and laser rocket engines), in furnaces of steam boilers and other power facilities;
3. Heat transfer in various gas discharges and in intensive shock waves, combustion fronts and laser deflagration waves;
4. Heat transfer in thermal protection materials, in combustion in semitransparent porous matrices, in laser and plasma engineering, in electronic engineering, etc.

CRCC is the most general type of heat transfer; in this general case of heat transfer in a moving medium, all the three mechanisms act. However (depending on medium temperature, velocity, density, geometry and optical and physical properties), it is possible for one or two mechanisms to dominate. In these limiting cases, heat transfer may be designated as "convective", "conductive," "radiative," "conductive-convective," "radiative-conductive" and "radiative-convective" heat exchanges.

When solving problems of CRCC in whichever mechanisms of the heat transfer are important, a useful first step is to determine the influence of radiation on the moving medium parameters. If this influence is small, one can simplify the problem, treating the medium motion and the radiation heat transfer separately. The problems are then solved successively by the methods for a moving media and for radiation heat transfer—the combined action of the mechanisms being taken into account using the additivity principle. In the case of such weak interaction account may be taken of the interdependency of the processes using perturbation methods. If radiative transfer influences strongly the medium's thermal state and its motion, the additivity principle is inapplicable. This case is the most complicated one in the CRCC problem.

To describe the CRCC processes mathematically and estimate how important is each of the heat transfer mechanisms, the following systems of equations are used [Pai, (1966)]; [Bond et al. (1965)]; [Siegel and Howell, (1972)]; [Ozisik, (1973)]; [Sparrow, and Cess, (1970)]:

1. **The continuity equation.**

$$\frac{\partial \rho}{\partial t} + \frac{\partial \rho u_i}{\partial x_i} = 0, \qquad i = 1, 2, 3, \tag{1}$$

where ρ is the medium density, $\mathbf{V} = \{u_1, u_2, u_3\}$ is the flow-velocity vector, u_i is x_i-velocity component and t is the time. In (1) and thereafter, we use tensor designations, the summation is over repeating indices.

2. **The momentum equation.**

$$\rho \frac{du_i}{dt} = F_i - \frac{\partial p}{\partial x_i} + \frac{\partial \tau_{ij}}{\partial x_i} + \frac{\partial p_{R,i,j}}{\partial x_j}, \tag{2}$$

where

$$\frac{d}{dt} = \frac{\partial}{\partial t} + u_j \frac{\partial}{\partial x_j}$$

$$\tau_{ij} = \mu\left(\frac{\partial u_i}{\partial x_j} + \frac{\partial u_j}{\partial x_i} - \frac{2}{3}\delta_{ij}\frac{\partial u_k}{\partial x_k}\right)$$

is the viscous stress tensor, F_i is the x_i-component of the force F_i acting on Unit volume (for example, the gravity force $F_i = \rho g_i$ where g_i is the x_i-component of the free-fall acceleration vector); p is the gas-dynamic pressure, μ is the dynamic viscosity coefficient; δ_{ik} is the Kronecker delta symbol ($\delta_{ik} = 1$ for i = k and $\delta_{ik} = 0$ for $i \neq k$); $pR,i,j = \frac{1}{c}\int_{4\pi} \omega_i \omega_j \, J \, d\omega$ (i, j = 1,2,3) is the radiation stress tensor determined when integrating the integral radiation intensity $J = \int_0^\infty J_\nu \, d\nu$ over the solid angle $\Omega = 4\pi$, here, J_ν is the spectral radiation intensity, ν is the radiation frequency, ω_i and ω_j are direction cosines for the vector $\vec{\omega}$, which characterizes the radiation propagation direction. In the evaluation of the significance of radiation pressure, a useful dimensionless parameter is R, the ratio of mean radiation pressure of isotropic black-body at a temperature T to gas dynamic pressure

$$R = \frac{4\pi}{3c}\tilde{\sigma}T^4/p,$$

where $c = 3 \cdot 10^8$ m/s is the velocity of light, and $\tilde{\sigma}$ (Joule/s · m^2 k^4) is the Stefan-Boltzmann constant. At $p = 10^5$ Pa, $T = 2 \cdot 10^4$ K, $R \approx 1{,}27 \cdot 10^{-3} \ll 1$. In most problems of CRCC (at temperatures lower than 20,000 K), one can neglect the radiation stress tensor.

3. **Energy conservation equation.**

$$\rho c_p \frac{dT}{dt} = -\frac{\partial}{\partial x_i}(q_{c,i} + q_{R,i}) + \frac{dp}{dt} + S + \mu\Phi \tag{3}$$

where

$$\Phi = 2\sum_{i=1}^{3}\left(\frac{\partial u_i}{\partial x_i}\right)^2 - \frac{2}{3}\left(\sum_{i=1}^{3}\frac{\partial u_i}{\partial x_i}\right)^2$$

$$+ \left(\frac{\partial u_2}{\partial x_1} \mathrm{pl} \frac{\partial u_1}{\partial x_2}\right)^2 + \left(\frac{\partial u_3}{\partial x_2} + \frac{\partial u_2}{\partial x_3}\right)^2 + \left(\frac{\partial u_1}{\partial x_3} + \frac{\partial u_3}{\partial x_1}\right)^2$$

is the viscous energy dissipation function for a Newtonian fluid, S is the energy released or external energy input per unit volume of the gas per unit time; $q_{c,i}$ is the x_i-component of the conductive heat flow according to the Fourier law $q_{c,i} = -\lambda \frac{\partial T}{\partial x_i}$, λ is the heat conductivity coefficient; $q_{R,i}$ is the x_i-component of integral radiation flow density

$$q_{R,i} = \int_{4\pi} J\vec{\Omega}\mathbf{n}_i d\mathbf{\Omega}$$

where n_i is the unit vector in the x_i-direction.

4. **Transport equation for selective radiation.**

$$\frac{1}{c}\frac{\partial J_\nu}{\partial t} + \frac{\partial J_\nu}{\partial s} + (\kappa_\nu + \sigma_\nu)J_\nu = j_\nu + \frac{\sigma_\nu}{4\pi}\int_{4\pi} p_\nu(\vec{\Omega}', \vec{\Omega})J'_\nu d\Omega', \tag{4}$$

where $J_\nu = J_\nu(s, \vec{\Omega}, t)$; $J'_\nu = J_\nu(s, \vec{\Omega}', t)$ is the spectral intensity of radiation in the medium; s is the coordinate along the ray in the $\vec{\Omega}$-direction; $k_\nu = k_\nu(s)$, $\sigma_\nu = \sigma_\nu(s)$ are the volume spectral coefficients of absorption and scattering, $p_\nu(\vec{\Omega}', \vec{\Omega})$ is the spectral phase function, $j_\nu = j_\nu(s)$ is the volume spectral coefficient of the radiation emission. According to Kirchhoff's law, j_ν is proportional to the Plank function

$$j_\nu = \kappa_\nu J_{b\nu} \tag{5}$$

where $J_{b\nu}$ is the black-body spectral intensity (Plank's intensity). When solving problems of CRCC concerned with nonrelativistic motion of a medium, the unsteady term on the left-hand side of Equation (4) can be ignored.

The systems of equations for CRCC is closed by the state equation which, for the perfect medium, is of the form

$$p = \rho \frac{R_0}{M} T, \quad \textbf{(6)}$$

where M is the molecular weight and $R_0 = 8314$ J/K · kmol is the universal gas constant.

To solve CRCC problems for multiphase, multicomponent and multitemperature media, one should formulate similar equations for each component and include models, where appropriate, for the turbulent motion of the radiating gas. The necessary steps to solve particular problems are:

- Development of relationships for the thermodynamic, thermophysical, transfer and optical properties of the media;
- Formulation of initial and boundary conditions, including the specification of thermophysical, physical-chemical and optical properties of surfaces;
- Solution of the problem for gas flow with internal heat sources;
- Solution of selective radiation problems and determination of the radiation flow fields and their divergence.

If the radiation stress tensor is negligibly small compared with the viscous stresses, then the radiational energy influences the medium motion described by Equations (1) and (2) only by its influence on the temperature field. The relationship between various heat exchange mechanisms is obtained on the basis of the energy conservation equation (3), written in dimensionless form

$$\rho c_p \frac{dT}{dt} = \left(-\frac{1}{PrRe} \frac{\partial}{\partial x_i} q_{c,i} \right) - \frac{1}{Bo} \frac{\partial q_{R,i}}{\partial x_i}$$

$$+ Ec \frac{dp}{dt} + \frac{S}{Bo^*} + \frac{Ec}{Re} \mu \Phi, \quad \textbf{(7)}$$

where ρ, c_p, μ, λ are referred to their values at characteristic temperature T_0 (accordingly ρ_0, c_{p0}, λ_0); and T, p, S, u are referred to T_0, $\rho_0 u_0^2$, S_0, u_0; t and x_i are referred to L/u_0 and L, where L is the typical space length.

Heat transfer mechanisms are characterized by the dimensionless similitude criteria of Prandtl ($Pr = c_{p0}\mu_0/\lambda_0$), Reynolds ($Re = \rho_0 u_0 L/\mu_0$), Boltzmann ($Bo = \rho_0 u_0 c_{p0} T_0/\bar{\sigma} T_0^4$), Eckert ($Ec = u_0^2/c_{p0} T_0$) and the generalized Boltzmann criterion ($Bo^* = \rho_0 u_0 c_{p0} T_0/LS_0$). The Boltzmann criterion gives an estimation of the relationship between convective, conductive and radiation heat transfer

$$Bo = 4 \cdot N \cdot Re \cdot Pr/\tau_0 = 4 \cdot N \cdot Pe/\tau_0, \quad \textbf{(8)}$$

where $N_0 = \lambda_0/4\bar{\sigma} T_0^3 l_0$ is the radiation-convective parameter (to compare heat conductivity and radiation contributions into the process); $Pe = Re \cdot Pr$ is the Pecklet number (to compare heat conductivity and convection), $\tau_0 = L/l_0$ is the typical optical thickness, l_0 is the mean free path to characterize radiation heat transfer. There are problems in estimately the latter value since it characterizes the photon mean free path averaged over the entire frequency spectrum while the spectral mean free paths vary by a few orders of magnitude. The approximations of optically thin ($\tau_0 \ll 1$) and optically thick ($\tau_0 \gg 0$) media are widely used in CRCC analyses since they make it possible to avoid the solution of the transport equation (4). In particular, for $\tau_0 \ll 1$:

$$\frac{\partial q_{R,i}}{\partial x_i} \approx 4\bar{\sigma} k_p T^4, \quad \textbf{(9)}$$

which corresponds to radiation energy losses. In this last equation, k_p is the mean Planck absorption coefficient. For $\tau_0 \gg 1$, radiation flow can be represented in the Fourier law form

$$q_{R,i} = -\lambda_R \frac{\partial T}{\partial x_i}$$

where $\lambda_R = 16\bar{\sigma} T^3/3k_R$ is the radiative conduction coefficient, with k_R as the mean Rosseland coefficient.

If the medium is optically thin for certain wave lengths and thick for others, the above-mentioned approximations for the radiation flux and its divergence can not apply. Instead, the transport equation (4) must be solved. For a highly-scattering medium, the integro-differential form of Equation (4) must be solved and this makes the problem more complicated. The standard way to simplify the problem is to use assumptions on diffusional or δ-like character of the scattering and then to solve the problem employing diffusion approximation. This approximation appears to be even more justified in this case than in the nonscattering case, where it can also be recommended for application.

CRCC problems have been intensively studied theoretically and experimentally since the mid-50s and are central to the achievements in aerodynamics and plasma-dynamics, astrophysics, in nuclear reactor theory, shock waves and high temperature physics. The classical problems of CRCC are problems of viscous, heat conductive radiating gas flow in shock and boundary layers and of Couette flow in radiation magnetogasdynamics.

All CRCC regimes appear in the problems of laser wave deflagration, which are analyzed for optic plasmatron and for laser thruster development. For a small laser beam, laser-supported plasma moves to meet the radiation due to the heat conductivity mechanism, giving rise to motion of the surrounding gas. When laser beam size increases, the heat conductivity mechanism for plasma motion is replaced by a radiational one due to the absorption of radiation in the cold gas surrounding the plasma.

References

Bond, J. W., Watson, K. M., and Welch, J. A. (1965) *Atomic Theory of Gas Dynamics,* Addison-Wesley Reading.

Ozisik, M. N. (1973) *Radiative Transfer and Interaction with Conduction and Convection,* A Wiley-Interscience Publication.

Pai, S. I. (1966) *Radiation Gas Dynamics,* Springer Verlag.

Siegel, R. and Howell, J. R. (1972) *Thermal Radiation Heat Transfer,* Mc. Graw-Hill Book Company.

Sparrow, E. M. and Cess, R. D. (1970) *Radiation Heat Transfer,* Brooks/Cole Publishing Company.

Leading to: *Coupled radiation and convection*

S.T. Suzzhikov

COVALENT BONDING (see Atom; Molecule)

COWPER STOVES (see Regenerative heat exchangers)

CRACKING (see Pyrolysis)

CRAMER'S RULE

Cramer's rule is a simple method of solving n linear equations in n unknowns.

$$\begin{cases} a_{11}x_1 + a_{22}x_2 + \cdots + a_{1n}x_n = b_1 \\ a_{21}x_1 + a_{22}x_2 + \cdots + a_{2n}x_n = b_2 \\ \vdots \qquad\qquad \vdots \qquad\qquad \vdots \\ a_{n1}x_1 + z_{n2}x_2 + \cdots + a_{nn}x_n = b_n. \end{cases} \qquad (1)$$

If the *determinant* of system (1)

$$D = \begin{vmatrix} a_{11} & a_{12} & \cdots & a_{1n} \\ a_{21} & a_{22} & \cdots & a_{2n} \\ \vdots & \vdots & \ddots & \vdots \\ a_{n1} & a_{n2} & \cdots & a_{nn} \end{vmatrix} \qquad (2)$$

is not zero, then system (1) has a unique solution

$$x_k = \frac{D_k}{D},$$

where D_k is the determinant obtained from the determinant of system D by replacing the elements $a_{1k}, a_{2k}, \ldots, a_{nk}$ of the kth column by the corresponding free terms $b_1, b_2, \ldots, b_n$ or

$$D_k = \sum_{i=1}^{n} A_{ik}b_i \qquad k = 1, 2, \ldots, n, \qquad (3)$$

where A_{ik} is a cofactor of the element a_{ik} of determinant D.

Thus, the solution of a linear system of Equation (1) in n unknowns reduces to the calculation of the (n + 1)th determinant of order n. The number of operations required to solve the system of equations (1) with the help of Cramer's rule is thus proportional to (n + 1). For a sufficiently large n, the solution of system (1) with the use of Cramer's rule is computer-time consuming and presents a considerable challenge to practical calculations of heat transfer problems.

V.N. Dvortsov

CRE (see Coal Research Establishment)

CREEPING FLOW (see Cross flow; Flow of fluids)

CRITICAL CONCENTRATION (see Thickeners)

CRITICAL DEPOSITION VELOCITY (see Liquid-solid flow)

CRITICAL FLOW

Following from: *Compressible flow; Gas-liquid flow*

Critical, or choked, flow is not only an interesting academic problem but is also important in many practical applications, such as in power generation and in chemical process industries where, without a precise knowledge of critical flow behaviour, safety or performance of a system may be compromised. Experimental data may be available for a specific application but reliance often has to be placed upon theoretical models which, in turn, must be validated against well-qualified data.

Single-phase choking is well understood [see, for example, Roy (1988)] but, with the introduction of a phase change, the behaviour of fluid becomes far more complex [for a detailed review of two-phase critical flow models and data, see Elias and Lellouche (1994)]. The occurrence of choking in a system is conventionally defined as the maximum mass flow rate as a function of downstream pressure. To illustrate this, consider **Figure 1** where a large reservoir is connected by a flow path of arbitrary geometry to another reservoir whose thermodynamic state can be controlled precisely.

Reducing P_1 promotes the flow of fluid from A to B at a rate which increases with pressure drop until the velocity at some point in the connector achieves the local sonic velocity.

A choked plane forms at this location and further reductions in downstream pressure have no effect on conditions upstream as the rarefaction waves travel at the local sound speed and are stalled at the choked plane. However, further reductions in P_1 will increase pressure drop across the choked plane—where the pressure gradient is now mathematically indeterminate—although physically the pressure drop occurs over a finite distance, resulting in a large pressure gradient. The ratio of the critical pressure P_c at the choked plane to the inlet pressure P_0 is known as the *critical pressure ratio* (P_c/P_0). The geometry of the flow path has a direct bearing on the flow. For a converging/diverging *nozzle,* the choked plane forms at the minimum flow area. For example, in an isentropic flow in a *De Laval nozzle,* the critical pressure ratio is given by:

$$\frac{P_c}{P_0} = \left(\frac{2}{\gamma + 1}\right)^{\gamma/(\gamma-1)} \qquad (1)$$

For air, with $\gamma = 1.4$, the critical pressure ratio is 0.528 and for steam ($\gamma = 1.3$), 0.546. The maximum flow rate is given by:

$$M_c = A_c\sqrt{\gamma P_0 \rho_0}\left(\frac{2}{\gamma + 1}\right)^{(\gamma+1)/2(\gamma-1)}, \qquad (2)$$

where A_c is the area of the choked plane, which in our example is the nozzle throat (the minimum flow area), and ρ_0 is the fluid density, corresponding to P_0, T_0.

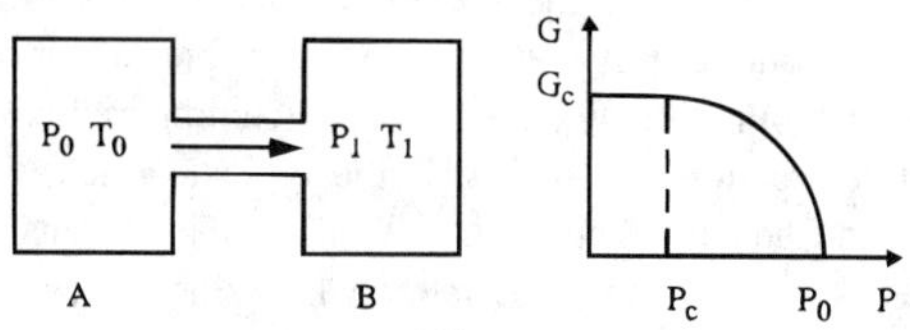

Figure 1. Flow between two arbitrarily large reservoirs.

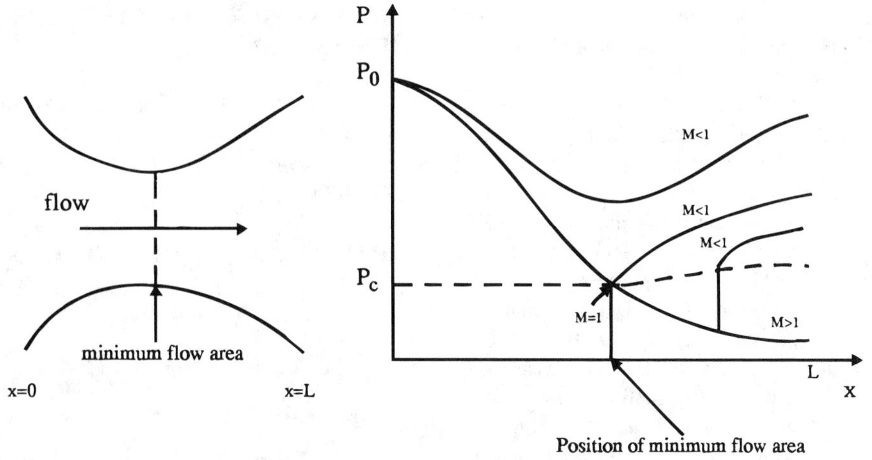

Figure 2. Axial pressure profiles in a De Laval nozzle.

For steady, 1-D horizontal flow, the *Euler equation* reduces to:

$$u du + \frac{dP}{\rho} = 0 \tag{3}$$

and introducing the *sonic velocity* $c^2 = dP/d\rho$ gives,

$$\frac{d\rho}{\rho} = -M^2 \frac{du}{u} \tag{4}$$

where M is the Mach number, defined by the ratio of flow velocity to local sonic velocity. Invoking continuity of mass then gives:

$$\frac{du}{u} = \frac{1}{M^2 - 1} \frac{dA}{A}. \tag{5}$$

Thus for subsonic velocities, a decrease in flow area results in an increase in flow velocity whereas for supersonic velocities, the converse is true. De Laval nozzles are often used in steam turbines, where a *condensation shock* may form downstream of the throat if the static pressure falls sufficiently for thermodynamic conditions to cross the *Wilson line* (a line approximately parallel to the saturation line and 115 kJ/kg below it) on the steam/water h-s diagram.

Therefore, repeating the experiment with the two reservoirs now connected by a De Laval nozzle will give the characteristic family of axial pressure profiles as in **Figure 2**.

Generalising to *two-phase critical flow*—where it is now assumed that reservoir A contains liquid at or near saturation conditions—it can be seen that for a sufficiently long flow path, the static pressure of the fluid accelerating through the connector will eventually fall to a level where flashing to vapour begins. Bubble nucleation and growth rely on heat transfer at the vapour/liquid interface, which introduces a time delay (typically ~1ms) in the development of voids, similar in many respects to the case of a condensation shock described above. The degree of sustainable superheat prior to void formation depends upon the availability of nucleation sites on the walls and in the bulk of the fluid. Shin and Jones (1993) have considered this problem and have introduced a critical flow model which included the effect of wall nucleation and bubble growth.

Once flashing to vapour occurs, the presence of vapour bubbles reduces the average density of the fluid, and hence the mass flow rate. It also has an impact on local sound speed, which exhibits a dependence upon both pressure and frequency as well as the character of the pressure disturbance itself i.e., a pressure pulse or a continuous wave [see Chen et al. (1983) for a detailed discussion].

A theoretical criterion for determining critical flow in two-phase systems can be derived from the method of characteristics and the previously-mentioned mathematical indeterminacy at the choked plane. For a detailed discussion, see Giot (1981). Various

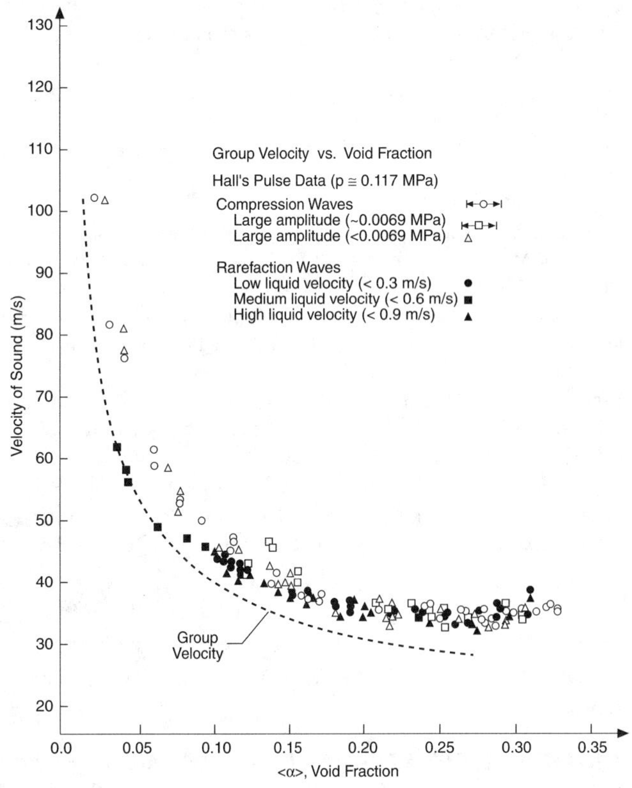

Figure 3. Sonic velocity vs. void fraction.

models for critical flow have been developed over many decades; early attempts include the Homogenous Equilibrium Model (HEM) through to more sophisticated models, including a number of full two-fluid six equation models. However, none of the currently-available critical flow models, to this author's knowledge, have been able to account for the full measure of critical flow parameters.

In the absence of a sufficiently accurate critical flow model, trends in the data may be identified to allow extrapolation from available experiments. Holmes and Allen (1995) have identified a number of data trends in two-phase critical flow for the purposes of Pressurised Water Reactor safety studies. Amongst the most important are that increasing inlet stagnation pressure leads to generally increasing choked flow rates, although evidence existed that this increase is not always monotonic and—for a given inlet stagnation pressure—the variation in fluid density and sound speed with quality combines to produce a monotonically decreasing mass flow rate with increasing quality. Also, sharp-edged inlet geometries generate *vena contracta,* reducing the area of the choked plane and hence mass flow rates, whereas well-rounded or gradually converging inlets maximise flow rates. The effects of inlet geometry are more marked for saturated inlet conditions and shorter flow paths. Exit geometry may also affect flow rates for short flow paths, where delayed flashing can shift the choked plane downstream of the minimum flow area in diverging nozzle, resulting in a choked plane with a larger surface area. For short flow paths, where the effects of delayed flashing dominate mass flow, the correlating parameter is flow path length, whereas longer flow paths allow equilibration and wall friction dominates so that the L/D ratio is the appropriate correlating parameter. Choking in complex flow paths, such as safety valves, may differ from design values owing to flow separations and the presence of noncondensable gases and particulates which reduce flow rates by promoting earlier flashing to vapour than would be experienced in a pure liquid.

References

Chen, L-Y, Drew, D. A., and Lahey, Jr. R. T. (1983) *An Analysis of Wave Dispersion, Sonic Velocity and Critical Flow in Two-Phase Mixtures.* NUREG/CR-3372, July.

Elias, E. and Lellouche, G. S. (1994) *Two-Phase Critical Flow.* International Journal of Multiphase Flow, Annual Reviews in Multiphase Flow 1994, edited by G. Hetsroni.

Giot, M. (1981) *Critical Flow. Thermohydraulics of Two-Phase Systems for Industrial Design and Nuclear Engineering.* Hemisphere Publishing Corporation.

Holmes, B. J. and Allen, E. J. (1995) *A Review of Critical Flow Data for Pressurised Water Reactor Safety Studies.* Critical Reviews of Multi-Phase Science and Technology, to be published.

Roy, D. N. (1988) *Applied Fluid Mechanics.* Ellis Horwood Ltd.

Shin, T. S. and Jones, O. C. (1993) *Nucleation and Flashing in Nozzles-1 A Distributed Nucleation Model. Int. J. Multiphase Flow,* 19, 6, 943–964.

B.J. Holmes

CRITICAL FLOW RATE, IN ORIFICES (see Orifice flowmeters)

CRITICAL HEAT FLUX, CHF (see Boiling; Boiling water reactor, BWR; Burnout, forced convection; Post dryout heat transfer; Nuclear reactors; Rod bundles, heat transfer in)

CRITICAL HEAT FLUX IN BOILING LIQUID METALS (see Liquid metals)

CRITICAL HEAT FLUX IN COILS (see Helical coil boilers)

CRITICALITY (see Nuclear reactors)

CRITICAL POINT, DRYING CURVE (see Drying)

CRITICAL POINT, THERMODYNAMICS

Following from: *Thermodynamics*

Figure 1 shows a plot of the relationship of the pressure p in a pure substance to its molar volume, $\tilde{v}$, for various temperatures, T, while **Figure 2** shows a projection of the same behaviour with pressure and temperature as the coordinates and volume as a parameter.

If a pure liquid is contained in a sealed tube in equilibrium with its vapour at a temperature T_1 then, in a gravitational field, the top

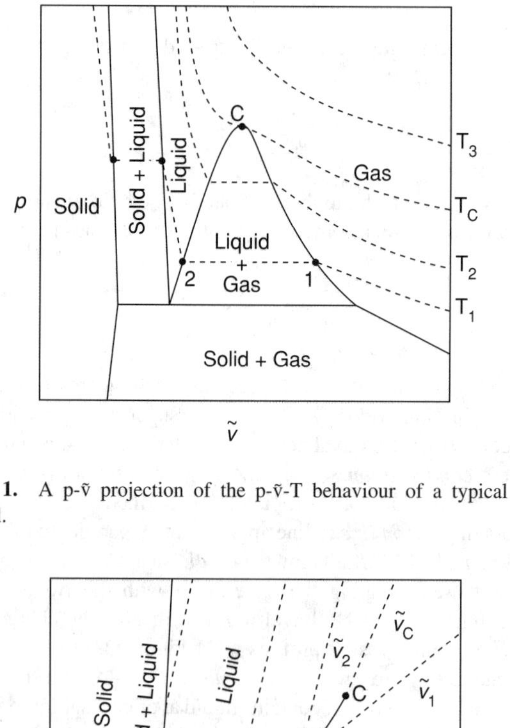

Figure 1. A p-$\tilde{v}$ projection of the p-$\tilde{v}$-T behaviour of a typical pure material.

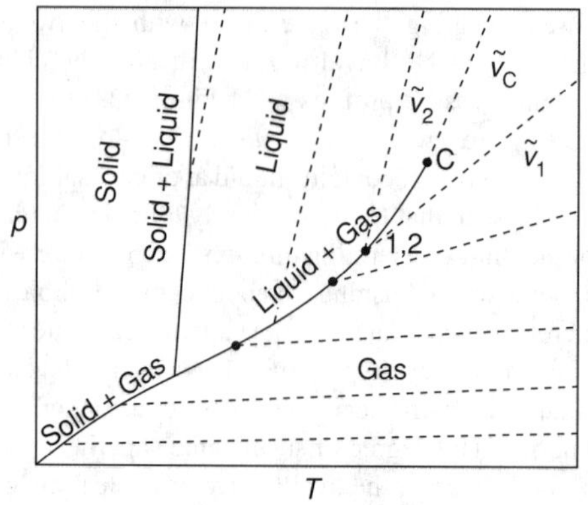

Figure 2. A p-T projection of the p-$\tilde{v}$-T behaviour of a typical pure material.

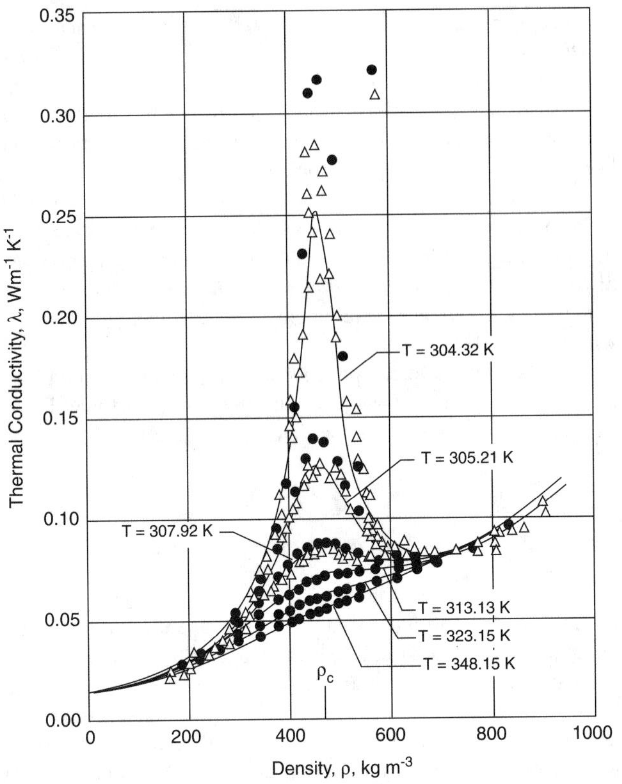

Figure 3. Thermal conductivity of carbon dioxide near the critical point.

of the liquid is indicated by an interface (meniscus). Such a situation corresponds to the isotherm T_1 in **Figure 1** where point 2 represents the system at the vapour-pressure of the liquid and corresponds to the liquid molar volume, whereas point 1 corresponds to the larger molar volume of the vapour phase at the same pressure and temperature. In **Figure 2,** both points 1 and 2 lie on the vapour-pressure curve which separates liquid and vapour phases. If tube temperature is increased to T_2 in **Figure 1**, vapour pressure is increased along 1c of **Figure 2**. At the same time, the molar volume of the liquid phase is increased and that of the vapour decreased, as shown in **Figure 1**. If this process is continued, the liquid phase and the vapour phase eventually become indistinguishable at some temperature, which is the situation indicated by point c in **Figures 1** and **2**. At this temperature, the *critical temperature*, T_c, the meniscus disappears and the molar volumes of the two phases are equal to $\tilde{v}_c$. Critical temperature may also be defined as the highest temperature at which the liquid phase of a substance can exist.

As can be seen from **Figure 1** along the critical isotherm T_c, the relationship between fluid pressure and its molar volume shows a point of inflection at the molar volume $\tilde{v}_c$, the *critical volume,* and the corresponding *critical pressure* p_c, so that the conditions

$$\left(\frac{\partial p}{\partial \tilde{v}}\right)_{T_c=0} \text{ and } \left(\frac{\partial^2 p}{\partial \tilde{v}^2}\right)_{T_c} = 0 \left(\frac{\partial^3 p}{\partial \tilde{v}^3}\right)_{T_c} < 0$$

define the **critical point**. This is the unique thermodynamic state for which, at temperature T_c, molar volume is $\tilde{v}_c$ and pressure, p_c. It is necessary only to prescribe two of these critical state parameters since the third is then automatically determined.

The critical state parameters T_c, $\tilde{v}_c$ and p_c are characteristic of each pure substance and must be determined experimentally. An up-to-date and extensive compilation of the critical state parameters of pure substances is currently being prepared by the Subcommittee on Thermodynamic Data of the IUPAC Commission on Thermodynamics [Young (1995); Ambrose (1995)].

Many of the other physical properties of pure fluids exhibit special behaviour near the critical point, which is the subject of current intensive investigation [Sengers and Levelt-Sengers (1986)]. For example, the heat capacity of fluid becomes infinitely large at the critical point as does thermal conductivity, whereas thermal diffusivity becomes zero.

Figure 3 shows experimental results for the thermal conductivity of carbon dioxide near the critical point. Some of these effects, a result of large-sale fluctuations in the density of fluid, extend over a wide range of temperature and density around the critical point. For example, the effect upon isobaric heat capacity and thermal conductivity can be more than 10%, even 50K, above the critical temperature at the critical density.

References

Ambrose, D. (1995) Article to appear in *Pure and Applied Chemistry,* citation not yet available.

Bett, K. E., Rowlinson, J. S. and Saville, G. (1975) *Thermodynamics for Chemical Engineers,* Athlone Press, London.

Sengers, J. V. and Levelt-Sengers, J. M. H. (1986) *Ann. Rev. Phys. Chem. 37,* 189.

Young, G. J. (1995) Article to appear in *Pure and Applied Chemistry,* citation not yet available.

Leading to: *Equations of state; Critical properties*

W.A. Wakeham

CRITICAL PRESSURE (see Critical point, thermodynamics; Pressure)

CRITICAL PRESSURE RATIO (see Critical flow)

CRITICAL SEDIMENTATION POINT (see Sedimentation)

CRITICAL STATE (see Critical point, thermodynamics)

CRITICAL SURFACE TENSION (see Contact angle)

CRITICAL TEMPERATURE (see Critical point, thermodynamics)

CRITICAL TEMPERATURE, FOR SUPERCONDUCTIVITY (see Superconductors; Supercritical heat transfer)

CRITICAL TRANSITION VELOCITY (see Liquid-solid flow)

CRITICAL ZONE (see Thickeners)

CROCCO'S THEOREM (see Compressible flow)

CROCCO TRANSFORMATION (see Boundary layer heat transfer)

CROSS CORRELATION (see Correlation analysis)

Following from: *External flow, overview*

Flow of a fluid normal to objects or groups of objects such as cylinders is referred to as crossflow. It is to be differentiated from *longitudinal flow* along the axis of symmetry, the term *inclined crossflow* being used to describe the intermediate condition. A crossflow is usually considered an **External Flow**, though for objects inside a duct, or groups of objects, there will be an overall pressure gradient and the flow may have some features of an *internal flow.* The objects could be *bluff* or *streamlined*, i.e., the cylinders need not be circular.

The flow of a *perfect fluid* (potential flow) past a cylinder bifurcates at the front edge of the body, $\phi = 0°$, where static pressure is a maximum, accelerates to the pressure minimum at 90°, and decelerates in the presence of the adverse pressure gradient, re-uniting at $\phi = 180°$. For a *real fluid*, the flow is substantially influenced by viscous effects, and the **Streamlines** are concentrated in a **Boundary Layer** near the wall (**Figure 1**). The solid wall also exerts *drag* on the fluid. Such a flow is characterised by the ratio of inertia to viscous forces, or **Reynolds Number**, Re,

$$\mathrm{Re} \equiv \frac{\rho u D}{\eta} \qquad (1)$$

where ρ is fluid density, η is fluid viscosity and u is a mean or bulk fluid velocity. D is a diameter or other suitable length, such as a hydraulic diameter.

For a very small Re or **Creeping Flow**, the streamlines are symmetric and drag is mainly due to skin-friction. In the range $1 \le \mathrm{Re} \le 40$, symmetry is lost and *boundary-layer separation*

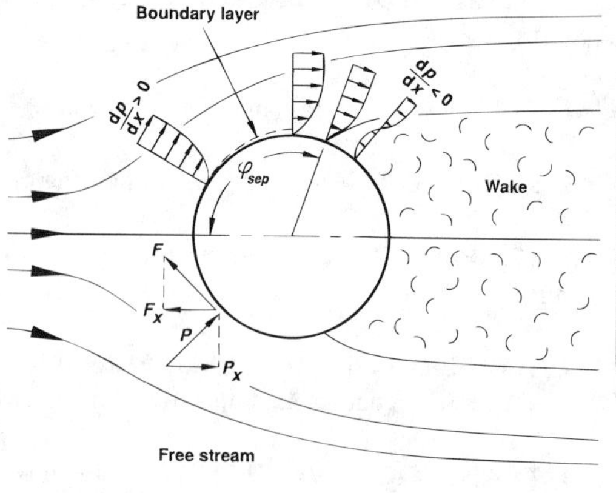

Figure 1. Crossflow past a cylinder.

occurs at $\phi = 82°$. Downstream, a *wake* forms with two stationary **Vortices** being observed at the rear of the cylinder. The flow is **Laminar**. For Re > 40, the flow becomes periodic with the vortices being shed alternately at a frequency, f, corresponding to a **Strouhal Number**, Sr = fD/η of 0.2. As Re further increases, the behaviour becomes less coherent and turbulent eddies are observed in the wake. At the *critical Re,* $\mathrm{Re_c} = 2 \times 10^5$, the separation point suddenly moves downstream, $\phi > 90°$, and the boundary layer becomes *turbulent*. There may be a separation bubble. The presence of *free-stream turbulence* or *surface-roughness* will decrease $\mathrm{Re_c}$. In the *supercritical* range, a turbulent periodic behaviour is apparent with Sr around 0.3.

Drag is due to two components: a) *skin-friction* or the viscous shear-force, F; and b) *form drag* due to the normal force, P, i.e., **Pressure**. The overall drag coefficient, c_D, is defined as:

$$c_D = \frac{F_x + P_x}{\frac{1}{2}\rho u^2 A_p} \qquad (2)$$

where F_x and P_x are integrated values projected in the main-flow direction (see **Figure 1**) and A_p is the area (usually projected normal to the flow for a bluff body, and parallel to the flow for a streamline body). **Figure 2** shows c_D vs. Re for a single cylinder. At low Re, drag is due mainly to skin-friction and c_D decreases with Re. By Re = 200, drag is almost entirely pressure-related and c_D is fairly constant; the slight rise in the range $2 \times 10^3 - 2 \times 10^5$ is a result of wake turbulence and narrowing of the wake due to upstream movement of the separation-point. The so-called drag crisis occurs at $\mathrm{Re_c}$, with c_D falling abruptly as the wake visibly narrows.

This description is qualitatively true for other similar geometries such as spheres. (See **Spheres, Flow Around and Drag**) For other shapes, there are significant differences in both flow and pressure fields. For streamlined bodies such as a rectangular fin (flat plate) aligned with the flow, form drag and hence, c_D, will be much reduced. Conversely, for a baffle-plate placed across the flow, pressure will dominate, the wake will be of uniform width, and c_D will be a relatively constant value of around 2 over a wide Re-range. Other effects such as blockage, free-stream turbulence, surface roughness, and secondary-flows in complex geometries such as finned tubes, will significantly alter the nature of the flow.

For flow past *banks of objects*, there is in an overall pressure gradient. The *friction coefficient,* f, is typically defined as:

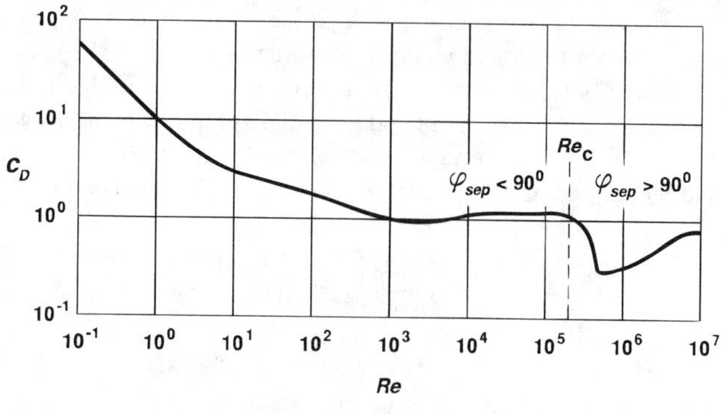

Figure 2. Drag coefficient vs. Re for a cylinder.

$$f = \frac{\tau}{\frac{1}{2}\rho u^2} = \frac{A_c}{A}\frac{\Delta\bar{p}}{\frac{1}{2}\rho u^2} = \frac{r_h}{L}\frac{\Delta\bar{p}}{\frac{1}{2}\rho u^2}, \quad (3)$$

where $\tau = (F_x + P_x)/A$ is an equivalent stress due to drag on *all* surfaces in the bank (which may or may not include walls, entrance effects, etc.). $\Delta\bar{p}$ is the mean pressure drop along distance L; A_c is the *minimum* free cross-sectional area; A is the total (unprojected) surface area in contact with the fluid; and $r_h = A_cL/A$ is a hydraulic radius [Kays and London (1984)]. The term radius is actually a misnomer which has persisted for historical reasons; since for a circular cylinder $r_h = r/2$. Numerous variations on f exist, e.g., the **Euler Number**, Eu (see **Tube Banks**, **Crossflow Over**). Charts of f, c_D, or Eu as a function of Re, present important information to the fluids engineer and are available for a wide variety of geometries and flow conditions.

Reference

Kays, W. M. and London, A. L. (1984) *Compact Heat Exchangers,* McGraw-Hill, New York.

Leading to: *Crossflow heat transfer; Tubes, crossflow over; Tube banks, crossflow over*

S.B. Beale

CROSSFLOW HEAT TRANSFER

Following from: *Crossflow*

When a fluid flows across a solid object or ensemble of solids at a different temperature, crossflow heat transfer results. Heat transfer is a function of **Reynolds Number**,

$$Re = \frac{\rho u D}{\eta} \quad (1)$$

and **Prandtl Number**,

$$Pr = \frac{\eta c_p}{\lambda} \quad (2)$$

where ρ is the density, u is a bulk velocity, η is viscosity, λ is the fluid conductivity, and c_p is specific heat. D is a characteristic length, such as a diameter. Another popular choice for crossflow heat exchangers is a hydraulic diameter, D_h.

At very low flow rates, heat transfer is by **Conduction** alone. If the objects are inclined to the vertical, *natural or mixed-convection* may be important. Under these circumstances, the engineer may also be interested in maintaining *thermal stratification*. As Re is increased, *forced-convection* becomes the dominant mode of heat transfer. The *local heat transfer coefficient*, α, is defined by,

$$\dot{q} = \alpha(T_w - T_b) \quad (3)$$

where $\dot{q}$ is the local wall heat flux, T_W is the wall temperature, and T_b is a reference bulk temperature. For single objects, T_b is chosen as the free-stream temperature. For banks of objects, a variety of references are used. α is nondimensionalized in terms of a local **Nusselt Number**, Nu,

$$Nu \equiv \frac{\alpha D}{\lambda} \quad (4)$$

Figure 1(a) and **(b)** show Nu distributions around a circular cylinder in the Re ranges 4×10^3 to 5×10^4 and 3.98×10^4 to 4.26×10^5, respectively. It can be seen that at low Re, Nu is a maximum at $\phi = 0°$, where skin-friction is zero. In this region, the flow is similar to that occurring at the stagnation point for flow normal to a flat-plate. Nu decreases to a minimum near the separation point, at the side of the cylinder, and then increases in the *wake.* Most heat transfer occurs through the front half of the cylinder. As Re increases, heat transfer in the latter half increases due to wake-turbulence. At $Re = 2 \times 10^5$, a *turbulent thermal boundary layer* is established and the Nu distribution is quite complex, with twin minima occurring at around 90° and 140° probably due to laminar-to-turbulent transition in the boundary-layer and

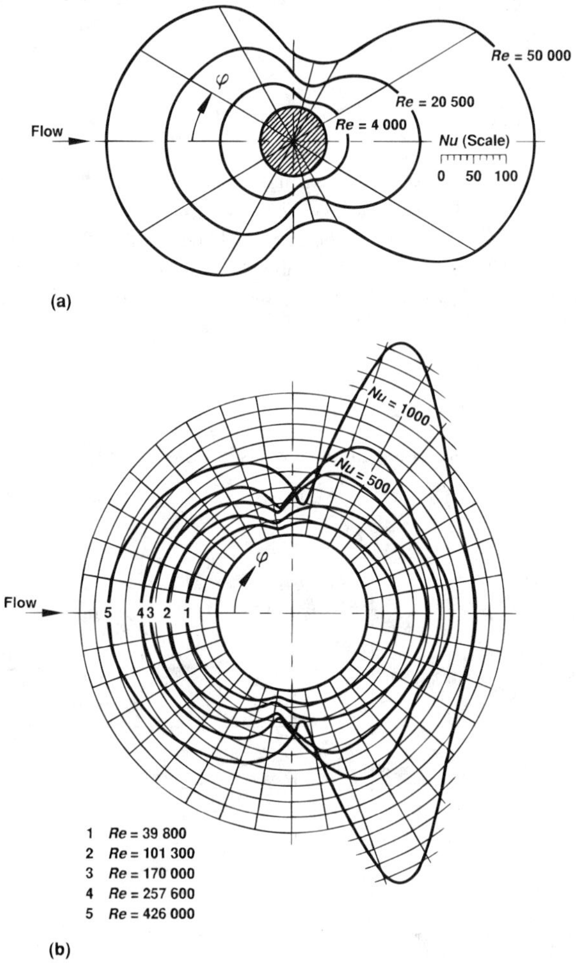

Figure 1. Distribution of local heat transfer coefficient around a circular cylinder for flow of air. *Sources:* a) Lohrisch (1929) and b) Schmidt and Wenner (1941), by permission of V.D.I. Verlag.

boundary-layer separation, respectively. At high Re, there is a large Nu maximum in the turbulent boundary-layer around $\phi = 110°$. Heat transfer at the rear stagnation-point is by now as much as at the front stagnation point.

Overall heat transfer is expressed in terms of an *average heat transfer coefficient*, $\bar{\alpha}$, defined by means of a *rate equation*,

$$\dot{Q} = \bar{\alpha} A \Delta T_M \tag{5}$$

where $\dot{Q}$ is the total rate of heat transfer, A is the total heat transfer area and ΔT_M is an average or effective temperature difference between the solid-wall and the bulk of the fluid. For situations involving the use of extended surfaces or *fins*, a modified rate equation is employed. (See **Tube Banks, Crossflow Over.**) $\bar{\alpha}$ is nondimensionalized either as an *average Nusselt number*, $\overline{Nu}$, according to Equation (4) or as an average **Stanton Number**, $\overline{St}$

$$\overline{St} = \frac{\bar{\alpha}}{\rho c_p u}. \tag{6}$$

Re and Pr are often evaluated at the bulk temperature, T_b, though some authors advocate the use of a *mean film temperature* $(T_w + T_b)/2$. $\overline{Nu}$ and $\overline{St}$ are frequently correlated according to a power-law relationship,

$$\overline{Nu} = (a + cRe^m)Pr^n \tag{7}$$

where c and m are Re-dependent and a accounts for natural convection, if present. Use of $n = {}^1/_3$, based on the *Colburn-Chilton analogy* is widespread, though empirical correlations may involve different values of m. (See **Analogy Between Heat, Mass and Momentum Transfer.**) **Figure 2** shows $\overline{Nu}$ and $\overline{St}$ as a function of Re and Pr. A Pr-independent *heat transfer factor*, j, is defined as,

$$j = \overline{St}Pr^{1-n} \tag{8}$$

and may be considered the heat transfer analogue of the friction factor, f/2.

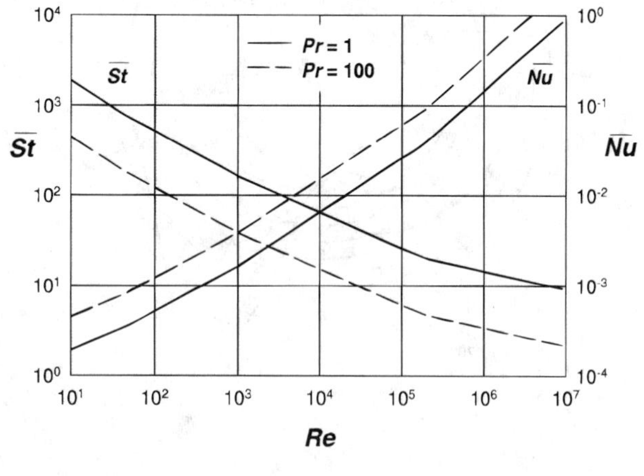

Figure 2. Average heat transfer for flow around a circular cylinder expressed in the form of Nusselt and Stanton numbers. *Source:* Žukauskas and Žiugžda. (1985).

Temperature variation of fluid properties affects both f and j. This may be accounted for by writing,

$$j = j'\left(\frac{Pr}{Pr_w}\right)^p \approx j'\left(\frac{\eta}{\eta_w}\right)^p \tag{9}$$

where j′ denotes a temperature-independent (adiabatic) value and Pr_w is evaluated at the wall temperature, T_w. f is treated in a similar manner. Temperature-independent values of f′ and j′ appear in the literature for a large number of geometries. These measures of overall performance are of much interest to the heat transfer engineer.

The geometry of the heat transfer surface will substantially alter the mechanisms of fluid flow and heat transfer in crossflow. Other influencing factors include: the effect of thermal boundary conditions (constant $\dot{q}_w$ vs. T_w), the influence of containing ducts or other blockage, free-stream turbulence, as well as the use of roughened surfaces or fins to enhance heat transfer.

References

Lohrische, W. (1929) Forschungsarbeiten auf dem Gebeite des Ingenieurwesens. 322, 46. (In German).

Schmidt, E. and Wenner, K. (1941) Forschung auf dem Gebeite des Ingenieurwesens. 12, 2, 65–73. (In German).

Žukauskas, A. and Žiugžda, J. (1985) Heat Transfer of a Cylinder in Crossflow. Hemisphere, New York. (Translator E. Bagdonaite, Editor G. F. Hewitt).

Leading to: *Immersed bodies, heat and mass transfer; Tube banks, single-phase heat transfer in; Tubes, single-phase heat transfer to in crossflow; Tubes and tube banks, boiling heat transfer on; Tube banks, condensation heat transfer in; Tubes, condensation on outside in crossflow*

S.B. Beale

CROSS SECTIONS (see Nuclear reactors)

CROSS SPECTRUM (see Spectral analysis)

CRUDE OIL (see Oil refining)

CRYOGENIC FLUIDS

It is generally agreed that **cryogenic fluids** are those whose boiling points (bp) at atmospheric pressure are about 120K or lower, although liquid ethylene with its boiling point of 170K is often included. A list of the cryogenic fluids, together with some selected properties, is given in **Table 1**. Detailed properties are available commercially on computer disc. (Cryodata Inc.)

Perhaps the most important and widely-used fluids are *liquefied natural gas or LNG* (bp = boiling point about 120 K), *liquid oxygen* (bp 90.2K) and *liquid nitrogen* (bp 77.3 K).

Table 1. Some properties of cryogens at their normal boiling point[a]

	He^4	n-H_2	D_2	Ne	N_2	CO	F_2	Ar	O_2	CH_4	Kr	Xe	C_2H_4
Normal boiling point (K)	4.22	20.4	23.7	27.1	77.3	81.7	85.0	87.3	90.2	111.6	120.0	165.0	169.4
Liquid density (kg/m^3)	125	71.0	163	1205	809	792	1502	1393	1141	423	2400	3040	568
Liquid density/vapour density	7.4	53	71	126	175	181	267	241	255	236	270	297	272
Enthalpy of vaporisation (kJ/kg)	20.42	446	301	86	199	216	175	161	213	512	108	96	482
Enthalpy of vaporisation (kJ/kg-mole)	80.6	899	1211	2333	5565	6040	6659	6441	6798	8206	9042	12,604	13,534
Volume of liquid vaporised by energy input of 1W-hr (cm^3)	1410	114	74	35	22	21	14	16	15	17	14	13	13
Dynamic viscosity of liquid (μNsec/m^2)	3.3	13.3	28.3	124	152	—	240	260	195	119	404	506	170
Surface tension (mN/m)	0.10	1.9	≈3	4.8	8.9	9.6	14.8	12.5	13.2	13.2	5.5	18.3	16.5
Thermal conductivity of liquid (mWm^{-1}K^{-1})	18.7	100	≈100	113	135	—	—	128	152	187	94	74	192
Volume of gas at 15°C released from 1 volume of liquid	739	830	830	1412	681	806	905	824	842	613	689	520	475

[a] Pressure of 1.01325 bar.
Source: Cryogenic Engineering, ed. B.A. Hands, Academic Press (1986).

The availability of cryogenic fluids forms an essential part of the infrastructure of a modern industrialised and civilised society. One of the major reasons for using liquid cryogens is to allow transport and storage as liquid at atmospheric pressure, rather than as high-pressure gas in thick-walled vessels, although there is an energy penalty involved in **Refrigeration**. However, the distillation of *liquid air* (air separation) enables the production of very high-purity oxygen and nitrogen. Plants producing up to several hundred tonnes per day and more of oxygen are commonplace, sometimes connected permanently to a chemical plant or steel works. Liquid nitrogen—formerly a by-product of the process—is now a product in its own right, being used principally as a convenient source of refrigeration, especially in the frozen food industry.

The other important by-product of air separation is *liquid argon*, which again can be produced at a very high purity. For welding, it is increasingly being stored as liquid at the factory rather than being delivered in high-pressure cylinders.

All cryogenic fluids except **Helium** and **Hydrogen** behave as 'normal' fluids, their common distinguishing features in general being a low specific heat and enthalpy of vaporisation. All gaseous cryogens are odourless and all liquid cryogens are colourless apart from **Oxygen**, which is pale blue, and **Fluorine**, which is pale yellow. They are all diamagnetic except oxygen, which is quite strongly paramagnetic.

With the exception of oxygen, all the gases are asphyxiants, and even oxygen will not support human life in concentrations greater than about 60%. Fluorine and oxygen are powerful oxidisers even in liquid form. Some cryogens are flammable; hydrogen is especially delicate to handle.

Hydrogen is an unusual fluid in that the molecule exists in two forms known as ortho and para, with somewhat different properties. The ratio of ortho to para is determined by conventional thermodynamics and is dependent on temperature. There are also different forms of isotopes (deuterium and tritium).

An explanation of the behaviour of the hydrogen molecule requires a knowledge of quantum mechanics and will not be discussed here. At low temperatures, equilibrium hydrogen (e-H_2) is entirely para. At room temperature, the ortho:para ratio is 3. The equilibrium state at room temperature is often known as normal hydrogen or *n-hydrogen*. The transition from the ortho to the para state involves a heat of conversion—which can be greater than the enthalpy of vaporisation—so that the vaporisation rates of hydrogen are often much larger than expected. It is for this reason that a catalyst is often included in a hydrogen liquefier to ensure that only para hydrogen is present in the liquid.

Helium is the one cryogenic fluid which can be claimed to be unique. Because of its low molecular weight and chemical inertness, quantum mechanical effects are important. There are two isotopic forms: the natural form He^4, which has a nucleus consisting of two protons and two neutrons; and the comparatively rare man-made form He^3, with only one neutron. The two isotopes have markedly different properties due to their different nuclear spins. He^3 is not considered here.

Below 2.2K, He^4 becomes *'superfluid'*, and is often known as HeII, the 'normal' liquid being known as 'HeI'. The locus of the HeI/HeII transition is known as the *'lambda line'* from the shape of the curve of specific heat as a function of temperature. The phase diagram of He^4 is shown in **Figure 1**, in which features of particular interest are the absence of a triple point and the fact that the liquid can only be solidified under pressure (greater than about 26 bars).

The temperature of the normal/superfluid transition depends somewhat on pressure. One end of this boundary forms with solid HeI and HeII the 'upper lambda point' (at 1.77K and 30.2 bars). The other end of the line (at 2.18K, 0.005 bar) where vapour, HeI and HeII coexist is known as the 'lower lambda point.'

HeI behaves as a conventional liquid (except when near the λ line) but requires much more care in handling than other cryogenic

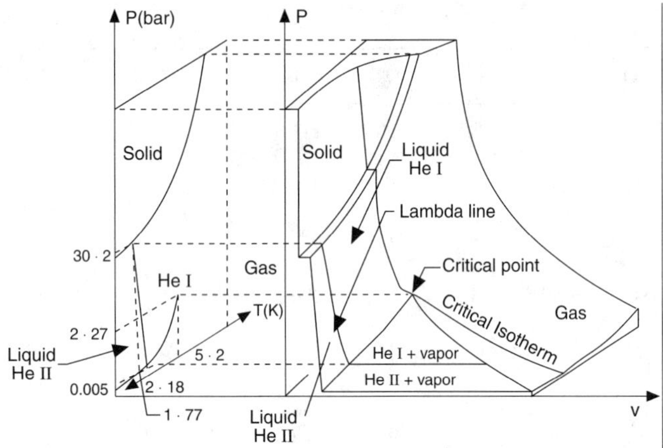

Figure 1. From Cryogenic Engineering (1986) by permission of Academic Press Ltd.

fluids, principally because of its extremely low latent heat of vaporisation. HeII is quite different, having a variety of properties quite different from those of any other liquid. It will, for instance, climb up over the edge of a container and drip off the bottom; it has a small or zero viscosity and a very large thermal conductivity. Flow velocity through fine capillaries is independent of the pressure head and is greater in tubes of smaller diameter. Flow may be induced by a temperature gradient in the absence of any pressure gradient. A consequence of the very high thermal conductivity is that below the λ point, boiling ceases and the liquid becomes "quiescent;" although the rate of heat transfer remains very high. Vinen has published a brief but useful review of the properties of superfluid helium.

References

Cryogenic Engineering, ed. B.A. Hands (1986) Academic Press.

Cryodata Inc. PO Box 558 Niwot, CO 80544 U.S.A.

Vinen, W. F. 'Physical Properties of Superfluid Helium—A General Review', *Proc. Workshop on Stability of Superconductors in Helium I and Helium II,* 43–51, Saclay, France. (International Institute of Refrigeration, 1981).

Leading to: *Cryogenic plant*

B.A. Hands

CRYOGENIC PLANT

Following from: *Cryogenic fluids; Liquefaction of gases*

A common feature of a cryogenic plant is the separation and/or **Liquefaction of Gases** at process conditions which may be at elevated pressures, but always involving very low temperatures. There are many industrial cryogenic processes which operate at temperatures in the region of −165°C to −195°C at their coldest point, with some operating as low as −269°C. Consequently, the conservation of cold becomes a dominant feature in the design of such processes, which focuses on highly efficient heat exchange. However, a typical cryogenic process has many elements to it, the cryogenic section being only a part of the whole flowscheme. **Figure 1** illustrates a typical cryogenic plant which consists of a pretreatment section, a cryogenic section and a compressor/expander section which provides refrigeration for the process. In many instances, feed compression may be required as in the case of air separation. Another aspect of cryogenic plant is the use of aluminium and stainless steel for the cold sections of the plant to avoid embrittlement problems encountered with carbon steel.

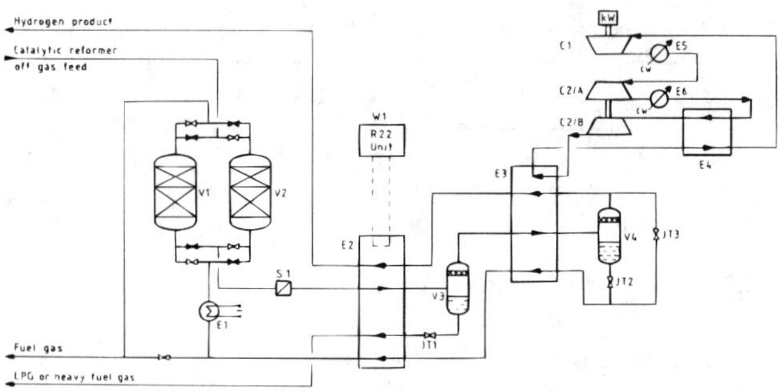

Figure 1. A flow sheet of a hydrogen recovery unit treating refinery off-gas.

Table 1. Freezable impurities present in cryogenic plant feed Gas

Feed gas	Typical impurities present
Air	H_2O, CO_2, traces of hydrocarbons
Natural gas	H_2O, CO_2, COS, Hg, heavy hydrocarbons, aromatics, CH_3OH
NH_3 purge gas	H_2O, NH_3
NH_3 plant synthesis gas	H_2O, CO_2
Petrochemical plant off-gases	H_2O, CO_2, H_2S, aromatics, CH_3OH
Refinery off-gases	H_2O, CO_2, H_2S, COS, NH_3, HCl NH_4Cl, HCN, arsenic, aromatics, heavy hydrocarbons

The prepurification section is usually needed upstream of a cryogenic plant because most feed gases will contain constituents that may freeze inside or even corrode the cryogenic equipment and will therefore require removal. Wet hydrocarbons can form hydrates at temperatures above 0°C (typically 5–15°C) and in such cases, water removal may be necessary. Freezable components include H_2O and CO_2 and NH_3. These are usually removed from the feed gas by absorption, adsorption, chilling, permeation, reversing heat exchangers, or combinations of these in a prepurification section.

Table 1 gives a list of typical gases treated by cryogenic units. Typical impurities are listed for each of these gases. In general, these impurities are removed upstream to less than 1 ppm. However, in some cases the liquids formed in the process dissolve the freezable components to a significant extent, making it possible to allow several hundred ppm to enter the cryogenic unit. Each case has to be considered carefully by the cryogenic unit designer, who uses reliable solubility and equilibrium data to determine the levels acceptable in the cryogenic unit. The impurity concentration in the

Table 2. Pretreatment methods used to treat freezable components

Component	Typical concentrations	Pretreatment used
H_2O	0–2000 v.p.m.	Adsorption, freezing*
	> 2000 v.p.m.	Chilling + adsorption. freezing*
CO_2	0–5000 v.p.m.	Adsorption, freezing*
	>5000 v.p.m to 5%	Absorption + adsorption. freezing*
	≤5%	Absorption, permeation
H_2S	0–5000 v.p.m.	Adsorption, reaction
	>5000 v.p.m.	Absorption, reaction, permeation
HCl	0–100 v.p.m.	Adsorption, absorption
COS, CS_2	0–50 v.p.m.	Reaction, adsorption
	>50 v.p.m.	Absorption, adsorption
CH_3OH	0–500 v.p.m.	Adsorption
	>500 v.p.m.	Adsorption, chilling
NH_3	0–2000 v.p.m.	Adsorption, absorption
	>2000 v.p.m.	Absorption, chilling
Heavy hydrocarbons and aromatics	>5000 v.p.m.	Adsorption, absorption, chilling

*By using reversing heat exchangers.

feed gas determines the nature of the pretreatment process used. **Table 2** shows a very general guide to pretreatment processes used for various impurity levels; each case must be carefully analysed to select the proper scheme. Impurities such as HCl, NH_4Cl, Hg and arsenic compounds are generally found in very low concentrations and catalytic methods are often used to remove them.

The clean, dry gas is admitted to the cryogenic part of the plant which cools, condenses and separates the desired components in the gas at low temperatures. In order to produce low temperatures, it is imperative that all parts of the cryogenic unit are designed as efficiently as possible (see **Liquefaction of Gases**). **Exergy** analysis, Tomlinson et al. (1990), shows how important process efficiency is in reducing power consumption and improving product yields.

Table 3. Types of heat exchangers used in cryogenic service

	Shell and tube	Wound coil	Plate-fin
1. Maximum single unit size (M × M × M)	2 dia. × 25 approx	4.3 dia × 25	1.0 × 1.4 × 6.5
2. Maximum design pressure (kPa)	>20 000	>20 000	10 000
3. Maximum number of streams	2 (3 in exceptional cases)	5–7	up to 7
4. Temperature limits (°C)	−269 to +350 or higher	−269 to + 350 or higher	−296 to +65
5. Construction material	Stainless steel/copper/ brass	Stainless steel/A1/ Cu	A1*
6. Typical approach temperature (°C)	6–10	3–5	2–4
7. Area for heat transfer per unit volume (m^3/m^3)	40–80	20–40	800–1200

*Recently, stainless steel designs have emerged for around 2000 kPa pressure.

Both reciprocating and rotating compression equipment is used in feed compression and refrigeration cycles in a cryogenic plant. The major difference between compressor equipment for cryogenic plant compared with noncryogenic plant is in the lubrication system. Many gas compressor preceding cryogenic processes are oil-free on the process side. If oil-lubricated compressors are used, then oil-removal systems are included after the machines. For these types of reciprocating compressors, piston rings are made of composite materials, which include graphite, thereby promoting lubrication. Packing glands are purged with dry gases to form a seal between the oil-lubricated crankcase side and the process end. In centrifugal machinery, a dry-seal system is provided to prevent process gas being contaminated with oil so that the metal bearings are allowed thorough lubrication while the process gas is kept dry.

Expansion turbines are used extensively to provide efficient refrigeration. These require dry operation for the process gas in a similar manner to ‘dry’ centrifugal compressors, and therefore employ similar lubrication and seal systems.

Reciprocating expanders are also used today in helium liquefiers, where flows are relatively small. Also, design of rotary expanders for low molecular weight gases to achieve a high efficiency is difficult without using very high speeds.

One of the most important aspects of a good cryogenic plant design is the effective use of heat exchangers. This is clearly brought out by exergy analysis during the conceptual design stage. Cryogenic plants generally utilise three types of heat exchangers: shell-and-tube, wound-coil, and plate-fin. The most widely used is the plate-fin. To appreciate why this is so, the three types of heat exchanger are compared in **Table 3** in terms of area per unit volume, maximum design pressure, practical approach temperatures, number of streams handled in one unit and materials of construction. The materials of construction have cryogenic service in mind. Another effective heat exchanger that can be applied in cryogenic service is the etched plate compact heat exchanger. This is particularly suitable where pressures exceed 10 MPa.

A cutaway drawing of a plate-fin heat exchanger is shown in **Figure 2**.

Distillation is extensively used in cryogenics to produce pure components from mixtures of gases. There are some significant differences in cryogenic plant mass transfer equipment which aim to make the units as compact as possible. Consequently, sieve trays with very low tray spacing (80 to 100mm) are extensively used in air separation and other areas. Structured packing is also seeing a significant increase in applications in cryogenic service. **Table 4** gives a summary of mass transfer equipment used in cryogenic plant.

To improve efficiency of cryogenic processes, process intensification is always seriously examined during process selection. An excellent example where this has succeeded is in the use of plate-fin heat exchangers to carry out simultaneous heat and mass transfer. **Figure 3** shows a section of plate-fin heat exchanger used for distillation purposes. This is particularly suitable for a situation where a lower molecular weight gas contains a few percent of

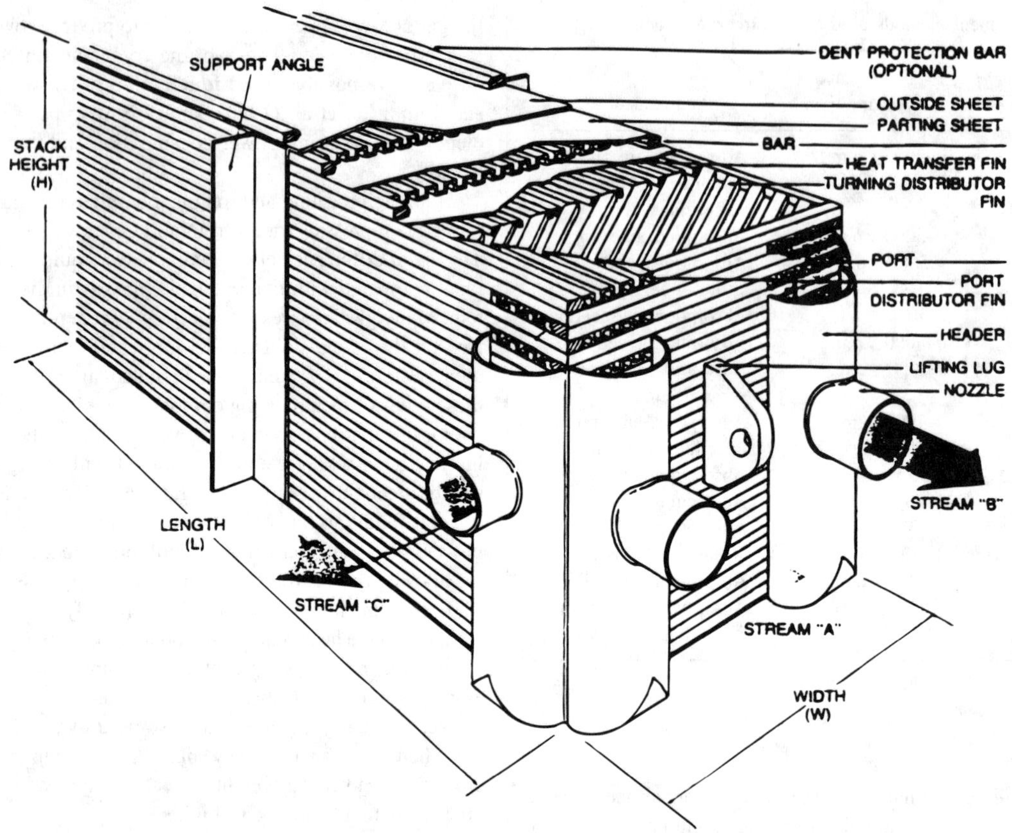

Figure 2.

Table 4. Mass transfer equipment summary

Method	Applications
Loose packing vertical in towers	Various types. Not used frequently in cryogenics. Uses mainly limited to argon recovery and rare gas recovery.
Structured packing	Various types. Increasingly used in air separation columns.
Valve and bubble cap trays	Various types. Used in hydrocarbons processing where good turndown is required.
Sieve trays (crossflow and annular)	Used in a variety of cryogenic applications. Lends itself to design for very low tray spacing.
Spray towers and falling films	Seldom used in cryogenics.
Plate-fin refluxing exchangers	Becoming very popular to facilitate simultaneous heat and mass transfer.

heavier components. The partly-cooled vapour may enter at point A, any condensate is knocked out in V1, the vapour then enters the reflux exchanger, E1, at B, and is cooled by refrigerant DE and product XY. The exchanger is always mounted vertically so that the feed stream cools in an upward direction to C. As it cools, condensate forms. Instead of the condensate passing up with the main gas flow, it runs back down the heat exchanger to the bottom and B. As it runs back, it is warmed by the upcoming vapour stream which tends to boil-off "light ends". Therefore, the liquid that emanates from the base B contains only small amounts of "light ends" and most of "the heavy ends". This liquid is approaching the equilibrium condition of the vapour entering at B and *not* the condition at the cold end of the exchanger C. This is the important factor of the reflux exchanger, which makes it more efficient than straight cooling and partial condensation. Several equilibrium stages can be catered for in a single long core, at the same time providing refrigeration at each theoretical stage.

The designer has to cater for heat transfer and distribution problems for plate-fins exchangers, as well as mass transfer. In counter-flow of vapour and liquid an appropriate flooding correlation is required.

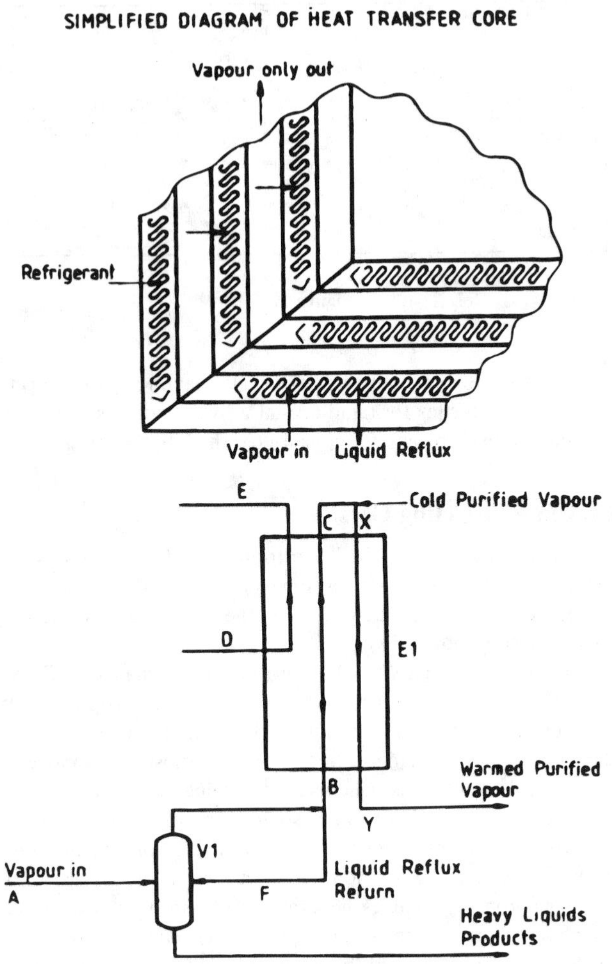

Figure 3. A plate-fin heat exchanger used as a mass transfer device.

References

Blakey, P. (1981) 'Cryogenic gases' *Chemical Engineering* (Oct 5) 113–24.

Clarke, M. E. and Gardener, J. B. (1976) 'Refrigeration with expansion turbines', *Contemp Phys.,* 17(6), 507–28.

Diehl, J. E. and Koppany, C. R. (1969) 'Flooding velocity correlation for gas liquid counterflow in vertical tubes'. *Chem. Eng. Prog. Symp. Series,* 65(92), 77.

Diery, W. (1984) 'The manufacture of plate-fin heat exchangers at Linde'. *Linde Reports Sci. Tech. 37.3.*

Gregory, E. J. (1987) 'Heat Exchangers'. In *Cryogenics Engineering,* ed. B. A. Hands), Chap. 8, 193 Academic Press, London.

Haselden, G. G. (1971) *Cryogenic Fundamentals.* Academic Press, London.

Holm, J. (1985) 'Energy recovery with turboexpander processes'. *C.E.P.* (July), 63.

Isalski, W. H. (1989) *Separation of Gases,* Oxford Science Publications.

Isalski, W. H. (1993) Refrigeration B/1, *Kempe's Engineers Yearbook,* Benn.

Keens, D. (1980) 'Technology of LNG plants still undergoing a steady evolution'. *Oil and Gas Journal* (Apr. 7) 84.

Tomlinson, T. R. et al. (1990) Exergy Analysis in Process Development, *The Chemical Engineer,* 18 and 25 October.

W.H. Isalski

CRYOGENIC PUMP (see Vacuum pumps)

CRYOGENIC USE OF STEEL (see Steels)

CRYOSCOPIC CONSTANT (see Solubility)

CRYOSTATS

***Following from:** Liquefaction of gases*

A cryostat (or *dewar*) is a vessel designed to store or transport liquefied gases (cryogens) at low temperatures. Details of design and construction materials depend on the cryogens, the operating temperature, and the duty cycle. (See **Cryogenic Fluids.**)

A range of techniques is used to achieve the required thermal insulation between the cryogen and its surroundings. Powder or foam insulation can be used economically for relatively high-temperature cryogens, but Sir James Dewar's invention of the double-walled, vacuum-insulated vessel has been one of the most important contributions to this technology.

The high-vacuum jacket around the cryogen reservoir effectively eliminates thermal conduction and convection. Thermal radiation is reduced by using low-emissivity materials, and more recently by using multilayer superinsulation. A single reservoir without actively-cooled shielding is sufficient for cryogens with normal boiling points above approximately 50 K. However, liquid helium (4.2 K) and liquid hydrogen (20 K) are usually stored in vessels surrounded by a jacket filled with liquid nitrogen (at 77 K), or by multiple radiation shields to reduce heat load from the surroundings. These shields are typically cooled by enthalpy of the exhaust gas or by mechanical coolers.

Cryostats have a wide range of applications. They are often used to contain other equipment that has to run at low temperatures, such as:

- superconducting magnets designed to produce high magnetic fields for medical imaging, research applications, or energy storage;
- superconducting quantum interference device (SQUID) magnetometers;
- mechanical coolers which can now reach temperatures below 4 K;
- specialised refrigerators working at temperatures below 1 K.

N. Balshaw

CRYSTAL GROWTH (see Crystallization)

CRYSTALLIZATION

Crystallization is the process of forming solid material from a liquid solution or melt, where the solid material formed has a crystalline (as opposed to amorphous) structure. A crystallization process generally has the following characteristics:

- The feed material is either in solution or is a liquid above the melting point of the solid phase. If in solution, there may be more than one solvent present.
- There may be dissolved or solid impurities present. Some impurities may have very similar properties to the solute (especially for side-products from organic reactions). During crystallization, impurities may remain in solution, crystallize separately, or incorporate in some way into the product crystals.
- The product material is solid, and present as particles in a range of sizes.
- The product is generally surrounded by mother liquor.
- The waste stream from the process is liquid, containing both residual dissolved product and impurities.

The general advantages of crystallization as a process are:

- High purification can be obtained in a single step.
- Produces a solid phase which may be suitable for direct packaging and sale.
- Operates at a lower temperature and with lower energy requirements than corresponding distillation separations.
- Plant can be simple and easy to construct and maintain.
- May be more economic than alternative separation processes.

The general disadvantages are:

- Generally only purifies one component.
- Yield is limited by phase equilibria.
- Process kinetics are more complex and less well-understood than some alternatives; obtaining detailed kinetic parameters involves complex experimental procedures.

Crystallization is used industrially in a wide variety of processes. Typical applications include bulk chemicals, such as salt, sugar and fertilizers; high value-added products such as specialty chemicals and pharmaceuticals; and difficult separations such as ortho- and para-xylene.

Crystallization processes relating to a single crystal and to multiple crystals in vessels are discussed below while typical *equipment* used for crystallization are dealt with in a separate article on **Crystallizers**.

Single Crystals

A crystal is a solid regular lattice of atoms, ions or molecules, formed by replicating a unit cell.

These lattices can be categorised by symmetry into a number of crystal systems: regular, tetragonal, orthorhombic, monoclinic, trigonal, triclinic and hexagonal. Over 80% of elements and simple inorganic materials crystallize in the regular or hexagonal systems; complex organics favour orthorhombic and monoclinic systems. The shape or habit of a crystal is defined by the faces of the crystal, which can align in different ways with the crystal lattice. The overall shape of a crystal is defined by the rate at which the various faces grow; the fastest growing faces disappear, leaving the slowest growing faces to dominate. Lattices can also have a range of defects. These can form sites for rapid crystal growth and, in some cases, are the dominant means for crystal growth.

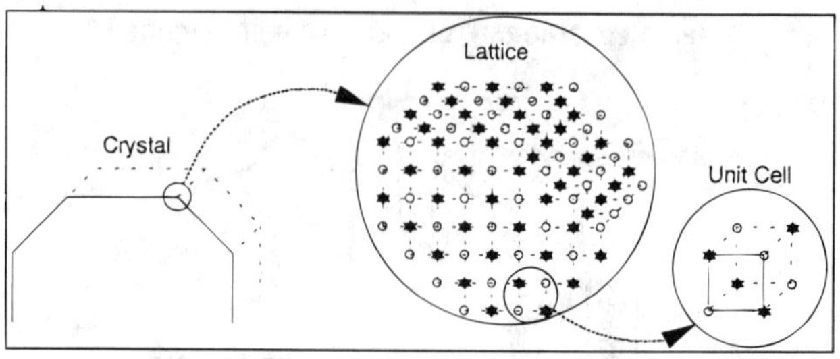

Figure 1.

Processes Affecting Crystals

The driving force for both the formation of new crystals and the growth of existing ones is *supersaturation.* This arises from the concentration of solute exceeding the equilibrium (saturation) solubility concentration.

Thermodynamically, the driving force is the change in Gibbs' free energy, $\Delta\mu = \mu - \mu^* = RT \ln \gamma/\gamma^*$ where γ is the activity coefficient. However, this driving force is difficult to determine so the concentration difference, $\Delta c = c - c^*$, is most commonly used in practical correlations. Uncrystallized solute molecules start by being dispersed randomly in a solution or melt. *Nucleation* is the process whereby new crystals are formed. *Primary nucleation* occurs in crystal-free solution; its exact mechanism is not understood but probably involves the formation of semistructured clusters which rearrange to form crystal nuclei. Ideal *primary homogeneous nucleation* can be derived thermodynamically by considering the formation of liquid droplets from a vapour phase:

$$J = A \exp\left(-\frac{\beta\sigma^3\Omega^2}{k^3T^3(\ln S)^2}\right), \quad (1)$$

which indicates that the nucleation rate J is governed by the interfa-

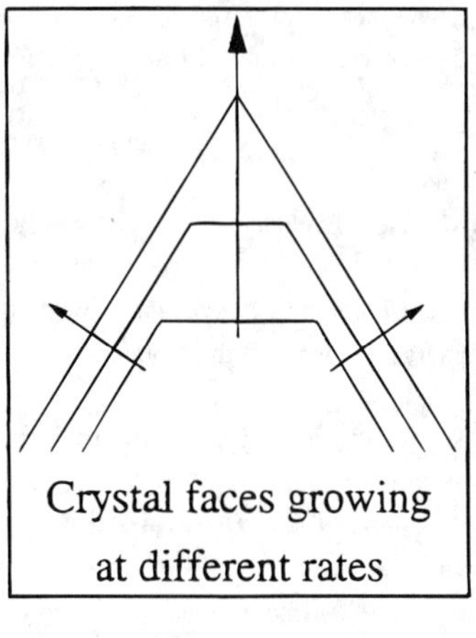

Figure 2.

cial tension σ, the supersaturation ratio $S = c/c^*$, and the temperature T, with the constant A, shape factor β, molecular volume Ω and Boltzmann's constant k all being constant for a given system. This equation can in theory, be applied to the formation of idealised solid particles in a liquid phase. The equation has two practical problems: the interfacial tension is very difficult to measure (it is usually easiest to get it *from* the nucleation rate) and primary homogeneous nucleation is rare in industrial systems unless solutions and vessels are very clean. In practice, *primary heterogeneous nucleation* occurs in crystal-free solution, where crystals form on particles in suspension (e.g., dust) or *secondary nucleation*, where new crystals are formed from existing crystals in solution. As a result, pragmatic nucleation correlations of the form $B = k_n \Delta c^b$ (primary) or $B = k_b \Delta c^b M_T^m N^n$ (secondary) are often found. Measurement of nucleation rates is difficult, and usable values are rarely found in the literature; nucleation is the most difficult parameter to characterise for most systems.

Crystal *growth* occurs when solute molecules in solution diffuse to the surface of the crystal, become adsorbed onto the surface and are then incorporated into the crystal lattice. For some systems, incorporation of solute into the crystal is easy and growth is limited by diffusion to the crystal surface through bulk solution or boundary layer. For other systems, surface integration of solute is rate controlling and growth on a flat crystal face is difficult; growth mainly occurs on stepped or kinked edges. For extreme cases, growth is strongly dependent on dislocations in the crystal.

At low supersaturation levels, crystal growth occurs but primary nucleation is not significant; at higher supersaturation levels, primary nucleation rates increase dramatically. This leads to the concept of a metastable zone in which growth dominates, and a labile zone in which primary nucleation dominates. Typical metastable zone widths range between 1–2°C to 30–40°C; inorganic compounds generally have lower widths than organic ones. This is an important concept, and it can be useful to construct such a diagram for a system and then plot the feed, crystallizer and product temperature/composition points.

Crystal growth rates can be measured by laboratory experiments and are also found in the literature, although there is no good summary of available data. Growth correlations are typically found in the form:

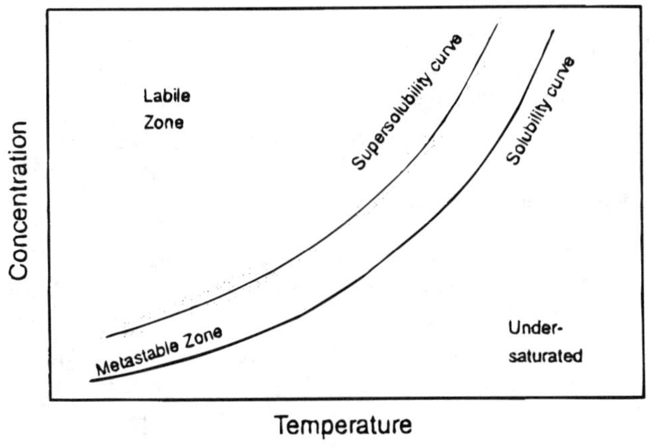

Figure 3.

$$G = k_g \Delta c^g e^{-E/RT}$$

Caution should be exercised when using data of this form, as impurities can have order-of-magnitude effects on growth rates.

Crystals can also join together to form *agglomerates;* this affects a number of bulk crystal properties including shape, purity, strength, size and packing density. Agglomeration is more usually seen in processes producing small (<50μm) particles.

Impurities can be incorporated into crystals in a number of ways. Surface impurities are left when residual mother liquor on the surface of the crystals evaporates, leaving behind any dissolved impurities. *Inclusions* of mother liquor may be formed in crystals, especially at high growth rates. *Occlusions* are voids formed between individual crystals, usually in agglomerates.

In addition to the growth of crystals, there are also processes where small or large fragments can be attritted or broken off crystals, acting as new nuclei for crystal growth. This reduces the size of large crystals, increases the number of smaller ones, and thus contributes to secondary nucleation. At supersaturation levels below the metastable zone limit, secondary nucleation is the dominant mechanism for formation of new crystals. Another factor to consider is exactly what crystal is forming. For many systems—especially organic ones—more than one structural arrangement is possible, leading to the crystallization of different *polymorphs* under different conditions. Each polymorph will have a different solubility, stability, etc. The formation of an unstable polymorph is usually undesirable. Unfortunately, the stable polymorph at a particular temperature has the lowest solubility and the slowest growth rates. Solvates can exhibit similar behaviour, although the growth unit is different in these cases.

Small crystals (<5μm), usually formed by precipitation, exhibit an additional effect—size-dependent solubility. The highly curved solid-liquid interface has a higher energy associated with it, and the solubility of very small crystals increases. This leads to a ripening (or ageing) process where smaller crystals held in suspension near the solubility concentration tend to dissolve and larger crystals grow.

Crystals are bounded by their slowest-growing faces. Different faces, especially in organic systems, can have different electronic, structural and chemical characteristics. This can lead to impurities becoming adsorbed to different extents on different faces, causing changes in relative growth rate and hence overall crystal shape or *habit*. Different growth conditions can also lead to this effect. It is sometimes possible to tailor additives to change crystal habit; more commonly perhaps, impurities cause unwanted changes.

Crystals in Vessels

In solution crystallization, there are a range of crystal sizes present in the vessel. The crystal size distribution can be expressed as number or mass-based, and as a continuous or discrete distribution, usually as the number n of crystals of a particular size L or size range ΔL. From this bulk parameters can be calculated:

Number of crystals $\quad N_c = \sum_{i=0} n_i \quad N_c = \int_0^\infty n_L dL \quad$ (2)

Mass of crystals $\quad M = f_v \rho \sum_{i=0} n_i \overline{L_i^3} \quad M = f_v \rho \int_0^\infty n_L L^3 dL \quad$ (3)

Number-mean size $\mu_n = \dfrac{\sum_{i=0} n_i \overline{L_i}}{\sum_{i=0} n_i} \qquad \mu_n = \dfrac{\int_0^\infty n_L L dL}{\int_0^\infty n_L dL}$ **(4)**

Mass-mean size $\mu_m = \dfrac{\sum_{i=0} n_i \overline{L_i^4}}{\sum_{i=0} n_i \overline{L_i^3}} \qquad \mu_m = \dfrac{\int_0^\infty n_L L^4 dL}{\int_0^\infty n_L L^3 dL}$ **(5)**

Reference is also commonly made to the moments of distribution:

n^{th} moment $m_n = \sum_{i=0} n_i \overline{L_i^n} \qquad m_n = \int_0^\infty n_L L^n dL$ **(6)**

Size distributions are usually summarised in terms of their mass-mean size and coefficient of variation CV. For a Gaussian distribution, CV = standard deviation of size/mean size × 100%, and on a cumulative undersized or oversized plot,

$$CV = |L_{16\%} - L_{84\%}|/2L_{50\%}.$$

An important concept in the analysis of crystallizer behaviour is that of *population balance.* The number of crystals of a particular size or size range is a balance of the formation and removal rate; for a simple system:

$$\frac{\partial n}{\partial t} + \frac{\partial (Gn)}{\partial L} + D_L - B_L + n\frac{d(\log V)}{dt} = -\sum_k \frac{n_k \dot{Q}_k}{V} \quad \textbf{(7)}$$

This is most commonly applied to steady-state continuous crystallizers.

For a more complete analysis of population balance and its application to crystallizers, refer to Randolph & Larson (1988) or Nývlt (1992).

Nomenclature

A Constant
B Nucleation rate (kg^{-1}/s)
B_L Birth rate of crystals size (m^{-4}/s)
b Supersaturation exponent
c Concentration (kg/kg)
CV Coefficient of variation
D_L Death rate of crystals (m^{-4}/s)
E Activation energy (J/mol/K)
f_v Volume shape factor
G Growth rate (m/s)
g Supersaturation exponent
J Primary nucleation rate
k Boltzmann's constant (J/K)
k_b Secondary nucleation constant
k_g Growth rate constant
k_n Primary nucleation constant
L Crystal size (m)
M Mass of crystals (kg)
m_n n^{th} moment of distribution
M_T Slurry density (kg/kg)
N_c Number of crystals
n Number density of crystals (m^{-4})
$\dot{Q}$ Flow rate (m^3/s)
R Gas constant (J/mol/K)
S Supersaturation ratio
T Absolute temperature (K)
V Volume of crystallizer (m^3)
Ω Molecular volume (m^3)
x Mole fraction
β Shape factor
γ Activity coefficient
μ Gibbs free energy (J/mol/K)
μ_m Mass-mean size (m)
μ_n Number-mean size (m)
σ Surface tension (J/m^2)
ρ Density (kg/m^3)

References

Mullin, J. W. (1993) *Crystallization* 3rd Edition, Butterworth-Heinemann, ISBN 0-7506-1129-4.

Myerson, A. S. (Editor) (1992) *Handbook of Industrial Crystallization,* Butterworth-Heinemann, ISBN 0-7506-9155-7.

Nývlt, J. (1992) *Design of Crystallizers,* CRC Press, ISBN 0-8493-5072-7.

Randolph, A. D. and Larson, M. A. (1988) *Theory of Particulate Processes,* 2nd Edition, Academic Press Inc, ISBN 0-12-579652-8.

Söhnel, O. and Garside, J. (1992) *Precipitation: Basic Principles and Industrial Applications,* Butterworth-Heinemann, ISBN 0-7506-1107-3.

SPS Crystallization Manual, Separation Processes Service, Harwell Laboratory, Didcot, Oxon, UK.

Leading to: *Crystallizers*

N.J. Hallas

CRYSTALLIZATION FOULING (see Fouling)

CRYSTALLIZERS

Following from: *Crystallization*

There is a wide variety of equipment used to carry out the crystallization process, called crystallizers. Such equipment can be classified into four broad types:

1. Bulk solution crystallizers. Crystals are suspended in solution for a significant time while nucleation and growth occurs.
2. Precipitation vessels. Feed streams entering the vessel generate high supersaturation levels (by chemical reaction, drowning or salting out), very rapidly forming large numbers of small crystals.
3. Melt crystallizers forming multiple crystals. The bulk (typically > 90%) of the solution or melt forms crystals either

in suspension or on a cooled surface. Impurities remain in the small amount of uncrystallized mother liquor.

4. Melt crystallizers forming large high-purity single crystals. Crystals form very slowly from high-purity melts, yielding large, pure and defect-free crystals. These are typically used for semiconductor manufacture.

All these type of equipment have aspects in common:

- A region where supersaturation is generated to drive the crystallization.
- A region where crystals are in contact with supersaturated solution for crystal growth. In some cases, crystals are present throughout the vessel, suspended by some form of agitation; in other cases, the crystals occupy only part of the vessel, typically as a fluidised bed.

Selection/Design

The selection of the appropriate crystallizer for a particular task will depend upon the feed material available, the properties of the system and the product requirements of the customer. A typical design sequence involves:

- Basic data collection.
- Selection of supersaturation generation method.
- Choice of batch or continuous operation.
- Choice of specific equipment type.
- Bench and pilot scale tests.
- Full scale design.

A full procedure is too complex for this article; however some key aspects are outlined below. For further details, refer to any of the books given in the reference list.

Supersaturation generation

Five main methods can be used to generate supersaturation:

- Cooling, using the vessel walls, internal coils, or by pumping mother liquor through an external heat exchanger. This is used when solubility changes significantly with temperature and when the feed stream is near saturation at a high temperature.
- Evaporation, by heating the mother liquor or reducing the pressure to form a boiling zone at the top of the vessel. This can be used for a wide range of systems, although it is more energy-intensive than cooling.
- Reaction, where feed streams enter and mix resulting in a chemical reaction generating the product, usually at high levels of supersaturation.
- Drowning out, where a miscible solvent is added resulting in a mixture in which the product is less soluble. This has similar characteristics to reaction crystallization.
- Salting out, where a salt with a common ion is added to precipitate the product from solution. Again, this has similar characteristics to reaction crystallization.

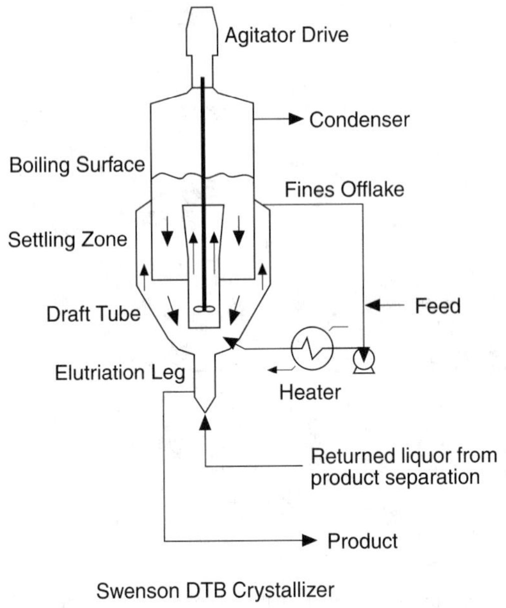

Figure 1.

Other techniques, e.g., pressure change, are occasionally found.

Crystallizer mode

Crystallizers can be designed to operate in either batch or continuous mode (and, rarely, combinations of the two).

Batch crystallization is generally easier to control and is more flexible. It can operate over a wide range of conditions.

Continuous crystallizers produce a consistent product and are generally smaller and more energy efficient than batch equipment for the same production rate. Thus, continuous crystallizers are favoured for high-production rate systems. However, they operate over only a narrow range of conditions, so more process knowledge is generally required to make sure they produce the required product specification.

Standard equipment

For *solution* crystallizers, the simplest equipment is an agitated, cooled vessel. Although simple, it is far from optimal in terms of hydrodynamics, with poor crystal suspension. A draft tube and baffles are often added to improve suspension characteristics, and this leads to designs such as the Swenson Draft Tube Baffled (DTB) and Oslo-Krystal crystallizers.

The Swenson DTB has the main recirculation provided by a propeller inside a draft tube, with a settling zone to allow fines to be removed and dissolved, and a product elutriation leg, where large crystals are extracted against an upward flow carrying small crystals back into the vessel. Both evaporative and cooling versions are found.

The Oslo-Krystal unit has a fluidised bed of crystals, suspension and agitation being provided by an external circulation loop of either crystal magma or relatively crystal-free solution.

For further information on industrial solution crystallizers, refer to Myerson (1992), Mullin (1993) or the SPS Crystallization Manual.

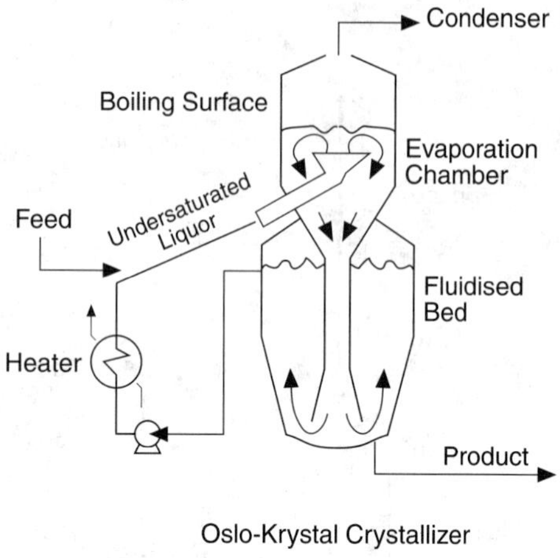

Figure 2.

Melt crystallizers fall into three general categories. The first has crystals in suspension within the vessel; crystals form on a cooled wall and are removed by a scraper. A melting zone can be added to increase purification of the crystals. The second has crystals forming on the outside of a cooled rotating drum or belt, with a scraper removing crystals directly as product. The third has crystals building up on a cooled wall; when crystallization is complete, residual mother liquor and impurities are drained off and the temperature raised to melt and recover the product. Product purity can be increased by "*sweating*" impurities out of the solidified crystals before recovery.

For further information on melt crystallizers, refer to Myerson (1992) or the SPS Crystallization Manual.

Precipitation processes are often carried out in agitated vessels of some kind, or by techniques such as impinging jets. Equipment selection is dependent upon system kinetics and product requirements. For further information, consult Söhnel and Garside (1992), Myerson (1992) or the SPS Crystallization Manual.

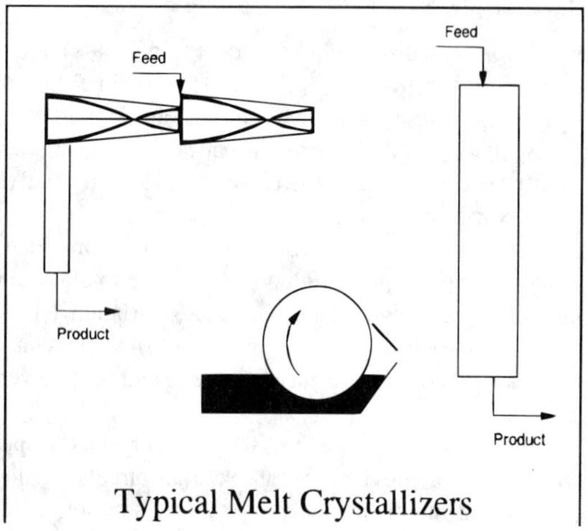

Figure 3.

Scale-up

Scale up is a very complex procedure for crystallizers, and several points should be noted. Growth rates, provided they are measured using liquor containing the correct impurities, are simple to scale-up from bench measurements. Nucleation rates, however, provide a *balancing complexity.* For continuous crystallizers, secondary nucleation rates dominate and these are a combination of several different mechanisms (e.g., crystal-crystal and crystal-wall collisions); each mechanism will scale differently. For batch crystallizers, primary nucleation rates are dependent upon hydrodynamic and thermal conditions which are easier to determine. However, when coupled with the inaccuracies in nucleation rate measurement, scale-up of nucleation rate is always difficult. If successful full-scale operation is highly dependent upon accurate scale-up, then several intermediate-sized tests may be needed. Alternatively, features such as fines or ultrasonic treatment may be used to provide some additional control over effective nucleation rate on the full scale plant.

Operation

Batch crystallizers

Batch crystallizers have the most control options available. Generally, the key parameter is the rate of generation of supersaturation expressed in terms of the cooling, evaporation or addition rate. The discussion below uses cooling rate as an example.

For the production of large crystals, the overall strategy is to form relatively few nuclei and then to grow these to product size under conditions which minimise further nuclei formation. Typically, this means cooling the solution (A) until it enters the metastable zone (B). Seeds can be added (G), or the solution cooled slowly until the labile zone is reached (C) and primary nucleation occurs (D). As the crystals grow (E) and the surface area increases, solute will be depleted increasingly rapidly, and the cooling rate can be increased to maintain the supersaturation level somewhere in the middle of the metastable zone (F).

Continuous crystallizers

Continuous crystallizers are more simple since they operate at steady-state. The only adjustable parameters are usually the feed

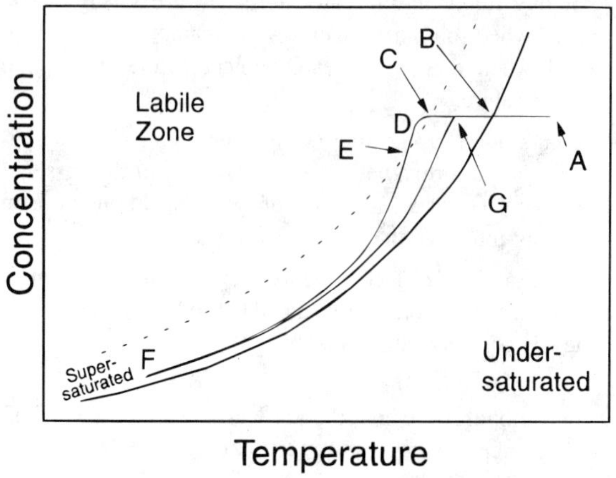

Figure 4.

rate (and hence residence time) and the supersaturation generation rate. If fines treatment or product classification is possible, then these are additional controls. As a result, continuous crystallizers tend to be easier to control, but can only be operated in a relative narrow range of conditions. It is also possible for them to operate in an oscillatory mode due to interactions between growth and nucleation rates, but this effect generally only appears in classifying crystallizers.

The size distribution from an ideal mixed-suspension, mixed-product removal (MSMPR) crystallizer, with size-independent growth and no growth dispersion, can be predicted by:

$$n_L = n^0 \exp(-L/G\tau)$$

where n^o is the population density of nuclei; L is the crystal size; G, the growth rate; and τ, the residence time. Thus plotting log (population density) against size for an MSMPR should yield a straight-line graph. The mass-mean size from such a distribution is $4G\tau$. Note that in the above equation, G is also a function of τ; increasing residence time does not proportionally increase the mean size since the growth rate would decrease.

A separate issue is the start-up of continuous crystallizers. It is generally recommended that at least 10 residence times are required for a continuous system to reach steady-state, though in practice some systems can take significantly longer. Thus, frequent start-ups and shutdowns of these types of equipment should be avoided.

Troubleshooting

Operating crystallizers do not always work perfectly. Some potential problems are listed below, but this is not exhaustive.

- Crystal mean size is too small or too many fines are present. Nucleation rate is too high.
- Crystal mean size is too large due to insufficient nucleation.
- Crystal size distribution is too wide.
- Variable size distribution from continuous crystallizer. Usually related to the use of fines treatment and/or product classification equipment.
- Formation of unwanted agglomerates.
- Formation of wrong polymorph or solvate; can also lead to problems with product caking or breaking down during storage.
- Encrustation on vessel surfaces due to local high supersaturation levels or low crystal concentrations.
- High impurity levels; either due to impurities becoming incorporated within the crystal or poor washing of residual mother liquor.

The cure for any given problem is too system-specific for a brief outline in this article; however some confirmatory tests and remedies are obvious from the above descriptions.

Nomenclature

G Crystal growth rate (m/s)
L Crystal size (m)
n_L Population density, size L
τ Residence time (s)

References

Garside, J. (1992) *Precipitation: Basic Principles and Industrial Applications,* Butterworth-Heinemann, ISBN 0-7506-1107-3.

Mullin, J. W. (1993) *Crystallization* 3rd Edition, Butterworth-Heinemann, ISBN 0-7506-1129-4.

Myerson, A. S. (Editor) (1992) *Handbook of Industrial Crystallization,* Butterworth-Heinemann, ISBN 0-7506-9155-7.

Nývlt, J. (1992) *Design of Crystallizers,* CRC Press, ISBN 0-8493-5072-7.

Söhnel, O., Söhnel, O. and Garside J. (1992) *Precipitation: Basic Principles and Industrial Applications,* Butterworth-Heinemann, ISBN 0-7506-1107-3.

SPS Crystallization Manual, Separation Processes Service, Harwell Laboratory, Didcot, Oxon, UK.

N.J. Hallas

CRYSTALS (see Crystallization)

CRYSTAL STRUCTURE ASYMMETRY (see Piezoelectricity)

CRYSTAL SUBLIMATION AND GROWTH (see Interfaces)

CSIRO (see Commonwealth Scientific and Industrial Research Organisation)

CUNNINGHAM COEFFICIENT (see Gas-solids separation, overview)

CURRENT VOLTAGE CHARACTERISTICS (see Electric arc)

CURVED FLOW (see Coanda effect)

CURVILINEAR CHANNELS (see Coiled tube, flow and pressure drop in)

CYANOGEN COMBUSTION (see Flames)

CYCLIC HYDROCARBONS (see Hydrocarbons)

CYCLOHEXANOL

Cyclohexanol ($C_6H_{11}OH$) is a colourless, oily liquid with a camphor-like odour. It is obtained by the hydrogenation of phenol. It is slightly toxic and narcotic in high concentrations. Its main uses are as a raw material in the production of nylon, esters and soap, as a solvent and in the textile industry.

Some of its physical characteristics are:

Molecular weight: 100.16
Melting point: 298.0 K
Boiling point: 434.3 K
Critical temperature: 624.3 K
Critical pressure: 3.75 MPa
Critical density: 307 kg/m^3

V. Vesovic

CYCLONE FURNACES (see Pulverised coal furnaces)

CYCLONE REYNOLDS NUMBER (see Gas-solids separation, overview)

CYCLONES (see Gas-solids separation, overview)

CYCLONE SEPARATOR (see Gas-solid flows; Vapour-liquid separators)

CYCLONE STOKES NUMBER (see Gas-solids separation, overview)

CYLINDER, INVISCID FLOW AROUND (see Inviscid flow)

CYLINDERS, FLOW OVER (see Vortices)

CYLINDRICAL CO-ORDINATES (see Co-ordinate system)

CYLINDRICAL POLAR CO-ORDINATES (see Conservation equations, single phase)

D

DALL TUBE (see Orifice flowmeters)

DALTON'S LAW (see Gas law)

DALTON'S LAW OF PARTIAL PRESSURES

Following from: *Thermodynamics*

Two Dalton's laws of partial pressure characterize the behaviour of nonreacting, ideal gas mixtures. One of them (1801) states that the pressure of gas mixture (p^{Σ}) under given conditions is equal to the sum of the partial pressures of components ($\Sigma_i\, p_i$), i.e., $p^{\Sigma} = \Sigma_i\, p_i$. It follows that $p_i = p^{\Sigma} x_i$, where x_i is the molar concentration of the given component in a mixture. According to the other law (1803), the solubility in a liquid of each component of the ideal gas mixture in equilibrium with this liquid is proportional to the partial pressure of the given component. Both laws reflect the fact that each of the gas mixture components is a separate independent gas in the nonreacting ideal gas approximation. In this case, the properties of a mixture are additive with respect to the components' properties (with the exception of configuration functions, which include entropy).

The partial pressure of a mixture component is the pressure of a corresponding individual gas at the mixture temperature, if this gas alone occupies the given volume of a mixture. In the absence of concentration gradients in a mixture, i.e., when the mixture is in equilibrium (the diffusion processes are absent), the spatial distribution of a partial pressure of each component is uniform.

When Dalton's law applies, the relationship between the partial pressure of a component (p_i) of gas mixture and its equilibrium solubility (the molar concentration x_i) in a liquid obeys quantitatively the **Henry Law** (1807). For each individual component, $p_i = K_i x_i$ where K_i is the Henry constant of the given component. The applicability of Dalton's law, as well as that of Henry's law, is at low pressures. As pressure grows, the forces of intermolecular interaction and the size of the molecules become influential; in this case, the pressure of a mixture loses its additivity with respect to partial pressures of components and the solubility of components in a liquid ceases to be linear with respect to these pressures. The latter circumstance may become apparent even in the region of low-gas solubilities, i.e., in that region of concentrations from which the Henry law has been just derived. The main reason for this is the dependence of the Henry constant on the total pressure of a gas mixture.

Just like Dalton's law for pressures, the additivity of volumes of nonreacting ideal gas mixture gives rise to **Amagat's Law**, which states that at a given pressure the total volume of a gas mixture is equal to the sum of the volumes of individual gases. In other words, the partial molar volume of a component is equal to the molar volume of a corresponding individual gas.

D.N. Kagan

DAMPING, OF HEAT EXCHANGER TUBES (see Vibration in heat exchangers)

DARCY EQUATION (see Porous medium)

DARCY'S LAW

Following from: *Porous media*

Our current ability to predict the flow of fluids through porous media, for example, in ground water movements, the recovery of oil or the design of filter beds originates from the experimental work performed by French engineer Henry Darcy (1856).

At the time, Darcy was concerned with developing the water works for the town of Dijon, and he needed information which would enable him to determine the size of the filter bed that should be installed to handle the daily throughput of water in the system. To provide this information, he performed a series of experiments in which water was passed downwards through vertical columns of sand at a controlled rate, and fluid pressures were measured at the top and bottom of the columns using mercury manometers. The results showed that there was a linear relationship between the flow rate per unit of cross-sectional area of the column, the column height, and the pressure differential when expressed as the differences in heights of equivalent water manometers above and below the sand. Historically (before the introduction of the SI system of units), the relationship was expressed in the form:

$$\frac{\dot{V}}{A} = \text{constant} \times \frac{(h_1 - h_2)}{l} \qquad \textbf{(1)}$$

where $\dot{V}$ is the volumetric flow rate, A is the cross-sectional area l is the thickness of the sand, and $(h_1 - h_2)$ is the water equivalent difference in manometer levels. The constant in this equation was found to be dependent upon the sizes of sand particles used in the experiment.

Other workers have subsequently shown that by introducing the concept of fluid potential and the physical properties of fluid, Darcy's original result can be extended to provide a relationship which is universally-applicable to any single-phase fluid, and for any flow direction. Expressed in SI nomenclature, it becomes:

$$\frac{\dot{V}}{A} = -K\,\frac{\rho}{\eta}\frac{d\Phi}{dl}, \qquad \textbf{(2)}$$

where η is the fluid viscosity; ρ is the fluid density; $d\Phi/dl$ is the potential gradient; and K is a property of the porous medium having the dimensions of $(\text{length})^2$ and known as *permeability.* This relationship has appropriately become known as **Darcy's Law**.

At the Darcy Centennial Hydrology Symposium in 1956, M. King Hubbert presented a paper in which he showed that Darcy's law has a theoretical foundation, and that the law can be derived from the Navier Stokes equation for a viscous fluid. He noted that Darcy's law has many analogies to Ohm's Law for the Conduction of Electricity and Fourier's expression for the conduction of heat. He pointed out, however, that whilst the relationship given in Equation (2) may appear to have similarities with the Poiseuille equation for laminar flow of fluids in straight cylindrical capillaries, expressed as:

$$\frac{\dot{V}}{\pi r^2} = \frac{r^2}{8}\frac{\rho}{\eta}\frac{d\Phi}{dl} \quad (3)$$

(where r is the tube radius), the fluid mechanics of flow through porous media are very different from those of flow through capillary tubes. (See **Fourier's Law**)

In porous media, inertia forces play an important role since each fluid particle moves through a tortuous path in which it is continuously being accelerated and decelerated; whereas in capillary flow, the fluid particles move in straight lines at a constant velocity. Thus, the similarities between Darcy's law and Poiseuille's equation are fortuitous. (See **Poiseuille Flow**.)

Darcy's law strictly applies to the flow of a single-phase fluid, in which case permeability is a property of the rock and is independent of the fluid flowing through the pores. When two or more phases flow through the rock at the same time however, each phase must flow through part of the total cross-section of pore space, thus reducing the flow area available to the other phases. As a result, the effective permeability of the individual phases is reduced. To accommodate this situation, Darcy's law has been modified by introducing a multiplication factor, known as *relative permeability,* k_i, for each phase.

$$\frac{\dot{V}}{A} = -Kki\,\frac{\rho}{\eta}\frac{d\Phi}{dl}. \quad (4)$$

The value of k_i will clearly depend upon the fraction of the pore flow area occupied by the other phases, and hence upon the saturation of the other phases. Typical relative permeability curves for two-phase oil and water flow through a sandstone rock are shown in **Figure 1**.

The relative permeability of one phase decreases as the saturation of the other phase increases, and vice versa. Note that the relative permeability of each phase becomes zero at a nonzero value of the phase saturation. This is because of the fluid pore structure, which prevents phase saturations below a certain level. Thus for two-phase flow, individual phase permeabilities depend on the nature of the other fluids as well as the structure of the rock matrix.

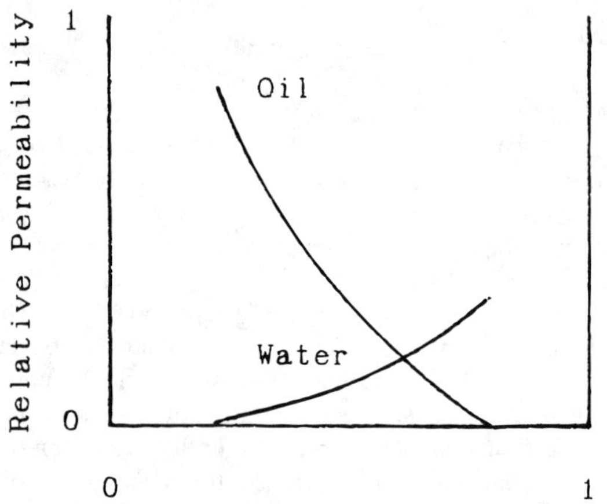

Figure 1. Example of water/oil relative permeability curves.

References

Darcy, H. (1856) *Les Fontaines Publiques de la Ville de Dijon.,* Victor Dalmont, Paris.

King, Hubbert M. (1956) Darcy's Law and the Field Equations of the Flow of Underground Fluids., *Trans. AIME,* 207: 222–239.

R.I. Hawes

DARCY NUMBER

There are two dimensionless groups given the name Darcy number.

$$\text{Darcy number 1, } Da_1 = \Delta p2D/\rho u^2L = 4f$$

where Δp is the pressure drop, D the diameter, ρ fluid density, u fluid velocity, and L the length over which the pressure drop is measured. f is the Fanning friction factor (see **Friction Factors**).

$$\text{Darcy number 2, } Da_2 = uL/D'$$

where D' is the permeability coefficient of a porous medium with units m^2/s (see **Darcy's Law**).

G.L. Shires

DATING OF ARCHAEOLOGICAL SAMPLES (see Thermoluminescence)

DDT, DEFLAGRATION TO DETONATION TRANSITION (see Explosion phenomena)

DEBORAH NUMBER

A **Non-Newtonian Fluid** is one for which stress is not linearly related to strain-rate. All non-Newtonian fluids are *elasticoviscous,* that is they combine elastic and viscous properties. When the time-scale of a flow t_f is much less than the relaxation time t_r of an elasticoviscous material, elastic effects dominate. When, on the other hand t_f is much greater than t_r, elastic effects relax sufficiently for viscous effects to dominate. The ratio t_f/t_r is a dimensionless number of particular significance in the study of flow of non-Newtonian fluids: depending on the circumstances, this number is called the Deborah number or the **Weissenberg Number**. The Deborah number, De, is named after the prophetess Deborah: ". . . the mountains flowed before the Lord . . . ," implying that all materials flow (are viscous) on a sufficiently long time-scale. Thus glass, normally considered to be elastic, flows on a time-scale of order centuries as the windows of medieval cathedrals testify. Similarly water, normally considered to be viscous, behaves elastically on a time-scale of order nanoseconds. The precise definition of De depends on the circumstances, though t_f is always taken to be a characteristic residence time. Thus, for flow at mean velocity U through a pipe of length L, $t_f = L/U$ and so $De = t_r U/L$. If the Deborah number is small, elastic effects can be neglected and the non-Newtonian fluid treated as a purely viscous material, albeit with a non-constant viscosity.

S.M. Richardson

DEBYE TEMPERATURE (see Thermal conductivity mechanisms)

DECANTATION (see Decanters)

DECANTERS

***Following from:** Liquid-solid separation*

The method of separation of a solid phase from a liquid by a repeating sequence of stages comprising dilution and gravity *sedimentation* is known as decantation. Most applications of the method involve continuous counter-current decantation (CCD), with the number of stages ranging from 2 and 10. The equipment selected for CCD may consist of multiple-compartment washing tray *thickeners* or a train of unit thickeners. Factors which make CCD a preferred choice of separation technique include the following: rapidly settling solids (settlement rate may be assisted by flocculation); a high ratio of solids concentration between the underflow and the feed; the need for high wash ratios during the dilution stage (greater than three times the liquor volume in the thickened underflow); a large quantity of solids to be processed (containing a significant fines fraction, which is otherwise difficult to concentrate).

Decanters is also a generic term used to cover the range of continuous sedimenting **Centrifuges** which utilise a scrolling mechanism for transport of solids through the machine and their subsequent discharge.

R.J. Wakeman

DECONVOLUTION, OPTICS (see Optical particle characterisation)

DEEP SHAFT (see Aeration)

DEFINITE INTEGRALS (see Integrals)

DEFLAGRATION (see Explosion phenomena)

DEFLAGRATION TO DETONATION TRANSITION, DDT (see Explosion phenomena)

DEGENERACY (see Boltzmann distribution)

DEGREES OF FREEDOM (see Boltzmann distribution; Phase rule; Thermodynamics)

DEHYDRATION (see Drying)

DE-ICING (see Ethylene glycol)

DE LAVAL NOZZLES (see Critical flow)

DELIQUESCENCE (see Hygroscopicity)

DELTA FUNCTION

The Dirac Delta Function $\delta(x - a)$ is an impulsive function defined as zero for every value of x, except for the point $x \neq a$ where it jumps to an infinitely large value. However, its graph encloses a unit area. It can be regarded as an idealization of a unit impulse. We define δ(x-a) through the following two properties:

1) For any x: $-\infty < x < +\infty$,

$$\delta(x - a) = 0 \text{ for every } x \neq a$$

2) $\int_{-\infty}^{+\infty} \delta(x - a)\, dx = 1.$

This function has the following important property: for any continuous function f(x),

$$\int_{-\infty}^{+\infty} \delta(x - a) f(x) dx = f(a),$$

that is, $\delta(x - a)$ applied to f(x) detects its value at $x = a$.

We can heuristically show the validity of this property using the following argument: Let us approximate $\delta(x - a)$ through the function $\delta_\varepsilon(x - a)$, such that:

$$\delta_\varepsilon(x - \alpha) = \begin{cases} 0, \text{ for all } x < a - \varepsilon/2 \\ 1/\varepsilon, \text{ for all } a - \varepsilon/2 < x < a + \varepsilon/2 \\ 0, \text{ for all } x > a + \varepsilon/2, \end{cases}$$

which approaches $\delta(x - a)$ as ε tends to zero.

Clearly, the area covered by $\delta_\varepsilon(x - a)$ is equal to one;

$$\int_{-\infty}^{+\infty} \delta_\varepsilon(x - a)\, dx = 1/\varepsilon \int_{a-\varepsilon/2}^{a+\varepsilon/2} dx = 1.$$

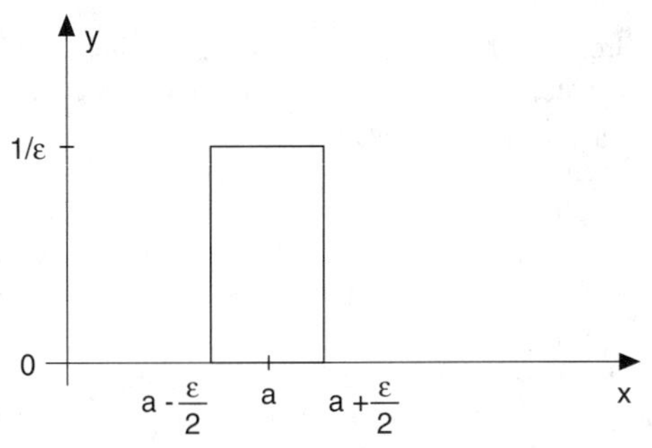

Figure 1. The function $\delta_\varepsilon(x - a)$ an approximation to the Delta function.

Furthermore, let F(x) be the primitive of f(x) (that is $F'(x) = f(x)$), then:

$$\int_{-\infty}^{+\infty} \delta_\varepsilon(x - a)f(x)\,dx = (1/\varepsilon)\int_{a-\varepsilon/2}^{a+\varepsilon/2} f(x)\,dx$$

$$= (1/\varepsilon)[F(a + \varepsilon/2) - F(a - \varepsilon/2)];$$

as ε goes to zero, the last expression defines the derivative of F(x) at x = a, which is precisely f(a).

The function δ(x − a) has a number of important applications in mathematical physics, in particular the solution of differential equations. In fact, it belongs to a class of generalized functions called *distributions*.

References

Schwartz, L. (1973) *Théorie des Distributions,* Hermann, Paris.

Leading to: *Differential equations*

E.L. Ortiz

DENSITY GAS MODEL (see Free molecule flow)

DENSITY, HOMOGENEOUS (see Multiphase flow)

DENSITY MEASUREMENT

Following from: *Density of gases; Density of liquids*

The measurement of the density of a substance is, in principle, extremely simple since from the definition of density, one merely needs to determine the mass of the material contained in a given volume. In practice, the measurement is not so simple, particularly if the thermodynamic state at which the density is required departs significantly from ambient temperature and pressure. The technique employed for measurements also depends upon the phase of the substance.

Measurements in Liquids

Measurement of the density of a liquid at atmospheric pressure and temperatures near to ambient can readily be conducted with a specific-gravity bottle to an accuracy of better than ±0.1%. The device merely consists of a small glass bottle with an extremely accurately-calibrated volume, which is weighed empty and then filled with the liquid. Obviously, such a device becomes impractical in this form for measurements at high pressures and/or temperatures far removed from ambient.

Measurements of liquid densities at moderate pressures are now most easily performed with the aid of vibrating U-tube densimeters (Wood, 1989). Such devices are commercially-available for use at pressures up to 40 MPa over a wide temperature range. The principle of the technique is that a thin steel or quartz tube bent into the shape of a 'U' (as shown in **Figure 1**) is set into oscillation perpendicular to the plane of the U. The frequency of oscillation of the tube is determined by its mass which, in turn, is related to the mass of liquid contained within it. Since the frequency of oscillation is rather easily measured with high precision, accuracies in the measurement of liquid density of 1 part in 10^5 are claimed.

For operation at higher pressures, these devices are unstable owing to the hydrostatic distortion of the thin-walled tube. Considerably greater efforts must be expended to make measurements of liquid densities under such conditions. It is often necessary to resort to methods that measure the volume of a fixed mass of sample as a function of pressure (Whalley, 1975; Dymond and Malhotra, 1988). Padua et al. (1994) have recently described an alternative method making use of an oscillating body.

Measurements in Gases

In the gas phase, some of the techniques for measurement of density are quite different owing to the property of a gas to expand and to fill the space made available to it. Thus, at moderate pressures the gas under investigation is contained in a well-defined volume and pressure and temperature are measured and/or controlled. Measurements are then made by varying one quantity and determining the effect on the second. Thus, one may vary the volume and examine the resulting pressure change with the temperature constant throughout. This is the method most often employed for the determination of the second *virial coefficient.* (See **Density of Gases.**) If the volume in the quantity is maintained constant and pressure is measured as a function of temperature, one has the so-called isochoric system which is often employed also at higher pressures.

It follows from this general description of the principles of measurements that precise and accurate measurements of a temperature and pressure are required, but modern techniques allow these measurements quite readily. The determination of the volumes of vessels involved in this process is often reduced to a gravimetric

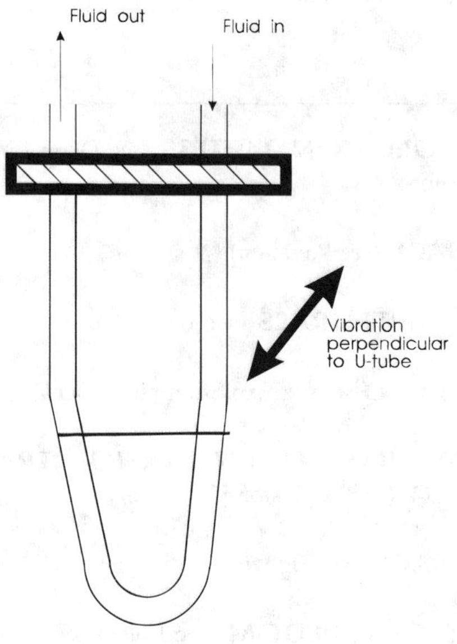

Figure 1.

measurement armed with the knowledge of the density of a liquid, such as mercury, used to fill the vessels.

Accuracy in direct gas density measurements of one part in 10^5 are possible, but require extreme care and patience (Saville, 1975; Brielles et al., 1975; Malbrunot, 1975).

References

Wood, R. H. (1989) *Thermochimica Acta* 154, 1.

Whalley, E. (1975) *The Compression of Liquids,* in Le Neindre, B., Vodar, B., eds., *Experimental Thermodynamics of Non-Reacting Fluids,* IUPAC, Butterworths, London, Chapter 9.

Dymond, J. H. and Malhotra, R. K. (1988) *Int. J. Thermophys.* 9, 941.

Padua, A. A. H., Fareleira, J. M. N. A., Calado, J. C. G., and Wakeham, W. A. (1994) *Int. J. Thermophys.* 15, 229.

Saville, G. (1975) *Measurements of p-V-T Properties of Gases and Gas Mixtures at Low Pressure,* in Le Neindre, B., Vodar, B., eds., *Experimental Thermodynamics of Non-Reacting Fluids,* IUPAC, Butterworths, London, Chapter 6.

Brielles, J., Dédit, A., Lallemand, M., Le Neindre, B., Leroux, Y, Verneuse, J., and Vidal, D., ibid., Chapter 7.

Malbrunot, P., ibid., Chapter 8.

W.A. Wakeham

DENSITY OF GASES

Following from: *Equations of state*

In most treatments of heat and mass transfer in engineering, it is usual to treat fluids as continua. Thus, it becomes possible to introduce the concept of a fluid (liquid or gas) density at a point. On this basis, the mass density of a gas, ρ, is simply the mass of a gas contained in a macroscopic volume whereas the amount-of-substance density, $\tilde{\rho}$, is the amount (number of moles) of a gas in the same volume. The density of a gas is a strong function of temperature and pressure. For very low pressures, every gas conforms to the **Perfect Gas** equation

$$p = \tilde{\rho}\tilde{R}T$$

Where R is the universal gas constant and T, the absolute temperature. But there are significant deviations from this simple behaviour as the pressure is increased.

One of the simplest means of representation of the behaviour of real gases qualitatively is the **Van der Waals' Equation of State**, expressed as:

$$p = \frac{\tilde{\rho}\tilde{R}T}{(a - b\tilde{\rho})} - \tilde{\rho}^2 a$$

in which a and b are constant characteristic of a particular substance which account for the attractive forces between molecules and their finite size. Since the van der Waals' equation reveals a critical point, the constants a and b can be related to the critical state parameters

$$T_c = 8a/27\tilde{R}b, \qquad p_c = a/27b^2$$

and

$$\tilde{\rho}_c = 1/3b.$$

These relationships do, in fact, provide a very crude means of estimating the density of a gas from the van der Waals' equation of state if the critical state variables are known for the substance. More accurate equations of state, such as that of Redlich and Kwong or Benedict, Webb and Rubin, have been developed which may be used in a similar manner for the estimation of the density of gases. However, none of these simple equations of state represent the behaviour of any real substances over a wide range of conditions so that one must always have recourse to measurements, at least to define the parameters in an equation of state.

For some fluids, very accurate equations of state exist which describe the density, as well as other thermodynamic properties of a gas, over a wide range of conditions. These equations are based on a set of critically-evaluated experimental data. Equations used to describe these data are often quite complicated [e.g., de Reuck and Craven (1993)].

The density of a gas is sometimes expressed in terms of a *virial expansion,* which emerges from a statistical mechanical treatment of the gas. Here,

$$Z = \frac{p}{\tilde{\rho}RT} = 1 + B\tilde{\rho} + C\tilde{\rho}^2 + \cdots$$

where B, C . . . are known as virial coefficients which have a known relationship to the forces between molecules. For moderately low densities, equations of this form can be used to calculate gas densities if B and C are available, which is sometimes the case (Dymond and Smith, 1980).

Other means of estimating the density of gases are available, such as those based upon **Corresponding States** procedures (Reid et al.).

References

Bett, K. E., Rowlinson, J. S., and Saville, G. (1975) *Thermodynamics for Chemical Engineers,* Athlone Press, London.

de Reuck, K. M. and Craven, R. J. B. (1993) *International Thermodynamic Tables of the Fluid State*—12. Methanol (Blackwell Scientific, London).

Dymond, J. H. and Smith, E. B. (1980) *The Virial Coefficients of Pure Gases and Mixtures. A Critical Compilation* (Clarendon Press, Oxford).

Reid, R. C., Prausnitz, J. M., and Sherwood, T. K. (1977) *The Properties of Gases and Liquids,* 3rd ed. (McGraw-Hill, New York).

Leading to: *Density measurement; Critical point; Thermodynamics*

W.A. Wakeham

DENSITY OF LIQUIDS

Following from: *Equations of state*

Fluids are often regarded as continua in most treatments of heat and mass transfer in engineering. This treatment makes it possible to introduce the concept of fluid (liquid or gas) density at a point. On this basis, the mass density of a liquid, ρ, is simply the mass of a liquid contained in a macroscopic volume, whereas the amount-of-substance density, $\tilde{\rho}$, represents the amount (number of moles) of a liquid contained in the same volume. The density of a liquid is a relatively strong function of temperature at constant pressure, but a relatively weak function of pressure at constant temperature. This is because, in the liquid phase, the molecules that comprise the liquid move around largely in the well region of the **Van der Waals Forces** between them. Thus, increases in molecular energy cause an expansion as the molecules stray further out into the relatively-weak attractive region of the potential. However, as the pressure is increased, the tendency for the molecules to be pushed further together is opposed by the strong repulsive forces between molecules.

A particularly simple **Equation of State** for the description of the density of liquids is due originally to van der Waals, expressed as:

$$p = \frac{\tilde{\rho}RT}{(1 - b\tilde{\rho})} - \tilde{\rho}^2 a,$$

which recognises the existence of attractive forces between molecules at large distances and repulsive ones at short distances. As discussed under **Density of Gases**, the parameters a and b in this equation can be estimated from critical constants. However, this is not a particularly satisfactory method since it can lead to large errors.

For several pure fluids, empirical equations of state of very high accuracy are available. These equations have been fitted to a set of experimental data that have been critically evaluated and are thermodynamically consistent with other thermodynamic quantities (de Reuck and Craven, 1993). For the majority of fluids, such equations are not available and one must then have recourse to methods of density estimation. These methods can be based upon refinements of the van der Waals equation of state, such as that due to Redlich and Kwong or Benedict, Webb and Rubin [Bett et al., (1975)]. However, all of these methods require a knowledge of some experimental information on the fluid in question. In the absence of such methods, entirely empirical procedures for the extraction of the liquid phase density, such as those listed by Reid et al. (1977), must be employed.

For liquid mixtures, it is sometimes sufficiently accurate for many purposes to assume that there is no volume of mixing, so that the density of the mixture can be written as:

$$\tilde{\rho}_{mix} = \Sigma \tilde{x}_i \tilde{\rho}_i,$$

where the $\tilde{x}_i$ are the amount-of-substance fractions. For systems with strong specific interactions, this is often a poor approximation and more sophisticated methods are required (Rowlinson and Swinton, 1982).

References

Bett, K. E., Rowlinson, J. S., and Saville, G. (1975) *Thermodynamics for Chemical Engineers,* Athlone Press, London.

de Reuck, K. M. and Craven, R. J. B. (1993) *International Thermodynamic Tables of the Fluid State*—12. Methanol, Blackwell Scientific, London.

Reid, R. C., Prausnitz, J. M., and Sherwood, T. C. (1977) *The Properties of Gases and Liquids,* 3rd ed., McGraw-Hill, New York.

Rowlinson, J. S. and Swinton, F. L. (1982) *Liquids and Liquid Mixtures,* 3rd Ed., Butterworths, London.

Leading to: *Critical point; Thermodynamics; Density measurement*

W.A. Wakeham

DENSITY, OF THE ATMOSPHERE (see Atmosphere, physical properties of)

DENSITY-WAVE OSCILLATIONS (see Instability, two-phase)

DEPARTMENT OF THE ENVIRONMENT, DoE

2 Marsham Street
London SW1P 3EB
UK
Tel: 0171 276 3000

DEPARTURE FROM FILM BOILING (see Leidenfrost phenomena; Rewetting of hot surfaces)

DEPARTURE FROM NUCLEATE BOILING, DNB (see Boiling; Burnout, forced convection; Nuclear reactors)

DEPHLEGMATOR

A dephlegmator is a device arranged for the partial condensation of a multicomponent vapour stream. The vapour stream flows vertically upwards and the condensate (condensed vapour) runs back down under the influence of gravity. The vapour stream and condensate thus move counter-currently and are in direct contact with each other. In addition to heat transfer between the vapour stream and cooling medium, mass is transferred between the rising vapour and falling condensate. Vapour leaving the device has become concentrated in the more volatile components, while the condensate is richer in the less volatile components. In the industrial rather than laboratory contexts, many writers use the word 'dephlegmator,' only if the device being described is a main unit in the process plant.

If an overhead condenser is used in continuous counter-current distillation in an industrial setting, it is often designed to receive a multicomponent vapour which flows vertically upwards and is condensed only partially. The condensate forms a reflux of liquid to the distillation column and the remaining vapour has become enriched in the more volatile components. However, the overhead condenser is not a key process unit, but is merely part of a distillation column. Many writers do not, therefore, consider the rather uncommon word 'dephlegmator' appropriate to this subordinate and commonly-encountered device. In its stead, they use 'backward-return condenser' or less informatively, 'partial condenser' or 'reflux condenser.'

In contrast, a particular instance where the importance of the dephlegmator as a heat and mass transfer device is clear (the word is thus used without dispute) is in the separation and recovery of ethene from a cracked gas feed, which contains a significant proportion by volume of light components (hydrogen, carbon monoxide and methane). The heavy component of the feed, ethene, is separated from the light components in a dephlegmator, and the ethene-rich condensate is passed on to a distillation column to remove any remaining methane. An earlier method of recovering ethene involved fractionation of the entire feed stream in a conventional adiabatic distillation column. The introduction of the dephlegmator, constituting a preseparation stage, resulted in vast savings in energy costs. A substantial part of the heat removed in the dephlegmator is transferred to the coolant at a higher temperature than was possible when the fractionation was done entirely in the conventional column. (See also **Condensers**; **Distillation**)

A.C. Lintern

DEPOSITION (see Liquid-solid flow)

DEPOSITION OF PARTICLES (see Dispersed flow)

DERIAZ TURBINES (see Hydraulic turbines)

DESALINATION

Following from: *Mass transfer; Water*

The production of fresh water from sea water or brackish water by reduction of the mineral content is called desalination or saline water reclamation. Creation of potable water by distillation was described a thousand years ago by the Persian chemist Abu Mansur Muwaffak (Kobert 1890) and the technique is doubtless much older. Today, desalination is used to provide not only drinking water but water for farming and industry, and a wide variety of methods is available.

Sea water usually contains between 30 and 40 grams of minerals per kilogram (Beaton 1986), mostly common salt, NaCl, and subterranean waters from almost zero to 70 g/kg, the former representing fresh water **Aquifers** and the latter some conate waters associated with oil deposits. In many parts of the world water with mineral concentrations significantly greater than 1 g/kg is drunk by humans; some ruminant animals can tolerate levels of 10 g/kg. Industrial equipment is much more sensitive in this respect than are living creatures, boilers requiring mineral contents as low as 0.01 g/kg.

Desalination Processes

The following Table summarises the main desalination processes.

Thermal processes	**Distillation**
	Freezing
Membrane processes	**Reverse osmosis**
	Electrodialysis
Chemical processes	**Ion exchange**

(See also separate entries about these topics.)

Distillation

Distillation is the oldest method, and was used on ships in the 17th century. The first land-based unit, built in England in 1912, consisted of a brine boiler with internal steam heating coils and a separate water-cooled condenser. Scale formation on heat transfer surfaces was a major problem and this system was eventually displaced by flash distillation.

Figure 1 is a simplified diagram illustrating the principle of multistage flash distillation. Vaporisation is brought about by pressure reduction between stages and the boiling heat transfer surface is thereby eliminated. Energy is conserved by using the heat given up by the condensing water vapour to heat the brine. The largest plant of this design in Saudi Arabia, produces almost a million tonnes a day of pure water.

A more recent, though less widely-used design, is the vertical tube evaporator (illustrated diagramatically in **Figure 2**). The multiple vertical tubes, which have special fluted surfaces, provide good, controlled film heat transfer, and being heated by vapour condensation from the previous effect, are not subject to excessive temperature. These factors reduce the risk of scaling.

A third type of desalination plant commonly used for moderate sizes (i.e., several tonnes per day) is illustrated in **Figure 3**. This is a horizontal tube distillation plant with vertical stacking; horizontal arrangements are also used.

Freezing

Pure water can be obtained from brine by freezing, and in the 1970's a great effort was made to develop an economic system. Unfortunately, the separation of ice from the residual brine and other engineering problems made this impossible and the system was not commercially-adopted.

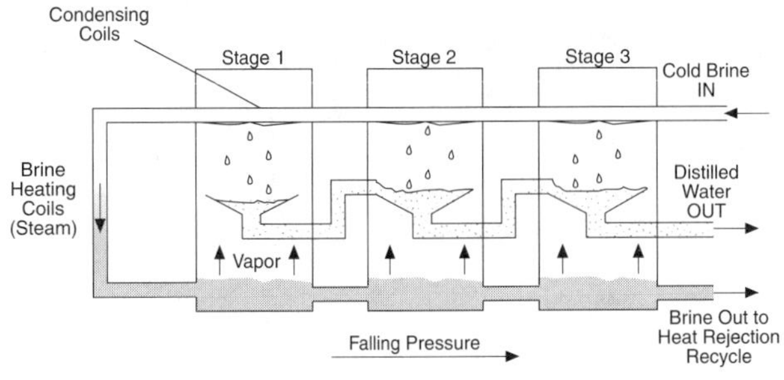

Figure 1.

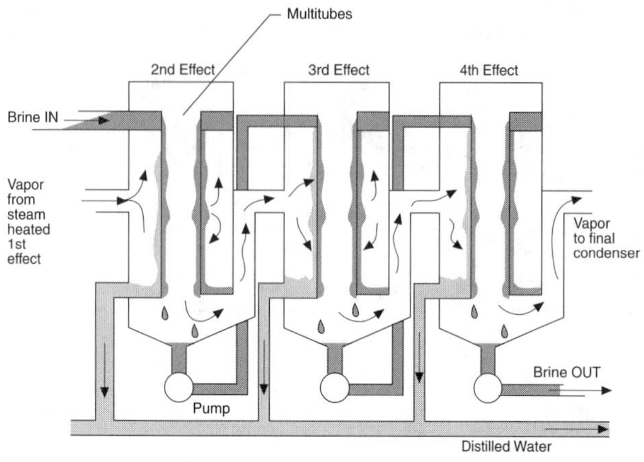

Figure 2.

Reverse Osmosis

If pure water and a salt solution are separated by a semipermeable membrane, water tends to diffuse through the membrane into the salt solution, raising its pressure above that of the water. If excess pressure is now applied to the salt solution, the process is reversed and pure water flows from the salt solution through the membrane in the opposite direction; this is called reverse osmosis (see **Membrane Processes**). The application of this technique to desalination has been studied intensely in the 1950's and 1960's, using mainly cellulose acetate membranes. Today, reverse osmosis

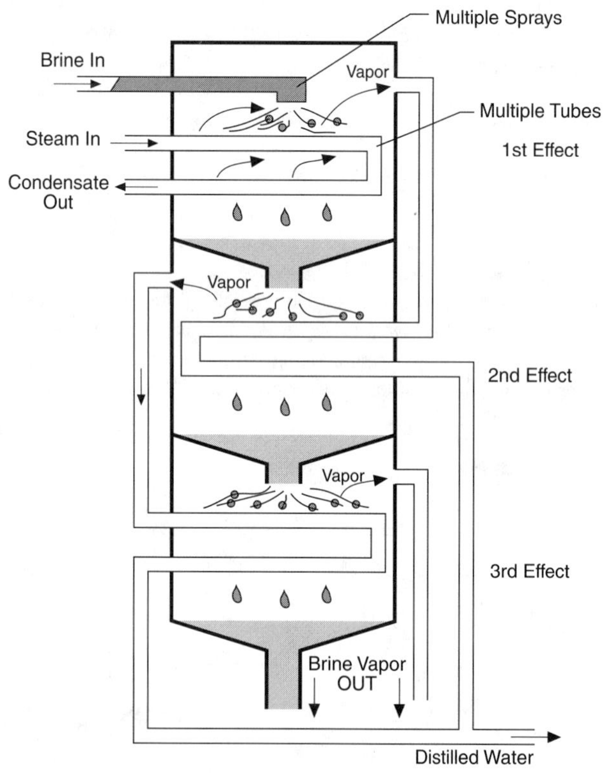

Figure 3.

employing polymeric materials is used commercially for the distillation of brackish water (pressure 2–3 MPa) and sea water (pressure 6–8 MPa). The largest brackish water reverse osmosis plant in Riyadh, Saudi Arabia produces over one hundred tonnes per day of fresh water.

Electrodialysis

If an electric current is passed through a salt solution, the ions migrate to the electrodes, seeking the opposite charge. Using special permeable membranes, the ions can be redistributed into selected volumes which are swept by two separate fluid streams. In this way, a stream with a high concentration of ions and another of almost pure water can be produced. Electrodialisis is used today commercially as a means of desalination of brackish waters.

Ion Exchange

Ion exchange is a standard procedure for industrial water treatment, particularly when dissolved solids are less than 0.5 g/kg (500 ppm). The equipment consists of a porous bed of organic resin particles, which exchange ions on contact with the salt solution. Both catonic and anionic materials are used (see **Water Preparation**).

Energy Requirements of Desalination

It is interesting to compare global estimates of the energy required to produce one kilogram of pure water by desalination with the amount of energy required to evaporate one kilogram of water at atmospheric pressure (i.e., 2.257×10^6 J). Single-effect distillation, as used in pioneering days, required 100% of this amount. Modern multistage flash distillation plants consume about 10%, freezing and reverse osmosis, about 2% and electrodialysis, less than 1%. Energy usage is not, of course, the only factor which determines the choice of a desalination system. For further reading, see Spiegler and Laird (1980).

References

Kobert, R. (1890) *Historische Studien aus dem pharmakologischen Institute der Kaiserlichen Universitat*, Dorpat, Halle.

Beaton, C. F. (1986) Ch 5.5.13 HEDH

Spiegler, K. S. and Laird, A. D. (1980) *Principles of Desalination*

G.L. Shires

DESALINATION, FLASH EVAPORATION FOR (see Evaporators)

DESALINATION OF OIL (see Electrostatic separation)

DESICCANTS (See Molecular sieves)

DESICCATION (see Drying)

DESIGN BASIS ACCIDENT (see Reflood)

DESTRUCTION OF SURFACES (see Melting)

DETERGENTS (see Surface active substances)

DERTERMINANTS (see Cramer's rule)

DETERMINISTIC CHAOS (see Non-linear systems)

DETONATION (see Explosion phenomena; Vapour explosions)

DEUTERIUM (see Neutrons)

DEUTERIUM OXIDE (see Heavy water)

DEUTSCH-ANDERSON EQUATION (see Gas-solid flows)

DEVIATORIC STRESS (see Newtonian fluids; Non-newtonian fluids)

DEVOLATISATION OF COAL PARTICLES (see Flames)

DEWAR (see Cryostats)

DEWATERING (see Drying)

DEW POINT (see Humidity measurement)

DIAMETER, HYDRAULIC (see Hydraulic diameter)

DIAPHRAGM GAUGE (see Pressure measurement)

DIE (see Casting of metals; Extrusion, plastics)

DIE-CASTING (see Casting of metals)

DIELECTRIC HEATING (see Dryers; High frequency heating; Joule heating)

DIELECTROPHORETIC FORCES (see Augmentation of heat transfer, single phase)

DIESEL ENGINES (see Internal combustion engines)

DIESEL FUEL (see Oils)

DIFFERENTIAL CONDENSATION CURVE (see Condensation curve)

DIFFERENTIAL EQUATIONS

A differential equation (DE) is a relationship between a function required (dependent variable), its derivatives of different orders, and independent variables. A DE describing a macroscopic physical process in continuum medium can be derived in two ways. The first is called phenomenological and based on mathematical representation of the corresponding physical law (or laws) for an elementary volume in a vicinity of a point in the space of independent variables, the representation being normalized by the value of the volume, followed by a formal transformation to the point limit. The other, statistical way involves a probabilistic approach to phenomena concerned with individual medium particles and deals with characteristics averaged over an ensemble of particles or process realisations.

Differential equations can be classified into ordinary differential equations containing derivatives with respect to one independent variable, and partial differential equations containing partial derivatives of the required function with respect to several independent variables. The order of the DE is the order of the highest derivative entering it. Every function substituted into a DE and reducing it to an identity is called its solution. The relationship $L(\varphi) = f(t, x, y, z)$ is called a linear DE when the sum of two partial solutions, φ_1 and φ_2, of the homogeneous DE $L(\varphi) = 0$ is its solution too, i.e., $L(\varphi_1 + \varphi_2) = 0$, where $L(\varphi)$ is the linear operator of differentiation.

To complete the mathematical description of a nonsteady physical process, represented by a given DE, the initial values of the required function must be specified in the whole domain of its description at the initial time (for a partial DE). Also, the values of the function or (and) its derivatives of the order less than n on the domain boundary (or on its part), or alternatively relationships describing the physical processes of interest on the boundary, must be specified for the whole time interval. In the case of steady physical process, only boundary values need to be specified. The initial and boundary values form the boundary conditions, and the set of DEs and corresponding boundary conditions, the boundary problem.

The order of a set of DEs (SDE) is the sum of orders of all DEs entering the SDE. Generally, the number of equations and additional relationships between them have to be equal to a number of unknown functions.

The Cauchy problem for a second order partial DE consists of finding a solution in some neighbourhood of the smooth surface Γ, where the values of the function and its derivative along some nontangent direction $\vec{l}$ are specified.

$$\varphi|_\Gamma = \varphi_0(x), \qquad \left.\frac{\partial\varphi}{\partial l}\right|_\Gamma = \varphi_1(x).$$

Functions φ_o and φ_l are the Cauchy data, and surface Γ is the Cauchy surface.

Normally, the boundary conditions have to be specified so that the boundary problem would have a unique solution. Sometimes, it is necessary to prove the theorems of existence and uniqueness for a given problem. For the general description of the boundary problem in the form of equation $D\varphi = \Phi$, where D is the operator of the given boundary problem and Φ is the set of right-hand sides of the DE and boundary conditions, the uniqueness theorem is equivalent to the existence of reverse transformation D^{-1} and the existence theorem, which refer to the coincidence of the operator D results domain with the results domain of right hand side Φ. The problems of solution existence and uniqueness are especially important for nonlinear DE's.

For boundary problems, the important question is accuracy, besides the existence and uniqueness. Small variations on boundary conditions or DE coefficients, which can always take place due to some inaccuracy of physical quantity measurements, may lead to variations in boundary problem solution too.

Partial DEs of the second order form

$$\sum_{j,k=1}^{m} A_{jk}(x)\frac{\partial^2\varphi}{\partial x_j \partial x_k} + \Phi\left(x_1, \ldots, x_m, \varphi, \frac{\partial\varphi}{\partial x_1}, \ldots \frac{\partial\varphi}{\partial x_m}\right) = 0$$

are classified according to the real characteristic values λ_i of the matrix (A) of the highest coefficients, which can be supposed to be symmetrical (it can always be achieved) and are obtained from the equation

$$\det(A - \lambda I) = \begin{vmatrix} A_{11} - \lambda & A_{12} & \cdots & A_{1m} \\ A_{21} & A_{22} - \lambda & \cdots & A_{2m} \\ \vdots & \vdots & \ddots & \vdots \\ A_{m1} & A_{m2} & \cdots & A_{mm} - \lambda \end{vmatrix} = 0.$$

The equation is of the type (α, β, γ) in the given point x, if among characteristic values λ_i there are α positive, β negative and γ zero values (α + β + γ = m). Three types of such partial DEs take particular roles in the theory of fluid flow and heat and mass transfer. The type (m, 0, 0) = (0, m, 0) is called elliptic (for instance, the Laplace and Helmholtz equations); the type (m − 1, 0, 1) = (0, m − 1, 1), is parabolic (for instance, the diffusion equation); and the type (m − 1, 1, 0) = (1, m − 1, 0) is hyperbolic (for instance, the wave equation). Due to the fact that the majority of physical problems leads to such equations, the theory for this type of equation is more developed than the theory for other partial DE types.

The characteristic surfaces (characteristics) of a partial DE of the second order are defined by nontrivial solutions of the first order equation

$$\sum_{j,k=1}^{m} A_{jk}(x)\frac{\partial\omega}{\partial x_j}\frac{\partial\omega}{\partial x_k} = 0.$$

The characteristics are invariant with respect to independent variable transformations. Elliptic equations do not have real characteristics. The parabolic type DE characteristics are planes x_m = const (x_m is a coordinate corresponding to γ = 1). The Cauchy data on characteristics are linked by some relationship. If it is violated, the Cauchy problem for a hyperbolic DE with boundary conditions on a characteristic has no solution. The boundary conditions for elliptic DE's, are distinguished from those for hyperbolic DE's. For elliptic DE's, the problem is to find a function in the whole definition domain under the given function (or its normal derivative) value along the whole boundary (Dirichlet, Neumann or mixed problems). For hyperbolic DE, the problem is to find a function in the whole domain where its value, and the value of its derivative along some direction - nontangential to the boundary surface - is defined on part of, not the whole, boundary (Cauchy problem). For a DE of parabolic type, boundary conditions for the parabolic direction ($A_{mn} = 0$) are defined on the characteristic $x_m = const$. Depending on conditions, the same general physical process can be described by DEs of different types. For instance, a heat conduction process in solid body with constant physical properties and with sharp variation of conditions at its surface can be, at very short times (of the order of nanoseconds), of a wave character. It can be described by a hyperbolic DE, of the rectangular coordinate system, in the form

$$\tau\frac{\partial^2 T}{\partial t^2} + \frac{\partial T}{\partial t} = \kappa\left(\frac{\partial^2 T}{\partial x^2} + \frac{\partial^2 T}{\partial y^2} + \frac{\partial^2 T}{\partial z^2}\right) \equiv \kappa\Delta T,$$

where τ is the heat relaxation time and is the κ − thermal diffusivity. For moderate conditions, when a vector of heat flux density can be described by the Fourier law $\vec{q} = -\lambda$ **grad**T, the equation is of the parabolic type and takes the form

$$\frac{\partial T}{\partial t} = \kappa\Delta T.$$

Under steady-state boundary conditions, the steady temperature distribution (if it exists) is described by the elliptic Laplace equation

$$\Delta T = 0.$$

Cases where analytical solutions exist for partial DE and SDE are very rare, and related mainly to linear problems and to conditions when a solution has a self-similar character—i.e., when an independent variable can be expressed as a complex of initial independent variables, leading to a reduction of the independent variables' number.

Among the analytical methods of partial DE solution, the following are the most widely employed: separation of variables method, where the solution is (if it is possible) in the form of a product of two or more functions of different independent variables (thus the problem is reduced to solving two or more DEs, depending on the reduced set of independent variables); methods of integral transforms, including Laplace, Fourier and other transforms; method of Green's functions; method of asymptotic expansions (mainly matched asymptotic expansions); variational method (for instance, the Bubnov-Galerkin method), where the solution of a problem—described by a partial DE and boundary and initial conditions—is obtained by solving a minimization problem for the corresponding functional.

In recent years, numerical methods have been widely-employed for the solution of problems described by partial DE and SDE; these include finite-differences, finite elements and boundary elements.

Leading to: *Bessel function; Characteristics, method of; Chebyshev equation; Green's function; Laplace transforms*

I.G. Zaltsman

DIFFERENTIAL PRESSURE FLOW-METERS

Following from: *Flow metering*

The operation of a differential pressure flow meter device depends on measurement of the pressure differential caused by an arrangement installed in a pipeline, or by elements of the pipeline. A flow meter consists of a primary converter which builds up differential pressure, a differential manometer and the connecting pipes between them.

Flow meters in the form of contraction devices (see **Orifice Flow Meter**) are the most common. Here, differential pressure is created by converting part of the potential energy of the flow into kinetic energy by changing the cross-section of the pipeline. The difference (Δp) between the pressure upstream of the contraction device (pin) and the pressure downstream of the device (p_{out}) is related to the flow rate $\dot{M}$ by $\dot{M} = \alpha S\sqrt{2\rho\Delta p}$ where S is the cross sectional area of the channel upstream of the meter and α is a characteristic coefficient.

The second group of flow meters depends on the differential pressure caused by a local hydraulic resistance in the pipeline. The flow condition of a such a resistance is usually laminar in order to obtain a linear dependence $\Delta p = 128\dot{V}\eta l/\pi d^4$. Such a dependence exists if the length of the capillary tube bank (**Figure 1a**) or of flat thin orifices is $l = (200 - 300)d$ whereas d is the hydraulic diameter. Such converters are usually used in measuring very small flow rates. As an alternative to capillary tubes, primary converters may be used in which the resistance is produced by a porous septum (**Figure 1b**). A linear dependence $\dot{V}(\Delta p)$ is realized in this case for $l \leq 0.25$ mm.

The operation of centrifugal flow meters (elbow meters) is based on the fact that in curvilinear sections of the pipeline (knees), differential pressure occurs due to the effect of a centrifugal force which in turn, depends on the flow rate (**Figure 1c**). For a pipeline with diameter D and with a knee having a radius of curvature R, the flow rate is determined by the relation $\dot{V} = \alpha\ \pi D^2/4\ (R/D\ \rho\Delta P)^{0.5}$. As a large body of experiments show, $\alpha = 1 \pm 0.04$ for $Re > 2 \times 10^5$. The accuracy of measurement of R and D must not be less than $\pm 0.5\%$ and $\pm 0.15\%$, respectively. Pressure taps are located on the diameter which coincides with the bisector of the central angle of knee turn and must be strictly perpendicular to the inner surface of the pipe. A straight section not less than 25D long must be placed ahead of the knee. The main advantage of such flow meters is the absence of any obstructions in the flow. They are mainly used for measuring water flow rate.

Flow meters with pressure devices, which include the Pitot tube (**Figure 1d**) and Prandtl tube (**Figure 1e**) and a device which averages of the impact pressure along the flow section (**Figure 1f**), measure the impact pressure which is the sum of the dynamic, $p_d = 0.5\rho u^2$, and static, p_s, pressures. The static pressure is measured either on the pipeline wall, or with the help of a static-pressure tube, or with Prandtl's tube (Figure 2). The differential manometer measures $\Delta p = p_d - p_s$, and the velocity at the point of probe location is defined from the relationship $u = k(2\Delta P/\rho)^{0.5}$. The main design dimensions of the Prandtl pipe are given in **Figure 2a**. For such a pipe, $k = 1 \pm 0.0025$. It consists of two concentric pipes: the central one has an open-end facing into the flow and takes up the impact pressure; on the outer pipe, a static pressure connection in the form of 6–8 holes measuring 0.1–0.2 d in diameter is located at a distance of 6 d from the probe nose. The area of the probe midsection must not exceed 2% of the flow section of the pipeline. With tubes measuring impact pressure, a local velocity is measured; to obtain the value of the flow rate, various methods are used: the method of integration of the velocity profile; the measurement of average velocity $\bar{u}$ at the point of the velocity profile where the local velocity is equal to the average velocity (in a developed turbulent flow in a smooth round pipe, the value of $\bar{u}$ is obtained at a distance of 0.768R from pipe centre); measurement of the velocity on the pipe axis u_0 coupled with the use of the relation between $\bar{u}$ and u_0 for a developed turbulent flow in hydraulically-smooth round pipes the ratio $\bar{u}/u_0$ is a function Re as shown in Table 1)

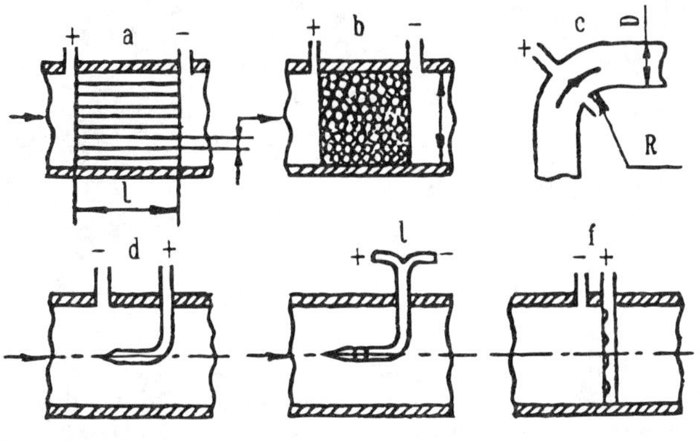

Figure 1. Differential pressure measurement devices.

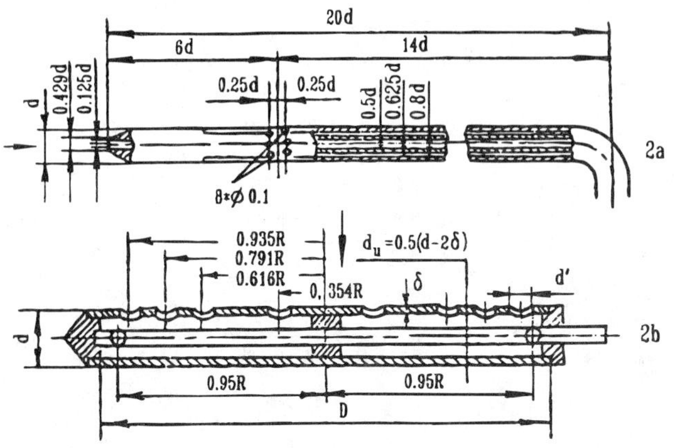

Figure 2. Prandtl tube (a) and pressure averager tube (b).

Table 1. Ratio of mean velocity ($\bar{u}$) to axial velocity (u_0) for fully developed turbulent flow in a smooth pipe

Re	4×10^3	10^4	$5\ 10^4$	10^5	$5\ 10^5$	10^6
$\bar{u}/u_0$	0.791	0.800	0.812	0.817	0.837	0.846

All these measurements call for a sufficiently-long, straight section of the pipeline $l = (25 - 40)$ D ahead of the place of measurement.

In pressure averagers (**Figure 2b**), an averaged over-the-pipeline radius value of dynamic head is measured rather than the local value. The primary converter consists of a pipe located along the pipeline diameter, which has a number of holes facing the flow. The outer diameter of the probe d should be within the limits of $d = (0.02 - 0.05)$ D, the wall thickness $\delta \leq 0.1d$, and the diameter of the exhaust holes of impact pressure tap $d_{exh} = (2 - 3)\delta$. There is a central tube which has tabbings (normal to the flow direction) located at a distance of 0.95R from the pipe axis; this allows measurement of the average impact pressure. For $D \geq 50$ mm, Re $\geq 2 \times 10^5$ and a long enough straight section of the pipeline ahead of the measuring section, the pipe coefficient $k = 1$. With the help of such a pipe, $\bar{u}$ is measured and the flow rate is derived from the equation

$$\dot{V} = \frac{\pi}{4} D^2 \sqrt{\frac{2\Delta p}{\rho}}.$$

The static pressure on the wall of the pipeline is tapped at a distance of 2d upstream of the pressure averager pipe.

Reference

Fluid Meters. Their Theory and Application. (1971) Report of ASME, Research Committee on Fluid Meters, New York, Publishers by ASME.

***Leading to:** Orifice meter*

Yu. L. Shekhter

DIFFERENTIAL PRESSURE TRANSDUCERS (see Pressure measurement)

DIFFRACTION (see Optics)

DIFFUSER (see Enlargement, flow and pressure change in)

DIFFUSION

***Following from:** Mass transfer*

Diffusion is a process leading to equalization of substance concentrations in a system or establishing in a system an equilibrium concentration distribution that results from random migration of the system's elements.

Three types of diffusion are distinguished, viz. molecular, *Brownian*, and turbulent. *Molecular diffusion* occurs in gases, liquids, and solids; both diffusion of molecules of extraneous substances (impurities) and self-diffusion are observed. Molecular diffusion occurs as a result of thermal motion of the molecules. It proceeds at a maximum rate in gases, at a lower rate in liquids, and at a still lower rate in solids—these differences being accounted for by the nature of thermal motion in these media.

In a gaseous phase, molecules possess a certain mean velocity depending on the temperature, but their motion is chaotic and in colliding, they change the direction of this motion. However, on the whole, the molecules of the substance migrate at a velocity much lower than the mean velocity of the molecular free motion. The higher the pressure, the denser is the molecule packing, the less is the free-path length, and the slower is the diffusion. The same occurs as molecule mass and size increase. Conversely, elevation of temperature causes an increase in the free-path length, a decrease in the number of collisions, and growth of free-motion velocity. These factors all lead to a speed-up of diffusion.

In liquids, molecular diffusion occurs by jumps of the molecules from one position to another; this arises when the energy of the molecule is high enough to rupture the bonds with the neighboring molecules allowing the molecule to move. On average, the jump does not exceed an intermolecular spacing, and since in a liquid this is much less than in a gas, the diffusion is substantially lower. Since a liquid is virtually incompressible, the diffusion rate is independent of pressure. Elevation of temperature increases intermolecular spacings and the velocity of vibrations and jumps of molecules, which enhances diffusion.

Gases contained in solids diffuse as ions or atoms migrating through interstitials of the crystal lattice, and the same is observed for atoms and ions with a radius much smaller than that of the ion or atom of the base substance constituting the solid. Diffusion of solid impurities occurs by interchange of sites of atoms and vacancies (unoccupied sites of crystal lattice), by migration of atoms through interstitials, by a simultaneous cyclic migration of several atoms, by a direct interchange of sites of two neighboring atoms, etc. Each displacement requires imparting to a particle a definite amount of energy (activation energy). Therefore, diffusion is extremely sensitive to temperature elevation, which manifests itself in its exponential dependence on temperature. Nevertheless, even at high temperature, diffusion in solids is much slower than in liquids.

So far, the above discussion has focused on the so-called pure concentration diffusion proceeding under the effect of concentration gradient (or chemical potential) in a medium unaffected by external factors. However, it is known that temperature gradient gives rise to a thermal diffusion, pressure gradient to pressure diffusion, an electrical field to electrical diffusion of charged particles, and so on. These diffusion types are beyond the scope of this discussion.

Molecular diffusion

In the general case, at a constant temperature and pressure, the molar diffusion flux $\dot{n}$ of substance A is proportional to the molar concentration gradient dC_A/dy and a one-dimensional formulation is described by an equation called Fick's law

$$\dot{n}_A = -D_A \frac{dC_A}{dy},$$

where D_A is the diffusion coefficient. In a binary system, the flow of one component must be balanced by the counter-flow of the other component

$$\dot{n}_B = -D_B \frac{dC_B}{dy},$$

but since $C_A + C_B = const$, then

$$\left|\frac{dC_A}{dy}\right| = \left|\frac{dC_B}{dy}\right| \quad \text{and} \quad D_A = D_B = D_{AB}.$$

D_{AB} is said to be the interdiffusion coefficient.

In order to describe a unidirectional diffusion of A molecules in a multicomponent mixture of ideal gases, the Stefan-Maxwell equation

$$\frac{dY_A}{dy} = \sum_{j=A}^{n} \frac{d_A C_j}{C_T^2 D_{A_j}} (u_j - u_A),$$

based on the kinetic theory of gases is used, where Y_A is the mole fraction of component A, $C_T = p/RT$, the total concentration (density) of mixture; $C_A = pY_A/RT$, $c_j = pY_j/RT$, D_{A_j}, the interdiffusion coefficient for a pair A, j; and u_j and u_A are the diffusion rates for the respective components of the pair.

Brownian diffusion

If fine particles (no more than a few microns in size) are placed in a stationary gas or liquid at rest, they randomly migrate in the bulk, the motion not decaying and being independent of the medium chemical properties. This phenomenon has come to be known as Brownian motion. It is brought about by the absence of a strict compensation of momentum in the molecule-particle collisions, i.e., the pressure pulsation affecting the particle. The suspended particles migrate independent of each other along intricate zigzag trajectories, changing direction up to 10^{14} times a second.

According to the general principles of statistical mechanics, the mean square value of displacement projection $\Delta\bar{x}^2$ of a particle on an arbitrarily-chosen axis is proportional to the observation time t

$$\Delta\bar{x}^2 = 2D_B t, \qquad (1)$$

where D_B is the diffusion coefficient of the Brownian particle.

For spherical particles of radius r, D_B is determined by the Stokes-Einstein equation:

$$D_B = \frac{kT}{6\pi\eta r},$$

where k is the Boltzmann constant; T, the temperature; K, η, the medium viscosity.

Equation (1) has been obtained under the assumption that particle displacements in any direction are equiprobable and, consequently, the mean value of the products of particle displacement in non-overlapping time intervals t_1 and t_2 is zero: $\overline{\Delta x_{t_1} \Delta x_{t_2}} = 0$. It holds true if we neglect the particle inertia, which is allowed for high enough t's.

If gas or liquid contains in its volume not a single particle but a great number of them, then, by virtue of the statistical character of the process, ultimately the particles appear to be uniformly distributed throughout the volume.

Turbulent diffusion

In turbulent flows or in artificially-stirred media (e.g., by stirrers or packings), mass transfer occurs on a macro level when the finite gas or liquid volumes are displaced. However, under these conditions, mass transfer may also be proportional to the concentration gradient. This fact is indicative of turbulent (eddy) diffusion.

In a turbulent flow, eddies continuously form, break up, disappear and appear once again. Large-scale eddies may include small-scale ones. The element of substance transferred is not a single molecule of the substance but some quantity of it that depends on the eddy size. The intensity of a turbulent transfer is considerably higher than that resulting from molecular motions. Treating a non-stationary diffusion of the substance from a point source (point 0) yields a simple relation

$$\bar{y}^2 = 2D_E t,$$

where $\bar{y}^2$ is the mean square value of particle displacement from the longitudinal axis running through point 0; D_E, the eddy diffusion coefficient; and t, the time. It has been proven that D_E depends on the time of motion for the particle volume considered.

The total mass transfer (e.g., in mixing) is a result of the joint effect of molecular and eddy diffusion. It is assumed in calculations that both processes are additive, i.e., the coefficients of eddy D_E and molecular D_M diffusion are added: $D = D_E + D_M$.

Diffusion in Electrolyte Solutions

A substance in electrolyte solutions, particularly dilute ones, exists in the form of ions (cations and anions). The theory of salt diffusion in dilute aqueous solutions is quite well-established, and the diffusion coefficient under an inifinite dilution is determined by the *Nernst-Heckell equation*

$$D_{AB}^0 = \frac{RT}{Fa^2} \frac{1/n_+ + 1/n_-}{1/\lambda_+^0 + 1/\lambda_-^0},$$

where D_{AB}^0 is the diffusion coefficient; (m^2/s) T, the temperature; K, Fa, the Faraday number; n_+ and n_-, the cation and anion valencies; λ_+^0 and λ_-^0, the limiting ionic conduction of the cation and the anion at a given temperature, $m^2/ohm \cdot mol$.

With increasing solution concentration, D_{AB}^0 first increases and then grows.

In mixed electrolyte systems with a light cation, e.g. H^+, the latter may migrate leaving behind the anion, and the absence of current is due to another heavier, cation lagging. In these systems, a unidirectional diffusion proceeds due to a joint action of electrical and concentration gradients

$$\dot{m}_+ = \frac{\lambda_+^0}{Fa^2}\left[-RT\frac{dc_+}{dy} + Fa\, c_+ \frac{dE}{dy}\right],$$

$$\dot{m}_- = \frac{\lambda_-^0}{Fa^2}\left[-RT\frac{dc_-}{dy} + Fa\, c_- \frac{dE}{dy}\right],$$

$$\sum \dot{m}_+ = \sum \dot{m}_-, \qquad \sum \dot{m} = 0.$$

In a multiion system, it is important to allow for ion interaction when several ionic conductions greatly differ from other ionic conductions, e.g., H^+ and OH^-, for H^+ $\lambda_+^0 = 349.8\ cm^2/\Omega\ mol$, for OH^- $\lambda_-^0 = 197.6\ cm^2/\Omega\ mol$, while for the rest ions of multiion systems ionic conductions are under 100.

Diffusion in porous bodies

There exist three mechanisms, viz., *molecular or bulk diffusion, Knudsen diffusion,* and *surface diffusion.* These can all come into action simultaneously in the same system.

Molecular diffusion is predominant in solids with large pores, whose size is much more than the free-path of the diffusing gas molecules. In this case, diffusion can be described as presented above.

Knudsen diffusion occurs in gas-filled solids with small pores, or under low pressure when the mean free-path of molecules is more than the pore size and the molecules collide with the walls more often than between themselves. Molecule reflection from the walls is normally diffuse, i.e., the molecules recoil in all directions, and diffusion of molecules along pores depends precisely on these collisions. The roles Knudsen and molecular diffusions perform are commensurable within a certain range of pore sizes and gas pressures.

Surface diffusion is observed during adsorption of a diffusing substance by a solid. Since the equilibrium surface gas concentration increases with an increase in partial pressure of the adsorbed species, a surface concentration gradient of a diffusing substance appears in the surface layer of a pore. Under certain conditions, this may enhance the total flow of a diffusing component.

For molecular diffusion in porous solids, use is made of the effective diffusion coefficient, arbitrarily related to the concentration gradient of a substance diffusing normal to the external surface of the body

$$D_{eff}^{M} = D_M\theta/\xi,$$

where D_M is the molecular diffusion coefficient, θ is a free (open) porosity and ξ, the so-called *tortuosity factor* determined in general experimentally in each specific case.

In order to calculate the flux density due to the Knudsen diffusion, it is also advisable to use some effective coefficient D_{eff}^{K}

$$D_{eff}^{K} = \frac{8\theta^2}{3\xi S_g\rho_s}\left(\frac{2RT}{\pi\bar{\mu}}\right)^{0.5}$$

$$= 19.400\frac{\theta^2}{\xi S_g\rho_s}\left(\frac{T}{\bar{\mu}}\right)^{0.5},$$

where θ and ξ are the values of open porosity and the tortuosity factor; ρ_s and S_g, the density of a porous solid and its specific surface; and $\bar{\mu}$, the molecular mass of a diffusing substance.

In a transition region in which both molecular and Knudsen diffusions are essential, an overall effective diffusion coefficient is recommended

$$D_{eff} = \left(\frac{1}{D_{eff}^{M}} + \frac{1}{D_{eff}^{K}}\right)^{-1}.$$

The above refers only to a single-component diffusion. Multicomponent diffusion in porous solids has been inadequately investigated and data of any reliability for its calculation are not available.

In considering the role of surface diffusion, it is generally assumed that the mass flux is proportional to the concentration gradient, and the absorbed layer of this sustance is extremely thin and does not change the pore cross section. Under these assumptions, the mass flux is:

$$\dot{m} = -\left(\frac{1}{D_{eff}^{M}} + \frac{1}{D_{eff}^{K}}\right)^{-1}\frac{dc}{dy} - D_{sp}\frac{d(S_g\rho_a c_{sur})}{dy},$$

where D_{sp} is the surface diffusion coefficient and c_{sur}, the surface concentration of adsorbed substance (the product $S_g\rho_s c_{sur}$ is the quantity of the adsorbed substance in a unit volume of porous mass). Assuming that an adsorbed layer and a gaseous phase in pores are at equilibrium and that the adsorption isotherm is linear, i.e., $S_g\rho_s c_{sur} = K_c$, then

$$\dot{m} = -\left[\left(\frac{1}{D_{eff}^{M}} + \frac{1}{D_{eff}^{K}}\right)^{-1} + DK_{sp}\right]\frac{dc}{dy}.$$

This equation describes the diffusion flux in a porous solid when all the three diffusion mechanisms come into action simultaneously.

References

Sherwood, T., Pigford, P., and Wilke, C. (1975) *Mass Transfer,* McGraw Hill, New York.

Leading to: *Ficks law of diffusion; Diffusion coefficient; Thermal diffusion*

I.L. Mostinsky

DIFFUSION COEFFICIENT

Following from: *Diffusion*

Diffusion coefficient is the proportionality factor D in *Fick's law* (see **Diffusion**) by which the mass of a substance dM diffusing in time dt through the surface dF normal to the diffusion direction is proportional to the concentration gradient **grad** c of this substance: dM = −D **grad** c dF dt. Hence, physically, the diffusion coefficient implies that the mass of the substance diffuses through a unit surface in a unit time at a concentration gradient of unity. The dimension of D in the SI system is a square meter per second.

The diffusion coefficient is a physical constant dependent on molecule size and other properties of the diffusing substance as well as on temperature and pressure. Diffusion coefficients of one substance into the other are commonly determined experimentally and presented in reference tables. Here, examples of self-diffusion and interdiffusion (binary diffusion) coefficients in a gaseous and liquid media are given in **Tables 1, 2**, and **3**.

As is obvious from comparing the data of **Tables 1** and **2** with those of **Table 3**, the diffusion coefficients in a gaseous and a liquid phases differ by a factor of 10^4–10^5, which is quite reasonable considering that diffusion is the movement of individual molecules through the layer of molecules of the same substance (self-diffusion) or other substances (binary diffusion in which the molecules of two substances interdiffuse). The number density of molecules in liquid is also very much higher and their mobility is lower, which

Table 1. Self-diffusion coefficient D_A of some gases at T = 273 K and p = 0.1 MPa

Gas	D_A, cm^2/s	Gas	D_A, cm^2/s	Gas	D_A, cm^2/s
H_2	1.604	O_2	0.192	H_2O	0.276
He	1.386	N_2	0.155	CH_4	0.188
Ar	0.157	CO_2	0.106	NH_3	0.192

Table 2. Interdiffusion coefficient D_{AB}

Diffusing gas A	Medium B	P, mmH_2O	T, K	D_{AB}, cm^2/s
H_2	He	760	298	1.64
Ar	He	751	288	0.703
N_2	CO_2	760	288	0.158
H_2O	air	760	273	0.219
H_2	N_2	760	288	0.743
H_2O	CO_2	760	273	0.146

Table 3. Diffusion coefficient of gases in liquids

Gas	Liquid	T, K	D,10^{-5}cm/s
Air	water	293	2.5
CO_2	water	298	1.92
H_2	water	298	4.50
NH_3	water	285	1.64
CO_2	ethanol	298	3.42
CO_2	heptane	298	6.03

implies a much lower diffusion coefficient. In solids, diffusion is still slower.

If experimental data are lacking, the diffusion coefficient can be calculated.

Diffusion in gases. For ideal gases, the diffusion coefficient does not depend on substance concentration. In accordance with the kinetic theory of gases, the mean free-path length l of molecules is inversely proportional to the mean cross-sectional area of the molecule S and the number density of the molecules n in a mixture. The latter is inversely proportional to the space occupied by the mixture, i.e., T/p, where T is the temperature and p, the pressure. The mean velocity of the molecules u is proportional to $(T/\bar{\mu})^{1/2}$, where $\bar{\mu}$ is the molecular mass. Thus, in the case of interdiffusion of gases with the same molecular mass μ or self-diffusion, the expression is:

$$D \sim ul \sim \frac{1}{n}\sqrt{\frac{T}{\bar{\mu}}} \sim \frac{T^{3/2}}{pS\bar{\mu}^{1/2}}.$$

If molecules of types A and B interact (binary diffusion), then the interdiffusion coefficient is;

$$D_{AB} = \text{const}\,\frac{T^{3/2}(1/\bar{\mu}_A + 1/\bar{\mu}_B)^{1/2}}{pS_{AB}} \qquad (1)$$

where S_{AB} is the mean value of the cross-sections of molecules of both types. The kinetic theory of gases makes it possible to determine the constant in Equation (1), assuming the molecules are spherical and their cross-sections are equal to cross-sections of these spheres.

Sutherland has made a correction to Equation (1), taking into account the forces of intermolecular attraction which influence the free-path length of molecules; thus,

$$D_{AB} = \text{const}\,\frac{T^{3/2}}{pS_{AB}}\left(\frac{1}{\bar{\mu}_A} + \frac{1}{\bar{\mu}_B}\right)^{1/2}\frac{1}{1 + \dfrac{C_{AB}}{T}}$$

where C_{AB} is the *Sutherland coefficient.*

Contemporary kinetic theory takes into account the intricate character of molecular interaction; molecules repel one another when they are close and attract one another at a distance. Many researchers have studied the potentials of this molecular interaction, but great recognition has been won by the so-called *Lennard-Jones potential*

$$\varphi(r) = 4\varepsilon\left[\left(\frac{\sigma}{r}\right)^{12} - \left(\frac{\sigma}{r}\right)^{6}\right],$$

where $\varphi(r)$ is the potential energy; r, the distance between the centers of molecules; ε and σ are Lennard-Jones' interaction constants determined for many gases and summarized in tables.

For dilute gas mixtures, and assuming that the molecular collisions are only binary and elastic, that the motion of colliding molecules is in terms of classical mechanics, that quantum effects are absent, and, finally, intermolecular forces act only along the center line, the expression

$$D_{AB} = \frac{1.883\ 10^{-20}\ T^{3/2}(1/\bar{\mu}_A + 1/\bar{\mu}_B)^{1/2}}{p\sigma_{AB}^2\Omega}$$

for the diffusion coefficient in a binary mixtures is obtained by Bird, Hirshfelder and Curtiss. Here, $\Omega = f(kT/\varepsilon_{AB})$ is the collision integral and k is the Boltzmann's constant. The interaction parameters ε_{AB} and σ_{AB} are determined for the binary system from the appropriate constants for pure substances:

$$\frac{\varepsilon_{AB}}{k} = \left(\frac{\varepsilon_A}{k}\cdot\frac{\varepsilon_B}{k}\right)^{1/2}, \qquad \sigma_{AB} = \frac{\sigma_A + \sigma_B}{2}.$$

If the data on ε and σ are not available, they can be estimated using the well-known critical parameters for a given substance:

$$\frac{\varepsilon}{k} = 0.75\ T_c, \qquad \sigma = \frac{5}{6}\cdot 10^{-8}V_c^{1/3},$$

where T_c and V_c are the critical temperature (K) and the critical molar volume (cm^3/mol) respectively.

Wilke and Lee noted the coefficient $1.885 \cdot 10^{-2}$ is not constant in reality and depends on molecular masses of the diffusing gases:

$$\left[1.533 + 0.352\left(\frac{\bar{\mu}_A + \bar{\mu}_B}{\bar{\mu}_A \cdot \bar{\mu}_B}\right)^{0.5} \cdot 10^{-20}\right].$$

This accurate definition results in a better agreement between experimental and computed data.

A correlation formula obtained by *Fuller, Schetter and Gittings* by means of computer-aided correlation of 340 experimental points, expressed as:

$$D_{AB} = \frac{1.01\ 10^{-4}\ T^{1.75}(1/\bar{\mu}_A + 1/\bar{\mu}_B)^{1/2}}{p[(\Sigma\ V_A)^{1/3} + (\Sigma\ V_B)^{1/3}]^2}$$

has come to be widely known, where ($\Sigma\ V_A$) and ($\Sigma\ V_B$) are the values derived from summation of atomic diffusion volumes for each component of the binary mixture, i.e., molecules A and B. The values for some atoms and simple molecules are presented in **Table 4**.

In order to calculate the diffusion coefficient in multicomponent systems, Wilke used the Maxwell-Stefan equation to derive the expression

$$D'_A = \frac{1 - Y_A}{Y_B/D_{AB} + Y_C/D_{AC} + \cdots},$$

where D'_A is the diffusion coefficient of the component A in the mixture with B, C, . . . , Y_A; Y_B, Y_C are the molar fractions of the appropriate components; and D_{AB} and D_{AC} are the diffusion coefficients in the AB and AC binary systems, respectively.

Diffusion in liquids. As has been noted, diffusion in liquids encounters greater resistance and the diffusion coefficients for liquids lower than 10^4 to 10^5 times.

One of the earliest equations for determining the diffusion coefficient in dilute solutions was the *Stokes-Einstein equation*, based on the model of motion of a spherical particle of diffusing substance A in a viscous liquid continuum B

$$D_{AB} = \frac{kT}{b\pi r_0 \eta_B},$$

where r_0 is the particle (molecule) radius and η_B, the liquid viscosity. The constant b depends on the size of diffusing molecules: b = 6 for molecules larger than those of the base substance; b = 4 for identical molecules; and b can be less than 4 for smaller molecules.

Assuming that the molecule diameter $2r_0 = (\tilde{v}/N_0)^{1/3}$, where $\tilde{v}$ is the molar volume of a diffusing substance and N_0 is the Avogadro number, then

Table 4. Diffusion volumes of atoms and simple molecules

Substance	$\tilde{v}$	Substance	$\tilde{v}$	Substance	$\tilde{v}$
C	16.5	CO_2	26.9	CO	18.9
H	1.98	H_2	7.07	NH_3	14.9
O	5.48	O_2	16.6	H_2O	12.7
N	5.69	N_2	17.9	SO_2	41.1

$$D_{AB} = \frac{2kT}{b\pi\eta_B}\left(\frac{N_0}{\tilde{v}_A}\right)^{1/3}.$$

Comparison of this formula with experimental data has shown that in most cases, the discrepancy is moderate and reaches 40% only in some cases.

In 1955, *Wilke and Chang* have suggested a more general formula based on extensive experimental investigations, but involving many empirical values as well

$$D_{AB} = 7.4\ 10^{-8}\left[(\varphi\bar{\mu}_B)^{1/2}\frac{T}{\eta_B\tilde{v}_A^{0.6}}\right],$$

where D_{AB} is the interdiffusion coefficient in an infinitely-dilute solution, cm^2/s; φ, the parameter of association of solvent B; $\bar{\mu}_B$, the molecular mass of substance B; $\tilde{v}_A$, the molar volume of solute A at a boiling point under normal conditions, cm^3/mol, η_B, the substance viscosity, Ns/m^2; and T, the temperature, K.

Introduction of the association parameter into the formula is brought about by the fact that associated molecules behave like large-size molecules and diffuse at a lower rate; the degree of association varying with mixture composition and with molecule types. Therefore, Wilke and Chang presented the values for most widespread solvents: for water $\varphi = 2.6$; methanol, 1.9; ethanol, 1.5; benzene, ester, heptane and nonassociated solvents, 1.

A semiempirical formula suggested by *Scheibel*,

$$D_{AB} = 8.2\ 10^{-8}\ T\left[1 + \left(\frac{3\tilde{v}_B}{\tilde{v}_A}\right)^{2/3}\right]\eta_B^{-1}\tilde{v}_A^{-1/3},$$

is worthy of attention. In some cases, it appears to be more exact than the preceding one; but for $v_A/v_B \leq (1 - 2)$, the deviation from experiment becomes important and the following relations are recommended:

$$D_A = 25.2\ 10^{-8}T\eta_B^{-1}\tilde{v}_B^{-1/3}\ \text{(water is the solvent)};$$

$$D_A = 18.6\ 10^{-8}\ T\eta_B^{-1}\tilde{v}_B^{-1/3}\ \text{(benzene is the solvent)};$$

$$D_A = 17.5\ 10^{-8}\ T\eta_B^{-1}\tilde{v}_B^{-1/3}\ \text{(other solvents)}.$$

Reddy and Doraiswamy have suggested the equation

$$\frac{D_{AB}\eta_B}{T} = K_{RS}\frac{\bar{\mu}_B^{1/2}}{(\tilde{v}_A\tilde{v}_B)^{1/3}},$$

where K_{RS} varies depending on the ratio of molar volumes: $K_{RS} = 8.5\ 10^{-8}$ for $\tilde{v}_A/\tilde{v}_B > 1.5$ and $K_{RS} = 10^{-7}$ for $\tilde{v}_A/\tilde{v}_B < 1.5$. Comparison of this equation with 96 experimental points has shown good agreement, the spread of points being about +15%.

All formulas suggested above for calculating the diffusion coefficient hold true for low-viscosity liquids. For a high-viscosity solvent, they are in great error and therefore inapplicable.

The temperature effect on the diffusion coefficient has been poorly studied so far. Within a narrow temperature range—from 10 to 20°C—the temperature dependence of the diffusion coefficient can be assumed to be linear

$$D = D_{298}[1 + \alpha(T - 298)],$$

where D_{298} is the diffusion coefficient at $T = 298K$, $\alpha = 20\eta_{298}^{1/2}\,\rho^{-1/3}$, η_{298}, the solvent viscosity at $T = 298K$, Ns/m, ρ, the solvent density, g/cm^3.

Theoretically, this must be an exponential dependence of the type $D = AT\exp(-E/RT)$.

The experimental data of Wilke and Chang give available evidence that the activation energy varies from 12.6 to 28.1 kJ/mol.

The dependence of the diffusion coefficient on concentration of diffusing substance, strictly speaking, is a consequence of the fact that diffusion flow depends on the difference (gradient) of the thermodynamic potential of the system rather than concentration, i.e., the formula must allow for activity of the diffusing substance. Hence, at V_B = const,

$$D = D_0 \frac{d\ln a}{d\ln c} = D_0\left(1 + \frac{d\ln \nu}{d\ln c}\right),$$

where D_0 and D are the diffusion coefficients, respectively, in an infinitely-dilute solution and in a solution with finite concentration c; a and c, the activity and the concentration of diffusing substance; and ν, the activity coefficient of this substance.

The semiempirical formulas presented above are more exact than the theoretical ones because the latter were derived making assumptions. Nevertheless, to avoid an appreciable error it is advisable to make calculations by several formulas concurrently and to compare the results.

In electrolyte solutions, salts dissociate and diffuse as ions and molecules depending on the degree of dissociation. The theory of salt diffusion is elaborated mainly for dilute solutions in which the degree of dissociation is close to one. Thus, the diffusion coefficient for a simple salt that is infinitely diluted can be found using the *Nernst-Heckell equation*

$$D_{AB} = \frac{RT}{Fa^2}\,\frac{1/n_+ + 1/n_-}{1/\lambda_+^0 + 1/\lambda_-^0},$$

where D_{AB} is the diffusion coefficient, defined as the proportionality factor between the molecular flow of dissolved salt and the gradient of its molecular concentration, cm^2/s; T, the temperature, K; Fa, the Faraday number, n_+ and n_-, the cation and anion valences; λ_+^0 and λ_-^0, the limit (under an infinite dilution) ionic conductions of cation and anion, cm^2/Ω mol.

In electrolyte solutions, the diffusion coefficient substantially depends on the concentration of diffusing substance. If its concentration is no more than 2N, the formula

$$D = D_0 \frac{V}{x_1 V_1}\frac{\eta_1}{\eta}\left(1 + m\frac{d\ln \nu}{dm}\right)$$

is suggested by Gordon, where D_0 and D are the diffusion coefficients for an infinitely dilute solution and a molar solution, respectively; V, the solution volume; V_1, the partial volume of the solvent; x_1, the number of solvent moles in volume V; ν, the molarity of the solution; η and η_1, the viscosity of solution and solvent; v, the molar activity coefficient.

References

Hirschfelder, J. O., C. F. Curtiss, C. F. and Bird, R. B. (1954) *Molecular Theory of Gases and Liquids,* Wiley, New York.

Sherwood, T. K., Pigford, R. L., and Wilke, C. R. (1975) *Mass Transfer,* McGraw Hill, New York.

I.L. Mostinsky

DIFFUSION COEFFICIENT OF GASES (see Kinetic theory of gases)

DIFFUSION FLAMES (see Flames)

DIFFUSION IN ELECTROLYTE SOLUTION (see Diffusion)

DIFFUSION LAW (see Condensation of multicomponent vapours)

DIFFUSION PUMP (see Vacuum pumps)

DILATANCY (see Viscosity)

DILATANT FLUIDS (see Non-Newtonian fluids)

DILATION OF GRANULAR MATERIAL (see Granular materials, discharge through orifices)

DILUTANT FLUIDS (see Fluids)

DIMENSIONAL ANALYSIS

Dimensional analysis is a method of reducing the number of variables required to describe a given physical situation by making use of the information implied by the units of the physical quantities involved. It is also known as the "*theory of similarity*".

Physical Quantities, Units and Dimensions

In observing the physical world, we make use of different physical concepts such as size, distance, time, temperature, etc., and to make quantitative deductions, we must adopt independent and reproducible methods of measuring these physical quantities. "Measurement" in essence means the comparison of the unknown quantity with a known reference, as in the measurement of distance using a ruler, or the balancing of an unknown weight against known weights on a chemical balance. For this purpose, agreement on a known "unit" for each physical quantity is needed; for example, the French Bureau of Standards keeps a block of platinum whose mass is defined to be one kilogramme, and known "weights" used for chemical balances are ultimately "calibrated" by comparison with this primary reference—usually by a series of intermediate comparisons.

In fact, it is not necessary to have an independent primary reference for every physical quantity of interest since some physical quantities are defined in terms of others. For example, "velocity" is defined as rate of change of position, or as length travelled in a given time; thus if length is measured in metres and time in seconds, velocity can be measured in metres per second. Units based directly on primary references are called fundamental or *basic* units, while those defined in terms of other units are called compound or *derived* units. (See also **SI Units.**)

Given such a definition of one quantity in terms of others, a choice of basic primary references is thus possible. In the above case for example, astronomers prefer to take the velocity of light and the terrestrial year as primary references, so that length is measured in the derived units "light-years."

Some physical quantities are defined through physical laws. Such a law is simply the expression of a general conclusion from a set of experimental observations—often expressed in the form of a mathematical equation. For example, **Fourier's Law** of heat conduction: "The heat flux ($\dot{q}$) due to conduction through a solid is proportional to the temperature gradient," may be written:

$$\dot{q} = -\lambda \, dT/dx$$

where T is the temperature; x, the distance, and the "constant" of proportionality λ is called the **Thermal Conductivity**. Experiment shows that λ is not, in fact, a constant but varies with both temperature and the nature of the solid; it is, in fact, a new physical variable defined by the physical law, just as velocity was defined above.

There are situations where the new physical variable defined by a physical law does turn out to be constant. For example, Joule's experiments on the conversion of mechanical work (W) into heat (Q) are summarized by the equation:

$$W = J \cdot Q,$$

which defines J as the "mechanical equivalent of heat". Joule found that J was independent of the particular conversion process involved and always had the same value ($J = 4.184 \times 10^7$ *erg/calorie*).

Such a constant is called a "universal constant" (other examples are Planck's constant and Boltzmann's constant), and when a law of this kind is available, there is further flexibility in choosing the set of fundamental units:

a. Keep the original system, with heat and work both based on arbitrary references (e.g., calorie and erg), and accept J as a new physical quantity (with units erg/calorie) which is relevant in any situation where work is converted to heat or vice versa.

b. Drop one of the original primary references (e.g., calorie) and use the law to define its units in terms of the other quantities involved by making the universal constant unity (i.e., a pure number). In this example, this implies that heat and work are the same kind of physical quantity (both are forms of energy). This is called a *coherent* choice of units.

It may be thought that a coherent choice is always the preferred choice, but this is not necessarily the case.

To summarize, a given physical system is described in terms of a set of physical variables of various types. To make quantitative observations, units of measurement for each variable are required. For this purpose, a subset of physical variables must be chosen, to which arbitrary reference units are assigned, then definitions or physical laws are used to derive units for the remaining variables.

The basic physical quantities for which arbitrary references are assigned are referred to as "dimensions," and the following notation is used:

[L] = metre means "the unit for dimension length is the metre"

v = [L/T] means "velocity v has dimensions length/time"

Example: Heat transfer to fluid flowing through a pipe. The heat transfer coefficient (α) between the fluid and pipe-wall will possibly depend on fluid properties: density (ρ), viscosity (η), specific heat (c_p), thermal conductivity (λ), and also on the fluid mean velocity (u), the length (l) and diameter (D) of the pipe, and the temperature difference (ΔT) between the wall and the fluid.

It is reasonable to take mass [M], length [L], time [T], heat [H] and temperature [Θ] as basic dimensions.

Then, for example, $\alpha = [H/L^2T\Theta]$ and the indices for all the variables can be set out in the form of a table called the *dimensional matrix:*

	α	D	l	u	ΔT	ρ	η	λ	c_p
M	0	0	0	0	0	1	1	0	−1
L	−2	1	1	1	0	−3	−1	−1	0
T	−1	0	0	−1	0	0	−1	−1	0
H	1	0	0	0	0	0	0	1	1
Θ	−1	0	0	0	1	0	0	−1	−1

The value of a physical variable is always written as a number of units. For example, density $\rho = \bar{\rho}Kg/m^3$ using the SI system of units. Of course, a change in the primary references, or basic units, induces a change in the corresponding numbers, $\bar{\rho}$, $\bar{\alpha}$, etc., for given values of the variables. This is referred to as a unit-transformation. Thus, if two alternative sets of basic units U_{i1}, U_{i2} are available for the basic dimensions:

$$[D_i] = U_{i1} \quad \text{or} \quad [D_i] = U_{i2}, \quad i = 1, 2, \ldots, k,$$

then for a typical physical variable v_j:

$$v_j = \bar{v}_{j1}\left[\prod_{i=1}^{k} U_{i1}^{m_{ij}}\right] = \bar{v}_{j2}\left[\prod_{i=1}^{k} U_{i2}^{m_{ij}}\right], \qquad j = 1, 2, \ldots, n,$$

where m_{ij}, i = 1, 2, . . . k, j = 1, 2, . . . n are the entries in the dimensional matrix, and it follows that:

$$\frac{\bar{v}_{j1}}{\bar{v}_{j2}} = \prod_{i=1}^{k}\left(\frac{U_{i2}}{U_{i1}}\right)^{m_{ij}} = c_j. \qquad (1)$$

This equation formally defines a unit-transformation.

Dimensionless Form of Physical Laws

Clearly the outcome of any set of experiments cannot depend on the particular choice of basic units for the variables, which implies that any mathematical equation representing a valid physical law must be invariant to a unit-transformation. Such an equation is said to be *dimensionally homogeneous.* Thus if: $f(v_1, v_2, \ldots v_n) = 0$ represents a physical law, then:

$$f(c_1v_1, c_2v_2, \ldots c_nv_n) = 0$$

where $c_1, c_2 \ldots c_n$ satisfy Equation (1) for some set of ratios U_{i1}/U_{i2}, $i = 1, 2, \ldots k$. This condition is a constraint on the form of an equation representing a physical law, which allows it to be expressed in terms of a reduced number of variables.

Example: Consider the simple fluid flow equation:

$$\frac{1}{2}\rho_1u_1^2 + g\rho_1z_1 + p_1 = \frac{1}{2}\rho_2u_2^2 + g\rho_2z_2 + p_2.$$

Using SI units, each variable can be written in the form $\rho_1 = \bar{\rho}_1$ kg/m^3 etc., where $\bar{\rho}_1$ etc. are pure numbers. Then each term in the equation has the units of pressure (kg/m.s^2), and the units cancel from the equation, confirming that it is indeed dimensionally homogeneous, and leaving a similar equation with $\bar{\rho}_1$ replacing ρ_1 etc.

Using a different set of basic units, these would also cancel giving a relation between a different set of numbers $\bar{\rho}_1'$, $\bar{u}_1'$ etc. Notice that if only the two sets of numbers $\bar{\rho}_1, \bar{u}_1, \ldots$ and $\bar{\rho}_1', \bar{u}_1', \ldots$, were given, it would be difficult to tell which units were used; so the equation must also be invariant to changes in the variables within a given system of units, provided that the variables are scaled in the same way as if changes have been made in the system of units (i.e., are subjected to a unit transformation).

This can be seen more clearly if the equation is divided by its first term, to give:

$$1 + 2\left(\frac{gz_1}{u_1^2}\right) + 2\left(\frac{p_1}{\rho_1u_1^2}\right) = \left(\frac{\rho_2u_2^2}{\rho_1u_1^2}\right)\left(1 + 2\left(\frac{gz_2}{u_2^2}\right) + 2\left(\frac{p_2}{\rho_2u_2^2}\right)\right)$$

Here, the units cancel within each bracketed group of variables, which are thus **Dimensionless Groups** or pure numbers, and the equation has been reduced to a relationship between five groups, rather than the nine original variables. Different physical situations giving the same values for the five groups are said to be "similar."

More generally, any dimensionally homogeneous equation can be reduced to dimensionless form and "similar" solutions can be exploited.

Complete Sets of Dimensionless Groups

It is clear that the members of a given set of variables can be combined to form dimensionless groups in various different ways. In general, we have

$$\Pi = \prod_{j=1}^{n} \nu_j^{p_j} \qquad (2)$$

and the indices p_j, $j = 1, 2, \ldots n$ can be chosen to make the product Π dimensionless. From the dimensional matrix, $\nu_j = [\Pi_{i=1}^{k} D_i^{m_{ij}}]$, $j = 1,2, \ldots n$, so substituting this in (2) and equating the resulting index for each D_i to zero yields the set of equations:

$$M \cdot p = 0 \qquad (3)$$

where p is the column vector with elements p_j. Now (3) has $(n - r)$ linearly independent solutions $p^{(1)}, p^{(2)}, \ldots p^{(n-r)}$, where r is the rank of the matrix M, and any other solution is a linear combination of these solutions. This means that a set of $(n - r)$ independent dimensionless groups can be formed, such that none of these can be formed by combination of the other groups in the set, but any group not in the set can be formed by combination of groups in the set. Such a set of dimensionless groups is called a **complete set,** and clearly *any* physical law must be expressible as a relation between members of this set.

In order to find the complete set, a subset of linearly independent columns of M must be found, then the columns permuted so that these are the first r columns and M can be written $[M_1, M_2]$. Then, using (3):

$$p^{(j)} = \begin{bmatrix} \bar{p}^{(j)} \\ e_j \end{bmatrix}, \qquad [M_1, M_2]\begin{bmatrix} \bar{p}^{(j)} \\ e_j \end{bmatrix} = 0,$$

$$M_1\bar{p}^{(j)} = -M_2e_j, \qquad j = 1, 2, \ldots, (n - r),$$

where e_j is a unit vector, with the jth element unity and the remaining elements zero. Then each $\bar{p}^{(j)}$ generates one group of the complete set.

A more intuitive way to describe this procedure is to select "units" for each basic dimension as a combination of variables describing the system, using the same number of variables as there are basic dimensions. Then using these units for each of the remaining variables generates the required set of dimensionless groups.

Example: Heat transfer to fluid flowing through a pipe. Using the dimensional matrix given earlier, lengths can be measured in pipe-diameters D, and temperatures with ΔT as the unit. For mass, the mass of unit volume of the fluid, ρD^3 can be used, and for heat, the capacity of this volume for unit temperature-rise, $\rho D^3c_p\Delta T$. Finally, for time, D/u can be used.

For the remaining variables:

$$\alpha = [HL^{-2}T^{-1}\Theta^{-1}] = \Pi_1(\rho D^3c_p \cdot \Delta T)(D)^{-2}(D/u)^{-1}(\Delta T)^{-1},$$

$$\text{whence } \Pi_1 = \alpha/\rho c_pu$$

$$1 = \Pi_2 \cdot D \qquad \Pi_2 = l/D$$

$$\eta = \Pi_3(\rho D^3)(D)^{-1}(D/u)^{-1} \qquad \Pi_3 = \eta/\rho Du$$

$$\lambda = \Pi_4(\rho D^3c_p\Delta T)(D)^{-1}(D/u)^{-1}(\Delta T)^{-1} \qquad \Pi_4 = \lambda/\rho c_puD$$

Here, units for all basic dimensions (effectively finding M_1^{-1} as a solution to $M_1 \cdot M_1^{-1} = I$) have been formed, which indicates that M is of full rank. However, note that none of the four groups obtained involve ΔT, which implies that α does not, in fact, depend on ΔT (or that other possibly relevant variables have not been considered, such as the coefficient of cubic expansion).

Had ΔT been omitted from the table, the rows for H and Θ would have been identical, except for a change of sign, showing that all variables only involve H and Θ as a ratio H/Θ, so units for H and Θ cannot be separately formed. Replacing the separate rows by a row for H/Θ (in fact identical to that for H), a unit can be defined $[H/\Theta] = \rho D^3 c_p$, and the same four dimensionless groups as before can then be obtained.

In general, more complicated groupings of the original choice of basic dimensions may have to be used to obtain a matrix with linearly independent rows, which thus define a new reduced set of basic dimensions, necessary and just sufficient to define the units for all the remaining variables. Again, this is called a complete set of basic dimensions.

This leads to the famous "**Π-Theorem**:" "For a physical system described by n physical variables using a complete set of r basic dimensions, the laws governing the system can be expressed as mathematical relations among at most $(n - r)$ dimensionless groups of variables."

Buckingham enunciated this theorem in 1914 **without** the all-important qualification that the basic dimensions form a complete set, and, of course, was unable to prove it. This remained a challenge until the correct form was enunciated and proved by Langhaar (1951).

Applications of Dimensional Analysis

The most obvious advantage of putting physical laws in dimensionless form is that it reduces the number of independent variables needed to describe the situation. Thus, for example, in planning an experimental investigation of heat transfer to fluid in a pipe, the form of the function: $\Pi_1 = \varphi(\Pi_2, \Pi_3, \Pi_4)$ can be investigated, rather than $\alpha = f(D, L, u, \Delta T, \rho, \eta, \lambda, c_p)$. Moreover, to vary Π_2, Π_3 and Π_4, the most convenient parameters can be chosen. Thus a rig can be built using only a single diameter pipe, and temperatures measured at several points along the length to obtain the effect of varying Π_2, while varying the flow-rate (for a given fluid) gives the effect of varying Π_3. It is in fact better to use $\Pi_4' = \Pi_3/\Pi_4 = c_p\eta/\lambda$, which involves only fluid properties, rather than Π_4, and to investigate variation of Π_4' by choosing a range of different fluids. Note again that it is easier to find a range of fluids to cover a range of values for Π_4', rather than separate ranges for the individual properties c_p, η, λ, which tend to vary together.

In this example, the analysis itself indicated that one variable (ΔT) was irrelevant—or that other factors were being ignored. It also showed that the effect of varying pipe-diameter can be deduced from experiments on a single pipe.

The latter is a special case of exploiting "similar" solutions, which is perhaps better illustrated by the following example:

Example: Wind-tunnel testing of aircraft. Assume that the drag force F on the aircraft is a function of the density (ρ), viscosity (η) and speed of sound (u_s) of the air, and of the velocity (u), wing-span (l) and other dimensions ($l_1, l_2, l_3 \ldots$) of the aircraft. Dimensional analysis yields:

$$\frac{F}{ul\eta} = f\left(\frac{u}{u_s}, \frac{ul\rho}{\eta}, \frac{l_1}{l}, \frac{l_2}{l}, \frac{l_3}{l}, \cdots\right)$$

To eliminate effects of the shape factors (l_1/l, ... etc.), a scale model is built, geometrically similar to the prototype aircraft, so that these have the same value for model and prototype.

To vary the other two groups independently, the air velocity u can be altered, but otherwise it is necessary either to build several models of different sizes or vary the air properties—in practice, a single model is used and either air temperature or air pressure is varied.

However, if the wind-tunnel uses only atmospheric air and only one model is available, only a partial solution is possible. If in fact u is well below u_s, then drag does not depend on u_s, and hence only one group (the **Reynolds Number** $ul\rho/\eta$) is important. Remember however that the model size (l_m) is much smaller than the prototype (l_p), so the velocity (u_m) required in the wind-tunnel will be higher than the prototype velocity of interest (u_p); hence it is the **Mach Number** (u_m/u_{sm}) of the *model* which limits the range of validity of the tests. In general, of course, conditions can be chosen so that:

$$\frac{u_m}{u_{sm}} = \frac{u_p}{u_{sp}}, \quad \frac{u_m l_m \rho_m}{\eta_m} = \frac{u_p l_p \rho_p}{\eta_p},$$
$$\frac{l_{1m}}{l_m} = \frac{l_{1p}}{l_p}, \quad \frac{l_{2m}}{l_m} = \frac{l_{2p}}{l_l} \cdots \tag{4}$$

which yields $F_p/u_p l_p \eta_p = F_m/u_m l_m \eta_m$. These are then similar solutions, and conditions (4) are often called "*similarity conditions.*"

It is quite common for similarity conditions to be incompatible, making it impossible to model actual conditions in all respects on a different scale.

Finally, the larger the number of basic dimensions for a given set of variables, the smaller the number of dimensionless groups in the complete set, and the simpler the resulting system.

Now if a given physical law is relevant, a noncoherent choice of units requires addition of the universal constant as a relevant physical variable. This is not necessary for a coherent choice, but then there is one less basic dimension and the number of groups is the same in each case.

On the other hand, if a physical law is not relevant the universal constant is not needed, but a coherent choice of units will reduce the number of basic dimensions and create an extra dimensionless group.

This is illustrated by the heat transfer example, where generation of heat by fluid friction was ignored. A coherent choice of heat unit would not have indicated that ΔT was irrelevant, and hence would have generated five groups.

References

Buckingham, E. (1914) "On Physically Similar Systems: Illustrations of the Use of Dimensional Equations", *Phys. Rev., 4,* 345.

Langhaar, H. L. (1951) "*Dimensional Analysis and the Theory of Models*", (John Wiley, New York).

Leading to: *Dimensionless groups*

R.W.H. Sargent

DIMENSIONAL MATRIX (see Dimensional analysis)

DIMENSIONAL STABLE ANODES, DSAS (see Electrolysis)

DIMENSIONALLY HOMOGENEOUS EQUATIONS (see Dimensional analysis)

DIMENSIONLESS GROUPS

A dimensionless group is a combination of dimensional or dimensionless quantities having zero overall dimension. In a system of coherent units, it can therefore be represented by a pure number.

The value of dimensionless groups for generalising experiemental data has been long recognised. Over a hundred years ago, G. G. Stokes demonstrated the importance of the group now known as **Reynolds Number** as a criterion for similarity of fluid behaviour under different conditions (Stokes, 1966). It was Helmholtz who showed the significance of the groups now known as **Froude Number** and **Mach Number.** Initially, the dimensionless groups did not have specific names, and the first to attach names was M. G. Weber in 1919, when he allocated the titles Froude, Reynolds and Cauchy to groups.

The naming of numbers is an informal process, and there are several cases where the same dimensionless group has been given more than one name, e.g., Froude and Boussinesq, Bond and Eötvös. On the other hand, some savants have been honoured by a multiplicity of numbers; Lagrange for instance has three, and Damköhler, no less than five. There does however seem to be one law which, though unwritten, is broken only rarely; that is, that dimensionless titles are awarded posthumously.

The following table lists the dimensionless groups most relevant to heat and mass transfer.

Table 1. Dimensionless groups

Archimedes number	$Ar = gL^3\rho_L(\rho_S - \rho_L)/\eta^2$	
Bond number	$Bo = gL^2(\rho_L - \rho_G)/\sigma$	also called Eötvös number, Eo
Biot number, heat transfer	$Bi = \alpha L/\lambda_S$	
Biot number, mass transfer	$Bi_m = \beta L/\delta_S$	
Clausius number	$Cl = u^3L\rho/\lambda\Delta T$	
Darcy number	$Da = \Delta p2D/\rho u^2L$	alternative = 4f
Deborah number	$D = t_r/t_o$	relaxation time/observation time, also called Weissenberg number
Eckert number	$Ec = u^2/c_p\Delta T$	also called Brinkman number
Euler number	$Eu = \Delta p/(\rho u^2/2)$	
Fanning friction factor	$f = \Delta pD/2\rho u^2L$	
Fourier number, heat transfer	$Fo = \kappa t/L^2$	
Fourier number, mass transfer	$Fo_m = \delta t/L^2$	
Froude number	$Fr = u/(gL)^{1/2}$	or u^2/gL also called Boussinesq number, Bo
Galileo number	$Ga = L^3g/\nu^2$	
Graetz number	$Gz = uD^2/\kappa L$	$Gz = RePrD/L$
Grash of number	$Gr = \beta gL^3\Delta T/\nu^2$	
Jakob number	$Ja = c_p\Delta T/h_{LG}$	also called phase change number
J factor, heat transfer	$j_H = Nu/(RePr^{1/3})$	$= St\ Pr^{2/3}$
J factor, mass transfer	$j_M = Sh/(Re\ Sc^{1/3})$	$= St_m\ Sc^{2/3}$
Kutateladze number	$K = h_{LG}/cp\Delta T$	
Lewis number	$Le = \kappa/\delta$	$= Sc/Pr$
Mach number	$Ma = u/u_{sonic}$	
Nussett number	$Nu = \alpha L/\lambda$	
Peclet number	$Pe = uL/\kappa$	$= Re\ Pr$
Prandtl number	$Pr = c_p\eta/\lambda$	$= \nu/\kappa$
Rayleigh number	$Ra = g\beta'\Delta TL^3/\nu\kappa$	$= Gr\ Pr$ (β' is coefft. of thermal expansion)
Reynolds number	$Re = uL\rho/\eta$	
Schmidt number	$Sc = \nu/\delta$	
Sherwood number	$Sh = \beta L/\delta$	
Stanton number, heat transfer	$St = \alpha/\rho uc_p$	$= Nu/(Re\ Pr)$
Stanton number, mass transfer	$St_m = \beta/u$	$=Sh/(Re\ Sc)$
Strouhal number	$St = f'L/u$	(f' is frequency)
Weber number	$We = \rho u^2L/\sigma$	

References

Catchpole, J. P. and Fulford, D., Dimensionless Groups *Ind. Eng. Chem.* 58(3):46, 1966 and 60(3):71 (1968.)

Stokes, G. G. (1966) *Mathematical and Physical Papers* 2nd ed, vol. 3.

G.L. Shires

DIMERS (see Molecule)

DIOXINS (see Incineration)

DIPHENYL

Diphenyl ($(C_6H_5)_2$) is a white crystalline solid insoluble in water. It is obtained from benzene or toluene and is extremely toxic. Its main usage is as one of the components of *heat-transfer fluids* capable of operating at temperatures around 700 K.

Below are some of its physical characteristics:

Molecular weight: 154.21
Melting point: 342.4 K
Boiling point: 528.4 K
Critical temperature: 789 K
Critical pressure: 3.85 MPa
Critical density: 323 kg/m^3

V. Vesovic

DIPOLE MOMENT (see Piezoelectricity)

DIRAC DELTA FUNCTION (see Delta function)

DIRECT CONTACT CONDENSERS (see Condensation of pure vapour; Condensation, overview; Condensers; Heat exchangers)

DIRECT CONTACT EVAPORATORS (see Evaporators)

DIRECT CONTACT MASS TRANSFER (see Agitated vessel mass transfer; Extraction, liquid/liquid)

DIRECT CONTACT HEAT EXCHANGERS

Following from: *Heat exchangers; Direct contact heat transfer*

Introduction

Direct contact heat exchangers have been used by heat transfer practitioners for more than one hundred years. In fact, the success of the industrial revolution has much to do with their initial use by James Watt in creating the needed vacuum for efficient steam engines. In 1900, Hausbrand's book, "Evaporation, Condensing and Cooling Apparatus," published information dealing with several types of direct contactors including barometric condensers. Despite this early start, the development of a true understanding of their nature lagged and still lags behind the understanding of surface-type heat exchangers. Nevertheless, they are widely used as open-feed water heaters in power plants, open-evaporative cooling towers, barometric condensers throughout the petroleum industry, and in gas (air) separation plants. Still another use is in absorption refrigeration plants. Other applications are in rotary retorts, drying processes, etc. Thus, knowledge of them as alternatives to conventional regenerators or recuperators is necessary to economically optimize systems that include heat exchange.

Direct contact heat exchange takes place between two process streams. The streams can include combinations such as gas-solid, gas-liquid, liquid-liquid, liquid-solid, or solid-solid streams. For obvious reasons, gas-gas systems cannot be achieved directly; however, two direct contactors can be used in series where a third stream extracts heat from one gas stream and transfers it to another. Thus, direct contactors can be used for almost all systems; but, the complexity of multiple component systems may overcome their economic advantage over surface type heat exchangers.

Advantages and Disadvantages in Utilizing Direct Contactors

The exchange of heat between two fluid streams can, in general, be accomplished using either direct contact or surface-type heat exchangers. There are, however, several limitations to the use of direct contactors. First, if two fluid streams are placed in direct contact, they will mix, unless the streams are immiscible. Thus, stream contamination will occur depending on the degree of miscibility. The two streams must also be at the same pressure in a direct contactor, which could lead to additional costs. The advantages in utilizing a direct contactor include the lack of surfaces to corrode or foul, or otherwise degrade the heat transfer performance. Other advantages include the potentially superior heat transfer for a given volume of heat exchanger due to the larger heat transfer surface area achievable and the ability to transfer heat at much lower temperature differences between the two streams. Still another advantage is the much lower pressure drop associated with direct contactors as compared to their tubular counterparts. A final advantage is the much lower capital cost as direct contact heat exchangers can be constructed out of little more than a pressure vessel, inlet nozzles for the fluid streams, and exit ports. Of course, it is sometimes advantageous to provide internals, as will be discussed later.

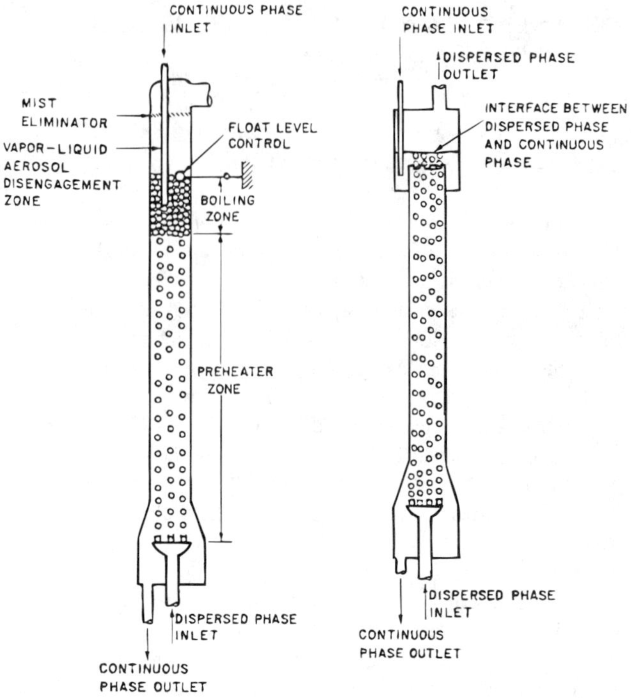

Figure 1. Schematics of spray columns for evaporation and for sensible heating of the dispersed lighter phase.

Varieties of Direct Contact Heat Exchangers

A typical direct contactor provides heat transfer between two fluid streams. The processes include the simple heating or cooling of one fluid by the other; cooling with the vaporization of the coolant; cooling of a gas-vapor mixture with partial condensation; cooling of a vapor or vapor mixture with total condensation; and cooling of a liquid with partial or complete solidification. Most of the direct-contact applications can be accomplished with the following devices: a) *Spray columns*, b) *Baffle tray columns*, c) Sieve tray or *bubble tray columns*, d) Packed columns, e) Pipeline contactors, and f) Mechanically agitated contactors.

Figures 1–6 illustrate the general configurations of a) through f), respectively. Except for the turbulent pipe contactor, all of the devices are counter-current devices and depend upon the relative buoyancy of the dispersed phase through a continuous phase. While the figures illustrate a less-dense dispersed phase being introduced at the bottom of the column, it is possible for the dispersed phase to be denser and introduced at the top, with the configuration internals appropriately revised.

The *turbulent pipe contactor* is a parallel-flow device and has the limits of efficiency of all such systems, whether they be direct contact or surface-type heat exchangers. That is, the maximum temperature achieved by the cool stream is that of the mixing cup temperature.

The size of the turbulent pipe contactor is dictated by the relative mass flow rate and the nature of the turbulence. Turbulence promoters can be installed to enhance the turbulence and, thereby, reduce

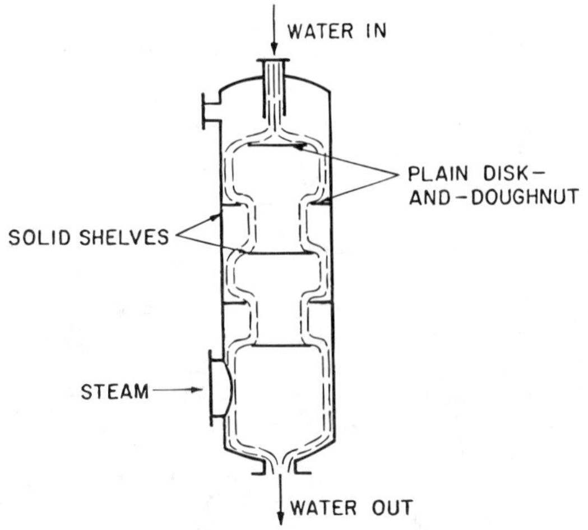

Figure 2. Schematic of a disk and donut baffle tray column for use as a steam condenser (Jacobs and Nadig, 1987).

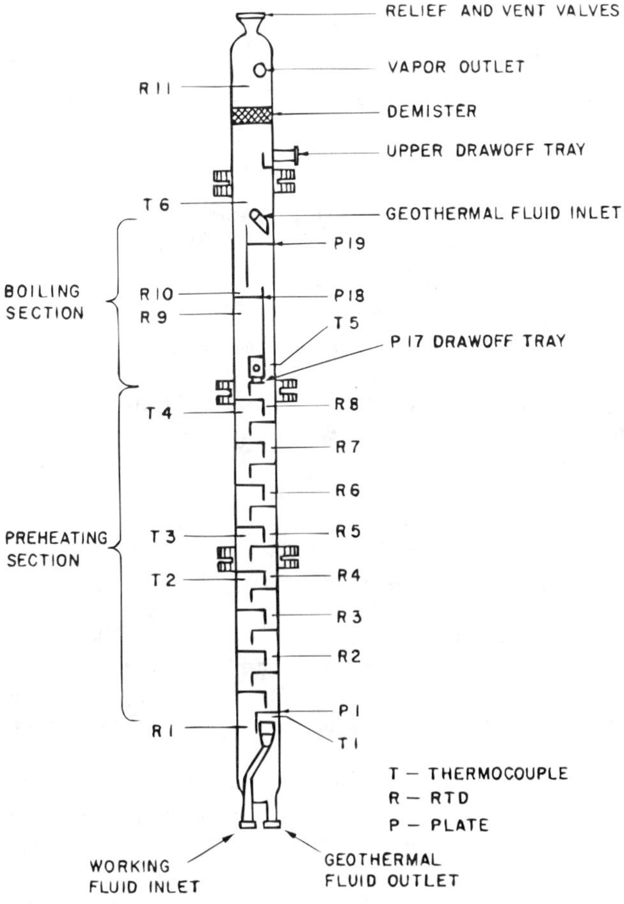

Figure 3. Schematic of a sieve tray column used for extracting heat from geothermal brine (Jacobs and Eden, 1986).

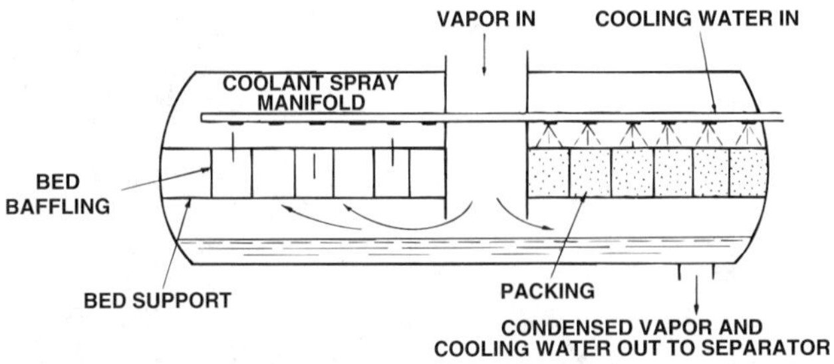

Figure 4. A possible configuration of a packed bed condenser (Jacobs and Eden, 1986).

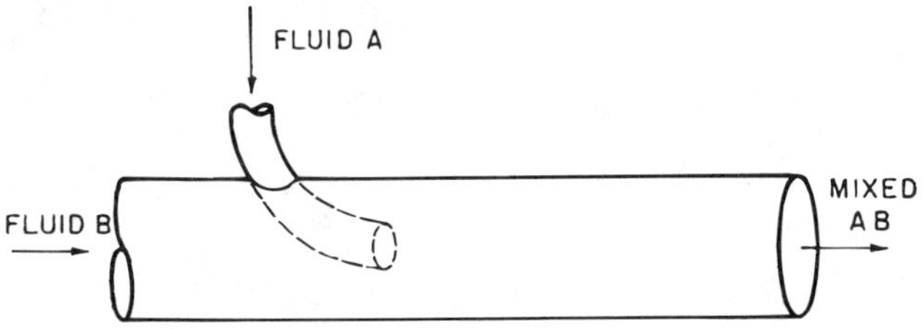

Figure 5. Turbulent pipe contactor.

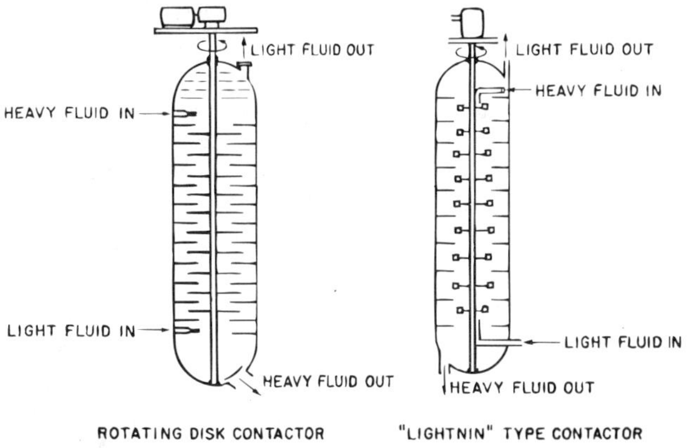

Figure 6. Typical mechanically agitated towers (Treybal, 1966).

the length of contactor necessary to essentially obtain the mixing cup temperature. If separation of the streams is desired, the contactor must be followed by a separation device such as a settler, a cyclone separator, or other mechanisms. While the turbulent pipe conductor is very inexpensive, if separation is desired, the cost of the settler will in all probability dictate the economics of the process.

The remaining apparatus all have the heat transfer take place between a continuous phase and a clearly defined disperse phase in the form of drops, bubbles, jets, sheets, or thin supported films in the case of packed beds. Heat exchangers with mechanical agitators (**Figure 6**), while often superior as heat or mass transfer equipment, are more difficult to design as the dispersed phase may have a wide range or drop or bubble sizes. Thus, empirical data from the manufacturer to establish performance is necessary. Further,

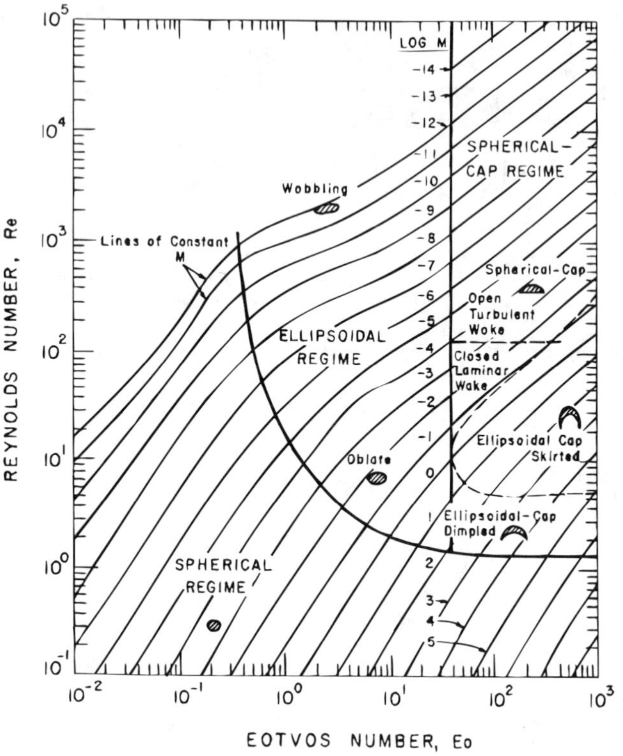

Figure 7. Drop characterization map (Grace 1983).

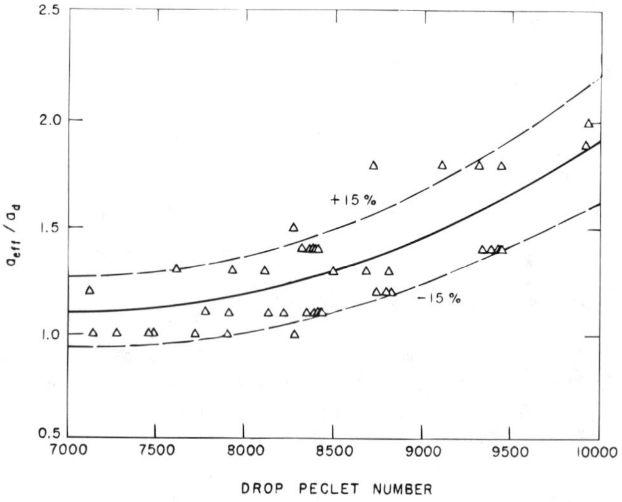

Figure 8. Effective thermal diffusivity to molecular diffusivity as a function of drop peclet number (Jacobs and Eden, 1986).

problems may result in seals at the penetration point of the drive shafts. Special designs may therefore be necessary.

Baffle tray columns may have similar problems in defining the nature of the curtain of the dispersed phase. Depending on flow rates and battle design, the dispersed phase may be a sheet, a series of rivlets or defined streams, which can break up into drops. If the

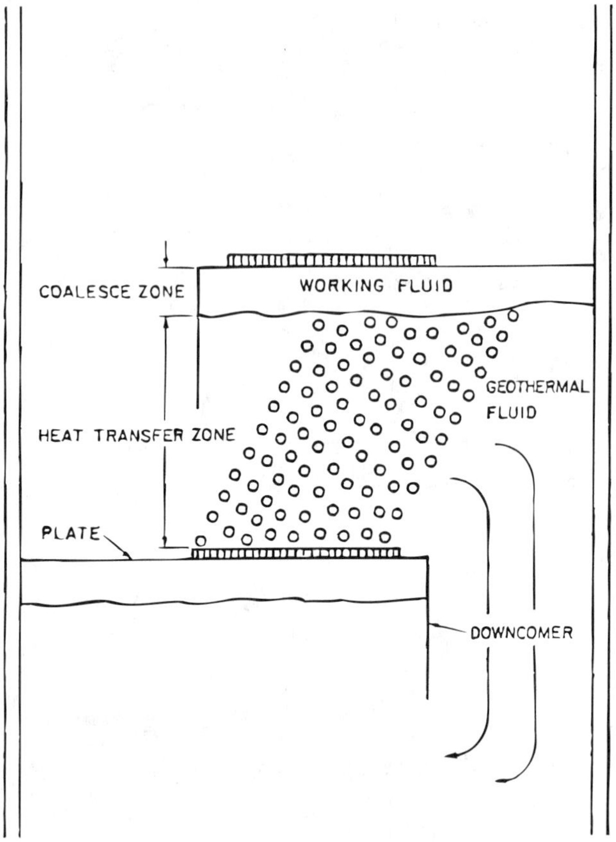

Figure 9. Schematic of a tray in a sieve tray column (Jacobs and Eden, 1986)

baffles are, in fact, trays with serrated or notched rims, the dispersed phase can be designed to be a series of well-defined streams and the heat transfer is more easily analyzed. The baffles/trays then result in mixing of the dispersed phase and enhance the internal-to-the-dispersed phase mixing.

The spray column shown in **Figure 1** is an open column whose only internals are the inlet nozzles for the dispersed and continuous phase. Ideally, such columns are capable of pure counter-flow operation, with the dispersed phase made up of nearly uniform diameter drops. While it is possible to design the dispersed phase inlet nozzle to achieve the desired characteristics, providing a uniform flow in the continuous phase is more difficult. Great care must be taken or maldistribution of the continuous phase may lead to diminished heat transfer. Thus, the design of continuous phase inlet nozzles are sometimes proprietary, or patented.

The bubble column or sieve tray column (see **Figure 3**), enhances the internal heat transfer coefficient by repeatedly reforming the drops at each tray. Proper tray or baffle design can lead to shorter columns, and potentially small heat exchanger volume for the same service. Their major disadvantage is fouling, corrosion or blockage of some of the holes in the sieve tray. Details of design methods and references to recent improvements are given by Jacobs (1988) and Jacobs (1995a, 1995b).

References

Hausbrand, E. (1933) *Condensing and Cooling Apparatus,* 5th Ed, Van Nostrand, New York

Jacobs, H. R. (1988) Direct Contact Heat Transfer for Process Technologies. *ASME Journal of Heat Transfer,* Vol. 110, pp. 1259–1270.

Jacobs, H. R. (1995a) Direct Contact Heat Exchangers, *Heat Exchanger Design Handbook.*

Jacobs, H. R. (1995b) Direct Contact Heat Transfer, *Heat Exchanger Design Handbook.*

H.R. Jacobs

DIRECT CONTACT HEAT TRANSFER

Following from: *Heat transfer*

Direct contact heat transfer is generally defined as heat transfer between two or more mass streams without the presence of an intervening wall. The mass streams can be co-current, counter-current or even cross-flow. The streams can be immiscible or miscible or partly so. Typical two-stream direct contactors include: liquid-liquid, liquid-vapor, liquid-solid, gas-solid, or even solid-solid. Common systems that have been studied extensively include: water-air, water-stream and water-organic liquid. Also, there have been extensive studies of fuel drops in an oxidizing gaseous environment. The possibilities are legion.

Direct contact heat transfer can take place across the interface between two continuous fluid streams, such as a gas flowing over a thin liquid film or between a disperse spray and a gaseous or vapor stream into which the spray is injected. The former might involve hot gas quenching or fuel vaporization and combustion, while the latter might involve the condensation of the vapor on droplets within the spray. Still another example is the cooling of fine drops of a liquid undergoing solidification, as in the manufacture of glass beads or metal shot.

In many cases of direct-contact heat transfer, chemical reactions can be taking place between the mass streams, and one of the gas streams may totally be consumed by the other. There can, of course, be simple sensible heat transfer as in the case of two immiscible liquids.

When the mass streams include at least one fluid, that stream can be either laminar or turbulent; in many applications, turbulence needs to be avoided as it can lead to problems in either characterization of the mass streams or bulk fluid dynamical changes.

A further characteristic of direct contact processes is that the fluid streams must locally be at the same pressure. Many industrial direct contactors thus develop relative flows of the mass streams by imposing an external body force. The most common of these is the effect of gravity or centrifical forces on two fluids of different density, although electrical or magnetic fields can be imposed on some fluids to achieve the desired effect of relative fluid motion.

Direct contact heat transfer has received considerable attention since the 1970's, although the field lacks the maturity associated with other heat transfer. Specific reviews of direct contact heat transfer include those of Sideman (1966), Sideman and Moalem-Maron (1982), Jacobs (1988) and Jacobs (1995). Jacobs has defined direct-contact heat transfer as being associated with either continuous streams or dispersed phases. In the former, the heat transfer can be of steady-state while for the latter, the heat transfer is always transient when one considers the individual drops, bubbles or particles of which the dispersed phase may be composed. When on of the fluids is composed of a dispersed stream, the bulk flow may appear to be undergoing a steady-state energy transfer, as in the case of a spray column or a sieve tray column; however, the individual fluid elements are undergoing transient heating. It is thus necessary to carry out a combined Eulerian-Lagrangian analysis of the flow and heat transfer. It is this characteristic of direct-contact heat transfer that makes it more difficult to model than surface-type heat exchange. All of the difficulties of modeling multiphase flow are present, together with the complications of heat exchange, and define the interfacial phenomena. Nevertheless, the advantages of potentially much higher heat transfer rates, the ability to transfer the heat at much lower temperature differences between the streams and the potentially lower cost makes direct-contact heat transfer extremely attractive.

To learn more about direct contact heat transfer, refer to the literature reviews cited above and to the articles on **Direct Contact Heat Exchangers, Dryers, Cooling Towers, Condensers, Quenchers, Multiphase Flow,** etc.

References

Sideman, S. (1966) Direct Contact Heat Exchange Between Immiscible Liquids, *Advances in Heat Transfer,* Academic Press, New York, NY, 207–286.

Sideman, S. and Moalem-Maron, D. (1982) *Direct Contact Condensation, Advances in Heat Transfer,* Academic Press, New York, NY, 228–276.

Jacobs, H. R. (1988) Direct Contact Heat Transfer for Process Technologies, *ASME Journal of Heat Transfer,* Vol. 110, 1259–1270.

Jacobs, H. R. (1995) Direct Contact Heat Transfer, To be published in the *Heat Exchanger Design Handbook* (HEDH) Update, Begell Press.

Leading to: *Direct contact heat exchangers*

H.R. Jacobs

DIRECT INVERSION OPTICAL TECHNIQUE (see Optical particle characterisation)

DIRECT NUMERICAL SIMULATIONS, DNS (see Turbulence models)

DIRICHLET CONDITIONS (see Conduction)

DIRICHLET'S PROBLEM (see Green's function)

DISCRETE ORDINATE APPROXIMATION (see Radiative heat transfer)

DISCHARGE COEFFICIENT (see Weirs)

DISK AND DOUGHNUT BAFFLES (see Shell-and-tube heat exchangers)

DISK TYPE CENTRIFUGE (see Centrifuges)

DISK TYPE STEAM TURBINE (see Steam turbine)

DISORDER (see Entropy)

DISPERSED FLOW

Following from: *Multiphase flow; Two-phase flows*

Dispersed flow is characterized by the flow where one phase is *dispersed* in the other *continuous* phase. This flow configuration is observed in all types (gas-liquid, gas-solid, liquid-liquid and liquid-solid) of two-phase flows as shown in **Figure 1**.

In **Gas-solid** and **Liquid-Solid Flows**, the dispersed phase is always in solid phase because solid particles never coalesce with each other. On the other hand, in **Gas-Liquid** and **Liquid-Liquid Flows**, the dispersed phase is determined mainly by the flow rates of both phases since the interface between both phases is deformable, and the dispersed phase coalesces and finally becomes the continuous phase as flow rate increases. For example, in gas-liquid flow the gas phase is dispersed (bubbly flow, **Figure 1b**) when the gas flow rate is small compared with liquid flow rate. On the other hand the liquid phase is dispersed (annular dispersed **Figure 1c** and droplet flow **Figure 1d**) when the liquid flow rate is small compared with gas flow.

However, even in the gas-solid and liquid-solid flows, solid particles agglomerate when the solid flow rate increases as shown in **Figure 2**. In this case, the group of solid particles behave like in the continuous phase and therefore, such flow configuration is called *plug flow* (sometimes, actually plugs the pipe) and is distinguished from dispersed flow.

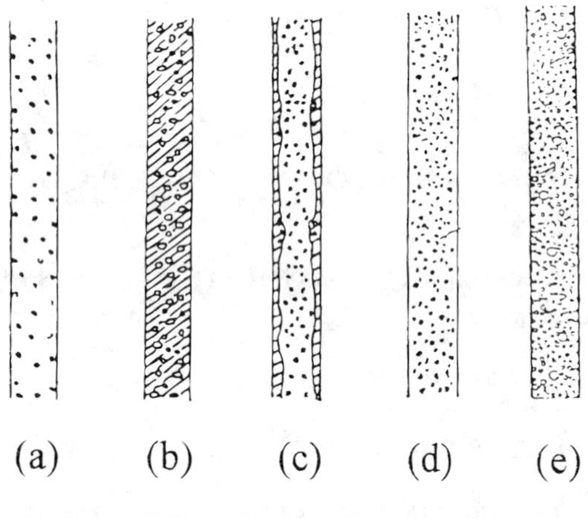

Figure 1. Dispersed flows, (a) gas-solid, liquid-solid, (b) bubbly, (c) annular dispersed, (d) droplet, (e) liquid-liquid.

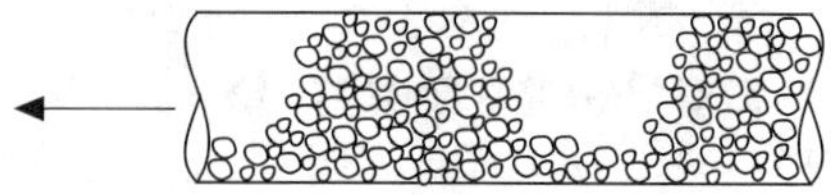

Figure 2. Plug flow in gas-solid and liquid-solid flow.

One of the most important features common to all types of dispersed flows is that mass, momentum and energy transfer between the phases are carried out from each particle (here, particle means solid particle, bubble, droplet in gas and liquid) to the surrounding continuous phase. Therefore, the mechanisms of mass, momentum and energy transfer from a single particle basically control the interaction between phases, although, of course, multi-particle effects must be considered. The correlations for phase interactions are usually based on that for a single particle, with some modification, due to the multiparticle effect of volume fraction or mass fraction of the dispersed phase.

Among the interaction terms between phases, the most important one is the *drag force* acting on the particle, since this reflects two-phase flow effects in determining the flow fields of the dispersed and continuous phases. Drag force, F_D, is given in terms of drag coefficient, C_D, which is defined by:

$$F_D = C_D A \frac{1}{2} \rho_c (u_c - u_d) | u_c - u_d |, \qquad \textbf{(1)}$$

where ρ is the density; u, velocity; suffixes c and d denote the continuous and dispersed phases; and A is the projected area of a particle in its flow direction.

For gas-solid and liquid-solid flows, the correlation of drag coefficient for a single spherical particle is used although in actual systems, the solid particle is not necessarily spherical. Schiller & Naumann (1933) have expressed it as:

$$C_D = 24(1 + 0.15\,Re_d^{0.687})/Re_d \qquad Re_d \le 1000$$

$$= 0.43 \qquad Re_d \ge 1000 \qquad \textbf{(2)}$$

where Re_d is the **Reynolds Number** based on the particle diameter. In dispersed flow, the effect of the multiparticle system is given by Bouilard (1989) as:

$$C'_D = C_D(1 - \varepsilon_d)^{-2.7} \qquad \textbf{(3)}$$

where C'_D is the drag coefficient of multiparticle system and ε_d is the volume fraction of the dispersed phase.

For gas-liquid and liquid-liquid flows, the effect of the deformation of particle (bubble or droplet) on the drag coefficient must be considered, particularly for larger particles. For smaller particles, the effects of the deformation are small and Equation (1) is approximately applied, although some modification is needed to consider the viscosities of both phases (Hadamard, 1911). However, as the particle diameter increases, the deformation becomes appreciable and the effects of deformation on drag coefficient become predominant. The drag coefficient is given by Harmathy (1960) as:

$$C_D = 0.57\sqrt{\frac{g\Delta\rho D_p^2}{\sigma}}, \qquad \textbf{(4)}$$

where Δρ is the difference in density between the phases, D_p, the particle diameter and σ the interfacial surface tension. This drag coefficient gives the constant *terminal velocity* for deformable particle (bubble or droplet), which is given by:

$$u_\infty = 1.53\left(\frac{\sigma g \Delta\rho}{\rho_c^2}\right)^{1/4}. \quad \textbf{(5)}$$

The drag coefficient is taken from the maximum of the values given by Equations (2) and (4). The effects of the multi-particle system on gas-liquid and liquid-liquid systems have been studied by Ishii, and Chawla (1979). For smaller particles the drag coefficient of which is given by Equation (2), the drag coefficient of the multiparticle system is given by:

$$C'_D = C_D(1 - \varepsilon_d)^{-n} \quad \textbf{(6)}$$

n = 1 for bubble in liquid; n = 1.75 for drop in liquid; n = 2.5 for drop in gas.

For larger particles, the drag coefficient of which is given by Equation (4),

$$C'_D = C_D\left\{\frac{1 + 17.67(f(\varepsilon_d))^{6/7}}{18.67 f(\varepsilon_d)}\right\}^2 \quad \textbf{(7)}$$

$$f(\varepsilon_d) = (1 - \varepsilon_d)^m \quad \textbf{(8)}$$

m = 1.5 for bubble in liquid; m = 2.25 for drop in liquid; m = 3 for drop in gas.

In applying the drag coefficients for dispersed flow discussed above, there are two approaches, i.e., *Eulerian* and *Lagrangian.*

In the Eulerian approach, both continuous and dispersed phases are averaged in appropriate time and/or space domains. The resulting basic equations of mass, momentum and energy conservations are expressed in Eulerian coordinate. As a result of averaging, the dispersed phase is also treated as a kind of continuous fluid. Depending upon the model which regards two-phase flow as a single mixture fluid or two separate fluids, there are two types of formulations called mixture model and two-fluid model [Ishii (1975), Delhaye (1968)]. Such a Eulerlian approach and basic equations in dispersed flow are the same as those in other types of two-phase flows.

For example, the momentum conservation equation based on two-fluid model is:

$$\frac{\partial(\varepsilon_k\rho_k)}{\partial t} + \mathrm{div}(\varepsilon_k\rho_k\overline{u_k}\,\overline{u_k}) = -\varepsilon_k \,\mathrm{grad}\, \overline{P_k} + \mathrm{div}(\varepsilon_k\overline{\tau_k})$$

$$+ \varepsilon_k\rho_k g + M_k \qquad (k = d, c) \quad \textbf{(9)}$$

where suffix k denote dispersed (k = d) and continuous (k = c) phases and the upper bar denotes averaged value. In Equation (9), M_k represents the momentum transfer term between phases. Using the drag coefficient in a multiparticle system (C'_D derived above), this is given by:

$$M_d = \frac{3}{4} C'_D \frac{1}{D_p} \varepsilon_d\rho_c |\overline{u_c} - \overline{u_d}|(\overline{u_c} - \overline{u_d}) \quad \textbf{(10)}$$

$$M_c = -M_d. \quad \textbf{(11)}$$

Using this momentum equation with mass and energy equations in Eulerian approach, the averaged behavior of heat, mass and momentum transfer in dispersed flow can be analyzed.

In the Lagrangian approach, the motion of each particle is calculated in Lagrangian coordinate, whereas the continuous phase

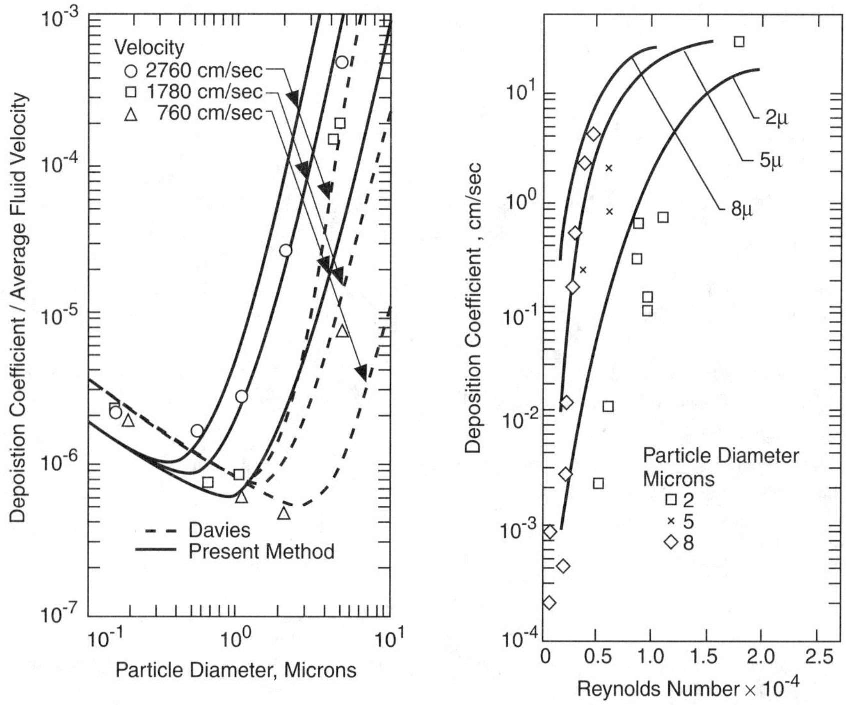

Figure 3. Deposition coefficient of solid particle in gas [Beal S.K. (1970)]. Nuclear Science and Engineering, 40, p. 8, with permission.

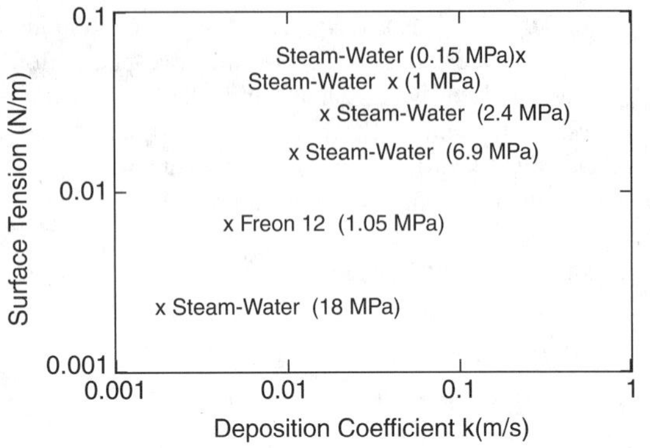

Figure 4. Deposition coefficient of droplet [Whalley et al. (1974)].

is treated in Eulerian coordinate. Using the drag coefficient of particles in a multiparticle system, the trajectory of each particle motion is calculated using the equation of motion of particle:

$$m_p \frac{du_d}{dt} = -C_d'A \frac{1}{2} \rho_c |u_c - u_d| (u_c - u_d) + f_d \tag{12}$$

where, m_p is the mass of single particle and f_d is the force acting upon the particle, such as gravity, pressure force and collision between particles, etc. Here, it should be noted that velocity u_c and u_d are local instantaneous (not averaged) velocity of particle and continuous phase. When the flow of continuous phase is turbulent, u_c is not given by the solution of averaged momentum equation (Equation 9). In order to get u_c, appropriate **Turbulence Models**, such as direct numerical simulation (DNS), large-eddy simulation (LES), two-equation model (k-ε, k-kL models), etc. are needed. This Lagrangian approach reflects physical mechanisms of dispersed flow more precisely. However, there are numerous dispersed particles in actual dispersed flow so that simulations of all particles are almost impossible with the present capacity of large-scale computers. Therefore, only simulation of sample particles is carried out and, using some statistical assumptions, the flow behavior and heat and mass transfer of dispersed flow are evaluated. Such Lagrangian approach is carried out particularly in the area of gas-solid dispersed flow [Yuu et al., (1978)].

Along with Lagrangian analysis of particle behavior in dispersed flow, many experimental studies have been also carried out for various parameters of dispersed flow. Among these, *deposition* rate of particles on the pipe wall is quite important and common to all types of dispersed flow. The deposition rate, $\dot{m}_D$, is defined as the mass flux of dispersed particle deposited on the wall per unit of time and unit area. This quantity is related to the mass concentration of dispersed phase in the continuous phase, C. The coefficient relating deposition rate and concentration is called deposition coefficient (k_D) and defined by:

$$\dot{m}_D = k_D C. \tag{13}$$

There are many correlations of deposition coefficient for gas-solid and gas-liquid (droplet or annular droplet) flows.

In real gas-solid flow, particle diameter is relatively small. Therefore, the particle follows Brownian motion (particle diameter less than 1 micron) and turbulent motion of continuous phase (less than 10 micron). Experiments and analyses have been carried out and correlations have been derived [Beal S.K. (1970)]. A typical result of deposition coefficient of gas-solid flow is shown in **Figure 3**.

In gas-liquid flow, particle (droplet) diameter is relatively large (10 to 100 micron) and droplet may deform due to hydrodynamic force. In this case, the mechanisms of deposition are quite complicated. Therefore, the derived correlations are mainly based on experimental data of droplet deposition. The earliest and standard correlation for deposition coefficient has been derived at Harwell Laboratory, UK [Whalley et al. (1974)], shown in **Figure 4**. More recently, the correlation considering turbulence droplet interaction has been presented by McCoy & Hanratty (1977).

References

Bouilard, J. X., et al. (1989) *AIChE J., 35,* 908.

Delhaye, J. M. (1968) *Equations Fondamentales des Ecoulements Diphasiques,* CEA-R-3429.

Beal, S. K. (1970) *Nuclear Sci. & Eng., 40,* I.

Ishii, M. (1975) *Thermo-fluid Dynamic Theory of Two-Phase Flow,* Eyrolles Paris.

Ishii, M. and T. C. Chawla, (1979) ANL 79-105.

Hadamard, J. (1911) Compt. Rend. Acad. Sci. Paris, *52,* 1735.

Harmathy, T. Z. (1960) *AIChE J. 6,* 281.

McCoy, D. D. and Hanratty, T. J. (1977) *Int. J. Multiphase Flow, 3,* 319.

Schiller, L. and Naumann, A. (1933) Z.V.D.I., *77,* 318.

Whalley, P. B. et al. (1974) *Proc. 5th Int. Heat Transfer Conf.* Versailles, Vol. 4, p. 290.

Yuu, S., et al. (1978) *AIChE J., 24,* 509.

A. Serizawa and I. Kataska

DISPERSED FLOW, IN NOZZLES (see Nozzles)

DISPERSED LIQUID FLOWS (see Liquid-liquid flow)

DISPERSION OF PARTICLES (see Particle transport in turbulent flow; Phase separation)

DISPERSION RELATIONSHIPS, FOR WAVES IN FLUIDS (see Waves in fluids)

DISPLACEMENT CHROMATOGRAPHY (see Chromatography)

DISPLACEMENT THICKNESS, OF BOUNDARY LAYER (see Boundary layer)

DISSIPATION OF HEAT FROM EARTH'S SURFACE (see Environmental heat transfer)

DISSIPATIVE SYSTEMS (see Non-linear systems)

DISSOLVED AIR FLOTATION, DAF (see Flotation)

DISSOLVED SOLIDS (see Water preparation)

DISTILLATION

Following from: *Mass transfer*

Introduction

Distillation is the most widely-used method of *separating fluid mixtures* on a commercial scale, it is thus an important part of many processes in the oil and chemical industries.

Many of the tall, thin towers which may be seen in an oil refinery or chemical plant are distillation columns. The most common column diameter is about 2.5 m, but 6 m diameter is commonplace and towers of 12 m dia have been built. Column heights may be as much as 30 m.

The advantages of distillation are: a) High purity products; b) Economies of scale; c) Well-established technology and competitive supply of equipment; d) Use of low temperature, low cost energy; e) Well suited for energy integration into the surrounding process.

It is generally-accepted that if it is possible to achieve a separation by distillation, then distillation will be the most economical method to use. Unless azeotropes are formed (see below), this means that nearly all mixtures where all components have a molecular weight between 140 and 40 will be separated by distillation. Low molecular weight fluids with a critical temperature below 50°C may not be condensed by cheap cooling water, and the additional costs associated with refrigerated or cryogenic distillation may mean that another separation process will be cheaper, at least for small-scale operation. Large molecular weight materials may thermally decompose or polymerise at their boiling temperature even when distilled under a high vacuum. Within the 40 to 140 molecular weight range, distillation is used. In 1992, Darton estimated the world-wide throughput of distillation columns as **Oil Refining,** 5 billion tonnes per year and *petrochemicals,* 130 million tonnes per year.

This article describes how distillation columns work, what they contain and how they are designed.

Vapour Liquid Equilibria

The separation of a mixture by distillation depends on the difference between the compositions of a boiling liquid mixture and the vapour mixture in equilibrium with the liquid. For example, the equilibrium line in **Figure 1a** correlates the mole fraction of benzene in the vapour, y, in equilibrium with x, the mole fraction of benzene in the liquid for a binary mixture of benzene and toluene. Benzene is more volatile than toluene (i.e., it has a higher vapour pressure at the same temperature) thus at equilibrium, y is greater than x. A distillation column with a boiler at the bottom and a condenser at the top provides a means of counter-current contact between the rising vapour and the descending liquid, such that at all levels the benzene moves from the liquid into the vapour and the toluene moves from the vapour into the liquid. Thus, benzene concentrates at the top of the column and toluene at the bottom.

For some mixtures, a constant boiling mixture or *azeotrope* exists where the composition of the vapour is the same as that of the liquid. For example, the equilibrium line of benzene and ethanol shown in **Figure 1b** has an azeotrope at 0.55 mol fraction of benzene. This is a low-boiling point azeotrope, i.e., it has a boiling temperature lower than that of both benzene and ethanol. Separation is limited by the composition of the azeotrope.

The prediction of vapour-liquid equilibria *ab initio* from the molecular structure of the mixture components is not yet possible because a) there is no complete molecular theory of liquids; and b) the equations of state for mixtures of vapours are still essentially empirical. The determination, correlation and prediction of vapour liquid equilibrium have been studied for at least a century and is the subject of specialist texts (Walas, 1985, and Reid et al., 1977). Collections of experimental data are available (Gmehling and Onken, 1977). The brief outline of the subject which follows includes the definition of parameters used in designing distillation columns.

For an ideal liquid mixture in contact with a low pressure vapour, the equilibrium compositions may be predicted from *Raoult's and Dalton's Law,* so that for a component, i,

$$y_i = x_i \frac{P_{vti}}{\pi}. \qquad \textbf{(1)}$$

For non-ideal liquids and higher pressures (approaching but less than the critical pressure), it is necessary to introduce additional terms so that Equation (1) becomes

$$\emptyset_i y_i = \frac{\gamma_i x_i P_{vti}}{\pi}. \qquad \textbf{(1a)}$$

The vapour phase fugacity coefficient $\emptyset_i$ may be estimated from a suitable equation of state. The liquid phase activity coefficient is more difficult to predict but if some experimental data are available, methods derived from the **Gibbs-Duhem Equation** are available for predicting or correlating changes in activity coefficients with composition. **Activity Coefficients** may be predicted from the molecular groups of the components by methods such as the UNIFAC method, now widely-used (Fredenslund et al. 1977). A review of methods of predicting vapour liquid equilibrium (V.L.E.) is presented by Prausnitz (Prausnitz, 1981).

The two most commonly-used ways of exploiting V.L.E. in column calcuations are those of the Equilibrium Constant or K-value, and relative volatility, α. The K-value provides a linear relationship for each component between its concentrations in the vapour and liquid, that is y = Kx. Thus from Equations (1) or (1a)

$$K_i = \frac{P_{vti}}{\pi} \quad \text{or} \quad \frac{\gamma_i P_{vti}}{\emptyset_i \pi}.$$

A component's K-value thus varies with both temperature and pressure. In general, either temperature or pressure is specified. The procedure for calculating the composition of, say, the vapour in equilibrium with a known composition multicomponent liquid is based on adjusting (by trial) the other variable, such that $\Sigma yi = 1.0$ where $y_i = K_i x_i$. In distillation column calculations, the variable is usually temperature so that the equilibria calculations yield a column temperature profile. The complex nature of this calculation means that nowadays all multicomponent column designs are done by means of computer-based numerical methods [see, for example, Prausnitz et al. (1980)].

Relative volatility is defined as the ratio of the K-value of one component to that of another, that is:

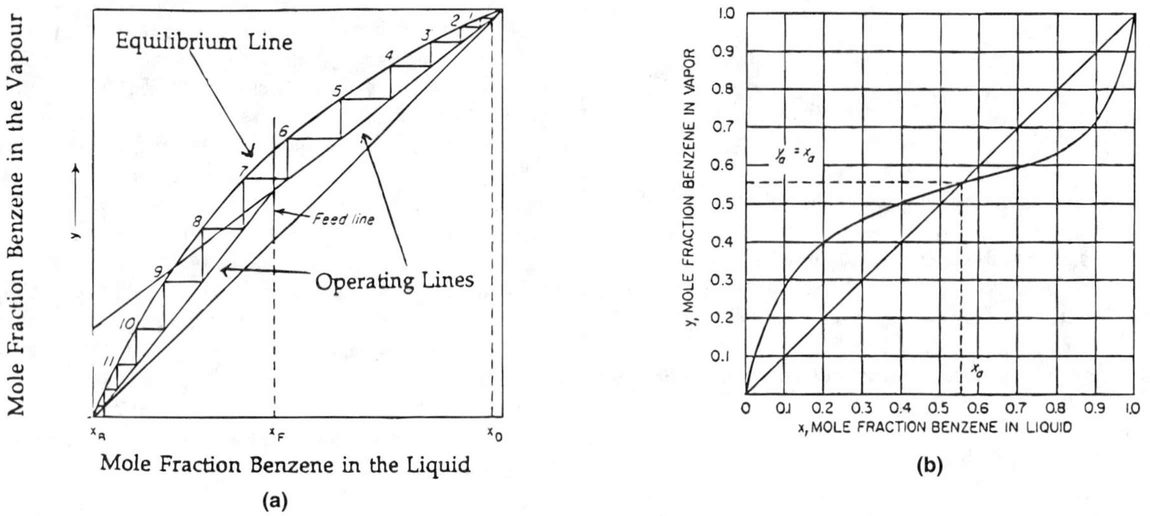

Figure 1.

$$\alpha_{ij} = \frac{K_i}{K_j} = \frac{\gamma_i P_{vti}/\text{Ø}_i}{\gamma_j P_{vtj}/\text{Ø}_j}.$$

For ideal mixtures, $\alpha_{ij} = P_{vti}/P_{vtj}$. Both vapour pressures P_{vti} and P_{vtj} depend strongly on temperature, but their ratio, α_{ij}, is often relatively constant throughout the column. Relative volatility provides an estimate of the difficulty of a particular separation, i.e., $\alpha = 10$ for easy separation, $\alpha = 1.1$ for difficult separation.

The dependence of vapour pressures on temperature provides the link between the difficulty of a particular separation and the difference in normal boiling points of the components, i.e., 20°C difference for easy separation; 1°C difference, difficult separation. A difficult separation requires a large column and a large energy input.

Design of Distillation Columns

The procedure is introduced by first considering a binary mixture; **Figure 2** shows a distillation column. Some of the liquid from the condenser at the top of the column, L_c, is returned as reflux. The *reflux ratio* is defined as the ratio of the liquid returned to the column divided by the liquid removed as product, i.e., $R = L_c/D$. **Figure 2** also shows the column as a series of *theoretical plates.* A *theoretical plate* is defined as a vapour-liquid contacting device such that the vapour leaves it in equilibrium with the liquid which leaves it. The first stage of column design is to calculate a) the column reflux ratio; and b) the number of theoretical plates.

The theoretical plates in **Figure 2** are numbered from the top of the column, and streams leaving a plate (in equilibrium) have the number of that plate. Thus, stream V_2 leaving plate 2 has vapour concentration y_2, and L_2 has liquid concentration x_2. If y_2 and x_2 refer to benzene in the benzene-toluene binary mixture, then y_2 and x_2 are each located at the same point on the equilibrium curve of **Figure 1a**.

A mass balance may be used to obtain a relationship between the concentrations of any component in the streams passing each other counter-currently between the plates and the product streams

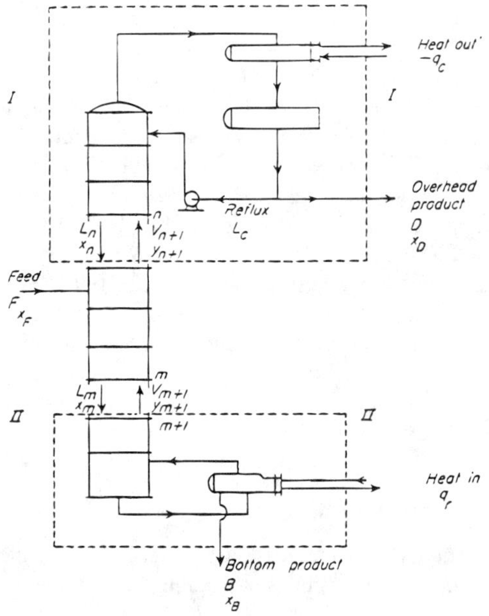

Figure 2.

leaving the ends of the column. Thus for the column section above the feed in **Figure 2**, between plates 1 and 2,

$$\text{Overall Mass Balance } V_2 = L_1 + D; \tag{2}$$

$$\text{Component (e.g., benzene) Balance } y_2V_2 = x_1L_1 + x_DD. \tag{2a}$$

In general, a heat balance is also required to allow for changes in L and V as the composition changes, but in most cases, it is reasonable to assume that the molar latent heats of all components of the mixture are approximately equal and that as much liquid evaporates as vapour condenses. That is, L and V are assumed to be constant between the top of the column and the feed, and constant (but different) between the feed and the bottom of the column. Thus for the section above the feed, the mass balances becomes:

$$V = L + D$$

$$y_{n+1}V = x_nL + x_DD$$

$$y_{n+1} = x_n \frac{L}{V} + x_D \frac{D}{V}. \quad \textbf{(3)}$$

Below the feed, in a similar way, the following equations may be derived

$$L = V + B$$

$$x_mL = y_{m+1}V + x_BB$$

$$y_{m+1} = x_B \frac{B}{V} - x_m \frac{L}{V}. \quad (4)$$

Note that for all sections of a particular column that (x_DD/V) and (x_BB/V) remain constant, the component mass balance **Equations 3** and **4** maybe drawn as straight lines (Operating Lines) below the Equilibria Line (**Figure 1**). Each has a slope equal to its' (L/V).

The flow rate ratios are related to the reflux ratio so that **Equation 3** may be written as:

$$y = \left(\frac{R}{R+1}\right)x + \left(\frac{1}{R+1}\right)x_D. \quad \textbf{(5)}$$

In general, the calculation of the number of theoretical plates required for a given separation at a given reflux ratio proceeds as follows: from a known vapour composition leaving a plate (say plate 1 where $y_1 = x_D$), use the theoretical plate concept and V.L.E. data to calculate the composition of the liquid leaving the plate (say x_1); then use a mass balance (say **Equation 3**) to calculate the composition of the vapour leaving the plate below (y_2 from plate 2). Repeat the calculation using theoretical plate-mass balance to get the composition of vapour leaving the next plate (say y_3 from plate 3) until the composition matches that of the feed. Then adjust the mass balance equations and deal with the section of the column below the feed.

These same principles are applied for multicomponent mixtures, and for changes in flow rates caused by taking off product at intermediate positions (e.g., sidestreams) or multiple feeds. They are well-described in many text books (Coulson and Richardson, 1978 and Treybal, 1981). The case of a binary mixture may be used to explain the meaning of Total Reflux and Minimum Reflux Ratio. Note that for a binary mixture, the process may be represented graphically as in **Figure 1** and the number of theoretical plates obtained by stepping off plates between the Operating Lines and the Equilibrium Line (known as the *McCabe - Thiele method*).

Total Reflux is the operating condition where vapour and liquid are passing each other in the column but no product is removed (i.e., $D = 0$ and $R = L/D = \infty$). The slopes of the Operating Lines are then $L/V = 1.0$, that is by mass balance, between each plate $y = x$.

At total reflux, the number of theoretical plates required is a minimum. As the reflux ratio is reduced (by taking off product), the number of plates required increases. The *Minimum Reflux Ratio* (R min) is the lowest value of reflux at which separation can be achieved even with an infinite number of plates. It is possible to achieve a separation at any reflux ratio above the minimum reflux ratio. As the reflux ratio increases, the number of theoretical plates required decreases.

The *Optimum Reflux Ratio* (R_0) is that at which the total cost of the distillation is a minimum, taking into account the capital cost of the column (which depends on the number of theoretical plates) and running cost, which depends on the reflux ratio. Note that the capital costs of the reboiler and condenser also depend on the reflux ratio. Thus, usually for very low energy costs $R_0/R_{min} = 1.3$, and for high energy cost $R_0/R_{min} = 1.1$ (It is best not to use $R_0/R_{min} < 1.1$ to allow for possible errors in the V.L.E data).

Column Internals, Trays and Packing

Examples of *trays* and *packings* are shown in **Figures 3a** and **3b**. Trays (or plates) consist of horizontal plates across which the descending liquid flows through streams of the ascending vapour, which is distributed across the column by various sorts of perforations (holes, valves, bubblecaps) in the plates.

Packings provide surfaces over which the liquid flows while the vapour passes between these surfaces.

The number of real trays required (N) is related to the number of theoretical trays (NTP) by the concept of tray efficiency, $E_0 =$ (NTP)/N.

The height of the column occupied by trays is then Z = N. (T.S.) where (TS) is the tray spacing, which is usually 300 mm, 450 mm, or 600 mm except in cryogenic distillation where (TS) is 100 to 150 mm.

The height of the column containing packing is usually calculated by Z = (NTP) (HETP) where (HETP) = Height of Packing Equivalent to One Theoretical Plate.

Column Diameter Pressure Drop and Flooding

To keep the column diameter (and cost) as small as possible, columns are designed to operate at the maximum permissible vapour velocity. Except for those cases where pressure drop is controlling [see Strigle R.F. (1994)], this is usually at about 80% of the flooding velocity. (See **Flooding and Flow Reversal.**) At flooding vapour velocity, the pressure gradients and drag forces overcome the gravitational forces on the liquid and it ceases to descend the column. Flooding may be correlated in terms of the Load Factor (C_S) and the Flow Parameter (X) [see **Figures 4a** and **4b**] where

$$C_s = U_s\left(\frac{\rho_v}{\rho_L - \rho_v}\right)^{1/2} \quad \text{and} \quad X = \left(\frac{L_w}{V_w}\right)\left(\frac{\rho_v}{\rho_L}\right)^{1/2}.$$

The Load Factor C_S is very approximately proportional to $G/(\sqrt{\rho_v})$ and this, in part, explains the important practical observation that throughput—expressed as a mass flow per unit column cross section area—goes up with pressure. The mass of liquid descending is similar to that of vapour ascending. Typical values of liquid flow per unit area are given in the following table.

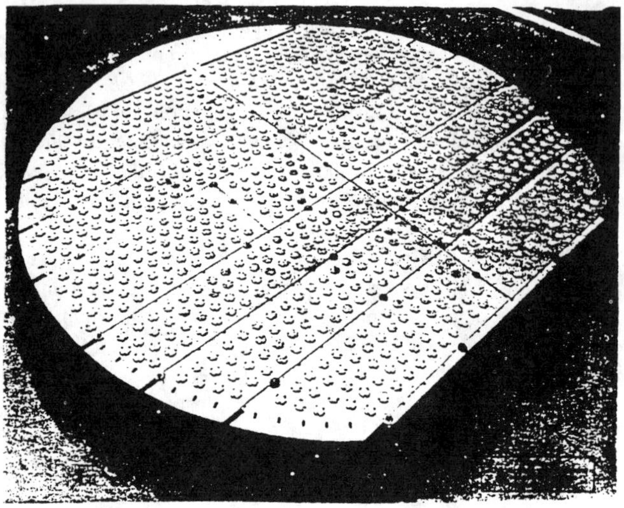

valve tray

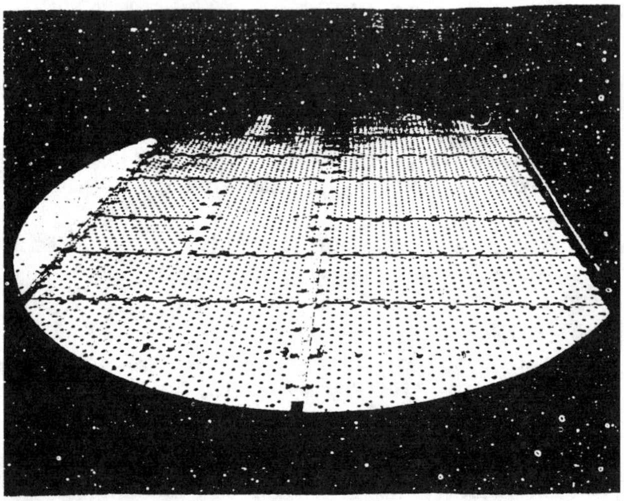

sieve tray

Metal Saddle Shape (IMTP)

Pall Ring

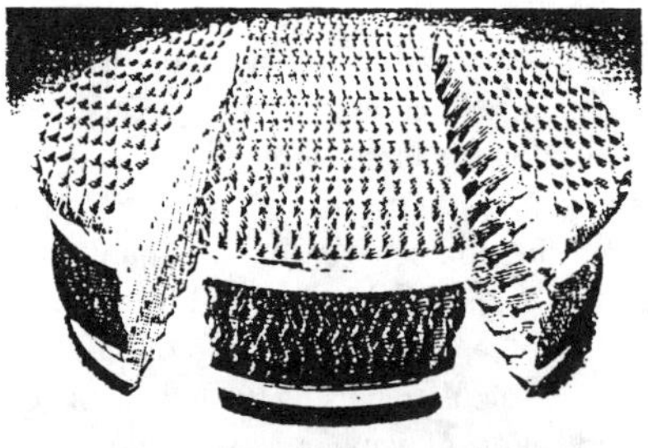

Intalox Structured Packing

Figure 3.

System	Pressure (bar)	Liquid Flow/Area ($Kg/m^2 \cdot h$)
Butanes	6	2000
Benzene	1	1000
Styrene	0.15	500

Tray Efficiency and HETP

A great deal has been written on tray and packing mass transfer in trays and packings. Tray mass transfer is interpreted in terms of a) Point Efficiency, E_{og}, b) Murphree Tray Efficiency, E_{mv}; and c) Overall Column Efficiency, Eo. Point efficiency refers to mass transfer at a point on the tray and is related to single-tray or Murphree Efficiency, taking into account the flow pattern on the tray. Overall column efficiency is then obtained by considering concentration (and flow patterns) throughout the column. Reference can be made to Lockett, 1986. Packing mass transfer is interpreted in terms of vapour and liquid mass transfer coefficients, and sometimes by the transfer unit concept [see, for example, Kister, 1989 and Strigle, 1994]. Here, only general comments can be made. Tray Efficiency does not change much with the type of tray or tray spacing, but varies with operating pressure being lower for

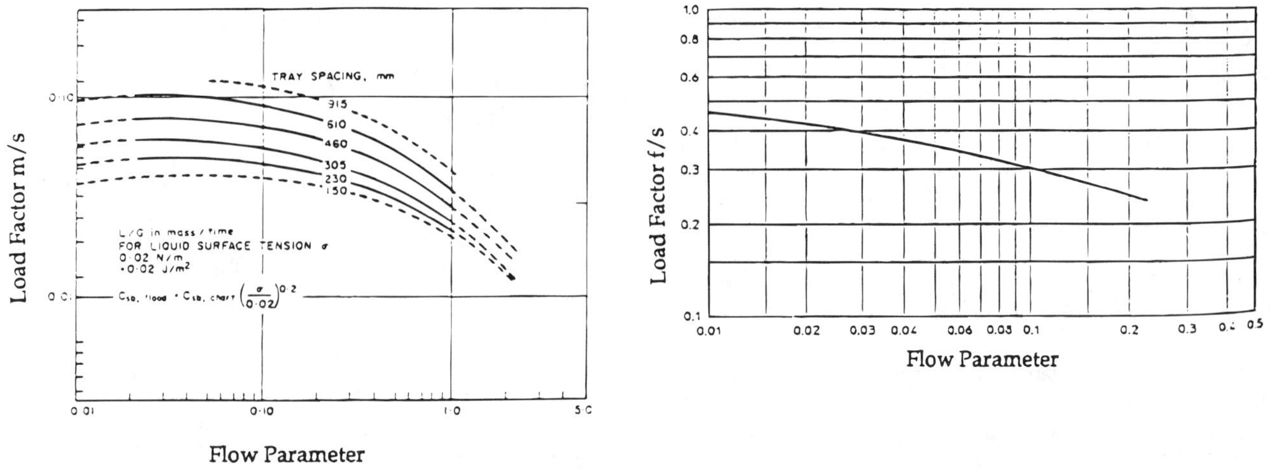

Figure 4.

vacuum distillation than for pressure distillation. This reflects the changes in liquid rate mentioned above (0.5 bar, Eo approx 0.5; 1.0 bar, Eo approx 0.7; 6 bar, Eo approx 0.9).

Packing HETP might be expected to go down with pressure but, in practice, does not change much with a system for a given packing. However, it changes with packing size, which determines the dry area per unit volume. For example, for random packings, HETP (m) approximately equals dp/60 (where dp is packing size in mm). A high quality of the initial distribution of liquid and vapour is essential in large-diameter columns, (even in those larger than 0.5 m). Without this scale up, failures have occurred. In general, a column with more theoretical trays for a given height will require a larger diameter, that is closer tray spacings or high area packings flood at a lower throughput. The design of distillation columns is still essentially empirical, in part based on avoiding earlier failures. (For example, there is no theoretical basis so far for two-phase flow on trays or in packings, nor is there a theory for multicomponent mass transfer on cross-flow trays).

Theory and practice are interpreted by Porter and Jenkins in a paper which explains why it is relatively easy to correlate distillation experience even in terms of the wrong theory.

Process Design (Sequencing Distillation Columns)

A simple continuous distillation column separates a feed into two products (top product and bottom product); thus, complete separation in one column is possible only for a binary mixture. For a three-component mixture (say ABC), a complete separation may be achieved in either one of two alternative sequences each of two columns, (that is A/BC, B/C or AB/C, A/B). The process design problem is to choose which of these sequences is the optimum from the point of view of cost or of suitability for energy integration. For a mixture of n components, the number of columns required in each alternate sequence is always (n-1), but the number of possible sequences goes up rapidly with the number of components, e.g., n = 3, S = 2; n = 4, S = 5; n = 6, S = 42, n = 10, S = 3,500). Choosing the optimum sequence is thus a complex problem which has received much attention. It is solved either by sophisticated numerical procedures to minimise the number of complete design and costings to be performed, or by oversimplifying rules of thumb which sometimes contradict each other. The best known rules are: a) Do the easiest separation first and the hardest last; b) separate in order of volatilities (the direct sequence); c) favour 50/50 splits; and d) remove the most plentiful component first. A simple equation proposed by Porter and Momoh has been found to identify the best sequence for energy integration and an example of its use is given by Porter, Jenkins and Momoh, 1987.

Nomenclature

B	Molar flow of bottom product (kg mol/h)
C_S	Vapour Load Factor = $u_s\,(\rho_v/(\rho_L - \rho_v))^{1/2}$(m/s)
D	Molar Flow of top product (distillate) from the condenser (kg mol/h)
E_{og}	Point Efficiency on a Tray
E_{mv}	Single Tray Murphree Efficiency
E_o	Overall column tray efficiency
G	Mass flow of vapour per unit column cross-sectional area (kg/s.m^2)
HETP	Height of Packing Equivalent to a Theoretical Plate (m)
K	Equilibrium Ratio (y/x)
L	Molar liquid flow in a section of the column (kg mol/h)
L_c	Molar liquid flow to the top of the column (kg mol/h)
L_w	Mass liquid flow in a section of the column (kg/h)
N	Number of real trays (or plates) in a distillation column
NTP	Number of Theoretical Trays (or plates) in a column
n	Number of components in the mixture
P_{vti}	Vapour pressure of component i at temperature t. (bar)
R	Reflux ratio (L_c/D)
R_{min}	Minimum Reflux Ratio
R_o	Optimum Reflux Ratio

S	Number of possible sequences to separate a multicomponent mixture
TS	Tray Spacing (m)
u_s	Superficial (i.e., empty tower) vapour velocity (m/s)
V	Molar vapour flow in a section of the column (kg.mol/h)
V_w	Mass vapour flow in a section of the column (kg/h)
V.L.E.	Vapour Liquid Equilibria (refers to vapour and liquid concentrations)
x	Mol. fraction of a component in the liquid
y	Mol fraction of a component in the vapour
Z	Height of the column which contains trays or packing. (m)
α_{ij}	Relative volatility of component i to that of j
γ	Activity coefficient in the liquid
Ø	Fugacity coefficient in the vapour
π	Total pressure of the system (bar)
ρ_v, ρ_l	Densities of vapour, liquid kg/m^3
i, j	Subscripts refering to components i, j, etc.

References

Coulson, J. M. and Richardson, J. F. (1978) *Chemical Engineering,* Vol 2, Third Edition, Pergamon Press Oxford.

Fredenslund, Aa, Gmehling, J. and Rasmussen, P. (1977) *Vapour-Liquid Equilibria using UNIFAC* (Elsevier).

Gmehling, J. and Onken, U. *Vapour-Liquid Data Collection,* DECHEMA Chemistry Data Series, Vol. 1, Part 1 (1977), Part 2a (1977), Part 2b (1978) (Together with W. Arlt) and subsequent volumes.

Kister, Henry Z. (1989) *Distillation Operations,* McGraw-Hill, New York.

Lockett, M. J. (1986) *Distillation Tray Fundamentals* Cambridge University Press.

Porter, K. E. and Jenkins, J. D. (1979) *The Interrelationship between Industrial Practice and Academic Research in Distillation and Absorption.* I.Chem.E Symposium Series *56* 5.1/1.

Porter, K. E., Momoh, S., and Jenkins (1987). *Some Simplified Approaches to the Design of the Optimum Heat-Integrated Distillation Sequence* I.Chem.E. Symposium Series 104, A449.

Prausnitz, J. M. (1981) Calculation of Phase Equilibria for Separation Processes *Trans. I. Chem. E, 59,* 3.

Prausnitz, J. M. et al. (1980) *Computer Calculations for Multicomponent Vapour-Liquid and Liquid-Liquid Equilibria,* Prentice-Hall Englewood Cliffs, N.J.

Reid, R. C., Prausnitz, J. M., and Sherwood, T. K. (1977) *Properties of Gases and Liquids* (McGraw Hill).

Strigle, Ralph F. (1994) *Packed Tower Design and Applications: Random and Structured Packings,* Gulf Publishing Company, Houston.

Treybal, R. E. (1981) *Mass Transfer Operations*, McGraw Hill.

Walas, S. M. (1985) *Phase Equilibria in Chemical Engineering* 1st Ed, Butterworth, London.

K.E. Porter

DISTILLATION REBOILERS (see Tubes and tube bundles, boiling heat transfer on)

DISTRIBUTIONS (see Delta function)

DISTURBANCE WAVES, IN ANNULAR FLOW (see Annular flow)

DISYMMETRY OF SCATTERED LIGHT (see Optical particle characterisation)

DITTUS-BOELTER CORRELATION (see Spiral heat exchangers; Tubes, condensation in)

DITTUS-BOELTER EQUATION (see Supercritical heat transfer; Tubes, single phase heat transfer in)

DNB, DEPARTURE FROM NUCLEATE BOILING (see Boiling; Burnout, forced convection)

DOE (see Department of the environment)

DONNEN EFFECT (see Membrane processes)

DOPING (see Semiconductors)

DOPPLER ANEMOMETRY (see Anemometers, laser Doppler; Doppler shift; Lasers)

DOPPLER BROADENING (see Radiative heat transfer)

DOPPLER BURST (see Anemometers, laser Doppler)

DOPPLER EFFECT (see Optical particle characterisation)

DOPPLER GLOBAL VELOCIMETRY (see Tracer methods)

DOPPLER SHIFT

The change in frequency of a wave observed at a detector whenever the source or detector is moving relative to one another is known as the Doppler shift. This frequency change is named after the Austrian physicist Christian Doppler who predicted its existence in 1842. The existence of the Doppler shift was first verified in experiments associated with sound waves. The Doppler shift principle has found practical application in acoustical and optical environments relating to remote-sensing, high-energy physics, astrophysics, spectroscopy, radar and meteorology.

The Doppler shift principle is used in modern *acoustic flow meters* to estimate the flow velocity in a pipeline. A pair of ultrasonic transmitters beam to receivers placed across the flowline; one transmitter beams upstream and the second downstream. The detected differences in the times of travel are then related to the flow velocity.

Optical Doppler shift due to light scattered by small particles suspended in a flowing medium forms the basis of modern laser based *Doppler anemometry.*

Leading to: *Anemometers, laser Doppler; Lasers*

C.J. Bates

DOUBLE EXPOSURE HOLOGRAPHY (see Film thickness determination by optical methods)

DOUBLE FLASH METHODS (see Optical partical characterisation)

DOUBLE-PIPE EXCHANGERS

Following from: *Heat exchangers*

Double-pipe exchangers is the generic term covering a range of jacketed 'U' tube exchangers normally operating in counter-current flow of two types:

True Double Pipes: Single inner pipe formed into a removable 'U' or element, with either longitudinal fins or centralising supports. Typical standard sizes, layout and overall dimensions are shown in **Figure 1** and finned geometric data are in **Table 1**.

Multitubular Hairpins: A removable bundle of U-tubes with either shell-&-tube type segmental baffles, rod type supports or longitudinal fins. A typical layout is shown in **Figure 2**, which shows longitudinal fins. These units are available as standard designs ranging from 75mm (3″ N/B) to 406mm (16″N/B) pipe sizes. Larger units up to 762mm (30″ N/B) are also available from some manufacturers (e.g., Brown Fintube). The larger units comply fully with TEMA–1988 and standard industry shell-&-tube specifications. Typical dimensions are shown in **Figure 3**. (See Figure 3 on page 326.) Geometric data are given in **Table 2**, based on the use of 19.7 mm (3/4″) OD tubes arranged on 23.8125 mm (15/16″) triangular pitch.

Unlike TEMA-type BEU units, the bundle or element is removed from the U-tube end of the exchanger, which means the tube sheets have to slide through the shells. Special closures are required with a combination of split flanges, split rings and sealing rings as shown in **Figure 2**.

The shells are supported in bracket assemblies designed to cradle both shells simultaneously. The brackets are configured to permit the modular assembly of many hairpin sections into an exchanger bank for inexpensive future expansion. This construction also allows the shells to be fully supported without welding the brackets to the shells, minimising the transfer of thermal stresses to the pipework.

The multitubular hairpin units should be more economic than TEMA-type shell-&-tube units when one or more of the following criteria apply:

1. Temperature cross or very close temperature approach. Often one counter-current hairpin unit will replace two multipass shell-&-tube units in series.
2. High tube-side design pressure (>5,000 kPa). The smaller diameter compact hairpin unit will have thinner flanges & closures, hence a lower cost. Present mechanical design technology has allowed the production of dependable, removable bundle hairpin multitubular units at tube-side design pressures of up to 82,500 kPa.
3. Alloy tubeside materials. Same as (2), hence a lower cost.
4. Where heat transfer augmentation is the best solution. Hairpin exchangers are normally single-pass so tube-side heat transfer enhancement devices, such as twisted tape turbulators, cores, cored helixes and helixes, are used to increase the velocity and/or film coefficient. A recent development at Brown Fintube has improved the design of units with horizontal tube-side boiling and dry exit vapour. Shell-side longitudinal fins can be beneficial when the tube-side film coefficient is significantly higher (>2 times) than the shell-side.
5. Where differential expansion is a problem. Multitubular hairpin units can be used as alternatives to TEMA floating-head types because the minimum U-bend dimension is much larger than a TEMA-type BEU unit, such that flexible mechanical tube cleaning devices can be used.

Thermal design is normally carried out by the manufacturer, but you can use shell-and-tube exchanger programmes by specifying a single-pass BEM unit with twice the nominal hairpin length.

The tube-side method is identical to a one-pass BEM unit.

The shell-side is also identical when using segmental baffles. Rod-type baffles are often used with triangular pitch, so the shell-and-tube 'Phillips rod-baffles' method cannot be used as this is only for square pitch. True longitudinal flow can be simulated with triangular pitch by specifying triple segmental baffles on maximum pitch. If longitudinal fins are used, a tube-side type correlation is applied with an equivalent diameter in place of the tube outside diameter. Also, the finned area is not as effective as bare tube area so a fin efficiency must be used. It is possible to carry out such calculations by hand and the method is adequately covered in various textbooks, including Process Heat Transfer [Hewitt et al. (1994)].

The HTFS shell-&-tube program TASC will have a multitubular hairpin and double-pipe option installed by 1996.

References

Brown Fintube UK, PO Box 790, Wimborne, Dorset, BH21 5YA, UK.

Brown Fintube Company, 12602 FM 529, Box 40082, Houston, Texas 77240, USA

TEMA (1988) *Standards of Tubular Exchanger Manufacturers Association,* 7th Edition New York.

Hewitt, G. F., Shires, G. L., and Bott, T. R. (1994) *Process Heat Transfer,* CRC Press, Boca Raton, FL.

HTFS: *Heat Transfer and Fluid Flow Service,* 392.7 Harwell, Didcot, Oxfordshire, OX11 ORA, UK.

A.R. Guy

DOUBLING TIME (see Nuclear reactors)

External Split Flange Design – for ease of maintenance

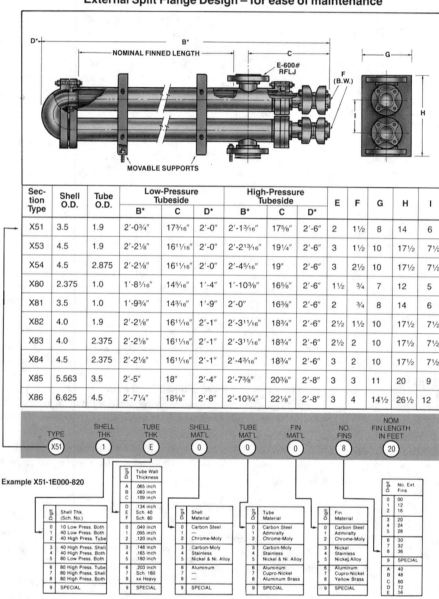

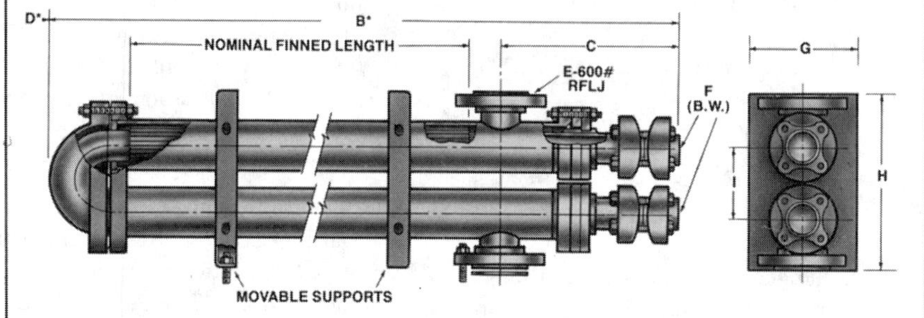

Section Type	Shell O.D.	Tube O.D.	Low-Pressure Tubeside B*	Low-Pressure Tubeside C	Low-Pressure Tubeside D*	High-Pressure Tubeside B*	High-Pressure Tubeside C	High-Pressure Tubeside D*	E	F	G	H	I
X51	3.5	1.9	2′-0¾″	17 3/16″	2′-0″	2′-1 3/16″	17⅝″	2′-6″	2	1½	8	14	6
X53	4.5	1.9	2′-2⅛″	16 11/16″	2′-0″	2′-2 13/16″	19¼″	2′-6″	3	1½	10	17½	7½
X54	4.5	2.875	2′-2⅛″	16 11/16″	2′-0″	2′-4 5/16″	19″	2′-6″	3	2½	10	17½	7½
X80	2.375	1.0	1′-8 1/16″	14 5/16″	1′-4″	1′-10⅜″	16⅝″	2′-6″	1½	¾	7	12	5
X81	3.5	1.0	1′-9¾″	14 3/16″	1′-9″	2′-0″	16⅜″	2′-6″	2	¾	8	14	6
X82	4.0	1.9	2′-2⅛″	16 11/16″	2′-1″	2′-3 11/16″	18¾″	2′-6″	2½	1½	10	17½	7½
X83	4.0	2.375	2′-2⅛″	16 11/16″	2′-1″	2′-3 11/16″	18¾″	2′-6″	2½	2	10	17½	7½
X84	4.5	2.375	2′-2⅛″	16 11/16″	2′-1″	2′-4 3/16″	18¾″	2′-6″	3	2	10	17½	7½
X85	5.563	3.5	2′-5″	18″	2′-4″	2′-7⅜″	20⅜″	2′-8″	3	3	11	20	9
X86	6.625	4.5	2′-7¼″	18⅝″	2′-8″	2′-10¾″	22⅛″	2′-8″	3	4	14½	26½	12

TYPE	SHELL THK	TUBE THK	SHELL MAT'L	TUBE MAT'L	FIN MAT'L	NO. FINS	NOM FIN LENGTH IN FEET
X51	1	E	0	0	0	8	20

Example X51-1E000-820

Digit	Shell Thk. (Sch. No.)
0	10 Low Press. Both
1	40 Low Press. Both
2	40 High Press. Tube
3	40 High Press. Shell
4	40 High Press. Both
5	80 Low Press. Both
6	80 High Press. Tube
7	80 High Press. Shell
8	80 High Press. Both
9	SPECIAL

Digit	Tube Wall Thickness
A	.065 inch
B	.083 inch
C	.109 inch
D	.134 inch
E	Sch. 40
F	Sch. 80
0	.049 inch
1	.095 inch
2	.120 inch
3	.148 inch
4	.165 inch
5	.180 inch
6	.203 inch
7	Sch. 160
8	xx Heavy
9	SPECIAL

Digit	Shell Material
0	Carbon Steel
1	—
2	Chrome-Moly
3	Carbon-Moly
4	Stainless
5	Nickel & Ni. Alloy
6	Aluminum
7	—
8	—
9	SPECIAL

Digit	Tube Material
0	Carbon Steel
1	Admiralty
2	Chrome-Moly
3	Carbon-Moly
4	Stainless
5	Nickel & Ni. Alloy
6	Aluminum
7	Cupro-Nickel
8	Aluminum Brass
9	SPECIAL

Digit	Fin Material
0	Carbon Steel
1	Admiralty
2	Chrome-Moly
3	Nickel
4	Stainless
5	Nickel Alloy
6	Aluminum
7	Cupro-Nickel
8	Yellow Brass
9	SPECIAL

Digit	No. Ext. Fins
0	00
1	12
2	16
3	20
4	24
5	28
6	30
7	32
8	36
9	SPECIAL
A	40
B	48
C	60
D	72
E	56

*Add nominal tube length to these dimensions for overall length "B" and length to remove hairpin dimension"D"

BROWN FINTUBE UK, PO BOX 790, WIMBORNE, DORSET, ENGLAND BH21 5YA.
TEL: 01258-840776 FAX: 01258-840961

Figure 1. Double-pipe exchanger dimensions.

Table 1. Double pipe hairpin section data

Shell Pipe O.D.		Inner Pipe O.D.		Fin Height		Fin Count	Surface Area Per Unit Length	
mm.	in.	mm.	in.	mm.	in.	(max.)	sq.m./m	sq.ft./ft
60.33	2.375	25.4	1.000	12.7	0.50	24	0.692	2.27
88.9	3.500	48.26	1.900	12.7	0.50	36	1.07	3.51
114.3	4.500	48.26	1.900	25.4	1.00	36	1.98	6.51
114.3	4.500	60.33	2.375	19.05	0.75	40	1.72	5.63
114.3	4.500	73.03	2.875	12.70	0.50	48	1.45	4.76
141.3	5.563	88.9	3.500	17.46	0.6875	56	2.24	7.34
168.3	6.625	114.3	4.500	17.46	0.6875	72	2.88	9.44

Table 2. Multitube hairpin section Data

	Shell O.D.		Shell Thk.		Tube Count	Surface Area for 6.1 m (20 ft.) Nominal Length	
Size	mm.	in.	mm.	in.	19 mm	sq.m.	sq.ft.
03-MT	88.9	3.500	5.49	0.216	5	3.75	40.4
04-MT	114.3	4.500	6.02	0.237	9	6.73	72.4
05-MT	141.3	5.563	6.55	0.258	14	10.5	113.2
06-MT	168.3	6.625	7.11	0.280	22	16.7	179.6
08-MT	219.1	8.625	8.18	0.322	42	32.0	344.3
10-MT	273.1	10.75	9.27	0.365	68	52.5	564.7
12-MT	323.9	12.75	9.53	0.375	109	84.7	912.1
14-MT	355.6	14.00	9.53	0.375	136	107.	1159.
16-MT	406.4	16.00	9.53	0.375	187	148.	1594.
18-MT	457.2	18.00	9.53	0.375	241	191.	2054.
20-MT	508.0	20.00	9.53	0.375	304	244.	2622.
22-MT	558.8	22.00	9.53	0.375	380	307.	3307.
24-MT	609.6	24.00	9.53	0.375	463	378.	4065.
26-MT	660.4	26.00	9.53	0.375	559	453.	4879.
28-MT	711.2	28.00	9.53	0.375	649	529.	5698.
30-MT	762.0	30.00	11.11	0.4375	752	630.	6776.

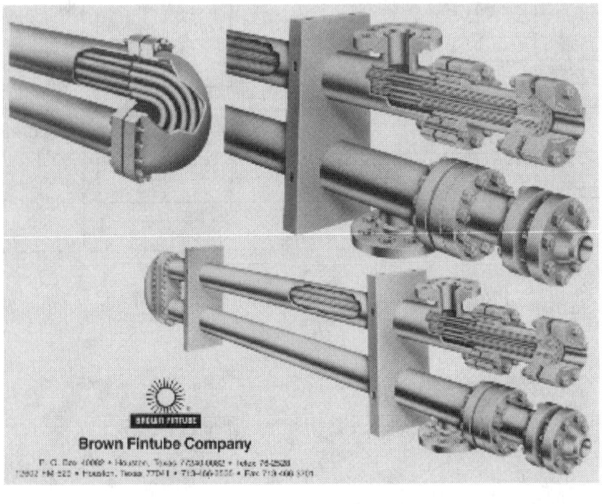

Figure 2. Multitubular hairpin exchanger.

Separated Head Multitube with Taper-Lok® High Pressure Tubeside

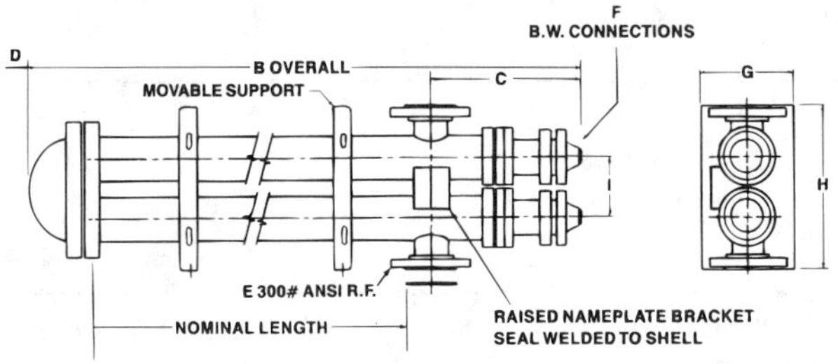

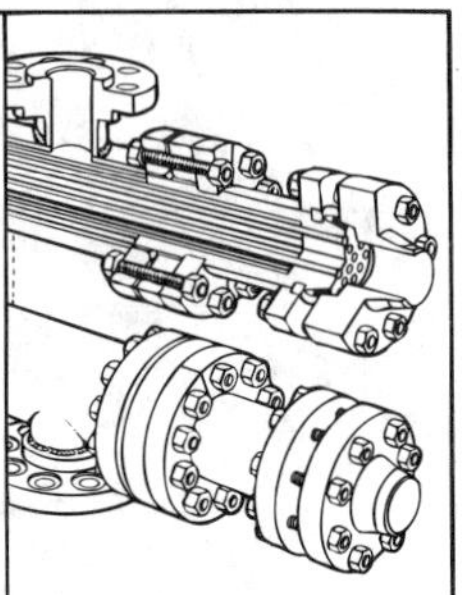

Nominal Shell Diameter	"B*"	"C"	"D*"	"E"	"F"	"G"	"H"	"I"	Surface Square Feet (1)	Dry (1) Weight Pounds
4	2′–4⅝″	19¼″	1′–6″	3″	2″	10″	17½″	7½″	75	1250
5	2′–7″	20⅛″	1′–10″	3″	3″	11″	20″	9″	113.2	1650
6	2′–10 5/16″	21 9/16″	1′–9″	4″	3″	14½″	26½″	12″	180	2400
8	3′–7″	27¼″	2′–0″	6″	4″	16⅝″	30″	13⅜″	342	3900
10	4′–5¾″	30¾″	2′–4″	8″	4″	22″	39½″	17½″	564.6	6200
12	4′–10″	34″	2′–5″	8″	6″	24″	44″	20″	912	8150
16	5′–7¾″	38½″	3′–0″	10″	8″	30″	56″	26″	1503	12,550

*Add nominal tube length to these dimensions for overall length "B" and length to remove hairpin "D."
(1) Surface areas and weights based on 20′ nominal length and ¾″ O.D. tubes.

Taper-Lok® Closure– This patented pressure energized closure employs the Taper-Lok® seal ring which remains visible during assembly, assuring proper ring alignment and sealing. Excellent in cycling, vibrating or high temperature services.

Separated Head Multitube with Taper-Lok® (1500 PSIG) Tubeside

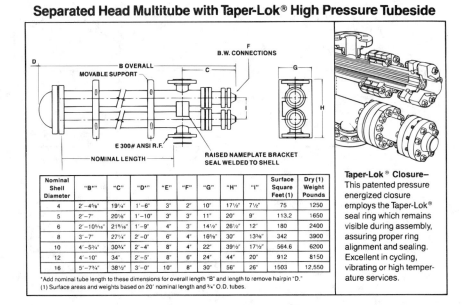

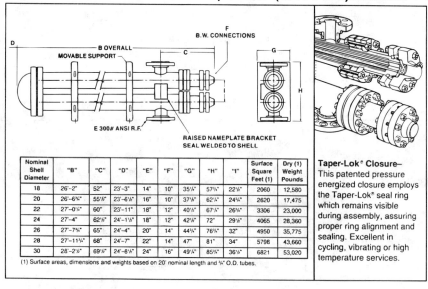

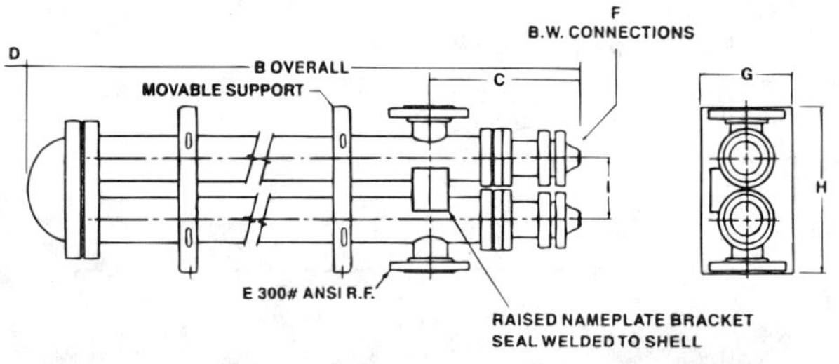

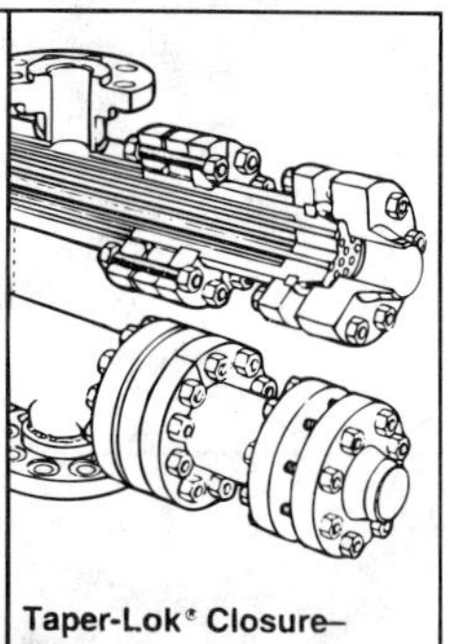

Nominal Shell Diameter	"B"	"C"	"D"	"E"	"F"	"G"	"H"	"I"	Surface Square Feet (1)	Dry (1) Weight Pounds
18	26′-2″	52″	23′-3″	14″	10″	35¼″	57¾″	22½″	2060	12,580
20	26′-6¾″	55½″	23′-6½″	16″	10″	37½″	62¼″	24¾″	2620	17,475
22	27′-0¼″	60″	23′-11″	18″	12″	40½″	67¼″	26¾″	3306	23,000
24	27′-4″	62½″	24′-1½″	18″	12″	42½″	72″	29½″	4065	28,360
26	27′-7¾″	65″	24′-4″	20″	14″	44¾″	76¾″	32″	4950	35,775
28	27′-11¾″	68″	24′-7″	22″	14″	47″	81″	34″	5798	43,660
30	28′-2½″	69½″	24′-8½″	24″	16″	49¼″	85¾″	36½″	6821	53,020

(1) Surface areas, dimensions and weights based on 20′ nominal length and ¾″ O.D. tubes.

Taper-Lok® Closure– This patented pressure energized closure employs the Taper-Lok® seal ring which remains visible during assembly, assuring proper ring alignment and sealing. Excellent in cycling, vibrating or high temperature services.

Figure 3. Multitubular hairpin dimensions.

DOWTHERM

Following from: *Heat transfer media*

Table 1.

Substance	Data	Property	Temperature, °C −150	−100	−75	−50	−25	0	20	50	100	150	200	250
			Temperature, K 123.15	173.15	198.15	223.15	248.15	273.15	293.15	323.15	373.15	423.15	473.15	523.15
DOWTHERM A	Chemical formula* Molecular weight: 166	Density, ρ_l (kg/m^3)	S	S	S	S	S	S	1 060	1 036	995	951	906	858
	Melting point: 12°C	Specific heat capacity, $c_{p,l}$ (kJ/kg K)	S	S	S	S	S	S	1.574	1.660	1.800	1.947	2.087	2.219
	Boiling point: 257.1°C Critical temperature: 497°C	Thermal conductivity, λ_l [(W/m^2)/(K/m)]	S	S	S	S	S	S	0.141	0.137	0.132	0.125	0.119	0.113
	Critical pressure: 3.134 MPa	Dynamic Viscosity, η_l(10^{-5} Ns/m^2)	S	S	S	S	S	S	380	215	100	58	39	28
*Mixture $(C_6H_5)_2O$ (73.5%); $(C_6H_5)_2$ (26.5%)														
DOWTHERM J	Chemical formula: $C_{10}H_{14}$ Molecular weight: 134	Density, ρ_l (kg/m^3)	S	S	S	917	897	888	872	842	801	754	V	V
	Melting point:— Boiling point: 181°C	Specific heat capacity, $c_{p,l}$ (kJ/kg K)	S	S	S	1.650	1.713	1.772	1.830	1.924	2.093	2.278	V	V
	Critical temperature: 383°C	Thermal conductivity, λ_l [(W/m^2)/(K/m)]	S	S	S	0.137	0.135	0.134	0.133	0.130	0.126	0.122	V	V
	Critical pressure: 2.837 MPa	Dynamic viscosity, η_l (10^{-5} Ns/m^2)	S	S	S	410	225	140	90	62	36	22	V	V

From: Chapter 5.5.10 of the *Heat Exchanger Design Handbook*, 1986. Physical properties of liquids at temperatures below their boiling point by C. F. Beaton Hemisphere Publishing Corporation. With permission.

DRAFT TUBE MIXER (see Fluidics)

DRAG (see Crossflow; Immersed bodies, flow around and drag; Tube, crossflow over)

DRAG COEFFICIENT

Following from: *Immersed bodies, flow around and drag*

Drag coefficient is a dimensionless factor of proportionality between overall hydrodynamic force vector $\bar{F}$ on a body in a liquid or gas flow and the product of reference area S of the body (commonly at midship section) and velocity head q

$$\bar{F} = C_d S \frac{\rho}{2}(\bar{v} - \bar{v}_s)|\bar{v} - \bar{v}_s|/2 = C_d S \frac{\rho u^2}{2}$$

where $q = \frac{\rho}{2}(\bar{v} - \bar{v}_s)|\bar{v} - \bar{v}_s|2 = \frac{\rho u^2}{2}$, $\bar{v}$ and v_s are the velocity vectors of the fluid and the body, $u = |\bar{v} - \bar{v}_s|$ is the relative velocity of the body, ρ the liquid (gas) density, S the midship section area of the body, and C_d the drag coefficient.

This relation follows from similarity theory and is extensively used in engineering for simplified calculation of the force acting on a body or a particle in liquid or gas in which it moves.

In practice, drag coefficient is calculated in most cases using empirical relations generalizing experimental data. The most widely studied case is the sphere. **Figure 1** graphs the dependence of *drag coefficient for a sphere* and a cylinder in cross-flow on the **Reynolds Number** Re = ρuD/η, where D is the sphere (cylinder) diameter, η the viscosity of liquid, and $u = |\bar{v} - \bar{v}_s|$. The drag coefficient decreases drastically from extremely high values at small Re numbers, to unity and lower at Re > 10^3. For Re < 0.2, Stokes has derived a theoretical formula for drag coefficient for a sphere:

$$C_d = \frac{24}{Re}, \qquad Re < 0.2.$$

Here, a purely viscous nonseparating flow occurs. The drag is governed by a high molecular friction of the liquid, the effect of which extends far upstream. With increasing Re number, inertial forces begin to predominate over viscosity forces and a laminar boundary layer is originated. Now, viscous forces are manifested only in this fairly thin layer. Flow beyond the boundary layer is virtually not affected by viscosity. Flow separation in the stern (point S in **Figure 1**) also occurs. As Re grows, the area of separation increases and attains the highest values at Re ~ 10^3; the drag coefficient in this case no longer diminishes and even slightly increases, remaining close to 0.4 for the range $2 \times 10^3 < Re < 2 \times 10^5$.

In the $0.2 < Re < 2 \times 10^3$ range, an approximation formula for calculating a drag coefficient for a sphere is:

$$C_d = \frac{21.12}{Re} + \frac{6.3}{\sqrt{Re}} + 0.25. \qquad (1)$$

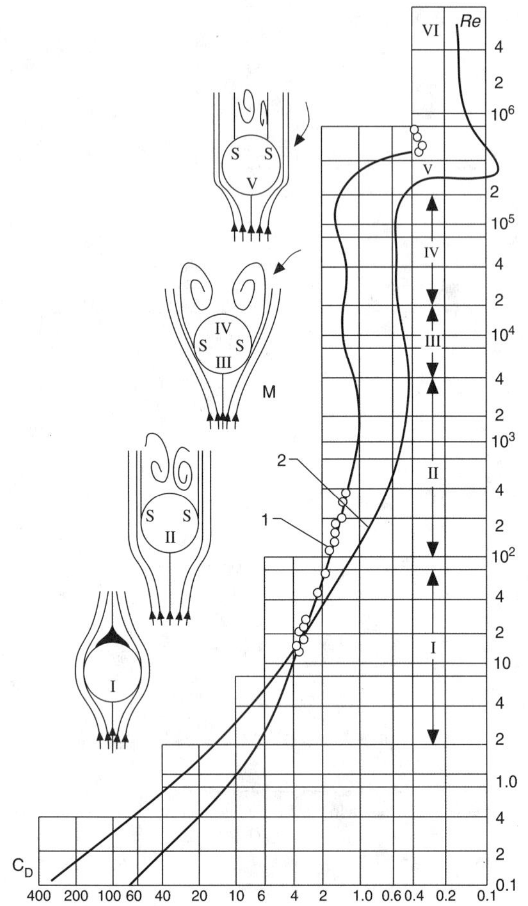

Figure 1. Drag coefficient for cylinders (1) and spheres (2) as a function of Reynolds number (Re).

If Re continues to increase, the situation arises (at Re $\sim 2 \times 10^5$) when the laminar boundary layer becomes partially turbulent in the nonseparating flow region of the sphere. The velocity profile in the turbulent boundary layer is fuller and better resists a positive pressure gradient. The area of separation is sharply displaced toward the stern, thereby drastically decreasing the drag coefficient. A self-similarity regime sets in and with further enhancement of the Re number, drag coefficient remains unchanged.

At high gas velocitiy, the drag coefficient also depends on the **Mach number** Ma = u/a, where a is the velocity of acoustic waves in the gas. At Ma < 1, a formula approximating a vast body of experimental data is:

$$C_d = C_{d_0} \frac{1.0 - 0445\,\mathrm{Ma} + 4.84\,\mathrm{Ma}^2 - 9.73\,\mathrm{Ma}^3 + 6.93\,\mathrm{Ma}^4}{\sqrt{1 + 1.2\,\mathrm{Ma}\,C_{d_0}}}$$

where C_{do} is calculated using Equation (1), has gained acceptance.

Drag coefficient is strongly affected by a body's shape. It is taken into account via the sphericity coefficient, which is the ratio of the sphere surface area of the same volume as the body relative to the body's surface area. For a tetrahedron, this is 0.67; for a cube, 0.806; and for octahedron, 0.85. Introduction of the sphericity coefficient in reality means the changing from an irregular shape

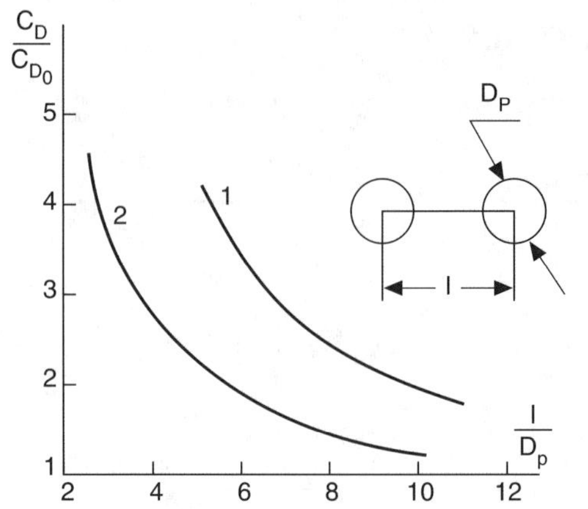

Figure 2.

of the body to some equivalent spherical shape, with sphere diameter taken as a reference dimension for determining the Re number and the midship section area.

Bodies of irregular shape, e.g., those of great length or twisted ones, move by tortuous trajectories and rotate, substantially changing the drag coefficient.

Where there are two spheres in close proximity, the drag coefficient is increased as shown in **Figure 2.**

Y.V. Polezhaev and I.V. Chirkov

DRAG FORCE (see Aerodynamics)

DRAG FORCE ON PARTICLES (see Dispersed flow)

DRAG INDUCED FLOW (see Couette flow)

DRAG ON A PARTICLE (see Gas-solids separation, overview)

DRAG ON PARTICLES AND SPHERES (see Particle transport in turbulent fluids)

DRAGON REACTOR (see Gas-graphite reactors)

DRAG ON SOLID SPHERES AND BUBBLES (see Stokes law for solid spheres and spherical bubbles)

DRAG REDUCTION (see Non-Newtonian fluids)

DRIFT FLUX (see Drift flux models; Void fraction)

DRIFT FLUX MODELS

Following from: *Void fraction*

In the article on **Void Fraction**, a description was given of the one-dimensional flow method and the *drift flux* j_{GL} was defined as the flux of the gas phase relative to a plane moving at the total superficial velocity U.

A prime assumption of the one-dimensional method was that the local velocity and void fraction were constant across the channel; in fact, they can vary considerably. Thus, the flow parameters have to be averaged over the cross-section and models incorporating these averages are often referred to as *drift flux models*. Following Zuber and Findlay (1975), we may define average and weighted mean values of the local parameters. Let F be any one of the local parameters (for example, u_G, U, U_L). An average value for F over a channel cross-section A can be defined as follows:

$$\langle F \rangle = \frac{1}{A}\int_A F dA, \tag{1}$$

and a weighted mean value for F may be also defined:

$$\bar{F} = \frac{\langle \varepsilon_G F \rangle}{\langle \varepsilon_G \rangle} = \frac{\frac{1}{A}\int_A \varepsilon_G F dA}{\frac{1}{A}\int_A \varepsilon_G dA}. \tag{2}$$

Expressions for average gas velocity and the weighted mean gas velocity are obtained as follows:

$$\langle u_G \rangle = \langle U \rangle + \langle u_{GU} \rangle \tag{3}$$

$$\bar{u}_G = \frac{\langle U_G \rangle}{\langle \varepsilon_G \rangle} = \frac{\langle \varepsilon_G u_G \rangle}{\langle \varepsilon_G \rangle} = \frac{\langle \varepsilon_G U \rangle}{\langle \varepsilon_G \rangle} + \frac{\langle \varepsilon_G u_{GU} \rangle}{\langle \varepsilon_G \rangle} \tag{4}$$

where u_{GU} is the *drift velocity* of the gas phase relative to the total flow velocity U at a given point in the channel (note for the one-dimensional model, $j_{GL} = \varepsilon_G u_{GU}$). Equation 4 can be written as:

$$\bar{u}_G = C_0 \langle U \rangle + \frac{\langle \varepsilon_G u_{GU} \rangle}{\langle \varepsilon_G \rangle} \tag{5}$$

where

$$C_0 = \frac{\langle \varepsilon_G U \rangle}{\langle \varepsilon_G \rangle \langle U \rangle} = \frac{(1/A)\int_A \varepsilon_G U dA}{(1/A^2)\int_A U dA \int_A \varepsilon_G dA}. \tag{6}$$

Similarly, the weighted mean liquid velocity is given by:

$$\bar{u}_L = \frac{\langle U_L \rangle}{\langle 1 - \varepsilon_G \rangle} \frac{\langle U - U_G \rangle}{\langle 1 - \varepsilon_G \rangle} = \frac{\langle U \rangle - \langle U_G \rangle}{\langle 1 - \varepsilon_G \rangle}, \tag{7}$$

and the *slip ratio* is given by:

$$S = \frac{\bar{u}_G}{\bar{u}_L} = \frac{\langle 1 - \varepsilon_G \rangle}{1/(C_0 + \langle u_{GU}\varepsilon_G \rangle / \langle \varepsilon_G \rangle \langle U \rangle) - \langle \varepsilon_G \rangle}. \tag{8}$$

For the case of no local velocity difference between the phases ($u_{GU} = 0$):

$$S = \frac{\langle 1 - \varepsilon_G \rangle}{1/C_0 - \langle \varepsilon_G \rangle} \tag{9}$$

thus for $C_0 \neq 1$, the slip ratio is not unity, even though the phases are travelling at the same velocity at each point in the channel. This arises because of differences between the distribution of velocity and void fraction. If u_{GU} is constant across the channel, then Equation 8 can be converted into a void-quality relationship of the form:

$$\langle \varepsilon_G \rangle = \frac{x\rho_L}{C_0[x\rho_L + (1 - x)\rho_G] + \rho_L \rho_G u_{GU}/\dot{m}} \tag{10}$$

where x is the quality (fraction of the total mass flow which is vapour); ρ_L and ρ_G, the liquid and gas phase densities; and $\dot{m}$, the mass flux.

In fully-developed bubble and/or slug flow, C_0 is usually of the order of 1.1–1.2. It is often convenient to correlate data for void fraction in terms of the parameters C_0 and u_{GU}. An extensive empirical correlation in this form is that of Chexal and Lellouche (1991).

References

Chexal, B. and Lellouche, G. (1991) *Void Fraction Correlation for Generalised Applications*. Nuclear Safety Analysis Centre of the Electric Power Research Institute, Report NSAC/139.

Zuber, N. and Findlay, J. A. (1965) Average Volumetric Concentration in Two-Phase Flow Systems. *J. Heat Transfer* 87, 453–468.

G.F. Hewitt

DRIFT VELOCITY (see Drift flux models)

DROPLET DEPOSITION AND ENTRAINMENT, IN ANNULAR FLOW (see Annular flow)

DROPLET/LIQUID SEPARATION (see Hydrocyclones)

DROP FORMATION (see Drops)

DROPS

Following from: *Gas-liquid flow; Liquid-liquid flow*

Drops are isolated liquid volumes kept intact by surface tension force. Drops can have sizes ranging from a tenth of a micron to several millimeters. Very small drops, for which gravity force is not significant, and drops in a zero gravity situation have a closely spherical form. The shapes of drops may be changed by external forces.

Drop Formation

Drops are formed from jet and sheet splitting, by larger drop breaking down, by drops impacting on a wall and by vapor condensation. Drops are also formed as a result of liquid spraying as gas bubbles burst at interfaces, by splashing, by liquid entrainment by gas flows over interfaces and by boiling liquid passing over a surface a temperature higher than saturation temperature.

Jet splitting

The problem of a jet splitting as it flows from an orifice has been examined by Rayleigh. The intensity of jet splitting increases with the dynamic interaction between jet and gas. At some distance from the orifice, unstable waves begin to appear. Along the jet's length, wave amplitude increases as a result of disturbances caused by the relative motion of liquid and gas. Jet splitting takes place when the velocity difference between phases exceeds a critical value (**Figure 1a**). The amplitudes of short and long waves grow with the difference in velocities. Ryleigh has observed that wave growth rate increases and reaches its maximum at an optimal wave length λ_{opt}, which is roughly 3.5 times the jet diameter. With increasing outflow velocity, the form of fluctuation is changed from symmetric to nonsymmetrical (**Figure 1b**). The length of the unstable waves and drop sizes also decrease.

Sheet splitting

The sheet from a centrifugal nozzle breaks down with the formation of two groups of waves. The first group of waves moves along the direction of the jet and seeks to transform a sheet into rings. The second group of waves extends to sheet circumference and seeks to separate the liquid in the jet fan spreading out from the nozzle center. As a result of wave motion, drops are formed. The smallest drops are blown away from the film surface under elevated pressures.

Drop splitting in gas flow

Drop flow is affected by inertial, frictional and Archimedean forces. Unlike hard particles, surface deformation and splitting of liquid drops are possible in gas flow. Conditions for the onset of drop splitting can be estimated from stability analysis of spherical interface in flow. Interface stability in the back and front parts of a drop is characterized by the use of the **Bond number**; instability arises when the Bond number is more than the critical value:

$$Bo = \frac{4D^2\rho_L g}{\sigma} > Bo_c = 4\pi^2.$$

Interface instability in the upper and lower parts of the drop is characterized by the use of the **Weber number**. Instability occurs when the Weber number is more than the critical value:

$$We = \frac{2D\rho_G u^2}{\sigma} > We_c = 2\pi.$$

With the increasing Weber number, a variety drop splitting mechanisms in a gas flow can be observed, as follows (**Figure 2**).

1. Drop splitting to smaller drops takes place at $We \approx We_c$;
2. Drop collapse with preliminary flow of a "bag" formed by a thin liquid film pulled on the ground (splitting of the "parachute" type) is completed at $We_c < We < 20$, with sufficient time for drop flow and with low viscous liquid. The film breaks down into small drops and on the ground breaks down into big drops;
3. Drop splitting with formation of a "bag" ("parachute") and jet occurs at $20 < We < 30$;
4. Drop splitting with formation of "bags" and fibres drowned on the flow takes place in the Weber number range of 30 to 60;
5. Drop splitting resulting in the liquid film or liquid fibres collection being blown away from the drop surface occurs when the Weber number ranges from 60 to 1000;
6. Explosive splitting is realized at larger Weber numbers. At very high flow velocities, the drop breaks down at once into numerous small drops.

If there is no drop slipping relative to flow, splitting can occur as a result of resonant interaction between the free fluctuation of the drop and a small-scale turbulent vortex, with the drops having close sizes.

Drop splitting upon impact with a wall

If the surface has a temperature near the saturation level and is coated by liquid film, the impact of drops causes splitting of liquid. If the surface which the drop comes into contact with has a small size, spraying and cutting of drops are possible. If surface temperature is higher than the temperature at which contact of

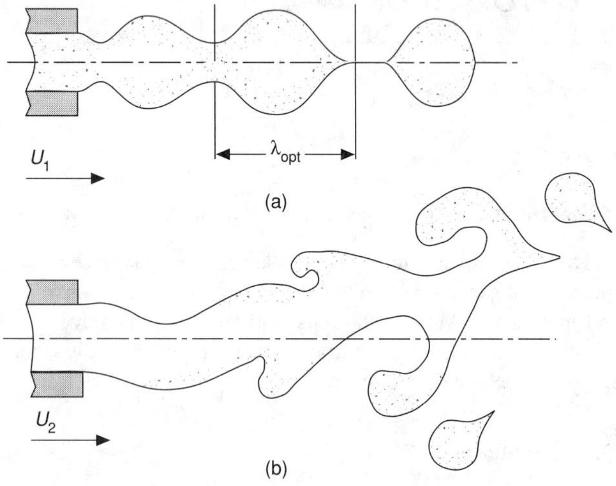

Figure 1. Jet breakup.

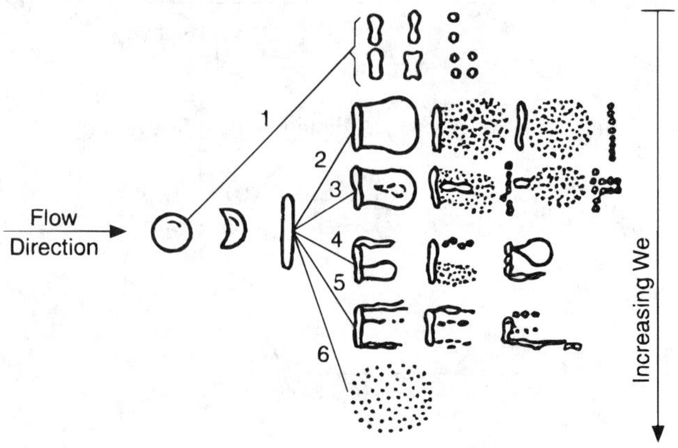

Figure 2.

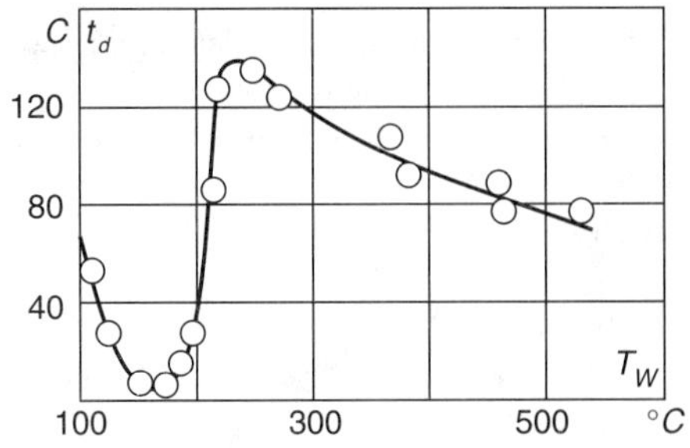

Figure 3. Drop evaporation time as a heated surface.

liquid with the wall is terminated, also called the **Leidenfrost temperature** (see **Leidenfrost Phenomena**), a continuous vapor film appears between the drop and the wall. At low We numbers, the drop can recoil from the wall coated by vapor film and be split. At large We numbers, the drop can spread over the vapor film and break down into numerous small drops.

In drop flow, coagulation, evaporation and condensation of drops can take place.

Coagulation of Drops

The coagulation (or coalescence) of drops can take place when drops collide with each other. The result of drop collision is not unique and depends on many factors. At the moment of collision, drops can break down or coalesce. A gas or vapor layer can be conserved between the drops that are close to each other. With increasing drop velocity, the thickness of the gas layer decreases and the probability of collision increases. At some velocity range, drop coagulation takes place.

Evaporation and Condensation on the Surfaces of Drops

The vapor-drop system is unstable: drops which have sizes less than a critical value evaporate and disappear; drops which have sizes more than critical value grow as a result of condensation on their surfaces. The critical drop radius r_c is coupled to vapor supercooling $T_s - T_G$ by the following equation:

$$r_c = \frac{2\sigma T_s}{h_{LG}\rho_L(T_s - T_G)},$$

where σ is the surface tension; T_s, the saturation temperature; h_{LG}, the latent heat of evaporation; and ρ_L, the liquid density. If there are a great number of drops with sizes more than critical value in volume, vigorous condensation on drops takes place. The noncondensable gas or vapor mixture leads to the concentration and partial pressure gradients near the drops; this causes a change in the saturation temperature of the diffusive vapor layer. With noncondensable gas, the growth velocity of drops is less than that for condensation of pure vapor because of thermal and partial pressure gradients. Condensation and evaporation velocities increase when gas flows around the drops.

Drop evaporation near a heated surface

The time of full drop evaporation t_d depends on the heated surface's temperature T_w. The dependence t_d of water drop evaporation time, in a volume 5 10^{-8} m^3 placed on a heated surface, on temperature is shown on **Figure 3**. Three dependencies can be distinctly found. Drop evaporation time decreases with an increase of T_w at a small wall superheated relative to saturation temperature. Then drop evaporation time increases sharply in the narrow range of temperature T_w and at last, the value t_d is reduced again with the high wall temperature. The first downward branch of dependence $t_d(T_w)$ corresponds to drop boiling; the second upward branch corresponds to the transition to a spheroidal condition, at which the drop is separated from the wall by vapor film; the third, again downward, branch corresponds to a developed spheroidal condition. The function $t_d(T_w)$ presents approximately the dependence $q(\Delta T)$ under condition of pool boiling of liquid.

Reference

Hewitt, G. F. and Hall-Taylor, N. S. (1970) *Annular Two-Phase Flow.* Pergamon Press.

***Leading to:** Dropsize measurement*

Yu.A. Kuzma-Kitchta

DROP SHAPES (see Contact angle)

DROPSIZE MEASUREMENT

***Following from:** Drops*

Methods used for drop size measurement can be classified under four main groups: *photographic methods, optical methods, sampling methods, electrical contact methods* and acoustic methods.

Photographic Methods

These methods are widely-used and they allow a wide range of information to be obtained are size, form and concentration in a given flow field. Drop trajectories and break-down and coagulation phenomena can also be studied. High-speed cameras and objectives with high resolution are utilized and stroboscopic light sources are often used to light the zone being investigated. Localized lighting is often obtained using light guides.

Optical Methods

The optical methods which have gained ground are: method of *scattering indicatrix,* method of small angles, polarisation method of scattering light, method of light attenuation and methods related to Laser-doppler anemometry.

The method of scattering indicatrix is used for measuring drops within the radius range of 0.01 to 100 μm. When a light beam is incident on an individual particle, the particle becomes a second source of radiation, scattering light all around in a way characterized by the scattering indicatrix—which is defined as the distribution of light scattered over all solid angles relative to the direction of the incident beam.

The angular distribution of light scattering from the particle depends on its radius. If the particle radius r_d is considerably less than the wave length of the incident light λ, the scattering indicatrix is symmetric relative to the plane that is perpendicular to the incident beam. The scattering indicatrix changes considerably for larger particles ($r_d \geq \lambda$). In this situation, it is not symmetric and scattering in the direction of the incident beam predominates.

Typical indicatrices as calculated for different diameters of water drops are presented in **Figure 1**. **Figure 1a** shows light scattering by particles with very small radius $\bar{r}_d = r_d/\lambda$ (where $\bar{r}_d$ is the dimensionless drop radius). The curve $I_2(\beta)$ represents the component of scattering light in the plane parallel to the observed plane and the curve $I_1(\beta)$, the plane perpendicular to it. **Figure 1b** shows the growing asymmetry of light scattering in the plane normal to the incident beam as drop size increases. The ratio of scattered light intensities measured for two angles relative to the plane perpendicular is the incident beam is a function of r_d and can be used for determining r_d. **Figure 1c** represents the change in the scattering indicatrix with an increase in particle size. The asymmetry grows, the tongue of the scattering indicatrix becomes narrow and is extended and light reflection is attenuated (at $\bar{r}_d \approx 1$, reflected light is equal to 1/10 of the transmitted light; at $\bar{r}_d \approx 10$, the ratio is 1/10000).

The intensity of light scattering by a particle at different angles to the incident beam direction is determined by the equation

$$I_r' = I_0 \frac{8\pi^4}{z^2\lambda^2} r_d^6 \frac{m^2 - 1}{m^2 + 2} (1 + \cos^2 \beta),$$

where r_d is the drop radius; m, the refractive index of the particle

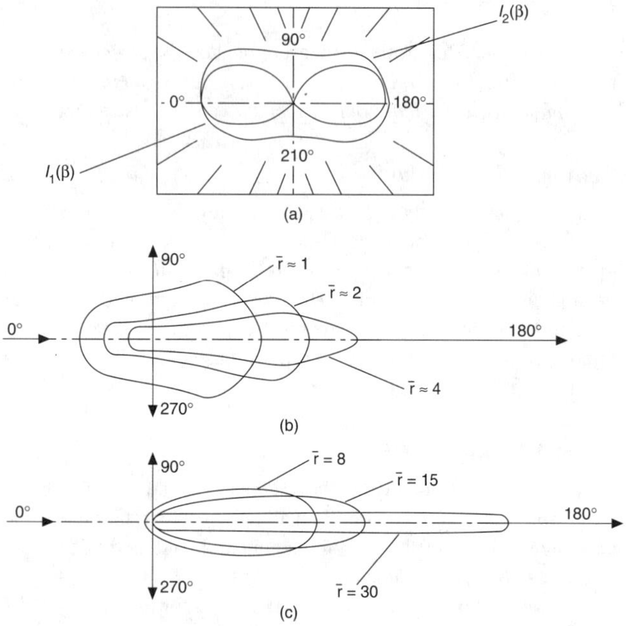

Figure 1. Typical scattering indicatrixes for various sizes of water droplet.

material; λ, the wave length of the incident light; z, the distance from the particle to the detection device; I_0, the incident light intensity; and β, the angle to incident beam direction. If there are n particles with different sizes with the volume sampled V, the expression for I_r reduces to the form

$$I_r = I_r' \frac{nV}{\lambda^2},$$

where

$$r_d^6 = \frac{1}{n}\int_{r_{min}}^{r_{max}} f(r) r^6 dr; \qquad n = \int_{r_{min}}^{r_{max}} f(r) dr,$$

and f(r) is the distribution function of particle size.

Secondary scattering need not be accounted for until the distance between particles is less than their radius; this corresponds to a range of liquid fractions of $0 < \varepsilon_L < 6\%$. At high liquid fractions ($\varepsilon_L \approx 20$–30%), it is necessary to take into account secondary scattering.

The *method of small angles* is used for determining particle sizes ranging from 1 to 300 μm. It is based on measurement of the scattering indicatrix of radiation at small angles. The limitations of this method are two-fold, as follows:

1) Particle concentration in the translucent volume cannot be too high; otherwise the effect of secondary scattering becomes significant.

2) The thickness of the translucent volume is limited by the value $\delta = \ln I_0'/I_0 \leq 0.3$, where I_0' is the transferred light intensity without scattering particles and I_0, light intensity transferred through the investigated volume with scattering particles.

The *polarization* of scattered light in the visible part of the spectrum, for range of drop radius $r_d = 0.05$–0.025 μm and observation angle $\beta = 90°$, changes smoothly with r_d. Drop radius is determined by measuring the intensity of the polarized light. As a result of the presence of particles, the intensity maximum for the polarized light is shifted in the direction of angles more than 90°.

According to *Buger's law*, light beam intensity I on outlet from dispersed medium is bounded to the intensity I_0 on inlet by dependence

$$I = I_0 \exp(-kl),$$

where k is the index of light beam attenuation and, l is the path length of the light beam.

Index k takes account of the scattering (k_s) and absoption (k_a) of light in medium

$$k = k_s + k_a.$$

For $\lambda = 0.5$ μm, light absorption by water particles is very small and it can be supposed that $k \approx k_s$. The equation of light attenuation in the dispersed medium stress has the form

$$I = I_0 \exp(-k_\lambda \pi r^2 nl),$$

where $k_\lambda \pi r^2$ characterizes the effective scattering surface for each particle with radius r and n is the number of particles per unit volume. This method measures, as the same point in a two-phase medium, the attenuation of radiation with wavelengths λ_1 and λ_2 and stress determine the ratio $k_{\lambda_1}/k_{\lambda_2}$.One can then find the particle size using the relationships $\bar{r}_1/\bar{r}_2 = \lambda_1/\lambda_2$ and $k_\lambda = f(\bar{r})$.

A number of methods for drop size determination based on *laser/doppler anemometry* leave been developed [see Hewitt (1982) for a review]. These methods use a variety of principles, which include:

1. Measurement of "visibility", which is the ratio of the maximum to minimum value of the fluctuating signal in the doppler bust.
2. Measurement of the amplitude of the doppler bust from a given drop.
3. Measurement of the phase behaviour of the doppler signal ("phase doppler anemometry").

Though quite expensive, the last of these methods appears to be finding the widest use.

Sampling Methods

The sampling method is realized by using glass plate that is coated with a layer of viscous liquid and is inserted briefly into the flow. Drop sizes are determined from photographs or with the aid of a microscope. In an alternative version of this method, the plate is covered by magnesium oxide. Pits are formed in the soft magnesium layers by the drop impact and their size is related to drop size.

Water drops can also be collected and inserted into oil for determination of their sizes with the aid of a microscope. The drops can also be frozen when they are inserted into a low temperature liquid. In this case, the particle size distribution can be determined with the help of sieves.

Electrical Contact Method

Wicks and Dukler have proposed a method that is based on recording the drop contact frequency between an inline pair of needles where tips have an adjustable distance. The electric contact between needles arises if the drops touch them simultaneously. The counter is used to monitor closure frequency. The distance between the needles is change and the closure frequency measured again. Drop size distribution can be determined from the obtained results, using probability theory. In the case of high velocity flow, the amplitudes of useful pulses are small and the effect of false pulses, which are created by charged drops that do not close the electrodes, is important. Circuits with amplitude limitation are used to exclude the effect of these pulses. Useful pulses that have amplitudes less than critical value are eliminated. This method can be used at a flow velocity of more than 40–50 m/s. At low velocity, the liquid drop can adhere to the electrodes and the accuracy of measurement decreases sharply.

A circuit with high-frequency correction has been applied to measure the drop distribution at very high flow velocities (up to 180 m/s). Upon contact of the electrodes with the drop, the current increases in the first winding of a transformer. The pulse number is determined by a counter. This method allows the use of different designs of hours. The drop size distribution function is determined by the following equation:

$$f(D_d) = \frac{FD_d}{\pi} \int_1^{\bar{D}_{km}} \sqrt{\bar{S} - 1}\, n(D_d, S)\, d\bar{S},$$

where $\bar{S} = S/D_d$ is the distance between the electrodes; D_d, the drop diameter; f, the closure frequency; S, the cross-section area of the investigated volume; $\bar{D}_{km} = D_{km}/D_k$, where D_{km} is the maximum drop diameter.

A comparison of the different methods for drop size measurement shows that their usability depends on flow conditions. For example, at high velocities (150–180 m/s), the optical and electric methods give similar results. At small velocities, the methods of Wicks-Dukler, optical method and method of prints give compatible results.

References

Hewitt, G. F. and Hall-Taylor, N. S. (1970) *Annular Two-Phase Flow.* Pergamon Press, Oxford.

Hewitt, G. F. (1982) Measurement of Drop and Bubble Size. Chapter 10.2.2.6 of the *Handbook of Multiphase System* (Ed. G. Hetsroni). McGraw Hill, New York.

Yu.A. Kuzma-Kichta

DROPS, MASS TRANSFER TO AND FROM (see Mass transfer)

DROP SPLITTING (see Drops)

DROP TOWERS (see Microgravity conditions)

DROPWISE CONDENSATION

Following from: *Condensation of pure vapour; Condensation, overview*

Dropwise condensation occurs when a vapour condenses on a surface not wetted by the condensate. For nonmetal vapours, dropwise condensation gives much higher heat transfer coefficients than those found with film condensation. For instance, the heat transfer coefficient for dropwise condensation of steam is around 10 times that for film condensation at power station condenser pressures and more than 20 times that for film condensation at atmospheric pressure. In circumstances where the filmwise coefficient is of similar magnitude to that on the cooling side, a change of mode to dropwise condensation offers a potential improvement in overall coefficient by a factor of up to around 2.

Clean metal surfaces are wetted by nonmetallic liquids and film condensation is the mode which normally occurs in practice. *Nonwetting agents,* known as *dropwise promoters,* are needed to promote dropwise condensation. Successful industrial application of dropwise condensation has been prevented by promoter breakdown, often associated with surface oxidation. Polytetrafluorethylene (ptfe, "teflon") provides an excellent nonwetting surface, but it has not been possible to produce sufficiently thin durable surface layers. (See also **Wettability**.)

Various promoters have been identified and used successfully in laboratory tests, mainly with steam and also a few organic fluids having relatively high surface tension. A typical example is dioctadecyl disulphide, which has given lifetimes of hundreds or thousands of hours in laboratory investigations. Dropwise condensation of steam has been observed on chromium and gold surfaces which are very smooth, without use of a promoting agent. It seems probable, however, that nonwetting impurities were present.

The high heat transfer coefficients obtainable with dropwise condensation are very susceptible to reduction by the presence in the vapour of *noncondensing gas.* In the absence of significant vapour velocity, very small gas concentrations lead to appreciable lowering of the heat transfer coefficient. This was largely responsible for wide discrepancies between early published values.

During dropwise condensation, the bare surface is continually exposed to vapour by coalescences between drops and by the sweeping action of the falling drops as they are removed from the surface by gravity or vapour shear stress. "Primary" drops are formed at nucleation sites on the exposed surface (typical nucleation site densities are 10^7 to 10^8 sites/mm^2). The primary drops grow by condensation until coalescences occur between neighbours. The coalesced drops continue to grow and new ones form and grow at sites exposed through coalescences. As the process continues, coalescences occur between drops of various sizes while the largest drops continue to grow until they reach their maximum size, when they are removed from the surface by gravity or vapour shear stress. The diameter ratio between the largest and smallest drops during dropwise condensation of steam is around 10^6.

Le Fevre and Rose (1966) have noted that three factors are involved in the mechanism of heat transfer through a single drop. These are: conduction in the liquid (important for relatively large drops), interphase matter transfer at the vapour-liquid interface (important for very small drops) and curvature of the vapour-liquid interface (important for the smallest drops).

By using an approximate expression for the distribution of drop sizes, together with their equation for the heat-transfer rate through a drop of given size, Le Fevre and Rose (1966) obtained the relation between the mean heat flux for the surface and the vapour-to-surface temperature difference. This showed, in agreement with experiment and in contrast to film condensation, that the heat transfer coefficient for dropwise condensation increases with increasing vapour-to-surface temperature difference.

Theory and experiment also indicate that the heat transfer coefficient decreases with decreasing pressure. Although in closed form and in principle applicable to any fluid, the expression giving the heat transfer coefficient is lengthy. An empirical equation in good agreement with both theory and experiment for dropwise condensation of pure, quiescent *steam* is:

$$\frac{\alpha}{\mathrm{kW/m^2K}} = \theta^{0.8}\left(5 + 0.3\,\frac{\Delta T}{K}\right) \qquad (1)$$

where α is the vapour-to-surface heat-transfer coefficient, ΔT is the vapour-to-surface temperature difference and θ is the Celsius temperature of the vapour. Discussion of the theory and comparisons with experimental data are given in Rose (1988, 1994).

There is conflicting evidence for the effect of the thermal conductivity of the condenser surface material on the heat-transfer coefficient resulting from non-uniformity of the surface heat flux. The balance of evidence suggests that this is only important at very low vapour-to-surface temperature difference where the condensing side resistance would probably be negligible in practice (see Rose (1994)).

Reviews of dropwise condensation heat transfer have been made by Le Fevre and Rose (1969), Rose (1988), Tanasawa (1991), Marto (1994) and Rose (1994).

References

Le Fevre, E. J. and Rose, J. W. (1966) A Theory of Heat Transfer by Dropwise Condensation. *Proc. Third Int. Heat Transfer Conf.* 2, 362–375.

Le Fevre, E. J. and Rose, J. W. (1969) Dropwise Condensation. *Proc. Symp. Bicentenary of James Watt Patent,* Univ. Glasgow, 166–191.

Marto, P. J. (1994) Vapor Condenser, in McGraw Hill *Year Book of Science and Technology,* 428–431, McGraw Hill.

Rose, J. W. (1988) Some Aspects of Condensation Heat Transfer Theory. *Int. Communications in Heat and Mass Transfer,* 15, 449–473.

Rose, J. W. (1994) *Dropwise Condensation.* Heat Exchanger Design Update, 2.6.5, 1–11, Begell House.

Tanasawa, I. (1991) *Advances in Condensation Heat Transfer,* Advances in Heat Transfer, 21, 55–139, Academic Press.

J.W. Rose

DROPWISE PROMOTERS (see Dropwise condensation)

DROWNING OUT (see Crystallizers)

DRUM TYPE STEAM TURBINE (see Steam turbine)

DRY-BULB TEMPERATURE

Following from: *Humidity measurement*

The dry-bulb temperature is the temperature of a perfectly-dry surface when exposed to convective heating of a humid gas stream. This temperature is higher than that of a wetted surface exposed to the same gas conditions of humidity, temperature and velocity. If the surface receives only convective heating from the gas, being isolated from other heating sources through conduction or radiation, then the dry-bulb temperature is the temperature of the gas itself. A dry-bulb thermometer is one of two temperature sensors which make up a dry- and wet-bulb *hygrometer* or *psychrometer* used to measure the moisture content (or humidity) of a moist gas. In a properly-designed psychrometer, both sensors are shielded from thermal radiation and exposed to a flow of the moist gas at a sufficient rate to ensure fully-developed forced-convective heat transfer. Normally, a gas velocity of 5 m s^{-1} is adequate [Wiederhold (1995)].

Reference

Wiederhold, P. (1995) Humidity Measurements, Chapter 42 of *Handbook of Drying Technology,* (Ed. A.S. Mujumdar), 2nd ed. Marcel Dekker, New York and Basel.

Leading to: *Wet-bulb temperature*

R.B. Keey

DRY CONTAINMENTS, FOR NUCLEAR REACTORS (see Containment)

DRYERS

Following from: *Drying*

A dryer (or drier) is a machine or apparatus used to remove moisture. It may be a small laboratory oven taking a few grams of moist material or a large industrial unit handling tonnes of wet feed per hour. An industrial dryer is never a stand-alone unit; it is part of a drying system which includes the feeder, the heaters and product collectors besides the actual drying section itself. The **Drying** of natural products has been accomplished from the beginning of time by the use of wind and sun as agents to drive off moisture. Because of the vagaries of weather, alternative methods have been gradually adopted, such as drying indoors in specially-heated and-ventilated rooms. Today, we still talk of *drying chambers,* which are rooms shrunk to the size of a cabinet or vessel. The variety of materials to be dried and the corresponding range of dryer designs available defies simple classification. However, there are three principal factors that define the nature of a dryer:

1. The method of material conveyance through the drying section;
2. The method of heating the material;
3. The pressure and temperature of operation.

These factors will be considered in turn.

Method of Conveying the Feed

More than any other factor, the method of conveying the feed through the dryer governs the outward appearance of a dryer and limits its operating parameters. While free-flowing granules can be handled in a variety of ways, more awkward materials like loose fibres, which can tangle together, and very wet sticky feeds require special techniques. The following table illustrates the range of conveying methods adopted industrially.

Table 1. Methods of conveying goods to be dried

Method	Typical Dryers	Typical Materials
1. Material not transported	Tray dryers	Wide range
	Bin dryers	Particles and grains
2. Material falls by gravity	Rotary dryers	Free-flowing solids
3. Material pushed by blades	Screw-conveyor dryers	Wet sludgy and sticky solids
4. Material moved on trucks	Tunnel dryers	Wide range
5. Material pulled over rolls	Hot-cylinder dryers	Webs, sheets and boards
6. Material held on band	(Unperforated) Band dryers	Discrete articles
	Perforated-band dryers	Particles
7. Material shaken on band	Vibrated-band dryers	Slightly sticky particles
8. Material suspended in gas	Spouted-bed dryers	Wide-range-sized particles
	Fluidised-bed dryers	Small-range-sized particles
	Pneumatic-conveying dryers	Uniform easy-dry particles
9. Material thrown into gas	Spray dryers	Pumpable slurries/ solutions

Adapted from Keey (1978)

Figure 1 illustrates various batch-drying methods, while **Figure 2** shows examples in which solids remain undisturbed or are cascaded through a continuously-worked dryer.

Most modern dryers are operated continuously, or semicontinuously over the working day, as a continuous dryer will require less labour, fuel and floor space than a batch dryer of the same capacity. However, batch drying would be chosen whenever the production rate is small (under 200 kg h^{-1}), or a large number of products have to be handled in the same unit, or whenever large bulky items have to be dried under extensive and complex schedules, for which the drying of porcelain sanitary ware and sawn timber boards provide examples.

Heating Methods

Dryers may be indirectly- or directly-heated. The heating medium in an indirect dryer is separate from the carrier gas. Steam may be used in conjunction with convective heating coils or supplied to a jacket surrounding the drying chamber. In *freeze-drying* units, hot oil is circulated around a heated platen which supports the drying material. For many direct dryers, a heated airflow passes straight through the drying chamber and the dried product is discharged or separated from the outlet air stream. The air may be indirectly heated though a heat exchanger. Alternatively, the dryer can be directly-fired or supplied with flue gas from a burner. If such a gas is used as a drying medium and the exhaust from the dryer is partially recycled to the intake, then the atmosphere within the drying chamber can become inertized. This scheme is often adopted should there be an explosion hazard with a dried product that is finely divided. On the other hand, care has to be taken in obtaining a clean flue gas with direct firing to avoid contamination of the product. With particularly sensitive products, it is wiser to not to use direct firing. Even without direct firing, many installations for foods and light-coloured products will need an air filter to keep out dirt and other specks of foreign matter.

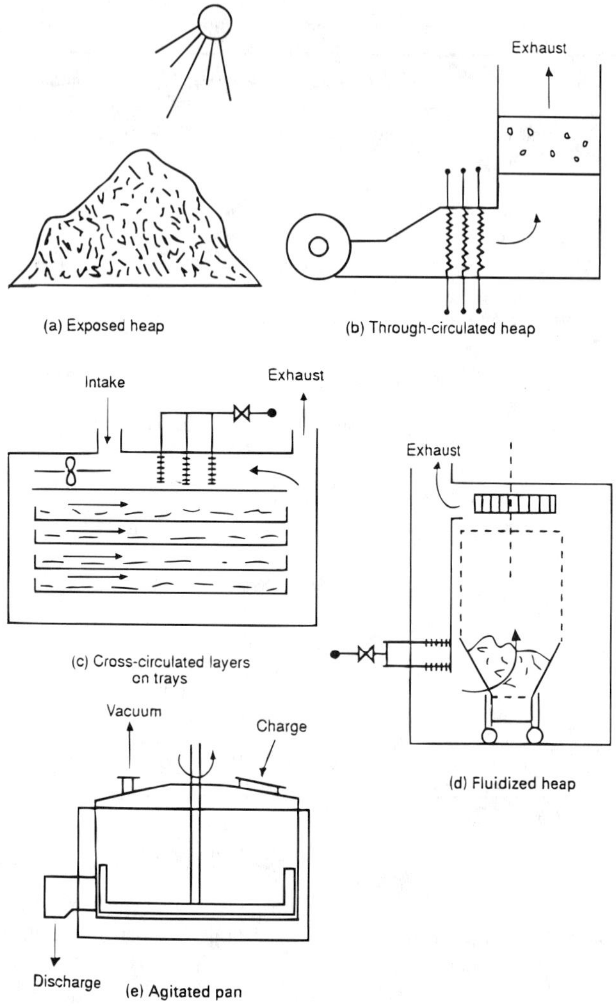

Figure 1. Some batch-drying methods. (*a*) Exposed heap; (*b*) through-circulated heap; (*c*) cross-circulated layers on trays; (*d*) fluidised heap; (*e*) agitated pan. *After Keey (1992).*

There are various modes of heating the moist material. **Convection** is perhaps the commonest. In convective heating, the carrier gas for the evaporated moisture is preheated before passing over or through the material, and the drying conditions are readily controlled by the temperature and humidity of the warm gas. The temperature of the solid is always less then the gas temperature and once the solids have warmed up, they may remain close to **Wet-bulb Temperature** for a significant period of time in the early stages of drying. When bulky porous materials are dried, such as wound textile bobbins and board timber, another quasi-steady temperature appears intermediate between the **Wet- and Dry-bulb Temperatures** within the material. This temperature is thought to be associated with the movement of an evaporative front into the material (Keey, 1978). An example of a continuously-worked convective dryer is shown in **Figure 3**. The material to be dried is placed in shelves of trays which are stacked on trolleys and slowly moved through a drying tunnel. Heaters are placed in the air space above the train of trolleys, and internal fans circulate the air around or over the drying material. Some heat is received by radiation from the heating coils and by gas radiation, as well as by conduction through the contact of the goods with the supporting base. These additional heat transfer mechanisms can significantly boost convection in cases where the solids are supported on trays or bands. Likewise in rotary cascading dryers, heat transfer to the raining curtain of particles is not solely by convection with the crossflowing air stream; gas radiation is important, and some heat can be transferred from the hot cyclindrical wall as the particles rest in the lower quadrants of the revolving drum.

If the material to be dried is very thin or wet, then heat may be suppled by way of **Conduction**. All the heat now passes through the material itself, from hot surfaces supporting or confining it, so the material's temperatures are higher than in convective heating. For this reason, thermally-sensitive materials, if in the form of slurry, would be dried under vacuum to reduce operating temperatures. **Figure 4** shows a twin-drum dryer which can accept a lump-free slurry. The rotating drums drag the slurry around their slowly-revolving peripheries while knives peel off the dried product, which falls into a conveying duct. With single-drum units, the wet paste may be splashed onto the surface of the drum, or the drum may dip into a pool of feed material, as is the commoner practice. Thin materials, in the form of sheets, can be drawn over and under a series of internally-heated cylinders to provide discontinuous thermal contact. The drying section of a paper-making machine is designed on this basis. In this instance, besides the conductive heat transfer, there is some adsorption of moisture by pressing felts and additionally, local thermal boosting by radiators is sometimes adopted for moisture-profile adjustment.

Energy may be supplied by various forms of *electromagnetic radiation,* whose wavelengths varies between those of solar radiation to those of microwaves (0.2 m to 0.2 μm). Longer-length radiation within this waveband barely penetrates beyond the exposed surface of the material, which normally only absorbs significant incident radiation at certain wavelengths. *Radio-frequency (RF)* and *microwave* energy, however, can penetrate a material significantly and may be regarded in some cases as providing volumetric heating. The cost of electrical energy compared with other energy sources confines the application of electrical heating methods to modest throughputs of high-value material or to finishing operations achieving a more uniformly-dried product. In general, the energy output of the various electromagnetic emitters is restricted. For example, *infra-red* emitters are limited to about 10 kW per unit, and where a particular power source greater than this is required, a number of individual emitters would be installed. However, the use of a number of emitters gives flexibility in heating arrangements; further flexibility can be achieved by varying individual voltages on the heaters, depending on the local moisture content. Two further attractive features of the use of shorter-wave RF and microwave energy are the volumetric and selective nature of heating. The latter property arises because most materials to be dried are nonmetallic, and are thus good electrical insulators with a low-loss factor, whereas moisture such as water has a high-loss factor. Moisture may evaporate *in situ.* In general, it is easier and cheaper to build a RF heater than a microwave unit, as complex arrangements are needed in a continuous process with microwave sources to allow the material being dried to be conveyed through the unit without excessive leakage of radiation at the inlet and outlet ports. There is some advantage in combining RF heating with convective drying, as *dielectric heating* (providing 10 to 20%

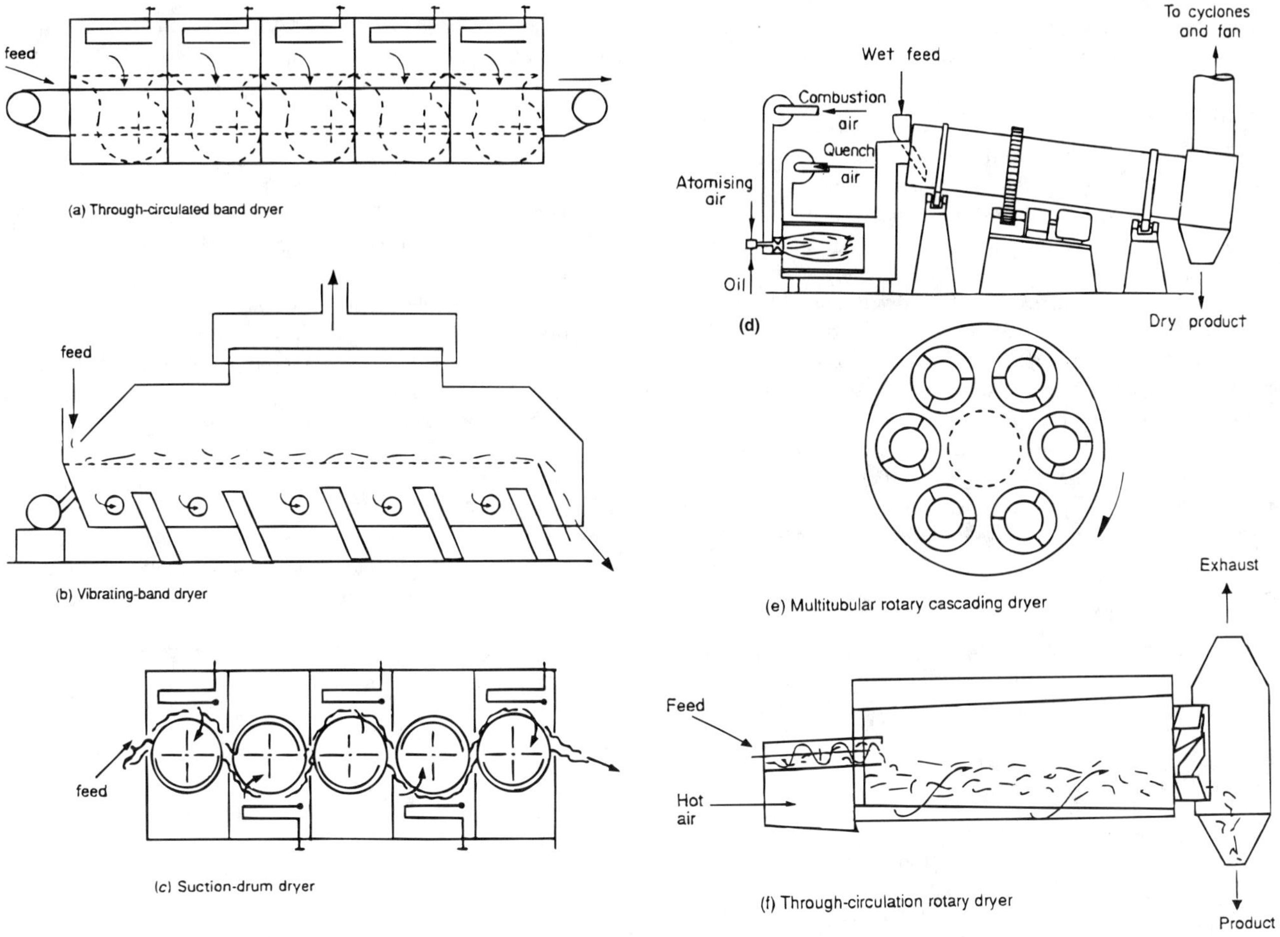

Figure 2. Continuous-drying methods with undisturbed or cascading solids. (*a*) Through-circulated band dryer; (*b*) vibrated-band dryer; (*c*) suction-drum dryer; (*d*) rotary cascading dryer; (*e*) multitubular rotary cascading dryer; (*f*) through-circulation rotary dryer. *After Keey (1992).*

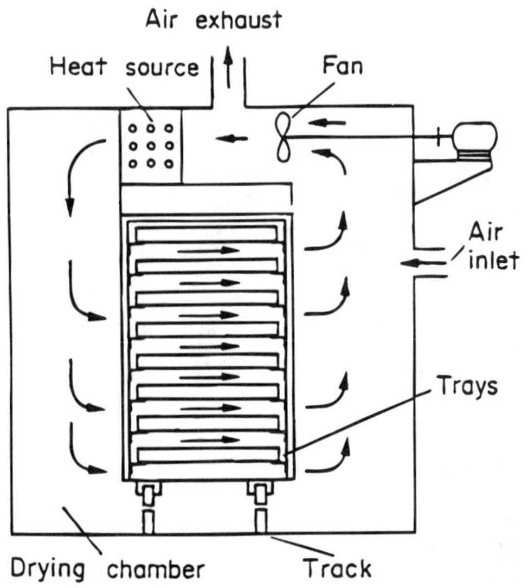

Figure 3. A drying tunnel for trucked trays. *After Sloan (1967).*

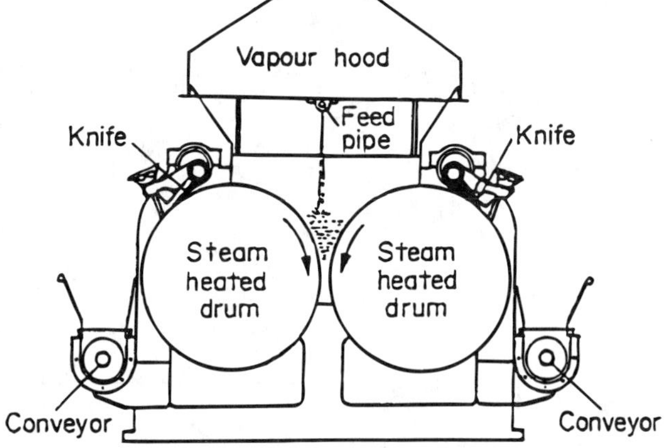

Figure 4. A twin-drum dryer for thin slurries. *After Sloan (1967).*

of the total) can be used to raise moisture temperature, and thus enhance the convective drying rate.

Pressure and Temperature of Operation

The thermal sensitivity of the material fixes the temperature-time limits for safe drying. For many materials, the rate of thermal degradation follows an Arrhenius relationship, and the maximum permissible working temperature falls exponentially with an increase in holding time. Polymers may be safely dried in a fluidised-bed dryer, for example, in which the dwell-time of the particles may be of the order of 10 to 20 min while in a static bed, with its inherently slower drying rates, drying conditions would have to be very mild and the associated dryer large and costly. *Spray dryers* for the manufacture of milk products can be operated with an air-inlet temperature of the order of 200°C since the residence time of particles in the drying chamber is very short, of the order of 20s or less. Nevertheless, it is sometimes necessary to dry thermolabile materials under vacuum to reduce process temperatures. *Freeze-drying* can only be undertaken below the triple-point pressure of 630 Pa. Sensitive pharmaceutical products are dried in agitated vacuum pans. Degrade of timber for high-quality end use can be reduced by seasoning under vacuum, rather than by a lengthier kiln-drying schedule at atmospheric pressure.

Dryer Selection

The selection of a complete drying installation includes many considerations other than the drying characteristics of the wet feedstock. These factors include the storage and delivery of the feed material, any equipment for performing it or blending back dried fine particles, the means of conveying the material as it dries, the equipment for collecting the dried product, and ancillary plants for the supply of heat, vacuum or refrigeration (Keey, 1992). Past experience in operating equipment will be a guide in the case of an existing product or drying process, and careful consideration of past choices normally reveals some deficiencies which can be rectified. Simple bench tests can reveal considerable semiquantitative information about a material's drying behaviour under proposed drying conditions, leading to the elimination of some types of dryer. There have been many attempts to provide first guides for the selection of a dryer for a particular job. Simple decision trees are set out by van't Land

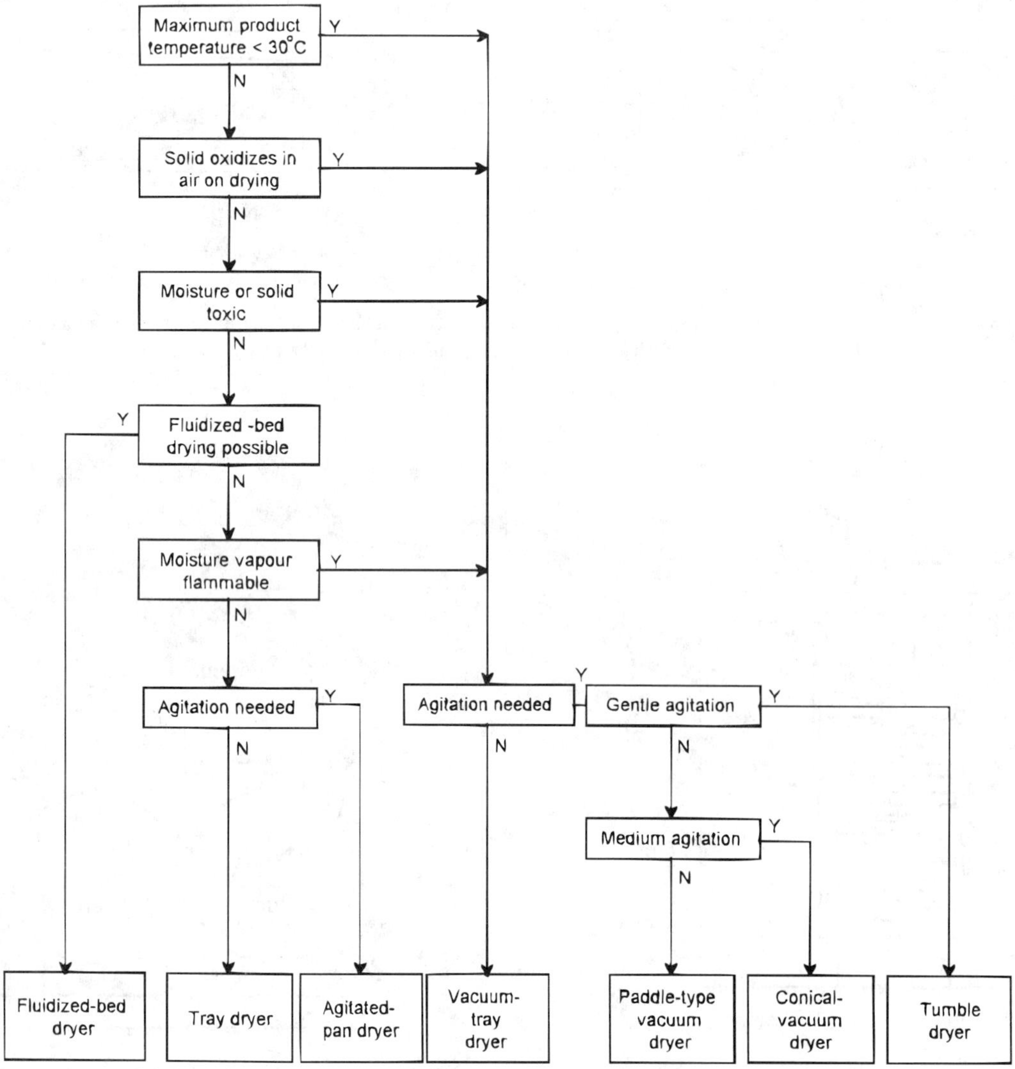

Figure 5. Decision tree for the selection of a batch dryer. *After van't Land (1984).*

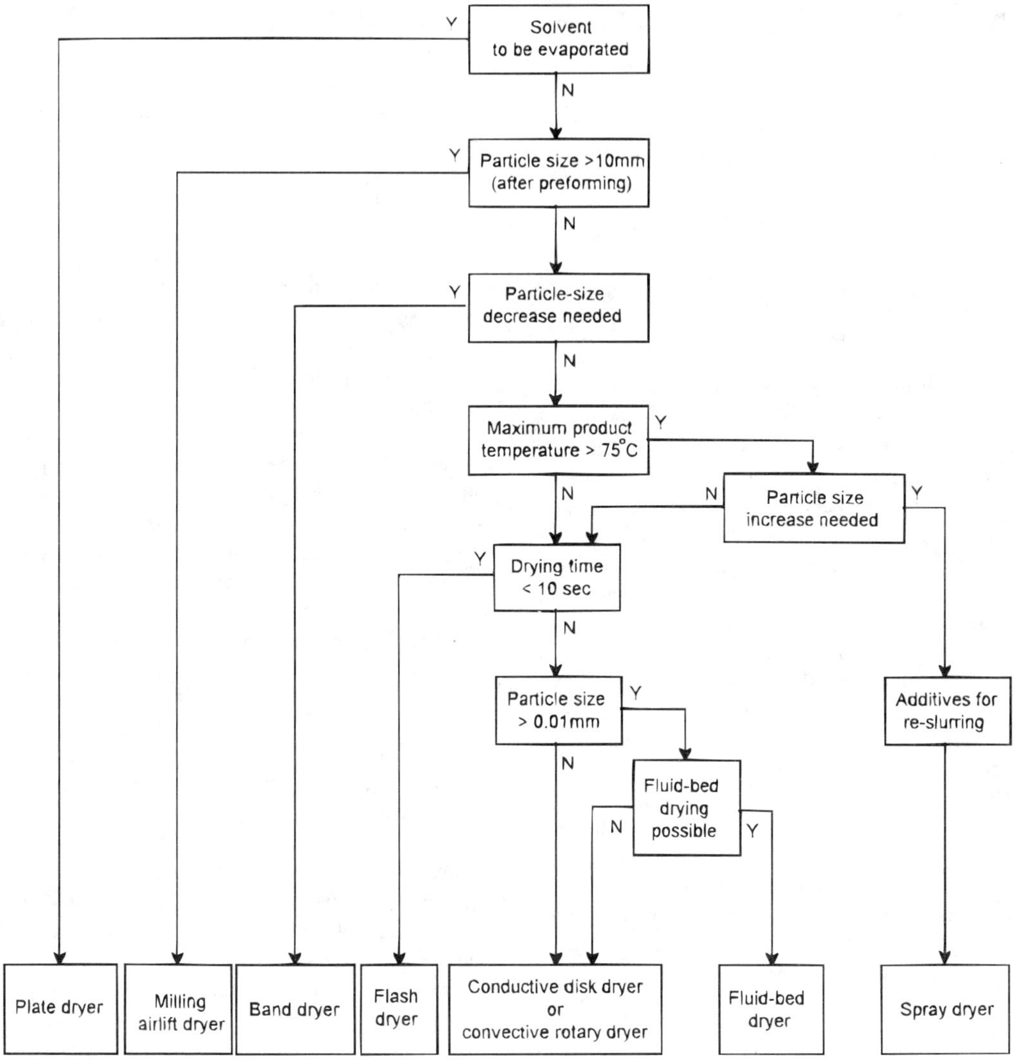

Figure 6. Decision tree for the selection of a continuous dryer. *After van't Land (1984).*

(1984) for batch and continuous dryers, and these are reproduced in **Figure 5** and **6**. More advanced methods based on expert systems are currently being developed (Kemp, 1994).

Such methods provide general indications and frequently, special considerations apply. When high-valued materials are to be dried, the costs of drying may relate more to product losses or dust emissions than to costs of investment and energy use. Design of the dryer may then be secondary to the careful specification of appropriate gas-cleaning technology. The dominance of material losses on processing is also related to the safe handling of toxic materials and the prevention of hazards in drying solvent-wet solids. To avoid losses, sophisticated drying assemblies can often be justified, such as a vacuum batch chamber for drying pharmaceutical granules by microwave heating. Other possibilities include the design of hybrid equipment in which several unit operations can be carried out in sequence in the same vessel. Thus, an agitated *vacuum dryer* might be used for crystallisation and filtration before drying or blending and granulation afterwards.

Cook and DuMont (1991) have provided a description of modern drying practice, including methods of improving the performance of existing dryers, besides the selection and commissioning of drying units.

References

Cook, E. M. and DuMont, H. D. (1991) *Process Drying Practice,* McGraw-Hill, New York.

Keey, R. B. (1978) *Introduction to Industrial Drying Operations,* Pergamon, Oxford.

Keey, R. B. (1992) *Drying of Loose and Particulate Materials,* Hemisphere, New York.

Kemp, I. C. (1994) *A new algorirthmn for dryer selection,* Proc. 9th Internat. Drying Symp., Gold Coast, 1–4 Aug.

Sloan, C. P. (1967) *Drying systems and equipment,* Chem. Eng., 74(14), 169–200.

van't Land, C. M. (1984) *Selection of industrial dryers,* Chem. Eng., 91(5), 53–61.

R.B. Keey

DRYING

Following from: *Mass transfer*

Drying has been defined as the process whereby moisture is vaporised from a material and is swept away from the surface, sometimes under vacuum, but normally by means of carrier gas which passes through or over the material (Keey, 1992). Commonly, drying is conceived as the removal of water into a hot airstream, but drying may encompass the removal of any volatile liquid into any heated gas. For drying, so defined, to take place, the moist material must obtain heat from its surroundings by convection, radiation or conduction, or by internal generation such as dielectric or inductive heating; the moisture in the body evaporate and the vapour is received by a carrier gas. This drying process is sketched in **Figure 1**.

Drying has a number of close synonyms. *Dehydration* is the process of depriving a material of its water or the loss of water as a constituent. The term is often used in food-drying operations to describe processes which strive to expel moisture but retain other volatile constituents in the original material, and which are responsible for valuable aromatic and flavouring properties. *Desiccation* implies a more thorough removal of water. It is applied in the drying of foodstuffs to indicate almost complete dehydration of these materials for preservation. The term is also commonly used to describe the thorough removal of moisture from gases.

While heat may be used to drive off moisture from a wet substance, moisture can be severed from its host material by the action of pressure gradients. This process is known as *dewatering,* and is normally used as a precursor to the drying of very wet materials when the moisture-solid bonding is not strong. Dewatering may be undertaken by mechanical means, such as pressing or centrifuging. These operations are not considered in this encyclopedia since combined heat and mass transfer processes are not involved.

Water can also be removed by *osmotic dehydration.* Foods can be treated with concentrated solutions of salt or sugar to attain substantial removal of water with limited solute uptake. The process has been described in terms of coupled diffusion of the water and solute [Raoult-Wack et al. (1989)]. Sludges have been dewatered through the action of an external direct-current field; this is known as *electro-osmosis* (Yoshida and Yukawa, 1992). These osmotic techniques also lie outside the realm of heat and mass transfer processes.

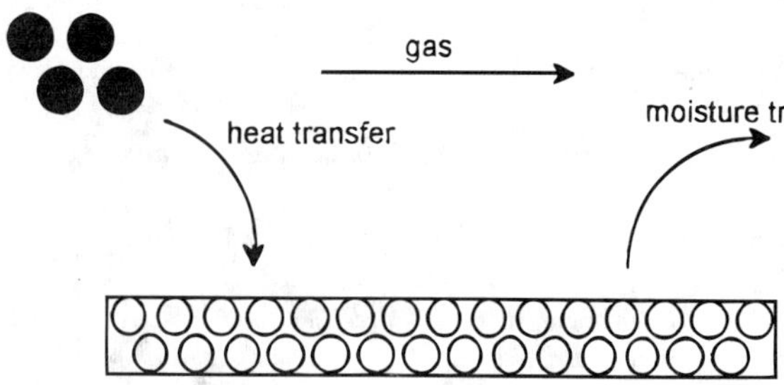

Figure 1. Convective drying process.

Drying is an energy-intensive operation of some significance. Estimates of energy demand range between 7 and 15% of a nation's industrial energy consumption (Keey, 1992), but a more recent survey for the United Kingdom in 1990 suggests the figure may be as high as 20%, an increase from the 12% obtained in a similar survey in 1978 (Oliver and Jay, 1994). This difference may reflect the changing pattern of industrial activity in that country over the period.

Although the drying medium is normally considered to be air, there are advantages in using other media. If the solid being dried forms a combustible powder or the moisture itself is a flammable solvent, then the use of an inertized or inherently inert gas is advisable. Drying in steam has the added advantages of lower energy use and higher heat transfer rates. Above the so-called *inversion-point temperature,* drying in steam is faster than drying in perfectly dry air at the same temperature. The evolution of moisture is uniform, and this feature is observed for example in the *superheated-steam drying* of wood under vacuum, a process used commercially for the production of high-quality seasoned board timber with minimum degrade caused by the development of drying stresses. A closed vacuum or high-pressure vessel is not necessary for steam drying. By allowing air in the drying chamber to be displaced by water vapour as the vessel warms up and moisture evolution begins, the steam will remain in the dryer and no complex sealing arrangements are needed for the intake and discharge of solids. Steam at 100°C has only 55% of the density of air at the same temperature and thus will remain trapped inside the chamber. The patented technique is known as *airless drying,* and the arrangements for batch operation are shown in **Figure 2**. If the vented steam can be used for other purposes, such as the production of

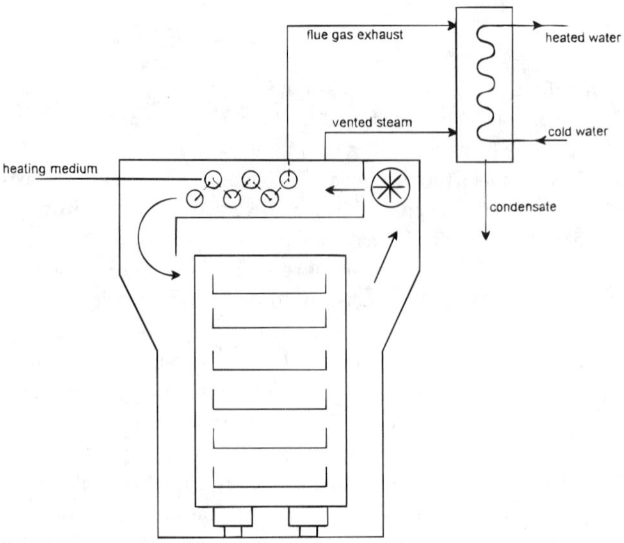

Figure 2. An airless drying system with heat recovery. *After Stubbing (1994).*

hot water, then the airless drying system can show considerable thermal economies over conventional drying in air.

Drying only takes place if the wet material contains more moisture than the equilibrium value for its environment. The earliest ideas on convective drying implied that liquid moisture diffuses to the exposed surface of a wet body where it evaporates, the vapour diffusing through the boundary layer into the bulk of the surrounding air. This view is clearly unsatisfactory, except for drying of homogeneous materials in which the moisture is effectively dissolved. Mechanisms of moisture movement are generally more complex. Most materials are composed of subentities, such as particles and fibres, which may be loose or held in some kind of matrix. The number and nature of the voids between these entities and the pores within them govern the quantity of moisture retained by and the extent of bonding to the solids. If the openings form a capillary network, the material is said to be *capillary-porous.* A capillary-porous material may be *nonhygroscopic*: that is, the moisture held within the body exerts its full vapour pressure. This is a limiting case found in some coarse, nonporous mineral aggregates. Moisture is simply trapped between the particles. As the space between the particles becomes more confined, vapour pressure is lowered according to the *Kelvin equation*

$$p_k = p_o \exp\text{-}\{2\sigma\nu/d_pRT\}, \tag{1}$$

where p_k is the capillary-vapour pressure, p_o is the saturation vapour pressure, σ is the surface tension, ν is the molar liquid volume and d_p is the capillary size.

Hygroscopicity may be related to this capillary condensation in the voids, but normally it stems from the structure of the primary entities with their finer passageways and their ability to hold moisture in a variety of ways. Material may not be simply capillary-porous, but be composed of complex arrangements of capillaries, vessels and cells, as seen with materials of vegetable origin. Some of these may be described as being both capillary-porous and colloidal, composed of a matrix colloidal by nature but developing a pore structure upon drying. Colloidal slurries of inorganic particles lose volume as moisture is driven off until the mass becomes consolidated. Colloidal material of biological origin, by contrast, does not shrink until condensed moisture in the intermiscellar spaces has been emptied. This condition is called the *fibre-saturation point* in woody material drying. Thereafter, as cellular moisture is expelled, the stuff shrinks as the cells shrivel.

The ratio of the moisture-vapour pressure to the saturation value at the same temperature is called *relative humidity* ψ. The lower the relative humidity, the more strongly is the moisture bound to the host material. The free energy ΔG required to release unit molal quantity of this moisture is given by

$$\Delta G = -RT \ln \psi \tag{2}$$

for an isothermal, reversible process without change of composition. Isothermal variation of the equilibrium moisture content (which is a function of this free-energy change) with relative humidity yields a *moisture isotherm,* and it is the desorption isotherm which is of interest in drying. Moisture isotherms are normally sigmoid in shape when plotted over the whole relative humidity range, but often a simple exponential expression can be fitted over a more limited range at higher relative humidities. The general shape of the isotherm reflects the nature of the moist material, as illustrated in **Figure 3**. An exception to this kind of behaviour is that of inorganic crystalline solids which have multiple hydrates. With these materials, relative humidity falls in stepwise fashion with loss of moisture as each hydrate disappears.

When pores in a solid are of molecular size, moisture can only be held therein by *volume filling* such that the adsorbate, although highly compressed, is not considered to be a separate phase. The chemical potential of the adsorbent varies with the amount adsorbed, unlike the behaviour at higher moisture contents when a separate adsorbate phase exists—with equilibrium between phases—and the chemical potential remains invariant. The fractional filling of these micropores is a complex exponential function of free-energy sorption, $RT \ln\psi$. With larger pore sizes, moisture can sorb in molecular layers on the host material. Consideration of multimolecular adsorption leads to the equation

$$\frac{X}{X_1} = \frac{C\psi}{(1 - k\psi)\{1 + (C - k)\psi\}} \tag{3}$$

for the equilibrium moisture content X at a relative humidity ψ, where X_1 is the moisture content for a complete monolayer, k is the exp ($\pm$ $\Delta H/RT$), where ΔH is the constant enthalpy difference in adsorptive heats between the first and successive molecular layers of moisture, and C is a coefficient. When k is zero, the equation reduces to that for monomolecular adsorption. Over the range $0 < k < 1$, this equation can describe the moisture-sorption behaviour of materials, which appears to reach a finite moisture content as relative humidity approaches unity. This quantity is sometimes known as the *maximum hygroscopic moisture content.* This three-coefficient equation (with C, k and X_1 as the adjustable parameters) has been tested for sorption of water vapour on 29 materials at room temperature over a wide range of relative humidity (from 0.07 to 0.97) and for some of these materials, over a narrower range of temperatures between 45 and 75°C (Jaafar and Michalowski, 1990). In most cases, experimental data could be fitted to within $\pm 8\%$ up to a relative humidity of 0.7, and in some instances over the whole humidity range. To cope with the hygroscopic behaviour at high relative humidities with colloidal material, which swells with increasing moisture content, Schuchmann et al. (1990) have recommended that $-\ln(1 - \psi)$ be chosen as the dependent variable rather than ψ itself in the correlation. It is unwise to extrapolate sorption correlations beyond their tested

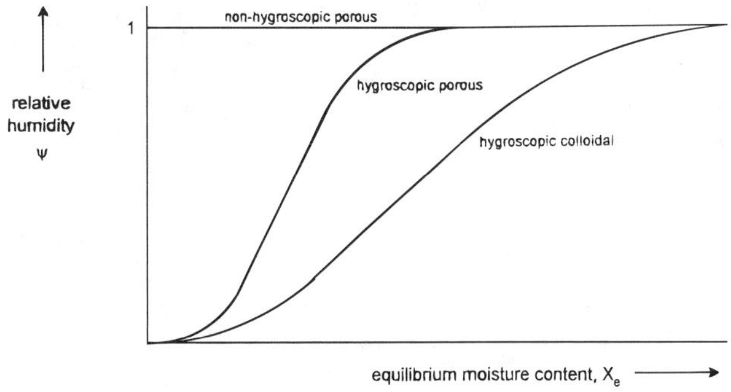

Figure 3. Variation of relative humidity with equilibrium moisture content for different materials. *After Keey (1978).*

range of relative humidities, owing to changes in hygroscopic behaviour at extremes in this range compared with that at intermediate values.

The manner in which a material dries out depends not only on its structure but also on its physical form. The drying of small wood chips is controlled essentially by moisture-vapour transport through the boundary layer; veneers and thin slats of the same wood by the dry fraction of the exposed surface; while the drying of board timber, by the internal moisture-transport mechanisms within the timber itself. Early experiments on drying materials in sample trays in an air stream have noted that initially, the drying rate was almost the same as that of a free liquid surface under the same conditions and remained relatively constant as the material dried out (Keey, 1972). This period of drying is followed by one in which the drying rates fell off sharply as moisture content was reduced to the equilibrium value even though the drying conditions remain unchanged. This marked difference in behaviour has led to the division of drying into the *constant-rate period* and *falling-rate period,* respectively. The "knee" in the drying curve between these two periods is known as the *critical point.* Sometimes, these periods are referred to as *unhindered drying* and *hindered drying,* respectively, to indicate whether the material itself plays a controlling role in restricting moisture loss. Appearance of the initial period can be masked by the induction effects at the start of drying as a moist solid warms up or cools to a dynamic equilibrium temperature, which is the **Wet-bulb Temperature**, if the surface is only heated convectively. This surface temperature is maintained as long as the surface is sufficiently wet to effectively saturated it. An example of a drying curve is shown in **Figure 4**.

The reasons for the appearance of a drying period of constant, or near-constant, drying rates are complex, particularly as it is unlikely that there exists a film of liquid moisture at the surface except in rare circumstances (van Brakel, 1980). Indeed, a constant-rate period can be observed if the dimensions of the wet and dry patches on the surface are sufficiently small compared with the thickness of the boundary layer. The requirement is that moisture-vapour pressure at the surface maintain the saturation value at the

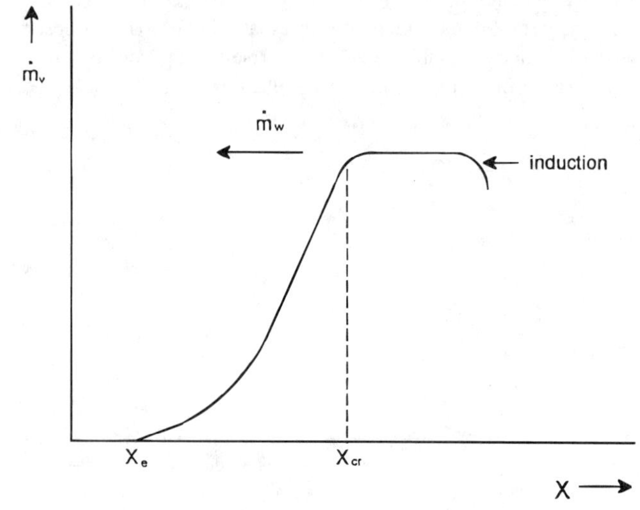

Figure 4. An example of a drying curve.

mean surface temperature, and thus the rate of moisture loss over unit exposed surface ($\dot{m}_W$) is:

$$\dot{m}_W = \beta(p_S - p_G), \tag{4}$$

where p_G is the partial pressure of the moisture vapour in the bulk of the gas and p_S is the value of the gas adjacent to the moist surface. In drying calculations, it is more useful to use humidities (ratios of the mass of moisture vapour to that of dry gas), and the above expression transforms to

$$\dot{m}_W = \beta_Y \phi (Y_S - Y_G) \tag{5}$$

where β_Y is a mass transfer coefficient based on the humidity difference; ϕ is the humidity potential coefficient which effectively "corrects" for the introduction of a linear humidity driving force (Keey, 1978); and Y_S and Y_G are the humidities at the surface and in the bulk of the gas, respectively. The falling-rate period begins when moisture movement within the solid can no longer maintain the rate of evaporation, or if moisture content at the surface falls below the maximum hygroscopic value and the partial pressure of the moisture vapour also dwindles. Clearly, a constant-rate period may never be observed in colloidal material drying that reaches unit relative humidity only at very high moisture contents.

To a first approximation, the drying kinetics in the falling-rate period may be regarded as of the first-order, and thus the drying rate is directly proportional to the difference between the mean moisture content of the wet material (X) and its equilibrium value (X_e):

$$-\frac{dX}{dt} \propto (X - X_e) \tag{6}$$

Integration of this equation yields the time Δt to dry from a moisture content of X_1 to X_2:

$$-\frac{\Delta t}{dt} = \frac{1}{a} \ln \frac{X_1 - X_e}{X_2 - X_e} \tag{7}$$

where a is a coefficient which may be proportional to a moisture "diffusivity." However, the validity of this equation is no test that moisture movement is by diffusion as this expression correlates drying time for a number of materials dried in static and fluidised beds, in which diffusional processes are unlikely (van Brakel, 1980). The appearance of first-order kinetics is sometimes referred to as the *regular regime* of drying.

There have been a number of attempts to provide a fundamental theoretical framework to describe drying and the development of moisture-content and temperature profiles in materials being dried. These include the use of irreversible thermodynamics to define the appropriate transport potentials (Luikov, 1966); theories based on moisture-vapour diffusion and capillary transport of liquid in porous media (Krischer and Kast, 1978); and volume-averaging the equations of continuity, mass and energy conservation which apply to each of the discontinuous phases within a moist porous body (Whitaker, 1980). While these approaches have found some limited success in describing heat and mass transfer processes under certain idealised conditions, these theories are limited in application by the assumptions made to get numerical solutions.

In practice, more empirical approaches have been found which are useful in describing drying behaviour of actual material under industrial conditions. One such method is based on the concept of the *characteristic drying curve* (Keey, 1978). This is a generalised drying curve obtained from laboratory tests under constant drying conditions with a sample material. It is a plot of the drying rate, normalised with respect to its maximum value (m_W) in the constant-rate period, against a characteristic moisture content—defined as the ratio of the freely evaporable moisture content ($X - X_e$) to the free moisture content at the critical point ($X_{cr} - X_e$). These nondimensional parameters thus become

$$f = \dot{m}_V/\dot{m}_W \quad (8)$$

and

$$\Phi = (X - X_e)/(X_{cr} - X_e), \quad (9)$$

respectively. An example of a characteristic drying curve is shown in **Figure 5**.

A unique characteristic drying curve is only found if the ratio of the exposed surface to the material volume is maintained constant or has a unique value at a given characteristic moisture content if the material shrinks. Strictly speaking, such a curve is only found at extremes of drying intensity, when the material's moisture content is essentially uniform (low-intensity drying) or when there is a sharp front between dried-out material close to the surface and a still-wet interior (high-intensity drying). In practice, however, over the range of practical drying conditions, characteristic drying curves appear when drying a variety of particulate and loose materials, provided the particle size is below 20mm (Keey, 1992). There is normally no simple relationship between the parameters f and Φ, although there are some special cases. First-order kinetics correspond to the identity $f = \Phi$, with drying of permeable materials approximating this behaviour. When drying is controlled by the dry fraction of exposed surface (with thin sheets, for example), then $f = \Phi^{2/3}$. Drying of impermeable materials like heartwood timber corresponds approximately to $f = \Phi^2$. If no critical point is observed in the drying experiment, a characteristic curve can be drawn up based on normalising the moisture content with respect to the ostensible start of the falling-rate period, provided a simple algebraic relationship can be fitted to the characteristic curve. The particular advantage of the characteristic drying curve concept is that a simplified equation can be written to describe the rate of drying at any place within a dryer if the humidity potential ($Y_W - Y_G$) is known, where Y_W is the saturation humidity at the wet-bulb temperature:

$$\dot{m}_v = f\{\dot{m}_{oW}\} = f\{\beta_Y\phi(Y_W - Y_G)\}. \quad (10)$$

Examples of the use of this expression to describe industrial drying processes are given by Keey (1978, 1992).

The intensity of drying (I) is given by the ratio of the maximum unhindered drying rate (m_W) to the maximum moisture-transfer rate through the material assuming a diffusion-like process:

$$I = \frac{\dot{m}_W}{D(X_0 - X_e)/b} \quad (11)$$

where D is a moisture diffusion coefficient, X_0 is the initial moisture content and b is the effective thickness or radius of the material. This suggests that the product $m_W b$ is a useful property; it is called the *flux parameter,* F. If ln F is plotted against moisture content X, a separate curve is found for each initial moisture content X_0 in the penetration period as the moisture content and temperature profiles develop within the material. In the regular regime, there is a common curve independent of initial moisture content and drying flux.

Under certain circumstances, drying curves can appear with both negative and positive gradients in drying rate as moisture is lost. The drying of layers of soluble dyestuffs can show discontinuities due to the build up and cracking of surface crusts, particularly if the crust is removed periodically. The drying of porous bodies containing a mixed volatile solvent may also show periods of falling and rising rates due to selective evaporation of the moisture with changes in relative volatility as the composition alters.

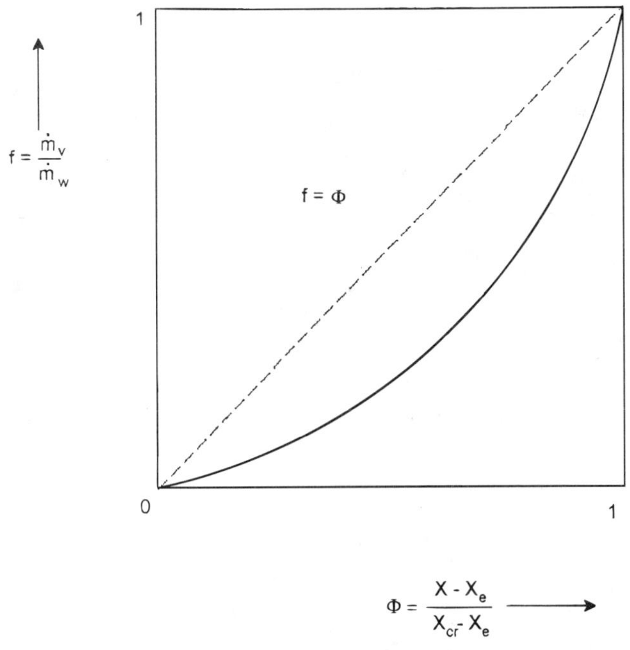

Figure 5. An example of a characteristic drying curve.

References

Jaafar, F. and Michalowski, S. (1990) Modified BET equation for sorption/desorption Isotherms, *Drying Technol.* **8**(4), 811–827.

Keey, R. B. (1978) *Introduction to Industrial Drying Operations,* Pergamon, Oxford.

Keey, R. B. (1992) *Drying of Loose and Particulate Materials,* Hemisphere, New York.

Krischer, O. and Kast, W. (1978) *Die wissenschaftlichen Grundlagen der Trocknungstechnik,* 3rd edn, Springer-Verlag, Berlin Heidelberg New York.

Luikov, A. V. (1966) *Heat and Mass Transfer in Capillary-Porous Bodies,* Pergamon, Oxford.

Stubbing, T. J. (1994) Airless Drying, in *Proc. 9th Internat. Drying Symp.*, Gold Coast, Qld, 1–4 Aug.

van Brakel, J. (1980) Mass Transfer in Convective Drying, Chapter 7 in *Advances in Drying*, 1, (Ed. A. S. Mujumdar), Hemisphere, Washington.

van Brakel, J. (1980) Mass Transfer in Convective Drying, Chapter 7 in *Advances in Drying*, 1, (Ed. A. S. Mujumdar), Hemisphere, Washington.

Whitaker, S. (1980) Heat and Mass Transfer in Granular Porous Media, Chapter 2 in *Advances in Drying*, 1, (Ed. A. S. Mujumdar), Hemisphere, Washington.

Leading to: *Dryers*

R.B. Keey

DRYING CHAMBERS (see Dryers)

DRYOUT (see Burnout, forced convection; CANDU nuclear power reactors; Heat pipes; Post dryout heat transfer)

DUAL-PURPOSE HEAT PUMPS (see Heat pumps)

DUBOIS' BODY SURFACE (see Body, human, heat transfer in)

DUCTILE FRACTURE (see Fracture of solid materials)

DUCTS, NON-CIRCULAR, FLOW IN (see Channel flow)

DUFOUR EFFECT (see Thermal conductivity)

DUNE FLOW (see Gas-solid flows)

DUST, AS AN AIR POLLUTANT (see Air pollution)

DUSTS (see Aerosols)

DYE LASERS (see Lasers)

DYNAMICAL SIMILARITY (see Flow of fluids)

DYNAMIC INSTABILITIES IN TWO-PHASE SYSTEMS (see Instability, two-phase)

DYNAMIC PRESSURE

Following from: *Bernoulli equation; Velocity head*

The dynamic (or velocity) pressure describes the kinetic energy per unit volume within a given streamtube. If u is the velocity and ρ the density, the dynamic pressure is given by $\rho u^2/2$. If the flow within a pipe, channel or duct is uniform throughout a given cross-section, this definition of the dynamic pressure applies to the entire cross-section. However, if the velocity is nonuniform, a correction factor (α) similar to that identified in the **velocity head**, must be applied to account for the variation in the kinetic energy from one streamtube to another. In this case, the dynamic pressure appropriate to the entire section is:

$$\frac{\alpha \bar{u}^2}{2g} \quad \text{where} \quad \alpha = \frac{1}{\bar{u}^2}\frac{\int_A u^3 dA}{\int_A u dA}$$

where A is the cross-sectional area and ū is the mean velocity. If the **Bernoulli Equation** is described in terms of pressures rather than energy heads, the steady-flow solution yields:

$$P + \frac{1}{2}\rho u^2 + \rho g z = \text{constant}$$

where the terms on the left are, respectively, the static pressure, the dynamic pressure and the potential or position pressure. The sum of these terms is usually referred to as the *stagnation pressure.*

C. Swan

DYNAMICS (see Flow of fluids)

DYNAMIC WAVES (see Kinematic waves)

E

EA (see Electricity Association)

EBULLIOSCOPIC CONSTANT (see Solubility)

ECKERT ENTHALPY (see Adiabatic wall temperature)

ECKERT, ERG

Born on 13 December 1904 in Prague, Czechoslovakia, Ernst Rudolf Georg Eckert attended the German Institute of Technology in Prague where he received the Diploma Ingenieur and Dr. Ing. He held the chair of thermodynamics at his alma mater and was also a section chief for the Aeronautical Research Institute in Braunschweig, Germany. At the end of World War II, he came to the U.S. where he first worked at Wright-Patterson Air Force Base and later at the NASA Lewis Flight Propulsion Laboratory. In 1951, he joined the University of Minnesota where his career continues today as Regents Professor Emeritus in the Mechanical Engineering Department. In Minnesota, he founded the Heat Transfer Laboratory and built up a significant research group working on many different aspects of heat transfer.

Eckert's contributions in heat transfer cover almost the entire field from radiation to buoyancy driven convection to high speed flows to many other areas as well. His early books (in the 1940s) and research established a quantitative, analytical and experimental base for the engineering science of heat and mass transfer. He has received a number of awards for outstanding research from technical societies as well as honorary doctorates. He has helped establish several journals in the area of heat transfer and was a founding father of the International Heat Transfer Conferences. He is recognized as one of the most outstanding contributors to heat transfer analysis, experiments and understanding in the international community for well over a half a century. He is also known for his major contributions to international cooperation in the research community and for his support of many individuals who have gone on to leadership positions in the heat transfer community.

R.J. Goldstein

ECKERT NUMBER

The Eckert number (Ec), first named in the early 1950s, is a dimensionless quantity useful in determining the relative importance in a heat transfer situation of the kinetic energy of a flow. It is the ratio of the kinetic energy to the enthalpy (or the dynamic temperature to the temperature) driving force for heat transfer

$$\mathrm{Ec} = \frac{U^2}{c_p \Delta T}$$

where U is an appropriate fluid velocity (e.g., outside the boundary layer or along the centerline of a duct), c_p is the specific heat at constant pressure and ΔT is the driving force for heat transfer (e.g., wall temperature minus free stream temperature). For small Eckert number (Ec << 1) the terms in the energy equation describing the effects of pressure changes, viscous dissipation, and body forces on the energy balance can be neglected and the equation reduces to a balance between conduction and convection.

The Eckert number when multiplied by the Prandtl number (Pr) is also a key parameter in determining the viscous dissipation of energy in a low speed flow.

$$\mathrm{Ec} \cdot \mathrm{Pr} = \frac{U^2 \eta}{\lambda(\Delta T)}$$

where η and λ are the dynamic viscosity and the conductivity, respectively, of the fluid. The parameter Ec · Pr (sometimes called the *Brinkman number*) is essentially the ratio of the kinetic energy dissipated in the flow to the thermal energy conducted into or away from the fluid.

When (Ec · Pr) << 1, the energy dissipation can be neglected relative to heat conduction in the fluid. For large (Ec · Pr) the energy dissipated is an important parameter in the heat transfer process and the kinetic energy can play a significant role in determining the temperature distribution in the flow and the overall heat transfer.

R.J. Goldstein

ECONOMIC PENALTIES OF FOULING (see Fouling)

EDF (see Electricité de France)

EDDIES (see Burners)

EDDIES IN TURBULENT FLOW (see Turbulence)

EDDY CORRELATION METHOD OF SURFACE HEAT TRANSFER (see Environmental heat transfer)

EDDY VISCOSITY (see Boundary layer heat transfer; Turbulence models)

EFFECTIVE DIFFUSIVITY METHOD (see Condensation of multicomponent vapours)

EFFECTIVENESS - NTU METHOD (see Mean temperature difference)

EFFECTIVENESS OF HEAT EXCHANGER (see Tube banks, single phase heat transfer in)

EFFICIENCY AND STEAM TURBINES (see Steam turbine)

EFFICIENCY, IN TURBINES (see Turbine)

EFFICIENCY OF CYCLES (see Carnot cycle)

EFFICIENCY OF HEAT EXCHANGERS (see Waste heat recovery)

EFFICIENCY OF POWER CYCLES (see Rankine cycle)

EFFICIENCY OF PROCESSES (see Exergy)

EFFLUENT TREATMENT (see Ion exchange)

EIGEN FUNCTIONS (see Eigen values)

EIGEN VALUES

Eigen value of the operator L is the value of the parameter λ (complex or real) for which the equation $Lu = \lambda u$ has a nonzero solution. The appropriate nonzero solutions are called *eigen functions* of the operator L, corresponding to eigen value λ. L nonlinear operators are usually considered as differential, integral, etc. A set of eigen values is a discrete spectrum of the operator L. Eigen functions belonging to different eigen values are linearly independent.

The *Hermitian (conjugate) linear operators* [for instance, the differential operator involved in the stationary equations of heat transfer and diffusion $L = \mathrm{div}(k \cdot \mathrm{grad}T)$] play an important part in solving problems of heat transfer. If the operator L is self-conjugate, then all its eigen values are real. Eigen functions corresponding to different eigen values are mutually orthogonal. If a self-conjugate operator L has a purely discrete spectrum, then it has a complete orthonormal sequence of eigen values.

The expansion of functions into a series in terms of the orthonormal sequence of eigen functions (the **Fourier Series**) is of paramount importance in solving problems of hydrodynamics and heat transfer (in analyzing computational algorithms, in particular).

V.N. Dvortsov

EINSTEIN EQUATION FOR MIXTURE VELOCITY (see Bubble flow)

EKMAN-LAYERS (see Rotating disc systems, applications; Rotating disc systems, basic phenomena)

ELASTICOVISCOUS FLUIDS (see Deborah number; Non-Newtonian fluids; Weissenberg number)

ELBOW FLOW METERS (see Differential pressure measurement)

ELECTRET (see Piezoelectricity)

ELECTRICAL COALESCERS (see Electrostatic separation)

ELECTRICAL CONDUCTIVITY

Electrical conductivity primarily refers to the capability of physical bodies to conduct electric current and is secondarily a quantitative measure of this property. Conductivity is attributed to *free-charge carriers*. The conductive property of a substance is characterised by its conductivity, σ, defined according to **Ohm's Law** by the expression $j = \sigma E$. Here, j is the electric current density and E is the strength of the electric field. The magnitude $\rho = 1/\sigma$ is called *resistivity.*

In metals, free electrons are the charge carriers. Among all metals, silver has the least resistivity ($\rho = 1.6 \times 10^{-8}$ Ohm/m at 300 K). Special high resistance alloys (used, e.g., for making electric heaters in electric furnaces) have ρ values up to 1.5×10^{-6} Ohm/m at 300 K. The dependence of ρ on temperature, T, has the form $\rho_T/\rho_{T0} = 1 + \alpha(T - T_0)$; here ρ_T, ρ_{T0} are the resistivity at T, T_0, respectively and α is the temperature resistance factor. For most metals, $\alpha = (3 \div 5) \times 10^{-3}\ K^{-1}$; for high resistance alloys, α is substantially lower (e.g., for constantan, $\alpha = 10^{-5}\ K^{-1}$).

At temperatures close to absolute zero, the conductivity of some metals abruptly falls to zero. This phenomenon is called low-temperature (classical) superconductivity and is the converse of high-temperature superconductivity, which arises at significantly higher temperatures (80 K and higher) in nonmetals of special compositions.

The conductivity of *semiconductors* is attributed to electron transition in a conduction band (electron conductivity in n-type semiconductors) or in a valance band (hole conductivity in p-type semiconductors). The intrinsic conductivity of semiconductors is attributed to the movement in opposite directions of the same quantities of electrons and holes. Practically the most significant is conductivity due to donor-type or acceptor-type additive, which provides high electron or hole conductivity, correspondingly. Semiconductor resistivity very strongly decreases with temperature: $\rho_T/\rho_{T0} = \exp(A/t)$, where A is a coefficient depending on a semiconductor's properties.

The conductivity of electrolytes (such as solutions of acids, alkalis and salts in water and other dissolvents along with molten salts) is attributed to positive and negative ions. The resistivity of electrolytes decreases with temperature (in contrast to metals): $\rho_T/\rho_{T_0} = 1 - \alpha(T - T_0)$. For most electrolytes, the values of α and ρ_{T0} depend on solution concentration.

The conductivity of gases and plasma is attributed to free electrons. The input of ions to conductivity is usually small. The two types of gas conductivity are: unself-maintained, which occurs due to gas ionization by external factors (such as X-radiation); and self-maintained, when ionization is due only to internal processes in a gas or plasma. Self-maintained conductivity of low-temperature (less then 10,000 K) plasma initially rises very strongly with temperature, then this rise significantly decelerates. To increase low-temperature plasma conductivity, alkali metals or their compounds are seeded into it.

Y.S. Svirchook

ELECTRICAL CONDUCTIVITY OF BUBBLY MIXTURES (see Bubble flow)

ELECTRICAL CONTACT METHOD (see Dropsize measurement)

ELECTRICALLY CHARGED PARTICLES (see Ionising radiation)

ELECTRICALLY DRIVEN SHOCK TUBES (see Shock tubes)

ELECTRICAL POWER GENERATION FROM GEOTHERMAL ENERGY (see Geothermal systems)

ELECTRICAL RESISTANCE STRAIN GAUGES (see Electrostatic separation)

ELECTRICAL RESISTIVITY OF PARTICLES (see Electrostatic separation)

ELECTRICAL SEPARATION (see Electrostatic separation)

ELECTRIC ARC

An electric arc is a form of a self-maintained gas discharge—i.e., a discharge which does not need an external gas ionization source for continuous burning. An electric arc burns between two electrodes: positive (anode) and negative (cathode). If an electric arc is fed from an (AC) power source with a given frequency, then the cathode and anode replace each other at the same frequency. The term "arc" is due to the fact that a sufficiently long discharge between the horizontal electrodes has an arc shape, caused by free-convective vertical gas motion. A long electric arc can be divided into three areas: a conducting column, the properties of which at some length apart from the electrodes are independent of physical phenomena near the electrodes; and two areas near the electrodes, namely, the near-anode and near-cathode areas. In near-electrode areas, a noticeable increase of electric field strength usually occurs compared with an electric arc column. Voltage drops in these areas are called cathode and anode voltage drops. Their values usually don't exceed 10 volts.

In an electric arc column, gas is heated to a high temperature and its electrical conductivity is attributed mainly to thermal ionization processes. At pressures higher then atmospheric pressure, gas in an electric arc column is usually in local thermodynamic equilibrium state.

An electric arc which burns in a large gas volume and isn't affected by external factors (e.g., by gas flow or applied magnetic field) is called a free-burning arc. Such an arc usually rapidly and randomly moves and changes its shape. In special devices, particularly in plasmotrons, it is possible to have a stationary electric arc (e.g., arc burning in a narrow, cylindrical, insulating channel) or to arrange its motion in an ordered fashion. Such electric arcs are called stabilized arcs.

The dependence of electric arc voltage on its current is called *current-voltage characteristic* (CVC). CVCs are classified into the static CVC, which is based on stationary current and voltage values and dynamic CVC's, which connect the corresponding instantaneous values.

The CVC of most direct current (DC) electric arcs is such that a current rise leads to a voltage decrease (drooping characteristic, see **Figure 1**, curve 1) or to a constant voltage (independent characteristic). Thus, an electric arc doesn't follow **Ohm's Law** and represents a nonlinear element of an electric circuit. To keep a stable electric arc burning, an additional resistor is connected in series with an arc to increase a power source's own CVC slope (see **Figure 1**: curve 2 is a CVC of a power source without resistor; curve 3 is a CVC of a power source with resistor). Point A corresponds to unstable electric arc burning because with an occasional increase of current I_a on a magnitude of ΔI, a positive potential difference, ΔV, arises which causes further current increases until point B is reached. This corresponds to stable arc burning at current I_b. An additional resistor substantially decreases the energy efficiency of an electric arc device. To avoid this disadvantage, special power sources are sometimes used. Certain stabilized electric arcs have rising CVCs; in this case, it is possible to substantially decrease resistor magnitude or to entirely remove it from a feeding circuit.

For alternating current (AC) electric arcs, current-time dependence during each half-period is near sinusoidal; voltage-time dependence usually has a near-rectangular shape, with characteristic sharp voltage peak at the point of origin (so-called ignition peak). A dynamic AC CVC has a loop shape that indicates a hysteresis phenomenon caused by thermal inertia of the electric arc column. A CVC, plotted by the effective values of current and voltage, has the same shape as a DC arc under the same conditions. That is why for stable AC arc burning, an induction coil is connected to a circuit in series with arc (more seldom a resistor is used). An advantage of an induction coil over a resistor is that the coil has a low resistance and consequently doesn't influence an electric arc device's efficiency. On the other hand, this leads to a significant decrease in power factor.

An electric arc is a powerful, highly-concentrated source of heat and light. These electric arc properties determine the main areas of its application. Electric arcs are widely used in various

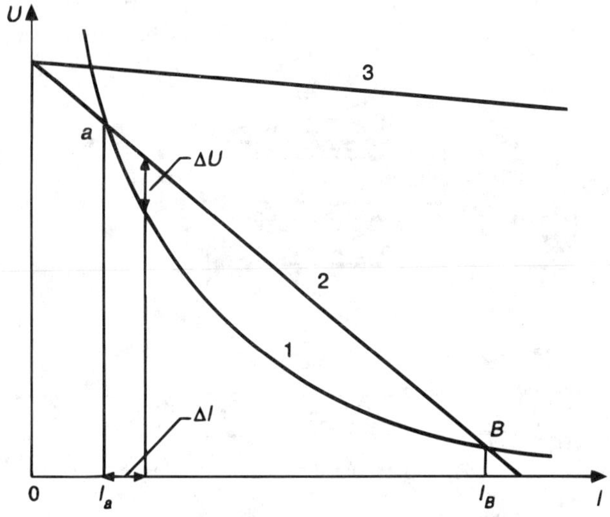

Figure 1. Current - voltage characteristics for electric arcs (1 - "drooping" characteristic, 2 - CVC for power source without resistor, 3- CVC with resistor).

welding devices, in steel-melting arc furnaces and in plasmotrons. Arc light sources are used in various lighting devices (e.g., in floodlights). In cinematographic projection equipment, high-pressure *xenon arc* lamps are used. The light spectrum of a xenon electric arc is close to sunlight, which is why such lamps provide "white" light and correct color transmission.

Leading to: *Electric arc heater*

Y.S. Svirchook

ELECTRIC ARC HEATER

Following from: *Electric arc; Plasma arc furnace*

An electric arc heater (*plasmotron*) is a low-temperature plasma generator in which an arc discharge is used as a heat release element. An electric arc is a self-maintained gas discharge characterized by:

- low cathode voltage drop (less then 20 V);
- high current density (10^2 A/cm^2 in the arc column and 10^3–10^7 A/cm^2 in near-electrode zones);
- temperature range of 5–50 *kK* and ionization degree of 1–100%.

In a plasmotron, such gases as air, inert gases, water vapor, natural gas, can be used as a working fluid. Heating of the gas is accomplished by conductive, radiative and convective heat exchange between the arc and a gas. The electric arc is practically the only method of stationary heating of any gas to a temperature over 4 *kK*.

Depending on the purpose, various types of plasmotrons have been developed. These include linear (**Figure 1a,b**), coaxial; combined, multiarc and other types with both direct current (*DC*) and alternating current (*AC*). Plasmotron powers range from hundreds of W to tens of MW; arc currents, from several A to several kA; mass flow rates, from fractions of g/s to several kg/s; pressures, from 0.1 to 20 MPa. Plasmotrons show high arc-burning stability,

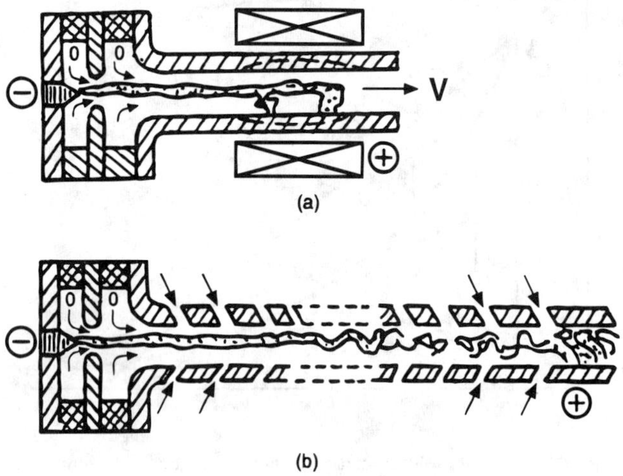

Figure 1. Linear (a) and co-axial (b) plasmatrons.

are relatively small and the process of gas heating can be easily and automatically controlled. This is because an electric arc is a low inertia Ohmic heater.

The main characteristics and parameters of a plasmotron are as follows: current-voltage arc characteristic (CVC); thermal efficiency; specific erosion of electrodes, which determines a plasmotron's lifespan; and heat flux. One of the main physical processes in a plasmotron is the by-passing phenomenon, i.e., unwanted electrical arc-wall and arc-arc (in a loop) contacts (breakdowns). Large-scale by-passing between an arc, placed along a channel axes, and a wall determines arc mean length and the maximum achievable electric power in the plasmotron of the type depicted in **Figure 1a**, sometimes called plasmotrons with self-aligning arc length. Large-scale by-passing also determines arc CVC, frequency of voltage pulsations ($\sim 10^4$ Hz), etc. Small-scale by-passing in near-electrode zones affects the erosion rate and the dynamics of arc movement on electrode surfaces.

The main elements of a plasmotron are an anode, a cathode, a gas injection chamber and an arc channel. As a rule, an anode works in a form of water-cooled, hollow metal cylinder, which is part of the arc chamber. An arc moves into an inner surface of the chamber due to electrodynamic and aerodynamic forces (**Figure 1a,b**). Specific erosion of a copper cathode in an air medium for an arc current range of between 100–1,000 A, is equal to $10^{-10} - 10^{-9}$ kg/s; in an argon, it is less by 2–3 orders of magnitude.

Two cathode types are mainly used, namely, the bar and the tube. Bar cathodes fabricated from refractory materials with high emission (W, C, etc.) are successfully used in the case of an inert gas (argon, helium, nitrogen, etc.) for an arc current range of 5–10 kA. A tungsten bar cathode, flush-pressed into a copper, water-cooled casing, effectively operates in an inert media at $I \sim 1$ kA, $p = 0.1$ MPa and shows a minimal specific erosion of 10^{-13} kg/c. This is due to atom recirculation, that is, the return of cathode vaporized atoms to the cathode surface in the form of ions. This phenomenon reveals the wide opportunities for low-erosion cathode development.

In nitrogen or oxidizing media with current values ≤ 200 A, it is worthwhile to use cathodes made of hafnium or zirconium and flush-pressed into a casing. Tube cathodes of copper, bronze, cast iron, etc. operate well in oxidizing and other media at the current range of 100–1,000 A and at a specific erosion of $\sim 10^{-9}$ kg/s. A method of decreasing erosion in a high-current tube cathode (e.g., tungsten cathode in argon medium) is the use of an auxiliary low-current arc, which provides preliminary generation of charged particles in a near-cathode zone. The specific erosion of such cathodes at the 3–8 kA current range doesn't exceed $(2–5) \times 10^{-12}$ kg/c.

The electric field strength of an arc in a laminar gas flow is $E \leq 10$ V/cm; in a turbulent flow, the strength is several times higher. That is why for the development of powerful linear plasmotrons it is worthwhile to use a turbulent arc. These plasmotrons require protection of the arc chamber walls from high heat fluxes. One of the effective means of doing this is by cold gas injection through the wall, which substantially decreases heat fluxes into the wall.

In the power balance of a plasmotron, the part determined by plasma radiation is appreciable. The spectrum of radiation determined by plasma temperature ($10^3 - 10^5$ K) corresponds mainly to the optical wavelength range. It consists of lines, a continuum and molecular bands.

At high arc currents and gas pressures, heat losses to the arc chamber walls—attributed to arc radiation—can be comparable with convective heat losses or can even be higher. One way to decrease radiative heat losses is to place a porous interelectrode insert and to inject plasma-forming gas through it. In a porous material, a regeneration of radiative and convective heat losses take place and these are returned to the main flow.

Another approach is based on arc turbulization by gas flow or magnetic field. In this case, an appreciable arc radiation decrease occurs.

In experimental plasmotron development and in associated theoretical investigations generalization of experimental data based on similarity criteria has a special importance. These criteria include both those well known from classical gas dynamics (**Mach**, **Reynolds**, **Prandtl numbers**, etc.) and also criteria peculiar to electric arc plasmotrons. Among the latter, the most significant criteria are the electric field strength criteria $\sigma ED^2/I$ or $Su = \sigma VD/I$; the power criterion which shows a relation between *Joule heat* release and flow heat power, $S_I = I^2/(\sigma \rho vH\ D^3)$; the magnetic interaction criterion $S_B = IB/(\rho v^2 D)$, which is important in the investigation of plasmotrons with magnetic control of an arc; and the criterion $S_r = 4\pi\varepsilon D^2/(\mu H)$ shows the input of radiation power to the process of heat transfer in a plasma. Based on similarity criteria, the use of semiempirical methods for calculating plasmotron electrical and heat characteristics has been developed.

Plasmotrons are used for direct reduction of metals from ores, for metal cutting, for workpiece surface strengthening, for overlaying of corrosion-resistant, refractory and another coatings, for a synthesis of new materials, for the destruction and utilization of toxic wastes, as test facilites for aerospace purposes, etc.

M.F. Zhukov

ELECTRIC CONTACT METHOD, FOR FILM THICKNESS MEASUREMENT (see Film thickness measurement)

ELECTRIC FURNACES (see Electric (Joule) heaters)

ELECTRICITÉ DE FRANCE, EDF

2 rue Louis Murat
75384 Paris cedex 08
France
Tel: 1 40 42 22 22

ELECTRICITY ASSOCIATION, EA

30 Millbank
London SW1P 4RD
UK
Tel: 0171 344 5700

ELECTRIC (JOULE) HEATERS

***Following from:** Joule heating*

An electric (Joule) heater is a device in which heat release in an electrical conductor due to electric current flow (**Joule Heating**) is used for heating solid, liquid and gaseous substances. It is more convenient to call such devices *electric furnaces*, in which electrical heaters are the heat release units. Electric furnaces have the following main advantages compared with burner furnaces:

1) The possibility of concentrating large releases of heat energy in a small volume.

2) The possibility of providing high heating uniformity, or given nonuniformity, by appropriate arrangement of heat sources (heaters).

3) The relative simplicity of electric power control (consequently, temperature) and the automation of the electric furnace's temperature regime.

4) The convenience of mechanization and automatization of product or materials charging/discharging processes that substantially facilitate the inclusion of electric furnaces in processing assemblies.

5) Electric furnaces can be easily sealed, preventing the oxidization of heated products or materials by using a protective atmosphere or vacuum.

6) The ecological purity of an electric furnace.

The main fields of electric furnaces application are in metallurgy, machine-building, chemical, food, structural industries, laboratory furnaces and domestic electrical heated devices.

Depending on the heating method, electric furnaces are divided into two types: indirect heating furnaces, in which heat is released from a heater and then transferred to a heated body by convection and/or thermal conduction, and direct heating furnaces, where a body is directly connected in an electrical circuit and is heated by current flowing through it. Most electric furnaces are of the first type. In the case of convective heating, heat transfer occurs by gas, liquid substance (liquid metal, molten salt, etc.) or by fluidized bed (heat from the heater is transferred to a processed product by numerous fine hard bodies, which are in continuous motion and in contact with the heater and the heated product). Indirect electric furnaces are divided into two types; namely, electric furnaces for thermal processing of products and materials and electric furnaces for melting.

Electric furnaces are also classified according to:

1) Operating time, either periodic or continuous (methodical) electric furnaces;

2) Operating temperature. Under this classification are low-temperature electric furnaces (less than 1,000 K) in which a major part of the heat is transferred to a product by convection; moderate-temperature electric furnaces (up to 1,700 K) where heat transfer is predominantly by radiation; high-temperature electric furnaces (more than 1,700 K) which include furnaces that can not operate with metal heaters without a protective atmosphere or vacuum).

3) Atmospheric volume during operation. Electric furnaces may be furnaces with oxidizing medium (usually air), with controlled atmospheres or with vacuum.

The presence in electric furnaces of zones with high temperatures requires, on the one hand, materials which can withstand such temperatures and on the other, materials which provide heat insulation for the zones from both the other parts of the electric furnace and the ambient medium. As applied in electric furnaces, such materials are divided into the following types: 1) the refractory; 2) heat-insulating; 3) heat resistant; and 4) materials for heaters.

Refractory materials

Refractoriness is the property of a material which enables it to withstand high temperatures. It is determined by comparing the deformation of a material undergone after heating with a special standard geometry test specimen and is measured in K. Besides refractoriness, materials must have sufficient thermal stability, i.e., the capability to sustain without failure a sufficient number of sharp thermal changes.

Refractory materials must be durable at temperatures near the heating source, be sufficiently strong and have a low thermal conductivity. The last requirement often is not provided, that is why thermal protection (so-called lining) of moderate- and high-temperature electric furnaces consists of at least two layers: refractory and thermal protection. The second layer is free from mechanical loading, but limits thermal losses to an acceptable level. In case of excessive mechanical loading of a refractory material, support mountings made from heat-resistant steel are provided.

Refractory materials are based on the application of the following oxides: SiO_2 (silica, refractoriness up to 2,000 K); Al_2O_3 (alumina, up to 2,100 K); MgO (magnesia, up to 3,100 K). The most widely used in electric furnaces are refractory products made of chamotte (formed from clays with large amounts of alumina). This material is sufficiently heat-resistant, has a relatively small linear thermal expansion coefficient and can sustain sharp temperature fluctuations. The critical operating temperature of chamotte is 1,700 K.

Another widely-used refractory material is dinas (made of quartzite). The distinct feature of dinas is its mechanical strength at high temperatures. The disadvantage is its inclination to crack during sharp temperature changes. The critical operating temperature of dinas is 1,900 K.

Magnesia-based refractory materials have good heat-resisting qualities but do not have sufficient strength and thermal stability at high temperatures.

Besides the above materials, refractory products made of zirconium dioxide (ZrO_2), zircon ($ZrO_2 + SiO_2$), carborundum (SiC), graphite, etc. are also used in electric furnaces.

In low- and moderate-temperature electric furnaces, porous refractory products are often used, particularly light chamotte and foam chamotte. They have less refractoriness and mechanical strength, but have much less thermal conductivity compared with the usual chamotte; therefore, they can be simultaneously used as refractory and heat insulating materials.

Heat-insulating materials

One of the most widely-used heat-insulating material is diatomite. It is composed of dried-up remains of very fine aquatic plants (diatom), consisting mainly of SiO_2 and pierced through by extremely fine pores which cause low density and thermal conduction. The maximal operating temperature of diatomite is 1,200 K. It is used in the form of bricks or as a fill.

Asbestos and derivative products are also good heat insulators. Its maximal operating temperature is up to 800 K. Asbestos, however, has a high hygroscopicity and presents potential health hazards. Heat-insulating wools (glass, mineral, etc.) also enjoy wide usage.

Lining design and the accuracy of its performance have an influence on heat loss magnitude, electric energy consumption, run-up time, life span, weight, overall dimensions and cost of an electric furnace. A lining is characterized by a value of accumulated heat, Q, during electric furnace run-up to a stationary regime, and a value of heat loss through the lining to an ambient air, W. A choice of lining (materials and thickness) is made by comparing these two values over a continuous operating time, t. One of the possible criteria for the optimal choice of a lining is a minimal value of $Q/t + W$. Lining thermal calculation is performed taking into account the temperature dependence of thermal conductivity.

Heat-resistant materials

These materials must meet the following requirements: 1) Sufficient heat resistance, i.e., an ability to endure prolonged operation at high temperatures without substantial surface oxidizing; 2) Sufficient high-temperature strength, i.e., an ability to maintain a mechanical strength at high temperatures; 3) Sufficient high-creep limit, i.e., small residual deformations under mechanical loading at high temperatures; 4) Characteristic stability of heaters during prolonged operations; 5) Adaptability to machining and welding for the production of various parts and devices.

The main heat-resistant materials are metals, in particular steel because they are adaptable and have sufficient mechanical strength and a relatively low creep. Chrome-nickel steel is the most widely used in electric furnaces.

Materials for heaters

Materials for heaters must have the following qualities:

1) Heat-resistance;

2) Sufficient high-temperature strength, providing an absence of heater deformation under its own weight during operation;

3) High electric resistivity. If this is not the case, the length of heaters is relatively large and difficulties of its placement in electric furnace appear;

4) Low temperature coefficient of electric resistivity. For practically all heater materials, this coefficient is positive; therefore during electric furnace run-up, the consumed electric current can exceed its rated value by more than 10 times. This fact has to be taken into account when designing an electric furnace power system. In some cases during the electric furnace run-up period, it is necessary to lower the voltage;

5) Characteristic stability of heaters during operations. For some materials, electrical resistivity increases over time, therefore the consumed power decreases. If it is necessary to sustain a constant power then a controlled step-up transformer is used;

6) Constant size. Some heaters have high thermal expansion and this leads to structural and operational inconveniences;

7) Adaptability to heater fabrication processes. It is necessary to make metal wires and strips for heaters. It is also desirable to have a knowledge of their welding. For nonmetal materials, it is

necessary to mould and press them to have heaters of a required shape.

Materials for heaters, which satisfy the above requirements to a great extent are nickel-chrome alloys (nichromes). The heat-resistance of nichromes is due to their very strong films of chrome oxide Cr_2O_3, which has a higher melting temperature than the alloy and endures well cyclical thermal loads. Nichromes can withstand operational temperatures reaching 1,500 K. Ferrochromenickel and ferrochromealuminium alloys are cheaper then nichromes but have operational disadvantages, such as brittleness, low mechanical strength, etc. In high-temperature electric furnaces, heaters made of molybdenum, tungsten, tantalum and niobium are used, but due to their intense oxidation in air they require a protective atmosphere.

Among nonmetallic materials for heaters, the most widely-used is silit. It is fabricated from carborundum (silicide of carbone, SiC). The heaters are made in the form of rods and tubes. Their operating temperature reaches 1,800 K. For operations in an oxidizing atmosphere with temperature of up to 2,000 K, heaters made of disilicide of molybdenum are used. In electric furnaces with a neutral atmosphere or a vacuum, heaters made of carbides of refractory metals (zirconium, niobium, tantal, hafnium) are used. The most refractory among them is hafnium carbide (T_{melt} = 4160 K).

Metal heaters are usually fabricated in the form of wire spirals and zigzags (made from wire or strip). Also in abundance are closed-tube heaters. Such heaters are made in a form of a tube of high-temperature steel, with the heaters (usually of spiral form) located at tube axes and having hermetically-sealed leads. The space between the spirals and the tube wall is filled by materials with high thermal conductivity and good electric insulation. The main advantage of a tube heater is the absence of electric potential on the tube surface, which ensures safe operation where the electric feed is from a standard electric network, for example in domestic heating devices. Nonmetal heaters are usually made in the form of rods and tubes.

Vacuum electric furnaces

Use of a vacuum instead of an inert medium is often more economic. Moreover, many technological processes and scientific researches can be conducted only in a vacuum. The main disadvantage of vacuum electric furnaces is their high cost (including vacuum equipment). Heaters for vacuum electric furnaces are usually made of the same materials as those for electric furnaces. In the case of electric furnaces with a relatively rough vacuum, it is necessary to take into account a decrease in electrical strength of the residual gas, according to the Paschenig law.

V. Svirchook

ELECTRIC POWER RESEARCH INSTITUTE, EPRI

PO Box 10412 Palo Alto
CA 94303
USA
Tel: 415 855 2000

ELECTROCATALYSTS (see Electrolysis)

ELECTROCHEMICAL CELLS

Following from: *Electrolysis*

When two electronically-conducting materials (*electrodes*) are placed in a solution containing ions (*electrolyte*), an electrochemical cell is formed and the potential for *redox reactions* to occur at the interfaces between the electrodes and electrolyte exists. Redox reactions are those reactions which involve a change in oxidation state, that is a gain or loss of electrons. If reactions are allowed to occur by electrically connecting the electrodes outside the solution, current will flow—as electrons in the electrodes (and external connections) and as ions in the solution. At one electrode, the *anode,* electrons are accepted and oxidation takes place and at the other electrode, the *cathode,* electrons are discharged and reduction takes place. In the absence of current flow, the "potential" for reactions to occur can be measured as a voltage. This voltage, called the equilibrium *cell potential,* is related to the overall **Free Energy**, ΔG, of the cell reactions made up of the sum of the reactions at both electrodes. It is expressed as

$$\Delta G = -nFE_o$$

where n is the number of electrons transferred and F is a constant (**Faraday's Constant**).

In any electrochemical cell, the amount of reduction at the cathode is equal to the amount of oxidation at the anode. This is regardless of whether the cell is generating electrical energy, as for example in a battery or **Fuel Cell**, or is consuming electrical energy in the generation of specific chemicals (as an *electrolytic cell*). These two types of cells are distinguished by the polarity of the electrodes; for example, in the case of a battery the anode is at negative potential relative to the cathode and vice-versa for an electrolytic cell. It is common usage to define electrochemical cells as the summation of two *half-cells,* that is the two halves of the cell complete with electrolyte and electrode. Thus, the equilibrium potential is the sum of two half-cell potentials.

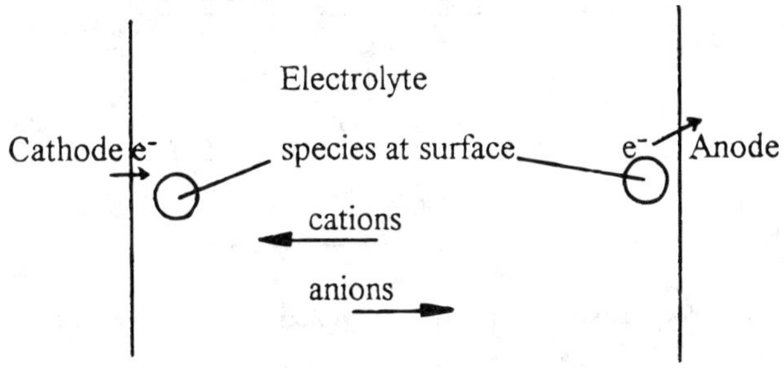

Figure 1. Oxidation and reduction in an electrochemical cell.

Leading to: *Electroplating; Fuel cells*

K. Scott

ELECTROCHEMICAL METHODS (see Shear stress measurement)

ELECTROCHEMICAL THEORY OF CORROSION (see Corrosion)

ELECTROCHEMISTRY (see Faraday's laws)

ELECTRODE (see Electrochemical cells)

ELECTRODEPOSITION (see Electroplating)

ELECTRODIFFUSION METHOD (see Velocity measurement)

ELECTROFLOTATION, EF (see Flotation)

ELECTROHYDRODYNAMIC AUGMENTATION (see Augmentation of heat transfer, two phase; Tube banks, condensation heat transfer in)

ELECTRODIALYSIS

Electrodialysis is a process for the separation of electrolyte from a solvent, typically water. The process is widely used in the **Desalination** of water and process solutions. It uses a direct electrical current to transport ions through sheets of ion-exchanger membranes and is operated in a unit with at least three compartments, as shown in **Figure 1**. The terminal compartments house an anode and a cathode, between which a potential difference is applied to drive the ions through the electrolyte solutions and the membranes.

Two types of membranes are used: one which is preferentially permeable to the transport of anions (anion-selective), and one which is preferentially permeable to cations (cation-selective). Membranes are arranged alternately between the electrodes, forming individual compartments (or cells). The solution to be desalinated is held in one compartment and during current flow, anions move through the anion exchange membrane in the direction of the anode into adjacent compartment while cations move in the opposite direction into an adjacent compartment on the other side. Thus, overall, the solution becomes depleted in ions in one compartment and solutions in adjacent compartments become enriched in ions. In practice, solutions flow through the compartment to allow continuous operation and several hundred cell pairs (one concentrated and one diluted solution) are used.

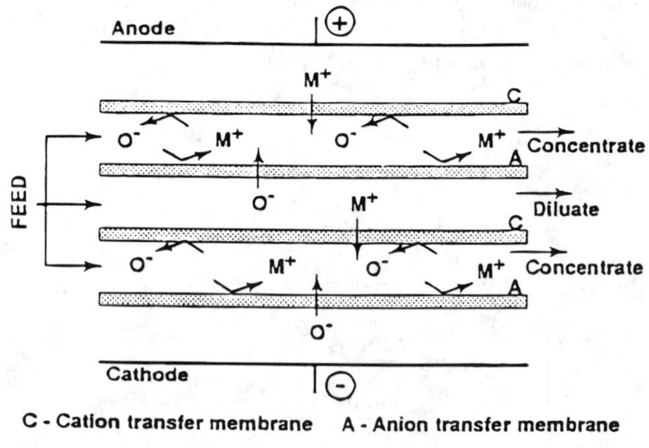

Figure 1. An electrodialysis unit showing alternate anion and cation exchange membranes.

The membranes generally used in electrodialysis are copolymers of styrene and divynylbenzene. Ion exchange characteristics are typically introduced by sulphonation in the case of cation-exchange, and by substitution with quaternary ammonium groups in the case of anion-selective materials. The degree of cross-linking of polymers determines the amount of water absorbed into the membrane after activation, creating channels through which ions can diffuse. The fixed charge groups—essentially electrostatically—repel ions of the same charge (e.g., anion in the case of sulphonate) and thus impart the appropriate ion-selective exchange characteristic.

Transport of ions through membranes is defined in terms of the transport number for a particular ion, which can be as high as 0.98 for some anion exchange membranes and is greater than 0.9 for cation-exchange membranes. A transport number less than 1 means that a proportion of the current is carried by ions of the opposite charge in the wrong direction. This behaviour controls the current efficiency of the process, which ideally should be 100%, i.e., 1 F of electricity allows the transport of 1 mol of salt in each cell pair.

Leading to: *Desalination; Membrane processes*

K. Scott

ELECTROLYSIS

Electrolysis is the process which occurs in an electrochemical cell when electrons pass from the anode to the cathode via an external circuit connecting the electrodes. For this to occur, there must be a mechanism for charge transport in the electrolyte between the electrode. This charge transport is due to the movement of *ions* in the electrolyte. These ions, on approaching the electrodes, undergo electrolysis and either accept electrons from the cathode or release electrons to the anode. In this way chemical change occurs at both electrodes.

The extent of chemical change depends on the charge passed, according to **Faraday's Laws** of electrolysis. The rate of chemical change depends upon current density, $j = I/S$ (Am^{-2}), at the electrode and is given by:

$$\text{Rate}(\text{mol m}^{-2}\,\text{s}^{-1}) = I/nFS \qquad \textbf{(1)}$$

where n is the number of electrons in the electrode reaction and F is the *Faraday constant.* The rate of chemical change and thus current density is determined by the kinetics of the electrode reaction(s).

Under conditions of electrolysis, the cell is operating away from its equilibrium (reversible) potentials determined from **Thermodynamics**. Certain electrode reactions are very fast and depart very little from the equilibrium potential. Such reactions are frequently referred to as reversible (see **Figure 1**). Other electrode reactions are inherently slow and require a potential, E, significantly greater in magnitude than the equilibrium potential to achieve a reasonable current density. This potential is called the *overpotential,* η ($=E-E_e$), and the electrode is the said to be *polarised.* Such reactions are referred to as irreversible (**Figure 2**). As overpotential is increased in magnitude (more negative for cathodic processes, more positive for anodic processes) current density increases, typically exponentially at high overpotentials. The relationship between current density and electrode potential is the subject of electrode kinetics.

There are a wide variety of electrochemical cells in practice: batteries where electrical energy is produced from the electrode reactions, electrolytic cells where chemical change is derived from an applied potential and fuel cells where electrical energy is produced continuously by the supply of a fuel. Electrolysis can involve different types of processes including reactions in the liquid phase, reactions in the solid phase, located on the surface of the electrodes, and reactions involving the gas phase. An important example of gas phase processes is the electrolytic generation of gas, e.g.,

$$2Cl^- - 2e^- \rightarrow Cl_2 \quad (2)$$

$$2H_2O - 4e^- \rightarrow O_2 + 4H^+ \quad (3)$$

$$2H_2O + 2e^- \rightarrow H_2 + 2OH^-. \quad (4)$$

The anodic production of chlorine and the cathodic formation of H_2, and thus hydroxide ion, is the basis for the *chlor-alkali industry* utilising a sodium chloride electrolyte. For this process, the anode and cathode reactions in the cells are separated by a diaphragm or **Membrane** to prevent chlorine gas reacting with the hydroxide (of sodium). This technique of separating the anode and cathode reactions during electrolysis is common. The appropriate separator, by necessity, must allow transport of suitable ions through its structure. In the electrolysis of sodium chloride for chlorine production, the separator should ideally enable transport of sodium ion alone, as shown in **Figure 2**, to form sodium hydroxide with hydroxide ion generated at the cathode (reaction 3).

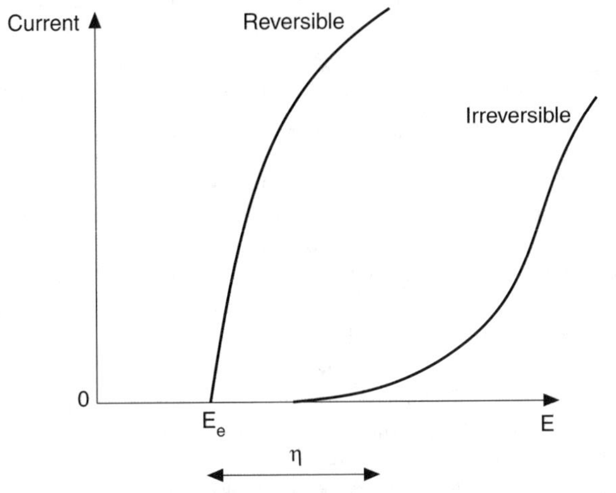

Figure 1. Current potential curve at a polarised electrode.

In all cells, electrolysis involves reactions at both electrodes and the overall movement of ions in the solutions maintains a neutrality of charge in the electrolyte. Generally therefore, electrolysis processes are written as overall reactions of two-electrode processes, e.g., in *chlor-alkali electrolysis* the reaction

$$2NaCl + 2H_2O \rightarrow 2NaOH + Cl_2 + H_2 \quad (5)$$

represents the electrolysis of sodium chloride in water to give sodium hydroxide, chlorine gas and hydrogen.

For electrolysis to occur, a source of energy is required to move ions in electrolytes and to overcome overpotentials at the electrodes. This energy is either supplied by an external power source, as in

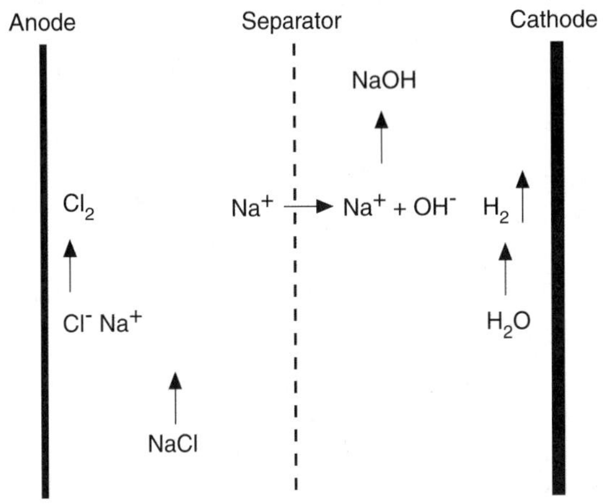

Figure 2. Electrolysis of NaCl solution to form chlorine and sodium hydroxide. Other examples of electrolysis, or more specifically "half-cell reactions," include:

- Simple electron transfer, e.g., anodic oxidation of Ce (III) ions

$$Ce^{3+} \rightarrow Ce^{4+} + e^-.$$

- Metal deposition, e.g., nickel plating

$$Ni^{2+} + 2e^- \rightarrow Ni$$

- Surface film transformation, e.g., in lead acid batteries

$$PbO_2 + 4H^+ + SO_4^{2+} + 2e^- \rightarrow PbSO_4 + 2H_2O$$

- Anodic dissolution, e.g., of iron

$$Fe \rightarrow Fe^{2+} + 2e^-$$

- Gas reduction, e.g., oxygen in porous gas diffusion electrodes used in fuel cells

$$O_2 + 4H^+ + 4e^- \rightarrow 2H_2O$$

electrolytic processes (e.g., recharging of lead/acid batteries) or is obtained from available "chemical" or free energy of the reaction (e.g., discharging of batteries). It is essentially transformed into heat in the electrolyte and is directly related to the internal resistance of the cell R and the applied current, i.e., I^2R (Joules s^{-1}).

In electrolysis, this energy requirement can be reduced by factors which lower internal resistance (small distance between the electrode, high-electrolyte conductivity) and which reduce overpotential, e.g., by the use of *electrocatalysts* or alternative electrode materials. Electrocatalysts provide alternative reaction pathways for the inherently kinetic, slow step of the electrode process and enable reaction at higher current densities closer to equilibrium potential. Many electrolyses have benefited from the use of electrocatalysts. For example, in chlorine cells anode design has been revolutionised in the 1960s by the use of coated titanium electrodes (so-called *dimensional stable anodes* or DSAs) as replacements to carbon. The coating, based on ruthenium oxide, valve metals, precious metals and transition metals, gave significant reductions in overpotential (< 50 mV for chlorine generation). Such materials are widely used for many commercial electrolyses.

***Leading to:** Electrochemical cells; Faraday's laws*

K. Scott

ELECTROLYTE (see Electrochemical cells)

ELECTROLYTE FLOW MEASUREMENT (see Electromagnetic flowmeters)

ELECTROLYTE SOLUTION, DIFFUSION IN (see Diffusion)

ELECTROLYTE SOLUTIONS, ADSORPTION FROM (see Adsorption)

ELECTROLYTIC CELL (see Electrochemical cell)

ELECTROLYTIC REACTIONS (see Faraday's laws)

ELECTROMAGNETIC FLOWMETERS

***Following from:** Flowmetering*

The electromagnetic flowmeter is one of the most successful nonintrusive types of flowmeter. The basic concept that an emf is induced in a conducting liquid moving through a magnetic field, in accordance with **Faraday's Law** of electromagnetic induction, has been known since the early part of the last century. However, it was not until the 1950s that industrial applications became a reality. Some of the earliest applications were concerned with measuring the *flow rate of blood,* Kolin (1960).

The primary device consists of an electrically-insulating metering tube, typically within a nonmagnetic stainless-steel tube. The choice of liner material depends mainly on the liquid, which can range from slurries or molasses to chemically-aggressive mixtures or molten metals. *Conductivity thresholds* as low as 5 micro-Siemen/cm are common, with some manufacturers claiming considerably lower values. When measuring the flow rate of an *electrolyte,* a further problem is that because of the relatively low ion mobility, conductivity is likely to be affected by the flow, making it a function of velocity.

One or more pairs of electrodes, diametrically-opposed, are set into the wall such that the diameter on which they lie, the magnetic field and the flow direction are mutually orthogonal. If the typical electrodes, e.g., dome-headed screws, are replaced by relatively large plates, the larger measuring areas can sometimes reduce the effect of nonstandard flow patterns on the meter. Such electrodes, however, are more prone to fouling, introducing additional errors. The secondary device processes the signal from the electrodes, typically millivolts. It provides outputs which can be in volts, milliamps or pulses per second. Usually, a reference signal from the primary device—which is proportional to the magnetic flux—is compared in the secondary device with the flow signal.

The use of direct current as a means of exciting the magnet has been quickly discarded because of the difficulty of electrochemical and polarisation emfs appearing at the electrodes, and for many years excitation by alternating current was the norm. However, a 'transformer' effect can exist since the conducting liquid, the electrodes and leads effectively form a single-turn loop, giving rise to a quadrature signal which can be of the same order of magnitude as the metering signal. It should be possible to remove the quadrature signal but complications arise when, for example, liquid conductivity is not uniform or earthing difficulties exist. Since the mid-1970s most manufacturers have used square or trapezoidal-wave excitation, operating usually at a few Hertz and minimising most of the above difficulties. An additional benefit is that the zero is continually reestablished, avoiding the earlier problems associated with zero drift [Baker (1982)]. The impedance of the liquid path should ideally be at least two orders of magnitude less than that of the input of the secondary device. The situation has improved since the days of the voltameter thermionic valve and germanium transistor. Modern electronic components make it possible for input impedance of the secondary device to be several orders of magnitude greater than that of its predecessors. One of the most recent developments is to position the electrodes outside a dielectric liner, the system operating by capacitive coupling, with no contacts within the metering tube.

Theoretically, a uniform magnetic field can only be achieved by using an infinite magnet. This means that, for practical purposes, certain regions of the cross-section containing the electrodes will produce larger signals than others at the same velocity. Some of the earliest recorded experimental work was that by Williams (1930). He realised that since the velocity profile of a flowing liquid was rarely flat, the emf existing across the pipe diameter deriving from the central velocity would be effectively shunted by those from the slower moving flows near the pipe wall. This would give rise to circulating currents, and the measured signal would be lessened by an Ohmic drop. As it turns out, as Shercliff (1962) explains, because the induced currents flow only in the plane normal to the meter axis, the above effect is self-cancelling if the velocity distribution is axisymmetric. Shercliff (1962) has developed the very important 'weight-function' map, **Figure 1**, which assisted in the prediction of the effects of asymmetry, especially relevant in short-pattern meters. Hartmann (1937) has pointed out that a similar 'shorting' takes place at the ends of the necessarily finite meter, where the magnetic intensity falls to zero. This effect has been minimised by the designs of the pole pieces.

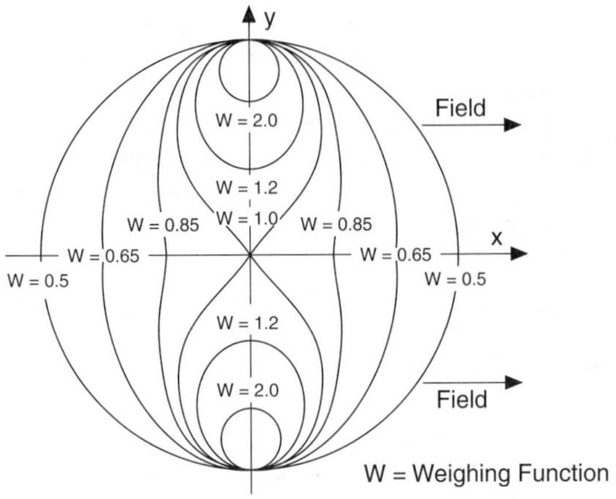

Figure 1. Electromagnetic flow meter.

Installation requirements are given in ISO 6817 (1992) and BS 5792 Part 1 (1993) Standards. The meter should be installed in a straight pipe at a distance of at least 10 times the nominal diameter (10DN) from any upstream disturbance and 5 DN downstream. The flow must be swirl-free and the meter should not be larger or more than 3% smaller than the connecting pipework. To prevent the formation of gas bubbles on the electrodes, these should normally be installed horizontally. In ideal conditions, such meters should be capable of measuring flow rate with an uncertainty of 0.2%.

References

Baker, R. C. (1982) *Electromagnetic flowmeters.* in: Scott, R. W. W. (Ed). London and New Jersey; Applied Science Publishers Ltd.

British Standards Institution (1993) *Measurement of conductive liquid flow in closed conduits - Method Using Electromagnetic Flowmeters.* BS 5792, Part 1, London.

Hartmann, J. (1937) Hg-dynamics 1, *Math-Fys.* Medd. 15. No. 6. (Royal Danish Academy of Science and Letters).

International Organisation for Standardization. (1992) *Measurement of conductive liquid flow in closed conduits - method Using electromagnetic flowmeters.* ISO 6817.

Kolin, A. (1960) Circulatory system; methods; Blood flow determination by the electromagnetic method. in: Glasser, O. (Ed). *Medical Physics,* 3: 141. Chicago I11: Year Book Medical Publishers.

Shercliff, J. A. (1962) *The Theory of Electromagnetic Flow Measurement.* Cambridge University Press.

Williams, E. J. (1930) The Induction of emfs in a moving fluid by a magnetic field and its application to an investigation of the flow of liquids. *Proc. Phys. Soc.,* London, 42: 466.

J.S. Humphreys

ELECTROMAGNETIC RADIATION (see Bremsstrahlung; Ionizing radiation)

ELECTROMAGNETIC SPECTRUM

Following from: *Electromagnetic waves*

Electromagnetic Waves occur with frequencies ranging from 10^4 to 10^{24} Hz and have wavelengths from 10^{-16} to 10^4 metres. The common feature of all types of electromagnetic waves is that their speed in a vacuum is 3.00×10^8 m/s.

Radio waves make up the long wavelength, low-frequency end of the electromagnetic spectrum. The longest wavelengths (up to 10 km) are used for underwater communication (e.g., submarines). Long-wave radio stations use 1–2 km wavelengths (150–300 kHz) while medium-wave broadcasts use frequencies of up to 1.5 MHz (200 m wavelength). Short-wave radio frequencies continue up to 11–12 MHz (wavelengths down to a few metres). FM radio and TV transmissions use frequencies in the range of 60–900 MHz (wavelengths down to 300 cm).

Radar and *Microwaves* use the frequency range 1–20 GHz (wavelengths down to 1 cm).

Infrared radiation covers the wavelength region between 1 μ and 1 mm (frequencies between 10^{11} and $\sim 3 \times 10^{14}$ Hz).

Visible light occupies a small, but important, part of the electromagnetic spectrum. Frequencies between 4 and 8×10^{14} Hz and wavelengths decreasing from 7 to 4×10^{-7} m, correspond to light with colors ranging from red, orange, yellow, green, blue and indigo light. Visible light, together with infrared and ultraviolet radiation, is emitted following electron transitions in atoms or molecules. Such transitions may be stimulated by heat or by the application of an electric current.

Ultra-violet rays have higher frequencies, from a few $\times\ 10^{16}$ Hz, and shorter wavelengths, down to 10^{-8} m, than visible light.

X-rays are emitted following electron transitions between the K, L, M and N atomic electron shells. X-ray wavelengths cover the range 10^{-8} to 10^{-11} m, with frequencies from 10^{16} to 10^{19} Hz. K X-rays are those produced by electron transitions to the K-shell; L X-rays are those produced by electron transitions to the L-shell and so on. Since these electron transitions involve the release of well-defined and precise quantities of energy, it is common to characterise X-rays by their energies and the normal units used are electron-volts. X-ray energies cover the range from a few eV to 100 keV.

Gamma rays are emitted by nuclear transitions and are usually emitted at higher energies than X-rays, with a range of a few keV to several MeV. Gamma rays have the shortest wavelength, down to 10^{-16} m, and the highest frequency, up to 10^{24} Hz, in the electromagnetic spectrum.

Characterising X-rays and gamma rays by energy implies that these types of radiation can be thought of as small energy packets. These energy packets are called *quanta* or *photons,* and in some ways are particle-like. This dual nature of electromagnetic radiation extends right through the electromagnetic spectrum. Some features of electromagnetic radiation, such as interference or diffraction, are best explained with reference to electromagnetic waves whereas others, such as some X and gamma ray interactions are easier to envisage as particle-like interactions. Planck has shown that individual photons or quanta of electromagnetic radiation carry an energy proportional to frequency. The constant of proportionality is known as **Planck's Constant** and has the value 6.63×10^{-34} J.s.

Leading to: *Gamma rays; Radiative heat transfer*

T.D. MacMahon

ELECTROMAGNETIC WAVES

A moving magnet is known to produce an electric field: this is the principle of the dynamo. It is also known that a moving electric charge produces a magnetic field: this is the principle behind an electromagnet. An oscillating electric charge (in a transmitting aerial, for example) has an associated oscillating electric field which induces a magnetic field. The magnetic field will, in turn, create a new electric field which will then induce a new magnetic field and so on, giving rise to self-supporting electromagnetic oscillations known as electromagnetic waves, which may propagate through empty space.

The laws governing the mutual induction of electric and magnetic fields were established by *James Clerk Maxwell* [see, for example, Ohanian (1989)]. From his equations, Maxwell predicted that electromagnetic waves would propagate with a speed $c = 1/\sqrt{\mu_0 \varepsilon_0}$, where μ_0 and ε_0 are, respectively, the *permeability* and *permittivity* of vacuum. Inserting numerical values of these constants leads to $c = 3.00 \times 10^8$ m/s, which Maxwell recognised as the measured **Speed of Light** in vacuum. He therefore, deduced that light waves are electromagnetic waves.

In common with other wave phenomena, electromagnetic waves have characteristic frequency (f) and wavelength (λ) whose product equals speed (c) : $c = \lambda f$. *Radio waves* may be generated by an oscillating electric charge on an aerial. A typical frequency for FM radio waves would be 100 MHz which, from the above relationship, corresponds to a wavelength of 3 metres. A long-wave radio frequency of, say, 200 kHz corresponds to a wavelength of 1.5 km. *Visible light* covers the wavelength region 4 to 7×10^{-7} m. The complete electromagnetic spectrum is described in the entry **Electromagnetic Spectrum**. In a propagating electromagnetic wave, the electric and magnetic field directions lie in the plane perpendicular to the direction of motion and are themselves at right angles to each other. If the electromagnetic wave is generated by oscillating electric charges in a vertical aerial, then the electric field will remain in the vertical direction and the magnetic field will be in a horizontal direction perpendicular to the horizontal direction of propagation. Such a wave would be called a *plane polarised wave.* To receive such a polarized wave at a receiving aerial, that aerial would also have to be vertical.

Electromagnetic waves carry both energy and momentum and can exert pressure on surfaces on which they fall. If, in an electromagnetic wave, the electric and magnetic fields are represented by the vectors **E** and **B** (**E** and **B** are perpendicular to each other and lie in the plane perpendicular to the direction of propagation of the wave) then the energy flux (**S**) in the wave is given by:

$$\mathbf{S} = \frac{1}{\mu_o}\mathbf{E} \times \mathbf{B}.$$

The energy flux vector **S** lies in the direction of propagation and will have units of watts per square metre. Whenever an electromagnetic wave strikes a body and is absorbed by it, the wave will exert a force on the body and transfer momentum to it. The force per unit area, or pressure, is given by S/c. Since c is a very large number, the resulting pressure is very small. For example, the average energy flux from sunlight on the Earth is about 1.4×10^3 W/m^2; this exerts a pressure on the Earth of 6×10^8 N which is, fortunately, much smaller than the gravitational force between the sun and the earth of about 4×10^{22} N. Note that when an electromagnetic wave is totally reflected from a body, the momentum transfer is double that when the wave is totally absorbed.

Reference

Ohanian, H. C. (1989) *Physics,* 2nd Edition. W. W. Norton & Company, New York.

Leading to: *Bremsstrahlung; Cerenkov radiation; Electromagnetic spectrum; Gamma rays*

T.D. MacMahon

ELECTROMAGNETIC WAVES, ABSORPTION AND SCATTERING (see Mie scattering)

ELECTROMAGNETISM (see Electromagnetic waves)

ELECTRON ENERGY LEVELS (see Ionization)

ELECTRON GAS (see Semiconductors)

ELECTRONIC THEORIES, FOR CATALYSIS (see Catalysis)

ELECTRONS (see Ionizing radiation)

ELECTRON SPIN RESONANCE SPECTROSCOPY (see Spectroscopy)

ELECTRON VOLT (see Ionizing radiation)

ELECTRO-OSMOSIS (see Drying)

ELECTROPHORETIC FORCES (see Augmentation of heat transfer, single phase)

ELECTROPLATING

Following from: *Electrochemical cells*

Electroplating is a method for the formation of a thin coating, typically of metal, alloy or composite 1–75 μm onto a suitable substrate (metal, alloy, polymer, ceramic or composite). The substrate to be plated forms the cathode of an **Electrochemical Cell**; the anode of the cell is preferably of pure metal which is electroplated. In this way the cathode reaction and anode reaction (metal dissolution) of a metal-metal ion couple $M-M^{n+}$

$$M^{n+} + ne^- \rightarrow M$$

$$M - ne^- \rightarrow M^{n+}$$

can be controlled to maintain a constant bath composition, which is important for the reproduction of plated workpieces.

The composition of the electroplating bath (cell) varies widely according to the required properties of the plate and the metal deposit. Metal can be electroplated from a simple aquo ion or from a metal complex; for example, cyanide. *Complexing ions* can serve several purposes: to prevent passivity of the anode; to make plating potentials more negative and prevent chemical reactions between substrate and the plating ion (e.g., Cu^{2+} with Fe); and to improve the throwing power. The throwing power is a measure of the uniformity of metal deposit, which is an important factor with objects of complex shape having variable contours and recesses. In addition to metal ion and electrolyte, other components (organic additives, brightness, levellers, structure modifiers and wetting agents) may be added to the bath to help impart specific properties or to avoid poor deposits. Preparation of the article prior to electroplating is extremely important to ensure a uniform, adherent deposit.

The process of *electrodeposition* occurs in two stages; the formation of nuclei of the new metal phase and their growth into crystals with the characteristic lattice, until a few a atomic layers are formed, and then the gradual build-up of the electrodeposit layer itself.

There are various approaches to electroplating articles: jig or rack mounting (for large objects); barrel plating (for large batches of small objects); individual mounting; wire mounting and continuous cathode transfer. A range of metals (Sn, Cu, Ni, Cr, Zn, Cd, Pb, Ag, Au and Pt) are plated as well as several alloys (e.g., Ni-Fe, Pd-Ni) and newer composites of metals with polymers (*polymets*), ceramics (*cermets*) and other dispersed solids, e.g., SiC or diamond. Newer techniques in electroplating include the use of programmed current waveforms (e.g., pulsing) and periodic current reversal to produce dense, non-porous deposits.

K. Scott

ELECTROSTATIC ATOMIZERS (see Atomization)

ELECTROSTATIC EFFECTS (see Gas-solids separation, overview)

ELECTROSTATIC EXTRACTION (see Extraction, liquid/liquid)

ELECTROSTATIC FIELDS (see Augmentation of heat transfer, single phase)

ELECTROSTATIC PRECIPITATION (see Gas-solid flows; Electrostatic separation)

ELECTROSTATIC PRECIPITATORS (see Gas-solids separation, overview)

ELECTROSTATIC SEPARATION

Following from: *Gas solids separation*

The separation of particulates in a fluid can be efficiently accomplished using electrical forces, the main criterion being the fluid must have insulating properties so an electric field can be superimposed across it. Electrostatic separation is done by initially charging the particulates or contaminants and then moving and concentrating them at the electric field boundary.

Electrical separation was first demonstrated by Hohlfield in 1824 by clearing a smoked-filled jar with an electrified point. In 1884, Oliver Lodge attempted to collect lead fume using an elementary form of electrostatic precipitator, energised from a Voss machine driven by a steam engine. It was not until the early 1900's that satisfactory means of electrical energisation enabled various workers to install plants to collect sulphuric acid mists. During the 1920's, the use of electrostatic precipitators became firmly established across all industries for the collection of both solid and liquid particulates.

Industrial Electrostatic Precipitation [White (1963)]

In industrial applications, there are generally two arrangements:

a. The horizontal-flow plate type—where a series of parallel plates, spaced up to 400 mm, form the gas passages and the discharge electrodes are insulated and hanging centrally between them.
b. The tube type precipitator—where the gases pass upwards through vertical tubes measuring up to 250 mm dia., with the discharge electrode taking the form of a coaxial element.

Because of its higher potential separation efficiency, having several stages/precipitator fields in series, the plate precipitator has largely replaced the tube type for dry particulates. Nowadays precipitators capable of treating 1,000 m^3/s of gas, removing at least 99.8% of the particulates from power plants are required to meet current legislation. These units are energised from silicon rectifiers having microprocessor-based thyristor control devices to optimise maximum voltage, and hence performance, at all times.

Where precipitators are used for mist/droplet collection, the vertical flow unit is ideally suited since the collected product is self-draining. Efficiencies in excess of 99% can be obtained from a single tube length.

Basic Principle of Operation [Rose and Wood (1956)]

The discharge electrode energised by a high negative potential, up to 70 kV depending on plant configuration, causes local ionisation of the gas molecules as a result of high electrical stress on the sharp-edged electrode. The positive ions are immediately collected by the electrode while the negative ions escape into the space charge region, moving towards the positively-charged collector. As the gas-borne particles pass through the space charge area, they receive a negative saturation charge in around 0.1 second and, as charged particles, move under the influence of the field to be collected on earthed plates.

Removal of Particulates

In order for the process to continue effectively, particulates which arrive at the collectors need to be periodically removed.

Typically this is done by simple impact rapping, which shears the agglomerated product, which then falls into the hoppers. The frequency/intensity of the rapping blows is regulated by the thickness and type of deposition. To maximise corona production, the discharge electrodes are similarly cleaned to maintain their sharp edges, and hence their emission characteristic.

For gases which are close to dew point or dusts which are particularly adhesive, wet-type precipitators are used where the deposited material is water-washed from the electrode system in the form of a slurry.

Effect of particle size

For particles having a diameter in excess of 2 microns, the charge is acquired by ionic collision, whereas for particles less than 0.2 microns, the charge is acquired by diffusion processes. A typical fractional efficiency relationship, [McEvoy (1986)] is illustrated in **Figure 1**. The dip in the 0.5 micron range follows classical charging theory; then once diffusion charging becomes effective, efficiency begins to recover.

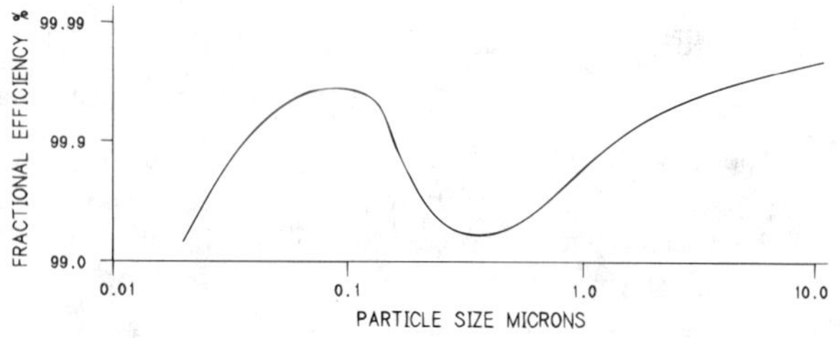

Figure 1. Typical fractional efficiency curve.

Electrostatic precipitator sizing

In spite of precipitators being widely used throughout industry for almost a century, the size of unit to meet a specific duty still cannot be derived from first principles. Sizing is based on evaluated precipitation constants, as determined from similar processes. Initial work by Deutsch (1922) has developed an exponential relationship between efficiency and contact time, i.e., plant size. More recently, because of the higher efficiencies now required, to correctly size the plant it is necessary to modify the Deutsch relationship. The equation currently used is that of Matts and Ohnfeldt (1963), incorporating the precipitation constant w_k, which is derived/calculated from practical measurements.

$$\text{Efficiency} = 1 - e^{-\left(\frac{w_k A}{\dot{v}}\right)^n}$$

n_i = a factor ~0.5

where

$\dot{v}$ = Gas Flow m^3/s

A = collector plate area m^2

w_k = Precipitation constant for the particulate m/s

Importance of gas distribution

Because of the above exponential relationship, it is important that gas distribution throughout the plant is uniform. In practice, a 15% RMS standard of deviation is accepted as the norm [Lloyd (1988)]. Operating velocities are typically in the range 1.0–2.0 m/s.

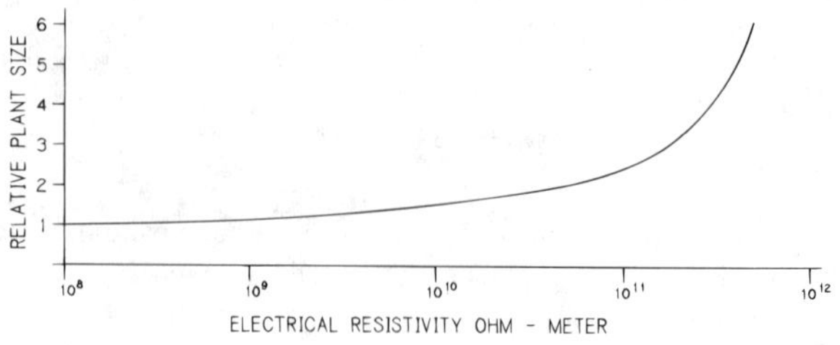

Figure 2. Effect of particle resistivity on plant size.

Effect of particle resistivity

The most important physical property of the particle, with regard to precipitation, is its electrical resistivity. As resistivity increases above 10^9 ohm-m, the precipitation constant w_k decreases, resulting in increasing plant size requirements. A typical relationship between relative plant size and electrical resistivity is shown in **Figure 2**.

Efficiency problems of existing plant can be overcome when handling particulates having resistivities > 10^{10} ohm-m resulting from fuel changes. This can be achieved by chemical conditioning methods, such as sulphur trioxide and ammonia injection, or by modifying the energising wave form to intermittent DC/pulse modulation or by pulse charging techniques.

In summary, the advantages of electrostatic separation are:

a. Separation efficiencies of 99.9%+ are attainable.

b. High efficiency possible for particles of 0.01 micron.

c. Designs available for temperatures up to 850°C.

d. Operational pressure range, minus 0.2 to plus 20 bar.

e. Low overall pressure loss, typically <20 mm wg.

f. Low power requirement <500 $W/m^3/s$ for 99.5% efficiency.

g. Material can be recovered in its original form.

h. Equipment has low maintenance and high reliability.

With the empirical nature of precipitation and the large number of factors which affect sizing, it is recommended that anyone contemplating using this technique approach recognised manufacturers and authorities for technical advice.

Electric coalescers [Darby (1971)]

In the hydrocarbon industry, electric coalescers, although slightly different in form to electrostatic precipitators, are used to remove suspended liquid droplets from many feed stock and product lines. With coalescers, however, friction and contact rather than ionic particle charging occurs. In moving to the field boundary, increased concentration and reduced surface tension results in particle coalescence. When sufficient mass is achieved at the boundary, it will fall out of the mainstream fluid under the influence of gravity. Again, field contact time is important to achieve effective separation. While electrostatic precipitators are usually DC-energised, both AC and DC energisation can be found with coalescers, which are often applied for desalination of crude oil and hydrofiners on diesel and similar fuels.

References

Darby, K. (1971) The use of electrostatic forces for the separation of suspended materials in gases and liquids. *Symposium on Electrochemical Engineering*. University of Newcastle upon Tyne. April.

Deutsch, W. (1922) *Ann d physik.* 68: 335.

Lloyd, D. A. (1988) *Electrostatic Precipitator Handbook.* Adam Hilger. Bristol & Philadelphia.

Matts, S. and Ohnfeldt, P. O. (1963) *Flakt Review.* 64.

McEvoy, L. M. et al. (1986) The collection of fine particulates in power plant electrostatic precipitators. *EPA/EPRI Symposium, New Orleans, February.*

Rose, H. E. and Wood, A. J. (1956) *An Introduction to Electrostatic Precipitation in Theory and Practice.* Constable & Co. London.

White, H. J. (1963) *Industrial Electrostatic Precipitation.* Addison Wesley and Pergamon Press, New York.

K.R. Parker

ELLIPTIC DIFFERENTIAL EQUATIONS (see Differential equations)

ELLIPTIC EQUATIONS (see Green's function)

ELUTION (see Ion exchange)

ELUTION CHROMATOGRAPHY (see Chromatography)

ELUTRIATION (see Classifiers)

EMERGENCY CORE COOLING SYSTEM, ECCS, (see Blowdown; Reflood; Rewetting of hot surfaces)

EMISSIVE POWER (see Emissivity; Radiative heat transfer)

EMISSIVITY

Following from: *Radiative heat transfer*

Emissivity (ε) is a measure of the ability of media to emanate thermal radiation (i.e., electromagnetic radiation in the wavelength range 10^{-1} to 10^2 μm) relative to radiation emanation from an ideal "black body" a similar temperature.

In the general case, emissivity is determined using *radiation spectral intensity* $I_\omega(\vec{r}, \vec{\Omega})$—the electromagnetic radiation energy propagating inside a unit solid angle in the $\vec{\Omega}$-direction through a unit area at a unit time in a unit spectral range.

For reference, black body spectral intensity $I_{b,w}(T)$ is used which is given by:

$$I_{b,\omega}(T) = 11.9086(\omega/T)^3/[\exp(1.4388\omega/T) - 1]. \quad (1)$$

Here, the units of $I_{b,w}(T)$ are W/(cm^2·micron).

The effective directional spectral intensity of a surface at distance $\vec{r}$ from the surface is a result of the directional intensity of self-radiation $I^e_\omega(\vec{r}, \vec{\Omega})$ and reflected radiation $I^r_\omega(\vec{r}, \vec{\Omega})$ (**Figure 1**)

$$J_\omega(\vec{r}, \vec{\Omega}) = I^e_\omega(\vec{r}, \vec{\Omega}) + I^r_\omega(\vec{r}, \vec{\Omega}). \quad (2)$$

In local thermodynamic equilibrium, the effective directional spectral intensity $J^e_\omega(\vec{r}, \vec{\Omega})$ is given by:

$$J^e_\omega(\vec{r}, \vec{\Omega}) = \alpha_\omega(\vec{r}, \vec{\Omega}) I_{b,\omega}[T(\vec{r})], \quad (3)$$

where $\alpha_\omega(\vec{r}, \vec{\Omega})$ is the directional absorptivity.

The spectral intensity of the reflected radiation is represented as:

$$I^r_\omega(\vec{r}, \Omega) = \int_0^{2\pi}\int_0^{\pi} I_\omega(\vec{r}, \vec{\Omega}')f_\omega(\vec{r}, \vec{\Omega}, \vec{\Omega}')\cos\Theta' \sin\Theta' \, d\Theta' \, d\varphi' \quad (4)$$

where $I_\omega(\vec{r}, \vec{\Omega}')$ is the spectral intensity of the radiation incident on a surface element $\vec{r}$ in the Ω'-direction, and $f_\omega(\vec{r}, \vec{\Omega}, \vec{\Omega}')$ is the spectral distribution function of the reflected radiation. For the particular cases of mirror-like and isotropically reflecting surfaces, the angle-dependence of the distribution function for the reflected radiation is supposed to be taken in the form of the δ-function, or constant. The directional spectral emissivity of any real surface at fixed temperature is always less than the intensity of radiation emanated by a black-body surface at the appropriate temperature and wave number.

The ratio

$$\varepsilon_\omega(\vec{r}, \vec{\Omega}) = I^e_\omega(\vec{r}, \vec{\Omega})/I_{b,\omega}[T(\vec{r})]. \quad (5)$$

is called *spectral emissivity.*

If the Kirchhoff law holds, then $\varepsilon_\omega(\vec{r}, \vec{\Omega}) = \alpha_\omega(\vec{r}, \vec{\Omega})$.

The *spectral hemispherical emissivity* is of the form

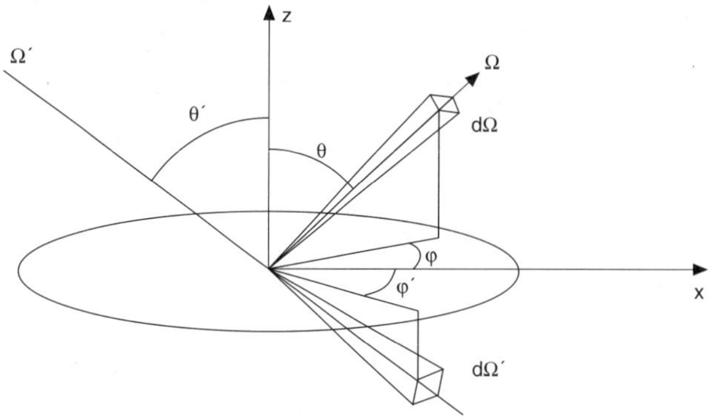

Figure 1. Emission of thermal radiation from a surface.

$$\varepsilon_\omega(\vec{r}) = \frac{\int_0^{\pi/2}\int_0^{2\pi} I^e_\omega(\vec{r}, \vec{\Omega})\cos\Theta\, d\Theta\, d\varphi}{\pi I_{b,\omega}[T(\vec{r})]}. \quad (6)$$

The *total hemispherical emissivity* is:

$$\varepsilon(\vec{r}) = \frac{\int_0^{\infty} \varepsilon_\omega(\vec{r}) I_{b,\omega}[T(\vec{r})]\, d\omega}{\int_0^{\infty} I_{b,\omega}[T(\vec{r})]\, d\omega}. \quad (7)$$

The spectral emissivity of the effective surface radiation is determined in a similar way when replacing I^e_ω by $J_\omega(\vec{r}, \vec{\Omega})$ in Equations 5 and 6.

The *directional spectral intensity* from a volume element is defined by the relation

$$J_\omega(\vec{r}, \vec{\Omega}) = I^e_\omega(\vec{r}) + \frac{\sigma_\omega(\vec{r})}{4\pi}\int_{4\pi} I_\omega(\vec{r}, \vec{\Omega}')p_\omega(\vec{r}, \Omega, \Omega')d\Omega' \quad (8)$$

where $I^e_\omega(\vec{r})$ is the self-radiation intensity. Both $J_\omega(\vec{r}, \vec{\Omega})$ and $I^e_\omega(\vec{r})$ have the units of energy per unit volume per unit time into a unit solid angle in a unit spectral range: $\sigma_\omega(\vec{r})$ and $p_\omega(\vec{r}, \vec{\Omega}, \Omega')$ are the spectral scattering coefficient and spectral phase function of the radiation incident to a unit volume characterized by $\vec{r}$ in the Ω'-direction. The self-radiation energy is emanated isotropically, therefore $I^e_\omega(\vec{r})$ depends on the location of the elementary volume only. If the approximation of local thermodynamic equilibrium is valid, then the intensity of self-radiation I^e_ω is expressed in terms of the spectral intensity of the black-body radiation according to Kirchhoff's law

$$I^e_\omega(\vec{r}) = \kappa_\omega(\vec{r}) I_{b,\omega}[T(\vec{r})], \quad (9)$$

where $\kappa_\omega(\vec{r})$ is the *volumetric spectral absorption coefficient.*

In a given medium volume, the spectral emissivity of the beam segment $[s_0, L]$ in a fixed direction (see **Figure 2**) is determined when solving the transfer equation. This solution can be expressed formally as:

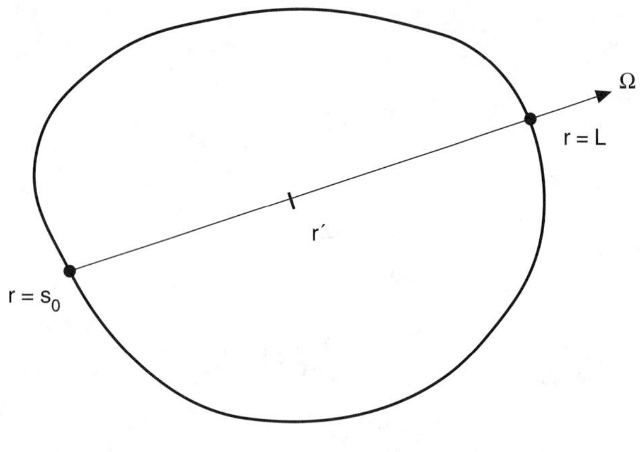

Figure 2. Spectral emission from a volume element.

$$J_\omega(r = L, \vec{\Omega}) = \int_{r=s_0}^{r=L} [\kappa\omega(r') + \sigma_\omega(r')]G_\omega(r', \vec{\Omega}') \times \exp\left\{-\int_{s_0}^{r'} [\kappa_\omega(r'') + \sigma_\omega(r'')]dr''\right\}dr', \quad (10)$$

where

$$G_\omega(r', \Omega') = \frac{\kappa_\omega(r')}{\kappa_\omega(r') + \sigma_\omega(r')} I_{b,\omega}[T(r')] + \frac{1}{4\pi}\frac{\sigma_\omega(r')}{\sigma_\omega(r') + \kappa_\omega(r')}\int_{4\pi} p_\omega(r, \vec{\Omega}, \vec{\Omega}')I_\omega(r, \vec{\Omega}')d\Omega'.$$

It is assumed that the radiation does not reach the boundary $r = s_0$. When there is no light scattering and the medium properties are constant on the entire segment (i.e., the optical path is uniform) the spectral radiation intensity is expressed as:

$$J_\omega(r = L, \vec{\Omega}) = I_{b,\omega}(T)\{1 - \exp[-\kappa_\omega(L - s_0)]\}. \quad (11)$$

The hemispherical (or total) *emissive power* which is a radiation flux density on the base of hemispherical uniform volume of the *R* radius is the most widely-used quantity and is of the form

$$E_\omega(R) = \int_0^{2\pi}\int_0^{\pi/2} J_\omega(r = R, \vec{\Omega}) \cos\Theta\, d\Theta\, d\varphi = \pi I_{b,\omega}(T)[1 - \exp(-\kappa_\omega R)]. \quad (12)$$

The spectral emissive power of a plane layer characterizes the radiation flux density on the boundary of a uniform plane layer of H thickness

$$E_\omega(H) = \pi I_{b,\omega}(T)[1 - 2E_3(\kappa_\omega H)], \quad (13)$$

where $E_3(t)$ is an integro-exponential function of the third order. It has been shown by numerous investigations that the emissive power of uniform volumes of various geometries can be approximately calculated using the hemispherical emissive power $E(R)$ when an equivalent radius is given.

Like surface spectral emissivity, the concept of a spectral emissivity for a hemispherical volume is introduced

$$\varepsilon_{\omega,sph} = E_\omega(R)/\pi I_{b,\omega}(T) = 1 - \exp(-\kappa_\omega R), \quad (14)$$

as well as that of a plane layer

$$\varepsilon_{\omega,pl} = E_\omega(H)/\pi I_{b,\omega}(T) = 1 - 2E_3(\kappa_\omega H) \quad (15)$$

and for other volumes. The physical meaning of spectral emissivity is a ratio of the spectral radiation flux density on the volume boundary to that of a black-body.

In the case of small optical thickness $R\kappa_\omega \ll 1$ the following approximate relation for the spectral emissivity of a uniform flow holds

$$\varepsilon_{\omega,\mathrm{total}} \approx \kappa_\omega R. \quad (16)$$

The total and group emissive power (directional, hemispherical, plane layer, etc.) are obtained by integrating the spectral emissive power over the entire range of wave numbers or over an isolated spectral interval, respectively

$$J = \int_0^\infty J_\omega d\omega, \qquad J_{\Delta\omega} = \int_{\Delta\omega} J_\omega d\omega. \quad (17)$$

In a similar way, total and group emissivities are defined, e.g., for a semispherical volume and plane layer

$$\varepsilon = \int_0^\infty \varepsilon_{\omega,\mathrm{sph}} d\omega/\tilde{\sigma}T^4, \qquad \varepsilon_{\Delta\omega} = \int_{\Delta\omega} \varepsilon_{\omega,\mathrm{sph}} d\omega/\tilde{\sigma}T^4, \quad (18)$$

To calculate total and group emissive power for volumes of various geometries, one needs to know the volumetric spectral absorption coefficient $\kappa_\omega(\vec{r})$ which is, as a rule, a very complicated function of the radiation wave number with variations of an order of magnitude and with rapidly-oscillating spectral dependence due to monatomic and molecular absorption.

Due to the complicated spectral dependence of the absorption coefficient, there are essential difficulties in calculating group and total emissive power even where there is no light scattering. For gases frequently used in engineering practice (air, CO_2 and H_2O), detailed experimental and theoretical results are classified in tabular and graphic data on spectral hemispherical emissive power [Siegel and Howell (1972); Ozisik, (1973); Ludwig, Malkmus, and Reardon et al. (1973)].

If a line structure manifests itself only weakly in the absorption coefficient spectrum, emissive power can be calculated efficiently by group approximation. The spectral range under investigation is divided into a number of intervals (spectral groups) inside of which each of the spectral coefficient is assumed to be independent of the radiation wave number. In this case, the total emissive power is represented as a sum

$$E(r) = L, \vec{\Omega}) = \sum_{l=1}^{N\Delta\omega} I_{b,\Delta\omega,l}(T)\{1 - \exp[-k_{\Delta\omega,l}(L - s_0)]\} \quad (19)$$

where $N_{\Delta\omega}$ is the number of spectral groups, $\kappa_{\Delta\omega,l}$ is the mean absorption coefficient of each interval, $\Delta\omega_l$, and $I_{b,\Delta\omega l} = \int_{\Delta\omega l} I_{b,\omega}(T)\, d\omega$. This approach gives satisfactory results for the line spectrum as well if conditions for full line overlapping are implemented and the spectrum practically becomes continuous.

To describe the spectrum line structure, the group approach may also be used. But within each spectral group, the absorption coefficient is no longer a constant value and models of the molecular line bands are used [Siegel and Howell (1972); Ozisik (1973); Ludwig, Malkmus, Reardon et al. (1973); Rodgers and Williams (1974); Goody (1964); Penner (1959); Edwards and Menard (1964); Tien (1968)]. When taking into account statistical models, the total emissive power is represented as:

$$E(r) = L, \vec{\Omega}) = \sum_{l=1}^{N\Delta\omega} I_{b,\Delta\omega l}(T) \times \left\{1 - \exp\left[-\sum_{i=1}^{Nb,l} W_i(L - s_0)\right]\right\},$$

$$W_i(L - s_0) = \sqrt{W_{L,i}^2 + W_{D,i}^2 - (W_{L,i}W_{D,i}/\tau_i)^2},$$

$$\tau_i = k_{i,l}(L - s_0), \quad (20)$$

$$W_{L,i} = \tau_i/, \sqrt{1 + \tau_i/4a_{L,i}},$$

$$W_{D,i} = 1{,}7a_{D,i}\sqrt{\ln[1 + (0{,}589\tau_i/a_{D,i})^2]},$$

$$a_{L,i} = \frac{\gamma_{L,i}}{d_i}, \qquad a_{D,i} = \frac{\gamma_{D,i}}{d_I},$$

where $N_{b,l}$ is the number of spectral line bands in a spectral range $\Delta\omega_l$; $W_i(L - s_0)$ is the equivalent thickness of i-th spectral line band for a Voigt contour, $W_{L,i}$; $W_{D,i}$ is the equivalent thickness for the Lorentz and Doppler contour; $k_{i,l}$ is the averaged absorption coefficient in the i-th line band of a spectral range l; $\gamma_{L,i}$ and $\gamma_{D,i}$ are the line half-width widened by collision (the Lorentz profile) and the Doppler widening mechanisms; and d_i is the line density in the i-th band. In obtaining Equation 20, a statistical model with exponential distribution of line intensities is used as well as methods that approximate equivalent length. For other approximation methods see Ludwig et al. (1973), Goody (1964), Penner (1959), Edwards and Menard (1964) and Tien (1968).

There are many methods to group lines in a band; the simplest one is to combine all the lines of the gas component into one band. Then $N_{b,l}$ in Equation 20 is the number of gas mixture components which have lines in the range $\Delta\omega_l$.

The numerical and experimental results on $k_{l,i}$ and d_i are given as tables by Ludwig et al. (1973) (CO, NO, CN, OH, HCl, HF, CO_2, H_2O) and in the form of approximations by Siegel and Howell (1972) and Tien (1968) (CO_2, CH_4, H_2O, CO).

The method of line-by-line integration is also sometimes used, with the spectral range divided into thousands and tens of thousands of intervals to describe in detail the absorption in each line. Such laborous calculations are performed to obtain basic results which are compared with the results obtained by approximate methods.

In the case of scattering, the problem of determining emissive power becomes more complicated since it is necessary to find the scattering coefficients $\sigma_\omega(\vec{r})$ and the phase functions $p_\omega(\vec{r}, \vec{\Omega}, \vec{\Omega}')$. For this the Mie theory, geometrical optics approximation and the Rayleigh theory [Bohren and Huffman, (1983)] are used as well as solutions of the integro-differential radiation transfer equation. The general effect of radiation scattering on emissive power is to diminish it, and this usually occurs if the albedo is sufficiently high

$$\lambda(\vec{r}) = \frac{\sigma_\omega(\vec{r})}{\kappa_\omega(\vec{r}) + \sigma_\omega(\vec{r})} > 0{,}95.$$

The determination of emissive power in a nonequilibrium medium hinges on solving kinetic equations which determine the population of excited states of radiating particles and which calculate the probabilities of radiation transition from highly-excited energetic states to lower ones. The relation between emissive power and volumetric absorption coefficients as in Equation 9 is inapplicable here.

The total amount of spectral, group or total radiation energy emitted from a volume of a medium per unit time in all directions over the solid angle 4π is the total emissive power.

References

Bohren, C. F. and Huffman, D. R. (1983) *Absorption and Scattering of Light by Small Particles,* A Wiley-Interscience Publication. John Wiley & Sons, Inc.

Edwards, D. K. and Menard, W. A. (1964) Comparison of models for correlation of total band absorption. *Applied Optics.* 3:621–625.

Goody, R. M. (1964) *Atmospheric Radiation. I. Theoretical Basis.* Oxford, At the Clarendon Press.

Ludwig, C. B., Malkmus, W., Reardon, J. E. et al. (1973) *Handbook of Infrared Radiation from Combustion Gases.* Washington: NASA SP—3080: 486.

Ozisik, M. N. (1973) *Radiative Transfer and Interaction with Conduction and Convection.* A Wiley-Interscience Publication.

Penner, S. S. (1959) *Quantitative Molecular Spectroscopy and Gas Emissivities.* Addison-Wesley, Reading, MA.

Rodgers, C. D. and Williams, W. (1974) Integrated absorption of a spectral line with the Voigt Profile. *JQSRT.* 14(4)319–323.

Siegel, R. and Howell, J. R. (1972) *Thermal Radiation Heat Transfer.* McGraw-Hill, New York.

Tien, C. L. (1968) Thermal radiation properties of gases. in *Advances in Heat Transfer.* 5. Academic Press, New York.

S.T. Surzhikov

EMISSIVITY OF PLANTS AND ANIMALS (see Environmental heat transfer)

EMULSIFYING AGENT (see Emulsions)

EMULSIONS

An emulsion is formed when two nonsoluble liquids (e.g., an oil and water) are agitated together to disperse one liquid into the other, in the form of drops. Emulsions can either be oil-in-water (O/W) or water-in-oil (W/O), depending on whether the continuous phase is the water or the oil, respectively. Drop sizes normally vary from 1 μm to 50 μm. When the agitation stops, if the drops coalesce and the two phases separate under gravity, the emulsion has been temporary. To form a stable emulsion, an *emulsifying agent* must be added to the system.

Sometimes, the formation of an emulsion is the deliberate outcome of a manufacturing process. This is the case, for example, in the production of mayonnaise, where ground mustard seeds are normally added to act as an emulsifying agent. Other times, the formation of an emulsion is totally undesirable. An example is the case of the oil industry where emulsification of oil and brine is common. It may occur in the oil reservoir itself or while flowing through pipelines, mechanical devices, such as pumps, and gas separators.

Formation and Stability of Emulsions

Controlling factors in the formation of an emulsion are: mechanical energy, agitation time, temperature, volumetric ratio between the two phases, degree of dispersion of the internal phase and presence of impurities or surfactants. The material of the shearing plates for the homogeniser used in the emulsification process also influences the type of emulsion formed, e.g., oil-wetted plates strongly favour W/O emulsions [J. T. Davies, 1964].

There are many ways of producing an emulsion and it is usually achieved by applying mechanical energy through agitation, normally by using an *homogeniser.* Initially, the interface between the two phases is deformed and large droplets are formed. These droplets are subsequently broken up into smaller droplets by the continuing agitation. Impurities or *surfactants* present in the system adsorb at the interfaces of the droplets, lower the interfacial tension, and thereby facilitate coalescence. However, the surfactant film formed at the interface of the droplet also tends to resist coalescence. A detailed review of the principles of emulsion formation has been published recently by P. Walstra (1993), where droplet break-up in laminar and turbulent flow is discussed and quantitative relations are presented.

The stability of an emulsion is dependent on the magnitudes of the previously-mentioned opposing effects and is affected by: interfacial viscosity, electric charge on drops, droplet size and concentration, and viscosity of the continuous phase. Aging of an emulsion may also affect its stability as the nature of the interfacial film, which helps to keep it stable, can change with time.

Choice of Emulsifiers

The choice of surfactants for a particular process depends on the restrictions established for that particular application, e.g., in the food industry, emulsifying agents must be edible. Another factor to be considered in the choice of a stabilising agent is whether the desired type of emulsion is an O/W or W/O, as the stabilising agent largely determines which phase is the continuous one [Bancroft (1913)]. The phase in which the surfactant is more soluble will become the continuous phase.

Emulsions are often used in the most diverse fields, e.g., food industry, pharmaceutical products and manufacture of lubricants.

Some examples of emulsifiers are soaps, proteins, starch and gelatine.

Separation of Emulsions

Methods normally used to break emulsions are:

- Gravity settling—Settling of emulsions is more rapid when the drop size is larger and when the continuous phase viscosity is lower. For faster separation, heat can be applied to reduce the viscosity of the continuous phase and sometimes to reduce the effectiveness of the surfactant.
- Centrifugation—Faster separation by increasing the centripetal acceleration force.

- Electrical coalescence—The application of an electrical current (direct or alternating) causes the internal phase droplets to coalesce.
- Chemical methods—Coalescence can be achieved by the addition of suitable chemicals. For instance, by adding electrolytes the charge at the droplets' interfaces may be neutralised and coalescence can result.

A combination of the above methods may also be chosen: heat to modify the continuous phase, chemistry to modify the emulsion, and electricity to finalise the separation.

References

Bancroft, W. D. (1913) *J. Phys. Chem.* 17:514.

Davies, J. T. (1964) in *Recent Progress in Surface Science*, (Danielli, J. F., Pankhurst, K. G. A. and Riddiford, A. C. eds). 2, Academic Press, New York and London.

Walstra, P. (1993) *Chem. Engng. Science* 48, (2)333–349.

M.A. Mendes-Tatsis

ENDOTHERMIC REACTIONS (see Thermodynamics)

ENEA (see Comitato Nazionale per la Ricerca e per lo Sviluppo dell'Energia Nucleare e delle Energie Alternative)

ENEL (see Ente Nazionale per l'Energie Elettrica)

ENERGY ACCOMMODATION COEFFICIENT (see Accommodation coefficient)

ENERGY CONSERVATION (see Exergy; Heat pumps)

ENERGY EFFICIENCY

Introduction

Successive UK governments have been aware of the importance of energy efficiency and have introduced major initiatives to improve the acceptance of energy-saving technologies and techniques——in industry, commerce and the public and domestic sectors. Despite early successes of these initiatives, the relatively low cost of energy during the 1980's resulted in energy efficiency receiving low priority within many organisations. Thus, at the end of the decade, numerous barriers to the adoption of energy-saving measures were still clearly apparent. In March 1989, the Department of the Environment's Energy Efficiency Office (EEO) launched its Energy Efficiency Best Practice programme. This coincided with increased national and international concern about environmental damage associated with energy use and, in particular, awareness of the need to reduce fossil fuel consumption.

Best Practice Programme

The aim of the Best Practice programme is to stimulate adoption of energy-efficient, good practices in terms of technologies, techniques, processes and energy management. It seeks to be of value to all energy users, covering every aspect of energy efficiency through four interrelated programme elements:

- Energy Consumption Guides;
- Good Practice;
- New Practice;
- Future Practice.

Emphasis is placed not only on specific energy-efficient technologies, but also on design considerations, management techniques, operating practices, education and training, and staff motivation.

Benchmarking: Energy Consumption Guides

Energy Consumption Guides allow organisations reviewing their energy use for the first time to undertake a 'benchmarking' exercise—comparing their own consumption with that of other operators in the same industry or building-type—thus establishing where, and how much, potential saving is available to them. This series of guides allows comparisons to be made within a broad range of industries and building types: from iron foundry cupolas to the liquid milk sector of the dairy industry; from domestic housing to schools and offices. The information is gathered from a representative sample of organisations within the appropriate sector. Each guide also contains an Action Plan of practical, achievable energy-saving techniques. These may range from low-cost good housekeeping measures to capital projects or plant modifications.

Stimulating take-up of good practice

The Good Practice element of the programme is designed to provide energy users with detailed and up-to-date information about existing energy-efficient measures. It promotes examples of low-risk, proven techniques which have already achieved significant energy cost savings. An extensive series of case studies offers concise, specific examples of the application of these successful measures. Each case study project has the potential to stimulate national energy cost savings, whilst achieving individual project paybacks acceptable to the appropriate sector(s). Complementing the case studies are Good Practice Guides which detail the best practice currently associated with a particular industry, process or building type.

New practice: Support for novel energy efficiency measures

The role of the New Practice element is to stimulate confidence in novel energy efficiency measures. The programme seeks to offset some of the risks involved in the wider application of such measures by providing detailed and objective information on new or improved techniques, or on novel applications of existing technologies. Successful projects are promoted through New Practice reports and profiles, seminars and site visits.

A longer-term approach to energy efficiency

The Best Practice programme supports basic R & D into energy efficiency measures for industrial applications through its Future Practice element. The EEO works closely with the Department of the Environment's Energy Related Environmental Issues (ENREI), which similarly supports buildings-related projects. Projects are again selected on the criteria of novelty, energy savings, environmental benefits and payback period. Findings are published through Future Practice profiles and reports.

The success of R & D projects can only really be measured when the developed technologies are adopted first as New Practice and ultimately as established Good Practice. Such acceptance has, for instance, been attained by compact heat exchangers developed under a number of Future Practice projects (see ETSU 1991, 1992 and 1993). Their adoption has already resulted in savings of around 1 million GJ/year in the UK. Furthermore, all the major UK heat exchanger manufacturers have now developed, or are in the process of developing, units of this type. Increasing availability will lead to their wider use in heat recovery and heat transfer applications in the future.

Compact heat exchangers

Compact heat exchangers are characterised by their high 'area density,' that is, high ratio of heat transfer surface area to heat exchanger volume. **Table 1**, cited from ETSU (1994) compares the area densities of a range of compact heat exchangers with that of a typical shell-and-tube exchanger.

The resultant small overall size of compact heat exchangers has meant that their initial development and use has been in the aerospace, road transport and marine sectors. More recently, there has been increasing interest in the concept of process intensification, stimulated by the availability of compact units and by environmental and other constraints. Printed circuit heat exchangers (PCHEs), welded plate and plate-fin exchangers (PFHEs) have established good markets and are now challenging conventional types of heat exchanger. Their value is now being explored in the traditional process industries, particularly in the chemical, pharmaceutical and food and drink sectors.

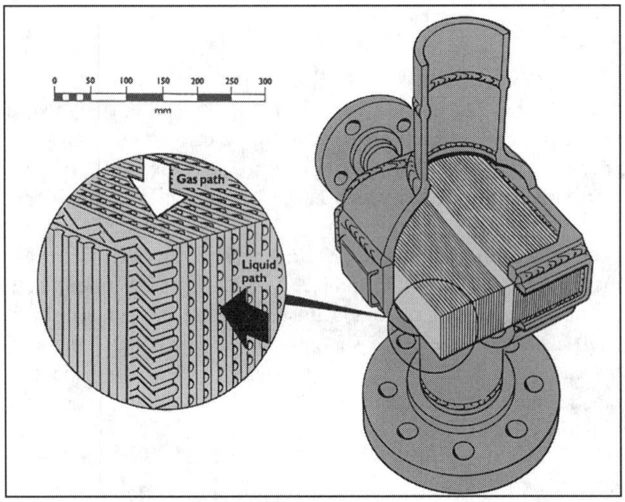

Figure 1. Printed circuit heat exchanger showing gas and liquid paths through the core.

Table 1. Area densities of compact and shell-&-tube heat exchanger

Compact heat exchanger type	Area density (m^2/m^3)
Liquid-liquid compact heat exchanger	> 300
Gas-liquid compact heat exchanger	> 700
Laminar flow heat exchanger	> 3,000
Micro heat exchanger	> 10,000
Conventional shell & tube exchanger (19mm dia tubes)	100 (typical)

Advantages of compact heat exchangers

Improved effectiveness

A major advantage of most compact designs is their greater efficiency or effectiveness. Effectiveness is expressed as the ratio of actual heat transfer to maximum possible heat transfer. It is a function of the heat capacity of fluid streams, the overall heat transfer coefficient and the area of the heat transfer surface.

Higher effectiveness allows the use of closer approach temperature differences between fluid streams. This can lead to significant energy cost savings in process heating and cooling duties since power requirements, particularly of plants such as refrigeration compressors, can be reduced.

Smaller volume

The smaller volume of compact units for a given heat transfer duty (compared with most shell-and-tube exchangers) provides benefits that extend well beyond the heat exchanger itself. Support structures and piping needs are reduced; the exchanger may be installed in a more convenient location (particularly important for 'green field' sites); less maintenance space is required for removal of the heat exchanger core.

Lower capital cost

When the total installed cost is considered, compact heat exchangers tend to be significantly cheaper than their conventional counterparts. This is particularly true when process requirements demand that the exchanger is made from an expensive material, such as nickel or titanium. Here, the cost per kg of raw material dominates the cost of the exchanger, and often the ancillaries.

Conclusion

Considerable advances have thus been made in the development and use of compact heat exchangers—incorporating new materials,

Table 2. Compact heat exchanger types

Type	Nature of fluids
Gasketed plate and frame	Liquids; two-phase[1]
Brazed plate	Liquids; two-phase
Welded plate and frame	Liquids; two-phase[2]
Plate-fin	Gases; liquids; two-phase
Printed circuit	Gases; liquids, two-phase
Welded stacked plate	Gases; liquids; two-phase
Laminar flow	Liquids; two-phase[3]
Compact shell and tube	Liquids; two-phase

[1]Two-phase includes boiling and condensation.
[2]One flow-path may have welded plates; the other relating gaskets.
[3]This and other types may be constructed using polymers.

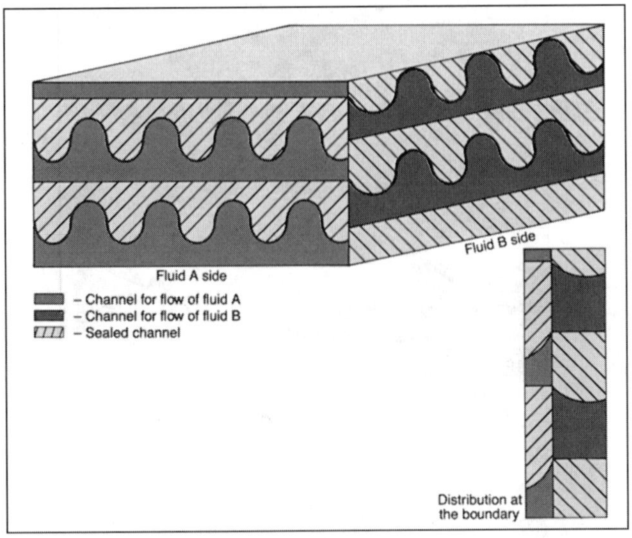

Figure 2. Flow distribution in a polymer film heat exchanger matrix.

construction methods and process integration techniques. Much of the work of overcoming the barriers to their acceptance has been supported by the Best Practice programme. This and other work of the Department of the Environment's EEO is managed by the Energy Technology Support Unit (ETSU) at Harwell, Oxfordshire (for industrial projects) and the Building Research Energy Conservation Support Unit (BRECSU) at Watford (for projects relating to buildings).

References

FPF 12. *Testing of Printed Circuit Heat Exchangers.* ETSU 1991.

FPF 29. *Investigation of a Novel Compact Heat Exchanger Surface.* ETSU 1992.

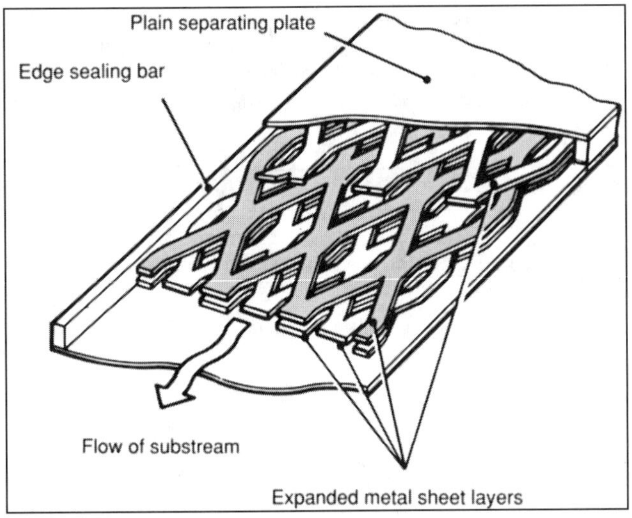

Figure 3. Schematic of new porous matrix exchanger surface.

FPF 40. *Design Data for Compact Polymer Film Heat Exchangers.* ETSU 1993.

GPG 89. *Guide to Compact Heat Exchangers.* ETSU 1994.

S. Marriott

ENERGY EFFICIENCY BEST PRACTICE PROGRAMME (see Energy efficiency)

ENERGY FRACTURE CRITERION (see Fracture mechanics)

ENERGY, RENEWABLE (see Renewal energy)

ENERGY SPECTRUM OF TURBULENCE (see Turbulence)

ENERGY STORAGE (see Solidification)

ENERGY SUPPLY (see Power plants)

ENERGY TECHNOLOGY SUPPORT UNIT, ETSU

Harwell, Didcot
Oxon OX11 ORA
UK
Tel: 01235 821000

ENGINEERING SCIENCES DATA UNIT, ESDU

ESDU International plc
27 Corsham Street
UK
London N1 6UA
Tel: 0171 490 5151

ENHANCED OIL RECOVERY

Following from: *Oils*

During the production of oil from a reservoir, oil moves towards the producer well through the pores of the sedimentary rocks in which it is contained. Historically, the pressure gradients needed to promote this movement have been provided by the release of the natural energy stored in the reservoir fluids, underlying aquifer and reservoir rocks—a process known as *primary recovery.* However, the amount of this natural energy is somewhat limited and the average reservoir pressure declines as oil is produced. As a result, the rate of oil production falls and the quantity of oil that can be recovered is small; typically less than 25% of the total amount of oil in the reservoir.

The oil recovery factor can be substantially increased by injecting fluids, either gas or water, into the reservoir to replace the oil that has been extracted, thus maintaining reservoir pressure. By using these techniques, the amount of oil recovered can be increased to about 50–60%. These were called *secondary processes* since they could restore production to fields that had reached their primary production limits. Pressure maintenance by water injection is now conventional practice, and it is used in the majority of new oil field developments. After the water injection scheme has reached its economic limit, the 40–50% of the oil that remains in the reservoir has two forms:

1. Some of the oil remain as droplets or ganglia trapped within the rock pores in regions that have been swept by the waterflood, **Figure 1**. As much as 30% of the oil originally in these zones may be left in this way. The mechanism that causes this oil to become trapped and the forces that prevent the ganglia from being moved both depend upon the wetting characteristics of the rock and the oil/water interfacial tensions. These interfacial forces are generally several orders of magnitude higher than the viscous forces imparted by the flowing water. To move this trapped oil, therefore, it would be necessary either to reduce substantially the interfacial tension or to increase the viscous forces. Increasing the viscous forces by several orders of magnitude is not a practical option, however, since the pressures involved would be enough to fracture the rock matrix.
2. A significant fraction of the reservoir remains unswept by the waterflood. Some of this unswept oil is in regions where the amount of oil that could be recovered makes it uneconomic to drill more wells; but much of it is in areas that have been bypassed by the waterflood. A number of factors account for this poor sweep efficiency:

- The density of water is higher than that of oil and gravity forces cause the water to flow preferentially through the lower regions of the reservoir,
- Heterogeneities within the reservoir cause fluids to flow preferentially through the high-permeability layers, **Figure 2**.
- The streamlines which define the paths taken by the fluids as they flow from an injector to a producer well vary in length, **Figure 3**. Fluids flowing through the innermost streamtubes reach the producer well much earlier than those flowing through the outer streamtubes. As a result, high water cuts can occur before the oil in the outermost streamtubes has been fully displaced.

For each of these mechanisms, sweep efficiency is dependent upon the oil/water viscosity ratio. If the oil is more viscous than the water, sweep efficiency is low and vice versa.

By injecting suitable materials into the reservoir, the interfacial behavior and/or the viscous ratios can be changed. Interfacial tensions can be reduced by adding a *surfactant* to the injection water or by injecting a gas that is miscible with the oil. Viscosity ratios can be changed by adding a high molecular weight polymer to increase the viscosity of the injection water or by heating the reservoir to reduce the viscosity of the oil. These methods were originally called tertiary processes since they were used after secondary recovery had been completed, but they are more familiarly known as **enhanced oil recovery** processes. They fall into three main categories:

A fourth category includes some more speculative processes, such as the injection of microorganisms to recover oil.

Figure 1. Oil ganglia trapped in the interstices of the rock matrix.

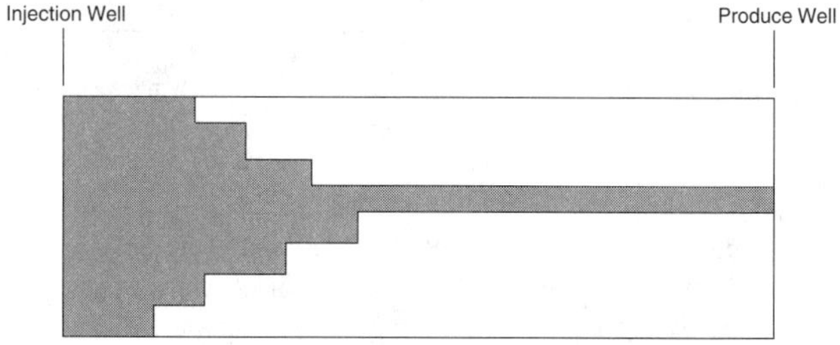

Figure 2. Example of sweep pattern occurring in heterogeneous reservoir having a high-permeability layer.

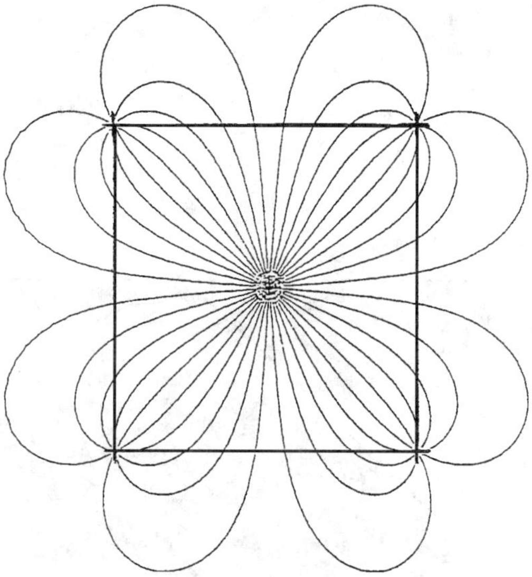

Figure 3. Example of streamline pattern for a single injector surrounded by four producer wells.

Table 1 Three main categories of enhanced oil/recovery processes

	Method	Target oil
Gas injection	Miscible Hydrocarbon	Residual
	Carbon Dioxide	Residual
	Nitrogen	Residual
Chemical Processes	Surfactant	Residual
	Polymer	By-passed
	Caustic	Residual
	Polymer Gels	By-passed
Thermal Processes	Steam Injection	Residual/By-passed
	Hot Water	Residual/By-passed
	In Situ Combustion	Residual/By-passed

Each of these processes is effective for a limited range of reservoir conditions so each application must be specifically designed for a particular situation.

R.I. Hawes

ENHANCEMENT OF FILM CONDENSATION HEAT TRANSFER (see Condensation, overview)

ENHANCEMENT OF HEAT TRANSFER (see Augmentation of heat transfer, single phase; Rotating duct systems, orthogonal, heat transfer in; Tube banks, condensation heat transfer in; Tube, single-phase heat transfer to in crossflow; Wavy flow)

ENHANCEMENT OF MASS TRANSFER (see Wavy flow)

ENIAC (see Computers)

ENLARGEMENT, FLOW AND PRESSURE CHANGE IN

Following from: *Channel flow; Pressure drop, single-phase; Pressure drop, two-phase*

The flow of fluid through an enlargement (increase in pipe diameter) results in a decrease in velocity and consequently, a pressure rise. If the contraction is sharp or sudden, the behaviour of single-phase flow is as shown in **Figure 1** and involves a recirculation region. This starts at the enlargement and extends about three outlet pipe diameters downstream. The dissipation of energy caused by this recirculation region means that not all the kinetic energy is converted to a pressure rise, and reversible and irreversible components of pressure drop must be considered. If the enlargement is being used to convert kinetic energy into pressure, it is necessary to employ a more gradual change in diameter to eliminate or minimise losses. This gradual increase in diameter is known as a *diffuser.* For diffusers which have a linear increase in diameter, an included angle of less than 7° is required to avoid separation of the flow and thus minimise losses.

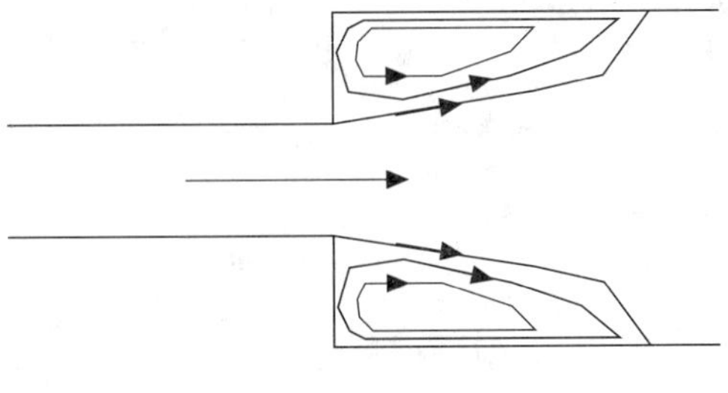

Figure 1. Flows structure for single-phase flow through a sudden contraction.

As for a **Contraction**, it is not possible to use a momentum balance to calculate pressure change through a generalised enlargement since there is an unknown reaction force from the walls to be accounted for. An energy balance gives a computable expression for the pressure drop. The exception is an abrupt enlargement, where the resultant forces on the annular downstream-facing wall can be assumed (with reasonable accuracy) to be equal to the pressure at the end of the upstream pipe times the area of the annulus. For turbulent flow, the total pressure change is then

$$(p_d - p_u)_{tot} = \frac{\dot{m}_1^2}{2\rho}[1 - S^2], \qquad (1)$$

where P_d and P_u are the downstream and upstream pressures, respectively, $\dot{m}$ is the mass flux, ρ, the density and s the area ratio between upstream and downstream pipes. The reversible pressure change is:

$$(p_d - p_u)_{rev} = \frac{\dot{m}_1^2}{2\rho}[2S - 2S^2]. \qquad (2)$$

The irreversible portion is determined by difference.

For a diffuser, methods which calculate growth of the **Boundary Layer** are required; for example, those similar to Ghose and Kline (1978) which uses a momentum integral approach. If the diffuser angle is >45°, it can be taken as a sudden enlargement.

Gas/liquid flows through sudden enlargements also involve a recirculation region downstream of the step. The length of the recirculation zone is similar to that for single-phase flow [Chouikhi et al. (1983)]. Application of a separated flow model equivalent to that used for single-phase flow has been found to do poorly in predicting pressure change [Schmidt (1993)]. However, recent work by Schmidt, which allows for a pressure on the annulus of the downstream-facing wall different from that at the upstream pipe outlet as well as other factors, gives good predictions.

For diffusers with annular two-phase flow, separation of the boundary layer has been inferred even at angles of 5° and depends on gas and liquid flow [Azzopardi (1992)]. **Figure 2** shows the effect of diffuser angle on pressure rise. The lower rise at 15° is probably due to the start of separation of the boundary layer.

References

Azzopardi, B. J. (1992) Gas-liquid flows in cylindrical venturi scrubbers: boundary layer separation in the diffuser section. *The Chemical Engineering Journal.* 49:55–64.

Chouikhi, S. M., Patrick, M. A., and Wragg, A. A. (1983) Two-phase turbulent wall transfer processes downstream of abrupt enlargements in

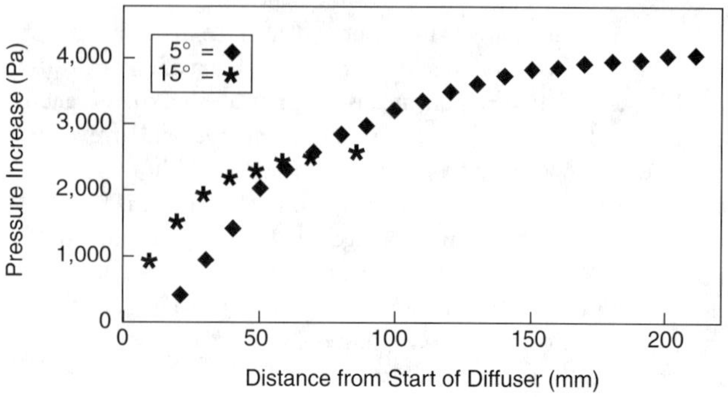

Figure 2. Measured pressure profiles in diffusers showing the effect of diffuser angle on pressure rise.

pipe diameter. *Proc. Int. Conf. on Physical Modelling of Multiphase Flow,* Coventry. April 19–21. Pub BHRA. 53–66.

Ghose, S. and Kline, S. J. (1978) The computation of optimum pressure recovery in two-dimensional diffusers, *Trans A.S.M.E., Journal of Fluids Engineering.* 100:419–426.

Schmidt, J. (1993) *Berechnung und Messung des Druckabfalls uber plötzliche scharfkantige Rohrerweiterungen und -verengungen bei Gas/Dampf-Flüssigkeitsströmung.* VDI-Forschungsheft.

B.J. Azzopardi

ENSEMBLE AVERAGES (see Ergodicity)

ENTE NAZIONALE PER l'ENERGIA ELETTRICA, ENEL

Direzione Studi e Ricerche
via G B Martini 3
CP 386, 00198 Rome, Italy
Tel: 6 850 91

ENTHALPY

***Following from:** Thermodynamics*

Enthalpy, denoted by H, is a convenient energy concept defined from properties of a system:

$$H - U + pV \quad \text{and} \quad dH - dU + d(pV) \tag{1}$$

where U is internal energy, p is pressure and V is volume, and is thus also a property of the system. The term U + d(pV) appears often in equations resulting from the *First Law of Thermodynamics.* Internal energy, and hence enthalpy, can only be quantified relative to an arbitrary reference value. In most applications, however, it is the changes in enthalpy that are important, so the values of the reference state chancel out. The convenience of enthalpy as a thermodynamic concept is illustrated by the definitions of the molar heat capacities of a closed system of constant volume, $\tilde{c}_v$, and constant pressure, $\tilde{c}_p$, in a reversible process:

$$\tilde{c}_v = \left(\frac{\partial \tilde{u}}{\partial T}\right)_v \quad \text{and} \quad \tilde{c}_p = \left(\frac{\partial \tilde{h}}{\partial T}\right)_p \tag{2}$$

where h, u and v are the molar enthalpy, molar internal energy and molar volume, respectively.

For solids and liquids, enthalpy change resulting from a change in pressure and temperature can be calculated from

$$\tilde{h}_2 - \tilde{h}_1 - \int_{T_1}^{T_2} \tilde{c}_p dT + \int_{p_1}^{p_2} \left[\tilde{v} - T\left(\frac{\partial \tilde{v}}{\partial T}\right)_p\right] dp. \tag{3}$$

Enthalpy and other thermodynamic properties of gases and vapours are usually expressed in terms of their deviation from ideal behaviour. *Residual enthalpy,* $\Delta\tilde{h}^R$, is defined as the difference between the actual enthalpy and the ideal enthalpy. For an isothermal process,

$$\Delta\tilde{h}^R - \int_0^p \left[T\left(\frac{\partial \tilde{v}}{\partial T}\right)_p - \tilde{v}\right] dp. \tag{4}$$

Accurate experimental data should be used in evaluating this expression, but approximate values (±2%) can be obtained by using the **Compressibility Factor** $Z = p\tilde{v}/RT$, which is plotted in compressibility charts as a function of reduced temperature and pressure. Residual enthalpy can be expressed in terms of the compressibility factor for a change at constant temperature and composition:

$$\Delta\tilde{h}^R - RT^2 \int_0^p \left(\frac{\partial Z}{\partial T}\right)_p \frac{dp}{p}. \tag{5}$$

It follows from the definition of enthalpy and the First Law of Thermodynamics that the change in enthalpy in a closed system at constant pressure in a reversible process is equal to the heat input. The change in enthalpy in a chemical reaction at constant pressure is then the heat of reaction $\Delta\tilde{h}_r$, which in most reactions result mainly from changes in bond strength over the course of a reaction. As enthalpy is a state function, the change in enthalpy is independent of the path of change, so that $\Delta\tilde{h}_r$, for any reaction can be calculated from a suitable combination of $\Delta\tilde{h}_r$ values for standard reference reactions.

The heat of formation of a chemical compound is a special case of the heat of reaction, where the compound is the only product from reactants comprising its component elements. The heat of combustion is the enthalpy change when a substance is oxidised with molecular oxygen. Enthalpy change associated with a phase change $\Delta\tilde{h}_{ph}$ in a pure substance, known as latent heat, is calculated from the **Clapeyron Equation**:

$$\frac{dp^*}{dT} = \frac{\Delta\tilde{h}_{ph}}{T\Delta\tilde{v}} \tag{6}$$

where p* is the saturation pressure and $\Delta\tilde{v}$ is the change in molar volume between the two phases. In the special case of vaporisation (or sublimation), where the vapour behaves as an ideal gas and the volume of liquid (or solid) is small compared to that of the vapour,

$$\frac{dp^*}{dT} = \frac{p\Delta\tilde{h}_{ph}}{RT^2}. \tag{7}$$

Enthalpy changes associated with changes in concentration, such as the heat of dilution, mixing or crystallisation, can be evaluated from enthalpy concentration charts.

M. Winkler

ENTHALPY, EFFECTIVE FOR SURFACE DESTRUCTION (see Ablation)

ENTHALPY METHOD (see Solidification)

ENTHALPY OF VAPORISATION (see Latent heat of vaporization)

ENTRAINMENT OF DROPLETS (see Wavy flow)

ENTRAINMENT OF DROPS, IN ANNULAR FLOW (see Annular flow)

ENTRANCE LENGTH EFFECTS (see Tube banks, crossflow over)

ENTRANCE REGION HEAT TRANSFER, IN TUBES (see Tubes, single phase heat transfer in)

ENTRAPMENT PUMPS (see Vacuum pumps)

ENTROPY

Following from: *Thermodynamics*

Entropy is a macrophysical property of thermodynamic systems which has been introduced by *Clausius.* It is completely transferred from one system to the other during *reversible processes,* while it always increases during *irreversible processes* in closed systems. The *First Law of Thermodynamics* claims that energy is conserved. This means that nonthermal energy lost by a system (for example, through friction) must reappear in a system or its surrounding in the form of thermal energy. The definition of entropy is obtained from the **Carnot Cycle** of heat engines. The efficiency of reversible heat engines working between two absolute temperatures, T and T − ΔT, is

$$\frac{W}{Q_{in}} = \frac{\Delta T}{T}, \tag{1}$$

where W is the work done and Q_{in} is the heat absorbed by the engine. The state function entropy, denoted by S, is then defined in terms of

$$dS - \frac{dQ}{T}. \tag{2}$$

According to the *Second Law of Thermodynamics,* only irreversible processes are possible in nature. Thus, the entropy change of a system and its surrounding is positive and tends to zero as the system approaches reversibility.

$$\Delta S_{tot} = \Delta S_{sys} + \Delta S_{sur} \geq 0. \tag{3}$$

For a closed system of constant composition,

$$dU - TdS - pdV. \tag{4}$$

This equation contains only properties of the system and is therefore a process-independent fundamental property relation of the system. It can also be derived that

$$dH - TdS + Vdp \tag{5}$$

and hence

$$\left(\frac{\partial H}{\partial S}\right)_p = T; \qquad \left(\frac{\partial H}{\partial p}\right)_s = V \tag{6}$$

$$T\left(\frac{\partial S}{\partial T}\right)_V = c_v; \qquad T\left(\frac{\partial S}{\partial T}\right)_p = c_p. \tag{7}$$

The entropy change of an ideal gas can be written as:

$$d\tilde{s} = R \ln\frac{V_2/N_2}{V_1/N_1} + \tilde{c}_v \ln\frac{T_2}{T_1} \tag{8}$$

with N as the number of moles present, $\tilde{s}$ is entropy per mole and $\tilde{c}_v$ is specific heat capacity per mole. The value of $d\tilde{s}$ is 0 if the system is separated into two identical parts (i.e., V/N is constant). However, entropy of mixing is generated if two different gases at the same temperature and pressure are mixed i.e., (N is constant and $V_2 = 2V_1$ for each component). Before removing the separating wall, the molecules of each species occupy only half the available space and are more ordered than in the final state, where they are randomly distributed over the total volume. From a microscopic view, entropy is therefore a measure of the potential number of microstates within the same macrostate, i.e., the *disorder* of the system. These ideas have been expressed mathematically by L. Boltzmann and J. W. Gibbs in terms of a *thermodynamic probability* Ω, which is the number of ways microscopic particles can be distributed among the states accessible to them.

$$\Omega = \frac{n!}{(n_1!)(n_2!)(n_3!)\cdots}. \tag{9}$$

with n as the total number of particles and n_1, n_2, etc. being the number of particles in states 1, 2, etc. Because of the large number of particles contained in thermodynamic systems and the randomness of their position at any time, statistical means have to be applied, giving rise to the expression *Statistical Thermodynamics.* The connection postulated by Boltzmann between entropy and thermodynamic probability is

$$S = k \ln \Omega \qquad (10)$$

with k as the *Boltzmann constant R/N,* and

$$\Delta S = k \ln \frac{\Omega_2}{\Omega_1} \qquad (11)$$

with Ω_1 and Ω_2, the probability for states 1 and 2.

If applied to information theory, entropy is the logarithm of the missing yes/no statements to provide a complete set of information.

M. Winkler

ENTROPY OF VAPORISATION (see Latent heat of vaporisation)

ENVIRONMENTAL CONCERNS (see Heat pumps)

ENVIRONMENTAL HEAT TRANSFER

Following from: *Heat transfer*

Introduction

Agriculturalists, foresters and ecologists are interested in surface heat balance because it is a major factor in the productivity of vegetation and the irrigation of crops. In hydrology, heat balance is important for the water balance of surfaces and water resources. In climatology, the surface heat balance determines surface temperature and for individual organisms, heat balance determines the energy needs of homoiothermic animals and the water requirements of terrestrial plants and animals. (See **Physiology and Heat Transfer.**)

Heat Balance is the key concept in *environmental heat transfer.* The heat balance equation is simply the statement of the *First Law of Thermodynamics* for a particular system. The usual convention adopted in this field is that fluxes toward the surface are positive on the left hand side of the equation:

$$\Sigma \text{ Inputs} = \Sigma \text{ Outputs} + \text{Storage}.$$

Heat balance of extensive surfaces

For uniform, extensive surfaces, **Heat Transfer** through the boundary layer may be treated as quasi one-dimensional, with transport being driven by gradients of concentration, temperature and velocity normal to the surface. Rates of heat transfer are expressed in terms of the **SI unit** of energy flux density, W m^{-2}. The alternative for climatological time scales is MJ m^{-2} d^{-1}. Surface conditions in nature are rarely uniform, but the solution to the heat balance can be recognised as the equilibrium temperature and vapour pressure at the surface. For example, for a dry surface, the difference between surface temperature and air temperature is proportional to the supply of heat by radiation and inversely proportional to the resistance to transport across the **Boundary Layer** above the surface.

Net radiation

For natural surfaces, the main driving term of the heat balance is *Net Radiation*, R_n,—the sum of four terms representing the components of the radiation microclimate of the surface:

$$R_n = S(1 - \rho) + (L_d - L_u) \qquad (1)$$

where S and ρS are the incident and reflected *solar radiation,* respectively; ρ is the *reflection coefficient* for solar radiation; L_d *downward thermal radiation;* and L_u the *upward thermal radiation* emitted by the surface. (See also **Solar Energy**.) L_u may be estimated from *Stefan's law,* provided the surface temperature T_s is known:

$$L_u = \varepsilon\sigma T_s^4 \qquad (2)$$

where σ (= 56.7 × 10^{-9} W m^{-2} K^{-4}) is the *Stefan-Boltzmann constant* and ε is the **Emissivity** of the surface. (See **Radiative Heat Transfer**.)

Natural surfaces may be approximated to black bodies in the thermal infrared (ε = 1), since measured values of ε *for plant leaves* are in the range 0.94–0.99 [Monteith and Unsworth (1990)]. Similar values have been measured for animal coats [Hammel (1955)]. Net radiation can therefore be approximated to:

$$R_n = S(1 - \rho) + (L_d - \sigma T_s^4). \qquad (3)$$

Solar irradiance is determined by illumination geometry and atmospheric attenuation, i.e., by season and weather. Obviously, S = 0 at night while daytime peak insolation is about 1,000 and 850 W m^{-2} in the tropics and in temperate latitudes, respectively. L_d is also determined by weather, but ground surface and vegetation cover have a strong influence on net radiation, via ρ and T_s. Taking an extreme example, ρ is about 20% for bare tundra but 80 to 85% for fresh snow [Lewis and Callaghan (1975)]. Even in strong sunshine, the high reflectivity of snow makes net solar radiation S(1 − σ) smaller than the (negative) net thermal radiation (L_d − L_u), and therefore R_n is negative. The resulting negative feedback in the microclimate tends to preserve snow-cover in spring, despite high insolation.

Dissipation terms

Strictly, we are concerned with the total heat (enthalpy) transfer from surfaces, but conventionally the components are treated separately, ignnoring the effects of height changes within the turbulent boundary layer. The dissipation terms in the heat balance are usually regarded as the *sensible heat flux* to the air above the surface, g_s; the *latent heat flux* carried by water evaporating from the surface g_E (equal to the product of the evaporation mass flux E with the specific enthalpy of evaporation h_{lg}, i.e., $g_E = h_{lg}E$); the *ground heat flux,* g_G; and heat storage in the surface layer CdT/dt:

$$R_n = g_s + g_E + g_G + C\frac{dT}{dt} \quad (4)$$

where C and dT/dt are the thermal capacity and the instantaneous rate of change of temperature of the surface layer (air plus a vegetation canopy), respectively. Over periods of several days, the storage term is negligible [Monteith and Unsworth (1990)] and is omitted from the remaining analysis.

Ground heat flux

Ground heat flux depends on the **thermal conductivity**, λ, and volumetric specific heat of the substrate (density $\times$ specific heat, $\rho' C_s$). The diurnal and annual cycles of solar heating result in the propagation into the ground of temperature waves, with damping depth δ given by:

$$\delta = \left(2\frac{\kappa}{\omega}\right)^{0.5} \quad (5)$$

where κ is the thermal diffusivity of the soil, equal to $\lambda/(\rho' c_s)$, and ω is the angular frequency of the temperature signal. (See **Thermal Diffusion**.) The *thermal properties of soil* depend on composition and water content, with values of δ for the daily cycle ranging from about 5 cm for peat soils; 7–14 cm for clay; and 7–18 cm for sandy soils [Van Wijk & De Vries (1964); Monteith & Unsworth 1990].

The other determinant of ground heat flux is net radiation. Fluxes below dark surfaces, such as a tarmac, are high because they absorb most solar radiation; but under vegetation, g_G is moderated by shade and becomes a relatively constant fraction of R_n above the canopy. The heat balance equation can then be simplified to:

$$(R_n - g_G) = g_S + g_E. \quad (6)$$

Sensible and latent heat fluxes

Sensible and latent heat fluxes are usually considered together because they are both transported by atmospheric **Turbulence** and because they 'compete' to dissipate the radiant heat input. The simplest case is for a dry surface—a tin roof, road surface or desert—where all the radiant heat must be used to heat the air. In this case,

$$(R_n - g_G) = g_S = \rho_a c_p \frac{(T_s - T_a)}{r_a} \quad (7)$$

where $\rho_a c_p$ (J m^{-3} K^{-1}) is the volumetric specific heat of air; T_a is the air temperature; and r_a (s m^{-1}) is the *aerodynamic resistance* for turbulent transport between the surface and the air. r_a depends mainly on wind speed and surface roughness, but it is also influenced by atmospheric stability [Monteith & Unsworth (1990)].

For wet surfaces, the evaporative heat flux g_E may be expressed as:

$$g_E = h_{lg}\frac{(c_{s(T_s)} - c_a)}{r_a + r_c} \quad (8)$$

where r_c is the surface (canopy) resistance and $c_{s(Ts)}$ and c_a are the saturated concentration of water vapour at the surface temperature and the atmospheric concentration, respectively. For surfaces with free water, such as lakes, reservoirs and vegetation covered by dew or recent rain, $r_c = 0$; but for vegetated surfaces, r_c is usually larger than r_a and therefore limits the flux of water vapour and the associated latent heat flux.

Potential evaporation

For dry surfaces, it is easy to solve the heat balance equation to estimate surface temperature. In contrast, the solution for wet surfaces is complicated by *stomatal control of water loss from plants* and the approximately exponential temperature dependence of the saturation vapour pressure over water.

A number of algorithms have been developed to predict water loss from natural surfaces, but most employ empirical relations which are invalid when used away from their original context. The scientific basis for the prediction of evaporation was first proposed by Penman (1948), who manipulated the heat balance equation to eliminate the unknown term, surface temperature.

The *Penman Equation* estimates *Potential Evaporation* E_p from a surface as the combination of two terms, which represent radiative energy supply and the evaporative demand of the atmosphere;

$$E_p = \frac{\Delta}{(\Delta + \gamma)}\frac{R_n}{h_{lg}} + \frac{\gamma}{(\Delta + \gamma)}\frac{(c_{s(T_a)} - c_a)}{f(u)} \quad (9)$$

where $c_{s(Ta)}$ is the saturation vapour pressure at the air temperature and f(u) is a function of wind speed with appropriate units. The weighting terms $\Delta/(\Delta + \gamma)$ and $\gamma/(\Delta + \gamma)$ depend on the *psychrometer 'constant'* γ, equal to 0.066 kPa K^{-1} at sea level, and Δ, the slope of saturation vapour pressure versus temperature relation at air temperature. Paw U (1992) has pointed out that Penman's original derivation involved significant errors because of the approximations necessary to permit analytical solution and hand calculation. However, McArthur (1990: 1992) argues that the principles of Penman's work remain valid, and that approximations may be avoided when computing E_p.

The Penman equation remains the standard basis for the estimation of potential evaporation from weather measurements, both for irrigation scheduling and as a measure of climate. Monteith (1965) [Monteith and Unsworth (1990); Jones (1992)] has subsequently adapted the Penman equation to include stomatal resistance, while Jarvis and McNaughton (1986) have shown that the atmosphere interacts with surface extent, roughness and canopy resistance to modify heat balance and evaporation at the landscape scale. Jarvis and McNaughton have identified coupling between the surface and the atmosphere as a key factor in the heat balance and evaporation in natural environments.

At one extreme are well-coupled conditions; when wind speed is high and/or the vegetation tall and rough, when the boundary layer resistance is small and the canopy temperature is close to air temperature. Well-coupled surfaces lose water at the imposed evaporation rate, determined by canopy resistance and atmospheric saturation deficit, with little effect of net radiation. At the other extreme are poorly-coupled conditions, over smooth surfaces and/or at low wind speeds. Then water is lost at the equilibrium evaporation rate, which is driven by net radiation.

Measurement of surface heat balance

There are three basic methods for measurement of surface heat balance: the *profile* and *Bowen Ratio* methods, both of which are valid only over extensive uniform surfaces, and *eddy correlation*.

The *profile method* depends on an understanding of the vertical structure of the turbulent boundary layer. The vertical profiles of wind speed, atmospheric temperature and water vapour concentration are measured in the constant flux layer [e.g., Biscoe et al. (1975)], then the turbulent diffusivity is estimated from the wind profile, with adjustment for atmospheric stability. The sensible and latent heat fluxes from the gradients of temperature and humidity and that of turbulent diffusivity.

The *Bowen Ratio* β is the ratio of sensible to latent heat fluxes, g_S/g_E. In principle, the Bowen ratio method only requires measurements of net radiation, g_G and the temperature and humidity at two heights [Monteith and Unsworth (1990)], but may also be applied to analyze profile measurements. The method depends on rearrangement of the heat balance equation so that there is only one unknown. Expansion of the Bowen Ratio by substitution for g_S and g_E from Equations 7 and 8 gives:

$$\beta = \frac{g_s}{h_{lg}E} = \frac{\rho c_p(T_1 - T_2)}{h_{lg}(c_1 - c_2)} \quad \textbf{(10)}$$

Substitution for g_s in the heat balance equation therefore gives:

$$E = \frac{(R_n - g_G)}{h_{lg}(1 - \beta)}. \quad \textbf{(11)}$$

Eddy correlation is simple in principle [e.g., Leuning et al. (1982)], but its routine application to measurements in the constant flux layer has been possible only since the development of portable computers, sonic anemometers [e.g., Kaimal et al. (1974)] and rapid response sensors. The net flux of sensible heat is computed as the time integral of the instantaneous vertical velocity, u_z, multiplied by the volumetric specific heat and temperature:

$$g_S = \int u_z(\rho c_p T)\, dt. \quad \textbf{(12)}$$

Similarly, the flux of latent heat is given by:

$$g_E = \int u_z(h_{lg}c_a)dt. \quad \textbf{(13)}$$

Eddy correlation therefore allows measurement of both sensible and latent heat fluxes, provided the sensors are fast enough to follow the full spectrum of turbulence.

Heat transfer from individual organisms

Within the constraints of this article, it is impossible to cover both heat transfer from extensive surfaces and that from individual organisms. Those interested in the heat balances of leaves, fruits and animals can refer to the texts of Monteith and Unsworth (1990) and Jones (1992).

References

Biscoe, P. V., Clark, J. A., Gregson, K., McGowan, M., Monteith, J. L., and Scott, R. K. (1975) Barley and its environment. I. Theory and practice. *Journal of Applied Ecology,* 12:227–257.

Hammel, H. T. (1955) Thermal properties of fur. *American Journal of Physiology,* 182:369–376.

Jarvis, P. G. and McNaughton, Stomatal control of transpiration: scaling up from leaf to region. *Advances in Ecological Research,* 15:1–49.

Jones, H. G. (1992) *Plants and Microclimate.* 2nd ed. Cambridge University Press, Cambridge.

Kaimal, J. C., Newman, J. T., Bisberg, A., and Cole, K. (1974) An improved three component sonic anemometer for investigation of atmospheric turbulance. Flow, its Measurement and Control in Science and Industry. *Instrument Society of America,* 1: 349–345.

Leuning, R., Denmead, O. T., and Lang, A. G. R. (1982) Effects of heat and water vapour transport on eddy covariance measurements of CO2 fluxes. *Boundary-layer Meteorology,* 23:209–222.

Lewis, M. C. and Callaghan, T. V. (1975) Tundra. Chapter 13, *Vegetation and the Atmosphere* (Ed. J. L. Monteith) Academic Press, London.

McArthur, A. J. (1990) An accurate solution to the Penman equation. *Agricultural and Forest Meteorology,* 51:87–92.

McArthur, A. J. (1992) The Penman form equations and the value of Delta: a small difference of opinion or a matter of fact? *Agricultural and Forest Meteorology,* 57:305–308.

Monteith, J. L. (1965) Evaporation and environment. *Symposia of the Society for Experimental Biology,* 19:205–234.

Monteith, J. L. and Unsworth, M. H. (1990) *Principles of Environmental Physics,* Second edition. Arnold, London.

Penman, H. L. (1948) Evaporation from open water, bare soil and grass. *Proceedings of the Royal Society of London,* A. 194:120–145.

Paw, U. K. T. (1992) A discussion of the Penman form equations and comparisons of some equations to estimate latent energy flux density. *Agricultural and Forest Meteorology,* 57:297–304.

Van Wijk, W. R. and De Vries, D. A. (1963) Periodic temperature variations. In *Physics of Plant Environment* (Ed. W. R. Van Wijk) North Holland, Amsterdam.

Leading to: *Atmosphere; Physiology and heat transfer*

J.A. Clark

ENVIRONMENTAL POLLUTION (see Air pollution; Power plants)

ENZYMATIC REACTION KINETICS (see Biochemical engineering)

EPRI (see Electric Power Research Institute)

EQUATION OF MOTION (see Flow of fluids)

EQUATION OF STATE

Following from: *Thermodynamics*

The equation of state relates the pressure p, volume V and temperature T of a physically homogeneous system in the state of *thermodynamic equilibrium* f(p, V,T) = 0. The equation called the thermic equation of state allows the expression of pressure in terms of volume and temperature p = p(V,T) and the definition of an elementary work $\delta A = p\delta V$ at an infinitesimal change of system volume δV. As distinguished from thermic equations, the caloric equation of state specifies the dependence of the internal energy of the system E on volume V and temperature T (either p and V or p and T).

The equation of state is a fundamental characteristic of a substance which makes possible the application of the general principles of thermodynamics and hydrodynamics to particular physical objects.

Equations of state can not be derived from thermodynamic relations alone but rely also on either experimental measurements or on theoretical calculations by the methods of statistical physics, using various models of interparticle interactions in a system. The First Law of Thermodynamics suggests that there exists a caloric equation of state while the Second, a relation between the thermodynamic and the caloric equations of state in the form $(\partial E/\partial V)_T = T(\partial p/\partial T)_V - p$. The thermodynamic relationship allows the determination of the thermic and caloric equations of state, if the thermodynamic potential, specified in the form of a function of appropriate variables, is known. Thus, for instance, for Helmholtz free energy F as a function of volume V and temperature T, we have

$$p = -\left(\frac{\partial F}{\partial V}\right)_T, \qquad E = -T^2\frac{\partial}{\partial T}\left(\frac{F}{T}\right)_V.$$

A simple example of the equation of state for gases is the Clapeyron-Mendeleev equation $p\nu = RT$, where R is the gas constant and ν is the volume of one mole. For real gases, and considering the interaction between particles, virial equations of state $p\nu = RT(1 + B(T)/\nu + C(T)/\nu^2 + \ldots)$ are widely used, where B(T), C(T), . . ., are the second, third, etc. virial coefficients. These coefficients are the functions of temperature alone and, depend on the forces of two, three or more particles interacting in the system. This equation of state for real gases is the most theoretically-substantiated one, but the calculation of higher order virial coefficients is limited by the difficulties presented by multiple integrals. With a rise in density, the character of the convergence of the virial expansion also deteriorates drastically. In dense systems, nonadditivity of interparticle interaction becomes significant which, when considered, changes drastically the values of the higher order virial coefficients.

A qualitatively accurate picture of a phase diagram of real gases yields the *van der Waals equation of state* $(p + a/\nu^2)(\nu - b) = RT$, in which a and b are constants whose values are determined from the adequate description of experimental data. This equation considers both the existence of attractive forces between the gas molecules, which brings about a decrease in pressure, and the presence of repulsive forces, which appears when the distances between molecules are small, preventing an infinite compression of the gas. The van der Waals equation of state has made it possible for the first time to obtain a thermodynamically-consistent description of a phase gas-liquid transition, which terminates at the critical point with parameters p_c, ν_c, and T_c—where the difference between the liquid and gaseous states disappears. If the equation of state is represented in a reduced form, i.e., in dimensionless parameters p/p_c, ν/ν_c, T/T_c, then the equations of state of various substances differ only slightly (the so-called law of corresponding states) within the limits of a definite group (simple liquids, alkali metals) over a reasonably large near-the-critical area.

The weak dependence of liquid structure on temperature at constant density is the decisive factor in creating models of equations of state for a liquid.

Molten metals, ionic systems and dielectrics, having substantially different attractive forces, and this testifies to the small role these forces play in forming the equilibrium properties of a liquid. These facts reveal the leading contribution of repulsive potential, while attractive forces and accounting for the temperature effects in a liquid bring about only minor corrections. A simple potential of interaction of solid spheres with radius a is widely used as a model for the description of repulsive forces in the system:

$$\varphi(r) = \begin{cases} \infty, & \text{if } r \geq a \\ 0, & \text{if } r > a \end{cases}$$

which characterizes the limit state of a highly-compressed and hot liquid. The model for hard spheres, being the simplest nontrivial model of the equation of state of a liquid, is widely used in **Integral Equations**, numerical *Monte-Carlo methods* and in molecular dynamics. Using different methods, the calculation of a system of particles with the potential of solid spheres is made over the entire range of a liquid state. This can be represented with high accuracy by the Pade analytic Carnahon-Starling approximation

$$p = \frac{RT}{V}\frac{1 + \nu + \nu^2 - \nu^3}{(1 - \nu)^3},$$

which has a free parameter $\nu = 4\pi a^3 n/3$ (n is the number density of the particles) that characterizes the degree of solid sphere packing. It must be noted that numerical calculations of the potential of solid spheres, with density approximating the dense packing of spheres $\nu \leq 1$, have shown a tendency towards the appearance of a long-range order in the system which is associated with crystallization.

Considerable progress in describing the thermodynamic properties of a liquid phase have been attained with the help of a soft sphere model. This model uses a power potential of repulsion

$$\varphi(r) = \begin{cases} \varepsilon(a/r)^n, & \text{if } r \leq a \\ 0, & \text{if } r > a \end{cases}$$

which is also the basis for carrying out numerical calculations by the Monte-Carlo method and by the molecular dynamics method. The resulting calculations of the equation of state of soft spheres is described for any $4 \leq n \leq 12$ by the approximating formula

$$p(V, T) = \frac{RT}{V}\left[1 + \frac{n(n + 4)}{18}\eta^{n/\rho}\left(\frac{\varepsilon}{kT}\right)^{1/3}\right],$$

where $\eta = a^3n/\sqrt{2}$ and ε and n are the additional free parameters of the model. The power potential of repulsion, like the potential of hard spheres, brings about crystallization—whose characteristics depend strongly on the value of n.

A modified van der Waals' model based on the combination of an approximate formula for the potential of soft spheres and an approximated average field considerably improves the description of the thermodynamic properties of a dense gaseous phase and a liquid. Greater accuracy in describing experimental data is achieved by using the results of calculations for the thermodynamics of a system of soft spheres and introducing adjusting parameters into the average field model. The modified van der Waals' equations of state can be equally useful for calculating the thermodynamic properties of liquids and dense gaseous phase. They have asymptotics of an ideal gas, which makes it possible to use them successfully for expressing a wide-range of semiempirical equations of state. These models can also define the equilibrium liquid-vapour line and the critical point. But as a whole, they are quantitative and do not give a detailed description of the fine characteristics of a liquid; in particular, the dual correlation function g(r) which defines all equilibrium properties of a liquid phase for a given potential of interparticle interaction $\varphi(r)$.

The density of a liquid near the melting point curve differs slightly (by 10–20%) from that of a crystal, which explains the success of equations of state for describing solid-state models. The notion of liquid molecules localizing in a small area in space under the action of an average field of neighbors forms the basis for the lattice model, or for the unconfined space theory. Such an approach, however, is applied only for a strongly-compressed liquid at low temperatures when kinetic energy forms a small fraction of total energy, a condition which, in general, does not correspond to a liquid state. A combination of solid-state and gaseous notions brings about a multistructural model, according to which an actual liquid is an equilibrium mixture of gaseous and solid bodies, whose properties can be described by a dual distribution function. Similar notions are used in a cell model where the liquid is divided into molecular cells, which are filled depending on thermodynamic conditions. Such models describe only schematically the thermodynamic properties of a real liquid.

Models of hard and soft spheres properly take into account the basic qualitative characteristics of multiparticle interactions in a dense liquid. This makes it possible to use them as a zero approximation model in calculating equations of state. The details of real potential of interparticle interaction are carried out within the framework of the thermodynamic perturbation theory, the chaotic phase approximation or on the basis of the variational method of perturbation theory. The variational method is most convenient for calculating the thermodynamics of particular liquids over a wide range of parameters and is especially effective in describing the properties of a dense liquid phase. Gibb's-Bogoliubov's inequality is used in finding the free energy F(V,T) of a system of particles with an arbitrary potential of interaction $\varphi(r)$

$$F \le F_0(\lambda) + \frac{n}{2}\int [\varphi(r) - \varphi_0(r)]g_0(r\lambda)d^3r,$$

where F_0 and g_0 are the free energy and a dual correlation function of the original system with interparticle interaction potential $\varphi_0(k)$, respectively. The equation of state of an arbitrary system is defined by averaging the perturbation over the structural characteristics of an initial approximation and further minimizing a free parameter of this approximation λ according to $(\partial F/\partial\lambda)_{V,T} = 0$. It is good practice to take a system of solid spheres as the zero approximation, for there are analytical expressions defining the free energy and dual correlation functions. This allows the explicit calculation of the thermodynamic characteristics of an arbitrary potential of the interparticle interaction. The radius of solid spheres is, in this case, the minimizing parameter λ. But at large densities, use of the solid spheres system as the initial one brings about essential errors, caused by the strong repulsion of particles at small distances. The choice, *as the zero approximation model,* of "soft" power potential $\varphi(r) = C/r^{12}$ is usually favoured here. Use of the soft and solid spheres model for the original approximation results in equations of state of inert liquids or molecular gases and simple metals that are of high accuracy.

The development of the variational method of perturbation theory for calculating the equation of state of liquid metals has led to the use of the results obtained from the Monte-Carlo method for a single-component plasma as the zero approximation. The model of a single-component plasma is based on numerical simulation of a simple system of point-like particles with Coulomb interaction, located, for its stability, on a homogeneous background of a balance charge with an opposite sign. In this case, there exists an analytical expression for F_0, with the Coulomb nonideality parameter $\Gamma = e/kTr_{ws}$ as the minimizing quantity, and the average interparticle distance $r_{ws} = (3/4\pi n)^{1/3}$. Results of calculating the equations of state have shown that for liquids with slight repulsion (alkali metals, solt melts, alloys), a single-component plasma model is best suited. It should be noted that the final expression for the equation of state of liquid metals—in addition to the configurational term determined by the variational method, the expression for energy—must include terms according to the Coulomb interaction energy and the kinetic energy of particles of the system, whose total value at high temperatures becomes significant. In simple liquids, the same holds true when accounting for electron excitation at high temperatures.

Additional information on equations of state are given in the article **PVT Relationships**.

To form equations of state for a solid body, phonon and electron components are distinguished and are considered separately to a large degree. A distinguishing feature of solid bodies is their atoms which oscillate about the nodes of a crystal lattice. A quasi-harmonic model of a solid body is based on the representation of a crystal lattice in the form of 3N oscillators. The most widespread interpolation of the Debye equation for a solid body has been obtained in the approximation of a sound spectrum of heat variations, which yields the following expression for heat components of energy and pressure:

$$E(V, T) = 3RTD(\theta_D(V))/T, \qquad p(V, T) = \gamma(V)E(V, T)/V,$$

where $D(x) = 3/x^3 \int_0^x t^3dt/(e^t - 1)$ is the Debye function, θ_D is the Debye temperature dependent on volume and $\gamma = -dl\ n\ \theta_D/d\ ln\ V$ is the Grüneisen coefficient. The Debye function $D(\theta_D/T)$ ensures in this case a low-temperature $c_v \propto T^3$ and a high-temperature $c_v \cong 3R$ asymptotes for the heat capacity of a lattice. In the Debye model, the solid body is taken to be homogeneous and isotropic, thermal excitations of a crystal boil down to sound waves. External pressure in this case only influences the boundary temperature, while the spectrum itself remains harmonic.

Calculations for an electron spectrum are usually made within the framework of the zone model. Here the crystal presents an ideal periodic structure with immobile cores in the cell nodes and each electron moves in a periodic self-consistent potential, which allows expressing the wave function of the whole system through single-electron Bloch's functions. The distinction between the models of equations of state reduces to the methods of calculating single-electron functions within the elementary atomic cells, and to the methods of joining the conditions at their boundaries. In the widely-used Wigner-Seitz model of spherical cells, a complete self-consistent problem is replaced by the solution of the Chariri-Fock equation for a single cell with Bloch's boundary conditions. In doing this, zones of allowed states for electrons arise. The association of pressure with wave functions on the cell boundary makes it possible to determine the equation of state of the system.

Under normal conditions, the approach of spherical cells is too schematic; besides, a considerable contribution is made by the effects of exchange and correlation interactions. For this reason, methods of adjoined flat or spherical waves, linear orbitals, etc., have been developed which accounted for the real shape of an elementary crystal cell. Exchange-correlation effects are accounted for in local density approximation.

A pseudo-potential model of a solid body is based on a true potential interaction of an electron with multicharge ion onto a simplified, single-particle. A *consecutive theory of pseudo-potentials accounting for two-, three-, etc., ionic interaction through conduction electrons is a rigorous expansion into a series in terms of the parameter of electron-ion interaction.* The application of a pseudo-potential approach at high degrees of compression is limited by the effect of ion core overlapping.

The increase in temperature brings about the appearance—both in gas and in liquid—of free charges, with the long-range character of the Coulomb interaction making significant its contribution into the equation of state over a large area of thermodynamic parameters. In a rarefied plasma, a separate description of states of discrete and continuous spectra—which define the internal structure of atoms and ions as well as the character of the behaviour of free electrons—is practiced. Such an approximation forms the basis of a "chemical" model of plasma in which the number of particles of $\{N_j\}$ kind is determined from the conditions of **chemical equilibrium**

$$\sum \mu_j = 0, \qquad (\partial F(V, T, \{N_j\})/\partial N_k)_{V,T} = 0,$$

where μ_j is the chemical potential of particles, and all the hypotheses specifying structure and interaction are involved in the expression for total free energy. This expression includes the kinetic energy of ions and electrons of a continuous spectrum, the contribution of the discrete spectrum and various corrections for interparticle interaction. At extremely high temperatures of the plasma or with strong rarefaction, the energy contribution of a photon gas $F_l = (4\sigma/3c)VT^4$ (σ is the *Stefan-Boltzmann constant* and c is the velocity of light).

References

Godwall, B. K., Sikka, S. K., and Chidambaram, R. (1983) *Phys. Rep.* 102.

Ross, M. 1985. *Rep. Prog. Phys.*, 48.

Leading to: *PVT Relationships; Van der Waals equation*

V.A. Bushman and V.E. Fortov

EQUILIBRIUM (see Thermodynamics)

EQUILIBRIUM STATE (see Vapour-liquid equilibrium)

EQUILIBRIUM STATES, THERMODYNAMIC (see Free energy)

EQUILIBRIUM TEMPERATURE (see Adiabatic wall temperature)

EQUILIBRIUM VAPOUR PRESSURE, CHANGE WITH TEMPERATURE (see Clapeyron-Clausius equation)

ERGODICITY

A *random process* for which *ensemble averages* are identical to time or spatial averages is said to be ergodic. (See **Stochastic Process** for definitions of averages.) Both types of averaging are acceptable ways of describing a random process.

Ensemble averages are directly linked to a probabilistic formulation of random processes, i.e., to all realisable states to each of which can be ascribed or derived a unique probability of occurrence. Most current statistical theories use ensemble averaging, especially so in the theory of **Turbulence**. However, in practice, such averages are difficult to measure and recourse is made to the measurement of time or spatial averages instead. Meaningful comparison between theory and measurement thus relies on the process being ergodic. It is valid for a stationary random process [see e.g., Beran (1968)]; however, in all other cases 'ergodicity' is generally an hypothesis to be validated by experiment.

For readable accounts see Beran (1968) and Yaglom (1962); more in depth discussions are to be found in Truesdell (1961) and in the basic article by Birkhoff (1931).

References

Beran, M. J. (1968) *Statistical Continuum Theories,* Interscience Publishers (JW).

Birkoff, G. D. (1931) *Proc. Natl. Acad. Sci.* 17:656.

Truesdell, C. (1961) Ergodic theories. *Proc. Intern. School Phys.* "E. Fermi" XIV Course: 21. Academic Press, New York.

Yaglom, A. (1962) *An Introduction to Mathematical Stationary Random Functions.* Prentice-Hall, Englewood Cliffs, N.J.

M.W. Reeks

ERGODIC PROCESSES (see Ergodicity)

EROSION

Erosion is the wearing away of mass from the surface of a body (barrier) by a flow of a fluid containing particles. Under natural conditions, the motion flow of particles is associated with the mobility of the carrier phase whether gas or liquid. Driven by inertia or other forces, the particles in a flow fall or onto the surface of the body and may cause damage. In contrast to a chemical or a thermal phenomenon (see **Ablation**), this is called *erosion* destruction.

Until 1930, studies on erosion were mainly experimental in nature, directed toward fluid and their break-up processes. More recently, the problem of wearing of steam turbine rotor blades arose, with water drops appearing in the steam circuit due to condensation. No less serious were problems involving pulverized coal in gas-turbine locomotives as well as problems in the fields of catalytic cracking, hydraulic power engineering, aviation and cosmonautics. Of particular interest was air vehicle flight through atmospheric formations (rain, snow, hail, dust) where erosion made optical and radio location devices, heat fairings and elements of control system, inoperative in a short time. Studies on the breakdown of control surfaces in solid propellant rocket engineering even spawned a new term—"dust knife."—which refers to the formation of narrow zones with increased particle concentrations in carrier flows.

As knowledge of high-speed heterogeneous flows spread in scientific and technical literature, the notion of an erosion barrier (analogous to sound barrier in aviation) appeared. This concept implied that no modification of construction materials and coatings could ensure the required level of stability and availability of a unit unless the entire structure of a heterogeneous flow is deliberately changed.

Erosion destruction is the sum of many elementary processes, which complicate the interpretation of experimental data and the construction of a general physical model of this phenomenon. Several typical conditions for the interaction of particles with a streamlined body can be distinguished in a heterogeneous flow, depending on the size and velocity of particles and also on their concentration.

Flow and collisions of isolated (independent) particles

The impact of an isolated particle into a surface type the mechanical motion of a body encountering resistance. Unlike classical problems of hydrodynamics and ballistics, a more intricate relationship between resistance and the instantaneous velocity of a particle V_p exists. For a simple case of a particle colliding in a direction normal to a semi-infinite homogeneous plate, the momentum equation is expressed as:

$$m_p \frac{dV_p}{d\tau} = -S_p\left(H_0 + C_D \frac{\rho_0 V_p^2}{2}\right). \qquad (1)$$

Here, m_p and S_p are the mass of a particle and the projection of its contact surface with the barrier onto a plane perpendicular to the velocity vector V_p (**Figure 1**); H_0 and ρ_0 are the dynamic hardness and the density of the barrier material; C_D is the inertia resistance coefficient, which depends mainly on particle shape. For a sphere, $C_D \cong 1$; for a cylinder with a flat face, $C_D \cong 2$; for a cylinder with a conical lip, $C_D \cong 2 \sin^2 \Theta/2$ where θ is the vertex angle of the cone.

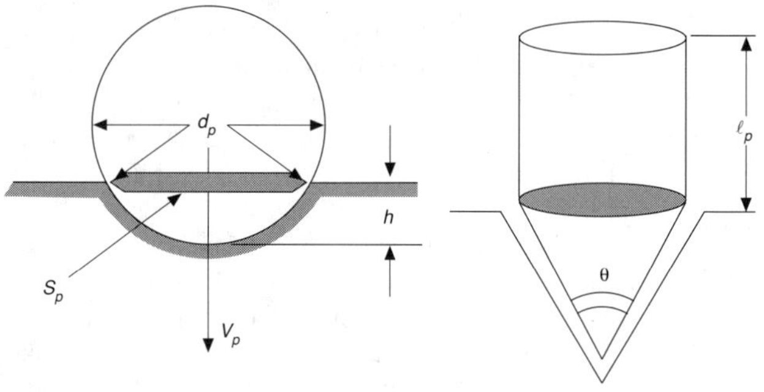

Figure 1. Penetration into a solid surface by particles impinging normally onto it.

Figure 1 shows the penetration process of a spherical particle with diameter d. Depending on the depth of penetration or the depth of crater h, the projection area of a contact spot S_p varies within the range of $(h/d_p) < 0.5$ as:

$$S_p = \pi h(d_p - h). \qquad (2)$$

Denoting the density of the particle material by ρ_p leads to solution of Equations 1 and 2:

$$\left(\frac{h}{d_p}\right)^2 = \frac{2}{3}\frac{\rho_p}{\rho_0}\frac{1}{C_D} \ln\left(1 + C_d \frac{\rho_0 V_p^2}{4H_0}\right). \qquad (3)$$

For small values of $(C_d\, \rho_o V_p^2/4H_0)$ the logarithmic term is given approximately by these values and:

$$h/d_p \cong \sqrt{\rho_p/(6H_0)}V_p. \qquad (4)$$

Equation 4 can be conveniently used for the solution of a reverse problem, i.e.: the determination of the dynamic hardness of the barrier material H_0 from measured values of striker velocity V_p and crater depth h. The dynamic hardness H_0, in contrast to static hardness H_B, is not a constant parameter of the material; however, for most metals, hardness H_0 grows only slightly with striker velocity at room temperatures, thus $H_0 \sim V_p^n$, where $n = 0.02 - 0.04$. Within the range of $10 < V_p < 1000$ m/s, dynamic hardness H_0 exceeds the Brunel hardness by no more than $1.5 - 2$ times.

Figure 2 shows the general results of studies on collisions between iron (curve 1) and copper (curve 2) balls with a massive lead plate, and also of iron, copper and lead balls with plates of the same material (curve 3). Comparing the theoretical dependence (Equation 3 or 4) with the experimental data, a qualitative agreement is noted: the theoretical model for the penetration of an undistorted (solid) particle into a barrier is not effective at high levels of striker energy $(\rho_0 V_p^2)/H_0$ since it does not describe the typical "hump" and a "trough" on the curves for iron and copper balls, which strike against a lead barrier. But more important, it does not explain why all the three experimental curves merge over the range of high velocities V_p.

In erosion destruction, pressure develops in the zone of particle contact with the barrier, which eventually exceeds the yield

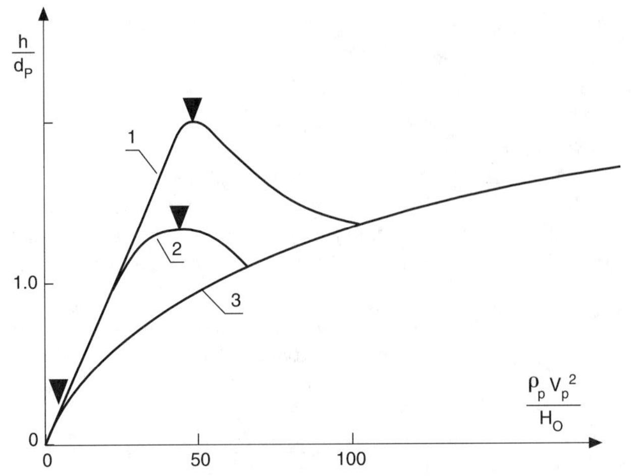

Figure 2. Results for impaction of iron (1) and copper balls with a lead plate.

point of the material. This provides a basis for the construction of a hydrodynamic analogy with a liquid jet directed into a basin filled with liquid. However, the liquid jet model poorly describes the geometrical features of compact (for instance, spherical) particles.

The limiting factors cited above have spurred the search for other approaches in constructing theoretical models of the process of particle collisions with barriers. As in the energy approach, the unknown quantity is the amount of ablated mass of the barrier material rather than the barrier depth. By reducing the determining parameters and replacing the mechanical characteristics of the material with thermodynamic ones, a unified interpretation of the experimental data has been achieved—both in single and in repeated impacts of particles on the barrier.

Figure 3 shows the experimental data for the effect of impact velocity of steel particles on the dimensional mass loss rate of lead

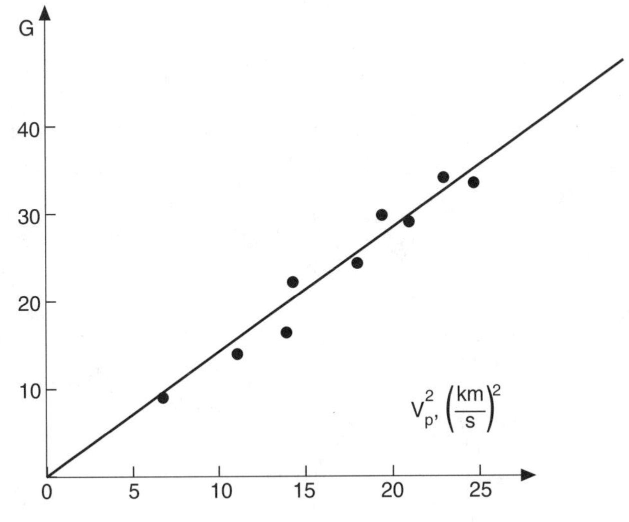

Figure 3. Dimensionless mass loss as a function of V_p^2.

barrier G. This important characteristic of erosion is defined as the ratio of mass loss of the barrier m_{er} to the mass of a striking particle m_p: $G = m_{er}/m_p$. The dimensionless mass loss rate is directly proportional to kinetic energy over a wide range of collisions velocity V_p

$$G = V_p^2/2H_{er}(1). \quad \textbf{(5)}$$

The proportionality factor has an enthalpy dimension and is called the effective enthalpy of erosion destruction $H_{er}(1)$. The index in (1) shows the type of the process: single or independent collisions. Although the ablated mass m_{er} can exceed by many times the mass of the striking particle m_p, it does not fully correspond to the entire volume of the crater. Thus in case of lead barriers, only 15% of the crater volume is formed due to mass loss. In brittle barriers, this can increase up to 50%. Most of the crater is formed due to compressive (radial) strains or the displacement of the substance beyond the original surface of the barrier (the so-called "collar"). It must be noted that the displaced-to-ablated mass ratio is not constant, especially within the range of low and moderate impact velocities V_p.

Steady-state flow of repeated collisions

The relationship between independent particle collisions and repeated collisions in a steady-state flow can be described with the help of the *covering coefficient* K_p. This is defined as the ratio of the midship sections of all incident particles $\Sigma_{i=n} S_{pi}$ to the area of the exposed surface S_0

$$K_p = \sum_{i=n} S_{pi}/S_0 \quad \textbf{(6)}$$

If all the particles are spheres fixed diameter d_p and density ρ_p, the covering coefficient K_p is proportional to a mass flux of particles $\dot{m}_p$:

$$K_p = \frac{3}{2}\dot{m}_p/(\rho_p d_p). \quad \textbf{(7)}$$

The flow of independent particle collisions corresponds to $K_p \ll 1$. But if $K_p \gg 1$, then incident particles cover the entire surface of the body closely and repeatedly.

For a higher flux of particles erosion destruction loses the typical characteristics of local action; craters are absent. The denser the flow of particles and the less the size of each of them, the more complete and stronger is the analogy between erosion and ablation (see **Ablation**); or in the general case, between the action of single-phase gas flow and the interaction of a body with a heterogeneous dispersed medium. Nonetheless, the fundamental difference between homogeneous and heterogeneous flows cannot be forgotten. The latter has a great or zonal depth of interphase interaction with the body (**Figure 4**). Not only the shock wave before the barrier (1), but also the boundary layer (2), the body's surface and the subsurface layer have no clear geometrical interpretation. The flow of high-velocity particles moves, as a rule, along trajectories which intersect the surface of the body (**Figure 4**), with the surface itself being highly irregular which makes the application of the

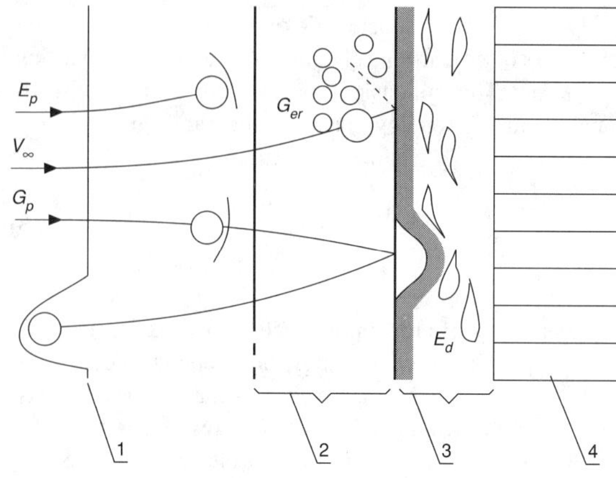

Figure 4. Multiple particle/surface interactions.

notion of wall temperature T_w difficult. The so-called "regenerated" layer (3) in which many longitudinal transverse discontinuities are formed due to the repeated passage of shock waves, is moved to a considerable depth underneath bottom of the craters. Not only the mechanical but also the thermophysical characteristics of this layer can be far from those of the original material (4). Under these conditions, the only correct approach is the use of the energy balance equation for relatively large elementary volumes, which embraces all three areas in **Figure 4**:

$$E_p = E_r + E_d + G_{er}\Delta Q_{er} + G_p\Delta Q_p + E_b. \quad \mathbf{(8)}$$

The supplied energy E_p is the sum of the kinetic energy of incident particles $G_pV_p^2/2$ and the change in the heat content $G_pC_p(T_p - T_w)$. The reflected part of energy flow E_r depends on the capability of particles to recoil. The dissipated energy E_d is identified with the energy of elastic waves inside the body, which transfers into heat and is absorbed by the barrier substance. Analogous with the thermal description (see **Ablation**) for the steady-state flow with velocity G_{er}, it can be assumed that $E_d \cong G_{er}C_0(T_w - T_0)$, where C_0 is the mean heat capacity of the barrier substance over the temperature range T_0 to temperature T_w on the exposed surface. The two next terms in the balance (8) are the energy consumption for the destruction of the barrier $G_{er}\Delta Q_{er}$ and the destruction of particles $G_p\Delta Q_p$; finally, the last term E_b corresponds to a blocked part of the initial energy of the flow of particles. The blocking effect arises due to the formation of a layer of reflected and ejected particles over the exposed surface.

Effective enthalpy of erosion destruction mentioned above can now designated as the coefficient to be used before the mass loss rate when combining the second and the third terms in Equation 8:

$$E_d + G_{er}\Delta Q_{er} = G_{er}[C_0(T_w - T_0) + \Delta Q_{er}] = G_{er}H_{er}.$$

Experiments have shown that with a rise in impact velocity V_p, the energy of reflected particles decreases rapidly, and as a consequence decreases the contribution of E_r. Conversely, the role of blocking E_b can rise with increasing impact velocity V_p (at constant concentration of particles in the heterogeneous flow z_p). Finally, the analysis of experimental data given in **Figure 2** shows that the contribution of the term $G_p\Delta Q_p$ is limited to a comparatively narrow range of collision velocity.

Schematic diagrams of the process of erosion development with $\Delta Q_p = 0$ and $E_b = 0$ are given in **Figure 5**. Depicted by the solid line is the functional dependence of the dimensional mass loss rate of erosion or the intensity of erosion G on the impact velocity V_p.

Taking into account the capability of particle recoil from the barrier surface demands that a correcting factor η be introduced in Equation 5

$$G = \eta V_p^2/(2H_{er}), \quad \mathbf{(9)}$$

where $\eta = 0$ for $V_p \leq V_{cr}$ and $\eta \to 1$ for $V_p \geq V_p^*$. The critical velocity V_{cr} is the boundary which determines the transition from elastic to plastic deformation of the barrier material. The threshold velocity value V_p^* corresponds to the onset of so-called *flow* or *hydrodynamic* interaction between the colliding bodies and the surface. It is a known fact that such an interaction usually sets in when the pressure at the point of contact exceeds three or four times the yield point of the barrier material. Hence, it follows that the threshold value of velocity must exceed the critical by about two times ($V_p^* > \sqrt{3}V_{cr}$). In view of this, note that following analytical relation for η can be written as

$$\eta = 1 - \exp\left[\frac{V_p - V_{cr}}{0.5V_{cr}}\right] \quad \text{for} \quad V_p \geq V_{cr}. \quad \mathbf{(10)}$$

Figure 6 summarizes the experimental data on the effect of impact velocity of solid spherical particles (with a diameter $d_p = 1.58$ mm and made from tungsten carbide $\rho_p = 14500$ kg/

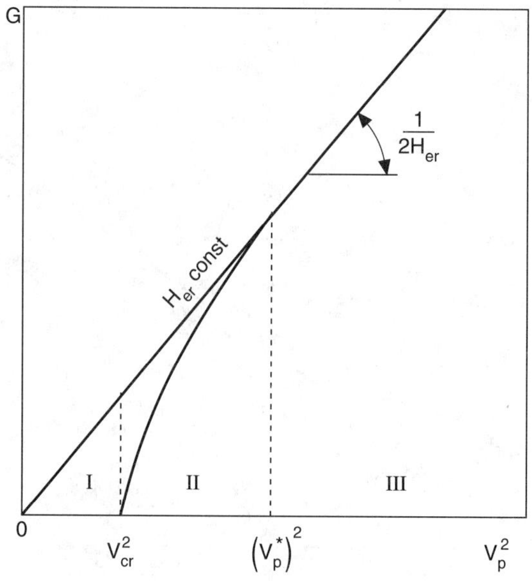

Figure 5. Idealised relationship between G and V_p.

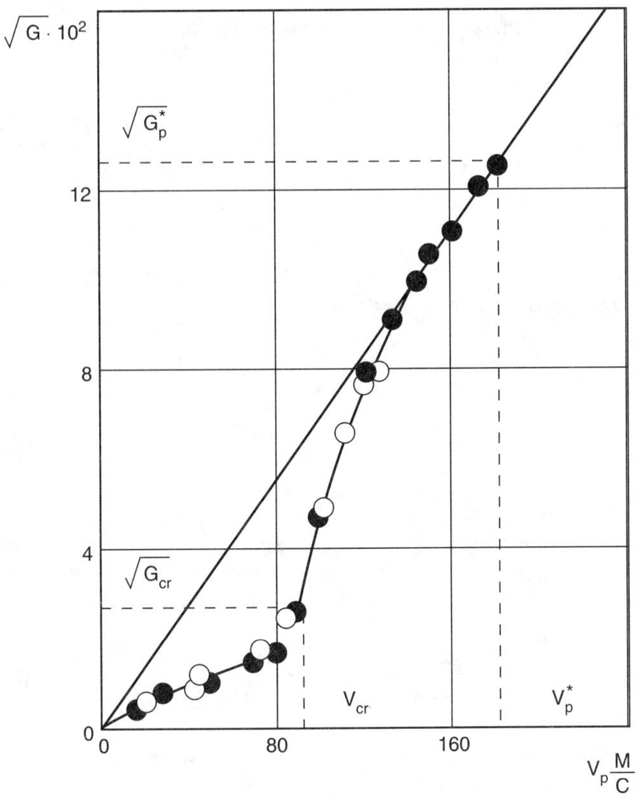

Figure 6. Actual data for G as a function of V_p for a tungsten carbide/aluminum system.

m^3) on the erosion intensity of an aluminum barrier. As can be seen, the idealised model of the process (**Figure 5**) does not fully correspond to the experimental data presented in **Figure 6**. A measurable process of destruction begins in a flow of particles at rather low impact velocities $V_p < V_{cr}$. However, the intensity of the initial destruction (up to $V_p < V_{cr}$) is so small that it is of no practical interest ($G \leq 0.001$).

A sharp bend in the dependence of G on V_p (**Figure 6**) occurs when the velocity reaches its critical value $V_p = V_{cr}$ is a convenient starting point for creating the design diagram for determining this criterion. An essential change in the resistance C_d, and also in the pressure at the point of contact of a spherical particle, occurs when particle penetration increases to $h/d_p \geq 0.25$. By comparing **Equation 4**, with the condition that $H_0 \cong 2H_B$, the critical value of the velocity can be approximately estimated:

$$V_{cr} \cong \sqrt{\frac{3H_0}{8\rho_p}} \cong \sqrt{\frac{3H_B}{4\rho_p}} \cong 0.2\sqrt{H_{er}\rho_0/\rho_p},\ [\text{m/s}]. \qquad \textbf{(11)}$$

Thus in the absence of blocking ($E_b = 0$), the intensity of erosion destruction is described by Equations 9, 10 and 11, with the only empirical constant in these equations being effective enthalpy H_{er}. As a rough estimation of the quantity H_{er}, the following relationship which relates H_{er} and the thermodynamic parameters for composite and heat shielding materials is recommended:

$$H_{er} \cong \frac{\varphi n}{3}\,[C_0(T_w - T_0) + \Delta Q_m],$$

where φ is the fraction of reinforcing fibers and n is the number of longitudinal cross bonds in the composite material (for homogeneous materials $\varphi = n = 1$). The melting heat ΔQ_m and heat capacity C_0 define the thermodynamic enthalpy for most stable (reinforcing) components of the material.

Nonsteady flow of repeated collisions

Experiments have indicated that even at constant impact velocity V_p, the action of a flux of particles on the barrier is not an arithmetic sum of single actions. Every preceding particle does not merely knock out a certain mass from the barrier but, by loosening them up, prepares the underlying layers for destruction. Damage studies have been carried out which provide a basis for the use of models of endurance failure, but these have still to account for the experimental results.

The result of the "loosening up" process is that there is an "incubation period" (at a given particle flux and velocity) for erosion rate to become constant. There is an analogy between this incubation period of erosion and the process of nonsteady thermal destruction (see **Melting**), during which almost half of the supplied energy goes into the substance heating and the other half is only absorbed in phase transformations. According to this analogy, there is a threshold value of the total kinetic energy of the particles impacted

$$\dot{M}_p^* \frac{V_p^2}{2} = a^* \qquad \textbf{(12)}$$

which on being reached makes the process of erosion stable, i.e., the derivative (slope) of the dependence $\dot{m}_{er}(\dot{m}_p)$ tends to a constant value, which does not depend on the mass of the particles which are subsequently deposited. In equation 12, $\dot{M}_p^*$ is the mass of particles (per unit surface area) which need to be deposited before the incubation period ends:

$$\frac{d\dot{m}_{er}}{d\dot{m}_p} \xrightarrow{\dot{m}_p \to \dot{m}_p^*} G = \eta\,\frac{V_p^2}{2H_{er}}.$$

The parameter a* is the second fundamental characteristic of erosion destruction (the first being H_{er}) of the barrier substance and is independent of impact velocity, mass, density of paticles, etc. The results of the experiments show that a* changes only slightly from one class of materials to another: $10^5 < a^* < 10^6$ J/m^2.

Figure 7 is a comparison of the calculated and experimental data for the dependence of the intensity of erosion of a composite material on the covering coefficient K_p, under the action of a flow of particles with diameter $d_p = 0.5 \times 10^{-3}$ m with an impact velocity of 1,800 (curve 1), 2,400 (curve 2), 3,000 (curve 3), 3,650 m/s (curve 4). Also shown by a dotted line are the results of the calculation of a threshold value of the covering coefficient K_p^* for different impact velocities V_p, obtained with the use of Equations 7 and 12.

In case $\dot{M}_p V_p^2/2 < a^*$, the intensity of erosion depends on the number of particles that have fallen out or on the covering coeffi-

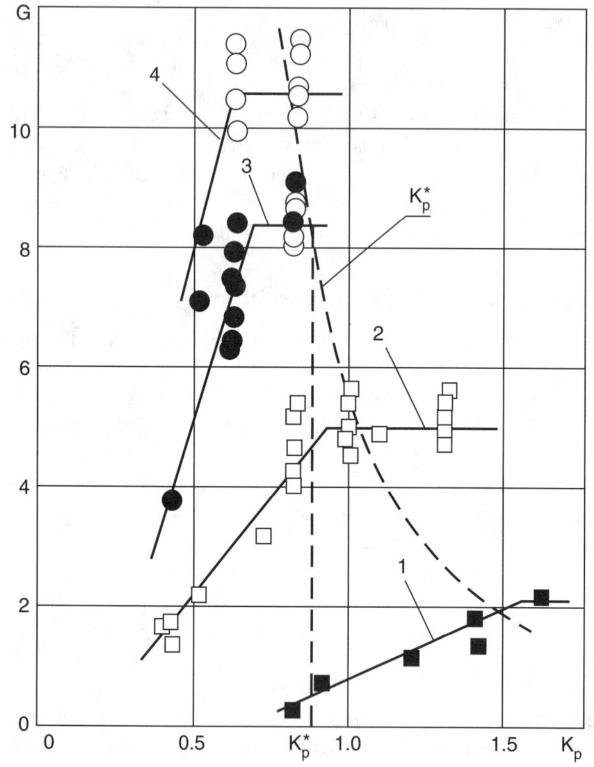

Figure 7. G as a function of K_p for a typical composite erosion.

cient (**Figure 7**). This defines the relation between effective enthalpy H_{er} and the quantity $H_{er}(1)$ (**Equation 1**) introduced earlier. The mass flux of a single particle when falling, is $\dot{M}_p(1) = m_p/S_p$, which in the case of a spherical particle yields $\dot{M}_p(1) = \frac{2}{3}\rho_p d_p$. Using Equations 5 and 12 yields

$$H_{er}(1)/H_{er} = 1 + (1/n),$$

where

$$n = \frac{\dot{M}_p(1)}{M_p^*} = \rho_p d_p V_p^*/3a^*.$$

Accordingly, the critical velocity of the start of destruction with a single impact V_{cr} (Equation 1) will differ from the corresponding steady-state value V_{cr}:

$$V_{cr}(1)/V_{cr} = \sqrt{H_{er}(1)/H_{er}} = \sqrt{V_{cr}}[3a^*/\rho_p d_p]^{0.25}.$$

This rule is indicative of the presence of the scale effect: critical velocity V_{cr} (Equation 1) increases in proportion to $d_p^{-0.25}$ with a decrease in particle size d_p.

All the above results apply to the *case of the impact of particles in a direction normal to the barrier surface.* At present, there is only an elementary theoretical explanation for the process of collisions at an oblique angle. The experimental results disagree with predictions based on the law of conservation of momentum.

YuV. Polezhaev

ERROR FUNCTION

An error function is defined by the integral

$$\text{erf}\, x = \frac{2}{\sqrt{\pi}} \int_0^x e^{-t^2}\, dt, \qquad (1)$$

and it occurs frequently in engineering problems; e.g., in heat conduction problems. The error function represents the area under the Gaussian function $2e^{-t^2}/\sqrt{\pi}$ from t = 0 to t = x, so that erf ∞ = 1. The complementary error function is:

$$\text{erfc}\, x = 1 - \text{erf}\, x = \frac{2}{\sqrt{\pi}} \int_x^\infty e^{-t^2}\, dt. \qquad (2)$$

The error function erf x is a monotonically increasing odd function of x; i.e., erf $(-x) = -$erf x and erf $x_1 \le$ erf x_2 whenever $x_1 \le x_2$. Its *Maclaurin series* (for small x) is given by:

$$\text{erf}\, x = \frac{2}{\sqrt{\pi}}\left(x - \frac{x^3}{3\cdot 1!} + \frac{x^5}{5\cdot 2!} - \frac{x^7}{7\cdot 3!} + \cdots\right), \qquad (3)$$

and for large values of x, the asymptotic expansion is:

$$\text{erf}\, x = 1 - \frac{e^{-x^2}}{\sqrt{\pi}x}\left(1 - \frac{1}{2x^2} + \frac{1\cdot 3}{(2x^2)^2} - \frac{1\cdot 3\cdot 5}{(2x^2)^3} + \cdots\right) \qquad (4)$$

where erf x may be approximated as

$$\text{erf}\, x \approx 1 - \frac{1}{\sqrt{\pi}x} e^{-x^2}. \qquad (5)$$

There exist extensive tabulations of erf x [see Abramowitz and Stegun (1965), for example].

Reference

Abramowitz, M. and Stegun, I. (1965) *Handbook of Mathematical Functions,* Dover Publications, New York.

H.G. Khajah

ESA (see European Space Agency)

ESDU (see Engineering Sciences Data Unit)

ETHANE

Ethane (C_2H_6) is a colourless and odourless gas that is slightly heavier than air. It occurs as a constituent of natural gas, which is the main raw material for its production. It forms an explosive and flammable mixture with air at low concentrations. It is used in the petrochemical industry as a feedstock for the production of ethylene and vinyl chloride.

Some of its physical characteristics are:

Molecular weight: 30.07
Melting point: 89.88K
Normal boiling point: 184.52K
Normal vapor density: 1.355 kg/m^3 (@ 273.15K; 1.0135MPa)
Critical temperature: 305.33K
Critical pressure: 4.872MPa
Critical density: 206.6 kg/m^3

References

Beaton, C. F. and Hewitt, G. F. (1989) *Physical Property Data for the Design Engineer.* Hemisphere Publishing Corporation, New York.

Friend, D. G., Ingham, H., and Ely, J. F. (1991) Thermophysical properties of ethane. *J. Phys. Chem. Ref. Data,* 20:275.

Hendl, S. et al. (1994) The transport properties of ethane. I viscosity. *Int. J. Thermophys.* 15:1.

Vesovic, V., et al. (1994) The transport properties of ethane. II thermal conductivity. *Int. J. Thermophys.* 15:33.

V. Vesovic

ETHANOLAMINES (see Natural gas)

ETHENE, SEPARATION OF (see Dephlagmator)

Table 1. Ethane: Values of thermophysical properties of the saturated liquid and vapor

T_{sat}, K	184.6	200	220	240	260	270	280	290	300
P_{sat}, kPa	102	217	492	967	1713	2211	2808	3517	4356
ρ_l, kg/m^3	543.9	524.0	496.1	465.1	429.0	407.9	383.1	351.7	303.2
ρ_g, kg/m^3	2.06	4.17	9.04	17.49	31.70	42.26	56.58	77.44	114.5
h_l, kJ/kg	−666.9	−628.8	−577.3	−522.4	−462.7	−430.2	−394.8	−354.6	−303.2
h_g, kJ/kg	−177.5	−161.2	−142.5	−127.7	−118.8	−117.8	−120.5	−129.5	−154.0
$\Delta h_{g,l}$, kJ/kg	489.4	467.7	434.8	394.7	343.9	312.4	274.3	225.1	149.2
$c_{p,l}$, kJ/(kg K)	2.43	2.50	2.64	2.84	3.18	3.47	3.97	5.09	9.98
$c_{p,g}$, kJ/(kg K)	1.34	1.48	1.71	2.02	2.50	2.91	3.61	5.19	12.74
η_l, μNs/m^2	165.8	136.7	108.2	86.3	68.3	60.2	52.3	44.2	34.6
η_g, μNs/m^2	—	6.02	6.68	7.47	8.54	9.28	10.25	11.64	14.23
λ_l, mW/(K m)	166.5	150.7	132.0	114.8	99.1	91.8	85.1	79.4	75.1
λ_g, mW/(K m)	—	—	—	16.0	19.7	22.2	25.6	31.1	44.7
Pr_l	2.42	2.27	2.16	2.13	2.19	2.27	2.44	2.83	4.60
Pr_g	—	—	—	0.94	1.08	1.22	1.45	1.94	4.05
σ, mN/m	15.8	13.3	10.1	6.85	4.28	3.14	2.00	1.14	0.43
$\beta_{e,l}$, kK^{-1}	2.33	2.57	3.01	3.72	5.06	6.31	8.64	14.4	43.3

The values of thermodynamic properties are from Friend et al. (1991), transport properties from Hendl (1994) and Vesovic (1994) and surface tension from Beaton and Hewitt (1989).

ETHANOL

Ethanol (C_2H_5OH) is a colourless, volatile liquid that is miscible with water. It used to be known simply as *alcohol,* which originally comes from an Arabic word and over time evolved to indicate 'purity of highest degree.'

Naturally-occurring ethanol is produced by fermentation of sugar or starch. Industrial production is based primarily on catalytic hydration of ethylene. Ethanol vapor forms an explosive and flammable mixture with air at low to medium concentrations. It is toxic if ingested in large quantities.

Ethanol has widespread use as a solvent, and as a raw material in the organic synthesis of dyes, pharmaceuticals, detergents, cosmetics, ethylamines, ethylacrylate and ethylacetate.

Some of the physical properties of ethanol are:

Molecular weight: 46.1
Melting point: 158.65 K
Normal boiling point: 351.45 K
Critical temperature: 516.25 K
Critical pressure: 6.390 MPa
Critical density: 280 kg/m^3

Reference

Beaton, C. F. and Hewitt, G. F. (1989) *Physical Property Data for the Design Engineer,* Hemisphere Publishing Corporation, New York.

V. Vesovic

Table 1. Ethanol: Values of thermophysical properties of the saturated liquid and vapor

T_{sat}, K P_{sat}, kPa	351.45 101.3	373 226	393 429	413 753	433 1 256	453 1 960	473 2 940	483 3 560	503 5 100	513 6 020
ρ_l, kg/m^3	757.0	733.7	709.0	680.3	648.5	610.5	564.0	537.6	466.2	420.3
ρ_g, kg/m^3	1.435	3.175	5.841	10.25	17.15	27.65	44.40	56.85	101.1	160.2
h_l, kJ/kg	202.5	271.7	340.0	413.2	491.5	576.5	670.7	722.2	837.4	909.8
h_g, kJ/kg	1 165.5	1 198.7	1 225.5	1 247.2	1 264.4	1 275.3	1 269.0	1 259.0	1 224.6	1 190.3
$\Delta h_{g,l}$, kJ/kg	963.0	927.0	885.5	834.0	772.9	698.9	598.3	536.7	387.3	280.5
$c_{p,l}$, kJ/(kg K)	3.00	3.30	3.61	3.96	4.65	5.51	6.16	6.61		
$c_{p,g}$, kJ/(kg K)	1.83	1.92	2.02	2.11	2.31	2.80	3.18	3.78	6.55	
η_l, μNs/m^2	428.7	314.3	240.0	185.5	144.6	113.6	89.6	79.7	63.2	56.3
η_g, μNs/m^2	10.4	11.1	11.7	12.3	12.9	13.7	14.5	15.1	16.7	18.5
λ_l, (mW/m^2)/(K/m)	153.6	150.7	146.5	141.9	137.2	134.8	129.1	125.6	108.0	79.11
λ_g, (mW/m^2)/(K/m)	19.9	22.4	24.5	26.8	29.3	32.1	35.3	37.8	43.9	50.7
Pr_l	8.37	6.88	5.91	5.18	4.90	4.64	4.28	4.19		
Pr_g	0.96	0.95	0.96	0.97	1.02	1.20	1.31	1.51	2.49	
σ, mN/m	17.7	15.7	13.6	11.5	9.3	6.9	4.5	3.3	0.9	0.34
$\beta_{e,l}$, kK^{-1}	1.41	1.60	1.90	2.41	3.13	4.18	6.06	7.56	16.0	

Thermophysical properties from Beaton and Hewitt (1989).

ETHYLENE

Ethylene (C_2H_4) is a colourless and sweet-smelling gas. It occurs naturally in plant tissue, but for industrial purposes, it is obtained by thermal cracking of naphtha or low alkanes. It forms an explosive and flammable mixture with air at low to medium concentrations.

It is used in the petrochemical and polymer industries as a raw material. The principal reactions include polymerization (production of polyethylene), oxidation (production of ethylene oxide, ethylene glycol, acetylaldehyde, etc.), halogenation (production of halogenated ethylenes), alkylation (production of ethylbenzene, ethyltoluene, etc.) and hydration (production of ethanol). It is also used in agriculture for ripening fruits.

Some of its physical properties are:

Molecular weight: 28.054
Melting point: 104K
Critical temperature: 282.35K
Critical pressure: 5.0408MPa
Normal boiling point: 169.35K
Critical density: 214.2 kg/m^3
Normal vapor density: 1.26 kg/m^3 (@ 273.15K; 1.0135MPa)

References

Beaton, C. F. and Hewitt, G. F. (1989) *Physical Property Data for the Design Engineer,* Hemisphere Publishing Corporation, New York.

Holland, P. M., Heaton, B. E., and Hanley, H. J. M. (1983). A correlation of the viscosity and thermal conductivity data of gaseous and liquid ethylene. *J. Phys. Chem. Ref. Data.* 12:917.

Jacobson, R. T. et al. (1988) *International Thermodynamic Tables of the Fluid State,* 10. Ethylene. Blackwell Scientific Publications, Oxford.

V. Vesovic

Table 1. Ethylene: Values of thermodynamic, transport and surface tension properties of the saturated liquid and vapor

T_{sat}, K P_{sat}, kPa	169.4 102	190 296	210 673	230 1321	240 1774	250 2331	260 3005	270 3813	280 4782
ρ_l, kg/m^3	567.9	537.3	504.8	467.6	446.3	422.2	393.6	356.6	291.5
ρ_g, kg/m^3	2.09	5.65	12.35	24.20	33.09	45.00	61.59	86.80	139.5
h_l, kJ/kg	−659.1	−608.8	−557.9	−503.4	−474.2	−442.8	−408.5	−368.6	−311.5
h_g, kJ/kg	−176.9	−158.6	−145.0	−137.2	−136.5	−138.8	−145.7	−160.7	−199.9
$\Delta h_{g,l}$, kJ/kg	482.1	450.2	412.9	366.2	337.7	304.0	262.8	207.9	111.6
$c_{p,l}$, kJ/(kg K)	2.39	2.47	2.61	2.85	3.05	3.37	3.95	5.40	18.21
$c_{p,g}$, kJ/(kg K)	1.30	1.42	1.61	1.94	2.22	2.66	3.48	5.59	24.82
η_l, μNs/m^2	175.5	135.6	107.5	85.0	75.0	65.4	56.0	46.1	33.2
η_g, μNs/ms2	6.03	6.76	7.55	8.54	9.16	9.94	11.0	12.5	16.1
λ_l, mW/(K m)	186.7	164.1	143.9	124.9	115.6	106.3	97.2	88.9	83.5
λ_g, mW(K m)	9.7	11.1	12.8	15.3	17.0	19.4	23.2	30.8	58.6
Pr_l	2.25	2.04	1.95	1.94	1.98	2.08	2.27	2.80	7.24
Pr_g	0.81	0.87	0.95	1.09	1.20	1.36	1.64	2.27	6.81
σ, mN/m	16.6	12.8	9.37	6.23	4.78	3.46	2.18	1.16	0.25
$\beta_{e,l}$, kK^{-1}	2.53	2.92	3.51	4.60	5.54	7.06	9.96	17.8	97.5

The values of thermodynamic properties are from Jacobson et al. (1988), transport properties from Holland et al. (1983) and surface tension from Beaton and Hewitt (1989). A correlation of the viscosity and thermal conductivity data of gaseous and liquid ethylene.

ETHYLENE GLYCOL

Ethylene Glycol (CH_2OHCH_2OH) is a colourless, odourless, sweet-tasting liquid that is soluble in water. It is obtained by the oxidation of ethylene then followed by hydration, although there are a number of other industrially viable processes for its production. It is moderately toxic. Mixed with water, it drastically reduces the freezing point of water; so its main uses are as *anti-freeze, de-icing agent, heat-transfer fluid,* solvent and extractant. It is also used in the production of polyester and of paints.

Its physical properties are:

Molecular weight: 62.07	**Critical temperature:** 645 K
Melting point: 260.2 K	**Critical pressure:** 7.7 MPa
Boiling point: 470.4 K	**Critical density:** 360 kg/m^3

V. Vesovic

ETSU (see Energy Technology Support Unit)

EUCLIDEAN SPACE (see Conservation equations, single phase)

EULER CORRELATION (see Thermal conductivity mechanisms)

EULER EQUATION (see Bernoulli equation; Conservation equations, single phase; Critical flow; Hydraulic turbines; Inviscid flow)

EULER EFFICIENCY (see Hydraulic turbines)

EULER FORMULA (see Exponential function)

EULERIAN APPROACH (see Dispersed flow)

EULERIAN BALANCES (see Conservation equations, single phase)

EULERIAN DESCRIPTION (see Streamline flow)

EULERIAN DESCRIPTION OF MOTION (see Conservation equations, single phase)

EULERIAN INTEGRAL SCALES (see Particle transport in turbulent fluids)

EULERIAN SPECIFICATION (see Flow of fluids)

EULER NUMBER

Euler number, Eu, represents the nondimensional relationship between pressure drop and momentum in fluid flow

$$Eu = \Delta p/\rho u^2$$

where ρ is fluid density and u, fluid velocity. Stated differently, Eu represents the static pressure loss divided by two velocity heads.

An alternative form of Euler number is sometimes used, which represents 2 × *fanning friction factor.*

G.L. Shires

EURATOM (see European Atomic Energy Agency)

EUROPEAN ATOMIC ENERGY AGENCY, EURATOM

200 rue de la Loi
10499 Brussels
Belgium
Tel: 235 11 11

EUROPEAN SPACE AGENCY, ESA

8–10 rue Mario NIKIS
75738 Paris cedex 15
France
Tel: 42 73 76 54

EVAPORATION COEFFICIENT (see Accommodation coefficient; Sublimation)

EVAPORATION FROM EARTH'S SURFACE (see Environmental heat transfer)

EVAPORATION OF DROPLETS (see Interfaces)

EVAPORATION OF DROPS (see Drops)

EVAPORATIVE COOLING

Following from: *Evaporators; Heat protection, sublimation*

Evaporative cooling is one of the most efficient techniques for thermal protection (see **Ablation** and **Sublimation**). It is based on the application of two mechanisms of heat absorption: phase conversion of a heat protective coating and the injection of gaseous products of decomposition of the coating into the boundary layer surrounding a body with a high-velocity and high-temperature gas flow around it. With increased incoming flow enthalpy H_e, the role of the second mechanism grows and becomes predominant.

Consider the heat balance on a destructing surface (**Figure 1**). The convective heat flux $\dot{q}_0$ (equal to the product of (α/c_p) and the enthalpy drop across the boundary layer $(H_e - H_w)$ where α is the heat transfer coefficient) leads to a rise in the body's surface temperature T_w. Unlike melting, sublimation and evaporation, do not have a constant temperature of phase transition. This is determined by the joint solution of the system of two equations: the kinetics of sublimation and mass balance on the destructing surface (see **Sublimation**). Assuming that the temperature of the surface

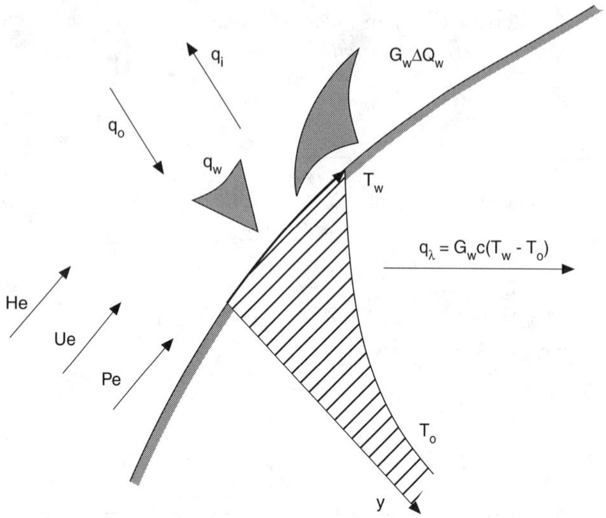

Figure 1. Heat balance on a destructing surface.

has stabilized, i.e., it has reached the maximum value $T_{w,max}$, which depends only on flux pressure p_e:

$$T_w \rightarrow T_{w,max} = [\Delta Q_v \bar{\mu}_v]/\{R[b - \ln(p_e \bar{\mu}_\Sigma)]/\bar{\mu}_v(1 - \gamma)]\}.$$

Then the heat balance at the destructing surface can be written as (see **Ablation**):

$$\dot{q}_0 = (\alpha/c_p)(H_e - H_w) = \epsilon\sigma T_w^4 + G_w[\bar{c}(T_w - T_0) + \Delta Q_v] + \dot{q}_i$$

$$\{\cdots\} = \left\{\begin{matrix}\text{re-}\\ \text{emission}\end{matrix}\right\} + \left\{\begin{matrix}\text{heating and}\\ \text{evaporation}\end{matrix}\right\} + \left\{\begin{matrix}\text{injection}\\ \text{effect}\end{matrix}\right\}. \qquad \textbf{(1)}$$

Here, $\varepsilon\sigma T_w^4$ is the heated surface radiation which can be neglected at sufficiently large values of mass loss rate G_w; $\bar{c}$ is the mean heat capacity of the condensed phase of the matter in the temperature range T_w up to the initial value T_0; and $\dot{q}_i$ is the thermal effect of the injection, which can be presented as the known function of mass loss rate.

Consider two possible approximations of the injection effect. The first is a linear one which, strictly speaking, holds only for moderate values of mass loss rate G_w:

$$\dot{q}_i = \dot{q}_0 - \dot{q}_w = [(\alpha/c_p) - (\alpha/c_p)_w](H_e - H_w) = \gamma G_w(H_e - H_w). \qquad \textbf{(2)}$$

Substituting Equation 2 into 1 yields

$$\gamma\bar{G}_w = \frac{\gamma H}{1 + \gamma H}, \quad \bar{G}_w = G_w/(\alpha/c_p)_0, \quad H = \frac{H_e - H_w}{H^*}. \qquad \textbf{(3)}$$

Here, γ is the injection factor ($\gamma < 1$) dependent on the ratio of molecular masses of the incoming flow gas mixtures to gaseous products of evaporation; $H^* = c(T_w - T_0) + \Delta Q_w$ stands for the total quantity of heat absorbed by a unit mass of the substance while evaporating.

The second approximation of the injection effect has no upper limit for the value of injection intensity G_w:

$$\dot{q}_i = \dot{q}_0(1 - \psi), \quad \psi = (\dot{q}_w/\dot{q}_0) = 1/[3(\gamma\bar{G}_w)^2 + \gamma\bar{G}_w + 1]. \qquad \textbf{(4)}$$

Solving Equations 1 and 4 together results in

$$\gamma\bar{G}_w = \sqrt[3]{\gamma H/3}(1 - 0.2/\sqrt[3]{(\gamma H)^2} - 1/9). \qquad \textbf{(5)}$$

Figure 2 presents the comparative results obtained from calculating Equations 3 and 5. Also presented are the experimental data obtained from the destruction of Teflon ($H^* = 2750$ kj/kg), glass-plastic with epoxide binder ($H^* = 12500$ kj/kg), quartz-ceramics ($H^* = 15000$ kj/kg) and graphite ($H^* = 34000$ kj/kg) in a high-temperature air flow. Inspite of the rather large distinction between the values of H* and the known specific features of the destruction mechanisms, these results with both the prediction data and with each other.

Thus it has been shown that the injection effect becomes decisive when $H > 1.5$. In reality, however, the destruction mechanism of a heat protective material (evaporation, sublimation) can be complicated by a number of subsidiary effects which make it impossible to use the above relations for small and moderate values of H. To these complicating factors are attributed, first of all, the chemical reactions of separate components of the heat protective materials with each other and with the components of the incoming gas flux.

For example, **Figure 3** presents a model of sublimation of glass-graphite materials in a vacuum and in air flow. The term "glass-graphite materials" covers a wide class of glass-plastics comprised of a glass-like refractory die (reinforcement) and organic binding with a large yield of coke residues (e.g., phenolformaldehyde resin). When heated up to temperatures higher than 1,000°C, this composite heat protective material transforms into a mixture of glass (SiO_2) and carbon (C). There can be chemical interaction between them, both at the body surface (with a preliminary stage of sublimation of quartz glass and steam dissociation):

$SiO_2(s) \rightleftharpoons SiO_2(g)$	glass sublimation
$SiO_2(g) \rightleftharpoons SiO + O_2$	steam dissociation
$C + O \rightleftharpoons CO$	carbon burning

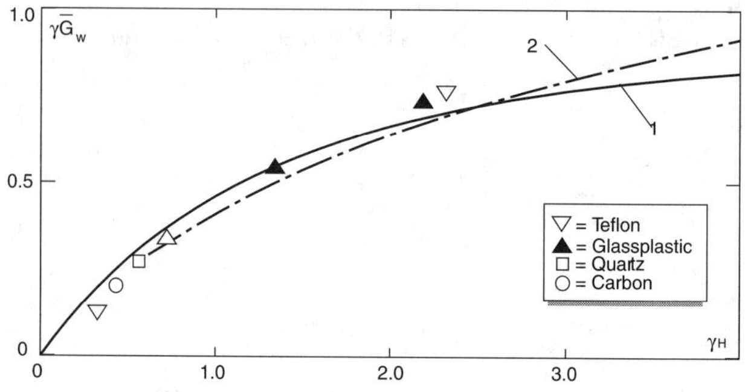

Figure 2. Mass loss rate from various materials.

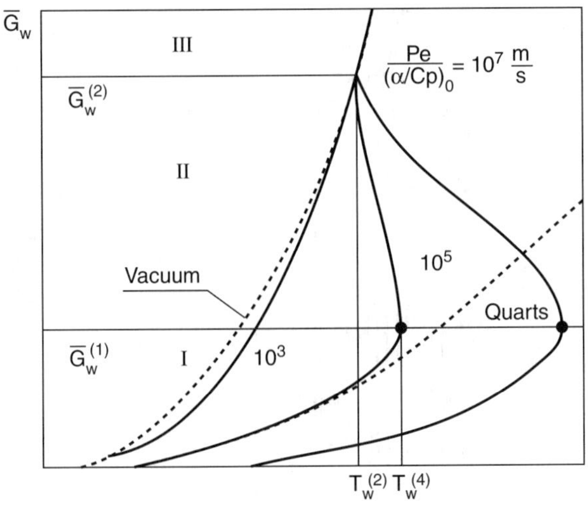

Figure 3. Sublimation of glass/graphite materials.

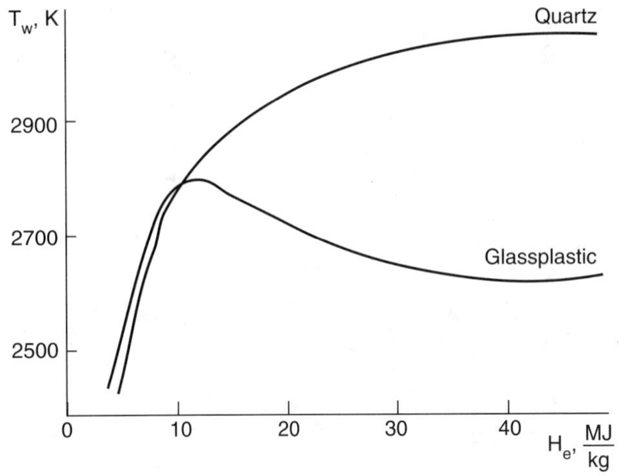

Figure 4. Surface temperature as a function of air enthalphy for quartz and quartz plastic.

and also heterogeneous interaction inside the heated layer, driven by the kinetics of reaction described by the **Arrhenius Equation**. In any case, the carbon is efficient in capturing oxygen atoms which, according to the law of mass action, sharply increases the degree of dissociation:

$$p_{SiO}/p_{SiO_2} = K_p(T_w)/\sqrt{p_{O_2}}.$$

This, in turn, intensifies glass evaporation nonequilibricity since it diminishes the possibility of a reverse process (condensation) taking place (see **Sublimation**).

At fairly small values of mass loss rate G_w, sublimation of a glass-graphite material in the air is almost the same as sublimation of a uniform quartz glass. Oxygen partial pressure p_{o_2} at the destructing surface is maintained at a sufficient level due to diffusion from the incoming flow.

An increase in the destruction rate to a certain threshold $\bar{G}_w(1)$ sharply changes the oxygen mass balance. Diffusion no longer provides the required amount of oxygen, and near the surface p_{o_2} starts to drop. As a result, the degree of SiO_2 molecule dissociation quickly increases, which intensifies evaporation velocity, the value of which approaches the maximum possible magnitude:

$$G_w \rightarrow G_{w,max} = \frac{p^H_{SiO_2}(T_w)}{\sqrt{2\pi(R/\bar{\mu}_{SiO_2})T_w}}.$$

In this case, the process of glass-graphite material evaporation in a dense air flow does not differ greatly from sublimation in a vacuum.

Figure 4 shows the dependence of the destructing glass-graphite material's surface temperature T_w on the enthalpy of the air flow H_e. Clearly, the difference between glass-graphite material and quartz glass runs to several hundred degrees. The presence of a viscous melt film at the destructing surface can correct the process model considered above; in particular, at the expense of decreasing the values of H^* and $T_{w,max}$ (see **Melting**).

Evaporative cooling has much in common with the systems of transpiration cooling (see **Transpiration Cooling**). A fundamental difference between them lies in the fact that evaporative coolant is self-regulating; i.e., the flow rate of a coolant is determined by the intensity of an external heat transfer whereas in the systems of transpiration cooling, this parameter is *a priori* fixed.

Yu.V. Polezhaev

EVAPORATORS

Following from: *Boiling; Falling film heat transfer; Nucleate boiling*

The process of *evaporation* is used widely in the chemical and process industry, and for a variety of purposes. These include the concentration of solutions (often as a precursor to crystallisation of the solute), revaporisation of liquefied gases, refrigeration applications (cooling or chilling), and generation of pure and mixed vapors for process applications. The term *evaporators* is usually reserved vapors for the first of these applications, namely the evaporation of the solvent from a solution in order to concentrate the solution. Evaporators may be classified into *falling film evaporators* (in which evaporation takes place from the film interface with no nucleate boiling at the wall), *nucleate boiling evaporators* (in which wall nucleate boiling occurs over part or all of the heat transfer surface), *flash evaporators* and *direct contact evaporators.*

Falling Film Evaporators

The process of falling film evaporation is illustrated schematically in **Figure 1**. A liquid film falls down the channel wall and evaporates, with the vapour either flowing co-currently with the film (**Figure 1 (a)**) or counter-current to it (**Figure 1 (b)**). Heat transfer in such systems is discussed in the article on **Falling Film Heat Transfer.** If counter-current flow is used, then there is a limit on the vapour velocity which can be generated before the liquid

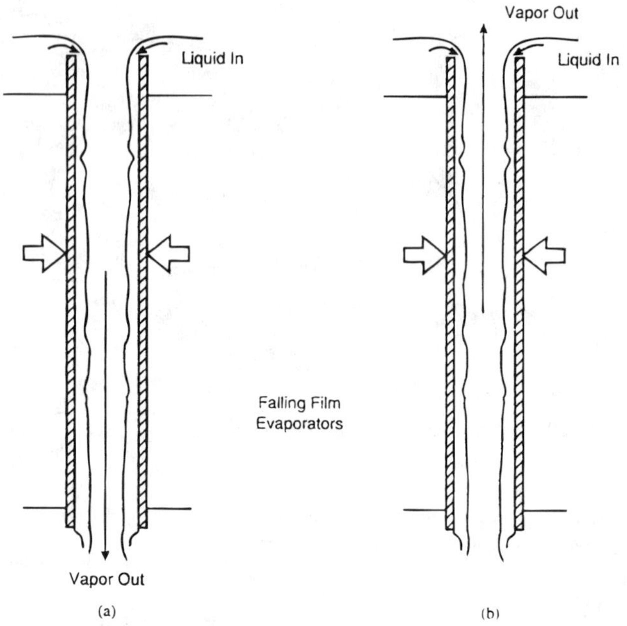

Figure 1. Modes of falling film evaporation.

begins to be carried upwards (see article on **Flooding and Flow Reversal**). In falling film evaporators, evaporation occurs directly at the liquid-vapour interface with no nucleate boiling at the wall; this means that lower temperature driving forces can be used and that the extensive fouling often associated with nucleate boiling of crystallising solutions is avoided. Falling film evaporators can take a number of different forms; an evaporator in which evaporation occurs on the inside of vertical tubes is shown in **Figure 2**. Note that particular care has to be taken to ensure uniform distribution of the liquid at the top of the tubes. In other forms of falling film evaporators, the liquid film may fall over banks of horizontal tubes, over plates and over coils.

Nucleate Boiling Evaporators

In evaporators with nucleate boiling, vapour generation is initiated by wall nucleation (though in much of the evaporator, nucleation may be supressed). Such evaporators are really a subclass of boiling systems in general, and the heat transfer processes are those described in the article on **Forced Convection Boiling**. The types of evaporators in this category include:

1. *Climbing Film Evaporator.* Here, there is a short region of nucleate boiling near the evaporator entrance but the main part of the evaporator has climbing film flow in the tubes with no nucleate boiling.
2. *Short-tube Vertical Evaporator.* Here, relatively short tubes may be used with an internal circulation device to aid circulation and minimise fouling. A typical system is shown in **Figure 3**.
3. *Basket-type Evaporator.* This is a variant of the short-tube type in which the whole tube bank may be easily removed for cleaning.

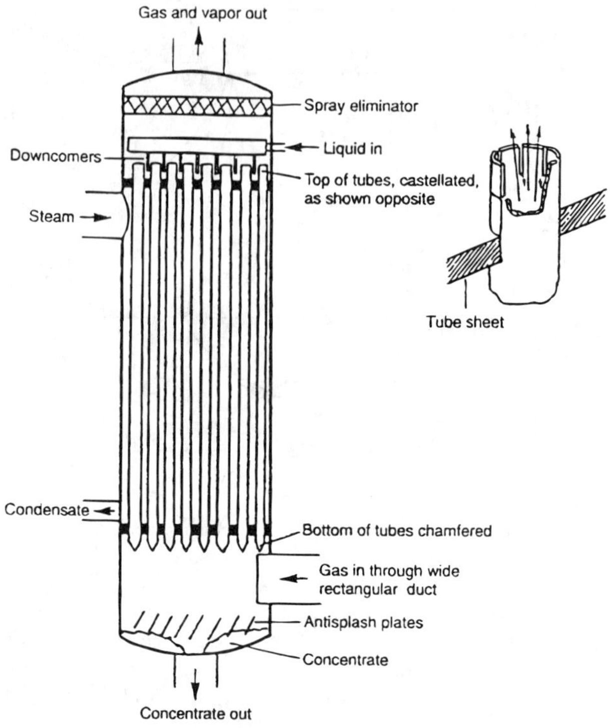

Falling-film evaporator with bottom entry of gas

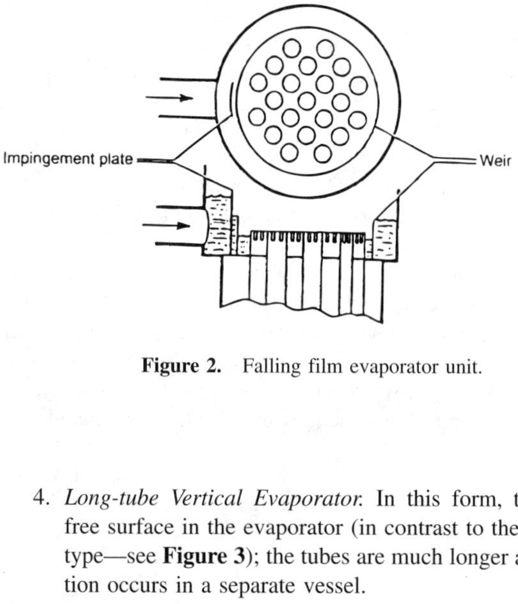

Figure 2. Falling film evaporator unit.

4. *Long-tube Vertical Evaporator.* In this form, there is no free surface in the evaporator (in contrast to the short-tube type—see **Figure 3**); the tubes are much longer and separation occurs in a separate vessel.
5. *Plate Evaporators.* Here, evaporation occurs in channels separated by plates of one form or another. The *plate heat exchanger* can be adapted to act as an evaporator and this also applies to the **Plate-fin Heat Exchanger**. Evaporators with *spiral plates* are also employed.
6. *Horizontal Tube Shell-Side Evaporator.* This is one of a number of types in which evaporation takes place outside surfaces in crossflow, with the heating on the inside of the tubes.

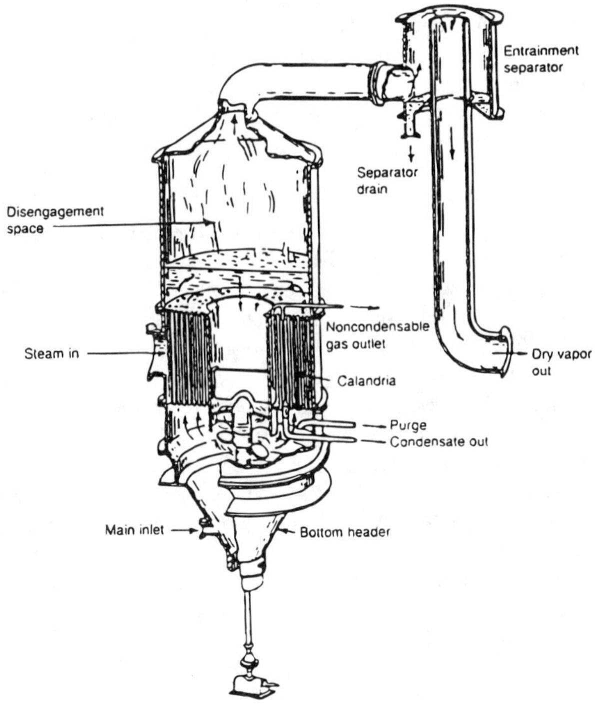

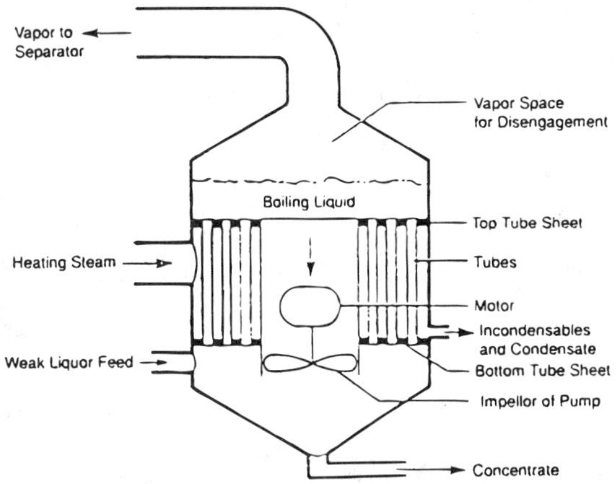

Figure 3. Short-tube vertical evaporator.

Flash Evaporation

In flash evaporation, the liquid is preheated at pressure and then flashed through a restriction into a vessel at lower pressure where vapour is formed in the process of restoring the liquid to its saturation temperature. Perhaps the best known case of flash evaporation is that used in seawater desalination plants, where salt water is evaporated to produce distilled water in a series of stages at different temperatures as illustrated schematically in **Figure 4**. (See also **Desalination)**

Direct Contact Evaporators

In this type of evaporator, a hot gas is injected into a pool of liquid and causes it to evaporate, the vapour being carried away

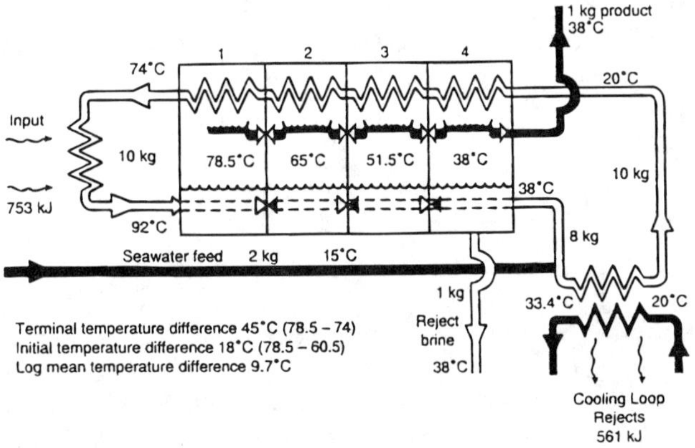

Figure 4. Multistage flash evaporation for desalination.

with the efflux gas. Such evaporators are relatively cheap and useful for the concentration of corrosive fluids, viscous liquids and slurries which may be difficult to handle in more conventional heat exchanger. A convenient way of providing the hot gas is to use a *submerged combustion* system in which the hot gas is produced *in situ* by the combustion of gas or fuel oil in a combustion chamber mounted in the centre of the vessel containing the liquid. (See also **Direct Contact Heat Exchangers)**

References

Hewitt, G. F., Shires, G. L., and Bott, T. R. (1994) *Process Heat Transfer.* Begell House, New York and CRC Press, Boca Raton, FL.

Smith, R. A. (1983) Evaporators. Section 3.5 of *Heat Exchanger Design Handbook.* Hemisphere Publishing Corporation, New York.

Smith, R. A. (1986) *Vaporisers: Selection, Design and Operation.* Longmans Scientific and Technical. Wiley, New York.

***Leading to:** Evaporative cooling*

G.F. Hewitt

EVENT TREES (see Risk analysis techniques)

EXERGY

Exergy is a quantity used by process engineers in analysing energy flows in industrial processes to improve designs and to minimise total energy usage.

The *First Law of Thermodynamics* states that in any process energy is conserved, and for a steady state flow process

$$H_0 - H_1 = Q + W. \quad (1)$$

A real process must comply with **Equation** 1, but compliance does not guarantee that the process is actually feasible.

The Second Law of Thermodynamics stipulates that energy transformations in which entropy is reduced are not possible. From the definition of entropy,

$$S_0 - S_1 \geq \frac{Q}{T_0} \quad \textbf{(2)}$$

and by substitution in **Equation** 1,

$$T_0(S_0 - S_1) \geq H_0 - H_1 - W. \quad \textbf{(3)}$$

The maximum work available from a process is therefore

$$W = (H_1 - H_0) - T_0(S_1 - S_0). \quad \textbf{(4)}$$

The maximum available work defined by **Equation** 4 is the *stream availability,* more commonly termed *exergy.* This is the amount of work or energy that can be obtained from a reversible steady-state flow process. With real irreversible processes, driving forces (such as temperature differences between cooling and warming streams) result in a net increase in total entropy. Exergy is lost and the equivalent potential for doing work is lost.

For a steady-state process which has feed-and-product streams and exchanges energy with its surroundings, the work input to the process is given by the *Gouy-Stodola equation:*

$$W = \Sigma\, Ex_{prod} - \Sigma\, Ex_{feed} + T_0 \Delta S_{irr} \quad \textbf{(5)}$$

The term $T_0\Delta S_{irr}$ is the 'lost work' and is the amount of potential work lost from the process. In a reversible process, this value is zero.

Carnot showed that if heat produces work by driving a reversible heat engine, then the maximum amount of work that can be delivered is:

$$W = Q\left(1 - \frac{T_{cold}}{T_{hot}}\right). \quad \textbf{(6)}$$

Similarly,

$$Ex = W_{max} = Q\left(1 - \frac{T_0}{T}\right) \quad \textbf{(7)}$$

as exergy is the maximum work that can be obtained via a reversible process, from a temperature T, when the cold sink is the surroundings at temperature T_0. The amount of work obtained is always less than the heat input and the ratio is termed the 'Carnot efficiency.'

If T is less than T_0, **Equation** 7 gives the minimum (reversible) work input to 'pump' heat from a low to high temperature, such as in a refrigeration cycle.

A difference in pressure or chemical concentration can produce work. In general, any system not in equilibrium with its environment can produce work via some sort of engine and the maximum quantity of work equals the initial exergy value of the system.

The concept of reversible process, upon which exergy is based, is useful because it gives a yardstick for comparing the actual performance of real processes. Clearly, 'lost work' should be minimised to improve overall process energy efficiency because it represents wastage—either of work input to the process or work that could have been usefully extracted from the process. Recognition of this is crucial for good process plant design.

Lost work is reduced and energy is saved by operating processes more closely to thermodynamic reversibility. An evaluation of process irreversibility and inefficiency is termed exergy analysis. Exergy analysis plays a key role in providing an understanding of how to minimise overall energy consumption [Kotas (1985), Sussmann (1980)].

Exergy analysis, as typically applied to a large-scale process plant, involves evaluation of lost work for each part of the process and examination of where the largest losses lie. It identifies areas of the overall process where further design work and optimisation will be of most value and the scope for improvement of energy utilisation. Results of exergy analyses have often led to the adoption of energy-efficient technology, especially in processes with high-energy consumption such as cryogenics and distillation [Gaggioli (1980)].

A conventional exergy analysis can be a painstaking exercise and recently, short-cut techniques have been developed to help designers examine process inefficiency via simple equations and checklists [Linnhoff (1983), Tomlinson et al. (1990), Steinmeyer (1992)]. These techniques can generate ideas to aid in process selection prior to fixing details of the process and can even enable process energy requirements to be estimated prior to the selection of process technology.

Exergy is useful as it gives both ideal and practical targets for energy conversion processes. However, exergy is only one aspect of energy, and there is no direct relation between exergy and the economic value of energy. It is up to the engineer to decide how to interpret the results of an exergy analysis, but used correctly it can be a very powerful tool for optimising process flowsheets and individual process equipment.

References

Gaggioli, R. A. ed. (1980) Thermodynamics Second Law Analysis. American Chemical Society. Washington, D.C.

Kotas, T. J. (1985) *The Exergy Method of Thermal Plant Analysis,* Butterworth. London.

Linnhoff, B. (1983) New concepts in thermodynamics for better chemical process design, *Chem. Eng. Res. Dev.* 61(4)207.

Steinmeyer, D. (1992) Save energy, without entropy. *Hyd. Proc.* 71(10):55.

Sussmann, M. V. (1980) *Availability (Exergy) Analysis.* Mulliken House. Lexington, MA.

Tomlinson, T. R., Finn, A. J., and Limb, D. I. (1990) Exergy analysis in process development, *The Chemical Engineer.* No. 483 & 484.

EXHAUST EMISSION LEVELS (see Internal combustion engines)

EXOSPHERE (see Atmosphere)

EXOTHERMIC REACTIONS (see Thermodynamics)

EXPANSION BELLOWS (see Expansion joints)

EXPANSION, FLOW THROUGH AND PRESSURE DROP (see Enlargement, flow and pressure change in)

EXPANSION JOINTS

***Following from:** Mechanical design of heat exchanger*

An expansion joint is a specially-designed component of compact dimensions that allows differential movement between two adjacent components and maintains the pressure envelope. Sometimes a packed joint is used, (see **Mechanical Design of Heat Exchangers Figure 1** for a packed floating head) but these are limited to low pressure and low severity duty. More common are the metallic *expansion bellows* used in the shell of fixed tube sheet exchangers or at the floating end of single-pass floating head exchangers of the S or T type (**Figure 1**).

Specialist suppliers produce three types of metallic bellows, with their typical characteristics given in **Table 1**. Thick-walled bellows are invariably used in the shell of fixed tube sheet exchangers and are fabricated from a plate of the same material and thickness as the shell barrel; the form is shown in **Figure 2**. Thin-walled bellows, shown in **Figure 3**, have convolutions that are cold-formed, either by rolling or hydraulic forming. For high pressures, bellows with multi-ply construction are used. This design of bellows is more flexible than the thick-wall type, although it is more susceptible to damage during exchanger construction and operation. **Figure 3** shows essential restraining bolts which keep the unit rigid during handling, but must be removed to allow axial thermal movement. Hot-formed medium wall bellows offer a compromise between thick and thin wall types.

Tube sheet design standards such as TEMA (1988) allow the fitting of bellows in fixed tube sheet exchangers. They are fitted when tube sheet calculation indicates a tube sheet thickness that is uneconomical. This can occur when the metal temperature difference between shell and tubes is greater than 100°C for carbon steel construction or 50°C when the shell is carbon steel and the tubes are stainless steel. With high temperature differences, it is sometimes not possible to satisfy the tube sheet calculations with any tube sheet thickness; obviously, bellows are then considered.

The bellows themselves are designed and constructed according to a code, the most comprehensive being the Standards of the Expansion Joint Manufacturers Association (1993). The Heat Exchanger Design Handbook (1994) has a chapter giving more information on exchanger bellows and in addition, provides recommendations on specification and operation.

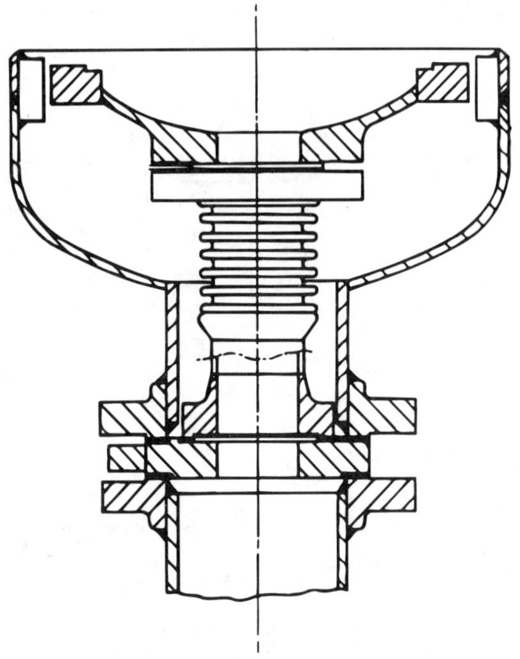

Figure 1. Floating-head bellows.

References

The Standards of the Tubular Exchanger Manufacturers Association (1988). New York.

The Standards of the Expansion Joint Manufacturers Association (1993). New York.

Heat Exchanger Design Handbook (1994). Begell House Inc., New York.

M. Morris

Table 1. Range of bellows classifications

Bellows	Thin wall	Medium wall	Thick wall
Wall thickness, mm	0.5 to 2	2 to 4.5	4 to 13+
Convolution height, mm	25 to 75	50 to 65	75 to 150
Materials include	Stainless steel Monel Inconel Incoloy	CrMo alloy steel Stainless steel	As shell
Manufacture	Cold rolling Hydraulic forming	Hot rolling	Pressing and welding
Comments	Restraining rings often needed Flexible External protection needed	Compromise between thin and thick wall	Rugged but stiff Usually a maximum of two or three convolutions because of difficulty of equalizing movement

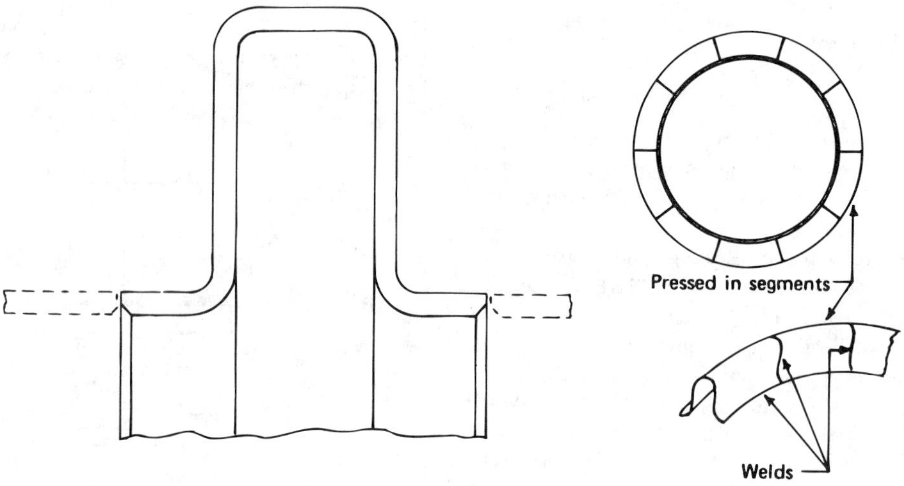

Figure 2. Thick-walled bellows.

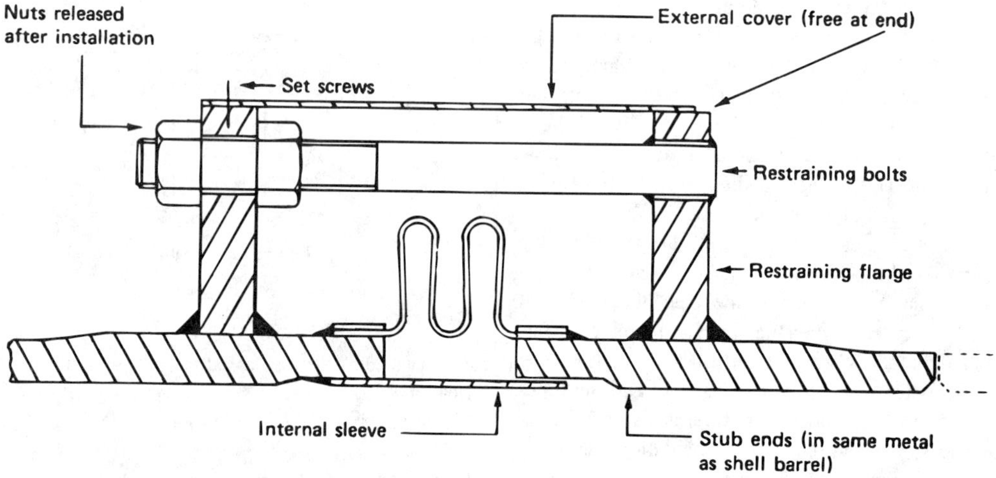

Figure 3. Thin-walled bellows.

EXPLOSION PHENOMENA

An explosion is an interaction between molecules which takes place at such a rate, and with such increase in volume and energy release, that it produces a major disturbance in adjacent matter.

The interaction can be chemical (e.g., **Combustion**) or physical (e.g., rapid phase transitions); single (i.e., vapour, liquid or solid explosions) or multiphase (e.g., mist or dust explosions); controlled (e.g., blasting) or uncontrolled (e.g., runaway reactions). Explosions can be unconfined (e.g., UVCE = unconfined vapour cloud explosion) or confined (in this case, their effect strongly depends on the extent of venting). Because of safety and hazard implications, the last category attracts a high level of attention relative to its frequency of occurrence. Most of these incidents are from combustion of a gaseous phase, which represents a suitable model for general description.

Gaseous explosions burn with a laminar, turbulent or detonative flame. The first two are *deflagration* waves in which the flame propagates to unburned reactants by diffusion; in a *detonation* the flame is constantly ignited by adiabatic shock compression. Depending on initial conditions, temperatures will range from 1,000–4,000 °K, pressures from 1 to 1,000 bar, while *flame speed* across these three categories rises from 1 to 10^4 $m.s^{-1}$.

Explosion ignition only occurs if suitable reactants are premixed within their flammability/explosion concentration range in the presence of an ignition source. Heat released by the source will start and increase the rate of the exothermic reaction until *autoignition* occurs; the autoignition temperature is only a function of the reactivity of the reactants. A minimum ignition energy is required to ensure that the size and rate of growth of the flame will exceed critical limits, so that a stable flame will be established rather than extinguished by heat dissipation into unburned surrounding reactants.

The *flame speed,* U_f, of each gas mixture is approximately related to its burning velocity, U_o, by the expansion factor, E, so that

$$U_f = U_o E.$$

For a system of reactants, burning velocity is at a maximum at the *stoichiometric mixture ratio,* (see **Combustion**) it increases with temperature and reduces with increased pressure. E is related to the additional volume produced by conversion of reactants into products, which tends to be high for complex and high-density reactants. Especially in confined explosions, this volume increase behind the flame will accelerate flame speed.

The volume of burnt gases is a product of U_o and the flame area. Hence, flame speed is further enhanced by an increase of flame area from the minimal plane front, A_p, to a curved flame. A_f. Also, initially spherical flames will slow down once partial confinement restricts further growth in all directions. However, boundary drag and shear increase flame area when flames travel through ducts, channels, pipes or even along walls or obstacles; confined or obstructed explosions therefore tend to accelerate faster than unconfined ones. In turn, shear and boundary drag also promote **turbulence**, expressed by a further factor, F_t, that leads to increased flame area and flame speed. Collectively,

$$U_f = U_o E(A_f/A_o)F_t.$$

When by the combined influence of these factors flame speed can rise above the local velocity of sound, it ceases to be controlled by mass and heat transfer processes and *deflagration-to-detonation transition* (DDT) takes place. The shock wave then initiates the fast reaction, and compression waves from the fast reaction maintain the shock front ahead. The critical factor is the strength of the weakest bond in the fuel molecule, which requires a minimum shock temperature for rupture and reaction initiation.

Detonation can also be initiated by shock impact from detonators, other explosives, or decompressions. When either the initiating shock or the DDT are too energetic an overdriven detonation may temporarily be established. Eventually, however, the detonation wave will settle down to its characteristic velocity.

The main damage from explosions is caused by the short-duration impact wave and shock pressures and the long-duration underpressures in the expansion wave. Light and heat are secondary factors.

References

Bowden, F. P. and Yoffe, Y. D. (1985) *Initiation and Growth of Explosions in Liquids and Solids.* Cambridge Science Classics.

Lewis, B. and von Elbe, G. (1987) *Combustion, Flames, and Explosions of Gases.* 3rd Edition. Academic Press Inc.

Nettleton, M. A. (1987) *Gaseous Detonations, Their Nature, Effects and Control.* Chapman and Hall Ltd.

H. Michels

EXPONENTIAL FUNCTION

An exponential function e^z is an elementary transcendental function defined for any value of z (real or complex) by

$$e^z = \lim_{n\to\infty}\left(1 + \frac{z}{n}\right)^n, \quad (1)$$

where e is the base of natural logarithms.

The exponential function is a solution of the differential equation $\omega' = \omega$ and can be presented in the entire open complex plane (i.e., in the entire complex plane except when point $z = \infty$) as:

$$e^z = 1 + \frac{z}{1!} + \frac{z^2}{2!} + \cdots + \frac{z^n}{n!} = \sum_{n=0}^{\infty} \frac{z^n}{n!}. \quad (2)$$

The exponential function e^z (where $z = x + iy$) is an entire transcendental function and is an analytic continuation of the function e^x from a real axis into a complex plane by the *Euler formula*: $e^z = e^{x+iy} = e^x(\cos x + i \sin y)$.

The exponential function e^z can accommodate all complex values except for 0. The equation $e^z = a$ has an infinite number of solutions for any complex number $a \neq 0$. The solution is found by the formula $z = \ln a + i \arg a$, where ln a is a logarithmic function, reverse of the exponential.

The exponential function is periodic with a purely imaginary period $2\pi i$: $e^{z+2\pi i} = e^z$.

The theorem of addition $e^{z_1+z_2} = e^{z_1} \cdot e^{z_2}$ is valid for the function e^z.

The derivative of the exponential function e^z is coincident with the function $(e^z)' = e^z$.

The integral of the function e^z is $\int e^z\, dz = e^z + C$.

For real numbers of x, the value of the function e^x increases faster than any degree x for $x \to \infty$ and for $x \to -\infty$, it tends to zero faster then any other degree 1/x, i.e., for any natural $n > 0$

$$\lim_{x\to\infty} \frac{e_x}{|x|^n} = \infty; \qquad \lim_{x\to-\infty} |x|^n e^x = 0. \quad (3)$$

Exponential functions are often met in problems of radiation heat transfer. A crucial element in such problems are the integral exponents $E(t) = \int_0^1 u^{n-2} \exp(-t/u)\, du$ for flat layers, and $Du(t) = \int_0^1 \exp(-t/\sqrt{1-u^2})\,(\sqrt{1-u^2})^{n-2} du$ for cylindrical geometry.

V.N. Dvortsov

EXPONENTIAL SUMS

In the analysis of decay processes, an empirical function f(x) is approximated on a real interval [a, b] by a finite sum of the form

$$\sum_{\nu=1}^{n} \alpha_\nu e^{\beta_\nu x}$$

where α_ν and β_ν are real numbers. This is a problem in nonlinear approximation theory. For a fixed integer $n > 0$, the unit interval [0, 1] is considered with an equidistant partition of length 1/2n;

$$0 \le x_0 < x_1 < \cdots < x_{2n} \le 1.$$

If, at these $2n + 1$ points, the values of the function to be approximated are known, then $f(x_k) = f_k (k = 0,1,\ldots,2n)$ and the following system of nonlinear equations is obtained:

$$\sum_{\nu=1}^{n} \alpha_\nu e^{\beta_\nu x_k} + (-1)^k \lambda = f_k \quad (1)$$

where λ accounts for the maximum error in the approximation. Since $x_k = k/2n$, then $z\nu = e^{\beta\nu/2n}$ and Equation 1 can be written as:

$$\sum_{\nu=1}^{n} \alpha_\nu z_\nu^k + (-1)^k \lambda = f_k. \quad (2)$$

The nonlinear Equation 2 for the unknowns α_ν, z_ν ($\upsilon = 1, \ldots, n$) and λ (note that $\beta_\nu = 2n \ln z_\nu$) can then be solved. The Newton iteration method is applied to improve on this first approximation.

References

Braess, D. (1986) *Nonlinear Approximation Theory,* VI. Springer-Verlag, Berlin.

Meinardus, G. (1967) *Approximation of Functions: Theory and Numerical Methods*, 10. Springer-Verlag, Berlin.

H.G. Khajah

EXTENDED SURFACE HEAT TRANSFER

Introduction

Extended surfaces have fins attached to the primary surface on one side of a two-fluid or a multifluid heat exchanger. Fins can be of a variety of geometry—plain, wavy or interrupted—and can be attached to the inside, outside or to both sides of circular, flat or oval tubes, or parting sheets. Fins are primarily used to increase the surface area (when the heat transfer coefficient on that fluid side is relatively low) and consequently to increase the total rate of heat transfer. In addition, enhanced fin geometries also increase the heat transfer coefficient compared to that for a plain fin. Fins may also be used on the high heat transfer coefficient fluid side in a heat exchanger primarily for structural strength (for example, for high pressure water flow through a flat tube) or to provide a thorough mixing of a highly-viscous liquid (such as for laminar oil flow in a flat or a round tube). Fins are attached to the primary surface by brazing, soldering, welding, adhesive bonding or mechanical expansion, or extruded or integrally connected to tubes. Major categories of **extended** surface heat exchangers are *Plate-fin* (**Figure 1**), and *Tube-fin* (**Figure 2**, individually finned tubes - **Figure 2a** and flat fins on an array of tubes - **Figure 2b**) exchangers. Note that shell-and-tube exchangers sometimes employ individually finned tubes—low finned tubing (similar to **Figure 2a** but with low height fins) [Shah (1985)].

Basic heat transfer and pressure drop analysis methods for extended and other heat exchangers have been described by Shah (1985). An overall design methodology for heat exchangers has also been presented by Shah (1992). Detailed step-by-step procedures for designing extended surface plate-fin and tube-fin type counterflow, crossflow, parallelflow and two-pass cross-counterflow heat exchangers have been outlined by Shah (1988).

In this entry, the theoretical and experimental/analytical nondimensional heat transfer coefficients (**Nusselt Number, Nu,** or *Col-*

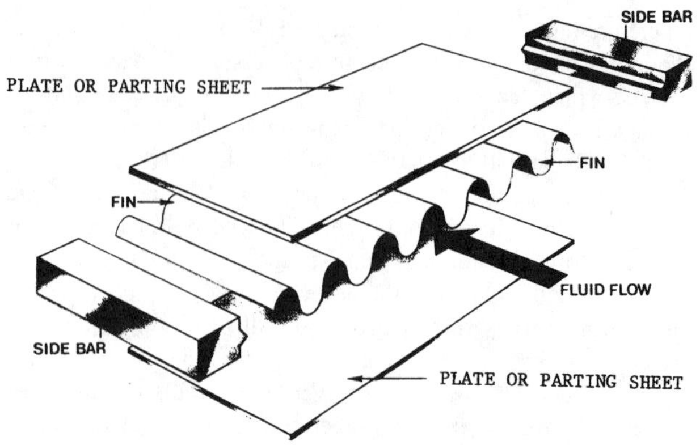

Figure 1.

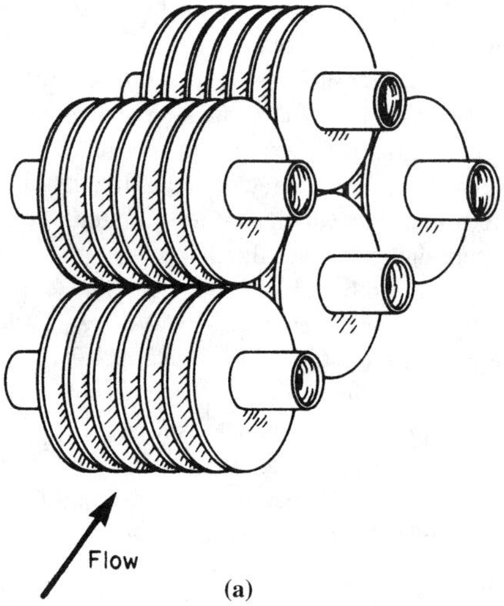

(a)

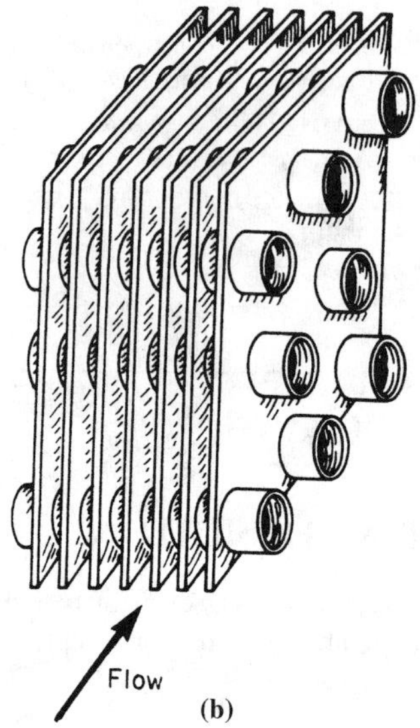

(b)

Figure 2.

burn factor, j) and the **Fanning Friction Factor** for some important extended surface geometries are summarized and a table of fin efficiencies for some important extended surfaces is provided.

Fin efficiency and extended surface efficiency

The concept of fin efficiency accounts for the reduction in temperature potential between the fin and the ambient fluid due to conduction along the fin and convection from or to the fin surface, depending on fin cooling or heating situation. The *fin temperature effectiveness* or *fin efficiency* is defined as the ratio of the actual heat transfer rate through the fin base divided by the maximum possible heat transfer rate through the fin base, which can be obtained if the entire fin is at base temperature (i.e., its material thermal conductivity is infinite). Since most real fins are "thin," they are treated as one-dimensional (1-D), with standard idealizations used for analysis [Huang and Shah (1992)]. This 1-D fin efficiency is a function of fin geometry, fin material thermal conductivity, heat transfer coefficient at the fin surface and fin tip boundary condition; it is not a function of the fin base or fin tip temperature, ambient temperature or heat flux at the fin base or fin tip. Fin efficiency formulas for some common plate-fin and tube-fin geometries of uniform fin thickness are presented in **Table 1** [Shah (1985)]. These results are not valid when the fin is thick or is subject to variable heat transfer coefficients or variable ambient fluid temperature, nor for fins with temperature depression at the base [see Huang and Shah (1992) for specific modifications to the basic formula or for specific results]. In an extended surface heat exchanger, heat transfer takes place from both the fins ($\eta_f < 100\%$) and the primary surface ($\eta_f = 100\%$). In this case, the total heat transfer rate is evaluated through a concept of *total surface effectiveness* or *surface efficiency* η_o defined as:

$$\eta_o = \frac{A_p}{A} + \eta_f \frac{A_f}{A} = 1 - \frac{A_f}{A}(1 - \eta_f) \qquad \textbf{(1)}$$

where A_f is the fin surface area, A_p is the primary surface area and $A = A_f + A_p$. In Equation 1, the heat transfer coefficients of finned and unfinned surfaces are idealized to be equal. Note that η_o is always required for the determination of thermal resistances for heat exchanger analysis [Shah (1985)].

Heat transfer and flow friction characteristics

Accurate and reliable surface heat transfer and flow friction characteristics are key input for exchanger heat transfer and pressure drop analyses, or the rating and sizing problems [Shah, (1985), (1992)]. The nondimensional surface heat transfer characteristic is usually presented in terms of the **Nusselt Number**, **Stanton Number** or *Colburn factor* vs. **Reynolds Number**.

The nondimensional surface pressure drop characteristic is usually presented in terms of the *Fanning friction factor* vs. **Reynolds Number**. Some important analytical solutions and empirical correlations for some important extended surfaces are summarized below.

Analytical solutions

Analytical solutions for developed and developing velocity/temperature profiles in constant cross-section noncircular flow passages are important for extended surface (plate-fin) heat exchangers. Fully developed laminar flow solutions are applicable to highly compact plate-fin exchangers with plain uninterrupted fins, developing laminar flow solutions to interrupted-fin geometries, and turbulent flow solutions to not-so-compact extended surfaces.

Fully developed laminar flow analytical solutions are presented in **Table 2** for specified ducts for three important thermal boundary conditions [Shah and London (1978); Shah and Bhatti (1987)]. The following observations may be made from this table:

1. There is a strong influence of flow passage geometry on Nu and fRe. Rectangular passages approaching a small aspect ratio exhibit the highest Nu and fRe.
2. Three thermal boundary conditions (denoted by the subscripts H1, H2, and T) have a strong influence on the Nusselt numbers. Here, H1 denotes constant axial wall heat flux with constant peripheral wall temperature, H2 denotes constant axial and peripheral wall heat flux and T denotes constant wall temperature.
3. As $Nu = \alpha D_h/\lambda$, a constant Nu implies the convective heat transfer coefficient α is independent of the flow velocity and fluid **Prandtl Number**.
4. An increase in α can be best achieved either by reducing D_h or by selecting a geometry with a low aspect ratio, rectangular flow passage. Reducing the hydraulic diameter is an obvious way to increase exchanger compactness and heat transfer, or D_h can be optimized using well-known heat transfer correlations based on design problem specifications.
5. Since fRe = constant, $f \propto 1/Re \propto 1/u_m$. In this case, it can be shown that $\Delta p \propto u_m$. Many additional analytical results for fully developed laminar flow ($Re \le 2{,}000$) are presented in Shah and London (1978) and in Shah and Bhatti (1987). The entrance effects, flow maldistribution, free convection, fluid property variation, fouling and surface roughness all affect fully developed analytical solutions. In order to account for these effects in real plate-fin plain fin geometries having fully developed flows, it is best to reduce the magnitude of the analytical Nu by at least 10% and to increase the value of the analytical fRe by 10% for design purposes.

The transition regime ($2{,}000 < Re < 10{,}000$) correlations for f and Nu can be found in the work of Bhatti and Shah (1987).

Fully-developed, turbulent flow Fanning friction factors are given by Bhatti and Shah (1987) as

$$f = A + B\,Re^{-1/m}, \qquad \textbf{(2)}$$

where

$A = 0.0054$, $B = 2.3 \times 10^{-8}$, $m = -2/3$ for $2100 \le Re \le 4000$

$A = 0.00128$, $B = 0.1143$, $m = 3.2154$ for $4000 \le Re \le 10^7$.

Equation 2 is accurate within ±2% [Bhatti and Shah (1987)]. The fully developed, turbulent flow Nusselt number correlation for a

Table 1. Finn efficiency for plate-fin and tube-fin geometries of uniform fin thickness

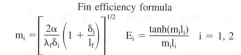

Geometry	Fin efficiency formula $m_i = \left[\frac{2\alpha}{\lambda_f \delta_i}\left(1 + \frac{\delta_i}{l_f}\right)\right]^{1/2}$ $E_i = \frac{\tanh(m_i l_i)}{m_i l_i}$ $i = 1, 2$
 Plain, wavy, or offset strip fin of rectangular cross-section	$\eta_f = E_1$ $l_1 = \frac{b}{2} - \delta_1 \qquad \delta_1 = \delta_f$
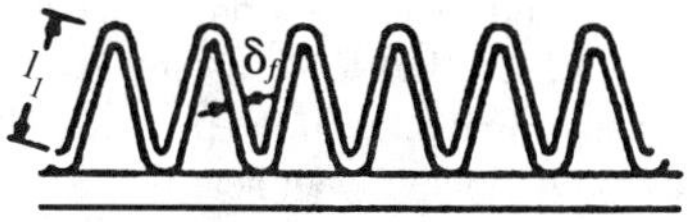 Triangular fin heated/cooled from one side	$\eta_f = \dfrac{\alpha A_1(T_0 - T_a)\dfrac{\sinh(m_1 l_1)}{m_1 l_1} + q_e}{\cosh(m_1 l_1)\left[\alpha A_1(T_0 - T_a) + q_e \dfrac{T_0 - T_a}{T_1 - T_a}\right]}, \ \delta_1 = \delta_f$
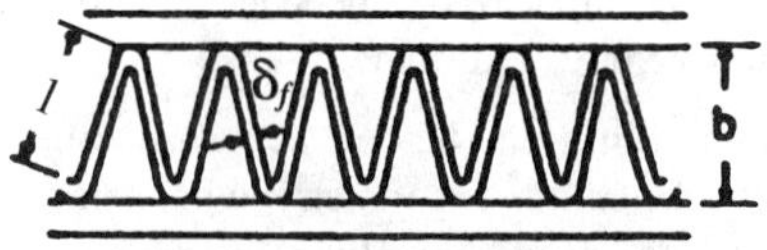 Plain, wavy, or louver fin of triangular cross-section	$\eta_f = E_1$ $l_1 = l/2, \qquad \delta_1 = \delta_f$
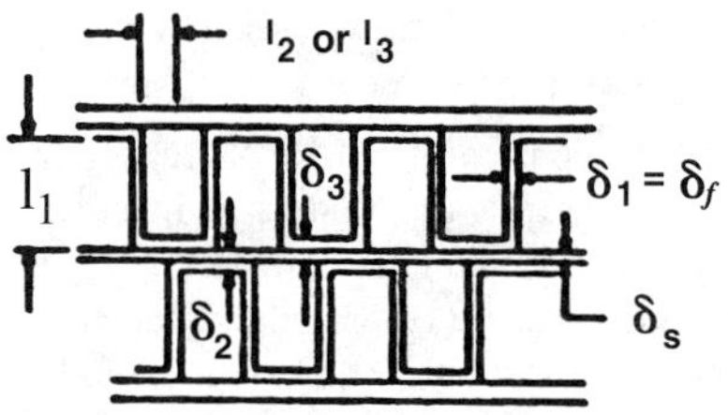 Double sandwich fin	$\eta_f = \frac{E_1 l_1 + E_2 l_2}{l_1 + l_2} \frac{1}{1 + m_1^2 E_1 E_2 l_1 l_2}$ $\delta_1 = \delta_f \qquad \delta_2 = \delta_3 = \delta_f + \delta_s$ $l_1 = b - \delta_f + \frac{\delta_s}{2} \qquad l_2 = l_3 = \frac{p_f}{2}$
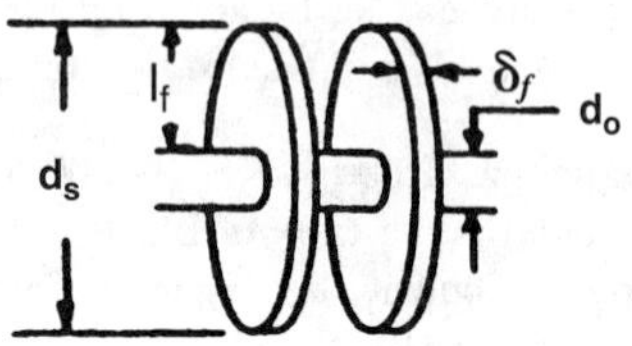 Circular fin	$\eta_f = \begin{cases} a(ml_e)^{-b} & \text{for } \Phi > 0.6 + 2.257(r^*)^{-0.445} \\ \dfrac{\tanh \Phi}{\Phi} & \text{for } \Phi \le 0.6 + 2.257(r^*)^{-0.445} \end{cases}$ $a = (r^*)^{-0.246} \qquad \Phi = ml_e(r^*)^{\exp(0.13ml_e - 1.3863)}$ $b = \begin{cases} 0.9107 + 0.0893r^* & \text{for } r^* \le 2 \\ 0.9706 + 0.17125 \ln r^* & \text{for } r^* > 2 \end{cases}$ $m = \left(\frac{2\alpha}{\lambda_f \delta_f}\right)^{1/2} \qquad l_e = l_f + \frac{\delta_f}{2} \qquad r^* = \frac{d_e}{d_o}$
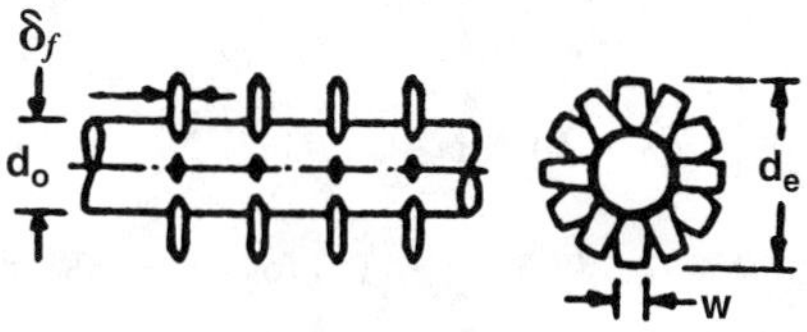 Studded fin	$\eta = \frac{\tanh(ml_e)}{ml_e}$ $m = \left[\frac{2\alpha}{\lambda_f \delta_f}\left(1 + \frac{\delta_f}{\omega}\right)\right]^{1/2} \qquad l_e = l_f + \frac{\delta_f}{2} \qquad l_f = \frac{(d_e - d_o)}{2}$

Table 2.

Geometry ($L/D_h > 100$)	Nu_{H1}	Nu_{H2}	Nu_T	fRe
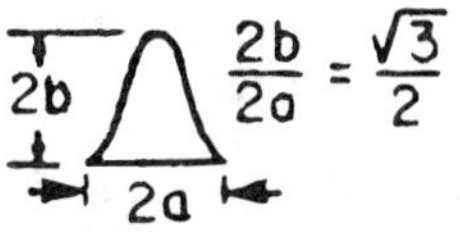 $\frac{2b}{2a} = \frac{\sqrt{3}}{2}$	3.014	1.474	2.39	12.630
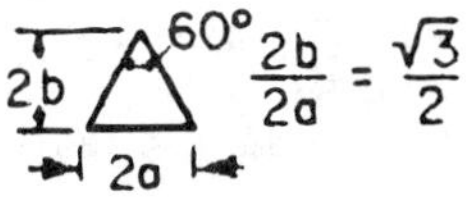 $60°$, $\frac{2b}{2a} = \frac{\sqrt{3}}{2}$	3.111	1.892	2.47	13.333
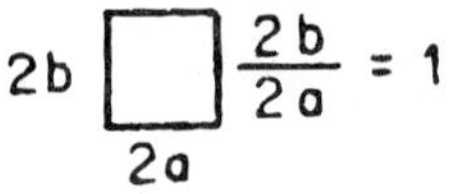 $\frac{2b}{2a} = 1$	3.608	3.091	2.976	14.227
	4.002	3.862	3.34	15.054
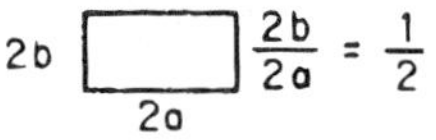 $\frac{2b}{2a} = \frac{1}{2}$	4.123	3.017	3.391	15.548
	4.364	4.364	3.657	16.000
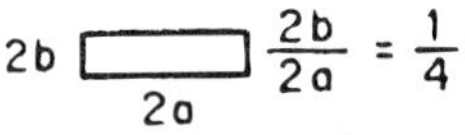 $\frac{2b}{2a} = \frac{1}{4}$	5.331	2.94	4.439	18.233
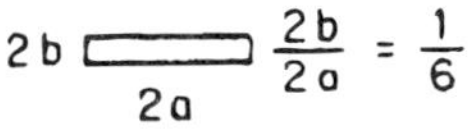 $\frac{2b}{2a} = \frac{1}{6}$	6.049	2.93	5.137	19.702
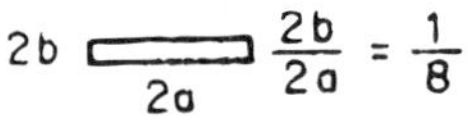 $\frac{2b}{2a} = \frac{1}{8}$	6.490	2.94	5.597	20.585
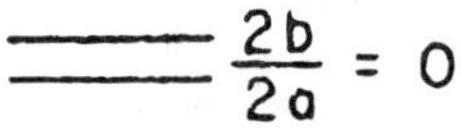 $\frac{2b}{2a} = 0$	8.235	8.235	7.541	24.000

circular tube is given by Gnielinski, as reported in Bhatti and Shah (1987), as:

$$Nu = \frac{(f/2)(Re - 1000)Pr}{1 + 12.7(f/2)^{1/2}(Pr^{2/3} - 1)}, \tag{3}$$

which is accurate within about ±10% with experimental data for $2{,}300 \le Re \le 5 \times 10^6$ and $0.5 \le Pr \le 2{,}000$. For higher accuracies in turbulent flow, refer to the correlations by Petukhov et al. reported by Bhatti and Shah (1987).

A careful observation of accurate experimental friction factors for all noncircular smooth ducts reveals that ducts with laminar $fRe < 16$ have turbulent f factors lower than those for a circular tube, whereas ducts with laminar $fRe > 16$ have turbulent f factors higher than those for a circular tube; [see Shah and Bhatti (1988)]. Similar trends have been observed for the Nusselt numbers. If one is satisfied within ±15% accuracy, Equations 2 and 3 for f and Nu can be used for noncircular passages, with the hydraulic diameter as the characteristic length in f, Nu and Re; otherwise refer to Bhatti and Shah (1987) for more accurate results.

For hydrodynamically and thermally developing flows, the analytical solutions are boundary condition dependent (for laminar flow heat transfer only) and geometry-dependent (see Shah and London (1978), Shah and Bhatti (1987) and Bhatti and Shah (1987) for specific solutions).

Experimental correlations

Analytical results presented in the preceding section are useful for well-defined constant cross-sectional extended surfaces with essentially unidirectional flows. Flows encountered in enhanced extended surfaces are generally very complex, having flow separation, reattachment, recirculation and vortices. Such flows significantly affect Nu and f for specific exchanger surfaces. Since no analytical or accurate numerical solutions are available, the information is derived experimentally. Kays and London (1984) and Webb (1994) have compiled most of the experimental results reported in open literature. Empirical correlations for some important extended surfaces are summarized.

Plate-fin extended surfaces

Offset Strip Fins. This is one of the most widely used, enhanced fin geometries (**Figure 3**) in aircraft, cryogenics and many other industries that do not require mass production. This surface has one of the highest heat transfer performance relative to the friction factor. Extensive analytical, numerical and experimental investigations have been conducted over the last 50 years. The most comprehensive correlations for j and f factors for offset strip-fin geometry are provided by Manglik and Bergles (1995), as follows:

$$j = 0.6522\, Re^{-0.5403}\left(\frac{s}{h}\right)^{-0.1541}\left(\frac{\delta_f}{l_f}\right)^{0.1499}\left(\frac{\delta_f}{s}\right)^{-0.0678}$$

$$\times\left[1 + 5.269 \times 10^{-5}\, Re^{1.340}\left(\frac{s}{h}\right)^{0.504}\left(\frac{\delta_f}{l_f}\right)^{0.456}\left(\frac{\delta_f}{s}\right)^{-1.055}\right]^{0.1} \tag{4}$$

$$f = 9.6243\, Re^{-0.7422}\left(\frac{s}{h}\right)^{-0.1856}\left(\frac{\delta_f}{l_f}\right)^{0.3053}\left(\frac{\delta_f}{s}\right)^{-0.2659}$$

$$\times\left[1 + 7.669 \times 10^{-8}\, Re^{4.429}\left(\frac{s}{h}\right)^{0.920}\left(\frac{\delta_f}{l_f}\right)^{3.767}\left(\frac{\delta_f}{s}\right)^{0.236}\right]^{0.1} \tag{5}$$

where

$$D_h = 4\, A_o/(A/l_f) = 4shl_f/[2(sl_f + hl_f + \delta_f h) + \delta_f s]. \tag{6}$$

Geometrical symbols in Equation 6 are shown in **Figure 3**.

These correlations predict the experimental data of 18 test cores within ±20% for $120 \le Re \le 10^4$. Although all the experimental data for these correlations have been obtained for air, the j factor takes into consideration minor variations in the Prandtl number, and the above correlations should be valid for $0.5 < Pr < 15$.

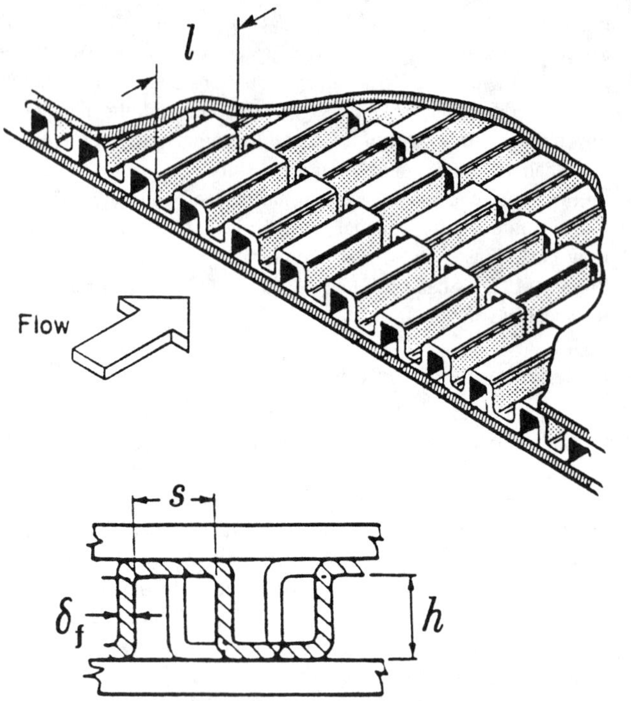

Figure 3.

Louver Fins. Louver or multilouver fins are extensively used in the auto industry due to their mass production manufacturability and hence, lower cost. It has generally higher j and f factors than those of the offset strip-fin geometry; also, the increase in the friction factors is usually higher than the increase in the j factors. However, the exchanger can be designed for higher heat transfer and the same pressure drop compared with offset strip-fins by a proper selection of exchanger frontal area, depth and fin density. Published literature on and correlations for louver fins have been summarized by Webb (1994) and Cowell (1995) while flow and heat transfer phenomena have been discussed by Cowell (1995). Because of the lack of systematic studies on modern louver fin geometries in the open literature, no correlation can be recommended for design purposes.

Tube-fin extended surfaces

Two major types of tube-fin extended surfaces are: a) individually-finned tubes, and b) flat fins (also sometimes referred to as plate fins) with or without enhancements/interruptions on an array of tubes as shown in **Figure 2**. An extensive coverage of published literature on and correlations for these extended surfaces are provided by Webb (1994), Kays and London (1984) and Rozenman (1976). Empirical correlations for some important geometries are summarized below.

Individually-Finned Tubes. This fin geometry, helically-wrapped (or extruded) circular fins on a circular tube as shown in **Figure 2a**, is commonly used in process and waste heat recovery industries. The following correlation for j factors is recommended by Briggs and Young [see Webb (1994)], for individually-finned tubes on staggered tube banks.

$$j = 0.134\ Re_d^{-0.319}(s/l_f)^{0.2}(s/\delta_f)^{0.11} \tag{7}$$

where l_f is the radial height of the fin, δ_f is the fin thickness, p_f is the fin pitch and $s = p_f - \delta_f$ is the distance between adjacent fins. Equation 7 is valid for the following ranges: $1100 \le Re_d \le 18{,}000$, $0.13 \le s/l_f \le 0.63$, $1.01 \le s/\delta_f \le 6.62$, $0.09 \le l_f/d_o \le 0.69$, $0.011 \le \delta_f/d_o \le 0.15$, $1.54 \le X_t/d_o \le 8.23$; fin root diameter d_o between 11.1 and 40.9 mm; and fin density N_f $(=1/p_f)$ between 246 and 768 fins per meter. The standard deviation of Equation 7 from experimental results has been computed at 5.1%.

For friction factors, Robinson and Briggs [see Webb (1994)] recommended the following correlation:

$$f_{tb} = 9.465\ Re_d^{-0.316}(X_t/d_0)^{-0.927}(X_t/X_d)^{0.515} \tag{8}$$

Here, $X_d = (X_t^2 + X_l^2)^{1/2}$ is the diagonal pitch and X_t and X_l are the transverse and longitudinal tube pitches, respectively. The correlation is valid for the following ranges: $2000 \le Re_d \le 50{,}000$, $0.15 \le s/l_f \le 0.19$, $3.75 \le s/\delta_f \le 6.03$, $0.35 \le l_f/d_o \le 0.56$, $0.011 \le \delta_f/d_o \le 0.025$, $1.86 \le X_t/d_o \le 4.60$, $18.6 \le d_o \le 40.9$ mm, and $311 \le N_f \le 431$ fins per meter. The standard deviation of Equation 8 from correlated data was 7.8%.

The extensive work on low-finned tubes has been assessed by Rabas and Taborek (1987), Ganguli and Yilmaz (1987) and Chai (1988). A simple but accurate correlation for heat transfer, given by Ganguli and Yilmaz (1987), is

$$j = 0.255\ Re_d^{-0.3}(d_e/s)^{-0.3} \tag{9}$$

A more accurate correlation for heat transfer is given by Rabas and Taborek (1987). By comparing existing data in literature and various correlations, Chai (1988) has arrived at the best correlation for friction factors:

$$f_{tb} = 1.748\ Re_d^{-0.233}\left(\frac{l_f}{s}\right)^{0.552} \cdot \left(\frac{d_o}{X_t}\right)^{0.599} \cdot \left(\frac{d_o}{X_l}\right)^{0.1738} \tag{10}$$

This correlation is valid for $895 \le Re_d \le 713{,}000$, $20 \le \theta \le 40$, $X_t/d_o < 4$, $N_r \ge 4$, and θ is the tube layout angle in degrees. It predicts 89 literature data points within a mean absolute error of 6%; the range of actual error is from −16.7 to +19.9%.

Flat Plain Fins on a Staggered Tube bank. This geometry, as shown in **Figure 2b**, is used in the air-conditioning/refrigeration industry as well as in applications where the pressure drop on the fin side prohibits the use of enhanced flat fins. An in-line tube bank is generally not used unless very low fin-side pressure drop is the essential requirement. Heat transfer correlation for **Figure 2b** flat plain fins on staggered tube banks with four or more tube rows has been provided by Gray and Webb [(see Webb 1994)] as follows:

$$j_4 = 0.14\ Re_d^{-0.328}(X_t/X_l)^{-0.502}(s/d_0)^{0.031} \tag{11}$$

For the number of tube rows N from 1 to 3, the j factor is lower and is given by:

$$\frac{j_N}{j_4} = 0.991[2.24\ Re_d^{-0.092}(N/4)^{-0.031}]^{0.607(4-N)} \tag{12}$$

Gray and Webb hypothesized that the friction factor consists of two components: one associated with the fins and the other associated with the tubes, as follows:

$$f = f_f \frac{A_f}{A} + f_t\left(1 - \frac{A_f}{A}\right)\left(1 - \frac{\delta_f}{p_f}\right) \quad \textbf{(13)}$$

where

$$f_f = 0.508\ Re_d^{-0.521}(X_t/d_0)^{1.318} \quad \textbf{(14)}$$

and f_t is the friction factor associated with the tube defined the same as f_t. In equation form, it was expressed in terms of the **Euler Number**, $Eu = 4\ f_{tb} = f_t\pi d_o/[N(X_t - d_o)]$, by Zukauskas and Ulinskas (1983). **Equation 13** correlated with 90% of the data for 19 heat exchangers within ± 20%. The range of dimensionless variables of **Equations 13 and 14** are: $500 \leq Re \leq 24{,}700$, $1.97 \leq X_t/d_o \leq 2.55$, $1.7 \leq X_l/d_o \leq 2.58$, and $0.08 \leq s/d_o \leq 0.64$.

Conclusion

The subject of extended surface heat transfer is very extensive and is difficult to condense in a few pages. This attempt to summarize some important typical results, both analytical and experimental, is but an introduction to the subject. Key references are provided below for further exploration of the subject.

Nomenclature

- A total heat transfer area (primary + fin) on one fluid side of a heat exchanger, A_p- primary surface area, A_f- fin surface area, m^2
- A_o minimum free flow area on one fluid side of a heat exchanger, m^2
- b plate spacing, $h + \delta_f$, m
- D_h hydraulic diameter of flow passages, m
- d_o tube outside diameter, m
- f Fanning friction factor, $\rho\Delta P D_h/(2L\dot{m}^2)$, dimensionless
- f_{tb} Fanning friction factor for the tube bank outside, $\rho\Delta P/(2N\dot{m}^2)$, dimensionless
- h height of the offset strip fin (see Fig. 3), m
- j Colburn factor, $NuPr^{-1/3}/Re$, dimensionless
- l_f offset strip fin length or fin height for individually finned tubes, m
- $\dot{m}$ mass velocity, kg/m^2s
- N number of tube rows
- N_f number of fins per meter, 1/m
- Nu Nusselt number, $\alpha D_h/\lambda$, dimensionless
- Pr fluid Prandtl number, dimensionless
- p_f fin pitch, m
- Re Reynolds number, $\dot{m}D_h/\eta$, dimensionless
- Re_d Reynolds number, $\rho u_m d_o/\eta$, dimensionless
- s distance between adjacent fins, p_f- δ_f, m
- u_m mean axial velocity in the minimum free flow area, m/s
- X_d diagonal tube pitch, m
- X_l longitudinal tube pitch, m
- X_t transverse tube pitch, m
- α heat transfer coefficient, W/m^2K
- δ_f fin thickness, m
- η_f fin efficiency, dimensionless
- η_o extended surface efficiency, dimensionless
- λ fluid thermal conductivity, W/mK
- η fluid dynamic viscosity, Pa · s
- ρ fluid density, kg/m^3

References

Bhatti, M. S. and Shah, R. K. (1987) Turbulent and transition convective heat transfer in ducts. *Handbook of Single-phase Convective Heat Transfer.* ed. by S. Kakaç, R. K. Shah and W. Aung, 4: 166 pages. John Wiley. New York.

Chai, H. C. (1988) A simple pressure drop correlation equation for low-finned tube crossflow heat exchangers. *Int. Commun. Heat Mass Transfer.* 15:95–101.

Cowell, T. A., Heikal, M. R., and Achaichia, A. (1995) Flow and heat transfer in compact louvered fin surfaces, *Exp. Thermal and Fluid Sci.* 10:192–199.

Ganguli, A. and Yilmaz, S. B. (1987) New heat transfer and pressure drop correlations for crossflow over low-finned tube banks. *AIChE Symp. Ser.* 257(83):9–14.

Huang, L. J. and Shah, R. K. (1992) Assessment of calculation methods for efficiency of straight fins of rectangular profiles. *Int. J. Heat and Fluid Flow.* 13:282–293.

Kays, W. M. and London, A. L. (1984) *Compact Heat Exchangers.* 3rd ed. McGraw-Hill, New York.

Manglik, R. M. and Bergles, A. E. (1995) Heat transfer and pressure drop correlations for the rectangular offset strip fin compact heat exchanger. *Exp. Thermal and Fluid Sci.* 10:171–180.

Rabas, T. J. and Taborek, J. (1987) Survey of turbulent forced-convection heat transfer and pressure drop characteristics of low-finned tube banks in crossflow. *Heat Transfer Eng.* 8(2):49–62.

Rozenman, T. (1976) Heat transfer and pressure drop characteristics of dry cooling tower extended surfaces. I: *Heat Transfer and Pressure Drop Data.* Report BNWL-PFR 7-100; II: *Data Analysis and Correlation. Report BNWL-PFR 7-102.* Battelle Pacific Northwest Laboratories. Richland, WA.

Shah, R. K. (1985) Compact heat exchangers. in *Handbook of Heat Transfer Applications.* 2nd Ed. ed. by W. M. Rohsenow, J. P. Hartnett, and E. N. Ganić. 4, III: 4–174 to 4–311. McGraw-Hill, New York.

Shah, R. K. (1988) Plate-fin and tube-fin heat exchanger design procedures. in *Heat Transfer Equipment Design.* ed. by R. K. Shah, E. C. Subbarao and R. A. Mashelkar. 255–266. Hemisphere Publishing Corp., Washington, DC.

Shah, R. K. (1992) Multidisciplinary approach to heat exchanger design. *Industrial Heat Exchangers,* ed. by J-M. Buchlin. Lecture Series No. 1991:04. von Kármán Institute for Fluid Dynamics. Belgium.

Shah, R. K. and Bhatti, M. S. (1987) Laminar convective heat transfer in ducts. *Handbook of Single-phase Convective Heat Transfer.* ed. by S. Kakaç, R. K. Shah and W. Aung, pages, 3: 137 pages, John Wiley, New York.

Shah, R. K. and Bhatti, M. S. (1988) Assessment of correlations for single-phase heat exchangers. *Two-Phase Flow Heat Exchangers: Thermal-Hydraulic Fundamentals and Design.* ed. by S. Kakaç, A. E. Bergles, and E. O. Fernandes. 81–122. Kluwer Academic Publishers, Dordrecht. The Netherlands.

Shah, R. K. and London, A. L. (1978) Laminar flow forced convection in ducts, Supp. 1 to *Advances in Heat Transfer.* Academic Press. New York.

Webb, R. L. (1994) *Principles of Enhanced Heat Transfer.* John Wiley. New York.

Zukauskas, A. and Ulinskas, R. (1983) Banks of plain and finned tubes. *Heat Exchanger Design Handbook.* 2: 2.2.4. Hemisphere. New York.

R.K. Shah

EXTENDED SURFACES (see Augmentation of heat transfer, single phase; Shell-and-tube heat exchangers)

EXTENSIONAL FLOW (Non-Newtonian fluids; Viscosity)

EXTENSIONAL VISCOSITY (see Non-Newtonian fluids)

EXTERNAL FLOWS: OVERVIEW

Following from: *Laminar flow; Turbulent flow; Flow of fluids*

External flows are defined as those motions of a fluid medium which occur when the medium flows around a body immersed in it. External flows may be addressed by the methods of fluid mechanics.

In the general case, the motion of a fluid medium in a certain volume can be specified if, at any moment of time t, the fluid velocity field of the medium $\vec{V}(\vec{x}, t)$ at any point $\vec{x}$ of the volume can be determined (or calculated with the required accuracy). At present, this problem has been solved only for some particular cases. To a large extent, this difficulty is associated with defining boundary conditions. When a fluid medium flows around bodies immersed in a fluid, the zone of influence of the body on the flow can extend over a considerable distance both up and down the flow. Inside the fluid medium, sharp boundaries of the flow can arise (for instance, cavities in cavitation or shock waves in retarding supersonic flows).

To describe the motion of the fluid medium, an appropriate mathematical model for this phenomenon must be chosen. As a rule, only the most essential properties of the medium are considered because the wider the context of the problem, the more difficult it is to close the mathematical model, to carry out physical analysis, and in the long run, to correlate theory with experiment. In many cases, it is the correct choice of a simple model that guarantees progress in the solution of a posed problem.

For gases, two models of fluid flow are used: molecular and viscous; for liquids, only a viscous model is used.

The molecular flow of gases is considered typical when the mean free path length λ of molecules exceeds the characteristic linear dimension of a streamlined body d. The **Knudsen Number**, $Kn = \bar{\lambda}/d$, is the dimensionless characteristic of the rarefied gas medium. The motion of gases at any Knudsen number is described by the *Boltzmann equation.* The equations of classical hydrodynamics follow the Boltzmann equation as the limiting case for the Knudsen number $Kn \to 0$.

The fundamental difference between the molecular and the viscous models of fluid flow lies in the statement of boundary conditions on the surface of streamlined bodies. Instead of conditions of equality of velocity and temperature of the gas at a solid wall (typical for viscous models), molecular models leave scope for a gas to slip and for the temperature to jump. In strongly-rarefied gases, the notions of temperature or velocity have to be interpreted differently than in the case of higher pressure systems.

Consider now the flow about a body at small Knudsen numbers. We assume that the continuous medium model can be applied to a fluid—which implies a continuous distribution of the substance and of the physical characteristics of its state and motion in space. An important property which differentiates a liquid or gaseous medium from a solid medium is their fluidity. Offering a noticeable counteraction to compressive strain, the fluids poorly resist shear deformation, i.e., a mutual slip of the layers of the medium.

The lower the tangential stresses that arise in motion, the less the relative velocity of mutual slip. This is what distinguishes fluids from granular solid media, whose slip is accompanied by the appearance of a friction force which is only slightly dependent on the slip velocity. There is practically no constant component of the resistance force in the fluid.

A complete set of equations which describes the motion of fluids, approximated as a continuous medium, is called the *Navier-Stokes equations.* An analysis of these equations reveals a set of dimensionless numbers or similarity criteria which describe external flow around a compact body with a geometrical shape and a length scale d. The description of the external flow includes the velocity field of fluid particles $\vec{v}(\vec{x}, t)$, temperature distribution $T(\vec{x}, t)$ and density $\rho(\vec{x}, t)$ of the medium in the entire volume under study.

Depending on a particular statement of the problem, the set of similarity criteria can be wider or narrower but, as a rule, it includes the **Reynolds Number** $Re = \rho vd/\eta$, **Mach Number** $Ma = v/a$, the **Prandtl Number** $Pr = \eta C_p/\lambda$ and the ratio of heat capacities at constant pressure and constant volume c_p and c_v (or an adiabatic exponent) $\gamma = c_p/c_v$. The notation used here is that v and a are the characteristic values of flow velocity and of sound velocity in the fluid while ρ, η, λ are density, viscosity and thermal conductivity, respectively. To this set of similarity criteria may sometimes be added the temperature factor, or the ratio of temperature on the wall (of the surface of a streamlined body) T_w to the temperature of a free-stream flow T_∞. Accounting for the instability of the flow would require consideration of the **Strouhal Number** and accounting for mass or volume forces, the **Froude Number**.

The influence of the various similarity criteria on external flow varies from case to case. For a closer analyses of these influences, see the articles on **Aerodynamics, Boundary Layers, Compressible Flow, Drag Coefficient, Inviscid Flow, Turbulent Flow, Free Convection, Jets** and **Waves in Fluids.** Here, only the fundamental features of flows around bodies as a function of free-stream velocity are considered; to be more exact, these cases are governed by only two similarity criteria, i.e., the Reynolds and Mach numbers. It is interesting to note that there exist rather wide ranges of variations for both criteria, within which their particular values play no part in the description of flow. These are the self-similarity ranges of Reynolds numbers from 10^3 to 10^6 and of Mach numbers from 5 to ∞ (the numerical values are given for air flows). The discussion below focuses on flows with low values of Reynolds and Mach numbers. The assumption of a simple, axi-symmetric geometry (a sphere or a cylinder) as the shape of the streamlined bodies in all flows considered here does not detract from the general applicability of the concepts presented.

Visualisation of a supersonic gas flow, with high values of Re and Ma, around the sphere (**Figure 1**) makes it possible to divide the entire field of flow into a number of areas for each of which a mathematical model of the local phenomena may be used.

In **Figure 1**, the uniform (homogeneous) incoming flow retains its free-stream velocity in the zone to the left of the bow shock wave 1. Conversely, at the right of jump 1, the flow undergoes considerable deformations, and the range of velocity variations is rather wide. In the area of the front stagnation point on the surface of body (2), the flow is subsonic and only on line 3 does the flow accelerate anew up to Ma = 1. This line is called the sound line and the point of its intersection with the body surface—sound point 4—is important for determining the intensity of heat transfer, both in laminar and in turbulent flows, in the **Boundary Layer**.

The sphere and the cylinder belong to a class of poorly-streamlined bodies. This means that the boundary layer leaves their surface at the mid- (or the largest) cross-section of the body, where the flow is accelerated to maximum velocity and local pressure p reaches a minimum (see **Inviscid Flow**). The separated or free boundary layer forms a shear flow 6, which separates the stagnant zone 7 in the bottom part of the body from an inviscid flow 5 zone expanding in Prandtl-Mayer waves.

When the free boundary layers 6 come together, a trailing vortex neck is formed behind the body which divides the entire trailing vortex into a short-range zone (denoted by B) and a long-range zone (denoted by C in **Figure 1**). The long-range trailing vortex (9) is a typical isobaric turbulent jet (see **Jets**), around which a suspended rear shock wave 8 is formed.

For a flow with high values of Mach and Reynolds numbers, the shock wave and the boundary layer can be treated as infinitely-thin, discontinuous areas, which allows the use of an inviscid flow model (see **Inviscid Flow**). In this case, the velocity field in the area A can be calculated adequately up to the separation point of the boundary layer.

It is well-known that in high-velocity gas flows, the stagnation temperature behind the shock wave can reach rather high values. This, in turn, brings about a considerable difference in the gas composition behind the shock wave and in the free-stream flow. Dissociation of molecules, ionization of atoms, numerous physico-chemical interactions between the components of the gas mixture and the material of the streamlined body occur (see **Ablation**).

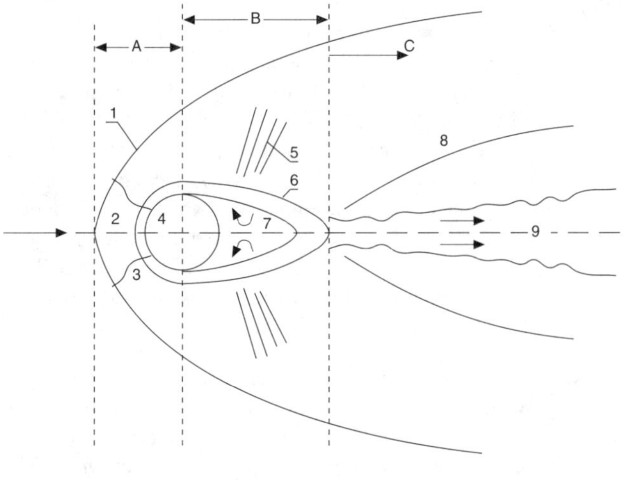

Figure 1. Requires in flow round a cylinder or sphere.

All these phenomena reflect on the accuracy of predicting the thermodynamic and transfer properties of the medium (see, for instance **Thermal Conductivity**), which must be known in order to apply the inviscid flow model. In particular, the adiabatic exponent $\gamma = c_p/c_v$ varies considerably with increasing temperature in an air-type gas mixture so that the error in calculating the pressure may be quite large.

In this connection, a modified Newton formula for the calculation of local pressure on the surface of the body with an arbitrary geometry evokes interest because, in essence, it does not require any knowledge about the mechanisms of real physico-chemical processes behind the bow shock wave. When solving the problem involving the resistance of bodies in a fluid flow, Newton believed that all the fluid particles are the same, fill in the space ahead of the body and do not interact with one another. The particles strike against the body surface quite inelastically, i.e., the particles lose the momentum component normal to the body surface which is the cause of the surplus pressure exerted on the nose and the windwood parts of the body surface (**Figure 2**). As for body surface parts which find themselves in the wind shadow (dashed in **Figure 2**), Newton assumed that the gauge pressure there is zero.

If an element of the body surface with area dS is inclined to the velocity vector of the free-stream flow at angle θ, then the mass flow rate of the gas from which momentum is lost is $\rho_\infty V_\infty dS \sin\theta$ and the normal ("lost") component of velocity is $V_\infty \sin\theta$. According to the law of momentum conservation, $(p - p_\infty)dS = [\rho_\infty V_\infty dS \sin\theta][V_\infty \sin\theta - 0]$, where the gauge pressure on the surface of the body with an angle of slope θ is determined by the relation called the Newton formula:

$$\frac{p - p_\infty}{\rho_\infty V_\infty^2} = \sin^2\theta.$$

The Newton resistance model postulates that the direction of fluid particles motion in the flow around the body varies instantaneously. In the general case, this assumption is not fulfilled. Disturbances caused by the presence of the body propagate in a subsonic flow well upstream. In supersonic flows, however, the disturbance region is limited by a shock wave (**Figure 1**). At large Mach numbers ($Ma_\infty \to 5$), the thickness of the shock layer is small and the Newton scheme is realized more or less adequately. In the modern version, Newton's formula is modified by replacing the denominator ($\rho_\infty V_\infty^2$) by the stagnation pressure p_0' (at $Ma_\infty \to 5$ the difference of p_0' from $\rho_\infty V_\infty^2$ does not exceed 5%).

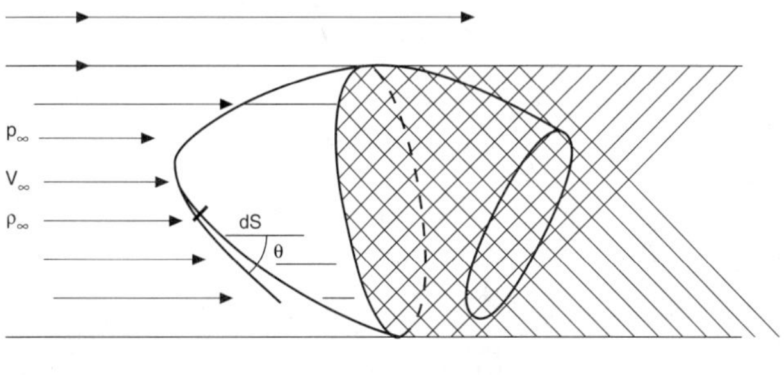

Figure 2. Bases of Newton method for force on a body in a flow.

From Newton's formula for pressure distribution, the effect of the geometry of the body on drag coefficient can be defined easily (see **Drag Coefficient**); to be more precise, the formula gives only that part of the drag coefficient which is not related to viscous friction. For a spherical segment or a cylindrical sector with central angle ω (**Figure 3**) the respective expressions are:

$$C_D = 2\left(1 - \frac{1}{2}\sin^2\omega\right) \quad \text{and} \quad C_D = 2\left(1 - \frac{1}{3}\sin^2\omega\right).$$

Hence, the drag coefficient of the sphere ($\omega = \pi/2$) is $C_D = 1$; of the cylinder, $C_D = 4/3$; and of the flat-end face perpendicular to the flow, $C_D = 2$. For the cone with a half-angle at the apex θ, the drag coefficient is $C_D = 2\sin^2\theta$.

The Newton formula also permits defining the positions of sound points in the surface of bodies whose shapes have no breaks (i.e., of smooth bodies). Gas motion that follows the shock wave from the critical point along the surface obeys the laws of adiabatic expansion, when the local pressure p is related to the local Mach number as

$$\frac{p}{p'_0} = 1/\left[1 + \frac{\gamma + 1}{2}\mathrm{Ma}^2\right]^{\gamma/(\gamma-1)}.$$

Thus the ratio of the pressure p_* on the sound line, to the stagnation pressure p'_0 when Ma = 1, within a broad range of γ variation from 1.05 to 1.4, will be:

$$\frac{P_*}{P'_0} = \left(\frac{\gamma + 1}{2}\right)^{-\gamma/(\gamma-1)} = 0.525 \text{ to } 0.605.$$

Since by the Newton formula the p/p'_0 ratio is proportional to $\cos^2\omega$, the range of variation of the central angle ω_* at the sound point is limited, $\omega_* = 36° - 41°$.

A slight dependence of the coordinate of the sound point on the adiabatic exponent γ allows the determination, with sufficient accuracy, the distribution of velocity of a subsonic flow in the vicinity of the stagnation point of a spherical or cylindrical blunt body. So, the velocity gradient on the longitudinal coordinate X, directed along the blunted body generatrix, can be calculated as follows:

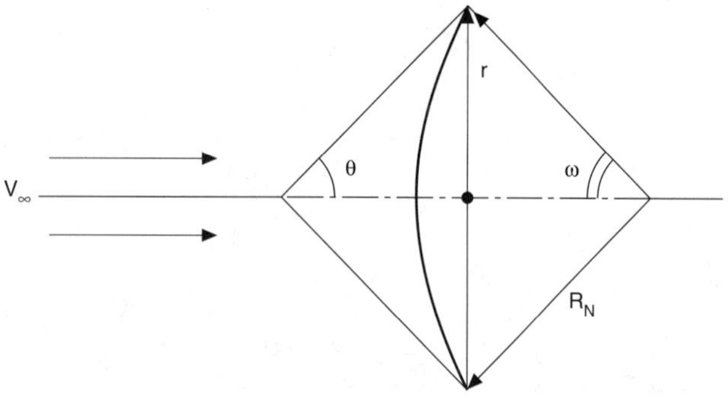

Figure 3. Force on a spherical segment or cylindrical sector.

$$\left.\frac{dU_e}{dx}\right|_{x=0} \cong \frac{a(T'_0)}{R_N\omega_*}.$$

This parameter determines the effect of the dimension of the body R_N on heat transfer in a laminar boundary layer. Here, a is the velocity of sound at the stagnation temperature T'_0.

We will now consider a further limiting case of flow around a body, i.e., the region of low free-stream velocities, to be more precise, the region of low Mach numbers $\mathrm{Ma} \le 0.3$ and low Reynolds numbers $\mathrm{Re} \le 1$.

The condition of low subsonic flow velocity allows the assumption that variations of density at various points in space filled with a fluid are not significant, i.e., the fluid is practically incompressible. However, the subsonic range of velocities prevents, as was noted above, the flow from being subdivided into free-stream and distorted flows. Therefore, exact solutions of the Navier-Stokes equations have only been obtained for rather specific conditions, whose applied value is limited. Thus, the well-known Stokes solution has been obtained for a single sphere in an infinite medium. When a spherical particle moves near the wall or some other solid surface, the surface affects the structure of flow formed by the sphere. For instance, a correction for the resistance force has the form [1 + 0.56(d/h)], where d is the sphere diameter and h is the distance from the flat surface along which the particle moves.

Flows for which the Reynolds number is rather small are called creeping motions. Interesting data have been obtained as a result of studying the fall of spherical particles under gravity. This case of a free motion of a body has been more thoroughly investigated (see **Drag Coefficient**). The stages characteristic of a fluid motion around spherical particles, and the trajectories of the fall in spheres, can be separated as follows:

1. At small Reynolds numbers ($\mathrm{Re} \le 0.5$), a smooth continuous flow around spheres is observed. No vortices arise in the bottom region. The trajectories of particle fall are strictly vertical. This strictly corresponds with a laminar flow.
2. At Reynolds numbers greater than 10, the formation of initially unstable and then stable vortices is observed, which adhere to the spheres. The trajectories of fall warp gradually and acquire an angle of slope to the vertical (transition streamline flow).
3. At Reynolds numbers greater than 100, the vortices begin to separate and at $\mathrm{Re} \cong 500$, separated vortices become regular. Leaving the bottom part of the body alternately, the vortices turn alternately to the right and to the left. Such a sequence of vortices is called the Karman vortex street. The vortex street moves with a velocity less then the velocity of flow V_∞ incoming on the body. The trajectory of particle fall is a twisting curve (turbulent, streamlined flow). At this stage, the concept of fall velocity becomes arbitrary and the length of the trajectory can differ considerably with the height of the fall.

All the above values of characteristic Reynolds numbers depend on body shape and the degree of turbulence of the main flow in which motion takes place. In streamline flow over an assembly of particles, their motion can change considerably if the mean distance between the particles is less than some limiting distance. It is significant that the lesser the Reynolds number, the stronger is the interference of neighbouring particles. Thus, the drag coefficient of spherical particles increases three times if Re $= 2 \times 10^2$ and only 1.5 times if Re $= 10^4$ at the same interparticle distance l/d = 5 (here, l is the distance between the centres of neighboring spherical particles).

For group motion of particles of various shapes, the problem of interaction becomes more involved, but reliable experimental data are not available.

Leading to: *Immersed bodies; Tubes, cross flowover; Tube banks, cross flowover; Rod bundles; Rotating disc systems; Wedge flow; Coanda effect.*

YuV. Polazhaev

EXTERNAL JACKET (see Agitated vessel heat transfer)

EXTINCTION, PHOTO (see Mie scattering)

EXTRA STRESS (see Non-Newtonian fluids)

EXTRACT PHASE (see Extraction, liquid/liquid)

EXTRACTION, LIQUID-LIQUID

Following from: *Mass transfer*

Liquid-liquid (or solvent) extraction is a counter-current separation process for isolating the constituents of a liquid mixture. In its simplest form, this involves the extraction of a solute from a binary solution by bringing it into contact with a second immiscible solvent in which the solute is soluble. In practical terms, however, many solutes may be present in the initial solution and the extracting 'solvent' may be a mixture of solvents designed to be selective for one or more solutes, depending upon their chemical type.

Solvent extraction is an old, established process and together with distillation constitute the two most important industrial *separation* procedures. The first commercially-successful liquid-liquid extraction operation was developed for the petroleum industry in 1909 when Edeleanu's process was employed for the removal of aromatic hydrocarbons from kerosene, using liquid sulphur dioxide as solvent. Since then many other processes have been developed by the petroleum, chemical, metallurgical, nuclear, pharmaceutical and food processing industries.

Whereas distillation effects a separation by utilising the differing volatilities of the components of a mixture, liquid-liquid extraction makes use of the different extent to which the components can partition into a second immiscible solvent. This property is frequently characteristic of the chemical type so that entire classes of compounds may be extracted if desired. The petroleum industry takes advantage of this characteristic of the process and has used extraction to separate, for example, aromatic hydrocarbons from paraffin hydrocarbons of the same boiling range using solvents such as liquified sulphur dioxide, furfural and diethylene glycol. In general, extraction is applied when the materials to be extracted are heat-sensitive or nonvolatile and when distillation would be inappropriate because components are close-boiling, have poor relative volatilities or form azeotropes.

The simplest extraction operation is single-contact batch extraction in which the initial feed solution is agitated with a suitable solvent, allowed to separate into two phases after which the solvent containing the extracted solute is decanted. This is analagous to the laboratory procedure employing a separating funnel. On an industrial scale, the extraction operation more usually involves more than one extraction stage and is normally carried out on a continuous basis. The equipment may be comprised of either discrete mixers and settlers or some form of column contactor in which the feed and solvent phases flow counter-currently by virtue of the density difference between the phases.

Final settling or phase separation is achieved under gravity at one end of the column by allowing an adequate settling volume for complete phase disengagement.

Any one extraction operation gives rise to two product streams: the extracted feed solution, more usually termed the *raffinate phase,* and the solvent containing extracted solute termed the *extract phase.* This nomenclature is unique to liquid-liquid extraction processes and will be used from hereon.

Choice of solvents

No single criterion can be used to assess the suitability of a solvent for a particular application and the final choice is invariably a compromise between competing requirements. Thus not only should the solvent be selective for the solute being extracted but it should also possess other desirable features such as low cost, low solubility in the feed-phase and good recoverability as well as being noncorrosive and noninflammable. Furthermore, interfacial tension between the two phases should not be so low that subsequent phase disengagement becomes difficult and the density difference between the phases should be large enough to maintain counter-current flow of the phases under the influence of gravity.

Of these factors, the first to be considered is the selectivity of the solvent, or the ease with which it extracts the desired solute from the feed stream. This is most readily understood by considering a simple ternary system consisting of a solution of solute C in a solvent A (the feed solution) and an extracting solvent B, which is designed to extract C from A. A simple single-stage extraction is shown on conventional triangular coordinates in **Figure 1a**. Here,

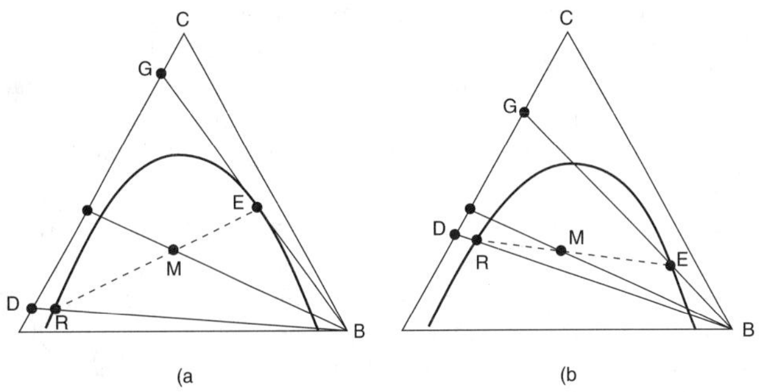

Figure 1. Extraction of solute C from A using solvents B and B′.

a mixture of A and C of composition F is mixed with a pure solvent B in such proportions as to give an overall composition M. This lies inside the miscibility curve and so the mixture will separate into two separate phases, R and E, joined by an equilibrium tie line RE. If the solvent B is now stripped from each phase, the solvent-free composition of R is given by D and the solvent-free composition of E by the point G. Both solvent-free compositions lie, of course, on the side of the triangle AC and it will be seen that the initial feed solution of composition F has been separated into two solutions D and G, which have low and high concentrations of C, respectively.

If this operation is now repeated using another solvent B′, the corresponding concentrations may be as shown in **Figure 1b**. In this instance, the solvent-free concentrations D and G are closer to the initial feed concentration F, and the separation of C is not as good as in the first case. It will be noted that the two solvents B and B′ are associated with equilibrium tie lines of very different slopes, and effects of this nature may be quantified by defining a solvent selectivity β_{CA}, analogous to relative volatility in distillation, such that:

$$\beta_{CA} = x_{CB}x_{AA}/x_{CA}x_{AB}.$$

Since X_{CB}/X_{CA} is the partition coefficient of the system, m,

$$\beta_{CA} = mx_{AA}/x_{AB} \qquad \textbf{(1)}$$

In most instances, β varies widely with concentrations; and for practical purposes, a solvent should be selected that gives high values of β in excess of unity and satisfies the other criteria listed above.

The extraction of aqueous solutions is usually carried out using organic solvents or mixtures thereof. In recent years, interest has developed in the possibilities of using a second aqueous phase loaded with a suitable polymer so the extracted solute does not come into contact with organic solvents. This is of particular interest to the pharmaceutical and foodstuffs industries [see Verrall (1992) and Hamm (1992) for details of such aqueous-aqueous systems.]

Phase Equilibria

The first step in the design of any extraction process is the determination of the equilibrium relationships between the feed solution and the proposed solvent. This enables the suitability of the solvent to be assessed in terms of its selectivity, as well as the calculation of the numbers of extraction stages required for any set of flow conditions and degree of separation.

Equilibrium data are usually determined directly in the laboratory since such measurements are more accurate than values calculated from predictive equations. Equilibria may be represented graphically on either triangular or rectangular coordinates, and a full discussion of the determination and representation of liquid-liquid equilibria has been presented by Treybal (1963). The correlation of equilibrium data is best achieved in terms of activity coefficients calculated from laboratory equilibrium measurements. A large number of semi-empirical equations are available for this purpose, but two models have found wide acceptance: the NRTL and the UNIQUAC equations for nonelectrolytes. In the absence of experimental data, it is not possible to determine the parameters of these equations and one must turn to purely predictive models, such as the regular solution and the UNIFAC models. All these procedures, as well as correlations for electrolyte solutions, have been discussed in detail by Newsham (1992) and this source should be consulted for further information.

Contactors

Contactors or extractors are specialised items of equipment designed to bring the feed and solvent phases together in such a manner that rapid transfer of the solute takes place from one phase to the other, followed by subsequent phase separation. In practice, efficient extraction involves four separate requirements:

a. The initial dispersion of one phase into the other in the form of droplets.
b. The maintenance of a fine dispersion in order to provide a large interfacial area for diffusion from one phase to the other.
c. The provision of an adequate holding or retention time for an acceptable level of diffusion to take place.
d. Final separation of the dispersion into raffinate and extract phases.

Numerous contactors have been described in the literature and the characteristics of the principal types have been summarized by Pratt and Stevens (1992). These authors also discussed the selection, design and scale-up of industrially-relevant contactors.

In its simplest form, a contactor merely consists of a *stirred tank* in series with a settling chamber through which the two phases flow (**Figure 2a**). Such arrangements are termed *mixer-settlers* and

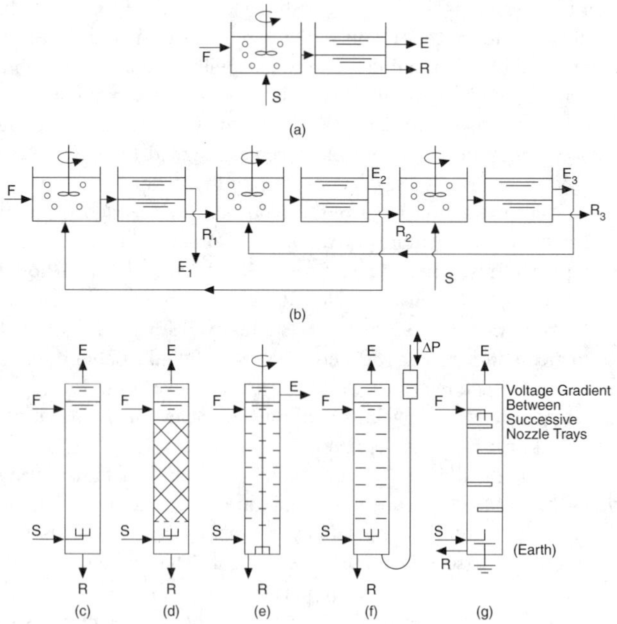

Figure 2. Contactor arrangements. (a) Single-stage mixer-settler. (b) Counter-current multiple contact using mixer-settlers. (c) Spray column. (d) Packed column. (e) Rotating disc column. (f) Air-pulsed plate column. (g) Electrostatic column. (F = feedstream; S = solvent; R = raffinate; E = extract. The feedstream is assumed to be the heavier phase throughout).

a number of units may be assembled in cascade to give the required degree of extraction. A typical assembly with counter-current phase flows is shown in **Figure 2b**. Such devices can become uneconomical when a high level of extraction is called for because of the multiplicity of units employed. Each calls for separate stirrers and instrumentation for interface control in each settler, and it is more usual to employ some form of 'column' contactor in which only one settling chamber is involved, irrespective of the degree of extraction required.

The simplest column contactor is the *spray tower* (**Figure 2c**) in which one phase is dispersed into the other and overall flows are counter-current through the column. Such units are inefficient and are of little interest outside the laboratory. If however some form of ordered or random packing, such as raschig rings, is introduced into the tower, extraction efficiency is increased several-fold. Such *packed columns* (**Figure 2d**) are an important item of industrial equipment. The packing not only reduces the gross back-mixing evident in the spray tower but also serves to establish a controllable droplet size distribution, as well as inducing additional turbulence inside and outside the droplets so diffusion from one phase into the other proceeds more rapidly [Batey and Thornton (1989)].

In column contactors described so far, energy available for droplet dispersion (and hence, the interfacial area available for solute transfer) is derived solely from the density difference between the phases; so the physical properties of the system set a limit to achievable extraction efficiency. This limitation may be overcome by supplying additional energy to the contactor and this concept has given rise to a large variety of so-called mechanical columns. Columns with coaxial rotating members of various designs have been described in the literature and the so-called *rotating disc contactor* illustrated in **Figure 2e** is a good example. This is basically a spray column with a central rotating shaft bearing a series of flat discs that rotate between fixed annular baffles. The shear forces set up produce very small droplets of dispersed phase and a correspondingly large interfacial area for mass transfer. Whilst such a unit gives good extraction efficiencies, the rotating shaft involves seals or bearings within the column and is therefore unsuitable for processing toxic or corrosive liquids.

This limitation relative to corrosive liquids may be overcome if mechanical energy is introduced in the form of reciprocatory, rather than rotary, motion. Such contactors are known as *pulsed columns,* and the reciprocatory motion may be applied either to the plates in the column or to the process fluids themselves. The latter procedure is now virtually universal. A typical pulsed column is illustrated in **Figure 2f** and consists, in essence, of a column shell fitted with a number of fixed perforated plates or sieve trays. The pulse may be imparted to the process fluids by attaching a cylinder closed by a reciprocating piston to the base of the column, or more usually by applying a sinusoidally-varying air pressure to a vertical standpipe connected to the base of the column (**Figure 2f**). Such a device is known as an air-pulsed column (Thornton 1954) and has the advantage that the process fluids are isolated from the pulsing mechanism by a pocket of air or inert gas. Such contactors have found wide application in the nuclear reprocessing industries. Column contents are usually pulsed sinusoidally at a frequency within the range 1–3 cycles per second and with an amplitude, measured in the column, of 12 mm or less. The perforated plates typically have a free area of 25% and are drilled with 3 mm diameter holes on a triangular pitch; the spacing between successive plates is 50 mm. Such a plate geometry does not allow the dispersed phase droplets to pass through readily except under the influence of the pulse, and very small droplets giving rise to large interfacial areas are readily obtained by varying the pulse frequency and/or amplitude. The extraction efficiency of the unit is therefore easily varied by changing the pulse characteristics and very high rates of extraction may be obtained with these columns. From an industrial viewpoint, mixer-settlers, rotating disc and pulsed columns have been employed successfully in a wide range of situations and extensive performance data are available in published literature.

Mechanical energy is not the only method of producing small droplets and thereby large interfacial areas for solute transfer. On the laboratory scale, both sonic and electrical energy have been employed successfully. Thus a sonic generator in contact with process fluids in a spray-type column gives rise to good droplet dispersions and high extraction rates [Thornton (1953)].

A promising procedure developed in recent years is *electrostatic extraction,* wherein electrical energy is employed to effect dispersion of one phase into the other by charging the dispersed phase entry nozzle to a high potential relative to a secondary electrode downstream in the column [Thornton and Brown (1966); Stewart and Thornton (1967)]. The droplets formed at the entry nozzle are now very small and carry an electrical charge so that they are accelerated at high velocity towards the secondary electrode. Furthermore, since the droplets carry a charge they oscillate rapidly due to the lowered interfacial tension, and contactors can be designed with very small contact times coupled with high extraction rates comparable with those usually associated with pulsed-plate columns. Rapid extraction coupled with low retention times in the contactor are particularly appropriate to processes in which the solute is of biological origin, or is otherwise unstable during extraction operation.

In practice, such electrostatic contactors comprise a column equipped with a number of insulated nozzle trays with a potential gradient between successive trays (**Figure 2g**). Designs for large-scale units have not yet been investigated in any detail, but the use of radial nozzle trays for larger diameter columns has been proposed [Thornton (1989)]. For a general discussion of electrostatic extraction, see Thornton (1976) and Weatherley (1992).

Factors in contactor design

The design of a column-type contactor basically involves the calculation of two geometrical parameters viz., the height of column necessary to give the required degree of extraction and the diameter to handle the necessary flow rates under prescribed operating conditions. In the design of stage-wise units, such as mixer-settlers, the corresponding parameters are power input and residence time of the mixing chamber and the residence time of the settler necessary for satisfactory phase setting.

The fundamental quantities usually employed to describe the hydrodynamic behaviour of column-type contactors are the phase flow rates, fractional holdup of the dispersed phase and the characteristic mean velocity of the droplets. If the fractional voidage of the column is denoted by ε, the quantities are related by the holdup equation:

$$V_d/x + V_c/(1 - x) = \varepsilon\overline{V}_o(\phi) \quad (1)$$

The term (ϕ) on the right hand side of **Equation 1** takes into account the reduction in mean velocity of a multiplicity of droplets by comparison with the velocity of a single droplet in infinite media. This so-called hindered rising term can frequently be represented by a function of holdup ($1 - x$) so long as the droplet size is small ($Re \ngtr 2$) and is independent of the phase flow rates—conditions frequently met in mechanical contactors. On this basis, the holdup equation takes the form:

$$V_d/x + V_c/(1 - x) = \varepsilon\overline{V}_o(1 - x) \quad (2)$$

A plot of $[V_d + (x/1 - x)V_c]$ versus $x(1 - x)$ is linear with a slope of $\varepsilon\bar{V}_o$ and enables characteristic velocity to be determined from flow rate and holdup measurements. Progressive increases in either or both phase flow rates finally result in flooding of the contactor, which is manifested by the appearance of a second interface at the opposite end of the column to the main interface. This condition corresponds to maximum values of the flow rates beyond which holdup remains constant, and can be found by differentiating **Equation 2** [Thornton and Pratt (1953)].

$$dV_c/dx = 0 \qquad V_{df} = 2\varepsilon\overline{V}_o x_f^2(1 - x_f) \quad (3)$$

$$dV_d/dx = 0 \qquad V_{cf} = \varepsilon\overline{V}_o(1 - x_f)^2(1 - 2x_f) \quad (4)$$

Eliminating εV_o between **Equations 3 and 4** and solving for x_f yields

$$x_f = [(L^2 + 8L)^{1/2} - 3L]/4(1 - L) \quad (5)$$

where L represents the ratio V_{df}/V_{cf} and subscript f denotes values at the flood point. Thus the phase superficial velocities at the flood point, together with the associated dispersed phase holdup, may be determined from **Equations 3–5** once the value of the droplet characteristic velocity $\bar{V}_o$ is known; the latter is readily obtained by plotting experimental holdup measurements in accordance with **Equation 2** and measuring the slope of the linear plot. By this means flood point data may be obtained from holdup measurements and vice versa [Thornton and Pratt (1953); Thornton (1956)]. It is important to note that **Equations 2–5** are only appropriate for situations where mean droplet size is small ($Re \ngtr 2$) and is constant up to the flood point. Furthermore the assumption is made that the effective buoyancy force between the phases is proportional to the difference in densities of the mixture and the dispersed phase ($\rho_m - \rho_d$). By the mixture law, this is equal to $(\rho_c - \rho_d)(1 - x)$, thus accounting for the $(1 - x)$ term on the right hand side of **Equation 2**. At higher Reynolds numbers, it is possible, in principle, to derive equations analogous to **Equations 2–5** provided that satisfactory equations can be formulated to describe droplet motion in the column. This is not always possible at present and numerous semi-empirical expressions have been proposed in lieu of **Equation 2** [Pratt and Stevens (1992)]. It should be noted that **Equations 3–5** cannot be used to predict flooding rates in packed columns since droplet coalescence sets in prior to the flooding point.

Relationships can be established between the Sauter mean droplet size, d_{vs}, and the characteristic velocity $\bar{V}_o$ by introducing the concept of column impedance, I. Thus in the case of a spray tower of height H, let the time taken for a droplet to move through this distance be t. The terminal velocity of the droplet, U, is then equal to H/t. In a contactor such as a perforated plate-column, the same size droplets will take longer than time *t* to pass through a height H because they will suffer a small but finite delay, Δt, at each plate. In a column of N plates, the total delay will amount to $N\Delta t$ and the velocity of the droplets relative to the continuous phase will be given by $H/(t + N\Delta t)$. In the limit as holdup tends to zero, this velocity approaches the characteristic velocity $\bar{V}_o$, and taking the ratio of spray and plate column velocities yields the expression

$$U/V_0 = 1 + N(\Delta t/t) = 1 + I \quad (6)$$

where I is the column impedance and is defined as the fractional increase in time of passage of a single droplet with respect to an empty spray column. The terminal velocity U can be written in terms of mean droplet size and the physical properties of the systems using published drag coefficient data, thereby establishing the link between droplet size and $\bar{V}_o$ for a known value of I via **Equation 6**. Values of I range from zero to some finite value, depending upon the characteristics of the contactor and the properties of the system. For the application of this concept to pulsed plate-columns and the use of laboratory holdup measurements to characterise the behaviour of such contactors, see Batey et al. (1987).

Diffusion of a solute from one liquid phase to another is a complex process governed by molecular and/or eddy diffusional mechanisms. Mass transfer flux is proportional to the instantaneous concentration driving force, and the ratio of these quantities is called the mass transfer coefficient and may be defined in terms of either the continuous or the dispersed phase driving force. Thus if solute A is transferring from phase c to phase d, the flux N_A is given by:

$$N_A = k_c(c_c - c_{ci}) = k_d(c_{di} - c_d) \quad (7)$$

In practice, it is not feasible to measure the two interfacial concentrations c_{ci} and c_{di} in a practical extraction system and so, since the interfacial concentrations are assumed to be in equilibrium, diffusion from phase c to d can be considered to be diffusion of solute through two phases or resistances in series. No resistance will be offered by the interface itself because the chemical potentials in each phase will be equal. On this basis, the so-called overall coefficients K_o can be defined:

$$N_A = K_{od}(c_d^* - c_d) = K_{oc}(c_c - c_c^*) \quad (8)$$

where all the concentrations are now known. For a linear equilibrium curve, the relationships between the individual phase or film coefficients and the overall coefficients are readily shown to be [Treybal (1963)]

$$I/K_{od} = I/k_d + m/k_c \quad (9)$$

$$I/K_{oc} = I/k_c + I/mk_d \quad (10)$$

Numerous mathematical models have been proposed for calculating values of the individual phase coefficients k, but these make assumptions regarding the hydrodynamic characteristics of the phase in question [Skelland (1992); Javed (1992)]. Knowledge of turbulence levels and flow patterns in the vicinity of the interface is still far from complete and this limits the use of models for predicting k values. Contactor design is therefore based upon experimental measurements of the mass transfer coefficients, using the actual type of contactor and extraction system in question.

For a detailed consideration of design procedures for column contactors and mixer-settler devices, see the comprehensive account by Pratt and Stevens (1992).

Reference must be made, however, to the complications arising when the **Marangoni Effect** or spontaneous interfacial turbulence is present. Many solutes promote intense turbulence at the interface as they diffuse from one phase to the other [Perez de Ortiz (1992)]. The level of turbulence cannot be predicted from first principles and the consequences can only be assessed from extensive pilot plant studies. Thus, for example, extraction efficiency of countercurrent column contactors is usually expressed in terms of the height of a transfer unit or HTU [Treybal (1963)]. The overall HTU based upon the continuous phase driving force is defined as:

$$(HTU)_{oc} = V_c/K_{oc}a \quad \textbf{(11)}$$

where the specific surface area a is equal to 6 $\varepsilon x/d_{vs}$, so that

$$(HTU)_{oc} = (V_c/6)[d_{vs}/K_{oc}\varepsilon x]_M \quad \textbf{(12)}$$

Marangoni effects can influence mean droplet size, d_{vs}, through enhanced interdroplet coalescence rates; the holdup x, by virtue of a correspondingly increased value of $\bar{V}_o$ in **Equation 2** and the overall mass transfer coefficient K_{oc} through enhanced turbulence at the droplet interface [Thornton et al. (1985); Javed et al. (1989)]. Moreover, values of K_{OC} become time-dependent and decrease as the droplet interface ages. See Thornton (1987) for a further discussion of this problem.

The terms in square brackets with subscript M in **Equation 12** are therefore all dependent upon the level of interfacial turbulence induced by the solute. The extent to which these quantities are modified by Marangoni phenomena is not yet quantifiable. It is therefore always desirable to study the characteristics of any proposed extraction system in the laboratory before proceeding to the design stage.

Industrial extraction operations

Details of relevant process chemistry and extraction operations in the hydrometallurgical, nuclear, pharmaceutical and food industries are provided by Thornton (1992).

Nomenclature

- a Specific interfacial area
- c Solute concentration
- d_{vs} Sauter mean droplet diameter
- H Height of column
- $(HTU)_{oc}$ Height of an overall transfer unit based on the continuous phase
- I Column impedance defined by Equation 6
- k Single-phase or film mass transfer coefficient
- K_o Overall mass transfer coefficient
- m Equilibrium distribution or partition coefficient
- N_A Transport flux of solute A
- U Droplet terminal velocity in infinite media
- V Superficial phase velocity
- $\bar{V}_o$ Droplet characteristic velocity i.e., mean droplet velocity when V_c = o and $V_d \rightarrow$ o.
- x Fractional holdup of dispersed phase. With subscripts concentration (wt fraction)
- β_{CA} Selectivity of solvent for solute C from an A–C solution
- ε Fractional voidage of column
- ρ Phase density

Subscripts

- c Continuous phase
- d Dispersed phase
- i At interface
- AA A in A-rich solution
- AB A in B-rich solution
- CA C in A-rich solution
- CB C in B-rich solution

Superscript

- * At equilibrium

References

Batey, W., Thompson, P. J., and Thornton, J. D. (1987) Column impedance and the hydrodynamic characterisation of pulsed plate-columns. Extraction '87. *Instn. Chem. Engrs. Symposium Series.* 103:133.

Batey, W. and Thornton, J. D. (1989) Partial mass transfer coefficients and packing performance in liquid-liquid extraction. *I & EC Research.* 28:1096.

Hamm, W. (1992) *Science and Practice of Liquid-liquid Extraction* (Ed. Thornton). Clarendon Press. Oxford. 2, Ch 4.

Javed, K. H., Thornton, J. D., and Anderson, T. J. (1989) Surface phenomena and mass transfer rates in liquid-liquid systems. *AIChE Journal.* 35(7):1125.

Javed, K. H. (1992) Science and Practice of Liquid-liquid Extraction (Ed. Thornton). Clarendon Press, Oxford. 1, Ch 4.

Newsham, D. M. T. (1992) Ibid. 1, Ch 1.

Perez de Ortiz, E. S. (1992) Ibid. 1, Ch 3.

Pratt, H. R. C. and Stevens, G. W. (1992) Ibid. 1, Ch 8.

Skelland, A. H. P. (1992) Ibid. 1, Ch 2.

Stewart, G. and Thornton, J. D. (1967) Charge and velocity characteristics of electrically charged droplets. *Instn. Chem. Engrs. Symposium Series.* 26: 29, 37.

Thornton, J. D. (1953) Improvements in or relating to columns for liquid-liquid extraction processes. Brit. Pat. 737,789 (filed 1953, published 1955).

Thornton, J. D. and Pratt, H. R. C. (1953) Flooding rates and mass transfer data for rotary annular columns. *Trans. Instn. Chem. Engrs.* 31:289.

Thornton, J. D. (1954) Recent developments in pulsed column techniques. *Chem. Eng. Progr. Symposium Series. 50* (No. 13): 39. See also Thornton, J. D. Improvements in or relating to liquid-liquid extraction columns. Brit. Pat. 756,049 (filed 1954, published 1956).

Thornton, J. D. (1956) Spray liquid-liquid extraction columns. Prediction of limiting holdup and flooding rates. *Chem. Eng. Sci.* 5:201.

Thornton, J. D. and Brown, B. A. (1966) Liquid-fluid extraction process. Brit. Pat. 1,205,562 (filed 1966, published 1970).

Thornton, J. D. (1976) Electrically enhanced liquid-liquid extraction. *Birmingham University Chemical Engineer.* 27(No.1):6.

Thornton, J. D., Anderson, T. J., Javed, K. H., and Achwal, S. K. (1985) Surface phenomena and mass transfer interactions in liquid-liquid systems. *AIChE Journal.* 31(7):1069.

Thornton, J. D. (1987) Interfacial phenomena and mass transfer in liquid-liquid extraction. *Chemistry and Industry* (SCI London). 16 March: 193.

Thornton, J. D. (1989) Euro. Pat. Specn. 0356030 AZ (filed 1989).

Thornton, J. D. (1992) *Science and Practice of Liquid-liquid Extraction* (Ed. Thornton). Clarendon Press, Oxford. 2.

Treybal, R. E. (1963) *Liquid Extraction.* McGraw Hill, NY.

Verrall, M. S. (1992) Science and practice of liquid-liquid extraction (Ed. Thornton). Clarendon Press, Oxford. 2, Ch 3.

Weatherley, L. R. (1992) Ibid. 2, Ch 5.

J.D. Thornton

EXTRACTORS (see Extraction, liquid/liquid)

EXTRUDATE SWELL (see Non-Newtonian fluids)

EXTRUSION PLASTICS

Plastics extrusion refers to the process whereby a *polymer*—usually in the form of pellets or powder—is mixed, heated, melted, compressed and pumped through a shaped *die* to form a continuous stream of product of the desired cross-sectional shape. A typical extrusion system is shown in **Figure 1**. Typical extrusions include pipe, sheet, film, coated wire and cable, monofilaments, parison for blow-molding, foams, fibres and various profiles [Hensen (1988)]. Twin-screw extruders are common in applications requiring superior mixing or improved heat and mass transfer [White (1990)].

In performing its task, the extruder functions as a solids pump, a heat exchanger and melting device [Davis (1993)], a melt pump, a chemical reaction or mass transfer vessel [Davis (1993); Deason (1983)] and an intensive mixer or homogenizer. Heat transfer is important in all extruder applications, but becomes critical in cases such as reactive extrusion [Davis (1993)] and foam extrusion. In practice, there is little difficulty in melting polymer granules since, besides heating through the barrel, significant amounts of mechanical energy are added to the system through the rotation of the screw. However, removing heat is not so straightforward. Polymer viscosity increases sharply with decreasing temperature and for this reason, some of the energy removed through cooling from the barrel is unavoidably added back to the system in the form of viscous heating.

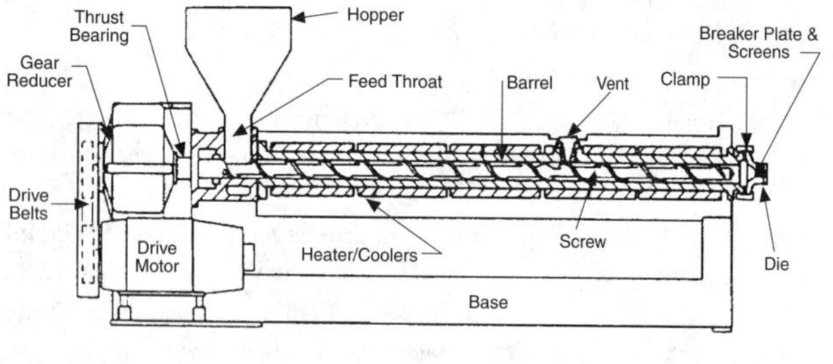

Figure 1. Illustration of a Single Screw Plasticating Extruder (from "Computer Modeling for Extrusion", K. T. O'Brien (Ed.), Carl Hanser Verlag, Munich, (1992).

Of all components of an extruder, the single most important is the screw. Its effect on performance is so great that, in many respects, extruder design is equivalent to screw design. Screw configuration varies depending on the particular application. **Figure 2** shows various typical *plasticating screw* designs.

It is convenient for analysis to divide an extruder screw in three "zones": The *Feed or Solids Conveying Zone* is designed to preheat the polymer granules and convey them to subsequent zones. In the *Transition or Melting or Plasticating Zone,* the polymer granules are melted and the resulting melt compressed. It has been shown in most extrusion applications (with the exception of twin-screw and pin-barrel extruders or certain PVC extrusions [Rauwendaal (1986)]) that during melting the polymer particles are pressed together to form a contiguous solid bed which is pushed against the trailing flight flank [Rauwendaal (1986); Tadmor and Klein (1970)]. In the *Metering or Pumping Zone,* the screw channel is full of molten polymer which is pumped towards the die. The motion of the screw relative to the barrel induces a drag flow which, because of the geometry of the screw, has the net effect of pushing the polymer forward. This is resisted by the pressure gradient operating in the opposite direction, and the net melt flow rate through the metering zone is the sum of these two terms [Tadmor and Klein (1970); Pearson and Richardson (1983); Rao (1986)].

The basic theoretical understanding of plasticating extrusion was developed in the 1960s. Models for solids conveying, melting of polymer granules and melt flow in the metering section have been systematically presented in the classical textbook of Tadmor and Klein (1970). Software for extruder screw design became available in the early seventies and have continued to evolve ever since [Pearson and Richardson (1983); Rao (1986)]. A variety of general purpose and specialized (CAD) codes can be used for the rigorous simulation of the extrusion process [Vlachopoulos (1992)

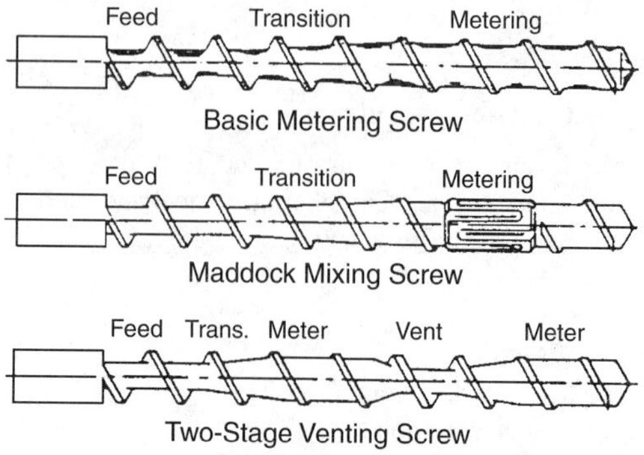

Figure 2. Plasticating Screw Designs (with permission from "Computer Modeling for Extrusion', K. T. O'Brien (Ed.), Hanser, Munich, 1992).

and other references therein]. Such simulations help designers gain a better understanding of flow and heat transfer in an extruder, identify reasons for possible product defects and develop better physical models for the processes involved.

References

Davis, W. M. (1993) Heat transfer in extruder reactors. *Reactive Extrusion.* M. Xanthos, (Ed.). Hanser, Munich.

Denson, C. D. (1983) Stripping operations in polymer processing. *Advances in Chemical Engineering. 12:61.*

Hensen, F. (Ed.). (1988) *Plastics Extrusion Technology.* Hanser, Munich.

Pearson, J. R. A. and Richardson, S. M. (Eds.). (1983) Computational analysis of polymer processing. Elsevier.

Rao, N. S. (1986) *CAD of Plasticating Screws.* Hanser, Munich.

Rauwendaal, C. (1986) *Polymer Extrusion.* Hanser, Munich.

Tadmor, Z. and Klein, I. (1970) *Engineering Principles of Plasticating Extrusion.* Van Nostrand Reinhold.

Vlachopoulos, J., Silvi N., and Vlcek, J. (1992) PolyCAD: a finite element package for polymer flow. *Computer Modelling for Extrusion and Other Continuous Polymer Processes.* O'Brien, K. T. (Ed.). Hanser, Munich.

T.A. Papathanasiou

F

FACHINFORMATIONSZENTRUM KARLSRUHE, FIZ

7514 Eggenstein-Leopoldshafen 2
Karlsruhe, Germany
Tel: 7247 8080

FAHRENHEIT TEMPERATURE SCALE

Following from: *Temperature*

The Fahrenheit degree (named after German physicist G.D. Fahrenheit, 1686–1736) is the unit of temperature (designated by °F) which is equal to 1/180 part of temperature interval between the points of ice melting (32 °F) and of water boiling (212 °F) at normal atmospheric pressure. Thus, $T_C = (T_F - 32)/1.8$, where T_C is the temperature in °C, and T_F is the temperature in °F (see **Figure 1**).

M.P. Orlova

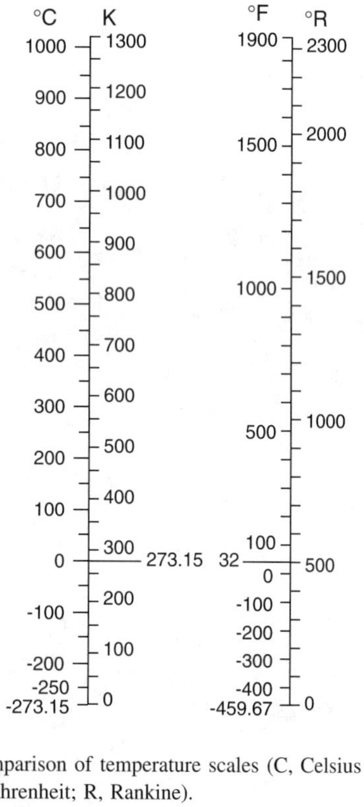

Figure 1. Comparison of temperature scales (C, Celsius or Centigrade; K, Kelvin; F, Fahrenheit; R, Rankine).

FALKNER-SKAN EQUATION (see Wedge flows)

FALLING FILM EVAPORATORS (see Evaporators)

FALLING FILM, COMBINED HEAT AND MASS TRANSFER

Following from: *Falling film flow; Falling film heat transfer; Falling film mass transfer*

Heat transfer in a falling film depends on the film hydrodynamics, mass transfer and the direction of heat flow.

Evaporation of a falling film

Film surface evaporators concentrate heat sensitive materials. They operate at low flux densities (<50 Kw/m^2), low boiling point elevations and temperature differences (<10 °C). Laminar film conditions can be predicted using **Equation 1** and suitable boundary conditions.[1]

$$U(y)\frac{\delta T}{\delta x} = \alpha \frac{\delta^2 T}{\delta y^2} \tag{1}$$

Predicted Nu_x values follow the general trend for laminar heating for both constant heat flux and temperature conditions but the actual values are 5% to 50% (corresponding x values 10^{-3} to 1) less than the laminar heating prediction. Data[2] have a +50% higher value than the evaporation prediction due to the enhancement effect of surface waves.

Prediction of laminar-wavy evaporation[3] has been attempted with fair results.[2,4]

For turbulent flow, evaporation prediction at both constant temperature and constant wall flux has been attempted based on similar methods employed for heating. Most use a constant film thickness model but recently a variable thickness model has been attempted. The basic **Equation 2** is applied plus appropriate boundary conditions.[5]

$$U^+\frac{\delta \theta}{\delta \bar{x}} = \frac{\delta}{\delta \eta}\left[\left(1 + \frac{\epsilon_H}{T}Pr\right)\frac{\delta \theta}{\delta \eta}\right] \tag{2}$$

Various turbulent evaporation models are detailed in **Table 1**. The modified *Van Driest model* appears to be the most promising particularly when incorporating a damping factor.

Schnabel and Schlunder[6] have proposed the following correlation from data:

$$h^*_{\infty lam} = 1.4287\, Re^{-1/3}; \quad Re < 400 \text{ laminar} \tag{3}$$

$$h^*_{\infty turb} = 0.00357\, Re^{0.4}\, Pr^{0.65}; \quad 3200 < Re \text{ turbulent} \tag{4}$$

$$h^*_{\infty trans} = [(h^*_{\infty lam})^2 + (h^*_{\infty turb})^2]^{1/2}; \quad 400 < Re < 3200 \tag{5}$$

The entrance length for evaporation is always longer than that for heating and is important with small Re and Pr. At large values of these parameters, it can be neglected. Variable film thickness mod-

Table 1. Models for turbulent falling films with evaporation and interfacial shear

Authors	Relevant equations	Comparison with data
Dukler[7]	Deissler model $y^+ \le 20$ $\epsilon_m/\nu = 0.125^2 u^+ y^+[1 - \exp(-0.125^2 u^+ y^+)]$ Von Karman model $y^+ > 20$ $\epsilon_m/\nu = 0.4^2\,(du^+/dy^+)^3/(d^2u^+/dy^{+2})^2$	$\bar{h}_x$ predicted +25 to +100% above data[9,10]
Domanskii[8]	Von Karman-Nikuradse model $\epsilon_m/\nu = 0$ $0 \le y^+ \le 5$ $\epsilon_m/\nu = y^+/5$ $5 \le y^+ \le 30$ $\epsilon_m/\nu = y^+/2.5$ $30 < y^+$	δ^+ $\overline{\text{vs}}$ Re same as Dukler[7] $\bar{h}_x$ $\overline{\text{vs}}$ Re higher than data[9,10]
Murthy[11]	Sleicher model $\epsilon_m/\nu = 0.091^2\,y^{+2}$ $y^+ < 50$	δ^+ $\overline{\text{vs}}$ Re poor agreement $\bar{h}_x$ $\overline{\text{vs}}$ Re agrees within ±15% with data[2,12]
Mills[13]	Van Driest model $y/\delta \le 0.6$ $\epsilon_m/\nu = -0.5 + 0.5[1 + 0.64y^{+2}\,[1 - \exp(-y^+/26)^2]^{1/2}$ Lamourell model $y/\delta > 0.6$ $\epsilon_m/\nu = 6.47 \times 10^{-4}(\rho g^{1/3}\,\nu^{4/3}/\sigma\delta^{+2/3})\,(\delta^+ - y^+)^2\,Re^{1.678}$	δ^+ $\overline{\text{vs}}$ Re +15% above data $\bar{h}_x$ $\overline{\text{vs}}$ Re agrees with data[2,14]
Beccari[15]	Sleicher model $y^+ \le 26$ Von Karman model $y^+ > 26$	no comparison
Hubbard[16]	Van Driest model $y/\delta \le 0.6$ Chung model $y/\delta > 0.6$ $\epsilon_m/\nu = 8.13 \times 10^{-17}[(\nu g)^{2/3}/Cb\bar{u}^2]$ $Re^{2n}[1 + b(\tau_b/\tau_w^0)]^2\,(\delta^+ - y^+)^2$	$\bar{h}_x$ $\overline{\text{vs}}$ Re lower slope than data[2]
Faghri[1,14]	Van Driest model $y/\delta \le 0.6$ ϵ_m/ν = constant $0.6 < y/\delta \le 1$	$\bar{h}_x$ $\overline{\text{vs}}$ Re agrees with data at Pr 1.77 but falls as Pr rises
Murthy[17]	Sleicher model $y^+ \le 30$ Reichardt model $y^+ > 30$ $\epsilon_m/\nu = 0.667y^+(2 + y^+/\delta^+)[3 + 4(y^+/\delta^+) + 2(y^+/\delta^+)^2]$	δ^+ $\overline{\text{vs}}$ Re agrees with data[11,17] $\bar{h}_x$ $\overline{\text{vs}}$ Re higher than data[2] at $Pr = 5$
Wassner[18]	Domanskii model $\tau/\tau_w = 1 - y/\delta$	$\bar{h}_x$ $\overline{\text{vs}}$ Re 0.7 lower than,[9] better agreement with data
Mostofizadeh[19]	Van Driest-Nikuradse model $\epsilon_m/\nu = -0.5 + 0.5[1 + 4[0.14 - 0.08(1 - y/\delta)^2$ $- 0.66(1 - y/\delta)^4]^2\delta^2[1 - \exp(-y^+/26)]^2\tau/\tau_w]^{1/2}$ $\epsilon_H/\nu = 1.875(\epsilon_m/\nu)\exp[-0.9/(0.64\,Pr$ $\epsilon_m/\nu) - Pr^{-0.1}]\,[\exp(1.2564y/\delta - 1)]$	δ^+ $\overline{\text{vs}}$ Re good agreement $\bar{h}_x$ $\overline{\text{vs}}$ Re good agreement at $Pr = 5.16$ under at $Pr = 1.75$
Yih & Liu[20]	Faghri model τ/τ_w modified to include interfacial shear	$\bar{h}_x$ $\overline{\text{vs}}$ Re good agreement with data. $Pr = 5.5$ higher than data $Pr = 2.0$

els have proven to give good predictions, but require extensive data and are too tedious for practical engineering calculations.

Condensation

Film condensation on a vertical surface results in the liquid at the top flowing in laminar mode, then changing to wavy and finally turbulent flow as it proceeds down the surface. Practical film steam condensation heat transfer coefficients range from 7 to 50 Kw/m^{-2} °K. There are a number of heat transfer resistances, but pure vapour condensation resistance mainly lies in the liquid film itself. Interfacial shear normally only has a significant effect on heat transfer for condensation inside tubes. The complexity of the mixed hydrodynamic and thermal conditions requires complex experimental calculations to obtain accurate output.

Laminar film flow (i.e. $Re < 10$) has been handled theoretically by *Nusselt*.[21] Data are consistently about 25% above these predictions due to waves on the surface. Stability analyses[22] show that, in practical situations, laminar condensate films are unstable and waves will appear. Predictions for the wavy laminar condition[23] give better results. Data[23] has given the result:

$$h_m^* = p\,Re^q \tag{6}$$

	p	q	Range
McAdams[24]	1.882	−1/3	<1800
Kutateladze[23]	1.01	−0.22	$40 < Re < 2000$
Gogonin[25]	1.39	−0.293	$20 < Re < 400$

For fully developed turbulent flow condensation, modelling methods follow heating/evaporation models and are detailed in **Table 2**.

Generally, if the vapour velocity is high (>60 ms^{-1}) interfacial shear must be considered in the model. There is a marked effect of condensation on flow regime transitions. Generally, Re_c between laminar wavy and turbulent is taken to be 1000.[29] Colburn[34] devel-

Table 2. Models for turbulent falling film condensation with interfacial shear

Turbulence models	Relevant equations	Comparison with data
Seban[26] Rohsenow[27] Shekriladze[28] Ueda[29] Dobran[30]	Von Karman-Nikuradze c.f. Domanskii Table 1	δ^+ v Re and $\bar{h}_\infty$ fair agreement with data[29,33,34]
Dukler[7] Razavi[31]	Deissler model c.f. Table 1	δ^+ vs Re, $\bar{h}_\infty$ higher than data, better agreement of momentum exchange included
Kutateladze[23]	Prandtl two and three layer model	δ^+ vs Re agrees Nusselt $\bar{h}_\infty$ agrees with own data
Kunz[32] Hubbard[16] Blangetti[33] Mostotizadeh[19] Yih & Liu[20]	Van Driest model c.f. Table 1	Best predictions with models employing a damping factor

oped methods of calculation for heat transfer with condensation which were extended and corrected by Soliman.[35]

$$h_x = 0.030\ Pr^{0.65}\ \tau_w^{1/2}\lambda\rho^{1/2}/\eta \qquad \textbf{(7)}$$

Nomenclature

b, C, p, q	constant
g	gravitational acceleration m/s^2
h	heat transfer coefficient W/m^{2}K
h_m^*	average dimensionless heat transfer $(hx/\lambda)\ (\nu^2/g)^{1/3}$
λ	thermal conductivity W/mK
Pr	Prandtl number ϵ_m/ϵ_H
q_w	heat wall flux W/m^2
Re	Reynolds number Uδ/v
U	velocity m/s
U^+	dimensionless velocity
$\overline{U}$	frictional velocity $(\tau_w/\rho)^{1/2}$
T	temperature °K
x	axial distance m
y	radial distance m
y^+	dimensionless yU^*/ν radial distance
α	thermal diffusivity m^2/s
$\epsilon_{m,H}$	eddy diffusivity momentum, heat m^2/s
δ	film thickness m
$\bar{\delta}$	frictional film thickness $\delta/(\nu^2/g)^{1/3}$
δ^+	dimensionless film thickness $U\delta/\nu$
θ	dimensionless temperature $\Delta T/(q_w\ \delta/\lambda)$
η	dynamic viscosity kg/ms
ν	kinematic viscosity m^2/s

References

1. Faghri, A. (1976) PhD Thesis, Univ. Calif. Berkley.
2. Chun, K. R. and Seban, R. A. (1971) *J. Heat Trans., TASME,* 93, 391.
3. Hirshburg, R. I. and Florschuetz, L. W. (1982) *J. Heat Trans., TASME,* 104, 452, 459.
4. Struve, H. (1969) *VDI,* Forschungsh, 534.
5. Yih, S. M. and Chen, C. H. (1982) *Proc. Int. Heat. Trans. Conf.,* 7, 3, 125.
6. Schnabel, G. and Schlunder, E. U. (1980) *Verfahrenstechnik,* 14, 79.
7. Dukler, A. E. (1960) *AIChE Symp.* Ser., 56, 1.
8. Domanskii, I. V. and Sokolov, V. N. (1967) *J. Appl. Chem. USSR,* 40, 56.
9. Portalski, S. (1963) *Chem. Eng. Sci.,* 18, 787.
10. Dukler, A. E. and Berglin, O. P. (1952) *CEP,* 48, 557.
11. Murthy, V. N. and Sarma, P. K. (1973) *J. Chem. Eng. Japan,* 6, 457.
12. Wilkie, W. (1962) *VDI,* Forschungsh, 490.
13. Mills, A. F. and Chung, D. K. (1973) *Int. J. Heat Mass Trans.,* 16, 694.
14. Seban, R. A. and Faghri, A. (1976) *J. Heat Trans., TASME,* 98, 315.
15. Beccari, M., Di Pinto, Santori, Biasi, Prosperetti, Tozzi (1975) *Int. J. Multiphase Flow,* 2, 357.
16. Hubbard, G. L., Mills, A. F., and Chung. D. K. (1976) *J. Heat Trans., TASME,* 98, 319.
17. Murthy, V. N. and Sarma, P. K. (1977) *Can. J. Chem. Eng.,* 55, 372.
18. Wassner, L. (1980) *Warme-und Soffubertragung,* 14, 23.
19. Mostofizadeh, C. and Stephen, K. (1981) *Warme-und Soffubertragung,* 15, 93.
20. Yih, S. M. and Liu, J. L. (1983) *AIChEJ,* 29, 903.
21. Nusselt, W. (1910) *VDI,* Zeitschrift, 54, 1154.
22. Marshall, E. and Lee, C. Y. (1963) *Int. J. Heat Mass Trans.,* 16, 41.
23. Kutateladze, S. S. (1963) *Fundamentals of Heat Transfer,* 15, AP., N.Y.
24. McAdams, W. H. (1942) *Heat Transmission, 2nd Ed.,* McGraw-Hill, N.Y.
25. Gogoin, I. I., Dorokhov, A. R. and Sosunov, V. J. (1981) *Heat Transfer—Soviet Research,* 13, 51.
26. Seban, R. A. (1954) *TASME,* 76, 299.
27. Rohsenow, W. M., Webber, J. H. and Ling, A. T. (1956) *TASME,* 78, 1637.
28. Shekriladze, I. and Mestvirishvili, S. (1973) *Int. J. Heat Mass Trans.,* 16, 715.
29. Ueda, T., Kubo, J. and Inove, M. (1974) *Proc. Int. Heat Trans. Conf.,* 5, 3, 304.
30. Dobran, F. (1983) *Int. J. Heat Mass Trans.,* 26, 1559.
31. Razovi, M. D. and Damle, A. S. (1978) *Trans. Inst. Chem. Eng.,* 56, 81.
32. Kunz, H. R. and Yerazunis, S. (1969) *J. Heat Trans., TASME,* 91, 413.
33. Blangetti, F. and Schlunder, E. U. (1978) *Proc. Heat Trans., Conf.,* 6, 2, 457.
34. Colburn, A. P. (1934) *Tran. AIChE.,* 30, 187
35. Soliman, M., Schaster, J. K. and Berenson, P. J. (1968) *J. Heat Trans.,* 90, 267.

P.L. Spedding

FALLING FILM FLOW

A falling film is the gravity flow of a continuous liquid film down a solid tube having one free surface. Other geometries and conditions are possible variables,[1] but normal practice is with a vertical tube and countercurrent gas flow. Falling films are used in industry for distillation, absorption, reactions, condensation, etc.[1] Its advantages are that liquid residence time is small, transfer rates are high for comparatively small pressure losses and it possesses a large interface of simple geometry. Design and operation of the equipment need careful attention to eliminate flooding or poor wetting phenomena.[2–3] (See also **Flooding and Flow Reversal**; **Capillary Action**.)

Immediately under the liquid distributor is a non-uniform entrance region followed by the main film. Both depend on liquid distributor design, Re, K_F, We and Fr.[4]

$$Re = 4\,\dot{M}/\pi\eta d = 4\Gamma/\eta = aK_F^{0.1} \quad (1)$$

$$K_F = \rho\sigma^3/g\eta^4 \quad (2)$$

where $\dot{M}$ is the mass flow rate, η the fluid viscosity, d the tube diameter, Γ the mass flow rate per unit of circumference, ρ is the fluid density and σ the surface tension.

The entrance region can be under stable accelerating or unstable retarding conditions depending on the film thickness, δ being under or over the distributor entrance setting. The entrance region persists until the growing underlying boundary layer reaches the free surface so frictional and gravitational forces are in equilibrium. Predictions for the entrance region using boundary layer equations show reasonable agreement with data.[5–6]

Flow regimes in the main film have been detailed.[4,7,8] c.f. **Table 1**.

A falling film consists of a base film with waves on the top of it. The wave structure is complex.[10] Small amplitude waves can be overtaken and overridden by larger waves, but the sublayer base film virtually becomes constant when a > 1600. The hydrodynamics are dissimilar for the various flow regimes giving some basis for the wide variation in reported data. *Film thickness* measurements have been reviewed.[4] Prediction models are given in **Table 2**.

Table 1.[7] Film flow regimes

"a" Values Eqn (1)	Flow regime	Sublayer state
0–1.22	smooth stratified	laminar
1.22–2.88	ripple (sine) wave	laminar
2.88–140; Fr → 1.0	roll (gravity) wave	laminar
140–1600; We → 1.0	capillary (surge) wave	laminar
1600–2400	capillary (surge) wave	turbulent
>2400	overriding (ring) swell wave	turbulent

A plot of Nusselt number against film Reynolds number can be used to correlate data. Film and wave data are available for falling film flow in the presence of a gas stream.[2,4,22] Hydrodynamic stability theory[23] and the universal velocity profile[24] for turbulent flow have been used to give reasonable agreement with data.[25] A general model has been suggested[26] which predicted data[27] within ±5%. The effect of gas flow,[13,28] dampening by surfactants[29,30] and wall roughness[7] have all been investigated. The onset of waves is predicted successfully by theory[13] for vertical flow but is not as reliable at other angles.[31] Pressure drops for falling film flow in a vertical tube have been reported for various fluid flow conditions.[2,4,32–34]

References

1. Sack, M. (1967) Falling-film shell-and-tube heat exchangers, *Chem. Eng. Prog.,* 63, 55–61.
2. Feind, K. (1960) Stromungsuntersuchungen bei gegenstrom von rieselfilmen und gas in lotrechten rohren, *VDI-Forschungsh,* 481, 32.
3. Malewski, W. (1968) Wave structure and mass transfer in falling films with ripples, *Chem. Ing. Tech.,* 4, 201–218.
4. Fulford, G. D. (1964) The flow of liquids in thin films. *Advances in Chemical Engineering,* T. B. Drew, J. W. Hopes and J. V. Vermeulen, eds., Vol. 5, 115–237.
5. Lynn, S. (1960) The acceleration of the surface of a falling film, *AIChEJ,* 6, 703–705.
6. Cerro, R. L. and Whitaker, S. (1971) Entrance region flows with a free surface, *Chem. Eng. Sci.,* 26, 785–798.
7. Brauer, H. (1956) Flow and heat transfer at falling liquid films, *VDI Forschungsh,* 457, 40.
8. Uehara, H., Naksoka, T., Egashira, S., Tagucki, Y. (1989) Body forced convection condensation on a vertical smooth surface, *Nippon Kikai Gakkai Ranbunshu B-hen,* 55, 442–9.
9. Yih, S. M. (1986) Modelling heat and mass transport in falling liquid films in *Handbook of Heat and Mass Transfer,* N. P. Cheremisinoff, ed., Vol. 2, C5, 111–210.
10. Chu, K. J. and Dukler, A. E. (1974, 1975) Statistical characteristics of thin, wavy films, *AIChEJ,* 20, 695–706, 21, 583–593.
11. Nusselt, W. (1910) Die oberflachen kondensation das wasserdampfes, *VDI-Zeitschrift,* 54, 1154–1178.
12. Jackson, M. L. (1955) Liquid films in viscous flow. *AIChEJ,* 1, 231–240.

Table 2[9]**.** Average film thickness correlation (no gas flow) in the form of $\delta(\text{meter}) = a(\nu^2/g)^{1/3}(4\Gamma/\eta)^b$

Author	Region	a	b	Basis	Velocity
Nusselt (2, 11)	Laminar	0.91	1/3	Theory	(12)
Kapitza (13; 14)	Wavy laminar	0.8434	1/3	Theory	(13)
Lukach et al. (15)	Wavy laminar	0.805	0.368	Experiment	
Brotz (16)	Turbulent	0.0682	2/3	Experiment	
Brauer (7)	Turbulent	0.2077	8/15	Experiment	
Feind (2)	Turbulent	0.266	1/2	Experiment	
Zhivaikin (17)	Turbulent	0.141	7/12	Experiment	
Ganchev et al. (18)	Turbulent	0.1373	7/12	Theory	
Kosky (19)	Turbulent	0.1364	7/12	Experiment	
Takahama and Kato (20)	Turbulent	0.2281	0.526	Experiment	
Mostofizadeh (21)	Turbulent	0.1721	0.562	Theory	

13. Kapitza, P. L. (1964) Collected papers by P. L. Kapitza, Macmillan, N.Y., 662.
14. Penev, V., Krylov, V. S., Boyadjiev, C., and Vorotilin, V. P. (1972) Wavy flow of thin liquid films. *Int. J. Heat Mass Trans.,* 15, 1395–1406.
15. Lukach, Y. Y., Radchenko, L. B., and Tananayiko, Y. M. (1972) Determination of the average thickness of a film of water during gravitation of flow along the exterior surface of vertical polymeric pipes. *Int. Chem. Eng.,* 12, 517–519.
16. Brotz, W. (1954) Uber die Vorausberedinung der absorptions geschwineig von Gayen instromenden flussig kectsschichten, *Chem. Ing. Tech.,* 26, 470–478.
17. Zhivaikin, L. Y. (1962) Liquid film thickness in film-type units. *Int. Chem. Eng.,* 2, 337–345.
18. Gauchev, B. G., Kozlov, X., and Lozovetskig, V. (1972) *Heat Transfer Soviet Research,* 4, 102.
19. Kosky, P. G. (1971) Thin-liquid films under simultaneous shear and gravity forces. *Int. J. Heat Mass Trans.,* 14, 1220–1224.
20. Takahama, H. and Katos. (1980) Longitudinal flow characteristics of vertically falling liquid films, *Int. J. Multiphase Flow,* 6, 203–211.
21. Mostofizadeh, C. (1980) PhD Thesis, Univ. Stiuttgart.
22. Anshus, B. E. and Goren, S. L. (1966) A method of getting appropriate solutions to the Orr-Sommerfeld equation for flow on a vertical wall, *AIChEJ,* 12, 1004–1008.
23. Roy, R. P. and Jam, S. (1989) A study of thin water film flow down an inclined plate without and with countercurrent air flow. *Exp. Fluids,* 7, 318–328.
24. Dukler, A. E. and Bergelin, O. P. (1952) Characteristics of flow in falling liquid film, *Chem. Eng. Prog.,* 48, 557–563.
25. Portalski, S. (1963) Studies of falling liquid film flow film thickness on a smooth vertical plate, *Chem. Eng. Sci.,* 18, 787–804.
26. Yih, S. M. and Liv, J. L. (1983) Prediction of heat transfer in turbulent falling liquid films with or without interfacial shear, *AIChEJ,* 29, 903–909.
27. Ueda, T. and Tanaka, T. (1974) Studies of liquid film flow in two phase annular and annular-mist flow regions, *Bull. JSME,* 17, 603–613.
28. Hewitt, G. F. (1961) Analysis of annular two-phase flow application of the Dukler analysis to vertical upward flow in a tube, *AERE-R3680.*
29. Emmert, R. E. and Pigford, R. L. (1954) Interfacial resistance-gas absorption in falling liquid film, *Chem. Eng. Prog.,* 50, 87–93.
31. Tailby, S. R. and Portalski, S. (1961) The optimum concentration of surface active agents for the suppression of ripples, *Trans. Inst. Chem. Eng.,* 39, 328–336.
31. Binnie, A. M. (1957) Experiments on the onset of wave formation on a film of water flowing down a vertical plane, *J. Fluid Mech.,* 2, 551–553.
32. Chien, S. F. (1961) An experimental investigation of the liquid structure and pressure drop of vertical downward annular two-phase flow. PhD Thesis, Univ. Minnesota.
33. Hewitt, G. F., King, I., and Lovegrave, P. C. (1961) Holdup and pressure drop measurements in the two-phase annular flow of air-water mixtures, *AERE-R 3764.*
34. Thomas, W. J. and Portalski, S. (1958) Hydrodynamics of counter-current flow in wetted-wall columns, *IEC,* 50, 1081–1088.

Leading to: *Falling film heat transfer; Falling film mass transfer; Falling film, combined heat and mass transfer*

P.L. Spedding

FALLING FILM HEAT TRANSFER

Following from: *Falling film flow; Heat transfer*

The rate of heat transfer in a falling film depends on whether heat goes to or from the wall alone or is accompanied by mass transfer. The transfer rate also is affected by hydrodynamics of the film.

Heat Transfer from the Wall Alone to or from a Falling Film

For fully developed laminar liquid flow

$$U(y)\frac{\partial T}{\partial x} = \frac{\lambda}{\rho c_p}\frac{\partial^2 T}{\partial y^2} \qquad (1)$$

applies. Different boundary conditions and velocity profiles used gave the results in **Table 1**.

Where the following symbols are used

Table 1. Models for laminar falling film with heat to or from the wall alone

Contact time	Method	Resulting relations
short	Leveque velocity profile. ΔT between T_W and T_{AV}	$Nu_x = \frac{\alpha_x \delta}{\lambda} = 0.678\ \bar{x}^{-1/3}$ $\bar{x} = (x\kappa)/(\delta^2 U_{max})$
short	entrance effect included[1]	$Nu_x = 0.0942\ Re\ Pr(\delta/L) + 1.88$
long $X > 0.5$ m	fully developed flow[2,3]	$Nu_\infty = 1.88$ $\bar{\alpha}_\infty = \frac{\alpha_\infty}{\lambda}\left(\frac{\nu^2}{g}\right)^{1/3} = 2.066\ Re^{-1/3}$
long	fully developed flow constant wall flux[2,3]	$Nu_\infty = 2.059$ $\bar{\alpha}_\infty = 2.262\ Re^{-1/3}$

Table 2. Models for turbulent falling film with heating from the wall with interfacial shear

Model	Relevant equations	Comparison with data[3,8]
Van Driest[7]	$\epsilon_m/\nu = -0.5 + 0.5\ (1 + 0.64(y^+)^2(T/T_w)\ [1 - \exp[-(y^+)\ (T/T_w)^{1/2}/25.1]]^2 f^2)^{1/2}$ $y/\delta \le 0.6$. ϵ_m/ν = constant; $0.6 \le y/\delta \le 1$	δ^+ $\overline{vs}$ Re follows data Nu_x $\overline{vs}$ Re above data
Van Driest[3]	same as above ϵ_m/ν = constant; $0.6 \le y/\delta \le y_c/\delta$	δ^+ $\overline{vs}$ Re follows data Nu_x $\overline{vs}$ Re above data
Van Driest[6]	same as above, T/T_W modified to include interfacial shear	δ^+ $\overline{vs}$ Re $\pm$ 5% data Nu_x $\overline{vs}$ Re, 0 to +40% data

Table 3.

Γ/η	a	b	c	d
$<615\ Pr^{-0.646}$	1.88	0	0	0
$615\ Pr^{-0.646}$ to 400	0.0614	8/15	0.344	0
400 to 800	0.00112	6/5	0.344	0
>800	0.0066	14/15	0.344	0
<800	0.00105	6/5	0.344	1 cooling
>800	0.00621	14/15	0.344	1 cooling

Nomenclature

- c_p heat capacity J/kg °K
- g gravitational acceleration m/s^2
- α heat transfer coefficient W/m^{2}K
- λ thermal conductivity W/mK
- Nu Nusselt Number $\alpha\delta/K$
- Nu_x Nusselt number at distance x
- Nu_∞ asymptomatic fully developed Nu
- Pr Prandtl Number ϵ_m/ϵ_H
- Re Reynolds Number $U\delta/\nu$
- T temperature °K
- U velocity m/s
- x axial distance m
- y radial distance m
- κ thermal diffusivity m^2/s
- Γ mass flow rate per unit periphery kg/ms
- δ film thickness m
- $\epsilon_{m,H}$ eddy diffusivity, momentum, heat m^2/s
- η viscosity kg/ms
- ν kinematic viscosity m^2/s
- ρ density kg/m^3

Subscripts

- w at wall
- AV average

The local fully developed heat transfer rates were correlated by Wilkie[8] for heating and cooling.

$$Nu_\infty = a(\Gamma/\eta)^b Pr^c (\eta/\eta_w)^d \quad (4)$$

where the values of the constants are given in **Table 3**.

The asymptomatic heat transfer coefficient was predicted[9] thus

$$h^*_\infty = (0.165\ Re^{0.16} - 0.4)Pr^{0.34}(Pr/Pr_w)^{0.25} \quad (5)$$

An analysis of film heating led to the correlation[10] which followed data within ± 15% for all but the fully developed turbulent film where a spread of +25% was obtained

$$h^*_\infty/(\eta/\eta_w)^{0.25} = i[(\Gamma/\eta)^j Pr^k[(\nu^2/g)^{1/3}/L]^l]^m \quad (6)$$

where the values of the constants are given in **Table 4**.

References

1. Nusselt, W. (1910) *VDI, Zeitschrift*, 54, 1154.
2. Yih, S. M. and Huang, P. G. (1980) *J. Chin. Inst. Chem. Eng.*, 11, 71.
3. Faghri, A. (1976) PhD Thesis, Univ. Calif. Berkley.
4. Kapitza, P. L. (1964) *Collected Papers of P. L. Kapitza*, Macmillan, N.Y., 662.
5. Bays, G. S. and McAdams, W. H. (1937) *IEC*, 29, 1240.
6. Yih, S. M. and Liu, J. L. (1983) *AIChEJ*, 29, 903.
7. Limberg, H. (1973) *Int. J. Heat Mass Trans.*, 16, 1691.
8. Wilkie, W. (1962) *VDI, Forschungsh*, 490.
9. Gimbutis, G. (1974) *Proc. Int. Heat Trans. Conf.*, 5, 2, 85.
10. Schnabel, G. and Schlunder, E. U. (1980) *Verfahrenstechnik*, 14, 79.

***Leading to:** Falling film, combined heat and mass transfer*

P.L. Spedding

FALLING FILM MASS TRANSFER

***Following from:** Falling film flow; Mass transfer*

Mass transfer in a falling film is complicated by changing characteristics of the film, end effects and the need for an exact heat balance. Spedding and Jones[1] outlined how experimental technique can nullify the effect of these parameters. Mass transfer can be formulated in terms of overall, liquid-side or gas-side coefficients with or without chemical reaction. Generally soluble solutes are gas-side controlled and sparingly soluble solutes are liquid-side controlled.

Liquid-Side Coefficients

For laminar falling film flow, the Higbie penetration theory[2] has been applied to short contact times and film lengths (used to suppress rippling):

$$\beta_L = 0.73((D_L/L)^{1/2}(\nu g)^{1/6}Re^{1/3} \quad (1)$$

Table 4.

Γ/η	Condition	Constant wall Flux i	Constant wall Temp i	j	k	l	m
$<615\ Pr^{-0.646}$	laminar-entrance	1.1	0.912	1/3	1	1	1/3
$<615\ Pr^{-0.646}$	laminar-film	1.43	1.3	−1/3	0	0	0
$615\ Pr^{-0.646}$–300	transition-film	0.0425	0.0425	0.2	0.344	0	0
>300	turbulent-film	0.0136	0.0136	0.4	0.344	0	0

$$N_A = N_A^1/2\pi(R + \delta)L = 2(C_e - C_o)(3D_L\Gamma/2\pi\delta L)^{1/2} \quad (2)$$

For long falling film contact times and lengths, several solutions have been obtained[3] the simplest being:

$$\beta_L = (U_{avL}\delta/x)\ln[(C_{eb} - C_1)/(C_e - C_0)] \quad (3)$$

where $(C_e - C_1)/(C_e - C_0) = 0.7857 \exp(-5.1213*x) + 0.1001 \exp(-39.318*x) + 0.03599 \exp(-105.64*x) + 0.01811 \exp(-204.75*x) + \cdots$ (4)

Brauer[4] correlated an equation:

$$\beta_L\delta/D_L = 3.4145 + 0.267(1.5*x)^{-1/2}/[1 + 0.2(1.5*x)^{-1/2}] \quad (5)$$

These equations represent appropriate data nicely provided ripples, entrance acceleration and end effects are eliminated. The use of surface active agents to suppress rippling should be discouraged, since they add to the mass transfer resistance.[5] Nevertheless, short falling film and wetted spheres have been used to measure diffusion coefficients of a number of gases.

The onset of laminar wavy flow results in a seven-fold increase in mass transfer, but only a 20–30% increase in heat transfer.[5] Various explanations, such as interwave areas of reverse flow and convective mass flux normal to the interface due to wave action, have been advanced to explain these phenomena. However, there is substantial disagreement on the subject. A number of modifications to the surface renewal theory have been suggested for wavy laminar flow with limited success,[6] hence empirical relations have been used. The extensive relations of Carrubba:[7]

$$\frac{\epsilon_m}{D_L} = G_1 - (G_1^{10} + G_2^{10})^{0.1} + G_2 - (G_2^{10} + G_5^{10})^{0.1} + G_5 - (G_5^{10} - G_6^{10})^{0.1} + G_6 \text{ for } 12 \le \Gamma/\eta \le 70 \quad (6)$$

$$\frac{\epsilon_m}{D_L} = G_3 - (G_3^{10} + G_4^{10})^{0.1} + G_4 - (G_4^{10} - G_5^{10})^{0.1} + G_5 - (G_5^{10} + G_6^{10})^{0.1} + G_6 \text{ for } 70 \le \Gamma/\eta \le 400 \quad (7)$$

$$G = \left[\left[\left[a\left(\frac{\Gamma}{\eta}\right)^b Sc\right]^c + p^f\right]^i - j\right](y/\delta)^o(1 - y/\delta)^q \quad (8)$$

	a	b	c	p	f	i	j	o	q
G_1	1.24×10^{-3}	1.6	1	0	0	1	0	2	0
G_2	2.13×10^{-4}	1.6	4	4.94	4	1/4	4.94	1.5	0
G_3	1.58×10^{-2}	1	1	0	0	1	0	2	0
G_4	2.72×10^{-3}	1	4	4.94	4	1/4	4.94	1.5	0
G_5	2.5×10^{-4}	1.5	4	3.94	4	1/4	3.94	0	2.5
G_6	7.52×10^{-4}	1.5	1	0	0	1	0	0	3

or those of Brauer:[8]

$$\beta_L = 2.02(D_L/0.066)(g^2/\nu)^{1/3}(\Gamma/\eta)^{1/3} \qquad \Gamma/\eta < 100 \quad (9)$$

$$\beta_L = 3.605(D_L/0.066)(g^2/\nu)^{1/3}(\Gamma/\eta)^{1/3} \qquad \Gamma/\eta > 100 \quad (10)$$

These data plus that of Hikita et al.[9] fit these relations satisfactorily.

A large number of models have been suggested for turbulent flow (Re > 1600) many of which also have been used for gas-side coefficients as well.

In general **Turbulence** is a complex and, as yet, not fully understood process. Therefore, prediction of turbulent mass transport has proved to be difficult. Models based on film or **Boundary Layer** concepts present a gross oversimplification of the mass transfer. Eddy diffusion or mixing length models require some empiricism for solutions to be obtained to the time-averaged Navier-Stokes equations. There has been a wide variation between the theories[10] and data.[11] Nevertheless, some models have proved useful under limited conditions. The analogy between heat and mass transfer of Friend and Metzner:[12]

$$Sh = \frac{\sqrt{f/2}\,Re\,Sc}{1 + 11.8\sqrt{f/2}(Sc - 1)/Sc^{1/3}} \quad (11)$$

gave a result within a standard deviation of 9.4% of data. Also, the general solution[13] of the Reynolds equations gave reasonable (±10%) agreement with data. Models based on the surface renewal concept[14] have proved to be more useful, since they are based on the stochastic replacement of fluid elements at the interface due to turbulent eddies and gave a more fundamental insight into the mechanisms involved. Some practical results of the concept give a form similar to the transport analogies.

$$Sh = S\,Re^w\,Sc^z \quad (12)$$

Model[14]	S	W	Z	Data
Ruckenstein	0.0096	0.9	0.33	good agreement
Thomas-Fan	0.0292	0.8	0.33	−33% of data
Thomas-Fan	0.0292	0.8	0.5	+50% of data
Pinczewski Sideman	0.0102	0.9	0.3	good agreement
Harriott Hamilton	0.0096	0.913	0.346	good agreement

Herron[15] checked the Pinczewski-Sideman model against data for falling film and other geometries and reported good agreement ±10%. Such models are useful regardless of the form of the mass transfer relation. More specifically, data in the form:

$$\beta_L = 1\,Re^m\,Sc^n\,Ga^r \quad (13)$$

have been suggested (**Table 1**) of which those of Carrubba[7] seems to give the most acceptable result.

Gas-Side Coefficients

Hikita and Ishimi[16] used the transport equations

$$U_G\frac{dC_G}{dx} = \frac{1}{r}\frac{d}{dr}\left(rD_G\frac{dC_G}{dr}\right) \quad (14)$$

for cocurrent flow and the velocity profile,

$$U_G = U_m\left(2 - \frac{U_i}{U_m}\right) - 2U_m\left(1 - \frac{U_i}{U_m}\right)\left(\frac{r}{r_i}\right)^2 \quad (15)$$

used with appropriate boundary conditions, to solve the laminar gas liquid film condition. This is an eddy diffusivity[17] approach for turbulent conditions and simultaneous heat and mass transfer.[18] The theory showed good agreement with data for all three cases. The turbulent condition produced a Gilliland-Sherwood type[19] equation (12). The data of Kast[20] and Braun and Hiby[21] have been used to develop relations for countercurrent and cocurrent flow

$$Sh_{Gm} = 0.015\ Re_G^{0.75}(\Gamma/\eta)^{0.16}Sc^{0.44}[1 + 5.2(L/d)^{-0.75}] \quad \textbf{(16)}$$

$$Sh_{Gm} = 0.18\ Re_G^{0.4}(\Gamma/\eta)^{0.16}Sc^{0.44}[1 + 0.4(L/d)^{-0.75}] \quad \textbf{(17)}$$

which parallel independent data within ±10%.

Gas Absorption with Chemical Reaction

A detailed treatment of the subject is given elsewhere.[22,23] Most of the work has been concentrated on processes of industrial interest using short falling film columns.

Nomenclature

a,b,c,f,i,j,l,m,n,o,q,r,s,w,z	constants
C_o, C_i	average inlet, outlet gas concentration mol/m^3
C_e	solubility of gas mole/m^3
$D_{L,G}$	molecular diffusivity liquid, gas m^2/s
d	diameter m
f	friction factor
G	constant
G_n	Galileo number gL^3/ν^2
g	gravitational constant m/s^2
β_L	average liquid side mass transfer coefficient m/s
$\overline{\beta}_L$	dimensionless liquid side mass transfer coefficient $(\beta_L/D)(v^2/g)^{1/3}$
L	absorption length m
N_A	average absorption rate per unit area mol/m^{2}s
N_A^1	average absorption rate mol/s
Re	Reynolds number $(4\Gamma/\eta)$
r	tube inside radius m
r_i	tube radius to interface m
Sh	Sherwood number $\lambda\delta/D$
Sc	Schmidt number ν/D
U_{av}	average phase velocity m/s
U_m	maximum phase velocity m/s
U_i	interfacial velocity m/s
$U_{G,L}$	gas, liquid velocity m/s
x	axial length m
*x	axial film distance xD/δ^2V_{max}
y	radial length m
ϵ_m	eddy diffusivity momentum m^2/s
δ	film thickness

Table 1. Constants and exponents of the equation $\beta_L = 1\ Re^m\ Sc^n\ Ga^r$

Author[14]	Range of Re	1	m	n	r	Other variables range
Brotz	Re < 1200	0.725	1/3	1/2	1/2	
	1200 < Re < 2360	0.0209	5/6	1/2	1/2	
	2360 < Re	0.076	2/3	1/2	1/2	
Kamei and Oishi	2000 < Re	0.018	0.7	0.444	0.083	282 < T < 232
Brauer	$5.4K_F^{1/10} \leq Re \leq 0.072K_F^{1/3}$	0.0346	1/3	0	0	
	$0.072K_F^{1/3} \leq Re \leq 1600$	0.1293	1/5	0	0	
	1600 < Re	0.00235	2/3	0	0	
Hikita et al.	30 ≤ Re ≤ 150	0.011	0.5	0.62	0.04	283 < T < 318
	150 < Re ≤ 2000	0.106	0	1/2	0	
Banerjee et al.	340 < Re < 10,900	0.0001344	0.933	1/2	0	
Johannisbauer	60 < Re < 300	0.408	0.4	0.5	−0.05	60 < Sc < 1100
	300 < Re < 1600	1.28	0.2	0.5	−0.05	$10^8 < Ga < 10^{15}$
	1600 < Re < 10,800	0.069	0.6	0.5	−0.05	
Lamourelle and Sandall	1300 < Re < 8300	0.0002593	0.839	1/2	0	
Carrubba	48 ≤ Re ≤ 280	0.00812	0.4667	0.5	0	for $Sc \geq 2.13 \times 10^5/Re^{1.6}$
	48 ≤ Re ≤ 280	3.7522	−1/3	0	0	for $Sc \leq 2.13 \times 10^5/Re^{1.6}$
	280 ≤ Re ≤ 1600	0.04396	0.1667	0.5	0	for $Sc \geq 7.28 \times 10^3/Re$
	280 ≤ Re ≤ 1600	3.7522	−1/3	0	0	for $Sc \leq 7.28 \times 10^3/Re$
	1600 ≤ Re	0.0007575	0.7167	0.5	0	for $Sc \geq 4.59 \times 10^5/Re^{2.5}$
	1600 ≤ Re	3.7522	−1/3	0	0	for $Sc \leq 4.59 \times 10^5/Re^{2.5}$
Bakopoulos	Same as Carrubba except that at 1600 ≤ Re and for $Sc \geq 4.59 \times 10^5/Re^{2.5}$ included a term $\{1+[11.3(L/d)^{-0.5}]\}^{0.5}$					287 < T < 313
Koziol et al.	170 < Re < 335	1.668	0.39	0.5	−0.1667	844 ≤ Sc ≤ 1085
	335 < Re < 1080	3.882	0.24	0.5	−0.1667	$1.7 \times 10^{13} \leq Ga \leq 2.7 \times 10^{13}$
	1080 < Re < 2513	0.0008923	0.71	0.5	0	
Yih and Chen	49 ≤ Re ≤ 300	0.01099	0.3955	1/2	0	282 ≤ T ≤ 232
	300 < Re ≤ 1600	0.02995	0.2134	1/2	0	148 ≤ Sc ≤ 981
	1600 < Re ≤ 10,500	0.0009777	0.6804	1/2	0	

Γ mass flow rate per unit periphery kg/ms
η viscosity kg/ms
ω kinematic viscosity m^2/s

References

1. Spedding, P. L. and Jones, M. T. (1988) *Chem. Eng. J.,* 37, 165–176.
2. Higbie, R. (1935) *Trans. Inst. Chem. Eng.,* 31, 365.
3. Olbrich, W. E. and Wild, J. D. (1969) *Chem. Eng. Sci.,* 24, 25.
4. Brauer, H. (1981) Fortschritte der Verfahrenstechnik, *VDI-Verlag,* 19, 81.
5. Emmert, R. E. and Pigford, R. L. (1954) *CEP,* 50, 87.
6. Banerjee, S., Rhodes, E. and Scott, D. S. (1967) *Chem. Eng. Sci.,* 22, 43.
7. Carrubba, G. (1976) PhD Thesis, Tech. Univ. Berlin.
8. Brauer, H. (1958) *Chem. Ing. Tech.,* 30, 75.
9. Hikita, H., Nakanishi K. and Kataoka, J. (1959) *Kagaku Kogaku,* 23, 459.
10. Launder, B. E., and Spalding, D. B. (1972) *Lectures in Mathematical Models of Turbulence,* AP, London.
11. Blom, J. (1970) *Int. Heat Trans. Conf.,* 4.
12. Friend, W. L. and Metzner, A. B. (1958) *AIChEJ,* 4, 393.
13. Wassel, A. J. and Cotton, I. (1973) *Int. J. Heat Mass Trans.,* 16, 1547.
14. Sideman, S. and Pinczewski, W. (1975) *Turbulent Heat and Mass Transfer at Interfaces in Topics in Transport Phenomena,* C. Guffinger, ed., Halstead, 74–207.
15. Herron, W. (1993) *Mass Transfer From Surfaces of Various Geometries,* PhD Thesis, QUB.
16. Hikita, H. and Ishimi, K. (1976) *J. Chem. Eng. Japan,* 9, 362.
17. Hikita, H. et al. (1978) *J. Chem. Eng. Japan,* 11, 96.
18. Hikita, H. et al. (1979) *Can. J. Chem. Eng.,* 57, 578.
19. Gilliand, E. R. and Sherwood, T. K. (1934) *IEC,* 26, 516.
20. Kast, W. (1966) *Chem. Tech.,* 18, 152.
21. Braun, D. and Hiby, J. W. (1970) *Chem. Ing. Tech.,* 42, 345.
22. Danckwerts, P. V. (1970) *Gas-liquid Reactions,* McGraw Hill, N.Y.
23. Astarita, G. (1967) *Mass Transfer with Chemical Reaction,* Elsevier, Amsterdam.

Leading to: *Falling film, combined heat and mass transfer*

P.L. Spedding

FALLING RATE PERIOD, DRYING CURVE (see Drying)

FANNING EQUATION (see Friction factors for single phase flow)

FANNING FRICTION FACTOR (see Friction factors for single phase flow; Heat pipes; Tubes, single phase flow in)

FANNO FLOW (see Compressible flow)

FARADAY MHD GENERATOR (see Magnetohydrodynamic, MHD, power generators)

FARADAY'S LAWS

Following from: *Electrolysis*

Faraday's Laws of electrolysis played a great part in putting *electrochemistry* on a sound quantitative scientific base. Two experimental discoveries embody these laws i) the amount of chemical change during an electrochemical reaction is proportional to the amount of electricity passed, ii) for the same amount of electricity passed, the amount of chemical change is proportional to the molecular weight of the substance divided by the number of electrons, n, required to change one molecule. This number of electrons is usually 1 or 2. The amount of electricity passed is the electrical charge, Q, defined as the integral of current passed with respect to time, t, i.e.

$$\text{Charge } Q = \int I dt$$

Faraday's Laws can be expressed in the equation

$$\text{Amount (mol)} = m = \frac{Q}{nF}$$

where F is Faraday's constant of proportionality. If the current in a given amount of time, t, is constant, then the amount of chemical change is

$$m = \frac{It}{nF}$$

Faraday's Laws apply separately to the reactions occurring at both electrodes in an electrochemical cell, i.e. to the formation of both oxidation and reduction products, and they apply equally well to *galvanic (spontaneous) reactions* and to *electrolytic (non-spontaneous or driven) reactions.* Furthermore Faraday's Laws apply when more than one reaction takes place at an electrode. For example, in the *electroplating* of a metal, M, at a cathode a secondary (side) reaction can be the formation of hydrogen gas

$$2H^+ + 2e^- \rightarrow H_2$$

The total amount of material produced (in mols) by a charge Q is

$$m_M + m_{H_2} = \frac{Q_M}{n_M F} + \frac{Q_{H_2}}{n_{H_2} F}$$

where

$$Q = Q_M + Q_{H_2}$$

The Faraday's constant is equivalent to the charge associated with a unit amount of electrons and is equal to the product of the fundamental charge on a single electron, Q_e, and the Avogadro constant, N_A, i.e.

$$F = N_A \cdot Q_e$$

$$= 6.023 \cdot 10^{23}\,(\text{mol}^{-1}) \times 1.602 \times 10^{-19}\,(\text{C})$$

$$= 96485\ \text{C mol}^{-1}$$

K. Scott

FAST NEUTRONS (see Liquid metal cooled fast reactor)

FAST REACTORS (see Liquid metal cooled fast reactor; Nuclear reactors)

FAULT TREES (see Risk analysis techniques)

FAXEN FORCES (see Particle transport in turbulent fluids)

FEEDBACK CONTROL (see Process control)

FEEDBACK, DYNAMIC (see Nyquist stability criterion and Nyquist diagrams)

FEEDFORWARD CONTROL (see Process control)

FEEDWATER HEATERS

Following from: Boilers

Introduction

A feedwater heater is used in a conventional power plant to preheat boiler feed water. The source of heat is steam bled from the turbines, and the objective is to improve the thermodynamic efficiency of the cycle. The most common configuration of feedwater heater is a shell and tube heat exchanger with the feedwater flowing inside the tubes and steam condensing outside. (See **Boilers** and **Shell and Tube Heat Exchangers**.)

Temperature Profiles

Figure 1 depicts the temperature profiles for a high-pressure feedwater heater which receives superheated steam extracted from a high-pressure turbine.

If sufficient superheat is available, it is possible to make use of the large temperature difference by specifying a separate section within the heater in which desuperheating occurs with a dry wall. This gives a higher heat flux than if condensation occurs, and also allows the possibility of raising the feedwater outlet temperature above that of the steam saturation temperature. The steam condenses almost isothermally, and the condensate is subcooled below the saturation temperature.

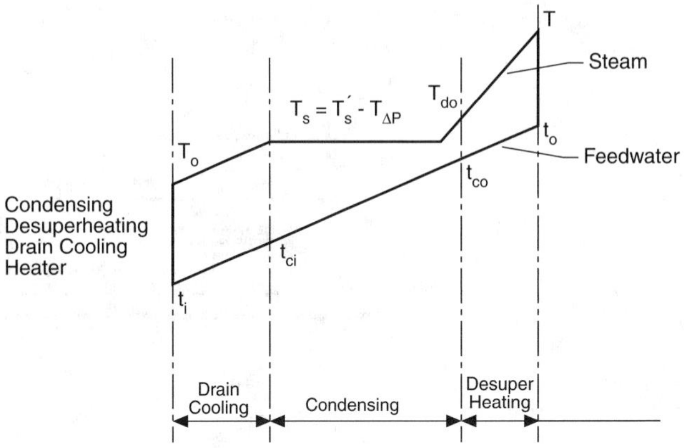

Figure 1. Temperature profiles for a high pressure feedwater heater.

In the subcooling zone heater surface is assigned to extract heat from the condensate (drains) from the condensing zone.

A heater may have neither a desuperheating zone nor a drain cooling zone.

Feedwater Heater Geometries

Figure 2 shows, in schematic form, the general arrangement of a three-zone heater. The shell contains a bundle of tubes (normally U-tubes). Two tube passes are almost always used. The feedwater inlet and outlet nozzles are connected to a channel on one side of the tube plate.

In the condensing zone, the tubes are supported by plates or grids of rods. The desuperheating and drain-cooling zones are contained within the shell by a shroud or wrapper, and are usually well baffled to both support the tubes and promote a satisfactorily high shellside heat transfer coefficient. Sometimes other types of a baffle support, based on some form of grid or array of rods, are used to minimise the risk of tube vibration.

High pressure units are sometimes of the "header-type" construction. This is a specialised design in which the feedwater inlet and outlet headers take the form of separate cylindrical vessels which penetrate into the heater shell. Each tube is individually welded onto the headers, and the headers are welded to the shell. There are usually four tube passes.

Feedwater heaters can be located either horizontally or vertically. The horizontal orientation is more common, but vertical heaters are sometimes preferred.

A feedwater heater must be equipped with a vent to allow removal of non-condensing gases.

Thermal Design Considerations

Thermal design of a feedwater heater requires an economic optimisation of many factors, including material and operating costs.

Two publications which describe feedwater heaters, and their design, in some detail are those of BEAMA (1968) and HEI (1984). These documents provide performance charts which can be used to estimate the surface area requirement. However, a computer program is required to achieve an optimised design. The paper by Clemmer and Lemezis (1965) presents a design logic which is

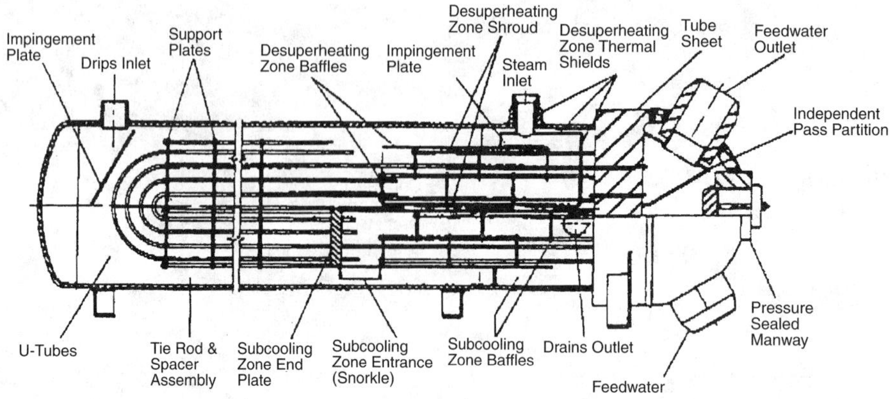

Figure 2. Typical arrangement of a three zone feedwater heater, (From *Process Heat Transfer*, 1994, CRC Press.)

suitable for implementation in a computer program. Further background information can be found in the publication by EPRI (1984).

Special attention must be paid to avoidance of (a) wet-wall conditions in the desuperheating section, in order to avoid erosion/corrosion problems and (b) excessive pressure drop in the drain cooler, which could cause flashing, and consequent tube damage.

Pressure loss in the desuperheating zone causes a reduction in the saturation temperature of the steam condensing zone. This in turn causes a reduction in the temperature difference in the condensing zone. Design of the two zones is therefore a compromise between the need to maintain a high heat transfer coefficient in the desuperheating zone, while avoiding an excessive reduction in the overall mean temperature difference.

References

BEAMA (1968) *Guide to Design of Feedwater Heating Plant,* The British Electrical and Allied Manufacturers' Association Ltd., London.

EPRI (1984) *Symposium on State-of-the-art Feedwater Heater Technology,* Report No. CS/NP-3743, EPRI, Palo Alto, California.

Clemmer, A. B. and Lemezis, S. (1965) Selection and Design of Closed Feedwater Heaters, *ASME Paper 65-WA/PTC-5,* ASME. Winter Annual Meeting, Chicago, November 7–11, (1965) ASME, Vol. 79, No. 7, 1494–1500.

HEI (1984) *Standards for Closed Feedwater Heaters, 4th edition,* Heat Exchange Institute, Cleveland, Ohio.

Hewitt, G. F., Shires, G. L., and Bott, T. R. (1994) *Process Heat Transfer,* CRC Press.

J.M. McNaught

FERMI-DIRAC DISTRIBUTION (see Boltzmann distribution)

FERROELECTRICITY (see Piezoelectricity)

FERTILE ISOTOPES (see Nuclear reactors)

FERTILE MATERIAL (see Liquid metal cooled fast reactor)

F-FACTOR METHOD (see Mean temperature difference)

FIBRE FILTERS (see Gas-solids separation, overview)

FIBRE OPTICS

Following from: Optics

The technology of fibre optics refers to the transmission of information, encoded within a light signal, along a suitable waveguide. The transmission medium is normally an optical fibre fabricated from silica. Transmission along a fibre relies upon a difference in **Refractive Index** between the core and cladding. Provided the refractive index of the core is higher than the that of the cladding total internal reflection can occur. (See also **Reflectivity**.)

When the incidence angle (θ_0) of the incoming light is below a limiting value known as the maximum acceptance angle, light follows the path shown by θ_1, θ_2 in **Figure 1**. The refractive indices of the fibre core, cladding and surrounding air are n_1, n_2, and n_0 respectively. At larger incident angles some light escapes (θ_3). A more detailed description of this topic is given by Dunlop and Smith (1989). For singlemode fibre the core diameter is much smaller than the cladding diameter (9μm cp 125 μm). This allows only the fundamental (LP_{01}) mode to propagate at normal operating wavelengths. *Multimode fibre* contains a larger core diameter (for example 50 μm) and allows higher order modes (LP_{11}. . . .) to propagate.

The transmission capacity of a an optical fibre is bounded by *modal bandwidth, chromatic dispersion* and *attenuation.* In a multimode fibre, the time taken for the fundamental and higher order modes to travel along the fibre differs as the path lengths vary. This gives rise to modal dispersion and so limits bandwith to the order of 1000 MHz.km. This phenomenon is absent in singlemode fibres which are partially limited by chromatic dispersion. Chromatic dispersion is wavelength dependent and is given by the sum of the waveguide and material dispersions for the fibre. Because these vary with wavelength in an opposite sense a zero

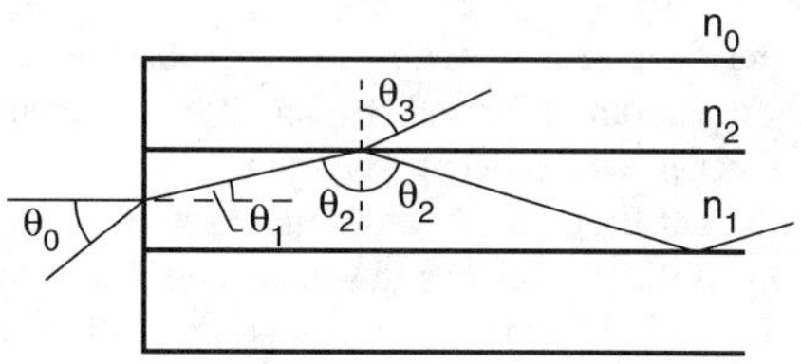

Figure 1. Total internal reflection in clad optical fibre.

dispersion point occurs close to 1300 nm for conventional step index fibre.

Fibre attenuation has fallen from 20 dB/km (1970) to 0.2 dB/km (1985). The loss mechanisms arise from three absorption characteristics. Firstly, ultra violet resonances increase significantly below 1000 nm. This is associated with electronic structure of the silica lattice atoms. Secondly, an infra-red absorption increases beyond 1500 nm, caused by oxide bond vibration. This leaves the basic loss mechanism known as *Rayleigh scattering*. For high silica based fibres, this is the dominant loss mechanism in the 1300 to 1600 nm region. The Rayleigh scattering loss, L_R (dB/km) is expressed as:

$$L_R = A\lambda^{-4}$$

where A is the scattering coefficient with values in the range 1–1.6 dB/km-μm^{-4}, λ is wavelength.

The most significant extrinsic loss is that due to OH. The overtone absorption bands occurring due to water in silica have combination bands at 1.24μm. When this consideration is combined with the intrinsic loss behaviour the favoured transmission windows become 1260–1360 nm and 1480–1580 nm.

Bulk silica has a tensile strength of approximately 14 GPa. However, the strength of silica based fibres is dominated by extrinsic flaws in the fibre surface. The distribution of flaws on the surface may be quantified and so the probability F of a fibre failing at or below a stress, S_i is given by the *Weibull equation:*

$$F = 1 - \exp\{-L/L_o(S_i/S_o)^m\}$$

where L_o is a standard length, S_o and m are the Weibull shape and scale constants respectively. The depth of the flaws grows under applied strain and in the presence of agents which increase the stress corrosion susceptibility constant (n) of the fibre. Note that as the dimensionless unit n decreases the rate of flaw growth increases. For a fibre with original inert strength S_i, experiencing a constant stress S the time to failure, t_f is given by:

$$t_f = \{S_i^{n-2}B/S^n\}$$

B is a material constant derived during inert strength measurements.

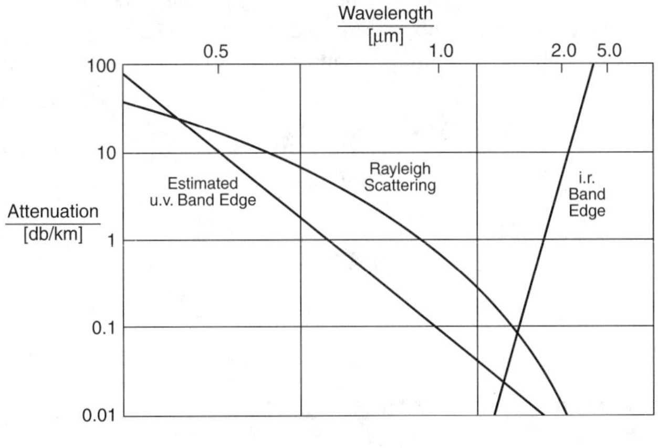

Figure 2. Extrinsic loss behaviour from Gowar, J. (1983), Optical Communication Systems, Prentice Hall, Hemel Hampstead, UK, with permission from the publisher.

Typically S_i,, under inert conditions is 14 GPa, B is 10^{10} Pa^2s and n is 20.

The installation of fibre into communications networks requires fibre to be coated with buffering materials to allow cabling into robust structures which can be pulled into underground ducts or directly buried. Bark and Lawrence (1987) discuss generic cable designs. Alternatively, groups of fibres can be blown into preinstalled tubing (see Cassidy and Hornung, 1984).

References

Dunlop, J. and Smith D. G. (1989) Optical Fibre Communications, Chapter 12, *Telecommunications Engineering,* Van Nostrand Reinhold International.

Gowar, J. (1983) *Optical Communication Systems,* Chapter 3, Prentice-Hall International.

Bark, P. R. and Lawrence, D. O. (1987) Fibre Optic Cables: Design, Performance Characteristics, and Field Experience, Chapter 5, *Optical Fibre Transmission,* E. E. Basch, ed. Howard W Sams & Co.

Cassidy S. A. and Hornung S. (1984) *A Fibre Unit for Blown Fibre Cable IEE Colloquium on Implementation and Reliability of Optical Fibre Cable Link,* London.

D.J. Stockton

FIBRE SATURATION POINT (see Drying)

FICK'S LAW (see Body, human, heat transfer in; Ion exchange)

FICK'S LAW, GENERALISED (see Irreversible thermodynamics)

FICK'S LAW OF DIFFUSION

***Following from:** Diffusion*

In a material composed of two or more chemical species in which there are spatial inhomogeneities of the composition, there is a driving force for inter-diffusion of the various molecular species so as to render the composition of the material uniform. In a mixture of just two species, the diffusive flux of each molecular species is proportional to the gradient of its composition. This proportionality is known as Fick's Law of Diffusion and is, to a limited extent, the mass transfer analogue of *Newton's Law of Viscosity* and **Fourier's Law of Heat Conduction** (Bird et al., 1960).

The mathematical formulation of Fick's Law must be performed with great care, because there are a number of ways of expressing the composition of a material and because it is necessary to prescribe the frame of reference with respect to which the flux of a particular species is measured. For this reason, it is first necessary to identify the molar average velocity in a mixture of n species as

$$\mathbf{v}^* = \sum_{i=1}^{n} c_i\mathbf{v}_i / \sum_{i=1}^{n} c_i = \sum_{i=1}^{n} c_i\mathbf{v}_i/c$$

where $\mathbf{v}_i$ is the mean velocity of molecules of species i with respect to a fixed co-ordinate axis. Here, c_i is the amount-of-substance density of species i and c the total amount-of-substance density for the mixture. This definition means that cv^* is the total amount of substance that flows through unit area perpendicular to $\mathbf{v}^*$ in unit time.

According to Fick's Law for a binary system, the diffusive amount-of-substance flux of species 1 relative to a reference frame moving with the molar average velocity of the two species is given by

$$J_1 = -cD_{12}\nabla x_1$$

where x_1 is the amount-of-substance fraction of species 1 and D_{12} is the *inter-diffusion coefficient* of the two species, sometimes known as the *mutual diffusion coefficient.*

It is rather difficult to arrange in practice to study diffusion relative to the molar average velocity of the two species so it is usually studied relative to a reference frame fixed in the laboratory. As a consequence, the total amount-of-substance flux observed in this reference frame differs from that given by Fick's Law. This can arise whenever there is a net flow of a species from one point to another, for example, when a liquid evaporates into a gas and diffuses through it or in a case of adsorption of a gas on to a solid surface. It may also arise when there is a change of volume of the mixture upon mixing since, then, the molar average velocity is not constant.

In *multicomponent mixtures* the process of diffusion is much more complicated since there can be coupling of the gradients of some species to the fluxes of others (Bird et al., 1960).

References

Bird, R. B., Stewart, W. E., and Lightfoot, E. N. (1960) *Transport Phenomena,* John Wiley, New York.

Tyrrell, H. J. V. and Harris, K. (1984) *Diffusion in Liquids*, Butterworths, London.

Cussler, E. L. (1984) *Diffusion: Mass Transfer in Fluid Systems,* Cambridge University Press, Cambridge.

***Leading to:** Mass transfer*

W.A. Wakeham

FIGURE OF MERIT FOR NUCLEAR REACTOR COOLANTS (see Coolants, reactor)

FILIPPOV EQUATION, FOR THERMAL CONDUCTIVITY OF SOLUTIONS (see Thermal conductivity mechanisms)

FILM BOILING (see Boiling; Post dryout heat transfer)

FILM BOILING COLLAPSE (see Rewetting of hot surfaces)

FILM CONDENSATION (see Condensation, overview)

FILM CONDUCTANCE METHOD, FOR FILM THICKNESS MEASUREMENT (see Film thickness measurement)

FILM COOLING

***Following from:** Heat protection; Transpiration cooling*

Film cooling is used in many applications to reduce convective heat transfer to a surface. Examples are the film cooling of gas turbine combustion chambers, vanes and blades which are subjected to high heat transfer from combustion gases [Metzger et al., (1993)]. Gas which is cooler than the freestream is passed onto the external surface via small slots or rows of holes within the surface. The aim is to introduce the coolant into the boundary layer without significantly increasing turbulence and entraining additional hot freestream gas. There are three temperatures in this problem: the freestream temperature, the coolant temperature and the wall temperature. For incompressible flow with constant fluid properties, the heat transfer rate to the surface, $\dot{q}$, may be expressed in the following form.

$$\dot{q} = h(T_{aw} - T_w) \quad (1)$$

h is a heat transfer coefficient. T_{aw} and T_w are the adiabatic wall temperature, which is now different to the freestream temperature, and wall temperature respectively. Alternatively, the relationship may be given in terms of the known temperature differences as,

$$\dot{q} = \alpha(T_g - T_w) + \beta(T_g - T_c) \quad (2)$$

α and β are solely functions of the flowfield. The film cooling effectiveness, η is thus given as,

$$\eta = \frac{T_g - T_{aw}}{T_g - T_c} = -\frac{\beta}{\alpha} \quad \text{also} \quad h = \alpha \quad (3)$$

(as can be seen if $\dot{q}$ in **Equation 1** is made equal to zero)

Simple modelling of the process assumes that the coolant injection does not disturb the boundary layer and as a result entrainment takes place as for a turbulent boundary layer. The gas temperature within the boundary layer is determined by an enthalpy balance for coolant and entrained mainstream giving the local adiabatic wall temperature. Due account is taken of mass addition and non uniform temperature through the boundary layer to give predictions for film cooling effectiveness at a distance x downstream from a slot of width s of the form below.

$$\eta = \frac{1.9\mathrm{Pr}^{2/3}}{1 + 0.329\frac{c_g}{c_c}\xi^{4/5}E} \quad (4)$$

$$\xi = \frac{\rho_g u_g x}{\rho_c u_c s}\left[\frac{\mu_c}{\mu_g}\mathrm{Re}_s\right]^{-1/4}, \quad \text{where} \quad \mathrm{Re}_s = \frac{\rho_c u_c s}{\mu_c}$$

ρ, c, μ and Pr are density, specific heat at constant pressure, viscosity and **Prandtl Number** respectively. The suffices c and g refer to coolant and freestream. E is a factor which gives an increase in entrainment associated with angled injection. This formula gives satisfactory predictions for slots and low injection rates as shown in **Figure 1**.

Further models which are applicable to higher injection rates employ wall jet entrainment. The coolant from discrete holes often penetrate the boundary layer at high injection rates and may increase the heat transfer above the value in the absence of film cooling. Film cooling holes may be shaped so as to diffuse the coolant as it enters the boundary layer and multiple rows of holes are often employed. Due to the wide range of geometries used in film cooling, a large number of empirical correlations are present in the literature. Coolant exit mass flux, momentum flux, and velocity ratios with respect to the local freestream values are parameters which have been used in correlating experimental results. Freestream pressure gradients, turbulence and wall curvature and roughness are all factors which influence film cooling performance. An article by Harnett (1985) gives a comprehensive review of this work and builds on an earlier article by Goldstein (1971). LeFebvre (1983) deals with combustion chamber film cooling.

Boundary layer numerical computations are also employed to predict the cooling process. For example, Crawford et al. (1980) distribute the coolant locally within the boundary layer according to prescribed rules dependent on the injection rate and angle. The local mixing length is also altered to take into account enhanced turbulence.

In recent years, the entire flowfield from the coolant plenum has been computed using the three dimensional, Navier-Stokes equations [e.g. Fougeres and Heider (1994) and Garg and Gaugler (1994)]. The separation in the coolant holes is simulated as is the complex vortex structures which develop in the coolant jet as it turns on encountering the mainstream. Deficiencies in the turbulence modelling limit these predictions, however, such methods do indicate the way forward in design.

References

Crawford, M. E., Kays, W. M., and Moffat, R. J. (1980) Full Coverage Film Cooling on Flat, *Isothermal Surfaces: A Summary Report on Data and Predictions,* NASA CR-3219.

Fougeres, J. M. and Heider, R. (1994) *Three-Dimensional Navier-Stokes Prediction of Heat Transfer with Film Cooling,* ASME Paper 94-GT-14, The Hague.

Garg, V. K. and Gaugler, R. E. (1994) *Prediction of Film Cooling on Gas Turbine Airfoils,* ASME Paper 94-GT-16, The Hague.

Goldstein, R. J. (1971) Film Cooling, *Advances in Heat Transfer,* T. F. Irvine, Jr. and J. P. Hartnett), eds., Vol. 8, Academic, New York.

Hartnett, J. P. (1985) Mass Transfer Cooling, *Handbook of Heat Transfer Applications,* 2nd Edition, W. M. Rohsenow, J. P. Hartnett and E. N. Ganic, eds., McGraw Hill.

LeFebvre, A. H. (1983) *Gas Turbine Combustion,* Hemisphere, New York.

Metzger, D. E., Kim, Y. W. and Yu, Y. (1993) *Turbine Cooling: An Overview and Some Focus Topics,* Proc 8th Int. Symp. Transport Phenonema in Thermal Engineering, Seoul.

T.V. Jones

FILM FLOW RATE MEASUREMENT

Following from: *Annular flow*

In annular two-phase flow, part of the liquid flows as a liquid film on the channel wall, and part as droplets entrained in the gas core of the flow. The distribution between film and core is an important system variable and affects many other parameters (dryout, heat transfer coefficient, pressure gradient, etc). In vertical flows in round tubes, the film flow rate is constant around the periphery since the flow is axisymetric. For horizontal flows, the film flow rate may vary around the periphery of the tube and, in this case, there is considerable interest in measuring *local* flow rates as a function of circumferential position.

Average film flow rate can be measured by sucking off the liquid film through a porous section of tube as illustrated in **Figure 1**. The entrained droplets have sufficient momentum to carry them beyond the porous wall section; a small amount of the gas phase is collected in between waves on the film, but the liquid collection rate becomes independent of the gas collection rate over a wide range once all of the film has been captured. However, at high liquid flow rates, the results are not so clear cut.

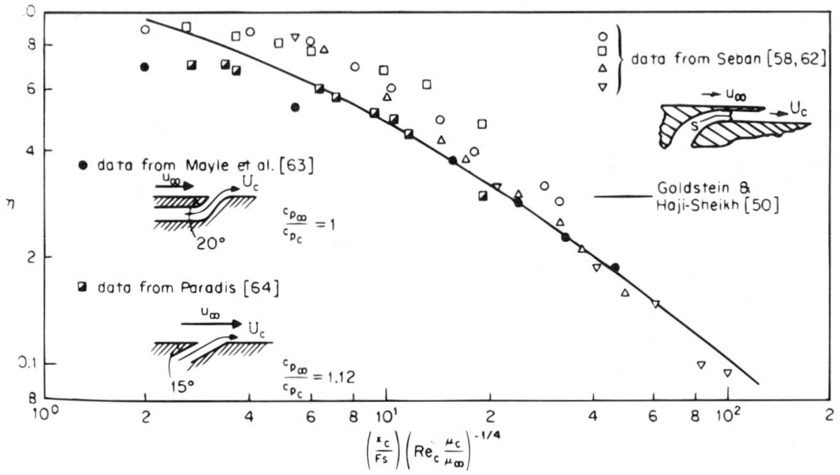

Figure 1. A comparison of the predicted effectiveness with experimental results from Hartnett (1985).

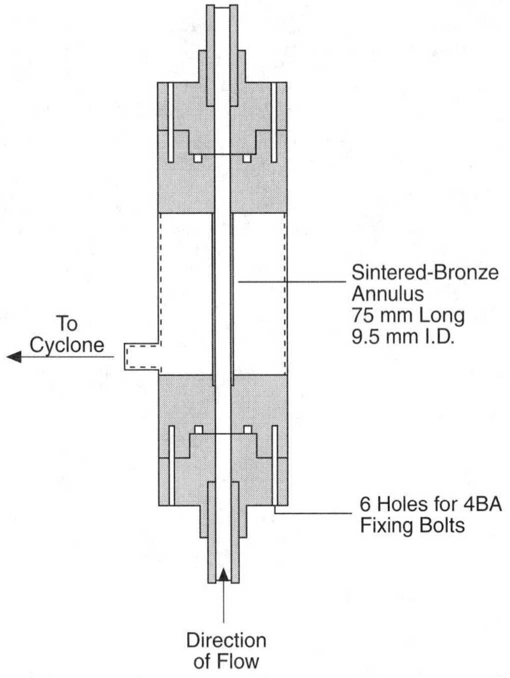

Figure 1. *Film suction method* for determination of film flow rate.

To measure *local film flow rate* a strip of porous wall can be used (bounded by longitudinal fins), and the position of this can be rotated around the circumference to give distribution of film flow rate in, say, horizontal annular flow. An alternative method of measuring local film flow rate was described by Coney and Fisher (1976) and is illustrated in **Figure 2**.

Here, part of the liquid film is separated by guide fins and, in the separated region, potassium chloride solution is pumped in at a constant rate. The salt solution mixes with the film liquid and the concentration of the salt in the mixed film is determined using miniature conductivity probes as shown. The effectiveness of the mixing was checked by having two successive probes.

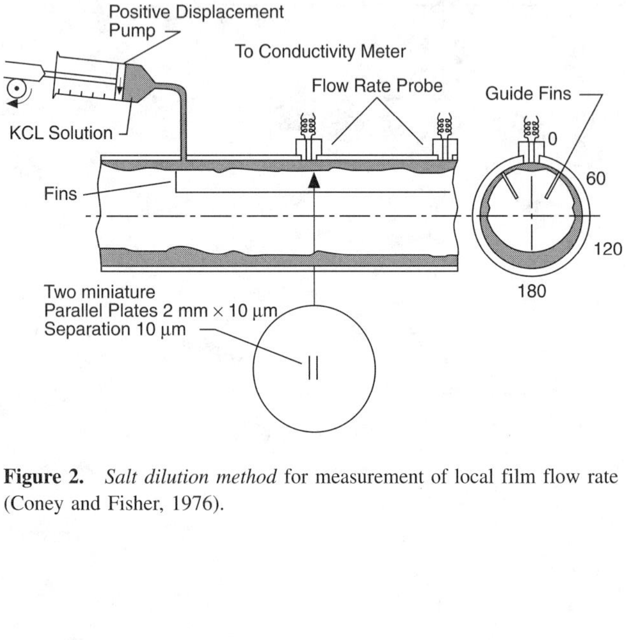

Figure 2. *Salt dilution method* for measurement of local film flow rate (Coney and Fisher, 1976).

Reviews of film flow rate measurement are given by Hewitt (1978)(1982).

References

Coney, M. W. E and Fisher, S. A. (1976) *Instrumentation for Two-phase Flow in Use or Under Development at the Central Electricity Research Laboratories,* paper presented at the meeting of the European Two-Phase Flow Group, Erlangen, Germany.

Hewitt, G. F. (1978) *Measurement of Two-phase Flow Parameters,* Academic Press, London.

Hewitt, G. F. (1982) Measurement techniques, Chapter 10 of *Handbook of Multiphase Systems,* G. Hetsroni ed., McGraw-Hill, New York.

G.F. Hewitt

FILM FLOW REGIMES (see Falling film flow)

FILM INVERSION (see Helical coil boilers)

FILM METHOD (see Condensation of multicomponent vapours)

FILM THICKNESS DETERMINATION BY OPTICAL METHODS

Following from: Optics

A film of gas or liquid has a *refractive index* which differs from that of the bulk fluid and the surface to which it is attached. This property may be used to determine the thickness of the film remotely provided that its refractive index is known. For a gas, *Gladstone and Dale's law* gives $(m - 1) = K\rho$ where m is the refractive index, ρ is the density and K is a constant depending on the constituents. Thus the variation with temperature or pressure is known.

The thickness enters through the *phase* of the propagating wave, given by mkL where L is distance travelled and $k = 2\pi/\lambda$ is the *wave number.* The phase of a wave may be measured using **Interferometry**. If the surface is opaque the interference is observed between waves reflected from the front face of the film and that which is transmitted and reflected from the surface. For normal incidence the difference in phase between the two waves is 2mkt, where t is the thickness. If the film thickness varies a set of interference fringes is observed, the fringe separation being given by the distance over which the phase difference changes by one wavelength of the light. (*Newton's rings* are an example of this kind of interference.) If the surface is transparent the same measurement may be made in transmission, except that in this case a separate reference wave must be employed. Transmission can also be used if the measurement is made along the film. Here the variation in path length, or refractive index, can be obtained as a function of position within the film. This has proved useful, for example, for measuring temperature variation in boundary layers.

Interference fringes are distorted by the presence of other phase objects such as windows or lenses. These effects can be overcome by the use of *holography.* If a hologram is taken of an object and the reconstructed image is superimposed on the original, then any small changes in the object show up as interference fringes. These are independent of the other phase objects which are constant. In this way undistorted interferometry can be carried out in real time. A similar process is *double exposure holography.* Here two holograms of the same object are taken on the same plate. On reconstruction interference fringes are seen indicating the differences that have occurred between the two exposures. (See also **Holograms** and **Holographic Interferometry**.)

If the film is opaque variations in surface height can be measured by *profiling.* In this technique fringes are projected onto the surface and their pattern reveals changes in height. *Moiré fringes* are often employed for this purpose. Provided that the underlying surface is flat the thickness variation is obtained from the height changes.

Further Reading

Sandhu, S. S. and Weinberg, F. J. (1972) *J. Phys. E: J. Sci. Inst.,* 5, 1018–1020.

Williams, D. C. (ed.) (1993) *Optical Methods in Engineering Metrology,* Chapman and Hall, London.

A.R. Jones

FILM THICKNESS (see Falling film flow)

FILM THICKNESS IN ANNULAR FLOW (see Annular flow)

FILM THICKNESS MEASUREMENTS

Following from: *Annular flow*

The methods of film thickness measurement in annular flow can be classified into two groups: methods involving contact with the fluid and methods without contact.

The methods with contact are: quick-closing valve method; methods based on measurement of conductance or conductance and capacitance of the film; electrical contact method.

The methods without contact are: methods based on measurement of the absorption of light, β, or γ beams; optical methods; fluorescence method.

Quick Closing Valve Method

The liquid volume V is measured in a channel with length L by quickly closing valves at the start and end of the length. The mean film thickness $\delta = V/\pi DL$ is determined. The complexity of method consists of the realization of instantaneous and simultaneous closure of the valves. It is usually assumed that the part of the liquid volume associated with liquid drops is negligible.

Methods Based on Measurement of Film Conductance

In the conductance method, the conductance of the liquid film is measured between electrodes flush with the channel wall. An electrolyte is often added to liquid to make it more conductive. The measurement section must be electrically insolated. A wide variety of electrode designs have been employed (see Hewitt, 1982). The closer the electrodes, the more localized the measurement but the more restricted the linear range of the relationship between film thickness and conductance. Less commonly, for non-aqueous systems, the capacitance of the film is measured.

Electrical Contact Method

Here, a needle is moved in a direction normal to the liquid film and the distance from the wall is determined when electrical contact between the needle and the film appears. If there are the waves on the film surface, short-time contacts with the wave crests arise. An uninterrupted contact exists if the needle touches the wave trough. The film can stick to the needle and the needle can influence on the flow. The electric contact method is not usually suitable for small film thickness ($\delta \leq 0.1$ mm).

The Light Absorption Method

In this method, a light beam is passed through the film. The light absorption depends on film thickness. The intensity of transmitted radiation (determined with aid of a photo-diode or photomultiplier) decreases not only due to absorption. If the film surface is wavy, the light can scatter, refract or reflect. These effects must be considered and may decrease the usefulness of the method.

Radiation Absorption Methods

In these methods, β or γ beams are passed through the film. The transmitted intensity is given by:

$$I = I_0 e^{-k\rho y},$$

where I_0 = intensity of incident radiation, ρ = liquid density, k = absorption coefficient, y = layer thickness. The method allows measurement of the mean film thickness. There are obvious problems in many systems of gaining interrupted access of the beam to the film.

Optical Methods (Shadow Methods, Photography)

When investigating the liquid film flow on outer cylinder surface, the thickness can often be measured using shadowgraphy or photography.

Fluorescence Spectroscopy Method

This method is based on using a light beam with a given wave length that passes through the liquid film. A fluorescent substance is added to the circulated liquid and the incident light stimulates a fluorescence with a different wave length. The quantity of emitted fluorescent light increases with increasing of film thickness. This value is measured with aid of a photomultiplier and spectrometer. The method allows highly localized measurements.

Reference

Hewitt, G. F., Hall-Taylor, N. S. (1970) *Annular Two-Phase Flow,* Pergamon Press, Oxford.

Hewitt, G. F. (1978) *Measurement of Two-Phase Flow Parameters*, Academic Press, London.

Hewitt, G. F. (1982) Measurement Techniques. Chapter 10 of *Handbook of Multiphase Systems*, G. Hetsroni, ed. McGraw Hill Book Company, New York.

Yu.A. Kuzma-Kichta

FILM-WISE CONDENSATION (see Surface condensers)

FILTERING CENTRIFUGES (see Centrifuges)

FILTERS

***Following from:** Filtration*

Filters are apparatuses for cleaning liquids and gases from suspended particles by filtration. The main component of a filter is a filter membrane which lets liquids and gases pass and removes suspended particles which settle on the surface or in the pores of the membrane.

The simplest filter (**Figure 1**) is a vessel separated by a filter membrane into two sections, dirty and clean. A pressure difference, giving rise to fluid flow through the filter membrane, is produced between them.

Filter membranes are made of cotton, wool, synthetic, glass, and metal fibers and cloths, of ceramic, cermet, synthetic, and other porous materials shaped as plates, pipes, and elements of other shapes, and of granular bulk materials of the same nature.

Depending on their mode of operation, filters are divided into *batch-type filters* and *continuous filters*. In turn, filters of both groups vary in the method of producing pressure difference (pressure or vacuum), the geometry of filter membranes (plane, tubular, etc.) and the type and the material of the filter membrane.

In batch-type filters, filtration proceeds simultaneously over the entire surface of the filter membrane, covering it with a layer of captured particles and clogging up the pores. Then, on achieving a certain limiting pressure difference the filter is switched off, regenerated, and put into operation once again.

Continuous filters are divided into sections, where the same operations proceed uninterruptedly and independently, but with some phase shift permitting the degeneration of each section one by one without switching off the filter as a whole.

Although the basic principle of operation is the same, filters are commonly classified into liquid filters for suspension separation and gas filters for aerosol separation and cleaning of dustladen gases.

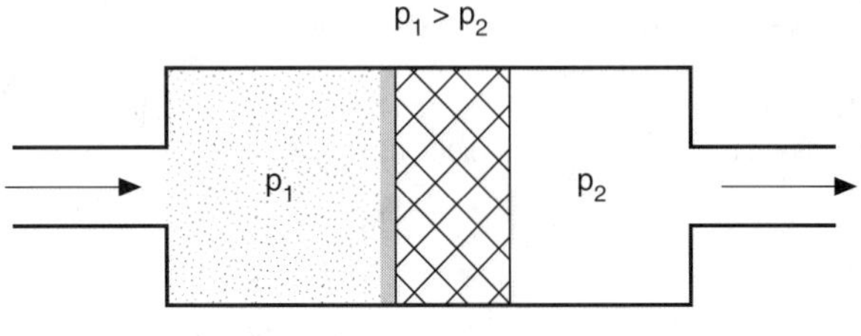

Figure 1. Simple filter system.

Liquid filters can be both batch-type and continuous. Filtration of suspensions gives rise to a sediment which, on the one hand, favors a finer cleaning and, on the other hand, increases pressure drop and reduces filter capacity. The latter circumstance has a severe effect on separation of fine-dispersed particles or easily deformed flocculi. In this case the filter capacity can be raised making an addition in a suspension of a subsidiary substance in the form of relatively coarse solid particles. The substance (a "filter aid") must possess a low bulk density and be chemically inert in relation to the liquid filtered and the sediment. The most commonly used is diatomite and, in food industry, cellulose. The quantity of the subsidiary substance needed is generally not large, and it may be recovered by washing.

Filtration can be performed under both hydrostatic pressure, pressure that is produced by pumping the suspension, or by a vacuum on the filtrate side.

The hydrostatic pressure is low and, as a rule, does not exceed a few meters of water gauge, and is used in a filter with easily filtering materials or at low filtration rates. The main advantage of these filters—known as Nutsche filters—is simple design and low cost. They have a perforated bottom with various filtering materials, from paper to sand and gravel, stacked on it. The suspension is fed from the top while the cleaned liquid (filtrate) is discharged into the lower chamber. The sediment is removed by either liquid back-flushing or mechanically, the filtration membrane is often merely replaced by a new one.

In order to enhance filter capacity, the pressure drop may be increased across the filtration membrane. If this is achieved by increasing the pressure on the suspension side (on the other side of filtration membrane the pressure approaches atmospheric in this case), the filters are considered to operate under pressure.

Pressurized filters, which run under periodic duty, include chamber and frame filter presses. In a frame filter press, the suspension is fed at a pressure up to 7 MPa into spaces between alternating plates and frames (**Figure 2**). The ribbed-surface plates serve as a support for the filtration membrane and afford filtrate drainage. Solid particles are collected in the frame plane as a compact mass which is removed when the plates are drawn apart.

The operation of a chamber filter press is similar to that of a frame filter press, but the working cavities are made by recessions in the plates. In a Karver filter press operating on this principle, the sediment before ejection is pressed at 42 MPa pressure, dried, solidified, and afterwards automatically ejected when the plates are drawn apart. A variety of filter press designs are used in industry.

Commonly encountered systems are the *plate* and *pipe filters* in which the filtering membrane is made of solid porous materials such as cermet cloth fastened on a strong screen membrane with a skeleton or on a grooved plate. An example is the pipe filter. This is also a batch-type filter: the pipe pores plugged with sediment are washed by water back-flushing and air-blown.

Continuous filters can also operate under pressure, but in this case removal of sediment on the pressure side of filtration membrane presents a problem which is solved much more readily in vacuum filters with atmospheric pressure on the pressure side. The drum-type vacuum filter, which has gained wide acceptance, contains, as the main component, a drum whose internal space is separated into three sections. This makes it possible to run in one

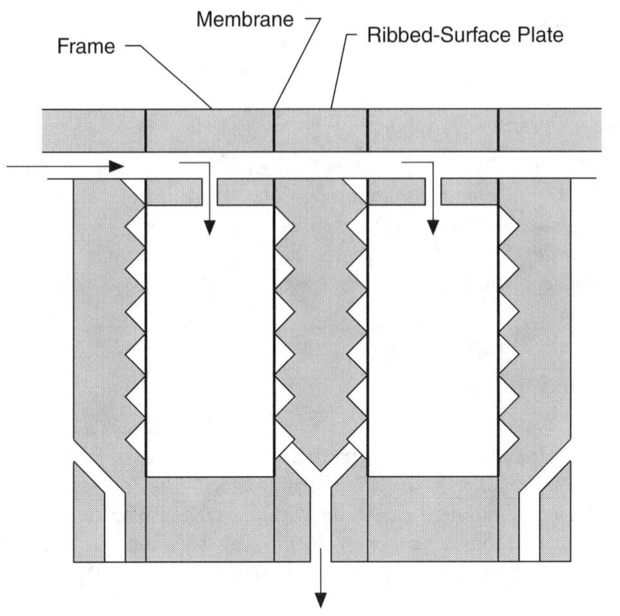

Figure 2. Frame filter press.

rotation of the drum through the filtration stages, washing the sediment with water, drying with air, and cake removal.

There are also disk, band, rotary, and other types of vacuum filters of batch type in which the sediment is either washed with water or removed mechanically.

The pressure difference can be also exerted by the centrifugal effect. The filters based on this principle are hollow rotating drums, the internal surfaces of which are covered with a filtration material. The suspension fed is centrifuged on the surface of the filtering material, the liquid flows through it, and is discharged out of the drum either by grooves or through perforations, while the sediment is left on the filtration membrane and later is removed mechanically. The subsidiary filtering substances described above (filter aids) may be used in *Centrifugal filters*.

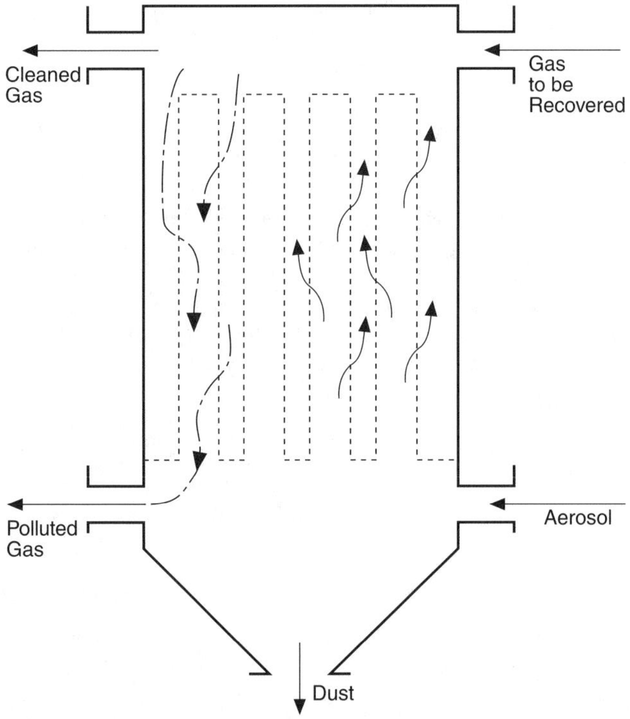

Figure 3. Multi-sectional bag filter.

Gas filters aimed at gas cleaning are, as a rule, continuous. They are designed for large volumetric flow rates of gas when it contains dust up to tens of grams per cubic meter. For the gases to conform to environmental requirements (see **Air Pollution**), it is necessary to achieve no less than a 99% dust collection. Multi-element cloth filters in housings fit best of all for the purpose. Filtration elements may take the form of a plane broad cloth resting on a support grid, but more often they have the form of cylindrical bags stretched on a grid or grid-free. Their diameter varies within wide ranges, but does not exceed 600 mm. The bags from 125 to 300 mm in diameter and from 2.5 to 3.5 m long (high) are commonly used. In grid-free *bag filters,* dustladen gas is supplied inside the bag (fixed from either above or below) and inflating it. The collected dust settles on the inside of the bag. In the supported filters, filtration may proceed inwardly, with sedimentation on the outward side of the bag. The number of elements in one housing may exceed 10,000.

Filters are regenerated by mechanically shaking off the dust layer and/or back blowing by the cleaned gas. The latter technique is more efficient. The dust shaken off is collected in a hopper from which it is removed mechanically or pneumatically.

Sectional design of filters makes it possible to regenerate it section by section without switching it off. A schematic diagram of this multisectional bag filter with shaking off and back blowing is presented in **Figure 3**.

Cleaning smaller volumes of industrial gases contaminated with reactive condensed substances is carried out in fibrous or grain filters, as well as, dry or wet filters. In the latter case, filters may operate under self-cleaning regime, which is also the case when mist droplets are captured.

Felt made of special ultra thin fibers, in which filtration membranes is supported of a frame structure, is applied for fine and superfine gas cleaning. The filtration membrane is placed in the frame structure as a tape between pi-shaped frames. Corrugated separators are mounted between the adjacent layers. These filters

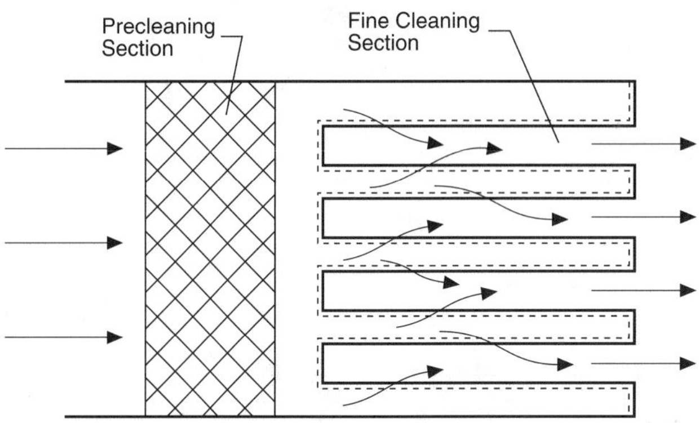

Figure 4. Combined gas filter.

permit the cleaning of gases with the 99.99% efficiency and higher; however, as a rule, they require precleaning. Combined filters (**Figure 4**) are designed to account for this.

Special cases are air filters aimed at cleaning dust from atmospheric air supplied to a room or technological apparatus, where ingress of dust is not allowable. A characteristic property of these filters is a higher rate of filtration, up to 3 m/s, and, hence, the risk of carrying away the particles already settled.

Air is commonly purified either by periodically replaced coarse fiber filtration membranes made of wire, or coarse synthetic fibers up to 250 μm in diameter, or by screens wetted in oil and mounted in the form of cassettes, or by oil self-cleaning filters with filtration membranes as a plane woven cloth. A filter of this type is shown in **Figure 5**. An endless tape moves in a vertical plane, passes through the filtration zone, collects dust on its surface, and then passes through an oil bath, where the cloth is washed of dust and oiled once again. The dust settles on the bath bottom as slime and is removed as the oil is recovered.

For fine air cleaning, filters of the same type are used as those for a fine and superfine gas cleaning described above.

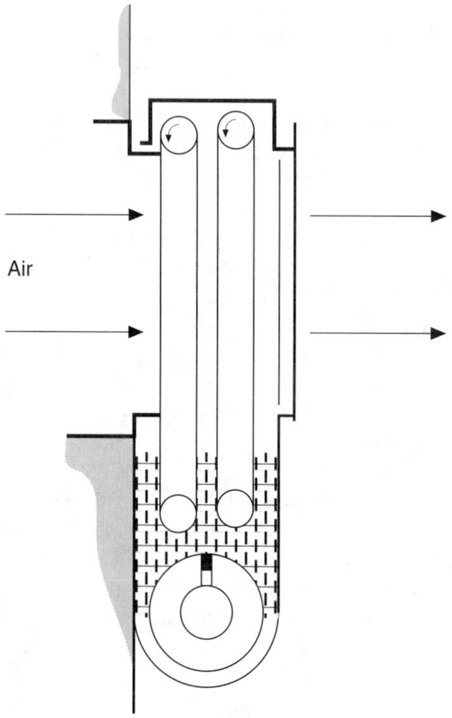

Figure 5. Self-cleaning air filter.

References

Perry, R. H. and Chilton, C. H. eds. (1973) *Chemical Engineer's Handbook*, 5th ed., Mc-Graw Hill, New York.

I.L. Mostinsky

FILTRATE (see Filtration)

FILTRATION

Following from: *Liquid-solid separation; Gas-solid separation*

Filtration is the separation of suspended impurities from liquid or gas by passing the fluid through a porous membrane that retains the particles on its surface or in its pores.

Since separation of liquid-solid suspensions is often different from separation of dust-laden gases, including aerosols, we shall discuss them separately.

Liquid Filtration

In liquid filtration, the suspension passes through a filtration membrane; the suspended particles remain on its surface and in its pores, while the clarified liquid, called *filtrate,* is collected behind the filtration membrane. The separated solid phase, the sediment, forms a continuously growing layer on the filtration membrane surface. It commonly consists of randomly lying particles of different shapes.

Fresh portions of liquid are passed through this layer and then through the filtration membrane. The liquid flow velocities, as a rule, are low, and the flow is laminar and described by the equation

$$\frac{dV}{Fdt} = \frac{\Delta p}{\eta\left(\alpha \frac{M}{F} + r\right)}, \qquad (1)$$

where V is the filtrate volume, m^3, F the filtration surface, m^2, t the time, s, Δp the pressure difference, Pa, η the filtrate viscosity, N s/m^2. The terms in parenthesis represent the *specific pressure drop*—i.e. the pressure drop for a superficial velocity of 1 m/s—for the sediment and filtration membrane respectively.

The total mass of dry sediment is determined as

$$M = mV = \frac{\rho C}{1 - \omega C} V, \qquad (2)$$

where m is the mass of dry sediment per unit filtrate volume, kg/m^3, ρ the filtrate density, kg/m^3, C the mass concentration of particles in suspension, kg/kg, ω is the ratio of wet to dry sediment mass. α is the specific pressure drop of sediment per unit of dry mass, r the specific pressure drop per unit surface of the filtration membrane.

Integration of Eq. (1) at Δp = const yields

$$\frac{tF}{V} = \frac{\eta\alpha m}{2\Delta p}\frac{V}{F} + \frac{\eta r}{\Delta p} \qquad (3)$$

or for specific conditions at a constant rate of filtration

$$\frac{V}{Ft} = \frac{\Delta p}{K_2} + C', \qquad (4)$$

where K_2 and C' are the constants that are characteristic for these specific conditions.

The above equations are not valid for separation of finely dispersed suspensions of low concentration by loose fibrous filtration partitions when the particles penetrate deep into the filter bed and the sediment is not formed on the filtration membrane surface.

The average specific pressure drop α of the sediment depends on pressure drop:

$$\alpha = \alpha'(\Delta p)^s \tag{5}$$

where α' is a constant value depending on particle shape and size and s is a coefficient of sediment compresibility which varies from 0 to 1. If the sediment consists of solid particles, increasing pressure does not change their size and packing density, i.e. $s = 0$, and the pressure drop of filtration membrane is low relative to that of the sediment layer, then Eq. (1) takes the form

$$v = \frac{dV}{Fdt} = \frac{\Delta pF}{\eta\alpha' M}, \tag{6}$$

i.e. at a constant pressure difference Δp the instantaneous rate of filtrate flow is inversely proportional to the quantity of sediment. Where the sediment consists of easily deformable particles, s approaches 1 and

$$v = \frac{dV}{Fdt} = \frac{F}{\eta\alpha' M}, \tag{7}$$

i.e. the instantaneous rate of filtrate flow is independent of pressure. These are extreme cases. Actually, s varies from 0.1 to 0.8. Hence, in filtration of a suspension with solid, granulated or crystalline particles, increasing pressure difference brings about a nearly proportional growth in the flow rate of the suspension being filtered. If flocculent or clay sediments are predominant in the suspension, then increasing pressure difference has almost no effect on the flow rate. Some sediments are compressed, and the rate of filtrate flow is reduced, if a certain critical pressure drop is exceeded.

A special filtering substance is sometimes added to slimy suspensions to improve filtration, and clarification is carried out at a constant rate of filtrate flow (not at $\Delta p = $ const) because otherwise a large amount of filtrate-contaminating impurity slips through at the beginning and a fast growth of the layer and its pressure drop loss is observed afterwards. It has been shown in practice that in general it is more advantageous to start filtration at a low pressure difference, gradually increasing it as the sediment layer grows. Centrifugal pumps with a steep characteristic operate in much the same fashion.

Equation (1) implies that the rate of filtrate flow is inversely proportional to filtrate viscosity η. Therefore, in order to increase filter capacity it is advisable either to heat the suspension or to add substances to reduce η.

Particle size plays an important role in filtration. Filtration of fine suspensions causes an increase in α' and a fall in the filtration rate. Therefore, it is a good practice to coarsen fine particles, e.g. by coagulation by either heating the suspension or adding coagulants.

In order to achieve the maximum capacity of *batch-type filters* one must know to what thickness the sediment layer should be brought and at what moment regeneration should be started. Such optimization of filter operating regime has been carried out for (1) $\Delta p = $ const, (2) $v = $ const, and (3) variable v and Δp. Here it was assumed that the complete cycle of filter operation that is a vessel with a filtration membrane and filtrate drain consists of six operations: preparation of the filter, feeding of suspension, filtration, sediment washing, air blowing, and discharging. Filtration, washing, and blowing are basic operations, the rest are auxiliary ones. Thus, the operation period consists of two components $t = t_b + t_{aux}$. The optimization has shown that for $\Delta p = $ const and a substantial pressure drop of filtration membrane the filter capacity reaches the maximum when the relation between t_b and t_{aux}

$$t_b = t_{aux}\left(1 + \sqrt{\frac{2\eta r^2}{\Delta p r_0 x_0}}\right) \tag{8}$$

is valid and for $v = $ const the relation

$$t_b = t_{aux}\left(1 + \sqrt{\frac{\eta r^2}{\Delta p r_0 x_0}}\right) \tag{9}$$

is valid, where η is the viscosity of suspension liquid phase, N s/m^2, and r_0 are the permeabilities of the filter and the sediment respectively, m^{-2}, x_0 the ratio of the sediment volume to the filter volume, Δp the pressure difference, N/m^2.

Comparison of optimum values of $(t_b - t_{aux})$ for $\Delta p = $ const and $v = $ const yields $\sqrt{2}$, i.e. 40% difference.

Optimization of the $v = $ const regime changing over to $\Delta p = $ const was much more complicated, and the results, as was expected, appeared to depend on the duration of operation under each regime, but on the whole were within the 40% indicated above.

Gas Filtration

The main distinction between filtration of polluted gases and that of liquid suspensions follows from the large (of the order of 10^3) difference in densities of the carried medium (gas) and dust particles. A dust-gas system can be stable only in the case of fine particles of 10 μm in size and smaller. This is known as an aerosol. Coarser particles, up to 100 μm in size and larger, can be present in real conditions, for instance, in industrial gas exhausts, but more rough and primitive filtration devices such as cyclones and settlers are used more frequently than filters for their separation.

Due to the great difference of gas and solid particle densities the latter, even for a high mass concentration, are at a sufficiently large distance from each other that they can be assumed noninteracting between themselves.

In what follows we shall first discuss filtration of aerosols containing solid particles (filtration of mists will be considered later).

Aerosol filtration

Aerosol filtration involves two stages, a stationary one and a nonstationary one. The first stage includes particle sedimentation on or in a clean or cleaned-by-regeneration filtration membrane, where the role of the dust layer formed is not significant. Trapping efficiency and pressure drop do not change in time and are determined only by the properties of filtration membrane and settling particles and by the gas flow parameters. The duration of this stage depends on the specific rate of filtrate flow, the dust content of the flow (concentration of solid particles in it), and the extent of dust trapping.

The second, nonstationary, stage begins at the moment when the sediments of the trapped dust clog up pores producing on the surface of filtration membrane a layer of such thickness that it has an appreciable effect on the pressure drop Δp and the efficiency of particle trapping η. The former parameter is growing progressively and the latter, as a rule, first rises and then may drop. A rise in η is attributed to the appearance of an additional filter bed, while a reduction, to the local "pressing through" of the dust as a result of increase in Δp. Variation of Δp and η versus time is depicted in **Figure 1**.

Theoretically, filtration of aerosols can be calculated only at the first stage, and even then only by modeling filtration membrane as single-row fibers separated by a space large enough in relation to a particle size. Sedimentation of particles on the fiber (a single cylinder) is schematically shown in **Figure 2**. In an aerosol flow, the particles follow the streamlines. However, if the latter bend in the filtration zone, the particle path deflects from the streamlines and under inertia forces (coarse particles), tangency (finer particles), or electrical attraction (charged particles) the particles settle out on the fiber surface. These effects also include the Brownian motion of highly dispersed particles and sedimentation of coarse particles by gravity.

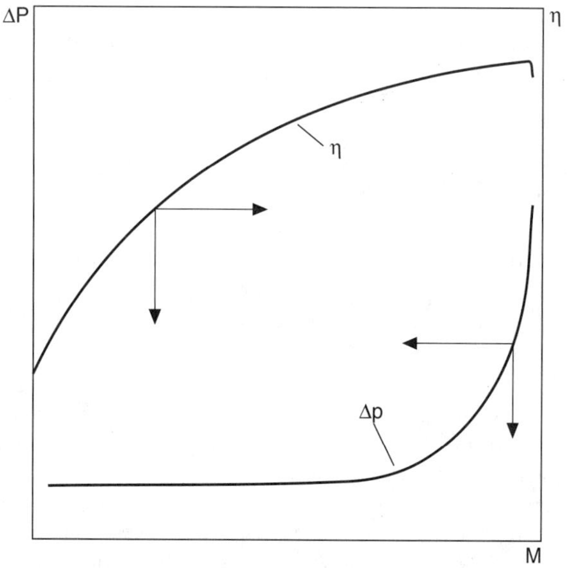

Figure 1. Variation of particle trapping efficiency (η) and pressure drop (Δp) with amount of material trapped on a filter.

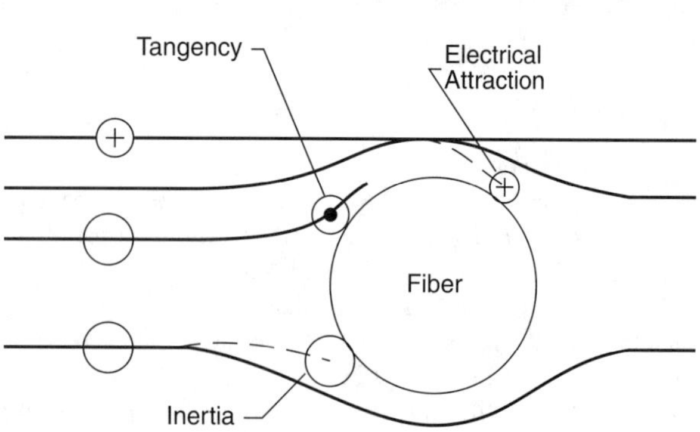

Figure 2. Mechanisms of particle trapping in flow over a cylinder.

Considering that the effects work independently, Devies derived the overall efficiency of sedimentation of particles of a definite size on the cylinder (so called fractional efficiency) under inertia, diffusion, and tangency

$$\eta = 0.16\left[R + (0.25 + 0.4R)\left(St + \frac{2}{Pe}\right) - 0.0263\left(St + \frac{1}{Pe}\right)^2\right], \quad (10)$$

where St and Pe are the Stokes and Peclet numbers, $R = r/a$, r and a are the radii of a particle and a cylinder (a fiber) respectively. Other formulae are also available for calculation of fractional of efficiency of particle trapping from aerosol flow by fibers. These include the Friedlander, Torgeson and Kirsch-Stechkina-Fuchs formulae. In order to calculate the efficiecy of polydispersed particle trapping one must know their size spectrum and do an appropriate summation. However, we should keep in mind a selective breakthrough that shows up as an anomalous increase in the breakthrough of 0.2–0.3 μm particles. With increasing fiber thickness and flow velocity the anomaly shifts toward coarser particles.

Mist filtration

In contrast to suspension and *aerosol filtration,* mist filtration does not give rise to a filter bed produced from the trapped substance, but the liquid clogged in the filtration membrane produces a so-called capillary effect. It involves spreading of clogged droplets with subsequent merging into coarser droplets or production of liquid films on fibers, accumulation of liquid at the sites of fiber interlacing, capillary condensation and adhesion of neighboring fibers due to capillary forces. All this affects the structure of filter beds, hindering gas flow and raising the pressure drop and, thereby, droplet breakthrough as a result of liquid forcing through the filtration membrane.

A particular emphasis should be laid on filtration in air oil filters. They use as the filtration membrane thick (25–100 mm) layers of coarse fibers (from 30 to 250 μm) and wire meshes, punched sheets, paper, and corrugated board impregnated with oil. The efficiency of trapping of the dust particles out of the air flow by the oil film, is higher than that without oil.

Filtration through granular filtration membranes

The mechanism of particle trapping by aerosol flow filtration through a layer of granular material is similar in outline to that through fibrous material; inertia, diffusion, tangency, and gravity effects work simultaneously much in the same way. The breakthrough $K = 1 - \eta$ is described by the formula of the type

$$K = \exp(-k\delta Pe^{k_1/d_g}), \quad \mathbf{(11)}$$

where δ is the thickness of granular layer, Pe the Peclet number,

d_g the grain diameter, and k and k_1 are empirical constants with k_1 varying within the range $-1/3$ to $-2/3$.

Investigations have established that at low flow velocities (up to 0.05 m/s), particles smaller than 0.5 μm in diameter settle out mainly under the action of diffusion, while the coarser particles settle by gravity. Selective breakthrough is inherent in granular as well as fibrous filtration membranes, but the greatest breakthrough is due to the coarse particles, from 0.4 to 0.9 μm in size. Increasing velocity results in increasing breakthrough and with the released particles becoming smaller.

Gas flow in a granular layer pores involves multiple breakup of the main flow and variation of velocity in value and direction. Some ("flow-through") channels have velocities which are substantally higher than the average one and inertia effects are predominant; also, there are stagnation zones with practically zero velocity in which particle sedimentation is due to diffusion and gravity.

At low flow velocities the stream is laminar and the relation between the average velocity and the pressure difference is linear and described by Darcy's equation. As the velocity increases, the laminar flow is disturbed, the importance of inertia effects grows as a result of multiple jet convergence and divergence, and the proportionality factor between velocity and Δp reduces.

The pressure difference for a layer of grains of arbitrary shape may be calculated for higher gas velocities by the formula

$$\Delta p = \lambda \frac{\delta U^2 \rho_g}{2 d_e \Phi_s^{3-n}} \frac{(1-\epsilon)^{3-n}}{\epsilon^3}, \qquad (12)$$

where λ and n are respectively the pressure drop coefficient of the layer and an exponent both of which are the functions of the Reynolds number, δ the layer thickness, U the average flow velocity related to the entire filtration surface, ρ_g gas density, d_e the equivalent grain diameter, ε the porosity, and Φ the coefficient of the grain shape: $\Phi_s = 1$ for spheres, $\Phi_s = 0.8$ for shingle and sand, and $\Phi_s = 0.5$ for coal and coke. For laminar flow $n = 1$ and formula (12) transforms into the *Kozeny-Karman equation*

$$\Delta p = \frac{180 \delta U \eta_g (1-\epsilon)^2}{d_e^2 \Phi_s^2 \epsilon^3}. \qquad (13)$$

where η_g gas viscosity, N s/m^2.

For calculating Δp of granular layers for the laminar and turbulent gas flow use can be made of the Ergun equation.

$$\Delta p = \frac{150(1-\epsilon^2)}{\epsilon^3} \frac{\eta_g U \delta}{d_e^2} + 1.75 \frac{1-\epsilon}{\epsilon^3} \frac{\rho_g U^2 \delta}{d_e} \qquad (14)$$

that does not involve a shape coefficient.

Specific empirical formulas are available in the literature for many specific grain types.

Formulas 11–14 pertain to regenerated filtration membranes; as they are clogged by trapped particles, Δp increases.

References

Perry, R. H. and Chilton, C. H. eds., (1973) *Chemical Engineer's Handbook*, 5th ed., McGraw-Hill, New York.

Leading to: *Filters*

I.L. Mostinsky

FIN EFFICIENCY (see Extended surface heat transfer; Tube banks, single phase heat transfer in)

FIN EFFICIENCY FACTOR (see Coupled conduction and convection)

FINITE DIFFERENCE METHODS (see Tau method)

FINITE ELEMENT METHODS (see Tau method)

FINNED TUBES (see Air cooled heat exchangers; Heat recovery boilers; Tube banks, condensation heat transfer in; Tube banks, crossflow over; Tube banks, single phase heat transfer in)

FINS (see Augmentation of heat transfer, single phase; Augmentation of heat transfer, two phase; Coupled conduction and convection)

FINS, CIRCUMFERENTIAL OR HELICAL (see Augmentation of heat transfer, two phase)

FIN TEMPERATURE EFFECTIVENESS (see Extended surface heat transfer)

FIRE BALLS (see Fires)

FIRE POINT (see Flammability)

FIRES

Following from: *Combustion*

Introduction

The discovery of how to produce fire was a significant step in the development of mankind and is still a process which we all experience personally from a very early age. A fire may be defined as "uncontrolled combustion" in order to differentiate it from well controlled combustion as obtained, for example, inside a furnace or car engine.

Physical Processes

Fires involve a large number of physical and chemical processes:

- Chemistry: The reaction of the fuel with the oxidant (normally air). It may be possible to express the overall reaction as a simple chemical reaction e.g.

$$CH_4 + 2O_2 \rightarrow CO_2 + 2H_2O$$

but, in fact, the burning process is usually very complex and involves a whole series of intermediate reactions.

- Mass Transfer: If the fuel is solid or liquid it will be evaporated or volatilised to release a combustible gas which, after burning, is transported away as combustion products or soot.
- Heat Transfer: The combustion of the fuel produces heat. Most of the heat initially goes into heating up the combustion products and, particularly, the inert gases such as nitrogen present in the air (this is why fuel burnt in pure oxygen produces much higher temperatures eg in an oxy-acetylene torch). The heat is then transferred to the surrounding environment by convection and, principally, radiation.
- Buoyancy driven flows: Fires usually depend upon buoyancy driven flows to remove the hot combustion products from the source of the fire, a process which at the same time draws in fresh air. The movement of smoke and combustion products generated by a fire is usually dominated by buoyancy as the driving force.

Types of Fire

Fires can be categorised through the type and physical state of fuel which is being burnt, the location of the fire, and its duration. A few examples are:

Cellulosic fires

Involving solid cellulosic fuels (e.g. wood). Most fires inside rooms fall into this category. Heat output is typically of the order of 1MW and flame temperatures about 600–800°C.

Pool fires

Involving liquid fuels, such as oil, in an open pool. Heat output is directly related to size and may be over 100MW. Typical flame temperatures are about 1100°C for most hydrocarbon fuels.

Jet fires

Produced when gaseous or liquid fuel, released as a jet from a pressurised pipe or container, is ignited. The shape of the flame is normally dominated by the initial momentum of the fuel. Heat output is directly related to the release rate of the fuel. Heat fluxes inside the flames can be significantly higher than inside pool fires due to the greater gas velocities and hence convection coefficient. Flame temperatures can also be higher than in pool fires by 100 to 200 °C.

Forest fires

A type of cellulosic fire but very different to a room fire both in size and nature.

Fire balls

Produced when a large volume of gaseous fuel is suddenly released and ignited or when an established cloud of fuel is suddenly ignited. These fires can be very large but are of short duration, only lasting until the fuel is used up.

Fire Modelling—Objectives

Fire modelling covers a wide range of disciplines with differing objectives. Workers involved with fires inside buildings may wish to model the rate of growth of a fire, its heat output and rate of generation of toxic combustion products. Alternatively, they may not be interested in the fire source so much as the movement of the smoke and its effect upon the evacuation of personnel or the speed with which the fire is detected.

Workers involved with areas where flammable liquids and gases are processed, e.g. offshore platforms, are interested in the shape of the fire, the heat flux to objects engulfed by it and the heat flux from *radiation* to objects or people nearby. The efficiency of fire walls and extinguishing systems such as water sprays is also a focus of interest.

In the nuclear field, designers need to demonstrate that transport containers could survive being engulfed in a pool fire. In this case, the focus of attention is not so much on the fire itself as the performance of the container in a fire environment.

Modelling Methods

Because the objectives of fire modelling depend upon the application, a wide range of methods have been developed, each with advantages and disadvantages.

At the most basic level, basic physical principles and laws can sometimes be applied to obtain a reasonable measure of quantities such as heat output, heat fluxes, rate of heating and mass flow rates. For example, the heat flux to an object engulfed by a pool fire can be determined, assuming that the convective heat flux is small using the standard equation for radiation heat transfer:

$$\dot{Q} = \sigma\epsilon(T_f^4 - T_s^4)$$

Using reasonable estimates for the quantities involved:

- Surface emissivity, ϵ 0.9
- Effective emissivity of flame 1.0
- Temperature of fire 1373K
- Temperature of surface 373K (say)

gives a heat flux of 180kW/m^2.

Where the situation is too complex to apply basic physical principles alone, engineering correlations have been developed based upon measured experimental data. Correlations have been developed covering a wide range of fire phenomena from flame size and shape to burning rates, heat fluxes and gas velocities.

Usually the correlations assume a given fuel or type of fuel, geometry and location. Where the situation of interest is within the range for which a correlation is valid, these correlations, because they are based upon experimental results, usually provide an accurate modelling tool.

In situations such as room fires, there are many variables, such as ceiling height, room area and size of openings which each have an influence upon the fire. However, the basic mechanisms and flow patterns are well understood (e.g. cold air entering through the lower part of the doorway, reaching the fuel which is burning, a plume of smoke and combustion products rising above the fire and forming a layer beneath the roof). In this situation, a mixture of basic physical equations and correlations can be used to model the heat and mass transfer between the effectively homogeneous "zones" of the room (e.g. cold air layer, fire plume and hot layer). Once a closed system of equations has been generated, these zone models can be used to model both steady state conditions and the transient development of a fire. Many of these zone models are well established and validated, particularly for applications such as room fires.

Occasionally it is necessary to model fires in conditions for which there are no available correlations or established zone models. This may occur, for example, when novel designs of building are developed (e.g. buildings with large atria). Under these conditions, the most reliable modelling is obtained by using a **Computational Fluid Dynamics (CFD)** computer code to represent the fire and flow of air and smoke. In a sense, this is just an extension of the zone models to many thousands of separate "zones", with the heat and mass transfer between neighbouring zones given by basic physical equations. Although CFD is the most general of the modelling methods it still currently requires significant time to set up the model and analyse the output and also needs very considerable computing power to solve the many thousands of separate equations.

References

Drysdale, D. D. (1985) *Introduction to Fire Dynamics,* John Wiley and Sons, New York.

The SFPE handbook of Fire Protection Engineering, NFPA, (1988).

Fire Safety Journal, Elsevier Applied Science

Blackshear, P. L. (1974) *Heat Transfer In Fires: Thermophysics, Social Aspects Economic Impact,* John Wiley and Sons, New York.

Leading to: *Flammability; Flames*

C.J. Fry

FIRE TUBE BOILERS (see Tubes and tube banks, boiling heat transfer on)

FIRST LAW OF THERMODYNAMICS (see Carnot cycle; Combustion; Entropy; Exergy; Flow of fluids; Heat; Irreversible thermodynamics; Specific heat capacity; Thermodynamics)

FIRST NORMAL STRESS DIFFERENCE COEFFICIENT (see Non-Newtonian fluids)

FISSILE MATERIAL (see Liquid metal cooled fast reactor)

FISSION (see Liquid metal cooled fast reactor)

FISSION PRODUCTS (see Nuclear reactors)

FISSION REACTION (see Nuclear reactors)

FIXED BED REGENERATORS (see Regenerative heat exchangers)

FIXED BEDS

Following from: *Porous media*

Fixed beds are used in the process industries as catalytic chemical reactors, adsorption/desorption beds, thermal regenerators, heat storage devices and pebble bed heaters. In all of these operations, the transfer of heat and mass takes place between the fluid and solid phases and the transfer can be either steady state or transient. The solid phase can be in several forms: randomly dumped packings, ordered mono-sized particles and monolithic blocks of various geometrical shapes. The fixed bed geometry itself is normally cylindrical and the flow of the fluid through the bed is parallel to the axis of the cylinder. Radial flow through annular fixed beds is also used when low pressure drop restrictions are specified.

The major design parameters are the pressure drop across the fixed bed, and the heat and mass transfer coefficients between the fluid and the surface of the solid phase. Diffusion of heat and mass into the interior of the solid phase can be a significant mechanism of transfer, but it is common to employ lumped transfer coefficients at the surface to account for the internal diffusion, and to use average solid temperatures and concentrations in the design calculations. Pressure drop across a fixed bed is calculated from empirical formulae: the most common formula being that proposed by Ergun in 1952, i.e.

$$\frac{\Delta p}{L} = \frac{c_1 \rho u^2(1-\epsilon)}{d\epsilon^3} + \frac{c_2 \eta u(1-\epsilon)^2}{d^2\epsilon^3} \tag{1}$$

where Δp is the pressure drop, L is the length of the fixed bed, d is the equivalent diameter of the particle, defined as the equivalent volume sphere (= 6 × volume/surface area), ϵ is the *bed voidage* (porosity = free volume/total volume), u is the superficial velocity based on flow through an empty fixed bed, η is the fluid viscosity and c_1 and c_2 are correlation values obtained by regression of experimental data. Universal values of c_1 and c_2 do not exist, see Heggs (1983), although, for randomly packed spherical particles, the values obtained by MacDonald et al. (1979) ($c_1 = 1.8$ and $c_2 = 180$) provide reasonable predictions of the pressure drop over a wide Reynolds number range: 0.1 to 10000. Otherwise, it is essential that they are evaluated from experimental data of the fixed bed under consideration. The measurement of the bed voidage ϵ, is crucial, because Equation 1 is very sensitive to this parameter. The value of the *equivalent volume sphere,* d, is straight forward for mono-sized particles, but in many process systems a mixture of particle sizes, and occasionally shapes, is used for the fixed bed. A suitable weighted mean of the equivalent volume sphere must be used in Equation 1, Hawkins (1993).

For the calculation of the pressure drop across *annular fixed beds,* it is necessary to take into account the direction of the radial flow and the effects of the fluid momentum changes, i.e.

$$\Delta p = \frac{\dot{M}_\ell}{\rho}\left\{\begin{aligned} &\frac{2(D_o - D_i)\dot{M}_\ell}{D_i D_o \epsilon^2}\left[\frac{c_1(1-\epsilon)}{d\epsilon} \pm \frac{(D_i - D_o)}{D_i D_o}\right] \\ &+ \frac{c_2 \eta (1-\epsilon)^2}{d^2\epsilon^3}\ln\left(\frac{D_o}{D_i}\right) \end{aligned}\right\} \tag{2}$$

where $M_l = \dot{M}/2\pi L$, where $\dot{M}$ is the total mass flow rate and L the height of the annulus, D_i and D_o are the inner and outer diameters of the annular fixed bed. The positive sign is for flow inwards and the negative sign is for flow outwards. **Equation 2** assumes that the flow is evenly distributed along the length and that the inlet and outlet mainfolds do not effect the total pressure drop of the annular system. However, flow mal-distribution can occur in annular systems due to the pressure profiles in the inlet and outlet manifolds, Heggs et al. (1994b); this will result in larger overall pressure drops than the value predicted by Equation 2. Two further relevant physical properties of the fixed beds are required for transfer processes: surface area, A_v, per unit bed volume

$$A_v = 6(1 - \epsilon)/d \quad \textbf{(3)}$$

and the bulk density—mass per volume of the system,

$$\rho_b = \rho_s(1 - \epsilon) \quad \textbf{(4)}$$

Heat transfer in fixed beds can be either a steady state or a transient process and various mathematical models have been proposed to describe transfer of heat between the fliud flowing through the fixed bed and the packing of the fixed bed. In some process applications there is transfer from the fluid to and across the wall of the container of the fixed bed. The definitive paper of Amundson (1956) contains almost all of the models needed to describe the transfer of heat in fixed beds. Without exception, it is necessary to solve coupled differential equations, either ordinary or partial. If chemical reactions or sorption processes are taking place in the fixed bed, then the system of equations becomes non-linear. Space precludes a detailed discussion of all the possible models and solutions. Only the transient non-reacting and non-sorptive systems will be discussed here. This corresponds to several applications: heat storage devices, pebble bed heaters, cooling of desorption beds, start-up and shut-down of process equipment and pipelines. The level of sophistication of the calculation is directly related to the number of parameters required for the evaluation of the solution pertaining to the problem.

The simplest representation of the transient heating or cooling of a fixed bed corresponds to the assumption that the film heat transfer coefficient between the fluid and the solids is infinite and the temperatures throughout the fixed bed are identical, Heggs (1994a). If the fixed bed is initially at a uniform temperature, T_o, and the inlet fluid temperature to the bed is changed to T_{fi} by some forcing function, fn(t), then the time for the fixed bed and its container to equilibrate at the inlet fluid temperature is given by

$$t_{sat} = \frac{L}{v}\left(1 + \frac{\rho_b c_s}{\epsilon(\rho c_p)_f} + \frac{(A_c \rho c)_w}{A_{cb}\epsilon(\rho c_p)_f}\right) + t_{fn} \quad \textbf{(5)}$$

where v is the interstitial velocity, $= u/\epsilon$, and A_{cb} and A_{cw} are the cross-sectional areas of the empty fixed bed and container wall respectively, and t_{fn} is the time for the forcing function of the inlet fluid temperature to reach the value T_{fi}. If the forcing function is a step change, then t_{fn} is zero. The first term in the brackets of **Equation 5** is the residence time of the fluid in the fixed bed, the second and third terms represent the times for the packing and the container wall to equilibrate. Normally it is assumed that the fixed bed is adiabatic with respect to the container wall, so that the third term in **Equation 5** is neglected. This is often the cause of the underprediction of the time for equilibrium.

The time to saturation is also delayed when the fluid to solids heat transfer coefficients are finite. In this representation of the transfer processes, the temperatures of the phases are different during the transient period of operation and it is impossible to present an explicit formula to predict the time of heating or cooling. The equations describing the system are as follows: Heat balance on the fixed bed

$$\frac{\partial T_f}{\partial t} + v\frac{\partial T_f}{\partial z} + \frac{\rho_b c_s}{\epsilon \rho_f c_{pf}}\frac{\partial T_s}{\partial t} = 0 \quad \textbf{(6)}$$

this balance ignores any thermicity effects of chemical reaction or the sorption process—no heat of adsorption or desorption. This equation becomes the pseudo-homogeneous system, if $\alpha \to \infty$ and the container wall effects are ignored. For finite values of the heat transfer coefficient, a rate equation is required to couple the temperatures between the fluid and solid phases, as follows:

$$\rho_b c_s \frac{\partial T_s}{\partial t} = \alpha A_v(T_f - T_s) \quad \textbf{(7)}$$

Initial condition:

$$t \leq 0, \quad 0 \leq z \leq L, \quad T_g = T_s = T_0 \quad \textbf{(8)}$$

Boundary condition:

$$t \geq 0, \quad z = 0, \quad T_f = fn(t) \quad \textbf{(9)}$$

The outlet fluid temperature, T_{fo}, of the above system of **Equations 5** to **8** is a function of two dimensionless groups: the **Number of Transfer Units**

$$NTU = \alpha A_v L/\dot{m}c_{pf} \quad \textbf{(10)}$$

and the *utilisation factor*

$$U_t = \dot{m}c_{pf}(t - L/v)/\rho_b c_s L \quad \textbf{(11)}$$

The following graph, **Figure 1**, shows the response of the dimensionless outlet fluid temperature, θ_f $(=\{T_{fo} - T_o\}/\{T_{fi} - T_o\})$, over the range of NTU values:0.01 to ∞ for U_t from 0 to 2.0. The forcing function for these results is a step change in the fluid inlet temperature.

The delay caused by the finite heat transfer between the fluid and the solid depends upon the value of NTU and, if this goes to infinity, then the vertical line at $U_t = 1.0$ corresponds to **Equation 5** without the wall term. The literature abounds with correlations for the prediction of the fluid to solid heat transfer coefficient, but which should be used to predict the value of the coefficient remains a bone of contention. Foumeny and Ma (1991) list a number of correlations for randomly dumped fixed beds of spherical particles and recommend the use the equation proposed by Abou-Ziyan (1988).

$$\alpha d/\lambda_f = 0.4722\ Re_p^{0.747} Pr_f^{0.33} \quad \textbf{(12)}$$

where the Reynolds number is defined as

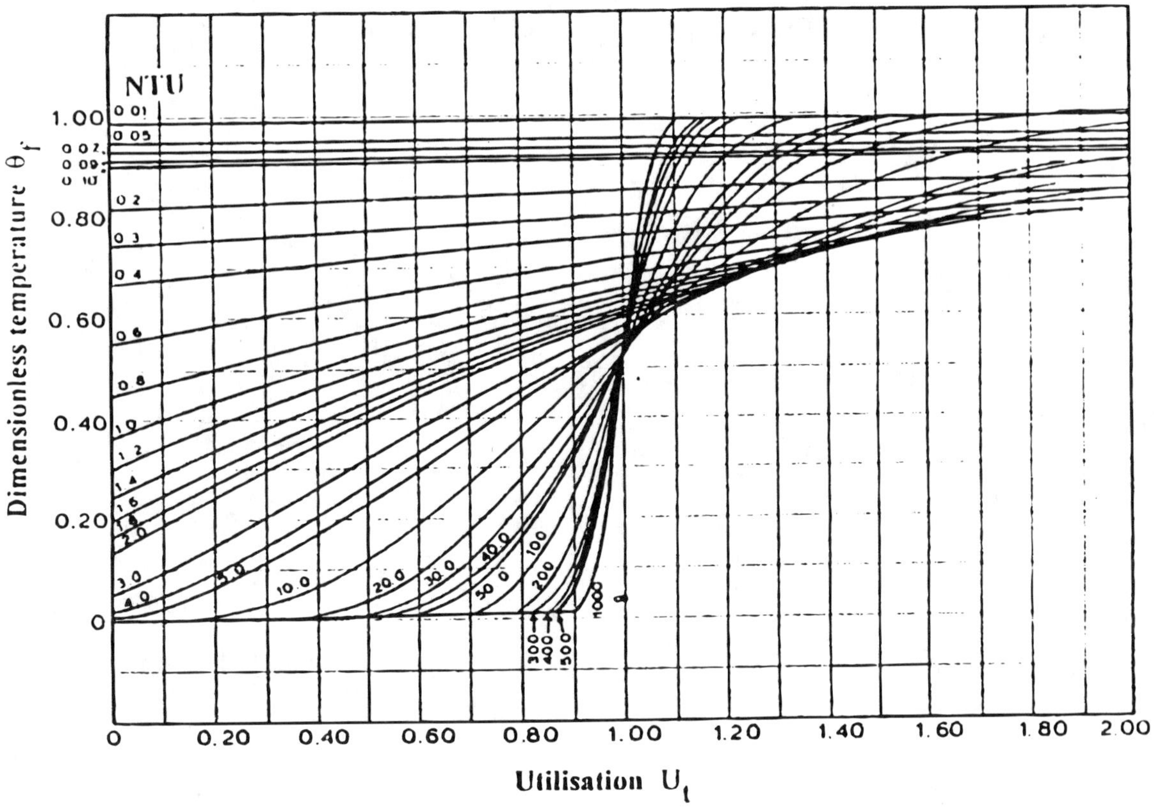

Figure 1. Dimensionless outlet fluid temperature response.

$$Re_p = 2\dot{m}d/3(1-\epsilon)\eta \quad \textbf{(13)}$$

The response of annular fixed beds can be predicted by the same method as above if the dimensionless groups NTU and U_t are redefined as follows:

$$NTU_s = \alpha A_v \pi (D_o^2 - D_i^2)L/\dot{M}c_{pf} \quad \textbf{(14)}$$

$$U_{ta} = \dot{M}c_{pf}(t - t_{res})/\rho_b c_s \pi (D_o^2 - D_i^2)L \quad \textbf{(15)}$$

$$t_{res} = (D_o^2 - D_i^2)/4D_i v_i, \text{ the residence time} \quad \textbf{(16)}$$

The effects of other mechanisms of transfer in the fixed beds can be accommodated by using a lumped heat transfer coefficient, $\alpha_z \ell$:

$$\frac{1}{\alpha_\ell} = \frac{1}{\alpha} + \frac{d}{10\lambda_s} + \frac{6(1-\epsilon)}{\dot{m}c_f Pe_d} \quad \textbf{(17)}$$

where the second term accounts for conduction in a spherical particle and the third term accounts for axial dispersion as the fluid flows through the fixed bed. The **Peclet Number**, Pe_d can be predicted from the following empirical equation:

$$\frac{1}{Pe_d} = 0.5 + \frac{\epsilon}{1.5Re_d Pr} \quad \textbf{(18)}$$

where Re_d is the **Reynolds Number** based on particle diameter and Pr is the **Prandtl Number**.

References

Abou-Ziyan, Z. Z. H. (1988) Heat and Momentum Transfer in Porous Material used for Thermal Energy Storage, PhD Thesis, University of Leeds, UK.

Amundson, N. R. (1956) Solid-Fluid Interactions in Fixed and Moving Beds, *Ind. Eng. Chem.*, 49(1)26–43.

Foumeny, E. A. and Ma, J. (1991) Design Correlations for Heat Transfer System, Chapter 11, *Heat Exchange Engineering: Volume 1, Design of Heat Exchangers,* Ellis Horwood Limited, London, 159–178.

Hawkins, A. E. (1993) The Shape of Powder-Particle Outlines, RSP Series—*Materials Science and Technology Series No. 1,* John Wiley and Sons, London.

Heggs, P. J. (1994a) Heat Transfer in Particulate Systems—The Infamous Film Heat Transfer Coefficient, Heat Transfer 1994—*Proceedings of the Tenth International Heat Transfer Conference, Vol. 1,* IChemE, Rugby, UK, 461–466.

Heggs, P. J., Ellis, D. I. and Ismail, M. S. (1994b) The Modeling of Fluid Flow Distributions in Annular Packed Beds, *Gas Separation and Purification,* 8, 4, 257–264.

Heggs, P. J. (1983) Fixed Beds, Section 2.2.5, *Heat Exchanger Design Handbook,* 2, Hemisphere Publishing Corporation, N.Y., 2.2.5-1–5.

MacDonald, I. F. El-Sayed, M. S., Mow, K., and Dullen, F. A. L. (1979) Flow Through Porous Media—the Ergun Equation Revised, *Ind. Eng. Chem. Fund,* 18, 198.

P.J. Heggs

FIXED TUBE SHEET EXCHANGERS (see Shell-and-tube heat exchangers)

FIZ (see Fachinformationszentrum Karlsruhe)

FLAMEOUT (see Combustion chamber)

FLAMES

***Following from:** Combustion; Fires*

Flame involves the **Chemical Reaction** between one chemical substance called a **Fuel**, and another chemical which is an oxidiser (or oxidant). In special cases, the fuel and the oxidant may be combined within the same chemical molecule and this is the case in some propellants and explosives. The chemical reaction between the fuel and oxidant is called **Combustion**; it is accompanied by the release of heat and usually by the emission of light in the visible region of the spectrum. In the case of a premixed hydrocarbon flame burning in air, the light emitted is normally blue if the mixture is fuel lean, and it gives the location of the flame and in particular, because of its greater intensity, the position of the flame front. However, if the mixture entering the flame is fuel-rich a yellow soot-producing flame is produced which is termed a luminous flame.

Most flames result from highly exothermic reactions giving *flame temperatures* about 2200K although flames may be capable of burning down to about 1300K, depending on the fuel-air ratio. Certain flames can be sustained below this temperature and are termed "cool" flames, but here only partial combustion occurs. Typical flames result from the combustion with air of a gaseous fuel such as natural gas, commercial and industrial liquid fuels, usually termed fuel oils, which are burned as a spray, or by pulverised coal particles suspended in air as is the case in a power station boiler.

Different types of flames can result from the way in which the fuel and oxidant are mixed in a burner and by their flow rates. *Premixed* gas flames can arise from fuel gas and air being mixed prior to entering the burner, or if they mix after leaving the burner they are called *diffusion flames.* The gas flow rate may be relatively low, in which case, the incoming gaseous flow of fuel and air is laminar, as is the flame. With high gas flows, they may be turbulent. Thus, flames can be laminar premixed, laminar diffusion, turbulent premixed or turbulent diffusion flames as indicated in **Table 1**. The transition from laminar to turbulent flame takes place as the flow velocity increases as shown in **Figure 1** (Hottel and Hawthorne, 1949). In addition, they can also be categorised into stationary flames or propagating (travelling) flames, the former being the most widely used in domestic or industrial burners, the latter being involved in explosions.

Table 1. Different ways of mixing fuel and oxidant

Mixing arrangement \ Flow	Low gas velocity	High gas velocity
Premixed	Premixed laminar flame	Turbulent premixed flame
Diffusive	Laminar diffusion flame	Turbulent diffusion flame

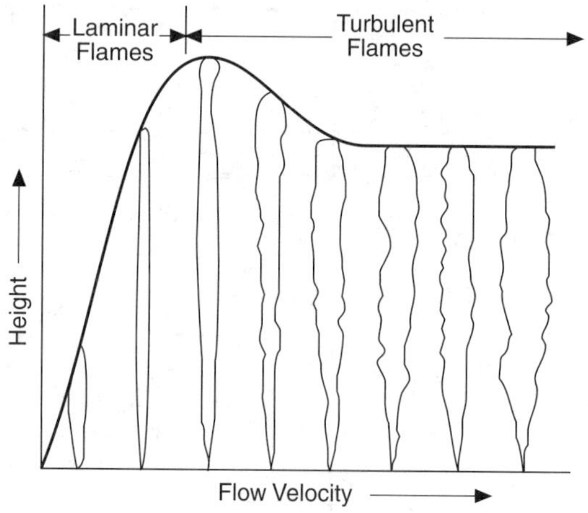

Figure 1. Change in flame type with increase in gas velocity (after Hottel and Hawthorne, 1949).

The type of flame that has been most studied is the laminar premixed flame of a gaseous fuel and oxidant, usually air, because it is the simplest flame and exhibits characteristics common to many other systems. Typical is the *Bunsen burner flame,* a type of flame widely used in gas fires, gas cookers and central heating units.

The Bunsen burner flame is shown in **Figure 2** (Gaydon and Wolfhard, 1970), but for research purposes, a flat flame is used using a special burner which produces uniform flow, as shown in **Figure 3(a)**. The Bunsen burner, however, illustrates both the premixed flame and the principle of the diffusion flame. The inner core is the reaction zone of a premixed flame, but the flame is fuel-rich, so the products of incomplete combustion burn in the outer core as a diffusion flame with the surrounding air. The exact nature of the flame is determined by the *fuel to air (mixture) ratio.* If there is excess fuel, it would be termed rich and the flame would be a yellow luminous one. If there is excess air (or oxygen), it would be termed lean. If it has exactly the correct amount of fuel

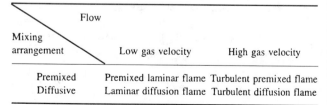

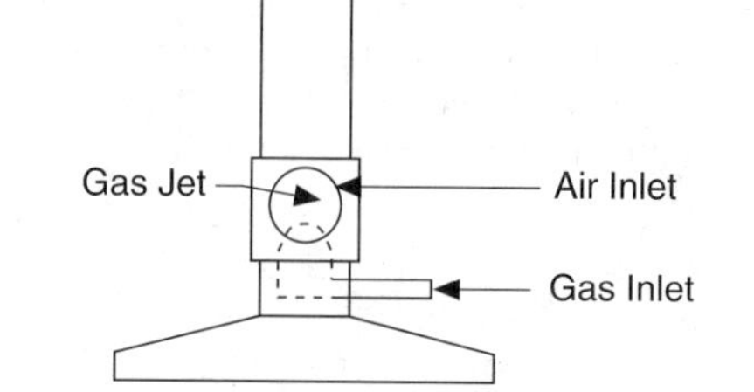

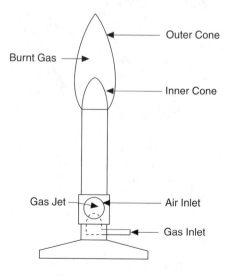

Figure 2. Bunsen burner.

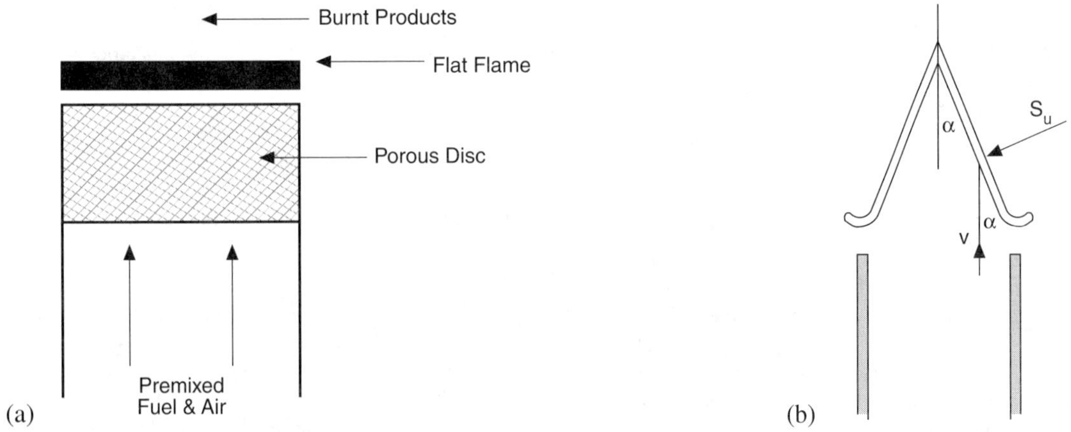

Figure 3. (a) Flat flame burner illustrating definition of burning velocity. (b) Flame cone angle and showing definition of burning velocity $S_u = v \sin \alpha$.

and air, it would be termed stoichiometric. Overall the combustion products would be represented by the stoichiometric equation, which in the case of methane (the major constituent of natural gas) is:

$$CH_4 + 2O_2 = 2CO_2 + 2H_2O - \Delta H_c$$

where $-\Delta H_c$ is the heat released by combustion, known as the heat of *combustion or calorific value* (cv). The term *stoichiometric* refers to the situation where combustion is complete with no unused fuel or no unused oxidant. The region of incomplete combustion in the flame shown in Figure 2 represents only partial combustion of the fuel resulting in carbon monoxide and hydrogen which subsequently burns with the secondary air to give CO_2 and H_2O. This two stage combustion can be represented by the reactions, for example for methane, by

$$CH_4 \rightarrow\rightarrow \{CO, H_2\} \rightarrow CO_2, H_2O$$

In general, all hydrocarbon flames, whether rich or lean, go through such a stage with CO and H_2 being formed in the first main reaction zone, and the second stage combustion of the CO and H_2 initially formed is characterised by the emission of blue light; this blue emission being a key feature of the combustion of all carbon (and hydrocarbon) containing fuels. This process is more pronounced in slightly fuel-rich flames and is termed *afterburning.*

A premixed flame of a particular fuel-air combination is characterised by three main parameters, the *burning velocity, flame temperature* and *flammability limit,* which are also determined by the pressure, temperature and, of course, mixture ratio. Premixed fuel air mixtures have a characteristic burning velocity and this enables flames to be stabilised on a burner as shown in **Figures 2**, **3(a)** and **3(b)** if the flow of gas mixture equals the laminar burning velocity. The burning velocity is simply as defined for a flat laminar flame as in **Figure 3(a)**, that is the approach velocity gives the burning velocity (relative to the unburned gas, S_u) and usually given as m/s. For a conical flame, the laminar flame is as defined in **Figure 3(b)**. In the case of laminar diffusion flames the fuel and the oxidant only meet at the burner mouth (i.e. they are not premixed) and mix by diffusion processes as the flame burns, as illustrated in **Figure 4**. In this case, the fuel gas and oxidant gas streams are slots giving a flat flame, but analogous axisymmetric flames can be obtained by the use of concentric tubes with the fuel usually entering via the inner tube.

Flames will only burn if they are within the flammability limit, that is the composition of the fuel-oxidant mixtures that will sustain a stable flame. There are two types of limits associated with the propagation of a laminar flame. The first is associated with the chemical reactive capability of the mixture to support a flame, i.e. the flammability limit. The second is associated with gas flow influences. Typical values are methane where the lower and upper flammability limits are 5 and 14 mol %, the stoichiometric ratio would be 9.47 mol %. In the case of n-heptane the limits would be 1 and 6 mol % respectively with a stoichiometric ratio of 1.87 mol %.

The combustion of liquid fuels or pulverised coal (or pulverised fuel, both abbreviated to pf) are both widely used in industrial burners, especially for large scale boilers used to generate steam

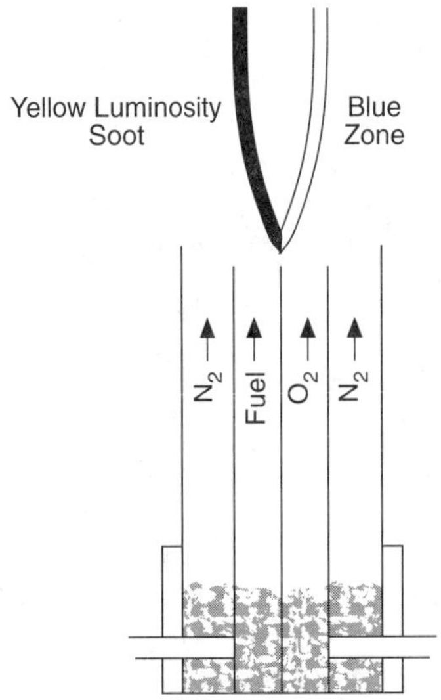

Figure 4. Typical type of laminar diffusion burner and flame.

for power generation. Industrial flames are generally turbulent in nature, and, for convenience and safety reasons, involve diffusion flames where the fuel and air are injected separately for safety reasons and the development of the length of the diffusion flame after the burner exit occurs with increased gas flow velocity as shown in **Figure 1**. With increasing gas velocity, there is a transition from laminar to turbulent diffusion combustion, although some diffusion mixing takes place, much results from turbulent interaction (turbulent diffusion).

Flames of liquid fuels (Williams, 1990) can range from blue, premixed like flames, though to highly luminous flames akin to coal flames. For liquid fuels to burn, they must be completely vaporised to give a vapour which burns in the same way as a gaseous flame. This is termed *"homogenous" spray combustion.* For more involatile fuels, the partially volatilised fuel burns as a spherical flame surrounding each droplet as shown in **Figure 5**, this is called *"heterogeneous" spray combustion.* An example of the first mode of combustion is in the combustion of aviation kerosine in an aircraft gas turbine where the fuel is largely vaporised after injection as a spray into the combustion chamber, some larger droplets however burn heterogeneously and tend to give smoke. The second mode is termed spray combustion where combustion takes place heterogeneously. This takes place in industrial furnaces and boilers and in diesel engines.

In pulverised fuel combustion, the coal particles, typically 100 μm in diameter, involve the following stages.

The coal particle enters the hot combustion chamber and heats up which results in pyrolysis of some of the more reactive components (typically 50%) of the coal, a process termed *devolatilisation.* The devolatilisation products burn in the first part of a pf flame with a yellow gas phase flame. Then the resultant carbonaceous char burns more slowly in a heterogeneous way involving particle surface reactions of the type

$$C_{(solid)} + O_2 = CO_2$$

leaving residual mineral ash if combustion goes to completion. The flame produced is of a luminous, highly radiating cloud of red hot char and mineral ash particles.

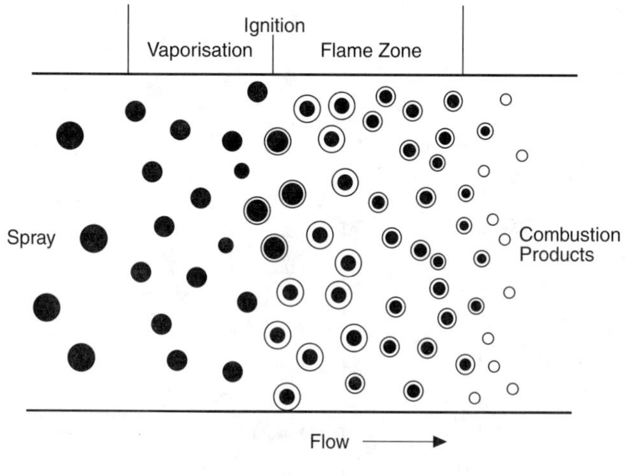

Figure 5. A diagrammatic model of idealised heterogeneous spray combustion.

Whilst the most commonly used flames are those of hydrocarbons (natural gas, oil, coal) with air, numerous other combinations should be noted. Of particular interest are:

1. High temperature flames produced by oxygen-fuel combinations such as *acetylene* (ethyne) with oxygen for welding and cutting torches, and with *natural gas* for the rapid melting of metals. The former produces flame temperatures (T_f) of about 3300K and the latter about 2700K. These high temperatures result because of the absence of the diluting nitrogen present in air and the high energy release from energetic (highly exothermic) fuels.

   In principle, the combustion of *hydrogen with fluorine* produces the hottest flame (T_f 4300K). Powdered metals, e.g. *aluminum,* also produce high flame temperatures which are approximately controlled by the boiling point of the oxide produced which, in the case of aluminium, is 3800K. They can be used for cutting through thick metal sheets such as a safe, or through concrete walls. Metal tubes can also be used and the use of a steel tube with oxygen flowing through it forms the basis of the thermic lance, also used for cutting purposes.

2. *Propellants:* Rockets are propelled by the combustion of solid or liquid propellants, that is, by propellants that can be stored. Propellants of liquid fuels may consist typically of liquid hydrogen (LH_2) plus liquid oxygen (LOx) (T_f = 3100K) used in the space shuttle, or liquid *nitrogen dioxide* (NO_2) with *liquid hydrazine* (T_f = 3000K) used in smaller rockets. A solid propellant may be used, e.g. *ammonium perchlorate* and a rubber polymer binder. The perchlorate sublimes to form a gas ($NH_3 + HClO_4$) and the highly oxidising perchloric acid reacts rapidly with the NH_3 and the binder especially at high pressures. The object in all cases is to produce high temperature flames so that thrust can be maximised, although other factors can come into play.

   These propellants may contain the fuel and the oxidant in one molecule (monopropellants) or they may be mixed before or during combustion (bipropellants).

3. Self-decomposition flames: Certain flames can be burned without the addition of an oxidiser because the fuel decomposes exothermically, although from such systems only relatively low flame temperatures (T_f) are achieved. Two well-known examples are that of *acetylene (ethyne)* and that of *hydrazine.* In the case of the former, the combustion reaction is:

   $$C_2H_2 \rightarrow C_{(s)} + H_{2(g)}$$

   which gives a flame temperature of 1650K.
   For hydrazine, the self decomposition reaction is

   $$H_2H_4 \rightarrow 0.5\ N_2 + 0.5\ H_2 + NH_3$$

   giving a flame temperature of 1800K.

   Since hydrazine is a storable liquid, it, or its methyl derivatives $(CH_3)_2N_2H_2$, is used in rocketry.

4. Unusual flames: Flames may be produced by the combustion of air or oxygen with a variety of unusual fuels such as

ammonia (NH_3, giving products of $N_2 + H_2O$), *carbon disulphide* (CS_2, giving $CO_2 + SO_2$). *Cyanogen* (C_2N_2) and another uncommon fuel *carbon subnitride* give some of the highest flame temperatures, namely 4800K and 5260K respectively, because the products, CO and N_2, do not dissociate significantly.

Whilst many flames burn in unconfined situations, e.g. in a gas fire or in a gas cooker, most industrial flames, and many flames used in domestic environments such as central heating boilers, are burned in a partially confined situation, namely a combustion chamber. In certain exceptional circumstances, usually during malfunction, explosions can occur. Thus, during ignition, if too much fuel has entered the combustion chamber, or if the rate at which fuel enters the combustion chamber is too great, then the rate of energy generated and volume of combustion products generated is too great for them to escape through the flue. In such circumstances, the expanding flame can result in the build up of pressure and an explosion can occur. In general, such confined combustion processes result in a rapid increase in pressure which can damage the combustion chamber and indeed can result in destruction of the equipment. For this reason, in many industrial combustion processes an explosion (or pressure) relief is built into the combustion chamber (or process vessel) to overcome the problems caused by the high build-up of pressure and the subsequent explosion.

In the extreme case, a confined explosion can result because of the combustion of a flammable mixture in a completely closed system. The most common but safe example is that found within the combustion chamber of a reciprocating piston motor vehicle during its operation. Here, spark ignites the mixture, the flame expands and undergoes transition to a confined explosion, although actually in this case it is only partially confined because of the movement of the piston. Examples of truly confined explosions occur in the case where one has an explosion resulting from a leak of natural gas in a room, or of flammable process mixtures such as hydrocarbon vapour in a chemical plant explosion. Details of this are dealt with in another section. (See **Internal Combustion Engines**; **Explosion Phenomena**.)

References

Gaydon, A. G. and Wolfhard, H. G. (1970) *Flames: Their Structure, Radiation and Temperature,* Third Edition, Chapman and Hall Ltd., London.

Glassman, I. (1977) *Combustion,* Academic Press, New York.

Hottel, H. C. and Hawthorne, W. R. (1949) *Third Symposium on Combustion and Flame and Explosion Phenomena,* 254, Williams & Wilkins, Baltimore.

Williams, A. (1990) *Combustion of Liquid Fuel Sprays,* Butterworths, London.

Leading to: *Fires; Flammability; Combustion chambers; Furnaces*

A. Williams

FLAME SPEED, OR VELOCITY (See Explosion phenomena)

FLAME TEMPERATURES (see Flames)

FLAMMABILITY

Introduction

The flammability of a substance is a measure of its ability to burn. It would be convenient if all substances could be categorised as either flammable or non-flammable but, unfortunately, a significant proportion of substances fall into the category of burning under certain conditions. By burning, one generally means self sustained combustion. Thus, the heat generated when a substance is oxidised must be sufficient to overcome any heat losses and heat up fresh fuel to its ignition temperature. The ability of a substance to do this will depend upon its condition, geometry, and environment. For example, flammability is normally considered with reference to burning in fresh air, but many substances which are not normally flammable will burn easily in an environment of pure oxygen.

Flammability of Gases

Gases are the easiest substances to define in terms of flammability. For a gas/air mixture, if it is flammable at all, then there exists a lower limit of fuel concentration below which self-sustained burning does not occur. This is called the *lower flammability* or *lower explosion limit* (LFL or LEL). For hydrogen, for example, the LFL is about 4% by volume. Even this basic limit can be affected by geometry with the LFL for downward propagating flames, for example, being different to that for upward propagating flames.

The flammable gas needs air to burn and, as the fuel concentration increases, the concentration of air is decreased until a point is reached where burning again cannot be sustained. This point is termed the *upper flammability* or *upper explosion limit* (UFL or UEL). It should be noted that these various limits are normally quoted for a gas mixed with pure air. The presence of other additional gases or particulates or aerosols (such as water sprays) can significantly affect the flammability of the fuel.

Flammability of Liquids

It is normally the vapour given off from liquids which burns rather than the liquids themselves. For a fire to be initially ignited, the concentration of the vapour needs to be above the LFL. This is strongly affected by the temperature of the liquid (hence the rate at which it is evaporating), as well as the ambient conditions and geometry. The liquid temperature necessary to produce ignition is termed the *flash point.* Whether a fire, once started, can be sustained will depend upon the rate of feedback of the heat from the flames to the liquid fuel. The liquid temperature required to generate sufficient vapour for this is termed the *fire point.*

Flammability of Solids

Solids burn by the heat from the flames causing volatile gases to be given off from the surface of the material. Whether a fire can be sustained is therefore very dependant upon the geometry and conditions. In general, situations where heat losses are minimised and the ratio of surface area to solid mass is greatest are the most favourable to sustained burning. Various standards and tests exist to demonstrate whether burning will be sustained in various geometries, e.g. propagation up a vertical wall, propagation from a lighted cigarette on a seat.

References

The SFPE Handbook of Fire Protection Engineering, NFPA, (1988).

C.J. Fry

FLAMMABILITY LIMIT (see Flames; Flammability)

FLANGES (see Mechanical design of heat exchangers)

FLASH DISTILLATION (see Desalination)

FLASH EVAPORATORS (see Evaporators)

FLASHING FLOW

The term flashing flow is reserved for the flow with dramatic evaporation of liquid due to a drop of pressure P. The process of production of the vapour phase is usually accompanied by massive thermodynamic and mechanical nonequilibrium by virtue of a difference in temperature and velocity of both phases. The rate of evaporation in a fluid volume during the flashing flow changes according to two factors the number of nuclei and the superheat of liquid. Thermodynamic nonequilibrium, which plays the most important role in the flashing flow reflects the necessity of liquid being superheated above its saturation temperature $T_s(P)$, to encourage the production of vapour bubbles. The vapour phase can be created on nuclei like particles of a gas dissolved in the liquid or solid particles. With the emergence of vapour bubbles on these nuclei, there begins a process of heterogeneous flashing. In the case of pure and gas-free liquid, the vapour phase can be produced at the occurrence of fluctuations of density in the liquid due to temperature fluctuations. That type of flashing is referred to as homogeneous flashing and takes superheating of dozen or several dozen of degrees. The initial difference between temperature of the liquid T_l, and temperature of the vapour T_v, namely $\Delta T = T_l - T_v > 0$, is a driving force of the growth of vapour bubbles and thus for the increase in void fraction.

Several stages of the **Bubble Growth** can be distinguished. In the first early stage, as the vapour bubble is small, the inertia force of the liquid surrounding the bubble has to be taken into account and the Rayleigh-Lamb equation is appropriate. The second stage is called the thermal stage when bubble growth is determined entirely by the heat flux from the liquid to the vapour due to the superheat ΔT

$$\dot{q} = \alpha \Delta T, \qquad (1)$$

where α is the heat transfer coefficient referring to the transfer of heat from the superheated liquid to the interfacial area (e.g. to the vapour bubble in the early stage of the flashing flow). In practical applications, when surface tension is neglected, it can be assumed that the entire heat flux from the superheated liquid to the interfacial area is used for the production of the mass flux of the vapour phase per unit of the volume

$$\dot{m} = \rho \frac{dx}{dt} = \frac{\dot{q}}{h_{fg}}, \qquad \text{kg/m}^2\text{s}, \qquad (2)$$

where x denotes the dryness fraction, ρ density of the two-phase mixture being produced and h_{fg} the heat of evaporation. For the two-phase flow through a channel, the dryness fraction can be averaged across the channel, being a function of the distance z from the origin of the coordinate system and of time t. In the case of flashing flow, the dryness fraction x has a lower value than the dryness fraction $\bar{x}$ referred to the equivalent equilibrium state. The latter is determined by means of two thermodynamic parameters, e.g. pressure P and enthalpy h.

$$\bar{x} = \frac{h - h'(P)}{h_{fg}(P)}, \qquad (3)$$

where h′P) is enthalpy of saturated liquid corresponding to the line of $\bar{x} = 0$.

Flashing flow was investigated in detail on the basis of the well-documented Moby Dick experiments described by Réocreux[1] and Bouré et al.[2] The Moby Dick experiments were carried out in a channel consisting of a straight portion followed by a conical expander provided with a 7° included-angle divergence. Measurements of pressure and void fraction, as functions of longitudinal distance, were reported. A typical pressure profile is shown in **Figure 1** which represents runs at the same upstream conditions and mass-flow rate, but at different back pressures. The diagram also plots the variation of the void fraction, measured by beta-ray absorption.

A distinct feature of flashing flow is that the liquid is in a metastable region, that is, at temperature T_l higher than the saturation temperature $T_s(P)$; $T_l > T_s$. The transition which the liquid undergoes during flashing flow with a pressure drop bears much resemblance to the isothermal transition as shown in **Figure 2**. Heterogeneous flashing in a channel can be initiated at a superheat of 2–3°C like that found in the Moby Dick experiments. Practically, it can be established that in the early stage, when the flashing flow starts becoming bubbly in structure, the temperature of the vapour is equal to its saturation temperature.

A theoretical description of flashing flow based on the one-dimensional relaxation model[3] was proposed by Bilicki et al.[4] The assumption was made that the flow turns into two-phase at a state where the liquid becomes superheated by ΔT. The value of superheat at the flashing point was determined from experiments and ranged from 2.66 to 2.91°C. So far, no theory exists which would make it possible to determine ΔT for heterogeneous nucleation, although such a possibility was considered, among others, in Skripov et al.[5] The above mentioned mathematical model of flashing flow draws on the set of one-dimensional equations of conservation of mass, momentum and energy:

$$\frac{D\rho}{Dt} + \rho \frac{\partial w}{\partial z} = -\frac{1}{\rho w A} \frac{\partial A}{\partial z} \qquad (4)$$

$$\frac{Dw}{Dt} + \frac{1}{\rho} \frac{\partial P}{\partial z} = -\frac{C\tau}{\rho A} \qquad (5)$$

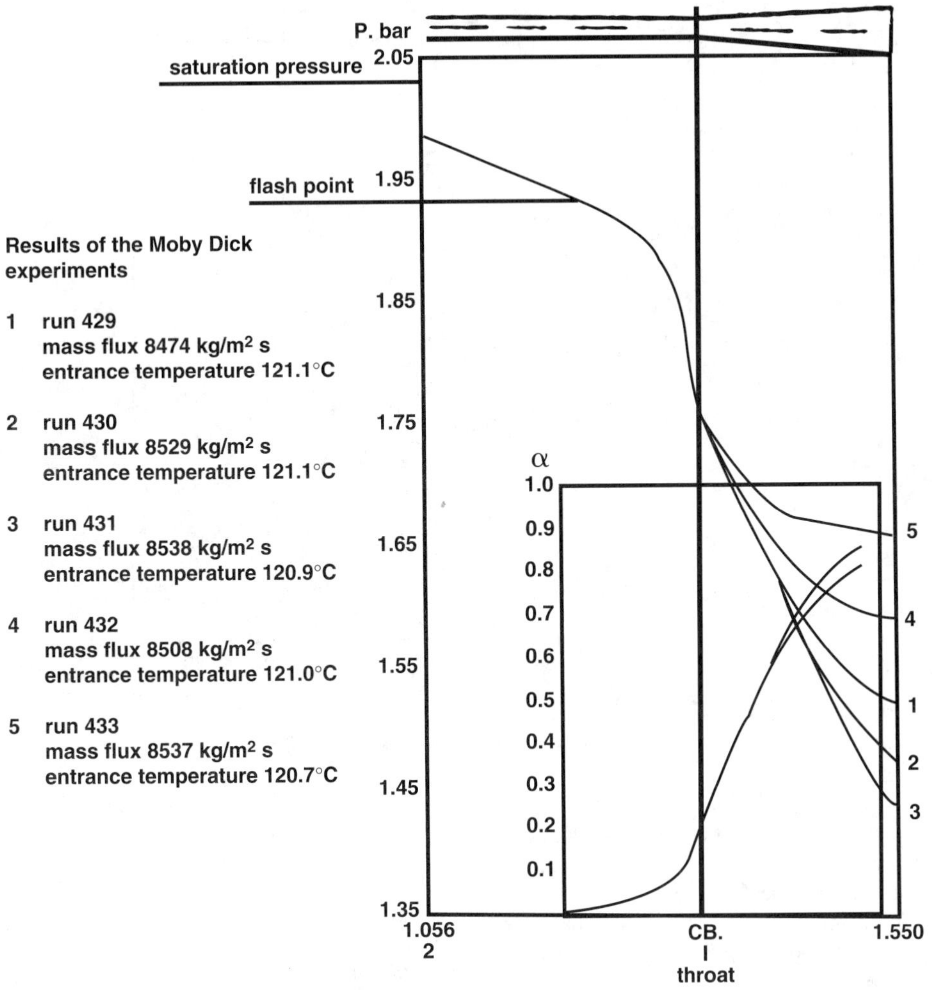

Figure 1. The Moby Dick experiments investigating flashing flow.

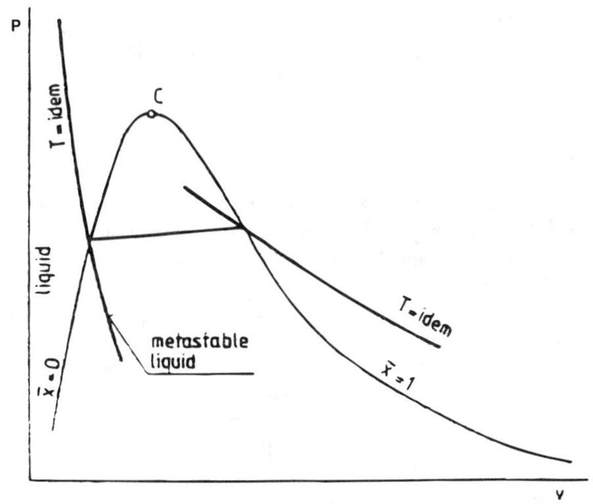

Figure 2. The liquid in the metastable region.

$$\frac{Dh}{Dt} - \frac{1}{\rho}\frac{DP}{Dt} = \frac{C\tau w}{\rho A} \tag{6}$$

$$\text{with} \quad \frac{D}{Dt} \equiv \frac{\partial}{\partial t} + w\frac{\partial}{\partial z} \tag{7}$$

where w denotes the barycentric velocity of the two-phase mixture, τ is the shear stress given by a closure equation, C the circumference of the channel, A its cross-section area. The above set of equations is supplemented with the rate equation which describes nonequilibrium state

$$\frac{Dx}{Dt} = -\frac{x - \bar{x}}{\theta_x} \tag{8}$$

where θ_x is the relaxation time, x denotes the actual dryness fraction as different from its equilibrium value. The state equation must also be added

$$h = h(v, P, x) = xh''(P) + (1 - x)h_l[P, T_l(P, v, x)], \tag{9}$$

where v is the specific volume of the two-phase mixture during the flashing flow, h_l enthalpy of the liquid in the metastable region.

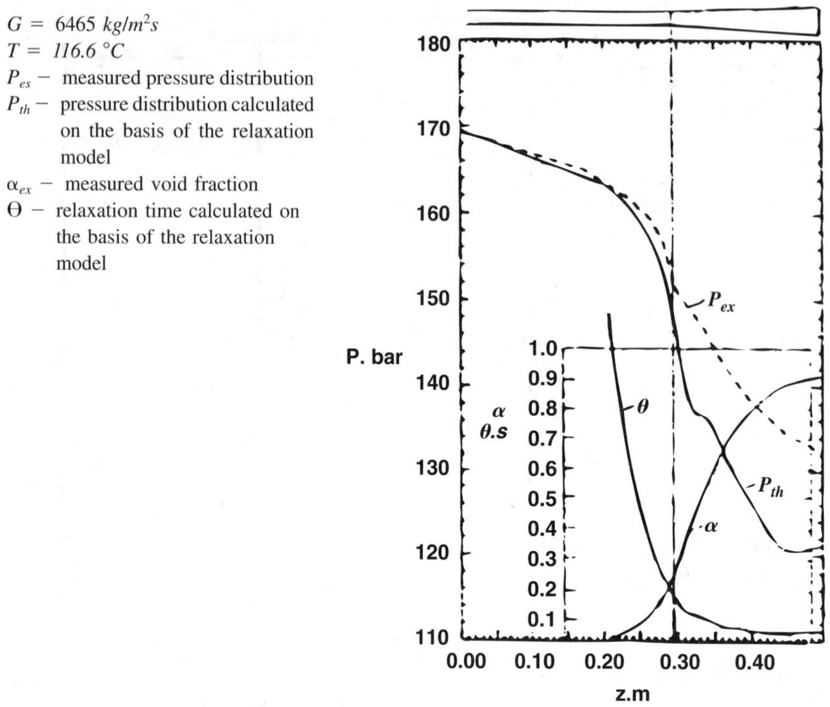

Figure 3. Experimental results for run 401 of the Moby Dick experiments[1] and theoretical predictions[3].

The presented model of flashing flow assumes knowledge of the relaxation time and superheat at the flashing point. Both quantities should be determined from experiments as we do not have at our disposal any method for their theoretical determination. In **Figure 3** there is presented a comparison of pressure profiles obtained theoretically and from the Moby Dick experiments.

A different model of flashing flow was recently proposed by Riznio and Ishii.[6] This model relies on the wall nucleation theory, bubble growth model and drift flux transport model.

In the preceding years flashing flow was linked with critical flow.[6] The papers of[7–13] illustrate the variety of approaches to this problem.

References

1. Réocreux, M. L. (1974) Contribution a l'étude des debits critiques en écoulement diphasique eau-vapeur. Ph.D. thesis, Université Scientifique et Medicale de Grenoble.
2. Bouré, J. A., Fritte, A. A., Giot, M. M. and Réocreux, M. L. (1976) Highlights on two-phase critical flow. *Int. J. Multiphase Flow,* 3, 1–22.
3. Bilicki, Z., Kestin, J., and Pratt, M. M. (1990) A reinterpretation of the Moby Dick experiments in terms of nonequilibrium model. *J. Fluids Engineering, Trans. ASME,* 112, 112–117.
4. Bilicki, Z. and Kestin, J. (1990) Physical aspects of the relaxation model in two-phase flow, *Proc. Royal Soc.,* Lond. A, 428, 379–397.
5. Skripov, V. P., Sinitsyn, E. N., Pavlov, P. A., Ermakov, G. V., Muratov, G. N., Bulanov, N. V. and Baidakov, V. G. (1988) *Thermophysical Properties of Liquids in the Metastable (Superheated) State.* Gordon and Breach Science Publishers New York, London.
6. Riznic, J. R. and Ishii, M. (1989) Bubble number density and vapor generation in flashing flow. *Int. J. Heat Mass Trans.* 32, 1821–1833.
7. Ardron, K. H. and Furness, R. A. (1976) A study of the critical flow models used in reactor blowdown analysis. *Nucl. Eng. Des.* 39, 257–266.
8. Jones, Jr. O. C., and Saha, P. (1977) Non-equilibrium aspects of water reactor safety, *Proc. Symp. on the Thermal and Hydraulic Aspects of Nucl. Reactor Safety, Vol. 1 Liquid Water Reactors* (Edited by O. C. Jones, Jr. & S. G. Bankoff). ASME, New York.
9. Weismann, J. and Tentner, A. (1978) Models for estimation of critical flow in two-phase systems. *Prog. Nucl. Energy,* 2, 183–197.
10. Abdollahian, D. Healzer, J. and Janseen, E. (1980) *Critical Flow Data Review and Analysis. Part I—Literature Survey,* SLI-7908-1, revised 11/80, November.
11. Jones, Jr. O. C. (1979) Flashing inception in flowing liquids, *Non-equilibrium Two-phase Flow* (Edited by J. C. Chen and S. G. Bankoff), 29–34. ASME, New York.
12. Shin, T. S. and Jones, O. C. (1993) Nucleation and flashing in nozzles-1. A distributed model. *Int. J. Multiphase Flow,* 19, 943–964.
13. Blinkov, V. N., Jones, O. C. and Nigmatulin, B. I. (1993) Nucleation and flashing in nozzles-2. Comparison with experiments using a five-equation model for vapor void development. *Int. J. Multiphase Flow,* 19, 965–986.

Z. Bilicki

FLASH SMELTING (see Smelting)

FLAT PLATE, BOUNDARY LAYER ON (see Boundary layer)

FLAT PLATE COLLECTOR (see Solar energy)

FLAT PLATE, COMBINED RADIATION AND CONVECTIVE HEAT TRANSFER TO (see Coupled radiation and convection)

FLASH POINT (see Flammability)

FLOATING HEADER EXCHANGERS (see Shell-and-tube heat exchangers)

FLOCCULATION (see Liquid-solid separation; Sedimentation; Viscosity)

FLOODING AND FLOW REVERSAL

Following from: *Gas-liquid flow*

At low gas velocities, it is possible for *counter current two-phase flow* to occur with the gas going upwards, and the liquid falling as a film on the channel wall. At the other extreme, when the gas velocity is high, the liquid flows upwards in the film, and we have **Annular Flow**. The transition between these two extreme cases is illustrated in **Figure 1**. As the gas flow is increased, the system passes from one of falling film flow (a) through the "flooding" transition at which liquid begins to travel upwards (b), to simultaneous upward and downward flow (c and d), to climbing film flow (e). When the gas flow is now reduced, a point is reached at which liquid begins to creep below the injection point, and this is termed "*flow reversal*". The flooding transition is extremely important in reflux condensation, in falling film mass transfer equipment and in many aspects of reactor safety. Flooding is also important in governing the transition between **Plug Flow** and **Churn Flow** (see **Gas-Liquid Flow**).

The onset of flooding is extremely sensitive to the entrance geometry. This is illustrated by the results shown in **Figure 2**.

Obviously, the liquid and gas must be separated in some way at the bottom of the channel. If this is done smoothly through a porous wall tube section, then the flooding gas velocities are relatively high and depend on tube length, as shown. If, on the other hand, the liquid exits through a sharp-ended tube (as illustrated in **Figure 2**), then the flooding phenomenon is governed by the exit zone and is not dependent on tube length; furthermore, the flooding velocity is significantly lower than that obtained with smooth entrance and exit. These results are further illustrated by more recent data obtained by Govan et al. (1991) and shown in **Figure 3**, which show that, even with a taped (bell-mouthed) exit for the liquid, the data are significantly below those for porous wall sections.

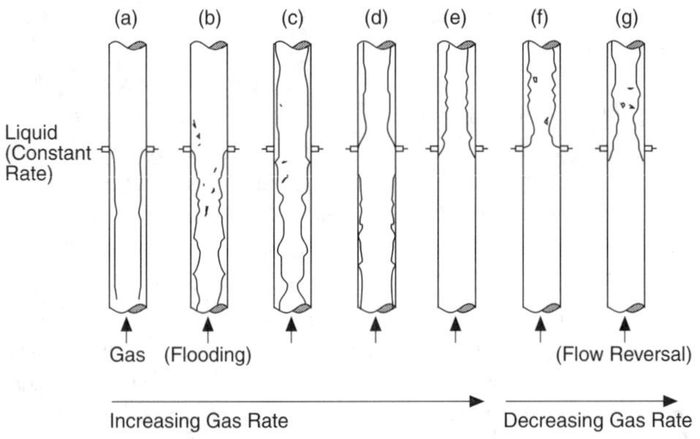

Figure 1. Flooding and flow reversal.

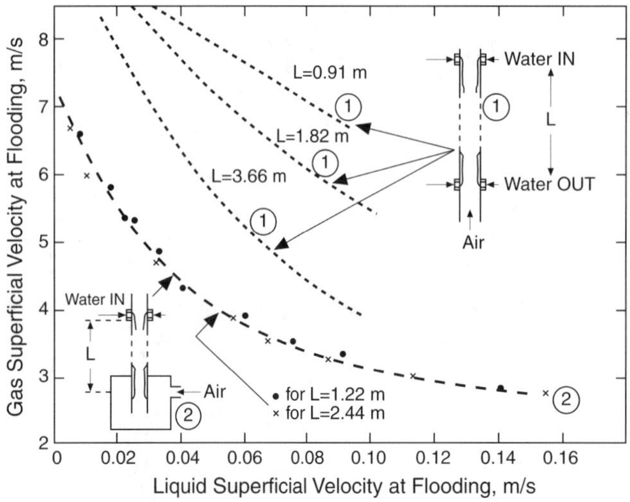

Figure 2. Flooding data for air-water flow in vertical tubes showing effect of entrance conditions (Hewitt, 1982).

In **Figure 3**, the data are plotted in the non-dimensional parameters:

$$U_G^* = U_G \rho_G^{1/2}[gD(\rho_L - \rho_G)]^{-1/2} \quad (1)$$

$$U_L^* = U_L \rho_L^{1/2}[gD(\rho_L - \rho_G)]^{-1/2} \quad (2)$$

where U_G and U_L are the superficial velocities of the gas and liquid phases, ρ_G and ρ_L their densities, g the acceleration due to gravity and D the tube diameter.

The most widely used correlation for flooding is that due to *Wallis* (1961), which is as follows:

$$\sqrt{U_G^*} + m\sqrt{U_L^*} = C \quad (3)$$

For sharp-edged entry (see **Figure 2**), values of m = 1 and C = 0.75 appear to fit the data. For smooth inlet and outlet conditions, where there is an influence of tube length of flooding velocity, Jayanti and Hewitt (1992) fitted the data with C = 1 and m given by:

$$m = 0.1928 + 0.01089\left(\frac{L}{D}\right) - 3.754 \times 10^{-5}\left(\frac{L}{D}\right)^2 \text{ for } \frac{L}{D} \le 120 \quad (4)$$

$$m = 0.96 \quad \text{for} \quad \frac{L}{D} > 120 \quad (5)$$

where L is the distance between the liquid inlet and the liquid outlet.

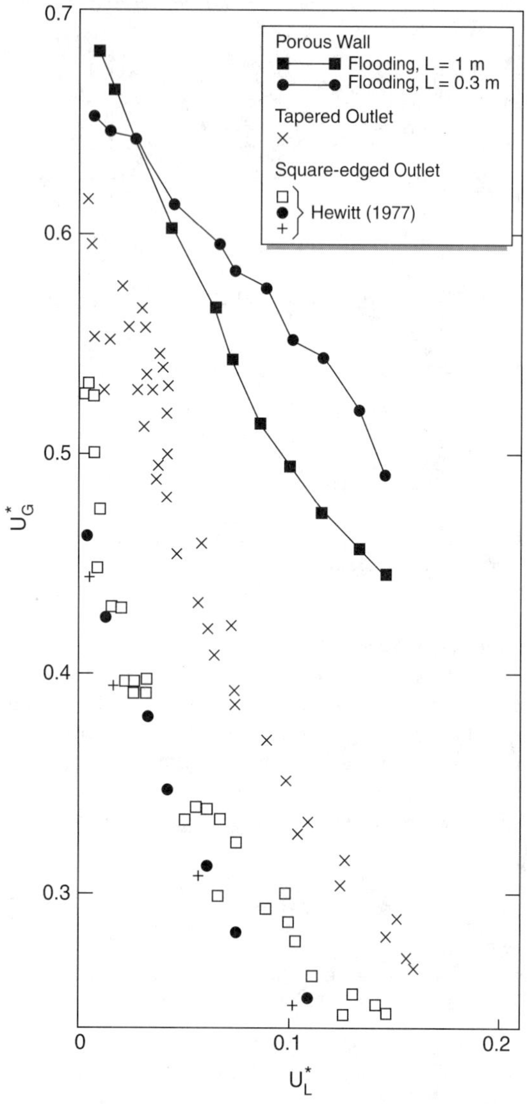

Figure 3. Comparisons of flooding data for sharp-edged, bell-mouthed (taped) and porous wall liquid exit (Govan et al., 1991).

There is a general consensus that flooding is caused by the formation of large waves on the interface. Very often, the precise mechanism is clouded by entrance effects. Thus, if a *vena contracta* is formed at the inlet for the gas phase, this generates a low pressure zone which can lead to the creation of a spurious wave. This is the probable reason why non-smooth inlets and outlets give such different results to those obtained with smooth conditions. However, in the case of a smooth inlet and outlet (see **Figure 2**), then it seems that, for smaller diameter tubes at least (i.e. tubes less than approximately 5 cm in diameter), the flooding mechanism is one in which a coherent large wave is formed at the liquid outlet; this wave is then swept up the tube to beyond the inlet, this process leading to the onset of **Churn Flow** above the injector. For larger diameter tubes, there is some evidence that the large waves formed are not coherent, and that entrainment of the wave tips (giving rise to droplet transport) occurs before the waves themselves are transported, and that this gives rise to upwards transport of liquid in the form of droplets which deposit above the injector position. For larger-diameter tubes, it seems probable that an alternative correlation for flooding velocity may be more appropriate. This is due to *Pushkina and Sorokin* (1969) and is in terms of the Kutateladze number K as follows:

$$K = U_G\rho_G^{1/2}[g\sigma(\rho_L - \rho_G)]^{-1/4} = 3.2 \qquad (6)$$

The *flow reversal* transition has been formed to be relatively insensitive to liquid flow rate, and to occur at an approximately constant gas superficial velocity. For smaller diameter tubes, the flow reversal transition is correlated approximately by the expression:

$$U_G^* = U_G\rho_G^{1/2}[gD(\rho_L - \rho_G)]^{-1/2} = 1 \qquad (7)$$

and for tube diameters greater than about 5 cm, Equation 6 may be used for both flow reversal and flooding.

References

Govan, A. H., Hewitt, G. F., Richter, H. J. and Scott, A. (1991) Flooding and churn flow in vertical pipes. *Int. J. Multiphase Flow* 17, 27–44.

Hewitt, G. F. Flow Regimes. Chapter 2.1 of *Handbook of Multiphase Systems,* G. Hetsroni ed., McGraw-Hill Book Company, New York, ISBN 0-07-028460-1.

Jayanti, S. and Hewitt, G. F. (1992) Prediction of the slug-to-churn flow transition in vertical two-phase flow. *Int. J. Multiphase Flow,* 18, 847–860.

Pushkina, O. L. and Sorokin, Y. L. (1969) Breakdown of liquid film motion in vertical tubes. *Heat Transfer Sov. Res.,* 1, 56–64.

Wallis, G. B. (1961) *Flooding velocities for air and water vertical tubes.* UKAA Report, AEEW-R123.

G.F. Hewitt

FLOTATION

***Following from:** Liquid-solid separation*

Flotation is a process in **Liquid-Solid Separation** technology whereby solids in suspension are recovered by their attachment to gas (usually air) bubbles, usually with the objective of removing the solids from the liquid. The particles most effectively removed are in the size range from 10 to 200 μm. The particle-bubble aggregates that are formed have a density less than the suspension itself; they rise to the surface and are removed. In the minerals industry, *selective froth flotation* is used to concentrate specific species of particles in a finely ground ore. "Flotation" is a generic term for a number of processes known as "*adsorptive bubble techniques*", Lemlich (1972); these are classified into foaming and non-foaming separation methods. Foaming methods require the addition of surfactants to generate a relatively stable foam or froth which then acts as a carrier fluid during particle removal. The properties most important in determining the success of a flotation process are solid hydrophobicity, bubble to particle size ratio, and the extent of turbulence in the fluid.

Most inorganic, and many organic, particles have *hydrophilic surfaces,* and as such they are not floatable. To attach a gas bubble to these solids therefore requires the displacement of a water film from their surfaces. This is achieved using surfactants (known as *collectors*) to render the particle surface *hydrophobic.* Collectors are usually long chain hydrocarbon molecules containing polar groups; these adsorb on to the surface at the charged group, with the hydrocarbon chain being presented to the aqueous phase. (See **Surface Active Substances**.)

Other reagents used in flotation are *frothers* to promote a smaller bubble size and a more stable froth, and *modifiers* or *conditioners* to adjust the pH of the suspension and assist attachment of bubbles to specific particle species in the suspension. Conditioners used include coagulants and polymer flocculants.

All flotation systems require a source of gas bubbles. In *natural flotation,* release of gas in the form of bubbles from, for example, a fermentation is the cause of scum at the surface of the liquid. *Aided flotation* is a natural flotation improved by blowing air bubbles into the suspension, and is useful for the separation of greases dispersed in a turbid liquid. In most process industry flotations, the bubbles are formed by three main techniques: (i) mechanically, using an agitator combined with an air injection system or by pumping the air through a porous plate or nozzle, to form bubbles that measure 0.2 to 2 mm in diameter (*mechanical flotation* or *froth flotation*); (ii) by gas nucleation from a solution to form bubbles of about 40 to 70 μm in diameter, induced by subjecting the suspension to a vacuum (*vacuum flotation*), or by saturating water with the gas under pressure and injecting it into the suspension (*dissolved air flotation*), or by supersaturating the suspension under pressure and then relieving the pressure (*microflotation*); and, (iii) by electrolysis of the aqueous phase (*electroflotation*) to produce bubbles <50 μm with minimal turbulence.

As general purpose liquid-solid separation technologies, the two flotation techniques which have found increasing value in recent years are *dissolved air* (DAF) and *electro-flotation* (EF). EF has the advantage that no pressurized recycle is required; the flow handled by the EF unit comprises the feed only. The recycle employed with DAF increases the flow into the unit. In EF there is no need for recycle pumps, compressors or saturators, but EF does require transformer-rectifier systems which can be relatively expensive. Both DAF and EF may be used either for clarification of liquids or for thickening or concentrating of solids suspensions. Power consumptions and air flows are dependent on many factors, but comparable data (per m^3 of water treated) may be given for water treatment applications: aided flotation (grease removal) requires about 5 to 10 Wh m^{-3} and 100 to 400 Nl m^{-3}, froth flotation 60 to 120 Wh m^{-3} and 10,000 Nl m^{-3}, and DAF (clarification) 40 to 80 Wh m^{-3} and 15 to 50 Nl m^{-3}.

The microscopic mechanisms that occur in flotation are complex, making it impossible to describe in mathematical detail or to develop fundamental design equations. According to Fuerstenau (1976), most flotation systems with a high degree of turbulence approximate to perfect mixers and the recovery of solids is then given by:

$$\text{Recovery} = \frac{c_0 - c}{c_0} = \frac{k\bar{t}}{1 + k\bar{t}}$$

In those dissolved air flotation systems where turbulence is low and in electroflotation cells, hydrodynamic conditions are probably closer to plug flow. In practice, it is difficult to achieve plug flow conditions, and a reasonable approximation to describe recovery by flotation is:

$$\frac{c_0 - c}{c_0} = 1 - e^{-k\bar{t}}$$

In these equations, c is the solids concentration, c_0 is the solids concentration in the feed, and $\bar{t}$ is the mean retention time of the suspension. Comparing the above equations shows that theoretically a plug flow flotation cell is more efficient than a perfectly mixed system (for the same k and $\bar{t}$); to put this another way, a smaller plug flow tank will give the same recovery as a larger perfectly mixed system (Jowett and Sutherland (1979)).

Stevenson (1986) has outlined practical methods for the determination of design and scale-up data for flotation systems. (See also **Foam Fractionation**.)

References

Fuerstenau, M. C., (Ed.) (1976) *Flotation, A. M. Gaudin Memorial Volume,* American Institute of Mining Engineers, New York.

Jowett, A. and Sutherland, D. (1979) A simulation study of the effect of cell size on flotation costs, *The Chemical Engineer,* 603–607.

Lemlich, R. (1972) *Adsorptive Bubble Separation Techniques,* Academic Press, London.

Stevenson, D. G. (1986) Chapter 5, *Solid/Liquid Separation Equipment Scale-up,* (D. B. Purchas and R. J. Wakeman, eds.), Uplands Press and Filtration Specialists, London.

R.J. Wakeman

FLOW ACROSS CYLINDERS, TUBES (see Tube, crossflow over)

FLOW EXCURSION INSTABILITIES (see Instability, two-phase)

FLOW METERING

Following from: Flow of fluids

Flow metering is the measurement of the quantity of a substance flowing through a given flow section in unit time. Flow metering is an unseparable part of any technological process flow chart; it is required to achieve maximum efficiency of automatization of production. It can also be widely used in scientific investigations.

The existing flowmeters can be classed into:

- Devices based on hydrodynamic methods: variable pressure differential, variable level, streamline flow, vortex methods and others;
- Devices with continuously moving bodies: tachometric, power, etc.;
- Devices based on various physical phenomena: thermal, electromagnetic, acoustic, optical. etc.

The devices of the first group have gained wide acceptance. Among them are the differential pressure flowmeters (see **Differential Pressure Flowmeters**), whose operation depends on the differential pressure across a device which changes the flow section of the pipe-line (**Figure 1**). The flowmeter includes: a primary transformer which creates a pressure differential; a differential pressure gauge which measures this pressure differential and connecting pipes. The most common type of primary transformers are restrictions (see **Orifice meter**): other types include a standard diaphragm, a standard nozzle, a pipe, Venturi nozzle, double diaphragms. Half-circle and quarter circle nozzles are used in measuring small flow rates.

Variable-level flowmeters are devices based on the dependence between the flow rate and the height of the level in the vessel into which a liquid is constantly supplied and, at the same time, flows out from the hole in the bottom or in a side wall. A vessel with an orifice which has an area S serves as a primary transformer. The quantity measured is the height of the level of a liquid z in the vessel, and the flow rate is defined from the relationship $\dot{V} = \alpha\, s\sqrt{2gz}$ and does not depend on the density of the liquid (**Figure 2**), where α is the flow rate coefficient.

For measuring the flow rate of the liquid in open channels (trays) *slit flowmeters* are widely used which are miniature weirs in the wall of a vessel into which a liquid is constantly supplied. The flow rate is defined by the height of the level of a liquid above the lower edge of the slit. The characteristic of such a flowmeter depends on the shape of the profile of the slit cross section: for a rectangular slit b_1 wide and b_2 high, the flow rate is $\dot{V} = 0.667 b_1 b_2 \sqrt{2gz}$; profiles $b_1(b_2)$ have been developed on applying which a linear relationship $\dot{V}(z)$ is realized.

In streamline flow flowmeters, the primary transformer takes up the dynamic pressure of the flow and moves under its action by the value dependent on the flow rate. The most widely used are the constant pressure defferential flowmeters in which a streamlined body moves vertically and a counterforce is produced by the weight of the body. These include **Rotameters (Figure 3)**, and also floating and piston-type (slide valve) flowmeters. Rotameters are made in the form of a vertical conical (conicity, 0.001–0.01), upwardly divergent glass pipe, on which the graduations are made; a float, on whose upper rim inclined ribs are made, moves inside the pipe. The float comes upward and rotates under the action of the flow,

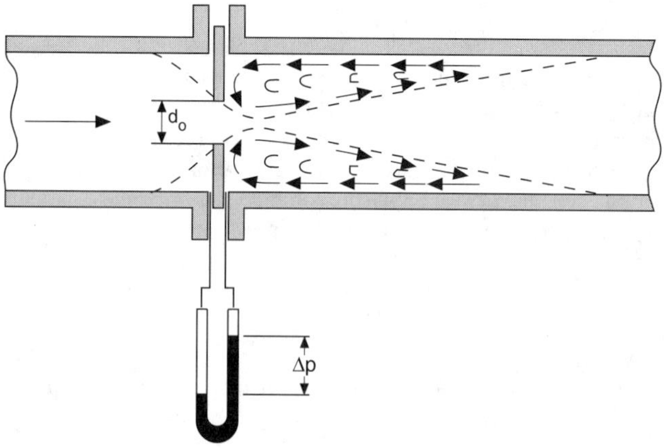

Figure 1. Differential pressure (orifice) flowmeter.

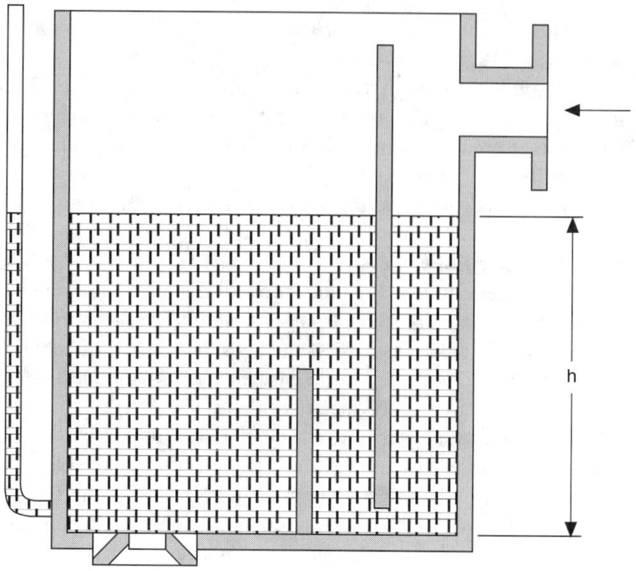

Figure 2. Variable level flowmeter.

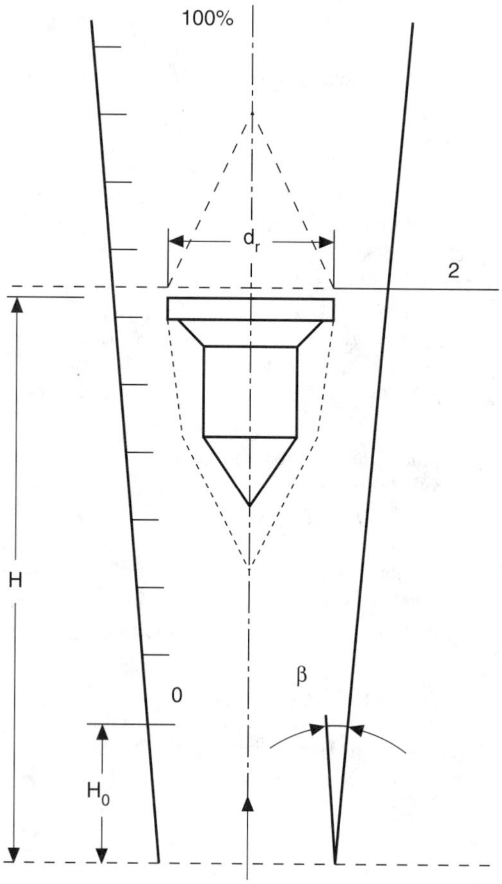

Figure 3. Principle of rotameter.

the rotation ensures the centring of the float in the middle of the flow. The volume flow rate of the liquid is defined by the height z of the float lift. It depends on the float characteristics (its volume V_f, midsection area $S_f = (\pi/4)d_f^2$ and on the density of the float material ρ_f), the density ρ of the liquid measured and is proportional to the area of the circular gap $S_c(z)$ between the walls of the pipe and the float, $\dot{V} = \alpha S_c\sqrt{2gV_f(\rho_f - \rho)/S_f\rho}$. For small taper angles of the pipe, the relationship $S_c(z)$ is practically linear, in particular conditions of measurement $\dot{V} = Az$, where A is defined by a preliminary calibration. The float-type flowmeters operate similarly. In piston-type flowmeters under the action of dynamic head the piston moves in a bush with specially shaped windows, through which the liquid flows out with a flow rate $\dot{V}$.

In *vortex flowmeters,* the frequency of variations of pressure or velocity which occur in the cross flow over a body (cylinder, prism, plate) (**Figure 4**) are determined and depend on flow rate. The frequency f is related to the mean streamlining velocity u and the size of the body d by the Strouhal number $Sr = du^{-1}f$. For a flow section area $S = (\pi/4)D^2$ the flow rate is defined by the relation $\dot{V} = SdSr^{-1}f$. The proportionality between $\dot{V}$ and f is ensured for $Sr = const$, which is realized when the cylinder is streamlined over the range $10^4 \leq Ro \leq 2 \times 10^5$ ($Ro = \omega d/u$ is the Rossby number). This ensures a wide range of the measured flow rate $\dot{V}_{max}/\dot{V}_{min} = 20$, but $\dot{V}_{min}$ is limited by the conditions of steady vortex formation (for instance, for water $u > 0.2$ m/s). Primary transformers with $d/D = 0.15 - 0.2$ are usually used (D is the diameter of the pipe-line). The pressure pulsations are transformed into an electric signal with the help of piezo ceramic pressure pickups. The error of flow rate measurement is estimated at 0.5–1.5%.

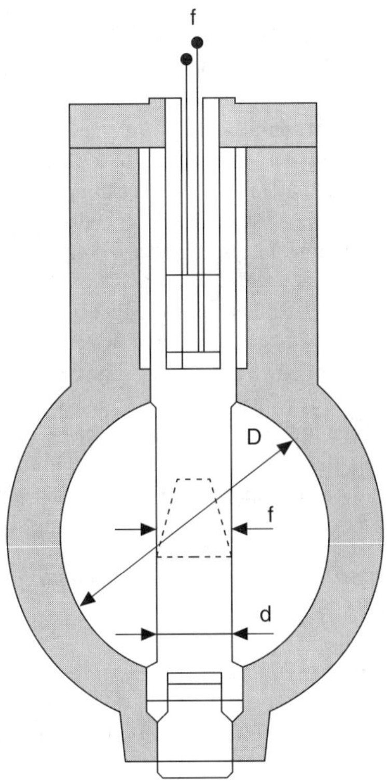

Figure 4. Vortex flowmeter.

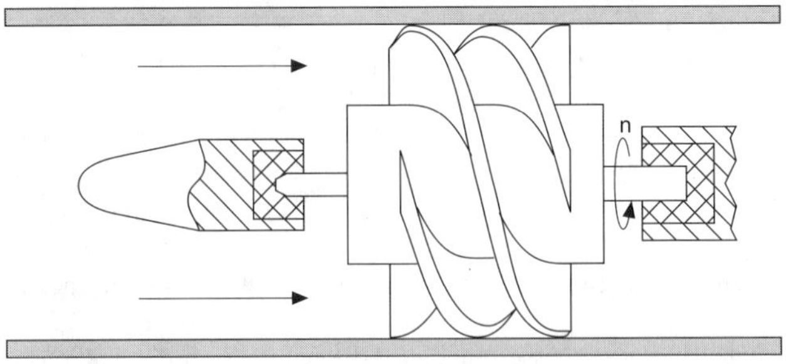

Figure 5. Turbine flowmeter with rotation in line of flow.

The *tachometric flowmeters* have a rotating element whose velocity of measurement is proportional to the volume flow rate.

Flowmeters in the form of a small turbine have found a wide utility, the speed of the turbine rotation being determined by the number of the electric pulses in unit time measured by a frequency meter (see **Anemometers, Vane**). Turbine flowmeters are designed either in the form of an axial small turbine (**Figure 5**) which has propeller blades with a variable helix angle, or in the form of a tangential small turbine (**Figure 6**) with flat radially arranged blades. In *ball-type flowmeters* a ball moving around the circuit due to the flow swirling with the help of a propeller guide is used as a moving element.

In *power flowmeters,* the value of a parameter is measured, which characterizes action of a force on the flow, the action being proportional to the mass flow rate. The flow is accelerated by the force. Depending on the character of acceleration, the flowmeters are classed as turbo-power (**Figure 7**) in which the flow is swirled either due to an external action (a rotor with an electric drive) or with the help of a fixed auger, Coriolis flowmeters in which Coriolis acceleration occurs due to the force action, and gyroscopic flowmeters in which the gyroscopic moment is measured. In the flowmeter

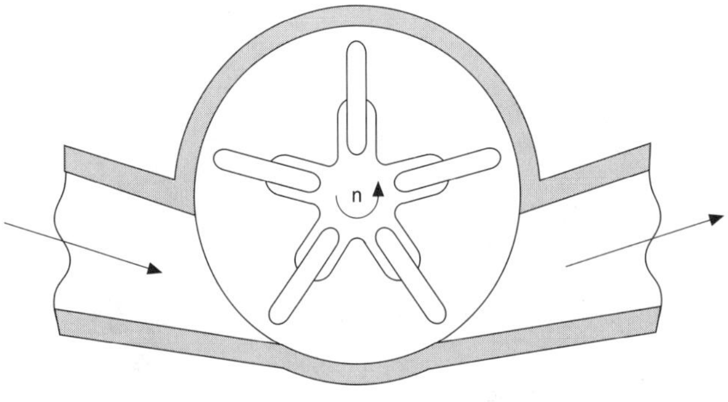

Figure 6. Turbine flowmeter with rotation normal to line of flow.

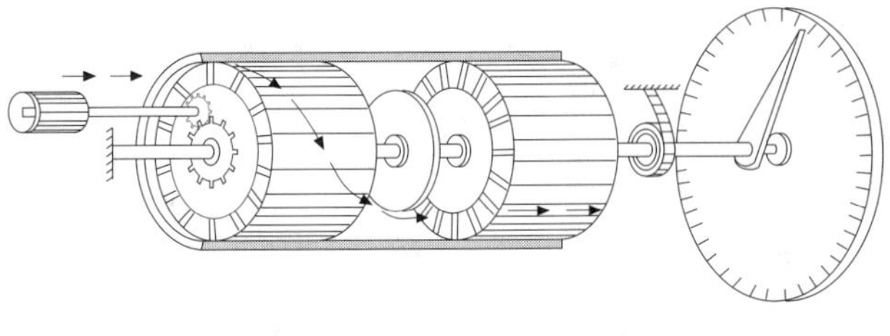

Figure 7. Turbo-power flowmeter.

shown in Figure 7, the torque on the rotating shaft is measured and is proportional to mass flow rate.

Thermal flowmeters (**Figure 8**) depend on the flow rate of the quantity of heat received by the flow of a liquid from a heater. A heater (usually an electric one) is placed into the flow on the pipe-line section and its power W is measured, and the difference in the temperatures of the flow $\Delta T = T_{out} - T_{in}$ at the inlet and at the outlet. Then the mass flow rate $\dot{M} = KW/c_p\Delta T$ is proportional to the power of heating W with ΔT kept constant. The coefficient K depends on heat losses into the surroundings, the nonuniform distribution of velocity along the pipe-line cross section, etc., a preliminary calibration is therefore made. When the flowmeter is manufactured and calibrated carefully, it can give an accuracy of measurement of the flow rate of $\pm(0.3\text{–}0.5)\%$ and can be used as a standard for testing and calibrating other flowmeters. The thermoanemometric method (see **Hot-wire and Hot-film Anemometers**) measures local velocity by determining the temperature attained by a hot wire or hot film fed with a constant current. This local velocity can then be related to the mean velocity using known relationships.

Electromagnetic Flowmeters (**Figure 9**) are designed to measure the flow rate of a liquid with conductance, as a rule, not less than 10^{-3} Ohm/m. Their operation relies on the interaction between the moving current-conducting liquid and the transverse magnetic field. In this case, an electromagnetic force E induced in a liquid, is proportional to the magnetic induction of the transverse field B, to the volume flow rate V of the liquid and to the distance D between the electrodes (located along the normal both to the velocity vector of the liquid and to the vector of the magnetic field indensity). D is typically equal to the inner diameter of the pipe-line. The current generated (E) is given by $E = 4B\dot{V}/\pi D$. The electromagnetic flowmeter has the advantages of independence of viscosity and density of the substance, the absence of pressure loss, scale linearity, high speed response the possibility of measiring in agressive, abraisive and highly viscous liquids. However, it cannot be applied for measuring flow rates of gas, vapours and dielectrics.

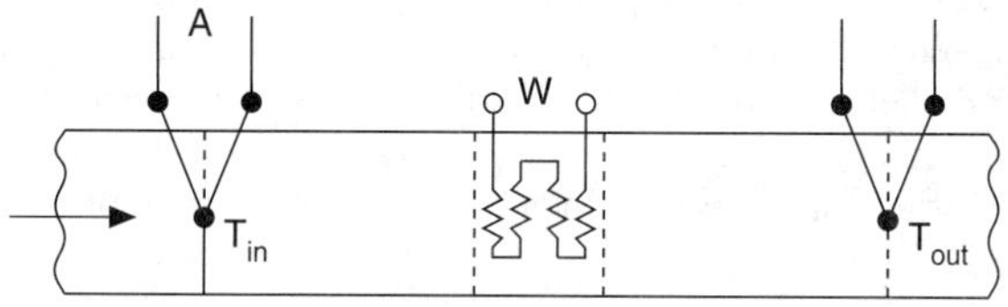

Figure 8. Thermal flowmeter.

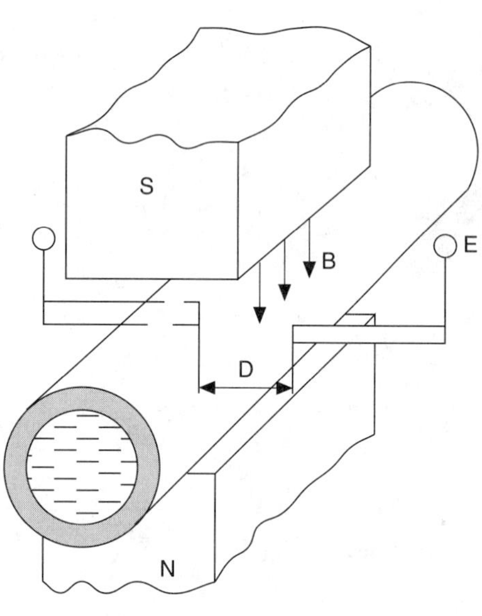

Figure 9. Electromagnetic flowmeter.

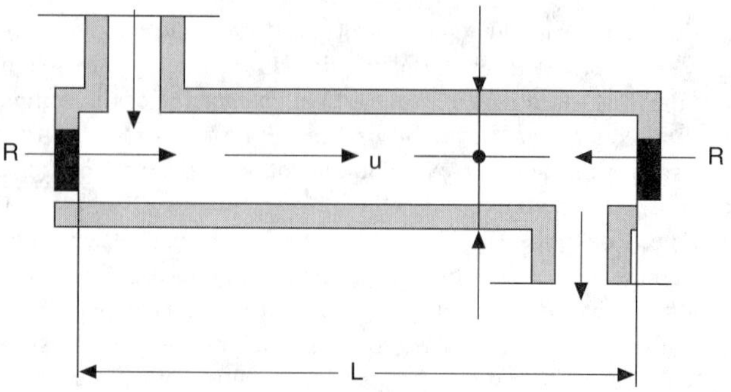

Figure 10. Acoustic flowmeter.

Acoustic (ultrasound) *flowmeters* (**Figure 10**) are based on measuring the difference in the time Δt for the passage of acoustic waves over a distance L in the direct of flow and counter to the direction of flow respectively. The sound is emitted and received by the radiators detector devices R. The time Δt is proportional to L and to the mean mass velocity u and inversely proportional to the square of the velocity a of propagation of acoustic oscillations in the measured medium $\Delta t = 2Lu/a^2$. Accounting for the fact that high-frequency oscillations (0.1 . . . 10 MHz) are applied for measuring very small time intervals, various methods have been developed. Most extensively employed is the doppler shift method in which the difference of pulse frequency rate along and against the flow is measured. The volume flow rate can be defined from the relationship $\dot{V} = (\pi/8)\,\Delta t a^2 D^2/L$. The flowmeter consists of a pipe section in whose end faces disc piezoceramic (for instance, from barium titanium $BaTiO_3$) elements are installed which are switched on alternately to radiation or to reception. Ultrasound flowmeters are particularly useful for measuring the flow rate of non-conducting liquids, of petroleum products and agressive media. The accuracy of measurement is estimated at 0.5–1%.

Optical flowmeters use the dependence on the flow rate of the substance of one or the other optical effect in the flow. The most generally employed are laser doppler flowmeters (see **Anemometers, Laser Doppler**) which rely on measuring the difference of frequencies Δf of incident and reflected light, which is due to the interaction between the light beam and the particles moving in the liquid. In this case the local velocity of the flow in some point of the cross section of the pipe is measured; if the relationship between the local and mean velocity is known, the flow rate may be determined.

Leading to: *Differential pressure flowmeters; Electromagnetic flowmeters; Mass-flowmeters; Rotameters*

References

Fluid Meters: Their Theory and Application (1971) New York, ASME.

Yu.L. Schehter

FLOW OF FLUIDS

When a fluid is viewed on a molecular, that is on a small, scale, properties of the fluid have an extremely non-uniform spatial distribution. *Fluid mechanics,* also called *fluid dynamics* or *hydrodynamics,* is normally concerned with the behaviour of the fluid on a large scale. As a result, the molecular structure of the fluid need not be taken into account explicitly and the *continuum hypothesis* can be invoked so that the fluid can be assumed to have a continuous structure. That is, there are volumes called *material points* in which properties of the fluid are constant. The analysis of fluid flows is greatly facilitated by distinguishing between *kinematics* and *dynamics.* A kinematical description of a motion does not consider the forces which cause the motion. A dynamical description, on the other hand, does consider them.

Consider a flow relative to a given reference frame. Let **x** denote the position of a material point X at time t, and $\mathbf{x} + \Delta\mathbf{x}$ the position of the same material point X at time $t + \Delta t$. The velocity **u** of X is defined by:

$$\mathbf{u}(\mathbf{x}, t),\ \mathbf{u}^{\dagger}(X, t) = \lim|_{\Delta t \to 0}\left[\frac{1}{\Delta t}\Delta\mathbf{x}\right] \tag{1}$$

Note that this defines two quantities: $\mathbf{u}(\mathbf{x}, t)$ is the *Eulerian* specification of velocity and gives the spatial distribution of velocity as a function of time; $\mathbf{u}^{\dagger}(X, t)$ is the *Lagrangian* specification of velocity and gives the velocity of material points of a fluid as a function of time. The Lagrangian specification is useful in certain contexts, for example in flows of elasticoviscous **Non-Newtonian Fluids** in which the stress $\boldsymbol{\tau}$ in a material point X at time t depends on the flow in the neighbourhood of X at all times $t^* \leq t$. In general, however, use of a Lagrangian specification is cumbersome and an Eulerian specification is used instead.

Although a motion may be steady in an Eulerian sense, so that the velocity at each position is independent of time, a material point X of a body may accelerate. The acceleration **a** of X is defined by:

$$\mathbf{a} = \lim|_{\Delta t \to 0}\left[\frac{1}{\Delta t}(\mathbf{u}(\mathbf{x} + \Delta\mathbf{x}, t + \Delta t) - \mathbf{u}(\mathbf{x}, t))\right] \tag{2}$$

Thus it follows that:

$$\mathbf{a} = \frac{\partial\mathbf{u}}{\partial t} + \mathbf{u}\cdot\nabla\mathbf{u} = \frac{D\mathbf{u}}{Dt} \tag{3}$$

where D/Dt denotes the *substantial derivative.*

It can be shown that:

$$\mathbf{u}(\mathbf{x} + \Delta\mathbf{x}) = \mathbf{u}(\mathbf{x}) + \frac{1}{2}\Delta\mathbf{x}\cdot\mathbf{e}(\mathbf{x}) - \frac{1}{2}\Delta\mathbf{x} \wedge \boldsymbol{\omega}(\mathbf{x}) \tag{4}$$

where **e** denotes the *strain-rate* given by:

$$\mathbf{e} = \nabla\mathbf{u} + (\nabla\mathbf{u})^T \tag{5}$$

and $\boldsymbol{\omega}$ the *vorticity* given by:

$$\omega = \nabla \wedge \mathbf{u} \tag{6}$$

Thus, the velocity field $\mathbf{u}(\mathbf{x} + \Delta\mathbf{x})$ can be decomposed into three components: a *translation* $\mathbf{u}(\mathbf{x})$, a *straining* $(1/2)\ \Delta\mathbf{x} \cdot \mathbf{e}(\mathbf{x})$ and a *rotation* $-\ (1/2)\ \Delta\mathbf{x} \cdot \boldsymbol{\omega}(\mathbf{x})$.

An *irrotational flow* is one for which $\boldsymbol{\omega} = \Delta \wedge \mathbf{u} = 0$. It can be shown that there exists a *scalar potential* ϕ for an irrotational flow such that:

$$\mathbf{u} = \boldsymbol{\nabla}\phi \tag{7}$$

Subject to some usually minor restrictions, an **Inviscid Flow** that is flow of fluid of zero viscosity, which starts as an irrotational flow always remains an irrotational flow. For this reason, an inviscid flow is often analysed using a scalar potential and called a *potential flow.* In contrast, a *solenoidal flow* is one for which $\boldsymbol{\nabla} \cdot \mathbf{u} = 0$. It can then be shown that there exists a *vector potential* $\boldsymbol{\Psi}$ for a solenoidal flow such that:

$$\mathbf{u} = \boldsymbol{\nabla} \wedge \boldsymbol{\Psi} \tag{8}$$

Conservation of mass for an incompressible fluid can be shown to mean that the flow is solenoidal. For this reason, an incompressible flow is often analysed using a vector potential. Furthermore, if the flow is two-dimensional, all but one component of $\boldsymbol{\Psi}$ vanish and **u** can be expressed in terms of a **Streamfunction** ψ. Thus, if **u** is two-dimensional and planar, that is in an appropriate rectangular or Cartesian coordinate system it only has an x-component and a y-component but no z-component, $\boldsymbol{\Psi} = \psi\mathbf{i}_z$ and hence:

$$\mathbf{u} = \frac{\partial\psi}{\partial y}\mathbf{i}_x - \frac{\partial\psi}{\partial x}\mathbf{i}_y \tag{9}$$

where $\mathbf{i}_x$, $\mathbf{i}_y$ and $\mathbf{i}_z$ denote unit vectors in the x-direction, y-direction and z-direction, respectively. The streamfunction ψ is constant along a **Streamline,** which is a line instantaneously at every point aligned with the local flow direction. As a result, streamlines are close together in those parts of a flow which are moving fast and far apart in those which are moving slowly.

There are usually two kinds of forces acting on a fluid:

- long-range forces, such as gravitational forces, which generally vary only slowly if at all over the fluid;
- short-range forces, such as viscous forces, which have a molecular origin and are, as a result, generally negligible unless there is physical contact between parts of the fluid. Long-range forces are called body forces or volume forces. Short-range forces can be approximated by forces on the surface of each part of the fluid: such surface forces lead to the concept of *stress* $\boldsymbol{\tau}$ in a fluid, and in particular to components called **Shear Stress** and **Normal Stress.**

It is assumed that mass, linear momentum, angular momentum and energy are conserved. The *conservation equations* resulting from these **Conservation Laws** are:

$$\frac{D\rho}{Dt} + \rho\boldsymbol{\nabla}\cdot\mathbf{u} = 0 \tag{10}$$

$$\rho\frac{D\mathbf{u}}{Dt} = \rho\mathbf{g} - \boldsymbol{\nabla}p + \boldsymbol{\nabla}\cdot\boldsymbol{\tau} \tag{11}$$

$$\boldsymbol{\tau} = \boldsymbol{\tau}^T \tag{12}$$

$$\rho \frac{DU}{Dt} = -\nabla \cdot \dot{\mathbf{q}} - p\nabla \cdot \mathbf{u} + \boldsymbol{\tau}:\nabla \mathbf{u} \tag{13}$$

where ρ denotes density, **u** velocity, **g** gravitational acceleration, p pressure, **τ** stress, U specific internal energy, $\dot{\mathbf{q}}$ heat flux and t time. The mass conservation equation (10) is often referred to as the *continuity equation.* The linear momentum equation (11) is often referred to as the *equation of motion.* The angular momentum conservation equation (12) is in a form appropriate to a *non-polar fluid,* that is a fluid for which the angular momentum is just the moment of the linear momentum. The energy conservation equation (13) is appropriate to a system in which there is no heat generation by chemical or nuclear reaction and is an expression of the *first law of thermodynamics.*

Note that, for incompressible fluids, that is fluids of constant density ρ, p is not the thermodynamic pressure, which is in fact undefined for such fluids. Furthermore, the gravitational acceleration **g** can be eliminated from (11) by defining a modified pressure P:

$$P = p - \rho \mathbf{g} \cdot \mathbf{x} \tag{14}$$

where **x** denotes spatial position. Then (11) becomes:

$$\rho \frac{D\mathbf{u}}{Dt} = -\nabla P + \nabla \cdot \boldsymbol{\tau} \tag{15}$$

In certain problems such as those involving free surfaces, however, pressure appears explicitly in the boundary conditions and gravity cannot, therefore, be eliminated. If the specific internal energy U is independent of the kinematics of the flow, then classical thermodynamic arguments can be applied. It can then be shown that (13) becomes:

$$\frac{DS}{Dt} + \frac{1}{\rho T}\nabla \cdot \dot{\mathbf{q}} - \frac{1}{\rho T}\boldsymbol{\tau}:\nabla \mathbf{u} = 0 \tag{16}$$

where S denotes specific entropy and T absolute temperature. The term $(1/\rho T)\ \boldsymbol{\tau}\cdot\nabla \mathbf{u}$ is called the *irreversible dissipation.* It can be shown using the *second law of thermodynamics:*

$$\frac{DS}{Dt} \geq 0 \tag{17}$$

that, for an incompressible fluid:

$$\boldsymbol{\tau}:\nabla \mathbf{u} \geq 0 \tag{18}$$

that is the irreversible dissipation is non-negative.

The conservation equations (10), (11), (12) and (13) govern the flow of any fluid. They describe how the density ρ, velocity **u** and specific internal energy U vary for a flowing fluid. They do this, however, by reference to the otherwise unknown pressure p, heat flux $\dot{\mathbf{q}}$ and stress **τ** (though conservation of angular momentum does interrelate some components of **τ**). Equations are clearly needed for p, $\dot{\mathbf{q}}$ and **τ** if well-posed problems are to be formulated. These equations must be obtained by phenomenological and molecular arguments corroborated by experiments. The equations will, in general, differ for different materials. Indeed, it is in these equations that the nature of a given material is manifested. The equations are called *constitutive equations* or *equations of state* because they describe the constitution or state of a material. Perhaps the most commonly used constitutive equations are those for:

- an *incompressible fluid,* that is one for which density ρ is constant so that (10) becomes:

$$\nabla \cdot \mathbf{u} = 0 \tag{19}$$

- a *Newtonian fluid* for which stress **τ** is related to strain-rate **e** given by (5) thus:

$$\boldsymbol{\tau} = \eta \mathbf{e} \tag{20}$$

where η denotes *viscosity:* then (11) becomes:

$$\rho \frac{D\mathbf{u}}{Dt} = \rho \mathbf{g} - \nabla p + \eta \nabla^2 \mathbf{u} \tag{21}$$

which are called the *Navier-Stokes equations*;

- a *Fourier fluid* for which heat flux $\dot{\mathbf{q}}$ is related to temperature T thus:

$$\dot{\mathbf{q}} = -\lambda \nabla T \tag{22}$$

where λ denotes *thermal conductivity:* then for a Newtonian fluid (13) becomes:

$$\rho c \frac{DT}{Dt} = \lambda \nabla^2 T - p\nabla \cdot \mathbf{u} + \frac{1}{2}\eta \mathbf{e}:\mathbf{e} \tag{23}$$

where c denotes specific heat.

Other constitutive equations exist: for an inviscid fluid, $\eta = 0$ and so $\boldsymbol{\tau} = 0$; for a **Non-Newtonian Fluid** **τ** is given by a rather more complex equation than (20).

The combined conservation-constitutive equations (10) or (19), (21) and (23) determine the behaviour of a Newtonian Fourier fluid within a flow domain. In order to determine the flow fully, however, additional conditions must be specified. *Initial conditions* are required on ρ, **u** and T in order that the equations can be solved. Specification of initial conditions usually poses few problems in practice. Indeed, no initial conditions are needed if the flow is steady in an Eulerian sense, that is if $\partial\rho/\partial t = 0$, $\partial\mathbf{u}/\partial t = \mathbf{0}$ and $\partial T/\partial t = 0$. Because the conservation-constitutive equations apply within but not at the edge of the flow domain, *boundary conditions* are also needed to determine ρ, **u** and T completely. At a rigid, impermeable wall, the normal component of the fluid velocity **u** at the wall equals the normal component of the wall velocity $\mathbf{u}_w$:

$$\mathbf{u} \cdot \mathbf{n} = \mathbf{u}_w \cdot \mathbf{n} \tag{24}$$

where **n** denotes the unit normal to the fluid at the wall. If the wall is at rest in the reference frame being used, then **u·n** vanishes at the wall. Boundary condition (24) is referred to as the *no flow-through condition.* If the fluid is inviscid, then the fluid can slip over the wall and the tangential components of **u** cannot be specified. If, on the other hand, the fluid is viscous, for example it is Newtonian, then molecular arguments corroborated by experiments show that the tangential components of **u** at the wall equal the tangential components of $\mathbf{u}_w$, so that there is no slip at the wall:

$$\mathbf{u} \wedge \mathbf{n} = \mathbf{u}_w \wedge \mathbf{n} \qquad (25)$$

Combining (24) and (25) yields:

$$\mathbf{u} = \mathbf{u}_w \qquad (26)$$

Boundary condition (26) is referred to as the *no slip condition.* In distinction to (24), which is entirely kinematic in origin, (25) is dynamic in origin so that (26) is partly kinematic and partly dynamic in origin. Arguments for the validity of the no slip condition are molecular and empirical in nature. Molecular arguments are based on notions of thermodynamic equilibrium at a wall and hence equality of the velocities of the molecules of the fluid and those of the wall. Empirical arguments are based on the extensive experimental corroboration of the consequences of assuming that there is no slip at a wall.

Boundary conditions on temperature are based on two notions. The first is that the heat flux must be continuous normal to the wall, in order that energy is conserved. Thus, for a Fourier fluid and a Fourier wall:

$$\lambda \nabla T \cdot \mathbf{n} = \lambda_w \nabla T \cdot \mathbf{n} \qquad (27)$$

where λ_w denotes the thermal conductivity of the wall. The second notion is that of thermodynamic equilibrium, which implies that the fluid temperature T at the wall is the same as the wall temperature T_w:

$$T = T_w \qquad (28)$$

It appears to be impossible to develop general methods for obtaining exact solutions of equations (10) or (19), (21) and (23) subject to appropriate boundary and initial conditions. There is, therefore, a need for a systematic method for solving them approximately, either analytically or numerically using computers. A description of the methods involved in the latter, called **Computational Fluid Dynamics**, is beyond the scope of this article (instead see Roache (1972) and Crochet, Davies and Walters (1984)). A rational basis exists for the development of analytical methods by the use of order-of-magnitude arguments based on dimensionless formulations. Most flows admit a characteristic length L_c, a characteristic speed U_c and a characteristic temperature T_c. Thus, for example, for flow in a long narrow pipe, L_c might be the pipe radius, U_c the mean fluid velocity along the pipe and T_c the temperature of the pipe wall. Define dimensionless variables (distinguished by a superscript asterisk) as follows:

$$\mathbf{u} = U_c\mathbf{u}^*, \quad p = \rho U_c^2 p^*, \quad \mathbf{g} = g\mathbf{g}^*, \quad t = L_c U_c^{-1} t^*,$$

$$T = T_c T^*, \quad \nabla = L_c^{-1}\nabla^*, \quad \nabla^2 = L_c^{-2}\nabla^{2*}, \quad \nabla^2 = L_c^{-2}\nabla^{2*} \qquad (29)$$

where $g = |\mathbf{g}|$. Then for an incompressible fluid (21) and (23) become respectively:

$$\frac{D\mathbf{u}^*}{Dt^*} = \frac{1}{Fr}\mathbf{g}^* - \nabla^* p^* + \frac{1}{Re}\nabla^{2*}\mathbf{u}^* \qquad (30)$$

$$Pe\frac{DT^*}{Dt^*} = \nabla^{2*}T^* + Br\, \mathbf{e}^*{:}\mathbf{e}^* \qquad (31)$$

where:

- the **Froude Number**. **Fr**, is given by:

$$Fr = U_c^2/gL_c \qquad (32)$$

  which may be interpreted as the ratio of inertial forces to gravitational forces;
- the **Reynolds Number**, **Re**, is given by:

$$Re = \rho U_c L_c/\eta \qquad (33)$$

  which may be interpreted as the ratio of inertial forces to viscous forces
- the **Peclet Number**, **Pe**, is given by;

$$Pe = \rho c U_c L_c/\lambda \qquad (34)$$

  which may be interpreted as the ratio of convection to conduction;
- the *Brinkman number, Br,* is given by:

$$Br = \eta U_c^2/2\lambda T_c \qquad (35)$$

  which may be interpreted as the ratio of heat generation by viscous dissipation to conduction (note that the factor of two in (35) is entirely arbitrary).

There are two main reasons for making (21) and (23) dimensionless. The first is as follows. Suppose that two flow problems have the same relative geometry, that is they are *geometrically similar* (an example is flow in two pipes of different size but of the same length-to-radius ratio), the same dimensionless boundary and initial conditions and the same values of the dimensionless groups *Fr, Re, Pe* and *Br.* Then the dimensionless solutions of the two problems are identical and the flows are said to be *dynamically similar.* The second reason concerns making rational simplifications to the flow equations based on order-of-magnitude arguments, in order to obtain approximate solutions. To see this, consider simplifications to the linear momentum conservation equation, that is the dimensionless Navier-Stokes equations (30). What generally matters is the relative importance of the inertial term $D\mathbf{u}^*/Dt^*$ and the viscous term $1/Re\, \nabla^{2*}\mathbf{u}^*$, in which the magnitude of the Reynolds number Re plays a central role.

If Re is very small, that is if $Re << 1$, it follows from (30) that the inertial term should be negligible compared with the viscous term and hence that the Navier-Stokes equations should become approximately:

$$\mathbf{0} = \rho\mathbf{g} - \nabla p + \eta\nabla^2\mathbf{u} \tag{36}$$

These are *Stokes' equations.* If for a given flow Re is low enough that inertial effects can be neglected, that flow is called a *slow flow,* a *creeping flow* or a *Stokes flow.* (An interesting feature of solutions of Stokes' equations is that, if an arbitrary cylindrical cross-section is projected normally from one flat wall to another parallel flat wall, the streamlines of the flow about the cylinder are identical to those for inviscid potential flow about the same cylinder: this is called *Hele-Shaw flow.*) For steady flow at speed U round a stationary solid sphere of radius R in an unbounded incompressible Newtonian fluid of density ρ and viscosity η, it can be shown that the magnitude F of the force exerted on the sphere is given by:

$$F = 6\pi\eta UR \tag{37}$$

provided the flow is slow, that is provided:

$$Re = 2\rho UR/\eta << 1 \tag{38}$$

The expression for F in (37) is called **Stokes' Law** experimentally, it is found to be valid for Re < 0.1 or so. In dimensionless form, Stokes' law is:

$$C_d = 24/Re \tag{39}$$

where the *drag coefficient,* C_d, which is a dimensionless drag force, is given by:

$$C_d = F/\left(\frac{1}{2}\rho U^2\pi R^2\right) \tag{40}$$

It can be shown that the analogous steady flow past a long stationary solid (right circular) cylinder in an unbounded incompressible Newtonian fluid does not exist. This is referred to as *Stokes' paradox.* Of course, it is not a real paradox, since it is clearly possible for fluid to flow past a cylinder at arbitrarily small Re: it merely means that the problem has been over-constrained by making at least one assumption too many.

If Re is very large, that is if Re >> 1, it follows from (30) that the viscous term should be negligible compared with the inertial term and hence that the Navier-Stokes equations should become approximately:

$$\rho\frac{D\mathbf{u}}{Dt} = \rho\mathbf{g} - \nabla p \tag{41}$$

These are *Euler's equations* which govern the flow of an inviscid fluid. For a steady incompressible inviscid flow, it can be shown from (41) that:

$$\frac{1}{2}|\mathbf{u}|^2 - \mathbf{g}\cdot\mathbf{x} + \frac{1}{\rho}p = C \tag{42}$$

where **x** denotes position and C a constant. This is **Bernoulli's Equation**: it may be interpreted physically as stating that, for steady flow of an incompressible inviscid fluid, the sum of the kinetic energy per unit mass ($1/2\ |\mathbf{u}|^2$), the potential energy per unit mass ($-\mathbf{g}\cdot\mathbf{x}$) and the pressure energy per unit mass ($1/\rho$ p) is constant. Equivalent, if more complicated, forms of Bernoulli's equation apply for an unsteady or compressible flow. For steady inviscid flow round a stationary solid sphere in an unbounded incompressible Newtonian fluid, it can be shown that the magnitude F of the force exerted on the sphere is given by:

$$F = 0 \tag{43}$$

Thus, an incompressible inviscid fluid moving at constant velocity round a stationary sphere exerts no drag force on the sphere. This is referred to as *d'Alembert's paradox.* Experimental evidence shows that, for flow of a real fluid, the viscosity of which never vanishes, (43) is false, no matter how large Re may be. The reason for this is that Re >> 1 implies that flows are effectively inviscid and hence approximately irrotational almost, but not quite, everywhere. Viscous effects, which lead to drag forces, are always important near and downstream of walls in **Boundary Layers** and *wakes,* respectively.

If Re is neither very high nor very low, that is if Re ~ 1, viscous and inertial forces should both be comparable and hence both be significant. Provided Re is not very large, there will be **Laminar Flow**: if Re is very large, there will be **Turbulent Flow**. Two main simplifications are possible if the flow is laminar. One is when there is *fully-developed flow.* The other is when there is *boundary layer flow.*

It happens that, in some common laminar flows, the presence of walls means that the acceleration term $\rho\ D\mathbf{u}/Dt$ in (21) is negligible even though Re is not small: such a flow is said to be *locally fully-developed,* or *fully-developed* if $\rho\ D\mathbf{u}/Dt = 0$. A locally full-developed flow is governed by Stokes' equations (36). Laminar **Channel Flow** is fully-developed except near the inlet and outlet, provided the channel is sufficiently long. **Poiseuille Flow** is pressure-induced channel flow in a long duct, usually a pipe. The volumetric flow rate $\dot{V}$ of a fluid of viscosity η through a pipe of radius R and length L is related to the *pressure drop* Δp by:

$$\dot{V} = \Delta p\pi R^4/8\eta L \tag{44}$$

This is the *Hagen-Poiseuille equation.* Pressure-driven flow is to be distinguished from drag-induced flow such as **Couette Flow** between two parallel flat plates a constant distance apart: one plate remains stationary, the other moves in its own plane. Provided the plates are long compared with the distance between them, the flow is fully-developed almost everywhere. If an incompressible, Newtonian liquid of density ρ and viscosity η falls in a layer of thickness h on a flat vertical plate of infinite extent, the liquid flows down the plate as a result of the gravitational acceleration. Beside the liquid is a gas of negligible density and viscosity and uniform pressure. Analysis of such **Falling Film Flow** shows that the volumetric flow rate $\dot{V}$ per unit width is given by:

$$\dot{V} = \rho g h^3/3\eta \tag{45}$$

A high Re flow is effectively inviscid almost everywhere. Near a wall, however, in what is called a boundary layer, viscous effects are always important. The reason for this is that there is no slip of a viscous fluid at a rigid impermeable wall, however small its viscosity may be. Analysis of such high Re flows is based on the assumption that there are two distinct regions in the flow:

- a region away from the walls in which viscous effects are negligible;
- a thin region immediately adjacent to the walls in which viscous effects are not negligible.

In the region away from the walls, the Navier-Stokes equations simplify to Euler's equations. In the boundary layer region adjacent to the walls, inertial effects and viscous effects are both significant: because it is thin, however, the transverse velocity is much smaller than the axial velocity, axial derivatives of velocity components are much smaller than transverse derivatives of those same components and the transverse pressure gradient is much smaller than the axial pressure gradient. For flow at velocity U past a semi-infinite flat plate, it can be shown that, if the axial velocity $u_x = \partial\psi/\partial y = Uf_\beta$, where:

$$f = \psi/\sqrt{2\eta x U/\rho} \quad \textbf{(46)}$$

$$\beta = y\sqrt{U\rho/2\eta x} \quad \textbf{(47)}$$

(see **Figure 1**) then f satisfies *Blasius' equation*:

$$f_{\beta\beta\beta} + ff_{\beta\beta} = 0; \qquad f(0) = 0, \qquad f_\beta(0) = 0; \qquad f_\beta(\infty) \to 1 \quad \textbf{(48)}$$

where a subscript denotes differentiation. The solution, obtained numerically, is shown in **Figure 2**.

Any laminar flow becomes unstable if Re or some other equivalent dimensionless number becomes sufficiently large. The initial instability is usually a **Secondary Flow** superimposed on the basic laminar flow. When Re is sufficiently large, the flow becomes turbulent. A **Turbulent Flow** is always a high Re flow. Two main classes of turbulent flow must be distinguished:

- a wall flow, for example flow in a pipe;

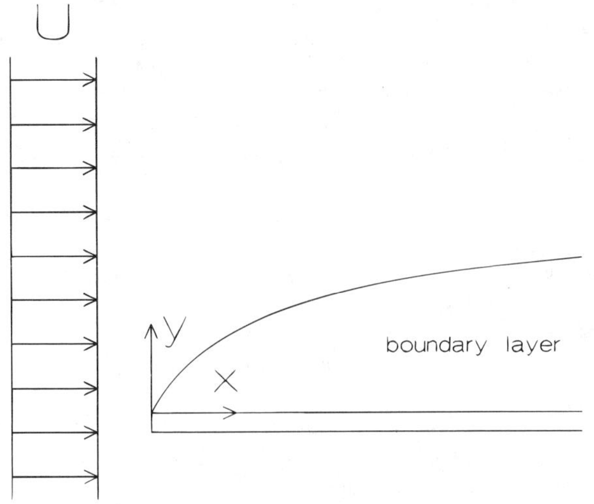

Figure 1. Flow past flat plate.

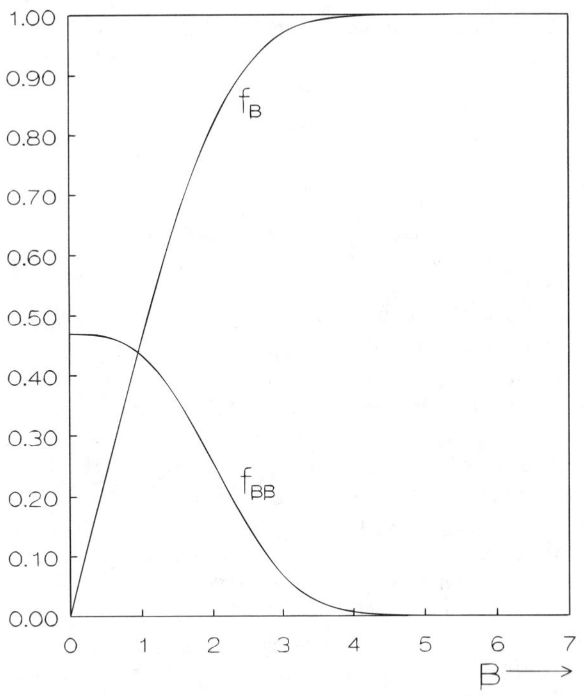

Figure 2. Solution of Blasius' equation.

- a free or **External Flow**, for example flow of a jet.

The details of the transition from laminar to turbulent flow are not well understood. In a turbulent flow, the flow variables such as velocity **u** and pressure p fluctuate in space and time in an apparently random manner. The fluctuations, which are often called eddies, are, as far as is known, completely predictable from the conservation equations, given appropriate boundary and initial conditions. Furthermore, it appears that many turbulent flows are dominated by large-scale motions which are not random. Because of the apparent randomness of turbulent flows, statistical flow descriptions are appropriate and average value of a flow quantity such as **u** or p are used. A volume average cannot be used, however, because a turbulent flow comprises fluctuations over a range of length-scales. Similarly, a time average cannot be used because a turbulent flow can be unsteady on average. Instead, an *ensemble average* is used in which an average is taken over many repetitions of an apparently identical flow, that is one with as nearly the same boundary and initial conditions as possible. For a stationary turbulent flow, that is a turbulent flow which is steady on average, the ergodic hypothesis is that, under certain fairly general conditions, the ensemble average of each flow variable is the same as its time average. For such a flow, a time average can be used instead of an ensemble average: indeed such time averages are what are commonly used in experimental measurements. Decomposition of each flow variable into an ensemble average component (denoted by an overbar) and a fluctuating component (denoted by a prime) permits definition of the *relative turbulent intensity* I_r given by:

$$I_r = [\overline{|\mathbf{u}'|^2}]^{1/2}/|\bar{\mathbf{u}}| \quad \textbf{(49)}$$

In a typical turbulent flow, $I_r \sim 0.1$.

The mass and linear momentum conservation equations (19) and (21) for an incompressible Newtonian fluid can be averaged to give:

$$\nabla \cdot \bar{\mathbf{u}} = 0 \tag{50}$$

$$\rho \frac{D\bar{\mathbf{u}}}{Dt} = -\nabla \bar{p} + \eta \nabla^2 \bar{\mathbf{u}} + \nabla \cdot (-\rho \overline{\mathbf{u}'\mathbf{u}'}) \tag{51}$$

which are *Reynolds' equations.* Note that (50) for $\bar{\mathbf{u}}$ is the same as (19) for **u** but (51) for $\bar{\mathbf{u}}$ and $\bar{p}$ differs from (21) for **u** and p: there is an extra term involving the so-called *Reynolds' stress*—$\rho\overline{\mathbf{u}'\mathbf{u}'}$ in (51). Its presence in (51) illustrates the closure problem in turbulence; there are more unknowns than equations—a closure problem always arises when a non-linear system is averaged since averaging loses information. In order to eliminate the closure problem in turbulence, a turbulence model is needed for the Reynolds stress. Such a model cannot, however, be obtained from the flow equations: it must be obtained using other arguments.

There are many general texts on fluid mechanics. Four excellent ones are Batchelor (1967), Denn (1980), Landau and Lifshitz (1959) and Streeter (1961); another is Richardson (1989). A good understanding of fluid mechanics is greatly enhanced by observing real flows, that is by *flow visualisation.* An outstanding collection of photographs is to be found in Van Dyke (1982). Many more are to be found in Tritton (1977). Good introductions to the mathematics underlying fluid mechanics are given in Leigh (1968) and Truesdell (1977).

It is as yet impossible to prove whether or not the flow equations for an incompressible Newtonian fluid have a solution and, if so, whether the solution is unique for arbitrary initial and boundary conditions. The standard text on this subject is Ladyzhenskaya (1969). A more recent one is Temam (1977); a rather more readable account is in Shinbrot (1973). The existence and uniqueness of a solution of the flow equations do not guarantee that the flow corresponding to the solution can actually occur in practice: the flow must also be stable. The stability of flows is discussed in Chandraskehar (1961) and Drazin and Reid (1981).

A comprehensive treatment of slow flow is to be found in Happel and Brenner (1973). An excellent text, with particularly good coverage of the paradoxes which arise in slow and inviscid flow is Van Dyke (1964). The standard text on inviscid flow is Lamb (1932). A very good, more recent, one is Lighthill (1986). A masterly discussion of wave motions in fluids is to be found in Lighthill (1978). A very readable account of locally fully-developed flow and boundary layer flow is given in White (1974). The standard text on boundary layer flow is Schlichting (1968); an extremely comprehensive one is Rosenhead (1963). An excellent introductory text on turbulence is Tennekes and Lumley (1972), while more advanced ones are Bradshaw (1976) and Hinze (1959). Turbulence models are discussed in Launder and Spalding (1972).

A good and comprehensive text on incompressible flows is Panton (1984). A standard and excellent text on *compressible flow* is Courant and Friedrichs (1948). A very useful introduction to one-dimensional compressible flow, for example in ducts, is given by Daneshyar (1976).

A good text on the mechanics of isolated bubbles, drops and particles is Clift, Grace and Weber (1978). A comprehensive text on **Multiphase Flow** is Hetsroni (1982), while a standard account of multiphase flows and many related topics is Levich (1962).

References

Batchelor, G. K. (1967) *An Introduction to Fluid Dynamics,* Cambridge University Press, Cambridge.

Bradshaw, P. (editor) (1976) *Turbulence,* Springer, Berlin.

Chandrasekhar, S. (1961) *Hydrodynamic and Hydromagnetic Stability,* Clarendon Press, Oxford.

Clift, R., Grace, J. R., and Weber, M. E. (1978) *Bubbles, Drops and Particles,* Academic Press, New York.

Courant, R. and Friedrichs, K. O. (1948) *Supersonic Flow and Shock Waves,* Wiley Interscience, New York.

Crochet, M. J., Davies, A. R. and Walters, K. (1984) *Numerical Simulation of Non-Newtonian Flow,* Elsevier, Amsterdam.

Daneshyar, H. (1976) *One-dimensional Compressible Flow,* Pergamon, Oxford.

Denn, M. M. (1980) *Process Fluid Mechanics,* Prentice-Hall, Englewood Cliffs.

Drazin, P. G. and Reid, W. H. (1981) *Hydrodynamic Stability,* Cambridge University Press, Cambridge.

Happel, J. and Brenner, H. (1973) *Low Reynolds Number Hydrodynamics,* Noordhoff, Leyden.

Hetsroni, G. (editor) (1982) *Handbook of Multiphase Systems,* Hemisphere, New York.

Hinze, O. (1959) *Turbulence,* McGraw-Hill, New York.

Ladyzhenskaya, O. A. (1969) *Mathematical Theory of Viscous Incompressible Flow,* Gordon and Breach, New York.

Lamb, H. (1932) *Hydrodynamics,* Cambridge University Press, Cambridge.

Landau, L. D. and Lifshitz, E. M. (1959) *Fluid Mechanics,* Pergamon, Oxford.

Launder, B. E. and Spalding, D. B. (1972) *Mathematical Models of Turbulence,* Academic Press, London.

Leigh, D. C. (1968) *Nonlinear Continuum Mechanics,* McGraw-Hill, New York.

Levich, V. I. (1962) *Physicochemical Hydrodynamic,* Prentice-Hall, Englewood Cliffs.

Lighthill, M. J. (1978) *Waves in Fluids,* Cambridge University Press, Cambridge.

Lighthill, M. J. (1986) *An Informal Introduction to Theoretical Fluid Mechanics,* Clarendon Press, Oxford.

Panton, R. L. (1984) *Incompressible Flow,* Wiley, New York.

Richardson, S. M. (1989) *Fluid Mechanics,* Hemisphere, New York.

Roache, P. J. (1972) *Computational Fluid Dynamics,* Hermosa, Albuquerque.

Rosenhead, L. (editor) (1963) *Laminar Boundary Layers,* Clarendon Press, Oxford.

Schlichting, H. (1968) *Boundary-Layer Theory,* McGraw-Hill, New York.

Shinbrot, M. (1973) *Lectures on Fluid Mechanics,* Gordon and Breach, New York.

Streeter, V. (editor) (1961) *Handbook of Fluid Dynamics,* McGraw-Hill, New York.

Temam, R. (1977) *Navier-Stokes Equations,* North-Holland, Amsterdam.

Tennekes, H. and Lumley, J. L. (1972) *A First Course in Turbulence,* M. I. T. Press, Cambridge.

Tritton, D. J. (1977) *Physical Fluid Dynamics,* Van Nostrand Reinhold, New York.

Truesdell, C. A. (1977) *A First Course in Rational Continuum Mechanics*, Academic Press, New York.

Van Dyke, M. (1964) *Perturbation Methods in Fluid Mechanics*, Academic Press, New York.

Van Dyke, M. (1982) *An Album of Fluid Motion*, Parabolic Press, Stanford.

White, F. M. (1974) *Viscous Fluid Flow*, McGraw-Hill, New York.

Leading to: *Aerodynamics; Boundary layer; Channel flow; External flow, overview; Flow metering; Free molecule flow; Inviscid flow; Porous media; Turbulent flow; Waves in fluids*

S.M. Richardson

FLOW PATTERNS (see Liquid-liquid flow; Three phase, gas-liquid-liquid, flows)

FLOW REGIME INDUCED INSTABILITIES, TWO-PHASE SYSTEMS (see Instability, two-phase)

FLOW REGIME RELAXATION INSTABILITIES (see Instability, two-phase)

FLOW REGIMES IN BUBBLE FLOW (see Bubble flow)

FLOW REVERSAL (see Churn flow)

FLOW SEPARATION (see Tube, crossflow over)

FLOW SPLITTING (see T-junctions)

FLOW VISUALISATION (see Photoluminescence; Tracer methods)

FLUID DYNAMICS (see Flow of fluids)

FLUID FILLED THERMOMETERS (see Temperature measurement, practice)

FLUIDICS

Fluidics represents a family of process equipment which may be used to intensify a range of heat and mass transfer processes. The common theme connecting all fluidic items is the absence of packings and of moving parts within the equipment. The technology was developed initially to meet the needs of the UK nuclear industry to pump radioactive liquors using equipment of assured reliability. These applications are described by Etherington (1984).

The *vortex* represents the basis of all Fluidic heat and mass transfer equipment. The *vortex chamber* (**Figure 1**) is used as an intense short residence time mixer. Flows are introduced to the chamber tangentially at the periphery. Under conservation of angular momentum, the fluid accelerates towards the outlet port on the centre-line (see Stairmand (1990)). This arrangement offers homogenisation of the fluids within a residence time of several milliseconds. The *vortex mixer* has been used for precipitation, liquor blending and in-line gas-liquid mixing.

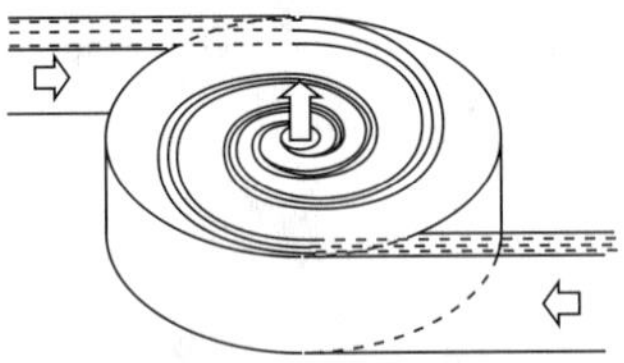

Figure 1. Vortex mixer.

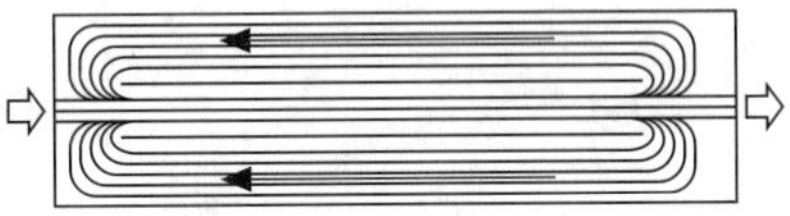

Figure 2. Draft tube mixer.

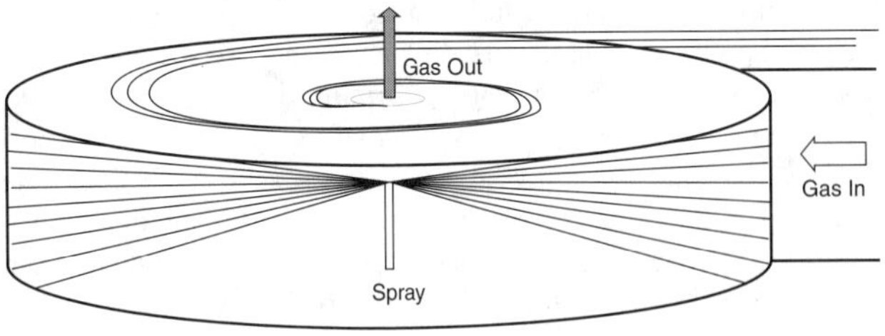

Figure 3. Gas-liquid contactor.

The *draft tube mixer* (**Figure 2**) is frequently used downstream of the vortex mixer to provide an extended residence time. A jet flow enters a central tube and also entrains liquor from the outside of the tube. Thus, a torroidal vortex is set up with a flow greatly larger than the feed jet. This creates a uniform mixing over residence times of order 10's of minutes. The duties to which the draft tube mixer has been applied are precipitate ripening, gas disolution and solvent extraction. Heat transfer may be effected by the use of an internal coil outside the draft tube, or by a jacket.

The vortex chamber is also used for *gas-liquid contacting* duties such as scrubbers, strippers and distillers. A radial spray head on the centre-line of the chamber (**Figure 3**), creates a stream of droplets which hit the walls, ceiling and floor of the chamber. The droplet diameter is typically 0.3 mm and the velocity 15 m/s. Gas is introduced tangentially at the chamber periphery, establishes a vortex flow pattern, and exits on the centre-line. This creates a counter-current contact between the liquor and gas at droplet Reynolds numbers of order 10^3. Gas mass transfer rates are therefore high and the equipment is typically a factor of 5 smaller than conventional gas-liquid contactors for gas-side limited systems. In the case of some liquid-side limited systems, the performance of the fluidic gas-liquid contactor can be maintained by recycling the liquor. In addition to its application as a scrubber, the unit has been used for stripping organics from process liquors and, in a staged form, for carrying out distillation duties.

References

Etherington C. (1984) Power fluidics technology and its application in the nuclear industry, *Nucl. Energy*, 23(4), 227–235.

Stairmand J. W. (1990) Flow patterns in vortex chambers for nuclear duties, *Nucl. Energy*, 29(6), 413–418.

J.W. Stairmand

FLUIDISED BED GASIFICATION (see Gasification)

FLUIDIZED BED

Following from: Gas-solids flow

A fluidized bed is a state of a two-phase mixture of particulate solid material and fluid, which is widely used in many modern technologies for efficient implementation of various physical and chemical processes. Fluidized beds have been used in technological processes such as: cracking and reforming of hydrocarbons (oil), carbonization and gasification of coal, ore roasting, Fischer-Tropsch synthesis, polyethylene manufacturing, limestone calcining, aluminuium anhydride production, granulation, vinil-chloride production, combustion of waste, nuclear fuel preparation, combustion of solid, liquid and gaseous fuels, drying, adsorption, cooling, heating, freezing, conveying, storing and thermal treating of various particulate solid materials.

The term "fluidized bed" is unavoidably connected to the term "particulate solid material". Particulate materials are mechanical mixtures of multitude of solid particles. Natural particulate materials originate from many long-term natural processes: heating, cooling, thermal dilatation, coliding, crushing, chopping up, atmospheric changes, river erosion and erosion caused by sea waves. Many technological processes also produce particulate solid material: grinding, chopping up, milling, evaporation, crystalization, spraying and drying. Particulate materials can also be of organic (plant) origin: fruits and seeds.

Particulate materials most commonly consist of solid particles with a range of shape and size. The majority of inorganic particulate solid materials found in nature have an extremely wide range of particle sizes. Such materials are called polydisperse materials. By certain technological processes, it is possible to produce particles with practically the same shape and size. Organic particulate materials found in nature (fruits and seeds) consist of particles of similar shape and size. Such materials are called monodisperse materials.

The geometrical, physical and aerodynamical properties of particulate solid materials all affect the onset of fluidization, and the characteristics, behaviour and the main parameters of fluidized beds. The most important solid properties are:

- particle density (not taking porousity into account),
- skeletal (true) density,
- bulk density—mass per unit volume of fixed bed,
- porosity (or void fraction) of the fixed bed—ratio of volume of space between the particles and the volume of the fixed bed,
- mean equivalent particle diameter—particle characteristic dimension,
- particle shape,
- particle size distribution—probability distribution of particle distribution due to their size,
- free fall (or terminal) velocity—velocity of falling particle at which gravitational, Archimedes and drag forces are in equilibrium.

For the exact definition of the term "fluidized bed," it is not sufficient to say that the fluidized bed is a state of the two-phase mixture of the particulate solid material and the fluid. Between two limiting states of the mixture—fluid percolation in the vertical direction through a fixed bed of particulate solids and the free fall of the particles through the stagnant fluid due to the gravitational force, a variety of different states of the solid-fluid two-phase mixture exist. The common characteristic of all these states in vertical, upward or downward, flow (of fluid, particles or both, in the same or opposite directions) is the existence of fluid-to-particle relative velocity and drag force. The various states of solid particle-fluid two-phase mixtures differ from each other by the following characteristics:

- the solid particles can be stagnant, floating or moving chaotically,
- the solid particles movement can be in a prefered direction or chaotic—one phase or both can be in movement,
- the flow direction can be vertical or horizontal,
- the fluid phase can be in cocurrent or countercurrent flow,
- the movement of the solid material can be free or limited by some kind of mechanical device (a perforated plate for example),
- the density or concentration of the mixture may differ greatly from one state to another.

Possible solid particle fluid mixture states are: fixed bed, stationary fluidized bed, fluidized bed with particle feeding at the bottom and overflow at the free surface of the bed, or vice versa, vertical conveying in the dense bed, low density vertical and horizontal conveying, downward particle movement in the dense bed with cocurrent fluid flow, and low density conveying downwards. Special cases of the above states are the moving bed and spouted bed. Dense phase, nonfluidized solid flow, in which particles move en bloc, with little relative velocity, has been refered to as moving-bed flow, packed bed flow or slip-stick flow. The voidage is close to the minimum fluidization value. Vertical down flow (in stand-pipes) is often used with the fluid moving faster than solids. Upflow of nonfluidized particles is not common. The spouted bed is a combination of a jet-like upward-moving dilute fluidized phase surrounded by a slow downwards moving bed through which gas percolates upward. The use of such systems is limited to a few physical operations with large particles. Transition boundaries between these states of solid material-fluid mixture are defined by the well-known Zenz diagram.

The fluidized state occurs when a fixed bed of the particulate material is penetrated in the vertical direction with fluid at sufficient velocity to break up the bed. In a fixed bed, the particles are immobile, leaning on one another at numerous contact points and applying forces to one another. Gravity forces—particle weight and the weight of the whole bed—are spread in all directions through the particle contact points. When the critical velocity (minimum fluidization velocity) is reached, the solid particles start floating, moving chaotically and colliding. Mutual contacts between the particles are of short duration and the forces between them are weak; the particulate solid material is then in the fluidized state. In the fluidized state, particles are in constant, chaotic movement, and their mean particle distance grows with increasing fluid velocity causing the bed height to rise. The pressure drop in the fluid phase across the bed is constant and equal to the bed weight over unit surface of the bed cross section. This value is reached at the minimum (incipient) fluidization velocity.

When the bed is fluidized with liquids, we have the case of the "homogenous" fluidization. Gas fluidization leads to so-called "heterogenous" fluidization. At gas velocities just above the minimum fluidization velocity, bubbles form and the fluidized bed can be treated as if it consists of two phases: bubbles, in which there are virtually no particles and a particulate (emulsion) phase, which is in a condition similar to that of the bed at the minimum fluidization velocity. Bubbles which form near the distribution plate, rise up the bed, grow and coalesce, producing bigger bubbles which sometimes break up into smaller bubbles. On the bed surface, bubbles eruptively burst, ejecting the particles far from the bed surface. Such bubble behaviour makes particle circulation in the bed very intensive. Behind the bubble, in its trail, particles move upwards. Around the bubbles and between them, and especially near the walls, particles move downwards. Bubble movement thus promotes intensive gas and particle axial mixing in the fluidized bed.

The chaotic movement of the particles in the fluidized bed is the main reason for the fact that the various fluidized bed characteristics are similar to those of liquids, which is why this state of two-phase fluid-solid mixture got its name. The free surface of the fluidized bed is horizontal, but of irregular shape due to the bubbles bursting; however it makes a clear, distinctive boundary between the bed of high concentration and the space above it (freeboard), in which particle concentration decreases exponentially. Bodies of greater density sink in the fluidized bed, and those of lower density float or chaotically move near the surface. In the fluidized state, particulate solid material flows out through the openings of the vessel. Just as in the liquids, mixing of two particulate materials is intensive and homogenous. Heat transfer is also intensive, maintaining a homogenous temperature field during the heating or cooling processes or when the heat is generated by fuel combustion in the bed. Fluidized beds obey the laws of hydrostatics.

The fluidized bed characteristics listed above enable its use in various devices for efficient implementation of physical and chemical processes.

The fluidized state, occuring between the filtration of the fixed bed and the pneumatic conveying regime, includes three different regimes: stationary *bubbling fluidized bed,* turbulent fluidized bed and the regime of fast fluidization. Recently, many technological processes are being carried out in the last two of the above mentioned regimes.

The bubbling fluidized bed best fits the already mentioned characteristics. The most outstanding difference between this and the other two regimes is the existence of large bubbles and a clearly outlined free surface. The bed is very nonhomogenous due to the presence of the bubbles and the pressure drop across the bed oscillates in time.

As the fluidization velocity increases, large bubbles break up into several smaller ones. When the break-up process overcomes the coalescence of the bubbles, oscillations of the pressure drop become smaller. This is the moment when turbulent regime occurs, with no big bubbles in the bed. In the particulate phase of the bed, which is becoming more homogenous, smaller voids exist in the form of the channels and jets, and particles form clusters. Neither the gaseous nor the emulsion phases can be said to be continuous. Mixing of particles becomes more intensive, and the interaction of phases is stronger. The free surface of the bed is very irregular and not clearly outlined. A lot of particles and clusters are being carried off the bed surface and then fall in to the bed again. The maximum velocity of the turbulent regime is the so-called "transport velocity," at which the fast fluidization regime occurs. At velocities above the transport velocity, particles are carried out of the system by the fluid and the bed can only be maintained by feeding new particulate material into it.

At velocities higher than transport velocity, depending on the flow rate of the solid particles returning to the bed, fast fluidization with higher or lower concentration occurs. In the fast fluidization regime, particles move upwards in clusters through the middle of the bed cross section. Near the walls, particles move downward. In this regime, particle mixing is even more intensive in both axial and radial directions. Particle concentration in the freeboard decreases exponentially in the upward direction.

All mentioned regimes of fluidization as well as the two boundary states—fixed bed and pneumatic conveying—have found their use in various devices and technological processes. Solid fuel (coal) combustion is the obvious example. Boilers with grates use fixed beds, fluidized bed combustion boilers use bubbling fluidized beds or turbulent regime (in this case with fly ash recirculation), circulating fluidized bed boilers use the fast fluidization regime and pulverized coal combustion boilers work in the pneumatic conveying regime.

Parameters that describe macroscopic, overall, behaviour of the fluidized bed in the bubbling regime are: minimum fluidization velocity, pressure drop across the bed, bed height increase and particle elutriation.

The easiest way of detecting the transition from the fixed bed regime to the fluidized bed regime is to measure the pressure drop across the bed as a function of the fluid velocity. A curve of characteristic shape is obtained (shown on **Figure 1**).

When *minimum fluidization velocity* ($-v_{mf}$) is reached, the Δp_p vs v_g curve bends and the *pressure drop* remains constant although

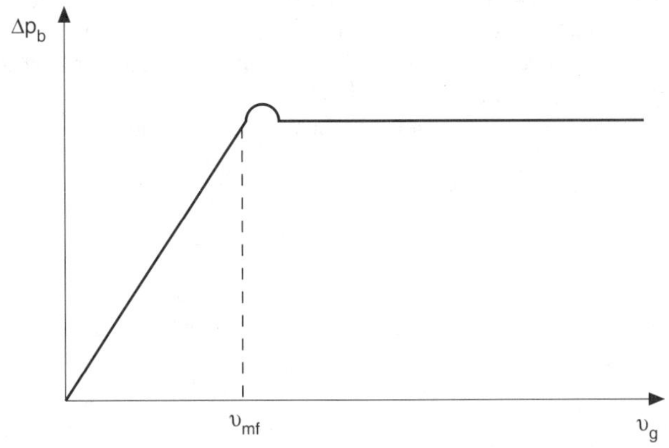

Figure 1. Variation of fluidized bed pressure drop with gas velocity.

the fluid velocity increases. Pressure drop across the fluidized bed becomes equal to the ratio of the overall particle weight and the bed cross section:

$$\Delta p_b = (1 - \epsilon_{mf})(\rho_p - \rho_g)gH_b. \quad (1)$$

Pressure drop across a fixed bed is practically proportional to gas velocity and can be calculated from the well-known Ergun relation (see **Fixed Beds**):

$$\Delta p_b = H_b \frac{150(1-\epsilon)^2}{\epsilon^3} \cdot \frac{\mu_g v_g}{(\Phi_s d_p)^2} + H_b \frac{1.75(1-\epsilon)}{\epsilon^3} \cdot \frac{\rho_g v_g^2}{\Phi_s d_p}. \quad (2)$$

When these two expressions are equalled, the Wen and Yu relation for minimum fluidization velocity is obtained:

$$Re_{mf} = \frac{d_p v_{mf} \rho_g}{\mu_g} = (33.7^2 + 0.0408 A_r)^{0.5} - 33.7,$$

$$A_r = g d_p^2 \rho_g (\rho_p - \rho_g)/\mu_q^2. \quad (3)$$

Minimum fluidization velocity has the same physical meaning as the free fall velocity of the particle—that is the velocity at which all particles in the bed are floating. More accurate values for the minimum fluidization velocity for the particular particulate material can be obtained only by means of direct measurement.

The surface of the fluidized bed is not precisely horizontal. Rising through the bed, the bubbles grow and burst reaching the surface, eruptively ejecting particles into the space above the bed (the freeboard). Particles of greater size return (fall back) into the bed, but lots of smaller particles are carried away from the bed by the flowing gas. For that reason, the fluidized bed surface is very disturbed and there is no sudden change in particle concentration. In the region above the free surface, there is the area called the splash zone, in which the particle concentration is changed from very high value in the fluidized bed to the low value characteristic of the area far from the bed surface.

Although the surface of the bubbling fluidized bed cannot be easily defined, the fact is that the bed height increase of the fluidization velocity. Increase of the bed height, i.e. the increase in bed volume, is closely related to the fluidized bed porosity and the volume fraction occupied by bubbles. If we suppose, according to the Davidson two-phase model of the fluidized bed, that all particles are outside bubbles, and always in the state of the minimum fluidization, i.e. that the porosity of the particulate (emulsion) phase is equal to ϵ_{mf}, than the increase in the bed height depends only on the volume fraction occupied by bubbles:

$$\frac{H}{H_{mf}} = \frac{1}{1-\delta_b} = \frac{1-\epsilon_{mf}}{1-\epsilon}. \quad (4)$$

Calculation of the bed height, porosity of the bed or the bubble volume fraction, hasn't been solved satisfactorily yet. Empirical correlations of type:

$$\epsilon - \epsilon_{mf} = C(v_g - v_{mf})^n \quad (5)$$

don't yield good results, and the constants C and n need to be determined experimentally for each specific particulate material. A relation used more often is Todes' equation:

$$\epsilon = \epsilon_{mf}\left(\frac{Re_p + 0.02\,Re_p^2}{Re_{mf} + 0.02\,Re_{mf}^2}\right)^{0.21}. \quad (6)$$

A recent approach to this problem uses the relations for the volume fraction occupied by the bubbles, and the corresponding relations for bubble diameter and velocity.

When the bubbles burst at the bed surface two processes take place: at the surface true local gas velocity can be much higher than the mean fluidization velocity, so the larger particles can also be thrown out far from the bed surface; only particles with low free fall velocity will be elutriated far from the bed surface. Due to these two processes, not only single particles are being thrown from the bed surface, but also particle clusters. Moving upwards, these clusters crumble and disintegrate. Some clusters continue their upward movement, and some start to fall back into bed. Particles separated from the clusters, or single particles thrown out of the bed, move upwards though some of them (the larger ones) begin to fall back into the bed. Only the smallest particles, whose free fall velocity is less than the gas velocity will be elutriated from the bed. The number of upward moving particles decreases as the height above the bed increases. Particle concentration and the density of the two-phase mixture decreases exponentially. For each particular fluidized bed and fluidization conditions, a maximum height exists, above which only particles whose free fall velocity is less than the gas velocity, can be found. In the case of a monodisperse material, above this height the particle concentration is zero. This height is called TDH—Transport Disengaging Height. In spite of thorough experimental research no reliable relation for TDH calculation has been developed yet. One of the relations currently in use is the Geldart equation:

$$TDH = 1200 \cdot H_{mg} \cdot Re_p^{1.55} A_r^{-1.1} \quad (7)$$

obtained for particles with $d_p = 75\ \mu m$–$2000\ \mu m$.

The net mass flow rate of particles with mean diameter d_{pi} at the distance h from the bed surface can be expressed by the relation:

$$F_i = F_{i\infty} + (F_{i0} - F_{i\infty})\exp(-a_f h) \quad (8)$$

where $F_{i\infty}$ is the particle flow rate above the TDH:

$$F_{i\infty} = E_{i\infty} y_i \quad (9)$$

and $E_{i\infty}$ is elutriation constant for which one of the empirical relations obtained is the following:

$$\frac{E_{i\infty}}{\rho_f v_f} = 23.7 \exp\left(-5.4\frac{u_t}{v_f}\right). \quad (10)$$

F_{i0} is the particle mass flow rate at the bed surface:

$$F_{i0} = E_0 y_i \quad (11)$$

Elutriation constant at the bed surface is:

$$E_0 = 0.3\ D_B d_p^{-1.17}(v_g - v_{mf})^{0.66}. \quad \textbf{(12)}$$

In order to acquire deeper knowledge and better understanding of the characteristic properties of the fluidized bed, it is necessary to research in more detail the local structure of the fluidized bed and the microprocesses taking place in it: bubble generation and growth, movement and mixing of particles and gas.

Bubbles in the heterogenous fluidized beds appear either immediatelly or just after the minimum fluidization velocity is reached. Bubble occurence is of statistically random nature. There is no place on the distribution plate predisposed for bubble occurence. Bubbles initially take a spherical shape, but as they grow they take the shape typical for gas bubbles in liquids—with a concave bottom. The irregularly and randomly spaced bubbles move mostly upward. The presence of other bubbles and particle circulation cause bubbles to move latterally, which leads to coalescence of bubbles. Moving upward, bubbles grow mainly by coalescence, causing the number of bubbles to decrease towards the bed surface. Large bubbles may be unstable so the bubbles can also break up before reaching the surface. Studying the bubbles assumes learning about: bubble generation, rise velocity, growth and coalescence, break up and bursting at the bed surface. All these processes are very complex and still inadequatelly explained. It has been noticed that, due to the particle size, bubbles can be "fast" and "slow", depending on their velocity, being higher or lower than the *minimum fluidization velocity,* i.e. velocity of the gas flowing through the bed emulsion phase. In fluidized beds with small particles (materials of the group B by the Geldart (1973) classification), bubbles are fast. In the beds with large particles (materials of the group D by Geldart classification), bubbles are slow. Fast bubbles pass through the fluidized bed with very low mass exchange between the gas in the bubble and the gas in the emulsion phase. At higher bubble velocities, a so-called "cloud" forms around the bubble, in which gas from the bubble circulates, while the gas from the emulsion phase travels around the bubble. Fluidized beds with "slow" bubbles are much more useful for chemical processes, since the bubbles are intensively washed out by the gas from the emulsion phase, so the whole of the gas flow can take part in the reaction. With the fast bubbles, gas passes through the bed not taking part in reactions with the particles.

Bubble rise velocity can be calculated from the following relation:

$$v_{B\infty} = 0.711(gD_B)^{0.5} \quad \textbf{(13)}$$

which is similar to the relation for the bubbles in liquids. One of the simple relations for calculating the size of the bubble at different distances of the distribution plate is Rowe's relation:

$$D_B = (v_g - v_{mf})^{0.5}(h + h_0)^{3/4}g^{-0.25}. \quad \textbf{(14)}$$

Upward gas flow through the bubble keeps the particles at the "roof" of the bubble, and the inflow of gas at the "bottom" turbulizes the particles behind the bubble and drags them up. This "wake" behind the bubble has a volume about 1/3 that of the bubble. Moving upwards, bubbles drag particles from the emulsion phase in their "wake". Because of the overall stationary state of the fluidized bed, upward movement of the particles causes downward movement in other areas of the bed. That is how quasi-stationary circulating particle flow in the emulsion phase of the fluidized bed is caused. The intensity, shape, character and number of these circulating flows depend on numerous parameters, but primarily on the size and the shape of the fluidized bed. That is the one of the basic reasons why it is not possible to obtain similarity between small beds in laboratories and the large industrial devices. Chaotic particle movement, and the organized circulation of particles caused by the bubbles are the main reason for particle mixing, uniformity of the temperature field in the fluidized bed, intensive heat transfer and favourable conditions for chemical reactions.

When gas mixing in the emulsion phase is taken into consideration, three processes must be considered: molecular gas diffusion, turbulent mixing and the directed gas movement as the result of the particle circulation. Bubbles, in their wake, along with particles also drag the gas upwards. In their downward movement, particles can significantly disturb the gas flow. Very intensive downward particle movement can even cause gas to move downward.

Particle mixing in the vertical direction is much more intensive than the mixing in the lateral direction. Using the analogy with the molecular diffusion and by defining the apparent particle diffusion coefficient, experimentally obtained values of this coefficient in the axial direction range from 100 to 1000 cm^2/s, and in the lateral direction only 10 to 50 cm^2/s. Gas and particle mixing is very important for processes in the fluidized beds. For example, in fluidized bed combustion of solid fuels, the following processes depend on the gas and particle mixing: mixing of the fuel and inert particulate material, even fuel distribution in the bed, the place and manner of feeding fuel, the number of feed points, ash behaviour in the bed, heat transfer to the immersed surfaces, combustion of volatiles, erosion of the heating surfaces, etc.

For gas-gas and gas-particle chemical reactions, gas mixing processes in the fluidized bed are also important. If we suppose that the gas passes through the bed in bubbles, either very little or none at all, takes part in chemical reactions, then the gas mixing in the emulsion phase is the most important process concerning the chemical reactions. Apparent gas diffusion coefficient in the axial direction is 0.1 to 1 m^2/s and in lateral direction several degrees of magnitude lower, namely, 10^{-3} to 10^{-4} m^2/s.

Data on gas mixing in the fluidized beds are scarce and differ a lot from experiment to experiment. There is still no established opinion on the effects fluidization velocity, particle and bed size influence on the mixing processes. Yet, it is known that the mixing is more intensive in large fluidized beds and at the higher fluidization velocities.

Another process important for carrying on processes in the fluidized beds is gas exchange between bubbles and the emulsion phase. In this exchange, a very important role is played by "washing" out of the bubbles with gas, a process which is very different in the "slow" and the "fast" bubbles. Uniformity of the temperature and concentration fields is always pointed out as the basic property contributing to the extensive implementation of chemical reactions in fluidized beds. In reality, in large scale chemical reactors, it sometimes happens that these conditions are not fulfilled. If gas and particle mixing and the gas exchange between bubbles and the emulsion phase are not of the sufficient intensity, large temperature variations can occur in the bed. Hot spots and uneven distribution of the reacting gases, fuel particles and oxygen are present. The most frequent reason for these drawbacks is the nonuniform gas distribution due to the wrong choice of the distribution plate and its poor construction which causes nonuniform fluidization and bad particle and gas mixing.

Heat Transfer in Fluidized Beds

Heat transfer in the fluidized bed is, apart from the particle and gas mixing, the most important process contributing to the intensity of the physical and chemical processes. In fact, several different processes can be distinguished: particle-gas heat transfer, heat transfer between different points in the bed, heat transfer between the fluidized bed particles and the larger particles floating in the bed and the heat transfer to the submerged surfaces in contact with the bed. All of these heat transfer processes are very intensive in fluidized beds.

In the case of uniform fluidization, the temperature difference between points in the bed does not exceed 2–5°C, with mean bed temperatures of several hundred, even 1000°C. Gas temperature, when leaving the bed, is practically the same as the particle temperature. These facts tell us of the great capability of the solid particles to exchange heat with the fluidizing gas. Intensive heat transfer is, first of all, a consequence of the large specific heat transfer surface (3000 to 45000 m^2/m^3), although heat transfer coefficients to the particles in the bed are relatively small, 6–25 $W/m^{2}°C$. The large heat capacity of the solid particles also makes the temperature difference between gas and particles small. Gas temperature follows the particle temperature.

Gas to particle heat transfer coefficients can be calculated from the Gelperin and Einstein relation:

$$Nu_p = 1.6\ 10^{-2}(Re_p/\epsilon)^{1.3}\mathbf{Pr}^{0.33} \quad \text{for} \quad Re_p/\epsilon < 200 \tag{15}$$

and

$$Nu_p = 0.4(Re_p/\epsilon)^{2/3}\mathbf{Pr}^{0.33} \quad \text{for} \quad Re_p/\epsilon > 200 \tag{16}$$

$$Pr = C_g \nu_g \rho_g/\lambda_g.$$

Despite the small values of the gas-to-particle heat transfer coefficient, even at a short distance from the distribution plate, the gas and particles temperature are practically equal. Five to ten particle diameters from the distribution plate, the temperature difference between gas and particles has decreased around 100 times. The temperature of gas in the bubbles also very quickly becomes equal to the particle temperature. Some ten millimeters from the plate is enough for this equalization to occur. If, in an inert fluidized bed, there are active particles (usually larger ones) which chemically react with the fluidizing gas, releasing heat (as in the solid fuel combustion), very complex processes of heat and mass transfer between these particles and the bed take place. At the begining, these larger particles heat up in the contact with the inert bed material. Simultaneously, evaporation and devolatilization take place. These processes depend on the intensity of heat transfer between the bed as a whole and these particles. When chemical reactions between gas and particles begin (combustion of the char), particle temperature is higher and the reverse process of heat exchange between active particles and fluidized bed takes place. The combustion process is limited by the mass transfer process, i.e. the diffusion of the reacting gas to the surface of the active particle. The mechanisms of mass transfer towards the active particles are molecular diffusion and the convective transport, while inert bed material disrupt these processes merely by being there. Because larger active particles mainly inhabit the emulsion phase, mass transfer intensity increases with the increase of the size of inert bed material (i.e. with v_{mf}) and decreases with increase of the active particle diameter.

One of the most recent relations for calculating mass transfer coefficient β is that of La Nauze and Jang:

$$Sh = \frac{\beta d_p}{D_G} = 2\epsilon_{mf} + 0.69(Re/\epsilon)^{0.5}Sc^{0.33} \tag{17}$$

who proposed two mass transfer mechanisms: "packets" (clusters) of particles carry fresh gas from the bulk bed towards the active particle, the movement of these packets is controled by the bubble flow; the other mechanism is classical convective mass transfer by gas which percolates through the emulsion phase with a velocity equal to v_{mf}.

Heat transfer between the active particle and the fluidized bed is controlled by three mechanisms: gas convection, particle convection and radiation. Depending on the active particle size and temperature, the mechanisms mentioned above do not have the same contribution to the overall heat transfer. These processes are also involved in the heat transfer between the bed and the immersed surfaces. Heat transfer by radiation becomes significant only when temperature differences exceed 400–500°C. For active particle size less or equal to the inert material particle size, the main transfer process is convective heat transfer; for larger particles, heat transfer due to the collision and contact with the fluidized bed particles may be important. Heat transfer coefficients α_p for active particles or immersed surfaces have a maximum for optimal fluidization velocity.

For active particle size close to the inert material particle size, relations (15) and (16) can be used. For larger active particles, Agarwall's relation:

$$Nu_p = \frac{\alpha_p d_p}{\lambda_g} = 2.0 + 1.8\ Re_p^{0.5}Pr^{0.33} \tag{18}$$

is suggested.

When the fluidization velocity increases from v_{mf} to the optimal value, heat transfer coefficients for the surfaces in contact with the fluidized bed also increase, due to the increase of heat transfer by particle contacts. For velocities greater than the optimal, the main contribution to the heat transfer is the gas convection because of the decrease of particle concentration, i.e. the increase of the bed porosity. Optimum velocity can be calculated from the Todes relation:

$$Re_{opt} = \frac{A_r}{18 + 5.22\ A_r^{0.5}}. \tag{19}$$

The factor with the greatest influence on the heat transfer, apart from the fluidization velocity, is the particle size. Particle size influences the change of the relative contribution of various mechanisms in the overall heat transfer. In the fluidized bed with small (< 0.1 mm) particles, convection by particles account for 90% of the overall heat transfer, while in the beds of large particles (> 1 mm) only 20% of the heat transfer is done by particle convection. Particle heat capacity is also important for the amount of heat transfered by particle convection. The maximum heat transfer coefficient is often calculated from the Zabrodski relation:

$$\alpha_{max} = 35.8\rho_p^{0.2}\lambda_g^{0.6}d_p^{-0.36}. \qquad \textbf{(20)}$$

In the literature, numerous relationships can be found for calculating the heat transfer to immersed surfaces of different shape: vessel walls, single horizontal or vertical pipe, tube bundles with smooth and finned tubes in corridor or stagered arrangement. Parameters that influence the heat transfer to these surfaces are: height and dimensions of the bed, bubble size, tube diameter, arrangement and position of tubes, tube distance, quality and shape of the surface. Existing relationships for calculating the heat transfer coefficient do not include all these parameters and for that reason, experimental data is greatly scattered and the accuracy of the formulae proposed is in the range up to ±50%.

References

Davidson, J. F. and Harrison, D. (1963) *Fluidized Particles*. Cambridge University Press, Cambridge, England.

Grace, J. R. (1982) Fluidization. Chapter 8 of *Handbook of Multiphase Systems*, McGraw Hill, New York.

Geldart, D. (1973) Types of gas fluidization. *Powder Technology*, 7, 285–292.

S. Oka

FLUIDIZATION (see Gas-solid flows; Fluidized bed)

FLUID MECHANICS (see Flow of fluids)

FLUIDS

The word 'fluid' is the generic title used to encompass the set of materials which cannot support a tangential or shear stress without flow. It therefore includes gases and liquids that may be both **Newtonian** and **non-Newtonian.** It follows from this definition that, for such a fluid material, it is possible to write a constitutive equation that relates the stress tensor τ to the rate of strain tensor E so that (Richardson, 1989)

$$\underline{\underline{\tau}} = \eta \underline{\underline{E}}$$

where η itself may depend upon the rate of strain tensor $\underline{\underline{E}}$ (see **Flow of Fluids**). In the case where η is a constant, this is the constitutive equation of a Newtonian fluid and η is the **Viscosity** this class of fluids includes all gases and most liquids. However, non-Newtonian fluids are important in many circumstances and display a wide variety of different behaviour since, for some fluids, the viscosity increases with increasing rate of strain (*dilutant fluids*) while for other materials, the viscosity decreases with increasing rate of strain (*pseudoplastic fluids*). Finally, there are materials where the behaviour depends upon time such as *thixotropic fluids,* where the viscosity shows a limited decrease with time under a suddenly applied constant shear stress, and others (*visco-elastic*) fluids, where the material partially returns to its original form when the applied stress is removed.

The physical properties of fluids enter the equations of conservation of fluid mechanics through constitutive equations such as that given above for the viscosity and through **Fourier's Law of Heat Conduction,** and an **Equation of State** for the **Density** in terms of the **Pressure** and **Temperature.** These constitutive equations may sometimes be deduced from statistical mechanics, such as the equation of state for a perfect gas, but more often must be founded upon experimental observations. Furthermore, the physical properties of the fluids themselves are most usually determined empirically since statistical mechanical theory can provide no more than guidance to their evaluation. Direct measurement of the physical properties of all fluids of interest is not possible owing to the shear magnitude of the task, so that very often it is necessary to make use of predictive procedures to evaluate properties that have varying degrees of reliance on rigorous theory.

References

Richardson, S. M. (1989) *Fluid Mechanics,* Hemisphere, New York.

Bird, R. B., Stewart, W. E., and Lightfoot, E. N. (1960) *Transport Phenomena,* Wiley, New York.

Reid, R. C., Prausnitz, J. M., and Sherwood, T. K. (1977) *The Properties of Gases and Liquids,* 3rd ed. McGraw-Hill, New York.

Leading to: *Cryogenic fluids; Density of liquids; Density of gases; Specific heat capacity; Thermal conductivity; Viscosity.*

W.A. Wakeham

FLUE GASES (see Chimneys)

FLUMES (see Weirs)

FLUORESCEIN (see Fluorescence)

FLUORESCENCE

Following from: *Photoluminescence*

In general fluorescence is the emission of light which occurs within about one hundred nanoseconds of excitation. Longer lived emission is termed **Phosphorescence.** Many minerals and biological molecules fluoresce when irradiated with ultraviolet light. The term is derived from the mineral *fluorspar* which is naturally fluorescent and which was used as a flux (latin *fluere*-flow) in smelting.

In molecular photochemistry, fluorescence is defined as the emission of light associated with a transition between states of the same spin multiplicity. The radiative or natural lifetime of molecular fluorescence is typically between one and one hundred nanoseconds, but, as discussed by Turro (1991), competing decay processes often result in an observed lifetime less than this. *Resonance fluorescence,* where absorption and emission wavelengths are the same, occurs in low pressure gases where collisional deactivation is inefficient. In condensed phases, fluorescence usually occurs from the lowest vibrational level of the emitting state. Molecular fluorescence usually arises from the transition from the first excited singlet to the ground state. The emission spectrum is often a rough mirror image of the lowest energy absorption band but shifted up to a few tens of nanometers to higher wavelength (See **Photoluminescence** for details of absorption and emission processes and spectra).

Spectrofluorimetry is a very sensitive analytical technique and fluorescent compounds are used as molecular "probes" and "labels". *Fluorescein* and *Rhodamine* are examples of highly fluorescent dyes which emit in the green and orange-red spectral regions. Fluorescence from organic compounds is discussed by Krasovitskii and Bolotin (1988).

References

Turro, N. J. (1991) *Modern Molecular Photochemistry,* University Science Books, California.

Krasovitskii, B. M. and Bolotin, B. M. (1988) *Organic Luminescent Materials,* VCH, Weinheim.

P. Douglas

FLUORESCENCE METHOD, FOR FILM THICKNESS MEASUREMENT (see Film thickness measurement)

FLUORESCENCE PHOTOGRAPHY (see Photographic techniques)

FLUORINE

Fluorine—(L. and F. *fluere,* flow, or flux), F; atomic weight 18.998403 ± 0.000001; atomic number 9; melting point—219.62°C (1 atm.); boiling point—188.14°C (1 atm); density 1.696 g/l (O°C, 1 atm.); specific gravity of liquid 1.108 at boiling point; valence I.

In 1529, Georgius Agricola described the use of fluorspar as a flux, and as early as 1670 Schwandhard found that glass was etched when exposed to fluorspar treated with acid. Scheele and many later investigators, including Davy, Gay-Lussac, Lavoisier, and Thenard, experimented with hydrofluoric acid, some experiments ending in tragedy. The element was finally isolated in 1886 by Moisson after nearly 74 years of continuous effort. Fluorine occurs chiefly in *fluorspar* (CaF_2) and *cryolite* (Na_2AlF_6), but is rather widely distributed in other minerals. It is a member of the halogen family of elements, and obtained by electrolyzing a solution of potassium hydrogen fluoride in anhydrous hydrogen fluoride in a vessel of metal or transparent fluorspar. Modern commercial production methods are essentially variations on the procedures first used by Moisson. Fluorine is the most electro-negative and reactive of all elements. It is a pale yellow, corrosive gas, which reacts with practically all organic and inorganic substances. Finely divided metals, glass, ceramics, carbon, and even water burn in fluorine with a bright flame.

Until World War II, there was no commercial production of elemental fluorine. The atom bomb project and nuclear energy applications, however, made it necessary to produce large quantities. Safe handling techniques have now been developed and it is possible at present to transport liquid fluorine by the ton. Fluorine and its compounds are used in producing uranium (from the hexafluoride) and more than 100 commercial fluorochemicals, including many well-known high-temperature plastics. Hydrofluoric acid is extensively used for etching the glass of light bulbs, etc. Fluorochloro hydrocarbons are extensively used in air conditioning and refrigeration. It has been suggested that fluorine can be substituted for hydrogen wherever it occurs in organic compounds, which could lead to an astronomical number of new fluorine compounds. The presence of fluorine as a soluble fluoride in drinking water to the extent of 2 ppm may cause mottled enamel in teeth, when used by children acquiring permanent teeth; in smaller amounts, however, fluorides are said to be beneficial and used in water supplies to prevent dental cavities. Elemental fluorine is being studied as a rocket propellant as it has an exceptionally high specific impulse value. Compounds of fluorine with rare gases have now been confirmed. Fluorides of xenon, radon, and krypton are among those reported. Elemental fluorine and the fluoride ion are highly toxic. The free element has a characteristic pungent odor, detectable in concentrations as low as 20 ppb, which is below the safe working level. The recommended maximum allowable concentration for a daily 8-hr time-weighted exposure is 0.1 ppm.

Handbook of Chemistry and Physics, CRC Press

FLUTED TUBES (see Tube banks, condensation heat transfer in)

FLUX METHODS (see Furnaces; Radiative heat transfer)

FLUX PARAMETER (see Drying)

FLYING VEHICLES, AERODYNAMICS OF (see Aerodynamics)

FMEA, FAILURE MODES AND EFFECTS ANALYSIS (see Risk analysis techniques)

FOAM FRACTIONATION

Following from: *Foams; Liquid-liquid separation; Liquid-solid separation*

Foam fractionation is one of the adsorptive bubble separation techniques (*adsubble techniques*). (See Lemlich (1972)). It operates through the selective **Adsorption** of a portion of one or more dissolved (or perhaps finely colloidal) components of a liquid mixture at the surfaces of **Bubbles**, usually of air or nitrogen, that rise through the mixture and then overflow as foam. Adding a surface-active collector may permit the adsorption of an otherwise surface inactive colligend via *chelation, counterionic attraction,* or otherwise. (See Lemlich (1993), and Matis and Mavros (1991).)

The solid lines of **Figure 1** illustrate continuous foam fractionation in the simple mode. Alternatively, the stripping mode is obtained by elevating the feed inlet. The enriching mode is obtained by returning some collapsed foam (foamate) to the top of the column as external reflux. Also, spontaneous or deliberate coalescense within the ascending foam can furnish internal reflux. Foam fractionation is analogous to fractional **Distillation** with entrainment of liquid. (See also **Flotation**)

Equations 1 and **2** apply to the continuous simple mode taken as a single theoretical (equilibrium) stage.

$$c_Q = c_W + \dot{V}_G A_G \Gamma_W / \dot{V}_Q \tag{1}$$

$$c_W = c_F - \dot{V}_G A_G \Gamma_W / \dot{V}_F \tag{2}$$

A_G is the ratio in the foam of bubble surface to bubble volume, c_F is the concentration of the component in the feed, c_W is the concentration in the bottom product, and c_Q is the concentration in the top product. $\dot{V}_F$, $\dot{V}_G$, and $\dot{V}_Q$ are the volumetric flowrates of feed, gas, and top product respectively. Γ_W is the solute surface excess (which is effectively the concentration on the bubble surface) in equilibrium with c_W.

For stripping, enriching, or combined operation, the ascending stream at any level can be conveniently viewed as consisting of bubble surface plus entrained liquid in mutual equilibrium. Effective operating lines can then be obtained and transfer units or theoretical stages calculated (Lemlich, 1972, 1993).

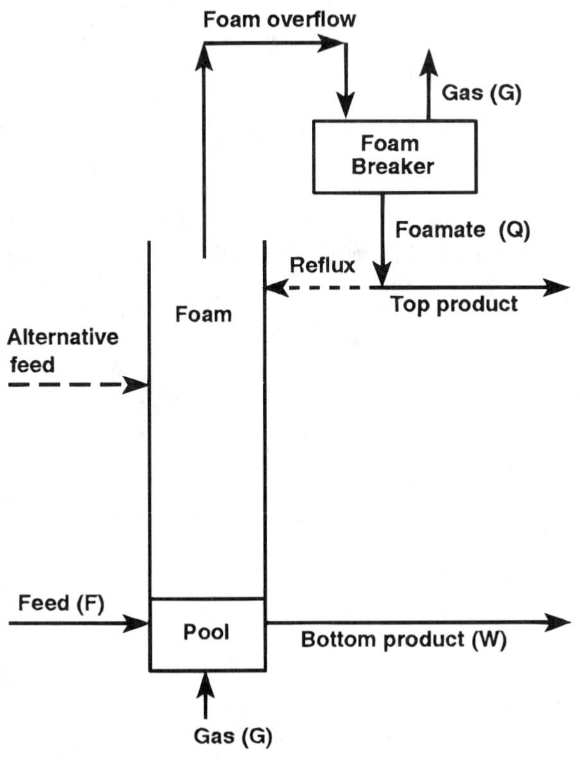

Figure 1. Continuous foam fractionation.

Since surface capacity is limited, foam fractionation is best suited to low c_F. Low superficial gas velocity favors foam drainage and hence enrichment, but limits throughput. The volumetric fraction, ϵ, of liquid in the foam at any predetermined level can be estimated conductimetrically with semi-theoretical **Equation 3** (Lemlich, 1985).

$$\epsilon = 3K - 2.5K^{4/3} + 0.5K^2 \tag{3}$$

K is the electrical conductivity of the foam divided by that of the liquid. **Equation 3** has been tested over the entire range of foam and dispersion; that is, for $0 \le \epsilon \le 1$. At extremely low ϵ, **Equation 3** approaches Lemlich's limit of $\epsilon = 3K$.

Bubble sizes can be roughly estimated visually, or indirectly by light scattering and transmission. From bubble sphericity to polyhedricity, $6/D_{32} \le A_G < 6.6/D_{32}$, where D_{32} is the Sauter mean bubble diameter. For reasonably stable homogenous foam of low ϵ ascending in plug flow through a column of uniform cross-section A_C, $\dot{V}_Q$ is roughly directly proportional to $\dot{V}_G^2/A_C D^2$, and also depends on liquid density, liquid viscosity, surface viscosity, and gravity. D is the bubble diameter, best averaged in some still arguable manner. (See also **Optical Particle Characterisation**)

As an alternative to internal sparging, bubbles can be generated through the release of dissolved gas, or by **Electrolysis,** or by external Venturi action as microbubbles. Also, reactive gases, columns of nonuniform cross-section, plate columns, and individual fractionators connected countercurrently have been investigated.

Adsorption at equilibrium is governed by the classical Gibbs relationship, Equation 4.

$$d\sigma = -\tilde{R}T \sum \Gamma_i d \ln a_i \tag{4}$$

$\tilde{R}$ is the universal gas constant, T is the absolute temperature, Γ_i is the solute surface excess of the i-th component, a_i is the activity of i-th component, and σ is the surface tension. **Equation 4** simplifies to **Equation 5** for a nonionic surfactant in pure water at concentration c_i below critical micelle.

$$\Gamma_i = -\frac{1}{\tilde{R}T} \frac{d\sigma}{d \ln c_i} \tag{5}$$

For the major surfactant in a foam, Γ_i is roughly constant because the bubble surfaces are essentially saturated. Typical values are of the order of 3×10^{-9} kmol/m^2 for a molecular weight of several hundred.

If a surfactant is the collector of a trace colligend, Γ_i for the latter will be directly proportional to its c_i at equilibrium if c_i is sufficiently low. For collection via counterionic attraction, the coefficient of linearity for adsorption of a trace polyvalent *colligend* ion is generally many times that of a monovalent ion. Too much collector can decrease the separation due to the formation of *micelles* which compete for colligend.

The bursting bubbles from foam fractionation and other adsubble techniques can inject a fine aerosol into the atmosphere. This can be a consideration if toxic, pathogenic, or otherwise noxious substances are involved.

References

Lemlich, R. Ed. (1972) *Adsorptive Bubble Separation Techniques,* Academic Press, New York.

Lemlich, R. (1985) Semitheoretical Equation to Relate Conductivity to Volumetric Foam Density, *Ind. Eng. Chem. Process Des. Dev. 1985,* 24, 686–687.

Lemlich R. (1993) Foam Fractionation, 296–312 in *Encyclopedia of Chemical Processing and Design, Vol. 23,* J.J. McKetta and W.A. Cunningham, eds., Marcel Dekker, New York (1985) reprinted in *Unit Operations Handbook, Vol. 1,* J.J. McKetta ed., 523–539, Marcel Dekker, New York, 1993.

Matis, K.A. and Mavros, P. (1991) Recovery of Metals by Ion Flotation from Dilute Aqueous Solutions, *Separ. Purif. Meth. 1991,* 20, 1–48. Foam/Froth Flotation: Part II. Removal of Particulate Matter, *ibid.,* 163–198.

R. Lemlich

FOGGING (see Condensers)

FOGGING IN CONDENSERS (see Condensation of multicomponent vapours)

FOKKER-PLANK EQUATION (see Stochastic process)

FORCED CONVECTION BOILING

Following from: Boiling

Boiling is most often understood as a phase transition from a liquid to a vapor state involving the appearance of vapor bubbles on a hot surface. In this respect, forced convection boiling and pool boiling have much in common. However, forced convection imparts a number of specific features to the conditions of bubble production and breakaway into the bulk of the liquid. The structures of vapor-liquid mixtures resulting from boiling and mixing of liquid and vapor phases also differ appreciably from each other.

Forced convection intensifies these processes compared to free motion accompanying pool boiling. **Figure 1** shows variation of the bulk flow temperature $\overline{T}_b$ and the wall temperature T_w in a heated channel with boiling heat transfer when a subcooled liquid with $\overline{T}_b < T_s$, where T_s is the saturation temperature, is supplied

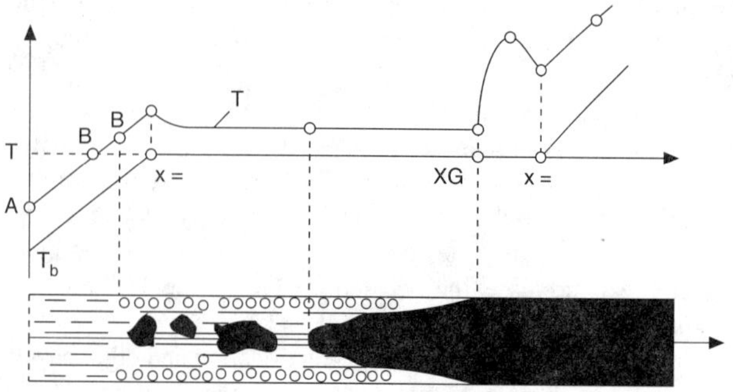

Figure 1. Regions of heat transfer in forced convective boiling (vertical tube shown horizontally for diagramatic purposes).

to the inlet. It also shows the change of the basic flow regimes along vapor-generation channel.

T_w is lower than T_s in the zone AB. Therefore, an ordinary convective heat transfer occurs between the wall and the liquid in this zone. The same is observed at the section BB′, where the wall superheat $\Delta T = T_w - T_s$ is insufficient to activate nucleation centers. The first vapor bubbles appear on the wall at the point B′. The degree of the wall overheating needed for incipience of boiling depends on local values of heat flux density $\dot{q}$, mass velocity of liquid in $\dot{q}$ and its subcooling $\Delta T_{sub} = T_s - \overline{T}_b$. We note that despite overheating of the liquid layers near the hot wall, the bulk flow temperature $\overline{T}_b$ at the point B′ remains lower that T_s. As a result the so called "surface boiling" or "subcooled boiling" is observed.

The subcooled boiling zone extends up to point C where $\overline{T}_b$ becomes equal to T_s, and the vapor quality of the flow $x = 0$. Here, $x = (h - h_L)/h_{LG}$, where h is the flow enthalpy, h_L the saturated liquid enthalpy on the saturation line, and h_{LG} the latent heat of vaporization. The zone of saturated liquid boiling follows, where $\overline{T}_b = T_s$ and $x > 0$. Initially, the vapor bubbles in subcooled boiling (B′C) do not break away from the wall or slip along it. At the condition of net vapor generation, bubbles leave the wall and are condensed in the flow of subcooled liquid; after this point, an ever increasing quantity of vapor is accumulated in the flow core. Eventually (for $x > 0$) the condensation process ceases.

In the neighborhood of point D, where the fraction of the channel cross section occupied by vapor is fairly large, an annular flow arises with the liquid film flowing by the channel wall and a vapor core being in the center (see **Annular Flow**). Within this regime, heat transfer occurs directly from the wall to the interface and, eventually, this process becomes so efficient that the wall temperature is insufficient to sustain nucleate boiling which is then suppressed. As the vapor-liquid mixture flows to still higher vapor quality region the quantity of liquid in the flow decreases and at a certain boundary vapor quality x_h (point E) wall dryout sets in, i.e., there is no longer any liquid-to-heat-transfer-surface contact, and the wall temperature rises. A transition occurs to dispersed, or the fog-type, flow of the mixture (zone EG).

The description of change of regimes with the growth of vapor quality x set forth above is slightly simplified. Actually the region B′E covers the bubbly (1) slug (2) churn (3) and annular-dispersed (4) flow regimes in a vertical channel (depicted in **Figure 2b**) and bubbly (1), plug (2), stratified (3), wave (4), slug (5), and annular-dispersed (6) flows in horizontal channels (**Figure 2a**). The wider variety of regimes in horizontal channels is due to gravity accounting for flow stratification.

For the most part, the above flow regimes are also observed in the channels of more complex geometry, such as annular and curvilinear channels and assemblies of fuel rods. Determination of the most probable flow regime can be conveniently done using the so-called regime charts (see **Gas-Liquid Flow**). The most commonly encountered ones are the Baker and Taitel-Dukler diagrams for horizontal flows and the Hewitt-Roberts and Oshinowo-Charles diagrams for upward and downward vertical flows respectively. It should be noted that the regime can be appreciably affected by the conditions of mixture injection to the channel and the presence and intensity of heat input on the wall. Therefore, the diagrams cannot be considered as universal and can be used only tentatively.

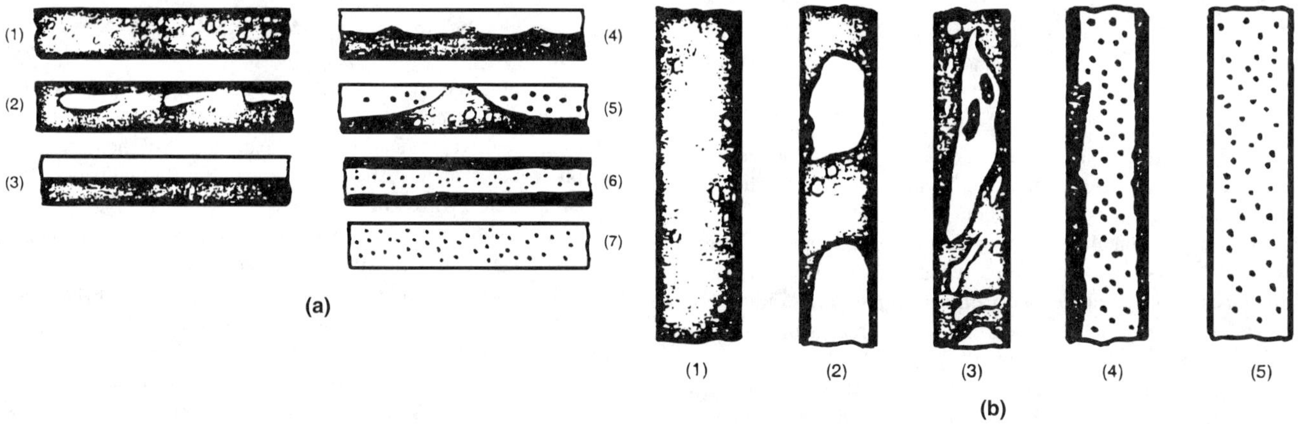

Figure 2 a & b. Flow regimes in horizontal and vertical channels.

Heat transfer in forced flow boiling is determined by both transfer of heat accumulated by vapor in the bubbles being broken off and by liquid convection.

Universal design formulas fitting all the regimes are not available for heat transfer. Commonly, individual relations are used for each region. Thus, for subcooled boiling the heat flux $\dot{q}$ removed from the hot wall is represented as a sum of two components, vis. a convective flux $\dot{q}_c$ and that of boiling with convective motion $\dot{q}_{cb}$

$$\dot{q} = \dot{q}_c + \dot{q}_{cb}. \qquad (1)$$

The component $\dot{q}_c = \alpha_L(T_s - \overline{T}_b)$, where α_L, the liquid heat transfer coefficient, is calculated by the formulas for single-phase convection heat transfer (see **Forced Convection**). The component $\dot{q}_{cb} = \alpha_{cb}(T_w - T_s)$, where α_{cb} is the heat transfer coefficient that is used to determine the nucleate boiling flux $\dot{q}_{cb}$. It approaches the heat transfer coefficient in pool boiling α_{pb} calculated with respect to $\dot{q}_{cb}$ rather than the total heat flux q. It is commonly assumed that $\alpha_{cb} = (0.7 - 0.8)\alpha_{pb}$, where $\alpha_{pb} = Aq_{cb}p^m$ (see **Pool Boiling**).

In the subcooled boiling region ($T_b < T_s$) the combined effect of nucleate boiling and forced convection is as illustrated in **Figure 3**. The dependence of heat transfer coefficient on the velocity of the liquid U_L with a single-phase convection without boiling $\alpha = \alpha_L$ is plotted as straight line 3. As the heat flux density grows, $\dot{q}_1 > \dot{q}_2 > \dot{q}_3$ heat transfer enhances and the dependences for α shift upward. It is also obvious that at low mixture velocities U_L has only a small effect on heat transfer, and curves 1 and 2 run horizontally with α close to the appropriate heat transfer coefficients in pool boiling α_{pb}. Conversely, at high U_L its effect on α turns out to be determining and curves 1 and 2 approach straight line 3.

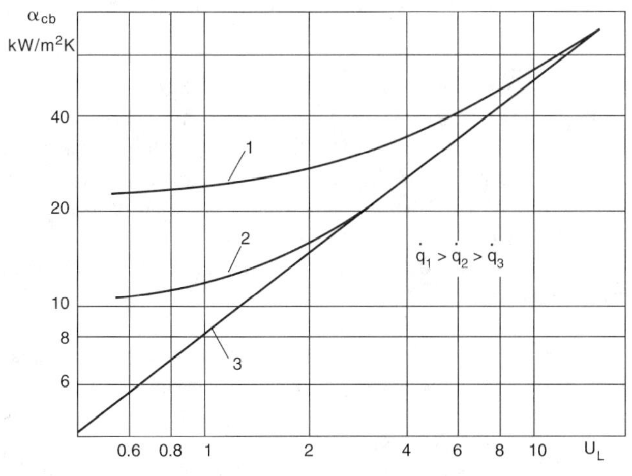

Figure 3. Variation of heat transfer coefficient with velocity and heat flux in the subcooled boiling region.

Quantitatively the heat transfer coefficient α_{cb} in the region of joint effect of nucleate boiling and forced convection is well described by Kutateladze's formula

$$\alpha = \alpha_L[1 + (\alpha_{cb}/\alpha_L)^2]^{1/2}, \qquad (2)$$

where, as before, $\alpha_{cb} = (0.7 - 0.8)\alpha_{pb}$.

In the developed boiling region use is often made of the approach formulated by Rohsenow and extended by Chen

$$\alpha_{tp} = \alpha_{mic} + \alpha_{mac}, \qquad (3)$$

where $\alpha_{mic} = \alpha_{pb}S_c$ determines the contribution of microconvection or nucleate boiling and $\alpha_{mac} = \alpha_L F_c$ determines the contribution made by macroconvection or forced convection.

Chen presents the curves for S_c and F_c which best describe experimental data of a great many researchers for water and some organic liquids. These curves are presented in **Figure 4**, where F_c is graphed as a function of Martinelli's parameter X

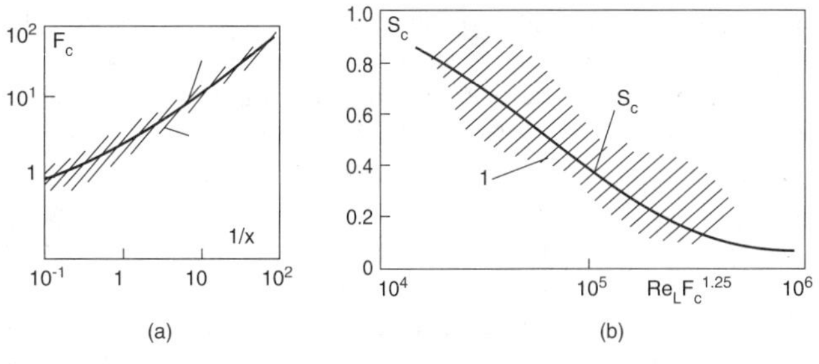

Figure 4. Factors in the Chen correlations.

$$X = \left[\frac{(dp/dz)_L}{(dp/dz)_G}\right]^{1/2},$$

where $(dp/dz)_L$ and $(dp/dz)_G$ are the frictional pressure gradients for the liquid and gas phases respectively flowing alone in the channel. The Reynolds number Re_L, which is needed for the calculation, is determined from the liquid flow rate.

The experimental data are within the dashed region **Figure 4**.

The factors such as material of the heat transfer surface, its roughness, contact angle (see **Contact Angle**), and fouling layer thickness on the heat transfer surface may exert an additional effect on the heat transfer coefficient α_{tp}.

When suppresion of nucleate boiling is observed (i.e., heat from the heating wall is transmitted by heat conduction through a thin liquid film to the interface, where vaporization occurs), the heat transfer coefficient α_{tp} can be determined by

$$\alpha_{tp}/\alpha_L = 3.5(X)^{-0.5}, \quad \textbf{(4)}$$

where α_L is calculated from the liquid phase velocity. Alternatively, the value of α_{tp} can be calculated from **Equation 3** with $\alpha_{mic} = 0$.

When the film is thin and when heat transfer through the film is governed by conduction rather than by turbulent convection, α_{tp} can be calculated from:

$$\alpha_{tp} \approx \lambda_L/\delta_L, \quad \textbf{(5)}$$

where λ_L is the thermal conductivity of liquid phase and δ_L the thickness of liquid film (see **Annular Flow**). This approach yields good results in estimating heat transfer in boiling of molten metals (high λ_L).

In the dispersed flow regime (zone FG in Figure 1) heat from the heating wall is first transmitted to the vapor and then from the vapor to the evaporating droplets. Within the framework of this two-stage process, the thermal resistance is mainly concentrated in transmission of heat to the vapor, and the heat transfer coefficient in pipes and channels is calculated by *Miropolskii's formula*

$$Nu_G = 0.023(Re_H Pr_G)^{0.8} Y,\ Y = 1 - 0.1\left(\frac{\rho_L}{\rho_G} - 1\right)^{0.4}(1 - x)^{0.4},$$

where $Nu_G = \alpha_{tp} d/\lambda_G$, $Re_H = \dot{m}d/\eta_G[x + \rho_G/\rho_L(1 - x)]$, ρ_L and ρ_G are the densities of liquid and vapor phase respectively, η_G and λ_G the dynamic viscosity and thermal conductivity of the vapor phase and d the channel diameter.

As was noted above, at vapor quality $x = x_b$ (Figure 1) the wall is dried out, which involves drastic deterioration of heat transfer and elevation of temperature of heat release surface. At sufficiently high heat flux densities burnout (transition heat flux) may also occur in other sections of the vapor-generating channel (zone B′ C D E in Figure 1). However, the nature of burnout turns out to be different. It will be associated not with the drying of the wall liquid film, but with coagulation of vapor bubbles into a continuous vapor film separating the wall from the main flow (see **Burnout (forced convection)**).

The overall pressure drop in the channel, where a two-phase mixture flows Δp_{tp}, is the sum of three components (see **Pressure Drop, Two-Phase Flow**)

$$\Delta p_{tp} = \Delta p_f + \Delta p_{ac} + \Delta p_h. \quad \textbf{(6)}$$

Here Δp_f is the pressure loss due to friction, Δp_{ac} the component due to the flow acceleration (of liquid and vapor phases) owing to the change of vapor quality, pressure or change of the flow cross section of the channel, Δp_h the pressure drop brought about by overcoming the hydrostatic pressure.

Either the homogeneous model or the separated flow model is used most often to describe two-phase flows and calculate pressure losses. In the first model the two-phase flow is treated as a homogeneous medium with averaged parameters (the velocity of the gaseous phase u_G and of the liquid u_L are equal, $1/\rho_H = x/\rho_G + 1 - x/\rho_L$). The most effective description is furnished by the homogeneous model for bubbly and dispersed flow regimes which are characterized by a fairly uniform distribution of the dispersed phase in a carrier medium (the liquid or vapor flow respectively). This model is also efficient for high pressures when the densities of liquid and vapor phases approach each other.

The separated flow model allowing for the difference in phase velocities, and their force, and energy interaction is most efficient in describing flows with extended interface boundaries, viz. stratified (in a horizontal channel), annular, annular-dispersed, wave, and other flows.

The components Δp_{ac} and Δp_h for the homogeneous flow model are calculated by integrating the equations:

$$\left(\frac{dp}{dz}\right)_{ac} = \dot{m}^2 \frac{d}{dz}\left(\frac{1}{\rho_H}\right) \quad \textbf{(7)}$$

and

$$\left(\frac{dp}{dz}\right)_h = \rho_H g \sin\theta, \quad \textbf{(8)}$$

where θ is the angle of inclination of the channel axis to the horizontal. Δp_f is obtained for the homogeneous model by integrating:

$$\left(\frac{dp}{dz}\right)_f = \xi \frac{1}{d}\frac{\dot{m}^2}{2}\left[1 + \psi x\left(\frac{\rho_G}{\rho_L} - 1\right)\right], \quad \textbf{(9)}$$

where the resistance coefficient $\xi = f(Re = \dot{m}d/\eta_L)$.

The presence of bubbles at the wall causes an increase in friction which is taken into account using the factor ψ. Rather cumbersome calculation formulas or the corresponding graphs are available for determining ψ. Generally ψ is a function of $\dot{m}$, pressure, and geometric characteristics of the channel.

For a discussion of the separated flow model, see the article **Pressure Drop, Two-Phase Flow**.

Boiling of liquid in channels is often accompanied by fluctuations of flow characteristics, which in engineering practice leads untimely to failure of equipment and significantly hampers its running. Fluctuations of parameters are inherent in some regimes, e.g. slug and plug ones, by virtue of their nature; in other cases they are an undesirable side effect which is to be suppressed (see **Flow Instabilities**). Static and dynamic instabilities of channels with boiling heat transfer agent are distinguished. The static (Ledinegg) instability is related to the fact that, at a constant input of thermal power, the same pressure drop in a vapor-generating channel may correspond to different combinations of flow rate and vapor quality. This may lead to a spontaneous reduction of flow rate of the liquid, growth of vapor quality, and development of off-design, often emergency, thermal conditions in the channel.

Dynamic instability most frequently manifests itself as density waves in which the channel exhibits pulsations of flow velocity with a certain frequency. This is due to the fact that the response of the system to variation of inlet parameters comes with a certain phase shift. The appearance of density waves is also promoted by pulsating heat release.

The pulsations can be also caused by compressed volumes (e.g. reservoirs filled with gas, vapor, or vapor-liquid mixture) as a result of acoustic effects. Stabilization of parameters in vapor-generating channels should receive primary consideration.

The behavior of vapor-generating channels in unsteady regimes, particularly following drastic increases of heat flux, has been inadequately investigated. In this case, the fluid dynamics, the heat transfer coefficient, and the conditions for achieving burnout may differ, sometimes substantially, from those implemented under steady regimes.

References

Butterworth, D., and Hewitt, G. eds. (1977) *Two-Phase Flow and Heat Transfer,* Oxford Univ. Press.

Collier, J. G. and Thome, J. (1994) *Forced Convective Boiling and Condensation* (3rd Edn.) Oxford University Press, Oxford.

Leading to: *Augmentation of heat transfer, two phase; Tubes and tube banks, boiling heat transfer in; Burnout, forced convection*

Yu.A. Zeigarnik

FORCED DRAFT AIR COOLED HEAT EXCHANGERS (see Air cooled heat exchangers)

FORCED VORTEX (see Vortices)

FOREST FIRES (see Fires)

FORM DRAG (see Crossflow)

FOSSIL FUEL FIRED BOILERS

Following from: *Boilers*

The term is generally used to describe **Boilers** in which steam is generated by firing *coal, oil or gas* in a furnace whose envelope consists of water-cooled tubes. Combustion products may be ducted over the tube banks providing superheat, reheat, economising and combustion air-heating duties. This type of equipment is also known as a *furnace-fired boiler* to distinguish it from ones in which solid fuel is burned on a grate or stoker or in a fluidised bed, these being the other major classes of fossil fuel fired boiler.

The boilers can be classified in a variety of ways depending on the firing scheme, as above; the duty, whether for electricity generation (utility boiler) or industrial heat and/or power purposes; the geometrical arrangement, tower, two-pass, **Coiled Tube, Water-tube**, *Shell* and many others; or by the type of generation, recirculation, natural circulation, **Once-through** etc.

Combustion systems and the furnace geometry are chosen to suit the type of fuel being burnt. Coal composition varies so widely that different concepts need to be adopted dependent on the carbon/volatile ratio, the ash content and the calorific value (CV). For example, most bituminous and sub-bituminous coals mined in the northern hemisphere have high enough CV and low enough ash contents to be suitable for wall firing whereas many southern hemisphere bituminous coals from India and South Africa have ash contents up to 45% and may require fluidised beds to sustain combustion.

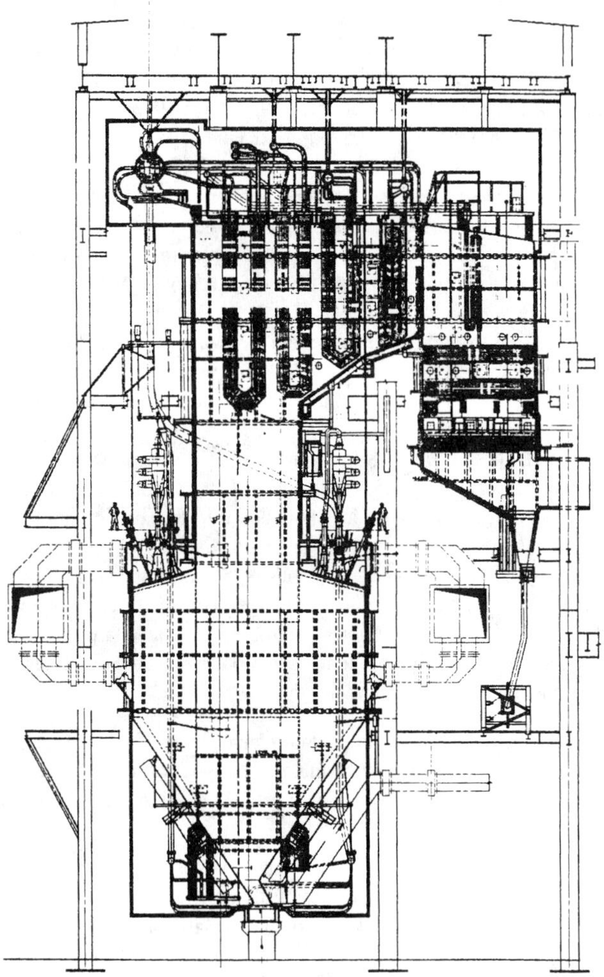

Figure 1. Babcock downshot-fired utility boiler (courtesy of Babcock Energy Ltd).

In contrast, anthracite and some bituminous coals have such a low volatile content that downshot firing has to be adopted (as shown for example in **Figure 1**).

A.E. Ruffell

FOSSIL FUELS (see Fuels)

FOULING

The fouling of heat exchangers may be defined as the accumulation of unwanted deposits on heat transfer surfaces. The foulant layer imposes an additional resistance to heat transfer and the narrowing of the flow area, due to the presence of deposit, results in an increased velocity for a given volumetric flow rate. Furthermore, the deposit is usually hydrodynamically rough so that there is an increased resistance to the flow of the fluid across the deposit surface. Therefore, the consequences of fouling are, in general, a reduction in exchanger efficiency and other associated operating problems including excessive pressure drop across the exchanger. It is only in recent years that the problem of heat exchanger fouling has attracted scientific and theoretical treatment and many aspects remain to be investigated. Fouling of heat exchangers has been reviewed (Bott 1990, Hewitt et al. 1994, Bott 1995).

In some heat exchangers fouling occurs rapidly, in others the equipment may operate for long periods, perhaps several years, before a problem becomes apparent. Much depends on the particular fluid and the conditions under which the exchanger operates. The nature of deposits is extremely variable. In some examples, deposits are hard, tenacious and difficult to remove. Other accumulations are soft and friable that lend themselves to removal.

It is usual to find that deposits are made up of different components. The deposit associated with cooling water for instance, may include: corrosion products, particulate matter, crystals and living biological material.

The extent of each of the components in the deposit will depend on many factors including the origin of the water, its treatment and the processing conditions. It is possible that one component is dominant, e.g. scale formation or corrosion. Because of this extremely variable quality of deposits, it has become common practice to consider different fouling mechanisms in the development of techniques to mitigate the problem. Six mechanisms have been identified that give rise to fouling problems in heat exchangers. They include:

1. *Crystallisation fouling* often referred to in water systems as *scaling,* involves the formation of crystals form solution on the surface or deposition of crystals formed in the bulk liquid or in the laminar sub-layer.
2. *Particulate fouling,* as the name suggests, depends on the arrival of discreet particles at the transfer surface. Particles may be small (i.e. <1 μm) or may be large (i.e. several mm). Particulate fouling is common in both liquid and gas systems.
3. *Biological fouling* may be defined as the growth of living matter on heat exchanger surfaces. The phenomenon usually associated with water systems, e.g. cooling water, involves both micro-organisms and macro-organisms. The former may include bacteria, fungi or algae while the latter includes mussels and barnacles.
4. Chemical reactions at or near the surface may give rise to what is often called *chemical reaction fouling.* The effect of heat on a process fluid as it passes through the exchanger, may accelerate chemical reactions, e.g. cracking or polymerisation reactions that can give rise to deposition on the surface. In some instances the metallic surface of the heat exchanger acts as a catalyst thereby aiding the fouling process. Chemical reaction fouling is usually associated with liquids but it may also occur in vapour or gas streams.
5. Some heat exchange materials of construction are subject to corrosion from the aggressive nature of the fluids or impurities in the fluids in contact with the surface. The result may be a protective layer, but more likely, the corrosion process produces a thick corrosion layer. *Corrosion fouling* may be observed in both liquid and gas systems.
6. In cooling operations, it is possible to freeze a liquid being processed at the cold surface. The frozen layer represents a resistance to heat transfer. For instance, *freezing fouling* may occur during the production of chilled water.

Although the different mechanisms have been identified as leading to fouling, it is possible to consider fouling in an idealised, general way. If the deposit thickness due to fouling is plotted against time, three idealised curves may be visualised as shown on **Figure 1**.

Curve A represents a straight line relationship, i.e. deposit thickness is proportional to time. Curve B is a falling rate curve, i.e. the rate of deposition declines with time. Curve C is the usual form of the relationship between deposit thickness (or thermal fouling resistance) and time; it is an exponential or asymptotic curve. For a short time the heat exchanger fouls relatively slowly, but after a period of time the rate of deposition rapidly accelerates to be followed by a fall in the rate of deposition. Eventually the deposit thickness remains constant. The thickness of a deposit is a measure of the resistance of the deposit to the transfer of heat so that similar curves would result if foulant thermal resistance was plotted against time.

The curves on **Figure 1** are shown to develop from the origin. In some examples of fouling, an initiation or induction period is required before fouling begins. Induction periods may be extremely

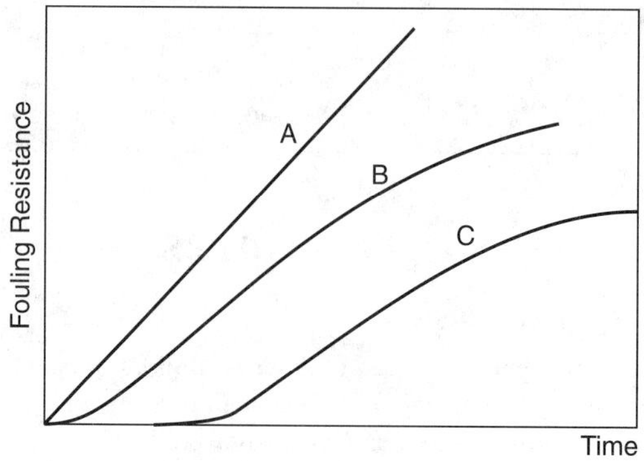

Figure 1. Idealised fouling curves.

short, i.e. a few seconds, or they may involve several weeks or months. The extent of the induction period depends on the nature of the heat exchanger surface, i.e. the material of construction and the surface roughness, together with the properties of the fluid (and any impurities) passing over the surface.

The presence of a deposit on the surfaces of a heat exchanger give rise to two major problems:

1. The efficiency of the heat exchanger is reduced in respect of heat transfer as a result of the thermal resistance of the deposit. In general, the thermal conductivity of deposits is very much lower than metals so that even a thin layer can cause significant thermal resistance. The fact that the surface of the foulant layer is rough compared to the original metal surface will, in general, increase the heat transfer due to the increased turbulence generated by the roughness elements. To some extent, this offsets the effects of increased thermal resistance across the heat exchanger. However, the benefits, due to the presence of the fouling layer, are usually relatively small compared with the restrictions to heat flow imposed by the insulating properties of the foulant.

   The overall effect on heat transfer may be summarised by the following statement:

   Change in overall heat transfer coefficient is some function of

   - Change due to the thermal resistance of the foulant layer
   - Change due to the surface roughness of the deposit
   - Change due to the increased velocity for a given volumetric flow rate, resulting from the restrictions on flow area imposed by the presence of the deposit.

   Although the dependence has been cited regarding the heat exchanger as a whole, it is strictly applicable to local conditions. Still, the incidence and quality of a deposit is likely to vary with its location in the exchanger.

2. Due to the roughness of the deposit and the restrictions of the flow area that are responsible for enhanced turbulence and heat transfer compared to the condition for the same flow rate, the pressure drop also increases. Relatively the layers on the inside of a tube can, for instance, give rise to substantial increases in pressure drop. For example a 1mm layer on the inside of a heat exchanger tube with a nominal internal diameter of 18mm increases the velocity for a given volumetric flow rate by a factor or $18^2/16^2 = 1.27$.

Since pressure drop is a function of the square of the velocity, the pressure drop with the deposit in place, will be increased by a factor of 1.27^2 or around 1.6 times, i.e. a 60% increase. These calculations neglect the effect of deposit roughness so that, if it were possible to take the additional contribution to pressure drop into account numerically, the pressure drop through the exchanger under fouled conditions may approach double that for a clean exchanger.

Reduced heat transfer and increased pressure drop under fouled conditions can have significant implications for energy utilisation on the process plant—namely reduced heat recovery that may have to be made good using primary energy, and increased energy requirements for monitoring the fluid through the exchanger. Both effects have implications in terms of cost of operation. The *economic penalties of* heat exchanger *fouling* on the financial performance include other effects that are due to the presence of deposits that may not always be recognised at the design stage, or during subsequent operation.

During the initial stages of heat exchanger selection when the basic concepts of the design are being considered, the problem of potential fouling will be a major concern. It is usual, after careful thought, to include additional thermal resistance over and above that for the clean conditions, to allow for the anticipated deposit accumulation. The result is an increase in heat transfer area for a given heat load and temperature change requirements. The increase in heat transfer area (i.e. an increase in the size of the heat exchanger for a given duty) represents an increase in capital cost. The choice of the anticipated increase in heat transfer resistance due to the fouling is crucial to the ultimate cost of the heat exchanger.

Where excessive fouling is anticipated and a particular heat exchanger is vital to the process, it may be prudent to duplicate the heat exchanger in addition to allowing for deposit accumulation in the basic design, so that production can be maintained without the need to shut down the process.

Duplication and oversizing, to allow for fouling, represents substantial additional capital costs. Where special materials of construction are required, due to the nature of the fluids passing through the heat exchanger, the additional cost may be considerable.

The cost of fouling

Unless the problem of fouling is recognised at the design stage and adequate steps taken to ensure that the problem is contained, serious operating and maintenance difficulties may result that add to the cost of production. Emergency shut down may be necessary because the heat exchanger rapidly loses heat transfer efficiency or more likely, the flow cannot be maintained due to the excessive pressure drop generated by the presence of the deposit. A further consequence of the high pressures necessary to drive the fluid through the exchanger may be failure of joints and packings, and increased wear and tear on the associated pumps. Furthermore the presence of deposits may encourage corrosion of the underlying metal with the need for early replacement.

All the factors directly attributable to fouling represent increased maintenance costs that may be considerable in some cases. Where fouling is experienced, sooner or later, it will be necessary to clean the heat exchanger. Not only will the cleaning process involve labour costs, it may require special equipment, particularly if chemical cleaning is involved. Additional circuitry involving pumps, tanks, pipelines and valves may be required, chemicals purchased and the cleaning process may produce effluents that require treatment before discharge.

Conventional *cleaning techniques* such as tube drilling, surface brushing and high pressure water jetting will also involve some financial investment in addition to the cost of the associated labour. Unless adequate precautions are taken, the cleaning process may result in damage to the heat exchanger. In some instances where deposits are tenacious and difficult to remove, damage may be inevitable.

Additional maintenance and cleaning means that the process will have to be shut down to allow access unless standby equipment has been provided. As a consequence, there is a period of lost production which represents a loss of return on the capital investment in the process equipment as a whole, and reduced profitability. Lost production as a result of unplanned shut down due to any cause including fouling, can have a substantial effect where there is keen competition.

It also has to be remembered that continual difficulties with the operation of a process plant, especially heat exchangers subject to excessive fouling, will affect the morale of the labour force. It may be particularly so where production bonuses and incentives are involved. The outcome of these concerns may be a general lowering of a sense of responsibility, poor housekeeping and slack attitudes. The costs are difficult to quantify, but there will be a cost penalty associated with these personnel problems.

The problem of fouling has led to remedial technologies that may include on-line cleaning or the use of additives. The purpose is to prevent fouling occurring or to reduce its effects on the performance of the heat exchanger in question. Mitigation by whatever means will involve additional costs either in terms of capital for the associated equipment or the purchase of treatment chemicals.

The cost of heat exchanger fouling can be considerable although much depends on the nature of the fluids being handled, the design of the heat exchanger and the conditions under which it is operated.

The *economic penalties* described above may be summarised:

1. Increased capital costs i.e. additional heat transfer area, mitigation and cleaning equipment.
2. Additional energy requirement to allow for reduced energy recovery.
3. Labour costs associated with additional maintenance, cleaning and mitigation.
4. Cost of any antifoulant chemicals.
5. Lost income resulting from lost production.
6. Equipment replacement costs.
7. Additional costs associated with low labour morale.

Fouling Mechanisms

The underlying mechanism involved in the accumulation of deposits on surfaces of heat exchangers may generally be considered to involve three stages:

1. The foulant or the agents or impurities (e.g. bacteria, solid particles or corrosive agents) that lead to deposit formation, approach the surface from the bulk through the viscous sublayer adjacent to the heat exchanger surface, across which the fluid is flowing. The principles of mass transfer apply.
2. At the surface, adhesion can take place involving a number of factors including the interaction of surface forces, chemical reactions and structural orientation.
3. Once deposited on the surface, the material may be subject to forces that compact or weaken the deposit. The quality of the deposit is likely to be time dependent.

These three stages will be influenced by a number of system parameters, but principally these are associated with fluid flow, heat and mass transfer. As a consequence, the process whereby a surface becomes fouled and the deposit maintained, is complex resulting from the interaction of a number of factors. Although it is extremely rare for a single mechanism to apply, it is useful to examine separately the mechanisms briefly mentioned earlier in order to have an appreciation of the effects of the system variables involved.

Crystallisation

The deposition or formation of crystals on a surface, sometimes referred to as scaling, is often associated with aqueous systems, e.g. cooling water circuits. (See also **Crystallisation**.)

Before crystallisation can occur, it is necessary to have conditions of supersaturation, i.e. dissolved solid concentrations above the saturation solubility at a particular temperature. The supersaturation provides the "driving force" for precipitation to occur. In general, the degree of supersaturation involved is quite small; its location will depend very much on the temperature distribution within the heat exchanger between the bulk fluid and surface temperatures.

Two solubilities of salts in water are recognised. Normal solubility salts (e.g. NaCl or Na_2SO_4) display increased solubility as the solution temperature is raised, so that when a saturated solution is cooled supersaturation can occur followed by precipitation. Fouling from normal solubility salts is likely when these solutions are cooled. Inverse solubility salts (e.g. $CaCO_3$, $Mg_2(SO_4)$), on the other hand, have reduced solubility as the temperature is raised. Consequently, as the temperature of a saturated solution increases, supersaturation and precipitation will occur. Fouling or scaling due to inverse solubility salts is probable when solutions of these salts are heated. Furthermore, the solubility of inverse solubility salts in general is quite low, so that relatively small temperature changes are likely to cause fouling problems. The problem is very prevalent in cooling water systems due to the so-called hardness salts usually present in the water. There are two contributory factors:

1. As the cooling process takes place the temperature of the water rises. If the water is already saturated even a small temperature rise leads to precipitation.
2. In cooling water systems, the technique used to remove unwanted heat is to evaporate some of the water in a cooling tower or spray pond. The loss of water vapour concentrates the dissolved salts so that saturation is soon achieved with the consequent increased possibility of precipitation of sparingly soluble salts. To offset this concentration effect, some of the saturated water is removed in the so-called "blow down" discharge, to be replaced with fresh "make up" water.

The scale produced on heat exchanger surfaces as a result of crystallisation from water is often tenacious and difficult to remove, although in some instances, it is soft, resembling a sludge that is more easily removed from the surface. The condition of the deposit very much depends on the conditions, particularly of temperatures that prevail close to the solid surfaces. The colour of scale depends very much on the system involved. In purer forms, the colour is near white, but colour can be imparted to the scale from trace compounds in the system, e.g. oxides of iron that may give black (magnetite) or red (haematite) colours.

Controlling scale formation in boilers is particularly important, since the accumulation of hardness salts on the heat transfer surfaces (e.g. the inside of water tubes) can lead to catastrophic failure. The presence of the scale can lead to excessive metal temperatures and consequently rupture.

Particle Deposition

Particle accumulation on heat exchanger surfaces is a common fouling phenomenon. A wide range of particle sizes may be involved in the deposition process. The origin of the particles varies. In some instances (e.g. water and crude oil), the particulate matter may be inherent in fluid stream.

In combustion systems, particulate matter (e.g. the mineral matter in coal) may be released as the combustible components are burnt. The mineral particles go forward with the flue gasses and are deposited on the heat exchanger surfaces. It is also possible in combustion systems, that, if the combustion process is not adequately controlled, incomplete combustion occurs and particles of unburnt carbon are carried in the gas stream. The high temperatures prevailing in combustion systems may mean that some particulate matter is in the form of liquid droplets that solidify as they are cooled. Agglomeration and chemical reaction of the particles with gaseous components in the flue gas can add to the complexity of the fouling process. (See also **Combustion** and **Flames**.)

In some process streams, temperature effects lead to chemical reactions which produce large molecules (e.g. polymers) as particulate matter capable of being deposited on surfaces.

Biological Growth on Surfaces

Biofouling of heat exchanger surfaces is generally falls in two groups: macrofouling and microfouling. In general, aqueous environments are involved such as might be found in a cooling water system. Water originates from a natural source such as a river or the sea. Generally macrofouling is associated with living creatures such as mussels and barnacles. It is also possible for vegetation such as sea weed to become attached and grow on surfaces. Macrofouling is often experienced—although not exclusively—in the use of sea water. Microfouling is the result of the attachment and growth of microorganisms on surfaces. As far as heat exchangers are concerned, the micro organisms involved are usually bacteria. Other microorganisms that may be found in cooling water systems include fungi and algae. Algae need a source of light. In nature this is sunlight; so these microbes are to be found in exposed regions, for example the basin below a cooling tower. Fungi, on the other hand, may be found on the internal structure of a cooling tower. The structure of the tower can be damaged and destroyed by fungal activity.

Although algae and fungi may not grow on heat exchanger surfaces, they may still be the source of deposits on the heat transfer surfaces. With the passage of time, either living or dead algae and fungi may become detached upstream of the heat exchanger only to deposit in the downstream equipment. Under the conditions, this organic material may provide a suitable nutrient for supporting bacterial activity. In general, the surface of heat exchangers in cooling water systems is ideal for microbial growth. The aqueous medium contains nutrients, it is aerated due to passage through a cooling tower or spray pond, and the temperature is near the optimum for maximum biological activity, i.e. in the range 20–50°C.

The slime layer associated with fouling from micro-organisms is characterised by a gelatinous, hydrated material often nearly transparent, but more usually coloured by micro-organisms, or by other contaminants in the water (e.g. oxides or hydroxides of iron). Another common feature of biofilms is that they are uneven and deformable under the action of the water flow. Under certain conditions filamentous structures grow out from the surface and are free to oscillate within the flowing water. Biofilms are notoriously difficult to remove from surfaces, but if they spontaneously slough off the surface as does happen during operation, the lumps of biofilm can cause blockages downstream. Very thick biofilms with serious implications for heat transfer and pressure drop are possible under favourable conditions.

The presence of a biofilm on a metal surface may promote corrosion. The local conditions under the biofilm are likely to be different from those in the bulk flow, this is particularly applicable to the pH. The activity of the living material is likely to lower the local pH that could initiate corrosion. Furthermore, under a biofilm, due to the restrictions on ingress of bulk fluid, galvanic cells may be established that enhances perhaps only locally, the rate of corrosion.

Chemical Reaction Fouling

The accumulation of deposits may occur as the result of chemical reactions at or near a heat exchanger surface. Because chemical reactions are usually enhanced by raised temperatures, *chemical reaction fouling* may be experienced where a reactive fluid is being heated. Under these circumstances, the slow-moving layers of fluid near the heat transfer surface are subject to relatively high temperatures.

Fouling due to the incidence of chemical reactions can occur in a wide range of process streams and over a wide range of temperature, i.e. from near ambient to temperatures associated with combustion conditions. Processes where chemical reaction is responsible for deposit formation, include food processing (e.g. the pasteurisation of milk), oil refining (e.g. preheating crude oil prior to primary distillation), and chemical manufacturing that involves polymerisation reactions. Even in combustion, the impurities contained in the original fuel (e.g. coal or combustible waste) may give rise to chemical reactions in conjunction with acid gases such as SO_2 and SO_3 contained in the flue gases. Incomplete combustion may lead to the generation of soot particles that deposit on heat transfer surfaces. Free radicals may be responsible for chain reactions in organic liquids that ultimately lead to deposit formation. These examples of chemical reaction demonstrate the wide range of substances that may be regarded as being deposits from chemical reactions. Consequently, the deposits may be soft and friable or hard and difficult to remove depending on the chemical reactions involved and the conditions.

Corrosion of Surfaces

Unlike other fouling mechanisms, in *corrosion fouling* the heat transfer surface itself degenerates to form a layer of lower thermal conductivity than the original metal. The resistance of many metals and alloys to corrosion is due to the presence of a thin layer of oxide on the surface that restricts the flow of electrons and ions usually necessary for corrosion to occur. A protective oxide layer may be regarded as controlled but desirable corrosion. At the same time it also represents a thermal resistance. In general, the thickness of the protective layer is such that the heat flow is not greatly impaired. If however, conditions are such that the corrosion of the surface is extensive, then the oxide layers (and probably hydroxide layers in aqueous systems) represent a fouling problem. It is also possible that corrosion is facilitated by the removal of protective oxide layers by aggressive chemical agents, e.g. acid attack.

In combustion systems, inorganic salts deposited on heat transfer surfaces may be subject to such high temperatures that they become molten. The liquid condition may provide pathways for electron transfer that accelerates corrosion beneath the deposit.

Corrosion fouling is very much dependent upon the material of construction from which the heat exchanger is fabricated. The problem of corrosion fouling can be eliminated by the correct choice of construction material, but in general, corrosion resistant alloys are expensive. Circumstances may be such that the high costs involved cannot be entertained. Other techniques must then be employed to restrict the incidence of corrosion.

Freezing Fouling

Freezing fouling may occur where the temperature in the region of the heat transfer surface is reduced to below the freezing point of the fluid being processed. The deposition of wax from waxy hydrocarbons by cooling is often considered to represent freezing fouling, but it is probably better defined as crystallisation fouling. A good example of freezing fouling is the production of chilled water. If the temperature of the heat exchanger surface on the water side is at or below 0°C, then it is likely that a layer of ice will form on the surface. The thickness of the ice deposit will be very dependent on the magnitude of the temperature distribution between the coolant on the one side of the exchanger and the water on the other. The elimination of freezing fouling may be achieved by the choice of coolant temperature, so that the surface in contact with the liquid from which heat is being extracted is slightly higher than the freezing point of the liquid.

Mixed Mechanisms of Fouling

Although six mechanisms of fouling have been briefly described, it is rare for practical heat exchanger fouling to be the result of a single mechanism. In most process streams where fouling occurs, two or probably more mechanisms are involved. It is possible that one mechanism may be dominant and, from a practical standpoint, the other mechanisms present can be ignored when remedial action is being considered. In cooling water systems, it is likely that, in addition to micro-organisms, the circulating water will contain dissolved solids, suspended particulate matter and, perhaps, also aggressive chemicals. The accumulated deposit on the equipment surfaces may therefore contain micro-organisms, particles, scale and products of corrosion. The gelatinous nature of the biofilm may aid the development of the foulant layer by capturing particles as they collide with its surface. Concentration effects may occur near the film that encourages crystal formation, and the changed conditions underneath the deposit may enhance corrosion.

In fouling associated with combustion, the fouling on heat exchangers may be due to particle deposition, chemical reactions, and corrosion as described earlier.

It will be clear from these two examples that the process of fouling may be extremely complex necessitating as it does, a rather empirical approach to its understanding and investigation.

Factors that affect the incidence of fouling

A number of system variables affect the incidence of fouling on heat exchange surfaces, but three generally carry more importance than all others including: fluid temperature and the associated temperature distribution, stream velocities and—as would be expected—the concentration of all foulant, or foulant precursors that are contained in the fluid streams. Variables of less general importance, but which may assume significance in certain examples, include: pH, material of construction and associated surface condition. Attention to the magnitude of these variables associated with all mechanisms of fouling, can go a long way to mitigating particular fouling problems, although it has to be said that certain problems may not respond as well as others.

Guidelines associated with temperature that are useful for the initial design and subsequent operation of heat exchangers suggest:

Low temperatures will:

1. reduce the affects of chemical reaction and corrosion since rates of reaction are generally temperature sensitive; high temperatures favour accelerated reactions;
2. reduce the activity of micro- and macro-organisms;
3. reduce the opportunity for supersaturation conditions to occur for inverse solubility salts.

High temperatures will:

1. reduce the incidence of biofouling where the temperature is above that for optimum growth;
2. avoid conditions that could lead to freezing fouling;
3. reduce the opportunity of supersaturation conditions to occur with dissolved salts with normal solubility.

In addition to temperature effects that influence the fouling process, temperature may also be a factor in the long term retention of the deposit on the surface. Over a period of time it is more than likely that a particular deposit will age. The aging process may be influenced by temperature. The effects could be beneficial or detrimental to the continued operation of the heat exchanger. It is possible that, due to internal chemical reactions, the deposit becomes more tenacious and difficult to remove. For other encrustations, the effect of changed temperature distribution as the deposit grows in thickness, planes of weakness and inconsistencies in the deposit lead to fracture and spalling.

The temperature effects are, in general, associated with the temperature distribution across the heat exchanger. For a given temperature difference between the hot and cold streams within the equipment, the growth of deposit (usually on both sides): will affect the distribution of that total temperature driving force so that the metal dividing wall separating the fluids; will experience a changing temperature. The deposits themselves will also be affected in terms of their respective temperatures. For large temperature differences and thick deposits, there is likely to be a considerable temperature difference across the deposit with implications for the quality of the deposit. For instance, the chemical reactions involved when the deposit is relatively thin, may be quite different from those associated with thick deposits. Such conditions, for instance on the flue gas side of coal combustion equipment, may give rise to stratified deposits and chemical transformations as time passes.

Although the temperature of the streams within a heat exchanger will be specified, there is some flexibility open to the designer to modify the temperature distribution. By investigating changes in velocity that affect the thermal resistance in the respective streams, it is possible to change beneficially, the various interface temperatures. The changes in velocity have implications in their own right.

Some comments on the effects of velocity have already been made namely the effects on temperature distribution and pressure drop. The latter is closely linked to fluid shear: increasing velocity increases fluid shear at the interface between a solid surface and the fluid flowing across it. High shear forces may result in foulant removal, that tends to maintain a static fouled condition, i.e. near the asymptotic or equilibrium fouling condition. Under these circumstances the velocity controls the deposit thickness. Increasing velocity may appear attractive for minimising the effects of deposits, but for a particular deposit, the necessary velocity may be unacceptably high leading to high pumping costs and possibly problems of erosion. It also has to be remembered that increasing velocity will increase turbulence, so that where the fouling process is mass transfer controlled, deposition is facilitated. In biological fouling, for instance higher velocities, although leading to enhanced removal opportunities, will also facilitate nutrient transfer to the living matter colonising the particular surface.

The choice of velocity therefore, is very much a compromise depending on the particular system under consideration. As a rough guide velocities of the order of 2m/s for liquid flows in tubes will have some controlling effect without excessive pumping costs. On the shell side mean fluid velocities for liquids across the tube bank of around 1m/s may be regarded as a suitable guide. In gas systems much higher velocities are possible but it is difficult to provide reliable guidelines.

In general, the higher the concentration of foulant or deposit precursor, the greater the fouling of surfaces is likely to be, since the mass transfer driving force, i.e. the concentration gradient towards the target surface is enhanced. It is usually not in the gift of the heat exchanger designer or operator, to influence the concentration of foulant precursors in the stream handled by the exchanger. Often the potential fouling problem is not recognised at the design stage; it may only be discovered during subsequent operation as trace constituents of the flow stream. It may be possible to limit the deposit precursor, for example particulate matter or unreacted species, by improved control of processes upstream from the exchanger. In certain exceptional circumstances it may be necessary to reduce, or remove altogether, the components responsible for the fouling process.

Fouling Resistance Concept

The usual method of allowing for the incidence of fouling in heat exchanger design is to employ resistances that account for the fouling on both sides of the exchanger. Sometimes these fouling reactions are referred to as "*fouling factors*". The latter description is not altogether satisfactory since the term "factor" is usually applied to a multiplier.

The thermal resistance due to fouling is additive as illustrated in the following equation which sums all the thermal resistance between the two fluids. If the **Overall Heat Transfer Coefficient** is U_1 then:

$$1/U_1 = R_1 + R_2 + R_w + 1/\alpha_1 + D_1/D_2 \cdot 1/\alpha_2$$

where R_1 and R_2 are the fouling resistances associated with fluid streams 1 and 2 respectively.

R_w represents the thermal resistance of the metal wall separating the two fluids. In general, this resistance is quite small and often it can be neglected altogether due to the high thermal conductivity of metals. α_1 and α_2 are the heat transfer coefficients at the metal wall for fluids 1 and 2 respectively, and D_1 and D_2 are the inner and outer diameter of the tube through which fluid 1 is passing. Fluid 2 passes over the outside of the tube. It will be seen that, in reality, the equation is not mathematically sound except for steady state conditions, i.e. when the deposit thickness on both sides of the exchanger remains constant. Under these conditions it is likely that the asymptotic fouling resistance has been attained. The earlier discussion has shown that fouling development is transient so that the steady state condition does not apply.

In order to help designers and others, tables of fouling resistances are published, e.g. the TEMA (Tubular Exchanger Manufacturers' Association) fouling resistances (Chenoweth 1990). The data are classified according to the fluid and process and the figures are based on the experience of recognised experts in the field. Although the information is a useful guide, it has to be treated with caution in the light of the earlier discussion on the influence of temperature, velocity and foulant precursor concentration. In general, the tables do not specify any of the variables so that it becomes difficult to relate them to a particular set of conditions. It also has to be understood that the published fouling resistances are only applicable to shell and tube heat exchangers and may not be used in the design of plate heat exchangers for instance. Furthermore, it has to be remembered that by taking into account the anticipated fouling resistance, a clean (newly on stream) heat exchanger will overperform. To compensate, adjustments to the fluid flow rate(s) will be made that could encourage the fouling process so that the fouled condition prediction is self-fulfilling.

Wherever possible data on fouling resistances relating to the actual process streams and the conditions of velocity and temperature pertaining to the particular design should be used for assessment purposes. Unfortunately these data are not generally to hand. The choice, then, becomes one of experience and judgement with guidance from published figures. In this connection, it has to be appreciated that the increased capital cost of a heat exchanger over and above the clean condition to allow for fouling, may very well depend upon the arbitrary choice of fouling resistance.

Mitigation of Fouling

In order to control the incidence of fouling a wide range of on-line mitigation techniques may be employed, but they generally fall into two groups namely mechanical methods and the use of chemical additives.

Mechanical methods as the name suggests, use physical methods of removal. Some examples are given in the following table:

Table 1. Mechanical control techniques

Technique	Application
Circulation of sponge rubber balls	Inside of power station condenser tubes
Brush and cage systems	Inside of tubes in shell and tube heat exchangers
Soot blowers	Outside of tubes in combustion spaces
Sonic vibration	Tubular exchangers
Vibrating springs activated by the fluid flow	Inside of tubes

Some examples of the use of additives are given in the following table:

It will be seen that wherever possible the use of the additive is complementary to the shear effects produced by the fluid velocity across the heat exchanger surface. Other additives impart changes either to the depositing particles or the surface so that the particles are "held off" the surface.

In general the concentration of the additive is relatively small, i.e. up to 100 mg/l but in many applications the concentration may be as low as 1 mg/l. In order to be cost effective the concentration of additive must be kept as low as possible.

Table 2. Chemical additive application

Fouling problem	Additive
Scaling or crystallisation	Dispersant to prevent attachment. Crystal modifier to weaken the deposit structure.
Particle deposition	Dispersant to prevent deposition. Flocculant followed by settling or filtration. Modifiers to reduce the strength of deposits in combustion systems
Biological growth	Biocide to kill the living matter. Biostat to reduce biological activity. Dispersant to prevent attachment to surfaces.
Chemical reaction	Antioxidants to prevent oxidation. Reaction chain terminators to remove free radicals.
Corrosion	Corrosion inhibitors to restrict the opportu - nity for corrosion reactions Chemicals to maintain protective oxide layers. Chemicals that react with acid combustion gases.
Freezing	Crystal modifiers to weaken the deposit structure.

References

Bott T. R. (1990) *Fouling Notebook,* Instn. Chem. Engrs.

Hewitt G. F., Shires G. L. and Bott T. R. (1994) *Process Heat Transfer,* CRC Press, Boca Raton.

Bott T. R. (1995) *Fouling of Heat Exchangers,* Elsevier, Amsterdam.

Cheroweth, J. M., (1990) Final report of the HTRI/TEMA Joint Committee to Review the Fouling Section of the TEMA Standards. *Heat Trans. Eng.,* 11, 1. 73–107.

T.R. Bott

FOULING FACTORS (see Fouling)

FOULING RESISTANCE (see Fouling; Wilson plot)

FOURIER-BESSEL SERIES (See Bessel's series)

FOURIER EQUATION (see Conduction)

FOURIER FLUID (see Flow of fluids)

FOURIER INTEGRAL (see Harmonic analysis)

FOURIER, BARON JEAN BAPTISTE JOSEPH (1768–1830)

French mathematician and physicist famous for his pioneer work on the representation of functions by trigonometric series, Fourier was born at Auxere on March 21, 1768, the son of a tailor. He became a teacher in mathematics in 1784 at the military school there. He taught at the Ecole Normale at Paris from its founding in 1795, where

JEAN BAPT. JOS. FOURIER 1768–1830

Figure 1.

his success soon led to the offer of the Chair of Analysis at the Ecole Polytechnique. In 1807 he was made a member of the academy of sciences.

Fourier's masterpiece was his mathematical theory of heat conduction stated in "Theorie Analytique De La Chaleur" (1822), one of the most important books published in the 19th century. It marked an epoch both in the history of pure and of applied mathematics, for in it, Fourier developed the theory of the series known by his name and applied it to the solution of boundary-value problems in partial differential equations. This work brought to a close a long controversy, and henceforth it was generally agreed that almost any function of a real variable can be represented by a series involving the sines and cosines of integral multiples of the variable. Fourier died in Paris on May 16, 1830.

U. Grigull, A. Sandner and J. Straub

FOURIER'S LAW

***Following from:** Heat transfer*

An empirical relationship between the conduction rate in a material and the temperature gradient in the direction of energy flow, first formulated by Fourier in 1822 [see Fourier (1955)] who concluded that "the heat flux resulting from thermal conduction is proportional to the magnitude of the temperature gradient and opposite to it in sign". For a unidirectional conduction process this observation may be expressed as:

$$\dot{q}_x = -\lambda \frac{dT}{dx}$$

where the vector $\dot{q}_x$ is the heat flux (W/m^2) in the positive x-direction, dT/dx is the (negative) temperature gradient (K/m) in the direction of heat flow (i.e. conduction occurs in the direction of decreasing temperature and the minus sign confirms this thermodynamic axiom) and the proportionality constant λ is the **Thermal Conductivity** of the material (W/mK). Fourier's Law thus provides the definition of thermal conductivity and forms the basis of many methods of determining its value. Fourier's Law, as the basic rate equation of the conduction process, when combined with the principle of conservation of energy, also forms the basis for the analysis of most **Conduction** problems.

References

Fourier. J. (1955) *The Analytical Theory of Heat,* Dover Publications, New York.

***Leading to:** Conduction; Thermal conductivity*

T.W. Davies

FOURIER LAW, GENERALISED (see Irreversible thermodynamics)

FOURIER NUMBER

***Following from:** Conduction*

The Fourier Number (Fo) is a dimensionless group which arises naturally from the non-dimensionalisation of the conduction equation. It is very widely used in the description and prediction of the temperature response of materials undergoing transient conductive heating or cooling. The significance of the Fourier Number may be exemplified by considering a simple case of one-dimensional conduction. **Figure 1** shows an instantaneous temperature profile at some time t during the cooling of a plate, which was initially at temperature T_i throughout, has one side insulated ($\dot{q}_x = 0$) and the other side exposed to some cooling environment. In the absence of any internal energy conversion processes the conduction equation for this system (see **Conduction**) simplifies to:

$$\frac{\partial^2 T}{\partial x^2} = \frac{1}{\kappa}\frac{\partial T}{\partial t}$$

where κ is the thermal diffusivity of the material ($\lambda/\rho c_p$). The solution of this equation yields the required information on the spatial temperature distribution at any time t, as represented by the curve in **Figure 1**.

When a problem is posed in dimensionless form the resulting solution is widely applicable to other problems in the same class. With transient conduction problems a dimensionless form of the conduction equation is easily obtained by scaling the dependent and independent variables using some convenient and constant system parameters. Choosing the initial, constant, temperature (T_i) and the plate width (L) a dimensionless temperature ($\theta = T/T_i$) and dimensionless position ($X = x/L$) can be defined so that the conduction equation is transformed to:

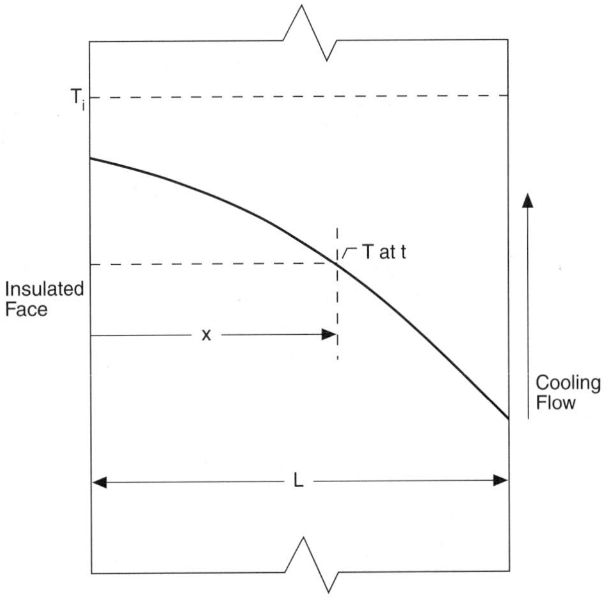

Figure 1. One-dimensional conduction in a cooled plate.

$$\frac{\partial^2\theta}{\partial X^2} = \frac{L^2}{\kappa}\frac{\partial\theta}{\partial t}$$

To complete the non-dimensionalisation process a dimensionless time. $\kappa t/L^2$, the **Fourier Number**, is introduced so that the completely dimensionless form of the conduction equation becomes:

$$\frac{\partial\theta}{\partial X^2} = \frac{\partial\theta}{\partial Fo}$$

Solutions to transient conduction problems are often presented with a dimensionless temperature expressed as a function of Fourier Number, (see Rohsenow et al. (1985)).

References

Rohsenow, W. M., Hartnett, J. P. and Ganic, E. N. (1985) *Handbook of Heat Transfer,* 2nd Ed., McGraw-Hill, New York.

Leading to: *Transient conduction*

T.W. Davies

FOURIER SERIES

Following from: *Series expansion; Harmonic analysis*

The Fourier series of the function f in terms of a system of functions $\{\varphi_n(t)\}$ orthonormal on the interval [a, b] is an infinite series

$$\sum_{k=0}^{\infty} d_k\varphi_k(t), \qquad \textbf{(1a)}$$

whose coefficients are determined by

$$d_k = \int_a^b f(t)\varphi_k(t)dt, \qquad k = 0, 1, \ldots \qquad \textbf{(1)}$$

and are called the Fourier coefficients of the function f. The function f is in the general case supposed to be integrable over the interval [a, b]. More often than not, a system of trigonometric functions used as the orthonormal system of functions. Thus, a Fourier series in a trigonometric system is usually what we mean by Fourier's series.

If an interval of expansion is given $-\pi < t < \pi$, then the Fourier series for a real function f(t), is an *infinite trigonometric series*

$$\frac{1}{2}a_0 + \sum_{k=1}^{\infty}(a_k\cos(kt) + b_k\sin(kt)) \equiv \sum_{-\infty}^{\infty} c_k e^{ikt} \qquad \textbf{(2)}$$

whose coefficients are defined by

$$a_k = \frac{1}{\pi}\int_{-\pi}^{\pi} f(\tau)\cos(k\tau)d\tau; \qquad b_k = \frac{1}{\pi}\int_{-\pi}^{\pi} f(\tau)\sin(k\tau)d\tau \qquad \textbf{(3)}$$

$$c_k = \overline{c_k} = \frac{1}{2}(a_k - ib_k) = \frac{1}{2\pi}\int_{-\pi}^{\pi} f(\tau)e^{-ik\tau}\,d\tau;$$

$$k = 0, 1, \ldots \qquad \textbf{(4)}$$

Here a_k, b_k, are real, and $\overline{c_k}$ are, generally speaking, complex numbers. The Fourier coefficients a_k, b_k of the function f(t) for $k \to \infty$ tend to zero.

In case of expansion of an even function ($f(t) = f(-t)$) into a Fourier series, it has only cosines, and of odd function ($f(t) = -f(-t)$) it has only sines.

If a *trigonometric series* (2) converges to f(t) uniformly, then the coefficients are necesserily the Fourier coefficients (3) of f(t).

Among the trigonometric polynomials of order n

$$T_n(t) = A_0 + \sum_{k=1}^{n}(A_k\cos(kt) + B_k\sin(kt)) \qquad \textbf{(5)}$$

the least value for the root-mean-square error

$$\frac{1}{\pi}\int_{-\pi}^{\pi}[f(t) - T_n(t)]^2\,dt \qquad \textbf{(6)}$$

is reached when as $T_n(t)$ a partial sum of a Fourier series, for the function f

$$S_n(t) = \frac{a_0}{2} + \sum_{k=1}^{n} a_k\cos(kt) + b_k\sin(kt) \qquad \textbf{(7)}$$

a system of trigonometric functions is complete (closed) and Parseval's equation holds for it

$$\frac{a_0^2}{2} + \sum_{k=1}^{\infty}(a_k^2 + b_k^2) = \frac{1}{\pi}\int_{-\pi}^{\pi} f^2(t)dt \qquad \textbf{(8)}$$

If a function f(t) on a finite interval satisfies the Dirichlet's conditions (i.e., has a finite number of extremums and is continuous everywhere, besides the finite number of points in which in can have discontinuities of the first kind), then Fourier series of function f converges for all t, during which it converges to f(t) at continuity points, and to a half-sum $1/2[f(t - 0) + f(t + 0)]$ at discontinuity points.

This statement is extended into arbitrary function of bounded variation (i.e., into functions representable in the form $f = f_1 - f_2$, where f_1 and f_2 are bounded and decreases on [a, b]). If the function f(x) is continuous and has an bounded variation, then a Fourier series converges to this function uniformly (the Dirichlet-Jordan test).

A specific property of Fourier series, associated with peculiarities in its behaviour (an excess of partial sums of Fourier series over the exact value of the function) in the vicinity of discontinuity points of the function f(t) is called Gibb's phenomenon. Gibb's phenomenon is associated with nonuniform convergence, limits the application of Fourier series in the vicinity of discontinuity points and requires additional effort in order to ensure reliable approximate computations in this field (the summation over arithmetic means).

V.N. Dvortsov

FOURIER TRANSFORMATION (see Harmonic analysis)

FOUR STROKE CYCLE (see Internal combustion engines)

FRACTAL ATTRACTOR (see Non-linear systems)

FRACTIONATION (see Foam fractionation)

FRACTURE MECHANICS

Fracture mechanics deals with the behavior of materials and structural members when cracks and failures develop in them, and with the development of methods for calculating crack growth rate and reducing residual strength. Fracture mechanics determines the critical size of cracks and cavities in a material admissible at given loads, and the time of crack growth from a certain initial size up to the critical one.

A material fails by either rupture or shear. The theoretical estimation of ultimate stress of the material for a simultaneous rupture or shear of the neighboring atomic crystal planes yields the quantity of the order of $K/2\pi$ or $G/2\pi$ respectively, where K is the compression modulus and G is the shear modulus. The real strength of crystal solids is about one or two orders of magnitude lower. This fact was clarified by A.A. Griffith in 1921 in terms of an energy approach. Comparing the increment in the body elastic energy and the crack surface energy with increasing crack length, he derived the stability criterion of the crack in a brittle material exposed to tensile stress

$$\sigma_c = (Eg_{1c}/(\pi a))^{1/2},$$

where E is Young's modulus, a the crack length, and g_{1c} the constant known as the critical velocity of energy release. At a stress higher then σ_c the crack in the body becomes unstable and catastrophically grows. Thus, the low strength of materials is accounted for by Griffith's cracks of micron size.

The *energy fracture criterion (Griffith's criterion)* is a necessary, but insufficient condition for the crack growth. In particular, if the curvature radius at the crack tip exceeds by far an atomic radius, the stress concentration appears insufficient for the rupture of atomic bonds. This crack grows at higher stresses than follows from the energy balance. Measuring the work of developing a new crack surface, we arrive, even for brittle materials, at the values at least an order of magnitude higher than the specific surface energy. The analysis shows that the crack development is practically always brought about by plastic deformation at its tip. The work of plastic deformation is so great the energy fracture criterion is determined by this value rather than the specific surface energy.

Fracture is significantly affected by various stress concentrators such as pits and grooves in the material and the environmental factors, e.g. temperature and chemical processes, in corrosion. It has been revealed that the characteristics of material strength such as yield limit and ultimate stress are adequate for calculating structural members, but insufficient in case there is the probability of crack nucleation.

The analysis of the stress field in an elastic or an elastoplastic body with a crack reveals that in the vicinity of the crack tip stresses grow. The stress field around the crack tip is expressed in the generalized form in terms of the so called stress intensity factor. For instance, for a through-thickness cleavage crack 2a in length in an infinite elastic plate

$$\sigma_{ij} = K_I(2\pi r)^{-1/2}f_{ij}(\theta), \qquad \text{where } K_I = \sigma(\pi a)^{1/2}.$$

Here τ and θ are the polar coordinates. The stress field at the crack tip is completely determined given the stress intensity factor K_I. Attainment of the critical value K_{Ic} causes failure. If K_{Ic} is known from the results of tests of the specimen with a preset size of a crack, it is possible to calculate the strength of the same material with cracks of any size or to determine the limit of the admissible crack size at a given load. K_{Ic} is a measure of a material's crack resistance known as a fracture toughness in the planar-strained state. The lower the fracture toughness, the lower is the admissible crack size in a material.

High-strength materials commonly posses a low fracture toughness. Fracture in these materials is studied by the methods of *linear elastic fracture mechanics* (LEFM). The materials with a low yield limit are commonly characterized with a high viscosity. The failure of these materials leads to a large-size plastic zone at the crack tip in comparison to the crack size and LEFM cannot be used in this case. The concept of crack opening is used for these materials. The crack is assumed to propagate if plastic deformation at the crack tip achieves the maximum admissible value. The deformation at the crack tip can be also expressed in terms of opening which is a measurable quantity.

A subcritical slow growth of a cavity or an incipient crack can occur under cyclic loading (a fatigue crack) or as a result of corrosion cracking under stress and by other mechanisms. The rate of growth of fatigue and corrosion cracks and, hence, the time of fracture are determined by the stress intensity factor.

An important role in physics and fracture mechanics is played by cold brittleness, which consists in transition of materials from a plastic to a brittle state and involves a drastic decrease in fracture toughness. The cold-brittle metals capable of turning brittle at low temperatures include in the first place metals and alloys with a bcc crystal lattice, e.g. iron, α—Fe-based steels, tungsten, and molybdenum, and with a hcp cristal lattice, e.g. cadmium and magnesium. Metals with a fcc crystal lattice such as copper, aluminium, and nickel and austenitic steels, do not show any signs of cold brittleness due to a weak dependence of yield limit on temperature and due to a large difference between yield limit and ultimate stress.

When structural members are under load, specifically at elevated temperatures, there is a temporal dependence of strength: a long-term load application reduces the strength of solids the more, the longer the time of loading. In the moderate temperature range the relation between the life (the time until fracture) t_D and the stress σ is described by the empirical equation

$$t_D = Ae^{a\sigma},$$

where A and a are the temperature-dependent constants. The expression

$$t_D = a_1\sigma^{-m}$$

is used at higher temperatures for high-temperature alloys. The widely known thermofluctuation concept of strength treats the fracture of solids as a successive rupture of atomic bonds due to thermal motion of atoms. The role of stresses in this model is to endow the bonds in the dissipation processes, reversible in an unloaded body, a preferred orientation resulting in accumulation of ruptures. The energy of thermal fluctuations is expended on overcoming the potential barriers in a crystal lattice. The life of the material under loading is described by

$$t_D = t_0 \exp[(U_0 - \gamma\sigma)/kT],$$

where U_0 is the magnitude of the energy barrier at $\sigma = 0$, the activation volume γ depends on local overstresses, and the preexponential factor t_0 is of the order of $10^{-12} - 10^{-13}$ s.

The material's strength in a microsecond range of load durations is studied by analizing cleavages. On reflection of the shock compression momentum from the free surface of the body it generates tensile stresses which may give rise to a failure or to a cleavage in the body. Under conditions of specimen loading by plane shock waves deformation is one-dimensional and the stressed state in cleavage approaches the state of three-dimensional tension. A short-term loading restricts the spread of information on failure development in separate portions of the specimen. In distinction from static rupture, failure under these conditions is not initiated in a single weakest site, but in a large number of sites, it is scattered throughout the body and develops through crack and void growth and merging. The development of cleavages until complete fracture of the body into fragments may last for a relatively long time compared to the loading time. The process can stop, depending on the duration of initial loading, at different phases: nucleation of microdiscontinuities on inclusions, their growth and joining together, development of the main crack, and separation of the cleavage element. The resistance to fracture in cleavage is commonly 1.5—3-fold higher than the real rupture stress under quasistatic conditions.

Leading to: *Fracture of solid materials*

G.I. Kanel and V.E. Fortov

FRACTURE OF SOLID MATERIALS

Following from: *Fracture mechanics*

Fracture of solid materials may arise as a result of the applied stress or purely as a result of the corrosive action of the environment. Referring only to stress induced fracture the two main modes are *ductile* and *brittle.* Most metals can withstand a large deformation, of the order of several percent, without failing. As the deformation increases past the yield point, microvoids are formed, as shown in **Figure 1**.

At any given instant, the fraction of microvoids, f, may be taken to define a "damage factor", increasing the true stress.

Fracture finally takes place when the microvoids coalesce and results in the dimpled appearance of the fracture surfaces. The dimples are equiaxed, in case of pure tension, or elongated when in the presence of shear, as is normally found near free surfaces where fracture follows from the slip along the planes of maximum shear stress. The ligaments between large clusters of microvoids may also tear, forming sharp ridges.

Metals are normally built up from crystals of atoms in the face-centered cubic, hexagonal close-packed or body-centred cubic structures. Of these, the face-centred cubic are the most ductile and always fail by the void coalescence mechanism after yield. Such metals include gold, copper and alloys such as brass and aluminium alloys.

Metals with the other crystal structures do not possess the easy dislocation motion of face-centred alloys and under low temperature or high strain rate they may fail by cleavage, e.g. separation occurs along well-defined crystal planes. The cleavage is not related to the total plastic deformation of the specimen.

From an engineering or macroscopic standpoint, brittle fracture is said to occur in a tensile specimen if no permanent deformation is observed in the two specimen halves, which may be fitted back together so that the appearance is that of the original specimen.

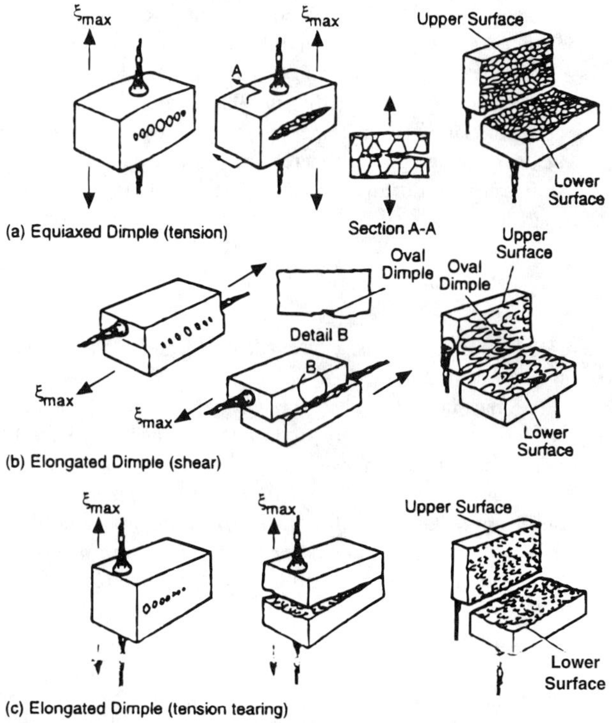

Figure 1. Appearance of fracture surfaces from microvoid coalescence. From ASM Metals Handbook, 8th Edition, Vol. 9, ASM International, Metals Park, Ohio (1994) with permission.

The fracture mechanism may be cleavage, i.e. separation of planes within a grain, since cracks are associated with limited crystallographic slip and hence small strains, but the most characteristic mechanism is intergranular fracture. The fracture surfaces reveal the outline of the grains that have separated due to the presence of weak or brittle phases between them or to environmental effects such as stress corrosion or hydrogen embrittlement.

Brittle fracture is the main mode of failure of glass and ceramics. Fracture starts from a small flaw—typically a few microns—and develops into a fracture mirror before entering into a mist characterised by the presence of fine radial ridges. This is followed by a hackle, with longer, better defined radial lines. Finally, a major crack or crack sweep through the rest of the surface.

Failure depends on the size of the starter flaw, on the applied stress and on the *fracture toughness* of the material, in accordance with the general expression

$$Y\sigma\sqrt{\pi a} = K_c$$

where Y is a dimensionless parameter depending on geometry, σ is the nominal applied stress, a the characteristic size of the flaw and K_c the fracture toughness. For a ceramic, K_c may be of the order of 5 $MPa\sqrt{m}$ when the stress is normal to the plane of the flaw. In a steel, it may be as high as 150 $MPa\sqrt{m}$ so that, for the same applied stress, the flaw needed to start a brittle fracture would be of the order of millimetres rather than of microns.

When a crack is present, a material that would otherwise fail in a ductile manner may fail without any prior plastic deformation provided that the crack size is sufficiently large or that external constraints inhibit the free plastic deformation. This occurs in large sections and in welded structures with a high level of residual stresses.

References

Derby, B., Hills, D. A, and Ruiz, C. (1992) *Materials for Engineering*, Longman.

Fellows, J. A. ed. (1974) *Metals Handbook* Vol. 9, Fractography and Atlas of Fractographs, Amer. Soc. Metals.

C. Ruiz

FRACTURE TOUGHNESS (see Fracture of solid materials)

FRANCIS TURBINES (see Hydraulic turbines)

FRAUNHOFER DIFRACTION (see Optical particle characterisation)

FREDHOLM INTEGRAL EQUATIONS (see Integral equations)

FREE CONVECTION

Following from: *Convective heat transfer*

Free convection, or *natural convection*, is a spontaneous flow arising from non-homogeneous fields of volumetric (mass) forces (gravitational, centrifugal, Coriolis, electromagnetic, etc.):

$$\Delta\vec{F} = \Delta(\rho\vec{g}) = \Delta\rho\vec{g} + \rho\Delta\vec{g}.$$

If density variation $\Delta\rho$ is caused by spatial non-uniformity of a temperature field, then a flow arising in the Earth gravitational field is called *thermal gravitation convection*. The density variability may also result from non-uniform distribution of concentration of any component in a mixture or from chemical reactions, difference in phase densities or from surface tension forces at the phase interface (in this case concentration diffusion or convection is implied).

Free-convective flows may be laminar and turbulent. A flow past a solid surface, the temperature of which is higher (lower) than that of the surrounding flowing medium, is the most widespread type of free convection. Figures 1 and 2 schematically illustrate characteristic examples of free convection. At the beginning of heating of a vertical surface (x = 0) (**Figure 1a**) a laminar boundary layer is formed. The layer thickness grows along the flow direction and at a certain distance, corresponding to x_{c1}, the fluid flow becomes unstable changing within the range from x_{c1} to x_{c2} from laminar to turbulent. To this character of flow structure variation there corresponds the change in the coefficient of heat transfer α_x which in the case of the developed turbulent FC remains constant along the plate length where the characteristics of thermal

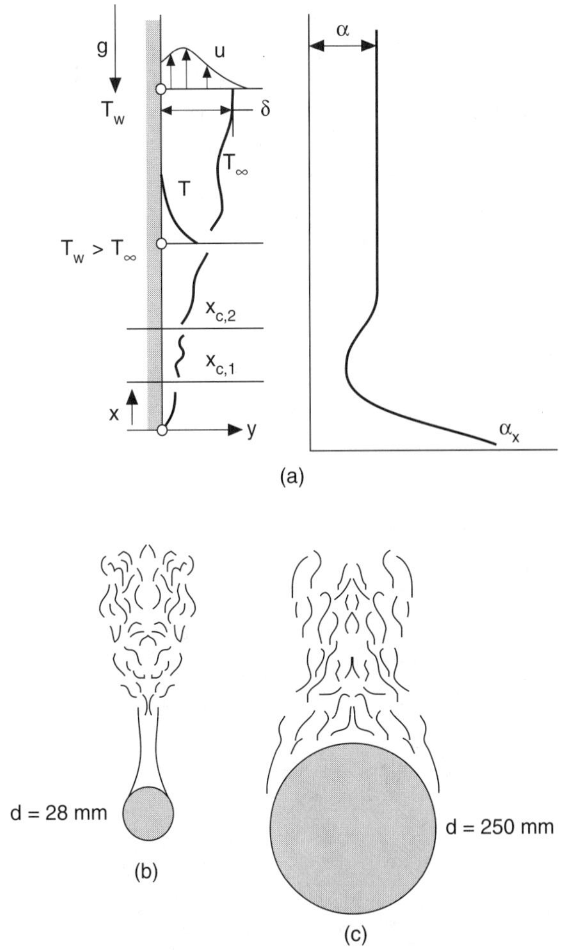

Figure 1. Development of free convection boundary layer on surfaces.

turbulence become statistically equal. The pictures of FC development in the flow past a hot sphere or horizontal cylinder are qualitatively similar (**Figures 1b,c**). On bodies of large diameters (**Figure 1c**) a turbulent boundary layer develops thus forming an ascending turbulent thermal plume in a trailing edge. From hot bodies of small diameters a laminar thermal plume ascends (**Figure 1b**) which at some distance from a body becomes turbulent. In narrow and closed cavities a FC flow is much more complex (**Figure 2**), due to the interaction between near-wall fluid flows formed on the heat exchanging surfaces. On heating one vertical wall (temperature T_h) and cooling the other (temperature T_{cl}) the modes with a common fluid flow are possible through the entire cavity that involve local secondary flows near vertical walls as is exemplified in **Figure 2a** by FC in a square cavity. A flow in narrow slots between parallel vertical plates is formed in the form of periodic circulations (**Figure 2b**). In a horizontal fluid layer between cold upper (T_{cl}) and hot lower (T_h) walls, the fluid flow has a cellular form with hexagonal cells (*Benard cells*) in the center of which fluid ascends from the hot surface to the cold one whereas in the periphery, it descends (**Figures 2c,d**). Such a form of the fluid flow was first observed by Benard in 1901. With increasing heat flux the cells are destroyed and the flow converts to a turbulent one.

In the theoretical analysis of FC flows and heat transfer the laws of momentum, mass and energy conservation at certain boundary conditions are used. The Boussinesq approximation of "weak" thermal convection is widely applied, i.e. density deviations from a mean value are considered to be negligible in all the equations, except for the equation of motion where they are taken into account in the buoyancy force term. For small temperature drops in a flow the relation $\rho(T)$ may be considered linear

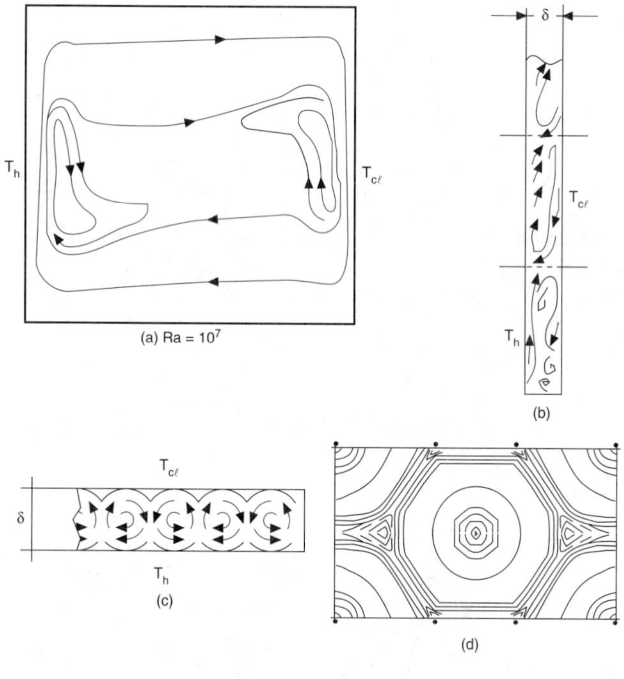

Figure 2. Free convection in cavities.

$$\rho = \rho_0[1 + \beta(T_0 - T)],$$

where ρ_0 is the fluid density at temperature T_0, $\beta = -[\partial\rho/\partial T]_p/\rho$ is the volumetric coefficient of thermal expansion.

Numerical values of β are usually small (water: $\beta = 1.5 \times 10^{-4}$, air: $\beta = 3.5 \times 10^{-3}$ at T = 273 K), therefore the density variation is taken into account only in those cases where it affects the gravitational forces. The Boussinesq approximation correlates the coefficient of volumetric expansion of a medium β with the gravity acceleration g; they enter into the governing equations only as a product. Physical substantiation of the Boussinesq approximation is based on the smallness of accelerations in FC flows as compared with the acceleration due to gravity.

Comparison with vast experimental material indicates the fact that the Boussinesq approximation well reflects the main specific features of *thermal gravitation convection* in a wide class of real convective flows.

As is shown by experimental data, in many cases of FC the main variations of the characteristics of thermal and hydrodynamic fields are concentrated in relatively narrow boundary layers near the heat transfer surface where viscous forces are commensurable with inertial and volumetric forces. The smallness of a boundary layer thickness as compared with characteristic dimensions of bodies allows one to introduce additional simplifications into the equations of motion and heat transfer.

The concept of a boundary layer is far more complex for FC than for forced convection, because thermal and hydrodynamic problems cannot be treated separately due to the fact that the fluid flow is completely determined by heat transfer. The main motive force (the difference between wall and surrounding temperatures) noticeably manifests itself only in a thin near-wall zone. This region of a temperature field with the thickness δ_T is called a thermal boundary layer.

The difference of temperature in a boundary layer creates a volumetric buoyancy force which causes motion. At the surface, the fluid is stationary (the "no-slip" condition). With distance from a wall the velocity u gradually grows to a maximum and then, under the effect of viscous friction, it vanishes (**Figure 1a**). Beyond the limits of this dynamic boundary layer there is a region of inviscid (potential) flow. The distance along the normal from the wall to the place, where the velocity differs from zero by 1 per cent of the value of u_{max}, is taken as the dynamic boundary layer thickness δ.

When $\delta_T < \delta$, the motion outside the thermal layer, where the buoyancy force is absent, is determined by the forces of dynamic and turbulent interaction between separate fluid layers.

When $\delta < \delta_T$, outside the dynamic boundary layer and within the thermal layer δ_T the flow may be considered as potential.

A flow in a boundary layer makes a main contribution into the transfer processes, whereas the induced outer flow is secondary and provides only higher order correction. This is the manifestation of the secondary effect of a boundary layer on the flow in the surrounding medium.

It follows from the dimensional analysis that a relative boundary layer thickness δ/x has the order of $Gr^{-0.25}$ where $Gr = g\beta(T_w - T_\infty)x^3/\nu^2$. At very large Grashof numbers characteristic of practical applications of the FC boundary layer theory, the boundary layer thickness is usually very small compared to the body size. Comparatively thick boundary layers take place for media with small Prandtl numbers (Pr) and with small differences between the body and surrounding temperatures.

In the Boussinesq approximation for an incompressible fluid and a steady-state regime, the equations of momentum, mass and energy conservation for laminar FC in a plane boundary layer are as follows

$$u\frac{\partial u}{\partial x} + v\frac{\partial u}{\partial y} = g\beta(T - T_\infty) + \nu\frac{\partial^2 u}{\partial y^2},$$

$$\frac{\partial u}{\partial x} + \frac{\partial v}{\partial y} = 0, \tag{1}$$

$$u\frac{\partial T}{\partial x} + v\frac{\partial T}{\partial y} = a\frac{\partial^2 T}{\partial y^2}.$$

The system of equations (1) allows the determination of the both velocity components (u, v) and of the temperature field (T) for various boundary conditions.

To generalize the solution results or experimental data as well as to reduce the quantity of problem parameters, similarity theory is used.

Some problem parameters are substituted by their combinations, the so-called generalized variables. Their structure depends on the form of differential operators entering into equations (1). We shall reduce the equations to the dimensionless form. It is convenient to use the quantities entering into the unambiguity conditions (boundary conditions) as the reduction scales. As a linear scale we shall take some characteristic dimension of a body L, for a temperature it is convenient to use, for instance, the relation $\theta = (T - T_\infty)/(T_w - T_\infty)$, where T_w is the body surface temperature, T_∞ is the surrounding temperature, T is the local temperature. The characteristic velocity may be obtained from the comparison of volumetric and viscosity forces $u_0 = \beta g\Delta TL^2/\nu$ or from the estimates of the type $u_0 = L/\tau_0$, where τ_0 is the time scale.

The dimensionalization yields

$$U\frac{\partial U}{\partial X} + V\frac{\partial U}{\partial Y} = \theta + Gr^{-1/2}\frac{\partial^2 U}{\partial Y^2},$$

$$U\frac{\partial \theta}{\partial X} + V\frac{\partial \theta}{\partial Y} = Gr^{-1/2}Pr^{-1}\frac{\partial^2 \theta}{\partial Y^2}. \tag{2}$$

The Grashof number $Gr = \beta g\Delta TL^3/\nu^2$ is the main governing criterion and the most important characteristic of FC heat transfer. It is the measure of the relation between the buoyancy forces in a non-isothermal flow and the forces of molecular viscosity. It also determines the mode of medium motion along the heat transfer surface. In its physical meaning, it is similar to the Reynolds number for a forced flow.

At small Gr numbers a FC flow is absent and heat transfer is carried out by molecular thermal conduction. In particular, in a horizontal layer (**Figure 2c**) this takes place at $Ra_\delta = Gr_\delta Pr = \beta g(T_h - T_c)\delta^3/\nu a < 1708$. When $Ra_\delta = 1708$, the stability of a horizontal layer is disturbed and a FC fluid flow develops in the form of Benard cells (**Figures 2c,d**). At $Ra_x = Gr_x Pr \approx 10^9$ on a vertical plate there takes place the transition from a laminar to turbulent flow (**Figure 1a**).

Free convection heat transfer, similar to that under forced convection, is characterized by the Nusselt number $Nu = \alpha L/\lambda$. This is usually an unknown quantity since it involves the heat transfer coefficient α which should be found. Thus, the dimensionless form of the heat transfer coefficient, Nu, depends on the dimensionless numbers Pr, Gr and the coordinate $X = x/L$

$$Nu = f(Pr, Gr, X).$$

In the theory of a *FC boundary layer* a wide use is made of integral relations obtained by averaging the motion and energy equations over the boundary layer thickness. For stationary conditions, with dissipation and compression work being neglected, these equations have the form

$$\left.\begin{aligned} \frac{d}{dx}\int_0^\delta u^2 dy &= -\nu\left(\frac{\partial u}{\partial y}\right)_0 + g\beta\int_0^{\delta_T} \vartheta\, dy; \\ \frac{d}{dx}\int_0^{\delta_{min}} u\vartheta\, dy &= -\alpha\left(\frac{\partial \vartheta}{\partial y}\right)_0. \end{aligned}\right\} \tag{3}$$

The system of equations (3) is not suitable for use in approximate calculations. Multiplying the boundary layer equations by the velocity and integrating we obtain the balance of mechanical energy

$$\frac{1}{2}\frac{d}{dx}\int_0^\delta u^3 dy = -\nu\int_0^\delta\left(\frac{\partial u}{\partial y}\right)^2 dy + g\beta\int_0^{\delta_{min}} u\vartheta\, dy.$$

Another system of integral relations may be obtained by successive multiplication of the boundary layer equations by the ordinate y powers with subsequent integration. For example, the equations of the first moment are

$$\frac{d}{dx}\int_0^\delta yu^2 dy - \int_0^\delta uv\, dy = g\beta\int_0^{\delta_T} y\vartheta\, dy,$$

$$\frac{d}{dx}\int_0^{\delta_{min}} yu\vartheta\, dy - \int_0^{\delta_{min}} u\vartheta\, dy = a\vartheta_c.$$

The most widely used method for processing the results of calculations and experiments is the application of the exponential function between the similarity criteria

$$Nu = C\, Gr^m Pr^n, \tag{4}$$

where C, m, and n are the constant dimensionless numbers. If, in logarithmic coordinates all the points fall on a straight line, this forms the basis of the practical method for constructing the exponential function. If the test points fall on a curve, then the single line is substituted by a segmented line. For separate segments of such a curve, the values of C, m, n are different.

To expand the applicability region of the relation of type (4), it may be presented in the form of the sum

$$Nu = C_1Gr^{m1}Pr^{n1} + C_2Gr^{m2}Pr^{n2} + C_3Gr^{m3}Pr^{n3} + \cdots$$

If the physical properties of a medium depend on temperature, then equations determining the form of these relations should be among the main equations. In this case, the similarity criterion should be treated as the arguments of correlations. Here the application of the generalized analysis is impossible and one has to restrict oneself to approximate solutions. In particular, if thermophysical characteristics can be represented by exponential functions of temperature, an additional parametric criterion, introduced into relation (4), is presented in the form of the surrounding medium-to-wall temperature ratio, viz.: $(T_\infty/T_w)^{l_1}$. Physical parameters should be referred to one of the two characteristic temperatures. This method is applicable to gases.

The dependence of liquid heat transfer on the heat flux direction and temperature difference are approximately allowed for by the introduction of an additional multiplier $(Pr_\infty/Pr_w)^{l_2}$ into the similarity equation. For fluid heating $Pr_\infty/Pr_w > 1$; for cooling $Pr_\infty/Pr_w < 1$. The ratio Pr_∞/Pr_w the more is different from zero the higher is the temperature head. The variability of physical parameters may be taken into account by parametric simplexes of the type λ_∞/λ_w, η_∞/η_w, $c_{p\infty}/c_{pw}$, c_p/c_{pw}, etc. as well as by the introduction of the temperature, which is determining for the given process.

Nusselt suggested the averaging of the physical parameters by the equation

$$\overline{\varphi} = \frac{1}{\vartheta_w}\int_{T_\infty}^{T_w} \varphi dT, \qquad \vartheta_w = T_w - T_\infty,$$

and to calculate the determining temperature as log-mean one

$$\overline{T} = \vartheta_w/\ln(1 + \vartheta_w/T_\infty). \tag{5}$$

When $\vartheta_w << T_w$, relation (5) is presented in the form of the power series of T_∞/T_w. If we restrict ourselves to the first term of the series, then $\bar{T} = T_\infty$; in the case of restriction to two terms $T = (T_\infty + T_w)/2$. In the FC problems the determining temperature is often chosen in the form of a linear combination of wall and surrounding medium temperatures:

$$\overline{T} = aT_w + bT_\infty,$$

where a and b are the coefficients varying from 0 to 1: $(a + b) = 1$.

The determining linear size is usually taken to be that which to a greater extend corresponds to a physical essence of the process (e.g. the plate height, cylinder or sphere diameter, gap or boundary layer thickness, etc.). The other dimensions enter into the similarity equation in the form of simplexes $P_{Lk} = L_k/L$ (the width and thickness of a plate, the height of a vertical or length of a horizontal cylinder, gap height). In a number of cases, a combination of heterogeneous physical quantities entering into the unambiguity conditions (the length scale in the asymptotic theory $L/Gr^{1/4}$, the linear dimension in the case of jet convection) is taken as the determining linear size. To universalize computational relations and eliminate parametric criteria, a common characteristic dimension is introduced. As an example, we shall give: $1/L = 1/a + 1/b$ for a horizontal plate; πD for a horizontal cylinder; $\pi D/2$ for a sphere; $D_{hyd} = 4S/P_l$ is the hydraulic diameter for a horizontal channel of an arbitrary cross-section (S is the cross-section area, P_l is the wetted perimeter).

For applied problems in the calculation of heat transfer from surfaces of an arbitrary shape in an infinite fluid the equation

$$\overline{Nu} = C\, Ra^n_{L\infty}(Pr_\infty/Pr_w)^{1/4}$$

is suggested, or in the dimensional form

$$\overline{\alpha} = A(\vartheta_w/L^{1/n-3})^n.$$

where the quantities C, A and n depend on Ra_L $(= L^3gp(T_\infty - T_w)\beta/\eta^2)$ and the body shape. A mean boundary layer temperature is taken to be the determining temperature. Correction factors are introduced for inclined and horizontal surfaces.

A specific feature of laminar FC on a vertical plate at a constant wall temperature (**Figure 1a**) is the fact that it allows a self-similar solution if a new variable is introduced into equations (1) in the form

$$\eta_s \equiv \frac{y}{x}\left(\frac{Gr_x}{4}\right)^{1/4}.$$

The boundary conditions are

$$Y = 0: \quad u = v = 0, \quad T = T_w$$

$$Y \Rightarrow \infty: \quad u \to 0, \quad T \to T_\infty.$$

Having represented the stream function and the dimensionless temperature as

$$\Psi(x, y) \equiv f(\eta_s)\left[4\nu\left(\frac{Gr_x}{4}\right)^{1/4}\right]; \qquad \theta = \frac{T - T_\infty}{T_w - T_\infty},$$

obtain equations (1) in the form

$$f''' + 3ff' - 2(f')^2 + \theta = 0$$

$$\theta'' + 3\, Prf\theta' = 0,$$

where $f'(\eta_s) = df/d\eta_s$.

The boundary conditions are

$$\eta_s = 0: \quad f = f' = 0; \quad \theta = 1$$

$$\eta_s \to \infty: \quad f' \to 0; \quad \theta \to 0.$$

Local heat transfer rate at a distance x from the plate edge may be determined by the formula $Nu_x = \alpha x/\lambda = (Gr_x/4)^{1/4}\, H(Pr)$ (where $H(Pr) = 0.75Pr^{1/2}/(0.609 + 1.22\, Pr^{1/2} + 1.238\, Pr)^{1/4}$) valid for $0 \le Pr < \infty$. The presented correlation reflects two noteworthy physical facts:

1. In the case of laminar FC, the coefficient of heat transfer along a vertical surface varies according to the law $\alpha(x) = Ax^{-1/4}$.

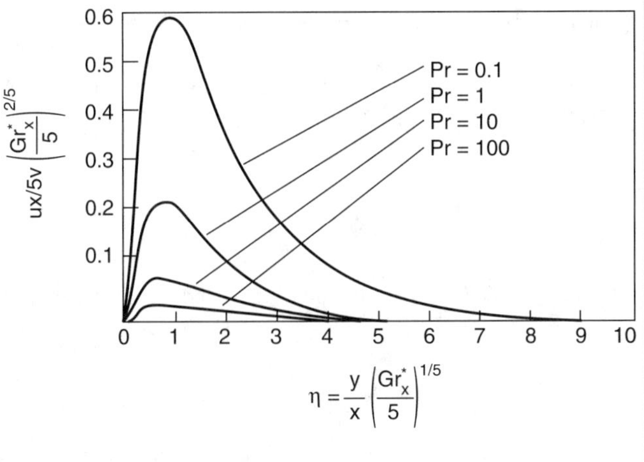

Figure 3. Dimensionless velocity profiles in free convection boundary layers on flat plates with constant heat flux.

2. In the limiting cases of $Pr \to 0$ and $Pr \to \infty$, the dependence of the dimensionless coefficient of heat transfer on the Prandtl number has different characters, viz. $Nu(x) = 0.600\ Pr^{1/2}\ Gr_x^{1/4}$ when $Pr \to 0$ and $Nu(x) = 0.503\ Pr^{1/4}\ Gr_x^{1/4}$ when $Pr \to \infty$.

The case of large Prandtl numbers corresponds to a very high viscosity and consequently to a slow flow usually called a creeping flow. For such flows the inertia terms in the equation of motion may be neglected and the relation for the Nusselt number Nu has the form F(GrPr). The case of $Pr \to 0$ corresponds to small viscosity thus allowing the neglection of viscous effects in the equation of motion and the relation for the Nusselt number acquires the form $F(Gr\ Pr^2)$.

The mean Nusselt number on a plate with a length x = L is

$$\overline{Nu_L} = \frac{\overline{\alpha}L}{\lambda} = \frac{4}{3}\left(\frac{Gr_x}{4}\right)^{1/4} H(Pr)$$

or $\overline{Nu_L} = 4/3\ Nu_{x=L}$.

The constancy of heat flux on a wall ($\dot{q}_w = const$) that corresponds to constant heat supply to the heat transfer surface (e.g. in electric heating devices, in the elements of radioelectronic equipment) is a boundary condition important in practice. This case can be easily realized in practice by heating a thin metal foil of a constant thickness with an electric current, therefore this is often used in experiments. In these problems, the wall temperature T_w is an unknown quantity. By virtue of the above, a modified Grashof number calculated from the heat flux on a wall, viz.: $Gr_x^* = g\beta\dot{q}_w x^4/\lambda\nu^2$ is taken to be the determining dimensionless parameter instead of the traditional $Gr_x = g\beta(T_w - T_\infty)x^3/\nu^2 = Gr_x^*\ Nu$. Here the temperature difference entering into the ordinary Grashof number is replaced by the multiplier $\dot{q}_w x/\lambda$. The wall temperature grows along the flow as $(T_w - T_\infty) \sim x^{1/5}$ and, consequently, the dimensionless coefficient of heat transfer, the Nusselt number, at $\dot{q}_w = const$ depends on Gr_x^* as $Nu_x = A_1(Pr)Gr_x^{*1/5}$ which compares to the relation $Nu_x = A_2(Pr)Gr_x^{1/4}$ for $T_w = const$ (as will also be seen by substitution of $Gr_x = Gr_x^*/Nu$). For the given case of $\dot{q}_w$ = const in **Figures 3** and **4** examples are presented of velocity and temperature distributions in a laminar boundary layer at various

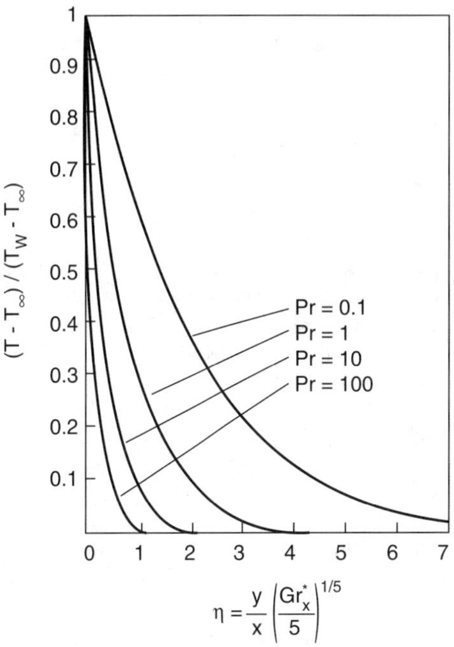

Figure 4. Dimensionless temperature profiles in free convection boundary layers on a flat plate with constant heat flux.

Prandtl numbers constructed based on the results by Sparrow and Gregg (*J. Heat Transfer.* 1956, v. 78, p. 435). These theoretical results are in a good agreement with experimental data.

Many FC flows occurring in nature and technology are turbulent, i.e. they have irregular pulsatory character. As compared with a large amount of theoretical and experimental studies of turbulence in forced turbulent flows, turbulence with FC is studied considerably less. However, the basic mechanisms of a turbulent flow are quite similar. Their main difference is in the fact that in FC flows the values of averaged velocities are smaller, while the levels of disturbances are higher than in forced flows. The flow field is related to the temperature field and to study turbulence with FC one requires simultaneous diagnostics of the both fields. This relation greatly complicates both the theoretical analysis and measurements. Experimental data on local heat transfer with developed *turbulent FC* on vertical and inclined surfaces at $\dot{q}_w = const$ within the range of $Gr_x^*\ Pr = 2 \times 10^{13} - 10^{16}$ are described by Wlitt and Ross (*J. Heat Transfer.* 1975, v. 97, p. 549) as

$$Nu(x) = 0.17(Gr_\infty^*\ Pr\ \cos\gamma)^{0.25},$$

where γ is the angle of surface inclination to the vertical plane. It follows from the equation given that the coefficient of heat transfer with turbulent FC is independent of x. Some data indicate the possibility of the presence of a weak dependence α(x).

For engineering calculations of heat transfer from bodies of different geometry and orientation in space, a variety of dimensionless empirical relations have been suggested.

For a vertical plate surface (**Figure 1a**) at Rayleigh numbers $Ra_L = Gr_L\ Pr$, varying within the range from 10^4 to 10^{13} and covering both the laminar and turbulent flow zones, a length-mean Nusselt number is equal to

$$\overline{Nu_L} = \left\{0.825 + \frac{0.387\ Ra_r^{1/6}}{[1 + (0.492/Pr)^{9/26}]^{8/27}}\right\}^2.$$

This equation may be used for liquid metals ($Pr < 0.1$) when Ra_L is substituted by $Gr_L\ Pr^2$ according to the above considered case of limiting values of Pr. This expression for $\overline{Nu_L}$ may be applied to the determination of mean heat transfer from a vertical cylinder with a height H, if the boundary layer thickness is much smaller than the cylinder diameter D, i.e. $D/H \geq 35/Gr^{1/4}$. The effects caused by the body curvature are especially substantial at small and moderate Grashof numbers.

For a horizontal cylinder (**Figures 1b,c**) mean heat transfer may be determined by the formula

$$\overline{Nu_L} = \left\{0.60 + \frac{0.387\ Ra_D^{1/6}}{[1 + (0.559/Pr)^{9/26}]^{8/27}}\right\}^2.$$

which is valid for $Ra_D < 10^{12}$.

When a cylinder is inclined at some angle φ to the horizontal, there appears an axial velocity component and the flow becomes three-dimensional. The growth of the boundary layer with cylinder inclination weakens convective heat transfer. Heat transfer near the lower end of an inclined cylinder of a finite length is determined by relations characteristic of the flow along the cylinder. In the upper portion of the cylinder the flow approaches the case of that around a horizontal cylinder. For small inclination angles this effect is insignificant. For example, $\overline{Nu}$ decreases by 8 per cent with φ growing from 0° to 45°. When the cylinder axis approaches the normal heat transfer from a cylinder decreases sharply.

The correlation for calculating averaged heat transfer from a sphere to a surrounding medium is presented in the following form

$$\overline{Nu} = (2^{0.816} + 0.15\ Ra_D^{0.277})^{1/0.816}.$$

for $Pr = 0.7 - 6$, $Ra_D = 10^{-6} - 10^4$, $\overline{T} = (T_w + T_\infty)/2$.

Heat transfer from a horizontal flat surface greatly depends on its position (upwards or downwards) relative to the direction of buoyancy force and also on body dimensions. The greatest coefficient of heat transfer with a free flow near a horizontal plate should be expected at the places of maximum flow rate, i.e. at the plate ends. At rather large plate dimensions the average coefficient of heat transfer ceases to depend on the end effect. The mean heat transfer coefficient may be determined by the formulas

$$\overline{Nu_L} = 0.54\ Ra_L^{0.25} \qquad (10^4 \leq Ra_L \leq 10^7)$$

$$\overline{Nu_L} = 0.15\ Ra_L^{1/3} \qquad (10^7 \leq Ra_L \leq 10^{11})$$

If the hot heat transfer surface is facing down, the resultant flow occurs from the center to the edges and the averaged values of the coefficient of heat transfer are calculated by the expression

$$\overline{Nu_L} = 0.27\ Ra_L^{0.25} \qquad (10^4 \leq Ra_L \leq 10^{10}).$$

King (*Mech. Eng.* 1932, v. 54, p. 347) obtained a practically important result and suggested a general equation for approximate determination of heat transfer from a body of an arbitrary shape that was found during generalized studies of heat transfer from plates, cylinders, bars, spheres and bodies of other geometric forms. This equation is similar in form to the formula for laminar flow near vertical surfaces. This formula may be used in the absence of more definite data for a body of the given form. The King formula is

$$\overline{Nu} = 0.6\ Ra^{1/4} \quad \text{when } 10^4 < Ra < 10^9.$$

The characteristic body dimension L is found from the relation

$$\frac{1}{L} = \frac{1}{L_h} + \frac{1}{L_v},$$

where L_h and L_v are the body dimensions along the horizontal and vertical lines. Thus, for a vertical plate it is equal to the height, and for a sphere it is equal to the radius.

In a fluid confined between two vertical surfaces (**Figures 2a,b**) heat transfer at small Grashof numbers (approximately up to 2000) is mainly performed by heat conduction and the Nusselt number is equal to 1. At large Grashof numbers $10^6 < Pr < 10^9$, $1 < Pr < 20$ for the cavity aspect ratio $H/\delta = 1 - 40$ the averaged heat transfer between the vertical surfaces (the horizontal sides of the cavity are thermally insulated) is described by the relation

$$\overline{Nu_\delta} = 0.046\ Ra_\delta^{1/3}.$$

In a fluid between horizontal surfaces at small Grashof numbers ($Gr < 1700$) the heat conduction mode is set and the Nusselt number is equal to unity both in the case when the heated surface is below and in the case when it is above. At large Grashof numbers corresponding to the turbulent flow mode which begins when $Gr \approx 5 \times 10^4$ there is no dependence on δ and mean heat transfer between the upper heated and lower cooled plates is described by the relation

$$\overline{Nu} = 0.069\ Gr^{1/3}Pr^{0.407},$$

which is valid for $0.02 < Pr < 8750$, $3\ 10^5 < Ra < 7\ 10^9$.

Study of the dynamics of a thermal FC plume over the heated elements is a special branch of the FC theory. The two idealized models (**Figure 5**) are considered: in the form of a two-dimensional plume (2) from a linear heat source (1) or in the form of an axisymmetric plume (2) from a point source (1). The plumes arise as a result of continuous heat supply. If heat is released only during a short period of time then the thus originating flow is called a thermic.

In practice the problem of the interaction between ascending flows having free boundaries and other flows and surfaces is often encountered. In particular, this refers to the cooling of elements of electronic equipment when the flows produced by heat sources placed at different places interact. Therefore, it is important to position these sources in such an order that maximum heat removal can be obtained. Elements with heat sources may be located on a surface which is usually thermally insulated and the resultant flow is caused by the interaction of flows formed by sources arranged in different places (**Figure 6**). In many of the production processes related to heating one has also to deal with the interaction of flows

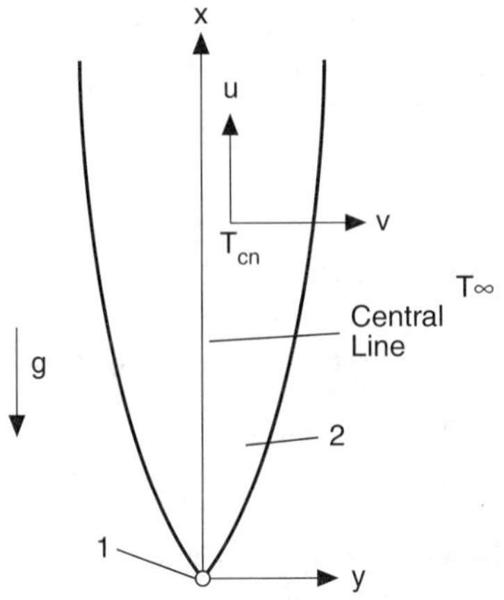

Figure 5. Thermal plume formation in free convection.

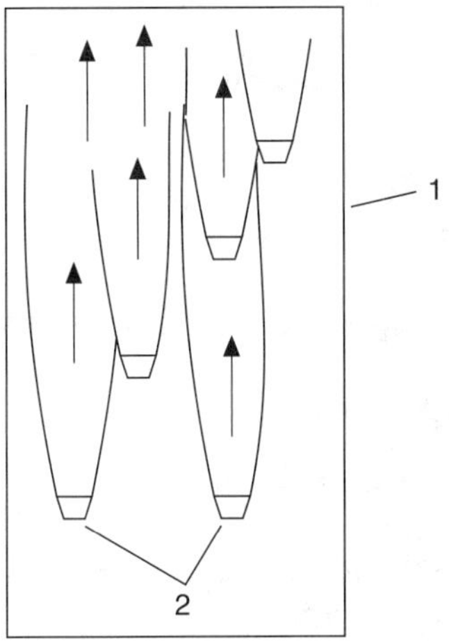

Figure 6.

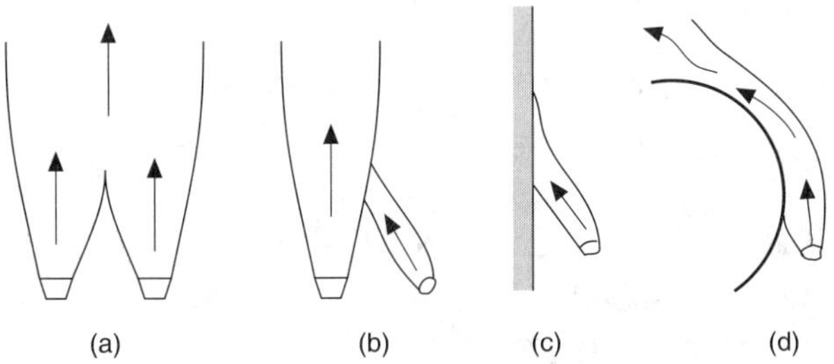

Figure 7.

created by a system of heated elements, in particular, by water-cooling towers, pipe-lines for transportation of hot liquid. **Figures 7a** and **b** illustrate the interaction of plumes with equal and different heat supplies and **Figures 7c** and **d** show the effect of vertical and curvilinear surfaces on a flow in a plane plume adjacement to them.

References

Gebhart, B. (1973) *Natural Convection Flows and Stability: Advances in Heat Transfer, v. 9,* Academic Press.

Jaluria, Y. (1980) *Natural Convection. Heat and Mass Transfer.* Pergamon Press.

Leading to: *Archimedes number*

O.G. Martynenko and A.F. Polyakov

FREE CONVECTION BOUNDARY LAYERS (see Boundary layer heat transfer)

FREE ENERGY

Following from: *Thermodynamics*

Gibbs free energy, G, and *Helmholtz free energy,* A, are both extensive thermodynamic properties defined as:

$$G = H - TS \tag{1}$$

and

$$A = U - TS \tag{2}$$

where T is temperature, S entropy and U internal energy.

Differentiation of these equations and substitution of the fundamental equations

$$dU = TdS - PdV \tag{3}$$

$$dH = TdS + VdP \tag{4}$$

where P is pressure and V volume leads to:

$$dA = -SdT - PdV \tag{5}$$

$$dG = -SdT + VdP \tag{6}$$

indicating that T and P are the natural variables associated with G, and T and V those associated with A.

The main utility of the free energy functions is in the determination of *equilibrium states.* For a closed system maintained at a fixed pressure and temperature in which the only work done is that of expansion of the system against the surrounding pressure, the equilibrium state is the one in which the Gibbs Free Energy assumes it lowest value. If the system is maintained at fixed volume and temperature and no work is done, it is the Helmholz Free Energy which assumes its lowest value. These conditions apply whether the system is single or multicomponent, single or multiphase.

G. Saville

FREE JETS (see Impinging jets)

FREE MOLECULE FLOW

Following from: *Flow of fluids*

The idea that materials may be treated as continua allows one to formulate equations of conservation for mass, momentum and energy in which all variables are continuous functions of space. (See **Conservation Equations.**) Thus, with the aid of the continuum hypothesis, it is possible to speak of the density of a material at a point in space whereas, from the molecular viewpoint, there actually may be no molecule at the point so that the concept of density has no meaning. The idea that materials may be treated as continua is founded upon the fact that in any element of volume that is small on a practical scale, there are a very large number of molecules (approximately 10^{16} in a cubic millimetre). Thus, for many purposes, it is possible to find a sufficiently small volume still containing a sufficiently large number of molecules that the discrete molecular nature of matter does not reveal itself.

However, under certain circumstances, the continuum hypothesis is inappropriate. Such circumstances occur when the distance between the molecules or, more correctly, the *mean free path* that they travel between collisions with other molecules, λ, is comparable with some physical dimension of the container of the flow channel, d ($\lambda/d \sim 1$). Naturally, this most often arises when the density of the gas is very low (so that the mean free path is large) and when the gas interacts with solid surfaces with a small-scale structure such as a porous solid or a capillary tube. In such circumstances, the gas molecules may interact as frequently with the solid surface as they do with other molecules and one has what is called a transition regime. As the density of the gas is reduced further, the collisions of molecules with walls completely dominate the processes and one reaches the free-molecule or *Knudsen regime* when $\lambda/d >> 1$. The ratio λ/d is termed the **Knudsen Number**, Kn.

For flow of a gas in the free-molecule or Knudsen regime through a tube of circular cross-section of radius R and length L, it is possible to show that the amount-of-substance flow through the tube is (Kauzmann, 1966)

$$\dot{M} = \frac{2\pi R^3(p_1 - p_2)}{M\bar{u}L}$$

under the influence of a pressure difference (p_1–p_2) where $\bar{u}$ is the mean velocity of the gas molecules and M is the molecular mass.

$$\bar{u} = (8RT/\pi M)^{1/2}$$

It should be noted that this flow is independent of the viscosity of the gas and of its density and is proportional to the third power of the radius of the tube, unlike for viscous flow where it is proportional to the fourth power.

In the transition regime, the behaviour is naturally intermediate between the continuum and free-molecule behaviour and is generally described in terms of slip at the walls so that the normal boundary condition of continuum mechanics, that there is no relative motion between the wall and the fluid, is abandoned. There are no exact means to treat free-molecule or transition flow since it is necessary to consider the nature of collisions between the molecules and the wall explicitly. Nevertheless, some progress has been made with relatively simple models, including the *Dusty Gas Model,* that is particularly appropriate for the treatment of *porous media* where the phenomenon of free-molecule flow often occurs (Cunningham and Williams, 1980).

References

Cunningham, R. E. and Williams, R. J. J. (1980) *Diffusion in Gases and Porous Media,* Plenum, New York.

Kauzmann, W. (1966) *Kinetic Theory of Gases,* Benjamin, New York.

W.A. Wakeham

FREE SETTLING (see Sedimentation)

FREE TURBULENT FLOWS (see Turbulent flow)

FREE VORTEX (see Hydrocyclones; Vortices)

FREEZE DRYING (see Dryers)

FREEZING (see Solidification)

FREEZING FOULING (see Fouling)

FREEZING POINT DEPRESSION (see Solubility)

FREONS (see Refrigerants, properties)

FRESNEL'S FORMULAS (see Reflectivity)

FREYN CHEQUERWORK REFRACTORIES (see Regenerative heat exchangers)

FRICTIONAL PRESSURE DROP (see Chezy formula)

FRICTION COEFFICIENT (see Crossflow)

FRICTION FACTOR (Hydraulic resistance)

FRICTION FACTORS FOR SINGLE PHASE FLOW IN SMOOTH AND ROUGH TUBES

Following from: *Pressure drop, single-phase*

Fanning Equation

For a single phase, fully developed flow in a pipe, the shear stress at the fluid-solid boundary is balanced by the pressure drop (see **Figure 1**). A one-dimensional force balance equation of this flow can be written as:

$$S\Delta p = \tau_W A \quad (1)$$

where S is the pipe cross-sectional area and A is the pipe surface area. Here, τ_w is the wall shear stress which is dependent upon the following parameters

- fluid velocity
- fluid properties, namely, density and viscosity
- pipe diameter
- surface roughness of the interior pipe wall

The first two parameters are due to the nature and the flow characteristics of the fluid itself. The last two depend on the physical geometry of the pipe. The stress can be expressed as

$$\tau = \frac{f\rho u^2}{2} \tag{2}$$

where f is the *Fanning friction factor.*

Friction Factor (f)

The friction factor, f, is a dimensionless factor that depends primarily on the velocity u, diameter D, density ρ, and viscosity η. It is also a function of wall roughness which depends on the size ϵ, spacing ϵ′ and shape of the roughness elements characterised by ϵ″. ϵ and ϵ′ have the dimension of length whereas ϵ″ is dimensionless. Since the friction factor is dimensionless, the quantities that it depends upon should appear in the dimensionless form. In this case, the terms u, D, ρ and η can be rearranged as uDρ/η which is the **Reynolds Number, Re**. For the characteristic *roughness factors* (ϵ and ϵ′), it may be made dimensionless by dividing these terms by D (the term ϵ/D is called the *relative roughness*). Hence, the friction factor can be written in a general form as:

$$f \propto \left(\mathrm{Re}, \frac{\epsilon}{D}, \frac{\epsilon'}{D}, \epsilon''\right) \tag{3}$$

From this we see that the friction factor of pipes will be the same if their Reynolds number, roughness patterns, and relative roughness are the same. For a smooth pipe, the roughness term is neglected and the magnitude of the friction factor is determined by fluid Reynolds number alone.

Fanning Friction Factor

The friction factor is found to be a function of the Reynolds number and the relative roughness. Experimental results of Nikuradse (1933) who carried out experiments on fluid flow in smooth and rough pipes showed that the characteristics of the friction factor were different for laminar and turbulent flow. For laminar flow (Re < 2100), the friction factor was independent of the surface roughness and it varied linearly with the inverse of Reynolds number. In this case, the friction factor of the Fanning equation can be calculated using the Hagen-Poiseuille equation (see **Poiseuille Flow**).

$$f = \frac{16}{\mathrm{Re}} \qquad \mathrm{Re} < 2{,}100 \tag{4}$$

For turbulent flow, both Reynolds number and the wall roughness influence the friction factor. At high Reynolds number, the friction factor of rough pipes become constant, dependent only on the pipe roughness. For smooth pipes, Blasius (1913) has shown that the friction factor (in a range of 3,000 < Re < 100,000) may be approximated by:

$$f = \frac{0.079}{\mathrm{Re}^{0.25}} \qquad 3{,}000 < \mathrm{Re} < 100{,}000 \tag{5}$$

However, for Re > 10^5, the following equation is found to be more accurate:

$$f = \frac{0.046}{\mathrm{Re}^{0.2}} \tag{6}$$

and this was used by Taitel and Dukler (1976).

Karman-Nikuradze Equation

Nikuradse (1933) measured the velocity profile and pressure drop in smooth and rough pipes where inner surface of rough pipes were roughened by sand grains of known sizes. He showed that the velocity profile in a smooth pipe is given by:

$$\frac{u}{u^*} = \frac{1}{\kappa} \ln \frac{u^* y}{\nu} + B \tag{7}$$

where κ (Von Kármán constant) and B are 0.4 and 5.5.

u^* is the friction velocity ($u^* = \sqrt{\tau_w/\rho}$)

$\nu = \eta/\rho$ is the kinematic viscosity

The friction factor can be related to mean flow velocity and Reynolds number by employing equation (1) and the relationship

$$f = 2\tau_W/\rho u^2 \tag{8}$$

The friction factor for smooth pipe is then expressed as:

$$\frac{1}{\sqrt{f}} = 0.86 \ln(\mathrm{Re}\sqrt{f}) - 0.8 \tag{9}$$

which is called the Karman-Nikuradse equation.

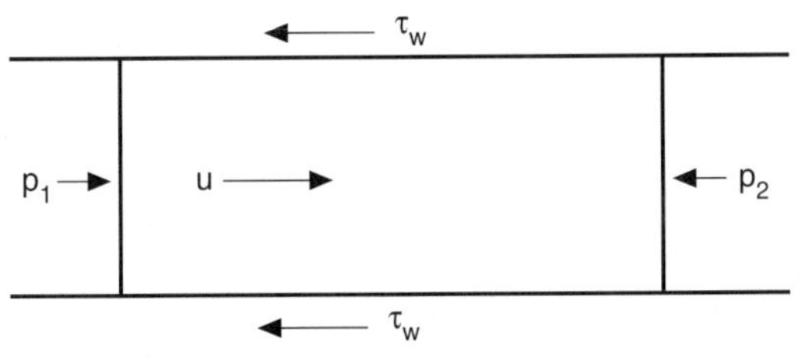

Figure 1. Forces acting on the fluid during single phase steady flow in a pipe.

For the rough pipes, the velocity distribution is defined as:

$$\frac{u}{u^*} = \frac{1}{\kappa} \ln \frac{y}{\epsilon} + B' \quad \textbf{(10)}$$

Here the constant B′ depends on the geometric characteristic of the roughness elements. Nikuradse classified the characteristics of the rough surface into three regimes based on the value of the *dimensionless characteristic roughness,* $u^*\epsilon/\nu$, where ϵ is the equivalent roughness height. The three roughness regimes are as follows:

1. Dynamically smooth: $0 \leq u^*\epsilon/\nu \leq 5$
2. Transition: $5 < u^*\epsilon/\nu \leq 70$
3. Completely rough: $u^*\epsilon/\nu > 70$

For the completely rough regime, the value of B′ is equal a constant of 8.48. The friction factor for rough pipes can be expressed in a form similar to that for smooth pipe as:

$$\frac{1}{\sqrt{f}} = 1.14 - 0.86 \ln(\epsilon/D) \quad \textbf{(11)}$$

Colebrook-White Formula

Friction factor of commercial pipes can be calculated using equation (5) if the pipe roughness is in the completely rough region. In the transition region where the friction factor depends on both Reynolds number and the relative roughness (ϵ/D), the friction factor of the commercial pipe is found to be different from those obtained from the sand roughness used by Nikuradse (see (**Figure 2**). This may be because the roughness patterns of commercial pipes are entirely different from, and vary greatly in uniformity compared to the artificial roughness. However, the friction factor of the commercial pipe in this zone can be calculated using an empiricism equation which is known as the Colebrook-White formula:

$$\frac{1}{\sqrt{f}} = -0.86 \ln\left[\frac{\epsilon/D}{3.7} + \frac{2.51}{Re\sqrt{f}}\right] \quad \textbf{(12)}$$

The formula can also be used for the smooth or fully rough pipes

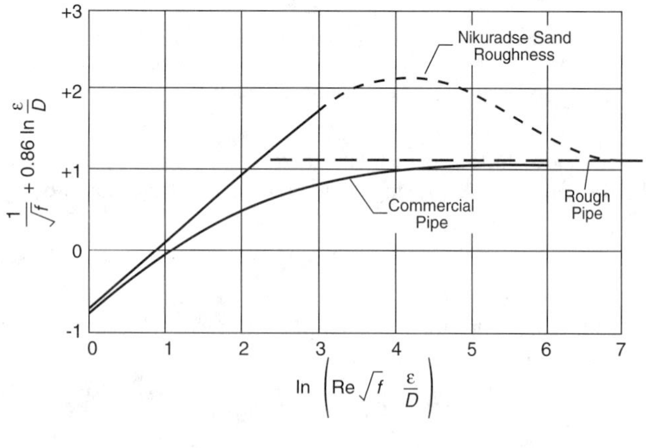

Figure 2. The difference between the Nikuradse sand roughness and the commercial roughness.

Table 1. Average roughness of commercial pipes, Streeter and Wylie (1983)

Material	ϵ mm
Cast iron	0.26
Galvanised iron	0.15
Asphalt cast iron	0.12
Commercial steel or wrought iron	0.046
Drawn tubing	0.0015
Glass	smooth

where it gets similar values to the Karman-Nikuradse equation when $\epsilon \to 0$ or $Re \to \infty$.

Moody Chart

In engineering applications there is a wide range of pipe wall roughness due to the different materials and methods of manufacture used to produce commercial pipes. Although the Colebrook-White formula can be used to calculate the value of the friction factor accurately from given value of the relative pipe roughness, the use of the formula is not practicable because of the complicated structure of the equation itself. Moody (1944) used the Colebrook-White formula to compute the friction factor of commercial pipes of different materials and summarised the data in the graph showing the relationship between friction factor, Reynolds number and relative roughness (**Figure 3** which is known as the *Moody Chart* or *Diagram*). Typical values of the roughness size of different pipe material are given in **Table 1**.

It is important to note that the value of friction factor obtained from the Moody Chart is equal four times of the Fanning friction factor.

$$f_{Moody} = 4f_{Fanning} \quad \textbf{(13)}$$

A useful explicit equation that applies to turbulent flow ($10^4 > Re > 4 \times 10^8$) in both smooth and rough pipes has been presented by Chen (1979):

$$\frac{1}{\sqrt{4f}} = -2.0 \log\left\{\frac{\epsilon}{3.7065D} - \frac{5.0452}{Re} \log\left[\frac{1}{2.8257}\left(\frac{\epsilon}{D}\right)^{1.1098} + \frac{5.8506}{Re^{0.8981}}\right]\right\} \quad \textbf{(14)}$$

References

Vennard J. K. and Street R. L, *Elementary Fluid Mechanics,* 5th Ed., Jonh Wiley and Sons Inc, USA.

Streeter V. L. and Wylie E. B. (1985) *Fluid Mechanics,* McGraw-Hill Inc, USA.

Nikuradse, J. (1950) Stromungsgesetze in rauhen rohren, *VDI-Forschungsheft,* 361, 1933, see English Translation NACA TM 1292.

Blasius, H. (1913) *Forschungsarbeiten auf dem Gebiete des Ingenieurwesens,* 131.

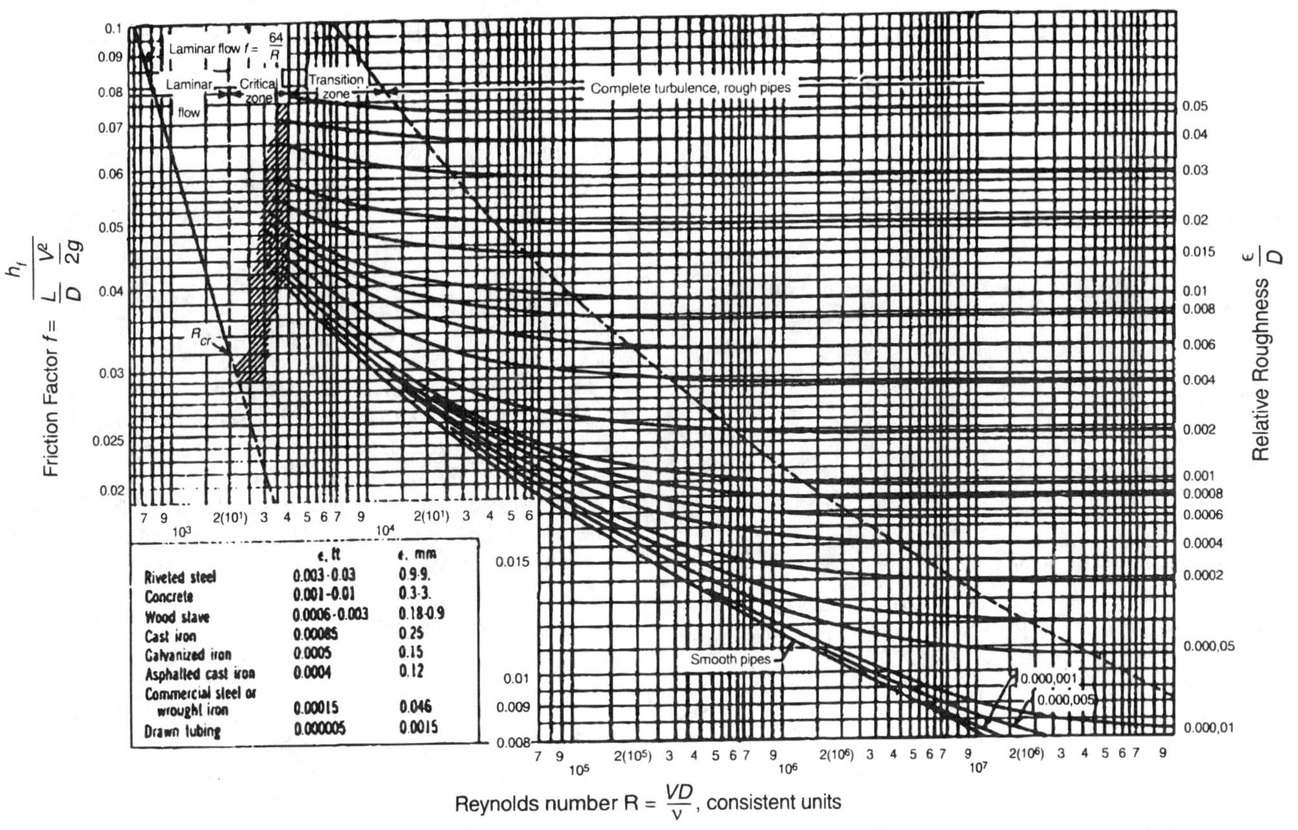

Figure 3. Moody diagram. (Adapted from Streeter V.L. et al. (1985) with permission.)

Taitel, Y. and Dukler, A. E. (1976) A model for predicting flow regime transition in horizontal and near horizontal gas-liquid flow, *AIChE J,* vol 22, pp 47–55.

Moody, L. F. (1944) Friction Factors for Pipe Flow, *Trans. A.S.M.E.,* vol. 66.

Douglas, J. F. Gasiorek, J. M. and Swaffield, J. A. (1985) *Fluid Mechanics,* 2nd Ed., Pitman Publishing Ltd.

Chen, N. H. (1979) An explicit equation for friction factor in pipes, *Int. Eng. Chem. Fundam.,* 18(3), 296.

S. Srichai

FRICTION VELOCITY

Following from: *Boundary layers*

The friction velocity (also known as the *shear-stress velocity*), u*, of a flow is defined by the relation:

$$u_*^2 = \tau/\rho \quad (1)$$

where τ is the shear stress and ρ the fluid density. This quantity has the dimensions of velocity and is frequently encountered in the study of **Boundary Layers.** For turbulent flow it is approximately constant in the region near to a wall, and in the atmosphere, this region extends over the lowest several metres. The flow velocity in the boundary layer is commonly expressed in terms of the friction velocity and the height, z, as:

$$u = \frac{u^*}{\kappa}\ln(z/z_0) \quad (2)$$

where z_0 is the roughness length. κ is an empirical constant known as the *von Kármán constant* and is found to have a value of $\kappa \approx 0.35$ for all cases of turbulent flow. The derivation of this formula is given in, for example, Holton (1979) and Streeter and Wylie (1983).

References

Holton, J. R. (1979) *An Introduction to Dynamic Meteorology,* Academic Press, New York.

Streeter, V. L. and Wylie, E. B. (1983) *Fluid Mechanics,* McGraw-Hill, Singapore.

M.J. Pattison

FRIEDEL CORRELATION (see Pressure drop, two phase flow)

FRINGE MODEL, FOR LDA (see Anemometers, laser Doppler)

FROSSLING-MARSHALL EQUATION (see Mass transfer)

FROSTING

Following from: *Condensation; Freezing*

Frost is a material composed of ice and air. Frosting is the vapour-to-solid phase change process in which airborne water vapour becomes frost on a cooled surface. Frost forms on cold surfaces when the surface temperature is below 0°C and below the dew point of the water vapour contained in a surrounding body of air. Supercooled water droplets first form on the cold surface. These droplets subsequently freeze to form a thin ice layer and frost begins to grow on this ice layer. Hayashi et al. (1977) defined three chronological periods of frost formation.

One dimensional growth period

In this period, ice crystals grow in a direction perpendicular to the cold surface to form frost with widely-spaced distinct ice needles.

Three dimensional growth period

Branches grow on the above ice needles until the upper frost layer becomes almost flat. During this period, part of the water vapour flux from the environment causes increased frost depth and part enters the frost layer by vapour diffusion, causing densification to occur.

Quasi-steady growth period

The outer frost surface reaches a temperature of 0°C and fluctuates close to this temperature. In this period, heat transfer rates remain almost constant as the frost layer thickens because increased thermal resistance due to increased frost depth is offset by resistance reductions due to frost density increases. In this period, density is believed to increase due to soakage and subsequent freezing of liquid water from the surface entering the frost layer.

Frost depth increases approximately in proportion to the square root of time prior to the quasi-steady growth period. (Schneider (1978)). Hayashi et al.(1977) classified frost into types A, B, C and D, with type A frost being more needle-like and less dense at all stages of chronological development and Type D being most dense. Type A frost generally forms with high vapour concentration differences between the air and the cold surface and at low surface temperatures.

Sensible heat transfer through a growing frost layer is supplemented by the release of latent heat of evaporation as water vapour becomes ice within and at the surface of the frost layer. The thermal resistance of the frost layer formed on heat exchange equipment may be calculated by knowing frost depth and the frost effective conductivity. At a particular depth position in the layer, the local frost effective thermal conductivity is defined as

$$\lambda = \dot{q}/(dT/dr)$$

where $\dot{q}$ includes sensible and latent heat transfer.

Frost effective conductivity may take values between that of air, 0.024 W/m K and that of ice, 2.3 W/m K. A generalised correlation of frost thermal conductivity has been presented by Dietenberger (1983) who considers frost conductivity to depend on: frost density; local frost temperature; cold wall temperature; air temperature; air relative or specific humidity and ambient total air pressure. Conductivity is particularly sensitive to frost density.

Frost density is not uniform through the frost layer thickness. Density is influenced by the conditions under which the frost forms and may have values from 50 kg/m^3 for Type A frost at the earliest stages of development up to the value for ice, 917 kg/m^3, for very mature frost.

Calculation of frost layer thermal resistance requires mathematical modelling using numerical techniques. A finite-difference transient 1-dimensional model for heat transfer and frost growth has been developed by Monaghan and Oosthuizen (1990) which is capable of calculating heat transfer, frost depth and frost mass or density over time when the environmental conditions and wall temperature are given but varying. For single tubes in a steady environment and for rows of tubes in an unsteady outdoor environment, Monaghan and Oosthuizen's model was found to be within +/− 25% in prediction of overall thermal conductance in a wide range of conditions.

Tests of typical refrigeration heat exchangers under frosting conditions have been carried out by Kondepudi and O'Neal (1989). In these tests, Kondepudi and O'Neal found that: rate of frost mass accumulation and pressure drop increased with air humidity, cold refrigerant temperatures and higher fin densities; and latent energy contributed about 30%–40% of total heat transfer.

References

Dietenberger, M. A. (1983) Generalised Correlation of the Water Frost Thermal Conductivity, *Int. J. of Heat Mass Transfer,* 26, 607.

Hayashi, Y., Aoki, A., Adachi, S. and Hori, K. (1977) Study of Frost Properties Correlating with Frost Formation Types, *J. Heat Transfer,* 99, 239.

Kondepudi, S. N. and O'Neal, D. L. (1989) The Performance of Finned Tube Heat Exchangers under Frosting Conditions, *ASME Winter Annual Meeting,* San Francisco.

Monaghan, P. F. and Oosthuizen, P. H. (1990) Mathematical model of frost growth on a single cylinder in steady crossflow. *In Heat Transfer 1990,* Proc. of 9th Int. Heat Transfer Conference, IHTC9, Jerusalem, ISRAEL, *3,* 115–120, Hemisphere Publishing Corp., Washington, USA.

Schneider, H. W. (1978) Equation of the Growth Rate of Frost Forming on Cooled Surfaces, *Int. J. Heat Mass Transfer, 21,* 1019.

P.F. Monaghan

FROTH (see Foam fractionation)

FROTH FLOW (see Churn flow)

FROTH FLOTATION (see Flotation)

FROUDE NUMBER

Froude number, Fr, is a dimensionless group which occurs frequently in the study of hydraulic phenomena involving a free surface.

$$Fr = u^2/gL$$

where u is velocity, g acceleration due to gravity and L height. It represents the ratio of momentum force to gravitational force.

For physical modelling, for example in naval architecture, Froude number is maintained constant by reducing the liquid velocity in the model in proportion to the square root of the linear dimension.

Often the term Froude number is used for the ratio $u/(gL)^{1/2}$. This velocity ratio is also called **Boussinesq Number.**

G.L. Shires

FROUDE NUMBER, EFFECT ON JET IMPINGEMENT (see Impinging jets)

FUEL ASSEMBLIES (see Boiling water reactor, BWR; Liquid metal cooled fast reactor)

FUEL CELLS

Following from: *Electrochemical cells*

The discoverer of the fuel cell was W. R. Grove who in 1839 reported the working of a hydrogen-oxygen system in dilute sulphuric acid.

A fuel cell is an electrochemical device which converts the "*chemical*" energy of a fuel into electrical energy. The "*chemical*" energy is associated with the free-energy change of an electrochemical reaction from which one can write.

$$\Delta G^0 = -nFE_r^0$$

where n is the number of electrons transferred, F is the Faraday Constant and E_r^0 is the reversible potential of the cell (see **Faraday's Law**).

A fuel cell will, in principle, continuously supply direct electrical energy when continuously supplied with fuel and an "oxidant", usually oxygen. The simplest of fuel cells is the $H_2 - O_2$ system. In this, hydrogen gas is oxidised at the anode and oxygen is reduced at the cathode to form water overall

$$H_2 + {}^1/_2\,O_2 = H_2O$$

The performance of fuel cells is best illustrated in its cell potential versus current density response (see **Figure 1**). This behaviour is largely determined by the kinetics of the electrode reactions and the internal resistance of the cell. The major kinetic factor is associated with the oxygen electrode, which requires much more negative potentials than the reversible potential of −1.23 V (w.r.t. reversible hydrogen electrode) to drive the reaction at significant current

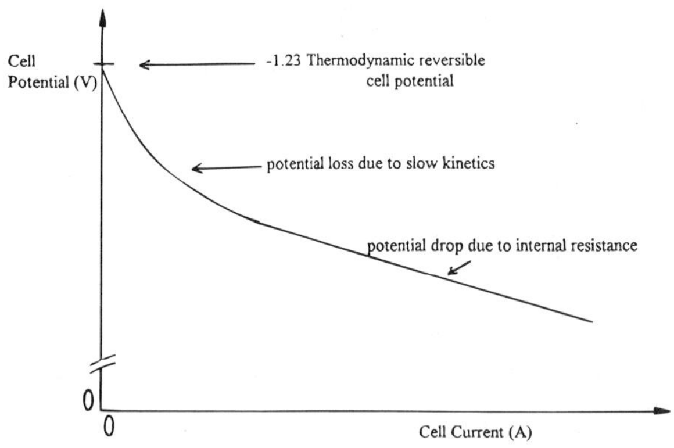

Figure 1. Schematic cell potential vs current behaviour for hydrogen-oxygen fuel cell.

densities. With catalytic electrodes based on dispersed platinum on carbon, the loss of voltage can be limited to values of 0.4 V at current densities around 0.1 A cm^{-2}

Hydrogen is the most common fuel utilised in most fuel cells although other fuel such as hydrazine, methane and methanol are possible. There are several types of fuel cell, which can be classified according to the "electrolyte"—alkaline (usually KOH), acid (mainly phosphoric acid), molten carbonate (mixture of Li and K), solid oxide (typically yttria-stabilised zirconia) and solid polymer (proton-conducting membrane).

Alkaline fuel cells usually utilise potassium hydroxide at a concentration of 30%, which is the approximate optimum value for maximum conductivity. The two half cell reactions are

anode $\quad H_2 + 2OH^- \rightarrow 2H_2O + 2e^-$

cathode $\quad O_2 + 2H_2O + 4e^- \rightarrow 4OH^-$

Alkaline fuel cells operate at low temperatures, typically 60–80°C, and at low pressures (1–2 bar).

The phosphoric acid fuel cell is the most technically advanced for large scale power generation. The electrolyte is a thin layer of ortho-phosphoric acid absorbed into a solid matrix (SiC/carbon composite). The electrodes are teflon bonded porous carbon structures using dispersed platinum based electrocatalysts. Operating temperatures are around 200°C which serves to improve the kinetics of oxygen reduction. Operating pressures are typically 4.5—5 bar for air/hydrogen systems, hydrogen being supplied usually by the reforming of methane, the primary fuel. This, or similar primary fuels, is the usual source of hydrogen for fuel cells.

Molten carbonate fuel cells, which can operate with either carbon monoxide or hydrogen as fuel, use an electrolyte immobilised in a porous inorganic matrix. Temperature of operation is 650°C and the anode is a porous nickel or Ni/Cr alloy and the cathode is porous NiO. Solid oxide fuel cells operate at 1000°C and are based on ceramic oxide electrolytes.

The solid polymer fuel cell is based on a solid polymeric proton conducting membrane as the electrolyte sandwiched between two platinum-catalysed porous carbon electrodes. Operating temperatures are approximately 100°C at their highest, otherwise damage to the membrane is risked. Applications are in space, underwater, standby power and transportation. The principle fuel used is hydrogen although methanol powered cells are attracting significant interest in view of the fuels superior energy density.

K. Scott

FUEL-COOLANT INTERACTION, FCI (see Vapour explosions)

FUEL RODS (see Boiling water reactor, BWR)

FUELS

Fuels are materials that create usable energy through chemical, nuclear or electrochemical reactions. The evolution of energy is usually controlled so that it may be used for heating or converted into some other form of energy, such as electricity. Most commonly used fuels are derived from hydrocarbon materials and release their energy in the form of heat during the process of **Combustion.** The most important fuels, not derived from hydrocarbons, include the nuclear fuels, uranium and plutonium, and liquid hydrogen for use in fuel cells and rocket engines. Fuels may be solid, liquid or gaseous and can be classified as primary or secondary. A *primary fuel* occurs naturally in a form that is suitable for immediate use, such as coal, whereas a *secondary fuel* is derived from a naturally occurring material by some form of processing that has radically altered it, such as coal gas from coal. *Carbonaceous fuels* may be further classified as fossil or non-fossil. *Fossil fuels* are derived from the decay of organic matter over millions of years and are the most extensively used group of fuels around the world.

The three major fossil fuels are *coal, petroleum* and *natural gas.* Coal has been used extensively throughout the industrialised world since the beginning of the industrial revolution. It can be mined from deep under the ground or deposits close to the surface can be strip-mined. The latter technique is used commonly around the world and provides the most economic source of coal. Coal requires very little processing before use. Some washing may be required to remove rock followed by crushing as necessary for the particular use. The major use for coal is combustion to provide electricity or heat, but it is also used to make *coke* for the production of iron and steel and for some production of chemicals. Coal occurs in various forms depending on its geological source and degree of coalification or rank. Of the three major fossil fuels coal has the highest carbon-to-hydrogen ratio and the lowest calorific value. The calorific value (on a mineral matter free basis) varies with rank from about 23 MJ/kg for the lower rank coals to about 34 MJ/kg for the higher rank and bituminous coals. Despite its solid form, coal is traded extensively around the world and still finds the greatest use of all fossil fuels for the generation of electricity. Its great advantage is that there are known deposits available for many hundreds of years and the deposits are found all around the world in all continents, hence giving a competitive world market and stable prices.

The use of petroleum has increased significantly since the 1950s with massive increases in the use of transport fuels, which are almost exclusively supplied from petroleum, and the growth of the petrochemical industry. Petroleum, or crude oil, is extracted, as a liquid, from wells on both land and sea. (See **Oils**.) As with coal, there are different forms of crude from different parts of the world. However, in all cases the crudes are processed (i.e. refined) to give a variety of liquid products which find different applications. As mentioned above, many of these are for transport fuel, such as petrol, diesel and aviation spirit. But some fractions, such as gas oil and fuel oil, are used for combustion for heating and the generation of electricity. Fuels, such as gas oil, are very convenient for industrial heating purposes since they are clean liquids which can be easily transported, stored and pumped. However, the cost of such fuels compared to others may make them less attractive and the economic usage of a particular fuel will depend very much on individual circumstances. Petroleum-derived fuels are attractive for use since they are in the liquid form and they have a lower carbon-to-hydrogen than coal and higher calorific values ranging from about 45–48 MJ/kg. There are sources of petroleum around the world and new finds are being made. However, there is still a concentration of resources in the Middle East, which could, in certain circumstances, influence the petroleum market and prices.

Natural Gas resources have a recent history of usage. Natural gas is mainly **Methane** and has the lowest carbon-to-hydrogen ratio of the fossil fuels and the highest calorific value of about 55 MJ/kg, although this value will be lower if the methane content in the gas is lower. Natural gas is a highly convenient and "clean" form of fossil fuel requiring minimal treatment before use, and is used for combustion for both heat and electricity generation. It is very applicable to local use where a network of distribution pipelines exists. However, it is a difficult fuel to transport over distances as a world commodity in the same way as coal and petroleum and requires expensive pipelines or very specialised shipping facilities.

The major fossil fuels mentioned above represent a very large part of the fuel usage throughout the world. However, other energy sources are used, such as nuclear and hydro, for the generation of electricity and these can be major sources in some local areas and countries. (See **Hydropower** and **Nuclear Reactors**.) There are also emerging sources of fuels which include renewables such as biomass, where the main sources are: wood; municipal solid wastes; agriculture and industrial wastes; methane from landfill gas; and ethanol. (See **Alternative Energy Sources**.) There are also existing or developing technologies which convert one particular fuel into another. This could be coal into liquids or gases, biomass into liquids or gases; and gases into liquids. Conversion has the advantage of meeting local needs, converting a fuel into a more transportable form (such as gas to liquid) or as a solution to environmental problems. For example, the *gasification of coal* yields better methods for the removal of *pollutants* such as sulphur and nitrogen, and enables the coal to be used more efficiently in combined cycles.

For the engineer or scientist there is a great deal of literature on the composition, properties and usage of different fuels. Some references are given below which may prove useful in finding information.

References

Technical Data on Fuel, 7th Edition, J. W. Rose and J. R. Cooper eds., (1977) Scottish Academic Press.

Fuels and Fuel Technology, 2nd Edition, W. Francis and M. C. Peters, Pergamon (1980).

Fuel Science and Technology Handbook, J. G. Speight ed., Chemical Industries Series, No. 41, (1990) Dekker.

Encyclopedia of Chemical Technology, 4th edition, 11 and 12, J. I. Kroschwitz ed., (1993) Wiley-Interscience.

M. Cloke

FUELS, PROPERTIES OF (see Furnaces)

FUEL TO AIR RATIO (see Flames)

FUGACITY

The fugacity of a component in a mixture is defined as:

$$f_i = (P\tilde{y}_i)\exp[(\mu_i - (\mu_i)_{pg})/\tilde{R}T]$$

where μ_i is the chemical potential of component i in the mixture, $(\mu_i)_{pg}$ is the chemical potential of the hypothetical perfect gas at the same pressure, P, temperature, T, and composition (mole fraction) of the mixture, $\tilde{y}$. $\tilde{R}$ is the universal gas content.

Fugacity has the dimensions of pressure. In many ways, it behaves in the same way as does partial pressure in a *perfect gas mixture,* although one must be careful not to take the analogy too far. It is most frequently used when discussing the thermodynamic properties of real gases and rather less so when considering liquids.

If fugacity is used to represent non-ideality in both gas and liquid phases, then the condition for equilibrium becomes equality of fugacity for each component in both phases.

G. Saville

FUGACITY COEFFICIENT (see Thermodynamics)

FULLER-SCHETTER-GITTINGS EQUATION, FOR DIFFUSION COEFFICIENT IN GASES (see Diffusion coefficient)

FUNCTIONS (see Delta function)

FURNACE FIRED BOILER (see Fossil fuel fired boilers)

FURNACE MODELS (see Furnaces)

FURNACES

Introduction

A furnace is a device in which heat is generated and transfered to materials with the object of bringing about physical and chemical changes. The source of heat is usually combustion of solid, liquid or gaseous fuel, or electrical energy applied through resistance heating (Joule heating) or inductive heating. However, solar energy can provide a clean source of high temperature if focussed onto a small area. This was recognised over two hundred years ago by Lavoisier who built a large mobile "magnifying glass" system, **Figure 1**, to bring about the combustion of metals in a sealed glass container and subsequently the demise of the phlogiston theory.

Furnaces employing combustion produce a hot gas which transfers heat to the material by radiation and convection. Solids are heated by direct contact, but fluids are usually heated indirectly, being carried inside pipes within the furnace. Alternatively, a **Regenerative Heat Exchanger** may be used to transfer heat from the combustion gases. Indirect heating has the advantage of avoiding contamination by combustion products.

There are two principle categories of indirect heating furnace. The first is that of **Boilers**, where the heat is used to generate steam for power generation or process plant use. The second is that of furnaces designed to heat process fluids other than water. Some of the latter category are conventional heaters used simply to increase the fluid temperature, others are process heaters used to bring about physical and chemical changes in the products, for example distillation, and pyrolysis of hydrocarbons or catalytic steam-gas reforming of synthetic natural gas.

Process Heaters

In the petrochemical industry process furnaces, or *process heaters,* are usually fired by oil or gas. They are designed to ensure that the fluid receives the right amount of heat and has the required residence time within specified temperature limits. This is achieved by appropriate disposition of tubes and careful control of firing rate and fluid flow. The following **Figure 2** shows three typical geometries.

Furnace Fuels

The following tables give the *properties of* gaseous, liquid and solid *fuels* used in furnaces. Gross calorific value includes the heat released by the condensation and cooling of combustion products, and the combustion air/fuel ratio is the stoichiometric value based on theoretically perfect combustion (see **Combustion**). In practice, excess air is required to ensure complete combustion. The excess is usually about 10% of the stoichiometric value for premixed

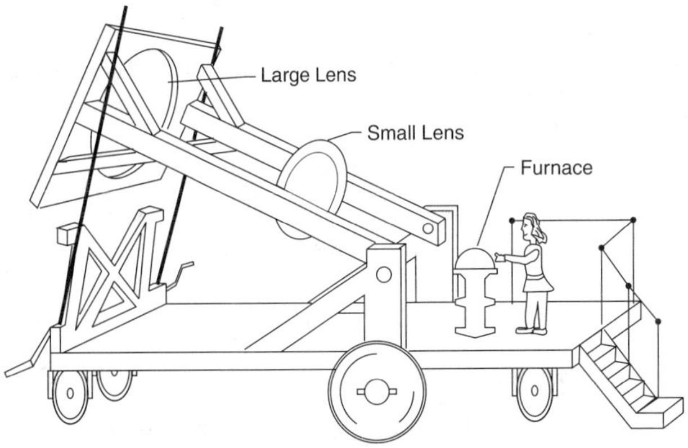

Figure 1. Lavoisier's solar powered furnace c 1774.

Table 1. Properties of selected gaseous fuels[a]

Fuel	Composition (% by volume)												Calorific value (MJ/kg)		Combustion air
	CO_2	N_2	CO	H_2	CH_4	C_2H_6	C_3H_8	C_4H_{10}	C_5H_{12}	C_2H_4	C_3H_6	C_4H_8	Gross	Net	(kg/kg)
Pure gases															
Carbon monoxide			100										10.10	10.10	2.46
Hydrogen				100									142.0	120.0	34.27
Methane					100								55.48	49.95	17.20
Ethane						100							51.88	47.45	15.90
Propane							100						50.35	46.33	15.25
n-Butane								100					49.55	45.73	14.98
n-Pentane									100				49.01	45.33	15.32
Ethylene										100			50.28	47.11	14.81
Propylene											100		48.91	45.75	14.81
Butylene												100	48.46	45.30	14.81
Fuel gases															
North Sea gas	0.2	1.5			94.5	3.0	0.5	0.2	0.1				53.5	48.2	16.6
Göningen gas	0.9	14.0			81.8	2.7	0.4	0.1	0.1				42.3	38.1	13.1
Synthetic natural gas	2.0		0.1	0.7	95.2		2.0						52.3	47.2	16.2
Commercial propane						1.5	91.0	2.5			5.0		50.3	46.3	15.2
Commercial butane					0.1	0.5	7.2	88.0			4.2		49.6	45.8	15.0
Water gas	4.7	4.5	41.0	49.0	0.8								16.5	15.1	4.0
Blast furnace gas	17.5	56.0	24.0	2.5									2.49	2.45	0.61
Coal gas	4.0	6.6	18.0	49.4	20.0							2.0	30.3	27.2	8.4
Producer gas	5.0	54.5	29.0	11.0	0.5								4.55	4.34	1.12
Lurgi crude gas	25.6	1.8	24.4	37.3	10.3					0.3	0.3		13.1	11.8	3.5
Lean reformer gas	16.7		2.2	46.4	34.7								30.9	27.3	8.9
Rich reformer gas	21.0		1.0	17.0	61.0								30.2	27.1	9.2

[a]From the *Heat Exchanger Design Handbook*, Hemisphere Publications. With permission.

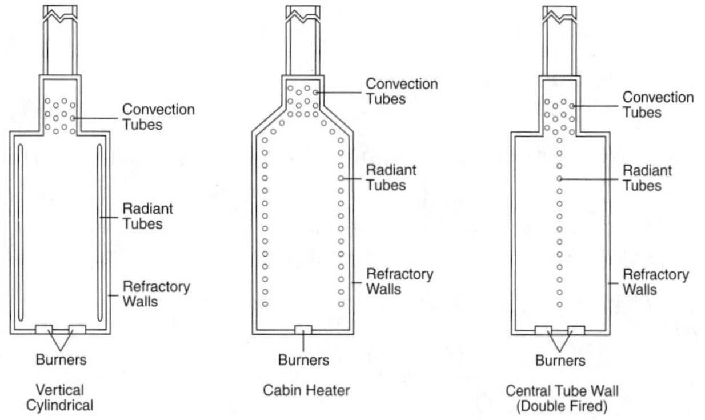

Figure 2. Process heaters. (From the *Heat Exchanger Design Handbook*, Hemisphere Publications. With permission.)

gaseous fuels, 20% for distillate oil and pulverised coal, and 30% for residual oil, although lower values may be achieved with efficient burners.

Furnace Heat Balance

The rate of heat released in a fuel fired furnace is given by the product of the mass rate of feed of fuel, $\dot{M}_f$, and the calorific value Δh_f, i.e.

$$\dot{Q}_f = \dot{M}_f \Delta h_f \tag{1}$$

The heat generated by combustion appears initially as sensible heat in the gaseous products of combustion, generated at the rate $\dot{M}_g$. If this were an adiabatic process (i.e. no heat transfer), the gas would attain the adiabatic flame temperature, T_f, given by

$$\dot{Q}_f = \dot{M}_g c_{pg}(T_f - T_0) \tag{2}$$

where T_o is the inlet air/fuel temperature and c_{pg} an appropriate averaged specific heat capacity of the gases for the range T_o to T_f.

However, as illustrated in **Figure 3**, part of the heat generated passes to the tubes containing the process fluid at rate $\dot{Q}_g$ and part is lost through the furnace walls at rate $\dot{Q}_1$. The remainder escapes at rate $\dot{Q}_p$ as waste heat of the combustion products leaving the furnace. The overall heat balance therefore is

$$\dot{Q}_f = \dot{Q}_g + \dot{Q}_l + \dot{Q}_p \tag{3}$$

Because of this the actual gas temperature, T_g, reached in the furnace is less than the adiabatic flame temperature, T_f. Neglecting wall losses an approximate value of T_g can be obtained from

$$\frac{\dot{Q}_f - \dot{Q}_g}{\dot{Q}_f} = \frac{T_g - T_0}{T_f - T_0} \tag{4}$$

For example, in a typical oil fired furnace the value of T_f is 2,200 K and the value of T_g is 1,300 K (Shires, Hewitt and Bott, 1994).

Table 2. Properties of selected liquid and solid fuels[a]

Fuel	Composition (% by mass, as fired)							Calorific value (MJ/kg)		Combustion air (kg/kg)
	C	H	O	S	N	Ash	Moisture	Gross	Net	
Liquids										
Kerosene	85.8	14.1		0.1				46.5	43.5	14.7
Gas oil	86.1	13.2		0.7				45.6	42.8	14.4
Light fuel oil	85.6	11.7	0.1	2.5	0.08	0.02		43.5	41.1	14.0
Medium fuel oil	85.6	11.5	0.15	2.6	0.12	0.03		43.1	40.8	13.9
Heavy fuel oil	85.4	11.4	0.2	2.8	0.15	0.05		42.9	40.5	13.8
Methanol	37.5	12.5	50.0					22.7	19.9	6.5
Ethanol	52.2	13.0	34.8					30.2	27.2	9.1
Solids (coals)										
Anthracite	78.2	2.4	1.5	1.0	0.9	8.0	8.0	29.7	28.9	9.8
Low volatile bituminous	77.4	3.4	2.0	1.0	1.2	8.0	7.0	30.6	29.7	10.1
Medium volatile bituminous	75.8	4.1	2.6	1.2	1.3	8.0	7.0	30.8	29.8	10.2
High volatile bituminous	71.6	4.3	3.8	1.7	1.6	8.0	9.0	29.5	28.4	9.7
Lignite	56.0	4.0	18.4	0.6	1.0	5.0	15.0	21.5	20.2	7.1

[a]From the *Heat Exchanger Design Handbook*, Hemisphere Publications. With permission.

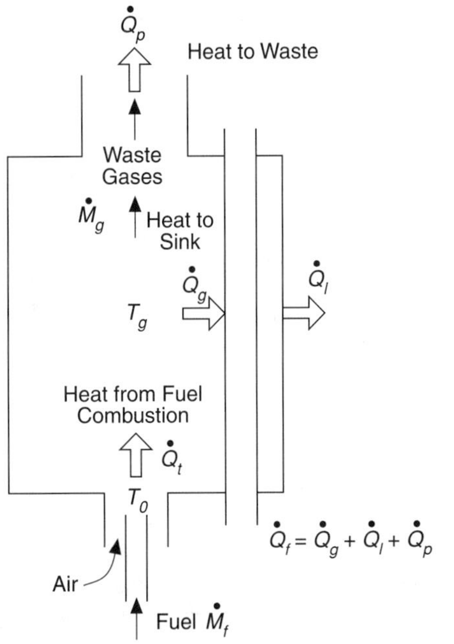

Figure 3. Furnace heat balances.

The rate of heat transfer, $\dot{Q}_g$, from the hot gases to the tubes is a function of the gas temperature as well as the radiation properties of the gases, the geometry of the furnace and tubes, and the emissivity of the tube surface. In applying simple furnace models iteration is therefore necessary.

Furnace Radiation

In most process heaters, the major part of the heat transfer from the hot gases to the tubes is by radiation. To calculate the radiative component it is necessary to know the effective emissivity, ϵ_g, of the combustion gases (typical value 0.25). This is dependent on the ratio of the partial pressures of CO_2 and H_2, the temperatures of the gas and the radiation source and the effective size of the radiating gas cloud. The latter is represented by the term pL_o, the product of partial pressure and effective length of the furnace—a term first introduced by H. C. Hottel. For details of the procedure see Hottel and Sarofim, 1967, and Hewitt, Shires and Bott, 1994.

The rate of heat transfer to the furnace product is also a function of the geometry of the tube banks and the fraction of the furnace surface area covered by them.

If the effective area of the tubes, A_1, is only a small fraction of the total area, A_t, (**Figure 4a**), the surface receives blackbody radiation (see **Radiative Heat Transfer**) and the rate of heat transfer is independent of gas emissivity, ϵ_g, i.e.

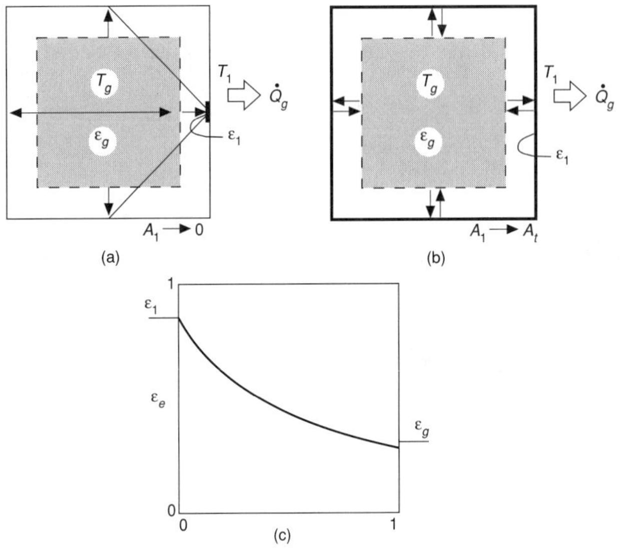

Figure 4. Effect of heat sink area on effective emissivity. (a) Small heat sink area; (b) large heat sink area; (c) example based on Equation 7 with $\epsilon_g = 0.3$ and $\epsilon_1 = 0.85$.

$$\dot{Q}_{gA_1\rightarrow 0} = A_1\epsilon_1\sigma(T_g^4 - T_1^4) \quad (5)$$

where ϵ_1 is the effective emissivity of the receiving surface and T_1 its temperature and σ is the Stefan-Boltzmann constant, 5.670×10^{-8} W/m²K⁴. On the other hand, if A_1 is almost equal to the total surface area, A_t, then

$$\dot{Q}_{gA_1\rightarrow A_1} = A_1\left(\frac{1}{1/\epsilon_1 + 1/\epsilon_g - 1}\right)\sigma(T_g^4 - T_1^4) \quad (6)$$

Intermediate ratios are covered by the speckled surface equation (Hottel and Sarofim, 1967).

$$\dot{Q}_g = A_1\left\{\frac{1}{1/\epsilon_1 + C(1/\epsilon_g - 1)}\right\}\sigma(T_g^4 - T_1^4) \quad (7)$$

where $C = A_1/A_t$.

Effects of Tube Bank Geometry

In practice the area, A_1, receiving radiation is a tube bank, **Figure 5**, which intercepts only a fraction of the incident radiation; some passes through to the refractory wall and is re-irradiated. For a detailed analysis see Hottel and Sarofim, 1967.

The fraction of radiation intercepted by a single row of tubes having a pitch to diameter ratio B (equals p/d) is given by

$$F = 1 - \frac{1}{B}\left\{(B^2 - 1)^{1/2} - \cos^{-1}\left(\frac{1}{B}\right)\right\} \quad (8)$$

Figure 6 shows F as a function of B for one and two rows of tubes.

The effective emissivity based on the projected area using Equation 7 and allowing for re-irradiation from the refractory backing is

$$\epsilon_{eff} = \frac{1}{\{1/F(2 - F) + (B/\pi)(1/\epsilon_1 - 1)\}} \quad (9)$$

This is illustrated in **Figure 7** for a typical tube surface emissivity of 0.85.

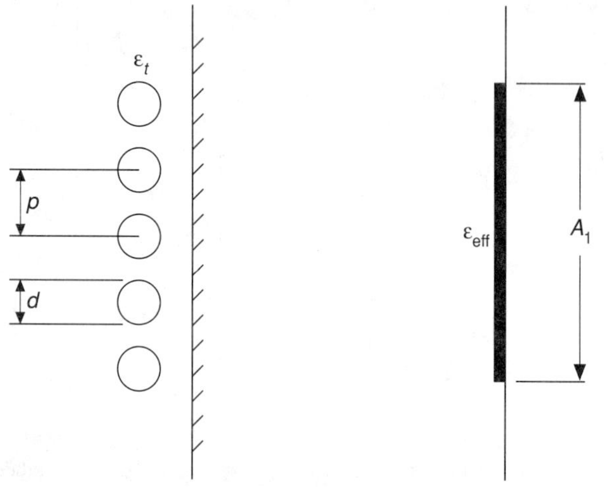

Figure 5. Plane surface equivalent of tube bank heat sink; B = p/d.

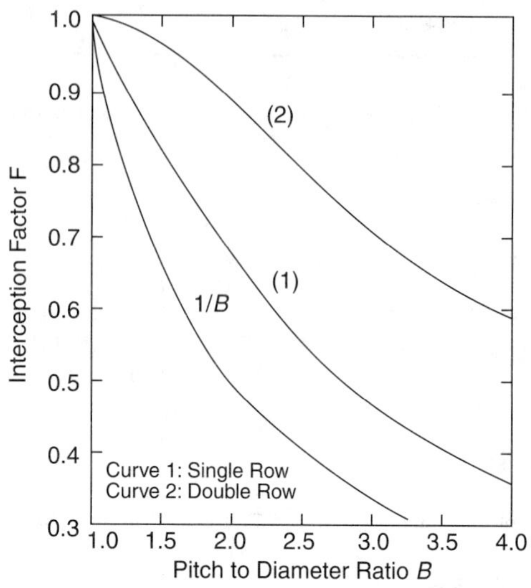

Figure 6. Fraction of incident radiation intercepted by tubes. (From the Heat Exchanger Design Handbook, Hemisphere Publications. With permission.)

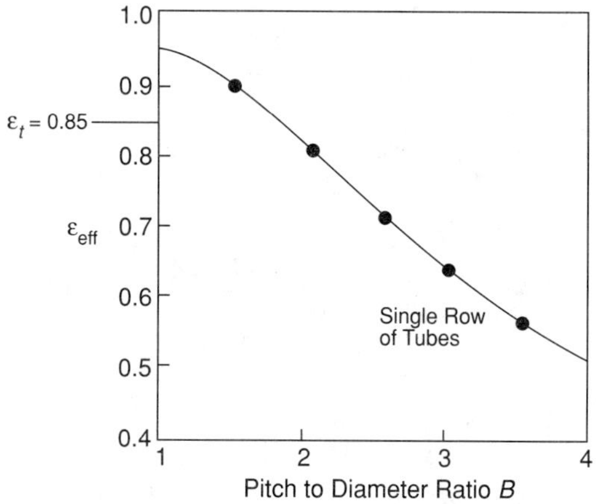

Figure 7. Effective emissivity of equivalent plane surface.

The relationship between the rate of heat transfer, $\dot{Q}_g$, from the gas to the tubes and the gas temperature, T_g, is obtained by combining Equations 6, 7 and 8. This in turn can be combined with Equation 3 to determine the furnace thermal characteristics. An example of this procedure is described in Hewitt, Shires and Bott, 1994.

Furnace Models

The complete mathematical description of a practical furnace is complex, combining aerodynamics, chemical reactions and heat transfer, and computer programs are normally used for detailed solutions. There are two basic types of approach; *zone methods* and *flux methods.*

Zone methods are employed when the heat release pattern from the flame is known or can be calculated independently. Conceptually, the furnace and its walls are divided into discrete zones, the effective exchange areas between zones are determined, and radiative heat transfer corresponding to the prescribed heat release pattern is calculated.

In flux methods, instead of dividing the space into zones the radiation arriving at a point in the system is itself divided into a number of characteristic directions, representing averages over a specified solid angle. Flux methods are well suited for use in combination with modern methods of prediction of fluid flow and mixing. Simultaneous solutions of the radiative heat transfer equations using flux methods and turbulent flow models are feasible.

For further information on these two methods see Beer, 1974, and Afgan and Beer, 1974, where examples of their application can be found.

Well Stirred Furnace Model

As a first approach to the estimation of furnace performance the well stirred furnace model is relatively simple and quick to use. One of the first versions was introduced by Lobo and Evans, 1939, and was used by Kern, 1986. An improved version expressed in non-dimensional terms was introduced by Hottel and Sarofim, 1967. This was reveiwed by Hottel, 1974 and subsequently described by Truelove, 1983, and Hewitt, Shires and Bott, 1994.

The general performance equation for furnaces derived from this model is

$$\left(\frac{\dot{Q}_g}{d}\right)(D'd) + (T_1')^4 = \left(1 - \frac{\dot{Q}_g'}{d}\right) + L_r'\left(1 - \frac{\dot{Q}_g'}{d} - T_e'\right) \quad \textbf{(10)}$$

where

$$\dot{Q}_g' = \frac{\dot{Q}_g}{\dot{Q}_f}\frac{(T_f - T_0)}{T_f} \text{ is reduced furnace density} \quad \textbf{(11)}$$

$$D' = \frac{\dot{Q}_f}{\sigma g_{rc} T_f^3 (T_f - T_0)} \text{ is reduced firing density} \quad \textbf{(12)}$$

$$T_1' = T_1/T_f \text{ is reduced sink temperature} \quad \textbf{(13)}$$

$$L_r' = \frac{U_r A_r}{g_{rc}\sigma T_f^3} \text{ is reduced heat loss} \quad \textbf{(14)}$$

and

$$g_{rc} = A_1/\{1/\epsilon_{eff} + C(1/\epsilon_g - 1)\} + \alpha A_c/4\sigma T_g^3 \quad \textbf{(15)}$$

where d is a factor to account for imperfect mixing, approximately equal to 1.2, T_e is external temperature, U_r is overall heat transfer coefficient for the refractory wall, A_r is the area of the refractory surface, A_1 is the area of tube subject to radiative heat transfer, A_c is the area of tube subject to convective heat transfer and α the average convective heat transfer coefficient for A_c.

Figure 8a shows graphically the *well stirred furnace model* performance prediction for zero wall losses, and **Figure 8b** for the same conditions but with typical wall losses included. Comparison of these figures shows that wall losses have a very significant effect when tube temperatures are high. As the firing rate, represented by D′d, is reduced the efficiency, represented by $\dot{Q}'/d$, reaches a peak then falls, eventually approaching zero as the major part of the heat is lost through the walls.

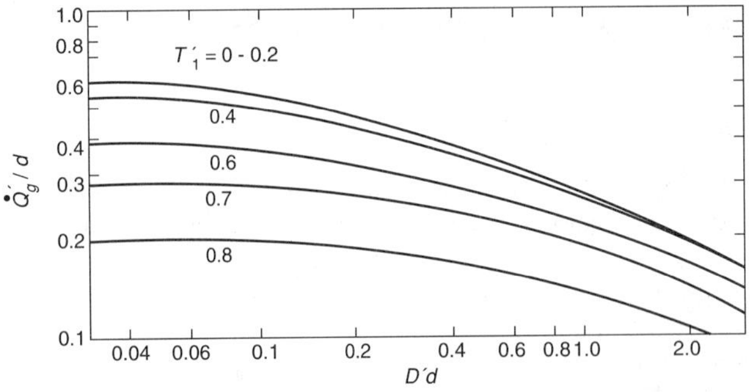

Figure 8(a). Performance curves for stirred reactor furnace with negligible wall losses. (From the Heat Exchanger Design Handbook, Hemisphere Publications. With permission.)

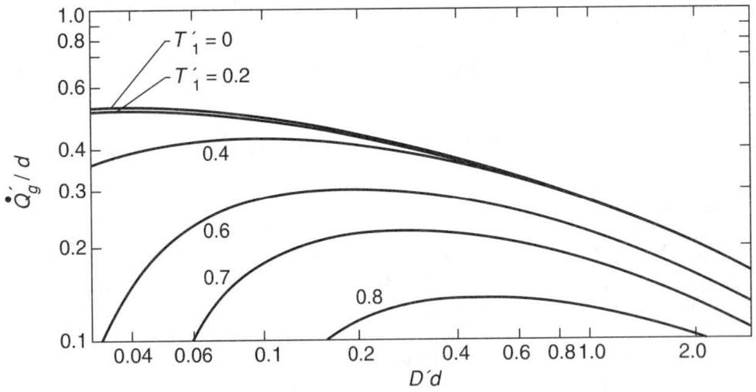

Figure 8(b). Performance curves for stirred reactor furnace with wall losses; $L_r' = 0.02$, $T_e' = 0.125$. (From the Heat Exchanger Design Handbook, Hemisphere Publications. With permission.)

References

Afgan, N. H. and Beer, J. M. (1974) *Heat Transfer in Flames,* Scripta Book Co., Wiley, New York.

Beer J. M. (1974) Methods of calculating radiative heat transfer from flames in combustors and furnaces, *Heat Transfer in Flames,* Scripta Book Co., Wiley, New York.

Hewitt, G. F. Shires, G. L., and Bott T. R. (1994) *Process Heat Transfer,* CRC Press.

Hottel, H. C. (1974) First estimates of industrial performance; the one—gas—zone model re-examined, *Heat Transfer in Flames,* Scripta Book Co., Wiley, New York.

Hottel, H. C. and Sarofim, A. F. (1967) *Radiative Heat Transfer,* McGraw Hill, New York.

Kern, D. O. (1986) *Process Heat Transfer,* McGraw Hill, New York.

Lobo, W. E. and Evans, J. E. (1939) *Trans AIChE,* 35–743.

Truelove, J. S. (1983) Furnaces and combustion chambers, *Heat Exchanger Design Handbook,* Hemisphere Publishing, New York.

Leading to: *Kilns; Plasma-arc furnaces*

G.L. Shires

FUSED SILICA OPEN TUBULAR COLUMNS, FSOT (see Gas chromatography)

FUSION, NUCLEAR FUSION REACTORS

Following from: *Power plant*

A *nuclear fusion reaction* is the interaction of two atomic nuclei to form a heavier nucleus. Fusion of light nuclei (atomic wt. $A_i < 50$) is generally exothermic, especially where helium-4 is formed. Examples are:

$$p + p \rightarrow D + e^+ + \nu + 0.9\ \text{MeV};\ [4.0 \times 10^{-22}] \qquad \mathbf{1(a)}$$

$$p + D \rightarrow {}^3He + \gamma + 5.5\ \text{MeV};\ [2.5 \times 10^{-4}] \qquad \mathbf{1(b)}$$

$${}^3He + {}^3He \rightarrow {}^4He + 2p + 12.9\ \text{MeV};\ [5.0 \times 10^3] \qquad \mathbf{1(c)}$$

$${}^{12}C + p \rightarrow {}^{13}N + \gamma + 1.9\ \text{MeV};\ [1.4 \times 10^0] \qquad \mathbf{2(a)}$$

$$D + D \rightarrow {}^3He + n + 3.3\ \text{MeV};\ [5.4 \times 10^1] \qquad \mathbf{3(a)}$$

$$D + T \rightarrow {}^4He + n + 17.6\ \text{MeV};\ [9.6 \times 10^3] \qquad \mathbf{3(b)}$$

Here p, D and T stand for proton, deuteron and triton (isotopes of hydrogen); n for a neutron, e^+ is a positron, ν a neutrino, and γ a photon; 1 MeV = 1.6×10^{-13}J; [ ] is the normalized cross-section in (KeV) $m^2 \times 10^{-28}$ [see Bahcall and Pinsonneault (1992)]. For fusion to occur, the interacting nuclei must penetrate or surmount the coulomb barrier. The fusion reactions 1(a) to 1(c) occur in *stars,* providing 97% of the solar luminosity. Reaction 2(a) starts the C-N cycle whereby ^{12}C and ^{14}N catalyse the fusion of hydrogen to helium. Because of the higher coulomb barrier, it predominates in higher temperature stars. Reactions 3(a)(b) are used in terrestrial fusion reactors. 3(b) provides the highest rates; the tritium is produced by neutron reactions in lithium:

$$n + {}^6Li \rightarrow {}^4He + T + 4.8\ \text{MeV} \qquad \mathbf{4(a)}$$

$$n + {}^7Li \rightarrow {}^4He + T + n - 2.8\ \text{MeV} \qquad \mathbf{4(b)}$$

Fusion Reactors are assemblies of nuclei undergoing fusion reactions. The *stars* are natural fusion reactors, held together by gravity, with their temperature and pressure sustained by fusion energy. *Hydrogen bombs* are transient fusion reactors, where the fusion fuel is compressed and heated by radiation from a nuclear fission explosion. Teller (1987) reports that they can be cylindrical devices, a foot or so in diameter and a few feet in length, and may produce an energy equivalent to a few hundred thousand tonnes of high explosive (1 tonne of H.E. is 4.2×10^9J).

Inertial Confinement Reactors are conceived as containable fusion explosions of up to about 3×10^8J. The fuel is compressed by radiation provided by converging pulsed energy from lasers or possibly particle accelerators. The most fusion energy, so far reported (1993), from a single event is about 50J, achieved at Lawrence Livermore National Laboratory with the pulsed laser, NOVA (up to 50kJ of 0.35 μm light in 10^{-9} s), focussed on to a small sphere (D ~ 0.3mm) containing D-T fuel. [See references in IAEA (1993) vol.3.]

Magnetic Fusion Reactors use magnetic fields to contain fusion fuel at pressure up to about 10 bar, heated by particle beams and, ultimately, by the 3.5 MeV helium nuclei produced by reaction 3(b). The principal system researched, the Tokamak (a Russian acronym), uses a large current flowing in the fuel, inside a metal torus, to provide the confining magnetic fields; externally-generated fields stabilize the configuration. Fusion power lasting 2s peaking at 1.7 MW has been obtained in the Joint European Torus (JET) at Culham, UK, using 3.1 MA current to confine dilute D-T fuel. When optimized, both JET and the USA's Tokamak Fusion Test Reactor are expected to produce fusion power up to the 10 MW level, using >10 MW power inputs from external sources.

Thermonuclear fusion reactions occur when the velocities of the colliding nuclei arise from high temperature of the fuel. The reaction rate R_{12} is given by

$$R_{12} = n_1 n_2 < u\sigma_R(u) > m^{-3}s^{-1} \qquad \mathbf{(5)}$$

where n_1 and n_2 are the number densities of two reacting species, $\sigma_R(u)$ is the reaction cross-section for a relative velocity u, and the average < > is taken over the Maxwellian distribution of u. **Figure 1** shows $< u\ \sigma_R\ (u) >$ as a function of temperature. The values provide a criteria for a net energy output from a self-sustained thermonuclear D-T fusion reactor, namely:

$$nT\tau_E > 4 \times 10^{28}\ m^{-3}Ks \qquad \mathbf{(6)}$$

Here $n = n_1 + n_2$, T is the temperature of about 10^8 K, and τ_E is the energy confinement time (a measure of the thermal insulation of the fuel) defined as the thermal energy of the fuel divided by the power used to sustain it. Nearly $10^{28}\ m^{-3}$Ks has been reached in tokamaks. Solar fusion power is thermonuclear. Up to about half the fusion power in JET is thermonuclear, the rest arises from fusion reactions of suprathermal particles used for heating, see IAEA (1993).

Heat transfer in fusion reactors is by *radiation,* by *thermal conduction* and by *convection.* The atoms of the hot (>10^6K) fuel are all ionized and form *plasma,* the highly-conducting, high-temperature state of matter, see Dendy (1993). Fully ionized plasma

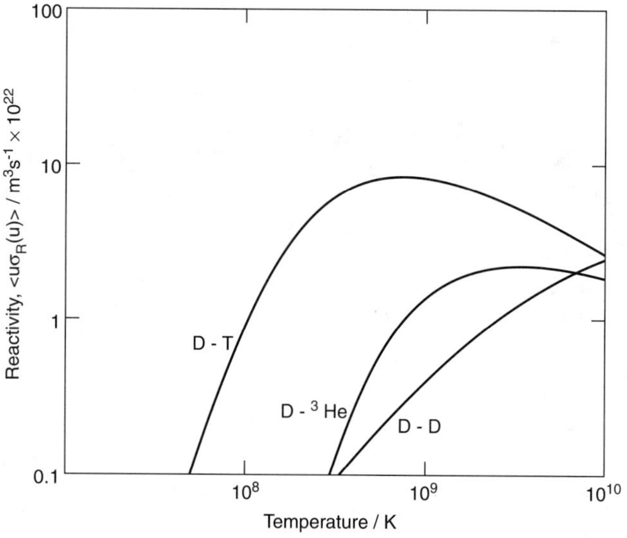

Figure 1.

radiates energy through electron scattering on ions of charge Ze and number density n_i at a rate P_B

$$P_B = 1.4 \times 10^{-28} n_i^2 Z^3 T^{1/2}, \qquad Wm^{-3} \qquad \textbf{(7)}$$

with photon energies up to a few k_B T, where k_B is the Boltzmann constant and Z the atomic number. In laboratory-scale plasma the radiation escapes. Stars are easily large enough to re-absorb the radiation, so that the radiant energy diffuses out at the rate $\dot{q}_R$

$$\dot{q}_R = -4(ac)(3\kappa\rho)^{-1} T^3 (dT/dr), \qquad Wm^{-2} \qquad \textbf{(8)}$$

where a is the radiation density constant, c the speed of light, κ the mass absorption coefficient and ρ the mass density. Radiative diffusion dominates in the inner region of the sun. In the outer regions ($r/r_\odot > 0.71$), the heat is transferred by convective cells.

The thermal conductivity of field-free plasma is due to the electrons and is approximately

$$\lambda = 6 \times 10^{-10} T^{5/2}/(Z \ln \Lambda), \qquad W/mK \qquad \textbf{(9)}$$

where ln Λ is the so-called coulomb logarithm, a slowly varying function of n, T and Z given in Wesson (1987), having a value of about 10. This conduction occurs in inertial confinement plasma ($T \sim 10^7$–10^8K, $n \sim 10^{30}$–10^{31} m^{-3}), and along lines of force in magnetic fields. The conductivity normal to a strong magnetic induction B is much reduced from (9), so that magnetic fields provide thermal insulation as well as confining plasma pressure. Heat and mass transfer due to electron and ion collisions across the confining magnetic fields of a tokamak are interdependent, are dependent on geometry and electric fields, and are predicted by *neoclassical theory,* see Wesson (1987). The thermal conductivity is due to the ions and is about

$$\lambda_\perp = 10^{-42} (\epsilon A_i/T)^{1/2} Z^2 n_i^2 (\ln \Lambda)/B^2, \qquad W/mK, \qquad \textbf{(10)}$$

where $\epsilon \equiv r/r_M$ is ratio of minor to major radius of the torus, and B is the induction arising from the confining electric current. A semi-empirical expression due to Bohm for *convective thermal* and *plasma diffusivity* across the magnetic field is

$$\delta \sim D = 5.4 \times 10^{-6}\, T/B, \qquad m^2s^{-1} \qquad \textbf{(11)}$$

The experimentally observed values of δ and D are generally intermediate between (10) and (11), and are in the range 1–100 m^2/s in large (~3m bore) tokamaks, where, typically, $n \sim 10^{20}m^{-3}$, $T \sim 10^8$K $B \sim 3T$, and the currents are < 7MA.

Envisaged *industrial fusion reactors* have a central fusion reactive zone of plasma, surrounded by a lithium-containing blanket about 1m thick, where the 14.1 MeV fusion neutrons transfer energy by collisions with atoms to a heat transfer fluid used to raise steam to power turbo-alternators. The tritium fuel is produced by reactions 4 in the blanket. The helium exhaust system has also to handle the heat convected out with the exhaust, and is a key factor in the designs.

References

Bahcall, J. N. and Pinsonneault, M. H. (1992) Standard Solar Models and solar neutrinos. *Rev. Mod. Phys.,* 64, 885.

Teller, E. (1987) Fusion Devices Explosive, *Encyclopedia of Physical Science and Technology,* 5, 723, Academic Press.

IAEA (1993), *Plasma Physics and Controlled Nuclear Fusion Research,* Vols. 1 & 3 (IAEA, Vienna).

Wesson, J. (1987) *Tokamaks,* Clarendon Press, Oxford.

Dendy, R. (1993) *Introduction to Plasma Physics,* CUP.

R.S. Pease

FUSION REACTORS (see Fusion, nuclear fusion reactors)

G

GALERKIN METHOD (see Integral equations)

GALILEO NUMBER

Galileo number, Ga, is a dimensionless group representing a ratio of forces present in the flow of viscous fluids.

$$Ga = gL^3\rho^2/\eta^2$$

where g is acceleration due to gravity, ρ density of the fluid, η viscosity of the fluid, and L the appropriate linear dimension.

Note that

$Ga = (\rho u^2L^2)(g\rho L^3)/(\eta uL)^2$

$=$ (momentum force)(gravitational force)/(viscous force)^2

or Ga = Re $\times$ gravitational force/viscous force

Galileo number is **Archimedes Number** with the particle density equal to that of the fluid.

G.L. Shires

GALILEO PRINCIPLE (see Wind tunnels)

GALVANIC REACTIONS (see Faraday's laws)

GAMMA FUNCTION

The gamma function, one of the special functions introduced by L. Euler (1927), is the extension of the factorial into fractional and complex values of the argument and can be obtained as a solution of the equation

$$\Gamma(z + 1) = z\Gamma(z)$$

The function was first difined (Euler) by the integral

$$\Gamma(z) = \int_0^\infty e^{-t}t^{z-1}dt$$

for the values of a complex argument z with positive real part (Re z > 0). It is widely used in analytical solutions of equations of mathematical physics by the integral transformation method, in particular, when applying the Laplace transform to the function written (approximated) as a power series in time.

For a whole value of an argument

$$\Gamma(n + 1) = 1\cdot 2\cdot 3\cdot \ldots \cdot n = n!$$

For a fractional value of an argument $0 \le x \le 1$

$$\Gamma(x + 1) = \sum_{i=0}^{5} a_i x^i + \epsilon(x),$$

where $a_0 = 1$; $a_1 = -.57486$; $a_2 = -.95124$; $a_3 = -.69986$; $a_4 = -.42455$; $a_5 = -.10107$; $|\epsilon(x)| \le 5 \times 10^{-5}$.

Other useful properties of the Gamma function, which reflect various expansions into series, asymptotic and approximating expressions, etc. are given in handbooks on special functions.

Out of the related special functions we shall note a polygamma function

$$\psi^{(n)}(z) = \frac{d^{n+1}}{dz^{n+1}} \ln \Gamma(z), \qquad n = 0, 1, \ldots$$

and an *incomplete gamma function*

$$\gamma(\alpha, x) = \int_0^x e^{-t}t^{\alpha-1}dt \qquad (\text{Re } \alpha > 0),$$

which have a number of known asymptotic and other properties.

References

Abramowitz, M. and Stegun, I. A. (1964) *Handbook of Mathematical Functions,* National Bureau of Standards, Appl. Math. Series-55.

I.G. Zaltsman

GAMMA RAY ABSORPTION TECHNIQUE (see Void fraction measurement)

GAMMA RAYS

Following from: *Electromagnetic spectrum; Electromagnetic waves; Ionising radiation*

Gamma rays constitute the highest frequency, shortest wavelength end of the **Electromagnetic Spectrum**. They are emitted following transitions between excited states, or between an excited state and the ground state, of atomic nuclei. Atomic nuclei are often left in excited states following a radioactive decay process such as alpha decay or beta decay. The emitted gamma ray energy is equal to the change in excitation energy of the nucleus, apart from a small and usually insignificant recoil energy.

High energy gamma rays can travel significant distances through solid material and, in doing so, they interact with electrons, or nuclei, of the material producing ionisation. Gamma rays are therefore included with other types of radiation under the title **Ionising Radiation**. There are three main types of gamma ray interaction with matter:

Compton Interactions

These involve the scattering of the gamma ray by a free electron. The laws of conservation of energy and momentum determine the relationship between the scattered gamma ray energy, the electron energy and the scattering angles to the energy of the initial gamma ray. The probability of Compton scattering per atom of absorber material depends on the number of electrons available as scattering targets, and therefore increases linearly with the atomic number (Z) of the material. The probability of *Compton scattering* generally falls off as the gamma ray energy increases. The angular distribution of the scattered gamma ray is such that there is a strong preference for forward scattering at high gamma ray energies ($\sim$10 MeV), while at 1 keV the scattering approaches isotropic.

The Photoelectric Effect

In the photoelectric process an incoming gamma ray undergoes an interaction with an absorber atom in which the gamma ray completely disappears. In its place, an energetic photoelectron is ejected by the atom from one of its bound shells. The interaction is with the atom as a whole and cannot take place with free electrons. The photoelectric process is the predominant mode of interaction for low energy gamma rays. The *photoelectric effect* is strongly enhanced in materials of high atomic number. In gamma ray spectroscopy the sharp peaks (photopeaks) in the spectra arise from photoelectric interactions in the detector. These photopeaks may be used to identify the radioisotopes which emitted the gamma rays.

Pair Production

In the field of an atomic nucleus a gamma ray energy above 1 MeV can interact to produce an electron-positron pair. In this process the gamma ray energy is converted into the mass of the electron-position pair. The electron mass corresponds to an energy of 511 keV so pair production requires gamma rays of at least 2 $\times$ 511 keV = 1.022 MeV energy. Any energy in excess of this appears as kinetic energy of the electron-positron pair which move off in opposite directions.

Reference

Glenn F. Knoll (1989) *Radiation Detection and Measurement* 2nd Edition, John Wiley and Sons.

T.D. MacMahon

GANGLIA (see Enhanced oil recovery)

GAS BONDED PHASE CHROMATOGRAPHY, GBC (see Gas chromatography)

GAS CHROMATOGRAPHY

Gas chromatography (GC) is divided into two classes, *gas liquid* (partition) and *gas solid* (adsorption) termed *GLC* and *GSC*, respectively. GLC is subdivided into two modes, namely *packed column,* low performance (liquid stationary phase, SP, on a solid-inert support) and *capillary* or open tubular, *column,* high performance (liquid stationary phase, on the inner surface, physically adsorbed or chemically bonded) chromatography.

Of the two classes GLC is far more versatile than GSC, the latter only being usable for the separation and analysis of permanent, natural and synthetic gas mixtures and very volatile, gaseous organics, because of the adsorptive nature of the solid (sorbent) phase while the former can be applied to analyse organics over a wide range of volatilities (volatile solvents to C_{60} hydrocarbons).

In GLC applications, the capillary column mode has largely superseded the packed column mode, since the advent of *fused silica open tubular (FSOT)* columns in 1979 and the ability to apply and employ *chemically-bonded* SP (*stationary phase*) (*GBC* is a term that can be used to describe *gas bonded-phase chromatography*).

Chromatographic Theory

For all chromatographic separations, the solute is distributed (partitioned) between the liquid SP or solid sorbent and the mobile-gas phase (MP) and the equilibrium involved for GLC is:

$$K_d = c_L/c_G \quad (1)$$

where K_d is the distribution (equilibrium) coefficient, c_L and c_G are the molar concentrations in SP (L) and MP (G), respectively. The movement of the solute down the column under these conditions whereby K_d is directly related to c_L/c_G is termed linear chromatography. The ratio of the time spent by (or amount of) the solute in the two phases is given by k′, the partition ratio (or capacity factor)

$$k' = \frac{c_L \cdot V_L}{c_G \cdot V_G} \quad (2)$$

$$= K_d/\beta \quad (3)$$

where β is the phase ratio, V_G/V_L

GSC is governed by a process of adsorption involving an equilibrium between the solute in the solid sorbent phase and the gas phase.

The theory applied to packed and capillary columns stems from the 1950's, see for example Ettre (1965), and is required to account for

1. relative rates of migration solutes
2. rate of band broadening (BB) during migration.

The original theory applied was the *Plate Theory* whereby the movement of solute is considered as a series of stepwise transfers from one theoretical plate to the next (equilibrations of solute between SP and MP at each plate) which accounted for the variables influencing migration rates but not band broadening during migration.

The more complete treatment is provided by the *Kinetic Theory* using Plate Theory terminology which describes migration and band broadening quantitatively, and this theory is based on random mechanisms for the migration of solute molecules through a column.

The number of theoretical plates, N, and the height equivalent to a theoretical plate, H (in cm), are related to the length of a column, L (in cm) by

$$N \cdot H = L \tag{4}$$

Using error theory and Gaussian distribution applied to the spread of velocities of particles about a mean velocity of the most average particle H is equated with the variance per unit length of the column

$$H = \sigma^2/L \text{ and thus } N = L^2/\sigma^2 \qquad \textbf{(5) and (6)}$$

where σ^2 the variance is the square of the standard deviation σ. An expression for N can be derived in terms of the chromatographic variables, retention time t_r, and base width, w_b, by eliminating L and σ.

$$N = 16\left(\frac{t_r}{w_b}\right)^2 \tag{7}$$

or

$$5.54\left(\frac{t_r}{w_{0.5}}\right)^2 \tag{8}$$

where $w_b = 1.7_{w_{0.5}}$

The mechanisms of BB involve mass transfer processes in and out of the SP and in the MP during migration and are described in equations for packed and capillary columns by several terms which contribute to the overall H

$$H = A + \frac{B}{\bar{u}} + C_s \cdot \bar{u} + C_m \cdot \bar{u} \tag{9}$$

where A term accounts for the contribution of the packing (eddy diffusion), B is the coefficient for longitudinal diffusion important when the MP is a gas, C_S and C_m are the coefficients for resistance to mass transfer in and out of the SP and in the MP, respectively, and u is the mean average linear velocity of the MP. The equation reduces to three terms to describe the contributions to BB for capillary columns since $A = 0$ for no packing, and further to two terms $B/\bar{u} + C_m \cdot \bar{u}$ if the contribution of $C_s \cdot \bar{u}$ is negligible, i.e. when the film thickness of SP is a minimum, $d_f = 0.1$ μm and the column radius is small, $r_c = 0.2$ mm.

Corrections can be applied to the equation to account for compression of the gas at the inlet (i) due to $P_i > P$ ambient and decompression due to the pressure drop down the column $P_i > P_o$. The pressure drop in the column is inversely proportional to the column diameter, d_c, and thus wide bore columns will experience a lower ΔP than narrow bore columns.

For packed columns the packing and the limitation in d_p (30 μm) serve to act as a barrier to flow (permeability); thus, the pressure drop is so severe that the effective column length is restricted to about 5 m, whereas capillary columns have a high permeability and columns of several orders of magnitude greater in length can be used.

The factors which influence the efficiency of a column measured in terms of H depend on $\bar{u}$, d_p (particle size-packed), d_f (film thickness of SP) η (gas viscosity), r_c and the column temperature, T_c. H decreases with a decrease, $\bar{u}$, d_p, d_f, η and r_c and with an increase in T_c.

Chromatography Parameters

The parameters that can be determined directly from the chromatogram are the retention time, (t_r) and the base width (w_b) or width at half height ($w_{0.5}$). Using **Equation 7** the number of theoretical plates, N can be determined to give a measure of column efficiency dependent on k′ for the solute analysed. By obtaining t_m for an unretained solute, the adjusted retention time t'_r can be determined and the number of effective theoretical plates obtained by replacing t_r by t'_r in **Equation 7**.

The retention volume V_r can be determined by measuring the volumetric flow rate F_r at the column exit using the equation

$$V_r = F_r \cdot t_r \tag{10}$$

The retention volume, V_r^o, can be corrected for the effect of gas compression and thus

$$V_r^o = jF_r \cdot t_r \tag{11}$$

where j is the James-Martin compressibility factor, see James and Martin (1952).

The specific retention volume V_g is dependent only on the density of the SP, and the column temperature and has been extensively used to characterise solutes on columns with SP of widely differing polarities.

The resolution R_s of two solutes A and B (slower moving solute) is defined in terms of the retention times and base widths as follows

$$R_s = 2\left(\frac{t_r - t_r}{w_b + w_b}\right) \tag{12}$$

and complete separation is defined at $R_s = 1.5$. R_s is proportional $\sqrt{N}$ and N increases by a factor of $\sqrt{2}$ when the column length is doubled.

The selectivity, α, is a measure of the capacity of a column to separate two closely eluting solutes and is defined by the ratio

$$\frac{K'_B}{K'_A} = \frac{k'_B}{k'_A} = \frac{t'_r}{t'_r} \tag{13}$$

The difficulty in separation of two solutes A and B increases as α approaches unity.

Retention Index, I(x)

The *Retention Index* (Kovats) was introduced to characterise different SP and *solute functionality* and is based on the retention parameters (t'_r or k′) of an homologous series, viz n-alkanes on any SP for which I(x) is $100 \times C_n$. Depending on solute functionality (non-polar and polar) the k′ of a solute will be shifted on a polar SP compared to a non-polar SP and ΔI is a measure of the functionality—polarity, polarisability or bonding activity of the solute. The characterisation of any SP is based on the polarity, P, obtained additively from the shifts in ΔI for selected solutes which are representative of the range of functionalities encountered (*Rohrschneider* and *McReynolds constants*).

For isothermal conditions, the relationship between log t_r or k' and C_n (or $100 \times C_n$) is linear while for temperature programmed conditions, the relationship is directly linear (equal chromatographic spacing between successive components, c_n and c_{n+1} of an homologous series, (e.g. n-alkanes or fatty acid methyl esters as examples).

SP and Column Fabrication

With packed columns a very wide range of SP has been employed because of the limitations in the separation capability of packed columns. With the advent of *fused silica* (capillary) *open tubular columns* which are capable of high resolution the range of SP has narrowed dramatically such that only 5 or 6 phases (polydimethyl siloxane based or polyethylene glycol) are required to cover the entire range of solute functionality and volatility anticipated. As an example, a non-polar polydimethyl siloxane SP of d_f 0.1 μm chemically-bonded to 25 m × 0.2 mm FSOT column will allow separation of a very wide range of solutes in a complex mixture according to their boiling points. Moreover, such columns are capable of chiral selectivity and thus can be used to separate enantiomers. Other features of SPs currently in use include high temperature stability and chirality for specialised applications.

For an introduction to the theory of chromatographic separations with reference to gas chromatography, see Bartle (1993).

References

Bartle, K. D. (1993) Chapter 1, *Gas Chromatography—a Practical Approach* (Ed. P. J. Baugh), Oxford University Press.

Ettre, L. S. (1965) *Open Tubular Columns in Gas Chromatography,* Plenum Press, New York.

James, A. T. and Martin, A. J. P. (1952) *Biochem. J.* 50, 679.

P.J. Baugh

GAS CLEANING (see Gas-solids separation, overview)

GAS COOLED NUCLEAR REACTORS (see Advanced gas cooled reactor)

GAS CYCLE REFRIGERATION (see Refrigeration)

GAS DISPERSION (see Agitated vessel mass transfer)

GAS DYNAMICS (see Aerodynamics)

GAS ENTRAINMENT (see Plunging liquid jets)

GASEOUS COOLANTS, FOR NUCLEAR REACTORS (see Coolants, reactors)

GASEOUS FUELS (see Fuels)

GASES, DIFFUSION IN (see Diffusion coefficient)

GASES, THERMAL CONDUCTIVITY OF (see Thermal conductivity; Thermal conductivity mechanisms)

GASES, TRANSPORT PROPERTIES OF

Table 1. Gases, transport properties

Substance	Data	Property (at low pressure)	−150 °C / 123.15 K	−100 °C / 173.15 K	−50 °C / 223.15 K	0 °C / 273.15 K	25 °C / 298.15 K	100 °C / 373.15 K	200 °C / 473.15 K	300 °C / 573.15 K	400 °C / 673.15 K	500 °C / 773.15 K	1000 °C / 1273.15 K	1500 °C / 1773.15 K	2000 °C / 2273.15 K
Air	Chemical formula: Molecular weight: 28.96	Specific heat capacity, $c_{p,g}$ (kJ/kg K)	1.019	1.008	1.005	1.005	1.005	1.005	1.022	1.043	1.051	1.089	1.189	1.239	1.269
	Normal density (at 0°C, 101.3 kPa): 1.29 kg/m³ Boiling point: −194.25°C	Thermal conductivity, λ_g[(W//m²)(K/m)]	0.011	0.016	0.020	0.024	0.026	0.031	0.039	0.044	0.050	0.056	0.079	0.106	0.128
	Critical temperature: −140.55°C Critical pressure: 3.769 MPa	Dynamic viscosity, η_g(10^{-5} Ns/m²)	0.853	1.17	1.46	1.72	1.82	2.17	2.57	2.93	3.25	3.55	4.79	5.78	6.81
Carbon Dioxide	Chemical formula: CO_2 Molecular weight: 44.01	Specific heat capacity, $c_{p,g}$ (kJ/kg K)	S	S	0.775	0.816	0.846	0.934	1.001	1.063	1.114	1.156	1.269	1.319	1.352
	Normal density (at 0°C, 101.3 kPa): 1.96 kg/m³ Boiling point: −78.5°C	Thermal conductivity, λ_g[(W/m²)/(K/m)]	S	S	0.011	0.015	0.016	0.022	0.030	0.038	0.045	0.052	0.083	0.108	0.132
	Critical temperature: 31.04°C Critical pressure: 7.382 MPa	Dynamic viscosity, η_g(10^{-5} Ns/m²)	S	S	1.13	1.37	1.49	1.82	2.22	2.59	2.93	3.24	4.59	5.70	6.57
Carbon Monoxide	Chemical formula: CO Molecular weight: 28.01	Specific heat capacity, $c_{p,g}$ (kJ/kg K)	1.038	1.038	1.038	1.038	1.038	1.038	1.055	1.080	1.105	1.130	1.235	1.285	1.315
	Normal density (at 0°C, 101.3 kPa): 1.25 kg/m³ Boiling point: −191.49°C	Thermal conductivity, λ_g[(W/m²)/(K/m)]	0.011	0.015	0.019	0.023	0.025	0.030	0.037	0.043	0.049	0.055	0.083	0.107	0.128
	Critical temperature: −139.99°C Critical pressure: 3.498 MPa	Dynamic viscosity, η_g(10^{-5} Ns/m²)	0.81	1.12	1.40	1.66	1.78	2.11	2.51	2.87	3.20	3.52	4.90	5.89	6.92

Table 1. Gases, transport properties—*continued*

Substance	Data	Property (at low pressure)	−150	−100	−50	0	25	100	200	300	400	500	1000	1500	2000
		Temperature, °C / Temperature, K	123.15	173.15	223.15	273.15	298.15	373.15	473.15	573.15	673.15	773.15	1273.15	1773.15	2273.15
Helium	Chemical formula: He Molecular weight: 4.00 Normal density (at 0°C,	Specific heat capacity, $c_{p,g}$ (kJ/kg K)	5.200	5.200	5.200	5.200	5.200	5.200	5.200	5.200	5.200	5.200	5.200	5.200	5.200
	101.3 kPa): 0.18 kg/m^3 Boiling point: −268.94°C Critical temperature:	Thermal conductivity, λ_g[(W/m^2)/(K/m)]	0.083	0.104	0.124	0.143	0.150	0.174	0.205	0.237	0.270	0.302	0.423	0.538	0.587
	−267.96°C Critical pressure: 2.29 MPa	Dynamic viscosity, $\eta_g(10^{-5}$ Ns/m^2)	1.09	1.35	1.63	1.89	1.96	2.28	2.67	3.06	3.41	3.75	5.19	6.61	7.84
Hydrogen	Chemical formula: H_2 Molecular weight: 2.02 Normal density (at 0°C,	Specific heat capacity, $c_{p,g}$ (kJ/kg K)	11.85	13.00	13.50	14.05	14.34	14.41	14.41	14.41	14.41	14.55	15.51	16.95	17.36
	101.3 kPa): 0.09 kg/m^3 Boiling point: −252.77°C Critical temperature:	Thermal conductivity, λ_g[(W/m^2)/(K/m)]	0.073	0.113	0.141	0.171	0.181	0.211	0.249	0.285	0.321	0.367	0.519	0.680	0.836
	−239.92°C Critical pressure: 1.316 MPa	Dynamic viscosity, $\eta_g(10^{-5}$ Ns/m^2)	0.488	0.618	0.734	0.841	0.892	1.04	1.22	1.39	1.54	1.69	2.38	2.86	3.28
Nitrogen	Chemical formula: N_2 Molecular weight: 28.01 Normal density (at 0°C,	Specific heat capacity, $c_{p,g}$ (kJ/kg K)	1.058	1.046	1.038	1.038	1.038	1.038	1.047	1.068	1.089	1.114	1.223	1.273	1.306
	101.3 kPa): 1.25 kg/m^3 Boiling point: −195.80°C Critical temperature:	Thermal conductivity, λ_g[(W/m^2)/(K/m)]	0.012	0.017	0.021	0.024	0.026	0.031	0.037	0.042	0.047	0.052	0.072	0.104	0.125
	−146.9°C Critical pressure: 3.396 MPa	Dynamic viscosity, $\eta_g(10^{-5}$ Ns/m^2)	0.836	1.14	1.41	1.66	1.78	2.09	2.47	2.82	3.14	3.42	4.61	5.56	6.36
Oxygen	Chemical formula: O_2 Molecular weight: 32.00 Normal density (at 0°C,	Specific heat capacity, $c_{p,g}$ (kJ/kg K)	0.927	0.909	0.903	0.909	0.913	0.934	0.963	0.992	1.026	1.051	1.122	1.164	1.197
	101.3 kPa): 1.43 kg/m^3 Boiling point: −182.97°C Critical temperature:	Thermal conductivity, λ_g[(W/m^2)/(K/m)]	0.011	0.016	0.020	0.024	0.026	0.032	0.039	0.045	0.052	0.058	0.086	0.115	0.139
	−118.38°C Critical pressure: 5.09 MPa	Dynamic viscosity, $\eta_g(10^{-5}$ Ns/m^2)	0.955	1.31	1.63	1.92	2.03	2.43	2.88	3.29	3.67	4.03	5.59	6.92	8.20

From Chapter 5.5.11 of the Heat Exchanger Design Handbook, 1987, Transport properties of superheated gases by Clive F. Beaton, Hemisphere Publishing Corporation.

GASES, RADIATION PROPERTIES OF (see Radiative heat transfer)

GAS FILTRATION (see Filtration)

GAS FLOW NUMBER (see Agitated vessel mass transfer)

GAS-GAS EJECTORS (see Jet pumps and ejectors)

GAS-GRAPHITE REACTORS

***Following from:** Nuclear reactors*

Experimental piles apart, gas-cooled graphite moderated reactors began with the plutonium producing *Windscale pile,* located in Cumbria in the North West of England at a site now called Sellafield. Because of the urgency with which plutonium was required for military purposes, a simple reactor design was chosen, i.e. an unpressurised core with a once-through coolant. The obvious choice of water as coolant; in use at the time in the USA, was rejected for the more densely populated UK, because of the potential risk of loss of coolant leading to fuel melt-down. Instead, atmospheric air was chosen as the coolant but, in order to remove the large quantity of heat, the aluminium cladding was made with longitudinal fins. A disadvantage of air is its exothermic reaction with graphite. Consequently when a particular *Wigner energy release* increased the graphite moderator temperature to a point where the heat of reaction overtook the cooling effect, the moderator went on fire. This happened in 1958, eight years after the pile went critical.

When the UK moved towards nuclear power generation at *Calder Hall* on the same site, it was natural at the design stage in 1952 to opt for gas-cooling again. Here the **Magnox** reactor used carbon dioxide as coolant, as did the subsequent **Advanced Gas-Cooled Reactor**. Carbon dioxide is much less reactive than air and its reaction can be inhibited by additives.

Helium is non-reactive and has been favoured for high and very high temperature gas-cooled reactors, the forerunner being the European DRAGON experimental project at Winfrith in the South of England. The advantage of non-reactivity is counterbalanced by increased cost, increased leakage and problems of sticking of moving parts. Designs for full scale versions have been proposed in USA, Europe, Japan and the former USSR and considerable developmental work on fuel and core has been carried out. As well as having a graphite core, the fuel is usually in intimate contact with graphite in the form of a matrix of graphite coated spheres or as a pebble bed.

Gas-cooled reactors with secondary **Containment**, as is now standard, have advantages in safety over liquid-cooled reactors. Firstly, when loss of coolant or of coolant pressure occurs there is little change in reactivity and there is much smaller change in heat transfer. Natural convection is sufficient to prevent melting of the cladding.

Leading to: *Advanced gas cooled reactor; Magnox power station*

D. Wilkie

GASIFICATION

Solid and liquid carbonaceous materials can be gasified by reaction with steam, or oxygen, or mixtures of both, to make a gas containing hydrogen, carbon oxides and methane that is suitable for use as a fuel gas, or for chemical synthesis. If air is used for gasification, the nitrogen present dilutes the product gas. The composition of the gas varies according to the reactants, e.g. oil, or coal, the process conditions, and the type of reactor used.

Coal Gasification

For the *gasification of coal,* three main types of reactor are used: moving bed, fluidised bed and entrained flow reactors, but the basic chemical reactions are the same. Gasification involves the combustion of some of the carbon in the coal with oxygen.

$$C + O_2 \rightarrow CO_2 \qquad (1)$$

This provides the heat required for the endothermic reactions between steam and carbon dioxide with the remaining carbon.

$$C + H_2O \rightarrow CO + H_2 \qquad (2)$$

$$C + CO_2 \rightarrow 2CO \qquad (3)$$

In moving bed reactors, lump coal is charged to the top of a fuel bed and it is gasified as it moves down the bed counter-current to the gasifying medium, a mixture of steam and oxygen, that is introduced into the bottom of the fuel bed. The oxygen is completely consumed by reaction with residual carbon in the ash, which is then discharged from the reactor. As the hot gases from the combustion zone move upwards through the bed, steam and carbon dioxide are decomposed by reactions (2) and (3) until the temperature falls to about 1200°C, when the gasification rate becomes negligible and equilibrium is approached in the water gas shift reaction (4)

$$H_2O + CO \rightleftharpoons CO_2 + H_2 \qquad (4)$$

At the top of the fuel bed, the hot gases pyrolyse and dry the incoming coal and leave the reactor, together with tar, oil, etc, at temperatures of 500°C–600°C. Methane is formed in the top of the fuel bed by the reaction of hydrogen with the coal substance

$$C + 2H_2 \rightarrow CH_4 \qquad (5)$$

The *Lurgi gasifier,* developed in the 1930s for the gasification of lignite, is the best example of a moving bed reactor. It produces a gas containing about 12% methane, 16% carbon monoxide, 35% hydrogen and 30% carbon dioxide, by volume, on a dry basis.

In fluidised bed reactors, the coal is crushed and charged to a bed of partly gasified coal, that is fluidised by the gasifying medium (steam and oxygen). The good heat and mass transfer that occurs enables most of the carbon in reactive coals, such as lignite, to be gasified at temperatures below the fusion point of the ash (1100°C). The gas produced contains about 80% of carbon monoxide and hydrogen with a small amount of methane (2%) and very little tar. The *Winkler process,* developed in the 1920s is still in commercial use at some locations.

Entrained flow reactors operate at high temperatures (1500°C) and high pressures with short residence times; the powdered-coal is entrained and gasified in co-current flow by the gasifying medium. Ash in the coal forms a liquid slag. The product gas, which contains mostly carbon monoxide and hydrogen and no tar or methane, leaves at the reaction temperature, and large heat recovery systems are required. Gasifiers developed by Shell and Texaco are in commercial use. In the former, dry pulverised coal is burned in oxygen, whereas, in the latter a slurry of coal in water is burned in oxygen.

Oil Gasification

Heavy and residual oils are gasified commercially using partial oxidation processes; the oil is atomised into a stream of steam and oxygen, the amount of oxygen being insufficient for complete combustion. The process is very similar to the gasification of coal in an entrained flow reactor. Indeed, the Shell and Texaco partial oxidation processes for the gasification of oil were the basis for the design of their coal gasification reactors. A typical gas composition obtained from heavy oil is 45% carbon monoxide, 45% hydrogen and 4% carbon dioxide.

The most effective way of gasifying light oils that can be purified to reduce sulphur, is by reaction with steam over a nickel catalyst, when the following reactions occur

$$C_nH_m + nH_2O \rightleftharpoons nCO + \left(\frac{m}{2} + n\right)H_2 \qquad (6)$$

$$CO + H_2O \rightleftharpoons CO_2 + H_2 \qquad (7)$$

$$CO + H_2 \rightleftharpoons CH_4 + H_2O \qquad (8)$$

Reactions (7) and (8) are close to equilibrium at the outlet of the catalytic reactor, the gas composition being largely determined by the temperature and pressure that prevails. At temperatures below 500°C the formation of methane is favoured at the expense of hydrogen. Thus a distillate oil gasified in an adiabatic catalytic reactor with an outlet temperature of 500°C will produce a gas that contains 60% methane and 17% hydrogen, the overall reaction being exothermic.

At temperatures above 500°C, more hydrogen and less methane is produced and the reactions become strongly endothermic, the process usually being carried out in an externally fired tubular reformer with the catalyst contained within the heated tubes. At an exit temperature of 850°c the gas contains 53% hydrogen and 17% methane. Reforming processes of this type are widely used in the petrochemical industry.

References

Elliot M. A. (1981) *The Chemistry of Coal Utilisation* Wiley Interscience, Chapter 24

Ullmann's Encyclopaedia of Industrial Chemistry (1989) Volume 12 Gas Production. VCH, Verslagsgesellschaft

J. Lacey

GASIFICATION OF COAL (see Fuels; Gasification)

GASKETS IN HEAT EXCHANGERS

Following from: *Mechanical design of heat exchangers*

The gasket is a key component in a flange assembly. As discussed in **Mechanical Design of Heat Exchangers**, flanges are composed of three sub-components. The complex interaction between these sub-components under bolting up and operating conditions determines the successful operation of the flange. However, it is through the gasket that any leakage occurs and for this reason it receives most attention.

A wide range of gasket materials and types are used, most of them to national standards which specify quality and dimensions. Gasket materials are described in the *Heat Exchange Design Handbook,* 1994, of which the following are the most important:

a. **Rubber**. Natural rubber was the first material used; synthetic rubbers now is widely available in many grades.
b. **Compressed asbestos fiber jointing (CAF)**. This was the most common material used for low pressure service, although alternatives are now sought as manufacturers and operators try to minimise their use of asbestos. The gasket consists of asbestos fiber bound together with an elastomeric binder, sometimes with a steel mesh, and can be used up to 550°C.
c. **Compressed synthetic fiber jointing**. This new range of gasket materials utilise *Kevlar,*™ a registered trademark of Du Pont, and glass fibre with elastomeric binders and is an attempt to replace CAF.
d. **Graphite foil**. The material comprises almost 100% graphite laminated to a metal core to give it strength for handling and service. It has a wide temperature range.
e. **Metal jacketed asbestos**. A popular gasket for exchangers as it can be manufactured with integral pass partition bars for multipass heads. A section is shown in **Figure 1** and shows the asbestos millboard core surrounded with a metal jacket which can be produced in a wide range of materials.
f. **Spiral wound**. These gaskets are made from stainless steel or high alloy metal wound in a spiral formation with a nonmetallic filler. These are suitable for high integrity gaskets and can be produced in a wide range of materials. **Figure 2** shows a gasket with the addition of solid rings to provide a compression stop and a location stop against the bolts. The internal ring can be fitted with partition bars as e) above.
g. **Solid metal joints**. Used for high pressure applications, they are available in a range of sections, e.g. flat or octagonal, or as proprietary designs. The flange surface finish is critical in achieving a leak free gasket. The finish should be selected to suit the particular gasket used, and the gasket supplier involved if necessary. A "gramophone" finish is one of the most useful finishes.

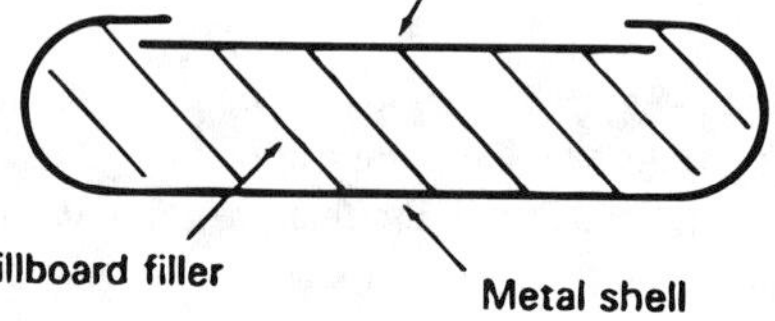

Figure 1.

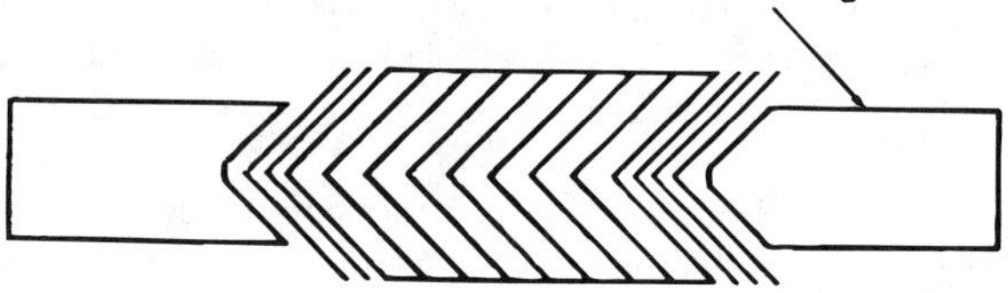

Figure 2.

The gasket is contained in a flange facing. A fully confined facing has machined steps inside and outside the gasket which gives complete location and prevention of gasket blowout or extrusion. This facing has the highest integrity, partial or unconfined facings are also used for lower pressure/integrity.

REFERENCES

Heat Exchange Design Handbook (1994) Begell House Inc., New York.

M. Morris

GAS LAW

Following from: *Perfect gas*

Ideal Gas Law

The p-v-T behaviour of ideal gases can be determined through the kinetic theory of gases assuming that i) the gas consists of a large number of equi-sized molecules which move randomly in all directions, ii) all collisions between the molecules and the molecules and container walls are elastic, iii) the volume of the molecules is negligible, iv) attractive forces between the molecules are negligible and v) the kinetic energy of the molecules is proportional to the absolute temperature. A direct result of this model is the Ideal Gas Law

$$p\tilde{v} = \tilde{R}T \quad \text{or} \quad pV = N\tilde{R}T \tag{1}$$

which includes the gas laws postulated by Charles, Boyle and Avogadro. In equation (1), $\tilde{v}$ is the molar volume (m^3/mol), V is the volume (m^3), N is the number of moles and $\tilde{R}$ is the Universal Gas Constant, $\tilde{R} = 8.314$ J/(mol K).

In a mixture of ideal gases, the momentum transfer between each molecular species and the container walls will be independent of the presence of other molecules. The total pressure can then be determined by *Dalton's Law*

$$p_{tot} = \sum p_l \tag{2}$$

as the sum of all partial pressures. The *Ideal Gas Law* can be applied to each component of a gas mixture as if it alone were present.

Non-Ideal Gases

At low temperature and high pressure, the assumption of negligible molecular volume and inter-molecular attraction becomes increasingly approximate. The deviation is expressed by the compressibility factor

$$z = \frac{p\tilde{v}}{\tilde{R}T} \tag{3}$$

Typical effects of temperature and pressure on the compressibility factor are shown in **Figure 1**.

Various prediction methods for the p-v-T behaviour of non-ideal gases have been recommended in the literature. The most common method is the *Van-der-Waals Equation*

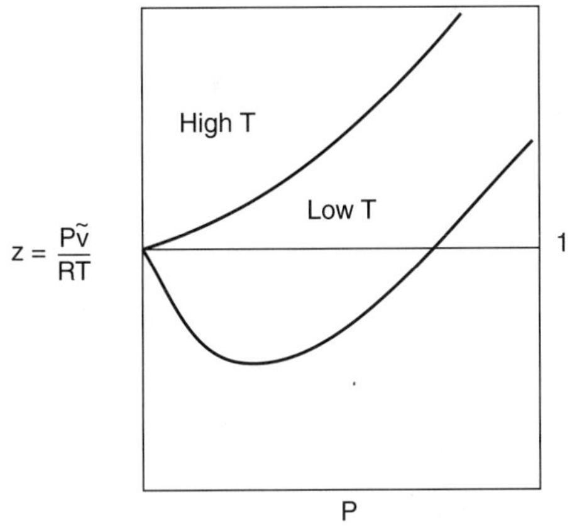

Figure 1. Compressibility factor.

$$\left(p + \frac{a}{\tilde{v}^2}\right)(\tilde{v} - b) = \tilde{R}T \tag{4}$$

The value of a corrects the Ideal Gas Law for the intermolecular attraction, the value of b for the finite molecular volume. The Van-der-Waals constants vary between gases and are listed below. A generalized, but less accurate formulation of the Van-der-Waals Equation may be obtained from the *Law of Corresponding States*

$$\left(p_r + \frac{3}{v_r^2}\right)(3v_r - 1) = 8T_r \tag{5}$$

with the reduced properties p_r, v_r and T_r being the actual pressure, volume and temperature divided by the appropriate critical value. Critical properties are also given below.

Other equations of state have been proposed to improve the prediction of the Van-der-Waals Equation. The most common are: The *Redlich-Kwong Equation* which is recommended for hydrocarbons

$$p = \frac{\tilde{R}T}{\tilde{v} - b} - \frac{a}{T^{0.5}\tilde{v}(\tilde{v} + b)} \tag{6}$$

the *Peng-Robinson Equation*

$$p = \frac{\tilde{R}T}{\tilde{v} - b} - \frac{a}{\tilde{v}^2 + 2b\tilde{v} - b^2} \tag{7}$$

the *Berthelot Equation*

$$p = \frac{\tilde{R}T}{\tilde{v} - b} - \frac{a}{T\tilde{v}^2} \tag{8}$$

the *Dieterici Equation*

$$p = \frac{\tilde{R}T}{\tilde{v} - b} e^{-a/\tilde{v}\tilde{R}T} \tag{9}$$

the *Virial Equation*

$$z = \frac{p\tilde{v}}{\tilde{R}T} = 1 + \frac{V}{\tilde{v}} + \frac{C}{\tilde{v}^2} + \frac{D}{\tilde{v}^3} + \cdots \tag{10}$$

The constants in Equations 6–10 can be found in Tables. They can also be estimated from the critical properties and the knowledge that the critical point is the point of inflection of the isotherms in the pressure-volume diagram.

Law of Corresponding States

Two fluids are in corresponding states if they have the same reduced properties p_r, v_r and T_r. Fluids in corresponding states deviate from ideal behaviour to the same extent, i.e. have the same **Compressibility Factor z**.

Generalized Compressibility Plots

According to the law of Corresponding States, p-v-T data for all gases can be obtained from a plot of compressibility factor as

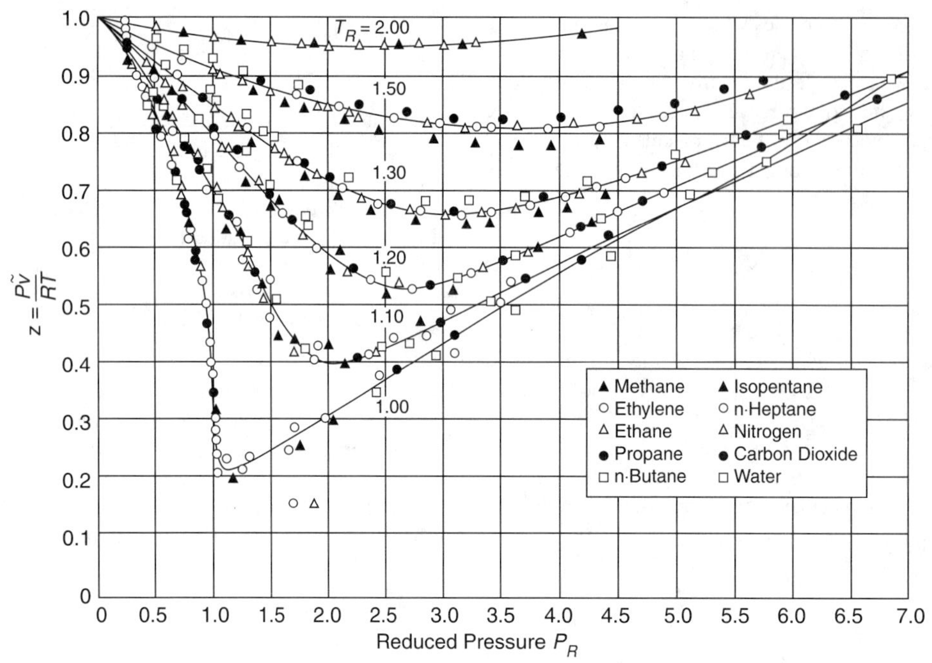

Figure 2. Generalized compressibility plot.

a function of reduced properties. The most common diagram is shown in **Figure 2**.

Gas Mixtures

For reduced equations of state and compressibility plots, pseudocritical constants of gas mixtures are used. The critical temperature T_c of a gas mixture can be obtained with good accuracy from *Kay's Law*

$$T_{c,mix} = \sum y_l T_{c,i} \tag{11}$$

with y_l being the mole fraction of component i in the gas mixture.

The critical pressure p_c is more accurately determined using the *Prausnitz and Gunn method*

$$p_{c,mix} = \frac{\tilde{R} T_{c,mix} \sum y_l z_l}{\sum y_l \tilde{v}_{c,i}} \tag{12}$$

For more information refer to Reid, Prausnitz and Sherwood.

References

Reid, R., Prausnitz, J. and Sherwood, T. (1977) *The Properties of Gases and Liquids,* McGraw-Hill Book Company.

H. Müller-Steinhagen

GAS LAW CONSTANTS

Following from: *Gas law*

The following table lists the constants a and b for the equation of state according to Van-der-Waals and the critical properties of the most common compounds.

Table 1.

		Van-der-Waals constants		Critical constants		
Gas	Formula	a ($N\ m^4\ mol^{-2}$)	b ($m^3\ mol^{-1}$)	T_c (K)	P_c ($N\ mm^{-2}$)	$\tilde{v}_{zc}$ ($mm^3\ mol^{-1}$)
Acetylene	C_2H_2	0.446	$5.15 \cdot 10^{-5}$	309.2	6.29	$112 \cdot 10^3$
Air				132.5	3.77	$88.0 \cdot 10^3$
Ammonia	NH_3	0.423	$3.72 \cdot 10^{-5}$	405.6	11.3	$72.4 \cdot 10^3$

Table 1.—*continued*

Gas	Formula	Van-der-Waals constants a (N m^4 mol^{-2})	Van-der-Waals constants b (m^3 mol^{-1})	Critical constants T_c (K)	Critical constants P_c (N mm^{-2})	$\tilde{v}_{zc}$ (mm^3 mol^{-1})
Argon	Ar	0.137	$3.23 \cdot 10^{-5}$	151.0	4.86	$75.3 \cdot 10^3$
Benzene	C_6H_6	1.83	$11.5 \cdot 10^{-5}$	561.7	4.83	$25.6 \cdot 10^3$
Butane (n−)	C_4H_{10}	1.47	$12.3 \cdot 10^{-5}$	426.0	3.65	—
Carbon dioxide	CO_2	0.365	$4.28 \cdot 10^{-5}$	304.3	7.40	$95.8 \cdot 10^3$
Carbon monoxide	CO	0.151	$4.00 \cdot 10^{-5}$	134.2	3.55	$90.1 \cdot 10^3$
Chlorine	Cl_2	0.659	$5.64 \cdot 10^{-5}$	417.2	7.72	$124 \cdot 10^3$
Cyanogen	$(CN)_2$	0.779	$6.93 \cdot 10^{-5}$	401.2	5.98	—
Ethane	C_2H_6	0.554	$6.47 \cdot 10^{-5}$	305.3	4.95	$143 \cdot 10^3$
Ethylene	C_2H_4	0.453	$5.73 \cdot 10^{-5}$	282.9	5.12	$127 \cdot 10^3$
Ethyl ether	$(C_2H_5)_2O$	1.76	$13.5 \cdot 10^{-5}$	467.8	3.60	$282 \cdot 10^3$
Helium	He	0.0034	$2.38 \cdot 10^{-5}$	5.3	0.229	$57.8 \cdot 10^3$
Hydrogen	H_2	0.0248	$2.67 \cdot 10^{-5}$	33.3	1.30	$64.6 \cdot 10^3$
Hydrogen bromide	HBr	0.452	$4.44 \cdot 10^{-5}$	363.2	8.52	
Hydrogen chloride	HCl	0.373	$4.09 \cdot 10^{-5}$	324.6	8.28	$86.9 \cdot 10^3$
Hydrogen cyanide	HCN	1.095	$8.25 \cdot 10^{-5}$	456.7	5.39	$135 \cdot 10^3$
Hydrogen iodide	HI	0.782	$5.31 \cdot 10^{-5}$	424.2	8.32	—
Hydrogen sulphide	H_2S	0.449	$4.30 \cdot 10^{-5}$	373.6	9.01	$11.9 \cdot 10^3$
Methane	CH_4	0.229	$4.30 \cdot 10^{-5}$	190.7	4.64	$99.4 \cdot 10^3$
Methanol	CH_3OH	0.967	$6.71 \cdot 10^{-5}$	513.2	7.98	$117 \cdot 10^3$
Methyl chloride	CH_3Cl	0.758	$6.51 \cdot 10^{-5}$	416.3	6.67	$136 \cdot 10^3$
Nitric oxide	NO	0.136	$2.80 \cdot 10^{-5}$	179.2	6.59	$58 \cdot 10^3$
Nitrogen	N_2	0.141	$3.92 \cdot 10^{-5}$	126.1	3.40	$90.1 \cdot 10^3$
Nitrous oxide	N_2O	0.384	$4.43 \cdot 10^{-5}$	309.7	7.27	$98 \cdot 10^3$
Oxygen	O_2	0.138	$3.19 \cdot 10^{-5}$	154.4	5.04	$74.4 \cdot 10^3$
Propane (n)	C_3H_8	0.879	$8.47 \cdot 10^{-5}$	370.0	4.26	$200 \cdot 10^3$
Propylene	C_3H_6	0.850	$8.30 \cdot 10^{-5}$	365.5	4.56	$180 \cdot 10^3$
Sulphur dioxide	SO_2	0.681	$5.65 \cdot 10^{-5}$	430.4	7.88	$123 \cdot 10^3$
Water	H_2O	0.554	$3.30 \cdot 10^{-5}$	647.3	22.15	$55.7 \cdot 10^3$

H. Müller-Steinhagen

GAS LIQUID CHROMATOFRAPHY, GLC (see Gas chromatography)

GAS-LIQUID CONTACTING (see Fluidics)

GAS-LIQUID FLOW

Following from: *Two-phase flow overview*

Of the four types of **Two-Phase Flow** (gas-liquid, **Gas-Solid**, **Liquid-Liquid** and **Liquid-Solid**), gas-liquid flows are the most complex, since they combine the characteristics of a deformable interface and the compressibility of one of the phases. For given flows of the two phases in a given channel, the gas-liquid interfacial distribution can take any of an infinite number of possible forms. However, these forms can be classified into types of interfacial distribution, commonly called *flow regimes* or *flow patterns.* Detailed discussions of these patterns are given by Hewitt (1982), Whalley (1987) and Dukler and Taitel (1986). The regimes encountered in vertical flows are illustrated in **Figure 1**. They include **Bubble Flow**, where the liquid is continuous, and there is a dispersion of bubbles within the liquid; **Slug** or **Plug Flow** where the bubbles have coalesced to make larger bubbles which approach the diameter of the tube; **Churn Flow** where the slug flow bubbles have broken down to give oscillating churn regime; **Annular Flow**

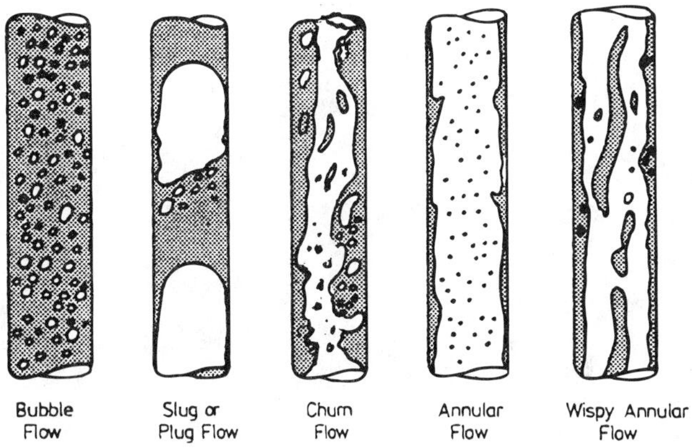

Figure 1. Flow patterns in vertical flow.

where the liquid flows on the wall of the tube as a film (with some liquid entrained in the core) and the gas flows in the centre; and **Wispy Annular Flow** where, as the liquid flow rate is increased, the concentration of drops in the gas core increases, leading to the formation of large lumps or streaks (wisps) of liquid.

Flow regimes in horizontal flow are illustrated in **Figure 2**. Here, as gravity acts normally to flow direction, separation of the flow occurs. The respective flow regimes are **Stratified Flow**, where the gravitational separation is complete; *stratified-wavy flow;* **Bubble Flow**, where the bubbles are dispersed in the liquid continuum (though there is some separation due to gravity as illustrated); *annular dispersed flow,* which is similar to that in vertical flow, though there is asymmetry in the film thickness due to the action of gravity; and a variety of *intermittent flows.* This latter category includes **Plug Flow**, in which there are large bubbles flowing near the top of the tube; *semi-slug flow,* where very large waves are present on the stratified layer; and **Slug Flow**, where these waves touch the top of the tube and form a liquid slug which passes rapidly along the channel. Pipe inclination is an important parameter in determining flow regimes, and flows in inclined pipes and in other geometries are discussed by Hewitt (1982). It is necessary to predict regimes as a basis for carrying out calculations on two-phase flow, and the usual procedure is to plot the information in terms of a flow regime map. Many of these maps are plotted in terms of primary variables (superficial velocity of the phases or mass flux and quantity, for instance), but there has been a great deal of work aimed at generalising the plots, so that they can be applied to a wide range of channel geometries and physical properties of the fluids. A generalised map for vertical flows is shown in **Figure 3**, and is due to Hewitt and Roberts (1969) (see Hewitt, 1982).

This map is plotted in terms of the superficial momentum fluxes of the two phases $\rho_L U_L^2$ and $\rho_G U_G^2$. A generalised flow pattern map for horizontal flow is that of Taitel and Dukler (1976) (see Dukler and Taitel, 1986), and is illustrated in **Figure 4**. This is plotted in terms of the following parameters:

$$X^2 = \frac{(dp_F/dz)_L}{(dp_F/dz)_G}$$

$$F = \sqrt{\frac{\rho_G}{\rho_L - \rho_G}} \frac{U_G}{\sqrt{Dg \cos \alpha}}$$

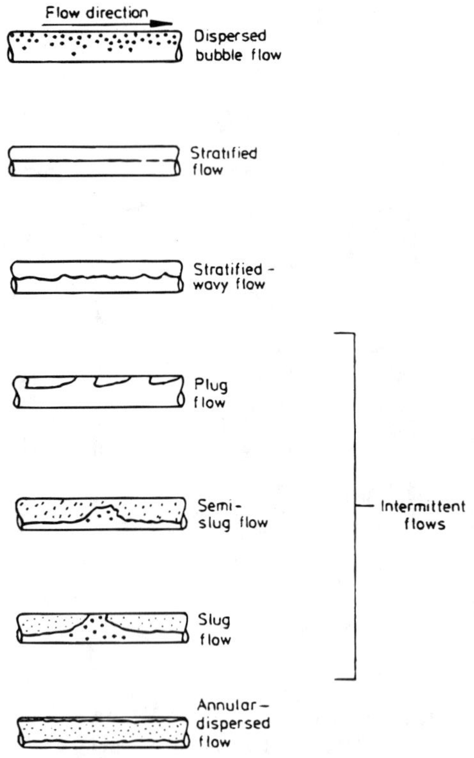

Figure 2. Flow regimes in horizontal two-phase flow.

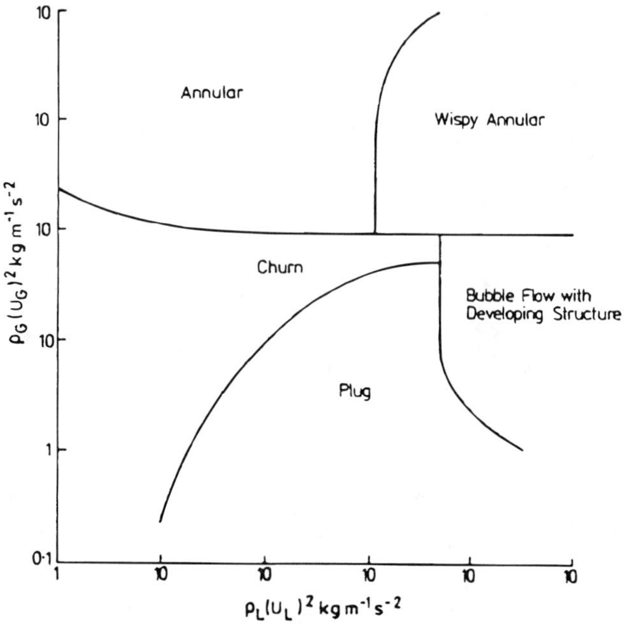

Figure 3. Flow pattern map obtained by Hewitt and Roberts (1969) for vertical two-phase co-current upwards flow in a vertical tube.

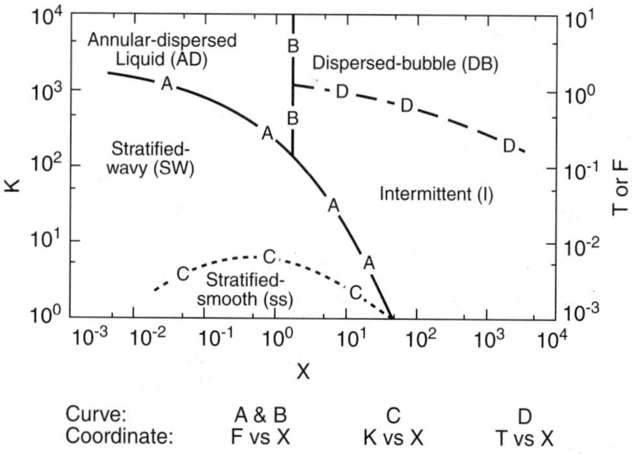

Figure 4. Flow pattern map for horizontal co-current flow obtained by Taitel and Dukler (1976). (See Dukler and Taitel, 1986).

$$K^2 = F^2\,Re_L = \frac{\rho_G U_G^2}{(\rho_L - \rho_G) D g \cos\alpha}\,\frac{D U_L}{\nu_L}$$

$$T = \left[\frac{(dp_F/dz)_L}{(\rho_L - \rho_G) g \cos\alpha}\right]^{1/2}$$

where $(dp_F/dz)_L$ and $(dp_F/dz)_G$ are the pressure gradients for the liquid phase and gas phase respectively, flowing alone in the channel, ρ_L and ρ_G are the phase densities, U_L and U_G are the superficial velocities of the phases, D the tube diameter, ν_L the liquid kinematic viscosity, g the acceleration due to gravity, and α the angle of inclination of the channel.

Taitel et al. (1980) also produced a flow pattern map for vertical flow, but this has met with less widespread use. Following similar approaches, Barnea (1987) has produced a unified model for flow pattern transitions for the whole range of pipe inclinations.

The subject of transitions from one flow pattern to another is still one of active debate. The maps given here should be used for general guidance only and, if there is a crucial interest in any specific transition, then closer study should be made. in what follows below, a brief commentary is given on some of the transitions covered, with some recent references where appropriate.

Flow pattern transitions in vertical flow

The most important transitions are as follows:

1. Bubble-Plug transition. It has been traditional to regard this transition to occur as a result of *bubble coalescence* leading to gradual bubble growth and the formation of large Taylor-type bubbles which occupy the whole pipe cross-section. Typically, the transition to slug flow occurs when the void fraction is around 25–30%. In highly turbulent flows, break-up of the bubbles may be postulated to occur (though this is rarely seen in actual bubble flows) to offset the progression of the coalescence. However, recent information seems to indicate that this view of the transition may be quite wrong. It seems more likely that void waves are formed in the flow, and that, within these waves, the bubbles become closely packed and are better able to coalesce, leading to plug flow. Evidence for this is provided, for instance, by Beisheuvel and Gorissen (1990).
2. Plug-annular flow transition. This has been an area of major controversy. Probably the main difficulty has been a semantic one. Taitel et al. (1980) define a "churn flow" which is essentially a developing plug or slug flow. However, churn flow as defined here does exist in fully developed flow, and has the following unique characteristics:
    a. The regime is entered from slug flow by the formation of flooding-type waves (see **Flooding and Flow Reversal**), and these persist as a characteristic of the regime throughout. Such waves are absent in both slug flow and annular flow but are formed repeatedly in the churn flow regime and transport the liquid upwards (Hewitt et al. 1985 and Govan et al. 1990).
    b. In between successive flooding waves, the flow of the liquid phase in the film region near the wall reverses direction, and is eventually entrained by the next upward-moving wave.

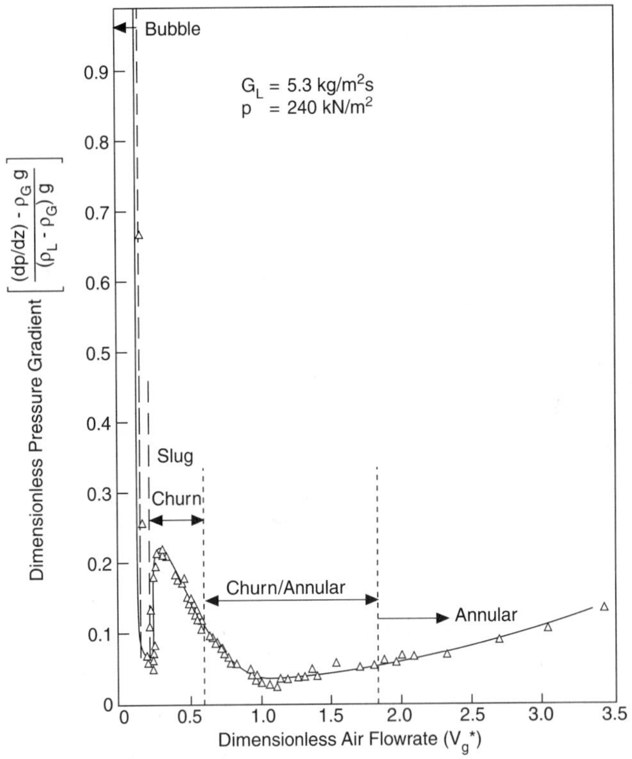

Figure 5. Data for pressure gradient in fully developed air-water flux in a vertical tube (Owen, 1986).

The onset of churn flow is accompanied by a sharp increase in pressure gradient, as is illustrated by the results obtained by Owen (1986) and illustrated in **Figure 5**. A detailed evaluation of the plug to churn flow transition in terms of this flooding mechanism is given by Jayanti and Hewitt (1992).

3. Churn to annular flow transition. As the gas velocity is increased after the churn flow regime has been entered, the pressure gradient initially decreases and the passes through a minimum value (see **Figure 5**). The flooding waves (and their associated intensive gas-liquid interactions promote large pressure gradients, and as they disappear, the pressure gradient reduces.

   Eventually, however, the pressure gradient increases again as the gas flow rate increases. Properly speaking, one might define the onset of true annular flow as that corresponding to the point at which there is no flow reversal within the liquid film. This might correspond approximately to the pressure drop minimum (though not accurately so). Another definition might be the flow reversal point (see **Flooding and Flow Reversal**). It is clear that, though both churn and annular flow have the characteristic of having a liquid layer at the wall and a gas core in the centre of the pipe, their flow behaviour is quite different. The definition of the exact point of transition is, nevertheless, quite difficult.

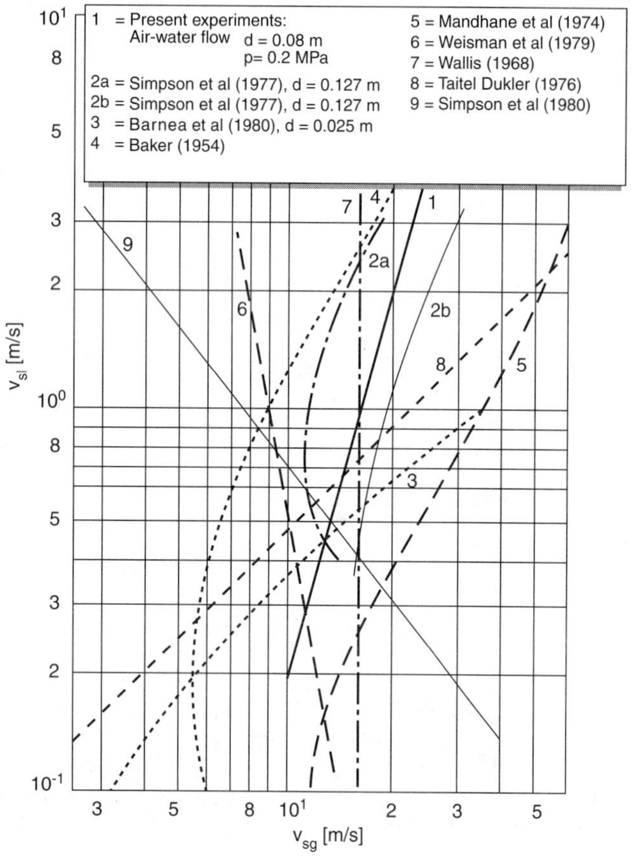

Figure 6. Comparison of relationships for slug/annular transition in horizontal flow with data of Reiman.

4. Annular to wispy-annular transition. This is supposed to occur approximately at a critical liquid momentum flux (see **Figure 3**) though, again, identification of the transition is to some extent subjective.

Flow pattern transitions in horizontal flow

Again, there has been an intensive interest in flow pattern transitions for flow in horizontal pipes, mainly arising out of the importance of such phenomena in the offshore pipeline area. A commentary on some of the important transitions in such flows is as follows:

1. Stratified-slug transition. The Taitel and Dukler (1976) model ascribed this transition to the onset of Kelvin-Helmholz instability and applied an inviscid form of this instability to the prediction of the transition. However, it was found that this did not agree well with the actual transition and it was necessary to introduce an arbitrary correction function to bring the data into line with the predictions. A closer approximation is presented in the analysis of Lin and Hanratty (1986) who present a theory taking account taking of viscous effects. However, the prediction of this transition still remains an area of uncertainty.
2. Slug-annular transition. In the Taitel and Dukler (1976) model, this transition was assumed to occur if the equilibrium liquid level at the onset of the Kelvin-Helmholz instability was less than half the channel diameter. However, this does not appear to fit the transition well, nor are there any generally satisfactory models. **Figure 6** shows a comparison of the model predictions from various approaches and it will be seen that there is not only a discrepancy between these in magnitude, but also in trend!
3. Slug-dispersed bubble transition. In the Taitel and Dukler and related maps, this transition is ascribed to the capability of the turbulence in the liquid phase to suspend the bubbles. When the bubbles cannot be suspended, then they agglomerate to form the gas bubble regions in slug flow. In actual fact, however, the mechanisms of bubble entrainment and behaviour in slug flow are extremely complex and need much further investigation.

References

Barnea, D. (1987) A unified model for predicting flow-pattern transitions for the whole range of pipe inclinations, *Int. J. Multiphase Flow,* 13, 1–12.

Biesheuvel, A. and Gorissen, W. C. M. (1990) Void fraction disturbances in a uniform bubbly fluid, *Int. J. Multiphase Flow* 16, 211–231.

Dukler, A. E. and Taitel, W. (1986) Flow Pattern Transitions in Gas-Liquid Systems: Measurement and Modelling, Chapter 1 of *Multiphase Science and Technology, Vol. 2* (Ed. G. F. Hewitt, J. M. Delhaye and N. Zuber), Hemisphere Publishing Corporation.

Govan, A. H., Hewitt, G. F., Richter, H. J. and Scott, A. (1991) Flooding and churn flow in vertical pipes, *Int. J. Multiphase Flow,* 17, 27–44.

Hewitt, G. F. (1982) Chapter 2, *Handbook of Multiphase Systems* (Ed. G. Hetsroni), Hemisphere Publishing Corporation, New York.

Hewitt, G. F., Martin, C. J. and Wilkes, N. S. (1985) Experimental and modelling studies of annular flow in the region between flow reversal and the pressure drop minimum. *Physico-Chemical Hydrodynamics,* 6, 43–50.

Lin, P. Y. and Hanratty, T. J. (1986) Prediction of the initiation of slugs with linear stability theory, *Int. J. Multiphase Flow,* 12, 79–98.

Taitel, Y., Barnea, D. and Dukler, A. E. (1980) Modelling flow pattern transitions for steady upward gas-liquid flow in vertical tubes, *AIChE J,* 26, 345–354.

Whalley, P. B. (1987) *Boiling and Condensation and Gas-Liquid Flow,* Clarendon Press, Oxford.

Leading to: *Annular flow; Bubble flow; Cavitation; Dispersed flow; Drops; Falling film flow; Flooding and flow reversal; Instability, two-phase; Interfaces; Plug flow; Plunging liquid jets; Pressure drop, two-phase; Slug flow; Stratified flow; Wispy annular flow*

G.F. Hewitt

GAS-SOLID FLOWS

Following from: *Multiphase flow*

A gas-solid flow is characterized by the flow of gases with suspended solids. This type of flow is fundamental to many industrial processes such as *pneumatic transport, particulate pollution control, combustion of pulverized coal, drying of food products, sand blasting, plasma-arc coating and fluidized bed mixing.* The dynamics and thermal history of particles in gases also affect the performance of rocket motors using metallized fuels, the quality of some pharmaceutical products and the design of advanced techniques for materials processing.

Various features of gas-solid flows can be best understood by considering the motion and thermal history of particles in a gas flow field. The equation of motion for a spherical particle of diameter D accelerating in a gas stream is

$$m\frac{du_p}{dt} = \frac{1}{2}\rho_g C_D \frac{\pi D^2}{4}(u_g - u_p)|u_g - u_p| + mg \qquad (1)$$

where m is the particle mass, g is the acceleration due to gravity, ρ_g is the gas density, C_D is the drag coefficient and u_p and u_g are the particle and gas velocity respectively. Dividing through by the particle mass yields

$$\frac{du_p}{dt} = \frac{18\eta}{\rho_p D^2}\frac{C_D Re_p}{24}(u_g - u_p) + g \qquad (2)$$

where ρ_p is the material density of the particle and Re_p is the particle **Reynolds Number** based on the relative velocity, particle diameter and gas properties. The factor $C_D Re_p/24$ is the ratio of the *drag coefficient* to *Stokes drag* and represented by the factor f. This factor is a function of the particle Reynolds number but may also depend on other parameters such as particle shape, Mach number based on the relative velocity and turbulence level (Clift et al., 1978). For small particle Reynolds number ($Re_p \leq 1$) the factor approaches unity which corresponds to Stokes flow. (See also **Stokes Law** and **Spheres, Flow Around and Drag**)

The coefficient of the drag term in the particle motion equation has dimensions of reciprocal time and leads to the definition of the *particle velocity response time.*

$$\tau_V = \frac{\rho_p D^2}{18\eta} \qquad (3)$$

This time is a measure of the responsiveness of a particle to a change in gas velocity. For example, the time required for a particle released from rest in a gravity-free environment to achieve 63% of the gas velocity in Stokes flow is one velocity response time. Also the distance a particle would penetrate into a stagnant gas before stopping is $u_0\tau_V$ where u_0 is the initial particle velocity. This distance is called the *stopping distance*. The *terminal velocity* a particle would achieve falling through a stagnant gas is $g\tau_v$.

Neglecting radiative heat transfer and assuming there is no change of phase of the particle material, the equation for particle temperature assuming a uniform internal temperature is

$$mc\frac{dT_p}{dt} = Nu\lambda\pi D(T_g - T_p) \qquad (4)$$

where c is the specific heat of the particle material, Nu is the **Nusselt Number** and T_p and T_g are the particle and gas temperatures respectively. Dividing through by the mass and specific heat yields a coefficient for the heat transfer term which is the reciprocal of the *thermal response time;* namely,

$$\tau_T = \frac{\rho_p c D^2}{6Nu\lambda} \qquad (5)$$

which is a measure of the thermal responsiveness of the particle to a change in gas temperature. At low Reynolds numbers, the Nusselt number is 2. Empirical expressions are available for higher particle Reynolds numbers and other effects (Clift et al., 1978). For particles in a gas, the thermal response time and velocity response time are of the same order of magnitude.

The response times are used in the definition of *Stokes number* which is the ratio of the particle response time to a time characteristic of the flow system. For example, consider the gas-solids flow through a venturi geometry shown in **Figure 1**. A time which would be representative of the fluid residence time in the venturi would be

$$\tau_F = \frac{L}{u_T} \qquad (6)$$

where L is the throat diameter and u_T is the flow velocity in the throat region. The Stokes number based on the velocity response time is:

$$St = \frac{t_V}{t_F} = \frac{\tau_V u_T}{L} \qquad (7)$$

For Stokes numbers much less than unity, the particles have sufficient time to maintain near velocity equilibrium with the gas and the gas-solids flow could be regarded as a single phase flow with modified density. On the other hand, if the Stokes number is large compared to unity, the particle motion is unaffected by the change in gas flow velocity through the venturi.

The same definition of Stokes number applies to thermal response time as well. For low Stokes numbers, the particles maintain near thermal equilibrium with the gas. At large Stokes numbers, the particles have no time to respond to the changes in gas temperature.

A basic definition in gas solids flows is *dilute* and *dense* flows. A dilute flow is a gas-particle flow in which the particle motion is controlled by the drag and lift forces on the particle. In a dense flow, on the other hand, the particle motion is controlled primarily by particle-particle collisions. A dilute flow would correspond to the conditions where the stopping distance is less than that between particle-particle collisions. This definition is important in under-

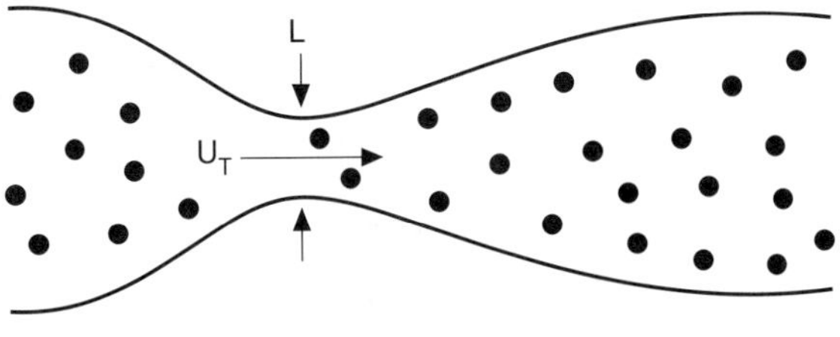

Figure 1.

standing and modeling the velocity and thermal fields in a gas-solid flow.

Another important feature of gas-solid flows is *coupling* which is the interaction between phases. If the gas affects the motion and temperature of the particles but the particles do not change the gas velocity or thermal flow fields, then flow is *one-way coupled.* On the other hand if there is a mutual interaction between phases, the flow is *two-way coupled.* As an example of coupling, consider hot particles being transported by a cold gas in a duct as shown in **Figure 2**. Assuming one-way coupling, the particle temperature decreases toward the gas temperature while the gas temperature remains constant. With two-way coupling, the gas temperature increases due to heat exchange with the particles and the rate of particle temperature is reduced. Actually all flows are two-way coupled, but for low particle concentrations, the effect of the particles on the gas field may be negligible. Therefore, the assumption of one-way coupling would be justified thereby simplifying the analysis of the flow system.

Pneumatic transport is an important example of gas-solid flows because of its wide use in industry to transport metal particles, grains, ores, cement, coal and other products not susceptible to damage by contact with the pipe walls. The main advantage of pneumatic transport is the flexibility of line location and the capability to tap the line at arbitrary locations. The gas velocity for vertical transport has to exceed the settling (terminal) velocity of the particles to maintain transport.

For horizontal pneumatic transport, various flow patterns are identified which depend on several factors such as flow velocity and particle loading. The various flow patterns are shown in **Figure 3**. *Homogeneous flow* occurs when the velocity is sufficiently high to keep the particles in suspension. *Dune flow* begins as the velocity is lowered and particles begin to settle out on the wall forming a dense flow region with a pattern like sand dunes. The velocity at which particles begin to settle out is called the *saltation* velocity. Further reduction in velocity leads to *slug flow* where there are alternate regions where particles fill the pipe and where they are in suspension. The flow behaves like slug flow of the gas-liquid flow regimes. Finally, with further reduction in velocity the particles become packed in the pipe and form a *packed bed* while the gas moves through the interstitial region between the particles. Even though the particles fill the pipe there may still be a slow motion of the particle bed.

A typical variation in pressure drop with flow velocity is shown in **Figure 4**. At high velocities (in the homogeneous flow regime), the pressure drop varies with the nearly the square of the velocity as with a single phase flow. The pressure drop for the particle-laden flow is higher because the particles lose momentum on contact with

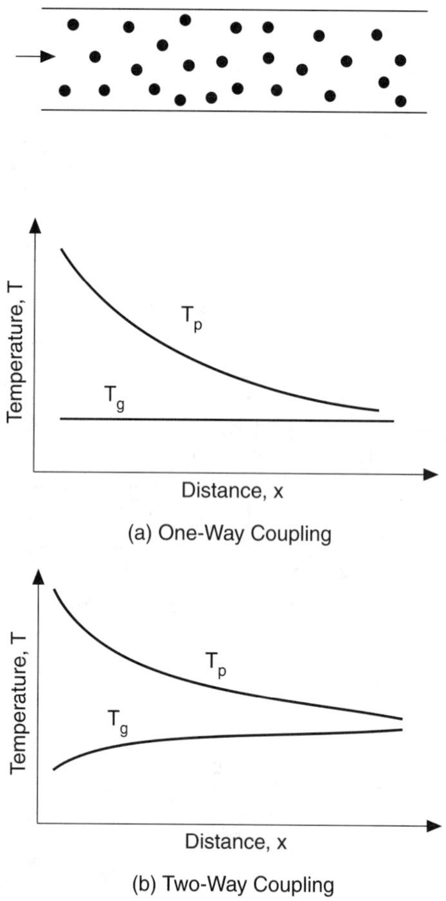

Figure 2.

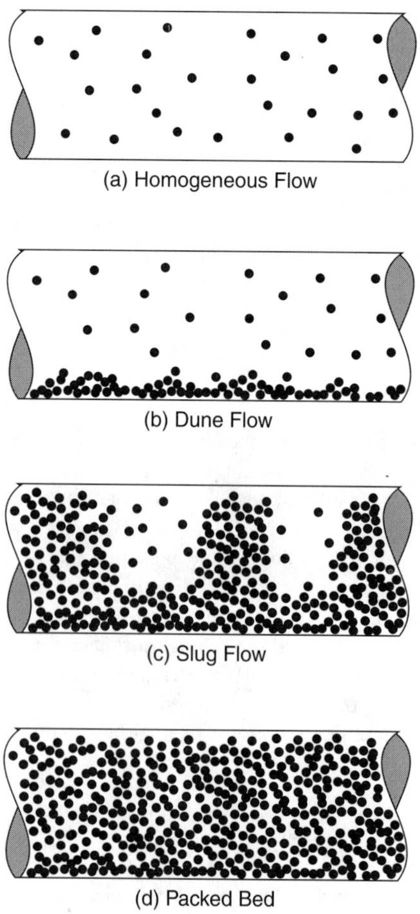

Figure 3.

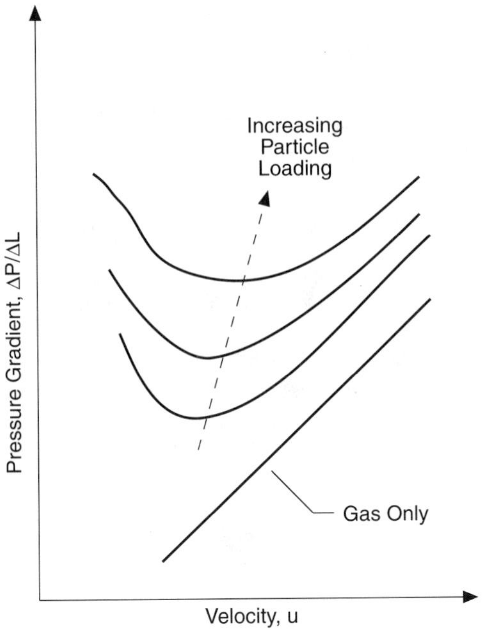

Figure 4.

the wall. The force applied to the fluid over a length ΔL of pipe by the drag on the particles is

$$F_p \sim nA\Delta L 3\pi\eta Df(u_g - u_p) \quad \textbf{(8)}$$

where n is the number of particles per unit volume and A is the cross-sectional area of the pipe. Equating this force to the augmented pressure gradient in the pipe due to the presence of the particles yields

$$\frac{\Delta p}{\Delta L} \sim \frac{\rho_d' f}{\tau_V}(u_g - u_p) \quad \textbf{(9)}$$

where ρ'_p is the mass of particles per unit volume or the *apparent* or *bulk* particle density. Thus one notes that the pressure loss increases with increased particle concentration and gas-particle velocity difference (more momentum loss at the wall). The increase in pressure gradient corresponds to a two-way coupling effect. There are several empirical formulations available in the literature to estimate the pressure drop due to the particulate phase.

As the velocity is reduced to the saltation velocity, the cross-sectional area for the flow is reduced because of the deposits on the wall. The effective flow velocity is increased and the pressure loss is higher. To avoid deposition, higher pressure loss and possible plugging, the pneumatic system should be designed with velocities exceeding the saltation velocity of the slug flow regime. Various empirical expressions for saltation velocity and friction factors for pressure drop due to the solid phase can be found in Klinzing (1981).

The **Fluidized Bed** is another important example of a gas-solids flow and is a key element in many chemical processes, particularly coal gasification, combustion and liquification. (Azbel and Cheremisinoff, 1983). Fluidized beds are also used for roasting ores and for the disposal of organic, biological and toxic wastes. In essence, the fluidized bed consists of a vertical cylinder loaded with particles and supplied with a gas through a distributor plate in the bottom of the cylinder. As the gas flow is increased the bed goes through several flow regimes (Hetsroni, 1982) as shown in **Figure 5**. At low flow rates, there is no significant motion of the particles as the gas passes through a packed bed. With increasing gas flow a point is reached where the particles are just supported by the hydrodynamic forces which is called *particle fluidization.* Further increase in gas velocity leads to the formation of bubbles (region of low particle concentration) which move upward in the bed and enhance mixing. With more gas flow, the bubbles grow to fill the tube and the slug flow regime is realized and is similar to the slug flow pattern in vertical gas-liquid flows. As the gas flow rate is further increased clusters of particles move about the field in an irregular fashion. Finally, at the highest gas flow rate, clusters move up the tube and out, with some downward motion near the wall, so the particles have to reintroduced at the bottom. This condition is called *fast fluidization.* Thus, the fluidized bed has a range of flow regimes from dense to dilute flows. The fluidized bed is attractive as a reactor because the gas is exposed to a large solids surface area to enhance surface reactions and heat transfer. Also the bed operates at a nearly uniform temperature which provides control of the reaction rates. Depending on the application, heat exchanger tubes may be located in the bed itself (Howard, 1983). The flow in a fluidized bed is very complex and difficult to scale up from bench scale to prototype operation.

Another important industrial example of gas-solid flows is the removal of particulates from exhaust gases for pollution control. The two most common devices for removal of particulates from a gas stream are the *cyclone separator* and the *electrostatic precipitator.* The cyclone separator is simply a device which separates the solids from the gas by centrifugal acceleration. Standard designs are available in engineering handbooks. The performance of the cyclone separator is quantified by the cut size, which is the particle size above which all the particles are collected and below which

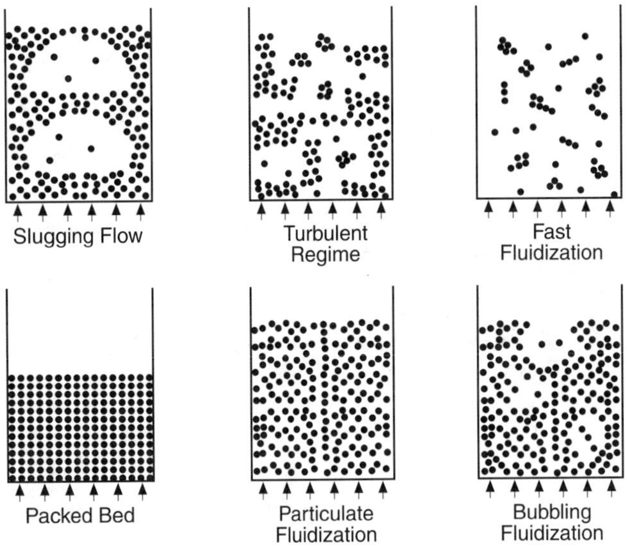

Figure 5. From Hetsroni, ed. (1982), *Handbook of Multiphase Systems*, Hemisphere Publishing Corp.

are carried out by the exhaust gases. The cyclone separator is widely used because it is inexpensive and robust. (See **Cyclones**).

The electrostatic precipitator operates by charging the particles and applying a Coulomb force to move the particles toward the collecting surface. It is capable of removing smaller particles from a gas stream than the cyclone separator and operates at a higher efficiency. It is commonly used to remove flyash from coal-fired power plants. The particulate-laden flow passes through an array of vertically suspended metal plates as shown **Figure 6**. The cross-sectional configuration is shown on the same figure. The ribs protruding into the flow provide mechanical rigidity for the large plates. High voltage wires produce a corona at the wires and an electric field between the wires and the walls. For a negatively charged wire, electrons generated by the corona travel along the electric lines of force and accumulate on the particles yielding negatively charged particles. The resulting Coulomb force on the particle moves the particle toward the wall (collection surface).

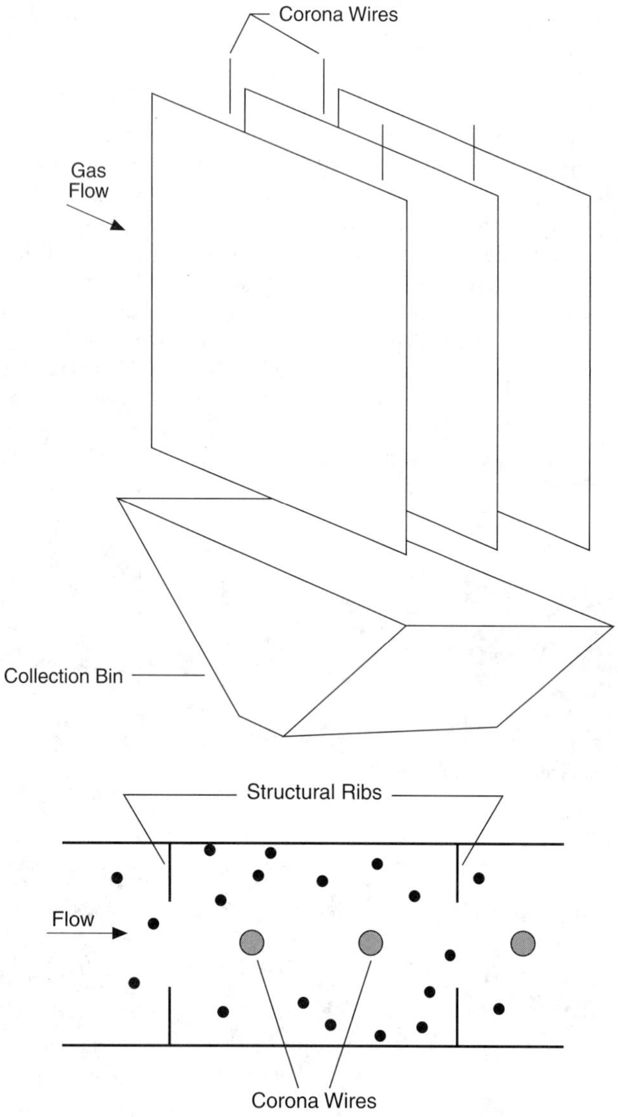

Figure 6.

Periodically the walls of the precipitator are "rapped" and the particles fall into a collection bin below the precipitator (Oglesby and Nichols, 1978).

The drift velocity toward the wall is given by

$$u_{drift} = \frac{q}{m}\frac{\tau_V E}{f} \tag{10}$$

where E is the electric field intensity and q/m is the charge to mass ratio on the particle. For very small particles the drag factor f must account for noncontinuum effects because the particle size may be comparable to the mean free path of the gas. A higher drift velocity implies an improved collection efficiency (particles collected/particles entering). The modified *Deutsch-Anderson equation* for efficiency

$$E = 1 - \exp\left[-\left(\frac{u_{drift}A}{u}\right)^k\right] \tag{11}$$

where A is the plate surface area, u is the gas velocity through the precipitator and k is an empirical constant is often used to predict efficiency but cannot be used with confidence to extrapolate performance predictions under different operating conditions. The effect of the charged particles on the electric field distribution and the influence of turbulence on particle dispersion currently preclude the capability of making an accurate predictions of collection efficiency. (See also **Electrostatic Separation**)

The particle-laden jet is another important example of a gas-solids flow. Gas-particle jets are fundamental to the operation of a furnace burning pulverized coal as well as sand blasting equipment and plasma-arc coating devices. A jet conveying pulverized coal issuing into a corner-fired furnace is shown schematically in **Figure 7**. Intense radiative heat transfer to the entering coal particles quickly causes devolatization where the volatiles in the coal (methane, hydrogen, etc.) are driven off as gaseous fuel to support combustion in the furnace. The remaining char particles burn more slowly as they pass through the furnace and react with the water vapor and other compounds. The mixing and dispersion of the

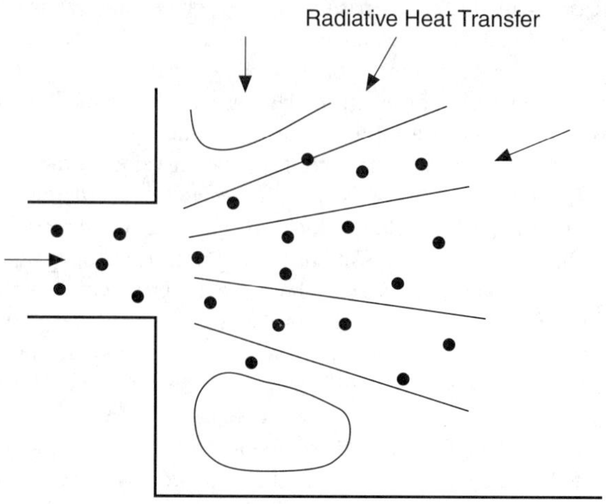

Figure 7.

particles by the jet is important to the efficient operation of the system (Smoot and Smith, 1985).

The production of powdered foods and other products through *spray drying* is a further example of gas-solids flows. Some examples of spray dried products include powdered milk, laundry detergent and pharmaceutical powders. Spray drying is particularly attractive for drying products which are heat sensitive. Slurry droplets are sprayed into a chamber, dried by hot gases and the resulting particles are collected as products. A typical counterflow spray dryer is shown in **Figure 8**. Here the droplets are sprayed downward from the top of the chamber as the hot gases are introduced from below The gases are introduced with a tangential velocity component which produces a swirling motion as they pass upward and out through the top. The dried particles fall through the port at the bottom where they are further treated or packaged as the final product.

Many factors have to be considered in the design of the spray dryer. The thermal coupling cools the drying gases as they pass upward through the particles thereby reducing the drying effectiveness. The gas temperatures cannot be too high to prevent burning the powder or detracting from the taste of a food product. Also the particles have to be sufficiently dry so as not to accumulate on the walls and lead to the danger of fire in the dryer.

Considerable progress has been made in the last ten years in developing numerical models for gas-solid flows (Crowe, 1991). The availability of large memory machines and high speed processors have enabled model developers to simulate sufficient detail of gas-solid flows to produce numerical models adequate for many industrial needs. Two approaches have emerged; the trajectory and two-fluid approach. In the trajectory approach, the particle field is generated by solving for the particle trajectories and the velocity and thermal history along the trajectories in the gas flow field. The local heat and momentum transfer to the gas are used update the gas field calculations and thereby include two-way coupling effects. This method is especially attractive for dilute flows. The two-fluid approach is to treat the particulate phase as a second fluid with an effective viscosity, thermal conductivity and diffusion coefficients. This approach has found application for dense flows such as fluidized beds. The continued development of numerical models will lead to ever improving simulations to complement design and support the operation of gas-solids systems.

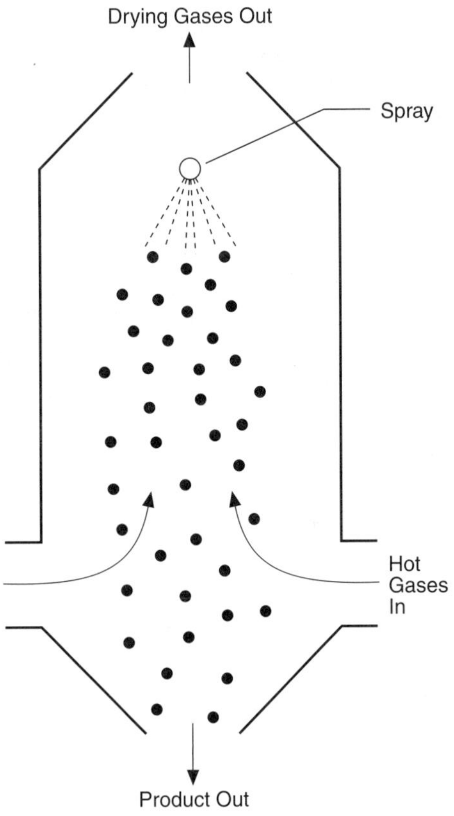

Figure 8.

References

Azbel, D. S. and Cheremisinoff, N. P. (1983) *Fluid Mechanics and Unit Operations,* Ann Arbor Science Publishers.

Clift, R., Grace, J. R. and Weber, M. E. (1978) *Bubbles, Drops and Particles,* Academic Press.

Crowe, C. T. (1991) The State of the Art in the Development of Numerical Models for Dispersed Phase Flows, *Proc. Intl. Conf. on Multiphase Flows—'91 Tsukuba,* Vol.3, pp. 49–60.

Hetsroni, G. (ed.) (1982) *Handbook of Multiphase Systems,* Hemisphere Publishing Corp.

Howard, J. R. (ed.) (1983) *Fluidized Beds Combustion and Applications,* Applied Science Publishers.

Klinzing, G. E. (1981) *Gas-Solid Transport,* McGraw-Hill.

Masters, K. (1985) *Spray Drying Handbook.* G. Goodwin, Ltd, London.

Oglesby, Jr., S. and Nichols, G. B. (1978) *Electrostatic Precipitation,* Marcel-Dekker, inc.

Smoot, L. D. and Smith, P. J. (1985) *Coal Combustion and Gasification,* Plenum Press, NY.

Leading to: *Aerosols; Fluidised beds; Gas-solids separation, overview*

C.T. Crowe

GAS-LIQUID INTERFACES (see Interfaces; Surface and interfacial tension)

GAS-LIQUID-SOLID FLOWS (see Multiphase flow)

GAS OILS (see Oils)

GASOLINE ENGINES (see Internal combustion engines)

GASOLINES (see Oils)

GAS RADIATION (see Parallel plates, radiative heat transfer between)

GAS REACTORS (see Nuclear reactors)

GAS SOLID CHROMATOGRAPHY, GSC (see Gas chromatography)

GAS-SOLIDS SEPARATION, OVERVIEW

***Following from:** Gas-solid flows; Phase separation*

Particulate solids are separated from gas streams for a variety of reasons. In some instances, the interest is in the recovery of solids as a product, e.g. following a milling operation for which a combined classification and separation is often required (Prasher, 1987). In other cases, the emission of fine particulate solids and dust from a unit operation may be excessive and therefore, a reduction and control of particulate level is required for the protection of subsequent process equipment or for the environmental emission (Strauss, 1975, and Theodore and Buonicore, 1976). This is illustrated in **Figure 1** for the case of advanced power generation from coal, where it is desirable to expand the hot gases in a gas turbine before exhaust to atmosphere to enhance the efficiency of power generation. Typical particulate effluent from a pilot-scale fluidised bed combustor is shown, together with upper limits of allowable emission to atmosphere, and of the inlet dust tolerance of high performance gas turbines (Henry et al., 1982). In this section, an overview of the principles of gas-solids separation is given. Details of various technologies are addressed in the next three sections.

The most commonly used techniques for the separation of particulate solids from gases are *inertial separators,* **Electrostatic Separators** and **Filters**. The physical phenomena involved in the separation of solids from gases are influenced by a number of important factors such as the properties of the gas, and gas-particle and particle-particle interactions. These factors are briefly reviewed

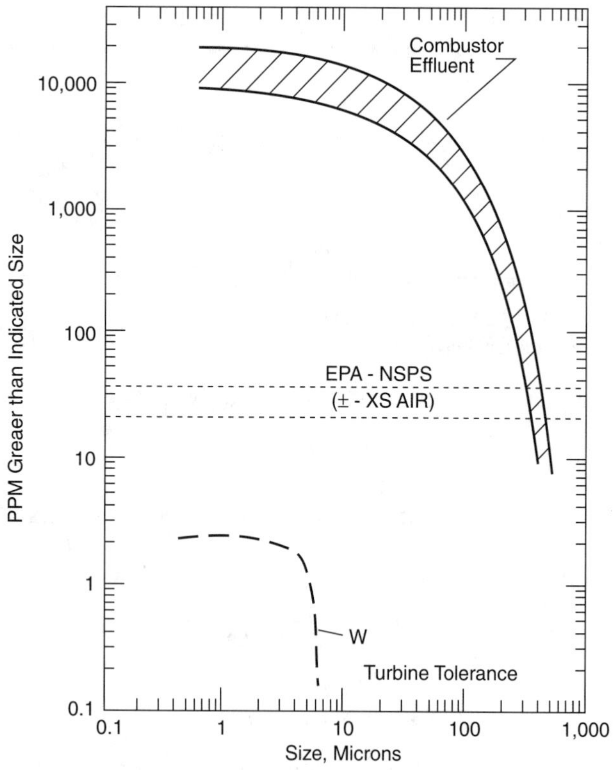

Figure 1. Industrial process and environmental requirements for gas-solids separation.

here first before addressing the principles of various types of gas-solids separators.

Gas Properties

Over the range of pressures commonly used in industrial gas-solids separation processes, departures from ideal gas behaviour are negligible. Therefore the density of a gas at pressure P and absolute temperature T can be approximated by

$$\rho_g = \frac{Mw}{\tilde{v}} = \frac{PMw}{\tilde{R}T} \tag{1}$$

where Mw is the mean molecular weight of the gas $\tilde{v}$ the molar volume, and $\tilde{R}$ the universal gas constant.

Elementary kinetic theory gives a first approximation for the effect of temperature and pressure on gas viscosity. **Viscosity** is predicted to be independent of pressure, and proportional to the square root of absolute temperature.

In addition to these macroscopic properties, the **Mean Free Path** of gas molecules, λ, is of interest for removal of fine particles. Elementary kinetic theory predicts that λ is inversely proportional to density, so that

$$\lambda \alpha \frac{T}{P} \tag{2}$$

Gas-particle Interaction

Almost any particle removal process requires the particles to migrate relative to the gas, so that the drag of the gas on the particle is of prime interest. The particle **Reynolds Number** is defined as

$$Re_p = \frac{ud_p\rho_g}{\eta} \tag{3}$$

where d_p is the particle diameter and u the velocity of the particle relative to the gas and η the gas viscosity. In most applications, Re_p remains small. The drag force, F_D, can then be estimated from **Stokes' Law** with a correction to allow for "slip effects" which arise when the particle diameter is comparable to the mean free path of gas molecules:

$$F_D = \frac{3\pi\eta ud_p}{C} \tag{4}$$

where C is the slip correction factor or "*Cunningham coefficient*" defined as

$$C = \frac{\text{Drag on particle in continuum flow at same Re}}{\text{Drag on particle in presence of slip}} \tag{5}$$

C can be estimated empirically from Davies' correlation (1945):

$$C = 1 + Kn\left[2.514 + 0.8\exp\left(\frac{-0.55}{Kn}\right)\right] \tag{6}$$

where Kn is the **Knudsen Number**:

$$Kn = \frac{\lambda}{d_p} \quad (7)$$

Equation 6 is applicable to spherical particles, and Beard (1976), and Clift et al., (1978) summarise modifications for non-spherical particles. Some representative values for C are given by Clift et al., (1981). It is worth noting that, whereas C only departs significantly from unity for submicron particles at ambient conditions and at elevated pressure, slip effects are significant for particles several microns in diameter at elevated temperatures and ambient pressure.

For Reynolds numbers larger than, say, 0.1, **Equation 4** is better replaced by the general form:

$$F_D = \frac{\pi \rho_g u^2 d_p^2 C_D}{8} \quad (8)$$

Where C_D is an empirical function of Re_p. Of the many forms suggested for C_D (Re_p), that due to Schiller and Nauman (1933), is probably most accurate for $Re_p < 800$:

$$C_D = \frac{24}{Re_p}(1 + 0.15\, Re_p^{0.687}) \quad (9)$$

When Re_p is sufficiently large that **Equation 8** must be used rather than Equation 4, then slip effects can be ignored, unless u is significant by comparison with the speed of sound in the gas. However, this is not normally the case in industrial gas-solids separation processes.

For a particle settling freely under its own weight in a gas, the drag force counterbalances the immersed weight of the particle. For low Re_p, terminal velocity then follows from **Equation 4** as

$$u_t = \frac{Cgd_p^2(\rho_p - \rho_g)}{18\eta} \quad (10)$$

where ρ_p is the particle density. Normally $\rho_p \gg \rho_g$, so that Equation 1 can be written

$$u_t = \frac{Cg\rho_p d_p^2}{18\eta} \quad (10a)$$

Increasing temperature, through its effect on gas viscosity, increases fluid-particle drag, reduces settling velocity, and generally makes removal of particles from gases more difficult. The effect of pressure is smaller, and acts through the slip effect so that it is only significant for particles typically smaller than about 1 μm; in general, the effect of increasing pressure is again to make particle removal more difficult. *Electrostatic precipitators* can be an exception to this rule because of the effect of pressure on the electrical properties of a gas. Increasing the pressure, within the range of interest here, widens the gap between the corona-starting and sparkover voltages. Therefore, the effect of increased drag on the particle at high pressure and high temperature may be compensated or even overcome by increasing the electrostatic field intensity (see below).

In certain types of filter the *Brownian diffusivity* of particles in the gas is of concern. This is usually best evaluated by the *Stokes-Einstein equation* (Clift et al., 1981):

$$D_{AB} = \frac{Ck_BT}{3\pi\eta d_p} \quad (11)$$

where k_B is Boltzmann's constant, 1.380622×10^{-23} JK^{-1}. Taking the effect of temperature on viscosity into account, for given particle size

$$D_{AB} \alpha CT^{1/2} \quad (11a)$$

so that the Brownian diffusivity increases with increasing temperature. Pressure has a weaker effect, through the slip correction factor, C, increasing pressure decreases $D_{AB.}$ It follows from the above that, if the efficiency of a gas-solids separation process is to be investigated, it is more important to match the process temperature than pressure. Pressure is important for fine particles, so that devices such as electrostatic precipitators intended to collect micron-sized particles would be tested at process temperature and pressure; electrical properties also dictate that both pressure and temperature should be matched.

Particle-particle Interaction

In some devices for particle collection, most notably electrostatic precipitators and some types of filter, it is essential for the collected particles to form relatively large agglomerates which can then be removed easily. Particle-particle cohesive forces arising from electrostatic, *van der Waals and capillary effects* are then important.

Electrostatic effects act over the longest range and may therefore be effective in separating particles from gases. However, they can only contribute to the cohesiveness of a deposited particulate if the constituent particles carry both positive and negative charges. Coury and Clift (1984) have shown that fly-ash from fluidised combustion can carry such charges, and be highly cohesive as a result. However, it is not known whether this phenomenon occurs more generally nor, for example, whether ash which has been raised above the fusion temperature carries significant charges. More generally, particles are likely to carry charges of the same sign, so that interparticle Coulombic repulsion acts against cohesion. Leakage of this charge requires a conductive path to an earthed surface, and this depends in particular on the surface characteristics of the particles. If the particles have for some reasons low surface resistivity, then electrical contact between particles can be good and the charge can leak away relatively rapidly. This is particularly important in electrostatic precipitation.

The other two types of force act over shorter ranges, and therefore represent genuine cohesion. Whereas electrostatic forces result from overall surplus of deficit of electrons, van der Waals forces arise from the attraction between atomic or molecular dipoles. Because they result from processes occurring on a molecular scale, van der Waals forces act over much shorter ranges than electrostatic forces. The strength of van der Waals cohesion depends on the value of the *Hamaker constant,* a characteristic property of the material comprising the particles. The Hamaker constant is effectively independent of pressure, and the dependence on temperature is not well understood but is probably weak. Hence the environment affects van der Waals cohesion mainly through its effect on surface properties. Dahneke (1972) showed that local deformation of the contacting surfaces can increase substantially the strength of inter-particle cohesion.

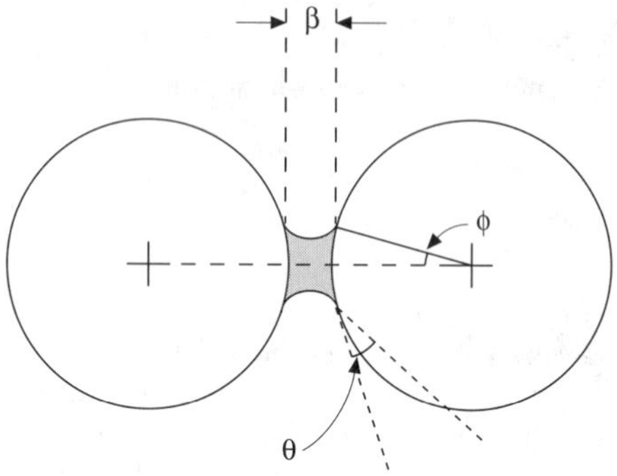

Figure 2. Pendular liquid bridge between solid particles.

Capillary forces, which result from a liquid film on the surface of the particles, are orders of magnitude stronger than van der Waals forces (See also **Capillary Action**). **Figure 2** shows schematically a "pendular bridge" between two idealised particles of diameter d_p. For zero particle separation ($\beta = 0$) and a fully-wetting liquid (contact angle, $\theta = 0$) the attractive force between the particles is (Fisher, 1926):

$$F_c = \frac{\pi d_p \sigma}{\left[1 + \tan\left(\frac{\varphi}{2}\right)\right]} \quad \textbf{(12)}$$

where ϕ is the angle defining the size of the bridge and σ is the tension at the gas/liquid interface. According to **Equation 12**, the cohesive force increases as the liquid bridge becomes smaller, to a limiting value for $\phi \rightarrow 0$ of $\pi d\sigma$. This theoretical result is not applicable to real particles with rough surface (Cheng, 1970); for particles which contact at points of asperity the cohesive force increases with liquid film thickness (Coughlin et al., 1982).

Principles of Gas-solids Separation Devices

Inertial Separators cover devices in which the main property used in recovering particles is their density, so that they are removed by centrifugal action. Deliberate changes in the direction of gas flow causes the particles trajectories to deviate from the gas streamlines, thus concentrating and separating the particles from the gas. Inertial separators are varied in design. Most separators in this category use passive mechanical separation with induced centrifugal motion, such as cyclones. There are however separators in which the centrifugal motion is induced by a rotating propeller. These devices are often used for classification purposes, and are covered in **Classifiers**.

Inertial separators can be used in a wide range of pressure and temperature conditions. Their performance is satisfactory for coarse particles, but as the particle size decreases much below 10 μm the collection efficiency deteriorates very rapidly. Of the various types of design, the *reverse flow cyclone* with tangential entry is the most common type of inertial separators due to its compactness, simplicity of construction and operation, and high throughput. The inlet can be made either truly tangential or wrapped around the body, commonly referred to as "tangential entry" and "volute entry" or "scroll inlet", respectively. The former has a higher collection efficiency, but at the expense of higher pressure drop. There are, of course, other designs using axial entry with vanes installed within the cyclone to induce rotation, direct through-flow of gas without reversing, and blowdown, where a small fraction of the gas is allowed to flow with the collected solids (Strauss, 1975).

Performance analysis of cyclones has been approached on both phenomenological as well mechanistic levels. On a phenomenological level, conventional dimensional analysis leads to the definition of groups which can be used to "scale" cyclone performance for effects of size, throughput and process conditions (see Abrahamson, 1981). The collection efficiency, E, for particles of diameter d_p and density ρ_p from a gas of density ρ_g and viscosity η is given by a relationship of the form

$$\eta = \text{fn}\left[\frac{d_p^2 \rho_p \dot{V}}{\eta D^3}, \frac{\dot{V}\rho_g}{\eta D}\right] \quad \textbf{(13)}$$

while the pressure drop across the cyclone, ΔP, follows a relationship with the general form

$$\frac{\Delta P D^4}{\rho_g Q^2} = \text{fn}\left[\frac{\dot{V}\rho_g}{\eta D}\right] \quad \textbf{(14)}$$

In Equations 13 and 14, $\dot{Q}$ is the volumetric flowrate of gas through the cyclone, and D is the cyclone barrel diameter. Therefore $\dot{Q}/D^2$ represents a characteristic gas velocity, such as the mean inlet gas velocity, as the inlet dimensions are fixed relative to the barrel diameter for a given design. The two independent dimensionless groups in Equation 13 can be regarded as the *cyclone Stokes number,* St_c, and *Reynolds number,* Re_c.

$$St_c = \frac{d_p^2 \rho_p \dot{V}}{\eta D^3} \quad \textbf{(15)}$$

$$Re_c = \frac{\dot{V}\rho_g}{\eta D} \quad \textbf{(16)}$$

The Stokes number describes the tendency of the particle trajectories to deviate from gas streamlines, and the Reynolds number describes the gas flow condition within the cyclone. The above analysis is only valid for low particle concentrations, where the probability of collection of any individual particle is unaffected by other particles present. For industrial cyclones, the gas flow in the cyclone is turbulent and Re_c is very large. It is therefore commonly assumed that the collection efficiency is not affected by the gas flow pattern, and the effect of Re_c is neglected. This leads to the commonly used scaling laws (see for example Strauss, 1975):

$$E = \text{fn}[St_c] \quad \textbf{(17)}$$

$$\frac{\Delta P D^4}{\rho_g \dot{V}^2} = \text{constant} \quad \textbf{(18)}$$

Equations 17 and **18** have been shown to work well for low particle loadings and for particles carrying low levels of electric charge

(Giles, 1982). However, for the cases where particle/particle interactions are significant, including electrical effects, the above correlations are not satisfactory and E is less sensitive to St_c.

The effect of temperature and pressure is in principle reflected in the density and viscosity of the gas. However, care must be taken in the use of **Equation 17** in this case because temperature and pressure can also modify Re_c, which it can then alter the flow condition, thus producing an effect which is not taken into account in **Equation 17**.

The alternative approach to analysis of cyclone performance is by mechanistic modelling, which is necessary to account for the effect of pressure and temperature, as well as for predicting the effects of changing cyclone geometry. Clift et al. (1991) have recently reviewed a number of widely used models of cyclones, i.e. those due to Leith and Licht (1972), Muschelknautz (1970), Dietz (1981) and Mothes and Löffler (1988). It emerges from their analysis that the models of Dietz (1981) and Mothes and Löffler (1988) provide the best agreement with a very wide range of experimental data reported in the literature.

Electrostatic Separators cover devices in which the main separating effect is migration of electrically charged particles in an imposed electric field. These devices have a high collection efficiency for a wide range of particle sizes, and are particularly suitable for submicrometre solids. They have at the same time a high throughput and low pressure drop, which are the main attributes for their wide use in the power generation industry. Operation under extreme conditions of temperature (i.e. above about 800°C) is difficult unless it is accompanied by high pressures, as it will be described below. Apart from the problems of integrity of materials of containment and construction, further considerations limit operation at elevated temperature and pressure. Insulating materials are required to prevent excessive current leakage between the electrodes. Breakdown of the gas between the electrodes must also be avoided. The minimum potential gradient causing ionic breakdown of a gas generally decreases with increasing temperature but increases with pressure. Therefore, at elevated temperature, there is generally a minimum pressure, dependent on gas composition, below which a precipitator cannot be operated. The electric properties of the particulate are also relevant: if it is highly resistive, then collected dust retains its charge and reduces the efficiency of collection of further particulates. It is also necessary for the collected dust to be removed from the precipitating electrode. This is normally achieved intermittently by some mechanical action such as "*rapping*" and the detached particles are allowed to settle to the base of the equipment. For this to be possible, the dust must detach as agglomerates with high terminal settling velocity; thus some cohesiveness of the collected dust is necessary to avoid excessive re-entrainment on electrode cleaning.

The process of electrostatic precipitation involves three stages:

a. charging of suspended particles by corona discharge;
b. migration and deposition of the charged particles under the influence of an applied electrostatic field;
c. removal of the collected material from the collecting electrode and transfer to a suitable receptacle outside the precipitator.

In industrial-scale electrostatic precipitation, stages (a) and (b) are generally combined and the electrostatic field also produces the necessary corona discharge for particle charging. The principles of operation are described by Strauss (1975) and by Böhm (1982). It is important to establish a stable corona having adequate current to provide complete charging of the particles, without getting sparkovers. The onsets of corona voltage and sparkover voltage depend on the geometry of the electrodes, type of gas and the operating temperature and pressure. The corona-starting voltage increases slowly as the pressure is increased, for both positive and negative corona. The effect of pressure on sparkover voltage is more prominent. The sparkover voltage increases rapidly as the pressure is increased above atmospheric, but reaches a maximum and declines, to coincide eventually with the corona-starting voltage (see Robinson, 1971). The pressure at which the corona-starting and sparkover voltages coincide is called the "critical pressure". The interaction of two opposing effects is considered responsible for the existence of the critical pressure point as the pressure is raised (at constant temperature).

a. shorter mean-free paths impede ionisation by collision, and so tend to raise the sparkover level;
b. enhanced photo-ionisation and reduced ion diffusion tend to facilitate streamer propagation from the anode across the gas.

As the pressure increases, the initially dominant first effect gives way to the second, the streamer develops across the gas, and at the critical pressure, spark breakdown ensues.

The effect of temperature has been studied by various authors and the results of the earlier work have been summarised by Robinson (1971). Briefly, the earlier works indicated that, for temperatures up to about 1073 K (800°C), the current-voltage relations for positive corona were a function of relative gas density only, but for negative corona the density and temperature had independent influences for temperatures above 823 K (550°C). At temperatures higher than 800°C, thermal ionisation was considered to become high enough to play a significant role and eventually lead to breakdown. In this case a positive corona was predicted to be required for most types of gases, as the space charge produced by thermal ionisation would generally be positive due to sweeping out of highly mobile electrons.

In more recent years, Bush et al. (1977 and 1979) investigated the range of temperature and pressure in which stable corona discharge may be obtained, and established current-voltage characteristics in a particle-free electrostatic precipitator for dry air, a simulated combustion gas and a substitute fuel gas at temperatures up to 1366K (1093°C) and pressures up to 35.5 bar. Their results show that excessive current and sparkover due to thermal ionisation, anticipated previously, are not encountered within the range tested. The results also reveal the tendency for the positive sparkover voltage to exceed that of the negative sparkover for temperatures above 533 K and low relative air densities (below about 2), a trend which had been observed by earlier workers. However, experience with dust-laden gases has consistently shown that positive coronas are inherently less stable than negative, and also yield a lower collection efficiency due to lower corona currents.

The results of Bush et al. show that the critical pressure increases with temperature. The critical pressure for negative corona discharge is much higher than for positive. It is worth noting that, due to the increase of critical pressures with temperature, the range of pressure for stable corona becomes wider at high temperatures. Thus precipitation could be made more efficient by applying higher voltages as temperature and pressure increase together. However, these conclusions are based on dust-free gases, and the performance evaluations reported so far cannot confirm with confidence their extension to dust-laden gases.

The bulk of the particles passing through the charging zone acquire electric charges of the same polarity as the discharge electrode; i.e. negative for negative corona. Two distinct mechanisms are involved in the charging process:

a. Field charging, caused by bombardment of the particles by ions migrating under the influence of the overall electric field.
b. Diffusion charging, resulting from attachment to the particles of ions which contact them in the course of their random movement through the gas.

Field charging is the dominating mechanism for large particles, typically greater than 1.0 μm. Its rate depends on the field intensity as well as on ion concentration. Diffusion charging is dominant for very fine particles, smaller than 0.1 μm. In industrial precipitators the particle residence time in the charging zone is large, and the particles reach their saturation charge level, which may be estimated from **Equation 19** due to Cochet (1956 and 1961). This equation includes the effect of diffusion charging in the field charging process, hence taking into account both mechanisms

$$q = \pi\epsilon_o d_p^2 E_o \left[\left(1 + \frac{2\lambda}{d_p}\right)^2 + \frac{2}{1 + \frac{2\lambda}{d_p}} \times \frac{\kappa_p - 1}{\kappa_p + 2}\right] \quad \textbf{(19)}$$

where q is the charge acquired by a particle of diameter d_p in the field strength is E_o, κ_p is the relative dielectric constant of the particle, ϵ_o is the permittivity of free space, and λ is the gas mean free path.

Under the influence of the electric field, charged particles migrate towards the collecting electrode. The migration velocity, ω, is commonly calculated from the quasi-steady force balance between Stokes' drag and the applied electrostatic force:

$$F_e = qE_p \quad \textbf{(20)}$$

where E_p is the precipitating electric field. Hence,

$$\omega = \frac{qE_pC}{3\pi\eta d_p} \quad \textbf{(21)}$$

where C is the slip correction factor. In practice, ω is determined empirically because of various complicating factors such as the presence of particle size and shape distributions, turbulence and ionic wind.

The collection efficiency of an electrostatic precipitator is related to the migration velocity by the celebrated Deutsch (1922) model. Details of this model and other technological issues such as the grade efficiency and particle collection and discharge are given in the section on the electrostatic precipitators.

Filters cover all devices in which the gas to be cleaned passes through a porous or permeable medium which collects and retains particles carried by the gas. In a *membrane-type filter,* the passages through the medium are comparable to or smaller than the particulate. Filtration then occurs by mechanical obstruction, and the particulate builds up as a filter "cake" on the upstream face of the filter. The filtration efficiency is essentially absolute, and virtually all the dust is retained on the upstream surface of the filter. In other types of filter, such as *granular bed* or *fibre filters,* the passages through the medium are typically large compared to the particulate to be collected. The particulate is therefore collected initially within the bed by the individual granules or fibres. This process is called depth filtration, and it continues until the passages on the upstream side of the filter narrow down, leading to bridging over the openings where a cake starts to form. In the initial stage of filtration here, the efficiency of the filter and retention of the particulate by the medium is of concern, while in the later stage of filtration, where a cake has formed, the efficiency is high and the increase in the pressure drop across the filter is of concern as in the membrane-type filters.

Filters require regular cleaning as the level of dust builds up on the filter. Various methods of cleaning are used depending on the filter type. Membrane filters are sometimes used just once and are then replaced or they are cleaned *in situ* by some intermittent mechanical process such as vibration or reverse pressure pulse or reverse gas flow rate. A similar cleaning process is applied to those filters which may operate initially by depth filtration, but rely on cake formation as the main filtration mechanism. The granular type filters, on the other hand, employ beds of unbounded filter elements, which are usually removed continuously or intermittently for regeneration of the filter medium. In some cases, the elements are cleaned *in situ,* as for a membrane filter.

The range of materials used as filter medium is very wide indeed; it includes natural and man-made fibres, and porous sintered polymer, metal and ceramic sheets and tubes. The main filter medium requirements are:

a. durability of the medium at process conditions;
b. the ability of the dust to form a cake on the medium;
c. filter performance during the initial stage of cake formation following cleaning;
d. ability of the medium to withstand mechanical stresses during cleaning;
e. form of cake detached during cleaning;
f. "blinding" of the filter medium either by depth filtration or by permanent adhesion of cake to medium.

Appropriate choice of filter medium depends on the process conditions and type of dust material. High temperature applications impose a great constraint on the choice of filter medium. Recent developments in this field have been addressed in a symposium on gas cleaning at high temperatures (see Clift and Seville, 1993).

The process of filtration is analysed in terms of three fundamental aspects of primary collection of dust from the gas, retention or rebound of dust from the collector, and the effect of collected dust on the filter structure.

For primary collection, the important mechanisms in general are: (i) deposition by diffusion; (ii) gravitational settling; (iii) inertial deposition; (iv) direct interception; (v) electrostatic deposition. Magnetic and thermal deposition may also occur depending on the application. The collection efficiency of each individual mechanism is conventionally presented in terms of single fibre or granule collection efficiency, E, defined by

$$E = \frac{\text{number of dust particles collected}}{\text{number of particles in approach volume}} \quad \textbf{(22)}$$

with E a function of dust and granule or fibre size and charge, dust density, gas properties, and bed voidage. For a particular dust size in a given filter, the penetration is defined as:

$$f = \frac{\text{Penetrating dust concentration}}{\text{Challenging dust concentration}} \quad \textbf{(23)}$$

The single particle collection efficiency depends on type and structure of filter medium. Furthermore, its relationship with the penetration is established through mathematical modelling of filtration process. For granular type filters, the single particle collection efficiency of the first four mechanical collection mechanisms has been summarized by Ghadiri et al. (1993). The electrical deposition mechanism is given by Coury (1983), and the magnetic collection by Birss and Parker (1981). For depth filters using fabric medium, various correlations for E have been summarized by Strauss (1975) and Löffler (1971). For granular bed filters, the relationship between E and f has been outlined by Ghadiri et al. (1993) and by Clift et al. (1981). Applications to fluidised bed filters at elevated temperatures have been discussed by Ghadiri et al. (1986). For fabric filters, the relationship between E and f has been analysed by Schweers and Löffler (1994), taking account of the distribution of inter-fibre spacings.

When a dust particle makes contact with a filter element, it is essential that it is retained rather than rebounding, as it may lead to the re-entrainment of the particle. Retention is dominated by short-range adhesive forces. The analysis of retention was initiated by Dahneke (1971), and developed further by Stenhouse and Freshwater (1976), Hiller and Löffler (1978), Clift (1983) and more recently by Ning (1995). The approach is based on an energy balance between the kinetic energy of the approaching particle, E_i, and the detachment energy, E_d. In addition to E_i, the particle acquires further energy in approaching the collector due to the very short van der Waals forces, E_v. The energy at the instant of rebound is then $e^2(E_i + E_v)$, where e is the coefficient of restitution; the kinetic energy lost is due to the dispersion of elastic waves as well as plastic deformation of the particle or the collector. In order to detach from the surface, the particle energy must exceed the detachment energy, E_d. Thus the particle adheres if

$$d_p^3 v_i^2 < \frac{12}{\pi \rho_p}\left[\frac{E_d}{e^2} - E_v\right] \quad \textbf{(24)}$$

The above inequality indicates a critical impact velocity above which a particle will rebound from the filter element. The detachment energy E_d is the sum of energies due to van der Waals forces, and electrostatic and capillary attractions if they are present. Plastic deformation enhances cohesion because of the larger contact area than otherwise possible with elastic deformation alone, and this has recently been modelled by Ning (1995).

Cake formation is essential in those filter types whose medium has large passages and the initial stage of filtration is by depth filtration. Interparticle cohesion is essential for cake formation to occur. The process starts with particle chain formation, leading to arching over the openings of the filter, hence providing the foundation for build up of a cake. If the interparticle cohesion is low or the span of the passages is too wide, the arch may be weak and it may fail under stresses due to the pressure drop in the cake. In this case, the dust particles will break into the filter, i.e. pinhole formation, which causes long term "blinding" of the filter. Ghadiri et al. (1989) showed that pinhole formation can be reduced by electrical enhancement of the interparticle cohesion. Coury (1983) has shown that particles that carry substantial electrical charges of both signs, such as fly-ash, readily form a cake, where uncharged and uncohesive particles, such as coal char, fail to form a cake under otherwise identical conditions. Cake formation is therefore strongly influenced by interparticle forces including van der Waals, electrostatic and capillary forces.

Filter cleaning is another important aspect of filter operation. It should be carried out in such a way that the cake is detached completely from the filter, i.e. without leaving patches of particles still adhering to the filter. At the same time, the cake coherence should be preserved so that the particles are not re-dispersed in the gas stream. Current practice in filter cleaning is largely empirical. However, much of recent work has centred on developing an understanding of the mechanisms involved from which some operational guidelines have emerged (see Clift and Seville, 1993).

References

Abrahamson, J. (1981) *Progress in Filtration and Separation,* Vol. 2, ed. R. J. Wakeman, Elsevier, Amsterdam, pp. 1–74.

Beard, K. V. (1976) *J. Atmos. Sci.,* 33, pp. 851–864.

Birss, R. R. and Parker, M. R. (1981) High Intensity Magnetic Separation, in *Progress in Filtration and Separation,* Vol. 2, ed. R. J. Wakeman, Elsevier, Amsterdam, pp. 171–303.

Böhm, J. (1982) *Electrostatic Precipitators,* Elsevier, Amsterdam.

Bush, J. R., Feldman, P. L. and Robinson, M. (1977) *Development of a High-Temperature/High Pressure Electrostatic Precipitator,* EPA–600/7–77–132.

Bush, J. R., Feldman, P. L. and Robinson, M. (1979) High Temperature High Pressure Electrostatic Precipitation, *Air Pollution Control Ass.* 29, pp. 365–371

Cheng, D. C. H. (1970) *J. Adhesion,* 2, 82.

Clift, R. (1983) *Trans. Inst. Engrs. Aust.* ME8, 181.

Clift, R. Grace, J. R. and Weber, M. E. (1978) *Bubbles, Drops and Particles,* Academic Press, New York.

Clift, R., Ghadiri, M. and Thambimuthu, K. V. (1981) in *Progress in Filtration and Separation, Vol. 2,* ed. R. J. Wakeman, Elsevier, Amsterdam, pp. 75–124.

Clift, R., Ghadiri, M. and Hoffman, A. C. (1991) *A.I.Ch.E.J.* 37(2), pp. 285–289.

Clift, R. and Seville, J. P. K. (1993) *Gas Cleaning at High Temperatures,* Blackie Academic and Professional, London.

Cochet, R. (1956) *Compt. Rend.* 243, 243.

Cochet, R. (1961) *Colloq. Inter. Centre Nat. Rech. Sci. (Paris),* 102, 331.

Coughlin, R. W., Elbirli, B. and Vergara-Edwards, L. (1982) *J. Colloid and Interface Sci.* 87, 18.

Coury, J. R. (1983) Ph.D. Dissertation, University of Cambridge.

Coury, J. R. and Clift, R. (1984) in *Electrical and Magnetic Separation and Filtration Technology,* Koninklijke Vlaamse Ingenieurvereniging, Antwerp, p. 27.

Dahneke, B. (1971) *J. Colloid and Interface Sci.* 37, 342.

Dahneke (1972) *J. Colloid and Interface Sci.* 40, 1.

Davies, C. N. (1945) *Proc. Phys. Sci.* 57, 259.

Deutsch, W. (1922) *Ann. Physik,* **68**, 335.

Dietz, P. W. (1981) *Collection Efficiency of Cyclone Separators, A.I.Ch.E.J.* 27, pp. 888–892.

Fisher, R. A. (1926) *J. Agric. Soc.* 16, 492.

Ghadiri, M., Cleaver, J. A. S. and Seaton, R. (1989) Electrically Enhanced Gas Filtration Using Microsieves as Filter Media (1989) 1st European Symposium on Separation of Particles from *Gases, PARTEC,* Nuremberg, 19–21 April 1989.

Ghadiri, M., Seville, J. P. K. and Clift, R. (1986) The Use of Fluidised Beds to Filter Gases at High Temperatures, *I. Chem. E. Symp. Series No. 99,* pp. 351–361.

Ghadiri, M., Seville, J. P. K. and Clift, R. (1993) Fluidised Bed Filtration of Gases at High Temperatures, *Trans. I. Chem. E.* 71, Part A, pp. 371–381.

Giles, W. B. (1982) Cyclone Scaling Experiments, 4th Symp. on the Transfer and Utilisation of Particulate Control Technology, 11–15 Oct. 1982, Houston, USA.

Henry, R. F., Saxena, S. C. and Podolski, W. F. (1982) Particulate Removal from High-Temperature, High-Pressure Combustion Gases, *Argonne National Laboratory Report,* ANL/FE–82–11.

Hiller, R. and Löffler, F. (1978) *Deposition and Filtration of Particles from Gases and Liquids,* SCI, London, p. 81.

Leith, D. and Licht, W. (1972) The Collection Efficiency of Cyclone Type Particle Collectors—a New Theoretical Approach, *A.I.Ch.E. Symp. Ser.* 68 (126), pp. 196–206.

Löffler, F. (1971) *Collection of Particles by Fibre Filters in Air Pollution Control, Part I,* W. Strauss, Ed. Wiley-Intersciences, pp. 337–375.

Mothes, H. and, F. (1988) Prediction of Particle Removal in Cyclone Separators, *Int. Chem. Eng.* 28 (2), pp. 231–240.

Muschelknautz, E. (1970) Design of Cyclone Separators in the Engineering Practice, *Staub-Reinhalt. Luft,* 30 (5) pp. 1–12.

Ning, Z. (1995) Ph.D. Dissertation, Aston University, Birmingham.

Prasher, C. L. (1987) *Crushing and Grinding Process Handbook,* Wiley, Chichester.

Robinson, M. (1971) Electrostatic Precipitation in *Air Pollution Control, Part I,* W. Strauss, ed., Wiley-Intersciences.

Schiller, L. and Nauman, A. Z. (1933) *Ver. Deut. Ing.* 77, pp. 310–320.

Schweers, E and Löffler, F. (1994) Realistic Modelling of the Behaviour of Fibrous Filters through Consideration of Filter Structure, *Powder Technology,* 80 (3), pp. 191–206.

Stenhouse, J. I. T. and Freshwater, D. C. (1976) *Trans. I. Chem. E.* 54, 95.

Strauss, W. (1975) *Industrial Gas Cleaning,* 2nd ed. Pergamon Press, Oxford.

Theodore, L. and Buonicore, A. J. (1976) *Industrial Air Pollution Control Equipment for Particulates,* CRC Press, Cleveland, Ohio.

Leading to: *Centrifuges; Classifiers; Filtration; Electrostatic separator, Scrubbers*

M. Ghadiri and R. Clift

GAS STEAM TURBINE UNITS (see Gas turbine)

GAS THERMOMETER (see Thermometer)

GAS TRANSFER PUMPS (see Vacuum pumps)

GAS TURBINE

Following from: *Turbines; Power plants*

The gas turbine is a turbine in which potential energy of heated and compressed gas is converted into kinetic energy as a result of its expansion in the tubine blading. Subsequently, the energy is converted into mechanical work on the rotating shaft. The process of expansion of the gas in a gas turbine takes place in one, or more often several, stages. As a rule, the total number of stages in a gas turbine is not large (not more than 3—5). In this aspect, gas turbines differ greatly from steam turbines (see **Steam Turbine**) in which the number of stages may reach several dozen. Reaction stages have higher efficiency than impulse stages and are therefore those mostly used in gas turbines. Either combustion products of organic fuels (liquid or gaseous) or clean air or other clean gases (for example inert gases or their mixtures), heated in a special heat-exchanger heater, are used as the gas turbine working fluid. In the latter case, the clean air and other clean gases should not get mixed with the fuel combustion products. Mixtures of organic fuel combustion products with water or steam also can be used as gas turbine working fluids.

Gas turbines are usually components of gas turbine engines or gas turbine units. Therefore the term gas turbine often denotes both the gas turbine engine and the gas turbine unit. Here, only one term—a gas turbine unit—will be used whenever it does not result in the distortion of the meaning. In addition to the gas turbine, the simplest open-cycle gas turbine unit (see **Figure 1a**) consists of a compressor, 2, (see **Compressor**) with intake system, 1; combustion chamber, 3, (see **Combustion Chamber**); auxilliary systems (in particular the fuel feed system), 4. The air from the atmosphere passes from the intake system, 1, into the compressor, 2, where it is compressed and then delivered into the combustion chamber, 3, where the fuel is burnt. The combustion products (gas) enter the turbine, 5, and are expanded to a pressure close to atmospheric and after that they are exhausted into the atmosphere through the exhaust unit, 6. Some portion of the turbine power is used for driving the compressor, 2, for driving the auxiliary units (pumps, electric generators, mechanisms etc.) of the complete gas turbine unit and for overcoming hydraulic resistance in the gas turbine unit. The remaining power is used for driving the payload mounted on the turbine shaft, 7.

In all modern gas turbine units the process of heat input takes place at constant working fluid pressure in the combustion chamber, therefore these gas turbine units are defined as working through the cycle p = const.

Within the p = const gas turbine unit cycle the actual compression work in the "temperature-entropy" diagram (**Figure 1b**) is proportional to the area 1′ka′2, adiabatic work l_{k0} is measured by the area 1k′a′2′. Thus $l_k > l_{k0}$ and the difference $l_k - l_{k0}$ is proportional to the area 1k′k1′. The predicted expansion work l_{t0}

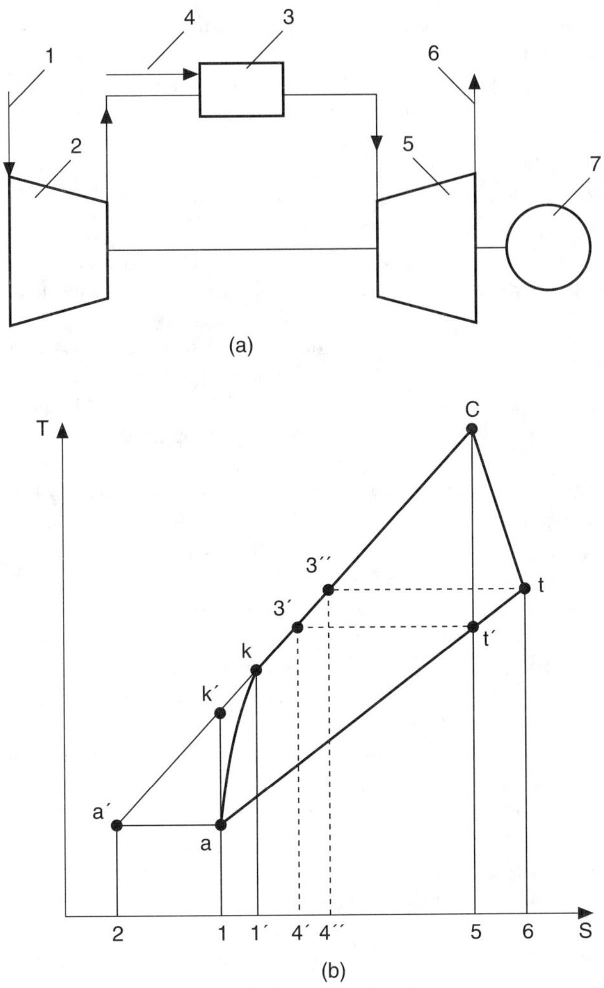

Figure 1. Components and T/S diagram for gas turbine units.

is proportional to the area 4′3′c5, the actual work of the turbine l_t is measured by the area 4″3″c5. Therefore $l_t < l_{t0}$ and the difference $l_{t0} - l_t$ is proportional to the area 4′3′3″4″. The specific useful work of the actual cycle is equivalent to the difference of areas 1′kc5 and 1at6. This difference is considerably smaller than the area akct, which is limited by the actual processes of the cycle (in the ideal cycle the work l_{eid} proportional to the area ak′ct′ is limited by the ideal processes).

The specific useful work of the simplest gas turbine unit actual cycle $l_e \approx \gamma RT_a [\vartheta(1 - 1/\chi)\eta_t - (\chi - 11)$, where R is the gas constant and γ is the adiabatic exponent for the process of compression in the compressor (or similar values for the expansion process in the turbine which are approximately assumed to be identical). T_a is the initial air temperature (corresponding to point "a", in **Figure 1b**); $\vartheta = T_c/T_a$; T_c is the initial temperature of the combustion products (gas) (point "c"); $\chi = \pi^{(\gamma-1)/\gamma}$ is a function of the pressure ratio $\pi = p_k/p_a$ in the cycle and η_t and η_k are the efficiencies of the turbine and the compressor. The gas turbine unit capacity is $N_e \approx l_e\dot{M}$, where $\dot{M}$ is the air (gas) flow rate through the compressor (the turbine). The simplest gas turbine unit cycle efficiency is

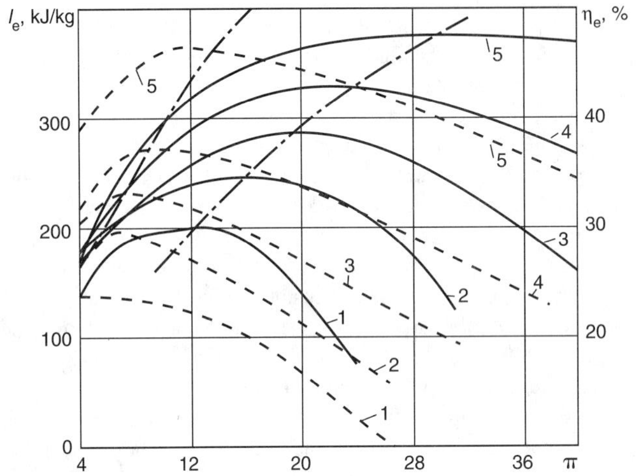

Figure 2. Variation of specific useful work l_e and unit efficiency η_e with $\pi = T_c/T_a$ at various T_c values.

$$\eta_e = [\vartheta(1 - 1/\chi)\eta_t - (\chi - 1)/\eta_k]/[\vartheta - 1 - (\chi - 1)/\eta_k].$$

The cycle efficiency η_e as well as specific work l_e increases with the growth of ϑ and with the efficiency of the turbine and the compressor. Approximate dependence of η_e (solid lines) and l_e (dashed lines) on π at T_c = 1000,1100,1200,1300,1500 K (curves 1, 2, 3, 4, 5) are given in **Figure 2**. Calculations were performed without taking into account losses connected with the turbine cooling. η_e and l_e grow considerably with the increase of T_c. From the figure it also follows that it is necessary to increase π with the increase of ϑ in order to employ the advantages connected with the increase of ϑ (or T_c). This is the current tendency of practical application of simple-cycle gas turbine units. The efficiency of such gas turbine units is not high (for example for stationary industrial or heavy duty gas turbine units for power, $\eta_e \approx$ 28–33%).

In the cycle of a *regenerative gas turbine* unit (**Figure 3**), the air entering through the intake system and compressed in the compressor, 2, passes into the heat enchanger, 6, where it is heated by gases issuing from the turbine, 5. The fuel flow rate, 4, in the combustion chamber, 3, can be decreased because the waste gas heat is used for heating the air. The advantage of schemes using heat regeneration is the possibility to maintain the efficiency of

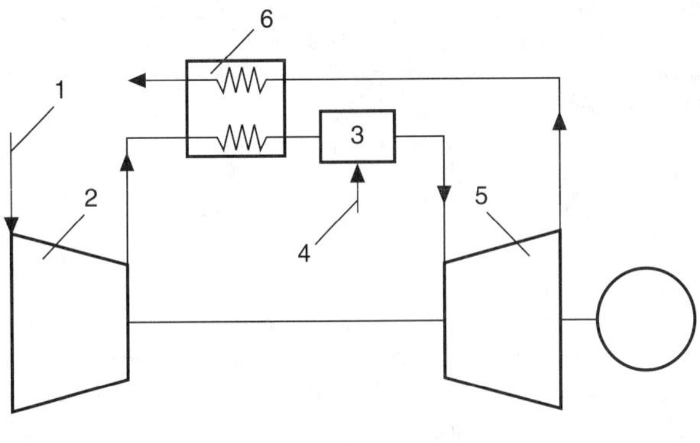

Figure 3. Regenerative gas turbine unit.

the unit practically constant and equal to the efficiency at design conditions in a wide range of loads (from 50 to 100%). The predicted efficiency of the turbine unit with regeneration at steady gas temperature T_c is higher than the efficiency of the simple-cycle gas turbine unit. Most gas turbine engines for wheeled vehicles and crawler-type vehicles are designed and constructed with the regenerative cycle because other possibilities of increasing the efficiency of these engines are rather limited.

For other types of gas turbine units, above all for stationary gas turbine units, regeneration is not applied in practice. This is connected with the large size of the heat-exchanger and with the necessity to use long and structually complicated piping systems, which causes additional hydraulic resistance. High specific power and efficiency (up to 50%) may be obtained in multicomponent gas-turbine units. In such plants, the value of π may be considerably higher (up to 22–55 and even 120) than in regenerative-type gas-turbine units. In simple-cycle gas-turbine units, the number of turbines and compressors is larger. However, they have no heavy and large-size regenerator and are less metal-intensive than regenerative-type gas-turbine units.

Closed-cycle gas-turbine units differ from the *open-cycle gas-turbine* units in construction. In addition to the regenerator, they also have a gas cooler at the exit which lowers the gas temperture to its initial level. In closed-cycle gas-turbine unit there is no combustion chamber. Instead, they have a heater (a furnace or a nuclear reactor) in which the working fluid (air, inert gases or their mixtures) do not mix with the combustion products of the fuel.

There are also combined units with gas and steam turbines—*steam-gas units* and *gas-steam units*—in which two working fluids, namely gas and steam, are used in one power-generating system. These units have better characteristics than modern steam-turbine systems and gas-turbine units. These combined gas and steam-turbine units make it possible to combine, in one thermal cycle, the high-temperature heat input (characteristic for gas-turbine units) and low-temperature (as in a steam-turbine unit condensor) heat removal. As a result, they provide conditions for obtaining high thermal efficiency of the cycle, thus high economy of energy generation. The efficiency of these combined units (steam-gas and gas-steam) may reach 50% at high values of the gas-turbine unit parameters; therefore, they are considered promising systems for power engineering. The high efficiency of turbine units with steam and gas turbines facilitate the solution of a very important problem of decreasing the amount of dangerous effluents with the aim of environmental protection.

The numerous combined turbine units in operation or under development differ by their heat schemes and the equipment used. They are classified mainly in accordance with the principle by which the fuel combustion heat is utilized.

If the main portion of heat is supplied with the fuel in the furnace of the boiler of the steam-turbine part these combined units are termed steam-gas units. If the main portion of heat is supplied with the fuel to the gas-turbine unit combustion chamber, these units are termed gas-steam units.

On the basis of the principle of interaction of working fluids, the combined systems are divided into two main groups: *separate circuit* units in which steam-water and gaseons working fluid move along separate paths and interact only across the heat exchanger surface, and *contact-type* units in which the combustion products mix with the steam-water working fluid prior to expansion in the gas-steam turbine.

Steam-gas units have separate circuits. There are steam-gas units of three types: those with a high head steam generator, those with a low head steam generator and those units using the waste heat of the gas turbine for heating the feed water in the steam-turbine unit. In the steam-gas units of the first type, the high head steam generator, is usually combined with the gas-turbine unit combustion chamber and the fuel burns at high pressure. In order to decrease the temperature of the waste gases, a gas-water heater is usually mounted downstream of the gas turbine. The heater partially replaces steam regeneration in the steam-turbine section of the unit. In steam-gas units with a low head steam generator, the waste gases from the gas-turbine unit pass into the furnace of a conventional boiler and are used for burning additional fuel, which may be not only liquid or gaseous but also solid, such as coal. There is also a gas-water heater downsteam of the low head steam generator. In such plants, the gas turbine unit is a sort of topping plant; this scheme is therefore used for reconstructing steam-turbine plants operating at lowered steam parameters. In the units of the third type, the exhaust gases from the gas-turbine unit are directed into the gaseous-feed water boiler where the heat of the exhaust gases is utilized. This heat may be sufficient to exclude regenerative feed water heaters. However, greater economic effect may be reached if the steam-gas system consists of multicomponent gas-turbine and steam-turbine units. In such a scheme, the condensate is heated in intermediate air coolers of the gas turbine unit and in the heater at the gas turbine unit exit. In order to create this type of combined system, it is not necessary to considerably change the boiler and the steam turbine regenerative system.

Gas-steam units may have either separate circuits or contact schemes. In gas-steam units with separate circuits, the fuel—or its main portion—is burnt in the gas-turbine unit combustion chamber. Sometimes the thermal cycle without fuel heat input to the steam-water working fluid is referred to as the *binary gas-steam cycle*. In some cases, a small amount of fuel (up to 15–20% of the total fuel flow rate of the gas circuit) is fed upsteam of the waste heat boiler. In the contact-type steam-gas units, water or steam is supplied to the high pressure circuit (downsteam of the compressor, into the combustion chamber or upsteam of the turbine). In gas-steam units with steam supply, a waste heat boiler is used for steam generation and utilizes a part of the turbine exhaust gases heat. This supply of water or steam increases the working fluid flow rate through the turbine as compared with the air flow rate through the compressor and thus increases its output. As the amount of power required for pumping the water is small, the capacity of the unit increases not less than 100 per cent. The disadvantage of the contact type gas-steam units is the requirement of chemical cleaning of the water which is exhausted with the gas.

One of the most promising designs is the steam-gas turbine which uses low calorific gas generated in the process of coal gasification as the gas-turbine unit fuel. The schemes of this type of units are essentially similar to the schemes of steam-gas units with high head steam generators and steam-gas units with low head steam generators. Their main structural difference is the inclusion of the gasification system with purification of the fuel gas generated of unburnt particles and sulphur.

Among other fields of application one should mention the application of driving stationary gas-turbine units in gas pipelines.

Superchargers and gas-turbine units comprise the gas-pumping system. Some of these system were created on the basis of existing aircraft and marine gas-turbine engines.

Gas-turbine units for power generation are used mainly as peak-load units, that is the units operating 500–1500 hours per year during peak loads in the power system. The main advantages of gas-turbine units which make them applicable in power engineering are their higher manoevrability (the possibility of quick start and reliable operation under cyclic conditions) as compared with steam-turbine units, simplicity of automation, compactness, and independence of water sources.

Gas-turbine units which are elements of steam-gas units and gas-steam units use gaseous and liquid fuel and have unit capacity from 100 to 600 mW. These units are especially widely used in the USA.

About 10 per cent of all gas-turbine units produced in the world are used in different kinds of ships. The *marine gas-turbine* units may be competitive with Diesel engines as far as their main economic characteristics are concerned. Both steam-gas and gas-steam units may be used as marine propulsion units. Specific fuel consumption of existing marine gas-steam units of about 40 mW capacity is not larger than 240 g/(kW hr). Also in use are combined marine propulsion systems consisting of a gas-turbine engine and a Diesel engine which are not combined by a common thermodynamic cycle.

Automotive gas-turbine engines are considered as a good alternative to piston-type internal combustion engines. Their further progress is closely linked with the development of ceramic blades and disks.

Gas-turbine engines are widely used in aviation. Their monopoly is connected with their high specific power, thrust and high efficiency.

Gas turbines and gas-turbine units are characterised by a great variety of designs (depending on their application). However, with the exception of automotive gas-turbine engines, all of them have one common feature; as a rule, they all need cooling. The air bled from the compressor outlet is generally used as a cooling medium. Practically the whole turbine flow passage is air cooled (see **Figure 4**). The cooling should be intensive because the thermal resistance of metal alloys is limited and ceramic materials for turbine blades are still in the process of development.

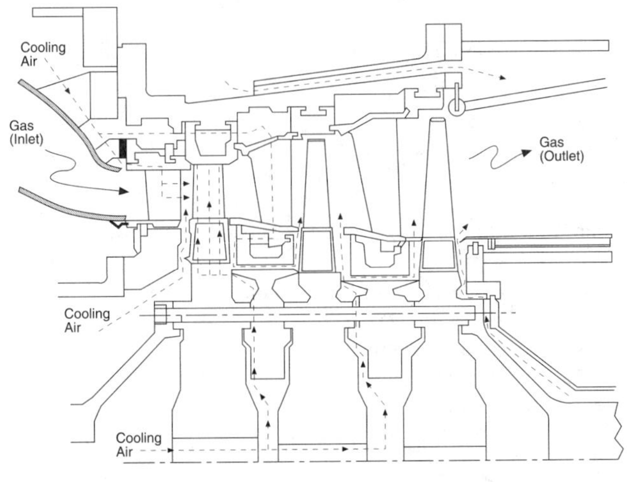

Figure 4. Cooling of gas turbine.

The general trend of research and development is to increase the *turbine blade cooling* efficiency which is determined by the depth of cooling $\theta = (T_g - T_w)/(T_g - T_a)$, where T_g is the temperature of the gas flowing over the blade, T_a is the coolant temperature and T_w is the temperature of the blade external surface. Dependences like those given in **Figure 5** are used to evaluate the cooling efficiency (g is the relative cooling air flow rate, which is determined as the ratio of the flow rate of the air used for cooling to the flow rate of the air at the compressor inlet). The blade with a porous sheath has the highest cooling efficiency (curve, 7); the cooling efficiency of the blade with film cooling is somewhat lower. Blades with film cooling (curve, 6) have several hundreds of small diameter holes (0.4 to 0.6 mm) in the thin-walled sheath. Curves 1 through 5 illustrate the cooling efficiency of blades cooled using different methods of convective cooling.

Intensification of heat transfer in the internal passages of cooled blades and reduction of additional losses due to cooling may increase the efficiency of gas-turbine units with cooled turbines.

Thermal conditions of gas-turbine blades differ somewhat from those of stationary surfaces across which a gas flows. They are characterized by a high turbulence level of the incoming flow, by surface curvature, pressure gradient and body forces. Transition from laminar to turbulent flow in the boundary layer at the turbine blade surface is very important for determining heat transfer. There are factors which either accelerate this process or retard it.

Figure 6 presents experimental data for the dependence of the dimensionless heat flux (Stanton number St) on the Reynolds number Re_x at the flat plate for various turbulence levels of the incoming flow Tu_∞.

An increase of Tu_∞ from 1.3 to 7.5% decrease the value of the Reynolds number at which turbulent flow is developed from $Re_x^* = 10^6$ to $Re_x^* < 10^5$.

As compared with the flat plate case, the flow around the gas-turbine blade is characterized by a high level of flow acceleration over the suction surface. This high flow acceleration retards transition from laminar to turbulent flow, that is, its action is opposite to the incoming flow turbulence level.

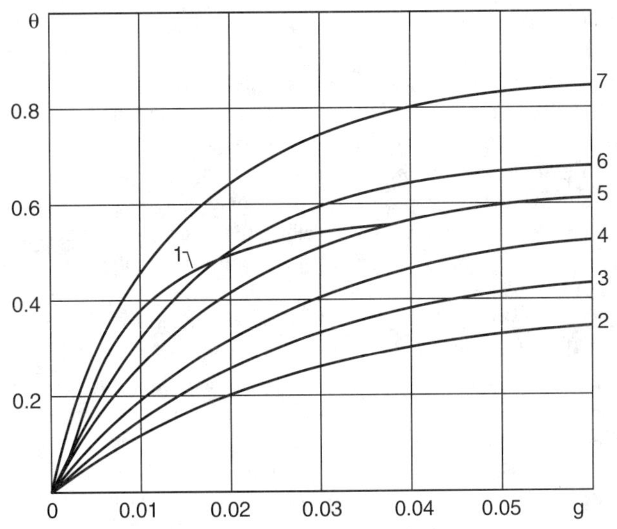

Figure 5. Depth of cooling (θ) as a function of relative cooling air flow rate g.

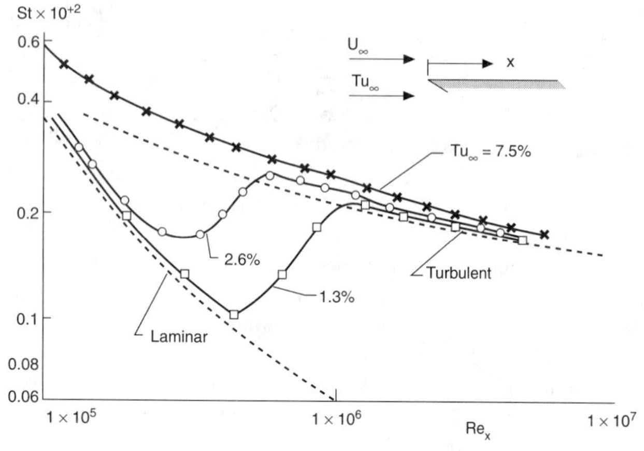

Figure 6. Effect of free stream turbulence level on heat transfer from a flat plate.

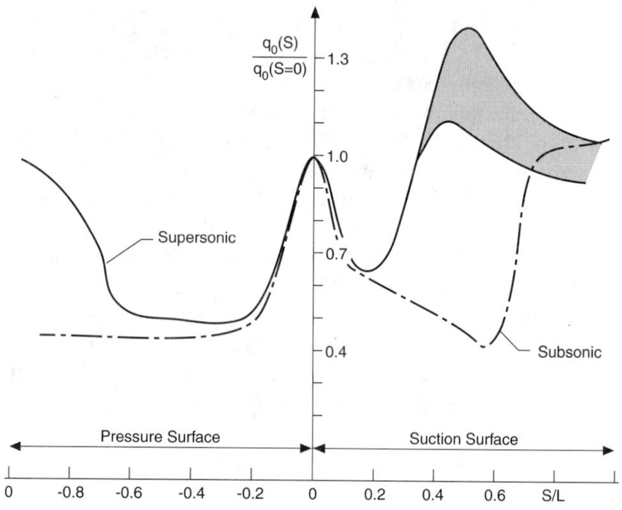

Figure 7. Comparison of heat flux profiles for supersonic and subsonic flows over turbine blades.

To describe accelerated flows, the dimensionless acceleration parameter $K = (\beta\nu)/U_e^2 = (\beta/U_e)/Re(1)$ is used where Re(1) is the unit Reynolds number $Re(1) = U_e/\nu$ $\beta = (dU_e/ds)$ is the gradient of velocity of the flow in the interblade channel and ν is the kinematic viscosity.

The acceleration parameter K may be defined without solving the boundary layer equation on the basis of known distributions of the flow gas dynamic parameters in the interblade channel.

The analysis of results of temperature measurements on the rotating blades shows that the determining factor is not the incoming flow turbulence level Tu_∞, but its local value Tu_e in a given interblade channel section. It may be assumed that

$$\frac{Tu_e}{Tu_\infty} = \frac{U_\infty/\sqrt{k_\infty}}{U_e/\sqrt{k_e}},$$

where k_e is the local value of turbulent kinetic energy, k_∞ is the same parameter in the incoming flow. Calculations show approximately 10 per cent decrease in k_e in the area from the leading edge to the trailing edge of the turbine blade. This allows the determination of the local turbulence intensity using the following relationship for inviscid flow: $Tu_e/Tu_\infty = U_\infty/U_e$.

Figure 7 presents the comparison of the heat flux q_0 distribution along the same profile for supersonic and subsonic flows. The variation of the heat flux along the relative coordinate S/L is complex and non-monotonic. From the analysis of the results obtained it follows that intensive turbulent pulsation of the flow around a supersonic blade accelerates transition to turbulent flow in the boundary layer at the blade surface. In the case of a subsonic blade, especially at the suction surface, this transition is retarded, which results in a considerable difference in the heat transfer level.

Suppression of turbulence in the boundary layer is observed only in the case when the acceleration parameter K is higher than 3×10^{-6}. If the free stream turbulence is not high, even relaminarization of already developed turbulent boundary layer is possible.

References

Horlock, J. H. (1966) *Axial Flow Turbines,* Butterworths, London, 1966.

Kostyuk, A. and Frolov, V. (1988) *Steam and Gas Turbines,* Moscow, Mir.

Heat Transfer and Fluid Flow in Rotating Machinery, (1987). Wen-jei Yang, ed., Washington etc., Hemisphere.

Ed. A. Manushin and Ju. V. Polezhaev

GAS TURBINE BLADE COOLING (see Film cooling)

GAS TURBINE HEAT RECOVERY BOILER (see Heat recovery boilers)

GAUGE PRESSURE (see Pressure)

GAUSSIAN DISTRIBUTION

The Gaussian distribution is a continuous distribution having the probability density function f(x) expressed as

$$f(x) = \frac{1}{\sigma\sqrt{2\pi}} e^{-\frac{1}{2}\left(\frac{x-\mu}{\sigma}\right)^2} \qquad (\sigma > 0, -\infty < x < \infty)$$

Where μ is the mean and σ is the standard deviation of the distribution. This distribution is also called the *normal distribution* and a random variable having this distribution is said to be normally distributed. The curve of f(x) is bell shaped and is symmetric with respect to μ as illustrated in **Figure 1**.

The corresponding probability distribution function F(x) is represented by the integral

$$F(x) = \frac{1}{\sigma\sqrt{2\pi}} \int_{-\infty}^{x} e^{-\frac{1}{2}\left(\frac{u-\mu}{\sigma}\right)^2} du \qquad (\sigma > 0, -\infty < x < \infty)$$

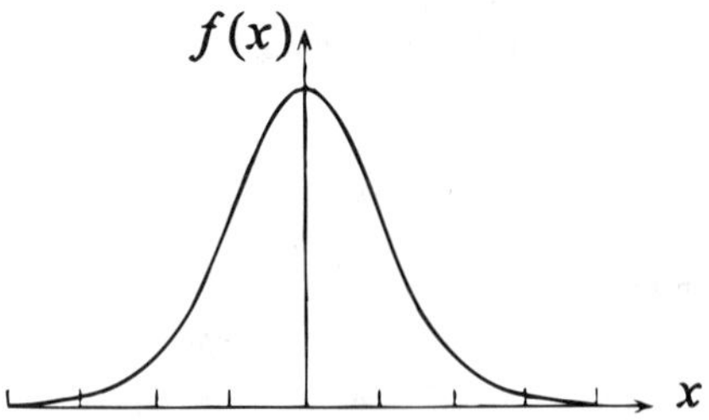

Figure 1. Curve of f(x).

which cannot be solved analytically but can be numerically evaluated for any x. Tabulated values of F(x) abound in standard textbooks on statistics. This is an important distribution because a host of random variables of practical interest such as the distribution of height and intelligence among males and females in an adult population, the lifetime of light bulbs and electric batteries, repeat measurements on a standard measurement, etc. are normal, approximately normal or can simply be converted into normal variables. The Gaussian distribution is also a useful approximation of more complicated distributions and it appears in the mathematical proofs of various statistical tests.

M.C. Tezock

GAUSSIAN ELIMINATION

A method of successive eliminations of unknowns for solving a system of linear equations. Consider the system

$$a_{i1}x_1 + a_{i2}x_2 + \cdots + a_{in}x_n = b_i \qquad (i = 1, 2, \ldots, n)$$

and let $A = (a_{ij})^n_{i,j=1}$. Gaussian elimination operates as follows. If the element (called pivot) $a_{11} \neq 0$, one eliminates x_1 from the last $(n - 1)$ equations by subtracting the first equation, multiplied by $m_{i1} = a_{i1}/a_{11}$, from the ith equation $(i = 2, \ldots, n)$. The last $(n - 1)$ equations form a system in the unknowns $x_2, x_3, \ldots, x_n$ given by

$$a^{(2)}_{i2}x_2 + a^{(2)}_{i3}x_3 + \cdots + a^{(2)}_{in}x_n = b^{(2)}_i \qquad (i = 2, 3, \ldots, n)$$

where $a^{(2)}_{ij} = a_{ij} - m_{i1}a_{1j}$ and $b^{(2)}_i = b_i - m_{i1}b_1$, $(i, j = 2, 3, \ldots, n)$.

Similarly, if $a^{(2)}_{22} \neq 0$ (again called pivot), using $m_{i2} = a^{(2)}_{i2}/a^{(2)}_{22}$, one can eliminate x_2 from the last $(n - 2)$ equations and get a system in the unknowns $x_3, \ldots, x_n$. The coefficients of this system are given by $a^{(3)}_{ij} = a^{(2)}_{ij} - m_{i2}a^{(2)}_{2j}$ and $b^{(3)}_i = b^{(2)}_i - m_{i2}b^{(2)}_2$, $(i, j = 3, \ldots, n)$.

If all the pivots $a_{11}, a^{(2)}_{22}, a^{(3)}_{33}, \ldots$ are nonzero, then the procedure continues. In the kth step, the last $(n - k)$ equations become:

$$a^{(k+1)}_{ik+1}x_{k+1} + \cdots + a^{(k+1)}_{in}x_n = b^{(k+1)}_i \qquad (i = k + 1, \ldots, n)$$

where $a^{(k+1)}_{ij} = a^{(k)}_{ij} - m_{i\,k}a^{(k)}_{kj}$ and $b^{(k+1)}_i = b^{(k)}_i - m_{i\,k}b^{(k)}_k$, $(i, j = k + 1, \ldots, n)$ with $m_{ik} = a^{(k)}_{ik}/a^{(k)}_{kk}$.

Gaussian elimination is performed in $(n - 1)$ steps. Collecting the first equation from each step one obtains the following upper triangular system:

$$a_{11}x_1 + a_{12}x_2 + \cdots + a_{1n}x_n = b_1$$

$$a^{(2)}_{22}x_2 + \cdots + a^{(2)}_{2n}x_n = b^{(2)}_2$$

$$\vdots$$

$$a^{(n)}_{nn}x_n = b^{(n)}_n$$

If $a^{(n)}_{nn} \neq 0$, there is a unique solution given as:

$$x_n = \frac{b^{(n)}_n}{a^{(n)}_{nn}} \quad \text{and}$$

$$x_k = \frac{b^{(k)}_k - \sum^n_{j=k+1} a^{(k)}_{kj}x_j}{a^{(k)}_{kk}}$$

$$(k = n - 1, n - 2, \ldots, 1)$$

Otherwise, if $a^{(n)}_{nn} = 0$, then there is no solution if $b^{(n)}_n \neq 0$, while there are infinitely many solutions if $b^{(n)}_n = 0$.

If for some k, the pivot $a^{(k)}_{kk} = 0$, then the elimination procedure cannot be continued unless an alternate pivot is chosen for the kth step. To this end, one looks for a nonzero coefficient of x_k in equations $k, k + 1, \ldots, n$, and if it is found in equation $j > k$, one interchanges equations j and k. This process is called *pivoting*.

The Gaussian elimination method can be carried out without pivoting if (i) matrix A is diagonally dominant i.e. $|a_{ii}| \geq \sum^n_{j=1,\, j\neq i} |a_{ij}|$ $(i = 1, 2, \ldots, n)$, or (ii) if A is symmetric and positive definite i.e. $A^\top = A$ and $x^\top A_x > 0$ for all vectors $x \neq 0$, ($\top$ denotes the transpose).

When applied to an $n \times n$ system, the Gaussian method takes $n(n^2 - 1)/3$ multiplications, $n(n - 1)/2$ divisions and $n(n^2 - 1)/3$ additions or subtractions.

References

Dahlquist, G and Björk, A. (1974) *Numerical Methods,* Prentice-Hall, N.J.

Golub, G. H. and Loan, C. F. van, (1983) *Matrix Computations,* North Oxford Acad.

M.K. El-Daou

GAUSSIAN MODEL OF PLUME (see Plumes)

GAUSSIAN QUADRATURE METHOD

The Gaussian quadrature method is an approximate method of calculation of a certain integral $I = \int^b_a y(x)\,dx$. By replacing the variables $x = (b - a)t/2 + (a + b)t/2$, $f(t) = (b - a)y(x)/2$ the desired integral is reduced to the form $\int^1_{-1} f(t)\,dt$.

The Gaussian quadrature formula is

$$\int_{-1}^{1} f(t)dt \simeq \sum_{i=1}^{n} A_i f(t_i) \qquad (1)$$

The cusps t_i of the Gaussian quadrature formula are the roots of a Legendre polynomial of degree n, $P_n(t)$. The *Legendre polynomial* has exactly n real and various roots in the interval $(-1, 1)$. The weights A_i of the Gaussian quadrature formula are defined by

$$A_i = \frac{2}{(1 - t_i^2)[P_n'(t_i)]^2}. \qquad (2)$$

Given in the table are the cusps and weights of the Gaussian quadrature formula for the first five values n.

Table 1.

n	Cusps	Weights	n	Cusps	Weights
2	±0.577350	1	4	±0.339981	0.652145
				±0.861136	0.347855
3	0	8/9	5	0	0.568889
	±0.774597	5/9		±0.538469	0.478629
				±0.906180	0.236927

Gaussian quadrature formula is exact for an arbitrary polynomial of degree not higher then $2n - 1$. The remainder term of Gauss's formula R_n for the integral $\int_a^b y(x)\,dx$ is expressed as follows

$$R_n = \frac{(b-a)^{2n-1} n!}{(2n+1)(2n)!^3} y^{(2n)}(\xi); \qquad a \le \xi \le b. \qquad (4)$$

The Gaussian quadrature method is applied when a subintegral function is smooth enough and a gain in the number of cusps is essential (for instance, in calculating multiple integrals as iterated integrals).

The Gaussian quadrature formula is widely used in solving problems of radiation heat transfer in direct integration of the equation of transfer of radiation over space. The application of Gauss's formula in this case works very well especially when the number of intervals of spectrum decomposition is great.

V.N. Dvortsov

GAUSS-OSTROGRADSKY FORMULA (See Integrals)

GAY-LUSSAC EXPERIMENT (see Internal energy)

GAY-LUSSAC's LAW

Following from: *Thermodynamics; Perfect gas*

One of the main (along with Charles', Boyle-Mariotte's and Avogadro's laws) empirical laws of ideal gases was established by J. Gay-Lussac in 1802. According to the Gay-Lussac Law, at constant (and low) pressure P the volume of the given gas mass M varies linearly with temperature: $V_t = V_0 + A\Delta T$, where V_t and V_0 are gas volumes at temperatures T and T_0, $\Delta T = T - T_0$, $A = V_0\alpha_p$ = const, $\alpha_p = V^{-1}(\partial V/\partial t)_p$ is the isobaric volume expansion coefficient, i.e. $V_T = V_0 + V_0\alpha_p\Delta T = V_0(1 + \alpha_p\Delta T)$. The quantity α_p for gas is found to be independent on gas nature and pressure, but dependent on temperature; in this case at $T_0 = 0°C$, $\alpha_p = 273.15^{-1}°C^{-1}$. The Gay-Lussac Law (GLL) describes the isobaric process of an ideal gas.

The GLL has played principal part in establishing the notion of absolute temperature and in deriving the universal equation of ideal gas state—the Clapeyron (Clapeyron-Mendeleyev) equation. At $T_0 = 0°C$ and $T = -273.15°C$ the ideal gas volume $V_t = 0$, i.e. the linear isobar of ideal gas vanishes at this temperature, thus intersecting the temperature axis. When temperature is counted from this thermodynamically minimally possible level, the GLL provides already not merely linear, but proportional dependence of gas volume on this absolute temperature T. The comparison, in this case, of the GLL with the other empirical ideal-gas laws (Charles', Boyle-Mariotte's and Avogadro's laws) allows a general equation of state of ideal gases $pV = M\tilde{M}^{-1}RT$, where $\tilde{M}$ is the molecular mass of gas, R is the universal gas constant (R = 8.314 J/mole K).

Like the other ideal-gas laws, including the *Clapeyron-Mendeleyev law,* the GLL, which is a particular case of the latter one, is valid at low pressures (far from critical one), when the influence of the actual size of particles and their forces of interaction is absent. The GLL, that was established by empirical method, and the notions obtained on its basis have been subsequently substantiated by using the molecular-kinetic theory. In particular, the physical meaning of absolute temperature T consists in the fact, that this quantity is proportional to the mean kinetic energy of translational motion of particles.

Of great importance in developing the universal equation of state of ideal gases and, in particular, in discovering *Avogadro's law,* according to which under equal conditions (p and T values) the number of moles in a unit of volume of any gas is the same (or 1 mole occupies the same volume, that is equal at normal pressure p = 0.1013 MPa and t = 0°C to the value of 22.4 litres) was the law established by Gay-Lussac in 1808. This law states, that the volumes of gases entering a chemical reaction relate to each other and to the volumes of gas products of a reaction as the ratios of small integers (for example, for the $2H_2 + O_2 = 2H_2O$ reaction this relation is 2:1:2).

Similarly to the GLL for volumes, the linearity of pressure of the given mass M of ideal gas at constant volume V establishes *Charles law* (1787), namely, $p_T = p_0(1 + \alpha_v T)$, where p_T and p_0 are gas pressures at temperatures T and T_0, and parameter α_v is the isochoric coefficient of pressure $\alpha_v = p^{-1}(\partial p/\partial T)_v = f(T)$, where for ideal gas $\alpha_p = \alpha_v$ and, accordingly, at T = 273.15°C $\alpha_v = 271.15^{-1}°C^{-1}$. Charles law describes the isochoric process of ideal gas.

D.N. Kagan

GENERAL CONFERENCE ON WEIGHTS AND MEASURES (see International temperature scale)

GENERALISED NEWTONIAN FLUIDS (see Non-Newtonian fluids)

GEOMETRIC SERIES (see Series expansion)

GEOMETRICAL THEORIES, FOR CATALYSIS (see Catalysis)

GEOTHERMAL ENERGY (see Alternative energy sources)

GEOTHERMAL HEAT PUMPS (see Heat pumps)

GEOTHERMAL SOURCES OF ELECTRICAL POWER (see Geothermal systems)

GEOTHERMAL SYSTEMS

Following from: *Alternative energy sources*

Holes drilled in the earth's crust can produce water as steam, as liquid or as mixtures of the two, and use can be made of this thermal source of electrical power generation, which has developed in twenty one countries during the present century. The world's 6300 MW electrical power capacity is provided by a number of different power cycles which have to contend with many difficulties in the technology, including solid deposition from the fluid, corrosion of plant and environmental pollution.

The continental crust has an average depth of 35 km, while the oceanic crust has 5 km of water plus 5 km of rock. The temperature rises with depth in the crust, commonly with gradients between 20 and 30°C/km but *thermal gradients* of several hundred degrees per kilometre depth are observed in some cases. Heat flows through the rock by conduction and more rapidly by fluid convection within rock fissures.

An economical geothermal field requires a source of heat from below, a permeable **Aquifer** holding the water or steam, a source of water recharge to make up for fluid losses and caprock to prevent uncontrolled loss of fluid into the atmosphere (Bowen, 1989). The temperature of geothermal fluids in hydrothermal convection systems should be higher than 180°C if they are to be used directly for power generation.

Hydrothermal sources producing geothermal fields with high temperature gradients are comparatively rare, however, as they are confined mainly to the earth's tectonic plate boundaries.

Since drilling is a large component of the capital cost of a geothermal power station, it is not often worth developing systems deeper than 2–3 km, and depths of 500–1500 m cover most operations.

There are two types of geothermal field. The first is the wet (or "liquid dominated") field which produces water under pressure at temperatures over 100°C. On reaching the surface, the pressure is reduced, and part of the water is "flashed" to steam, leaving a larger fraction as boiling water. The second is the dry (or "vapor dominated") field, which produces dry saturated, or superheated, steam at pressures higher than that of the atmosphere.

Lower temperature fluids are used for direct heating in buildings and industrial processes.

Electrical power was first generated from *geothermal energy* in 1904 at Larderello in Italy. Although the installed electrical generation capacity was about 130 MW by the time of the Second World War, no other country had, by then, exploited geothermal energy for power production. In the 1950s, the United States, Japan and New Zealand became more interested in the technology. The Wairakei power station was built in New Zealand between 1956 and 1963, with several units giving a total power rating of almost 200 MW. Development in the Geysers (California, USA) started with a unit rated at 11 MW commissioned in 1960, and the 22 MW Matsukawa plant inaugurated commercial electricity production in Japan in 1966. Mexico, Iceland and the Soviet Union also built plants about this period.

In the period up to 1985, there was a high growth in geothermal power exploitation following the oil crises in 1973 and 1979, in the Geysers and Imperial Valley in California, Central America, Japan and the Philippines—all located in the seismic zones of the world, where the geology supports flow of high temperature fluids close to the earth's surface. Countries in these regions which are heavily dependent upon imported oil have been particularly active in this growth, (see **Table 1**).

Several types of power generation cycle are used for geothermal systems. Steam extracted from dry wells, or after separation from wet wells, can be passed directly through a turbine and exhausted to atmosphere. This direct non-condensing cycle is the simplest and has the lowest capital cost. However, it uses twice as much steam as the condensing cycle for a given rate of electrical energy generation. If the field is liquid-dominated, the water can be flashed in a vessel operating at a pressure lower than that at which the main steam is introduced to the turbine. The flash steam is passed through the lower pressure stages of the turbine, and the unflashed hot water is discharged to waste, or may be used in industrial or district heating.

In other plants, power generation is achieved by passing fluid through well-head separators and then using the steam in a condens-

Table 1. Geothermal electric capacity (1992)

Country	MW Rating
USA	2979.2
Philippines	893.5
Mexico	725
Italy	635.2
New Zealand	286
Japan	270
Indonesia	142.7
El Salvador	105
Nicaragua	70
Iceland	50
Kenya	45
China	30.8
Turkey	20
ex-USSR	11
Guadeloupe	4.2
Azores	3
Greece	2
Romania	1.5
Argentina	0.7
Thailand	0.3
Zambia	0.2
TOTAL	6275.3

Source: Fridleifsson and Freeston, 1994.

ing turbine, using a tower cooled by natural ventilation. Small "portable" turbines are used directly at well-heads. Although they have lower efficiency, they are low cost and can be used to test the behaviour of a reservoir at the same time as drilling activities aimed at supplying heat to a central power plant. Although some plants use single flash systems, it is possible to use a second flash vessel to extract extra power. Flash steam systems can be rated at between 10 and 120 MW.

Binary cycle units employ low boiling point fluids in a heat exchanger which extracts the heat from the geothermal fluid. This allows more heat to be extracted because it rejects the geothermal fluid at lower temperature than the flash steam units. The technique allows use of geothermal fluids produced at lower temperatures, and restricts corrosion and scaling problems to the heat exchanger. The first binary cycle plant was commissioned in 1979 at East Mesa (Imperial Valley, California), and was rated at 10 MW. A number of other binary cycle plants since have been constructed, but this type of power generation system has not been used as often as the steam turbine units.

The efficiencies of geothermal power plants are low because of low fluid temperatures and pressures. Conventional fossil fuel plants are twice as efficient, so geothermal heat must be very cheap to compete with fossil fuels as a source of electrical power.

One of the problems encountered by the industry has been the drilling of "dry" wells, which are found to produce insufficient hot fluid to run a geothermal plant. In other cases, wells show a rapid decline in production rate. Much effort has been put into the modelling of geothermal reservoir performance to try to understand these problems. Nevertheless, it is to be expected that a borehole will have a useful life of only about 10 years, after which it may be deepened or new boreholes drilled to access other parts of the reservoir.

The chemistry of water/rock interaction has been the subject of extensive studies. Dissolution of rock components can cause changes in the permeability of the reservoir, and can produce highly saline geothermal fluids. Changes in temperature and pressure can subsequently cause precipitation of the dissolved solids (for example, silica or calcium salts).

This scaling phenomenon is particularly serious where water flashes to steam, and may result in premature blockage of the reservoir, or of the borehole walls and the surface equipment. The unwanted deposits can be removed from the borehole by periodic reaming operations, but in a few cases the scaling is so rapid and persistent that the well has to be abandoned.

Dissolved substances may corrode pipework and other equipment, and when discharged to the environment they may be an unacceptable source of pollution. Plants can use special reactors to remove dissolved solids before the fluid enters the main surface plant. Gas ejector discharges contain hydrogen sulphide, and the Geysers in California is an example where plant has been installed to extract this noxious component before discharge.

The fluids leaving the reservoir may also contain dust, which may erode turbines.

The more restrictive environmental legislation becomes, the more difficulty the geothermal power industry has in meeting these requirements. Hot water containing dissolved solids may be difficult to discharge into local water courses, and modern plants reinject the discharge water into the rock formation from which it came (as at the Geysers). Care has to be taken to locate this reinjection so that the cold water does not mix with the hot fluids in the reservoir too close to the extraction point, otherwise the temperature of the production fluid will be lowered.

The technology is well established, and geothermal plants are generally both efficient and reliable. A number of plants are planned and under construction world-wide but, like any alternative power source, further growth will depend on the oil price. There are a number of countries in which little or no exploitation of possible resources has taken place. In particular, the rate of exploitation in South America has been low.

This entry has not covered the use of geothermal energy for direct heating, and has therefore ignored such developments in countries such as Iceland, France and Japan, (Harrison et al., 1990).

Experiments have been carried out in the USA, in Europe and in Japan to extract geothermal energy in dry geological formations by artificially introducing water as a heat transfer medium. This "Hot Dry Rock" technology is still being developed, (Baria, 1990).

The ultimate process in geothermal energy exploitation could be the direct extraction of heat from a magma chamber. There are such chambers close to the surface of the earth, in volcanic regions (eg 1000°C at 4 km depth). The problems of drilling, and of designing a heat exchanger for these extreme circumstances have not been overcome. Nevertheless, the US Department of Energy has funded a study of this intense magma energy source.

References

Baria, R., (Editor) (1990) *Hot Dry Rock Geothermal Energy.* Robertson Scientific Publications, London.

Bowen, R. (1989) Geothermal Resources. Second Edition. Elsevier Applied Science, London and New York.

Fridleifsson, I. B., Freeston D. H. (1994) Geothermal Energy Research and Development, *Geothermics,* 23, 2, 175–214.

Harrison R., Mortimer N. D., Smarason O. B. (1990) *Geothermal Heating.* Pergamon Press, Oxford.

A.S.P. Green and R.H. Parker

GEYSERING INSTABILITIES (see Instability, two-phase)

G-FIN (see Air cooled heat exchangers)

GIBBS-DALTON LAW

This states that as the pressure of a real gas approaches zero, the **Fugacity** of each component in the mixture approaches its **Partial Pressure.**

This is often reinterpreted as meaning that as pressure approaches zero, a real gas becomes 'perfect'. Care must be taken when making this sort of reinterpretation since not all properties of real gases approach their 'perfect' values at zero pressure. The **Joule-Thomson Coeffient** is a notable example of this. It remains non-zero at $P = 0$, whereas the perfect gas value is identically zero. (See also **Perfect Gas**)

G. Saville

GIBBS-DUHEM EQUATION

There are several Gibbs-Duhem equations, although the name is most commonly associated with just one of them. They refer to the *properties of fluid (gas or liquid) mixtures* at fixed pressure and temperature. The equations all have the same form, namely:

$$\Sigma\, n_i d\tilde{f}_i = 0 \qquad \textbf{(1)}$$

where $\tilde{f}_i$ is a partial molar quantity for component i, n_i is the number of moles of i present in the mixture and the summation is over all components present in the mixture. $\tilde{f}$ can be any partial molar quantity, volume, enthalpy, entropy etc., but the one of greatest utility is the partial molar Gibbs free energy, usually called the **Chemical Potential,** μ, defined by:

$$\mu_i = (\partial G/\partial n_i)_{P,T,n_j} \qquad \textbf{(2)}$$

where G is the Gibbs Free Energy and n_j in the constraint list indicates that all amounts of substance are to be held fixed, except for n_i, when performing the differentiation.

The Gibbs-Duhem equation for this partial molar quantity, and the one most normally associated with the name, is then:

$$\Sigma\, n_i d\mu_i = 0 \qquad \textbf{(3)}$$

This equation shows that the chemical potentials in a mixture are not all independent and that there is a constraint equation which they must satisfy.

G. Saville

GIBBS ENERGY (See Chemical potential)

GIBBS FREE ENERGY (see Free energy; Multicomponent systems, thermodynamics of; Phase equilibrium)

GIBBS FUNCTION, SYNONYM FOR GIBBS FREE ENERGY (see Free energy)

GIBBS-HELMHOLZ EQUATIONS

The name Gibbs-Helmholtz is usually associated with the equation:

$$H = G - T(\partial G/\partial T)_P \qquad \textbf{(1)}$$

where H is enthalpy, G is Gibbs free energy and T temperature, with the implied assumption that the differentiation is to be carried out at constant composition. However, this is just one of a very large number of equations of similar form, which, at various times have been called Gibbs-Helmholtz equations. The equation given is, however, most appropriately called the Gibbs-Helmholtz equation since it was first recognised by both Gibbs and Helmholtz (independently).

The greatest utility of this equation lies in the presence of the temperature derivative of the Gibbs free energy. For example, the equilibrium constant in a chemical reaction is determined by the standard **Free Energy** change for the reaction. Thus, a knowledge of the temperature derivative of the free energy provides information relating the temperature derivative of the equilibrium constant. The equilibrium constant, K, may be written as:

$$\Delta g^\circ = -\tilde{R}T \ln K \qquad \textbf{(2)}$$

where Δg° is the standard Gibbs free energy change for the reaction at temperature T and $\tilde{R}$ is the universal gas constant.

Applying the Gibbs-Helmholtz equation to each component in the reaction gives:

$$\Delta h^\circ = \Delta g^\circ - T(\partial \Delta g^\circ/\partial T)_P \qquad \textbf{(3)}$$

and hence

$$(d \ln K)/(d[1/T]) = -\Delta h^\circ/\tilde{R} \qquad \textbf{(4)}$$

G. Saville

GIBBS POTENTIAL, SYNONYM FOR GIBBS FREE ENERGY (see Free energy)

GLADSTONE AND DALES LAW (see Film thickness determination by optical methods)

GLADSTONE-DALE EQUATION (see Interferometry; Visualization of flow)

GLASS-GRAPHITE MATERIALS (see Evaporative cooling)

GLASS-PLASTIC MATERIALS (see Evaporative cooling)

GLOBAL WARMING (see Greenhouse effect)

GLOW CURVES (see Thermoluminescence)

GLYCOL (see Natural gas)

ĠORTLER FLOWS (see Ġortler-Taylor vortex flows)

GÖRTLER-TAYLOR VORTEX FLOWS

Following from: *Secondary flows*

Where there is a flow over a concave surface, *centrifugal instabilities* can develop causing the formation of an array of **Vortices**. These vortices are aligned with their axes parallel to the main direction of flow with adjacent vortices rotating in opposite senses.

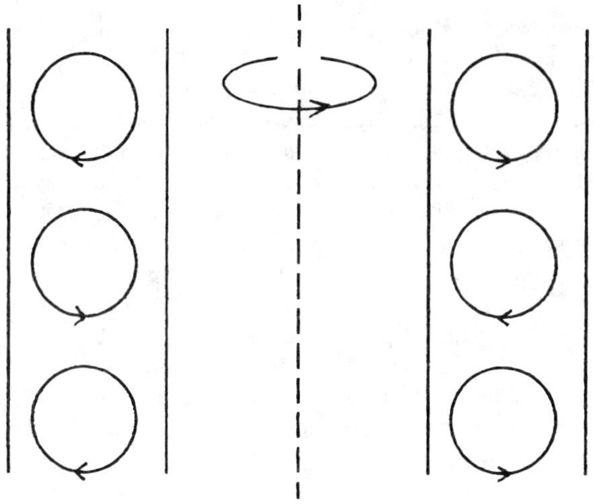

Figure 1. Taylor vortices between two rotating cylinders.

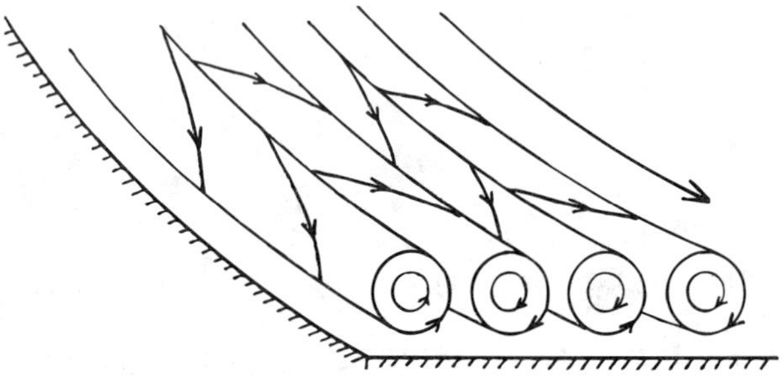

Figure 2. Vortices in the boundary layer of a concave wall.

These are laminar flows and are sometimes known as Görtler-Taylor (or Taylor-Görtler) vortex flows, but are often subdivided into *Taylor flows* and *Görtler flows.*

Taylor flows occur between two *rotating concentric cylinders.* At low rotation rates, there is **Couette Flow**, but for higher rotation rates (of one or both cylinders), Taylor vortices develop. **Figure 1** illustrates the arrangement of these vortices in an annulus. The point at which these vortices develop is dependent on the rotation rates of both the inner and outer cylinders and the width of the gap. The conditions over which such vortices are found are discussed in both Schlichting (1968), Tritton (1977) and Rosenhead (1963).

Görtler vortex flows are found where there is a flow over a *concave surface.* For sufficiently high flow speeds, centrifugal instabilities develop leading to the production of vortices in the boundary layer. This type flow are discussed in both Schlichting (1968), Rosenhead (1963) and Saric (1994) and is illustrated in **Figure 2**.

References

Rosenhead, L. (1963) *Laminar Boundary Layers,* Oxford University Press, London.

Saric, W. S. (1994) Görtler vortices, *Annu. Rev. Fluid Mech.,* 26, 379–409.

Schlichting, H. (1968) *Boundary-Layer Theory,* McGraw Hill, New York.

Tritton, D. J. (1977) *Physical Fluid Dynamics,* Van Nostrand Reinhold, Wokingham, UK.

Leading to: *Taylor instability*

M.J. Pattison

GOLD

Gold (Sanskrit *Jval;* Anglo-Saxon *gold*), Au (L. *aurum,* shining dawn); atomic weight 196.9665; atomic number 79; melting point 1064.43°C; boiling point 3080°C; specific gravity 18.88 (20°C); valence 1 or 3.

Known and highly valued from earliest times, gold is found in nature as the free metal and in tellurides; it is very widely distributed and is almost always associated with quartz or pyrite. It occurs in veins and alluvial deposits, and is often separated from rocks and other minerals by sluicing or panning operations. About two thirds of the world's gold output now comes from South Africa, and about two thirds of the total U.S. production comes from South Dakota and Nevada. The metal is recovered from its ores by cyaniding, amalgamating, and smelting processes. Refining is also frequently done by electrolysis. Gold occurs in sea water to the extent of 0.1 to 2 mg/ton, depending on the location where the sample is taken. As yet, no method has been found for recovering gold from sea water profitably. It is estimated that all the gold in the world, so far refined, could be placed in a single cube 60 ft. on a side.

Of all the elements, gold in its pure state is undoubtedly the most beautiful. It is metallic, having a yellow color when in a mass, but when finely divided it may be black, ruby, or purple. The Purple of Cassius is a delicate test for auric gold. It is the most malleable and ductile metal; 1 oz of gold can be beaten out to 300 ft^2. It is a soft metal and is usually alloyed to give it more strength. It is a good conductor of heat and electricity, and is unaffected by air and most reagents. It is used in coinage and is a standard for monetary systems in many countries. It is also extensively used for jewelry, decoration, dental work, and for plating. It is used for coating certain space satellites, as it is a good reflector of infrared and is inert.

Gold, like other precious metals, is measured in troy weight; when alloyed with other metals, the term *carat* is used to express the amount of gold present, 24 carats being pure gold.

The most common gold compounds are auric chloride ($AuCl_3$) and chlorauric acid ($HAuCl_4$), the latter being used in photography for toning the silver image. Gold has 18 isotopes; Au^{198}, with a half-life of 2.7 days, is used for treating cancer and other diseases. Disodium aurothiomalate is administered intramuscularly as a treatment for arthritis. A mixture of one part nitric acid with three of hydrochloric acid is called *aqua regia* (because it dissolved gold, the King of Metals). Gold is available commercially with a purity of 99.999 + %.

For many years the temperature assigned to the freezing point of gold has been 1063.0°C; this has served as a calibration point for the **International Temperature Scales** (ITS-27 and ITS-48) and the International Practical Temperature Scale (ITPS-48). In

1968, a new International Practical Temperature Scale (ITPS-68) was adopted, which demanded that the freezing point of gold be changed to 1064.43°C.

From Handbook of Chemistry and Physics, CRC Press

GOLD RECOVERY (see Ion exchange)

GOUY-STODOLA EQUATION (see Exergy)

GRAB SAMPLING (see Sampling)

GRAETZ LEO 1856–1941

A German physicist, born at Breslau on September 26, 1856, Graetz studied mathematics and physics at Breslau, Berlin and Strassburg. In 1881 he became assistant to A. Kundt at Strassburg and in 1883 he went to the University of München, where he became professor in 1908 and occupied the second chair for physics in parallel to Roentgen.

His scientific work was first concerned with the fields of heat conduction, radiation, friction and elasticity and after 1890 with problems of electromagnetic waves and cathode rays.

Graetz was a prolific technical writer, with 23 editions of his *Electricity and its Applications* and a five-volume work, *Handbook of Electricity and Magnetism,* which contributed to the wide dissemination of knowledge in electricity, which at that time was in its infancy. He died in München on November 12, 1941.

LEO GRAETZ 1856–1941

U. Grigull, H. Sandner and J. Straub

GRAETZ NUMBER

The Graetz number, G_z, is a nondimensional group applicable mainly to transient heat conduction in laminar pipe flow. It is defined as

$$Gz = \frac{UD^2}{\kappa x} \tag{1}$$

where U is the velocity of the fluid, D the diameter of the pipe, κ the fluid thermal diffusivity ($\lambda/\rho c_p$) and x the axial distance along the pipe.

Gz represents the ratio of the time taken by heat to diffuse radially into the fluid by conduction (sometimes called the "relaxation time"), D^2/κ, to the time taken for the fluid to reach distance x, x/U, i.e.

$$Gz = \frac{D^2/\kappa}{x/U} = \frac{UD^2}{\kappa x} \tag{2}$$

For small values of Gz (Gz $<$ 20) radial temperature profiles are fully developed, but for larger values thermal boundary layer development has to be taken into account.

Note that

$$Gz = \frac{D}{x}\,RePr \tag{3}$$

and Gz^{-1} is often used as a nondimensional form of axial distance in the representation of entrance effects on laminar flow heat transfer.

Graetz number is the reciprocal of Fourier number with time replaced by x/U, and many of the equations for transient heat conduction in laminar pipe flow are analogous to those of transient heat conduction in cylinders.

Reference

Hewitt G. F., Shires, G. L. and Bott, T. R. (1994) *Process Heat Transfer,* CRC Press, Boca Raton, FL, USA.

G.L. Shires

GRANULAR BEDS (see Gas-solids separation, overview)

GRANULAR FILTRATION MEMBRANES (see Filtration)

GRANULAR MATERIALS, DISCHARGE THROUGH ORIFICES

The mass flow rate $\dot{M}$ of a free flowing granular material through a circular orifice of diameter D_0 in the base of a cylindrical *bunker* might reasonably be expected to depend on the depth of the material in the bunker H, the diameter of the bunker D, the particle diameter d, the gravitational acceleration g, the density of the material ρ and the coefficient of friction μ.

Thus

$$\dot{M} = f(D_0, D, H, d, g, \rho, \mu) \quad \textbf{(1)}$$

For a conical or wedge-shaped *hopper,* the half angle α replaces the bunker diameter D.

It is found experimentally that neither the bunker diameter D nor the quantity of material in the bunker, as typified by the height H, has any significant effect on the flow rate. There are good theoretical reasons for both these observations as described by Nedderman (1993) and Nedderman et al. (1982) among many others. It is also found that, provided $d << D_0$, the effect of particle diameter is slight. Ignoring, therefore, the effects of H, D and d, dimensional analysis shows that the only permissible form is that

$$\dot{M} \propto \rho\sqrt{g}D_0^{5/2} \quad \textbf{(2)}$$

Early workers did not find the exponent of 5/2 on D_0, but Beverloo et al. (1962), plotting $\dot{M}^{2/5}$ vs. D_0, found a linear relationship but with an intercept which was proportional to the particle diameter d. They produced what is known as the *Beverloo correlation,*

$$\dot{M} = C\rho\sqrt{g}(D_0 - kd)^{5/2} \quad \textbf{(3)}$$

where C is roughly 0.58 but depends to some slight extend on the coefficient of friction. Grace and Raffle (1986) report a value as large as 0.64 for glass spheres. The parameter k depends on the particle shape, but takes a value close to 1.5 for spherical particles. It is of interest to note that Hagen devised a similar correlation some 100 years earlier, but this seem to have been forgotten in the interim.

It is important to use the correct density in Equation 3 and in similar correlations given below. The density of a granular material can be varied by compaction over a considerable range. However, experiments show that the mass flow rate is independent of the initial degree of compaction of the material. It appears that on initiation of discharge the material dilates to some density characteristic of the flowing material and it is this density that must be used in these correlations. This density seems to be close to the lowest stable density and can be measured by gently filling a container of known volume.

If the orifice is too small, particles may wedge in it, causing intermittent flow or even complete stoppage. Equation 3 should only be used if $D_0 > 6d$, and in this range the correction kd can never be a significant parameter.

The fact that the effective orifice diameter is $(D_0 - kd)$ is often called the empty annulus effect, since it seems to suggest that no particles pass through an annular zone of width 1/2kd. There is in fact no such empty annulus. Instead those particles near the edge of the orifice seem to be retarded by some process that is not yet fully understood.

For non-circular orifices the flow rate seems to be proportional to the group $A^*\sqrt{D_H}$, where the hydraulic mean diameter D_H is defined as $4A^*/P^*$ and A^* and P^* are the area and perimeter of the region remaining after an "empty annulus" of 1/2kd has been removed from the orifice. Casting Equation 3 into this form gives,

$$\dot{M} = \frac{4C}{\pi}\rho A^*\sqrt{(gD_H)} \quad \textbf{(4)}$$

For a long slot orifice of length L and width B (B << L) Equation 4 becomes,

$$\dot{M} = \frac{4\sqrt{2}C}{\pi}\rho\sqrt{g}(L - kd)(B - kd)^{3/2} \quad \textbf{(5)}$$

or, taking C = 0.58 and k = 1.5,

$$\dot{M} = 1.03\,\rho\sqrt{g}(L - 1.5d)(B - 1.5d)^{3/2} \quad \textbf{(6)}$$

A pre-Beverloo correlation by Rose and Tanaka (1959) gives the effect of hopper angle on the mass flow rate. The flow pattern in a discharging hopper can be classified as either *mass* or *core flow.* In mass flow all the material is in motion, whereas in core flow there exists a comparatively narrow flow channel between stagnant zones. We will denote the angle between the lower end of the stagnant zone boundary and the horizontal by ϕ_d as shown in **Figure 1**.

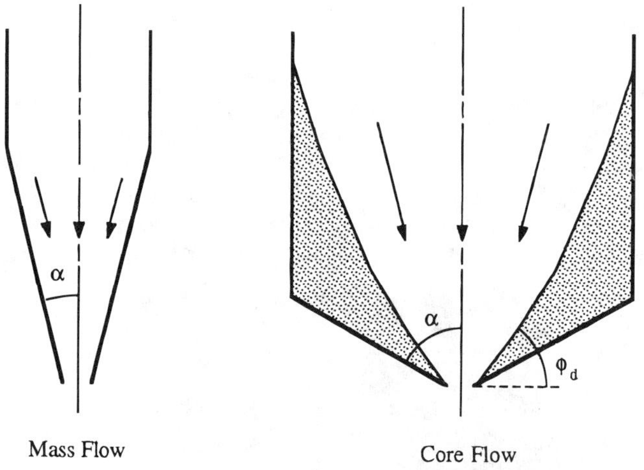

Figure 1.

Mass flow occurs in comparatively narrow hoppers whereas core flow occurs in wide-angled hoppers and cylindrical bunkers. Within core flow the mass flow rate is independent of the hopper angle and is therefore given by Equation 3. Within mass flow, the mass flow rate is a function of the hopper angle and Rose and Tanaka give the multiplicative factor $(\tan\alpha\tan\phi_d)^{-0.35}$. Incorporating this idea into the Beverloo correlation gives,

$$\dot{M} = CF\rho\sqrt{g}(D_0 - kd)^{5/2} \quad \textbf{(7)}$$

where

$$F = (\tan\alpha\tan\phi_d)^{-0.35} \quad \text{for } \alpha < 90 - \phi_d \quad \textbf{(8)}$$

$$F = 1 \quad \text{for } \alpha > 90 - \phi_d$$

Unfortunately it is not easy to predict the value of ϕ_d and a value of 45° is recommended by the Draft British Code of Practice for Silo Design [BMHB (1987)] in the absence of more reliable information.

It is seen that none of the correlations above contains any parameter which can easily be varied to control the flow rate. Only the orifice size can be used for this purpose, necessitating the use of a slide valve with the inevitable risk of jamming. Instead, the flow can be varied by injecting air into the hopper, giving a gauge pressure ΔP above the material. The absence of any term involving H in the Beverloo correlation shows that the mass flow rate is determined by conditions in the immediate vicinity of the orifice. If this is so, we can argue that in gravity flow the material is driven by its own weight ρg, but in air-augmented flow it is driven by the sum of its own weight and the pressure gradient at the orifice $(dp/dr)_0$. Thus one might expect to replace the term $\sqrt{g}$ in the Beverloo correlation by $\sqrt{(g + (dp/dr)_0/\rho)}$ giving

$$\dot{M} = CF\rho\left(g + \frac{1}{\rho}\left(\frac{dp}{dr}\right)_0\right)^{1/2}(D_0 - kd)^{5/2} \qquad \textbf{(9)}$$

The pressure gradient can be found from consideration of the percolation of air through the material and if the gas Reynolds number is low enough, **Darcy's Law** can be used, giving $(dp/dr)_0 = \Delta P/r_0$, where r_0 is the radius from the virtual apex to the edge of the orifice i.e. $D_0 = 2r_0 \cos\alpha$. In this case,

$$\dot{M} = Cf\rho\left(g + \frac{\Delta P}{\rho r_0}\right)^{1/2}(D_0 - kd)^{5/2} \qquad \textbf{(10)}$$

The results for higher Reynolds numbers are given by Thorpe (1984) and Nedderman, Tüzün and Thorpe (1983).

If a material is discharged from an unventilated hopper, air will have to enter through the orifice countercurrent to the discharging material. This will set up an adverse pressure gradient, causing a reduction in the mass flow rate. This can easily be evaluated from **Equation 9** if the permeability of the material is known.

The Beverloo correlation, and its modified forms given above are valid for coarse cohesionless materials. For cohesive materials, the flow rate tends to be irregular and unreproducible. This is because the cohesion is very sensitive to the previous treatment of the material and furthermore tends to increase during storage. Thus a material which flows reliably immediately after filling a hopper may flow irregularly, or not at all, after some period of storage.

Fine particles discharge at rates less that those predicted by the Beverloo correlation, even if cohesionless. The reason for this is that all materials dilate to some extend as they approach the orifice. Thus there must be relative motion between the material and the interstitial fluid. If the permeability is low, this will set up an adverse pressure gradient that retard the flow. Nonetheless such materials obey a modified form of the Beverloo correlation but with lower values of the parameter C. Verghese (1993) proposes the correlation,

$$C = 0.5\left(1 - \frac{0.0146}{d^2}\right) \qquad \textbf{(11)}$$

where d is measured in mm, but this correlation has been tested on a very limited range of materials and should therefore be used with caution.

References

Beverloo, W. A., Leniger, H. A. and Van de Velde, J. (1961) *Chem. Eng. Sci.* 15, 260.

BMHB (1987) *Silos; Draft Design Code.*

Grace, S. M. and Raffle, M. N. (1986) *Dept. of Chem. Engng. Part 2 Tripos Report,* University of Cambridge.

Nedderman R. M. (1992) *Statics and Kinematics of Granular Materials,* CUP.

Nedderman, R. M., Tuzun, U. and Thorpe, R. B. (1983) *Pow. Tech.,* 35, 69.

Nedderman, R. M., Tuzun, U., Savage, S. B. and Houlsby, G. T. (1982) *Chem. Eng. Sci.,* 37, 1597.

Rose, H. F. and Tanaka, T. (1959) *The Engineer (London),* 208, 465.

Thorpe, R. B. (1984) Ph.D. Thesis, University of Cambridge.

Verghese T. M. (1992) Ph.D. Thesis, University of Cambridge.

R. Nedderman

GRAPHICAL USER INTERFACE, GUI (see Computer programmes)

GRAPHITE BLOCK HEAT EXCHANGERS

Why Graphite?

Graphite is a corrosion resistant material of construction which gives no ion pick up, hence it is suited to applications mainly within the fine chemical pharmaceutical and fine chemical industries.

Why a Block Construction?

Although cylindrical (modular) and shell and tube *graphite heat exchangers* are still manufactured, the cubic block is much more compact, less wasteful of materials and more easily maintained. However, each type still has its particular niche, the cubic form being mainly used in conjunction with reactor vessels as condensers.

Condensing

For condensing applications, the sub assembly (see **Figure 1** on page 533) is usually made of a laminated construction so that slots can be made on the process side. A slot increases the available area for condensation by a factor of two for the same unit volume, which is particularly advantageous.

Liquid Duties

The header on either the process or the service side can be manufactured to contain a large number of baffles (e.g. 80 passes can be achieved in a unit having just 13.94m of heat transfer area). This is at the detriment of pressure drop and it is advisable not to have a liquid flow-rate of much more than 1m/s through the exchanger.

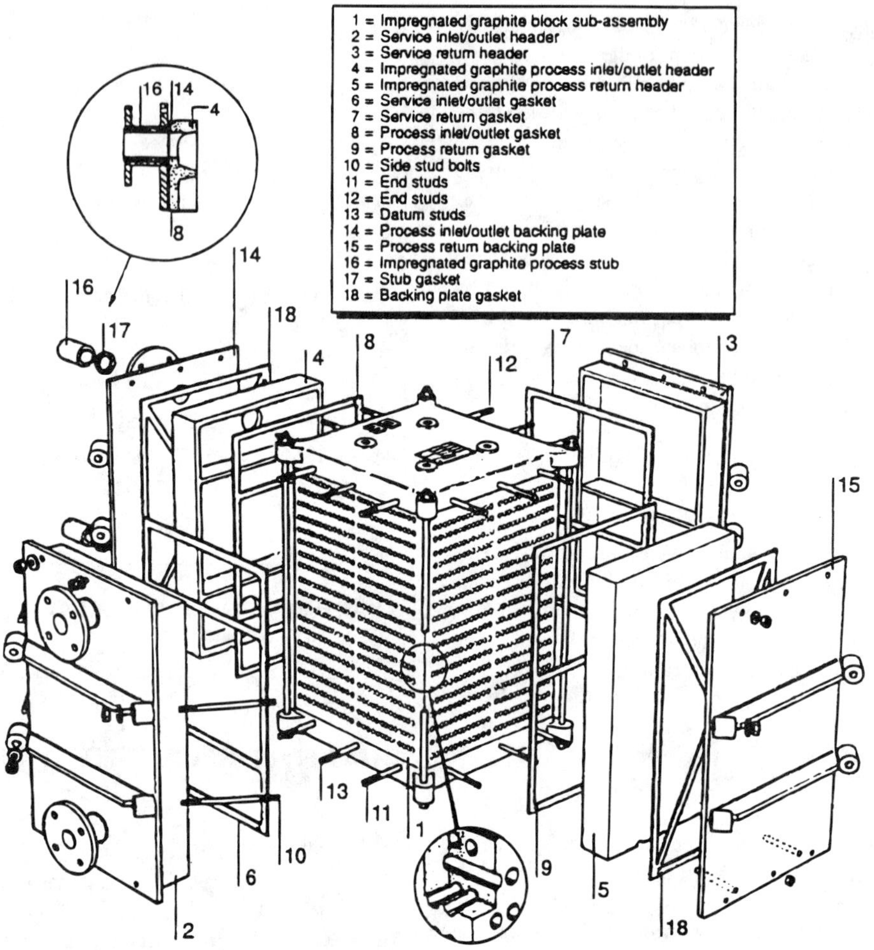

Figure 1. With permission from Secomak, Ltd., Stanmore, U.K.

Limitations

Graphite itself is a porous material, particularly when being used in the construction of heat exchangers and bursting discs, needs to be impregnated with a resin. There are two basic resins, furanic and phenolic based. These resins help form an impermeable barrier between process and service sides. Sometimes they are not compatible with chemicals, particularly highly oxidizing agents, caustic above 15% concentration and wet halogen gases; it is advisable to contact the manufacturer (for advice) on compatibility.

Advantages of Graphite Cubic Blocks

Graphite cubic blocks are made for a wide range of applications, for example:

1. To replace glass. Graphite is a more thermally efficient material of construction, typically using 1/2 to 1/3 the heat transfer area required by a glass unit. Graphite is more robust and can be subjected to greater pressure.
2. In sulphuric acid dilution. A small mixing chamber can be attached to the process inlet connection where sulphuric acid and dilution water are mixed and then instantly passed through the exchanger.
3. In vacuum pump exhausts. To condense vapors before and after vacuum pumps (Busch and Rietschle particularly).

The major advantage of a cubic unit is its compactness and ability to cope with a very large number of chemicals and duties up to 20 bar pressure and condensing duties at temperatures above 165°C. In addition, its form of construction makes it easily accessible for maintenance.

S.J. Oliver

GRAPHITE, USE IN HEAT EXCHANGERS (see Graphite block heat exchangers)

GRASHOF FRANZ 1826–1893

A German engineer, born July 11, 1826 at Düsseldorf, Grashof left school at the age of 15 to work as a mechanic, and then attended trade school in Hage and secondary school in Düsseldorf. From

1844 until 1847 Grashof studied mathematics, physics and machine design at the Berlin Royal Technical Institute. After a voyage of nearly three years which took him as far as the Dutch Indies and Australia he continued his studies at Berlin in 1852.

Grashof was one of the leaders in founding the Society of German Engineers (**Verein Deutscher Ingenieure**, **VDI**) and assumed an enormous load as author, editor, corrector and dispatcher. In 1863 Redtenbacher died and Grashof's name was so esteemed that the Technical University of Karlsruhe appointed him to be successor as superintendent of the engineering school. He also served as professor of applied mechanics and mechanical engineering and his lectures included strength of materials, hydraulics and theory of heat, in addition to general engineering.

After Grashof's death, October 26, 1893 at Karlsruhe the Society of German Engineers honored his memory by the Institution of the Grashof Commemorative Medal as the highest distinction that the society could bestow for merit in the engineering skills.

FRANZ GRASHOF 1826–1893

V. Grigull, H. Sandner and J. Straub

GRASHOF NUMBER

Grashof number, Gr, is a non-dimensional parameter used in the correlation of heat and mass transfer due to thermally induced natural convection at a solid surface immersed in a fluid. It is defined as

$$Gr = \frac{gl^3\xi\Delta T}{\nu^2} \tag{1}$$

where

g = acceleration due to gravity, m s^{-2}
l = representative dimension, m
ξ = coefficient of expansion of the fluid, K^{-1}
ΔT = temperature difference between the surface and the bulk of the fluid, K
ν = kinematic viscosity of the fluid, m^2s^{-1}

The significance of the Grashof number is that it represents the ratio between the buoyancy force due to spatial variation in fluid density (caused by temperature differences) to the restraining force due to the viscosisty of the fluid.

The form of the Grashof number can be derived by considering the forces on a small element of fluid of volume l^3.

The buoyancy force, F_b, on this element has the magnitude $gl^3\Delta\rho$ where $\Delta\rho$ is the difference in density between the element and the surrounding fluid. The order of magnitude of the viscous force, F_V, on the element is ηul where η is the fluid viscosity, and u the velocity of the element relative to the surrounding fluid. Hence,

$$\frac{F_b}{F_v} = \frac{gl^3\Delta\rho}{\eta ul} \tag{2}$$

The order of magnitude of the velocity u may be obtained by equating viscous and momentum forces, i.e.

$$\eta ul = \rho u^2 l^2 \tag{3}$$

or

$$u = \eta/\rho l \tag{4}$$

Substituting this value into the ratio of buoyancy to viscous forces

$$\frac{F_b}{F_v} = \frac{gl^3\rho\Delta\rho}{\eta^2} = \frac{gl^3\Delta\rho}{\nu^2\rho} \tag{5}$$

and using the relationship

$$\Delta\rho = \xi\rho\Delta T \tag{6}$$

$$\frac{F_b}{F_v} = \frac{gl^3\xi\Delta T}{\nu^2} \tag{7}$$

Since Reynolds number, Re, represents the ratio of momentum to viscous forces (see **Reynolds Number**) the relative magnitudes of Gr and Re are an indication of the relative importance of natural and forced convection in determining heat transfer. Forced convection effects are usually insignificant when $Gr/Re^2 >> 1$ and conversely natural convection effects may be neglected when $Gr/Re^2 << 1$. When the ratio is of the order of one, combined effects of natural and forced convection have to be taken into account.

Reference

Hewitt G. F., Shires, G. L. and Bott T. R. (1994) *Process Heat Transfer,* CRC Press, Boca Raton, FL, USA.

G.L. Shires

GRAVITATIONAL PRESSURE GRADIENT (see Multiphase flow; Two-phase flow)

GRAVITY DIE-CASTING (see Casting of metals)

GRAVITY EFFECTS ON BOILING (see Burnout in pool boiling)

GRAVITY FORCE (see Archimedes force)

GRAVITY SEPARATORS (see Vapour-liquid separators)

GRAVITY SETTLERS (see Liquid-solid separation)

GRAVITY WAVES (see Turbulence; Waves in fluids)

GRAY SURFACE (see Radiative heat transfer)

GREEK ALPHABET

Α	α	Alpha	Ν	ν	Nu
Β	β	Beta	Ξ	ξ	Xi
Γ	γ	Gamma	Ο	ο	Omicron
Δ	δ	Delta	Π	π	Pi
Ε	ϵ	Epsilon	Ρ	ρ	Rho
Ζ	ζ	Zeta	Σ	σ	Sigma
Η	η	Eta	Τ	τ	Tau
Θ	θ	Theta	Υ	υ	Upsilon
Ι	ι	Iota	Φ	ϕ	Phi
Κ	κ	Kappa	Χ	χ	Chi
Λ	λ	Lambda	Ψ	ψ	Psi
Μ	μ	Mu	Ω	ω	Omega

G.L. Shires

GREENHOUSE EFFECT

Following from: *Atmosphere; Environmental heat transfer*

The Basic Greenhouse Effect

We grow plants in greenhouses (glass-houses) because they provide a warmer environment than in the open air. The common explanation for the increase in temperature is that the glass is transparent to short-wave (solar) radiation and opaque to longwave (infrared) radiation. More radiant energy is trapped therefore temperature is increased. While this explanation turns out to be rather incomplete for glass-houses (it's actually suppression of convection which accounts for most of the increase in temperature), it is a very accurate explanation of why the surface of the Earth is kept warm by the presence of the so-called *greenhouse gases* in the atmosphere (see **Figure 1**).

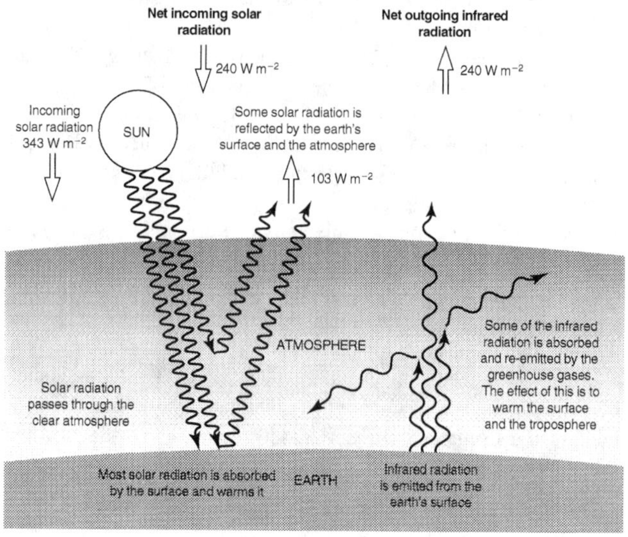

Figure 1. A simplified diagram illustrating the global long-term radiative balance of the atmosphere. Net input of solar radiation (240 Wm^{-2}) must be balanced by net output of infrared radiation. About a third (103 Wm^{-2}) of incoming solar radiation is reflected and the remainder is mostly absorbed by the surface. Outgoing infrared radiation is absorbed by greenhouse gases and by clouds keeping the surface about 33°C warmer than it would otherwise be (from IPCC, 1994 With permission).

While most planets exhibit a greenhouse effect, the magnitude of the warming depends on the composition of the planetary atmosphere, because it is only those components which absorb and re-emit infrared radiation which contribute to the effect. The atmosphere of Venus is composed almost entirely of carbon dioxide (CO_2), giving rise to a greenhouse effect of more than 500K and a surface temperature of around 750K. The atmosphere of Mars, also composed chiefly of CO_2 but much thinner than that of Venus, provides a greenhouse warming of around 10K. The Earth's atmosphere, with a greenhouse effect of around 30K, contrasts with the other planets in that water vapor is the most important greenhouse gas, followed by carbon dioxide, and that together these gases represent only a small fraction of the total mass of the atmosphere; the major components nitrogen and oxygen do not absorb infrared radiation and play no direct part in the Earth's greenhouse effect.

The Enhanced Greenhouse Effect

Changes in the atmospheric concentration of greenhouse gases will alter the greenhouse effect, leading to a change in climate as the atmosphere adjusts to a new equilibrium. Because marine and terrestrial vegetation plays such a large part in the exchange of greenhouse gases between the atmosphere and the surface, it is likely that past changes in climate occurring over time scales of thousands to millions of years, even when caused by external factors, have been modulated by changes in the marine and terrestrial biospheres. More recently, the increasing use of fossil fuels since the Industrial Revolution has demonstrably increased atmospheric concentrations of CO_2. Concentrations of methane (CH_4) and nitrous oxide (N_2O) have also increased due, at least in part, to human activities, and some widely-used synthetic chemicals such as the chlorofluorocarbons (CFCs) are new and powerful greenhouse gases.

In the absence of other changes, a doubling of CO_2 concentration (or equivalent increases in other greenhouse gases) would increase the Earth's surface temperature by around 1K. In practice, other complex changes would be expected to occur in the climate system. Some of the associated changes will tend to reduce the warming and some will tend to increase it, and the greatest uncertainty in calculating climate change under the enhanced greenhouse effect is due to uncertainty in these climate feedbacks, particularly those involving the amount and distribution of water vapor. Numerical models of climate generally predict that the net effect of climate feedbacks would be to increase the warming significantly above that by the simple calculation based on radiation (see **Figure 2**).

Atmospheric particulates (*aerosols*) derived from industrial emissions and from biomass burning exert a cooling influence on climate because they increase the amount of sunlight reflected to space. Though linked to the enhanced greenhouse effect through emissions (greenhouse gases often come from the same sources as the aerosols or their precursors, e.g. power station chimneys) atmospheric particulates are not themselves greenhouse agents.

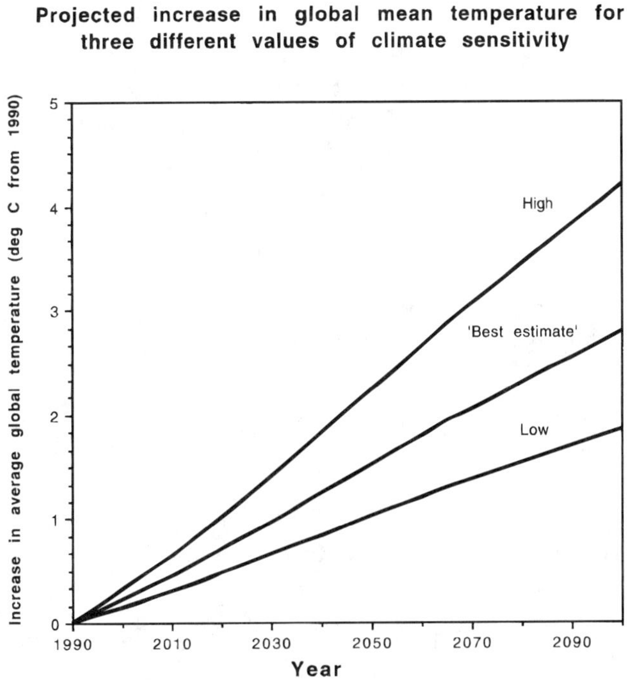

Figure 2. Future temperature rise will depend on future emissions of greenhouse gases. The diagram shows projected changes in mean surface temperature under the IS92 scenario of future emissions (see IPCC 1992). Climate sensitivity is an index of the equilibrium change in global mean temperature that would occur in response to a doubling of CO2 concentration. Low climate sensitivity = 1.5°C, best estimate = 2.5°C and High = 4.5°C (from IPCC, 1992. With permission). The calculations do not include the effect of atmospheric dust and particulates (aerosols).

History

The warming effect of greenhouse gases in the atmosphere was first recognized in 1827 by **Jean-Baptiste Fourier**. Around 30 years later John Tyndall measured the infrared absorption characteristics of water vapor and carbon dioxide, and suggested that reduced levels of atmospheric carbon dioxide may have been a cause of ice ages. In 1896, Svante Arrhenius estimated that doubling atmospheric concentrations of carbon dioxide would increase global average temperature by 5–6°C (not very different from current estimates), but it was about a half century before G. S. Callendar made a similar calculation for the effect of the burning of fossil fuels.

Such calculations received little attention outside of scientific circles and it was not until the mid-1970s that "the *Greenhouse Effect*" or "*Global Warming*" began to move into the public—and therefore political—spotlight, prompted by a variety of factors including: clear evidence of increasing atmospheric concentrations of carbon dioxide and of its link to human activities; results from computer models which suggested significant *regional* changes of climate change associated with an enhanced greenhouse effect; and, not least, media coverage which tended to associate the greenhouse effect with images of disaster and chaos, leaving the public (and governments) disturbed but not necessarily better informed.

To provide an authoritative assessment of current scientific understanding of the enhanced greenhouse effect the World Meteorological Organisation (WMO) and the United Nations Environmental Programme (UNEP) jointly established an Intergovernmental Panel on Climate Change (IPCC) in 1988. The IPCC reports in 1990 provided impetus to the negotiation of the *UN Framework Convention on Climate Change,* which was eventually signed in Rio de Janeiro, Brazil by over 150 countries in June 1992.

References

Basic physics of the greenhouse effect

Houghton, J. T. (1986) *The Physics of Atmospheres, 2nd* ed., Cambridge University Press, Cambridge.

The enhanced greenhouse effect and climate change

Climate Change 1990: The IPCC Scientific Assessment, J. T. Houghton, G. J. Jenkins and J. J. Ephraums, eds., Cambridge University Press, Cambridge.

Climate Change 1995: The Science of Climate Change Contribution of WGI to the Second Assessment Report of the Intergovernmental Panel on Climate Change, J. T. Houghton, L. G. Meira Filho, B. A. Callander, N. Harris, A. Kattenberg and K. Maskell, eds. Cambridge University Press, Cambridge.

Impacts of climate change

Climate Change 1990: The IPCC Impacts Assessment, W. J. McG. Tegart, G. W. Sheldon and D. C. Griffiths, eds., Australian Government Publishing Service, Canberra.

Climate Change 1995: Impacts, Adaptations and Mitigation of Climate Change: Scientific-Technical Analyses, Contribution of WGII to the Second Assessment Report of the Intergovernmental Panel on Climate Change, R. T. Watson, M. C. Zinyowera, and R. H. Moss, eds. Cambridge University Press, Cambridge.

Houghton, J. T. (1994) *Global Warming: the Complete Briefing,* Lion Publishing, Oxford

Greenhouse gases: emissions and strategies and options for reducing emissions

Climate Change 1990: The IPCC Response Strategies, Island Press, Washington, DC.

Climate Change 1995: Economic and Social Dimensions of Climate Change, Contribution of WGIII to the Second Assessment Report of the Intergovernmental Panel on Climate Change. J. Bruce, Hoesung Lee, and E. Haites, eds., Cambridge University Press, Cambridge.

The UN Framework Convention on Climate Change

The Earth Summit 1993: The United Nations Conference on Environment and Development (UNCED) S. P. Johnson, ed., International Environmental Law and Policy Series, Graham and Trotman/Martinus Nijhoff, London.

B.A. Callander

GREENHOUSE GASES (see Greenhouse effect)

GREEN'S FORMULA (see Integrals)

GREEN'S FUNCTION

Following from: *Differential equations*

Green's function is a function of many variables associated with integral representation of solution of a boundary promlem for a differential equation.

In the general case of a linear boundary problem with homogeneous boundary conditions

$$L\varphi(x) = f(x), \qquad x \in D, \tag{1a}$$

$$\Gamma_i\varphi(x) = 0, \qquad i = 1, 2, \ldots, I, \quad x \in S, \tag{1b}$$

where $\Gamma_i\ \varphi(x)$ are linear homogeneous functions of $\varphi(x)$ and its derivatives on the boundary S of domain D. An inverse transformation (if it exists) of the form

$$\varphi(x) = \int_D G(x, \xi)f(\xi)dv \tag{2}$$

uses Green's function $G(x,\xi)$ as a kernel for the given problem (Equation 1).

Equation 2 describes the solution as a superposition of elementary solutions which can be interpreted as point sources or power pulses $f(\xi)\ \delta(x,\xi)$ at the point $x = \xi$ (where $\delta(x,\xi)$ is the Dirac delta function).

The function $G(x,\xi)$ of the argument x must satisfy the homogeneous boundary condition (1b), and also the equation

$$LG(x, \xi) = 0 \qquad \text{for} \quad x \neq \xi \tag{3a}$$

and the condition

$$\int_D LG(x, \xi)dv = 1, \tag{3b}$$

or, as generalizeed function, the equation

$$LG(x, \xi) = \delta(x, \xi). \tag{4}$$

If the operator L is self-conjugate, Green's function $G(x,\xi)$ is symmetric, i.e., $G(x,\xi) = G(\xi,x)$. For a boundary problem for a linear ordinary differential equation

$$L\varphi \equiv a_n(x)\frac{d^n\varphi}{dx^n} + \cdots + a_1\frac{d\varphi}{dx} + a_0 = f(x) \tag{5}$$

the general solution on the section [a, b] can be presented in the form

$$\varphi = \int_a^b G(x, \xi)f(\xi)d\xi + \sum_{k=1}^{n} C_k\varphi_k(x), \tag{6}$$

where $\{\varphi_k\}$ is the functional system of solutions of a homogeneous equation $L(\varphi) = 0$, C_k are arbitrary constants obtaind from boundary conditions.

It often appears possible to determine Green's function so that a particular solution

$$\int_a^b G(x, \xi)f(\xi)dv$$

satisfies the given boundary conditions. Such Green's function must have a jump of (n − 1)th derivative for $x = \xi$

$$\left.\frac{\partial^{n-1}G}{\partial x^{n-1}}\right|_{x\to\xi^t} - \left.\frac{\partial^{n-1}G}{\partial x^{n-1}}\right|_{x\to\xi^-} = \frac{1}{a_n(\xi)}.$$

Further Green's function for linear differential equations with partial derivatives concerns.

Elliptic equations. The solution of *Dirichlet's problem* for the *Poisson equation*

$$\Delta\varphi = f(x), \qquad x \in D \tag{7a}$$

$$\varphi(x) = F(x), \qquad x \in S \tag{7b}$$

can be written with the help of Green's function $G(x,\xi)$ as

$$\varphi(x) = \int_D G(x, \xi)f(\xi)dv + \int_S F(s)\frac{\partial G}{\partial n}ds, \tag{8}$$

where n is the outer normal to the surface S. Green function for the given problem is represented in the form

$$G(x, \xi) = \begin{cases} \dfrac{1}{2\pi}\ln r + g(x, \xi) & \text{for } N = 2 \\ -\dfrac{1}{4\pi r} + g(x, \xi) & \text{for } N = 3, \end{cases}$$

where N is the problem dimensionality, r is the distance between the points x and ξ, $g(x,\xi)$ is a harmonic function of $(x,\xi) \in D$, chosen so that Green's function satisfies boundary condition (7b).

Parabolic equations. The solution of a boundary problem for the equation of thermal conductivity with homogeneous boundary conditions

$$L(\varphi) \equiv \frac{\partial \varphi}{\partial t} - a^2 \Delta \varphi = f(t, x), \qquad x \in D \tag{9a}$$

$$\Gamma_i \varphi(x) = 0, \qquad x \in S \tag{9b}$$

and the initial condition

$$\varphi(0, x) = F(x), \qquad x \in D,$$

where Γ_i are the linear boundary operators with coefficients which depend on t and x, can be written with the help of Green's function $G(t,x,\tau,\xi)$ as

$$\varphi(t, x) = \int_0^t \int_D G(t, x, \tau, \xi) f(\tau, \xi) dv \, d\tau + \int_D G(t, x, \tau, \xi) F(\xi) dv. \tag{10}$$

Green's function for the given problem as a function t, x satisfies the equation (3a) for $(t,x) \neq (\tau,\xi)$ and for $t > \tau \geq 0$, $x \in D$ condition (9b).

For instance, the solution of equation (9a) on the entire infinite space can be expressed in the form (10) with the help of Green's function

$$G(t, x, \tau, \xi) = -\left[\frac{1}{4\pi a^2 (t - \tau)}\right]^{n/2} \exp\left(-\frac{1}{4a^2} \frac{r^2}{t - \tau}\right) \tag{11}$$

$(t > \tau;\ n = 1,2,3)$, where r is the dinstance between the points x and ξ.

Hyperbolic equations. In a number of cases the solution of a two-dimentional Cauchy problem with boundary conditions specified on a boundary curve can be obtained employing an integral relation based on the Green-Riemann function which has a more complex character than in the case of elliptic and parabolic equations.

I.G. Zaltsman

GREGORIC EFFECT (see Augmentation of heat transfer, two phase)

GRID EFFECTS (see Rod bundles, heat transfer in)

GRIFFITH'S ENERGY FRACTURE CRITERION (see Fracture mechanics)

GROUND-COUPLED HEAT PUMP (see Heat pumps)

GROUND EFFECT (see Aerodynamics)

GROUND HEAT FLUX (see Environmental heat transfer)

GROUND SOURCE HEAT PUMP (see Heat pumps)

GROUND-TO-AIR HEAT PUMP (see Heat pumps)

GROWTH RING METHOD (see Solidification)

GUI, GRAPHICAL USER INTERFACE (see Computer programmes)

H

HABIT, CRYSTAL SHAPE (see Crystallization)

HAGEN-POISEUILLE EQUATION (see Heat pipes; Flow of fluids; Poiseuille flow; Tubes, single phase flow in; Turbulence)

HALL COEFFICIENT (see Magnetohydrodynamic, MHD, power generators; Magnetohydrodynamics, MHD)

HALL-TAYLOR EQUATION FOR ELECTRICAL CONDUCTIVITY (see Bubble flow)

HALOGENATED FLUOROALKANES (see Heat pumps)

HALOGENS

Halogens is the common name for five elements (fluorine, chlorine, bromine, iodine and astatine) that belong to group VII A of the Periodic Table. They are all characterised by having one electron missing in their outermost valance shell which makes them strongly electronegative and good oxidising agents. At room temperature and pressure, fluorine and chlorine are diatomic gases, bromine is a reddish-brown liquid, while iodine and astatine are solids. They are highly toxic, while fluorine, chlorine and bromine also present a fire risk.

V. Vesovic

HAMAKER CONSTANT (see Gas-solids separation, overview)

HAMPSON COIL (see Heat exchangers)

HANKEL FUNCTION (see Bessel functions)

HARDENING (see Steels)

HARMONIC ANALYSIS

Harmonic analysis is a branch of mathematics which includes theories of trigonometric series (**Fourier Series**), *Fourier transformations,* function approximation by trigonometric polynomials, almost periodic functions, and also generalization of these notions in connection with general problems of the theory of functions and functional analysis.

Each periodic function f(t) having a period T and satisfying Dirichlet's conditions (a discontinuity of the first kind, a finite and a countable number of extremums in a period) can be represented (expanded) in the form of a sum of an infinite number of sinusoidal functions

$$f(t) = A_0 + A_1 \sin(\omega t + \alpha_1) + A_2 \sin(2\omega t + \alpha_2) + \cdots$$

$$= A_0 + \sum_{n=1}^{\infty} A_n \sin(n\omega t + \alpha_n), \qquad \textbf{(1)}$$

where the frequency $\omega = 2\pi/T$, A_i and α_i are constant.

The components of expansion (1) are called the harmonic components (harmonics of the 1th, 2nd etc. kind), and the expansion itself, the harmonic analysis of the function f(t).

If we denote $x = \omega t$, then expansion (1) for the function f(x) with a period 2π has the form

$$f(x) = A_0 + \sum_{n=1}^{\infty} A_n \sin(nx + \alpha_n)$$

and after transformations

$$f(x) = \frac{a_0}{2} + \sum_{n=1}^{\infty} (a_n \cos nx + b_n \sin nx), \qquad \textbf{(2)}$$

which is called the Fourier series.

The coefficients in (2) are defined as

$$a_n = \frac{1}{\pi}\int_{-\pi}^{\pi} f(x)\cos nx\, dx$$

$$b_n = \frac{1}{\pi}\int_{-\pi}^{\pi} f(x)\sin nx\, dx \qquad n = 0, 1, 2, \ldots.$$

Series (2) converges to f(x) at points of its continuity, and to a half-sum $[f(x_i^-) + f(x_i^+)]/2$ at discontinuity points x_i.

The expansion of an even function $(f(-x) = f(x))$ into a Fourier series has only cosines, and of an odd function $(f(-x) = -f(x))$, only sines.

In general, the expansion of a function f(x) into a series according to any orthogonal system of functions $\{\varphi_n(x)\}$, defined on the interval [a,b],

$$f(x) = \sum_{n=0}^{\infty} c_n \varphi_n(x)$$

with coefficients defined as

$$c_n = \frac{1}{\lambda_n}\int_a^b f(x)\varphi_n(x)dx, \qquad n = 0, 1, 2, \ldots$$

is called the generalized Fourier series of a given function, and the coefficients c_n the generalized Fourier coefficients relative to expansion $\{\varphi_n(x)\}$.

For an arbitrary orthogonal on the interval [a,b] system of square integrable functions $\{\varphi_n(x)\}$, the least average deviation for a given m

$$\Delta_m = \int_a^b [f(x) - \sigma_m(x)]^2 dx$$

of the function f from an approximating polynomial

$$\sigma_m = \sum_{n=0}^{m} \gamma_n \varphi_n(x)$$

is obtained, if as σ_m is used a trancated series of a generalized Fourier series

$$S_m = \sum_{n=0}^{m} c_n \varphi_n(x).$$

Then

$$\Delta_m^{min} = \int_a^b f^2(x)dx - \sum_{n=0}^{m} \lambda_n c_n^2 \quad \text{(Bessel's identity)}.$$

If $\{\varphi_n\} = \{\sin nx; \cos nx\}$, then a minimum Δ_m is obtained, if the Fourier trancated series is taken and

$$\Delta_m^{min} = \int_{-\pi}^{\pi} f^2(x)dx - \pi\left\{\frac{a_0^2}{2} + \sum_{n=0}^{m} (a_n^2 + b_n^2)\right\}.$$

Similarly, each function f(x) defined on a number axis $(-\infty,\infty)$ which satisfies the Dirichlet conditions can be represented in the form of the Fourier integral

$$f(x) = \frac{1}{\sqrt{2\pi}} \int_{-\infty}^{\infty} F(s)e^{-isx}dx,$$

with the right-hand part being equal at discontinuity points to $[f(x^-) + f(x^+)]/2$.

The Fourier integral is also called the *Fourier transform* of function F(s) which is the *inverse Fourier transform* of f(x).

$$F(s) = \frac{1}{\sqrt{2\pi}} \int_{-\infty}^{\infty} f(x)e^{isx}dx.$$

A Fourier series and a Fourier integral relate each function f(x) and its spectrum (for a Fourier integral a set of harmonics is the spectrum). In some approaches (for instance, in turbulence theory) the study of functions by their spectra is used.

The notion of Fourier transformation can be easily generalized into the functions of many variables.

Harmonic analysis methods and similar methods associated with the expansion in terms of a complete orthogonal system of functions, are widely used in solving the problems of mathematical physics, associated with the solution of parabolic and elliptic partial differential equations.

Leading to: *Fourier series*

I.G. Zaltsman

HEADERS AND MANIFOLDS, FLOW DISTRIBUTION IN

Following from: *Flow of fluids; Channel flows*

Devices used to achieve the division of flows from a single source to many outlets are usually termed headers and manifolds. The former term is usually confined to heat exchangers and cases where the outlets are close together. Manifolds are more often involved in the distribution between pieces of equipment.

For single phase flow, the proportion of the flow emerging from each outlet depends on the resistance in the downstream pipework. **Figure 1** shows that when the resistance in the branches is low, there can be significant *maldistribution* which is noticeably decreased when the resistances are increased. Where it is important that the flow from each outlet is approximately uniform, additional resistances in the form of orifices (or ferrules) are provided. In cases such as fired boilers these ferrules are individually sized.

As well as the pressure drops in the downstream pipes, there are additional pressure drops associated with the junction, one for each outlet, which should be taken into account. In the modelling of these multi-branch systems, there have been two approaches. The first (Acrivos et al., 1959, Baruja and Jones, 1976) used the concept of *porosity* to represent a regular array of closely spaced branches. Such an approach might be suited to cases with a single flow direction but not for more complex configurations. Alternatively, the system is treated as a series of individual junctions. Care must be taken to account for any interaction between branches. McNown (1954) examined the effect on inter-branch distance on the pressure changes along the main pipe. For the branch pressure drop, Zenz (1962) suggests that the value for an individual junction be increases by 25%. However, in a systematic study, Scruton et al. (1988) showed that the branch pressure drop is well represented by the correlation of Gardel (1957) for inter-branch spacing of 1.5 to 15 times the branch diameter.

Separation of the phases occurs in the inlet or return headers of heat exchangers operating with the **Gas/Liquid Flows**. The liquid collects at the bottom. Hence tubes on the lower rows will be overloaded with liquid.

There is very little information available for the division of gas/liquid flows at manifolds. What few data exist, e.g. Collier (1976) and Coney (1980), show that the manifold acts as a series of

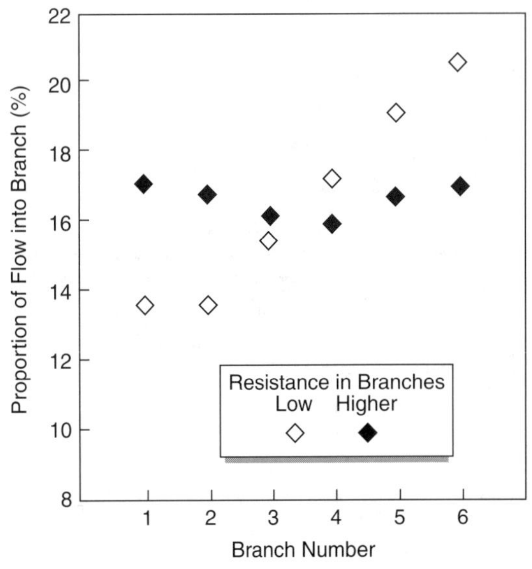

Figure 1. Effect of downstream resistance on maldistribution of single-phase flow between six branches.

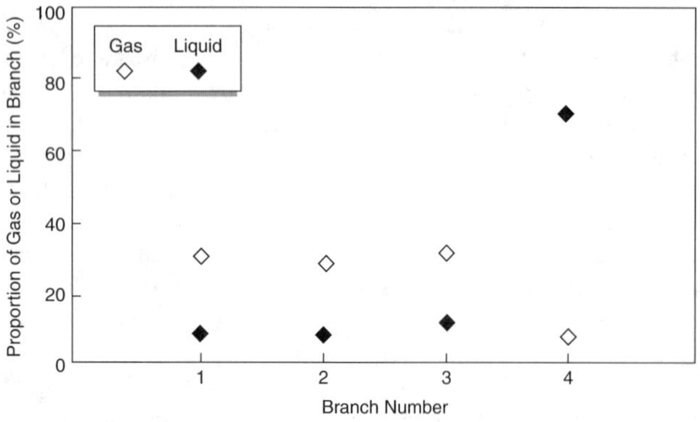

Figure 2. Division of gas and liquid at a four branch manifold with a vertical main pipe and horizontal branches—mass flux = 580 kg/m^2s; inlet quality = 16%.

junctions. However, because of the almost inevitable maldistribution of the phases this can mean that one phase accumulates at the last branch as in the case shown in **Figure 2**. Similar results have been reported with junctions in the horizontal plane. If the branches were connected to a series of condensers, the one connected to the final branch could seriously underperform. If the branches are placed very close together, there could be interaction between them but it is not known at which inter-branch spacing this becomes important as this parameter has not been researched.

References

Acrivos, A., Babcock, B. D., and Pigford, R. L. (1959) Flow distribution in manifolds, *Chem. Eng. Sci.*, 10, pp 112–124.

Baruja, R. A. and Jones, E. H. (1976) Flow distribution in manifolds, Trans ASME, *J. Fluids Eng.*, 98, pp 654–666.

Collier, J. G. (1976) Single-phase and two-phase behaviour in primary circuit components, *N.A.T.O Advanced Study Institute on Two-phase Flow and Heat Transfer,* Istanbul, Turkey.

Coney, M. W. E. (1980) Two-phase flow distribution in a manifold system, *European Two-Phase Flow Group Meeting,* Strathclyde.

Gardel, A. (1957) "Les pertes de charge dans les ecoulementes au travers de branchements en te", *Bulletin Technique de la Suisse Romande,* 9, pp 122–130 and 10, pp 143–148.

McNown, J. S. (1954) Mechanics of manifold flow, *ASCE Trans.,* 119, pp 1103–1142.

Scruton, B., Longworth, D., and Mays, C. J. (1988) Pressure losses for dividing single-phase flows in headers with multiple branches, *2nd National Heat Transfer Conference,* Birmingham.

Zenz, F. A. (1962) Minimise manifold pressure drop, *Hydrocarbon Processing and Petroleum Refiner,* 41 (12), pp 125–130.

***Leading to:** T-junctions*

B.J. Azzopardi

HEADS (see Mechanical design of heat exchangers)

HEAT

***Following from:** Thermodynamics*

Heat is less easily defined than work. In general terms, the temperature of a substance increases if heat is added to it. Therefore, heat was, for a long time, thought to be an invisible fluid named 'caloric' which flows along a temperature difference from one system to the other. We now know that this is false and that heat is not contained in a system but is manifested only as an interaction of the system with its surroundings as the system changes from one state to another. Like work, it can be considered as energy in transit. Heat is best defined through the *First Law of Thermodynamics:*

$$\Delta U = Mc_v\Delta T = Q - W = Q - \int pdV \quad \textbf{(1)}$$

$$\Delta H = Mc_p\Delta T = \Delta U + \int Vdp + \int pdV = Q + \int Vdp \quad \textbf{(2)}$$

where U is internal energy, M mass, T temperature, c_v and c_p specific heat capacity at constant volume and pressure respectively, and Q is heat, W work, p pressure, V volume, and H enthalpy.

Therefore,

$$Q = \Delta U \text{ for } V = \text{constant and } Q = \Delta H \text{ for } p = \text{constant} \quad \textbf{(3)}$$

If phase change occurs as a result of heat transfer, the latent heat of evaporation and/or the latent heat of solidification have to be included in **Equation 3**.

The sign convention is that heat transfer is positive into the system and negative out of the system.

H. Müller-Steinhagen

HEAT ACTUATED (see Heat pumps)

HEAT CAPACITY, OF AIR (see Air, properties of)

HEAT EXCHANGER NETWORKS (see Process integration)

HEAT EXCHANGERS

***Following from:** Heat transfer*

A heat exchanger is a device used to transfer heat between two or more fluids. The fluids can be single or two phase and, depending on the exchanger type, may be separated or in direct contact. Devices involving energy sources such as nuclear fuel pins or fired heaters are not normally regarded as heat exchangers although many of the principles involved in their design are the same.

In order to discuss heat exchangers it is necessary to provide some form of categorisation. There are two approaches that are normally taken. The first considers the flow configuration within the heat exchanger, while the second is based on the classification of equipment type primarily by construction. Both are considered here.

Classification of Heat Exchangers by Flow Configuration

There are four basic flow configurations

- Counter Flow
- Co-current Flow
- Crossflow
- Hybrids such as Cross Counterflow and Multi Pass Flow

Figure 1 illustrates an idealised counter flow exchanger in which the two fluids flow parallel to each other but in opposite directions. This type of flow arrangement allows the largest change in temperature of both fluids and is therefor most efficient (where efficiency is the amount of actual heat transferred compared with the theoretical maximum amount of heat that can be transferred).

In co-current flow heat exchangers, the streams flow parallel to each other and in the same direction as shown in **Figure 2**. This is less efficient than counter current flow but does provide more uniform wall temperatures.

Crossflow heat exchangers are intermediate in efficiency between counter current flow and parallel flow exchangers. In these units, the streams flow at right angles to each other as shown in **Figure 3**.

In industrial heat exchangers, hybrids of the above flow types are often found. Examples of these are combined cross flow/counter flow heat exchangers and multi pass flow heat exchangers. See for example **Figure 4**.

Classification of Heat Exchangers by Construction

In this section heat exchangers are classified mainly by their construction, Garland (1990), (see **Figure 5**). The first level of classification is to divide heat exchanger types into *recuperative* or regenerative. A *Recuperative Heat Exchanger* has separate flow paths for each fluid and fluids flow simultaneously through the exchanger exchanging heat across the wall separating the flow paths. A **Regenerative Heat Exchanger** has a single flow path which the hot and cold fluids alternately pass through.

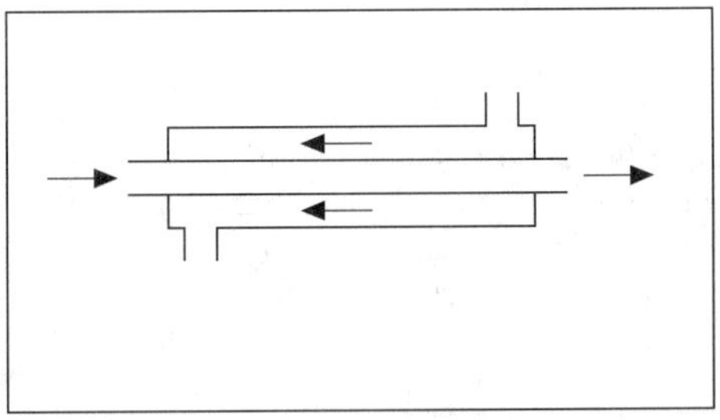

Figure 1. Counter current flow.

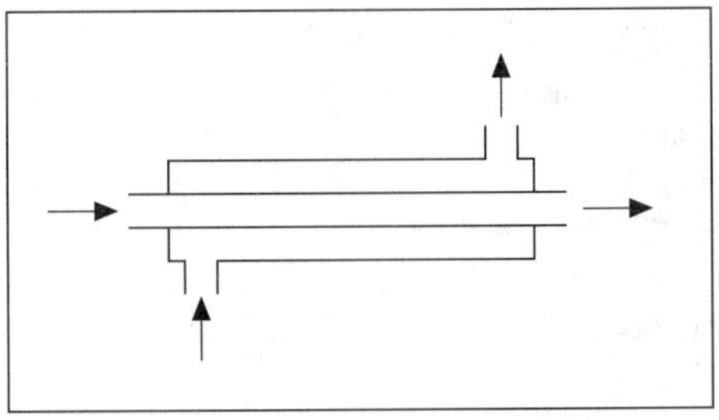

Figure 2. Co-current flow.

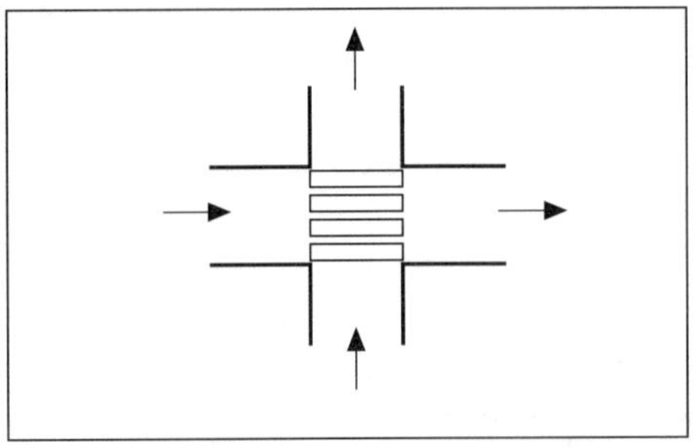

Figure 3. Crossflow.

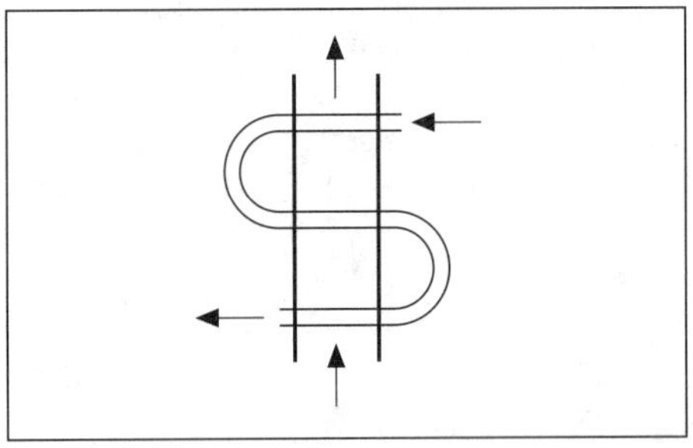

Figure 4. Cross/counter flow.

Regenerative heat exchangers

In a regenerative heat exchanger, the flow path normally consists of a matrix which is heated when the hot fluid passes through it (this is known as the "hot blow"). This heat is then released to the cold fluid when this flows through the matrix (the "cold blow"). Regenerative Heat Exchangers are sometimes known as *Capacitive Heat Exchangers.* A good overview of regenerators is provided by Walker (1982).

Regenerators are mainly used in gas/gas heat recovery applications in power stations and other energy intensive industries. The two main types of regenerator are Static and Dynamic. Both types of regenerator are transient in operation and unless great care is taken in their design there is normally cross contamination of the hot and cold streams. However, the use of regenerators is likely to increase in the future as attempts are made to improve energy efficiency and recover more low grade heat. However, because regenerative heat exchangers tend to be used for specialist applications recuperative heat exchangers are more common.

Recuperative heat exchangers

There are many types of recuperative exchangers which can broadly be grouped into indirect contact, direct contact and specials. Indirect contact heat exchangers keep the fluids exchanging heat separate by the use of tubes or plates etc.. Direct contact exchangers do not separate the fluids exchanging heat and in fact rely on the fluids being in close contact.

Heat Exchanger Types

This section briefly describes some of the more common types of heat exchanger and is arranged according to the classification given in **Figure 5**.

Indirect heat exchangers

In this type, the streams are separated by a wall, usually metal. Examples of these are tubular exchangers, see **Figure 6**, and plate exchangers, see **Figure 7**.

Tubular heat exchangers are very popular due to the flexibility the designer has to allow for a wide range of pressures and temperatures. Tubular heat exchangers can be sub-divided into a number of categories, of which the shell and tube exchanger is the most common.

A **Shell and Tube Exchanger** consists of a number of tubes mounted inside a cylindrical shell. **Figure 8** illustrates a typical

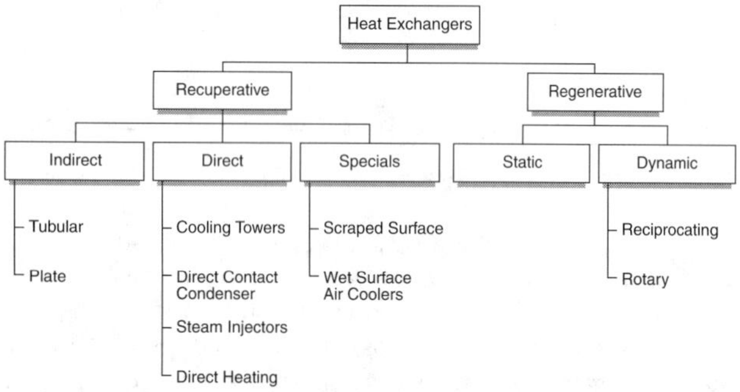

Figure 5. Heat exchanger classifications.

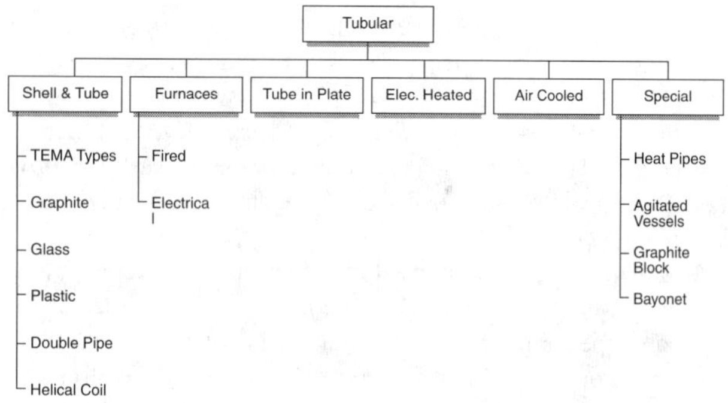

Figure 6. Tubular exchanger classification.

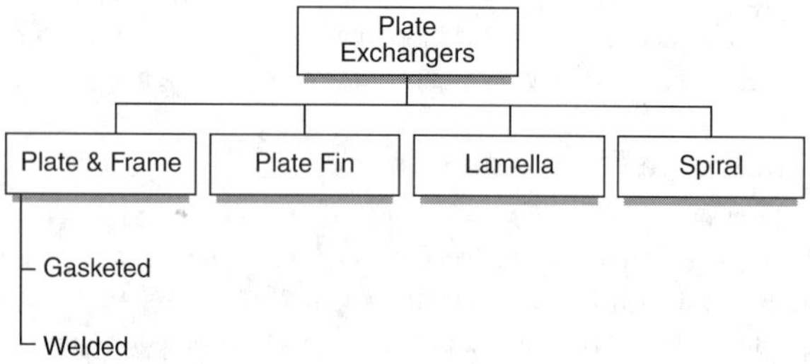

Figure 7. Plate exchanger classification.

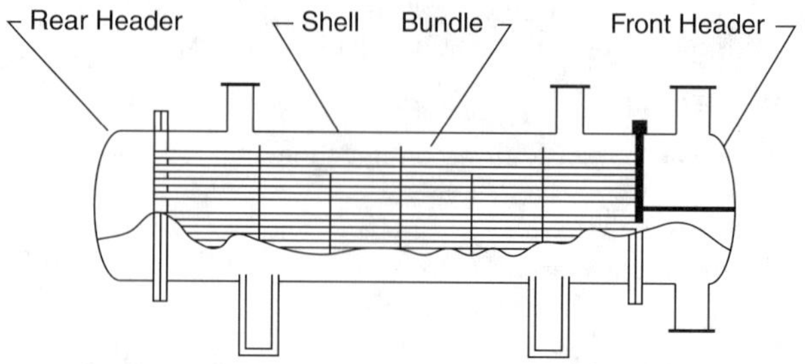

Figure 8. Shell and tube exchanger.

unit that may be found in a petrochemical plant. Two fluids can exchange heat, one fluid flows over the outside of the tubes while the second fluid flows through the tubes. The fluids can be single or two phase and can flow in a parallel or a cross/counter flow arrangement. The shell and tube exchanger consists of four major parts:

- Front end—this is where the fluid enters the tubeside of the exchanger
- Rear end—this is where the tubeside fluid leave the exchanger or where it is returned to the front header in exchangers with multiple tubeside passes
- Tube bundle—this comprises of the tubes, tube sheets, baffles and tie rods etc. to hold the bundle together.
- Shell—this contains the tube bundle

The popularity of shell and tube exchangers has resulted in a standard being developed for their designation and use. This is the **Tubular Exchanger Manufactures Association** (TEMA) Standard. In general shell and tube exchangers are made of metal but for specialist applications (e.g.: involving strong acids of pharmaceuticals) other materials such as graphite, plastic and glass may be used. It is also normal for the tubes to be straight but in some cryogenic applications helical or *Hampson coils* are used. A simple form of the shell and tube exchanger is the **Double Pipe Exchanger.** This exchanger consists of a one or more tubes contained within a larger pipe. In its most complex form there is little difference between a multi tube double pipe and a shell and tube exchanger. However, double pipe exchangers tend to be modular in construction and so several units can be bolted together to achieve the required duty. The book by E.A.D. Saunders (Saunders, 1988) provides a good overview of tubular exchangers.

Other types of tubular exchanger include:

- **Furnaces**—the process fluid passes through the furnace in straight or helically wound tubes and the heating is either by burners or electric heaters.
- Tubes in plate—these are mainly found in heat recovery and air conditioning applications. The tubes are normally mounted in some form of duct and the plates act as supports and provide extra surface area in the form of fins.
- Electrically heated—in this case the fluid normally flows over the outside of electrically heated tubes. (see **Joule Heating**)
- **Air Cooled Heat Exchangers** consist of bundle of tubes, a fan system and supporting structure. The tubes can have

various type of fins in order to provide additional surface area on the air side. Air is either sucked up through the tubes by a fan mounted above the bundle (induced draught) or blown through the tubes by a fan mounted under the bundle (forced draught). They tend to be used in locations where there are problems in obtaining an adequate supply of cooling water.

- **Heat Pipes, Agitated Vessels** and **Graphite Block Exchangers** can be regarded as tubular or could be placed under Recuperative "Specials". A heat pipe consists of a pipe, a wick material and a working fluid. The working fluid absorbs heat, evaporates and passes to the other end of the heat pipe were it condenses and releases heat. The fluid then returns by capillary action to the hot end of the heat pipe to re-evaporate. Agitated vessels are mainly used to heat viscous fluids. They consist of a vessel with tubes on the inside and a agitator such as a propeller or a helical ribbon impeller. The tubes carry the hot fluid and the agitator is introduced to ensure uniform heating of the cold fluid. Carbon block exchangers are normally used when corrosive fluids need to be heated or cooled. They consist of solid blocks of carbon which have holes drilled in them for the fluids to pass through. The blocks are then bolted together with headers to form the heat exchanger.

Plate heat exchangers separate the fluids exchanging heat by the means of plates. These normally have enhanced surfaces such as fins or embossing and are either bolted together, brazed or welded. Plate heat exchangers are mainly found in the cryogenic and food processing industries. However, because of their high surface area to volume ratio, low inventory of fluids and their ability to handle more than two streams, they are also starting to be used in the chemical industry.

Plate and Frame Heat Exchangers consists of two rectangular end members which hold together a number of embossed rectangular plates with holes on the corner for the fluids to pass through. Each of the plates is separated by a gasket which seals the plates and arranges the flow of fluids between the plates, see **Figure 9**. This type of exchanger is widely used in the food industry because it can easily be taken apart to clean. If leakage to the environment is a concern it is possible to weld two plate together to ensure that the fluid flowing between the welded plates can not leak. However, as there are still some gaskets present it is still possible for leakage to occur. Brazed plate heat exchangers avoid the possibility of leakage by brazing all the plates together and then welding on the inlet and outlet ports.

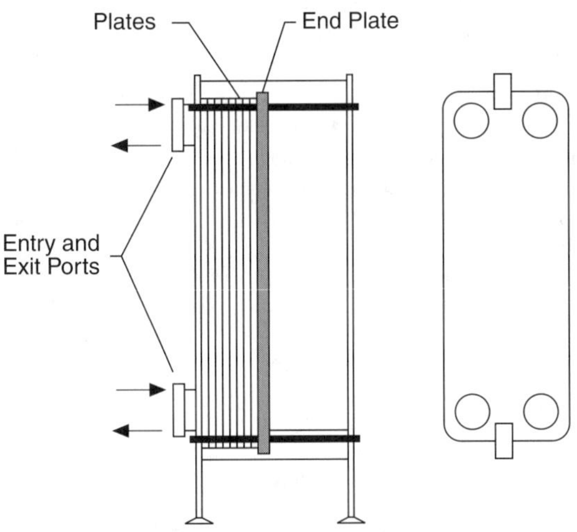

Figure 9. Plate and frame exchanger.

Plate Fin Exchangers consist of fins or spacers sandwiched between parallel plates. The fins can be arranged so as to allow any combination of crossflow or parallel flow between adjacent plates. It also possible to pass up to 12 fluid streams through a single exchanger by careful arrangement of headers. They are normally made of aluminium or stainless steel and brazed together. Their main use is in gas liquefaction due to their ability to operate with close temperature approaches.

Lamella heat exchangers are similar in some respects to a shell and tube. Rectangular tubes with rounded corners are stacked close together to form a bundle which is placed inside a shell. One fluid passes through the tubes while the fluid flows in parallel through the gaps between the tubes. They tend to be used in the pulp and paper industry where larger flow passages are required.

Spiral plate exchangers are formed by winding two flat parallel plates together to form a coil. The ends are then sealed with gaskets or are welded. They are mainly used with viscous, heavily fouling fluids or fluids containing particles or fibres.

Direct contact

This category of heat exchanger does not use a heat transfer surface, because of this, it is often cheaper than indirect heat exchangers. However, to use a direct contact heat exchanger with two fluids they must be immiscible or if a single fluid is to be used it must undergo a phase change. (See **Direct Contact Heat Transfer**)

The most easily recognisable form of direct contact heat exchanger is the natural draught **Cooling Tower** found at many power stations. These units comprise of a large approximately cylindrical shell (usually over 100m in height) and packing at the bottom to increase surface area. The water to be cooled is sprayed onto the packing from above while air flows in through the bottom of the packing and up through the tower by natural buoyancy. The main problem with this and other types of direct contact cooling tower is the continuous need to make up the cooling water supply due to evaporation.

Direct contact condensers are sometimes used instead of tubular condensers because of their low capital and maintenance costs. There are many variations of direct contact condenser. In its simplest form a coolant is sprayed from the top of a vessel over vapor entering at the side of the vessel. The condensate and coolant are then collected at the bottom. The high surface area achieved by the spray ensures they are quite efficient heat exchangers.

Steam injection is used for heating fluids in tanks or in pipelines. The steam promotes heat transfer by the turbulence created by injection and transfers heat by condensing. Normally no attempt is made to collect the condensate.

Direct heating is mainly used in dryers where a wet solid is dried by passing it through a hot air stream. Another form of direct heating is **Submerged Combustion.** This was developed mainly for the concentration and crystallisation of corrosive solutions. The fluid is evaporated by the flame and exhaust gases being aimed down into the fluid which is held in some form of tank.

Specials

The wet surface air cooler is similar in some respects to an air cooled heat exchanger. However, in this type of unit water is sprayed over the tubes and a fan sucks air and the water down over the tube bundle. The whole system is enclosed and the warm damp air is normally vented to atmosphere.

Scraped Surface Exchangers consist of a jacketed vessel which the fluid passes through and a rotating scraper which continuously removes deposit from the inside walls of the vessel. These units are used in the food and pharmaceutical industry in process where deposits form on the heated walls of the jacketed vessel.

Static Regenerators

Static regenerators or fixed bed regenerators have no moving parts except for valves. In this case the hot gas passes through the matrix for a fixed time period at the end of which a reversal occurs, the hot gas is shut off and the cold gas passes through the matrix. The main problem with this type of unit is that both the hot and cold flow are intermittent. To overcome this and have continuous operation at least two static regenerators are required or a rotary regenerator could be used.

Rotary regenerator

In a rotary regenerator cylindrical shaped packing rotates about the axis of a cylinder between a pair of gas seals. Hot and cold gas flows simultaneously through ducting on either side of the gas seals and through the rotating packing. (See **Regenerative Heat Exchangers**)

Thermal Analysis

The thermal analysis of any heat exchanger involves the solution of the basic heat transfer equation.

$$d\dot{Q} = \alpha(T_h - T_c)dA \qquad (1)$$

This equation calculates the amount of heat $d\dot{Q}$ transferred through the area dA, where T_h and T_c are the local temperatures of the hot and cold fluids, α is the local heat transfer coefficient and dA is the local incremental area on which α is based. For a flat wall

$$\alpha_w = \delta_w/\lambda_w \qquad (2)$$

where δ_w is the wall thickness and λ_w its thermal conductivity.

For single phase flow past the wall α for each of the streams is a function of Re and Pr. When condensing or boiling is taking place α may also be a function of the temperature difference. Once the heat transfer coefficient for each stream and the wall are known the overall heat transfer coefficient U is then given by

$$1/U = 1/\alpha_h + r_w + 1/\alpha_c \qquad (3)$$

where the wall resistance r_w is given by $1/\alpha_w$. The total rate of heat transfer between the hot and cold fluids is then given by

$$\dot{Q}_T = UA_T(T_h - T_c) \qquad (4)$$

This equation is for constant temperatures and heat transfer coefficients. In most heat exchangers this is not the case and so a different form of the equation is used

$$\dot{Q}_T = UA_T\Delta T_M \qquad (5)$$

where $\dot{Q}_T$ is the total heat load, U is the mean overall heat transfer coefficient and ΔT_M the mean temperature difference. The calculation of ΔT_M and the removal of the constant heat transfer coefficient assumption is described in **Mean Temperature Difference**.

Calculation of U and ΔT_M requires information on the exchanger type, the geometry (e.g. the size of the passages in a plate or the diameter of a tube), flow orientation, pure counter current flow or crossflow, etc. The total duty $\dot{Q}_T$ can then be calculated using an assumed value of A_T and compared with the required duty. Changes to the assumed geometry can then be made and U, ΔT_M and $\dot{Q}_T$ recalculated to eventually iterate to a solution where $\dot{Q}_T$ is equal to the required duty. However, in performing the thermal analysis a check should also be made at each iteration that the allowable pressure drop is not exceeded. Computer programs such as TASC from HTFS (Heat Transfer and Fluid Flow Service) perform these calculations automatically and optimise the design.

Mechanical Considerations

All heat exchangers types have to undergo some form of mechanical design. Any exchanger that operates at above atmospheric pressure should be designed according to the locally specified *pressure vessel design code* such as ASME VIII (American Society of Mechanical Engineers) or BS 5500 (British Standard). These codes specify the requirements for a pressure vessel, but they do not deal with any specific features of a particular heat exchanger type. In some cases specialist standards exist for certain types of heat exchanger. Two of these are listed below, but in general individual manufacturers define their own standards.

References

Garland, W. J. (1990) Private Communication.

Walker, G. (1982) *Industrial Heat Exchangers–A Basic Guide,* Hemisphere Publishing Corporation.

Rohsenow, W. M. and Hartnett, J. P. (1973) *Handbook of Heat Transfer,* New York: McGraw-Hill Book Company.

Saunders, E. A. D. (1988). *Heat Exchangers–Selection, Design and Construction,* Longman Scientific and Technical.

Tubular Exchanger Manufacturers, Association, (1988) (TEMA) Seventh Edition. *Shell and Tube Exchangers.*

American Petroleum Institute (API) 661: *Air Cooled Heat Exchangers for the Petroleum Industry.*

R.J. Brogan

Leading to: *Air cooled heat exchangers; Cooling towers; Direct contact heat exchangers; Double pipe heat exchangers; Feedwater heaters; Graphite block heat exchangers; Optimisation of heat exchanger design; Plate* *and fin heat exchanger; Plate and frame heat exchangers; Regenerative heat exchangers; Scraped surface heat exchangers; Shell and tube heat exchangers; Spiral heat exchangers*

R.J. Brogan

HEAT EXCHANGERS, MULTI-STREAM (see Plate fin heat exchangers)

HEAT EXCHANGERS, PLATE FIN (see Plate fin heat exchangers)

HEAT EXCHANGER STANDARDS, TEMA (see TEMA standards)

HEAT FLUX

Following from: *Heat transfer*

Heat flux (W/m^2) is the rate of thermal energy flow per unit surface area of heat transfer surface e.g. in a heat exchanger.

Heat flux is the main parameter in calculating heat transfer. A generalized classification distinguishes between heat fluxes by *convection*, heat *conduction*, and *radiation*. The heat flux vector is directed towards regions of lower temperature.

Convective heat flux is proportional to the temperature difference between solid, liquid, or gaseous media participating in heat transfer. A heat transfer coefficient serves as the proportionality factor.

Under heat conduction, the heat flux vector is proportional to and usually parallel to the temperature gradient vector. However, in anisotropic bodies the direction of the two vectors may not coincide.

In the case of simultaneous heat and mass transfer the effective heat flux may substantially, by several orders of magnitude, exceed the value due to heat conduction only.

The radiative heat flux is a flux of electromagnetic radiation and, in contrast to convection and heat conduction, may occur without any intervening medium, i.e. it can occur through a vacuum. For an idealized black body the radiation heat flux is described by Planck's law. The actual radiation flux values can be only lower than this idealised value.

Leading to: *Heat transfer coefficient; Heat flux measurements*

O.A. Gerashchenko

HEAT FLUX MEASUREMENT

Following from: *Heat flux*

Heat flux measurement devices (*heat flux meters*) are designed to obtain experimental information about the heat flux.

The majority of heat flux meters is based on the method of an extra wall which was first formulated by E. Schmidt at the beginning of the present century. Its essence is that the extra wall is located on the path of a measured heat flux. When this wall is permeated by the flux to be measured, temperature gradients and differences appear whose values are proportional to the flux. By measuring these gradients or differences, the flux can be determined.

The most successful group of sensing element designs, the functions of the extra wall material and temperature difference meter are combined. Thus, in the design of the heat flux meter with a three-layer extra wall, the intermediate layer is the intermediate thermoelectrode of a differential thermocouple whose e.m.f. is measured across the end layers.

To increase the sensitivity, separate thermometric elements contain in series a generated thermometric signal and in parallel a measured heat flux. The modern technology allows up to 2000 elements to be placed per 1 cm^2. Means are available to increase the packing density in order to elevate the sensitivity. Battery devices with an area of about 100 cm^2 allow effective measurement of heat fluxes, as low as 10^{-3} W/m^2. The upper limit of the measured heat flux density exceeds 10^7 W/m^2. Heat flux meters are thus available to cover the range 10^{-3} to 10^7 W/m^2 for temperatures 4–1000 K. At room temperature, the standard error is not beyond $\pm 1.5\%$ of the measured quantity. The error can approach $\pm 7\%$ at the boundaries of the mentioned temperature range.

O.A. Gerashchenko

HEAT FLUX METERS (see Heat flux measurement)

HEATING DEGREE DAYS (see Buildings and heat transfer)

HEATING OF BUILDINGS (see Buildings and heat transfer; Space heating)

HEATING, VENTILATION & AIR CONDITIONING, HVAC (see Air conditioning)

HEAT ISLAND (see Air pollution)

HEAT OF ADSORPTION (see Adsorption)

HEAT OF MELTING (see Melting)

HEAT OF VAPORISATION (see Latent heat of vaporisation)

HEAT PIPES

The heat pipe is a sealed system containing a liquid, which when vaporised transfers heat under isothermal conditions. The temperature of the vapor corresponds to the vapor pressure, and any temperature variation throughout the system is related directly to vapor pressure drop. The choice of liquid charge is related to the required operating temperature range of the heat pipe. This may vary from cryogenic conditions (well below O °C) to high temperature operation (above 600°C), in which case liquid metals are used (e.g. potassium, sodium or lithium).

The heat pipe has three major operating zones, namely evaporator, adiabatic section and condenser, see **Figure 1**. In the case of the elementary pipe design, liquid returns from the condenser via a wick structure. The wick is designed to provide a capillary pumping action, as described below. The heat pipe is a development of the *thermosyphon,* in which there is no wick structure and liquid is returned to the evaporator by gravity. Thus in the case of the thermosyphon the condenser region must be above the evaporator region, angle Ø in **Figure 1** being negative.

The heat pipe as we now know it was originated by Grover in Los Alamos for use in thermionic direct conversion devices. One of its main features, namely isothermalisation, is of major significance in this application. It is further possible to control the temperature of operation of the pipe by introducing a controlled pressure of inert gas, such as helium or argon. The vapor pressure of the liquid charge will be equal to that of the gas, provided operation is ensured to be of the nature illustrated by **Figure 2**. This pipe is referred to by Dunn and Reay as "gas-buffered" or "variable conductance" design.

Heat Pipe Operation Limits

The heat pipe has four major operating regimes, each of which sets a limit of performance in either heat transfer rate (axial or radial) or temperature drop. The limit for each regime is presented below for a simple cylindrical geometry heat pipe, as illustrated in **Figure 3**. These limits were catogorised by Busse and are as follows:

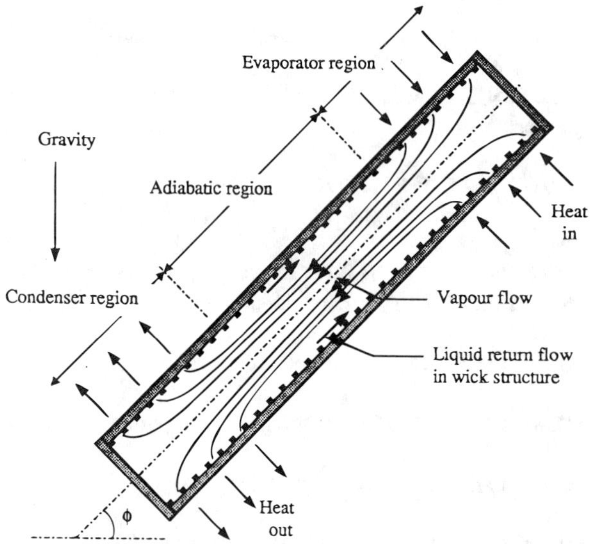

Figure 1. Heat pipe operation (for cylindrical geometry pipe).

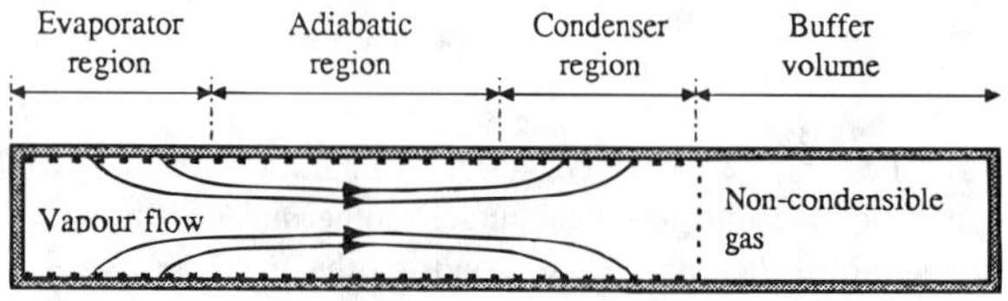

Figure 2. Variable conductance heat pipe.

Vapour Pressure or Viscous Limit

At low temperature range of operation of the working fluid, especially at start-up of the heat pipe, the minimum pressure at the condenser end of the pipe can be very small. The vapor pressure drop between the extreme end of the evaporator and the end of the condenser, represents a restriction in operation. The maximum rate of heat transfer under this restricted vapor pressure drop limit is given by:

$$\dot{Q}_V = \frac{D_v^2 h_{lg} p_v \rho_v}{64\, \eta_v l_{eff}} \tag{1}$$

where D_v is the diameter of the vapor passageway, h_{lg} is the enthalpy of vaporisation, p_v is the pressure, ρ_v the vapor density, and η_v the vapor dynamic viscosity.

NOTE: All vapor properties in Equation 1 refer to conditions at the closed end of the evaporator.

$$l_{eff} = \text{effective heat pipe length}$$

$$= \frac{l_e}{2} + l_a + \frac{l_c}{2}$$

where l_e is the length of the evaporator region, l_a the length of the adiabatic region and l_c the length of the condenser region.

Sonic Limit

At a temperature, above the vapor pressure limit, the vapor velocity can be comparable with sonic velocity and the vapor flow becomes "choked". The recommended maximum rate of heat transfer, to avoid choked flow conditions (i.e. sonic limit) is given by

$$\dot{Q}_S = 0.474\, A_v h_{lg} (p_v \rho_v)^{1/2} \tag{2}$$

where A_V is the area of the vapor passageway.

Entrainment Limit

The vapor velocity increases with temperature and may be sufficiently high to produce shear force effects on the liquid return flow from the condenser to the evaporator, which cause entrainment of the liquid by the vapor. The restraining force of liquid surface tension is a major parameter in determining the entrainment limit. Entrainment will cause a starvation of fluid flow from the condenser and eventual "*dry out*" of the evaporator.

The entrainment limit is given by

$$\dot{Q}_E = A_v h_{lg} \sqrt{\frac{\rho_v \sigma}{x}} \tag{3}$$

where σ is the surface tension of the liquid, x is the characteristic dimension of the wick surface ($\equiv 2r_\sigma$, where r_σ = effective radius of pore structure).

Circulation Limit

The driving pressure for liquid circulation within the heat pipe is given by the capillary force established within the wick structure, namely:

$$\Delta p_\sigma = \frac{2\sigma}{r_\sigma} \text{ (i.e. maximum capillary pressure)}$$

Circulation will be maintained provided:

$$\Delta p_\sigma \geq \Delta p_l + \Delta P_v + p_l g l \cos \phi \quad \textbf{(4)}$$

where Δp_l is the frictional pressure drop in liquid and Δp_V is the frictional pressure drop in the vapor.

For laminar flow conditions in the wick structure:

$$\Delta p_l = \frac{l_{eff}\dot{Q}\eta_l}{A_w K h_{lg} \rho_l} \text{ (i.e., \textbf{Darcy's Law})} \quad \textbf{(5)}$$

where $\dot{Q}$ is the rate of heat transfer, η_l the liquid viscosity, A_w the cross sectional area within the wick, K the permeability of the wick and ρ_l the liquid density.

The gravitational head ($\rho_l g 1 \cos \Phi$), may be positive or negative, depending on whether the pipe is gravity assist or working against gravity (see **Figure 1**).

In calculating the vapor pressure drop (Δp_v) it is important to ensure that the **Mach Number** $M < 0.2$ and incompressible flow conditions are assumed.

For laminar flow condition (i.e. $Re_v < 2000$) the *Hagen-Poiseuille equation* may be applied, thus

$$\Delta p_v = \frac{8\eta_v l\dot{Q}}{\rho_v h_{lg} \pi r_v^4} \quad \textbf{(6)}$$

where r_v is the radius of the vapor passageway.

For turbulent flow, the Fanning equation gives:

$$\Delta p_v = \frac{2fl\rho_v V_v^2}{D_v} \quad \textbf{(7)}$$

where l is the length of the vapor passageway, f is *Fanning friction factor* ($0.079/Re_v^{1/4}$ for $2100 < Re_v < 10^5$), V_v is the vapor velocity and D_v the diameter of the vapor passageway; and where Re_v is the Reynolds number for vapor flow.

NOTE: for laminar flow, i.e. $Re < 2100$ the *Fanning friction factor* quoted above is replaced by the *Hagen-Poiseuille* form, $f = 16/Re_v$. (See **Friction Factors for Single Phase Flow**)

The vapor pressure drop over the length of the evaporator plus the adiabatic region, for turbulent flow ($Re \geq 2000$) is given in ESDU 79012 as

$$\frac{\Delta p_{ve} + \Delta p_{va}}{\rho_v V_v^2} = \frac{0.158}{D_v Re_v^{1/4}}\left(\frac{l_e}{2.75} + l_a\right) + 0.93 \quad \textbf{(8)}$$

Boiling Limit

The temperature drop across the wick structure in the evaporator region increases with evaporator heat flux. A point is reached when temperature difference exceeds the degree of superheat sustainable in relation to nucleate boiling conditions. The onset of **Boiling** within the wick structure interferes with liquid circulation. This eventually leads to "*dry out*", which in the case of constant heat flux heating can cause "**Burn Out**" of the evaporator containment.

In the event of nucleate boiling the relationship between bubble radius and pressure difference sustainable across the curved surface is given by:

$$\Delta p = \frac{2\sigma}{R} \quad \textbf{(9)}$$

where R is the radius of the bubble.

The degree of superheat ΔT_s related to Δ_p is given by the *Clausius Clapeyron equation*

$$\frac{\Delta P}{\Delta T_s} = \frac{h_{lg}}{T_{sat}(v_v - v_l)} \quad \textbf{(10)}$$

where v_v is the specific volume of the vapor; and v_l is the specific volume of the liquid.

Since $v_l << v_v$, then

$$\Delta T_s = \frac{2\sigma T_{sat}}{\rho_v h_{lg} R} \quad \textbf{(11)}$$

Nucleation sites, at which bubbles first form, are provided by scratches or rough surfaces and by the release of absorbed gas. Dunn and Reay give the following impirical equation for the degree of superheat in a wick structure.

$$\Delta T_s = \frac{3.06\sigma T}{\rho_v h_{lg} \delta} \quad \textbf{(12)}$$

where δ is the thermal layer thickness (i.e. characteristic dimension) equal to 2.5×10^{-5} m for typical surfaces.

Using this characteristic dimension they have produced a table showing the degree of superheat for a range of candidate heat pipe working fluids, including ammonia, water and liquid metals (for high temperature operation). The liquid metals, having much higher surface tension give much higher degrees of superheat (e.g. 10°C $\Delta T_s < 54$ °C compared to $\Delta T_s \approx 2$°C for NH_3 and H_2O).

The associated evaporator heat flux with nucleate boiling is given by:

$$\dot{q}_b = \frac{\lambda_w \Delta T_s}{x} \quad \textbf{(13)}$$

where λ_w is the effective thermal conductivity of the wick (metal plus liquid) and x is the thickness of wick structure.

$$\lambda_w = \frac{\beta - \varepsilon}{\beta + \varepsilon}\lambda_l \quad (14)$$

where

$$\beta = \left(1 + \frac{\lambda_S}{\lambda_l}\right) \Big/ \left(1 - \frac{\lambda_S}{\lambda_l}\right) \quad (15)$$

and λ_s is the conductivity of the solid, λ_l the conductivity of the liquid and ε the porosity of the wick structure.

An alternative equation for the boiling limit is given in ETSU data sheet 79012 as

$$\dot{q}_b = \frac{T}{h_{fg}Z\rho_v}\left(\frac{2\sigma}{r_n} - \Delta p_\sigma\right) \quad (16)$$

where Δp_σ is the maximum capillary pressure provided by the wick (see above), r_n is the nucleate radius ($= 2 \times 10^{-6}$ m) and Z the thermal impedance of the wick.

The above limitations are seen to relate to temperature, according to working fluid, in the manner illustrated by **Figure 3**. The choice of working fluid must be such that the heat pipe is operated at a temperature well beyond the viscous limit, even at start up.

Thermosyphon

The *thermosyphon* differs from the heat pipe, in having no wick structure. The device can therefore only operate with the condenser above the evaporator with gravity-assist liquid flow return. Equations relating to the various limits of performance a two-phase closed thermosyphon are given in ESDU data sheet 81038. The viscous and sonic limits are the same as for wicked heat pipes and the equation for the boiling limit and counter-current flow limits are summarised below.

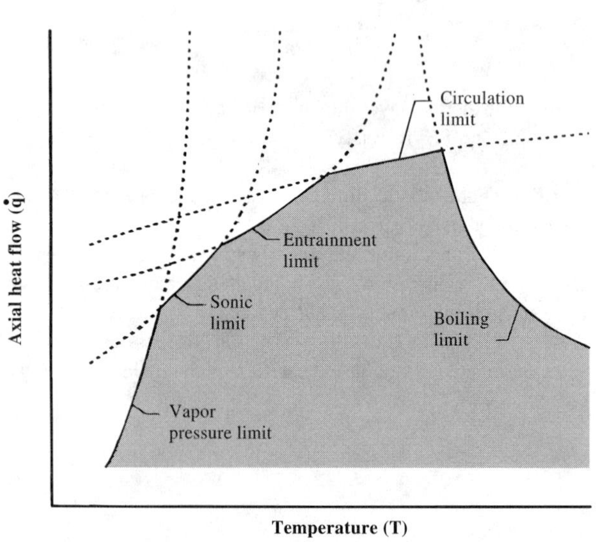

Figure 3. Heat pipe limits—operating envelope.

Boiling Limit

The boiling limit occurs when a stable vapor film is formed between the liquid and the evaporator wall. The maximum heat flux as given in ESDU 81038:

$$\dot{q}_b = 0.12\ h_{lg}\rho_v^{1/2}[\sigma g(\rho_l - \rho_v)]^{1/4} \quad (17)$$

Counter-Current Flow Limit

This condition relates to entrainment or flooding. The maximum heat transfer under this condition is given by

$$\dot{Q}_{max} = A_v h_{lg} f_1 f_2 f_3 \rho_v^{1/2}[\sigma_g(\rho_l - \rho_v)]^{1/4} \quad (18)$$

where f_1 is a function of the **Bond Number**, defined as

$$B_o = D_v\left[g\left(\frac{\rho_l - \rho_v}{\sigma}\right)\right]^{1/2} \quad (19)$$

f_1 can be found from ESDU 81038, but is seen to have a value of 4 at $B_0 = 1.0$ and a value of 8 at $B_0 = 10.0$.

The factor f_1 is a function of a dimensionless parameter K_p, which is defined as

$$K_p = \frac{p_v}{[\sigma g(\rho_l - \rho_v)]^{1/2}} \quad (20)$$

and $f_1 = K_p^{-0.17}$ if $k_p \leq 4 \times 10^4$

$f_2 = 0.165$ if $K_p > 4 \times 10^4$

The factor f_3 is a function of the inclination of the heat pipe. When vertical $f_3 = 1$. The varitation of f_3 with both angle of inclination of the pipe and Bond number is given in Figure 2 of ESDU 81038.

Figure of Merit

In selecting the working fluid for a heat pipe or thermosyphon it is necessary to ensure that the device operates within the above defined limits. The choice of working fluid very much depends on the thermophysical properties of the fluid as well as the mode of operation of the device. A figure of merit (Φ) may be used to establish the relative performance of a range of prospective working fluids. Values of Φ are given for a range of fluids in ESDU 80017 and a plot of Φ versus temperature illustrates the influence of the working fluid properties on maximum heat flux.

For capillary driven heat pipes

$$\Phi_l = \frac{h_{lg}\rho_l\sigma}{\eta_l} \quad (21)$$

see **Figure 4** for Φ_l versus Temperature for a range of fluids.

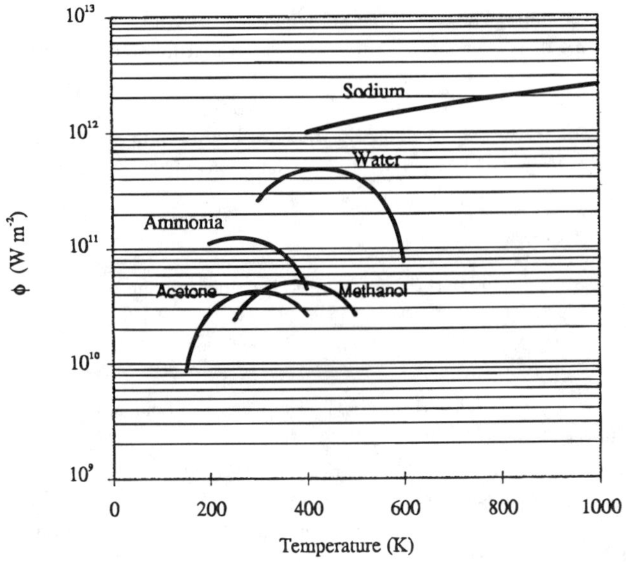

Figure 4. Figures of merit (Φ) for different working fluids in capillary driven heat pipes.

For a thermosyphon

$$\Phi_2 = \left[\frac{h_{lg}\lambda_l^3\rho_l^2}{\eta_l}\right]^{1/4} \tag{22}$$

Practical Design Consideration

The heat pipe may be used to transfer heat under near isothermal conditions and may also be used to effect temperature control, as illustrated by **Figure 2**. The use of a buffer gas to control vapor pressure and hence vapor temperature is seen to be a very effective method of temperature control. Both passive and active techniques are illustrated in Heat Pipes by Dunn and Reay. There is also the potential of enhanced heat pipe performance, when operating in the capillary limit regime, with use of composite wick structure design. (Ref. Dunn and Reay). A further advantage of the heat pipe is its application as a thermal transformer, see **Figure 5**.

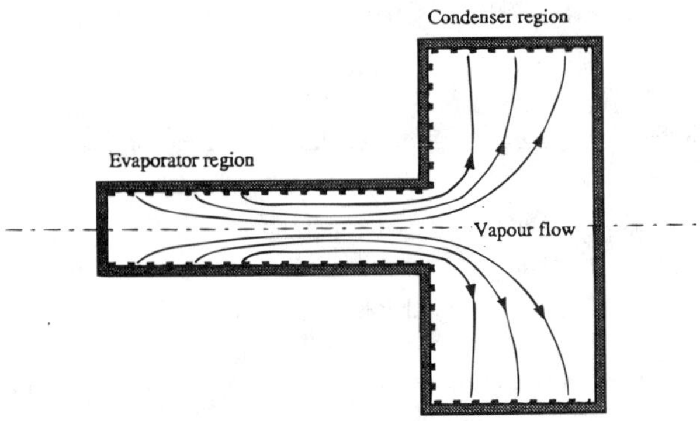

Figure 5. Thermal transformer.

Porous Element Boiler

The concept of vaporisation of a fluid in a heated porous element was developed firstly at Harwell by Dunn and Rice in the late 1960's for establishing a nuclear reactor design using this principle, and secondly at the University of Reading, leading to the successful submission of a PhD thesis by Rice (1971). Work at Reading lead to the use of the porous element heater for such applications as a fast response vapor diffusion vacuum pump, jointly developed with AERE Harwell and Edwards High Vacuum Ltd. The short residence time for liquid heating and evaporation was exploited in further work associated with pyrolytic chemical reactions.

The principle of vaporisation within a porous element, compared to vaporisation from a plain surface, is illustrated in **Figure 6** and 7. It is seen that stable boiling can only be achieved in a porous media if a uniform flow regime is established. The means for achieving this condition was brought about by the use of a dispenser region through which the liquid feed was fed into the element, see **Figure 8**. The dispenser also provided a thermal barrier to prevent subcooled boiling at inlet and porous alumina, with small pore size (typically 1–5 μm diameter pore) and low permeability, and produced high pressure drop compared to pressure drop across the heated porous media. Both flat plate and cylindrical geometry

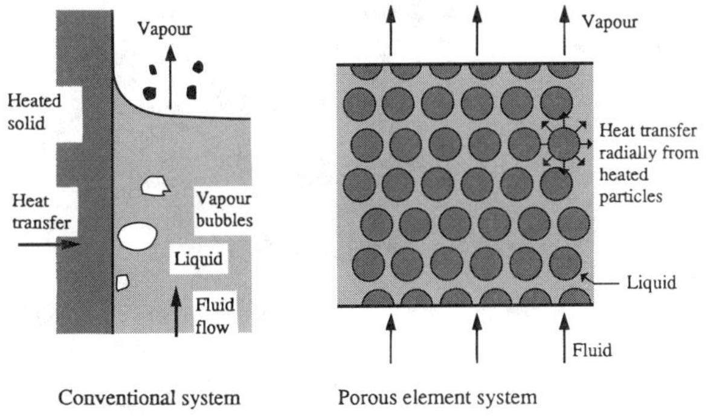

Figure 6. Principle of vaporisation within a porous element.

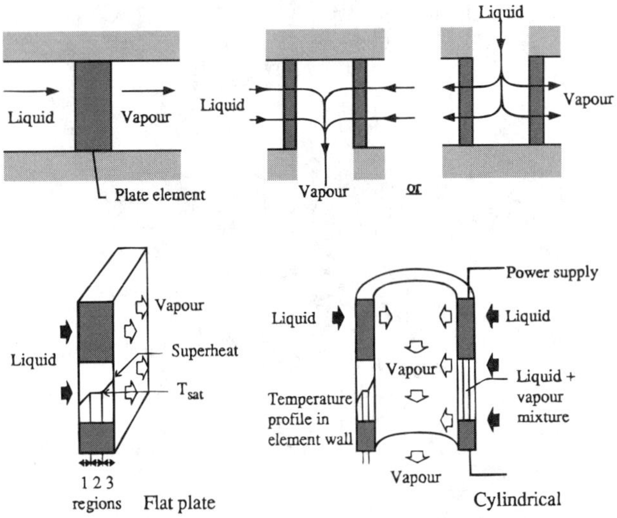

Figure 7. Porous element heating and vaporisation.

porous element boilers were constructed, with stable boiling and superheat in a single pass, see **Figure 8**.

The concept was developed using electrically heated porous elements, see **Figure 7**. Specific power ratings in excess of 1kW/cm^3 of element were achieved both when vaporising water and freon. In the case of freon, evaporation and superheat was achieved uniformly with a porous element in excess of 1m long.

It was originally conceived that the porous element boiler could be developed to provide a new concept of boiling water reactor design, see **Figure 9**. The porous element would consist of packed enriched UO_2 coated particles contained in a porous ceramic 'dispenser' tube. The reactor vessel would be fed with water through porous dispenser tubes. It was conceived that this reactor design would permit both boiling and superheating in a single pass through the porous element *"fuel rods."* An experimental *"in-pile"* steam generator was designed, as illustrated in **Figure 8**, in the hope that the concept may be demonstrated under nuclear heating conditions.

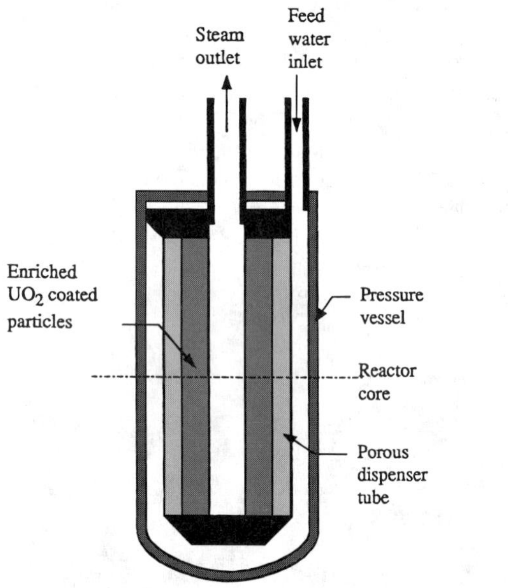

Figure 8. Schematic of in-pile porous element steam generator.

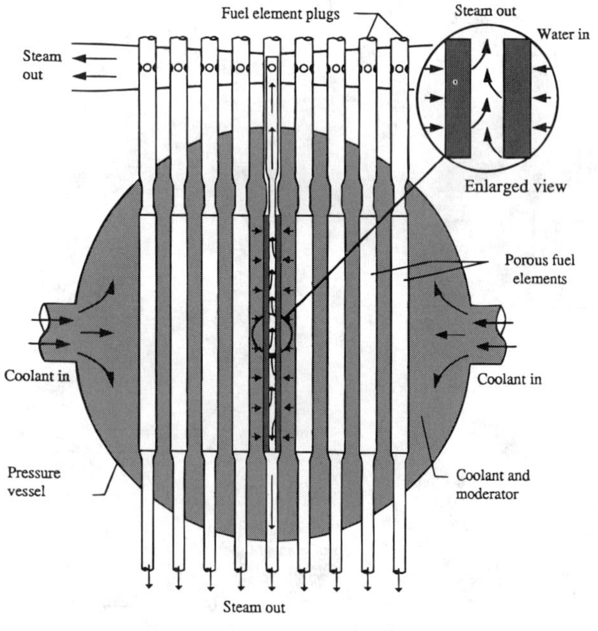

Figure 9. Porous element pressurised reactor.

References

Busse, C. A. (1973) Theory of ultimate heat transfer limit of cylindrical heat pipes, *Int. J. Heat and Mass Transfer,* 16, 169–186.

Chisholm, D. (1971) *The Heat Pipe,* Mills and Boon Ltd., London.

Cotter, T. P. (1965) *Theory of Heat Pipes,* LA 3246–MS, 26 March 1965.

Dunn, P. D. and Reay, D. A. (1994) *Heat Pipes,* 4th Edition, Pergamon.

Grover, G. M., U.S. Patent 3229759, Filed 1963.

Grover, G. M., Cotter, T. P., and Erickson, G. F., (1964) Structures of very high thermal conductance, *J Appl. Phys.,* 35, p. 1990.

Fulford, D., (1989) *Variable Conductance Heat Pipes,* PhD Thesis, University of Reading, U.K.

Heat Pipes—General Information in their Use, Operation and Design, ESDU data sheet 80013, Aug. 1980.

Heat Pipes—Performance of Capillary-driven Design, ESDU data sheet 79012, Sept. 1979.

Heat-Pipes—Properties of Common Small-pore Wicks, ESDU data sheet 79013, Nov. 1979.

Heat Pipes—Performance of Two-phase Closed Thermosyphons, ESDU data sheet 81038, Oct. 1981.

Rice, G., (1971) *Porous Element Boiler,* PhD Thesis, University of Reading, U.K.,

Rice, G., Dunn, P. D., Oswald, R. D., Harris, N. S., Power, B. D., Dennis, H. T. M., and Pollock, J. F. (1977) An industrial vapor vacuum pump employing a porous element boiler, *Proc. 7th Int. Vacuum Conf.,* Vienna.

Rice, G., Dunn, P. D., (1992) 'Porous Element Boiling and Superheating', *8th International Heat Pipe Conference,* Beijing, Sept. 1992, Publ. *Advances in Heat Pipe Science and Technology,* Ed. by M. A. Tangze—Int. Academic Publishers, ISBN 7-80003-272 1/T 9.

Thermophysical properties of heat pipe working fluids: operating range between −60°C and 300°C, ESDU data sheet 80017, Aug. 1980.

G. Rice

HEAT PROTECTION

Heat protection is a system of measures taken for preventing or decreasing the heat flow from the surrounding medium to the surface of the body being protected. Six basic techniques of heat removel or absorption are currently known:

- by thermal conductivity and heat capacity of a heat-absorbing layer of a substance,
- by releasing (re-radiating) absorbed heat to the surrounding medium,
- by convection (bleeding) of a coolant through a system of ducts in the subsurface layer,

- by coolant percolation through permeable wall with subsequent injection into the surrounding medium (see **Transpiration Cooling**),
- by sacrificial ablation of the surface layer (see **Ablation**),
- by producing an electromagnetic effect on the surrounding medium.

Under some operating conditions a combination of two or more techniques may be used for heat removal and absorption. The mechanisms and limits of applicability of methods most widespread in practice are set forth below.

Thermal conductivity/heat capacity

Heat protection systems based on thermal conductivity and heat capacity or, in other words, on accumulation of thermal energy are in essence low-temperature ones. They do not permit overheating of the surface layer above the temperature of physicochemical transformation. The capability of transferring energy depends on the thermal conductivity λ of the surface layer. Thus, by Fourier's law:

$$\dot{q}_0 = -\lambda\left(\frac{\partial T}{\partial n}\right)_{n=0}.$$

Where $\dot{q}_0$ is the heat flux into the surface layer and $(\partial T/\partial n)_{n=0}$ is the temperature gradient at the surface (n = 0). The maximum quantity of heat that can be absorbed by this system in time τ is determined by the expression

$$Q_\Sigma A_s = \int_0^\tau \dot{q}_0 A_s dt = M\int_{T_0}^{T_p} c dT = \bar{c}M(T_p - T_0),$$

where M is the mass of a substance, c and $\bar{c}$ are the true and the mean (over the temperature interval) heat capacities, T_p and T_0 are the limiting and the initial temperatures of the substance, A_s is the area of the heated surface.

The efficiency of this cooling technique is the higher, the higher the heat capacity of a material $\bar{c}$, its thermal conductivity λ and the limiting temperature (or the temperature of physicochemical transformation) T_p.

Table 1 presents the thermal characteristics of copper, tungsten, and graphite at T_0 = 20°C, while the melting point $T_p = T_m$ is taken as a limiting temperature.

Table 1. Heat-absorbing materials properties

Material	λ, W/mK	ρ, kg/m³	c, kJ/kgK	T_p, K
Copper	368	8950	0.37	1370
Tungsten	150	19,300	0.08	3640
Graphite	130	2190	1.68	4000

In order to determine the time of functioning of the heat-absorbing system τ_T and the mass needed M, it is necessary to solve the problem of heat propagation in the solid with a given distribution of heat flux $\dot{q}_0(\tau)$. We assume that the $\dot{q}_0(\tau)$ dependence can be approximated by the second-degree polynomial $\dot{q}_0(\tau) = b_2\tau^2 + b_2\tau + b_0$, where b_0, b_1, and b_2 are constant. Then the temperature of the heated surface will vary with time as

$$(T_w - T_0) = \frac{2\sqrt{\tau}}{\sqrt{\pi\lambda\rho c}} = \left[\frac{8}{15}b_2\tau^2 + \frac{2}{3}b_1\tau + b_0\right].$$

Hence, it follows, in particular, that for a time-constant heat flux $\dot{q}_0 = b_0$ the temperature of the body surface T_w reaches the limiting value T_p in the time interval

$$\tau_T = \frac{\pi}{4}\frac{\lambda\rho c}{\dot{q}_0^2}(T_p - T_0)^2.$$

The operating range $\tau_T^{3/2} = 3/4\,(T_p - T_0)/b_1\sqrt{\pi\lambda\rho c}$ results from a linear variation of heat flux $\dot{q}_1 = b_1\tau$.

In this time interval, the *heat wave* will propagate in the coating to the depth δ_T. We assume for definiteness that the depth of the heated layer δ_T corresponds to the distance from an outer (heated) surface to the isotherm $T = T_\delta$, where $\theta_\delta = (T_\delta - T_0)/(T_w - T_0)$ is a preset small value, for instance $\theta_\delta = 0.05$. In this case, the temperature of the outer surface veries monotonically, $\delta_T = k\sqrt{a\tau}$ with $k \cong \theta_\delta^{-0.3}$.

Since $\tau \le \tau_T$, δ_T or the dimensionless number $Bi = (\dot{q}_0\delta_t)/[\lambda(T_p - T_0)]$ cannot grow indefinitely.

It can be shown that it is not advisable to use heat-absorbing coatings, even metals, with a thickness over 25 to 50 mm. It is constraints imposed on the time τ_T and the needed thickness δ_T or mass M of the heat-absorbing layer that establish the applicability range of the thermal conductivity/heat capacity heat protection technique in the $\dot{q} - Q_\Sigma$ coordinate (**Figure 1**).

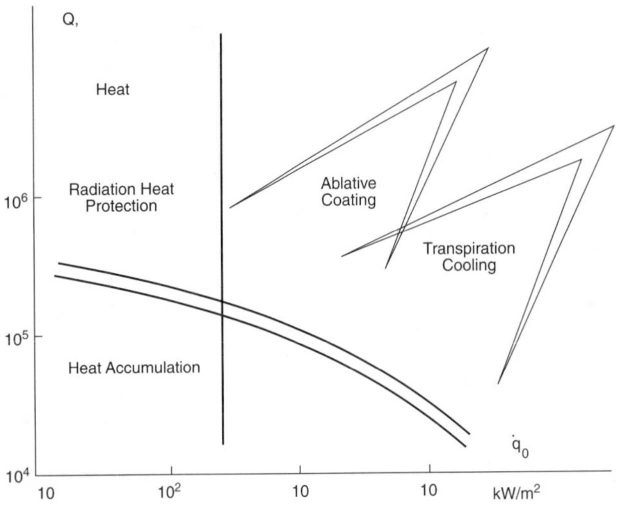

Figure 1. Regions of applicability of various heat protection methodologies.

Radiation cooling systems

Systems with partial release of the thermal energy supplied are also known as radiation cooling systems. The maximum quantity of energy that can be radiated by a body at a given temperature on a given wavelength is governed by the Planck law. **Figure 2** depicts the spectral density of radiation $E_{0\lambda}(T_w)$ for different temperatures of a black body surface. It is apparent that the maximum radiation shifts toward the region of short wavelengths in accordance with *Wien's displacement law*

$$\lambda_{max} T_w = 2897.6\ \mu mK.$$

The distribution of relative monochromatic radiation density is shown in **Figure 3**, where it can be seen that 80 per cent of radiation lies in the λ_T region varying from 2000 to 8000 μm K.

The total radiation energy emitted per unit time from a unit surface is

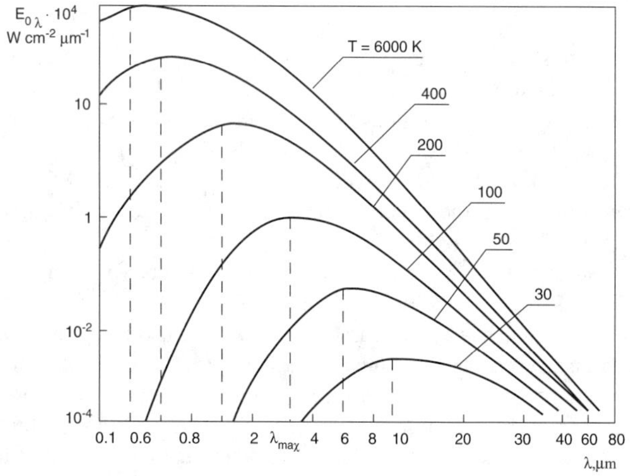

Figure 2. Radiation spectral density for a black body.

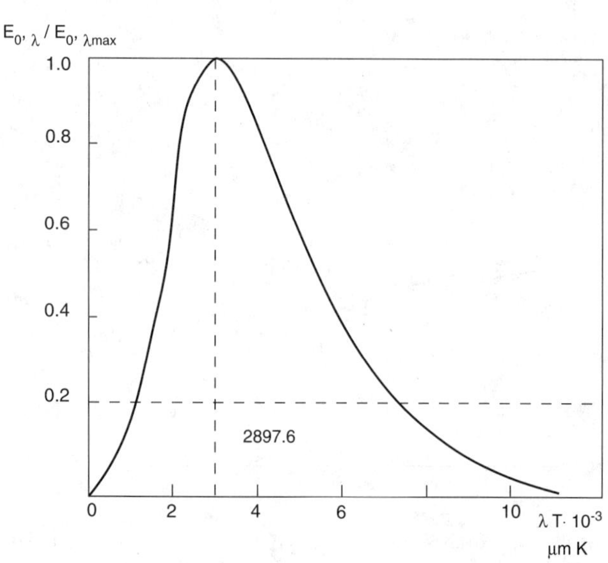

Figure 3. Distribution of relative monochromatic radiation density.

$$\dot{q}_R = \int_0^\infty E_{0\lambda}(T_w)d\lambda = \sigma T_w^4.$$

This relation is known as the Stefan-Boltzmann law for a black body. Here the constant $\sigma = 5.710^{-8}$ W/(m^{3}K^4).

Real substances are not black radiators. They radiate on any wavelength only a part of maximum possible energy $E_{0\lambda}(T_w)$ that is equal to $\varepsilon_\lambda E_{0\lambda}(T_w)$. The coefficient ε_λ is called spectral emissivity.

ε_λ should be distinguished from the integral emissivity

$$\varepsilon = \int_0^\infty \varepsilon_\lambda E_{0\lambda}(T_w)d\lambda \Big/ \int_0^\infty E_{0\lambda}(T_w)d\lambda.$$

The spectral emissivity factor of solids depends slightly on temperature T_w of the radiating surface, but only changes greatly with wavelength λ. The result is that the integral emissivity factor ϵ depends substantially on temperature, because the maximum in $E_{0\lambda}$ shifts, with growing T_w, toward the region of the short wavelength (Wein's displacement law).

The shape of ε_λ versus λ curves (**Figure 4**) fundamentally differs for polished metals or, in general case, conductors (curves 1 and 2) and oxides, or dielectrics (curves 4 and 5). At the room temperatures metals are characterized by a low integral emissivity factor, but surface contamination (oxidation, roughness) may equalize emissivity factors of metals and dielectrics (**Figure 4**, curve 3). With increasing temperature the integral emissivity grows for metals and drops for dielectrics (**Figure 5**).

The idea of equalizing the input convective heat flux $\dot{q}_0$ and the surface-radiated heat flux $\dot{q}_R = \varepsilon\sigma T_w^4$ is the basis of the radiation cooling method. If $\dot{q}_R = \dot{q}_0$, i.e. the substrate is perfectly heat-insulated, the surface temperature acquires an equilibrium value

$$T_w \to T_{eq} = \sqrt[4]{\dot{q}_0/(\varepsilon\sigma)}.$$

The radiation cooling method is used in particular on a shuttle spacecraft. Since T_w cannot exceed the temperature of phase or physicochemical transformation T_p, the applicability of this method is bounded by the heat flux per unit area $\dot{q}_0$ (**Figure 1**).

In principle, protection from radiation heat fluxes is also possible. Thus, in deep space, temperature-controlling coatings are used that possess a low emittance, or bulk absorption coefficient, in the visible spectral region in which solar radiation band largely lies and a high ε_λ in the IR region, where the coating radiates spontaneously (bearing in mind the low temperature of its surfaces).

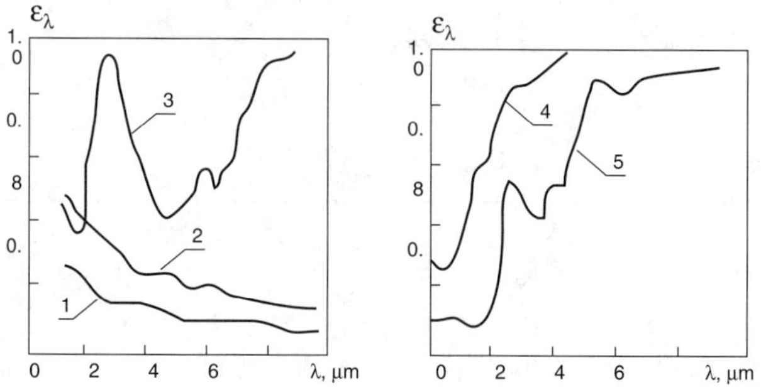

Figure 4. Typical curves for spectral emissivity.

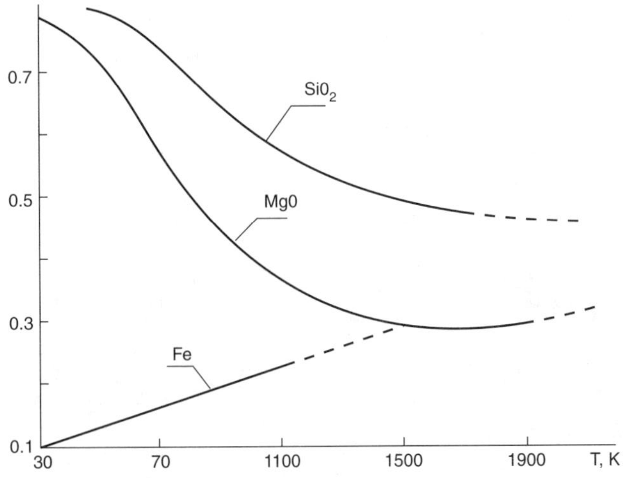

Figure 5. Variation of integral emissivity with temperature.

Convection cooling systems

Convective cooling systems have gained wide acceptance in engineering (**Figure 6**). Heat from a hot surface 3 is transferred to cooling liquid or gas pumped from a reservoir 1 by a special feeding system 2. Convective cooling systems may be classified as "closed" or "open"; in the open system, the coolant is discharged to the surroundings. In the closed system, the coolant flows in a closed circuit. A necessary component of closed systems is a heat exchanger (4) in which heat from the coolant is transferred to another heat transfer medium. In this case, the amount of coolant needed in the closed circuit does not depend on the duration of operation. On the whole, the applicability of this method is limited by a number of factors among which is the limiting temperature of the outer wall $T_w < T_p$. Under a stationary operating regime, the temperature of the outer heated wall is calculated using the set of equations

$$T_w - T_i = \dot{q}_0\delta/\lambda$$

$$\dot{q}_0 A_s = c\dot{M}(T_i - T_0),$$

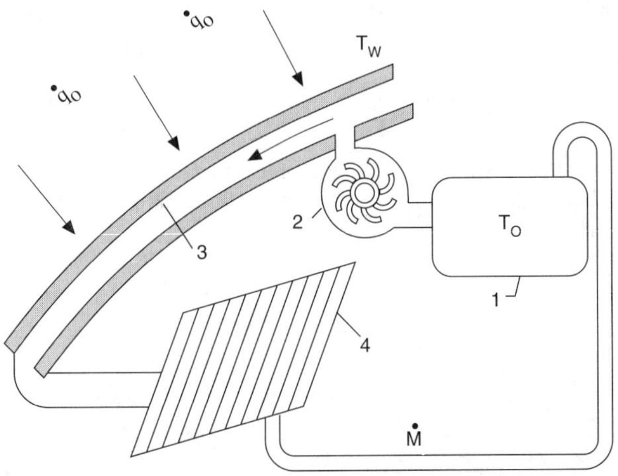

Figure 6. Convective cooling system.

where T_i and A_s are the temperature and the area of the inner surface of the heated wall, $\dot{q}_0$ the heat flux from the outside, δ the wall thickess, c and T_0 the coolant heat capacity and the temperature at the system outlet, and $\dot{M}$ the coolant flow rate. Among gaseous coolants the highest heat capacity (c = 14.5kJ(kg)) is characteristic of hydrogen, while amongst liquid coolants, fluids such as water and alcohol are the most widely used. Sodium or lithium melts can be used for cooling at high wall temperatures.

Extension of the range of permissible heat fluxes is possible by either increasing the coolant $\dot{M}$ flow rate or by raising its temperature T_i. The latter may bring about the boiling regime on the inner surface with additional heat being removed as latent heat of vaporization. A voluminous amount of literature discusses these problems, in particular as applied to cooling of units at thermal power stations and nuclear power plants.

Liquid-propellant rocket engines commonly employ a convective cooling system in which the fuel is used as a coolant before it enters the combustion chamber and there reacts with oxidizers.

Injection cooling systems

Systems with a permeable wall permeated by the coolant can be implemented as a film, transpiration, or jet curtain cooling (**Figure 7**). Injection of a gas or a liquid directly into the near-wall layer of a high-temperature free stream makes this layer thicker. The free stream is displaced from the surface protected which results in diminishing of heat transfer rate. This cooling technique is sometimes called active cooling. It is distinguished from ablation, first, by maintaining the geometry of the body being protected and, second, by the opportunity of controlling the coolant flow rate to maintain the surface temperature at the desired level.

Transpiration cooling is the easiest to realize. Its mechanism consists of two physical processes:

1. internal heat transfer during which the gaseous coolant is filtered through a porous (permeable) matrix (wall) and absorbs part of thermal energy delivered from the outside (see **Porous Medium**),
2. external heat transfer when the gaseous coolant leaves the wall and penetrates into the boundary layer diluting and displacing the high-temperature free stream fluid from the surface.

The latter process strongly depends on the regime of flow around the body, but it makes transpiration cooling highly efficient in relation to convective cooling. Sometimes this technique of heat protection is referred to as *mass transfer cooling*, which also emphasizes the role the transfer processes play in the boundary layer.

In a simplified formulation, external heat transfer is reduced by making an allowance for the injection effect. A gaseous coolant penetrating into the boundary layer is heated from the surface temperature T_w to a temperature close to gas temperature T_e on the outer layer boundary.

If the gas injected and the free stream are of the same composition, then it can be assumed, by analogy with thermal conductivity of solids, that absorption of heat in the boundary layer corresponds to the equation

$$\dot{q}_0 - \dot{q}_w = \dot{q}_b = \gamma G_w(h_e - h_w),$$

where $\dot{q}_0$ and $\dot{q}_w$ are the heat fluxes to impermeable and porous

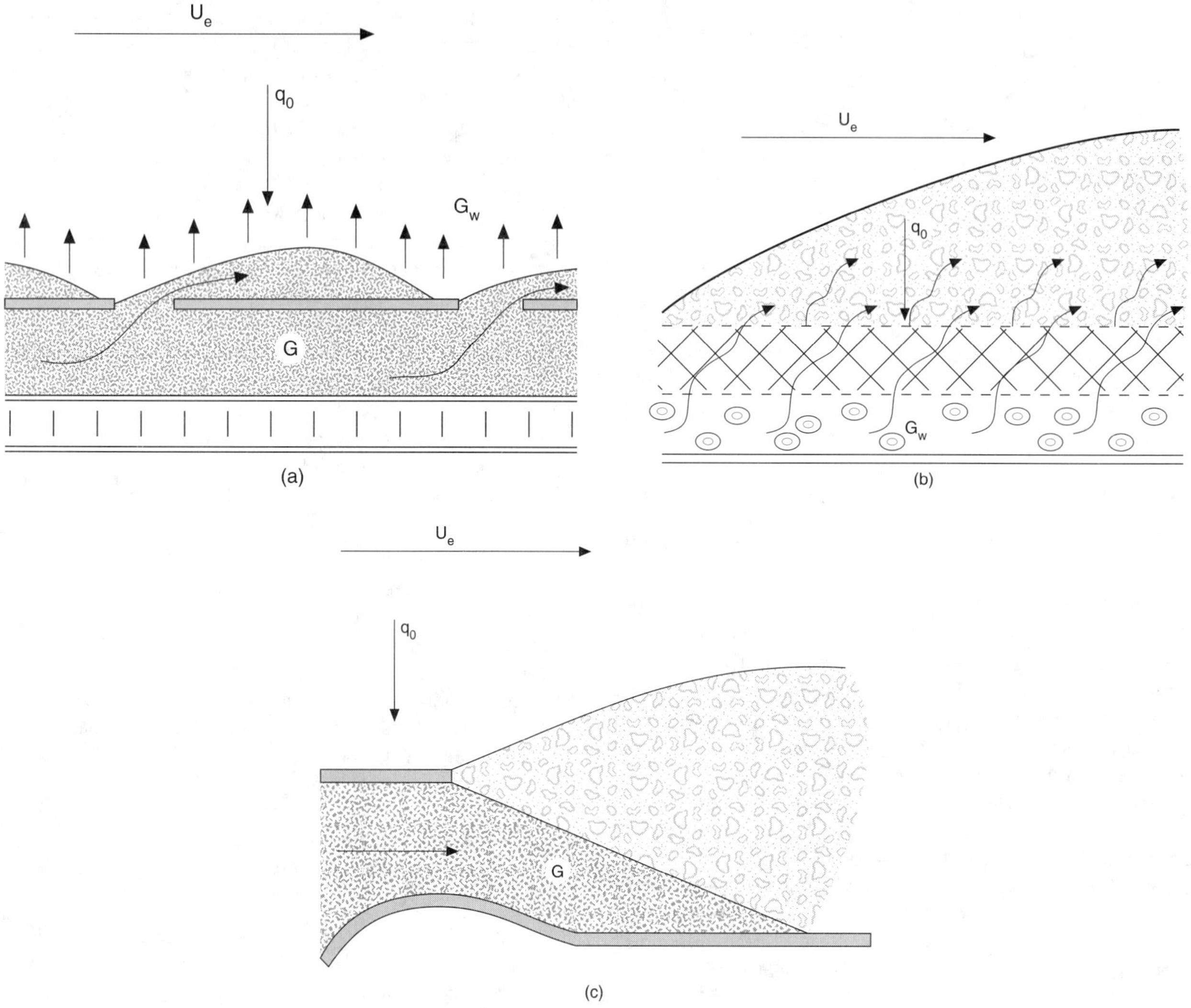

Figure 7. Injection cooling systems: (a) film cooling, (b) transpiration cooling, (c) jet curtain cooling.

surfaces respectively, G_w the specific flow rate (mass flow per unit area of surface) of the coolant, h_e and h_w the stagnation gas enthalpies on the outer boundary of the boundary layer and on the wall respectively.

The parameter γ, commonly referred to as the injection coefficient, must take account of incomplete heat absorption by the coolant due to the drift by the free stream at a temperature lower than T_e. In other words, this is the coefficient allowing for fluidity of the coolant due to convective and diffusion heat transfer in the boundary layer.

Making use of the traditional representation for the heat flux

$$\dot{q}_0 = (\alpha/c_p)_0(h_e - h_w), \qquad \dot{q}_w = (\alpha/c_p)_w(h_e - h_w)$$

yields for the injection effect a linear relation (**Figure 8**) where $\bar{G}_w = G_w/(\alpha/c_p)_0$ is the dimensionless flow rate of the coolant.

Some merely qualitative estimates indicate that the injection coefficient γ can be modified such that different physical properties of the coolant and the free stream are taken into account. To this end it is sufficient to introduce into it the ratio of the injected $\tilde{M}_v$ to the free $\tilde{M}_e$ gas molecular masses

$$\gamma = \alpha\left(\frac{\tilde{M}_e}{\tilde{M}_v}\right)^b.$$

Table 2 gives the values of injection coefficient for various gases for laminar regime flow in the vicinity of a blunt body stagnation point. These data can be represented approximately as:

$$\gamma = 0.6\left(\frac{\tilde{M}_e}{\tilde{M}_v}\right)^{0.24}.$$

However, the range of applicability of a linear approximation for the injection effect is, obviously, limited. First, even injection of homogeneous gases such as carbon dioxide gives rise to a substantial deviation from a linear dependence when the heat flux attains the level $(q_w/q_0) \leq 0.5$. Therefore, another approximation is required for calculating intense injection heat transfer. Preferable among many others is a quadratic dependence on the dimensionless coolant flow rate (**Figure 8**)

$$\dot{q}_w/\dot{q}_0 = [3(\gamma\bar{G}_w)^2 + \gamma\bar{G}w + 1]^{-1}$$

in which the injection coefficient γ is related, just as in the linear approximation, to the ratio of molecular masses $\tilde{M}_e/\tilde{M}_v$.

In contrast to blowing of homogeneous chemically inert gases, the real processes may involve changes in the thermodynamic and transfer properties, oxidizer suction from the outer flow, and its interaction with the coolant components. **Figure 8** presents the results of numerical solution of the equation for the laminar boundary layer on a glass-reinforced plastic ablative surface and a linear and a quadratic approximations of the injection effect are plotted.

It is important to note that an analogy of heat and mass transfer is also satisfactorily retained in injection, therefore, the ratio of mass transfer coefficients β_w/β_0 can be estimated using the same approximating relations as the ratio $\dot{q}_w/\dot{q}_0$, however, the injection coefficient is

$$\gamma_\beta = 0.7\left(\frac{\tilde{M}_e}{\tilde{M}_v}\right)^{0.55}.$$

In order to account for the effect of coolant injection in a turbulent boundary layer the formula

Table 2. Injection coefficients for various gases

Injected gas	Coefficient	The validity range for a linear approximation in the (q_w/q_0) value
Air	0.65	from −0.5 to 1.0
H_2	1.90	from 0 to 0.4
H_e	1.03	from 0 to 0.75
CO_2	0.67	from 0 to 0.8
SiO_2 vapors	0.50	from 0 to 1.3

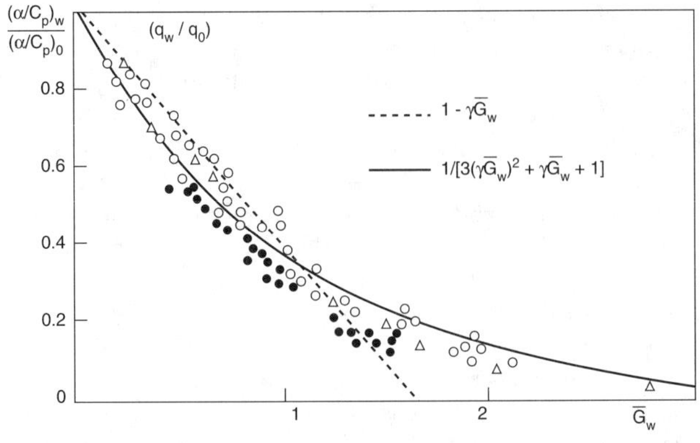

Figure 8. Effect of dimensionless injection flow rate on heat flux in transpiration cooling.

$$\dot{q}_w/\dot{q}_0 = \exp\left[-0.37\left(\frac{\tilde{M}_e}{\tilde{M}_v}\right)^{0.7}\bar{G}_w\right]$$

is recommended. Proceeding from the analysis of experimental data on the injection effect on heat transfer in the turbulent boundary layer on a plane plate, we can also recommend a linear approximation of the type used for the laminar flow. But $\gamma_T \cong 0.5\ \gamma_l$, i.e. the blowing efficiency is reduced twofold.

If a perforated surface is to be used choice of hole diameter and concentration of holes depends on the boundary layer thickness. It is required that the hole diameter be no more than the boundary layer thickness δ while the spacing between the neighboring holes be less than 5δ. As the experiments have shown, under the turbulent flow regime in the boundary layer perforation cooling is equivalent to transpiration cooling only at relative flow rates $\bar{G}_w \leq 0.5$. As the flow rate increases, or if the concentration of holes is low, perforation cooling is less efficient than transpiration cooling.

Let us compare the efficiency of transpiration and convective cooling. We assume an ideal situation in which the temperature difference in the wall being cooled is absent and the heat transfer coefficient between the coolant and the wall is infinitely large. Thus, we assume that the coolant temperature is equal to that of the outer surface. The heat balance in the convective cooling system can be reduced to

$$\dot{q}_0 A_w = \bar{c}\dot{M}(T_w - T_0) \quad \text{or} \quad q_0 = G_c(h_w - h_e).$$

Here $G_c = \dot{M}/A_s$ is the flow rate of the coolant per unit area of the protected surface. If, however, the transpiration cooling is used under the same heat conditions, then

$$q_0 = G_t(h_w - h_0) + q_b = G_t[(h_w - h_0) + \gamma(h_e - h_w)].$$

Here a linear approximation is used for estimating the injection effect. With consideration for equality of heat fluxes and the temperatures of the walls being cooled we obtain

$$G_t/G_e = \frac{1}{1 + \gamma[(h_e - h_w)/(h_w - h_0)]}.$$

The higher the temperature of the free stream T_e in relation to the wall protected T_w, the more efficient is transpiration cooling compared with convective cooling. The range of applicability of transpiration cooling is not in fact limited. As is shown in **Figure 1**, in the $Q_\Sigma - \dot{q}_0$ variables all the regimes to the right of or above the limiting curves can be realized only for permeable and ablative coatings. The last, sixth, method of heat protection based on an exposure of the surrounding medium to electromagnetic force is so far only within the scope of laboratory tests.

Leading to: *Ablation; Transpiration cooling; Film cooling; Evaporative cooling; Sublimation*

Yu.V. Polezhaev

HEAT PUMPS

Heat pumps are devices that operate in a cycle similar to the vapor-compression refrigerator cycle illustrated in **Figure 1**.

In its most basic form, a vapor-compression refrigeration system (see Van Wylen, 1985) consists of an evaporator, a compressor, a condenser, a throttling device which is usually an expansion valve or capillary tube and the connecting tubing. The working fluid is the refrigerant, such as Freon or ammonia, which goes through a thermodynamic cycle. (see also **Refrigeration**)

The thermodynamic cycle is shown schematically in **Figure 2** (Althouse et. al., 1982). Four important processes take place during the cycle. First, heat (Q_l) is transferred to the refrigerant in the evaporator from stations 4 to 1 in **Figures 1** and **2**, where both its pressure and temperature are lower than a thermal source such as air, water, or the ground. Evaporation of the refrigerant occurs from a liquid to a saturated vapor, theoretically at a constant pressure. In practice, however, a pressure drop is associated with fluid flow and heat transfer through the evaporator. Secondly, work (W) is done on the refrigerant as saturated vapor at low pressure and temperature enters the compressor and undergoes adiabatic compression (from 1 to 2). The result is a compressed refrigerant vapor at high pressure and temperature at the compressor outlet (point 2 in **Figures 1** and **2**). Third, heat (Q_h) is transferred from the hot vapor in the condenser (from 2 to 3), where its pressure and temperature are higher than a thermal sink that is at a higher temperature

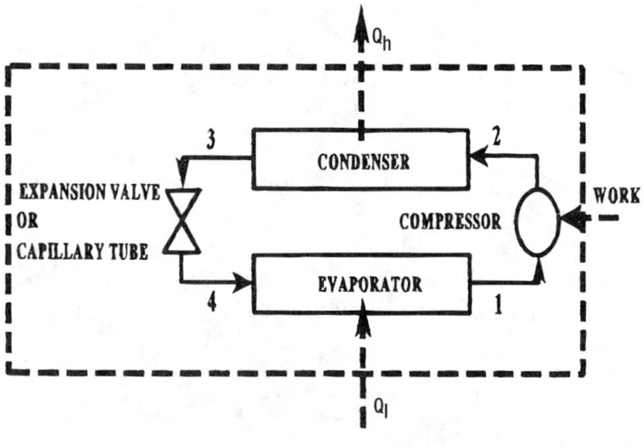

Figure 1. A simple vapor-compression refrigeration cycle.

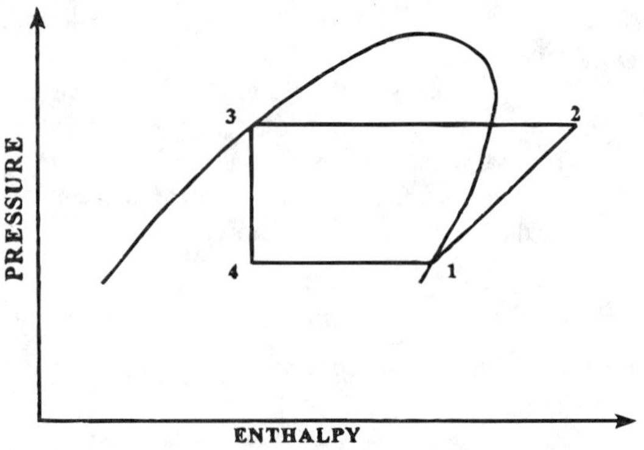

Figure 2. Thermodynamic cycle of the vapor-refrigeration cycle.

than the source. Condensation from a vapor to a saturated liquid occurs in the condenser, theoretically at constant pressure, but again in practice a pressure drop occurs for the same reasons as in the evaporator. The refrigerant leaves the condenser as saturated liquid. The last process before the refrigerant reenters the evaporator is the throttling of the refrigerant through the expansion valve or capillary tube from 3 to 4. During this process the pressure drop is adiabatic, resulting in a decreased refrigerant pressure and temperature. Usually the refrigerant enters as a liquid and leaves as a mixture of liquid and vapor.

The basic heat pump cycle is identical to the vapor-compression refrigeration cycle shown in **Figures 1** and **2**, the only difference between a heat pump and a refrigerator being their basic functions. A refrigeration system cools the external fluid flowing through the evaporator, whereas a heat pump heats the external fluid flowing through the condenser. The main difference between a refrigerator and a heat pump is in the manner of operation regarding cooling or heating. If the application is cooling then you would be interested in the cooling aspect, Q_l in **Figure 1**, occuring over the evaporator, and the cooling device will for example be called refrigerator, air conditioner, chiller, crycooler, etc. On the other hand if the application is heating, then you would be interested in the heating aspect, Q_h in **Figure 1**, occuring over the condenser, and the heating device will be called a heat pump.

Heat pumps are mostly used for heating water and air. Water can be heated for swimming pools and household purposes by using ambient air (a so called *air-to-water heat pump*) and air is usually heated during winter for space heating inside houses, buildings, factories, etc. also by using ambient air as the source (*air-to-air heat pump*). The ground can, however, also be used as a source for space heating. The heat pump will then be called a *ground-coupled, ground source* or *ground-to-air heat pump.* Another source of heat is water. Here, the evaporator is placed in a borehole, pond or lake for space heating. This type of heat pump is called a *water-to-air heat pump.* In the industry *water-to-water heat pumps* are used, for example, to produce hot and cold water simultaneously. These types of systems are also called *dual-purpose* systems.

Dual-purpose systems have been designed to be applied in both heating and cooling capacities at the same time. In a reverse cycle system, the functions of evaporator and condenser can be reversed. Both "hot" and "cold" can thus be delivered to the same thermal reservoir at different times. Dual systems are also called heat pumps, although the focus is not solely on their heating capacities.

Although the devices have different names, two features are common in both. First, the same basic refrigeration cycle takes place and second both are heat-pumping systems. Heat is "pumped" from a thermal source at low temperature to a thermal sink at a higher temperature. These heat-pumping systems or heat pumps have two major or advantages over conventional technology. First, depending on the application, more than one unit of heating per unit of energy input required can often be delivered, i.e. the *coefficient of performance* (COP) value is greater than one. Usually, for every one kilowatt of power required by the compressor more than one kilowatt of heating capacity is available at the condenser. In most practical applications the coefficient of performance is between two and six for a heat pump. To heat a swimming pool, the COP may be as high as six; whereas in a hot water system, where water is heated to a temperature of 55°C, the COP may be as low as three. The coefficient of performance (COP) of a heat pump is defined as

$$COP = Q_h \text{ (energy sought)}/W \text{ (energy that costs)}$$

$$= Q_h/(Q_h - Q_l)$$

A COP value of four would mean an energy saving of 75%. Heat pumps therefore offer the advantage of *energy conservation* and lowered costs compared to other methods of heating. Another advantage of heat pumps is that heat-pumping devices may be either *heat-actuated* or *work-actuated.* Heat-actuated heat pumps allow for use of lower-grade thermal energy, which in other cases might often remain unused. Heat pumps therefore may also help mitigate thermal *pollution* and *environmental concerns.*

The advantages of heat pumps have long been recognized. Research into improving performance, reliability, energy-efficiency, and environmental impact has been an ongoing concern for industrial, governmental, and academic organisations. Studies have centered on advanced cycle design for both heat- and work-actuated systems, improved components (including choice of refrigerant), and use in a wider range of applications. Specific areas of activity have included: a search for *refrigerant* replacements, advanced mobile air conditioning for transportation-applications, advanced vapor-compression technology, absorption heat pumps, ground-coupled (or geothermal) heat pumps and air cycle heat pumps. A few of these activities will now be discussed.

Replacement Refrigerants

In response to increasing concerns that certain chlorine-containing compounds, such as the fully *halogenated fluoroalkanes,* may be catalysing a decrease in stratospheric ozone levels has led to an active search for replacement refrigerants. The *ozone layer* is vital for mankind because it protects us from the dangerous ultraviolet rays emitted by the sun. *Chlorofluorocarbon* (CFC) refrigerants have been identified as major contributors in the ozone layer depletion problem. Ultraviolet radiation breaks CFC molecules apart, releasing its free chlorine as the CFC molecule reaches the upper level of the stratosphere. The free chlorine disrupts the delicate equilibrium of the ozone layer. Concern for the effect of CFCs on the ozone layer has previously led to an international meeting in Montreal, in September 1987. The *Montreal Protocol* on Substances that Deplete the Ozone layer was born out of this meeting. The Protocol was originally signed by forty-five nations. The Protocol was later amended after agreement was reached that CFCs should be completely phased out by the year 2000 and an intended phase-out of *hydrochlorofluorocarbon* (HCFC) refrigerants by 2020, with 2040 as the absolute deadline. This Protocol can, however, be expected to be amended again.

The search for replacement refrigerants has led to HCFCs and *hydrofluorocarbons* (HCFs). An HFC molecule contains no chlorine and poses no threat to the ozone layer. HCFC-22 is perhaps the most widely used refrigerant in heat pumps. It has a relatively low ozone-depletion potential, but since it contains chlorine, a replacement fluid is being sought. The important factor is to find replacement fluids with thermodynamic properties similar to the refrigerants being replaced. It is also desirable to match the enthalpy of vaporisation. If these parameters can be matched closely, the need for system redesign would be minimal. (See also **Refrigeration** and **Refrigerants**)

Ground-Coupled or Geothermal Heat Pumps

In the past ten years, home-owners all over the world have discovered that geothermal (ground source) systems are ideal for heating and cooling. In winter, water or other fluids circulating through a "loop" of underground pipe absorbs heat from the earth and carries it to the geothermal unit which extracts the heat at a higher temperature, and distributes it throughout the home. In summer, the unit extracts heat from your home and transfers it back to the circulating water in the underground loop system, where it is dissipated into the cooler earth. Loops are installed in two basic types: closed and open. Closed loops are buried in the earth or submerged in lakes or ponds. Open loops use ground water pumped from a well. The loop configuration will depend on the following factors: the subsoil geology of the land; the local cost of trenching and drilling; availability of quality ground water and availability of land area. If land area is limited, the closed loops can be inserted into vertical boreholes. U-shaped loops of pipe are inserted into the holes. The holes are then back filled with a sealing solution.

Air-Cycle Heat Pumps

Air is the ultimate refrigerant. It is non toxic, non-corrosive and does not harm the ozone layer or contribute to global warming. Leaks have no impact on the environment. No particular service precautions or refrigerant recovery procedures are needed. The price of air is obviously right—it is free. Air is readily available and delivery to the site is immediate. No recovery equipment is needed. From a design viewpoint, air is ideal, especially if the problem calls for very high or very low temperatures. Air does not change phase in the normal operating regimes and can be used over an exceptionally wide range of temperatures. Many thousands of air cycle systems are in use today. Virtually every jet propelled aircraft in production today, whether commercial or military, uses air cycle refrigeration. In addition, air cycle systems can operate over the broad range of temperatures to which they will be subject in the course of normal flight.

An electrically driven air cycle heat pump is a very simple device. The rotating group typically consists of a compressor and a turbine mounted on the same shaft as a high-speed motor. This assembly is the only moving part. The single stage, centrifugal compressor has a relatively low pressure ratio that is typically in the 1.4:1 range. The single-stage, radial inflow turbine wheel has a slightly lower pressure ratio. The motor is powered by the output of a variable frequency, variable voltage inverter so that the speed of the motor, compressor and turbine can change. Finally, one, two or three heat exchangers are needed, depending on the application and the concept.

Energy is required for compressing air. Mechanical energy is converted into thermal energy when air is compressed and the air becomes hotter. This is the source of heating. Correspondingly, when air is expanded and does work, thermal energy is converted to mechanical energy and the air becomes colder. This is the source of refrigeration. The compression process therefore consumes mechanical energy and the expansion process produces mechanical energy. By mounting the compressor and the turbine on the same shaft, the turbine helps drive the compressor. Unfortunately, the laws of physics are such that the turbine does not produce enough mechanical energy to drive the compressor by itself and therefore an electric motor is used to supplement the mechanical energy produced by the turbine.

Consider a system in which both heating and cooling are needed simultaneously. Perhaps a restaurant wants to heat 20 °C incoming city water up to 82 °C for dishwashing. Simultaneously, they would like to air-condition the kitchen. Assume that the outside air temperature is 25 °C. If outside air at atmospheric pressure is compared with a 1.8:1 pressure ratio, the discharge temperature will be about 89 °C. This is sufficient to heat the water to the required temperature using an appropriate heat exchanger. In heating the water, the compressed air is cooled to a lower temperature. If the compressed air is now expanded in the turbine back to atmospheric pressure, it will discharge at a temperature substantially below freezing—assuming that the air was dry and no condensation occurred. This is well below the temperature required for air conditioning and so the air will be mixed with room air before being discharged. The air remains breathable as there is no lubricating oil in the system to contaminate it. It is difficult to present a comparison with conventional equipment because the air cycle heat pump is not yet in production and it must be matched against a wide range of combinations of heating and cooling equipment. COP heating values, however, of between 1.9 and 3.6 are expected.

References

Althouse, A. D., Turnquist, C. H., and Bracciano, A. (1982) *Modern Refrigeration and Air Conditioning,* The Goodheart-Willcox Co., Inc.

Van Wylen, G. J. and Sonntag, R. E. (1985) *Fundamental of Classical Thermodynamics,* John Wiley and Sons, Toronto, 3rd edition.

J.P. Meyer

HEAT RECOVERY BOILERS

***Following from:** Boilers*

Boilers of this type generate steam by extracting heat from hot fluid, generally a gas, which is produced as a by-product of a chemical reaction and has no intrinsic value as a chemical, or which is in the form of exhaust from a gas-turbine or internal combustion engine. Heat recovery boilers are also often found in association with high temperature processes such as steel and glass making.

It is not uncommon for these boilers to receive additional heat by supplementary firing of other by-products or by burning gas or light oil to raise the boiler inlet gas temperature to a level commensurate with the desired steam conditions.

The convective elements of boilers fired by municipal industrial or other waste materials are often referred to as heat recovery boilers although, apart from their geometry, they are more akin to the rear end of a furnace-fired boiler. Perhaps the most common heat recovery boiler is now the *gas turbine heat recovery boiler.*

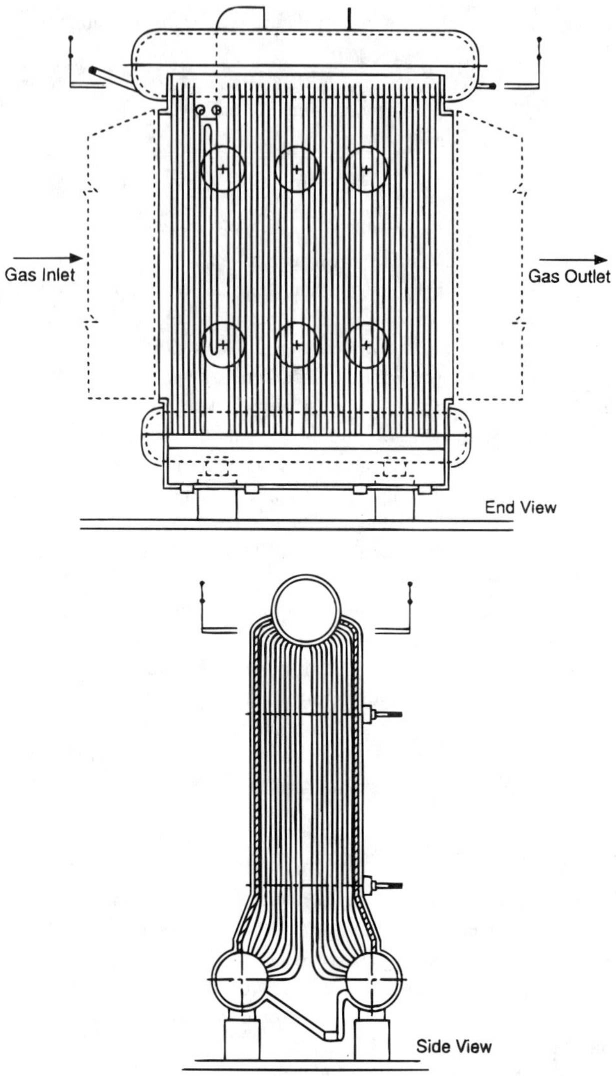

Figure 1. Tri-drum heat recovery boiler for float-glass plant. Drawing reproduced by permission of Babcock Energy Ltd., UK.

Most heat recovery applications are characterised by low heat capacity fluids and low pressures, which combine to limit the velocities over the tubes and result in low heat transfer coefficients. For this reason, when the gases are clean, it is almost universal to use finning on the outside of the tubes. Most boilermakers use serrated helical finning but some utilise solid finning despite the reduced thermal efficiency and weight penalty involved. (See **Augmentation of Heat Transfer**)

Boilers designed to recover heat from high temperature gases exhausted from glass making plant have to offer very low pressure drop to the gas stream and the tubes have to be arranged such that deposits, which are unavoidable as a result of the excess of sulphur trioxide in the gas, can be treated and removed regularly by sootblowing. Such a boiler design is known as the tri-drum, illustrated in **Figure 1**. Economiser and superheater surfaces may be incorporated before and after the evaporator tubing as required to increase

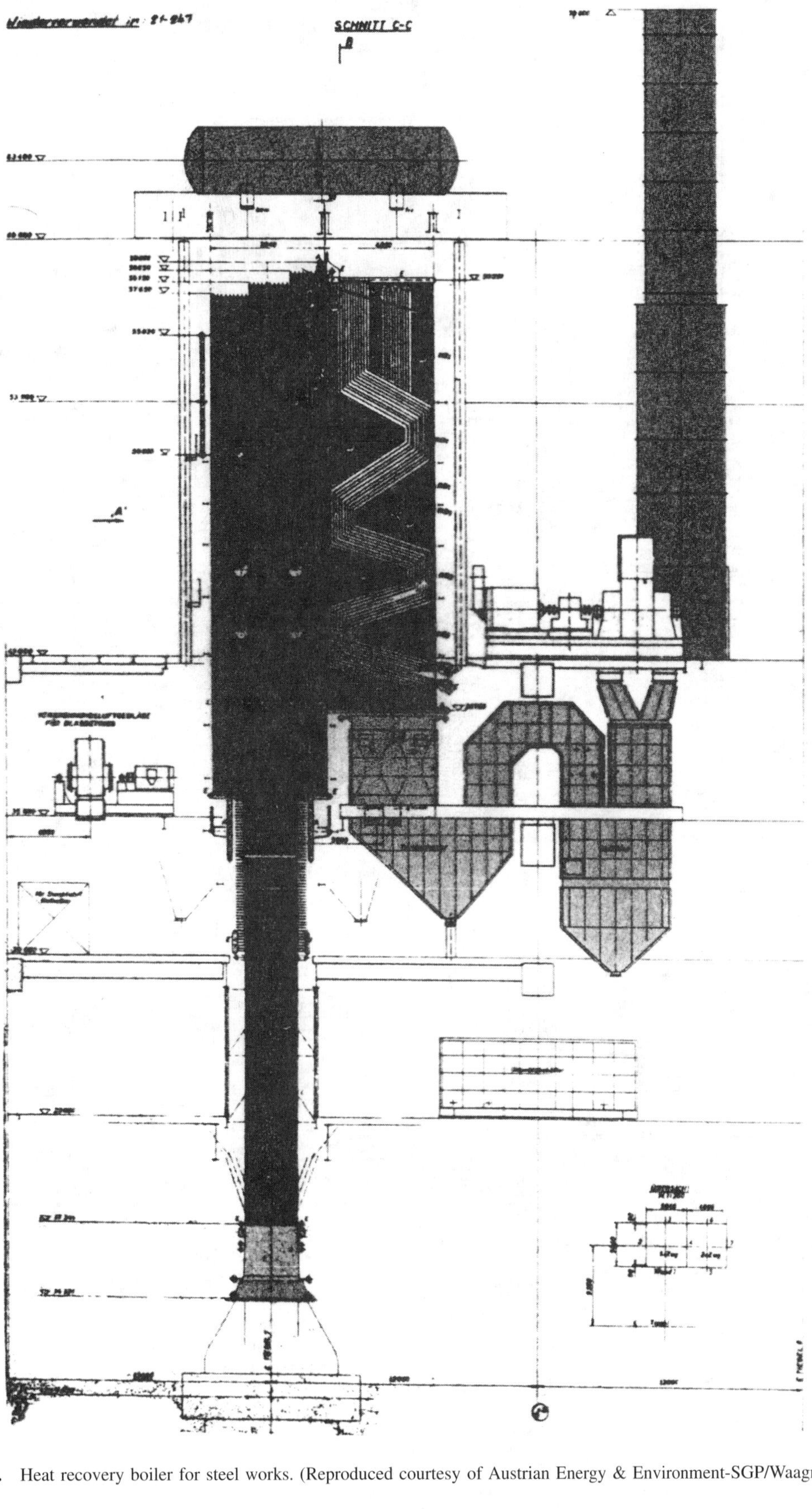

Figure 2. Heat recovery boiler for steel works. (Reproduced courtesy of Austrian Energy & Environment-SGP/Waagner Biro).

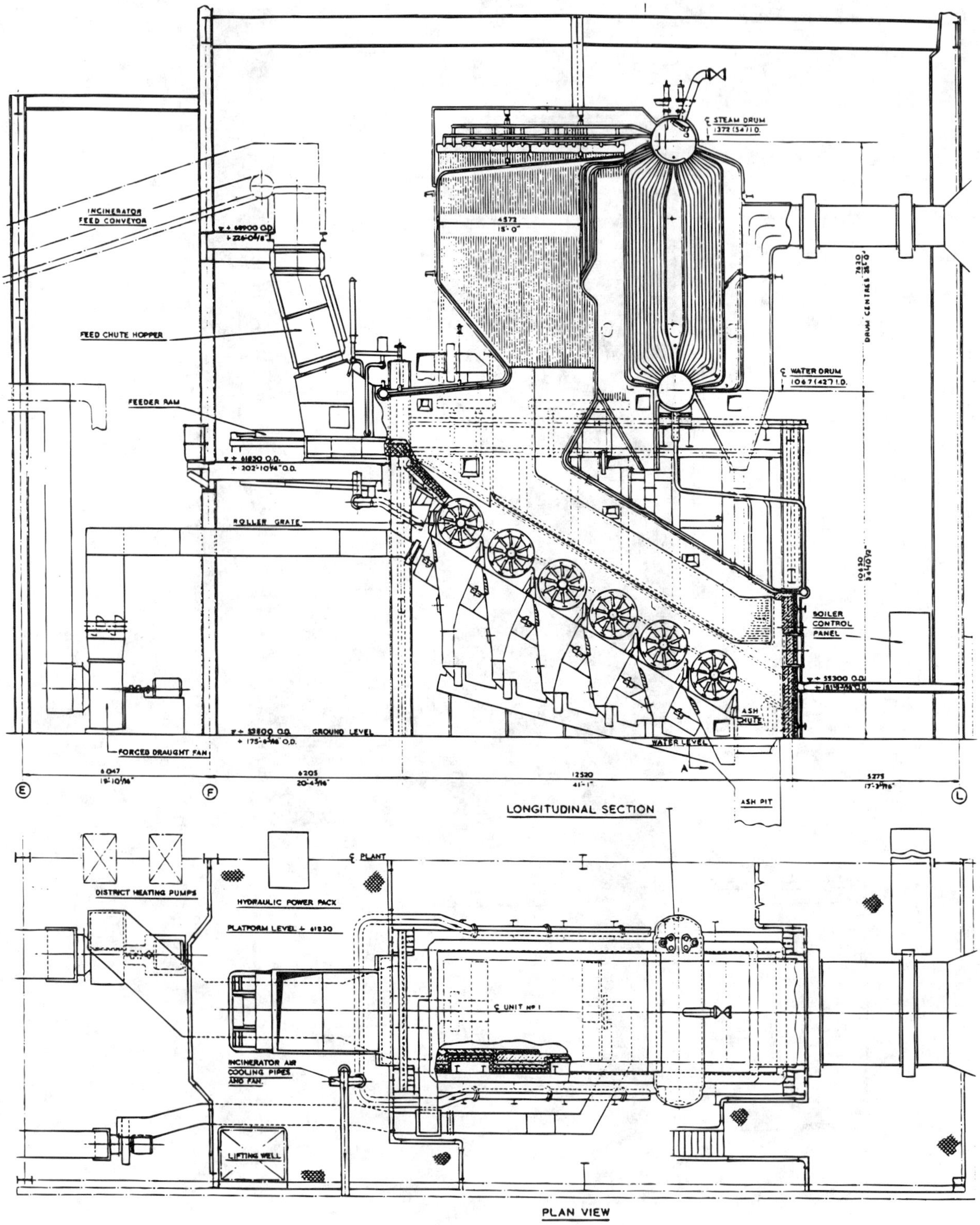

Figure 3. Bi-drum heat recovery boiler for waste heat incinerator. (Drawing reproduced by permission of Babcock Energy Ltd., UK).

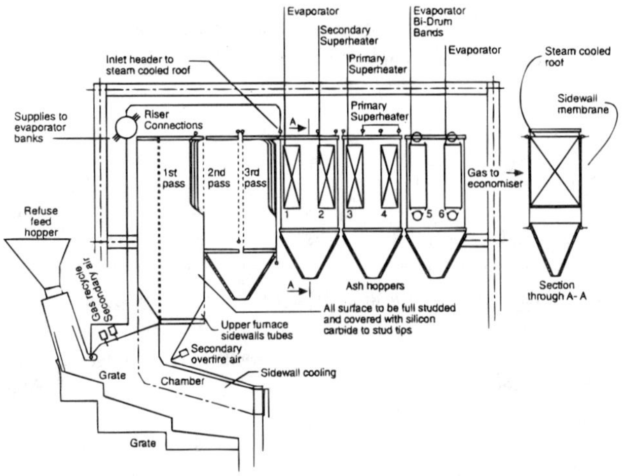

Figure 4. General arrangement of heat recovery boiler for MSW incinerator plant. (Drawing reproduced by permission of Babcock Ltd., UK).

boiler efficiency or optimise steaming rate. A large capacity steam drum is provided to accommodate variations in the steam production rate throughout the soot blowing cycles.

Boilers for recovering heat from steel-making plants are similar in concept to those above but, because of the wider swings in heat available from the source, these units are usually provided with auxiliary firing. A typical design is illustrated in **Figure 2** in which the principal features are the elongated duct to draw the hot gases away from the converter, the water-cooled square-section furnace with auxiliary burners forming the up-pass and the combined evaporator/economiser forming the parallel down-pass. Hoppers are provided to collect particulates carried over from the steel converter and the induced draft fan is used to draw hot gas through the heat recovery system. As in the preceding example, the large-capacity steam drum is noteworthy. Detailed layout of heating surfaces and choice of the balance between longitudinal and cross-flow areas is a matter of optimisation in conjunction with the auxiliary power of the fan.

Boilers to recover the heat of combustion from waste incinerators may range from simple bi-drums such as shown in **Figure 3** to combinations of these with banks of evaporator and superheater tubes mounted in water or steam-cooled ducts, the so called tail-end boiler as shown schematically in **Figure 4**.

The principal considerations in design of these boilers are support and accommodation of relative expansion, ensuring adequate circulation under all operating conditions and provision of means of removing deposits from tubes. It is usual to support heat recovery boilers from the bottom but this leads to difficulties in matching the expansion of the furnace with that of the boiler, so top supported units offer simpler and often cheaper solutions. Ensuring adequate circulation involves providing separate supplies and risers for each section of the boiler that is subject to significantly different different heat fluxes. Mechanical rapping gear is usually provided to ensure removal of deposits which fall into hoppers mounted below each section of the boiler.

A.E. Ruffell

HEAT RECOVERY NETWORK DESIGN (see Process integration)

HEAT STORAGE, SENSIBLE AND LATENT

***Following from:** Freezing; Melting*

In heat storage, use is made of the thermal capacity of solid or liquid materials, either by their sensible (specific) heat effect (heating/cooling cycles) or by their latent heat effect at a phase change (melting/freezing cycles). For heat storage, the important thermal characteristics are:

1. Heat capacity, MJ/m^3
2. Charge and discharge rates, $\dot{m}\ C_p\ (T_i - T_o)$, kW/m^3
3. Heat loss rate $\dot{Q}_{LOSS}$, kW/m^3
4. Thermal stratification, ΔT_{STR}

Heat Capacity

When use is made of the sensible heat effect a high volumetric specific heat (ρc_p) of the material is required. **Table 1** gives some

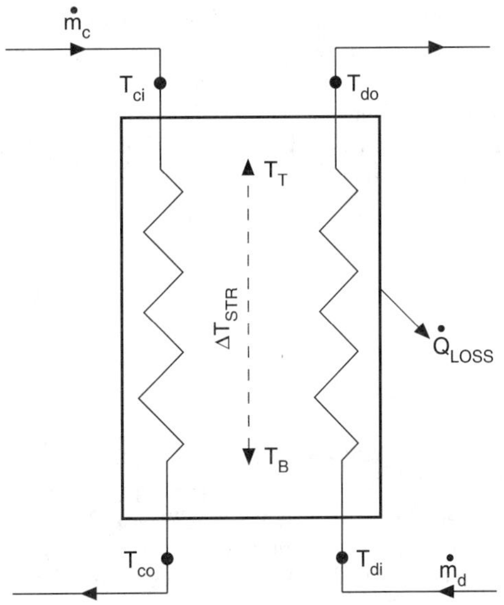

Figure 1. Principles of heat storage. Heat charge rate: $\dot{Q}_c = \dot{m}_c\ C_{pc}\ (T_{ci} - T_{co})$, heat discharge rate: $\dot{Q}_d = \dot{m}_d\ C_{pd}\ (T_{do} - T_{di})$, temperature in top of store T_T and bottom of store T_B.

Table 1. Sensible heat storage properties

	ρc_p ($MJ\ m^{-3}\ K^{-1}$)	λ ($Wm^{-1}\ K^{-1}$)	τ (s)
Concrete	1.9	1.3	80
Rock, stones	1.7	2.0	40
Steel	4	40	5
Water	4.2	0.6	360
Thermal oil	1.8	0.11	840

data for typical materials. Solid materials are often used in rock (packed) beds. As packing densities are often between 60% and 85% the volumetric capacity (ρc_p) is reduced proportionally. For higher temperatures the use of such packed beds as regenerators is common. For low temperature applications like home heating and solar energy, water storage is mostly used.

In latent heat stores the latent heat effect is important. At the phase change temperature the heat of solidification will be available when freezing a liquid. **Table 2** give some latent heat effects and the solidification temperature T_s for several materials. For some organic materials, there is a transition range ΔT_{tr}. Inorganic materials may show supercooling effects, which means that the heat discharges at a (unwanted) lower temperature. Also segregation can occur in inorganic solutions in water.

Charge and Discharge Rates

To load/unload a heat storage vessel a heat transfer fluid (like water, air, oil) takes up the heat either by direct contact or through a heat exchanger. Often the limiting heat transfer is on the heat storage material side. Conductivity (λ) in solid materials limits the heat transfer rate. In **Table 1** a typical discharge time τ based on transient heat conduction problem for a Fourier number $Fo = \kappa\tau/d^2$ equal to 0.5 has been given for a typical dimension d of 10 mm (See **Fourier Number**). The time τ gives the time for a 90% or more completed loading/unloading cycle. It increases proportional to d^2. Also for the solidification (unloading) part of a latent heat store this is the limiting heat transfer mechanism. For liquid heat stores and in melting, natural convection effects will increase the heat transfer rate. For organic latent heat materials with low conductivity-values finned pipes and embedded metal structures are used to improve the heat transfer.

Heat Loss Rate

Heat losses of the storage vessel to the surroundings can severely diminish the heat storage efficiency.

Thermal Stratification

In sensible heat stores the heat is loaded/unloaded in an axial direction. In this direction, a temperature gradient (ΔT_{STR}) will occur. In the optimal case, there is a rather sharp temperature front. However, due to axial dispersion of heat, and in liquids also due to convection flows, this front will be spread out. With large convection flows it can even completely disappear and result in a single, mixed temperature. Due to the dispersion the average temperature in the storage at the end of the loading will decrease. This lowers the storage efficiency.

Table 2. Latent heat storage properties

	$T_s(\Delta T_{tr})$ °C	ΔH kJ/kg
Lithium Chloride 3-Hydrate	8	253
Calcium Chloride 6-Hydrate	30	170
Disodium Phosphate 12-Hydrate	35	270
Sodium Thiosulfate 5-Hydrate	48	210
99% Eicosane	37	245
Paraffin wax 52/54	43–55	150
Micro-wax 85/88	72–93	180
Stearic Acid	68–74	191

References

Hoogendoorn, C. J. and Bart, G. C. J. (1992) Performance and modelling of lat. heat stores, *Solar Energy,* 48, 1, 53–58.

Ouden, C. den (1981) *Thermal Storage of Solar Energy,* Martinus Nijhoff Publishers, the Hague, NL.

Paykoc, E. and Kakac, S. (1987) *Solar Thermal Energy Storage; in Solar Energy Utilization,* H. Yüncü, ed., NATO ASI Series, Martinus Nijhoff Publishers, Dordr., NL.

Leading to: *Regenerators; Solar energy*

C.J. Hoogendoorn

HEAT TRANSFER

Heat transfer is a physical process of spontaneous, irreversile heat transport from hotter (i.e., having a higher temperature) bodies or parts of bodies to cooler ones. It is also a field of science that is concerned with the study and quantitative description of heat transfer laws.

Heat transfer as a physical process of heat transport in a space with nonuniform temperature distribution is characterized by the heat flux $\dot{q}$, W/m^2, i.e. the quantity of heat passing per unit time through a unit of arbitrary surface along the normal to it. If the surface is represented by the surface or the wall of the body which is in a state of heat transfer with other bodies, eg. between the wall and the fluid medium flowing over it, then we deal with the heat flux on the wall $\dot{q}_w$, which is one of the most important resultant characteristics of heat transfer. The quantity $\dot{q}_w$ which depends on the character and scale of temperature nonuniformities in the system (in the wall-fluid system this scale is given as temperature difference $\Delta T = T_w - T_f$, where T_f is the characteristic temperature of fluid, T_w the temperature of the wall), on the geometrical characteristics of the system, on the characteristics of fluid flow and by the physical properties of the bodies and media participating in heat transfer.

In nature, transport of heat is accomplished in three ways: molecularly, (as a result of thermal motion of molecules), by convection (displacement of macroscopic elements of the system in space), and by thermal radiation. The first two are observed only in the presence of material medium (i.e. an ensemble of microparticles such as molecules and atoms in a state of thermal collisions). If geometrically small elements of the medium possess the same properties as their macroscopic aggregate, then the medium is said to be a continuum. Solids and liquids fit well the concept of continuum. Gas can also be considered as a continuous medium if the Knudsen number $Kn = l/L < 10^{-3}$ (i.e. if the free path length l of molecules is small compared to the characteristic geometrical dimension L of the heat transfer system. For air $l \cong 2.27\ 10^{-5}T/p$, m, where T is the absolute temperature k and p the pressure, Pa. Heat transfer in continua is tackled using a phenomenological approach, making it possible to consider only the resultant characteristics of heat transfer for an entire ensemble of molecules. A distinction needs to be made between single phase and multiphase media. Examples of multiphase media are gas- and vapor-liquid mixtures and dust-laden gases. In certain cases, loose materials and dispersed media (multiphase media the condensed phases of which are in a finely dispersed state) can be regarded as continuous media with assigned effective properties.

Molecular heat transfer in a continuum is known as heat conduction. The phenomenological description of heat conduction is based on a fundamental hypothesis of heat transfer, Fourier's law,

$$\dot{q} = -\lambda \text{ grad } T, \qquad \textbf{(1)}$$

where λ, W/(mK), is the physical parameter (property) of the medium that is known as the coefficient of thermal conductivity. For many substances such as solids and ordinary liquids λ only depends slightly on temperature and in many cases can be considered a constant. In the case of gases, particularly, in states close to the saturation curve or the critical point, λ strongly depends on temperature and often this cannot be neglected. In isotropic bodies (media) the thermal conductivity does not depend on the heat propagation direction. Many widely encountered solid materials possess anisotropy (i.e. their heat conduction depends on the heat flux direction). Examples are reinforced concrete, fabrics, composite materials. For these media, Fourier's law is applied in a more general statement. It should be noted that heat transfer by heat conduction in most bodies, except carbon and metals, is inefficient and produces a high thermal resistance to heat flux. Therefore, materials with low λ are widely used for thermal insulation of dwellings, premises, and engineering objects. The system in which heat transfer occurs by heat conduction often consists of several bodies close together. Examples are multilayer walls, an iron, or a soldering iron contacting with heated items. In these cases, in addition to heat conduction inside each body of the system, the rate of heat transfer can be also essentially affected by direct contact heat transfer whose characteristics depend on physical processes responsible for heat transport via contacting body surfaces. In fluid media (liquids and gases) under terrestrial conditions heat transfer by heat conduction alone is rarely realized because the temperature field arising due to thermal expansion changes the medium density and gives rise to Archimedean forces (buoyancy forces) which cause free convection.

Convective heat transfer is characteristic of fluid media (fluids) which include liquids, gases, vapors and multiphase fluid media. Depending on which way the fluid flow is implemented, we distinguish between *forced convection*, which is maintained by supplying to the flow the needed quantity of mechanical energy from the outside, for instance, from the pump or a fan, and *free convection* as mentioned above. A forced flow under conditions of heat transfer appreciably affected by buoyancy is referred to as *mixed convection*. Heat transfer caused by the joint action of convective and molecular transfer is said to be convective heat transfer. For practical purposes, the most important case of heat transfer, is convective heat exchange between a flowing medium and its interface with another medium this interface being called a *heat transfer surface* or a *wall* in the case where the other medium is a solid. In heat exchangers, overall heat transfer, that is heat exchange between two flowing media across a solid separating wall, plays an essential role. For quantitative expression of convective heat transfer and overall heat transfer, a linear Newton-Richmann relation

$$\dot{q}_w = \alpha\Delta T = \alpha(T_w - T_f), \qquad \textbf{(2)}$$

$$\dot{q}_w = U\Delta T = U(T_{f1} - T_{f2})$$

is widely used, where α, W/(m^{2}K), and U are respectively the heat transfer and overall heat transfer coefficients. In contrast to α, the coefficient U takes into account not only the rate of fluid heat transfer by convection, but a thermal resistance of separating wall as well. Relations (2) are applied for both heat transfer surface elements (local heat transfer) and the entire surface on the average (average heat transfer). Average values are commonly used for the whole exchanger (see **Mean Temperature Difference**). It should be born in mind that the heat transfer coefficient is most frequently not a constant and may have an intricate dependence of process parameters, including $\dot{q}_w$ and ΔT. For boiling, vapor generation inside a tube, and intensive heating of a turbulent flow gas through a tube (especially at supercritical pressure) sharp reductions in heat transfer coefficient may be observed when q_w or ΔT increases. This process is known as *heat transfer crisis (burnout)*. In consideration of internal heat transfer in porous bodies, i.e., a convective heat transfer between a solid matrix and a fluid filtered through it, a quantitative relation

$$\dot{q}_v = \alpha_v(T_w - T_f) \qquad \textbf{(2a)}$$

is often used, where $\dot{q}_v$ is the heat flux per unit volume of the porous body, T_w the local temperature of the solid skeleton, and T_f the local temperature of fluid. The quantity α_v, W/(m^{3}K), is known as the volumetric heat transfer coefficient.

Convective heat transfer between a fluid and a wall is more intensive than for heat conduction alone owing to the additional convective heat transfer which is most intensive in the turbulent fluid flow regime where, due to turbulent mixing, the effective thermal conductivity of fluid sharply increases. Heat transmission deep into the fluid flow can be artificially intensified using various techniques of heat transfer augmentation, e.g. by making the wall rough, implementing secondary flows, and installation of promoters and porous elements (see **Augmentation of Heat Transfer**).

Convective heat transfer occurs both in external flow over bodies and in internal fluid flow. In the case of external flow over a body with fairly large geometrical dimensions and with high velocity, the resultant characteristics of heat transfer depend only on the processes occurring in a thin fluid layer, directly adjacent to the surface in the flow, that is called a **Boundary Layer**; owing to this such heat transfer is often called **Boundary Layer Heat Transfer**. In the case of internal flow which are often encountered in tubes and channels, the dynamic and thermal influence of the wall gradually, with increasing distance from entrance, extends to the entire fluid flow. Internal heat transfer is characterized by a dramatic variation of heat transfer at the entrance region of the channel and its subsequent stabilization far from entrance. Description of the mechanisms of heat transfer at entrance regions of channels is often referred to as the Graetz-Nusselt problem by the names of scientists who were the first to solve this problem for the case of laminar flow (Poiseuille flow) in a tube. Heat transfer in fluid flows having a free surface, such as flows in open beds and falling liquid films, form a special range of problems of great consequence for engineering.

In engineering convective heat transfer often involves physico-chemical transformations in the fluid. Of great interest is heat transfer in boiling (pool boiling and forced flow in pipes), evaporation, and film and dropwise condensation that are widely practised in power engineering and underlie many technologies. Convective heat transfer is commonly used in high-temperature aerospace and chemical engineering if chemical reactions proceed in the fluid flow and on the wall. The specific feature of these processes is an additional transmission of latent heat of transformation resulting from molecular and turbulent diffusion of chemical components of the medium. The specific features of hydrodynamics of conducting fluids in a magnetic field (see **Magnetohydrodynamics**) and liquids in which the laws of viscous friction differ from those of normal (rheological or non-Newtonian liquids) are responsible for the relevant characteristic properties of the mechanism of convective heat transfer.

Heat transfer by radiation occurs by the transmission of electromagnetic waves (photons) in the infrared, visible, and ultraviolet ranges. Such waves are constantly generated by substances as a result of molecular and atomic vibrations due to internal energy. Heat transfer by radiation often becomes dominant in heat transfer in optically transparent systems at temperatures above 1000°C. Radiative heat transfer involves the events of radiation, transport, and absorption of thermal energy. The resultant heat flux released or absorbed by the surface, which for simplicity we consider optically nontransparent, includes the radiation and absorption processes

$$\dot{q}_w = \dot{q} = \dot{q}_0 - \dot{q}_i = \dot{q}_n + R\dot{q}_i - \dot{q}_i, \qquad \mathbf{(3)}$$

where $\dot{q}_i$ is the density of radiant flux incident from the outside on the surface, $\dot{q}_0$ the effective radiation density of the surface itself consisting of both natural radiation $\dot{q}_n$ and the reflected part of incident radiation (R is the reflection factor). In contrast to heat conduction and convective heat transfer, no material medium is needed for radiative heat transfer to be accomplished because electromagnetic waves also can easily propagate in vacuum.

Combined heat transfer by radiation and heat conduction or by radiation, heat conduction, and convection is implemented in modern engineering objects. Convective and radiative-convective heat transfer is frequently closely interrelated with conductive heat transfer in the wall and, therefore, must be considered jointly. These processes are said to be conjugate.

Heat transfer as an independent field of science has been developing for over three hundred years, but it eventually became an independent science with its own methods, and terms early in the 20th century. The vigorous development of heat transfer science is linked with the development of power engineering, first thermal and in recent years atomic, as well as the development of chemical, aerospace, and cryogenic engineering. Protection of the environment is within the scope of this science. It is no exaggeration to say that in any sphere of industry, a proper implementation of heat transfer based on scientific knowledge of its mechanism ensures a positive effect and saving of fuel, energy, and resources. A powerful impetus to extension of the sphere of application of the heat transfer theory was given by computer engineering. We can state with assurance that an independent domain has been created that is aimed at developing the techniques of numerical heat transfer and the software for it. In accordance with classification of heat transfer processes by the predominating transmission mechanism the science of heat transfer falls into three independent divisions: the theories of heat conduction, convective heat transfer, and radiative heat transfer respectively.

What the three divisions have in common is the phenomenological approach mentioned above and extensive use of the methods of similarity and dimensional theory for analyzing, quantitative description (the so called generalized relationships or dimentionless numbers), and simulation of multiple-factor heat transfer processes. The similarity method enables the transformation of the functional dependence between the dimensional physical quantities that are essential for the process considered into a generalized dependence (a similitude relationship) between the dimensionless variables, i.e., similarity numbers or criteria the number of which is less than the number of original dimensional quantities by the number of quantities with independent dimensions (the Buckingham II-theorem) (see **Dimensional Analysis**). By tradition the most important and generally used similarity numbers are named after the scientists who made a significant contribution to the theory of heat and mass transfer. Such are the **Fourier** (Fo), **Biot** (Bi), **Reynolds** (Re), **Nusselt** (Nu), **Prandtl** (Pr), **Stanton** (St), **Peclet** (Pe), **Grashof** (Gr), **Rayleigh** (Ra), **Mach** (Ma), **Lewis** (Le), **Schmidt** (Sc), **Sherwood** (Sh), and other numbers. The similarity methods gained currency in the theory of heat conduction and convective heat transfer. Currently efforts are being made to use them in the field of radiative heat transfer.

The theory of turbulent convective heat transfer has made impressive headway owing to the method of statistical averaging of turbulent flow parameters elaborated by O. Reynolds late in the 19th century and applied by him to derive the differential equations of turbulent fluid flow, which were given his name, and the fundumentals of the theory of boundary layer which were formulated by L. Prandtl in 1904. Prandtl also put forward a fruitful model of turbulent transmission, viz. the mixing length model, which enabled a number of classic solutions to turbulent heat transfer problems to be obtained which are fairly consistent with the experimental data. The continued interest shown in improved models of turbulent transfer is accounted for by the fact that the Reynolds averaged equations are not formally closed.

References

Gröber, H., Erk, S., and Gridull, U. (1963) *Die Grundgesetze der Wärmeübertragung,* Springer, 3. Aufl. Berlin.

Kline S. J. (1965) *Similitude and Approximation Theory,* McGraw-Hill, New York.

Kays, W. M. and Crawford, M. E. (1980) *Convective Heat and Mass Transfer,* McGraw, New York.

Cebeci, T. and Bradshaw, P. (1984) *Physical and Computational Aspects of Convective Heat Transfer,* Springer, New York.

Siegel, R. and Howell, J. R. (1972) *Thermal Radiation Heat Transfer,* McGraw-Hill, New York.

Leading to: *Adiabatic wall temperature; Coupled conduction and convection; Coupled radiation and convection; Coupled radiation, convection and conduction; Convective* *heat transfer; Falling film heat transfer; Heat exchangers; Heat flux; Heat transfer coefficient; Radiative heat transfer;*

V.A. Kurganov

HEAT TRANSFER AND FLUID FLOW SERVICE

Heat Transfer and Fluid Flow Service, HTFS
Building 392.7, Harwell Laboratory
Didcot, 0X11 0RA,
UK
Tel: 01235 432908

HEAT TRANSFER COEFFICIENT

Following from: *Heat transfer; Heat flux*

Heat transfer coefficient is a quantitative characteristic of convective heat transfer between a fluid medium (a fluid) and the surface (wall) flowed over by the fluid. This characteristic appears as a proportionality factor α in the *Newton-Richmann relation*

$$\dot{q}_w = \alpha(T_w - T_f),$$

where $\dot{q}_w = -\lambda(\partial T/\partial n)_w$ is the heat flux density on the wall, T_w the wall temperature, T_f the characteristic fluid temperature, e.g. the temperature T_e far from the wall in an external flow, the bulk flow temperature T_b in tubes, etc. The unit of measurement in the international system of units (SI) (see **International system of units**) is $W/(m^2K)$, 1 $W/(m^2K)$ = 0.86 $kcal/(m^2h°C)$ = 0.1761 Btu/(h $ft^{2}°F$) or 1 $kcal/(m^2h°C)$ = 1.1630 $W/(m^2K)$, 1 Btu/(h $ft^{2}°F$) = 5.6785 $W/(m^2K)$. The heat transfer coefficient has gained currency in calculations of convective heat transfer and in solving problems of external heat exchange between a heat conducting solid medium and its surroundings. Heat transfer coefficient depends on both the thermal properties of a medium, the hydrodynamic characteristics of its flow, and the hydrodynamic and thermal boundary conditions. Using the methods of similarity theory, the dependence of heat transfer coefficient on many factors can be represented in many cases of practical importance as compact relations between dimensionless parameters, known as similarity criteria. These relations are said to be generalized or similarity equations (formulas). The **Nusselt number** $Nu = \alpha l/\lambda_f$ or the **Stanton number** $St = \alpha/(\dot{m}C_{pf})$ is used as a dimensionless number for heat transfer in these equations, where l is the characteristic dimension of the surface in the flow, $\dot{m} = \rho u$ the mass velocity of the fluid flow, λ_f and C_{pf} the fluid thermal conductivity and heat capacity. When solving the problems of heat conduction in a solid, the distribution of heat transfer coefficient α between the body and its surroundings is often given as a boundary condition. Here, it is useful to use a dimensionless independent parameter, the **Biot number** $Bi = \alpha l/\lambda_s$, where λ_s is the thermal conductivity of a solid and l its characteristic dimension. The dependence of the Nu and St numbers on the Re and Pr numbers plays an essential role in heat transfer by forced convection. In the case of fully developed heat transfer in a circular tube with laminar fluid flow the Nusselt number is a constant, namely Nu = 3.66 at a constant wall temperature and 4.36 at a constant heat flux (see **Tubes (single-phase heat transfer in)**). In the case of free convection, the Nu number depends on the Gr and Pr numbers. When the heat capacity of the fluid varies substantially the heat transfer coefficient is frequently determined in terms of enthalpy difference $(h_w - h_f)$. The concept of heat transfer coefficient is also used in heat transfer with phase transformations in liquid (boiling, condensation). In this case the liquid temperature is characterized by the saturation temperature T_s. The order of magnitude of heat transfer coefficient for different cases of heat transfer is presented in **Table 1**.

When analyzing internal heat transfer in porous bodies, i.e., convective heat transfer between a rigid matrix and a fluid permeating through it, use is often made of the volumetric heat transfer coefficient

$$\alpha_v = q_v/(T_w - T_f),$$

where q_v is the heat flux passing from the rigid matrix to the fluid in a unit volume of a porous body, T_w the local temperature of the matrix, and T_f the local bulk temperature of the fluid.

It should be emphasized that the constancy of α over a wide range of $\dot{q}_w$ and ΔT (other conditions being equal) is encountered only in the case of convective heat transfer when the physical properties of fluid change only slightly during heat transfer. Under convective heat transfer in a fluid with varying properties and in boiling, heat transfer coefficient may substantially depend on $\dot{q}_w$ and ΔT. In these cases an increase of heat flux may give rise to hazardous phenomena such as burnout (transition heat flux) and deterioration of turbulent heat transfer in tubes. If the $\dot{q}_w$ (ΔT) is nonlinear, it appears inappropriate to represent it in terms of the coefficient α when analyzing, for example, boiling stability.

An overall heat transfer coefficient

$$U = \dot{q}_w/(T_{f1} - T_{f2}) = \dot{q}_w/\Delta T_f,$$

where T_{f1} and T_{f2} are the temperatures of the heating and heated liquids, is used in calculations of heat transfer between two fluids through the separating wall. The U values for the most commonly used wall configurations are determined by the formulas

Table 1. Approximate values of heat transfer coefficient

Conditions of heat transfer	$W/(m^2K)$
Gases in free convection	5–37
Water in free convection	100–1200
Oil under free convection	50–350
Gas flow in tubes and between tubes	10–350
Water flowing in tubes	500–1200
Oil flowing in tubes	300–1700
Molten metals flowing in tubes	2000–45000
Water nucleate boiling	2000–45000
Water film boiling	100–300
Film-type condensation of water vapor	4000–17000
Dropsize condensation of water vapor	30000–140000
Condensation of organic liquids	500–2300

$$U = 1 \Big/ \left[\frac{1}{\alpha_1} + \sum_{i=1}^{n} \frac{S_i}{\lambda_i} + \frac{1}{\alpha_2}\right]$$

for a plane multilayer wall,

$$U = 1 \Big/ \left[\frac{1}{\alpha_1}\frac{D}{D_1} + \sum_{i=1}^{n} \frac{D}{2\lambda_1} \ln \frac{D_{i+1}}{D_i} + \frac{1}{\alpha_2}\frac{D}{D_2}\right]$$

for a cylindrical multilayer wall, and

$$U = 1 \Big/ \left[\frac{1}{\alpha_1}\frac{D^2}{D_1^2} + \sum_{i=1}^{n} \frac{(D_{i+1} - D_i)}{2\lambda_i}\frac{D^2}{D_i D_{i+1}} + \frac{1}{\alpha_2}\frac{D}{D_2}\right]$$

for a spherical multilayer wall.

Here D_1 and D_2 are the internal and external diameter of the wall, D the reference diameter by which a reference heat transfer surface is determined, S_i, D_i, D_{i+1}, and λ_i are the thickness, internal and external diameters, and the thermal conductivity of the i-th layer. The first and the third terms in brackets are said to be thermal resistances of heat transfer. In order to lower them the walls are finned and various methods of heat transfer augmentation are used. The second term in brackets is said to be the thermal resistance of the wall, which may greatly increase as a result of wall contamination, such as scale and ash build-up, or poor heat transfer between the wall layers. The values of α and U for a small element of heat transfer surface are called local ones. If they do not vary greatly then, in practical calculations of heat transfer on finite-size surfaces, we use the mean values of the coefficients and the heat transfer equation

$$Q = \overline{\dot{q}_w} A = \overline{U} A \overline{\Delta T_{Mf}} \quad \text{or} \quad \overline{\alpha} A \overline{\Delta T_{Mw}},$$

where A is the reference heat transfer surface, and $\overline{\Delta T_M}$ (ofen mean logarithmic $\overline{\Delta T_{LM}}$) temperature drop (see **Mean Temperature Difference**).

References

Jakob, M. (1958) *Heat Transfer,* Wiley, New York, Chapman and Hall, London.

Schneider, P. J. (1955) *Conduction Heat Transfer,* Addison-Wesley Publ. Co., Cambridge.

Adiutory, E. F. (1974) *The New Heat Transfer,* vols. 1,2, Ventuno Press, Cincinnati.

Leading to: *Mean temperature difference*

V.A. Kurganov

HEAT TRANSFER COEFFICIENT, IN COILED TUBES (see Coiled tubes, heat transfer in)

HEAT TRANSFER COEFFICIENT, IN POROUS MEDIA (see Porous media)

HEAT TRANSFER COEFFICIENT, TYPICAL VALUES (see Convective heat transfer)

HEAT TRANSFER ENHANCEMENT (see Augmentation of heat transfer, single phase; Augmentation of heat transfer, two phase; Rotating duct systems, parallel, heat transfer in)

HEAT TRANSFER ENHANCEMENT DEVICES (see Plate fin heat exchangers)

HEAT TRANSFER FLUIDS (see Diphenyl; Dowtherm; Ethylene glycol; Heat transfer media)

HEAT TRANSFER IN AGITATED VESSELS (see Agitated vessel heat transfer)

HEAT TRANSFER IN BOILING MIXTURES (see Multicomponent mixtures, boiling in)

HEAT TRANSFER IN BUBBLY FLOW (see Bubble flow)

HEAT TRANSFER IN FLUIDISED BEDS (see Fluidised bed)

HEAT TRANSFER IN PLUG FLOW (see Plug flow heat transfer)

HEAT TRANSFER IN POROUS MEDIA (see Porous medium)

HEAT TRANSFER MEDIA

Heat transfer media and refrigerants can be distinguished according to their range of application. The latter are capable of operation between −100°C and +150°C, the former between −50°C and +400°C. The great variety of such substances for industrial purposes is shown in **Table 1**.

A heat transfer medium in the most general sense may be present in the solid, liquid and/or vapor phase; it can be used to store heat

Table 1.

Refrigerants and heat carriers	Temperature (°C)
Liquified gas for refrigeration	−100 to 100
Various liquids for refrigerating and cooling baths	−100 to 100
Brines for cooling	
Aqueous solutions, inorganic compounds	−50 to 30
Aqueous solutions, organic compounds	−50 to 100
Organic compounds, anhydrous	−25 to 200
Petroleum oils	0 to 300
Synthetic heat carriers	0 to 400
High-temperature salts (HTS)	150 to 620
Molten metals	350 to 900
Gases	350 to 1000

in a reversible form and can be circulated within the installation, e.g. in pipes. With multiphase substances, a change of phase will occur during heat exchange with the surroundings, considerable amounts of enthalpy being exchanged. Whereas specific heat capacity c varies between 1.6 and 4.2 kJ/(kgK) for organic compounds or water respectively, the amount of vaporization/condensation enthalpy ranges from 80 to 2500 kJ/(kgK). The heat exchanged (dq) can be determined from the sensible heat ($c \cdot dT$) and the phase change enthalpy (r). For example with vaporization (t,p = const), the added heat for heating to boiling point and complete vaporization is found by $dq = c \cdot dT + r = dh + r$.

In order to calculate the heat exchange between a medium and a surface, the law of heat transmission has to be taken into account, for example as non-dimensional Nusselt number. For forced convection, the valid equation is Nu = f(Re, Pr), for free convection Nu = f(Gr, Pr), where $Nu = (\alpha \cdot D)/\lambda$, D signifying characteristic length. For the calculation of the heat transfer coefficient α[W/(m^2K)]. Therefore, the following temperature-dependent properties are indispensable:

1. density ρ [kg/(m^3)],
2. specific heat capacity c [kJ/(kg K)],
3. heat conductivity λ [W/(m K)] and
4. Kinematic viscosity ν [m^2/s] or dynamic viscosity η [Ns/m^2], where $\nu = \eta/\rho$.

From these are calculated the thermal diffusivity κ [m^2/s] [$\kappa = \lambda/(\rho c)$] and the Prandtl number $Pr = \nu/\kappa$

In successfully selecting a suitable heat transfer medium, a great many criteria besides the four properties mentioned have to be considered. Concerning the installation, these are permissible operating temperature and pressure range. At the same time, a multitude of marginal and secondary conditions have to be maintained, each of which is necessary but far from sufficient in itself.

The most important heat transfer medium may well be water, which exhibits thermodynamic properties favorable to heat transfer. Moreover it is nontoxic, physiologically harmless, nonflammable, mildly corrosive in a way that can easily be controlled, thermally stable, readily available, and cheap. These advantages are compromised by a freezing point which is far too high, at 0°C, and an accompanying increase in volume of 9%, as well as by a rapid increase of vapor pressure with temperature. Water, therefore, may be employed as a heat transfer fluid in a limited range of applications only. For higher temperatures, substances with lower vapor pressures are substituted, mainly mineral oils and synthetic heat carriers.

Criteria to be considered in chosing a heat carrier include: closed flash point, water content, neutralization value (for aquaeous acids), sulfur content chlorine content, and corrosive activity, for example with respect to copper. In addition the supplier has to provide information on various properties including: kinematic viscosity v at 0, 40 and 100°C; pour-point (DIN ISO 3016); density at 15°C; mass percentage of oxide ash; inflammability point in °C; mass fraction of coke residue in mineral oil based heat carriers and data on thermal stability. The properties required for the design of heat transmission installations have to be given, stating the test procedure for the temperature range intended. Thermal stability values of unused heat transfer liquids can be determined according to DIN 51 528 (01.94). In addition to the initial treatment of the samples the evaluation of the boiling range by gas-chromatography simulation is set out there. For the evaluation of used heat transfer media a similar testing procedure is being devised.

New products in this field include substitutes for refrigerants that destabilize the ozone layer and developments of suitable lubricants for compressors (R134a as a mixture with other substances is viable).

Leading to: *Dowtherm; Refrigerants, properties*

H.-J. Kilger

HEAT TRANSFER PROBE (see Shear stress measurement)

HEAT TRANSFER TO PIPELINES (see Pipelines)

HEAT TREATMENT OF STEELS (see Steels)

HEAT WHEELS (see Regenerative heat exchangers)

HEAVY WATER

Heavy water is the common name for *deuterium oxide* (D_2O). Its molecular weight is 20.03 and its physical properties are similar to those of water. It boils at 374.5 K and freezes at 277 K. Heavy water is found naturally in very low concentrations in water and its main usage is in nuclear reactors as a moderator.

V. Vesovic

HEAVY WATER REACTORS (see Nuclear reactors)

HEAT WAVES (see Heat protection)

HEDH (see Heat Exchanger Design Handbook)

HEDSTROM NUMBER (see Liquid-solid flow)

HELE SHAW FLOWS

Following from: *Flow of fluids*

This term describes the flow of a real fluid between two closely spaced parallel plates. **Figure 1** outlines a typical arrangement (usually referred to as a Hele Shaw cell) in which the fluid flow is generated by a pressure gradient applied across the ends of the plates, and the spacing d is sufficiently small to ensure that the viscous forces dominate (ie. viscous forces >> inertial forces).

If some form of "obstruction" is placed between the plates (ie. a vertical cylinder), then, provided the linear dimension of this object measured in the plane of the plates (L) is large in comparison to the spacing d, the flow behaves as if the local pressure gradient extended to infinity. As a result, the local velocity may be expressed in terms of the local pressure gradient:

$$u \approx \frac{-1}{2\eta}\frac{\partial P}{\partial x} z(d - z), \qquad v \approx \frac{-1}{2\eta}\frac{\partial P}{\partial y} z(d - z)$$

where η is the dynamic viscosity, and the Cartesian co-ordinates are aligned so that z is perpendicular to the plane of the plates. The condition necessary for this to occur is:

$$\frac{d}{L}\left(\frac{\rho d^3 |\nabla P|}{\eta^2}\right) << 1$$

The velocity components noted above define a two dimensional velocity field which is both irrotational and satisfies the zero normal flow condition at a solid boundary. As a result, the streamlines generated in a Hele Shaw cell are identical to the hypothetical streamlines associated with the two dimensional flow of an ideal fluid around a similar obstruction. The introduction of dye traces within a Hele Shaw cell thus provides a visual representation of an ideal fluid flow. Batchlor (1967) provides further investigation of this and other viscous dominated flows.

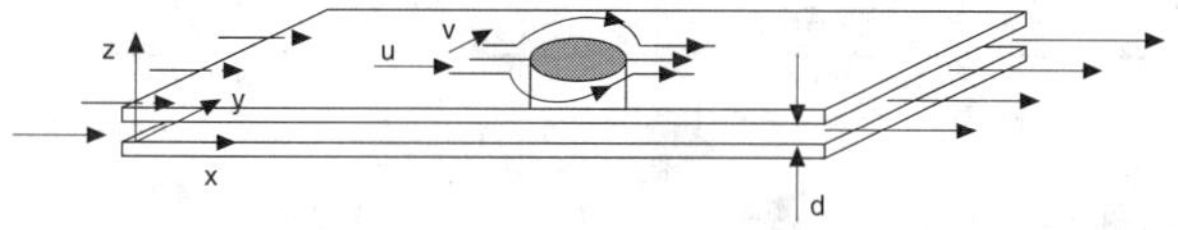

Figure 1. Hele Shaw cell.

References

Batchlor, G. K. (1967) *An Introduction to Fluid Mechanics,* Cambridge University Press, Cambridge.

C. Swan

HELICAL COIL BOILERS

Following from: Boilers

There are specific applications in which *helical coil heat exchangers* are advantageous over straight tube ones. The foremost of these is for high temperature applications where long straight tubes may pose severe mechanical problems due to thermal expansion which can be minimised in coiled tubes. Another advantage of coiled tubes is that they can be used to pack more surface area in a small volume and are therefore used for applications with small temperature difference or high volumetric heat rating. Coiled tubes also have better heat transfer coefficient and residence time distributions and are therefore used in *compact heat exchangers.* The main disadvantages of coiled tube heat exchangers are that they are more costly than the straight tube type to manufacture, they cannot be cleaned easily and are therefore not suitable for concentration or crystallization type of applications. **Figure 1** shows a coiled tube evaporator with five coils with boiling occurring on the shell-side.

A variant of the helical coil boilers is the **Spiral Plate Heat Exchanger** shown in **Figure 2**. Here, the spiral surfaces are separated by raised bosses and sealed with two end plates. The hot fluid enters at the center of the unit and flows towards the periphery in a spiral path. The cold fluid enters at the periphery and flows towards the center setting up a counter-current flow configuration. One of the advantages claimed for this design is that the swirling path of the fluids produces a scrubbing effect which reduces fouling. Also, the end plates can be removed to give full access to the flow passages for cleaning. However, this limits it to low pressure applications.

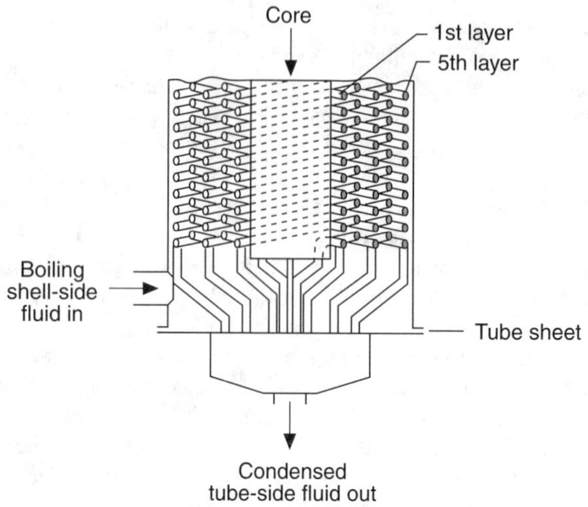

Figure 1. A helical coil evaporator. From Hewitt, Shires and Bott (1994). *Process Heat Transfer*, CRC Press.

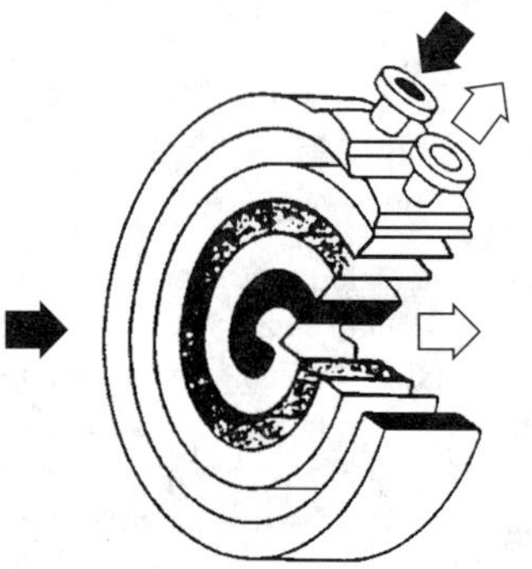

Figure 2. A spiral tube heat exchanger. From Hewitt, Shires and Bott (1994). *Process Heat Transfer*, CRC Press.

The swirling motion produced by coiling a tube increases the frictional pressure drop and the average heat transfer coefficient compared with those in a straight pipe of the same total (uncoiled) length. There are several empirical correlations to calculate these for single phase flow. These are usually in the form of an enhancement factor, over the straight tube value, correlated in terms of the flow Reynolds number and the tube-to-coil diameter ratio. The correlations of Ito (1959, 1969) and those of Schmidt (1967) are widely used to calculate the friction factor and the heat transfer coefficient, respectively, over a wide range of flow conditions. The calculation of these parameters for two-phase flow through coils is much less well-established, although some specific effects of coiling on the flow patterns (such as "*film inversion*") have been studied. An upper limit for these may be obtained by multiplying the two-phase straight tube values by the single phase enhancement factors implicit in the correlations of Ito and Schmidt mentioned above.

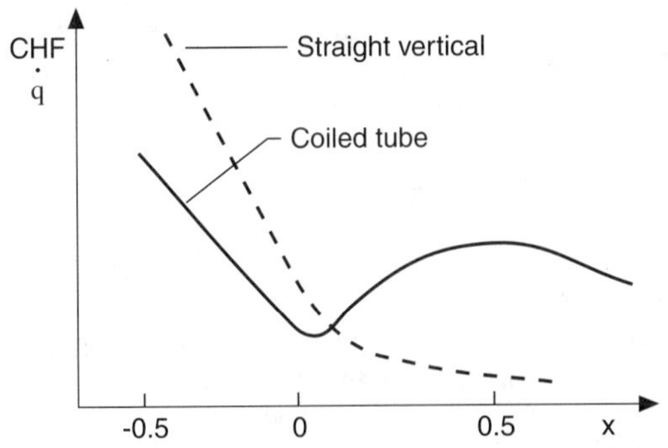

Figure 3. Schematic relation between critical heat flux and quality in straight tubes and coiled tubes.

A special feature of coiled tubes is that their *critical heat flux* (CHF) is less than that of straight tubes under subcooled flow conditions whereas it is significantly higher in the high-quality region (**Figure 3**). This led to the use of helical coil steam generators in the nuclear industry. The correlation of Berthoud and Jayanti (1990) can be used to calculate the CHF vs quality relation in the high quality region.

Some other applications of helical coil heat exchangers are discussed in Hewitt et al. (1994).

References

Berthoud, G. and Jayanti, S. (1990) Characterization of dryout in helical coils, *Int. J. Heat Mass Transfer,* 33(7), 1451–1463.

Hewitt, G. F., Shires, G. L., and Bott, T. R. (1994) *Process Heat Transfer,* CRC Press.

Ito, H. (1959) Friction factors for turbulent flow in curved pipes, *J. Basic Eng.,* 81, 123–134.

Ito, H. (1969) Laminar flow in curved pipes, *Z. Agnew. Math. Mech.,* 11, 653–663.

Schmidt, E. F. (1967) Heat transfer and pressure loss in spiral tubes, *Chem. Ing. Tech.,* 39, 781–789.

S. Jayanti

HELICAL COIL HEAT EXCHANGERS (see Helical coil boilers)

HELIUM

Helium—(Gr. *helios,* the sun), He; atomic weight 4.00260; atomic number 2; melting point below −272.2°C (26atm); boiling point −268.934°C; density 0.1785 g/l (0°C, 1 atm); liquid density 7.62 lb/ft^3 at boiling point; valence usually 0.

Evidence of the existence of helium was first obtained by Janssen during the solar eclipse of 1868 when he detected a new line in the solar spectrum; Lockyer and Frankland suggested the name *helium* for the new element; in 1895, Ramsay discovered helium in the uranium mineral *clevite,* and it was independently discovered in clevite by the Swedish chemists Cleve and Langlet about the same time. Rutherford and Royds in 1907 demonstrated that α particles are helium nuclei.

Except for hydrogen, helium is the most abundant element found throughout the universe. It has been detected spectroscopically in great abundance, especially in the hotter stars, and it is an important component in both the proton-proton reaction and the carbon cycle, which account for the energy of the sun and stars. The fusion of hydrogen into helium provides the energy of the hydrogen bomb.

The helium content of the atmosphere is about 1 part in 200,000. While it is present in various radioactive minerals as a decay product, the bulk of the world's supply is obtained from wells, for example in Texas, Oklahoma, Kansas and Swift River, Saskatchewan.

Helium has the lowest melting point of any element and has found wide use in cryogenic research, as its boiling point is close to absolute zero. Its use in the study of superconductivity is vital. Using liquid helium, Kurti and co-workers, and others, have succeeded in obtaining temperatures of a few microdegrees K by the adiabatic demagnetization of copper nuclei, starting from about 0.01 K.

Five isotopes of helium are known. Liquid helium (He^4) exists in two forms: He^4 I and He^4 II, with a sharp transition point at 2.174 K (3.83 cm Hg). He^4I (above this temperature) is a normal liquid, but He^4II (below it) is unlike any other known substance. It expands on cooling; its conductivity for heat is enormous; and neither its heat conduction nor viscosity obeys normal rules. It has other peculiar properties. Helium is the only liquid that cannot be solidified by lowering the temperature. It remains liquid down to absolute zero at ordinary pressures, but it can readily be solidified by increasing the pressure. Solid He^3 and He^4 are unusual in that both can readily be changed in volume by more than 30% by application of pressure. The specific heat of helium gas is unusually high. The density of helium vapor at the normal boiling point is also very high, with the vapor expanding greatly when heated to room temperature. Containers filled with helium gas at 5 to 10° K should be treated as though they contained liquid helium due to the large increase in pressure resulting from warming the gas to room temperature.

While helium normally has a 0 valence, it seems to have a weak tendency to combine with certain other elements. Means of preparing helium difluoride have been studied, and species such as HeNe and the molecular ions He^+ and He^{++} have been investigated. Helium is widely used as an inert gas shield for arc welding; as a protective gas in growing silicon and germanium crystals, and in titanium and zirconium production; as a cooling medium for nuclear reactors, and as a gas for supersonic wind tunnels. A mixture of 80% helium and 20% oxygen is used as an artificial atmosphere for divers and others working under pressure. Helium is extensively used for filling balloons as it is a much safer gas than hydrogen. While its density is almost twice that of hydrogen, it has about 98% of the lifting power of hydrogen. At sea level, 1000 ft^3 of helium lifts 68.5 lb.

One of the recent largest uses for helium has been for pressuring liquid fuel rockets. A Saturn booster such as used on the

Apollo lunar missions requires about 13 million cubic feet of helium for a firing, plus more for checkouts.

Handbook of Chemistry and Physics, CRC Press

HELIUM-NEON LASERS (see Lasers)

HELMHOLTZ FREE ENERGY (see Free energy; Multi-component systems, thermodynamics of)

HELMHOLTZ-TAYLOR INSTABILITY (see Plunging liquid jets)

HEMISPHERICAL EMISSIVITY (see Emissivity)

HENRY'S LAW

Henry's Law is concerned with the *solubility of a gas in a liquid* to form an ideal dilute solution. An ideal dilute solution is defined as one in which the **Activity Coefficient** of the solvent is unity. This will clearly be so at infinite dilution, but in practice, it often remains essentially unchanged during the addition of small amounts of solute, say a few mole per cent, and is therefore an adequate approximation for real dilute solutions.

If the non-ideality of the gas phase can be ignored, i.e. the pressure is sufficiently low, we find that the solubility of the gas in the liquid is directly proportional to its partial pressure in the gas phase. If the volatility of the solvent is low, this can be approximated as the total pressure and this is the form in which the Law is most usually used.

Henry's constant is usually defined as:

$$H = \tilde{x}_2/P$$

where $\tilde{x}_2$ is the mole fraction of solute gas in the liquid phase at pressure P. H is, of course, a function of temperature.

G. Saville

HEPTANE

n-Heptane (C_7H_{16}) is a colourless liquid that is insoluble in water. It is obtained by fractional distillation of petroleum. Heptane is not only flammable, but also moderately toxic if inhaled. It is used as a solvent, as an anaesthetic and in organic synthesis.

Molecular weight: 100.198
Melting point: 182.6 K
Normal boiling point: 371.6 K
Critical temperature: 540.61 K
Critical pressure: 2.736 MPa
Critical density: 234.1 kg/m^3

See **Table 1** below.

V. Vesovic

Table 1. Heptane: Values of thermophysical properties of the saturated liquid and vapor

T_{sat}, K	371.6	380	400	420	440	460	480	500	520	540.6
p_{sat}, kPa	101.3	130	219	349	529	721	1 094	1 513	2 046	2 736
ρ_ℓ, kg/m^3	614	606	585	563	540	512	484	448	397	234
ρ_g, kg/m^3	3.46	4.36	7.23	11.5	17.4	25.6	37.8	56.5	88.3	234
h_ℓ, kJ/kg	453.6	474.5	530.3	586.2	639.7	702.5	760.6	825.7	895.5	1 004.8
h_g, kJ/kg	765.3	786.2	823.4	860.6	897.8	937.4	972.3	1 009.5	1 035.1	1 004.8
$\Delta h_{g,\ell}$, kJ/kg	319.7	311.7	293.1	274.4	258.1	234.9	211.7	183.8	139.6	
$c_{p,\ell}$, kJ/(kg K)	2.57	2.61	2.72	2.82	2.93	3.03	3.19	3.39	3.85	
$c_{p,g}$, kJ/(kg K)	1.98	2.01	2.15	2.26	2.41	2.51	2.72	3.05	3.60	
η_ℓ, μNs/m^2	201	186	159	135	115	97	82	67	54	
η_g, μNs/m^2	7.3	7.7	8.3	9.0	9.7	10.7	12.1	14.0	17.6	
λ_ℓ, (mW/Km)	98.0	95.5	88.8	82.9	76.1	69.9	61.1	52.0	41.8	
λ_g, (mW/Km)	18.0	18.4	20.1	22.6	24.9	27.7	29.9	33.0	36.0	
Pr_ℓ	5.27	5.08	4.87	4.59	4.43	4.27	4.28	4.37	4.97	
Pr_g	0.80	0.84	0.89	0.90	0.94	0.97	1.10	1.29	1.76	
σ, mN/m	12.6	11.8	10.0	8.3	6.6	5.1	3.6	2.2	0.8	
$\beta_{e,\ell}$, kK^{-1}	1.6	1.7	1.90	2.2	2.7	3.2	4.2	6.8	24.5	

From Beaton C.F. and Hewitt G.F., *Physical Property Data for the Design Engineer*, Hemisphere Publishing Corp., New York, 1989, with permission.

HERMITIAN, CONJUGATE, LINEAR OPERATORS (see Eigen values)

HERRINGBONE CORRUGATIONS (see Plate fin heat exchangers)

HERSCHEL-BULKLEY FLUIDS (see Non-Newtonian fluids)

HETEROGENEOUS CATALYSIS (see Catalysis)

HETEROGENEOUS FLOW REGIME (see Liquid-solid flow

HETEROGENEOUS REACTIONS (see Chemical reaction)

HETEROGENEOUS SPRAY COMBUSTION (see Flames)

HETEROSPHERE (see Atmosphere)

HEXANE

n-Hexane (C_6H_{14}) is a colourless liquid that is insoluble in water. It is obtained by fractional distillation of petroleum. Hexane is classified as flammable and moderately toxic. At low concentrations in air it is also explosive, but under equilibrium conditions these concentrations are only achieved at temperatures below 2°C. Its main use is as a solvent in the extraction of oils from seeds. It is also used in motor fuels, as a polyolefins solvent, as polymerisation medium and paint diluent.

Molecular weight: 86.178	Critical temperature: 507.44 K
Melting point: 177.83 K	Critical pressure: 3.031 MPa
Normal boiling point: 341.88 K	Critical density: 232.8 kg/m^3

See **Table 1**.

V. Vesovic

Table 1. Hexane: Values of thermophysical properties of the saturated liquid and vapor

T_{sat}, K	341.88	370	385	400	415	430	445	460	475	507.44
p_{sat}, kPa	101.3	228	331	465	638	854	1 124	1 457	1 859	3 031
ρ_ℓ, kg/m^3	613	585	568	551	532	511	488	463	432	233
ρ_g, kg/m^3	3.3	7.0	10.0	14.1	19.5	26.7	34.7	48.6	66.2	233
h_ℓ, kJ/kg	395.4	465.2	507.1	546.6	586.2	632.7	676.9	725.7	774.6	930.4
h_g, kJ/kg	728.0	776.9	802.5	830.4	856.0	879.2	907.1	930.4	953.7	930.4
$\Delta h_{g,\ell}$, kJ/kg	332.6	311.7	295.4	283.8	269.8	246.5	230.2	204.7	179.1	
$c_{p,\ell}$, kJ/(kg K)	2.39	2.58	2.68	2.78	2.89	3.00	3.12	3.26	3.46	
$c_{p,g}$, kJ/(kg K)	1.91	2.07	2.18	2.28	2.39	2.51	2.66	2.92	3.32	
η_ℓ, μNs/m^2	202.2	158.9	145.7	128.5	113.6	99.9	87.4	75.6	64.3	
η_g, μNs/m^2	7.3	7.9	8.4	8.8	9.4	10.1	10.9	11.9	13.3	
λ_ℓ, (mW/m^2)/(K/m)	100.4	91.0	87	82	78	74.5	67.8	63.0	56.0	
λ_g, (mW/m^2)/(K/m)	16.7	18.8	20.4	22.4	24.2	26.2	28.5	30.4	33.8	
Pr_ℓ	4.81	4.51	4.49	4.36	4.19	4.01	3.99	3.83	3.97	
Pr_g	0.83	0.87	0.91	0.91	0.92	0.97	1.02	1.14	1.31	
σ, mN/m	13.33	10.57	9.13	7.71	6.35	5.02	3.78	2.66	1.63	
$\beta_{e,\ell}$, kK^{-1}	1.28	1.86	2.17	2.46	2.88	3.41	4.14	5.41	20.0	

From C. F. Beaton, G. F. Hewitt, 1989, *Physical Property Data for the Design Engineer,* Hemisphere Publishing Corporation, New York, with permission.

HIGH FREQUENCY HEATING

Radio frequency and *microwave* are sometimes used as alternatives to convective, conductive or radiant heat transfer for the proccessing of "non metals" (Langton, 1949). Industries making use of these techniques include, textiles, paper, food, plastic and chemicals. The applications are many and varied including drying, baking, defrosting, welding and polymerisation (Jones, 1987). Known as *high frequency* or *dielectric heating,* both are forms of electro-magnetic wave energy which share some characteristics but also have significant differences.

The perceived advantage of dielectric heating is based on the so called "volumetric" effect arising from the fact that the energy is absorbed directly in the body of the material rather than being transferred to it via a surface. The concept of "volumetric" heating needs to be qualified since there is in reality a limiting penetration depth which depends on the properties of the material being heated as well as on the wavelength of the energy source (Metaxas and Meridith, 1983). At the shorter wavelengths associated with microwave, penetration depth into a wet body is normally a few centimetres; at the longer wavelengths of radio frequency the depth can be a large fraction of a metre or a few centimetres depending on the ionic conductivity of the water in the material.

Dielectric process heating uses the frequency range from about 5 MHz to 5 GHz with radio frequency, RF, being normally defined as being less than 100 MHz. The definition of microwave usually is between 500 MHz and 5 GHZ. Within these ranges there are specific frequencies allocated for industrial, scientific and medical uses, the so called ISM bands. The most common of these are 13.56 and 27.12 MHz (wavelengths 22.4 and 11.2 metres respectively) for RF with 900 MHz and 2.45 GHz (wavelengths 0.35 and 0.13 metres respectively) being the permitted frequencies for microwave. These particular frequencies have been chosen in order to minimise the risk of interference with telecommunications by either the fundamental or by higher harmonics and have no particular significance as far as the resonance of the water dipole is concerned. The actual frequency within the "900 band" varies from country to country, depending on local regulations.

Heat Transfer

Dielectric heat transfer depends on a number of polarisation effects which take place over a broad band of frequencies. The most commonly described one is dipolar orientational polarisation (Hasted, 1973).

As can be seen from **Figure 1** which shows loss factor as a function of frequency, this is important at microwave frequency

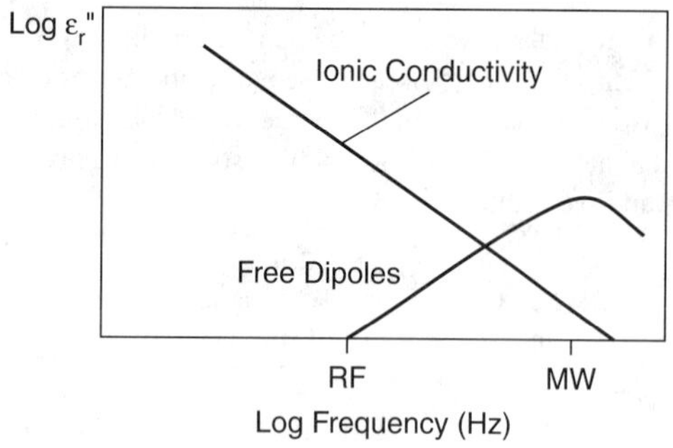

Figure 1.

but of relatively little significance at the lower, radio frequencies. The dominant mode in the RF range is space charge orientation, which in turn is dependent on the ionic conductivity of the material being processed. If the dielectric properties of a particular material are known it is possible, in theory, to choose the most appropriate frequency from those available in the ISM bands. In reality unless the dielectric loss factor, ε_r'' is very low most products can be dried or processed by either RF or microwave. The choice can then be made on other considerations such as the engineering required to make a satisfactory heat transfer applicator compatible with the process line requirements, ie product width, height and shape.

The heat transferred per unit volume of product is given by

$P = 2\pi f \varepsilon_0 \varepsilon_r'' E^2$ W/m where

f = frequency hertz

E = electric field strength V/m

ε_r'' = loss factor or relative permittivity

ε_0 = permittivity of free space (8.85×10^{12} farad/m)

Loss factor, ie the product of dielectric constant and loss tangent, varies with a number of parameters including frequency, moisture content and temperature. The relationships are often quite complex as for example in drying where as the temperature increases the moisture content falls. These relationships can be such that preferential heating and drying of the wetter areas takes place, in the right circumstances leading to "moisture profile correction".

When dealing with process heating, interpretation of published data, such as that by Von Hippel (1961) needs to be undertaken with care. For example the figures quoted for "water" often refer to deionised water, in this case the effect of having any level of ionic material present (as in most "real" water), will substantially increase the loss factor at RF but have relatively little effect at microwave frequency.

Mass Transfer Considerations

RF and microwave are alternatives to conventional heat transfer with several feature which need careful consideration. Internal mass transfer may be in liquid or vapor phase depending on the structure of the material, for example in certain capillary porous materials the internal generation of heat having first reduced the viscosity and surface tension of the bulk of the free water in the body will then cause a small quantity to be evaporated, raising the internal vapor pressure sufficiently for liquid phase flow to occur. In other materials where the pore structure is courser the more conventional mechanism of evaporation followed by vapor phase flow takes place. Perhaps the most significant aspect of this form of heating is that moisture gradients are avoided because the heat is transferred directly to the water and is not dependent on the thermal conductivity of the substrate and consequently the classic model of drying on a retreating wet front does not apply to this form of heating. Because the surface moisture content is normally at least as high as the average in the body the vapor pressure difference across the surface is such that relatively modest air flows are required.

Because the heating arises from the interaction between the high frequency field and the product there is no incidental heating of the oven metal work or the air. In processes which involve the evaporation of water it is important that recondensation is avoided. The minimum requirement is a flow of heated air over the product to act as a means of mass transfer from the surrounding atmosphere; this same air flow is used to maintain the metal work above the dew point.

Radio Frequency Power Sources and Applicators

Radio frequency power supplies are usually a "class C" oscillator based on a triode valve which has been built into a cavity type tank circuit. With new regulations in place increasing importance is being attached to the need avoid electromagnetic interference with other equipment, and therefore alternative generator types are being considered for some applications, such as the use of crystal driven linear amplifiers in conjunction with 50 ohm transmission lines.

Radio frequency applicators are essentially capacitors which contain the product requiring heating as the whole or a part of its dielectric. The simplest and most widely used is the through field or parallel plate electrode as shown in **Figure 2**.

When used for drying an air space is required above the dielectric to allow for the movement of the product through the machine and for ventilation of the water vapor. This then means an increase in voltage between the plate in order to maintain an adequate field strength in the product. It is therefore important to consider the relative dimensions of the dielectric and air space capacitors to give the desired heating effect without an electrical discharge taking place. For very thin materials such as paper it may be necessary to use an alternative electrode configuration.

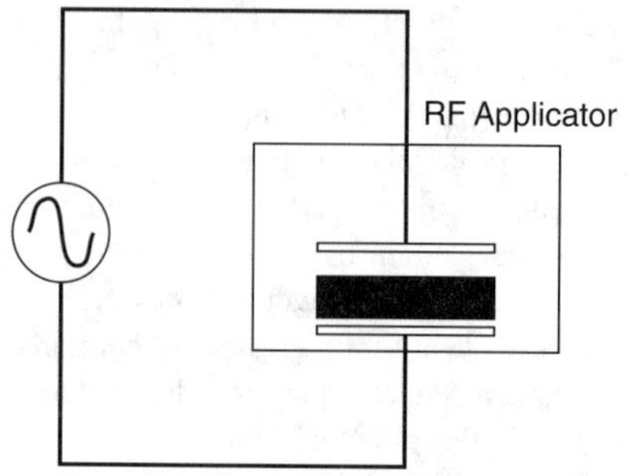

Figure 2.

Microwave Power Sources and Applicators

For industrial heating applications using microwaves the usual power source is the magnetron. The most common form of microwave heating applicator is the multimode cavity, similar in concept to the domestic oven. When used for continuous, conveyorised processing the design of the ports to allow the passage of product is critical in order to prevent the emission of microwave energy. The limits on the dimensions of these apertures can be very restrictive, typically 20 to 30 mm for 2.45 GHz and perhaps 80 to 100 mm at 900 MHz.

Industrial microwave heating has been used extensively in the rubber industry for curing and preheating prior to moulding. In the food industry it has been used for tempering, melting, cooking and drying. Recently microwave vacuum dryers have been developed for drying expensive, high quality temperature sensitive pharmaceuticals.

References

Hastel, J. B. (1973) *Aqueous Dielectrics,* Chapman and Hall, London.

Jones, P. L. (1987) Radio frequency processing in Europe, *J. Microwave Power,* 22–3, 143–153.

Langton, L. L. (1949) *Radio Frequency Heating Equipment,* Pitmans, London.

Metaxas, A. C. and Meridith R. J. (1983) *Industrial Microwave Heating,* Peter Perigrinus, London.

Von Hippel, A. (1961) *Dielectric Materials and Applications,* MIT Press, Cambridge Mass.

P.L. Jones

HIGH LEVEL LANGUAGE (see Computer programmes)

HIGH SPEED PHOTOGRAPHY (see Photographic techniques)

HIGH TEMPERATURE GAS REACTORS (see Nuclear reactors)

HIGH TEMPERATURE PLASMA (see Plasma)

HIGH TEMPERATURE REACTOR, HTR (see Nuclear reactors)

HINDERED DRYING (see Drying)

HINDERED SETTLING (see Sedimentation)

HOLDUP (see Three phase, gas/liquid/liquid, flows)

HOLDUP WAVES (see Wispy annular flow)

HOLES AND CONDUCTION ELECTRONS (see Semiconductors)

HOLLOW SCRUBBER (see Scrubbers)

HOLOGRAMS

Conventional photography represents two-dimensional (2D) records of three-dimensional (3D) scenes. Here the distribution of the light intensity passing through or reflected by the scene is registered on a photosensitive surface. By recording this way, information on the phase distribution of the light waves is not required. On the contrary, *holography* depends on recording the amplitude and the phase of the light waves. It is then possible to reconstruct the scene in three dimensions as well. This makes holography a special imaging method.

Because recording materials are only sensitive to light intensity, the phase information has to be transformed into a light intensity code. This can be done by using coherent light for object illumination and by adding to it a reference wave from the same light source as the object wave. Both waves, or light beams, interfere at the recording medium producing an interference pattern in which the local fringe density describes a function of the phase distribution and the gradient of darkness of the fringes is proportional to the intensity of the object wave. The recorded scene can be reconstructed if the photographic plate, which after development, will be called *hologram,* is illuminated with the reference wave. The information stored in the photo plate is encoded and can be shown as a true 3D image of the original scene. An observer cannot distinguish between the reconstructed and the original wave field.

Holograms can be classified according to the arrangement of the optical components in the holographic camera or according to the features of the developed photographic plate, as shown in **Figure 1**.

Classification of Holograms

With respect to the arrangement of the components

a. In-Line. A plane wave illuminates a particle or a particle field. A small amount of light is diffracted by the particles forming an object beam which is able to interfere with the undisturbed wave.

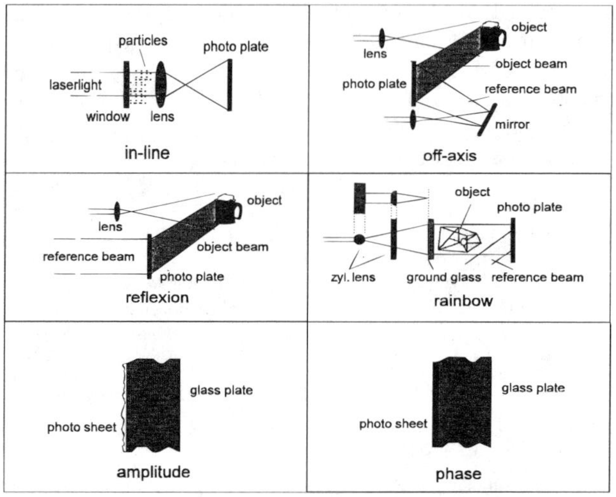

Figure 1. Types of holograms.

b. Off-Axis (Counter Light). A light beam illuminates a transparent medium containing the object to be holographed. The object beam is recorded simultaneously with a reference beam on a photo plate.
c. Off-Axis. An object being illuminated reflects part of the light to a photo plate. Simultaneously the plate is also exposed to a reference beam. The path length of both beams is almost equal.
d. Reflexion. The object beam meets the photo plate at the emulsion side while the reference beam incides on its glass side. It can be reconstructed with white light.
e. Rainbow. With the aid of two cylindrical lenses a laser light sheet is produced and sharply focussed onto a ground glass plate in order to obtain diffuse illumination of the object. The light travels through the partial transparent object and finally meets the photo plate. A plane wave is used as a reference. It can be reconstructed with white light.

With respect to the developed photo sheet

f. Amplitude. Once the recording has been made, the photo plate is developed and fixed. The photo emulsion shrinks a little but remains uniformly thick. The amplitude of the reconstructed wave results in a function of the darkness distribution in the photographic sheet.
g. Phase. This fixing process is replaced by bleaching the photo plate until total transparency. During this process the silver ions are removed from the photo emulsion producing an irregular thickness distribution on the photographic sheet which depends on the recorded intensity distribution. The reconstruction wave behind the plate results in phase modulation.

Application: Spray Characterization

Pulsed laser holography is a very suitable non-intrusive method to visualize disperse flows, when the particle size in the dispersed phase is bigger than 10 times the wavelength. Spray geometry, drop size and distribution, droplet velocities and trajectories can be measured and evaluated by means of digital image processing. The purpose of digital image processing is to reflect the main features of a picture more clearly and informatively than in the original and to judge the contents of an image quantitively by employing pattern recognition algorithms.

References

Chavez, A. and Mayinger, F. (1988) Single and double pulsed holography for characterization of sprays of refigerant R113 injected into its own saturated vapor, *Experimental Heat Transfer Fluid Mechanics and Thermodynamics,* 848–854.

Hariharan, P. (1984) *Optical Holography,* Cambridge University Press.

Mayinger, F. (1994) *Optical Measurements—Techniques and Applications,* Springer Verlag, Berlin, Heidelberg, New York.

Nishida, K., Nurakami, N., and Hiroyasu, H. (1978) Holographic measurements of evaporating diesel sprays at high pressure and temperature, *JSME Int. Journal,* 30–259, 107–115.

Leading to: *Holographic interferometry*

R. Fehle and F. Mayinger

HOLOGRAPHIC INTERFEROMETRY

Following from: *Holograms; Interferometry*

Holography allows various interferometric methods for measuring processes of heat and mass transfer to be used. Holographic Interferometry has displaced the Mach-Zehnder-Interferometry completely, because not only is it much cheaper to use, but is also much easier and convenient to handle. With Holographic Interferometry there is no need to machine or manufacture windows for test sections, mirrors and lenses of the optical components with special precision or accuracy, because imperfections are automatically balanced by the holographic two-step procedure.

Gabor (1948) invented a new method for recording and storing optical information which was called holography (see **Holograms**). Unlike photography which can only record the two-dimensional distribution of the radiation emitted by an object, holography can store and reconstruct three-dimensional pictures. The name holography comes from the ability of the method to record the totality (holos) of the information related to the wavefront of the light namely the amplitude, the wavelength and the phase position. It could only be used when the laser had been developed, which was 10 years later. A detailed description can be found in the literature: Ostrowsky (1980), Vest (1979).

Holographic Interferometry is a combination of interferometry and holography. **Figure 1** shows a simple holographic arrangement for the examination of transparent media. The main difference between classical and Holographic Interferometry is that the object beam is compared with itself.

Therefore, holographic and Mach-Zehnder-Interferometry differ from each other in such a way that the first is a two-step method (see **Figure 2**) and the second is a two-way method. In the former the undistorted object wave, called the comparison wave, is stored on a photographic plate and can be reconstructed after its development (then it is called hologram) by an illumination with the reference wave. The heat or mass transfer is then introduced, so that the momentary object wave, which is called measuring wave, experiences an additional phase shift. Now the measuring and the comparison wave interfere behind the hologram so describing the physical phenomenon which has been introduced and is to be investigated. (It must be pointed out, that macroscopic interference

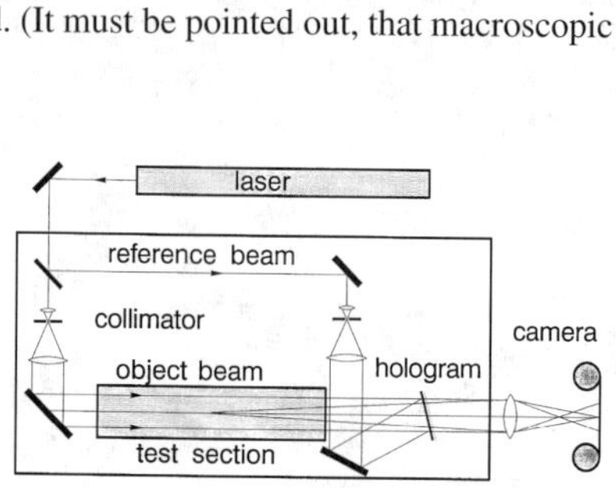

Figure 1. Holographic arrangement for the examination of transparent media.

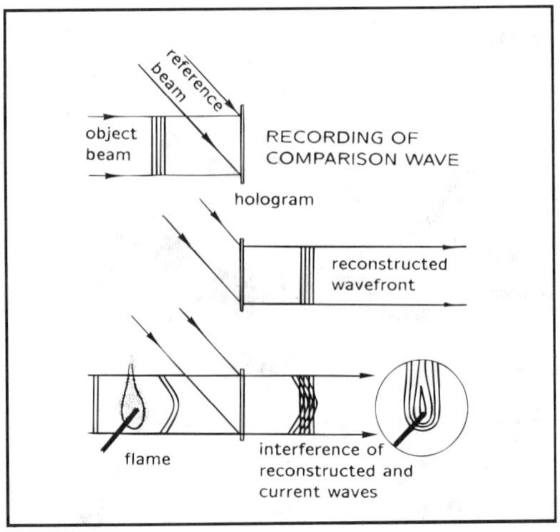

Figure 2. Real-time method: Superposition of measuring and comparison wave.

caused by the superposition of two object waves has to be clearly distinguished from the microscopic interference which takes place when the object and the reference wave are superimposed during the recording of the hologram.)

Double Exposure Technique

Several object waves—one after the other—can be recorded on one and the same hologram. When illuminating the hologram with the reference wave all are reconstructed simultaneously. This principle is used for the double exposure technique, see Mayinger (1994): In a first exposure the comparison and in a second exposure the measuring wave is recorded. From the interference pattern one can determine the differences between the comparison and measuring conditions caused, for example, by heat transfer.

Disadvantages of this method are that transient processes cannot be continuously observed and that the most favourable moment for the exposure with the measuring wave cannot be preselected, because the interference pattern only appears after the chemical process.

Real Time Method

This method allows a continuous observation of the process under investigation like Mach-Zehnder-Interferometry. After the first illumination in which the comparison wave is recorded on the photographic plate, it is developed and fixed. The chemical process can proceed in situ in a special housing made of glass, or the holographic plate can be removed, chemically treated and accurately repositioned afterwards.

Now the object wave of the unheated test section is reconstructed by illuminating the hologram with the reference wave. Simultaneously the test section is then irradiated by the measuring wave, however, now with the heat or mass transfer process switched on. Both object waves interfere behind the hologram. The continuously observable interference pattern is a result of the temperature or concentration field produced by the heat or mass transfer (see **Figure 3** on page 582).

An Interference Method for Simultaneous Heat and Mass Transfer

If the refractive index is simultaneously influenced by more than one parameter, for example by temperature and concentration, the interferogram cannot be evaluated directly. Attempts have been made to use the dependency of the refractive index on the wavelength of the light. This can be done by recording two interferograms originating from the light of two different wavelengths and from that to evaluate the temperature and the concentration field separately (Panknin, 1977).

Evaluation of Interferograms

See Mayinger (1994) and **Interferometry**.

References

Gabor, D. (1948) A New Microscopial Principle, *Nature,* 161, 1948; 1949, Microscopy by Reconstructed Wavefronts, *Proc. Roy. Soc.,* A 197, 1949; 1951 Microscopy by Reconstructed Wavefronts II, *Proc. Roy. Soc.,* A 197. 1951.

Mayinger, F. (1994) *Optical Measurements—Techniques and Applications,* Springer Verlag, Berlin, Heidelberg, New York.

Ostrowsky, Y. I. (1980) *Interferometry by Holography,* Springer Verlag, Berlin, Heidelberg, New York.

Panknin, W. (1977) Eine hol. Zweiwellenlängen-Interferometrie zur Messung überlagerter Temperatur- und Konzentrations-grenzschichten, Ph.D. Thesis, Tech. Univ. Hannover.

Vest, C. M. (1979) *Holographic Interferometry,* John Wiley, New York.

R. Fehle and F. Mayinger

HOLOGRAPHIC METHODS, FOR PARTICLE SIZING (see Particle size measurement)

HOLOGRAPHY (see Film thickness determination by optical methods; Holograms; Optical particle characterisation; Photographic techniques)

HOMEOSTASIS (see Physiology and heat transfer)

HOMOGENEOUS CATALYSIS (see Catalysis)

HOMOGENEOUS DENSITY (see Multiphase flow)

HOMOGENEOUS FLOW (see Gas-solid flows; Liquid solid flow)

HOMOGENEOUS FLOW, CONSERVATION EQUATIONS (see Multiphase flow)

HOMOGENEOUS MODEL (see Two-phase flows)

HOMOGENEOUS SPRAY COMBUSTION (see Flames)

HOMOGENISATION (see Agitation devices)

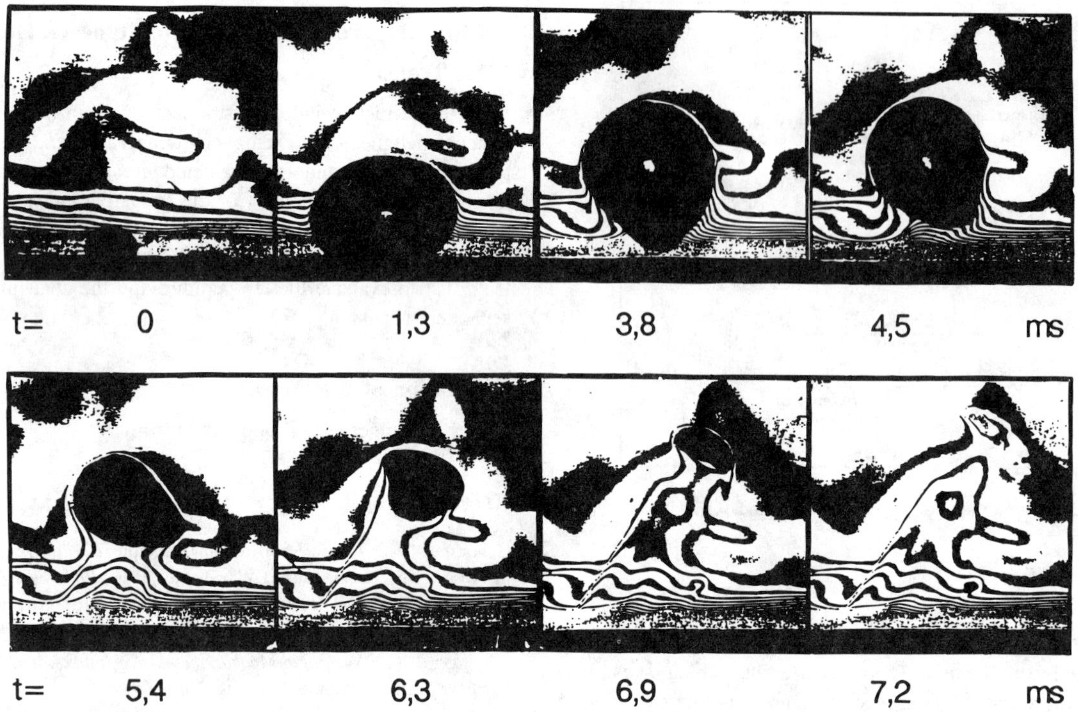

Figure 3. Growth and condensation of a steam bubble (p = 1bar, water temperature 8K below saturation, flow velocity w = 0.25 m/s, heat flux $\dot{q}$ = 9W/cm^2). (From Mayinger (1994) *Optical Measurements-Techniques and Applications*, Springer Verlag, Berlin, with permission).

HOMOGENISER (see Emulsions)

HOMOSPHERE (see Atmosphere)

HOOKEAN SOLID (see Non-Newtonian fluids)

HOOKE'S LAW (see Poisson's ratio)

HOPPERS, GRANULAR FLOW FROM (see Granular materials, discharge through orifices)

HORIZONTAL THERMOSYPHON REBOILER (see Reboilers)

HORIZONTAL TUBE SHELL SIDE EVAPORATOR (see Evaporators)

HOT DRY ROCK, HDR, GEOTHERMAL HEAT (see Alternative energy sources)

HOT FILM ANEMOMETERS (see Hot wire and hot film anemometers)

HOTTEL-WHILLIER-BLISS EQUATION (see Solar energy)

HOT-WIRE AND HOT-FILM ANEMOMETERS

Following from: *Velocity measurement*

The hot-wire anemometer, which is used to measure the instantaneous velocities of fluid flows, is based on the dependence of the sensor (gauge) heat transfer on the fluid velocity, temperature, and composition. A theoretical substantiation of this measurement method was set forth by King in his classic work in 1914. This method came into use for measuring turbulence characteristics following the development of a technique for compensation of the gauge's thermal lag.

The sensor in the hot-wire anemometer is a thin metal wire (see **Figure 1**) or a film made of a material with a high temperature coefficient of resistance such as tungsten, platinum, platinum-rho-

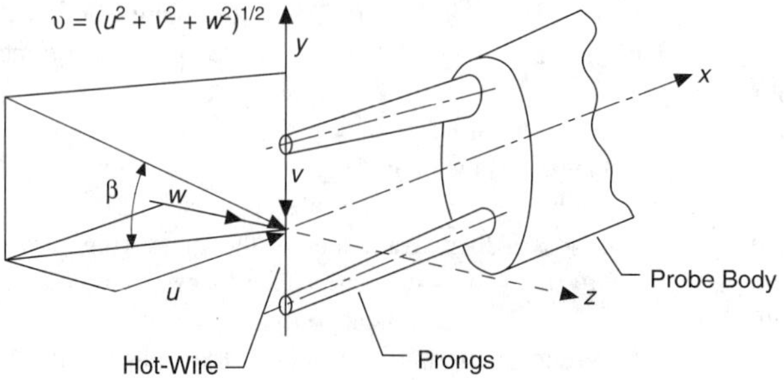

Figure 1. Sensor of hot wire anemometers.

dium, and platinum-iridium. The properties of materials used for manufacturing the sensors are presented in **Table 1**.

In these materials, resistance to a good accuracy depends linearly on temperature $R_w = R_0(1 + \alpha_\rho T_w)$, where R_0 is the wire resistance at T = 0°C and T_w is the wire transfer (0°C).

Usually the gauge wire is between 1 to 10 μm in diameter D and from 0.5 to 2 mm in length l. The wire is welded or soldered to stainless steel prongs with a diameter at the tips of 0.1 to 0.2 mm. In film gauges a 1 to 5 μm thick film is deposited on a cone-, wedge-, or cylinder-shaped substrate for measuring velocity or on a plane substrate for measuring shear stresses on the wall.

The gauge sensitivity to the flow velocity is provided by having the wire or film temperature T_w substantially higher than the flow temperature T_f, which is achieved by heating it by electric current. Measuring the velocity in gases is carried out with $(T_w - T_f)$ typically in the range 180 to 200 K and in water, from 20 to 40 K. Measurement of voltage drop across the wire allows, if the physical properties of the gauge material and the mechanism of heat transfer are known, the determination of local fluid velocity. The hot-wire anemometer operates in two regimes, viz. a constant-current regime (I_w = const) when the gauge voltage pulsations are attributable to temperature variation and, hence, the wire resistance, and the regime of constant temperature (T_w = const), maintained by the feedback system with a variable current heating the sensor. One or the other regime is used depending on the specific character of measurements.

The gauge is connected by a shielded coaxial cable to one of the arms of Wheatstone bridge, the opposite arm being connected to a variable resistance controlling the overheat factor $m = R_w/R_f$ (see **Figure 2**). In the constant- temperature hot-wire anemometer a feedback amplifier controls the bridge current so as not to upset the bridge balance and, consequently, to retain the gauge resistance and temperature constant and independent of the cooling rate. The instantaneous disbalance voltage of the bridge determines an instantaneous value of velocity pulsation. Contemporary instrument circuits afford, at an optimal alignment, a bandwidth up to 100 kHz.

The gauge thermal balance can be represented as

Table 1. Materials used in hot wire and hot film anemometers

Material	Tungsten	Platinum	Platinum-iridium	Platinum-rhodium
Electrical resistivity ρ, (10^{-6} Ω/cm)	5.5	10	32	18
Temperature coefficient of resistance α_ρ, (°C^{-1})	0.0035	0.0036	0.0008	0.0016
Specific heat C, (cal/g°C)	0.033	0.031	0.0032	0.0035
Thermal conductivity, λ, (W/cm°K)	1.78	0.69	0.255	0.501
Density, (g/cm³)	19.3	21.45	21.61	19.97
Maximum temperature, (°C)	300	800	750	850

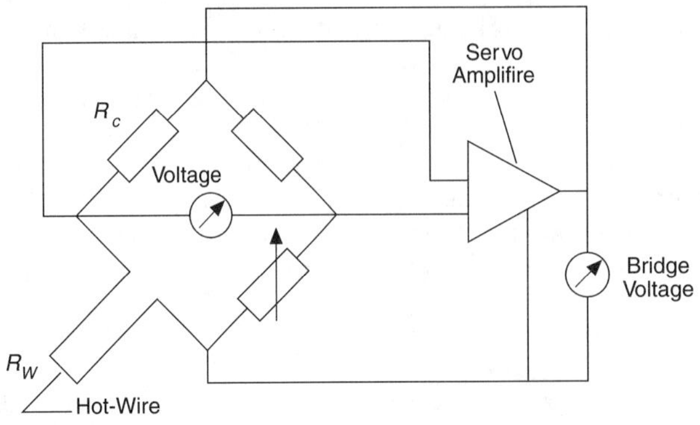

Figure 2. Wheatstone bridge circuit for hot wire anemometer.

$$Q_J = Q_{f.c} + Q_{fr.c} + Q_r + Q_{h.cond} + Q_{h.cap},$$

where Q_J is the Joule heat $I_w^2 R_w$, $Q_{f\cdot c}$ the heat removed due to forced convection, $Q_{fr\cdot c}$ and Q_r heat losses due to free convection and radiation, $Q_{h\cdot cond}$ heat removal to the sensor prongs, and $Q_{h\cdot cap}$ the change of heat content of the sensor with temperature change owing to turbulent pulsations of velocity.

Repeated measurements have shown that at mean velocities U > 0.2 m/s, gauge temperature T_w < 1000 K and the length-to-diameter ratio l/D > 200, the heat losses $Q_{fr\cdot c}$, Q_r, and $Q_{h\cdot cond}$ can be ignored, the more so if the gauge is individually calibrated. $Q_{f\cdot c}$ depends on the heat transfer coefficient α, the surface area A, the angle β subtended by the velocity vector direction and the normal to the sensor axis, and the temperature drop $(T_w - T_f)$.

The heat transfer coefficient depends on the physical conditions of heat transfer on the surface and at l/D > 200 and β = 0° can be expressed in dimensionless from

$$Nu = \frac{\alpha D}{\lambda} = f(Re, Pr, Kn, Ma, T_w/T_f).$$

where *l* is the length of the wire and D its diameter. When the effects of flow compressibility (Ma ≪ 1) and rarefaction (Knudsen number Kn < 0.015) can be neglected, the dependence

$$Nu = 0.42\,Pr^{0.2} + 0.57\,Pr^{0.33}Re^{0.5} \quad \textbf{(1)}$$

is commonly used which describes heat transfer from cylinders for $0.7 \le Pr \le 10^3$ and $10^{-2} \le Re \le 10^3$. In equation 1, the fluid physical properties are taken at $\bar{T} = 0.5(T_w + T_f)$.

Alternatively, for gas flows, the heat transfer is well described by the relation

$$Nu = (A + B\,Re^n)\left(\frac{\bar{T}}{T_f}\right)^{0.17},$$

in which the constants have the following values:

within $0.02 \le Re \le 44$ A = 0.24, B = 0.56, n = 0.45,

within $44 < Re < 140$ A = 0, B = 0.48, n = 0.51.

The limiting number Re = 44 is determined by origination of a vortex trail in the cylinder wake. The indicated range of Re numbers makes it possible to measure the velocity by the gauge with D = 5 μm wire within the $0.2 \leq U \leq 550$ m/s range.

The effect of compressibility is negligibly small at Ma < 0.3, but at high subsonic and supersonic velocities it must be allowed for. For supersonic velocities at $1.2 \leq Ma \leq 5$ generalization of a vast body of experimental data yielded the expression

$$Nu = 0.58\ Re^{0.5} - 0.795$$

holding for Kn < 0.015. The influence of Knudsen number has to be allowed for in rarefied gases.

The heat transfer equations of hot-wire anemometer under consideration describe a general character of the functional dependence between the power put out by a sensor and the flow over it. In practice the specific gauge connected to a hot-wire anemometer electronic circuit is individually calibrated to determine the quantitative relation of the voltage at the bridge output E and the flow velocity U

$$\frac{E^2 R_w}{(R_c + R_w)^2} = (A + BU^n)(T_w - T_f),$$

where R_w is the resistance of the heated wire at T_w, R_c the resistance connected to the bridge in series with the gauge.

Calibration is carried out in the flow with known parameters (temperature, pressure) using a reliable velocity standard. It should be pointed out that individual calibration of the gauge takes account of the effects due to heat losses to the wire prongs.

As follows from Equation 1, the constants $A = 0.42\pi l\lambda Pr^{0.2}$ and $B = 0.57 l\lambda(D/\nu)^{0.5}Pr^{0.33}$ are determined by the geometrical parameters of the gauge and physical properties of the medium. For diatomic gases B does not depend on temperature and A is proportional to the coefficient of thermal conductivity λ which is linearly dependent on T. In most practical cases n = 0.5 represents a good approximation as the power on Reynolds Number.

As a rule, in calibration and measurements the temperature difference $(T_w - T_f)$ remains the same and if T_f = const in both cases, the above equation is simplified to

$$E^2 = A_1 + B_1 U^{0.5}$$

and the constants A_1 and B_1 are determined from the results of calibration by the least squares method.

The hot-wire anemometer gauge is very sensitive to contamination by dust, smoke, and oil vapor particles which may be present in the flow. As is shown in practice, the change of the gauge characteristic owing to contamination is proportional to the time of operation; therefore, it is advisable to calibrate the gauge prior to and after measurements and to filter thoroughly the air in the volume being investigated so that no particles larger than 0.2D in size are found in the flow.

If the flow approaches the sensor at an angle β, then the effective rate of wire cooling is determined by the relation

$$U_{ef}^2 = U^2(\cos^2\beta + 0.04 \sin^2\beta)$$

which has been verified experimentally at l/D > 600.

The hot-wire anemometer, besides measuring an mean velocity, is an efficient instrument for diagnostics of turbulence characteristics, viz. intensities of temperature and velocity component fluctuations, coefficients of turbulent transfer of momentum and heat, spatial and temporal scales, etc.

Let U, V, and W be the relevant components of the velocity vector in the system of X, Y, Z coordinates (**Figure 1**) and their pulsation components be u, v, and w. The simplest case is represented by the vector of the averaged velocity which lies in the XY plane, while V and W equal zero. If the wire is arranged in the XY plane at right angle to U, it will measure the axial velocity component. In this case the relation between the instantaneous values of voltage pulsations at the bridge output and the velocity pulsation is $e = \partial E/\partial U\ u = -S_u u$, where $\partial E/\partial U$ is the derivative of calibration dependence on the measured mean velocity U. Velocity pulsations in turbulent flow are random and their mean value is 0. Therefore, commonly the root-mean-square value of the velocity pulsations is measured

$$\sqrt{\bar{u}^2} = \left[\frac{1}{T}\int_0^T u^2(t)dt\right]^{1/2}$$

using root-mean-square voltmeters

$$\sqrt{\bar{e}^2} = S_u\sqrt{\bar{u}^2}.$$

Transverse velocity components are measured using gauges with tilted wires. If the wire tilted to the gauge axis at an angle of 45° is arranged in the UV plane, then the voltage at the bridge output will depend on two components of velocity pulsations, $e_1 = -S_{u1}u - S_{v2}v$. When the gauge is rotated through 180°, the wire will be inclined at an angle of 135° to the U direction and the signal will measure $e_2 = -S_{u2}u + S_{v1}v$. By using the results of measurements by perpendicular and tilted wires in two positions and solving the set of equations

$$\bar{e}^2 = S_u^2\bar{u}^2,$$

$$\bar{e}_1^2 = S_{u1}^2\bar{u}^2 + 2S_{u1}S_{v1}\overline{uv} + S_{v1}^2\bar{v}^2,$$

$$\bar{e}_2^2 = S_{u2}^2\bar{u}^2 - 2S_{u2}S_{v2}\overline{uv} + S_{v2}^2\bar{v}^2$$

the values of $\sqrt{\bar{u}^2}$, $\sqrt{\bar{v}^2}$, and $\overline{uv}$ are obtained.

An X-shaped gauge with two crossed strings is often used instead of a single rotated string. In this case a two-channel apparatus is needed for simultaneous measurement of two signals. If both strings have identical sensitivity coefficients, which is achieved by careful manufacturing of the gauge and by selecting appropriate wire overheating, the root-mean-square sum and difference of instantaneous values of signals from the wires $\sqrt{(e_1 + e_2)^2} = 2S_u\sqrt{u^2}$, $\sqrt{(e_1 - e_2)^2} = 2S_v\sqrt{v^2}$ determine intensities of pulsations of longitudinal and transverse velocity components, and $\bar{e}_1^2 - \bar{e}_2^2 = 4S_uS_v\overline{uv}$, the component of turbulent stress $\overline{uv}$. However, it should be pointed out that the requirements for identical characteristics of the wire are extremely high. If $S_1 = (1 + n)S_2$, then using this technique for measuring $\sqrt{u^2}$ and $\sqrt{v^2}$ yields the error 100n%. An appropriate arrangement of gauge wires with respect to the mean velocity vector makes it possible to find $\overline{w^2}$ and $\overline{uw}$, $\overline{vw}$.

In nonisothermal flows the wire of hot-wire anemometer is simultaneously affected by temperature and velocity pulsations. Introducing the sensitivity coefficient $S_{\theta'''}$ according to $\sqrt{e^2} = S_\theta \sqrt{\theta^2}$ we can write $e = -S_uU + S_\theta\theta$ in the simplest case when the string is arranged perpendicular to the vector of the average velocity. Squaring and averaging yield

$$\bar{e}^2 = S_u\bar{u}^2 - 2S_uS_\theta\overline{u\theta} + S_\theta\bar{\theta}^2.$$

In order to determine $\overline{u}^2$, $\overline{\theta}^2$, and $\overline{u\theta}$, measurements are taken at several different values of $(T_w - T_f)$ for realization of different correlations between the sensitivity coefficients S_u and S_θ. The usual practice is to take 8 to 10 measurements of $\overline{e}^2$ at various $m = R_w/R_f$ and to treat the results by the least squares technique.

When measuring shear stresses on the wall τ_w the film gauge on the plane substrate is flush-mounted with the wall. The dependence between the power, put out by the film, and the shear stress τ_w is of the form

$$\frac{E^2R_f}{(R_c + R_f)^2} = A + B\tau_w^{1/3},$$

where R_f is the resistance of the hot film and the constants A and B are determined by precalibration in the flow given τ_w, e.g. stabilized isothermal flow in the pipe or in the boundary layer.

References

Lomass, C. G. (1986) *Fundamentals of Hot-Wire Anemometry,* Cambridge Univ. Press.

Bradshaw, P. (1971) *An Introduction to Turbulence and Its Measurement,* Pergamon Press, Oxford.

Yu.L. Shekhter

HTFS (see Heat Transfer and Fluid Flow Service)

HUGONIOT ADIABAT (see Adiabatic conditions)

HUMAN THERMOREGULATORY SYSTEM (see Physiology and heat transfer)

HUMID HEAT (see Humidity; Lewis relationship)

HUMIDITY

Humidity is the term used to denote the vapor content of a gas and usually refers to the air-water system. In all work on the related subjects of **Air Conditioning** and water cooling, it is customary to base definitions on unit mass of dry air (or other gas). Thus, the humidity of air is the mass of water vapor associated with unit mass of dry air. It is a dimensionless quantity (kg/kg or lb/lb) and, therefore, has the same numerical value whatever system of coherent units is employed.

On the assumption that the ideal **gas law** is applicable, the humidity of a gas may be expressed in terms of its total pressure (p) and absolute temperature (T), the molecular weights of the components and the partial pressure (p_v) of the vapor.

In unit volume of humid gas, the mass m_v of vapor of molecular weight M_v is given by:

$$p_v\,.1 = \frac{m_v}{M_v}RT$$

or

$$m_v = \frac{p_vM_v}{RT}$$

Similarly, the mass m_G of dry gas of molecular weight M_G is:

$$m_G = \frac{(p - p_v)M_G}{RT}$$

Thus the humidity, $\mathcal{H} = m_v/m_G = p_v/(p - p_v) \cdot M_v/M_G$

For the air water system, therefore $\mathcal{H} = p_v/(p - p_v) \cdot 18/29$ approximately.

For a gas, saturated with vapor, the corresponding saturation humidity $\mathcal{H}_o$ is:

$$\mathcal{H}_o = \frac{p_{vo}}{(p - p_{vo})} \cdot \frac{M_v}{M_G}$$

where p_{VO} is the saturated vapor pressure at the temperature T.

The *percentage humidity,* h

$$h = \frac{\mathcal{H}}{\mathcal{H}_o} \times 100$$

i.e.

$$h = \frac{p_v}{p_{vo}}\frac{(p - p_{vo})}{(p - p_v)} \times 100$$

The *percentage humidity* should not be confused with the *percentage relative humidity* ($\mathcal{R}$) which is the ratio of the partial pressure of the vapor to its saturation value at the same temperature ($\times$ 100) i.e.:

$$\mathcal{R} = \frac{p_v}{p_{vo}} \times 100$$

Thus

$$\mathcal{R} = h \cdot \frac{(p - p_v)}{(p - p_{vo})}$$

The percentage humidity and the percentage relative humidity are approximately equal only when:

a. p_{vo} (and p_v) are small compared with p

b. p_v is close to p_{vo}, i.e. the gas is nearly saturated.

Unlike the percentage relative humidity, the humidity and the percentage humidity are a function of the total pressure p.

For convenience, humidity may be shown as a function of temperature on a *Humidity Chart* (or *Psychrometric Chart*), for various values of the percentage relative humidity (see **Air Conditioning**). The region above the saturation curve (100% relative humidity) represents supersaturation conditions in which excess water is normally present as a mist composed of liquid droplets. The usual humidity chart applies to the air-water system at *atmospheric pressure;* charts can be constructed for any gas-liquid system at any pressure.

Other quantities which are frequently shown on a humidity chart are:

Humid volume, which is the volume occupied by unit mass of air, together with its associated water vapor.

Saturated volume, which is the humid volume of air when saturated with water.

Humid heat, which is the specific heat capacity of unit mass of air and associated water vapor ($= c_a + \mathcal{H}c_w$) (J/kgK) where c_a is the specific heat capacity of air, and c_w is the specific heat capacity of water vapor.

References

Coulson, J. M. and Richardson, J. F. (1996) *Chemical Engineering,* Vol. 1, 5th Ed., Pergamon Press, Oxford.

Penman, H. L. (1955) *Humidity,* Institute of Physics, London.

Treybal, R. E. (1980) *Mass Transfer Operations,* 3rd Ed., McGraw Hill Book Co.

Leading to: *Humidity measurement*

J.F. Richardson

HUMIDITY CHART (see Humidity)

HUMIDITY MEASUREMENT

Following from: *Humidity; Dry bulb temperature; Wet bulb temperature; Hygroscopicity*

The humidity of a gas may be measured by a variety of techniques. Some are based on the determination of the partial pressure of vapor in the gas (e.g. *dew point* determination); the relationship between humidity and partial pressure is given in the entry entitled **Humidity**. Measurements of wet and dry bulb temperatures involve the use of the *psychometric chart.* (see **Air Conditioning**). There are other methods which depend on the influence of humidity on the physical properties and behaviour of various substances.

Measurement of Dew Point

This is the determination of the temperature at which the partial pressure of the water vapor is equal to its saturation value. It involves gradually reducing the temperature of a mirror, or other highly polished surface, within the gas and determining the temperature at which condensation is first observed. This temperature is known as the dew point. The partial pressure of water in the gas is, therefore, that of saturated gas at this temperature. Therefore, the required value of humidity can be obtained from the saturation curve on the psychrometric chart.

The reliability of the measurement depends on the accuracy with which the first formation of dew can be detected. This may be done photometrically using a photo-resistive optical-sensing bridge. When the humidity is so low that no condensation occurs until the surface temperature has been reduced below 0°C, the moisture is deposited as a frost; this may be detected by methods based on alpha particle attenuation. Good control of the cooling of the mirror may be achieved by using a thermo-electric module in which heat is transferred away from the polished surface by the Peltier mechanism (see **Peltier Effect**).

As an alternative to reducing the temperature of the surface, this may be maintained constant (typically at 0°C) and the gas compressed until the partial pressure of the water vapor reaches the saturation value when moisture starts to be deposited on the surface. For a gas at a pressure p_1 with a partial pressure of water of p_{v1}, compression to a pressure p_2 will be necessary for moisture deposition to occur, where $p_2/p_1 = p_{vs}/p_{v1}$, p_{vs} being the saturation vapor pressure at the temperature of the surface.

Measurement of Wet and Dry Bulb Temperatures

If a stream of gas of known (dry bulb) temperature is passed rapidly over a small wet surface sufficiently rapidly for the condition of the air not to be significantly changed, a dynamic equilibrium is established with the heat transferred from the air being exactly balanced by the latent heat required to vaporise the water at the surface (See also **Wet Bulb Temperature** and **Dry Bulb Temperature**). If the air flow rate is such that forced convection is the dominant mechanism by which heat is transferred (conduction and radiation effects then negligible), the ratio of the coefficients of heat and mass transfer will be constant and the equilibrium temperature to which the surface falls will be the wet bulb temperature. For the air-water system, the *psychrometric ratio* is approximately unity and the wet bulb and adiabatic saturation temperatures are almost identical. Therefore, the adiabatic cooling lines on the psychrometric chart represent the compositions of all air-water vapor mixtures, not only with the same adiabatic saturation temperatures, but also with the same wet bulb temperatures. Thus, if the adiabatic cooling line corresponding to the measured wet bulb temperature is selected, the point on this line corresponding to the dry bulb temperature will indicate the humidity of the gas.

A wet and dry bulb *hygrometer* incorporates two thermometers, one with a bare (dry) bulb and the other (the wet bulb) covered with a porous fabric which is maintained saturated with water. Air is drawn rapidly ($>5 ms^{-1}$) over the thermometer bulbs and, when the temperature of the wet bulb falls to an equilibrium value, both temperature readings are taken.

Chemical/Physical Absorption Methods

A metered volume of gas is passed over a suitable absorbent whose increase in mass is determined. The method is very accurate, but somewhat laborious. Suitable absorbents for water vapor include phosphorus pentoxide dispersed on pumice and concentrated sulphuric acid.

Electrical Conductivity Measurements

The electrical resistance of thin films of hygroscopic materials, such as lithium chloride, is a function of temperature and of moisture content which itself depends on the humidity of the atmosphere to which it is exposed. In a lithium chloride cell, a skein of very fine fibres is wound on a plastic frame carrying two electrodes in contact with the fibres. The current flowing for a constant voltage difference applied across the electrodes gives a measure of electrical resistance. The instrument can be calibrated to give direct readings of humidity. An alternative system utilises a sensor in the form of a hydroxyethyl-cellulose film incorporating a matrix of conducting carbon particles which expands and contracts according to the moisture content of the atmosphere to which it is exposed. The electrical resistance is a function of the degree of compression of the matrix and the system can be calibrated to give direct readings of humidity.

Hair Hygrometer

Many cellulose materials, including hair, wool and cotton, are hygroscopic (see **Hydroscopicity**) and undergo changes in physical dimensions and shape according to the amount of moisture they have absorbed. In the hair hygrometer, the length of a hair or fibre is affected by the humidity of the surrounding atmosphere and this property can be exploited in instruments which can be constructed to give direct readings. There is a need to calibrate at frequent intervals because of drifting of the zero. Problems are acute when the instrument is used over a wide humidity range.

Thermal Conductivity Meters

The rate of heat loss from a heated wire depends on the thermal conductivity of the surrounding gas which, in general, is a function of its moisture content. A hot wire element, or a thermistor, can be used to fulfil the dual functions of heat source and temperature sensor. Thus, for a constant applied voltage, the temperature, and, hence, resistance of the element, will be a function of the humidity of the gas with which it is in contact.

Other Methods

A variety of other methods is available and each has its own field of application. The principles of operation include:

a. selective absorption of infra-red electromagnetic radiation.
b. measurement of heat of absorption on to a surface.
c. electrolytic hygrometry where the quantity of electricity required to electrolyse water adsorbed from the atmosphere on to a thin film of dessicant is measured.
d. piezo-electric hygrometry employing a quartz crystal with a hygroscopic coating on which moisture is alternately absorbed from a wet-gas and desorbed in a dry gas stream; the dynamics is a function of the gas humidity.
e. capacitance meters in which the electrical capacitance is a function of the degree of deposition of moisture from the atmosphere.
f. observation of colour changes in active ingredients, such as cobaltous chloride.

References

Hickman, M. J. (1970) *Measurement of Humidity,* 4th ed., HMSO, National Physical Laboratory, Notes on Applied Science, No. 4.

Meadowcroft, D. B. (1988) Chemical analysis—moisture measurement, *Instrumentation Reference Book,* Chapter 6, Nottingk, B. E., ed., Butterworth.

Wexler, A. (1965) ed., Humidity and Moisture. Measurements and Control in Science and Industry, Vol. 1, *Principles and Methods of Measuring Humidity in Gases,* Ruskin, R. E., ed., Reinhold, New York.

J.F. Richardson

HUMIDITY, OF EARTH'S ATMOSPHERE (see Atmosphere, physical properties of)

HUMID VOLUME (see Humidity)

HVAC, HEATING, VENTILATION & AIR CONDITIONING (see Air conditioning)

HYDRATE FORMATION (see Natural gas)

HYDRAULIC DIAMETER

***Following from:** Pressure drop (single phase)*

According to similarity theory, any geometrical parameter of the channel can be taken as an independent dimension. However, it is advisable to choose an independent dimension that takes into account the hydrodynamic characteristics of fluid flow. Therefore, as an independent dimension of channels with internal flow, use is made of an equivalent *hydraulic diameter* D_H characterizing the relation between the cross-sectional are S and the wetted perimeter P: $D_H = 4S/P$. D_H is chosen such that the ratio of pressure forces, acting on the flow cross section, to friction forces, applied along the channel perimeter of arbitrary shape, corresponds to the ratio of the forces in an equivalent circular pipe with the diameter $D = D_H$. Applicability of hydraulic diameter as a universal, independent dimension for channels of various geometrical shapes is limited by the following restrictions.

1. The thickness of the boundary layer, in which the greater part of the variation of velocity with distance from the wall occurs, should be much less than the *diameter* of the channel or the radius of curvature of the perimeter.
2. The thickness of the boundary layer and the wall shear stress should vary only slightly around the channel perimeter.

Using D_H makes it possible in some cases, within the restrictions indicated above, to calculate hydraulic characteristics of the flow in channels of various geometrical shapes by the formulas derived for a circular pipe. For free surface flows, the *hydraulic radius* r_H = S/P is commonly used as an independent dimension; for large flow width, this equals the mean depth of the fluid.

M.Kh. Ibragimov

HYDRAULIC GRADIENT LINE (see Static head; Velocity head)

HYDRAULIC JUMP (see Open channel flow)

HYDRAULIC RADIUS (see Hydraulic diameter)

HYDRAULIC REACTION TURBINE (see Turbine)

HYDRAULIC RESISTANCE

Following from: *Pressure drop, single phase*

There are two types of hydraulic resistance: friction resistance and local resistance. In the former case hydraulic resistance is due to momentum transfer to the solid walls. In the latter case the resistance is caused by dissipation of mechanical energy when the configuration or the direction of flow is sharply changed, by the formation of vortices and secondary flows as a result of the flow breaking away, by the centrifugal forces, etc. To categorise local resistances, we usually refer the resistances of adapters, nozzles, extension pieces, diaphragms, pipe-line accessories, swivel knees, pipe entrances, etc.

In defining a total resistance (pressure loss Δp_f) a conditional superposition is used

$$\Delta P_f = \sum \Delta P_{fr} + \sum \Delta P_l. \tag{1}$$

The friction resistance (pressure differential along the channels length) is calculated from Darcy's empirical formula

$$\Delta P_{fr} = \bar{f}\frac{l}{D_H}\frac{\rho\bar{u}^2}{2}, \tag{2}$$

where $\bar{f}$ is the Moody friction factor (4 times the Fanning friction factor - see **Friction Factor**), l and D_H = 4S/P are the length and the hydraulic diameter of the channel, ρ is the fluid density, and $\bar{u}$ is the mean velocity of flow.

In order to define the local hydraulic resistance (ΔP_l) the Weisbach formula is used

$$\Delta P_l = \zeta\frac{\rho\bar{u}^2}{2}, \tag{3}$$

where ζ is the coefficient of local resistance.

For flow in smooth channels the friction factor $\bar{f}$ depends on the flow conditions and is only a function of $Re = \bar{u}D_H/\nu$. For a laminar flow, the value for straight pipes is determined from the Poiseuille formula:

$$\bar{f} = C/Re. \tag{4}$$

The values of C depend on the shape of the section and are given in **Table 1**.

We can see from Equation 2 that in a laminar flow the pressure differential varies with the mean velocity of motion to the first power: a linear law of resistance (area I, **Figure 1**). In a turbulent flow the hydraulic friction resistance increases sharply (area II). Such a rise in the resistance is due to the heavy loss of energy associated with pulsating motion of turbulent vortices in the fluid flow. The value of $\bar{f}$ in a turbulent flow in a round pipe may be calculated from the Blasius formula for $5 \times 10^3 \le Re \le 10^5$

$$\xi = 0.3146/Re^{0.25} \tag{5}$$

and from the *Nikuradze formula* for $10^5 \le Re \le 4 \times 10^6$

$$\xi = 0.032 + 0.221/Re^{0.237} \tag{6}$$

The above formulas are valid for flow in channels with smooth walls with fully developed hydraulic and thermal conditions. In the inlet zone of the channel (up to $20D_H$ long) $\bar{f}$ has a higher value than that calculated by Equations 5 and 6. The friction factor is affected by variations of fluid physical properties caused by variations in temperature and by the action of buoyancy forces.

In *rough channels* the hydraulic resistance increases due to formation of vortices at the roughness elements leading to additional loss of flow specific energy. Three types of roughness can be distinguished:

1. Natural roughness, which is formed as a result of long operation of pipe-lines.
2. Sand roughness, characterized by a high density and various forms of nodules.
3. Artificial (or regular) roughness, when the elements of roughness have a particular geometrical shape and location.

Each type of roughness has its own specific character of variation of the resistance friction coefficient with Re. In the case of sand, roughness the ratio of the pipe radius r_0 to the mean protuberance height δ_r on the wall surface ($k = r_0/\delta_r$) is taken as the roughness parameter. Up to a certain value of Re, the resistance of the rough pipe varies in the same manner as for a smooth one (**Figure 2**) (in a laminar flow it varies according to Equation 4 (curve 1) in a turbulent flow, according to Equation 5 (curve 2). This is because at first the thickness of the laminar sublayer near the wall δ_{lam} exceeds the average height of the roughness protuberances. ($\delta_{lam} > \delta_r$). As Re increases further, δ_r becomes greater than δ_{lam}. This brings about an increase in the friction resistance of a rough pipe as compared to a smooth one above a certain transition number Re_{tr}, whose value depends on the roughness parameter: $Re_{tr} \cong 100k$. For $Re > Re_{tr}$ (self-similarity flow) a square law of resistance is observed, when the friction resistance coefficient depends only on the value of the parameter k (curve 3 in **Figure 2**): $\bar{f} \cong 0.1/k^{0.25}$. The value of $\bar{f}$ for pipes with commercial roughness can be evaluated from the Colebrook-White formula

$$1/\sqrt{\bar{f}} = 1.74 - 2\log[2k_s/D_h + 18.7/(Re\sqrt{\bar{f}})]. \tag{7}$$

Here k_s is the equivalent sand roughness, which for new pipes

Table 1 Factor relating friction factor and Re^{-1} in Equation 4

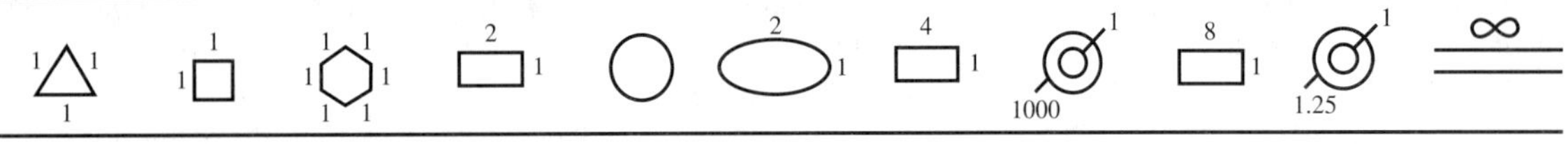

Section shape	triangle	square	hexagon	rectangle 2:1	circle	ellipse 2:1	rectangle 4:1	annulus 1000	rectangle 8:1	annulus 1.25	parallel plates
C in (4)	53.3	56.9	60.2	62.1	64.0	67.3	72.9	74.7	82.3	95.9	96.0

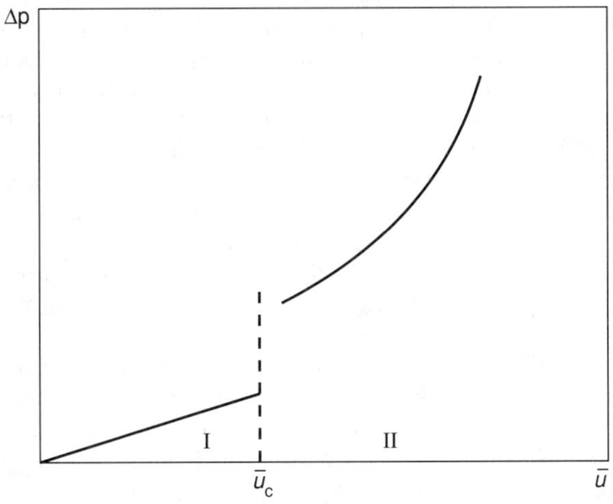

Figure 1. Variations of pressure ion with mean velocity.

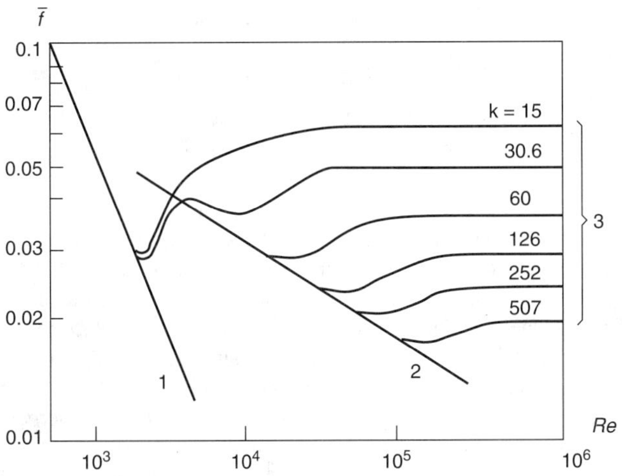

Figure 2. Variation of friction factor with Reynolds number.

drawn out of ferrous metals is about 0.01 mm and for new steel pipes is about 0.014 mm; after some years of operation, it increases up to about 0.2 mm. For old rusty pipes $k_s \cong 1 - 3$ mm and for new zinc-plated pipes 0.5 mm; for new astestos cement pipes it is 0.085 mm.

For artificial roughness, because of its diversity, there are no unique generalizing parameters for roughness. In such a case in order to determine the hydraulic resistance special calculation procedures can be used. Values of $\bar{f}$ for typical fittings, etc. are given in the book by Idel'chik (1992).

In smooth bends and in coiled pipes with $R/r_0 \geq 3$ we assume that $\Delta P_l = 0$, and the effect of centrifugal forces is taken into account by substituting the effective meaning of friction resistance coefficient in Equation 2: for a laminar flow

$$\bar{f} = \bar{f}_0(0.37\ D^{0.36}), \qquad 10^{1/6} < 2D < 10^3 \ldots; \qquad (8)$$

for a turbulent flow ($Re > 10^4$)

$$\bar{f} = \bar{f}_0(1 + 0.15D/Re^{3/4}) \ldots, \qquad (9)$$

where $\bar{f}$ is the friction resistance coefficient for a straight pipe; $D = 1/2\ Re\sqrt{r_0/R}$ is the Dean number, r_0 is the pipe radius, R is the of radius curvature.

References

Idel'chik, I. (1992) *Handbook of Hydraulic Resistance* (2nd Ed.) Begell House, New York.

Schlichting, H. (1979) *Boundary Layer Theory,* McGraw Hill, New York.

Heat Exchanger, Design Handbook (1983) vol. 1 and 2, Hemisphere Publishing Corporation.

M.Kh. Ibragimov and V.A. Kurganov

HYDRAULICS

***Following from:** Flow of fluids*

This is a general term which embraces all those subjects which are concerned with the dynamics of liquids (or hydrodynamics). In particular, it is especially applied to the study of practical engineering problems. As a result, although this general definition is perfectly valid, many aspects of hydraulic engineering are specifically related to the study of water flows.

In the broadest sense, hydraulics may be sub-divided into two areas. The first is concerned with the hydrodynamics of liquids apart from their motions (ie. the properties of fluids at rest), and is accordingly referred to as **Hydrostatics**. This includes the study of buoyancy and floatation, the calculation of forces on submerged bodies, and the measurement of pressures using manometers. Practical applications of these subject areas include the design of gates and barriers, the determination of pressures on dams and submerged containers, and the operation of hydraulic presses.

The second sub-division concerns the description of moving or flowing liquids. The terms *hydro-kinematics* is sometimes used to define this area. This sub-division includes flows which may be steady or unsteady, uniform or varied, laminar or turbulent, and with or without friction losses. Practical problems within this area include **Open Channel Flows**, pipe flows, estuarine dynamics, coastal hydraulics, wave mechanics, hydraulic structures, sediment transport, fluid loading, and hydraulic machinery including both **turbines** and pumps. For specific information concerning these areas the reader is directed to the following references.

References

Abbott, M. B. and Price, W. A. (1994) *Coastal, Estuarial and Harbour Engineers Reference Book,* Chapman and Hall, London.

Brater, E. F. and King, H. W. (1976) *Handbook of Hydraulics,* McGraw-Hill Inc., New York.

Francis, P. and Minton, J. D. R. (1984) *Civil Engineering Hydraulics,* Arnold, London.

Rouse, H. (1949) *Engineering hydraulics,* John Wiley and Sons, New York.

Vennard, J. K. and Street, R. L. (1976) *Elementary Fluid Mechanics,* John Wiley and Sons, New York.

Zippario, V. J. and Hasen, H. (1993) *Davis' Handbook of Applied Hydraulics,* McGraw-Hill Inc., New York.

Leading to: *Hydrostatics; Hydro power; Open channel flow; Torbine; Waves in fluids*

C. Swan

HYDRAULIC TURBINES

Following from: *Hydro power; Turbine*

A hydraulic turbine converts the potential energy of a flowing liquid to rotational energy for further use. In principle, there is no restriction on either the liquid or the use for the energy developed. However, in most cases, these are respectively water and electrical generation. Hence, hydraulic turbines have become synonymous with hydro electric power.

The rate of doing work (power) developed in a hydraulic turbine is:

$$\dot{W} = \Delta p \dot{V} \eta \quad (1)$$

where Δp is the drop in total pressure across the turbine, $\dot{V}$ is the volumetric flow rate and η the efficiency of the turbine. It is common practice to quote Δp in terms of the difference between the upstream and downstream total heads, called the turbine head which equals $\Delta p/g\rho$.

The basis of the design of the turbine hydraulic passages is the velocity diagrams at the entry and exit of the turbine rotating element (called the runner). These lead to the *Euler equation* for theoretical torque and to the theoretical *Euler efficiency* of the turbine. (See **Flow of Fluids**) Although elemental velocity triangles are employed for preliminary design of the hydraulic passages, for large turbines, model testing is necessary for verification of performance. Because of the cost and time involved in developmental model testing, more recently, a computerised finite element solution of the inviscid flow equations in the hydraulic passages, (see **Computational Fluid Dynamics**) cross correlated with general data from model test results, is employed for advanced design. In particular, the *efficiency of the hydraulic turbine* must be optimised and established for contractual purposes. The peak efficiency of a properly designed large hydraulic turbines can be as high as 95%, with typically every point of improved efficiency involving considerable monetary benefits in operation.

Testing cannot model the losses due to hydraulic friction. Hence model test efficiencies are converted to full scale values with established **Reynolds Number** based formulae. Confirmatory site efficiency testing of hydraulic turbines is possible using current meters, ultrasonic, salt velocity tracer, water column inertia (Gibson method) and thermodynamic methods to evaluate flow (IEC, 1991). However, because of cost and the measuring inaccuracy inherent in site testing, there is a general trend to rely solely on model test results.

Depending on the use, the amount of water and level difference available, the power of hydraulic turbines can be a few kilowatts up to hundreds of Megawatts. However, regardless of size, their performance can be equated through similarity laws; hence the applicability of tests on models to predict the performance of large turbines. From non dimensional considerations the similarity laws are:

$$\text{Head Coefficient} = \Delta p/(\rho u^2) \quad (2)$$

$$\text{Flow Coefficient} = \dot{V}/(uD^2) \quad (3)$$

$$\text{Power Coefficient} = \dot{W}/(\rho u^3 D^2) \quad (4)$$

where $\dot{W}$ is the power, ρ is the water density, u is the water velocity and, D is a characteristic diameter of the runner from which all other dimensions of the hydraulic passages follow.

Hydraulic turbines are classified according to specific speed. Specific speed is defined as the rotational speed (revolutions per minute) at which a hydraulic turbine would operate at best efficiency under unit head (one metre) and which is sized to produce unit power (one kilowatt). The equation for specific speed derived from non dimensional considerations is therefore:-

$$\text{Specific speed} = \text{Speed}.(\dot{W}/1000)^{0.5}/(\Delta p/\rho g)^{1.25} \quad (5)$$

The historical development of hydraulic turbines has culminated in two distinct types namely *impulse* (or constant pressure) and *reaction.* Reaction turbines are further divided into radial and axial flow and variable and fixed runner blade. In the impulse turbine, flow is directed through a nozzle to impact on a series of buckets attached to the periphery of the runner. The total transfer of energy is from the change of momentum of the fluid jet; there is no change in hydrostatic pressure once the fluid exits the jet. Impulse turbines (known as *Pelton turbines* after their inventor) are typically used for heads above 100 m and reasonably low flows. They can have up to six jets to better utilise larger flows, as shown in **Figure 1**.

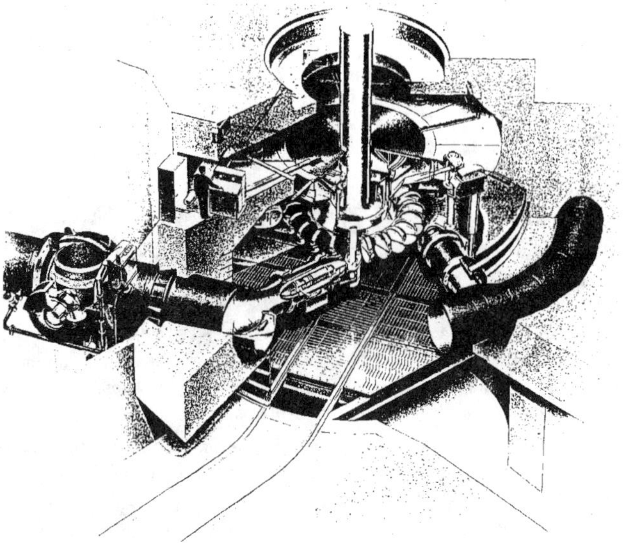

Figure 1. Four jet Pelton turbine (with permission of Kvaerner-Boving Ltd).

Reaction turbines for heads in the range 600 m to 30 m are known as *Francis turbines* (after their inventor). These are radial flow units in which the flow enters the runner radially and discharges axially, as illustrated in **Figure 2**.

The runner of a reaction turbine is equipped with blades which contain and direct the flow. Therefore, in addition to energy derived from the momentum changes of the fluid as it passes through the runner, it is also generated from the changes in hydrostatic pressure of the fluid within the runner passages. Below design heads of approximately 50 m axial flow turbines are used, known as *propeller units* because of their similarity to a ships propeller. A subsection of propeller units known as *bulb turbines* are used for heads below approximately 10 m. Small hydraulic turbines can be arranged with a horizontal shaft for ease of maintenance, but the larger units used for hydro-electric power installations are almost universally vertical. The exception is the bulb turbine which is only arranged horizontally.

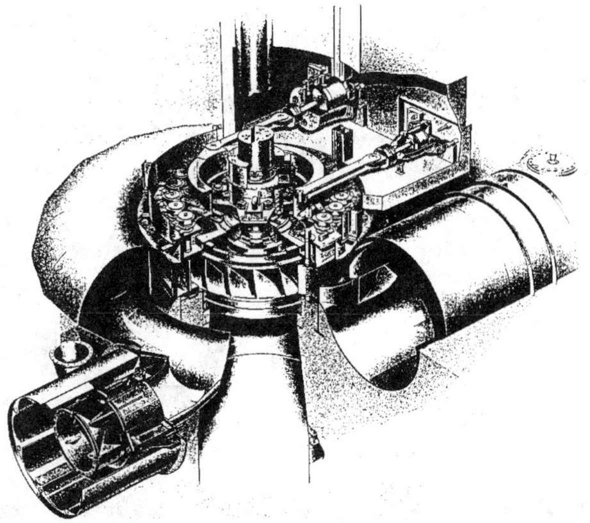

Figure 2. Francis turbine with pressure relief valve (with permission of Kvaerner-Boving Ltd).

Utilisation of the three basic turbine types are within the following ranges of specific speed (calculated from rpm, kW and m):-

Single jet Pelton: $3 < \text{Specific Speed} < 36$

Multiple jet Pelton: $36 < \text{Specific Speed} < 60$

Francis: $60 < \text{Specific Speed} < 400$

Propeller: $300 < \text{Specific Speed} < 1200$

Reversible pump turbines are a special type of reaction turbine. These change direction of rotation to operate both as a pump and a turbine and are used for pumped storage applications. Hydraulically, a reversible pump turbine is designed as a pump with only minor modifications to accommodate its role as a turbine.

The stability of operation and the internal hydraulic forces (both static and dynamic), are directly dependent on the velocity of flow through the turbine. For a given design head and flow velocity there is an unique specific speed. This leads to the relationship:-

$$\text{Specific Speed} = K/(\Delta p/\rho g)^{0.5} \qquad \textbf{(6)}$$

where K is a constant

There are strong commercial benefits in using as small a turbine (hence high flow velocities) as possible for any given application. However, this is restricted by the state of the art in respect of vibration and performance. In 1994, the generally accepted maximum value for K was about 2300.

In all three general groups of hydraulic turbine, flow is directed to the periphery of the runner of the turbine via a spiral casing and discharges from the runner through a draft tube. In reaction turbines, the rotation of the liquid commences as a free vortex in the spiral casing; it is directed through the fixed stay vanes and then through the adjustable turbine wicket gates, such that the angle of approach of the flow to the runner at the design conditions is precisely the runner blade angle (shockless entry). Flow through the reaction turbine to obtain the required power is regulated by the turbine wicket gates. Thus, shockless flow is only obtained at the output for best efficiency and at the design head. In a reaction turbine the draft tube is designed for maximum recovery of hydrostatic pressure. This is especially critical for low head turbines.

The efficiency and operational stability of turbines with low design heads is a strong function of the inlet approach angle, efficiency dropping rapidly with decrease in output and, to a lesser extent, with change in head. To maintain efficiency over the operating range, low head units often have adjustable runner blades, the runner inlet angle changing with wicket gate position and, if required, with operating head. These units are said to be double regulating and include the semi radial flow *Deriaz turbines,* axial flow *Kaplan turbines* (both named after their inventors) and bulb turbines.

Reaction turbine runners can suffer **Cavitation** at the blade inlet (due to off design flow conditions), in the runner hydraulic channels at part load operation and at the runner exit on the suction side of the runner blades. The latter is the most critical and is a function of the back pressure on the runner. Suction side cavitation in a hydraulic turbine is accordingly related to downstream (tailwater) level through the *Thoma Coefficient* defined as:

$$\sigma_{Th} = (p_a - p_{vap} - z.g.\rho)/\Delta p \quad \textbf{(7)}$$

where z is the height of the runner exit plane above tailwater level, p_a is atmospheric pressure and p_{vap} vapor pressure of the fluid. Height z is commonly called the setting of the turbine.

Typically, it is too expensive to set a large hydraulic turbine deep enough below tailwater level to completely eliminate cavitation and, in any particular application, an economic balance between cavitation repair and cost of excavation has to be established. For preliminary design, operating experience is used to establish an acceptable σ_{Th}. One such criterion is:

$$\sigma_{Th} = 7.54 \,.\, 10^{-5} \cdot (\text{specific speed})^{1.41} \quad \textbf{(8)}$$

For major installations, the cavitation performance of the hydraulic turbine should be established with model testing. Cavitation model testing of medium to low head hydraulic turbines is often conducted with **Froude Number** similarity to the prototype.

The wetted surfaces of hydraulic turbines are also prone to damage due to **Erosion** by transported silt and sand and corrosion from aggressive fluids. Damage is particularly problematical when silt erosion, corrosion and cavitation act in conjunction (synergistic effects). In applications where the silt content is extreme the hydraulic design may sacrifice efficiency for partial immunity from silt erosion (contouring of surfaces and thickening of runner blades for example).

Stainless steel, which has a far better resistance to cavitation and erosion than carbon steel, is extensively employed in susceptible areas.

The economic pressures to increase specific speed and hence flow velocities, for a particular head, have led to operational problems with flow induced vibrations from runner blades and stay vanes. These are particularly problematical when their forcing frequency coincides with the natural frequency of any other part of the mechanical, hydraulic or electrical system thus leading to resonance. Also, at part load operation reaction turbines suffer from draft tube pressure oscillations which can result in unacceptable power swings. Air admitted naturally to areas of low pressure or force fed from compressors can be effective in curing oscillations due to part load operation.

The speed of hydraulic turbines is regulated by a governor. The governor senses the speed of the turbine and adjusts the wicket gate opening to maintain speed within close limits. Speed sensing and the associated feed back control systems are typically digital. On all other than very small hydraulic turbines, the amplification from the digital governor to the wicket gates (and runner blades if required) is through a high pressure oil servomotor system. If the hydraulic turbine is operating on a large integrated network then its speed is controlled by the network and the governor is used to change output via its permanent speed droop. The governor feed back and gain have to accommodate the water column and generator inertias. The turbine and all equipment connected to it, both mechanically and electrically, are designed for runaway speed of the turbine (speed at zero torque) resulting from governor failure. To aid regulation, high to medium head turbines are often equipped with pressure relief valves. For the same reason, the jets of impulse turbines are often equipped with flow diverters. For security, isolation valves or hydraulic gates are commonly installed at spiral casing inlets.

References

Brekke, H. (1994) State of the Art in Pelton Turbine Design, *The International Journal on Hydropower and Dams,* Aqua Media International, 1–2.

Gummer, J. H. and Hensman, P. C. (1992) A Review of Stayvane Cracking in Hydraulic Turbines, *International Water Power and Dam Construction,* Reed Business Publishing, 44–8.

International Electrotechnical Commission (IEC) 41 (1991) Field acceptance tests to determine the hydraulic performance of hydraulic turbines, storage pumps and pump turbines, *Bureau Central de la Commission Electrotechnique Internationale,* Geneva, Switzerland.

Raabe, J. (1985) *Hydro Power, The Design, Use and Function of Hydromechanical Hydraulic and Electrical Equipment,* VDI, Verlag, Dusseldorf, Germany. ISBN 3 18 400616 6.

Vivier, L. (1966) *Turbines Hydrauliques et Leur Regulation,* Editions Albin Michel, Paris.

J.H. Gummer

HYDRAZINE COMBUSTION (see Flames)

HYDROCARBONS

Hydrocarbon is a generic name for a very large number of chemical compounds that consist entirely of hydrogen and carbon atoms. They are further subdivided into *aliphatic, aromatic* and *cyclic.* Aliphatic hydrocarbons consist of linear or branched chains and can be saturated (paraffins) or unsaturated (olefins, acetylenes etc.). Aromatic hydrocarbons consist of closed, cyclic, hexagonal rings of a distinctive chemical structure in terms of chemical bonds. The compounds benzene, napthalene and anthracene belong to this group. The name ‘aromatic’ derives from the distinctive odour associated with these compounds. They are in general chemically reactive compounds whose use would have been widespread if it was not for their carcinogenic effects. The third group, cyclic hydrocarbons, consists of compounds that contain three or more carbon atoms in a ring structure. They can be of the saturated (cycloparaffins) or unsaturated (cycloolefins, cycloacetylenes, etc.) variety. Cyclohexane is the most widely used chemical from this group. Hydrocarbons have widespread use as raw materials in the production of other organic compounds and as a primary, global energy source.

V. Vesovic

HYDROCHLORIC ACID

Hydrochloric acid is *hydrogen chloride* (HCl) in aqueous solution. It is a colourless, fuming liquid that comes in different industrial grades depending on the concentration of hydrogen chloride, which has a molecular weight of 36.46. The concentrated, fuming acid contains 35–38% hydrogen chloride and has a specific gravity of 1.19. It is a strong and corrosive acid which is highly toxic.

V. Vesovic

HYDROCYCLONES

***Following from:** Gas-solids separation, overview; Liquid-solid separation; Liquid-liquid separation*

The hydrocyclone is a simple piece of equipment that uses fluid pressure to generate centrifugal force and a flow pattern which can separate particles or droplets from a liquid medium. These particles or droplets must have a sufficiently different density relative to the medium in order to achieve separation.

The flow pattern in a hydrocyclone is cyclonic. This is induced by tangential injection of the liquid into a cylindrical chamber which causes the development of a vortex. The chamber has a restricted axial bottom outlet such that all of the liquid in the vortex can not escape via this outlet. Some of the liquid has to reverse its path and flow counter currently to an axial top outlet. This reverse flow continues to rotate and an air core develops due to lower pressure at the axis of rotation.

The conventional hydrocyclone consists of a cylindrical chamber connected to a conical body which leads to the bottom outlet at the apex of the cone. This discharges the "underflow". The reverse flow is located by a pipe which projects axially into the top of the chamber, is termed the "vortex finder", and discharges the "overflow". The spiral flows and general shape of a hydrocyclone are illustrated in **Figures 1** and **2**.

The terms "bottom" and "top" in this context are used only to simplify the description. A hydrocyclone can work upside down or with an inclined or horizontal axis.

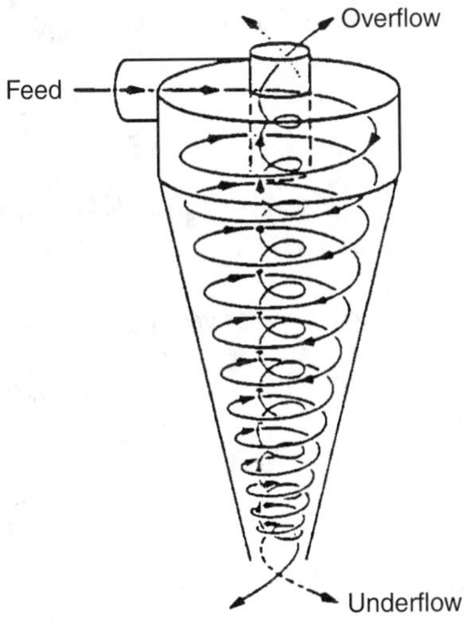

Figure 1. Schematic representation of the spiral flow.

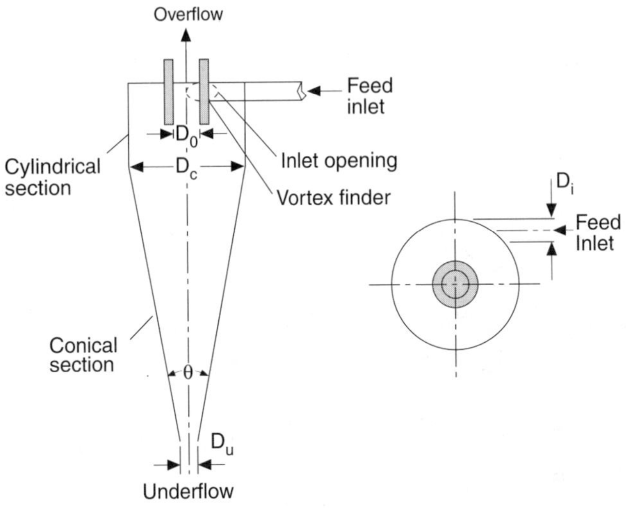

Figure 2. Principal features of a hydrocyclone.

The fluid velocity patterns that develop are illustrated in **Figures 3** and **4**.

The existence of both downward and upward flow streams means that there is a locus of zero vertical velocity (**Figure 3**). At the same time both streams have a tangential velocity; that in the outer vortex decreases with increase in radius; that in the inner vortex decreases with decrease in radius. Velocities can be represented by the equation ur^n = constant where $n = -1$ in the inner vortex and can have values ranging from +0.5 to +0.8 in the outer vortex (See **Vortex Flow**).

This means that the outer fluid is approaching a free vortex and is experiencing shear whereas the inner fluid is rotating as if it were a solid body, i.e. with constant angular velocity. The peak tangential velocity is not necessarily at the position of zero vertical velocity.

These flow patterns and relationships have been established experimentally by Kelsall (1952) and Bradley (1965). They have been substantiated more recently by the application of Computational Fluid Dynamics, Pressdee (1989).

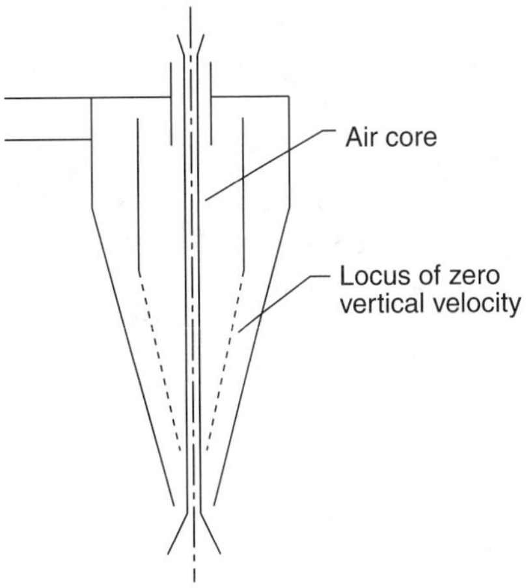

Figure 3. Schematic representation of the locus of zero vertical velocity and the air core.

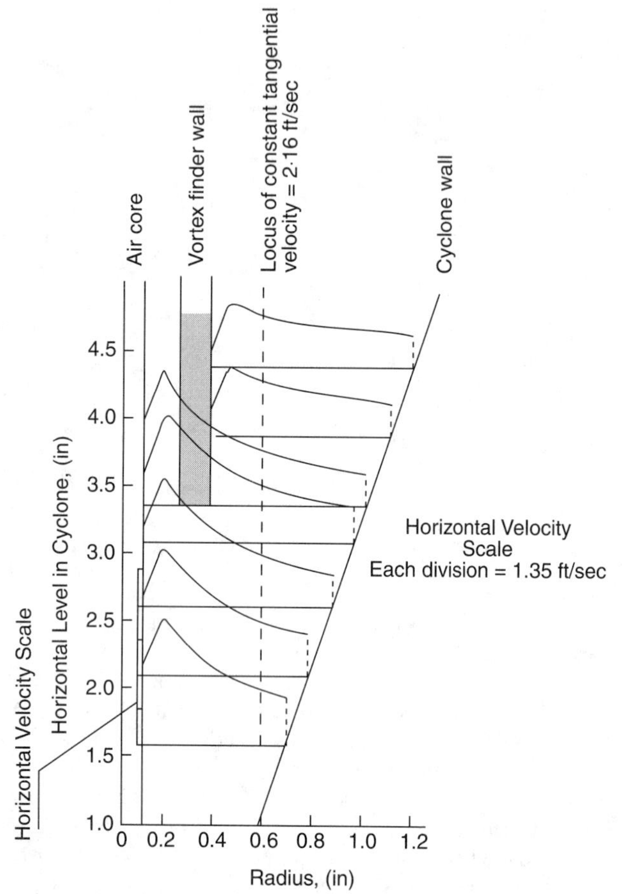

Figure 4. Tangential velocity distribution. Data of Kelsall, *Trans. Inst. Chem. Eng.* 30, 87 (1952).

The size and shape of the vortex finder, the shape of the chamber into which the feed liquid is injected and the geometry of the feed inlet all dictate the details of the flow pattern which will often include recirculatory streams (See **Figure 5**).

Due to this complexity many authors have produced correlations for both the separation efficiency and the pressure drop in hydrocyclones. Reviews have been published by Bradley (1965) and Svarovsky (1984).

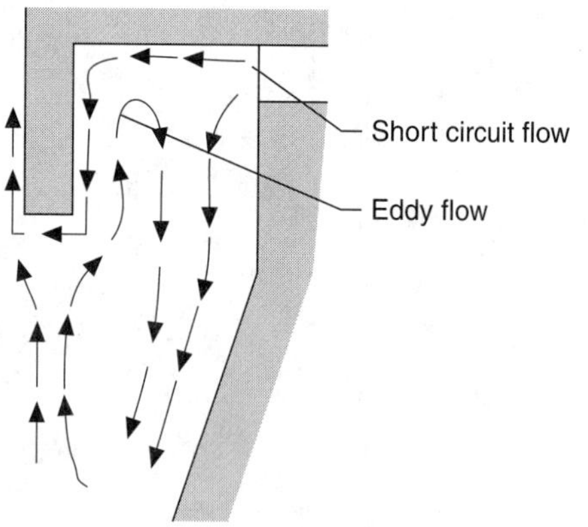

Figure 5. Schematic representation of the short circuit and eddy flows.

The centrifugal force that is generated is capable of separating particulate solids down to around 5 to 10 microns in size. The smaller the diameter of the cyclone the higher the force and the higher the efficiency of separation of small particles. Long cone designs also exist which claim to give good efficiency with particles of less than 5 microns. A small diameter has, however, a low fluid capacity and in most cases such cyclones have to be manifolded and connected in parallel to be of industrial use.

Applications of the hydrocyclone were originally in the mineral processing industry for solid-liquid separation, i.e., the de-watering of particulate solids suspended in a water stream. The mechanism of separation is, however, dependent on both the difference in specific gravity of the particles and the liquid medium as well as the size of the particles (See **Stokes' Law**). This means that the hydrocyclone is also a classifier if the range of size of particles present in the feed stream is such that large will pass to the outer vortex and small will be entrained in the inner vortex (See **Classifiers**). The mineral processing industry uses hydrocyclones in this way. It also uses them as heavy media separators where the liquid density is increased to be intermediate between a light mineral and a heavy mineral. The light then passes to the overflow, the heavy to the underflow.

Applications extended to the chemical and food industries. In the main these are de-watering, for example, in concentrating a crystal slurry prior to centrifugation or in the concentration of starch suspensions.

In the latter case, use is also being made of the shear which exists in the outer vortex to strip gluten from starch particles.

Because of this shear it was originally believed that hydrocyclones would have limited application to the separation of liquid droplets. Development work primarily at Southampton University, Thew (1986) has, however, succeeded in disproving this. By elongating the cone and using narrow angles as small as 1° to 2° the shear zone is minimized relative to the non shear inner zone and finely dispersed oil in water can be separated down to a few tens of ppm. These special forms of hydrocyclone are now successfully marketed for removing oil from water in the refinery and offshore oil industries (See **Liquid-Liquid Separation**).

The hydrocyclone is now an established simple, practical tool widely used in all process industries.

References

Bradley, D. (1965) *The Hydrocyclone,* Pergamon Press Limited, Oxford.

Bradley, D. and Pulling, D. J. (1959) *Transactions Institution of Chemical Engineers,* vol. 37, p. 34.

Kelsall, D. F. (1952) *Transactions Institution of Chemical Engineers,* vol. 30, p. 87.

Pressdee, A. W. (1989) *Process Industry Journal,* p. 29, June.

Svarovsky, L. (1984) *Hydrocyclones,* Holt Rinehart and Winston.

Thew, M. T. (1986) *The Chemical Engineer,* 427, p. 17, July–August.

D. Bradley

HYDROCHLOROFLUOROCARBON, HCFC (see Heat pumps; Refrigeration)

HYDROCHLOROFLUORCARBON-22 (see Heat pumps)

HYDROCRACKER (see Oil refining)

HYDRODYNAMIC ENTRANCE LENGTH, IN TUBES (see Tubes, single phase flow in)

HYDRODYNAMICS (see Flow of fluids)

HYDRODYNAMIC THEORY OF BOILING (see Burnout in pool boiling)

HYDROFLUORIC ACID

Hydrofluoric acid is *hydrogen fluoride* (HF) in aqueous solution. It is a colourless fuming, liquid that comes in different industrial grades depending on the concentration of hydrogen fluoride which has a molecular weight of 20.01. The concentrated, fuming acid contains about 70% hydrogen fluoride and is highly corrosive and toxic. It will corrode silica-containing material.

V. Vesovic

HYDROFLUOROCARBON, HFC (see Heat pumps; Refrigeration)

HYDROGEN

Hydrogen—(Gr. *hydro,* water, and *genes,* forming), H; atomic weight (natural) 1.0079; atomic weight (H^1)1.007822; atomic number 1; melting point −259.14°C; boiling point −252.87°C; density 0.08988 g/l; density (liquid) 70.8 g/l (−253°C); density (solid) 70.6 g/l (−262°C); valence 1.

Hydrogen was prepared many years before it was recognized as a distinct substance by Cavendish in 1766. It was named by Lavoisier. Hydrogen is the most abundant of all elements in the universe, and it is thought that the heavier elements were, and still are, being built from hydrogen and helium.

It has been estimated that hydrogen makes up more than 90% of all the atoms or three quarters of the mass of the universe. It is found in the sun and most stars, and plays an important part in the proton-proton reaction and carbon-nitrogen cycle, which accounts for the energy of the sun and stars. It is thought that hydrogen is a major component of the planet Jupiter and that at some depth in the planet's interior the pressure is so great that solid molecular hydrogen is converted into solid metallic hydrogen. In 1973, it was reported that a group of Russian experimenters may have produced metallic hydrogen at a pressure of 2.8 Mbar. At the transition the density changed from 1.08 to 1.3 g/cm^3. Earlier, in 1972, a Livermore (California) group also reported on a similar experiment in which they observed a pressure-volume point centered at 2 Mbar. It has been predicted that metallic hydrogen may may be metastable; others have predicted it would be a superconductor at room temperature.

On earth, hydrogen occurs chiefly in combination with oxygen in water, but it is also present in organic matter such as living plants, petroleum, coal, etc. It is present as the free element in the atmosphere, but only to the extent of less than 1 ppm, by volume. It is the lightest of all gases, and combines with other elements, sometimes explosively, to form compounds. Great quantities of hydrogen are required commercially for the fixation of nitrogen from the air in the Haber ammonia process and for the hydrogenation of fats and oils. It is also used in large quantities in methanol production, in hydrodealkylation, hydrocracking, and hydrodesulfurization. It is also used as a rocket fuel, for welding, for production of hydrochloric acid, for the reduction of metallic ores, and for filling balloons. The lifting power of 1 ft^3 of hydrogen gas is about 0.076 lb at 0°C, 760 mm pressure.

Production of hydrogen in the U.S. alone several amounts to several billion cubic feet per year. It is prepared by the action of steam on heated carbon, by decomposition of certain hydrocarbons with heat, by the electrolysis of water, or by the displacement from acids by certain metals. It is also produced by the action of sodium or potassium hydroxide on aluminum.

Liquid hydrogen is important in cryogenics and in the study of superconductivity as its melting point is only a few degrees above absolute zero. The ordinary isotope of hydrogen, ${}_1H^1$, is known as *protium.* In 1932, Urey announced the preparation of a stable isotope, *deuterium* (${}_1H^2$ or D) with an atomic weight of 2. Two years later an unstable isotope, *tritium* (${}_1H^3$), with an atomic weight of 3 was discovered. Tritium has a half-life of about 12.5 years. One atom of deuterium is found mixed in with about 6000 ordinary hydrogen atoms. Tritium atoms are also present but in much smaller proportion. Tritium is readily produced in nuclear reactors and is used in the production of the hydrogen bomb. It is also used as a radioactive agent in making luminous paints, and as a tracer.

Quite apart from isotopes, it has been shown that hydrogen gas under ordinary conditions is a mixture of two kinds of molecules, known as *ortho-* and *para*-hydrogen, which differ from one another by the spins of their electrons and nuclei. Normal hydrogen at room temperature contains 25% of the *para* form and 75% of the *ortho* form. The *ortho* form cannot be prepared in the pure state. Since the two forms differ in energy, the physical properties also differ. The melting and boiling points of *para*-hydrogen are about 0.1°C lower than those of normal hydrogen. (See also **Cryogenic Fluids**)

From Handbook of Chemistry and Physics, CRC Press

HYDROGEN AS ENERGY SOURCE (see Fuel cells)

HYDROGEN BOMBS (see Fusion, nuclear fusion reactors)

HYDROGEN CHLORIDE (see Hydrochloric acid)

HYDROGEN COMBUSTION (see Flames)

HYDROGEN FLUORIDE (see Hydrofluoric acid)

HYDROGEN/FLUORINE COMBUSTION (see Flames)

HYDROGEN IODIDE

Hydrogen Iodide (HI) is a non-flammable gas that in aqueous solution forms a strong acid. It is formed from hydrogen and iodine vapor over a catalyst. It is toxic and acts as an irritant to skin. Its main uses are in the pharmaceutical industry.

Molecular weight: 127.91	Critical temperature: 424 K
Melting point: 222.4 K	Critical pressure: 8.3 MPa
Boiling point: 237.6 K	Critical density: 976 kg/m^3
	(Normal vapor density: 5.71 kg/m^3 (@ 0°C, 101.3 kPa)

V. Vesovic

HYDROGEN PEROXIDE

Hydrogen Peroxide (H_2O_2) is a liquid with a molecular weight of 34.015 that freezes at 272.7K and boils at 423.4K. It is soluble in water and it comes in different industrial grades depending on the strength of the aqueous solution. It is a strong oxidising agent that is not only toxic but also flammable and explosive.

V. Vesovic

HYDROKINEMATICS (see Hydraulics)

HYDROMETALLURGY (see Ion exchange)

HYDROPHILE-LYOPHILE BALANCE, HLB (see Surface active substances)

HYDROPHILLIC/HYDROPHOBIC SURFACES (see Flotation; Surface active substances)

HYDRO POWER

***Following from:** Hydraulics; Power plant*

Hydro power is now almost exclusively used for the generation of electricity. In 1994, it accounted for approximately 18% of the world's electricity generation. Hydro power generation world-wide far outstripped any other *renewable resource* and it is likely to do so in the foreseeable future. Because of its very nature, hydro power is not evenly distributed world wide. For example in 1994, 95% of electricity generation in Brazil was from hydro whereas in England it only accounted for 0.1%. The largest remaining areas for economically feasible pure hydro power development are in China, Brazil and Russia respectively.

Pure hydro power exploits the potential energy of water in one of two ways: either by impounding it by means of a dam, and hence concentrating and releasing the potential energy at a chosen point, or by circumventing a region of large losses in potential energy, for example a waterfall, rapids or river horseshoe. In many instances, there is insufficient water at one site for economic development, in which case the water is diverted and collected from diverse sites by channels, pipelines and tunnels to a common point. Water thus collected or impounded is directed by pressure pipelines (called *penstocks*) to the power-houses hydraulic turbines. Pure hydro power projects are often staged, the outfall of one facility being the head water of the next. For low head applications power-houses are typically situated at the base of dams. Power-houses in high head facilities are often underground.

The water channels and pipelines have to be designed for the hydraulic transients resulting from the change of flow rate of the water. *Surge tanks,* overflows and pressure relief valves are used extensively. Hydro facilities have to accommodate maximum predicted floods and are normally equipped with spillways, often gated. Water flow in complex scheme is controlled with hydraulic valves and gates, which are also used for isolation during maintenance. Depending on climate and location, the facility may be equipped to deal with ice floes and large amounts of vegetation. Where waterborne silt is a problem upstream, settling beds are provided.

Pumped storage hydro projects have little or no natural water inflow. Water is pumped from a lower to a higher reservoir during periods of low electrical demand and released through turbines during periods of high demand. As such, pumped storage is the only practical method of substantial energy storage, permitting thermal power stations to operate at base load. The upper and lower reservoirs of pumped storage facilities are often completely man made. A large percentage of pumped storage facilities use reversible pump turbines. However, there are also many facilities where the pump and turbine share the same shaft which in turn is connected to the motor/generator. During turbine operation, the pump is dewatered and vise versa during pump operation. Alternatively, the two are isolated with mechanical couplings. For high head pumped storage schemes, multi stage pumps or pump turbines are used.

Because of the relatively fast response of hydro units, hydro power can be very effective as spinning reserve and peaking duty on a large generating system. In some instances hydro spinning reserve can be advantageously coupled with synchronous condenser operation of the associated hydro generators, for electrical system compensation. In a micro or macro form, relatively small hydro units can be installed to serve isolated communities, obviating the need for long and expensive transmission lines.

Power from a pure hydro facility is not constant and varies according to the availability of the water. Water flows tend to be seasonal (spring runoffs and monsoon) and, although the creation of large reservoirs alleviates this problem, they cannot completely eliminate it. Dependability of hydro power is also reduced if the scheme is multi purpose, providing water for irrigation or acting as flood control. In both cases the release of water could be dictated by factors other than the power demand (see also **Water**).

The initial capital outlay for a large hydro scheme can be substantial and difficult to arrange. The amount of construction work is usually considerable. Both of these factors give hydro projects a substantially longer lead time than an equivalent thermal generating facility. The development of large river schemes involving the flooding of substantial tracts of land upstream of the dam is disruptive to both the local environment and populace, especially if large communities have to be relocated. High head schemes are also environmentally contentious as they tend to be located in areas of natural beauty. Accordingly, it is increasingly difficult to obtain both approval and financing for the creation of new hydro power facilities and recent efforts have concentrated on the refurbishing of existing hydro developments.

References

Gummer, J. H., Barr D. M., and, Simms G. P. (1993) Evaluation Criteria for Upgrading Hydro Powerplants, *International Water Power and Dam Construction,* v. 45-12.

International Water Power & Dam Construction Handbook, (1994) Reed Business Publishing, London, ISBN 0617 00548 6.

Mosonyi, E. (1991) *Water Power Development,* Akademiai Kiado, Budapest, ISBN 9630542706.

Leading to: *Hydraulic turbines*

J.H. Gummer

HYDROSTATIC FORCES (see Hydrostatics)

HYDROSTATIC LIFT FORCE (see Archimedes force)

HYDROSTATICS

Following from: *Fluid properties*

Hydrostatics is defined as that branch of physics which has to do with the pressure and equilibrium of water and other liquids.

Liquids at rest present far simpler problems to solve than those of fluid dynamics, since individual fluid elements do not move relative to others in the fluid body, therefore shear forces are not involved and all pressure forces are normal to the fluid elements surfaces. Following the criteria that fluid elements should not move relative to one another, hydrostatics can be extended to systems of *relative equilibrium,* where elements have no relative motion although the body of liquid may be moving as a whole. The following will be briefly described: pressure variation in compressible and incompressible liquids; forces on planes; *stability of floating* and submerged *bodies;* relative equilibrium; and *manometry.*

Pressure Variation in Compressible and Incompressible Liquids

Consider a small liquid cylinder as shown in **Figure 1**.

δS the cross sectional area is very small therefore there is no pressure variation across it. As there are no shear forces the only forces acting on it are those due to pressure on the surfaces and gravity. The only forces on the sides of the cylinder are normal to the sides and therefore have no component along the axis. Resolving forces along the cylinder axis yields:

$$(p + \delta p)\delta S - p\delta S + \rho g \delta S \delta l \cos\theta = 0$$

If the difference in height between the ends of the cylinder $\delta z = \delta l \cos\theta$ then the above equation simplifies to $\delta p + \rho g \delta z = 0$. Taking the limit as $\delta z \to 0$ gives:

$$\frac{\partial p}{\partial z} = -\rho g \quad (1)$$

In an incompressible liquid this equation can be integrated giving $p + \rho g z =$ constant. If a datum level of z is taken at a pressure of p_a and the depth below this level is h, then the pressure p at this depth is:

$$p = p_a + \rho g h \quad (2)$$

This is the standard formula for pressure calculation in an incompressible liquid. Unless the variation of density with z is known, the pressure variation in a static compressible fluid cannot be found in the same way.

The three conditions for liquid equilibrium are therefore:

1. equal pressure over any horizontal plane
2. density must be the same over any horizontal plane
3. Equation 1 must hold.

Forces on Planes

Thrust is exerted on every part of any surface in contact with a fluid. It is important to be able to resolve a resultant force over an area and the position of this force. For a horizontal plane, the position of the resultant force is at the plane's centroid; the resultant force is simply the pressure times the area. For planes at an angle, the following method can be used to determine the resultant force vector.

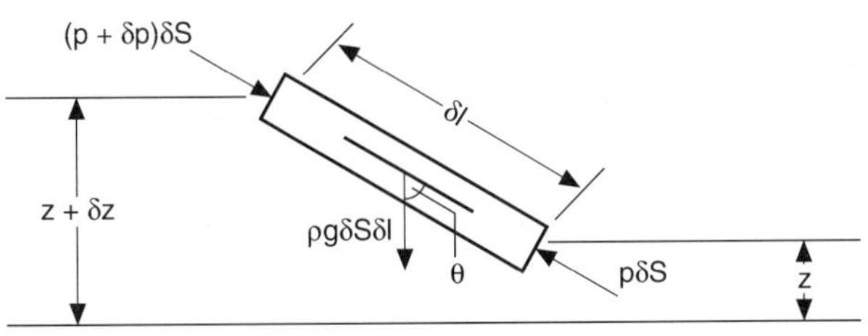

Figure 1. From Massey B.S. (1968), *Mechanics of Fluids*, Van Nostrand Reinhold (International) Co., Ltd., London, with permission.

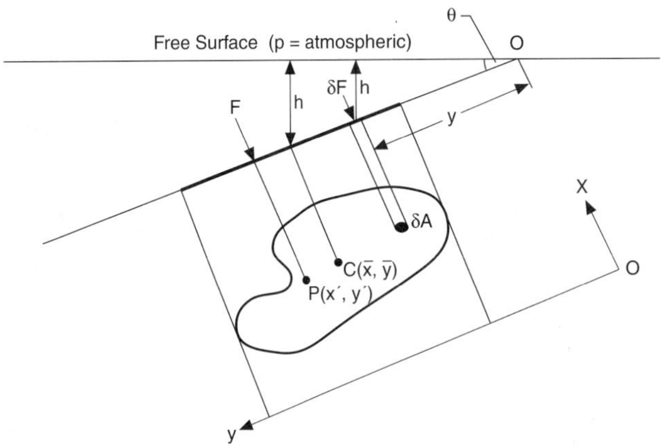

Figure 2. From Massey B.S. (1968), *Mechanics of Fluids*, Van Nostrand Reinhold (International) Co., Ltd., London, with permission.

Consider the plane in **Figure 2**. Each element of area δA is subjected to a force due to the liquid:

$$\delta F = p\delta A = \rho g h \delta A = \rho g y \sin\theta \delta A \qquad (3)$$

p is the gauge pressure. Integrating over the whole area gives:

$$F = \int_A \rho g y \sin\theta \cdot dA = \rho g \sin\theta \int_A y \cdot dA \qquad (4)$$

Where $\bar{y} = 1/A\int_A ydA$ (i.e. the "first moment of area") is zero is known as the centroidal axis. The other centroidal axis is at $\bar{x} = 1/A\int_A xdA$. Where these intersect is the "centroid" of area $(\bar{x},\bar{y})$. The centroid of volume is obtained by integrating with respect to volume rather than area i.e. $\bar{x} = 1/V\int_V xdV$, $\bar{y} = 1/V\int_V ydV$.

This gives:

$$F = \rho g \sin\theta A\bar{y} = \rho g A\bar{h} \qquad (5)$$

Therefore for an incompressible fluid, $\rho g\bar{h}$ is the pressure at the centroid. The total force exerted on the plane by the static fluid is the product of the area and the pressure at the centroid. The moment of the resulting force must equal the sum of the moments of the individual forces, therefore the line of action of F can be determined from the following equation:

$$Fy' = \int_A \rho g y^2 \sin\theta \cdot dA \qquad (6)$$

Substituting F from Equation 3 yields:

$$y' = \frac{\rho g \sin\theta \int_A y^2 \cdot dA}{\rho g \sin\theta \int_A ydA} = \frac{\int_A y^2 dA}{A\bar{y}} = \frac{2^{nd}\text{ moment of area}}{1^{st}\text{ moment of area}} \qquad (7)$$

Similarly for x', taking moments about Oy:

$$Fx' = \int_A \rho g xy \sin\theta dA = \rho g \sin\theta \int_A xydA \qquad (8)$$

$$x' = \frac{\int_A xy \cdot dA}{A\bar{y}} \qquad (9)$$

The centre of pressure (x',y') is always lower than the centroid of area except for horizontal surfaces. Curved surfaces require more complex mathematics but the resultant force vector can still be determined, see Streeter and Wylie (1983). Another method for determining the resultant force and line of action on a plane surface utilises a "pressure prism" concept, see Streeter and Wylie (1983).

Stability of Floating and Submerged Bodies

The resultant force acting on a body in a static fluid is called the "buoyant" or "buoyancy force", which always acts vertically upwards. It is the difference between the pressure forces exerted on its underside and upper side. This is equal to the gravity force of liquid that is displaced by the solid body, which forms *Archimedes' principle.* The centroid of buoyancy is therefore at the centroid of the displaced volume. All bodies floating in a static fluid have vertical stability as small vertical displacements set up restoring forces by increasing or decreasing the volume of fluid displaced.

Submerged bodies have rotational stability only if its centre of gravity is below the centre of buoyancy as shifts from equilibrium produce couples which return the body to equilibrium. If the centre of buoyancy and gravity coincide the body is in neutral equilibrium. The stability of floating objects can be determined by considering **Figure 3**.

If a floating body with centre of buoyancy Bo, and centre of gravity G, is moved slightly away from its stable position (a) as shown in (b), the new centroid of buoyancy B′ (which is the centroid of the displaced volume "abcd") causes an upward thrust setting up a couple which restores stable equilibrium only if M, where the vertical through B′ intersects the original centre line, called the "*metacentre*", is above G. If it is below, the body is unstable, and if G and M coincide, the body is in neutral stability. The distance MG is the "*metacentric height*".

Manometry

The measurement of pressure can be achieved by using suitable columns of liquids. Some simple manometers are shown in **Figure 4**.

The simple "piezometer tube" shown in **Figure 4a** gives the "gauge" pressure as $p = \rho gh$. It cannot be used for negative gauge pressures as air would be sucked into the tube. "U-tube" manometers can be used to measure pressure differences, or gauge pressure if one of the two vertical arms is open to atmosphere. By using fluids of different density various pressure ranges can be measured. Utilising two immiscible liquids in the arms, "micromanometers" can be constructed for the measurement of small pressure differences, see

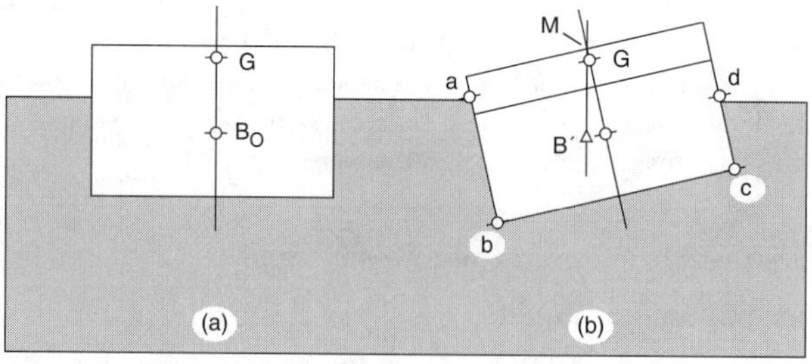

Figure 3.

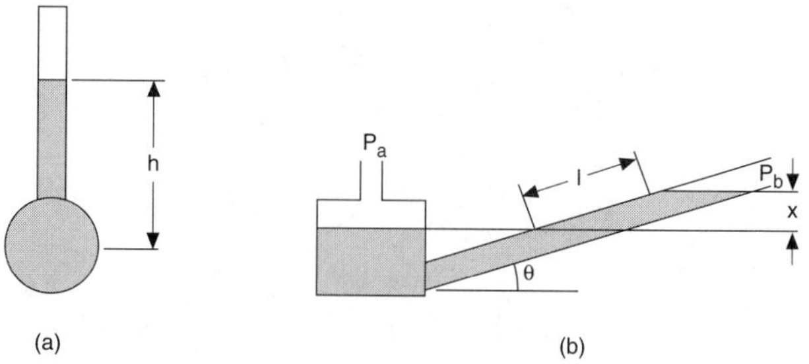

Figure 4.

Streeter and Wylie (1983). A simpler method for measuring small pressure differences is the "inclined manometer" shown in **Figure 4b**. The inclination forces the liquid to travel a larger distance l in order to achieve the same increase in height x therefore:

$$p_a - p_b = \rho g x = \rho g l \sin \theta \qquad \textbf{(10)}$$

If θ is small, there is considerable magnification of the meniscus' movement. For angles less than 5°, the position of the meniscus is difficult to determine.

For large pressure differences, U tube manometers can be connected in series. (See also **Pressure Measurement**.)

References

Massey, B. S. (1968) *Mechanics of Fluids,* Ch. 2, Sixth Edition, Van Nostrand Reinhold (International) Co. Ltd, (1989).

Streeter, V. L. and Wylie, E. B. (1983) *Fluid Mechanics,* Ch. 2, First SI Metric Edition. McGraw-Hill Book Co, 1986.

Leading to: *Archimedes force*

N. Hawkes

HYGROMETER (see Dry bulb temperature; Humidity measurement)

HYGROSCOPICITY

Hygroscopicity is the tendency of a solid substance to absorb moisture from the surrounding atmosphere. The process can take on a number of forms. Thus, with a porous solid such as activated carbon, water vapor will be physically adsorbed, both on the external surface and within the pores, to form a condensed layer. The process may initially take place at "active sites" from which spreading then occurs. With other solids, such as silica gel, the interaction at the surface may not be entirely of a physical nature and some loose chemical bonds may be established. Many cellulosic materials, including hair, cotton and wool, are hygroscopic and change their physical dimensions as a result of the take-up of water. Such materials may be used as the active elements in hygrometers (see **Humidity Measurement**).

The total amount of water which can be taken up by the hygroscopic material will be a function of the temperature and humidity of the atmosphere in which it is located and will ultimately be determined by the sorption isotherm of the system. In general, the rate of transfer of moisture will fall off progressively as equilibrium is approached, not only because the concentration driving force becomes smaller, but also because the overall diffusional resistance to mass transfer increases as the more easily accessible elements of surface reach their equilibrium state so that, in a porous solid, the vapor must then diffuse into the more remote pores.

A particular example of hygroscopic behaviour is *deliquescence* which is exhibited by many water-soluble solids, including inorganic salts (e.g. calcium chloride). At a given temperature, the vapor pressure of a saturated salt solution will be lower than that of pure water and, if it is less than the partial pressure in the atmosphere, moisture will be tranferred to the surface of the solids, part of which will dissolve to form a saturated solution. When all the solids have dissolved, the process will continue until the partial pressure of the now unsaturated solution equals that in the atmosphere. In some cases, the formation of a hydrate or higher hydrate of the salt may precede the formation of a liquid phase. Liquids (eg sulphuric acid) may also be deliquescent.

Leading to: *Humidity measurement*

J.F. Richardson

HYPERBOLIC DIFFERENTIAL EQUATIONS (see Differential equations)

HYPERBOLIC EQUATIONS (see Green's function)

HYPERSONIC FLOW (see Compressible flow)

HYSTERESIS OF CONTACT ANGLE (see Contact angle)

I

IAEA (see International Atomic Energy Agency)

ICE CONDENSER CONTAINMENTS, FOR NUCLEAR REACTOR (see Containment)

ICHEME (see Institution of Chemical Engineers)

ICING (see Frosting)

IDEAL DIODE LAW (see Solar cells)

IDEAL GAS (see Perfect gas; Thermodynamics)

IDEAL GAS LAW (see Gas law)

IDEAL MIXTURE (see Activity coefficient; Binary systems, thermodynamics of; Raoult's law)

IDEAL PLASMA (see Plasma)

IDEAL SOLUTIONS (see Thermodynamics)

IEA (see International Energy Agency)

IEE (see Institution of Electrical Engineers)

IFRF (see International Flame Research Foundation)

IGNITION, EXPLOSION (see Explosion phenomena)

IMECHE (see Institution of Mechanical Engineers)

IMMERSED BODIES, FLOW AROUND AND DRAG

***Following from:** External flows, overview*

External flows past objects encompass a variety of fluid mechanics phenomena. The character of the flow field depends on the shape of the object. Even the simplest shaped objects, like a sphere, produce rather complex flows. For a given shaped object, the flow pattern and related forces depend strongly on various parameters such as size, orientation, speed and fluid properties. By the concepts of dimensional analysis, the important dimensionless parameters are the **Reynolds Number** (ratio of inertia forces and viscous forces), the **Mach Number** (ratio of flow velocity and speed of sound) and sometimes the **Froude Number** (ratio of inertia forces and gravity forces). The study of flow around immersed bodies has a wide variety of engineering applications but in terms of heat and mass transfer, spheres are the most important. Thus most of this article deals with spheres.

There is a great similarity in the development of flow pattern at increasing Reynolds number between a sphere and a circular cylinder (or tube), except for the *vortex street* associated with the latter and other two-dimensional bodies, which is not formed for three-dimensional bodies, instead a *vortex ring* occurs, which for a sphere is formed at about $Re_D = 24$ (see **Figure 1**) and becomes unstable at about $Re_D = 200$ when it tends to move downstream of the body and is immediately replaced by a new vortex ring.

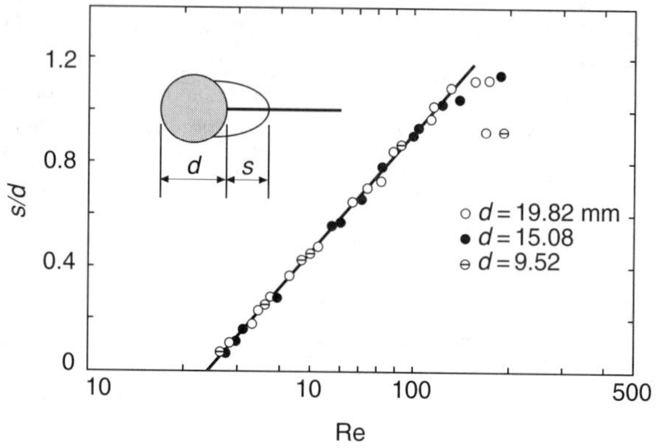

Figure 1. Observed lengths of the region of closed streamlines behind a sphere. From Taneda S. (1956)

No regular pattern of motion like the vortex street forms in the wake of a sphere (or of any three-dimensional body), although there is a general impression that vorticity is shed from the standing ring-vortex like a succession of distorted vortex loops not symmetrical around the central axis. (See **Vortex Shedding**) This flow process does not give rise to vibrations of the sphere. At large Reynolds number, flow over the front half of a sphere may be divided into a thin **Boundary Layer** region where viscous forces are dominant, and an outer region, where the flow corresponds to that of an inviscid fluid.

Pressure decreases over the front half of the sphere from the stagnation point onwards, thus having a stabilizing effect on the boundary layer, which remains laminar up to about $Re_D = 5 \cdot 10^5$. Beyond the minimum pressure point on the sphere surface, the boundary layer is subjected to an adverse pressure gradient and the fluid decelerates. Later, flow separation occurs. At low Re_D, the separation point is located at the rear stagnation point and with increasing Re_D moves forward and reaches $\varphi \approx 80°$ from the front stagnation point at $Re_D \approx 1000$. Pressure drag begins to dominate and the **Drag Coefficient** becomes almost independent of the Reynolds number until about $Re_D = 5 \cdot 10^5$, when the transition from **Laminar** to **Turbulent Flow** occurs before separation. As a result, the point of separation moves to the rear, making the wake smaller and abruptly reducing drag coefficient.

At very low Re_D, during the so-called *creeping flow*, inertia forces are assumed to be negligible; hence, the governing equations for the flow (*Navier-Stokes equations* and the continuity equation) are greatly simplified. Stokes succeeded in obtaining a solution for these equations and the drag force has been found to be

$$F_D = 6\pi\eta R U_\infty \quad (1)$$

where R is the radius of the sphere, U_∞ the freestream velocity and η the dynamic viscosity of the fluid. See also **Stokes' Law for Solid Spheres and Spherical Bubbles**. This relationship may be used up to $Re_D = 0.5$ with negligible error. This flow range is often called the *Stokes flow*. The drag coefficient is defined as:

$$C_D = \frac{F_D}{\frac{\rho U_\infty^2}{2} \cdot A} \tag{2}$$

where A is the projected area (equal to πR^2), which can be expressed as:

$$C_D = \frac{24}{Re_D} \tag{3}$$

where Re_D is

$$Re_D = \frac{U_\infty D}{\eta/\rho} \tag{4}$$

Equation 1 is known as Stokes' law. Sometimes Equation 3 is also referred to as **Stokes' Law**.

An extension of Equation 3 has been formulated using a method of successive approximations to the governing equations (still at low Re_D). This formula reads

$$C_D = \frac{24}{Re_D}\left(1 + \frac{3}{16} Re_D\right) \tag{5}$$

Equation 5 may be used up to $Re_D \approx 100$. At higher values of Re_D, it is necessary to rely on empirical expressions based on experiments. **Figure 2** provides a graph of C_D versus Re_D for a sphere.

When applied to particle mechanics, the following regimes are introduced:

$Re_D \leq 0.2$	**Stokes flow**	$C_D = \frac{24}{Re_D}$
$0.2 < Re_D \leq 500$	**Allen flow**	$C_D = \frac{18.5}{Re_D^{0.6}}$
$500 < Re_D < 10^5$	**Newton flow**	$C_D = 0.44$

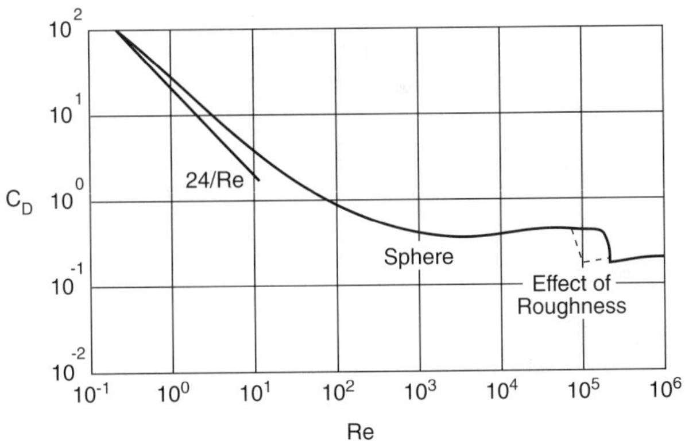

Figure 2. C_D as a function of Re_D for a sphere.

Several other expressions for drag coefficient may be found in other literature.

For blunt bodies like a sphere, an increase in surface roughness may cause a decrease in drag. The transition to turbulent boundary layer flow occurs at a lower Reynolds number than for a smooth sphere. One effect is the wake region behind the sphere becomes considerably narrower and overall drag is reduced.

Similar to a circular cylinder, the inviscid flow field around a sphere can also be determined analytically. The velocity components are:

$$u_r = -U_\infty\left\{1 - \left(\frac{r_o}{r}\right)^3\right\}\cos\theta \tag{6}$$

$$u_\theta = \frac{U_\infty}{2}\left\{2 + \left(\frac{r_o}{r}\right)^3\right\}\sin\theta \tag{7}$$

where θ is measured from the forward stagnation point, r is the radial coordinate, r_o is sphere radius and U_∞ is the freestream velocity. The maximum velocity occurs at $\theta = \pi/2$ and is

$$(u_\theta)_{max} = 1.5U_\infty \tag{8}$$

Experimentally however, the maximum value is about $1.3U_\infty$ and occurs upstream of $\theta = \pi/2$.

Velocity distribution in the laminar boundary layer over the front part of a sphere can be calculated using a series expansion technique. It requires, however, an experimentally determined pressure distribution if accurate values of skin friction and boundary layer thickness are to be achieved.

References

Batchelor, G. K. (1970) *An Introduction to Fluid Dynamics*. Cambridge University Press.

Munson, B. R., Young, D. F. and Okiishi, T. H. (1990) *Fundamentals of Fluid Mechanics*. J. Wiley & Sons.

Taneda S., (1956) Rep. Res. Inst. Appl. Mech., Kyushu Univ, 4, 99.

White, F. M. (1994) *Fluid Mechanics*. 3rd Edition. McGraw-Hill.

Leading to: *Drag coefficient; Immersed bodies, heat transfer and mass transfer; Tubes, cross flow over*

B. Sundén

IMMERSED BODIES, HEAT TRANSFER AND MASS TRANSFER

Following from: *Immersed bodies, flow around and drag*

This article describes heat and mass transfer from spheres. An immersed sphere is commonly understood to be a sphere made of solid material. But as far as heat transfer is concerned, fuel drops in an internal combustion engine or in the front of a nozzle of a furnace, or liquid drops in a spray drying oven, are considered as spheres of small size. In such cases, mass transfer takes place simultaneously with heat transfer. An understanding of heat transfer from single spheres is needed before predicting the thermal performance of clouds of spheres which are heated or cooled in a stream of fluid.

The distributions of local convective heat transfer coefficient around a sphere immersed in an air stream are shown in **Figure 1**.

The heat transfer coefficient decreases gradually from the front stagnation point to the minimum value at about 105° from the stagnation point. This decrease is due to the growth of the thermal boundary layer, which increases thermal resistance. Between 105° and 120°, a sharp increase in heat transfer coefficient occurs, after which the increase becomes less. The sudden increase of the local convective heat transfer coefficient following the minimum point may be explained by the fact that flow in the boundary layer undergoes a transition from **Laminar** to **Turbulent Flow** and that an intensive turbulent state exists on the downstream side of the sphere.

By using a series expansion technique for solving the laminar boundary layer equations, Frössling succeeded in obtaining the following relation for the local heat transfer coefficient

$$\frac{Nu}{\sqrt{Re}} = 1.8615 - 2.1477\left(\frac{x}{D}\right)^2 + 2.4609\left(\frac{x}{D}\right)^4 \quad (1)$$

where **Nusselt Number**, $Nu = \alpha D/\lambda$ (D sphere diameter, λ thermal conductivity of the fluid, α the heat transfer coefficient), **Reynolds Number**, $Re = U_\infty D/\nu$ (U_∞ is the freestream velocity and ν, the kinematic viscosity of the fluid) and x is the distance along the perimeter measured from the stagnation point. The coefficients in **Equation 1** are valid for naphtalene with a **Prandtl Number** equal to 2.53 and an experimental pressure distribution has been used in deriving it.

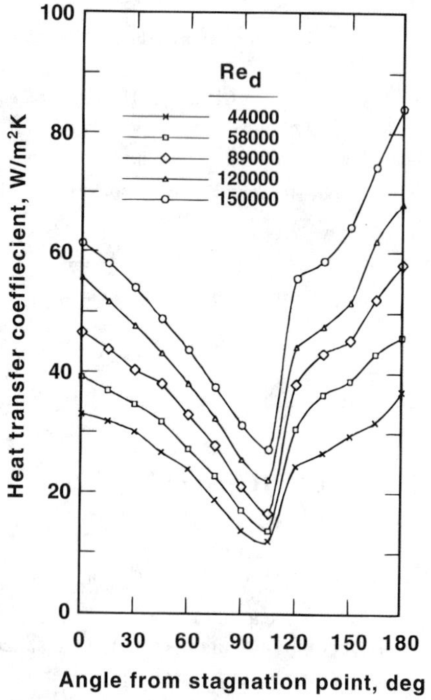

Figure 1. Local heat transfer coefficients for a sphere in an airstream.

From the engineering point of view, the average heat transfer coefficient is most important. This must be sought from experiments because of the difficulty of performing a theoretical analysis in the separated flow region on the downstream side of the sphere. Several experiments have been carried out and a variety of empirical formulae are available. A few will be presented here.

For air flowing in the Reynolds number interval $20 < Re_D < 150{,}000$, one formula reads

$$Nu_D = 0.33\, Re_D^{0.6} \quad (2)$$

which, for other gases, is modified to

$$Nu_D = 0.37\, Re_D^{0.6} Pr^{1/3} \quad (3)$$

In Equations 2 and 3, the thermophysical properties should be evaluated at the so-called film temperature (average of freestream temperature and surface temperature).

A common formula recommended for Reynolds numbers in the range of $3.5 \leq Re_D \leq 7.6 \cdot 10^4$ and Prandtl numbers in the range of $0.6 \leq Pr \leq 380$ is:

$$Nu_D = 2 + Pr^{0.4}\{0.4\, Re_D^{1/2} + 0.06\, Re_D^{0.67}\}\left(\frac{\eta_\infty}{\eta_w}\right)^{1/4} \quad (4)$$

where η_∞ and η_w are the dynamic viscosities at the freestream temperature and sphere surface temperature, respectively.

Freely-falling liquid drops present a special case of convective heat and mass transfer. In the Reynolds number range $1 \leq Re_D \leq 70{,}000$, a formula reads

$$Nu_D = 2 + 0.6\, Re_D^{1/2} Pr^{1/3}. \quad (5)$$

As is obvious, many formulas reduce to $Nu_D = 2$ as $Re_D \to 0$. This is the value for heat transfer by conduction in a spherical shell and can easily be found analytically.

For air flow across a sphere, recent measurements have been correlated as:

$$Nu_D = 2 + \left\{\frac{Re_D}{4} + 3 \cdot 10^{-4}\, Re_D^{1.6}\right\}^{1/2} \quad (6)$$

which is valid in the range $100 < Re_D < 2 \cdot 10^5$. At higher Reynolds number in the range of $4 \cdot 10^5 < Re_D < 5 \cdot 10^6$, the following formula has been suggested:

$$Nu_D = 430 + 5 \cdot 10^{-3}\, Re_D + 0.25 \cdot 10^{-9}\, Re_D^2 - 3.1 \cdot 10^{-17}\, Re_D^3. \quad (7)$$

For heat transfer from a sphere to a liquid metal, the following expression has been suggested:

$$Nu_D = 2 + 0.386(Re_D Pr)^{1/2}. \quad (8)$$

The range of application is $3.6 \cdot 10^4 < Re_D < 2 \cdot 10^5$.

The analogy between equations describing heat transfer and equations for isothermal mass transfer implies that there exists for every heat transfer situation a corresponding mass transfer situation, with analogous boundary conditions and with the same geometry.

The mass transfer coefficient is expressed by the dimensionless **Sherwood Number,** defined as:

$$\mathrm{Sh} = \frac{\beta L}{\delta} \tag{9}$$

where β is the mass transfer coefficient, L is the characteristic length and δ is the mass diffusivity.

If the Schmidt number, defined as

$$\mathrm{Sc} = \frac{\nu}{\delta} \tag{10}$$

is introduced, the formulae given previously can be used for mass transfer problems if Nu is replaced by Sh and Pr is replaced by Sc. (See **Analogy Between Heat, Mass and Momentum Transfer.**)

References

Eckert, E. R. G. and Drake, R. M. Jr. (1972) *Analysis of Heat and Mass Transfer.* McGraw-Hill.

Kreith, F. and Bohn, M. S. (1993) *Principles of Heat Transfer.* 5th edition. West Publ. Comp.

Schlichting, H. (1979) *Boundary Layer Theory.* 7th edition. McGraw-Hill.

B. Sundén

IMMERSED JETS (see Jets)

IMMISCIBLE LIQUIDS (see Interfaces; Liquid-liquid flow)

IMMISIBLE LIQUIDS, BOILING HEAT TRANSFER (see Multicomponent mixtures, boiling in)

IMPACT OF PARTICLES ON SURFACE (see Erosion)

IMPEDANCE METHOD FOR VOID FRACTIONS (see Void fraction measurement)

IMPELLER MIXERS (see Mixers)

IMPELLERS (see Agitation devices)

IMPINGEMENT SEPARATORS (see Vapour-liquid separation)

IMPINGING JETS

Following from: *Convective heat transfer; Flow of fluids*

Jets can be classified as submerged if they discharge into an ambient fluid of similar physical properties (e.g., air in air) and unsubmerged if the properties of the two fluids are quite different (e.g., water in air). Several configurations have been tested such as: jets issuing from orifices or round and slot nozzles different types of jet arrays; flat and curved impingement surfaces, normal and inclined impingement etc. Only the simple geometry of a single jet impinging on a flat surface is treated below. **Figure 1** shows the typical fluid-dynamic features of a submerged (top) and an unsubmerged (bottom) jet. For both, three main regions can be identified:

1. the free jet region, which extends from the nozzle up to a certain distance from the surface and can, in turn, be divided into a potential core, a developing and a fully developed zone;
2. the stagnation region, where the flow on the wall is accelerated by a streamwise stabilizing pressure gradient (the boundary layer thickness tends to keep constant);
3. the wall region, where the pressure gradient effect no longer holds and an even steep rise of turbulence can occur.

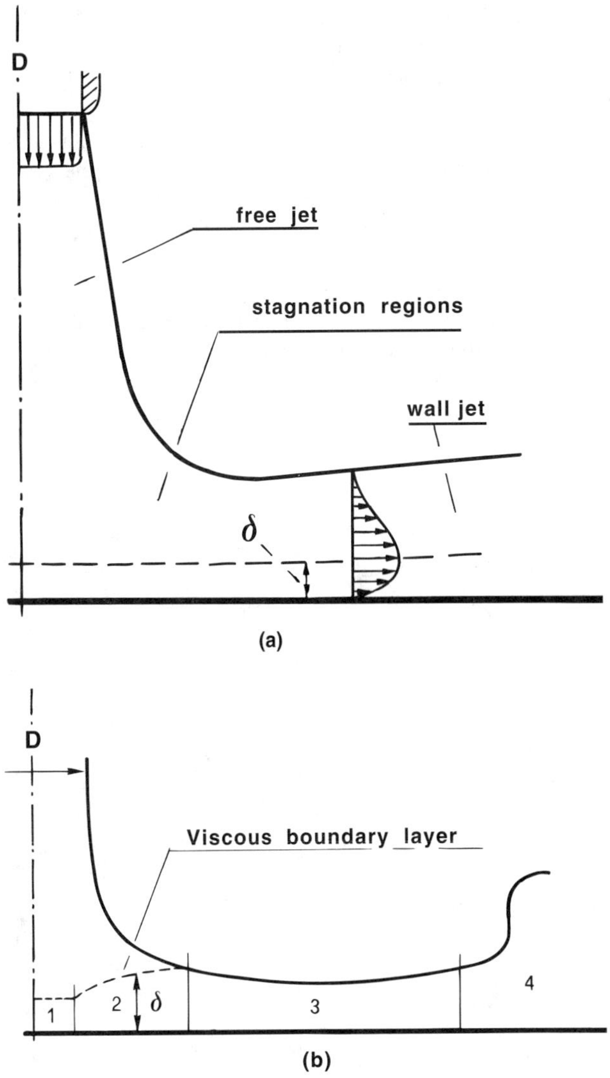

Figure 1. Submerged (top) and unsubmerged (bottom) jet shape.

After issuing from the nozzle, the submerged jet widens linearly with its length due to the exchange of momentum with the ambient fluid over the free boundaries. An outline of the trends of the development of the boundary layer and of the fluid velocity is given in **Figure 1a**.

The unsubmerged jet (liquid in gas) exibits a gas-liquid free boundary which bends close to the solid surface once the liquid starts spreading over the wall. The wall flow is subdivided into four regions as shown in **Figure 1b**.

1. a stagnation region with a constant boundary layer;
2. a zone where the viscous boundary layer enlarges up to the value of the liquid sheet thickness;
3. a fully viscous region;
4. the hydraulic jump with the occurrence of a sudden increase in liquid thickness and a slowing down of liquid velocity (**Figure 1b**).

The **Nusselt Number** is most commonly given as a function of **Reynolds Number** Re and **Prandtl Number**, Pr in the form $Pr^m Re^n$, the dimensionless distance from the stagnation point (line) on the surface, r/D, and the dimensionles nozzle to wall spacing z/D. Roughly speaking, Nu decreases with the increase of r/D and of z/D. Furthermore, the sensitivity of Nu to z/D is more pronounced for submerged jets than for unsubmerged ones. Nevertheless radial peaks, depending on the values of Re and z/D, occur both for submerged and unsubmerged jets and can be ascribed to the thermal boundary layer evolution and to the effect of turbulence. **Figure 2** from Lytle and Webb (1991) shows a sketch of this phenomenon. More recently, the influence of splattering [Lienhard et al. (1992)] and of the **Froude Number** [Di Marco et al. (1993)] have been accounted for. More detailed information on the treated matter and several correlations can be found in Faggiani and Grassi (1990) and Viskanta (1993).

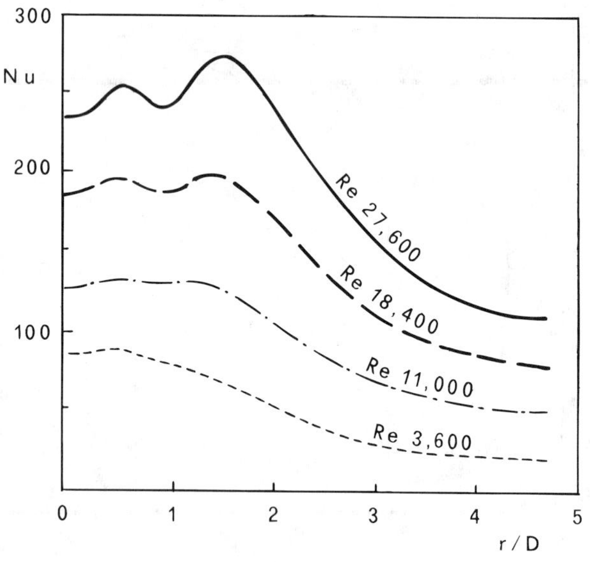

Figure 2. Nusselt number profile.

References

Di Marco, P., Grassi, W., Magrini, A. (1993) Heat transfer between a flat horizontal surface and an unsubmerged liquid jet at low velocity. *Eurotherm Seminar*. 32. Oxford, UK.

Faggiani, S., Grassi, W. (1990) Impinging liquid jets on heated surfaces. KN16. *9th Int. Heat Transfer Conf.* Jerusalem. Israel. 1.

Lienhard, J. H. (V), Liu, X., Gabour, L. A. (1992) Splattering and heat transfer during impingement of a turbulent liquid jet. *J. Heat Transfer, Trans.* ASME. 114.

Lytle, D., Webb, B., W. (1991) Secondary heat transfer maxima for air jet impingement at low nozzle-to-plate spacing. *Experimental Heat Transfer, Fluid Mechanics and Thermodynamics*. Elsevier, New York.

Viskanta R. (1993) Heat transfer to impinging gas and flame jets. *Experimental Thermal and Fluid Science*. vol. 6.

W. Grassi

IMPULSES (see Delta function)

IMPULSE TURBINES (see Hydraulic turbines)

IMPURITIES IN CRYSTALS (see Crystallization)

INCINERATION

Following from: *Combustion*

Introduction

At the present time, the western world produces about 10 tonnes of *waste* per person annually in the form of household waste, industrial waste and waste from activities such as energy production, agriculture, mining and sewage disposal. Incineration is the best way to dispose of much of this material in an environmentally-friendly manner. For example, clinical waste is still buried in many developing countries, only to be dug up by dogs with hazardous consequences: clearly, those countries have an urgent need for simple cheap incinerators. Incineration technology is still evolving rapidly throughout the world and further innovations can be anticipated in the next few years.

An important aspect of incineration combustion systems involves treatment of the gaseous, liquid and solid effluent in order to minimise their environmental effects. In addition, energy can be derived from many wastes and energy recovery is therefore an important part of incinerator design.

A typical incinerator consists of three components: the furnace chamber, the heat recovery boiler, and the flue gas treatment plant.

Incinerator Furnace Chamber

In a municipal solid waste (MSW) incinerator, waste material is usually burned in a refractory-lined furnace chamber as a moving bed on a grate with underfire air. The waste material is very inhomogeneous, and various techniques are used to mix the burning material on the grate as it progresses along the chamber. For example, a reciprocating grate or steps allowing the material to fall and disperse may be used. Other mixing techniques used include rotary kilns and fluidised beds.

Material introduced to the bed must dry first before it can be ignited. This drying stage is significant since municipal waste typically contains 30% water. This water content is the main factor influencing the calorific value of the waste. As the waste heats up, it loses volatiles which burn above the bed. When ignition temperature is reached, the material in the bed burns in the surrounding oxygen, emitting **Carbon Dioxide** and gradually pyrolysing to a char. Once the oxygen is used up, the carbon dioxide is reduced to **Carbon Monoxide,** which burns in the freeboard along with **Pyrolysis** products. Eventually, the char burns to an ash which drops into the ash pit. Usually the ash contains less than 5% carbon, and indeed the permitted carbon in ash may be controlled by legislation. Again it is emphasised that all these bed combustion processes take place while the bed is being mixed; at present, there is no good mathematical model of the total process and empirical correlations are used for bed design.

The overall air/fuel ratio in the furnace chamber is kept lean to ensure that there is sufficient air to minimise residual carbon in the ash, and secondary air is injected above the bed to ensure that solid and gaseous hydrocarbons are fully oxidised before they are passed to the flue gas treatment plant. Due to rapid variations in the local composition of waste, optimisation of air flow distribution poses a very difficult control problem and empirical techniques are usually used.

Heat Recovery Boiler

The heat recovery boiler consists of two stages. In the first stage, the gases are cooled by the radiation of heat to the walls whilst in the second, gases are further cooled by convective heat transfer to tubes located in the flow. In the radiant section, water in the wall tubes is boiled and passed to the boiler drum where the saturated steam and the water are separated. If superheated steam is required, then the first convection sections are used as superheater.

The main problems encountered in heat recovery boilers are related to corrosion since flue gases from the incineration can contain corrosive components, such as hydrochloric acid from the burning of PVC plastics. To minimise corrosion, wall temperatures must be kept relatively low and impingement of flame or particles on the wall must be avoided to prevent reducing conditions at the metal tube walls.

(See also **Heat Recovery Boilers** and **Waste Heat Recovery**.)

Flue Gas Treatment Plant

To minimise the emission of *pollutants,* the first stage of flue gas treatment is achieved in the furnace and boiler by ensuring that there is sufficient excess air and residence time to complete the combustion process.

The removal of pollutants such as particles and acid gases from the incinerator flue gases is accomplished in a flue gas treatment plant. This follows well-established chemical engineering design principles, and consists of components such as venturi and tower **Scrubbers, Heat Exchangers, Electrostatic** and bag **Filters, Fans** and **Pumps**.

Although *dioxins* can be destroyed in an incinerator, they can be formed in the flue gas treatment plant. The minimisation of dioxins may be achieved by ensuring that the optimum temperature/composition/time history of the flue gases is satisfied. Permitted levels of emission of all pollutants is tightly controlled by legislation.

Energy from Waste

As already mentioned, incinerators provide a vital stepping-stone in improving energy efficiency. A key point is that energy in the form of electricity is worth about eight times more than the same amount of energy as steam heat. Hence, it is important that the proportion of waste converted to electricity is maximised.

Since much of the heat produced by an incinerator during high summer is wasted, there are opportunities for **air conditioning** systems which use hot water to provide chilling during this period.

References

Nasserzadeh, V., Swithenbank, J. and Jones, B. (1993) "Effect of high-speed secondary air jets on the overall performance of a large MSW incinerator with a vertical shaft". *Combust. Sci. & Tech.* 92:389–422.

Nasserzadeh, V., Swithenbank, J., et al. (1993) "Three-dimensional mathematical modelling of the Sheffield Clinical Incinerator, using computational fluid dynamics and experimental data". *J. Inst. Energy.* 66:169–179.

J. Swithenbank

INCLINED TUBE BANKS (see Tube banks, crossflow over)

INCLINED TUBES (see Tube banks, single phase heat transfer in)

INCLUSIONS (see Crystallization)

INCOMPLETE GAMMA FUNCTION (see Gamma function)

INCOMPRESSIBLE FLUID (see Flow of fluids)

INDEFINITE INTEGRALS (see Integrals)

INDUCED DRAFT AIR COOLED HEAT EXCHANGERS (see Air cooled heat exchangers)

INDUCTION HEATING

Following from: *Furnaces*

When an electrically conductive body is placed in the region of a time varying magnetic field, electric currents are induced in the body causing thermal power generation in the body. This effect, known as induction heating, is widely used in industries ranging from the production of optical glass fibre to the heating of 25 tonne steel slabs [examples are given in BNCE (1984)]. The magnetic field is produced by a suitable arrangement of conductors, the induction coil, connected to a source which can provide the required time varying current in the coil. Electrical power supplied to the coil is thus converted to thermal power in the workpiece through the electromagnetic field, without physical electrical connection to the workpiece. Almost invariably, power sources used for induction heating provide an alternating current to the induction coil, the choice of frequency being critical to the particular heating application.

The induced 'eddy' current intensity is greatest at the surface of the workpiece and decreases towards its centre as a function of the ratio: thickness/skin depth. As the ratio increases, a greater proportion of the total power is dissipated near the surface, this phenomenon is known as the *skin effect*. The skin depth, δ, is defined as $\delta = \sqrt{2\rho/\omega\mu}$, where ρ is the electrical resistivity (Ωm) and $\omega = 2\pi f$ (rad/s) is the angular frequency of the coil current. The absolute magnetic permeability μ is $\mu_r\ \mu_0$, where $\mu_0 = 4\pi\ 10^{-7}$ (H/m); the relative permeability, μ_r, is a function of the applied magnetic field strength for magnetic materials and has the value 1 for nonmagnetic materials such as copper and aluminium.

Power generated in a workpiece and the induction heating efficiency can be derived for regular shapes, such as cylindrical rods or tubes and wide rectangular slabs, from analytical solutions to the diffusion equation of the induced current, supplemented by empirical factors. These derivations are given in Davies & Simpson (1979), Orfeuil (1987) and Davies (1990). The analytical solutions assume constant material properties throughout the workpiece, whereas resistivity and specific heat vary with temperature and the permeability of magnetic materials is a function of field strength and temperature, reducing to μ_0 above the Curie temperature ($\approx$ 750 °C for steel). Computer-based numerical solutions are now commonly used to take account of these variations, an early example being Gibson (1973).

For a solid circular billet of diameter d, and length L, heated in an enclosing circular coil of diameter D, length L_c and having N turns with a current of I amp/turn, the induced power P_W is approximately given by:

$$P_w = (N \cdot I \cdot K_c)^2 \cdot \frac{\pi \cdot d \cdot \rho}{\delta \cdot L} \cdot Q_{rod} \quad \text{(Watt)} \qquad \textbf{(1)}$$

where Q_{rod} is given in **Figure 1** as a function of d/δ, and K_c is dependent on the ratios d/D, d/δ and L/L_c. Orfeuil (1987) gives empirical values for K_c, which tend to unity as d/D and L/L_c approach unity. The power induced in hollow cylinders of wall thickness t is calculated with Q_{rod} in the above expression replaced by an equivalent flux factor Q_{cyl}, which is a function of t/d, d/δ and μ_r independently of δ. Davies (1990) shows graphs of Q_{cyl} for a range of these parameters.

Similarly, for a rectangular slab of length L, having width W, much greater than its thickness t, the induced power is:

$$P_w = (N \cdot I \cdot K_c)^2 \cdot \frac{2 \cdot W \cdot \rho}{\delta \cdot L} \cdot Q_{slab} \quad \text{(Watt)} \qquad \textbf{(2)}$$

Substitution of the Q factors by P_{rod} or P_{slab} from **Figure 1** gives the reactive power (VAR) in the workpiece, which is needed for the evaluation of the power factor of the coil.

The efficiency of conversion of the electrical power supplied to the coil into thermal power in the workpiece, known as the coil or electrical efficiency, η_c, is given by:

$$\eta_c = 100 \Big/ \left[1 + \frac{1}{K_A K_C^2} \cdot \frac{S_C}{S_W} \cdot \sqrt{\frac{\rho_C}{\rho_W \cdot \mu_W}} \cdot \frac{1}{Q} \right] \qquad \textbf{(3)}$$

where Q is the relevant flux factor, K_A is the space factor of the coil system and S_C/S_W is the ratio of the coil perimeter to that of the workpiece in the same plane. Harvey (1976) shows that coil efficiency can be significantly increased by the use of multilayer windings instead of the more conventional single-layer coil. These high-efficiency coils are now commonly used for heating nonferrous billets at mains frequency.

The overall efficiency of induction heating is $\eta_{supply} \cdot \eta_{thermal} \cdot \eta_c$. η_{supply} is typically 0.8–0.9 (per unit) and accounts for losses

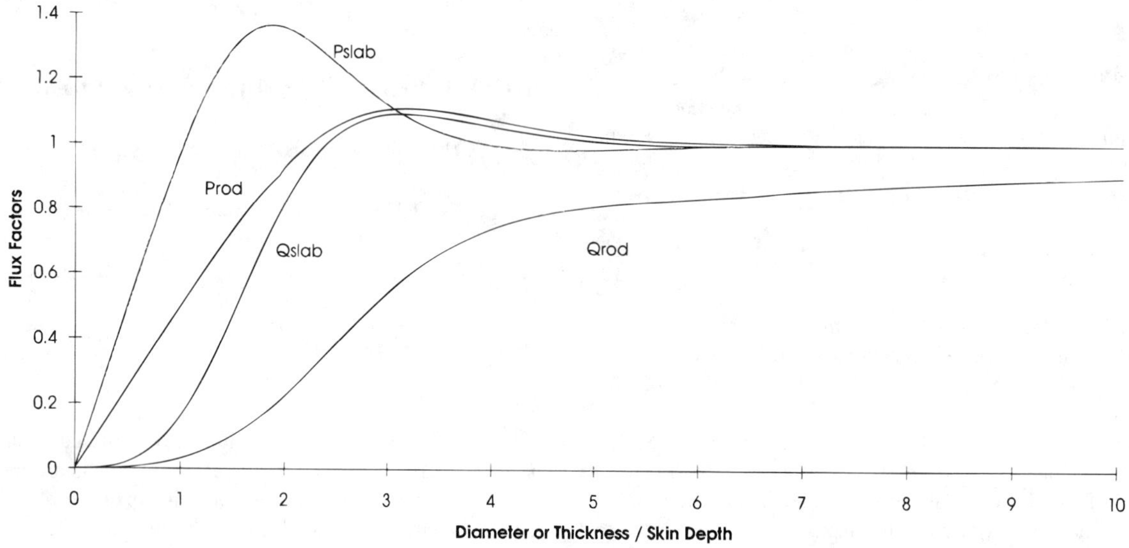

Figure 1. Flux factors Q and P for rods & slabs.

in cables, power factor correction capacitors and frequency conversion equipment; the thermal efficiency, $\eta_{thermal}$, represents thermal losses from the workpiece and is critically dependent on operating temperature, thermal insulation and method of operation of the heater. Typical values are in the range 0.7–0.9 (per unit).

Transverse flux induction heating is employed for heating continuous metal strips. In this mode, the magnetic field is directed at the broad face of the material rather than through its narrow cross-section, with the induced current flowing across the width of the strip. The advantages of the method include a higher efficiency, particularly for nonferrous strips, at much lower operating frequencies than are possible with conventional axial flux induction heaters. Ireson (1989) gives a useful overall account of the technique and its commercial realisation.

Apart from mains frequency installations, power supplies for modern induction heaters are derived from solid state frequency converters. Unit sizes up to 7 MW have been installed for metal melting at 1–3 kHz and 1 MW units are now available for frequencies up to 500 kHz, previously the domain of power vacuum tube triodes.

References

Davies, E. J. (1990) *Conduction and Induction Heating*. Peter Peregrinus Ltd. London.

Davies, E. J. and Simpson, P. G. (1979) *Induction Heating Handbook*. McGraw-Hill Book Company (UK) Limited. Maidenhead.

Gibson, R. C. (1973) SLEDDY, a computer programme for calculating the induction and other heating of metal slabs and long cylindrical billets. Report ECRC/MM16. *EA Technology*. Capenhurst, Chester.

Guide to induction heating equipment. (1984) *British National Committee for Electroheat (BNCE)*. 30 Millbank, London.

Harvey, I. G. (1976) The theory of multilayered windings for induction heating and their application to a 1 MW, 50 Hz, longitudinal flux billet heater. Paper II(a)4. *8th UIE Congress*. Liège.

Ireson, R. C. J. (1989) Induction heating with transverse flux in strip-metal process lines. *IEE Power Engineering Journal*. 3: (2). London.

Orfeuil, M. (1987) *Electric Process Heating*. Battelle Press. Columbus, Richmond, Ohio.

I.G. Harvey

INDUSTRIAL AERODYNAMICS (see Aerodynamics)

INDUSTRIAL FUSION REACTORS (see Fusion, nuclear fusion reactors)

INERT GASES (see Noble gases)

INERTIAL CONFINEMENT REACTORS (see Fusion, nuclear fusion reactors)

INERTIAL REFERENCE FRAMES (see Conservation equations, single phase)

INERTIAL SEPARATORS (see Gas-solids separation, overview)

INFINITE SERIES

An **infinite series** is an expression of the form $\sum_{k=1}^{\infty} a_k = a_1 + a_2 + \ldots + a_n + \ldots$, in which an infinite number of terms (complex and real numbers) are added. The sum of the first n terms of a series $S_n = \sum_{k=1}^{n} a_k$ is a partial sum of a series. If there is a finite limit S of partial sums for $n \to \infty$, then it is the sum of a series and the series is convergent ($S = \lim_{n\to\infty} S_n$); otherwise, the series is called divergent. The series is absolutely convergent if a series $\sum_{k=1}^{\infty} |a_k|$ converges.

If the terms of a series are presented by the functions of some variable, i.e., $a_k = a_k(x)$ then the series is called functional.

A series of functions $(a_1(x) + \ldots a_k(x))$ converges to a function (a sum) S(x) for each value of x, which has a finite limit

$$\lim_{n\to\infty} \sum_{k=1}^{n} a_k(x) = S(x).$$

A series converges uniformly on a set of values $x \in [a, b]$ to S(x) if a sequence of its partial sums S_n converges uniformly, i.e., for $\forall\epsilon > 0\ \exists N$ independent of $x \in [a, b]$, such that the condition $|S(x) - S_n(x)| < \epsilon$ is met for all $n \geq N$.

The most widely used criterion for uniform convergence is the Weierstrass M-test: if the terms of a functional series $\sum_{k=1}^{\infty} a_k(x)$ on the interval [a, b] satisfy the inequality $|a_k(x)| \leq M_k$, where M_k are the terms of some convergent series $\sum_{k=1}^{\infty} M_k$, then the series $\sum_{k=1}^{\infty} a_k(x)$ converges on [a, b] uniformly and absolutely.

To define uniform convergence of series of the form $\sum_{k=1}^{\infty} a_k(x)b_k(x)$, Abel's test for uniform convergence is used: if a series $\sum_{k=1}^{\infty} b_k(x)$ converges uniformly on the interval [a, b], and a function $a_k(x)$ forms a monotone bounded for any x and k sequence (i.e., $|a_k(x)| \leq M$), then the series $\sum_{k=1}^{\infty} a_k(x)b_k(x)$ converges uniformly on [a, b].

Uniformly-convergent series can be added term by term and multiplied by a bounded function. The series obtained as a result are also convergent. If separate terms of a uniformly-convergent series are continuous functions, then the sum of a series is also a continuous function. If the function $a_k(x)$ are integrable on the interval [a, b], a series consisting of the integrals converges uniformly; then this series can be integrated termwise, with the sum of the integrals being equal to the integral of the sum

$$\int_a^b S(x)dx = \sum_{k=1}^{\infty} \int_a^b a_k(x)dx. \tag{1}$$

If a function $a_k(x)$ has continuous derivatives $a_k'(x)$ and the series $\sum_{k=1}^{\infty} a_k'(x)$ composed of derivatives uniformly converges, then the series $\sum_{k=1}^{\infty} a_k(x)$ can be differentiated term by term, with $S'(x) = \sum_{k=1}^{\infty} a_k'(x)$.

The expansion of functions into infinite series is often employed in solving problems of hydrodynamics and heat transfer. With the help of series, the exact values of transcendental constants, functions and integrals are calculated and differential equations are solved.

Leading to: *Taylor series*

V.N. Dvortsov

INFINITE TRIGONOMETRIC SERIES (see Fourier series)

INFRA-RED DRYING (see Dryers)

INFRA-RED IMAGING (see Thermal imaging)

INFRA-RED PHOTOGRAPHY (see Photographic techniques)

INFRA-RED RADIATION (see Electromagnetic spectrum)

INFRA-RED SPECTROSCOPY (see Spectroscopy)

INGRESS (see Rotating disc systems, applications)

INJECTION (see Augmentation of heat transfer, single phase)

INLET EFFECTS IN CHANNEL FLOW (see Channel flow)

IN-LINE MIXERS (see Mixers, static)

IN-LINE TUBE BANKS (see Tube banks, crossflow over)

INSTABILITIES IN LAMINAR FLOW (see Laminar flow)

INSTABILITIES IN TWO-PHASE SYSTEMS (see Two-phase instabilities)

INSTABILITY (see Non-linear systems)

INSTITUTE OF ENERGY, IoE

18 Devonshire Street
London W1N 2AU
UK
Tel: 0171 580 0008

INSTITUTION OF CHEMICAL ENGINEERS, ICHEME

165–171 Railway Terrace
Rugby, CV21 3HQ
UK
Tel: 01788 578214

INSTITUTON OF ELECTRICAL ENGINEERS, IEE

Michael Faraday House
Six Hills Way
Stevenage
Herts SG1 2AY, UK
Tel: 01438 313311

INSTITUTION OF MECHANICAL ENGINEERS, IMECHE

1 Birdcage Walk
London, SW1H 9JJ, UK
Tel: 0–71 222 7899

INSULATION (see Heat protection)

INSULATORS, ELECTRICAL (see Semiconductors)

INTEGRAL CONDENSATION CURVE (see Condensation curve)

INTEGRAL EQUATIONS

An integral equation is a mathematical expression that includes a required function under an integration sign. Such equations often describe an elementary or a complex physical process wherein the characteristics at a given point depend on values in the whole domain and can't be defined only on the bases of the values near the given point (as in local-type problems described by differential equations). Most frequently, integral equations as well as integro-differential equations are found in such problems of heat and mass transfer as diffusion, the potential theory and radiation heat transfer. For instance, nondimensional radiation heat flux in a given point on diffusely reflecting inner surfaces of two parallel isothermal optical grey plates, $\dot{q}(x) = Q(x)/\sigma T_w^4$, dependent on fluxes and reflected from other points, is defined by the equation

$$\dot{q}(x) = (1 - \rho) + \frac{\rho\gamma^2}{2}\int_{-1/2}^{1/2} K(x, s)\dot{q}(s)ds,$$

where $K(x,s) = [(x - s)^2 + \gamma^2]^{-3/2}$; $\gamma = h/d$; $x = X/d$; d and h are the width of the plates and the distance between them; and T_w is surface temperature. Using an integral equation for local-type boundary value problems leads to more compact expressions than when differential equations are employed, including boundary conditions in the governing equation. This approach allows the possibility of using alternative solution methods such as the perturbation method or the approximate semianalytic averaging method.

Linear integral equations are of the form

$$A(x)y(x) + \int_D K(x, s)y(s)ds = f(x), \tag{1}$$

where function A is called a coefficient; K is a kernel; and f is a free term (right-hand side) of the equation while D is the domain of function y(s). If A and K are matrices while y and f are vectors, Equation 1 is a set of integral equations. If $f = 0$, Equation 1 is called homogeneous; otherwise it is a nonhomogeneous equation. Three kinds of linear integral equations may be distinguished depending on whether $A \equiv 0$ (first kind), $A \neq 0$ for all x (second kind) and $A = 0$ for some x (third kind). In the one-dimensional case, the first and second kind of equations are of the form

$$\int_a^b K(x, s)y(s)ds = f(x), \quad (2)$$

$$y(x) - \lambda \int_a^b K(x, s)y(s)ds = f(x). \quad (3)$$

If K(x,s) and f(x) are continuous functions, λ is a real number and a and b are constants, Equations 2 and 3 are called the *Fredholm integral equation* of the first and second kinds, respectively. The homogeneous integral equation always has a trivial (zero) solution. Those values of λ which provide nontrivial solutions of the integral equation are known as the **eigen values** of the equation (or the kernel) and the corresponding nontrivial solutions, the characteristic ones. In case of a symmetrical nondegenerate kernel, $K(x,s) = K(s,x)$, the integral equation has at least one characteristic solution and all its eigen values are real. For the kernel of the form $K(x,s) = K(x - s)$, the integral equation is of the convoluted type. If K(x,s) becomes zero for $s > x$, Equations 2 and 3 take the form

$$\int_a^x K(x, s)y(s)ds = f(x), \quad (4)$$

$$y(x) - \lambda \int_a^x K(x, s)y(s)ds = f(x) \quad (5)$$

and are called the *Volterra integral equations* of the first and second kind the linear Equation 3, which is not the Fredholm equation (i.e., when the integral operator is not continuous) is referred to as a singular integral equation for which solutions can be found by special methods. The equations below provide an example of nonlinear integral equations:

$$y(x) - \lambda \int_a^b K(x, s, y(s))ds = f(x),$$

$$y(x) - \lambda \int_a^b K(x, s)F(s, y(s))ds = f(x).$$

For continuous K and f, analytic solutions of Volterra integral equations of the second kind, Equation 5, can be obtained by means of operational calculus or the method of successive approximations. For the latter method, the number of iterations, which are necessary to obtain a solution with given accuracy, significantly depend on the choice of initial approximation $y_0(x)$. The solution for the Fredholm equation of the second kind (Equation 3) exists and can be constructed by means of the successive approximations method if parameter $|\lambda| < M^{-1}$, where M is such, that

$$\int_a^b |K(x, s)| ds \le M \quad \text{at} \quad a \le x \le b.$$

Projection methods can be developed for approximate solution of integral equations of the second kind with constant integration limits. The solution is often approached by summing a finite series

$$\tilde{y}(x) = \sum_{i=1}^{n} C_i y_i(x),$$

where y_i are known linear independent functions and C_i are free parameters which have to be defined. Depending on the approach to the approximate solution and definition of the free parameters, this method is classified as being a least squares method, the *Galerkin method* or the collocation method. Analytical solutions of integral equations of the second kind can be computed also by means of a resolvent R(x, s; λ) in the form

$$y(x) = f(x) + \lambda \int_D R(x, s; \lambda)f(s)ds,$$

where R is expressed as an infinite series of multiple integrals of kernel K. The Volterra equation of the first kind can be frequently reduced to the second kind of equation. The solution of the Fredholm equation of the first kind is difficult because small perturbations in K and f or errors in solving methods can give uncontollable errors. Solving this equation involves the use of regularization algorithms (i.e., the Tikhonov method, statistical regularization methods, etc.).

Numerical methods for solution of integral equation are based either on a discretisation of the calculation domain and then reducing the problem to solving a set of algebraic equations (linear for linear integral equation) for discrete values $\{y_j\}$, or on the methods described above; results are usually impracticable without the use of a computer.

I.G. Zaltsman

INTEGRALS

The problem of integrating a function or computing the value of an *integral* function arises in the solution of practical such as the problem of acceleration of a particle under the action of an applied variable force, or the problem of finding the area of a curvilinear trapezoid bounded by the graph of a given function and coordinate segments. The first problem, formulated in the general form, introduces the notion of an indefinite integral; the second, the notion of a definite integral.

Finding the indefinite integral of a function f(x) is the operational inverse of differentiation where an antiderivative function F(x) is sought, such that its derivative is equal to f(x), i.e., $F'(x) = f(x)$ and $dF(x) = f(x)dx$.

Hence, all the antiderivative functions of f(x) or its integrals can be expressed as

$$F(x) = \int f(x)dx + c,$$

where c is an arbitrary constant.

The antiderivatives of elementary functions can easily be found on the basis of formulas for derivatives of these functions. The most widely used are:

1. $\int a\,dx = ax + c;\ (a = \text{const});$
2. $\int x^\alpha\,dx = \dfrac{x^{\alpha+1}}{\alpha + 1} + c;\ (\alpha \ne -1)$
3. $\int \dfrac{dx}{x} = \ln|x| + c;$
4. $\int \dfrac{dx}{1 + x^2} = \arctan x + c;$

5. $\int \frac{dx}{\sqrt{1 - x^2}} = \arcsin x + c;$

6. $\int a^x\, dx = \frac{a^x}{\ln a} + c; \int e^x\, dx = e^x + c;$

7. $\int \sin x\, dx = -\cos x + c; \int \cos x\, dx = \sin x + c;$

8. $\int \frac{dx}{\sin^2 x} = -\cot x + c; \int \frac{dx}{\cos^2 x} = \tan x + c;$

9. $\int \sinh x\, dx = \cosh x + c; \int \cosh x\, dx = \sinh x + c;$

10. $\int \frac{dx}{\sinh^2 x} = -\coth x + c; \int \frac{dx}{\cosh^2 x} = \tanh x + c;$ etc.

A wide range of integrals is given in standard texts and handbooks.

Some other formulae can be obtained using the rules of integration, the simplest of which considers the property of operational linearity

$$\int af(x)dx = a \int f(x)dx \ (a = \text{const});$$

$$\int [f(x) + g(x)]dx = \int f(x)dx + \int g(x)dx.$$

In addition, methods of integration—such as a change of variables or integration by parts—are extensively used. The basis of the method change of variable or substitution, is the transformation of the independent variable. If

$$\int f(t)dt = F(t) + c,$$

then the integral

$$\int f(w(x))w'(x)dx$$

can be easily calculated by changing the variable to $t = w(x)$ according to

$$\int f(w(x))w'(x)dx = \int f(t)dt = F(t) + c = F(w(x)) + c.$$

For the method of integration by parts, the rule of differentiation of product functions applies. Let $u = f(x)$, $v = g(x)$ be functions with continuous derivatives. Then from $d(u\ v) = u\ dv + v\ dv$, it follows that

$$\int udv = uv - \int vdu.$$

Indefinite integrals of many even comparatively simple functions cannot be expressed in terms of elementary integrable functions by algebraic manipulation. Examples of this type include

$$\int e^{-x^2}dx, \quad \int \sin x^2 dx, \quad \int \frac{\sin x}{x} dx.$$

Functions whose indefinite integrals cannot be obtained by simple methods are called *transcendental functions.* They are, however, widely used in analysis. Among these are the sine and cosine integral functions, integral exponential function, probability integral, Fresnel's integrals, **Gamma Functions, Bessel Functions,** etc. Numerical and other techniques can be used to obtain values for such integrals and tables of values are given in standard texts.

The notion of a definite integral, on the other hand, brings about the problem of solving the area of a curvilinear trapezoid bounded by a graph in the *Cartesian coordinate* system of functions $f(x) > 0$ in the interval [a, b], which is the base of a trapezoid and the vertical intervals [a, f(a)] and [b, f(b)].

To calculate this area, the trapezoid is divided into n rectangles whose bases are the intervals $[x_i, x_{i+1}]$ of an arbitrary partitioning of the base [a, b] into n parts according to

$$a = x_0 < x_1 < x_2 < \cdots < x_i < x_{i+1} < \cdots < x_n = b,$$

and the heights are $f(\xi_i)$, where ξ_i is an arbitrary point on the interval $[x_i, x_{i+1}]$, $i = 0, 1, \ldots, n - 1$.

The sum of the areas of rectangles

$$S_n = \sum_{i=0}^{n=1} f(\xi_i)\Delta x_i, \qquad \Delta x_i = x_{i+1} - x_i$$

yields an approximate value of the area of a trapezoid. The exact value of the area can be obtained as a limit

$$S = \lim_{\max|\Delta x_i|\to 0} \sum_{i=0}^{n-1} f(\xi_i)\Delta x_i$$

for $n \to \infty$, with the maximum interval of partitioning tending towards zero.

Leaving aside the problem of area definition, a finite limit of the sum S_n is called the *definite integral* of the function f(x) over the interval [a, b] and is denoted by

$$\int_a^b f(x)dx.$$

If this limit exists, the function f(x) is called an integrable function in the interval [a, b]. The numbers a and b are called the lower and upper limits of integration, the sum S_n is called the integral sum.

Functions integrable over an interval [a, b] include continuous functions, bounded functions having only finite number of discontinuity points, and monotonic functions.

If the two functions f(x) and g(x) are integrable over the interval [a, b], then their sum, difference and products are also integrable. If a function is integrable over [a, b], then it is integrable on any of its parts $[\alpha, \beta] \in [a, b]$.

Properties of Definite Integrals

It follows from the definition that

$$\int_a^a f(x)dx = 0$$

$$\int_a^b f(x)dx = -\int_b^a f(x)dx.$$

It has been proven that if f(x) and g(x) are integrable on given intervals, then the following properties of definite integral are also valid:

a. $\int_a^b f(x)dx = \int_a^c f(x)dx + \int_c^b f(x)dx;$

b. $\int_a^b k\,f(x)dx = k\int_a^b f(x)dx$; k = const;
c. $\int_a^b [f(x) + g(x)]dx = \int_a^b f(x)dx + \int_a^b g(x)dx$;
d. for $b > a$ and $f(x) > 0$, $\int_a^b f(x)dx > 0$;
e. for $b > a$ and $f(x) > g(x)$, $\int_a^b f(x)dx > \int_a^b g(x)dx$;
f. if $m \le f(x) \le M$, $g(x) \ge 0$, $(g(x) \le 0)$ then $\int_a^b f(x)g(x)dx = \mu\int_a^b g(x)dx$; where $m \le \mu \le M$ (the generalized mean-value theorem);
g. if f(x) decreases monotonically in the range [a, b] and $f(x) \ge 0$, then $\int_a^b f(x)g(x)dx = f(a)\int_a^\xi g(x)dx$, where ξ is a value from the interval [a, b].
if $f(x) \ge 0$ and increases monotonically, then $\int_a^b f(x)g(x)dx = f(a)\int_a^\xi g(x)dx + f(b)\int_\xi^b g(x)dx$, (the second mean-value theorem);
h. $\int_a^b u\,dv = uv|_a^b - \int_a^b v\,du$, where $uv|_a^b = u(b)v(b) - u(a)v(a)$ (integration by parts formula);
i. if the product of a continuous function and its first derivative $g(t)\,g'(t) \in [a, b]$ for $t \in [\alpha, \beta]$, $g(\alpha) = a$, $g(\beta) = b$, then $\int_a^b f(x)dx = \int_\alpha^\beta f(g(t))g'(t)dt$.

If a function f(x) is integrable over [a, b], then it is also integrable over [a, x] where x is any value in the range [a, b]. The function

$$\Phi(x) = \int_a^x f(x)dx$$

will be continuous over [a, b], and at points where f(x) is continuous, $\Phi'(x) = f(x)$.

Thus, Φ(x) is an example of an antiderivative function f(x); any antiderivative function has the form

$$F(x) = \Phi(x) + c$$

or

$$F(x) = \int_a^x f(x)dx + C.$$

Having defined the constant c as

$$c = F(a) - \int_a^a f(x)dx = F(a),$$

yields

$$F(x) = F(a) + \int_a^x f(x)dx,$$

and finally, the main formula of integral calculus, the *Newton-Leibnitz formula*

$$\int_a^b f(x)dx = F(b) - F(a).$$

Where the integrand depends on several variables, the integrals which depend on the parameter are considered. These integrals are of the form

$$\phi(x) = \int_a^b f(x, y)dy.$$

Subject to some limitations on f, such integrals can be differentiated and integrated "under the integral", i.e.,

$$\phi'(x) = \int_a^b \frac{\partial f(x, y)}{\partial x}dy, \qquad \int_c^d \phi(x)dx = \int_a^b\int_c^d f(x, y)dx\,dy.$$

In cases of integrals with variable limits of integration a(x) and b(x), the differentiation formula has the form

$$\phi'(x) = \int_{a(x)}^{b(x)} \frac{\partial f(x, y)}{\partial x}dy + b'(x)f(x, b(x)) - a'(x)f(x, a(x)).$$

The notion of the definite integral is generalized to the case of an infinite interval of integration and also to some classes of unbounded functions, thus leading to improper integrals. Therefore, if the function f(x) is continuous on the interval [a, ∞] and there exists a limit

$$\lim_{b\to\infty}\int_a^b f(x)dx,$$

then it is called the integral with an infinite upper limit and is denoted by

$$\int_a^\infty f(x)dx.$$

If, in the interval [a, b], the function f(x) has one point of discontinuity C in which it is unbounded (otherwise, the definite integral exists in the previous sense), then the improper integral means

$$\lim_{\epsilon\to 0}\int_a^{C-\epsilon} f(x)dx + \lim_{\epsilon\to 0}\int_{C+\epsilon}^b f(x)dx,$$

if these limits exist.

The notion of integrals extends from the functions of a complex variable to vector-functions of many real variables (multiple integrals).

Note the integral relations for the functions of many variables used in the mechanics of continuous medium:

Green's formula:

$$\iint_D \left(\frac{\partial Q}{\partial x} - \frac{\partial P}{\partial y}\right)dx\,dy = \int_L (P\,dx + Q\,dy),$$

where L is the boundary of a plane area, and integration with respect to L is performed in a direction such that the area D is on the left;

Gauss'-Ostrogradsky's formula:

$$\iiint_V \left(\frac{\partial P}{\partial x} + \frac{\partial Q}{\partial y} + \frac{\partial R}{\partial z}\right)dx\,dy\,dz$$

$$= \iint_S (P\,dy\,dz + Q\,dz\,dx + R\,dx\,dy),$$

where the integral on the right-hand part is taken over the external side of surface S of volume V;

Stokes formula:

$$\iint_S \left\{\left(\frac{\partial Q}{\partial x} - \frac{\partial P}{\partial y}\right) dx\, dy + \left(\frac{\partial R}{\partial y} - \frac{\partial Q}{\partial z}\right) dy\, dz + \left(\frac{\partial P}{\partial z} - \frac{\partial R}{\partial x}\right) dz\, dx\right\} = \int_L (P\, dx + Q\, dy + R\, dz),$$

which is the generalization of Green's formula for the case of an arbitrary surface A, bounded by a closed contour L.

Green's, Gauss'-Ostrogradsky's and Stokes formulas are widely employed in formulating problems and theorems of hydrodynamics and convective heat and mass-transfer, and also in approximating the corresponding transfer equations in a controlled volume for numerical solution.

Stieltjes and *Lebesque integrals* are further generalizations of the notion of integrals.

The Stieltjes integral of a function f(x), continuous on the interval [a, b], with respect to a bounded monotonic function u(x) is defined as the following sum:

$$\sum_{i=0}^{n-1} f(\xi_i)[u(x_{i+1}) - u(x_i)],$$

where ξ_i are arbitrary points placed at intervals $[x_i, x_{i+1}]$ in an arbitrary partitioning of the interval [a, b], for $\Delta x_i = x_{i+1} - x_i \rightarrow 0$. This limit is denoted as:

$$\int_a^b f(x)du(x).$$

The Stieltjes integral also exists when the function u(x) can be represented in the form of a sum of two bounded monotonic functions.

$$u(x) = u_1(x) + u_2(x).$$

If the function u(x) has a derivative u′(x) bounded in the range [a, b] and integrable in the ordinary sense (in the sense of Riemann), then Stieltjes's integral is reduced to Riemann's integral

$$\int_a^b f(x)du(x) = \int_a^b f(x)u'(x)dx.$$

In order to determine Lebesque's integral of some function f(x) over the interval [a, b], the range of values of a variable y = f(x) is divided into small intervals (as distinct from Riemann's integral where the field x is divided)

$$\cdots < y_{-2} < y_{-1} < y_0 < y_1 < y_2 < \cdots$$

and an integral sum of the form

$$S = \sum_i \eta_i \mu(M_i)$$

is completed, where M_i is the set of points x in the interval [a, b], for which $y_{i-1} \le f(x) < y_i$ and $\mu(M_i)$ are values of the set M_i, $y_{i-1} \le \eta_i \le y_i$. The function f(x) is called integrable in the Lebesque sense if the series converges absolutely for $\Delta y_i = y_i - y_{i-1} \rightarrow 0$.

The limit of these sums is called the Lebesque integral and is denoted by $\int_E f(x)\, dx$.

In heat and mass transfer theory dealing with sufficiently smooth functions, the Lebesque integral is practically never used except for some statistical models of poorly determined phenomena, such as turbulent fluctuations of the parameters of a flow.

Leading to: *Bessel functions; Trapezoidal rule*

I.G. Zaltsman

INTEGRATION BY PARTS (see Integrals)

INTEGRODIFFERENTIAL EQUATIONS

An integro-differential equation is a mathematical expression which contains derivatives of the required function and its integral transforms. Such equations are typical for those processes where a quantity of interest (a required function) in each point is not unambiguously determined by its value near the point—as on processes described by differential equations—but also depends on the function distribution all over the domain. For instance, a radiation energy intensity along the ray $\vec{l}$ in the emitting, absorbing and scattering medium ($I\vec{l}$) is described by an integro-differential equation of radiation transfer, having the form

$$\frac{\partial I}{\partial l} = -K_l I + K_a I_p + \frac{K_s}{4\pi} \oint_{4\pi} I(\vec{l}')\Phi(\vec{l}, \vec{l}')d\Omega, \qquad (1)$$

where K_a, K_s are the absorbtion and scattering coefficients; $K_e = K_a + K_s$ is the attenuation coefficient; and Φ is the scattering phase function. The physical meaning of Equation 1 is that variation of intensity at a given point along a given direction depends on both the local processes of absorption, scattering (first term in write-hand side) the *self-emittance* of the medium (second term) and a process of radiation scattering converging to the given point from a whole volume. If only one independent variable is involved in an integro-differential equation, the latter is called an ordinary integro-differential equation. If an integro-differential equation includes derivatives of more than one independent variable, it is called a partial integro-differential equation.

The linear integro-differential equation is a relationship LD(y) + LI(y) = f, where LD is a linear differentiation operator; LI is a linear integration operator such that $LD(y_1 + y_2) = LD(y_1) + LD(y_2)$; $LI(y_1 + y_2) = LI(y_1) + LI(y_2)$; and f is an arbitrary function of the independent variables. A linear integro-differential equation with partial derivatives of the second order has a differential type of operator. For instance, the equation

$$\alpha \frac{\partial y}{\partial t} + (1 - \alpha)\frac{\partial^2 y}{\partial t^2} = \beta\left[\Delta y + \int_0^t y(x, y, z, \tau)\psi(t, \tau)d\tau\right] \qquad (2)$$

belongs to the hyperbolic type if $\alpha = 0$, $\beta = 1$; to the elliptic if $\alpha = 0$, $\beta = -1$; and to the parabolic type if $\alpha = 1$, $\beta \neq 0$. An integro-differential equation is singular if the kernel of integral transform tends to infinity at one or several points of the function domain. An example is the Prandtl equation for air circulation around the plane wing

$$\Gamma(z) + L(z)\int_{-1}^{1}\frac{\Gamma'(\xi)}{z-\xi}\,d\xi = f(z), \qquad (3)$$

which is also common for other fields of applied mathematics. Another example is one of the forms of heat conduction equation, which takes into account a finite speed of heat transport in the medium

$$\rho c_p\frac{\partial T}{\partial t} = \frac{\partial}{\partial x}\left(k\frac{\partial T}{\partial x}\right) - \int_0^t \exp\left(-\frac{t-t'}{\tau}\right)\frac{\partial^2}{\partial t'\partial x}\left(\frac{\partial T}{\partial x}\right)dt', \qquad (4)$$

where τ is the relaxation time. Analytical solutions of integro-differential equation can sometimes be achieved by reducing them to differential or integral equations and applying sufficiently worked out methods to solve the latter. In some cases, a method of separation of variables can be useful for reducing a partial integro-differential equation to an ordinary one (for instance, Equation 2). Semianalytical solutions of ordinary linear integro-differential equations containing an integral Volterra operator with a difference kernel can be obtained by the **Laplace Transform** method. Nowadays, numerical methods for solution of integro-differential equations are widely employed which are similar to those used for differential equations.

Z.G. Zaltsman

INTENSIFICATION OF HEAT TRANSFER (see Augmentation of heat transfer, single phase)

INTENSITY OF RADIATIVE ENERGY TRANSPORT (see Radiative heat transfer)

INTER-DIFFUSION COEFFICIENT (see Fick's law of diffusion)

INTERFACE HEAT TRANSFER COEFFICIENT (see Condensation of pure vapour)

INTERFACE MASS TRANSFER COEFFICIENT (see Condensation of pure vapour)

INTERFACE TEMPERATURE DROP (see Condensation of pure vapour; Condensation, overview)

INTERFACES

Gas-Liquid and Liquid-Liquid Interfaces

Intermolecular forces existing between molecules are responsible for the condensation of vapours to form the liquid state. Liquid surfaces tend to contract so that the surface to volume ratio is a minimum. The *surface* is in a state of *tension* which may be readily explained in terms of these forces [Fowkes (1962) and Wu (1969)]. The molecules in the bulk experience forces from neighbouring molecules which are, on average, equal in all directions. For a molecule in or near the surface, these forces will not balance and the molecule will experience a pull towards the bulk. Many molecules will leave the surface, which is consequently more sparsely occupied than the internal layers so that the average spacing is slightly greater than the minimum, giving rise to the contraction.

An alternative way of looking at it is that in order to bring a molecule from the bulk to the surface work must be done against the attraction. Hence, there is an excess **Free Energy** associated with the surface. It is usual to define the excess surface free energy as the work which must be done on the surface, isothermally and reversibly, to expand the area by unit amount. The same considerations hold for interfaces formed between *immiscible liquids*. Fowkes (1962) has developed a theory for interfacial tension in terms of the contributions of the polar and dispersion interactions. Many methods, both static and dynamic, are available for measuring surface and interfacial tensions. (See **Surface and Interfacial Tension**).

As a consequence of the tendency of liquid surfaces to contract, an excess pressure must exist in the interior of a bubble or a drop. If r_1 and r_2 are the principal radii of curvature of an interface x, y and it is expanded parallel to itself by a displacement dz, then the work done against the surface tension is $\sigma d(xy)$. Where σ is the excess free-surface energy. Also, if the pressure difference is $p = p - p_o$, then the work done by the pressure is p(xy)dz. For equilibrium,

$$p(xy)dz = \sigma d(xy). \qquad (1)$$

$$\text{Since } dx = (x/r_1)dz \text{ and } dy = (y/r_2)\,dz, \qquad (2)$$

$$p = \sigma\left(\frac{1}{r_1} + \frac{1}{r_2}\right) = \frac{2\sigma}{r} \qquad (3)$$

for a spherical drop and $4\sigma/r$ for a spherical bubble. This expression is known as the *Young-Laplace equation*.

Another consequence of the curvature of an interface is that outside vapour pressure is higher than over the corresponding flat surface. Energy must be expended to transfer liquid from a planar surface to a droplet since the free energy of the droplet will increase

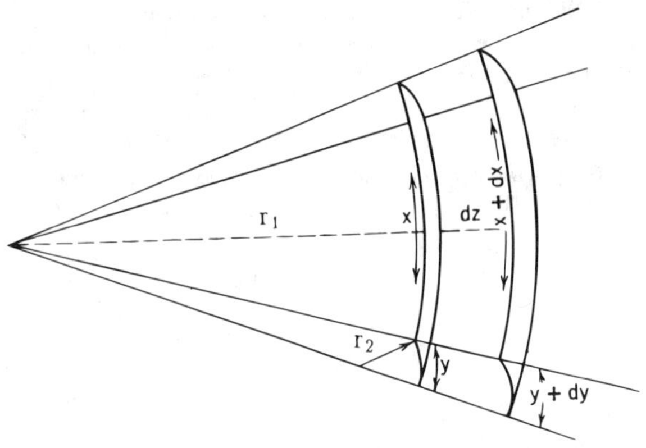

Figure 1.

as its surface area increases. Let the radius of the droplet be r and it is increased to r + dr by the transfer of dn moles of liquid from a plane liquid surface, whose vapour pressure is p_o, to the droplet where the vapour pressure is p_r. The surface free energy of the droplet will increase by $8\pi\sigma r dr$. Assuming the vapour behaves as a perfect gas, this free-energy change is also equal to dn RT ln (p_r/p_o). Hence,

$$dn\ RT\ \ln(P_r/p_o) = 8\pi\sigma r dr \quad \textbf{(4)}$$

where

$$dn = 4\pi r^2 dr(\rho/M) \quad \textbf{(5)}$$

then

$$RT\ \ln(p_r/p_o) = \frac{2\sigma M}{\rho r} = \frac{2\sigma V_m}{r} \quad \textbf{(6)}$$

where ρ is the density of the liquid, V_m is the molar volume of the liquid and M is the molar mass.

The above expression is known as the *Kelvin equation*. It explains many important effects which occur when small volumes of materials are involved. Examples of these phenomena are:

1. **Droplets in a vapour atmosphere**. Below a critical size, the difference in vapour pressure outside the droplet compared with a flat surface means that small droplets will evaporate. As the droplet decreases in size, the equilibrium vapour pressure increases further until the drop vanishes. Thus in a population of droplets of different radii surrounded by vapour, the large droplets grow at the expense of the smaller ones.
2. **Bubbles in a liquid**. In this case, the vapour pressure inside small bubbles is greater than that inside large ones. This explains the phenomenon of superheating. A liquid can only boil if vapour pressure is sufficiently large for bubbles to grow. If preexisting bubbles of suitably large radius are present (nuclei), the liquid can vapourize into the bubbles which then grow and initiate boiling. If no nuclei exist the temperature of the liquid must be increased until equilibrium vapour pressure in the smallest bubbles is great enough to allow them to grow.
   In general, superheating does not occur since bubbles arising from dissolved air or the presence of dust particles act as nuclei.
3. **Solid particles in a vapour atmosphere**. Small solid particles may be treated as spheres having an equivalent radius. Hence the vapour pressure over small crystals is greater than over large ones, and in the atmosphere of this vapour small crystals sublime and larger crystals grow. This is an example of *Ostwald ripening*. Small crystals also melt at temperatures lower than the bulk melting point.
4. **Crystals in saturated solution**. An equivalent expression can be written for the solubility of solid crystals in solution.

$$RT\ \ln\left(\frac{C_r}{C_\infty}\right) = \frac{2\sigma V}{r} \quad \textbf{(7)}$$

   where C_r and C_∞ are the solubilities of the crystals in the bulk, respectively. Dissolution and redeposition of material will take place and large crystals will grow at the expense of smaller ones.
5. **Capillary condensation**. This phenomenon is also described by the Kelvin equation. If a narrow capillary is completely wet by a liquid, i.e., the contact angle formed by the liquid on the solid capillary wall is zero, vapour will condense in the capillary at vapour pressure lower than that at which it would condense on the free liquid surface. The liquid in the capillary presents a concave interface where the vapour pressure is lower than over a flat surface so that the capillary continues to fill. The pores in a porous solid may be treated as capillaries having an effective value of the radius.

Work of Adhesion Between Liquids

Intermolecular forces acting within a liquid give rise to its surface tension and, in a similar way, the interactions across a liquid-liquid boundary result in the observed *interfacial tension*. Harkins (1952) has introduced the concept of work of cohesion and adhesion. If a column of liquid A of unit cross-sectional area is divided by a plane as illustrated in **Figure 2**, two new surfaces are created and work is done W_{AA}^{coh} against the forces of cohesion.

$$W_{AA}^{coh} = 2\sigma_{AV} \quad \textbf{(8)}$$

Similarly if liquid A is in contact with liquid B, an interfacial

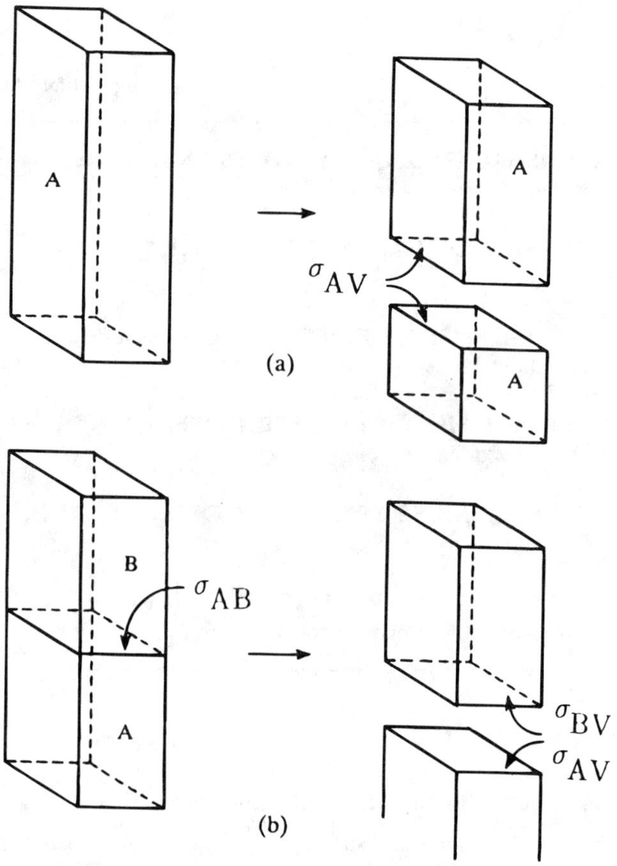

Figure 2.

tension σ_{AB} exists between them. If the two liquids are separated an amount of work W_{AB}^{adh} must be done and two surfaces having free energies σ_{AV} and σ_{BV} result.

Hence,

$$W_{AB}^{adh} = \sigma_{AV} + \sigma_{BV} - \sigma_{AB} \quad \text{Dupré equation.} \qquad (9)$$

Some values of the work of adhesion for some liquids against water are given in **Table 1**. It can be seen that values of interfacial tension usually lie between those of the individual surface tensions. However, the last three values depart from the rule. This is because these molecules contain a hydrophilic moiety which is attracted to the water and causes the molecules in the interface to reorient. The interfacial tension is thus reduced and there is a correspondingly high value of the work of adhesion.

Table 1. Some values of works of adhesion and cohesion

Liquid-air interface	Work of cohesion	Liquid-liquid interface	Work of adhesion
Octane	44	Octane-water	44
Octyl alcohol	55	Heptane-water	42
Heptanoic acid	57	Octyl alcohol-water	92
Heptane	40	Octylene-water	73
		Heptanoic acid-water	95

Spreading of One Liquid On Another

When two immiscible liquids such as oil and water are placed in contact, drops are formed which float on the surface as shown in **Figure 3**. For equilibrium,

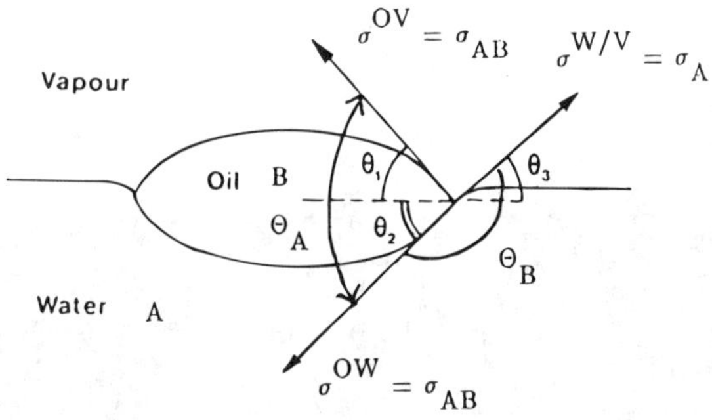

Figure 3.

$$\frac{\sigma_A}{\sin \Theta_A} = \frac{\sigma_B}{\sin \Theta_B} = \frac{\sigma_{AB}}{\sin(2\pi - \Theta_A - \Theta_B)}$$

As σ_A increases, Θ_A decreases and becomes equal to zero when

$$\sigma_A \geq \sigma_B + \sigma_{AB}$$

Under these conditions, the drop spreads spontaneously.

Table 2. Spreading coefficients at 20°C of liquids

Liquid, B	$S_{B/A}$	Liquid, B	$S_{B/A}$
On Water			
Isoamyl alcohol	44.0	Nitrobenzene	3.8
n-Octyl alcohol	35.7	Hexane	3.4
Heptaldehyde	32.2	Heptane (30°C)	0.2
Oleic acid	24.6	Ethylene dibromide	−3.2
Ethyl nonylate	20.9	o-Monobromotoluene	−3.3
p-Cymene	10.1	Carbon disulfide	−8.2
Benzene	8.8	Iodobenzene	−8.7
Toluene	6.8	Bromoform	−9.6
Isopentane	9.4	Methylene iodide	−26.5
On Mercury			
Ethyl iodide	135	Benzene	99
Oleic acid	122	Hexane	79
Carbon disulfide	108	Acetone	60
n-Octyl alcohol	102	Water	−3

In terms of the work of adhesion, this relation becomes

$$\sigma_A \geq 2\sigma_B + \sigma_A - W_{AB}.$$

The spreading coefficient, S, is defined by:

$$S = W_{AB} - 2\sigma_B$$

$$= \sigma_A - \sigma_B - \sigma_{AB}$$

$$\geq 0 \text{ for spreading to occur.}$$

The spreading coefficient may vary with time as the media become mutually saturated. Some values of spreading coefficients are given in **Table 2** and **Table 3** on page 616.

References

Fowkes, F. M. (1962) *J. Phys. Chem.* 66: 382.

Harkins, W. D. (1952) *The Physical Chemistry of Surface Films*. Reinhold. New York.

Wu, S. (1969) *J. Coll. Int. Sci.* 31: 153.

Leading to: *Marangoni effect; Surface and interfacial tension; Capillary action; Waves on interfaces*

A.I. Bailey

INTERFACIAL AREA (see Bubble flow)

INTERFACIAL FLOWS (see Marangoni effect)

Table 3. Initial versus final spreading coefficients

On Water

Liquid, B	σ_B	$\sigma_{B(A)}$	$\sigma_{A(B)}$	σ_{AB}	$S_{B/A}$	$S_{B(A)/A(B)}$	$S_{A/B}$	$S_{A(B)/B(A)}$
Isoamyl alcohol	23.7	23.6	25.9	5	44	−2.7	−54	−1.3
Benzene	28.9	28.8	62.2	35	8.9	−1.6	−78.9	−68.4
CS_2	32.4	31.8		48.4	−7	−9.9	−89	
n-Heptyl alcohol	27.5			7.7	40	−5.9	−56	
CH_2I_2	50.7			41.5	−27	−24	−73	

On Mercury

Liquid, B	σ_B	$\sigma_{B(A)}$	$\sigma_{A(B)}$	σ_{AB}	$S_{B/A}$	$S_{B(A)/A(B)}$	$S_{A/B}$	$S_{A(B)/B(A)}$
Water	72.8	(72.8)	448	415	−3	−40	−817	−790
Benzene	28.8	(28.8)	393	357	99	7	−813	−721
n-Octane	21.8	(21.8)	400	378	85	0	−841	−756

INTERFACIAL FRICTION FACTOR (see Wavy flow)

INTERFACIAL JUMP CONDITIONS (see Conservation equations, two-phase)

INTERFACIAL MOMENTUM TRANSFER (see Stratified gas-liquid flow)

INTERFACIAL RESISTANCE (see Thermal contact resistance)

INTERFACIAL SHEAR STRESS (see Stratified gas-liquid flow)

INTERFACIAL TENSION (see Interfaces; Marangoni effect)

INTERFERENCE (see Interferometry)

INTERFERENCE TECHNIQUES (see Photographic techniques)

INTERFEROMETRY

Interference

The effect produced by the superposition of two or more waves is called interference. Interference appears in the case of temporally and spatially coherent waves. Therefore, waves which are able to interfere have to be generated by a monochromatic, point light source (for example, lasers). Generation by means of separate light sources is not possible i.e., the light waves have to be generated by splitting up a single beam into two separate light waves. After passing along different optical paths, their superposition causes an interference pattern.

Originated by a single light source, waves have a limited connected length known as the coherent length. The coherent length is directly connected to the monochromatic quality of the light. As a result of the beam splitting, the light is subject to a path change. However, to obtain interference the path change must be smaller than the coherent length. Below, a description of Dual Beam Interferometry is provided.

Each of the two waves is precisely described by their amplitude and phase distribution. Both waves have different amplitudes and a constant temporal phase shift. To record the phase shift the phenomenon of interference is used. As a result of the phase shift a visible and recordable brightness and darkness distribution is created. Recorded in this way, the difference of the phase angle corresponds to a local difference of the optical paths. Due to the dependency of the interference effect on the ratio of path difference $\Delta(n \cdot d)$ to wave length λ, it is possible to use interferometry for determining path d (measurement of length), refraction n (refractometry) and wave length λ (spectroscopy), or to examine the modifications of these parameters.

In the field of heat and mass transfer, interferometry is mainly used for the determination of refraction fields which—due to their physical dependency—can later be converted into density, temperature or concentration fields [Ladenburg (1954)].

Basics of the Dual Beam Interferometry

The whole information about a phase influencing object (for example, a thermal boundary layer) is given by the distortion of an originally plane wavefront. Its structure can be made visible by one of the interference methods. Phase differences in a distorted wavefront relating to a reference wavefront cause changes in the intensity of light whereby they become visible. Two of the most important techniques used to generate interference are described below.

Normal dual beam interference (Mach-Zehnder). Figure 1 shows the interference pattern of natural, convective air cooling of printed circuit boards (PCBs) in a closed casing. An unstable thermal boundary layer exists at the cold top-wall of the casing, which interacts with the air flow rising up from the channels

Figure 1. Holographic interferogram of natural convective air cooling in a closed casing with five heated plates [Mayinger (1994)].

between the PCBs. It produces vortex flow and temperature oscillation. Interference lines are generated if the optical paths n · l of the reference and object beams differ in multiples S of the wave length λ. Maximum intensity occur when S is an integer whereas minimum intensity is caused by path lengths which are odd multiples of $\frac{\lambda}{2}$. From this, the refraction differences Δn along a test section l can be determined:

$$S \cdot \lambda = l \cdot \Delta n \quad (1)$$

In the example illustrated in **Figure 1** the refraction index within the test section was influenced only by changes in temperature. As first approximation the interference lines can be interpreted as a representing important differences in temperature, and with constant reference conditions as isotherms.

$$S \cdot \lambda = l \cdot \frac{dn}{dT} \Delta T \quad (2)$$

Differential interference. During this process instead of a plane reference wavefront, the same distorted wavefront is used; however, with a lateral shift. In the differential interferometer, measurement and reference waves pass the measurement field with a laterally-shifted distance y and interfere afterwards. This shifting y is known in terms of its amount and direction. The interference lines provide a scale for the gradient of the distorted wavefront. As in the case of normal interference, the phase difference is proportional to the temperature difference so that ΔT/Δy can be measured directly. Heat transfer experiments in particular require knowledge of the temperature gradient dT/dy, and differential interferometry has proved to be a suitable method [Merzkirch (1987)]. In general however, the accuracy of this measurement method is lower than that obtained with normal dual beam interference.

Types of Dual Beam Interferometer

In an interferometer the measuring and reference beams are generated by splitting one beam of a single light source, nowadays usually a laser. A classification can be made according to the *beam splitting method* used. Some types are listed below.

a. **Amplitude splitting**. Classical mirror-interferometers use semipermeable layers (mirrors) as beam-splitting elements. Usual measuring beam diameters (0.1 to 0.3 m.) require large and expensive interferometer mirrors with high surface quality.

Examples are the **Mach-Zehnder Interferometer,** which is suitable for transparent objects. It consists of two beam-splitting and two reflecting mirrors in a rectangle or parallelogram arrangement (**Figure 2a**). Furthermore, it is suitable for long models (4m) and investigation of temperature fields.

The *Jamin Interferometer* consists of two sloped mirrors whereby each mirror takes over the functions of beam splitting as well as beam reunification (**Figure 2b**). The shift between the measuring and reference beam is small. It can be used as a differential interferometer, but is also suitable for measuring refractive indices of gases.

The best known mirror-interferometer is the *Michelson Interferometer.* It is mainly used for measuring lengths and examining surfaces (**Figure 2c**). Furthermore, it is not suitable for measuring refractive indices of gases and liquids.

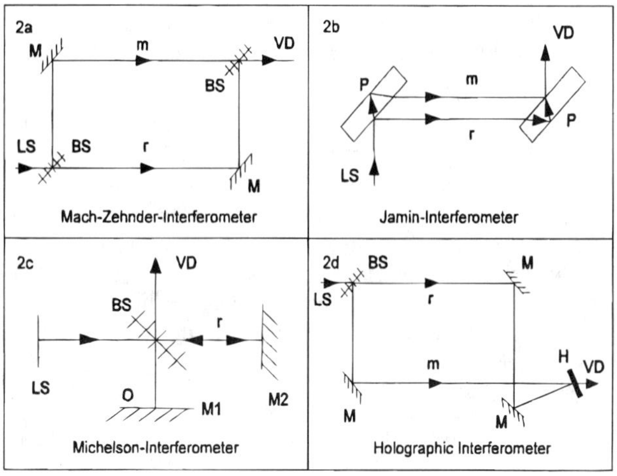

Figure 2. Different types of dual-beam interferometers. LS = light source, M = mirror, BS = beam splitter, VD = view direction, H = hologram, P = plane plate, O = object, m = measuring beam, r = reference beam.

b. **Splitting of the wavefront**. Lord Rayleigh developed a procedure to determine the refractive index by means of a path-shift compensating arrangement. This procedure has found widespread use in the field of measuring concentrations of two- and three-component mixtures of gases and liquids.

c. **Interference-holography**. The use of lasers with large coherent lengths as light sources in optical systems opened a wide field of application for holography. The possibility of storing the phase distribution of a light wave and reconstructing it later by means of holography led to several new interference techniques. An object wave can be stored on a hologram, later on released by illumination with the reference beam and superimposed on the present object wave. Thus, two waves passing the measuring object at different times interfere with each other so that the difference of comparison and measuring wave can be measured directly. In the field of transillumination interferometry, the Mach-Zehnder- and the Michelson-Interferometer have been displaced by the Holographic Interferometer (**Figure 2d**) which offers the advantage of largely eliminating imperfections in the optical components. Not only is it much cheaper to use, but it is also much easier and convenient to handle.

Ideal Interferometry

The characteristics of a real implementation of interferometry are discussed below by means of an idealized dual beam interferometer and an idealized phase object. In general, the real qualities are obtained by adding correction terms. In the case of an ideal interferometry the following assumptions, applicable to all types of interferometer,[2] are made.

- Strictly monochromatic light source: The light source is point shaped and is the origin of undamped spherical wavefronts, which are transformed into plane parallel wavefronts by the use of an ideal lens.
- No aberration: The beam paths of the interferometer contain no optical elements at all—particularly the phase influencing object included—which can cause a deflection of the beams

(aberration). The beams remain parallel during their passage through the object and no distortion is caused by the projection process.

- Interferometer equations in the case of ideal interferometry: In the interference pattern the difference $S_i \cdot \lambda$ of the optical paths at each point $P(x_i, y_i)$ of the beam cross-section is recorded. The modification of the optical path length of the measuring beam $n_m(x_i, y_i) \cdot l$ through the test section, compared to the optical path length $n_r(x_i, y_i) \cdot l$ of the corresponding reference beam (obtained by two-dimensional measuring methods, integrated along the path through the model), is:

$$S_i(x_i, y_i) \cdot \lambda = (n_r(x_i, y_i) - n_m(x_i, y_i)) \cdot l = \Delta n(x_i, y_i) \cdot l \quad (3)$$

The locations of a constant phase difference S_i in a two-dimensional object are, for example, points with a constant temperature difference $\Delta T_i = T(x_i, y_i) - T_{ref}$, if the gradient of the refractive index dn/dT is assumed to be constant. Consequently, an interference pattern can be considered as a field of isotherms and the interpretation requires at least the temperature T_{ref} of one point of the cross-section.

Interferograms, Test Media

Interferometry is frequently used in heat and mass transfer. **Figure 3** shows the interferogram of a tube bundle, where the change from laminar to turbulent flow occurs. Fluctuations in the boundary layer can be clearly seen.

The refractive index field can be determined from the interference pattern by Equation 3. The relation between refractive index n and density ρ is given by the *Lorentz-Lorenz equation,* which can be simplified for gases ($n \approx 1$) to the *Gladstone-Dale equation*:

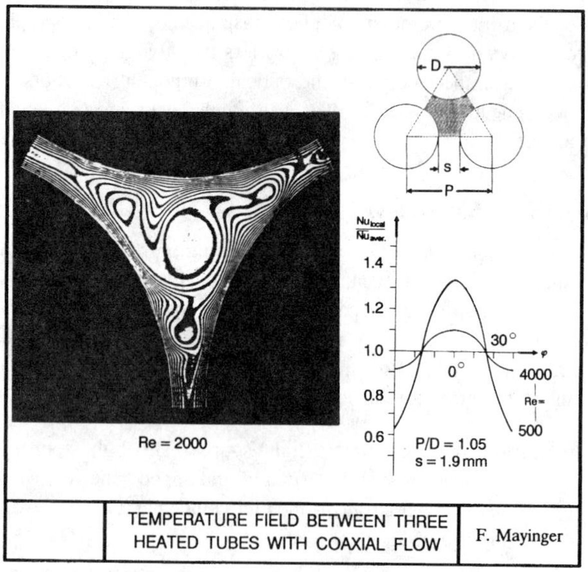

Figure 3. Temperature field between three heated tubes with coaxial flow [Mayinger (1974)].

Table 1. Refractive indices of gases (20°C, 760 Torr) and liquids (25°C)

	$\lambda_{Laser} = 0{,}6328 \cdot 10^{-6}$m	
	n [−]	dn/dT [1/K]
air	1,0002716	$0{,}927 \cdot 10^{-6}$
carbon dioxide	1,0004174	$1{,}424 \cdot 10^{-6}$
steam	1,0002337	$0{,}798 \cdot 10^{-6}$
water	1,3314	$0{,}985 \cdot 10^{-4}$
methyl alcohol	1,3253	$4{,}0 \cdot 10^{-4}$

$$N(\lambda) = \frac{2}{3}(n(\lambda) - 1) \cdot \frac{M}{\rho} \quad (4)$$

Here, N is the molecular refractivity (property constant) and M is the molecular weight. Including the equation of state for ideal gases, **Equations 3 and 4** result in:

$$T(x, y) = \left[S(x, y) \cdot \frac{2\lambda R}{3Npl} + \frac{1}{T_{ref}} \right]^{-1}. \quad (5)$$

If the temperature T_{ref} is known at any location, a specific temperature could be assigned to each interference line. If air is used as test fluid the temperature difference between two lines is only a few degrees. In the case of liquids, a linear dependency of the refractive index on temperature is usually assumed. So **Equation 3** can be transformed into:

$$S(x, y) \cdot \lambda = l \cdot \frac{dn}{dT} \cdot [T(x, y) - T_{ref}] \quad (6)$$

Refractive indices of gases and liquids are listed in **Table 1**. The sensitivity of interferometric methods for water is 100 times greater than for air.

With **Equation 7**, the heat transfer coefficient α is calculated from the temperature gradient at the wall $(dT/dy)_w$, determined from the temperature profile T(y) and the temperature difference $T_w - T_\infty$:

$$\alpha = -\frac{\lambda_{fluid} \cdot \left(\frac{dT}{dy}\right)_w}{T_w - T_\infty} \quad (7)$$

The interference lines in **Figure 3** are isotherms with a temperature difference of 2·3 K. The fringe density at the wall is proportional to the temperature gradient and hence, to the heat transfer coefficient. Therefore, the local **Nusselt Number** distribution can be estimated directly from the interference pattern. The evaluation of the interferogram in **Figure 3** is also presented as the ratio of local and average Nusselt number for different **Reynolds Numbers**.

References

Ladenburg, R. (1954) *Interferometry, in Physical Measurements in Gas Dynamics and Combustion.* A3. Princeton Univ. Press. Princeton, New Jersey.

Mayinger, F. and Panknin, W. (1974) *Holography in Heat and Mass Transfer.* 5th Int. Heat Transfer Conference. Tokyo.

Mayinger, F. and Wang, Z. (1994) Experiments on natural convective air cooling of a PCBs array in a closed casing with inclination; Kakaç, S., Yüncḧu, H. and Hijikata, K. (1994) *Proc. of the NATO Adv. Study Inst. on Cooling of Electr. Systems*. Kluwer Academic Publishers.

Merzkirch, W. (1987) *Flow Visualization*. Academic Press. 2nd edition. Orlando.

Tolansky, S. (1955) *An Introduction of Interferometry.* McGraw-Hill. New York.

R. Fehle and F. Mayinger

INTERMITTENT FLOW (see Gas-liquid flow)

INTERMITTENCY (see Boundary layer heat transfer)

INTERMOLECULAR FORCES (see Kinetic theory of gases)

INTERMOLECULAR PAIR POTENTIAL (see Van de Waal's forces)

INTERMOLECULAR POTENTIALS (see Reduced properties)

INTERNAL COILS (see Agitated vessel heat transfer)

INTERNAL COMBUSTION ENGINES

***Following from** Combustion*

The internal combustion (IC) engine has been the dominant prime mover in our society since its invention in the last quarter of the 19th century [for more details see, for example, Heywood, 1988]. Its purpose is to generate mechanical power from the chemical energy contained in the fuel and released through combustion of the fuel inside the engine. It is this specific point, that fuel is burned inside the work-producing part of the engine, that gives IC engines their name and distinguishes them from other types such as external combustion engines. Although **Gas Turbines** satisfy the definition of an IC engine, the term has been traditionally associated with *spark-ignition* (sometimes called *Otto, gasoline or petrol engines*) and *diesel engines* (or *compression-ignition engines*).

Internal combustion engines are used in applications ranging from marine propulsion and power generating sets with capacity exceeding 100 MW to hand-held tools where the power delivered is less than 100 W. This implies that the size and characteristics of today's engines vary widely between large diesels having cylinder bores exceeding 1,000 mm and reciprocating at speeds as low as 100 rpm to small gasoline two-stroke engines with cylinder bores around 20 mm. Within these two extremes lie medium-speed diesel engines, heavy-duty automotive diesels, truck and passenger car engines, aircraft engines, motorcycle engines and small industrial engines. From all these types, the passenger car gasoline and diesel engines have a prominent position since they are, by far, the largest produced engines in the world; as such, their influence on social and economic life is of paramount importance.

The majority of reciprocating internal combustion engines operate on what is known as the *four-stroke cycle* (**Figure 1**), which is subdivided into four processes: intake, compression, expansion/power and exhaust. Each engine cylinder requires four strokes of its piston which corresponds to two crankshaft revolutions to complete the sequence which lead to the production of power.

The intake stroke is initiated by the downward movement of the piston, which draws into the cylinder fresh fuel/air mixture through the port/valve assembly, and ends when the piston reaches bottom-dead-centre (BDC). The mixture is generated either by means of a carburetor (as in conventional engines) or by injection of gasoline at low pressure into the intake port through an electronically-controlled pintle-type injector (as in more advanced engines). Effectively, the induction process starts with the opening of the intake valve just before top-dead-centre (TDC) and ends when the intake valve (or valves in four-valve per-cylinder engines) closes shortly after BDC. The closing time of the intake valve(s) is a function of the design of the induction manifold, which influences the gas dynamics and volumetric efficiency of the engine, and engine speed.

The intake stroke is succeeded by the *compression* stroke which effectively starts at the intake valve closure. Its purpose is to prepare the mixture for combustion by increasing its temperature and pressure. Combustion is initiated by the energy released through the spark plug towards the end of the compression stroke and is associated with a rapid rise in the cylinder pressure.

The *power or expansion* stroke starts with the piston at TDC of compression and ends at BDC. At this point, the high temperature

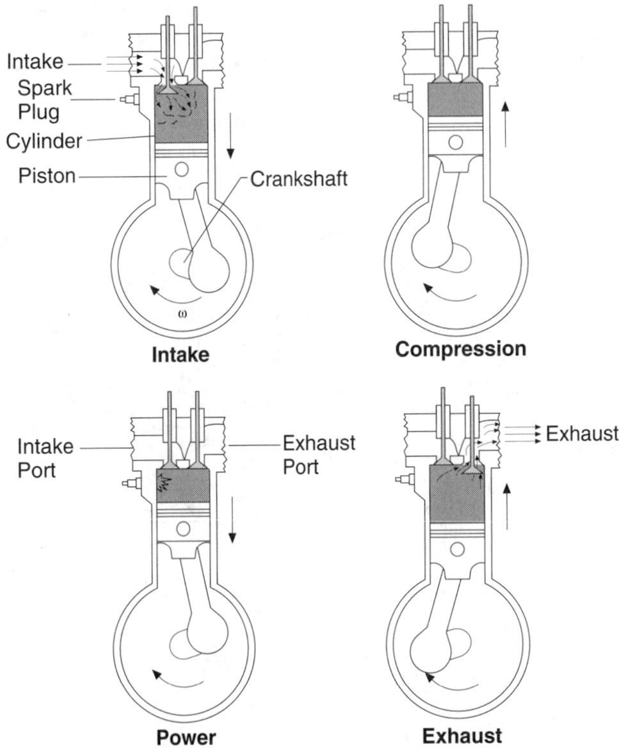

Figure 1. Four-stroke engine cycle.

and pressure gases generated during combustion push the piston down, thus forcing the crank to rotate. Just before the piston reaches BDC, the exhaust valve(s) opens and the burned gases are allowed to exit the cylinder due to the differential pressure between the cylinder and the exhaust manifold.

This *exhaust* stroke completes the engine cycle by evacuating the cylinder from burned, partially-burned or even unburned gases escaping the combustion process; the next engine cycle starts when the intake valve opens near TDC and the exhaust valve closes a few degrees crank angle later.

It is important to note that the properties of gasoline, in association with combustion chamber geometry, exert a significant influence on combustion duration, rate of pressure rise and *pollutant formation*. Under certain conditions, the mixture at the end gas may autoignite before the flame reaches that part of the cylinder, leading to *knock* which gives rise to high-intensity and frequency pressure oscillations.

The tendency of gasoline fuel to resist *autoignition* and thus prevent possible damage to the engine as a result of *knock,* is characterised by its *octane number.* Until recently, the addition of a small quantity of lead into the gasoline was the preferred method for suppressing knock but the associated health risks, combined with the need to use catalysts for reducing exhaust emissions, have necessitated the introduction of unleaded gasoline. This requires a reduction of the engine's compression ratio (ratio of the cylinder volume at BDC to the volume at TDC) in order to prevent knock with undesirable effects on thermal efficiency.

As already mentioned, the four-stroke cycle, also known as Otto cycle after its inventor Nicolaus Otto who built the first engine in 1876, produces a power stroke for every two crankshaft revolutions. One way to increase the power output of a given engine size is to convert it to a two-stroke cycle (**Figure 2**) in which power is produced during every engine revolution.

Because this mode of operation gives rise to increased power output—albeit not to the double levels expected from simple calculations—it has been extensively used in motorcycle, passenger car and marine applications with both spark-ignition and diesel engines. An additional advantage is the simple design of two-stroke engines since they can operate with side ports in the liner, covered and uncovered by piston motion, instead of the bulky and complicated overhead cam arrangement.

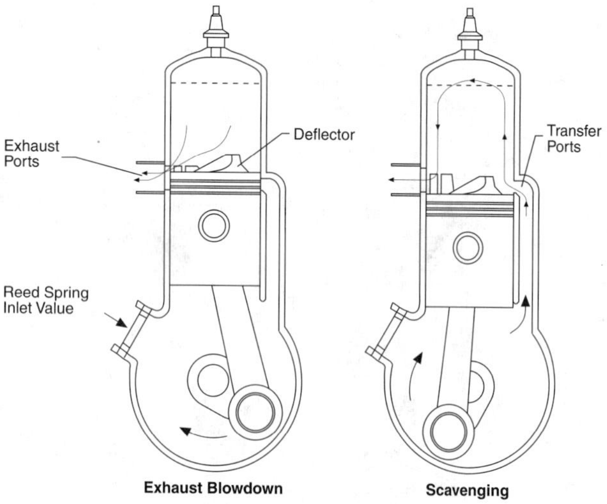

Figure 2. Two-stroke engine cycle.

In the two-stroke cycle, the *compression* stroke starts after the inlet and exhaust side ports are covered by the piston; the fuel/air mixture is compressed and then ignited by a spark-plug, similar to ignition in a four-stroke gasoline engine, to initiate combustion near TDC. At the same time, fresh charge is allowed to enter the crankcase before its subsequent compression by the downward-moving piston during the *power or expansion* stroke. During this period, burned gases push the piston until it reaches BDC, which allows first the exhaust ports and then the intake (transfer) ports to be uncovered. The opening of the exhaust ports permits the burned gases to exit the cylinder while partly at the same time the fresh charge, which has been compressed in the crankcase, enters the cylinder through the properly orientated transfer ports.

The overlapping of the induction and exhaust strokes in two-stroke cycle engines is responsible for some of the fresh charge flowing directly out of the cylinder during the scavenging process. Despite various attempts to reduce the magnitude of this problem by introducing a deflector into the piston (**Figure 2**) and directing the incoming charge away from the location of the exhaust ports, charging efficiency in conventional two-stroke engines remains relatively low. A solution to this problem is to introduce the fuel directly into the cylinder, separately from the fresh air, through air-assisted injectors during the period when both the exhaust and transfer ports are closed. Despite the short period available for mixing, air-assisted atomisers can achieve a homogeneous lean mixture at the time of ignition by generating gasoline droplets of less than 40 μm mean diameter, which vapourize very easily during the compression stroke.

Amongst the various types of internal combustion engines, the diesel or compression-ignition engine is renowned for its high efficiency, reduced fuel consumption and relatively low total gaseous emissions. Its name comes from the German engineer Rudolf Diesel (1858–1913) who in 1892 described in his patent a form of internal combustion engine which does not require an external source of ignition and where combustion is initiated by the autoignition of the liquid fuel injected into the high temperature and pressure air towards the end of the compression stroke.

The inherent efficiency advantages of the diesel engine stem from its lean overall mixture ratios, the high engine compression ratios afforded due to the absence of end-gas ignition (knock) and the greater expansion ratios. As a consequence, diesel engines in either the two-stroke or four-stroke configuration have been traditionally the preferred power plants for commercial applications such as ships/boats, energy-generating sets, locomotives and trucks and, over the last 20 years or so, passenger cars as well especially in Europe.

The low power-output disadvantage of diesel engines has been circumvented by the use of superchargers or turbochargers which increase the power/weight ratio of an engine through an increase of the inlet air density. Turbochargers are expected to become standard components of all future diesel engines, irrespective of application.

The operation of the diesel engine differs from that of the spark-ignition engine mainly in the way the mixture is formed prior to combustion. Only air is inducted into the engine through a helical or directed port and the fuel mixes with air during the compression stroke, following its injection at high pressure into a prechamber (indirect-injection diesel or IDI) or into the main chamber (direct-injection diesel or DI) just before combustion is started.

The need to achieve good fuel/air mixing in diesel engines is satisfied by high-pressure fuel injection systems which generate droplets of about 40 μm mean diameter. For passenger cars, the fuel injection systems consist of a rotary pump, delivery pipes and fuel injector nozzles which vary in their design according to the application; direct-injection diesel engines use hole-type nozzles while indirect-injection diesels employ pintle-type nozzles. Larger diesel engines use in-line fuel-injection pumps, unit injectors (pump and nozzle combined in one unit) or individual single-barrel pumps which are mounted close to each cylinder.

Over the last 20 years or so, the realisation that the resources of crude oil are finite and that the environment we live in is becoming more and more polluted, has urged governments to introduce laws which limit the *exhaust emission levels* of vehicles and engines of all types. Since their introduction in Japan and the USA in the late 60s and in Europe in 1970, emission regulations are consistently becoming more stringent and engine manufacturers are facing their toughest ever challenge with the standards agreed for 1996 onwards, which are summarised for passenger cars in **Table 1**. It is expected that the new standards to be imposed in Europe for the year 2000 will be even lower, following Californian levels which require zero emission levels after the turn of the century. However, it is uncertain whether existing engines will satisfy these limits despite the desperate attempts by engineers worldwide.

It is clear from **Table 1** that the major pollutants in spark-ignition engines are hydrocarbons (HC), carbon monoxide (CO) and oxides of nitrogen ($NO_x = NO + NO_2$) while in diesel engines, NO_x and particulates—which consist of soot particles formed during combustion of lubricating oil and hydrocarbons—are the most harmful.

At present three-way catalysts, which are a standard component of today's passenger cars equipped with spark-ignition engine running on unleaded gasoline, allow about 90% reduction of the emitted HC, CO and NO_x by converting them into carbon dioxide (CO_2), water (H_20) and N_2.

Table 1. European Emission Standards for 1996

	grams per kilometer		
Engine Type	CO	HC + NO_x	Particulates
GASOLINE ENGINES			
Certification Limits	2.2	0.5	—
Deterioration Factors (DF)	1.2	1.2	—
Certification Limits with DF Applied	1.833	0.416	—
IDI DIESEL ENGINES			
Certification Limits	1.0	0.7	0.08
Deterioration Factors (DF)	1.1	1.0	1.2
Certification Limit with DF Applied	0,909	0,700	0.066
DI DIESEL ENGINES			
Certification Limits	1.0	0.9	0.10
Deterioration Factors (DF)	1.1	1.0	1.2
Certification Limit with DF Applied	0.909	0.900	0.083

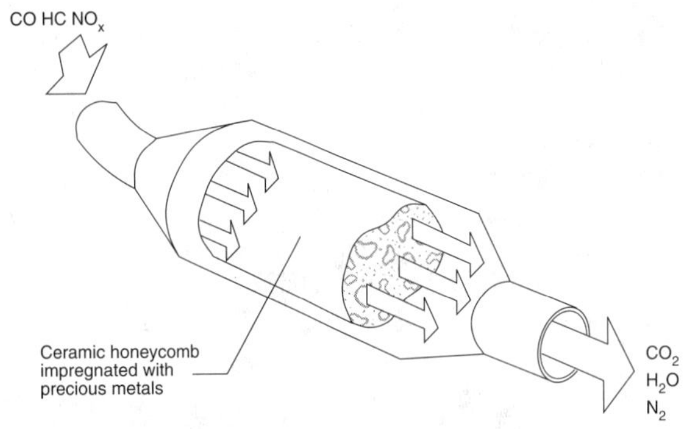

Figure 3. Model of a three-way catalytic converter.

Unfortunately these catalysts require stoichiometric (air-fuel ratio of ~14.5) engine operation, which is undesirable from both the fuel consumption and CO_2 emissions points of view. An alternative approach is the lean burn concept which offers promise for simultaneous reduction of fuel consumption and exhaust emissions through satisfactory combustion of lean mixtures with much higher than 20 air-fuel ratios. It is expected that the development of lean burn catalysts with conversion efficiencies over 60% may allow lean burn engines to satisfy future emissions legislation; this is an area of active research in both industry and academia. On the other hand, new diesel engines depend on two-way or oxidizing catalysts for reduction of exhaust particulates through conversion of HC into CO_2 and H_2O, and on exhaust gas recirculation and retarded injection timing for reduction of NO_x levels.

References

Arcoumanis, C. (ed.) (1988) *Internal Combustion Engines*. Academic Press.

Blair, G. P. (1990) *The Basic Design of Two-stroke Engines*. Society of Automotive Engineers.

Ferguson, C. R. (1986) *Internal Combustion Engines*. John Wiley & Sons.

Heywood, J. B. (1988) *Internal Combustion Engine Fundamentals*. McGraw Hill.

Stone, R. (1992) *Introduction to Internal Combustion Engines*. Macmillan Education Ltd. 2nd Edition.

Weaving, J. H. (ed.) (1990) *Internal Combustion Engineering: Science & Technology*. Elsevier Applied Science.

Leading to: *Combustion chamber; Octane number; Jet engines*

C. Arcoumanis

INTERNAL ENERGY

Following from: *Thermodynamics*

The *First Law of Thermodynamics* contains an explicit statement about the amount by which the internal energy U of a gas changes when work W or heat Q is received or given up by the system. It must be emphasized that contrary to Q and W, U is a state variable i.e, its value depends only on the state of the system and not on how this state was attained.

The *Gay-Lussac experiment* consists of two containers connected by a pipe and valve (**Figure 1**). Container 1 is filled with an ideal gas, container 2 is completely evacuated. The system is perfectly insulated from its surroundings.

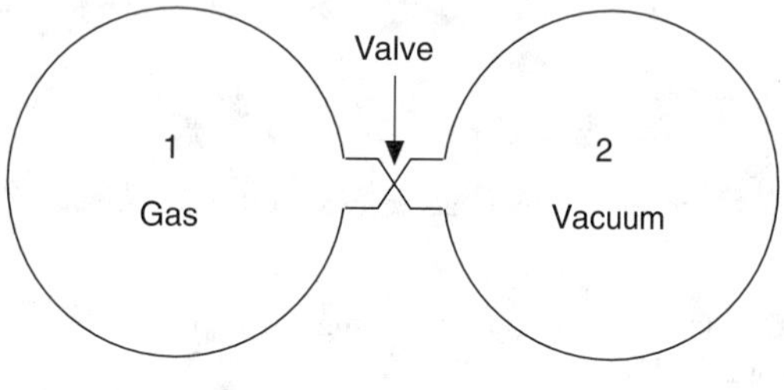

Figure 1.

The valve is opened and the gas confined in 1 expands into vacuum 2. Pressure and volume change while the temperature remains constant. Since no work or heat are exchanged with the surrounding, the internal energy will not change during this process. Thus, the internal energy of an ideal gas is only a function of its temperature.

$$dU = \left(\frac{\partial U}{\partial V}\right)_T dV + \left(\frac{\partial U}{\partial T}\right)_V dT = 0 + \left(\frac{\partial U}{\partial T}\right)_V dT \quad (1)$$

where V is volume and T, temperature. With

$$\frac{dU}{dT} = \left(\frac{\partial U}{\partial T}\right)_V = c_v, \quad (2)$$

the change in internal energy is obtained from

$$\Delta U = \int_{T_1}^{T_2} c_v dT. \quad (3)$$

Since for an ideal gas U is a function only of temperature, it follows from **Equation 2** that the specific heat capacity c_v for an ideal gas is independent of pressure and volume. Values of c_v are often expressed as polynomials in T, [see, for example, Reid, Prausnitz and Sherwwod].

For nonideal fluids, the following equation for pressure dependence of the internal energy can be derived from the fundamental relationship between enthalpy and internal energy:

$$\left(\frac{\partial U}{\partial p}\right)_T = \left(\frac{\partial H}{\partial p}\right)_T - p\left(\frac{\partial V}{\partial p}\right)_T - V \quad (4)$$

or

$$\left(\frac{\partial U}{\partial p}\right)_T = (\kappa p - \beta T)V \quad (5)$$

where κ is the isothermal compressibility and β is the volume expansivity. Equations 4 and 5 are usually only applied to liquids, with κ and β being very small if the fluid can be treated as incompressible, i.e, if it is not near the critical point.

Reference

Reid, R., Prausnitz, J., and Sherwood, T. (1977) *The Properties of Gases and Liquids*. McGraw-Hill Book Company.

H. Müller-Steinhagen

INTERNAL REBOILERS (see Reboilers)

INTERNATIONAL ATOMIC ENERGY AGENCY, IAEA

Wagramer Str 5
PF 100, 1400 Vienna
Austria
Tel: 1 206 00

INTERNATIONAL ENERGY AGENCY, IEA

2 rue Andre Pascal
75775 Paris cedex 16
France
Tel: 1 45 24 82 00

INTERNATIONAL FLAME RESEARCH FOUNDATION, IFRF

PO Box 10 000
1970 CA Ijmuiden
The Netherlands
Tel: 2514 93064

INTERNATIONAL TEMPERATURE SCALE

***Following from:** Temperature; SI Units; Thermodynamics*

The International temperature scale (ITS) is a combination of established numerical values of reference points, interpolation dependencies and techniques which ensure the unity of temperature measurements. It is developed on the basis of generalizations made by the consulting committee on thermometry and is affirmed by the General Conference on Weights and Measures.

The first ITS-27 approved by the 7th session of the General Conference on Weights and Measures in 1927 covered the sphere of determination from the normal boiling point of oxygen (−182.97°C) up to the gold solidifying point (1,063°C) and contained six reference points. Suggested as interpolation instruments between the reference points were the platinum resistance thermometer and the platinum-rhodium thermocouple (660°C − 1,063°C). For determining the temperature above the gold solidifying point, an optical thermometer and Wien's formula were recommended.

The international practical temperature scale (**IPTS**) of 1948 was the result of refinements in the values of reference points of **ITS-27**. In IPTS-68, "Kelvin" (K) was first defined as a unit of temperature. This scale can be used for the low temperature range down to the triple point of hydrogen of 13.81 K (−253.34°C). The number of main reference points was increased to 13. The estimated error of values of reference points was 0.01 K for the low-temperature range and reached 0.2 K at the gold solidifying point.

In 1976, a Preliminary Temperature Scale (EPT-76) was accepted for use which ensured unity of measurements in the temperature range of 0.5−30 K. ITS-76 coincided with IPTS-68 at the neon boiling point (~ 27 K).

ITS-90 was placed in service beginning Jan. 1, 1990. The unit of thermodynamic temperature T_{90}—kelvin (K)—was defined in ITS-90 as 1/273.16 part of the thermodynamic temperature of the triple point of water. The temperature expressed in degree Celsius (°C) is designated as t_{90} and is defined

$$t_{90}(°C) = T_{90}(K) - 273.15.$$

The ITS-90 has the following fields of definition: T_{90} is defined between 0.65 and 5.0 K from the saturated vapour pressure of helium isotopes ^3He and ^4He; T_{90} between 3.0 K and the triple point of neon (24.5561 K) is defined by the helium interpolation gas thermometer, calibrated to three experimentally-realized temperatures which have assigned values; T_{90} between the triple point of equilibrium hydrogen (13.8033 K) and the silver solidifying point (961.78°C) is defined by a platinum resistance thermometer, calibrated to reference points by the interpolation dependence of platinum resistance on the temperature between these points; T_{90} above the silver point is determined from reference points and from Plank's radiation formula.

The ITS-90 is based on 17 reference points, the 9th of which corresponds to the triple point of water where temperature values T_{90} = 273.16 K and t_{90} = 0.01°C. Differences between ITS-90 and IPTS-68 are shown in the **Figure 1**.

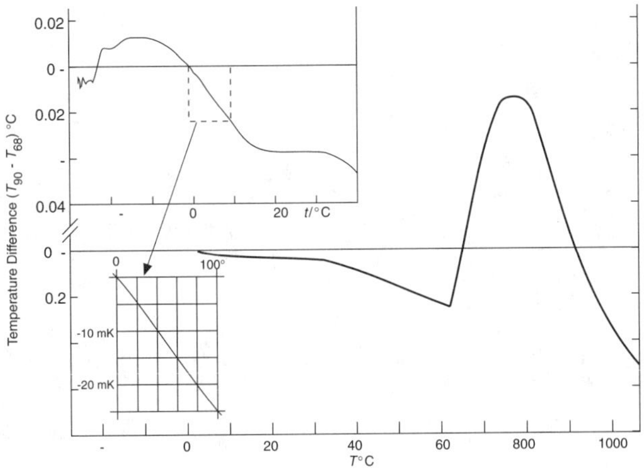

Figure 1. Temperature differences between ITS-90 and IPTS-68.

Ju.A. Dedickov

INUNDATION (see Condensation of pure vapour; Condensation, overview)

INUNDATION, EFFECT ON CONDENSATION (see Tube banks, condensation heat transfer in)

INVERSE ANNULAR FLOW (see Annular flow; Post dryout heat transfer)

INVERSION LAYER, EFFECT ON POLLUTION (see Air pollution)

INVERSION POINT (see Three phase, gas/liquid/liquid, flows)

INVERSION POINT TEMPERATURE (see Drying)

INVISCID FLOW

Following from: *Conservation equations, single-phase; Flow of fluids*

Inviscid flow is a schematic representation of the motion of mobile media (gaseous or liquid, and also solids under the rapid action of high pressures). It is the main theoretical model for many fields of modern technology. Calculation results obtained within the framework of this model are widely used in designing flying vehicles, rockets and their engines, turbines and compressors; in analyzing the motion of shells in the bore of a gun and their trajectories in the atmosphere; in designing the combustion and detonation of fuels and explosives; in determining the action of explosive waves on barriers; in describing high-speed collisions of solids; and in many other cases.

Studies of inviscid flows are carried out in gas dynamics, acoustics, electro- and magneto-gas dynamics, the dynamics of rarefied gases, plasma dynamics, etc. In the theory of rarefied gases and plasma a statistical description of the behaviour of fluid particles comprising the medium is used. In other cases, the flow may be considered to be within the scope of the continuous medium model with the use of small, volume-averaged values of mass, momentum and energy.

If the viscosity and thermal conductivity of a fluid are ignored, its velocity $\vec{v}$, pressure P and density ρ (at points in space where they are continuous) must be related by the following equations:

$$\rho \frac{d\vec{v}}{dt} = \rho\vec{F} - \text{grad } p,$$

$$\frac{1}{\rho}\frac{d\rho}{dt} = -\text{div } \vec{v},$$

$$\rho \frac{d}{dt}\left(U + \frac{v^2}{2}\right) = \rho\vec{F}\vec{v} - \text{div } p\vec{v} + \rho q.$$

The first equation, *Euler's equation*, relates the acceleration of a fluid particle (i.e., an elemental volume of fluid) to an external body force $\vec{F}$ and the pressure force applied on the side of the neighboring fluid particles. This equation is the generalization of Newton's second law (the law of conservation of momentum) as applied to the motion of fluid particles.

The second equation express of the law of mass conservation: the rate of change of density of a fluid particle is equal, with the sign reversed, to the rate of change of volume. The third equation is the law of energy conservation: a change in the internal and kinetic energy of a fluid particle is due to an action of impressed mass forces $\vec{F}$ and surface forces (pressure p), and to an inflow of heat with intensity q from an external source.

A strong heat release inside a moving gas or an inflow of heat from the outside can cause considerable changes in density. The compressibility of a medium is the principal feature of many problems which can be solved with the help the inviscid flow model. This, in turn, calls for a wider application of thermodynamics (see **Thermodynamics, Irreversible Thermodynamics, Multicomponent Systems, Thermodynamics of**) and the inclusion of the equation of medium state into the system of defining equations (see **Equations of State, PVT Relationships**).

In a number of problems, the equations for an inviscid flow must be supplemented with kinetic equations to derive the rates of internal physico-chemical transformations, chemical reactions between the components of a mixture, dissociation or ionization, and also for the excitation of internal degrees of freedom, etc.

The parameters of gas can undergo sudden changes on some surfaces inside the flow area. There are two types of discontinuity surfaces. If no flow passes through a discontinuity surface, then this is the tangential discontinuity surface (for instance, in parallel flow of two gases which differ in density, temperature and in velocity). A typical example of a discontinuity surface of the second type is a shock wave (see **Waves in Fluids**). In this case the flow of fluid particles intersects the discontinuity surface, and the equations for finite volume of gas are written in integral form. By geometrical criteria, problems solved with the help of an inviscid flow model are subdivided into internal and external problems. Internal flows refer to flows in nozzles, in blades of gas turbines, in aerodynamic wind tunnels, etc. (see **Nozzles, Turbines, Wind Tunnels**).

If a streamlined body is immersed in a liquid whose velocity at infinity is described by a uniform profile, then such a problem refers to a class of external inviscid flows (see **External Flows, Overview**).

The most important fluids, air and water, have significant (though rather low) coefficients of viscosity. But even for them, the inviscid flow model gives a good description of real processes at some distance from bounding solid surface. A more rigorous statement of the problem necessitates the use of the Navier-Stokes equations for fluid motion which, unlike Euler's equations, allow for internal friction or tangential stresses between neighboring layers and also for the molar thermal conductivity of the medium. Although, with the Reynolds number Re tending to infinity, the Navier-Stokes equations become at the limit the Euler ones, a drop in the order of the differential equations (from the fourth to the second order) occurs. Therefore, the solution of simplified equations will not satisfy the boundary conditions of a complete system of equations. It can also be supposed that in the limiting case of Re $\rightarrow \infty$, the solutions for the Navier-Stokes equations contain the assumption that the entire field of flow can be divided into two areas: a thin layer area (see **Boundary Layers**) which surrounds the body and propagates behind it in the form of a narrow wake; and an external flow area separated from the body by the said layer. It is a known fact that in a number of cases such a division is wrong; for instance, in the case of separation of the boundary layer.

The purpose of solving all the above-mentioned problems with an inviscid flow model is to determine the impact of the moving fluid on a streamlined body. Sometimes, the total field of gas-dynamic parameters (velocity, density, temperature) must be known in the entire flow or in some specified sections (at the nozzle cut, at the outer edge of the boundary layer, etc.).

Solutions to a system of equations of an inviscid flow are usually obtained with some simplification. The first simplification method reduces the number of independent variables. Cases of similitude and *self-similarity* of flows are of some interest. For instance, in describing strong explosions in half-space and also in the propagation of flame fronts or detonation caused by cylindrical (plane) sources, the dimensionless values of the gas parameters behind the shock wave can be described by a system of ordinary differential equations in one self-similar variable: $\xi = E_0 t^2/(\rho_1 r^{2+\nu})$ (where E_0 is the energy of explosion released immediately at the moment t = 0; ρ_1 is the initial density of the gas medium; r is the distance from the explosition site; $\nu = 3$ for spherical, $\nu = 2$ for cylindrical and $\nu = 1$ for plane shock waves).

Among stationary inviscid flows, the self-similarity model is typified by a homogeneous supersonic flow moving around an infinite thin cone (**Figure 1**). It is impossible to distinguish a typical linear dimension in this problem. Extending or compressing the flow field around the cone vertex an arbitrary number of times does not change the picture—i.e., it remains similar to itself. In an axially-symmetric flow around the cone, fluid flow is retarded first in a *conical shock wave* (OS) trailing at the cone vertex, and then in a compressed layer of gas present between a shock wave and the body; the flow is turned to a direction corresponding to the cone surface (the streamline "a").

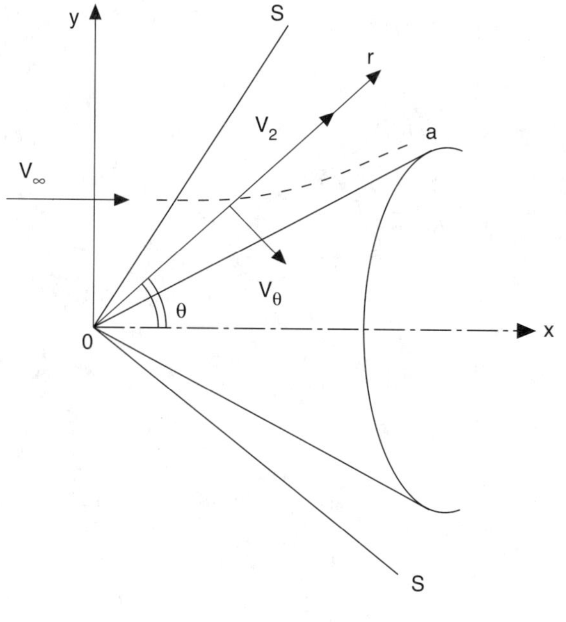

Figure 1. Homogeneous supersonic flow around an infinite thin cone.

If a shift is made from the Cartesian system x, y to a polar coordinate system r, θ and if we denote by v_r and v_θ the components of the velocity vector, then the only self-similar coordinate will be the polar angle θ. The equations of an inviscid flow are reduced in this case to a system of two ordinary differential equations:

$$\frac{\gamma - 1}{2}(2v_r + v_\theta \cot\theta + v'_\theta)[1 - (v_r^2 + v_\theta^2)$$

$$- v_\theta(v_r v'_r + v_\theta v'_\theta)] = 0, \quad v_\theta = v'_r,$$

where $\gamma = c_p/c_v$ is the ratio of specific heat capacities of the gas, and the prime means differentiation with respect to θ.

Self-similarity, in the broad sense, sometimes means the independence of dimensionless parameters characterizing the flow from similarity criteria (see **External Flows, Overview and Drag Coefficient**).

The most important results of theoretical analysis of inviscid flows have been obtained for stationary and irrotational flows. The acceleration of a fluid particle, entered into the Euler equation, can be written as follows:

$$\frac{d\vec{v}}{dt} = \frac{\partial\vec{v}}{\partial t} + \text{grad}\,\frac{\vec{v}}{2} + 2\vec{\omega} \times \vec{v},$$

where $\vec{\omega} = 1/2 \text{ curl } \vec{v}$ is the velocity vector curl. For a steady-state flow ($\partial\vec{v}/\partial t = 0$), when $\vec{\omega} = 0$ or $\vec{\omega} \parallel \vec{v}$, the Euler equation allows a general integral. Since the mass forces have a potential in a majority of cases—i.e., can be written as $\vec{F} = -\text{grad}\,\Phi$ where Φ is the mass force potential or the force function—the general integral of Euler's equation takes the form known as *Bernoulli's integral*:

$$\frac{v^2}{2} + \int \frac{dp}{\rho} + \Phi = \text{const.}$$

For a heavy incompressible fluid $\Phi = gZ$, and $\rho = \text{const}$,

$$gZ = \frac{p}{\rho} + \frac{v^2}{2} = \text{const},$$

where Z is the vertical coordinate.

For weightless, incompressible fluid,

$$p + \frac{\rho v^2}{2} = p_{00},$$

where p_{00} is the pressure of the stagnated flow (or impact pressure). The second term on the left-hand part of the equation is called velocity or dynamic head.

For the isothermal motion of compressible gas $p/\rho = \text{const}$, Bernoulli's integral is written as follows:

$$\frac{v^2}{2} + \frac{p_{00}}{\rho_{00}} \ln \frac{p}{\rho} = 0.$$

Finally, the equation of state of the adiabatic process $p/\rho^\gamma = \text{const}$ allows Bernoulli's integral to be written as:

$$v^2 = \frac{2\gamma}{\gamma - 1}\frac{p_{00}}{\rho_{00}}\left[1 - \left(\frac{p}{p_{00}}\right)^{\gamma/(\gamma-1)}\right]$$

(notation of this kind was suggested by De Saint Venant et Wantzel in 1839).

The idea of irrotational motion seems quite fruitful not only because Bernoulli's integral exists in this case but it also makes it possible to obtain a simple relation between the modulus of the fluid flow velocity vector and the pressure. Also, an important point here is that three unknown quantities—the projections of speed onto the coordinate axes, V_x, V_y and V_z—can can be expressed in term of one unknown function, the velocity potential φ(x, y, z).

The vector equation $\vec{\omega} = 1/2\ \textbf{curl}\ \vec{v} = 0$ in projections onto the coordinate axes is equivalent to the fact that there exists a certain function φ(x, y, z) whose partial derivatives with respect to x, y, and z are equivalent to the appropriate velocity components

$$V_x = \frac{\partial\varphi}{\partial x}, \qquad V_y = \frac{\partial\varphi}{\partial y}, \qquad V_z = \frac{\partial\varphi}{\partial z}.$$

If the mass force $\vec{F} \equiv 0$ is neglected, then a single scalar equation for velocity potential, instead of the Euler vector equation, can be written:

$$\left[1 - \frac{V_x^2}{a^2}\right]\frac{\partial^2\varphi}{\partial x^2} + \left[1 - \frac{V_y^2}{a^2}\right]\frac{\partial^2\varphi}{\partial y^2} + \left[1 - \frac{V_z^2}{a^2}\right]\frac{\partial^2\varphi}{\partial z^2}$$

$$- 2\frac{V_xV_y}{a^2}\frac{\partial^2\varphi}{\partial x\partial y} - 2\frac{V_xV_z}{a^2}\frac{\partial^2\varphi}{\partial x\partial z} - 2\frac{V_yV_z}{a^2}\frac{\partial^2\varphi}{\partial y\partial z} = 0,$$

where $a^2 = (dp/d\rho)$ is the sound velocity in gas.

Depending on whether the motion is subsonic or supersonic, the differential equation is an elliptic or hyperbolic one. Inviscid flows of an incompressible fluid form a large and important class because with the velocity of flow much lesser than the velocity of sound, the velocity potential equation takes the form of the Laplace linear equation, which is well-studied in mathematics:

$$\frac{\partial^2\varphi}{\partial x^2} + \frac{\partial^2\varphi}{\partial y^2} + \frac{\partial^2\varphi}{\partial z^2} = 0.$$

The set of its solutions forms the class of **Harmonic Functions**.

The solution of two-dimensional (plane) problems has been completed with the use of the theory of functions of complex variables. Any potential plane-parallel flow of inviscid gas can be described by a complex potential $W = \varphi - i\psi$, where φ(x, y) and ψ(x, y) are two harmonic, mutually conjugate functions

$$\frac{\partial\varphi}{\partial x} = \frac{\partial\psi}{\partial y} = V_x, \qquad \frac{\partial\varphi}{\partial y} = \frac{\partial\psi}{\partial x} = V_y.$$

The function ψ(x, y) has a simple hydrodynamic meaning: substituting into the streamline equation (see **Streamline**)

$$\frac{dx}{V_x} = \frac{dy}{V_y},$$

the value of the velocity projections expressed in terms of the function ψ yields

$$\frac{\partial \psi}{\partial x} dx + \frac{\partial \psi}{\partial y} dy = d\psi = 0;$$

hence, it follows that the function $\psi(x, y)$ preserves a constant value along the streamlines ψ = const (see **Stream Function**).

It is easy to verify that a family of isopotential lines $\varphi(x, y)$ = const and a family of streamlines $\psi(x, y)$ = const are mutually orthogonal, i.e., they intersect at a right angle (**Figure 2**).

Each of the two conjugate functions φ and ψ can be interpreted as the velocity potential or the current function. This procedure is usually used in designing the flows.

Introducing a complex potential W allows a decrease in the order of the equation: instead of an unknown function of two variables $\psi(x, y)$ the function of one complex variable W(Z) must be found, where $z = x + iy$ is in Cartesian coordinates and $Z = re^{i\alpha}$ is in polar coordinates.

By way of conclusion, some simple functions of a complex variable Z are considered below and a geometric interpretation of flows they describe is provided. Figure 2 shows a homogeneous flow with velocity V_∞ and inclined to the real axis of a physical plane at an angle α described by a linear function $W = V_\infty Z$. Source and run-off flows, described by logarithmic functions ($A > 0$ for the source and $A < 0$ for the run-off), are depicted in **Figure 3**. Finally, **Figure 4** shows a homogeneous flow with velocity V_∞ past a cylinder with radius a, obtained as the sum of a linear and a reverse function $W = |V_\infty|(Z + a^2/Z)$.

As the last example shows, the superposition principle is of prime importance for potential flows when the flow with the potential $\sum_{i=1}^{n} \varphi_i = \varphi$ is obtained out of several simple flows with potentials φ_i. The velocity vectors of integrable flows at each point are added vectorially.

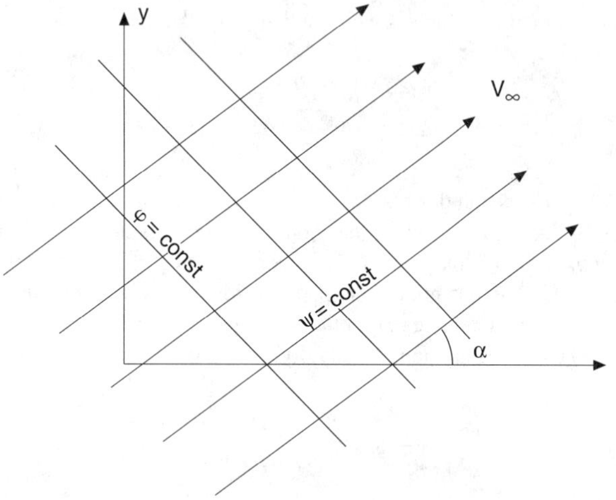

Figure 2. Orthoganality of isopotential lines ($\varphi(x, y)$ = constant) and streamlines ($\psi(x, y)$ = constant)

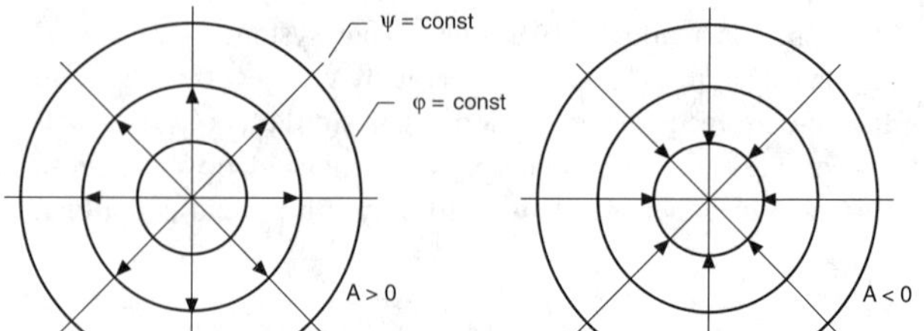

Figure 3. Source and run-off flows

This procedure manifests itself most eloquently for a plane potential flow. If the streamline configurations of two integrable plane flows are known then their superposition on one drawing forms a grid which makes it possible to draw the streamlines of the resultant flow. The flow around an infinite cylinder of radius a with velocity V_∞ (**Figure 4**) is the sum of the plane homogeneous flow parallel to the OX axis with complex potential $W_1 = |V_\infty|Z$, and of the velocity field dipole with a complex potential

$$W_2 = \frac{2\pi a^2 |V_\infty|}{2\pi Z} = \frac{a^2 |V_\infty|}{Z}.$$

The dipole is the limiting set of the source and the run-off when the distance r between them tends to zero, and the power Q increases up to infinity so that the product $m = Q \cdot r$ = const.

The flow picture in **Figure 4** consists of two areas: outside and inside the circle of radius a. The first area $|Z| \geq a$, is the external flow around the cylinder. The second area, $|Z| \leq a$, is the flow formed by the dipole inside the circle of radius a. The velocity modulus on the circle contour varies according to the sine law $|V| = 2|V_\infty| \sin \theta$. The points A ($\theta = \pi$) and B ($\theta = 0$) are called the front and rear critical points, respectively, or the branching points of the flow.

The velocity on the cylinder surface assumes its maximum value at points C and D, which correspond to the mid-section $\theta = \pm \pi/2$. It is interesting to note that the maximum velocity on the cylinder contour exceeds the velocity of the free-stream flow by a factor of 2

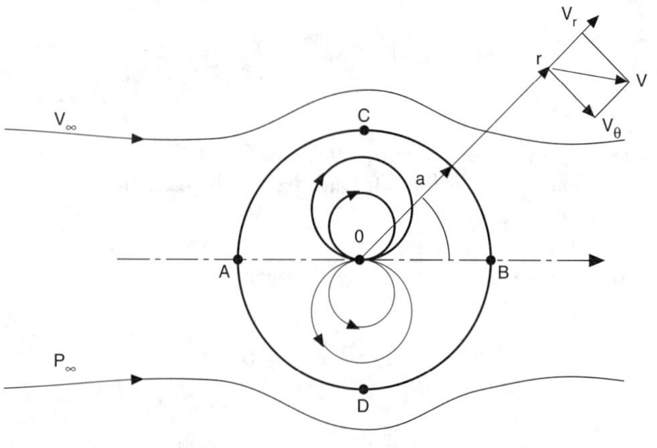

Figure 4. Flow around an infinite cylinder

$$|V_{max}| = 2|V_\infty|.$$

The distribution of static pressure on the cylinder surface can be easily obtained from the known distribution of velocity. For this purpose, Bernoulli's integral can be used and the result can be represented in dimensional form with the help of a pressure coefficient C_p:

$$C_p = \frac{P - P_\infty}{\frac{1}{2}\rho|V_\infty|^2} = 1 - \left(\frac{|V|}{|V_\infty|}\right)^2 = 1 - 4\sin^2\theta.$$

Such a representation of pressure is convenient because C_p is independent either of fluid density ρ, or of pressure or velocity at an infinite distance from the cylinder, p_∞ or V_∞. Thus, a universal form of notation for pressure distribution on a cylinder surface of any size under arbitrary conditions in a free-stream flow is obtained.

Experimentally, the measured pressure distribution starts to vary from the one calculated from inviscid flow theory even before the point of maximum velocity $|V_{max}|$ or minimum pressure p on the surface of the cylinder. The reason for this divergence is the impossibility of smooth (continuous) flow by a real fluid around a cylinder.

The cylinder and the sphere are two examples of poorly-streamlined bodies. After breaking down into two flows at the front critical point, the free-stream flow can merge again into one flow at point B only under certain conditions. Such conditions correspond to very low velocities $|V_\infty|$ of free-stream flow and, to be more precise, to Reynolds numbers less than unity. In the general case fluid particles on both sides of the cylinder or the sphere already in the vicinity of the mid-section tend toward the trajectories parallel to the flow axis. A stagnant zone with reduced pressure relative to p_∞ is formed between them in the wake of the cylinder. The pressure difference at points A and B is the source of the resistance the bodies experience while moving in fluid or gaseous media (see **Drag Coefficient**). The discrepancy between potential flow around the cylinder and the real process provides an explanation of D'Alembert's paradox. For well-streamlined bodies the stagnant zone is either absent completely or is reduced to a minimum, and the calculations of pressure according to an inviscid (potential) flow model agree well with the measurements.

The effect of the flow's three-dimensionality can be analyzed by comparing the distribution of velocity and pressure on the surface of a plane (cylinder) or a three-dimensional, axially-symmetric body, a sphere. Using the superposition principle, the potential for inviscid flow over a sphere is obtained by superposing a homogeneous flow with velocity $|V_\infty|$ onto the flow from a dipole:

$$\varphi = |V_\infty|R\left[1 + \frac{1}{2}\left(\frac{a}{R}\right)^2\right]\cos\theta.$$

Velocity distribution on the surface of a sphere is characterized by equations of the form:

$$V_\theta = -\frac{3}{2}|V_\infty|\sin\theta.$$

The maximum value of velocity is $(V_\theta)_{max} = 3/2\,|V_\infty|$. Comparing the results for the sphere and the cylinder, note that in three-dimensional flow the deformation is much lesser and the maximum acceleration of fluid particles on the cylinder's mid-section is almost 1.35 times higher than that of the sphere.

The distribution of pressure on the surface of the sphere has the form:

$$C_p = \frac{p - p_\infty}{\frac{1}{2}\rho|V_\infty|^2} = 1 - \frac{9}{4}\sin^2\theta.$$

Although the maximum reduction in pressure is not so great as on the cylinder, the break-off and resistance due to nonsymmetry of pressure at the critical points are also characteristics of the sphere.

Examples of flows around a sphere and a cylinder show that even at subsonic flow velocities (Mach numbers $Ma_\infty < 1$), the local values of velocity can be supersonic. The break-up of fluid particles that follows occurs in shock waves (see **Waves in Fluids**) and brings about a sharp increase in resistance, analogous to the so-called sound barrier in aviation.

Methods for solving the equations of motion and continuity equations for supersonic inviscid fluid flows have also been developed. The most widely used is the method of characteristics.

The characteristics of equations for supersonic flow, together with the gas velocity vector, form an angle at the given point equal to the angle of propagation of disturbances

$$\alpha = \pm\arcsin\frac{1}{Ma}.$$

At any point of space with supersonic flow, the velocity vector is directed along the bisectrix of the angle α contained by the characteristics. Projections of gas velocity from the normal to the characteristic at the given point is equal in magnitude to the local velocity of sound a.

Leading to: *Bernoulli equation; Vorticity; Stream function; Normal stress; Streamline*

Yu.V. Polezhaev and I.V. Chirkov

IoE (see Institute of Energy)

ION EXCHANGE

Ion exchange refers to the interaction of ionic species in aqueous solutions with adsorbent solid materials. It is distinguished from conventional **Adsorption** by the nature and morphology of the adsorbent material which in most cases is either a dynamic *polymer* matrix or an inorganic structure containing exchangeable functional groups. All modern ion exchange resins are polymeric structures, generally based either on styrene or an acrylic matrix. *Polystyrene sulphonic acid cation resins* are cross-linked copolymers of styrene (vinylbenzene), with divinylbenzene (DVB-containing pendant sulphonic acid groups capable of exchanging cations in solution over the entire pH range. *Acrylic cation resins* are made by copolymerising acrylic or methacrylic acid with DVB. These *resins* have a high capacity and will exchange cations in the alkaline pH range 7–14.

Polystyrene anion exchange resins are cross-linked copolymers of polystyrene-DVB, with pendant functional groups comprising primary, secondary and tertiary amines. Strong base resins operate over the entire pH range, whereas weak base resins exchange anions in the pH range 0–7. Polystyrene copolymers possess a gel type structure, swell and contract in aqueous solutions and are physically robust in most process operations. *Macroporous or macroreticular resins* have been developed which have a porous, sintered structure possessing free water channels within the matrix of about 1,000 Å diameter. Large molecules can travel freely within these pores and therefore, these materials are less prone to organic fouling. Inorganic materials are either naturally-occurring or synthetic materials. Typical examples are natural and synthetic *zeolites;* insoluble salts, e.g., hydrous zirconium phosphate; heteropolyacids, e.g., ammonium phosphomolybdate; natural clays, e.g., montmorillonite.

Ion Exchange Selectivity

The important properties of ion exchange materials are given below:

1. In order to preserve electroneutrality, ion exchange is stoichiometric and the capacity is independent of the nature of the counterion;
2. Ion exchange is nearly always a reversible process;
3. Ion exchange is a rate-controlled process, usually governed by diffusion in the solid phase or in the surrounding liquid film.

Simple binary cation exchange equilibrium can be expressed by the law of mass action. Assume that the ion exchanger is initially in the B form and the solution contains ions A.

$$z_A \overline{B^{z_B^+}} + z_B A^{z_A^+} \Leftrightarrow z_B \overline{A^{z_A^+}} + z_A B^{z_B^+} \quad (1)$$

Overbars denote the ionic species in the resin phase. The selectivity coefficient, $(K_c)_B^A$, can be defined in terms of the concentrations in the solution and resin phase.

$$(K_c)_B^A = \frac{(\bar{c}_A)^{zB}(c_B)^{zA}}{(\bar{c}_B)^{zA}(c_A)^{zB}} \quad (2)$$

It is usual to express the selectivity coefficient in terms of equivalent or mole fractions of the ionic species.

$$(K_c)_B^A\left(\frac{\bar{c}}{c}\right)^{zA-zB} = \frac{(y_A)^{zB}(x_B)^{zA}}{(y_B)^{zA}(x_A)^{zB}} \quad (3)$$

A typical equilibrium plot of univalent-univalent exchange, i.e., $z_A = z_B = 1$ is shown in **Figure 1**. In the industrially-important case of divalent-univalent exchange, i.e., $z_A = 2$, $z_B = 1$, the selectivity coefficient becomes:

$$(K_c)_B^A\left(\frac{\bar{c}}{c}\right) = \frac{y_A(1 - x_A)^2}{x_A(1 - y_A)^2} \quad (4)$$

The equilibrium curve is given in **Figure 2**. In this case, selectivity

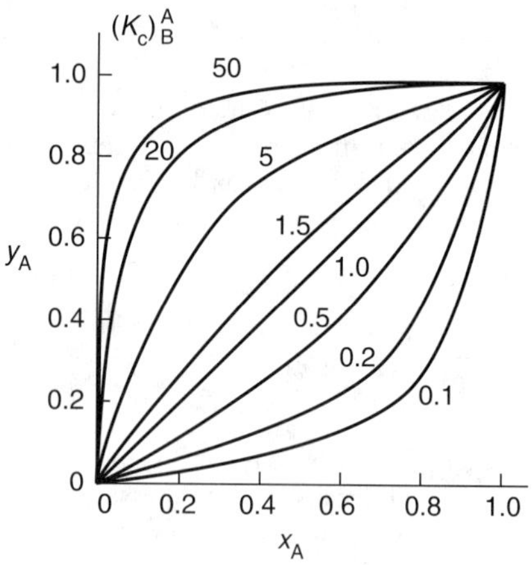

Figure 1. Equilibrium isotherm for uni-univalent exchange. From Rousseau R.W. (1987) *Handbook of Separation Process Technology*, John Wiley & Sons Inc, with permission.

is strongly dependent on the total ionic concentration of the solution phase.

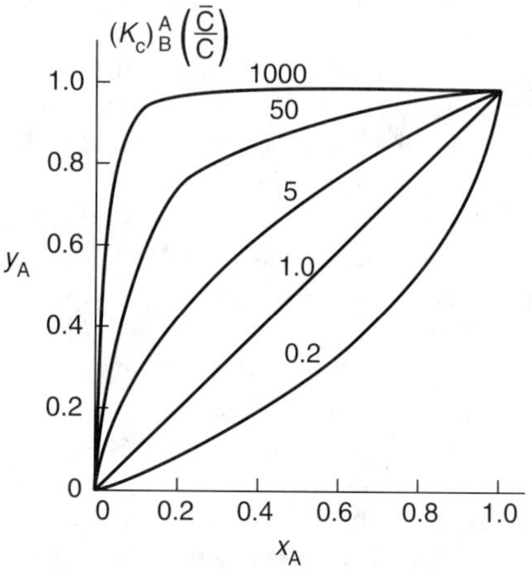

Figure 2. Equilibrium isotherm for di-univalent exchange. From Rousseau R.W. (1987) *Handbook of Separation Process Technology*, John Wiley & Sons Inc, with permission.

Ion Exchange Kinetics

Ion exchange is a rate-controlled process and the kinetics depends on the following five steps:

1. Diffusion of the counterions through the bulk solution to the surface of the ion exchanger;
2. Diffusion of the counterions within the solid phase;
3. Chemical reaction between the counterions and the ion exchanger;
4. Diffusion of the displaced ions out of the ion exchanger;
5. Diffusion of the displaced ions from the exchanger surface into the bulk solution.

The kinetics of ion exchange are governed by either a diffusion or mass action mechanism, depending on which is the slowest step. Diffusion of ions in the external solution is termed *liquid film control,* but is hydrodynamically ill-defined. The diffusion and transport of ions within the ion exchanger is termed *particle diffusion control*. Chemical reaction (step 3) is uncommon but can be rate-controlling in certain specialised cases.

Homogeneous diffusional mass transfer processes can be described by **Fick's Law**:

$$J_i = D \text{ grad } C_i.$$

J_i is the flux if the diffusing species i of concentration C_i and D is the diffusion coefficient. Fundamental treatment of the laws of ion exchange is given by Helfferich (1962). The concentration profiles within a spherical ion exchange bead are depicted in **Figure 3**.

Fractional conversion or attainment of equilibrium can be calculated under infinite solution volume conditions by the solution of Fick's Law for spherical geometry:

$$X = 1 - \frac{6}{\pi^2} \sum_{n=1}^{\infty} \exp\left(-\frac{\overline{D}\pi^2 n^2 t}{r_0^2} \right) \quad \textbf{(5)}$$

The mathematical treatment is quite different if the exchanger is treated as a solid phase and the mechanism is assumed to be a heterogeneous chemical reaction. The conceptual models to describe this case study are similar to those developed for noncatalytic fluid-solid reactions. *Shell progressive, shrinking core or ash-layer models* have been applied to ion exchange, especially if there is a complexing reaction. **Figure 4** gives a schematic representation of a partially-reacted ion exchange bead.

The following relationships have been obtained for each kinetic mechanism:

Liquid film diffusion control

$$t = \frac{a r_o c_{As}}{3 c_{A0} k_{mA}} X \quad \textbf{(6)}$$

Ash-layer diffusion control

$$t = \frac{a r_0^2 c_{As}}{D_e c_{A0}} \left[\frac{1}{2} - \frac{1}{2}(1 - X)^{2/3} - \frac{1}{3} X \right] \quad \textbf{(7)}$$

Chemical reaction control

$$t = \frac{r_0}{c_{A0} k_s} [1 - (1 - X)^{1/3}] \quad \textbf{(8)}$$

Conventional ion exchange reactions can often be explained by homogeneous diffusional kinetics. For example, simple ion exchange reactions encountered in water treatment, such as Na^+-H^+ or Cl^--OH^- with polyelectrolyte gels and macroreticular resins, can be fitted by diffusional theory. However, more complex ion exchange reactions involving complex ions or chelating ion exchange materials are not satisfied by diffusional relationships. For example, the sorption and desorption of copper by an iminodiacetic acid exchanger is better fitted by chemical reaction rate models. Also, sorption into inorganic ion exchangers is better explained by chemical reaction rate models. Slater (1991) has brought together most of the available ion exchange theory and discussed the interpretation and prediction of mass transfer data.

Applications of Ion Exchange

Industrial applications of ion exchange are extremely widespread and range from purification of low-cost commodities, such

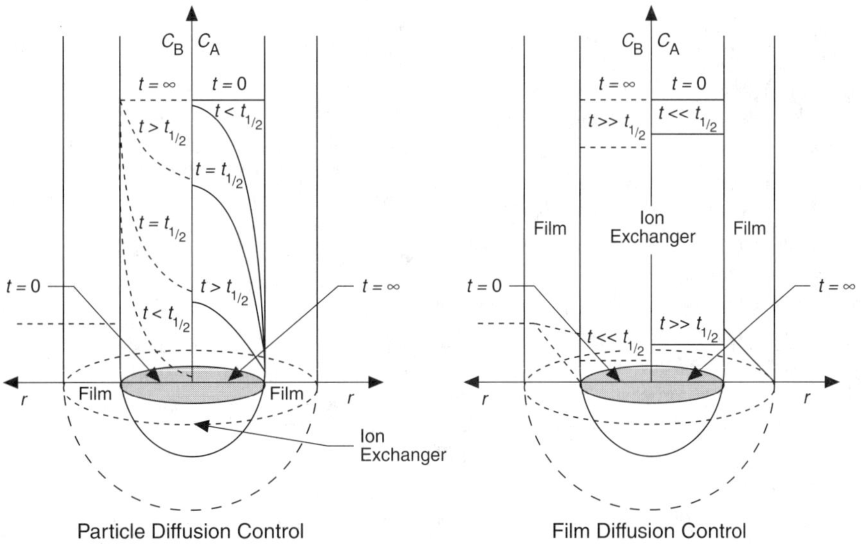

Figure 3. Radial concentration profiles at different times for ideal particle diffusion and film diffusion control. The right hand sides of the diagrams show the profiles of species A (initially in the resin) and the left sides, those of species B (initially in solution). From Helfferich, F (1962) *Ion Exchange*, McGraw Hill Co., with permission.

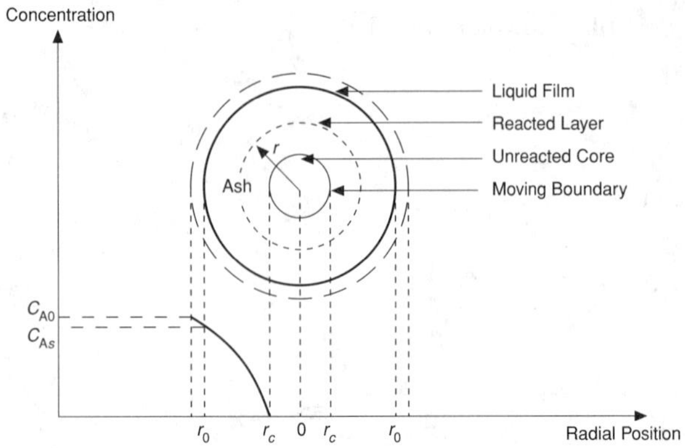

Figure 4. Schematic diagram of a partially-reacted resin bead. From Rousseau, R.W. (1987) *Handbook of Separation Process Technology*, John Wiley & Sons Inc, with permission.

as water, to the purification and treatment of high-cost pharmaceutical products as well as precious metals such as gold and platinum. Dorfner (1991) has reviewed the industrial applications of ion exchange and comprehensive references are to be found in his textbook.

The largest single application, measured in terms of ion exchange resin usage, is water treatment i.e., softening, demineralization for high pressure boilers and dealkalization. (See also **Water Preparation**). Enormous advances in ion exchange technology have occurred because of the relentless requirement for pure and ultrapure water. Other major industrial applications are the processing and decolorization of sugar solutions and the recovery of uranium from relatively low-grade mineral ore leach solutions. Ion exchange is also used in the fields of medicine, pharmaceuticals, chemicals processing, catalysis and laboratory analysis.

Water Treatment

The removal of salts and other ionic impurities from water is based primarily on the exchange of cations (Na^+, K^+, Ca^{2+}, Mg^{2+}, etc.) with the hydrogen form of a cation exchanger and the exchange of anions (Cl^-, HCO_3^-, CO_3^{2-}, SO_4^{2-}, etc.) with the hydroxide form of an anion exchanger. Usually, softening of water is achieved with cation exchangers and demineralization, with a mixed bed of cation/anion exchange resin. Natural groundwaters contain appreciable amounts of long-chain aliphatic acids, e.g., humic and fulvic acid and these tend to foul conventional anion exchange resins. Macroreticular polymeric anion exchange resins overcome this problem and offer relatively easy regeneration. Ultrahigh purity water for the nuclear power, semiconductor and pharmaceutical industries is produced using mixed bed ion exchange processes. Ion exchange is now widely used in *effluent treatment* and *pollution control.* Bolto and Pawlowski (1987) have reviewed these procedures in detail. The process strategy depends entirely on the waste to be treated, concentration of pollutants and flow rates. The treatment of mine drainage water, removal of ammonia, nitrates and pesticides from groundwater and the treatment of nuclear waste solutions are examples of typical applications.

Sugar Processing

The principal applications of ion exchange in the *purification* and treatment of sugar solutions, juices and syrups are:

1. Softening and demineralization of sugar juices to remove scale-forming elements prior to evaporation;
2. Decolorization with anion exchange resins;
3. Catalytic inversion of sucrose to fructose and glucose;
4. Glucose/fructose separation.

Ion exchange is predominantly used in the sugar beet industry since the syrup contains significant amounts of calcium and magnesium, and these can be exchanged with conventional cation exchangers in sodium form. Deionization of sugar syrups using both cation and anion exchange resins also reduces molasses and thus, sugar yield is increased. The inversion of sucrose can be catalyzed by cation exchange resins in the hydrogen form.

Pharmaceutical and Medical Uses

Some important uses of ion exchange are: processing of pharmaceuticals, use in artificial organs and analytical applications in medicine. Antibiotics, such as streptomycin, can be separated from fermentation broths with cation exchange resins under neutral pH conditions. Weak acid carboxylic ion exchange resins are usually employed in sodium form and regenerated with a mineral acid. Similarly, vitamin B_{12} can be recovered from microbial fermentation broths using weak acid ion exchangers. In medicine, ion exchange resins have also found use as preparative media and for various clinical applications.

Hydrometallurgy

Ion exchange has found widespread use in the recovery of metals from mineral leach solutions and from secondary waste solutions. The most important application is the recovery of uranium from low-level ore bodies in the mineral deposits of Australia, Canada, South Africa and the USA. Ion exchange has also been used for the recovery of thorium, rare earth elements, transition metals, transuranic elements, precious metals such as gold, silver and platinum, and base metals such as chromium.

Uranium is recovered from mineral leach solutions as an anionic sulphate or carbonate complex, depending on the nature of the ore body. Uranyl sulphate anions are sorbed preferentially at pH values in the range 0.5–1.5 on conventional strong base anion exchange resins and can be eluted using sulphuric acid (usually heated to improve regeneration). The uranium uptake is extremely favourable and for this reason, high enrichment of uranium is achieved in the eluant. Uranium is normally precipitated from the eluant with alkali. The process was traditionally carried out in a cascade of fixed bed columns operating on a "merry-go-round" principle, but in recent years sophisticated continuous counter-current ion exchange contactors have been developed and installed on mines in South Africa.

Gold can be recovered from low-concentration side-streams with activated carbon. Counter-current adsorption has also been applied to this process, although the regeneration step is more difficult and requires high temperature *elution* and thermal reactivation of the carbon. Recently, much work has been devoted to the development of novel ion exchange resins capable of recovering gold from cyanide liquors and amenable to chemical regeneration using conventional reagents at ambient temperatures.

Ion Exchange Equipment

A typical mixed-bed ion exchange column is shown in **Figure 5**. The inlet and bottom collector systems are major design features. Most processes use two, three or four columns operated in sequence to maintain a continuous flow of treated product solution. There have been great advances in the development of continuous counter-current ion exchange columns based on multi-stage fluidized bed contactors. These have been used on several uranium recovery plants in South Africa and a schematic installation is shown in **Figure 6**. Column diameters of up to 5 m are commonplace. The advantages of continuous counter-current ion exchange are realised for processes having very favourable equilibrium or treating high concentration solutions for the recovery of high value solutes, e.g., uranium and precious metals such as gold and platinum.

SYMBOLS

- a stoichiometric coefficient
- c concentration in solution phase
- $\bar{c}$ concentration in the particle phase
- c_{A0} concentration in bulk solution phase (**Figure 4**)
- c_{As} concentration at liquid/particle interface (**Figure 4**)
- D diffusion coefficient in solution phase
- $\bar{D}$ diffusion coefficient in particle phase
- D_e effective diffusion coefficient
- J mass flux of ions from solution to particle
- $(K_c)_B^A$ selectivity coefficient
- k_{mA} mass transfer coefficient of species A in the liquid film
- k_s rate constant based on surface area
- r_0 particle radius
- t time
- X extent of resin particle conversion
- x equivalent or mole fraction of ionic species in solution phase
- y equivalent or mole fraction of ionic species in particle phase
- z valency of ionic species

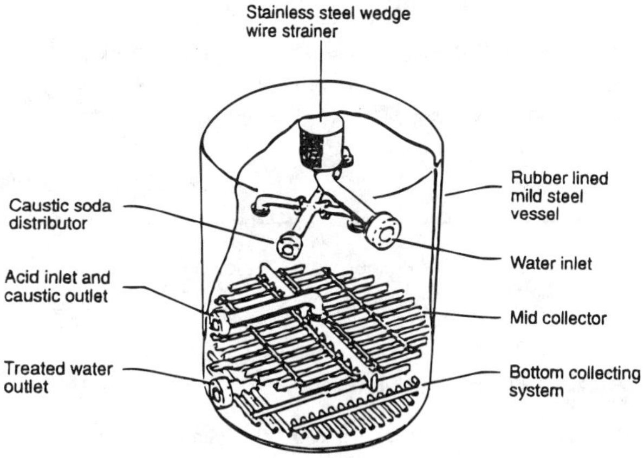

Figure 5. Schematic arrangement of a typical mixed bed ion exchange column. From Arden, T.V. (1968) *Water Purification by Ion Exchange*, Butterworth Heinemann, with permission.

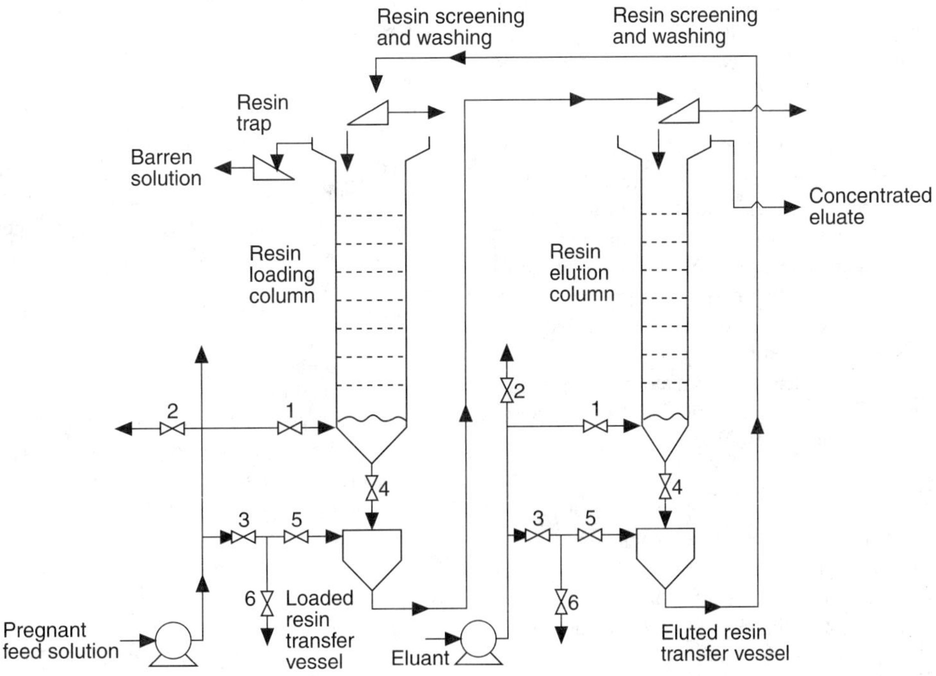

Figure 6. Schematic layout of a continuous countercurrent uranium plant. From Streat M and Naden D (1987) Ion Exchange and Sorption Process in Hydrometallurgy, S.C.I. London, with permission.

SUBSCRIPTS

A ionic species *A*
B ionic species *B*
i ionic species *i*

References

Bolto, B. A. and Pawlowski, L. (1987) *Wastewater Treatment by Ion Exchange*. E and F N Spon. London.

Dorfner, K. (1991) *Ion Exchangers*. Walter de Gruyter. Berlin and New York.

Helfferich, F. (1962) *Ion Exchange*. McGraw Hill. New York.

Slater, M. I. (1991) *The Principles of Ion Exchange Technology*. Butterworth-Heinemann. Oxford.

M. Streat

IONIC BONDING (see Atom; Molecule)

IONISATION

Ionisation is the removal of charge—usually electrons—from neutral atoms to produce positive ions or alternatively, the deposition of extra charge giving negative ions. The simplest examples of ionisation occur in gases due to high voltage breakdown, e.g., lightning or exposure to ultraviolet or **Ionising Radiation**. Similar processes ar also found in liquids and solids.

In the electronic system of an atom there are many possible *energy levels* or states that electrons may occupy. No two electrons may have exactly the same state and, allowing for the quantum parameter called spin for which there are two possibilities (up and down), this usually means that two electrons may occupy each possible energy level. The most strongly-bound states, corresponding to electrons in "orbits" closer to the nucleus, are filled first. An atom in its ground state has all of its electrons placed in the lowest possible available energy levels. In fact, the energy of an electron in a particular level is quoted as the *binding energy* or the work that would have to be done to remove it completely from the atom. The outer electrons of an atom are most easily removed and have smaller binding or ionisation energies. A general discussion of this area may be found in the work of Hecht (1994).

Any process that is able to transfer a large enough amount of energy, particularly to the outer electrons of an atom, may produce ionisation. Possibilities include: strong electric fields, **Electromagnetic Radiation, Ionising** (nuclear) **Radiation,** etc.

Ionisation by Electric Fields

In gases a large enough electric field strips electrons away from atoms. The electrons are then accelerated through the medium. If there is a sufficient path length, the electrons may gain enough kinetic energy to produce secondary electrons by impact ionisation, leading to an avalanche type breakdown. A very similar process may occur in solids, particularly semiconductor devices.

In certain molten materials and aqueous solutions the unit chemical cell is already dissociated into ions, e.g., sodium chloride into Na^+ and Cl^-. The ions in these materials will support conduction, but these ions already existed before the application of an electric field.

Ionisation by Electromagnetic Radiation

Electrons may be excited by the absorption of photons (quanta of electromagnetic energy). Photons of visible and ultraviolet light have energies similar to those required to liberate electrons from many materials.

Ionising Radiation

Various forms of nuclear and atomic radiation can be detected by their ability to ionise matter. *X-rays* and γ rays are high energy electromagnetic photons capable of penetrating some distance through matter and producing a trail of ionisation. α-particles, β-particles and other high energy particles associated with radioactive decay are stopped from absorbing materials leaving a track of ionisation.

Reference

Hecht, E. (1994) *Physics*. Brooks Cole.

Leading to: *Ionising radiation*

A.W. Campbell

IONISING RADIATION

Ionising radiation is radiation which, when passing through matter, interacts with the atoms in the matter producing free electrons and positive ions, i.e., ionisation of the matter. Since the production of free electrons and positive ions requires energy, the ionising radiation itself loses energy as it passes through matter. The rate of energy loss of the radiation will depend on its nature, its original energy and on the atomic number of the matter through which it is passing.

There are two main categories of ionising radiation:

1. *Electromagnetic radiation* which travels through matter at the speed of light. Such radiation can be thought of either as electromagnetic waves with a characteristic frequency or wavelength; or as individual energy packets called *photons,* which have a characteristic energy and zero mass. The photon energy E is related to electromagnetic wave frequency f by the equation

$$E = hf$$

where h is Planck's constant (6.626×10^{-34} J s). Photon energies are normally quoted in electronvolt (eV) units, where 1 eV $\equiv 1.602 \times 10^{-19}$ J. *X-rays* have energies up to about 100 keV while **Gamma Ray** energies extend up to several MeV.

2. *Electrically charged particles* such as electrons, alpha particles or other ions. Such particles are characterised by their mass, their electric charge and their kinetic energy.

Electrons carry a single unit, positive or negative, of electric charge (1.602×10^{-19} Coulomb) and have a mass (9.109×10^{-31} kg). When a high-energy electron is emitted from the nucleus of an atom undergoing radioactive decay, it is called a *beta particle*. Such beta particles can have kinetic energies up to a few MeV. *Alpha particles* are much more massive than electrons as they consist of two protons plus two neutrons (mass = 6.646×10^{-27} kg) and have two units of electric charge (3.204×10^{-19} Coulomb). Alpha particles also arise in the decay of radioactive isotopes and have typical energies of 4 or 5 MeV.

The ways in which the two categories of ionising radiation behave in their passage through matter differs considerably.

Photons (electromagnetic radiation) do not lose energy or cause ionisation in a uniform way as they pass through matter. They travel significant distances without causing any disturbance and then interact with an electron at some point giving up all, or a significant portion, of their energy to that electron. Photons with energies typical of X-rays or gamma rays will have sufficient energy to dissociate the electron with which they interact from its atom (i.e., cause **Ionisation**). The resulting high energy electron itself will often have sufficient energy to cause further ionisation.

Because of the way in which they interact with matter it is not possible to define a range for photons in matter. For any thickness of material, there is a finite probability that photons could pass through that thickness without interaction. One can say that if a flux of photons with intensity $I_0(m^{-2}\ s^{-1})$ impinges normally onto the surface of a slab of material of thickness x (m), then the flux of photons, I_x, transmitted without interacting in the slab can be given by

$$I_x = I_0 e^{-\mu x}$$

where μ is known as the *linear attenuation coefficient* and is a function of the photon energy and of the atomic number of the material (decreases with increasing photon energy, increases with increasing atomic number). **Lead** is therefore a good absorber of electromagnetic radiation.

Charged particles, on the other hand, lose energy continuously as they pass through matter. Their electric field causes excitation and ionisation of atoms along their path until they come to rest. It is thus possible to define a path length of charged particles in matter. Such a path length depends on the mass of the particle, its electric charge and the atomic number of the material through which it is passing. A 1 MeV electron (or beta particle) would have a range of about 2 mm in aluminum and less than 1 mm in lead. Because of their much greater mass and high electric charge, alpha particles come to rest much more quickly. Alpha particles of a few MeV are stopped completely by a single sheet of paper.

Reference

Knoll, G. F. (1989) *Radiation Detection & Measurement*. 2nd edition. John Wiley & Sons.

Leading to: *Gamma rays*

T.D. MacMahon

IONS (see Electrochemical cells)

IONS, TRANSPORT IN ELECTROLYTE (see Electrolysis)

IOTVOS NUMBER (see Bond number)

IRON

Iron—(Anglo-Saxon, iron), Fe (L. *ferrum*); atomic weight 55.847; atomic number 26; melting point 1535°C; boiling point 2750°C; sp. gr. 7.874 (20°C); valence 2, 3, 4, or 6.

The use of iron is prehistoric. Genesis mentions that Tubal-Cain, seven generations from Adam, was "an instructor of every artificer in brass and iron." A remarkable iron pillar, dating to about A.D. 400, remains standing today in Delhi, India. This solid shaft of wrought iron is about 7 1/4 m high by 40 cm in diameter. Corrosion to the pillar has been minimal although it has been exposed to the weather since its erection.

Iron is a relatively abundant element in the universe. It is found in the sun and many types of stars in considerable quantity. Its nuclei are very stable. Iron is found native as a principal component of a class of meteorites known as *siderites,* and is a minor constituent of the other two classes. The core of the earth, 2150 mi in radius, is thought to be largely composed of iron, with about 10% occluded hydrogen.

The metal is the fourth most abundant element, by weight, making up the crust of the earth. The most common ore is *hematite* (Fe_2O_3), from which the metal is obtained by reduction with carbon. Iron is found in other widely distributed minerals such as *magnetite,* which is frequently seen as *black sands* along beaches and banks of streams. *Taconite* is becoming increasingly important as a commercial ore.

Common iron is a mixture of four isotopes. Six other isotopes are known to exist. Iron is a vital constituent of plant and animal life, and appears in hemoglobin. The pure metal is not often encountered in commerce, but is usually alloyed with carbon or other metals. The pure metal is very reactive chemically, and rapidly corrodes, especially in moist air or at elevated temperatures. It has four allotropic forms, or ferrites, known as α, β, γ, and δ, with transition points at 770, 928, and 1530°C. The α form is magnetic, but when transformed into the β form, the magnetism disappears although the lattice remains unchanged. The relations of these forms are peculiar. *Pig iron* is an alloy containing about 3% carbon with varying amounts of S, Si, Mn, and P. It is hard, brittle, fairly fusible, and is used to produce other alloys, including steel. Wrought iron contains only a few tenths of a percent of carbon, is tough, malleable, less fusible, and has usually a "fibrous" structure. Carbon steel is an alloy of iron with carbon, with small amounts of Mn, S, P, and Si. Alloy steels are carbon steels with other additives such as nickel, chromium, vanadium, etc.

Iron is the cheapest and most abundant, useful, and important of all metals.

Handbook of Chemistry and Physics, CRC Press

IRREVERSIBLE PROCESSES (see Entropy; Thermodynamics)

IRREVERSIBLE THERMODYNAMICS

***Following from:** Thermodynamics*

Irreversible thermodynamics is a division of physics which studies the general regularities in transport phenomena (heat transfer, mass transfer, etc.) and their relaxation (transition from nonequilibrium systems to the thermodynamical equilibrium state). It is possible to use for this purpose, as in reversible thermodynamics (thermostatics), phenomenological approaches based on the generalization of experimental facts and statistical physics methods which establish the bonds between molecular models (microscopical structure, properties of molecules, intermolecular interaction) and macroscopical substance behavior.

The starting points of irreversible thermodynamics are the *first* and *second laws of thermodynamics* in local formulation. In the general case, the nonequilibrium continuum is a moving ν-component mixture of substances $A_1, A_2, \ldots A_\nu$, among which r chemical reactions run simultaneously

$$\sum_{i=0}^{\nu} \nu'_{ik} A_i \rightleftharpoons \sum_{i=0}^{\nu} \nu''_{ik} A_i; \quad 1 \le k \le r. \tag{1}$$

It is described by the fields of temperature $T(t, \mathbf{r})$, mass concentrations of components $\rho_i(t, \mathbf{r})$ and average mass flow speed $v(t, \mathbf{r})$. The local density of system $\rho(t, \mathbf{r})$ and mass fractions of components $\chi_i(t, \mathbf{r})$ are defined by the relations

$$\rho = \sum_{i=1}^{\nu} \rho_i; \quad \chi_i = \rho_i/\rho; \quad \sum_{i=1}^{\nu} \chi_i = 1. \tag{2}$$

The conservation equation of mixture mass and balance equations of component masses in the local coordinate system, moving with the mass centre speed $\mathbf{v}(t, \mathbf{r})$, can be written as:

$$\frac{d\rho}{dt} = -\rho \frac{\partial}{\partial \mathbf{r}} \mathbf{v}; \quad \rho \frac{d\chi_i}{dt} = -\frac{\partial}{\partial \mathbf{r}} \mathbf{j}_i + \sum_{k=1}^{r} m_i(\nu'_{ik} - \nu''_{ik}) R_k, \tag{3}$$

Here, $\mathbf{j}_i$ is the diffusion flow density of the ith component,

$$\sum_{i=1}^{\nu} \mathbf{j}_i = 0, \tag{4}$$

and R_k is the rate of the kth chemical reaction. The balance equation of specific internal energy u(t, r) has the following form:

$$\rho \frac{du}{dt} = -\frac{\partial}{\partial \mathbf{r}} \mathbf{q} - p \frac{\partial}{\partial \mathbf{r}} \mathbf{v} - \Pi\colon \frac{\partial \mathbf{v}}{\partial \mathbf{r}} + \sum_{i=1}^{\nu} \mathbf{F}_i \mathbf{j}_i, \tag{5}$$

where q is the heat flow density, p is pressure and Π is the viscosity part of pressure tensor

$$\mathrm{p} = p1 + \Pi,$$

$\partial \mathbf{v}/\partial \mathbf{r}$ is the speed deformation tensor, $\mathbf{F}_i$ is the mass force exerted on the ith component particles from the external fields. Equation 5 is the local formulation of the first law of thermodynamics for a nonequilibrium open system.

The second law of thermodynamics in the form of a local entropy balance is deduced by substituting Equations 3 and 5 with the thermodynamical relationship

$$ds = \frac{1}{T}\left(du + p dv - \sum_{i=1}^{\nu} \mu_i d\xi_i\right), \tag{6}$$

describing the variation of local specific entropy s(t, r) with time interval dt in the centre mass coordinate system of the continuum. In Equation 6, $v = 1/\rho$ is the specific volume, and $\mu_i(t, r)$ is the local specific chemical potential of ith component. In so far as du in Equation 6 contains the dissipative contributions (the two last terms on the right side of Equation 5); ds includes both the "equilibrium" component ds_e and the nonnegative term $ds_i \ge 0$; and conditioned by irreversibility:

$$ds = ds_e + ds_i; \quad ds_i \ge 0.$$

The local entropy balance equation follows from Equation 6:

$$\rho \frac{ds}{dt} = -\frac{\partial}{\partial \mathbf{r}} \mathbf{j}_s + \sigma, \tag{7}$$

where

$$\mathbf{j}_s = \frac{1}{T}\left(\mathbf{q} - \sum_{i=1}^{\nu} \mu_i \mathbf{j}_i\right) \tag{8}$$

is entropy flow density, and

$$\sigma = -\sum_{i=1}^{\nu} \mathbf{j}_i \left[\frac{\partial(\mu_i/T)}{\partial r} - \frac{1}{T}\mathbf{F}_i\right] - \mathbf{q}\frac{1}{T^2}\frac{\partial T}{\partial r} - \Pi\colon \frac{1}{T}\frac{\partial \mathbf{v}}{\partial \mathbf{r}} - \sum_{i=1}^{\nu} R_k \frac{1}{T} A_k \tag{9}$$

is entropy production. In Equation 9,

$$A_k = \sum_{i=1}^{\nu} m_i(\nu''_{ik} - \nu'_{ik})\mu_i$$

is the affinity of chemical reaction (Equation 1).

Equation 9 has the structure

$$\sigma = -\sum_{\alpha=1}^{n} J_\alpha X_\alpha, \quad \textbf{(10)}$$

where J_α are the flows ($\mathbf{J_i}$, $\mathbf{q}$, Π, R_k) and X_α are the matching thermodynamic forces. The second law of thermodynamics is reduced to

$$\sigma \geq 0. \quad \textbf{(11)}$$

Equality in Equation 11 corresponds to the thermodynamical equilibrium state—full or local, and nonequal—for the local entropy growth in irreversible processes. From the second law of thermodynamics (Equation 11) can be inferred the fact that flows in Equations 9 and 10 are in opposite direction to thermodynamic forces, $J_\alpha X_\alpha \leq 0$ and they disappear after the transition to thermodynamic equilibrium. Practice fully confirms these deductions.

In accordance with experience, flows and thermodynamic forces are bound by transfer laws $J_\alpha = F(X_1, \ldots, X_n)$; $1 \leq \alpha \leq n$; moreover if thermodynamic forces are absent, the flows are also equal to zero. The most simple relationship is linear,

$$J_\alpha = -\sum_{\beta=1}^{n} L_{\alpha\beta} X_\beta; \quad 1 \leq \alpha \leq n. \quad \textbf{(12)}$$

The assumption about linearity of transport laws is a basis of linear, irreversible thermodynamics. Quantities $L_{\alpha\beta}$ are called kinetic (transport) coefficients. They are properties of the system considered; it means they don't depend on thermodynamic forces and flows, but are functions only of the state parameters (temperature, pressure, mixture composition). Phenomenological irreversible thermodynamics neither establishes these dependencies nor indicates suitable limits to linear transport laws. However, it is possible to formulate a number of general statements about the structure and properties of the kinetic coefficient matrix $L_{\alpha\beta}$.

From Equations 9 and 10, it is clear that flows and thermodynamic forces have different tensor dimensions. Diffusion flows j_i and diffusion thermodynamic forces

$$\mathbf{d_i} = \frac{\partial(\mu_i/T)}{\partial \mathbf{r}} - \frac{1}{T}\mathbf{F_i}$$

are vectors, as well as heat flow $\mathbf{q}$ and the related thermodynamic force $(1/T)\partial T/\partial \mathbf{r}$. Reaction rates R_κ and affinities A_κ are scalars. *Some further scalar couples (J,X) are obtained from the last but one term in Equation 9 the normal and tangential viscosity tension contributions* are given by $\pi = (\Pi_{xx} + \Pi_{yy} + \Pi_{zz})/3$; $\mathring{\Pi} = \Pi - \pi 1$, and it also follows that

$$\Pi\text{:}\ \frac{\partial \mathbf{v}}{\partial \mathbf{r}} = \pi \frac{\partial}{\partial r}\cdot \mathbf{v} + \mathring{\Pi} S; \quad S = \frac{1}{2}\left(\frac{\partial \mathbf{v}}{\partial \mathbf{r}} + \frac{\partial \tilde{v}}{\partial \mathbf{r}}\right) - \frac{1}{3}\frac{\partial}{\partial \mathbf{r}}\cdot \mathbf{v}$$

(S is the shift rate tensor, $\partial/\partial \mathbf{r} \equiv$ div). Thus π and $(1/T)(\partial/\partial \mathbf{r}) \cdot \mathbf{v}$ are the scalar flow and the thermodynamic force, respectively; flow $\mathring{\Pi}$ and force S/T are second-range tensors. According to the Courier principle, only flows and thermodynamic forces with the same tensor dimensions may be related using linear correlations. Any kinetic coefficient $L_{\alpha\beta}$, relating a flow J_α and a thermodynamic force X_β with different tensor dimensions is identically equal to zero. The permissible linear correlations between flows and thermodynamic forces occurring in nonequilibrium, multicomponent reacting moving fluid (Equations 3–9) are terms limited to

$$\mathbf{j_i} = -\sum_{j=1}^{\nu} L_{ij}\mathbf{d_j} - L_{iq}\frac{1}{T^2}\frac{\partial T}{\partial \mathbf{r}}; \quad 1 \leq i \leq \nu; \quad \textbf{(13)}$$

$$\mathbf{q} = -\sum_{i=1}^{\nu} L_{qi}\mathbf{d_i} - L_{qq}\frac{1}{T^2}\frac{\partial T}{\partial \mathbf{r}}; \quad \textbf{(14)}$$

$$\mathring{\Pi} = -L\frac{1}{T}S; \quad \textbf{(15)}$$

$$\pi = -L_{pp}\frac{1}{T}\frac{\partial}{\partial \mathbf{r}}\cdot \mathbf{v} - \sum_{k=1}^{r} L_{pk}\frac{1}{T}A_k; \quad \textbf{(16)}$$

$$R_k = -L_{kp}\frac{1}{T}\frac{\partial}{\partial \mathbf{r}}\cdot \mathbf{v} - \sum_{l=1}^{r} l_{kl}\frac{1}{T}A_l; \quad 1 \leq k \leq r. \quad \textbf{(17)}$$

The correlations (Equation 12 and Equations 13–17) predict not only direct ($J_\alpha = L_{\alpha\alpha}X_\alpha$) but "overcrossed" irreversible processes when the flow of a certain physical characteristic is implemented by other natural, "nonrelated" thermodynamic forces ($J_\alpha = L_{\alpha\beta}X_\beta$; $\alpha \neq \beta$). Thus Equation 13, the *generalized Ficks law,* describes diffusion in a multicomponent mixture when diffusion flow $\mathbf{j_i}$ of component i is caused by the diffusion thermodynamic forces $\mathbf{d_j}$ (in particular, by concentration gradients $\partial \rho j/\partial \mathbf{r}$) of other components as well as temperature gradient. The quantities L_{ij} and L_{iq} represent the multicomponent diffusion and thermal diffusion coefficients.

The generalized Fourier law (Equation 14) describes a rise of heat flow at the expense of temperature gradient $\partial T/\partial \mathbf{r}$ (the transport coefficient of this direct process is related to the ordinary thermal conductivity $\lambda = L_{qq}/T^2$), and by diffusion thermodynamic forces $\mathbf{d}_i$. The latter process is called the diffusion thermo effect. It is in mutual overcross with thermal diffusion. The appearance of tangential viscous tensions, which is described by the generalized Newton-Stokes law (Equation 15), is a direct process (dynamic viscosity coefficient $\eta = [1/2]\ [L/T]$) and the overcross processes are absent here. Yet normal viscous tensions (at is clear from Equation 16) can arise both because of the direct process—which stipulates volume expansion of the flow and is formed by the volume viscosity coefficient $\kappa = L_{pp}/T$—and the overcross processes arising from nonequilibrium chemical reactions.

All the above irreversible processes have been observed in practice and are really linear until the fluid in which they proceed can be considered as a continuum.

It is impossible to say the same about the conditions of Equation 17. The linearity of conditions between chemical reaction rates and their affinities is broken already by small deflections from chemical equilibrium $A_k = 0$, $R_k = 0$.

Uniting Equations 11 and 12, it is possible to present entropy production in quadratic form

$$\sigma = \sum_{\alpha=1}^{n} \sum_{\beta=1}^{n} L_{\alpha\beta} X_\alpha X_\beta. \quad \textbf{(18)}$$

According to Equation 11, it is positively determined. Hence it follows that all the diagonal elements of the kinetic coefficients matrix forming direct irreversible processes are positive, $L_{\alpha\alpha} \geq 0$; all its principal minors are positive too. Moreover, important information on the structure of this matrix is derived from Onzager mutuality correlations, which establish equality of kinetic coefficients for overcrossed, irreversible processes:

$$L_{\alpha\beta} = L_{\beta\alpha}. \quad \textbf{(19)}$$

The derivation of accurate formulae for kinetic coefficient calculation from molecular data, as well as molecular kinetic justification of the second law of thermodynamics for nonequilibrium systems in the form of Equations 7–11, is feasible now only for dilute gases by applying the *Boltzman kinetic equation* and its modifications. Attempts to describe dense nonequilibrium system behaviour proceeding from reversible, equations of moleculular motion meets with problems that can not be completely overcome. Together with the kinetic equations method, the nonequilibrium statistical operator method and related approaches of linear reaction theory can be successfully employed.

In certain cases, the introduction of some critical nonequilibrity degree can suddenly increase an open system's regularity, which is followed by dissipative structure formation. Known examples are: the occurrence of Benard cells in a viscous liquid layer, heated from below in gravitational field, when the heat flux exceeds a critical value; the appearance of layers in a gas discharge column with the introduction of a critical current density; and so on. A richer variety of dissipative structures—both space and temporary—are found in nonlinear, noneqilibrium systems (for example, the Belousov—Zabotinski reaction). A comparatively new and swiftly-developing branch of irreversible thermodynamics is devoted to the study of the organization of phenomena in strong nonequilibrium, nonlinear open systems.

References

de Groot, S. R. and Mazur, P. (1962) *Nonequilibrium Thermodynamics*. North-Holland Publ. Co. Amsterdam.

Glansdorf, P. and Prigogine I. (1970) *Thermodynamic Theory of Structure, Stability and Fluctuations*. A division of John Wiley & Son. London/New York/Sidney/Toronto.

Nicolis, G. and Prigogine I. (1976) *Self-organization in Nonequilibrium Systems*. A. Wiley Interscience Publ. New York/London/Sidney/Toronto.

***Leading to:** Heat transfer; Mass transfer; Diffusion*

A.M. Semjonov

IRREVERSIBILITY (see Non-linear systems)

IRROTATIONAL FLOW (see Conservation equations, single phase; Flow of fluids; Inviscid flow)

ISENTROPIC EXPONENT (see Adiabatic conditions)

ISENTROPIC PROCESSES (see Rankine cycle)

ISOBAR

An isobar is a curve drawn through points of equal pressure. For example, it could be a line joining states of equal pressure in a graph representing thermodynamic processes. The term is commonly used in meteorology where an isobar is a curve joining points of equal atmospheric pressure on a given reference surface, e.g., sea level.

Note that isobar is also used in nuclear physics to describe nuclides having the same number of nucleons, but with different numbers of protons and neutrons.

G.L. Shires

ISOBARIC JET (see Jets)

ISO-BUTANE

ISO-Butane ($(CH_3)_2CHCH_3$) is a colourless gas, soluble in water. Although its physical properties are similar to that of n-Butane, it exhibits markedly different chemical behaviour. It is obtained by petroleum fractionation of natural gas or by isomerization of butane. It forms an explosive and flammable mixture with air at low concentrations. Its main uses are as a raw material in organic synthesis, for the production of synthetic rubber, and in the production of branched hydrocarbons of high octane grading.

Some of its physical characteristics are;

Molecular weight: 58.12	**Critical pressure:** 3.647 MPa
Melting point: 113.55 K	**Critical density:** 221 kg/m^3
Boiling point: 261.4 K	**Normal vapor density:** 2.59 kg/m^3 (@ 0°C, 101.3 kPa)
Critical temperature: 408.1 K	

(See **Table 1** on page 637.)

V. Vesovic

ISOGONAL MAPPING (see Conformal mapping)

ISOELECTRIC POINST (see Liquid-solid separation)

Table 1. Iso-butane: Values of thermophysical properties of the saturated liquid and vapor

T_{sat}, K	261.4	285	300	315	330	345	360	375	390	408.1
p_{sat}, kPa	101.3	233	355	553	805	1132	1540	2066	2697	3647
ρ_ℓ, kg/m^3	594	567	550	529	508	485	455	422	377	221
ρ_g, kg/m^3	2.87	6.33	9.81	14.6	21.2	30.0	42.2	59.7	87.3	221
h_ℓ, kJ/kg	232.5	286.1	323.3	360.5	397.7	437.3	481.5	528.0	574.5	697.8
h_g, kJ/kg	597.8	623.4	646.6	667.6	686.2	704.8	721.1	737.3	744.3	697.8
$\Delta h_{g,\ell}$, kJ/kg	365.2	337.3	323.3	307.1	288.5	267.5	239.6	209.3	169.8	
$c_{p,\ell}$, kJ/(kg K)	2.12	2.34	2.45	2.56	2.68	2.79	2.95	3.16	3.59	
$c_{p,g}$, kJ/(kg K)	1.53	1.69	1.81	1.94	2.09	2.28	2.55	3.00	4.18	
η_ℓ, μNs/m^2	240	190	145	116	101	86	74	61	44	
η_g, μNs/m^2	6.7	7.4	8.0	8.6	9.1	9.7	10.4	11.3	12.9	
λ_ℓ, mW/(Km)	100	92	87	83	80	77	73	68	61	
λ_g, mW/(Km)	11.6	14.4	16.3	18.3	20.4	22.7	25.2	27.9	31.2	
Pr_ℓ	5.09	4.83	4.08	3.58	3.38	3.12	2.99	2.83	2.59	
Pr_g	0.88	0.88	0.89	0.91	0.93	0.97	1.05	1.22	1.73	
σ, mN/m	14.1	11.4	9.7	8.1	6.5	5.0	3.6	2.3	1.1	
$\beta_{g,\ell}$, kK^{-1}	1.87	2.22	2.52	2.91	3.44	4.20	5.40	7.56	12.8	

Beaton, C. F. and Hewitt, G. F. (1989) *Physical Property Data for the Design Engineer.* Hemisphere Publishing Corporation. New York.

ISOCHORE

An isochore is a graph representing the state of a system using two variables, for example pressure and temperature, while the volume remains constant.

G.L. Shires

ISOENTHALPIC PROCESS

Following from: *Thermodynamics*

Isoenthalpic means constant enthalpy, and any material which passes through a system without a change of enthalpy has, by definition, passed through an isoenthalpic process.

For an open system at steady flow:

$$\Delta H = Q - W - \Delta KE - \Delta PE$$

If, in addition, the process is adiabatic (Q = 0) and there is no work transfer (W = 0), then if the sum of the changes in kinetic and potential energy are zero, the process is isoenthalpic. For instance, it is common to assume that a gas or vapour flashing through a valve is an isoenthalpic process. In such an example, the residence time and contact area available within the valve is so small that very little heat transfer can occur, so the process is approximately adiabatic. The valve does not transfer work to the surroundings and the inlet and outlet are at such similar elevations that changes of gravitational potential energy can be neglected. Finally, it is common to neglect the change in kinetic energy since the diameters of the inlet and outlet pipes can be, and often are, selected to minimise the change in velocity of the fluid.

The assumption that a process is isoenthalpic gives a simple method for determining the change in temperature of fluid flowing through the process, provided the upstream conditions and the downstream pressure are known, as follows:

Knowing the upstream conditions, the upstream enthalpy is known or can be calculated. Knowing that the downstream enthalpy is the same as the upstream enthalpy and knowing the downstream pressure, the temperature and condition of the downstream fluid can be determined.

N. Kirkby

ISO-OCTANE

Iso-octane ($(CH_3)_3CCH_2CH(CH_3)_2$ or 2,2,4-trimethylpentane) is an isomer of octane that boils at 372.4K, freezes at 165.75K and has a specific gravity of 0.69.

V. Vesovic

ISO-PROPANOL

Iso-propanol ($(CH_3)_2CHOH$) is a colourless liquid that is soluble in water. It is obtained from propylene. It forms an explosive and flammable mixture with air at low concentrations and it is moderately toxic. It is used as a raw material for the production of acetone and glycerol, as a solvent for alkaloids, resins, oils and cellulose derivatives and in medicine.

Some of its physical characteristics are:

Molecular weight: 60.1	**Critical temperature:** 508.75 K
Melting point: 194.15 K	**Critical pressure:** 5.370 MPa
Normal boiling point: 355.65 K	**Critical density:** 274 kg/m^3

(See **Table 1** on following page.)

V. Vesovic

Table 1. Iso-propanol: Values of thermophysical properties of the saturated liquid and vapor

T_{sat}, K	355.65	373	390	408	425	443	459	478	498	508
p_{sat}, kPa	101.3	200	380	580	925	1 425	2 025	3 039	4 052	5 369
ρ_ℓ, kg/m^3	732.3	712.7	683.0	660.0	630.1	597.4	566.0	514.8	460.5	288.0
ρ_g, kg/m^3	2.06	4.15	7.73	14.3	21.00	32.78	46.40	72.3	108.4	252
h_ℓ, kJ/kg	0.0	60.1	121.8	190.2	257.6	331.8	400.1	484.0		
h_g, kJ/kg	677.8	688.0	736.8	767.9	796.1	822.9	841.7	851.5		
$\Delta h_{g,\ell}$, kJ/kg	677.8	627.9	615.0	577.7	538.5	491.1	441.6	367.5	284.5	82.5
$c_{p,\ell}$, kJ/(kgK)	3.37	3.55	3.71	3.88	4.04	4.20	4.34	4.49		
$c_{p,g}$, kJ/(kg K)	1.63	1.71	1.80	1.94	2.15	2.37	2.83	3.97		
η_ℓ, μNs/m^2	502	376	295	230	184	147	122	93.5	72.5	
η_g, μNs/m^2	9.08	9.80	10.3	10.9	11.4	11.9	12.5	13.7	15.2	28.2
λ_ℓ, mW/(Km)	131.1	127.5	124.3	122.8	120.1	117.3	115.8	113.2	110.7	107.7
λ_g, mW/(Km)	19.8	22.2	24.6	27.1	29.3	31.8	34.9	38.9	42.3	47.1
Pr_ℓ	12.9	10.5	8.81	7.27	6.19	5.27	4.57	3.71		
Pr_g	0.75	0.75	0.75	0.78	0.84	0.89	1.01	1.40		
σ, mN/m	18.6	17.2	14.2	11.84	9.4	6.9	4.97	2.6	1.05	
$\beta_{g,\ell}$, kK^{-1}	1.41	1.81	2.27	2.65	3.20	3.95	5.10	7.76	37.3	

From: Beaton, C. F. and Hewitt G. F. (1989) *Physical Property Data for the Design Engineer.* Hemisphere Publishing Corporation. New York.

ISOTACH OR ISOVEL

Following from: *Velocity measurement*

Isotachs or isovels are lines linking points of constant velocity, and as such, are used to indicate velocity variation within a given section of the flow field. If **Figure 1** represents the cross-section of an open channel flow, the isotachs (indicated by the dotted lines) highlight the velocity reduction at the boundaries of the flow and the occurrence of the maximum velocity (point A) some distance beneath the water surface.

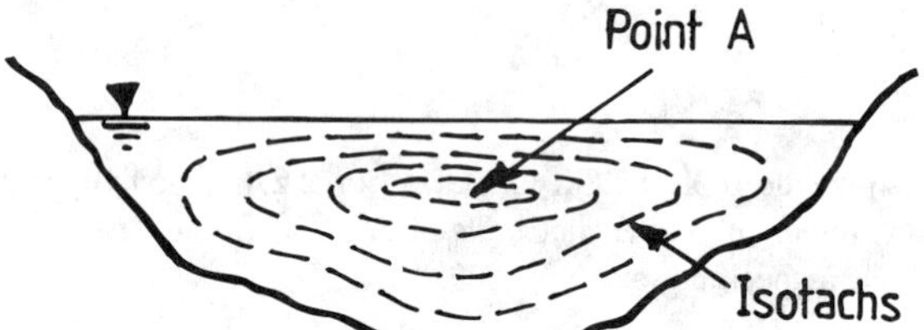

Figure 1. Open-channel flow.

C. Swan

ISOTHERM

A curve joining points of equal temperature is known as an isotherm.

G.L. Shires

ISOTHERMAL PROCESS

Following from: *Thermodynamics*

'Isothermal' means at constant temperature. In a strict sense, an isothermal process must be a reversible process because by definition, if every part of the system is at the same, constant temperature throughout the process, there can be no frictional or other irreversible effects giving rise to heat and causing local changes in temperature [see, for example, Rogers and Mayhew (1992)]. No real process can be perfectly isothermal, some come very close, especially if it is accepted that it is only the spatially-averaged temperature which must remain constant.

In processes operating on a single phase, heat transfer will result in a change in temperature unless exactly balanced by some other energy transfer, e.g., work, and this balance can be very difficult to achieve in practice. One solution might be to eliminate the heat transfer; but in reality, insulation can reduce heat transfer but cannot stop heat transfer completely. An alternative approach is to use systems which can accept some heat transfer without a change in temperature.

In two-phase systems, heat transfer can be accommodated without changing the temperature by altering the relative amounts of the two phases present. The most common example is a phase equilibrium in a pure substance at constant pressure. Ice in water is frequently used as a fixed point for temperature because this system remains at a constant temperature provided the pressure is constant and the rate of heat transfer is not sufficient to cause the system to depart significantly from equilibrium.

References

Rogers, G. F. C. and Mayhew, Y. R. (1992) *Engineering Thermodynamics: Work and Heat Transfer: SI units.* 4th Edition. Longman Group UK Ltd., Harlow, Essex.

N. Kirkby

ISOTOPES (see Atom; Atomic weight; Relative molar mass)

J

JACKETED VESSELS

***Following from:** Heat exchangers*

The prediction of the transient time required to heat and/or cool the contents of a jacketed vessel is dependent upon many variables: jacket configuration—plain jacket or dimpled jacket (with or without directional vanes) or a limpet coil; heat transfer medium in the jacket—isothermal, nonisothermal or pump-around fluid and an external heat exchanger; type of vessel, its geometrical dimensions and the type, speed and dimensions of the stirrer; and last but not least, the fluid in the vessel and its physical properties.

Heggs and Hills (1995) have developed a unifying set of equations for these predictions based on a number of assumptions: fluid in the vessel is perfectly mixed; an overall heat transfer coefficient applies and remains constant; flow rates are steady; physical properties remain fixed; heat losses or gains are negligible; the heat capacities of the jacket, vessel, stirrer are small relative to that of the liquid contents of the vessel; and the thermal response of the jacket and any external heat exchanger and pipework are instantaneous.

The **Overall Heat Transfer Coefficient** is given by:

$$\frac{1}{(UA)_j} = \frac{1}{(\alpha A)_m} + \frac{1}{(\alpha_f A)_m} + R_w + \frac{1}{(\alpha_f A)_p} + \frac{1}{(\alpha A)_p} \qquad (1)$$

where the film and fouling heat transfer coefficients α and α_f, are predicted from correlations and tables of empirical data: Kern (1965), Fletcher (1987) and Foumeny and Ma (1991). The symbol R_w is the conductive resistance of the vessel wall, and subscripts m and p apply to the outside and inside of the vessel respectively. (See also **Agitated Vessel Heat Transfer**, **Tank Coils** and **Fouling** for methods of calculating α and α_f.)

A proportionality factor, χ, is used in the unifying set of equations to account for the configuration and the means of affecting the transfer of heat, so that for an isothermal heating (condensing) or cooling (boiling) medium in the jacket (**Figure 1**) with $T_{m,1} = T_{m,2}$,

$$\chi_1 = (UA)_j. \qquad (2)$$

For a nonisothermal transfer medium, **Figure 1** with $T_{m,1} \neq T_{m,2}$,

$$\chi_2 = \dot{M}_m c_{pm}[1 - \exp(-(UA)_j/\dot{M}_m c_{pm})] \qquad (3)$$

For a pump-around fluid and an external heat exchanger, (see **Figure 2**)

$$\frac{1}{\chi_3} = \frac{1}{\varepsilon C_{min}} + \frac{1}{\chi_2} + \frac{1}{(\dot{M}c_p)_{int}} \qquad (4)$$

where ε is the thermal effectiveness of the external heat exchanger and $C_{min} = \min[(\dot{M}c_p)_m, (\dot{M}c_p)_{int}]$, and χ_2 is evaluated using the flow rate and specific heat capacity, $(\dot{M}c_p)_{int}$, of the pump-around fluid in **Equation 3**.

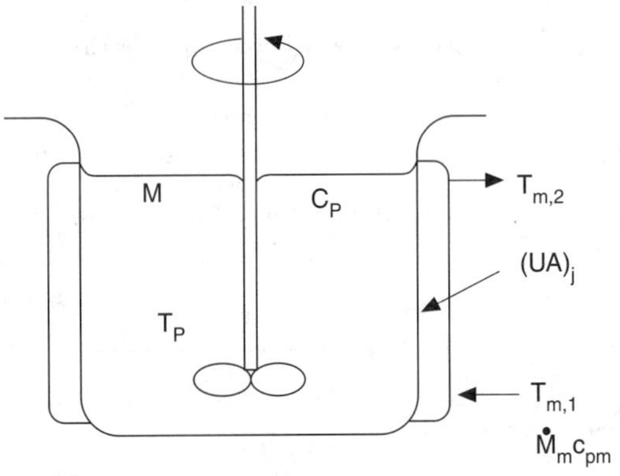

Figure 1. Schematic of a jacketed vessel.

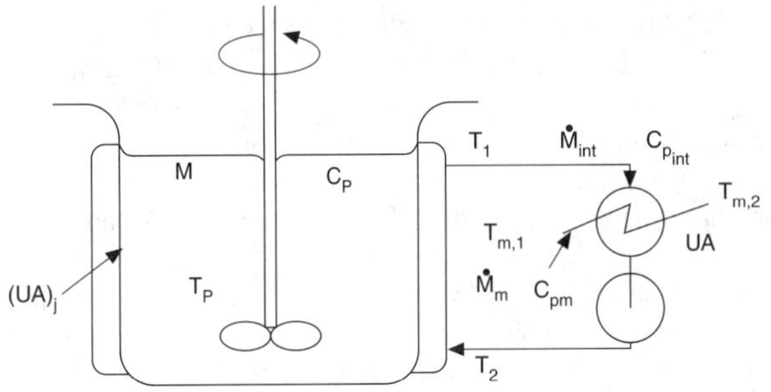

Figure 2. Jacketed vessel with a liquid pump-around and external heat exchanger.

The transient time of heating or cooling is given by:

$$t = \frac{Mc_p}{\chi} \ln\left\{\frac{T_{m,1} - T_{p,i}}{T_{m,1} - T_p}\right\} \qquad (5)$$

the contents temperature after a time t is given by:

$$T_p = T_{m,1} - (T_{m,1} - T_{p,i})\exp(-\chi t/Mc_p) \qquad (6)$$

the total heat requirements up to a time t is given by:

$$|Q| = Mc_p(T_{m,1} - T_{p,i})[1 - \exp(-\chi t/Mc_p)] \qquad (7)$$

total heat requirements to achieve a temperature T_p

$$|Q| = Mc_p(T_p - T_{p,i}) \qquad (8)$$

the rate of heat transfer at any instant of time t is given by:

$$|\dot{Q}| = \chi(T_{m,1} - T_{p,i})\exp(-\chi t/Mc_p). \qquad (9)$$

References

Heggs, P. J. and Hills, P. D. (1995) The design of heat exchangers for batch reactors. Ch. 18 in *Heat Exchange Engineering*. 4: 297–313.

Fletcher, P. (1987) Heat transfer coefficients for stirred batch reactor design. *The Chemical Engineer.* 435. April: 33–37.

Foumeny, E. A. and Ma, J. (1991) Design correlations for heat transfer systems. Ch. 11 in Heat exchange engineering: 1. *Design of Heat Exchangers.* Ellis Horwood Limited. London. 159–178.

Kern, D. Q. (1965) *Process Heat Transfer.* McGraw-Hill Book Co. New York.

P.J. Heggs

JAKOB, MAX 1879–1955

German physicist, born July 20, 1879 in Ludwigshafen. Jakob studied electrical engineering at the Technical University of München where he graduated in 1903. From 1903 until 1906 he was an assistant to O. Knoblauch at the Laboratory for Technical Physics. After working in the electrical industry Jakob joined the "Physikalisch-Technische Reichsanstalt" at Berlin-Charlottenburg in 1910, where he started his career in thermodynamics and heat transfer. He conducted a large amount of important works in these fields, covering such areas as steam and air at high pressure, devices for measuring thermal conductivity, the mechanisms of boiling and condensation, flow in pipes and nozzles, and others.

In 1936 he emigrated to the United States, where he became a research professor at the Illinois Institute of Technology and consultant in heat transfer for the Armour Research Foundation. In 1952, three years before his sudden death, he was awarded the Worchester Reed Warner Medal by the American Society of Mechanical Engineers.

His long years of research resulted in significant contributions to the literature of the profession; nearly 500 books, articles, reviews and discussions have been published.

Jakob, Max 1879–1955.

U. Grigull, H. Sandner and J. Straub

JAKOB NUMBER

Jakob Number or Jakob modulus, **Ja**, is a dimensionless group which occurs frequently in the study of **Boiling** and **Bubble Growth**.

$$Ja = c_p \Delta T / h_{LG}$$

where c_p is the specific heat capacity of the liquid, ΔT is the difference between the temperature of the liquid and saturation temperature, and h_{LG} is the latent heat of evaporation. Ja represents the ratio of sensible to latent heat involved in the boiling process.

Note that Jakob number is the reciprocal of **Kutateladze Number**.

Reference

Hewitt, G. F., Shires, G. L., and Bott, T. R. (1994) *Process Heat Transfer.* CRC Press.

G.L. Shires

JAMIN INTERFEROMETRY (see Interferometry)

JAR TEST (see Sedimentation)

JEFFREYS SHELTERING HYPOTHESIS (see Stratified gas-liquid flow)

JET ENGINE

***Following from:** Internal combustion engines*

A jet engine is an aircraft engine used to provide propulsion for a vehicle by ejecting a substance flow, i.e., creating a reactive force (thrust) which is applied against the vehicle. The jet (stream) can be continuous or discontinuous, gaseous or liquid, or in the form of ions, electrons, photons, etc. or separate solid particles. According to their design and the way the thrust is developed, jet engines are classified into two types: those using an outer medium (for instance *air-jet engines* or water-jet engines (ship engines)); and those which are independent of the outer medium, whose working substance is in the vehicle proper such as rocket engines (liquid-propellant), solid-propellant, ion-plasma jet, photon, etc.

Jet engines are characterised by the thrust R and the flow rate $\dot{M}$ (kg/s) of the working substance; $\dot{M}$ is the sum of the fuel flow $\dot{M}_f$ and the oxidant (air in air-jet engines) flow $\dot{M}_a$. Such engines are also characterised by their overall dimensions (length, diameter, midsection); overhaul period; reliability (mean time between failures); and by ecological characteristics such as noise, exhaust gas composition, radiation, etc.

The thrust of a jet engine is generally expressed in terms of the exhaust velocity W of the working substance, the pressure p_n at the nozzle cross-section at an area F_n and the flight velocity V in air with a pressure p_H:

$$R = \dot{M}_f[(1 + \beta)W - V] + F_n(p_n - p_H), \tag{1}$$

where for air-jet engines, β is the fuel mass-to-air flow ratio per second; for rocket engines, $\beta = 0$ and $V = 0$ because the fuel propels together with the vehicle. The jet engines can have a thrust of from 10^{-2} N in ion-plasma jet engines, to 10^7 N in liquid-propellant rocket engines and $10^3 - 3 \times 10^5$ N in air-jet engines.

The efficiency of converting the kinetic energy of a jet into useful work of engine propulsion is estimated by the flight efficiency

$$\eta_f = \frac{2W/V}{1 + W/V}. \tag{2}$$

The major parameters characterizing jet engines are

(a) Thermo- and gas dynamic performance:

Measures of performance include the specific impulse J_{sp} (thrust-to-working substance flow rate ratio, $(R/\dot{M})$ or the specific thrust R_{sp} (thrust-to-air flow rate ratio in air-jet engines, $R/\dot{M}_a$). J_{sp} is highest in photon engines (3×10^8). It is $10^4 \times 10^5$ in ion-plasma jet engines and 4 and 3×10^3 in liquid-propellant and solid-propellant rocket engines, respectively. In air-jet engines, R_{sp} varies from 10^2 to 1.5×10^3 Ns/kg.

Specific thrust is related to the useful work (L_c) of the thermodynamic cycle through which the engine generates jet kinetic energy

$$R_{sp} = \sqrt{2L_c + V^2} - V, \tag{3}$$

in this case, $V = 0$ for rocket engines.

Another important parameter is the specific consumption of the working substance C_R (in air-jet engines, of fuel), i.e., flow rate-to-thrust ratio. The total efficiency of the engine is given by:

$$\eta_0 = \frac{V}{C_R H_u} = \frac{J_{sp} V}{H_u} \tag{4}$$

where H_u is the heat of combustion. η_0 is equal to the product of the flight efficiency η_f and the efficiency of the cycle η_t

In rocket engines, $C_R = 1/J_{sp}$, i.e., $10^{-2} - 10^{-9}$; and in air-jet engines, $C_R \approx 0.3 \times 10^{-2}$ kg/sN.

(b) The design effectiveness:

A design objective is to increase the ratio of mid-section thrust (R_f) to the mass of the engine.

Operating Process and Main Parameters of Air-Jet Engines

The conversion of heat energy released in fuel combustion in air-jet engines into the kinetic energy of a jet issuing from the engine and developing a gas thrust occurs in the cycle at constant pressure (Brayton's cycle). An air stream at a velocity V, pressure p_H and temperature T_H taken in by the engine is initially compressed due to retardation in the air intake ($\Pi_v = p_{in}/p_H$) and further compressed due to mechanical energy supply in the compressor ($\Pi_c = p_c/p_{in}$). The air is heated by the fuel combustion in the combustion chamber to a temperature T_g and then is expanded first in the turbine ($\Pi_T = p_g/p_T$), which rotates the compressor and auxiliary units, and later in the nozzle ($\Pi_n = p_T/p_H$), accelerating up to a preset exhaust velocity W.

Figure 1 shows the effect of the parameters T_g/T_H and Π_c on a turbojet engine in ground conditions (H = 0 km, Ma = 0, i.e., for T_H = 228 K) and in flight conditions Ma = 2.2, H = 11 km relative to the $R_{sp} - C_R$ coordinates.

The maximum temperature of the gas in the T_g cycle (in gas-turbine engines, the temperature before the turbine) determines the efficiency of all types of air-jet engines (R_{sp} and C_R) first. At a temperature T_g lower than T_{gmin}, based on the equality of terms on the right-hand side in Equation 1, air-jet engine efficiency cannot be realized at all. The degree of pressure increase has only a slight effect on air-jet engines.

The rise in values of T_g in gas-turbine air-jet engines from 800°C (in 1940) up to 1,500°C (by 1990) has been realized by enhancing the high-resistance alloys used and, to a greater extent, by the introduction and improvement of methods for cooling the first-stage turbine blades, the turbine (rotor) blade and the nozzle guide vane. Rotor blades work under especially severe conditions; military air-jet engines have circumferential speeds of 500–600 m/s at a gas temperature of up to 1,500°C. The requirements of long-term strength and low-cycle thermal fatigue dictate the following temperatures of blade walls from modern alloys: for rotor blades, not higher than 850–950°C and for nozzle guide vanes, up to 1000–1100°C. Air bled from the compressor is passed through the blades to cool them from within.

Cooling efficiency depends on the complexity of the internal structure of the blade and on the ratio of air flow rates—cooling air vs. that flowing around the outside of the blade, $\delta = \dot{M}_c/\dot{M}_a$.

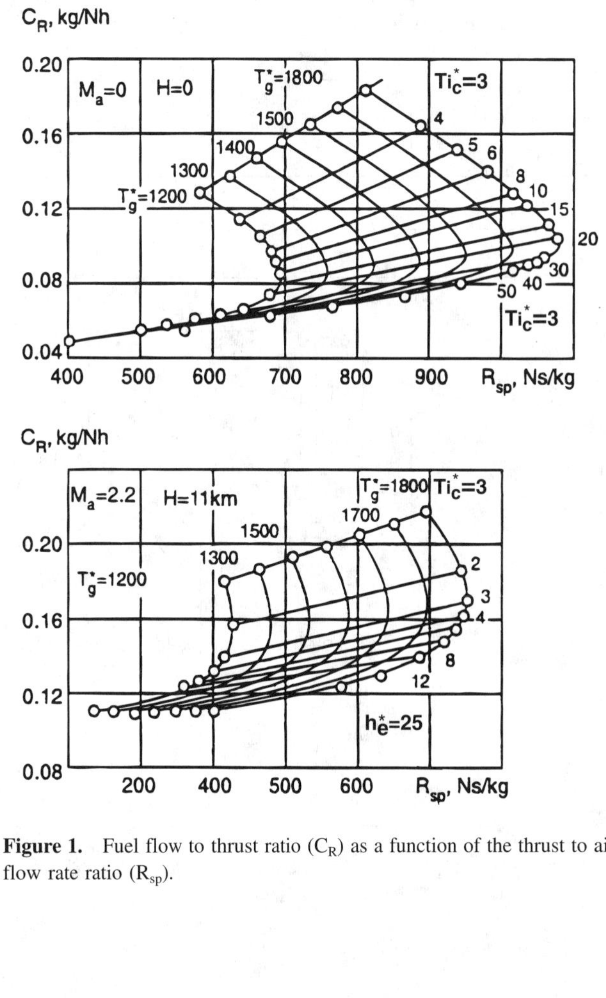

Figure 1. Fuel flow to thrust ratio (C_R) as a function of the thrust to air flow rate ratio (R_{sp}).

Figure 2 characterizes the efficiency of cooling $\theta = (T_g - T_{bl})/(T_g - T_c)$ of a series of blade cooling schemes. Thus, the value $\theta = 0.5$, which corresponds to equal temperature drops between gas flow and the blade and between the blade and cooling air, is characteristic of a high degree of heat exchange intensification inside the blade (intricate eddy flows and a specially developed heat transfer surface) because the gas velocity on the outside of the blade is 10 times higher than the air velocities inside it. The values of δ are usually in the range 2.5–5%. To increase T_g, intermediate air bled from the compressor is often cooled in a special heat exchanger with fuel cooling capability.

The entry of cooling air into the flowing duct, as a rule, diminishes the flow around the turbine blades, reducing their efficiency. But in this case, the useful effect of a gas temperature rise reduces the deterioration in efficiency.

The most efficient turbines are the high-temperature turbines made from super heat-resistant materials (ceramics, composites, etc.). This is the main trend in the development of future air-jet engines in which only blades made from silicon carbide and niobium are used, which operate at gas temperatures of about 1,300°C. The technological difficulties associated with the manufacture of a less than 1 mm thick ceramic blade training edge, however, make their efficiency worse than that of metal blades and, therefore, prevents the complete elimination of cooling mechanisms.

In air-jet engines of more intricate designs, by-pass engines (with or without mixing of flows from a turbofan, or an outer duct, or a turbocompressor, or an inner duct) with thrust augmentors are employed and the parameters R_{sp} and C_R are usually used. Turbojet by-pass engines are now the main type of aircraft engines used in both civil and military aviation. **Figure 3** shows how higher flight efficiencies (Equation 2) are obtained at the expense of larger air masses with lower exhaust velocities (a large value of the by-pass ratio $m = G_{aII}/G_{aI} = 3 - 6$) and accordingly, of smaller C_R when $H = 11$ km, $M = 0.8$ and $T_g = 1600$ K. In order to obtain minimum C_R for the given flight conditions and with a turbocompressor selected (T_g and $\Pi_{k\Sigma}$), the optimal combinations of by-pass ratio m and relative pressure increase in the fan Π_F—governed mainly by the requirement for equal conditions of flow pressures issuing from both ducts—can be determined. An increase in the by-pass

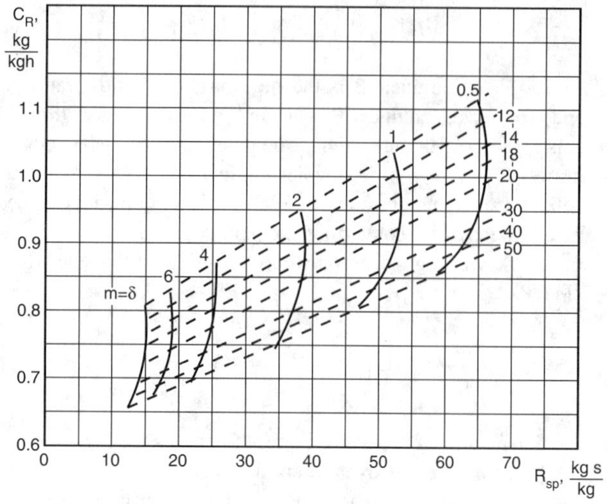

Figure 3. Effect of bypass ratio in overflow rate-to-thrust ratio.

Main Data	Engine Type				
	JT8D-11(15)	JT9D-3A	JT9D-7	JT9D-59/70D	F-100
Tg, K	1294~1340	1420	1500	1598...1643	1590...1672
δ, %	~1,0	1,96	2.1	3.4	~3.0
θ	~0.2	~0.3	~0.45	~0.5	~0.55
Cooling system development	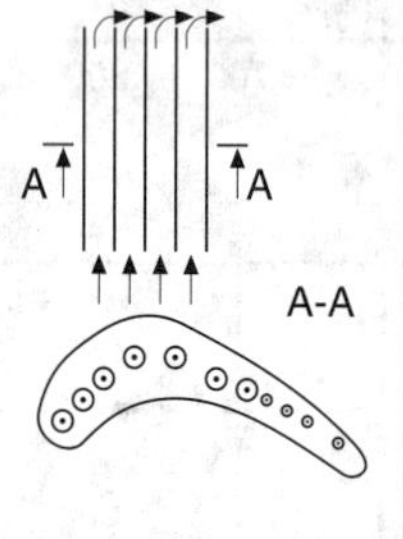 A A A-A Radial air flow ducts Multiflow - line schematic diagram of air flow	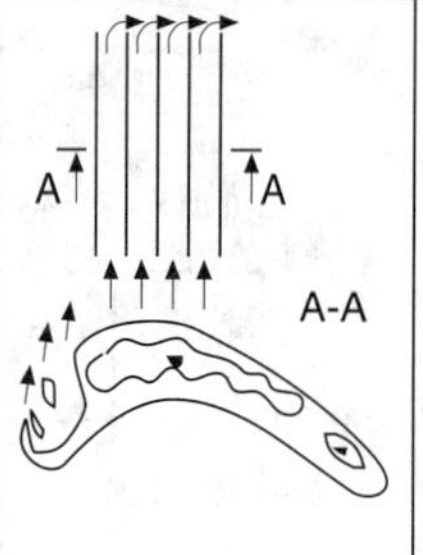 A A A-A Radial air flow ducts	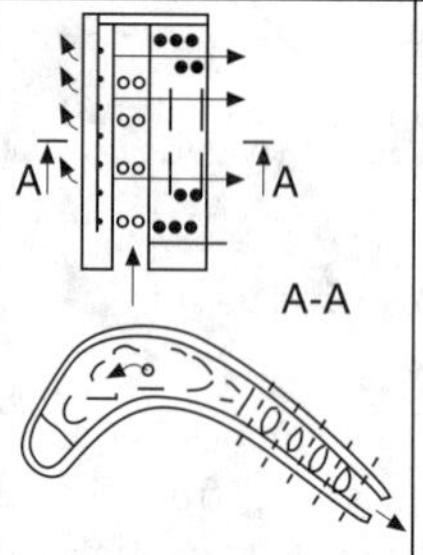 A A A-A Deflector blade with air jet stream	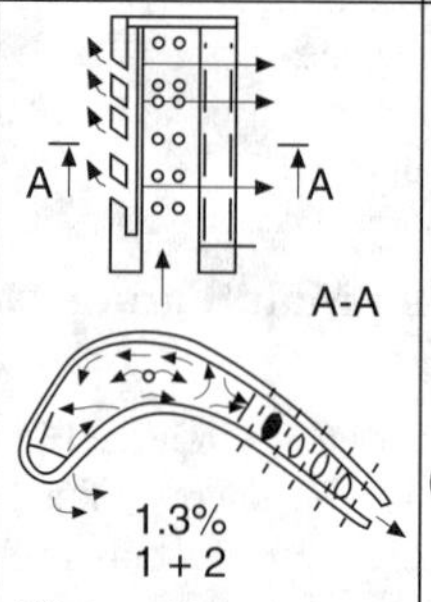 A A A-A 1.3% 1 + 2 Film cooling and cooling air twisting	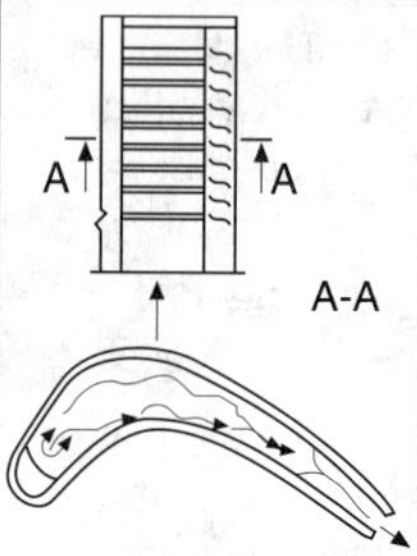 A A A-A 1. Deflector blade welded from two parts 2. Cooling air twisting 3. Development surface of inner finning

Figure 2. Cooling efficiency for various types of blade cooling in jet engines.

ratio is always accompanied by the need to reduce the Π_F, causing a decrease in specific thrust and a rise in the overall dimensions of an engine. It is the size of the latter in aircraft conditions (in the wing geometry and taking account of the total outer aerodynamic drag) that restricts the tendency for a decrease in m, starting with the reduction of C_R.

Figure 4 shows the effects of the various degrees of upgrading (with respect to specific thrust and fuel consumption) turbojet engines can undergo by feeding additional (after-burn) fuel into the air duct behind the turbine, i.e., the effect of gas temperature in the thrust augmentor of a turbojet engine, with $\Pi_c = 8$, $T_g = 1{,}400$ K.

Operating Characteristics of Air-Jet Engines

The conditions in which air-jet engines are operated determine the changes in their main parameters (the thrust and the fuel consumption) with velocity (Mach number), flight altitude (H) of the aircraft and the operation mode of the engine. Engine parameters change considerably at ground conditions with the temperature of the surrounding air.

The typical operating modes of air-jet engine operation are: *maximum mode* (with maximum permissible gas temperature and r.p.m.) where the time is usually limited and applications are during takeoff and acceleration of the aircraft; the *maximum continuous power mode* (80–85% thrust and 95% r.p.m.), with unlimited time and used for instance, in climbing; *cruising power mode* (80–85% thrust and this is usually the condition of highest efficiency), the most continuous and *idling power mode* (5–7% thrust), a holding mode. In some cases (as a rule, in turboshaft engines) a contingency power mode exists.

Air-jet engines with a thrust augmentor (afterburn systems) have additional modes: *maximum reheat power*, i.e., with a maximum thrust for limiting n, T_g and T_F (at take-off or at target interception); the mode of *partial thrust augmentation*, with a decreased n, T_F and even T_g and unlimited time of operation used for instance, in continuous supersonic flight and the *minimal afterburning mode*, defined by sustained combustion in the thrust augmentor and required for greater smoothness of control.

The change in the thrust with height, velocity of flight and mode is connected to the level of maximum gas temperature in

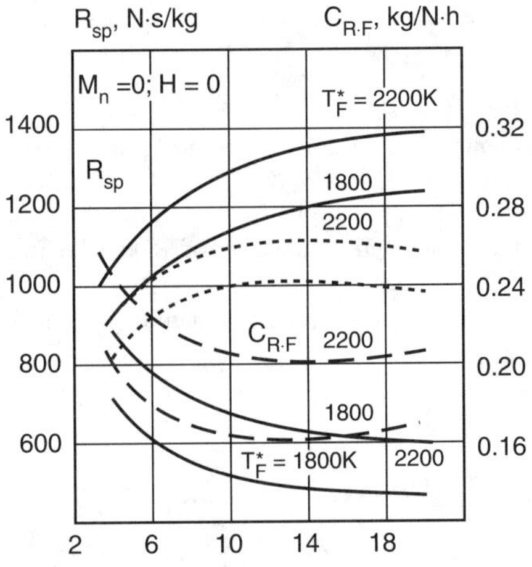

Figure 4. Effect of afterburn on engine performance.

the cycle, the temperature and pressure of the air which enters the engine, the values of pressure increase and with the efficiency of the elements under these conditions. These values, which enter into Equations 4 and 5, and the absolute values of air flow rate through the engine determine the thrust and the efficiency for the given M, flight altitude and rotor r.p.m. (n).

Air-jet engines operate under broad changes in environmental conditions, not only of the temperature and pressure of the feeding air but also of the degree of uniformity of the parameter fields W, P, T and their nonstationary characteristics. Such elements as air intake, the compressor and the combustion chamber have regions where their characteristics can be unstable (air flow separation, stall) and where either the operation can cease or damage due to the action of pulsations can occur. Agreement between the amount of fuel supplied under set conditions and the required charges in the geometry of the flowing duct (usually the guide vanes of the compressor and areas of the nozzle and air intake) is ensured by a system of computer based control which can take account of up to 12–14 optimal parameters.

Ram-jet engines are aircraft engines for high-speed flight in which the Brayton cycle is, as a rule, realized in flight conditions without a turbocompressor in the air intake-combustion chamber-nozzle system. Depending on the set flight velocity and on the characterstics of the main elements, ram-jet engines are classed as subsonic, super and hypersonic by the type of air intake; as having subsonic or supersonic velocities in the combustion chamber and as having continuous and pulsating action. Ram-jet engines are distinguished by a rapid increase in thrust with flight velocity due to a rise in pressure and in air density. But at flight velocities where the stagnation temperature of the air passing through the air intake into the combustion chamber approaches the combustion temperature (2,000–2,200 K), the warm-up decreases and the thrust diminishes drastically. Operation under conditions of supersonic combustion (such engines are called hypersonic ram-jet engines) considerably enlarges the application of ram-jet engines depending on flight Mach number. The real Ma for hypersonic ram-jet engines increases from 3–5 to 10–12.

References

Bathie, W. W. (1984) *Fundamentals of Gas Turbines.* John Wiley and Sons.

Cohen, H., Rogers, G. F. S., and Saravanamutoo, N. J. N. (1987) *Gas Turbines.* Longman Scientific and Technical.

Ferry, A. (1969) *Supersonic Turbojet Propulsion Systems and Components.*

Wilson, D. T. (1984) *Design of High-efficiency Turbomachinery and Gas Turbines.* MIT Press.

O. Favorsky

JET ENGINES, COMBUSTION CHAMBERS FOR
(see Combustion chamber)

JET FIRES (see Fires)

JET PUMP (see Boiling water reactor, BWR)

JET PUMPS AND EJECTORS

Following from: *Pumps*

Ejectors, or jet pumps, utilise the pressure energy of a high-pressure fluid stream to boost the pressure and/or flow of a low-pressure source. They can operate with either incompressible or compressible fluids as the primary (driving) and secondary (driven) flows. The main features of an ejector are shown in **Figure 1**. The figure also defines the subscripts used later for primary (1), secondary (2), etc.

The primary fluid is passed through a nozzle where the pressure energy is converted into kinetic energy. The high-velocity jet entrains the secondary fluid. The two streams mix in the mixing tube, leading to pressure recovery. Further static pressure is recovered in a narrow-angle diffuser downstream of the mixing tube.

Ejectors are generally inefficient devices. However, their simplicity and lack of moving parts make them worthy of consideration, particularly where a high-pressure stream of fluid is already available. **Table 1** summarises potential ejector applications.

Table 1. Summary of ejector applications

Primary fluid	Secondary fluid
Liquid	Liquid
	Gas
	Multiphase
Gas	Gas

Note that the term 'Jet Pump' is used to refer to a liquid—liquid ejector.

The most comprehensive source of design information for ejectors can be found in a series of Engineering Sciences Data Unit (ESDU) data items, Nos. 85032 and 84029. These are available on subscription as part of the ESDU Internal Flow series.

Liquid-Liquid (Jet Pumps)

Three key parameters for a jet pump are the pressure ratio, defined by

$$N = \frac{P_5 - P_2}{P_1 - P_5}$$

where P_1, etc. are total pressures, as indicated in **Figure 1**

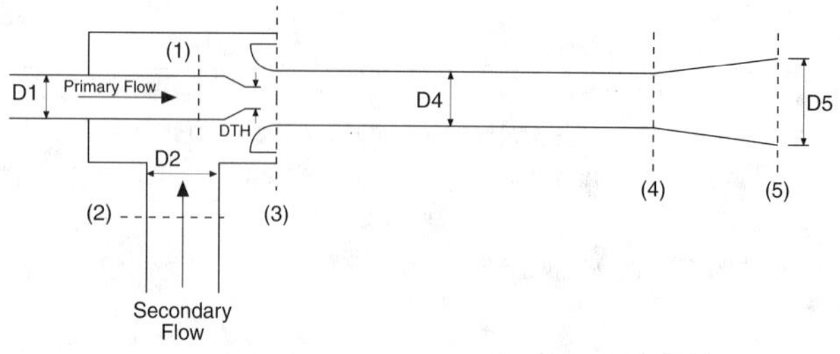

Figure 1. Main feature of a typical ejector.

The flow ratio

$$M = \dot{V}_2/\dot{V}_1$$

and the ratio of mixing tube-to-nozzle diameter (R). These are related through the equation

$$N = \frac{2R + \dfrac{2CM^2R^2}{1-R} - R^2(1+CM)(1+M)(1+K_m+K_d) - \dfrac{CM^2R^2}{(1-R)^2}(1+K_s)}{(1+K_p) - 2R - \dfrac{2CM^2R^2}{1-R} + R^2(1+CM)(1+M)(1+K_m+K_d)}$$

where C is the density ratio (secondary to primary). The loss coefficients K_p, K_s, K_m and K_d account for losses in the primary nozzle, secondary flow inlet, mixing chamber and diffuser, respectively. For high Reynolds number applications (above 2×10^5), values of 0.05, 0.1, 0.15 and 0.2 can be assumed for a well-designed jet pump.

The equation can be solved directly for N if C, M and R are known. Alternatively, either graphical (e.g., ESDU 85032) or equation solving methods can be used.

Ejector efficiency can be calculated from

$$\eta = M \times N.$$

Figure 2 shows a typical efficiency and performance curve for a jet pump, as a function of secondary flow rate. Note the typical shape, with a peak efficiency of around 35%.

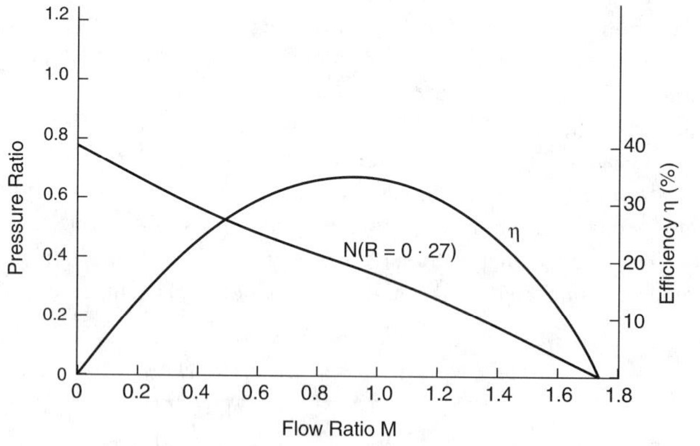

Figure 2. Typical jet pump performance curves.

For a particular application (i.e., given primary and secondary flows/pressures), an optimum value of R can be found by trial and error.

Once the ratios have been determined, the primary nozzle can be sized, from

$$D_n = \sqrt{\frac{4\dot{V}_1}{C_D\pi}\sqrt{\frac{\rho_1}{2(P_1 - P_2)}}}$$

For a well-designed nozzle, a value of 0.95 for discharge coefficient (C_D) can be used. The nozzle should be conical with a half-angle of 5–10°. A parallel section at the nozzle outlet (see **Figure 1**) is not critical to performance, but can improve the mechanical strength of the design.

Entry to the mixing tube needs to avoid large secondary flow losses: either a converging conical section or a bell-mouth entry should be used. A mixing tube length of 7–10 (mixing tube) diameters is recommended.

To reduce downstream pressure losses, the flow needs to be expanded downstream of the mixing tube to lessen flow velocities to a reasonable level. As this may involve a large area ratio, a narrow angle diffuser is required (typically 2–3° half angle).

When operating at low suction pressures and high flow ratios, cavitation can prove to be a problem with jet pumps. Detailed information is available in Cunningham et al. (1969) or ESDU 85032.

Gas-Gas Ejectors

For small pressure differences, gas-gas ejectors can be treated like liquid jet pumps. However, for higher pressure ratios, compressibility effects need to be taken into account. Above a critical pressure ratio between primary and secondary (around 1.8, depending on gas properties), flow in the primary nozzle reaches sonic velocity. Flow in the nozzle becomes independent of secondary pressure, and is given by:

$$\dot{M}_1 = C_D P_1 S_{TH} \sqrt{\frac{\gamma}{R_1 T_1}\left(\frac{2}{\gamma+1}\right)^{(\gamma+1)/(\gamma-1)}}.$$

where C_D is the discharge coefficient, S_{TH} the throat area, γ the ratio of specific heat at constant pressure to the specific heat at constant volume and R, the specific gas constant (See **Critical Flow**, **Jets** and **Nozzles**)

Downstream of the nozzle, flow will expand in a series of supersonic shocks until the pressures of both streams become equal and mixing occurs.

In some ejector designs, a converging-diverging nozzle is utilised to accomodate the expanding jet. An 'on-design' condition can be defined where the static pressures of primary and secondary flows are equal at the nozzle exit. However, work by Ashton, Green and Reade (1993) suggests that the use of a diverging section is not necessary for effective operation, at least at moderate pressure ratios. Performance can be then calculated by considering conservation of mass, momentum and energy in the mixing tube and diffuser. Owing to the complexity of the equations, these cannot be solved directly. A complex graphical method is available in ESDU 84029.

As with a jet pump, the key geometric factor in the design is the mixing tube diameter. Performance increases by reducing mixing tube diameter up to a point where the expanding supersonic primary jet almost fills the mixing tube before mixing can take place and choking occurs. A further decrease in mixing tube diameter (or any attempt to increase secondary flow) causes performance to decrease rapidly.

Design requirements for the nozzle, mixing tube and diffuser are similar to those for jet pumps.

Liquid-Gas Ejectors

These ejectors utilise liquid as the primary fluid: generally, gas-driving liquid is not a very effective arrangement due to differences in density (and hence momentum) between the streams. Performance can be generally characterised by an equation of the form

$$\dot{V}_G = B(\dot{V}_L - \dot{V}_{Lmin})$$

where G and L refer to the gas and liquid, respectively and V_{Lmin} is a minimum value of liquid flow below which no gas flow will occur. An expression for V_{Lmin} has been derived by Henzler (1980), given by:

$$\dot{V}_{Lmin} = 0.38\,\frac{D_m}{D_n}\left[\frac{\rho_L}{\rho_G}\right]^{0.09}\left[\frac{p_5}{p_2} - 1\right]^{-1/6}\sqrt{\frac{2(p_5 - p_2)}{\rho_L u_j^2}}$$

Given the different densities of the two streams, mixing duty tends to be more arduous than in a single-phase ejector, and generally longer mixing tubes than those for single-phase ejectors are used (typically 20–30 diameters). However, where mixing does occur, this is very intensive (the 'mixing shock'). This tends to result in significant energy losses. A spinner upstream of the primary nozzle is sometimes used to help disintegrate the jet and induce early mixing.

Ejectors for Mass Transfer

Gas-liquid ejectors can be very effective devices for mass transfer applications. They can be used either as stand-alone devices, or in combination with a contact vessel. They have the combined benefits of being able to draw in gas without the need for compression, and providing a very fine dispersion in the mixing tube.

Ejectors are characterised by cocurrent plug flow in the mixing tube, with very high energy dissipation rates (typically in the range 100–1,000 W/kg), but short residence times (less than 1 second). This leads to mass transfer coefficients typically 2–3 orders of magnitude greater than a typical stirred tank, making them particularly suited to absorption with rapid, competing chemical reactions where fast mixing is required to reduce byproduct formation.

Where longer residence time is required, ejectors are often combined with a contact vessel. In such cases, the ejector provides rapid initial mixing, a fine bubble dispersion and, if properly designed, good liquid mixing in the vessel. Gas can either be recycled, by using the suction characteristics of the ejector to draw gas back from the headspace of the vessel, or operated in once-through mode. The 'Buss Reactor,' successfully used for hydroginations, sulphonations, aminations, etc. is designed on the former principle [e.g., see van Dierendonck and Leuteritz (1988)].

References

Ashton, K. J., Green, A. J., and Reade, A. (1993) *Gas Production Improvements Using Ejectors.* SPE Paper 26684. Society of Petroleum Engineers.

Cunningham, R. G., Hansen, A. G., and Na, T. Y. (1969) *Jet Pump Cavitation.* Paper 69-WA/FE-29. ASME.

ESDU 84029. (1984) Ejectors and jet pumps. Design for compressible air flow. Engineering Sciences Data Unit

ESDU 85032. (1984) Ejectors and Jet Pumps. Design and performance for incompressible liquid flow. Engineering Sciences Data Unit.

Henzler, H. J. (1980) *Chem. Ing. Tech.* 47: 5: 659. (in German).

van Dierendonck, L. L. and Leuteritz, G. M. (1988) *Proc. Sixth Euro. Conf. on Mixing*. Pavia. Italy. 287.

A.J. Green

JETS

Following from: *External flows*

Jets are liquid or gas flows in a space filled with a fluid with different physical parameters, namely velocity, temperature, composition, etc. The distinctive features of these flows can be illustrated by gas jets flowing out into a space filled with gas at rest.

Figures 1 and **2** present two typical patterns of immersed jets. The first (Figure 1) corresponds to an isobaric jet in which static pressure is constant throughout an entire volume. The initial outflow velocity U_0 can be either sub or supersonic. The second pattern (Figure 2) is barrel-shaped in the initial section and corresponds to the supersonic efflux of an underexpanded jet when static pressure at the nozzle exit section p_a substantially exceeds the environmental pressure p_H. The $p_a/p_H = n$ ratio is the most important parameter characterizing gas expansion at the initial section of the jet. Within one or a few "barrels" the external boundary of the jet contour becomes more regular and monotonically expands. Beyond this point, the jet pattern is similar to an isobaric jet (cf. **Figures 1** and **2**). The initial sections of length x_0 of the two forms of jets differ from each other not only in external contours but also in the character of distribution of the most significant gas dynamic parameters.

The isobaric jet (**Figure 1**) has a core in the entrance region in which the axial gas velocity actually remains constant and equals U_0. Between the core and the external boundaries there arises a dynamic, thermal, or diffusion boundary layer where turbulent mixing of jet particles and the quiescent medium occurs. Beyond the initial section $x > x_0$, the boundary layers unite and axial flow velocity decreases inversely with x (the distance from the nozzle) for an axisymmetric jet, or with $x^{-0.5}$ for a plane jet. The external boundary of the jet is not a tangential discontinuity, i.e., shows no jumps in its parameters and, like the thickness of a boundary layer, can be determined only approximately. If the surface on which gas velocity is 0.1 of that of axial flow is taken as the jet boundary, the thickness of the isobaric jet, in fact, grows linearly with x. For $x > x_o$ the main section the proportionality factor is equal to 0.22, irrespective of jet shape, velocity and physical state. In the initial section ($x < x_o$) the proportionality factor depends on the ratio of densities ρ_0/ρ_H or temperatures (T_0/T_H), (the subscripts "0" and "H" denote the initial jet and environmental parameters).

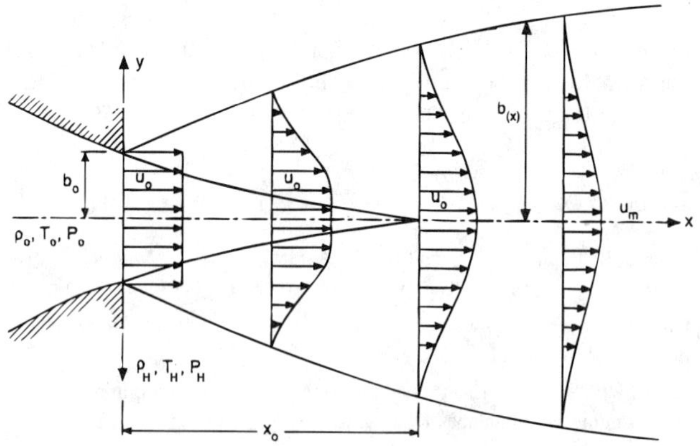

Figure 1. Subsonic isobaric jet.

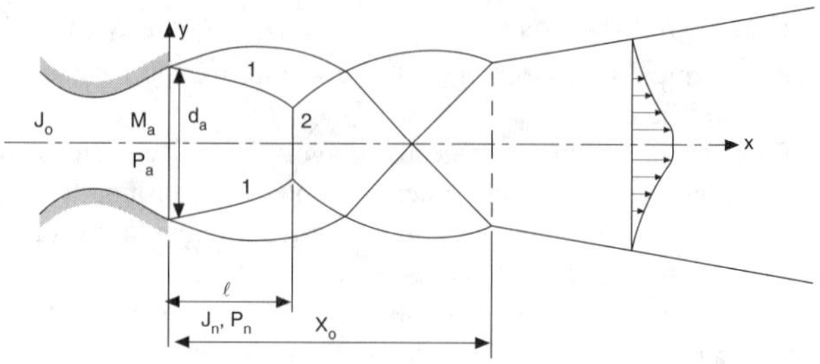

Figure 2. Supersonic jet.

Figure 3 demonstrates the effect of preheating $\theta = T_0/T_H$ on the length of the initial section $\bar{x}_0 = x_0/b_0$ for the axisymmetric (1) and plane (2) subsonic isobaric jets. **Figure 4** shows the axial velocity for the axisymmetric subsonic jet as a function of preheating θ.

The characteristic property of the isobaric jet is that the transverse velocity components are negligible relative to the longitudinal velocity in any jet section. The profiles of excess velocity, temperature and impurity concentration in cross-sections of both immersed jet and the jet in the cocurrent flow are similar in shape. **Figure 5** presents the velocity profiles at various cross-sections of a plane jet and **Figure 6** shows a universal profile of all the sections in dimensionless coordinates. To describe the velocity profiles in the main section ($x > x_0$) of the jet of any shape, Schlichting has derived the function f(η)

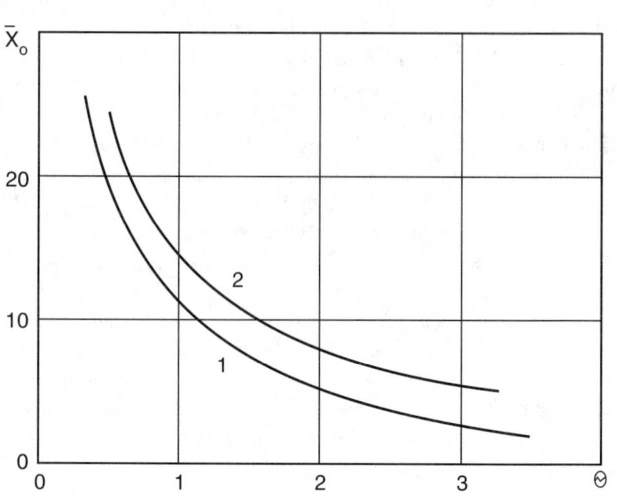

Figure 3. Effect of preheating on the length of the initial section of a jet (1, axisymmetric, 2 plane).

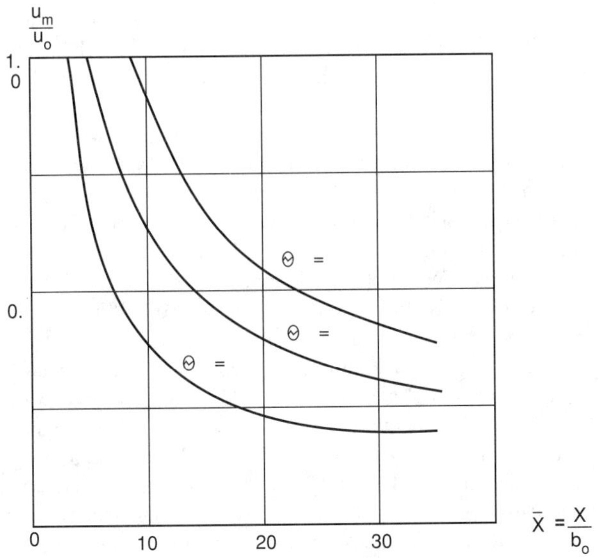

Figure 4. Axial velocity of a subsonic jet as a function of distance and preheat.

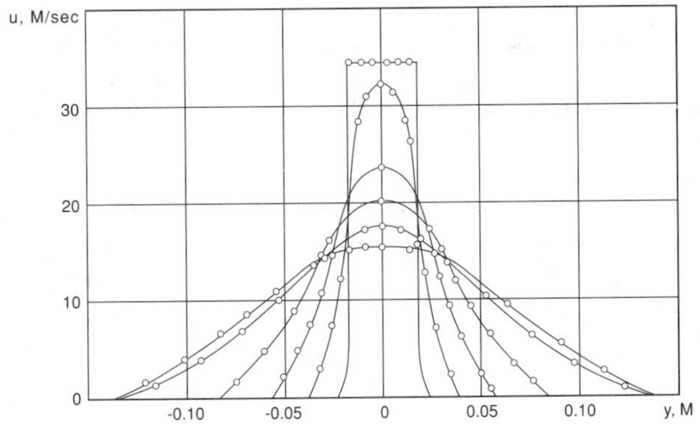

Figure 5. Velocity distribution in a jet.

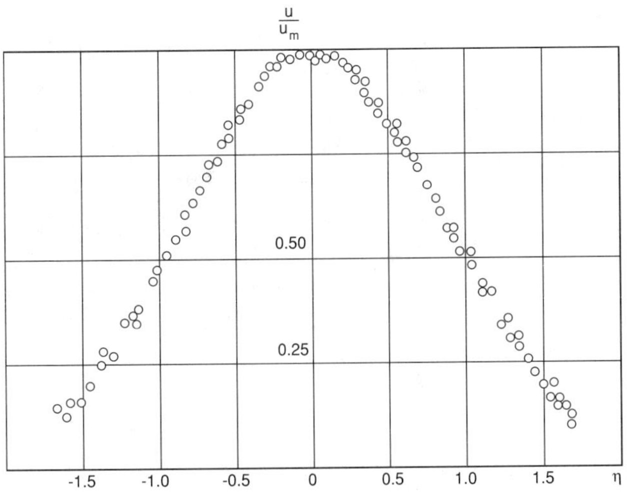

Figure 6. Universal profile of velocity in a jet.

$$\frac{u - u_H}{u_m - u_H} = f(\eta) = (1 - \eta^{3/2})^2.$$

Here $\eta = y/b$ is the distance from the jet axis to the radius at a given cross-section b. The function $f(\eta)$ holds for $Ma_0 \ll 1$ and for small diffences in density between the jet ρ_0 and the environment ρ_H, i.e., at $\beta = \rho_H/\rho_0 \approx 1$. However, experiments show that the universal nature of the velocity profiles sitll holds in the range $0.25 \leq \beta \leq 4$. For cross-sections in the main section ($x > x_0$) the dependence of excess temperature on excess velocity is

$$\frac{T - T_H}{T_m - T_H} = \left(\frac{u - u_H}{u_m - u_H}\right)^{Pr_t}$$

is valid. Here, Pr_t is the turbulent Prandtl number proportional to the ratio of heat released as a result of turbulent friction to that removed by turbulent mixing.

Experiments carried out at $\beta = 0.03 - 300$ show that $Pr_t = 0.8$ for axisymmetric jets and $Pr_t = 0.5$ for plane jets.

In jets flowing out of rectangular nozzles, flow sections can be distinguished in which the jet properties approach the appropriate analogous jets: the plane jet, near the nozzle and the axisymmetric jet, at long distances. Between them, the jet is specified by three coordinates. Initially, jet expansion toward the minor nozzle axis is more intense than toward the major axis, and at some distance from the nozzle the jet pattern changes orientation by 90°, but the ratio of the larger to the smaller side turns out to be below the original one. Depending on outflow conditions, the jet may become an axisymmetric one at the distance of 100h and over (h is the height of the shorter side of the nozzle). The mechanisms of velocity decay become the same as in an ordinary axisymmetric jet at different distances from the nozzle exit section in accordance with the nozzle aspect ratio. This characteristic distance can be estimated from experimental data using the dimensionless complex $x/\sqrt{F_0} >$ (3–6) (F_0 is the total area of the nozzle exit).

In supersonic, underexpanded ($n = (p_a/p_H) > 1$) axisymmetric jets the entrance region is characterized by an intricate flow pattern. With excessive pressure, the jet first expands and its velocity grows. Near the nozzle edge, there arises the beam of rarefaction waves that facilitate gas expansion in the jet from pressure p_a at the nozzle exit section to the pressure p_H of the ambient gas. Acceleration of flow involves the Prandtl-Meyer expansion, and the jet boundary remains linear until it crosses the first characteristic. Overexpansion due to radial gas effusion gives rise to an intercepting shock 1 (**Figure 2**). Depending on the inclination angle of the shock (which in turn, depends on the pressure ratio at the nozzle exit section), the intercepting shock is either reflected at a certain point lying on the axis (regular reflection) or forms a Mach disk (2 in. **Figure 2**). The Mach disk is the surface of a high intensity shock normal to the flow direction and interacting with the intercepting shock and the reflected shock wave at the triple point. The distance l from the nozzle exit section to Mach disk is proportional to the nozzle diameter d_a, the Mach number Ma_a, and the square root of the pressure ratio n. The reflected shock wave propagates in the outer region of the jet, making the gas flow once again from the center towards the periphery. When the reflected shock wave reaches the jet boundary the cycle is completed. Side oscillations may repeatedly occur along the flow before they damp altogether. The results from numerical methods demonstrate that at some length from the nozzle exit section, the flow is in good agreement with the model of gas expansion from a point source into vacuum.

Interaction of supersonic, underexpanded jet with a plane obstacle can be conventionally represented by three types of flow in accordance with the position of the nozzle relative to an obstacle. In the case when the obstacle is between the nozzle exit section and the point $H \leq 0.9x_m$, a stable interaction regime is observed (x_m is the distance from the nozzle exit section to the Mach disk in the case of a free, supersonic jet). As the distance from the obstacle to the nozzle exit section increases in the $(0.9–1.3)x_m$ interval, a peripheral supersonic flow, possessing a high stagnation pressure, pinches the central subsonic flow to give rise to a peripheral stagnation point. Closure of the central zone of flow brings about accumulation of gas in it and a shift of the central shock toward the nozzle, which increases pressure before the obstacle with subsequent opening of this zone, etc. With a further increase in the distance from the obstacle to the nozzle exit section, the flow is stabilized and a stable circulation zone is formed around the obstacle.

If the obstacle is positioned in the interval $(1.8–2.1)x_m$, an oscillation of the nozzle section wave structure arises once again, but the intensity of oscillations and the noise generated are much weaker than in the first instability regime. As the distance from the obstacle increases, the first barrel of the jet becomes undisturbed and a central compression shock, resulting from interaction of the second barrel with the obstacle, appears before the obstacle. The boundaries between the interaction regimes indicated above hold in the range $n = 2 - 20$, $Ma_a = 2 - 3$.

The changes in heat transfer characteristics from the jet to the obstacle in the above cases are considered next. If the distance is short $\overline{H} < \overline{H}_{1H}$ ($\overline{H}_{1H}$ is the distance from the nozzle exit section to the obstacle at which instability occurs) and the flow around the obstacle is stable, the maximum heat transfer coefficient α is at the stagnation point (curves 1–3, in **Figure 7**). With a greater distance from it α monotonically diminishes. As the pressure ratio increases and at $\overline{H}$ = *const*, α profiles become fuller near the stagnation point, remaining virtually constant in the center of the obstacle and growing substantially at the periphery. It should be noted that with variation of Ma_a the α value and its character of distribution change only slightly. A further increase in $\overline{H}$ within the stationary flow gives rise to a peripheral maximum α which shifts toward the stagnation point (curves 4–6 in **Figure 7**), as the obstacle is moved further from the nozzle. The same effect on α variation is exerted by n at $\overline{H}$ = *const*. The peripheral maximum in α arises due to the effect of turbulent pulsations on heat transfer in the inner mixing zone, originating from the triple point of the central compression shock. The peripheral maximum is also characteristic of unsteady flow. Interaction beyond the first barrel results in only one maximum at the stagnation point.

In the modeling of high-intensity, convective heat transfer in jets, determination of maximum Reynolds number at given values of pressure and temperature in a plenum chamber of a gas dynamic unit is of fundamental importance. This is reduced to an analysis of conditions under which the jets of finite diameter interact with blunt bodies of various shapes. It has been established that the maximum Re number, nonseparated flow around a hemispherical blunt body corresponds to the outflow regime at the nozzle section $Ma_a = 2.5$, the diameter of the body being no more than 0.9 of the nozzle diameter. The model can be enlarged about half as much again by using the cool gas wake with the same Mach number $Ma_a = 2.5$ at the nozzle exit section. The use of cool wakes makes it possible to sharply reduce the requirements for heat output of a test setup. However, all the conditions minimizing disturbances on the jet mixing boundary (shaping the nozzle edge, equalizing static pressures and Mach numbers in the exit section, smaller size of an edge) must be satisfied.

Short-duration and stationary jets differ greatly in both structure and parameter distribution.

A schematic representation of the initial stage of underexpanded, supersonic, short-duration jet is presented in **Figure 8**. Shock wave 1 appears in front of the jet gas, whereas secondary

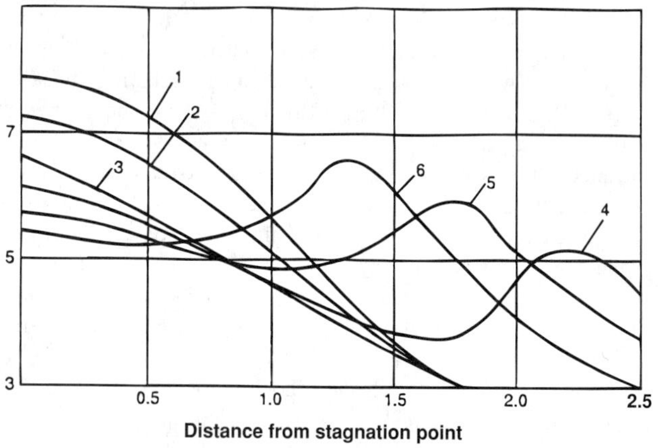

Figure 7. Variation of heat transfer coefficients with peripheral distance from stagnation point.

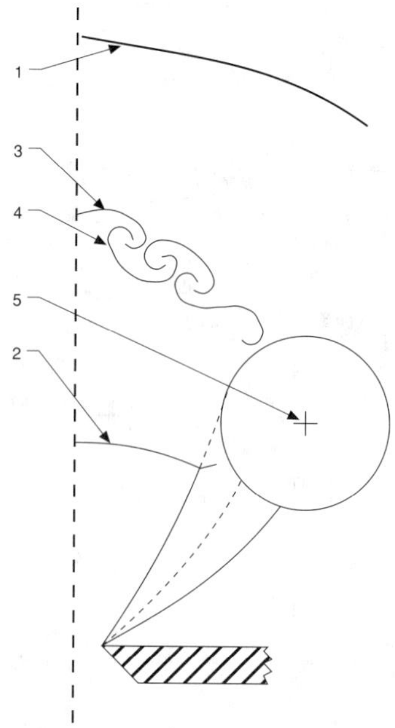

Figure 8. Schematic representation of underexpanded short-duration jet.

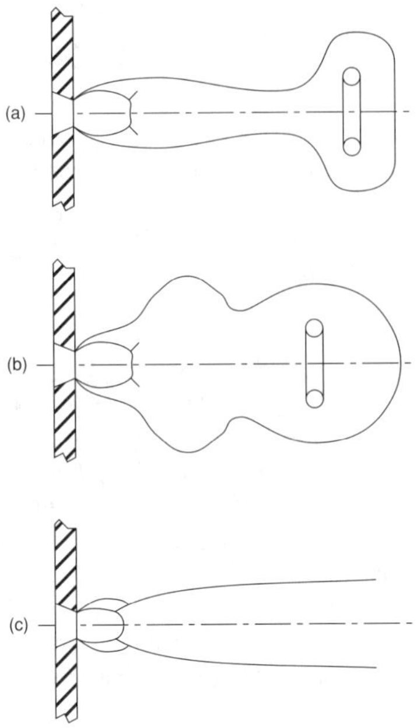

Figure 9. Transient behaviors of short-duration jets (a, cold gas, b, hot gas) compared with stationary jet (c).

shock wave 2 is formed in the jet itself. Three kinds of toroidal eddies are formed, with different causes underlying their formation. A vortex ring 3 emerges as a result of pulsed enhancement of pressure in the nozzle exit section at the initial moment of efflux and resembles in nature annular Wood vortices. Vortices of type 4 are formed due to instability of the forward front of contact discontinuity. Vortex 5 is formed as a result of swirling of the turbulent shear layer, which originates along the lateral boundary of the jet, and, during subsequent development of the jet, takes place at its "head." This increases by two or three times the cross sectional dimensions of the jet as compared to a stationary jet with the same parameters.

Just as in the case of the stationary jet, where the ratio of stagnation temperature to the environmental temperature and the Mach number may be such as to bring about quantitative changes in jet parameters, certain combinations of the Mach number and the temperature in short duration jets also cause structural changes in the flow. Around the jet, there may appear a region of backflow that is a swirling flow resembling a cocoon.

Figure 9 schematically depicts (a) the cold gas jet being formed, (b) the hot gas jet being formed, and (c) the stationary underexpanded jet. All three jets have identical pressure ratio.

V.V. Golub and Yu.V. Polezhaev

JETS, DEFLECTION NEAR SURFACES (see Coanda effect)

JETS, FREE (see Impinging jets)

JETS, LIQUID PLUNGING (see Plunging liquid jets)

JET SPLITTING, IN DROP FORMATION (see Drops)

JETS, SUBMERGED (see Impinging jets)

JETS, TWO PHASE (see Plunging liquid jets)

JETS, UNSUBMERGED (see Impinging jets)

JOINT CONDUCTANCE (see Thermal contact resistance)

JOINT RESISTANCE (see Thermal contact resistance)

JOULEAN DISSIPATION (see Magneto hydrodynamics)

JOULE HEATING

The use of electricity for heating purposes in the process industries is not widespread but it is increasing, particularly for special applications. The power may be purchased via the local electricity grid, or more often than not in a large chemical complex it will be generated by the recovery and transformation of waste heat via steam turbines and alternators.

The following are some of the advantages of electricity (a secondary form of energy) that need to be set against its relatively high cost compared with primary sources of energy such as coal and oil.

1. It is clean in operation
2. The energy is of constant quality
3. It is convenient and versatile
4. Control is relatively simple.

There are a number of different ways that the energy can be utilised, providing an opportunity to optimise cost and convenience. Common forms of electrical heating include:

1. *Resistance heating,* which involves passing an electric current through a resistance to generate heat. It is probably the most common method of using electrical power for process heating. The electric current may pass through an external resistance (indirect heating) or through the material to be heated (direct heating). It has been used for food processing [Skudder and Biss (1987)].
2. *Induction heating* utilises the transfer of energy from a coil to the workpiece via an alternating magnetic field. Traditionally, the technique has been used for metal heating, but in more recent times it has been applied to chemical reactors [Hobson and Day, 1985].
3. *Dielectric heating* involves subjecting the material to be heated—provided it contains suitable molecules capable of excitation—to the effects of an electric field alternating at radio or microwave frequencies. It has been used in polymerisation and curing processes.
4. *Infrared heating* depends on radiation effects. The principle is that an element (resistance-heated) radiates heat energy

that may be focused or reflected in the same way as light energy, and can therefore be directed as determined by the process requirements. The principles of **Radiative Heat Transfer** discussed elsewhere in this Encyclopaedia apply to this technique. Uses of infrared heating include drying of sheet material and spray-painted articles.

References

Hobson, L. and Day, J. (1985) Induction heating of vessels. *Int. J. Electr. Eng. Educ.* 2: 129.

Skudder, P. and Biss, C. (1987) Aseptic processing of food products using Ohmic heating. *Chem. Engr.* Feb: 26.

***Leading to:** Electric joule heater*

T.R. Bott

JOULE, JAMES (see Steam engines)

JOULE-THOMSON COEFFICIENT

When a gas in steady flow passes through a constriction, e.g., in an orifice or valve, it normally experiences a change in temperature. This is in part due to changes in kinetic energy, but there is another part contributed by the nonideality of the gas. If the upstream and downstream ducts are sufficiently large for kinetic energy to be negligible at these stations, upstream and downstream temperatures are measured far enough away from the disturbance created by the constriction and the system is adiabatic; the measured effect is due to the nonideality alone. From the *first law of thermodynamics,* such a process is isenthalpic and one can usefully define a Joule-Thomson coefficient as:

$$\mu_{JT} = (\partial T/\partial P)_H$$

as a measure of the change in temperature which results from a drop in pressure across the constriction.

For most real gases at around ambient conditions, μ is positive——i.e., the temperature falls as it passes through the constriction. For hydrogen and helium, it is negative and the temperature increases. At higher temperatures, for most gases, μ falls and may even become negative. μ can also become negative through application of pressure, even at ambient temperature, but pressures in excess of 200 bar are normally necessary to achieve this.

G. Saville

JOULE-THOMSON EFFECT (see Liquefaction of gases)

JURIN FORMULA, FOR CAPILLARY RISE (see Capillary action)

K

KADER'S FORMULA (see Tubes, single phase heat transfer in)

KAPLAN TURBINES (see Hydraulic turbines)

KARMAN-KOZEAY EQUATION (see Filtration)

KARMAN-NIKURADZE EQUATION (see Friction factors for single phase flow)

KAY'S LAW (see Gas law)

KELVIN (see International temperature scale)

KELVIN EQUATION (see Interfaces)

KELVIN EQUATION, FOR VAPOUR PRESSURE (see Capillary action)

KELVIN-HELMHOLTZ INSTABILITY

Following from: *Stratified gas-liquid flow; Two-phase flow*

The Kelvin-Helmholtz instability arises at the interface of two fluid layers of different densities ρ_g and ρ_l flowing horizontally with velocities u_g and u_l. By assuming that the flow is incompressible and inviscid, and applying a small perturbation it can be shown [Ishii (1982)] that the solution for the wave velocity is given by:

$$C = \frac{\rho_l u_l + \rho_g u_g}{\rho_l + \rho_g} \pm \left[C_\infty^2 - \rho_g \rho_l \left(\frac{u_g - u_l}{\rho_l + \rho_g}\right)^2\right]^{1/2} \quad (1)$$

where

$$C_\infty^2 = \frac{g}{k}\left(\frac{\rho_l - \rho_g}{\rho_l + \rho_g}\right) + \frac{\sigma k}{(\rho_l + \rho_g)}. \quad (2)$$

and k is the *wave number*, ie 2 π/wave length.

The displacement of the interface from the equilibrium configuration is proportional to exp $[ik(x - Ct)]$ and can therefore grow exponentially if the imaginary part of the wave velocity is nonzero. This will occur when:

$$\left[\frac{g}{k}\left(\frac{\rho_l - \rho_g}{\rho_l + \rho_g}\right) + \frac{\sigma k}{(\rho_l + \rho_g)}\right] < \rho_g \rho_l \left(\frac{u_l - u_g}{\rho_l + \rho_g}\right)^2, \quad (3)$$

where σ is the interfacial surface tension.

When rearranged this gives:

$$(u_g - u_l)^2 \geq \left(\frac{\rho_l + \rho_g}{\rho_g \rho_l}\right)\left[\sigma k + \frac{g}{k}(\rho_l - \rho_g)\right]. \quad (4)$$

For a system with finite depths h_l and h_g, modified densities of ρ_g coth kh_g and ρ_l coth kh_l should be used, leading to:

$$(u_g - u_l)^2 \geq [k^2\sigma + (\rho_l - \rho_g)g]\frac{h_g}{\rho_g}\left[\frac{\tanh(kh_g)}{kh_g} + \frac{\rho_g}{\rho_l}\frac{\tanh(kh_l)}{kh_g}\right] \quad (5)$$

For large wavelengths ($k \to 0$), the gravity term dominates and the stability criterion becomes:

$$(u_g - u_l)^2 \geq (\rho_l - \rho_g)g\frac{h_g}{\rho_g}. \quad (6)$$

References

Ishii, R. M. *Handbook of Multiphase Systems*. Ch 2.4.1, Hemisphere Publications, New York.

A. Hall

KELVIN, LORD, WILLIAM THOMSON (see Steam engines)

K-EPSILON MODEL OF TURBULENCE (see Boundary layer heat transfer)

KEROSENE (see Oil refining; Oils)

KETTLE REBOILERS (see Reboilers)

KEVLAR (see Gaskets in heat exchangers)

KILNS

A kiln is a device in which heat is applied to bring about physical and chemical changes in materials. It is, therefore, a type of **Furnace** and the two terms are sometimes used to describe very similar pieces of equipment. In general, the word "kiln" is applied to devices for thermal processing of nonmetallic solids, and is particularly associated with the ceramic, cement and lime industries. The three principle types are *verticle shaft kilns, rotary kilns* and *periodic kilns*.

Verticle Shaft Kilns

As shown in **Figure 1**, a vertical shaft kiln is a refractory-lined tower fitted with peripheral fuel burners and air inlets in the lower section, and gas exhaust ports in the upper. Material to be processed is introduced at the top and discharged through the base so that gases and solids are in counter-current flow and some preheating occurs at the top section. Kilns of this type have been used for

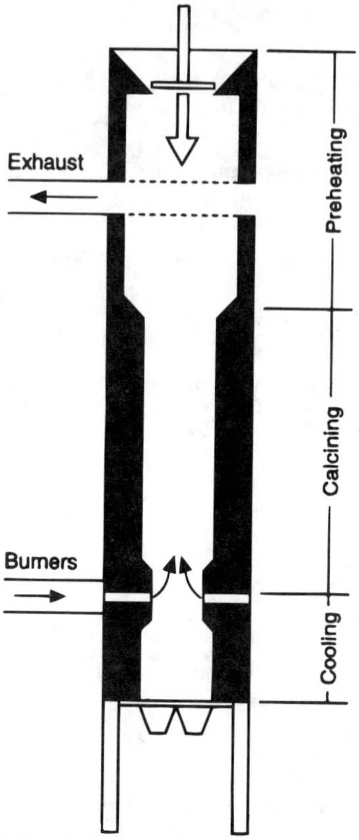

Figure 1. Vertical shaft kiln.

centuries for calcining limestone and in the manufacture of cement. They are still employed for the same purpose although in some modern plants, they have been replaced by fluidized-bed calcining plant.

Rotary Kilns

A rotary kiln, as shown diagramatically in **Figure 2**, is a long refractory-lined inclined tube slowly rotated to ensure the steady flow of solid material from the feeder at the higher end to the collector at the lower. A burner at the lower end produces hot gases which pass in counter-current flow to the moving solids. The fuel may be pulverized coal, gas or oil, or a combination of these.

The tube is fitted externally with steel hoops or "tyres" supported by rollers or "trunnions," and is rotated by an electric motor geared to a toothed ring. The rotational speed of cement or lime kilns is usually about 1 rpm and the inclination of the tube about 1 in 20. Improved heat transfer and longer residence time—which may be required for calcining some ores—can be obtained by reducing the inclination and increasing the rotational speed, but at the cost of lower throughput.

Rotary kilns are widely used for thermal processing of bulk materials such as cement and lime, for calcining or agglomeration of minerals, rocks and ores, clays and shales, and for incineration or pyrolysis of combustible wastes. They are produced in a range of sizes and units with diameters up to 6 m and lengths up to 200 m have been made.

Because of their poor thermal efficiency rotary kilns, particularly large ones, are often fitted with energy-saving devices. For example, the combustion air may be preheated by channeling it through the hot material leaving the kiln and the hot gases leaving the kiln may be diverted to preheated feedstock, or to raise steam in a waste heat boiler.

Periodic Kilns

This title covers a wide range of batch-type thermal processing kilns, all having the characteristic that the material passes through a cycle of heating, soaking and cooling. Basically there are two categories: one in which the material is stationary and the kiln temperature varies with time; the other in which the material passes through the kiln where temperature varies with position. An example of the first category is the small *muffle furnace,* often electrically heated, used by potters. An example of the second category is the *tunnel kiln*. The latter is a long tunnel with cold air entering at one end, being heated by burners in the central region, and leaving as hot gas at the other end. The material to be processed, usually ceramics, is carried by moving refractory-based cars going in the opposite direction to the gas and in their passage through the tunnel

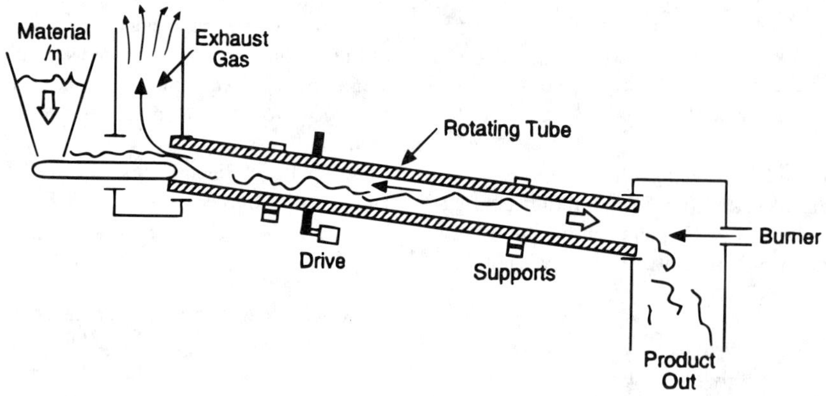

Figure 2. Rotary kiln.

their contents are subjected to the required three thermal phases, preheating, soaking and cooling.

G.L. Shires

KINEMATICS (see Conservation equations, single phase; Flow of fluids)

KINEMATIC SHOCKS (see Kinematic waves)

KINEMATIC TURBULENCE (see Particle transport in turbulent fluids)

KINEMATIC WAVES

***Following from:** Waves in fluids*

Kinematic waves (or *continuity waves*) occur when the one-dimensional flow rate of some quantity depends on the "density" of that quantity [Lighthill and Whitham (1995), Wallis (1969)]. For example, if the flow rate, $\dot{q}$ (cars per hour), on a motorway is a function of the number of cars per mile, n, the speed of a kinematic wave is:

$$U_W = \partial\dot{q}/\partial n. \tag{1}$$

The partial derivative indicates that other parameters, such as the width of the roadway or the radius of bends, which might influence $\dot{q}$, are kept constant. In another example, the volumetric flow rate per unit width of a viscous fluid, of thickness δ and viscosity η in the laminar regime down an inclined plane at an angle θ to the horizontal, in an environment of stationary gas of low density, is:

$$\dot{q} = \frac{\delta^3 g \sin\theta}{3\eta} \tag{2}$$

and the kinematic wave speed is:

$$U_W = \frac{\partial\dot{q}}{\partial\delta} = \frac{\delta^2 g \sin\theta}{\eta}, \tag{3}$$

which is three times the average velocity, $U = \dot{q}/\delta$.

Similarly, the flow per unit width in a broad river of constant slope and depth, y, is:

$$\dot{q} = \left(\frac{y^3 g \sin\theta}{2C_f}\right)^{1/2} \tag{4}$$

if the shear stress on the bottom of the river is related to the mean velocity by

$$\tau = C_f \frac{1}{2}\rho U^2, \tag{5}$$

with C_f being the coefficient of friction.

For this case, if C_f is constant, the kinematic wave speed is:

$$U_W = \frac{\partial\dot{q}}{\partial y} = \frac{3}{2}\sqrt{\frac{yg\sin\theta}{2C_f}} \tag{6}$$

or 3/2 times the average velocity.

Waves of finite amplitude are called *kinematic* (or continuity) *shocks*. For the motorway example, their speed is:

$$U_s = \frac{\dot{q}_1 - \dot{q}_2}{n_1 - n_2} \tag{7}$$

where subscripts 1 and 2 refer to conditions on either side of the shock and the algebraic sign gives the direction of propagation. A convenient graphical representation of U_w and U_s is as slopes of tangents or chords on a plot of $\dot{q}$ versus n. For a shock wave to be maintained as a sharp discontinuity, kinematic waves must run towards it on both sides. For example, if $U_{w1} < U_s < U_{w2}$, state 2 will be on the negative side of the shock.

Similarly, for a falling liquid film

$$U_s = \frac{\dot{q}_1 - \dot{q}_2}{\delta_1 - \delta_2} \tag{8}$$

and the shock will form when a thicker film is above a thinner film and rides over it, as does a surge wave in a river.

In two-phase flow, similar relationships apply, but they are subject to additional constraints. For incompressible flow of two components in a duct of constant area, for example, the constraint is that the overall volumetric flow rate, superficial velocity or volumetric flux, be unchanged across the wave. For gas-liquid flow, the wave speed is then

$$U_{=}(\partial U_{GS}/\partial\alpha)_{(U_{GS}+U_{LS})} \tag{9}$$

where the superscript S denotes "superficial" velocity.

In simple cases where relative motion can be described by a drift flux, j_{gf}, that is determined by a balance of forces—as in a bubble column or foam drainage system—the wave speed is:

$$U_W = U_{GS} + U_{LS} + \frac{\partial j_{gf}}{\partial\alpha} \tag{10}$$

where j_{gf} is the volumetric drift flux of bubbles in a coordinate system moving at the volumetric average velocity, and α is the void fraction. (See **Drift Flux Models**.) The corresponding speed of a kinematic shock wave is:

$$U_s = U_{GS} + U_{LS} + \frac{(j_{gf})_1 - (j_{gf})_2}{\alpha_1 - \alpha_2}. \tag{11}$$

When $U_{GS} + U_{LS} = 0$, as in a batch phase separation, j_{gf} becomes the volumetric flux of bubbles, U_{GS}, and the equations resemble those for single-phase flow. This is the basis of Kynch's (1952) theory of *batch sedimentation*.

Sometimes it is convenient to express wave speed in terms of forces on the components. For example, in stratified gas-liquid flow in an inclined pipe, the balance between buoyancy, interfacial and wall shear stresses may be expressed as:

$$F = f_{WL}\rho_L(\pi/4)^2\tilde{S}_L U_{LS}^2/2D\tilde{A}_L^3 - f_{WG}\rho_G(\pi/4)^2$$

$$\tilde{S}_G U_{GS}^2/2D\tilde{A}_G^3 - \frac{f_i\rho_G(\pi/4)^3\tilde{S}_i}{2D\tilde{A}_L\tilde{A}_G}\left(\frac{U_{GS}}{\tilde{A}_G} - \frac{U_{LS}}{\tilde{A}_L}\right)\left|\frac{U_{GS}}{\tilde{A}_G} - \frac{U_{LS}}{\tilde{A}_L}\right|$$

$$+ g(\rho_L - \rho_G)(\sin\theta) = 0 \quad \mathbf{(12)}$$

where $\tilde{A}_L = A_L/D^2$, $\tilde{A}_G = A_G/D^2$, $\tilde{S}_L = S_L/D$, $\tilde{S}_G = S_G/D$ and $\tilde{S}_i = S_i/D$ are dimensionless cross-sectional areas and wetted perimeters of the liquid stream, gas stream and interface. The symbols f_{WL}, f_{WG} and f_i represent the friction factors for the liquid on the wall, the gas on the wall and at the interface, respectively. The kinematic wave speed is then [Wu et al. (1987), Crowley et al. (1992)]:

$$U_W = \frac{\pi}{4}\frac{\partial F/\partial\tilde{A}_L}{\partial F/\partial U_{GS} - \partial F/\partial U_{LS}} \quad \mathbf{(13)}$$

where F is treated as a function of the three variables $\tilde{A}_L$, U_{LS} and U_{GS}.

Kinematic waves describe transients in highly-damped fluid systems, for which the contribution of inertia terms to the momentum balance is small (e.g., sedimentation or fluidization at low particle **Reynolds Number**). Otherwise, they usually correspond to the limit of "long" waves while smaller scale effects are influenced by other factors such as inertia and surface tension. A criterion for stability in the latter case is that kinematic waved do not override "*dynamic waves*" which predominate at the other extreme when frictional forces are neglected. For example, for turbulent flow of a liquid film down an inclined plane, the dynamic wave speed is $U + \sqrt{yg\cos\theta}$. Instability, leading to *roll-wave formation,* occurs when kinematic wave speed exceeds this value, i.e., when

$$\frac{1}{2}\sqrt{\frac{yg\sin\theta}{2C_f}} > \sqrt{yg\cos\theta} \text{ or } \tan\theta > 8C_f \quad \mathbf{(14)}$$

For instance, if $C_f = 0.005$, the critical slope is $\tan^{-1}$ (.04) or 2.3° and roll-waves will form on steeper slopes.

Kinematic waves may also be the mechanism whereby end-conditions can influence a flow, determining when a solution propagates up- or downstream, or causing a critical condition when the wave speed is zero, as in two-phase counter-current flow "flooding" (Wallis, 1969). (See also **Flooding and Flow Reversal**.)

References

Crowley, C. J. et al. (1992) Validation of a one-dimensional wave model for the stratified-to-slug flow regime transition, with consequences for wave growth and slug frequency. *Int. J. Multiphase Flow.* 18: 249–271.

Kynch, G. H. (1952) A Theory of Sedimentation. *Trans. Farad. Soc.* 48: 166–176.

Lighthill, M. J. and Whitham, G. B. (1955) On kinematic waves, I. Flood movements in long rivers, *Roy. Soc. Proc.* 229: 281–316.

Wallis, G. B., (1969) *One-dimensional Two-phase Flow.* McGraw-Hill. New York.

Wu, H. L. et al. (1987) Flow pattern transitions in a two-phase gas/condensate flow at high pressure in an 8″ horizontal pipe. In *Proc. 3rd Int. Conf. on Multi-phase Flow.* The Hague. The Netherlands. 13–21.

G.B. Wallis

KINETIC ENERGY OF FLUID FLOW (see Velocity head)

KINETIC THEORY OF GASES

The kinetic theory of gases is concerned with molecules in motion and with the microscopic and macroscopic consequences of such motion in a gas. Kinetic theory can be used to deduce some of the equilibrium properties of gases, but the methods of *statistical thermodynamics* are more powerful in that respect. The importance of kinetic theory lies in its ability to describe nonequilibrium phenomena such as the transport of heat or of momentum in a slightly nonuniform gas or the scattering of molecules by other molecules. Much of modern kinetic theory is due to the efforts of Maxwell, Boltzmann, Enskog and Chapman in the late 19th century and the early 20th century. In this article, emphasis will be placed on the results of these efforts; little attention will be given to the mathematical details involved [Chapman and Cowling (1970)].

The results of kinetic theory may be built up from the kinematics of binary molecular encounters to yield a description of the *transport properties of gases* at such densities that binary but not ternary collisions are important. For monatomic gases, this theory is both complete and easily applicable [Maitland et al. (1981)]; the microscopic and macroscopic results are presented below. There is no entirely satisfactory kinetic theory for gases at high densities or for liquids although the theory due to Enskog, which will be described briefly, is of some use. For polyatomic gases, the theory is made vastly more complicated by two facts. First, the forces between molecules depend not only upon intermolecular separation but also on the mutual orientation of the molecules. Second, polyatomic molecules have rotational and vibrational degrees of freedom which play an important role in determining the outcome of a collision between molecules. Although much of the relevant theory has been derived in a formal sense, it is, at present, extremely difficult to apply [Maitland et al. (1981)]. Consequently, simplified results will be discussed, from which the transport properties of polyatomic gases may be determined approximately [Reid et al. (1977)].

The kinematics of a binary collision between two spherically-symmetric molecules are easily derived. The forces between two such molecules act along the intermolecular axis and are a function of the separation r only. The potential energy U(r) stored in the *intermolecular forces* is therefore also a function only of the separation. The general form of this function is illustrated in **Figure 1**.

The kinetic energy of a pair of molecules can be separated into the sum of two independent terms; one relating to the motion of the centre of mass and the other, to the relative motion of the two molecules. Thus, in studying an encounter or collision between two molecules, it will be sufficient to consider only the relative motion in a body-fixed coordinate system with axes parallel to the space-fixed frame of reference and origin located at the centre of mass of the two molecules. In this frame, the relative trajectories of two molecules are always parallel. A typical case is illustrated

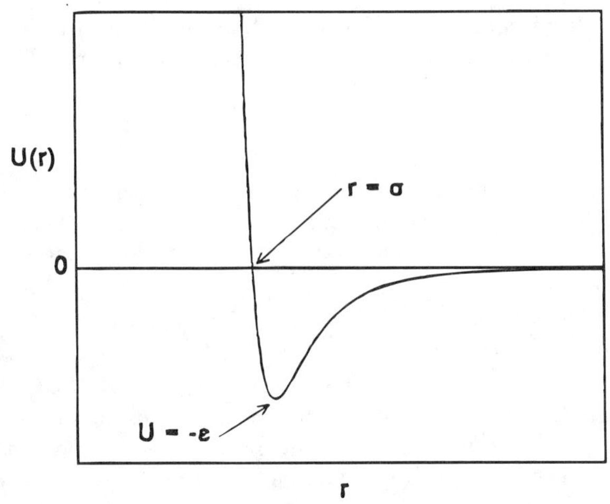

Figure 1. The intermolecular potential energy function.

in **Figure 2**. The initial and final trajectories are specified by the initial relative kinetic energy E, the impact parameter b, and the angle of deflection χ.

Both energy and angular momentum are conserved in the encounter and the angle of deflection [Guggenheim, (1960)] is

$$\chi(b, E) = \pi - 2b \int_{r_0}^{\infty} \frac{dr/r^2}{\sqrt{1 - (U/E) - (b^2/r^2)}}, \quad (1)$$

where r_0 is the distance of closest approach during the encounter:

$$r_0^2 = b^2/\{1 - (U/E)\}. \quad (2)$$

Although individual collisions cannot be studied experimentally, the *scattering of molecular beams* may be observed and is quantified in terms of scattering cross sections. In a typical experimental apparatus, two beams of particles with selected narrow velocity ranges are crossed and the scattered flux is measured as a function of the scattering angle χ. The differential scattering cross-section $\sigma(\chi,E)$ is defined as the ratio of the number of scattered particles

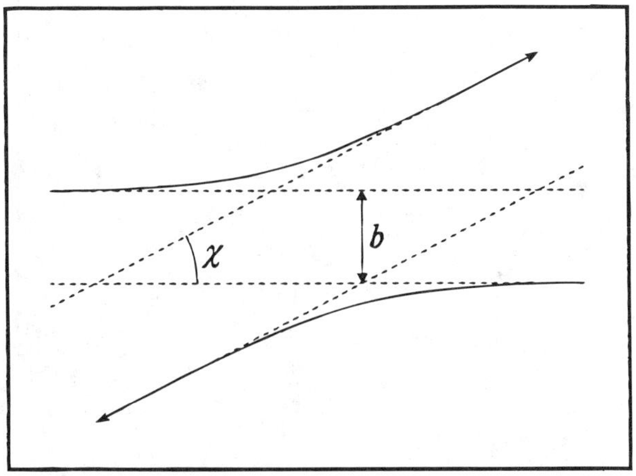

Figure 2. Trajectories of two molecules.

per unit time per unit solid angle to the number of incident particles per unit time per unit area. This is a function of relative initial kinetic energy and the angle of scattering. The differential cross-section is given by:

$$\sigma(|\chi|, E) = \left|\frac{b}{\sin\chi(d\chi/db)}\right|, \quad (3)$$

or, when more than one impact parameter leads to the same scattering angle, by a sum of such terms. Sometimes, only the integral cross-section Q(E) is measured:

$$Q(E) = \int_0^{\pi} \sigma(\chi,)2\pi \sin\chi \, d\chi. \quad (4)$$

These quantities may be used to infer information about the intermolecular potential U(r) [Maitland et al. (1981)].

For polyatomic molecules, the description is much more complicated and inelastic scattering is possible in which there is a change in the internal state of one or both molecules such that the initial and final relative kinetic energies differ. These complications will not be considered further here.

The kinetic theory of gases in bulk is described in detail by the famous *Boltzmann equation*. This is an integro-differential equation for the distribution function f(r,u,t), where f dxdydzdudvdw is the probable number of molecules whose centres have, at time t, positions in the ranges x to x + dx, y to y + dy, z to z + dz, and velocity components in the ranges u to u + du, v to v + dv, w to w + dw. For a gas in a state of thermodynamic equilibrium, this reduces to the **Maxwell-Boltzmann Velocity Distribution Function**, from which quantities such as mean speed, mean collision rate and mean free path may be determined. However, the Boltzmann equation may also be solved for cases in which small macroscopic gradients exist in either (bulk) velocity, temperature or composition. The solutions give the relation between the gradient and the corresponding flux in each case in terms of collision cross-sections. The coefficients of **Viscosity, Thermal Conductivity** and **Diffusion** are thereby related to intermolecular potential.

The Chapman-Enskog solution of the Boltzmann equation is mathematically complicated but it leads to comparatively simple results [Chapman and Cowling (1970)]. The transport coefficients are given in terms of a set of temperature-dependent collision integrals $\bar{\Omega}^{(l,s)}(T)$ which are themselves given in terms of a set of energy-dependent transport cross-sections $Q^{(l)}(E)$. Finally these, transport cross-sections are related to the angle of deflection $\chi(b,E)$. Specifically, the results are [Maitland et al. (1981)]:

$$\eta = \frac{5}{16} \frac{(\pi mkT)^{1/2}}{\bar{\Omega}^{(2,2)}(T)} \quad (5)$$

and

$$\lambda = \frac{25}{32} \frac{(\pi mkT)^{1/2} c_V}{\bar{\Omega}^{(2,2)}(T)}, \quad (6)$$

where m is the mass of one molecule, k is Boltzmann's constant, and c_v is the specific heat capacity at constant volume,

$$\Omega^{(l,s)}(T) = \frac{1}{(s+1)!(kT)^{s+2}} \int_0^\infty Q^{(l)}(E)\exp(-E/kT)E^{s+1}dE, \quad \textbf{(7)}$$

and

$$Q^{(l)}(E) = \frac{4\pi(1+1)}{(2l+1)-(-1)^l} \int_0^\infty (1-\cos^l \chi)b\, db. \quad \textbf{(8)}$$

Equations 5 and 6 are in fact only first-order approximations of the transport coefficients, but the necessary corrections to these formulae are usually less than 1%. It is worth noting that in this approximation, both η and λ depend upon the same collision integral and that the ratio λ/η in a monatomic gas is simply $5c_V/2$ (for monatomic gases, $c_V = 3k/2m$).

For rigid spherical molecules of diameter d, the integrals (**Equations 1, 7 and 8**) may be evaluated analytically with the results that $\cos(\chi/2) = (b/d)$ for all E, $Q^{(l)}(E) = \pi d^2$ for all E and $\bar{\Omega}^{(l,s)}(T) = \pi d^2$ for all T. The coefficients of viscosity and thermal conductivity for this model are therefore

$$\eta = \frac{5(mkT/\pi)^{1/2}}{16d^2}, \quad \textbf{(9)}$$

and

$$\lambda = \frac{25(mkT/\pi)^{1/2}c_V}{32d^2}. \quad \textbf{(10)}$$

For other intermolecular potentials, the integrals may be evaluated by quadrature [Maitland et al (1981)]. Equation 9 is often applied to obtain from viscosity data a rough estimate of the size of a molecule and hence, its mean free path.

The kinetic theory may be applied also to the transport coefficients of monatomic gas mixtures. The diffusion coefficient D_{12} in a binary mixture is:

$$D_{12} = \frac{3}{16\bar{n}} \frac{(2kT\pi\mu_{12})}{\bar{\Omega}_{12}^{(1,1)}(T)}, \quad \textbf{(11)}$$

where $\bar{n}$ is the number density of molecules, μ_{12} is the reduced mass $m_1m_2/(m_1 + m_2)$, and $\bar{\Omega}_{12}^{(1,1)}(T)$ is the collision integral evaluated through **Equations 1, 7 and 8** with U equal to the unlike intermolecular potential U_{12} between molecules of types 1 and 2. The exact results for η_{mix} and λ_{mix} are quite complicated but differ little from the following simple approximations:

$$\eta_{mix} = \sum_i \frac{\eta_i}{1 + \Sigma_{j\neq i}(x_j/x_i)\phi_{ij}} \quad \textbf{(12)}$$

and

$$\lambda_{mix} = \sum_i \frac{\lambda i}{1 + \Sigma_{j\neq i}(x_j/x_i)\phi_{ij}} \quad \textbf{(13)}$$

with

$$\phi_{ij} = (2\mu_{ij}/m_i)^{1/2}(\bar{\Omega}_{ij}^{(2,2)}/\bar{\Omega}_{ii}^{(2,2)}). \quad \textbf{(14)}$$

An important feature of the results is that the transport coefficients obey a principle of corresponding states amongst any set of substances whose intermolecular potentials may be written in the form:

$$u(r) = \varepsilon F(r/\sigma). \quad \textbf{(15)}$$

Here, ε and σ are scaling parameters for length and energy (defined in **Figure 1**) appropriate to a particular substance and F is a universal function. When **Equation 15** holds, the collision integrals are given by

$$\bar{\Omega}^{(l,s)}(T) = \pi\sigma^2\Omega^{(l,s)*}(T^*), \quad \textbf{(16)}$$

where $\Omega^{(l,s)*}(T^*)$ is a reduced collision integral determined by the form of the function F and $T^* = kT/\varepsilon$ is the reduced temperature. For nonattracting rigid spherical molecules of diameter $d = \sigma$, all of the reduced collision integrals are unity, whereas for real molecules, these integrals vary slowly with temperature as illustrated in **Figure 3**. Tables of reduced collision integrals are available for a number of model intermolecular potential functions which conform to **Equation 15** and scaling parameters are available for many substances [Reid et al. (1977)]. Universal correlations of the collision integrals $\Omega^{(1,1)*}(T^*)$ and $\Omega^{(2,2)*}(T^*)$ based on experimental data are also available, together with scaling parameters for a number of simple molecules [Maitland et al. (1981)]. Alternative correlations of transport coefficients in terms of the critical constants are available for cases in which suitable scaling parameters based on **Equation 15** are unavailable [Reid et al. (1977)].

As indicated above, the formal theory is much more complicated for polyatomic gases because of the possibility of inelastic collisions, and is very difficult to apply. Fortunately, it appears that the effects of inelastic collisions on the coefficients of viscosity and diffusion are not usually more than a few percent. Furthermore, it is an empirical fact that scaling parameters ϵ and σ can be found such that the viscosity and diffusion coefficients of polyatomic gases follow the same principle of corresponding states as do monatomic gases. Thus η and D_{12} may be predicted for pure polyatomic gases and for mixtures containing polyatomic components, provided that some data is available from which to determine the required values of ε and σ.

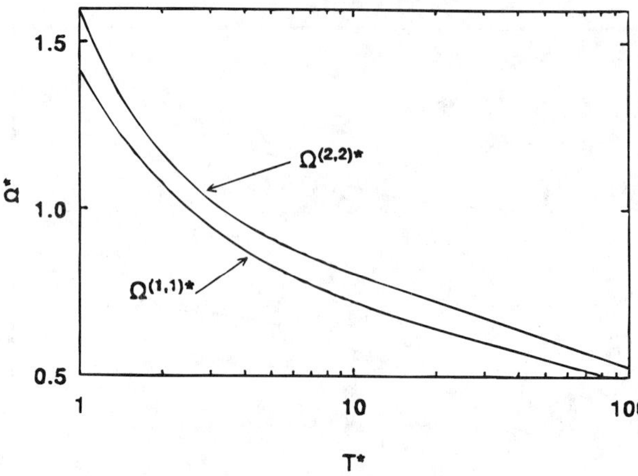

Figure 3. Reduced collision integrals determined from the principle of corresponding states. *Source:* Maitland et al. (1981).

The thermal conductivity of polyatomic gases is influenced considerably by inelastic collisions and cannot therefore be obtained with any accuracy from the monatomic theory or from the principle of corresponding states. A number of approximate treatments are available by which λ is represented as the sum of two terms: one (λ_{trans}) relating to the transport of translational energy and the other (λ_{int}), to the transport of internal molecular energy. In the simplest treatment, the ratio λ_{trans}/η is set equal to its value in a monatomic gas, (15R/4M) and the internal energy of the molecules is assumed to be transported at the same rate as the molecules themselves by the process of diffusion. The result of this treatment is the formula

$$\lambda/\eta c_V = \frac{5}{2}\Delta + (\rho D/\eta)(1 - \Delta) \quad \textbf{(17)}$$

in which $\Delta = 3k/2mc_v$ and D is the coefficient of self diffusion. The ratio ($\rho D/\pi$) is given by

$$\frac{\rho D}{\eta} = \frac{6}{5}\frac{\bar{\Omega}^{(2,2)}}{\bar{\Omega}^{(1,1)}} \quad \textbf{(18)}$$

and it turns out that this has an almost constant value of 4/3. **Equation 17** is based on the assumption that inelastic collisions are rare. If the opposite assumption is made, namely that translational and internal energy are exchanged easily in collisions, then the Eucken expression is obtained:

$$\lambda/\eta c_V = \frac{3}{2}\Delta + 1. \quad \textbf{(19)}$$

Usually, the ratio $\lambda/\eta c_v$ lies somewhere in between the values given by **Equations 17 and 19**. A more accurate treatment requires information on the frequency of inelastic collisions, but this is available from experiment in only a few cases. Thus, an approximation such as Equation 19 is often the best that can be done. Application of this formula requires η, which may be estimated using the monatomic theory or the principle of corresponding states, and c_v. Once pure component thermal conductivities have been estimated in this crude way, one might as well estimate λ_{mix} from **Equation 13**.

All of the kinetic theory discussed above applies only to gases at densities which are large enough that the mean free path is small compared with the dimensions of the container and, at the same time, small enough that only binary collisions are important. At very low densities, thermal conductivity and viscosity are smaller than those of the dilute-gas theory (**Equations 5 and 6**) and, ultimately, they can be reduced in proportion to the density. Otherwise, the transport coefficients are slowly varying function of ρ which, ignoring the very low density regime, extrapolate to the dilute-gas values as $\rho \to 0$. For gases at atmospheric pressure, the transport coefficients differ little from these 'zero-density' values but at higher pressures, significant variation is found. Unfortunately, there is no satisfactory theory to account for density dependence of the transport coefficients. The theory of Enskog [Hirschfelder et al. (1964)], is based on a model in which the molecules are treated as rigid elastic spheres and, although it gives results that agree with computer simulations of hard-sphere transport properties, the results as applied to real gases are disappointing. Various modifications have been proposed that go some way to alleviating this situation.

References

Chapman, S. and Cowling, T. G. (1970) *The Mathematical Theory of Non-Uniform Gases* (3rd ed). Cambridge University Press. London.

Guggenheim, E. A. (1960) *Elements of the Kinetic Theory of Gases*. Pergamon Press. Oxford.

Hirschfelder, J. O., Curtiss, C. F., and Bird, R. B. (1964) *Molecular Theory of Gases and Liquids* (Corrected Edition). Wiley. New York.

Maitland, G. C., Rigby, M., Smith, E. B., and Wakeham, W. A. (1981) *Intermolecular Forces, Their Origin and Determination*. Clarendon Press. Oxford.

Reid, R. C., Prausnitz, J. M., and Sherwood, T. K. (1977) *The Properties of Gases and Liquids* (3rd ed). McGraw-Hill. New York.

Leading to: *Boltzman distribution; Maxwell-Boltzmann distribution; Mean free path*

J.P.M. Trusler

KINETIC THEORY OF CHROMATOGRAPHY (see Gas chromatography)

KIRCHOFF'S LAW (see Radiative heat transfer)

KIRILLOV UNIVERSAL TEMPERATURE PROFILE (see Tubes, single phase heat transfer in)

KNOCK (see Internal combustion engines; Octane number)

KNUDSEN DIFFUSION (see Diffusion)

KNUDSEN-LANGMUIR EQUATION (see Sublimation)

KNUDSEN NUMBER

The Knudsen number K is usually defined for a gas as the ratio

$$K = l/L, \quad \textbf{(1)}$$

where l is the *mean free path* and L is any macroscopic dimension of interest. The importance of this number is that, when $K \ll 1$, the gas behaves as a continuum fluid on the length scale L. The regime in which $K \ll 1$ is known as the hydrodynamic regime while that in which $K \gg 1$ is known as the Knudsen regime. In the former, the gas obeys the *Navier-Stokes equations* of hydrodynamics while in the latter, *rarefied gas dynamics* apply.

References

Truesdell, C. and Muncaster, R. G. (1980) *Fundamentals of Maxwell's Kinetic Theory of a Simple Monatomic Gas*. Academic Press. New York.

J.P.M. Trusler

KNUDSEN REGIME (see Free molecule flow)

KNURLED SURFACES (see Augmentation of heat transfer, two phase)

KOLMOGOROV AND PRANDTL MODEL (see Turbulence models)

KOLMOGOROV EDDIES (see Burners)

KOLMOGOROV MICROSCALE (see Turbulence)

KOZEAY-KARMAN EQUATION (see Filtration)

KRYPTON

Krypton—(Gr. *kryptos,* hidden), Kr; atomic weight 83.80; atomic number 36; melting point −156.6°C; boiling point −152.30 ± 0.10°C; density 3.733 g/l (O°C); valence usually 0.

Discovered in 1898 by Ramsay and Travers in the residue left after liquid air had nearly boiled away, Krypton is present in the air to the extent of about 1 ppm. The atmosphere of Mars has been found to contain 0.3 ppm of krypton. It is one of the "noble" gases. It is characterized by its brilliant green and orange spectral lines. Naturally-occurring krypton contains six stable isotopes. Fifteen other unstable isotopes are now recognized. The spectral lines of krypton are easily produced and some are very sharp.

In 1960 it was internationally agreed that the fundamental unit of length, the meter, should be defined in terms of the orange-red spectral line of $Kr,^{86}$ corresponding to the transition $5p[O_{1/2}]$, $-6d[O_{1/2}]$, as follows: *1 m = 1,650,763.73 wavelengths (in vacuo)* of the orange-red line of Kr. This replaces the standard meter of Paris, which was defined in terms of a bar made of a platinum-iridium alloy.

Solid krypton is a white crystalline substance with a face-centered cubic structure which is common to all the "rare gases." While krypton is generally thought of as a rare gas that normally does not combine with other elements to form compounds, it now appears that the existence of some krypton compounds is established. Krypton difluoride has been prepared in gram quantities and can be made by several methods. A higher fluoride of krypton and a salt of an oxyacid of krypton also have been reported. Molecule ions of $ArKr^+$ and KrH^+ have been identified and investigated, and evidence is provided for the formation of KrXe or $KrXe^+$. Krypton clathrates have been prepared with hydroquinone and phenol. Kr^{85} has found recent application in chemical analysis. By imbedding the isotope in various solids, *kryptonates* are formed. The activity of these kryptonates is sensitive to chemical reactions at the surface. Estimates of the concentration of reactants are therefore made possible.

Krypton is used commercially with argon as a low-pressure filling gas for fluorescent lights. It is used in certain photographic flash lamps for high-speed photography.

Handbook of Chemistry and Physics, CRC Press

K-TUBE (see Shear stress measurement)

KUBELKA-MUNK APPROXIMATION, FOR COMBINED RADIATION AND CONDUCTION (see Coupled radiation and conduction)

KUTATELADZE NUMBER

Kutateladze number, K, is the reciprocal of **Jakob Number**.

$$K = h_{lg}/c_p \Delta T$$

Where h_{lg} is latent heat of evaporation, c_p is the specific heat capacity of liquid and ΔT is the difference between the temperature of the liquid and saturation temperature.

Note that there is a second Kutateladze number, Ku, which is related to electric arcs.

G.L Shires

L

LABILE ZONE (see Crystallizers)

LAGRANGE'S INTERPOLATION FORMULA

If n + 1 pairs (x_k, y_k), $(k = 0, 1, \ldots, n)$ of real or complex numbers are given, where $\{x_k\}$ are distinct, then there exists exactly one polynomial $P_n(x)$ of degree (at most n) such that

$$P_n(x_k) = y_k \qquad (k = 0, 1, \ldots, n). \tag{1}$$

One way of obtaining $P_n(x)$ is through the Lagrange interpolation formula [Davis (1975)]:

$$P_n(x) = \sum_{k=0}^{n} y_k L_k(x) \tag{2}$$

where

$$L_k(x) = \frac{F(x)}{F'(x_k)(x - x_k)} \quad \text{and} \quad F(x) = \prod_{k=0}^{n} (x - x_k). \tag{3}$$

When the interpolation points $\{x_k\}$ are equidistant, that is $x_k = x_0 + kh$ [Henrici (1964)], using $s = x - x_0/h$ (3) can be reduced to the form

$$L_k(s) = L_k(x_0 + sh)$$

$$= \frac{(-1)^{n-k}}{k!(n-k)!} \prod_{i=0, i \neq k} (s - i)$$

$$(k = 0, 1, \ldots, n).$$

If $y_k = f(x_k)$, with f(x) an (n + 1)-differentiable function in an interval [a, b] and with $x_k \in [a, b]$, then for all $x \in [a, b]$,

$$f(x) - P_n(x) = \frac{f^{(n+1)}(\xi)}{(n+1)!} F(x) \tag{4}$$

for some ξ, $x_0 \leq \xi \leq x_n$, [Davis (1975)]. If, furthermore, $|f^{(n+1)}|$ is bounded on [a, b] by a constant M and if $\{x_k\}$ are the zeros of the **Chebyshev Polynomial** of degree n + 1 defined on [a, b], (4) gives

$$|f(x) - P_n(x)| \leq M \frac{(b-a)^{n+1}}{(n+1)! 2^{2n+1}}.$$

The Lagrange formula for trigonometric interpolation is obtained from (2) with

$$L_k(x) = \prod_{i=0, i \neq k} \frac{\sin(x - x_i)/2}{\sin(x_k - x_i)/2}.$$

If the interpolation points are complex numbers $z_0, z_1, \ldots, z_n$ and lie in a domain D bounded by a piecewise smooth contour γ, and if f is analytic in D and continuous in its closure $\bar{D} = D \cup \gamma$, then the Lagrange formula has the form [Gaier (1987)]:

$$P_n(z) = \frac{1}{2\pi i} \int_\gamma \frac{F(\xi) - F(z)}{F(\xi)(\xi - z)} f(\xi) d\xi \tag{5}$$

with

$$f(z) - P_n(z) = \frac{F(z)}{2\pi i} \int_\gamma \frac{f(\xi)}{F(\xi)(z - \xi)} d\xi$$

If $\{z_k\}$ are the (n + 1) roots of $z^{n+1} - 1 = 0$, that is $z_k = e^{2\pi i k/n+1}$, then (2) takes the form:

$$P_n(z) = \frac{(1 - z^{n+1})}{n+1} \sum_{k=0}^{n} \frac{z_k f(z_k)}{z_k - z}$$

with

$$|f(z) - P_n(z)| \leq KE_n \log(n+1)$$

where K is a constant and E_n = infimum { maximum $|f(z) - p(z)|$; $z \in \bar{D}$}, the infimum is taken over all polynomials of degree n and the maximum over the closed domain $\bar{D}$.

REFERENCES

Davis, P. J. (1975). *Interpolation and Approximation*. Ch. 2 & 3. Dover, N.Y.

Gaier, D. (1987). *Lectures on Complex Approximation*. Ch. 2; §1 and §4. Birkhäuser. Boston. 1987.

Henrici, P. K. (1964). *Elements of Numerical Analysis*. Ch. 9 & 10. John Wiley. N.Y.

M.K. El-Daou

LAGRANGIAN APPROACH (see Dispersed flow)

LAGRANGIAN DESCRIPTION OF MOTION (see Conservation laws, single phase; Flow of fluids)

LANGRANGIAN INTERVAL SCALES (see Particle transport in turbulent flow)

LAMBERT'S COSINE LAW (see Radiative heat transfer)

LAMDA LINE (see Cryogenic fluids)

LAMELLA HEAT EXCHANGERS (see Heat exchangers)

LAMINAR BOUNDARY LAYER (see Boundary layer; Wedge flows)

LAMINAR FLOW

Following from: *Flow of fluids; Conservation equations, single-phase*

The velocity field in a drag-induced **Couette Flow** between parallel flat plates remains constant on plane surfaces which are parallel to the flat plates. Individual plane laminae are sheared and thus slide over each other, the velocity being purely axial everywhere. Thus an element of fluid with a rectangular section at time $t = 0$ in a plane spanned by the direction of flow and the direction of shear is distorted into a parallelogram at time $t > 0$ (see **Figure 1**). Similarly, the velocity field in a pressure-induced **Poiseuille Flow** in a long pipe of circular cross-section normal to its axis remains constant on cylindrical surfaces which are concentric to the axis of the pipe. Individual cylindrical laminae slide over each other, the velocity again being purely axial everywhere. Smooth flows of such kinds are called laminar flows [see Richardson (1989), Rosenhead (1963), Shapiro (1964) and White (1974)] or, in the older literature, **Streamline Flows** and generally occur at low values of an appropriate dimensionless number, usually a **Reynolds Number**. It is to be distinguished from a rough or erratic **Turbulent Flow**, which is a high Reynolds number flow.

The origin of turbulence appears to be instability of the associated low Reynolds number laminar flow. Indeed, all laminar flows seem to be susceptible to instabilities leading eventually to turbulence (see Drazin & Reid (1981) for a discussion of the instability of several basic laminar flows).

In a turbulent flow, flow variables such as velocity and pressure fluctuate in space and time in an apparently random manner. No such fluctuations are observed in a laminar flow, though it should be noted that a laminar flow need not be steady, nor necessarily very smooth.

The details of the transition from laminar to turbulent flow are not well understood. For flow by a wall, however, the main steps as the Reynolds number is increased appear to comprise:

- an initial, often two-dimensional, instability which leads to:
- a **Secondary Flow** which is generally three-dimensional and itself unstable which leads to:
- a succession of increasingly complex secondary flows which are again themselves unstable and which lead to:
- further, almost without exception three-dimensional, flows which are themselves unstable, and so on, until:
- intense, local, three-dimensional fluctuations are produced which grow both in size and in number, merge and eventually:
- the flow becomes fully turbulent.

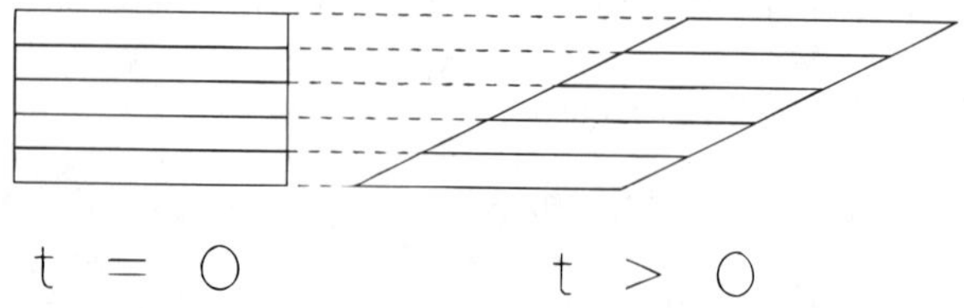

Figure 1. Laminar Couette Flow.

The exact point at which the flow ceases to be laminar is debatable and probably irrelevant.

It is important to note that an increase in Reynolds number occurs not only through an increase in speed. It also occurs through an increase in dimension or size. Thus, laminar flows tend to involve relatively slow motions of small objects. Turbulent flows, in contrast, tend to involve relatively fast motions of large objects. Some flows, particularly those associated with **Boundary Layers**, **Jets** and *wakes*, are laminar upstream and turbulent downstream. This is because the characteristic dimension involved in the Reynolds number is then the axial distance from the start of the flow. A typical example of this is smoke from a lighted cigarette. Near the cigarette, there is a smooth column of smoke: the flow is laminar. Further away, the column breaks down: the flow is turbulent.

References

Drazin, P. G. and Reid, W. H. (1981) *Hydrodynamic Stability*. Cambridge University Press. Cambridge.

Richardson, S. M. (1989) *Fluid Mechanics*. Hemisphere. New York.

Rosenhead, L. (editor). (1963) *Laminar Boundary Layers*. Clarendon Press. Oxford.

Shapiro, A. H. (1964) *Shape and Flow*. Heinemann. London.

White, F. M. (1974) *Viscous Fluid Flow*. McGraw-Hill. New York.

Leading to: *Couette flow; Poiseuille flow; Channel flow.*

S.M. Richardson

LAMINAR FLOW HEAT EXCHANGERS (see Coupled conduction and convection)

LAMINAR FLOWS, HEAT TRANSFER (see Convective heat transfer)

LAMINAR MIXED CONVECTION (see Mixed (combined) convection)

LANGEVIN EQUATION FOR BROWNIAN MOTION (see Stochastic process)

LANGMUIR ADSORPTION ISOTHERM (see Adsorption)

LANGMUIR-HINSHELWOOD MODEL, FOR HETEROGENEOUS CATALYSIS (see Catalysis)

LANGUAGES, COMPUTER (see Computers)

LAPLACE EQUATION (see Conduction)

LAPLACE EQUATION, FOR VELOCITY OF SOUND (see Adiabatic conditions)

LAPLACE TRANSFORM

***Following from:** Differential equations*

The Laplace transform is the basis of operational methods for solving linear problems described by differential or integro-differential equations. The Laplace transform involves the transformation of a complex function f(t) of the real argument t, which satisfies Gelder's condition, $|f(t + h)f(t)| \le A|h|^{\alpha}$, $\alpha \le 1$, $|h| < h_0$, the bounded growth condition, $|f(t)| < M\, e^{S_0 t}$ (S_0 is the index of growth); and the condition $f(t) = 0$ for $t < 0$. The Laplace transform has the form

$$F(p) = \int_0^{\infty} f(t)e^{-pt}dt.$$

Here f(t) is the function-original, F(p) is the Laplace transform of function f(t); and $p = s + i\sigma$, the complex variable.

For any original function f(t), the transform of F(p) can be defined in a half-plane Re $p > s_0$, where s_0 is the growth index of f(t). If the function f(t) is bounded, $|f(t)| < M$ for all t, then $s_0 = 0$.

To reconstruct the original of f(t) from the transform F(p), the inverse Laplace transform (transformation formula) exists which has the form

$$f(t) = \frac{1}{2\pi i}\int_{a-i\infty}^{a+i\infty} e^{pt}F(p)dp,$$

where the integral is taken along any vertical line on a complex plane for $a > s_0$ and is meant as the main Cauchy value.

Some of the properties of the Laplace transform are:

- Passage to the limit

$$\lim_{s\to\infty} F(p) = 0, \qquad \text{where} \quad s = \text{Re } p.$$

- Linearity: for any constant α and β

$$\alpha\, f(t) + \beta\, g(t) \overset{Lp}{\Rightarrow} \alpha\, F(p) + \beta\, G(p),$$

where f and g are the originals and F and Q are the respective transforms.

- Transform of a derivative

$$f'(t) \overset{Lp}{\Rightarrow} pF(p) - f(0)$$

$$f^{(n)}(t) \overset{Lp}{\Rightarrow} p^nF(p) - p^{n-1}f(0) - \cdots - f^{(n-1)}(0).$$

- Differentiation of a transform

$$F^{(n)}(p) \overset{Lp^{-1}}{\Rightarrow} (-1)^n t^n f(t).$$

- Transformation of an integral

$$\int_0^t f(t)dt \overset{Lp}{\Rightarrow} F(p)/p.$$

- Integration of a transform

$$\int_p^{\infty} F(p)dp \overset{Lp^{-1}}{\Rightarrow} f(t)/t.$$

- Time-shift theorem

$$f(t - \tau) \overset{Lp}{\Rightarrow} e^{-p\tau}F(p), \qquad \text{where } \tau > 0.$$

- Frequency-shift theorem

$$e^{p_0 t}f(t) \overset{Lp}{\Rightarrow} F(p - p_0).$$

Convolution theorem

$$F(p)G(p) \overset{Lp^{-1}}{\Rightarrow} \int_0^t f(\tau)g(t - \tau)d\tau.$$

These, and a number of other properties of the Laplace transform formulated as theorems and their corollaries, make up the body of the Laplace transform method.

Among other integral transformes related to the Laplace transform, and used in solving problems of mathematical physics by the operational method, are the Fourier transform, Mellin's transform, Hankel's transform (for cylindrical geometry) etc.

The operational methods (the Laplace transform or related methods) are especially useful in solving ordinary linear differential equations with constant coefficients and systems of such equations.

The method can also be used in solving integral and integrodifferential (Volterra's equations of the first and second kind) equations, and also of problems of mathematical physics described by nonstationary and stationary partial differential equations.

I.G. Zaltsman

LARGE EDDY SIMULATIONS, LES (see Particle transport in turbulent flow; Turbulence models)

LASER DEFLAGRATION WAVES (see Coupled radiation and convection)

LASER-DOPPLER ANEMOMETERS (see Anemometers, laser-Doppler; Lasers)

LASER-DOPPLER METHOD, FOR PARTICLE SIZING (see Particle size measurement)

LASER-DOPPLER METHODS, FOR DROPSIZE MEASUREMENT (see Dropsize measurement)

LASER-DOPPLER VELOCIMETRY, LDV (see Anemometers, laser-Doppler; Optical particle characterisation)

LASER ROCKET ENGINES (see Coupled radiation and convection)

LASERS

Following from: *Optical methods*

The term laser is an acronym for **l**ight **a**mplification by **s**timulated **e**mission and **r**adiation. The name therefore describes how a laser beam is generated. In essence, it consists of a substance which will emit light when excited, normally by an electric current. Lasers may be classified as being *pulsed* or *continuous wave* (CW); the latter produces hundreds of thousands of pulses every second and therefore looks continuous. Visible laser wavelengths range from 400 to 750 nm; within this range *Helium-neon* and *Argon-ion* lasers are most widely encountered. Other lasers operate below (ultraviolet) and above (infrared) the visible spectrum.

Gas lasers utilising a mixture of **Helium** (He) and *Neon* (Ne) produce a visible red light at a constant wavelength of 632.8 nm, with relatively low power outputs from a fraction of a milliwatt (mW) up to 100 mW. For higher power capacities and a range of wavelengths CW lasers based on **Argon** (Ar) are available. Small, compact, single-phase, air-cooled argon-ion lasers range from 25 mW up to 300 mW; as shown in **Figure 1** air passes over cooling fins attached to the tube in order to remove excess heat.

Higher power capacities of up to over 6 W are available from water-cooled, three-phase argon-ion lasers. These lasers produce a multitude of wavelengths in the blue and green regions of the visible spectrum. The gas is housed in a *plasma tube* constructed from a metal/ceramic combination. A line diagram of the components of a modern commercial argon-ion laser is shown in **Figure 2**. Each of the lasers shown in **Figures 1** and **2** have similar design features, the tubes are sealed, to maintain a vacuum, by high-quality windows. The cathode provides the electric current used to excite the gas and causes photons to be emitted. Further stimulation and amplification occur before the output laser beam is emitted via the output mirror. **Figure 2** shows that the plasma tube is surrounded by a water jacket which is used to cool the outer surface of the tube. An adjacent gas reservoir is used to ensure that the plasma tube is fully charged. The mirrors on either side of the *Brewster windows* reflect the light between them so that lasing will occur (the Brewster windows allow light of one polarisation to pass through without any reflection losses.) The mirrors are highly reflective so that only a few percent of the actual laser

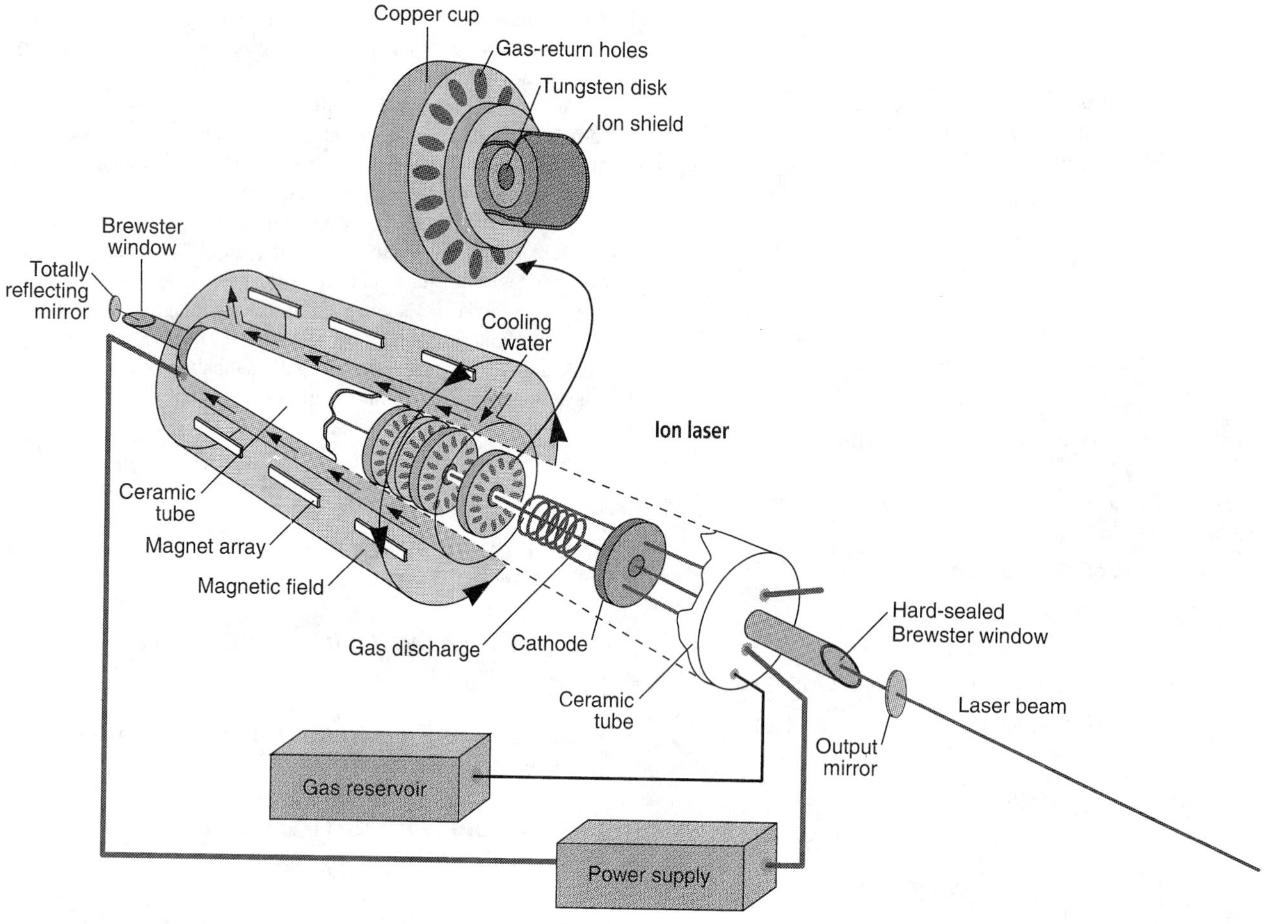

Figure 1. Typical air cooled argon-ion laser [Hecht (1992)].

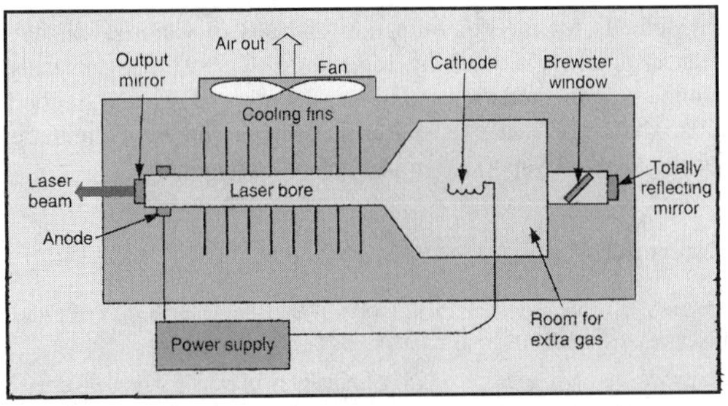

Figure 2. Typical water-cooled argon-ion laser [Hecht (1992)].

power available is used as the output laser beam. Wavelength selection may be achieved to provide single-line or multiline operation.

For two-dimensional *laser Doppler anemometry* systems (see **Anemometers, Laser Doppler**) advantage is taken of the green and blue wavelengths at 514.5 nm and 488 nm respectively; a mauve wavelength (476.5 nm) is used for three-dimensional laser Doppler systems.

The special properties of gas lasers which make them so well-suited for measurement purposes are:

a. Spatial and temporal coherence of the small diameter beam produced—typically 0.6 to 2 mm.
b. Long-term power stability.
c. Polarisation of the laser beam.
d. The plasma tube and associated optical cavity is sealed. The tube windows and the mirrors may be readily cleaned to maintain optimum output.
e. The intensity distribution of the beam is Gaussian.

Other high-power gas lasers are based upon a metal which has been vaporised, such as **Copper, Krypton and Gold**.

Dye lasers are made from a liquid which has been coloured by a dye. They produce beams of different wavelengths when excited by another laser.

Solid state and *chemical lasers* have found application for specific uses. Pulsed lasers are usually more powerful than CW lasers since their energy is concentrated in pulses whose length may be varied.

Semiconductor diode lasers able to operate at room temperatures, with lifetimes approaching or exceeding those of gas lasers are today, small, light-weight and inexpensive. These miniature lasers operate in the visible (700–800 nm), short (800–900 nm) and long (900–1600 nm) wavelengths. Small semiconductor lasers are used to highlight beams produced by lasers operating in the invisible infrared regions. These high-powered lasers are widely used in the medical environment. Modern applications of lasers cover an ever expanding range of scientific, medical and industrial uses. All lasers should be treated with caution irrespective of the environment in which they are used.

Reference

Hecht, J. (1992) Ion lasers deliver power at visible and UV wavelengths. *Laser Focus World*. December.

Leading to: *Anemometers, laser-Doppler*

C.J. Bates

LATENT HEAT FLUX (See Environmental heat transfer)

LATENT HEAT OF FUSION

Following from: *Melting*

During the process of melting, the solid and liquid phases of a pure substance are in equilibrium with each other. The amount of heat required to convert one unit amount of substance from the solid phase to the liquid phase—leaving the temperature of the system unaltered—is known as the latent heat of fusion. It is also equal to the enthalpy difference between the solid and liquid phases, ΔH_{SL}.

As a consequence, the latent heat of fusion depends upon the crystal form of the solid phase which should be strictly specified. Direct measurements of moderate accuracy of the latent heat of fusion is relatively straightforward with modern differential scanning calorimeters over a moderate range of conditions of temperature and pressure. In principle, the **Clausius-Clapeyron Equation** is available,

$$\Delta H_{SL} = -R\left(\frac{d \ln p}{d(1/T)}\right)_m$$

where the subscript m denotes along the melting line. This implies that information on the variation of the melting temperature with pressure is available; but this is seldom the case so it is not a practical route to evaluate ΔH_{SL}. Compilations of experimental results of ΔH_{SL} are available for a number of materials [Weast (1982)], but there is no reliable means of their estimation [Reid et al. (1977)]

References

Weast, R. C., ed. (1982) *Handbook of Chemistry and Physics*. CRC Press. Boca Raton, FL.

Reid, R. C., Prausnitz, J. M., and Sherwood, T. K. (1977) *The Properties of Gases and Liquids*. McGraw-Hill, New York.

W.A. Wakeham

LATENT HEAT OF VAPORISATION

Following from: *Saturated fluid properties*

A single-component system containing two phases (gas and liquid) equilibrium has one degree of freedom. It is thus completely specified by a single independent variable, usually the temperature. The remaining equilibrium state variables, *vapour pressure* p_s and the molar volumes of the two coexisting phases $\tilde{v}_s^g$ and $\tilde{v}_s^l$ follow from the combination of thermodynamic equilibrium criteria and the equation of state applied to the two coexisting planes. The latent heat of vaporisation ΔH corresponds to the amount of energy that must be supplied to the system to convert a unit amount of substance from the liquid to the vapour phase under conditions of equilibrium between the two phases. This transition thus always occurs at constant temperature and the corresponding vapour pressure, p_s. The second law of thermodynamics yields the relationship between the *heat of vaporisation* and the *entropy of vaporisation* as:

$$\Delta H_{LG} = T\Delta S_{LG} = T(S^g - S^l) \qquad (1)$$

The *enthalpy of vaporisation* is equal to the heat of vaporisation since the pressure is constant along the phase transition, so that

$$\Delta H_{LG} = (H^g - H^l)_s \qquad (2)$$

where the subscript s denotes saturation conditions.

The latent heat of vaporisation may be related to other thermodynamic quantities [Majer et al. (1989)]; for example, the equation

$$\Delta H_{LG} = T\int_{\tilde{v}_s^l}^{\tilde{v}_s^g} \left(\frac{\partial p}{\partial T}\right)_{\tilde{v}} d\tilde{v} \qquad (3)$$

relates it to the ($p\tilde{v}T$) behaviour of the fluid. A further relationship of some significance is the *Clapeyron equation*

$$\frac{\Delta H_{LG}}{T(\tilde{v}_s^g - \tilde{v}_s^l)} = \left(\frac{\partial p}{\partial T}\right)_s \qquad (4)$$

since integration with it yields an exact relationship expressing the dependence of the vapour pressure on temperature. From this result, an approximate relationship—valid in the limit of vanishing vapour pressure—can be derived

$$\Delta H_{LG} = -R\left(\frac{d\ln p}{d(1/T)}\right)_s \qquad (p_s \to 0), \qquad (5)$$

which is known as the **Clausius-Clapeyron Equation** and provides a means for estimating latent heat of vaporisation from vapour pressure data.

Heats of vaporisation of fluid mixtures are much more difficult to discuss simply because in a two-phase, multicomponent system, the number of degrees of freedom is equal to the number of components in the system. Vaporisation can occur in an infinite number of ways, characterised by the amount of mixture evaporated and changes in temperature and/or pressure. The various definitions are discussed in detail by Majer et al. (1989).

Methods for the determination of heats of vaporisation have been critically evaluated by Majer et al. (1989) who have also summarised methods for estimating quantity. Majer and Svoboda (1985) have given an extensive critical review and data compilation for the heats of vaporisation of organic compounds.

References

Majer, V., Svoboda, V., and Pick, J. (1989) *Heats of Vaporisation of Fluids*. Elsevier. Amsterdam.

Majer, V. and Svoboda, V. (1985) Enthalpies of vaporisation of organic compounds. A critical review and data compilation. *IUPAC Chemical Data Series No. 32*. Blackwell. Oxford.

W.A. Wakeham

LATENT HEAT STORAGE (see Heat storage, sensible and latent)

LAUGHING GAS (see Nitrous oxide)

LAVAL NOZZLE (see Nozzles)

LAW OF CORRESPONDING STATES (see Gas law)

LAW OF THE WALL (see Boundary layer heat transfer; Turbulence models; Turbulent flow; Universal velocity profile)

LDA (see Anemometers, laser-Doppler)

LEAD

Lead—(Anglo-Saxon *lead*), Pb (L. *plumbum*); atomic weight 207.2; atomic number 82; melting point 327.502°C; boiling point 1740°C; specific gravity 11.35 (20°C); valence 2 or 4. Long known, mentioned in Exodus, the alchemists believed lead to be the oldest metal and associated it with the planet Saturn. Native lead occurs in nature, but it is rare. Lead is obtained chiefly from *galena* (PbS) by a roasting process. *Anglesite* ($PbSO_4$), *cerussite* ($PbCO_3$), and *minim* (Pb_3O_4) are other common lead minerals.

Lead is a bluish-white metal of bright luster, is very soft, highly malleable, ductile, and a poor conductor of electricity. It is very resistant to corrosion; lead pipes bearing the insignia of Roman emperors, used as drains from the baths, are still in service. It is used in containers for corrosive liquids (such as in sulfuric acid chambers) and may be toughened by the addition of a small percentage of antimony or other metals.

Natural lead is a mixture of four stable isotopes: Pb^{204} (1.48%), Pb^{206} (23.6%), Pb^{207} (22.6%), and Pb^{208} (52.3%). Lead isotopes are the end products of each of the three series of naturally occurring radioactive elements: Pb^{206} for the uranium series; Pb^{207} for the actinium series; and Pb^{208} for the thorium series. Seventeen other isotopes of lead, all of which are radioactive, are recognized. Its alloys include solder, type metal, and various antifriction metals.

Great quantities of lead, both as the metal and as the dioxide, are used in storage batteries. Much metal also goes into cable covering, plumbing, ammunition, and in the manufacture of lead tetraethyl, used as an antiknock compound in gasoline. The metal is very effective as a sound absorber, is used as a radiation shield around X-ray equipment and nuclear reactors, and is used to absorb vibration.

White lead, the basic carbonate, sublimed white lead ($PbSO_4$), chrome yellow ($PbCrO_4$), red lead (Pb_3O_4), and other lead compounds were used extensively in paints; in recent years the use of lead in paints has been drastically curtailed to eliminate or reduce health hazards. Lead oxide is used in producing fine "crystal glass" and "flint glass" of a high index of refraction for achromatic lenses. The nitrate and the acetate are soluble salts. Lead salts such as lead arsenate have been used as insecticides, but their use in recent years has been practically eliminated in favor of less harmful organic compounds.

Care must be used in handling lead as it is a cumulative poison. Environmental concern with lead poisoning has resulted in a national program to reduce the concentration of lead in gasoline.

Handbook of Chemistry and Physics, CRC Press

LEAD BASED ADDITIVES TO PETROL (see Oils)

LEAD IN GASOLINE (see Internal combustion engines)

LEAD SHOT (See Prilling)

LEBESQUE'S INTEGRAL (See Integrals)

LE CHATELIER-BRAUN PRINCIPLE (see Chemical equilibrium)

LEDINEGG INSTABILITIES (see Instability, two-phase)

LEGENDRE POLYNOMIAL (see Gaussian quadrature method)

LEGENDRE POLYNOMIAL SERIES

Following from: *Series expansion*

The Legendre polynomial series is the decomposition of a function f(x) square integrable on the interval $(-1, 1)$ into a series according to the Legendre polynomial system $P_n(x)$ $n = 0, 1, \ldots$

$$f(x) = \sum_{n=0}^{\infty} c_n P_n(x), \tag{1}$$

the coefficients of series c_n are calculated by:

$$c_n = \frac{2n+1}{2} \int_{-1}^{1} f(x) P_n(x) dx. \tag{2}$$

The Legendre polynomial system is orthogonal to weight 1 and is complete on the interval $(-1, 1)$.

The orthogonality relationships for Legendre polynomials have the form

$$\int_{-1}^{1} P_n(x) P_m(x) dx = 0 \qquad n \neq m \tag{3}$$

$$\int_{-1}^{1} P_n^2(x) dx = \frac{2}{2n+1} \qquad n = 0, 1, \ldots.$$

Therefore, the Legendre polynomial series is a type of **Fourier Series** written in the system of orthogonal polynomials. The partial sums of a Legendre series bring the functions f(x) closer in the sense of a root-mean-square deviation and the condition $\lim_{n\to\infty} c_n = 0$ is satisfied.

The Fourier series, written in the form of Legendre polynomials within the interval $(-1, 1)$, is similar to the trigonometric Fourier series: convergence of both series takes place, which means that a Fourier-Legendre function f(x) converges at the points $x \in (-1, 1)$ when a trigonometric series of a Fourier function $F(\theta) = \sqrt{\sin\theta} \cdot f(\cos\theta)$ converges at the point $\theta = \arcsin x$.

Near the ends of the orthogonality interval, the properties of the Fourier and Legendre series are different since at points $x \in (-1, 1)$, the orthonormal Legendre polynomials increase infinitely.

If a function f(x) on the interval $(-1, 1)$ is continuous and satisfies a Lipschitz condition, then the Fourier and Legendre series converges to a function uniformly over the interval $(-1, 1)$.

The expansion in terms of a complete orthogonal system of adjoint Legendre functions $P_m^m(x), P_{m+1}^m(x), \ldots, P_n^m(x)$ also plays an important part in problems of radiation heat transfer.

V.N. Dvortsov

LEGENDRE TYPE POLYNOMIALS

Legendre type polynomials or spherical polynomials are polynomials orthogonal on the interval $[-1, 1]$ with unit weight.

Legendre polynomials are defined by:

$$P_n(x) = \frac{1}{n!2^n} \frac{d^n}{dx^n} (x^2 - 1)^n; \qquad n = 0, 1, \ldots. \tag{1}$$

The representation

$$P_n(x) = \sum_{k=0}^{[n/2]} \frac{(-1)^k (2n-2k)!}{k!(n-k)!(n-2k)!} x^{n-2k}; \qquad n = 0, 1, \ldots \tag{2}$$

is valid for Legendre polynomials.

Legendre polynomials are bounded on the interval $[-1, 1]$ by the solution of the differential Legendre equation

$$(1 - x^2)y'' - 2xy' + n(n+1)y = 0, \tag{3}$$

which appears in problems of heat conduction when solving the Laplace equation in spherical coordinates by the separation of variables method.

The following recurrent relations are valid for Legendre poynomials

$$(n + 1)P_{n+1}(x) = (2n + 1)xP_n(x) - nP_{n-1}(x)$$

$$(1 - x^2)P'_n(x) = nP_{n-1}(x) - xnP_n(x) \quad (4)$$

$$P'_{n+1}(x) - P'_{n-1} = (2n + 1)P_n(x)$$

Legendre polynomials have n different real roots. They all lie on the interval [−1, 1]. For Legendre polynomials the representation

$$P_n(x) = \frac{1}{\pi}\int_0^{\pi} (x \pm \sqrt{x^2 - 1}\cos\varphi)^n \, d\varphi$$

$$P_n(x) = \frac{1}{\pi}\int_0^{\pi} \frac{d\varphi}{(x \pm \sqrt{x^2 - 1}\cos\varphi)^{n+1}}, \quad |\arg x| < \frac{\pi}{2} \quad (5)$$

are valid.

Also expressed in terms of Legendre polynomials are adjoint functions, which are often used in soving problems of radiation heat transfer

$$P_n^m(x) = (1 - x^2)^{m/2} P_n^{(m)}(x), \quad m < n, \quad n = 0, 1, 2, \ldots . \quad (6)$$

V.N. Dvortsov

LEIBNITZ-NEWTON FORMULA (See Integrals)

LEIBNIZ RULE

Leibniz rule of calculus is to be found in most advanced texts in mathematics, such as Wylie and Barrett (1982) or Abramowitz and Stegun (1965). The rule is represented by the equation:

$$\frac{d}{dt}\left[\int_{a(t)}^{b(t)} \Phi(t, z)dz\right] = \int_{a(t)}^{b(t)} \frac{\partial \Phi(t, z)}{\partial t} dz + \Phi(t, b(t))\frac{db(t)}{dt} - \Phi(t, a(t))\frac{da(t)}{dt}.$$

This rule is used in systems where integrations need to be performed over a time-dependent domain of integration. It provides a convenient transformation from the integral between transient limits of a temporal derivative to the temporal derivative of an integral between transient limits and vice versa.

References

Abramowitz, M. and Stegun, I. A. (1965) *Handbook of Mathematical Functions*. Dover, New York.

Wylie, C. R. and Barrett, L. C. (1982) *Advanced Engineering Mathematics*. 5th Ed. McGraw-Hill Book Company.

P.F. Pickering

LEIDENFROST PHENOMENA

The term Leidenfrost phenomena is given to the body of phenomena observed when a small amount of liquid is placed or spilled on a very hot surface. It is named after the German medical doctor J. G. Leidenfrost (Leidenfrost, 1756).

When spilling liquid on a hot surface, one may note the presence of large and small masses moving rapidly about the surface without wetting it. The process is accompanied by 'dancing' of small droplets, disruption of large masses of liquid by bubbles breaking through, hissing and spitting when liquid contacts a cooler surface, and withal, slowing down of evaporation [Gottfried et al. (1966)].

The Leidenfrost temperature is defined as the plate temperature at which droplet evaporation time is the greatest. For water this temperature is 150 °C to 210 °C above saturation, depending on the surface and method of depositing the droplet. However, many researchers employ the term Leidenfrost phenomenon to describe the boundary between transition boiling and film boiling of a large liquid mass. The terms *rewetting, quenching, sputtering, departure from film boiling,* and *minimum film boiling,* are often synonymously used. (see **Boiling**.)

Several mechanisms have been proposed to determine the Leidenfrost temperature. The hydrodynamics approach holds that the separation of the liquid-vapor interface from the wall lasts only as long as the vapor generation rate is sufficient to maintain a stable vapor film, e.g., Berenson (1961). The thermodynamic approach assumes that the liquid can never exist beyond a "maximum liquid temperature," which depends on the liquid properties only. Thus, a heated surface whose surface temperature exceeds this limit cannot support liquid contact. The maximum liquid temperature can be determined either from the spinodal line, or from the kinetic theory of bubble nucleation in liquids, e.g., Blander and Katz (1975) and Lienhard and Karimi (1981).

Other explanations relying on system adsorption characteristics have been offered, e.g., Segev and Bankoff (1980) and Olek et al. (1988).

References

Berenson, P. J. (1961) Film boiling heat transfer from a horizontal surface. Trans. ASME. *J. Heat Transfer*. 83: 351–358.

Blander, M. and Katz, J. L. (1975) Bubble nucleation in liquids. *AIChE. J.* Vol. 21: 833–848.

Gottfried, B. S., Lee, C. J., and Bell K. J. (1966) The Leidenfrost phenomenon: film boiling of liquid droplets on a flat plate. *Int. J. Heat Mass Transfer*. Vol. 9: 1167–1187.

Leidenfrost, J. G. (1756) *De aquae communis nonnullis qualitatibus tractatus,* (A tract about some qualities of common water). Duisburg on Rhine. An original copy of this Treatise is in the Yale University Library.

Lienhard, J. H. and Karimi, A. (1981) Homogeneous nucleation and the spinodal line. Trans. ASME. *J. Heat Transfer*. Vol. 103: 61–64.

Olek, S., Zvirin, Y., and Elias, E. (1988) The relation between the rewetting temperature and the liquid-solid contact angle. *Int. J. Heat Mass Transfer.* Vol. 31: 898–902.

Segev, A. and Bankoff, G. (1980) The role of adsorption in determining the minimum film boiling temperature. *Int. J. Heat Mass Transfer.* Vol. 23: 623–637.

Leading to: *Rewetting of hot surfaces.*

S. Olek, Y. Zvirin and E. Elias

LENNARD-JONES POTENTIAL (see Diffusion coefficient; Thermal conductivity; Thermal conductivity mechanisms)

LENSES (see Optics)

LEVER RULE (see Saturated fluid properties)

LEWIS NUMBER

Following from: *Lewis relationship*

The Lewis Number (Le) is defined as the ratio of the **Schmidt Number** (Sc) and the **Prandtl Number** (Pr). The Lewis Number is also the ratio of thermal diffusivity and molecular diffusivity as is found from the definitions of Schmidt and Prandtl Number, as follows:

$$Le = \frac{Sc}{Pr} = \frac{\eta}{\rho D_{12}} \cdot \frac{\lambda}{c_p \eta} = \frac{(\lambda/\rho c_p)}{D_{12}} = \frac{\kappa}{D_{12}} \quad (1)$$

where $\kappa = (\lambda / \rho\, c_p)$ is the **Thermal Diffusivity**. The Lewis Number is important in determining the relationship between mass and heat transfer coefficients (the **Lewis Relationship**).

G.F. Hewitt

LEWIS RELATIONSHIP

Following from: *Heat transfer; Mass transfer*

Commonly-used relationships for heat and mass transfer in turbulent flow in channels take the form:

$$Nu = \frac{\alpha D}{\lambda} = 0.023\, Re^{0.8}\, Pr^{1/3} \quad (1)$$

$$Sh = \frac{\beta D}{D_{12}} = 0.023\, Re^{0.8}\, Pr^{1/3} \quad (2)$$

where Nu is the **Nusselt Number**; Sh, the **Sherwood Number**; α, the **heat transfer coefficient**; λ, the **Thermal Conductivity**; D, the tube diameter; β, the **Mass Transfer Coefficient**; and D_{12}, the **Diffusion Coefficient** for a binary mixture (this is replaced by effective diffusivity in the case of a multicomponent mixture). From these two equations, it follows that:

$$\frac{Sh}{Nu} = \frac{\beta D}{D_{12}} \frac{\lambda}{\alpha D} = \frac{\beta\lambda}{D_{12}\alpha} = \left(\frac{Sc}{Pr}\right)^{1/3} = Le^{1/3} \quad (3)$$

where Le is the **Lewis Number** (= Sc/Pr). From Equation 3, it follows that the ratio of mass transfer coefficient to heat transfer coefficient is given by:

$$\frac{\beta}{\alpha} = \frac{D_{12}Le^{1/3}}{\lambda} = \frac{1}{\rho c_p}\left[\frac{D_{12}^3 \rho^3 c_p^3}{\lambda^3} \cdot \frac{\lambda}{D_{12}\rho c_p}\right]^{1/3} = \frac{Le^{-2/3}}{\rho c_p} \quad (4)$$

which is known as the "Lewis Relationship". This relationship is particularly important in the *air-water system,* where the ratio $\alpha / \beta\, \rho_a\, s$ is known as the *psychrometric ratio* (b). Here, s is the *humid heat,* which is the amount of heat required to increase 1 kg of air plus its associated moisture by 1 Kelvin. For the air-water system, Le $\approx$ 1 and thus:

$$b = \frac{\alpha}{\beta \rho_a s} = \frac{c_p}{s} Le^{2/3} \approx \frac{c_p}{s} \quad (5)$$

since, for the air-water system with low concentrations of water vapour (as in the atmosphere), $c_p \approx s$ and $b = 1$. The implication of a unity value of b is that the *adiabatic saturation temperature* and the **Wet-bulb Temperature** are equal for the air-water system. Here, the adiabatic saturation temperature is defined as the temperature T_s reached by a gas stream containing a given amount of vapour, which is contacted adiabatically with a stream of liquid from where the vapour is derived, the temperature of the liquid at the entrance of the contactor also being at T_s.

Leading to: *Wet bulb temperature*

G.F. Hewitt

LEWIS RULE (see Multicomponent systems, thermodynamics of)

L-FIN (see Air cooled heat exchangers)

LIFT FORCE (see Aerodynamics)

LIGHT SCATTERING (see Optical particle characterisation)

LIGHT SCATTERING FROM PARTICLES (see Anemometers, laser-Doppler)

LIGHT HEATING OIL (see Oils)

LIGHT TRACK IMAGING (see Photographic techniques)

LIGHT WATER REACTOR SAFETY (see Rewetting of hot surfaces)

LIGNS ABSORPTION METHOD, FOR FILM THICKNESS MEASUREMENT (see Film thickness measurement)

LIMIT CYCLES (see Non-linear systems)

LIMPET COILS (see Tank coils)

LINEAR ATTENUATION COEFFICIENT (see Ionising radiation)

LINEAR ELASTIC FRACTURE MECHANICS (see Fracture mechanics)

LINEAR EQUATIONS (see Cramer's rule)

LINEARISED THEORY OF DIFFUSION (See Condensation of multicomponent vapours)

LINE AXIS CONCENTRATING COLLECTOR (see Solar energy)

LINE SETTLING (see Sedimentation)

LIQUEFACTION (see Natural gas)

LIQUEFACTION OF GASES

***Following from:** Cryostats*

Gases such as nitrogen, oxygen and methane require the use of very low temperatures to liquefy and store them at relatively low pressures. To achieve this, a whole range of cryogenic technologies has been developed to ensure the economical liquefaction of gases. The reason why such technology has become important is that storage of these gases is more economic and versatile in their liquid form.

There are several ways in which refrigeration can be supplied to a process to cool and/or condense a gas or mixture of gases. These can be grouped into the three fundamental principles used in commercial applications:

1. The Joule-Thomson effect
2. Compression/condensation and expansion of a pure component
3. Expansion turbines or engines.

Most processes in cryogenic technology use one or more of the above principles. Alternatively, they may use a mixed refrigerant as working fluid.

The refrigeration requirements for each cryogenic process are made up of the following elements:

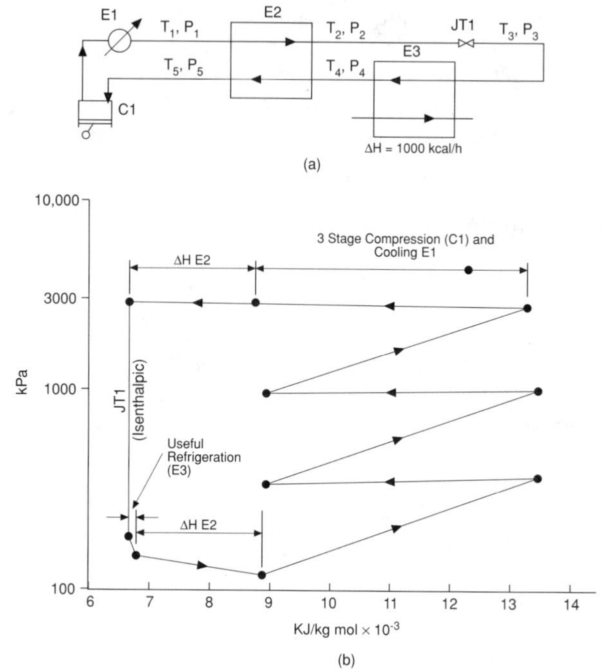

Figure 1. (a) The process scheme and (b) The P-H diagram for a closed Jowle-Thomson cycle.

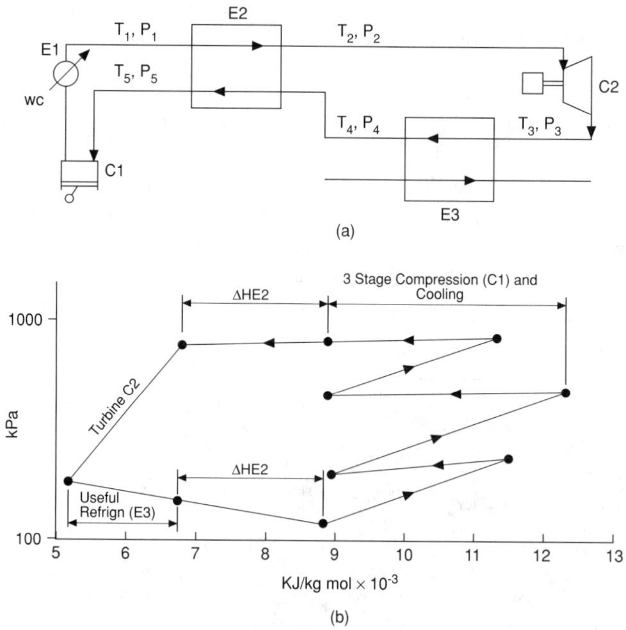

Figure 2. (a) The process scheme and (b) The P-H diagram for a closed-loop, compression-expansion cycle.

- heat ingress from the atmosphere
- irreversible thermodynamic losses through heat exchange
- the requirement to produce liquid from what is normally a gas.

Heat ingress is usually kept to a minimum by good insulation

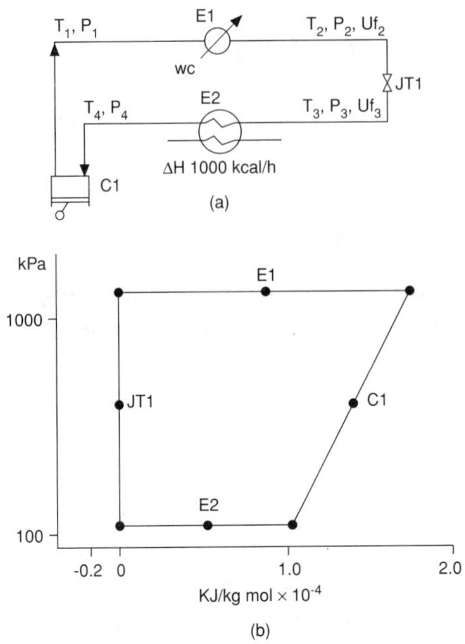

Figure 3. (a) The process schematic and (b) The P-H chart for the propane refrigeration cycle.

practices and is a small proportion of the refrigeration requirement in industrial liquefaction processes. Heat exchanger warm-end losses and other irreversible losses are minimised through the use of compact, high-efficiency heat exchangers. Such exchangers are plate-fin aluminium or stainless steel units, as well as wound coil or etched exchangers made in stainless steel or sometimes aluminium. They have economical temperature difference approaches of 2–3°C in contrast with normal shell and tube exchangers, which are economical at much higher temperature differences. (See also **Heat Exchangers**.)

The importance of good heat transfer in cryogenic processes cannot be overemphasised because of the significant cost of liquefaction of fluids, such as N_2, O_2 and CH_4. The references attached to this section give a more detailed analysis of the benefits of compact heat exchange.

The three basic methods of **Refrigeration,** used singly or in combination with each other, are compared in **Figures 1**, **2** and **3** where the cycle is represented above a pressure/enthalpy (P-H) chart, which illustrates refrigeration loads that can be produced by each cycle. The cryogenic plant designer must possess the skill to select the best fluid, mixture, technique or combination of techniques to achieve the optimum refrigeration process.

Table 1 compares the different methods from a specific power viewpoint whilst **Figure 4** provides a rapid assessment of the overall coefficient of performance of such liquefaction cycles as applied to different temperature levels. This serves as an easy method of estimating the approximate power input for such refrigeration cycles, where $\dot{Q}_R$ is the refrigeration duty and $\dot{Q}_{IN}$ is the power input.

References

Barron, R. F. (1985) *Cryogenic Systems* (2nd ed.) Oxford University Press.

Hands, B. A. (1986) *Cryogenic Engineering*, Academic Press. London.

Table 1. Comparison of three refrigeration cycles. The basis is to provide 1000 kcal h^{-1} refrigeration at different levels

Cycle type	Cycle fluid	Flowrate (Nm^3h^{-1})	Specific power consumption (kW/1000 kcal)	Required temperature of refrigerant (°)
Joule-Thomson (3000 kPa)	N_2	895	150	−40
Expander-compressor (Comp. to 800 kPa)	N_2	60	6	−40
	N_2	75	8	−80
	N_2	95	10	−120
	N_2	130	14	−160
Propane cycle (Comp., cond., expansion, and evap.)	C_3H_8	9.1	0.83	−40
	C_3H_8	8.4	0.57	−25
	C_3H_8	8.0	0.42	−14
	C_3H_8	7.7	0.32	−4.5
	C_3H_8	7.4	0.23	+5.5

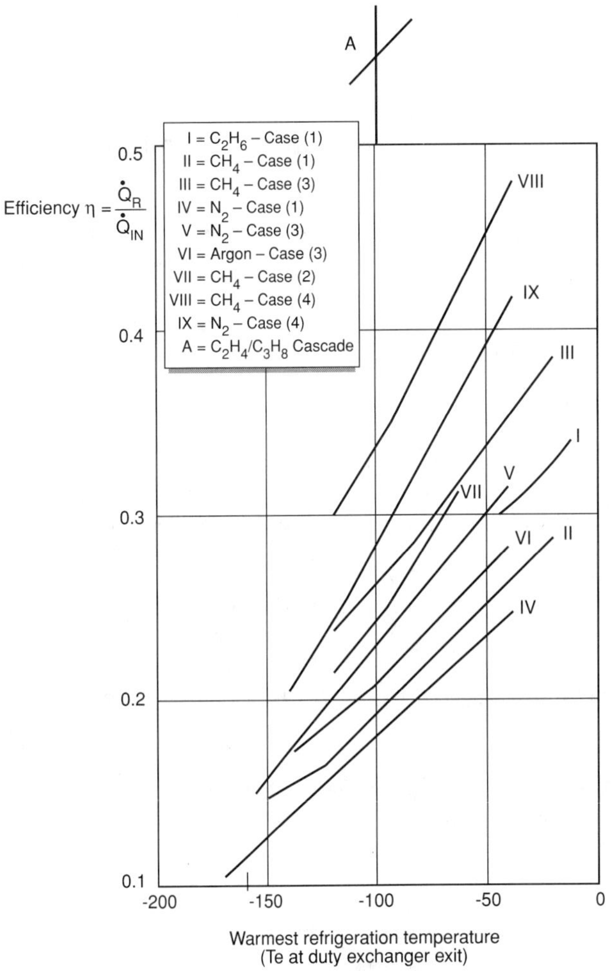

Figure 4. Performance coefficient for gas compression-expansion cycles.

Haselden, G. G. (1971) *Cryogenic Fundamentals*, Academic Press, London.

Isalski, W. H. (1989) *Separation of Gases*, Oxford Science Publications.

Isalski, W. H. (1993) *Kempe's Engineers Year Book*, Refrigeration B5/1, Benn.

Newton, C. L. (1976) Hydrogen Production and Liquefaction, *Chem. & Proc. Eng.* (Dec), p51.

Scott, R. B. Denton, W. H., and Nicholls, C.M. (1964) *Technology and Uses of Liquid Hydrogen*, Pergamon Press. Oxford.

Tomlinson, T. R. et al (1990) Exergy analysis in process development, *Chem. Eng.*, 11 & 25 October, p25.

Leading to: *Cryogenic plant*

W.H. Isalski

LIQUEFIED NATURAL GAS, LNG (see Cryogenic fluids; Natural gas)

LIQUEFIED PETROLEUM GAS, LPG (see n-Butane; Propane; Oil refining)

LIQUID AIR (see Cryogenic fluids)

LIQUID ARGON (see Cryogenic fluids)

LIQUID COOLANTS, FOR NUCLEAR REACTORS (see Coolants, reactor)

LIQUID FILTRATION (see Filtration)

LIQUID FUELS (see Fuels)

LIQUID FLUORINE (see Cryogenic fluids)

LIQUID-GAS EJECTORS (see Jet pumps and ejectors)

LIQUID HOLDUP (see Two-phase flows)

LIQUID IN GLASS THERMOMETERS (see Thermometer)

LIQUID IN STEEL THERMOMETERS (see Thermometer)

LIQUID-LIQUID EJECTORS (see Jet pumps and ejectors)

LIQUID-LIQUID EXTRACTION (see Extraction, liquid-liquid)

LIQUID-LIQUID FLOW

Following from: *Two-phase flow*

The flow of two *immiscible liquids* presents a wide variety of *flow patterns* ranging from relatively quiescent stratified flow to regimes in which intense mixing between the phases occurs.

The more recent studies are illustrated in **Figure 1** [Arirachakaran et al. (1989)]. As suggested by this figure the nature of liquid-liquid flows is highly complex, with full stratification occurring at low velocities and full dispersion at velocities. This makes prediction of these flows a real challenge.

Given the complexities of the flow, there has been a concentration of work on either fully-separated (stratified, and to a lesser extent, annular) flows on the one hand and fully-dispersed flows on the other. In the latter type of flow conditions under which there is an *inversion* between the continuous and the disperse phases is of great interest. Below the hydrodynamics of two-phase liquid flows and phase inversion are briefly discussed.

Hydrodynamics

Separated flows

This type of flow occurs when two liquids are in laminar flow. The parallel flow of two immiscible fluids in a two-dimensional channel is a classical problem discussed by Bird et al. (1960). Since the flows are laminar in both phases separated flows can be treated theoretically and solutions are in good agreement with experimental observations. In the case of flow in cylindrical tubes analytical solutions are not possible. However numerical solutions are being obtained that can be extended to turbulent flow by using suitable turbulence models. As the flow rates of the phases are increased, waves appear at the interface that, upon turbulence intensification, lead to dispersed flow.

Dispersed flows

With increased turbulence breakage of one phase into the other takes place, as shown in **Figure 1**. As dispersion is formed flow behaviour becomes dependent on dispersion viscosity. Water in oil dispersions behave as Newtonian fluid up to 10% by volume water. Above this fraction they show significant pseudoplastic behaviour;

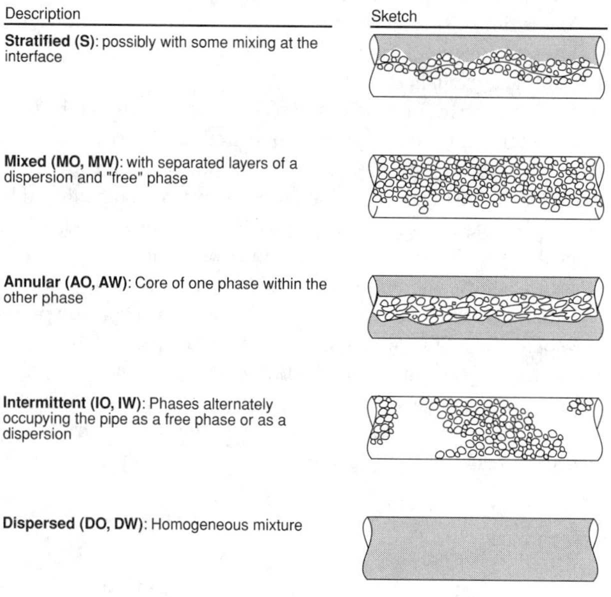
Description
Sketch
Stratified (S): possibly with some mixing at the interface
Mixed (MO, MW): with separated layers of a dispersion and "free" phase
Annular (AO, AW): Core of one phase within the other phase
Intermittent (IO, IW): Phases alternately occupying the pipe as a free phase or as a dispersion
Dispersed (DO, DW): Homogeneous mixture

Figure 1. Flow patterns defined by Arirachakaran et al. (1989).

as mass flow increases, viscosity decreases. It has been found that the smaller the drop size is, the higher the viscosity.

Models for fully-dispersed flows are discussed by Arirachakaran et al. (1989), who found good agreement between predicted and observed results in the fully-dispersed region provided the correct viscosity is used. There is some evidence that the presence of droplets can sometimes suppress the turbulence of the continuous phase.

Phase Inversion

In liquid-liquid systems, it is of crucial importance to know the conditions governing phase inversion. This is defined as the point at which the continuous phase becomes the dispersed one and vice versa. Thus, an inversion point would represent a change from an oil-in-water to a water-in-oil dispersion. The condition that triggers this change is the increase of dispersed phase hold-up. This is accompanied by an increase in dispersion viscosity. It has been observed that viscosity of the mixture reaches a peak near the inversion point, with the mixed viscosity often being much higher than that of the more viscous phase. As the hydrodynamic conditions of pipe flow and continuous (or batch) stirred tanks are different, conditions governing phase inversion in these two cases are not the same.

Phase inversion of dispersions in stirred tanks systems has received substantially more attention than that in pipe systems. Typical data for phase inversion in *stirred vessels* are illustrated in **Figure 2**. Conditions for phase inversion depend on the physical properties of the system and on hydrodynamic conditions. Therefore, for a given tank configuration and a given system the inversion point will depend on phase hold-up and stirring speed, as shown in **Figure 2**. The graph shows an upper and lower band for phase

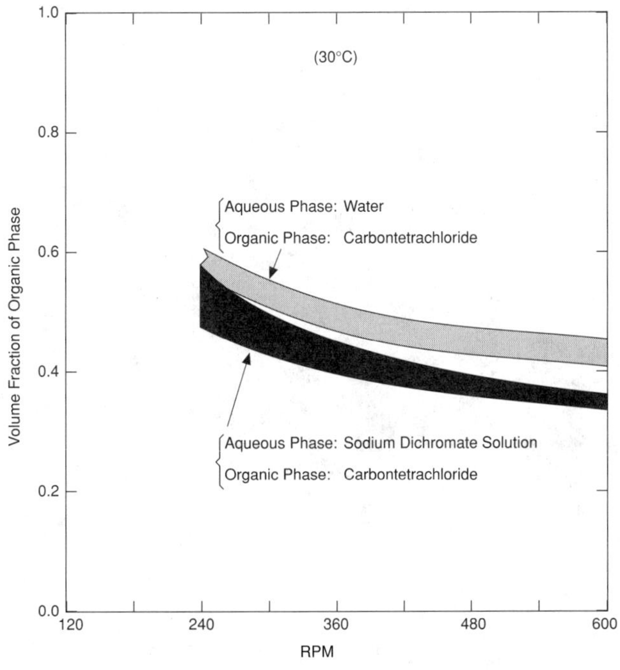

Figure 2. Inversion curves for binary immiscible liquid-liquid systems. *Source:* McClarey and Mansouri (1978).

inversion, rather than a single line. This means there is hysteresis. Thus on increasing stirrer speed to create inversion, a lower stirrer speed on lower volume fraction of the specified phase is required for reversion of the inversion process. In addition, the inversion band is not the same when starting from an oil-dispersed system or a water-dispersed one. It should be noted that if mass transfer is taking place, the conditions for phase inversion will most certainly change due to the presence of solute at the interface.

There is considerable controversy about the effect of wettability of the containing vessel (and also of the agitator) on phase inversion, but there are no general conclusions on this respect, although it has been suggested that wettability may only be significant at low stirring rates.

In pipe systems, the mechanisms of mixing leading to inversion are much more complex and, from the point of view of the experimentalist, uncontrolled. There is little information in the literature about phase inversion in pipes. Unlike the case of stirring speed, there seems to be no effect of mixture velocity or droplet size on phase inversion. Phase fraction and temperature, on the other hand, seem to be the key parameters.

References

Arirachakaran, S., Oglesby, K. D., Malinowsky, M. S., Shoham, O., and Brill, J. P. (1989) An analysis of oil/water flow phenomena in horizontal pipes. Paper presented at SPE Production Operation Symposium. (SPE 18836) March, Oklahoma City.

Bird, R. B., Stewart, W. S., and Ligutfoot, E. N. (1960) *Transport phenomena*. John Wiley and Sons.

Godfrey, J. and Hanson, C. (1982) Liquid-liquid systems, in *Handbook of multiphase systems*. (ed. Hetsroni, G.). McGraw-Hill Book Company.

McClarey, M. J. and Ali Mansouri, G. (1978) Factors affecting the phase inversion of dispersed immiscible Liquid-liquid mixtures. *A.I.Ch.E. Symp. Series 74*. 173:134–139.

Leading to: *Dispersed flow; Drops*

E.S. Perez de Ortiz

LIQUID-LIQUID INTERFACES (see Interfaces; Surface and interfacial tension)

LIQUID-LIQUID MASS TRANSFER (see Marangoni effect)

LIQUID-LIQUID SEPARATION (see Membrane processes)

LIQUID-LIQUID-SOLID FLOWS (see Multiphase flow)

LIQUID MEMBRANES (see Membrane processes)

LIQUID METAL BOILING HEAT TRANSFER (see Liquid metals)

LIQUID METAL COOLED FAST REACTOR

Following from: Nuclear reactors

In standard lightwater reactors, only 1 to 2% of the energy potential of natural uranium is used. In fact, essentially only the ^{235}U isotope is fissioned. The dominating ^{238}U isotope (99.28% in natural uranium) contributes only marginally. However, a two-step process can be envisaged first to convert ^{238}U into ^{239}Pu (by neutron capture) and successively to fission ^{239}Pu. In this manner one can make a much more efficient use of natural uranium. The first step of the process outlined above can be envisaged if the neutrons issued by fission (and which have energies around an average value of 2 MeV) are moderated to lower energies as little as possible.

In fact, the properties of nuclear reactions induced by neutrons of energies higher than few kiloelectronvolts ("fast neutrons") limit the neutrons which disappear by parasitic captures in the structures. Moreover, at these energies there is a favourable competition between fission and capture. The neutron balance in the core of this type of reactor (called fast reactor) is positive in the sense that neutrons are available: out of the average number of neutrons emitted per fission (2.4 for ^{235}U, 3 for ^{239}Pu), one neutron is needed to keep the chain reaction and one neutron is available to "breed" new fissile material if parasitic capture and captures in fissile materials are kept as low as possible. A similar "excess" of neutrons is not available in thermal reactors.

The need to keep the neutrons issued by fission as "fast" as possible makes it necessary to avoid the presence of light elements (like hydrogen) in the core, since neutrons loose higher amounts of energy when interacting with such nuclei. As mentioned above, the neutron capture probabilities are smaller for most materials for fast neutrons; as a consequence, the choice of structural materials for fast reactors is mostly related to their performance under irradiation and not to their nuclear properties. In the same energy domain, however, fission probabilities ("cross-sections") are small compared to those associated with thermal reactor neutron energies. Thus, there is the need for a high density of fissile nuclei per core unit volume and hence, for a high fissile material content in the nuclear fuel. Optimal values for large, fast power reactors (1500 MWe) are found when the Pu/U ratio in the fuel is in the range of 0.15 to 0.25. The typical value for the volume of a 1,000 MWe core is about 10 m^3 containing approximately 5 tons of Pu. The resulting high heat density needs a specific coolant fluid with very high thermal transfer capability. Liquid metals, and in particular, liquid Na have been unanimously chosen.

The Na thermal conductivity at 500°C, 67 W m^{-1} °C^{-1} is a hundred times higher than for water at 40°C. Na stays liquid over a large range of temperatures ($\sim$ 100°C to $\sim$ 900°C) and this allows high core outlet temperatures, favourable to increased thermodynamic efficiency. Despite these and other advantages of Na, some disadvantages have to be mentioned; in particular, the high chemical affinity for oxygen and water. The opacity of Na is an obstacle for easy in-core inspections, and the fact the Na is solid below 98°C, makes it necessary to keep Na temperature above 150°C in any configuration of the power plant. Since, Na becomes radioactive under irradiation, the formation of the radioactive isotope ^{22}Na (decay half-life $\sim$ 2.5 years) needs precautions. Finally, Na should be kept well purified to avoid corrosion.

The core of a typical fast reactor is made with fuel assemblies (hexagonal geometry), and can be surrounded by assemblies which contain the "fertile" material (^{238}U). The control of the chain reaction is provided by control rods in the core, which contain an absorber material (such as B_4C, natural or enriched in the isotope ^{10}B).

Typical fuel assemblies (e.g., in the case of the French fast reactor SUPER-PHENIX) are made of thin (4 to 5 mm) hexagonal

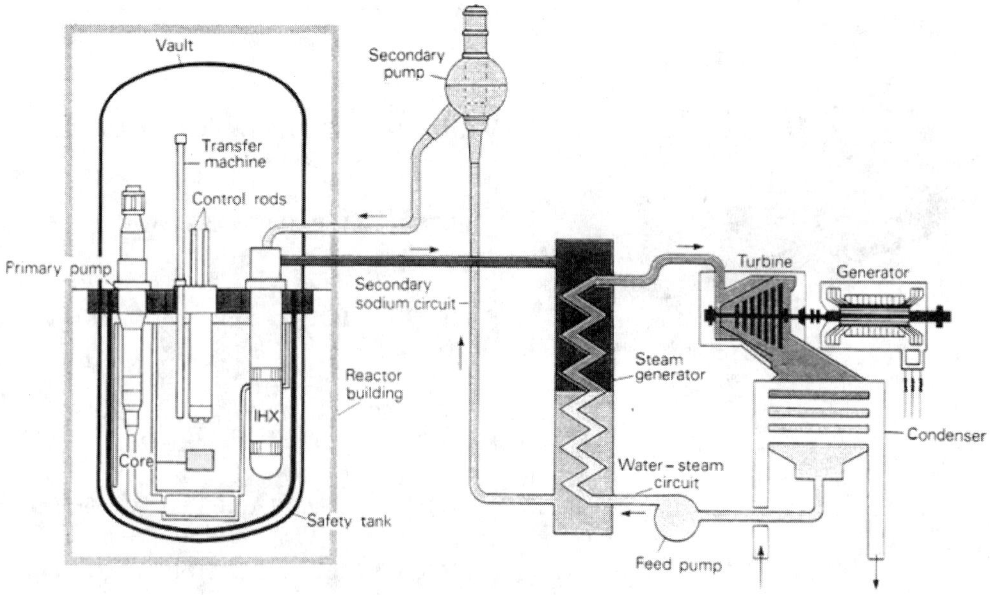

Figure 1. Schematic flow diagram of SUPER-PHENIX.

tubes in stainless-steel, which contain a regular lattice of fuel pins (271) with diameter less than 1 cm. Each pin is made up of fuel pellets of mixed UO_2-PuO_2 within a stainless-steel cladding, less than 0.5 mm thick. Pins are kept in place by an helicoidal spacer around the pin itself. The height of the fissile column is 100 cm. The Na enters from the bottom of the assembly with T = 395°C and has T = 545°C at the outlet. The maximum heat generation at the center of the core corresponds to a power of 460 KW/liter.

Economy motivates the increase of irradiation time as far as possible. A major constraint is represented by the metallurgical behaviour of materials (e.g., swelling). Present experience indicates the possibility of reaching fuel burn-ups of 100,000 MWd/ton (of heavy metal) and higher if appropriate choices are made for fuel-pin cladding materials (e.g., high-Ni SS) and for hexagonal tubes (e.g., ferritic SS).

Fuel types other than mixed oxides can be envisaged; in particular, mixed nitrides which could offer the advantage of higher density and better thermal conductivity. Metal alloys have also been used and are presently being actively studied in the USA, within the frame of a pyrometallurgical technology development for fuel reprocessing and fabrication.

As far as safety characteristics are concerned temperature increase gives rise to a reduction of core reactivity, partly by nuclear effect (Doppler effect) which acts without any delay. However, the ejection of Na from the center of the core can result in a reactivity increase and measures are taken to eliminate the risk of void formation (in particular, boiling). The high thermal inertia associated with the large mass of Na, which is kept in normal operation at approximately 300°C below the boiling point, is a further guarantee that boiling of Na, if ever attached, will be reached only after long-enough delay.

The risks related to Na fires and Na-water exothermic reactions are handled with appropriate design features of Na circuits, and several sophisticated detection instruments are provided. Fast reactors of different design, size and fuel type have been built and operated in the world for more than 40 years. The largest plant is SUPER-PHENIX in France, an industrial prototype of 1,200 MWe (see figure). A large amount of relevant experience has been gathered on the operation of this reactor type. Breeding properties have been successfully proved experimentally.

However, the present context of relatively low Uranium scarcity does not favour the industrial deployment of breeders before a few decades. But breeding capability is not the only attractive property of fast reactors. They are well-suited for using plutonium fuel and for burning it efficiently, if needed, to avoid stockpiling. Moreover, their excellent neutron economy indicated above offers possible solutions for the burning of long-lived radioactive wastes.

M. Salvatores

LIQUID-METAL HEAT TRANSFER

Following from: *Convective heat transfer*

Liquid Metals are a specific class of coolants. Their basic advantage is a high molecular thermal conductivity which, for identical flow parameters, enhances heat transfer coefficients. Another distinguishing feature of liquid metals is the low pressure of their vapors, which allows their use in power engineering equipment at high temperatures and low pressure, thus alleviating solution of mechanical strength problems.

The most widespread liquid metals used in engineering are alkali metals. Among them sodium is first and foremost, used as a coolant of fast reactors and a working fluid of high-temperature heat pipes. Potassium is a promising working medium for space power plants. In some cases eutectic Na−K and Pb−Bi alloys and Hg, Li, and Ga are also used.

The high thermal conductivity and, hence, low Prandtl numbers of liquid metals imply that heat transfer by molecular thermal conduction is significant not only in the near-wall layer, but also in the flow core even in a fully developed turbulent flow. The thickness of the thermal boundary layer proves to be substantially larger than the thickness of the hydrodynamic boundary layer (see **Single-phase Forced Convection Heat Transfer**).

The hydrodynamic characteristics of liquid metal flows (friction factor and the coefficient of local resistance) are calculated by conventional formulae.

Fully developed heat transfer to liquid metals in tubes may be calculated using the following generalized relations:

For the case $T_w = const$ (curve 1 in **Figure 1**)

$$Nu = 5 + 0.025\,Pe^{0.8}, \quad (Pe < 4\,10^3,\ Pr = 0.004 - 0.04) \qquad \textbf{(1)}$$

where $Nu = \alpha d/\lambda$ and $Pe = ud/\kappa$ where α is the heat transfer coefficient, d the tube diameter, λ the thermal conductivity and κ the thermal diffusivity. For the case $\dot{q}_w = const$ (curve 2 in **Figure 1**)

$$Nu = 7.5 + 0.005\,Pe, \qquad (300 \le Pe \le 10^4) \qquad \textbf{(2)}$$

At high values of Pe number Equations 1 and 2 approach each other (**Figure 1**).

An approximate calculation of mean heat transfer in the entrance region in the case of turbulent flow can be performed with Equations 1 and 2 by introducing a correction factor for this region

$$\varepsilon_e = 1.72(d/l)^{0.16}. \qquad \textbf{(3)}$$

where l is the tube length. The length of the thermal entrance region is 10 to 15d.

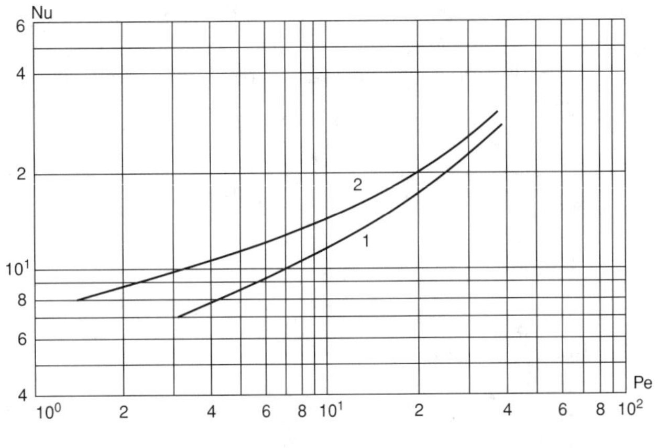

Figure 1. Single phase heat transfer relationships for liquid meters.

For tube bundles in longitudinal flow, the following relationships may be used:

$$Nu = 6 + 0.006\ Pe \qquad (30 \leq Pe \leq 4000,\ Re > 10^4) \qquad \mathbf{(4)}$$

and

$$Nu = 2Pe^{0.5} \qquad (50 \leq Pe \leq 7000), \qquad \mathbf{(5)}$$

where Pe is calculated from the free-stream velocity and the outside tube diameter. There relationships are valid for the range of pitch-to-diameter ratio $s/d = 1.2 - 1.75$ and may also be used for staggered and in-line tube bundles in crossflow.

The thermal properties of liquid metal depend only slightly on temperature. Taking into account the small transverse temperature difference in liquid metal flow due to the high thermal conductivity, the effect of nonisothermal conditions in heat transfer is not significant and, as a rule, is not considered.

Bundles of fuel rods in triangular or square arrays are often used in reactors with a liquid metal coolant. In this case, due to the nonuniform flow past the central, peripheral and angular fuel rods, the heat transfer to them is different. There are also variations in heat transfer around the perimeter of the fuel rod. Normally, heat transfer is determined above all by the Pe number. However, the pitch to diameter ratio of the rods in the bundle, the arrangement of the rods and the presence of plugs mounted for equalizing coolant flow rate over the bundle section have proved to be important. A uniform temperature distribution around the fuel rod perimeter, except for the above factors, depends on the ratio of the coefficient of thermal conductivity of the liquid metal to that of the rod enclosure. These factors assume particular importance in tightly-packed bundles.

Heat transfer for the cases indicated above is calculated using unwieldy empirical relations that are valid, as a rule, within a narrow range of parameters. Details of these are presented in handbooks.

Heat transfer by natural convection from a horizontal cylinder is described by the formula

$$Nu = C[GrPr^2/(1 + Pr)]^n \qquad \mathbf{(6)}$$

where $Nu = \alpha d/\lambda$, $Pr = c_p\ \eta/\lambda$ and $Gr = gd^3\rho^2\beta\Delta T/\eta^2$ where c_p is the specific heat capacity, η the viscosity, g the acceleration due to gravity, ΔT the temperature difference between the surface and the fluid and β the coefficient of volumetric thermal expansion. $C = 0.67$, $n = 1/4$ for $Gr = 10^2 - 10^8$, and $C = 0.35$, $n = 1/3$ for $Gr > 10^8$. For a vertical cylinder of height H and radius r, the equation:

$$Nu_H = 0.16\left[Ra_H \frac{r}{H}\right]^{0.3}, \qquad \mathbf{(7)}$$

may be used for

$$\left(Ra_H \frac{r}{H}\right) = 1.6 \times 10^6 - 4 \times 10^7, \qquad \frac{r}{H} = 0.39 - 1.50$$

where $Nu_H = \dfrac{\alpha H}{\lambda}$; $Ra_H = Gr_H\ Pr = g\beta\Delta TH^3/\nu\kappa$.

The relation

$$Nu = C(\varphi)Ra^{1/3}Pr^{0.074}, \qquad \mathbf{(8)}$$

where $C(\varphi)$ reduces from 0.069 to 0.049, with φ varying from 0 to 90°, is used to calculate heat transfer in a plane gap between two surfaces arranged at an angle φ to the horizontal in the $1.5 \times 10^5 \leq Ra \leq 2.5 \times 10^8$ range.

It has been established that heat transfer in liquid metals depends, to a high degree, on fouling resistances at the wall-liquid interface. These resistances appear due to chemical or electrochemical interaction between the wall material and a coolant to produce a surface layer of intermetallics, carbides and other compounds or solid solutions with reduced thermal conductivity. Mass transfer and deposition of corrosion products are also possible on the heat exchange surface. In all cases, much importance is attached not only to the chemical compatibility of the liquid metal coolant and the wall material, but also to the degree of metal purity. This is given a special attention with special in-line systems for metal purification used in many systems.

"Metal-metal" heat exchangers with a bilateral flow of single-phase coolant past the wall are calculated by conventional relations. However, the design of such apparatus—primarily shell-and-tube heat exchangers—has a specific feature. Due to the relatively low specific heat of liquid metals, heating of the coolant is comparable to, and in some cases appreciably exceeds, the value of the governing temperature difference. Thus, a correct allowance for the by-pass leakages of the coolant along the shell and the contribution of zones of deteriorated flow along the heat exchange surface are of crucial significance. In this case a "zone-by-zone" approach in designing heat exchangers has proven to be efficient.

One of the principal specific features of boiling of most of liquid metals—alkali metals above all—is that the superheat for the incipience of boiling may amount to tens, and in some cases, hundreds of degree. This is due to the good wetting of the solid metal surfaces by alkali metals, growing solubility of inert gases with temperature, and a small slope in the vapor pressure curve dp/dT_S. Thus, measures should be taken, whenever necessary, for reducing the incipient boiling superheat and ensuring the reproducibility of its value. Note that mercury, which poorly wets most technical surfaces, has extremely low incipient superheat.

Pool boiling of alkali metals at moderate heat loads is characterized by unstable boiling, i.e., spontaneous switching from a natural convection regime (curve 1 in **Figure 2**) to developed boiling (curve 2 in **Figure 2**) and vice versa (circles in **Figure 2**). This transition under the condition $\dot{q} = const$ is accompanied by wall temperature fluctuations; these are greatest at low pressure and low heat flux.

The fraction of the overall heat flux accounted for by bubble evaporation during growth on the heating surface is between 20–60%. The remainder of the heat is transferred by convection to the liquid bulk, or is removed with the superheated liquid surrounding the bubbles rising to the free surface. These effects are linked with the high thermal conductivity of liquid metals.

The heat transfer coefficient α_{pb} for nucleate pool boiling of alkali metals is described by an empirical formula that is typical for boiling of most of liquids under similar conditions

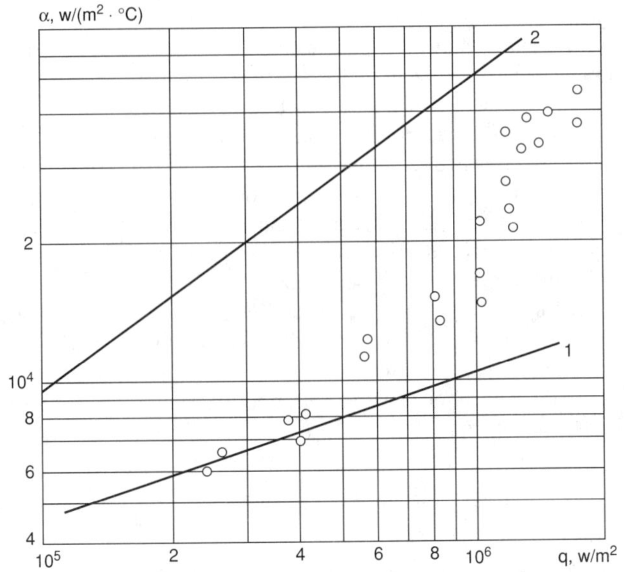

Figure 2. Pool boiling of liquid metals: spontaneous switching between natural connection (1) and developed boiling (2).

$$\alpha_{pb} = A\dot{q}^m p^n, \quad \mathbf{(9)}$$

where α_{pb} is expressed in W/m^2K, $\dot{q}$ in W/m^2, p in MPa. For sodium, A = 22.4, m = 0.67, n = 0.4 in the 5 to 30 kPa range and A = 7.55, m = 0.67, n = 0.1 in the 30 to 150 kPa range. For potassium, A = 6.35, m = 0.67, n = 0.1 in the 10 to 200 kPa range.

Critical heat fluxes $\dot{q}_{cr}$, which bring about the transition from nucleate to film boiling, are described for alkali metals by the empirical formula

$$q_{cr} = 0.7\lambda^{0.6}(p/p_{cr})^{1/6}, \quad \mathbf{(10)}$$

where λ is the coefficient of thermal conductivity and P_{cr}, the critical pressure.

Flow regimes of two-phase alkali metal flows are the same as those of ordinary liquids (see **Forced Convection Boiling**). However, owing to the high incipient boiling superheat and a high ratio of specific volumes of vapor and liquid phase (low operating pressures), the regions of bubble and slug flow regimes may correspond to an extremely narrow range of vapor quality or be missing altogether. The annular-dispersed flow regime (see **Annular Flow**) is predominant. Due to the high thermal conductivity, the temperature difference across the liquid film in annular flow (ΔT_{lf}) is small. Even at high evaporation rates, the temperature difference across the liquid-vapor interface ΔT_{ev} is small also. Thus, the overall difference between the wall and the saturation temperature $\Delta T_w = T_w - T_s = \Delta T_{lf} + \Delta T_{ev}$ is insufficient for incipience of vapor bubbles on the heating surface. This means that classic boiling with vapor bubbles growing on the wall is usually absent. Phase transition occurs by evaporation from the interface, to which heat is supplied by thermal conduction through the liquid film.

The heat transfer coefficient for forced convection boiling is determined as:

$$\alpha_{fcb} = \frac{\dot{q}}{\Delta T_{ev} + \Delta T_{lf}}, \quad \mathbf{(11)}$$

where ΔT_{ev} can be estimated by the Hertz-Knudsen correlation; ΔT_{lf}, by conventional relations for thermal conduction; and film thickness can be determined sufficiently accurately from the Martinelli-Lockart relation (see Forced Convection Boiling) for a given vapor quality. Typical values of heat transfer coefficients lie between 5×10^4 and 10^5 W/m^2K. It should be noted that α_{fcb} is independent of $\dot{q}$.

Burnout for a forced, two-phase liquid metal flow is commonly linked with the dryout of the near-wall liquid film. The boundary quality which gives rise to the dryout of the film x'_b depends on pressure and mass flow rate. It is fairly high and varies from 0.8 to 0.9 for pressures typical of alkali metals. A further increase in x'_b can be attained applying porous coatings, coiled tubes or other means facilitating retention on the wall of moisture deposited from the flow core. The length of the zone of the complete film dryout is $\Delta x \approx 0.05 - 0.1$.

Condensation of liquid metals readily wetting the surface obeys the same laws as condensation of conventional liquids does (see **Condensation**). As a rule, this is a film-type condensation. The high thermal conductivity of liquid metals leads to a drastic reduction of the contribution of liquid film thermal resistance to overall heat transfer resistance during condensation (for nonmetallic liquids it is the basic contribution). Simultaneously, the contribution of resistance at the vapor-film interface and, in particular, of diffusion resistance grows if noncondensable gases are present or chemical reactions proceed in vapor phase, which require that efforts be made to expel impurities from a vapor. Typical values of the heat transfer coefficient for condensation in the absence of noncondensable gases are from 8×10^4 to 10^5 W/m^2K.

Yu.A. Zeigarnik

LIQUID METALS

The main physical properties of liquid metals which differ from other fluids are:

- a high thermal conductivity
- a high melting point (except mercury and sodium-potassium)
- a high boiling point at atmospheric pressure
- a low electrical resistivity.

These characteristics are used for particular heat transfer applications, especially at high temperature. Sodium and some alloys are used to cool fast neutron nuclear reactors, which operate at low pressure and high temperature. Lithium or specific alloys are used to cool the fusion reactors blankets; magnetohydrodynamic phenomena are used in metallurgical processes.

Chemically, some liquid metals are highly reactive with oxygen. When used, special precautions should be taken to avoid contact with air, water, etc.

Basic information on heat transfer of liquid metals was compiled by Kutateladze (1959) and detailed data on sodium and sodium-potassium alloy were given by Foust (1976) including chemistry, physical properties, heat transfer correlations and industrial applications. Two other liquid metals handbooks were previously edited by Lyon (1952) and Jackson (1955).

Single-Phase Heat Transfer

For forced and natural convection the governing equations are the same as for any fluid, i.e., conservation of mass, momentum and energy.

Forced convection is governed by two dimensionless parameters: the **Reynolds Number (Re)** and the **Peclet Number (Pe)**. The specificity of liquid metals is in the low value of Prandtl number, so Pe << Re. The influence of Prandtl number on temperature distribution in a heated tube is shown in **Figure 1**.

Liquid metals heat transfer coefficients are high, a fact which could be advantageous but could also be a drawback. The forced convection heat transfer correlations for liquid metals depend on the Peclet number, while for classical fluids (water, air) they depend on the Reynolds number.

The heat transfer correlations for liquid metals in a heated tube have been reported by Foust (1976):

- For a uniform heat flux,

$$Nu = 7.0 + 0.025\ Pe^{0.8}. \quad (1)$$

- For a uniform wall temperature,

$$Nu = 5 + 0.025\ Pe^{0.8}. \quad (2)$$

Alternative expressions are given in the article on **Liquid Metal Heat Transfer**. Heat transfer correlations for liquid metals flow in annuli and between parallel plates have also been reported by Foust (1976).

Longitudinal flow and crossflow through rod or tube bundles have been extensively studied for nuclear reactor applications. The heat transfer correlations are reported by Foust (1976), showing the dependency on the pitch/diameter and on the geometrical arrangement of the bundle.

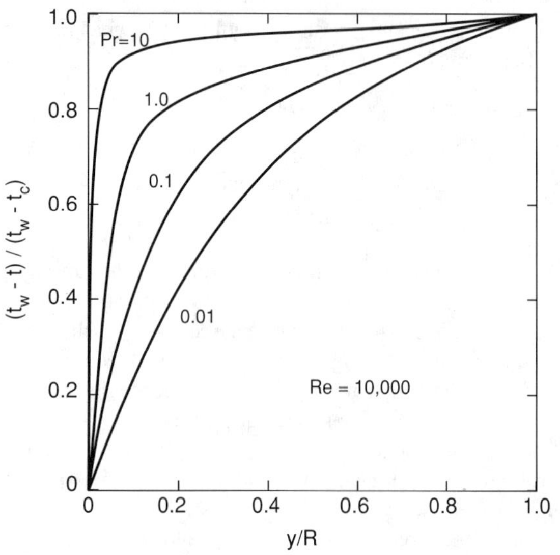

Figure 1. Effect of Prandtl number. *Source:* Foust (1976).

Natural convection is governed by two dimensionless parameters: the **Grashof Number (Gr)** and the **Boussinesq number (Bo)**. With the following definition

$Bo = (Gr)(Pr)^2 = (Ra)(Pr)$, liquid metals are characterized by Bo << Ra << Gr. Natural convection heat transfer correlations of liquid metals depend on the Boussinesq number while for classical fluids, they depend on Rayleigh number.

There is a large variety of natural convection configurations, but the basic one is natural convection along a heated plate. Sheriff (1979) has presented a review of the following cases:

for the vertical plate with a uniform wall temperature, the local Nusselt number is:

$$Nu = (0.57 \pm 0.03)(Gr_x Pr^2)^{0.25}. \quad (3)$$

and for the vertical plate with a uniform heat flux,

$$Nu = 0.73(Gr_x * Pr^2)^{0.20} \quad (4)$$

where $Gr_x = \beta g x^3 \Delta T/\nu^2$ and $Gr_x^* = \beta g x^4 \dot{q}/\nu^2$ and x is the vertical distance along the plate.

The configurations of downward-facing plate, upward-facing plate and the influence of inclination have also been reported by Sheriff (1979).

Heat Transfer During Boiling

Boiling of liquid metals has been extensively studied in the frame of nuclear reactors safety analyses [Kottowski, (1994)]. General features related to liquid metal boiling have been described by Dwyer (1976). Foust (1976) has reported various analytical calculations and experimental data, which tend to show a high level of superheat at boiling inception for sodium and potassium. However, this seems to be highly dependent on the purity of the liquid metals, the wetting of the surface, the heat-up kinetics (dT/dt) and the absence of microbubbles in the liquid. Most of the work performed since 20 years ago in subassembly configurations present a low level of superheat, which is clearly more representative of industrial applications.

Apart from superheat, effect of thermal conductivity and high saturation temperature, the boiling of liquid metals at normal pressure presents many analogies with boiling of water at normal pressure: high liquid-to-vapor density ratio, similar flow configurations (bubbly, slug, etc.). Friction pressure drop correlations developed for water are generally used for liquid metals.

The heat transfer coefficient under nucleate boiling is very high. A survey of pool boiling of liquid metals is proposed by Shah (1992).

One important feature for liquid metals boiling is the prediction of Critical Heat Flux (CHF). For water CHF depends strongly on the flow conditions. Several correlations have been established, mainly for sodium flow in nuclear reactor subassemblies. Kottowski (1982) proposes the following correlation for CHF:

$$\dot{q}_{CHF} = 0.216\, h_{lg}(1 - 2x)m^{0.807}\left(\frac{d}{l}\right)^{0.807} \quad (5)$$

for a hydraulic diameter d in the range: 0.002 to 0.010 m and for a heating length in the range: 0.3 to 1.5 m.

Further information on boiling of liquid metals is given in the section on **Liquid Metal Heat Transfer**.

Magnetohydrodynamics (MHD) in Liquid Metals

Liquid metals are good electrical conductors, and they interact with electromagnetic fields so that various MHD applications are possible. Analysis of MHD phenomena needs to solve simultaneously the Navier-Stokes and the Maxwell equations. Berton (1991) presents the basic dimensionless parameters deduced from the system of equations:

- the Hartmann Number: $Ha = BL\sqrt{\sigma/\rho\nu}$
- the hydrodynamic Reynolds Number, Re, and the *magnetic Reynolds number* $R_m = \mu_o \sigma u L$
- the Stuart number: $N = \sigma B^2 L/\rho u = Ha^2/Re$
- the **Prandtl Number**, Pr, and the *Batchelor number* (also known as the magnetic Prandtl number); $Ba = \nu\mu_o\sigma$.

where B is the magnetic flux in Teslas, μ_o is the permeability of the space ($4\pi \times 10^{-7}$ henry m^{-1}) and σ is the electrical conductance in Siemens.

Various applications of MHD in liquid metals have been developed in industrial processes: metallurgy, liquid metal cooled nuclear reactors, prototypes for fusion reactors, ship propulsion, power conversion. An overview of these applications has been included in the Proceedings of "Energy Transfer in MHD flows" Conference in Cadarache (1991).

Additional Symbols

Quantity	Symbol	SI Unit
Magnetic field	B	T
Critical pressure	p_c	Pa
Vacuum magnetic permeability	μ_o	H m^{-1}
Fluid magnetic permeability	μ	H m^{-1}
Electrical conductivity	σ	S m^{-1}

References

Berton, R. (1991) *Magnétohydrodynamique*. Ed. Masson.

Chen, J. C. and Bishop, A. A. (1970) Liquid-metal heat transfer and fluid dynamics. *ASME Winter Annual Meeting*. New-York, USA.

Dwyer, O. E. (1976) *Boiling Liquid Metal Heat Transfer*. ANS. 244 East Ogden Avenue. Winsdale, ILL.

Foust, O. J. (1976) *Sodium-NaK Engineering Handbook*. ch. 2. Gordon and Breach. Science Publishers, Inc.

Jackson, C. B. (1955) *Liquid Metals Handbook*. ch. II: 3 edition.

Kottowski, H. M. Liquid metal thermalhydraulics. INFORUM Verlags GmbH.

Kottowski, H., Saraterri, C. (1982) Convective heat transfer and critical heat flux at liquid metal boiling. *10th Liquid Metal Boiling Working Group*. Karlsruhe, Oct. 27, 29. Available from JRC Ispra.

Lyon, R. N. (1952) *Liquid Metals Handbook*. ch. 5. 2nd edition.

Shah, M. M. (1992) A survey of experimental heat transfer data for nucleate pool boiling of liquid metals and a new correlation. *Int. J. Heat and Fluid Flow*. 13.

Sheriff, N. and Davies, N. W. (1979) Liquid metal natural convection from plane surfaces: a review including recent sodium measurements. *Int. J. Heat and Fluid Flow*. 1.

J. Costa and D. Tenchine

LIQUID NITROGEN (see Cryogenic fluids)

LIQUID OXYGEN (see Cryogenic fluids)

LIQUID RING COMPRESSOR (see Compressors)

LIQUIDS, DIFFUSION IN (see Diffusion coefficient)

LIQUID-SOLID-GAS FLOWS (see Multiphase flow)

LIQUID-SOLID FLOW

Following from: *Two-phase flows*

Liquid-solid flow represents the flow of a liquid continuum carrying dispersed solid particles suspended and conveyed by the drag and pressure forces of the liquid acting on the particles. The aim of those *slurry* flows may be the transport of bulk-solids or physical or chemical processes between carrier liquid and solids. In reality such a flow comprises two very different flows: the total mixture flow characterized by the **Pipe Reynolds Number Re** and the relative flow between the solid particles and the carrier fluid characterized by the **Particle Reynolds Number Re_s**. The complete range of velocities is not possible with slurries as it is with pure liquids. The two superimposed phases influence each other and should be harmonized to flow without depositions or blockages.

Flow Behaviour, Flow Pattern and Flow Regimes

How the solid-particles behave in the mixture—whether they distribute evenly, and move suspended in the carrier-flow or segregate and deposit, depends as well on the solid properties (grain size, shape, density), on the properties of the carrier liquid (density, viscosity), on the operation parameters of pipe flow (velocity, pipe diameter, solid concentration) and on flow direction. Under some conditions, solid particles can change the rheologic behaviour of the slurry from Newtonian to non-Newtonian.

Rheologic Classification

With larger solid particles, the fluid and solid phases mostly retain their own identities because of the working inertial forces. Thus the increase of viscosity is relatively small, and the slurry flow behaviour then remains for any concentration Newtonian, like that of the Newtonian carrier fluid. Fine-grained slurries behave

Table 1. Limits for pseudohomogenous flow regime according to Weber (1978)

r_s/r_L —	1.5	2.0	2.5	3.0	4.0	5.0	6.0	7.0	8.0
$Re_s = 2$; $w_{so}/u_L = 0.0056$									
d_s μm	231	184	162	147	128	116	108	101	96
w_{so} cm/s	0.97	1.25	1.42	1.56	1.80	1.98	2.12	2.26	2.38
u_L m/s	1.73	2.23	2.53	2.78	3.21	3.53	3.78	4.03	4.25
$Re_s = 0.1$; $w_{so}/u_L = 0.00146$									
d_s μm	80.0	63.0	55.0	49.5	43.5	39.5	36.5	34.5	33.0
w_{so} cm/s	0.147	0.186	0.214	0.236	0.270	0.298	0.322	0.340	0.360
u_L m/s	1.00	1.27	1.46	1.62	1.85	2.04	2.20	2.33	2.47

likewise while the solids concentration remains < 25% because the distances between the suspended particles are still large enough to avoid intermolecular cohesive forces. By increasing the concentration, non-Newtonian flow behaviour will occur.

Flow Direction

A relatively low tendency towards *segregation* exists for vertical flow as a result of the symmetrical configuration of forces. Even for coarse material, a fairly uniform solid distribution can be expected in the pipe as long as the condition for conveyance is well-satisfied.

For horizontal conveyance gravity causes asymmetrical configuration of forces and segregation is always present, even when the conveyance condition for horizontal flow is well-satisfied. Horizontal slurry flows therefore show a solids concentration profile depending on the velocity and are called *settling slurries.*

Flow Regimes of Newtonian Slurries

Only very fine particles with $Re_s < 10^{-6}$, which can be conveyed by **Brownian Molecular Movement,** are kept in suspension without any **Turbulence**, the so-called *colloidal dispersions.*

Fine particles with $10^{-6} < Re_s < 0.1$ can be easily held in suspension by hydraulic forces, and this tendency is supported by a low solid density and by a nonspherical particle shape. Only little turbulence is needed to keep those particles homogeneously suspended; so liquid velocity can be low in this *homogeneous flow* regime.

Particles with $0.1 < Re_s < 2$ need some more turbulence and velocity to be held in suspension, but in the case of horizontal flow, completely uniform solid distribution cannot be reached. A certain degree of segregation is permitted. This type of suspension can exist at economically-feasible velocities and is called the *pseudo-homogeneous flow regime.*

To guarantee equal conditions for solids of any density relative to the concentration profile at these Re_s limits, the ratio of the settling velocity and the fluid velocity w_{so}/u_L must remain constant. Fluid velocities required for this condition can be obtained from **Table 1**.

For coarser particles with $Re_s > 2$, the segregation is greater and the mixture flow is more heterogeneous. In this *heterogeneous flow regime,* lower velocities can lead to conveyance by *saltation,* and finally to the so-called *critical deposition velocity,* where the solid particles begin to settle out. At this point the pressure drop of the mixture is minimum.

The rheologic behaviour and the flow regimes are listed in **Table 2**.

A more detailed impression of the complex relationship between particle size, friction velocity, pipe diameter and the flow pattern of the solid particles in a newtonian liquid-solid flow is given by the generalized phase diagram of Thomas (1962) in **Figure 1**. (d_s is particle size and δ boundary layer thickness, U_o friction velocity and ν kinematic viscosity.)

Flow Regimes of Non-Newtonian Slurries

Highly concentrated non-Newtonian slurries, so called nonsettling slurries are homogeneous, and do not need turbulence to prevent settling out while conveyed through a relatively short pipe because the settling process is very slow. However, for long distance transport, turbulent flow should be recommended. Here the transition velocity from laminar to turbulent flow is referred to as the *critical transition velocity.*

Basic Relationships

Basically a heterogeneous liquid-solid flow can also be treated as a homogeneous mixture flow, like a pseudo-liquid that obeys the usual equations of single-phase flows. Then suitable weighted average pseudo properties can be formed, based on the properties and conditions of the two single-phases. Those typical virtual properties are, for instance, mean velocity, mean density, mean consistency of the mixture and so on. To proceed with pressure drop calculation in this way indeed depends on grain size and flow direction.

Vertical upward liquid-solid flow

Solid particles can be conveyed upward when the transport condition is well-satisfied; that means when fluid velocity exceeds the terminal settling velocity of the solids:

$$u_L >> w_S \tag{1}$$

$$w_S = w_{S0}(1 - c_v)^\gamma \tag{2}$$

$$w_{so} = \sqrt{\frac{4}{3}\frac{d_{s50}}{c_D}\frac{(\rho_s - \rho_L)}{\rho_L}g} \tag{3}$$

Table 2. Classification of hydraulic flow regimes relying on Durand (1953)

Flow behaviour	Rheological behaviour	Concentration by volume	Reynolds number Re_s
homogeneous	Newtonian *)	<25%	≦0.1
pseudohomog.	Newtonian *)	<25%	$0.1 < Re_s < 2$
heterogeneous	Newtonian	<25%	>2

*)At concentrations > 30pc the slurry starts being non-Newtonian

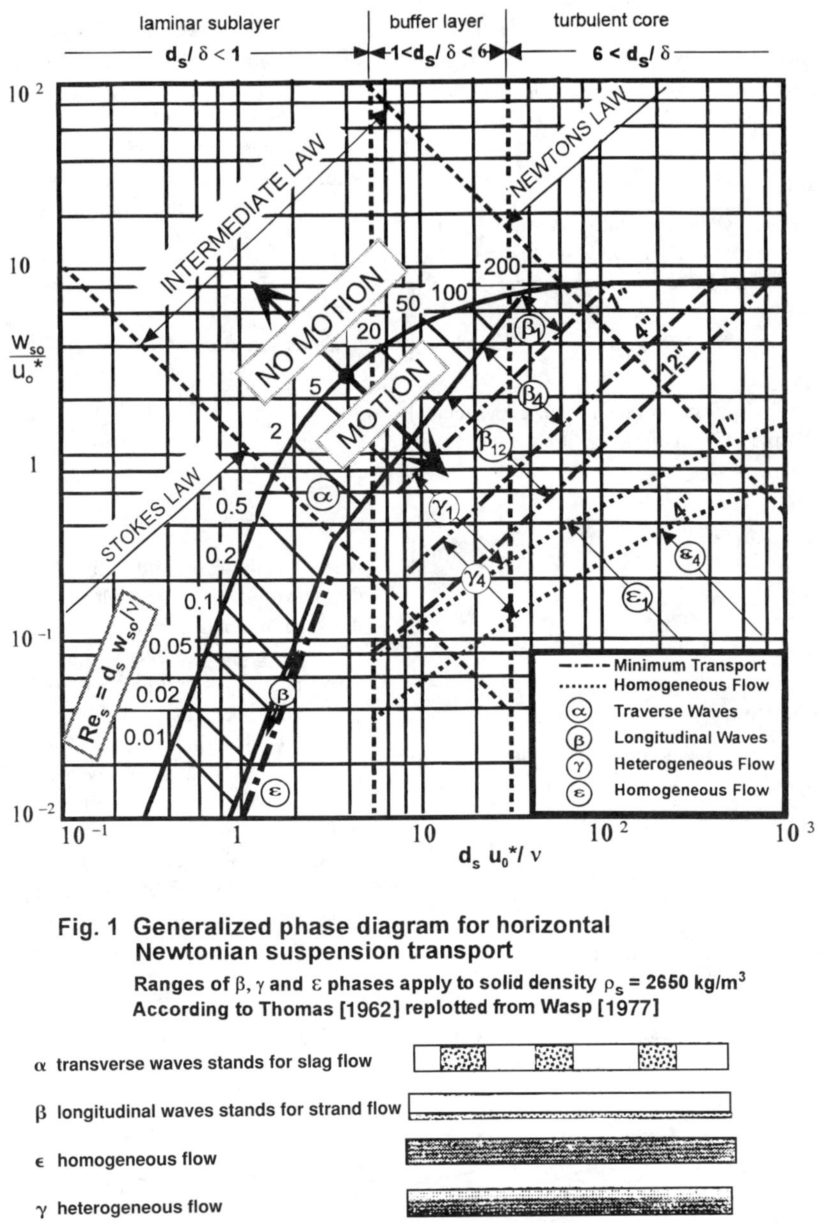

Figure 1. Generalized phase diagram for horizontal Newtonian suspension transport.

$$c_D = \frac{24}{Re_s} + \frac{4}{\sqrt{Re_s}} + 0.4 \qquad (4)$$

$$Re_s = \frac{w_{so} d_{s50}}{\nu_L} \qquad (5)$$

where w_s is the *hindered settling velocity* at higher concentration and/or at nonspherical particle shape; w_{so} is the single particle settling velocity; Re_s is the particle related Reynolds number; $c_v = V_s/(V_s + V_L)$ is the local solid concentration by volume; d_{s50} is the grain size at 50% passing sieve; ρ_s is solid density; ρ_f is liquid density; and ν_L is kinematic liquid viscosity.

Two hindering effects must be taken into account: the solids concentration and the particle shape. **Figure 2** gives the influence

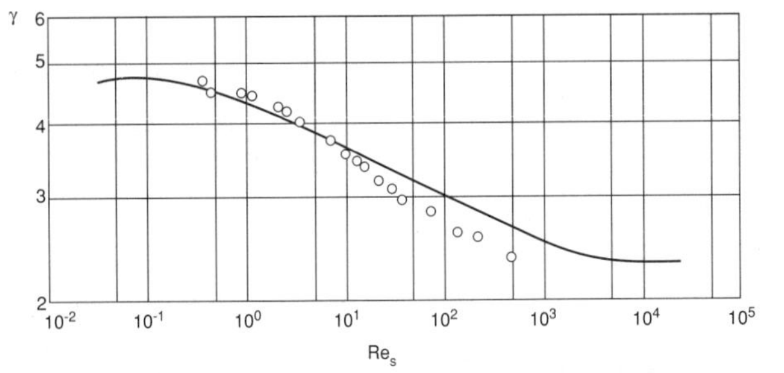

Figure 2. Influence of the concentration on settling velocity according to Maude and Whitmore (1958) replotted from Weber (1974).

of the concentration according to Maude and Whitmore (1958) and **Figure 3**, the shape which is described by the *sphericity* ψ, being the surface of a volume equivalent sphere related to the real surface of the particle.

Horizontal liquid-solid flow

Solid particles can be conveyed horizontally without *deposition* when the transport condition for horizontal flow is satisfied, which is true when fluid velocity exceeds the *critical deposition velocity:*

$$u_L > u_{crit} \tag{6}$$

The critical deposition velocity can be taken from experimental results according to Wasp (1977), correlating to the following equation:

$$u_{crit} = 3.525\, c_T^{0.234} \left(\frac{d_{s50}}{D}\right)^{1/6} \sqrt{2Dg\,\frac{\rho_s - \rho_L}{\rho_L}} \tag{7}$$

where D is pipe diameter.

Pressure Drop Calculation

The mentioned pseudo-liquid method can easily be applied to vertical and horizontal homogeneous and pseudo-homogeneous mixture flows, because there is only little slip between u_L and u_s, so that $u_m \approx u_L \approx u_s$ and $c_T \approx c_v$. The pressure drop should be calculated in this case by the following equation:

$$\Delta p = \lambda_L \frac{\rho_m}{2} u_m^2 \frac{\Delta L_{tot}}{D} + \rho_m g \Delta L_{vert} \tag{8}$$

where λ_L = to one-quarter of the **Fanning Friction Factor** of the pure liquid; $\rho_m = c_v\,\rho_s + (1 - c_v)\,\rho_L$ is the mean mixture density; $u_m = (\dot{V}_L + \dot{V}_s)/A$ is the mean mixture velocity; and $c_T = \dot{V}_s/(\dot{V}_s + \dot{V}_L)$ is the delivered or transport concentration by volume.

In the heterogeneous vertical flow regime, the solids are also homogeneously distributed but a considerable slip can exist. Therefore, the local concentration c_v has to be calculated for each velocity with respect to the delivered concentration and to the given solid mass flow rate before **Equation 8** can be applied.

$$c_v = \frac{u_m}{2w_{so}}\left[\frac{w_{so}}{u_m} - 1 + \sqrt{\left(1 - \frac{ww_{so}}{u_m}\right)^2 1 + 4c_T \frac{w_{so}}{u_m}}\right] \tag{9}$$

The pressure drop of horizontal liquid-solid flow can be calculated for any flow regime according to Weber (1986), by applying the generalized Durand equation:

$$\Delta p = \left[83^{1/m}\left(\frac{Dg}{u_m^2}\frac{s-1}{\sqrt{c_D}}\right)^{1.5/m^3} (1 - f)c_T + 1\right] \lambda_L \frac{\rho_{Ls}}{2} u_m^2 \frac{\Delta L_{hor}}{D} \tag{10}$$

where f is the fine solids (Re_s <2); $\rho_{Ls} = \rho_s\, fc_T + (1 - fc_T)\rho_L$ is the density of the carrier fluid enriched by "fines"; $s = \rho_s/\rho_{Ls}$ is the specific solid density; $m = 2 - (d_{s90}/d_{s10})^{-.04}$ is a correction factor for grain size distribution; d_{s90} is grain size at 90% passing sieve; and d_{s10} is grain size at 10% passing sieve.

It should be stressed that when applying the fine part of **Equation 10** the quantities ρ_{Ls}, and s are directly influenced. The quantities d_{s90}, d_{s50}, d_{s10} are influenced by separating the fine solids part and adding it to the carrier fluid, as shown in **Figure 4**. In consequence, the correction factor m is changing because of ρ_{Ls} and d_{s50} and also because of C_D.

For horizontal heterogeneous liquid-solid flow, the local solid concentration by volume related to the delivered concentration can be calculated by the following equation:

$$c_D/c_T = \sqrt{\rho_s/\rho_{LS}} \tag{11}$$

In **Figure 5** calculated results for an heterogeneous Newtonian slurry flow are compared with experimental results.

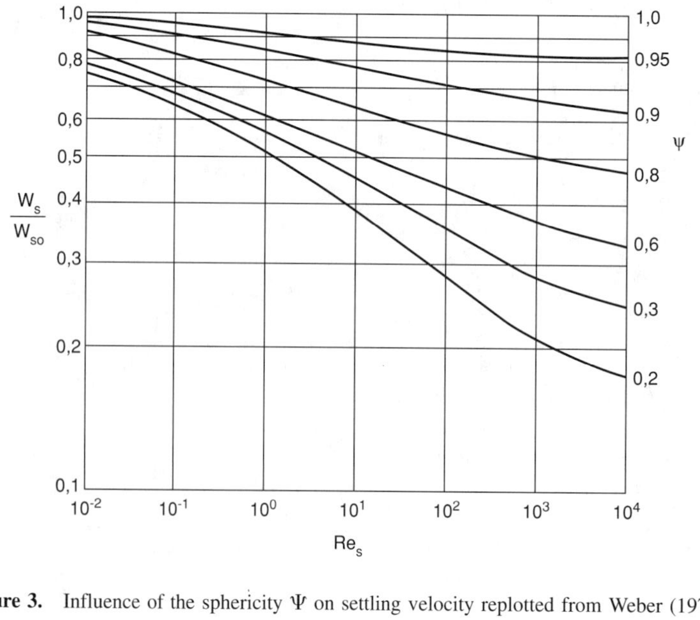

Figure 3. Influence of the sphericity Ψ on settling velocity replotted from Weber (1974).

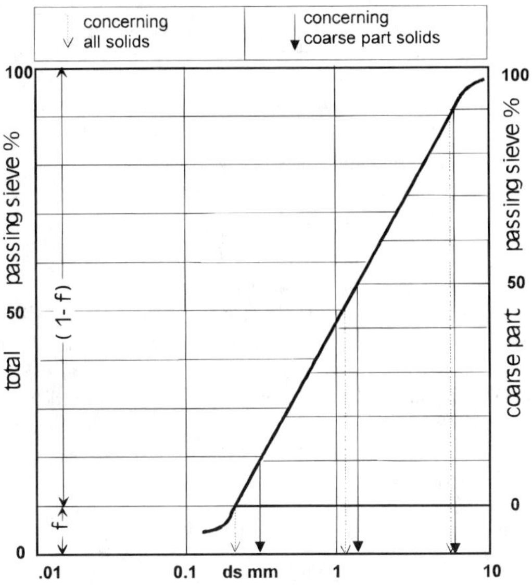

Figure 4. Effect of fine particles added to the carrier fluid on the remainder effective particle size.

Relationships for non-Newtonian slurries

Since a multitude of non-newtonian or nonsettling slurries are of the Bingham type, the rheological law of *Binghamian slurries* is relevant here:

$$\tau = \tau_o + \mu_B dv/dy \qquad (12)$$

where τ is the shear stress of the slurry; τ_O is the yield stress of the slurry; μ_B is the Bingham viscosity; and dv/dy is the velocity gradient.

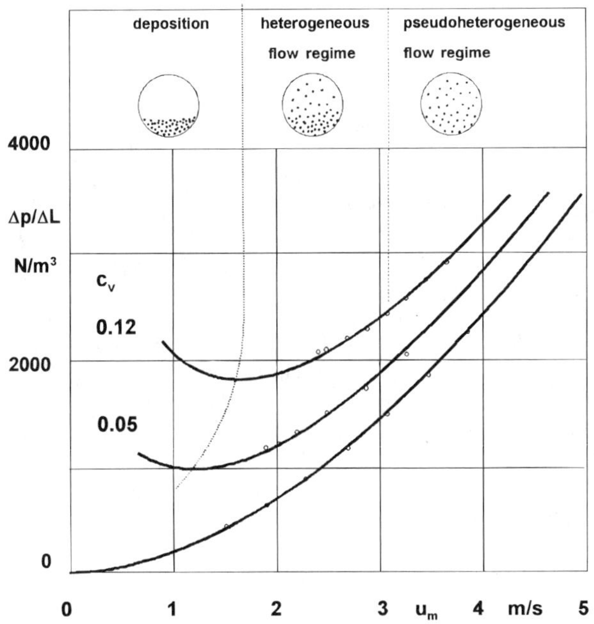

Figure 5. Heterogeneous Newtonian iron ore slurry flow with "fines" calculated by Equation 10. *Source:* Experimental data according to Report III (1973).

Laminar Binghamian slurry flow

For short pipe lengths, very economical low velocities within the subcritical laminar flow regime can be chosen. The critical transition velocity for binghamian slurries can be found using the following equation according to Durand and Condolios (1952):

$$Re_{crit} = 1000\left(1 + \sqrt{1 + \frac{He}{3000}}\right) \qquad (13)$$

where He $= \tau_O D^2 \rho_m/\mu_B{}^2$ is the Hedström number; $Re_{crit} = u_{mcrit} D\rho_m/\mu_B$ is the critical transition Reynolds number; and $u_{mcrit} = Re_{crit}\ \mu_B/(D\rho_m)$ is the critical transition velocity.

To ensure the stability of the laminar slurry flow, no particles should be greater than

$$d_{syield} = \frac{3\pi}{2} \frac{\tau_o}{(\rho_s - \rho_L)g}. \qquad (14)$$

Dedegil (1986) refers to this diameter, which is critical between settling and suspending. Otherwise, overcritical turbulent flow should be realized to avoid settling out of solids.

The pressure gradient of laminar Binghamian slurry may be calculated by the *Buckingham equation,* but as this can not be done explicitely, an approximate solution of it which neglects the four-power terms may be used instead, according to Wasp, (1977):

$$\frac{\Delta p}{\Delta L} = \mu_B\left(\frac{32 u_m}{D^2}\right) + \frac{16}{3}\frac{\tau_o}{D} \qquad (15)$$

Turbulent Binghamian slurries

Since in the case of turbulent Binghamian slurries the solid particles are pushed by turbulence eddies, in this turbulent flow regime the yield stress loses its effect. Therefore, the relationships of Newtonian slurries can be used, but the Reynolds numbers have to incorporate Binghamian viscosity. The friction factor for instance is $\lambda_L = \phi(Re_B)$ and the drag coefficient is $c_D = f(Re_{sB})$. This means the usual drag curve and the usual friction curve can be used. Only the Reynolds numbers, changed by the higher non-Newtonian consistencies, have to be applied.

References

Dedegil, M. Y. (1986) Drag coefficient and settling velocity of particles. *Proc. of Int. Symp. on Slurry Flows.* ASME. Dec. 07–12. Anaheim-CA. FED-Vol. 38. S. 9–15

Durand, R. (1953) Basic relationships of the transportation of Solids in pipes-experimental research. *Proc. Minnesota Int. Hydraulics Div.* ASCE 89–103. Sept.

Maude, A. D. and Whitmore, R. L. (1958) A generalized theory of sedimentation. *British Journal of Appl. Physics.* 9. Dec.

Thomas, D. G. (1962) Transport characteristics of suspensions: Part VI. Minimum transport velocity for large particle size suspensions in round horizontal pipes. *AlChE Journal.* 8. 3.

Wasp, E. J. et al. (1977) Solid liquid flow-slurry pipeline transportation. *Series on Bulk Materials Handling*. 1. No. 4.

Weber M. et al. (1974) *Strömungsfördertechnik*. Krausskopfverlag. Mainz.

Weber, M. (1978) Pseudohomogene Gemische, Teil B, aus: Hydraulischer Feststofftransport in Rohrleitungen. Ein praxisbezogener Einführungskurs. *Hydrotransport 5*. Hannover.

Weber, M. (1986) Improved Durand-equation for multiple application. *Int. Symposium on Slurry Flows*. ASME-VDI. Anaheim (Ca) USA.

Report III. (1973). *Saskatchewan Res. Council*. Experimental studies on the hydraulic transport of iron ore.

M. Weber

LIQUID-SOLID INTERFACES (see Contact angle)

LIQUID-SOLID SEPARATION

Following from: *Particle technology; Phase separation*

Liquid-solid separation involves the separation of two phases, solid and liquid, from a suspension. It is used in many processes for the: 1. recovery of valuable solid component (the liquid being discarded); 2. liquid recovery (the solids being discarded); 3. recovery of both solid and liquid; or 4. recovery of neither phase (e.g., when a liquid is being cleaned prior to discharge, as in the prevention of water pollution).

Any separation system design must consider all stages of pre-treatment, solids concentration, solids separation, and post-treatment. This encompasses a wide range of equipment and processes, summarized in **Table 1**. Pre-treatment is used primarily with difficult-to-filter slurries, enabling them to be filtered more easily. It usually involves changing the nature of the suspended solids by either chemical or physical means, or by adding a solid (filter aid) to the suspension to act as a bulking agent to increase the permeability of the cake formed during subsequent filtration. In solids concentration, part of the liquid may be removed by (gravity or centrifugal) thickening or hydrocycloning to reduce liquid volume throughput load on the filter.

Table 1. Components of the solid/liquid separation process

Pre-treatment		
Chemical	Physical	
Flocculation	Crystal growth	
Coagulation	Addition of filter aids	
pH adjustment	Freezing	
	Ageing	
Solids concentration		
Thickening	Clarification	Assisted separations
Gravity	—gravity	—magnetic
Centrifugal sedimenters	—centrifugal	—electric
Hydrocyclones		—diaelectric
Delayed cake filters		—acoustic
—high-shear crossflow		
Low-shear crossflow		
—microfiltration		
Solids separation		
Cake filter	Depth filters	
—pressure	Granular beds	
—vacuum	Cartridges	
—centrifugal	Precoat	
—gravity	Membranes	
Post-treatment		
Filtrate	Cake	
Polishing	Washing	
—microfiltration	—displacement	
—ultrafiltration	—reslurry	
(—reverse osmosis)	Deliquoring	
(—electrodialysis)	—gravity drainage	
	—displacement (gas blowing)	
	—mechanical expression	
	(thermal drying)	

A number of new 'assisted separation' techniques are making their way into the list of technical alternatives—these utilise magnetic, electrical or sonic force fields (or combinations) to provide more effective separation.

Solids separation involves a filter, types of which are classified in many different ways. For present purposes a division into those in which cakes are formed and those in which the particles are captured in the depth of the medium is adequate. Cake filters can be further divided into pressure, vacuum, centrifugal and gravity operations.

Post-treatment processes involve making improvements to the quality of the solid or liquid products. In the case of the filtrate, these operations are often referred to as polishing processes, and may involve micro or ultrafilters to remove finer substances. Further purification may involve removal of ionic and macromolecular species by, for example, reverse osmosis, ion exchange or electrodialysis. The relative position of these separation processes in the spectrum of the size of "particle" to be removed from the liquid is shown on **Figure 1**. Cake post-treatment processes include washing soluble impurities from the cake voids and removal of excess liquid from the voids. Thermal drying is often the final stage of liquid removal.

Solid/Liquid Separation in the Flowsheet

Solid/liquid separation is all too often designed as a 'stands alone' unit in a plant flowsheet. The performance of a solid/liquid separation device is sensitive to the history of the feed solution and, in particular, to the properties imparted to the suspension by its method of manufacture, e.g., on the shape, size and size distribution of the particles, which result from the operating conditions in the precipitator or crystallizer. A change in particle production conditions can affect the best choice of filter for a particular purpose. The economics and viability of producing the product is often affected by the amount of liquid removed in the post-treatment processes. For example, if the cake is to be transported, briquetted or pelletized, cake moisture content will need to be within a specified range; or if a bone dry product is required the thermal load on the dryers can be reduced by correct choice and operation of filter.

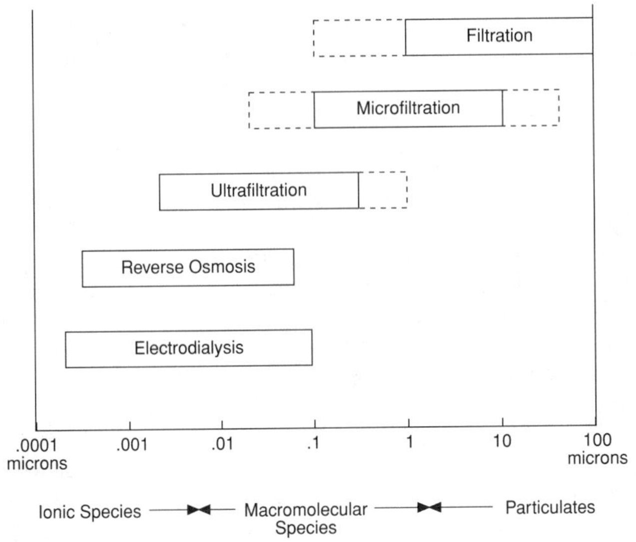

Figure 1. General techniques for contamination removal from liquids relative to the size of the species to be removed.

It is important, therefore, to consider simultaneously and in some detail those processes which are to feed suspension to the solid/liquid separations plant, and the subsequent processing of the solid or liquid products.

Particle and Liquid Properties in Solid/Liquid Separation

Three parameter types may be identified to fully describe a solid/liquid system. These are:

primary properties;

state of the system;

macroscopic properties.

Primary properties are those which can be measured independently of the other components of the system; specifically, they are the solid and liquid physical properties, the size, size distribution and shape of the particles, and the surface properties of the particles in their solution environment. The way the particle interacts with its surrounding fluid becomes important for smaller particles (notably when the particle size $< 10 - 20$ μm), since repulsive (surface), forces between the particles can become as significant as gravitationally-or hydrodynamically-induced forces. These factors decide whether the particles will, for example, settle slowly or quickly; whether they can be retained on some kind of septum or porous medium; or whether the resulting cake will be dry or sloppy.

The description of the state of the system (porosity or concentration, and the homogeneity and extent of dispersion of the particles) combines with primary properties to control the macroscopic properties, which are measured to investigate the application of a particular separation method. Such measurements may be the permeability or specific resistance of the filter bed or filter cake, the terminal settling velocity of the particles, or the bulk settling rate of the suspension.

Particle Shape

Liquid particles almost always approximate a sphere due to the distribution of surface forces around the particle. In contrast, solid particles are rarely either spherical or uniform. Certain classes of materials are essentially crystalline and may be made up from fairly uniform particles, each of which is, for example, cubic or rhombohedral. But even crystalline materials may be a mixture of shapes, especially if, as often happens industrially, breakage of the crystals occurs due to handling. Indeed, breakage can be caused within the separator itself and is a common problem in, for example, pusher centrifuges. The great majority of particles are of irregular shape: fibrous particles are common, but they may possess a wide range of length to diameter ratios; they may have smooth surfaces; they may be fibrillated, and so on. It is rare that the shape of the particles to be handled can be defined precisely.

Particle Size

Particles may vary in size from very fine or colloidal matter to coarse granular solids. Sometimes all the solids may be of the same material, i.e., of homogeneous composition or, as is often the case with effluent suspensions, the individual particles may have very different compositions. In general terms particle size has a significant effect on solid/liquid separation behaviour of the suspension. A knowledge of techniques for measuring size particles is therefore important to the process technologist.

In solid/liquid separation four reasons for measuring particles size can be identified [Scarlett and Ward (1986)]:

a. To measure and specify the quality of a liquid, which is the valuable product from a filtration process. In this case, the particles remaining in suspension are dilute in concentration, and are therefore difficult to filter. Often only the total concentration of solids is required as, for example, in water treatment processes. However, in operations such as the filtration of parental fluids or hydraulic fluids, the size (and occasionally shape) of the remaining solids is critical.
b. An extension of this requirement is to specify the performance of a filter medium in terms of its ability to retain particles of different sizes. This type of evaluation is usually associated with fluid polishing operations and the specification of a nominal pore size for a polishing medium, or with the performance assessment of separating devices such as sedimenting centrifuges.
c. In many operations the solid is the valuable product. It is rarely recovered in a completely dry state and is often processed further. Evaluation of this product is required for quality control and is not connected solely with the separation process. In this case, the method of evaluation is often dictated by the customer or by the standards accepted by the particular industry.
d. Occasionally, the requirement is to evaluate the solid in order to predict their probable behaviour in a separation process. This may be to enable an initial choice between different separation methods; to select or test a suitable pretreatment process or filter medium; to improve the efficiency of an existing machine, or to estimate the size of a new one. In any of these, the objective is predictive in nature, and the measuring technique must be selected more carefully than

for a quality control application. The measured size, for design purposes, should relate to the way the particles are handled in the process; but this is not essential for quality control purposes.

The Solution Environment

Interactions between the particle and the liquid in which it is suspended have the greatest influence when particles are smaller, particularly when their size is smaller than a few microns. Origins of interparticle repulsive forces lie in the distribution of solution ions around the charged surface of the particle, and the resultant electrical charge is dependent on the chemical species present at the surface. A potential energy of repulsion may extend appreciable distances from the particle surface, but its range may be compressed by increasing the electrolyte content of the solution. For practical purposes, the magnitude of net repulsive force between particles is represented by the zeta (ζ-) potential; and the following statements can be made about the influence of the ζ-potential in solid/liquid separation [Wakeman et al. (1988)]:

a. The net repulsive force increases with increasing magnitude of the ζ-potential.
b. Reducing the magnitude of the repulsive force causes the dispersion to become unstable, and generally more easily separated.
c. Repulsive forces can be reduced by either (1) adding a non-adsorbing electrolyte to the liquid to change the distribution of solution ions around the particle, or (2) altering the electrical charge on the surface of the particle by the specific adsorption of certain ions.

Around the isoelectric point of the suspension, the process engineer can expect:

1. faster settling rates;
2. more rapid filter cake formation; and
3. slightly higher moisture content cakes and sediments

due to the aggregation of particles in the suspension, where interparticle repulsion forces are very small. At the maximum or minimum ζ-potential, the engineer can expect:

1. slower settling rates;
2. slower cake formation rates, and;
3. slightly lower moisture content cakes and sediments

due to the existence of greater repulsive forces, which maintain the particles better dispersed throughout the liquid phase.

At pH's beyond those at which maximum or minimum ζ-potentials first occur, the process engineer can expect intermediate settling, filtration and expression rates and intermediate cake and sediment moisture contents. This is due to compression of the electrical double layer around the particle caused by high ionic strengths in the solution.

The Nature of the Fluid

Apart from the effect of the fluid at the fluid/particle interface, the density and viscosity of the fluid are most important in industrial filtration. Density is generally only significant where separation depends on a difference in density between the fluid and the particles, e.g., in thickeners or centrifugal sedimenters. Whatever difference in density exists must usually be accepted and cannot often be controlled to any significant degree; occasionally the influence of temperature may be important, or it may be possible to alter the density to a very limited extent by varying the amount of dissolved matter.

Viscosity has a more widespread effect and, at the same time, is more amenable to control since it is usually sensitive to temperature changes. The rate of filtration of liquids can be greatly accelerated in many instances by a relatively small increase in temperature, which causes a drop in viscosity.

Solid/Liquid Separation Equipment

There are many solid/liquid separation techniques which have established practical value for general application in the processing industries, and there are a few which are in their early stages of industrial exploitation. The dividing line between these two categories is arguable, particularly as the field is noted for innovation and rapid development. The area of solid/liquid separation techniques is broad; the following list indicates the diversity of equipment available to the engineer:

- *Gravity settlers,* e.g., clarifiers, deep thickeners, lamella separators, settling tanks, lagoons, thickeners;
- *Sedimenting centrifuges,* e.g., tubular bowl, skimmer pipe, disc, scroll discharge;
- Hydrocyclones, e.g., conical, circulating bed;
- Classifiers, e.g., hydraulic, mechanical, screens, sieve bends;
- Gravity filters, e.g., deep bed, Nutsche;
- Line filters, e.g., cartridges, strainers;
- Pressure filters, e.g., continuous pressure, diatomaceous earth, fibre bed, filter press, horizontal element, pressure Nutsche, vertical element, sand, sheet filter, tubular element;
- Filters with compression, e.g., belt press, membrane plate and frame, screw press, variable volume filter (e.g., tube);
- Vacuum filters, e.g., top/bottom fed drum, disc, leaf, belt, pan, table, precoat drum;
- Filter thickeners or crossflow filters, e.g., delayed cake, dynamic or high shear microfilters, low shear microfilters/ultrafilters;
- Filtering centrifuges, e.g., basket, pendulum, oscillating, tumbling, plough/peeler, pusher, worm screen;
- Flotation;
- Magnetic filters, e.g., low gradient (drum, grid or belt), high gradient.

Computer software is available [Tarleton and Wakeman (1991)] to aid in the selection of appropriate equipment for specific solid/liquid separation problems.

Interactions Between Separations Equipment and the Feed (Wakeman et al, 1988, and Tiller and Yeh, 1986)

Compressibility of particulate structures is a key factor in the behaviour of all solid/liquid separation equipment. Particulate

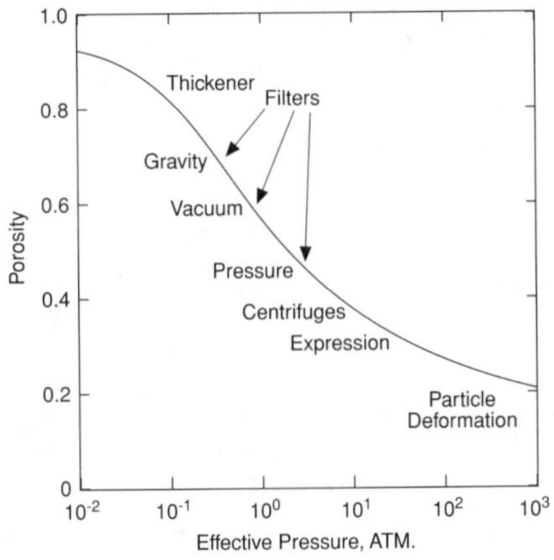

Figure 2. Separation operations in relation to pressure.

aggregation in suspensions determines the degree of compressibility, which is controlled by suitable pretreatment processes. As increasing thicknesses of deposits cover the separation surfaces, developing stresses continually compress the particulate bed. The principal sources of stress are (1) the unbuoyed weight in gravity in gravity thickeners, (2) centrifugal forces, (3) pump pressure converted into Darcian drag at the particle surfaces, and (4) surface forces developed by pressure-actuated impermeable membranes.

When stress is applied to a bed of flocculated particles, the bed is compressed by particle movement into open pores until no further movement into the interstices can occur. Any further deformation of the bed results in particle deformation or breakage. The porosity of the compressed floc is a function of the initial structure and particle shape——spherical particles generally have a compressed packing density (solidosity) of about 0.65. The difference between the initial solidosity resulting from flocculation and 0.65 represents the measure of compressibility that is used for correlation, design and scale-up purposes.

In sedimentation processes the solids are subjected to low compressive pressures of the order of 0.1 to 0.2 bar. At the other extreme involving high pressure expression the effective pressure can be up to hundreds of atmospheres. At low pressures, compression arises from squeezing of particles into unoccupied voids whilst at high pressures, particle deformation and crushing (where the coefficients of elasticity and ultimate strength are important) determine cake compaction. Separations equipment can be categorised according to the typical effective pressure involved in the process, as shown by **Figure 2**.

Particle size, size distribution, shape and degree of flocculation determine the solids packing density in solid/liquid mixtures, and hence the compressibility of the mixture. This is shown schematically in **Figure 3**. Particles larger than about 20 μm form beds which are essentially incompressible, the solidosity of which depends primarily on particle shape. Irregular particles form beds with larger porosities than those associated with spheres. Filter aids such as diatomaceous earths and expanded perlite are so irregular that they form beds whose solidosities range from 0.1 to 0.2, even though the primary particle sizes may be below 10 μm.

Stresses developed in the matrix of large particles during separation processes do not generally reach sufficient magnitude to disturb the structure.

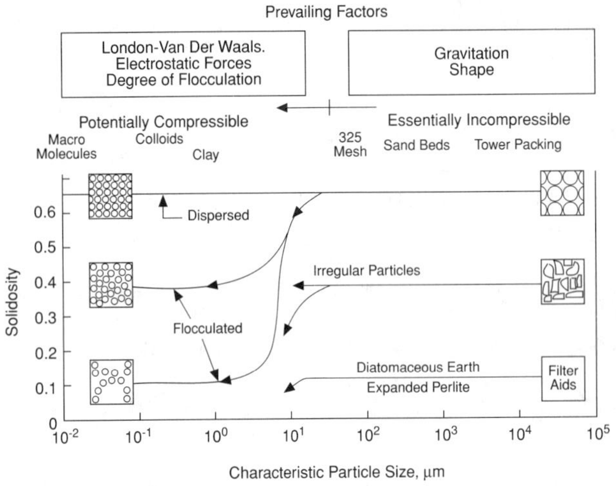

Figure 3. Relations between null stress volume fraction of solid and particle size, shape and degree of flocculation.

As the characteristic dimension of the particle decreases, the effects of interparticle forces increase relative to gravitational force. When attractive London-van der Waals forces predominate in comparison with electrostatic and gravitational forces, the particles tend to form aggregates. As the number of particles grows in an aggregate (or in a polymer-flocculated system) in suspension, internal porosity of the overall suspension increases. Although a highly-porous floc is distorted upon deposition, the unstressed solidosity of the cake or sediment may be as low as 0.1. On the right-hand side of **Figure 3**, beds formed from large particles are incompressible and have a packing density dependent on particle shape. On the left of the diagram, the solids volume fraction depends mainly on the degree on aggregation (or the state of dispersion) of the suspension.

References

Scarlett, B. and Ward, A. S. (1986) In *Solid/Liquid Separation Equipment Scale-up*. eds. D. B. Purchas and R. J. Wakeman. Uplands Press & Filtration Specialists.

Wakeman, R. J., Thuraisingham, S. T., and Tarleton, E. S. (1988) Proc. 391st EFCE Event Particle Technology in Relation to Solid/Liquid Separation. Technologisch Instituut-Koninklijke Vlaamse Ingenieursvereniging—The Filtration Society. Antwerp.

Tiller, F. M. and Yeh, C. S. (1986) In *Advances in Solid/Liquid Separation*. (Ed. H. S. Muralidhara). Battelle Press.

Tarleton, E. S. and Wakeman, R. J. (1991) Solid/liquid separation equipment simulation and design: p^c-SELECT—Personal computer software for the analysis of filtration and sedimentation experimental data and the selection of solid/liquid separation equipment. Separation Technology Associates. Exeter.

Leading to: *Filtration; Flotation; Centrifuges*

R.J. Wakeman

LIQUID, THERMAL CONDUCTIVITY OF (see Thermal conductivity; Thermal conductivity mechanisms)

LJUNGSTROM AIR PREHEATER (see Regenerative heat exchangers)

LOCA, LOSS OF COOLANT ACCIDENT, (see Blowdown)

LOCAL FILM FLOW RATE MEASUREMENT (see Film flow rate measurement)

LOCAL STRUCTURE (See Mixing)

LOCAL VOID FRACTION (see Void fraction)

LOCAL VOID FRACTION MEASUREMENT (see Void fraction measurement)

LOCKHART AND MARTINELLI CORRELATION (see Pressure drop, two phase flow)

LOGARITHMIC PROFILE, IN BOUNDARY LAYER (see Boundary layer)

LOG MEAN TEMPERATURE DIFFERENCE (see Mean temperature difference; Tube banks, single phase heat transfer in)

LONDON-VAN-DER-WAALS FORCES (see Liquid-solid separation)

LONG-TUBE VERTICAL EVAPORATOR (see Evaporators)

LORENTZ-LORENZ EQUATION (see Interferometry)

LOSCHMIDT CONSTANT

The Loschmidt constant, N_L, is defined as the number of particles (atoms or molecules) per unit volume of a perfect gas at STP (0°C and 760mmHg). The accepted value is

$$N_L = 2.687 \times 10^{25}\ m^{-3}$$

N_L is equal to the ratio of **Avogadro Number**, N_A to the molar volume, $\tilde{v}$ at STP.

$$\text{ie } N_L = \frac{N_A}{\tilde{v}} = \frac{6.022 \times 10^{23}\ mol^{-1}}{2.241 \times 10^{-2}\ m^3\ mol^{-1}}$$

$$= 2.687 \times 10^{25}\ m^{-3}$$

Note that Loschmidt constant is sometimes quoted as molar volume/molecular volume, which is the number of molecules per mole or **Avogadro Number**.

G.L. Shires

LOSS COEFFICIENT IN BENDS (see Bends, flow and pressure drop in)

LOSS OF COOLANT ACCIDENT, LOCA, (see Blowdown; Boiling water reactor, BWR; Reflood)

LOST WORK (see Thermodynamics)

LOUVER FINS (see Extended surface heat transfer)

LOWER EXPLOSION LIMIT, LEL (see Flammability)

LOWER FLAMMABILITY LIMIT, LFL (see Flammability)

LOW FIN TUBES (see Augmentation of heat transfer, single phase; Augmentation of heat transfer, two phase)

LOW FIN TUBING (see Shell-and-tube heat exchangers)

LOW PRESSURE DIE-CASTING (See Casting of metals)

LOW TEMPERATURE PLASMA (see Plasma)

LPG, LIQUEFIED PETROLEUM GAS (see n-Butane; Oil refining; Propane)

LUBRICATING BASE OILS (see Oils)

LUIKOV NUMBER (see Wet bulb temperature)

LUMINESCENCE (see Phosphorescence; Radiative heat transfer)

LURGI GASIFIER (see Gasification)

LURGI PROCESS (see Substitute natural gas)

LYOPHILLIC MOIETY (see Surface active substances)

LYOPHOBIC MOIETY (see Surface active substances)

M

McCABE - THIELE METHOD (see Distillation)

McREYNOLDS CONSTANT (see Gas chromatography)

MACH, ERNST 1838–1916

The Austrian physicist and philosopher, Ernst Mach, whose work in physics and in philosophy had a great influence on 20th-century thought, was born on February 18, 1838 at Turas in Moravia and educated in Wien. He was professor of physics at Graz from 1864 to 1867 and at Prag from 1867 to 1895, and professor of inductive philosophy at Wien from 1895 to 1901. He was made a member of the Austrian House of Peers in 1901 and died at München on February 19, 1916.

Mach was a thorough-going positivist and took the view, which most scientists now share, that no statement is admissible in natural science unless it is empirically verifiable. His criteria of verifiability were, however exceptionally rigorous: they led him not only to reject such metaphysical conceptions as that of the ether and that of absolute space and time but also to oppose the introduction of atoms and molecules into physical theory. Nevertheless it was his criticism along these lines of Sir Isaac Newton's system that made the way clear for Albert Einstein's Theory of Relativity. As a positivist, he regarded scientific laws as purely descriptive; and he held that the choice between alternative hypotheses covering the same facts was to be made on the grounds of economy.

Mach's name is associated with **Mach Number**, which expresses the speed of matter relative to the local speed of sound.

Figure 1.

U. Grigull, H. Sandner and J. Straub

MACHINE LANGUAGE (see Computer programmes)

MACH LINES (see Characteristics, method of)

MACH NUMBER

Mach number, Ma, is the ratio of velocity to the local velocity of sound in the medium.

$$Ma = u/u_{sonic}$$

See also **Aerodynamics**.

G.L. Shires

MACH NUMBER, IN NOZZLES (see Nozzles)

MACH WAVES (see Compressible flow)

MACH-ZEHNDER INTERFEROMETRY

***Following from:** Interferometry*

The Mach-Zehnder interferometer is a classical mirror-interferometer. For a long time it was the most common dual-beam-interferometer used to measure continuously refractive index distributions of transparent objects. Developed by Mach and Zehnder in 1892, it has been frequently used in heat and mass transfer and in gas dynamics for example, combustion, until it was superceded by the holographic interferometer.

Figure 1 shows a Mach-Zehnder interferometer with a specific phase object ("phase object" is used to denote the transparent object being studied which influences the phase of light passing through it) in its measuring beam. It consists of two reflecting and two beam-splitting mirrors in a parallelogram (can also be a rectangle) arrangement. First the interferometer has to be adjusted in such a way that the two beams have equal optical path lengths. When the physical process of interest, for example heat transfer, is introduced into the measuring beam an optical path difference between reference and measuring beam is produced. A superposition of the

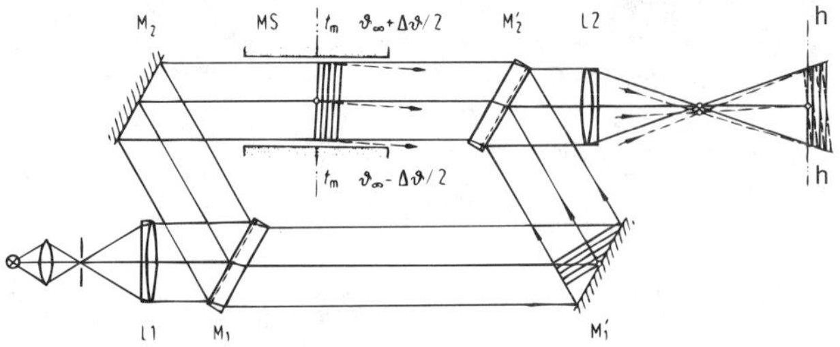

Figure 1. Parallelogram arrangement of a Mach-Zehnder-Interferometer, M_1, M_2' beamsplitting mirrors, M_2, M_1' reflecting mirrors, L1 and L2 lenses, MS test section with a constant temperature gradient (and so a constant refractive index gradient), $t_m - t_m$ adjustment plane, $t_i - t_i$ image plane.

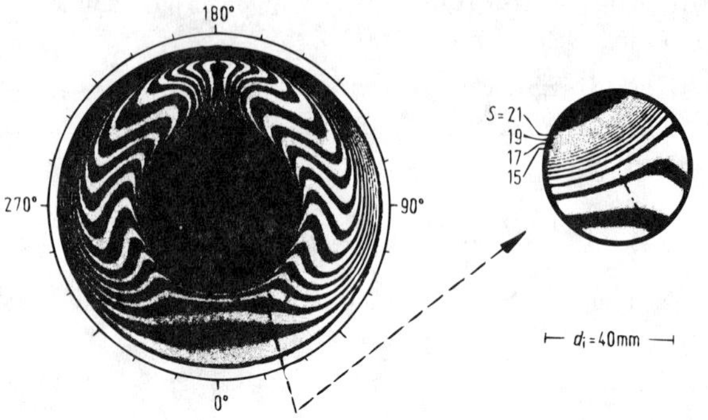

Figure 2. Interferogram of a horizontal annulus with infinite fringe field. $d_i = 40mm$; $d_a = 98mm$; $s = 29mm$; $\frac{s}{d_i} = 0, 73$;. Angle position 30°: $Nu_s = 4, 82$; $Gr_s = 7, 28 \cdot 10^4$; $\Delta\vartheta_\infty = 29, 7K$ [Photo according to Hauf (1966)].

two beams then generates an interference pattern in which the interference fringes correspond to lines of uniform difference of the refractive index.

Test section windows in the measuring beam, and corresponding plates compensating their effect in the reference beam, have to be manufactured with special precision and accuracy. The same goes for mirrors and lenses of all optical components because imperfections influence the beams differently. This makes the Mach-Zehnder interferometer very expensive.

Example

Figure 2 shows the interference pattern of a horizontal annulus filled with air when the inner cylinder is heated isothermally. The outer tube is cooled by water at constant temperature. When evaluating the interferogram the origin, i.e., the position of the inner wall, has to be determined. But the interference line close to the wall cannot be distinguished because of defraction effects. This problem is solved by extrapolating the temperature distribution—known from the interference pattern—from the wall temperature (measured by thermocouples) so the wall position and, as a result, the temperature gradient and the heat transfer coefficient can be determined.

References

Mach, L. (1892) Über einen Interferenzrefraktor. *Z. Instrumentenk.* 12:89–93.

Zehnder, L. (1891) Ein neuer Interferenzrefraktor. *Z. Instrumentenk.* 11:275–285.

Grigull, U. and Hauf, W. (1966) Natural convection in horizontal cylindrical annuli. *Proc. Inter. Heat Transfer Conf. 3rd.* II:182–195. Chicago.

Bennett, F. D. and Kahl, G. D. (1953) A generalized vector theory for the Mach-Zehnder-Interferometer. *J. Opt. Soc. Am.* 43:71–78.

Price, E. W. (1952) Initial adjustment of the Mach-Zehnder-Interferometer. *Rev. Sci. Instr.* 23:162

R. Fehle and F. Mayinger

MACLAURIN SERIES (see Error function; Power series; Taylor series)

MACRO-POROUS AND MACRO-RETICULAR RESINS (see Ion exchange)

MAGNESIUM

Magnesium—(*Magnesia,* district in Thessaly) Mg; atomic weight 24.305; atomic number 12; melting point 648.8 ± 0.5°C; boiling point 1090°C; specific gravity 1.738 (20°C); valence 2.

Compounds of magnesium have long been known. Black recognized magnesium as an element in 1755. It was isolated by Davy in 1808, and prepared in coherent form by Bussy in 1831. Magnesium is the eighth most abundant element in the earth's crust. It does not occur uncombined, but is found in large deposits in the form of *magnesite, dolomite,* and other minerals. The metal is now principally obtained in the U.S. by electrolysis of fused magnesium chloride derived from brines, wells, and sea water.

Magnesium is a light, silvery-white, and fairly tough metal. It tarnishes slightly in air, and finely divided magnesium readily ignites upon heating in air and burns with a dazzling white flame. It is used in flashlight photography, flares, and pyrotechnics, including incendiary bombs. It is one third lighter than aluminum, and in alloys is essential for airplane and missile construction. The metal improves the mechanical, fabrication, and welding characteristics of aluminum when used as an alloying agent. Magnesium is used in producing nodular graphite in cast iron, and is used as an additive to conventional propellants. It is also used as a reducing agent in the production of pure uranium and other metals from their salts. The hydroxide (*milk of magnesia*), chloride, sulfate (*Epsom salts*), and citrate are used in medicine. Dead-burned magnesite is employed for refractory purposes such as brick and liners in furnaces and converters. Organic magnesium compounds (Grignard's reaction) are important.

Magnesium is an important element in both plant and animal life. Chlorophylls are magnesium-centered porphyrins. The adult daily requirement of magnesium is about 300 mg/day, but this is affected by various factors. Great care should be taken in handling magnesium metal, especially in the finely divided state, as serious fires can occur. Water should not be used on burning magnesium or on magnesium fires.

Handbook of Chemistry and Physics, CRC Press

MAGNETIC FUSION REACTORS (see Fusion, nuclear fusion reactors)

MAGNETIC REYNOLDS NUMBER (see Liquid metals)

MAGNETO ACOUSTIC WAVES (see Magneto hydrodynamics)

MAGNETOHYDRODYNAMIC METHODS (see Velocity measurement)

MAGNETOHYDRODYNAMIC ELECTRICAL POWER GENERATORS

***Following from:** Magnetohydrodynamics*

MHD generators (MHD electrical power generators) are devices in which, according to the magnetohydrodynamics laws, a conversion of the energy of working fluid into electrical energy takes place. The principle of operation of MHD generators as well as conventional electrical generators is based on Faraday's induction law. In an electrically conducting fluid, moving at velocity $\vec{v}$ in a magnetic field $\vec{B}$, an electromotive force $(\vec{v} \times \vec{B})$ is induced. When electrodes connected to an external circuit are arranged to be flowed over by the fluid, electrical current of density $\vec{j}$ is produced by the electromotive force the current being limited by Ohm's law. **Figure 1** shows a schematic diagram of an MHD generator channel, constructed from electrode and insulation walls. In the channel, the fluid works against the electromagnetic body force $(\vec{j} \times \vec{B})$. A part of this work is used for electrical energy generation in the external circuit (load). Unlike a conventional turbogenerator, the MHD generator has no moving parts and this permits one to increase substantially the working fluid temperature.

The possibility of electrical current generation by a fluid moving in magnetic field was first pointed out by M. Faraday (1832). He made an attempt to measure electrical current induced by flow sea water in the river Thames estuary in the Earth's magnetic field. During the next century, various proposals on MHD energy conversion devices have emerged. The first experiments on electrical energy generation were performed by B. Karlovitz at the Westinghouse laboratory, U.S. (1938) using an MHD generator working with nonequilibrium plasma. At the Avco-Everette laboratory, U.S. the first highpowered MHD generator Mark-V worked on combustion products of liquid fuel and oxygen. The U − 25 MHD pilot plant using natural gas fuel was commissioned at the Institute of High Temperatures, U.S.S.R. (1971). A coal-fired MHD pilot plant (CDIF) was constructed in the U.S. (1984). In the U.S.S.R., a series of pulsed MHD generators (Pamir) with solid rocket fuel were developed for earthquake predictions and mineral deposits prospecting at the Kurchatov Atomic Energy Institute and other institutions (1973–1977). The pulsed MHD generator Khibiny (1976) was designed for geophysical research and used for feeding a huge current loop in sea water around the Rybachy peninsula in the Barenz Sea.

MHD electrical power generators have prospective uses in advanced high temperature energy production cycles and high energy pulsed electrical sources.

Depending upon the kind of working fluid, the following types of MHD generators are distinguished: open cycle MHD generators operated on combustion products of various fuels; closed cycle MHD generators working on noble gases, and liquid metal MHD generators.

According to the character of electrical field variation, MHD generators are divided into conduction and induction types. In the conduction type, the MHD generator electrical field is potential $(\vec{E} = -\text{grad } \varphi)$ and electrical current generated in the working fluid flows through a load (see **Figure 1**). In this type of MHD generator the electrodes should be arranged as shown in this figure. Conduction MHD generators can generate direct current or alternating current. In the induction type, the MHD generator's electrical

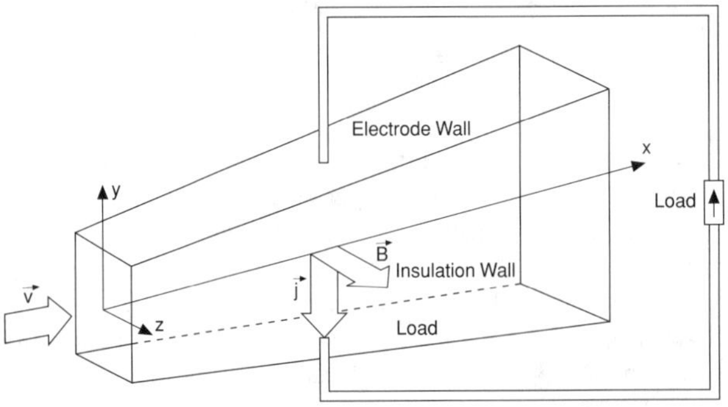

Figure 1. MHD generator channel.

field is excited by a transient magnetic field (rot $\vec{E} = -\partial\vec{B}/\partial t$) and the generated current can be short-circuited inside the working fluid. The magnetic field production the induction MHD generators can be of traveling magnetic field or transformer type. **Figure 2** is a schematic diagram of an MHD generator channel with a traveling magnetic field created by external conductors. There are many alternative geometrical configurations for MHD generator channels. In **Figures 1** and **2**, linear channels are shown in which working fluid flows in a rectilinear pattern. A rectilinear flow can occur also in coaxial and disk channels. A curvilinear flow takes place in a vortex MHD generator with a disk or coaxial channel and in helical and spiral MHD generators.

The basic electrical characteristics of MHD generators are power output $\dot{W} = \int \vec{j} \cdot \vec{E} dV$, which is generated in working fluid volume, and local electrical efficiency η, defined as a ratio of electrical power output density $\vec{j} \cdot \vec{E}$ to electromagnetic body force power density $(\vec{j} \times \vec{B}) \cdot \vec{v}$, $\eta = \vec{j} \cdot \vec{E}/(\vec{j} \times \vec{B}) \cdot \vec{v}$. In MHD generators, the values of $\vec{j} \cdot \vec{E}$ and $(\vec{j} \times \vec{B}) \cdot \vec{v}$ are negative.

Thermodynamic characteristics include enthalpy extraction coefficient η_N, which is determined as a ratio of the power output $\dot{W}$ to the inlet stagnation enthalpy flux, $\eta_N = \dot{W}/\dot{M}h_{0in}$, and isoentropic efficiency η_{oi}, defined as a ratio of the power output $\dot{W}$ to the difference of inlet and outlet stagnation enthalpies fluxes in the isoentropic expansion at given stagnation pressure ratio, $\eta_{oi} = \dot{W}/\dot{M}[h_{0in} - (h_{0out})_{is}]$. At any point in the energy conversion process, local isoentropic efficiency is determined as $\eta_0 = d\dot{W}\dot{M}(dH_0)_{is}$.

The values of η_0 and η are interconnected by relationship following from conservation equations. When wall friction and heat losses are negligible, this reduces to:

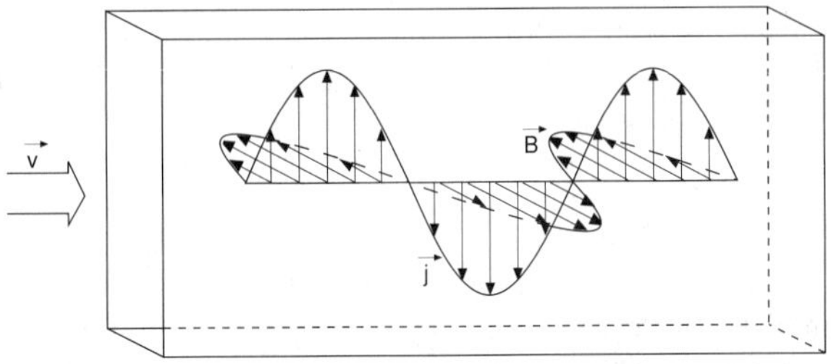

Figure 2. MHD generator channel with traveling magnetic field created by external conductors.

$$\eta_0 = \eta_0\left[1 + \frac{1}{2}(\gamma - 1)\mathrm{Ma}^2(1 - \eta)\right]^{-1}.$$

This formula reflects the influence of Joule dissipation upon η_0, which increases with decreasing gas static temperature.

The relation between local electrical characteristics and flow parameters in MHD generators is governed by Ohm's law. In a linear conduction MHD generator, in which the flow velocity is directed along x-axis ($\vec{v} = u\vec{e}_x$) and magnetic field is directed along z-axis, the components of the Ohm's law are written as

$$j_x = \sigma E_x - \beta j_y,\ j_y = \sigma(E_y - uB) + \beta j_x,\ j_z = \sigma E_z,$$

where σ is the electrical conductivity of working fluid and β is the Hall coefficient. In an MHD channel of rectangular cross-section with uniform flow, the z-component of current density equals zero ($j_z = 0$). In this case the electrical power density $\vec{j} \cdot \vec{E}$ is represented as

$$\vec{j} \cdot \vec{E} = -\sigma u^2 B^2 \eta(1 - \eta)(1 + j_x^2/j_y^2)^{-1}.$$

In this relationship, variables η and j_x/j_y are determined by the electrical scheme and load regime of MHD generator. The maximum power density is achieved when $\eta = 0.5$, $j_x/j_y = 0$, $B = B_{max}$, and $\mathrm{Ma} = \mathrm{Ma}_{opt}$. The optimum Mach number arises from the non-monotic dependence of the product σu^2 on velocity at fixed stagnation gas parameters. The maximum value σu^2 occurs at the following condition

$$\partial \ln \sigma/\partial \ln \mathrm{Ma} + 2\left(1 + \frac{1}{2}(\gamma - 1)\mathrm{Ma}^2\right)^{-1} = 0.$$

It follows from this condition that optimum flow is subsonic ($\mathrm{Ma}_{opt} < 1$) for atomic gases, and supersonic ($\mathrm{Ma}_{opt} > 1$) for molecular gases. For MHD generators designed for power plants, optimum Mach number Ma_{opt} has to be determined from the conditional maximum of $\sigma u^2 \eta(1 - \eta)$ at a fixed value of $\eta_0 = \eta_0(\mathrm{Ma}, \eta)$.

The dependence of electrical conductivity σ upon the working gas parameters is governed by the ionization of atoms and molecules and the movement of charged particles relative to neutrals. To increase the electrical conductivity, a small seeding flow of alkali metals (K, Cs, etc) with low ionisation potential is added to the gaseous working fluid. The electrical conductivity of sufficiently dense gases is determined by electron density n_e and mobility μ_e, $\sigma = n_e e \mu_e = n_e e^2 \tau/m_e$, where e and m_e are electron charge and mass, τ is mean time of electron-neutral collisions. The electron mobility enters in the definition of the Hall coefficient, $\beta = \mu_e B$.

At low degrees of ionization, when neutral particles provide major part of the effective cross-section for electron collision, the electrical conductivity is an exponential function of gas temperature. Such a strong dependence of electrical conductivity on temperature limits lower working temperature in MHD generator. The minimum useable outlet temperature of molecular gases is approximately equal to 2000 K. When noble gases are used in closed cycle MHD generator, it is possible to substantially decrease minimum working temperature due to nonequilibrium ionization.

In an MHD generator with nonequilibrium plasma the enhanced degree of ionization is sustained by the Joulean heating. The electrical conductivity depends on the electron temperature. To heat electrons it is necessary to have high values of the Hall coefficient ($\beta \geq 3$). However, at the condition $\beta > 1$, ionization instability emerges in the plasma. As the magnetic field increases, the plasma transfers to the state of ionization turbulence. One of the means of stabilization of turbulent plasma is the use of flow regimes with full ionization of alkali seeding. This approach is successfully realized in disk MHD generators with nonequilibrium plasma.

Electrical conductivity of liquid metals is practically constant and large enough for utilization of the induction scheme for an MHD generator. In induction MHD generators the main losses are caused by wall friction and eddy currents induced in electrodeless channels.

The most advanced and technically feasible MHD generators (in terms of research and development) are those of the conduction type using gaseous working fluids. The presence of the Hall effect leads to a variety of electrical schemes for such MHD generators, of which are described in what follows.

The Faraday MHD generator is characterized by an electrical scheme in which the load is supplied by the current flowing in the channel in the direction of induced electromotive force (along y-axis in **Figure 1**). At finite values of Hall coefficient ($\beta \geq 1$) the electrodes of Faraday MHD generator are segmented and switched to individual loads (see **Figure 3a**). The electrode segmentation prevents circulation of the longitudinal current in the electrodes and plasma ($j_x = 0$) and provides the establishment of an axial electrical field ($E_x < 0$). In a segmented MHD channel the power output and efficiency reach their maximum values. At small values of Hall coefficient ($\beta < 1$) the segmentation is not needed and their continuous electrodes switched to a single load are used. In a channel with continuous electrodes, the conditions $E_x = 0$ and

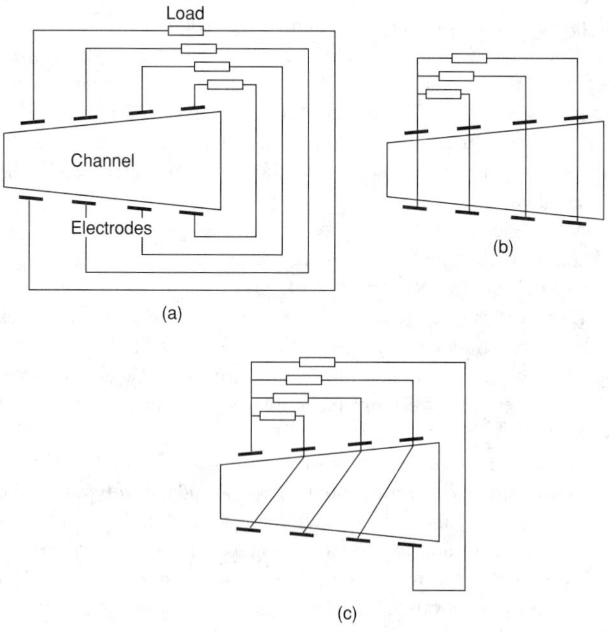

Figure 3. Types of MHD generator: (a) Faraday, (b) Hall, (c) Series type.

$j_x > 0$ are realised. For an ideal Faraday MHD generator (infinitely fine segmentation or absence of Hall effect) the electrical characteristics are the following

$$j_y = -\sigma uB(1 - K),\ j_x = 0,\ E_x = -\beta uB(1 - K),\ E_y = KuB,$$

where K is the load parameter, determined by the external circuit. In a Hall MHD generator the axial current j_x is supplied to the load, which is switched to the sections of segmented channel (see **Figure 3b**). In each section the opposite electrodes are shorted ($E_y = 0$). The power output of Hall generator is lower than that of Faraday channel at the same efficiency. The characteristics of the Hall channel reach the ideal segmented Faraday channel values when $\beta \rightarrow \infty$.

The series MHD generator has a prescribed direction of the electrical field vector in the channel. The ratio of the electrical field vector components defines the angle α between the equipotential lines and the axis of the channel

$$\alpha = -\arctan E_x/E_y.$$

This condition is provided by shorting of the electrodes of the segmented channel along the equipotential lines (see **Figure 3c**). The load current of series channel contains both components of electrical current j_x and j_y. A series channel is transformed to a Hall channel at $\alpha = \pi/2$. Both Hall and series channels can operate with a single load switched to inlet and outlet of the channel. For any regime of operation of the Faraday channel it is possible to configurate series channel having the same electrical characteristics by matching the equipotential angle α to the condition $j_x = 0$. But at the off-design conditions the characteristics of series channel with fixed α are worse than those of the Faraday channel.

Distributions of flow parameters over the channel cross section are nonuniform owing to skin friction and heat transfer at the walls. Major effects influencing electrical current flow in MHD channel are revealed when the equations of Ohm's law are averaged over the cross section. The Ohms law equations can be averaged if a certain model distribution of electrical parameters is assumed. In the case of the parallel equipotentials model, when the following assumptions are introduced

$$E_x = E_x(x),\ E_y = E_y(x, y),\ j_x = j_x(x, y, z),\ j_y = j_y(x, y, z),$$

$$u = u(x, y, z),\ B = B(x),\ \sigma = \sigma(x, y, z),\ \beta = \beta(x),$$

the averaged components of Ohm's law can expressed as

$$\langle\langle j_x\rangle\rangle = \langle\langle\sigma\rangle\rangle E_x + \beta\langle j_y\rangle_z,$$

$$\langle j_y\rangle =_x = (\langle\langle\sigma\rangle\rangle/G)(\langle E_y\rangle_y - \alpha\langle\langle u\rangle\rangle B) + \beta\langle\langle j_x\rangle\rangle;$$

here, double brackets denote averaging along both axes in the channel cross section, $\langle J_y \rangle_y$ does not depend on y because the normal component of current density at the insulator side wall is equal to zero, $\langle E_y \rangle_y$ does not depend on z because electrodes are assumed to be ideal conductors; the coefficients G and α are defined as

$$G = \langle\langle\sigma\rangle\rangle\langle\langle\sigma\rangle_x^{-1}\rangle_y(1 + \beta^2) - \beta^2,$$

$$\alpha = \langle\langle\sigma u\rangle_z\langle\sigma\rangle_z^{-1}\rangle_y\langle\langle u\rangle\rangle^{-1}.$$

G-factor reflects the two most important effects in a MHD channel with nonuniform flow, i.e. the Ohmic resistance of the electrode layers and the circulation of the Hall current which flows in a negative direction with respect to the x-axis in the core, where electrical conductivity is higher than the mean value, and in a positive direction with respect to the x-axis in the electrode layers. The latter effect is enhanced when the Hall coefficient increases. To suppress this effect, a curvilinear configuration of electrodes is used. At the surface of such electrodes the z-component of electrical current, which does not induce Hall, electromotive force, is substantial.

The other reason for Hall current circulation in the channel is shorting of the axial electrical field E_x by components of construction. For example, in a channel with continuous electrodes the condition $E_x \cong 0$ is imposed and in the plasma, a volume axial current is induced $j_x \sim \beta j_y$. In **Figure 4**, current stream lines (solid curves) and equipotential lines (broken lines) in continuous electrode channel are drawn for $\beta \approx 1$. Losses caused by Hall current lead to power output decreases by approximately $1 + j_x^2/j_y^2 \approx 1 + \beta^2$ times. Segmentation of electrodes allows an axial electrical field to be sustained of finite value E_x. At small ratio of electrode pitch Δl to distance between opposite electrodes Y ($\Delta l/Y \ll 1$) the condition $j_x \approx 0$ is fulfilled. However, at the surface of individual electrode, the Hall effect causes asymmetrical distribution of current density. When Hall coefficient β increases, the current concentrations at the electrode edges (upstream for the anode and downstream for the cathode) grow rapidly.

Nonuniform distribution of magnetic field can also be the cause of a short-circuited current in the channel. Usually the greatest nonuniformities of magnetic field occur at the inlet and the outlet of the channel. At nonuniform magnetic field distribution B(x) a variable electromotive force ($\vec{v} \times \vec{B}$) is induced which can produce current loops in the plasma volume. In **Figure 4** one can see the current loops in the end regions of the channel, where an exponential decrease of magnetic field is arranged. One of the ways of reducing the short-circuited current is through the loading of the end regions of the channel by means of segmented electrodes according to the variation of ($\vec{v} \times \vec{B}$) along the x-axis.

Magnetohydrodynamic interaction produces the additional nonuniformities of working fluid parameters in the flow. At the electrode wall, the electromagnetic body force $(\vec{j} \times \vec{B})_x$, which decelerates the flow, causes the deformation of the velocity profile in the boundary layer. Since the variation of $(\vec{j} \times \vec{B})_x$ across the boundary layer is relatively small, the velocity profile at the elec-

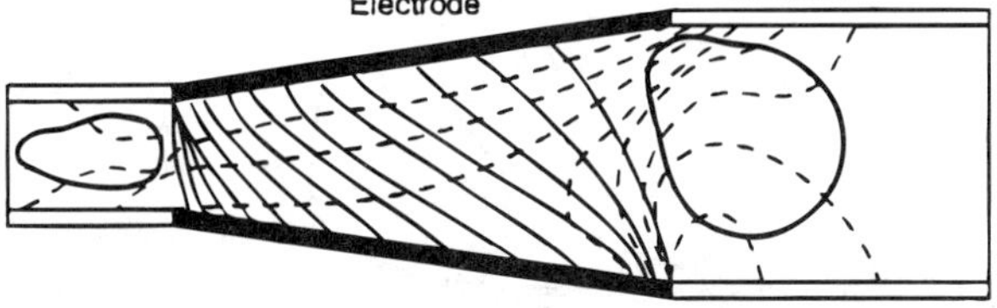

Figure 4. Current stream lines and equipotential lines in a continuous electrode channel.

trode wall becomes less steep. In an MHD channel with strong deceleration of the flow, boundary layer separation can occur. To prevent the boundary layer separation it is necessary to control the rate of deceleration of the MHD flow. At the insulation walls, where the electromagnetic body force $(\vec{j} \times \vec{B})_x$ is nonuniform, the opposite effect (Hartman effect) takes place. In **Figure 5** the deformation of the velocity profile is shown. Joulean heating flattens the temperature profile in the boundary layer and increases heat flux to the channel wall. When the transverse component of electromagnetic body force $(\vec{j} \times \vec{B})_y$ is substantial in an MHD channel, a secondary flow is generated. The secondary flow causes a nonuniform distribution of plasma parameters along the perimeter of the channel cross-section and redistribution of electrical field and current at the channel walls.

The principal influence upon the characteristics and reliability of the MHD generator is exerted by the regime of current discharge at the electrode surface. There are three main regimes of discharge: diffuse mode, microarcing and high current arcing. The existing theoretical and experimental data show that the transition from one mode of discharge to another one is determined by current density and thermal conditions at the electrode surface. In an open cycle, MHD generator the diffuse mode of discharge occurs at $T_W \geq 2000$ K and $j \leq 10^4$ A/m^2. At lower electrode temperature, arcing arises. Nonlinear arc phenomena result in nonuniform current and potential distribution at the electrodes and in interelectrode breakdown. The interelectrode arcs are affected by the transversal electromagnetic force $(\vec{j} \times \vec{B})_y$. When the force $(\vec{j} \times \vec{B})_y$ moves the arc into the interelectrode insulator, the ceramic material of insulator is locally overheated and can be destroyed. This mechanism is the one which is mainly responsible for failure of MHD channel constructions. In order to avoid the interelectrode breakdown, devices for current control in the electrodes circuits are used.

Mathematical modeling of flow in MHD channel is based on the equations of magnetohydrodynamics. A simple level of description of MHD flow is quasi-one-dimensional approximation, which employs the averaged equations of conservation and Ohm's law and takes into account all major effects influencing axial distribution of gasdynamic and electrodynamic parameters as well as integral characteristics of MHD generator. The quasi-one-dimensional equations are easily corrected with respect to experimental data and quite suitable to optimization procedures. In order to determine the optimal flow three variables B(x), η(x) and Ma(x) are varied along the channel. For accurate description of space structure of flow and conditions at the channel walls, the 2D and 3D MHD equations are solved, mainly by various numerical techniques.

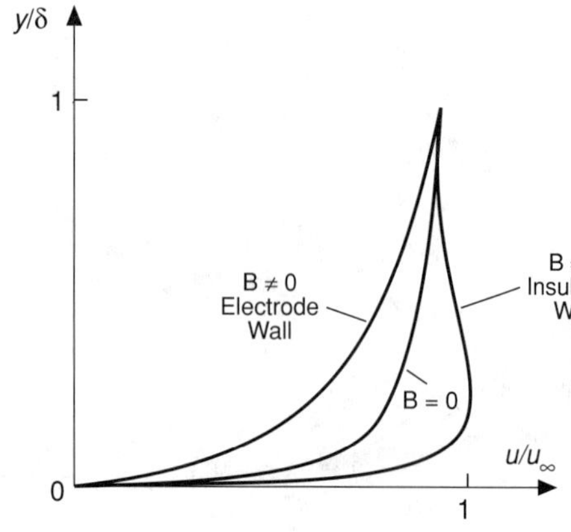

Figure 5. Deformation of the velocity profile.

References

Rosa R. J. (1968) *Magnetohydrodynamic Energy Conversion*, McGraw-Hill. New York.

Vatazhin A. B., Lyubimov G. A. and Regirer S. A. (1970) *Magnetohydrodynamic Flows in Channels*. Nauka, Moscow. (in Russian).

Patrick, M. and Shumyatsky, B. Ya., ed. (1979) *Open-cycle Magnetohydrodynamic Electrical Power Generation*. A Joint Publication U.S.A./U.S.S.R., Argonne National Laboratory, Argonne, IL., USA.

S.A. Medin

MAGNETOHYDRODYNAMICS

Magnetohydrodynamics is a branch of fluid dynamics which studies the movement of an electrically-conducting fluid in a magnetic field.

Faraday first pointed out an interaction of sea flows with the Earth is magnetic field (1832). In the beginning of the 20th century the first proposals for applying electromagnetic induction phenomenon in technical devices with electrically-conducting liquids and gases appeared. Systematic studies of magnetohydrodynamic (MHD) flows began in the 30s when the first exact solutions of MHD equations were obtained and experiments on liquid metal flows in MHD channels were performed [J. Hartmann and F. Lazarus (1937)]. The discovery of Alfven waves finalized the establishment of magnetohydrodynamics as an individual science [H. Alfven, (1942)].

Currently magnetohydrodynamics is applied in astrophysics and geophysics, fission and fusion, metallurgy and direct energy conversion, etc.

MHD applications fall within the conditions of the MHD approximation, according to which nonrelativistic, low frequency movement of an electrically-conducting fluid is considered when displacement and convection currents, electric body force and energy density of electric field can be neglected.

The MHD equations consist of the mass, momentum and energy conservation laws:

$$\frac{\partial \rho}{\partial t} + \operatorname{div} \rho\vec{v} = 0,$$

$$\rho \frac{\partial \vec{v}}{\partial t} + \rho(\vec{v}\nabla)\vec{v} = -\operatorname{grad} p + \operatorname{div} \hat{\tau} + (\vec{j} \times \vec{B}),$$

$$\rho \frac{\partial}{\partial t}\left(u + \frac{1}{2}v^2\right) + \rho(\vec{v}\nabla)\left(u + \frac{1}{2}v^2\right)$$

$$= -\operatorname{div} p\vec{v} - \operatorname{div} \vec{q} + \operatorname{div} \hat{\tau}\vec{v} + \vec{j}\vec{E},$$

Maxwell's equations and Ohm's law:

$$\text{curl}\ \vec{B} = \mu\vec{j},$$

$$\text{curl}\ \vec{E} = -\frac{\partial\vec{B}}{\partial t},$$

$$\text{div}\ \vec{B} = 0,$$

$$\text{div}\ \vec{E} = \frac{1}{\varepsilon}\rho_e,$$

$$\vec{j} = \sigma(\vec{E} + \vec{v} \times \vec{B}) - \beta(\vec{j} \times \vec{B})/B,$$

and equations of fluid properties:

$$u = u(p, \rho),\ \sigma = \sigma(p, \rho),\ \beta = \beta(p, \rho, B).$$

For many working fluids permittivity and permeability are equal to those of a vacuum ($\varepsilon_0 = 8.85410^{-12}$ Farad/m, $\mu_0 = 4\pi\ 10^{-7}$ Henry/m). The following quantities enter into MHD equations: ρ, p and u are fluid density, pressure and internal energy respectively, $\vec{\upsilon}$ is fluid velocity, $\vec{\tau}$ is the viscous stress tensor, $\vec{q}$ is the heat flux vector, $\vec{E}$ is the electric field, $\vec{B}$ is magnetic induction, $\vec{j}$ is current density, ρ_e is electric charge density, σ is fluid electrical conductivity, β is the Hall coefficient.

The conservation equations are related to Maxwell's equations and Ohm's law through electromagnetic body force $(\vec{j} \times \vec{B})$ and power density $\vec{j}\vec{E}$ of work done on the fluid by the electromagnetic field. $\vec{j}\vec{E}$ is related with power density of electromagnetic body force $(\vec{j} \times \vec{B})\,\vec{\upsilon}$ by the relationship

$$\vec{j}\vec{E} = (\vec{j} \times \vec{B})\vec{v} + j^2/\sigma,$$

where j^2/σ is the rate of Joulean dissipation.

Magnetohydrodynamics is characterized by dimensionl parameters which include, in addition to the conventional hydrodynamic parameters (Re, Pr, etc.), new ones containing electromagnetic variables: $A = \upsilon/\upsilon_a$ is the Alfven number where $\upsilon_a = B/\sqrt{\mu\rho}$ is the Alfven velocity; $Re_m = \mu\sigma\upsilon L$ is the magnetic Reynolds number; $K = E/\upsilon B$ is the parameter of electric field (or load parameter); $\beta = \omega\tau$ is the Hall coefficient being the ratio of electron cyclotron frequency and mean frequency of electron collisions with neutrals; $S = \sigma B^2 L/\rho\upsilon$ is the parameter of MHD interaction; $Ha = BL\rho\sqrt{\sigma/\eta}$ is the Hartmann number and some other parameters.

Boundary conditions used in MHD equations are formulated by traditional hydrodynamics and electrodynamics methods. External conditions represent the paths of electrical current and magnetic field lines outside the flow volume, and in particular, the configuration of external electrical circuit and the type of magnet system.

Exclusion of variables $\vec{j}$ and $\vec{E}$ from electrodynamic equations leads to an equation containing only one variable $\vec{B}$, called the equation of induction. In the case of $\sigma = const$ and $\beta = 0$ the equation of induction can be written as:

$$\frac{\partial\vec{B}}{\partial t} = \text{curl}(\vec{v} \times \vec{B}) + \nu_m \Delta\vec{B},$$

where $\nu_m = 1/\mu\sigma$ is the magnetic viscosity.

The first term in the right hand side of the equation determines the convective transport of magnetic field by fluid particles, the second term describes the diffusion of the magnetic field in the fluid. The relative role of convection and diffusion is determined by magnetic Reynolds number Re_m. At $Re_m = 0$ diffusion of magnetic field only takes place at finite velocity. In this case, magnetic viscosity ν_m determines either the characteristic time of magnetic field variation $t \sim l^2/\upsilon_m$ at distance l or the characteristic depth of magnetic field penetration $l \sim \sqrt{\upsilon_m t}$ during time t, defined in electrical engineering as the skin layer.

At $Re_m = 0$ fluid movement does not influence the applied magnetic field, which is induced by currents circulating outside flow volume.

At $Re_m = \infty$ the effect of magnetic field freezing is observed when magnetic flux through any closed liquid contour is conserved and all liquid particles—initially at the magnetic field line—continue to be on the line.

These properties of MHD flows are analogous to those of conventional hydrodynamics derived from theorems on vortex and circulation of velocity. In an ideally-conducting fluid ($\sigma = \infty$), when there is no dissipation of energy, small perturbations travel as nonattenuating MHD waves. For the plane wave of the type exp i(kx − wt) traveling in a uniform magnetic field, the linearized MHD equations are separated into two independent subsystems of equations which define Alfven and magnetoacoustic waves.

Alfven waves are characterized by transversal oscillations of perturbations of velocity υ_z' and magnetic field B_z' traveling with velocity

$$v_{ax} = \omega/k = \pm B_x/\sqrt{\mu\rho}.$$

In Alfven waves there is no density perturbation, so these waves can propagate both in compressible and incompressible fluids.

Magnetoacoustic waves include perturbations of density ρ' and also variables B_y', υ_x' and υ_y'. The solution of the dispersion equation reveals two types of magnetoacoustic waves, denoted as fast waves and a slow wave, respectively, which propagate at corresponding velocities

$$v_{+,-} = \frac{1}{2}\left[(\sqrt{a^2 + 2av_{ax} + v_a^2} + \sqrt{a^2 - 2av_{ax} + v_a^2})\right],$$

where a is the sonic speed.

In **Figure 1**, the phase velocity diagram shows the dependence of the wave velocity value upon the angle θ between the undisturbed magnetic field $\vec{B}$ and the wave vector $\vec{k}$.

For flows of ideally-conducting fluids, discontinuous solutions are possible. The relationships between flow parameters are obtained from the equations of mass, momentum and energy conservation and also from the boundary conditions for the electromagnetic field. Analysis of these relationships reveals four types of discontinuities.

Contact discontinuities. In a contact discontinuity, the normal component of velocity does not exist; velocity, pressure and magnetic fields are continuous. Density, temperature and also entropy have different values across the interface. Conditions for flow parameters at a contact discontinuity are written in the form:

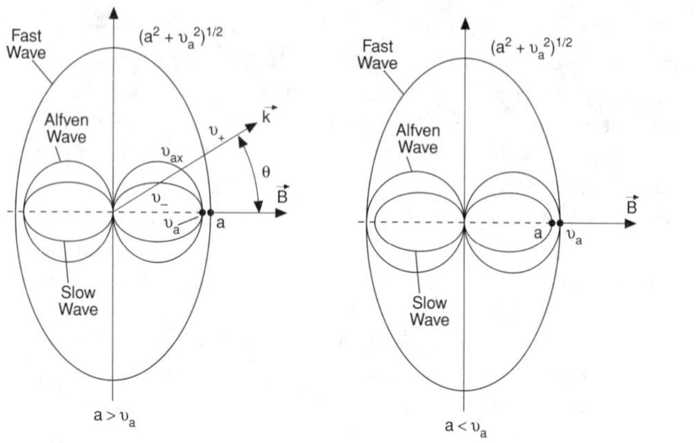

Figure 1. Phase velocity diagram.

$$v_n = 0;\ \{\vec{v}_r\} = 0,\ \{\rho\} \neq 0,\ \{p\} = 0,\ B_n \neq 0,\ \{\vec{B}_r\} = 0.$$

Here, the brackets denote the difference of values at the interface.

Tangential discontinuities. Conditions for tangential discontinuities are:

$$v_n = 0;\ \{\vec{v}_r\} \neq 0,\ \{\rho\} \neq 0,\ \{p + B_r^2/2\mu\} = 0,\ B_n = 0,\ \{\vec{B}_r\} \neq 0.$$

The velocity and magnetic field are tangent to the discontinuity surface and have arbitrary jumps in magnitude and direction. Density discontinuity is also arbitrary.

Rotational discontinuities. In rotational discontinuities, fluid thermodynamic properties and the normal component of velocity are continuous while the magnetic field vector rotates around the normal direction, being constant in magnitude:

$$\{v_n\} = 0;\ \{\vec{v}_r\} \neq 0,\ \{\rho\} = 0,\ \{p\} = 0,\ B_n \neq 0,\ \{\vec{B}_r\} = 0.$$

At the rotational discontinuity the jump of the tangential component of velocity is coupled to the jump of vector $\vec{B}_\tau$ by the equation $\{\vec{\upsilon}_\tau\} = \{\vec{B}_\tau\}/\sqrt{\mu\rho}$. The normal component of velocity is equal to the velocity of Alfven wave: $\upsilon_n = B_n/\sqrt{\mu\rho}$. The weak rotational discontinuity goes over into Alfven wave.

Shock waves. At the discontinuity surface, hydrodynamic parameters and magnetic field have jumps:

$$\{v_n\} \neq 0;\ \{\vec{v}_r\} \neq 0,\ \{\rho\} \neq 0,\ \{p\} \neq 0,\ B_n \neq 0,\ \{\vec{B}_r\} \neq 0.$$

It can be shown that vectors $\vec{\upsilon}_1, \vec{\upsilon}_2, \vec{B}_1$, and $\vec{B}_2$ lie in the plane normal to the discontinuity surface. The Hugoniot equation of MHD shock wave has the following form:

$$u_2 - u_1 + \frac{1}{2}(p_2 + p_1)\left(\frac{1}{\rho_2} - \frac{1}{\rho_1}\right) + \frac{1}{4\mu}\left(\frac{1}{\rho_2} - \frac{1}{\rho_1}\right)(b_{r_2} - B_{r_1})^2 = 0.$$

The condition that entropy must decrease ($S_2 > S_1$) determines the possible existence of only compression shocks ($\rho_2 > \rho_1$). As $B_{\tau 1}$ increases the degree of gas compression decreases at a given intensity of shock wave.

Two types of shock wave can occur: fast and slow shocks. In both types, the normal velocities are higher than υ_{+1} and υ_{-1} respectively in front of the shockwave and less than υ_{+2} and υ_{-2} behind the shock. The tangential component of magnetic field B_τ rises across a fast shock and drops across a slow shock. Weak MHD shock waves transform into the corresponding magnetoacoustic waves.

The following are possible transitions between the four discussed MHD discontinuities:

(a) Contact discontinuities may transform into tangential discontinuities and vice versa.

(b) Rotational discontinuities and shock waves may transform into tangential discontinuities and vice versa.

(c) Rotational discontinuities may transform into shock waves and vice versa.

The influence of an electromagnetic body force upon stationary MHD flows is distinctly demonstated by the solution of the Hartmann problem. This problem considers a flow of conducting viscous incompressible fluid between two parallel insulating planes in a uniformly applied magnetic field $\vec{B}_0 = B_0\vec{e}_z$. In fully-developed flow the quantities $\vec{\upsilon} = u\vec{e}_x$; $\vec{B} = B_x + \vec{e}_x + B_0\vec{e}_z$; $\vec{j} = j\vec{e}_y$ depend only on coordinate z in the direction normal to the planes placed at $z = \pm b$ and pressure $p = p(x) + p'(z)$ depends also on coordinate x in the direction of the velocity vector. The solution of the z-component of the momentum equation has the following form:

$$p + B_x^2/2\mu = p(x)$$

and reflects the balance between the z-component of the electromagnetic body force $(\vec{j} \times \vec{B})_z$ and the transverse pressure gradient (pinch effect).

The velocity distribution u(z) is determined by forces of viscous friction and $(\vec{j} \times \vec{B})_x$

$$\frac{u}{\langle u \rangle} = \left[1 - \frac{\cosh(\mathrm{Ha}\ z/b)}{\cosh \mathrm{Ha}}\right]\left[1 - \frac{\tanh \mathrm{Ha}}{\mathrm{Ha}}\right]^{-1},$$

where the Hartmann number $\mathrm{Ha} = B_0 b\sqrt{\sigma/\eta}$, $\langle u \rangle$ is average in z-direction velocity, which is interrelated with longitudinal pressure gradient dp/dx. In **Figure 2**, the velocity profiles (solid lines) at various values of the Hartmann number Ha are presented. When Ha increases the velocity profile becomes fuller, leading to a rise in shear force. This occurs due to nonuniform distribution of electromagnetic body force $(\vec{j} \times \vec{B})_x = \sigma(E_0 - uB_0)B_0$, where E_0 is the constant electric field determined by external circuit. It can be seen that the body force accelerates the near wall layers of the fluid relative to the core layers.

The solution of the equation of induction for boundary conditions $B_x(\pm b) = 0$, corresponding to the open circuit ($E_0 = \langle u \rangle B_0$), yields the distribution of $B_x(z)$:

$$\frac{B_x}{B_0} = \mathrm{Re}_m \frac{\sinh(\mathrm{Ha}\ z/b) - z/b \sinh \mathrm{Ha}}{\mathrm{Ha} \cosh \mathrm{Ha} - \sinh \mathrm{Ha}},$$

where $\mathrm{Re}_m = \mu\sigma\langle u \rangle b$. The induced magnetic field B_x is proportional to Re_m and produces a bend of magnetic field lines in the direction of fluid flow (connection effect). The broken curves in **Figure 2** present vector lines for the magnetic field for $\mathrm{Re}_m = 5$.

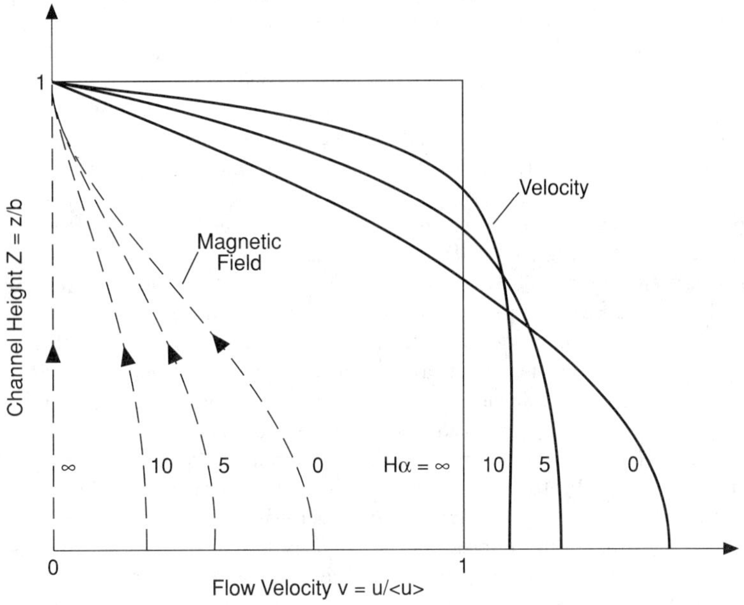

Figure 2. Velocity profiles in flow between parallel plates.

The analysis of the combined influence of electromagnetic body force ($\vec{j} \times \vec{B}$) and electrical power density $\vec{j}\vec{E}$ upon the flow of compressible fluid is performed by use of the equations of reverse action. These equations are derived from one-dimensional stationary MHD equations when magnetic and electrical field $\vec{B} = B\vec{e}_z$ and $\vec{E} = E\vec{e}_y$ are considered to be given quantities and effects of viscosity and thermal conductivity are negligible:

$$(Ma^2 - 1)\frac{du}{u} = -\frac{\sigma B^2}{pu}(u - u_1)(u - u_3)dx,$$

$$(Ma^2 - 1)\frac{d\,Ma}{Ma} = -\left(1 + \frac{\gamma - 1}{2}Ma^2\right)\frac{\sigma B^2}{pu}(u - u_2)(u - u_3)dx,$$

$$u_1 = \frac{\gamma - 1}{\gamma}\frac{E}{B},\ u_2 = \frac{1 + \gamma Ma^2}{2 + (\gamma - 1)Ma^2}u_1,\ u_3 = \frac{\gamma - 1}{\gamma}u_1.$$

These equations determine the velocity and **Mach number** variations in subsonic and supersonic flows in the presence of MHD interaction. As in conventional gasdynamics physical action with a given sign exerts opposite influences on subsonic and supersonic MHD flows. The effect of MHD upon the flow velocity is variable in direction and changes sign twice: at $u = u_1$, when the body force and power actions, influencing oppositely the velocity variation, are balanced, and at $u = u_3$, when the current density $j = -\sigma B(u - u_3)$ equal zero.

In **Figure 3**, the family of lines u(Ma) passing through an arbitrary point Ma_0, u_0 at the condition E/B = const is plotted. The arrows on the lines indicate the direction of u and Ma variations downstream. The broken lines separate regions of different variations of u and Ma. The line $u = u_3$ divides the plane $u - Ma$ in two parts: for $u < u_3$ the flow occurs in acceleration mode with consumption of energy from the external circuit ($jB > 0, jE > 0$),

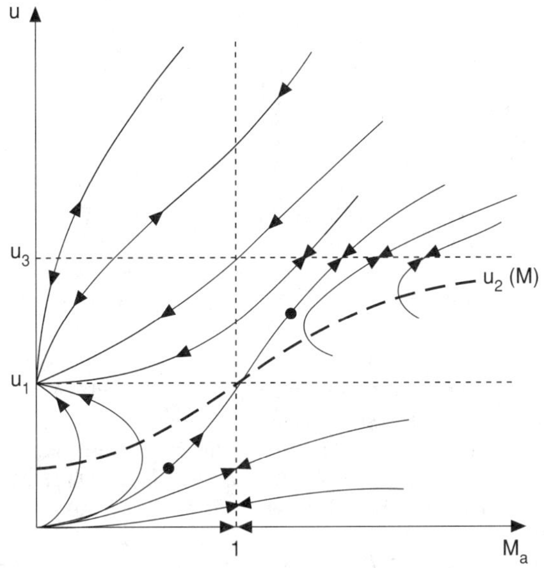

Figure 3. Flow velocity (u) as a function of Mach number (Ma) at E/B = const.

for $u > u_3$ the flow acts as a generator ($jB < 0, jE < 0$). The u/Ma diagram determines the properties and limit regimes of flow in MHD devices of constant cross-section. Thus, in an MHD generator, ($u > u_3$) subsonic flow is accelerated and in the limit mode is choked at the outlet ($Ma_2 = 1$), while supersonic flow is decelerated and is choked or asymptotically approaches minimum velocity u_3 (as $x \to \infty$). These limitations can be excluded by switching on an additional physical action, e.g. a geometric action by using an MHD channel of variable cross section. In MHD channels of constant cross-section, it is possible for the flow to pass smoothly through the sonic velocity. Such transition can occur when the sign of action is changed at Ma = 1. In **Figure 3**, there are two transitions which are possible for flow regimes described by u(Ma) lines passing through the points Ma = 1, u = u_1 and Ma = 1, u = u_3.

The influence of MHD interaction on volume structure of channel flow is determined by nonuniformities of current and potential distributions. These nonuniformities, in turn, are related to the viscous and thermal boundary layers at the walls and also with configuration of electrode and insulator wall elements and the form of the external electrical circuit. At the insulator walls, the Hartmann effect dominates. In compressible fluid at high velocities, the relative acceleration of wall layers causes a non-monotonic velocity profile with a maximum point inside the boundary layer. At the electrode walls, deceleration of flow is possible in the case when the body force is directed against the flow (e.g. in the generator regime). This effect increases the boundary layer thickness and causes boundary layer separation. The energy input into the boundary layer increases the temperature gradient, which consequentially increases the wall heat flux.

In cases when electromagnetic body force has a transverse component $(\vec{j} \times \vec{B})_y$, secondary flows can arise. For $Re_m \ll 1$, secondary flow occurs when the axial current density j_x is nonuniform over the channel cross-section and the transverse body force is not potential. From the equation for the velocity vortex ω_x an estimation for the occurrence of secondary flow is as follows:

$$\frac{\omega_x l}{u} \sim \frac{v}{u} \sim \frac{j_x BL}{\rho u^2} = S_x,$$

where l is the perimeter of the vortex cell. In an MHD generator, the condition $j_x > 0$ takes place. In this case, the negative pressure gradient $\partial p/\partial y$ in the channel cross-section, caused by $(\vec{j} \times \vec{B})_y$, induces a two-cell vortex structure in which the fluid moves in the boundary layer at the insulator wall in the positive direction of y-axis (toward the anode) and then to the middle of the anode. Fluid particles moving along the walls are gradually cooled and push electric current to the central part of the anode. A positive feedback exists between the anode current concentration, leading to the elevated local Joule heating, and the intensity of vortex flow. At sufficiently large values of the parameters of MHD interaction, an arcing occurs and overheating of the anode central part takes place.

Plasma flows in MHD channels undergo various types of instabilities. An analysis of the linearized system of MHD equations for $Re_m \ll 1$ shows the following main types of instabilities: acoustic, superheating and vortex MHD instabilities.

The acoustic instability is manifested in an enhancement of sonic waves under influence of current density fluctuations caused by perturbances of thermodynamic parameters and corresponding electrical conductivity. A positive contribution to the growth of short sonic waves is produced by two factors: electromagnetic body forces and Joule heating variations caused by conductivity fluctuations. The destabilising influence of these mechanisms is largest when the wave travels in the direction of the nondisturbed body force $(\vec{j}_0 \times \vec{B}_0)$, for the condition $(\partial\sigma/\partial T)_s > 0$ and where the damping action of magnetic friction is overcome.

The superheating instability is characterized by the growth of entropy disturbances caused by the positive feedback between fluctuations of entropy and Joule heating. This type of instability occurs for the condition $(\partial\sigma/\partial T)_p > 0$ and has the maximum increment at $\vec{k}\vec{j}_0 = 0$. The development of finite amplitude disturbances for electrical conductivity growing with temperature may result in moving localized regions with elevated temperature named T-layers.

Vortex instability is manifested by the growth of velocity vortex disturbances induced by the electrical conductivity and electromagnetic body force gradients in nonuniformly conducting fluid. For the condition $(\partial\sigma/\partial T)_p > 0$, two mechanisms of the instability initiation are possible: a type of Rayleigh-Taylor instability relating to departures from equlibrium of heavy liquid superimposed over a light liquid and the convective instability of a liquid layer heated on the underside.

In turbulent flows, the magnetic field dampens the vortex vector components normal to the field. The attenuation decrement of the velocity pulsations is estimated as $\sigma B_\perp^2/\rho$. The decrease in the velocity pulsations causes a decrease in turbulent shear stresses. The available experimental data on structure of turbulent MHD flows in channels show that with an increase of magnetic field, the pulsations of velocity, temperature and electric field decay, averaged velocity profiles are laminarized, the hydraulic resistance decreases (except when Hartmann effect prevails) and the transition from a laminar regime of flow to a turbulent one is made more difficult.

References

Kulikovsky, A. G. and Lyubimov, G. A. (1965) *Magnitohydrodynamics*, Addison-Wesley, Reading.

Landau, L. D. and Lifshitz, E. M. (1960) *Electrodynamics of Continuous Media*, Addison-Wesley, Reading.

Shercliff, J. A. (1965) *A Textbook of Magnetohydrodynamics*, Pergamon Press, Oxford.

Leading to: *Magnetohydrodynamic electrical power generators*

S.A. Medin

MAGNETOHYDRODYNAMICS IN LIQUID METALS (see Liquid metals)

MAGNOX (see Nuclear reactors)

MAGNOX POWER STATION

Following from: *Nuclear reactors*

Magnox nuclear power stations were introduced in the UK with the design and building of Calder Hall at Sellafield in Cumbria.

This plant, claimed to be the first nuclear power station in the world, currently generates 60 MWe in each of its four reactors. It first supplied power to the national grid in October 1956. Its sister station at Chapelcross near Annan in Scotland is also still in operation.

Fuel elements used in such stations consist of uranium rods one inch in diameter canned in Magnox, an alloy of magnesium with low neutron absorption and high thermal conductivity.

The initial can design used closely-spaced fins in transverse position to coolant flow direction, an improvement on the longitudinal fins adopted for the Windscale pile. Longitudinal fins suffered from the fact that increasing the number or size of fins in order to increase heat transfer restricts the flow at the base of the fins, which reduces heat transfer [Fortescue and Hall (1957)]. Experimental studies using metal-coated insulating material enabled the determination of heat transfer coefficient distribution over transverse fins [Harris and Wilson (1960)]. Flow visualization studies revealed the formation of vortices between fins thus helping to achieve acceptable heat transfer rates.

When larger commercial versions of Calder Hall were built at Bradwell, Dungeness, Hinkley Point, Sizewell, Hunterston and Trawsfynnid in the UK, at Tokai Murai in Japan and Latina in Italy, the transverse fins were displaced by polyzonal fins of various designs, e.g., spiral fins in quadrants formed by straight splitter fins. Experimental work was carried out on these designs, including optimisation of the number and angle of fins. Although average heat transfer rates were higher than for transverse fins, these designs suffered from longitudinal and transverse variations in heat transfer, caused by interruption of spiral flow by splitters forming the quadrants and by blockage produced by braces added to prevent fuel element distortion.

Reactor operation is dictated by maximum can temperature and hence by minimum heat transfer coefficient. The location of this minimum coefficient is dependent on manufacturing variability and could occur at different positions along the can, although usually slightly downstream of a brace. The probabilistic treatment of this problem has been developed by Wilkie (1962).

An improvement over the polyzonal design has been the four-zone "herringbone" design introduced in France. This type is now used in all magnox reactors. But it also suffers from longitudinal and transverse variations in heat transfer caused by fin deformation during reactor operation. (See also **Augmentation of Heat Transfer, Single-phase**.)

References

Fortescue, P. and Hall, W. B. (1957) *J. Brit. Nucl. Energy Conf.* Vol 2. 83.

Harris, M. J. and Wilson, J. T. (1960) *Heat Transfer and Fluid Flow Investigations on Large-scale Transverse Fins.* Paper 7 at the Joint I. Mech. E. and Brit. Nucl. Energy Soc. Symposium on the use of secondary surfaces for heat transfer with clean gases. London.

Wilkie, D. (1962) *The Probability of Obtaining a Low Stanton Number on a Polyzonal Fuel Element.* UK Atomic Energy Authority TRG Report 217 (w).

D. Wilkie

MAGNUS FORCE

Following from: *Flow of fluids*

A fluid flow perpendicular to the axis of a *rotating cylinder* produces a force perpendicular to both the axis and the flow direction (**Figure 1**). This force is known as the Magnus Force and can be explained by the *Bernoulli effect.* In the case shown in the figure, the fluid moves faster over the upper surface of the cylinder than over the the lower surface, which results in a lower pressure at the upper surface. This phenomenon is discussed in, for example, Streeter and Wylie (1983) and Massey (1989).

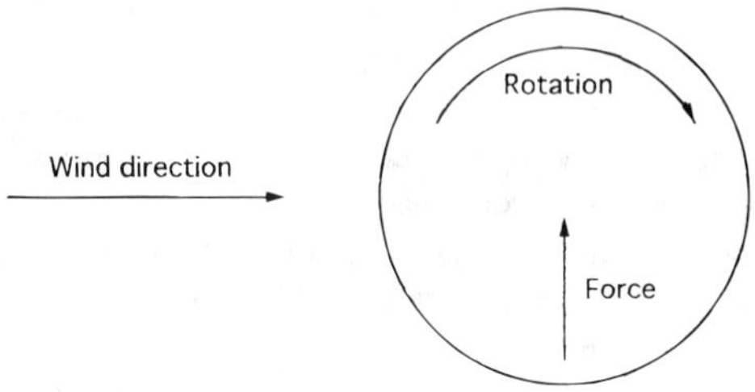

Figure 1. Force on a rotating cylinder.

References

Massey, B. S. (1989) *Mechanics of Fluids.* Van Nostrand Reinhold, London.

Streeter, V. L. and Wylie, E. B. (1983) *Fluid Mechanics.* McGraw-Hill. Singapore.

M.J. Pattison

MAGNUSSEN AND HJERTAGER MODEL (see Combustion)

MAINFRAME COMPUTERS (see Computers)

MALDISTRIBUTION OF FLOW (see Headers and manifolds, flow distribution in)

MALVERN, SCATTERING, METHOD FOR PARTICLE SIZING (see Particle size measurement)

MANOMETERS (see Pressure measurement)

MANOMETRY (see Hydrostatics)

MARANGONI EFFECT

Following from: *Interfaces*

It has been observed that the process of mass transfer across the interface of certain systems produces spontaneous interfacial convection. When these *interfacial flows* are driven by local changes in interfacial tension the phenomenon is called the Marangoni effect. Marangoni phenomena cover different degrees of interfacial turbulence, from microscale structured convection that requires special optical systems to be detected (roll cells) to violent eruptions that can be observed without any optical help. This interfacial convection has multiple effects in *liquid-liquid mass transfer* processes, the most relevant of which are: (1) increase in mass transfer coefficients over predicted values; (2) interfacial deformations that alter the interfacial area; (3) changes in the coalescence rate due to either an increase or a decrease in the rate of drainage of film trapped between colliding drops; and (4) changes in conditions for drop breakage due to their effect on the stability of the ligaments present in deformed drops. The effects on drop coalescence and breakage affect drop size distribution in liquid-liquid dispersions and therefore the performance of liquid-liquid contactors. Substantial research has been devoted to the study of the conditions that lead to the Marangoni phenomena and to the quantification of their effects on mass transfer rate and dispersion characteristics.

The mechanism of Marangoni convection can be explained using **Figure 1**. **Figure 1a** shows a schematic representation of the transfer of a solute S from phase A into phase B at steady state. The transfer may be accompanied by heat effects, in which case, temperature profiles are also present. While interfacial tension over the whole interface is uniform, the interface is quiescent. However, mechanical disturbances in the system may bring an element of fluid from the bulk of one of the phases to the interface, producing a local change in concentration and temperature (**Figure 1b**). The resulting local change in interfacial tension will generate radially spreading flow from points of low interfacial tension. Depending on the physical properties of the system and the magnitude of the disturbance these movements may either die out or be sustained. In the first case the system will return to its initial state, while in the other it will become unstable. In some simple cases conditions required for the system to become interfacially convective can be predicted qualitatively. However, it is from mathematical analysis that the stability criteria have been established for different types of mass transfer mechanisms and interfacial geometry. The references listed below contain extensive and comprehensive reviews of the subject and its effect on contacting equipment performance.

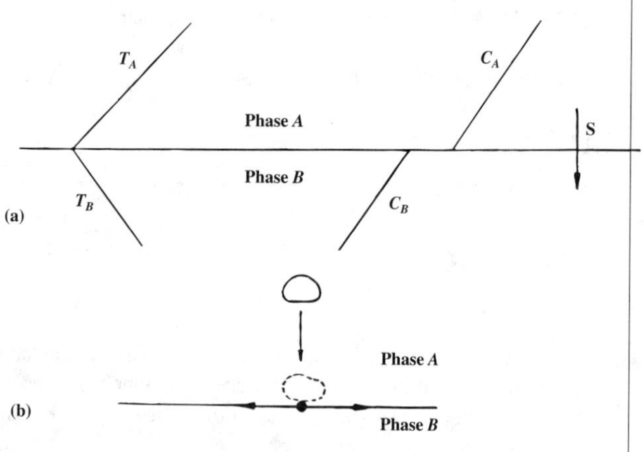

Figure 1. Mechanism of Marangoni flows.

References

Berg, J. C. (1972) Interfacial phenomena in fluid phase separation processes. *Recent Developments in Separation Science*. Vol. II. CRC Press. Cleveland. Ohio.

Sawistowski, H. (1971) Interfacial phenomena. *Recent Advances in Liquid-Liquid Extraction* (ed. C. Hanson). Pergamon Press. Oxford.

Ortiz, E. S. P. de. (1992) Marangoni phenomena. *Science and Practice of Liquid-Liquid Extraction*. Vol. 1 (ed. J. D. Thornton). Clarendon Press. Oxford.

E.S. Perez de Ortiz

MARGULES EQUATION (see Activity coefficient)

MARINE FUEL OILS (see Oils)

MARINE GAS TURBINES (see Gas turbine)

MARIOTTE LAW (see Boyle's, Boyle-Mariotte, law)

MASSACHUSETTS INSTITUTE OF TECHNOLOGY, MIT

Cambridge
MA 02139
USA
Tel: 617 253 1000

MASS ACTION LAW

Following from: *Chemical reaction; Chemical equilibrium*

The two related definitions of the Law of Mass Action are:

1. At constant temperature the product of active masses on one side of a chemical equation, when divided by the product of active masses on the other side of the chemical equation, is a constant, regardless of the amounts of each substance present at the beginning of the action.
2. At constant temperature the rate of reaction is proportional to the concentration of each kind of substance taking part in the reaction. [Weast (1982)]

The second definition is often used in stating reaction rate laws for single, homogeneous, unidirectional chemical reactions (e.g., rate $= k\ C_A\ C_B$, where k is the rate constant and C denotes concentrations). The rates of forward and reverse chemical reactions in a reversible system can be expressed in similar terms, leading to the first definition above.

The Law of Mass Action has been described as a restrictive condition expressing the indestructibility of matter during chemical reactions [Rushbrook (1962)]. There are relatively few references in modern texts to this law, which relates to the generally accepted indestructibility of matter under conditions of changing equilibria. Considering the general homogeneous, reversible chemical reaction

$$A + B \rightleftarrows C + D,$$

the Law of Mass Action requires the rate of forward reaction to be proportional to the concentrations of A and B. Defining these as C_A and C_B respectively, the rate of the forward reaction can be written as

$$r_{forward} = k_f C_A C_B.$$

Similarly, the rate of the reverse reaction is expressed as

$$r_{reverse} = k_r C_C C_D,$$

where C_C and C_D denote the concentrations of species C and D, respectively. In these equations, k_f and k_r denote the forward and reverse (temperature dependent, concentration independent) reaction rate constants; the temperature dependence of k_f and k_r are described by the Arrhenius equation.

At equilibrium, $r_{forward} = r_{reverse}$ and

$$k_f C_A C_B = k_r C_C C_D,$$

which can be used to define the *equilibrium constant* for this particular reaction:

$$K = \frac{C_C C_D}{C_A C_B} = \frac{k_f}{k_r}$$

More generally, if the chemical reaction has stoichiometric coefficients a,b,c and d such that

$$aA + bB \rightleftarrows cC + dD,$$

at equilibrium

$$K = \frac{C_C^c C_D^d}{C_A^a C_B^b} = \frac{k_f}{k_r}$$

[Moore (1963)]. The concept finds ready application in mass action equilibria of sorption and ion exchange processes Perry and Chilton, (1984).

References

Atkins P. W. (1986) *Physical Chemistry.* OUP. Oxford. UK.

Moore W. J. (1963) *Physical Chemistry.* Prentice Hall. Englewood Cliffs. New Jersey.

Perry R. H. and Chilton C. H. (Eds.). (1984) *Chemical Engineers' Handbook.* McGraw-Hill. New York. USA.

Rushbrooke G. S. (1962) OUP. Oxford. UK.

Weast, (1982) (63rd Edition) *Handbook of Chemistry and Physics.* CRC Press. Boca Raton. Florida. USA.

Leading to: *Arrhenius equation*

A.A. Herod and R. Kandiyoti

MASS FLOW METERS

Following from: *Flow metering*

By measuring mass rather than volume, many problems associated with fluid expansion can be avoided. Most mass flow meter designs only satisfy specialised applications. Three classes of mass meter techniques can be found.

'Hybrid' meters use two meters, one measuring velocity and the other kinetic energy. Many possibilities exist, but most are specialised curiosities.

Two principles of 'Inferential' meters can be found, the most popular being the combination of a volume meter and a densitometer. This technique is popular and uses proven technology, but it suffers from the uncertainties associated with two measurements.

A critical flow (sonic) nozzle accelerates gas to the speed of sound within the nozzle. At this point, mass flow is proportional to the gas properties and upstream pressure. This provides very accurate measurement (and control) of mass flow of gas, albeit with a high pressure drop.

'True' mass meters utilise four principles. The first makes use of the momentum of the fluid by imparting swirl to the fluid, then the force generated by removing it is measured. A driven rotor followed by a fixed measuring rotor is one design example. Other momentum measuring techniques are also used.

The second principle uses **Differential Pressure**, usually produced by forcing unbalanced flow through **Orifices** or **Venturis**. Problems involving the provision of a steady unbalanced flow however allow very few practical flow meters.

Thermal meters measure the energy removed from a heated element. Heat loss and temperature rise techniques, although dependent on the heat capacity of fluids, provide good metering results, especially for low flows of gas.

The fourth true mass flow meter principle depends on the *Coriolis force*. This force is produced by a moving body subjected to an angular acceleration. In this case, the moving body is the flowing fluid while the angular acceleration is provided by a vibrating element oscillating within the fluid. The resulting Coriolis force reacts on the vibrating element in the same direction of the flow. This causes the vibrating element to bend slightly. This bending is detected by the resultant phase differences. Normally a tube is anchored at each end and vibrated in the centre. Sensors between the anchors detect the phase difference, which is proportional to mass flow. Twin tubes increase noise rejection, and bending the tube into loops or a 'U' shape improves sensitivity. A computer provides control, makes corrections and calculates mass flow and density. Future designs may replace the vibrating tube with 'tuning fork arrangements.'

The Coriolis meter design is the first practical, accurate, true mass flow meter with a variety of uses for many different fluids.

R. Paton

MASS MEDIAN DIAMETER, MMD (see Atomization)

MASS SPECTROSCOPY (see Spectroscopy)

MASS TRANSFER

Mass transfer is the transport of a substance (mass) in liquid and gaseous media. Depending on the conditions, the nature, and the forces responsible for mass transfer, four basic types are distinguished: (1) diffusion in a quiescent medium, (2) mass transfer in laminar flow, (3) mass transfer in the turbulent flow, and (4) mass exchange between phases.

The simplest case is mass transfer in a medium at rest in which the driving force is the difference of concentrations in adjacent regions of the medium and the mechanism is molecular diffusion. The substance flows, by virtue of the statistical character of molecule motion, from a high concentration region to a low concentration one tending to equalization of concentration throughout the entire volume. This mass transfer is described by an equation known as Fick's law which, when applied to a binary mixture, has the form

$$\dot{m} = -D_{AB}\,\tilde{M}a\,\frac{dC_A}{dy}, \tag{1}$$

where $\dot{m}$ is the flow of substance A, kg/m^2s, in the reverse direction to the concentration gradient of this substance dC_A/dy ($kmole/m^3.m$), $\tilde{M}$ is the molecular weight of component A (kg/kmole) and D_{AB} (m^2/s) is the interdiffusion coefficient of substance A in substance B and is determined by the physical properties of these substances. D_{AB} has been determined experimentally for many gas pairs and can also be calculated using a molecular-kinetic theory. D_{AB} is known to depend on temperature and pressure. The diffusion coefficient in gases under normal conditions is of the order of 10^{-4} m^2/s, while in liquids it is about five orders lower (see **Diffusion** and **Diffusion Coefficient** for more details).

Actually, in addition to concentration gradient, temperature and pressure gradients which affect mass transfer via thermal and pressure diffusion may come into play. These effects are most significant in gas mixtures with a widely varying molecule size, e.g. He – Cs. Thermal diffusion underlies one of the methods of separation of uranium isotope U^{235}.

In laminar flow of gaseous and liquid mixtures, calculation of mass transfer does not present particular difficulties. For instance, for a plate in the stream of incompressible liquid, the set of equations 2 describing the velocity and concentration fields has the form

$$\frac{\partial\bar{u}}{\partial x} + \frac{\partial\bar{v}}{\partial y} = 0,$$

$$\bar{u}\,\frac{\partial\bar{u}}{\partial x} + \bar{v}\,\frac{\partial\bar{v}}{\partial y} = \nu\,\frac{\partial^2\bar{u}}{\partial y^2}, \tag{2}$$

$$\bar{u}\,\frac{\partial C}{\partial x} + \bar{v}\,\frac{\partial C}{\partial y} = D\,\frac{\partial^2 C}{\partial y^2},$$

where x and y are the longitudinal and transverse coordinates, u and v the longitudinal and transverse components of velocity, ν and D the coefficients of kinematic viscosity and molecular diffusion, respectively, and C the local concentration of a substance, $C = f(x,y)$.

Under the boundary conditions $\bar{u} = \bar{v} = 0$ and $C = C_1$ at $y = 0$, $\bar{u} = \bar{u}_0$ and $C = C_0$ at $x < 0$ or $y = \infty$ the solution to equation system 2 yields

$$Sh_x = \frac{\beta x}{D} = 0.332\,Re_x^{1/2}Sc^{1/3}, \tag{3}$$

where β is the local mass transfer coefficient at the distance x from the leading edge of the plate ($\beta = [\dot{m}/\tilde{M}(C_1 - C_0)]$, $Re_x = x\bar{u}_0/\nu$ is the Reynolds number, and $Sc = \nu/D$ the Schmidt number. Note that there is a departure from the linear dependence of $\dot{m}$ or $(C_1 - C_0)$ at high rates of mass transfer (see Sherwood et al, 1975).

If the total length of the plate is L, then the length-averaged mass transfer coefficient is found from the equation

$$Sh_L = \frac{\bar{\beta}L}{D} = 0.664\,Re_L^{1/2}Sc^{1/3}, \tag{4}$$

where $Re_L = Lu_0/\nu$.

Mass transfer fundamentally changes in transition to a turbulent flow. Its vortex flow characteristics lead to a large-scale transport of fluid. This transport commonly has rates which are orders of magnitude higher than molecular ones and promotes a faster equalization of the concentration field and, given a substance source, rapid propagation of the substance over the flow cross section. Since a rigorous theory of turbulence is lacking, it is desirable to describe the flow itself and heat and mass transfer in it by a set of equations similar to 2 for the laminar flow using averaged velocity values and replacing ν and D by their effective values conforming to the conditions in the turbulent flow, i.e., by the coefficients of "eddy viscosity" and "turbulent diffusion" (see **Diffusion** and **Diffusion Coefficient**). The molecular diffusion also occurs in the turbulent flow, e.g. between and inside eddies. Its role is enhanced (relative to turbulent transport) as the channel surface is approached and becomes predominant near it. It is commonly believed that the molecular D_M and turbulent D_T diffusion coefficients are additive, i.e., $D = D_M + D_T$.

Since in a developed turbulent flow the substance, energy, and momentum transport occurs via large-scale eddies, the transport rate is considered identical and D_T, a_T, and ν_T are about equal. (This is a triple analogy between the transport of the substance, energy, and momentum.) This makes it possible to use empirical dimensionless equations describing heat transfer for calculation of mass transfer.

Mass exchange between a gas and a droplet is commonly encountered in engineering. For a medium at rest the solution can be written thus

$$Sc = \frac{\beta d_d}{D} = 2, \tag{5}$$

where β is the mass transfer coefficient, d_d the droplet diameter, and D the diffusion coefficient of the exchanged substance in the gas. When the drop moves with respect to a medium in the $Re_d \le 200$ range the Frössling-Marshall formula

$$Sh = 2 + 0.6\,Re_d^{1/2}Sc^{1/3} \tag{6}$$

(where $Re_d = d_p\,u/\nu$ with u and ν being the velocity of the gas relative to the drop and the gas kinemetic viscosity respectively) is quite consistent with the experiment. This formula also holds in the case of evaporation of droplets in a gas stream provided the evaporation rate is small or moderate.

The rate of mass transfer from the surface of liquid film (e.g. evaporation) flowing on the inner surface of the tube toward the central gas flow can be calculated using an empirical formula

$$Sh = \frac{\beta d}{D} = 0.023\ Re_d^{0.83} Sc^{0.44}, \quad (7)$$

where d is the diameter of the tube (cf. $Nu = 0.023\ Re_d^{0.8}\ Pr^{0.4}$ for heat transfer).

One more example of mass transfer is diffusion of some substance, such as A from one moving medium to another through interface (two-film theory). If it is assumed that there is no concentration jump on the boundary, i.e., $C_{ABi} = C_{AEi}$ (Fig. 1a), then the flow of substance A can be represented in the form

$$\dot{m} = \beta_{AB}\,\tilde{M}_A\,(C_{AB} - C_{ABi}) \quad (8)$$

$$= \beta_{AE}\,\tilde{M}_A(C_{AEi} - C_{AE}) = K_A\,\tilde{M}_A\,(C_{AB} - C_{AE}),$$

where β_{AB}, β_{AE}, and K_A are the coefficients of mass transfer for substance A in media B and E and the overall mass transfer coefficient, respectively, and C_A the concentration of substance A at the points indicated in **Figure 1**.

Hence,

$$\frac{1}{K_A} = \frac{1}{\beta_{AB}} + \frac{1}{\beta_{AC}}, \quad (9)$$

i.e., the total resistance to mass transfer is a sum of resistances in each medium.

In many cases, the concentrations of the transfering substance are not identical at the interface in the two respective media. For instance, if B is a liquid phase and E is gas, and an equilibrium on the boundary obeys Henry's law $C_{ABi} = H\,C_{AEi}$ (**Figure 2**), then

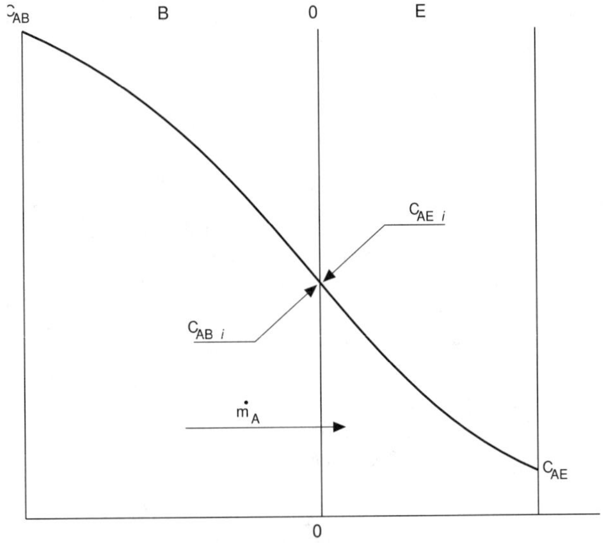

Figure 1. Mass transfer of component A between media B and E with no concentration jump at the interface.

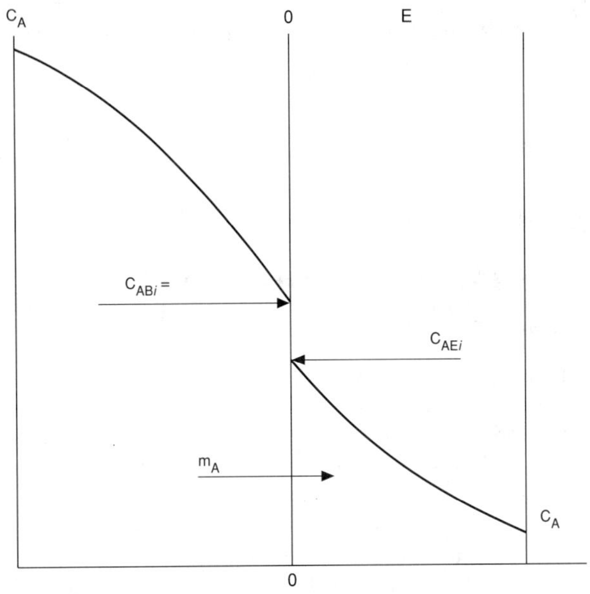

Figure 2. Mass transfer of component A between media B and E with a concentration jump at the interface.

$$\dot{m} = \beta_{AB}\,\tilde{M}_A\,(C_{AB} - C_{ABi}) = \beta_{AE}\,\tilde{M}_A(C_{AEi} - C_{AE}) \quad (10)$$

$$= K_B\,\tilde{M}_A(C_{AB} - HC_{AE}) = K_E\,\tilde{M}_A\left(\frac{C_{AB}}{H} - C_{AE}\right).$$

Whence

$$\frac{1}{K_B} = \frac{1}{\beta_{AB}} + \frac{H}{\beta_{AC}} = \frac{H}{K_E}. \quad (11)$$

Given the flow parameters for both films, β can be determined using, e.g., the above formulas.

Mass transfer, mostly in combination with heat transfer, is widely employed in industry, in chemical process equipment, metallurgy, power engineering, and so on. The equipment includes fractionating towers, absorbers and extractors, driers and cooling towers, combustion chambers, heterogeneous catalysis apparatuses, and many others.

References

Sherwood, T. K., Pigford, R. L., and Wilke, C. R. (1975) *Mass Transfer.* McGraw Hill, New York.

Spolding, D. B. (1963) *Convective Mass Transfer.* Arnold Publ. London.

Leading to: *Ablation; Adsorption; Agitated vessel mass transfer; Chromatography; Desalination; Diffusion; Distillation; Dying; Extraction, liquid/liquid; Falling film mass* *transfer; Membrane processes; Multicomponent systems thermodynamics; Phase equilibrium*

I.L. Mostinsky

MASS TRANSFER COEFFICIENTS

***Following from:** Mass transfer*

The process of **Mass Transfer** across an interface, or across a virtual surface in the bulk of a phase, is the result of a chemical potential driving force. This driving force is more usually expressed in terms of **Concentrations** of the species, or **Partial Pressures** in the case of gas phases. The rate of transfer of a given species per unit area normal to the surface, i.e., the species flux, depends on some of the physical properties of the system and on the degree of **Turbulence** of the phases involved. In general the relationship between the flux and these parameters is not easily developed from fundamentals of mass transfer, so that mass transfer coefficients have been defined that lump them all together. These definitions are of the form:

$$\text{Flux} = \text{coefficient.(concentration difference)}$$

In the case of species crossing an interface, there are several expressions for the flux based on different driving forces. The interfacial flux, $\dot{m}$, can be expressed in the four following ways depending on the concentration driving force used:

$$\dot{m} = \beta_G(p - p_i) = \beta_L(c_i - c)$$

$$= \beta_{OG}(p - Hc) = \beta_{OL}[(p/H) - c]$$

where $\dot{m}$ is the mass flux; β, the mass transfer coefficient; and the subscripts L and G indicate the gas and liquid phases. The first two equation define the single-phase gas and liquid mass transfer coefficients. Since the interfacial concentrations p_i and c_i are usually unknown, the overall mass transfer coefficients β_{OG} and β_{OL} defined by the two last equations, are more commonly used. In these equations H is the equilibrium distribution coefficient of the solute between the two phases at equilibrium. Since the interfacial flux must be the same irrespective of the driving force used to express it, the four numerical coefficients are different and have different units. This is also the case when dimensionless driving forces, such as molar or mass fractions, are used. The mass transfer coefficients depend on the diffusivity of the solute and the hydrodynamics of the phases. They can be calculated using expressions derived from fundamentals of mass transfer, in the case of laminar flow, or from empirical correlations.

References

Skelland, A. H. P. (1974) *Diffusional Mass Transfer.* John Wiley.

E.S. Perez de Ortiz

MASS TRANSFER, ELECTROCHEMICAL, PROBE (see Shear stress measurement)

MASS TRANSFER UNDER REDUCED GRAVITY (see Microgravity conditions)

MATHEMATICAL MODELLING (see Computational fluid dynamics)

MATTE (see Smelting)

MAXI-COMPUTERS (see Computers)

MAXIMUM HEAT FLUX (see Burnout in pool boiling)

MAXIMUM HYGROSCOPIC MOISTURE CONTENT (see Drying)

MAXIMUM LIQUID TEMPERATURE (see Rewetting of hot surfaces)

MAXWELL-BOLTZMANN DISTRIBUTION

***Following from:** Kinetic theory of gases*

The distribution of molecular velocities in a gas, established first by Maxwell and later proved rigorously by Boltzmann, is given by a function F and is today known as the Maxwell-Boltzmann velocity distribution function. Since this probability function depends upon the specified velocity u, F = F(u) and is defined such that F(u) dudvdw gives the probability that a molecule selected at random will, at any instant, have a velocity u with Cartesian components in the ranges u to u + du, v to v + dv, and w to w + dw.

The Maxwell-Boltzmann velocity distribution function refers specifically to a gas which is at rest (in the sense that no macroscopic flow exists) and in a state of thermodynamic equilibrium. Subject to these assumptions, the distribution law states that

$$F(u) = f(u)f(v)f(w) \tag{1}$$

$$= \left(\frac{m}{2\pi kT}\right)^{3/2} \exp(-mc^2/2kT),$$

where m is the mass of one molecule, k is the Boltzmann's constant, and $c = |u|$ is the speed of the molecule. Note that F is given as the product f(u)f(v)f(w) and that the velocity components in different directions are therefore uncorrelated. In other words, the probability of the molecule possessing a specified velocity u in the x direction is not influenced by the values of v and w for that or any other molecule. The function f is thus a velocity distribution function for motion in a specified direction, and is given by

$$f(q) = \left(\frac{m}{2\pi kT}\right)^{1/2} \exp(-mq^2/2kT), \tag{2}$$

where q represents one of (u,v,w).

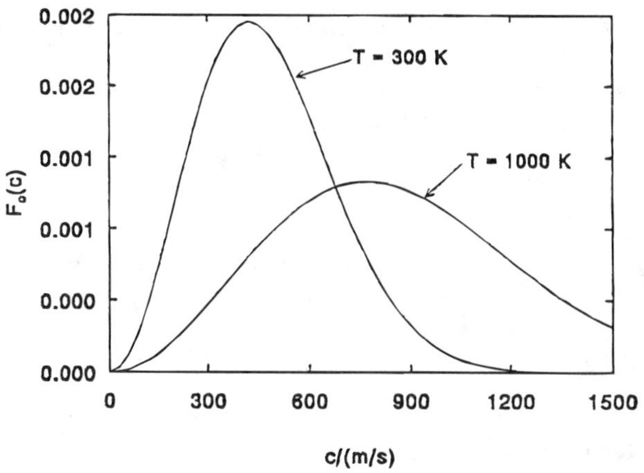

Figure 1. Maxwell-Boltzmann distribution of molecular speeds for nitrogen gas.

The distribution function $F_c(c)$ for molecular speed, irrespective of direction, is easily found by:

$$F_c(c) = 4\pi c^2 \left(\frac{m}{2\pi kT}\right)^{3/2} \exp(-mc^2/2kT). \quad \textbf{(3)}$$

The behaviour of the function $F_c(c)$ is illustrated in **Figure 1** for nitrogen gas at temperatures of 300 K and 1000 K. Note that f and F_c are normalised distribution functions which, upon integration with respect to their argument on [0, ∞], yield unity. Various 'average' molecular speeds may be obtained easily from the distribution function F_c as follows:

The most probable speed (maximum F_c):	$\hat{c} = (2kT/m)^{1/2}$
The mean speed:	$\langle c \rangle = \int_0^\infty F_c(c)c\,dc = (8kT/\pi m)^{1/2}$
The root-mean-square speed:	$\langle c^2 \rangle^{1/2} = (\int_0^\infty F_c(c)c^2 dc)^{1/2} = (3kT/m)^{1/2}$

These measures of average speed may be compared with the *speed of sound* in the perfect gas: $(kT\gamma/m)^{1/2}$. For nitrogen at 300 K, the speed of sound is 353 m/s; $\hat{c}$ = 420 m/s, $\langle c \rangle$ = 478 m/s; and $\langle c^2 \rangle^{1/2}$ = 516 m/s: all are proportional to $\sqrt{T}$.

References

Kennard, E. H. (1938) *Kinetic Theory of Gases*. McGraw-Hill. New York.

J.P.M. Trusler

MAXWELL EQUATION (see Multicomponent systems thermodynamics)

MAXWELL EQUATION FOR ELECTRICAL CONDUCTIVITY (see Bubble flow)

MAXWELL FLUIDS (see Non-Newtonian fluids)

MAXWELL MODEL FOR ACCOMMODATION COEFFICIENT (see Accommodation coefficient)

MAXWELL RELATIONS (see Thermodynamics)

MAXWELL'S EQUATIONS (see Electromagnetic waves; Magnetohydrodynamics)

MAXWELL-STEFAN EQUATIONS

***Following from:** Diffusion*

The Maxwell-Stefan (M-S) equations [Maxwell (1866), Stefan (1871)] describe the process of diffusion, where diffusive fluxes, J_i, of species through a plane, across which no net transfer of moles occurs, depend on all (n-1) independent driving forces in a mixture of n species. Their predictions, in particular that a species need not diffuse in the direction of its own driving force, have been confirmed by reliable experimental studies, e.g. Duncan & Toor, 1962. A recent text describing their formulation, development and applications is Taylor and Krishna, 1993, who quote them as,

$$\frac{\tilde{y}_i}{RT} \nabla_{T,P}\mu_i = \sum_{\substack{k=1 \\ k\neq i}}^{n} \frac{\tilde{y}_i J_k - \tilde{y}_k J_i}{\tilde{c}\mathcal{D}_{ik}} \quad \textbf{(1)}$$

In ideal mixtures at constant temperature and pressure, the left hand side simplifies to $\nabla\tilde{y}_i$ or $d\tilde{y}_i/ds$ for one space dimension. The $\mathcal{D}_{ik}$ diffusivities are then the widely available *binary diffusivities*, $\mathcal{D}_{ik}$. Further simplification for a binary gives **Fick's Law**. Equation 1 has been used to describe many processes, notably **Distillation** and **Condensation**. For a film model of interfacial transport, Krishna & Standart, 1977, show an exact solution. The M-S equations are often written in matrix form, explicit in diffusive flux. Here braces and square brackets denote (n-1) column and square matrices, respectively, and normal matrix products are implied,

$$(J) = -\tilde{c}[M]^{-1}[\Gamma](\nabla\tilde{y}) = -\tilde{c}[D](\nabla\tilde{y}) \quad \textbf{(2)}$$

$$M_{ii} = \frac{\tilde{y}_i}{\mathcal{D}_{in}} + \sum_{\substack{k=1 \\ k\neq i}}^{n} \frac{\tilde{y}_k}{\mathcal{D}_{ik}} \quad M_{ik} = -\tilde{y}_i\left[\frac{1}{\mathcal{D}_{ik}} - \frac{1}{\mathcal{D}_{in}}\right]$$

$$\Gamma_{ik} = \delta_{ik} + \tilde{y}_i \left.\frac{\partial(\ln \gamma_i)}{\partial \tilde{y}_k}\right|_{T,p,\tilde{y}_k\neq\tilde{y}_i}$$

The second term in **Equation 2** is the generalised form of Fick's Law and the composition dependence of the diffusion coefficients [D] is clear through the stated relationship with the M-S equations. **Equation 2** is the origin of the *Linearised Theory* approach of Toor, 1964, and Stewart & Prober, 1964, which can be applied generally to film, surface renewal and boundary layer models of mass transfer. It also shows the dependence on activity coefficient, γ, in non-ideal mixtures. The $\mathcal{D}_{ik}$ then become composition dependent, but may in some cases be estimated from binary diffusivities, $\mathcal{D}^{o}$, at infinite dilution, Taylor & Krishna, 1993. The Maxwell-Stefan equations are the correct description of multicomponent mass transfer and their usage is steadily broadening as experimental evidence supporting their better applicability mounts.

References

Duncan, J. B. and Toor, H. L. (1962) *An Experimental Study of Three Component Gas Diffusion,* AIChEJ, *8,* 38–41.

Krishna, R. and Standart, G. L. (1976) *A Multicomponent Film Model Incorporating an Exact Matrix Method of Solution to the M-S Equations,* AIChEJ, *22,* pp 383–389.

Maxwell, J. C. (1866) On the dynamic theory of gases, *Phil. Trans. Roy. Soc.*, 157, 49–88.

Stefon, J. (1871) Über das Gleichgewicht und die Bewegung, insbesondere die Diffusion Von Gasmengen, Sitzungsber. Akad. Wiss. Wien, 63, 63–124.

Stewart, W. E. and Prober, R. (1964) *Matrix Calculation of Multi-component Mass Transfer in Isothermal Systems,* Ind Eng Chem Fundam, *3,* pp 224–235.

Taylor, R. and Krishna, R. (1993) *Multicomponent Mass Transfer,* Wiley Interscience.

Toor, H. L. (1964) *Solution of the Linearised Equations of Multi-component Mass Transfer,* AIChEJ, *10,* 448–465.

Leading to: *Condensation of multi-component vapours*

D.R. Webb

McCABE-THIELE METHOD (see Distillation)

MEAN FREE PATH

Following from: *Kinetic theory of gases; Maxwell Boltzmann distribution*

In kinetic theory, the mean free path is defined as the mean distance travelled by a molecule between collision with any other molecule. For a dilute gas composed of hard spherical molecules of kinds A and B, the mean time τ_{AB} between successive collisions of a given A molecule with B molecules is given by

$$\tau_{AB} = \frac{1}{\pi d_{AB}^2 \bar{n}_B \langle c_{AB} \rangle}. \quad (1)$$

Here, d_{AB} is the mean diameter of molecules A and B, $\bar{n}_B$ is the number density of B molecules, and $\langle c_{AB} \rangle$ is the mean relative speed of molecules A and B. According to the Maxwell-Boltzmann velocity distribution law, molecular velocities are not correlated and $\langle c_{AB} \rangle$ is therefore given in terms of the mean speeds of the individual molecules by $\langle c_{AB} \rangle = \sqrt{\langle c_A \rangle^2 + \langle c_B \rangle^2}$ where, for molecules of type i with mass m_i, $\langle c_i \rangle = (8kT/\pi m_i)^{1/2}$, where k is Boltzmann's constant and T is the absolute temperature.

The mean free path l_{AB} travelled by a given A molecule between successive collisions with B molecules is simply $\langle c_A \rangle \tau_{AB}$ or

$$l_{AB} = \left(\frac{m_B}{m_A + m_B}\right)^{1/2} \frac{1}{\pi d_{AB}^2 \bar{n}_B}. \quad (2)$$

Real molecules interact through intermolecular forces which vary smoothly with distance. Consequently, there is no unique counterpart of the hard-sphere collision cross-section πd_{AB}^2 and no unique definition of what constitutes a collision. Often, d_{AB} is replaced by the separation σ_{AB} at which the A–B intermolecular potential energy crosses zero. Alternatively, d_{AB} may be determined from viscosity data by comparison with the theoretical predictions for hard spheres (Kennard, 1938). These formulae are easily specialised to the case where A and B are identical and, if d is estimated from viscosity data, then

$$l = \frac{32}{5\pi} \frac{\eta}{\rho \langle c \rangle} \approx \frac{2\eta}{\rho \langle c \rangle}. \quad (3)$$

where η is viscosity and ρ density.

References

Kennard, E. H. (1938) *Kinetic Theory of Gases.* McGraw-Hill. New York.

J.P.M. Trusler

MEAN PHASE CONTENT (see Multiphase flow)

MEAN TEMPERATURE DIFFERENCE

Following from: *Heat exchangers; Heat transfer coefficient*

Figure 1 shows a generalised **Heat Exchanger** in which heat is transferred between two streams, (stream 1 and stream 2).

The rate of heat transfer, $\dot{Q}$, between the streams may be expressed as a function of the area available for the transfer of heat, A, the overall heat transfer coefficient, U, and a mean temperature difference, ΔT_m, such that

$$\dot{Q} = AU\Delta T_m \quad (1)$$

Estimating the rate of heat transfer for a given design of heat exchanger requires techniques for estimating the **Overall Heat Transfer Coefficient** and techniques for estimating the mean temperature difference.

Techniques for estimating mean temperature difference are based upon the following assumptions:

a. the heat exchangers have only two streams;

b. heat exchange with the surroundings is negligible;

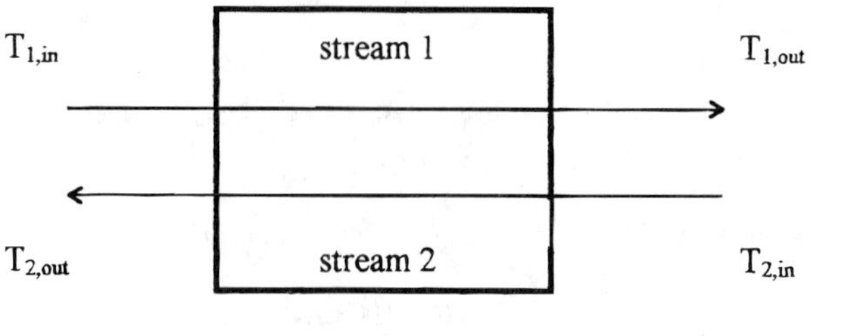

Figure 1. Generalised two stream heat exchanger.

c. there is a linear relationship between specific enthalpy and temperature for both streams (i.e. constant specific heat capacities);

d. the overall heat transfer coefficient between the stream is constant throughout the heat exchanger;

e. where a heat exchanger consists of multiple parallel paths, the flowrates and heat transfer areas in each path are identical.

The above assumptions are most likely to be met when both streams are single-phase fluids (i.e. all liquid or all gas) and where the temperature changes are small such that the specific heat capacities and other properties of the fluids stay constant throughout the heat exchanger. The approach could be applied to heat exchangers involving boiling or condensing but only under circumstances where there are no significant changes in overall heat transfer coefficient. Heat exchangers involving the onset of boiling or condensation or the dryout transition are therefore not suitable for treatment using the traditional mean temperature difference approach. Such heat exchangers will need to be analysed using techniques which make allowance for changes in heat transfer coefficient.

With the above assumptions, the rate of heat transfer in any geometry of heat exchanger can, in principle, be calculated. The simplest case is pure counter current flow. Here the mean temperature difference can be expressed in terms of the inlet and outlet temperatures of each stream.

$$\Delta T_m = \frac{(T_{1,in} - T_{2,out}) - (T_{1,out} - T_{2,in})}{\log_e \dfrac{(T_{1,in} - T_{2,out})}{(T_{1,out} - T_{2,in})}} = \Delta T_{LM} \quad (2)$$

For a heat exchanger with counter current flow, the mean temperature difference is known as the log mean temperature difference, ΔT_{LM}. The log mean temperature difference is the maximum mean temperature difference that can be achieved in any geometry of heat exchanger for any given set of inlet and outlet temperatures. For any other type of heat exchanger, the mean temperature difference can be expressed as

$$\Delta T_m = \Delta T_{LM} F \quad (3)$$

where F is always less than or equal to 1. Estimating the mean temperature difference in a heat exchanger by calculating the log mean temperature difference and estimating F is known as the *F factor method.*

F varies with geometry and thermal conditions. The thermal conditions are defined by parameters such as the overall heat transfer coefficient, U, the area available for heat transfer, A, the mass flow rates of the two streams $\dot{M}_1$ and $\dot{M}_2$, the specific heat capacities of the two streams c_1 and c_2, and the temperature change in each stream $(T_{1,in} - T_{1,out})$ and $(T_{2,in} - T_{2,out})$.

For any given geometry, F is often presented as a function of two non-dimensional parameters, R, the ratio of the thermal capacities of the two streams and P, (sometimes known as effectiveness, E) the ratio of the achieved heat transfer rate to the maximum possible heat transfer rate. These parameters are typically defined as

$$R = \frac{\dot{M}_1 c_1}{\dot{M}_2 c_2} = \left|\frac{(T_{2,in} - T_{2,out})}{(T_{1,in} - T_{1,out})}\right| \quad (4)$$

and

$$P = \frac{\text{achieved heat transfer rate}}{\text{maximum heat transfer rate}} = \frac{(T_{in} - T_{out})_{larger}}{(T_{1,in} - T_{2,in})} \quad (5)$$

where

$(T_{in} - T_{out})_{larger}$ is the larger of $(T_{1,in} - T_{1,out})$ and $(T_{2,in} - T_{2,out})$

Unfortunately, there are no standard definitions of R and P. Users of the F method should always check the definitions used by any supplier of F value information and apply identical definitions when using this information in heat transfer calculations. This variation of definition can lead to problems when comparing data from different authors.

Values of F are frequently presented as graphs showing the relationship between F and P for a range of values of R. **Figure 2** shows a typical relationship. The figure shows that F always tends to 1 as the amount of heat transferred reduces to zero. The figure also shows that for any given value of R, there is typically a maximum achievable value of P. The value of F changes rapidly as P approaches its maximum value. Because of this sensitivity, heat exchanger designs are rarely developed near the maximum value of P and are typically restricted to conditions which give values of F greater than 0.8.

The F factor can be used for both design and rating calculations. For design calculations, the mass flowrates, specific heat capacities and required temperature changes will be specified. R and P can therefore be calculated directly. The design engineer will need to check that the required value of P is less than the maximum value of P for the specified value of R. The value of F can then be found from the graph. The combination of the log mean temperature difference and F gives the mean temperature difference and with an estimate of the overall heat transfer coefficient, the required heat transfer area can be established. If the required value of P is greater than the maximum value for the specified value of R, a different type of heat exchanger must be considered until a feasible design is found. A simple counter current heat exchanger will always be able to achieve any design requirement but may be physically impractical.

For rating calculations, the geometry of the heat exchanger and its heat transfer area, the mass flowrates and the specific heat capacities of the streams and hence the overall heat transfer coefficient and the inlet temperatures will be defined. It will therefore be possible to calculate R but not P. The rating is estimated using an iteration which may start by guessing a heat transfer rate and calculating the exit temperatures, the log mean temperature difference and P and then using the figure to estimate F. The resulting heat transfer rate is then calculated from F, the log mean temperature difference, the overall heat transfer coefficient and the heat transfer area. The guessed and calculated heat transfer rates are compared and the guessed value is adjusted until convergence is achieved.

Published relationships for F in graphical form are available for most geometries of shell and tube heat exchanger and a range of geometries of crossflow heat exchanger (see guide to references

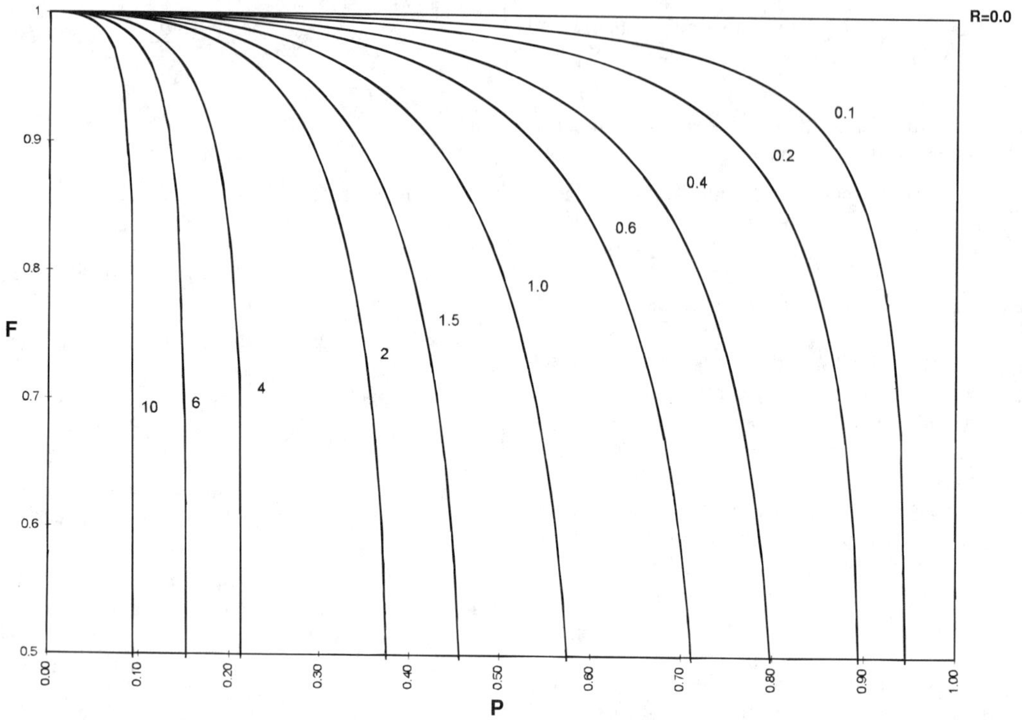

Figure 2. Typical relationship between F and P for various values of R (Based on Single E-shell with any even number of tube passes).

at the end of this section). The size of the graphical presentations rarely allows values of F to be estimated to better than two significant figures. This accuracy of estimation is consistent with the overall accuracy of the mean temperature difference approach and the lack of compliance with the underlying assumptions. Attempts to improve the accuracy in the estimation of F values are therefore unlikely to produce significant benefits for heat exchanger designers.

An alternative method of presenting mean temperature difference information is known as the *effectiveness—N_{TU}* method. This method is based upon exactly the same initial assumptions. The heat transfer behaviour are presented as a relationship between effectiveness, E, (defined in a similar way to P), the ratio of the thermal capacities of the streams, R, and the number of heat transfer units, N_{TU} which as calculated from the expression

$$N_{TU} = \frac{UA}{(\dot{M}c)_{smaller}} \quad \textbf{(6)}$$

where $(\dot{M}c)_{smaller}$ is the smaller of $(\dot{M}c)_1$ and $(\dot{M}c)_2$

Unfortunately, the parameters E, R and N_{TU} are again not consistently defined and any user should check the definition used by the supplier of data and apply those definitions when using the data.

Effectiveness—N_{TU} information is typically presented graphically as the relationship of effectiveness against N_{TU} for various values of R. **Figure 3** shows a typical relationship. This shows that the effectiveness tends to zero as the N_{TU} tends to zero and the effectiveness tends to a maximum value as N_{TU} becomes large.

Users of the effectiveness N_{TU} technique are not required to calculate the log mean temperature difference when carrying out design or rating calculations. For design calculations, E and R can be calculated from the mass flowrates, specific heat capacities and inlet and outlet temperatures. The value of N_{TU} can be read from the graph for the chosen design of heat exchanger and used to calculate the required surface area. In rating calculations, the surface area, mass flowrates and specific heat capacities can be used to calculate R and N_{TU}. The value of E can be read from the graph and used to calculate the rate of heat transfer in the heat exchanger. It can therefore be argued that the effectiveness—N_{TU} method can be used for rating calculations without the need for an iteration. In reality, since heat transfer coefficient will change with temperature, it is still likely that an iterative calculation will be required.

Effectiveness—N_{TU} relationships are generally presented graphically and may be used to produced estimates to two significant figures. Again, because of deviations from the underlying assumptions, this level of precision in the calculations is consistent with the precision of the overall method. An attempt has been made to produce a single algebraic expression with a number of variable coefficients to represent a range of common heat exchangers, *ESDU, 93012*. Approximately 90 geometries were represented by an expression with 14 variable coefficients. The curve fitting approach matches the exact relationships to better than 2%. By incorporating the algebraic expression into a computer program, a range of geometries and designs can be readily accessed. The accuracy of the computer calculations does not, however, bring any increase in the accuracy of the overall method.

F values or effectiveness—N_{TU} relationships are obtained by making simplifying assumptions about the geometry of a heat exchanger and then carrying out a process of integration. The integration can either be carried out algebraically (most designs of shell and tube heat exchanger) or using finite element methods (most designs of crossflow heat exchanger). In all but the simplest of

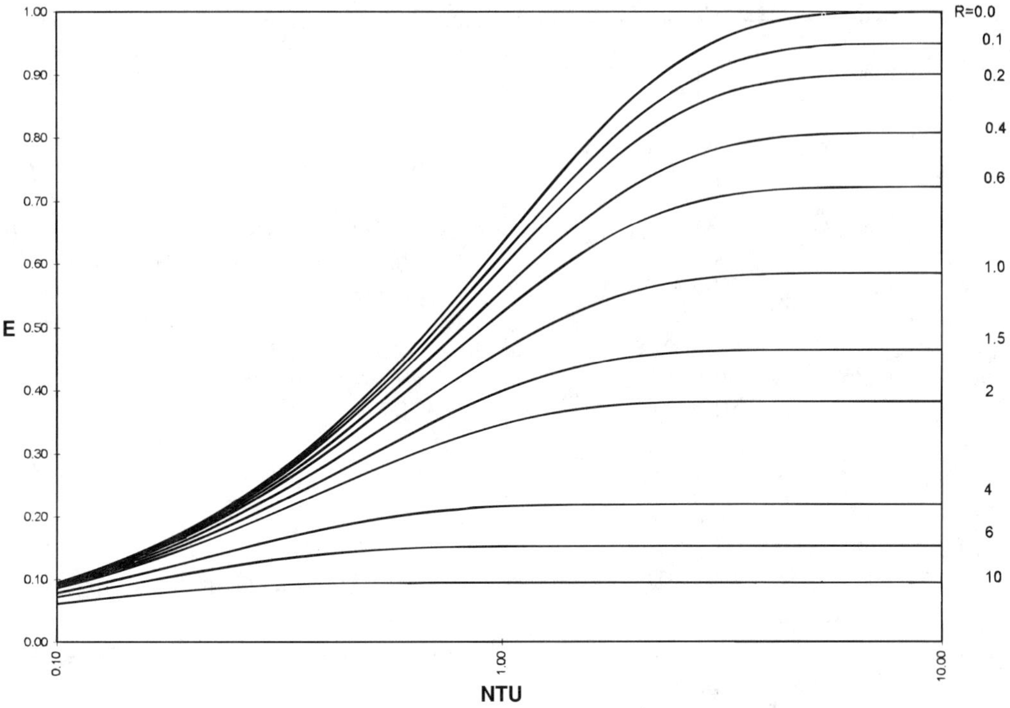

Figure 3. Typical relationship between E and N_{TU} for various values of R (Based on Single E-shell with any even number of tube passes).

geometries, the process of integration is complex. For shell and tube exchangers, the resulting algebraic expressions require care in application to avoid error. For example, for the single E-shell with any even number of passes, the expression linking NTU with E is

$$N_{TU} = \frac{-\log_e\left[\dfrac{2 - E(1 + R - \eta)}{2 - E(1 + r + \eta)}\right]}{\eta} \quad \textbf{(7)}$$

where

$$\eta = \sqrt{(1 + R^2)}$$

and

$$F = \frac{\log_e\left[\dfrac{1 - ER}{1 - E}\right]}{N_{TU}(1 - R)} \quad \textbf{(8)}$$

For crossflow heat exchangers, particularly with more than one pass, the finite element integration often also involves iteration as well. Users of mean temperature difference techniques are therefore advised to use the graphical presentations of these integrations or curve fits to the data.

Table 1 gives references to sources of graphical data for various types of heat exchanger.

Table 1. Sources of graphical data for various types of heat exchanger

Geometry of heat exchanger	F Factor data	Effectiveness—N_{TU} data
Counter current flow		ESDU, 85042 Kays and London (1984)
Co-current flow	Taborek (1983)	ESDU, 85042 TEMA (1978) Kays and London (1984)
Shell and Tube		
Single E-shell, even tube passes	Kern (1950) TEMA (1978) Perry (1973) Taborek (1983)	ESDU, 85042 Ten Broeck (1938) TEMA (1978) Kays and London (1984)
Two E-shells in series	TEMA (1978) Perry (1973) Taborek (1983)	ESDU, 85042 Kays and London (1984)
Three E-shells in series	TEMA (1978) Perry (1973) Taborek (1983)	ESDU, 85042 Kays and London (1984)
Four E-shells in series	TEMA (1978) Perry (1973) Taborek (1983)	ESDU, 85042 Kays and London (1984)
Five E-shells in series	TEMA (1978) Taborek (1983)	ESDU, 85042
Six E-shells in series	TEMA (1978) Perry (1973) Taborek (1983)	ESDU, 85042
E-shell with three tube passes		ESDU, 85042

Table 1.—*continued*

Geometry of heat exchanger	F Factor data	Effectiveness—N_{TU} data
Two E-shells in series, parallel		ESDU, 88021
Three E-shells in series, parallel		ESDU, 88021
Four E-shells in series, parallel		ESDU, 88021
G-shell with two tube passes	TEMA (1978) Taborek (1983)	ESDU, 85042
J-shell with one tube pass		ESDU, 85042
J-shell with two tube passes	TEMA (1978) Taborek (1983)	ESDU, 85042 Kays and London (1984)
J-shell with four tube passes		ESDU, 85042
J-shell with infinite tube passes		ESDU, 85042
Two J-shells in series (various configurations)		ESDU, 88021
F-shell	Taborek (1983)	ESDU, 86018, ESDU, 88021
Crossflow		
Single pass	Perry (1973) Taborek (1983)	ESDU, 87020 Kays and London (1984)
Two pass—counter current flow		ESDU, 87020 Kays and London (1984) Pignotti (1984)
Two pass—co-current flow		ESDU, 87020 Kays and London (1984) Pignotti (1984)
Three pass—counter current flow		ESDU, 87020 Pignotti (1984)
Three pass—co-current flow		ESDU, 87020 Pignotti (1984)
Single pass—discrete tube rows	Perry (1973) Taborek (1983)	ESDU, 87020
Two pass—discrete tube rows	Taborek (1983)	ESDU, 87020
Three pass—discrete tube rows		ESDU, 87020
Four pass—discrete tube rows		ESDU, 87020
Serpentine—counter current flow	Taborek (1983)	ESDU, 87020
Serpentine—co-current flow		ESDU, 87020

References

ESDU, 85042. Effectiveness—N_{TU} relationships for the design and performance rating of two-stream heat exchangers, ESDU, Data Item, 85042, December 1985.

ESDU, 86018. Effectiveness—N_{TU} relationships for the design and performance rating of two stream heat exchangers, ESDU, Data Item, 86018, July 1986, Amended July 1991.

ESDU, 87020. Effectiveness—N_{TU} relationships for the design and performance evaluation of multi-pass crossflow heat exchangers, ESDU, Data Item 87020, October 1987, Amended November 1991.

ESDU, 88021. Effectiveness—N_{TU} relationships for the design and performance evaluation of additional shell and tube heat exchangers, ESDU, Data Item 88021, November 1988, Amended July 1991.

ESDU, 91036. Algebraic representations of effectiveness—N_{TU} relationships, ESDU, Data Item 91036, November 1991.

Kays, W and London, A. L. (1984), *Compact Heat Exchangers. Third Edition.* McGraw-Hill.

Kern, D. Q (1950), *Process Heat Transfer, First Edition,* McGraw-Hill.

Perry (1973), *Chemical Engineers' Handbook,* McGraw-Hill.

Pignotti, A. (1984), Matrix formalism for complex heat exchangers, Trans ASME, *Journal of Heat Transfer,* Vol. 106, pp. 352–360.

Taborek, J (1983), *Heat Exchanger Design Handbook, Section 1.5,* Hemisphere Publishing Corporation.

TEMA (1978), *Standards of Tubular Exchanger Manufacturers Association, Sixth Edition.*

J. Ward

MEAN TEMPERATURE DRIVING FORCE

(see Condensation curve)

MECHANICAL DESIGN OF HEAT EXCHANGERS

***Following from:** Heat exchangers*

The **Shell and Tube**, the **Air Cooled** and the **Plate-type Exchanger** are the three most commonly used types of exchangers in the chemical and process industries. With increasing effort in recent years to reduce weight and size and increase efficiency, other types of exchangers are increasingly used. However the mechanical (and thermal) design of these alternative exchangers tends to be of a proprietary nature which may explain why many clients prefer the tried-and-tested shell and tube exchanger type which still predominates in most plants.

The general principles of the mechanical design of the following types of exchangers are given in the *Heat Exchanger Design Handbook*, 1994, and full descriptions of each, are given under the corresponding entries in this encyclopedia.

- a. **Shell and Tube Exchangers**
- b. **Air Cooled Exchangers**
- c. **Plate Type Exchangers**
- d. **Plate Fin Type Exchangers**
- e. **Double Pipe Exchangers**
- f. **Graphite Block Exchangers**
- g. **Spiral Plate Type Exchangers**
- h. **Direct Contact Exchangers**
- i. **Heat Pipes**

The shell and tube exchanger basically consists of a number of connected components, some of which are also used in the construction of other types of exchangers. The pressurised components of the shell and tube exchanger are designed to be in accordance with a pressure vessel design code such as ASME VIII, 1993, or BS5500, 1994.

To meet the relevant regulations (see **Pressure Vessels**) the pressurised components of alternative types of exchangers must meet at least the principles of a relevant pressure vessel design code.

A pressure vessel design code alone cannot be expected to cover all the special features of heat exchangers. To give guidance and protection to designers, manufacturers and purchasers, a supplementary code is desirable. A universally accepted code for shell and tube exchangers is TEMA, 1988, which although designed to supplement ASME VIII, can be used in conjunction with other pressure vessel codes. TEMA specifies minimum thicknesses, corrosion allowances, particular design requirements, tolerances, testing requirements, aspects of operation, maintenance and guarantees. (See also **TEMA Standards**)

One of the most useful functions of TEMA is to provide a simple three-letter system that completely defines all shell and tube exchangers with respect to exhanger type, stationary end head, rear end head and shell side nozzle configuration. This system is shown in **Figure 1**. The first letter defines the stationary end head, the middle letter defines the shell type and the last letter the rear end type.

In specifying TEMA (Fig 1) the purchaser will choose one of the three classes:

Class R for "generally severe requirements of petroleum and related processing applications"

Class C for "generally moderate requirements of commercial and general process applications"

Class B for "chemical process service."

Heat transfer equipment may be designated by type or function it performs, such as chiller, condenser, cooler reboiler, etc. The choice of shell and tube type is determined chiefly by factors such as the need for the provision for differential movement between shell and tubes, the design pressure, the design temperature, and the fouling nature of the fluids rather than the function. More information on the choice of types, their main features and their design, is given in Saunders, 1988.

A common type of shell and tube exchanger is the fixed tubesheet type. This is shown in **Figure 2**, and has the TEMA designation AEM. The main components of the exchanger shown in **Figure 2** feature in most shell and tube exchangers and are given a reference number which relates to the component descriptions below.

The following components perform a function mainly related to fluid flow.

1. **Tubes** The usual outside diameter range for petroleum and petrochemical applications is 15 to 32mm, with 19 and 25 being the most common. Tubes may be purchased to minimum or average wall thickness. The thickness tolerances for minimum wall tubes are minus zero, plus 18% to 22% of the nominal thickness, while those of average wall tubes are plus and minus 8% to 10% of the nominal wall thickness. Tube thickness must be checked against internal and external pressure but the dimensions of the most commonly used tubes can withstand appreciable pressures. The most common tube length range is 3600 to 9000mm for removable bundles and 3600 to 15000mm for the fixed tube type. Removable bundle weights are often limited to 20 tons. TEMA specifies minimum tube pitch/outside diameter ratios and minimum gaps between tubes.
2. **Channel partition plates** For exchangers with multiple tube passes, the channels are fitted with flat metal plates which divide the head into separate compartments. The thickness of these plates depends on channel diameter but is usually 9 to 16mm for carbon and low alloy steels and 6 to 13mm for the more expensive alloys. Except for special high pressure heads, the partition plates are always welded to the channel barrel and also to the adjacent tubesheet or cover if either of these components is in turn welded to the channel. If the tubesheet or cover is not welded to the channel, the tubesheet or cover is grooved and the edge of the partition plate sealed by a gasket embedded in the grooves.
3. **Shell baffles** Shell cross baffles have the dual purpose of supporting the tubes at intervals to prevent sag and vibration, and also of forcing the shell side fluid back and forth across the bundle, from one end of the exchanger to the other. Segmentally single cut baffles are the most common, however, thermal or pressure drop may dictate baffles of more complicated shape. Split backing ring and pull through floating head exchangers have a special support type baffle adjacent to the floating head to take the weight of the floating head assembly. TEMA specifies the minimum baffle thickness, the maximum unsupported tube length, the clearances between tubes and holes in the baffles and between shell inside diameter and baffle outside diameter. Two shell pass exchangers, (see **Figure 1** shell types F, G or H), require a longitudinal baffle, which for F type exchangers is welded to the stationary tubesheet. Leakage of the shell side fluid between the shell and the longitudinal baffle edges must be minimised. When removable bundles are used, this leakage gap is sealed by flexible strips or packing devices. **Figure 3** shows a typical flexible strip.
4. **Tie rods** Tie rods and spacers are used to hold the tube bundle together and to locate the shell baffles in the correct position. Tie rods are circular metal rods screwed into the stationary tubesheet and secured at the farthest baffle by lock nuts. The number of tie rods depends on shell diameter and is specified by TEMA.

The following components perform a function mainly related to pressure and fluid containment. Their design is carried out in accordance with the relevant pressure vessel code, see **Pressure Vessels**.

5/6. **Shell barrel and channel barrel** TEMA specifies minimum barrel thicknesses depending on diameter, material and class. Most barrels larger than 450mm internal diameter are fabricated from rolled and welded plate. The shell barrel must be straight and true as a tightly fitting tube bundle must be inserted and particular care has to be taken in fabrication. Large nozzles may cause "sinkage" at the nozzle/shell junction due to weld shrinkage and temporary stiffeners may be needed.

7/8. **Dished heads and flat heads** Small diameter, low pressure dished heads are sometimes cast but most dished heads are fabricated from plate and are of semi-ellipsoidal, tori-

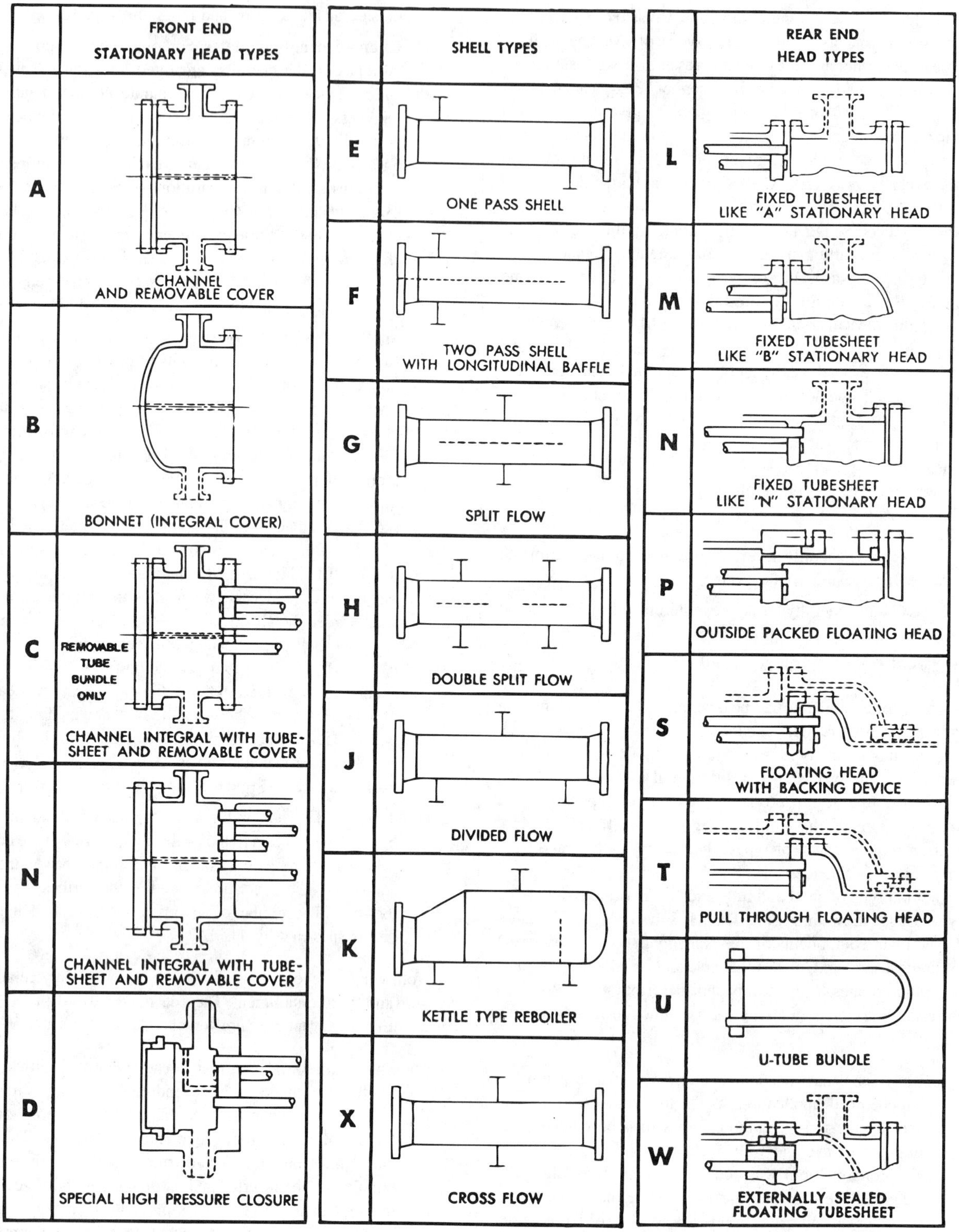

Figure 1. TEMA Designation System. ©1988 by Tubular Exchanger Manufacturers Association.

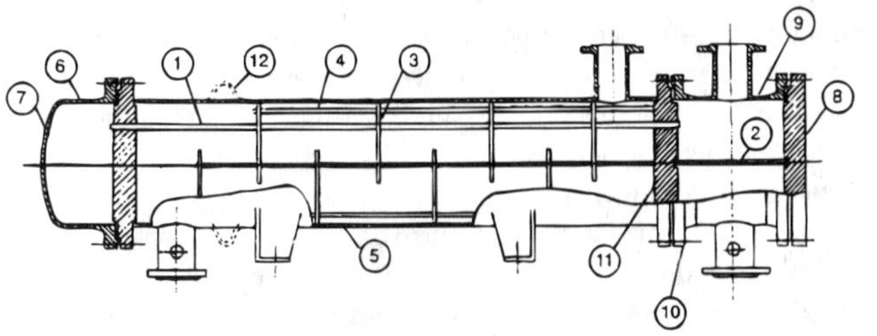

Figure 2. Fixed tubesheet exchanger.

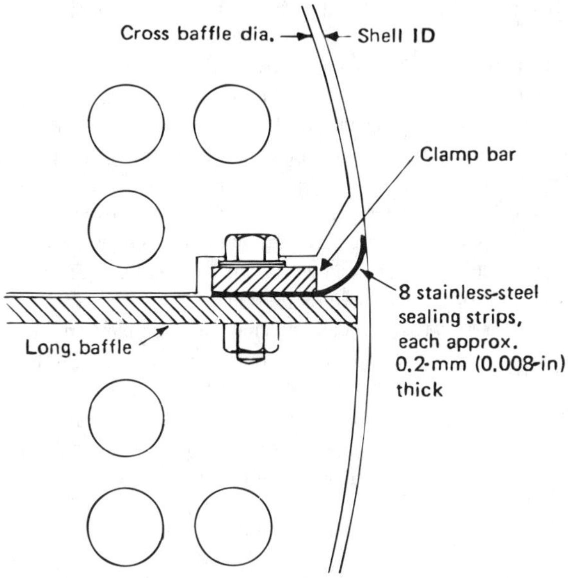

Figure 3. Longitudinal baffle.

spherical or hemispherical shape. The minimum thickness of dished heads is the same as for adjacent barrels. Tube cleaning with a welded channel bonnet (TEMA front end B) would require the breaking and remaking of the channel nozzle flanges to enable the channel to be removed. A flat head (TEMA front end A) avoids this and allows the pipework to remain in place.

9. **Nozzles** Most nozzles are sized to match the adjacent schedule piping. The openings in the barrels require reinforcement in accordance with the relevant pressure vessel code which in turn will limit the maximum size of nozzle opening. **Figure 4** shows a typical nozzle in moderate service, with reinforcement provided by a reinforcement plate and with a weld neck nozzle flange.

10. **Flanges** Three types of flanges are found in shell and tube exchangers, namely, Girth flanges for the shell and channel barrels; internal flanges in the floating head exchanger to allow disassembly of the internals and removal of the tube bundle; and nozzle flanges where the flange and gasket standards, the size and pressure rating will be set by the line specification. **Figure 5** shows three types of flanges. The weld neck flange type, which has a tapered hub with a smooth stress transition and accessibility for full non-destructive examination, provides the highest integrity of

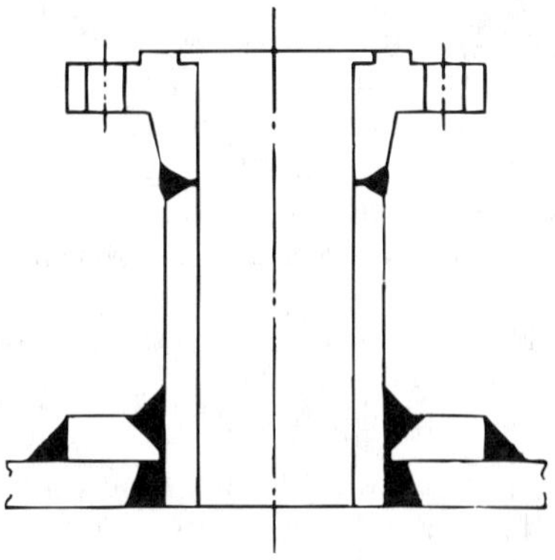

Figure 4. Welds neck nozzle.

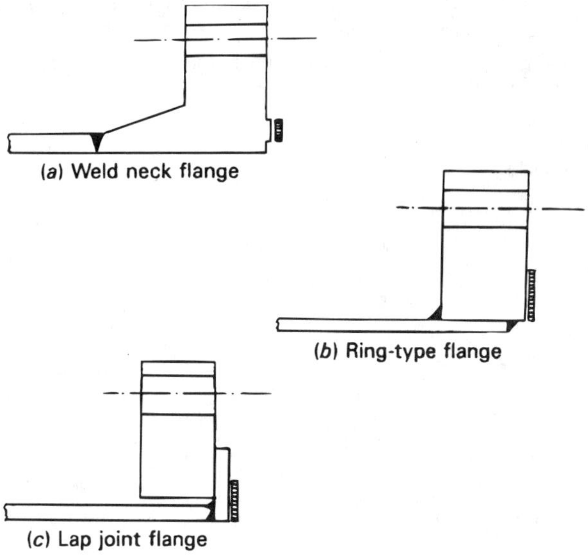

Figure 5. Flange types.

the three types. A flange consists of three sub-components, the flange ring, the gasket and the bolting. The successful operation of the flange depends on the correct choice, design and assembly of these sub-components. The Heat Exchange Design Handbook contains two chapters discussing these factors.

11. **Tubesheets** Tubesheets less than 100mm thick are generally made from plate material. Thicker tubesheets, or for high integrity service, are made from forged discs. Clad plate is commonly used where high alloy material is required for process reasons. A clad tubesheet consists of a carbon or low alloy backing plate of sufficient thickness to satisfy the pressure vessel design code, with a layer of the higher alloy material bonded onto it by welding or by explosion cladding. TEMA gives design rules to calculate the tubesheet thickness, which give similar but not identical results to the rules in ASME and BS5500. It also specifies tolerances for tube hole diameter, ligament width and for drill drift. Different methods are available for the attachment of the tube end to the tubesheet. The most common method is roller expansion where the force produced by

an expanding tool deforms the tube radially outward to give a mechanical seal. In explosive expansion a charge is placed inside the tube within the tubesheet thickness. It is more expensive than roller expansion but can produce tighter joints. Welded tube joints can be produced at the "outer" face of the tubesheet or downhole at the "inner" face of the tubesheet. The success of the tube end joints is highly dependent on the correct choice of type and the experience of the manufacturer. This is discussed in detail in Saunders, 1988.

12. **Expansion bellows** These may be required in the shell of a fixed tubesheet exchanger or at the floating head of single tube pass floating head exchangers. They are discussed in more detail in **Expansion Joints.**

Other important heat exchanger components include those in the floating head assemblies, supports and rectangular headers in air cooled exchangers. These and other components are described in the Heat Exchanger Design Handbook.

References

ASME VIII Division 1. *ASME Boiler and Pressure Vessel Code.* (1993) Rules for the construction of pressure vessels. ASME New York.

BS5500 British Standard for the Specification for Unfired Pressure Vessels. (1994) BSI London.

E. A. D. Saunders, (1988) *Heat Exchangers: Selection, Design and Construction.* Longman, London.

Heat Exchanger Design Handbook, (1994) Begell House Inc, New York.

TEMA Standards of the Tubular Exchanger Manufacturers Association. (1988) TEMA New York

Leading to: *Expansion joints; Gaskets in heat exchangers*

M. Morris

MEISSNER EFFECT (see Superconductors)

MELT FILMS (see Melting)

MELTING

Melting is the process of changing a solid substance into a liquid accompanied by heat absorption. The reverse of melting, solidification, occurs, in equilibrium conditions, at the same temperature T_m and with the evolution of the same amount of heat ΔQ_m which is absorbed in melting. Some of the substances with a complicated molecular structure such as glass, resin and polymers, which are supercooled liquids at room temperature, soften gradually on heating and have no definite melting temperature T_m. Melting is only associated with a certain weakening of interatomic bonds and therefore ΔQ_m is only three to four per cent heat of substance sublimation ΔQ_v, which corresponds to a complete break of the bonds.

The heat of melting is related to melting temperature. The molar entropy of melting for metals is $\Delta H_m/T_m = \Delta S_m \cong 9.7$ kJ/(k) for monovalent interatomic bonds, but for covalent crystals changing to a metal state, this ratio can increase three times.

The model of melting crystalline substances in a high-temperature gas flow is widely used to establish general laws of conjugated heat transfer problems, in particular, as applied to problems of heat protection (see **Heat Protection, Ablation**).

If the parameters of a gas flow (pressure, velocity and temperature) do not change significantly with time, then we can clearly distinguish two periods in the character of heating melting substances (**Figure 1**). At first, for time $\tau \leq \tau_T$, the temperature of the heated surface grows from the initial value T_0 to the phase transition (melting) temperature $T_m = T_d$ where T_d is the temperature at which surface destruction begins.

Beginning at $\tau = \tau_T$, a melt film is formed on the surface of the crystalline substance, its viscosity, however, is so small that it may be carried away instantaneously by the gas flow under the action of shear stresses (friction, pressure gradient). By the term "instantaneously" we mean that the temperature drop over the melted film is always less than the temperature drop $(T_m - T_0)$ over the heated layer of the solid, therefore the contribution of sublimation is negligibly small and the entire ablation occurs in a molten state.

In what follows, a model is presented which aims to represent all the relevant parameters and mechanisms (including melting) in the ablation of a surface under the influence of a gas flow at constant (high) temperature. Generalization of the results of the theoretical analysis is obtained by the use of dimensionless coordinates. One of the dimensionless groups used is a parameter describing the thermal efficiency of ablation or destruction:

$$m = C(T_m - T_0)/(\Delta Q_m).$$

where C is the specific heat capacity of the solid material and ΔQ_m is the heat of phase transitions. Presented schematically in **Figure 2** is the heat destruction and heating of the body at time $\tau > \tau_T$. Defining $S(\tau)$ as the thickness of the layer of the substance already removed from the surface, z is a dimensionless space coordinate with respect to the moving interface between the gas and liquid (melted) phases:

$$z = [y - S(\tau)]/\sqrt{\dot{\kappa}\tau_T}.$$

where $\dot{\kappa}$ is the thermal diffusivity of the solid material (= $\lambda/\rho c$

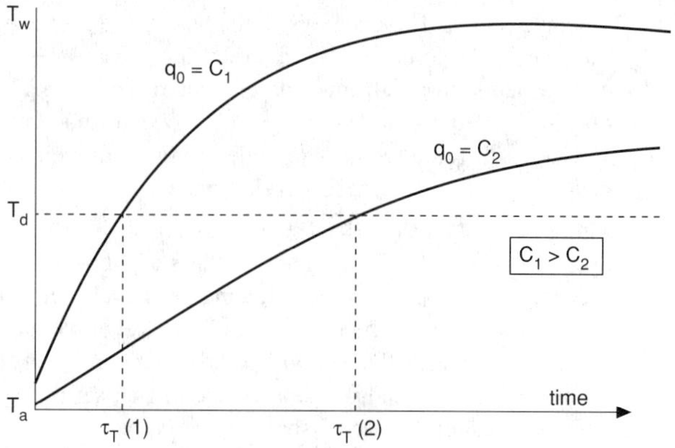

Figure 1. Change of temperature with time in the melting and destruction of a material.

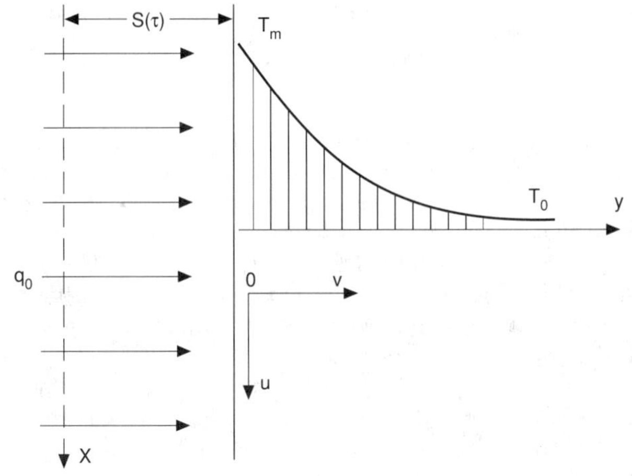

Figure 2. Temperature profile in solid material during melting.

where λ is the thermal conductivity and ρ the density). The value of λ_T may be calculated from:

$$\tau_T = \left[\frac{\pi}{4}(\lambda\rho C)(T_m - T_0)^2\right]/\dot{q}_0^2.$$

where $\dot{q}_o$ is the heat flux to the surface. We can define a dimensionless time coordinate as:

$$t = \tau/\tau_T - 1.$$

Figure 3 shows how coordinate z changes with t for the isotherm $\theta_b = (T_\delta - T_0)/(T_m - T_0) = 0.1$ for various values of m. By changing the given value of θ_b = const, we obtain the temperature field inside the destructing body.

It is of interest to correlate the process of development of the temperature field with the change of the dimensionless carried-

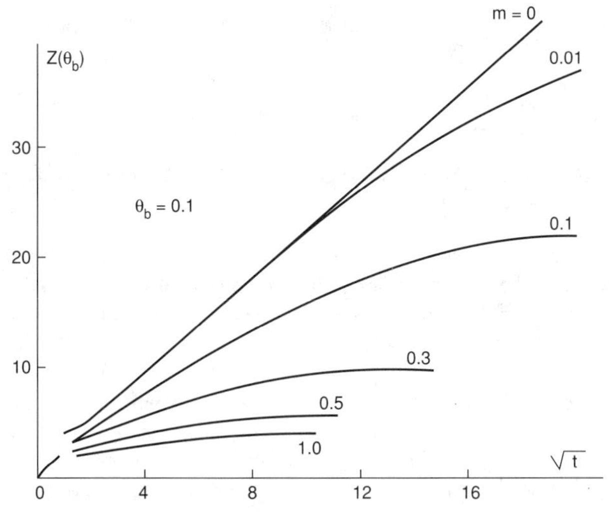

Figure 3. Variation of z with $\sqrt{t}$ for a given dimensionless temperature isotherm θ.

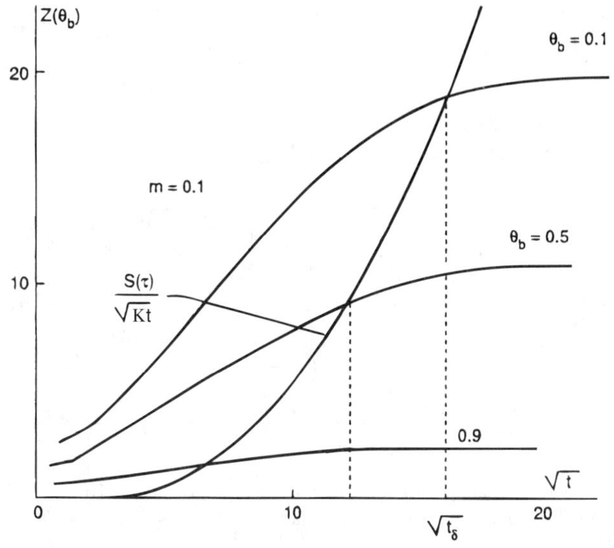

Figure 4. Comparison isotherms and melt layer thickness.

away layer thickness $S(\tau)/\sqrt{\kappa\tau_T}$. **Figure 4** gives the coordinates of three isotherms for a fixed value of the parameter m (= 0.1) and also gives dimensionless thickness of the carried away layer, all as a function of dimensional time t. We notice that with an accuracy of 10–15% the coordinate of the corresponding isotherm approaches its set (stationary) state at the moment of its intersection with the curve $S(\tau)/\sqrt{\kappa\tau_T}$. We may choose the onset time t_δ of the "quasi-stationary" value of depth of the heated layer to correspond to $\delta_T = z\ (\theta_b = 0.1)$.

The stationary temperature profile is an exponential function of the coordinate z and the parameter m:

$$\frac{T - T_0}{T_m - T_0} = \theta(z, t \to \infty) \to \bar{\theta}(z) = \exp\left[-\frac{\sqrt{\pi}mz}{2(m+1)}\right].$$

Under these quasi-stationary conditions, there is a constant value of ablation rate $\bar{\upsilon}_\infty$:

$$\frac{dS}{d\tau}(t \to \infty) \to \bar{v}_\infty = \frac{m}{m+1}\frac{\dot{q}_0}{\rho c(T_m - T_0)}$$

$$= \frac{\dot{q}_0}{\rho c(T_m - T_0) + \rho\Delta Q_m}.$$

The stationary temperature profile makes possible the establishment of a simple relationship for the amount of heat required for heating the subsurface layers of the destructing body, this amount of heat playing an important role in the energy balance on the surface of destruction:

$$\dot{q}_\lambda = -\lambda \left.\frac{\partial T}{\partial y}\right|_{y=S(\tau)} = \rho c \bar{v}_\infty (T_m - T_0).$$

Within the framework of this simple one-parameter model of destruction we can obtain the dependence on m of the times for the onset

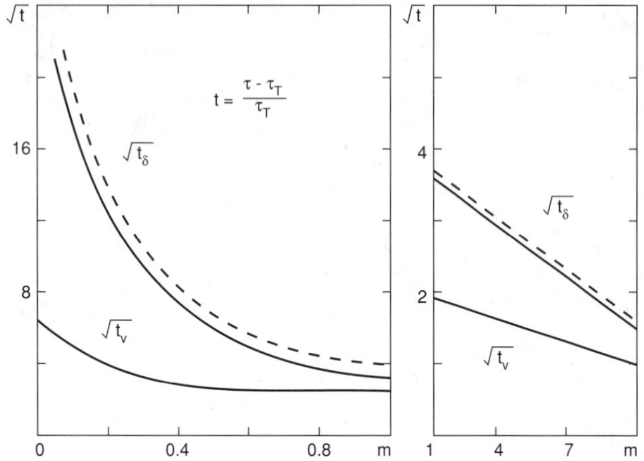

Figure 5. Variation of times for the onset of a quasi-stationary ablation rate (t_υ and quasi-stationary depth of heated layer (t_δ) with m.

of quasi-stationary values of ablation velocity t_υ and of the heated layer depth, t_υ and t_δ respectively (**Figure 5**). Here, we define a quasi-stationary value of ablation velocity ($dS/d\tau$) as one that differs from the stationary one $\bar{v}_\infty$ by no more than 10%. The coordinate of the isotherm $\theta_b = 0.1$ is defined as the depth of heated layer.

In contrast with crystalline substances which have a specific melting temperature T_m and a heat of phase transition ΔQ_m, amorphous substances change from a solid to a liquid state gradually without practically and additional heat absorption. They have a high melt viscosity which depends exponentially on temperature:

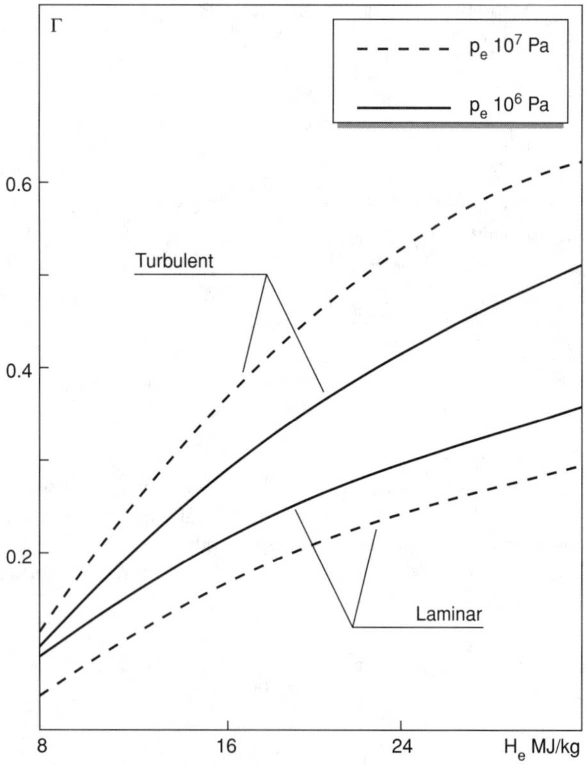

Figure 6.

$$\eta(T) = \exp\left(\frac{a}{T} + b\right).$$

Therefore, not only is the temperature of transition from a solid to a liquid state unknown in advance but also the temperature of the outer surface. The fraction Γ of the heat load which goes into sublimation (gasification) of the solid material depends on the thermal characteristics of the body over which the hot gas flows. The resultant characteristics of ablation of amorphous materials is the sum of the action of two opposite processes: the higher the density or pressure of the incident flow the faster the melt film is carried away, but also the heat flux also grows which brings about an increase in the outer surface temperature and in the evaporation rate.

A careful analysis of the equation of the melt film motion over the surface of amorphous bodies allows us to reveal an interesting rule. It turns out that the ratio of the film mass loss to the evaporated substance flow rate depends mainly on the parameter $p_e/(\alpha/c_p)_0^2$, where p_e is the stagnation pressure, α the heat transfer coefficient and c_p the gas specific heat capacity. The body dimensions, the enthalpy of the stagnated flow and other determining quantities, together affect the ablation characteristics of amorphous substance of the quartz glass type by not more than ±30%.

In a wide range of incident flow pressures (from 1 kPa to 1 MPa) almost identical characteristics of heat ablation efficiency are realized for a laminar flow in the boundary layer (**Figure 6**), since $(\alpha/c_p)_0^L \cong \sqrt{p_e}$. In a turbulent boundary layer in which $(\alpha/c_p)_0^T \cong p_e^{0.8}$, a noticeable redistribution of ablation in the direction of sublimation will occur, and the share of melt on the surface will decrease the more substantially the higher p_e (**Figure 6**). (see also **Ablation**).

Yu.V. Polezhaev

MELTING HEAT (See Melting)

MEMBRANE POLARISATION (see Membrane processes)

MEMBRANE PROCESSES

***Following from:** Diffusion; Liquid-solid separation; Mass transfer*

Membrane processes cover a group of *separation processes* in which the characteristics of a membrane (porosity, selectivity, electric charge) are used to separate the components of a solution or a suspension. In these processes the feed stream is separated into two: the fraction that permeates through the membrane, called the *permeate*, and the fraction containing the components that have not been transported through the membrane, usually called the *retentate*. The size of the components to be separated and the nature and magnitude of the driving force provide criteria for a classification of the membrane separation processes, as shown in the table. It should be noted that the boundaries between some of the processes, such as *reverse osmosis* and *ultrafiltration*, are arbitrary.

Of the processes listed in **Table 1**, reverse osmosis and ultafiltration are the most widely used industrially. Abundant examples of their application can be found in water **Desalinisation**, the dairy industries and the separation of organic solutes from aqueous solutions. Membrane processes do not require heating, which makes the process suitable for the treatment of thermos-labile products. In addition the relatively low capital and operating costs involved make membrane processes an appealing alternative to more conventional separation processes, particularly when dealing with dilute solutions. In the next sections the principles and characteristics of the processes listed in Table 1 will be described briefly.

Osmosis

Osmosis is the process of transfer of the solvent of a solution across a membrane separating two liquid solutions of different concentrations. The transfer takes place from the phase in which the chemical potential solvent is higher. A classical example is the transfer of water to a sugar solution. **Figure 1(a)** shows two compartments separated by a water permeable membrane. One contains the solution, while the other holds pure water. Initially the level of the solution is placed at the same level as that of the water. Since the chemical potential of pure water is higher than that of water in the solution, water transfers across the membrane to the solution, increasing the level on the solution side and thus creating a hydrostatic head (**Figure 1b**). When the process is at equilibrium the difference in the hydrostatic pressure between the two phases is that required to make the chemical potential of water in both phases the same. In the special case of this example, when the pure solvent is on one side of the membrane this pressure difference is the *osmotic pressure,* Π. If solutions are at both sides of the membrane, the pressure difference gives the osmotic pressure difference $\Delta\,\Pi$. For very dilute solutions the osmotic pressure is given by

Table 1. Classification of membrane separation processes [Bowen (1991)]

Name of process	Driving force	Separation size range
Microfiltration	Pressure gradient	10 − 0.1 μm
Ultrafiltration	Pressure gradient	< 0.1 μm − 5 nm
Reverse osmosis (hyperfiltration)	Pressure gradient	< 5 nm
Electrodialysis	Electric field gradient	< 5 nm
Dialysis	Concentration gradient	< 5 nm

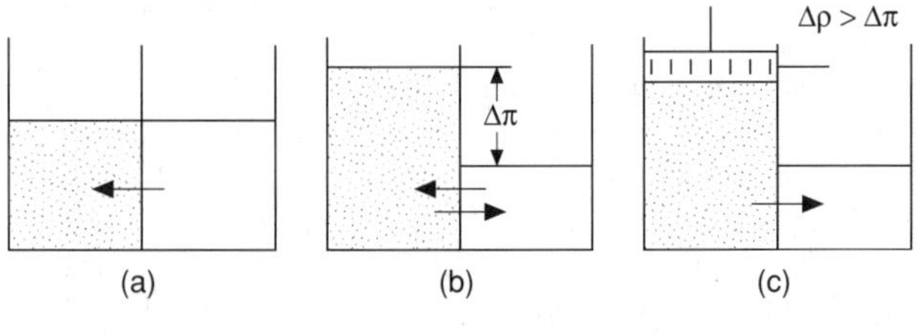

Figure 1. Osmosis and reverse osmosis, (a) osmotic flow, (b) osmotic equilibrium and (c) reverse osmosis.

$$\Pi = \beta c_i RT \tag{1}$$

where β is a constant that takes into account the dissociation of the solute, c_i is the molar concentration of solute i, R the gas constant and T the absolute temperature. The constant β is in general equal to the number of ions produced by the dissolved solute; thus for electrolytes that dissociate fully such as Nalco and CAC_2 it is equal to 2 and 3 respectively, while $\beta = 1$ for molecules that do not dissociate. When the solute has a large molecular mass the osmotic pressure of the solvent ceases to depend linearly on the concentration of the solute and a polynomial approximation must be used. For concentrated solutions as well as non-ideal solutions the osmotic pressure must be determined experimentally or estimated using activities instead of concentrations.

Reverse Osmosis

The term reverse osmosis is applied to the separation of water from solutions by transfer though a semipermeable membrane that ideally should only be permeable to water. If as shown in **Figure 1(c)** a mechanical pressure is applied to the compartment where the level is higher, the chemical potential of the solvent on the solution side will increase, forcing the transfer of the solvent through the membrane into the pure solvent phase. This pressure difference between the phases, ΔP, creates the driving force for the transfer and is the principle on which the reverse osmosis and ultrafiltration processes are based. In order to achieve the transfer, ΔP must be greater than the difference between the osmotic pressures $\Delta\Pi$. The rate of solvent permeation across the membrane is measured as its linear velocity normal to the membrane, J_1. Ideally the membrane should be only permeable to the solvent. However this is very difficult to achieve and membranes have some permeability to the solute. The rate of permeation of the solvent is given by:

$$J_1 = K_1(|\Delta P| - |\Delta\Pi|) \tag{2}$$

where K_1 is the *permeability coefficient* of the membrane and ΔP is the transmembrane pressure,. The permeability coefficient depends on the membrane resistance, the resistance to transfer caused by deposits on the membrane and the diffusion resistance in the boundary layer at the retentate side of the membrane. Depending on the thickness of the solid deposits and the hydrodynamic conditions of the retentate the two latter resistances may be negligible. In fact membrane equipment are operated under conditions that tend to minimise these two resistances in order to increase J.

The difference in solute concentration between the retentate and the permeate (in practice membranes have some permeability to the solute) creates a driving force for solute transfer across the membrane. The flux of solute, J_2, is given by:

$$J_2 = K_2(c_R - c_P) \tag{3}$$

where K_2 is a constant that includes the characteristics of the membrane, and c_R and c_p are the concentration of the solute in the retentate and permeate respectively.

Equations 2 and 3 indicate that the driving forces for solvent and solute transport are independent. Therefore by increasing transmembrane pressure it is possible to increase the rate of solvent transfer without increasing the concentration of solute in the retentate. This in fact leads to an increase in the solute partition coefficient between retentate and permeate. However the increase in solute concentration in the retentate, and in particular at the membrane wall, may lead to an increase in osmotic pressure important enough to reduce J_1. This effect is called *membrane polarisation* and will be discussed in connection with ultrafiltration.

Ultrafiltration

Ultrafiltration is also a membrane separation process driven by pressure. Its main difference from reverse osmosis is the type of membrane used. While reverse osmosis membranes are non-porous and therefore permeable only to very small molecules, ultrafiltration membranes are micro-porous and permeable to small solute species in addition to the solvent. Ultrafiltration is therefore used to keep large solute species in the retentate. The effect of osmotic pressure is therefore less important in this process so that the transmembrane pressures required are not as high as for reverse osmosis. The solvent flux obtained with ultrafiltration is also given by equation 2, but the membrane permeability K_1 now depends on the characteristics of the ultrafiltration membrane and the diffusional resistances specific to this process. As a result of solvent transfer the retentate becomes more concentrated and a concentration gradient of solute builds up in the direction of transfer with the higher concentration at the membrane wall. This phenomenon, called membrane polarisation, creates a concentration driving force for solute diffusion in the direction opposite to solvent transfer. At equilibrium the diffusional flux of solute back to the retentate phase equals the convective flux towards the membrane:

$$J(c - c_p) = -D(dc/dy) \qquad \textbf{(4)}$$

where J is the velocity of the permeate phase through the membrane, c_p is the concentration of solute in the permeate, D is the diffusivity of the solute in the retentate and y the distance in the direction normal to the membrane. As the separation proceeds the concentration gradient becomes steeper due to solute accumulation next to the membrane. As the solutes retained in ultrafiltration are larger than in reverse osmosis, their diffusivities are lower and thus accumulation near the membrane wall is more likely in ultrafiltration than in reverse osmosis.

Equation 2 allows comparison of the response of reverse osmosis and ultrafiltration to transmembrane pressure. As ΔP is increased, solute accumulation at the membrane wall on the retentate side is higher in ultrafiltration than in reverse osmosis due to the lower diffusivity of the solute. This accumulation of solute leads to membrane polarisation as well as solute precipitation on the membrane (gelification), which in turn decrease K_1 and increase $\Delta\Pi$. The process reaches a region in which the permeate flow rate becomes independent of pressure.

Microfiltration

The separation of particles of micron and sub-micron levels can effectively be performed using membrane filters. The suspended particles for which the process is industrially used include colloids, micro-organisms and emulsion droplets. There are two different configurations for the microfiltration operation: (1) dead-end microfiltration, and (2) cross-flow microfiltration. In the first, the membrane plane is normal to the feed flux, while in the second, it is tangential. The advantages and disadvantages of the type of configuration depend on the characteristics of the feed. Highly concentrated suspensions are not suitable for dead-end treatment since the separated particles rapidly accumulate on the membrane, increasing the resistance to filtration by forming a cake and/or clogging the membrane. In cross-flow filtration, the flow is parallel to the membrane and the drag forces close to the membrane wall reduce the amounts of particulate material deposited on the membrane. Therefore the first is used for dilute suspensions while the latter can deal with concentrated ones, such as slurries.

Microfitration is a pressure driven process and as such the flux of filtrate is given by **Equation 2**. In this case $\Delta\Pi$ can be neglected since it is not relevant to this process.

Dialysis

Dialysis is a process in which the solute, usually an electrolyte, transfers across the membrane driven by the difference in concentration between the two sides of the membrane. Two conditions have to be fulfilled for the process to be effective: (1) the concentration on the permeate side has to be kept low so that the driving force remains as high as possible, and (2) the osmotic pressure must be low, and remain low during the process, so that a counterflux of solvent does not result in feed dilution. Since the combination of these two conditions is not likely to occur, the use of selective membranes has become general practise so that the partition coefficient of the solute between the two phases can be substantially increased by using the *Donnen effect.* This is achieved by adding to the feed a salt with one of its ions in common with the solute and the other one rejected by the membrane. If for example NaCl is the solute to be separated, the addition of NaX (where X is the ion rejected by the membrane) will lead to the following expression for the NaCl partition coefficient at equilibrium:

$$([NaCl]_I/[NaCl]_{II})^2 = 1 + [NaX]_I/[NaCl]_{II} \qquad \textbf{(5)}$$

where the square brackets indicate concentrations. This equation show that the larger the concentration of NaX, the greater the separation enhancement. (See also **Dialysis**)

Electrodialysis

Electrodialysis is a process in which ion-selective membranes are used together with an electric field normal to the membrane phases. The basic principles of electrodialysis are better explained with reference to **Figure 2**. An electrodialysis stack consists of parallel compartments separated alternately by cation-exchange and anion-exchange membranes. The feed is introduced into each compartment, so that aided by the electric field and selected by the membranes, the cations and the anions are transferred in opposite directions to the neighbouring compartments. In this way a demineralised solution leaves compartments 2 and 4 and the concentrated one compartments 1, 3 and 5.

Depending on the type of membrane used, ions can be selected according to their valence. Therefore this process can be used for fractionating ions of different valence. The main application of

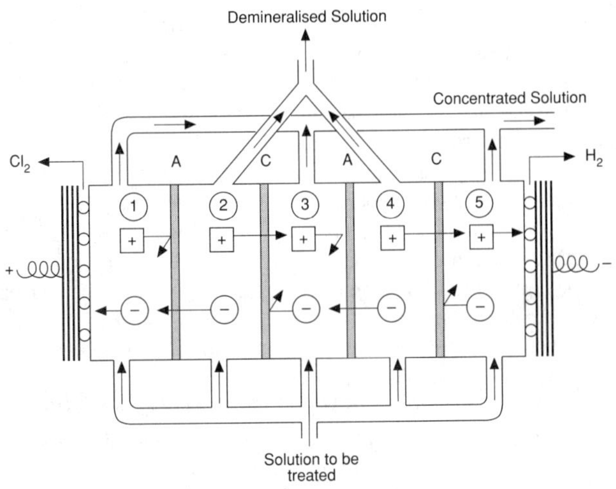

Figure 2. Schematic flow-diagram for an electrodialysis stack.

electrodialysis is in demineralisation in general and in the desalinisation of brackish water. (See also **Electrodialysis**)

Pervaporation

Pervaporation is the combination of the selective separation and transfer of a component across the membrane and its evaporation on the permeate side. In order to achieve evaporation the pressure on the permeate side must be such that the partial pressure of this component is lower than its saturation vapour pressure. Therefore although the driving force for transfer is the difference in activity of the transferred species, this is the result of applying a vacuum on the permeate side.

Pervaporation is more expensive than other membrane processes due to the required supply of heat to produce the evaporation. It is, therefore, used in the separation of components from mixtures that are difficult to treat, such as azeotopic solutions and isomer mixtures, for which conventional processes would be more costly.

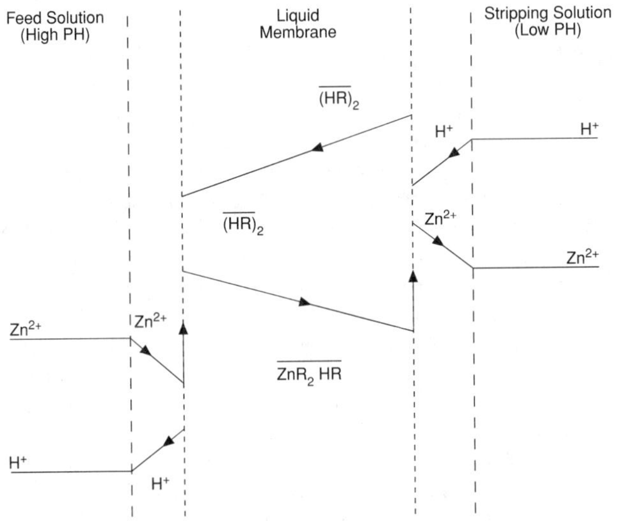

Figure 3. Concentration profiles in the extraction of zinc with a liquid membrane.

Liquid Membranes

A liquid membrane is a liquid phase that separates two fluid phases of different composition. In most applications the liquid membrane is an organic phase placed between two aqueous phases. The principles involved in liquid membrane separations are those of solvent extraction, rather than membrane separation. The solute is first selectively extracted by the liquid membrane, it then transfers across the liquid film driven by its concentration gradient and when it reaches the other side of the liquid membrane it is stripped by the third phase.

There are two types of mechanisms of transfer to and from the membrane: (1) Physical transfer, and (2) carrier mediated transfer. The first is based on the solubility of the solute in the membrane, while the second requires the presence of a selective reactant in the organic phase called the carrier. In this case the process is based on the reversibility of the chemical reaction. The extraction of Zn^{2+} from an aqueous feed with a carrier RH dissolved in the organic membrane, will be used as an example of carrier mediated extraction. The overall reaction between the cation and the carrier can be represented by:

$$Zn^{2+} + 3RH \Leftrightarrow ZnR_2.HR + 2H^+ \quad \textbf{(6)}$$

where bars indicate species in the organic phase. Reaction 6 is reversible. At pH 3 the forward reaction takes place while at low pH values the reverse reaction predominates. A suitable liquid membrane system for the separation of a membrane containing RH should have a high pH value in the feed phase and a low one in the stripping phase. **Figure 3** shows a schematic drawing of the concentration gradients in the feed, the membrane and the stripping phase with the corresponding concentration profiles.

The main advantage of liquid membranes over conventional solvent extraction is that by contacting the three phases simultaneously, the solute partition coefficient between the two aqueous

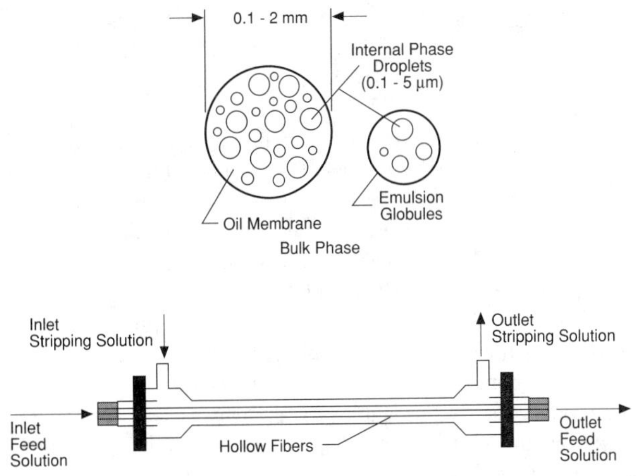

Figure 4. Liquid membranes: (a) emulsion liquid membrane, (b) supported liquid membrane.

phases may be several orders of magnitude greater than the one that can be obtained with conventional solvent extraction.

There are two ways in which a liquid membrane can be formed leading to two configurations: (1) emulsion liquid membranes and (2) supported liquid membranes. These are illustrated in **Figures 4a** and **4b**. The emulsion liquid membrane is obtained by adding a surfactant into the organic phase in order to stabilise a water in oil emulsion. The emulsion is then dispersed in the feed so that the organic phase, which is the continuous phase of the dispersed globules, becomes the membrane Supported liquid membranes are obtained by impregnation of porous fibres of hollow fibre modules.

References

Bowen, W. R. (1991) Membrane separation processes, in *Chemical Engineering,* Vol. 2, 4th edition, Coulson, J. M. and Richardson, J. F., Pergamon Press, Oxford.

Rautenbach, R. and Albrecht, R. (1989) *Membrane Processes,* John Wiley and Sons, Chichester.

Way, J. D., Noble, R. D., Flynn, T. M., Sloan, E. D. (1982) Liquid membrane transport: a survey, J. *Memb. Sci.,* 12, 239–259.

***Leading to:** Desalination; Dialysis electrodialysis*

E.S. Perez de Ortiz

MEMBRANES, ION EXCHANGE (see Dialysis; Ion exchange; Membane processes)

MEMBRANE TYPE FILTERS (see Gas-solids separation, overview)

MENDELEEV-CLAPEYRON EQUATION (see Nozzles)

MERCURY

Mercury—(Planet *Mercury*), Hg (*hydrargyrum,* liquid silver); atomic weight 200.59; atomic number 80; melting point −38.842°C; boiling point 356.58°C; specific gravity 13.546 (20°C); valence 1 or 2.

Known to ancient Chinese and Hindus, found in Egyptian tombs of 1500 B.C., mercury is the only common metal liquid at ordinary temperatures. It only rarely occurs freely in nature. The chief ore is *cinnabar* (HgS). Spain and Italy produce about 50% of the world's supply of the metal. The commercial unit for handling mercury is the "flask," which weighs 76 lb. The metal is obtained by heating cinnabar in a current of air and by condensing the vapor.

Memory is a heavy, silvery-white metal; a rather poor conductor of heat, as compared with other metals, and a fair conductor of electricity. It easily forms alloys with many metals, such as gold, silver, and tin, which are called *amalgams.* Its ease in amalgamating with gold is made use of in the recovery of gold from its ores. The metal is widely used in laboratory work for making thermometers, barometers, diffusion pumps, and many other instruments. It is used in making mercury-vapor lamps and advertising signs, etc. and is used in mercury switches and other electrical apparatus. Other uses are in making pesticides, mercury cells for caustic soda and chlorine production, dental preparations, antifouling paint, batteries, and catalysts.

The most important salts are mercuric chloride, $HgCl_2$ (corrosive sublimate—a violent poison), mercurous chloride, Hg_2Cl_2 (calomel, occasionally still used in medicine), mercury fulminate, $Hg(ONC)_2$, a detonator widely used in explosives), and mercuric sulfide HgS, vermillion, a high-grade paint pigment). Organic mercury compounds are important. It has been found that an electrical discharge causes mercury vapor to combine with neon, argon, krypton, and xenon. These products, held together with van der Waals' forces, correspond to HgNe, HgAr, HgKr, and HgXe.

Mercury is a virulent poison and is readily absorbed through the respiratory tract, the gastrointestinal tract, or through unbroken skin. It acts as a cumulative poison since only small amounts of the element can be eliminated at a time by the human organism. Since mercury is a very volatile element, dangerous levels are readily attained in air. Air saturated with mercury vapor at 20°C contains a concentration that exceeds the toxic limit many times. The danger increases at higher temperatures. *It is therefore important that mercury be handled with care.* Containers of mercury should be securely covered and spillage should be avoided. If it is necessary to heat mercury or mercury compounds, it should be done in a well-ventilated hood. Methyl mercury is a dangerous pollutant and is now widely found in water and streams. The U.S. National Bureau of Standards has redetermined the triple point of mercury and found it to be −38.84168°C.

Handbook of Chemistry and Physics, CRC Press

MERKEL'S EQUATION (see Cooling towers)

MESOSPHERE (see Atmosphere)

METALS

Metals are often described as having a high lustre, being malleable and ductile, and having good electrical and thermal conductivity. Many metals do have all these properties, but many non-metallic substances can also exhibit one or more of them. The most characteristic property of a metal is its high electrical and thermal conductivity. Moreover, these two properties are proportional to one another at constant temperature (*Wiedemann-Franz-Lorenz law*), (Smallman, 1970).

Over eighty elements in the periodic table are metals, characterised by having electrons in the outer shell which can be easily removed. This contributes the component of their electrical and thermal conductivity; the second major component of heat transfer within metals is the energy carried by lattice vibrations, called phonons, Klemens (1969) and Edwards (1990).

The atoms in a solid metal are packed together in a highly regular manner. The arrangement follows a specific pattern, and the structure of the metal is characterised by the unit of the pattern, known as a cell. In the majority of metals the atoms are packed so that they occupy the minimum volume; these are the body-centered cubic (bcc), face-centered cubic (fcc) and close-packed hexagonal (cph) arrangements. Other packing arrangements are orthorhombic, rhombohedral, tetragonal, etc.

Emission and absorption of heat by radiation is governed by temperature and surface finish. The latter can significantly affect radiation to and from metals due to surface oxidation, (Roberts and Miller, 1961).

Selected thermal data for commonly used metallic elements are presented in **Table 1**, [Lide (1991/92) and Smithells (1992)]. The defining equations for thermal conductivity, λ, and coefficient of linear expansion, α, are as follows:

$$\dot{Q} = \frac{\lambda(T_2 - T_1)At}{\delta} \quad (1)$$

where $\dot{Q}$ is the rate of heat flowing a distance δ through a body with cross-sectional area A in time t.

$$l_{T2} = l_{T1}[1 + \alpha(T_2 - T_1)] \quad (2)$$

where l_{T1} and l_{T2} are lengths at temperatures T_1 and T_2 respectively.

References

Edwards, P. P. (1990) Chapter 7, *Advances in Physical Metallurgy* J. A. Charles and G. C. Smith, eds., The Institute of Metals, London.

Klemens, P. G. (1969) Chapter 1, *Thermal Conductivity.* R. P. Tye, ed., Academic Press, London and New York.

Lide, D. R., ed. (1991/92) *Handbook of Chemistry and Physics,* CRC Press, Boca Raton, FL.

Roberts, J. K. and Miller, A. R. (1961) *Heat and Thermodynamics,* Blackie, London.

Smallman, R. E. (1970) *Modern Physical Metallurgy,* Butterworth, London.

Smithells, C. J. (1992) *Smithells Reference Handbook.* E. A. Branches and G. B. Brook, eds., Butterworth-Heinemann.

Table 1. Thermal data for commonly used metallic elements

	Thermal Conductivity, λ, W/mK	Specific Heat Capacity, c, J/kgK	Coeff. of Linear Exp. K^{-1} ($\times10^6$)	Density kg/m³	Melting Point, T_m °C
Aluminium	237	897	23.1	2700	660
Chromium	94	449	4.9	7150	1863
Cobalt	100	421	13.0	8860	1495
Copper	401	385	16.5	8960	1085
Gold	318	129	14.2	19300	1064
Iron	80	449	11.8	7870	1538
Lead	35	129	28.9	11300	328
Molybdenum	138	251	4.8	10200	2623
Nickel	91	444	13.4	8900	1455
Silver	429	235	18.9	10500	962
Tin	67	228	22.0	7280	232
Titanium	22	523	8.6	4500	1670

Handbook of Chemistry and Physics, CRC Press.

Leading to: *Steels*

R.L. Keightley and K.W. Tupholme

METHANE

Methane (CH_4) is a colourless, odourless and tasteless gas that is lighter than air. It is the main constituent of natural and coal gas and is a by-product in the decay of organic matter. It forms an explosive and flammable mixture with air at low concentrations. Its main uses are in the petrochemical industry as a raw material for the production of **Methanol**, ammonia, acetylene, chlorinated methanes and carbon black. It is also used as a fuel in the form of **Natural Gas**. (See Table 1 below)

Table 1. Methane: Values of thermophysical properties of the saturated liquid and vapour

T_{sat}, K	111.7	120	130	140	150	160	170	180	190
P_{sat}, kPa	102	192	368	642	1041	1593	2328	3283	4515
ρ_l, kg/m³	422.5	410.1	394.3	377.2	358.1	336.1	309.6	275.2	200.4
ρ_g, kg/m³	1.83	3.27	6.00	10.18	16.37	25.46	39.10	61.43	125.2
h_l, kJ/kg	−909.5	−880.3	−844.2	−806.8	−767.3	−725.0	−678.2	−623.6	−530.7
h_g, kJ/kg	−399.6	−386.6	−373.2	−362.7	−356.0	−354.2	−359.7	−377.9	−449.7
Δh_{gl}, kJ/kg	510.0	493.7	471.0	444.1	411.4	370.7	318.4	245.7	81.1
$c_{p,l}$, kJ/(kg K)	3.48	3.54	3.64	3.80	4.06	4.47	5.23	7.22	92.3
$c_{p,g}$, kJ/(kg K)	2.23	2.31	2.44	2.64	2.94	3.44	4.43	7.19	111.5
η_l, μNs/m²	115.9	97.1	79.9	66.4	55.4	46.0	37.6	29.7	18.7
η_g, μNs/m²	4.48	4.82	5.27	5.74	6.28	6.90	7.68	8.84	12.4
λ_l, mW/(K m)	185.9	172.7	157.6	143.2	129.0	114.7	100.4	87.6	98.8
λ_g, mW/(K m)	13.2	14.5	16.2	18.1	20.2	22.9	26.6	34.1	88.6
Pr_l	2.17	1.99	1.84	1.76	1.74	1.79	1.96	2.44	17.5
Pr_g	0.75	0.77	0.79	0.84	0.91	1.04	1.28	1.86	15.6
σ, mN/m	13.2	11.5	9.28	7.22	5.31	3.58	2.06	0.81	0.01
$\beta_{e,l}$, kK^{-1}	3.47	3.75	4.23	4.93	6.02	7.84	11.3	21.0	487

Thermodynamic properties from Angus, S., deReuck, K. M. and Armstrong, B. (1978); transport properties from Friend D. G., Ely J. F. and Ingham H. (1989) and surface tension from Beaton, C. F. and Hewitt, G. F. (1989).

Molecular weight: 16.043
Melting point: 90.7K
Normal boiling point: 111.63K
Normal vapour density: 0.72 kg/m^3
(@273.15K; 1.0135MPa)

Critical temperature: 190.55K
Critical pressure: 4.595MPa
Critical density: 162.2 kg/m^3

References

Angus, S., deReuck, K. M. and Armstrong, B. (1978) *International Thermodynamic Tables of the Fluid State-5, Methane.* Pergamon Press, Oxford.

Beaton, C. F. and Hewitt, G. F. (1989) *Physical Property Data for the Design Engineer.* Hemisphere Publishing Composition, New York.

Friend, D. G., Ely, J. F. and Inghams, H. (1989) *Phys. Chem. Ref. Data,* 18, 583.

V. Vesovic

METHANOL

Methanol (CH_3OH) is a colourless, polar liquid that is miscible with water. It used to be known as *wood alcohol* due to the fact that it was first produced by wood distillation. Industrially, methanol is produced by the catalytic synthesis of a pressurized mixture of **Hydrogen, Carbon Monoxide** and **Carbon Dioxide** or by steam reforming of **Natural Gas**. It is flammable and at low to medium concentration in air it forms an explosive mixture. It is also toxic and if ingested in large quantities can lead to blindness. It is widely used as a fuel, as a solvent for celluloses and dyes, as a gasoline additive and as raw material in organic synthesis, especially for the production of formaldehyde and acetic acid. (See Table 1 below)

Molecular weight: 32.04
Melting point: 175.6K
Normal boiling point: 337.6K

Critical temperature: 512.6K
Critical pressure: 8.1035MPa
Critical density: 275.5 kg/m^3

Table 1. Methanol: Values of thermophysical properties of the saturated liquid and vapour

T_{sat}, K	337.7	370	390	410	430	450	470	490	510
P_{sat}, kPa	102	320	586	1006	1637	2543	3794	5473	7750
ρ_l, kg/m^3	748.2	714.4	691.1	665.2	635.5	600.4	557.1	497.4	374.5
ρ_g, kg/m^3	1.22	3.67	6.62	11.40	18.99	30.83	48.55	80.36	165.7
h_l, kJ/kg	105.7	202.0	267.1	337.1	412.9	495.8	588.0	695.9	853.5
h_g, kJ/kg	1207	1232	1243	1248	1244	1235	1227	1194	1079
Δh_{gl}, kJ/kg	1101	1030	975.7	910.4	831.5	739.2	638.6	498.1	225.5
$c_{p,l}$,kJ/(kg K)	2.83	3.14	3.36	3.63	3.96	4.41	5.11	6.90	34.6
$c_{p,g}$, kJ/(kg K)	4.44	5.07	5.59	6.36	7.43	8.27	8.01	11.88	59.6
η_l, μNs/m^2	—	221	176.1	140.7	112.8	91.2	73.9	60.2	42.9
η_g,μNs/m^2	11.1	12.3	13.0	13.9	14.8	15.8	17.2	19.7	25.5
λ_l,mW/(K m)	191.5	182.2	178.9	171.4	165.0	159.5	153.9	148.2	142.5
λ_g,mW/(K m)	18.3	22.8	25.7	29.1	33.1	38.5	45.7	57.9	95.1
Pr_l	—	3.81	3.31	2.98	2.71	2.52	2.45	2.80	10.4
Pr_g	2.69	2.73	2.83	3.04	3.32	3.39	3.01	4.04	15.9
σ, mN/m	18.7	16.0	13.9	11.8	9.65	7.28	4.88	2.48	0.27
$\beta_{e,l}$, kK^{-1}	1.34	1.58	1.82	2.17	2.71	3.60	5.35	10.85	119.03

Thermodynamic properties from deReuck, K. M. and Craven, R. J. B. (1993); transport properties and surface tension from Beaton C. F. and Hewitt G. F. (1989).

References

Beaton C. F. and Hewitt G. F. (1989) *Physical Property Data for the Design Engineer,* Hemisphere Publishing Corporation, New York.

deReuck K. M. and Craven R. J. B. (1993) *International Thermodyanic Tables of the Fluid State −12, Methanol,* Blackwell Scientific Publications, Oxford.

V. Vesovic

METHOD OF CHARACTERISTICS (see Compressible flow)

METHYLAMINE

Methylamine (CH_3NH_2) is a colourless, ammonia-like smelling gas which is soluble in water. It is obtained by a catalytic reaction between **Methanol** and ammonia at high temperature. It is flammable, very toxic and at low to medium concentration it forms an explosive mixture. Its main use is in the paint and polymer industries, although it is also used in the manufacture of insecticides and fungicides.

	Critical temperature: 430 K
Molecular weight: 31.06	Critical pressure: 7.46 MPa
Melting point: 179.7 K	Critical density: 222 kg/m^3
Boiling point: 266.8 K	Normal vapor density: 1.39 kg/m^3 (@ 0°C, 101.3 kPa)

V. Vesovic

METHYLCHLORIDE

Methylchloride (CH_3Cl) is a colourless, sweet smelling gas that is slightly soluble in water. It is obtained by either the chlorination of methane or the reaction of hydrochloric acid with methanol. It forms a flammable mixture with air at low concentrations and is highly toxic. Its usage as a raw material for the production of CFCs has dramatically decreased in recent years, but it is still used in the production of silicones and synthetic rubber.

	Critical temperature: 416.3 K
Molecular weight: 50.49	Critical pressure: 6.68 MPa
Melting point: 175.45 K	Critical density: 363 kg/m^3
Boiling point: 248.9 K	Normal vapor density: 2.25 kg/m^3 (@ 0°C, 101.3 kPa)

V. Vesovic

METRE (see Speed of light)

METZNER-OTTO CONSTANTS FOR IMPELLERS (see Agitated vessel heat transfer)

MICELLAR CATALYSIS (see Catalysis)

MICELLES (see Foam fractionation)

MICHAELIS-MENTEN RELATIONSHIP (see Biochemical engineering)

MICHELSON INTERFEROMETRY (see Interferometry)

MICRO-COMPUTERS (see Computers)

MICROFILTRATION (see Membrane processes)

MICROFIN TUBES (see Augmentation of heat transfer, two phase)

MICROGRAVITY CONDITIONS

Until recent years, scientific experiments have always been carried out in laboratories on Earth and therefore experimental results were always obtained under gravitational conditions. In some cases, gravity does not affect the phenomenon or process under study, but in other cases it may be so important that it may mask other phenomena present in the system. Recently, new opportunities for research have developed with the space age, and experiments in reduced gravity conditions have become possible. There are three main benefits which may arise from carrying out experiments under microgravity. First, pure phenomena can be investigated without the influence of gravity. Second, while investigating some systems, new phenomena may arise which may be of importance but which have been masked under the effect of gravity. Third, new manufacturing processes may be developed which take advantage of the new environment. Examples of the last might be the production of pure crystals and of new metal alloys in space.

One of the difficulties encountered with this new type of experimental environment is that scientists are not used to the new concepts under which experiments and investigators work. Therefore, new experimental methods and techniques need to be devised and applied.

Microgravity Levels

Although it is normally assumed that in Space the gravitational level is zero, this is not strictly correct; hence the use of the terminology "*microgravity*" or "*reduced gravity*" instead of "*zero-gravity*".

Microgravity conditions may be simulated to a certain extent in Earth laboratories, but can be obtained, in decreasing order of time duration, in various platforms such as:

- Satellites
- Orbiting space stations
- Rockets
- Aircraft (parabolic flights)
- Drop Towers

Experiments may be carried out in any of the above but in each one different microgravity levels will be achieved (**Figure 1**) depending on the absolute rotation and angular acceleration of the platform, the forces acting on it and the non-uniformities of the external fields.

The choice of which platform to use depends fundamentally on the requirements of the experiments, in terms of the microgravity duration and level, and also on whether experimental conditions need to be modified during the microgravity period, in which case the investigator needs to be present. Another important factor is economics, as, doing an experiment in a space station is much more expensive than in a parabolic flight.

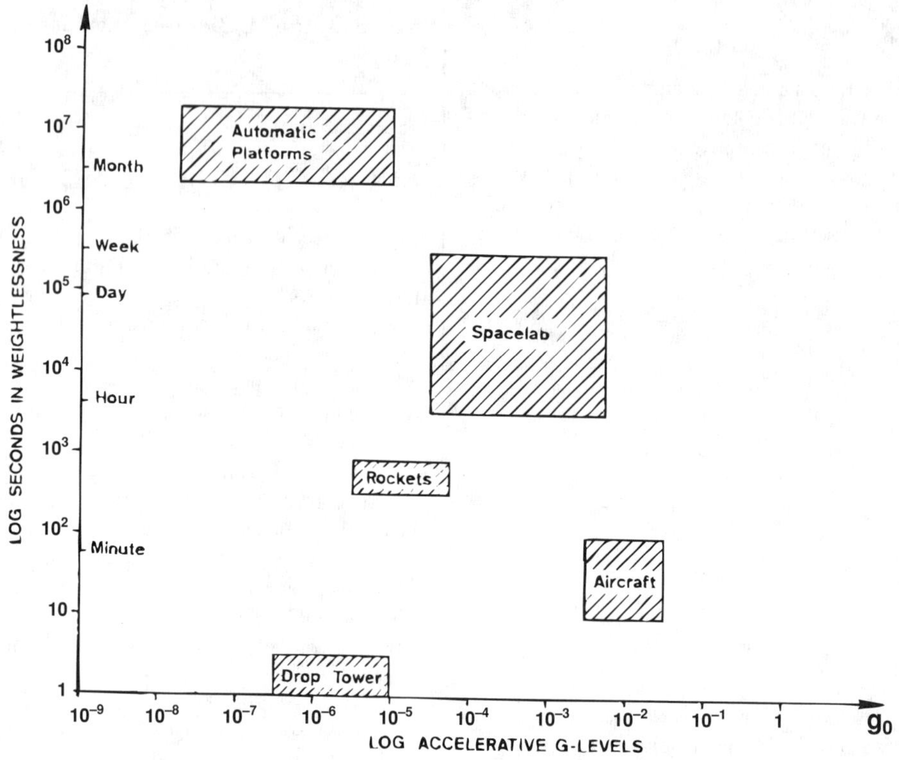

Figure 1. Nominal gravitational levels as a function of duration achievable with the main microgravity platforms.

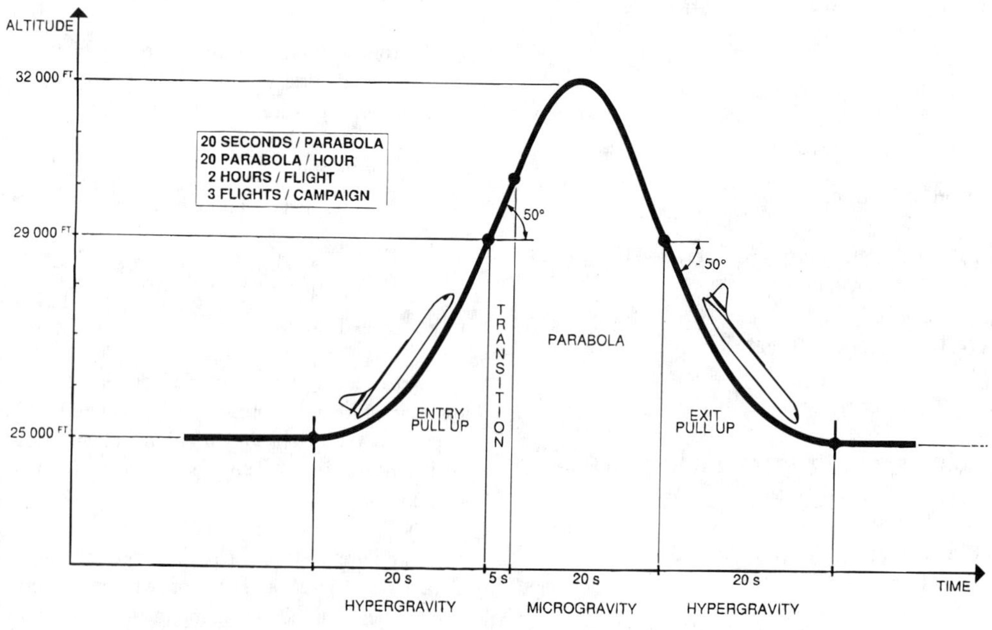

Figure 2. Parabolic flight trajectory (ESA/CNES Caravelle Aircraft).

AIRCRAFT	LOCATION	ESA CAMPAIGNS	DATES	NUMBER PARABOLAE	MATERIAL SCIENCES				LIFE SCIENCES			TOTAL
					Fluid Physics	Combustion Physics	Material Proc./Surf. Phys.	Technology	Human Physiology	Biology	Technology	
KC-135	HOUSTON	1st	Dec. 84	44	11	-	2	-	3	2	-	18
		2nd	Jun. 85	51	14	-	-	-	7	2	-	23
		3rd	Mar. 86	75	6	1	1	-	2	-	-	10
		4th	Apr. 87	75	-	7	-	1	2	-	-	10
		5th	Oct. 87	75	2	1	-	2	11	1	-	17
		6th	Aug. 88	75	6	1	1	2	4	2	-	16
CARAVELLE	BRETIGNY	Demo	Feb. 89	30	-	-	-	2	1	-	-	3
		7th	Mar. 89	92	2	1	-	2	4	1	-	10
		8th	Oct. 89	92	1	1	1	1	2	2	2	10
		9th	Apr. 90	90	2	1	-	-	2	-	5	10
		10th	Jul. 90	90	2	-	1	-	2	-	3	8
		11th	Oct. 90	87	2	-	-	1	2	-	3	8
		12th	Feb. 91	90	2	-	-	2	5	-	2	11
		13th	Jun. 91	93	4	-	2	1	2	1	-	10
		14th	Sep. 91	75	1	-	1	3	1	-	2	8
		15th	Mar. 92	90	5	1	-	1	3	-	-	10
		16th	Jan. 93	92	6	1	1	-	2	2	-	12
		17th	Nov. 93	93	2	1	-	-	5	3	-	11
Total				1409	68	16	10	18	60	16	17	205

Figure 3. ESA parabolic flight campaigns.

Parabolic Flights

One of the most economic ways of carrying out experiments under microgravity is in parabolic flights, where around 30 parabolae may be performed, per flight, and for each parabola 20–30 seconds of microgravity may achieved (**Figure 2**). An important advantage of this mode of experimentation is that the investigator flies with the experiment and may therefore alter conditions at different stages of the experiment.

Parabolic flight campaigns have been run, for example, by the European Space Agency (ESA), since December 1984. Up to November 1993, ESA run a total of 17 campaigns, where 205 experiments were performed during 1409 parabolae (**Figure 3**). This is approximately equivalent to 8 hours of microgravity which is, in turn, equivalent to approximately 5.3 orbits around the Earth.

During the ESA parabolic flights experiments (**Figure 3**) have been carried out in the fields of Material Sciences (Fluid Physics, Combustion Physics, Material Processing/Surface Physics and Technology) and Life Sciences (Human Physiology, Biology and Technology).

Example

An example of results from a Fluid Physics experiment performed under the microgravity conditions achieved in parabolic flights is shown in **Figure 4**. During mass transfer from the drop, containing acetylacetone, to the outer water phase, interfacial convection is observed, although not as intense (**Figure 5**) and not lasting for as long as under gravitational conditions [Mendes-Tatsis and Perez de Ortiz (1992)]. On Earth, two types of convection occur during the transfer of acetylacetone into water: gravitational convection (caused by density gradients) and Marangoni convection (caused by interfacial tension gradients). Under microgravity, interfacial tension gradients become the only reason for the convection observed and therefore pure Marangoni convection can be observed.

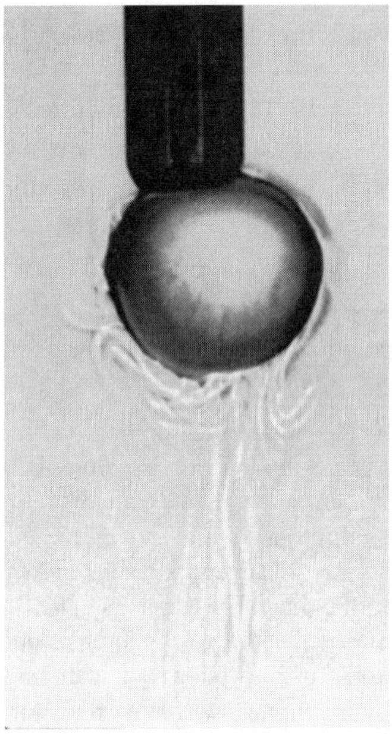

Figure 4. Transfer of acetylacetone from the drop into water (gravitational conditions, 7 seconds after formation of drop).

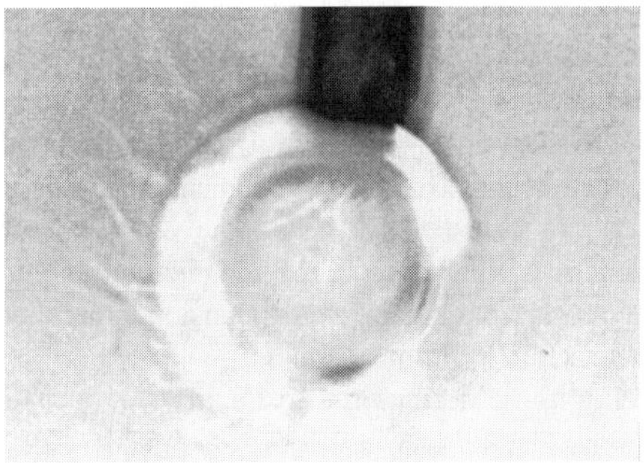

Figure 5. Transfer of acetylacetone from the drop into water (microgravity conditions, 7 seconds after formation of drop).

Another interesting occurrence in microgravity is the extension of interfaces, which can also be observed in **Figures 4** and **5**. Under similar experimental conditions, the diameter of the drop under microgravity is larger than under gravitational conditions.

Acknowledgements

The author is grateful to the European Space Agency for providing some of the data and some of the figures included here.

Reference

Mendes-Tatsis, M. A. and Perez de Ortiz, E. S. (1992) Proc. R. Soc. Lond. A 438, 389–396.

M.A. Mendes-Tatsis

MICROWAVE DRYING (see Dryers)

MICROWAVE HEATING (see High frequency heating)

MICROWAVES (see Electromagnetic spectrum)

MIE SCATTERING

Following from: *Optics*

The Mie theory is a theory absorption and scattering of plane electromagnetic waves by uniform isotropic particles of the simplest form (sphere, infinite cylinder) which are in a uniform, isotropic dielectric infinite medium. Though the initial assumptions of the Mie theory are idealized its results are widely used when solving problems of radiation heat transfer in light scattering media.

The basic aim of the theory is the calculation of efficiency coefficients (factors) for absorption (Q_a), scattering (Q_s) and extinction (Q_e). The ratio of σ_i, the cross-section for the appropriate process, to the particle profected area

$$Q_i = \frac{\sigma_i}{\pi r^2},$$

defines the efficiency coefficients (factors) Q_i, where r is the particle radius. The cross-section σ_i is the ratio of the energy flux absorbed, scattered or (in sum) extinguished by a particle to the incident energy flux density (i.e. to the energy of undisturbed electromagnetic waves per unit area oriented normally to the wave front). The cross-section is of area dimension while the efficiency coefficients factors are dimensionless.

According to the definition of extinction

$$\sigma_e = \sigma_a + \sigma_s, \qquad Q_e = Q_a + Q_s \tag{1}$$

The mathematics of the Mie theory for application to interaction with a spherical object can be divided into the following steps:

1. Introduction of the spherical coordinate system with the particle center as an origin.
2. Plane electromagnetic wave expansion in vector spherical functions.
3. Expansions of a spherical electromagnetic field and field inside the particle in spherical vector functions.
4. Determination of the coefficients of function expansions into series in vector functions: the coefficients are obtained for the field inside a particle and for the scattering field being solved.
5. The coefficients of scattering Q_s and attenuation efficiencies Q_e are calculated by integrating the Pointing vector (expressed in terms of the electric and magnetic field expansions) with respect to angle and space variables.

In Bohren and Huffman (1983), Deirmendjian (1969) and Van de Hulst (1957) the detailed procedure is given to obtain the resulting relations of the Mie theory. The relations for scattering and extinction are of the form

$$Q_s = \frac{2}{x^2}\sum_{j=1}^{\infty}(2j+1)(|a_j|^2 + |b_j|^2), \tag{2}$$

$$Q_e = \frac{2}{x^2}\sum_{j=1}^{\infty}(2j+1)\mathrm{Re}\{a_j + b_j\}, \tag{3}$$

where $x = 2\pi r/\lambda$ is the diffraction parameter, Re is the real part of the sum of the complex numbers:

$$a_j = \frac{\Psi_j(x)[\Psi_j'(m_\lambda x)/\Psi_j(m_\lambda x)] - m_\lambda \Psi_j'(x)}{\xi_j(x)[\Psi_j'(m\lambda x)/\Psi_j(m_\lambda x)] - m_\lambda \xi_j'(x)}, \tag{4}$$

$$b_j = \frac{m_\lambda \Psi_j(x)[\Psi_j'(m_\lambda x)/\Psi_j(m_\lambda x)] - \Psi_j'(x)}{m_\lambda \xi_j(x)[\Psi_j'(m\lambda x)/\Psi_j(m_\lambda x)] - \xi_j'(x)}, \tag{5}$$

a_j, b_j are the expansion coefficients, (the Mie coefficients) expressed in terms of the Riccatty-Bessel functions $\Psi_j(t)$ and $\xi_i(t)$ which are expressed in terms of the Bessel functions of non-integer order

$$\Psi_j(t) = \sqrt{\frac{\pi t}{2}}\, J_{j+1/2}(t), \tag{6}$$

$$\xi_j(t) = \sqrt{\frac{\pi t}{2}}\, J_{j+1/2}(t) + (-1)^{ni} J_{-n-1/2}(t), \tag{7}$$

$$i = \sqrt{-1},$$

the stroke Equations 4 and 5 means the differentiation with respect to an argument, $m_\lambda = n_{1,\lambda} + in_{2,\lambda}$ is the complex refractive index of a particle material in the surrounding medium, $n_{1,\lambda}$ is the refractive index, $n_{2,\lambda}$ is the absorption index associated with the spectral absorption coefficient by the relative $\kappa_\lambda = 4\pi n_{1,\lambda}/\lambda$.

The coefficient of absorption efficiency is determined after finding Q_e and Q_s with (1) taken into account. The dependence of Q_e on x for water drops is given in **Figure 1**. For particle radii commensurable with a wave length ($r = 1\ \mu m$) the typical spectral peculiarities are manifested, namely, the interference structure (large scale oscillations), ripple (irregular fine structure), and the weak spectral dependence region.

The Mie series (2), (3) are poorly converging series, especially for diffraction parameter $x > 20$. Numerous investigations are devoted to this problem which resulted in effective computational

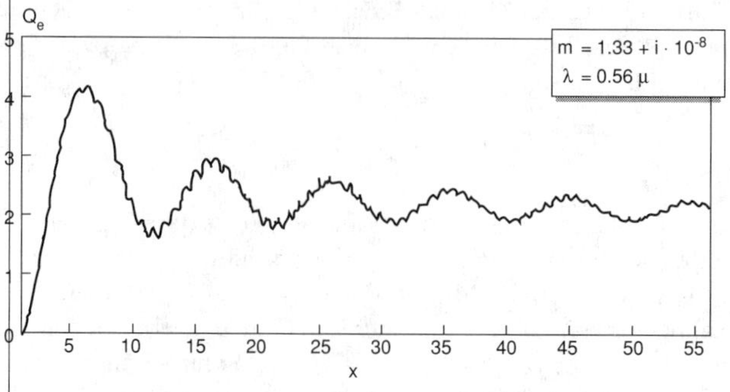

Figure 1. Extinction efficiency (Q_e) as a function of diffraction parameter x ($= 2\pi r/\lambda$).

algorithms using formulas of direct and inverse recursion (the expression of the series subsequent terms in terms of the preceding ones) as well as computational programs [Bohren and Huffman (1983)].

Since the problem is azimuthally symmetrical the phase function p_{gl}, according to Mie theory depends on the latitude angle θ between the scattering direction and that of undisturbed wave front propagation. The averaged cosine of a scattering angle

$$\mu = \frac{1}{2}\int_{-1}^{1} p_\lambda(\mu)\mu d\mu,$$

(where $\mu = \cos\theta$) characterized the elongation degree of the phase function and is determine using the Mie coefficient

$$\bar{\mu} = \frac{4}{x^2 Q_s}\sum_{j=1}^{\infty}\frac{j(j+2)}{j+1}\text{Re}\{a_j a_{j+1}^* + b_j b_{j+1}^*\} + \frac{2j+1}{j(j+1)}\text{Re}\{a_j b_j^*\}. \quad \textbf{(9)}$$

Here * means the complex conjugate quantity.

For small values of diffraction parameter, $x \ll 1$, one can retain just the first summands of the Mie series which corresponds to the Rayleigh law for scattering by particles the size of which are essentially less than the radiation wave length. For large $x \geq 20$ the efficiency coefficients are found from geometrical optics. The intermediate region for the diffraction parameter variation is called the "Mie scattering region."

The polarization characteristics of scattered radiation are rarely taken into account in the radiation heat transfer theory. However, if necessary they can be computed according to the Mie theory [Bohren and Huffman (1983)].

Volume spectral absorption coefficients involved in the radiation transfer equation are determined by formulas

$$\kappa_\lambda = \pi N_0 \int_0^\infty Q_a(m\lambda, x) r^2 N(r) dr, \quad \textbf{(10)}$$

$$\sigma_\lambda = \pi N_0 \int_0^\infty Q_s(m\lambda, x) r^2 N(r) dr,$$

where N_0 is the total number of particles in unit volume, N(r) is the particle distribution (along the radius) function. For a monodisperse system of particles (of radius r_0):

$$\kappa_\lambda = \pi r_0^2 Q_a(m\lambda, x),$$

$$\sigma_\lambda = \pi r_0^2 Q_s(m\lambda, x).$$

References

Bohren, C. F. and Huffman, D. R. (1983) *Absorption and Scattering of Light by Small Particles.* A Wiley Interscience Publication, John Wiley & Sons, Inc.

Deirmendjian, D. (1969) *Electromagnetic Scattering on Spherial Polydispersions,* Elsevier, New York.

Van de Hulst, H. C. (1957) *Light Scattering by Small Particles,* Wiley, New York.

S.T. Surzhikov

MIE SERIES (See Mie scattering)

MIE THEORY (see Optical particle characterisation, Mie scattering)

MIKIC, ROHSENOW AND GRIFFITH EQUATION, FOR BUBBLE GROWTH (see Bubble growth)

MINI-COMPUTERS (see Computers)

MINIMUM FILM BOILING TEMPERATURE (see Leidenfrost phenomena; Rewetting of hot surfaces)

MINIMUM FLUIDISATION VELOCITY (see Fluidised bed)

MIROPOLSKII FORMULA, FOR POST DRYOUT HEAT TRANSFER (see Forced convective boiling)

MIST ELIMINATORS (see Vapour-liquid separation)

MISTS (see Aerosols)

MIT (see Massachusetts Institute of Technology)

MIXED (COMBINED) CONVECTION

Following from: *Convective heat transfer*

Mixed (combined) convection is a combination of forced and free convections which is the general case of convection when a flow is determined simultaneously by both an outer forcing system (i.e. outer energy supply to the fluid-streamlined body system) and inner volumetric (mass) forces, viz. by the non-uniform density distribution of a fluid medium in a gravity field. The most vivid manifestation of mixed convection is the motion of the temperature stratified mass of air and water areas of the Earth that the traditionally studied in geophysics. However, mixed convection is found in the systems of much smaller scales, i.e. in many engineering devices. We shall illustrate this on the basis of some examples referring to channel flows, the most typical and common cases. On heating or cooling of channel walls, and at the small velocities of a fluid flow that are characteristic of a laminar flow, mixed convection is almost always realized. Pure forced laminar convection may be obtained only in capillaries. Studies of turbulent channel flows with substantial gravity field effects have actively developed since the 1960s after their becoming important in engineering practice by virtue of the growth of heat loads and channel dimensions in modern technological applications (thermal and nuclear power engineering, pipeline transport).

In a mathematical description of mixed convection in the equations of motion (the Navier-Stokes equations), both the term characterizing the pressure head loss dp/dx and the term characterizing mass forces ρg are retained. The simplest case allowing an analytical solution of the problem refers to the steady-state laminar flow at a distance from the tube inlet with a constant heat flux on a wall (q_w = const). Figure 1 shows the calculated velocity and temperature distributions both in an ascending flow, i.e. when the directions of forced and free near-wall convections coincide (case "a"), and in a descending flow in heated tubes, i.e. when the directions of forced and free near-wall convections are opposite (case "b"). The curves refer to different values of the Rayleigh number $Ra_A = g\beta D^4 (dT/dx)/\nu^2 = 4\ Gr_q/Re = (4/Re)(g\beta q_w D^4/\lambda\nu^2)$. In the case of an ascending fluid flow, velocity profiles peak near the wall giving an M-like shape which becomes sharper with the growth of the effect of thermogravitation convection until instability of the laminar flow occurs before the calculated values of the axis velocity have vanished (**Figure 1a**, $Ra_A = 625$). In a descending flow the stability is disturbed rather quickly, as indicated by the velocity profiles (**Figure 1b**) which even at relatively small growth of the bouyancy effect acquire zero values of the velocity gradient on the wall. A remarkable property of temperature distributions should be noted for all the cases considered (**Figures 1a,b**); even with strong deformation of velocity distribution, the temperature profiles only differ slightly from the profiles for pure forced convection. The shown specific features of velocity and temperature shown in Figure 1 are reflected on the corresponding variation of the relative heat transfer value presented in the left portion of Figure 2 where Nu_0 means the Nusselt number for a "purely" forced flow.

The variation of heat transfer with Gr_q in turbulent and laminar mixed convection is considerably different (**Figure 2**). In the case of a descending flow (b) (Nu/Nu_0), falls with Gr_q in laminar mixed convection due to the retardation near a wall, but with turbulent mixed convection (Nu/Nu_0) grows as the heat load increases due

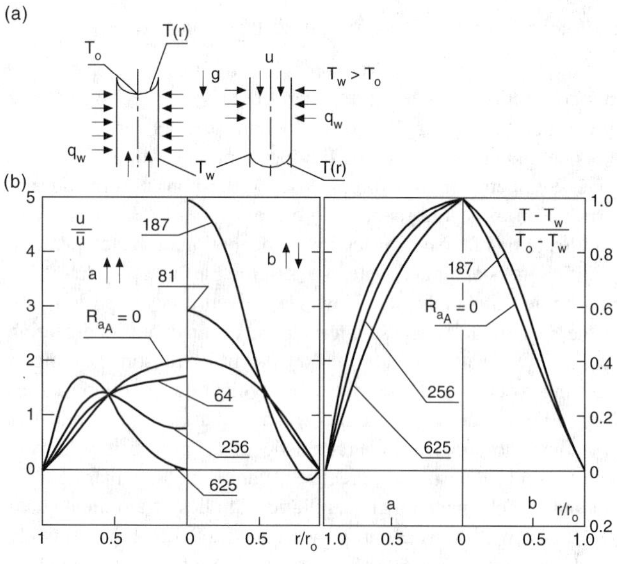

Figure 1. Velocity and temperature profiles in mixed convection.

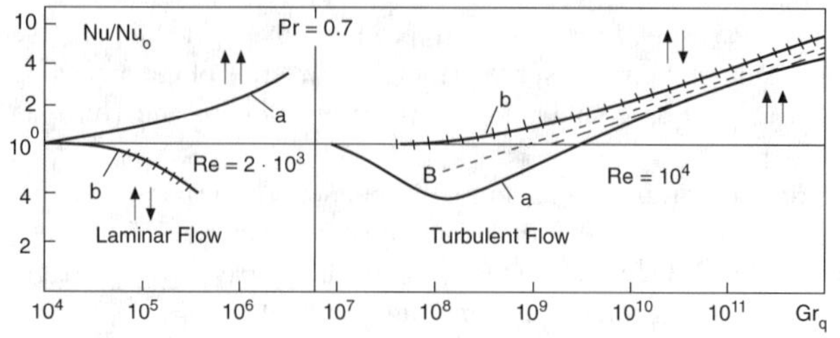

Figure 2. Variation of Nusselt number Nu (= $\alpha D/\lambda$ where α = least transfer coefficient, D = tube diameter and λ = thermal conductivity) with Gr_q (= $g\beta\dot{q}_w D^4/\lambda w$).

to additional flow turbulization. For an ascending turbulent flow the behaviour of the curve (Nu/Nu_0) − Gr_q is non-monotonic. First, heat transfer deteriorates with increasing Gr_q. This is caused by the attenuation of turbulent momentum and heat transfer as a result of the effect of buoyancy forces on the shear stress profile and hence as the generation of turbulence. With further increase in the Grashof number, heat transfer begins to grow, thus reflecting an intensive development of the free convection effect on the flow as a whole as also takes place in the laminar mode (case "a"). In turbulent flow, the velocity profiles also acquire a typical M-like shape, but the quantitative characteristics, as well as the mechanism of momentum and heat transfer, are quite different for turbulent and laminar mixed convection. In the region of a strong effect of free convection the mode of thermal turbulent convection with the dependence of the Nusselt number on the Grashof number, characteristic to turbulent free convection $Nu = A(Pr)Gr^{1/4}$ and excluding the effect of a geometry parameter on heat transfer, is established. Line B in **Figure 2** is constructed by the relation for the Nusselt number with developed free convection on a vertical plate. Under normal conditions (in particular, for air at atmospheric pressure and temperatures close to room ones, the data of which are given in **Figure 2**) the effect of the gravity field on a forced turbulent flow manifests itself only at relatively small Reynolds numbers, of the order of 10^4 for tubes, and with channels of rather large dimensions equal with characteristic equivalent diameters of the order of 1 m and more. However, for media with strong density variation, e.g. under near-critical conditions where density varies from the values typical of gases to the values characteristic for a liquid, mixed convection is realized in tubes of small diameters (of the order of several millimeters) and at large values of the Reynolds number (see **Figure 3** for a water flow $p/p_c = 1.1$ in a diameter 3 mm tube at $Re = (2 - 3) \times 10^4$).

In a turbulent flow in the presence of heat transfer, the gravity field leads not only to the manifestation of large-scale free convection covering the entire flow as in the laminar mode, but also to the manifestation of local effects on turbulence that radically change the character of heat transfer (see **Figure 2**). One may judge the character of the manifestation of the effect of the buoyancy forces on turbulence from the velocity distributions given in **Figure 4.** These velocity profiles are plotted in the near-wall region in terms of the universal parameters $u^+ = u/u^*$ and $\gamma^- = u^*\rho y/\eta$ where u^* is the friction velocity ($\sqrt{\tau_w/\rho}$) and y^+ the distance from the wall. The data are for a horizontal channel. The channel can either be heated from above giving stable stratification or from below giving unstable stratification. Points 1 and 2 represent the profile near

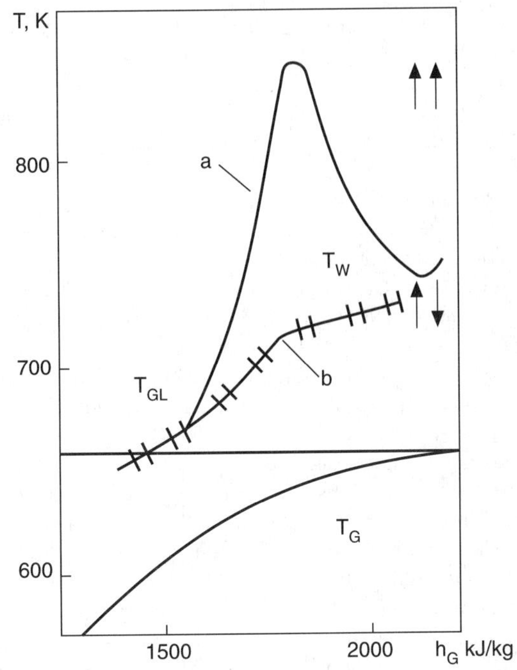

Figure 3. Temperature profiles as a function of fluid enthalphy (h_{GC}) for ascending (a) and descending (b) flow in a 3 mm tube at (p/pc) = 1.1. (T_w = wall temperature, G_{ct} = mean fluid tempeature).

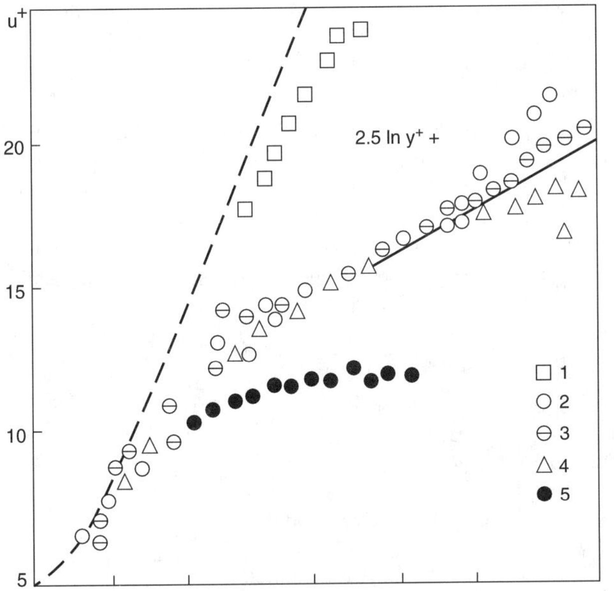

Figure 4. Dimensionless velocity profiles in mixed convection.

the bottom wall for a case with stable density distribution (stable stratification) and from points 4 and 5 show the profile with unstable density distribution (unstable stratification); those results may be compared with the universal velocity distribution in an equilibrium wall flow (points 3). Points 2 and 4 indicate to a small effect of density stratification on turbulence. With strong stable stratification, turbulent transport of fluid particles is retarded and a turbulent flow, which occurred under isothermal conditions, laminarizes at the same Reynolds number and the velocity distribution in this case (points 1) approaches the distribution typical of a laminar flow (a dashed line). With a strong unstable stratification (points 5) the buoyancy forces additionally turbulize the flow and in the limited case the wall logarithmic law for a velocity is transformed to the "−1/3" − law, viz.: $u \sim y^{-1/3}$ (line 6) found when studying the near-Earth atmospheric turbulence. The data indicate that in physical studies and mathematical simulation of mixed convection, one should simultaneously take into account both the global effect of gravitation on the flow as a whole and its direct effect on turbulence.

References

Petukhov, B. S., and Poliakov, A. F. (1988) *Heat Transfer in Turbulent Mixed Convection,* Hemisphere Publishing Corp., N.Y.

A.F. Poliakov

MIXER-HEAT EXCHANGERS (see Mixers, static)

MIXERS

Following from: *Mixing*

Impeller mixers are simple but versatile equipment used widely in the process industries to mix a liquid with another liquid, a gas or solids using forced convection (flow). A variety of impeller designs is available but their classification is not unique. Clearly the viscosity of the mix for which the impeller is intended is an important consideration and can be used as a first broad basis for classification. Low viscosity fluids are usually processed by small impellers but highly viscous fluids require comparatively bulkier impeller designs. A typical example is the distinct physical difference between a propeller and a helical ribbon impeller. Other bases, relating also to flow, are the way in which impellers generate flow (by shear stress or pressure) and the types of flow pattern they produce (tangential, radial, or axial). Paddle, turbine and propeller impellers, for example, transmit momentum by direct pressure of the blades on the liquid but a rotating disk impeller instead shears the liquid. A marine type mixing propeller will generate axial flow whereas a flat blade turbine generate a radial flow and some designs produce mixed radial and axial flows. An example of this classification is illustrated in **Figure 1** and further details are given by Starbacek and Tausk (1965) and Zlokarnik and Judat (1988). From a hydrodynamic point of view, impeller mixers resemble pumps in which a power input, $\dot{W}$ is expended for the circulation of the liquid (pumping capacity, $\dot{V}$) and the generation of shear as a result of the hydrodynamic head, Z developed giving:

$$\dot{W} = \rho \dot{V} Z$$

where ρ is the density of the liquid. Consequently, impeller mixers can also be classified on the basis of their typical power input requirement (see **Table 1**) and ability to produce at a constant

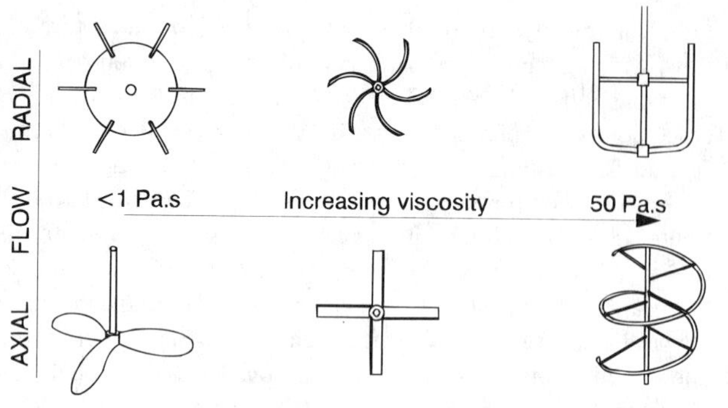

Figure 1. An example of a flow based classification of impellers.

Table 1. Power based classification of impellers

Low Power (0.2 kw/m^3): Propellers, Turbines, Paddles
blending of low viscosity liquids dispersing gases in low viscosity liquids suspending fine solids in low viscosity liquids
Very High Power (4 kw/m^3): Anchors, Helical Ribbons
blending of high viscosity liquids, pastes suspending solids in viscous fluids

power input the required balance of flow circulation and shear as illustrated in **Figure 2** and described further by McDonough (1992).

For example, dispersion of immiscible liquids and of solid agglomerates in liquids will be more effectively carried out in high shear devices whereas the blending of liquids will require axial flow impellers which are capable of producing high flow. Clearly, the power and flow-head characteristics of an impeller mixer are basic design and operation information. Because of the large number of variables involved in an impeller mixer system and the way in which the flow is induced, it is theoretically not yet possible to predict these characteristics (although much progress has been made in recent years with computational fluid dynamics). Experiments become necessary and are carried out usually with liquids in well baffled vessels (to prevent vortexing which hinders mixing) on a laboratory scale for application to a large scale.

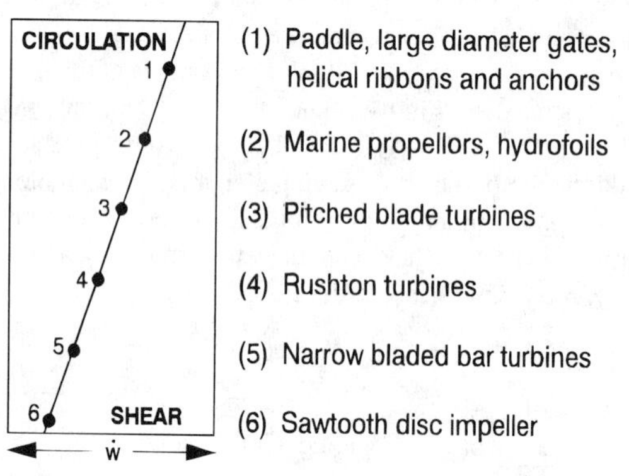

Figure 2. Impeller mixers flow-head (shear) characteristics.

Typically, experiments are carried out in cylindrical vessels filled with liquid to a height equal to the diameter of the vessel. The principal variables become: the impeller diameter, D_i and its rotational speed, ω (rev/s), the vessel diameter and the physical properties of the liquid (its density, ρ and viscosity, η). For each impeller-vessel system there will also be a number of dimensions defining the geometrical arrangement. Measurements of power, $\dot{W}$, and flow circulation, $\dot{V}$, are taken over a range of conditions and **Dimensional Analysis** principles are used to correlate the data which are expressed as *power number,* $N_W = \dot{W}/\rho\omega^3 D_i^5$ and *circulation rate number,* $N_V = \dot{V}/\omega D_i^3$ as functions of the **Reynolds Number,** $Re = \rho\omega D_i^2/\eta$ and the impeller vessel geometrical ratios which are numerous. In addition to these basic measurements, a tracer can be dropped in the liquid and a mixing time, t, evaluated and expressed as a mixing time number, $N_t = \omega t$. It too is a function of the Reynolds number and impeller-vessel geometrical ratios.

Some key observations can be made on these functions as illustrated in **Figure 3**. As expected, the Reynolds number which controls the nature of the flow is the dominant variable. We observe a unique relationship between N_W and Re in both the laminar (N_W = constant/Re) and turbulent (N_W = constant) regimes. The same holds with the mixing time and flow circulation numbers showing the beneficial effect of turbulence on mixing, but the geometry of the impeller in relation to the vessel also has an influence particularly on the pumping capacity-shear characteristics. In general, for a given power level, high circulation and low shear are observed with large impeller diameter running at low speeds. Small diameter impellers running at high speeds usually produce low circulation but intense shearing. An axial flow turbine shows typically:

$$(\dot{V}/Z)_W \alpha D_i^{8/3}$$

The literature abounds with correlations and graphs which relate N_W, N_V, N_t to Re and geometrical ratios [see for example Nagata (1975) and Uhl and Gray (1975]. When using these data, care must be taken in identifying the applicable geometrical ratios. This is very important when considering scale-up of equipment. Of course, scaling up cannot be based on geometry alone, kinematic (relating

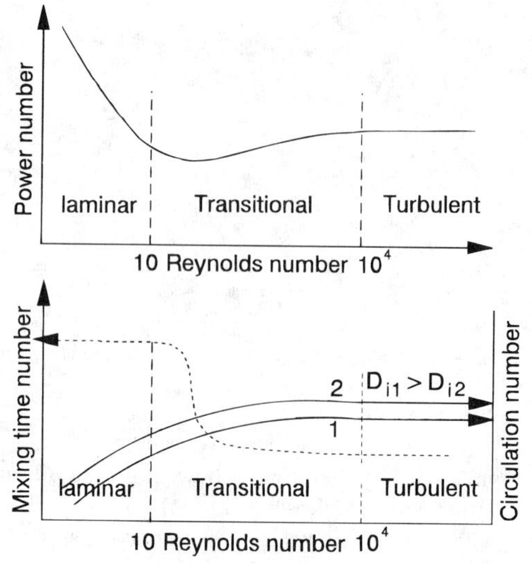

Figure 3. General flow characteristics of impeller mixers.

Table 2. Scale-up criteria for impeller mixers

Dynamic Similarities	Kinematic Similarities
Reynolds Number **Froude Number** **Weber Number** Power per unit volume	Impeller tip speeds volumetric flow per shear
Geometric Similarities are Usually Satisfied	
Impeller diameter/tank diameter Impeller diameter/tank clearances	

to velocities and flow patterns) and dynamic (relating to forces) similarities must also be taken into account. A number of scale-up criteria can result (see **Table 2**) but unfortunately not all similarities hold as they usually conflict with each other and a consideration of the important physical aspects of the operation become necessary. Even then, an intermediate scale-up prior to full scale design may be necessary. These considerations are clearly very important and are addressed in most of the references cited. All the treatment discussed so far applies strictly to a single Newtonian liquid. It forms however a fundamental basis for analysing mixing in the more complex situations where often all that is required is an impeller mixer delivering the correct balance of flow and shear at minimum energy input.

References

McDonough, R. J. (1992) *Mixing for the Process Industries,* Van Nostrand, Reinhold, New York.

Nagata, S. (1975) *Mixing: Principles and Application,* John Wiley & Sons, Inc., New York.

Sterbacek, Z. and Tausk, P. (1965) *Mixing in the Chemical Industry.* Pergamon Press, Oxford.

Uhl, V. W. and Gray, J. B. (1966) *Mixing Theory and Practice,* Academic Press.

Zlokarnik, M. and Judat, H. (1988) Stirring, in *Ullmann's Encyclopedia of Industrial Chemistry, 5th Ed.* Vol. B2.

H. Benkreira

MIXERS, STATIC

Following from: *Mixing*

Static mixers provide a means of achieving homogenization of gases, liquids and viscous materials without the use of moving mechanical parts. In its simplest form, the materials are passed through a fixed geometric structure which repeatedly splits the stream of material into numerous parts as it passes through and then reunifies them with a different part of the stream. The mixer housing is generally pipe-shaped and is mostly supplied flanged so that it can be easily installed in line as part of a continuous process.

A simple static mixer offers many advantages:

- Static mixers do not require a separate energy supply. The pumps or blowers, delivering the materials to be mixed, supply all the energy required.
- Pressure drop is small so energy consumption is low.
- They have no moving parts, so they require little maintenance and down time is minimised
- They require minimal and operational cost investment.
- Performance is predictable, uniform and consistent. Homogeneity, expressed as a deviation from the mean, is quantifiable.
- They are compact and require little space.
- Shear forces set up in static mixers are generally small, so the product is treated gently during processing.
- Sealing problems are eliminated.
- They are suitable for quick response on-line proportional control dosing systems to provide representative samples.
- Differences in concentration, temperature and velocity are equalized over the cross-section of the flow.

Since 1970 basic static mixers have undergone a lot of development which has led to a very wide range of applications in different forms. They are no longer used only for simple blending, but also where heat and mass transfer operations or chemical reactions are involved. Such processes take advantage of the ability to provide not only good blending, but also good heat transfer, uniform residence time, and where two or more phases are involved intimate dispersal and contact.

Specialist forms of static mixers are now used in many industrial sectors: petroleum, natural gas and refineries; petrochemicals; chemicals; polymer production and plastics processing, pulp and paper; cosmetics and detergents; foods; water and wastewater treatment; energy and environmental protection.

There are now at least 20 manufacturers, some of whom produce a limited variety of fairly basic designs, but the market leaders have now developed a wide range of specialist designs, based on the original concept, but adapted and refined to suit particular specialist purposes. Each manufacturer adopts a different nomenclature for his own designs. The following relate to the products of the Sulzer-Chemtech Division of Sulzer (UK) Ltd at Farnborough, Hampshire, but similar products can be found from other manufacturers.

The Sulzer range starts with two types of the basic static mixer for turbulent and laminar flow, the SMV and SMX, respectively. The SMV mixing elements are made in the form of corrugated plates which form open intersecting channels. Apart from their use for mixing low-viscosity liquids, gases and dispersing insoluble liquids, they are used for contacting gases with liquids and as a mass transfer device. The SMX is similar but has mixing elements in the form of a lattice of intermeshing and intersecting bars and is used for higher viscosity material.

Where static mixers are used for the physical absorption of gases into liquids and for gas/liquid chemical reactions they give safer operation and reduce the inventory of material. The reaction takes place quickly. An example is the dissolution of chlorine into water and its subsequent reaction with alkenes.

Another product based upon the same static mixing principle is mixer packing (SMVP) for use in bubble extraction and reaction columns. This has proved to have good hydrodynamic properties with strong cross mixing, little back mixing and high capacity.

Static mixers have also been developed in various ways to make them suitable for use as heat exchangers for viscous liquids. The simplest *mixer-heat exchanger* is the monotube version in which the tube containing the geometric mixing structure and product is enclosed within another pipe, so that the heating or cooling medium is fed into the space between the inner and outer pipes. This device generally remains limited to the transfer of small amounts of heat or the throughput of small amounts of product, otherwise the pipe lengths or pressure drop become too great.

For larger throughputs an alternative is multitube heat exchangers. Here the product is divided into parallel streams and mixing occurs only within the part streams; there is no radial mixing.

A more advanced development is a mixer in which the mixing elements are formed from hollow tubes, which contain the heat transfer medium. This then becomes a mixer-reactor (SMR) in which high heat transfer coefficients are obtained with a large internal heat transfer surface. This allows highly exothermic reactions to be closely controlled at the correct temperature. Equally it can be used for temperature control of viscous products, giving very quick response.

The SMR is designed for use in continuous processes and there are many possible configurations. One is the simple "once-through" product flow (plug-flow) suitable for applications such as cooling solutions, increasing viscosity of thermoplastic melts, adjusting melt temperature, continuous heat treatment and reaction control.

The SMR can also be configured to operate with product recirculation (loop). This is suitable for continuous high rate exo- or endothermic chemical reactions, either single or multi-phase.

A third common configuration is a combination of loop and plug flow in order to achieve a rapid preliminary or main reaction, perhaps with further conversion down stream by admixed additives.

The SMR and SMXL (mixer reactor for laminar flow) are suitable for temperature-sensitive materials, where simultaneous reactions are required, or where temperature-controlled reactions require a uniform residence time and long mixing lengths. Static mixers are available in a wide variety of materials for different applications: stainless steel; exotic metals for corrosive materials; fibreglass reinforced plastic; polypropylene and so on. Mixing elements in various plastics such as PVDF, PP and ETFE are available installed in glass pipes (NPS, 40, 25, 15mm) for laboratory experiments. Other designs are available in sizes up to eight metres or more in diameter.

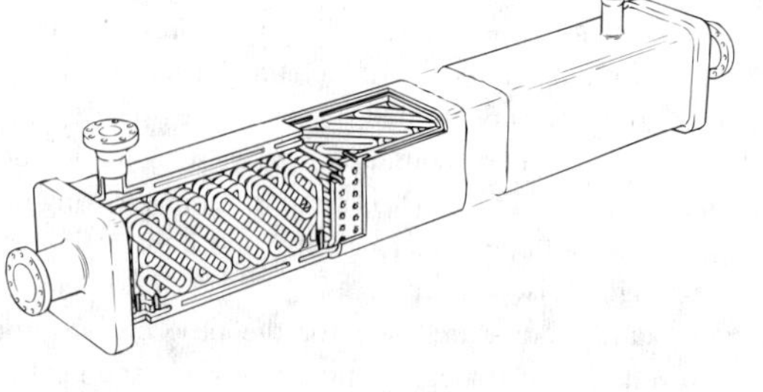

Figure 1. Schematic diagram of the SMR mixer-reactor. The product flows in the square channel, in which the mixing elements are arranged, constructed from tubing through which the heat transfer medium flows.

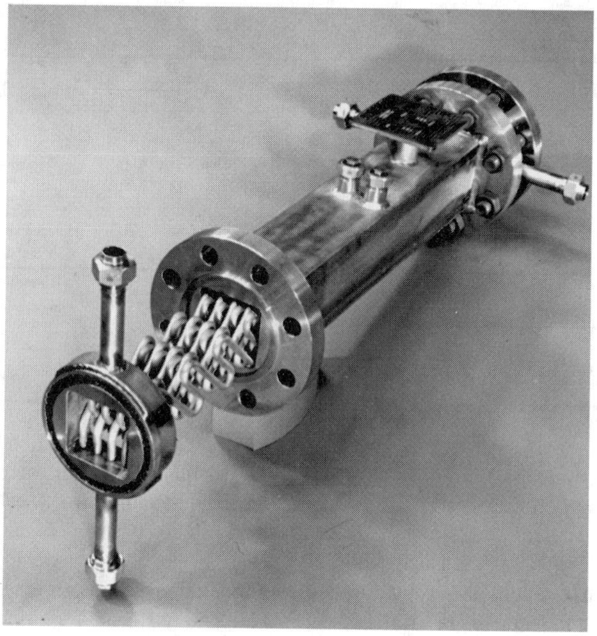

Figure 2. An SMR mixer-reactor module, showing the mixing partly drawn from its casing. A series of modules can be assembled in loops or other configurations according to the process requirement.

Pilot tests are often necessary when a continuous process is being developed, for example, carrying out temperature controlled reactions. Scale up to large production rates is without risk if the results of such test results are available. Most suppliers of static mixers are able to undertake such tests in their own pilot plants. Alternatively, a small range of mixers are generally available from many manufacturers for trial purposes. On this basis, the suppliers are generally able to provide performance guarantees.

References

Gerstenberg, H., Schuhr, P., Steiner, U. R. (1982). *Chemie-Ingenieur-Technik 54* 6, pp551–53.

Heierle, A. (1980): *Chemie-Technik 9* pp. 83–85.

Lynn, S., Oldershaw, C. F. (1984): *Heat Transfer Engineering 5* 1–2, pp. 85–92.

Müller, W. (1982): *Chemie-Ingenieur-Technik 54* 6, pp. 610–11.

Schneider, G. (1990): Institution of Chemical Engineers Symposium Series 121, paper 8, pp. 109–119. *Static Mixers as Gas/Liquid reactors.*

Streiff, F. A., (1986): *Wärmeubertragung bei der Kunstsoffaufbereitung.* VDI-Verlag, pp. 241–275.

Sulzer *Mixing Processes, Prospectus* 23.27.06.20.

A.W.A. Reeve

MIXER SETTLERS (see Extraction, liquid-liquid)

MIXING

Introduction

Mixing can be defined as an operation which reduces the degree of non-uniformity of all properties of a system, single or multiphase with one or many components. It is carried out in equipment (**Mixers**) which rearrange the components of the system into a state where they are, ideally, uniformly distributed in the system and completely dispersed to their ultimate, smallest pieces. The relative importance of distribution and dispersion to overall mixing is illustrated in **Figure 1**. The size of these ultimate pieces may be that of a molecule, a drop, a bubble or a particle depending on the nature of the components and phases of the system. As well as homogenising a system and its properties, **Figure 1** shows that mixing also creates large interfacial areas between the components of a mixture and thus promotes and enhances mass and heat transfer and chemical reaction. It is therefore an important unit operation in all processing industries, as described by Harnby et al. (1992) and Sterbacek and Tausk (1965), and one which controls the quality and consistency of the final products whether they issue from a simple blending operation or a complex chemical reaction. In most applications perfect mixing is not feasible nor necessary and it is sufficient that the degree of mixing achieved meets the requirements of a process or a product. A method of evaluating the degree of mixing achieved is, however, critical for assessing and controlling the performance of a process and the ensuing quality of the product.

Assessment of Mixing

The true measure of the standard of mixing in a batch or in a run is the degree of uniformity of the product. This requires consideration of *gross uniformity, texture and local structure* of a sample withdrawn from the product. **Sampling** in itself can be difficult, particularly when solids are involved. In practice, the scale of examination of the sample should be just enough to indicate whether the mixing has been sufficient and thus will vary for different processes. Gross uniformity is a crude measure of how overall a minor component or phase is distributed in a mixture. Texture describes non-uniformity of composition in a mixture. It is a measure of the scale and intensity of segregation of the minor component or phase in the mixture and may be apparent through the presence of regions of unmixed or poorly mixed material. Local structure is the way in which the ultimate pieces are arranged, either clumped together as agglomerates, large bubbles and drops or dispersed individually. Local structure is therefore a measure of the degree of dispersion in a mixture and require a scale of scrutiny sufficiently fine to allow resolution of these ultimate entities. The physical meaning of these three quantitative measures of mixing derives directly from the concepts of distribution and dispersion illustrated in **Figure 1**.

The mathematical definition of these indices can easily be derived (see Tadmor and Gogos, 1979) using the size and number distribution of the dotted pieces illustrated in Figure 1. The evaluation of the degree of mixing is not always assessed from measurements carried out on the product but also from on line monitoring, at one or more points in the mixer, of concentration, conductivity, light or noise adsorption and other properties as described comprehensively by Shah (1992). Whichever technique is used, the aim of the measurements is usually to obtain a mixing time (i.e. the time necessary to achieve the desired degree of mixing) from which the mixing rate (i.e. the rate at which uniformity is approached) can be obtained and to correlate the data against the variables of the mixer system, particularly the power input and thus obtain a mixer effectiveness.

Since the number of design and operation variables in a mixer are large, **Dimensional Analysis** is used to correlate the data, and the principles of similarity are used to scale up equipment. In most cases it is difficult to satisfy all the scale-up criteria (beside the essential geometric similarity) and a consideration of the physical factors of the mixer is necessary. In the mixing of fluids in agitated vessels for example, the **Reynolds Number** plays an important part since it is a measure of the relative importance of the inertial and viscous forces, i.e. it indicates the onset and level of turbulence in the flow. In the mixing of gas-liquid or immiscible liquid-liquid systems, in addition to the Reynolds number, the **Weber Number** must also be considered since it represents the ratio of applied to surface tension forces, which control the mixing and mass transfer characteristics of bubbles and drops. These and other aspects of mixing have been reviewed by Sweeney (1978) and recently more comprehensively by Shah (1992).

Mixing Mechanisms

Three basic mechanisms, either singly or in combination, can be present in a mixing operation; molecular diffusion, eddy diffusion and bulk or convective flow. Molecular diffusion is spontaneous, driven by the gradients of concentration and temperature and can be found in the mixing of miscible liquids of low viscosities and all gases. In most practical operations it is a slow process and eddy diffusion must be superimposed on it to speed up mixing. Eddy diffusion results from **Turbulence** in the flow of fluids and requires greater energy input. In the mixing of fluids thus, **Viscosity** is the dominant property of the system resisting mixing and the lower it is the easier the turbulent conditions can be achieved and the quicker the mixing operation. Examples of mixing systems which use flow turbulence are the pumping of liquids in pipes,

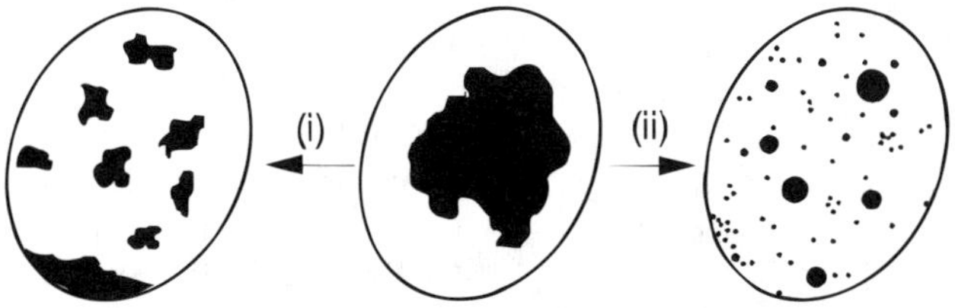

Figure 1. Distributive (i) and dispersive (ii) mixing.

their mechanical agitation in vessels, the jetting of a liquid into another liquid and the airlifting of liquids as reviewed in details by Harnby et al (1992) and Sterbacek and Tausk (1965).

Much knowledge has been gained on turbulent mixing of low viscosity fluids and suspensions using experimental techniques such as **Laser Doppler Anemometry** to map flow circulation patterns and velocities and levels of flow turbulence in various mixing systems and mixer designs. With the advent of fast and powerful computers, theoretical computational fluid dynamics analyses, as described in the proceedings of the 8th European Conference on Mixing (1994), are increasingly reproducing experimentally observed flow distributions, though much work is still needed to overcome the intrinsic complexities of mixer design, fluids rheological behaviour and multiphase features.

All research points to the need of providing the correct balance between the levels of turbulence and flow circulation rates to ensure effective mixing of particular systems. A top to bottom axial-flow circulating propeller, for example, is very well suited for homogenising miscible liquids and keeping solid sediments suspended in large tanks. Gas-in-liquid dispersions, on the other hand, are better achieved using radial-flow turbine type impellers which produce high shear zones near the impeller blade tips to break the incoming gas into fine bubbles. Very fine bubbles indicate very good mixing but usually require greater coalescing times, hence, yield poor mass transfer. Turbulence levels induced by the impeller must be tuned to produce optimal size bubbles. Even higher shear can be obtained with fast moving sawtooth-disc stirrers which are particularly suited for emulsification and dispersion of liquids over a wide range of viscosities.

When turbulence is not experienced bulk or convective flow is required for mixing, and can be achieved by rearranging the materials without deforming them or by deforming the materials in the laminar shear or elongational flows associated with systems of very high viscosities. The no-deformation rearrangement mode clearly cannot yield any dispersive mixing; it is distributive in nature and can be either ordered, as for example in the in-line static mixer for blending polymer melts, food pastes and other similar high viscosity materials or random, as in a V-blender for mixing free-flowing solid-solid systems which segregate when sheared, vibrated or fluidised and can only be effectively mixed by bulk convective flow. The laminar flow rearrangement mode of mixing is governed by the extent of strain undergone by the pockets of the minor component or phase when distributive mixing is the only concern. When dispersive mixing is required in laminar flow, the stresses within the field of flow become important to break down solid agglomerates, for example, and clearly the application of power and the forces brought to bear become of greater significance. Mixers for such duties are bulky and hold much less material. Examples of equipment are vessels agitated with anchors and helical ribbons for the processing of high viscosity materials, single or twin extruders, blade and Banbury mixers and mixing rolls for kneading, shearing and tearing pastes, polymer melts and cohesive solid systems.

Two fundamental observations on mixing in laminar flow should be made. The first is with regard to the ability of elongation of creating larger interfacial area than shearing for the same strain. Clearly, when designing mixers extensional flow features should be incorporated. However, while it is easy to generate high flows by shear, in practice extensional flows are difficult to obtain. The second observation concerns the importance of orientation of the sheared or elongated layers during mixing. For a given strain, the largest interfacial area is usually achieved when the layers are perpendicular to the plane of shear or elongation. Application of such simple observations yields large improvement in mixing in extruders, such as when pins, barrier flights and cavities are implanted in the flow field to disrupt shear, reorient and elongate layers to improve distribution and dispersion.

References

Harnby, N., Edwards, M. F., and Nienow, A. W. (1992) *Mixing in the Process Industries,* Butterworth Heinemann Ltd., Oxford.

Eigth European Conference on Mixing (1994). I Chem E Symposium Series No 136, Cambridge, 21–23 Sept 1994.

Shah, Y. T., (1992) *Design Parameters for Mechanically Agitated Reactors, Advances in Chemical Engineering,* Vol. 17, Academic Press Inc.

Sterbacek, Z. and Tausk, P. (1965) *Mixing in the Chemical Industry,* Pergamon Press, Oxford.

Sweeney, E. T., (1978) *An Introduction and Literature Guide to Mixing, BHRA Fluid Engineering Series,* Vol. 5, Cranfield.

Tadmor, Z and Gogos, C. G. (1979) *Principles of Polymer Processing,* John Wiley & Sons, Inc., New York.

Leading to: *Agitation devices; Mixers; Mixers static.*

H. Benkreira

MIXING GAS-LIQUID (see Jet pumps and ejectors)

MIXING IN ROD BUNDLES (see Rod bundles, flow in)

MIXING LENGTH (see Boundary layer heat transfer)

MIXING LENGTH HYPOTHESIS (see Boundary layer)

MIXING LENGTH MODELS (see Turbulence models)

MIXING OF PARTICLES IN FLUIDISED BEDS (see Fluidised bed)

MIXING REYNOLDS NUMBER (see Agitated vessel mass transfer; Agitation devices)

MIXTURE CONSERVATION EQUATIONS (see Multiphase flow)

MIXTURES (see Activity coefficient; Binary systems, thermodynamics of; Gibbs-Duhem equation)

M J BOX COMPLEX METHOD (see Optimization of exchanger design)

MODEL BANDWIDTH (see Fibre optics)

MODELING TECHNIQUES (see CANDU nuclear power reactor)

MODERATING RATIO (see Nuclear reactors)

MODERATORS (see Nuclear reactors)

MODULUS OF THE SOLID (see Non-Newtonian fluids)

MOIRÉ FRINGES (see Film thickness determination by optical methods; Visualization of flow)

MOISTURE ISOTHERM (see Drying)

MOISTURE MEASUREMENT

Following from: *Humidity measurement*

Moisture measurement has found wide use in scientific research and production, and during the past decades many moisture measurement techniques have been introduced, among which the classic combination of **Dry and Wet Bulb Thermometers** is one of the most convenient and reliable. However, in order to improve its measurement accuracy it is necessary to study the relation between the measured wet bulb temperature T_w and the adiabatic saturation temperature T* of moist air.

The adiabatic saturation temperature is the equilibrium temperature of moist air under adiabatic conditions, and it is very difficult to measure. In practice the wet bulb temperature is often used as a substitute for the adiabatic saturation temperature although the measured wet bulb temperature is quite different from the adiabatic saturation temperature, because it is determined under nonadiabatic contitions.

A dry/wet bulb thermometer is shown in **Figure 1**. When moist air flows along the wet bulb surface, a very thin boundary layer of saturated most air forms around it. The airstream (with a mass flow $\dot{m}_a$.) flowing near the saturated boundary layer will absorb a certain amount of vapour (with a mass flow $\dot{m}_v$) diffusing from the saturated layer. Because of insufficient heat and mass exchange the airflow cannot reach the saturated state and it will have a temperature higher than the wet bulb temperature. Since there is heat and mass exchange between the boundary layer and the airflow,

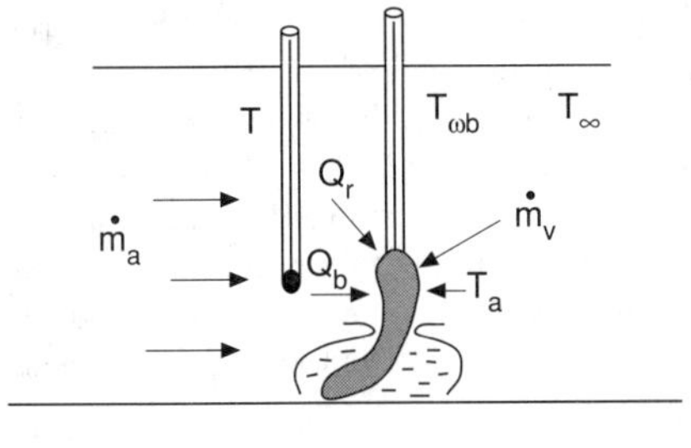

Figure 1. Dry/wet bulb thermometer.

the temperature in the saturated boundary layer cannot be treated as the adiabatic saturation temperature; that is, the measured wet bulb temperature is not the adiabatic saturated temperature. The measured wet bulb temperature is affected by air velocity, wet bulb diameter, the heat exchange with surroundings and the thermal state of the moist air.

Experiments (Zhao Yuzhen (1992)) have been conducted under the following conditions: dry bulb temperature from 16°C to 88.3°C; air velocity from 0 to 40m/s, and wet bulb diameters of 6, 8, 10, 11 and 12mm. The experimental curves are shown in **Figure 2**. For the unshielded dry/wet bulb thermometer, the relative deviation $(T_{wb} - T^*)/(T - T_{wb})$, is shown with a solid curve drawn through the data points. For the shielded dry/wet bulb thermometer data are also plotted in **Figure 2**. and correlated with the dashed curve. It can be seen that for the unshielded wet bulb, as ln(ReA/D^2) increases T_{wb} tends to T*; at ln (ReA/D^2) = 16.2, the deviation is zero, that is T_{wb} = T*, and when ln (ReA/D^2) > 16.5, the deviation become constant.

The explanation of this phenomenon is as follows. When ln (ReA/D^2) is comparatively small, the evaporating vapor is transferred mainly by the pressure difference of vapor in the boundary layer, so it spreads slowly. For the unshielded wet bulb thermometer, the latent heat of vaporization of water at the wet bulb is supplied through radiation as well as convection, so the partial pressure and the corresponding saturation temperature of the vapor around the wet bulb are higher, thus resulting in the wet bulb temperature being higher than the adiabatic saturated temperature. When ln (ReA/D^2) increases, the vapor diffusion not only depends on the pressure differences, but also upon the influence of the flow, so the partial pressure of the vapor arund the wet bulb falls, and its saturated temperature becomes lower and tends to the adiabatic saturated temperature. If ln (ReA/D^2) > 16.2 vapor is removed instantly by flowing air, the partial pressure of vapor around the wet bulb remains stable and so does the wet bulb temperature. It can be seen from **Figure 2** that for an unshielded wet bulb, when ln (ReA/D^2) > 16, the wet bulb temperature is much closer to the adiabatic saturated temperature.

From the solid line in **Figure 2**. the following expression is obtained in which the dependence of the relative deviation on Re and wet bulb diameter D is correlated as:

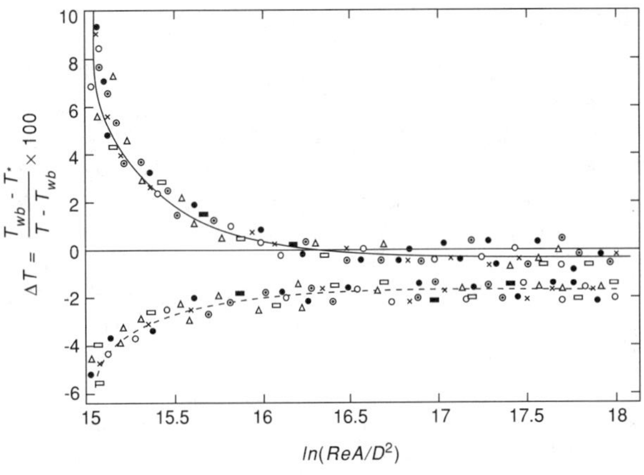

Figure 2. Experimental results.

$$\frac{T_{Wb} - T^*}{T - T_{Wb}} 100 = 746.9 - 126.87\left[\ln\left(\frac{Re\ A}{D^2}\right)\right] + 7.24\left[\ln\left(\frac{Re\ A}{D^2}\right)\right]^2 - 0.1395[\ln(Re\ A/D^2)]^3 \quad (1)$$

Zhao Yuzhen (1991) recommended a correlation for the following conditions: dry bulb temperature at $5 \sim -15°C$; wind velocity varied from o to 40m/s and wet bulb diameters of 2, 5, 6, 7, 5, 8mm.

$$\frac{T_{wb} - T^*}{T - T_{wb}} 100 = 119.1 - 21.6\left[\ln\left(\frac{Re\ A}{D^2}\right)\right] + 1.191\left[\ln\left(\frac{Re\ A}{D^2}\right)\right]^2 - 0.0186[\ln(Re\ A/D^2)]^3 \quad (2)$$

James L. Threlkeld (1970), obtained two correlations: When the dry bulb temperature equals that of the surroundings, the relative deviation $(T_{wb} - T^*)/(T - T_{wb})$ is:

$$100\frac{T_{wb} - T^*}{T - T_{wb}} = \frac{Le(1 + h_c/h_r) - 1}{1 + B/k^*} \quad (3)$$

For the shielded dry/wet bulb thermometer the following formula is employed:

$$100\frac{T_{wb} - T^*}{T - T_{wb}} = \frac{Le - 1}{1 + B/k^*} \quad (4)$$

where

h_r = radiation heat transfer coefficient
h_c = convection heat transfer coefficient
B = coefficient of wet bulb
$k^* = \frac{c_p}{r}$

Kent (1985), recommends a new type dry/wet bulb thermometer, which can be used for the measurement of high temperature moist air. With dry bulb temperature 200°C and wet bulb temperature 99°C and an uncertainty of less than 2% is claimed.

References

Threlkeld, J. L. (1970) Thermal Environmental Engineering 2nd ed. Prentice-Hall, Englewood Cliffs, N.J.

Kent, A. C. (1985) An Aspirated Humidity and Energy Meter for High Temperature Moist Air. Proceedings of Moisture and Humidity.

Zhao Yuzhen and Jiang Baocheng, (1992) Experimental Study of the Effects of Wind Speed, Radiation and Wet Bulb Diameter on Wet Bulb Temperature. Expermental Thermal and science.

Zhao Yuzhen and Jiang Baocheng, (1991) The Influence on Wet Bulb Temperature of Wind Velocity, Radiation and Wet Bulb Diameter at Low Temperature. Journal of Engineering Thermophysics of China.

Leading to: *Dry bulb temperature; wet bulb temperature*

Z. Yuzhen

MOISTURE METER (see Humidity measurement)

MOLALITY OF A SOLUTION (see Solubility)

MOLAR MASS

Following from: *Mole*

The molar mass (symbol M, SI unit $kg \cdot mol^{-1}$) is defined as the mass per unit amount of substance of a given chemical entity. In keeping with the definition of the mole, the chemical entity in question should always be specified. Usually this will be a recognised atom, molecule or ion, but any collection, or indeed fragment, of such species may be specified. The molar mass of an element E may be obtained from tables such as those in this Encyclopaedia; usually these give the dimensionless '**Atomic Weight**' $A_r = M(E)/(10^{-3}\ kg \cdot mol^{-1})$. For a molecule, the molar mass is obtained by summing the molar mass of the constituent atoms. In the case of atomic or molecular ions, allowance should strictly be made for the surplus or deficit of electrons.

The term '*molecular weight*' is commonly attached to the quantity M even though M is a mass (not a weight) of 1 mol (not one molecule) of a particular substance. Often the same term is incorrectly used for the dimensionless *relative molar mass.* In view of this confused situation, use of the term molecular weight should be abandoned in favour of the molar mass or the **Relative Molar Mass**, whichever is meant. Similarly, the term '*molecular mass*' properly means the mass of one molecule and is therefore not a synonym for molar mass.

J.P.M. Trusler

MOLD CONSTANT (see Casting of metals)

MOLE

The mole (symbol mol) is the SI unit for the quantity amount of substance (symbol n). It is defined as the amount of substance that contains as many of the specified entities as there are ^{12}C atoms in 0.012 kg of ^{12}C. It is essential to specify the entities in question, such as atoms, molecules, ions, molecular fragments, electrons, protons, etc., or any defined collection of such entities. It is therefore possible to specify the amount of substance such as 1 mol of SO_4^{2-}, 1 mol of $2H^+$, or 1 mol of H_2SO_4. It is also correct to specify, for example, 1 mol of $^1/_2\ Cl_2$ or 1 mol of $CH_3\cdot$.

The actual number of entities present in amount of substance n is of course equal to Ln where $L = 6.0221367 \times 10^{23}\ mol^{-1}$ is Avagadro's constant. (**Avogadro Number**). Similarly, the mass of material present in amount of substance n is Mn where M is the molar mass of the specified entities. Thus, the amount of substance can frequently be determined from mass measurement.

J.P.M. Trusler

MOLECULAR DIFFUSION (see Diffusion)

MOLECULAR FLOW OF GAS (see External flows, overview)

MOLECULAR INTERACTIONS (see Kinetic theory of gases; Mean free path)

MOLECULAR MASS (see Molar mass)

MOLECULAR PARTITION FUNCTION (see Boltzmann distribution)

MOLECULAR SCATTERING (see Kinetic theory of gases)

MOLECULAR SPECTROSCOPY (see Spectroscopy)

MOLECULAR SPEEDS (see Maxwell-Boltzmann distribution)

MOLECULAR SIEVES

The term molecular sieves describes microporous media capable of separating molecules on the basis of size. It can relate to certain carbons and silicas, as well as porous gels/resins for *polymer separations,* but molecular sieving is best illustrated by the *aluminosilicate zeolites.* The first separations were on chabazite, a natural mineral, but now *synthetic zeolites* (e.g. zeolites A, X Y) are widely used.

Zeolites have unique 3-D molecular architectures based on linking $[AlO_4]^{5-}$ and $[SiO_4]^{4-}$ tetrahedra. Their structures contain cations and water molecules, located in regular channels and cavities. Two examples are in **Figures 1 (a) and (b)**.

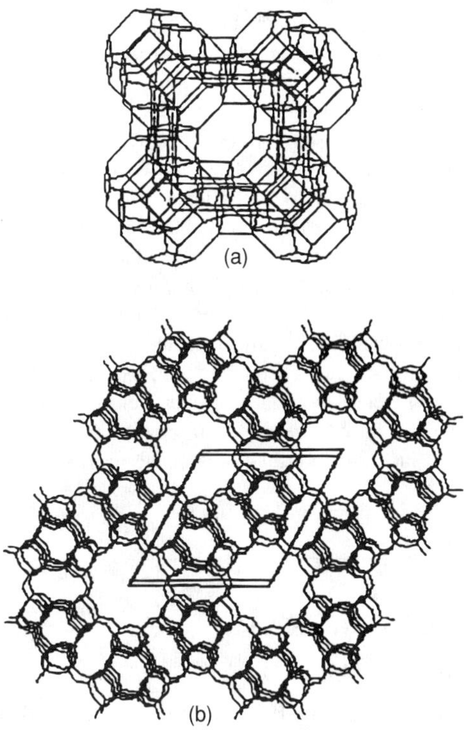

Figure 1. (a) Synthetic zeolite A (b) synthetic zeolite L.

In these schematic diagrams, a line represents an oxygen atom, and an intersection Si/Al. Entry to internal cavities/channels is via pore openings whose size is defined by the number of oxygen atoms forming the constricting "window". Removal of water enables zeolites to selectively *adsorb* (or "sorb") gaseous molecules as a function of window size. Synthetic zeolite 4A takes in, e.g. CO, H_2O, NH_3, N_2, CH_4, C_2H_6, CO_2 and excludes larger molecules (4A represents 4 Å the restricting dimensions).

Synthetic zeolite X has larger pores (7.8 Å) and can accept e.g. SF_6, C_6H_6, naphthalene. The mechanism of molecular sieving is a complex interaction to which zeolite structure, cation content and Si/Al composition all contribute. Permeation of gases through zeolites is a diffusion process so temperature controls gas separations. Additionally, molecular polarizability/dipolar nature will contribute to some separations.

Zeolites are widely used in industry as *desiccants* (air supplies, refrigerators, double glazing units, vehicle braking, large scale drying of H_2,O_2, liquid propane gas, ethylene, propylene, natural gas). They also perform many useful separations [e.g. CO_2 from natural gas and air, "*sweetening*" of gases (removal of S,N. compounds), i-n paraffins, benzene/toluene/xylene, oxygen/nitrogen enrichment from air]. They also separate liquids (alcohol from water, p-xylene and ethylbenzene from their isomers, sugars). Zeolites are widely used as *catalysts* [cracking, hydrocracking, selectoforming, methanol to gasoline (MTG)] and in some cases, products selectivity can be controlled by molecular sieving (*shape selective catalysis*).

References

Breck, D. W. (1984) Zeolite Molecular Sieves, Robert. E. Krieger, Florida.

Dyer, A. (1988) *An Introduction to Zeolite Molecular Sieves*, John Wiley, Chichester.

Ruthven, D. M. (1984) *Principles of Adsorption and Adsorption Processes*, John Wiley, New York.

***Leading from:** Adsorption*

A. Dyer

MOLECULAR WEIGHT (see Molar mass; Relative molar mass)

MOLECULE

***Following from:** Atom*

A molecule may be defined as a finite collection of atoms, possibly bearing a net electrical charge, between which there exists some kind of binding. Thus a molecule may be as simple as H_2 or H_2^+ or as complex as a polymer chain or a protein chain. The most common kind of chemical binding in molecules is that of the *covalent bond,* in which the atoms share electrons and thereby lower the overall energy of the molecule relative to that of the separated atoms. Such molecules usually have some degree of stability. *Ionic bonding* is common in extended arrays of atoms but much less so in small isolated molecules. However, many large molecules have three dimensional conformations which are stabilised in solution by interactions between ionic groups within the molecule. Other kinds of intramolecular forces, such as *van der Waals forces* and, especially, hydrogen bonding, play an important part in determining the conformation, and hence chemical and biochemical activity, of large molecules. Sometimes two molecules which do not react chemically with each other nevertheless combine or associate by means of hydrogen bonding or van der Waals forces to form a *dimer* which may itself be considered as a molecule. Such dimers usually dissociate readily.

J.P.M. Trusler

MOLLIER DIAGRAM

Following from: *Thermodynamics*

Richard Mollier (1863–1935) spent most of his working life at the Technische Hochschule in Dresden studying the properties of thermodynamic media and their effective representation in the form of charts and diagrams. His major contribution was in popularizing the use of enthalpy. In 1904, Mollier devised the first enthalpy—entropy chart still most closely associated with his name. However, he published a number of other enthalpy-based charts and in recognition of his work, the US Bureau of Standards recommended in 1923 that all such charts should be known as Mollier diagrams.

Mollier's H-S diagram (Enthalpy v Entropy) was a logical extension of the T-S diagram (Temperature v Entropy) first proposed by Gibbs, retaining the advantages of T-S diagrams but introducing several new advantages. A typical H-S Mollier diagram for a thermodynamic fluid such as steam is shown in **Figure 1**.

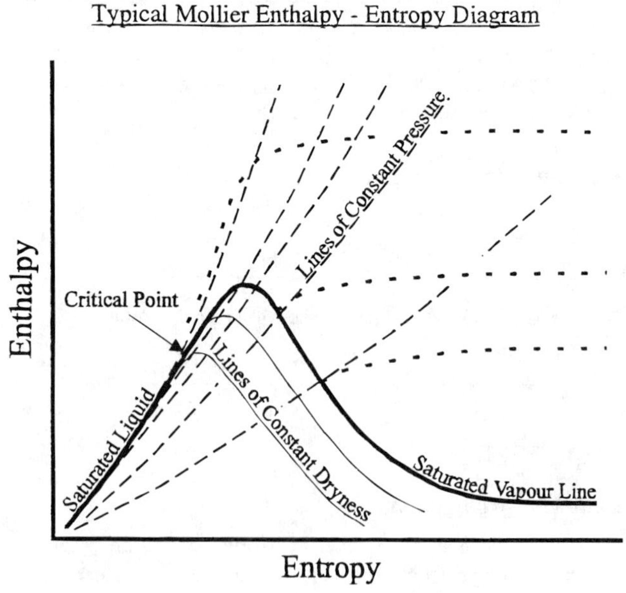

Figure 1.

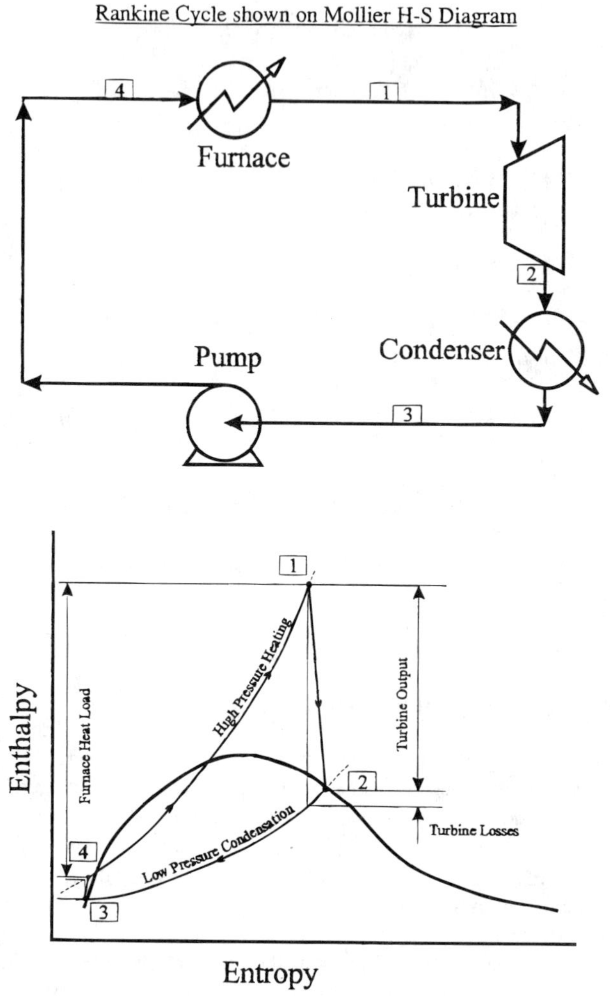

Figure 2.

The advantages of such a diagram are that *vertical lines represent reversible processes and horizontal lines represent lines of constant energy.* Power generation and refrigeration cycles are most conveniently represented on H-S diagrams because work can be calculated directly from vertical distances as opposed to areas on T-S and P-V diagrams. Further the inefficiencies due to irreversibility in real processes are shown clearly on an H-S diagram as shown in **Figure 2**.

N.F. Kirkby

MOLYBDENUM

Molybdenum—(Gr. *molybdos,* lead), Mo; atomic weight 95.94 ± 0.01; atomic number 42; melting point 2617°C; boiling point 4612°C; specific gravity 10.22 (20°C); valence 2, 3, 4?, 5?, or 6.

Before Scheele recognized molybdenite as a distinct ore of a new element in 1778, it was confused with graphite and lead ore. The metal molybdenum was prepared in an impure form in 1782 by Hjelm. Molybdenum does not occur native, but is obtained principally from *molybdenite* (MoS_2). *Wulfenite* ($PbMoO_4$), and *powellite* ($Ca(MoW)O_4$) are also minor commercial ores. Molybdenum is also recovered as a by-product of copper and tungsten mining operations. The metal is prepared from the powder made by the hydrogen reduction of purified molybdic trioxide or ammonium molybdate. The metal is silvery white, very hard, but is softer and more ductile than tungsten. It has a high elastic modulus, and only tungsten and tantalum, of the more readily available metals, have higher melting points. It is a valuable alloying agent, as it contributes to the hardenability and toughness of quenched and tempered steels. It also improves the strength of steel at high temperatures. It is used in certain nickel-based alloys, such as the "Hastelloys,®" which are heat-resistant and corrosion-resistant to chemical solutions. Molybdenum oxidizes at elevated temperatures.

The metal has found recent application as electrodes for electrically heated glass furnaces and forehearths. The metal is also used in nuclear energy applications and for missile and aircraft parts. Molybdenum is valuable as a catalyst in the refining of petroleum. It has found application as a filament material in electronic and electrical applications. Molybdenum is an essential trace element in plant nutrition. Some lands are barren for lack of this element in the soil. Molybdenum sulfide is useful as a lubricant, especially at high temperatures where oils would decompose. Almost all ultra-high strength steels with minimum yield points up to 300,000 psi(lb/in.^{2}) contain molybdenum in amounts from 0.25 to 8%.

Handbook of Chemistry and Physics, CRC Press

MONOCHROMATIC LIGHT (see Optics)

MONODISPERSE AEROSOLS (see Aerosols)

MONOD MODEL OF CELL GROWTH (see Biochemical engineering)

MONTE CARLO METHOD (see Non-linear systems; Optimization of exchanger design; Radiative heat transfer)

MONTE CARLO MODELLING, OF TURBULENCE (See Turbulent flow)

MONTREAL PROTOCOL (see Heat pumps; Refrigeration)

MOODY CHART (see Friction factors for single phase flow)

MOODY, OR WEISBACH, FRICTION FACTOR (see Tubes, single phase flow in)

MOSSBAUER SPECTROSCOPY (see Spectroscopy)

MOTOR GASOLINE (see Oil refining; Oils)

MOULD (see Casting of metals)

MOVING BOUNDARY PROBLEMS (see Solidification)

MTBE (see Oils)

MUFFLE FURNACE (see Kilns)

MULTICOMPONENT MIXTURES, BOILING IN

Following from: *Boiling*

Frequently, in industrial processes, liquid mixtures of two or more components have to be evaporated in order to separate them from one another. It is known from experiments that heat transfer coefficients during evaporation of mixtures can be substantially smaller than those of the pure components of the mixture. On the other hand, marked improvements of heat transfer have been noted if one of the components of the mixture is surface-active. Mixtures of organic or inorganic liquids, however, contain surface-active components only in certain cases (soaps, addition of wetting agents), so that a decrease of the heat transfer coefficient is inevitable in comparison with that of the pure components.

Bonilla and Perry (1941) were the first to investigate a large number of binary mixtures of organic liquids and of water with organic liquids. Detailed discussions of the phenomena and experimental results are given by Collier (1972), van Stralen and Cole (1979) and Stephan (1992). As an example, heat transfer coefficients for ethanol-water mixtures are plotted in **Figure 1** for a pressure of 1 bar and a heat flux of 10^5 W/m^2. As is recognized, the heat transfer coefficients, $\alpha = \dot{q}/\Delta T$, of the mixture, where $\dot{q}$ is the heat flux and ΔT the difference between wall and saturation temperature, are clearly smaller than the values α_{id} that would be obtained if one were to interpolate linearly between the heat transfer coefficients of the pure components. One recognizes also a clear decrease of the heat transfer coefficient in the region, in which vapor and liquid compositions $\tilde{y} - \tilde{x}$ are strongly different, as is seen from a comparison with the upper curve in **Figure 1**.

As shown by experiments, the heat transfer decreases with the difference in concentration more strongly at high pressures than at low pressures. This can be explained by the fact that the number of vapor bubbles being formed per surface unit increases with the pressure and there is less surface available for the mass flow rate in the liquid space.

One notices several peculiarities during the boiling of *oil-refrigerant-mixtures,* which can be encountered frequently in the evaporators of refrigeration plants, because the refrigerant always drags along some lubricating oil from the compressor into the evaporator. According to **Figure 2**, depending upon the type of oil and the heat flux, small amounts of oil can lead to a slight decrease of the heat transfer coefficient, and amounts up to 3% by mass fraction can even lead to an increase, resulting in a better heat transfer than that of the pure refrigerant. In general, one can conclude that mass fractions of oil over 5% lead to large reductions in heat transfer.

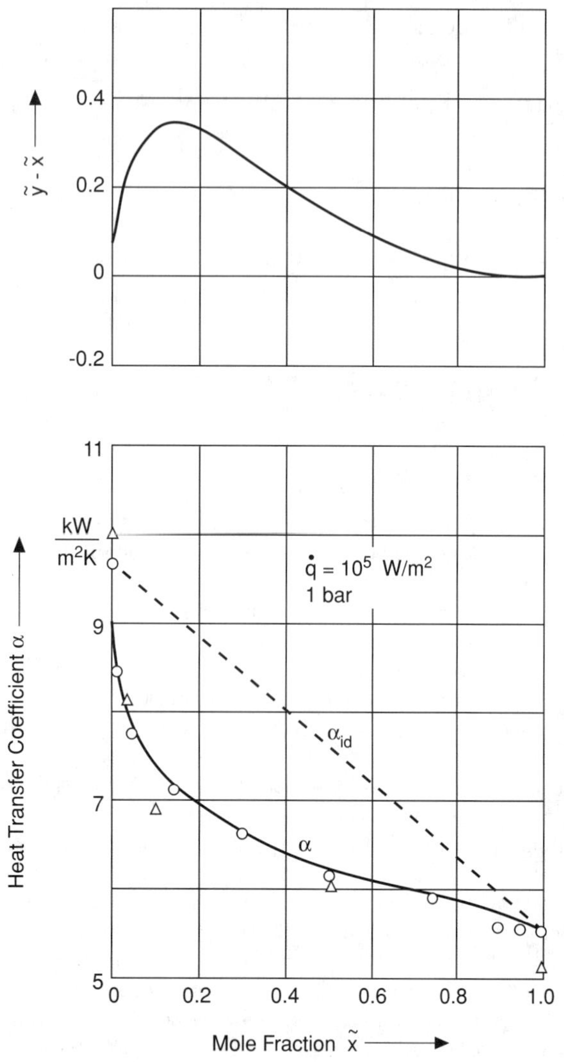

Figure 1. Heat transfer coefficients for boiling ethanol-water mixtures. $\tilde{y}$ is the mole fraction of ethanol in the vapor and $\tilde{x}$ is the mole fraction of ethanol in the liquid.

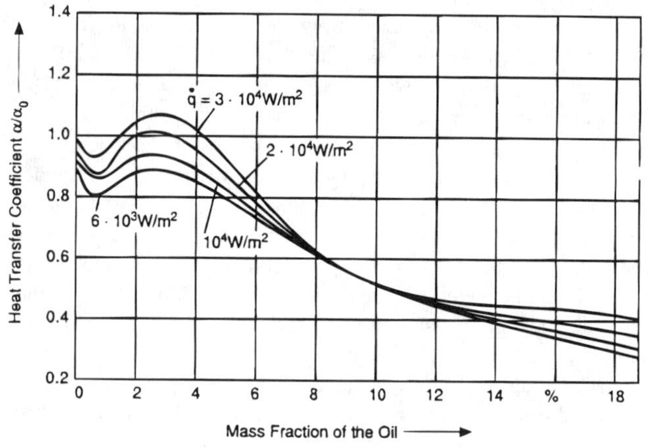

Figure 2. Ratio of the heat transfer coefficients α/α_0. α is the heat transfer coefficient of the oil-refrigerant mixture. α_o that of the oil-free refrigerant R 114 at 1.285 bar

One should, therefore, take suitable measures, such as in the use of an oil separator after the compressor, to ensure that the oil content in the evaporator is always below a mass fraction of 5%.

Investigations of the heat transfer in mixtures with more than three components remain unknown, and until now the heat transfer of only a few ternary mixtures has been measured. Basically, the findings for binary mixtures were confirmed.

Essentially two methods have proven reliable for the reproduction of heat transfer data. One starts from empirical correlations for pure substances. Such correlations usually contain dimensionless parameters, which can now be formed with the properties of the mixtures. By use of an additional term, allowance is made for the decrease of heat transfer resulting from the obstruction of bubble growth by diffusion:

$$\alpha = \alpha_0 f \tag{1}$$

In it the heat transfer coefficient α_0 can be calculated from the equation established for pure substances, (**see Boiling**). One must, however, enter the properties of the mixture.

A considerable amount of the reduction of heat transfer in mixtures, compared to pure substances, is conditioned by the change in the thermal properties, whereas the reduction factor f makes a comparatively small contribution. It lies between 0.8 and close to unity for most hydrocarbon mixtures and the mixtures of hydrocarbons with water. In order to avoid the lengthy computation of the property values of the mixture one prefers simple correlation procedures. Such a procedure starts from the fact that, for the transfer of a definite heat flux to the mixture, a larger wall superheat $\Delta T = T_w - T_s$ is required than for the evaporation of pure substances. The saturation temperature T_s in this case is the boiling temperature of the mixture with the mean composition x of the liquid. In order to calculate the superheat, an "ideal" wall superheat ΔT_{id} is defined by

$$\Delta T_{id} = \sum_{i=1}^{k} \tilde{x}_i \Delta T_i, \tag{2}$$

in which the temperature differences ΔT_i between the wall and saturation temperatures result from the heat transfer coefficients α_i of the pure substances with the heat flux $\dot{q}$ of the mixture according to $\Delta T_i = \dot{q}/\alpha_i$ and can be calculated from pure component-equations. The actual driving temperature difference ΔT is different from the ideal one

$$\Delta T = \Delta T_{id} + \Delta T_E \quad \text{or} \quad \Delta T = \Delta T_{id}(1 + \theta) \tag{3}$$

with $\theta = \Delta T_E/\Delta T_{id}$. The supplementary term depends mainly on the difference between vapor and liquid composition, and is always positive because of the reduction of the heat transfer in the mixture. Experiments with binary mixtures yielded the simple linear relation

$$\theta = K_{12}|\tilde{y} - \tilde{x}|, \tag{4}$$

where K_{12} is a positive number, which is approximately independent of the composition. One can interpret K_{12} as a binary interaction parameter that must be determined for a given mixture and pressure.

In the pressure range between 1 and 10 bar, the pressure dependence of K_{12} could be reproduced approximately by the empirical equation

$$K_{12} = K_{12}^0(0.88 + 0.12\ p/p_0) \tag{5}$$

with $p_o = 1$ bar and $p \leq 10$ bar. The value K_{12}^0 is different for every mixture, but is independent of pressure. Values K_{12}^0 for different binary mixtures are given by Stephan (1992). A mean value for K_{12}^0 for all these mixtures is approximately 1.4.

For a mixture of K components, we have instead of Equation 4

$$\theta = \left|\sum_{i=1}^{K-1} K_{iK}(\tilde{y}_i - \tilde{x}_i)\right|. \tag{6}$$

Here K_{ik} are interaction parameters that must be ordered in a sequence according to increasing boiling points of the pure substances.

For a ternary mixture, Equation 6 reduces to

$$\theta = |K_{13}(\tilde{y}_1 - \tilde{x}_1) + K_{23}(\tilde{y}_2 - \tilde{x}_2)|, \tag{7}$$

which contains the associated binary mixtures as limiting cases. If one assumes $\tilde{y}_1 = \tilde{x}_1 = 0$, then one reduces the ternary to the binary mixture of components 2 and 3, whereas for $\tilde{y}_2 = \tilde{x}_2 = 0$, the associated binary mixture consists of components 1 and 3. One recognizes from this that the coefficients K_{iK} with $K_{iK} = 0.14$ as a mean value, are identical with the values for binary mixtures. Heat transfer coefficients calculated according to this method reproduce quite well the previously known test data for ternary mixtures.

Solutions of Solids in Liquids

The boiling point rises in an isobaric solution of solids in liquids. The driving temperature difference between wall and saturation temperature thus decreases with the amount of the dissolved solids. The vapor consists of the pure solvent. In the vicinity of the wall, the solvent must overcome a mass transfer resistance in order to get from the liquid to the bubble surface. This impedes bubble growth, and, in comparison with the pure solvent, reduces the heat transfer, exactly as in the case of the binary and multi-component mixtures of liquids.

Experiments with aqueous solutions of sucrose, sodium chloride, sodium hydroxide, and ammonium nitrate in the pressure range between 1 and 16 bar lead to the empirical correlation

$$\frac{\alpha}{\alpha_w} = \left(\frac{\rho_L}{\rho_{Lw}}\right)^{0.816} \left(\frac{\rho_G \Delta h_{LG} c_{pw}}{\rho_{Gw} \Delta h_{LG,w} c_p}\right)^{-0.716} \left(\frac{\rho_L - \rho_G}{\rho_{Lw} - \rho_{Gw}}\right)^{-1.0795} \cdot \left(\frac{\eta}{\eta_w}\right)^{-0.1} \left(\frac{\sigma}{\sigma_w}\right)^{-1.2135} \left(\frac{\lambda}{\lambda_w}\right)^{0.284}. \tag{8}$$

The index L signifies liquid, G gas, and w water. The property values $\rho_L, c_p, \eta, \lambda$ of the solution and the surface tension σ are to be evaluated at the mean temperature $T_m = (T_w + T_s)/2$ and the mass fraction x of the solution.

Boiling of Immiscible Liquids

In a binary mixture consisting of two immiscible liquid phases and the vapor phase, the boiling temperature is clearly determined, according to the Gibbs phase rule, by prescribing the pressure. The boiling temperature of the mixture is less than that of the pure components. If, for example, one mixes water with perchloroethylene, two immiscible liquid phases are formed, whose boiling temperature at a pressure of 1 bar lies at 87.8°C, whereas pure water boils at about 100 °C and pure perchloroethylene at about 121 °C. If, therefore, one wishes to evaporate a liquid that would decompose at its boiling temperature, one can add an immiscible liquid or its vapor and thus lower the boiling temperature. The heat transfer to such immiscible liquids is determined to a great extent by which of the two liquid phases touches the heating surface. If, for example, a mixture of water/perchloroethylene fills a container with a horizontal heating surface at the bottom, then the lower phase contains mainly the heavier and less volatile perchloroethylene, whereas the upper phase consists primarily of water. At sufficiently large heat fluxes, either nucleate or film boiling occurs, depending upon the heat flux.

In spite of this very complicated phenomenon, the heat transfer during the boiling of immiscible liquids is determined to a great extent by the properties of the liquid at the heating surface. Measurements could be reproduced quite well as long as there was film boiling at the wall with an equation, which starts with the known equations for film boiling and allows for the influence of radiation and the convective heat transport in the case of a subcooled liquid:

$$\alpha = \alpha_f + 0.88\alpha_r + 0.12\alpha_c\theta. \tag{9}$$

α_f is the heat transfer coefficient for film boiling, which is calculated from

$$\alpha_f = 0.41\left[\frac{\lambda_G^3 \rho_G(\rho_L - \rho_G) g \Delta h}{\eta_G (T_w - T_s)\sqrt{\dfrac{\delta}{g(\rho_L - \rho_G)}}}\right]^{1/4} \tag{10}$$

with $\Delta h = \Delta h_{LG} + 0.95 c_{vG}(T_w - T_s)$. The quantity α_r is the heat transfer coefficient for radiation, $\alpha_r = \dot{q}_r / (T_w - T_s)$. The heat transfer coefficient α_c is obtained from one of the known equations for heat transfer during turbulent free convection in the vapor film. A subcooling factor $\theta = (T_s - T_b) / (T_w - T_s)$ plays a role, if the temperature T_b in the core of the liquid is below the boiling temperature. The properties are those of the phase at the heated surface. A condition for the use of these equations is that the fluid layers are thicker than 4.5 cm. The above equations are not valid if there is nucleate boiling at the wall. In this case, the heat transfer with boiling immiscible liquids has not yet been sufficiently researched.

Boiling of Mixtures in Forced Flow

As in the case of pure substances, during the boiling of mixtures in forced flow in tubes or channels, there is a decrease in saturation temperature because of the pressure drop along the path of flow. In addition, because the more volatile component is converted into vapor first, the liquid becomes richer in the less volatile component.

In many industrial processes where the tube diameter is not very small, and, thus, the pressure drop not very large, the rise in boiling temperature, because of the increase in the less volatile component in the liquid, offsets the decrease in boiling temperature as a consequence of the pressure drop. It is thus possible that the saturation temperature even rises downstream. In a two-fluid heat exchanger, the driving temperature gradient decreases. In general, the heat transfer coefficient also decreases.

Figure 3 shows the course of the vapor temperature T_G in the core of an annular flow of an evaporating liquid film which is directed upward in a vertical evaporator tube. The course of the phase interface temperature T_I on the vapor side of the liquid film is also plotted in the figure. The mixture evaporated was ethanol/water. The concentration of the higher boiling component in the liquid, in most cases, causes an increase in viscosity, which likewise contributes to a reduction in the heat transfer coefficient.

In the case of boiling of pure substance convective boiling is in general governed by two mechanisms (see **Boiling**). In the extreme case when the heat flux is too low to support nucleate boiling, vapor is generated within the stream from minute nuclei and by evaporation on the vapor liquid interface. We then have "pure convective boiling". In most practical applications, however, heat flux and wall superheat are high enough for the onset of nucleate boiling. Nucleate and pure convective boiling are superimposed according to their relative magnitudes.

The onset of nucleate boiling (see **Nucleate Boiling**) is given by

$$\dot{q}_{0NB} = \frac{2\sigma T_s \alpha_c}{r_{cr} \rho_G \Delta h_v}, \qquad (11)$$

wherein the critical radius r_{cr} for usual drawn tube materials $r_{cr} = 0.3 \cdot 10^{-6}$ m is recommended. α_C is the convective heat transfer coefficient. It can be obtained from an equation for single-phase heat transfer during turbulent tube flow (see **Convective Heat Transfer**) for example from

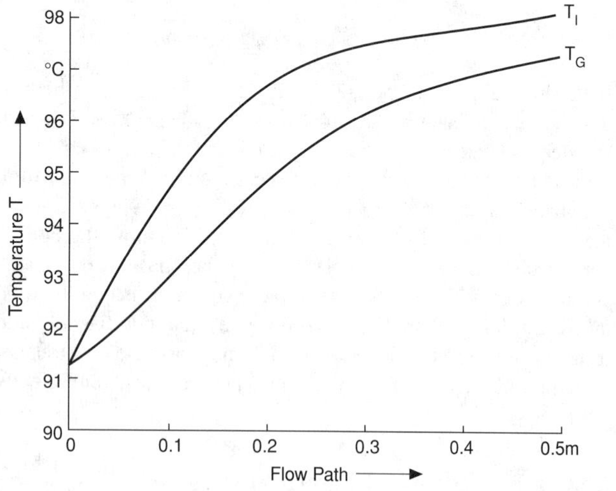

Figure 3. Temperature T_G of the vapor and T_I at the phase interface of a boiling ethanol/water mixture with upwards directed annular flow in a vertical tube of 37 mm inner diameter. At inlet: mass flow rate $\dot{M} = 0.1$ kg/s, quality $x^* = 0.05$, mole fraction of ethanol $\tilde{x} = 0.041$. Heat flux $\dot{q} = 2 \cdot 10^6$ W / m^2.

$$Nu = \alpha_c d/\lambda_L = 0.023\, Re^{0.8} Pr^{0.4} \qquad (12)$$

with the **Reynolds Number** $Re = \dot{m}_L d / \eta_L$ and the **Prandtl Number** Pr of the liquid.

Until recently two models of convective boiling were most often used, one by Chen (1963) and one by Shah (1976). Neither model presents a satisfactory solution, as attested by numerous publications. A new model based on an asymptotic addition of the two boiling components was recently presented by Steiner and Taborek (1992) and tested over 13000 data points in vertical convective boiling, mostly of pure substances. It applies to mixtures, as well. One must, however, enter the properties of the mixture. As in the case of pure substances heat fluxes must be below the critical heat flux where film boiling or dry-out of the heated surface occurs. For steam-water no dry-out will occur up to vapor qualities of $x^* = 0.99$. The convective boiling heat transfer coefficient α_{2Ph} is composed of a nucleate boiling and a forced convection contribution

$$\alpha_{2Ph} = [(\alpha_{B,o} F_{nbf})^3 + (\alpha_C F_{2Ph})^3]^{1/3}. \qquad (13)$$

Here the nucleate boiling heat transfer coefficient $\alpha_{B,o}$ is the value at normalized conditions $\dot{q}_0, p_r, o$, for example $\dot{q}_0 = 20000$ W/m^2, $p_r o = 0.1$ derived from the appropriate equation for pure substances in **Nucleate Boiling**.

The nucleate flow boiling correction factor is given by

$$F_{nbf} = \left[2.816 p_r^{0.45} + \left(3.4 + \frac{1.7}{1 - p_r^7}\right) p_r^{3.7}\right] \cdot \left(\frac{\dot{q}}{\dot{q}_0}\right)^n \left(\frac{d}{0.01\text{m}}\right)^{-0.4} \left(\frac{Ra}{1\mu\text{m}}\right)^{0.133} f(M) \qquad (14)$$

with $p_r \leq 0.95$ and $n = 0.8 - 0.1 \exp(1{,}75\, p_r)$ for all fluids except cryogenics. For cryogenics we have $n = 0.7 - 0.13 \exp(1.105\, p_r)$. The residual correction f(M) is a function of the molar mass, see **Figure 4**. An equation representing the values of **Figure 4** can be found in the paper of Steiner and Taborek (1992).

The nucleate boiling correction term $\alpha_{B,0} F_{nbf}$ in **Equation 13** can only be used if the wall superheat or heat flux is above a

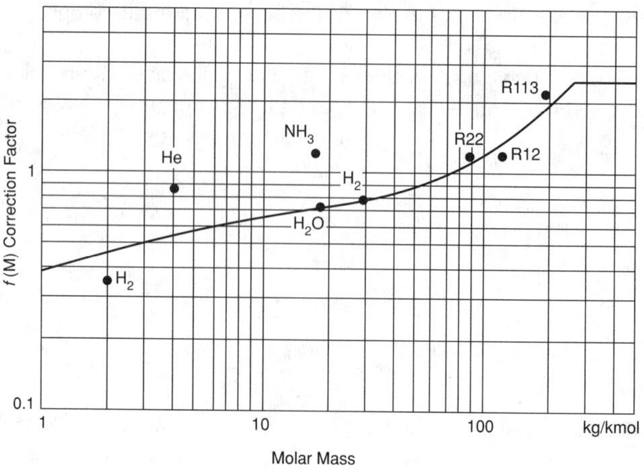

Figure 4. Correction factor F(M) as a function of molecular mass.

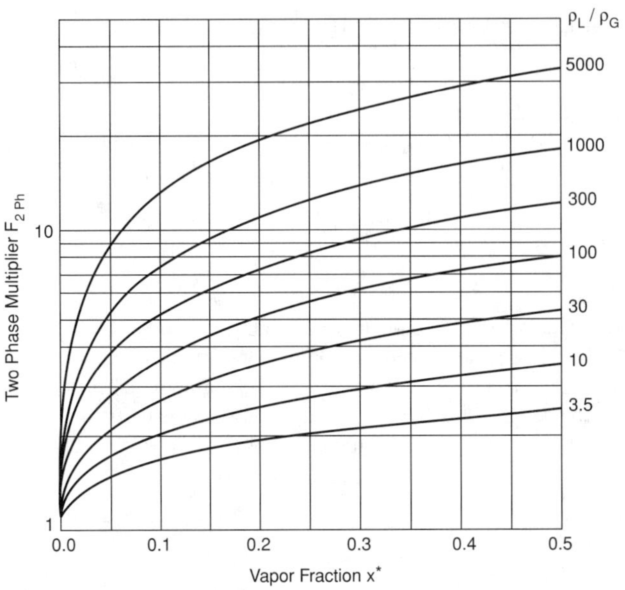

Figure 5. Two-phase flow multiplier F_{2Ph} as a function of vapor fraction x^*. Validity restricted if x^*_{cr} is reached.

certain minimum value $\dot{q}_{oNB}$ required for the onset of nucleate boiling, **Equation 11**.

The local convective heat transfer coefficient α_C in Equation 13 can be calculated from one of the standard equations for convective heat transfer, such as **Equation 12**. F_{2Ph} is two-phase multiplier accounting for the enhancement of heat transfer in the liquid vapor mixture. It is a function of the vapor fraction, x^* $\dot{q}$ and the ratio of liquid to vapor density, ρ_L/ρ_G **Figure 5**. An equation representing the values of **Figure 5** is also given in the paper of Steiner and Taborek (1992).

Material from Stephan, K. (1992) Heat Transfer in Condensation and Boiling, by permission of Springer Verlag GmbH.

References

Bonilla, C. F. and Perry, C. W. (1941) Heat Transmission to Boiling Mixtures. *Am. Inst. Chem. Eng. J.* 37 685–705.

Chen, J. C. (1966) A Correlation for Boiling Heat Transfer to Saturated Liquids in Convective Flow, *Ind. Engng. Chem. Process Design Developm.* 5 322–329.

Collier, J. G. (1972) *Convective Boiling and Condensation,* Mac Graw Hill, London.

Shah, M. M. (1976) A New Correlation for Heat Transfer During Boiling Flow Through Pipes, *ASHRAE Trans.,* 82 66–86.

Steiner, D., Taborek, J. (1992) Flow Boiling Heat Transfer in Vertical Tubes Correlated by an Asymptotic Model, *Heat Transfer Eng.* 13 43–69.

Stephan, K. (1992) *Heat Transfer in Condensation and Boiling,* Springer, Berlin.

van Stralen, S. and Cole, R. (1979) *Boiling Phenomena,* Vol. 1 and 2, Hemisphere, Washington.

K. Stephan

MULTICOMPONENT MIXTURES, DIFFUSION IN (see Fick's law of diffusion)

(see Fick's law of diffusion)

MULTI-COMPONENT SYSTEMS THERMODYNAMICS

***Following from:** Thermodynamics; Mass transfer*

Multi-component systems are the mixtures (the solutions) which consist of several pure (individual) substances (components). The initial amount each of the substances in the ν-component system may be represented in terms of the component masses m_i $(1 \le i \le \nu)$. Another way, which is widely used in thermodynamics, is to represent the system composition in terms of the number of moles of a component, $N_i = m_i/M_i$, where M_i is the molecular mass of the component.

The thermodynamic behavior of the ν-component system is described by the First Law and the Second Law of thermodynamics. If a system under consideration is an open one, then, not only energy exchange in the form of work or heat may take place but also mass exchange. When there are no external force fields the joint equation of the First and the Second Laws may be written in the form:

$$dU \le TdS - pdV + \sum_{i=1}^{\nu} \mu_i dN_i, \tag{1}$$

where U, S and V are the total internal energy, entropy and volume of the system; μ_i is a chemical potential. An equality sign is valid in Equation 1 for equilibrium processes or, in other words, for reversible or quasistatic processes; an unequality sign relates to the case when nonstatic processes or dissipative effects take place.

Equation 1 defines as the internal energy U as a thermodynamic potential with regards to independent variable S, V, $N_1, \ldots N_\nu$. If the set of the independent variables is not convenient to use, then after the Lejandre transformation of equation 1 it is possible to introduce the other thermodynamic potentials such as an enthalpy $H(S, p, N_1, \ldots, N_\nu) = U + pV$; a Helmholtz energy $A(T, V, N_1, \ldots, N_\nu) = U - TS$; a Gibbs energy $G(T,p,N_1 \ldots, N_\nu) = H - TS$. For the last case Equation 1 may be rearranged into the form

$$dG \le -SdT + V\,dp + \sum_{i=1}^{\nu} \mu_i dN_i. \tag{2}$$

which allows the calculation of any thermodynamic function for a multi-component system under equilibrium. For this case the Gibbs energy should be known as a function of its own independent variables T, p, $N_1 \ldots, N_\nu$, which is more convenient for practical applications. For equilibrium conditions it follows from the Equation 2 that:

$$S = -\left(\frac{\partial G}{\partial T}\right)_{p,N_1,\ldots,N_\nu}; \quad V = -\left(\frac{\partial G}{\partial p}\right)_{T,N_1,\ldots,N_\nu}; \tag{3}$$

$$\mu_i = -\left(\frac{\partial G}{\partial N_i}\right)_{T,p,N_{j\neq i}}.$$

A chemical potential μ_i of the component i may be calculated by differentiating any of the similar thermodynamic potentials, with respect to the number of moles of the chosen component:

$$\mu_i = \left(\frac{\partial U}{\partial N_i}\right)_{S,V,N_{j\neq i}} = \left(\frac{\partial H}{\partial N_i}\right)_{S,p,N_{j\neq i}} = \left(\frac{\partial A}{\partial N_i}\right)_{T,V,N_{j\neq i}}. \quad \textbf{(4)}$$

Also from Equations 1 and 2, exact correlations may be derived (the Maxwell equations), which are the joint thermodynamic parameters of the multi-component system. These correlations may be used for describing the behavior of the system under consideration if limited experimental data is available.

To describe the composition of a multi-component system, concentrations of the components are used: in terms of mass $\rho_i = m_i/V$ as well as of moles $\tilde{\rho}_i = N_i/V$, which define the content of the component per unit volume. Fractions of the components are also commonly used: thus mass fraction $\chi_i = m_i/m$ and mole fraction $\tilde{\chi}_i = N_i/N$, where m and N are the total mass and the total number of moles in the system respectively. System compositions in terms of molar quantities $\tilde{\rho}_i$ or $\tilde{\chi}_i$ are more convenient to use in chemical thermodynamics or in the thermodynamics of solutions.

It is easy to calculate the mass compositions when the mole composition values are known:

$$\rho_i = \tilde{\rho}_i M_i; \qquad \chi_i = \tilde{\chi}_i M_i/M, \quad \textbf{(5)}$$

where an equivalent molecular mass of the mixture M is defined as

$$M = m/N = \rho/\tilde{\rho} = \sum_{i=1}^{\nu} \tilde{\chi}_i M_i.$$

Extensive thermodynamic functions such as V, H, G, S, etc. which appear in Equations 1–3 depend not only on thermodynamic parameters of the system but on the amount of the substance included in the system. As such, the extensive values can not be considered thermodynamic properties. On the contrary, the intensive thermodynamic functions—such as specific volume v, enthalpy h, Gibbs energy g, etc.—which have been obtained dividing the extensive functions by the total mass or the total number of moles the system, are exactly defined by the same parameters which characterize a thermodynamic state of the system. In this sense, the intensive functions may be considered as thermodynamic properties of the system. Below we will use molar thermodynamic values defined for any extensive function Ψ in the form

$$\tilde{\psi} = \Psi/N = f(T, \tilde{\rho}_1, \ldots, \tilde{\rho}_\nu) = \varphi(T, p, \tilde{\chi}_1, \ldots, \tilde{\chi}_\nu). \quad \textbf{(6)}$$

To obtain the corresponding specific (mass-based) values, the molar functions of the form (6) have to be divided by molecular mass of the mixture defined by Equation 5.

By definition the chemical potential μ_i appearing in Equations 3 and 4 is an intensive thermodynamic function as are other functions of the type

$$\bar{\psi}_i = \left(\frac{\partial \Psi}{\partial N_i}\right)_{T,p,N_{j\neq i}} = f_i(T, \tilde{\rho}_1, \ldots, \tilde{\rho}_\nu)$$

$$= \varphi_i(T, p, \tilde{\chi}_1, \ldots, \tilde{\chi}_\nu), \quad \textbf{(7)}$$

which used to be called partial thermodynamic functions: for example, partial volume $\bar{v}_i$ and entropy $\bar{s}_i$. It follows from equation 3 that the chemical potential μ_i is equal to the partial Gibbs's energy $\bar{g}_i$ of the ith component.

Any thermodynamic functions, whether extensive or specific, may be determined according to the additive rule:

$$\Psi = \sum_{i=1}^{\nu} N_i \psi_{mi}; \qquad \tilde{\psi} = \sum_{i=1}^{\nu} \tilde{\chi}_i \bar{\psi}_i. \quad \textbf{(8)}$$

Both molar and specific thermodynamic functions of the form defined in Equation 6 and partial functions defined as in Equation 7 satisfy the same thermodynamical correlations as the corresponding extensive functions. The correlations between the partial functions and the mixture composition given as $\tilde{\chi}_1, \ldots, \tilde{\chi}_\nu$ may be obtained from the Gibbs—Duhem equation

$$\sum_{i=1}^{\nu} \tilde{\chi}_i d\bar{\psi}_i = 0. \quad \textbf{(9)}$$

Any variations of the partial functions caused by a small change of the mixture composition satisfy Equation 9 if T and p stay constant.

Equation 8 does not imply that properties of any multi-component mixture may be calculated if these properties were known for the components. This is possible only for the ideal mixtures (solutions). In such cases, the partial functions with exception of these containing entropy (i.e., volume, internal energy, enthalpy, heat capacity, etc.) will be equal to the corresponding specific functions calculated for the pure components at given T and p (Amagat's Law). In general this may be written as:

$$\bar{\psi}_i^{id}(T, p, \tilde{\chi}_1, \ldots, \tilde{\chi}_\nu) = \tilde{\psi}_i(T, p). \quad \textbf{(10)}$$

(It is implied that the aggregate state should be the same for each of the components and the mixture). For an ideal mixture, the partial entropy s_i as well Gibbs energy $\bar{g}_i$ contains terms additional to those indicated in Equation 10:

$$\bar{s}_i^{id}(T, p, \tilde{\chi}_1, \ldots, \tilde{\chi}_\nu) = \tilde{s}_i(T, p) - R \ln \tilde{\chi}_i, \quad \textbf{(11)}$$

$$\bar{g}_i^{id}(T, p, \tilde{\chi}_1, \ldots, \tilde{\chi}_\nu) = \tilde{g}_i(T, p) + RT \ln \tilde{\chi}_i. \quad \textbf{(12)}$$

These terms take into account the entropy increase due to the irreversible mixing of the components:

$$\Delta\tilde{s}^m = -R \sum_{i=1}^{\nu} \tilde{\chi}_i \ln \tilde{\chi}_i. \quad \textbf{(13)}$$

Equation 12 is a form of **Raoult's Law**.

For mixtures whose components are perfect gases only, the Gibbs energy of the components $g_i(T,p)$ in Equation 12 may be calculated as

$$\tilde{g}_i^{pg}(T, p) = G_i^0(T) + RT \ln(p/p_0), \quad \textbf{(14)}$$

where $p_0 = 101325$ Pa is the standard atmospheric pressure and $G_i^0(T)$ is the standard Gibbs energy of the pure substances.

The perfect gas mixtures obey not only Amagat's and Raoult's Laws but also **Dalton's Law** which gives the correlations:

$$p^{pg} = \sum_{i=1}^{\nu} p_i; \qquad p_i = \tilde{\rho}_i RT = \tilde{\chi}_i p. \qquad \textbf{(15)}$$

Real mixtures do not obey the Amagat's and Raoult's Laws. The deviations from the ideal solution behavior are described by the *excess thermodynamic functions*

$$\tilde{\psi}^E = \sum_{i=1}^{\nu} \tilde{\chi}_i \bar{\psi}_i^E(T, p, \tilde{\chi}_1, \ldots, \tilde{\chi}_\nu); \qquad \bar{\psi}_i^E = \bar{\psi}_i - \bar{\psi}_i^{id}. \qquad \textbf{(16)}$$

The excess volume υ^E, enthalpy h^E, entropy s^E define the volume, heat and entropy of mixing effects accordingly. As a result of the mixing effects, the system changes volume and a heat of mixing may may need to be added or removed from the system, even at constant p and T. Also the entropy change may be greater or less than Δs^m calculated from Equation 13. These effects not observed in ideal mixtures, but they are important in a qualitative and quantitative sense to describe the behavior of real mixtures. The effects are due to molecular interactions, which are essentially different in pure substances and in mixtures where the interactions of different kinds of molecules take place.

Activity coefficients (γ_i) are commonly used in the thermodynamics of solutions instead of excess partial Gibbs energies $\bar{g}_i^E$. The two quantities are related by the defining equation:

$$\gamma_i(T, p, \tilde{\chi}_1, \ldots, \tilde{\chi}_\nu) = \exp(\bar{g}_i^E/RT). \qquad \textbf{(17)}$$

If the mixture is an ideal then for all the components $\gamma_i = 1$.

From Equations 12, 16 and 17 it is possible to relate $\bar{g}_i$ to activity coefficient:

$$\bar{g}_i(T, p, \tilde{\chi}_1, \ldots, \tilde{\chi}_\nu) = \tilde{g}_i(T, p) + RT\ln(\tilde{\chi}_i\gamma_i), \qquad \textbf{(18)}$$

The excess Gibbs energy $\tilde{g}^E$ is also related to activity coefficient by:

$$\tilde{g}^E(T, p, \tilde{X}_1, \ldots, \tilde{\chi}_\nu) = RT \sum_{i=1}^{\nu} \tilde{\chi}_i \ln \gamma_i. \qquad \textbf{(19)}$$

If the dependence $\tilde{g}^E$ on $\tilde{x}_1, \ldots \tilde{\chi}\nu$ is known then the activity coefficients γ_i for all the components may be calculated.

The volumetric, heat and the other mixing effects may be deduced from Equation 19 if the equations similar to (3) are used:

$$\tilde{v}^E = (\partial \tilde{g}^E/\partial p)_{T,p,\tilde{\chi}_1,\ldots,\tilde{\chi}_\nu}; \qquad \tilde{h}^E = [\partial(\tilde{g}^E/T)/\partial(1/T)]_{p,\tilde{\chi}_1,\ldots,\tilde{\chi}_\nu}.$$

Similar equations may be used to calculate from Equation 17 any of the excess partial thermodynamic functions of the components:

$$\bar{v}_i^E = RT(\partial \ln \gamma_i/\partial p)_{T,\tilde{\chi}}; \qquad \bar{h}_i^E = R[\partial \ln \gamma_i/\partial(1/T)]_{p,\tilde{\chi}}, \text{ etc.}$$

In order to describe thermodynamic properties of real gases mixtures, equations of state given as functions of pressure on temperature, molar density and mixtures composition are used:

$$p/\tilde{\rho}RT = z(T, \tilde{\rho}, \tilde{\chi}_1, \ldots, \tilde{\chi}_\nu) \qquad \textbf{(20)}$$

(if a mixture includes ideal gases only, then $z = 1$). Statistical mechanics allows the derivation of the exact function (20) having correlated it with a potential energy of the molecular interactions. Such equations are valid for gases of moderate density. For high-density gases and liquids, there are many empirical equations of state. Many of them represent the modifications of the famous **van der Vaals Equation**. Such equations may often describe both the gaseous and the liquid phases of pure substances and their mixtures.

If the equation of state (20) is available then it is possible to calculate any of the thermodynamic functions of the mixture in terms of the variables T, $\tilde{\rho}$, $\tilde{x}_1, \ldots, \tilde{x}\nu$. For example,

$$\tilde{s} = \sum_{i=1}^{\nu} \tilde{\chi}_i[S_i^0(T) - R\ln\tilde{\chi}_i] - R\ln(\tilde{\rho}RT/p_0) - R\left[q + T\left(\frac{\partial q}{\partial T}\right)_{\tilde{\rho},\tilde{\chi}}\right]; \qquad \textbf{(21)}$$

$$\tilde{h} = \sum_{i=1}^{\nu} \tilde{\chi}_i H_i^0(T) + (z-1)RT - RT^2\left(\frac{\partial q}{\partial T}\right)_{\tilde{\rho},\tilde{\chi}},$$

etc., where $S_i^0 = dG_i^0/dT$, $H_i^0 = G_i^0 + TS_i^0$ are the standard entropy and enthalpy, and

$$q(T, \tilde{\rho}, \tilde{\chi}) = \int_0^{\tilde{\rho}} [z(T, \tilde{\rho}, \tilde{\chi}) - 1]\frac{d\tilde{\rho}}{\tilde{\rho}}. \qquad \textbf{(22)}$$

In many cases it is more convenient to use, instead of a chemical potential $\mu_i = \bar{g}_i$, a **Fugacity** f_i of the i-component. Based on equations 12, 14, 15, a fugacity may be introduced as an analog of a partial pressure p_i (15):

$$f_i = p_0 \exp([\bar{g}_i(T, p, \tilde{\chi}) - G_i^0(T)]). \qquad \textbf{(23)}$$

For the mixture comprising the perfect gases the fugacity of a component equals to its partial pressure:

$$f_i^{pg} = p_i = p\tilde{\chi}_i.$$

If a pure substance fugacity is defined in a manner similar to (23) as

$$f_i^+ = p_0 \exp([\tilde{g}_i(T, p) - G_i^0(T)]), \qquad \textbf{(24)}$$

where $\tilde{g}_i$ is a molar Gibbs energy of the pure i-substance, then according to the definition (18) it is possible to obtain a relationship which links the fugacities (23) and (24) in the real mixture to the activity coefficients (17):

$$f_i(T, p, \tilde{\chi}_1, \ldots, \tilde{\chi}_\nu) = f_i^+(T, p)\tilde{\chi}_i\gamma_i. \qquad \textbf{(25)}$$

Taking into account (25) Raoult's Law (12) may be rearranged into the form of the Lewis Rule: $f_i^{id} = f_i^+\tilde{X}_i$.

The fugacity coefficient $\varphi_i = f_i/p\tilde{X}_i$ can be evaluated from equation of state (20) taking account of Equation 21:

$$\ln \varphi_i(T, \rho\tilde{\chi}) = z - 1 - \ln z + q + \left(\frac{\partial q}{\partial \tilde{X}_i}\right)_{T,\tilde{\rho},\tilde{X}_{k\neq i}} - \sum_{j=1}^{\nu} \tilde{X}_j \left(\frac{\partial q}{\partial \tilde{X}_j}\right)_{T,\tilde{\rho},\tilde{X}_{k\neq j}} . \quad (26)$$

In order to calculate the thermodynamic functions (21), (22), (26) of the mixture at given $T, p, \tilde{X}_1, \ldots, \tilde{X}_\nu$, it is necessary to solve (as a rule numerically) the nonlinear Equation (20) with regards to the mixture density $\tilde{\rho}(T, p, \tilde{X}_1, \ldots, \tilde{X}_\nu)$ using a defined route in the application of the correlations which are to be used for the purpose. If the equation of state (20) describes both a gas and a liquid phase, then in the vapor-liquid phase there exist several routes. The minimal of them relates to the gaseous and the maximal one to the liquid phase, all the other routes have no physical sense and could describe none of the stable states of the system.

In summary, there are two methods of describing the thermodynamic behavior of multi-component systems': one of them is based on excess functions and another on the equation of state. Both of them are equivalent, but each has its own advantages. The first one has been found best for mixtures all the components of which, at the given p and T, are in the same aggregate state as the mixture one. The method is difficult to use in describing the phase equilibrium and critical phenomena in solutions. The second method, however, is invalid in describing solid solutions and melting processes. Even the most modern equations of state are less accurate than the experimental data they aim to describe.

References

Prigogine, I., Defay, R. (1954) *Chemical Thermodynamics.* Longmans, London.

Reid, R. C., Prausnitz, J. M., Poling, B. E. (1987). *The Properties of Gases and Liquids.* McGraw-Hill Book Co.

Walas, S. M. (1985). *Phase Equilibrium in Chemical Engineering.* Butterworth Publishers.

Leading to: *Binary systems, thermodynamics of; Phase equilibrium*

M.Yu. Boyarsky and A.M. Semjonov

MULTICOMPONENT VAPOUR CONDENSATION (see Dephlegmator)

MULTIFLUID MODELS (see Two-phase flow)

MULTIMODE FIBRE (see Fibre optics)

MULTIPHASE FLOW

A multiphase flow is defined as one in which more than one phase (ie. gas, solid and liquid) occurs. Such flows are ubiquitous in industry, examples being gas-liquid flows in evaporators and condensers, gas-liquid-solid flows in chemical reactors, solid-gas flows in pneumatic conveying, etc. This introductory article attempts to give an overview, with more detailed material appearing on each individual type of multiphase flow in separate entries.

In multiphase flows, solid phases are denoted by the subscript S, liquid phases by the subscript L and gas phases by the subscript G. Some of the main characteristics of these three types of phases are as follows:

Solids

In a multiphase flow, the solid phase is in the form of lumps or particles which are carried along in the flow. The characteristics of the movement of the solid are strongly dependent on the size of the individual elements and on the motions of the associated fluids. Very small particles follow the fluid motions, whereas larger particles are less responsive.

Liquids

In a multiphase flow containing a liquid phase, the liquid can be the continuous phase containing dispersed elements of solids (particles), gases (bubbles) or other liquids (drops). The liquid phase can also be discontinuous, as in the form of drops suspended in a gas phase or in another liquid phase. Another important property of liquid phases relates to *wettability.* When a liquid phase is in contact with a solid phase (such as a channel wall) and is adjacent to another phase which is also in contact with the wall, there exists at the wall a *triple interface,* and the angle subtended at this interface by the liquid-gas and liquid-solid interface is known as the **Contact Angle**.

Gases

As a fluid, a gas has the same properties as a liquid in its response to forces. However, it has the important additional property of being (in comparison to liquids and solids) highly compressible. Notwithstanding this property, many multiphase flows containing gases can be treated as essentially incompressible, particularly if the pressure is reasonably high and the **Mach Number** with respect to the gas phase, is low (e.g., < 0.2).

Types of Multiphase Flow

The most common class of multiphase flows are the **Two-Phase Flows**, and these include **Gas-Liquid Flow, Gas-Solid Flow, Liquid-Liquid Flow** and **Liquid-Solid Flow**. The reader is referred to the general overview article on **Two-Phase Flows** and to the individual articles on each of the respective two-phase flow types.

Three-phase flows are also of practical significance, and examples are as follows:

1. *Gas-liquid-solid flows:* this type of system occurs in two-phase fluidised bed and gas lift chemical reactors where a gas-liquid reaction is promoted by solid catalyst particles suspended in the mixture.
2. *Three-phase, gas-liquid-liquid flows:* mixtures of vapours and two immiscible liquid phases are common in chemical engineering plants. Examples are gas-oil-water flows in oil recovery systems and immiscible condensate-vapour flows in steam / hydrocarbon condensing systems.

3. *Solid-liquid-liquid flows:* An example here would be that of an immiscible liquid-liquid reaction, in which a solid phase is formed, that separates out in the system.

Multiphase flows are not restricted to only three phases. An example of a *four phase flow* system would be that of direct-contact freeze crystallisation in which, for example, butane liquid is injected into solution from which the crystals are to be formed, and freezing occurs as a result of the evaporation of the liquid butane. In this case, the four phases are, respectively, butane liquid, butane vapour, solute phase and crystalline (solid) phase.

Basic Quantities In Multiphase Flows

The *mean phase content* (ε_i) of the ith phase is defined as the time-averaged volume fraction of that phase in a section of the channel or as the time-averaged area fraction of the phase in a given cross-section (the two definitions may be taken as equivalent in most practical situations). Specifically, the mean phase content of the gas phase (ε_G) is often termed the **Void Fraction**. The volume flux (or *superficial velocity*) U_i of a phase is defined as:

$$U_i = \frac{\dot{V}_i}{S} \tag{1}$$

where $\dot{V}_i$ is the volume flow rate of the phase (m^3/s) and S is the channel cross-sectional area (m^2). The total superficial velocity U is given by:

$$U = \sum_{i-1}^{n} U_i \tag{2}$$

where n is the total number of phases present. The *average* phase velocity (u_i) of the ith phase is given by:

$$u_i = \frac{U_i}{\varepsilon_i} = \frac{\dot{V}_i}{S\varepsilon_i} \tag{3}$$

and the *flow quality* x_i of the ith phase is defined as:

$$x_i = \frac{\dot{m}_i}{\sum_{i=1}^{n} \dot{m}_i} \tag{4}$$

where $\dot{m}_i$ is the mass flux of the ith phase (given by $\dot{M}_i/S$, where $\dot{M}_i$ is the mass rate of flow the phase through the channel). A *multiphase density* ρ_{MP} may be defined as the mass of the multiphase mixture per unit channel volume, and this is given by:

$$\rho_{MP} = \sum_{i=1}^{n} \varepsilon_i \rho_i \tag{5}$$

where ρ_i is the density of the ith phase.

Conservation Equations For Homogeneous Flow

The simplest approach for representation of multiphase flows is to treat them as homogeneous mixtures in which the velocities of all the phases are identical and equal to the homogeneous velocity u_H. This is given by:

$$u_H = U = \frac{\dot{m}}{\rho_H} = \frac{\dot{M}}{S\rho_H} \tag{6}$$

where $\dot{m}$ is the total mass flux, $\dot{M}$ the total mass rate of flow and ρ_H is the *homogeneous density* given by:

$$\rho_H = \frac{\dot{m}}{U} = \frac{1}{\sum_{i=1}^{n} (x_i/\rho_i)} \tag{7}$$

The *homogeneous conservation equations* for mass, momentum and energy are stated as follows (detailed derivations are given by Hewitt, 1983):

$$\frac{\partial}{\partial z}(U\rho_H S) + S\frac{\partial \rho_H}{\partial t} = 0 \tag{8}$$

$$\frac{\partial \dot{m}}{\partial t} + \frac{1}{S}\frac{\partial(\dot{m}^2 S/\rho_H)}{\partial z} = \frac{\partial p}{\partial z} - g\rho_H \sin\alpha - \frac{\tau_o P}{S} \tag{9}$$

$$\rho_H\left(\frac{\partial e}{\partial t} + U\frac{\partial e}{\partial z}\right) = \frac{\dot{q}P}{S} + \dot{q}_V + \frac{\partial p}{\partial t} \tag{10}$$

where z is the axial distance, t time, p pressure, g the acceleration due to gravity, α the angle of inclination of the channel, τ_o the wall shear stress, P the channel periphery, $\dot{q}$ the wall heat flux, $\dot{q}_V$ the internal heat generation rate in the fluid per unit volume and e the energy convected per unit fluid mass, given by:

$$e = h + \frac{u^2}{2} + gz\sin\alpha \tag{11}$$

where h is the specific enthalpy given by:

$$h = \mu + \frac{p}{\rho_H} \tag{12}$$

where μ is the specific internal energy.

For steady state flow in a constant cross-section duct, the momentum equation (Equation 9) reduces to:

$$-\frac{dp}{dz} = \frac{\tau_o p}{S} + \dot{m}^2\frac{d(1/\rho_H)}{dz} + g\rho_H \sin\alpha \tag{13}$$

where the three terms of the right hand are, respectively, the *frictional pressure gradient,* the *accelerational pressure gradient* and the *gravitational pressure gradient.*

Conservation Equations For Separated Multiphase Flows

For the *separated flow model* for multiphase flows, the phases are considered to be flowing in separated zones of the channel, each phase having its own velocity, as illustrated in **Figure 1**.

To develop equations for this case, it is possible to write the conservation equations for each separate phase, considering the interaction between that phase and the channel wall and also its interactions with the adjacent phases. This approach for two-phase flows is described in detail in the article on **Conservation Equations, Two-Phase**. A common practice is to add the conservation equations for the respective phases together, and this leads to the *mixture conservation equations.* Details of the approach for multiphase systems are given by Lahey and Moody (1977) and by Hewitt (1983). Here, for the sake of brevity, we will merely state the mixture conservation equations which, for mass, momentum and energy are respectively:

$$\frac{\partial}{\partial t}(\rho_{MP}S) + \frac{\partial}{\partial z}(\dot{m}S) = 0 \quad (14)$$

$$-\frac{\partial p}{\partial z} - g\rho_{MP}\sin\alpha - \frac{\sum_{i=1}^{n}\tau_{io}P_{io}}{S} = \frac{\partial \dot{m}}{\partial t} + \frac{1}{S}\frac{\partial}{\partial z}\left(\dot{m}^2 S\sum_{i=1}^{n}\frac{x_i^2}{\rho_i\varepsilon_i}\right) \quad (15)$$

$$S\frac{\partial}{\partial t}\left(\sum_{i=1}^{n}\rho_i h_i\varepsilon_i\right) + \frac{\partial}{\partial z}\left(S\sum_{i=1}^{n}\dot{m}x_i h_i\right) = \dot{q}P + \dot{q}_v S - \frac{\partial}{\partial z}\left(\dot{m}^3 S\sum_{i=1}^{n}\frac{x_i^3}{\rho_i^2\varepsilon_i^2}\right) - g\dot{m}S\sin\alpha - \frac{\partial}{\partial t}\left(\dot{m}^2\sum_{i=1}^{n}\frac{x_i^2}{\rho_i\varepsilon_i}\right) + S\frac{\partial p}{\partial t} \quad (16)$$

where h_i is the enthalpy of the ith phase. for steady state flow in a constant cross-section duct, Equation 15 reduces to:

$$-\frac{dp}{dz} = \frac{\tau_o P}{S} + \dot{m}^2\frac{d}{dz}\left(\sum_{i=1}^{n}\frac{x_i^2}{\rho_i\varepsilon_i}\right) + g\rho_{MP}\sin\alpha \quad (17)$$

where the three terms on the right hand side of the equation are respectively the *frictional, accelerational* and *gravitational pressure gradients.*

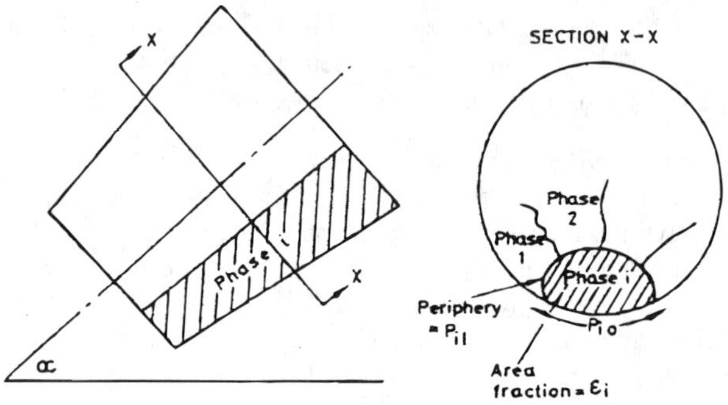

Figure 1. Separated flow models.

References

Hewitt, G. F. (1983) Multiphase fluid flow and pressure drop: Introduction and fundamentals. Chapter 2.3.1 of the *Heat Exchanger Design Handbook*, Hemisphere Publishing Corporation, New York.

Lahey, R. T. and Moody, F. J. (1977) The *Thermal Hydraulics of a Boiling Water Nuclear Reactor.* American Nuclear Society.

Leading to: *Gas-liquid flow; Gas-solid flow; Liquid-liquid flow; Liquid-solid flow*

G.F. Hewitt

MULTIPHASE DENSITY (see Multiphase flow)

MULTI-STAGE TURBINES (see Turbine)

MULTI-START HELICALLY COILED TUBE BOILER (see Coiled tube boilers)

MURPHREE EFFICIENCY (see Distillation)

MUTUAL DIFFUSION COEFFICIENT (see Fick's law of diffusion)

N

NAPHTHA (see Oil refining)

NAS (see National Academy of Sciences)

NASA (see National Aeronautics and Space Administration)

National Academy of Sciences, NAS

2101 Constitution Avenue
Washington, DC 20418
U.S.A.
Tel: 202 334 2000

National Aeronautics and Space Administration, NASA

Langley Research Centre
Hampton, Virginia
U.S.A.
Tel: 804 864 1000

National Engineering Laboratory, NEL

East Kilbride
Glasgow G75 OQU
U.K.
Tel: 013552 20222

National Institute of Standards and Technology, NIST

Gaithersburg, Maryland 20899
U.S.A.
Tel: 301 975 2000

National Research Council of Canada, NRC

Division of Energy Research and Development
Building M50
Montreal Road, Ottawa ON KIA OR6
Canada
Tel: 613 993 9102

NATURAL CIRCULATION LOOPS, THERMOSYPHONS

Natural circulation loops (thermosyphons) are flow systems heated from below and cooled from above, such that the heat sink is higher than the heat source. This specific configuration creates a density gradient which generates the driving force. Thermosyphons appear in geophysical and geothermal systems and have been used in many applications in diverse energy conversion systems, such as solar heating devices, absorption refrigerators, reboilers in chemical industries and cooling of various engines. One of the most important uses of thermosyphons is in emergency core cooling of nuclear reactors. This subject gained more interest following the recovery of the reactor after the *Three Mile Island* (TMI) *accident* in 1979, when it was demonstrated that natural circulation was the only effective way to remove the decay heat.

Natural circulation flows are often divided into single- and two-phase loops. Reviews on thermosyphons were written by Zvirin (1981) and Greif (1988). Summaries of recent advances appear in D'Auria and Vigni (1990) and Knaani and Zvirin (1993).

Theoretical methods have been developed in order to simulate various loops, derive scaling laws for experiments and explain physical phenomena including stability characteristics. The mathematical models are based on the coupled conservation equations, rendering the problem non-linear. The continuity equation in one-dimensional models yields the result that the velocity, v, is a function of time only (and an unknown constant for steady state). The temperature distribution, T, is obtained in terms of v by solving the energy equation. For two-phase loop sections the quality, x, is obtained and for double-diffusive loops the salinity, S, is found from the diffusion equation. The momentum equation is integrated along the closed loop, to yield v. Analytical solutions exist for simple loops. Numerical methods are needed for more complex ones and for transient calculations. Stability features have been obtained by linear stability analysis as well as by finite amplitude methods; numerical solutions are used for both. (See also **Instability, Two-phase**).

Data on natural circulation exist in the literature for the whole range of scales, from large working systems to laboratory experiments. The former include nuclear reactors, (e.g. post-accident TMI and many tests on others), solar energy systems and reboilers. The latter are usually simple geometry loops to study various phenomena and to validate computer codes.

The experimental and theoretical investigations have produced the information needed to understand, predict and simulate the behaviour of thermosyphons. The interaction of the participating physical forces is complex and non-linear; gravity, friction and inertia depend on heat and mass transfer characteristics. This leads to several interesting features of the convective flows.

In general, steady state loop flows (SF) are established for a certain range of **Rayleigh Numbers**, Ra, above some thresholds and below critical instability limits. These flows can be reached either from a perturbation of a rest state (conductive solution) or by coast-down from forced flow. For two-phase thermosyphons, the curve of flow rate vs. loop inventory exhibits a local maximum. Transients leading to SF can be monotonic or oscillatory. For a range of Ra, the SFs are unstable (growing oscillations). This can lead to bifurcation (multiple SF s), long term periodic flow and chaotic behavior. These phenomena have also been observed in systems with parallel loops and thermosyphons with throughflows.

Finally, there does not yet exist a general set of heat transfer and friction correlations for natural circulation loops, and in theoretical and numerical studies forced flow correlations are often used, with some loss of accuracy. Other approximations have also been made in cases where there is a lack of more accurate information, such as linear profiles of void fraction.

References

D'Auria, F. and Vigni, P., (eds.) (1990) Proc. Eurothem Seminar Nr. 16: Natural Circulation in Industrial Applications, Pisa, Italy.

Greif, R. (1988) Natural Circulation Loops. *J. Heat Transfer.* Vol. 110, 1243–1258.

Knaani, A. and Zvirin, Y. (1993) Bifurcation Phenomena in Two-Phase Natural Circulation Loops. *Int. J. Multiphase Flow.* Vol. 19, 1129–1151.

Zvirin, Y., (1981) A Review of Natural Circulation Loops in Pressurized Water Reactors and other Systems. *Nucl. Engng. & Des.*, Vol. 67, 203–225.

Y. Zvirin and A. Knaani

NATURAL CIRCULATION REBOILERS (see Tubes and tube banks, boiling heat transfer on)

NATURAL CONVECTION (see Free convection)

NATURAL FREQUENCY OF TUBE VIBRATION (see Vibration in heat exchangers)

NATURAL GAS

The major constituents of natural gas are paraffinic hydrocarbons, principally **Methane** with smaller quantities of **Ethane, Propane** and **Butane; Nitrogen, Carbon Dioxide** and hydrogen sulphide may also be present. Natural gas is found in many parts of the world in sedimentary rock basins, where it was probably formed from the decay of large accumulations of organic matter that were covered by sediment and subjected to intense heat and pressure over long time periods. Gas fields occur where the gas has migrated and collected in porous strata beneath an anticline formed by an impervious layer of rock. The gas is found at pressures of 10 to 200 bar, and both oil and water may be present in the strata.

Natural gas is extracted from the reservoir through multiple wells, the rate of extraction depending on the rate of flow through the rocks, and other factors. Gas withdrawn from most geological formations is saturated with water vapour at the prevailing temperatures and pressures and steps must be taken to avoid the formation, downstream of the well head, of solid hydrates (by the reaction of methane and water), which, at the elevated pressures can occur at temperatures well above the freezing point of water. An inhibitor such as **Methanol** is injected into the gas at the well head to avoid *hydrate formation.* The subsequent treatment depends on the type of gas, but in most cases involves the use of primary separators to remove liquids and any particulate matter. There is also a need to avoid the formation, in the transmission pipes, of liquid hydrocarbons by *retrograde condensation*, as the pressure falls. The well head pressure is therefore reduced to the working level, liquids are removed and the gas is finally dried to adjust the water vapour dew point by direct contact with *glycol.*

Wet gases that contain substantial amounts of liquid hydrocarbons are treated to reduce the propane and higher hydrocarbon content for economic reasons and to facilitate transmission. This can be achieved by cooling the gas to −35°c and scrubbing it with a refrigerated wash oil. The wash oil is then regenerated by steam stripping to recover the liquid hydrocarbon 'condensates.'

Sour gases are treated to reduce the carbon dioxide and hydrogen sulphide contents to the required levels and to meet statutory limits. A number of processes using chemical absorbents such as solutions of *ethanolamines*, or *potassium carbonate* are used; the gas is contacted with the absorbent in a packed tower to remove the unwanted compounds, and the absorbent is regenerated by heating.

Natural gas is transmitted through an extensive network of *high pressure pipelines*, of up to 1.0 m, or more diameter, that operate at 70 bar pressures. Recompression stations at intervals of 60 tm may be required on long distance pipelines.

Natural gas is also transported in large quantities in the liquid form at −160°c by ship, eg. from Alaska etc., to Japan. The natural gas is purified using solid absorbents to remove the last traces of carbon dioxide, water vapour and sulphur compounds prior to cooling and **Liquefaction** in a cascade refrigeration cycle. The liquid, LNG, is stored either above ground, or in ground insulated tanks at the terminals, a variety of equipment being used for re-vaporisation of the liquid prior to use.

References

Barenblatt, G. I., Entov V. M. and Ryzhik, (1990) *Theory of Fluid Flows Through Natural Beds.* Flower Academic Publishers.

Bereez, E., and Balla-Achs, M. (1983) Studies in Inorganic Chemistry No. 4. *Gas Hydrates.* Elsevier.

Selby, R. C. (1985) *Elements of Petroleum Geology.* Freeman and Co.

The Gas Engineers Handbook, (1966). The Industrial Press.

J.A. Lacey

NATURAL GAS COMBUSTION (see Flames)

NAVIER-STOKES EQUATIONS (see Conservation equations, single phase; Combustion; Couette flow; Flow of fluids; Newtonian fluids; Particle transport in turbulent fluids; Poiseuille flow; Turbulence models)

NEA (see Nuclear Energy Agency)

NEAR INFRA-RED SPECTROSCOPY (see Spectroscopy)

NEGATIVE CATALYSIS (see Catalysis)

NEGATIVE PRESSURE CONTAINMENT, FOR NUCLEAR REACTORS (see Containment)

NEL (see National Engineering Laboratory)

NEON

Neon—(Gr. *neos*, new), Ne; atomic weight 20.179; atomic number 10; melting point −248.67°C; boiling point −246.048°C (1 atm); density of gas 0.89990 g/l (1 atm, 0°C); density of liquid at boiling point 1.207 g/cm^3; valence 0.

Discovered by Ramsay and Travers in 1898, neon is a rare gaseous element present in the atmosphere to the extent of 1 part in 65,000 of air. It is obtained by liquefaction of air and separated from the other gases by fractional distillation. Natural neon is a mixture of three isotopes. Five other unstable isotopes are known. It is a very inert element; however, it is said to form a compound with fluorine. It is still questionable if true compounds of neon exist, but evidence is mounting in favor of their existence. The following ions are known from optical and mass spectrometric studies: Ne^+, $(NeAr)^+$, $(NeH)^+$, and $(HeNe)^+$. Neon also forms an unstable hydrate. In a vacuum discharge tube, neon glows reddish orange. Of all the rare gases, the discharge of neon is the most intense at ordinary voltages and currents.

Neon is used in making the common neon advertising signs, which accounts for its largest use. It is also used to make high-voltage indicators, lightning arrestors, wave meter tubes, and TV tubes. Neon and helium are used in making gas lasers. Liquid neon is now commercially available and is finding important application as an economical cryogenic refrigerant. It has over 40 times more refrigerating capacity per unit volume than liquid helium and more than three times that of liquid hydrogen. It is compact, inert, and is less expensive than helium when it meets refrigeration requirements.

Handbook of Chemistry and Physics, CRC Press

NERNST-HECKELL EQUATION (see Diffusion)

NERNST-HECKELL EQUATION FOR DIFFUSION IN ELECTROLYTE SOLUTIONS (see Diffusion coefficient)

NERNST'S LAW (see Thermodynamics)

NETWORK MODELS (see Buildings and heat transfer)

NEUMANN CONDITIONS (see Conduction)

NEUMANN FUNCTIONS (see Bessel functions)

NEUMANN'S SOLUTION (see Solidification)

NEURAL NETWORKS (see Computers)

NEUTRON CROSS-SECTION (see Nuclear reactors)

NEUTRON DECAY (see Neutrons)

NEUTRON INTERACTIONS (see Neutrons)

NEUTRONS

Neutrons, together with protons, form the fundamental constituents of atomic nuclei. Neutrons have a mass of 1.675×10^{-27} kg, just slightly greater than that of protons (1.673×10^{-27} kg). Neutrons have a zero electric charge, whereas protons have an electric charge equal in magnitude to that of an electron. Neutrons are found in the nuclei of all atoms with the exception only of hydrogen whose nucleus normally consists of one proton. A few atoms of hydrogen contain nuclei with one proton and one neutron and even fewer have one proton with two neutrons. These rarer forms of hydrogen are known as *deuterium* and *tritium* and are called isotopes of hydrogen. Within stable nuclei, such as deuterium, the neutron remains stable indefinitely. Outside the nucleus, however, free neutrons are found to be unstable and decay by beta decay. In this process the neutron changes into a proton and the small reduction in mass produces a high energy electron (beta particle) and an *antineutrino* (an uncharged and essentially massless particle).

$$n \rightarrow p^+ + e^- + \tilde{\nu}$$

The half life of free neutrons is approximately 10 minutes. This means that, if at some initial time there are N free neutrons, there will be N/2 free neutrons after 10 minutes and N/2 neutrons will have decayed to protons in that time.

Neutron Interactions

Having no electric charge, neutrons do not cause ionization as they pass through matter. Neutrons, however, interact readily with atomic nuclei since there is no coulomb barrier to overcome. For low energy neutrons, the most likely interaction is radiative capture. In this process, the capture of a neutron by a nucleus is followed by the emission of gamma rays resulting from the conversion of the neutron's binding energy to excitation energy of the compound nucleus. In most nuclei this amounts to about 6–8 MeV. The probability of radiative capture occurring increases, in most nuclei, as the neutron energy decreases in the range below ~0.1 eV. At higher energies, the capture probability (or cross-section) varies in a more complex way, exhibiting a resonance structure. At energies in the 100 keV − 1 MeV range, the neutrons have sufficient energy to eject other particles such as protons or alpha particles from nuclei.

Elastic scattering reactions are those where the neutron shares its kinetic energy with a nucleus, conserving the total kinetic energy. Inelastic scattering occurs with high energy neutrons when part of their kinetic energy is transferred to excitation energy of the nucleus; kinetic energy is then not conserved.

An important neutron-induced reaction is *nuclear fission.* This occurs in very heavy nuclei such as ^{235}U when the capture of a neutron leads to an excited compound nucleus in which the balance between the coulomb forces and the strong nuclear forces is such that fission of the nucleus is the most likely outcome. The fission process itself releases further neutrons (about 2.5 per fission) leading to the possibility of a self-sustaining chain reaction. The neutron thus plays a central role in the production of nuclear power and the subject of nuclear reactor physics is concerned mainly with the behaviour of neutrons and their interactions with matter. **Nuclear Reactors** are controlled by adjusting the positions of neutron absorbers within the reactor core.

Leading to: *Nuclear reactors*

T.D. MacMahon

NEWCOMEN ATMOSPHERIC ENGINE (see Steam engines)

NEWTON FLOW (see Immersed bodies, flow around and drag)

NEWTONIAN FLUIDS

Following from: *Fluids*

A Newtonian fluid is one where there is a linear relationship between *stress* and *strain-rate*. The ratio of stress to strain-rate is the **Viscosity** of the fluid. A fluid where there is no linear relationship between stress and strain-rate is called a **Non-Newtonian fluid**.

For a Newtonian fluid the stress (strictly, the *deviatoric stress*) $\boldsymbol{\tau}$ is given by:

$$\boldsymbol{\tau} = \eta \mathbf{e} \tag{1}$$

where $\mathbf{e}$ denotes the strain-rate given in terms of the velocity $\mathbf{u}$ by:

$$\mathbf{e} = \nabla \mathbf{u} + (\nabla \mathbf{u})^T \tag{2}$$

and η denotes the viscosity of the fluid, which is independent of $\mathbf{e}$, though it may depend on temperature T and perhaps on pressure p.

Because $\mathbf{e}$ is a *symmetrical tensor*, it follows from (1) that $\boldsymbol{\tau}$ too is a symmetrical tensor:

$$\boldsymbol{\tau} = \boldsymbol{\tau}^T \tag{3}$$

Angular momentum conservation then implies that Newtonian fluids are *non-polar* and hence, that the angular momentum is just the moment of the linear momentum.

Most fluids under normal circumstances are Newtonian so the study of Newtonian fluids comprises the major part of the subject of fluid mechanics. Linear momentum conservation then leads to the *Navier-Stokes equations.*

Leading to: *Non-Newtonian fluids*

S.M. Richardson

NEWTON-LEIBNITZ FORMULA (See Integrals)

NEWTON-RICHMANN RELATIONS (see Heat transfer; Heat transfer coefficient)

NEWTON'S FORMULA FOR PRESSURE DISTRIBUTION AROUND A BODY (see External flows, overview)

NEWTON'S LAW OF COOLING

Following from: *Heat transfer*

This relationship was derived from an empirical observation of convective cooling of hot bodies made by Isaac Newton in 1701, who stated that "the rate of loss of heat by a body is directly proportional to the excess temperature of the body above that of its surroundings." Accordingly, the temperature of a hot object (T_1) which is cooling down as a result of exposure to a convective flow at $T_2 < T_1$, would vary as:

$$\frac{dT_1}{dt} \propto (T_1 - T_2) \tag{1}$$

If the energy loss from the hot body to the cooler fluid is replenished by a heat flux $\dot{q}$ such that T_1 remains constant then the steady state version of Newton's Law of Cooling can be expressed as

$$\dot{q} \propto (T_1 - T_2) = \alpha(T_1 - T_2) \tag{2}$$

This rate equation is universally used to define the **Heat Transfer Coefficient** (α) for all convective flows (free, forced, single/multiphase, etc.) involving either heating or cooling. It should be noted that in some cases (α) is temperature dependent and then $\dot{q}$ is not a linear function of the driving force (T_1-T_2). It should also be noted that the defining driving force varies from system to system (boundary layer flows, tube flows, etc), but the complexity of any particular process is usually reflected in the formulation of the expression for (α), whose value depends upon the nature and properties of the flow system and ranges from 10 W/m^{2}K for **Natural Convection** between air and a vertical plate to 100,000 W/m^{2}K for dropwise **Condensation** of saturated water vapour at a vertical plate.

The study of convective heat transfer is ultimately concerned with finding the value of the heat transfer coefficient, as defined by Newton's Law of Cooling, in terms of the physical parameters of the convection system.

Leading to: *Heat transfer coefficient*

T.W. Davies

NEWTON'S LAW OF VISCOSITY (see Fick's law of diffusion; Non-Newtonian fluid heat transfer)

N-HYDROGEN (see Cryogenic fluids)

NICKEL

Nickel—(Ger, *Nickel*, Satan or "Old Nick's and from *kupfernickel*, Old Nick's copper), Ni; atomic weight 58.71; atomic number 28; melting point 1453°C; boiling point 2732°C; specific gravity 8.902 (25°C); valence 0,1,2,3.

Discovered by Cronstedt in 1751 in kupfernickel (*niccolite*), nickel is found as a constituent in most meteorites and often serves as one of the criteria for distinguishing a meteorite from other minerals. Iron meteorites, or *siderites*, may contain iron alloyed with from 5 to nearly 20% nickel. Nickel is obtained commercially from *pentlandite* and *pyrrhotite* of the Sudbury region of Ontario, a district that produces a major part of the world supply. Other deposits are found in New Caledonia, Australia, Cuba, Indonesia, and elsewhere.

Nickel is silvery white and takes on a high polish. It is hard, malleable, ductile, somewhat ferromagnetic, and a fair conductor of heat and electricity. It belongs to the iron-cobalt group of metals and is chiefly valuable for the alloys it forms. It is extensively used for making stainless steel and other corrosion-resistant alloys such as Invar™, Monel®, Inconel®, and the Hastelloys®. Tubing made of a copper-nickel alloy is extensively used in making desalination plants for converting sea water into fresh water. Nickel is also now used extensively in coinage and in making nickel steel for armor plate and burglar-proof vaults, and is a component in Nichrome®, Permalloy®, and constantan. Nickel added to glass gives a green color. Nickel plating is often used to provide a protective coating for other metals, and finely divided nickel is a catalyst for hydrogenating vegetable oils. It is also used in ceramics, in the manufacture of Alnico magnets, and in the Edison® storage battery. The sulfate and the oxides are important compounds.

Natural nickel is a mixture of five stable isotopes; seven other unstable isotopes are known. Exposure to nickel metal and soluble compounds (as Ni) should not exceed 1 mg/M^3 (8-hr time-weighted average—40-hr week). Nickel carbonyl exposure, however, should not exceed 0.007 mg/M^3, and is considered to be a very toxic material. Nickel sulphide fume and dust is recognized as having carcinogenic potential.

Handbook of Chemistry and Physics, CRC Press

NIKURADZE FORMULA FOR FRICTION FACTOR (see Hydraulic resistance)

NIKURADZE MEASUREMENTS IN TUBES (see Tubes, single phase flow in)

NIST (see National Institute of Standards and Technology)

NITRIC ACID

Nitric acid (HNO_3) is a yellowish, fuming liquid with a molecular weight of 63.01. It is readily miscible with water, highly corrosive and is a strong oxidising agent that will attack most metals. It boils and decomposes at 359.2K, its freezing point is 231.6 K while its specific gravity is 1.5. It comes in different industrial grades depending on the strength of the aqueous solution.

V. Vesovic

NITRIC OXIDE

Nitric Oxide (NO) is a colourless, non-flammable gas that readily reacts with oxygen to form nitrogen dioxide at room temperature. It is highly toxic and is obtained by the high temperature oxidation of ammonia. Its main usage is as an inert atmosphere in electrical systems.

	Critical temperature: 180 K
Molecular weight: 30.01	Critical pressure: 6.48 MPa
Melting point: 109.6 K	Critical density: 517 kg/m^3
Boiling point: 121.4 K	Normal vapor density: 1.34 kg/m^3 (@ 0°C, 101.3 kPa)

V. Vesovic

NITROGEN

Nitrogen—(L. *nitrum*, Gr. *nitron*, native soda; genes, *forming*), N; atomic weight 14.0067; atomic number 7; melting point −209.86°C; boiling point −195.8°C; density 1.2506 g/l; specific gravity liquid 0.808 (−195.8°C), solid 1.026 (−252°C); valence 3 or 5.

Nitrogen was discovered by Daniel Rutherford in 1772, but Scheele, Cavendish, Priestley, and others about the same time studied "burnt or dephlogisticated air," as air without oxygen was then called. Nitrogen makes up 78% of the air, by volume. The atmosphere of Mars, by comparison, is 2.6% nitrogen. The estimated amount of this element in our atmosphere is more than 4000 billion tons. From this inexhaustible source it can be obtained by liquefaction and fractional distillation. Nitrogen molecules give the orange-red, blue-green, blue-violet, and deep violet shades to the aurora. The element is so inert that Lavoisier named it *azote*, meaning without life, yet its compounds are so active as to be most important in foods, poisons, fertilizers, and explosives. Nitrogen can also be easily prepared by heating a water solution of ammonium nitrite.

Nitrogen, as a gas, is colorless, odorless, and a generally inert element. As a liquid it is also colorless and odorless, and is similar in appearance to water. Two allotropic forms of solid nitrogen exist, with the transition from the α to the β form taking place at −237°C. When nitrogen is heated, it combines directly with magnesium, lithium, or calcium; when mixed with oxygen and subjected to electric sparks, it first forms nitric oxide (NO) and then the dioxide (NO_2); when heated under pressure with a catalyst with hydrogen, ammonia is formed (*Haber process*). The ammonia thus formed is of the utmost importance as it is used in fertilizers, and it can be oxidized to nitric acid (*Ostwald process*).

The ammonia industry is the largest consumer of nitrogen. Large amounts of the gas are also used by the electronics industry, which uses the gas as a blanketing medium during production of such components as transistors, diodes, etc. Large quantities of nitrogen are used in annealing stainless steel and other steel mill products. The drug industry also uses large quantities. Nitrogen is used as a refrigerant both for the immersion freezing of food products and for transportation of foods. Liquid nitrogen is also used in missile work as a purge for components, insulators for space chambers, etc., and by the oil industry to build up great pressures in wells to force crude oil upward. Sodium and potassium nitrates are formed by the decomposition of organic matter with compounds of the metals present. In certain dry areas of the world these saltpeters are found in quantity. Ammonia, nitric acid, the nitrates, the five oxides (N_2O, NO, N_2O_3, NO_2, and N_2O_5), TNT, the cyanides, etc. are but a few of the important compounds.

Handbook of Chemistry and Physics, CRC Press

NITROGEN DIOXIDE

Nitrogen Dioxide (NO_2) is a red-brown, non-flammable gas that at room temperature can also exist in a liquid form. It is highly toxic and is classified as one of the major *air pollutants.* It is obtained by the oxidation of nitric oxide. Its main uses are as an oxidising and nitrating agent and in the production of nitric acid.

Molecular weight: 46.01
Melting point: 261.9 K
Boiling point: 294.3 K
Critical temperature: 431.4 K
Critical pressure: 1.013 MPa
Critical density: 271 kg/m^3
Normal vapor density: 2.05 kg/m^3(@ 0°C, 101.3 kPa)

V. Vesovic

NITROGEN DIOXIDE/HYDRAZINE COMBUSTION (see Flames)

NITROUS OXIDE

Nitrous Oxide (N_2O) is a colourless, asphyxiant gas that is commonly known as *laughing gas.* It is only slightly soluble in water and is produced by the thermal decomposition of ammonium nitrate. It forms explosive mixtures with air. Its main uses are as an anaesthetic and as a propellant.

Molecular weight: 44.01
Melting point: 182.3 K
Boiling point: 184.7 K
Critical temperature: 309.58 K
Critical pressure: 7.159 MPa
Critical density: 452 kg/m^3
Normal vapor density: 1.96 kg/m^3(@ 0°C, 101.3 kPa)

V. Vesovic

NMR, SUPERCONDUCTING MAGNETS FOR (see Superconducting magnets)

NOBLE GASES

Noble or *inert gases* is the common name for six gaseous elements (helium, neon, argon, krypton, xenon and radon) that belong to group VIII A of the Periodic Table. They are all characterised by having full outermost electron shells which makes them chemically unreactive under most conditions. The name 'noble' derives from their lack of chemical affinity for other elements, although only **Helium, Neon** and **Argon** are truly inert since no chemical reactions with other elements have been observed. The elements of this group are monatomic and, because of their simplicity, the physical properties of the first five members of the series are known relatively accurately.

V. Vesovic

NOISE, FANS (see Air cooled heat exchangers)

NOMENCLATURE

Quantity	Symbol	Coherent SI unit
Absorptivity (radiation)	α	—
Amount of substance	N	mol
Molar flow rate	$\dot{N}$	mol/s
Molar mass velocity (= $\dot{N}/A_c$)	$\dot{n}$	mol/m^2s
Angle		
Plane	$\alpha,\beta,\gamma,\theta,\phi$	rad
Solid	ω	sr
Area		
Cross-sectional	S	m^2
Surface	A	m^2
Coefficient of cubic expansion		
$(1/v)(\partial v/\partial T)_p$	β	K^{-1}
Concentration		
Mass (= M/V)	c	kg/m^3
Molar (= N/V)	$\tilde{c}$	mol/m^3
Contact angle	θ	rad.
Density		
Mass (= M/V)	ρ	kg/m^3
Molar (= N/V)	$\tilde{\rho}$	mol/m^3
Diameter	D	m
Diffusion coefficient	D	m^2/s
Diffusivity	δ	m^2/s
Thermal (= $\lambda/\rho c_p$)	κ	m^2/s
Dryness fraction (quality)	x	—
Emissivity (radiation)	ε	—
Energy	E	J = kg m^2/s^2
Enthalpy (= U + pV)	H	J = kg m^2/s^2
Specific, molar	$h,\tilde{h}$	J/kg, J/mol
Of phase change at constant p:		
Melting	$h_{sl},\tilde{h}_{sl}$	J/kg, J/mol
Sublimation	$h_{sg},\tilde{h}_{sg}$	J/kg, J/mol
Evaporation	$h_{lg},\tilde{h}_{lg}$	J/kg,J/mol
Entropy	S	J/K = kg m^2/s^2 K
Specific, molar	$s,\tilde{s}$	J/kg K, J/mol K
Force	F	N = kg m/s^2
Fraction		
Mass, of species i	$x_i y_i$	—
Mole, of species i	$\tilde{x} y_{\tilde{i}}$	—
Phase	ϵ	—
Frequency	f	Hz, s^{-1}
Frequency circular	ω	rad/s
Gas constant	$\tilde{R}$	J/mol K
Molar (universal)		= kg m^2/s^2 mol K

Quantity	Symbol	Coherent SI unit
Specific, of species i	R_i	J/kg K = m^2/s^2 K
Gravitational acceleration	g	m/s^2
Gibbs function (H − TS)	G	J = kg m^2/s^2
Specific (h − Ts)	g	J/kg = m^2/s^2
Molar ($\tilde{h}$ − T$\tilde{s}$)	$\tilde{g}$	J/mol = Kg m^2/s^2 mol
Heat		
Quantity of	Q	J = kg m^2/s^2
Rate	$\dot{Q}$	W = kg m^2/s^3
Flux ($\dot{Q}$/A)	$\dot{q}$	W/m^2 = kg/s^3
Heat capacity	c	J/K = kg m^2/s^2 K
Specific, at constant volume or pressure	c_v, c_p	J/kg K = m^2/s^2 K
Molar at constant volume or pressure	$\tilde{c}_v$, $\tilde{c}_p$	J/mol K = kg m^2/s^2 mol K
Ratio c_p/c_v	γ	—
Heat transfer coefficient	α	W/m^2 K = kg/s^3 K
Overall	U	W/m^2 K = kg/s^3 K
Internal energy	U	J = kg m^2/s^2
Specific	u	J/kg = m^2/s^2
Molar	$\tilde{u}$	J/mol = kgm^2/s^2mol
Joule Thompson coefficient [$(\delta T/\delta P)_h$]	μ_{JT}	m^2K/N = ms^2K/kg
Length	l(L)	m
Diameter	d(D)	m
Radius	r(R)	m
Breadth	b(B)	m
Height	z(Z)	m
Thickness	δ	m
Liquid film thickness	δ	m
Liquid holdup	ε_L	—
Mass	M	kg
Flow rate	$\dot{M}$	kg/s
Mass velocity or mass flux ($\dot{M}$/A)	$\dot{m}$	kg/m^2s
Flux of species i	$\dot{m}_i$,	kg/m^2s
Mass transfer coefficient	β	m/s
Molar mass	$\tilde{M}$	kg/kmol
Pressure	p	Pa = N/m^2 = Kg/ms^2
Drop	δp	Pa = N/m^2 = kg/ms^2
Reflectivity	ρ	—
Shear stress	τ	Pa = N/m^2 = kg/ms^2
Surface tension	σ	N/m = kgs^2
Temperature		
Absolute	T	K
Difference or interval	ΔT	K
Logarithmic mean difference	ΔT_{LM}	K
Mean temperature difference	ΔT_M	K
Thermal conductivity	λ	W/mK = kg m/s^3K
Time	t	s
Velocity	u	m/s
Component in Cartesian coordinates x,y,z	u,v,w	m/s
View factor (geometric or configuration factor)	ϕ_{12}	sr
Viscosity		
Dynamic (absolute)	η	Pas = N s/m^2 = kg/ms
Kinematic (= η/ρ)	ν	m^2/s
Void fraction	ϵ_G	—
Volume	V	m^3
Volumetric flow rate	$\dot{V}$	m^3/s
Specific, molar volume	v,$\tilde{v}$	m^3/kg
Work	W	J = kg m^2/s^2
Rate (power)	$\dot{W}$	W = kg m^2/s^3
Wavelength	λ	m
Subscripts and Superscripts		
Solid or saturated solid	S (s)	
Liquid or saturated liquid	L(l)	
Gas or saturated vapor	G(g)	
Change of phase at constant p:		
Melting	SL (sl)	
Sublimation	SG (sg)	
Evaporation	LG (lg)	
Critical state	c	
Initial	o	
Inlet	in,	
Outlet	out,	
At constant value of property	p,v,t. etc	
Mean	—	
Molar (per unit of amount of substance)	~	
Stagnation	o	
Wall	W(w)	

G.L. Shires

NON-CIRCULAR DUCTS, FLOW IN (see Channel flow)

NON-CONDENSIBLE GASES (see Condensation, overview; Surface condensers)

NON-CONDENSING GASES (see Dropwise condensation)

NON-HOOKEAN SOLIDS (see Non-Newtonian fluids)

NON-HYGROSCOPIC MATERIAL (see Drying)

NON-IDEAL PLASMA (see Plasma)

NON-INTRUSIVE FLOW MEASUREMENT (see Electromagnetic flowmeters)

NON-LINEAR SYSTEMS

Introduction

A system is defined to be non-linear if the laws governing the time evolution of its state variables depend on the values of these variables in a manner that deviates from proportionality. Non-linearity is widely spread in nature (Nicolis, 1995). At the microscopic level the equations of motion of a system of particles under the effect of their own collisions, or the equations describing the interaction of radiation with matter are nonlinear; at the macroscopic level, the equations describing the evolution of the conserved variables {x} of a one-component fluid exhibit the universal "inertial" non-linearity ∇. Vx where v is the fluid velocity—itself part of the set of the variables {x}—and V the gradient operator; likewise, the composition variables of a chemically reactive mixture obey typically a set of non-linear equations in which the principal source of non-linearity is to be found in the law of mass action, linking the reaction velocity to the products of concentrations of the species involved.

In recent years it has been realized that quite ordinary systems obeying simple non-linear laws can give rise spontaneously to behaviours of considerable complexity associated with abrupt transitions, a multiplicity of states, rhythmic activity, pattern formation or a random-looking evolution which is referred to as *deterministic chaos.* Thermal convection in a fluid layer heated from below, Taylor vortex flow, turbulence, the generation of coherent light by a laser, chemical oscillations or the formation of a flame front provide some well-established examples of this ubiquitous property of non-linear systems, which is also referred to as *self-organization.* The self-organization paradigm has proven to be a powerful tool for analyzing complex systems outside the traditional realm of physical sciences, notably biological systems and systems encountered in environmental science. (See for example **Turbulence**)

Non-linearity, Instability and Bifurcation

Non-linearity may remain "inactive" or, on the contrary, lead to qualitative changes of behavior depending on the values of the *control parameters* describing the way a system has been initially prepared or is being permanently solicited by the external world. In a *conservative* system. (e.g., a system whose relevant observables obey the laws of mechanics), these may stand for the initial amount of energy communicated to it or for the action of an external force deriving from a potential. In a *dissipative* system (a system whose relevant observables generate during their evolution a strictly positive entropy production) a typical control parameter is the distance from thermodynamic equilibrium or from a phase transition point.

The awakening of the system's nonlinearities does not happen gradually but involves a succession of explosive events, in the form of *instabilities.* Specifically, when the constraints exerted by the environment reach certain thresholds small perturbations of the environment or small, spontaneously arising fluctuations, become amplified, leading the system out of its basic state and pushing it towards a new regime. This transition resembles a *bifurcation:* when the initial state becomes unstable, it is replaced not by a unique regime but, generally speaking, by a multitude of stable regimes that are accessible simultaneously (**Figure 1**). Nothing in the initial preparation of the system enables one to know which particular regime will actually be chosen. Only chance, in the form of a critical variation which is going to prevail at the propitious moment, will decide which particular branch will be followed. This makes the system *sensitive to the parameters* controlling the position of the bifurcation point, since two macroscopically indiscernible systems, submitted to the same constraints, may follow entirely different paths.

Bifurcation is far from being a unique event. As the constraints are varied, a system typically undergoes not just a single transition

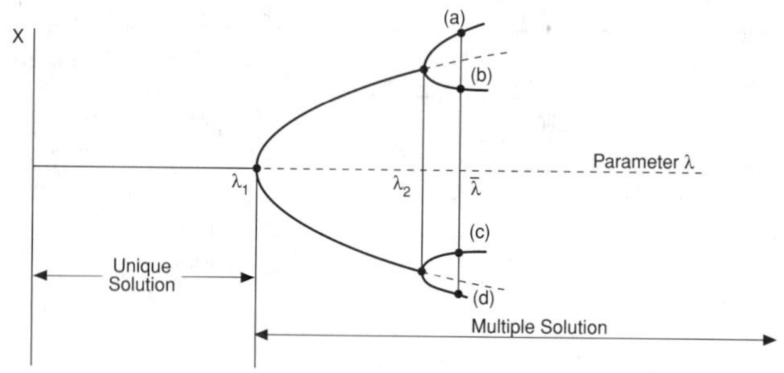

Figure 1. Typical bifurcation diagram. As the control parameter λ is varied, a reference state loses its stability and two new branches of solutions emerge at $\lambda \geq \lambda_1$. These branches, in turn, lose their stability beyond a secondary bifurcation point $\lambda = \lambda_2$, etc.

but a whole sequence of transition phenomena, the characteristics of which depend on the nature of the non-linearities present. In many cases, these transitions culminate in a regime which, despite its deterministic origin, is characterized by an irregular evolution of the variables in space and time resembling in many respects a game of chance. One refers to this phenomenon as *deterministic chaos.*

Modeling Non-linear Systems

In their vast majority, natural systems are composed of a large number of interacting elementary subunits. At the microscopic level, this interaction is manifested through the inter- or intramolecular forces. At the mesoscopic and macroscopic level, the interparticle forces do not come into play in an explicit manner but are manifested indirectly through, for instance, the transfer of matter, energy, mechanical work or information between a subsystem and its surroundings.

As a rule, because of the interactions, each subunit undergoes complex non-linear dynamics. At the microscopic level, this resembles in many respects a noise process, of which **Brownian Motion** is a typical example, but actually this complexity finds its origin in an extreme form of deterministic chaos characteristic of systems involving a large number of degrees of freedom. At the macroscopic level complexity arises through the mechanism of bifurcations and a milder form of deterministic chaos characteristic of systems involving a small number of macrovariables describing the thermodynamic state of the system. Finally, at the mesoscopic level, an intermediate view is adopted in which the central quantity is the probability distribution of the macrovariables. *Complexity* is manifested here through a qualitative change in the properties of this probability (multi-modality, critical fluctuations, etc.) across a transition point leading to multiple states or to chaos. A considerable amount of effort has been devoted to bridge the gap between these various levels of description, [Prigogine (1980); Nicolis and Prigogine (1989)] but our understanding of this fundamental problem is still far from being complete.

Motions of particles and their interactions at the microscopic level are governed by the fundamental laws of physics, essentially those of mechanics, quantum mechanics and electrodynamics. Progress in *chaos theory* has allowed one to reformulate on a new basis the connection between the complex dynamics induced by these interactions and the foundations of statistical mechanics, including the origin of *irreversibility*, through the study of the spectrum of the Liouville-like operators governing the evolution of probability densities. The advent of supercomputers has made it possible to envisage a complementary approach, in which the corresponding equations are written out explicitly for all the particles involved and solved on the computer through the microscopic simulation techniques of molecular dynamics or *Monte Carlo procedures* (Mareschal & Holian, 1992). The complex phenomena characteristic of non-linearity emerge, then, from the collective behavior in the ensemble—for instance, by performing averages of a certain property over the particles or over successive runs differing slightly in their initial conditions.

The mesoscopic approach is based on phenomenological equations describing an intermediate level. An example of this class is **Chemical Reaction** kinetics where the elementary steps are restricted to reactive collisions without explicit consideration of vibrational and rotational states of the molecules. At this level, the evolution laws are formulated as **Stochastic Processes** governed by Master equations or equations of the Fokker-Planck type which are accessible to the tools of *probability theory* [Nicolis & Prigogine (1989); van Kampen (1981)].

The third, macroscopic level of description is the **Thermodynamic** level. Here the individuality of processes disappears in the mathematical formulation and one resorts to the *constitutive relations* of thermodynamics linking phenomenologically the thermodynamic potentials to the state variables and the fluxes of the various processes present to the thermodynamic forces such as temperature or chemical potential gradients (De Groot and Mazur, 1962) Molecular dynamics reveals that linear phenomenological laws of this second kind, with state-dependent coefficients, remain surprisingly robust even under very strong constraints. Still, far away from equilibrium, a purely thermodynamic characterization of the states of a system is not sufficient. In this range the most relevant mode of approach becomes the analysis of the phenomenological rate equations (reaction-diffusion equations, Navier-Stokes equations, etc.) derived from the balance equations of mass, momentum and energy supplemented with the constitutive relations, using the methods of stability, bifurcation and chaos theories. These methods are briefly summarized in the next two subsections.

Characterization of Non-linear Systems

Phase space, attractors

A useful tool in classifying and comparing different types of dynamic behavior that are likely to arise within a system when the conditions to which it is submitted are varied is afforded by *qualitative* analysis, based on the geometric view of non-linear systems (Guckenheimer and Holmes, 1983).

Consider a system described by a finite set of observables such as temperature, chemical composition, flow velocity, pressure, etc. One embeds its evolution into the abstract space spanned by all these variables. In this *phase space*, an instantaneous state of the system is represented by a point, and as time goes by, the point in question follows a curve, called the phase trajectory. By following the trajectories emanating from different initial states, one obtains a *phase portrait* which provides a valuable qualitative idea of the system's potentialities. As a general rule, for every natural system obeying the second law of thermodynamics (dissipative system), the phase trajectory will converge, after a certain time, towards an object in phase space whose dimension is strictly smaller than that of the phase space itself, and which is referred to as the *attractor.*

In view of the foregoing, understanding the types of behavior generated by a system amounts to classifying all the attractors that can be realized in a phase space. The simplest element one can embed in a space is the point. Consequently, one can imagine point attractors (**Figure 2a**), whose existence means that after a sufficiently long period of time, any initial state will tend towards a regime that will no longer evolve once it is established: the system will be in a stationary state.

One level higher than a point in the hierarchy of geometric forms is the line. We can thus imagine attractors in the form of a closed curve towards which different possible histories of the system will tend (**Figure 2b**); these attractors are referred to as *limit cycles.* In this type of behavior, events are repeated in a regular and reproducible way: one has here an archetype of periodic phenomena, of which biological rhythms are a particularly important example.

A further level upwards in the hierarchy of complexity are *strange attractors* (**Figure 2c**). Contrary to the cases illustrated in **Figures 2a** and **2b** where the behavior became stable and regular once the system was fixed on the attractor, one now finds that the system performs an aperiodic and apparently erratic movement on the attractor, which has already been referred to as deterministic chaos. This behavior results from two opposite tendencies: an

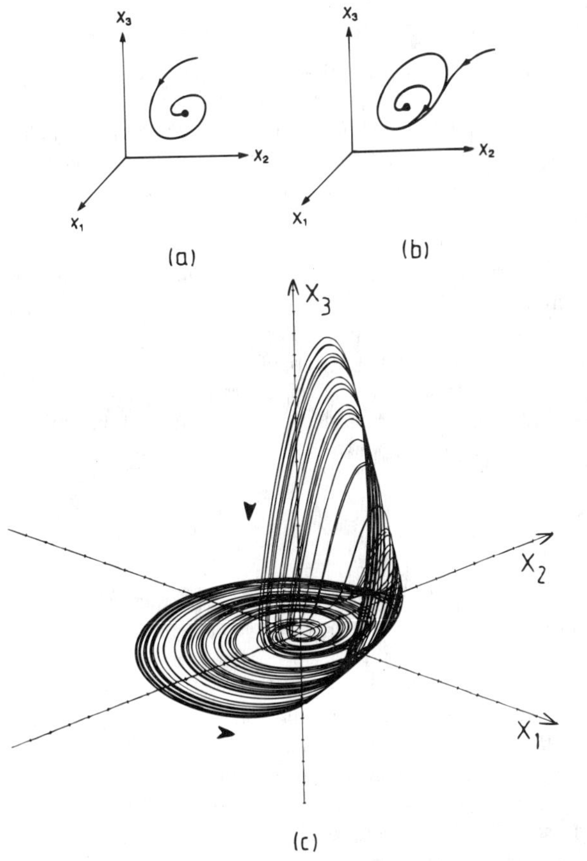

Figure 2. Fixed point (a), limit cycle (b), and chaotic (c) attractors.

instability arising along certain directions of the attractor (horizontal arrow in **Figure 2c**) coexisting permanently with a stabilizing trend which prevents the trajectories from escaping and re-injects them into the attractor (vertical arrow in **Figure 2c**). In order to reconcile these two antagonistic tendencies, the system must free itself from the constraints of Euclidean geometry which dictated the form of the trajectories of the examples in **Figures 2a** and **2b**. A *fractal attractor*—that is, a set of points constituting an object somewhere between a surface and a three-dimensional volume, whose dimensionality (generally not an integer) is thus greater than in Euclidean geometry—appears as the result of this radical change (Ott 1993).

The existence of an unstable element in the dynamics of deterministic chaos means that the system evolves in a radically different way if its initial state is slightly different, even if it obeys exactly the same laws. This property of *sensitivity to initial conditions* leads to an unexpected consequence. Consider a system operating in the regime of deterministic chaos, and suppose that its initial state is given by the value of a certain variable at the instant zero. Due to the fact that the precision of any physical measurement is limited, two slightly different states, represented by two points in the phase space separated by a distance less than the precision of the measurement, will be indiscernible to the observer. And yet, after a certain lapse of time, these two states will end up becoming distinct and projecting themselves on distant parts of the attractor. It follows that it is no longer meaningful to make long-term predictions concerning the future of the underlying system.

Analytic Approach to Non-linear Systems

A major limitation hindering the *quantitative* study of non-linear systems is that, as a rule, exact non-trivial solutions of non-linear evolution equations are not available. Nevertheless, judicious application of perturbation theory combined with geometric techniques leads to the unexpected conclusion that in the vicinity of transition phenomena the dynamics of a non-linear system is dramatically simplified, in the sense that it is dominated by a limited number of key variables, in terms of which all others can be expressed, obeying furthermore universal evolution equations (Guckenheimer and Holmes, 1983). For instance, in a system involving a finite number of variables and operating in the vicinity of the first bifurcation depicted in **Figure 1**—referred to as pitchfork bifurcation—there exists a single combination z of the initial variables, obeying to the equation

$$\frac{dz}{d\tau} = (\lambda - \lambda_c)z - uz^3 \qquad \textbf{(1)}$$

in which τ is a scaled time, λ the control parameter, λ_c its critical value and u a real parameter whose value depends on the detailed structure of the system. **Equation 1** is known as *normal form*, and the one-dimensional subspace of the initial phase space on which z is defined is known as the *center manifold*. The variable z itself is referred to as the *order parameter*.

More intricate situations arise in the presence of interaction between instabilities generating higher order bifurcation phenomena and, possibly, chaotic dynamics. In such cases one can still guarantee that the part of the dynamics that gives information on the bifurcating branches takes place in a phase space of reduced dimensionality. The explicit construction of the normal forms becomes, however, much more involved and their universality can no longer be guaranteed. Still, in the regime of chaotic dynamics universal properties of a new type may emerge. (Ott, 1993)

The above developments open the tantalizing possibility of modelling the great diversity of complex phenomena observed in real-world multivariable non-linear systems in terms of a limited number of variables or, equivalently, in terms of low-dimensional attractors. What is more, these results can be extended to spatially distributed systems involving at the outset an infinite number of degrees of freedom. The type of normal form depends, in this case, not only on the nature of the instability (e.g., non-oscillatory or oscillatory) but also on whether or not a *symmetry-breaking* process generating a characteristic length previously absent in the system is taking place. As an example, in the vicinity of an oscillatory instability without spatial symmetry breaking, one arrives at an equation known as the complex Landau-Ginzburg equation (Coullet & Gil, 1988)

$$\frac{\partial z}{\partial t} = (\lambda - \lambda_c)z + (1 + i\alpha)\nabla^2 z - (1 + i\beta)|z|^2 z \qquad \textbf{(2)}$$

where z is now a complex-valued order parameter and the particular values of α and β depend on the detailed structure of the system at hand. Similar equations can be derived for symmetry-breaking instabilities or for interfering instabilities involving the interaction between oscillatory and symmetry-breaking modes. These equations have been studied intensely in recent years and have provided the interpretation of a large body of experimental data in such different contexts as fluid mechanics, optics, chemistry and materials science.

References

Coullet P. and Gil, L. (1988) Normal form description of broken symmetries, *Solid State Phenomena* 3–4, 57–76.

De Groot S. and Mazur, P. (1962) *Nonequilibrium Thermodynamics*, North-Holland, Amsterdam.

Guckenheimer J. and Holmes, Ph. (1983) *Nonlinear Oscillations, Dynamical Systems and Bifurcation of Vector Fields*, Springer, Berlin.

Mareschal M. and Holian B. (eds). (1992) *Microscopic Simulations of Complex Hydrodynamic Phenomena*, Plenum, New York.

Nicolis G. and Prigogine, I. (1989). *Exploring Complexity*, Freeman, New York.

Nicolis G. (1995) *Introduction to Nonlinear Science*, Cambridge University Press, Cambridge.

OH E, (1993) *Chaos in Dynamical Systems*. Cambridge University Press, Cambridge.

Prigogine, I. (1980) *From Being to Becoming*, Freeman, San Francisco.

Van Kampen, N. (1981) *Stochastic Processes in Physics and Chemistry*, North-Holland, Amsterdam.

G. Nicolis

NON-NEWTONIAN FLUID HEAT TRANSFER

Following from: *Non-Newtonian fluids*

Most slurries, suspensions, dispersions, solutions of polymeric materials and melts exhibit complex flow behaviour which cannot be described by *Newton's law of viscosity* $\tau = \eta\gamma$ where τ is the shear stress, γ is the shear rate and the constant of proportionality η is the material property called **Viscosity**. Convective heat transfer to such fluids depends on the fluid rheology, geometric configuration of the flow domain as well as the flow regime (e.g. laminar, turbulent etc). The apparent viscosity of **Non-Newtonian Fluids**, $\eta_a = \tau/\gamma$, is not a material property (as is the case for **Newtonian Fluids**) but may depend on the rate of shear and previous flow history of the fluid. *Viscoelastic fluids* display properties of both fluids and elastic solids. The following chart presents a rheological classification of non-Newtonian fluids encountered in the chemical, food, petrochemical, detergent, printing inks, coatings, etc. industries.

Convective heat transfer to such fluids depends on the rheology of the fluid. It is affected by viscous dissipation due to the very high viscosities coupled with high shear rates, and the temperature-dependence of the apparent viscosity as well as by thermal conductivity, possible chemical reactions, etc. Due to space limitations, we confine our attention to steady convective heat transfer to *pseudoplastic or dilatant fluids* described by the well known *Ostwald-de-Waele power law model* ($\tau = K\gamma^n$ or $\eta_a = K\gamma^{n-1}$; $n < 1$ for pseudoplastic fluids and $n > 1$ for dilatant fluids). Further, we will examine only internal flow through smooth, straight conduits of circular and noncircular cross-section. Since most non-Newtonian fluids are highly viscous, laminar flow is often encountered in industrial applications. External flows involving non-Newtonian fluids are generally of less practical interest than internal flows. Time-dependent fluids which undergo significant shearing before entry into the channel or duct are affected by the time-dependency only over short times of deformation. Also, except in the hydrodynamically developing entrance region of the duct or channel fluid elasticity has little influence on the flow since the elastic stresses do not change in the flow direction.

Since most non-Newtonian fluids have high viscosities the hydrodynamic entrance length beyond which the flow becomes independent of axial distance is shorter than the thermal entrance length. Therefore we focus on the case of fully developed velocity profiles, considering a number of geometric configurations, e.g. circular pipe, parallel plates, rectangular ducts, cylindrical annuli and several non-circular cross section ducts. We further restrict attention to rectilinear ducts of uniform cross-section. Effects of viscous dissipation, variable apparent viscosity, effect of buoyancy, chemical reactions, external flow, turbulence, etc. have been discussed in the literature, e.g. Mashelkar (1988), Lawal and Mujumdar (1989), and Etemad and Mujumdar (1994), among others. The interested reader is referred to the excellent textbook by Skelland (1967) for a thorough analysis of non-Newtonian flow and heat transfer.

Forced Convection Channel Flows

Laminar flow

Since many non-Newtonian fluids can be adequately considered as purely viscous over flow ranges of interest, for the sake of convenience (and with some loss of generality) the correlations and analytical as well as numerical results presented here assume a power law model. For visoelastic fluids elastic effects become of significance only when secondary flows appear in the flow domain. Thus, for a fully developed laminar tube flow there is no mechanism whereby elasticity can play a role. However for rectangular ducts the secondary flows caused by normal stress differences acting on the fluid boundaries cause enhanced heat transfer due to elasticity effects. The convective heat transfer coefficient expressed as the local Nusselt Number, Nu_x, depends on the duct geometry and the thermal boundary condition applied. The two most commonly used thermal boundary conditions are:

a. Uniform temperature (T boundary condition),
b. Uniform heat flux condition (often referred to as H2 boundary condition to denote uniformity in axial as well as peripheral directions).

Further, the flow may be hydrodynamically and thermally developing in the streamwise direction. For a hydrodynamically devel-

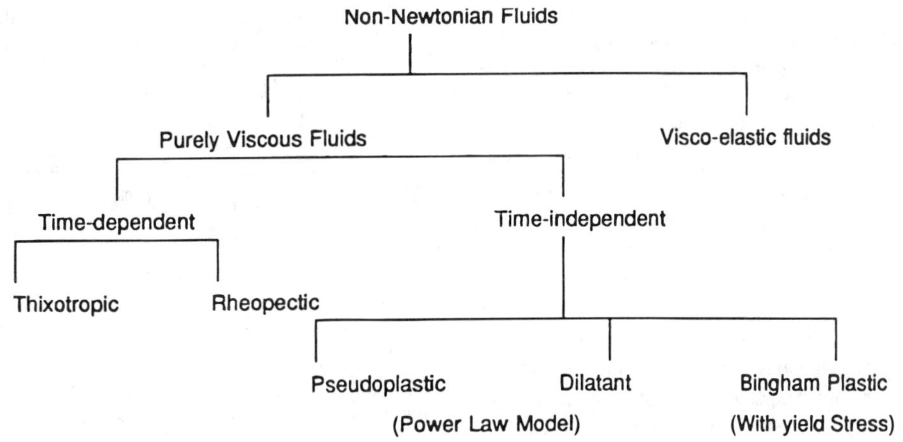

oped flow the local **Nusselt Number** in the thermal entrance region can be obtained analytically using the following relations:
for $Gz > 25\pi$

$$Nu_{H2,x} = 1.41[(3n + 1)/4n]Gz^{1/3} \tag{1}$$

$$Nu_{T,x} = 1.16[(3n + 1)/4n]^{1/3}Gz^{1/3} \tag{2}$$

for $Gz > 33\pi$, where **Graetz Number**, Gz, is defined as

$$Gz = \dot{m}c_p/\lambda x$$

For laminar flow through rectangular ducts the Nusselt number is a function of the aspect ratio as well as n. For a fixed aspect ratio the influence of n is less than 10 per cent over $0.5 \leq n \leq 1.0$ which covers most fluids of practical interest.

The variation of the local Nusselt number with axial distance for a simultaneously developing flow of a power law fluid with Re = 500 and Pr = 10 though a circular tube is shown in **Figure 1a** and **1b**. Both Re and Pr are based on the apparent viscosity of the fluid Figure 1a shows the variation of $Nu_{T,x}$ with dimensionless axial distance X_{th} for uniform wall temperature for three values of n viz. n = 0.5 (pseudoplastic) 1.0 (Newtonian) and 1.25(dilatant). The effect of n on the fully-developed value can be seen to be minimal. **Figure 1b** shows the corresponding Nusselt number distributions for the H2 thermal boundary condition. The Nusselt numbers are higher for the uniform heat flux boundary condition.

Figures 2a and **2b** present computed Nusselt number distributions for a square duct. Although numerically different (but of the same order) the results display trends similar to those obtained for a circular tube (**Figure 1a** and **1b**). These results were computed by solving the equations of continuity, momentum and energy over the domain of interest. Note that the Nusselt number is dependant on the Prandtl number as well, which is not a fluid property for non-Newtonian fluids unlike Newtonian fluids.

Table 1 below compares the fully-developed Nusselt numbers (Nu_T and Nu_{H2}) for twelve different cross-sectional geometries. For noncircular ducts the Reynolds number is based on the hydraulic diameter of the duct cross-section. These results were obtained by Etemad (1995) by a finite element solution of the governing conservation equations for Re = 500 and Pr = 10. It is interesting to note the influence of rounding the corners of a square duct; for example rounding the square duct with a radius one-sixth the side of the duct results in an increase of Nu_T from 3.19 to 3.43 for n = 0.5. Heat transfer and pressure drop results for power law fluids flowing in various noncircular ducts, including effects of temperature-dependent viscosity as well as viscous dissipation are available in Etemad (1995). A discussion of these aspects is beyond the scope of this text.

Turbulent flow

Several empirical correlations are available to estimate the fully developed Nusselt number. One may use the following correlations due to Gnielinski for constant temperature as well as constant flux conditions under turbulent flow conditions (Hartnett (1994):

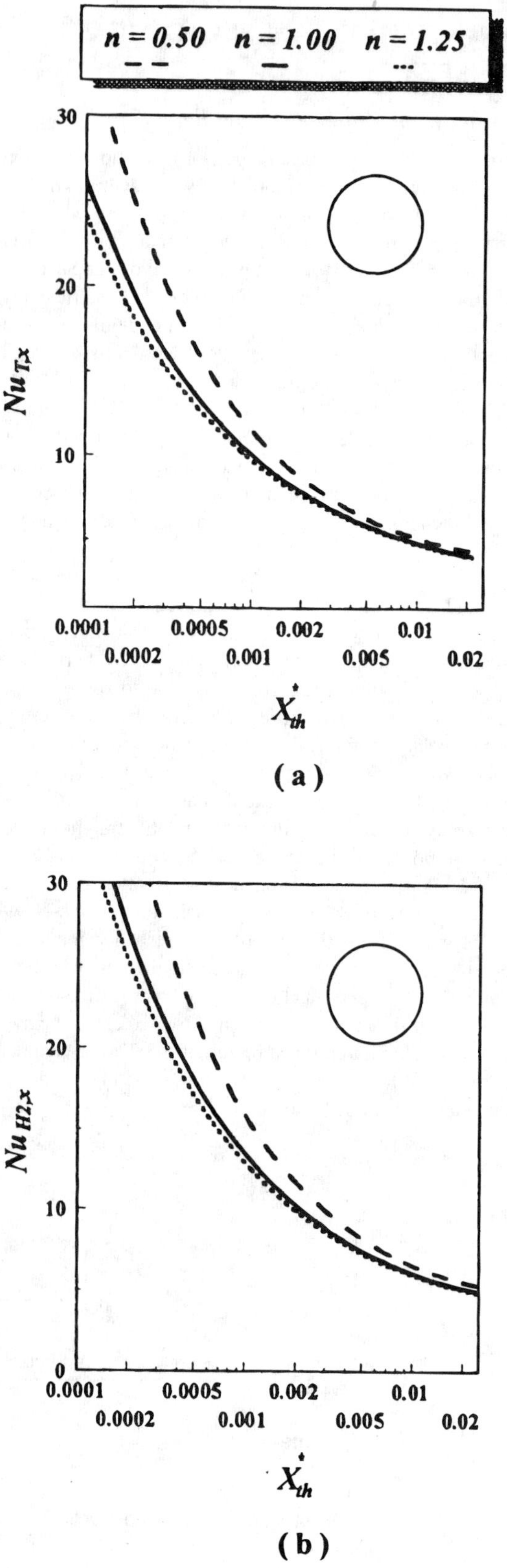

Figure 1. Simultaneously developing flow and heat transfer for Re = 500, Pr = 10. a) Isothermal wall, (2) Constant heat flux wall.

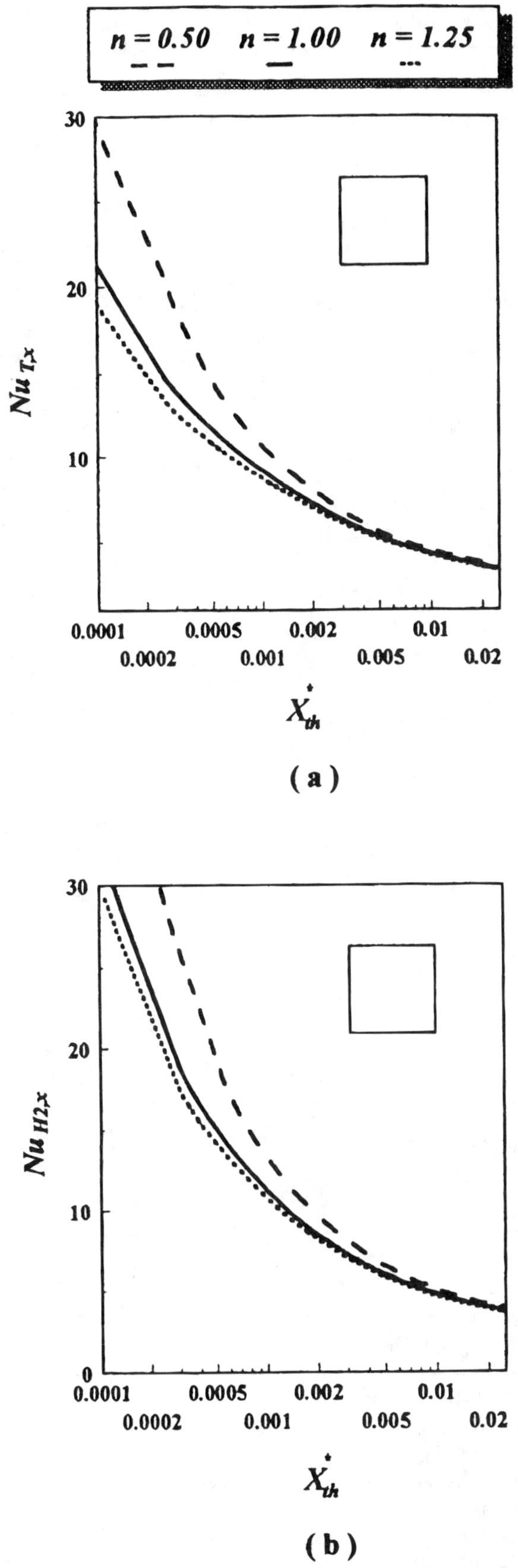

Figure 2. Simultaneously developing flow and heat transfer in square duct for Re = 500 and Pr = 10. (a) Isothermal wall, (b) Constant heat flux at wall.

$$Nu = \frac{\frac{f}{2}(Re - 1000)Pr}{1 + 12.7\left[\left(\frac{f}{2}\right)^{1/2}(Pr^{2/3} - 1)\right]} \tag{3}$$

where the friction factor f is estimated by

$$f = 0.0791(Re_j)^{-0.25} \tag{4}$$

for both round and rectangular ducts. Here

$$Re_j = [\rho U D_h]/[\eta(a + b)] \tag{5}$$

For circular tube, a = 0.25 and b = 0.75. For rectangular ducts, the values of a and b are given below in **Table 2** (Hartnett, 1994). The estimates are within ±20% of measured data. No general correlations exist for heat transfer in turbulent flow of a non-Newtonian fluid. For the flow of dilute suspensions of fibres in water in circular tubes substantial reductions in turbulent flow heat transfer have been reported in the literature at Reynolds numbers up to 2×10^4. However, above this value the heat transfer rate begins to rise sharply.

Closing Remarks

Relatively little literature exists on predictive correlations useful for engineering design involving non-Newtonian fluids. For space limitations no information is included on external flow heat transfer viz. flow over objects of various shapes. Also excluded is discussion of such important effects as temperature or composition dependent apparent viscosity, viscous dissipation, visco-elasticity, mixed convection, etc. The reader is referred to the extensive reviews by Metzner (1965), Shenoy and Mashelkar (1982), Lawal and Mujumdar (1989) and Etemad (1995), among others. The textbook by Skelland (1967) is strongly recommended for a tutorial review of the subject matter.

Notation

a	geometric constant in Kozicki generalized Reynolds number
A	area of duct cross-section
b	geometric constant in Kozicki generalized Reynolds number
c_p	Specific heat of fluid
D_h	hydraulic diameter 4A/P
f	Fanning friction factor
Gz	Graetz number $\dot{m}\ c_p/\lambda x$
$\dot{m}$	mass flow rate
n	power law index
Nu_x	axially local Nusselt number
$Nu_{T,x}$	axially local Nusselt number for T boundary condition
$Nu_{H2,x}$	axially local Nusselt number for H2 boundary condition
P	perimeter of the duct cross-section
Pr	Prandtl number $\eta\ c_p\ (u_e/D_h)^{n-1}$
Re	Reynolds number $\rho U^{2-n} D_h^{n/\eta}$
Rej	Reynolds number $\rho U D_h/(a + b)\eta$

Table 1. Fully developed Nusselt numbers for various duct cross-sectional geometries and power law indices

Geometry	Nu_T			Nu_{H2}		
	n = 0.50	n = 1.00	n = 1.25	n = 0.50	n = 1.00	n = 1.25
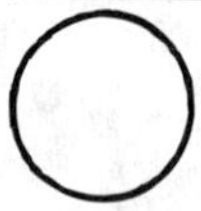	3.950	3.659	3.590	4.744	4.363	4.272
	3.190	2.979	2.925	3.310	3.090	3.032
	3.270	3.047	2.991	3.461	3.221	3.160
	3.431	3.188	3.129	3.764	3.488	3.417
	3.600	3.388	3.350	3.150	3.021	2.998
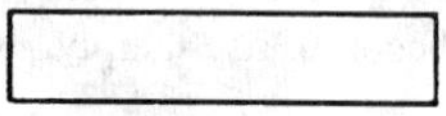	4.922	4.831	4.817	2.717	2.924	2.996
	7.950	7.541	7.442	8.758	8.235	8.109
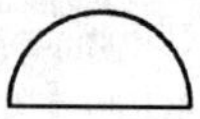	3.480	3.318	3.265	3.038	2.920	2.880
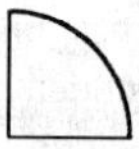	3.206	3.060	3.013	3.121	2.980	2.937
	2.936	2.820	2.789	2.510	2.430	2.408
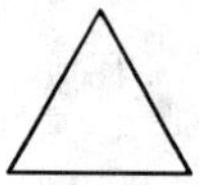	2.594	2.503	2.478	1.951	1.896	1.880
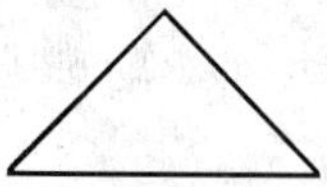	2.409	2.350	2.335	1.370	1.351	1.344

Table 2.

Aspect Ratio	a	b
1.0	0.2121	0.6771
0.8	0.2155	0.6831
0.6	0.2297	0.7065
0.4	0.2659	0.7571
0.2	0.3475	0.8444

u_e inlet velocity
x axial distance
X_{th} x/D_h Pr · Re

Greek symbols

η viscosity of newtonian fluid
η_a apparent viscosity of non-Newtonian fluid
τ shear stress
γ shear rate (velocity gradient for one dimensional flow)
ρ density of fluid
λ thermal conductivity of fluid

Acknowledgement

The author is grateful to Dr. S. Gh. Etemad for his valuable help in the preparation of this contribution.

References

Cho, Y. I. and Hartnett, J. P., (1987) Non-Newtonian fluids, Chapter 2 in *Handbook of Heat Transfer Applications*, 2nd Ed. (Ed. Rohsenow, Hartnett, J. P. and Ganic), Kingsport Press.

Etemad, S. Gh., Mujumdar, A. S., Huang, B., (1994) Viscous dissipation effects in entrance region heat transfer for a power law fluid flowing between parallel plates, *International Journal of Heat and Fluid Flow*, vol. 15, no. 2, 122–131.

Etamad, S. Gh., Mujumdar, A. S., Huang, B, (1994). Laminar forced convection heat transfer of a non-Newtonian fluid in the entrance region of a square duct with different boundary conditions, 10th International Heat Transfer Conference, Brighton, U.K.

Etemad, S. Gh., (1995). Ph.D. Thesis, McGill University, Montreal, Quebec, Canada.

Hartnett, J. P., (1994), Single phase channel flow forced convection heat transfer, Keynote Lecture, 10th International Heat Transfer Conference, Brighton, U.K.

Irvine, T. F. and Karni, J., (1987) Non-Newtonian fluid flow and heat transfer, Chapter 20 in *Handbook of Single Phase Convective Heat Transfer* (Ed. Kakac, S. Shah, R. K. Aung, W.), 20-1–20-57.

Lawal, A. and Mujumdar, A. S., (1989) Laminar duct flow and heat transfer to purely viscous non-Newtonian fluids, *Advances in Transport Processes*, vol. VII (Ed. A. S. Mujumdar and R. A. Mashelkar). Wiley, New York, 353–442.

Mashelkar, M. A. (1988) Convective heat transfer in laminar internal flows design considerations for heat exchanges handling non-Newtonian fluids, *Heat Transfer Equipment Design* (Ed. R. K. Shah, E. C. Subbarao and R. A. Mashelkar). Hemisphere Publishing Corporation, N.Y. 719–746.

Metzner, A. B., (1965) Heat transfer to non-Newtonian fluids, *Advances In Heat Transfer*, vol. 2, (Ed. Hartnett, J. P. and Irvine, T. F.) Academic Press, New York, p 357–397.

Shenoy, A. V. and Mashelkar, R. A., (1982) Thermal convection in non-Newtonian fluids, *Advances in Heat Transfer*, vol. 15, Academic Press, N.Y., 143–225.

Skelland, A. H. P. (1967) *Non-Newtonian Flow and Heat Transfer*, John Wiley & Sons, New York.

A.S. Mujumdar

NON-NEWTONIAN FLUIDS

Following from: Fluids

A **Newtonian Fluid** is one where there is a linear relationship between *stress* and *strain-rate:* the ratio of stress to strain-rate is the viscosity of the fluid. A *Hookean solid* is one where there is a linear relationship between *stress* and *strain:* the ratio of stress to strain is the *modulus* of the solid. Many materials have intermediate properties between those of a Newtonian fluid and a Hookean solid. If the properties are predominantly solid-like, the materials are called *non-Hookean* and the materials are described as *viscoelastic.* If they are predominantly fluid-like, they are called *non-Newtonian* and the materials are described as *elasticoviscous.*

A non-Newtonian fluid has, therefore, a simultaneously elastic and viscous nature. In fact, all fluids are non-Newtonian on an appropriate time-scale, though for many common fluids such as air and water the time-scale is extremely short. When the time-scale of a flow t_f is much less than the *relaxation time* t_r of an elasticoviscous material, elastic effects dominate. This typically happens when there are abrupt changes in flow geometry. When on the other hand t_f is much greater than t_r, elastic effects relax sufficiently for viscous effects to dominate. This typically happens when there are not abrupt changes in flow geometry. The ratio of t_f to t_r is a dimensionless number of particular significance in the study of flow of non-Newtonian fluids: depending on the circumstances, this number is called the **Deborah Number** or the **Weissenberg Number**. If the Deborah or Weissenberg number is small, elastic effects can be neglected and the non-Newtonian fluid treated as a purely viscous material, albeit with a non-constant viscosity.

The mass and linear momentum conservation equations which govern the flow of any fluid are respectively (see Richardson 1989):

$$\frac{\partial \rho}{\partial t} + \nabla\cdot(\rho \mathbf{u}) = 0 \tag{1}$$

$$\rho\frac{\partial \mathbf{u}}{\partial t} + \rho\mathbf{u}\cdot\nabla\mathbf{u} = \rho\mathbf{g} - \nabla p + \nabla\cdot\boldsymbol{\tau} \tag{2}$$

where ρ denotes the density of the fluid, u velocity, p pressure, $\boldsymbol{\tau}$ *deviatoric stress*, g gravitational acceleration and t time. Note that it is assumed that τ is a symmetrical tensor, thus ensuring that angular momentum is conserved. Determination of the flow of any fluid requires solution of these conservation equations, subject to appropriate boundary and initial conditions, together with *constitutive equations* for the fluid. It is in the constitutive equations that the nature of a given fluid is manifested. For a Newtonian fluid $\boldsymbol{\tau}$ is given by:

$$\boldsymbol{\tau} = \eta\mathbf{e} \tag{3}$$

where **e** denotes the strain-rate given by:

$$\mathbf{e} = \nabla\mathbf{u} + (\nabla\mathbf{u})^T \tag{4}$$

and η denotes the viscosity of the fluid, which is independent of **e** (though it may, of course, depend on temperature T and perhaps on pressure p). A non-Newtonian fluid is one for which Equation 3 does not hold or one for which Equation 3 holds but only with η dependent on **e**.

A flow in a long duct such as a pipe is susceptible to rather simpler analysis than a more general flow because it is essentially a *shear flow* (which may loosely be regarded as one in which the velocity varies in a direction perpendicular to the direction of flow) as opposed to an *extensional flow* (which may loosely be regarded as one in which the velocity varies in the direction of flow). In a shear flow, the relevant rheological or flow property is *viscosity* η and $\boldsymbol{\tau}$ is given by Equation 3. A fluid for which (3) holds but η depends on **e** is called a *generalised Newtonian fluid.*

The viscosity η is a scalar; the strain-rate **e** is a tensor. If η depends on **e**, then it can be shown (see Richardson (1989) and

Bird, Armstrong and Hassager (1977)) that η must be a function of the three scalar invariants I_e, II_e and III_e of **e** which are given by:

$$I_e = \mathrm{tr}(\mathbf{e}) \qquad \textbf{(5)}$$

$$II_e = \frac{1}{2}\left[\mathrm{tr}(\mathbf{e}^2) - (\mathrm{tr}(\mathbf{e}))^2\right] \qquad \textbf{(6)}$$

$$III_e = \det(\mathbf{e}) \qquad \textbf{(7)}$$

where tr( ) and det( ) denote trace and determinant, respectively. For an incompressible fluid:

$$I_e = 0 \qquad \textbf{(8)}$$

A shear flow is one for which:

$$III_e = 0 \qquad \textbf{(9)}$$

Thus II_e is the only non-zero invariant for a shear flow of an incompressible fluid. The shear-rate γ is defined by:

$$\gamma = \sqrt{II_e} \qquad \textbf{(10)}$$

Hence the constitutive equation for the stress $\boldsymbol{\tau}$ of an incompressible generalised Newtonian fluid is:

$$\boldsymbol{\tau} = \eta\{\gamma\}\mathbf{e} \qquad \textbf{(11)}$$

For a Newtonian fluid, η is independent of I_e, II_e and III_e and hence of γ.

Many different semi-empirical relationships for $\eta(\gamma)$ exist [(see Bird, Armstrong & Hassager (1977), Skelland (1967) and Wilkinson, (1960)]. One of the most widely used is the *power-law fluid:*

$$\eta = \eta_c\gamma^n \qquad \textbf{(12)}$$

where η_c denotes consistency or unit shear-rate viscosity and n the power law exponent. A Newtonian fluid is one for which $n = 0$. Many non-Newtonian fluids are *shear-thinning* or *pseudoplastic*, that is the viscosity η decreases with increasing shear-rate γ (see **Figure 1**). The magnitude of the shear stress always increases with increasing shear-rate, however, so the product $\eta\,\gamma$ always increases with increasing γ. Thus $-1 < n \leq 0$; typically, n is between -0.8 and 0. If $n = 0$, the viscosity is, of course, independent of shear-rate; if $n = -0.8$, an increase in γ by a factor of 10 causes a decrease in η by a factor of about 6. For a *shear-thickening* or *dilatant* fluid, in contrast, η increases with increasing γ (see **Figure 1**).

A relationship which is applicable to fluids which have a yield stress is the *Bingham plastic* or *Bingham fluid:*

$$\eta = \begin{cases} \infty & \text{if } |\boldsymbol{\tau}| \leq \tau_y \\ \eta_P + \dfrac{\tau_y}{\gamma} & \text{if } |\boldsymbol{\tau}| \geq \tau_y \end{cases} \qquad \textbf{(13)}$$

(see **Figure 1**) where η_p denotes the plastic viscosity and τ_y the yield stress. If $|\boldsymbol{\tau}| \leq \tau_y$, that is the stress in the fluid is less than its yield stress, the fluid does not deform and so $\gamma = 0$ but $\eta = \infty$ since $|\boldsymbol{\tau}| \neq 0$ in general. In contrast if $|\boldsymbol{\tau}| > \tau_y$ the fluid deforms and γ is linearly related to $|\boldsymbol{\tau}|$. If $|\boldsymbol{\tau}| \gg \tau_y$ then $\eta \simeq \eta_p$. For a Newtonian fluid, $\tau_y = 0$.

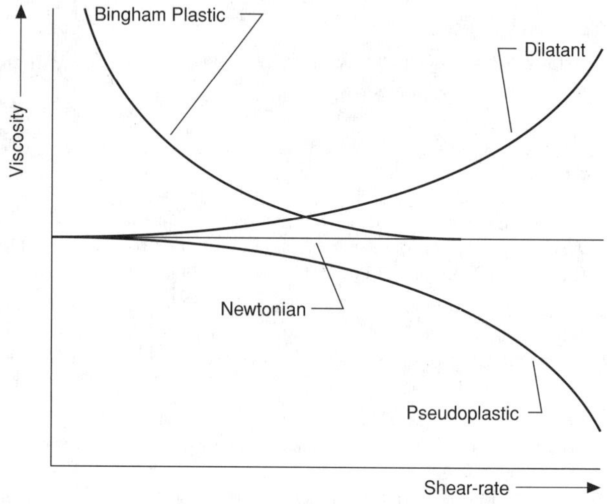

Figure 1. Purely viscous fluids.

A somewhat more complicated relationship, which combines features of the power-law fluid and Bingham plastic, so that it is often called a *yield power-law fluid*, is the *Herschel-Bulkley fluid:*

$$\eta = \begin{cases} \infty & \text{if } |\boldsymbol{\tau}| \leq \tau_y \\ \dfrac{1}{\gamma}(\eta_P\gamma)^{n+1} + \dfrac{\tau_y}{\gamma} & \text{if } |\boldsymbol{\tau}| \geq \tau_y \end{cases} \qquad \textbf{(14)}$$

which reduces to the Bingham plastic if $n = 0$ and to the power-law fluid if $\tau_y = 0$ and η_p^{n+1} is identified with η_c.

A large number of other generalised Newtonian fluids exist (see Bird, Armstrong & Hassager (1977), Skelland (1967), Tanner (1985) and Wilkinson (1960)). These can, for example, involve the introduction of a characteristic time or a limiting viscosity at high or low shear-rates and hence generally involve further parameters. Thus the Newtonian fluid involves one parameter, η; the power-law fluid involves two, η_c and n; the Bingham plastic involves two, η_p and τ_y; the Herschel-Bulkley fluid involves three, η_p, τ_y and n.

In some fluids, the viscosity η varies with time t when the shear-rate γ is held constant. Two different types of behaviour can be distiguished [see Skelland (1967) and Wilkinson (1960)]. For a *thixotropic fluid*, η decreases with increasing t; for a *rheopectic fluid*, η increases with increasing t (see **Figure 2**). The reason for the variation of η with t is associated with changes in the structure of the fluid. When straining of the fluid breaks down that structure, for example by destroying local linkages within the fluid, η decreases with increasing t and the fluid is thixotropic. A similar argument can be advanced for the decrease in η with increasing γ for a pseudoplastic fluid. In contrast, when straining builds up the structure, for example by causing local alignment in the fluid, η increases with increasing t and the fluid is rheopectic. Again, a similar argument can be advanced for the increase in η with increasing γ for a dilatant fluid. There is, therefore, an analogy between

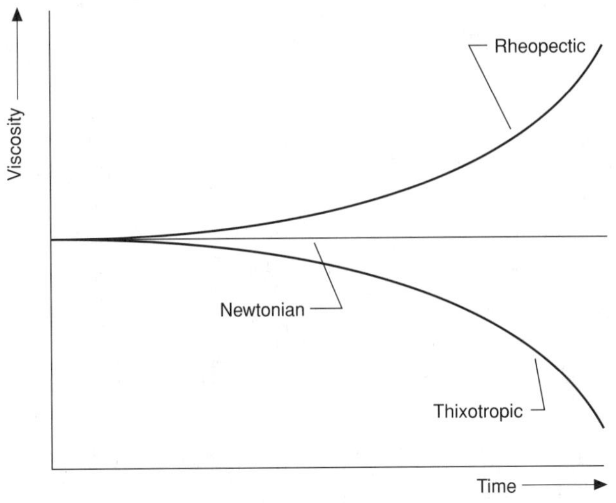

Figure 2. Time-dependent fluids.

thixotropy and pseudoplasticity on the one hand and between rheopexy and dilatancy on the other. Nevertheless, variation of viscosity with time is different from, and should not be confused with, variation with shear-rate.

Constitutive equations for non-Newtonian fluids which are not purely viscous are considerably more difficult to develop and, because of their complexity, are considerably more difficult to use. The constitutive equations in general define what is called the *extra stress* $\boldsymbol{\tau}^E$: the deviatoric stress $\boldsymbol{\tau}$ in **Equation 2** is related to $\boldsymbol{\tau}^E$ by:

$$\boldsymbol{\tau} = \boldsymbol{\tau}^E - \frac{1}{3}\,\mathrm{tr}(\boldsymbol{\tau}^E)\mathbf{I} \tag{15}$$

where $\mathbf{I}$ denotes the unit tensor. There are five main principles which can be used to deduce equations for $\boldsymbol{\tau}^E$. In a purely mechanical theory (which thus ignores thermal effects and so on) for a purely viscous fluid, the first three of these principles are:

- translational invariance: $\boldsymbol{\tau}^E$ depends on $\nabla\mathbf{u}$ and not on $\mathbf{u}$;
- local action: $\boldsymbol{\tau}^E$ at a given spatial position $\mathbf{x}$ depends only on the flow in the neighbourhood of $\mathbf{x}$,
- reference-frame indifference or rotational invariance: it can be shown (see Leigh, 1968) that this means that $\boldsymbol{\tau}^E$ depends on $\mathbf{e}$ given by (4) and not on vorticity $\mathbf{w}$ given by:

$$\mathbf{w} = \nabla\mathbf{u} - (\nabla\mathbf{u})^T \tag{16}$$

It can then be shown [(see Richardson (1989) and Leigh (1968)] that:

$$\boldsymbol{\tau}^E = A\mathbf{I} + B\mathbf{e} + C\mathbf{e}^2 \tag{17}$$

where A, B and C are constants depending on I_e, II_e and III_e given by (5), (6) and (7), respectively. For an incompressible fluid, the term A $\mathbf{I}$ can be subsumed into pressure p. Generalised Newtonian fluids are thus fluids for which $B = \eta$ and $C = 0$. A fluid for which $B = \eta$ and $C \neq 0$ is called a *Reiner-Rivlin fluid.* For elasticoviscous fluids, a fourth principle is added:

- determinism: $\boldsymbol{\tau}^E$ for a given material point X at time t depends on the deformation of X at all times $t^* \le t$ and not $t^* > t$. A typical constitutive equation might then be:

$$\boldsymbol{\tau}^E(X, t) = \int_{-\infty}^{t} m(t, t^*)\mathbf{D}(X, t^*)dt^* \tag{18}$$

where m( ) denotes a memory function and **D**( ) a deformation measure. An elastic material has a perfect memory in the sense that, however it is deformed, it will always try to revert to its original shape. A viscous material, in contrast, has no memory and never tries to revert to its original shape. An elasticoviscous material has a fading memory; it remembers the recent past better than the distant past. A typical functional form for m( ) is:

$$m(t, t^*) = \exp(-(t - t^*)/t_r) \tag{19}$$

where t_r denotes a relaxation time. Constitutive equations can be written in integral form, such as Equation 18, or in differential form. A widely used one in differential form is the *Oldroyd fluid:*

$$\boldsymbol{\tau}^E + t_{r1}\frac{\mathcal{D}\boldsymbol{\tau}^E}{\mathcal{D}t} = \eta\mathbf{e} + \eta t_{r2}\frac{\mathcal{D}\mathbf{e}}{\mathcal{D}t} \tag{20}$$

where t_{r1} and t_{r2} denote relaxation times and:

$$\frac{\mathcal{D}}{\mathcal{D}t} = \frac{D}{Dt} + (\cdot\mathbf{w} - \mathbf{w}\cdot) + a(\cdot\mathbf{e} + \mathbf{e}\cdot)$$
$$+ b\,\mathrm{tr}(\cdot\mathbf{e})\mathbf{I} + c\,\mathrm{tr}(\cdot\mathbf{e})\mathbf{e} \tag{21}$$

where a, b and c are constants. Thus $\mathcal{D}/\mathcal{D}t$ is a generalisation of the substantial derivative D/Dt given by:

$$\frac{D}{Dt} = \frac{\partial}{\partial t} + \mathbf{u}\cdot\nabla \tag{22}$$

D/Dt allows for translation following a given material point X: $\mathcal{D}/\mathcal{D}t$ allows for rotation and stretching as well as translation following X. A special example of the Oldroyd fluid is the *Maxwell fluid* for which $t_{r2} = 0$ and $t_{r1} = t_r$. For very small deformations, it can be shown that:

$$\frac{\mathcal{D}}{\mathcal{D}t} \simeq \frac{\partial}{\partial t} \tag{23}$$

and hence that, for a Maxwell fluid:

$$\boldsymbol{\tau}^E + t_r\frac{\partial\boldsymbol{\tau}^E}{\partial t} \simeq \eta\mathbf{e} \tag{24}$$

For a nearly steady flow, that is one for which $|\partial/\partial t| << \frac{1}{t_r}$:

$$\boldsymbol{\tau}^E \simeq \eta\mathbf{e} \tag{25}$$

which is the constitutive equation for a Newtonian fluid. In contrast for a rapidly changing flow, that is one for which $|\partial/\partial t| >> \frac{1}{t_r}$:

$$\frac{\partial \boldsymbol{\tau}^E}{\partial t} \simeq \frac{\eta}{t_r}\mathbf{e} = G\mathbf{e} \qquad \textbf{(26)}$$

where G denotes the *shear modulus.* Thus Equation 26 can be integrated to yield:

$$\boldsymbol{\tau}^E \simeq G\int_{-\infty}^{t} \mathbf{e}dt^* \qquad \textbf{(27)}$$

But the time-integral of the strain-rate **e** is just the strain so Equation 22 is the constitutive equation for a Hookean solid. Thus the Maxwell fluid can be seen to possess both viscous and elastic characteristics.

The study of the flow of non-Newtonian fluids is called *rheology:* measurements are made in *rheometers* (see Walters, 1975). In addition to viscosity, or strictly *shear viscosity,* η, three other parameters are of interest. They are *extensional viscosity* η_{ext}, *first normal stress difference coefficient* Ψ_1 and *second normal stress difference coefficient* Ψ_2. This is because, in addition to **Shear Stress** which is normally important for Newtonian fluids, **Normal Stress** is also important for non-Newtonian fluids. For a simple shear flow (see **Figure 3**):

$$\mathbf{u} = \gamma y \mathbf{i}_x \qquad \textbf{(28)}$$

where $\mathbf{i}_x$ denotes a unit vector in the x-direction. The stress $\boldsymbol{\tau}$ comprises nine components. For a Newtonian fluid, seven of these (τ_{xz}, τ_{zx}, τ_{yz}, τ_{zy}, τ_{xx}, τ_{yy} and τ_{zz}) vanish for a simple shear flow. For a non-Newtonian fluid in contrast, only four of these (τ_{xz}, τ_{zx}, τ_{yz} and τ_{zy}) vanish. Of those which do not vanish:

$$\eta = \tau_{xy}/\gamma = \tau_{yx}/\gamma \qquad \textbf{(29)}$$

$$\Psi_1 = (\tau_{xx} - \tau_{yy})/\gamma^2 \qquad \textbf{(30)}$$

$$\Psi_2 = (\tau_{yy} - \tau_{zz})/\gamma^2 \qquad \textbf{(31)}$$

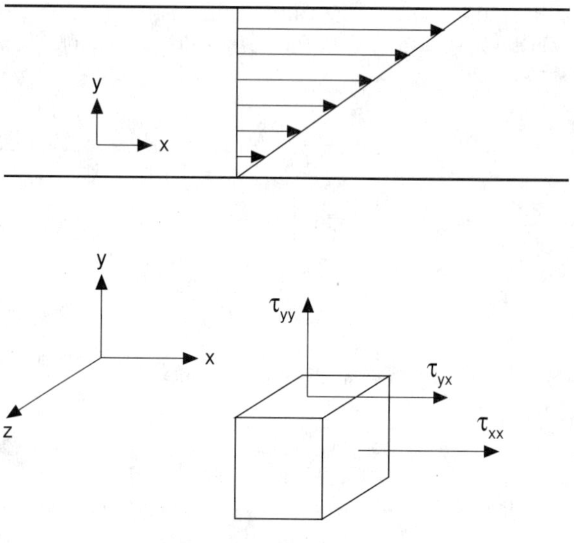

Figure 3. Simple shear flow.

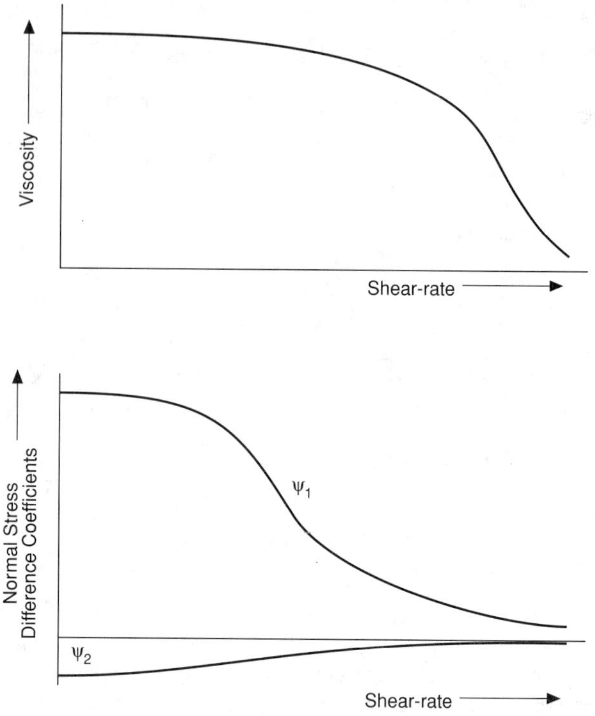

Figure 4. Pseudoplastic fluids.

For a pseudoplastic fluid such as a molten polymer, η is positive and decreases as γ increases; Ψ_1 is positive and decreases as γ increases and Ψ_2 is negative and increases as γ increases (see **Figure 4**).

For a simple extensional flow with extension-rate ϵ (see **Figure 5**):

$$\mathbf{u} = \epsilon z \mathbf{i}_z - \frac{1}{2}\epsilon r \mathbf{i}_r \qquad \textbf{(32)}$$

where $\mathbf{i}_z$ and $\mathbf{i}_r$ denote unit vectors in the z-direction and r-direction, respectively. By definition:

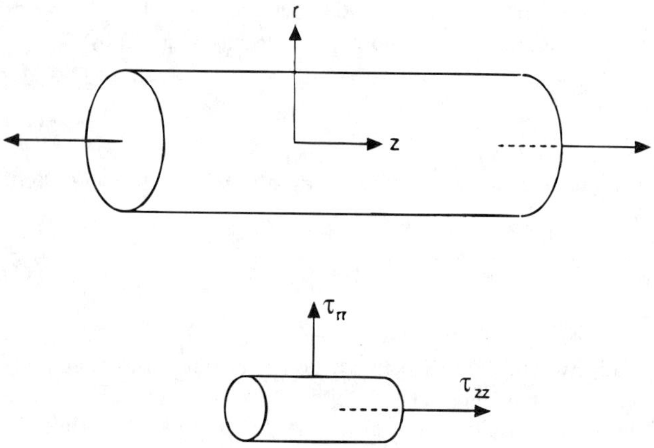

Figure 5. Simple extensional flow.

$$\eta_{ext} = (\tau_{zz} - \tau_{rr})/\epsilon \tag{33}$$

For a Newtonian fluid:

$$\eta_{ext} = 3\eta \tag{34}$$

For a pseudoplastic fluid, η_{ext} is greater than 3 η and increases as ϵ increases: η_{ext} can be several orders of magnitude larger than η at high ϵ.

Non-Newtonian fluids exhibit a number of effects not shown by Newtonian fluids. A Newtonian fluid issuing at high Reynolds number from an orifice of diameter D will tend to form a jet of diameter 0.87D whereas a non-Newtonian fluid issuing in the same way, though normally at low Reynolds number because of the higher viscosity involved, will tend to form a jet of typical diameter 3D. This so-called *extrudate swell* has important consequences, for example in the formation of droplets. *Rod-climbing*, as the name suggests, means that a non-Newtonian fluid will, because of normal stress effects, tend to climb up a rotating rod, whereas a Newtonian fluid will tend to move down the rod because of centrifugal effects. This has important, consequences, for example in the stirring of reactors. *Drag-reduction* is the significant reduction in friction factor, typically by two orders of magnitude, for turbulent flow of certain non-Newtonian fluids such as polymer solutions in water compared with that of the corresponding Newtonian fluid such as water. Unfortunately, the reduction in friction factor is accompanied by a similar reduction in heat transfer coefficient, so there is no obvious application of this effect to flows in heat exchangers.

References

Bird, R. B., Armstrong, R. C. and Hassager, O. (1977) *Dynamics of Polymeric Liquids: I—Fluid Mechanics*, Wiley, New York.

Leigh, D. C. (1968) *Principles of Continuum Mechanics*, McGraw-Hill, New York.

Richardson, S. M. (1989) *Fluid Mechanics*, Hemisphere, New York.

Skelland, A. H. P. (1967) *Non-Newtonian Flow and Heat Transfer*, Wiley, New York.

Tanner, R. I. (1985) *Engineering Rheology*, Clarendon, Oxford.

Walters, K. (1975) *Rheometry*, Chapman & Hall, London.

Wilkinson, W. L. (1960) *Non-Newtonian Fluids*, Pergamon, London.

Leading to: *Non-Newtonian fluid heat transfer*

S.M. Richardson

NON-NEWTONIAN SLURRIES (see Liquid-solid flow)

NON-POLAR FLUIDS (see Newtonian fluids)

NORMAL DISTRIBUTION (see Gaussian distribution)

NORMAL STRESS

Following from: *Inviscid flow*

In a rectangular or Cartesian coordinate system (x,y,z), the *stress tensor* $\boldsymbol{\tau}$ is given by:

$$\begin{aligned}\tau = {} & \tau_{xx}\mathbf{i}_x\mathbf{i}_x + \tau_{xy}\mathbf{i}_x\mathbf{i}_y + \tau_{xz}\mathbf{i}_x\mathbf{i}_z \\ & + \tau_{yx}\mathbf{i}_y\mathbf{i}_x + \tau_{yy}\mathbf{i}_y\mathbf{i}_y + \tau_{yz}\mathbf{i}_y\mathbf{i}_z \\ & + \tau_{zx}\mathbf{i}_z\mathbf{i}_x + \tau_{zy}\mathbf{i}_z\mathbf{i}_y + \tau_{zz}\mathbf{i}_z\mathbf{i}_z\end{aligned} \tag{1}$$

where i_x, i_y and i_z denote unit vectors in the x-direction, y-direction and z-direction, respectively. The components of τ may be interpreted thus:

- τ_{xx} is the x-component of force per unit area exerted across a plane surface element normal to the x-direction;
- τ_{xy} is the y-component of force per unit area exerted across a plane surface element normal to the x-direction; and so on. (Precisely analogous interpretations can be given to components of stress in any orthogonal coordinate system).

  Each component with identical suffices (τ_{xx}, τ_{yy} and τ_{zz}) is called a *normal stress* since it represents a force in a direction per unit area normal to that direction. Each other component with different suffices (τ_{xy}, τ_{yx}, τ_{xz}, τ_{zx}, τ_{yz} and τ_{zy}) is called a **Shear Stress**.

S.M. Richardson

NORMAL STRESS DIFFERENCE (see Non-Newtonian fluids)

NOZZLES

Following from: *Channel flow*

Nozzles are profiled ducts for speeding up a liquid or a gas to a specified velocity in a preset direction. Nozzles are used in the rocket and aircraft engineering to produce jet propulsion, in intensive shattering and spraying technologies, in jet devices and ejectors and in gas dynamic lasers and gas turbines (see **Gas turbine**). Nozzles are the basic components of wind tunnels (see **Wind Tunnels**), and they allow the design of systems of intensive (jet) cooling, such as in electrical engineering.

Nozzles are extremely diverse in geometry (**Figure 1**): round (axisymmetric) and two-dimensional, annular and tray, with oblique and right-angle exit sections, etc. The diversity of nozzle contours enables one to obtain a high degree of outflow uniformity both in absolute value and the divergence angle of the velocity vector, which is of prime importance for increasing jet propulsion and for modeling flow around aircraft and rockets.

Axisymmetric and two-dimensional nozzles of the simplest shape are smoothly converging and then diverging ducts (see **Figure 2**. Known as a Laval nozzles, they were named after a Swedish

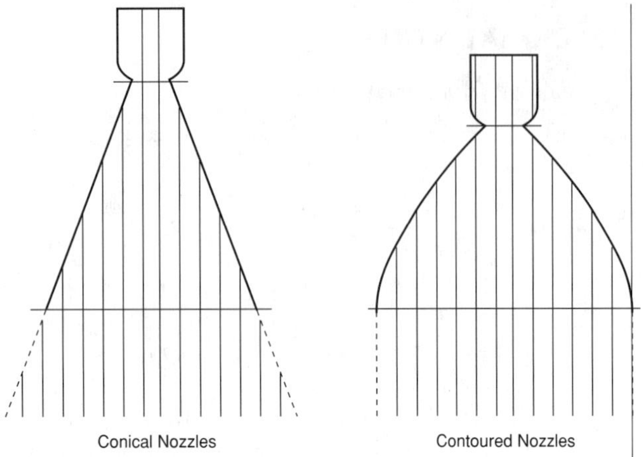

Figure 1a. Round Laval nozzles.

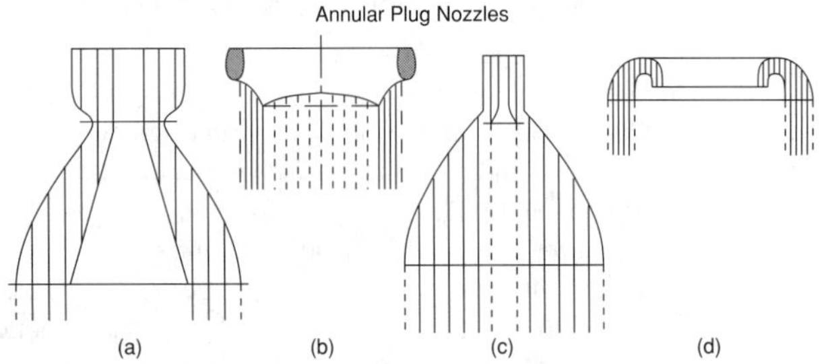

Figure 1b. Conical nozzles with pin shaped plug (b) Aerospike nozzles with pin shaped plug, (c) Tray nozzles, (d) Tray nozzles with backward flow.

engineer who was the first to design them in 1889 for generating supersonic water vapor jets to rotate an impeller in a steam turbine.

We shall first discuss the principal laws governing an adiabatic flow of an ideal gas in the Laval nozzle. In order to describe the gas state we make use of the Mendeleev-Clapeyron equation

$$p = \rho \frac{R}{\bar{M}} T$$

and the adiabatic equation $p/\rho^{\gamma} = \text{const}$, where p, ρ, and T are the gas pressure, density, and temperature, R the universal gas constant, $\bar{M}$ the gas molecular mass, and γ the adiabatic exponent which, for an ideal gas, is determined as the ratio of heat capacities at a constant pressure and volume $\gamma = c_p/c_v$ with $c_p - c_v = \frac{\gamma - 1}{\gamma} = \frac{R}{\bar{M}}$. The sound velocity is determined as

$$a^2 = \frac{dp}{d\rho} = \gamma \frac{p}{\rho} = \gamma \frac{R}{\bar{M}} T = (\gamma - 1)c_p T.$$

Let the nozzle axis be rectilinear and horizontal and the deflection of the stream velocity vector from the axis be negligible. In each section x = const (**Figure 2**) the gas velocity, pressure, density, and temperature take constant values (a one-dimensional flow model).

Then the mass, momentum, and energy conservation laws (see **Conservation Laws**) take the form

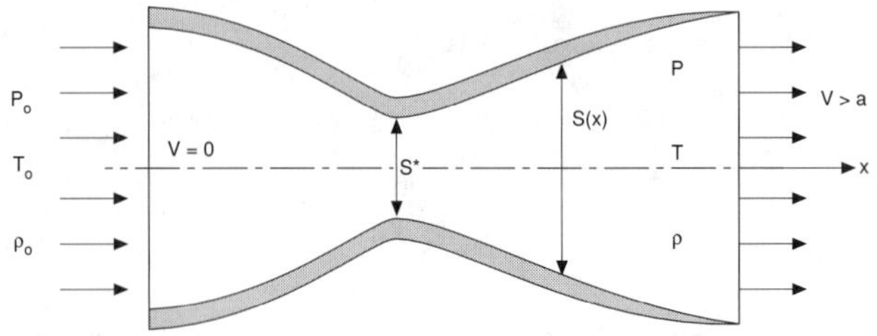

Figure 2. Flow of a gas through a Laval nozzle.

$$\frac{d}{dx}(\rho v S) = 0, \quad (1)$$

$$v\frac{dv}{dx} = -\frac{1}{\rho}\frac{dP}{dx}, \quad (2)$$

$$\frac{d}{dx}\left(\frac{v^2}{2} + u\right) = \frac{d}{dx}\frac{P}{\rho} \quad (3)$$

for a nonfrictional and non-heat-conducting flow). Here v(x) is the velocity, S(x) the cross section area of the nozzle, and $u = \int c_v dT = c_v T$ the internal gas energy related to enthalpy by $h = \int c_p dT = c_p T = (c_v + R/\bar{M})T = u + p/\rho$. The integral of Equation 1 is the constant gas flow rate $\dot{M} = \rho v S = \text{const}$. The integral of Equation 2 is known as the Bernoulli integral. For the adiabatic process it is

$$\frac{v^2}{2} + \int \frac{dp}{\rho} = \frac{v^2}{2} + \frac{\gamma}{\gamma - 1}\left(\frac{p}{\rho}\right) = \frac{\gamma}{\gamma - 1}\frac{p_0}{\rho_0} = \text{const},$$

where p_0 and ρ_0 are the values for the gas brought to rest, i.e., the "stagnation" values. The integral of the energy conservation equation $v^2/2 + h = h_0 = \text{const}$ makes it possible to introduce the concept of an overall enthalpy of gas flow h_0 which corresponds to the total energy of the gas brought to rest (the stagnation enthalpy). Accordingly, $T_0 = h_0/c_p$, or the total gas temperature is linked with p_0 and ρ_0 by the adiabatic equation $p_0/\rho_0^{\gamma} = \text{const}$ and by the equation describing the state of the ideal gas a $p_0 = \rho_0 R/\bar{M} T_0$.

Correlating gas parameters in different nozzle sections, we can derive from the energy conservation law

$$T_0 = T + \frac{v^2}{2c_p} = T\left(1 + \frac{v^2}{2c_p T}\right) = T\left(1 + \frac{\gamma - 1}{2} Ma^2\right). \quad (4)$$

Here Ma = v/a is the dimensionless ratio of the flow velocity to the sound velocity referred to as the Mach number (see **Mach Number**). Since

$$\frac{p}{p_0} = \left(\frac{\rho}{\rho_0}\right)^{\gamma} = \left(\frac{p}{p_0}\right)\left(\frac{T}{T_0}\right)\left(\frac{\rho}{\rho_0}\right)^{\gamma - 1},$$

then

$$\frac{p}{p_0} = \left(\frac{\rho}{\rho_0}\right)^{\gamma} = \left(\frac{T}{T_0}\right)^{\gamma/(\gamma - 1)}.$$

Thus, pressure and density monotonically decrease with increasing local value of the Mach number

Table 1. Stagnation values in gas flows

Relative parameter	Mach Number in the Gas Flow 1	2	3	4
T_0/T	1.2	1.8	2.7	4.2
p_0/p	1.9	8	38	155
ρ_0/ρ	1.6	4.5	13.5	38

$$p = \frac{p_0}{\left(1 + \frac{\gamma - 1}{2}Ma^2\right)^{\gamma/(\gamma-1)}}, \qquad \rho = \frac{\rho_0}{\left(1 + \frac{\gamma - 1}{2}Ma^2\right)^{1/(\gamma-1)}}. \tag{5}$$

Table 1 presents the relative values of gas temperature, pressure, and density for various Mach numbers Ma and $\gamma = 1.4$ (the value of γ for air). As is seen from **Table 1**, at moderate Mach numbers the temperature drops approximately as Ma, the pressure drops as $Ma^{3.5}$, and the density drops as $Ma^{2.5}$. As Ma increases, the exponents grow tending to 2, 7 and 5 for $\gamma = 1.4$.

In contrast to pressure and density, the temperatures of gases and gaseous mixtures cannot diminish indefinitely when they are accelerated in the nozzle. This is because condensation i.e., a transition of the gas molecules to a condensed state occurs. Even for dry air, where drops of water cannot be formed, liquid oxygen droplets show up in the flow at a temperature ~ 90 K and nitrogen begins to condense at T = 77 K. To avoid this the air in the nozzle inlet is heated to a temperature T_0 that depends on the required value of the Mach number.

Figure 3 graphs the dependence of the minimum temperature T_0 and pressure p_0 on the Ma number of the air stream which during expansion, does not bring about oxygen condensation. The dependence on pressure p is due to the change of the gas condensation (saturation) temperature with pressure along the saturation curve (see **Vapor Pressure**). An account for this phenomenon requires a joint (solution) of the set of equations

$$p_i = x_i p = p_i^H(T) = \exp\left(-\frac{\Delta Q_{i,v}\bar{M}}{RT} + C_0\right),$$

$$T = T_0\Big/\left(1 + \frac{\gamma - 1}{2}Ma^2\right),$$

$$p = p_0\Big/\left(1 + \frac{\gamma - 1}{2}Ma^2\right)^{\gamma/(\gamma-1)}.$$

The laws governing the contour or the cross section area S(x) of the nozzle can be derived by combining Equations 1 and 2. Eliminating density and pressure yields an equation for the Laval nozzle

$$\frac{dS}{S} = \frac{dv}{v}(Ma^2 - 1).$$

In a subsonic flow (Ma < 1) an increase in velocity (dv > 0) can

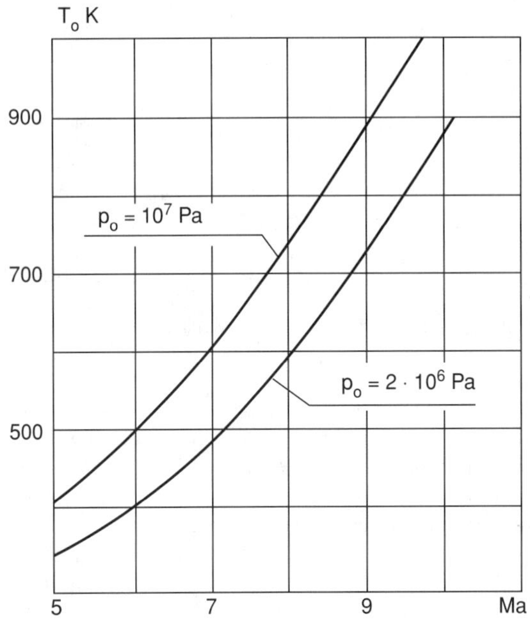

Figure 3. Temperature T_0 to which air entering a nozzle must be preheated in order to prevent condensation of oxygen at the throat.

be achieved only by diminishing of the cross section area dS < 0. Conversely, expanding the duct, we cause the gas flow velocity to decrease. However, after the sound velocity in the flow (M > 1) has been attained, the law of variation of the cross section area is reversed. The minimum of function S(v) corresponds to the nozzle throat S* at which Ma = 1 (**Figure 4**).

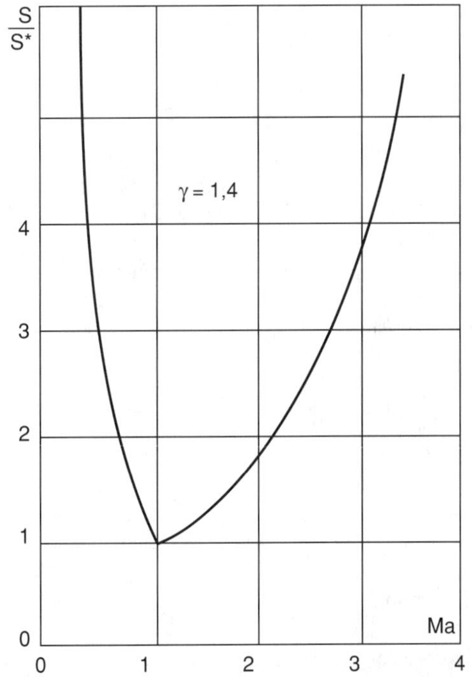

Figure 4. Variation of Mach number with S/S* in a nozzle.

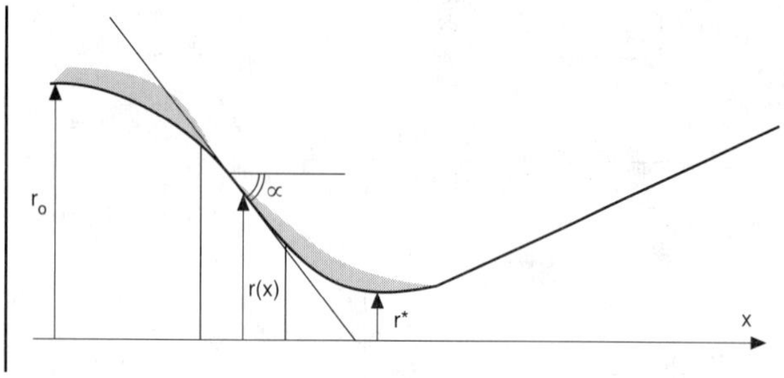

Figure 5. Nozzle profile parameters.

The one-dimensional inviscid gas model is insufficient for the accurate determination of the nozzle contour, i.e., find the law of variation of the cross section area S(x). In addition, in order to provide a highly uniform flow at the nozzle exit section, the angles and radii of the convergent and divergent section of the nozzle must be carefully selected to match one another. There exists an easy technique of shaping the nozzle subsonic section. It is based on using a conical flow with an inclination angle α of the wall (**Figure 5**). As a rule, $\alpha < 24°$. The radii r_0 and r^* must be also specified as initial data for the inlet section and a throat. In the entrance section a cylindrical prechamber smoothly matches the conical section. Here the law of profiling is

$$r(x) = \frac{2r_0}{\pi}\arctan[z + 0.34z^3 + 0.135r^5 + 0.052z^7 + 0.103z^{12}],$$

where $z = \pi/2/x/r_{00} \tan\alpha$.

The length of this section is determined by the range of variation of the cylindrical coordinate x

$$\frac{1.666r_0}{\pi\tan\alpha} \le x \le \frac{5.2r_0}{\pi\tan\alpha}.$$

After the conical section the flow must be changed into a cylindrical one, the length of the third section being

$$0 \le x \le \frac{5.2r^*}{\pi\tan\alpha}$$

and the contour being described by the function

$$r(x) = r^* + x\tan\alpha$$

$$-\frac{2r^*}{\pi}\arctan[z + 0.33z^3 + 0.13z^5 + 0.04z^7 + 0.008z^{12}],$$

where $z = \pi x \tan\alpha/2r^*$.

The minimum admissible ratio of the inlet section radius to the throat radius $n = r_0/r^* = 2.015$ yields the contour of the subsonic section without using a conical section. However, a low value of contraction ratio n does not guarantee suppresion of flow fluctuations at the nozzle outlet.

A universal contour can also be recommended for the supersonic section with a conical flow with an angle $\alpha < 15°$. In this case the law of profiling is

$$r(x) = 1 + x\tan\alpha - A\tan\alpha\arctan\left(\frac{x}{A} + 0.03\left(\frac{x}{A}\right)^4\right),$$

$$A = \frac{2}{\pi}\left(\cot\alpha + 3.686\sin(\alpha/2) - \frac{0.48}{\sin(\alpha/2)} + 0.0048\right).$$

Numerical calculations demonstrate that the nozzle so designed is advantageous within a wide range of gas composition and with an adiabatic exponent in the range $1.13 < \gamma < 1.67$.

In engineering, great emphasis is placed on problems related to particle acceleration in the nozzle and, in particular, finding an optimized profile or, similarly, optimum distribution throughout the nozzle length of gas dynamic parameters (gas velocity or pressure).

Gas-particle flows in nozzles

The one-dimensional equation of motion of a solid and accelerating particle motion in an ideal gas flow is as follows

$$\frac{dv_p'}{dx} = \frac{3}{4}C_d\frac{\rho_0}{\rho_p d_p}\left(1 - \frac{\gamma - 1}{\gamma + 1}\lambda^2\right)^{1/(\gamma-1)}\frac{(\lambda - v_p')^2}{v_p'}. \quad \textbf{(6)}$$

Here v_p, ρ_p, and d_p are the velocity, density, and diameter of a particle, C_d the drag coefficient of a particle, $v_p' = v_p/a_{cr}$, $\lambda = v/a_{cr}$.

The dimensionless gas velocity λ depends on the nozzle geometry which allows us to consider it as a parameter which may be varied with the aim of designing nozzles with high performance characteristics.

It is required to find the gas velocity distribution which ensures the particle speeds up from v_{p0}' to the required value of v_p' in the minimum length $(x - x_0)$. This variational problem consists in seeking the minimum of the function

$$x - x_0 = \theta[\lambda(v_p')] = \int_{v_{p0}'}^{v_p'} F[v_p', \lambda(v_p')]dv_p'.$$

Assuming C_d = const and introducing new variables $x' = C_d\frac{\rho_0}{\rho_p d_p}x$ and $z = \frac{\gamma - 1}{\gamma + 1}\lambda^2$ reduces the solution of this problem to taking an integral

$$x' - x_0' = \frac{1}{(\gamma - 1)^2}\int_{z_0}^{z}\frac{\gamma^2 z^2 - (\gamma - 1)^2}{z(1 - z)^{(2\gamma-1)/(\gamma-1)}}dz.$$

For $\gamma = n - 1/n - 2$, where $n = 3,4,5,\ldots$ the solution can be obtained in an analytical form

$$x' - x_0' = \frac{2}{3}\left[\left(\frac{\gamma}{\gamma - 1}\right)^2\left(\frac{1}{(n - 1)(1 - z)^{n-1}} - \frac{1}{(n - 2)(1 - z)^{n-2}}\right.\right.$$

$$\left.\left. + \ln\frac{1 - z}{z}\sum_{i=1}^{n-1}\frac{(n - 1)(n - 2)\cdots(n - 1)z^i}{i!i(1 - z)^i}\right)\right]_{z_0}^{z}, \quad \textbf{(7)}$$

$$v_p' = \gamma\lambda - \frac{\gamma + 1}{\lambda}. \quad \textbf{(8)}$$

Speeding up the particle to the specified velocity in the minimum length obviously requires the maximum aerodynamic force on the particle. This force, in the model postulated in which C_d = const, depends only on the particle and gas velocities, which follows from the equation of motion (Equation 6). It is not difficult to separate out two limiting cases when a particle with the velocity v'_p does not at all interact with the gas flow ($dv_p/dx = 0$). The first is observed in equilibrium flow of particles and gas $\lambda = v'_p$, and the second occurs at the maximum gas velocity $\lambda = \sqrt{\gamma + 1/\gamma - 1}$ when its density diminishes and becomes zero. Within the range, $v'_p < \lambda < \sqrt{\gamma + 1/\gamma - 1}$, there is a certain v'_p-depending value of λ^* at which a particle acceleration dv'_p is maximum. In the optimized nozzle it must be maximum throughout the particle path. Fulfilling these conditions yields Equation 7.

We now consider the operation of this optimized nozzle under off-design regimes. All the factors leading to off-design regimes, except for an isentropic exponent, are incorporated in a generalized complex $m = C_d\rho_0/\rho_p d_p$. It has a dimension which is the reciprocal of length and as a scale factor it is involved in a dimensionless coordinate of solution 7. Optimization is achieved only for a given, quite specific, value $m = m^*$. In this case the particle possesses, in each nozzle section, the velocity for which the flow parameters provide the maximum acceleration. We denote this velocity by $(v'_p)^*$.

If the equality $m = m^*$ does not hold, for instance, due to deviation of particle size or gas density in the prechamber from the estimated values, this nozzle is no longer optimized. For $m > m^*$ particle velocities are higher than $(v'_p)^*$ and lie within the range $(v'_p)^* < v'_p < \lambda$. For $m < m^*$ the particle begins to lag behind $v'_p < (v'_p)^*$ and, failing to gather velocity at the initial portion of the path, it will steadily deviate from the optimum speadup parameters. Thus, a polydisperse flow is characterized by a broadened range of particle velocities.

We note one more specific feature of the solution obtained. Equation (8) relates those instantaneous dimensionless velocities of gas and particles which make possible the most rapid speedup of particles and implies that $\lambda \geq \sqrt{\gamma + 1/\gamma - 1} > 1$ holds for each $v'_p \geq 0$. In other words, an optimized nozzle is supersonic everywhere, while actual nozzles have a subsonic section and a supersonic section $\lambda < \sqrt{\gamma + 1/\gamma - 1}$. Theoretically the subsonic section can be made arbitrarily short by using steep walls and changing to a multinozzle structure (an assembly of nozzles). In actual fact, this can be advisable, for instance, for avoiding particle overheating by the gas flow since the particle temperature rise is predominantly in the subsonic section.

The profile of the optional nozzle is closely linked with the distribution of particle velocity in the acceleraton of a polydisperse flow. If relatively low particle velocities that correspond to the entrance section of the nozzle are needed, the particle velocity spectrum in the polydisperse flow will be strongly blurred due to a fairly broad range of particle velocities and high gradients of their variation along the flow. To equalize the velocities, a constant gas velocity section is needed in which the velocities of particles of various size will be equalized and approach the gas velocity. Here, the Re numbers of the particles, become smaller because the difference in velocity between the gas and the particles is small. C_d can no longer be taken as constant and should be calculated from:

$$C_d = \frac{1}{4\eta}(\eta + b\sqrt{\eta} + C), \qquad \eta = \frac{v - v_p}{v}, \qquad b = \frac{25.2}{\sqrt{Re_p}},$$

$$C = \frac{84.48}{Re_p}, \qquad Re_p = \frac{\rho v d_p}{\mu} \qquad x'' = \frac{\rho x}{\rho_p d_p}$$

which allows for the dependence of the drag coefficient on Re (see **Drag Coefficient**).

The particle motion equation is written as

$$\frac{dv_p}{dx''} = \frac{3}{4} C_d \frac{(v - v_p)^2}{v_p}$$

or, after transformations,

$$dx'' = \frac{16}{3} \frac{\eta - 1}{\eta(\eta + b\sqrt{\eta} + c)} d\eta.$$

This equation has an analytical solution

$$x'' = \frac{32}{3}\left[\frac{\alpha^2 - 1}{\alpha(\alpha - \beta)}\ln(z - \alpha) - \frac{\beta^2 - 1}{\beta(\alpha - \beta)}\ln(z - \beta) - \frac{1}{\alpha\beta}\ln z\right]_{z_0}^{z},$$

where

$$\alpha = -\frac{b}{2} + \sqrt{\left(\frac{b}{2}\right)^2 - C} = -\frac{3.981}{\sqrt{Re_p}},$$

$$\beta = -\frac{b}{2} - \sqrt{\left(\frac{b}{2}\right)^2 - C} = -\frac{21.219}{\sqrt{Re_p}}, \qquad z = \sqrt{\eta}.$$

The solution allows us to find the length of the section with constant parameters of the gas flow at which the particle is speeded up from velocity η_0 to velocity η.

The physics of the flow of a medium in a nozzle is substantially more complex if viscosity becomes significant, if the gas is chemically reactive, electromagnetic fields appear, or the concentration of the disperse phase is high.

Yu.V. Polezhaev and I.V. Chirkov

NOZZLES, HYDRAULIC LOSSES IN CRITICAL FLOW (see Pressure drop, single phase)

NRC (see National Research Council of Canada)

NSSS, NUCLEAR STEAM SUPPLY SYSTEM (see Steam generators, nuclear)

NTU, NUMBER OF TRANSFER UNITS (see Cooling towers; Mean temperature difference)

NUCLEAR COUPLED INSTABILITIES (see Instability, two-phase)

NUCLEAR FISSION (see Liquid metal cooled fast reactor; Neutrons; Nuclear reactors)

NUCLEAR FUEL (see Advanced gas cooled reactor, AGR; Boiling water reactor, BWR; Magnox power station)

NUCLEAR FUEL, FAST REACTOR (see Liquid metal cooled fast reactor)

NUCLEAR FUSION REACTORS (see Fusion, nuclear fusion reactors)

NUCLEAR MAGNETIC RESONANCE SPECTROSCOPY (see Spectroscopy)

NUCLEAR REACTORS

Following from: *Power plant*

Introduction

A nuclear reactor is an engineered system for allowing a controlled nuclear chain reaction to take place to provide a source of heat. The level of heat production may be set at any level from almost zero in reactors designed to study reactor physics, to very intense sources of heat for electric power production. Since the *fission reaction*, which is the basis of the *chain reaction* and provides the source of heat, was discovered in 1938, there have been an enormous number of reactor designs, most of which have never been built, involving diversity in the form of the fuel, in the heat transfer medium to take heat away from the active region and in the layout of the engineering structures and components. Applications range from electricity production to district heating, from space propulsion to industrial heat, from semiconductor production to medical radioisotopes and from peaceful civilian applications to military weapon material production.

This entry is concerned with applications which generate large amounts of heat and therefore present problems of heat transfer both in normal operation and in upset conditions. The high capital cost of nuclear reactors has tended to push power density to levels as high as possible to optimise the revenue or other benefits of the plant and has necessitated large programmes of experimental and theoretical development in heat transfer. The entry will describe the basic elements of reactor physics, describe the most important types of reactor and then list some of the issues of heat and mass transfer which have to be addressed.

Reactor Physics

In boilers using fossil fuel, the heat derives from chemical reactions which involve the electrons in the outer region of the atoms. In a nuclear reactor, the reactions take place within the nucleus of the atoms and the energy release per event is about 2×10^7 bigger than in a chemical reaction such as the oxidation reaction of combustion.

Nuclei can react with neutrons in one of three ways, scatter, capture or fission; it is a question of chance which reaction will occur. The probability of occurrence of each reaction is measured by its *neutron cross section* which can be thought of as the average diametric area of the nucleus as seen by a travelling neutron. However, these cross-sections are strong functions of the relative velocity between the neutron and the nucleus and have different values for the different types of reaction.

In scatter, the nucleus does not change but its motion may be changed in direction and velocity or in vibrational amplitude and frequency if it is bound in a lattice. At the same time, the neutron velocity and its direction will be changed in a manner analogous to the way in which a billiard ball slows down and changes direction when it collides with another ball. In capture, the neutron is absorbed into the target nucleus, the nucleus increases its mass by one unit and may become radioactive. In the case of absorption into *Uranium 238*, and after two rapid stages of radioactive decay, *Plutonium 239* is formed. This is called *breeding* since fissile isotopes (those capable of undergoing fission) are produced to replace those being used up in fission. Isotopes which can be used for breeding are called *fertile isotopes* and the most important ones are Uranium238, giving Plutonium239, Plutonium240 giving Plutonium241 and Thorium232 giving Uranium233. Neutrons can be captured to some degree by almost all the materials which go to make up the reactor as well as in the fuel materials.

In *fission*, the target nucleus absorbs the neutron and then splits into normally two unequal size pieces, the *fission products*. The two pieces move apart at high velocity and it is this kinetic energy which provides most of the nuclear heat. In addition, between two and three neutrons are emitted at high velocity from the fission process. These take part in further neutron reactions including fission, and hence a chain reaction can occur. If on average one, and only one, neutron leads to a further fission, the reactor is said to be in a steady state at constant power. If there is not a perfect balance, the power level will either increase or decrease and a balance is established by moving absorbers into the reactor core or by adjusting the net leakage of neutrons from the core. A reactor in which the balance is maintained at a constant level is said to be critical.

Only isotopes of **Uranium, Plutonium** and other higher man-made isotopes are able to support fission. Properties of the most important fissile isotopes are given in **Table 1**. *Cross-sections* are given both for thermalised neutrons (2200m/s), appropriate for *thermal reactors* and for fast neutrons, appropriate for *fast reactors*.

Neutrons are emitted from the fission process with an energy of about 2.5MeV. In the fissile isotopes, the cross-section for fission is highest at the low neutron energies typical of neutrons scattering in thermal equilibrium with the surrounding atoms. Hence materials called *moderators* are provided within the lattice to slow the neutrons down by elastic scatter. The most effective moderators are those having nuclei with a low mass to maximise the energy loss per collision, with a high cross-section for scattering and with a low cross-section for absorption to avoid excessive neutron losses. The most suitable ones are hydrogen in the form of water, deuterium in the form of heavy water, carbon in the form of graphite and beryllium. **Table 2** gives data on the key properties of moderators. The *moderating ratio* is the ratio of energy loss to absorption and

Table 1. Cross section in *barns* (10^{-28} m^2) and neutron yields of fissile isotopes for both thermal and fast reactors

Isotope	Fission cross section for thermal reactor	Capture cross section for thermal reactor	Neutrons per fission for thermal reactor	Fission cross section for fast reactor	Capture cross section for fast reactor	Neutrons per fast fission for fast reactor
U235	584	98	2.43	1.2	0.2	2.51
U233	526	49	2.49	1.9	0.1	2.52
Pu239	735	284	2.87	1.6	0.1	2.95
Pu241	1,028	373	2.93	1.5	0.2	3.03

Table 2. Moderator properties

Moderator	Scatter cross section	Capture cross section	No. of collisions from fission to thermal energy	Moderating ratio
Water	44.3	0.66	18	72
Heavy water	10.4	1.2E-3	25	12,000
Graphite	4.7	3.2E-3	114	170

should be high. Light water is not the best moderator in this sense but it has been used more than the others because of its ready availability and low cost.

Criticality in a reactor is a question of obtaining a balance between fission, absorption and leakage from the sides of the reactor core. Increasing the size of the reactor decreases the surface area to volume ratio and hence reduces the fraction of neutrons which leak out compared to those that undergo fission or absorption in the volume of the core. From this is derived the concept of critical size.

Fission and absorption are accompanied by very high *radiation* levels from which people and equipment have to be protected by massive shielding. Radiation also causes structural materials to become radioactive and causes radiolasis in water. The gases from thc lattcr may have significant effects on heat transfer.

Fast and Thermal Reactors

The use of moderators reduces the neutron energy to levels where the fission cross-section is very high. One of the consequences of this is that the amounts of fissile isotopes as indicated by their enrichment can be quite low (few %). A critical reactor in which all reactions take place at high neutron energies can still be obtained if the enrichment is high enough. Such a reactor is known as a fast reactor since fission takes place before *thermalisation*; fast reactors are built without moderator. The advantage of this mode of operation is that the number of neutrons from each fission is higher than in low energy or thermal fission and a higher proportion of the spare neutrons can be absorbed in fertile materials. More fissile material can be produced than is burnt in fission; the stock of fissile material for use in further reactors increases with time and is recovered by reprocessing. Since this allows for ultimate use of the fertile U238, which constitutes 99.3% of uranium mined from the earth, the potential power production per tonne of uranium ore can be increased by a factor of over 60 by use of fast reactors as compared to thermal reactors.

In the early stages of development of the nuclear power programme, the benefits of breeding were thought to be so important that reducing the *doubling time*, the time taken to produce enough surplus-bred plutonium to allow the commissioning of a second fast reactor, was a prime design consideration. This called for compact reactor cores with a very high power density. To cope with the heat transfer requirements without introducing significant moderation, *liquid metals* (usually Na, Na/K and in some proposals Pb or Bi) have been used as coolants.

Main Reactor Types

Fuel

In most reactors, the fuel elements are in the form of metallic tubes filled with fuel material either in the form of metal rods or ceramic pellets. The most common ceramic is uranium dioxide or a mixture of uranium and plutonium oxides. Other ceramics are nitrides and carbides. Ceramics were chosen in preference to metallic forms because of their superior properties under irradiation, particularly in their dimensional stability. They do, however, have the disadvantage of low thermal conductivity, resulting in high centre temperatures, and high specific heat resulting in a large amount of stored energy which can produce problems in some emergency loss of coolant conditions.

High temperature gas cooled reactors have all-ceramic cores. The fuel is in the form of small (approx. 0.5mm) spheres coated with layers of graphite and silicon carbide. These spheres are normally embedded in a graphite matrix which in turn is moulded into rods or spheres and covered in a further layer of graphite.

Water cooled reactors

These are the most widely adopted designs for power production with over 400 in operation in 1994. Ordinary water is used as both moderator and coolant and the same water serves both purposes. From **Table 2** it is seen that the absorption cross-section of water is relatively high compared with the other moderators and this leads to a close spaced lattice of fuel rods to reduce the water content. High water velocities are required to remove the heat. Typical fuel rods have a diameter of around 10 mm with UO_2 pellets in a cladding of *Zircaloy*, an alloy of *zirconium.*

There are two forms of water reactors, **Boiling Water Reactors** (BWR) and *Pressurised Water Reactors (PWR).* BWRs operate at a pressure of 7 MPa and generate steam by boiling within the reactor core. The core is contained within a steel vessel which also contains steam separators and dryers in its upper region. Steam passes directly to the turbine. Water may circulate through the core by natural circulation, or, more usually be directly or indirectly pumped. Some designs, for example, have jet pumps around the core which are driven by high pressure pumps taking only a fraction of the total core flow outside the vessel.

In a PWR, the pressure is raised to 15MPa by an electrically heated pressuriser to prevent boiling in the reactor core at its operating temperature of around 310C. Water is pumped into the pressure vessel, down an annular downcomer, up through the core and out to a *steam generator.* The latter may be of the inverted U-tube type or be a once-through steam generator. Variants of the PWR may be of the integral type with the steam generators and pumps, if provided, within the pressure vessel.

Water reactors are provided with *safety systems* designed to prevent overheating of the core and release of radioactive material from the core, in case of reasonably conceivable accidents short of failure of the main pressure vessel. In the unlikely event of failure of one of the main coolant pipes in a PWR, for instance, most of the core water inventory could be lost in a few seconds. Large accumulators are provided to reflood the core before it overheats. There are also pumped systems designed to supply sufficient water flow to the vessel at high and low pressure. To provide adequate reliability allowing for failure of these systems to operate on demand, they are duplicated up to four times to provide redundancy. Analysis of heat and mass transfer during emergency conditions presents some very challenging problems.

Demonstrable safety requires that consideration is given to failure of all the safety systems. A **Containment** structure is provided to contain the steam released from the most serious possible failure of the pipe systems and to contain the fission products which may be released from overheated fuel. These are large buildings, in some cases (particularly for BWRs and some integral PWRs) provided with a pressure suppression system, and with further provision for the ultimate removal of heat. Many designs, particularly the smaller ones, make use of natural convection and heat conduction to avoid the need for reliable redundant emergency electricity supplies.

Gas reactors

Gas reactors have a graphite moderator and use gas for heat removal to a steam generator normally of the once-through type. Most of the operating gas cooled power reactors are in the UK. The early ones use a finned *Magnox* cladding and CO_2 coolant. The natural uranium fuel is in the form of metal rods 2.5 cm in diameter. Graphite has a small neutron absorption cross-section but the atoms of carbon are less efficient than the hydrogen in water for neutron slowing down. The spacing between each fuel element is about 20 cm with a 10 cm channel for coolant. The **Advanced Gas Cooled Reactor (AGR)** is a later design still using CO_2 but at a temperature permitting the production of superheated steam at a temperature and pressure typical of modern coal fired plant. This requires the use of stainless steel cladding and enriched ceramic fuel. The fuel element is in the form of a cluster of pins of smaller size than the Magnox reactors and the cladding is ribbed to improve heat transfer.

A further advance in gas reactor technology is the *High Temperature Reactor (HTR).* The coolant gas is Helium and the fuel is the all-ceramic type described above. It is possible to reach temperatures of 950°C enabling HTRs to be used as a heat source for some high temperature industrial processes

Heavy water reactors

These usually have heavy water moderator and usually use a separate flow of heavy water as coolant. The properties of heavy water permit the use of natural UO_2, clad in Zircaloy as fuel. The fuel bundles and coolant are contained in an array of pressure tubes insulated from the main body of the moderator by a gas gap and a second tube known as the calandria tube. There are header arrangements at each end of the pressure tubes which are normally horizontal. There are systems for the injection of water in emergencies. There is a second heavy water reactor type which has boiling light water coolant with vertical calandria tubes. Steam is separated from the circulating water in a steam drum and then passes directly to the turbine.

Water cooled graphite moderated reactors

Like the heavy water reactors, these are pressure tube reactors. The pressure tubes are vertical and very long. Boiling takes place within them and there are large external steam drums. The stations which have been built do not have a containment structure but do have a pressure suppression pool underneath the reactor itself.

Fast reactors

These use highly enriched fuel and, in most designs, liquid metal coolant. This allows them to operate at near atmospheric pressure, eliminating many of the problems of pressurised loss of coolant accidents which have to be considered in other designs. There are both loop type and integral type designs in which the core, the coolant pumps and the primary heat exchangers are contained in a single vessel filled with liquid metal. Since it is undesirable for there to be any chance of water interacting with the primary coolant in the event of a steam generator leak, there is an intermediate heat transfer loop to link the primary heat exchangers with the steam generators usung liquid metal as the heat transfer medium.

Other reactor concepts

Over the years of reactor development, there have been a very large number of combinations of fuel type, coolant, moderator, pressure vessel or pressure tube, operating temperature and pressure and power output. Reactors have been designed for use on land, to power ships and submarines and in space to provide electric power as well as to provide the driving force for rocket propulsion. There have been reactors operating with a liquid salt fuel, which was directly pumped to a heat exchanger, and reactors with gaseous fuel for use in space.

Heat and Mass Transfer Issues—Normal Operation

In normal operation, the prime concern is to remove the heat generated from the fuel with a margin for failure large enough to allow for expected transients in the local power level. Excessive margins result in economic penalty through the provision of, for example, unnecessary pumping power or unnecessary downrating of the plant. Failure means failure of the fuel cladding by melting, chemical reaction with the coolant or physical weakening of the material. A good knowledge of the heat transfer coefficient under all operating conditions, coupled with an ability to calculate the resulting fuel and cladding temperature including the effects of the gap between the fuel and the cladding, is required. The gap clearance changes with irradiation due to dimensional changes in the fuel and cladding and due to the build up of fission product gases within it, which changes its conductivity. With liquid coolants, it is necessary to avoid a crisis in heat transfer associated with the phase change to vapour. In non-boiling systems, this is the point of *departure from nucleate boiling* (DNBR) and in boiling water reactors it is the *critical heat flux* (CHF) in both cases where the cladding surface becomes dry and is only cooled by water vapour as opposed to water or a water/steam mixture.

Heat and Mass Transfer Issues—Emergency Conditions

It is necessary to determine what would happen to the reactor, and particularly to its fuel, in the event of failure of the heat transfer systems, whether by electrical failure or by mechanical failure of equipment or rupture of the piping. Failure of large diameter pipes, such as the main coolant ducts in a PWR which have a diameter approaching 1m, results in massive turbulence and rapid ejection of much of the primary coolant. In making the reactor safety case, it is normally assumed that all the fluid is ejected, a very pessimistic assumption that may lead to over-design of safety injection systems. Following the cooling which occurs during this rapid **Blowdown** phase, the fuel may dry completely and experience a rapid heat-up from the decay heat. Subsequent introduction of water causes the fuel to quench with an advancing quench front along its length. Efforts are being made (through 1994) to develop the methods and supporting data needed for a best estimate approach, rather than a pessimistic one. The calculation problem in gas reactors is less severe since there is no phase change.

Prolonged failure of the coolant systems may allow the fuel to melt and form a molten pool, either within the reactor vessel or piping system or below it, having melted through the vessel wall. This introduces new issues in heat transfer as water is introduced to cool the molten pool. There will also be release of gaseous fission products and radioactive aerosols which are carried through the duct systems by steam flows and will be deposited at various points within the reactor vessel and piping and, for the material which escapes through the rupture, within the containment. It is important to know how much of this material is available from any possible failure in the containment system.

I. Gibson

Leading to: *Advanced gas cooled reactor, AGR; Blowdown; Boiling water reactor, BWR; Containment; Fusion, nuclear fusion reactor; Liquid metal cooled fast reactor; Magnox power station*

NUCLEAR REACTOR SAFETY ANALYSIS (see Vapour explosions)

NUCLEAR STEAM GENERATORS (see Steam generators, nuclear)

NUCLEATE BOILING, EVAPORATORS (see Evaporators)

NUCLEATION SITES (see Boiling)

NUCLEONS (see Atom; Relative molar mass)

NUKIYAMA, BOILING CURVE (see Boiling)

NUCLEATE BOILING

Following from: *Boiling*

When boiling occurs on a solid surface at low superheat, bubbles can be seen to form repeatedly at preferred positions called *nucleation sites.* Nucleate boiling can occur in **Pool Boiling** and in **Forced-Convective Boiling**. The heat transfer coefficients are very high but, despite many years of research, empirical correlations for the coefficients have large error bands. Much of the difficulty arises from the sensitivity of nucleate boiling to the microgeometry of the surface on a micron length scale and to its wettability; it is difficult to find appropriate ways of quantifying these characteristics. There is still disagreement about the physical mechanisms by which the heat is transferred so phenomenological models for nucleate boiling at present do no better, and often worse, than the empirical correlations. An empirical correlation of wide application has been given by Gorenflo (1991), based on the general scaling of fluid thermal and transport properties with reduced pressure p/p_c and reduced temperature T/T_c. (See **Reduced Properties**) Recent reviews of the voluminous research literature on mechanisms in boiling include those by Dhir (1990) and Fujita (1992). This article describes the features of nucleate boiling on which there is broad agreement and indicates the areas of disagreement and further development.

The approach to modelling of nucleate boiling at low wall superheats has been to try to understand separately how many nucleation sites are active at a specified superheat, how bubbles grow and depart and how they influence heat transfer. We shall see that the processes are in fact linked, that wall superheat cannot be specified by a single value and that the flow conditions of the bulk liquid in pool boiling or convective boiling have some influence when nucleation sites are widely spaced in the so-called 'isolated-bubble' regime. First, however, we consider an idealised situation, the conditions for equilibrium of a small spherical vapour bubble of radius r_e in pure, uniformly superheated liquid and the consequences of departures from equilibrium. 'Superheat' and 'subcooling', which occur so frequently in the descriptions of boiling, are defined relative to the saturation temperature $T_{sat}(p_o)$ corresponding to the system pressure p_o, being the condition for equilibrium between liquid and vapour at an interface of zero curvature, **Figure 1**. A spherical bubble of finite radius r has an interface of curvature 2/r and this has two effects: (1) for mechanical equilibrium of the bubble interface there must be an excess internal pressure of $2\sigma/r_e$ to resist the collapsing membrane stress caused by the surface tension σ; (2) the vapour pressure for a given interfacial temperature is decreased (Kelvin equation)

$$p/p_{sat} = \exp(-2\sigma vM/r\tilde{R}T) \tag{1}$$

where σ is surface tension, v specific volume of the liquid, M molecular weight and $\tilde{R}$ the universal gas constant.

There is a similar effect with exponent of opposite sign for the vapour pressure in equilibrium with a droplet of liquid. The effect is negligible for radii greater than about 10nm.

From (1) and (2) the vapour pressure must be greater than p_o by $2\sigma/r_e$ and the interface must be superheated, **Figures 1** and **2(a)**. In a uniformly superheated liquid maintained at constant pressure the equilibrium of the bubble is unstable against any disturbance. A decrease in radius leads to a requirement for a higher vapour pressure for equilibrium; this cannot be provided so collapse continues. An increase in radius leads to a requirement for a lower vapour pressure and therefore a lower interfacial superheat. The resulting temperature gradient from the bulk liquid to the interface drives the heat flow that provides the latent heat for continued growth, **Figure 2(b)**. A radial pressure difference is required to drive the motion of the liquid but this declines as growth proceeds, also the

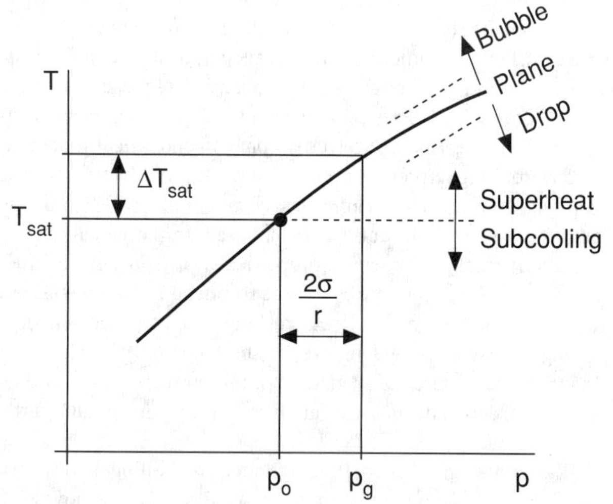

Figure 1. Equilibrium at plane and curved interfaces.

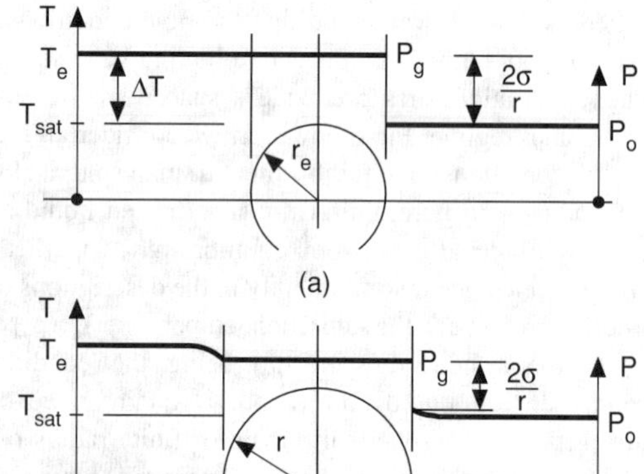

Figure 2. Unstable equilibrium and growth of a bubble nucleus.

interfacial temperature approaches the saturation temperature as $2\sigma/r$ becomes negligible. Then the rate of growth of the bubble is controlled by the rate of heat transfer, which can be modelled approximately by transient conduction in the liquid:

$$4\pi r^2\left(\frac{\lambda_l \Delta T_{sat}}{\sqrt{\pi\kappa_l t}}\right) = h_{lg}\frac{d}{dt}\left(\frac{4}{3}\pi r^3 \rho_g\right) \tag{2}$$

$$r = \frac{2}{\sqrt{\pi}} Ja\sqrt{\kappa_l t}, \qquad \text{where Jakob Number } Ja = \frac{\rho_l c_l \Delta T_{sat}}{\rho_g h_{lg}} \tag{3}$$

where λ_l is thermal conductivity of the liquid, κ_l thermal diffusivity of the liquid, h_{lg} latent heat of evaporation, ρ_g density of the vapour ρ_l density of the liquid and c_l specific heat capacity of the liquid.

In *homogeneous nucleation* the unstable nuclei from which growth commences are supposed to be formed by the random fluctuations in local energy in the superheated liquid. The number distribution of clusters of high-energy molecules (i.e. bubbles) depends on their work of formation, including a contribution by surface free energy (surface tension). Some of the clusters will be above the critical size for unstable equilibrium, some below. By combining the expression for cluster size distribution with a model for rate of growth or collapse, the net rate of bubble nucleation can be predicted, Skripov (1974), Blander and Katz (1975). The rate is extremely sensitive to temperature, increasing by many orders of magnitude over a very small range of temperature so that an effective homogeneous nucleation temperature T_n can be calculated. Lienhard (1976) obtained an approximate generalisation of the analyses in the form

$$\frac{T_n - T_{sat}}{T_c} = 0.905 - \frac{T_{sat}}{T_c} + 0.095\left(\frac{T_{sat}}{T_c}\right)^8 \tag{4}$$

For system pressures well below the critical pressure, the homogeneous nucleation temperature is approximately 0.91 T_c, whatever the value of p_o, corresponding to very high superheats $T_n - T_{sat}$. These superheats can be achieved in very carefully-controlled experiments but generally bubbles nucleate at superheats far smaller than those predicted by homogeneous nucleation models, particularly at solid walls.

When water boils at atmospheric pressure on a heated metal wall, bubbles appear at wall superheats of around 10K, compared to the superheat of 216K required for homogeneous nucleation. Similarly low superheats are required for the boiling of most other liquids on heated solid walls. Superheats approaching the homogeneous nucleation values can be achieved only by subjecting the liquid-wall system to prolonged periods of high pressure at low temperature (subcooling) or sometimes in the initiation of boiling of extremely well-wetting liquids such as fluorocarbons (Bar-Cohen, 1992). Wetting can be quantified by measurement of the **Contact Angle** θ between the liquid-vapour interface and the surface of the solid but the contact angle can exhibit hysteresis, its value depending on the direction and rate of motion of the contact line between liquid, vapour and solid, **Figure 3**. Microscopic examination, combined with the sensitivity of the boiling superheats to the previous temperature-pressure history, wettability and surface finish of the heated wall, provides strong evidence that the preferred sites for bubble formation are cavities with dimensions of a few

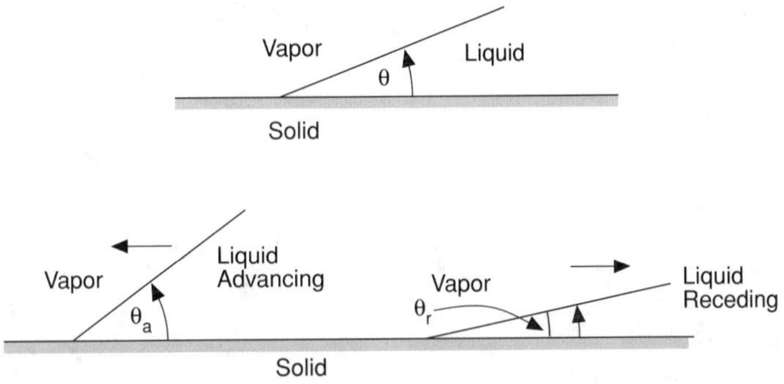

Figure 3. Definition of contact angle Q; hysteresis.

microns or smaller, that trap and stabilise liquid-vapour interfaces at far larger radii of curvature than those of the nuclei in homogeneous nucleation. Thus 'nucleate boiling' is a misnomer, since new clusters of vapour phase do not have to be created. Instead bubbles form repeatedly from tiny reservoirs of continuously-maintained vapour. The processes inside such small cavities cannot be observed directly.

The supposed mechanisms by which liquid-vapour interfaces are stabilised are summarised in **Figure 4**. Most systems in which boiling occurs are initially filled with cold liquid so the liquid-vapour interface must be stabilised when subcooled liquid first enters the cavity, i.e. when the vapour pressure p_g is less than the system pressure p_o. This requires reversal of the curvature of the interface. This could occur at a region in the cavity that is so poorly wetted that the local contact angle greatly exceeds 90° (**Figure 4a**) However, contact angles measured on large plane surfaces generally range from nearly zero for cryogenic and fluorocarbon liquids on clean metals to around 70° for water on poorly cleaned stainless steel Reversal of interfacial curvature when θ is much less than 90° requires a re-entrant geometry, (**Figure 4b**). The presence of trapped or dissolved non-condensible gas can have a large effect on the stability of the interface under subcooled conditions by increasing the total pressure in the reservoir of gas plus vapour so that it exceeds the system pressure; the effect may be time-dependent as soluble gas diffuses between the interface and the interior of the liquid. When the temperature of the wall surrounding the cavity is increased, the vapour pressure increases until the curvature of the liquid-vapour interface reverses again and the trapped vapour becomes unstable to growth by evaporation (**Figure 4c**). The excess pressure and the corresponding superheat for equilibrium may go through several local maxima before the vapour finally emerges from the cavity and 'nucleates' the growth of a visible bubble, (**Figure 4(d)**. The highest of these superheats determines the wall superheat for the inception of bubble production. Maintaining production may then be possible at a lower superheat that only has to overcome the local maximum at the mouth of the cavity, position 6 in **Figure 4(d)**. This model explains the sensitivity of the onset of boiling of well-wetting fluids to pre-boiling conditions and the hysteresis between boiling curves for increasing and decreasing heat flux, (**Figure 5**).

As an embryo bubble emerges from a cavity it encounters a large negative temperature gradient in the liquid surrounding it, resulting from the efficient heat transfer driven by the motion of previous bubbles produced by the cavity itself or by adjacent nucleation sites. This gradient has been modelled by transient or steady conduction into the liquid. It reduces the effective superheat at the interface of a spherical bubble at the mouth of a cavity (**Figure 6**) and limits the size range of cavities that can be active. When combined with information about the size distribution of cavities actually present on a surface (which may be difficult to obtain) and the further assumption that the wall superheat is uniform, this model should define the number of nucleation sites active at any superheat. Increasing the superheat should activate progressively smaller cavities, causing the steep gradient of the nucleate boiling curve. However, the model is over-simplified and does not take into account the inherent patchiness of nucleate boiling heat transfer which, in some circumstances, can lead to large local variations from the mean value of the wall superheat (Kenning, 1992).

Cavities which are stable traps for subcooled vapour prior to boiling may not be the only nucleation sites for bubbles once boiling has been established. Rather shallow cavities which are poor vapour traps may be 'seeded' with vapour from bubbles growing at more stable sites (Judd and Chopra, 1993). Small bubbles bursting through the liquid layers under larger bubbles may produce clouds of tiny bubbles that act as secondary nucleation sites that are not associated with surface cavities (Mesler, 1992), by a process for which there is as yet no quantifiable model. Because of the various mechanisms by which nucleation sites can be created and interact, it is not possible to specify the number of active nucleation sites without also considering bubble motion and the localised processes of heat transfer.

The mechanism of growth of a bubble in uniformly superheated liquid, described previously, is modified when nucleation occurs at a solid wall. Growth as a perfect hemisphere (**Figure 7a**) is prevented by the difficulty of displacing liquid from the solid boundary so a microlayer of liquid is left under the base of the bubble (**Figure 7b**). The curvature at the periphery of the bubble depends on the local viscous and inertial stresses. It is sometimes sharp enough to give the appearance of a contact angle between the bubble and the wall but there is no triple contact line so the properties of the wall can exert no influence. The thickness of the microlayer at the bubble boundary can be estimated from viscous boundary layer theory without detailed consideration of the bubble shape. As it grows, the bubble displaces liquid so by the time it reaches a point at distance R from the nucleation site in time t the liquid at R has been in motion for time t and the boundary layer

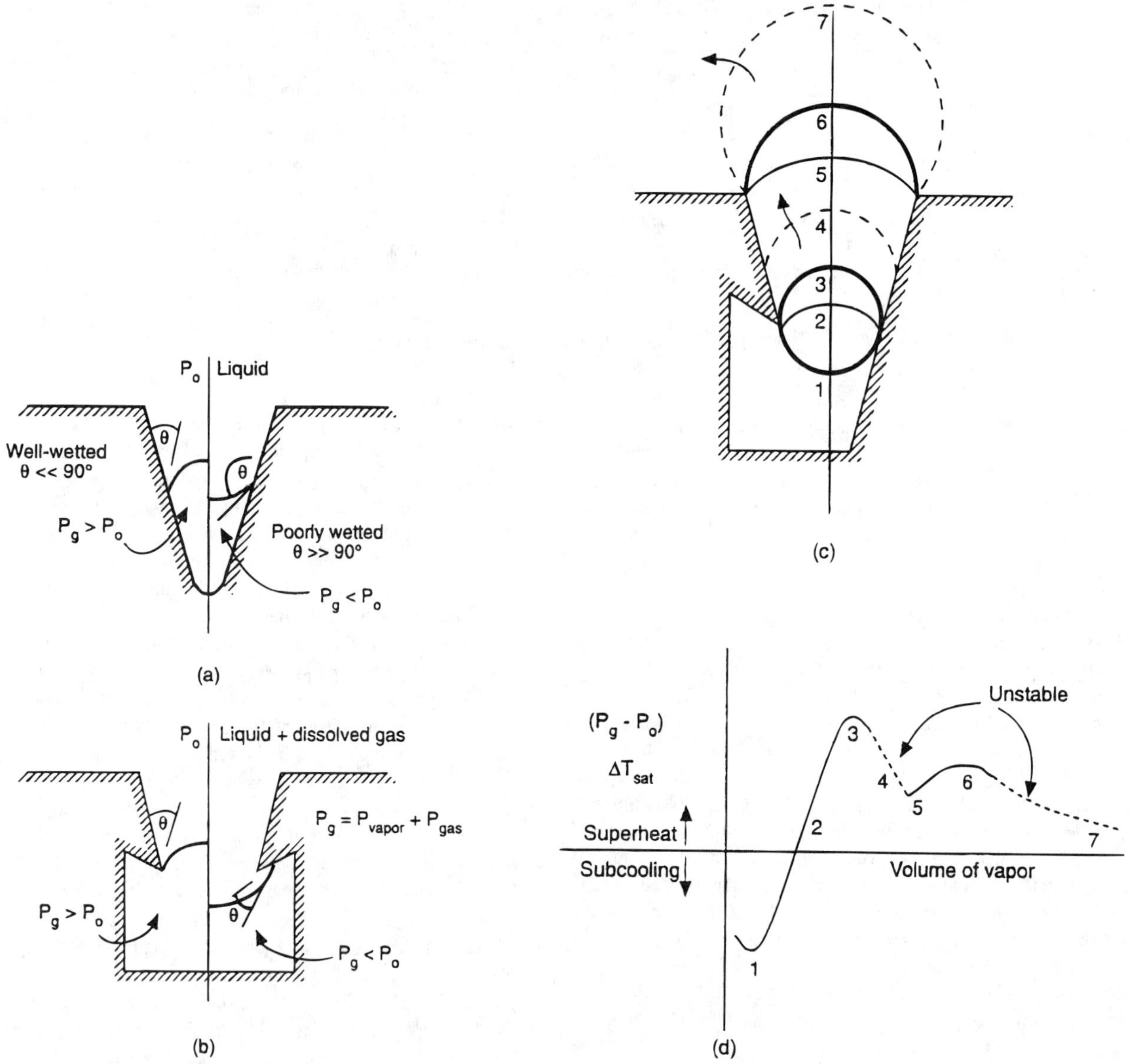

Figure 4. Nucleation at a wall cavity.

of slow-moving liquid that is overtaken by the bubble is of thickness δ_{Ro}, where

$$\delta_{Ro} \sim \sqrt{\nu_l t} \quad \textbf{(5)}$$

where ν_l is the kinematic viscosity of the liquid.

The bubble grows by transient conduction of heat to its interface, as in **Equations 2 and 3** but modified by the temperature gradient in the liquid near the wall, and by additional conduction through the microlayer so approximately

$$R \propto \sqrt{t} \quad \textbf{(6)}$$

From **Equations 5 and 6** the initial thickness of the microlayer under a growing bubble increases approximately linearly with radius to a thickness ranging from a few microns for small, fast-growing bubbles to tens of microns under slow-growing bubbles in pool boiling at low wall superheats. Once formed, the microlayer decreases in thickness by evaporating into the bubble as heat is conducted from the superheated wall across the thin microlayer. As the bubble sticks further out from the wall into liquid that is less superheated, or even subcooled, its rate of growth decreases and it starts to move away from the wall under the combined influence of hydrodynamic and hydrostatic forces. In saturated pool boiling on a horizontal wall the bubble lifts off vertically and the periphery of the base of the bubble moves back towards the nucleation site, (**Figure 7c**). Initially it moves over wall that is still covered by the microlayer but at small radii it may encounter a region where the microlayer has evaporated

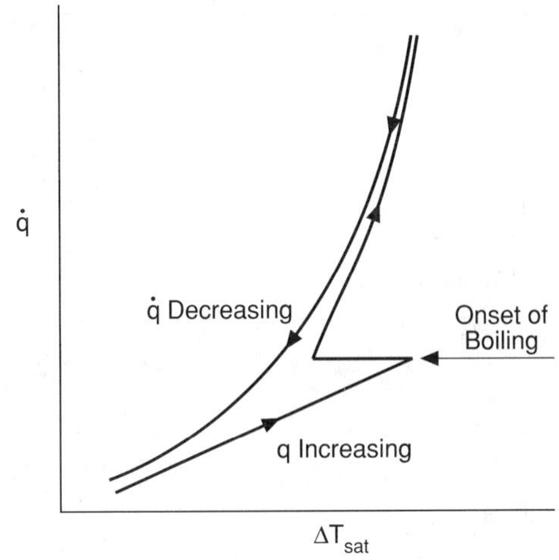

Figure 5. Boiling curve hysteresis.

to dryness and then it would be appropriate to refer to a dynamic advancing contact angle at the base of the bubble. Cooper and Chandratilleke (1981) have presented nondimensional functions to describe the evolution from near-hemispherical to near-spherical shape during the growth of bubbles under various idealised conditions but analytical models for bubble growth often assume inaccurately that the bubble is a truncated sphere. Correlation of the departure size by a balance between buoyancy and surface tension forces with a static contact angle θ, (**Figure 7d**) gives no more than the right order of magnitude for the radius:

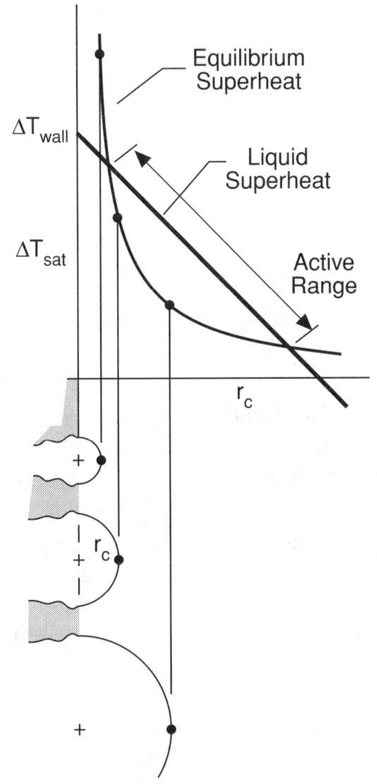

Figure 6. Active size range of nucleation sites.

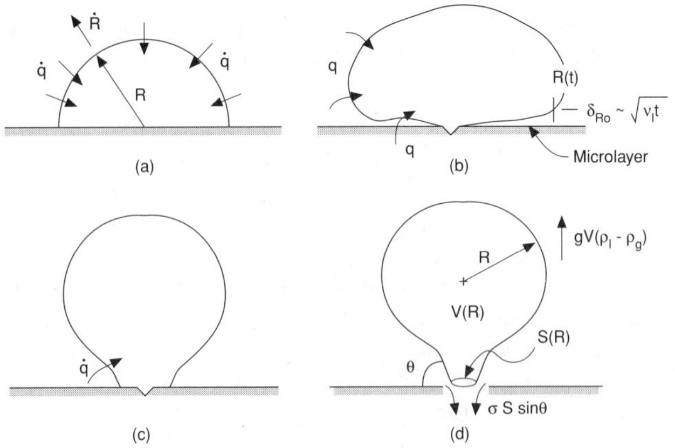

Figure 7. Bubble growth and detachment.

$$\frac{4}{3}\pi R^3 f_1(\theta) g(\rho_l - \rho_g) = 2\pi R f_2(\theta)\sigma \sin\theta, \tag{7}$$

$$R \sim 0.1\,\theta \sqrt{\frac{\sigma}{g(\rho_l - \rho_g)}}, \quad \theta \text{ in degrees.}$$

Improvements in the understanding of bubble departure are to be expected from numerical modelling that takes account of changes in bubble shape and the associated liquid inertia that can drive bubbles away from the wall, even against a hydrostatic buoyancy force. In subcooled boiling the bubbles recondense, either after moving away from the superheated wall at low subcooling, or in close proximity to the wall at large subcooling of the bulk liquid. (See also **Bubble Growth**)

The overall mechanism of heat transfer must involve heat removal from the wall, followed by transport into the interior of the bulk liquid. In nucleate boiling, the bubbles somehow greatly reduce the thermal resistance that occurs close to the wall in heat transfer to a single-phase liquid. The mechanisms of heat removal from the wall, summarised in **Figure 8**, are generally supposed to be:

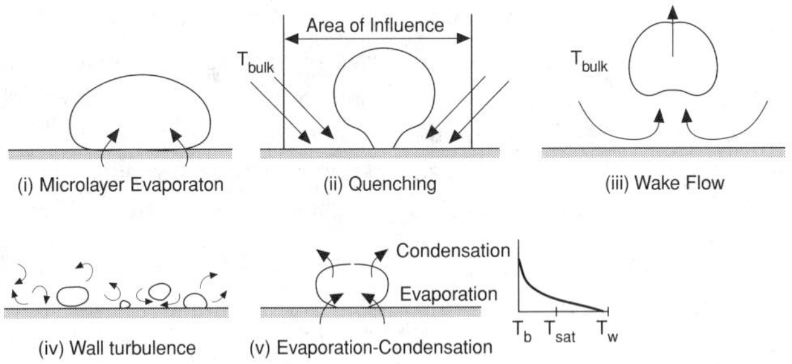

Figure 8. Mechanisms of heat transfer in nucleate boiling.

1. conduction across the very thin microlayers under growing bubbles;
2. quenching by relatively cold bulk liquid moving towards the wall as bubbles round off and detach, modelled by transient conduction into the liquid from 'areas of influence' on the wall about four times the maximum contact area of the bubbles;

3. further localised convective cooling by the motion of bulk liquid in the wakes of departing bubbles;
4. a general increase in turbulence in the liquid close to the wall.

Heat is transferred into the bulk liquid by the motion of bubbles away from the wall (latent heat transport), which may also carry some superheated liquid round each bubble, or by turbulent transport in the liquid. In subcooled boiling there may be a 'heat-pipe' effect of vapour evaporating at the base of bubbles and recondensing where the bubbles are in contact with the subcooled bulk liquid, (**Figure 8v**).

Mechanisms (1), (2) and (3) are concentrated round the nucleation sites and fluctuate as bubbles grow and depart so there must be some unsteady lateral conduction of heat in the wall, (**Figure 9**). Only a wall made of a material with infinite thermal diffusivity can have a uniform, steady superheat. In experiments using very thin, electrically heated walls the local variations in temperature are accentuated and can be measured by observing a layer of thermochromic liquid crystal on the back of the wall, Kenning(1992) and Kenning and Yan (1995). In pool boiling of water at low heat fluxes such measurements confirm that there is strong cooling by microlayer evaporation (1); they show that mechanisms (2) and (3) are less effective than the transient conduction 'quenching' model suggests and that they operate on a wall area of influence that is no bigger than the maximum projected areas of the bubbles (**Figure 10**); the general level of convective cooling (4) is several times the level expected for single phase convection. The localised cooling round the nucleation sites interacts with the processes of bubble nucleation and growth (**Figure 11**). The waiting between bubbles depends on the rate of recovery of the local wall superheat after the departure of a bubble. This recovery may be interrupted by cooling by bubbles growing at adjacent sites or by fluctuations in the general convective cooling so that sites produce bubbles intermittently. The spatial variations in wall superheat affect the rate of microlayer evaporation that helps to drive bubble growth. The model for nucleation site activity based only on the mean wall superheat, summarised in **Figure 6**, cannot represent the intricacies of the real nucleation processes on thin walls. On thicker walls the variations in superheat should be smaller but they can only be measured at a few locations by microthermometers. However, the variations can be modelled numerically on a supercomputer and preliminary studies suggest that they influence nucleate boiling, even on a wall of high thermal conductivity such as copper (Sadasivan et al., 1994). This sort of study should improve our understanding of nucleate boiling but the fundamental difficulties of specifying

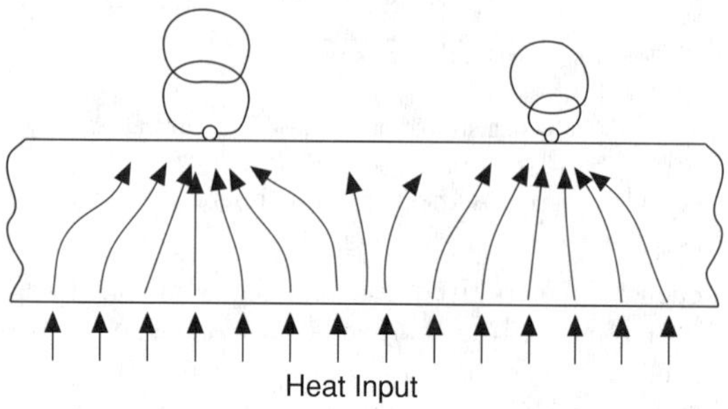

Figure 9. Lateral conduction in the wall.

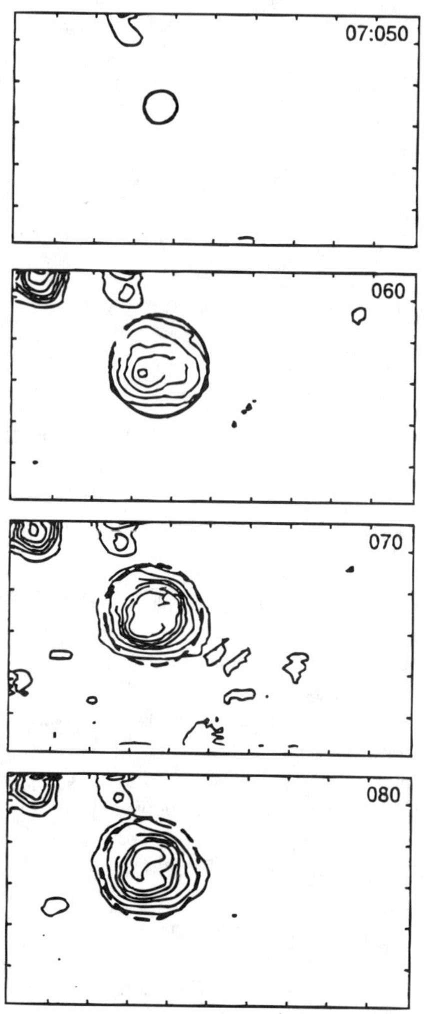

Figure 10. Wall cooling during bubble growth 50–60: growth to maximum radius; 60–70: detachment; 70–80: rising bubble.

the microgeometry and internal wetting characteristics of the nucleation sites will remain.

As the heat flux and the mean wall superheat are increased in saturated pool boiling, the active nucleation sites become so numerous that their bubbles start to coalesce a short distance from the wall. There is a transition to 'fully developed' nucleate boiling, in which the wall is covered by a liquid-rich 'macrolayer' less than 1mm thick through which thin stems of vapour are connected to an overlying cloud of large 'mushroom' bubbles (**Figure 12**). Heat transfer is assumed to occur by conduction across the unsteady macrolayer causing evaporation at the bases of the mushroom bubbles, and by evaporation at the wall into the vapour stems feeding the bubbles. The fluctuating solid-liquid-vapour contact lines at the bases of the stems may be zones of efficient heat transfer. Wayner (1992) has described the processes of flow and heat transfer in liquid films so thin that they are influenced by van der Waals forces. There is still debate about the mechanisms of heat transfer in fully-developed nucleate boiling. The boiling curve (heat flux vs. mean wall superheat) loses the sensitivity to the orientation of the wall that is evident at lower heat fluxes (Nishi-

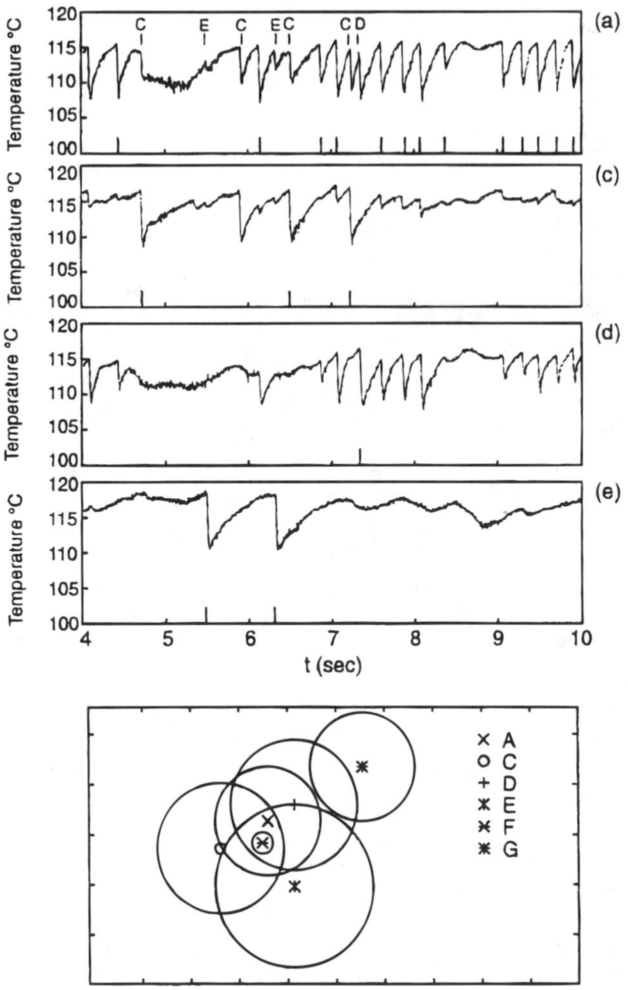

Figure 11. Interaction between nucleation sites; influence of sites C, D, E on site A.

kawa et al., 1984) but it is still sensitive to the surface condition of the wall, so there is no discontinuity in the curve. It is unclear what role is played by the individual nucleation sites as the heat flux is increased and the macrolayer gets thinner. Nucleate boiling breaks down when the macrolayer can no longer be replenished with liquid at a sufficient rate, or when local dry spots are stabilised by the resulting local increase in wall superheat, (see **Burnout in Pool Boiling**).

In forced-convective boiling the heated walls form confining channels through which liquid is forced by an externally-applied

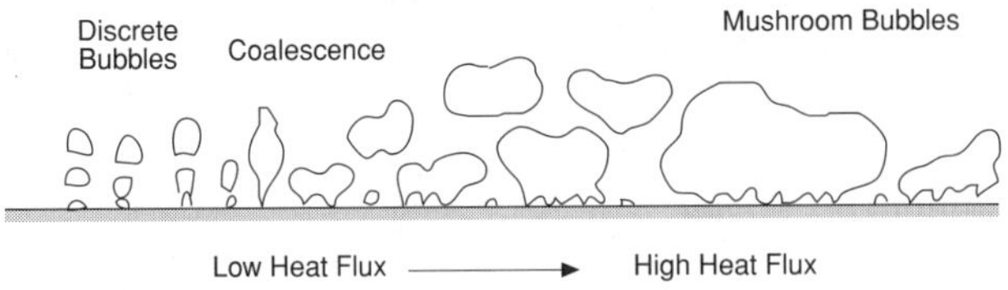

Figure 12. Transition from partial nucleate boiling to fully-developed nucleate boiling.

pressure gradient. The conditions that have received most experimental attention are flow inside vertical and horizontal tubes and flow outside bundles of horizontal tubes. Most experiments involve uniform electrical heating, which does not always represent well the boundary conditions for boiling in heat exchangers, where the source of heat is a hot fluid. The liquid is usually subcooled when it enters the heated region. Vapour is first generated by nucleate boiling; the wall must be superheated to a value that depends on its microgeometry and wettability in order to activate nucleation sites, as in pool boiling. This superheat may be generated by increasing the heat flux, by decreasing the liquid flow rate or by decreasing the system pressure (or perhaps by a combination of all three in industrial systems). In uniformly heated systems boiling is initiated near the downstream end of the heated channel, where the wall temperature is highest and the pressure is lowest, giving the highest wall superheat. With further increases in heat flux, for instance, the initiation point moves upstream and flow boiling develops on its downstream side through regions of subcooled boiling in which the vapour bubbles condense at or close to the wall, bubbly flow in which bubbles move into the bulk flow (even if it is still slightly subcooled , i.e. the averaged thermodynamic quality is still negative), then the flow regimes corresponding to higher vapour fractions described in the articles on **Forced Convection Boiling and Two-phase Flow**. A typical wall temperature profile along a uniformly heated channel is shown in **Figure 13**. The wall superheat is approximately constant in the subcooled nucleate boiling region; as the quantity of flowing vapour increases, the wall superheat decreases as other mechanisms of heat transfer come into effect. (Wall superheats generally refer to time-space average values, since little work has been done in flow boiling on the local variations in superheat that have been shown to be important in pool boiling.) Like pool boiling, flow nucleate boiling of well-wetting liquids can exhibit hysteresis, which can modify the axial distributions of wall superheat (Wadekar, 1993).

Design correlations usually treat flow boiling as a combination of nucleate boiling and convection. Care is required because nucle-

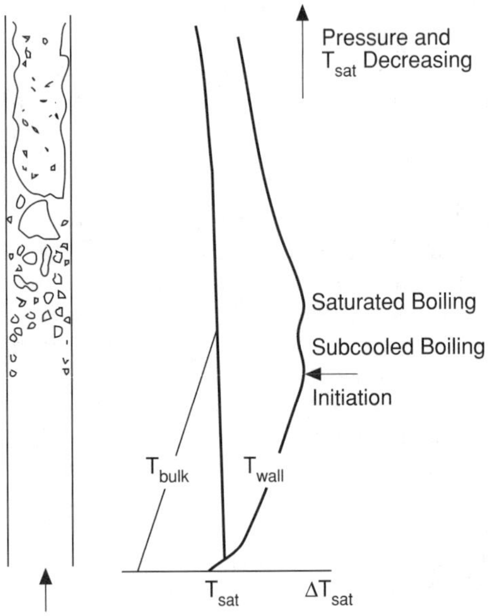

Figure 13. Temperature changes along a uniformly-heated channel.

ate boiling is expected to be a nonlinear function of the wall superheat and convection may be driven either by the wall to bulk temperature difference in subcooled flows or by the wall superheat in saturated flows; it is safer to combine heat fluxes at a given wall superheat, rather than heat transfer coefficients. The nucleate boiling contribution is often based on information obtained from pool boiling experiments so it is subject to the usual difficulties of specifying surface conditions; this makes it difficult to obtain accurate correlations or even to choose between correlation schemes. One such scheme is the simple addition of the nucleate boiling and liquid convective heat fluxes, which seems to work reasonably well for large nucleate boiling fluxes at large liquid subcooling (e.g. del Valle and Kenning, 1985), although the nucleate boiling flux depends on the subcooling. This is not surprising because subcooling has a large effect on bubble behaviour, reducing size but increasing frequency. This simple scheme does not work when the mass fraction x of vapour becomes significant in saturated flow boiling. The presence of the vapour increases the convective heat flux by mechanisms that depend on the flow regime. The velocity and turbulence of the liquid near the wall may be increased, sliding bubbles may continuously create thin liquid microlayers analogous to the transient microlayers in pool boiling (Cornwell, 1990), the flow may oscillate in the slug-churn flow regime or liquid may flow on the wall as a thin but highly disturbed film in the annular flow regime. These effects may be represented approximately by multiplying the single-phase liquid heat transfer coefficient by an enhancement factor F that is a function of the local quality x and the fluid properties to obtain the convective heat flux $\dot{q}_{fc} = F\dot{q}_l$. The changes in flow conditions at the wall should have an effect on the nucleate boiling heat flux $\dot{q}_{nb}$ at a given wall superheat. The nuclei are exposed to larger temperature gradients, which from **Figure 6** may suppress the activity of some sites, and hydrodynamic forces cause bubbles to detach at smaller sizes than in pool boiling. On the other hand, nucleation may be aided by the seeding of unstable sites as bubbles slide along the wall or by the entrainment of microbubbles created at liquid-vapour interfaces in the interior of the two-phase flow. The details of these processes are not understood but it is assumed that they can be combined in a suppression factor S that is a function only of the local flow conditions and which reduces the basic nucleate boiling heat flux. Chen (1966) introduced a correlation scheme for heat $\dot{q}$, of the form

$$\dot{q} = S\dot{q}_{nb} + F\dot{q}_l \qquad \textbf{(8)}$$

of which there have been many subsequent developments (see article on **Boiling**). However, some experimental data are better represented by setting $\dot{q}$ equal to whichever is larger of $\dot{q}_{nb}$ or $F\dot{q}_l$, (Kenning and Cooper, 1989).

This article has so far dealt with nucleate boiling on surfaces that are nominally smooth. We have seen that nucleation depends on microscopic cavities that are accidental consequences of the method of manufacture of the surface. Prolonged service may modify the nucleation characteristics by corrosion or by deposition of corrosion products or dissolved solids. For non-fouling service, tubing is commercially available with 'enhanced' surfaces designed to provide large but stable nucleation sites, sometimes combined with short fins to extend the surface area. Thome (1990) provides a detailed description of boiling on enhanced surfaces. (see **Augmentation of Heat Transfer, Two Phase**)

References

Bar-Cohen, A. (1992) Hysteresis phenomena at the onset of nucleate boiling, Proc. Engineering Foundation Conf. on Pool and External Flow Boiling, Santa Barbara, 1–14.

Blander, M. and Katz, J. L. (1975) Bubble nucleation in liquids, *AIChE Journal 21*, 833–848.

Chen, J. C. (1966) Correlation for boiling heat transfer to saturated fluids in convective flow, *Ind. Eng. Chem. Process Design and Development* 5, 322–329.

Cooper, M. G. and Chandratilleke, T. (1981) Growth of diffusion-controlled vapour bubbles at a wall in a known temperature gradient, *Int. J. Heat Mass Transfer* 24, 1475–1492.

Cornwell, K. (1990) The influence of bubbly flow on boiling from a tube in a bundle, *Int. J. Heat Mass Transfer* 33, 2579–2584.

del Valle, V. H. and Kenning, D. B. R. (1985) Subcooled boiling at high heat flux, *Int. J. Heat Mass Transfer* 28, 1907–1920.

Dhir, V. K. (1990) Nucleate and transition boiling under pool and external flow conditions, Proc. 9th Int. Heat Transfer Conf., Jerusalem, 1, 129–156.

Fujita, Y. (1992) The state-of-the-art nucleate boiling mechanism, Proc. Engineering Foundation Conf. on Pool and External Flow Boiling, Santa Barbara, 83–98.

Gorenflo, D. (1991) Behaltersieden, *VDI-Warmeatlas*, 6th ed., VDI-Verlag, Dusseldorf.

Judd, R. L. and Chopra, A. (1993) Interaction of the nucleation processes occurring at adjacent nucleation sites, *J. Heat Transfer* 115, 955–962.

Kenning, D. B. R. and Cooper, M. G. (1989) Saturated flow boiling of water in vertical tubes, *Int. J. Heat Mass Transfer* 32, 445–458.

Kenning, D. B. R. (1992) Wall temperature patterns in nucleate boiling, *Int. J. Heat Mass Transfer* 35, 73–86.

Kenning, D. B. R. and Yan, Y. (1996) Pool boiling heat transfer on a thin plate: features revealed by liquid crystal thermography, *Int. J. Heat Mass Transfer* 30, 3117–3137 published June 1996.

Lienhard, J. H. (1976) Correlation for the limiting liquid superheat, *Chem. Eng. Science* 31, 847–849.

Mesler, R. B. (1992) Improving nucleate boiling using secondary nucleation, Proc. Engineering Foundation Conf. on Pool and External Flow Boiling, Santa Barbara, 43–48.

Nishikawa, K., Fujita, Y. and Ohta, H. (1984) Effect of surface configuration on nucleate boiling heat transfer, *Int. J. Heat Mass Transfer* 27, 1559–1571.

Sadasivan, P., Unal, C. and Nelson, R. A. (1994) Nonlinear aspects of high heat flux nucleate boiling heat transfer, Los Alamos National Laboratory Reports TSA-6-94-R105, R106.

Skripov, V. P. (1974) *Metastable Liquids*, Wiley, New York.

Thome, J. R. (1990) *Enhanced Boiling Heat Transfer*, Hemisphere, New York.

Wadekar, V. (1993) Onset of boiling in vertical upflow, Heat Transfer-Atlanta, AIChE Symposium Series 295, 89, 293–299.

Wayner, P. C. (1992) Evaporation and stress in the contact line region, Proc. Engineering Foundation Conf. on Pool and External Flow Boiling, Santa Barbara, 251–256.

Leading to: *Bubble growth*

D.B.R. Kenning

NUMERICAL HEAT TRANSFER

Introduction

Numerical heat transfer is a broad term denoting the procedures for the solution, on a computer, of a set of algebraic equations that approximate the differential (and, occasionally, integral) equations describing conduction, convection and/or radiation heat transfer.

The usual objective in any heat transfer calculation is the determination of the rate of heat transfer to or from some surface or object. In conduction problems, this requires finding the temperature gradient in the material at its surface. In convection problems, the temperature gradient in a fluid flowing over a surface is needed to find the heat flux at that surface. In both cases, the determination of the complete temperature distribution in the region of interest is needed as a first step, and in convection one must also find the velocity distribution. Thus, a full solution of the energy equation and perhaps also the equations of motion is required. These are partial differential equations, possibly coupled.

Radiation is somewhat different, involving surfaces separated (in general) by a fluid which may or may not participate in the radiation. If it is transparent, and if the temperatures of the surfaces are known, the radiation and convection phenomena are uncoupled and can be solved separately. If the surface temperatures are not specified, but are to be found as part of the solution, or if the fluid absorbs or emits radiant energy, then the two phenomena are coupled. In either case, the solution of the algebraic and/or integral equations for radiation is required in addition to that of the differential equations for convection and conduction.

The equations describing heat transfer are complex, having some or all of the following characteristics: they are non-linear; they comprise algebraic, partial differential and/or integral equations; they constitute a coupled system; the properties of the substances involved are usually functions of temperature and may be functions of pressure; the solution region is usually not a simple square, circle or box; and it may (in problems involving solidification, melting, etc.) change in size and shape in a manner not known in advance. Thus analytical methods, leading to exact, closed form solutions, are almost always not available.

Two approaches are possible. In the first, the equations are simplified—for example by linearisation, or by the neglect of terms considered sufficiently small, or by the assumption of constant properties, or by some other technique until an equation or system of equations is obtained for which an analytical solution *can* be found. It could be said that an exact analytical solution will be obtained for an approximate problem. The solution will, to some extent, be in error, and it will not normally be possible to estimate the magnitude of this error without recourse to external information such as an experimental result.

The second approach is to use a numerical method. To do this, the continuous solution region is, in most methods, replaced by a net or grid of lines and elements. The solution variables—temperature, velocity, etc.—are not obtained at all of the infinite number of points in the solution region, but only at the finite number of nodes of the grid, or at points within the finite number of elements. The differential equations are replaced by set of linear (or, rarely, non-linear) algebraic equations, which must and can be solved on a computer.

There is also an error in this approach: for example, if derivatives are replaced by finite differences, only an approximate value for the derivative will be obtained. In principle, if the problem is well-posed and if the solution method is well-designed, this error will approach zero as the grid is made increasingly fine. In practice, a fine but not infinitesimal grid must be used. The error can, in general, be estimated and it can also be reduced at the price of increased effort (meaning increased computer time). It could now be said that an approximate solution will be obtained for an exact problem.

In this entry, emphasis will be given to methods for conduction and convection problems, with only a brief mention of radiation. Attention will be limited to incompressible fluids except when buoyancy is important, in which case the Boussinesq approximation will be made. The ideas described can often be used in more general situations, but other methods are also available, especially for high speed flow, which are beyond the present scope.

The Governing Equations for Conduction and Convection

For an incompressible Newtonian fluid, the continuity, momentum and energy equations are

$$\nabla \cdot \mathrm{u} = 0 \tag{1}$$

$$\rho \frac{Du}{Dt} = \nabla \cdot \{\eta \nabla u + \eta (\nabla u)^T\} - \nabla p + \rho g \tag{2}$$

$$\rho c_p \frac{DT}{Dt} = \nabla \cdot (\lambda \nabla T) + q''' \tag{3}$$

where $(\ )^T$ denotes vector transpose, ρg, is a buoyancy (body force) term, and q''' denotes the rate of internal generation of heat per unit volume. These equations represent a system of five equations in three dimensions, or four equations in two dimensions. For conduction in solids, DT/Dt in (3) is replaced by $\partial T/\partial t$, and (1) and (2) are not relevant.

The thermodynamic and transport properties ρ, C_p, λ and η of most materials are in general dependent on temperature and (particularly in the case of ρ) also on pressure. It is often possible to neglect this dependency, with considerable simplification to the computational procedure. If buoyancy is important, density must be considered as a variable and an equation of state is required. A common assumption for low speed flow (small Mach number, together with other restrictions) is to invoke the *Boussinesq approximation*, in which the variation of density is neglected except in the body force term of the momentum equations, namely the last term of (2). For a Boussinesq fluid, it is assumed that $\rho = f(T)$. The usual assumption is a linear relationship:

$$\rho = \rho_o\{1 - \beta(T - T_a)\} \tag{4}$$

where β is the volumetric expansion coefficient, and the subscript denotes a reference state. However water near 4°C, and some other substances, may require a more complex equation.

Equations 1–3 are written in terms of what are called the *primitive variables:* ρ, **u**, T and p. Because of difficulties which can arise with the specification and implementation of pressure boundary conditions, they are sometimes transformed into equations for the *derived variables* vorticity $\boldsymbol{\zeta}$ and vector potential $\boldsymbol{\psi}$, or their two-dimensional counterparts ζ and the stream function ψ, together in either case with ρ and T. The relationships between the primitive and derived variables are:

$$\mathbf{u} = \nabla \times \boldsymbol{\psi}, \qquad \boldsymbol{\zeta} = \nabla \times \mathbf{u}. \tag{5}$$

Vorticity and vector potential are thus related by

$$\nabla^2 \boldsymbol{\psi} = \boldsymbol{\zeta} \tag{6}$$

and taking the curl of the momentum equation (2) leads to the vorticity transport equation

$$\rho \frac{\partial \boldsymbol{\zeta}}{\partial t} - \nabla \times (\rho \mathbf{u} \times \boldsymbol{\zeta}) = \mu \nabla^2 \boldsymbol{\zeta} + \nabla \rho \times \mathbf{g}. \tag{7}$$

As well as the elimination of pressure and the need to solve the continuity equation (which is automatically satisfied by the introduction of the vector potential or stream function), this formulation has, in two dimensions, the advantage of requiring the solution of just three equations 3, 6 and 7 instead of four. In three dimensions, however, the number is actually increased from five to seven, but the elimination of pressure is still sometimes seen as making the transformation worthwhile.

Equations 2 and 3 are written in what is called the *flux* or *gradient* formulation. With the help of Equation 1, they can also be written in the *conservation* form

$$\rho\left\{\frac{\partial \mathbf{u}}{\partial t} + \nabla\cdot(\mathbf{uu})\right\} = \nabla\cdot\{\eta\nabla\mathbf{u} + \eta(\nabla\mathbf{u})^T\} - \nabla p + \rho g \tag{2a}$$

$$\rho c_p\left\{\frac{\partial T}{\partial t} + \nabla\cdot(\mathbf{u}T)\right\} = \nabla\cdot(\lambda\nabla T) + q''' \tag{3a}$$

which is mathematically (and physically) equivalent, but which is sometimes felt to offer a computational advantage because this formulation explicitly recognises the principles of conservation of momentum and energy.

Non-dimensional equations

It is almost always advantageous to transform the dimensional equations into a non-dimensional form, by scaling the dimensional variables with respect to known *reference* or *characteristic quantities*. First, if a parametric solution is needed—to study the effects of changes in density, velocity, etc. over a range of values—then the number of independent parameters is usually reduced by such a transformation. Second, useful physical insight can often be garnered in the process. If characteristic quantities are selected for length, velocity, pressure, time, etc., the equations can be rewritten in terms of dimensionless variables, and the coefficients in the new equations involve parameters such as Re, Pr and Ra (or Gr).

For example, let $\theta = (T - T_0)/(T_1 - T_0)$, where T_0 and T_1 are two specified temperatures (perhaps a wall temperature and a free stream temperature), and use U, L and L/U as the characteristic velocity, dimension and time scale of the problem. Then the energy and vorticity transport Equations 3 and 7 become

$$\frac{\partial \theta}{\partial t} + \nabla\cdot(\mathbf{u}\theta) = \frac{1}{RePr}\nabla\cdot(\nabla\theta) + Q \tag{8}$$

$$\rho \frac{\partial \boldsymbol{\zeta}}{\partial t} - \nabla \times (\mathbf{u} \times \boldsymbol{\zeta}) = \frac{1}{Re}\nabla^2\boldsymbol{\zeta} + \frac{Gr}{Re^2}\nabla \times \theta. \tag{9}$$

in which, for simplicity, properties have been taken as constants (other than ρ in the body force term, for which (4) has been used), Q denotes the non-dimensional heat generation rate

$$q''' L/\rho\, c_p U\, (T_1 - T_0).$$

Boundary conditions

These equations must be accompanied by boundary conditions appropriate to the particular problem. For transient problems, an initial condition is required, and in all problems boundary values for all variables are needed. In heat transfer, three types of condition on T are encountered:

- specified temperature: $T = T_w$ at a boundary, where the wall temperature T_w is known; this is called a Dirichlet condition;
- specified heat flux: $-\lambda(\partial T/\partial n)_w = q_w$, where n is normal to the surface, and the wall heat flux q_w is known; this is called a Neumann condition;
- specified heat transfer coefficient α: $-\lambda\,(\partial T/\partial n)_w = \alpha(T_w - T_i)$; this is called a mixed condition.

A variation of the second condition occurs when the heat flux is by radiation from a source of known temperature T_s: $-\lambda\,(\partial T/\partial n)_w = \alpha F(T_s^4 - T_w^4)$ in which F is a view factor and σ is the Stefan-Bolzmann constant. When either the second or third conditions apply, the wall temperature T_w must be found as part of the solution.

In convection problems, boundary values of **u** and/or its derivatives are also required. Typically, if the flow in the x-direction is parallel and fully developed, $\partial u/dx = v = w = 0$. At the free surface of a liquid, tangential stresses are often taken to be zero and $\partial u/dx = 0$. On any solid surface, the no-slip condition must be applied (the tangential velocity components are equal to those of the surface) and in the absence of suction or blowing, the normal velocity component is zero. If derived variables are used, boundary conditions for $\boldsymbol{\zeta}$ and $\boldsymbol{\psi}$ may be found from those for **u**.

Turbulence modelling

Most flows of practical importance are turbulent, which adds a considerable burden to their computation. In principle, turbulent flow must obey the Navier-Stokes equations. However, their use to calculate it is impractical because of the huge demands which would be made on computer memory and speed. A common alternative is to use the two equation $k - \epsilon$ turbulence model (Rodi, 1993), which starts with the time-averaged equations of motion in which all variables are represented as the sum of a mean and a fluctuating component. Using tensor notation (summation over repeated suffixes), these equations are

$$\frac{\partial \bar{u}_i}{\partial x_i} = 0 \tag{10}$$

$$\frac{\partial \bar{u}_i}{\partial t} + \bar{u}_j \frac{\partial \bar{u}_i}{\partial x_j} = -\frac{1}{\rho_r}\frac{\partial \bar{p}}{\partial x_i} + \frac{\partial}{\partial x_j}\left(\nu \frac{\partial \bar{u}_i}{\partial x_j} - \overline{u_i' u_j'}\right) + g_i \frac{\rho - \rho_r}{\rho_r} \tag{11}$$

$$\frac{\partial T}{\partial t} + \bar{u}_i \frac{\partial T}{\partial x_j} = \frac{\partial}{\partial x_i}\left(\lambda \frac{\partial T}{\partial x_i} - \overline{u_i' T'}\right) + S \tag{12}$$

where overbars denote time-averaged quantities and primes denote turbulent fluctuations from the mean. S is a volumetric source term. The terms such as $\overline{u_i' u_j'}$, when multiplied by the density, are called Reynolds stresses, and are unknown quantities. The *closure problem* of the $k - \varepsilon$ turbulence model is based upon relating the Reynolds stresses to the mean quantities by

$$-\rho\overline{u_i' u_j'} = \eta_t\left(\frac{\partial \bar{u}_i}{\partial x_j} + \frac{\partial \bar{u}_j}{\partial x_i}\right) - \frac{2}{3}\rho k \delta_{ij} \tag{13}$$

where η_t, is the turbulent or eddy viscosity and $k = 0.5\,\overline{u_i u_j}$ is the kinetic energy of the turbulence. In turn, η_t is given by

$$\eta_t = \frac{C_\mu \rho k^2}{\varepsilon} \tag{14}$$

where ϵ is the rate of dissipation of turbulent energy

$$\varepsilon = \frac{\eta_t}{\rho}\left(\frac{\partial u_i'}{\partial x_j}\right)\left(\frac{\partial u_i'}{\partial x_j}\right)$$

and is seen to be proportional to the fluctuating vorticity. Finally, transport equations are constructed for k and ε:

$$\rho\frac{Dk}{Dt} = \frac{\partial}{\partial x_j}\left(\frac{\eta_t}{\sigma_k}\frac{\partial k}{\partial x_j}\right) + \underbrace{\eta_t\left(\frac{\partial u_i}{\partial x_j} + \frac{\partial u_j}{\partial x_i}\right)\frac{\partial u_i}{\partial x_j}}_{P} + \underbrace{\beta g_i \frac{\eta_t}{\sigma_t}\frac{\partial T}{\partial x_t}}_{G} - \rho\varepsilon \tag{15}$$

$$\rho\frac{D\varepsilon}{Dt} = \frac{\partial}{\partial x_j}\left(\frac{\eta_t}{\sigma_\varepsilon}\frac{\partial \varepsilon}{\partial x_j}\right) + \frac{\rho C_{\varepsilon 1}\varepsilon}{k}(P + G)(1 + C_{\varepsilon 3}R_f) - \frac{\rho C_{\varepsilon 2}}{k}. \tag{16}$$

The quantity σ_t is the turbulent Prandtl number; its value lies in the range 0.5–0.9 depending upon the nature of the flow. The first terms on the right hand sides of (15) and (16) represent diffusive transport of k and ε respectively. The term labelled P in (15) is usually designated "production": it represents the transfer of kinetic energy from the mean to the turbulent motion. The buoyancy term G represents an exchange between the turbulent kinetic energy k and potential energy. The last term in (15) is viscous dissipation. In (16), the second and third terms on the right denote generation of vorticity by vortex stretching and viscous destruction of vorticity, respectively. R_f is the flux Richardson number, defined as $-G/P$.

The various constants in these equations are commonly set to the values given in the following table.

$C\mu$	$C_{\varepsilon 1}$	$C_{\varepsilon 2}$	$C_{\varepsilon 3}$	σ_k	σ_e
0.09	1.44	1.92	~0 or ~1*	1.0	1.3

*For vectical or horizontal layers, respectively.

Solution Methods for Conduction and Convection

The most common solution methods are *finite difference* and *finite volume* methods on the one hand, and *weighted residual methods* on the other. Among the latter, the most common are *finite element* and *spectral* methods. The approach used depends upon whether the flow is steady or unsteady, and on whether primitive or derived variables are being used. A necessarily brief description of some common methods follows.

Finite difference methods

To illustrate a finite difference solution, consider the equation

$$\frac{\partial T}{\partial t} + u\frac{\partial T}{\partial x} + v\frac{\partial T}{\partial y} = \alpha\left(\frac{\partial^2 T}{\partial x^2} + \frac{\partial^2 T}{\partial y^2}\right) \tag{17}$$

with suitable initial and boundary conditions, where $\alpha = \lambda/\rho c_p$ is the thermal diffusivity of the material. For a conduction problem, u and v would be zero. For a convection problem there would be the need also to solve the equations of motion for u and v; these equations might be coupled to (17) if buoyancy is significant. We will consider (17) as it stands, assume that u and v are known, and suppose at first that T is also known at all boundaries of the solution region.

Discretisation

A rectangular grid is superimposed over the solution region, a portion of which is shown in **Figure 1**. The intention now is to

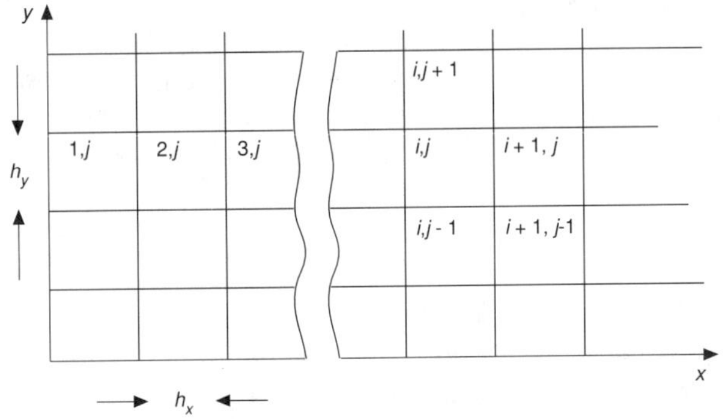

Figure 1. Portion of a computational grid, showing the numbering system.

find approximations to the values of the solution variables at the nodes of the grid.

The vertical grid lines are numbered i = 1, 2, 3 . . . M along the x-axis and the horizontal lines are numbered j = 1, 2, 3 . . . N along the y-axis. A superscript n is used to indicate time: $t = t_0 + nk$, where t_0 is the initial time (usually set to zero) and k is the size of the time step. Thus the temperature at the point (i,j) and at time n is denoted by T^n_{ij}.

FTCS scheme

The equations to be solved contain terms such $\partial T/\partial t$, $u\,\partial T/\partial x$ and $\partial^2 T/\partial y^2$. Using the simplest difference operators, namely forward in time and central in space, these terms may be represented by

$$\frac{T^{n+1}_{i,j} - T^n_{i,j}}{k} + O(k), \qquad u^n_{i,j}\frac{T^n_{i+1,j} - T^n_{i-1,j}}{2h_x} + O(h_x^2)$$

and

$$\frac{T^n_{i,j+1} - 2T^n_{i,j} + T^n_{i,j-1}}{h_y^2} + O(h_y^2) \tag{18}$$

respectively, where h_x and h_y are the *grid* or *mesh sizes* in the x and y directions respectively (e.g. de Vahl Davis, 1986). It is assumed for simplicity that these quantities are constant (although not necessarily equal); often, this is not the case. Expressions for the truncation errors O(k), etc. can be found from a Taylor series expansion of the finite difference approximations.

With these and similar finite difference approximations, and neglecting the truncation errors, (16) may be written

$$\left(\frac{T^{n+1}_{i,j} - T^n_{i,j}}{k} + u^n_{i,j}\frac{T^n_{i+1,j} - T^n_{i-1,j}}{2h_x} + v^n_{i,j}\frac{T^n_{i,j+1} - T^n_{i,j-1}}{2h_y}\right) = \tag{19}$$

$$\alpha\left(\frac{T^n_{i+1,j} - 2T^n_{i,j} + T^n_{i-1,j}}{h_x^2} + \frac{T^n_{i,j+1} - 2T^n_{i,j} + T^n_{i,j-1}}{h_y^2}\right).$$

Equations like (19) may be written for every one of the (M-2)(N-2) internal mesh points. This is known as the forward time-central space (FTCS) scheme. (19) is a *finite difference approximation* (FDA) to (17), the partial differential equation (PDE).

Explicit solution

One way to solve (19) is to rearrange it to express $T^{n+1}_{i,j}$ as a function of the values of T at time n. Since the initial condition provides values for T at all mesh points at time t = 0 (i.e. n = 0), the new or n = 1 values of T may be calculated *explicitly* at each internal mesh point. Once this is complete, the values of T at n = 2 may be found, and so on. This is the easiest way of solving (19) and, for equations as simple as this, it is adequate. It does not, however, always work.

Consistency, convergence and stability

There are three important characteristics of any finite difference scheme: consistency, convergence and stability. A scheme is said to be *consistent* if the truncation errors such as those appearing in (18) tend to zero as the grid and time steps tend to zero, so that the FDA approaches the PDE. Clearly, those in (18) satisfy this requirement, but there are some methods—notably, the DuFort-Frankel scheme—which are not unconditionally consistent. Such methods must be used with considerable care.

Convergence goes one step further: it requires the *solution* of the FDA to approach the *solution* of the PDE as the step sizes go to zero. Convergence does not always follow from satisfying consistency, and it is in fact not normally possible to determine directly whether a method will be convergent.

The third characteristic of a scheme is its *stability.* A scheme is stable if any errors such as round-off errors in the computer decay to zero (more strictly, remain bounded) as the calculations progress, i.e., as $n \to \infty$. Stability can be examined by several methods including the von Neumann or Fourier Series method and the Matrix method. For the FTCS scheme (19) for solving (17), it can be shown that stability requires $\alpha k/h^2 \le 0.5$, where $h = \min(h_x, h_y)$. Thus if the mesh is fine (small h) or if α is large, a small value of k and therefore a long computation time will be required.

A further limitation on stability can exist; the Courant or CFL condition, which is that uk/h must be no greater than unity at any mesh point. The physical significance of this condition is that the fluid should not move a distance greater than one mesh length during one time step, and it applies to explicit schemes for hyperbolic flows, e.g., for flows with little or no diffusion.

Although a direct determination of convergence is not usually possible, the Lax equivalence theorem (Richtmeyer and Morton, 1967, p. 45) provides some help: "Given a properly posed linear initial value problem and a finite difference approximation to it that satisfies the consistency condition, stability is the necessary and sufficient condition for convergence". Unfortunately, most convection heat transfer problems are *not* linear, and the Lax theorem at best provides guidance to necessary but not always sufficient conditions.

The word "convergence" is also used in a different sense from that discussed above. In an iterative procedure for the solution of the algebraic FDAs, the error in each estimate of the solution must decrease with each successive iteration. If it does, the procedure is said to be convergent; if it does not, the procedure is of no use.

Coupled systems for convection problems

The FTCS method can be immediately and easily extended to convection problems using the derived variable formulation, equations (6) and (7). The main complication arises from the need to determine boundary conditions on vorticity. A common formula is that known as the Woods method (Roache, 1972).

Boundary conditions on the vector potential (or, in two dimensions, the stream function) must be obtained from those on velocity; see Hirasaki and Hellums, (1968, 1970).

From some given initial state, the FDA of (7) may be used to obtain new values of vorticity. The FDA of (6) is then solved for the vector potential or stream function, and new values for the velocity components obtained from the FDA of (5). The calculations for the time step are completed by using (19) to obtain new values of temperature.

Implicit solution

The limitation on the time step k imposed by the stability criterion led to the development of *implicit* methods which require more work per time step, but allow the use of larger values of k and therefore fewer time steps. Implicit methods involve the construction of FDAs in which some or all of the spatial derivatives are evaluated at time level n + 1, rather than at n.

For example, in the Crank-Nicolson method, the FDAs of $\partial T/\partial x$ and $\partial^2 T/\partial x^2$ are written

$$\frac{1}{2}\left(\frac{T^n_{i+1,j} - T^n_{i-1,j}}{2h_x} + \frac{T^{n+1}_{i+1,j} - T^{n+1}_{i-1,j}}{2h_x}\right)$$

and

$$\frac{1}{2}\left(\frac{T^n_{i+1,j} - 2T^n_{i,j} + T^n_{i-1,j}}{h_x^2} + \frac{T^{n+1}_{i+1,j} - 2T^{n+1}_{i,j} + T^{n+1}_{i-1,j}}{h_x^2}\right) \quad (20)$$

respectively. These averages of the FDAs at n and n + 1 can be interpreted as central difference approximations to the derivatives at an intermediate time n + 1/2.

The application of (20) and the corresponding FDAs in the y direction will introduce unknown values of T at time level n + 1 to the equations to be solved. In two dimensions there will be five unknowns in each equation, namely the values of T at the grid point (i,j) and at the four surrounding grid points. In three dimensions there will be seven unknowns.

The algebraic equations to be solved thus constitute a system. $T^{n+1}_{i,j}$ can no longer be written explicitly in terms of known quantities. All the new values of T at time level n + 1 must be found by a solution of this system of equations, i.e., *implicitly.*

The solution may be found iteratively, using for example the Gauss-Seidel or successive over relaxation (SOR) methods for large linear systems, or it may be found directly, using an elimination method. Gaussian elimination or a similar method is usually impractical for the very large systems which result from the use of a fine mesh, and even SOR can be extremely demanding. The *alternating direction implicit* (ADI) method is frequently employed to overcome this difficulty.

In ADI, the derivatives are written implicitly, i.e. at time n + 1, in just one of the space directions, while those in the other direction(s) are written explicitly, i.e., at time n. Thus there are only three unknowns in each equation; for example in the x-direction, they are the time n + 1 values of T at the points i + 1, i and i − 1. The system of equations is therefore tridiagonal. For such a system, a highly efficient elimination algorithm known as the Thomas algorithm (see, e.g., Roache, 1972) is available. The values of T thus computed are regarded as intermediate values. The equations are now written implicitly in the next coordinate direction, and explicitly in the other direction(s), using the intermediate values just computed. The direction in which implicit differencing is applied cycles through the two (or three) coordinates in turn. One time step is regarded as having been completed when the two (or three) intermediate steps have been taken. It is interesting to note that each intermediate ADI step is unstable but that the complete sequence of two (or, in three dimensions, three) steps is unconditionally stable.

Upwind differences and false diffusion

A stability limitation on the FTCS method has been noted arising from the use of central differences for the spatial derivatives. In highly convective flows, an additional, very severe restriction also occurs, associated with the value of a Reynolds (or Peclet) number based on the local velocity and the mesh size. If, in (19), $Re_h = uh/\alpha > 2$, unphysical oscillations in time and/or space will develop, leading eventually to a total breakdown of the solution. To overcome this problem, the idea of *upwind* differences has been developed: one-sided differences in which the direction of differencing is determined by the sign of the corresponding velocity. Thus, for $u > 0$, $\partial u/\partial x$ is approximated by $(u_i - u_{i-1})/h_x$. This device is very effective in preventing the instability but introduces a truncation error which is proportional to $uh(\partial^2 u/\partial x^2)$ and is thus like a diffusion term. In fact, for small values of viscosity and large values of velocity, this "false" diffusion can swamp the true diffusion, preventing an accurate solution from being obtained. Second or higher order upwinding, and other differencing schemes, have been developed to overcome this deficiency. Among such methods is the QUICK scheme (Quadratic Upstream Interpolation for Convective Kinematics) and other related methods, originated by Leonard (1988).

Boundary conditions

The implementation of Dirichlet boundary conditions is straight forward; indeed, it happens automatically when an FDA is constructed at a mesh point adjacent to a boundary. Neumann and mixed boundary conditions, involving an unknown boundary value but a known boundary gradient, are more complex. Consider the boundary point (1,j) shown in **Figure 1**, and suppose that the condition

$$\frac{\partial T}{\partial x} = aT_{i,j} + b \quad (21)$$

must be imposed there. There are several ways in which this may be achieved. The simplest is to replace the derivative $\partial T/\partial x$ by the first order forward FDA $(T_{2,j} - T_{1,j})/h_x$, so that (21) may be written

$$T_{1,j} = \frac{T_{2,j} - bh_x}{1 + ah_x}. \quad (22)$$

After the values of T at internal mesh points have been computed, the boundary values can be found from expressions such as (22). An improved accuracy may be obtained by using a second order forward FDA of the derivative. Better still is to use a central FDA at the point (1,j). This requires the introduction of a hypothetical mesh point outside the solution region, the value of T at which is obtain from the boundary condition. The boundary value of T can then be computed as part of the solution procedure.

Weighted residual methods*

The characteristic feature of finite difference methods is that values are sought and obtained for the unknown quantities (velocity, temperature, etc.) at the node points of a computational mesh. The variation of these quantities between node points is implied by the

* The outline given in this section draws heavily on the text of Fletcher (1991).

order of the approximations used, but in practice, no use is made of this knowledge.

In weighted residual methods, on the other hand, the assumption is made that the solution can be represented by an appropriate combination of analytical functions over the whole of the solution region. In one dimension, for example, a solution of the form

$$T = \sum_{i=1}^{I} a_i(t)\phi_i(x) \tag{23}$$

would be assumed, where $a_i(t)$ are coefficients which in general may be time-dependent and which must be evaluated during the solution process, and where $\phi_i(x)$ are *trial functions* whose form might be chosen, at least in part, to be compatible with some anticipated features of the solution. The challenge is to choose appropriate functions ϕ and then to find the coefficients a such that (23) satisfies the given differential equation and its boundary and initial conditions as well as possible at all points in the solution domain.

Suppose that the differential equation to be solved is written

$$L(\bar{T}) = 0 \tag{24}$$

where L is a differential operator, and the overbar denotes the exact solution. If an approximate solution T is inserted into (24) it will not, in general, evaluate to zero but to some *residual R:*

$$L(T) = R \tag{25}$$

The coefficients $a_i(t)$ are found by requiring that the integral of the *weighted residual* over the solution region is zero;

$$\iiint W_m(x, y, z)R \, dx \, dy \, dz = 0. \tag{26}$$

This represents a system of M equations, $m = 1, 2, 3, \ldots M$, from which the coefficients a_i may be obtained. If these coefficients are (as implied here) functions of t, the system is comprised of ordinary differential equations. In steady problems, a system of algebraic equations is obtained.

With different choices of weighting functions W_m, different weighted residual methods are obtained. In the *finite volume* method, the solution region is divided into M subdomains D_m within which $D_m = 1$ and beyond which $D_m = 0$. In the *collocation* method, $W_m = \delta(x - x_m)$ where δ is the Dirac delta function; substitution into (26) then leads to the conclusion that finite difference methods are a form of collocation method. In the *least squares* method, $W_m = \partial R/\partial a_m$, from which follows the requirement that the integral of R^2 over the solution region will be a minimum. Finally, in the *Galerkin* method, the weighting functions are chosen from the same family as the trial functions themselves, and these are usually linear or low order polynomial functions.

We will now examine the finite volume and Galerkin methods in a little more detail.

Finite volume method

Consider the problem of solving Laplace's equation

$$\frac{\partial^2 \bar{\phi}}{\partial x^2} + \frac{\partial^2 \bar{\phi}}{\partial y^2} = 0 \tag{27}$$

in a two-dimensional region which, as shown in **Figure 2**, is divided into a number of quadrilateral elements. The directions of the computational grid lines have been deliberately chosen not to be coincident with the directions of the axes to show the power and generality of the method.

We insert (26) into (25), with $W = 1$, and apply Green's theorem, replacing the integral over the area ABCD with an integral around the contour ABCD, to obtain

$$\iint_{ABCD} 1\left(\frac{\partial^2 \phi}{\partial x^2} + \frac{\partial^2 \phi}{\partial y^2}\right) dx \, dy = \int_{ABCD} \mathbf{H} \cdot \mathbf{n} \, ds = 0 \tag{28}$$

where

$$\mathbf{H} \cdot \mathbf{n} \, ds = \frac{\partial \phi}{\partial x} dy - \frac{\partial \phi}{\partial y} dx. \tag{29}$$

Each derivative in (28) must be approximated on each of the four edges of the subdomain ABCD. For example,

$$\left.\frac{\partial \phi}{\partial x} dy\right|_{AB} = \left[\frac{\partial \phi}{\partial x}\right]_{j,k-1/2} \Delta y_{AB} \tag{30}$$

where

$$\left[\frac{\partial \phi}{\partial x}\right]_{j,k-1/2} \approx \left(\frac{1}{S_{A'B'C'D'}}\right) \iint \left(\frac{\partial \phi}{\partial x}\right) dx \, dy = \left(\frac{1}{S_{A'B'C'D'}}\right) \int \phi \, dy$$

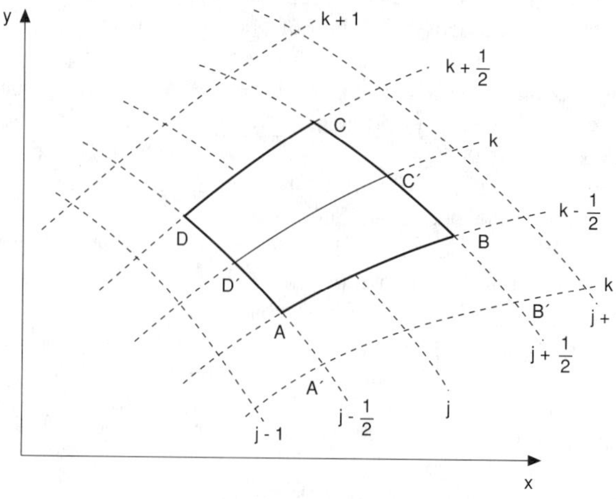

Figure 2. V finite volume mesh.

and in turn

$$\int_{A'B'C'D'} \phi \, dy \approx \phi_{j,k-1}\Delta y_{A'B'} + \phi_B \Delta y_{B'C'} + \phi_{j,k}\Delta y_{C'D'} + \phi_A \Delta y_{D'A'}.$$

If the mesh is not too distorted, $\Delta y_{A'B'} \approx -\Delta y_{C'D'} \simeq \Delta y_{AB}$ etc., so that

$$\left[\frac{\partial \phi}{\partial x}\right]_{j,k-1/2} = \frac{(\phi_{j,k-1} - \phi_{j,k})\Delta y_{AB} + (\phi_B - \phi_A)\Delta y_{k-1,k}}{S_{AB}} \text{ etc.}$$

After further algebraic manipulation, we obtain

$$\begin{aligned}
\phi_{j,k} = \{&0.25(P_{CD} - P_{DA})\phi_{j-1,k+1} \\
&+ [Q_{CD} + 0.25(P_{BC} - P_{DA})]\phi_{j,k+1} \\
&+ 0.25(P_{BC} - P_{CD})\phi_{j+1,k+1} \\
&+ [Q_{DA} + 0.25(P_{CD} - P_{AB})]\phi_{j-1,k} \\
&+ [Q_{BC} + 0.25(P_{AB} - P_{CD})]\phi_{j+1,k} \\
&+ 0.25(P_{DA} - P_{AB})\phi_{j-1,k-1} \\
&+ [Q_{AB} + 0.25(P_{DA} - P_{BC})]\phi_{j,k-1} \\
&+ 0.25(P_{AB} - P_{BC})\phi_{j+1,k-1}\}/ \\
&(Q_{AB} + Q_{BC} + Q_{CD} + Q_{DA})
\end{aligned} \quad \textbf{(31)}$$

where

$$Q_{AB} = (\Delta x_{AB}^2 + \Delta y_{AB}^2)/S_{AB},$$

$$P_{AB} = (\Delta x_{AB}\Delta x_{k-1,k} + \Delta y_{AB}\Delta y_{k-1,k})/S_{AB}$$

etc., and ϕ_A, X_A, y_A, etc. are evaluated as the averages of the respective four surrounding nodal values. Equation 31 is in a form convenient for iterative solution by successive over-relaxation.

It is now apparent that, in implementation, the finite volume method is very similar to the finite difference method. However, the FVM, being derived from an integral form of the governing equations, has better conservation properties than the FDM. It is also capable of handling computational domains of a complex shape more easily. It should also be noted that derivative boundary conditions can be easily implemented by direct substitution into one of the approximations (29).

SIMPLE methods

An important and popular class of methods originated with the SIMPLE algorithm (Semi-Implicit Method for Pressure-Linked Equations) due to Patankar and Spalding (1972) and described in Patankar (1981). They are FVMs, with the discretisation of the primitive variable equations performed on a staggered grid as illustrated in **Figure 3**. The control volume (CV) associated with the point (i,j) is shown shaded. The different variables are identified with different points in the control volume: pressure, temperature and any other scalar with the centre of the CV, and the u and v velocity components with the centres of the upstream and downstream faces of the CV in the x and y directions respectively.

The basis of the method is the use of discretised momentum equations to advance from a known (or initial) state at time or iteration number n to estimate new velocities at time or iteration number n + 1. These estimates will not, in general, be accurate, as they do not satisfy continuity. Equations for corrections to the values of the pressure and the velocities are then employed, and the corrected pressure used as the starting point for the next cycle of calculations.

Improvements and extensions to SIMPLE, known as SIMPLER, SIMPLEC, SIMPLEX etc. have been developed since the original publication of the method in 1972.

Galerkin finite element method

The Galerkin FEM is probably the most widely used of the "true" FEMs (in contrast to the FVM), although related methods including collocation, least squares and boundary elements are also used. To illustrate it, consider the Poisson equation

$$\frac{\partial^2 \bar{w}}{\partial x^2} + \frac{\partial^2 \bar{w}}{\partial y^2} + 1 = 0 \quad \textbf{(32)}$$

with $\bar{w} = 0$ on the boundaries of a square region. In the Galerkin FEM, the approximate solution of (32) is written directly in terms of the nodal unknowns, i.e.

$$w = \sum_{i=1}^{I} w_i \phi_i(x, y) \quad \textbf{(33)}$$

where $\phi_i(x,w)$ are interpolating or approximating functions of some assumed form. Normally, linear or quadratic forms are used.

Consider the portion of the solution region shown in **Figure 4**. In this two-dimensional example, the local solution is interpolated

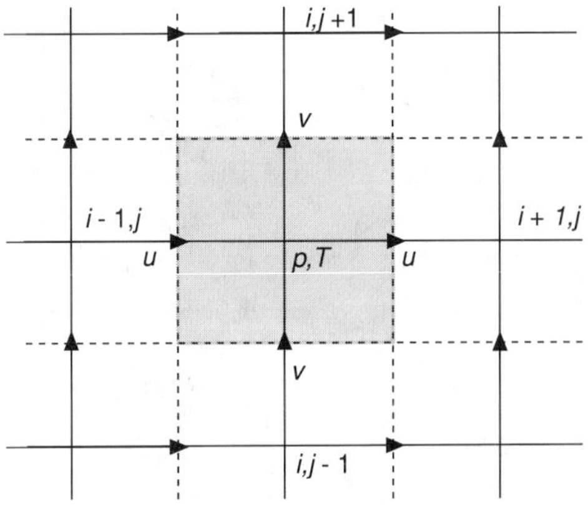

Figure 3. Main grid (solid lines) and staggered grid (broken lines). The horizontal and vertical arrows indicate the grid points for the u and v velocity components.

or approximated, separately, in each of the four elements A, B, C and D which surround the (j,k) node.

A coordinate system (ξ, η) such that $-1 \le \xi, \eta \le 1$ is introduced for each element, as illustrated for element C in **Figure 4**. Then a suitable bi-linear form for each approximating function ϕ_i is

$$\phi_i(\xi, \eta) = 0.25(1 + \xi_i\xi)(1 + \eta_i\eta) \quad \textbf{(34)}$$

Substitution of (33) into (32), application of the Galerkin formulation (26) and some manipulation yields

$$\mathbf{BW} = \mathbf{G} \quad \textbf{(35)}$$

where **W** is the vector of unknown nodal values w_i, an element of **B** is given by

$$b_{m,i} = -\int_{-1}^{1}\int_{-1}^{1}\left[\frac{\partial\phi_i}{\partial x}\frac{\partial\phi_m}{\partial x} + \frac{\partial\phi_i}{\partial y}\frac{\partial\phi_m}{\partial y}\right]dx\,dy, \quad \textbf{(36)}$$

and an element of **G** is given by

$$g_m = \int_{-1}^{1}\int_{-1}^{1}\phi_m\,dx\,dy \quad \textbf{(37)}$$

The parameter m ranges over all the internal nodes, i.e., $2 \le j \le NX - 1$, $2 \le k \le NY - 1$. The integrals in (36) are best evaluated over each element, individually, using the element-based coordinates (ξ, η). Only the four elements around node m make non-zero contributions to the integrals in (36) and (37).

For a uniform grid, with $h_x = h_y = h$, the following is eventually obtained:

$$\begin{aligned}
&(w_{j-1,k+1} - 2w_{j,k+1} + w_{j+1,k+1})\\
&+ 4(w_{j-1,k} - 2w_{j,k} + w_{j+1,k})\\
&+ (w_{j-1,k-1} - 2w_{j,k-1} + w_{j+1,k-1})\\
&+ (w_{j+1,k-1} - 2w_{j+1,k} + w_{j+1,k+1}) \quad \textbf{(38)}\\
&+ 4(w_{j,k-1} - 2w_{j,k} + w_{j,k+1})\\
&+ (w_{j-1,k-1} - 2w_{j-1,k} + w_{j-1,k+1})\\
&+ 6h^2 = 0.
\end{aligned}$$

Equation (38) represents a system of linear equations which can be solved by, for example, successive overrelaxation.

As a matter of interest, the finite difference representation of (32), using central difference approximations, would be

$$w_{j-1,k} + w_{j+1,k} + w_{j,k-1} + w_{j,k+1} - 4w_{j,k} + h^2 = 0, \quad \textbf{(39)}$$

which is a five point formula, compared with (38) which is a nine point formula.

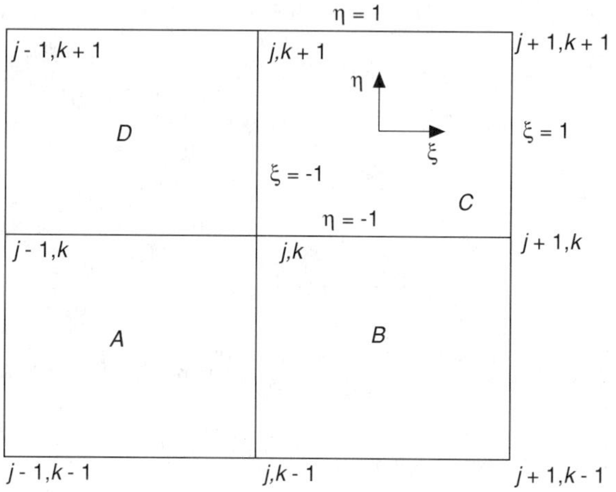

Figure 4. Global and local notation.

Spectral method

The spectral method uses the same form of approximate solution as the traditional Galerkin method. Like the latter, the trial functions W_m and weighting functions Wm are non-zero throughout the computational domain. On the other hand, the unknown coefficients a_i in (23) cannot be identified with nodal unknowns.

In the spectral method, the trial and weighting functions are orthogonal:

$$\iiint \phi_i(x, y, z)\phi_m(x, y, z)dx\,dy\,dz \neq 0 \text{ when } m = i$$
$$= 0 \text{ when } m \neq i.$$

Fourier Series, Legendre Polynomials and **Chebyshev Polynomials** are examples of orthogonal functions often used in spectral methods.

Radiation

Whereas conduction and convection heat transfer are described by partial differential equations, the basic governing equations for radiation are integral equations. The integral nature of the equations arises from the fact that it is necessary to sum the radiant energy exchanges over surfaces, the radiant intensity at which depends upon the relative orientation to, and separation from other surfaces. When combined modes of heat transfer occur, integral and differential equations are both relevant.

Under idealised conditions—uniform surface temperatures, constant properties, etc.—radiation calculations may be complex because of a particular geometry, but are not difficult and can be undertaken using classical methods. Under more realistic conditions, the challenge is somewhat greater.

The complexities which are present in heat transfer problems were mentioned at the start of this entry. For radiation in particular, they are compounded by a geometry which is generally difficult and by the fact that the material properties relevant to radiation are dependent on the properties of the radiation itself (especially wave length) as well as on the nature and surface characteristics of the materials.

The integral nature of the equations has demanded different approaches to their solution. As outlined by Edwards (1985) and Chung (1988), these include the methods of discrete ordinates, Monte Carlo, invariant imbedding, zone analysis and finite element imbedding.

In the method of discrete ordinates, a set of fixed directions is chosen to approximate the solid angle integrals over all necessary directions by a sum over the chosen discrete set of directions.

In the Monte Carlo method, on the other hand, a sequence of random directions from any starting point is chosen and the paths and histories of "photons" taking those initial directions are followed. For a given photon, its scattering, transmission, reflection and/or absorption are determined based on the geometry of the problem and the properties of the materials involved. A statistical analysis of the behaviour of tens or hundreds of thousands of photons leads to an accurate result. The method can be used for complete radiation calculations, or for the more limited task of determining radiation view factors in complex geometries which can then be used in conjunction with more conventional methods of calculation.

Invariant or differential imbedding has been used in astrophysics calculations by Chandraskhar (1960). The method is based on the principle of invariance of the emergent radiation from a semi-infinite plane-parallel layer or slab to the addition or subtraction of layers of arbitrary optical thickness. Discrete ordinates are used to convert integrals to summations. The method is ideal for layers or semi-infinite half-spaces, but has difficulties in more complicated geometries.

In zone analysis [Ozisik (1973); Siegel and Howell (1981)], the surfaces of an enclosure are subdivided into zones over which the temperatures, radiation properties and incident and emitted radiant intensities are assumed to be uniform. The accuracy of the results depends on the fineness of the subdivisions and the validity of the assumptions of uniformity.

For specularly transmitting slabs, and for reflecting or diffusely transmitting and reflecting slabs whose bidirectional properties are known from, for instance, differential imbedding, a multislab system may be built up by finite element imbedding. Interference effects may be included as well as polarisation and directional effects. Finite element methods may also be used to find radiation view factors between surfaces, and to solve problems of combined conduction-convection-radiation heat transfer.

References

Anderson, J. D. (1994) *Computational Fluid Dynamics: The Basics with Applications*, Mc-Graw-Hill, New York.

Chandraskhar, S. (1960) *Radiative Heat Transfer*, Dover, New York.

Chung, T. J. (1988) Integral and integro-differential systems, Chapter 14 of *Handbook of Numerical Heat Transfer* (eds. W. J. Minkowycz, E. M. Sparrow, G. E. Schneider and R. H. Pletcher), Wiley, New York.

de Vahl Davis, G. (1986) *Numerical Methods in Engineering and Science*. Allen & Unwin, London.

Edwards, D. K. (1985) Gas radiation transfer, Part 6 of Chapter 14 of *Handbook of Heat Transfer Fundamentals* (eds. W. M. Rohsenow, J. P. Hartnett and E. N. Ganic), 2nd Edition, McGraw-Hill, New York.

Fletcher, C. A. J. (1991) *Computational Techniques for Fluid Dynamics*, Vols. I and II, 2nd Ed., Springer-Verlag, Berlin, Heidelberg.

Hirasaki, G. J. and Hellums, J. D. (1968) A general formulation of the boundary conditions on the vector potential in three dimensional hydrodynamics, *Quart. Appl. Math., 26*, 331–342.

Hirasaki, G. J. and Hellums, J. D. (1970) Boundary conditions on the vector and scalar potentials in viscous three-dimensional hydrodynamics, *Quart. Appl. Math., 28*, 293–296.

Leonard, B. P. (1988) Elliptic systems: finite-difference method IV, Chapter 9 of *Handbook of Numerical Heat Transfer* (eds. W. J. Minkowycz, E. M. Sparrow, G. E. Schneider and R. H. Fletcher), Wiley. New York.

Ozisik, M. N. (1973) *Radiative Transfer.* Wiley-Interscience, New York.

Patankar, S. V. (1980) *Numerical Heat Transfer and Fluid Flow*, Hemisphere/McGraw-Hill, New York.

Patankar, S. V. and Spalding, D. B. (1972) A calculation procedure for heat, mass and momentum transfer in three-dimensional parabolic flows, *Int. J. Heat Mass Transfer*, 15, 1787–1806.

Reddy, J. N. and Gartling, D. K. (1994) *The Finite Element Method in Heat Transfer and Fluid Dynamics*, CRC Press, Boca Raton, FL.

Richtmeyer, R. D. and Morton, K. W. (1967) Difference methods for initial-value problems, *Interscience*, New York.

Rodi, W. (1993) *Turbulence Models and Their Application in Hydraulics—A State of the Art Review*, 3rd Ed., A. A. Balkema, Rotterdam/Brookfield.

Siegel, R. and Howell, J. R. (1981) *Thermal Radiation Heat Transfer.* 2nd Ed., Hemisphere, Washington.

Torrance, K. K. (1985) Numerical methods in heat transfer, Chapter 5 of *Handbook of Heat Transfer Fundamentals* (eds. W. M. Rohsenow, J. P. Hartnett and E. N. Ganic), McGraw-Hill, New York.

G. de Vahl Davis

NUMERICAL ANALYSIS (see Computational fluid dynamics)

NUMERICAL METHODS OF HEAT EXCHANGER DESIGN (see Tube banks, single phase heat transfer in)

NUSSELT CONDENSATION MODEL (see Tubes, condensation in)

NUSSELT EQUATIONS (see Condensation of pure vapour; Condensation, overview)

NUSSELT, WILHELM (1882–1957)

A German engineer, born on November 25, 1882 at Nürnberg, Nusselt studied machinery at the Technical Universities of Berlin-Charlottenburg and München where he graduated in 1904 and conducted advanced studies in mathematics and physics. He became an assistant to O. Knoblauch at the laboratory for technical physics in München and completed his doctoral thesis on the conductivity of insulating materials in 1907, using the "*Nusselt Sphere*" for his experiments. From 1907 to 1909 he worked as an assistant of Mollier in Dresden, qualifying himself for a professorship with a work on heat and momentum transfer in tubes.

Figure 1.

In 1915, Nusselt published his pioneer paper: "The Basic Laws of Heat Transfer" in which he first proposed the dimensionless groups now known as the principal parameters in the similarity theory of heat transfer. Other famous works were concerned with the film condensation of steam on vertical surfaces, the combustion of pulverized coal and the analogy between heat and mass transfer in evaporation. Among the primarily mathematical works of Nusselt, the well known solutions for laminar heat transfer in the entrance region of tubes, for heat exchange in cross-flow and the basic theory of regenerators should be mentioned.

Nusselt was professor at the Technical Universities of Karlsruhe from 1920 to 1925 and at München from 1925 until his retirement in 1952. He was awarded the Gauss-medal and the grashof commemorative medal. Nusselt died in München on September 1, 1957.

U. Grigull, M. Sandner and J. Straub

NUSSELT NUMBER

Nusselt number, Nu, is the dimensionless parameter characterising convective heat transfer. It is defined as

$$Nu = \frac{\alpha L}{\lambda}$$

where α is convective heat transfer coefficient, L is representative dimension (e.g. diameter for pipes), and λ is the thermal conductivity of the fluid. Nusselt number is a measure of the ratio between heat transfer by convection (α) and heat transfer by conduction alone (λ/L).

Convective heat transfer relationships are usually expressed in terms of Nusselt number as a function of **Reynolds Number** and **Prandtl Number**.

Reference

Hewitt, G. F., Shires, G. L. and Bott, T. R. (1994) *Process Heat Transfer*, CRC Press, Boca Raton, FL.

G.L. Shires

NUSSELT SPHERE (see Nusselt Wilhelm)

NUTSCHE, GRAVITY FILTERS (see Liquid-solid separation)

NYQUIST STABILITY CRITERION AND NYQUIST DIAGRAMS

The Nyquist stability criterion has been used extensively in science and engineering to assess the stability of physical systems that can be represented by sets of linear equations. The mathematical foundations of the criterion can be found in many advanced mathematics or *linear control theory* texts such as Wylie and Barrett (1982), D'Azzo and Houpis (1975) or Willems (1970). The criterion has found popularity partly because of its method of application, which is through a relatively straight-forward geometrical construction. Consider the simple *dynamic feedback* system represented in the frequency-domain by the following block diagram, **Figure 1**. This system response δy can be related to the external forcing function δy_{ext} by the frequency-domain equation,

$$\delta y = \Phi(s)\delta y_{ext} = \frac{G(s)}{1 + G(s)H(s)}\delta y_{ext} \quad (1)$$

The quotient $\Phi(s)$ can usually be written so that no singularities appear in the denominator. The system is stable, for a bounded forcing function, if the real parts of the zeros of the denominator have negative real parts. This is because if transformed back into time-domain it is these which form the indices of the exponentials

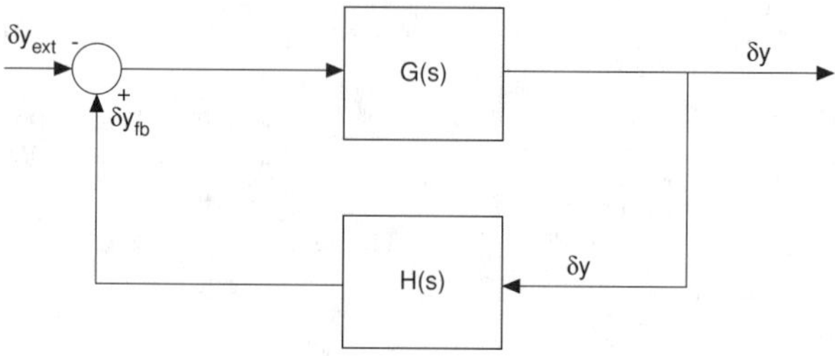

Figure 1. Block diagram of a simple feedback system.

in the solution. The stability of a linear system is thus determined by examining the signs of the real parts of the roots of the characteristic equation. For the simple feedback system described here the characteristic equation is,

$$E_c(s) = 1 + G(s)H(s) = 0 \quad \textbf{(2)}$$

The Nyquist criterion considers the variation of the characteristic equation with variation of s over a closed contour C which is formed by a semicircle in the right-hand half of the s-plane and the imaginary axis. Provided that the characteristic equation is analytic everywhere on the closed contour then the net variation of the argument of $E_c(s)$ can be related to the number of poles P and zeros Z of $E_c(s)$ inside C by the following equation,

$$Z - P = \frac{1}{2\pi}\,[\text{net variation of } \arg(E_c(s)) \text{ around C}] \quad \textbf{(3)}$$

In geometric terms the number of clockwise encirclements of the origin of the mapping of $E_c(s)$, for a single traverse of s in the s-plane around the contour C, is equal to the difference between the number of zeros and the number of poles of $E_c(s)$ inside C. If the semicircle, or Nyquist "D-contour", is then enlarged to an infinite size, so as to take up the entire right-hand half of the s-plane, the method can be used to check if the characteristic equation has any roots in the right-hand half of the plane. As mentioned earlier it is normally the case that $\Phi(s)$ is rearranged so as to remove any poles from $E_c(s)$. It is assumed that this has been done and hence $P = 0$. If any roots lie in the right-hand half of the plane then they have positive real parts and the system is, therefore, unstable.

For Laplace transforms of most physical systems, the application of the criterion is simplified by the fact that the mapping of s into $E_c(s)$ along the semicircular portion of C collapses onto the origin as the radius of the semicircle tends to an infinite size. (See **Laplace Transformations**) Furthermore, the mappings along the positive and negative branches of the imaginary axis of s into $E_c(s)$ are symmetric about the real axis. For Laplace transformed systems the application of the Nyquist criterion amounts simply to plotting the locus produced by the mapping of the characteristic equation in its plane (the w-plane), for variation of s along the positive branch of the imaginary axis in the s-plane. In algebraic terms the locus in the w-plane is formed by evaluating,

$$w = E_c(i\omega) \quad \text{for} \quad 0 \le \omega \le \infty \quad \textbf{(4)}$$

An example of the application of the criterion is to the stability

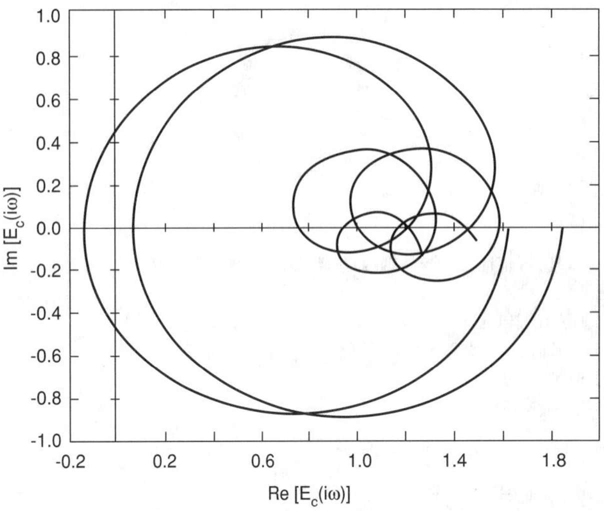

Figure 2. Nyquist diagram from a boiling channel stability analysis.

analysis of boiling channels (Lahey and Podowski, 1989). A sample Nyquist diagram which shows two loci calculated for a boiling channel is given below. The locus that encircles the origin shows an unstable system and the locus which does not, shows a stable system. (See also **Instability, Two-phase**)

To avoid any confusion, it is useful to note that the Nyquist criterion is often applied to the open loop transfer function, G(s)H(s). Stability of the closed loop system is then determined by encirclement of the point $(-1,0)$ in the w-plane.

References

D'Azzo, J. J. and Houpis, C. H. (1975) *Linear Control System Analysis and Design: Conventional and Modern*, McGraw-Hill, USA.

Lahey Jr., R. T. & Podowski, M. Z. (1989) On the analysis of various instabilities in two-phase flows, in *Multiphase Science and Technology*, Eds. G. F. Hewitt, J-M. Delhaye & N. Zuber, vol. 4, pp 183–370, Hemisphere.

Willems, J. L. (1970) *Stability Theory of Dynamical Systems*, Nelson, UK.

Wylie, C. R. and Barrett, L. C. (1982) *Advanced Engineering Mathematics*, 5th Ed. McGraw-Hill Book Company.

P.F. Pickering

O

Oak Ridge National Laboratory, ORNL

Energy Division
PO Box 2008
Oak Ridge TN 37831-6206
U.S.A.
Tel: 615 576 5454

OCCLUSIONS (see Crystallization)

OCEAN THERMAL ENERGY CONVERSION

Following from: *Alternative energy sources*

Ocean thermal energy conversion (OTEC) is essentially a low temperature **Rankine Cycle** using the warm surface waters of the tropical and subtropical oceans as a heat source and the colder water from 500–1,000 m deep as the heat sink. With the surface water at a maximum of 27°C and the deeper water in the 5–10°C range, the maximum Carnot efficiency of the cycle is about 7 percent and the practical net station efficiency about 1.5–2 percent. Hence enormous quantities of water need to be circulated and large quantities of heat transferred even for plants in the 5–10 Mwe (net) range.

The OTEC concept was first proposed by the French physicist J. D'Arsonval in 1881. Two basic cycles have been considered. The closed cycle uses a low boiling working fluid (ammonia with 0.1–0.2 percent water is the clear choice on thermodynamic and thermal-hydraulic grounds) which is vaporized by warm sea water in the vaporizer, the vapor expanded in a turbine connected to an electrical generator, the exhaust vapor condensed by the cold sea water in the condenser, and the condensate returned to the vaporizer by a feed pump. The heat exchangers must be very large, but the turbine is small and the power cycle technology straightforward. Shell and tube, plate, and plate fin exchanger designs have been tested for this service.

The open cycle uses the water itself as the working fluid, a small fraction of the warm water being flashed to vapor at about 2,500 Pa absolute in a large vacuum chamber, the vapor then being expanded through a turbine before being condensed by cold water in a direct contact condenser at about 1,000 Pa. All of the equipment must be very large to accommodate the low density vapor and the technology has not been fully developed. A variant of the open cycle uses a surface condenser to condense the water vapor to produce potable water.

A number of OTEC pilot plants and test facilities have been built and operated, and there is a low but continuing level of interest, especially for isolated tropical islands with few other energy sources.

Reference

Avery, W. H. and Wu, C. (1994) *Renewable Energy from the Ocean: A Guide to OTEC,* Oxford University Press, New York.

K.J. Bell

OCTANE

n-Octane (C_8H_{18}) is a colourless liquid that is insoluble in water. It is obtained by fractional distillation of petroleum. Octane is flammable and moderately toxic. It is used for the **Octane Number** calibration of antiknock properties of fuels. It is also used as a solvent and in organic synthesis.

Molecular weight: 114.224	Critical temperature: 568.8 K
Melting point: 216.35 K	Critical pressure: 2.486 MPa
Normal boiling point: 398.8 K	Critical density: 232 kg/m^3

Reference

Beaton, C. F. and Hewitt G. F. (1989) *Physical Property Data for the Design Engineer,* Hemisphere Publishing Corporation, New York.

V. Vesovic

OCTANE NUMBER

Following from: *Internal combustion engines*

In a petrol engine, under certain operating and ambient conditions, *knock* may occur which can cause damage to the engine. Its occurrence depends as well on engine and fuel properties. A measure of the anti-knock performance of a fuel in a given engine is given by its *octane number;* the higher the octane number, the higher the resistance to knock. A scale has been deviced in which automotive fuels are assigned an octane number, which is based on two hydrocarbons defining the ends of the scale: normal heptane (n-C_7H_{16}) which has a value of zero and iso-octane (C_8H_{18}: 2, 2, 4-trimethylpentane) with an octane number of 100. Blends of these two hydrocarbons define the knock resistance e.g. a blend of 10% n-heptane and 90% iso-octane by volume has an octane number of 90.

The octane number of a fuel is determined in a standard test engine (single-cylinder, variable compression ratio CFR engine developed under the auspices of the Cooperative Fuel Research Committee in 1931) by means of either the research method (ASTM D-2699 ; BS 2637) or the motor method (ASTM D-2700 ; BS 2638). The corresponding research octane number (RON) and motor octane number (MON) are obtained under different test conditions which are summarised in **Table 1** [Heywood (1988) and Owen and Coley, (1990)]. Typical octane numbers for automotive fuels are given in **Table 2**. A worldwide summary of octane ratings is published regularly by the Associated Octel Co. Ltd, London.

Table 1. Octane: Properties of the saturated liquid and vapor

T_{sat}, K	398.8	415	435	455	475	495	515	535	555	568.8
P_{sat}, kPa	101.3	156	252	386	569	809	1 127	1 535	2 052	2 486
ρ_ℓ, kg/m^3	611	595	575	553	529	502	470	432	373	232
ρ_g, kg/m^3	3.80	5.67	8.98	13.7	20.4	29.9	44.0	65.0	105	232
h_ℓ, kJ/kg	514.1	558.2	609.4	667.6	725.7	786.2	851.3	921.1	990.9	1 088.6
h_g, kJ/kg	814.1	844.3	883.9	923.4	965.3	1 004.8	1 046.7	1 081.6	1 109.5	1 088.6
$\Delta h_{g,\ell}$, kJ/kg	300.0	286.1	274.5	255.8	239.6	218.6	195.4	160.5	118.6	
$c_{p,\ell}$, kJ/(kg K)	2.50	2.61	2.74	2.89	3.03	3.18	3.36	3.60	4.23	
$c_{p,g}$, kJ/(kg K)	2.11	2.19	2.30	2.42	2.56	2.73	2.96	3.30	4.80	
η_ℓ, μNs/m^2	203	174	149	126	107	91	75	61	48	
η_g, μNs/m^2	7.4	7.9	8.5	9.2	10.0	11.1	12.5	14.7	19.4	
λ_ℓ(mW/m^2)/(K/m)	98	93	88	83	78	73	68	62	56	
λ_g, (mW/m^2)/(K/m)	18.4	20.3	22.5	24.3	26.7	29.1	31.6	34.4	38.6	
Pr_ℓ	5.18	4.88	4.64	4.39	4.16	3.96	3.71	3.54	3.63	
Pr_g	0.85	0.85	0.87	0.92	0.96	1.04	1.21	1.41	2.41	
σ, mN/m	11.9	10.5	8.9	7.3	5.8	4.4	2.9	1.8	0.55	
$\beta_{e,\ell}$, kK^{-1}	1.62	1.79	2.05	2.39	2.88	3.60	4.82	6.96	15.9	

The values of thermophysical properties have been obtained from Beaton, C.F. and Hewitt, G.F. (1989).

Table 1. Operating conditions for research and motor methods

Condition	Research method	Motor method
Inlet temperature	52°C	149°C
Inlet pressure	Atmospheric	
Humidity	0.0036–0.0072	kg/kg dry air
Coolant temperature	100°C	
Engine speed	600 rev/min	900 rev/min
Spark advance	13°btdc (constant)	19–26°btdc (varies with compression ratio)
Air/fuel ratio	Adjusted for maximum knock	

Table 2. Typical octane numbers*

Fuel	RON	MON
Four-star Leaded	97.7 (97.0 min)	86.6 (86.0 min)
Unleaded premium	96.0 (95.0 min)	85.3 (85.0 min)
Super Plus Unleaded	98.2	87.0

*Technical inspections courtesy of Esso Petroleum Co.

References

Heywood, J. B. (1988) *Internal Combustion Engine Fundamentals*, McGraw Hill.

Owen, K. and Coley, T. (1990) *Automotive Fuels Handbook*, Society of Automotive Engineers.

D. Arcoumanis

OECD (see Organisation for Economic Co-operation and Development)

OHM'S LAW

Ohm's Law is the relationship between the potential difference applied across an electrical conductor and the current flowing in the conductor. This relationship is

$$V = RI$$

where R is the electrical *resistance,* the constant of proportionality. The unit of resistance is the ohm, Ω, that being the resistance between two points of a conductor when a potential difference of 1 volt, produces a current of 1 amp between the two points. An alternate definition of the ohm is the resistance of conductor in which a current of 1 amp generates heat at the rate of 1 watt, i.e.

$$\dot{Q} = I^2R$$

The reciprocal of resistance is called the *conductance,* G; the unit of conductance is the *siemen,* S. The resistance, and thus conductance, depend upon the material, the cross-section, s and the length, l, of the conductor. When the resistance is calculated on the basis of unit length and unit cross-section, the quantity called *resistivity,* ρ, is defined

$$\rho = R\ s/l$$

The inverse of the resistivity is the *conductivity,* $\kappa = 1/\rho$. For solid substances and single component liquids, resistivity is a characteristic of a particular substance. A highly conducting solid such as copper has a resistivity of 1.725×10^{-8} Ωm (at 20°C) whereas a poorer conducting solid, such as carbon has a resistivity of 7.10^{-6} Ωm.

The resistivity of all pure metals increases with temperature whereas the resistivity of carbon, insulating materials and electrolytes decreases with temperature. The resistance of electrolytes depends largely on the ionic components and their concentrations in solution. Generally over moderate concentration ranges, resistivity decreases with increasing concentration but reaches a minimum value at a particular, moderately high concentration. For example, aqueous solutions of sodium hydroxide have minimum restivities at a concentration of approximately 6 kmol m^{-3} (at approximately 20°C).

Leading to: *Electrical conductivity*

K. Scott

OHNESORGE NUMBER (see Atomization)

OIL FILM TECHNIQUE (see Visualization of flow)

OIL GASIFICATION (see Gasification)

OIL REFINING

Following from: *Oils*

Oil refining is the process whereby *crude oil* is split—refined—into commercially useful products. **Distillation** is the primary means of separating the constituents, which may be sold directly, or be used as feedstock for further processes. These secondary processes may involve separation by extraction or may use catalysts to change the chemical species such that a further range of products are produced.

The objective of refining crude oil is to meet the marked demand in the most economical manner. The nature of the market (for example, whether there is strong demand for *motor gasoline* or for *kerosene*) and the relative values of the individual products (their marginal values) largely dictate the mix of refinery processes that are used. The relative values of products differ, with the high value materials typically occurring in the mid-boiling range materials such as motor gasoline, kerosene and diesel fuel. Values are influenced by geographical location, market profile and by the seasons.

Whereas a few refineries process a single source crude, most process a range of crude oils, the choice being dependent partly on the price of the crude cargo, and partly on the ability to most economically meet the product range demand.

Figure 1 illustrates the flow scheme of processing units that might be found in a complex refinery, although not all refineries necessarily operate with all the units shown.

Crude oil entering the refinery is first distilled in the crude distillation unit, operating at nominally atmospheric pressure with a crude preheat furnace temperature in the range 370–380°C, any higher temperature causing excessive thermal cracking. The residue from the atmospheric distillation stage is then redistilled under a vacuum of typically 10–50mb absolute to recover heavier distillates using a heater temperature up to approximately 400–440°C depending on crude type and distillate demand.

Liquid petroleum gases (LPG) are either sold directly, or converted to heavier high octane products for motor gasoline. **Propane** is converted by catalytic polymerisation, and **butane** by isomeration or alkylation, to produce high octane liquid products for motor gasoline blending. *Naphtha* is catalytically converted (reformed) to convert naphthenes into higher octane aromatic components, also for motor gasoline. (see also **Hydrocarbons**) Gas oil is used, after sulphur removal, for diesel fuel and heating oil.

Distillates from the vacuum distillation unit are used variously as hydrocracker or catalytic cracker feedstock, both these units giving products ranging from LPG gases to heavy gas oils. Heavy distillates from vacuum distillation are also used as direct feed for the manufacture of lubricating oils in which case the distillates are further processed to remove aromatic components and wax. The residue from vacuum distillation is normally used as fuel, road bitumen or petroleum coke.

Additional processes are used to remove sulphur compounds from both liquid and gaseous streams. Between approximately 2–8% of the feed on an energy basis, depending on refinery complexity, is used to provide fuel to the refinery.

Further Reading

Nelson, W. L. *Petroleum Refinery Engineering,* McGraw Hill.

D.W. Reay

OIL REFRIGERANT MIXTURES, BOILING HEAT TRANSFER (see Multicomponent mixtures, boiling in)

OIL PRODUCTION (see Enhanced oil recovery)

OILS

Following from: *Petroleum*

Oil is the generic name given to the products from petroleum refining that are liquid at ambient conditions, other products being heavy residues and *bitumen* which are solid at ambient conditions, and small quantities of gas and coke. Other mineral oils may be derived from coal.

Oil products are primarily characterised by their boiling range, with their specific properties determining their suitability for different applications. These may be finished products in their own right, after treatment to remove sulphur, etc., or as feedstock for further processing. Different oils need to have different properties and meet different quality criteria depending upon their ultimate use and, often, geographic destination. The nomenclature and specific properties needed reflect these requirements.

The lightest oils, termed *gasolines,* have a boiling range of C5 to approximately 90°C and are used primarily as a component of *motor gasoline,* or *petrol.* Gasoline itself is not a high octane material, but is a mixture of several components which provide the desired properties of volatility and **Octane Number** etc. Thus, naphthas, boiling at approximately 90–160°C, are catalytically reformed to convert naphthenes into aromatics which contribute high octane gasoline components (See **Hydrocarbons**). Other high octane blending constituents are obtained from converting lighter materials (C3–C5 components) into different chemical species.

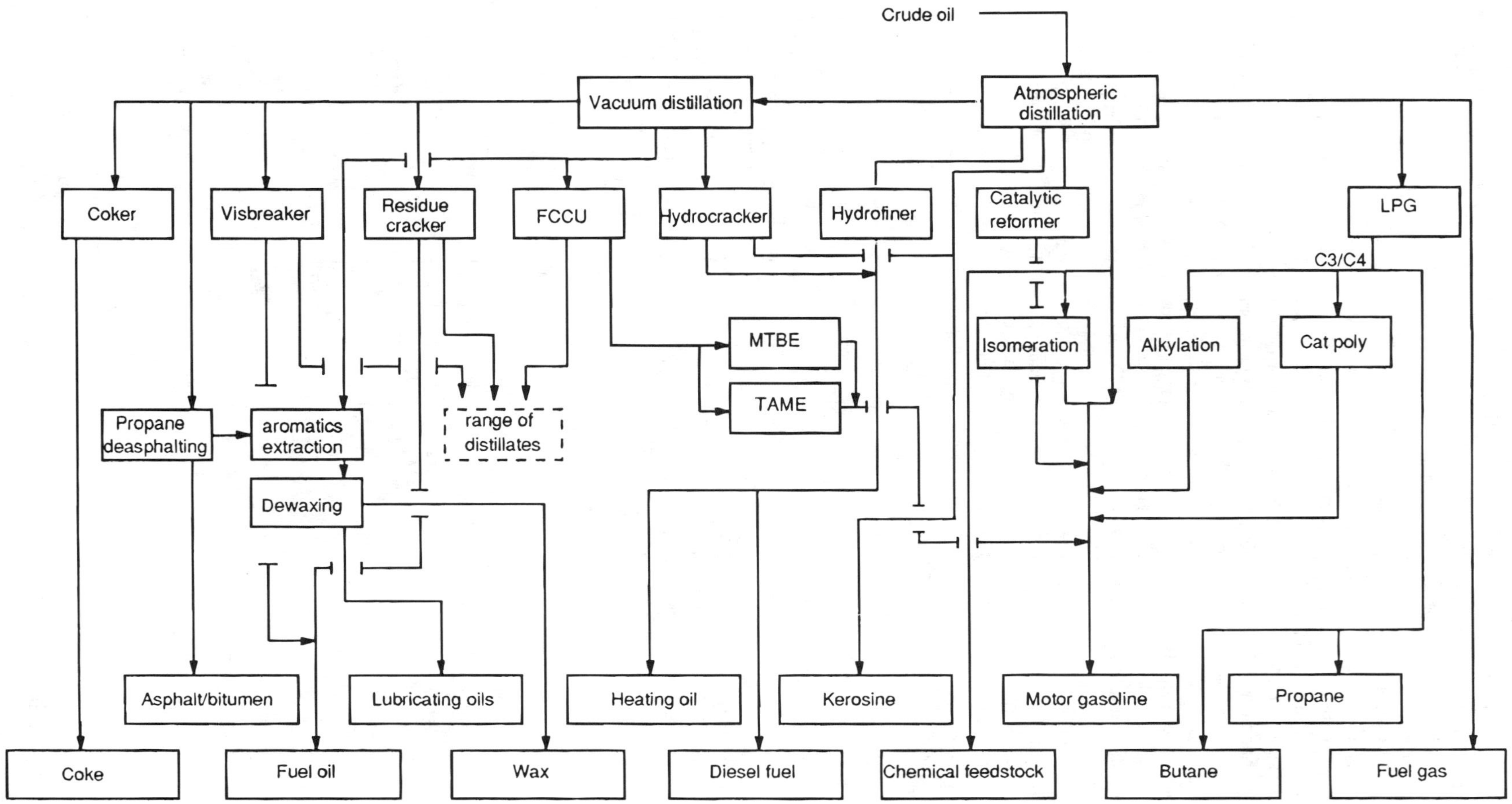

FIGURE 1. The interaction of major processing steps in a complex oil refinery.

Typically, such products will be alkylates, isomerates, butyl and methyl ethers etc, the latter two compounds commonly known as TAME and MTBE, respectively. These components in particular have significantly reduced the use of *lead based additives* for enhancing octane number as a result of environmental pressures but with the attendant increase in processing costs. Straight run (unreformed) naphthas are also used extensively as chemical plant feedstock to produce a wide range of plastics.

Kerosene boils in the range 160–240°C and is primarily used as aviation fuel (Jet A-1). Although this fuel is required to meet numerous stringent property specifications, one of the critical properties is freezing point which effectively limits the highest boiling point components that it may contain. Fluid density is approximately 700kg/m^3. In some locations, kerosene is used as a burning fuel, in which case rather less stringent property specifications apply.

Gas oils boil up to approximately 360°C, with different types having different end points. The lightest products are used as *diesel fuel* and *light heating oil,* the heaviest used as components for industrial and *marine fuel oils;* heavy marine fuels also contain crude oil residues. Density and sulphur content are important properties, as are those properties affecting flow. In particular diesel, fuel needs to meet a *cloud point specification* which is an indication of the wax content and thus, the ability to remain fluid in cold climates. Different specifications are used for different geographical locations and seasons.

Oils boiling up to approximately 500–550°C are used as *lubricating base oils,* from which, after extensive processing and treatment to control properties such as oxidation stability, colour, wax content etc., a wide variety of finished lubricating oils are formulated. These oils are normally classified in terms of a viscosity range, an important property being the variation of viscosity over a given temperature range. To some extent this is a function of crude oil origin but processing criteria also has an effect.

The heaviest materials are oils of extremely high viscosity, virtually solid at ambient conditions, and are used for bituminous based products.

References

Physical Principles of Oil Production. International Human Resources Development Corporation, Boston, USA ISBN 0-934634-07-6.

Leading to: *Oil refining*

D.W. Reay

OIL WATER SEPARATION (see Hydrocyclones)

OIL WELLS (see Enhanced oil recovery)

OLDROYD FLUIDS (see Non-Newtonian fluids)

ONCE-THROUGH BOILERS

These boilers, in contrast to recirculation or natural circulation units, are characterised by continuous flow paths from the evaporator inlet to the superheater outlet without a separation drum in the circuit. They are almost exclusively used for steam production in connection with utility electricity generation, and have been the most popular design in Europe and Scandinavia for many years.

A number of different detailed designs are offered by Siemens KWU (owners of the **Benson Boiler** patent), Sulzer and, in recent years, Japanese and American companies. All once-through boilers incorporate relatively small bore evaporator tubes (usually about 25mm bore) which are generally arranged in a spiral fashion to form the furnace envelope. Two principal options are adopted for the section of the boiler after the furnace. Either a two-pass arrangement with a vestibule cage and downpass may be used (as shown in **Figure 1**) or a tower concept with the superheat, reheat and economiser sections above the furnace (as shown in **Figure 2**). The walls and roof of the boiler may consist of tubes forming part of the evaporator or superheater circuits and inside these are mounted banks of tubes acting as the principal superheat, reheater and economiser circuits.

Once-through boilers are generally associated with high pressure operation and the feed water enters at high sub-critical (>180

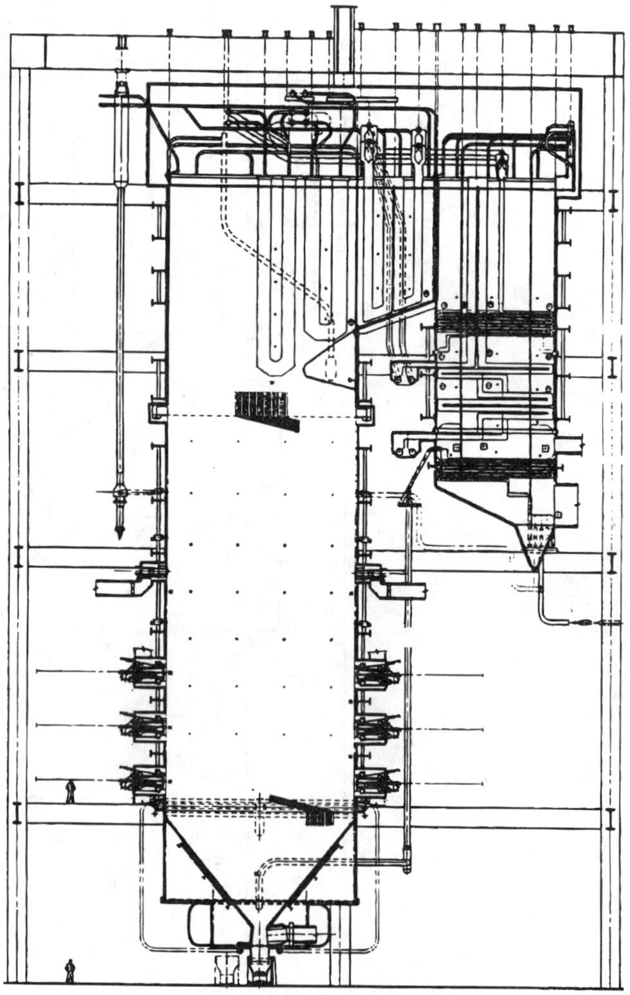

Figure 1. A typical Babcock two-pass once-through utility boiler. (Courtesy of Mitsui Babcock Energy Ltd.)

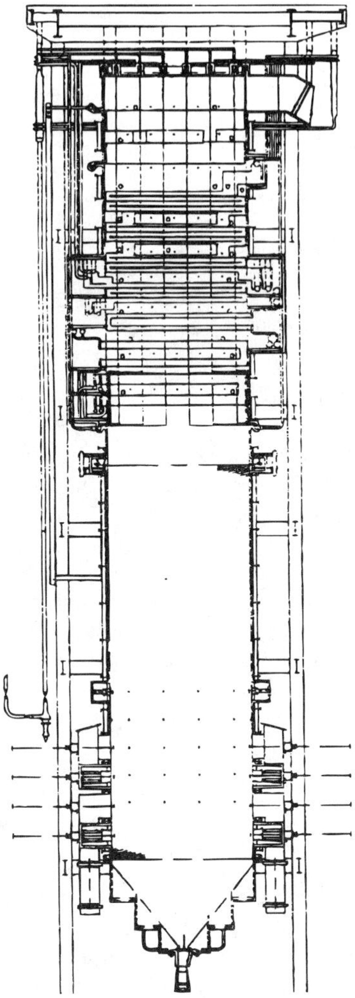

Figure 2. A typical Babcock once-through tower boiler. (Courtesy of Babcock Energy Ltd.)

bar) or supercritical pressure whilst superheated steam leaves at a pressure some 20–30 bar lower. Sliding pressure operation is adopted to accommodate requirements of part-load running.

Furnaces may be fired by burners mounted in the front wall or by opposed firing with burners, normally, in the front and rear walls or by tangential firing achieved by slot-burners mounted in the corners to create a circulating flow which is claimed to have advantages for suppressing pollutant formation.

Because the water is evaporated to high quality in a once-through boiler, it is particularly important to guard against dryout occurring in high heat flux zones or to take other precautions against the phenomenon being associated with burnout. In spiral furnace boilers operating at sub-critical pressures, where dryout would take place at about 40% quality, it is possible to arrange for this position to coincide with an area of reduced heat flux above the primary combustion zone, such that the temperature rise in the tubing at dryout is limited. An alternative solution is to use rifled-bore tubing which, by creating centrifugal forces, causes more of the liquid phase to remain in contact with the tube wall, thus delaying dryout to higher qualities and/or enabling lower water velocities to be adopted.

A.E. Ruffell

ONCE THROUGH EVAPORATORS (see Tubes and tube bundles, boiling heat transfer on)

ONCE THROUGH STEAM GENERATORS (see Steam generators, nuclear)

ONSET OF NUCLEATE BOILING (see Rod bundles, heat transfer in)

OPACITY (see Optical particle characterisation)

OPEN CHANNEL FLOW

Following from: *Hydraulics*

In contrast to pipe flows, open channel flows are characterised by a free surface which is exposed to the atmosphere. The pressure on this boundary thus remains approximately constant irrespective of any changes in the water depth and the flow velocity. These free-surface flows occur commonly in engineering practice, and include both large-scale geophysical flows (rivers and estuaries) and small-scale man-made flows (irrigational channels, drainage channels and sewers). Although many of the traditional examples are of primary interest to the civil engineer, the underlying theory of open channel hydraulics is appropriate to any free-surface flow. In general, these flows may be laminar or turbulent, steady or unsteady, and uniform or varied. However, as in the case of pipe flow, this general class of problems may be subdivided into two distinct groups. The first involves significant changes in the water depth over relatively short channel lengths. These are classified as "rapidly varying flows," and are largely unaffected by shear forces. In contrast, the second group involves less rapid changes in the water depth (occurring over longer distances), and are classified as "gradually varying flows." This latter group may be significantly affected by shear forces.

Most elementary open channel hydraulics is based upon three underlying assumptions: (a) the fluid is homogeneous and incompressible; (b) the flow is steady; and (c) the pressure distribution is hydrostatic at all control sections. Although there are important exceptions, notably the inhomogeneity caused by the entrainment of air in a high-speed flow or the unsteadiness associated with the propagation of flood waves or tidal bores, these assumptions are widely applicable and lead to important simplifications of the conservation equations. In particular, if the pressure distribution is assumed hydrostatic at all control sections, this implies that the *streamlines* are straight, parallel, and approximately horizontal, and that there are no pressure gradients due to the curvature of the flow. In this case the hydraulic gradient line, or *piezometric line,* is the same for all streamtubes and is co-incident with the free surface. This result accounts for the wide application of the energy line —hydraulic gradient line as a means of describing an open channel flow.

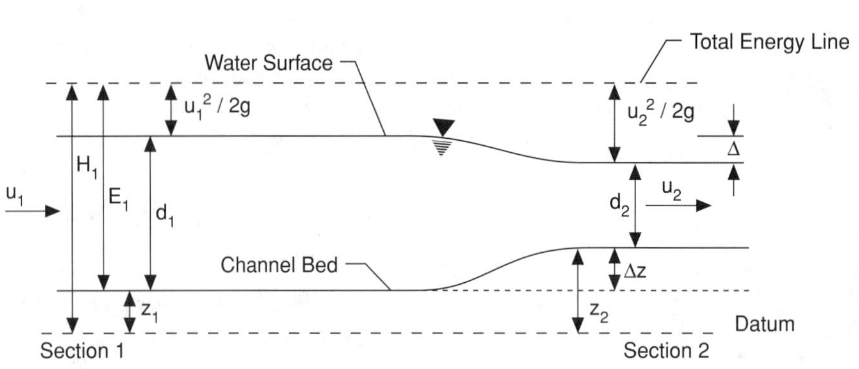

Figure 1a. Transitional flow.

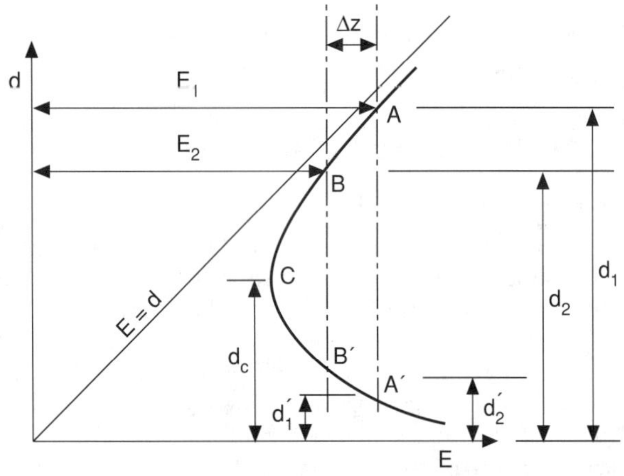

Figure 1b. Specific energy curve (not to scale).

To emphasize the importance of the unconstrained free-surface boundary, a transitional flow involving a step in a rectangular channel is considered (**Figure 1a**).

If the volume flow per unit width is $\dot{V}$, mass conservation defines the velocity (u) in terms of the water depth (d) such that $\dot{V} = u_1 d_1 = u_2 d_2$. Furthermore, if the energy head (E) is expressed in terms of the specific head (or "specific energy") measured relative to the channel bed, E may be defined in terms of d and $\dot{V}$ alone:

$$E = d + \frac{\dot{V}^2}{2gd^2}$$

where the final term is an alternative representation of the velocity head. **Figure 1b** concerns the variation of E and d for a given value of $\dot{V}$, and demonstrates that if E and $\dot{V}$ are fixed there are (in general) two potential solutions for d. In the present case if there are no energy losses, **Bernoulli's Theorem** gives $E_2 = E_1 - \Delta z$, and thus the state (E_2, d_2) could be represented by point B or point B′ on the specific energy curve.

The depths d_2 and d_2' both represent physically realistic solutions, and are often referred to as "alternate depths" corresponding to two different flow regimes. Although the specific energy in each case is the same (E_2), point B corresponds to a deep slow flow whereas point B′ describes a shallow fast flow.

Point C on the specific energy curve (**Figure 1b**) defines the common boundary of these two flow regimes, and represents the so-called **critical flow.** This state, which is usually defined in terms of a critical depth (d_c), represents the minimum specific energy for a given volume discharge. If the water depth increases above the critical depth ($d > d_c$) the specific energy also increases due to the flow-work associated with the hydrostatic pressure. In contrast, if the water depth reduces below the critical depth ($d < d_c$) the kinetic energy (or velocity head) accounts for the increase in the specific energy. The critical flow may also be interpreted as producing the maximum flow for a given specific energy.

If $dE/dy = 0$ at the critical limit, it follows that the critical depth (d_c) and the critical velocity (u_c) are given by:

$$d_c = \frac{2}{3}E, \qquad u_c = \sqrt{gd_c}$$

These definitions allow the classification of the flow regimes noted above. If $d > d_c$ (or $u < u_c$) the regime is described as subcritical (or *subundal*) flow; whereas if $d < d_c$ (or $u > u_c$) supercritical (or *superundal*) flow is said to occur. A close analogy exists between these definitions of an open channel flow and the distinction of subsonic or supersonic flow in a compressible fluid. Indeed, this analogy can be further extended since the critical velocity u_c defines the speed of a surface wave in water of depth d_c. As a result the **Froude Nunber** (Fr), defined by $Fr^2 = u/(gd)$, defines the ratio of the free stream velocity to the surface wave velocity. In the context of open channel flows $Fr < 1$ implies sub-critical flow, $Fr > 1$ supercritical flow, and $Fr = 1$ critical flow. This approach is directly analogous to the **Mach Number** (M) description of a compressible flow. This defines the ratio of the gas velocity to the sonic velocity, such that $M < 1$ implies subsonic flow and $M > 1$ supersonic flow.

Returning to the transition problem (**Figure 1a**), the development of the water surface over the downstream step is dependent upon the "accessibility" of the two flow regimes. The specific energy curve (**Figure 1b**) provides guidance in this respect. If the upstream state is defined by A, and the discharge per unit width is constant, any changes must take place along the E-d curve shown on **Figure 1b**. To move from A to B is clearly possible, but to move from B to B′ requires a reduction in the specific energy below E_2, and thus cannot be justified in the present transition.

As a result, the specific energy curve suggests that if the upstream flow is subcritical (point A), an increase in the bed elevation will produce a reduction in the water depth from d_1 to d_2. Not only is this result somewhat unexpected, the accessibility of the various flow regimes is dependent upon the upstream condition. For example, if the initial conditions were described by A′ (rather than A) the upstream regime would be supercritical. In this case, a similar accessibility argument suggests that an increase in the bed elevation would produce an increase in the water depth from d_1' to d_2', with the final energy state represented by B′ on the E-d curve. This notion of the accessibility of the open channel flow regimes is analogous to a similar process in thermodynamic theory, in which the accessibility of various gas states is not only dependent upon the change in the energy level, but also on the required entropy change.

We have already noted that the distinction between subcritical and supercritical flow is dependent upon the velocity of a surface wave or disturbance. This has important implications for the control of any open channel flow, and in particular the estimation of water levels for a given volume discharge. If a flow is supercritical a disturbance at the water surface is unable to travel upstream (relative to a stationary observer) because the velocity of the flow exceeds the wave velocity ($Fr > 1$). As a result, all supercritical flows are controlled from upstream, and may be considered "blind" to any changes which arise downstream. In contrast, subcritical flows ($Fr < 1$) are such that a surface disturbance can either travel upstream or downstream, and as a result these flows are typically controlled from downstream.

In its simplest form, a control structure is designed to change the water depth to (or through) the critical depth (d_c), so that the discharge is fixed relative to the depth. In practice, most control structures accelerate a subcritical flow, through the critical regime, to produce a shallow fast supercritical flow. The most common examples of such structures include sluice gates and **Weirs** (**Figures 2a** and **2b**).

To include more than one effective control within an open channel, the supercritical flow produced by an upstream control must be reconverted to a subcritical flow. This is usually achieved by a *hydraulic jump* (or stationary bore) in which the characteristics of the subcritical flow are determined by a second downstream control. These events are associated with large energy losses, and are often used as an effective means of dissipating unwanted kinetic energy downstream of an overflow (spillway) or underflow (sluice gate) structure. The hydraulic jump is in many respects analogous to a shock wave arising within a compressible flow. For example, whereas the hydraulic jump provides a transition from supercritical to subcritical flow, the shock wave involves a transition from supersonic to subsonic flow. In both cases there is a critical velocity below which these transitions cannot occur, and both processes involve an increase in entropy. Indeed, in the case of a hydraulic jump the increase in entropy per unit mass is proportional to the cube of the depth change, whereas in a shock wave this increase is proportional to the cube of the pressure difference (provided this is small). Specialist texts describing open channel flow are given by Chow (1959), Henderson (1966), Francis and Minton (1984) and Townson (1991).

References

Chow, V. Y. (1959) *Open Channel Hydraulics*. McGraw Hill Inc., New York.

Francis, J. D. R. & Minton, P. (1984) *Civil Engineering Hydraulics*. Arnold, London.

Henderson, F. M. (1966) *Open Channel Flow*. Macmillan, London.

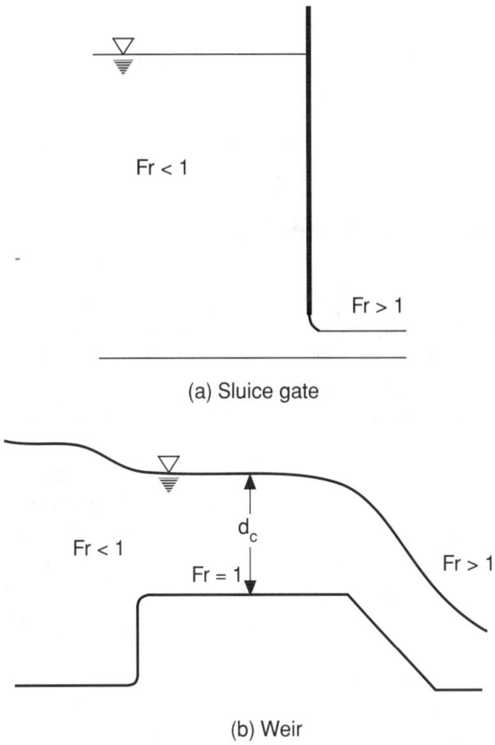

Figure 2a,b. Open channel control structures.

Townson, J. M. (1991) *Free Surface Hydraulics*. Unwin Hyman, London.

Leading to: *Weirs*

C. Swan

OPEN COLUMN CHROMATOGRAPHY (see Gas chromatography)

OPEN CYCLE GAS TURBINE (see Gas turbine)

OPEN CYCLE MHD GENERATORS (see MHD electrical power generators)

OPEN SYSTEM (see Thermodynamics)

OPERATING SYSTEMS (see Computers)

OPTICAL KINEMATIC METHODS (see Velocity measurement)

OPTICALLY THICK MEDIA, COMBINED RADIATION, CONVECTION AND CONDUCTION IN (see Coupled radiation, convection and conduction)

OPTICALLY THIN MEDIA, COMBINED RADIATION, CONVECTION AND CONDUCTION IN, (see Coupled radiation, convection and conduction)

OPTICAL METHODS, FOR DROPSIZE MEASUREMENT (see Dropsize measurement)

OPTICAL PARTICLE CHARACTERISATION

Following from: Optics

Optical methods are widely used to determine physical properties because they are usually non-perturbing and do not involve taking samples. The techniques available may be broadly classified in terms of the ratio of the particle size to the wavelength of the probing light, expressed through the *particle size parameter* $x = \pi D/\lambda$. If the particles are sufficiently large *photography* may be used. With care, particles as small as 10 μm ($x \approx 50$) can be photographed directly. However, for smaller particles, large magnification is required, which inevitably results in lack of depth of field. Use of a microscope enables particles down to about 1 μm to be observed, but here the depth of field is so restricted that only particles sampled onto a glass slide can be measured. This difficulty can be partly resolved by the use of *holography* which can reconstruct the distribution of particles in space. The resulting image can be studied in any plane, for example, by a digital television system. Reconstruction of images of particles down to about 0.1 μm has been claimed. A number of instruments are available to perform automated image analysis. Both methods need short pulse light sources (lasers for holography) to freeze the particle motion. Particle velocity and trajectory can be obtained by *double flash methods.* (See also **Photographic Techniques, Holograms, Holographic Interferometry** and **Interferometry**)

The above methods have the advantage that the image can be seen and the size and shape obtained directly. However, analysis can be very time-consuming. Difficulties also arise for small particles, and at high concentrations where shadowing can occur. The alternative is to use *light scattering* methods. These can, in principle, be applied to any particle size depending on the technique employed, and analysis can be automated and rapid. However, the analysis is indirect and relies upon a suitable theory to interpret the observations. Because of its convenience, the most widely used is *Mie theory,* but this only describes scattering by a sphere. In reality almost all solid particles are not spherical and neither are many liquid drops. Consideration must be given, therefore, to the errors that the spherical assumption introduces.

A light wave incident upon a particle can undergo two processes; scattering and absorption. The former can be thought of as being due to reflection, refraction and diffraction. The fraction of scattered light power to that incident is the *scattering efficiency* (Q_{sca}). An *absorption efficiency* (Q_{abs}) is similarly defined and the total loss due to both processes is the extinction for which $Q_{ext} = Q_{sca} + Q_{abs}$. Propagation through a cloud of particles is described by the *Beer-Lambert law*

$$\tau = \exp(-K_{ext}L) \tag{1}$$

where L is the path length and

$$K_{ext} = \int_0^\infty n(D)AQ_{ext}dD \tag{2}$$

is the extinction coefficient. A is the cross-sectional area of the particle and n(D) is the particle size distribution function defined such that

$$N = \int_0^\infty n(D)dD \tag{3}$$

is the number of particles per unit volume.

τ is the *transmissivity.* The opposite of this $(1 - \tau)$ is the loss of light, or the *opacity.*

The principle of measuring extinction is illustrated in **Figure 1**. A parallel beam of light transluminates the particle cloud. Unscattered light is brought to a focus at the centre of a small aperture and passes through to a detector. The aperture only transmits scattered light if the scattering angle is less than θ. The definition of extinction efficiency assumes that all scattered light is excluded from the measurement. As the particles become larger, scattering is increasingly concentrated into very narrow angles around the forward direction. Measurement of transmitted light has to be restricted to extremely small angular ranges in order that true extinction is obtained, and a sufficiently small aperture must be chosen to ensure that this is so.

A second difficulty arises in dense particle clouds due to multiple scattering. This is found to happen to $\tau < 0.9$. In this case, the Beer-Lambert law ceases to be true unless, again, all scattered light is excluded. For finite detector apertures, the scattered light collected is added back and must be taken into account unless it is a very small fraction of the losses. For dense clouds, multiple scattering theory becomes necessary to describe the result.

The basic set up for measuring scattering is shown in **Figure 2**. In this case, the size of the volume within the particle cloud that is seen by the detector is defined by the aperture and the depth of field of the lens along θ. The aperture of the lens determines the angular range over which scattered light is collected. An alternative scheme is to collect parallel light in the fashion of **Figure 1**, but with the lens and aperture rotated through the angle θ. In this case, the aperture determines the angular range.

Light scattering depends upon the particle size, refractive index, shape and concentration. In principle all these parameters can be measured, though the most effort has been directed to determination of size. The technique to be used depends upon the size. The very

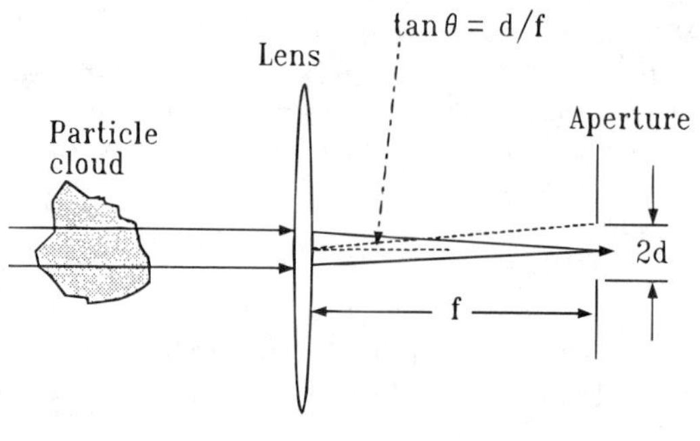

Figure 1. Schematic diagram showing extinction measurement.

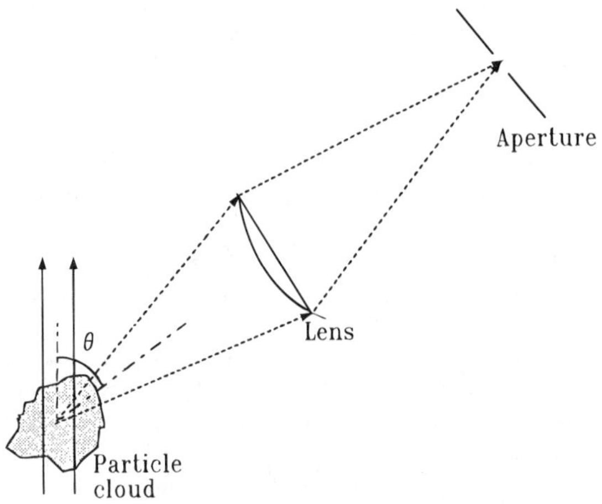

Figure 2. Schematic diagram showing scattering measurement.

smallest particles ($x \ll 1$) fall into the *Rayleigh scattering* régime where the variation of scattering with angle or wavelength is independent of size. The scattered intensity is proportional to NV^2 where V is the particle volume. For absorbing particles, the extinction is proportional to NV. The ratio of these yields V, and N is also be found. For non-absorbing particles, it is necessary to measure the equivalent refractive index of the particle cloud which is also proportional to NV. Typical examples of such particles may be smokes or the very early stages of nucleation and condensation.

Another method appropriate to small particles is *photon correlation spectroscopy.* Particles suspended in a fluid are subjected to collisions with molecules. If they are sufficiently small, this results in **Brownian Motion.** This random velocity causes a frequency broadening of the incident light due to the *Doppler effect.* The increase in bandwidth results in a reduction in temporal coherence and a consequent decrease in the autocorrelation when the scattered signal is compared with itself after increasing time delays. The decay is exponential with a rate directly related to the diffusion coefficient, which is inversely proportional to the particle size. In turbulent flow the technique is complicated by the random velocity fluctuations, though it has been succesfully applied to soot in flames.

As the particle size increases, so does the ratio of forward to backscatter and the extinction becomes size-sensitive. For sizes less than about 1 μm, these vary in a simple monotonic way. The ratio of scattered intensity at two symmetric angles (commonly 45° and 135°) can be used, which is the *disymmetry.* The alternative measurement of the variation of extinction with wavelength is the *spectral extinction method.*

For particles greater than 1 μm, forward scatter techniques are most widely used. This is because it is recognised that these are insensitive to shape and refractive index. They are also simple, which is especially true for particles above about 4 μm where *Fraunhofer diffraction* can be used to describe the scattering. It is preferable, however, to use Mie theory and to supply a refractive index. The particles are illuminated by a parallel laser beam and the scattered light is detected in the focal plane of a receiving lens, commonly by ring photodiodes, as seen in **Figure 3**. Unscattered light is brought to focus at O and light scattered through the angle θ at P. The angular distribution of the scattering is characteristic

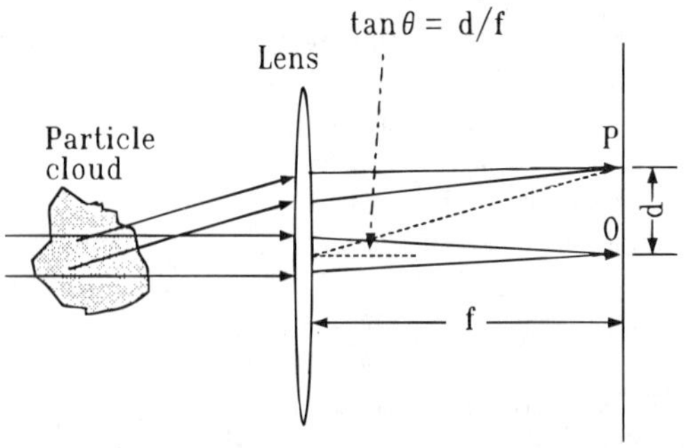

Figure 3. Schematic diagram showing diffraction measurement.

of the particle size distribution. By using a range of lens focal lengths, it is claimed that a size range from 1 μm to 1800 μm can be covered. The method is normally limited to transmissivities greater than 50% due to multiple scattering, though ways are being developed to deal with this.

All the above methods examine clouds of large numbers of particles which have a distribution of sizes. These have been recovered by matching the scattering measurements against predicted results for a range of assumed distributions. Typically, two parameter distributions are used for simplicity. The most common of these are the *Rosin-Rammler* and log-normal distributions. However, there is no guarantee that these represent the true situation, especially if the actual distribution is multi-disperse. Thus, attention is now switching to "model independent" techniques of *direct inversion.* Perhaps the most widely used of these is that of Phillips and Twomey.

The number density of the particles is readily obtained if the size distribution is known. This is most readily achieved by measurement of the transmissivity. For large particles ($x > 10$) it is found that $Q_{ext} \cong 2$, a constant. Then it can also be demonstrated from Equations 1 and 2 that

$$\tau \cong \exp(-3f_v L/D_{32}) \quad \textbf{(4)}$$

where f_v is the particle volume fraction and D_{32} is the *Sauter mean diameter.* However, the measurement of transmissivity has the difficulty already outlined above that forward scatter must be excluded. Also, if the concentration is low it becomes difficult to measure a small change in a high intensity. In that circumstance, it is preferable to measure the scattered intensity at some large angle (as in **Figure 2**) where there is no interference from the incident light. This technique is *nephelometry.* The intensity is proportional to the particle number concentration multiplied by the volume of the test space. This latter component must be found either from geometry or by calibration. There is the added complication that the collected power depends upon the measurement angle and the aperture of the lens.

The alternative to making measurements on clouds of particles is to examine them individually; sometimes referred to as *particle counting.* In the simplest instruments of this type, particle laden fluid is drawn though an enclosure containing the light beam and collecting optics. The use of mirrors ensures that scattered light is collected over a very large solid angle. The size is obtained from a measurement of the received power. In this way particles can be measured down to 0.1 μm diameter. Ultimately there is a limit due to noise caused by scattering from gas molecules. Because the particles arrive at random they are governed by *Poisson statistics* and the error in count is proportional to the square root of the number. This means that to ensure high accuracy in the size distribution very large numbers of particles have to be counted. The smallest concentration that can be tolerated is determined by the allowable count time and the flow rate. Typical flow rates are of the order 0.03 $m^3\ min^{-1}$. The maximum concentration is governed by the requirement that there should only be one particle at a time in the test space. Typically, this is of the order $10^{11}\ m^{-3}$.

When the light source is a laser a difficulty arises due to the Gaussian beam profile. A small particle passing through the centre of the beam, where the intensity is highest, will result in the same intensity as a large particle passing through the outer parts of the beam. This is the *trajectory problem.* There are three ways to resolve this. One is to filter the beam so that it has a uniform intensity profile; the so-called *top hat profile.* The second is to provide a second small diameter laser beam which acts as a pointer to the centre of the main beam. The two may be discriminated using either differences in wavelength or polarisation. The third method is *deconvolution,* in which a calibration leads to a matrix inversion procedure to recover the true size distribution.

Particle counting yields a temporal average, whereas cloud methods give a spatial average. In order to convert the temporal average into a spatial one, the particle velocities are needed. If the beam profile has a variation which is known, then the particle may be timed between two points on the profile. The alternative is to use *laser Doppler velocimetry* (LDV). (See also **Anemometers, Laser Doppler**). The principle of this is that two laser beams cross to produce an interference pattern at the test space. A particle crossing this pattern generates a signal which varies sinusoidally in time, the frequency of which is simply related to the velocity. The sense of the velocity may be established by frequency shifting one of the beams so that the interference fringes move through the test space.

The general form of the Doppler signal is

$$I = I_1 + I_2 \cos(\omega t - \phi)$$

The particle size can be established from the mean scattered intensity (I_1), as for particle counters. However, methods which rely on the measurement of an absolute quantity have a number of disadvantages, including the need for calibration with particles having known properties. Also, in many systems, access is only available via windows. If these become soiled, their transmissivity is reduced, affecting both the incident and scattered intensities. For these reasons, relative methods are preferable. One such variable is the ratio I_2/I_1, which is the *visibility.* While this has been shown to work satisfactorily in a limited range of circumstances, it is normally restricted to particles which are smaller than the fringe spacing because the visibility is not monotonic beyond that. Also, if the particle does not pass exactly through the centre of intersection of the laser beams, the visibility is distorted. In dense systems scattering along the paths of the incident beams can result in an imbalance of their intensities, which causes a reduction in visibility not related to particle size.

The third term in the Doppler signal is the phase (ϕ). It has been found that this increases linearly with particle size over a wide range. The particle can be larger than the fringe spacing, which means that small test volumes can be retained giving high resolution and enabling high concentrations to be measured. There is one complication due to the fact that a phase change of $2n\pi$, where n is an integer, cannot be discriminated. This problem has been resolved over a limited range by the use of multiple detectors, each with a small angular offset. Typically, three detectors are used. This technique is *phase Doppler anemometry* (PDA). Currently, it is claimed that the size range 1 μm to 10 mm can be studied at concentrations up to $10^{12}\ m^{-3}$. (For further information see **Anemometers, Laser Doppler).**

Particle counting methods have excellent spatial resolution, but the size distribution function is built up over a period of time. The concentration is established by knowledge of the area of the test space through which the particles flow and the volumetric flow rate. On the other hand, cloud methods yield a size distribution almost immediately, and so are capable of measuring rapid temporal fluctuation. However, they have very poor spatial resolution. Further, by measurement close to the forward direction, cloud methods can be made insensitive to unknown shape and refractive index. Particle counters are sensitive to both of these, and phase Doppler instruments normally incorporate a technique for indicating particle non-sphericity.

The fact that scattering depends upon shape and refractive index implies that it can be used to measure these parameters. There is a growing interest in these areas. Shape can be determined provided that it can be described in terms of simple variables—such as axial ratio. This has been applied, for example, to the sizing and counting of fibres. The full complex refractive index has been measured, and in combustion studies the real refractive index of drops has been used remotely to indicate their temperature.

Further Reading

Allen, T. (1990) *Particle Size Measurement* 4th ed., Chapman and Hall, London.

Bohren, C. E. and Huffman, D. R. (1988) *Absorption and Scattering of Light by Small Particles,* Wiley-Interscience, New York.

Heitor, M. V. Stårner, S. H. Taylor, A. M. K. P. and Whitelaw, J. H. "Velocity, size and turbulent flux measurements by laser Doppler velocimetry". *Instrumentation for Flows with Combustion,* A. M. K. P. Taylor, ed., pp 113–250, Academic Press, London.

Jones, A. R. (1993) Light scattering for particle characterisation, *Instrumentation for Flows with Combustion,* A. M. K. P. Taylor, ed., pp 323–404, Academic Press, London.

Kerker, M. (1969) *The Scattering of Light,* Academic Press, New York.

A.R. Jones

OPTICAL PROBES, FOR LOCAL VOID FRACTION (see Void fraction measurement)

OPTICAL PYROMETERS (see Temperature measurement, practice)

OPTICAL TECHNIQUES, FILM THICKNESS (see Film thickness determination by optical methods)

OPTICAL TECHNIQUES OF FLOW VISUALISATION (see Tracer methods)

OPTICS

Optics is a branch of science that deals with light propagation in vacuum and in various media. By "light," it is usually understood a radiation visible to the human eye (spectral range 400–700 nm), although the laws of optics are valid for a much larger wavelength range. From the point of view of classical physics, light is transverse electromagnetic waves; the propagation velocity of such waves (light velocity) in the vacuum, c, is constant in any coordinate system and represents one of the fundamental constants. The velocity of light in a medium, v, is always less than in a vacuum. The ratio $c/v = n$ is called the *refractive index*. From the point of view of quantum physics, light is a flow of "light quantums," or photons. Radiation with a single wavelength (λ) which is described by an infinite sinusoid, is called *monochromatic*. Near-monochromatic radiation is generated by some types of lasers. Nonmonochromatic radiation can be represented as a superposition of monochromatic waves . If the phase of monochromatic vibration is the same in each point of a certain surface, then this surface is called a *wave surface*, or *wave front*. Usually, the form of wave front varies when the wave transmits through a medium and/or optical system.

An important concept in optics is *radiation coherence*, which means a conservation of constant phase difference of electric (or magnetic) field at two different points in the radiation field during a time period Δt. Spatial and temporal coherence should be distinguished. If a pair of points is chosen on a wave front surface (which at time instant $t = 0$ is, by definition, a locus of equiphase points) they are said to be spatially coherent during time interval Δt_s. If a pair of points is chosen along the line of radiation propagation, they are said to be temporarily coherent during time interval Δt_t. A value $c\Delta t_t$ is named the coherence length.

Polarization of monochromatic radiation is characterized by the spatial-temporal behaviour of the electric field vector, $\vec{E}$.

If the Cartesian components of vector $\vec{E}$ have a form:

$$\vec{E}_x = \vec{e}_x E_x \sin \omega t, \qquad \vec{E}_y = \vec{e}_y E_y \sin(\omega t + \varphi)$$

then a vector $\vec{E} = \vec{E}_x + \vec{E}_y$ rotates in a transversal plane with a rate of angular motion, ω, relative to a center of coordinates and its end describes an ellipse (elliptic polarization, **Figure 1**). At $\varphi = \pi/2$ and $E_x = E_y$, an ellipse degenerates to a circumference (circular polarization); at $\varphi = 0$ it degenerates to a straight line (linear polarization). Natural ("white") light is nonpolarized. It represents a collection of light waves with all possible $\vec{E}$ vector directions, simultaneously existing and replacing each other at random. Partially polarized light is characterized by a predominant vibration direction, but not exclusively. After transmittance of natural light through some crystals, it polarizes partially or completely. Polarization also takes place after light reflection or refraction at a boundary of two dielectrics under the condition of oblique incidence of a light ray on the boundary.

Typically, light propagation is described by the wave equation, which follows from electromagnetic field theory (Maxwell's equations). This rigorous approach is the basis of wave (physical) optics. However, many very important results from the practical point of view can be obtained more simply with the use of a geometrical optics approach.

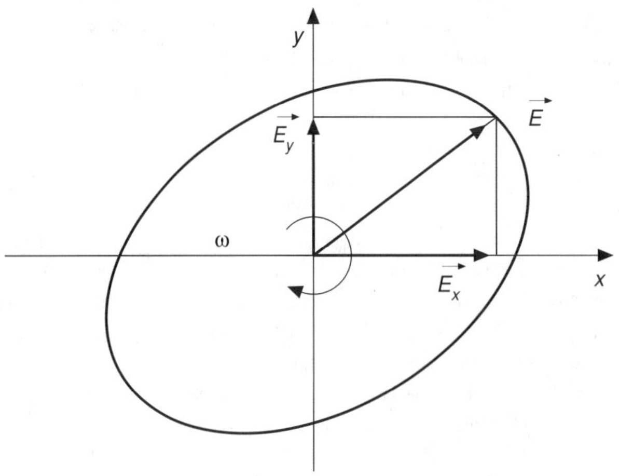

Figure 1. Elliptic polarization.

Geometrical optics is a name of system of optics in which the limiting case of $\lambda = 0$ is considered . In this case, the optics laws can be formulated in terms of geometrical concepts. Geometrical optics deals with infinitely fine light beams, namely light rays, along which light propagates. A boundary between light and shadow is considered as absolutely sharp. Light intensity changes occur due only to change of the light beam cross-section. The main laws of geometrical optics are as follows:

1. Law of rectilinear light propagation: in a uniform medium the light rays are straight lines.
2. Law of independent action of light beams: action produced by one beam doesn't depend on the presence of another beam.
3. Law of light reflection: the incident ray, reflected ray and the normal to the reflecting surface all lie on the same plane, and the angles between these rays and the normal are equal, i.e. an angle of incidence is equal to an angle of reflection.
4. Law of light refraction: if ray transists from one medium to another then it changes its direction (refracts). The incident ray, refracted ray and a normal to a separating surface the two media at the point of ray incidence all lie on the same plane. The angle of incidence, γ, and angle of refraction, θ, (respective to the normal) are connected to each other by equation $\sin \gamma / \sin \theta = n_{\theta\gamma}$, here $n_{\theta\gamma}$ is an index of refraction of the second medium with respect to the first. If it is assumed that $n_{\theta\gamma} = -1$ then it leads to the law of reflection $\gamma = -\gamma$. Thus, any equation for a calculation of a refracting system can be applied to a reflecting system.

Commonly, an index of refraction depends on the wavelength. This dependence is called dispersion; it explains light decomposition to a colour spectrum after light transmittance through a prism.

An optical system for image formation or for object illumination consists of a set of lenses and mirrors. A lens is called thin if its maximal thickness is significantly less than the curvature radii of its surfaces. A straight line which passes through the lens center and is normal to both lens surfaces is called the main optical lens axis. A point where all rays parallel to the main optical axes converge is called the *focus* and the length between focus and lens center is called *focal length*, f. An equation of a thin lens is $-1/a_1 + 1/a_2 = 1/f$; here a_1, a_2 are the lengths from lens center to an object and its image, respectively. The value of a is assumed positive if the direction with respect to the lens center coincides with the light propagation direction, and negative in the opposite case (**Figure 2**). For off-axes point A (object AB correspondingly) image formation of any two rays amongst three special rays are used. Ray 1 is parallel to the main optical axes, and after refraction in a lens, it passes across a rear focus F_2. Ray 2 passes across a center 0 of a thin lens without refraction. Ray 3 passes across a front focus F_1 and after refraction it passes parallel to the main optical axes. The point where these rays intersection, A′, is an image of point A. If the rays do not intersect at A′ (this case corresponds to object location between lens and its front focus) then their extensions intersect to the left of a lens and the A image is a virtual one.

An optical system is called ideal if an image of a point also represents the point. Such an image is called stigmatic. Real optic systems have errors, or aberrations, attributed to violation of stigmatic image conditions. The most important are spherical aberration (attributed to nonideal shapes of lenses and mirror surfaces leading to deviation from a stigmatic image), coma and astigmatism (both attributed to violation of the stigmatic image of a point not lying on a main optical axes), distortion (violation of proportions at the object image) and chromatic aberration in nonmonochromatic illumination (colored halos attributed to dispersion of the index of refraction).

During light propagation in a medium, there is light intensity decrease along a propagation direction, attributed to two phenomena, namely light absorption and light scattering. Decrease of monochromatic radiation intensity, during light propagation through an uniform layer with thickness δ is described by Buger's law: $I = I_O \exp[-(k_a + k_s)\delta]$; here I_O is intensity of incident radiation, k_a, k_s are the absorption and scattering indexes, respectively, which usually depend on the wavelength. The value $k_e = k_a + k_s$ is called the extinction index. In contrast with light scattering, absorption is connected with the transfer of a part of the light energy to the medium. The absorption and refraction indexes depend on only physical-chemical material properties and are called its optical constants. Light scattering is attributed to two phenomena:

a. deflection from rectilinear light propagation in a medium with nonuniform distribution of index of refraction (e.g. in a gas medium with nonuniform density distribution);
b. light reflection and diffraction in a heterogeneous medium (e.g. in emulsions and suspensions).

A medium with large quantity of fine inhomogeneities is called turbid. The product $K = k_a\delta$ is called optical thickness of a layer. If $K \ll 1$ then a layer is *optically thin*, i.e. light propagates through it practically without absorption. For large K a layer is termed *optically thick*. Light scattering by fine particles attributed to the diffraction phenomenon is described by Mie theory and is called **Mie Scattering**.

Wave Optics

In wave optics, the law of independent action of light beams is usually invalid. If coherent beams superimpose then an interference occurs, which simply appears as alternating light and dark bands or rings on a screen. A phase difference of two coherent waves with the same polarization, arriving to a point of observation, A, depends on initial phases of the waves and on a length difference, Δl, between point A and wave sources S_1 and S_2 (**Figure 3**). If the initial wave phases are the same and Δl is a multiple of wavelength, then the waves arrive to point A with the same phase, field strengths E_1 and E_2 are added and light intensity, which is proportional to $(E_1 + E_2)^2$, correspondingly increases. If l is a multiple of odd number of half-waves, λ/2, then the waves arrive to A with the opposite phases and the field strengths are subtracted. At $E_1 = E_2$ light intensity in point A is equal 0. At all points B equidistant from S_1 and S_2, an interference leads to light intensity increase.

If a parallel (collimated) beam of monochromatic light is incident with inclination on a wedge-shaped transparent plate (**Figure**

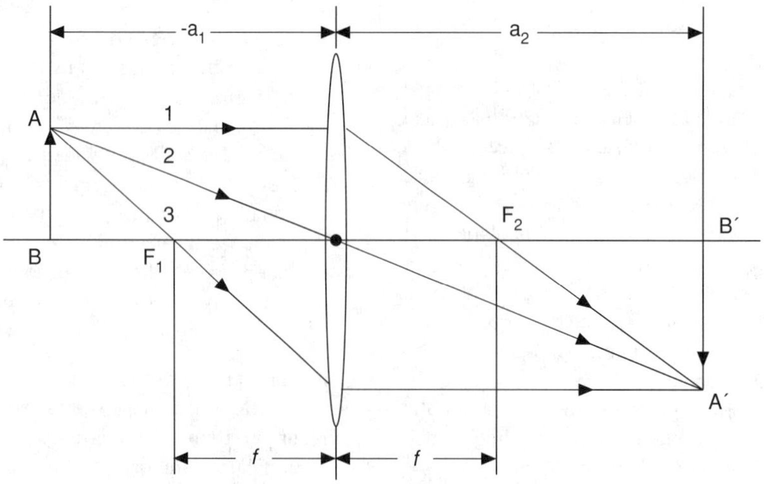

Figure 2. Optical system for image formation or for object illumination.

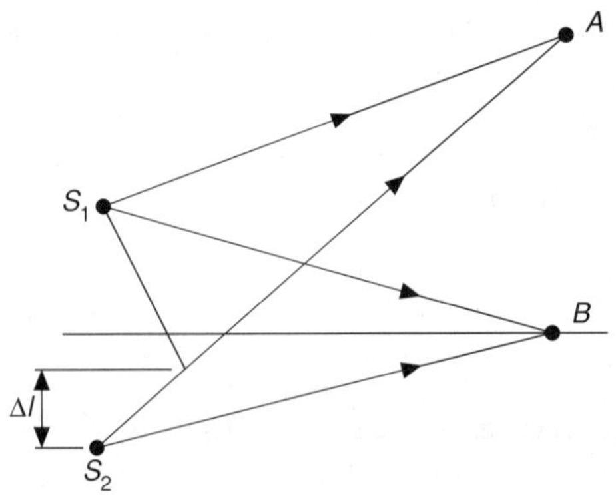

Figure 3. Phase difference of coherent waves with the same polarization.

4), then due to interference of the beams reflected from front and rear plate faces, an interference picture appears in the form of alternating light and dark bands parallel to wedge edge. If the difference between BA and BCDA is multiple to an odd number of half-waves then point A lies on a light band; otherwise, it lies on a dark one. These bands are called the uniform thickness bands, because they correspond to a collection of the points where the thicknesses of wedge are the same. Distortion of a straight shape of the bands reveals defects of plate faces. If a plate with parallel faces is illuminated by a monochromatic diverging light beam, then alternating light and dark concentric rings (or its parts) occur on when the reflected light is projected onto a screen. They also appear as a result of interference of the beams, reflecting from two plate faces and are called uniform inclination bands.

Based on interference phenomenon, optical methods and instruments (Michelson, **Mach-Zender**, Twaiman-Green and other interferometers) are widely used for determining the quality of optical elements, of optical and other materials media homogeneity, for precise measuring of distances and for many other purposes. The most convenient are the laser interferometers in which a laser is used as a source of light.

Diffraction

In wave optics, the law of rectilinear light propagation in a homogeneous medium is invalid. If a plane monochromatic wave falls on an aperture of an arbitrary shape in an opaque screen, then for a small distance behind the screen the cross-section shape of the light beam replicates the aperture shape, i.e. there is a sharp boundary between light and shadow. As the distance increases, and the light spot enlarges, its shape deforms and the boundary between light and shadow becomes less and less sharp. This is due to the light diffraction phenomenon. Light intensity distribution in beam cross-section behind a screen can be approximately calculated by the application of the Fresnel-Huygens principle. According to this principle, each point of wave front is a coherent source of a semispherical elementary wave. An envelope of these waves taking into account their interference represents a new wave front in the next moment. If the condition of that the diffraction zone is remote from the beam source is valid (Fraunhofer diffraction), $L \gg D^2/\lambda$, where L is the distance from aperture to the screen, D is a characteristic aperture size, then the distribution curve of light intensity in a diffraction pattern does not depend on the value of L. Diffraction in a near zone is called Fresnel diffraction. In the wave diffraction of a plane monochromatic wave by a round aperture, the remote zone diffraction pattern forms a central light circle, surrounded by alternating dark and light rings.

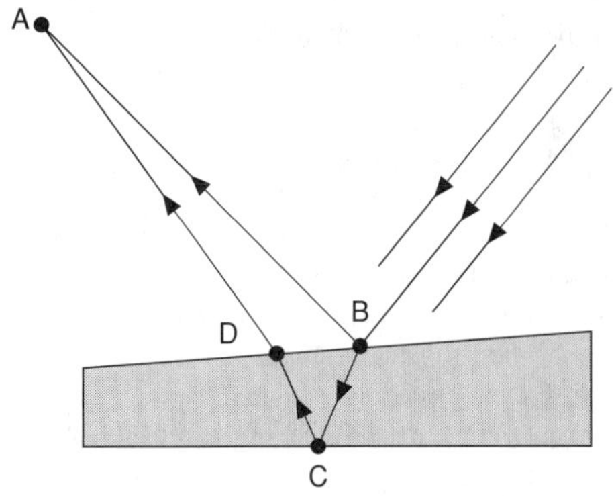

Figure 4. Wedge-shaped transparent plate.

Figure 5 shows relative local intensity, $\bar{I} = I/I_O$, versus relative radius, $\bar{r} = \pi Dr/\lambda L$, and also the total radiation power, $\dot{W}$, passing through a circle with radius $\bar{r}$. The value (φ of $\bar{r}$ corresponding to the central circle is determined by the equation $\varphi = 2.44\lambda/D$ and 84 percent of total radiation energy passes through this circle. In diffraction of a parallel monochromatic light beam by a narrow long slit, a set of equidistant light and dark bands are formed on a screen, and a distance between the bands is proportional to the wavelength. An assembly of a large number of narrow parallel equidistant slits is called diffraction grating. When nonmonochromatic radiation is incident on a diffraction grating, spectral dispersion of the radiation occurs; thus such a grating is used as a dispersing element in many spectral instruments.

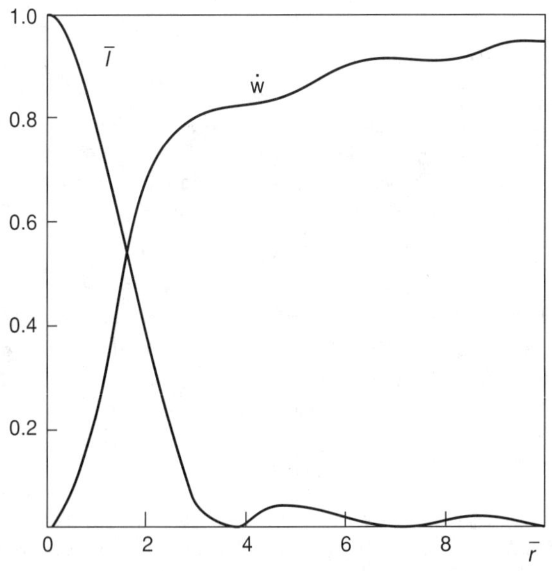

Figure 5. Light intensity distribution on a screen irradiated by monochromatic light which has passed through a small aperture at some distance from the screen.

Leading to: *Mie scattering; Speed of light; Optical particle characteristics; Refractive index*

Y. Svirchook

OPTIMIZATION OF EXCHANGER DESIGN

Following from: *Heat exchangers*

The optimization of a heat exchanger design can be viewed at three different levels: 1) the identification of a heat exchanger design that meets the process specifications (described below) at the lowest initial cost; 2) the identification of a heat exchanger design that will meet process specifications and operate most satisfactorily over the lifetime of the plant; 3) the identification of a system of heat exchangers and auxiliary components that will meet plant process specifications with minimum total cost to the process (including utilities and lost production).

The distinctions among these three levels of optimization can be best understood if we list the required and desired criteria of the ideal heat exchanger:

1. The exchanger must achieve the required changes in the thermal conditions of the process streams within the allowable pressure drops; this is what is meant by "meeting process specifications."
2. The heat exchanger construction must withstand the mechanical stresses of manufacturing, transport, installation, operation under normal and foreseeable operating conditions (including emergency shut downs) and maintenance, as well as minimizing effects of corrosion and fouling.
3. The heat exchanger must be maintainable with minimum downtime, including cleaning and repair or replacement of short-lived components such as gaskets and tubes.
4. The heat exchangers should be flexible enough to meet process specifications under reasonable changes in process conditions, such as normal fouling transients and seasonal and diurnal changes in service stream temperatures.
5. Subject to the above, the exchanger should cost as little as possible (installed).
6. The exchanger may need to meet special requirements of length, diameter, weight, or inventory standards, especially in retrofit applications.
7. Other requirements may apply in special situations, such as manufacturing time, or experience and/or capability of operating and maintenance personnel.

Another vital but often overlooked factor is the uncertainty associated with every step in the exchanger design (and of course characteristic to some degree of every piece of engineering equipment). Palen and Taborek (1969) showed that the best proprietary method for shell and tube exchangers available at the time could only predict overall heat transfer rate within about ±30 percent and shell side pressure drop within about ±40 percent for a large sample of carefully tested exchangers under minimum fouling conditions and using fluids with well-known properties. While there have been some improvements since, the uncertainties for heat exchangers in normal plant service are probably no better and may well be worse than these values.

The design of a heat exchanger that meets process requirements at lowest first cost (Level 1 above) explicitly satisfies criteria 1 and 5 (though transport and installation cost differences are sometimes ignored) and implicitly those aspects of criterion 2 that are governed by the ASME Unfired Pressure Vessel Code or other applicable codes. Criteria 3, 6, and 7 may be met if the purchasor has specified *a priori* those features which he deems important in each case; these specifications operate as initial conditions or constraints on the optimization process. Unfortunately, personnel doing the specification and bid evaluation for heat exchangers all too often lack the plant operating experience or input to include these requirements.

Optimization at this level can be handled by various case-study methods, with the objective function being the cost of the heat exchanger. A cost estimation program is required. If the vendor is doing the optimization, the cost program can be quite precise; if the customer is doing the optimization, the cost program can represent at best only an estimate of what the quoted cost will be because of variations in manufacturing capabilities (and therefore costs) among manufacturers.

Palen et al. (1974) demonstrated the use of the *M. J. Box Complex Method* (1965) in the optimization of shell and tube heat exchangers, using a rating program somewhat simpler than the best *Stream Analysis Method* available at that time. Even with the computing power available (CDC6400), the optimization program for a realistic case ran in under 10 seconds. With present computing capabilities, there is no need to use anything less that the best available rating method. It is also true that very little improvement (<1 percent) in objective function was realized in the last half of the runs, though at the end, the design was rationalized to standard dimensions (e.g., tube diameter was set to 3/4 inch, rather than 0.7 inch.). Given the irremovable uncertainties in the basic rating method, a wide range of plausible solutions would fall in the uncertainty band of the objective function.

The second level of optimization (as described in the first paragraph) requires consideration of all of the criteria listed. Concentration upon first cost as the objective function may result in selection of basic design features that preclude maximum operational efficiency (e.g. fixed tube sheet vs. removable bundle in an application where mechanical cleaning of the shell side will result in a cleaner heat exchanger and longer/higher efficiency operation between cleanings than with chemical cleaning only). Unfortunately, many of these considerations can only be expressed qualitatively and do not lend themselves to decision making solely on quantitative computation. Both lack of real-world operational experience and time pressure on the designer militate against proper weight being given to these often substantial cost factors.

Criterion 4 can only be considered in the global context of the entire process design and economic and market forecasting, requiring input from a variety of company personnel and necessarily involving high degrees of uncertainty. But it is only in these circumstances that one can move to Level 3 optimization. Examples of heat exchanger systems where very high economic value (even the technical feasibility of the process) ride on this level of optimization include crude oil preheat trains and feed-effluent heat recovery systems for chemical reactors. In another application, central station heating/cooling systems are moving towards this need.

Great progress has been made in this area, notably through the use of "*Pinch Technology,*" now increasingly referred to as "*Pinch Analysis*" (Linnhoff (1994)), to synthesize heat exchanger networks and integrate them with the rest of the process system. A number of organizations offer computer programs to retrofit existing systems or to design new ones using this approach. (See also **Process Integration**).

Heat recovery (or more precisely, efficient use of heat within the process, primarily by using hot process streams to heat cold process streams within the plant) originally provided the incentive for this kind of optimization, with less attention paid to capital cost. Recently, attention and capability have expanded to include capital minimization and waste stream reduction.

These are powerful computational techniques, but they still must be used with great care to consider the non-quantitative factors (e.g., maldistribution of a stream among parallel paths in an exchanger can render it incapable of meeting a very close approach temperature difference) and the uncertainties in the basic design methods and especially in future operating conditions. Some investigations have been done on the effect of uncertainties on system design using *Monte Carlo methods* [(Al-Zakri and Bell (1981); Uddin and Bell (1988))]. Until these approaches can be fully developed and integrated into practical design guidance, large-scale systems need to incorporate redundancy and resilience: stand-by capability, trim heating and cooling, and flexible piping layouts with adequate instrumentation and control.

References

Al-Zakri, A. S., and Bell, K. J., (1981) Heat transfer: estimating performance when uncertainties exist, *Chem. Eng.* Prog. *77,* 7, 39.

Box, M. J. (1965) A new method of constrained optimization and a comparison with other methods, *Computer J. 8,* 42.

Linnhoff, B. (1994) Use pinch analysis to knock down capital costs and emissions, *Chem. Eng.* Prog. *90,* No. 8, 32.

Palen, J. W., and Taborek, J., (1969) Solution of shell side flow pressure drop and heat transfer by stream analysis method, CEP Symp. Series No. 92, *65,* 53.

Palen, J. W., Cham, T. P., and Taborek, J. (1974) Optimization of shell and tube heat exchangers by case study method, AIChE Symp. Series No. 138, *70.*

Uddin, A. K. M. Mahbub, and Bell, K. J., (1988) Effect of uncertainties upon the performance of a feed-effluent heat exchanger system, *Heat Trans. Eng. 9,* No. 4, 63.

K.J. Bell

ORBITING SPACE STATIONS (see Microgravity conditions)

Organization for Economic Cooperation and Development, OECD

2 rue Andre Pascal
75775 Paris cedex 16
France
Tel: 1 45 24-82 00

ORGANISMS, HEAT TRANSFER FROM (see Environmental heat transfer)

ORIFICE BAFFLES (see Shell-and-tube heat exchangers)

ORIFICE FLOWMETERS

***Following from:** Differential pressure flowmeters*

Measurement of the flow rates of liquids, gases and vapours with orifice meters has found wide use both in industrial and in scientific measurements. A restriction fulfilling the function of a primary converter, is installed in a pipe-line and produces in it a local change of a flow section. The method depends upon the fact that an increase in velocity and kinetic energy of the flow behind the restriction as compared with the parameters upstream of it, brings about a decrease in a static pressure p_{out} downstream of the restriction with respect to the pressure p_{in} upstream of it. The differential pressure $\Delta p = p_{in} - p_{out}$ depends on the fluid flow velocity and can serve as a flow rate measure (see **Differential pressure flowmeter**).

The following designs of orifice are used: the standard orifice (**Figure 1a**), the double orifice (**Figure 1b**), orifice with an input cone (**Figure lc**), and an orifice with a double cone (**Figure 1d**). Closely related to orifices, but offering advantages of better downstream pressure recovery and (sometimes) range, in exchange for increased cost, are a whole variety of other restriction designs. These include standard nozzles (**Figure 1e**), "half-circle" nozzles (**Figure 1g**), "quarter-circle" nozzles (**Figure 1i**), cylindrical nozzles (**Figure 1k**), Venturi tube (**Figure 1l**), Venturi nozzle (**Figure 1m**), and Dall's tube (**Figure 1n**).

The most widely used standard restrictions are orifice plates converging nozzles and Venturi nozzles (see **Figure 2**). The orifice plate is a thin disk with a hole with diameter d and area S, located in line with the pipe-line whose diameter is D. The nozzle is made in the form of an insert with an orifice smoothly contracting at the inlet and ending with a cylindrical part. The nozzle profile allows us to realize a smooth compression of a jet up to its minimum section, which ensures less overall loss of pressure than in the case of the orifice plate. The Venturi nozzle has minimum losses of pressure of all the restrictions due to the installation of a diffusor at the outlet, which recovers the pressure. The principle of measuring the substance flow rate of pressure differential is the same for all types of restrictions, and a quantitative relation between $\dot{M}$ and Δp is defined by the relation

$$\dot{M} = \alpha \, \epsilon \, S\sqrt{2\rho\Delta p}, \qquad kg/s.$$

where ρ is the fluid density upstream of the meter and ϵ is a

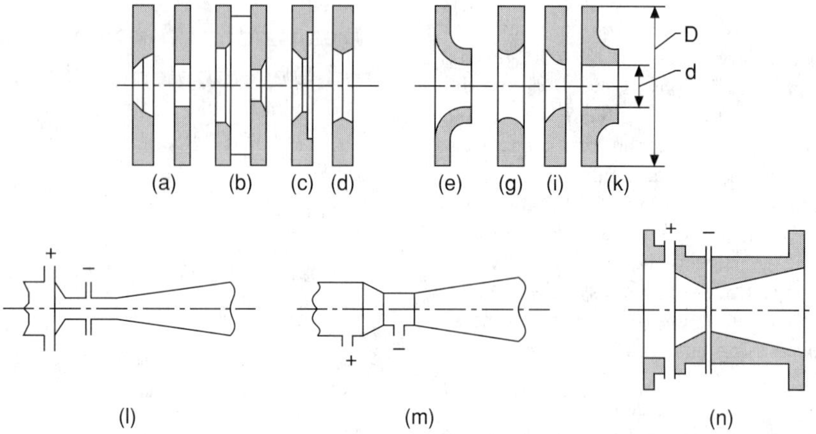

Figure 1. Types of restriction used in differential pressure flowmeter: (a)–(d), orifice plate, (e)–(k), nozzles; (l) Venturi tube; (m) Venturi nozzle; (n) Dall tube.

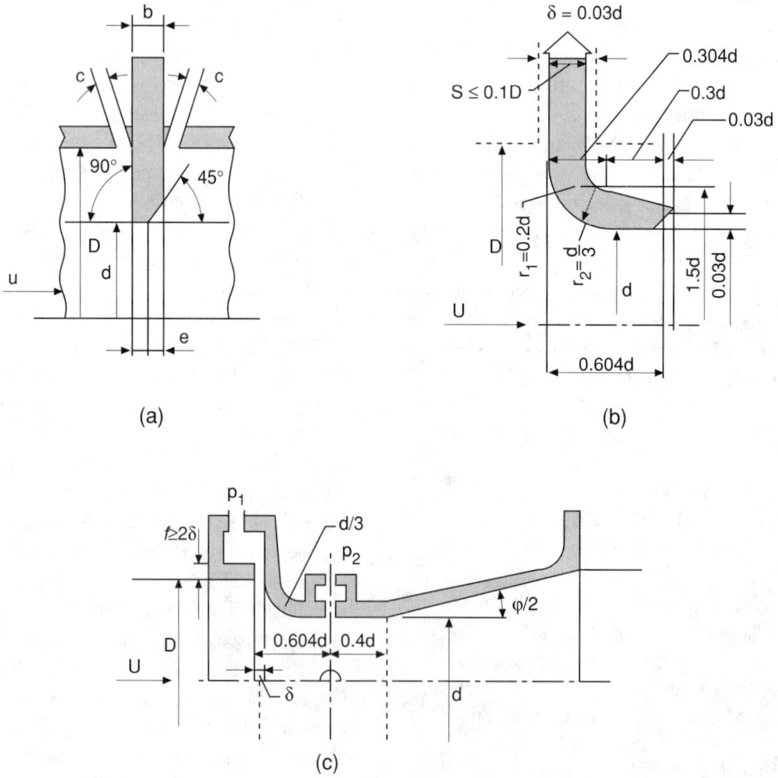

Figure 2. Details of the most widely used restrictions: (a) Orifice plate; (b) converging nozzle; (c) Venturi tube.

correction factor for compressibility (to be discussed below). The coefficient of propotionality α is called the "flow rate coefficient" or "discharge coefficient" and depends on the restriction type, on the degree of jet contraction (i.e., on the ratio of the flow area of the restriction to the cross sectional area of the pipe, where $m = d^2/D^2$), and on the Reynolds number $Re = \bar{u}\, D/\nu$ defined by the mean velocity $\bar{u}$ of the fluid.

For a given value of area ratio (m) α varies significantly with Re for small Re, but for Reynolds numbers higher than a boundary value Re_b ($Re > Re_b$) the flow in the restriction is self-similar and α is independent of Re and is constant for the given type of restriction. In measurements at low Re, double orifice plates are used, where an additional orifice with $m_1 > m_2$ is installed upstream of the main orifice, and the pressure taps are located immediately upstream of the additional and downstream of the main plate. Nozzles with the "quarter-circle" profile are also useful in this zone.

The values of boundary Reynolds numbers(Re_b) are given in **Table 1**.

The correction factor for the expansion of the medium ϵ being measured accounts for the changes in the substance density on its

Table 1. Values of boundary Reynolds number ($Re_b \times 10^{-4}$)

Area ratio (m)	0.05	0.10	0.20	0.30	0.40	0.50	0.60	0.65
Orifice	2.3	3.0	5.7	9.0	13.5	18.5	24.0	27.0
Nozzles	6.0	6.6	9.0	12.5	16.5	19.0	20.0	20.0
Double orifice plate	0.25	0.25	0.35	0.5	0.73	1.0	1.5	—
"Quarter-circle" nozzle	0.20	0.20	0.23	0.35	0.40	0.50	—	—

flowing through a restriction and depends on $\Delta p/p_{in}$, on the area ratio m, on the adiabatic exponent (γ) of the substance and on the type of the restriction. For standard diaphragms empirical equations can be used, for instance

$$\epsilon = 1 - (0.371 + 0.318\ m^2)[1 - \delta p^{1/\gamma}]^{0.935},$$

where $\delta p = (1 - \Delta p/p_{in})$. For standard nozzles and Venturi nozzles

$$\epsilon = \sqrt{\delta p^{2/\gamma} \frac{\gamma}{\gamma - 1} \frac{1 - \delta p^{(\gamma-1)/\gamma}}{\Delta p/p_{in}} \frac{1 - m^2}{1 - m^2 \delta p^{2/\gamma}}}.$$

For measuring the flow rate of a liquid $\epsilon = 1$.

Suppose we wish to measure flow of a substance with density ρ flowing in a tube of diameter D. We wish to determine the flow rate up to a maximum value $\dot{M}_{max}$ (kg/s) and the maximum pressure drop we wish to allow over the orifice is $\Delta p_{max}(p_a)$. In order to determine the diameter d, we invoke the relation

$$m\alpha = \frac{\dot{M}_{max}}{3.48 D^2 \epsilon \sqrt{\rho \Delta p_{max}}},$$

The relation between $m\alpha$ and α for standard restrictions is given in **Table 2**.

Using α determined from the table, we can calculate $m = (m\alpha)/\alpha$ and the diameter of the restriction: $d = Dm^{0.5}$. When metering hot fluids, it is sometimes important to take account of changes in orifice diamter as a result of thermal expansion; if the temperature of the operating medium flowing through the restriction is T and the diameter at 293K is d, the diameter is found from the relationship $d[1 + \beta(T - 293)]$, where β is the linear expansion coefficient of the material from which the restriction is manufactured.

Table 2. Relation between $m\alpha$ and α for standard restrictions

	α	
$m\alpha$	for orifice plates	for nozzles
0.05	0.601	0.987
0.10	0.609	0.990
0.15	0.621	0.993
0.20	0.637	0.999
0.25	0.665	1.006
0.30	0.676	1.015
0.35	0.696	1.025
0.40	0.718	1.038
0.45	0.742	1.050
0.50	0.768	1.067
0.55	0.797	1.085

In order to carry out the precision measurements, the flow rate coefficient α is often initially determined from the results of calibration by the weight or by the volume method.

The density of a substance flowing through a restrictor is determined from the pressure p_{in} measured directly at the inlet end face of the restriction, through separate holes which are not used for measuring Δp and the temperature T_{in} is determined with the help of a thermometer located at a distance of $l = (15\text{–}20)D$ from the front end face of the restriction.

The standard diaphragm is shown schematically in **Figure 2a**. The hole in the diaphragm must have a cylindrical form with a sharp edge at the inlet. The length of the cylindrical part is $0.005 \le l \le 0.02D$, and for $m > 0.5$, $l = b/3$. The diaphragm thickness should be $b \le 0.05D$, but not less than 2.5–3 mm. At the outlet the hole is made conical with an angle of half-opening 30–45°. The relative area of the orifice m should be in the range 0.05 – 0.64. The inlet and outlet planes of the orifice plate must be parallel, and the angle between the inlet plane of the orifice plate and a normal to the tube axis must not exceed 1°. Special attention in manufacturing must be given to treatment of the inlet edge: it must be free of chamfering and scratches. Pressure must be tapped with the help of holes which must be drilled in the immediate vicinity of the front and rear walls of the orifice plate, or through the use of circular chambers connected with the inner cavity of the pipeline by an annular slot or by several holes. In this case the pressure is averaged over the slot perimeter. The diameter of the orifice or the slot width $c \le 0.01D$, in doubled orifices it is 0.5D. At small Re, orifices with an inlet cone, with a double cone or with a rounded surface are used.

The standard nozzle is shown in **Figure 2b**. The nozzle consists of an inlet part formed by the arcs of the circles of radiuses $r_1 = 0.2d$ and $r_2 = 0.33d$, which transfer smoothly from one to the other, and of a cylindrical part of length $l = 0.3d$ at the outlet. The arc of radius r_2 transfers to the cylindrical part along the tangent, and the arc of radius r_1 is integrated with the inlet plane. The deviation of radiuses r_1 and r_2 from the nominal value must not exceed 10 percent. At the outlet, the cylindrical part of the nozzle orifice has a bore of diameter 1.06d and of length 0.3d to prevent the outlet edge from damages. The ellipticity of the nozzle orifice must not exceed 0.1 percent. Such nozzles have a coefficient of jet constriction close to unity, as a result of which the discharge coefficient is considerably higher than that of the orifice plate, which allows us to measure larger flow rates at the same differential pressure.

In the "quarter-circle" nozzle, the nozzle profile is formed by one arc, whose length is equal to 1/4 of the length of the circle of radius r, and a cylindrical part is absent. With the aim of ensuring constancy of the discharge coefficient α, the choice of the value of r is of fundamental importance. **Table 3** gives recommended values.

Table 3. Recommended radii for the quarter-circle nozzle

m	0.05	0.16	0.25	0.36	0.39	0.43	0.48
r/d	0.100	0.114	0.135	0.209	0.285	0.380	0.446

Nozzles with "half-circle" profile, rounded profile and combined nozzles are also used.

A restriction which has a diffusor at the end for recovering pressure is called a flow tube. To this class belong the Venturi tube and the Venturi nozzle. The use of these devices is also described in the article on **Venturi Meters**. The Venturi tube consists of an inlet cylindrical part with diameter D and of length $l = D$; of a confusor with a taper angle of $21 \pm 1°$ of length 2.7(D–d); a cylindrical throttle of length equal to its diameter d, and a diffusor with conicity of 7–8°, which has the diameter D at the outlet. The application of shortened Venturi's tubes with the outlet diameter less than D and the diffusor conicity of 14–16° with the diffusor length of (0.7–1)D is sometimes encountered.

The Venturi nozzle (**Figure 2c**) has an inlet part made according to the standard nozzle profile, but a somewhat lengthened cylindrical part (0.7–0.75d instead of 0.3d). The angle of flare of the diffusor is $\varphi = 12–16°$. The pressure p_{in} is tapped immediately ahead of the inlet plane, and the pressure p_{out} is tapped through four holes spaced around the circumference at a distance of 0.3d from the beginning of the cylindrical part. The application of shortened Venturi nozzles with a cut off diffusor is also encountered. Besides the above standardized geometries double Venturi nozzles and the Dall tube, which have low pressure losses, are also used. Of particular importance for increasing the accuracy of measurement of the flow rate are the conditions of mounting the restriction into the pipe-line system. This defines the necessity of straight sections in a pipe-line in front of the restriction, l_{in} and behind it l_{out}. Depending on perturbations introduced into the flow by the pipe-line elements, the minimum length of the straight part ahead of the restriction for orifice plates and nozzles is $l_{in} = (60–80)D$, and $l_{out} = (4–8)D$. For Venturi nozzles $l_{in} = (3–4)D$ and $l_{out} = (6–8)D$. The required lengths l_{in} and l_{out} increase with increasing m.

The restriction is connected with the differential pressure gauge by two tubes. In standard flowmetering devices, the diameter of these tubes is 8–12 mm and the length of the tubes must be minimal wherever possible. The lines must have a slope not less than 1:10 along the entire path. The tube ends must be smooth, without acute angles and hollows. Steel, copper brass and aluminium tubes are used. Tubes made from plastic materials can be used at pressures up to 0.6 MPa.

Differential pressure gauges are used for measuring pressure differential Δp in the restriction: U-shaped and cup-type manometers, inclined alcohol micromanometers, compensation micropressure gauges which allow the measurement of Δp with an accuracy of ± 0.1 Pa, liquid-sealed bell manometers and ring-balance manometers, diaphragm pressure gauges, bellows pressure gauges, etc. (see **Pressure Measurement**). The accuracy of measuring the flow rate on the employment of standard restrictions is estimated as 0.8–2 percent for liquids and 1–3 percent for gases. On carrying out an individual calibration of the restrictions, the accuracy increases significantly.

Reference

Fluid meters. their theory and application significantly (1971) Report of ASME, Research Committee on Fluid Meters, New York, published by ASME.

Yu L. Shekhter

ORNL (see Oak Ridge National Laboratory)

ORR-SUMMERFIELD STABILITY ANALYSIS (see Wavy flow)

OSCILLATIONS IN A STATIONARY FLUID (see Secondary flows)

OSMOSIS (see Membrane processes)

OSMOTIC DEHYDRATION (see Drying)

OSMOTIC PRESSURE (see Membrane processes)

OSTWALD DE WAELE POWER LAW MODEL (see Non-Newtonian fluid heat transfer

OSTWALD PROCESS (see Nitrogen)

OSTWALD RIPENING (see Interfaces)

OTEC (see Ocean thermal energy conservation system)

OTTO CYCLE (see Internal combustion engines)

OVERALL HEAT TRANSFER COEFFICIENT

Following from: *Heat transfer coefficient*

The overall heat transfer coefficient is employed in calculating the rate of heat transfer $\dot{Q}$ from one fluid at an average bulk temperature T_1 through a solid surface to a second fluid at an average bulk temperature T_2 (where $T_1 > T_2$). The defining equation is generally only applicable to an incremental element of heat transfer surface dA for which the heat transfer rate is $d\dot{Q}$, and the equation is strictly valid only at steady state conditions and negligible lateral heat transfer in the solid surface, conditions generally true enough in most practical applications. The defining equation is

$$d\dot{Q} = U(T_1 - T_2)dA \tag{1}$$

where U is referenced to a specific surface (see below).

In the particular situation of heat transfer across a plane wall of uniform thickness, U is related to the individual film heat transfer coefficients, α_1 and α_2, of the two fluids by the equation

$$U = \frac{1}{\dfrac{1}{\alpha_1} + \dfrac{\delta_w}{\lambda_w} + \dfrac{1}{\alpha_2}} \tag{2}$$

where δ_w is the thickness of the wall and λ_w is the thermal conductivity of the wall.

If there are fouling deposits on the wall, they have a resistance to heat transfer, R_1 and R_2, in units of m^2K/W, and these resistances must be added in (See **Fouling and Fouling Factors**)

$$U = \frac{1}{\dfrac{1}{\alpha_1} + R_1 + \dfrac{\delta_w}{\lambda_w} + R_2 + \dfrac{1}{\alpha_2}} \tag{3}$$

For the special but very important case of heat transfer through the wall of a plain round tube, the different heat transfer areas on the inside and outside surfaces of the tube need to be considered. Let dA_i be the inside incremental area and dA_o be the outside. Then (including fouling resistances R_{fi} and R_{fo} inside and out):

$$U_i = \frac{1}{\frac{1}{\alpha_i} + R_{fi} + \frac{r_i \ln(r_o/r_i)}{\lambda_w} + R_{fo}\frac{r_i}{r_o} + \frac{r_i}{\alpha_o r_o}} \quad (4)$$

where U_i is termed the "overall heat transfer coefficient referenced to (or based on) the inside tube heat transfer area", and r_i and r_o the inside and outside radii of the tube.

Alternatively, the overall coefficient may be based on the outside heat transfer area, giving

$$U_o = \frac{1}{\frac{r_o}{\alpha_i r_i} + R_{fi}\frac{r}{r_i} + \frac{r_o \ln(r_o/r_i)}{\lambda_w} + R_{fo} + \frac{1}{\alpha_o}} \quad (5)$$

where U_o is termed the "overall heat transfer coefficient based on the outside tube heat transfer area". Note that

$$d\dot{Q} = U_i(T_1 - T_2)dA_i = U_o(T_1 - T_2)dA_o. \quad (6)$$

These ideas may be extended to more complicated surfaces such as finned or composite tubes, but it is then necessary to add further resistance terms (and the area ratio corrections) for the fins or imperfect metal-to-metal contact.

Generally, in order to use these equations in heat transfer applications, the basic equation must be integrated:

$$A_T = \int_0^{\dot{Q}_T} \frac{d\dot{Q}}{U(T_1 - T_2)} \quad (7)$$

where A_T is the total area required to transfer $\dot{Q}_T$ and T_1, T_2 and sometimes U must be expressed as functions of the heat already transferred from one end up to a given point in the heat transfer device. This is the basic design equation for most heat exchangers.

References

Hewitt, G. F., Shires, G. L., and Bott, T. R. (1994) *Process Heat Transfer,* CRC Press, Boca Raton, Florida.

Incropera, F. P., and DeWitt, D. P. (1990) *Introduction to Heat Transfer, 2nd Ed.,* John Wiley & Sons, New York.

***Leading to:** Fouling; Fouling factors*

K.J. Bell

OVER POTENTIAL (see Electrolysis)

OXYGEN

Oxygen—(GR. *oxys,* sharp, acid, and *genes,* forming; acid former), O; atomic weight (natural) 15.99994; atomic number 8; melting point −218.4°C; boiling point −182.962°C; density 1.429 g/T(0°C); specific gravity liquid 1.14 (−182.96°C); valence 2.

For many centuries, workers occasionally realized air was composed of more than one component. The behavior of oxygen and nitrogen as components of air led to the advancement of the phlogiston theory of combustion, which captured the minds of chemists for a century. Oxygen was prepared by several workers, including Bayen and Borch, but they did not know how to collect it, did not study its properties, and did not recognize it as an elementary substance. Priestley is generally credited with its discovery, although Scheele also discovered it independently.

Oxygen is the third most abundant element found in the sun, and it plays a part in the carbon-nitrogen cycle, one process thought to give the sun and stars their energy. Oxygen under excited conditions is responsible for the bright red and yellow-green colors of the aurora. Oxygen, as a gaseous element, forms 21% of the earth's atmosphere by volume from which it can be obtained by liquefaction and fractional distillation. The atmosphere of Mars contains about 0.15% oxygen. The element and its compounds make up 49.2%, by weight, of the earth's crust. About two-thirds of the human body and nine-tenths of water is oxygen.

In the laboratory oxygen can be prepared by the electrolysis of water or by heating potassium chlorate with manganese dioxide as a catalyst. The gas is colorless, odorless, and tasteless. The liquid and solid forms are a pale blue color and are strongly paramagnetic. Ozone (O_3), a highly active allotropic form of oxygen, is formed by the action of an electrical discharge or ultraviolet light on oxygen. Ozone's presence in the atmosphere (amounting to the equivalent of a layer 3 mm thick at ordinary pressures and temperatures) is of vital importance in preventing harmful ultraviolet rays of the sun from reaching the earth's surface. There has been recent concern that aerosols in the atmosphere may have a detrimental effect on this ozone layer. Ozone is toxic and exposure should not exceed 0.2mg/M^3 (8-hr time-weighted average—40-hr work week.) Undiluted ozone has a bluish color. Liquid ozone is bluish black, and solid ozone is violet-black.

Oxygen is very reactive and capable of combining with most elements. It is a component of hundreds of thousands of organic compounds. It is essential for respiration of all plants and animals and for practically all combustion. In hospitals it is frequently used to aid respiration of patients. Its atomic weight was used as a standard of comparison for each of the other elements until 1961 when the International Union of Pure and Applied Chemistry adopted carbon 12 as the new basis. Oxygen has eight isotopes. Natural oxygen is a mixture of three isotopes. Oxygen 18 occurs naturally, is stable, and is available commercially. Water (H_2O with 1.5 percent O^{13}) is also available. Oxygen enrichment of steel blast furnaces accounts for the greatest use of the gas. Large quantities are also used in making synthesis gas for ammonia and methanol, ethylene oxide, and for oxy-acetylene welding. Air separation plants produce about 99 percent of the gas, electrolysis plants about 1 percent.

Handbook of Chemistry and Physics, CRC Press

OZONE DEPLETION (see Trichloroethane: Trichloroethylene)

OZONE, IN ATMOSPHERE (see Atmosphere)

OZONE LAYER (see Heat pumps)

P

PACKED BED (see Fixed beds; Gas-solid flows)

PACKED BEDS, WETTING (see Contact angle)

PACKED COLUMN CHROMATOGRAPHY (see Gas chromatography)

PACKED JOINTS (see Expansion joints)

PACKED SCRUBBER (see Scrubbers)

PACKED TOWERS (see Cooling towers)

PACKING, COLUMNS (see Distillation)

PACKING IN REGENERATORS (see Regenerative heat exchangers)

PAIR PRODUCTION (see Gamma rays)

PARABOLIC DIFFERENTIAL EQUATIONS (see Differential equations)

PARABOLIC EQUATIONS (see Green's function)

PARAFFIN

Paraffin or *alkane* is the common name for all saturated, *aliphatic hydrocarbons* of the chemical formula C_nH_{2n+2}. At room temperature and pressure the first four members of the normal series (**Methane**, **Ethane**, **Propane** and **Butane**) are gases, the next thirteen are liquids, while the rest are solids. The paraffins are the least toxic of all **hydrocarbons**. The name paraffin is also commonly used for an oily petroleum distillate which is a mixture of hydrocarbons. It is believed that the perceived chemical inertness of the compounds led to the name paraffin from *parum* (little) and *affinis* (related).

V. Vesovic

PARALLEL PLATES, RADIATIVE HEAT TRANSFER BETWEEN

Following from: *Radiative heat transfer*

Consider two parallel plane surfaces of infinite extent that have different uniform temperatures (**Figure 1**). The space between the surfaces is either a vacuum or contains a material that does not interact with radiative energy; the surface properties are independent of wavelength (*gray surfaces*). A useful *radiative exchange* result is the net transfer from a unit area of surface 1 across the separation space to surface 2. This is derived in standard heat transfer texts by flux or ray tracing methods, Siegel and Howell (1992). The **Emissivity** (which equals the *absorptivity* for a gray surface) of each surface is assumed independent of the angular direction of emission. Reflected energy, however, can be diffuse, specular (mirror-like), or can have an arbitrary angular distribution, and the same result is obtained. The surface properties can depend on temperature. For these conditions, after accounting for all exchanges between the parallel boundaries, the net energy flux (W/m^2) transferred by radiation from 1 to 2 is

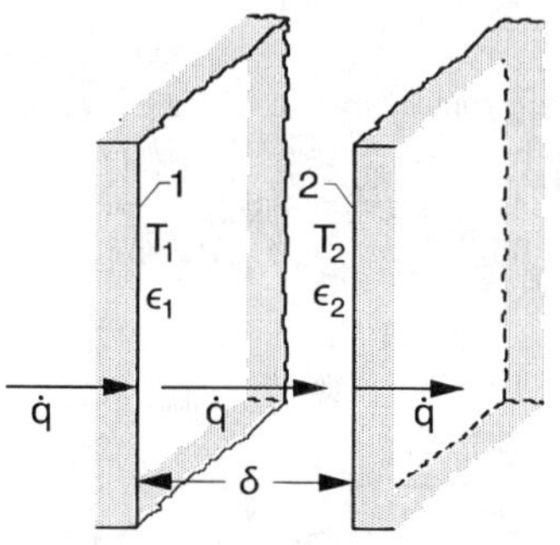

Figure 1. Radiative transfer across the space between two infinite parallel boundaries 1 and 2.

$$\dot{q} = \frac{\sigma(T_1^4 - T_2^4)}{\dfrac{1}{\epsilon_1(T_1)} + \dfrac{1}{\epsilon_2(T_2)} - 1} \tag{1a}$$

The ϵ_1 and ϵ_2 are emissivities, each selected at their surface temperature. The flux $\dot{q}$ is independent of the spacing between the surfaces.

The space between the plane surfaces is now filled with a gas that does not absorb, emit, or scatter radiation, but is heat conducting. If the spacing δ is narrow, or gravity is small, free convection is suppressed. Conduction occurs and is independent of radiation so the net flux transferred is

$$\dot{q} = \frac{\sigma(T_1^4 - T_2^4)}{\dfrac{1}{\epsilon_1(T_1)} + \dfrac{1}{\epsilon_2(T_2)} - 1} + \frac{\lambda}{\delta}(T_1 - T_2) \tag{1b}$$

Since the gas does not interact with radiation, convection can also be added to **Equation 1a** as an independent mode of heat transfer.

If the surface properties depend on wavelength, **Equation 1a** is written spectrally by using the blackbody function instead of σT^4. The net radiative transfer is obtained by integrating over all wavelengths to yield

$$\dot{q} = \int_{\lambda=0}^{\infty} \frac{e_{\lambda b,1}(\lambda, T_1) - e_{\lambda b,2}(\lambda, T_2)}{\dfrac{1}{\epsilon_{\lambda,1}(\lambda, T_1)} + \dfrac{1}{\epsilon_{\lambda,2}(\lambda, T_2)} - 1} d\lambda \tag{2}$$

where the ϵ are now functions of wavelength and temperature.

A technique for reducing radiative transfer is to place a series of highly reflecting thin parallel plates (*radiation shields*) in an evacuated space between two boundaries. As shown in **Figure 2**, there are N thin plates, all with gray properties and having the same property restrictions as for **Equation 1a**. Each thin plate, such as the n th plate, has emissivities, $\epsilon_{n,1}$ and $\epsilon_{n,2}$ on its sides;

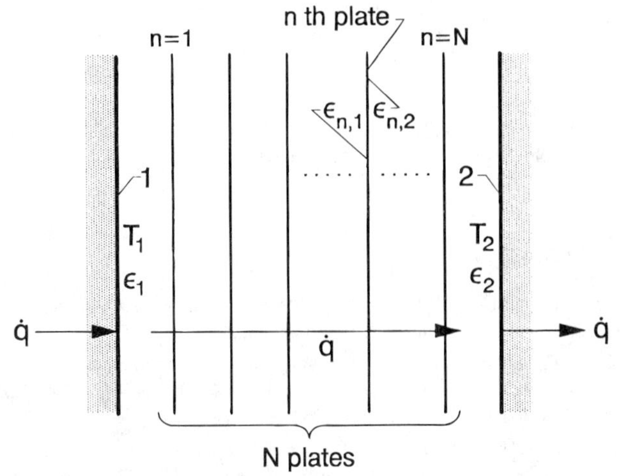

Figure 2. Radiative transfer between infinite parallel boundaries with N radiation shields between them.

for simplicity the ϵ are assumed independent of temperature. The net energy transfer is

$$\dot q = \frac{\sigma(T_1^4 - T_2^4)}{\frac{1}{\epsilon_1} + \sum_{n=1}^{N}\left[\frac{1}{\epsilon_{n,1}} + \frac{1}{\epsilon_{n,2}}\right] + \frac{1}{\epsilon_2} - (N+1)} \quad \textbf{(3)}$$

The complexity of having an *absorbing and emitting gas* between the plane boundaries is now added to **Equation 2**. The gas is mixed, as in a furnace, so it is at uniform temperature T_G (**Figure 3**). Energy is either being added to or removed from the gas to maintain T_G at steady state in the presence of the radiative interaction with the boundaries at T_1 and T_2. The net radiative energy leaving surface 1 is not equal to that removed at 2 (they are equal for the conditions of Equation 2). A control volume is considered bounded by opposing unit areas of the boundaries and extending across the space δ (**Figure 3**). The $\dot q_1$, $\dot q_2$, and $\dot Q_G$ are the energy quantities that must be supplied, by some means other than internal radiation, to the control volume boundaries on surfaces 1 and 2, and to the gas in the control volume, to maintain their steady temperatures. If any of these energies are negative, then energy is

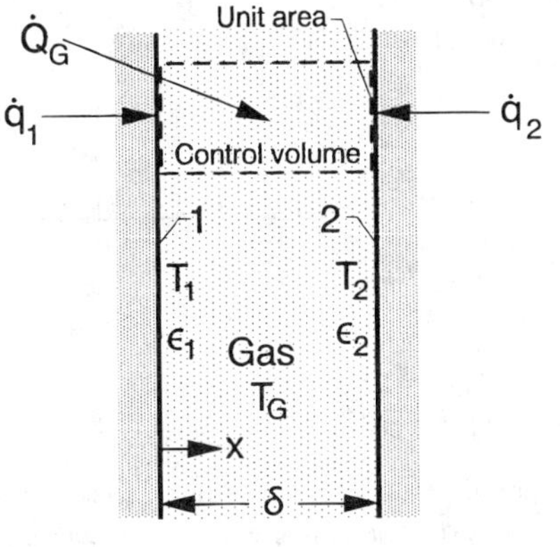

Figure 3. Infinite parallel boundaries with absorbing and emitting gas at uniform temperature T_G between them.

being removed from the corresponding surface (such as by external cooling of the boundary) or gas. From an overall energy balance, $\dot q_1 + \dot q_2 + \dot Q_G = 0$, so the energy added to the gas, for example by combustion, is $\dot Q_G = -(\dot q_1 + \dot q_2)$. Because of energy absorption in the gas, the directional behavior of radiation leaving a surface is significant since it determines the path length through the gas; for **Equations 4a** and **4b** both the reflected and emitted energies from the boundaries are diffuse. The energy fluxes supplied externally to surfaces 1 and 2 to compensate for their net radiative losses are

$$\dot q_1 = \int_{\lambda=0}^{\infty} \frac{\{\epsilon_{\lambda,1}\epsilon_{\lambda,2}\bar\tau_\lambda(e_{\lambda b,1} - e_{\lambda b,2}) + \epsilon_{\lambda,1}(1-\bar\tau_\lambda)[1 + (1-\epsilon_{\lambda,2})\bar\tau_\lambda](e_{\lambda b,1} - e_{\lambda b,G})\}}{1 - (1-\epsilon_{\lambda,1})(1-\epsilon_{\lambda,2})\bar\tau_\lambda^2}\, d\lambda \quad (4a)$$

$$\dot q_2 = \int_{\lambda=0}^{\infty} \frac{\{\epsilon_{\lambda,1}\epsilon_{\lambda,2}\bar\tau_\lambda(e_{\lambda b,2} - e_{\lambda b,1}) + \epsilon_{\lambda,2}(1-\bar\tau_\lambda)[1 + (1-\epsilon_{\lambda,1})\bar\tau_\lambda](e_{\lambda b,2} - e_{\lambda b,G})\}}{1 - (1-\epsilon_{\lambda,1})(1-\epsilon_{\lambda,2})\bar\tau_\lambda^2}\, d\lambda \quad (4b)$$

The mean spectral transmittance accounting for all directions of diffuse radiation through the gas is $\bar\tau_\lambda = 2E_3\,(a_\lambda\delta)$, where E_3 is the exponential integral function (tabulated in Siegel and Howell (1992) and in handbooks),

$$E_j(\beta) = \int_{\mu=0}^{1} \mu^{j-2} \exp\left(-\frac{\beta}{\mu}\right) d\mu \quad \textbf{(4c)}$$

with $j = 3$. The radiative behavior depends on the spacing δ between boundaries; **Equations 1a and 2** are independent of spacing.

A special case is for a gray gas ($a_\lambda = a$) between diffuse-gray parallel boundaries. **Equations 4a and 4b** then become

$$\dot q_1 = \frac{\epsilon_1\epsilon_2\bar\tau\sigma(T_1^4 - T_2^4) + \epsilon_1(1-\bar\tau)[1 + (1-\epsilon_2)\bar\tau]\sigma(T_1^4 - T_G^4)}{1 - (1-\epsilon_1)(1-\epsilon_2)\bar\tau^2} \quad \textbf{(5a)}$$

$$\dot q_2 = \frac{\epsilon_1\epsilon_2\bar\tau\sigma(T_2^4 - T_1^4) + \epsilon_2(1-\bar\tau)[1 + (1-\epsilon_1)\bar\tau]\sigma(T_2^4 - T_G^4)}{1 - (1-\epsilon_1)(1-\epsilon_2)\bar\tau^2} \quad \textbf{(5b)}$$

For a nonabsorbing (transparent) medium between the boundaries $\bar\tau = 1$, and **Equations 5a and 5b** each become the same as **Equation 1a** (there is a negative sign for $\dot q_2$ as its direction is opposite to $\dot q$). For a highly absorbing medium $\bar\tau \to 0$, which results in

$$\dot q_1 = \epsilon_1\sigma(T_1^4 - T_G^4); \qquad \dot q_2 = \epsilon_2\sigma(T_2^4 - T_G^4) \quad \textbf{(6a,6b)}$$

Since for a gray boundary $\epsilon = \alpha$, **Equations 6a and 6b** are each a local balance stating that the energy flux that must be supplied externally to a boundary to maintain its specified temperature is equal to its emission at its surface temperature minus absorption of energy emitted to it by the gas.

Now consider an absorbing and emitting stationary gas between parallel black surfaces with heat conduction in the gas. The gas attains a temperature distribution that depends on the local balance within it of radiant absorption, emission, and heat conduction. The energy balance in dimensionless form, **Equation 7a**, equates the gain by conduction to the loss by local radiant emission reduced by the radiant energy gained as a result of emission from the black boundaries and the other volume elements of the gas,

$$N_1 \frac{d^2t(\kappa)}{d\kappa^2} = t^4(\kappa) - \frac{1}{2}\left[E_2(\kappa) + t_2^4E_2(\kappa_\delta - \kappa) + \int_{\kappa^*=0}^{\kappa_\delta} t^4(\kappa^*)E_1(|\kappa - \kappa^*|)d\kappa^*\right] \quad \textbf{(7a)}$$

Equation 7a is an **Integrodifferential Equation** for the temperature distribution $t(\kappa)$; the solution depends on κ_δ, t_2, and N_1. The boundary conditions are $t(0) = 1$ and $t(\kappa_\delta) = T_2/T_1 = t_2$. The net transfer from boundary 1 to 2 by conduction and radiation is

$$\frac{\dot{q}}{\sigma T_1^4} = -4N_1 \left.\frac{dt}{d\kappa}\right|_{\kappa=0} + 1 - 2\left[t_2^4E_3(\kappa_\delta) + \int_{\kappa^*-0}^{\kappa_\delta} t^4(\kappa^*)E_2(\kappa^*)d\kappa^*\right] \quad \textbf{(7b)}$$

Numerical solutions, Viskanta and Grosh (1962), give results for $\dot{q}/\sigma T_1^4$ as in **Table 1**.

If the gas between the boundaries is a good absorber, the heat transfer can be obtained accurately by using a *diffusion approximation*. Radiation is represented as an effective heat conduction where, for a gray gas, the effective radiative conductivity depends on the local gas temperature to the third power. The energy equation is

$$\frac{d}{dx}\left[\left(\frac{16\sigma T^3}{3a} + \lambda\right)\frac{dT}{dx}\right] = 0 \quad \textbf{(8a)}$$

and the heat transfer by combined radiation and conduction is obtained from

$$\dot{q} = -\left(\frac{16\sigma T^3}{3a} + \lambda\right)\frac{dT}{dx} \quad \textbf{(8b)}$$

Table 1. Heat transfer between parallel black surfaces by combined radiation and conduction through a gray medium

Optical thickness, $\kappa_\delta = a\delta$	Temperature ratio, $t_2 = T_2/T_1$	Conduction-radiation parameter, N_1	Dimensionless energy flux, $\dot{q}/\sigma T_1^4$
0.1	0.5	0	0.859
		0.01	1.074
		0.1	2.880
		1	20.88
		10	200.88
1.0	0.5	0	0.518
		0.01	0.596
		0.1	0.798
		1	2.600
		10	20.60
1.0	0.1	0	0.556
		0.01	0.658
		0.1	0.991
		1	4.218
		10	36.60
10	0.5	0	0.102
		0.01	0.114
		0.1	0.131
		1	0.315
		10	2.114

The solution for a gray medium between plane boundaries is in Siegel and Howell (1992), and a spectral analysis is in Siegel and Spuckler (1994).

References

Siegel, R. and Howell, J. R. (1992) *Thermal Radiation Heat Transfer*, 3rd edition, Hemisphere Publishing Corporation, Washington DC.

Siegel, R. and Spuckler, C. M. (1994) Approximate solution methods for spectral radiative transfer in high refractive index layers, *Int. J. of Heat and Mass Trans.*, 37 (Suppl. 1), 403–413.

Viskanta, R. and Grosh, R. J. (1962) Effect of surface emissivity on heat transfer by simultaneous conduction and radiation, *Int. J. of Heat and Mass Trans.*, 5, 729–734.

Nomenclature

- a absorption coefficient, 1/m
- a_λ spectral absorption coefficient, 1/m
- E_j exponential integral function of order j
- $e_{\lambda b}$ spectral blackbody function, W/m^3
- N number of thin plates
- N_1 conduction-radiation parameter $\lambda a/4\sigma T_1^3$
- $\dot{Q}_G$ rate of energy addition to gas in control volume, W
- $\dot{q}$ heat flux, W/m^2
- T absolute temperature, K
- t dimensionless temperature T/T_1
- x coordinate between surfaces (origin at surface 1), m
- α absorptivity of surface
- δ spacing between parallel surfaces, m
- ϵ emissivity of surface
- κ optical coordinate ax; κ_δ, optical thickness $a\delta$
- λ wavelength, m; thermal conductivity, W/m · K
- σ Stefan-Boltzmann constant, W/m^2 · K^4
- $\bar{\tau}$ mean transmittance

Subscripts

- G gas
- n the nth radiation shield
- 1,2 the two parallel surfaces
- λ spectral quantity

R. Siegel

PARSONS, CHARLES (see Steam engines)

PARTIAL PRESSURE

The concept of partial pressure applies, exactly, only to perfect gas mixtures. i.e., ones in which each component gas follows the equation of state $PV = n\tilde{R}T$ for all pressures and temperatures. Then:

$$PV = n\tilde{R}T \quad \text{where} \quad n = \Sigma n_i$$

Where $\tilde{R}$ is the universal gas constant, T the absolute temperature and n_i is the amount of component i present.

From this one can define a partial pressure of component i as,

$$P_i = n_i P/n$$

or,

$$P_i = \tilde{x}_i P$$

where $\tilde{x}_i$ is the mole fraction of component i.

It is used, as an approximation, for real gases at low or moderate pressures. For permanent gases, the approximation is often acceptable up to atmospheric pressure, or even slightly above. For condensible gases, the approximation may be an order of magnitude worse.

G. Saville

PARTICLE CHARACTERISATION (see Particle technology)

PARTICLE CONCENTRATION MEASUREMENT (see Coulter counter)

PARTICLE COUNTING (see Optical particle characterisation)

PARTICLE DEPOSITION (see Fouling)

PARTICLE DIAMETER, EQUIVALENT VOLUME SPHERE (see Fixed beds)

PARTICLE DIAMETER, MEAN VALUES (see Sauter mean diameter)

PARTICLE FLOW, IN NOZZLES (see Nozzles)

PARTICLE IMAGE VELOCIMETRY, PIV (see Tracer methods; Visualization of flow)

PARTICLE-LIQUID SEPARATION (see Hydrocyclones)

PARTICLE-PARTICLE INTERACTIONS (see Gas-solids separation, overview)

PARTICLES, DRAG AND LIFT (see Particle transport in turbulent fluids)

PARTICLE SIZE MEASUREMENTS

Following from: *Particle technology*

In order to characterise a dispersion of particles in a fluid, the determination of the size distribution of the particles is of vital importance. In studying a collection of particles which have a practically spherical form (drops of a liquid in a gas flow and gas bubbles in a liquid), it is suffice to analyze the distribution of particles by their diameters. In the case of solid particles, we can speak about an effective diameter. The diameter of the particles X is an argument whose values vary over a range in a given collection of particles. The collection of particles can be characterised by dividing them into classes with an interval ΔX. The quantity ΔX is limited mainly by the method of measurement and by the characteristics of the measuring device (the scale factor of a microscope, the difference in size of the sieve mesh, the value of the angle by which the light beam is deflected by a particle, etc.).

Particles with diameter from $X_n - \Delta X/2$ up to $X_n + \Delta X/2$ form a class which corresponds to an average size of diameter X_n. Plotting the number of particles of a given class against X_n over the range of classes under study, we obtain an average curve of distribution of particles by diameter (a frequency curve):

$$f'(X) = \frac{di}{dX},$$

where i is the number of particles with diameter smaller than X.

Summing up the number of particles in all classes in succession and plotting as ordinate the sum of the particles with diameters smaller than the given one, we get a cumulative curve, which also characterizes the distribution of particles by diameter:

$$f(X) = i = \int_0^X \frac{di}{dX} dX.$$

Typical ranges of the particle sizes formed or applied in various two-phase flows in various processes are given in **Table 1**.

Table 1. Typical particle size ranges

Application	Size range
Combustion of liquid fuels	10–800 μm
Atomization drying	10–1000 μm
Annular two-phase flow	10–400 μm
Plasma jets with dispersive phase	0.1–100 μm
Smoke aerosols for measuring the velocity with the help of LDA	0.1–5 μm

The existing methods of measuring the particle size can be divided into three groups:

- direct methods, when a sample of particles is taken directly from the flow under study;
- solidification methods, which are used for size distribution determination for liquid sprays. Here, the liquid droplets are solidified by freezing. To facilitate freezing, a substitute liquid may be employed which has a lower latent heat of solidification or a higher melting point;
- indirect methods, among the most often used of these are photographic and holographic techniques, optical methods including light scattering and laser-Dopler anemometry, investigation of particle tracks on their collision with a screen coated with an immersion layer (for instance, carbon black) and techniques based on drop impact with the thermocouples.

Direct methods of particle measurement are only used in investigation of flows containing solid particles which are extracted from the required points of a flow with the help of sampling tubes. The particles are then investigated either with the help of microscopes or by successive sieving through sieves with different mesh sizes, the differences of these forming a series of size classes.

On studying the atomization of a liquid, it is sometimes replaced by, say a wax which is atomized at a higher temperature, the spray drops then solidify and can be collected as solid particles. In this case, to describe the distribution of solidified drops of wax by size, sieves are also used as well as measurements with the help of a microscope.

The distribution of particles by size is often studied by an indirect method based on the investigation of traces left by the particles (usually by drops) on a plate covered with a layer of carbon black, magnesium oxide or oil and the size of impact craters on the layer is related to drop size. The screen is exposed to the spray cone for a short time (usually fractions of a millisccond) with the help of a quick-acting mechanism and then is studied under the microscope.

Another indirect method of measuring drop size can be achieved through the use of a thermal probe which is a thermocouple heated by a current. Drops falling onto the junction evaporate taking heat away from the thermocouple. The measured electric signal depends on the drop radius. This dependence is determined from the calibration. The size of the drops measured is from 3 to 1180 μm. The probe is good for constant monitoring of the dispersed phase combustion chambers, steam boilers, etc. The resolving power of the probe is about 10^3 drops/s over the range of 3–30 μm.

However, optical methods of measuring particle size have the potential to allow us to obtain—without destroying the flow structure—complete information on the dispersed phase: the size, shape, concentration, and the relationship between velocity and size.

Optical methods can be subdivided into two groups. To the first group belong *photographic* and *holographic* methods, to the second, the methods of *light scattering* and of *laser-doppler anemometry.* Let us consider these methods in more detail.

Photographing small fast moving particles requires high intensity illumination of the particle ensemble, high quality of films and optics, and also short exposure time. The required illumination is determined by the size of the particles and by their velocity. It known from geometrical optics that the intensity of light scattered by small spheres (10—1000 μm) is

$$J = J_0F(\theta)d,$$

where J_o is the intensity of the incident light, $F(\theta)$ is a function describing the dependence on the direction of observation relative to the incident light and d is the particle diameter. The time of exposure τ is determined by the maximum velocity υ of motion of particles with radius r: $\tau \leq r/\upsilon$, i.e., the displacement of the particle over the time of exposure must be less then its radius.

Among a great many photographic methods used in studying two-phase dispersions the method in which the illumination is performed with the help of a "laser knife" (**Figure 1**) deserves more attention. The beam (2) of the laser (1) is shaped into a thin plane light beam by an optical system which is a combination of short-focus (3) and long-focus (4) spherical lenses moving on a common axis relative to each other coupled with a cylindrical lens

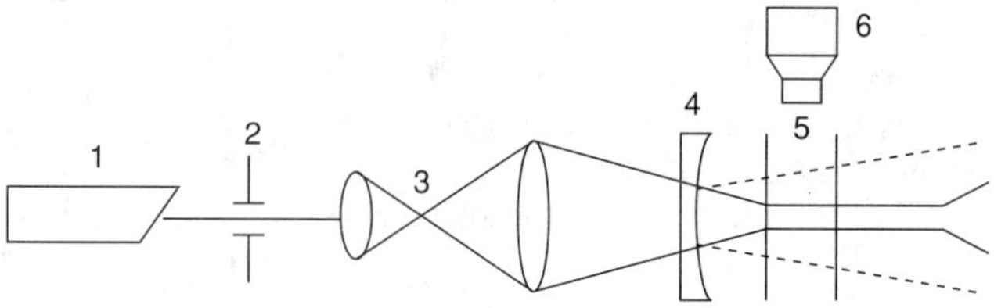

Figure 1. "Laser-knife" method for photography of dispersions.

(4). The "knife" can be displaced by changing the focal length of the telescope. It can be turned around its axis by changing the angle at which the cylindrical lens (4) is placed and set at sharp angles to the studied medium with the help of a prism. The process (5) under study is registered on a photographic camera (6). In order to obtain a sharp image of particles falling into the "light knife" area, a lens with a depth of focus larger than the transverse size of thc "light knife" is recommended to be used in the recording system. The resolving power of the device is determined by the quality of the photographic material and the lenses.

Holographic methods have been used successfully for registering a three-dimensional picture of development of fast-running processes in two-phase flows, spray cones and in burning. Holography is based on the registration of distribution of intensity in an interference pattern formed by a object (1,3,6,7) and by a reference (1,2,4,5) wave coherent with it (**Figure 2**). The hologram can be registered on some surface (two-dimensional recording) or in some volume (recording in a three-dimensional medium). In both cases, the main property of a hologram is preserved allowing it to transform a reference wave into a object (i.e. if a hologram is illuminated by a reference wave, then it creates the same amplitude-phase space distribution in its plane, as was created by an object wave on recording). A two-beam scheme, with an oblique reference beam which makes it possible to obtain a stereoscopic picture of particles, has received the widest acceptance in the diagnostics of two-phase flows. The limitation of particle dimension to a lower limit of up to 50–70 μm is mainly associated with the resolving power of recording materials. The main disadvantage of the method is the complexity of the experimental set-up and in the interpretation of data obtained; it is therefore often used for obtaining a qualitative picture of the flow.

Optical methods depending on the phenomenon of scattering of a plane monochromatic wave of light by the particles have gained wide spread acceptance. A curve, which characterizes the

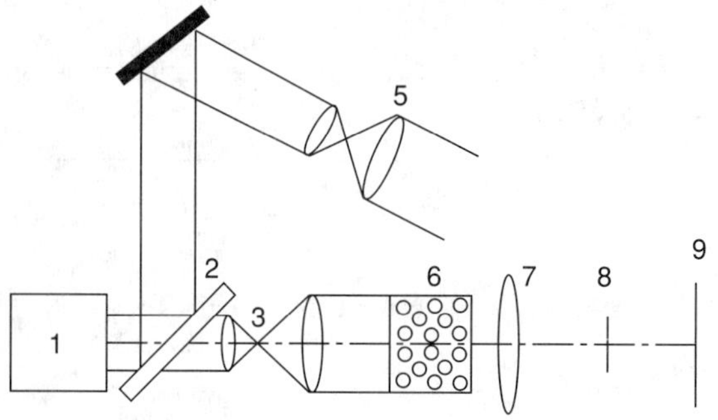

Figure 2. Holographic method for drop size determination.

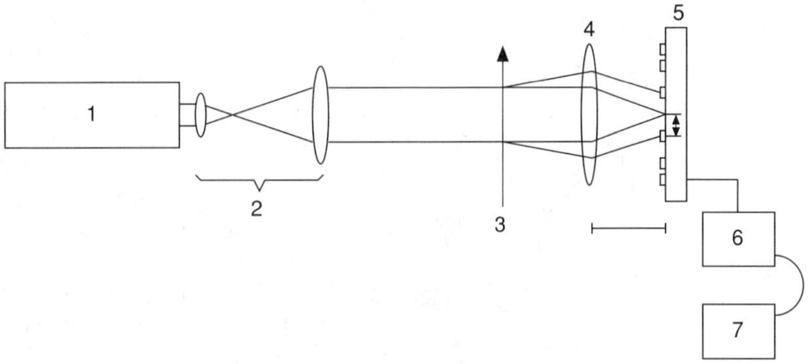

Figure 3. Malvern diffraction method.

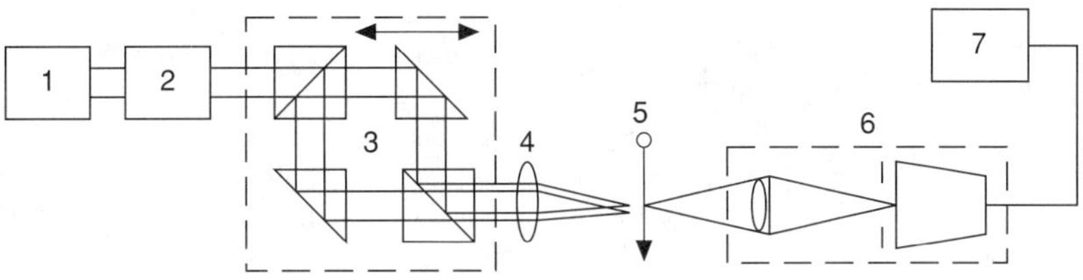

Figure 4. LDA measurement system.

intensity of scattered light in relation to the angle of observation (indicatrix of diffusion) can be used for determining the function of particle size distribution in the flow.

The methods of light scattering are used mainly for measuring the sizes of spherical particles over the range of from 0.02 to 2000 μm. The method of small angles enjoys wide application. It serves for measuring the size and the size distribution of spherical particles over the range of 3–2000 μm. The method of small angles is based on measuring the angular distribution of light scattered inside the cone of a small angle along the direction of the main light beam propogation.

Given in **Figure 3** is the schematic diagram developed by the Malvern company for measuring the particle sizes over the range of 0.5—1800 μm. The scheme combines the method of small-angle diagnostics of small particles deflecting light to large angles. Different particles (3) present in the laser light beam (1) diffract the beam at different angles, as a result of which diffraction rings of different diameters are formed in the focal plane of the Fourier lens (4). Each ring is associated with a certain dimension of particle. Recording the intensity of light in these rings with the help of specially developed multiple unit photodetector (5) and interpreting the signals by assuming a particular form of size distribution and applying computed processing (6,7) we can obtain the size particles distribution.

Particularly appealing for measuring the particle dimensions in a two-phase flow is the application of the laser-doppler anemometer (LDA) which makes it possible to also obtain in one experiment the information on the velocity of measured particles. In studying two-phases flows differentiating circuit of LDA are mainly used. For investigating two-phase flows an optical train with varying period of an interference field is required as well as the radio equipment for determining the parameters of doppler's signals. One of such is shown in **Figure 4**. Radiation from the laser (1) passes through a telescopic system (2) and falls into optico-mechanical divider (3), which allows the distance between the beams to be varied by to a displacement of one of the prisms, the beams being focused into an investigated point of flow (5), by means of lens (4). The procedure described allows the size of the interference field to be varied within the limits of 50–500 μm. The receiving part of LDA (6,7) is standard, and recording facilities allow us to analyze all the required parameters of the Doppler signal. More information on the application of the LDA technique to particle size measurement is given in the article **Optical Particle Characterisation**.

References

Azzopardi, B. I. (1979) Measurement of drop sizes, *Int. J. Heat Mass Trans*, 22, 1979.

Rinkyvichus, B. S. (1978) *Laser Anemometry*, Moscow, Energiya, 1978.

PARTICLES, RADIATIVE PROPERTIES OF (see Radiative heat transfer)

PARTICLE SIZE PARAMETER (see Optical particle measurement)

PARTICLES, RADIATIVE PROPERTIES OF (see Radiative heat transfer)

PARTICLES, SOLID IN LIQUID (see Liquid-solid flow)

PARTICLE TECHNOLOGY

A particle may be defined as a single entity comprising part of a solid or liquid discontinuous phase. According to this definition a particle may have any size. It is common to refer to a suspension of particles in a gas as an **Aerosol** and to particles in suspension in a liquid as a sol (*hydrosol* if the liquid is water). Clearly, when we are considering the stability of a suspension, particle size becomes important. For example, a suspension of $1\mu m$ (one micrometer = $1\mu m = 10^{-6}m$) particles in air may remain stable for many minutes, whereas $100\mu m$ particles will settle out in seconds. Similarly, flow-rates of particles from hoppers and bins, reaction rates of particles in reactors, separation efficiencies in filters and separators and most other aspects of particle behaviour depend on particle size.

Particle Technology Includes:

1. Methods of producing particles which meet a predetermined specification, expressed in terms of size distribution and shape distribution, by such processes as are appropriate including growth from the liquid phase and the gas phase.
2. Methods of producing composite particles by such processes as agglomeration and encapsulation.
3. The relationship between bulk properties of assemblies of particles and the measurable parameters of the individual particles.
4. The relationship between such properties as size, shape, their distribution and optimal range for the variables which control production, handling, processing and utilisation of particulate material.
5. Overall design of particulate processes. This could include the use of microprocessor and automatic control in general, in conjunction with on-line measurement devices.
6. Health and safety considerations of the production, handling, processing and utilisation of particulate materials including, for example, explosion risks and inhalation risks.

Leading to: *Gas-solids separation; Liquid-solids separation; Particle size measurement; Particle transport in turbulent fluids*

J.P.K. Seville and
IChemE Particle Technology Subject Group

PARTICLE TRACKING VELOCIMETRY (see Visualization of flow)

PARTICLE TRANSPORT IN TURBULENT FLUIDS

Following from: *Gas-solid flows; Particle technology*

Introduction

The transport of particles as solids, droplets or bubbles by a turbulent flow is a common enough feature in many natural and industrial processes; the mixing and combustion of pulverized coal in coal fired stations, and the dispersal of pollutants in the atmosphere and in rivers and estuaries are two of copious examples. In these sorts of flows the particles, while advected by the mean flow, are dispersed by the random motion of the **Turbulence**. In many cases of interest, the particle size and density difference (inertia) are sufficiently large that the particles do not follow either the variations in mean carrier flow or the turbulence, so unlike the transport of a passive contaminant, particle transport does not generally obey the heat mass transfer analogy; this is especially so in a turbulent boundary layer for example. Particle size and density can influence the transport in other ways. For instance, in many cases the mass loading of the dispersed phase is sufficiently high that it influences the motion of the carrier phase (two-way momentum coupling). To solve for the transport of an individual phase, it is necessary to solve for the transport of both phases coupled together. In the case of solid particles in a gas, this will occur when the volume fraction $\gtrsim 10^{-5}$ (Elghobashi, 1994). At higher volume fractions $\gtrsim 10^{-3}$ the motion of a particle will be influenced by the motion of its neighbours either directly through inter-particle collisions or indirectly through their influence on the particle drag law, i.e. four way coupling between phases may occur. Even if such coupling does not occur, at sufficiently high inertia particle-wall interactions (which determine whether a particle will bounce or adhere or resuspend) may dominate the transport.

While particle transport refers in general to the way particle-turbulence interaction influences all transport properties in either phase, the concern here is with the influence of the turbulence on the spreading and mixing of particles and ultimately on their deposition on containment walls.

There are two methods which are currently used to model particle transport. The first is the so-called *"trajectory" approach* where by solving their individual equations of motion, particles are tracked through a random flow field representative of the turbulence (see for example Berlemont et. al. 1990). The principal advantages of this method are that it is very easy to implement and relatively easy to make changes in the physics, e.g. in the particle equation of motion and in the nature of the particle wall interactions, Sommerfeld (1990). The disadvantage is that even in very dilute flows it requires many realisations of the flow field to obtain an adequate picture of the transport. In general, the method is restricted to fairly simple types of random flight models and the influence of large scales structure in complex flows is generally not accounted for. In addition, some approximation has to be made for the Lagrangian timescale of the fluid seen by the particles.

The second approach known as the *two-fluid approach* treats both carrier (continuous) and dispersed phases as two fluids co-existing and interacting with one another, modelling each fluid in terms of a set of "continuum" equations which represent the net conservation of mass, momentum and energy of either species within some elemental volume of the mixture. (See Elghobashi, 1994). The principal advantage of this approach over the trajectory approach is that it is computationally more efficient since it deals directly with the net transport property of interest. However, much controversy has surrounded the forms of these equations, not least of which concerns the type of averaging upon which they are based and the form of the constitutive relations that are necessary for their solution.

With the advent of super-computers *numerical simulation* of particles in random flow fields typical of real turbulence has been crucial in furthering our understanding and modelling of particle transport. The application of direct numerical simulation (DNS) tracks particles through a flow field, each realisation of which is a solution of the incompressible *Navier Stokes equation*. (See for example Squires and Eaton, 1991a,b, Elghobashi and Truesdell, 1992, and Mclaughlin, 1989.) The method is therefore relatively model free but limited to very low Reynolds number flow ~150. Despite these limitations, there is strong evidence that many of the main characteristics of real flows are being correctly predicted by DNS at least for moderate Reynolds numbers, especially in isotropic turbulence and turbulent boundary layers. The Reynolds number limitation has been overcome by *Large Eddy Simulation (LES)* resolving the 3 dimensional time dependent details of the largest scales of motion while using simple closure models for the smaller "sub-grid" scales. However, the method is only reliable with one way coupling, since in two way coupling it is the interaction between the sub-grid scales and the particles that dominates the coupling. Mention should also be made of simulations using so called *kinematic turbulence*. These are of two types: the first based on the Kraichnan's method of random Fourier modes (Fung et al., 1992); the second based on discrete vortex methods (Crowe et al., 1993) used to examine the dispersion by large-scale organized structures in wakes and shear layers. Unlike DNS each realisation of the flow field is not a solution of the Navier Stokes equations so it is suitable only for studying dispersion. Nevertheless, useful information that does not depend upon the detailed dynamics can be obtained.

The Particle Equation of Motion

Of crucial importance to any model of particle transport is the form used for the particle equation of motion. Unfortunately no equation of motion exists appropriate for all conditions. Analytic forms exist for only a restricted range of flow conditions. Nevertheless, these have proven useful in establishing empirical models that extend the range of applicability. As a useful starting point the general aerodynamic force $\underline{F}a$ acting on a particle moving in an unsteady flow field can be seen as the sum of several forces, i.e.

$$\underline{F}a = \underline{F}i + \underline{F}v + \underline{F}vi \tag{1}$$

where $\underline{F}i$ is the inertial inviscid force that would act in the absence of viscocity, $\underline{F}v$ is the viscous force in the absence of inertial forces, and $\underline{F}vi$ is the force due to the interaction of viscous and inertial effects. The inviscid force is composed of a force dependent upon the pressure gradient across the particle (and hence proportional to the local fluid acceleration) and an **Added Mass** term that accounts for the fact that a particle accelerating relative to the fluid is transferring momentum at a certain rate to the carrier flow: the increase in the apparent mass of the particle compared to the mass of the displaced fluid is commonly referred to as the added mass coefficient C_M. (For a sphere $C_M = {}^1/_2$.) In addition, a particle will experience a lift force normal to the direction of its relative velocity and the local vorticity of the flow. See Auton et al (1988) for the first correct derivation of $\underline{F}i$. Recent numerical simulations (Rivero et al., 1991) for small to moderate Reynolds numbers in laminar flow indicate that it is probably correct under a wide range of conditions.

The viscous force $\underline{F}v$ can be represented as the sum of two components:

$$\underline{F}v = \underline{F}v_s + \underline{F}v_h \tag{2}$$

where $\underline{F}v_s$ is the viscous force as if the flow were steady and $\underline{F}v_h$ an extra force derived from the diffusion of vorticity generated at the surface of the particle at a rate proportional to the particle's relative acceleration. Since the diffusion rate is finite, this means that the force is dependent on the history of the particle motion.

In one particular case of unsteady *Stokes flow*, the equation motion of a sphere has been derived exactly by Maxey and Riley (1983). For the low particle Reynolds number approximation $\underline{F}i$ is the same as that derived in Auton et al. (1988). The term corresponding to $\underline{F}v_s$, is due to Stokes steady drag while the remaining term is the history term $\underline{F}v_h$ commonly referred to as the *Basset history term*. Both terms differ from that in uniform steady flow in that they contain the influence of curvature in the ambient flow (evaluated at the centre of the particle); these "extra forces" are known as *Faxen forces*. The form originally derived by Maxey and Riley implicitly assumes that the particle is introduced into the flow with the same velocity as the carrier flow. In general, this cannot be true so an extra term must be included to account for shear stresses induced by the formation of an intense thin vortex layer arising from the mismatch of initial and flow conditions (Maxey, 1993).

The extent of viscous-inertial interaction measured by $\underline{F}vi$ had for some time been limited to the well known corrections due to Oseen (1927) and Proudman and Pearson (1957) for moderate particle Reynolds numbers, with higher Reynolds number forms based on measurement. All forms are generally expressed in terms of a **Drag Coefficient** C_D versus **Reynolds Number**, Re_p. In recent times significant advances have been made in our understanding of the influence of inertial forces in unsteady uniform flows and in steady uniform shear flows. Mei et al. (1991b) and Lovelanti and Brady (1993) have shown that at long times when inertial effects become important the kernel of $\underline{F}v_h$ (associated with the diffusion of locally created vorticity) decays faster than $t^{-1/2}$ based on simple diffusion theory. In a steady uniform shear a *sphere* for instance will experience a lift force (Saffman, 1965, 1968) as well as a rotating sphere in a uniform flow (Rubinow and Keller, 1961). The latter is important near a wall where depending on the conditions of impact a rebounding particle will have some induced spin. In an unbounded flow with kinematic viscosity ν and shear gradient G the frequently used form of a *lift* derived by Saffman is limited to conditions of low particle Reynolds number in which the *Saffman length* $(\nu/G)^1/_2 << $ the *Stokes length* $(\nu/|v_r|$ where v_r is the velocity of the particle relative to the local carrier flow velocity. In many flows of practical interest, especially in a turbulent boundary layer, these conditions are rarely satisfied. Very recently McLaughlin (1993) removed the restriction on Saffman length and extended the analytic forms to situations where the particle is near a wall. These lift predictions were consistent with those obtained numerically by Dandy and Dwyer (1990) who extended the range of lift for a stationary sphere to Reynolds numbers of about 100.

Transport in Simple Flows: Definitions, Scaling and Useful Concepts

We examine transport of particles with low Reynolds number (Stokes flow) in homogeneous stationary turbulence in which the mean flow is either uniform or a simple shear. These are somewhat idealized flows but help us to define some useful parameters and highlight some simple effects that influence transport in real flows. A useful measure of the influence of particle size and density (particle inertia) is the particle relaxation or response time τ_p defined for a sphere of radius a and density ρ_p as:

$$\tau_p = \frac{9\mu}{2a^2\left(\rho_p + \frac{1}{2}\rho_f\right)} \tag{3}$$

So if G is the typical gradient of the mean flow and $Tl_f^{(p)}$ the integral timescale of the turbulence measured along a particle trajectory, particles will follow the mean flow if $G\,\tau_p \ll 1$ and the turbulent fluctuations if $\tau_p/Tl_f^{(p)} \ll 1$. However in the case of the turbulence the value $Tl_f^{(p)}$ is not prescribed in any way, depending upon the value τ_p itself and the *Eulerian integral length scale* Le and *time scale* Te_f of the carrier flow turbulence. For instance in the absence of drift as the ratio τ_p/Te_f varies between 0 and ∞ the value of $Tl_f^{(p)}$ varies between $Tl_f^{(f)}$ (the fluid *Lagrangian integral timescale*) and Te_f (the *Eulerian integral timescale or eddy decay time*).

Transport in homogeneous turbulence

In uniform flows we consider motion in a frame of reference moving with the mean flow. Following Taylor's 1921 theory of diffusion by continuous movements, it is reasonable to assume that in an unbounded flow in the limit of diffusion times $>>$ both $Tl_f^{(p)}$ and τ_p the net spatial number density of particles decays in time according to **Fick's Law of Gradient Diffusion** with long time diffusion coefficients $D_{ij}(\infty)$ which denote the limiting values of $D_{ij}(t)$ where:

$$D_{ij}(t) = \langle v_i(t) y_j(t)\rangle \tag{4}$$

where $v_i(t)$ and $y_j(t)$ are the particle velocity and position at time t in the i and j directions and $\langle\ ..\ \rangle$ is a global ensemble average. In the case of a quasi-steady Stokes drag law one can show from the equation of motion that for long times the ratio of the particle to carrier flow (fluid element) long time diffusion coefficients is the ratio $Tl_f^{(p)}/Tl_f^{(f)}$ where $Tl_f^{(f)}$ is the carrier flow Lagrangian timescale. In the absence of drift, approximate calculations (Reeks, 1977) of $Tl_f^{(p)}/Tl_f^{(f)}$ for an isotropic random stationary Gaussian velocity field show that it is > 1 with a maximum value $Te_f/Tl_f^{(f)}$ for large particle that increases as the ratio of a structure parameter $\langle u'^2\rangle^{1/2}Te_f/Le$ (the parameter indicates the importance of persistent or coherent structures on the dispersion). That is, the long time particle diffusion is actually greater than that for an equivalent fluid element, a result also true of DNS isotropic decaying (Elghobashi and Truesdell, 1992) and forced turbulence (Squires and Eaton, 1991a) and confirmed by the experiments of Wells and Stock (1983). However, in the DNS study of decaying isotropic turbulence, the maximum value of this ratio occurred at an intermediate value of the ratio of τ_p/Te_f. This is further linked to the influence of structure on transport and to the trapping of particles within vortices: increasing particle inertia was found to produce a preferential bias in the particle trajectory towards regions of low vorticity and high strain rate (Squires and Eaton, 1991b). Similar biasing of the particle concentration towards regions of high velocity for particles settling under gravity produces a higher settling velocity than that in the absence of gravity (Maxey, 1987).

However a far more significant effect of gravitational settling is its influence on particle dispersion. If particles are settling out under gravity with some drift velocity v_g they see a shorter fluid timescale than those with zero drift because of the shorter time they spend in an eddy. So $Tl_f^{(p)}$ is smaller than its value for zero drift. If the transit time across any eddy due to the particles relative mean motion $>>$ the eddy decay time (i.e. $v_g >>$ the carrier flow rms velocity $\langle u'^2\rangle^{1/2}$ then this form for D_{ij} can be approximated by:

$$D_{ij}(\infty) = \int_0^\infty \langle u_i'(0,0)u_j'(\underline{v}_g s,)\rangle\, ds = \langle u'^2\rangle Le_{ij}/v_g \tag{5}$$

where Le_{ij} is the *Eulerian fluid length scale* appropriate for the i-j directions, one of these directions being in the direction of v_g. The influence of the mean motion on the diffusion coefficient is known as the crossing trajectory effect (Yudine, 1959); the implication is that diffusion is reduced normal and parallel to the direction of gravity. In the case of isotropic turbulence the ratio of these length scales is a $^1/_2$ implying a similar ratio of the corresponding particle **Diffusion Coefficients**, an effect also confirmed by the experiments of Ferguson and Stock (1994) and by DNS (Squires and Eaton, 1991a, and Elghobashi & Truesdell, 1992).

Finally, particle transport in homogeneous turbulence can be used to introduce the concept of particle pressure and to evaluate the so called equation of state. Pressure is used here in a general sense to denote the surface forces on an elemental volume due to the net momentum transferred across those surfaces from the turbulent motion of individual particles. So at equilibrium there will be a balance between the gradient of these surface forces and the net weight of the particles due to gravity. Using an approach similar to that used in classical thermodynamics to evaluate the virial of system of interacting molecules one can show (Reeks, 1991) that:

$$p_{ij} = (\langle \rho v_i v_j\rangle + \tau_p^{-1}\langle y_i(t)\cdot u_j^{(p)}(t)\rangle\langle\rho\rangle \tag{6}$$

where $-p_{ij}$ are the components of the particle stress tensor, $\langle p\rangle$ the net mass density of the particles and $u^{(p)}(t)$ the carrier flow velocity along a particle trajectory. The first term on the RHS is the contribution from the particle *Reynolds stresses* and the second term from the interfacial momentum transfer due to the turbulence, $\tau_p^{-1}\langle \rho u_i\rangle$ which is the gradient of a surface force at position $\underline{x}$ in the flow so that:

$$\langle \rho u_i\rangle = -\langle u_i^{(p)}(t) y_j(t)\rangle \frac{\partial\langle\rho\rangle}{\partial x_j} \tag{7}$$

A similar result is given by Simonin et al. (1993). The equation of state also implies a fundamental relationship between the quantities on the right handside of **Equation 6** and the particle diffusion coefficient $D_{ij}(\infty)$. Thus:

$$\frac{p}{\langle \rho \rangle} = \tau_p^{-1} D(\infty) = \langle v_i' v_j' \rangle + \tau_p^{-1} \langle y_i(t) u_j^{(p)}(t) \rangle \quad \textbf{(8)}$$

This relationship, valid in all turbulent flows, expresses a fundamental equivalence at equilibrium between the gradients of surface forces and diffusion fluxes.

Transport in a simple shear flow

Analysis based on Stokes drag of the long term *dispersion of particles* in a simple shear (Reeks, 1993) highlights certain deficiencies in the the traditional two fluid model assumption that the dispersed phase behaves as a **Newtonian Fluid**; i.e. at equilibrium in a bounded simple shear flow the shear stresses for example are directly proportional to the symmetric mean rate of shearing of the dispersed phase (the so-called *Boussinesq assumption*). In reality, each component of the particle stress is the sum of two components: a homogeneous component having a form identical to that in homogeneous turbulence i.e. explicitly independent of the shear; and a deviatoric component which depends upon the value of the shearing of *both* the carrier and dispersed phases. The deviatoric components are zero for fluid point motion but dominates over the homogeneous component as $G\tau_p \gg 1$. Furthermore, unlike the homogeneous stresses, the deviatoric stresses are always asymmetric (arising from the inherent asymmetry in the interfacial momentum transfer surface forcer). There are no violations of Cauchy's second law since the net surface couple generated on an elemental volume is precisely compensated by an equal and opposite body couple. However this asymmetry does influence the dispersion: with the streamwise diffusion coefficients increasing faster than the cross-streamwise components an elemental volume of particles is stretched into a long narrow ellipsoid that rotates until its major axis is along the streamwise direction (Hyland, McKee and Reeks, 1993).

Crossing trajectories will also influence the particle dispersion: so as in homogeneous flow, it is significantly reduced when the particles are settling under gravity. They can also play a part when at any point in shear the mean particle velocity is different from that of the carrier phase (Huang and Stock, 1994).

Transport including other aerodynamic forces

A thorough investigation of the particle transport using an equation of motion which includes added mass and history forces (see the Maxey and Riley equation) is currently restricted to kinematic simulation carried in a stationary random isotropic Gaussian homogeneous velocity fields, see Mei et. al, (1991). While particle RMS velocities are reduced, the particle dispersion coefficients in the long term are surprisingly little different from their values based on Stokes drag. This has some connection with the result that the particle stop distance is the same as its Stokes drag value. Wang and Stock (1988), and Mei et al, (1991) have also evaluated the influence of non-Stokes drag on particle dispersion in the same flow field: they found that the long term diffusion coefficient for large particles in the absence of gravity was slightly greater than the equivalent value based on Stokes drag. As to the dispersion in a simple shear only the influence of lift forces has so far been investigated. Ounis & Ahamdi (1991) have shown that the principle influence of lift is to increase the long term diffusion coefficient in the cross-streamwise direction compared to its Stokes drag value which is independent of the local shear. Hyland (1994) has confirmed this result and shown that the diffusion coefficient in the other directions are enhanced by the lift with similar trends in behaviour with increasing particle inertia. He has also considered the evolution of a point source of particles released at the centre of the shear, observing a similar stretching and rotation of an evolving ellipsoid but at a greater rate to that for Stokes drag. The dispersion is in general described by the telegraph equation which approaches the normal diffusion equation as $t \to \infty$.

Rational Theories for Transport in Real Flows

The simple analytic theories based on simple diffusion that we have used to quantify and analyze transport in the simple flows are no longer adequate in non uniform flows. However they form a basis for developing more general approaches. There are two approaches worthy mentioning, namely the approaches by Reeks (1992) and Simonin et al (1993) which in the end yield very similar results and generate the correct behaviour in the simple flows discussed above. Both use the *two-fluid model approach.* Thus in the case of a simple Stokes drag the net momentum equation for example would be found by averaging the instantaneous equations for the momentum based on the equation of motion. Using density weighted (Favre) averages you obtain:

$$\langle \rho \rangle \frac{D\bar{v}_i}{Dt} = -\frac{\partial}{\partial x_j} \langle \rho v_j' v_i' \rangle + \tau_p^{-1} \langle \rho \rangle (\langle u_i \rangle - \bar{v}_i) + \tau_p^{-1} \langle \rho u_i \rangle \quad \textbf{(9)}$$

where $\bar{\underline{v}}$ is the density weighted mean particle velocity and the instantaneous carrier flow velocity $\underline{u}$ is expressed as the sum of a mean $\langle u \rangle$ and fluctuating component $\underline{u}'$. The essential problem is then finding closed equations for the Reynolds stresses and the turbulent interfacial momentum transfer term $\langle \rho u'_j \rangle$. Reeks uses a pdf approach similar to that used in kinetic theory whereby the constitutive relations for the dispersed phase are derived from approximate solutions of a pdf equation representing the probability density that a particle has a certain velocity and position at time t (the analogue of the Boltzmann equation) with a term representing the net force experienced by a particle due to its interaction with random turbulent eddies. For a particle with velocity $\underline{v}$ and position $\underline{x}$ the component of this net force per unit volume is given by:

$$\tau_p^{-1} \langle u_i' W \rangle = -\left(\frac{\partial}{\partial v_j} \mu_{ji} + \frac{\partial}{\partial x_j} \lambda_{ji} + \gamma_i \right) \langle W(\underline{v}, \underline{x}, t) \rangle \quad \textbf{(10)}$$

where W $(\underline{v}, \underline{x}, t)$ is the instantaneous phase space density and the dispersion coefficients μ_{ij} and λ_{ij} and γ_i are dispersion coefficients dependent upon displacements in the velocity $\Delta\underline{v}$ $(\underline{v}, \underline{x}, t|0)$ and position $\Delta\underline{y}(\underline{v}, \underline{x}, t|0)$ of a particle starting out at time zero and arriving at $(\underline{v}, \underline{x})$ at time t The net interfacial momentum term $\langle \rho u'_i \rangle$ is then simply the value of $\langle u'_i W \rangle$ integrated over all particle velocities, namely

$$\langle \rho u_i' \rangle = -\langle u_i'(\underline{x}, t) \Delta y_j(\underline{x}, t|0) \rangle \frac{\partial \langle \rho \rangle}{\partial x_j} - \left\langle \Delta y_j(\underline{x}, t|0) \frac{\partial u_i'(\underline{x}, t)}{\partial x_j} \right\rangle \langle \rho \rangle \quad \textbf{(11)}$$

A similar result has been derived by Simonin et al 1993 by recognising that the fluctuating interfacial term is responsible for an extra fluid-particle drift over and above that due to the mean relative velocity $\langle \underline{u} \rangle - \underline{\bar{v}}$. It is the presence of this drift velocity in particle settling under gravity in homogeneous turbulence that causes the globally averaged settling velocity to be higher than in it is in still fluid (Maxey, 1987). The form derived for this drift term gives an expression which assuming an exponential decaying autocorrelation for the fluid is identical to the value derived by Reeks. The form is an obvious generalistation of the results derived for simple flows with suitable averages defined about a specific point in space, rather than a global average associated with the whole ensemble. We note that in addition there is an extra term which is zero in homogeneous flows equivalent to a body force which will tend to move particles from high to low regions of turbulence. In weakly inhomogeneous flows in the long time limit when the inertial acceleration term can be ignored the momentum equation reduces to a simple diffusion equation with value for D_{ij} etc given by the fundamental **Equation 8** and a drift velocity (derived from the body force and the gradient of the particle co variance). The latter is referred to as the *turbophoretic velocity* (Reeks, 1992) indicating that particles will drift from high to low regions of turbulence. This behaviour has been observed in experiment (Young and Hanratty, 1991). This has important implications for the behaviour of particles in turbulent boundary layers (Reeks et al., 1993).

References

Auton, T. R., Hunt, J. C. R., and Prud'homme, M. (1988) The force exerted on a body in inviscid unsteady non-uniform rotational flow, *J. Fluid Mechanics*, 197, 241–257.

Berlemont, A., Desjonqueres, P., and Gouesbet, G. (1990) Particle Lagrang ian simulation in turbulent flows, *Int. J. of Multiphase Flow*, 16, 19–34.

Crowe, C. T., Chung, J. N., and Troutt, T. R. (1993) Particle dispersion by organized turbulent structures, *Particulate Two-Phase Flow*, Rocco, M., ed., Butterworth.

Dandy, D. S. and Dwyer, H. A. (1990) A sphere in shear flow at finite Reynolds number: effect of shear on particle lift, drag, and heat transfer, *J. Fluid Mechanics*, 216, 381–410.

Elghobashi, S. (1994) On predicting particle-laden turbulent flows, *Applied Scientific Research*, 52, 309–329.

Elghobashi, S. and Truesdell, G. C. (1992) Direct simulation of particle dispersion in a decaying isotropic turbulence, *J. Fluid Mechanics*, 242, 655–700.

Ferguson, J. R. and Stock, D. E. (1994) Effects of fluid continuity on turbulent particle dispersion, *J. Fluid Mechanics*, Vol. submitted.

Fung, J. C. H., Hunt, J. C. R., Malik, N. A., and Perkings, R. J. (1992) Kinematic simulation of homogeneous turbulence by unsteady random Fourier modes, *J. Fluid Mechanics*, 136, 281–318.

Huang, X., Stock, D. E., and Wang, L.-P. (1993) Using the Monte-Carlo process to simulate two-dimensional heavy particle dispersion, *ASME/FED, Gas-Solid Flows*, 166, 153–167.

Hyland, K. E. (1994) The modelling of particle dispersion in turbulent fluid flows, Ph.D. Dissertation, University of Strathclyde, Glasgow.

Hyland, K. E., McKee, S., and Reeks, M. W. (1993) The dispersion of particles in a simple shear flow, *ASME/FED, Gas-Solid Flows*, 166, 177–182.

Lovalenti, P. M. and Brady, J. F. (1993) The hydrodynamic force on a rigid particle undergoing arbitrary time-dependent motion at small Reynolds number, *J. Fluid Mechanics*, 256, 561–605.

Maxey, M. R. (1987) The gravitational settling of aerosol particles in homogeneous turbulence and random flow fields, *J. Fluid Mechanics*, 174, 441–465.

Maxey, M. R. (1993) The equation of motion for a small rigid sphere in a nonuniform or unsteady flow, *ASME/FED, Gas-Solid Flows*, 166, 57–62.

Maxey, M. R. and Riley, J. J. (1983) Equation of motion for a small rigid sphere in a nonuniform flow, *Physics of Fluids*, 26, 883–889.

McLaughlin, J. B. (1989) Aerosol particle deposition in numerically simulated channel flow, *Physics of Fluids A*, 1, 1211–1224.

McLaughlin, J. B. (1993) The lift on a small sphere in wall-bounded linear shear flows, *J. Fluid Mechanics*, 246, 249–265.

Mei, R., Adrian, R. J., and Hanratty, T. J. (1991a) Particle dispersion in isotropic turbulence under Stokes drag and Basset force with gravitational settling, *J. Fluid Mechanics*, 225, 481–495.

Mei, R., Lawrence, C. J., and Adrian, R. J. (1991b) Unsteady drag on a sphere at finite Reynolds number with small fluctuations in the free-stream velocity, *J. Fluid Mechanics*, 233, 613–631.

Ossen, C. W. (1927) *Hydrodynamik*, Leipzig, 132.

Ounis, H. and Ahmadi, G. (1991) Motions of small particles in a turbulent simple shear flow field under microgravity condition, *Physics of Fluids A*, 3, 2559–2570.

Proudman, I. A. N. and Pearson, J. R. A. (1957) Expansions at small Reynolds number for the flow past a sphere and a circular cylinder, *J. Fluid Mechanics*, 2, 237–262.

Reeks, M. W. (1977) On the dispersion of small particles suspended in an isotropic turbulent field, *J. Fluid Mechanics*, 83, 529–546.

Reeks, M. W. (1991) On a kinetic equation for the transport of particles in turbulent flows, *Physics of Fluids A*, 3, 446–456.

Reeks, M. W. (1992) On the continuum equations for dispersed particles in nonuniform flows, *Physics of Fluids A*, 4, 1290–1303.

Reeks, M. W. (1993) On the constitutive relations for dispersed particles in nonuniform flows. 1: Dispersion in a simple shear flow, *Physics of Fluids A*, 5, 750–761.

Reeks, M. W., Swailes, D., Hyland, K. E., and McKee, S. (1993) A unifying theory for the deposition of particles in a turbulent boundary layer, *ASME/FED, Gas-Solid Flows*, 166, 109–112.

Rivero, M., Magnaudet, J., and Fabre, J. (1991) Quelques résultats nouvea ux concernant les forces exercées sur une inclusion sphérique par un écoule ment accéléré, *C. R. Acad. Sci. Paris*, 312, 1499–1506.

Rubinow, S. I. and J. B. Keller (1961) The transverse force on a spinning sphere moving in a viscous fluid, *J. Fluid Mechanics*, 11, 447–459.

Saffman, P. G. (1965) The lift on a small sphere in a slow shear flow, *J. Fluid Mechanics*, 22, 285–400.

Saffman, P. G. (1968) The lift on a small sphere in a slow shear flow-corrigendum, *J. Fluid Mechanics*, 31, 624.

Simonin, O., Deutsch, E., and Minier, J. P. (1993) Eulerian prediction of the fluid/particle correlated motion in turbulent two-phase flows, *Applied Scientific Research*, 51, 275–283.

Sommerfeld, M. (1960) Numerical simulation of particle dispersion in a turbulent flow, Numerical methods for multiphase flow: The importance of particle lift forces and particle/wall collision models, *ASME/FED, Gas-Solid Flows*, 91, 11–18.

Squires, K. D. and Eaton, J. K. (1991a) Measurements of particle dispersion obtained from direct numerical simulations of isotropic turbulence, *J. Fluid Mechanics*, 226, 1–35.

Squires, K. D. and Eaton, J. K. (1991b) Preferential concentration of particles by turbulence, *Physics of Fluids A*, 3, 1169–1178.

Taylor, G. I. (1921) Diffusion by continuous movements, *Proc. London. Math Soc.*, 20, 196–212.

Wang, L. P. and Stock, D. E. (1989) Numerical simulation of heavy particle dispersion time step and nonlinear drag consideration, *J. Fluids Engineering, Transactions of the ASME*, 114, 100–106.

Young, J. B. and Hanratty, T. J. (1991) Optical studies on the turbulent motion of solid particles in a pipe flow, *J. Fluid Mechanics*, 231.

Yudine, M. I. 1959. Physical consideration on heavy-particle dispersion, *Advances in Geophysics*, 6, 185–191.

M.W. Reeks

PARTICLE VELOCITY RESPONSE TIME (see Gas-solid flows)

PARTICLE WALL COLLISIONS (see Erosion)

PARTICLES IN LIQUID (see Brownian motion)

PARTICULATE FOULING (see Fouling)

PASQUILL CATEGORY (see Plumes)

PASSIVE CONTAINMENTS, FOR NUCLEAR REACTORS (see Containment)

PASSIVE PLUME (see Plumes)

PASSIVE SAFETY FEATURES (see Boiling water reactor, BWR)

PASSIVE SOLAR DESIGN OF BUILDINGS (see Solar energy)

PASSIVE SOLAR WATER HEATER (see Solar energy)

PAULI EXCLUSION PRINCIPLE (see Van der Waal's forces)

PEAK HEAT FLUX (see Burnout in pool boiling)

PDA (see Anemometers, laser-Doppler)

PECLET, JEAN CLAUDE EUGENE (1793–1857)

A French physicist, born February 10, 1793 at Besancon, Peclet, became one of the first scholars of the Ecole Normale at Paris, Gaylussac and Dulong being his teachers. Peclet was elected professor at the College de Marseille in 1816, teaching physical sciences there until 1827. He returned to Paris when nominated maitre de conferences at the Ecole Normale and was elected professor at the important Ecole Centrale des Arts et Manufactures. In 1840 he became inspecteur general de l'instruction publique and retired from this charge in 1852 to devote himself exclusively to teaching.

His publications were famous for their clarity of style, sharpminded views and well performed experiments. His famous book "Traité de la Chaleur et de ses Applications aux Arts et aux Manufactures" (Paris 1829) was distributed worldwide and had been translated into German.

Peclet continued lecturing until his death December 6, 1857 at Paris.

JEAN CLAUDE EUG. PECLET 1793–1857

U. Grigull, H. Sandner and J. Straub

PECLET NUMBER

Peclet number, Pe, is a dimensionless group representing the ratio of heat transfer by motion of a fluid to heat transfer by thermal conduction.

$$Pe = \frac{uL}{\kappa} = Re\ Pr$$

where u is fluid velocity, L is a characteristic dimension, and κ is thermal diffusivity of the fluid. Re is **Reynolds Number** and Pr is **Prandtl Number**.

Heat transported by the fluid per unit area is proportional to $u\rho c_p$ where ρ is density and c_p is specific heat capacity, while heat conducted per unit area is proportional to λ/L where λ is thermal conductivity. Hence

$$\frac{\text{heat transported}}{\text{heat conducted}} \approx \frac{u\rho c_p}{\lambda/L} = \frac{uL}{(\lambda/\rho c_p)} = \frac{\mu L}{\kappa}$$

Peclet number for mass transfer has the corresponding form

$$Pe_m = \frac{\mu L}{\delta} = Re\ Sc$$

where δ is diffusivity and Sc is **Schmidt Number**.

G.L. Shires

PECLET NUMBER EFFECT IN LIQUID-METAL HEAT TRANSFER (see Liquid-metal heat transfer)

PELTIER EFFECT

The Peltier effect, discovered by Jean Peltier in 1834, is an important *Thermoelectric Phenomenon* that relates to the energy transfer (positive or negative) that occurs, over and above **Joule Heating**, at the junction of two dissimilar materials when an electric current passes through it. When the junction is maintained at a given temperature, the Peltier effect results in the equivalent of a heat addition or heat removal—the Peltier heat—which is reversible and is proportional to the current. Thus,

$$\text{Peltier heat} = \pi I$$

where π is the Peltier coefficient and I is the electric current. Note that π depends on the materials forming the junction and the temperature.

The Peltier effect is one of the key phenomena (along with the *Thompson effect*) determining the emf generated in a thermocouple used for temperature measurement. For a thermocouple of materials A and B, with one junction at a constant temperature and the other at (absolute) temperature T,

$$\frac{d\varepsilon_{AB}}{dT} = \frac{\pi_{AB}}{T}$$

where ε_{AB} is the thermocouple emf generated at the junction of materials A and B. This equation can be used to calculate the Peltier coefficient for the combination of materials A and B.

The Peltier effect is used in *thermoelectric (Peltier) refrigerators* or *heat pumps*. Such devices provide heat removal, sometimes addition, in an easily controlled and reversible device without moving parts. They have been used in space vehicles as well as in a number of small commercial devices for controlling temperature or providing refrigeration, often using semiconductor materials as the thermoelectric elements.

R.J. Goldstein

PELTIER REFRIGERATOR (see Peltier effect)

PELTON TURBINES (see Hydraulic turbines)

PENG-ROBINSON EQUATION (see PVT relationships)

PENMAN EQUATION FOR POTENTIAL EVAPORATION (see Environmental heat transfer)

PENSTOCKS (see Hydro power)

PENTANE

n-Pentane (C_5H_{12}) is a colourless, highly volatile liquid that is soluble in water. It is obtained by fractional distillation of petroleum. It forms an explosive and flammable mixture with air at low concentrations. In high concentrations it acts as a narcotic. Its main use is in the production of gasoline for internal combustion engines. It is also used as a solvent, in extraction processes and in low-temperature thermometers.

Molecular weight: 72.151	Critical temperature: 469.6 K
Melting point: 143.4	Critical pressure: 3.369 MPa
Normal boiling point: 309.2 K	Critical density: 273.3 kg/m^3

[See **Table 1** on page 825]

V. Vesovic

PERFECT GAS

Following from: *Thermodynamics*

A perfect gas is a gas which follows the equation of state:

$$p\tilde{v} = \tilde{R}T,$$

where p denotes the pressure (N/m^2); $\tilde{v}$ the molar volume (m^3/mole); $\tilde{R}$ (= 8314 J/mole K or 8314 kJ/k mole K) the universal gas constant; and T absolute temperature, K.

If both sides of the equation are be divided by $\tilde{M}$ (the molecular mass of a given gas), the equation will be:

$$pv = RT, \tag{1}$$

where v denotes the specific volume (m^3/kg) and R the gas constant for the gas under consideration (J/kg K).

From the equation of state and from the equations of thermodynamics differentials

$$\left(\frac{\partial \tilde{u}}{\partial \tilde{v}}\right)_T = -p + T\left(\frac{\partial p}{\partial T}\right)_v \tag{2}$$

it can be shown that $(\partial\tilde{u}/\partial\tilde{v})_T = 0$, i.e. the internal energy of the

Table 1. Pentane: Values of thermophysical properties of the saturated liquid and vapor

T_{sat}, K p_{sat}, kPa	309.2 101.3	335 227	350 341	365 492	380 688	395 935	410 1249	425 1634	440 2103	469.6 3370
ρ_l, kg/m^3	610.2	582.9	566.0	548.0	528.9	507.9	484.1	456.5	423.5	280.9
ρ_g, kg/m^3	3.00	6.36	9.41	13.51	18.99	26.11	36.21	49.73	68.96	184.1
h_l, kJ/kg	319.8	383.8	423.3	458.2	504.7	546.6	588.5	637.3	686.2	846.7
h_g, kJ/kg	678.0	721.1	744.3	767.6	790.8	814.1	837.4	855.9	876.9	846.7
$\Delta h_{g,l}$, kJ/kg	358.2	337.3	321.0	309.4	286.1	267.5	248.9	218.6	190.7	
$c_{p,l}$, kJ/(kg K)	2.34	2.52	2.62	2.72	2.82	2.94	3.06	3.20	3.44	
$c_{p,g}$, kJ/(kg K)	1.79	1.96	2.05	2.16	2.28	2.48	2.66	2.96	3.37	
η_l, μNs/m^2	196	159	140	123	108	95	83	72	60	
η_g, μNs/m^2	6.9	7.6	8.1	8.5	9.0	9.5	10.2	11.1	12.4	
λ_l, mW/(K m)	107	98	93	88	83	79	75	71	69	
λ_g, mW/(K m)	16.7	19.3	21.0	22.8	24.8	26.7	29.0	31.7	34.9	
Pr_l	4.29	4.09	3.94	3.80	3.67	3.53	3.36	3.16	2.84	
Pr_g	0.74	0.77	0.79	0.81	0.83	0.88	0.94	1.03	1.19	
σ, mN/m	14.3	11.3	9.7	8.1	6.7	5.2	3.8	2.8	1.6	
$\beta_{e,l}$, kK^{-1}	1.40	1.93	2.21	2.52	2.92	3.54	4.48	5.95	15.5	

The values of thermophysical properties have been obtained from reference: C.F. Beaton, G.F. Hewitt (1989), *Physical Property Data for the Design Engineer,* Hemisphere Publishing Corporation, New York.

perfect gas does not depend on volume; this may be regarded as an independent property of a perfect gas.

Often it is assumed that at $p \to 0$, and at any temperature T, every real gas becomes identical to perfect gas. However, this is not necessarily true.

At moderate pressures as a theoretically proved equation of state for a real gas is the virial equation as follows:

$$\frac{p\tilde{v}}{\bar{R}T} = 1 + \frac{B}{\bar{v}} + \frac{C}{\bar{v}^2} + \cdots, \tag{3}$$

where B, C . . . are respectively the second, third etc. virial coefficients, which are only temperature dependent.

This equation can be rearranged as follows:

$$\frac{p\bar{v}}{\bar{R}T} = 1 + B'p + C'p^2 + \cdots, \tag{4}$$

where the primed coefficients B′, C′ etc. are also functions only of the temperature and can be expressed in terms of B, C. . . .

For instance

$$B' = B/\bar{R}T. \tag{5}$$

From Equation 4 it follows, that

$$\lim_{p\to 0} \frac{p\bar{v}}{\bar{R}T} = 1, \tag{6}$$

as it is for the perfect gas.

But some other properties of a real gas at $p \to 0$ differ from that of the perfect gas. For example from Equation 4

$$\bar{v} = \frac{\bar{R}T}{p} + B'RT + C'\bar{R}Tp + \cdots. \tag{7}$$

Since $\bar{R}T/p = \bar{v}_{perf.}$, the molar volume of a perfect gas, and taking into consideration Equations 5, from Equation 7 it follows

$$\bar{v} = \bar{v}_{perf.} + B' + C'\bar{R}Tp + \cdots. \tag{8}$$

From which

$$\lim_{p\to 0} \bar{v} = \bar{v}_{perf.} + B'.$$

The same is valid, for instance, for the Joule-Thomson coefficient α

$$\alpha = \left(\frac{\partial T}{\partial p}\right)_h = -\left(\frac{\partial \bar{h}}{\partial p}\right)_T / \bar{C}_p, \tag{10}$$

where $\bar{h}$ denotes the molar enthalpy and $\bar{C}_p$ molar heat capacity.

From the thermodynamic differential equation:

$$\left(\frac{\partial \bar{h}}{\partial p}\right)_T = \bar{v} - T\left(\frac{\partial \bar{v}}{\partial T}\right)_p \tag{11}$$

it follows that for a perfect gas $(\partial\tilde{h}/\partial p)_T = 0$ and $\alpha = 0$.

But for a real gas at $p \to 0$

$$\lim_{p\to 0}\left(\frac{\partial \tilde{h}}{\partial p}\right)_T = B - T\frac{dB}{dT},$$

which in general is not zero and therefore $\lim_{p\to 0} \alpha \neq 0$.

In thermodynamic and thermo-chemical calculations, the so called perfect gas thermodynamic functions are used: $\tilde{u}_0$; $\tilde{h}_0$; $\tilde{a}_0$

(molar Helmholtz energy); $\bar{g}_0$ (molar Gibbs energy.). For a real gas the respective functions are equal to the perfect gas functions at $p \to 0$.

***Leading to:** Gay-Lussac's law; Boyle-Marriotte law*

E.E. Shpilrain

PERFECT MIXTURES (see Fugacity)

PERFORATED TRAYS (see Distillation)

PERIODIC KILNS (see Kilns)

PERMANENT MOULDING (see Casting of metals)

PERMEABILITY (see Darcy's law; Porous medium)

PERMEABILITY COEFFICIENT (see Membrane processes)

PERMEABILITY OF VACUUM (see Electromagnetic waves)

PERMEATE (see Membrane processes)

PERMITTIVITY

Permittivity is the relationship between the electric flux density and the electric field strength of a capacitor with a certain dielectric material. i.e.

$$\frac{\text{Electric flux density}}{\text{Electric field strength}} = \frac{Q/a}{V/d} = \frac{Cd}{a}$$

where Q is the charge, V is the potential difference, d is the thickness of the dielectric in the direction of flux and a is the cross-section area perpendicular to the direction of flux. C is the capacitance of the dielectric. The permittivity of a material is defined relative to the permittivity of a vacuum (or air as a good approximation) that is the permittivity of free space.

$$\varepsilon_0 = Cd/a$$

ε_0 has the value 8.85×10^{-12} Fm^{-1}.

The relative permittivity, ε_r, is the ratio of the capacitance of a capacitor having a certain material as dielectric to that of a capacitor with vacuum as dielectric. This statement is embodied in the relationship

$$C = \frac{\varepsilon_o\varepsilon_r \cdot a}{d} \text{ Farads}$$

The absolute permittivity of a dielectric is therefore

$$\varepsilon = \varepsilon_o\varepsilon_r \ Fm^{-1}$$

Values of relative permittivity of common insulating materials are in the range 2–10 (e.g., bakelite, 4.5–5.5; glass, 5–10) and of air is 1.0006.

K. Scott

PERMITTIVITY OF VACUUM (see Electromagnetic waves; Permittivity)

PERTURBATION METHODS (see Asymptotic expansion)

PERVAPORATION (see Membrane processes)

PETROCHEMICALS (see Distillation)

PETROL (see Oils)

PETROL ENGINES (see Internal combustion engines)

PETROLEUM

***Following from:** Fuels*

Petroleum, or crude oil, consists of hydrocarbons covering a wide range of boiling points and molecular weights. Although compositional analysis is usually taken to C_{20}, crude oil is best visualised as a smooth continuous gradation of boiling point matter up to approximately C_{100}. The relative quantities of the many thousands of components vary according to the origin of the oil. Chemically, petroleum consists primarily of three dominant species—paraffins, naphthenes or cycloparaffins, and aromatics—within which there are numerous complex sub-species. (See also **Hydrocarbons** and **Paraffins**) In addition there are small quantities of sulphur, nitrogen and oxygen and trace quantities of heavy metals. There is some relationship between geographical location and the composition of the crude oil.

Petroleum is found as an accumulation of hydrocarbons in porous sedimentary rocks ranging in age from ten to several hundred million years old. It is believed that petroleum originates from anaerobic decomposition of constituents of marine life, plankton and algae leading to the formation of organic clays which, when subject to pressure, temperatures up to 300 °C and geological time, form petroleum.

In the petroleum industry, oil density is traditionally measured according to an inverse scale derived by the American Petroleum Institute, known as the *API gravity.* The relationship with specific gravity, measured at 60 °F (15.6 °C) is according to the following equation: Density = $(141.5/SG_{60}) - 131.5$. Crude oil density ranges from approximately 800.0 to 930.0 kg/m3 (45.3–20.6 degrees API)

Further Reading

Petroleum Engineering Handbook, H. B. Bradley—Editor in Chief, Society of Petroleum Engineers, Texas, USA. ISBN 1-55563-010-3.

Our Industry Petroleum, Published by British Petroleum Plc.

***Leading to:** Oils; Oil refining*

D.W. Reay

PETUKHOV AND KIRILLOV CORRELATION FOR HEAT TRANSFER (see Tubes, single phase heat transfer in)

PETUKHOV-POPOV CORRELATION (see Triangular ducts, flow and heat transfer)

PHASE (see Film thickness determination by optical methods)

PHASE DOPPLER ANEMOMETERS (see Anemometers, laser-Doppler)

PHASE DOPPLER ANEMOMETRY, FOR SIMULTANEOUS VELOCITY AND DROP SIZE MEASUREMENT (see Anemometers, laser Doppler)

PHASE DOPPLER ANEMOMETRY, PDA (see Optical particle characterisation)

PHASE EQUATIONS (See Conservation equations, two-phase)

PHASE EQUILIBRIA (see Extraction, liquid-liquid)

PHASE EQUILIBRIUM

***Following from:** Thermodynamics; Mass transfer*

Phase equilibrium conditions of a system comprising ν components (where $\nu \geq 1$) and φ phases (where $\varphi \geq 1$) results from the Second Law of Thermodynamics and may be expressed by following equalities: temperatures $T^{(\alpha)}$ (thermal equilibrium, in which there no heat flows); pressures $p^{(\alpha)}$ (mechanical equilibrium, i.e. the phases are not separated); chemical potentials $\tilde{\mu}_i^{(\alpha)}$ of each of the components (no mass transfer):

$$T^{(1)} = T^{(2)} = \cdots = T^{(\varphi)} = T; \quad (1)$$

$$p^{(1)} = p^{(2)} = \cdots = p^{(\varphi)} = p; \quad (2)$$

$$\tilde{\mu}_i^{(1)} = \tilde{\mu}_i^{(2)} = \cdots = \tilde{\mu}_i^{(\varphi)} = \tilde{\mu}_i; \qquad 1 \leq i \leq \nu. \quad (3)$$

In the presence of force fields, and/or surface tension, (which occurs at the phase interfaces) condition (2) needs to be modified.

The chemical potential of a pure substance is equal to the molar (specific) Gibbs energy at the given aggregate state $\tilde{\mu}^{(\alpha)} = \tilde{g}^{(\alpha)}(T,p)$. The chemical potential of a component in a mixture is equal to the partial Gibbs energy, which depends on the phase composition of the ν-component geterogeneous system:

$$\tilde{\mu}_i^{(\alpha)} = \tilde{g}_i^{(\alpha)}(T, p, \tilde{x}_1^{(\alpha)}, \ldots, \tilde{x}_\nu^{(\alpha)}),$$

where $\tilde{x}_i^{(\alpha)} = \tilde{N}_i^{(\alpha)}/\tilde{N}^{(\alpha)}$ is the mole fraction of the ith-component in phase α, where $\tilde{N}_i^{(\alpha)}$ is the number of moles of the component in phase α and $\tilde{N}^{(\alpha)}$ is the total number of moles of all components in the given phase (α).

$$\sum_{i=1}^{\nu} \tilde{x}_i^{(\alpha)} = 1, \qquad 1 \leq \alpha \leq \varphi. \quad (4)$$

In the ν-component, φ phase system there are $\nu\varphi$ component fractions $\tilde{x}_i^{(k)}$. Together with temperature T and pressure p, this gives a total of ($\nu\varphi$+2) variables defining the (equilibrium) system. However, the equilibria and summation expression (Equations 1–4), imply that the variables are not independent and only f variables were.

$$f = \nu - \varphi + 2 \quad (5)$$

are independent. The f value is called the number of the degrees of freedom (or the variants) of the system. Equation 5 expresses the Gibbs Phase Rule. As soon as $f \geq 0$, the number of the phases which are in the equilibrium can not exceed $\nu + 2$ in a ν-component system (if a pure substance, then $f \leq 3$); a system is called non–variant if $f = \nu + 2$.

For a pure substance, equilibrium of two phases (α and β, say) obeys the condition

$$g^{(\alpha)}(T, p) = g^{(\beta)}(T, p). \quad (6)$$

A differential equation of the equilibrium curve follows from (6) if the thermodynamic correlations

$$(\partial g/\partial T)_p = -s, \qquad (\partial g/\partial p)_T = v \quad (7)$$

are taken into account; the equation is the well-known **Clapeyron-Clausius Equation**:

$$dp_l^{(\alpha\beta)}/dt = (s_l^{(\beta)} - s_l^{(\alpha)})/(v_l^{(\beta)} - v_l^{(\alpha)}), \quad (8)$$

where $v_l^{(\alpha)}$, $v_l^{(\beta)}$, $s_l^{(\alpha)}$, $s_l^{(\beta)}$ are the molar (specific) volumes and entropies of the coexisting phases. The enthalpy change accompanying the isothermal-isobaric phase transformation α to β is determined by the difference in entropy or enthalpy of the coexisting phases, which are participating the process transformation:

$$r^{(\alpha\beta)} = T(s_l^{(\beta)} - s_l^{(\alpha)}) = h_l^{(\beta)} - h_l^{(\alpha)}.$$

Figures 1–3 presents the Gibbs energy dependency on phase temperature gaseous (g), liquid (l), solid (s) phases at three different pressure values. The correlations (Equation 7) explain a shape of the curves considering that the entropy should be less for those phases which have a more regular molecular structure; on the contrary a density for such a phase should be greater, as a rule. Consistent with Equation 6, the intersection points of the curves define the parameters of the phase transformations at equilibrium: sublimation (s → g), melting (s → l), evaporation (l → g). The curves formed by such points are given in **Figure 4** often referred

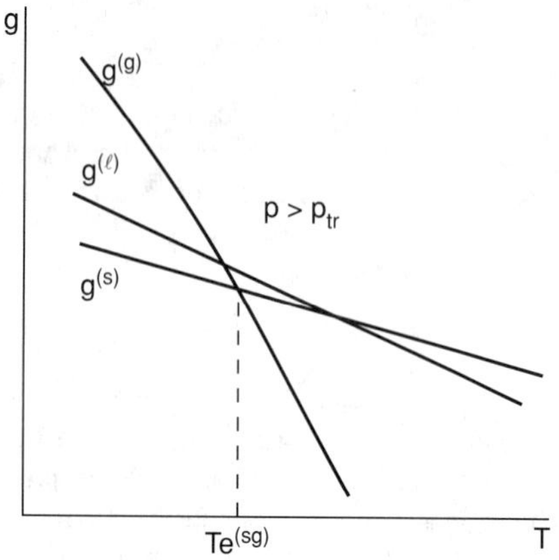

Figure 1. Phasic Gibbs energy as functions of temperature at pressures above the triple point.

to as a phase diagram. If all the lines intersect at the same point, then they give a triple point, with parameters p_{tr}, T_{tr}, at which three phases (solid, liquid, and gaseous) exit simultaneously. Triple points of the type s-l-g are inherent to all pure substances exept ^4He; because of quantum effects, the liquid phase of ^4He forms crystals at higher pressures. As a result, the phase diagram shape essentially differs in comparison with **Figure 4**. Also, this figure does not present the equilibrium curves and the triple points corresponding to phase transformations which occur between solid states due to allotropic modifications. The slope of the melting curve given in **Figure 4** is typical for the majority of the normal substances which obey the correlation $v^{(l)} > v^{(s)}$. However, for the case where $v^{(l)} < v^{(s)}$ in the vicinity of the triple point of some anomolous substances (for example water), the slope of the curve is negative ($dp_f^{(sl)}/dT < 0$). The vapor-liquid equilibrium curve starts at the triple point and finishes at the critical point T_c, p_c at which liquid and gaseous phases become identical. At super critical parameters $T > T_c$, $p > p_c$. There is no the phase transformation of type $l \rightarrow g$ and the substance is in high density state, which has sometimes been called the fluid state (f).

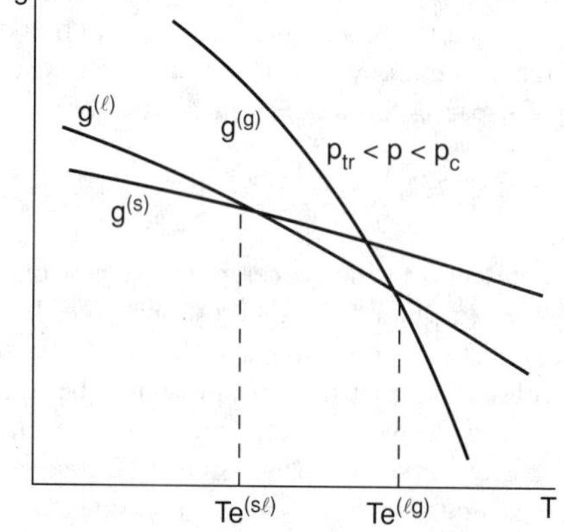

Figure 2. Phasic Gibbs energy as a function of temperature at a pressure between the triple and critical points.

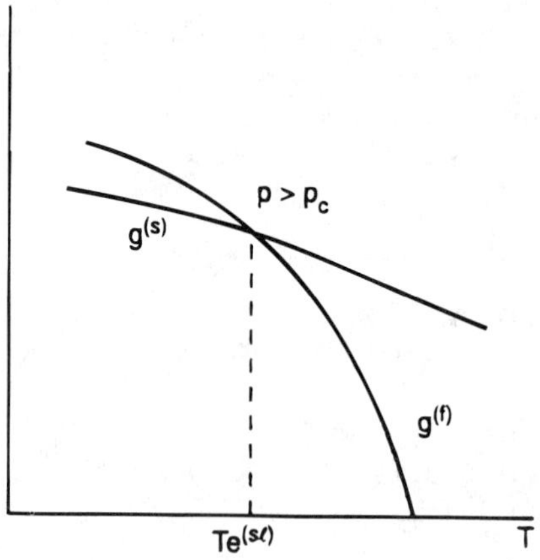

Figure 3. Phasic Gibbs energy as a function of temperature at pressure.

According to the Second Law of Thermodynamics, the stable states of substances at given T and p relate to minimal values of the Gibbs energy g(T,p). An unstable state can only exist transiently. Metastable states may exist indefinitely until some sufficiently strong external influence, causes the system to pass into the stable state. Stability of the metastable state diminishes as the parameters of the system depart more and more from the equilibrium state. The metastable states of a superheated liquid (which doesn't boil) and a supercooled liquid (which doesn't form crystals) as well as a supercooled vapor (which doesn't condense) are frequent in practice; these states are represented by lines above the lines which have minimum Gibbs energy as shown in **Figures 1** and **2**.

The thermodynamic properties ψ of the heterogeneous system—which may be defined as a first derivative of the Gibbs energy with respect to the parameters in Equation 7—specific (molar) volume, entropy, enthalpy, etc.— may be calculated by summing of the same property $\psi^{(\alpha)}$ related to the phase under equilibrium

$$\psi = \sum_{\alpha=1}^{\varphi} z^{(\alpha)}\psi^{(\alpha)}, \qquad \textbf{(9)}$$

where $z^{(\alpha)} = N^{(\alpha)}/N$ is a mass (or mole) fraction of phase α, N is the total number of moles in the system. Of course,

$$\sum_{\alpha=1}^{\varphi} z^{(\alpha)} = 1. \qquad \textbf{(10)}$$

Under equilibrium, for a pure substance, the fraction z^{α} may have any value between 0 and 1 at a given p and T subject to these values obeying Equation 10. An example, is the moisture fraction $z^{(l)}$ in the case of vapor-liquid equilibrium. Here, $z^{(l)}$ is referred to as **Quality**. This type of state of the heterogeneous systems is called *indifferent*. In such states, as well as at the critical point of pure substances—thermal expansion $(\partial v/\partial T)_p$, isothermal compressibility $(\partial v/\partial p)_T$, as well as an isobaric heat capacity $c_p = T(\partial s/\partial T)_p$—are infinitely large. Also, constant volume values of specific heat capacity $c_v = T(\partial s/\partial T)_v$, as well as, the derivative $(\partial p/\partial T)_v$ vary with a jump, when the substance passes from a one-phase into two-phase state; they stay defined in the two-phase region. **Figures 4 and 5** illustrate the described situation for pure substances with dependenties of specific volume and entropy on temperature and pressure. The heavy lines relate to borders of the coexisting phases regions, the weak lines are isotherms, and the broken ones are isochores.

If the total composition of the ν-component, φ-phase system is given, then it follows that:

$$\tilde{x}_i = N_i/N; \qquad N_i = \sum_{\alpha=1}^{\varphi} N_i^{(\alpha)};$$

$$N = \sum_{i=1}^{\nu} N_i = \sum_{\alpha=1}^{\varphi} N^{(\alpha)}; \qquad \sum_{i=1}^{\nu} \tilde{x}_i = 1, \tag{11}$$

and hence:

$$\sum_{\alpha=1}^{\varphi} z^{(\alpha)}\tilde{x}_i^{(\alpha)} = \tilde{x}_i; \qquad 1 \le i \le \nu. \tag{12}$$

The number of the equilibrium equations together with the material

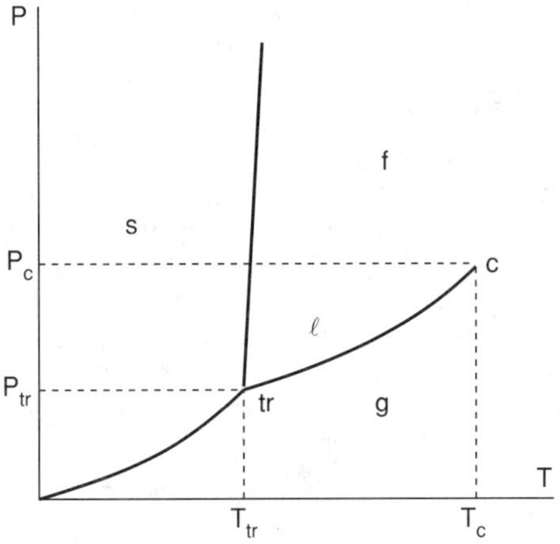

Figure 4. Phase diagram in terms of pressure and temperature.

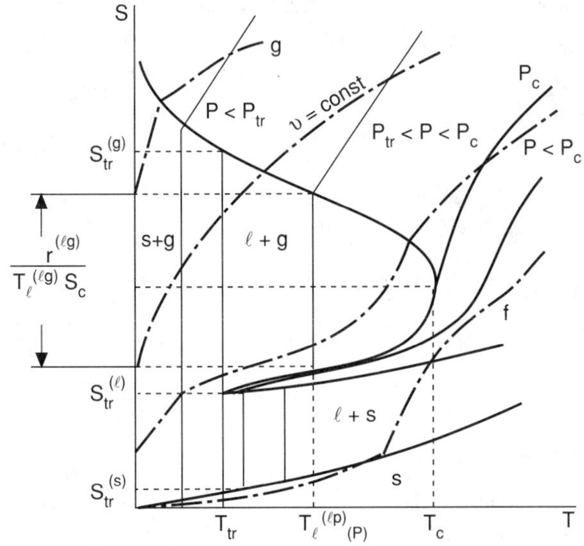

Figure 5. Phase diagram in terms of entropy and temperature.

balance equations are exactly equal to the number of the values $\tilde{x}_i^{(\alpha)}$, $z^{(\alpha)}$, which need to be defined.

The thermodynamic behavior of the multi-component heterogeneous systems is identical to that which has been described above for pure substances. In particular, appearance (or disappearance) of the phase causes a discontinuity in the temperature and pressure dependencies of those thermodynamic functions, which may be defined as the first derivatives of the Gibbs energy (v, s, h, etc.). As to the second derivatives ($(\partial v/\partial T)_p$, $(\partial v/\partial p)_T$, c_p, c_v, etc.), they change the values in a jumpwise manner. Indifferent thermodynamic states are inherent to azeotropic mixtures and non- or monovariant multicomponent mixtures.

The number of types of the phase equilibrium is far greater in multicomponent systems in comparison with pure substances. In particular, at certain conditions, some liquids or high-density gases may split into two phases, which have the same aggregate states but different compositions (liquid-liquid or gas-gas equilibria).

The types of phase equilibrium described above relate to the first order of phase change. By contrast, the second order of phase changes (ferromagnetic-paramagnetic, normal conductor— superconductor, normal—superfluid helium, etc.) are not accompanied by volume and heat effects; that is why, in such situations, there are no phase equilibrium states.

Reference

I. Prigogine, R. Defay, (1954) *Chemical Thermodynamics*, Longmans, London.

Leading to: *Vapour-liquid equilibrium*

M.Y. Boyarsky and A.M. Semjonov

PHASE INVERSION (see Liquid-liquid flow)

PHASE RULE

The conditions for there to be equilibrium between two more phases are that the **Pressure** P, **Temperature** T, and **Chemical Potential** μ_i, of each substance should have the same value in every phase. These conditions impose a restriction on the possible number of phases that can be in mutual equilibrium.

The Phase Rule is usually expressed in the form:

$$P + F = C + 2$$

where P is the number of phases present, C is the number of components and F is the number of *degrees of freedom*. The degrees of freedom refer to the values of pressure, temperature and mole fraction in a particular phase which can be varied at will. It is important that no other variable be treated as a degree of freedom. One common mistake is to treat an overall mole fraction as a degree of freedom.

A single component system has C = 1 and hence P + F = 3. If there is but a single phase present, F = 2 and the variables pressure and temperature can both be varied indepenently. If two phases are present, say liquid and gas, only pressure (or temperature) can be varied independently; i.e. we are moving along the vapour pressure curve. If three phases are present, nothing can be varied—we are at the triple point.

In a two component system, C = 2 and P + F = 4. If two phases are present, F = 2 and we can vary independently any two of pressure, temperature, mole fraction of, say, component 1 in phase 1, or mole fraction of component 1 in phase 2. A knowledge only of the overall composition and either P or T is not sufficient to specify the state.

G. Saville

PHASE SEPARATION, OVERVIEW

The best overview of phase separation is provided by the chart shown in **Figure 1** [Weast, 1972] on page 831. Most of this section will discuss what is on this chart and show how it can be used.

For *dispersions*, the most important characterizing parameters are the size and density of the dispersed phase and the density and viscosity of the continuous phase. The particle size is one of the coordinates of **Figure 1** on page 831. A single dispersed phase specific gravity of 2 is assumed. The behavior of these particles suspended in two different continuous phases, air and water, are displayed on this chart. Starting at the top of **Figure 1**, the sizes in various unit systems are given. Of particular importance, this chart shows the relation of particle sizes to several different screen sizes.

Particles that are familiar to all of us, of various sizes are given as examples for dispersions in both air and water. This helps us to relate the diameters of particles in general to the diameters of particles that we know. The range of particles that is displayed goes all the way from large, single molicules to gravel.

Methods of analyzing particles various sized are also suggested. These methods include all the methods also available for separating particles from the suspending fluid. Included are impingers, sieves, centrifuges, electrostatic precipitators, filters and gravity. (see also **Particle Size Measurement**)

Many porous or fluffy particles like ash, soot, or flocculent precipitates of various kinds have an effective density which is much closer to that of the continuous phase than that of the pure compound of which they are nominally composed. The average porosity of the particle and its diameter are the critical parameters in determining the relative velocity between these particles and the continuous phase.

Phase separation is a concern for a number of pairs of phases. These include:

- Liquid-Gas
- Liquid-Solid
- Gas-Solid
- Liquid-Liquid
- Solid-Solid Systems.

The separation of all these combinations of phases is discussed in detail in Perry (1973). Of the five combinations given above, the liquid-liquid and solid-solid systems are not mentioned in **Figure 1**, so a brief introduction to the separation of these systems will be included here. (See also **Vapour-Liquid Separation, Liquid-Solids Separation** and **Gas-Solids Separation**)

Liquid-Liquid Systems

Separation in liquid-liquid systems is most commonly accomplished by means of either gravity or a centrifuge. (See **Cyclones**) The key difference is whether the droplets which constitute the dispersed phase are large enough so that they rise or fall at a useful velocity. **Figure 1** shows how rapidly a particle of specific gravity 2 will settle in water. The settling velocity on Figure 1 is proportional to the density difference (See **Stokes' Law**), so we are speaking of very small settling velocities in a one g field when the particles are small.

The settling velocity can be greatly enhanced by use of a centrifuge, some of which have effective accelerations of several thousand g's.

Solid-Solid Systems

Separation of solids can be accomplished by a variety of processes working on density differences, size differences, electrostatic forces and magnetic forces. (See **Electrostatic Separation**). **Figure 2** on page 832, Perry (1973) shows the size ranges for which different kinds of sorting methods are appropriate. The chart is for mixtures of solids suspended in gases or liquids.

References

Weast, R. C. (1972) *Handbook of Chemistry and Physics*, CRC Press, Cleveland, OH, PF 230.

Perry, R. H. and Chilton, Cecil H. (1973) *Chemical Engineers Handbook*, Fifth Edition, Chapters 18, 19, 20, 21.

Leading to: *Cyclones; Electrostatic separation; Gas-solids separation; Liquid-solids separation; Vapour-liquid separation*

P. Griffith

PHASE SPACE (see Non-linear systems)

PHASE VELOCITY (see Speed of light)

PHASE VELOCITY, AVERAGE (see Multiphase flow)

PHENOL

Phenol (C_6H_5OH) is a white crystalline solid of distinctive odour. It used to be known as *carbolic acid*, as it was first isolated from coal tar. There are a number of processes for its production using cumene or benzene as a starting material. It is very toxic

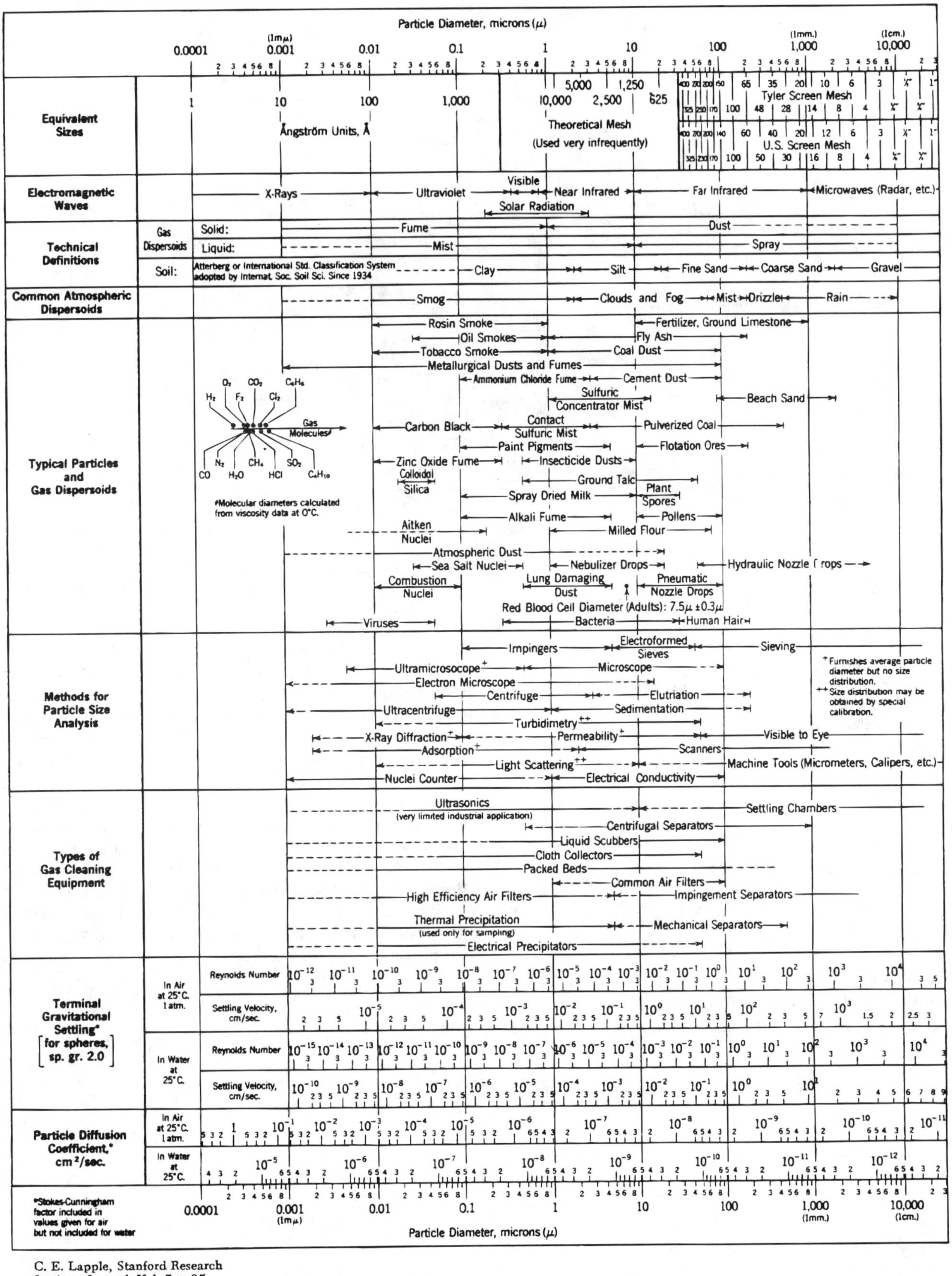

C. E. Lapple, Stanford Research Institute Journal, Vol. 5, p.95 (Third Quarter, 1961)

Figure 1. Characteristics of particles and particle dispersolids.

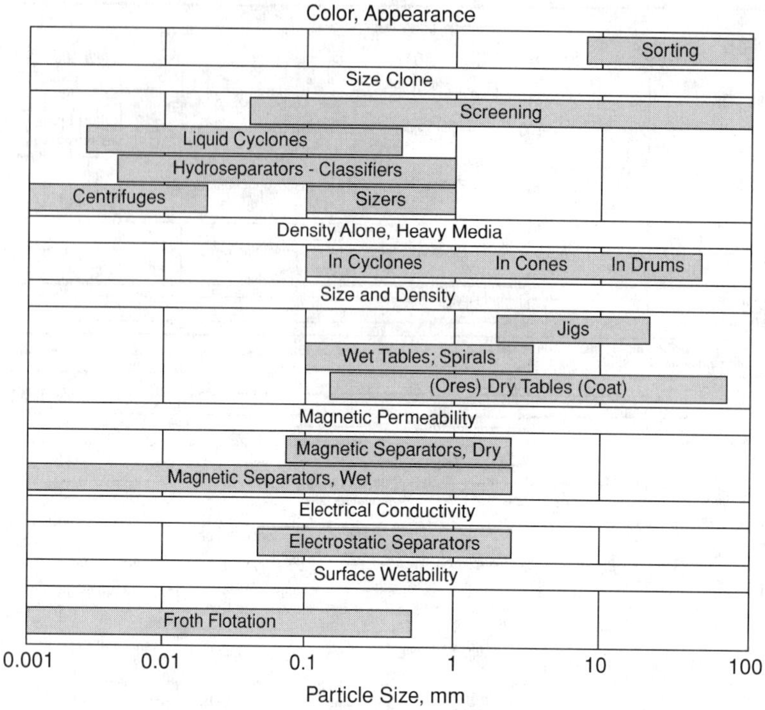

Figure 2. Particle size as a guide to the range of applications (Perry, 1973).

and contact with skin causes burns. Its main usage is in organic synthesis for the production of resins, pharmaceuticals, dyes and paints.

Molecular weight: 94.1	Critical temperature: 693.2 K
Melting point: 313.90 K	Critical pressure: 6.130 MPa
Normal boiling point: 454.95 K	Critical density: 435.7 kg/m^3

[see **Table 1** on page 833.]

V. Vesovic

PHENOMENOLOGICAL MODELS (see Pressure drop, two phase flow; Void fraction)

PHONONS, IN THERMAL CONDCUTIVITY OF SOLIDS (see Thermal conductivity mechanisms)

PHOSPHORESCENCE

Following from: *Photoluminescence*

In general phosphorescence is the long-lived emission of light which persists for longer than about one hundred nanoseconds after excitation. Shorter lived emission is termed **fluorescence**. The term is derived from the latin *phosphorus-light bringing*.

In molecular photochemistry, phosphorescence is defined as the emission of light associated with a transition between states of different spin multiplicity and usually arises from the transition from the first excited triplet to the ground state singlet. As discussed by Turro (1991) such transitions are "forbidden" and hence occur relatively slowly. Because of the long life of the triplet state the excited-state energy can be efficiently "quenched" in fluid solution it is usual to work at 77K in rigid organic glasses to observe phosphorescence. However in some cases, where molecules contain atoms other than those from the first row of the periodic table (e.g. halogenated polyaromatics and complexes of second or third row transition elements), room temperature phosphorescence can be detected. Oxygen is an efficient quencher of phosphorescence in fluid solution. Phosphorescence emission lies at lower energy than fluorescence (See **Photoluminescence** for further details.)

The term phosphorescence is often used in solid state photophysics to indicate a temperature dependent emission process involving recombination of trapped electrons and holes (see also **Thermoluminescence**).

References

Turro, N. J. (1991) *Modern Molecular Photochemistry*, University Science Books, California.

P. Douglas

PHOSPHORIC ACID

Phosphoric acid (H_3PO_4) is a crystalline solid of molecular weight of 97.995 when pure. It melts at 315.6 K and its specific

Table 1. Phenol: Values of thermophysical properties of the saturated liquid and vapor

T_{sat},K	455	480	505	530	555	580	605	635	665	693.15
P_{sat},KPa	101.3	216	404	693	1100	1650	2360	3410	4720	6129
ρ_l, kg/m^3	955	932	905	877	851	809	772	713	636	436
ρ_g, kg/m^3	2.60	5.36	9.88	16.9	27.2	41.9	62.9	101	170	436
h_l, kJ/kg	−153	−93	−41	20	80	147	206	290	399	514
h_g, kJ/kg	336	374	411	447	482	515	545	575	587	514
$\Delta h_{g,l}$, kJ/kg	489	467	452	427	402	368	339	285	188	
$c_{p,l}$,kJ/(kg K)	2.55	2.61	2.66	2.76	2.85	3.01	3.10	3.43	3.77	
$c_{p,g}$, kJ/(kg K)	1.63	1.74	1.85	1.95	2.06	2.22	2.44	2.94	5.10	
η_l, μNs/m^2	351	256	219	166	137	113	90.1	75.2	65.3	
η_g, μNs/m^2	12.8	13.5	14.3	15.2	16.2	17.3	18.6	20.2	23.4	
λ_l, mW/(K m)	175	170	166	162	157	154	149	141	130	
λ_g, mW/(K m)	28.9	31.6	34.8	38.4	42.4	46.8	51.4	56.2	64.3	
Pr_l	5.11	3.93	3.51	2.83	2.59	2.21	1.87	1.57	1.89	
Pr_g	0.72	0.74	0.76	0.77	0.79	0.82	0.88	1.07	1.86	
σ, mN/m	24.5	20.9	18.2	14.6	12.4	10.2	7.6	5.3	2.7	
$\beta_{e,l}$,kK^{-1}	0.91	1.05	1.27	1.34	1.83	2.41	3.25	4.72	15.6	

The values of thermophysical properties have been obtained from reference. C. F. Beaton, G. F. Hewitt—*Physical Property Data for the Design Engineer*, Hemisphere Publishing Corporation, New York, 1989.

gravity is 1.8. At lower strengths in aqueous solution it is a colourless liquid. It is corrosive and attacks ferrous metals. It is also toxic and an irritant to skin.

V. Vesovic

PHOSPHORUS

Phosphorus—(Gr. *phosphoros*, light bearing; ancient name for the planet Venus when appearing before sunrise), P; atomic weight 30.97376; atomic number 15; melting point (white) 44.1°C; boiling point (white) 280°C; specific gravity (white) 1.82, (red) 2.20, (black) 2.25 to 2.69; valence 3 or 5.

Discovered in 1669 by Brand, who prepared it from urine, phosphorus exists in four or more allotropic forms: white (or yellow), red, and black (or violet). White phosphorus has two modifications: α and β with a transition temperature at −3.8°C. Never found free in nature, it is widely distributed in combination with minerals. *Phosphate* rock, which contains the mineral *apatite*, an impure tri-calcium phosphate, is an important source of the element. Large deposits are found in the U.S.S.R., in Morocco, and in Florida, Tennessee, Utah, Idaho, and elsewhere. Phosphorus is an essential ingredient of all cell protoplasm, nervous tissue, and bones.

Ordinary phosphorus is a waxy white solid; when pure it is colorless and transparent. It is insoluble in water, but soluble in carbon disulfide. It takes fire spontaneously in air, burning to the pentoxide. It is very poisonous, 50mg constituting an approximate fatal dose. Exposure to white phosphorus should not exceed 0.1 mg/M^3 (8-hr time-weighted average—40-hr work week). White phosphorus should be kept under water, as it is dangerously reactive in air, and it should be handled with forceps, as contact with the skin may cause severe burns. When exposed to sunlight or when heated in its own vapor to 250°C, it is converted to the red variety, which does not phosphoresce in air as does the white variety. This form does not ignite spontaneously and it is not as dangerous as white phosphorus. It should, however, be handled with care as it does convert to the white form at some temperatures and it emits highly toxic fumes of the oxides of phosphorus when heated. The red modification is fairly stable, sublimes with a vapor pressure of 1 atm at 417°C, and is used in the manufacture of safety matches, pyrotechnics, pesticides, incendiary shells, smoke bombs, tracer bullets, etc.

White phosphorus may be made by several methods. By one process, tri-calcium phosphate, the essential ingredient of phosphate rock, is heated in the presence of carbon and silica in an electric furnace or fuel-fired blast furnace. Elementary phosphorus is liberated as vapor and may be collected under water. If desired, the phosphorus vapor and carbon monoxide produced by the reaction can be oxidized at once in the presence of moisture or water to produce phosphoric acid, an important compound in making super-phosphate fertilizers. In recent years, concentrated phosphoric acids, which may contain as much as 70 to 75% P_2O_5 content, have become of great importance to agriculture and farm production.

Phosphates are used in the production of special glasses, such as those used for sodium lamps. Bone-ash, calcium phosphate, is also used to produce fine chinaware and mono-calcium phosphate used in baking powder. Phosphorus is also important in the production of steels, phosphor bronze, and many other products. Trisodium phosphate is important as a cleaning agent, as a water softener, and for preventing boiler scale and corrosion of pipes and boiler tubes.

Handbook of Chemistry and Physics, CRC Press

PHOTOCHROMIC DYE TRACING

Following from: *Tracer techniques; Visualization of flow*

A photochromic dye is a material that will undergo a reversible change in its absorption spectrum when exposed to electromagnetic radiation. A powerful method of flow visualisation is provided through the use of such dyes. A non-intrusive "line marker" can be created within the liquid by exciting the dye solution with an electromagnetic radiation source, e.g. a laser. The line, or trace, will then follow the flow (fluid elements have been tagged) and its motion can be monitored (using high speed photography for instance) to reveal both qualitative and quantitative information relating to the fluid motion.

Chemically different dye types are available which will operate within different fluids, e.g. *Pyridine* for ethyl alcohol, Popovich and Hummel (1967), *Spiropyran* for aromatic solvents, Zolotorofe and Scheele (1970), or *Triarylmethane* for water, Fogwell and Hope (1987). The discovery of the latter dye for water flows allows studies to be conducted over a wide range of Reynolds numbers, and eliminates the safety considerations associated with the other dyes.

It is recommended that the dyestuff *Acid Violet 19 (monomethyl)* is used with water flows. In solution, the dye is coloured magenta. The solution can be bleached using a combination of potassium sulphite and sodium metabisulphate (to keep the pH between 5 and 6), Hope (1991). Irradiation with light of wavelength 308 nanometres (from a xenon-chloride excimer laser) will restore the colour. The coloured trace will then fade with time, the rate of which will depend on the bleach concentration, the solution temperature, the solution pH, and the overall ionic strength of the solution. The depth of trace penetration will depend on the laser power and dye concentration. Useful trace lengths of 200 millimetres have been produced. **Figure 1** shows an example of dye traces within a laminar flow. The trace contrast can be enhanced optically through the use of wavelength specific interference filters.

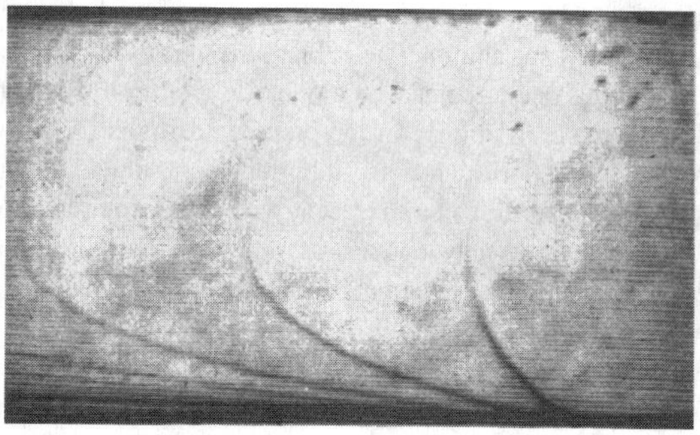

Figure 1. Photochromic dye traces within a laminar pipe flow.

References

Popovich, A. T. and Hummel, R. L. (1967) A new method for non-disturbing turbulent flow measurements very close to a wall, *Chem. Eng. Sci.*, 22.

Zolotorofe, D. L. and Scheele, G. F. (1970) Photochromic dye tracer measurements of small liquid velocities, *Ind. Eng. Chem. Fundam.*, 9.

Fogwell, T. W. and Hope, C. B. (1987) Photochromic dye tracing in water flows, *Experimental Heat Transfer*, 1.

Hope, C. B. (1991) *The Development of a Water Soluble Photochromic Dye Tracing Technique and its Application to Horizontal Two-Phase Flows*, PhD Thesis, Imperial College, University of London.

C.B. Hope

PHOTOCONVERSION (see Renewable energy)

PHOTO CORRELATION SPECTROSCOPY (see Optical particle characterisation)

PHOTODIODE ARRAYS (see Photographic techniques)

PHOTO ELECTRIC EFFECT (see Gamma rays)

PHOTOGRAPHIC METHODS, FOR DROPSIZE MEASUREMENT (see Dropsize measurement)

PHOTOGRAPHIC METHODS, FOR PARTICLE SIZING (see Particle size measurement)

PHOTOGRAPHIC TECHNIQUES

Following from: *Visualization of flow*

Photography provides two-dimensional records of three-dimensional scenes; the whole information about spatial objects can obtained by a totally different imaging process—*holography.* Modern imaging techniques often no longer use photosensitive layers to store images, but make use of the possibilities of electronic data processing. To process images by computer they have to be converted into a computer-compatible format, i.e. digitalized; in photographic models, scanners are used. To convert light signals into electrical signals, either *photomultiplier tubes, photodiode-arrays,* or *charge-coupled-devices (CCD) are used.* All three systems are photon counting devices, meaning the output signal is proportional to the number of photons received on the surface. With CCD-cameras, a two-dimensional resolution is achieved. They have to be connected with the computer system by analogue to digital converters.

Photographic Analysis of Motion

Neither fast nor very slow motion processes can be registered in detail by direct observation. However, various photographic techniques can evaluate such processes.

Time-lapse and time-stretch photography

Time-lapse photography allows analysis is of slow changes of an object for example the melting process by subsequent observation or measurement. With ν_A as recording frequency and ν_P as projection frequency in frames per second then $\nu_A < \nu_P$ in the case of time-lapse photography. When taking such pictures cinematographic cameras are normally used (super-8, 16-mm-, 35-mm-film). The camera is adjusted to single image recording and triggered by a control device at selected time intervals.

With *slow motion* or *time-stretch photography* very fast running processes can be analysed by evaluating single images of different phases of motion (Here $\nu_A > \nu_P$). For investigation of boiling phenomena for example, high speed cinematography is a helpful tool. All cinematographic cameras—except image converting cameras—use moving parts for separating the images. The specific camera construction depends on the recording frequency. Cameras with continuous film transport make it possible to record up to 10^4 frames per second. During exposure the image has to be driven like the film in order to compensate for the effect of film transport and avoid distortion; this problem is solved by optical components like the rotating prism. **Figure 1** shows photographs made by such a camera. With a drum camera recording frequencies up to 10^5 are possible and with rotating mirror cameras even 10^7 frames per second. With ultra high speed video cameras frequencies up to 2×10^7 are possible.

Short time lapse photography

For *high speed photography* exposure times may be as short as nanoseconds (10^{-9} s). With such short exposure times, achieved by illumination with sparks or flashes a sharp image of a specific phase of very fast running processes can be recorded. In the case of fast moving objects, flash and motion have to be synchronized and only limited sequences can be recorded.

Light track imaging

This technique depends upon small light sources or reflectors capable of moving with the object being photographed. For example reflecting particles can be added to a streaming fluid and their paths recorded on film by long time exposure. With this technique movement sequences show up very clearly.

Light emission techniques

Photography with radiation outside the visible range often provides information which is not accessible within it. Transformation of the spectral range to be registered can be carried out by electronic image converters. Often a signal amplifier is also used.

Infrared photography

Investigation of heated objects with temperatures between 250°C and 525°C is possible by recording their own radiation (heat radiation) on film sensitive to the infrared range. For recording radiation from objects with lower or higher temperatures, for example *thermographic cameras*, image converters, are needed.

Fluorescence photography

Evaluation of the natural spontaneous emission of photons from particles in gaseous systems is one of the oldest optical techniques for the determination of concentrations and temperatures. A major field of application of this technique is combustion. Low spectral signal intensities lead to limited time and

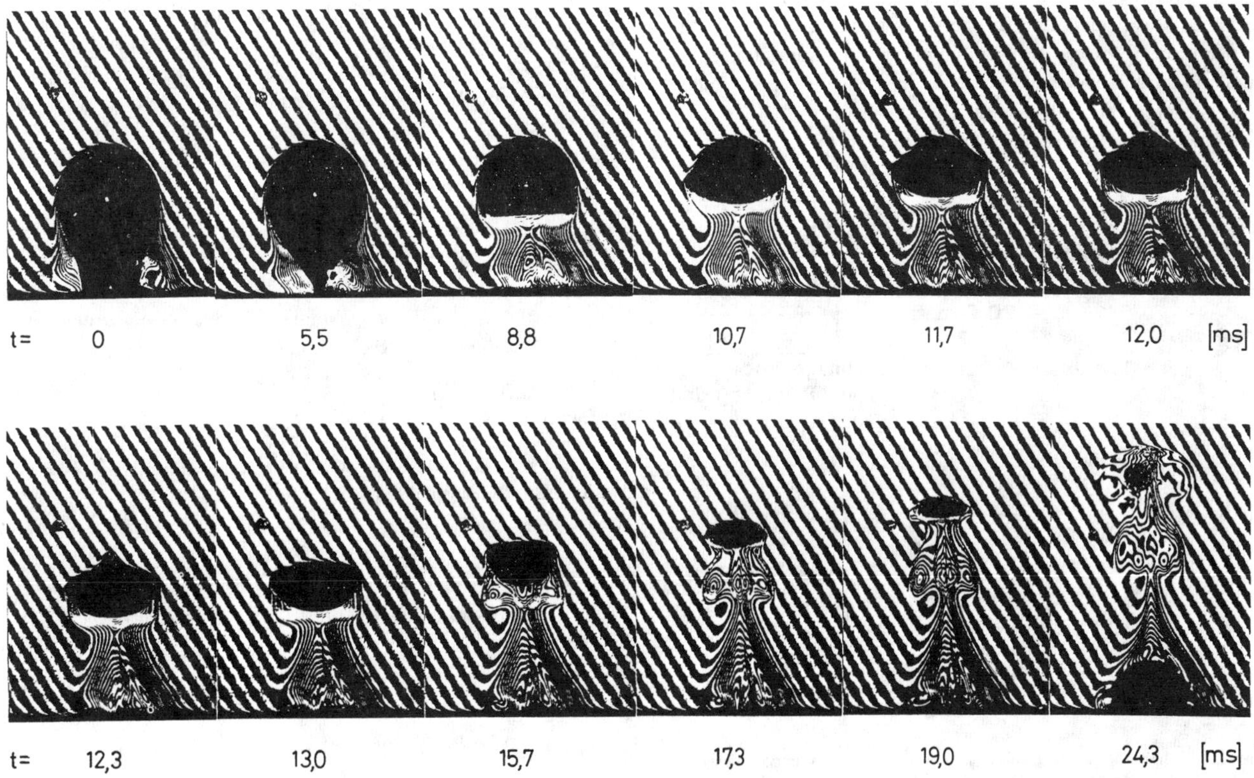

Figure 1. High speed cinematography for investigation of bubble condensation in subcooled ethanol with holographic interferometry (p = 0.5 bar: Δ T = 10,2 K; Ja = $(\rho_l \cdot c_p \cdot \Delta T)/\rho_g \cdot h_{lg}) = 30$).

spatial resolution and advanced detector technology is needed. As an example, **Figure 2** shows a system for self fluorescence investigations of high speed flames. The main components are the CCD video cameras with controller and image recording and processing units. The image intensifier of the camera provides the required increase in sensitivity, gateability and spectral shift to maximum sensitivity from the visible/red region to the UV. With shutter speeds reduced to 10 μs the dynamic structure of the flame can be investigated.

Optical Measurement Techniques and Specific Photographic Processes

Interference techniques

Interference techniques work with superposition of at least two light waves. The waves have optical paths which differ in length with the effect that superposition causes an interference pattern containing information about the process under investigation. The interference techniques have the advantage that photographic recording shows immediately the modifications of the refractive index field caused by diffusion, temperature gradients or flow in a two-dimensional field. (See also **Interferometry**)

Holography

In normal photography, the chosen point of view decides which perspective of the three-dimensional object is shown on the photograph. The basic idea of *holography* is to store the totality (holos) of a wavefront influenced by an object by adding a reference wave generated by the same coherent light source. The interference pattern is recorded on a film. Illuminating the chemically treated photo plate with the reference wave the object is reconstructed three-dimensionally (see **Figure 3**). (See also **Holography** and **Holographic Interferometry**)

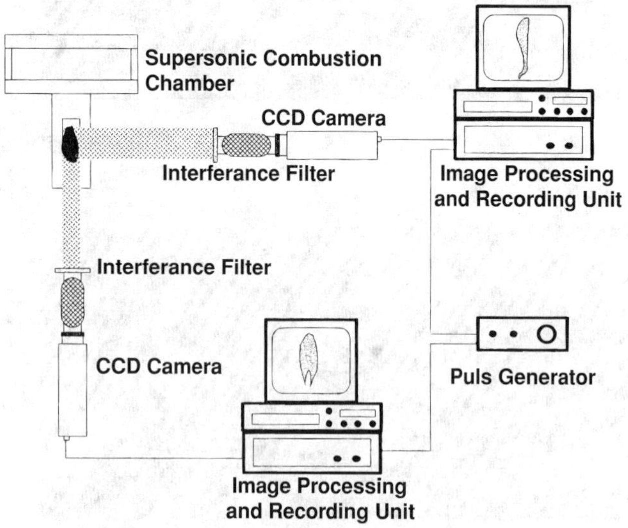

Figure 2. Schematic of setup of cameras for self fluorescence investigations of high speed flames, observed from two directions.

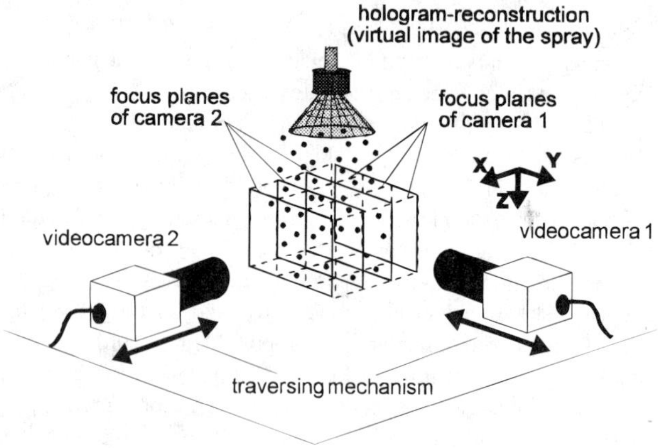

Figure 3. Scanning of a three-dimensional holographic reconstruction by using two videocameras and digital image processing.

Speckle-method

The *speckle-method* is an interferometric method which has found wide application since the laser has been developed. The granulation which could be seen when illuminating a diffuse reflecting surface by laser light is called "speckle". The single points of such a raw surface acts as coherent light sources of waves with different phases. These waves interfere and form a statistically irregular interference pattern. The fundamental characteristic of the speckles useful for this technique is that the speckle pattern follows a surface translation perpendicular to the optical axis. By analysing the movement of single parts the surface deformation can be determined. The conditions before and after the deformation are stored on a film by double exposure technique (*specklegram*).

Schlieren and shadowgraph methods

Schlieren and shadowgraph methods depend upon the fact that solid, liquid or gaseous objects influence a transmitted light beam by changing its direction according to refractive index differences relative to the homogeneous surrounding. The simplest schlieren method is the *shadowgraph* technique. (See entries on **Interferometry** and **Shadowgraph Technique**)

References

Gonzalez, R. C. and Wintz, P. (1977) *Digital Image Processing*, Addison-Wesley Publ. Co. Reading, MA.

Kingslake, R. (1965–1983) *Applied Optics and Optical Engineering*, Academic Press, New York.

Saxe, R.-F. (1966) *High-Speed Photography*, The Focal Press, London.

Walls, H. J. and Attridge, G. G. (1977) *Basic Photo Science, 2nd edition*, Focal Press, London.

Leading to: *Holograms; Holographic interferometry; Interferometry; Shadowgraph techniques*

R. Fehle and F. Mayinger

PHOTOGRAPHY (see Optical particle characterisation)

PHOTOLUMINESCENCE

This is the emission of light induced by the absorption of light. Two types of emission are usually considered: **Fluorescence**, which occurs within about one hundred nanoseconds of excitation; and **Phosphorescence**, which is longer lived.

Figure 1 shows the essential features of molecular photoluminescence as discussed by Turro (1991). Schematic potential energy curves for the ground state singlet, So, first excited *singlet state*, S_1, and first *triplet state*, T_1, are shown with associated vibrational energy levels. Transitions between states of the same spin multiplicity are *allowed* and occur rapidly; those between states of different multiplicity are *forbidden* and usually occur relatively slowly. Absorption (ABS) populates vibrational levels in S_1 and S_2. In condensed phases these generally relax within a few picoseconds via vibrational relaxation (VR) and internal conversion (IC) to the lowest vibrational level of S_1. At this point the relative rates of intersystem crossing (ISC), internal conversion (IC), and fluorescence (FL) determine the fluorescence and triplet quantum yields. If ISC occurs, then vibrational relaxation gives the lowest vibrational level of T_1 which generally decays via reverse ISC to S_0 at room temperature or phosphorescence at low temperatures.

Figure 2 shows the relative positions of absorption and emission bands. It can be seen that the degradation of the excitation energy leads to a *Stokes' shift*, i.e. fluorescence lies at lower energy than excitation, and that phosphorescence occurs at lower energy than fluorescence. Repopulation of S_1, by thermal excitation (TE) from T_1 or triplet-triplet annhiliation, can give rise to a weak but long lived "delayed fluorescence". Organic photoluminescent materials are discussed by Krasovitskii and Bolotin (1988).

The theory and applications of solid state luminescence are discussed by Kitai, (1993). The term phosphorescence is often used in solid state photophysics to indicate a temperature dependent emission process involving recombination of trapped electrons and holes (see also **Thermoluminescence**). Photoluminescence in solids arises from transitions involving conduction bands, valence bands, or localised energy levels at impurity or defect sites. Luminescent centres can be introduced into inorganic crystal or glass matrices by the addition of lanthanide or transition metal ion dopants. Emission and absorption bands may be narrow if the transitions involve only the luminescent centre, or broad-band if lattice vibrations are involved. In

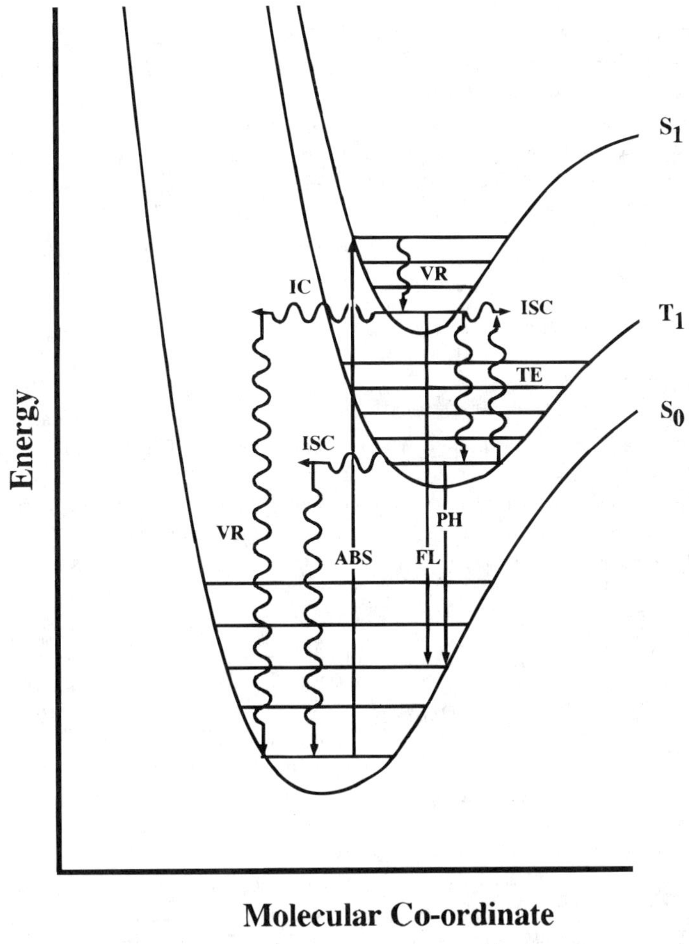

Figure 1. Processes in molecular photoluminescence.

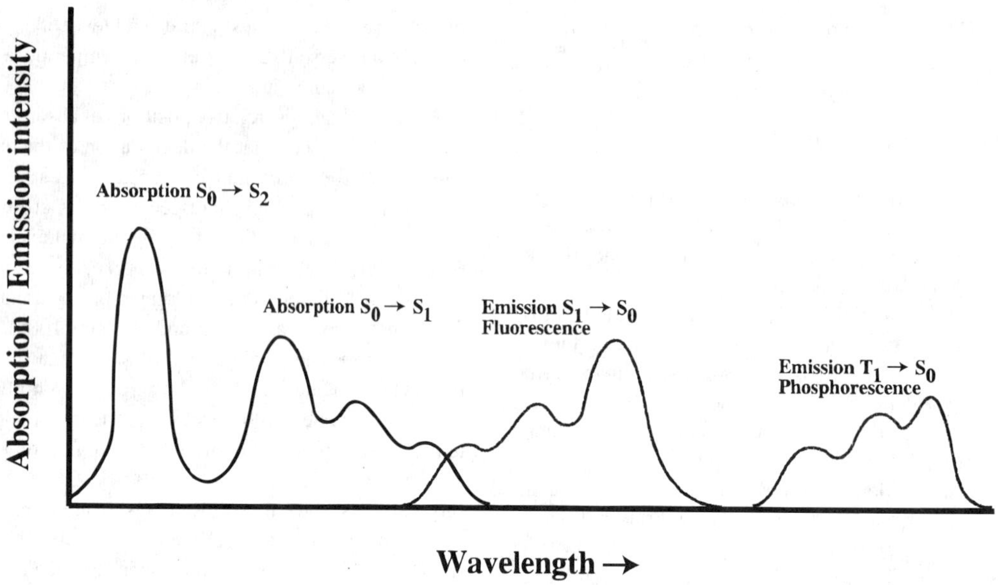

Figure 2. Typical absorption and emission spectra.

the latter case large Stokes' shifts may be observed as excitation energy is lost to lattice vibrations before emission. Applications of solid state luminescence include the ruby and neodymium lasers, and phosphors for fluorescent lamps.

Photoluminescence is an extremely sensitive analytical and imaging technique. Laser induced fluorescence from iodine vapour has been used in *flow visualisation* studies by Zimmermann and Miles (1983). The long lived emission from a suspended ZnS phosphor has been used by Nakatani and Yamadato (1983) to follow the flow of a glycerine/water mixture.

References

Kitai, A. H. (1993) *Solid State Luminescence Theory Materials and Devices*, Chapman and Hall, London.

Krasovitskii, B. M. and Bolotin, B. M. (1988) *Organic Luminescent Materials*, VCH, Weinheim.

Nakatani, N. and Yamada, Y. (1983) *Flow Visualisation III* (Ed. W. J. Yang), Hemisphere Publishing Corporation, 82–86.

Turro, N. J. (1991) *Modern Molecular Photochemistry*, University Science Books, California.

Zimmerman, M. and Miles R. B. (1983) *Flow Visualisation III* (Ed. W. J. Yang), Hemisphere Publishing Corporation, 460–462.

Leading to: *Fluorescence; Phosphorescence*

P. Douglas

PHOTOMULTIPLIERS (see Photographic techniques)

PHOTONS (see Electromagnetic spectrum; Ionizing radiation)

PHOTOVOLTAIC EFFECT (see Solar cells)

PHYSICAL ADSORPTION (see Adsorption)

PHYSIOLOGY AND HEAT TRANSFER

Living organisms require energy to sustain life and develop. Energy is derived from food which is *metabolised*, mostly into *adenosine triphosphate (ATP)* which can be readily extracted (*anabolism*). In these processes much of the available energy is released as heat, e.g. 60% during synthesis of ATP. Additional heat is produced when ATP-stored energy is transferred to the functional systems in the body. These energy losses leave, on the average, a mere 25% available for use by the functional systems (typical conversion efficiencies are: 65% for heart muscles, 30% for bicycling and 5% for respiration).

In all warm-blooded animals, life depends on *homeostasis*—the maintenance of fairly constant internal conditions. One of the chief manifestations of homeostasis is the maintenance of a constant body temperature. At rest these temperatures for mammals are within the narrow range of 35–39.5°C with species-specific levels. Most of the generated heat must be dissipated to the environment to maintain constant body temperature. Temperature control is achieved by the thermoregulatory system composed of 3 interconnected subsystems: sensory, control and regulatory. Thermal receptors, located throughout the body, form the major components of the sensory system. The control center is located in the hypothalamus which lies at the base of the brain. Regulatory mechanisms mediate heat production and, for the most part heat loss. Long-term adaptations to environmental changes are effected through the endocrine glands. Short-term responses of the *thermoregulatory system* to varying internal or external conditions, are produced by three basic mechanisms: metabolic, vasomotor and sudomotor which are depicted in **Figure 1**. Other thermoregulatory activities are behavioral rather than physiological. These include seasonal

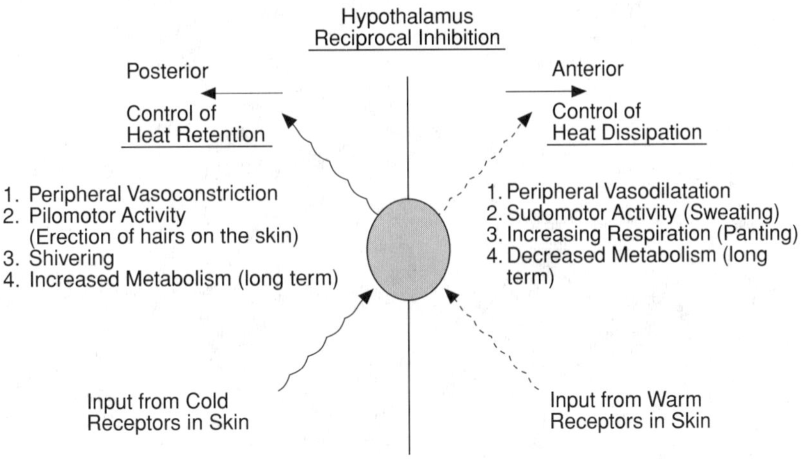

Figure 1. Schematic representation of mammalian thermoregulation.

migration to milder climates, seeking shelter, periodic basking in the sun, typical of poikilotherms (i.e., animals having a variable body temperature, "cold-blooded"), and, in the case of humans, design and utilization of garments, etc. A final reference is made to those animals, termed hibernators, who are usually homeotherms (i.e., animals having a relatively constant body temperature, "warm-blooded") e.g., ground squirrel. These animals possess the unique biological capability of lowering their body temperature to levels which are lethal to non-hibernating species.

References

Shitzer, A. and Eberhart, R. C. eds. (1985) *Heat Transfer in Medicine and Biology—Analysis and Applications*, Plenum Press, New York.

Whittow, G. C., ed. (1971) *Comparative Physiology of Thermoregulation*, Academic Press, New York.

Leading to: *Body, human, heat transfer*

A. Shitzer

PID, PROPORTIONAL-INTEGRAL-DERIVATIVE CONTROLLERS (see Process control)

PIEZOELECTRICITY

If stress is applied to certain crystalline materials such as quartz, equal and opposite charges appear on opposite faces of the sample. This phenomenom is called piezoelectricity. Conversely, if a potential difference is applied across the sample, it will change shape (as if it had compressed or stretched along a certain axis). This effect is used widely in manufacturing transducers and crystal oscillators. Devices such as gramophone pick-ups, ultrasonic transducers, accelerometers, microphones and load cells may all employ piezoelectric sensors to convert forces into electrical signals. The reverse effect is employed in the crystal control of electrical oscillator circuits that are used in clocks and precision measurement apparatus.

Piezoelectricity occurs almost exclusively in ionically bonded crystalline solids. In such materials there exists an ordered array of positive and negative ions. If an applied stress produces a strain or distortion of the crystal lattice that displaces one type of charge in a particular direction with respect to the other type an electric field or *dipole moment* will be produced. Simple crystals such as common salt which has a simple cubic arrangement do not exhibit this property. However, more complicated classes of crystal structure that lack a centre of symmetry are piezoelectric. The simplest structure of this type is sphalerite, or zinc blende (Zn S) (**Figure 1a**). The diagram (**Figure 1b**) shows that a shearing stress may result in compression of the cubic cell along one horizontal diagonal axis and extension along the other diagonal. As the zinc atoms a, b and c move apart the two sulphur atoms d and e will tend to move upwards. Also, as the three zinc atoms p, q and r are compressed together the two sulphur atoms s an t will again move upwards. The overall effect is a total upward displacement of the negatively charged ions in the diagram. Hence, the top of the specimen will assume a negative charge and the bottom will have an equal positive charge.

Piezoelectricity is also found in all *ferroelectric* materials. Detailed analysis of this phenomenom is found in Kittel (1971). These are crystalline structures that possess a permanent dipole moment. This means that for a number of reasons the centres of positive charges in the unit cells of the crystal structure are permanently displaced with respect to the centres of negative charge. If all these individual crystal cell dipoles are aligned, then the entire specimen will exhibit a permanent electric field. Such a system is called an *electret* (c.f. magnet). In practice, an electret will accumulate free charge from its environment that will after a short time neutralise the external fields (i.e. free negative charge will accumulate on the positive "pole" face etc.) If, however, the material is compressed along an axis parallel to the internal field, thereby altering the dipole separation, the internal field will change. The field change will be observed externally for a short time before the accumulated charge on the faces can change to neutralise the

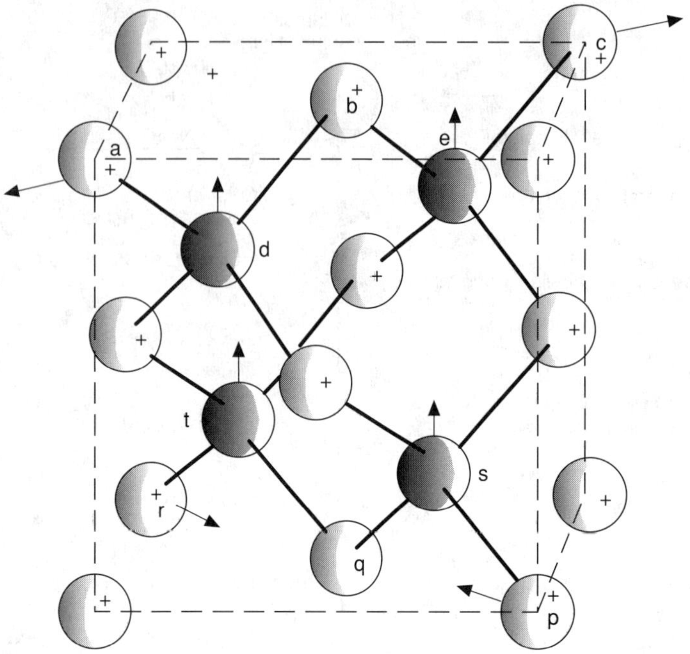

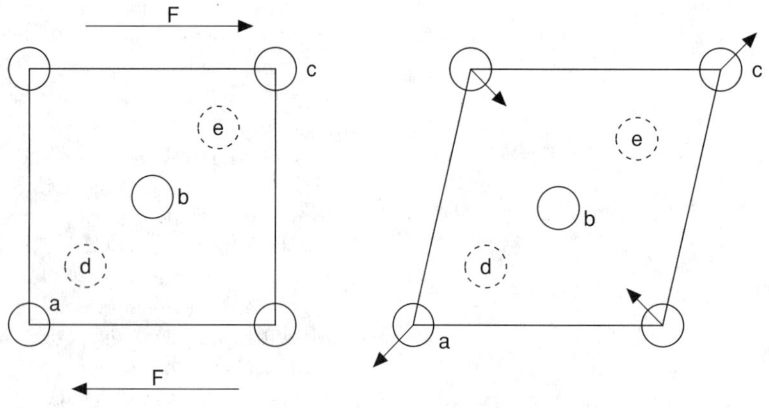

Figure 1.

new field. A fuller treatment of *ferroelectricity* is given by Feynman (1964).

References

Feynman R. P., Leighton R. B., Sands M. (1964) *The Feynman Lectures on Physics*, Addison Wesley.

Kittel C. (1971) *Introduction to Solid State Physics*, John Wiley.

A.W. Campbell

PIEZOMETRIC HEAD LINE (see Static head)

PIEZOMETRIC LINE (see Open channel flow)

PIG IRON (see Iron)

PINCH DESIGN METHOD (see Process integration)

PINCH POINT (see Process integration; Waste heat recovery)

PINCH TECHNOLOGY OR PINCH ANALYSIS (see Optimization of exchanger design)

PIPE FILTERS (see Filters)

PIPE JUNCTIONS (see T-junctions)

PIPELINES

***Following from:** Natural gas; Oils*

Two key values required to calculate pipeline temperature gradients are rarely known with certainty: the temperature and the thermal conductivity of the surrounding material. However, given reasonable estimates of these values, the equations described below allow calculation of flowing pipeline temperature changes to within 10%. This is adequate for most design purposes (e.g. process considerations and anti-corrosion coating design). For systems where more precision is needed (e.g. inventory calculation for loss control monitoring), temperature should be measured at regular intervals along the length of the pipeline and the equations "tuned" as operating experience is gained.

Although the method is suited to hand calculation, the following equations are more typically used as the basis for computer calculations where the pipeline is divided into a large number of segments. This approach can be extended to cover transients and to interface with pressure loss routines to provide iterative solutions for gas pipeline flows.

For a simple hollow cylinder, the conduction heat transfer coefficient based on diameter D is

$$\alpha_1 = \frac{2\pi\lambda_1}{D\,\ln(D_1/D_2)} \tag{1}$$

where is D_1 and D_2 are the outer and inner diameter respectively and λ_1 is the thermal conductivity of the material. The coefficient relating to the pipe surroundings (based on the outside diameter) has been found to be adequately predicted for engineering purposes by:

$$\alpha_n = \frac{2\pi\lambda_n}{D\,\ln(4h/D)} \tag{2}$$

Where h is the depth of cover to the pipe axis, D is the outside diameter of the wrapped pipe and λ_n the thermal conductivity of the surroundings.

The **Overall Heat Transfer Coefficient** for the pipe and any coatings or insulation is obtained by adding the inverse coefficients for each layer (based on the same diameter D):

$$\frac{1}{U} = \frac{1}{\alpha_1} + \frac{1}{\alpha_2} + \cdots + \frac{1}{\alpha_n} \tag{3}$$

The heat flow ratio, a, can then be calculated:

$$a = UD/\dot{M}c_p \tag{4}$$

The pipeline axial temperature profile can then be expressed as:

$$T_2 = (T_1 - T_a)e^{-aL} + T_a \tag{5}$$

Where T_2 is the temperature at a distance L from the inlet and T_1 is the inlet temperature. T_a is the asymptotic temperature or the temperature which the fluid approaches as it flows along the line. This is often wrongly assumed to be equal to the ground temperature where in fact it can be above or below ground temperature depending on the balance between the cooling effect as the product expands with falling pressure and heat gained through friction. These effects are represented through the **Joule-Thomson Coefficient**, μ_{JT} and, correcting for elevation change, Δz

$$T_a = T_g - (\mu_{JT}\Delta p + \Delta z g/c_p) \tag{6}$$

At typical pipeline conditions, light hydrocarbon gases can have Joule-Thomson coefficients, μ_{JT}, of the order of 0.4°C/bar resulting in asymptotic temperatures below ground temperature. For hydrocarbon oils the coefficient becomes negative with frictional heating effects outweighing cooling due to expansion. A coefficient of −0.04°C/bar would be typical for a heavy oil.

Typical thermal conductivities:

Material	λ(W/mK)
Steel	45
Epoxy coating	0.21
Concrete	1.4
Soil (landlines)	0.9 − 1.7
Soil (subsea—water saturated)	2.7

More complex systems involving multiple lines, trace heating etc. are addressed in King, 1980. Jee, 1992, deals with multiple pipelines within an outer carrier pipe and Maddox and Erbar, 1982, provide an approach for pipelines with two-phase flow.

References

Jee, T. P. (1992) *Assessing the Thermal Behaviour of Flowline Bundles*, IIR conference on Optimizing Design and Construction of Flowline Bundles, April.

King, G. G. (1980) *Geothermal Design of Buried Pipelines*, ASME Energy Conference and Exhibition, New Orleans.

Maddox, R. N. and Erbar, J. H. (1982) *Fluid Flow of Gas Conditioning and Processing*, Ch. 16, 3, Campbell Petroleum Series.

P. Harrison

PIPELINE SAMPLING (see Sampling)

PIPELINES, HIGH PRESSURE GAS (see Natural gas)

PISTON ENGINES (see Steam engines)

PI THEOREM (see Dimensional analysis)

PITOT STATIC TUBE (see Pitot tube)

PITOT TUBE

***Following from:** Bernoulli equation; Velocity head*

The *total pressure tube*, or Pitot tube, provides a common method of measuring the stagnation pressure within a pipe, channel or duct flow. In its simplest form this instrument consists of a symmetrical body such as a cylinder, cone, or hemisphere with a small hole or piezometric opening drilled along its central axis. If this is aligned with its central axis in the direction of the flow (**Figure 1a**) the fluid will accelerate around the upstream face with minimal energy losses, and a stagnation point arises at the piezometric opening. In this case the tapping point, typically connected to some form of manometer, provides a direct measure of the stagnation pressure, or total pressure, Ps. If a second piezometric opening records the static pressure in the undisturbed flow (P_o), the velocity of the flow may be inferred from the pressure difference (P_s-P_o).

The **Bernoulli Equation** states that for an incompressible fluid the dynamic pressure accounts for the difference between the stagnation pressure and the static pressure:

$$\frac{\rho u^2}{2} = P_s - P_o \tag{1}$$

where u is the velocity of the flow field.

An alternative arrangement, usually referred to as a *Pitot-static tube*, is indicated on **Figure 1b**. In this case the static pressure is recorded on the same instrument through a series of tapping points located at section B-B. This approach is often more convenient and, in particular, overcomes the difficulty experienced in curved flow where the transverse pressure gradient renders u indeterminate from (P_s-P_o). However, this latter approach suffers from the disadvantage that the static pressure P_o' measured at section B-B may be slightly less than that recorded in the free stream. This effect may be eliminated in the detailed design of the Pitot tube, or incorporated within an empirical coefficient (C_1) such that:

$$u = C_1\left(\frac{2(P_s - P_0')}{\rho}\right)^{1/2} \tag{2}$$

Since $P_o' < P_o$, C_1 is always less than 1.0. However, for most practical engineering purposes $C_1 = 1.0$ is appropriate for many conventional Pitot-static tubes.

Pitot tubes may also be used in compressible flows. If M_o defines the **Mach Number** of the undisturbed flow, the *Euler equation* appropriate to a subsonic flow ($M_o < 1$) yields:

$$\frac{u^2}{2} = c_p T_s\left[1 - \left(\frac{P_o}{P_s}\right)^{(\gamma-1)/\gamma}\right] \tag{3}$$

where P_s is obtained from the Pitot tube, P_o from the static tube (piezometer) and T_s is the temperature at the stagnation point which is usually measured using a thermocouple. In this equation c_p is the specific heat at constant pressure and γ is the ratio (c_p/c_v) or the adiabatic exponent.

For supersonic flow ($M_o > 1$) a normal shock wave will be located upstream of the stagnation point. In this case the free stream velocity (u) is given by:

$$\frac{P_s}{P_o} = \frac{\gamma + 1}{2} M_o^2 \left(\frac{(\gamma + 1)^2 M_o^2}{4\gamma M_o^2 - 2\gamma + 2}\right)^{1/(\gamma-1)} \tag{4}$$

$$\frac{u^2}{2}\left(1 + \frac{2c_p}{\gamma R M_o^2}\right) = c_p T_s \tag{5}$$

where R is the gas constant (or energy per unit mass per Kelvin).

***Leading to:** Differential pressure flow meters; Velocity measurement*

C. Swan

PITZER AND CURL RELATIONSHIP (see PVT relationships)

PLANCK'S FUNCTION (see Stefan-Boltzmann law)

PLANE POTENTIAL FLOW (see Inviscid flow)

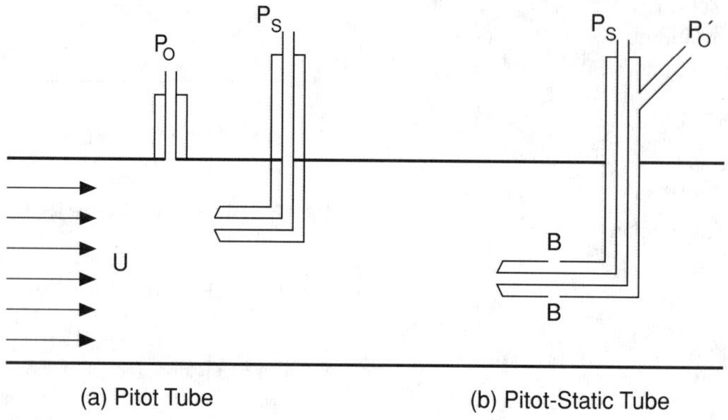

Figure 1. Flow measurement using (a) pitot tube, and (b) pitot-static tube.

PLASMA

Plasma is a gaseous, electrically neutral mixture of electrons, ions and atoms (molecules). The electrons of plasma dispose themselves presumably at vicinities of heavy positively charged ions. As a result, the electric fields exist in small domains of space. The dimension of the domain is known as the Debye radius

$$r_d = (\varepsilon_0 kT/e^2 n_e)^{1/2}$$

where ε_0 is the permittivity of the vacuum, k is the Boltzmann constant, T is the temperature, e is the charge of electron, n_e is the electron concentration. The Debye radius is always very small compared with the characteristic dimension of the plasma L, $r_d \ll L$.

In nature, plasma is ubiguitous (the sun, other stars, upper layers of planetary atmospheres and so on). Electrons and ions in metals and semiconductors also represent plasma-like media.

In laboratory conditions, plasma can be obtained by electric discharges in gases. A low pressure is usually a necessary condition for glow discharge in gases and high pressure results in arc discharges. The latter are characterized by large electric current (up to 10^5 Ampere), low voltage between electrodes and high brightness.

Plasma may arise in high temperature gases, for example in the fronts of strong shock waves.

Plasma is classified as ideal and nonideal. In *ideal plasma* the ratio of potential energy to the kinetic energy of the particles is very small. This ratio may be written as

$$\gamma = e^2 n_e^{1/3}/(4\pi\varepsilon_0 kT) < 1,$$

where $e^2 n_e^{1/3}$ is the potential energy of charged particle interaction at the average distance between them. The transition between ideal and non-ideal plasma occurs in the range $\gamma = 0.1$ to 1 (see **Figure 1** which also indicates the position of various examples of plasmas).

Plasma may be also classified as high temperature plasma and low temperature plasma. The assumed upper boundary of the low temperature plasma is $T = 10^5$ K. The wide variety of the applications of low temperature plasma include: light sources, arc discharge welding, plasma cutting of metals, hardening of surfaces and so on. The new and very promising direction of plasma technology is synthesising new chemical composites in plasma media. *Plasma rocket engines* find some special applications; propulsion is a result of outflow of plasma jets.

Hot plasma is being considered in various projects of controlled nuclear fusion (see **Fusion, Nuclear Fusion Reactors**).

The equilibrium state of plasma is completely defined by their thermodynamic parameters. For ideal equilibrium plasma the pressure is equal to the sum of the partial pressures of plasma components.

In the simplest, case plasma consists of the atoms, electrons and single charged ions, which densities are related by the Saha equation

$$n_e n_i/n_a = (2\Sigma_i/\Sigma_a)(2\pi m_e kT/\hbar^2)^{3/2}\exp(-I/kT),$$

where Σ_i and Σ_a are ion and atomic statistical sums, m_e is the electron mass, $\hbar$ is the Plank constant, I is the atomic energy of ionization.

Thermodynamic equilibrium is realized in closed systems. If plasma is affected by external perturbations—inhomogeneous heating (cooling), emitting charged particles to the environment or receiving them—that means that such a plasma is not a closed system and thermodynamic equilibrium does not exist.

The concept of local thermodynamic equilibrium is very important when the plasma state as a whole is nonequilibrium, but plasma properties in every spatial point are determined by the local values of temperature. Plasma nonequilibrium states are analyzed by methods of physical kinetics.

Plasma properties are determined by plasma composition and are very sensitive to temperature variations. It is very important

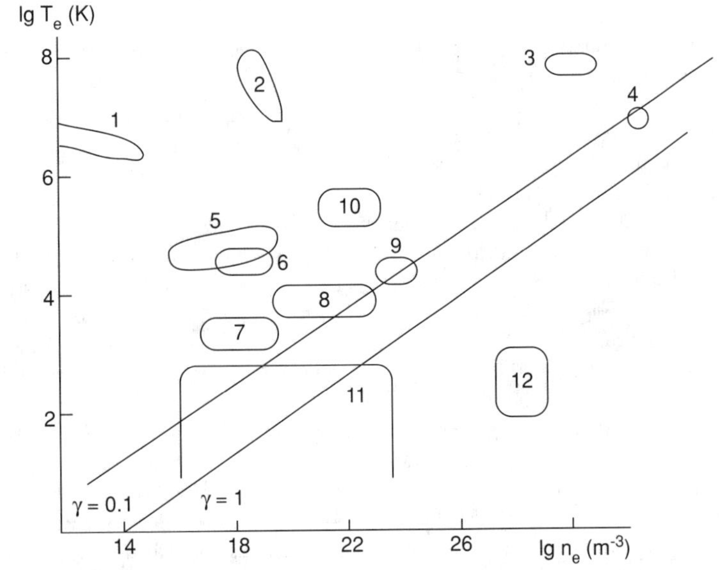

Figure 1. Transition between ideal and non-ideal plasmas.

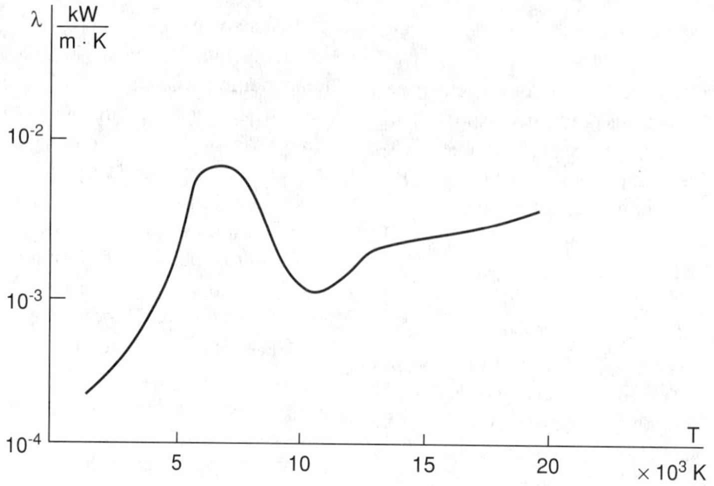

Figure 2. Plasma thermal conductivity as a function of pressure.

to note that the physical and chemical processes in plasma (such as dissociation and ionization) require a great deal of energy. Therefore, the contribution of the processes in plasma heat capacity, internal energy and enthalpy may be more closely compared to the contribution of particle motion.

Diffusion of plasma charged particles largely depends on the strong interaction between them and is defined by condition of quasi neutrality. This diffusion is known as ambipolar diffusion. The ambipolar diffusion coefficient D_a is largely depends on the slow motion of the heavy ions (if $T_e = T_i$, then $D_a = 2D_i$).

For *plasma thermal conductivity,* two processes are very important. The first process is the plasma particle motion; the role of every plasma component largely depends on its density and mobility. The second process is governed by the ambipolar diffusion and by transfer of energy of ionization. Electrons and ions diffuse to the low temperature region and recombine, releasing ionization energy. With the growth of temperature and degree of ionization the role of this mechanism becomes very important, but then decreases as conditions of full ionization are realized (**Figure 2**).

Plasma energy transfer is also realized by the emission and absorption of electromagnetic radiation. Each radiation process is efficient only in its own spectral interval and largely depends on the thermodynamic parameters of the plasma. As a result, the total emission and absorption coefficients vary sharply and irregularly with frequency (**Figure 3**). These characteristics lead to the failure of many simple approaches to the theory of these processes (for example the approximation of a "gray" gas).

The free path length of a photon l_v is closely connected with the spectral absorption coefficient K_v, $l_v = 1/K_v$. This value largely depends on frequency and may be large or small, $l_v \gtrless L$. As a result, the value of radiation flux cannot be considered as proportional to the temperature gradient. In consistent theory, the flow of radiation flux should be written as the integral over the whole plasma volume, accounting for the real temperature distribution in plasma volume. Such a characteristic as average emissivity of the plasma layer may be used only for very rough estimates (see **Radiation Heat Transfer**).

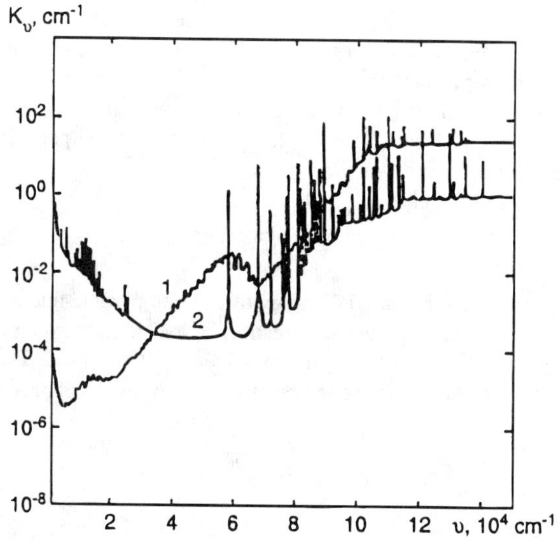

Figure 3. Total absorption (1) and emission (2) coefficients electromagnetic radiation from aplasma as a function of frequency ν.

An external electric field creates an electric current in the plasma. *Plasma conductivity* is defined by electron concentration, average length of the mean free path be and average thermal velocity $v_e = (8kT/\pi m_e)^{1/2}$. The contribution of ions to the electric conductivity may be neglected because $m_i \gg m_e$. The electrical conductivity may be defined by relation

$$\sigma = e^2 n_e l_e/(m_e v_e).$$

The conductivity of an ideal plasma at T = *const* decreases with increasing density. However, the plasma conductivity increases again with density when plasma nonideality becomes significant. At large densities, plasma conductivity can reach values close to metallic ones. This growth with increasing density may be explained by the lowering of the energy of atomic ionization in plasma, the emergence of clusters and another consequences of strong interaction between plasma particles.

The motion of charged particles is strongly influenced by external magnetic fields. In the absence of inter-particle collisions, plasma charges move along screwed trajectories whose axes are directed along the magnetic field. Plasma charge motion perpendicular to the magnetic field is inhibited. Therefore, the rate of transfer across the magnetic field is greatly suppressed (affecting diffusion, viscosity and thermal conductivity). Plasma may be isolated into specific regions if large enough magnetic fields are used. Magnetic confinement of hot plasma has been used in fusion projects.

If the **Knudsen Number** is small and local thermodynamic equilibrium is well realized, then the plasma motion is described by the equations of hydrodynamics and magneto-hydrodynamics.

It is difficult to formulate similarity criteria for plasma dynamics, if it necessary to account for radiation transfer. The uncertainty of the radiation Knudsen number $Kn_R = \langle l_v \rangle / L$ is the first reason of this difficulty arising from the sharp and irregular dependence of l_v versus v.

References

Krall, N. (1973) *Principles of Plasma Physics*, McGraw-Hill Book Co.

Fortov, V., Iakubov, I. (1989) *Physics of Nonideal Plasma*, Hemisphere Publ. Co., New York.

Biberman, L., Vorob'ev, V., Iakubov, I. (1987) *Kinetics of Non-Equilibrium Low-Temperature Plasma*, Plenum Publ. Co., New York.

Leading to: *Plasma arc furnace*

L.M. Biberman

PLASMA ARC FURNACE

Following from: *Plasma; Electric arc heaters; Furnaces*

The plasma arc furnace is a device to melt a substance by low-temperature plasma flow, typically created by an electric arc heater (plasmotron). The main field of application of the plasma furnace is electrometallurgy. The main advantages of a plasma furnace are:

1. high stability of the working process and an opportunity of continuous temperature adjustment by changing of the plasmotron electric conditions;
2. the opportunity of operation in practically any desired atmosphere;
3. high degree adoption of alloying additions;
4. a decrease of the impurities content and consequently an opportunity of smelting low-carbon steels and alloys;
5. an opportunity of smelting the nitrided steels by usage of gaseous nitrogen;
6. a relatively small pollution of ambient air.

Various direct current (DC) and alternating current (AC) plasmotrons are used in plasma furnaces. In large-scale plasma furnaces, several plasmotrons are used to provide more homogeneous heating.

There are three types of the plasma furnaces: plasma furnaces for melting in a ceramic crucible; plasma furnaces for melting in a crystallizer; plasma furnaces for melting in a scull.

The ceramic crucible plasma furnaces are used mainly for melting steel, nickel-based alloys and waste metals with alloying additions. Argon is usually used as the plasma forming gas. Plasma furnaces are used also for gray cast iron smelting. The main advantage of a plasma furnace in comparison with a usual blast cupola is in that the plasma furnace does not need intensive air blasts to support the burning process. Besides in the case of gray cast iron smelting in a plasma furnace, the consumption of electric power and the charge decrease.

Plasma furnaces with a crystallizer are used mainly for the metal refining process. By contrast to electroslag, vacuum-arc and electron-beam refining processes, the main technological means of action on the liquid metal is the gas phase. In particular, a reaction of nitrogen dissolving in liquid iron and its alloys runs faster than in alternative melting devices. Plasma furnaces with a crystallizer can be used in a wide pressure range (10^6—10^{-1} Pa). If necessary, pressure can be varied during a plasma furnace cycle.

A typical plasma furnace with crystallizer includes a remelting blank, two (or more) plasmotrons, a copper water-cooled crystallizer and a remelted ingot, all placed in a common chamber. The cylindrical blank (at the chamber top) and a crystallizer (at the chamber bottom) have mutual vertical axes. The plasma jets fall at an angle on a crystallizer bath and simultaneously flow about the blank, melting it. The melt drops into to a bath. The blank is driven downwards and is rotated to provide uniform melting. A solid ingot is extracted from the crystallizer by a power drive. The coordinated operation of blank drive and ingot extraction devices can provide a continuous remelting process.

The plasma furnaces for melting in a scull are designed to make steel castings, high-temperature alloys and refractory metals. These furnaces can operate in a wide pressure range. Their advantages are high metal purity, an opportunity of scrap charge and waste melting, a prolonged time of liquid metal bath holding for performing of such operations as alloying, mixing, taking a sample, etc.

The plasma furnaces are used also for ceramic melting. Their advantages are high purity of product and an opportunity to provide a ceramic casting process.

Y.S. Svirchook

PLASMA ROCKET ENGINES (see Plasma)

PLASMA THERMAL CONDUCTIVITY (see Plasma)

PLASMA TUBE (see Lasers)

PLASMOTRON (see Electric arc heater)

PLASTICS (see Extrusion, plastics)

PLATE AND FRAME HEAT EXCHANGERS

The original idea for the plate heat exchangers was patented in the latter half of the nineteenth century, the first commercially successful design being introduced in 1923 by Dr. Richard Seligman. The basic design remains unchanged, but continual refinements have boosted operating pressures from 1 to 25 atmospheres in current machines.

The plate and frame heat exchanger (see **Figure 1**) consists of a frame in which closely spaced metal plates are clamped between a head and follower. The plates have corner ports and are sealed by gaskets around the ports and along the plate edges. A double seal forms pockets open to atmosphere to prevent mixing of product and service liquids in the rare event of leakage past a gasket.

Recent developments have introduced the double wall plate. The plates are grouped into passes with each fluid being directed evenly between the paralleled passages in each pass.

An important, exclusive feature of the plate heat exchanger is that by the use of special connector plates it is possible to provide connections for alternative fluids so that a number of duties can be done in the same frame (Lane, 1966; Hargis et al., 1966; Marriott, 1971).

Plate Construction

Plates are made from a range of materials, for example the "Paraflow" plates are pressed from stainless steel, titanium, Hastelloy, Avesta 254 SMO, Avesta 254 SLX or any material ductile enough to be formed into a pressing. The special design of the trough pattern strengthens the plates, increases the effective heat transfer area and produces turbulence in the liquid flow between plates. Plates are pressed in materials between 0.5 and 1.2 mm thick and plates are available with effective heat transfer area from 0.03 to 3.5 m^2. Up to 700 plates can be contained within the frame of the largest Paraflow exchanger, providing over 2400 m^2 of surface area. Flow ports and associated pipework are sized

Figure 1. Plate heat exchanger. (With permission of APV p/c.)

in proportion to the plate area and control the maximum liquid throughput.

Gasket Material

As detailed in **Table 1**, various gasket elastomers are available which have chemical and temperature resistance coupled with good sealing properties. The temperatures shown are maximum, therefore possible simultaneous chemical action must be taken into account when selecting the most suitable material for a particular application.

Thermal Performance

Plate performance is determined by the plate geometry but it is not possible to estimate the film coefficient from the trough dimensions with some accuracy as can be obtained with a tube. The geometrical parameters involved such as plate gap, height, pitch and angle of the trough are too numerous for this to be possible but some work has been done on evaluating the effect of these variables (Kays and London, 1958; Maslov, 1965).

In a plate heat exchanger, the heat transfer can best be described by a Dittus-Boelter type equation:

$$Nu = A(Re)^n(Pr)^m\left(\frac{\eta}{\eta_w}\right)^x \quad (1)$$

Typical values of the constant and exponents are

$$A = 0.15 - 0.40,\ m = 0.30 - 0.45$$

$$n = 0.65,\ x = 0.05 - 0.20$$

where

$$Nu = \frac{\alpha d}{\lambda} \qquad Re = \frac{Vd\rho}{\eta} \qquad Pr = \frac{c_p\eta}{\lambda}$$

d is the equivalent diameter defined in the case of the plate heat exchangers as approximately 2 × the mean gap.

Table 1.

Gasket material	Approx. max operating temp °C	Application
Medium Nitrile	135	Resistant to fatty materials
EPDM	150	High temperature wide resistance for a range of chemicals
Resin cured butyl	150	Aldehydes, ketones
Flurocarbon	177	Mineral oils, fuels, vegetable and animal rubber base) oils

Where

α = *heat transfer coefficient*
λ = *thermal conductivity*
η = *viscosity*
c_p = *specific heat*
subscription w = wall

Typical velocities in plate heat exchangers for waterlike fluids in turbulent flow are 0.3–0.9 m/s but true velocities in certain regions will be higher by a factor of up to 4 due to the effect of the geometry of the plate design. All heat transfer and pressure drop relationships are, based on either a velocity calculated from the average plate gap or on the flow rate per passage.

Figure 2 illustrates the effect of velocity on pressure drop and film coefficient. The film coefficients are very high and can be obtained for a moderate pressure drop.

One particularly important feature of the plate heat exchanger is that the turbulence induced by the troughs reduces the Reynolds number at which the flow becomes laminar. Typical values at which the flow becomes laminar varies from about 100 to 400, according to the type of plate.

The friction factor is correlated with:

$$\Delta p = \frac{f \cdot L_p V^2}{2g \cdot d} \qquad f = \frac{B}{Re^y}$$

Where y varies from 0.1 to 0.4 according to the plate and B is a constant for the plate.

Fouling

In many applications, the heat transfer surface of the plate is less susceptible to fouling than a tubular unit. This is due to 4 principle advantages of the plate design:

1. There is a high degree of turbulence, which increases the rate of foulant removal and results in a lower asymptotic value of fouling resistance.

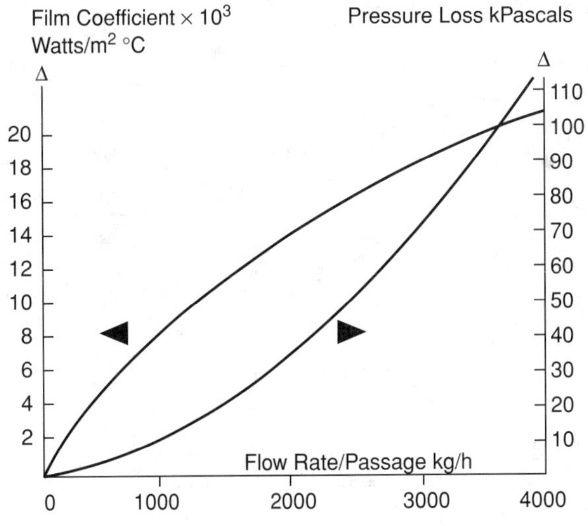

Figure 2.

2. The velocity profile across a plate is good. There are no zones of low velocity compared with certain areas on the shell side of tubular exchangers.
3. Corrosion is maintained at an absolute minimum by careful selection of use of corrosive resistant materials.
4. Materials used for pressing the plates have a very smooth surface.

The most important of these is turbulence. HTRI (Heat Trasfer Research Incorporated) has shown that for tubular heat exchangers, fouling is a function of low velocities and friction factor. Although flow velocities are low with the plate heat exchanger, friction factors are very high, and this results in lower fouling resistance. The effect of velocity and turbulence is plotted in **Figure 3**. The lower fouling characteristics of the plate heat exchanger compared to the tubular has been verified by HTRI's work (Suitor, 1976).

Tests have been carried out which tend to confirm that fouling varies for different plates, with the more turbulent type of plate providing lower fouling resistances.

Heat Transfer Coefficients

Higher overall heat transfer coefficients are obtained with the plate heat exchanger compared with the tubular for a similar loss of pressure because the shell side of tubular exchanger is basically a poor design from a thermal point of view. Considerable pressure drop is used without much benefit in heat transfer due to the turbulence in the separated region at the rear of the tube. Additionally, large areas of tubes even in a well-designed tubular unit are partially bypassed by liquid and low heat transfer area are thus created. Bypassing in a plate type exchanger is less of a problem and more use is made of the flow separation which occurs over the plate troughs since the reattachment point on the plate gives rise to an area of very high heat transfer.

For most duties, the fluids have to make fewer passes across the plates than would be required through tubes or in passes across the shell. Since in many cases a plate unit can carry out the duty with one pass for both fluids, the reduction in the number of required passes means less pressure lost due to entrance and exit losses and therefore more effective use of the pressure.

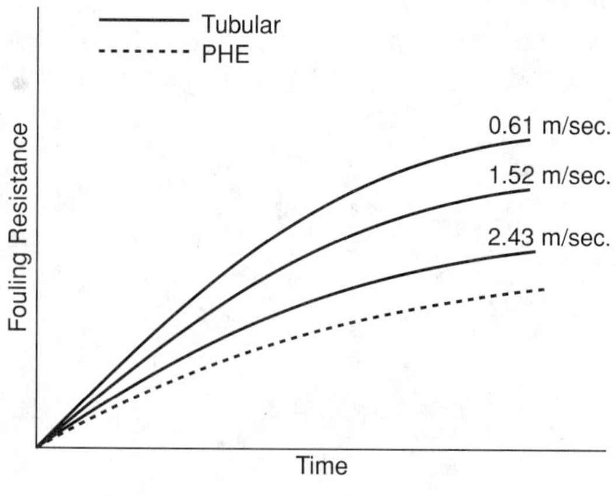

Figure 3. Effect of velocity and turbulence.

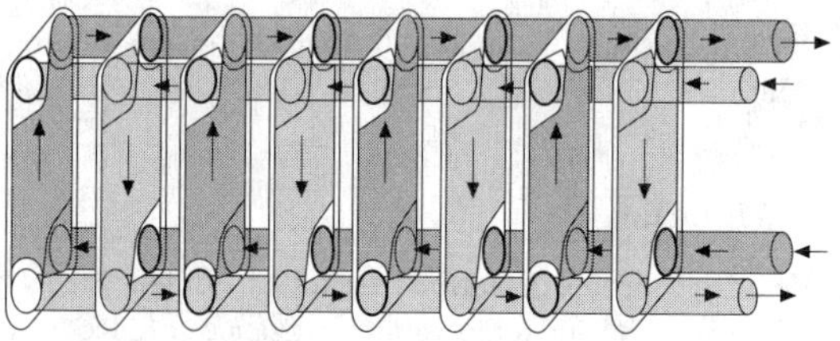

Figure 4. A diagrammattic view of a two way flow system.

Condensing

For condensing duties where permissible pressure loss is less than 7kPascals the tubular unit is most efficient. Under such pressure conditions only a portion of the length of a plate heat exchanger plate would be used and a substantial surface area would be wasted. However, when less restrictive pressure drops are available the plate heat exchanger becomes an excellent condenser, since very high heat transfer coefficients are obtained and the condensation can be carried out in a single pass across the plate.

Pressure Drop of Condensing Vapour

The pressure drop of condensing steam in the passage of plate heat exchangers has been investigated experimentally for a series of different Paraflow plates.

It is interesting to note that for a set of steam flow rates and given duty the steam pressure drop is higher when the liquid and steam are in countercurrent rather than co-current flow.

It can be shown that for equal duties and flow the temperature difference for countercurrent flow is lower at the steam inlet than at the outlet, with most of the steam condensation taking place in the lower half of the plate. The reverse holds true for cocurrent flow. In this case, most of the steam condenses in the top half of the plate, the mean vapour velocity is lower and a reduction in pressure drop of between 10–40% occurs. This difference in pressure drop becomes lower for duties where the final approach temperature between the steam and process fluid becomes larger.

The pressure drop of condensing steam is a function of steam flow rate, pressure and temperature difference. Since the steam pressure drop affects the saturation temperature of the steam, the mean temperature difference, in turn, becomes a function of steam pressure drop. This is particularly important when vacuum steam is being used, since small changes in steam pressure can give significant alterations in the temperature at which the steam condenses.

Gas Cooling

Plate Heat Exchangers also are used for gas cooling. The problems are similar to those of steam heating since the gas velocity changes along the length of the plate due either to condensation or to pressure fluctuations. Designs usually are restricted by pressure drop, therefore machines with low-pressure drop plates are recommended. A typical allowable pressure loss would be 3.5kPascals with low gas velocities giving overall heat transfer coefficients in the region of 300 W/m^{2}°K.

Evaporating

The plate heat exchanger can also be used for evaporation of highly viscous fluids when the evaporation occurs in plate or the liquid flashes after leaving the plate. Applications generally have been restricted to the soap and food industries. The advantage of thse units is their ability to concentrate viscous fluids of up to pascal seconds.

The unit is particularly suitable for high concentration especially as a finishing stage to a larger evaporator where the quantity of vapour is low and can be handled by the comparatively small ports of the plate (Jackson and Trouper, 1966).

Laminar Flow

One other field suitable for the plate heat exchanger is that of laminar flow heat transfer. It has been previously pointed out that the exchanger can save surface by handling fairly viscous fluids in turbulent flow because the critical Reynolds number is low. Once the viscosity exceeds 20–50 cP, however, most plate heat exchanger designs fall into the viscous flow range. Considering only Newtonian fluids since most chemical duties fall into this category, laminar flow can be said to be one of three types:

1. Fully developed velocity and temperature profiles (i.e. the limiting Nusselt case);
2. Fully developed velocity profile with developing temperature profile (i.e. the thermal entrance region); or
3. The simultaneous development of velocity and temperature profiles.

The first type is of interest only when considering fluids of low Prandtl number, and this does not usually exist with normal plate heat exchanger applications. The third is relevant only for fluids such as gases which have a Prandtl number of about one.

For type 2 correlations for heat transfer and pressure drop in laminar flow are in the form

$$Nu = e(Re \cdot Pr)^m \left(\frac{\eta}{\eta w}\right)^x \quad \textbf{(2)}$$

where

Nu = Nusselt number $(\alpha d/\lambda)$
Re = Reynolds number $(Vd\rho/\eta)$
Pr = Prandtl number $(c_p\eta/\lambda)$
η/η_w = Sieder Tate correction factor

(see ***Convective Heat Transfer***)
and

$$f = \frac{a}{Re}. \quad \textbf{(3)}$$

where f is the friction factor and a is a characteristic of the plate.

From this correlation it is possible to calculate the film heat transfer coefficient, for laminar flow. This coefficient, combined with that of the metal and the calculated coefficient for the service fluid together with the fouling resistance, are then used to produce the overall coefficient. As with turbulent flow, an allowance has to be made to the Log Mean Temperature Difference to allow for either end-effect correction for small plate packs and/or concurrency caused by having concurrent flow in some passes. This is particularly important for laminar flow since these exchangers usually have more than one pass. (See also **Heat Exchangers** and **Mean Temperature Difference; Overall Heat Transfer Coefficient**)

Comparing Plate and Tubular Exchangers

Ten points of comparison

1. For liquid/liquid duties, the plate heat exchanger will usually give a higher overall heat transfer coefficient and in many cases the required pressure loss will be no higher.
2. The effective mean temperature difference will usually be higher with the plate heat exchanger.
3. Although the tube is the best shape of flow conduit for withstanding pressure it is entirely the wrong shap for optimum heat transfer performance since it has the smallest surface area per unit of cross-sectional flow area.
4. Because of the restrictions in the flow area of the ports on plate units it is usually difficult (unless a moderate pressure loss is available) to produce economic designs when it is necessary to handle large quantities of low-density fluids such as vapours and gases.
5. A plate heat exchanger is more compact than a tubular and in many instances will occupy less floor space.
6. From a mechanical view point, the plate passage is not the optimum and gasketed plate units are usually not made to withstand operating pressures much in excess of 25 kgf/cm
7. For most materials of construction, sheet metal for plates is less expensive per unit area than tube of the same thickness.
8. When materials other than carbon steel are required, the plate will usually be more economical than the tube for the application.
9. When carbon steel construction is acceptable and when a closer temperature approach is not required, the tubular heat exchanger will often be the most economic solution since the plate heat exchanger is rarely made in carbon steel.
10. Until recently, applications for plate heat exchangers were restricted by the need for the gaskets to be elastomeric. Recent advances is design have introduced brazed plates and welded plates, thereby widening the range of applications.

References

Hargis, A. M. Beckman, A. T., and Loiacono, J. (1966) *The Heat Exchanger*, ASME Publication. PET, 21.
Jackson, B. W. and Troupe, R. A. (1966) Plate heat exchanger design by E NTU method chemical, *Engineer Prog. Symp Serv. 62 (No 64) 185*
Kays, W. M. and London A. L. (1958) Compact Heat Exchangers, McGraw-Hill, New York.
Lane, D. E. (1966) Design trends in plate heat exchangers, *Chemical Process Engineering Heat Transfer Summary*, 127, August.
Marriott, J. (1971) Where and how to use the plate heat exchangers, *Chemical Engineering*, 78 (no 8) 127.

Maslov, A. (1965) *Calculation of the Heat Exchange of Plate Apparatus on the Basis of Diagram (Graphs)*, Kholod ekh. 6, 25.
Suitor, J. W. (1976) *Plate Heat Exchanger Fouling Study, HTRI Report No. F-EX-18.*

B. Lamb

PLATE EVAPORATORS (see Evaporators)

PLATE FILTERS (see Filters)

PLATE-FIN EXTENDED SURFACES (see Extended surface heat transfer)

PLATE FIN HEAT EXCHANGERS

History

Aluminium alloy plate fin heat exchangers have been used in the aircraft industry for 50 years and in cryogenics and chemical plant for 35 years. They are also used in railway engines and motor cars.

Stainless steel plate fins have been used in aircraft for 30 years and are now becoming established in chemical plant.

Concept

The concept is shown in **Figure 1**. Corrugated metal fins are placed between flat plates. The structure is joined together by brazing (see later). The fins have the dual purpose of holding the plates together, thus containing pressure, and of forming a secondary (fin) surface for heat transfer. At the edges of the plates are bars which contain each fluid within the space between adjacent plates.

The heights of corrugations and bars may vary between plates, as shown. For a liquid stream we can use a low height corrugation, matching high heat transfer coefficient with lesser surface area while for a low pressure stream we can use a high corrugation height, matching low coefficient with higher surface area but also giving larger through area to achieve lower pressure drop. An industrial unit contains about 1000m^2 of surface per cubic metre.

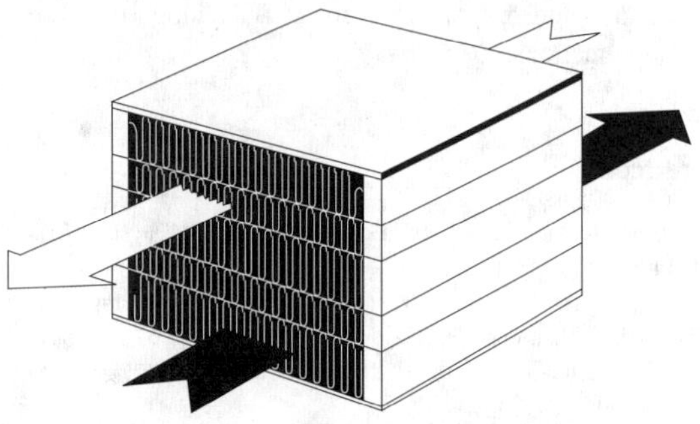

Figure 1. Basic construction of plate fin units (By courtesy of I.M.I. Marston Ltd.)

Aluminium units can be made up to 1.2 m × 1.2 m cross section and 6.2 m long.

Stainless steel units can be made up to 500 mm × 500 mm cross section × 1500 mm length.

Corrugations (Fins)

Corrugations are also made with *heat transfer enhancement devices*. Standard forms are shown in **Figure 2**. Characteristics are shown in **Figure 3**.

Plain corrugation is the basic form and is used normally for low pressure drop streams.

Perforated corrugation shows a slight increase in performance over plain corrugation, but this is reduced by the loss of area due to perforation. The main use is to permit migration of fluid across fin channels, usually in boiling duties.

Serrated corrugation is made by cutting the fins every 3.2 mm and displacing the second fin to a point half way between the preceding fins. This gives a dramatic increase in heat transfer.

Herringbone corrugation is made by displacing the fins sideways every 9.5 mm to give a zig-zag path. Performance is intermediate between the plain and serrated forms. The friction factor continues to fall at high Reynolds numbers, unlike the serrated, showing advantages at higher velocities and pressures.

The designer can, therefore, vary fin heights, fin pitch and fin thickness together with four standard fin types giving great versatility of design.

Construction

Plate-fin units are normally arranged for counter-flow heat exchange. Cross flow units are used for vehicle radiators and cross counter-flow is used for liquid subcoolers.

Figure 4 shows the typical layer arrangements for a *three-stream heat exchanger*. A *two-stream exchanger* can be constructed by using the first of the arrangements shown for the hot stream, alternating with the second arrangement shown for the cold stream.

In this way, the heat exchanger is built up to the appropriate height. It is then brazed together; headers and pipework are welded over the inlet and outlet parts of each stream to give a finished unit. Layers are normally arranged with alternating hot and cold streams, as below:

C H C H C H C H . . .

An alternative system, called double banking, is sometimes used, as below:

C H C C H C C H C . . .

A three-stream unit is made by using all the fin arrangements shown in **Figure 4**.

Figure 5 gives a drawing of a *five-stream heat exchanger*. This has "cut away" sections to show components, one of which shows the method used for stream distribution using an end entry inlet. When several cold streams are used in one unit the layers of each stream should be evenly disposed across the stack height.

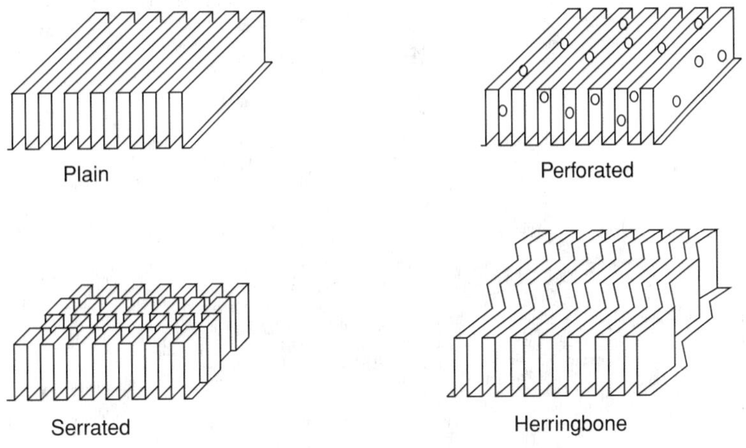

Figure 2. Types of corrugation (By courtesy of I.M.I. Marston Ltd.)

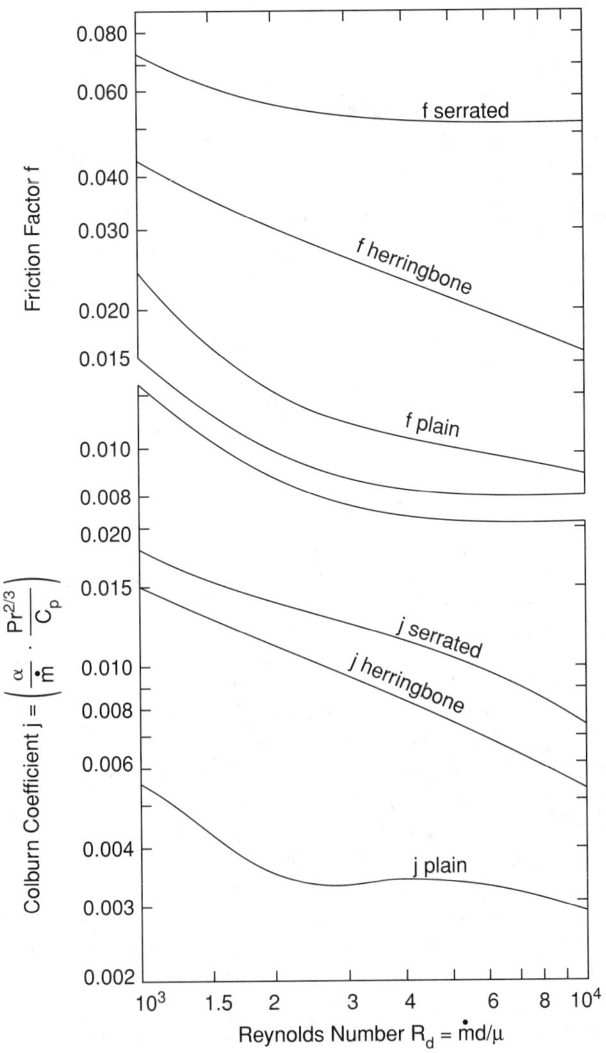

Figure 3. Characteristics of plate fin surfaces.

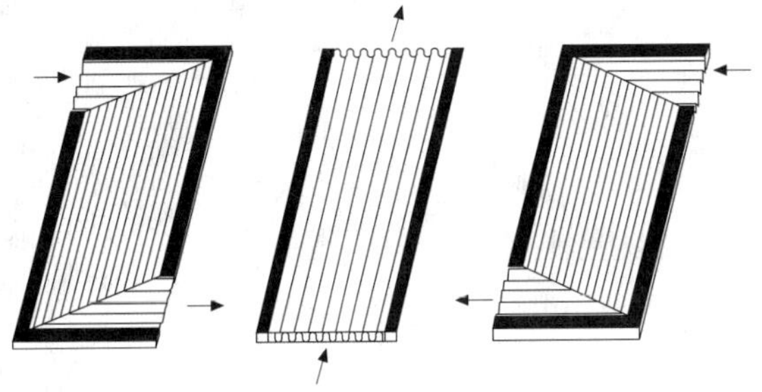

Figure 4. Layer arrangement for 3 stream units (By courtesy of I.M.I. Marston Ltd.)

Figure 6 on page 853 shows a method by which a stream which occupies layers at one end of a block can be taken out at part length and replaced by a second stream at the other end of the block. This can be repeated and in ethylene exchangers up to 5 streams occupy the same layers for different lengths.

A variation of the system allows part of a stream to be taken out (or added) at part length. The use of all these features permits a high degree of process intensification in a single block. The maximum number of streams in one block, so far, is ten, of which three had partway take-offs or additions.

Brazing and Materials

Aluminium units use material AS3003 in the exchanger block. Braze material is AS3003 + silicon. Plates are purchased with a thin film of braze metal on both sides. The unit is built and placed in a vacuum furnace. The braze takes place under vacuum and at a temperature of 580°C. The parts of the block are then firmly attached together.

AS5083 is used for headers and piping below 65°C. Above this temperature AS5454 is used.

Stainless steel units are made of AISI type 321. Braze material is essentially nickel and can be applied to the plates by spraying. Brazing takes place under vacuum at temperatures up to 1050°C.

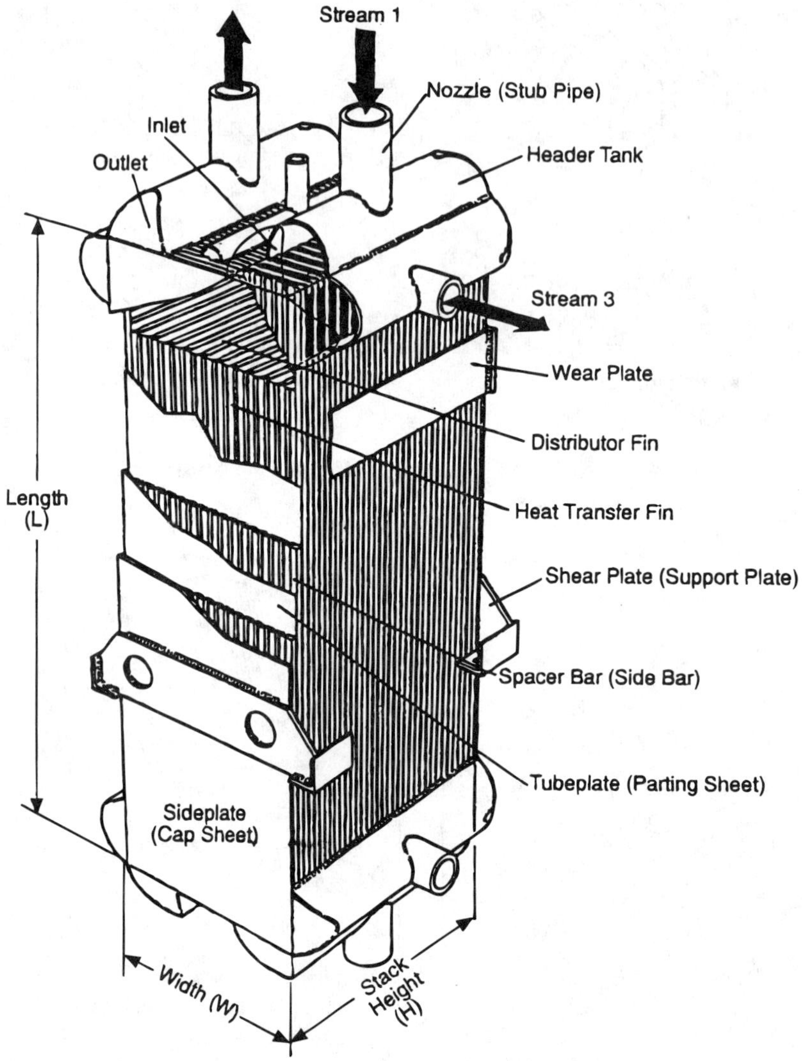

Figure 5.

Pressures and Temperatures

Aluminium units operate with design pressures up to 100 bars and at temperatures from absolute zero to 65°C. Above 65°C a change of header material will allow operation to 120°C with reduced design pressures. Stainless steel units are currently limited to 50 bars design pressure and temperatures up to 750°C.

E.J. Gregory

PLATE HEAT EXCHANGERS (see Heat exchangers)

PLATE THEORY OF CHROMATOGRAPHY (see Gas chromatography)

PLATE TYPE CONDENSERS (see Condensers)

PLATINUM

Platinum—(Sp. *platina*, silver), Pt; atomic weight 195.09; atomic number 78; melting point 1772°C; boiling point 3827 ± 100°C; specific gravity 21.45 (20°C); valence 1?, 2, 3, or 4. Discovered in South America by Ulloa in 1735 and by Wood in 1741. Platinum was used by pre-Columbian Indians.

Platinum occurs native, accompanied by small quantities of iridium, osmium, palladium, ruthenium, and rhodium, all belonging to the same group of metals. These are found in the alluvial deposits of the Ural mountains, of Columbia, and of certain western American states. *Sperrylite*($PtAs_2$), occurring with the nickel-bearing deposits of Sudbury, Ontario, is the source of a considerable amount of the metal. The large production of nickel offsets there being only one part of the platinum metals in two million parts of ore.

Platinum is a beautiful silvery-white metal, when pure, and is malleable and ductile. It has a coefficient of expansion almost equal to that of soda-lime-silica glass, and is therefore used to make sealed electrodes in glass systems. The metal does not oxidize

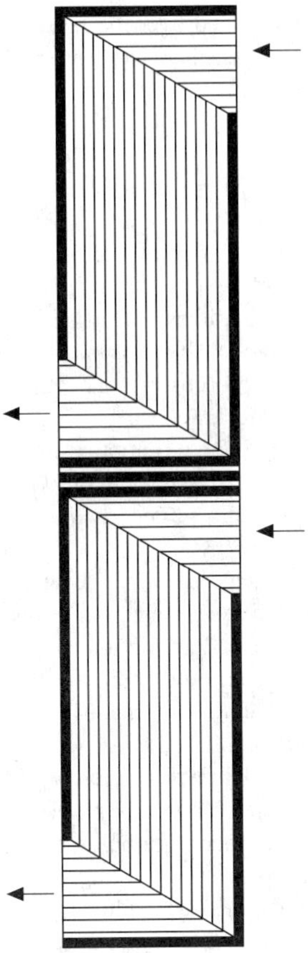

Figure 6. Re-entrant layers.

in air at any temperature, but is corroded by halogens, cyanides, sulfur, and caustic alkalis. It is insoluble in hydrochloric and nitric acid, but dissolves when they are mixed as *aqua regia*, forming chloroplatinic acid (H_2PtCl_6), an important compound.

The metal is extensively used in jewelry, wire, and vessels for laboratory use, and in many valuable instruments including thermocouple elements. It is also used for electrical contacts, corrosion-resistant apparatus, and in dentistry. Platinum-cobalt alloys have magnetic properties. One such alloy made of 76.7% Pt and 23.3% Co, by weight, is an extremely powerful magnet that offers a B-H (max) almost twice that of Alnico V. Platinum resistance wires are used for constructing high-temperature electric furnaces. The metal is used for coating missile nose cones, jet engine fuel nozzles, etc., which must perform reliably for long periods of time at high temperatures. The metal, like palladium, absorbs large volumes of hydrogen, retaining it at ordinary temperatures but giving it up at red heat.

In the finely divided state platinum is an excellent catalyst, having long been used in the contact process for producing sulfuric acid. It is also used as a catalyst in cracking petroleum products. There is also much current interest in the use of platinum as a catalyst in fuel cells and in antipollution devices for automobiles. Platinum anodes are extensively used in cathodic protection systems for large ships and ocean-going vessels, pipelines, steel piers, etc. Fine platinum wire will glow red hot when placed in the vapor of methyl alcohol. It acts here as a catalyst, converting the alcohol to formaldehyde. This phenomenon has been used commercially to produce cigarette lighters and hand warmers. Hydrogen and oxygen explode in the presence of platinum.

Handbook of Chemistry and Physics, CRC Press

PLATINUM RESISTANCE THERMOMETER (see Resistance thermometry)

PLESSET AND ZWICK EQUATION, FOR BUBBLE GROWTH (see Bubble growth)

PLUG FLOW (see Dispersed flow)

PLUG FLOW HEAT TRANSFER

Plug flow, also known as piston flow and slug flow, is a flow with a homogeneous velocity profile across the entire flow section: $u = w = \text{const}$ ($w = {}^{1}/^{\infty}A \int_A udA$ is the section average velocity). Examples of plug flow heat transfer are: the motion of the long bars of various profiles, wires, threads pulling through a heat-treatment zone (for instance, in hardening and annealing devices); the motion of granular bodies (the difference in velocity near the wall and in the core of the flow is in this case only 10–12 percent). A plug flow model can also be used to study heat transfer in a laminar flow of molten metals within the entrance region of a pipe (i.e., when $0 \le x < l_{eh}$ where l_{eh} is the hydrodynamic entrance region length and x the distance from the tube entrance). This is possible as a result of the low values of Prandtl number for molten metals ($Pr \cong 0.01 - 0.03$). The length to diameter ratio $l_{eh}/d_h \cong 0.03$ Re (here d_h is the hydraulic diameter, Re is the Reynolds number) considerably exceeds the length to diameter ration of the entrance region l_{et}/d_n. Note that thermal $l_{et}/d_h \cong 10^{-1}$ Pe ($Pe = ud_h/\kappa = Re\,Pr$ is the Peclet number), i.e., $l_{et}/l_{eh} \cong 3\,Pr \cong 0.03 - 0.1$. Therefore, the temperature field has a chance to stabilize as long as the velocity profile is at the initial stage of development and remains close to that for plug flow. The accepted model of plug flow allows us to present the flow/energy equation in the following form (it is assumed that the physical properties of the liquid, the density ρ, heat capacity c, and thermal conductivity λ, are constant, and that turbulence is absent):in the Cartesian system of coordinates

$$\frac{\partial T}{\partial x} = k\left(\frac{\partial^2 T}{\partial x^2} + \frac{\partial^2 T}{\partial y^2} + \frac{\partial^2 T}{\partial z^2}\right) + \frac{q_v}{\rho c};$$

and in a cylindrical system of coordinates

$$\frac{\partial T}{\partial x} = k\left(\frac{\partial^2 T}{\partial x^2} + \frac{1}{r}\frac{\partial}{\partial r}\left(r\frac{\partial T}{\partial r}\right) + \frac{1}{r^2}\frac{\partial^2 T}{\partial \varphi^2}\right) + \frac{q_v}{\rho c};$$

where $k = \kappa/u = d_n/Pe$ is the constant ($k = \lambda/(\rho c)$ is the thermal diffusivity), q_v is the volumetric heat generation rate. The equations are related to those which are considered in the heat conduction theory and can be solved by the analytical and numerical methods developed for heat transfer equations.

References

Carlslaw, G., Jaeger, D. (1964) *Heat Transfer in Solid Bodies*, Moscow, Nauka.

Petukhov, B. S. (1967) *Heat Transfer and Resistance in Laminar Fluid Flow Through Pipe*, Moscow, Energy.

V.A. Kurganov

PLUNGING JET (see Aeration)

PLUMES

Following from: *Chimneys; Jets*

Gas or aerosol released into the atmosphere at an approximately steady rate, for example from a chimney, will advect with the wind and take on an elongated shape reminiscent of a large feather (**Figure 1**). Such a cloud is known as a "plume". Plumes arise in a variety of circumstances, the essential features being the existence of a continuous, finite sized source of material and a preferential flow direction which elongates the cloud. The term is applied equally to underwater liquid releases, and although this article concentrates on atmospheric plumes the general principles are the same in the underwater case.

Sources of plumes in the atmosphere include discharges from a chimney, fires, and blow-down of a pressurised vessel (**Figure 2**) or pipe work through an accidental puncture. A number of features may introduce a preferential flow direction: the direction of the ambient wind; intrinsic momentum generated in a release from pressure; and buoyancy forces. A plume with high intrinsic momentum is a "jet" (in which the velocity decreases as one goes down stream) until a point is reached where the other factors dominate the directional flow. The term "plume" is sometimes used specifically for cases where these other mechanisms are dominant.

The simplest case is that of a steady plume, where discharge from a constant source has been occurring for long enough that the plume is well developed over the relevant range downstream from the source. In this case there is a conserved flux of the material emitted from the source. This is related to the principal plume variables by

$$\dot{M} = c(x)u(x)S(x)$$

where S(x) is an appropriate measure of the cross-sectional area at a distance x along the plume centre line from the source, c is a measure of the concentration, and u is a measure of the plume velocity. In different models c and u may be cross-sectional averages or, if self-similar profiles of concentration and velocity are assumed, centre line values. As one goes downstream, air is mixed into the plume causing the concentration to decrease and the area to increase. For plumes advecting with the wind, u(x) may be fairly constant or increase if the plume rises; for jets u(x) decreases dramatically.

The rate of mixing depends on the level of turbulence present. Turbulence is generated by the wind (atmospheric shear flow) and also by shear in any other fluid motion associated with the plume including buoyancy-driven flow or jet flow. In different situations, different turbulence generation mechanisms may dominate the mixing process; the most complex cases are where no single turbulence generation mechanism dominates. (see also **Turbulence**)

The above definition of a steady plume makes some implicit assumptions about atmospheric turbulence. In particular, that it is possible to define a concentration averaged over a sufficiently long time to even out fluctuations due to plume meander, but sufficiently short to make the idea of a fixed wind direction meaningful. Looking at visible plumes from chimneys, this assumption is intuitively plausible, but often requires careful consideration when applying the analysis to specific situations.

Perhaps the classic example of a steady plume description is the *Gaussian model* of a plume advecting horizontally with the wind at a height h. In this case mixing is purely due to atmospheric turbulence, and the concentration is assumed to have a self-similar Gaussian profile

$$c(x, y, z) = c(x)e^{(-y^2/2\sigma_y^2) - (z-h)^2/2\sigma_z^2}$$

The conserved flux is as given above with $S(x) = 2\pi\, \sigma_y(x)\, \sigma_z(x)$ and u the constant wind speed at height h. The parameters $\sigma_y(x)$ and $\sigma_z(x)$ are assigned empirical forms dependent upon atmospheric stability, which is usually quoted in terms of a *Pasquill category* (also empirical) in the range A to F or G, see for example Hanna and Drivas (1987), Panofsky and Dutton (1984). This model is simple but adequate for many purposes. It is often elaborated to allow for the presence of the ground (to ensure that the entire flux of material remains above the ground).

Such a model can be used to describe a *passive plume*. "Passive" means that the local flow is unaffected by the presence of the plume, which must have no buoyancy or momentum. It is not appropriate for dense or buoyant plumes, except in the very far field when dilution will eventually render either of these passive.

In discharges from chimneys, buoyancy is often an aid to getting the plume as high as possible into the atmosphere and enhancing dilution. Plumes from fires are also buoyant, and the plume rise may mitigate the immediate consequences of toxic combustion products, although not necessarily the long term environmental

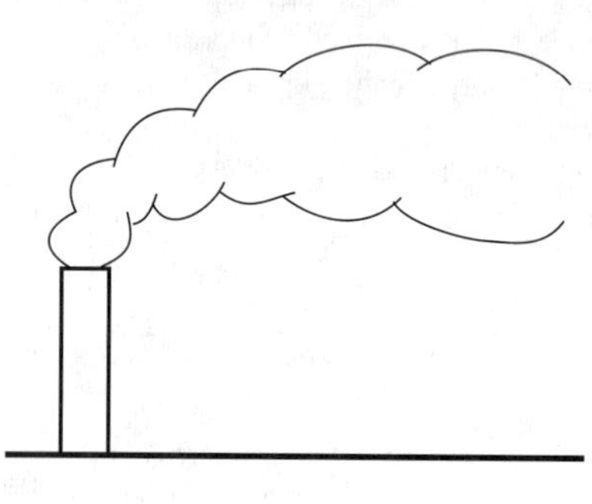

Figure 1. A chimney plume.

Figure 2. A ground level plume following a jet release.

Figure 3. A dense vertical jet and a descending plume, resulting in a dense plume at ground level.

impact. In the absence of wind, or where the speed of rise is much more rapid than the wind speed, such a plume will rise vertically. Its buoyancy determines its rate of rise, which in turn determines the level of turbulence, which induces mixing which dilutes the plume and reduces the buoyancy. Turner (1973) gives a clear analysis of this situation. Any variation in atmospheric density with height will also affect the behaviour of the plume.

If there is a wind then the plume will bend over. The trajectory and turbulent dilution rates are affected both by the wind flow and by buoyancy, as well as by atmospheric stratification and any momentum at the source. Simple analyses, similar to the vertical plume rise case, may apply in some cases where the buoyancy-generated turbulence dominates atmospheric turbulence, as discussed by Hanna, Briggs, and Hosker (1982). Relatively simple descriptions are also applicable to momentum-dominated plumes (**Jets**).

However, the general case of a (positively or negatively) buoyant airborne plume, possibly with initial momentum, in a turbulent atmosphere is more complicated. For example a dense jet may be emitted upwards and, as it loses momentum and becomes dominated by buoyancy forces return back to the ground (**Figure 3**), or it may dilute to the extent that it becomes passive while still airborne.

Dense plumes at ground level are often the expected result of *accidents to chemical plant*. Many substances may form dense clouds owing to high molecular weight, low temperature, or liquid aerosol content. (For example ammonia and LNG readily form heavy clouds despite having low molecular weight.) Such plumes have received much attention as the danger associated with them is enhanced by their density: they stay close to the ground and dilute more slowly as any mixing must overcome the stable density interface at the top of the cloud. (Although very close to the source mixing may be enhanced by turbulence generated from the slumping behaviour of the cloud.) The main parameter which affects the mixing rate is the **Richardson Number**—typically $Ri = g(\rho - \rho_a)H/(\rho_a u_*^2)$ for a cloud of height H and density ρ in an atmosphere of density ρ_a in which the velocity associated with turbulence is characterised by u^*. Thus a large cloud in a still atmosphere may be effectively dense, even if its density is only very slightly greater than that of the ambient air. The width of a ground-based dense plume may increase much more rapidly than the passive case as the plume slumps. Correspondingly the height increases more slowly or may even decrease for a limited distance downstream. Wheatley and Webber (1984) and Britter and McQuaid (1988) discuss heavy clouds in detail.

References

Britter, R. E. and McQuaid, J. (1988) Workbook on the dispersion of dense gases, *UK Health and Safety Executive Report 17/1988.*

Hanna, S. R., Briggs, G. A. and Hosker, R. P. (1982) *Handbook on Atmospheric Diffusion*, Technical Information Center of the U.S. Department of Energy DOE/TIC-11223.

Hanna, S. R. and Drivas, P. J. (1987) *Vapour Cloud Dispersion Models*, American Institute of Chemical Engineers, ISBN 0-8169-0403-0.

Panofsky, H. A. and Dutton, J. A. (1984) *Atmospheric Turbulence*, John Wiley and Sons, ISBN 0-471-05714-2.

Turner, J. S. (1973) *Buoyancy Effects in Fluids*, Cambridge University Press, ISBN 0 521 08623 ×.

Wheatley, C. J. and Webber D. M. (1984) Aspects of the dispersion of denser-than-air vapours relevant to gas cloud explosions, *Commission of the European Communities EUR 9592*, ISBN 92-825-4712-4.

D.M. Webber

PLUNGING LIQUID JETS

Following from: *Aeration*

The gas entrainment by a plunging liquid jet may occur in many problems of practical interest. A good understanding of the *air carryunder* and *bubble dispersion* process associated with a plunging liquid jet is vital if one is to be able to quantify such phenomena as sea surface chemistry, the meteorological and ecological significance of (breaking) ocean waves, the performance of certain type of chemical reactors, and the "**Greenhouse Effect**" (i.e., the absorption of CO_2 by the oceans). Indeed, the absorption of greenhouse gases into the ocean has been hypothesized to be highly dependent upon the air carryunder that occurs due to breaking waves. This process can be approximated with a plunging liquid jet (Monahan, 1991), (Kerman, 1984).

A number of prior experimental studies have been performed in which axisymmetric plunging liquid jets have been used to investigate the air carryunder process. These include the work of Lin and Donnelly (1966), Van De Sande and Smith (1973), and McKeogh and Ervine (1981). Recent experiments include those of Chanson and Cummings (1994) and Bonetto and Lahey (1993).

As shown in **Figure 1**, a converging nozzle oriented vertically produces an axisymmetric liquid jet. This jet impacts orthogonally a pool of water and, when a threshold velocity was exceeded, the plunging liquid jet causes significant air entrainment [Bonetto and Lahey, 1993]. In agreement with the observations of McKeogh and Ervine [1981], different two-phase jet characteristics were noted by Bonetto and Lahey [1993], depending on the turbulence intensity of the plunging liquid jet. For a laminar axisymmetric liquid jet (i.e., one having a turbulence intensity less than about 0.8%) the diameter of the entrained bubbles was in the range 15–300 μm. On the other hand, for liquid jet turbulence intensities of about 3%, the entrained air bubbles had diameters in the range of 1–3 mm. (See **Turbulence** and **Turbulent Flow**)

Two different methods were used for the measurement of void fraction in the two-phase jet: a KfK impedance probe and a DANTEC Fiber-optic Phase Doppler Anemometer (FPDA). The turbulence intensity of the liquid jet at the nozzle exit was found to be

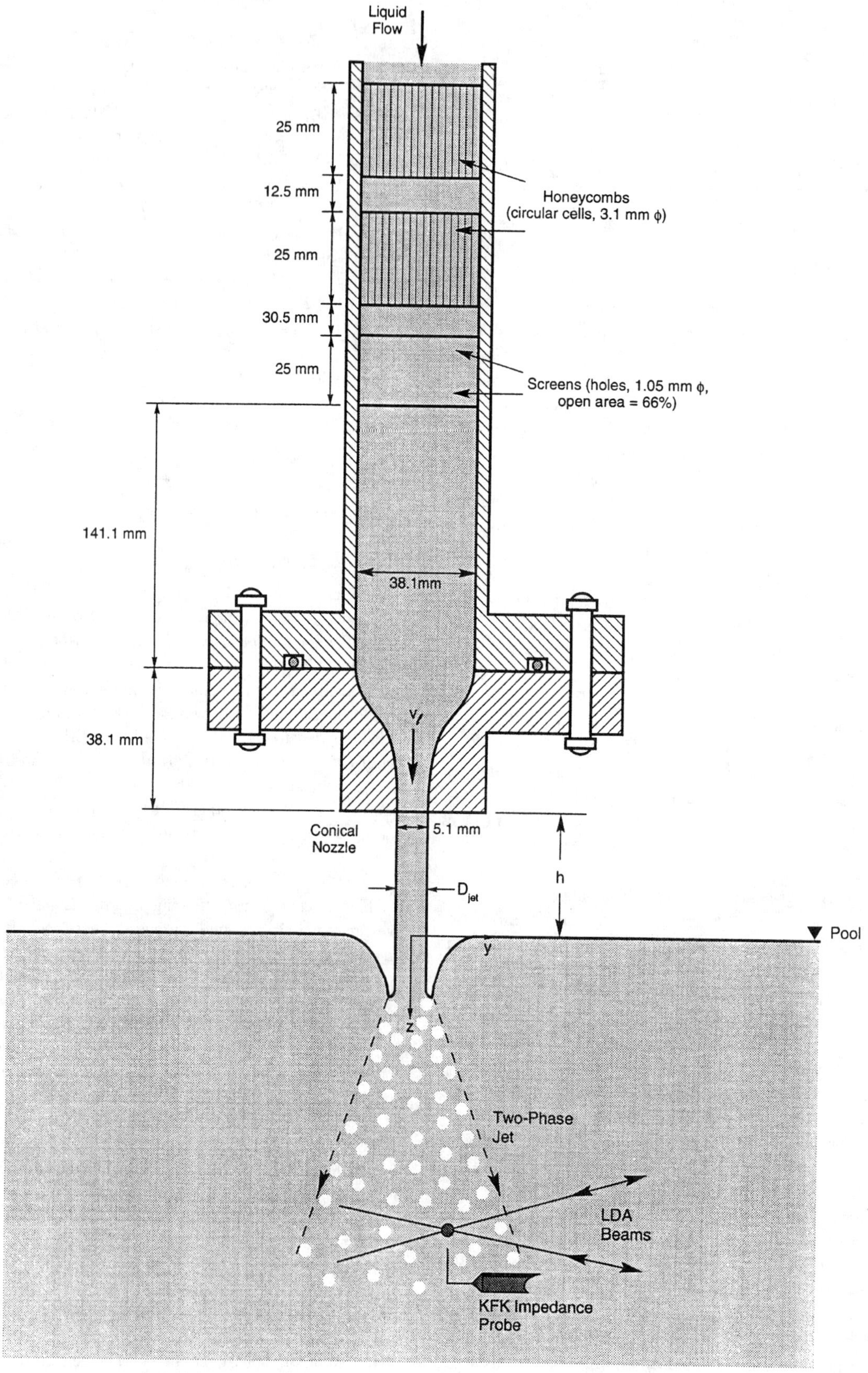

Figure 1. Schematic of a conical nozzle and plunging liquid/two-phase jets.

one of the most important parameters affecting jet roughness and the size of the bubbles entrained by the axisymmetric plunging liquid jet. As can be seen in **Figure 1**, an arrangement of honeycombs and screens was used to control the turbulence intensity of the flow entering the nozzle.

Figure 2 depicts a contour plot of the two-dimensional probability density function of the bubble diameters (D_b) and axial velocities (v_z) for an axisymmetric nozzle. Quantitatively the most probable value of peak no. 1 (liquid velocities) was at $D_b = 5$ mm, $v_z = 4.05$ m/s, and the most probable value of peak no. 2 (gas velocities) was at $D_b = 125$ mm, $v_z = 3.5$ m/s.

Figure 3 presents the mean liquid and gas velocities and the turbulent intensity of the liquid and gas velocities as a function of lateral position (y) for the liquid jet and for h = 8.5 mm, w_l = 1.8 kg/s and z = 31.0 mm. We see that at the edge of the spreading two-phase jet that the gas (bubble) velocity is negative, indicating buoyancy-driven countercurrent flow. It was also found that the two-phase jet was more dispersed than the corresponding single-phase flow case and that the turbulence intensity was higher. This turbulence enhancement is presumably due to bubble-induced turbulence. In this case the bubble-induced turbulence accounts for about 30% of the total turbulence level.

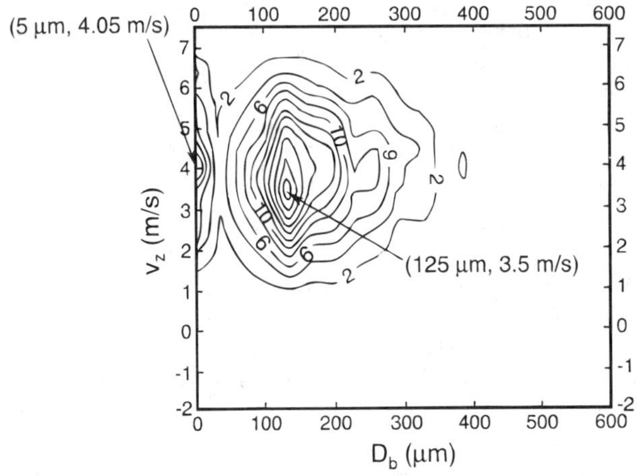

Figure 2. Contour plot of the two-dimensional probability density function-smooth jet (z = 35.1 mm, w_l = 0.143 kg/s; h = 9.0 mm; y = 0).

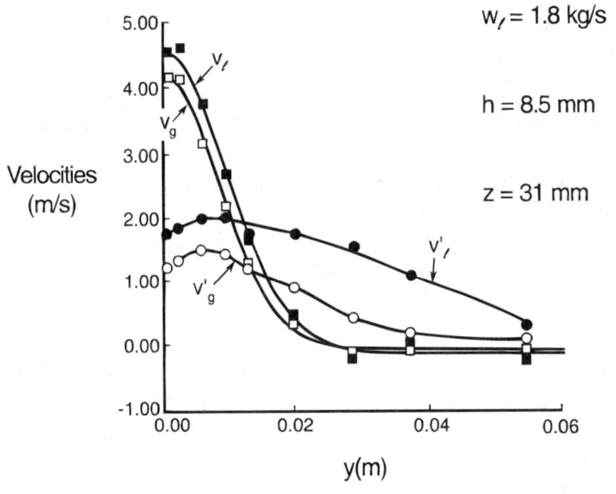

Figure 3. Mean and rms axial velocities for the gas and liquid phases.

As noted previously, when the liquid jet impacts the pool surface, air entrainment occurs around the jet's perimeter. In **Figure 4a** *the measured local void* fraction is presented as a function of y for z = 1.0 mm (i.e., with the probe 1 mm under the undisturbed liquid level) for a planar liquid jet [Bonetto and Lahey, 1994]. We see that the void fraction has a maximum at y approximately equal to 3 mm, and this peak disperses with z. Obviously, the air entrainment process is responsible for this effect. In the high-speed video visualization of these experiments it was rare to observe bubbles at the liquid jet's centerline for z < 10 mm. However, once the air was entrained, lateral dispersion of the gas phase occurred as z was increased. **Figure 4b** shows how the void peak was dispersed at z = 26.0 mm. We see that the maximum now occurs at y equal to 5.5 mm. Moreover, we see that there is significant void fraction at y = 0 (i.e., the jet's centerplane) because of the void dispersion process. **Figure 4c** shows the void fraction profile at z = 58.8 mm. Significantly, the curve now tends towards a maximum at the centerplane of the jet (y = 0). Again, this is a direct result of the void dispersion process in the two-phase jet.

For a turbulent (i.e., rough) liquid jet the entrained bubble sizes were of the order of $D_b \cong 3$ mm, and the slip ratio was close to calculated values based on the terminal rise velocity of a single bubble. Moreover, the turbulence intensity of the liquid jet had two components, one due to shear-induced turbulence and the other due to bubble-induced pseudo-turbulence. Both components of turbulence were of the same order to magnitude.

The mechanism that produces the gas entrainment currently is not well understood. Lezzi and Prosperetti [1991] have proposed that the instability responsible for the air entrainment was caused by the gas viscosity. In particular, they studied the linear stability of a vertical film of a viscous gas bounded by an inviscid liquid in uniform motion on one side, and by inviscid liquid at rest on the other side. They also obtained the marginal stability boundary with the gas gap width, d, as the control parameter (i.e., they numerically compute for a given d the range of wavelengths that makes the system unstable).

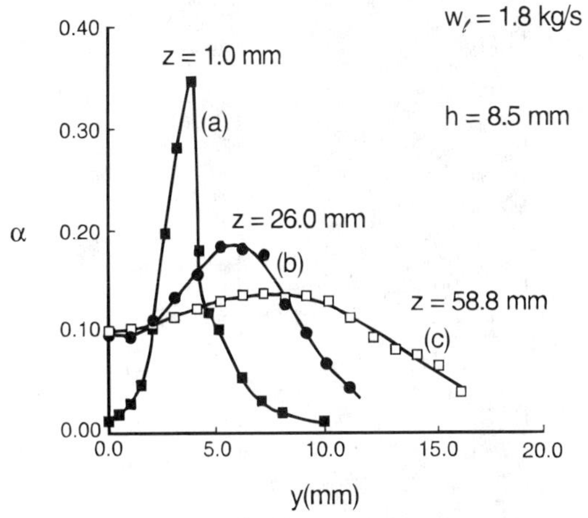

Figure 4. Void fraction as a function of the lateral position for different axial positions.

Bonetto et al. [1994] proposed that a *Helmholtz-Taylor instability* is responsible for the air entrainment. They assumed the liquid jet, the liquid in the pool and the gas in the gap to be inviscid fluids. A linear stability analysis was performed on the system, the perturbation being a sinusoidal wave on the liquid jet/gas gap interface. The celerity of the perturbation was calculated using the appropriate conservation equations. Also the sinusoidal perturbation imposed on the liquid jet/gas gap interface produces a sinusoidal perturbation on the gas gap/pool liquid interface with the same celerity and wave number and is out-of-phase (i.e. phase = 180°) **Figure 5** presents a schematic of the air entrainment process. Two interfacial waves of small amplitude and wave length λ_d grow as they move with celerity C. At a certain position they may touch each other entrapping a volume of air proportional to the shaded area. The shaded area, Λ, is given by $\Lambda = \lambda_d \delta$.

Figure 6 shows the volumetric flow rate of air, $\dot{Q}_a$, measured by McKeogh and Ervine [1981] as a function of the liquid jet velocity, v_z, for a liquid jet diameter of, $D_{jet} = 0.0051$m. The jet turbulence level in these experiments was 3% and the distance of the nozzle above the pool's surface was h = 0.03 m. The solid line corresponds to the theoretical predictions by Bonetto et al. (1994) for $\delta = 0.291$ mm. Significantly, no value of δ used with the results of Lezzi and Prosperetti (1991) agrees with McKeogh and Ervine experiments, thus it appears that Helmholtz-Taylor instability is the dominant air entrainment mechanism.

The spreading of a bubbly two-phase jet involves the interaction between the liquid turbulence and the bubbles. For the computation of jet dispersion it is important to appropriately model the turbulent intensity of the continuous phase. Rodi (1984) presented results using the classical k-ε model of Gibson and Launder (1976), and showed that k-ε models do not accurately predict jet spreading. Sini and Dekeyser (1987) solved the single-phase turbulent jet using Rodi's k-ε model. Significantly, it has been found that single-phase turbulent jet data can be used for the assessment of **Turbulence Models** because one does not have to constitute complicated turbulent closure laws, such as those required near solid (no slip) boundaries. Indeed, due to the absence of walls and the associated shear boundary conditions the turbulent jet is probably the simplest non-trivial case to analyze. Interestingly, the same conclusions can be reached for a two-phase turbulent jet.

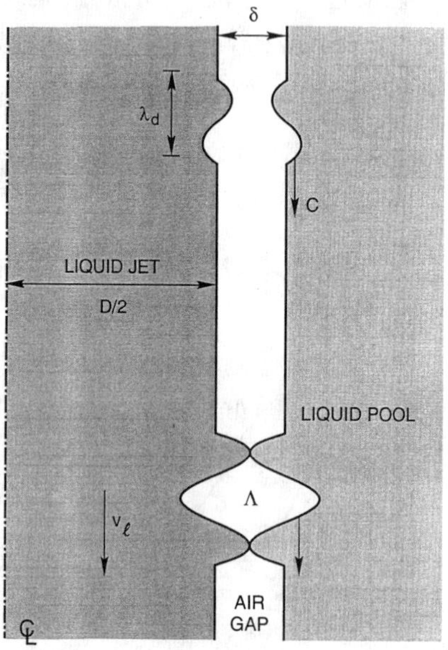

Figure 5. Mechanism responsible for air entrainment.

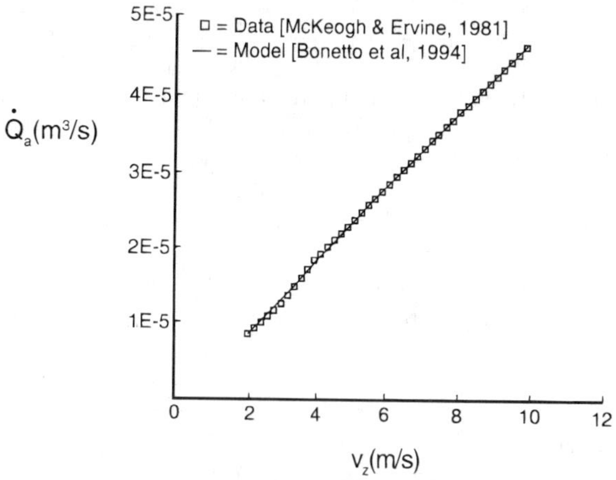

Figure 6. Entrained air volumetric flow rate as a function of liquid jet velocity.

Most researchers have analyzed liquid jets using a parabolic scheme instead of an elliptic one. Unfortunately, using a parabolic scheme one cannot compute the pressure distribution field accurately, more importantly one cannot compute recirculating flows (such as occur in vertical two-phase downflowing jets) and the downstream pressure has to be (arbitrarily) specified. Solving the partial differential equations as an elliptic system increases the complexity of the numerical problem but provides more detailed and accurate information on the flow field (pressure) than a parabolic scheme.

We note that the turbulence present in the liquid has two components in a two-phase jet. One component, the shear-induced turbulence, is due to viscosity and it is present in both single and two-phase flows. The other component is the bubble-induced turbulence due to slip between the bubbles and the surrounding liquid, and it only occurs in two-phase flows.

A state-of-the-art multidimensional two-fluid model obtained using ensemble averaging has been derived and was closed using cell average model [Arnold et al., 1989]. This approach provides equations for multiphase flows that are mechanistically based (as opposite to being empirical). The two-fluid model's conservation equations and the associated k-ε turbulence model were numerically integrated using a CFD code, PHOENICS. **Figure 7a** shows the computed v_z at z = 0 and **Figure 7b** at z = 31 mm. The open circles are experimental points. The spreading of the jet is well predicted. Significantly, the underprediction at the center-line is similar to that observed in single-phase flows (Rodi, 1984). **Figure 7c** shows the computed v_z at z = 59 mm and the trends are similar to **Figure 7b**. The agreement with the experimental data is quite good, indicated that suitable formulated two-fluid models are able to predict two-phase jet flows.

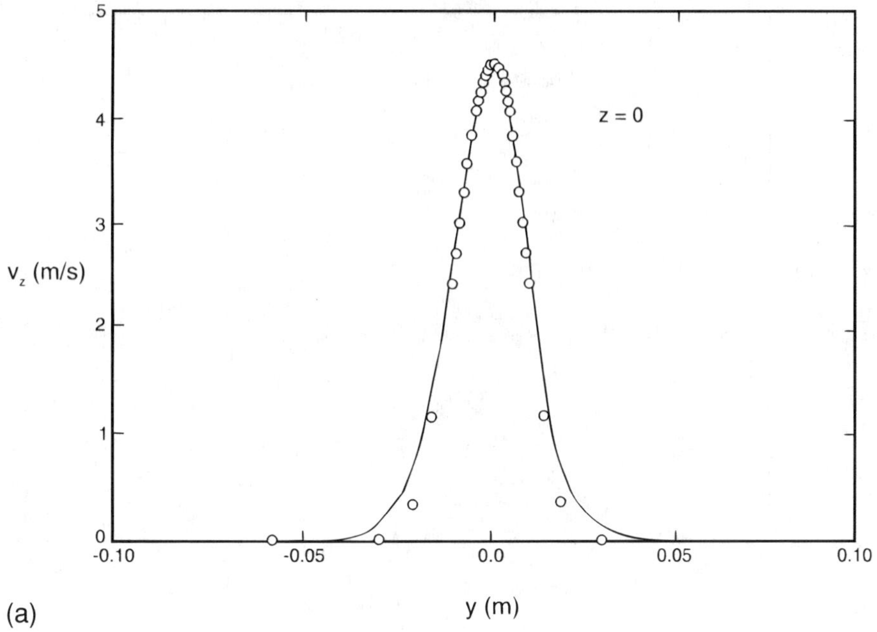

Figure 7a. Axial velocity as a function of the lateral coordinate (z = 0.0).

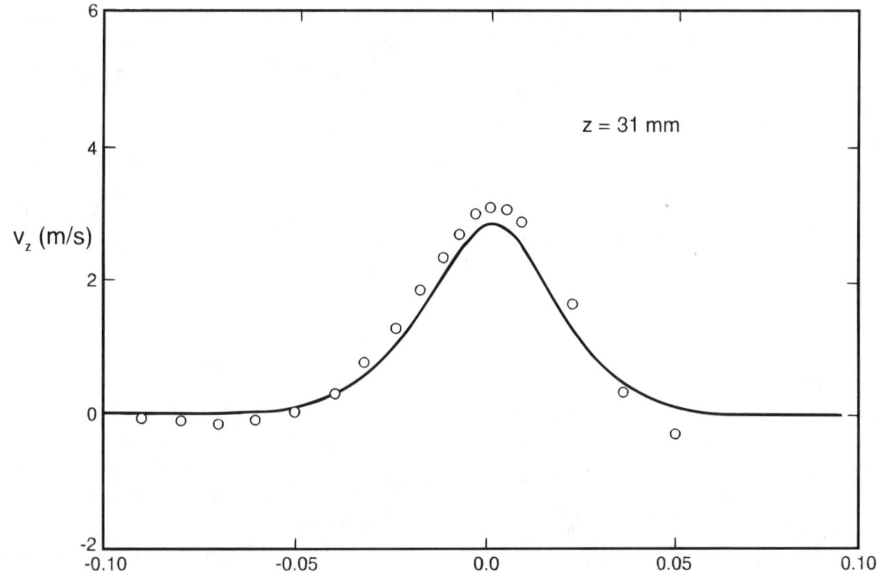

Figure 7b. Axial velocity as a function of the lateral coordinate (z = 31 mm).

References

Arnold, G., Drew, D. A., and Lahey, R. T., Jr. (1989) Derivation of constitutive equations for interfacial force and reynolds stress for a suspension of spheres using ensemble averaging, *J. Chem. Eng. Comm.*, 86, 43–54.

Bonetto, F. and Lahey, R. T., Jr. (1993) An experimental study on air carryunder due to a plunging liquid jet, *Int. J. Multiphase Flow*, 19, 2, 281–294.

Bonetto, F., Drew, D. A., and Lahey, R. T., Jr. (1994) The analysis of a plunging liquid jet—the air entrainment process, *J. Chem. Eng. Comm.*, 130, 11–29.

Bonetto, F. and Lahey, R. T., Jr. (1994) *Experimental Results of a Two-Phase Planar Jet*, in preparation.

Chanson, H., and Cummings, P. (1994) Modeling Air Bubble Entrainment by Plunging Breakers, Proceedings of the Int. Symp. Waves, 783–793, held in Univ. of Columbia, Vancouver, Canada, Aug. 21–24.

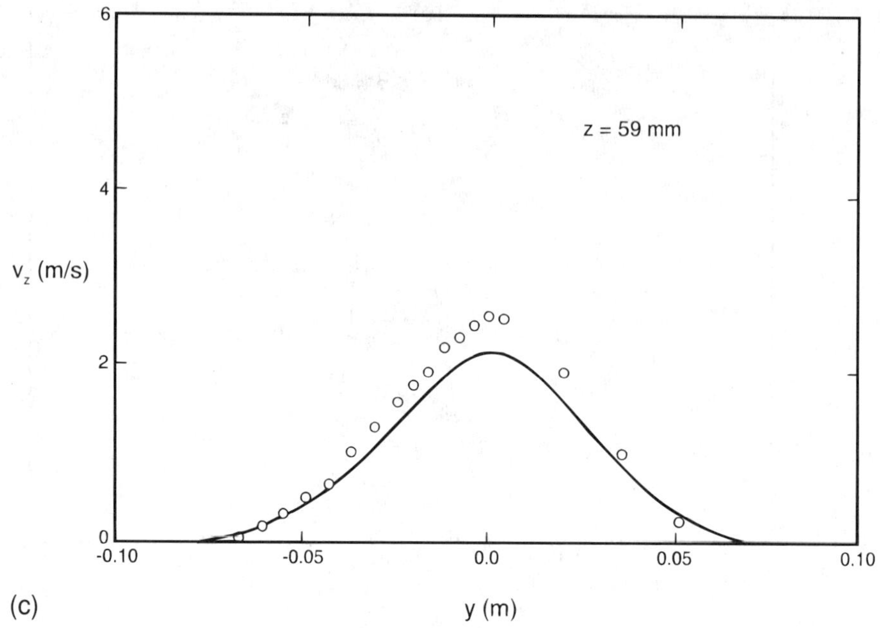

Figure 7c. Axial velocity as a function of the coordinate (z = 59 mm).

Delhaye, J. M. (1968) Equations Fondamentales des Ecoulement Diphasiques, Part I and II, *CEA-R-3429*, Centre d'Etudes Nucleaires de Grenoble, France.

Drew, D. A. and Lahey, R. T., Jr. (1979) Application of general constitutive principles to the derivation of multidimensional two-phase flow equations, *Int. J. Multiphase Flow*, 5, 423–264.

Ishii, M. (1975) *Thermo-Fluid Dynamic Theory of Two-Phase Flow*. Eyrolles.

Lezzi, A. M. and Prosperetti, A. (1991) The stability of an air film in a liquid flow, *J. Fluid Mech.*, 226, 319–347.

Lin, T. J. and Donnelly, H. G. (1966) Gas bubble entrainment by plunging laminar liquid jets, *AIChE Journal*, 12–3, 563.

McKeogh, E. J. and Ervine, D. A. (1981) Air entrainment rate and diffusion pattern of plunging liquid jets, *Chem. Eng. Sci.*, 36, 1161.

Park, J. W. (1992) *Void wave propagation in two-phase flow*, Ph.D. Thesis, Rensselaer Polytechnic Institute.

Rodi, W. (1984) Turbulence Models and their Application in Hydraulics, IAHR/AIRH Monograph.

Sini, J. F., and DeKeyser, I. (1987) Numerical prediction of turbulent plane jets and forced plumes by use of the k-ϵ model of turbulence, *Int. J. Heat Mass Transfer*, 30, 9, 1787–1801.

F. Bonetto and
R.T. Lahey, Jr.

PLUTONIUM

Plutonium—(Planet *pluto*), Pu; atomic number 94; isotopic mass Pu^{239} 239.13 (physical scale); specific gravity (α modification) 19.84 (25°C); melting point 641°C; boiling point 3232°C; valence 3, 4, 5, or 6.

Plutonium was the second transuranium element of the actinide series to be discovered. The isotope Pu^{238} was produced in 1940 by Seaborg, McMillan, Kennedy, and Wahl by deuteron bombardment of uranium in the 60-in. cyclotron at Berkeley, California. Plutonium also exists in trace quantities in naturally occurring uranium ores. It is formed in much the same manner as neptunium, by irradiation of natural uranium with the neutrons which are present. By far of greatest importance is the isotope Pu^{239}, with a half-life of 24,360 years, produced in extensive quantities in nuclear reactors from natural uranium:

$$U^{238}(n\gamma)U^{239} \xrightarrow{\beta} Np^{239} \xrightarrow{\beta} Pu^{239}$$

Fifteen isotopes of plutonium are known. Plutonium has assumed the position of dominant importance among the transuranium elements because of its successful use as an explosive ingredient in nuclear weapons and the place which it holds as a key material in the development of industrial use of nuclear power. One kilogram is equivalent to about 22 million kilowatt hours of heat energy. The complete detonation of a kilogram of plutonium produces an explosion equal to about 20,000 tons of chemical explosive. Its importance depends on the nuclear property of being readily fissionable with neutrons and its availability in quantity. The various nuclear applications of plutonium are well known. Pu^{238} has been used in the Apollo lunar missions to power seismic and other equipment on the lunar surface.

As with neptunium and uranium, plutonium metal can be prepared by reduction of the trifluoride with alkaline-earth metals. The metal has a silvery appearance and takes on a yellow tarnish when slightly oxidized. It is chemically reactive. A relatively large piece of plutonium is warm to the touch because of the energy given off in alpha decay. Larger pieces will produce enough heat to boil water. The metal readily dissolves in concentrated hydrochloric acid, hydroiodic acid, or perchloric acid with formation of the Pu^{+3} ion.

The metal exhibits six allotropic modifications having various crystalline structures. The densities of these vary from 16.00 to 19.86 g/cm^3. Plutonium also exhibits four ionic valence states in aqueous solutions: Pu^{+3} (blue lavender), Pu^{+4} (yellow brown), PuO^+ (pink?), and PuO^{+2} (pink orange). The ion PuO^+ is unstable in aqueous solutions, disproportionating into Pu^{+4} and PuO^{+2} the Pu^{+4} thus formed, however, oxidizes the PuO^+ into PuO^{+2}, itself being reduced to Pu^{+3}, giving finally Pu^{+3} and PuO^{+2}. Plutonium forms binary compounds with oxygen: PuO, PuO_2, and intermediate oxides of variable composition; with the halides: PuF_3, PuF_4, $PuCl_3$, $PuBr_3$, PuI_3; with carbon, nitrogen, and silicon: PuC, PuN, $PuSi_2$. Oxyhalides are also well known: PuOCl, PuOBr, PuOI.

Because of the high rate of emission of alpha particles and the element being specifically absorbed by bone marrow, plutonium, as well as all of the other transuranium elements except neptunium, are radiological poisons and must be handled with very special equipment and precautions. Plutonium is a very dangerous radiological hazard. Precautions must also be taken to prevent the unintentional formation of a critical mass. Plutonium in liquid solution is more likely to become critical than solid plutonium. The shape of the mass must also be considered where criticality is concerned.

Handbook of Chemistry and Physics, CRC Press

PLUTONIUM, BURNING IN FAST REACTOR
(see Liquid metal cooled fast reactor)

PLUTONIUM 239 (see Nuclear reactors)

PNEUMATIC TRANSPORT (see Gas-solid flows)

POD BOILERS (see Coiled tube boilers)

POISEUILLE EQUATION (see Contact angle)

POISEUILLE FLOW

Poiseuille flow is *pressure-induced* flow (**Channel Flow**) in a long duct, usually a pipe. It is distinguished from drag-induced flow such as **Couette Flow**. Specifically, it is assumed that there is **Laminar Flow** of an incompressible **Newtonian Fluid** of viscosity η induced by a constant positive pressure difference or pressure drop Δp in a pipe of length L and radius $R << L$. By a pipe is meant a right circular cylindrical duct, that is a duct with a circular cross-section normal to its axis or generator.

Because of the geometry, Poiseuille flow is analysed using cylindrical polar coordinates (r, θ, z) with origin on the centre-line of the pipe entrance and z-direction aligned with the centre-line (see **Figure 1**). Symmetry means that Poiseuille flow is swirl-free and axisymmetric. Thus, the only non-zero components of the velocity **u** are the radial component u_r and the axial component u_z: the angular component $u_\theta = 0$. Moreover, u_r and u_z are independent of θ, as is the pressure p. Because the pipe is long, Poiseuille flow is fully developed, that is the velocity **u** is independent of axial position z everywhere except near the entrance ($z = 0$) and exit ($z = L$) of the pipe, from which it follows that $u_r = 0$. Solution of the mass and linear momentum *conservation equations*, specifically the *Navier-Stokes equations*, with boundary conditions of no-slip at the pipe wall ($r = R$) and symmetry at the centre-line ($r = 0$) yields (see Richardson, 1989):

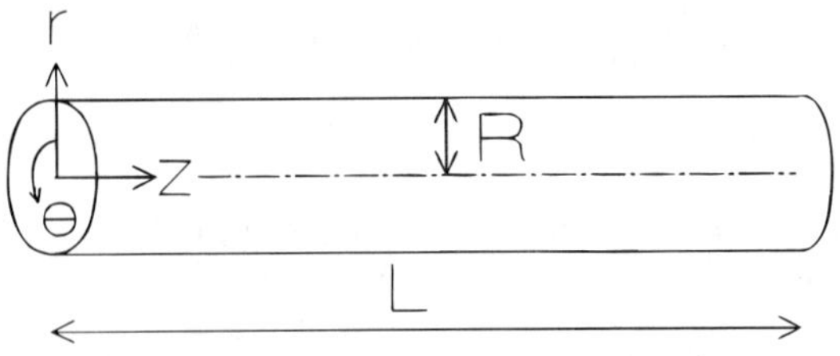

Figure 1. Poiseuille flow geometry.

$$u_z = \Delta p(R^2 - r^2)/4\eta L \quad (1)$$

Thus the axial velocity profile is parabolic (see **Figure 2**). The maximum axial velocity $u_{z_{max}}$ occurs at the centre-line ($r = 0$) and is given by:

$$u_{z_{max}} = \Delta p R^2/4\eta L \quad (2)$$

whereas the mean axial velocity $\bar{u}_z$ is given by:

$$\bar{u}_z = \Delta p R^2/8\eta L \quad (3)$$

from which it follows that $\bar{u}_z = \frac{1}{2} u_{z_{max}}$. The volumetric flow rate $\dot{V}$ through the pipe is given by:

$$\dot{V} = \Delta p \pi R^4/8\eta L \quad (4)$$

This is the *Hagen-Poiseuille Equation*, also known as *Poiseuille's Law*. Experimentally, Equation 4 is found to be corroborated provided the **Reynolds Number** (Re) given by:

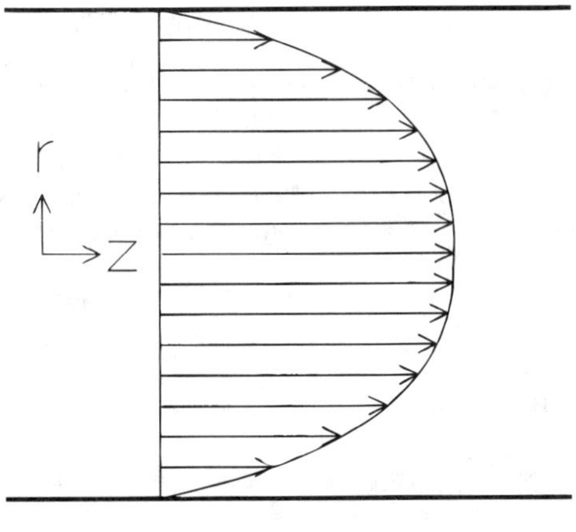

Figure 2. Axial velocity profile.

$$Re = 2\rho\bar{u}_z R/\eta \quad (5)$$

is less than some critical value $Re_c \simeq 2000$, so that there is laminar flow and not **Turbulent Flow**, though it should be noted that Re_c appears to be very much larger than 2000 if particular care is taken to minimise disturbances which might cause flow instabilities; and provided $L/R >> C\ Re$ where $C \sim 0.1$, so that the pipe is long enough for entrance and exit effects to be negligible and hence for **u** to be fully-developed.

Poiseuille flow is a shear flow with shear-rate γ given by:

$$\gamma = \left|\frac{du_z}{dr}\right| = \Delta pr/2\eta L \quad (6)$$

The *viscous dissipation rate* ϵ is given by:

$$\epsilon = \eta\gamma^2 \quad (7)$$

Dissipation of mechanical energy, that is conversion of mechanical energy (specifically pressure energy) into thermal energy by viscous action, increases the temperature T of the fluid. Assuming that the temperature T is fully developed, the solution of the energy conservation equation with boundary conditions of specified temperature T_w at the pipe wall ($r = R$) and symmetry at the centreline ($r = 0$) yields (see Richardson, 1989):

$$T - T_w = \Delta p^2(R^4 - r^4)/64\lambda\eta L^2 \quad (8)$$

where λ denotes thermal conductivity. Experimentally, (8) is found to be corroborated provided $L/R >> C\ Pe$ where $C \sim 0.1$ and the **Peclet Number** (Pe) is given by:

$$Pe = 2\rho c\bar{u}_z R/\lambda \quad (9)$$

where c denotes specific heat. The pipe is then long enough for entrance and exit effects to be negligible and hence for T to be fully-developed.

For Poiseuille flow in channels or ducts of non-circular cross-section, analogous expressions can be obtained for velocity and temperature [see Happel and Brenner (1973) and Shah and London (1978)].

References

Happel, J. and Brenner, H. (1973) *Low Reynolds Number Hydrodynamics*, Noordhoff, Leyden.

Richardson, S. M. (1989) *Fluid Mechanics*, Hemisphere, New York.

Shah, R. K. and London, A. L. (1978) Laminar flow forced convection in ducts. *Advances in Heat Transfer Supplement 1*, Academic Press, New York.

S.M. Richardson

POISEUILLE LAW (see Poiseuille flow)

POISSON EQUATION (see Conduction; Green's function)

POISSON'S RATIO

When a bar of a solid material is loaded in tension longitudinally, the longitudinal extension is accompanied by a lateral contraction. The ratio between the lateral contraction per unit length and the longitudinal extension also per unit length is known as the Poisson's ratio. In the elastic regime, this ratio is a material constant with a value of about 0.3 for most metals.

In the literature the Greek letter ν is commonly used as the symbol for the Poisson's ratio.

Consider an elemental cube with its sides parallel to the three coordinate axes x,y,z. Under a stress system σ_x, σ_y, σ_z the strains are ϵ_x, ϵ_y, ϵ_z. The relationship between strains and stresses in the elastic regime is given by *Hooke's laws*,

$$\epsilon_x = \frac{1}{E}[\sigma_x - \nu(\sigma_y + \sigma_z)]$$

$$\epsilon_y = \frac{1}{E}[\sigma_y - \nu(\sigma_x + \sigma_z)]$$

$$\epsilon_z = \frac{1}{E}[\sigma_z - \nu(\sigma_x + \sigma_y)]$$

where E is *Young's modulus* and v is the Poisson's ratio.

C. Ruiz

POISSON STATISTICS (see Optical particle characterisation)

POLARISATION (see Electrolysis; Reflectivity)

POLARISATION METHOD, FOR DROPSIZE MEASUREMENT (see Dropsize measurement)

POLARISED RADIATION (see Electromagnetic waves)

POLLUTANTS (see Burners; Combustion; Fuels; Incineration; Internal combustion engines)

POLLUTION (see Aeration; Air pollution; Burners; Chimneys; Combustion; Heat pumps; Prilling)

POLLUTION CONTROL (see Ion exchange)

POLYMER SEPARATION (see Molecular sieves)

POLYMETS (see Electroplating)

POLYMORPHS (see Crystallization)

POLYNOMIALS

An expression of the form

$$P_n(x) = a_0 + a_1x + a_2x^2 + \cdots + a_nx^n,$$

where at least $a_n \neq 0$, is called a polynomial of degree n. The

Fundamental Theorem of Algebra states that a polynomial of degree n has exactly n roots or zeros, which can be real or complex. Let us call them x_i, for i = 1, 2, . . . , n. Therefore, since $P_n(x)$ is equal to zero for each $x = x_i$, it can be written as:

$$P_n(x) = a_0 + a_1x + a_2x^2 + \cdots + a_nx^n$$

$$= a_0(x - x_1)(x - x_2) \cdots (x - x_n),$$

the last expression is called the factorization of $P_n(x)$.

Relations between the roots and the coefficients of a polynomial follow immediately expanding this last expression. For example, the sum of the roots is equal to $-a_1/a_0$.

From the Fundamental Theorem of Algebra it also follows that the representation of a polynomial $P_n(x)$ in terms of powers of x is unique. That is, a polynomial $P_n(x)$ is uniquely associated with a set of n + 1 coefficients a_i, for i = 0, 1, 2, , . . . , n. Polynomials of low degree receive special names: a linear polynomial is one of the first degree; quadratic is one of the second degree, cubic is one of the third degree.

Calculation of Numerical Values of a Polynomial

Notice that the calculation for a given value of the argument x of a quadratic polynomial $P_2(x) = a_0 + a_1 x + a_2 x^2$ requires two additions and three multiplications, one for the second term and two for the third. In general to calculate a value of a polynomial of degree n it is necessary to do n additions and n(n + 1)/2 multiplications. A considerable saving in computing time can be achieved using the so-called nested form of a polynomial. For $P_2(x)$ this is: $P_2(x) = (a_2 x + a_1)x + a_0$, the number of additions necessary to calculate a value of it for a given value of x remains equal to n = 2 but the number of multiplications is reduced from 3 to 2.

In general, for a polynomial of degree n, the nested form has the general expression

$$P_n(x) = ((\cdots((a_nx + a_{n-1})x + a_{n-2})x + \cdots + a_2)x + a_1)x + a_0,$$

using the nested form the number of multiplications is reduced substantially: from n(n + 1)/2 to simply n.

Other Types of Polynomials

Up to now we have referred only to polynomials in terms of powers of x, that is, we have used the basis of representation $\{1, x, x^2, \ldots, x^n\}$, but instead of that basis of representation we could use, for example, the trigonometric functions:

$$\{1, \cos x, \sin x, \cos 2x, \sin 2x, \ldots, \cos nx, \sin nx\},$$

and have then the trigonometric polynomial

$$a_0 + a_1 \cos x + b_1 \sin x + a_2 \cos 2x, + b_2 \sin 2x + \cdots$$

$$+ a_n \cos nx + b_n \cos nx.$$

Finite sections of **Fourier Series** are a special type of *trigonometric polynomials*. Similarly, we can define polynomials in terms of other representation basis: for example **Legendre or Chebyshev Polynomials** of degrees 0, 1, 2, . . , **Bessel functions** of orders 0, 1, 2, . . and many others. Such representations are useful in applications, for example for the solution of **Differential Equations**.

Reference

Fike, C. T. (1968) *Computer Evaluation of Mathematical Functions*, Prentice-Hall, New Jersey.

Leading to: *Fourier series; Legendre polynomial series; Chebyshev polynomials; Bessel series; Differential equations*

E.L. Ortiz

POLYSTYRENE ANION EXCHANGE RESINS (see Ion exchange)

POLYSTYRENE SULPHONIC ACID CATION RESINS (see Ion exchange)

POLYTROPIC INDEX (see Polytropic process)

POLYTROPIC PATH (see Polytropic process)

POLYTROPIC PROCESS

Following from: *Thermodynamics*

The term "polytropic" was originally coined to describe any reversible process on any open or closed system of gas or vapour which involves both heat and work transfer, such that a specified combination of properties were maintained constant throughout the process. In such a process, the expression relating the properties of the system throughout the process is called the *polytropic path*.

There are an infinite number of reversible polytropic paths between two given states; the most commonly used polytropic path is

$$T\frac{dS}{dT} = C_1 \quad \textbf{(1)}$$

where T is **Temperature**, S is **Entropy** and C_1 is a constant and is equal to zero for an adiabatic process. This path is equivalent to the assumption that the same amount of heat is transferred to the system in each equal temperature increment. In a reversible process following this polytropic path the heat and work transfer are as follows:

$$Q = C_1(T_2 - T_1) \quad \textbf{(2)}$$

and

$$W = (H_2 - H_1) - Q \quad \textbf{(3)}$$

where H is **Enthalpy**

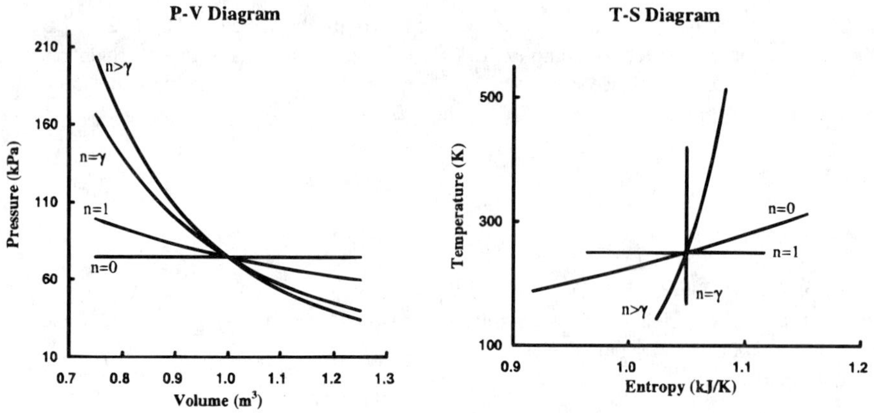

Figure 1. Examples of simple polytropic processes.

Pressure and volume, and pressure and temperature are related by the following expressions

$$\frac{dP}{P} - \frac{V}{P}\left[\left(\frac{C_p - C_1}{C_v - C_1}\right)\left(\frac{\partial P}{\partial V}\right)_T\right]\frac{dV}{V} = 0 \qquad (4)$$

and

$$\frac{dT}{T} - \left[\frac{P}{C_p - C_1}\left(\frac{\partial V}{\partial T}\right)_P\right]\frac{dP}{P} = 0 \qquad (5)$$

where C_p and C_v are heat capacity at constant pressure and volume respectively.

For an ideal gas this polytropic path simplifies to

$$PV^n = C_2 \qquad (6)$$

where C_2 and n are constants and n is called the *polytropic index*. The polytropic index characterises the process, as summarised in **Figure 1**, and is given by

$$n = (C_P - C_1)/(C_V - C_1). \qquad (7)$$

The equivalent pressure—temperature and work relationships are as follows:

$$\frac{P^{(n-1)/n}}{T} = C_3\ (= N\tilde{R}/C_2^{1/n}) \qquad (8)$$

and

$$W = \frac{n}{n-1}(P_2V_2 - P_1V_1). \qquad (9)$$

Many gas compression and expansion processes may be usefully approximated by a polytropic process. In each case the polytropic coefficient must be determined experimentally by measurement of the heat and work transfer and the initial and final states.

Bett, Rowlinson and Saville (1975) discuss the errors and inconsistencies which may arise when $PV^n = C_2$ is applied to non-ideal gases and vapours.

Reference

Bett K. E., Rowlinson J. S., and Saville G. (1975) *Thermodynamics for Chemical Engineers*, The MIT Press, Cambridge MA.

N.F. Kirkby

POOL BOILING OF LIQUID METALS (see Liquid-metal heat transfer)

POOL FIRES (see Fires)

POPULATION BALANCE, CRYSTALS (see Crystallization)

POROSITY (see Porous medium)

POROSITY MODEL OF BRANCHED PIPEWORK (see Headers and manifolds, flow distribution in)

POROUS BODIES, DIFFUSION IN (see Diffusion)

POROUS COATINGS, FOR INCREASING BURNOUT FLUX (see Burnout, forced convection)

POROUS MEDIA, HEAT TRANSFER IN (see Porous medium)

POROUS MEDIA, HYDRAULIC LOSSES IN (see Pressure drop, single phase)

POOL BOILING

Following from: *Boiling*

In pool boiling, vapour is generated at a superheated wall that is small compared to the dimensions of the pool of nominally stagnant liquid in which it is immersed. The motion of the liquid is induced by the boiling process itself (analogous to single-phase natural convection at a heated wall in an unbounded fluid) and the velocities are assumed to be low. These conditions are convenient for small-scale laboratory experiments and much of the understanding of boiling, such as the basic division into nucleate, transition and film boiling and studies of bubble nucleation and motion discussed in the article on **Boiling**, has been derived from pool boiling experiments. However, pool boiling is unusual in industrial equipment. Even if there is no forced flow of liquid past the heated wall, confinement of the liquid and close spacing of multiple heaters, as in kettle reboilers, **Figure 1**, means that conditions are closer to **Forced Convective Boiling**. The heat source is often a hot fluid separated from the boiling liquid by a thin metal wall, whereas electrical resistance heating is often used in pool boiling experiments. Consequently it is important to appreciate the special conditions of pool boiling experiments and to exercise caution in transferring the information they provide to large-scale industrial systems in which flow effects are generally significant. This article reviews the techniques that are used in pool boiling experiments. Pool boiling behaviour is described in more detail in the articles on **Boiling**, **Nucleate Boiling** and **Burnout** (**Pool Boiling**).

Pool boiling can be classified according to conditions in the pool, the geometry of the heated wall and the method of heating. These conditions influence the methods used to measure the primary variables of wall superheat and heat flux that are conventionally used to present boiling heat transfer performance as a "boiling curve".

In saturated pool boiling, or bulk boiling, **Figure 2(a)**, the pool is maintained at or slightly above the saturation temperature by interaction with the vapour bubbles rising from the superheated boiling surface. (Subsidiary heaters may be used to compensate for heat lost from the walls of the containing vessel.) The pool has a free surface at which the bubbles burst; the vapour space is usually connected to a condenser that returns liquid to the pool. The system pressure is controlled by the cooling applied to the condenser. In subcooled boiling, the pool temperature distant from the boiling surface is below the saturation temperature. There can be no escape of vapour from a subcooled pool, unless it is very shallow, so a heat sink must be provided by cooling regions on the walls of the vessel, **Figure 2(b)**. Alternatively, a subcooled experiment can be run for a short period without heat sink, relying on the thermal capacity of the cold pool. A subcooled pool cannot have a free surface in contact with its own pure vapour. Either the boiling vessel must be connected to a separate vessel in which the pressure is controlled, or there must be a gas space above the pool. Use of a cover gas leads to a concentration of dissolved gas which can influence boiling, particularly by improving the stability of nucleation sites and reducing the superheat required for their activation. Dissolved gas can be removed by a preliminary period of saturated boiling, either in the experimental vessel or in a separate vessel from which the experimental vessel is filled. The temperature-time-dissolved gas history can influence the subsequent boiling experiments, as described in the article on **Nucleate Boiling**, and may be different in industrial systems.

In both saturated and subcooled pool boiling, the operation of the heat sink requires a recirculatory flow in the pool that may interact with the boiling process in ways that depend on the geometry of the pool and of the superheated boiling surface. The shape of the vessel may be constrained by the need to observe the boiling process. Early experiments on pool boiling used heating surfaces that were thin horizontal wires of materials such as platinum, heated by the passage of direct electrical current. The electrical resistance of the wire provided a measure of its temperature, averaged over its length. Such experiments are useful to demonstrate some of the basic characteristics of boiling but they suffer from the disadvantage that the length scale of the bubbles is similar to that of the heater so that their behaviour is atypical of the extensive surfaces in industrial plant. Most experiments now use larger heaters in the form of horizontal cylinders with diameters in the range 10 to 20mm, horizontal plates of circular or rectangular shape and vertical or sloping rectangular plates, with dimensions in the range 5 to 100mm. Heaters much larger than this are rarely used because of the large power requirements resulting from the high heat fluxes in nucleate boiling. The small heaters interact with the recirculation of liquid in the pool through edge effects or because their dimensions are comparable with the critical wavelengths of interfacial instabilities in film boiling, **Figure 3**. The recirculatory flows that must return liquid right to the wall in nucleate boiling are rarely considered, except in the special case of vertical flow counter to the vapour flow.

Industrial plants frequently use heat transfer from hot single-phase or condensing fluids to drive boiling under conditions that approximate to controlled wall temperature, **Figure 4**. This is rarely done in pool boiling experiments because of the difficulty of measuring the wall temperature and the heat flux accurately. Electrical

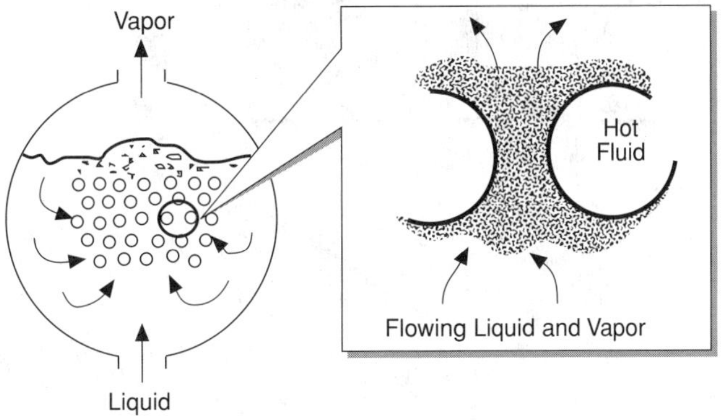

Figure 1. Kettle reboiler.

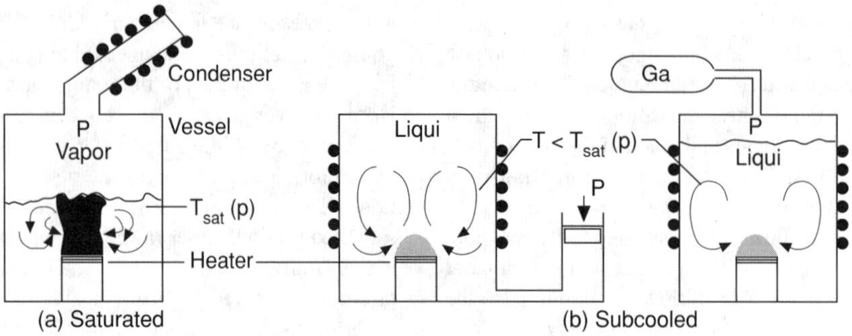

Figure 2. Pool boiling.

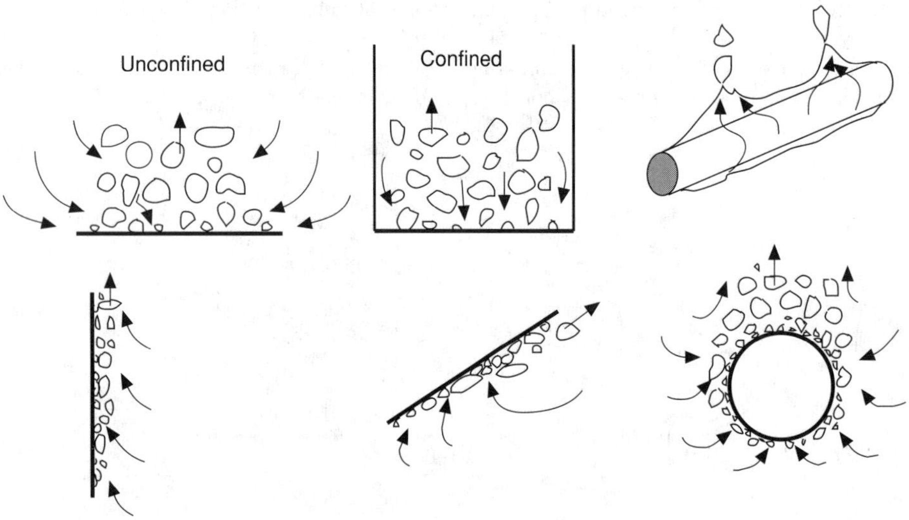

Figure 3. Influence of heater geometry.

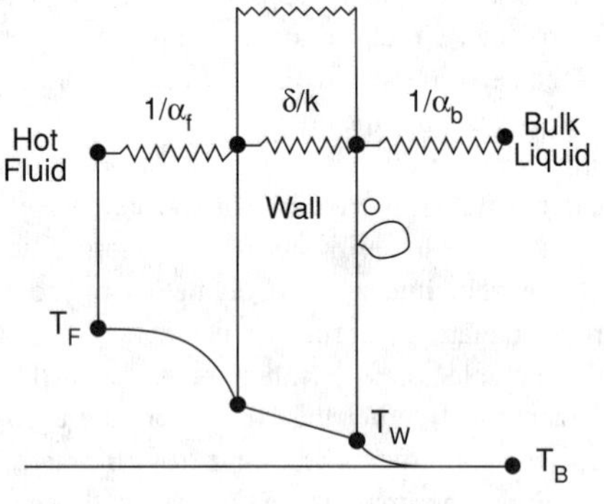

Figure 4. Heat supply from hot fluid.

resistance heating is generally used, so that experiments are performed at controlled heat flux. This can influence the boiling process, particularly in the departure from nucleate boiling and the transition region, **Figure 5**. Electrical heating is sometimes combined with sophisticated feedback control in order to operate in the unstable region of the boiling curve where dq/dΔTsat is negative. Electrical resistance heating is used in three ways:

a. Heaters in the form of cylindrical tubes or rectangular plates can be made of thin, electrically-conducting material. The uniform heat input is calculated from the current and the voltage across the heater. Temperature-measuring devices (thermocouples, thin resistance thermometers or thermochromic liquid crystal) are attached to the non-boiling surface of the wall, which is maintained adiabatic by thermal insulation or guard heating. The measurements are corrected for the parabolic temperature distribution across the wall to give the estimated temperature at the boiling surface, **Figure 6(a)**.
b. A variation on (a) is to deposit a very thin layer of electrically conducting material such as gold or tin oxide on an insulating substrate such as glass. The electrical heat input is generated at the boiling surface; the temperature is deduced from its electrical resistance or measured at the back of the substrate. When used with a heating layer that is so thin that it is transparent, this technique offers the special advantage that the boiling process can be observed through the wall; the disadvantage is that the wall effectively has the thermal properties of the substrate which, because it must be an electrical insulator, has a very low thermal conductivity (except for special materials such as sapphire).

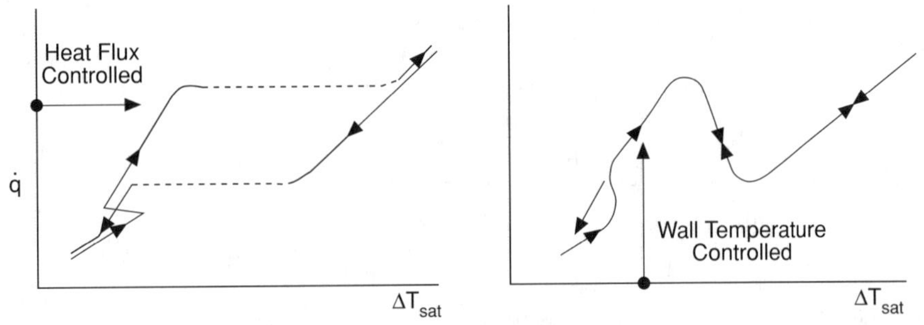

Figure 5. Effect of heating method on boiling curve.

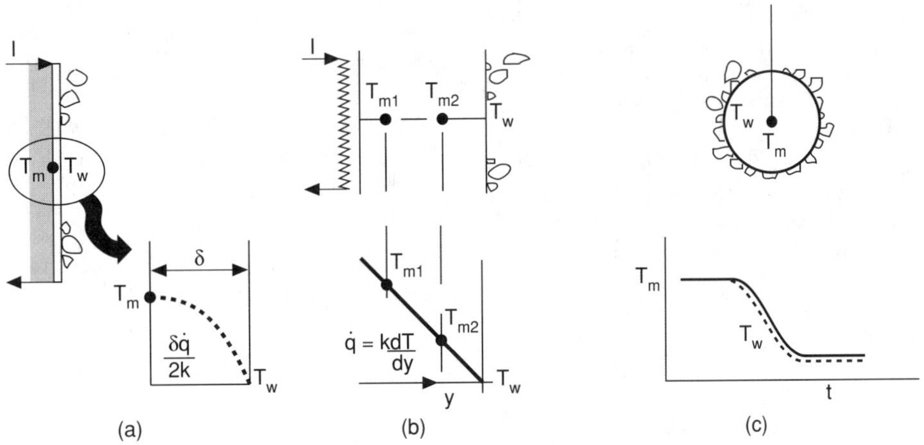

Figure 6. Heating methods and measurement of $\dot{q}$ and ΔT_{sat}.

c. Indirect electrical heating is used in conjunction with a thick wall of a good thermal conductor such as copper or aluminium. An electrically insulated resistance element is embedded in, or clamped to the back of, the wall. Thermocouples are embedded in the wall to measure the temperature gradient, from which are calculated the heat flux and the extrapolated temperature at the boiling surface, **Figure 6(b)**. This method can be used for heaters of circular or rectangular cross-section.

Electrical heating is difficult to arrange for other shapes of heaters, such as spheres. For these cases, a transient quenching method can be used in which the heater is treated as a calorimeter. Its temperature and rate of decrease of temperature are measured by embedded thermocouples when the preheated body is plunged quickly into the liquid pool, then held stationary, **Figure 6(c)**. The readings are corrected for the transient temperature gradients within the heater. This method is particularly convenient for the study of transition boiling.

The choice of heating method places constraints on the material and thickness of the heated wall, which may not match the conditions in industrial systems. The role of the bulk properties of the wall is overshadowed by the influence of the condition of its surface. The microgeometry and wettability of the boiling surface are known to have large effects on nucleate and transition boiling, as discussed in the articles on **Boiling** and **Nucleate Boiling**. In industrial plant they are dependent on the method of manufacture and on subsequent corrosion and fouling in ways that cannot be reproduced exactly (or even as yet quantified) in pool boiling experiments. Pool boiling surfaces are often subjected to artificial treatment in order to improve the reproducibility of the experiments but, for the reasons indicated above, there is considerable uncertainty in the application of the data to other systems.

D.B.R. Kenning

POROUS MEDIUM

Following from: *Flow of fluids*

A porous medium is a solid with voids distributed more or less uniformly throughout the bulk of the body.

The basic characteristic of this medium is *porosity*. The *bulk porosity* Π of a material is defined as the ratio of void volume V_v to body volume V_0, $\Pi = V_v/V_0$. Since the remaining portion V_s of the total volume of the material is in the form of a solid "skeleton", then

$$1 - \Pi = V_s/V_0.$$

For example, the porosity of porous materials with the skeleton formed by spherical particles with diameter d_p can be found from the relation

$$\Pi = 1 - N_P \frac{\pi d_p^3}{6},$$

where N_p is the number of particles per unit volume. These spheres can be arranged in various ways (**Figures 1a and b**). The cubic arrangement of spheres of the same diameter is characterized by a porosity of 0.476, while at a denser, rhombic, packing the porosity reduces to 0.259 (theoretically, this is the minimum porosity of packing of uniform spheres without deformation of the solid). The real porosity generally is estimated using its relation to density $\rho_\Sigma = \rho_0\,(1 - \Pi)$ or $\Pi = 1 - (\rho_\Sigma/\rho_0)$, where ρ_Σ and ρ_0 are the densities of the medium and of the solid material forming its skeleton respectively.

Permeability (or gas permeability) is the property which gives a measure of the gas flow through a porous medium exposed to a pressure difference. The superficial velocity V of fluid flow depends on permeability and pressure gradient in accordance with a modified *Darcy equation*

$$-\frac{dp}{dy} = \alpha\eta V + \beta\rho V^2.$$

Here, superficial velocity v is defined as the volumetric flow rate of the fluid per unit cross section of the medium. The coefficent α allows for friction losses that are characterized by the fluid viscosity η and the structure of a porous matrix. The coefficient of inertia β takes into account the losses associated with expansion, constriction, and bends in the pore channels; these losses are approximately proportional to ρV^2.

The modified Darcy equation is universal and describes isothermal liquid and gas flow in any porous solids without allowance for capillary forces. The influence of capillary forces (an increase in viscosity) is observed in water at the pore sizes $d_p < 1\ \mu m$. As a consequence, the ratio of water to gas flow rates for identical pressure difference may be either proportional to the ratio of their viscosities at high d_p or decrease by nearly a factor of 20 for porous channels of small diameter.

At low coolant velocity, the inertia term in the Darcy equation may be neglected, and the equation takes the form that is widely used in the theory of filtration

$$-\frac{dp}{dy} = \frac{\eta V}{K},$$

where $K = 1/\alpha$ is the Darcy permeability coefficient.

In this flow regime, the superficial velocity V or superficial mass flux $\dot{m}$ in porous material is proportional to pressure gradient. The higher the velocity, the stronger the influence of the quadratic term in the resistance law, while the dependence of flow rate on pressure gradient dp/dy gradually weakens and tends to a square root function

$$\dot{m} = \sqrt{-\rho(dp/dz)/\beta}.$$

Let us discuss in more detail a viscosity-inertia regime of filtration of a compressible ideal gas. Substituting gas mass flux for filtration velocity V

$$V = \frac{\dot{m}}{\rho} = \frac{\dot{m}RT}{p},$$

we transform the modified Darcy equation to the form

$$-p\frac{dp}{dy} = \alpha\eta\dot{m}RT + \beta\dot{m}^2RT.$$

If the pressure at the inlet to a porous plate is p_1 and at the outlet p_2, the temperature being constant throughout its thickness δ, then the solution to this equation is

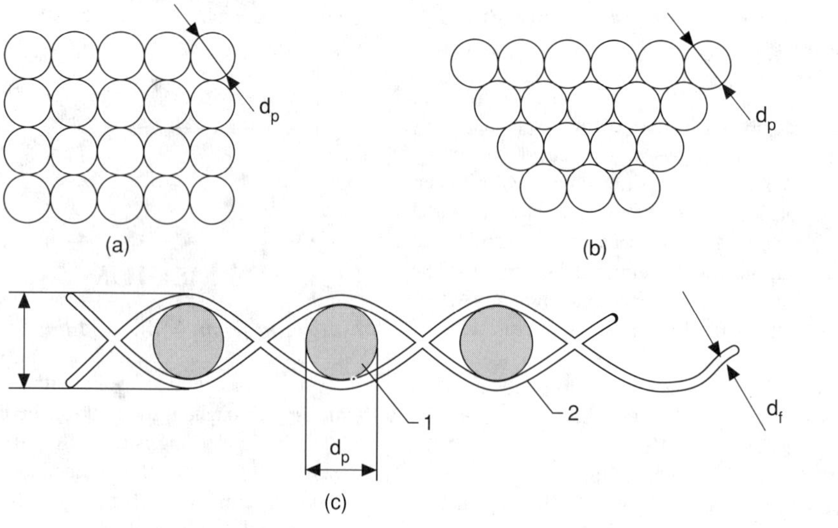

Figure 1. Configurations of porous media.

$$p_1^2 - p_2^2 = 2\alpha\eta\delta\dot{m}RT + 2\beta\delta\dot{m}^2RT$$

that, after simple manipulations, takes the form

$$\frac{p_1 + p_2}{2}(\Delta P) = \alpha\eta\delta\dot{m}RT(1 + Re),$$

where the Re number is calculated using the characteristic dimension of the porous medium determined as a ratio of the inertial to viscous resistance coefficients β/α

$$Re = \beta\dot{m}/(\alpha\eta) = (\beta/\alpha)\dot{m}/\eta.$$

In contrast to the flow in tubes, the choice of characteristic parameters for a porous wall is ambiguous. Even for powdered materials produced by sintering particles of the same shape and size determination of the pore channel diameter as a characteristic dimension involves difficulties and leads to a great scatter of experimental data being processed.

The authors of some papers make an attempt to relate porosity Π with the coefficients α and β in the modified Darcy law. In particular, a dependence of the type of

$$\alpha = 1.7 \times 10^2 \frac{(1 - \Pi)^2}{d_p^2\Pi^3}, \qquad \beta = 0.63\frac{1 - \Pi}{d_p\Pi^{4.72}}$$

is established for materials produced by sintering of spherical particles with diameter d_p, m.

Real materials virtually all have a distribution of grain size, which appreciably reduces porosity and pore size. For instance, the diameter of a pore formed by four spheres with diameter d_p is 0.156 d_p, by three d_p spheres and one 0.5 d_p sphere it is as low as 0.118 d_p. If, however, another 0.5 d_p sphere appears among the four d_p spheres, then the pore size reduces to 0.095 d_p. Porosity diminishes especially sharply in cases when fine particles are in pores formed by coarse grains.

To secure an adequate mechanical strength, porous materials are subjected to a special thermal treatment, i.e. sintering, at a temperature of the order of 0.8–0.9 of the melting point. The grains of powder, as a rule, possess an inherent residual porosity from 8 to 15 per cent. The overal porosity of metal porous bodies varies over the range 30 to 40 per cent.

Besides pure metals, various carbides, ceramics, and some other materials are used for production of porous bodies. In this case the porous structure is formed by adding an evaporable component to the material and subsequent firing the mixture. This method makes possible the production of materials with porosity up to 50–60 per cent.

Mesh porous matrices can be obtained from fibrous substances (**Figure 1c**). These materials possess some valuable advantages over powders, viz. a considerably higher strength, easy manufacture of materials with preset and uniform porosity depending on the type of original mesh, and, finally, the possibility of making products of any size.

The rate of intraporous convective heat transfer is commonly characterized by the volume *heat transfer coefficient* α_v. If we denote by T_s the local temperature of a porous wall and by T_g the coolant temperature at the same section (**Figure 2**), then the thermal state of a plane porous wall with constant thermal properties of the material and the coolant will be described in the steady state case by two equations, one

$$\lambda_\Sigma \frac{d^2T_s}{dy^2} = \alpha_v(T_s - T_g)$$

for the solid phase (the porous matrix) and the other

$$c_{pg}\dot{M}_g \frac{dT_g}{dy} = \alpha_v(T_s - T_g)$$

for the (gaseous) coolant.

We introduce a dimensionless temperature difference $\theta = \frac{T_s - T_g}{T_{sw} - T_{s0}}$ which accounts for a local normalized to the maximum temperature difference across the porous wall. In this case the set of two equations reduces to one equation with respect to θ

$$\frac{d^2\theta}{dy^2} + \frac{\alpha_v}{c_{pg}\dot{M}_g}\frac{d\theta}{dy} - \frac{\alpha_v}{\lambda_\Sigma}\theta = 0.$$

Its general solution is $\theta = C_1e^{k_1y} + C_2e^{k_2y}$, where the exponents k_1 and k_2 are the roots of the characteristic equation

$$k_{1,2} = -\frac{\alpha_v}{2c_{pg}\dot{M}_g} \pm \sqrt{\left(\frac{\alpha_v}{2c_{pg}\dot{M}_g}\right)^2 + \frac{\alpha_v}{\lambda_\Sigma}}.$$

The constants C_1 and C_2 are determined from the boundary condition on the inner surface and the heat balance in the porous wall. Here, however, there arises an uncertainty in the allowance for the fraction of heat energy which is supplied to the coolant before entering the porous wall. For instance, flowing along the inner

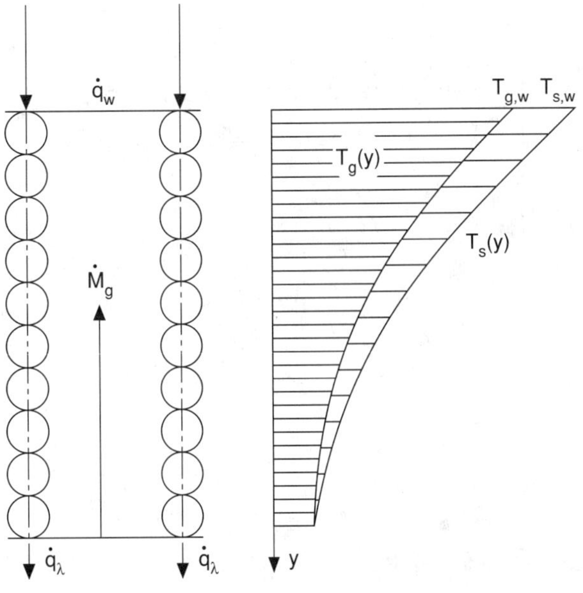

Figure 2. Heat flows in porous media.

surface of the wall, the coolant may be heated as if in a channel with suction through one of the channel surfaces.

Nevertheless, qualitative characteristics of coolant heating in a porous wall can be specified without elucidating the type of the boundary condition. **Figures 3** and **4** demonstrate how two reference parameters entering the equation for θ, namely $a = \alpha_v/\lambda_\Sigma$ and $b = \alpha_v/(c_{pg}\dot{M}_g)$, affect the efficiency of a porous system $\eta = (1 - \theta_w)$. **Figure 3** shows the effect of the porous wall thickness δ at a fixed $a = 0.3\ cm^{-2}$. **Figure 4** graphs the dependence of the dimensionless temperature on the set of parameters a and b for a wall thickness $\delta = 0.6$ cm. For a given value of b it is advisable to diminish the coefficient a by, for example, increasing the thermal conductivity of the solid phase, which brings about a flatter temperature profile. The coolant temperature at the outlet from the porous wall may turn out to lower than the temperature of the porous skeleton.

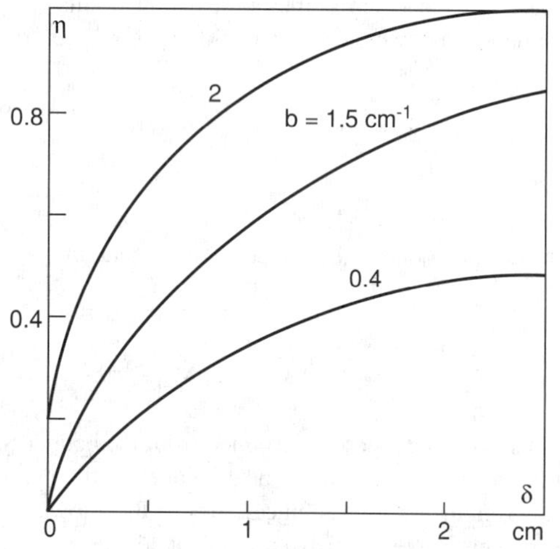

Figure 3. Efficiency of cooling of a porous medium as a function of thickness.

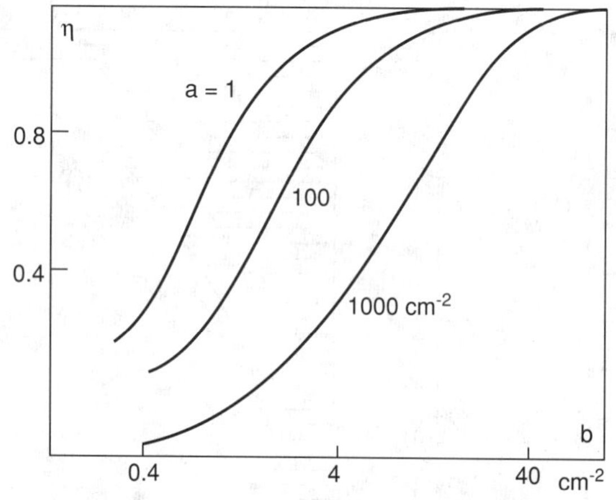

Figure 4. Efficiency of cooling of a porous medium as a function of $b = \alpha_v/c_{pg}\dot{M}_g$.

Temperature equilibrium in a porous body strongly depends on the thermal conductivity of the porous material λ_Σ and on the internal heat transfer α_v. In porous metals, thermal resistance to heat transfer is mainly observed in the region of contact of neighboring particles (**Figure 2**), where the smallest cross section area and the greatest inhomogeneity of a metal composition are observed. The quality of thermal contact depends on many practically irreproducible technological factors, including the shape and size of the original particles, material purity and composition, compacting pressure, sintering temperature and time. It is this irresponducibility that excludes accurate physical modeling of heat transfer and accounts for the spread of experimental data.

A large number of relationships have been developed for calculating the dependence of the thermal conductivity λ_Σ on porosity. However, none of them can describe the vast range of experimental data. Therefore, we confine ourselves to two approximating dependences of λ_Σ on porosity Π denoted by Roman numerals I and II in **Figure 5**. The great bulk of experimental data denoted by III in **Figure 5** lies between these two approximations, and they can be considered an upper and lower estimate of the thermal conductivity coefficient of real porous materials.

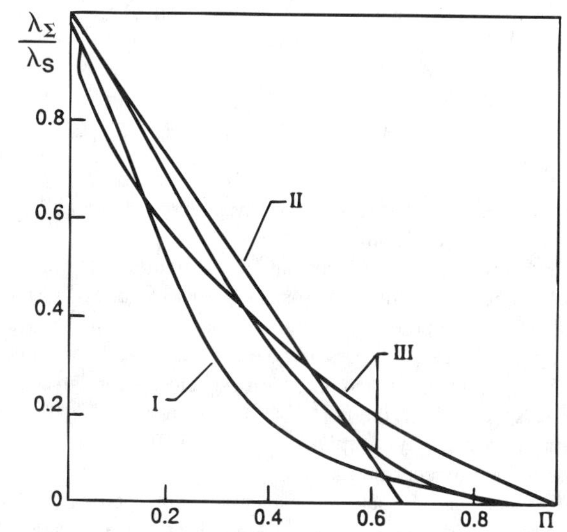

Figure 5. Variation of the thermal conductivity of a porous medium with porosity.

Curve I corresponds to a quadratic dependence

$$\lambda_\Sigma/\lambda_s = (1 - \Pi)/(1 + 11\,\Pi^2)$$

Curve II represents the simplest linear function

$$\lambda_\Sigma/\lambda_s = (1 - 1.5\Pi)$$

whose range of applicability is limited by the value $\Pi < 0.6$. As is seen from **Figure 5**, the discrepancy between curves I and II is often more than a factor of two. Other linear approximations, differing from II only by a numerical coefficient of porosity Π, that can vary from 2.1 to 1.5, are also encountered in the literature. It should be pointed out that λ_Σ measured for fibrous (mesh) and powdered (sintered from spheres) materials differs from the above expressions.

The effect of temperature on *thermal conductivity of porous materials* is ambiguous. For the materials with a high thermal conductivity of the solid phase (copper, tungsten) the dependence of λ_Σ on T is of the same character as for the relevant nonporous materials (see **Thermal Conductivity**). This is indicative of the absence of significant changes in the porous structure and the small contribution of radiant and convective components of heat transfer.

However, the effect of radiation or the kind of coolant gas may play an essential role for ceramic materials characterized by a low thermal conductivity of the matrix ($\lambda_\Sigma < 1$ W/mK) and a high porosity.

As for heat transfer in a fluid, we must take into account not only the molecular component $\lambda_g\Pi$ but also the possibility of a convective contribution which may substantially increase with increase pressure.

For fluid flow in pore channels it is not possible accurately to determine the heat transfer surface. Therefore, we use in this case not the traditional heat transfer coefficient per unit area of interface but the volume coefficient of heat transfer $\alpha_v = \dot{Q}/(V_0\Delta T)$, kW/m^3 K averaged over the entire volume of the porous medium with an average temperature difference ΔT.

Under stationary conditions the quantity of heat $\dot{Q}$ supplied to a coolant is calculated by its heating in a porous wall

$$\dot{Q} = c_{pg}\dot{M}_g(T_{gw} - T_{g0})$$

or by the specific internal heat release $\dot{q}_v$ as a result of ohmic heating

$$\dot{Q} = \dot{q}_v V_0,$$

$\dot{q}_v$ being totally determined by the current and the electrical resistance of a porous matrix. The main difficulty in experiments is measuring temperature difference $\Delta T = T_{gw} - T_{g0}$. The temperature difference cannot in actual fact be measured; therefore, the porous wall temperature is measured and the gas temperature obtained by heat balance.

Uncertainty in α_v is also introduced by errors in temperature measurements in the porous medium. Apart from common errors due to heat removal thermocouple junctions, there arise errors brought about by jamming of pores by the thermocouple, intense localized heat transfer between the coolant and the thermocouple, nonuniform heating of the thermocouple as a result of radiation from the matrix itself or from an external source under radiant heating. All the above factors are responsible for a substantial scatter of experimental data.

The results of experimental determination of the volume *heat transfer coefficient in porous media* are summarized in **Figure 6** (the numerals from 1 to 13 refer to the published papers). The comparison of the data obtained is hampered by different choice of characteristic dimensions for the Re and Nu numbers. In contrast to the surface heat transfer, the Nu number involves a squared characteristic linear dimension $Nu = \alpha_v l^2/\lambda$. Some researchers use as l the diameter of original particles d_p, others prefer the mean pore diameter, but most believe it more correct to take the ratio of coefficients of viscous α to inertial β resistance in the modified Darcy law.

The merit of the latter technique for the characteristic dimension choice is the fact that porosity Π is excluded from the similarity equation $Nu = f(Re)$. It follows from the comparison of the experimental data for various coolants such as air, alcohol and transformer oil, that their individual properties can be taken into account by introducing the Pr number to the same extent as by introducing the Re number. The product Re Pr = Pe is the Peclet number.

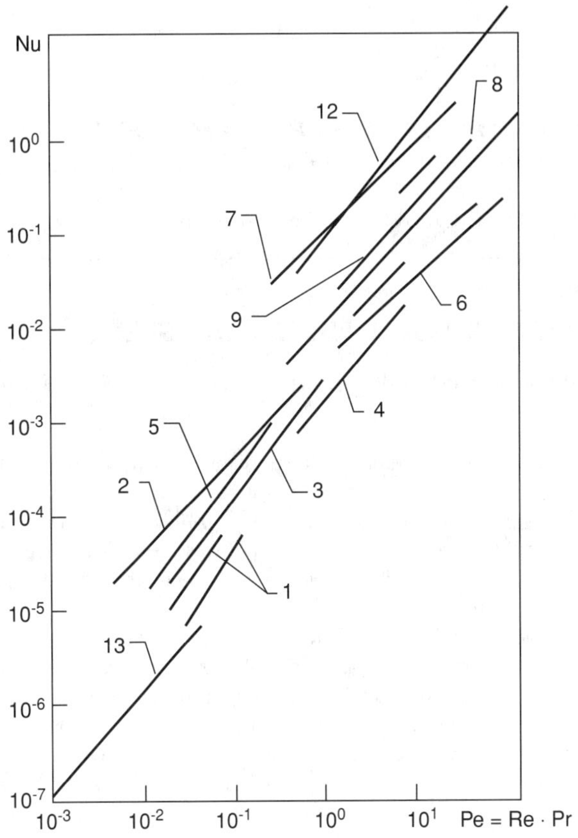

Figure 6. Published correlations for heat transfer in porous media.

The data in **Figure 6** make it possible to select various approximation dependences of the $Nu = A\,Pe^n$ type, where the exponent n on the Pe number in the papers mentioned varies from 0.65 to 1.84. Hence, any averaged empirical relation can be taken only as a first approximation. Choosing the β/α ratio as a characteristic dimension enables us to write the similarity relation for internal heat transfer as

$$Nu = 0.004\ Re\ Pr.$$

Account must be also taken of a limited thickness of the porous wall. The experimental results become independant of layer thickness only after 8 to 10 layers of spheres have been laid. However, the results are relatively insensitive to random fluctuations of the coolant flow rate $\dot{M}_g$.

Reference

Gnielinski, V. (1983) Fixed beds. Chapters 2.5.4 of *Heat Exhanger Design Handbook*, Hemisphere Publishing Corporation, New York.

Leading to: *Darcy's Law*

Yu.V. Polezhaev

POROUS WALL COOLING (see Transpiration cooling)

POSISTOR RESISTANCE THERMOMETERS (see Resistance thermometry)

POSITIVE CATALYSIS (see Catalysis)

POST-DRYOUT HEAT TRANSFER

Following from: *Boiling; Forced convection boiling; Burnout, forced convection*

Convective flow boiling downstream of the *critical heat flux* (CHF) location can be termed "post-dryout heat transfer". Conventionally, this regime is divided into a *transition boiling* region (unstable or partial film boiling) and a stable *film boiling* region. Transition boiling is defined as an unstable region with nucleate and film boiling occurring alternately, characterized by a decrease in the wall heat flux as the wall temperature increases. This definition of transition boiling is useful when dealing with a fluid that has undergone CHF with bulk subcooling or low vapor qualities. However, when CHF is characterized by film *dryout* (at high vapor qualities), the flow pattern is liquid droplets dispersed in a bulk vapor. In this case, bulk nucleate boiling is unlikely and the post-dryout heat transfer mechanisms are primarily vapor convection and some form of droplet-wall interaction. Hence, it is appropriate to consider post-dryout heat transfer as a consequence of the fluid condition at CHF rather than in terms of transition and film boiling as defined traditionally.

Figure 1 shows typical post-CHF flow conditions for the above two types of postulated CHF and post-dryout situations. If CHF occurs at low vapor fractions, then post-CHF behavior can be expected to be one of *inverse annular flow* followed by dispersed flow. If the flow rate is also low, then "film boiling" rather than convection will be the dominant heat transfer process. As quality and film thickness increase due to vaporization of the liquid, a decrease in heat transfer capability occurs as the wall temperature increases. This corresponds to the conventional definition of transition boiling.

If CHF occurs at high vapor fractions, the initiating mechanism is dryout of a liquid film at the wall, leading to dispersed drop flow in the post-dryout region. In this situation convection will dominate the heat transfer process, especially if the flow rate is high. Under these conditions, the heat removal capacity of the fluid can either decrease or increase as wall temperature increases, depending on the relative magnitudes of vapor convection and droplet-wall interaction. The usual negative-slope transition boiling portion of the classical boiling curve may not exist in this situation.

Inverse Annular Flow Regime

Post-dryout heat transfer with bulk subcooling or low vapor qualities occur with inverse annular flow, wherein a central core of liquid is isolated from the heat transfer surface by an annular vapor film. The heat is transferred from the wall to the vapor and subsequently from the vapor to the liquid interphase. If the liquid core is subcooled, a significant fraction of the heat flux is used for sensible heating of the liquid, the balance used to generate vapor at the liquid interphase. If the liquid core is saturated, heat transfer to the interphase is used directly for evaporation. The vapor annulus experiences higher acceleration than the liquid core, leading to substantial slip between the two phases. This velocity differential often causes instability and waves at the liquid interphase, sometimes resulting in entrainment of liquid drops into the vapor flow. These effects enhance heat transfer between the vapor and liquid phases, reducing the tendency for vapor superheating and keeping the vapor temperature close to the local saturation temperature.

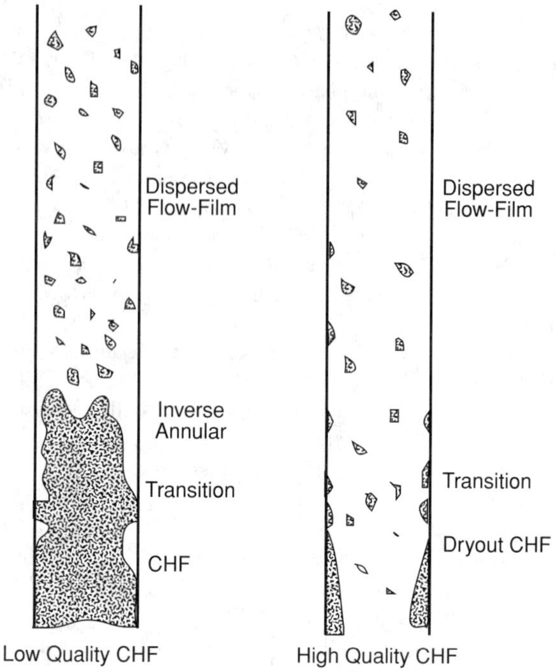

Figure 1. Flow patterns in post-dryout heat transfer.

Limited experimental measurements of post-dryout heat transfer in the inverse annular film regime have been obtained with refrigerants, cryogenic fluids and water (Bromley, et al., 1953, Fung, et al., 1979, Groeneveld and Gardiner, 1978, Quinn, 1965, Ragheb, et al., 1981). These experimental results indicate the following parametric effects on the post-dryout **Heat Transfer Coefficient** at the wall (α):

- α increases significantly with increases in bulk liquid subcooling, typically 2 to 5% per °C increase in subcooling,
- With positive equilibrium vapor **Quality** (i.e. x_e, the equilibrium or thermodynamic quality greater than zero) α decreases with increases in x_e at low mass fluxes but increases with x_e at high mass fluxes,
- The effect of mass flux ($\dot{m}$) is small or nonexistence for subcooled conditions and low mass fluxes; at higher flow rates and with net vapor quality α increases strongly with increasing mass flux,
- α increases with increasing system pressure, typically doubling in magnitude for a threefold increase in absolute pressure,

- In contrast to behavior in nucleate boiling, α is relatively insensitive to the absolute magnitude of the heat flux.

Some of these parametric effects are illustrated in **Figure 2**, from Groeneveld (1992).

Methods proposed for prediction of post-dryout heat transfer in this inverse annular flow regime are highly empirical. Groeneveld (1992) reviewed nine different correlations, noting that all were based on limited data and therefore are applicable only in specific ranges of operating parameters. A method that has a wider range of application and converges to predictions for the higher quality disperse regime was suggested by Groeneveld and Rousseau (1983):

$$\dot{q} = \text{convective heat flux}$$

$$= \text{greater of} \begin{array}{l} \text{(a) } \alpha_1(T_w - T_{sat}) \\ \text{(b) } \alpha_2(T_w - T_G) \end{array} \tag{1}$$

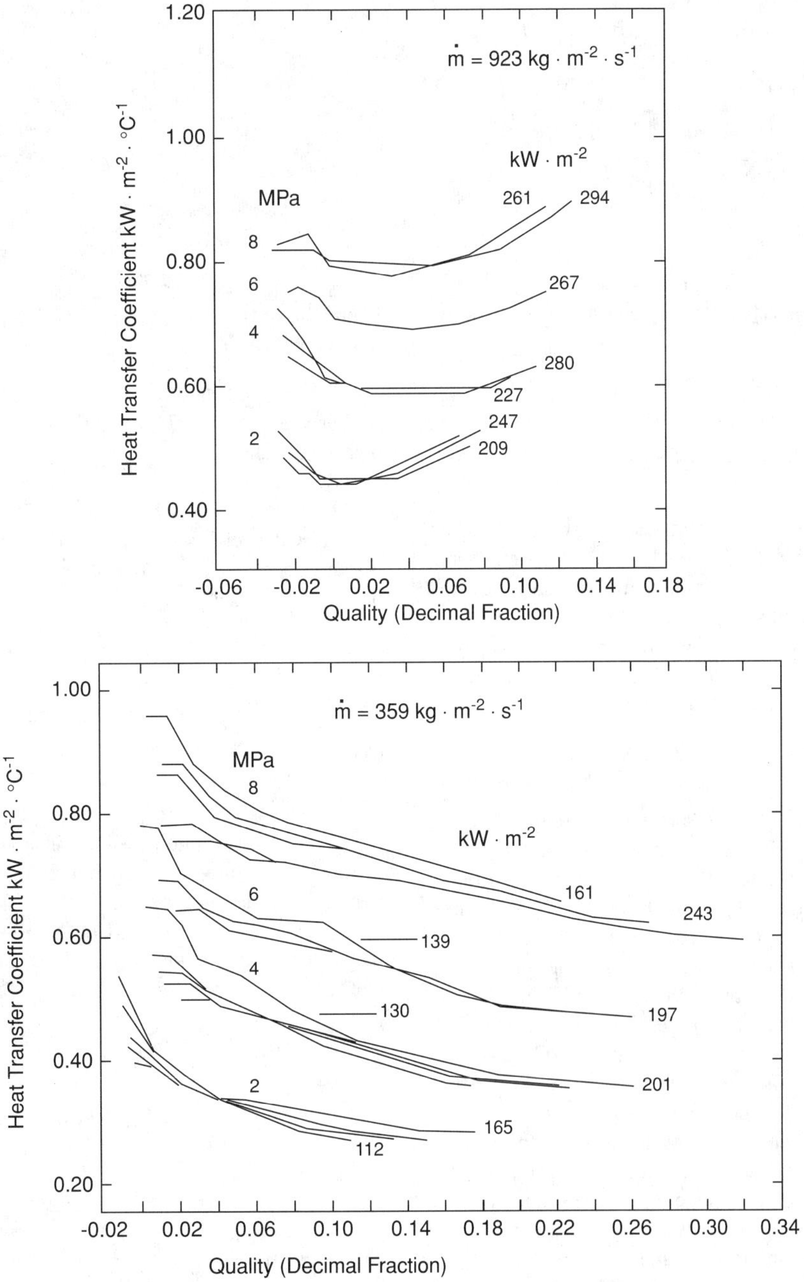

Figure 2. Parametric effects for inverse annular heat transfer, Groeneveld (1992).

where,
T_{sat} = saturation temperature
T_w = wall temperature
T_G = bulk vapor temperature

and,

$$\alpha_1 = \alpha_B(1 + 0.025\Delta T_{sub})\sqrt{(1 - x_a)} \quad (2)$$

$$\alpha_B = 0.425\left[\frac{g\lambda_G^3\rho_G(\rho_L - \rho_G)h_{LG}}{\eta_G\Delta T_s}\right]^{1/2}\left[\frac{g(\rho_L - \rho_G)}{\sigma}\right]^{1/4} \quad (3)$$

(Berenson 1961)

$$\alpha_2 = 0.0083\frac{\lambda_G}{D}\left\{\frac{\dot{m}D}{\eta_G}\left[x_a + \frac{\rho_G}{\rho_L}(1 - x_a)\right]\right\}^{0.88}(Pr)_G^{0.61} \quad (4)$$

$\Delta T_{sub} = T_{sat} - T_L$
$\Delta T_{sat} - T_W - T_{sat}$
x_a = actual quality

For inverse annular flow, the vapor phase is often close to thermodynamic equilibrium so that,

$$T_G \simeq T_{sat}$$

$$x_a \simeq x_e$$

The above approach considers only convective heat transfer from the wall. Radiation heat transfer is usually small by comparison, but may become significant at high temperatures. In such a case, the additional heat flux due to thermal radiation can be approximated as exchange between parallel infinite surfaces, assuming some effective emissivity for the wall and for the liquid interphase (See **Parallel Plates, Radiative Heat Transfer Between**).

Dispersed Flow Regime

In post-dryout heat transfer, the dispersed flow regime is more often encountered since it can occur over a wide range of vapor qualities, from as low as 0.2 to 1.0. Typical conditions for this regime are illustrated in **Figure 3**. Before dryout (CHF), fairly low wall superheats are sufficient to sustain convective nucleate boiling. Immediately downstream from the dryout point, the diminished heat transfer coefficient causes the wall temperature to rise to a much higher level, often approaching several hundred degrees Celsius superheat. As noted by several investigators, (Chen, et al., 1979, Groeneveld and Delorme, 1976, Evans, et al., 1985), thermodynamic non-equilibrium often occurs in this regime of two-phase flow, resulting in the actual vapor quality being less than the equilibrium quality (see **Figure 3**). The liquid phase temperature is normally close to the local saturation temperature. The need to predict the thermohydraulic conditions downstream from the point of dryout may be regarded as an entrance-region problem with the objective of calculating local pressure, equilibrium flow quality, actual flow quality, vapor superheat temperature, heat transfer coefficient at the wall, and either the wall heat flux or wall temperature.

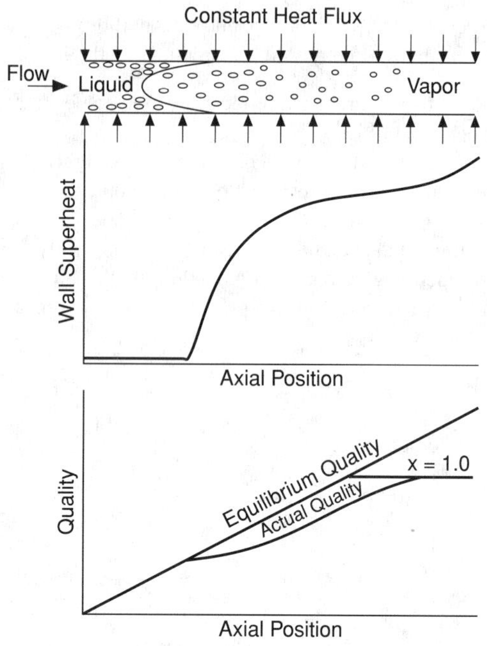

Figure 3. Thermal conditions typical of dispersed flow post-dryout heat transfer.

Early experimental investigations (e.g. Bennett et al., 1967) measured wall temperatures, mass flow, system pressure, inlet equilibrium quality, and wall heat flux, but without measurement of non-equilibrium vapor temperature (T_G) or actual quality (x_a). In concert with such initial experimental investigations, a simple equilibrium approach was taken to correlate the heat transfer results, based on two major assumptions:

- Thermodynamic equilibrium between the two phases at all axial locations,
- Heat transfer at the superheated walls primarily by turbulent vapor convection.

The first assumption resulted in taking all fluid temperatures at the local saturation temperature. Consequence of the second assumption was that the wall heat transfer could be represented by a turbulent convective **Nusselt Number**. This results in a simple model for predicting the post-dryout wall heat flux, given the wall temperature, or vice versa:

$$T_G = T_{sat}$$

$$\dot{q} = \alpha_G[T_w - T_{sat}] \quad (5)$$

where,
$\alpha_G = Nu_G\frac{\lambda_G}{D}$

Nu_G = vapour convection Nusselt number

Typical of such equilibrium correlations are those of Dougal-Rohsenow (1963) and Groeneveld (1969). These correlations calculate the wall heat transfer coefficient as a function of mass flux, equilibrium quality, and saturated vapor properties.

For example, the Dougall-Rohsenow correlation is,

$$\alpha_G = 0.023 \frac{\lambda_G}{D} \left[Re_G \left(x_e + \frac{\rho_G}{\rho_L} (1 - x_e) \right) \right]^{0.8} Pr_G^{0.4} \quad \textbf{(6)}$$

These models are simple and easy to use since they require only information on local conditions at the point of interest, i.e. local wall temperature and equilibrium quality. Compared with the experimental data available in the late 1970's, one finds that these equilibrium correlations predict reasonable trends, though often showing quantitative disagreements of 100 percent or more.

In recent years, experimental techniques improved to the point that researchers were able to measure superheated vapor temperatures as well as the wall temperature and local heat flux. Typical results as shown in **Figure 4**, verify that the state of thermodynamic non-equilibrium can exist in this heat transfer regime.

For such conditions, the various interactive mechanisms for heat transfer are illustrated in **Figure 5**. At high wall superheats, direct liquid contact on the hot surface is minimal and the wall rejects heat mainly by convection and radiation to the vapor and by radiation to the entrained droplets. If thermodynamic non-equilibrium can exist then the vapor could attain a temperature higher than the saturation temperature. In this situation, interfacial heat transfer between the superheated vapor and the droplet surface (which would be at saturation temperature) occur by both convection and radiation. The droplet in turn would be cooled by latent heat of evaporation. By most estimates, the radiative heat transfer is relatively small compared to convective heat transfer. Consequently, the degree of non-equilibrium that occurs is governed by the relative magnitudes of convective heat transfer from the hot wall to the vapor and from the vapor to the entrained droplets. The potential vapor superheat temperature (T_G) results from the dynamic balance between these two convective heat transfer pro-

cesses. Assuming that the equilibrium vapor quality is known at all axial locations from heat balance, the non-equilibrium actual vapor quality (x_a) is related to the vapor bulk superheat temperature (T_G) by the following relationship,

$$x_a = \frac{x_e h_{LG}(P)}{h_G(T_g, P) - h_G(T_{sat}, P)} \quad \textbf{(7)}$$

The experimental results of Unal et al. (1988) showed the parametric effects of operating parameters on the vapor superheat ($T_G - T_{sat}$) and on the temperature difference between wall and vapor ($T_w - T_G$). It is seen from **Figure 6** that the vapor superheat decreases with increasing mass flux, increases with increasing wall heat flux, and decreases with actual vapor mass flux. The temperature difference between wall and vapor ($T_w - T_G$) is relatively insensitive to these parameters, though showing a slight trend opposite to that of the vapor superheat.

A number of non-equilibrium models have been proposed in attempts to make better predictions of heat transfer in this dispersed flow regime. One class of models estimates the vapor superheat and wall heat transfer coefficient at any given axial location as functions of only local thermal hydraulic parameters at that location. The CSO model (Chen, Ozkaynak and Sundaram, 1979) utilizes momentum transfer analogy to calculate a convective heat transfer coefficient between the wall and vapor, together with empirical correlations for the non-equilibrium vapor temperature:

$$\dot{q}_{wG} = \alpha_{wG}(T_w - T_G) \quad \textbf{(8)}$$

$$\alpha_{wG} = \left[\frac{\dot{m}_G C_{pG}}{Pr_G^{2/3}} \right] \frac{f}{2} \quad \textbf{(9)}$$

where,

$$\frac{1}{\sqrt{f}} = 3.48 - 4 \log_{10} \left(\frac{\epsilon}{D} + \frac{9.35}{Re_G \sqrt{f}} \right) \quad \textbf{(10)}$$

$\dot{m}$	(kg/m²s)	X in	(kW/m²)	Ref.
□	20.7	0.65	26.8	Evans
+	19.7	0.65	23.0	Evans
○	19.4	0.59	23.0	Gottula
×	19.5	0.61	22.8	Gottula

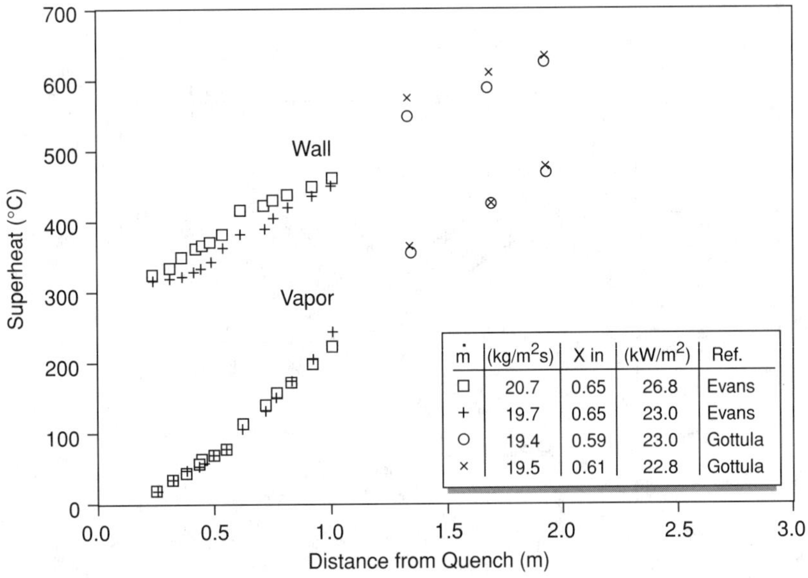

Figure 4. Experimental data showing thermodynamic non-equilibrium with superheated vapor, Chen and Costigan (1992).

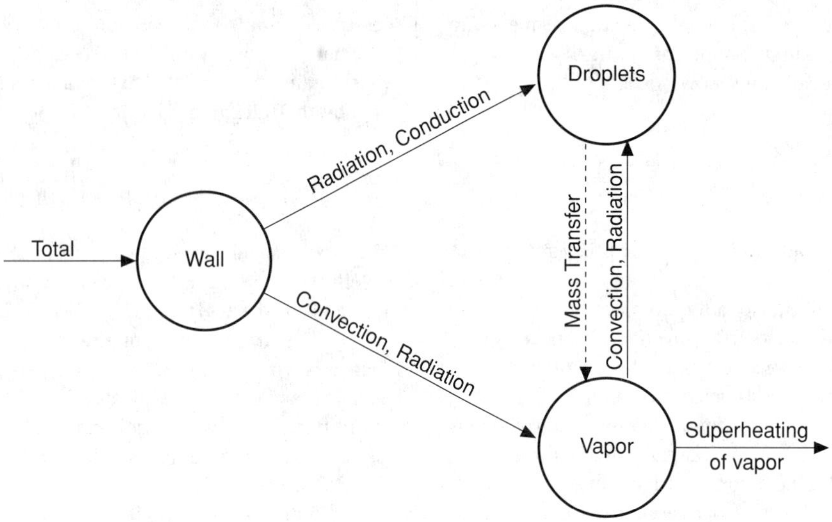

Figure 5. Exchange mechanisms in dispersed post-dryout heat transfer.

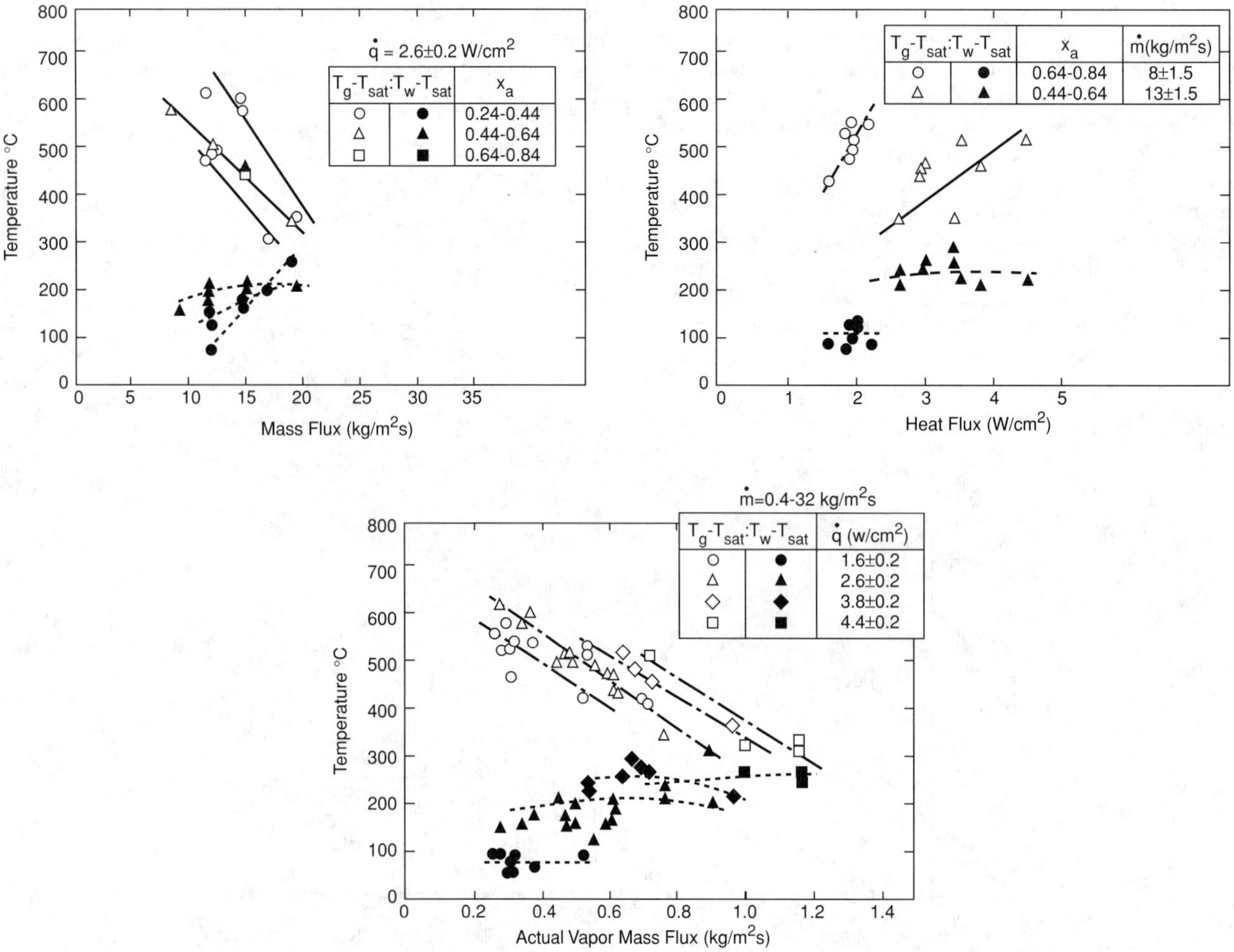

Figure 6. Parametric effects on superheat temperatures (Unal et al. 1988) (From *Journal of Heat Transfer*, 110, Aug. 1988, p. 725, with permission).

$$Re_G = \frac{D\rho_G\dot{m}}{\eta_G}\left(\frac{x_a}{\rho_G} + \frac{1 - x_a}{\rho_L}\right) \quad (11)$$

and x_a from **Equation 7** with,

$$\frac{T_G - T_{sat}}{T_w - T_G} = \frac{1 - \dfrac{x_a}{x_e}}{0.26}\left[1.15 - \left(\frac{P}{P_c}\right)^{0.65}\right] \quad (12)$$

Except for short distances immediately downstream of the dryout point, this wall-vapor convective heat flux represents the major contribution to total wall heat flux. Comparison of this model with experimental measurements show scatter in predicted vapor temperatures but reasonable predictions of wall heat flux.

A more realistic class of models recognizes that the degree of non-equilibrium at any axial location is historically developed by the upstream competitive heat transfer mechanisms (wall-to-vapor versus vapor-to-droplet). One way to treat this situation is to write the conservation equation for the vapor phase, with a volumetric vapor source term (Γ),

Γ = vapor source function

= rate of vaporization per unit mixture volume

$$\dot{m}dx_a = \Gamma dz \quad (13)$$

$$\dot{q}P_H dz = \dot{m}Ah_{LG}\,dx_e \quad (14)$$

where, P_H = heated perimeter and A = flow area

If the functional relationship for the source function Γ and the wall heat transfer coefficient α is known **Equations 13 and 14** can be integrated starting at the dryout CHF location to calculate the non-equilibrium actual flow x_a at any downstream location z. Various models have been proposed for estimating the vapor source function Γ and for the wall heat transfer coefficient α. An example is the correlation of Saha, et al. (1980):

$$\dot{m}dx_a = \Gamma dz \quad (15)$$

$$\Gamma = 6300\left(1 - \frac{P}{P_c}\right)^2\left[\left(\frac{\dot{m}x_a}{\epsilon_G}\right)^2\left(\frac{D}{\rho_G\sigma}\right)\right]^{0.5} \times \frac{\lambda_G(1 - \epsilon_G)(T_G - T_{sat})}{D^2h_{LG}} \quad (16)$$

$$\epsilon_G = 1 - \frac{1 - x_a}{1 + \dfrac{x_a(\rho_L - \rho_G)}{\rho_G} + \dfrac{\rho_L\bar{v}_{Lj}}{\dot{m}}} \quad (17)$$

$$\bar{v}_{Lj} = -1.4\left[\frac{g\sigma(\rho_L - \rho_G)}{\rho_G^2}\right]^{1/4} \quad (18)$$

σ = surface tension

$$\alpha = 0.0157\frac{\lambda_G}{D}Re_G^{0.84}Pr_G^{0.33}\left(\frac{L}{D}\right)^{0.04} \quad 6 < \frac{L}{D} < 60 \quad (19)$$

$$\alpha = 0.0133\frac{\lambda_G}{D}Re_G^{0.84}Pr_G^{0.33} \quad \frac{L}{D} > 60 \quad (20)$$

Alternate expressions for this source function Γ have been proposed by Webb and Chen (1982) and Unal, et al. (1991).

The above models account only for the direct convective heat transfer between the wall and vapor. While this is usually the dominant mechanism for post-dryout heat transfer in the dispersed flow regime, there are experimental indications that direct heat transfer from the wall to impinging liquid drops may be significant at short distances downstream of the dryout location, (Unal, et al., 1988, Cokmez-Tuzla, et al., 1993). Since models for this wall-liquid transfer mechanism are in early stages of development, current capability can only account for wall-vapor convection as described above. Radiative heat transfer needs to be considered also, but is found to be less than 10 percent of total heat transfer in most applications.

References

Berenson, P. J. (1961) Film boiling heat transfer from a horizontal surface, *J. of Heat Transfer*, 83, 351–358.

Bromley, L. A., LeRoy, N. R., and Robbers, J. A. (1953) Heat transfer in forced convective film boiling, *Ind. Eng. Chem*, 45, 2639–2646.

Chen, J. C., Ozkaynak, F. T., and Sundaram, R. K. (1979) Vapor heat transfer in post-CHF region including the effect of thermodynamic nonequilibrium, *Nuclear Eng. and Design*, 51, 143.

Chen, J. C., Sundaram, R. K., and Ozkaynak, F. T. (1977) A Phenomenological Correlation for Post-CHF Heat Transfer, NUREG-0237, U.S. Nuclear Regulatory Commission, National Technical Information Service, Springfield, VA, USA.

Chen, J. C. and Costigan, G. (1992) Review of Post-dryout heat transfer in dispersed two-phase flow, *Post Dryout Heat Transfer*, Multiphase Science and Technology, G. F. Hewitt, J. M. Delhaye, N. Zuber, eds. CRC Press, London, 1–37.

Dougall, R. S. and Rohsenow, W. M. (1963) Film Boiling on the Inside of Vertical Tubes with Upward Flow of the Fluid at Low Qualities, MIT Report No. 9079–26.

Evans, D., Webb, S. W., and Chen, J. C. (1985) Axially varying vapor superheats in convective film boiling, *J. of Heat Transfer*, 107, 3, 663–669

Fung, K. K., Gardiner, S. R. M. and Groeneveld, D. C. (1979) Subcooled and low quality flow film boiling of water at atmospheric pressure, *Nucl. Eng. and Design*, 55, 51–57.

Groeneveld, D. C. and Delorme, D. J. (1976) Prediction of thermal nonequilibrium in the post-dryout regime, *Nuclear Eng. and Design*, 36, 17.

Groeneveld, D. C. and Rousseau, J. C. (1983) CHF and Post-CHF heat transfer: An assessment of prediction methods and recommendations for reactor safety codes, Proceedings NATO Meeting on *Advances in Two-Phase Flow and Heat Transfer*, 1, 203–239, NATO ASI Series, Series E. No. 63, Nijhoff Publishers.

Groeneveld, D. C. (1992) A review of inverted annular and low quality film boiling, *Post-Dryout Heat Transfer*, G. F. Hewitt, J. M. Delhaye and N. Zuber eds., CRC Press, London, 327–366.

Groeneveld, D. C. (1969) An Investigation of Heat Transfer in the Liquid Deficient Regime, Report AECL-3281.

Groeneveld, D. C. and Gardiner, S. R. M. (1978) A method of obtaining flow film boiling data for subcooled water, *Int. J. Heat and Mass Transfer*, 21, 664–665.

Quinn, E. P. (1965) Forced Flow Transition Boiling Heat Transfer from Smooth and Finned Surfaces, GEAP-4786.

Ragheb, H. S., Cheng, S. C., and Groeneveld, D. C. (1981) Observations in transition boiling of subcooled water under forced convective conditions, *Int. J. of Heat and Mass Transfer*, 24, 7, 1127–1137.

Unal, C., Tuzla, K., Badr, O., Neti, S., and Chen, J. C. (1988) Parametric trends for post-CHF heat transfer in rod bundles, *J. of Heat Transfer*, 110, 721–727.

Unal, C., Tuzla, K., Cokmez-Tuzla, A., and Chen, J. C. (1991) Vapor generation rate model for dispersed drop flow, *Nuclear Eng. and Design*, 125, 161–173.

Webb, S. W., and Chen, J. C. (1982) Vapor generation rate in non-equilibrium convective film boiling, *Proc. 7th International Heat Transfer Conf.*, 4, 437–442, Munich.

Leading to: *Rewetting of hot surfaces*

J.C. Chen

POST DRYOUT HEAT TRANSFER (see Boiling; Forced convective boiling)

POTABLE WATER (see Water)

POTASSIUM

Potassium—(English, *potash*—pot ashes; L. *kalium*, Arab. *qali*, alkali), K; atomic weight. 39.0983 ± 0.0003; atomic number 19; melting point 63.25°C; boiling point 760°C; specific gravity 0.862 (20°C); valence 1.

Discovered in 1807 by Davy, who obtained it from caustic potash (KOH); this was the first metal isolated by electrolysis. The metal is the seventh most abundant and makes up about 2.4% by weight of the earth's crust. Most potassium minerals are insoluble and the metal is obtained from them only with great difficulty. Certain minerals, such as *sylvite, carnallite, langbeinite*, and *polyhalite* are found in ancient lake and sea beds and form rather extensive deposits from which potassium and its salts can readily be obtained. Potash is mined in Germany, New Mexico, California, Utah, and elsewhere. Large deposits of potash, found at a depth of some 3000 ft in Saskatchewan, promise to be important in coming years. Potassium is also found in the ocean, but is present only in relatively small amounts, compared to sodium.

The greatest demand for potash has been in its use for fertilizers. Potassium is an essential constituent for plant growth and it is found in most soils. Potassium is never found free in nature, but is obtained by electrolysis of the hydroxide, much in the same manner as prepared by Davy. Thermal methods also are commonly used to produce potassium (such as by reduction of potassium compounds with CaC_2, C, Si, or Na). It is one of the most reactive and electropositive of metals; except for lithium, it is the lightest known metal. It is soft, easily cut with a knife, and is silvery in appearance immediately after a fresh surface is exposed. It rapidly oxidizes in air and must be preserved in a mineral oil such as kerosene. As with other metals of the alkali group, it decomposes in water with the evolution of hydrogen. It catches fire spontaneously on water. Potassium and its salts impart a violet color to flames.

Nine isotopes of potassium are known. Ordinary potassium is composed of three isotopes, one of which is K^{40} (0.0118%), a radioactive isotope with a half-life of 1.28×10^9 years. The radioactivity presents no appreciable hazard. An alloy of sodium and potassium (NaK) is used as a heat-transfer medium. Many potassium salts are of utmost importance, including the hydroxide, nitrate, carbonate, chloride, chlorate, bromide, iodide, cyanide, sulfate, chromate, and dichromate.

Handbook of Chemistry and Physics, CRC Press

POTASSIUM CARBONATE (see Natural gas)

POTENTIAL FLOW (see Flow of fluids)

POWER FLOWMETERS (see Flow metering)

POWER LAW FLUIDS (see Non-Newtonian fluids)

POWER NUMBER (see Agitated vessel mass transfer; Agitation devices; Mixers)

POWER PLANTS

The adoption of electric power has a great significance for infrastructure and is a factor of primary importance for progress in science and technology and growth of labor productivity in all spheres of the economy. Output and structure of the production of primary energy resources are presented in **Table 1** by key figures of the world energy consumption in 1990 in relation to 1973 and 1985 (1 ton of coal equivalent (t.c.e.) corresponds to 7×10^6 kcal or 29 GJ).

Table 2 presents the structure of consumption of primary energy resources within the period from 1980 to 2020 (on an average) according to the forecast of the International Power Commission.

In 1980, power plants (PP) generated over 11600 TW-hr of electric energy.

The total annual consumption of electric energy in the world at the end of the 20th century is estimated to range from 13000 to 16000 TW-hr. This corresponds to an annual average growth of 2.5 to 3.0% between 1990–2000. By 2020, electric energy consumption is estimated at 25000 TW-hr. Installed capacity of power plants in 2000 is estimated at 3.3–3.7 TW Production of electric energy in the industrialized nations accounts for up to 35% of energy resources consumed. In the future, it is anticipated that coal will be the predominant fuel for power plants (up to 50%), less than 10% from renewables, and from 15 to 18% from nuclear energy.

A power plant is an assemblage of equipment, and apparatus used directly for generation of electric energy and also the buildings and structures needed for it. Power plants are classified into those

Table 1. World power supply

Indices of power supply	1973	1985	1990
Consumption of commercial resources, million t.c.e.	7438.1	9130.0	11333
Per capita consumption, kg coal equivalent	1923	1888	2142
Production of electric power, TW-hr	6126	9676	11659
Electric fuel coefficient, kW-hr/t	824	1060	1029
Structure of consumption of nonrenewable resources,%			
Contribution of solid fuel in consumption	46.7	40.8	40.2
Contribution of liquid fuel in consumption	30.3	32.9	28.8
Contribution of natural gas	20.7	22.5	21.9
Contribution of nuclear energy	0.3	1.9	6.8

Table 2. World energy supply

Power resource, Billion t.c.e (%)	1980	2000	2020
Solid fuel	2.6 (26.3)	4.1 (25.8)	6.6 (30.5)
Petroleum	3.8 (38.3)	4.9 (30.8)	4.4 (20.5)
Natural gas	1.9 (19.2)	2.7 (17.0)	3.7 (17.2)
Water power	0.6 (6.1)	1.0 (6.3)	1.4 (6.6)
Nuclear power	0.1 (1.0)	1.3 (8.1)	2.4 (11.3)
Nonstationary renewable energy surces	0.0 (0.0)	0.3 (1.9)	1.3 (6.0)
Noncommercial energy resurces	0.9 (9.1)	1.6 (10.1)	1.7 (7.9)
The total amount	9.9 (100)	15.6 (100)	21.5 (100)

using traditional and nontraditional energy resources. The former type includes thermal power stations (TPS), nuclear power plants (NPP), hydraulic power plants (HPP), and hydro pumped storage power plants (HPSPP). The latter type includes solar power plants (SPP), geothermal power plants (GTPP), wind power stations (WPS), tidal power plants (TPP), magnetohydrodynamic power plants (MHDPP), etc.

TPSs are the foundation of electric power industry, they generate electric energy by conversion of thermal energy released in burning of fossil fuel. Depending on the type of equipment they can be steam-turbine, gas-turbine, steam-gas, and diesel power plants. The basic items of equipment are boiler units, turbines, electric units, pumps, compressors, heat exchangers, electric switch-gears, etc. Steam-turbine TPSs are divided into condensation power plants (CPP) and combined heat and power generation plants (CHPP).

In an NPP the energy source is a nuclear reactor, in which thermal energy is generated as a result of a chain fission reaction of heavy elements. The removed heat is transferred from the reactor by a heat-carrying agent which is admitted to a steam generator or a turbine. Depending on the type of neutrons used, thermal and fast reactors are distinguished. The former are most commonly encountered. Unlike TPSs, NPSs are furnished with biological shield, equipment for nuclear fuel recharging, systems of special ventilation and emergency cooling, and other systems.

In the next century, electric energy will probably be generated in thermonuclear plants. The energy source at these plants is light nuclei fusion. Controlled thermonuclear fusion will make it possible for humankind to solve completely the problems of power supply.

In an HPP, hydraulic power is converted to electric energy. The HPP is a collection of water development works, power-generating plant, and machinery. The main components of an HPP in flat country are a dam across the river producing a concentrated drop of water level, the plant building in which hydroturbines are arranged, electric current generators, and other equipment. Navigation locks, water intake works for irrigation, water supply, and fish ladders are constructed if needed.

With pumped storage systems (HPSPPs) the system acts, during lower loading of the electric power systems, as a pumping plant consuming electric energy and pumping the water from a lower to an upper basin. With increasing electric power consumption in the system the water from the upper pool is passed via turbines to the lower pool. At this time the HPSPP works as a HPP, i.e., generates electric power. There are HPSPPs with 24-hour, weekly, and even seasonal power storage. Pumps and turbines or reversible hydraulic machines (pump-turbines) which can operate in turn as a pump or as a turbine are mounted in HPSPPs. An electric machine can also operate in a reversible regime, i.e., operate as either a motor or a generator.

In 1987, the annual output of electric power per capita on average was 2085 kW-hr all over the world, the maximum in Norway (24756 kW-hr), the minimum in Kampuchea (13 kW-hr). A total of 1790 GW was installed at TPSs in 1990. The most powerful blocks with a single-unit power 1365 MW and 1200 MW are installed in the USA (PP in Rockport) and in Russia (Kostroma TPS) respectively. In Western Europe the largest gas turbine, 135 MW, as a part of 600 MW steam-gas plant, is installed at the TPS in Amsterdam. In Japan TPS's work with 1000 MW steam-gas plant. In the USA the overall power of gas turbine and steam-gas plants exceeds 60 million kW (8% of installed power). Japan is planning to construct in mid-1990s a TPS with 2.6 million kW steam-gas plant (8 blocks). Siemens designed and constructed in Turkey a TPS with six steam-gas plants of 450 MW each.

Contribution of NPPs is about 12% of power, i.e., 334 GW as of the end of 1990. In 1990, 8.6 GW was put into operation and 3 GW was brought to a halt at NPPs all over the world. The most powerful NPPs are in Japan (9.0 GW), in Canada (9.4 GW), in France (5.5 GW). The installed power of NPPs in the USA is 106 GW, in France 55 GW, in Russia 20 GW, in Japan 32 GW, in Germany 24 GW. The most powerful nuclear block (1500 MW) is installed in Lithuania.

HPPs account for 24% of the total power, 550.5 GW in 1990. The biggest operative HPPs are in Venezuela (10.3 GW), in Brazil (12.6 GW, though not all the units in operation), in Grand Coulee (USA, 6.5 GW), and in the Sayano-Shushenskaya (6.4 GW), and Krasnoyarsk (6.0 GW) plants in Russia.

In 1990 there were 240 operative HPSPPs all over the world with the total power 70 GW. In addition, these were 16 HPSPPs, with total power of 13 GW, were under construction and 18, with total power 12 GW, were planned for construction. USA (36 HPSPPs, 15.1 GW), Japan (23 HPSPPs, 12.8 GW), Italy (32 HPSPPs, 11.8 GW), and Spain (36 HPSPPs, 8.3 GW) have the highest installed power.

In the latest forecasts for development of world power generation for the next 20–40 years, nontraditional energy sources are of minor importance. It is clear that, taking into account switching to systems using inexhaustible energy resourses such as thermonuclear, nuclear, and solar energy, there is no danger of energy shortage. However, fossil fuels will be of predominant importance in the world's energy balance up to the middle of the next century.

The total engineering potential of renewable energy sources is estimated to be 12 TW-yr a year (**Table 3**). **Table 4** presents an approximate cost of electric power from traditional and renewable energy sources.

It is commonly believed that nontraditional energy sources are advisable to use for decentralized energy supply. Until recently diesel electric power stations with the power from several kilowatts to several hundred kilowatts gained currency. However, even with the current expenditures the cost of electric power generated by them often turns out to be higher than from less powerful HPPs and WPSs. Heat supply from solar power plants even now can successfully compete with direct electric heating. If the cost of fossil fuel is doubled, solar power plants virtually appear to be more efficient than all the traditional heating systems.

Several powerful SPPs with the total power 145 MW are run in the USA. By 1995 the SPP power is planned to grow up to 590 MW and by 2000, up to 4000 MW. The highest single-unit powers are installed in the USA (15 MW), Ghana (6 MW), and Australia (2 MW). In addition, SPPs of small power are expected to increase in quantity on other countries too. In some states generation of power by SPPs increased considerably. For instance, in Israel SPPs generate 3.1% electric power.

Electric energy based on geothermal resources is produced in 16 countries, with the installed power in each country not exceeding a few tens or hundreds megawatts. There are extremely low values compared to the vast world geothermal energy resources. The total geothermal resources, including the three-kilometer continental shelf, are estimated at 4.1×10^{19} MJ of which 3.6×10^{15} MJ can be utilized by the present-day technologies of electric energy generation. This is equivalent to 1.2 TW of electric power used for 100 years. By the beginning of 1989 the total installed power of geothermal power plants amounted to 5.1 GW (233 units), including 2.02 GW in the USA, 0.89 GW in the Philippines, 0.645 GW in Mexico, and 0.519 GW in Italy. An annual growth in electric power output at these plants was 16.5% from 1978 to 1985. If this rate continued, then in 1990 the installed electric power must have attained 9.4 GW. Geothermal energy accounts for 20% of the entire electric power output in the Philippines and in Kenya, while in Mexico it is about 50%. The most powerful unit operates at a geothermal power plant in the USA (135 MW).

A wind power system (WPS) is a setup converting the kinetic energy of wind flow to electric energy. A WPS consists of a wind motor, electric current generator, automatic devices for controlling the wind motor and generator operation, and structures for their assembly and maintenance.

WPS's are used as low power electric energy sources in areas with strong winds, where the annual average wind velocity exceeds 5 m/s, and far away from the centralized electric power supply systems. Wind possesses a tremendous energy (26.6×10^{15} kW-hr), which constitutes 2% of all energy of solar radiation incident on the earth.

In recent 15 years more than 10000 WPSs with the power from 3 to 330 kW were constructed and run all over the world. The first WPS was put into operation in Great Britain (the power 3 MW), and another WPS in Denmark (2 MW).

Tidal energy attracts a considerable interest. It is commonly believed that technical reserves for electric energy generation at

Table 3. Engineering potential of renewable resources

Type of energy source	Technical potential, TW-yr per year	Potential realized in practice, TW-yr per year
Solar energy	5.0	1.0
Geothermal energy (thermal water and steam)	0.4	0.2
Wind	3.0	1.0
Tides	0.04	~ 0.013
Waves and sea currents	0.005	~ 0.002
Temperature gradient of the ocean	1.0	0 − 1.0
Organic wastes	2.5	1.0

Table 4. Comparison of costs of power plants

Type of electric station	Capital investment in 1985, dollars per kW	Cost of electric energy, cents per kW-hr with the coefficient of fuel cost rise	
		1	2
	A. Centralized electric power supply		
Coal TPS	1000–1300	5.5–7.0	6.8–8.3
NPP with light-hydrogen reactors	2000–2500	6.5–8.0	7.0–8.5
Solar power plant (SPP) of tower type	1500–3000	11.2–22.0	11.0–22.0
Gas turbine thermal power plant (GTPP)	1800–2600	5.0–7.5	5.0–7.5
Using the ocean thermal gradient	2200–2800	7.0–9.5	7.0–9.5
	B. Decentralized electric power supply		
Diesel electric station	800–1000	21.0–24.0	31.0–34.0
Small hydro plant (Mini-HPP)	1000–1400	5.0–7.0	5.0–7.0
Solar power plant (SPP) based on photoconverters	1000–15000	57.0–85.0	57.0–85.0
Wind power system (WPS)	1500–2000	11.0–18.0	11.0–18.0

tidal power plants (TPP) are about one-third of potential tidal energy. Thus, under Russian conditions technical resources of tidal energy are 250 billion kW-hr per annum. Sea tide electric power plants use tidal fluctuations of the sea level which, as a rule, occur twice in 24 hours.

At the end of the 1980s TPPs were constructed in a number of countries. In France a 240 MW TPP was constructed, in Russia, near Murmansk a pilot TPP was constructed. In addition, preliminary work was carried out on the possibility of full scale TPP construction in Russia. There is one TPP in the USA and a few in China. In Great Britain final plans and specifications are being prepared for construction of a tidal power plant with the power 7.2 GW. In Norway and Japan electric power stations using the sea wave energy have been run successfully.

The constant rise of prices for petroleum and natural gas gives impetus to search of new energy sources one of which is the energy of biomass. By composition it can be carbon-containing (plants, wood, seaweed, grain, paper, ets.) and sugar-containing (sugar beetroot, sugar cane, Chinese sugar cane). The sources of biomass are wood products, vegetable remnants, livestock breeding wastes, domestic garbage and industrial wastes, ets. Though the energy of biomass can meet only 6 to 10% of power demands of industrial states, its potential role is important because biomass is a renewable energy source.

Small gas turbines can use bioresources. At the end of 1992 in Great Britain, 24 biogas units operated, 8 were being constructed, and 18 were under design. It is expected that by 2000 the annual energy potential of this energy source will achieve 1 million t.c.e. In the state of Penjab, India, a 10 MW TPS has been constructed that will use straw as fuel. In 1987, by the data of the 14th Congress of the International Power Commission, the total of biomass used in power production was 1.8 billion t.c.e.

Further development of electric power plants is linked with using new cycles and working media. An analysis of various trends in developing electric power engineering shows that one of the most promising new power-generation technologies is the use of combined cycle steam-gas plants (SGP). In these plants, a gas turbine is used with the hot exhaust gases being used to generate steam which is passed to a steam turbine (often on the same shaft as the steam turbine). The greater part of operating SPGs in the world are the binary-type plants. Depending on the power ratio of steam to gas in the unit, on the initial temperature in combustion chamber, and the degree of air compression, the efficiency of such plants varies from 42% to 53%.

In recent years, considerable progress was achieved in solving the problem of cooling gas-turbine plant (GTP) elements. This allowed a substantial rise of the initial gas temperature in the combustion chamber, a change of power ratio of gas and turbine cycles and going over to a new version of SGP by the STIG cycle with admission of steam in the gas turbine. According to the data of the 14th Congress of the International Power Commission these plants are put into operation most actively in the USA, Japan, Western Europe, and other parts of the world. According to the data of the US Department of Energy as of the beginning of 1990 the total capacity of SGPs amounted to 5.3 GW. About 40 GW capacity is expected to be installed before 2000.

There exist real prospects of putting into operation of magnetohydrodynamic power plants and hydrogen-based power generation. The efficiency of TPS can be sharply raised by a combined cycle with an MHD topping plant. MHD power plants use, as a working medium, plasma produced by high-temperature (about 2700°C) fuel burning. An increases in the upper temperature limit of the working medium even now can give efficiency up to 50% and higher.

In addition to conventional elements installed at TPSs, MHD power plants possess compressor plants, high-temperature oxidizer heaters, combustion chambers, MHD channel, a superconduction magnetic system, inverter substantions, the cooling system of high-temperature elements, the system of additive inlet and outlet. These elements substantially raise the cost of the plant and complicate its operation. At the same time the higher efficiency of MHD power plants, the consequential reduction of environmental pollution, the possibility of producing high-power units, and their greater flexibility speak in favor of MHD power plant construction. Experience gained in these plants will form the basis for the adoption of high efficiency solid-fuel plants in power engineering.

A key problem in development of hydrogen-based power engineering is production of low-cost hydrogen. Different processes exist for this, such as using coal, water electrolysis, and plasma chemistry. It seem promising to use hydrogen in power-chemistry, power-metallurgy, and other systems. The economic benefit turns out to be the highest if both power and technological problems are solved simultaneously.

In the USA a pilot hydrogen power plant with a power of 1 MW was constructed in 1977. Then the demonstration station with a 4.5 MW hydrogen-air electrochemical generator began to be constructed. Tests of the different systems of this station were carried out in 1981. By the end of 1990s commercial stations of this type are planned to be put into practice. In Germany, an experimental hydrogen-oxygen steam generator with the thermal power 15 MW is being designed with steam parameters: temperature 850°C, pressure 8 MPa. Fuel systems for such power stations are still being improved. The power single units is rising. In Japan and USA 4 MW units have already been put into operation.

Electric power plants are a source of disturbance to environmental equilibrium. The interaction of energy generation and the biosphere in the majority of cases has a negative environmental consequence, primarily due to the generation of wastes such as noxious gases, solid and liquid pollutants, radioactive substances and waste heat (**Figure 1**). They pollute the atmosphere and water basins. Operation of electric power plants changes the regime of river run-off and withdraws valuable land from agricultural rotation. For instance, a TPS of the power 1 GW consumes 8 million tons of coal a year, ejecting about 10 million tons of CO_2 and hundreds of thousands of tons of ash. By the way, the dust radioactivity of TPS's is about twice as high as the radiation from all NPP. TPSs are not only the source of heavy pollution, but consume much oxygen from the air.

Currently more than 2.5 billion tons per annum of various substances are ejected in to the earth atmosphere. One of the most harmful components is sulfur dioxide SO_2. In 1970 about 90 million tons of SO_2 was ejected into the atmosphere, in 1980 the total ejections of SO_2 in developed countries amounted to 111.9 million tons incrising by 1990 by about 15%. Other harmful components in waste gases are nitrogen oxides. In practice, nitrogen oxide NO and dioxide NO_2 whose sum is denoted as NO_x pose a problem of atmospheric air protection. Globally the quantity of naturally formed nitrogen oxides by far exceeds its creation as a result of human activities. According to estimates made in 1990, 80 million tons of NO_x was generated annually. However, it should be taken into account that anthropogenic ejections of nitrogen oxides virtually grow twice every 20–25 years.

NPPs are environmently less dangerous compared to TPS, particularly TPP's using low-grade solid fuels with a high content of ash and sulfur and a high-sulfur content, but NPP's need strict observance of the radiation safety rules. Ensuring safety in nuclear power generation involves many technical aspects, but nuclear safety and removal of decay-heat in the active reactor zone are decisive due to their influence on general safety. Burial of radioactive waste and decommissioning NPPs present a number of grave problems.

In the coming years we cannot expect that nontraditional renewable energy sources can significantly improve the condition of the environment in a global scale because their share in the world energy production is small. The efficiency of technological measures taken to reduce the harmful effect of electric power plants,

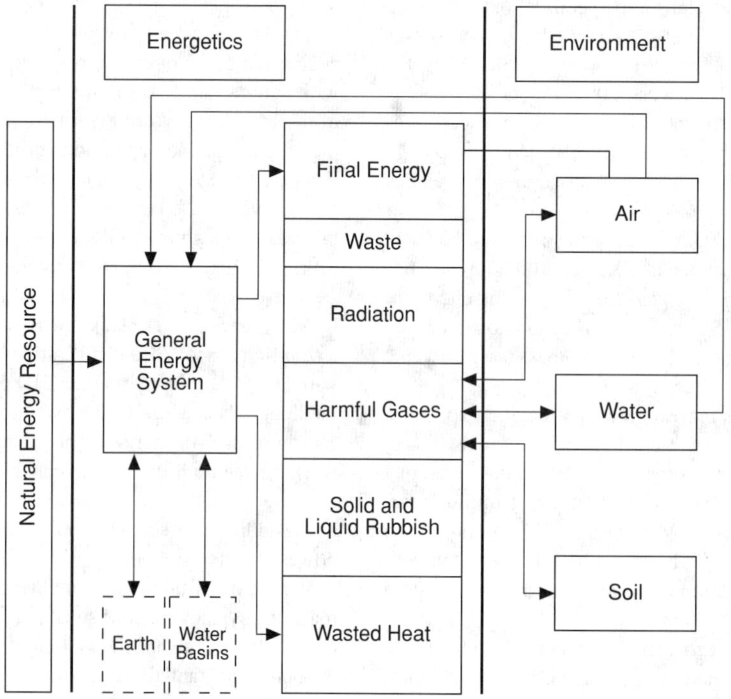

Figure 1.

using traditional energy resources, on the environment is substantially higher than going over to using nontraditional renewable energy sources.

The reduction of negative effect on the environment is possible owing to (1) improvement of the structure of traditional energy resources, (2) lowering of total utilization energy resources by improving efficiency, (3) improvement of engineering and technological design of electric power plants.

A more intensive use of renewable enegry resources for energy generation will also contribute to improvement of environment.

Leading to: *Fusion, nuclear tosion reactors; Gas turbine; Hydropower; Magnetohydrodynamic (MHD) power generation; Nuclear reaction; Renewable energy; Wind turbines*

Yu.N. Rudenko and
V.S. Polonsky

POWER SERIES

Following from: *Series expansion*

An infinite series of the form

$$\sum_{k=0}^{\infty} c_k(z - \zeta_0)^k \quad (1)$$

is called a "power series" expansion around the centre ζ_0 with constant "coefficients" c_k. The variable z and the constants ζ_0 and c_k may be real or complex numbers. For a given $z = z_0$, (1) becomes an infinite series of constant terms $c_n(z_0 - \zeta_0)^n$ which may or may not be convergent; in the first case, we denote the sum by $S(z_0)$. We say that R is the "radius of convergence" of (1) if this series converges for all z with $|z - \zeta_0| < R$ and diverges for all z with $|z - \zeta_0| > R$. Certainly, every series (1) converges at $z = \zeta_0$. Thus, if the series diverges for every $z \neq \zeta_0$ we put $R = 0$. On the other hand, if (1) is convergent for all values of z, we take $R = \infty$. If $R > 0$, we write

$$S(z) = \sum_{k=0}^{\infty} c_k(z - \zeta_0)^k, \qquad |z - \zeta_0| < R$$

The radius of convergence may be determined by taking either lim $|c_{n+1}/c_n|$ or lim $n\sqrt{|c_n|}$ as $n \to \infty$: if the limit is finite and equals $L > 0$ then $R = 1/L$, otherwise $R = \infty$. According to *Taylor's theorem*, every function f(z) which is differentiable in a domain D has a power series expansion of the form

$$f(z) = \sum_{k=0}^{\infty} \frac{f^{(k)}(\zeta_0)}{k!}(z - \zeta_0)^k$$

which is unique for every ζ_0 in D. This is called the *Taylor series* expansion of f(z) around the point ζ_0 and its radius of convergence is the largest number R such that all z with $|z - \zeta_0| < R$ lie within D. A Taylor series with centre $\zeta_0 = 0$ is called a *Maclaurin series.*

References

Hille, E. (1973) *Analytic Function Theory*, Vol.I, Chelsea Publishing Co., New York.

Kreyszig, E. (1983) *Advanced Engineering Mathematics*, John Wiley, New York.

Leading to: *Fourier series; Taylor series*

H.G. Khajah

POWER SPECTRUM (see Spectral analysis)

PRA, PROBABILITY RISK ASSESSMENT (see Risk analysis techniques)

PRANDTL, LUDWIG (1875–1953)

A German physicist famous for his work in aeronautics, Prandtl was born at Freising, Bavaria, on February 4, 1875. He qualified at Munchen in 1900 with a thesis on elastic stability, and was professor of applied mechanics at Gottingen from 1904 until his death there on August 15, 1953. In 1925 he became director of the Kaiser Wilhelm Institute for Fluid Mechanics. His discovery (1904) of the "**Boundary Layer**" which adjoins the surface of a body moving in a fluid led to an understanding of *skin friction* drag and of the way in which streamlining reduces the drag of airplane wings and other moving bodies. His work on wing theory,

LUDWIG PRANDTL 1875–1953

published in 1918–1919, which followed that of F. W. Lanchester (1902–1907), but was carried out independently, elucidated the flow over airplane wings of finite span.

Prandtl made decisive advances in boundary layer and wing theories, and his work became the basic material of aeronautics. He also made important contributions to the theories of *supersonic flow* and of *turbulence*, besides contributing much to the development of *Wind Tunnels* and other aerodynamic equipment. In addition, he devised the soap-film analogy for the torsion of noncircular sections and wrote on the theory of plasticity and of meteorology.

U. Grigull, H. Sandner and J. Straub

PRANDTL-MEYER RELATIONSHIP (see Compressible flow)

PRANDTL NUMBER

Prandtl number, Pr, is a dimensionless parameter representing the ratio of diffusion of momentum to diffusion of heat in a fluid.

$$\Pr = \frac{\nu}{\kappa} = \frac{\eta/\rho}{\lambda/\rho c_p} = \frac{\eta c_p}{\lambda}$$

where ν is kinematic viscosity and κ thermal diffusivity.

Prandtl number is a characteristic of the fluid only. For air at room temperature Pr is 0.71 and most common gases have similar values. The Prandtl number of water at 17°C is 7.56. Liquids in general have high Prandtl numbers, with values as high as 10^5 for some oils.

Reference

Hewitt, G. F., Shires, G. L., and Bott, T. R. (1994) *Process Heat Transfer*, CRC Press, Boca Raton, FL.

G.L. Shires

PRANDTL'S BOUNDARY LAYER (see Boundary layer)

PRANDTL'S FORMULA, FOR FRICTION FACTOR (see Tubes, single phase flow in)

PRANDTL'S MIXING LENGTH MODEL (see Turbulence models)

PRANDTL TUBE (see Differential pressure flowmeters)

PRECIPITATION (see Crystallizers)

PREMIXED FLAMES (see Flames)

PRESSURE

Following from: *Thermodynamics*

The pressure p of a fluid on a surface is defined as the normal force exerted by the fluid per unit area of the surface. If force is measured in Newtons and area in square meters, the unit of pressure is 1 N/m^2 or 1 *Pascal* (1 Pa). Other important units are 1 *bar* = 10^5 Pa, 1 atm = $1.0133 \cdot 10^5$ Pa, 1 *Torr* = $1.3332 \cdot 10^2$ Pa and 1 *psi* (pound per square inch) = $6.8948 \cdot 10^3$ Pa.

The total pressure exerted on a boundary wall is called the *absolute pressure*, whereas the pressure exerted by the atmosphere is called *atmospheric pressure*. *Gauge pressure* designates the difference between absolute and atmospheric pressure in a particular system and is normally measured by an instrument which has atmospheric pressure as a reference. Since a vertical column of a fluid with density ρ under the influence of gravity exerts a pressure at its base in direct proportion to its height h,

$$p = \frac{F}{A} = \frac{Mg}{A} = \frac{Ah\rho g}{A} = \rho g h \qquad (1)$$

pressure is also expressed as the equivalent height of a fluid column (e.g. mm water, mm mercury). The above definition of pressure is only valid for an area element sufficiently large so that the fluid may be treated as a continuum, i.e. as long as the average distance a fluid molecule travels between collisions is small compared with a side of the area element.

The *partial pressure* p_i of a component in a mixture of gases is the pressure this component would exert if it solely occupied the entire volume of the mixture at the given temperature. For ideal gases, it is defined by the Ideal Gas Law (See **Gas Law**)

$$p_i = \frac{N_i \tilde{R} T}{V} \qquad (2)$$

where N_i is the number of moles of component i, $\tilde{R}$ is the universal gas constant, T is the absolute temperature and V is the volume.

The *critical pressure* is the pressure at the critical state of a substance. At pressures higher than the critical pressure, no distinction can be made between liquid and vapour phases.

The pressure at which a liquid vaporises or a vapour condenses at a given temperature is called the *saturation pressure*.

H. Müller-Steinhagen

PRESSURE AVERAGES (see Differential pressure flowmeters)

PRESSURE DIE-CASTING (see Casting of metals)

PRESSURE DROP IN BENDS (see Bends, flow and pressure drop in)

PRESSURE DROP IN COILED TIBE (see Coiled tube, flow and pressure drop in)

PRESSURE DROP IN FLUIDISED BEDS (see Fluidised bed)

PRESSURE DROP MULTIPLIERS (see Pressure drop, two phase flow)

PRESSURE DROP OSCILLATIONS (see Instability, two-phase)

PRESSURE DROP, SINGLE-PHASE

Following from: *Channel flow; Conservation equations, single phase*

When a single-phase fluid flows in a hydraulic system which generally consists of elements such as constant cross section conduits; throttles (valves, diaphragms, filters, etc.); variable cross section channels, e.g. nozzles and reducers; bends, tanks, basins, heat exchangers, and so on, the fluid pressure changes both as a result of conversion of the kinetic energy to potential energy and vica versa and an irreversible conversion of part of the mechanical energy of the fluid to heat due to viscous friction forces in the flow. In actual fact, static pressure is most often constant over the flow cross section and changes only longitudinally, while the fluid flow can be considered as one-dimensional and characterized in each reference cross section by the mean parameters, i.e., the mean mass velocity $\dot{m} = \dot{M}/S$, where $\dot{M} = \int_S \rho u dS$ is the mass flow rate of the fluid, S the flow cross-section area, and by the bulk velocity $u_B = \dot{m}/\rho_B$, where $\rho(p, h_B)$ is the bulk density of the fluid determined depending on the local static pressure p and the bulk enthalpy $h_B = \frac{1}{S}\int \frac{\rho u}{\dot{m}} h\, dS$. Along with the fluid static pressure, i.e., the pressure which an instrument moving together with the flow would indicate, hydromechanics often deals with the total pressure p_o ($p_0 = p + \frac{\rho_B u_B^2}{2} + \rho_B gz$) that is a sum of static pressure p, dynamic pressure $\rho_B u_B^2/2$, and hydrostatic pressure $\rho_B gz$, where g is the gravitational acceleration, z the height of the flow axis above the ground reckoned from some fixed zero mark. The pressure difference in the initial (1) and end (2) reference flow cross sections is referred to as the pressure drop or hydraulic resistance and denoted as $\Delta p = p_1 - p_2$ or $\Delta p_0 = p_{01} - p_{02}$. Frequently $\Delta p \neq \Delta p_0$, therefore, one should mention without fail which pressures we are having to do with. In pipes and channels the length derivative dp/dx is called the pressure gradient and dp_0/dx the hydraulic slope. Determination of Δp and Δp_0 under given flow conditions is the most important problem of hydrodynamic analysis. Its results are used to find the parameters of a pump, a fan, or a compressor responsible for fluid motion or the draught needed in circuits with natural circulation.

Pressure drops are determined using the equations of continuity, energy, and fluid motion. We consider some of the solutions that are valid for one-dimensional or quasi-one-dimensional flows with a constant flow rate. The continuity equation for this flow has the form: $\rho_B u_B S$ = const or in a differential form

$$d\rho_B/\rho_B + du_B/u_B + dS/S = 0. \quad \textbf{(1)}$$

Using the energy equation for the stream

$$dh = d(\dot{Q} + \dot{Q}_f)/\dot{M} = du + pdv \ldots, \quad \textbf{(2)}$$

$$pdv = d(pv) + udu + gdz + dW_T/\dot{M} + dW_f/\dot{M} \ldots, \quad \textbf{(3)}$$

where $\dot{Q}$ is the rate of heat input per unit time into a flow of a fluid flowing at $\dot{M}$ kg/s, u the specific internal energy, $v = 1/\rho$ the specific volume, W_T the useful work, $W_f = \dot{Q}_f$ work againt friction forces in fluid fully converted to friction heat $\dot{Q}_f$, yields, for a flow doing no useful work, $dW_T = 0$, the relationships

(a) for the flow of incompressible fluid for which v is assumed constant and dv to be zero. The flows of dropping liquids, such as water and petroleum products, and gases at $\Delta v/v < 5\%$ (if T $\cong$ const at $\Delta p/p < 5\%$) are close to this case. The heat put into or removed at $d\dot{Q} \neq 0$ is mainly expended for changing the internal energy of fluid and does not affect the mechanical energy balance (3). Integrating (3) at p dV = 0 and averaging the integral quantities over the flow yields

$$\Delta p = p_1 - p_2 = \left(\alpha_2 \frac{\rho u_{B2}^2}{2} - \alpha_1 \frac{\rho u_{B1}^2}{2}\right) + g\rho(z_2 - z_1) + \Delta p_f \ldots.$$

Equation 4 is said to be the Bernoulli equation. The value $\Delta P_f = \rho W_f/\dot{M} > 0$ due to the work of friction in the flow is known as pressure loss or friction loss (hydraulic resistance). The first term on the right-hand side of Equation 4 is the dynamic pressure difference, the second term the hydrostatic pressure difference. The coefficient $\alpha = \frac{1}{S}\int_S (u/u_b)^3 dS$ where u is the local longitudinal component of the fluid velocity, takes into account the actual non-uniformity of the velocity profile and, consequently, that of kinetic energy distribution over the channel cross section and is referred to as the Coriolis coefficient. For fully developed turbulent flows in pipes $\alpha = 1.06 - 1.07$, for fully developed laminar flows $\alpha = 2.0$ in a circular pipe and $\alpha = 1.54$ in plane and annular pipes for $D_2/D_1 \to 1$. Equation 4 readily yields $\Delta p_0 = \Delta p_f$ in uniform ($u_2 = u_1$, $\alpha_1 = \alpha_2$) or homogeneous ($\alpha_1 = \alpha_2 = 1$) flows of incompressible fluid, i.e., the total pressure is reduced by the value of hydraulic losses.

(b) for flows of compressible fluids: v = v(P,T) (gases at $\Delta v/v > 5\%$)

$$\Delta p = p_1 - p_2 = \left(\alpha_2 \frac{\bar{\rho} u_{B2}^2}{2} - \alpha_1 \frac{\bar{\rho} u_{B1}^2}{2}\right) + g\bar{\rho}(z_2 - z_1) + \Delta p_f + \ldots, \quad \textbf{(5)}$$

where $\bar{\rho} = (p_2 - p_1)/\int_{p_2}^{p_1} \frac{dp}{\rho}$ is the effective density of fluid depending on the character of thermodynamic process, $\Delta p_f = \bar{\rho} w_f/\dot{M}$. An exact solution (5) can be obtained only for some particular cases. Thus, for an isothermal gas flow obeying the Clapeyron equation of state

$$\bar{\rho} = \rho_1(\varepsilon - 1)/\ln \varepsilon + \ldots, \quad \textbf{(6')}$$

where $\varepsilon = p_2/p_1$ is the backpressure ratio. When the gas flow is polytropic,

$$\bar{\rho} = \rho_1 \frac{n-1}{n} \frac{1-\varepsilon}{1-\varepsilon^{(n-1)/n}} \ldots, \qquad (6'')$$

where n is the polytropic exponent. Forthe adiabatic process $n \cong \gamma = c_p/c_v$, where γ is the isentropic exponent (see **Adiabatic conditions**). As follows from Equations 4 and 5, in cases where the hydrostatic pressure difference and friction loss are low in relation to the dynamic head the static pressure drops, $\Delta p > 0$, as a result of fluid acceleration, ($u_{B2} > u_{B1}$); in the case of stagnation $u_{B2} < u_{B1}$ the static pressure increases (is recovered) $\Delta p < 0$.

Conversion of the pressure energy to kinetic energy of the flow and vica versa is performed using channels (nozzles) of variable cross section (**Figure 1**) such as contractions ($du_B > 0$). These channels are as a rule short and therefore, the hydrostatic pressure difference can be neglected and the flow can be considered adiabatic. Solving equations 4 and 5 together with equation of continuity (1) for these conditions, we derive the relations

(a) for incompressible fluid

$$\Delta p = p_1 - p_2 = \Delta p_f + \alpha_1 \frac{\rho u_{B1}^2}{2}\left[\frac{\alpha_2}{\alpha_1}\left(\frac{S_1}{S_2}\right)^2 - 1\right]$$

$$= \Delta p_f + \alpha_2 \frac{\rho u_{B2}^2}{2}\left[1 - \frac{\alpha_1}{\alpha_2}\left(\frac{S_2}{S_1}\right)^2\right] = \ldots, \qquad (7)$$

in particular cases, for instance, when the fluid outflows from the reservoir (u_{B1}/u_{B2} and $S_1/S_2 \to 0$) or if a deep stagnation occurs in a diffuser such that u_{B2}/u_{B1} and $S_1/S_2 \to 0$, eq. (7) can be simplified. This equation is used to calculate the Venturi tubes (Fig. 1) employed for measuring the flow rate,

(b) in the case of compressible gas (p/ρ^γ = const) for a stream tube without friction the differential equations

$$\frac{du}{u} = \frac{1}{Ma^2 - 1}\frac{dS}{S}, \qquad \frac{dp}{p} = \frac{\gamma Ma^2}{1 - Ma^2}\frac{dS}{S}, \ldots \qquad (8)$$

are valid, where $Ma = u/u_{sound}$ is the Mach number, $u_{sound} = \sqrt{\gamma p/\rho}$ is the local velocity of sound. In the subsonic flow ($Ma < 1$) the flow velocity increases ($du > 0$) at $dS < 0$, i.e., in contracting channels, while flow stagnation occurs ($du < 0$) in expanding channels ($dS > 0$). In the supersonic flow ($Ma > 0$) the ratio du/dS changes sign. Therefore, in order to initiate supersonic flow a Laval nozzle is used that is a combination of the subsonic and supersonic contractors (**Figure 1**). Equation 8 implies the relations

$$\frac{u_2}{u_1} = \sqrt{1 + \frac{2}{\gamma - 1}\frac{1}{Ma_1^2}[1 - \varepsilon^{(\gamma-1)/\gamma}]} \ldots, \qquad (9)$$

$$\frac{S_1}{S_2} = \varepsilon^{1/\gamma}\sqrt{1 + \frac{2}{\gamma - 1}\frac{1}{Ma_1^2}[1 - \varepsilon^{(\gamma-1)/\gamma}]} \ldots, \qquad (10)$$

where $Ma_1 = u/u_{sound}$ is the Mach number in entrance section 1. Acceleration of the subsonic flow in a smoothly shaped contracting nozzle is possible up to

$$\varepsilon \geq \varepsilon_* = [2/(\gamma + 1)]^{\gamma/(\gamma-1)} \ldots. \qquad (11)$$

At $\varepsilon = \varepsilon^*$ the local sound velocity is achieved in the exit section and the critical outflow regime sets in under which the maximum possible gas flow rate is reached that is retained even at $\varepsilon < \varepsilon^*$.

Determination of pressure losses Δp_f in many cases is a most important element in hydrodynamic analysis (see **Hydraulic Resistance**).

Hydraulic loss in liquid filtration through a bulk porous layer can be determined by the formula

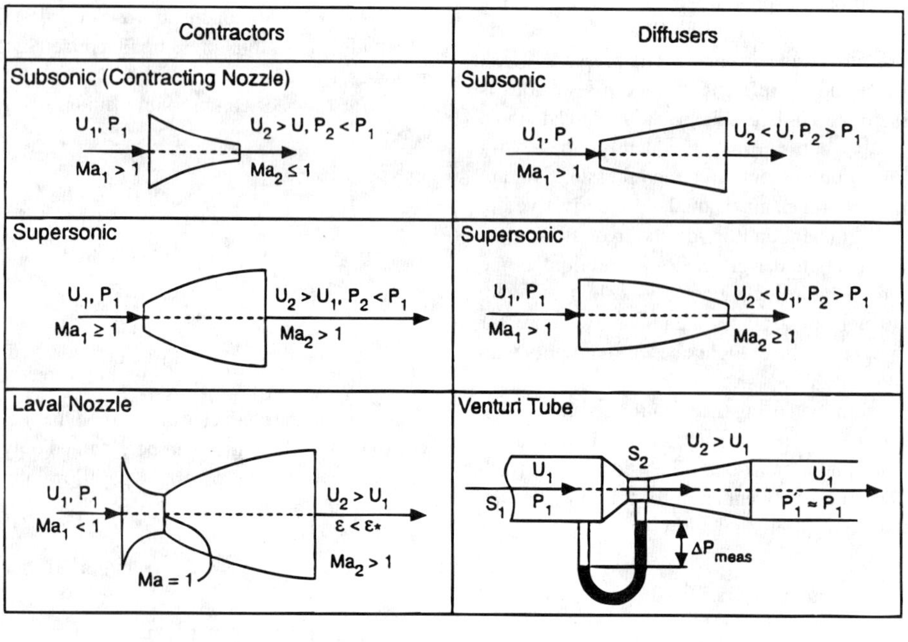

Figure 1. Flow through nozzles.

$$\Delta P = \xi_p \dot{m}^2/(2\rho\Pi)\,\frac{H}{d_p}\ldots, \qquad (12)$$

where $\dot{m}$ is the mass flow rate per unit area of the layer, Π the layer porosity, H the layer height, $d_p = \sqrt{32/\alpha_1\Pi}$ is the arbitrary diameter of the medium pores (α_1 is the viscous coefficient of resistance in the linear filtration region). (see also **Porous Medium**) Using **Figure 2**, the ξ_p values can be found as a function of the Reynolds number $Re_p = \dot{m}d_p/(\mu\Pi)$, $\alpha_1 = 10.5ad^{-2}\Pi^{-4.4}$, d is the mean diameter of particles. The approximate a and C values in **Figure 2** for bulk layers are presented in **Table 1**.

The pressure drop can be also determined using the equation of fluid motion (the momentum equations). This equation in the differential form is known as the Navier-Stokes equation and gained currency in theoretical solution to fluid dynamics problems. This approach is advantageous in analyzing pressure losses of fluids with variable density which flow in the channels with heat input and removal when there occurs an interconversion of thermal and mechanical energy which hinders analyzing Equations 2 and 3.

An integral equation of motion for fluid flow in the channel with impermeable walls depicted in **Figure 3** is of the form

$$P_1 + P_2 + F_g + R + \dot{M}(\beta_1 u_{b1} - \beta_2 u_{b2}) = 0, \ldots, \qquad (13)$$

where P_1 and P_2 are the total pressure forces at the inlet and outlet of the channel as shown in Figure 3. The vector polygon of forces corresponding to Equation 13 is also shown in **Figure 3**. If the pressure can be assumed constant over the flow section, then P = pS. The coefficient β in 13 is referred to as the momentum coefficient or the Boussinesq coefficient and allows for a nonuniform distribution of momentum over the flow section with ingomogeneous distribution of fluid velocity and density. For fully developed turbulent flows in pipes $\beta = 1.02 - 1.04$ and for stabilized laminar flow in circular pipes $\beta = 1.33$, in plane and annular pipes at $D_2/D_1 \to 1$, $\beta = 1.2$. The force R, i.e., the reaction of the channel walls, is a resultant of forces acting on the lateral surface of the flow, including the friction force on the wall. The same force but opposite in direction acts from the side of fluid on the channel section under consideration.

In the case of a straight uniform section tube often encountered in practice Equation 13 takes the form

$$\Delta p = p_1 - p_2 = \Delta p_\tau + \dot{m}(\beta_2 u_2 - \beta_1 u_1) + \Delta p_g. \qquad (14)$$

Using equation of continuity (1) for S = *const* yields

$$\Delta p = p_2 - p_1 = \Delta p_\tau + \Delta p_u + \Delta p_g \ldots, \qquad (15)$$

where $\Delta p_u = \dot{m}^2(\beta_2/\rho b_2 - \beta_1/\rho b_1)$ is the result of acceleration of

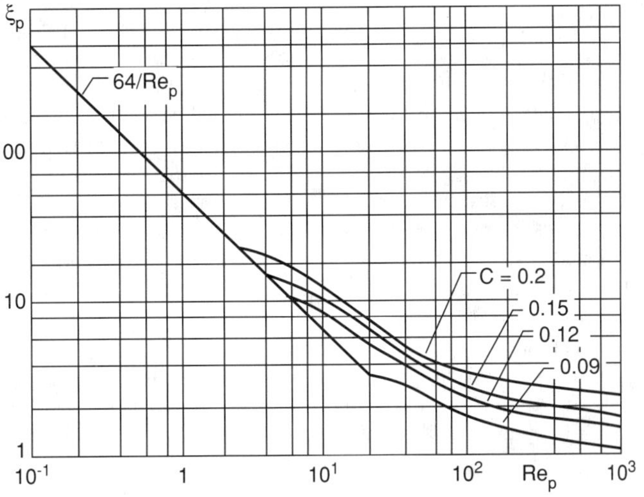

Figure 2. Resistance factor for flow in a porous medium.

the flow, $\Delta p_r = \int_1^2 \tau_w \frac{\Pi}{S} dx$ the friction drag (τ_w is the local tangential stress on the wall averaged over perimeter Π), $\Delta p_g = -g\cos\varphi \int_1^2 \bar{\rho}dx = -g\cos\varphi \int_1^2 \beta_\rho \rho_{bdx}$ is the hydrostatic pressure difference. Here φ is the angle formed by the vectors of velocity and gravitational acceleration, $\bar{\rho}$ the fluid density averaged over cross section, i.e., commonly $\bar{\rho} = \rho_b$ and $\beta_\rho = 1$. In the case of constant β and ρ

$$p_1 - p_2 = \Delta p_g + \Delta p_\tau = g\rho(z_2 - z_1) + \Delta p_\tau \ldots, \qquad (16)$$

i.e., the equation of motion coincides with Bernoulli Equation 5 and, hence, $\Delta p_f = \Delta p_\tau$. Δp, as well as Δp_f, in tubes and in channels are determined by the Darcy-Weisbach formula

Table 1.

Particle shape	a	C
Polished balls	1.0	0.09
Balls with a natural nonpolished liquid-impermeable surface	1.3 – 1.7	0.09 – 0.11
Sphere-shaped particles with a natural surface, rolled sand, plant seeds, etc.	2.3 – 3.3	0.12 – 0.14
Particles of irregular shape (lumps), hard coal, crushed stone	4.5 – 11.0	0.17 – 0.20
Porous particles, alumina, silica gel, activated carbon	3.7 – 7.6	0.17 – 0.20

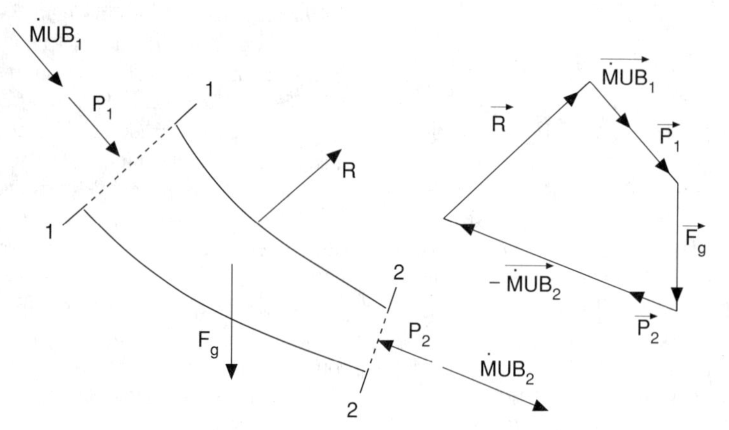

Figure 3.

$$\Delta p_\tau = \bar{f}\,\frac{\rho_b u_b^2}{2}\,\frac{1}{D_h}\ldots, \tag{17}$$

or

$$\frac{dp_\tau}{d(x/D_h)} = f\,\frac{\rho_b u_b^2}{2}, \tag{18}$$

where $\bar{f}$ and f are the average and the local friction factors respectively. We note that Δp_f, that is an energy characteristic of friction, and the dynamic characteristic Δp_T with interconversion of kinetic, potential, and thermal energies occurring in the flow have, generally, different values in the same way as the coefficients λ and f do. The components Δp_w and Δp_g can be expressed in Equation 15 by relations similar to 17 and 18, f_w and f_g are respectively the coefficients of internal and hydrostatic resistance. In case f_i, ρ_b, and u_b substantially vary over the tube length the differences Δp_i are determined by integration of equations of the type of Equation 18. In the case of flow in tubes of a fluid with variable physical properties Δp_T and f are experimentally determined often using a simplified version of eq. (14) in which it is assumed that $\beta_2 = \beta_2 = const = 1$, $\beta\rho = 1$,

$$\Delta p = p_1 - p_2 = \Delta p_\tau^0 + \dot{m}(u_{b2} - u_{b1}) + \Delta p_g^0 \ldots \tag{14}$$

This equation complies with a one-dimensional, or homogeneous, flow model, and the Δp_τ^0, Δp_g^0, f^0, etc. values are said to be one-dimensional. Actual values may differ from them (this can occur, e.g. in transcritical states of fluid), therefor, in engineering calculations one should use an equation of motion which was applied to obtain f, f_w, f_g. As an example we refer to Equation 14 for a flow of heated gas in a tube derived by Guggenheim. Writing it in a differential form and multiplying its both sides by p, we have

$$-p\,dp = \frac{f}{2}\dot{m}^2\frac{p}{\rho}\,d\left(\frac{x}{D_h}\right) + \dot{m}^2 p\,d\left(\frac{1}{\rho}\right)\ldots \tag{19}$$

Since for gas $p/\rho = RT$, then, discarding the intermediate manipulations, we obtain

$$\Delta p = p_1 - p_2 = \dot{m}^2\frac{R\bar{T}}{0.5(p_1 + p_2)}\left[\frac{f}{2}\frac{1}{D_h} + \frac{T_2 - T_1}{\bar{T}} + \ln\frac{P_1}{P_2}\right]\ldots \tag{20}$$

for f = *const*. Under flow regimes far from critical ($Ma \ll 1$) the values of T_1, T_2, and the pipe-averaged temperature $\bar{T}$ can be found from the thermal balance and the last term in brackets can be neglected in the first approximation.

Under the conditions when Equation 16 is valid, f can be determined by the Poiseuille, Blasius, and Colebrook-White formulas (see *Hydraulic resistance*). In the case of fluid heating or cooling when its viscosity and density cannot be assumed constant over the cross section, f can be evaluated by formulas for liquids for which only viscosity variation is essential

$$\bar{f} = \bar{f}_0(\eta_w/\eta_0)^n \ldots \tag{21}$$

When the flow is laminar, the mean length coefficient is determined over the length l, $\bar{f}_0$ by the Poiseuille formula at the Reynolds number Re_0 at the pipe inlet,

$$n = C\left(Pe\,\frac{D_h}{l}\right)^m\left(\frac{\eta_w}{\eta_0}\right)^{-0.062},$$

where at $Pe\,d_h/l < 1500$, $C = 2.3$, $m = -0.3$, at $Pe\,d_h/l > 1500$, $C = 0.535$, $m = -0.1$. The values of η_0 are taken for the pipe inlet, η_w for the wall temperature. If the flow is turbulent, $\eta_0 = \eta_b$, $\bar{f}_0 = \bar{f}_{0b}(Re_b)$ and $n = 0.33$ for heating and $n = 0.24$ for cooling of fluid. For gases

$$\bar{f} = \bar{f}_{0b}(\rho_w/\rho_b)^n \ldots \tag{22}$$

If the gas in the laminar flow is heated, $n = -1$, if cooled, $n = 0$. For fluid heated in the transcritical region of states $n = 0.4$.

In analyzing pressure loss in channels with permeable walls, e.g. heaters and manifolds, thermal pipes in evaporation and condensation zones, we use an equation of variable mass flow which for incompressible fluid takes the form

$$\frac{d\rho}{\rho} + \beta u_b du_b + u_b^2 d\beta + \beta u_b(u_b - \theta)\frac{d\dot{M}}{\dot{M}}$$

$$+ f\frac{u_b^2}{2} d\left(\frac{x}{D_h}\right) = 0. \ldots \quad (23)$$

Here θ is the projection of the velocity vector of separated or added fluid masses in the direction of the main flow, and β the Boussinesq coefficient which plays an active role in this equation.

Leading to: *Friction factors; Hydraulic diameter; Hydraulic resistance*

V.A. Kurganov

PRESSURE DROP, TWO-PHASE FLOW

Following from: *Conservation equations, two-phase; Multiphase-flow*

Pressure drop in two-phase flow is a major design variable, governing the pumping power required to transport two-phase fluids and also governing the recirculation rate in natural circulation systems. The conservation equations for two-phase flow are a subset of those for **Multiphase Flow**. Full derivation of the conservation equations is given in the article **Conservation Equations, Two-Phase**. For the purposes of this article, we shall use simplified forms of the momentum equations which apply to ducts of constant cross section and for steady state flow. Extensive background information on the subject of pressure drop is given by Hewitt (1982) and a review of pressure drop in orifices, valves, bends and fittings is given by Hewitt (1984).

Homogeneous Model

The simplest approach to the prediction of two phase flows is to assume that the phases are thoroughly mixed and can be treated as a single phase flow. This *homogeneous model* (see **Multiphase Flow**) will obviously work best when the phases are strongly interdispersed (i.e. at high velocities). For the homogeneous model, the pressure gradient (dp/dz) is given by:

$$-\frac{dp}{dz} = \frac{\tau_0 P}{S} + \frac{d(\dot{m}^2/\rho_H)}{dz} + g\rho_H \sin\alpha \quad (1)$$

where τ_o is the wall shear stress, P the tube peripherary, S the tube cross sectional area, $\dot{m}$ the mass flux, z the axial distance, g the acceleration due to gravity, α the angle of inclination of the channel to the horizontal and ρ_H the homogeneous density given by:

$$\rho_H = \frac{\rho_G \rho_L}{x\rho_L + (1 - x)\rho_G} \quad (2)$$

where ρ_G ρ_L are the gas and liquid densities and x is the quality (fraction of the total mass flow which is vapour).

The three terms on the right hand side of **Equation 1** may be regarded respectively as the frictional pressure gradient ($-dp_F/dz$), the accelerational pressure gradient ($-dp_a/dz$) and the gravitational pressure gradient ($-dp_g/dz$). Thus:

$$-\frac{dp}{dz} = -\frac{dp_F}{dz} - \frac{dp_a}{dz} - \frac{dp_g}{dz} \quad (3)$$

The frictional pressure gradient term in the homogeneous model is often related to a two-phase friction factor f_{TP} as follows:

$$-\frac{dp_F}{dz} = \frac{\tau_0 P}{S} = \frac{2f_{TP}\dot{m}^2}{D\rho_H} \quad (4)$$

where D is the tube diameter. f_{TP} may be related (via the normal single-phase friction factor relationships) to a two-phase Reynolds number defined as follows:

$$Re_{TP} = \frac{\dot{m}D}{\eta_{TP}} \quad (5)$$

where η_{TP} is a two-phase viscosity. There are some difficulties in defining the latter; a whole variety of forms being suggested in the literature. These are exemplified by the relationship:

$$\frac{1}{\eta_{TP}} = \frac{x}{\eta_G} + \frac{(1 - x)}{\eta_L} \quad (6)$$

where η_G and η_L are the gas and liquid viscosities respectively. In fact, though, the homogeneous model departs grossly from experimental data and simply readjusting the definition of viscosity has been found to be totally inadequate in bring agreement. Many authors have suggested empirical modifications of the friction factor to take account of the two-phase nature of the flow. Perhaps the most widely used of these corrections to the homogeneous model is the correlation of Beggs and Brill (1973), which corrects the homogeneous model for both flow regime and tube inclination. However, the preferred option has been to work using the *separated flow model* or, alternatively, *phenomenological models*. (see **Multiphase Flow** and **Two-Phase Gas-Liquid Flow**)

Separated Flow Model

Here, the phases are considered to be flowing separately in the channel, each with a given velocity and each occupying a given fraction of the channel cross section. The separated flow momentum equation reduces, for a duct of constant cross sectional area and steady flow, to:

$$-\frac{dp}{dz} = \frac{\tau_0 P}{S} + \dot{m}^2 \frac{d}{dz}\left[\frac{(1 - x)^2}{\rho_L(1 - \varepsilon_G)} + \frac{x^2}{\rho_G \varepsilon_G}\right] \quad (7)$$

$$+ g\rho_{TP} \sin\alpha$$

where ε_G is the void fraction and ρ_{TP} is given by:

$$\rho_{TP} = (1 - \varepsilon_G)\rho_L + \varepsilon_G\rho_G \quad (8)$$

Again, the three terms on the right hand side of **Equation 7** denote respectively the frictional, accelerational and gravitational pressure gradient terms. As will be seen from the above equations, calculation of the latter two terms (accelerational and gravitational) requires, in contrast to the homogeneous model, a value for ϵ_G the void fraction. Thus, to use the separated flow model, a void fraction correlation has to be invoked (see **Void Fraction**). The frictional pressure gradient $(-dp_F/dz)$ $(= \tau_o P/A)$ is widely calculated in terms of the ratio of the two-phase frictional pressure gradient to the frictional pressure gradients for the liquid phase flowing alone in the channel, $(dp_F/dz)_L$, the gas flowing alone in the channel, $(dp_F/dz)_G$, or the frictional pressure gradient for the total flow flowing in the channel with liquid phase properties, $(dp_F/dz)_{LO}$. *Pressure drop multipliers* are then defined as follows:

$$\phi_G^2 = \frac{dp_F/dz}{(dp_F/dz)_G} \tag{9}$$

$$\phi_L^2 = \frac{dp_F/dz}{(dp_F/dz)_L} \tag{10}$$

$$\phi_{LO}^2 = \frac{dp_F/dz}{(dp_F/dz)_{LO}} \tag{11}$$

The best known correlation in terms of such multipliers is that of *Lockhart and Martinelli* (1949) who ploted ϕ_G and ϕ_L as a function of the parameter X defined as follows:

$$X^2 = \frac{(dp_F/dz)_L}{(dp_F/dz)_G} \tag{12}$$

Different curves were suggested, depending on whether the phase-alone flows were laminar ("viscous") or turbulent, and the multipliers subscripted accordingly. The graphical correlation of Lockhart and Martinelli is shown in **Figure 1**.

Although still widely used, the Lockhart-Martinelli correlation has the disadvantage that it fails to predict adequately the effect of mass flux (and other parameters) and a whole variety of more sophisticated correlations have been produced as replacements (see Hewitt, 1982). Typical of these more recent correlations is that of Friedel (1979) whose correlation is given by:

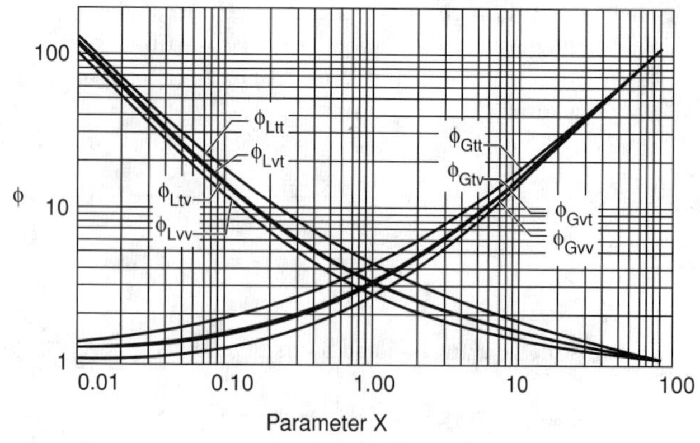

Figure 1. Correlation of Lockhart and Martinelli (1949).

$$\phi_{LO}^2 E + \frac{3.24\ FH}{Fr^{0.045} We^{0.035}} \tag{13}$$

where:

$$E = (1 - x)^2 + x^2 \frac{\rho_L f_{GO}}{\rho_G f_{LO}} \tag{14}$$

$$F = x^{0.78}(1 - x)^{0.24} \tag{15}$$

$$H = \left(\frac{\rho_L}{\rho_G}\right)^{0.91} \left(\frac{\eta_G}{\eta_L}\right)^{0.19} \left(1 - \frac{\eta_G}{\eta_L}\right)^{0.7} \tag{16}$$

$$Fr = \frac{\dot{m}^2}{gD\rho_H} \tag{17}$$

$$We = \frac{\dot{m}^2 D}{\rho_H \sigma} \tag{18}$$

where σ is the surface tension and f_{GO} and f_{LO} are the friction factors for gas and liquid single phase flows at the total mass flux. This correlation has a standard deviation of around 30% for single-component flows and about 40–50% for two-component flows.

Phenomenological Models

Prediction of pressure drop is, as will have been seen, subject to large error. This error can be reduced if proper account is taken of the actual nature of the two-phase flows, namely the flow pattern or flow regime (see **Two-Phase Gas-Liquid Flow**). Relationships are then developed for the various features of the flow and combined into an overall model which represents these features in order to predict the important design variables, such as pressure drop. Examples of such an approach are given in the articles on **Annular Flow**; **Bubble Flow**; **Stratified Flow Gas Liquid** and **Slug Flow**.

References

Beggs, H. D. and Brill, J. P. (1973) A study of two-phase flow in inclined pipes, *J. Petroleum Tch.*, 25, 607–617.

Friedel, L. (1979) Improved friction pressure drop correlations for horizontal and vertical two-phase pipe flow. European Two-Phase Flow Group Meeting, Ispra, Italy, paper E2.

Hewitt, G. F. (1982) Pressure drop. *Handbook of Multiphase Systems*, (Ch 2.2., G. Hetsroni,) ed., McGraw Hill Book Company, New York.

Hewitt, G. F. (1984) Two-phase flow through orifices, valves, bends and other singularities. Proceedings of the Eighth Lecture Series on Two-Phase Flow, University of Trondheim, 1984, 163–198.

Lockhart, R. W. and Martinelli, R. C. (1949) Proposed correlation of data for isothermal, two-phase, two-component flow in inpipes. *Chem. Eng. Prog.* 45, 39–48.

Leading to: *Instability, two-phase*

G.F. Hewitt

PRESSURE DUE TO RADIATION (see Electromagnetic waves)

PRESSURE EFFECTS ON BOILING (see Burnout in pool boiling)

PRESSURE GRADIENT (see Three phase, gas-liquid-liquid, flows)

PRESSURE GRADIENT, COMPONENTS OF IN MULTIPHASE FLOW (see Multiphase flow)

PRESSURE GRADIENT IN ANNULAR FLOW (see Annular flow)

PRESSURE INDUCED FLOW (see Poiseuille flow)

PRESSURE MEASUREMENT

***Following from:** Pressure; Pressure drop, single phase; Pressure drop, two-phase flow*

The measurement of pressure (and differential pressure between two points in a system) is of vital importance in studies of heat and mass transfer. A very wide range of techniques exists but the main ones, covered in this article, are the use of *manometers, Bourdon gauges* and *pressure transducers*. A good general source on pressure measurement is the book by Jones (1985). Specialised techniques relating to very high pressures are discussed by Beggs (1983).

Manometers

Using manometers, good accuracy is possible and, by the use of inclined manometers and micro-manometers, it is possible to cover a wide range of pressure (through the upper pressure limit for absolute pressure measurement is necessarily rather low). However, manometers are rather messy to use and have a poor response time (typically about 0.5 seconds).

Great care needs to be taken in ensuring that the lines connecting the manometer to the points at which measurements are to be made are filled with fluids of known density. Care also needs to be taken in the interpretation of manometer level differences in terms of the differences in pressure being measured. This is illustrated by the example shown in **Figure 1** where a manometer is used to measure pressure drop in a two-phase flow system; the diagram is somewhat simplified since, in such a system, purging of lines with liquid would be used to avoid ambiguity of line content (see Hewitt, 1978). A pressure balance can be carried out at level A:

$$p_1 + (z_2 - z_1)g\rho_c = p_2 + (z_4 - z_3)g\rho_c + (z_3 - z_1)g\rho_m \quad (1)$$

and rearranging we have:

$$p_1 - p_2 = (z_3 - z_1)g(\rho_m - \rho_c) + (z_4 - z_2)g\rho_c \quad (2)$$

If $p_1 = p_2$, then the manometric difference is given by:

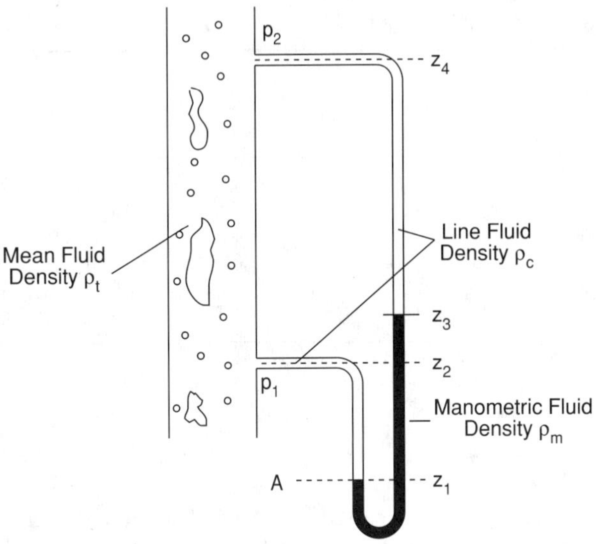

Figure 1. Measurement of differential pressure in a two phase flow using a manometer system.

$$(z_3 - z_1) = -(z_4 - z_2)\frac{\rho_c}{\rho_m - \rho_c} \quad (3)$$

thus, there is an "offset" on the manometer which depends on the distance between the tappings and the density (ρ_c) in the lines. In the absence of flow through the tube it follows that:

$$p_1 - p_2 = g\rho_t(z_4 - z_2) \quad (4)$$

and the manometric difference is:

$$(z_3 - z_1) = \frac{\rho_t - \rho_c}{\rho_m - \rho_c}(z_4 - z_3) \quad (5)$$

the manometer will have zero differential if the fluid in the line has the same density as the fluid in the tube.

A special form of the manometer is the **barometer** in which the atmospheric pressure is compared to vacuum using an inverted tube in a mercury bath as shown in **Figure 2**.

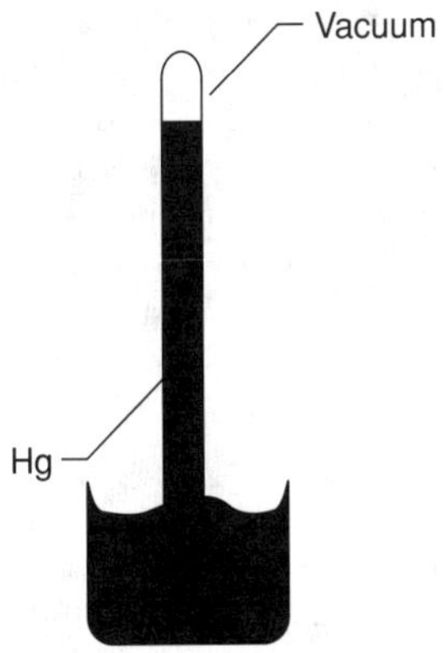

Figure 2. The mercury barometer.

Table 1. Classification of transducer type: with advantages and disadvantages

	Capacitance type	Strain gauge type	Reluctance type	Piezo-electric type	Potentiometric type	Magnetostrictive type	Eddy current type
Sensitivity % full scale	0.01	0.3	0.1	0.001	0.25–1	0.5	0.01
Response time μs	20	10–100	200	2	Poor	2	10
Stability	Excellent	Good	Fair	Variable	Moderate	Good	Poor
Maximum Temp °F (standard types)	720	600	600	325	—	—	—
Cost	High	High	High	V.high	Moderate	V.high	High
Commercial availability	V.good	Good	Good	Good	Good	Poor	Moderate

"Potentiometric"—diaphragm moves slidewire
"Magnetostrictive"—magnetic properties varying with force

Bourdon gauges

Bourdon gauges indicate pressure by the amount of deflection under internal pressure of an oval tube bent in the arc of a circle and closed at one end. Such gauges cover a wide range of conditions (vacuum up to 700 MPa), have good potential accuracy and are versatile. However, they have a rather poor response time (typically about 0.2 s). An alternative form of mechanical pressure indicator is the *diaphragm gauge* which depends on the deflection of a diaphragm when subjected to a difference of pressure between the two faces. The pressure range of such gauges is more restricted than that of more conventional Bourdon gauges; the *aneroid barometer* is a type of diaphragm gauge.

Pressure transducers

Pressure transducers depend on the (often very small) deflection of a diaphragm by the imposed pressure. The deflection is detected by various means and extremely fast response is possible. **Table 1** gives the classification of pressure transducer types, with their advantages and disadvantages.

Obviously, by mounting pressure transducers at two points in a system, their signals can be subtracted to give the differential pressure. Alternatively, a *differential pressure transducer* can be used in which two pressure tapping lines are attached to the respective points in the system and to two chambers separated by a diaphragm. The movement of the diaphragm indicates the differential pressure. Of the types of transducer itemised in **Table 1**, the strain-gauge and reluctance transducers are ideally suitable for differential pressure measurements but the capacitance and piezoelectric types are specifically unsuitable. There are problems in using differential transducers which are similar to those mentioned for manometers above. Care has to be taken that the lines are full of a fluid of known density and the problem of the "offset" may restrict the range of the instrument (this can be counteracted by using an intermediate manometer system—see Hewitt, 1978).

References

Hewitt, G. F. (1978) *Measurement of Two Phase Flow Parameters*, Academic Press, London. ISBN: 0–12–346260–6.

Jones, E. B. (1985) *Instrument Technology*, Vol. 1, *Mechanical Measurements*, Butterworth & Co. (Publishers) Ltd, London. ISBN: 0–408–01232–1.

Peggs, G. N. (1983) *High Pressure Measurement Techniques*, Applied Science Publishers, London. ISBN: 0–85334–189–3.

G.F. Hewitt

PRESSURE NOZZLES (see Atomization)

PRESSURE PRISMS (see Hydrostatics)

PRESSURE SUPPRESSION

Following from: *Containment; Nuclear reactors*

In a nuclear reactor, this term is applied to containment systems where measures are taken to minimise the containment pressure in the unlikely event of rupture of any of the reactor pipes. In a water reactor, such a rupture results in ejection of water and steam accompanied by flashing of the water as the reactor and containment pressures come to equilibrium. One approach is to design the containment to be large enough for this equilibrium pressure to kept to a reasonably low level. However containment structures contribute significantly to the overall cost of the plant and some designs, especially BWRs and integral PWRs, take steps to reduce this pressure by condensing much of the emitted steam. This can most easily be done by arranging that the path from the reactor cavity to the main body of the containment forces the steam to bubble into a large volume of water. **Figure 1** on page 893 illustrates the arrangement for the *Advanced Boiling Water Reactor (ABWR)* where the tank of water is situated in an annular region within the containment and below the level of the reactor. The water is cooled by circulating it through a residual heat exchanger which forms the ultimate heat sink in severe emergency conditions.

Alternative arrangements in other reactors have replaced the water tank by a rack of ice trays or by racks of gravel.

I. Gibson

PRESSURE SUPPRESSION CONTAINMENTS, FOR NUCLEAR REACTORS (see Containment)

PRESSURE-SWIRL NOZZLES (see Atomization)

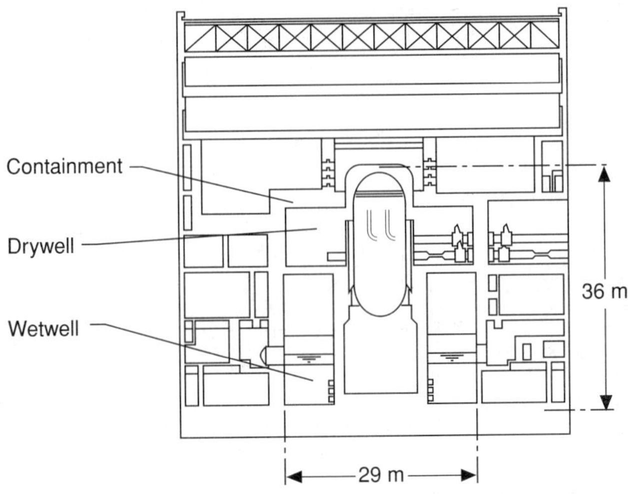

Figure 1. Advanced boiling water reactor containment.

PRESSURE TRANSDUCERS (see Pressure measurement)

PRESSURE VESSEL DESIGN CODES (see Heat exchangers)

PRESSURE VESSELS

A pressure vessel, as a type of unit, is one of the most important components in industrial and petro-chemical process plants. In the broad sense, the term pressure vessel encompasses a wide range of unit heat exchangers, reactors, storage vessels, columns, separation vessels, etc. (See also **Mechanical Design of Heat Exchangers**) Because of the risks that would be associated with any accidental release of contents, in many countries the production and operation of pressure vessels are controlled by legislation. This legislation may define the national standard to which the pressure vessel is to be designed, the involvement of independant inspection during construction, and subsequently the regular inspection and testing during operation. Some national pressure vessel standards such as ASME VIII, 1993, or BS5500, 1994, have effectively the status of defacto international standards.

The national legislation and/or standard generally define when a vessel is to be treated as a pressure vessel. A definition of minimum pressure (typically 5.10^4 N/m^2) will exclude low pressure tanks and a minimum of a few litres will exclude piping and piping components. Note that vessels operating at vacuum are often defined as pressure vessels to ensure that the design, construction etc. are of acceptable quality.

For design and construction purposes, the pressure vessel is generally defined as the pressure vessel proper including welded attachments up to, and including, the nozzle flanges, screwed or welded connectors, or the edge to be welded at the first circumferential weld to connecting piping. **Figure 1** on page 894 shows a typical pressure vessel envelope.

Several organisations are involved in the production and operation of a pressure vessel. These can be considered as follows.

a. The *Regulating Authority* is the authority in the country of installation that is legally charged with the enforcement of the requirements of law and regulations relating to pressure vessels
b. The *User* operates the plant and thus the pressure vessel. He is responsible to the regulating authority for the continued safe operation of the vessel.
c. The *Purchaser* is the organisation that buys the finished pressure vessel for its own use or on behalf of the purchaser.
d. The *Manufacturer* is the organisation that designs, constructs and tests the pressure vessel in accordance with the purchaser's order. Note that the design function may be carried out by the purchaser or by an independant organisation.
e. The *Inspecting Authority* is the organisation that verifies that the pressure vessel has been designed, constructed and tested in accordance with the order and with the standard.

Pressure vessels, as components of a complete plant, are designed to meet requirements specified by a team, typically comprising process engineers, thermodynamicists and mechanical engineers. The full design procedure is described in detail in Bickell and Ruiz (1967) and the interaction between the elements of the procedure is shown in **Figure 2**.

Operational Requirements

The first step in this design procedure is to set down the operational requirements. These are imposed on the vessel as part of the overall plant and include the following.

a. **Operating pressure**. As well as the normal steady operating pressure, the maximum maintained pressure needs to be defined. Regulations and/or standards will define how this maximum pressure is translated into vessel design pressure.
b. **Fluid conditions**. Maximum and minimum fluid temperatures will need to be specified and translated into metal design temperatures. Fluid physical and chemical properties will influence material choice and specific gravity will effect support design.
c. **External loads**. Loads to be considered include wind, snow, and local loads such as piping reactions and dead weight of equipment supported from the vessel.
d. **Transient conditions**. Some vessels may require an assessment of cyclic loads resulting from operational pressure, temperature, structural and accoustic vibration loading.

Functional Requirements

Next the functional requirements, which cover geometrical parameters, are defined. Some of these parameters are again defined by the plant design team whilst some are left to the discretion of the pressure vessel designer. The functional requirements include the following.

a. Size and shape of the vessel
b. Method of vessel support
c. Location and size of attachments and nozzles

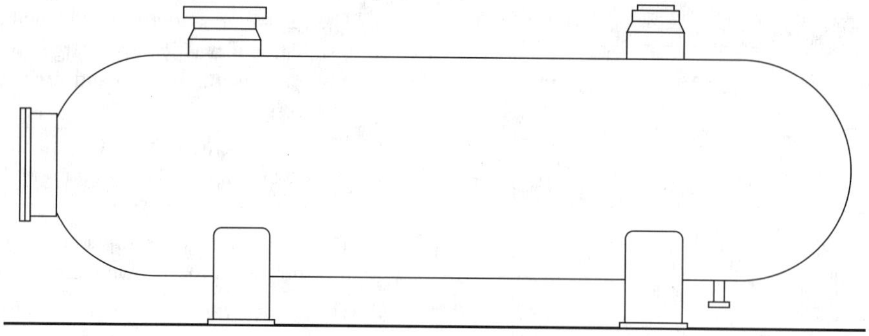

Figure 1. Pressure vessel envelope.

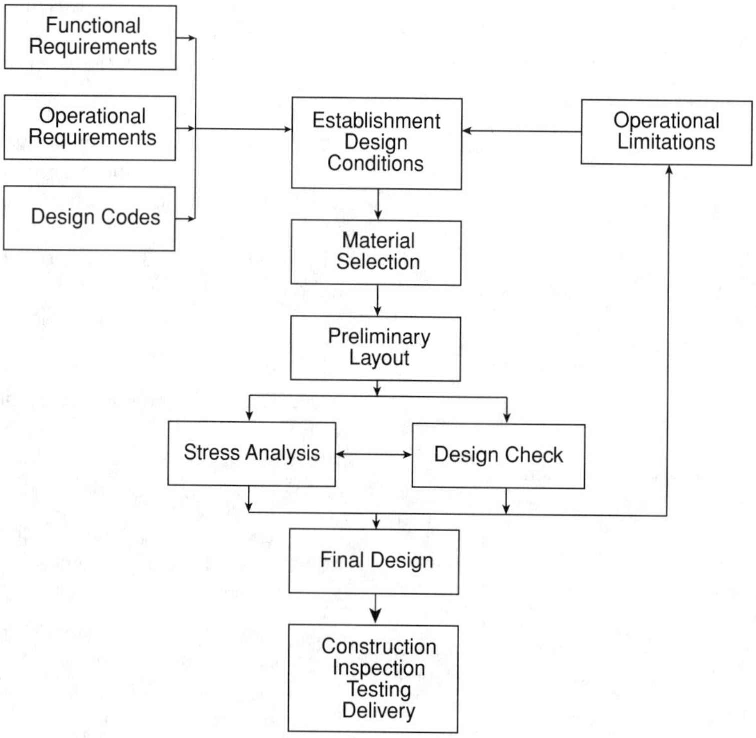

Figure 2. Pressure vessel design procedure.

Materials

Next the main materials are selected. Some national standards list acceptable materials with acceptable temperature ranges and design stresses. Design stresses are set using safety factors applied to material properties which include.

a. Yield strength at design temperature
b. Ultimate tensile strength at room temperature
c. Creep strength at design temperature

(See **Stress in Solid Materials; Fracture of Solid Materials**)

The standard will have selected the materials based upon the above material properties together with knowledge of the following properties that influence fabrication and operation.

a. Elongation and reduction of area at fracture
b. Notch toughness
c. Ageing and embrittlement under operating conditions
d. Fatigue strength
e. Availability

The range of materials used for pressure vessels is wide and includes, but is not limited to, the following:

a. Carbon steel (with less than 0.25% carbon)
b. Carbon manganese steel (giving higher strength than carbon steel)
c. Low alloy steels

d. High alloy steels
e. Austenitic stainless steels
f. Non-ferrous materials (aluminium, copper, nickel and alloys)
g. High duty bolting materials

Clad materials are accepted by national standards but often only the base material thickness can be used in design calculations.

Proprietary materials are used for special applications by agreement between the designer and purchaser although standard bodies will require evidence of previous successful applications before accepting as a material to be listed in the standard. (See **Metals**; **Steels**)

Design Rules

Figure 2 illustrates the overall pressure vessel design procedure. The design rules in standards will give minimum thicknesses or dimensions of a range of pressure vessel components. These thicknesses will ensure integrity of vessel design against the risk of gross plastic deformation, incremental collapse and collapse through buckling. The components covered by the design rules in standards are described in more detail in the **Mechanical Design of Heat Exchangers**.

The thicknesses determined by the relevant equations are minimal to which should be added various allowances, including allowances for corrosion, erosion, material supply tolerances and any fabrication thinning.

The preliminary thicknesses of components are generally obtained by using the relevant internal or external pressure equations of the standard. These thicknesses are then checked for the other loads that have been identified in the operational requirements.

Where components or loads are not covered by explicit equations in the standard additional analysis may be required and this is by agreement between the designer, purchaser and inspecting authority. An example of the assessment of additional analysis is Appendix A of BS5500, 1994, which identifies the general design criteria to be used in these circumstances. For further reading see Bickell and Ruiz, 1967.

Before construction starts, the manufacturer is often required to submit fully dimensioned drawings of the main pressure vessel shell and components for approval by the purchaser and inspecting authority. In addition to showing dimensions and thicknesses, these drawings include the following information:

a. Design conditions
b. Welding procedures to be applied
c. Key weld details
d. Heat treatment procedures to be applied
e. Non-destructive test requirements
f. Test pressures

The manufacturer is generally required to maintain a positive system of identification for the materials used in construction so that all material in the completed pressure vessel can be traced to its origin. The forming of plates into cylinders or dished ends will be either a hot or cold process depending on the material, its thickness and finished dimensions. The standard will define the allowable assembly tolerances and forming tolerances of cylinders and ends. These tolerances limit the stresses resulting from out-of-roundness and joint misalignment. Additional tolerances may be specified by the purchaser to allow for example, for the insertion of internals. The standards will usually show typical acceptable weld details for seams and attachment of components.

Depending on material and thickness at the weld joint, pre-heating and post-weld heat treatment may be required. Pre-heat is applied locally to the weld area but post-weld heat treatment is preferably applied to the complete vessel in an enclosed furnace.

Inspection and Testing

Each pressure vessel is inspected by the inspecting authority during construction. The standard specifies the stages from material reception through to completed vessel at which inspection by this authority is mandatory. The purchaser may require additional inspection, for example to check internals.

The manufacturer identifies the welding procedures required in the pressure vessel construction, together with test pieces that are representative of the materials and thicknesses used in the actual vessel. The production and testing of these test pieces are generally witnessed by the inspecting authority unless previously authenticated test pieces are available.

Welders have to pass approval tests which are designed to demonstrate their competence to make sound welds similar to those used in the actual vessel. These welder approvals are again authenticated by a recognised inspecting authority.

The national standard defines the level of non-destructive testing that is applied during the construction. This non-destructive testing is usually one or more of the following.

a. Magnetic particle or dye penetrant (for weld surface flaws)
b. Radiography (for weld internal flaws)
c. Ultrasonic (for weld internal flaws)

The degree of non-destructive testing depends upon material and thickness (i.e. upon the difficulty of welding). Some standards use a "joint factor" approach which allows a reduced amount of non-destructive testing if the designed thickness is increased. This joint factor is chosen and applied at the initial design stage.

Before delivery, most standards require a pressure test which is witnessed by the inspecting authority. Water is the preferred test fluid because of its incompressibility. If air is the only possible test fluid, special precautions have to be taken and consultations are needed with the inspecting authority and other relevant safety authorities. The test pressure is usually between 1.2 and 1.5 times the design pressure and this test pressure is gradually applied in stages and held for an agreed time to demonstrate the adequacy of the vessel.

Once delivered and placed into operation, the user picks up the responsiblity for safe service. Legislation will often require inspection at regular intervals during the vessel life and for some critical contents may require the involvement of the regulating authority.

References

ASME VIII Division 1, ASME Boiler and Pressure Vessel Code. (1993) Rules for the Construction of Pressure Vessels, ASME New York.

BS5500 British Standard for the Specification for Unfired Pressure Vessels. (1994) BSI London.

Bickell, M. B. and Ruiz, C. (1967) *Pressure Vessel Design and Analysis*, Macmillan, London.

M. Morris

PRESSURE WAVES (see Waves in fluids)

PRESSURISED WATER REACTORS, PWR (see Nuclear reactors; Reflood)

PRESTON TUBE (see Shear stress measurement)

PRILLING

Prilling is a method of producing reasonably uniform spherical particles from molten solids, strong solutions or slurries. It essentially consists of two operations, firstly producing liquid drops and secondly solidifying them individually by cooling as they fall through a rising ambient air stream. There is no agglomeration so the size distribution of the drops determines that of the product. The process is widely used in the fertiliser industry for making *ammonium nitrate*, calcium nitrate, urea and compound fertilisers containing N, P and K. A typical plant will produce 2,000 te/day or more. Prilling is also used in the explosives industry to produce a porous prill of ammonium nitrate which will absorb oil.

Prilling towers must be of sufficient height for the particles to be strong enough not to break on impact. Latent heat is transferred from the drop to the air as it falls, and if significant amounts of water are present evaporation also occurs, increasing the cooling effect on the drop. It is important for the temperature of the feed liquor to be as low as possible, just a degree or two above its solidification point. Higher temperatures require taller towers, as do larger particle sizes. Prilling towers in the fertiliser industry are typically over 50m high for a mean particle size of about 2mm. In the explosives industry the particle size is smaller, the feed wetter and towers of about 10m are used.

The drop producer has to make uniformly sized drops and distribute them uniformly across the width of the tower. Various designs are used. For simple solutions an array of downward-pointing sprays is suitable. For more difficult materials such as slurries, spinning discs and spinning perforated baskets are used. Attention has to be paid to the ease of clearing blockages: in the fertiliser industry the feeds are liquids or slurries which solidify in the range 130°C to 200°C. Little further processing is required after prilling. Cooling is however necessary to prevent caking. Screening of small quantities of oversize and undersize prills is also done. In the ammonium nitrate explosives industry where the molten feed contains about 5% H_2O further drying is essential to produce the porous structure for oil absorption.

NPK fertilizer products require an oil and dust anti-caking treatment. Other products do not, eg. in ammonium nitrate manufacture small quantities of salts of Mg or Al are added to the melt before prilling and these enhance the strength and anti-caking behaviour without further treatment.

In the mid 1960's a variation of the air prilling process was developed for manufacturing ammonium nitrate fertiliser. The drops fell through a rising cloud of finely ground mixed clay dusts and into a dense fluidised dust bed cooled by water radiators where the remaining heat was removed. The prilling operation was greatly intensified with much narrower towers, and tower heights of only 5–7m. The process was used successfully for nearly 30 years, but is now closed down.

Until the 1980s hot air from prilling towers was discharged directly to the atmosphere. It contained small amounts of fine solid particles and also some very fine fume. Recent *pollution* regulations require this air to be cleaned, which is done using irrigated fibre candle filters.

J.K. Walters

PRILLING TOWERS (see Prilling)

PRIMARY RECOVERY PROCESS (see Enhanced oil recovery)

PRINTED CIRCUIT HEAT EXCHANGER (see Reboilers)

PROBABILITY DENSITY FUNCTION, PDF (see Stochastic process; Tracer methods)

PROBABILITY THEORY (see Non-linear systems)

PROCESS (see Thermodynamics)

PROCESS CONTROL

The objectives of process control are generally either to maintain a process at a desired, constant operating condition (temperature, pressure, composition, etc.) in the face of disturbances or, less typically in conventional process applications, to force it to follow a desired trajectory with time. All process control configurations, whether manual, automatic, or computer-based, have three essential elements:

a. a measurement (often several);
b. a control strategy (embedded in a controller);
c. a final element for implementing the control action (a valve, heater or other variable input).

For example, in classical *feedback control*, one can correct for the effects of a disturbance by:

a. measuring some property of the system to be controlled (its temperature for example);
b. comparing it with its desired value (set point) and calculating what changes, if any, are required as inputs to the process;
c. making an appropriate adjustment to that/those inputs (a variable heater, for example).

Similarly, in *feedforward control*, one can anticipate and com-

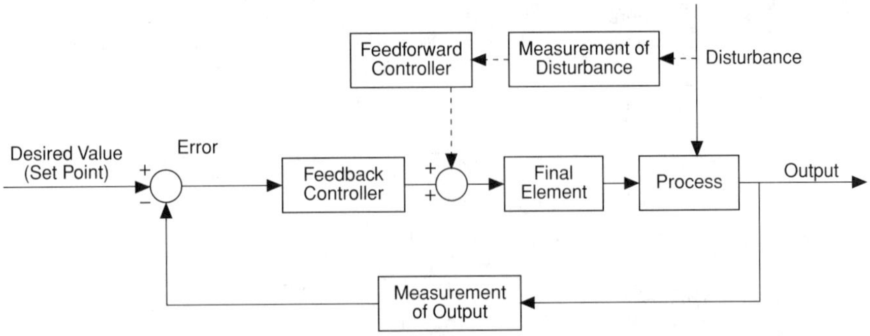

Figure 1.

pensate for the effects of a major disturbance to the process by first measuring the disturbance; this would then be followed by the other two steps described above. Both feedback and feedforward control can be used on the same process as illustrated diagrammatically in **Figure 1**.

In a conventional (analogue) control environment, the measurements, controller calculations and signals to the final elements are typically carried out by electrical signals, generally in the range of 4–20 mA; in more modern computer control systems, any analogue measurements are converted to digital form for processing and the resulting signal to the final element is reconverted to analogue form where necessary. Brief descriptions of digital control systems can be found in the books by Stephanopoulos (1984) and by Seborg et al. (1989), among others.

Measurements

A wide range of on-line measuring devices (sensors) exists in the process industries, the most common being those for flow rate, pressure, liquid level, temperature, pH, and other selective measures of chemical composition. There is a continuing development in this area; books by Nagy (1992), Noltingk (1985), and Clevett (1986) all provide summaries of many types of sensors currently available.

Controllers

Classical PID (*proportional-integral-derivative*) *controllers* still dominate feedback strategy in industrial applications. Here, the signal to the final control element, p, is related to its nominal value, p_s, and to the "error" signal, e, by the equation:

$$p = p_s + K_c[e + (\int e\, dt)/\tau_I + \tau_D\, de/dt]$$

where K_c, τ_I, and τ_D are the proportional, integral, and derivative constants respectively.

Final Element

The control action signal calculated by the controller is sent to the final control element, a device which implements the change of a suitable input to the system. This variable input is typically a flow rate (of fuel, air, coolant, reactant, etc.). Hence, an automatic control valve, either electrically or pneumatically operated, is usually appropriate (see Shinskey, 1988, for a discussion of types of valves).

Typical problems which lead to the poor performance of control systems include: inadequate/inaccurate measurements; long time delays in either the process or the measurements; varying or nonlinear nature of the system; the interaction of several quasi-independent control loops.

The major development over the past two decades has been the replacement of analogue controllers by digitally calculated control signals, even for conventional PID controllers. In addition, there has been considerable effort to develop more sophisticated model-based predictive controllers. In a conventional controller, the controller parameters are often adjusted ("tuned") on-line, although prior knowledge of the system to be controlled will provide some guidance in their choice. Model-based control, on the other hand, makes use of a model of the process (either fundamentally-based or empirical) to decide on the appropriate control strategy to be implemented to achieve the desired objective. In principle, these can lead to improved performance, but, in practice, their performance may be strongly dependent on the accuracy of the model used. Some simple examples are given by Seborg et al., 1989.

References

Clevett, K. J. (1986) *Analyzer Technology*, Wiley, New York.

Nagy, I. (1992) *Introduction to Chemical Process Instrumentation*, Elsevier, Amsterdam.

Noltingk, B. E. (1985) *Instrument Technology*, Butterworths, London.

Seborg, D. E., Edgar, T. F. and Mellichamp, D. A. (1989) *Process Dynamics and Control*, Wiley, New York.

Shinskey, E. (1988) *Process Control Systems*, McGraw-Hill, New York.

Stephanopoulos, G. (1984) *Chemical Process Control*, Prentice-Hall, Englewood Cliffs, NJ.

L. Kershenbaum

PROCESS HEATERS (see Furnaces)

PROCESS INTEGRATION

What is meant by Process Integration? In its broadest sense it could be taken to mean the achievement of more than one process

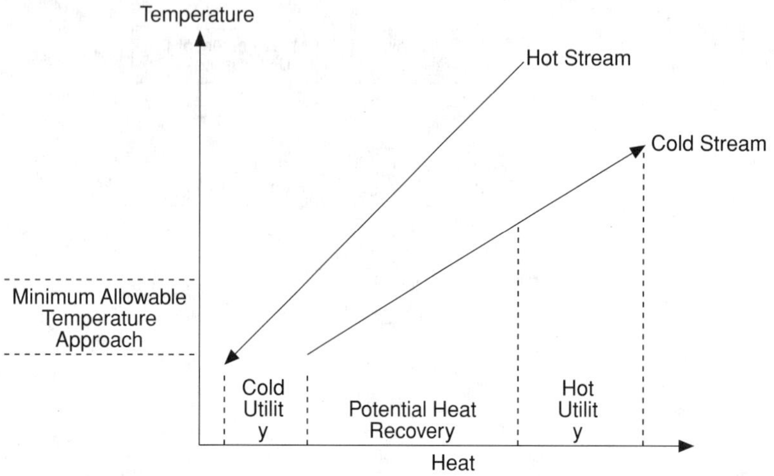

Figure 1. Heat recovery between one hot and one cold process stream.

objective in a single item of equipment. An example would be reactive distillation where both chemical reaction and product purification are achieved in a column. The simplest example is probably the heat recovery exchanger where the required cooling of one stream and the required heating of another are achieved in a single exchanger. This latter example leads to the most restricted sense of the term (and that is the sense relevant here) where it refers to the use of heat recovery for cost reduction in process plants.

Because of this emphasis on energy, process integration is often viewed as solely relating to *heat recovery network design*. This is a misconception. The technology, when properly applied, can have implications for reactor design, separator design and overall process optimisation in any plant in which energy is a significant factor.

T-$\dot{Q}$ Plots

A heat recovery problem involving just one hot and one cold stream can easily be solved using a *Temperature-Heat Load (T-$\dot{Q}$) Plot.*

The heat supply or demand associated with a stream is given by the relationship:

$$\dot{Q} = CP(TS - TT)$$

where, $\dot{Q}$ is the heat load, CP is the heat capacity flowrate, TS is the stream's "supply" temperature and TT is the stream's "target" temperature.

If the heat capacity flow-rate is independent of temperature, this relationship would appear as a straight line on a plot of Temperature against Heat Load.

Since, in heat recovery problems heat load differences are the important factor, the line may be placed in any position along the heat load axis. However, its position on the temperature axis is fixed.

When the line representing the hot stream is superimposed on that representing the cold stream such that they are separated by a specified minimum temperature difference (see **Figure 1**), the following observations can be made.

- where the lines overlap heat can be transfered from the hot stream to the cold stream at or above the specified minimum temperature difference. This 'overlap' indicates the scope for heat recovery.
- where the hot line "overshoots" the cold line there is insufficient temperature driving force to permit heat recovery and the heat must be rejected to a cold utility.
- the full heat content of the hot stream has now been accounted for, and any remaining cold stream heating duty—represented by the overshoot of the cold line—must be supplied through the use of hot utility.

The scope for heat recovery is dependent on the allowable temperature approach. As the minimum temperature approach is increased the lines move apart, the overlap reduces and the utility demands increase (**Figure 2**).

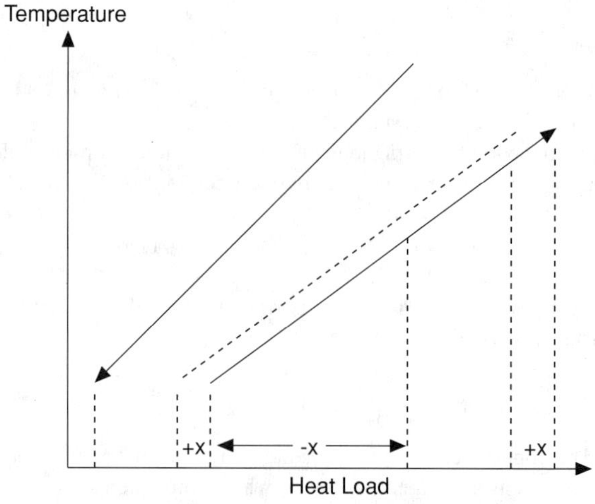

Figure 2. Effect of changing minimum allowable temperature approach.

Composite Curves

The approach described above can be applied to systems involving a multitude of hot and cold streams. Here, rather than deal with individual streams, overall supply and demand relationships are used. These relationships are called "Composite Curves" and were introduced by Huang and Elshout (1976). A methodology for generating such curves was presented by Umeda, Itoh and Shiroko (1978).

The Hot Composite Curve represents the overall heat supply within the processes as a function of temperature. It consists of a series on connected straight lines (**Figure 3**). Each change in slope represents a change in overall hot stream heat capacity flow-rate and is generally associated with either the "arrival" or "departure" of a stream from the temperature field. Where a stream arrives the overall heat capacity flow-rate increases and the slope of the Composite decreases.

The Cold Composite Curve represents the heat demands made within the process as a function of temperature.

Hot and Cold Composites can be superimposed in the manner described above for the two stream problem. Then, the overlap shows the scope for heat recovery within the process. The cold end overshoot shows the subsequent heat rejection to cold utility. The hot end overshoot shows the subsequent minimum quantity of heat required to operate the process.

Heat Recovery Pinch

With a two stream problem the close approach between the supply and demand lines occurs at the end of one of the lines. With Composite Curves the close approach often occurs at an intermediate point. This point was named the *Pinch* by Umeda, Itoh and Shiroko (1978).

The Pinch has significance for both the design of the heat recovery network and for the engineering of energy saving process modifications.

Consider the decomposition of the overall problem into two parts; an Above Pinch problem and a Below Pinch problem. Three key observations can be made.

- Below the Pinch the heat supplied by the process equates with that demanded by the process plus the minimum cold utility requirement.
- Above the Pinch the heat demanded by the process equates with that supplied from within the process plus the minimum hot utility requirement.
- Any transfer of heat across the Pinch results in both heat balances being affected with both hot and cold utility requirements increasing by the quantity transfered across the Pinch.

These observations have very significant implications. First, they lead to a significant simplification in heat recovery network design methodology. A heat recovery network design ensuring minimum utility usage results from following just one simple expedient: prevent heat flow across the Pinch. Second, as described by Umeda et al. (1979), the overall energy needs of a process can be reduced by introducing process changes that shift heat demands from above to below the Pinch or shift heat supplies from below to above the Pinch.

Pinch Design Method

The observation that the elimination of heat flow across the Pinch results in a network requiring minimum utility consumption resulted in the development of a powerful network design methodology called the *Pinch Design Method* (Linnhoff and Hindmarsh, 1983).

Heat flow across the Pinch can be eliminated by simply dividing the design problem into two parts: an above Pinch design and a below Pinch design. The results can then be merged in order to obtain an overall design.

The constraint that the designer now has to contend with is the minimum allowable temperature approach. This is most critical around the Pinch itself, where the temperature driving forces are smallest. The closest approach occurs at the Pinch itself and heat recovery matches must be selected such that the exchanger temperature profiles diverge as the streams move

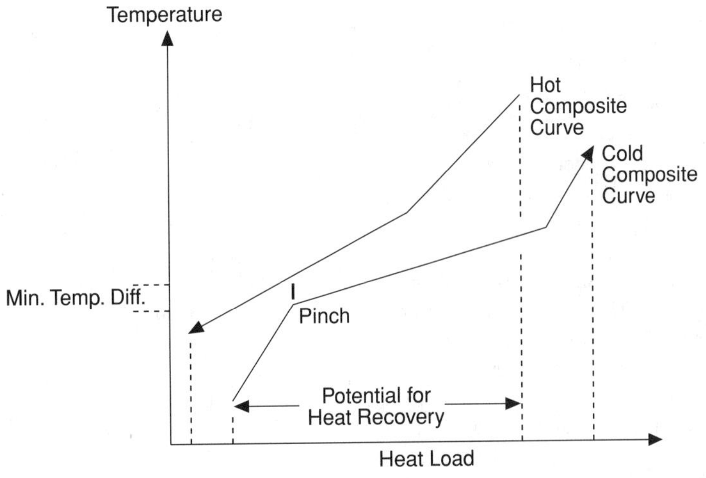

Figure 3. Multiple stream problem; superimposed composite curves.

away from the Pinch. This is achieved if the matches obey the following rule:

$$\text{CP out} \geq \text{CP in}$$

This rule generally applies solely to matches made at the Pinch. Temperature driving forces in regions away from the Pinch are often large enough to allow matches not conforming to this rule.

Another important factor affecting the cost of a heat recovery network is the number of heat exchangers used. It was recognised that this is minimised if the load on a match is maximised.

So, the Pinch Design Method consists of the following steps:

1. Divide the overall problem into two parts at the Pinch.
2. Develop solutions for the individual parts. In each case start the design at the Pinch, making matches conforming to the CP rule. Maximise the heat load on these matches, thereby removing a stream from the problem.
3. Complete the sub-problem design by making remaining heat recovery matches and by using utility units.
4. Merge the two designs to provide an overall solution.

The design achieved in this way will generally use more exchangers than necessary. This provides scope for further refinement. Exchangers can often be removed by shifting heat loads around the network using a technique called "loops and paths exploitation". The result is generally an infringement of the minimum temperature approach. The designer then has two options: the infringement can be accepted (in which case more area but fewer exchangers are used), or the temperature approach is restored through energy relaxation (in which case more energy but fewer exchangers are used).

Minimum Number of Units

As shown by Hohmann (1971) the minimum number of heat exchanger units required for a heat recovery network is given by:

$$N_{units} = Nh + Nc + Nu - Ns$$

where:
Nh = number of hot streams
Nc = number of cold streams
Nu = number of utility streams
Ns = number of separate heat balanced systems

Network Area

In addition to developing an understanding of the thermodynamics of network design Umeda et al. (1978) showed how the area needs of a network could be estimated prior to design. They recognised that, in the case where hot and cold streams have uniform film heat transfer coefficients, α, the network area was minimised if the network exhibits pure counterflow. They then developed a network structure that exhibited this property. This structure was subsequently the basis on which Linnhoff and Ahmad (1990) derived an equation for network area estimation:

$$A_{network} = \sum_{j}^{intervals} \left(\frac{1}{\Delta T_{LM}}\right)_j \sum_{i}^{streams} (\dot{Q}/\alpha)_i$$

In practice, differing process streams can be expected to exhibit differing film heat transfer coefficients. One factor that greatly influences the coefficient actually achieved is the stream's allowable pressure drop. A means of predicting area requirements on the basis of stream pressure drop has been developed by Polley and Panjeh Shahi (1991).

These algorithms only provide approximate estimations for network area. In practice, as pointed out by Umeda et al., pure counterflow cannot be achieved in network design without using an inordinately large number of heat exchangers. The use of fewer exchangers leads to poorer use of temperature driving force and consequently to larger network areas. This "penalty" can often be compensate by judicious selection of stream matches. Where there are marked differences in individual stream coefficients, matches between streams having similar values should be favoured. Then savings can be made through using higher temperature driving forces on matches having the lower overall heat transfer coefficients.

Network Capital Cost

A reasonable prediction of the heat exchanger capital cost of a network can be obtained from:

$$C_{network} = N_{units}\left\{Ca + Cb\left(\frac{A_{network}}{N_{units}}\right)^{Cc}\right\}$$

where, Ca, Cb and Cc are the cost factors normally applied in the estimation of the cost of a single heat exchanger.

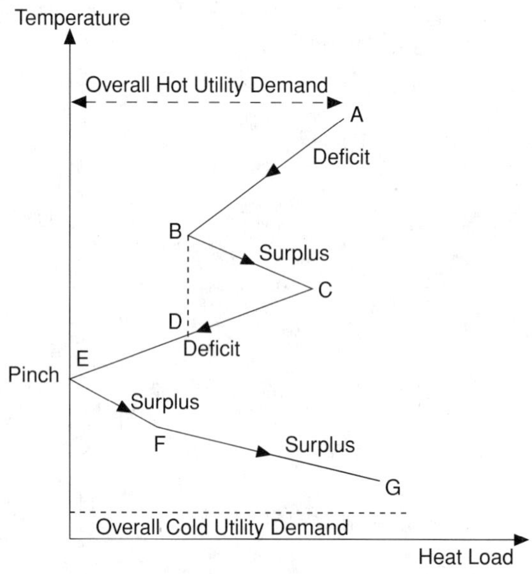

Figure 4. Heat demand and supply diagram.

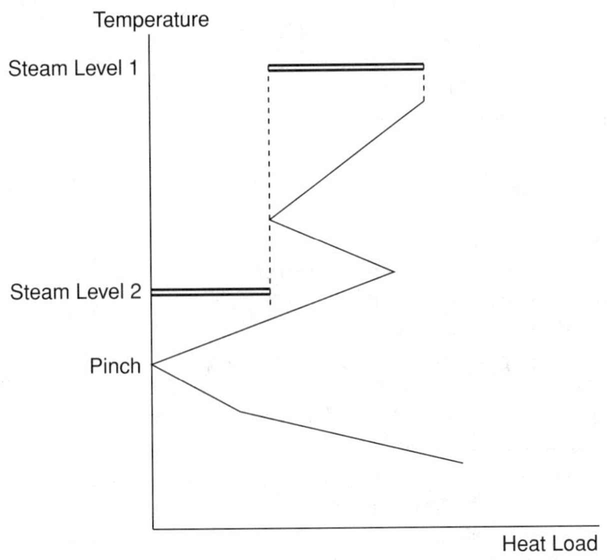

Figure 5. Selection of utility steam levels and loads.

Determining the Minimum Temperature Approach

As seen above, the minimum temperature approach has an effect on both network energy consumption and network area. The closer the temperature approach the lower the energy consumption but the higher the area required for the heat recovery (note: not necessarily total network area). This presents an optimisation problem. Umeda et al. (1978) proposed the following approach:

1. Develop composite curves
2. Set energy recovery level
3. Determine minimum area
4. Reduce total capital cost by simplifying network structure
5. Calculate total annual cost
6. Repeat steps 2 to 5 to determine optimum network.

The approach suggested by Linnhoff and Ahmad (1990) can be summarised as follows:

1. Develop composite curves
2. Set energy recovery level through specification of minimum allowable temperature approach
3. Determine minimum network area
4. Estimate minimum number of units
5. Estimate subsequent capital cost
6. Calculate total annual cost
7. Repeat steps 2 to 6 to determine optimum energy recovery level
8. Use the Pinch Design Method to subsequently design network at the identified minimum temperature approach.

The procedure suggested by Umeda et al. could be undertaken by hand but is best suited to computer application. That suggested by Linnhoff and Ahmad is a hand method. Since, it allows the designer to control the development of the network, it permits the engineer to both exercise art and to introduce practical constraints into the design. Consequently, it is possibly to be preferred. However, it should be recognised that Umeda's procedure could produce better results.

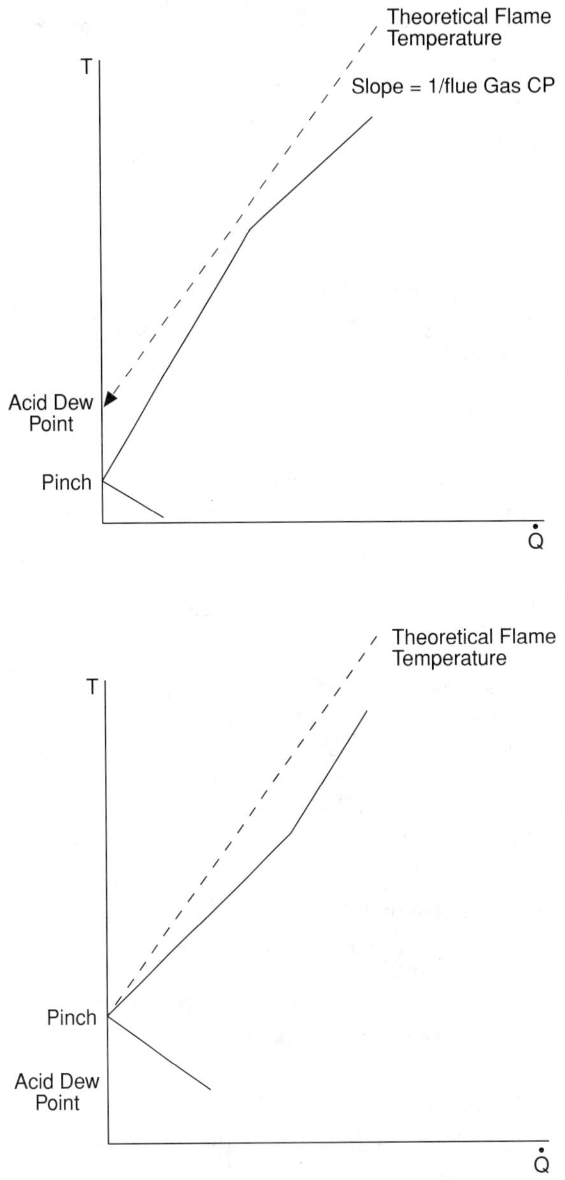

Figure 6. Minimum flue gas consumption. (a) Exhaust at acid dew point temperature. (b) Exhaust temperature constrained by process pinch.

Heat Demand and Supply Diagram

In 1982, Itoh, Shiroko and Umeda introduced the Heat Demand and Supply Diagram. The HDS diagram is prepared by plotting the horizontal distance between the Composite Curves as a function of temperature. As such it shows the variation of heat supply and demand within the process (**Figure 4**). This allows the designer

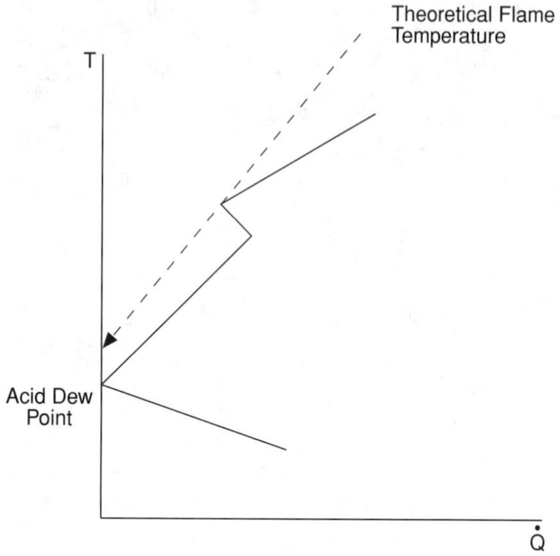

Figure 6 (continued). (c) Exhaust temperature constrained by process.

to determine the ways in which the residual heating and cooling duties can be met by available utilities. For instance, **Figure 5** shows how utility steam levels can be set; **Figure 6** shows how the quantity of flue gas necessary to operate a process can be determined; and, **Figure 7** shows how a process' heating needs can be satisfied through a mixture of steam and flue gas.

Developing Technology

Process integration is a wide ranging technology. Its application in retrofits has required the development of separate techniques (see Polley, Jegede and Panjeh Shahi, 1990). Developments that extend the technology to batch processes have been made by Kemp and Deakin (1989). Umeda, Niida and Shiroko (1979) showed how composite curves based on Carnot Factor rather than absolute temperature could be developed. Carnot factor curves have been used by Dhole and Linnhoff (1992) for the selection of refrigeration levels in cryogenic plants.

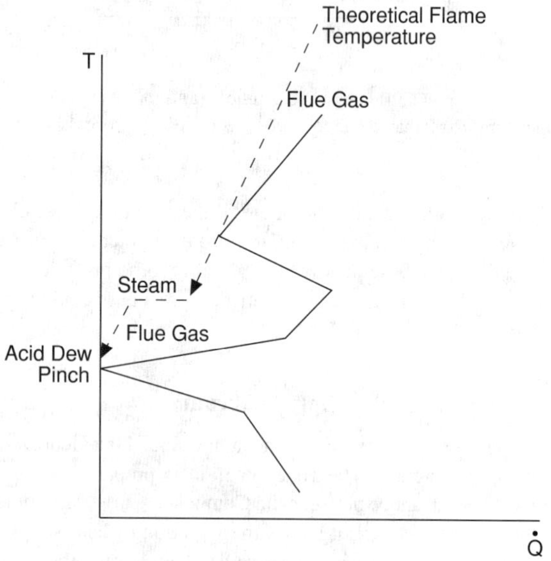

Figure 7. Selection of mixed utilities.

References

Huang, F. and Elshout, R. (1976) Optimizing the heat recovery of crude units, *Chem. Eng. Prog.*, 68–74, July 1976.

Umeda, T., Itoh, J., and Shiroko, K. (1978) Heat exchange system synthesis, *Chem. Eng. Prog.*, 70–76, July.

Umeda, T., Niida, K., and Shiroko, K. (1979) A thermodynamic approach to heat integration in distillation systems, *AIChеJ*, 25(3), 423–429.

Linnhoff, B. and Hindmarsh, E. (1983) The pinch design method for heat exchanger networks, *Chem. Eng. Sci.*, 38, 745–763.

Linnhoff, B. and Ahmad, S. (1990) Cost optimum heat exchanger networks, *Computers Chem. Engng.*, 14(7), 729–750.

Polley, G. T. and Panjeh Shahi, M. H. (1991) Interfacing heat exchanger network synthesis and detailed heat exchanger design, *Chem. Eng. Res. Dev.*, 69A, 445–457.

Hohmann, E. C. (1971) Optimum Networks for Heat Exchange, Ph.D. Thesis, University of Southern California.

Itoh, J., Shiroko, K., and Umeda, T. (1982) Extensive applications of the T-Q̇ diagram to heat integrated system synthesis, Int. Symp. on Process Systems Engineering, Kyoto, 1982, *Computers. Chem. Eng.*, 10, 59–66, 1986.

Polley, G. T., Panjeh Shahi, M. S., and Jegede, F. O. (1990) Pressure drop considerations in the retrofit of heat exchanger networks, *Chem. Eng. Res. Dev.*, 68A, 211–219.

Kemp, I. C. and Deakin, A. W. (1989) The cascade analysis for energy and process integration of batch processes, *Chem. Eng. Res. Dev.*, 67A, 495–525.

Linnhoff, B. and Dhole, V. (1992) Shaftwork targets for low temperature process design, *Chem. Eng. Sci.*, 47(8), 2081–2091.

G.T. Polley

PRODUCER GAS

***Following from:** Fuels*

Producer gas is a combustible gas manufactured by blowing a mixture of steam and air upwards through a bed of hot coke, or coal, such that the fuel is completely gasified. The gas obtained from coke consists mainly of a mixture of carbon monoxide and hydrogen with the nitrogen from the blast of air. When coal is used the gas will contain, in addition, tar and the gases liberated during the carbonisation of the coal in the fuel bed.

The process is carried out usually by charging the fuel by gravity from a hopper into a vertical, cylindrical, steel chamber, which is either lined with fire bricks, or it has an annular water-jacket from which steam required for gasification can be raised by heat transmitted from the fuel bed. The fuel bed is supported at the bottom by a grate/distributor, through which is introduced the blast, made by adding steam to the air supply such that it is saturated at a temperature of about 50°C.

A layer of ash is maintained at the bottom of the fuel bed; it serves to protect the grate and to distribute and pre-heat the blast. Immediately above the ash zone is a narrow combustion zone in which oxygen in the air reacts with carbon in the fuel to form carbon dioxide, thereby generating the heat to sustain the subsequent gasification reactions. The hot gases and steam move upwards into the reduction zone where endothermic reactions occur between carbon and carbon dioxide, and between carbon and steam to produce carbon monoxide and hydrogen. At the top of the bed the incoming fuel is dried and carbonised. Most of the sulphur in the coal appears as hydrogen sulphide in the product gas.

The hot gases leaving the producer are quenched and washed with water to remove dust and tar and, if required, purified to remove hydrogen sulphide. A typical producer gas obtained from coke contains 27% carbon monoxide, 12% hydrogen, 0.5% methane, 5% carbon dioxide and 55% nitrogen, by volume. It has a heating value of about 5,000 kJ/m^3. When coal is used as fuel the producer gas contains about 3% methane and 0.5% higher hydrocarbons.

References

Himus, G. W. (1972) The *Elements of Fuel Technology*, Leonard Hill, London.

Gas Making and Natural Gas, BP Trading Ltd.

J. Lacey

PROFILE METHOD OF SURFACE HEAT BALANCE (see Environmental heat transfer)

PROFILING, OPTICAL TECHNIQUE (see Film thickness determination by optical methods)

PROPANE

Propane (C_3H_8) is a colourless and non-corrosive gas. It is highly flammable and at low concentration it forms an explosive mixture with air. Its main uses are as a feedstock in the production of ethylene, and as a solvent, refrigerant and aerosol propellant. Propane is the major constituent of *liquefied petroleum gas* (*LPG*) and as such it is increasingly stored in liquefied form at low temperatures in specially adapted natural underground caverns.

Molecular weight: 44.096
Melting point: 85.47K
Normal boiling point: 231.18K
Normal vapour density: 2.01 kg/m^3
(@ 273.15K; 1.0135MPa)

Critical temperature: 369.9K
Critical pressure: 4.301MPa
Critical density: 220 kg/m^3

References

Buhner, K., Maurer, G. and Bender, E. (1981) *Cryogenics*, 21, 157.

Younglove, B. A. and Ely, J. F. (1987) *J. Phys. Chem. Ref. Data*, 16, 577.

Beaton, C. F. and Hewitt, G. F. (1989) *Physical Property Data for the Design Engineer*, Hemisphere Publishing Corporation, New York.

V. Vesovic

Table 1. Propane: Values of thermophysical properties of the saturated liquid and vapour

T_{sat}, K	231.2	250	270	290	310	330	350	360	365
P_{sat}, kPa	101	218	431	770	1276	1993	2974	3587	3932
ρ_l, kg/m^3	581.2	558.8	533.2	505.0	472.9	434.0	381.3	341.3	310.4
ρ_g, kg/m^3	2.41	4.93	9.42	16.63	27.95	46.01	77.96	107.6	133.4
h_l, kJ/kg	−535.2	−487.7	−438.6	−387.9	−333.7	−273.9	−204.6	−162.1	−135.0
h_g, kJ/kg	−104.0	−81.6	−58.6	−37.3	−18.8	−5.0	−1.7	−10.5	−22.8
$\Delta h_{g,l}$, kJ/kg	431.3	406.1	380.0	350.6	314.9	268.9	202.9	151.7	112.2
$c_{p,l}$, kJ/(kg K)	—	—	2.47	2.60	2.85	3.26	4.31	6.33	10.63
$c_{p,g}$, kJ/(kg K)	1.44	1.55	1.70	1.89	2.16	2.65	3.98	6.68	12.40
η_l, μNs/m^2	194.4	159.0	129.6	105.8	85.7	67.9	50.6	40.9	34.8
η_g, μNs/m^2	6.43	6.98	7.59	8.28	9.07	10.1	11.8	13.4	15.0
λ_l, mW/(K m)	128.6	116.9	105.8	95.7	86.4	77.4	69.0	66.0	65.8
λ_g, mW/(K m)	11.8	13.8	16.2	18.9	22.0	25.9	32.3	39.8	48.2
Pr_l	—	—	3.02	2.88	2.83	2.86	3.16	3.92	5.63
Pr_g	0.79	0.79	0.80	0.83	0.89	1.03	1.45	2.25	3.86
σ, mN/m	15.5	14.1	10.8	7.72	5.44	3.08	1.50	0.97	0.57
$\beta_{e,l}$, kK^{-1}	1.99	2.22	2.56	3.07	3.95	5.76	11.5	24.4	53.8

The values of thermodynamic properties have been obtained from reference Buhner et al. (1981), the values of transport properties have been obtained from reference Younglove and Ely (1987), while the values of the surface tension have been obtained from reference Beaton and Hewitt (1989).

n-PROPANOL

n-Propanol (C_3H_7OH) is a colourless liquid that is soluble in water. It occurs naturally as a by-product of vegetable fermentation. The main process for obtaining propanol is by hydroformulation of ethylene. It forms an explosive and flammable mixture with air at low concentrations and it has low toxicity. Its main uses are as a solvent and as a chemical intermediate.

Molecular weight: 60.1
Melting point: 184.15 K
Normal boiling point: 370.95 K

Critical temperature: 536.85 K
Critical pressure: 5.050 MPa
Critical density: 273 kg/m^3

V. Vesovic

Table 1. Propanol: Values of thermophysical properties of the saturated liquid and vapour

T_{sat}, K	373.2	393.2	413.2	433.2	453.2	473.2	493.2	513.2	523.2	533.1
P_{sat}, KPa	109.4	218.5	399.2	683.6	1089	1662	2426	3402	3998	4689
ρ_l, kg/m^3	732.5	711	687.5	660	628.5	592.0	548.5	492.0	452.5	390.5
ρ_g, kg/m^3	2.26	4.43	8.05	13.8	22.5	35.3	55.6	90.4	118.0	161.0
h_l, kJ/kg	0.0	65	139	222	315	433	548	691		
h_g, kJ/kg	687	710	733	766	802	860	904	955		
$\Delta h_{g,l}$, kJ/kg	687	645	594	544	486	427	356	264	209	138
$c_{p,l}$, kJ/(kg K)	3.21	3.47	3.86	4.36	5.02	5.90	6.78	7.79		
$c_{p,g}$, kJ/(kg K)	1.65	1.82	1.93	2.05	2.20	2.36	2.97	3.94		
η_l, μNs/m^2	447	337	250	188	148	119	90.6	70.0	61.4	53.9
η_g, μNs/m^2	9.61	10.3	10.9	11.5	12.2	12.9	14.2	15.7	17.0	19.3
λ_l, mW/(K m)	142.4	139.2	138.4	133.5	127.9	120.7	111.8	100.6	94.1	89.3
λ_g, mW/(K m)	20.9	23.0	26.2	28.9	31.4	34.7	38.0	43.9	47.5	53.5
Pr_l	10.1	8.40	6.97	5.14	5.81	5.82	5.50	5.42		
Pr_g	0.76	0.82	0.80	0.82	0.85	0.88	1.11	1.41		
σ, mN/m	17.6	16.15	14.42	12.7	10.77	8.85	6.35	4.04	2.6	0.96
$\beta_{e,l}$, kK^{-1}	1.33	1.57	1.91	2.43	3.15	4.14	5.82	9.19	13.6	

The values of thermophysical properties have been obtained from C. F. Beaton, G. F. Hewitt—Physical Property Data for the Design Engineer, Hemisphere Publishing Corporation, New York, 1989.

PROPYLENE

Propylene (C_3H_6) is a colourless gas. It is obtained by thermal cracking of ethylene. At low concentration it forms an explosive and flammable mixture with air, while at high concentrations it can cause asphyxiation and skin burns. It is used in the petrochemical industry as a fuel and alkylate and in the chemical industry for the production of polypropylene, isopropyl alcohol, propylene oxide and other chemicals.

Molecular weight: 42.08
Melting point: 87.9K
Normal boiling point: 225.46K
Normal vapour density: 1.91 kg/m^3
(@ 273.15K; 1.0135MPa)

Critical temperature: 365.57K
Critical pressure: 4.6646MPa
Critical density: 223.4 kg/m^3

[See **Table 1** on page 905]

V. Vesovic

PROTONS (see Atom)

PSA, PROBABILITY SAFETY ASSESSMENT (see Risk analysis techniques)

PSEUDO CRITICAL TEMPERATURE (see Supercritical heat transfer)

PSEUDO FILM BOILING (see Supercritical heat transfer)

PSEUDO HOMOGENEOUS FLOW (see Liquid-solid flow)

PSEUDOPLASTIC FLUIDS (see Fluids; Non-Newtonian fluids)

PSYCHROMETER (see Dry bulb temperature; Wet bulb temperature; Humidity measurement)

PSYCHROMETRIC CHART (see Air conditioning)

PSYCHROMETRIC RATES (see Lewis relationship)

PSYCHROMETRIC RATIO (see Humidity measurement; Wet bulb temperature)

PULSED COLUMNS (see Extraction, liquid-liquid)

Table 1. Propylene: Values of thermophysical properties of the saturated liquid and vapour

T_{sat}, K	225.5	255	270	285	300	315	330	345	360
P_{sat}, KPa	102	327	533	822	1212	1722	2375	3196	4220
ρ_l, kg/m^3	609.0	570.6	549.3	526.4	501.2	473.0	440.0	398.2	329.2
ρ_g, kg/m^3	2.36	7.06	11.26	17.24	25.62	37.39	54.41	80.85	133.8
h_l, kJ/kg	−543.0	−476.6	−440.9	−403.7	−364.6	−323.0	−278.0	−227.6	−162.0
h_g, kJ/kg	−103.9	−72.9	−58.4	−45.0	−33.4	−24.2	−18.8	−20.1	−40.2
$\Delta h_{g,l}$, kJ/kg	439.1	403.6	382.6	358.7	331.2	298.8	259.3	207.4	121.9
$c_{p,l}$, kJ/(kg K)	2.18	2.32	2.42	2.54	2.69	2.91	3.27	4.08	9.41
$c_{p,g}$, kJ/(kg K)	1.32	1.50	1.62	1.77	1.96	2.25	2.72	3.78	10.06
η_l, μNs/m^2	150.9	108.0	101.1	99.2	90.3	80.9	78.7	61.1	37.4
η_g, μNs/m^2	6.62	7.53	8.04	8.74	9.26	10.1	11.4	12.7	27.2
λ_l, mW/(K m)	119.1	104.0	98.6	93.6	90.9	88.0	83.3	76.1	56.0
λ_g, mW/(K m)	9.53	13.0	14.9	17.1	19.4	22.2	25.4	29.6	44.4
Pr_l	2.76	2.41	2.48	2.69	2.67	2.67	3.09	3.28	6.28
Pr_g	0.92	0.87	0.87	0.90	0.94	1.02	1.22	1.62	6.16
σ, mN/m	16.5	12.6	10.5	8.7	6.5	5.1	3.4	2.0	0.63
$\beta_{e,l}$, kK^{-1}	2.05	2.44	2.74	3.15	3.75	4.70	6.46	11.0	45.0

The values of thermodynamic properties have been obtained from S. Angus, B. Armstrong, K. M. deReuck—International Thermodynamic Tables of the Fluid State—7, Propylene, Pergamon Press, Oxford, 1980, while the values of transport properties and the surface tension have been obtained from C. F. Beaton, G. F. Hewitt—Physical Property Data for the Design Engineer, Hemisphere Publishing Corporation, New York, 1989.

PULSED LASERS (see Lasers)

PULSED THERMAL ANEMOMETERS (see Anemometers, thermal)

PULSED THERMOGRAPHY (see Thermal imaging)

PULVERSISED COAL COMBUSTION (see Gas-solid flows)

PULVERISED COAL FURNACES

Following from: *Furnaces*

Pulverised coal (*pulverised fuel*-pf) has been fired in rotary cement kilns (see **Kilns**) and boiler furnaces (see **Boilers**). The latter are basically boxes lined with tubes in which water is evaporated and contain a water/steam mixture. The coal is pulverised to a fine powder, usually so that 70% is less than about 75 μm in size, before being carried by part of the combustion air stream to the burners. These *coal burners* are usually mounted on one vertical wall, two opposing vertical walls or grouped one above the other at the four corners of the furnace, (see **Figure 1** and **Figure 2**). Corner burners fire tangentially into the furnace giving a single ball of flame in a central vortex. Other designs, for example using down-firing for low volatile coals and burners for low-grade high-moisture coals, have been discussed by Dryden (1975) and Lawn (1987).

In all the furnaces described so far hot dry ash falls to the "dry bottom" of the furnace chamber where it is removed. The exception to this is in cyclone fired boilers, which are usually of two types. In the *vertical cyclone furnace*, a variant of the tangentially fired design, larger coal particles are centrifuged out of the gas flow to burn on the refractory-lined walls of the lower part of the chamber, the slag running out of the "wet bottom" chamber. The other type of cyclone furnace uses individual cylindrical refractory-lined chambers in which the coal burns to give hot gases which are exhausted into the main furnace (see previous references).

Large pf furnaces have a fuel input of about 200 tons per hour of coal, (to provide a 500 MW electrical load), and chambers 35m high with a cross sectional area of about 300 m^2 are typical. The furnace designer has to ensure that there is the right amount of heat transferred from the flames to the wall tubes to evaporate the desired amount of water and to still have the correct gas temperature at the furnace exit. This must be done without any excessive local heat fluxes damaging the tubes, while at the same time complete combustion of the coal particles has to be achieved. There is the additional consideration that pollutants, e.g. oxides of nitrogen, must be kept to a minimum.

One of the problems specific to coal fired furnaces is the build up of *ash* or *slag* on the furnace walls leading to changes in temperature and emissivity. Raask (1985) has dealt at length with ash and slag deposits on furnace walls and the heat transfer properties of boiler deposits. He also describes measures to combat **Fouling** in boilers, (e.g. coal cleaning and blending, installation of soot-blowers and water-jets). Data on combustion efficiency, (typically over 98%), furnace exit gas temperatures of between about 1300 K and 1600 K and measured heat fluxes up to about 320 KWm^{-2}, (increasing by about 10% after soot-blowing), have been reported in large pf fired plant by Godridge and Read (1976).

Furnace designers make use of physical and/or mathematical models (see **Furnaces**). The latter use either heat balances or **Computational Fluid Dynamics**, CFD. In the first method the furnace is divided into regions or zones, see Hottel and Sarofim (1967) and Field et al. (1967), and a specific application to a pf fired furnace, also making use of a physical model to provide mass transfer information has been described by Cooper and Gibb (1984). The CFD method is based on the finite difference solution of momentum, enthalpy and species conservation equations. Applica-

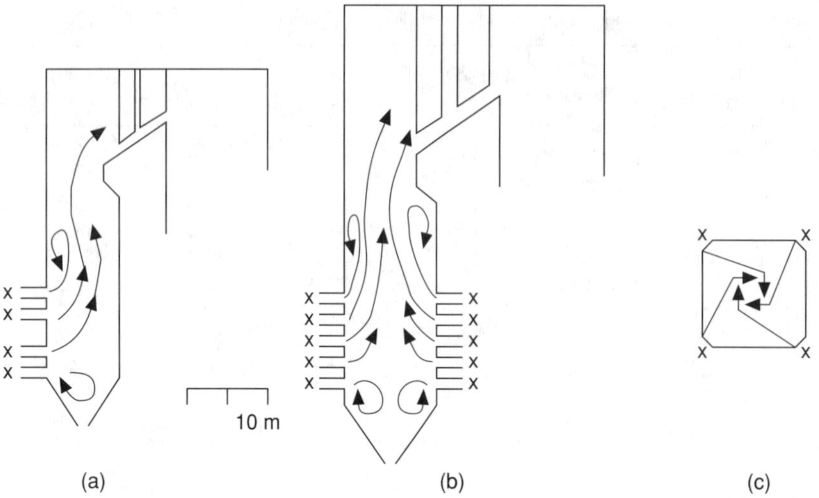

Figure 1. Water-tube boiler furnaces and gas flow patterns, (a) front-wall-fired furnace, (b) opposed-wall-fired furnace, (c) corner-fired furnace (horizontal section) x burners.

FIGURE 2. Wall-tubes and burner openings in a PF-fired water-tube boiler-furnace under construction. (Reproduced by permission of PowerGen.)

tion of the method to coal fired cyclone combustors has been reported by Boyson et al. (1986).

References

Boyson, F., Weber, R., Swithenbank, J., and Lawn, C. J. (1986) Modelling coal fired cyclone combustors, *Combustion and Flame*, 63, 73–86.

Cooper, S. and Gibb, J. (1984) High temperature heat transfer: furnace performance assessment, First U.K. Heat Transfer Conf., Leeds.

Dryden, L. G. C. Ed. (1975) *The Efficient Use of Energy*, I.P.C. Press, London.

Field, M. A., Gill, D. W., Morgan, B. B., and Hawksley, P. G. W. (1967) Combustion of Pulverised Coal, B.C.U.R.A., Inst. Energy, London.

Godridge, A. M. and Read, A. W. (1976) Combustion and heat transfer in large boiler furnaces, *Prog. Energy* and *Comb. Sci.*, 2, 83–95.

Hottel, H. C. and Sarofim, A. F. (1967) *Radiative Transfer*, McGraw-Hill, NY.

Lawn, C. J. Ed. (1987) *Principles of Combustion Engineering for Boilers*, Ch. 1, 3 and 5, Acad. Press, London.

Raask, E. (1985) *Mineral Impurities in Coal Combustion*, Hemisphere Pub. Corp. London.

A.M. Godridge

PULVERISED FUEL, PF (see Pulverised coal furnaces)

PUMPED STORAGE (see Hydro power)

PURIFICATION (see Ion Exchange; Water preparation)

PURIFICATION OF METALS (see Refining)

PUSHER CENTRIFUGE (see Centrifuges)

PUSHKINA AND SOROKIN CORRELATIONS, FOR FLOODING (see Flooding and flow reversal)

PVT RELATIONSHIPS

Following from: *Equations of state*

A PVT relationship is one of the forms of the equations of state (see **Equations of State**), which relates the pressure, molar volume V and the temperature T of physically homogeneous media in thermodynamic equilibrium.

Equation of state of a liquid for reasons of not sufficiently advanced qualitative theory is based either on experimental data immediately or on one of the variants of the law of corresponding states with a limited amount of information on a particular substance. In this case, while forming rational empirical equations the general ideas of the statistical theory of the equation of state or a character of density changes characteristic of liquids are usually considered. To describe the properties of liquids, the equations of state are used which define the PVT relationships both of liquid and gas in the form of a single analytical expression. The simplest are the so-called cubic equations of state corresponding to a cubic dependence of pressure on a specific volume (density) of a liquid and being a modification of van der Waals' equation. The intermolecular repulsive forces in them are included in the term similar to that in van der Waals' equation; the effect of the attractive forces in conveyed by the terms of different form. The most prominent in this group are:

The Redlich-Kwong equation of state is

$$p = \frac{\tilde{R}T}{\tilde{V} - b} - \frac{a}{T^{0.5}\tilde{V}(\tilde{V} + b)},$$

where $\tilde{V}$ is the molar volume, (m^3/mol) $\tilde{R}$ the universal gas constant (8.314J/mol k), T the temperature (k) and p the pressure (pa) whose two parameters a and b can be determined either by fitting the experimental data or related to the properties of a substance at the critical point:

$$a = 0.42748R^2T_c^{2.5}/p_c = \Omega_a R^2 T_c^{2.5}/p_c;$$

$$b = 0.08664RT_c/p_c = \Omega_b RT_c/p_c.$$

In the last case the values of the coefficients Ω_a and Ω_b for a particular substance can be refined from the experimental data, and, in equations generalized for a number of substances they are often represented in the form of functions of a reduced temperature $T_r = T/T_c$ and the *acentric* (Pitzer's) *factor* $\omega = (\log p_c/10p_s)T_r = 0.7$, p_c, p_s are the critical and saturation pressures.

The Soave equation of state is

$$p = \frac{\tilde{R}T}{\tilde{V} - b} - \frac{a\alpha}{\tilde{V}(\tilde{V} + b)},$$

where
$a = 0.42748RT_c^2/pc;$
$b = 0.08664RT_c/p_c;$
$\alpha = [1 + (1 - T_r^{0.5})(0.480 + 1.574\omega - 0.176\omega^2)]^2;$

and for highly polar substances $\alpha = 1 + (1 - T_r)(m + nT_r)$ with m and n being determined from the experimental data.

The Peng-Robinson equation of state is

$$p = \frac{\tilde{R}T}{\tilde{V} - b} - \frac{a\alpha}{\tilde{V}(\tilde{V} + b) + b(\tilde{V} - b)},$$

in which
$a = 0.45724R^2T_c^2/p_c;$
$b = 0.07780RT_c/p_c;$
$\alpha = [1 + (1 - T_r^{0.5})(0.37464 + 1.54226\omega - 0.26992\omega^2)]^2,$

is the one of two parameter cubic equations which most closely predicts the thermodynamic properties of a liquid. The compressibility of substances at the critical point according to this equation is $z_c = 0.307$. The error of this equation for individual substances is 2–3% except near the critical point and for highly polar substances. Cubic equations have found wide application for the description of properties of mixtures of substances when carrying out calculations in industry when a large number of iterative calculations must be performed, especially in defining the state of phase equilibrium of mixtures. For this purpose, procedures for determining the parameters of the equations of state of mixtures from the parameters of equations for pure substances have been developed. A considerable body of modifications of cubic equations are known which allow us to increase the accuracy of calculations both through various methods of determining the parameters of the equation (a and b) or by invoking more parameters.

Another trend in modifying the van der Waals equations of state starts from the assumption that the term which reflects the influence of intermolecular attractive forces is assumed to be the same as that in one of the cubic equations under study, and the term of the equation regarding for the repulsive forces is transformed. The development of this trend promises a greater accuracy in describing the properties of a substance in the liquid phase, where the intermolecular repulsive forces are the determining ones. Thus, the combination of the Carnahan-Starling equation approximating the interaction of particles with the potential of a rigid sphere and of the Redlich-Kwong equation brings about an equation of state of the form

$$p = \frac{\tilde{R}T}{\tilde{V}} \cdot \frac{1 + y + y^2 - y^3}{(1 - y)^3} - \frac{a}{T^{0.5}\tilde{V}(\tilde{V} - b)},$$

where $y = b_0\tilde{\rho}/4$, $b_0 = 2\pi\sigma^3/3$ is the second virial coefficient of rigid spheres, $\tilde{\rho}$ is the molar density and σ is the diameter of the molecules.

The more exact description of the PVT relationship can be obtained when applying multiconstant equations. The *Benedict-Webb-Rubin equation* of state is most generally used in practical calculations. Various sets of eight constants for a wide range of substances have been published repeatedly in literature. Starling, by adding additional terms, has come up with a new modification of this equation applied over the range of sufficiently low temperatures (not lower than $T_r = 0.3$) and for densities up to $3\rho_c$. The Benedict-Webb-Rubin-Starling equation of state has the form

$$p = \tilde{\rho}\tilde{R}T + \left(B_0\tilde{R}T - A_0 - \frac{C_0}{T^2} + \frac{D_0}{T^3} - \frac{E_0}{T^4}\right)\tilde{\rho}^2$$

$$+ \left(bRT - a - \frac{d}{T}\right)\tilde{\rho}^3 + \alpha\left(a + \frac{d}{T}\right)\tilde{\rho}^6$$

$$+ \frac{\tilde{c}\rho^3}{T^2}(1 + \gamma\rho^2)\exp(-\gamma\tilde{\rho}^2).$$

Its eleven constants can be either specific to a particular substance or generalized and expressed in terms of critical parameters of the substance and its acentric factor ω. In the latter case the PTV relationship is described less accurately. The rules of combining these constants in applying the equations to binary mixtures are also known.

When employing the law of corresponding states for representing the PTV relationships in a wide range of state parameters of liquid and gas, the method based on the *Pitzer and Curl relationship* has enjoyed the widest application

$$z = z^{(0)}(p_r, T_r) + \omega z^{(1)}(p_r, T_r),$$

where $z^{(0)}$ is the compressibility of "simple" liquid, $z^{(1)}$ is the correction for the deviation from the behaviour of "simple" liquid, and ω is the Pitzer acentric factor. Lee and Kesler have presented the most precise values of $z^{(0)}$ and $z^{(1)}$ in table form for a temperature range of $0.3 \le T_r \le 4$ and pressures of $0.01 \le p_r \le 10$, and have also suggested generalized equations for describing them. The error in describing the properties by this method does not exceed 2–3%, except for the near critical region ($T_r = 0.93 - 1.0$) and for highly polarized substances. In an effort to increase the reliability, modifications of the method recommended for particular classes of liquids have been developed.

The density (specific volume) of many liquids at atmospheric pressure or in a state of saturation has been measured experimentally and is given in handbooks. At the same time a number of methods of approximate determination of these properties for poorly studied liquids on the basis of a limited number of data have been developed.

A number of additive methods have been suggested for finding a molar volume of liquids. The idea behind these methods is that each chemical element or each type of chemical bonds is ascribed certain numerical values of the components whose sum allows us to calculate the molar volume. The error in calculated values for different liquids is 3–4% except for highly accociated liquids. A somewhat more accurate result (except for low boiling substances and polar nitrogen and fluorine containing compounds) for molar volume of boiling liquid $\tilde{V}_b$ can be determined by the *Tyn and Calus method*, if the value of the critical volume $\tilde{V}_c$ of the substance is known,

$$\tilde{V}_b = 0.285\ \tilde{V}_c^{1.048},$$

where $\tilde{V}_b$ and $\tilde{V}_c$ are expressed in cm^3/mole.

Some other methods for describing the temperature dependence of a specific volume of a saturated liquid have also been suggested. The most exact among them are evidently the methods which use various modifications of the *Rackett equation*

$$\ln(\tilde{V}_l/\tilde{V}_c) = (1 - T_r)^{2/7} \ln z_c,$$

where $T_r = T/T_c$ is the reduced dimensionless temperature, $z_c = p_c\tilde{V}_c/\tilde{R}T_c$ is the compressibility of the substance at the critical point. This equation makes it possible to describe a wide range of substances with an error less than 1.5%; however, much worse results are obtained when the method applied to quantum liquids and to substances whose molecules contain cyclic groups, to alcohols, nitriles, etc. Its modifications consist either in replacing the quantity z_c by the empirical constant peculiar to each substance or in using as the reference state the state at a certain temperature instead of the critical point, for which the specific volume of saturated liquid is well known. The value of z_c in this case can be calculated from the relationship

$$z_c = 0.29056 - 0.08775\omega,$$

where $\omega = (\log p_c/10\ p_s)T_r = 0.7$ is the Pitzer acentric factor.

In such a case, the accuracy of calculation of specific volumes of saturated liquid increases. Thus, for a majority of polar substances the error does not exceed 1%. In order to determine the pressure at which the saturated liquid is, we can make use of one of the generalized methods most of which are based on integration of the **Clapeyron-Clausius Equation** on condition that this or that additional assumption will be made. One of the most precise methods among them is the *Riedel-Plank-Miller method* to use which we must have an information on normal boiling temperature (expressed as $T_{b,r}$ which is the reduced value of this temperature) and the critical point. The saturation pressure in this method is calculated as

$$\ln p_{s,r} = -G[1 - T_r^2 + k(3 + T_r)(1 - T_r)^3]/T_r,$$

where
$G = 0.4835 + 0.4605h$;
$h = T_{b,r} \ln p_c/(1 - T_{b,r})$;
$k = h/G - (1 + T_{r,b})/[(3 + T_{b\cdot r})(1 - T_{b,r})^2]$

The error in calculating $p_{s,r}$ at pressures higher than 10 mm Hg for nonpolar liquids does not exceed 2–4% but is somewhat higher for polar liquid.

The density of a compressed liquid, i.e. of liquid at a pressure higher than the saturation pressure, within the temperature range from the triple point temperature up to temperatures somewhat exceeding the normal boiling temperature, changes only slightly with increase in the pressure. Its isothermal compressibility β_T in this case depends slightly on pressure. This fact is the basis for a whole family of empirical equations of state of a compressed liquid in which this or that form of dependence of isothermal compressibility or its reciprocal quantity $k_T = \beta_T^{-1}$, on pressure, is assumed as the basis. The most convenient among these is the secant bulk-modulus equation. Its linear variant, suggested by the *Tait* equation in 1888, is as follows

$$\frac{\tilde{V}_0 p}{\tilde{V}_0 - \tilde{V}} = k_{T,0} + mp,$$

where the quantities with subscript "0" refer to the initial pressure, makes it possible to describe a specific volume of a large number of liquids at pressures of several tens of MPa. The addition of terms containing p^2 and p^3, enables the field of application of these equations for instance for water, water solutions and liquid metals to be extended up to the pressures of several hundreds MPa. Extensive application has also been made of an equation which is erroneously called the Tait equation and approximates the linear secant bulk-modulus equation,

$$\frac{\tilde{V}_0 - \tilde{V}}{\tilde{V}_0} = C \log\left(\frac{B + p}{p}\right),$$

which corresponds to the implicit assumption that $\tilde{V}_0(\partial p/\partial \tilde{V})_T = k_{T,0} + m_1 p$. Also known are the variants generalizing this equation, in which the specific volume of the compressed liquid is related to its specific volume in the state of saturation ($\tilde{V}s$)

$$\frac{\tilde{V}_s - \tilde{V}}{\tilde{V}_s} = C \log\left(\frac{B + p}{B + p_s}\right),$$

the parameter as a function of C in this case being represented as a function of the Pitzer acentric factor, and B by the reduced temperature and the Pitzer factor.

An equation of the form

$$p = A(T)/\tilde{V}^2 + B(T)/\tilde{V}^8$$

is used for describing the properties of compressed liquids in a wider range of temperatures, for instance for organic liquids up to $T_r = 0.9$ and works well.

The application of the above equations is limited to the conditions under which liquid is the existing phase.

References

Reid, R., Prausnitz, J. M., and Sherwood, T. K. (1977) *The Properties of Gases and Liquids*, 3rd edition, McGraw-Hill.

Wales, S. M. (1985) *Phase Equilibria in Chemical Engineering*, Butterworth Publ.

Leading to: *Boyles Law; Van der Waals equation of state; Vapour pressure*

A.A. Alexandrov

PYRIDINE (see Photochromic dye tracing)

PYROLYSIS

Following from: *Fuels*

Heating organic materials in the absence of air/oxygen gives rise to chemical and physical changes. The oldest of chemical engineers, the *charcoal* maker, piled his *wood* in mounds covered with mud and heated ("burned") it in the relative absence of air. The great clearance of forests in north-western Europe and Britain, which reached its climax during the population expansion of the 11–12th centuries, is thought to have been accomplished by the hand of the charcoal-maker, himself. The grave effects of similar deforestation are today being felt in parts of Africa and Asia. Wood pyrolysis releases light gases like **Carbon Monoxide** and **Carbon Dioxide**, *light alcohols, aldehydes, ketones and organic acids. Tars* are the larger molecular mass volatile products of wood pyrolysis, which readily condense at ambient temperature. With a far higher carbon content than the parent wood (40–50% carbon vs. 75–90% carbon), and nearly negligible sulphur content, the solid residue ("char", "charcoal") has traditionally been the prized fuel of agrarian communities. The early steel manufacture of the 15–17th centuries in Europe, including the eventually successful attempts at casting steel guns directed personally by Henry VIII, were based on reducing iron oxides by carbon in charcoal; steel production using charcoal currently based on farmed eucalyptus wood-charcoal in modern Brasil appears to survive economically due to low wage structures.

To retain good ignition properties as a household fuel, the charcoal must retain some (ca. 10%) of its original volatile content; hence temperatures in this type of pyrolysis rarely exceed 400–450°C. As in all pyrolytic processes the primary products of the thermal breakdown of wood are reactive: the thermal cleavage of bonds gives rise to the formation of a wide array of free radicals. Smaller free radical are highly reactive, with nano- or micro-second half lives whilst larger free radicals, such as the triphenyl methyl free radical, are chemically stable in solution. Final product distributions of pyrolytic processes are therefore critically dependent on such reaction parameters as temperature, pressure, residence times of volatiles in the heated zone and degree of contact of tar vapours with heated solid surfaces. Without access to our terminology, the traditional charcoal-maker nevertheless piled his wood high, so as to extend the path of his evolving tars, to ensure maximum tar recondensation and secondary char formation. The expected product distributions in slow pyrolysis would involve 15–30% char and 35–45% tars and liquids (including an aqueous phase) with the remainder evolving as gases.

No longer considered economical or environmentally acceptable, the "destructive distillation of wood" was, until the last 60 years, a principal source of chemical feedstocks such as **Methanol** (wood alcohol), *acetone, acetic acid* and tars. The latter has historically been much used as isolation against penetration of water or humidity on structures as different as roofs and exterior walls of houses and wooden hulls of ships and boats. "Stockholm tar" is still widely marketed in the UK for similar routine household applications and for antiseptic use in horsecare. More recently, the pyrolysis of agricultural and forestry wastes and municipal solid waste has been studied as a likely process route for waste disposal and energy and chemical feedstock production. While some of these thermochemical processes are actively being researched, no industrial scale applications have as yet emerged. For a recent overview of the thermochemical processing of *biomass* see Bridgewater, 1994.

The widest current direct industrial application of pyrolysis is in *coke-oven* operation. This consists in heating, *coal,* in cellular ovens which may be 0.60–1 m in width, up to 15 m in length and several meters high. Coke oven design and reaction conditions (e.g. temperatures) may be changed to modify product distributions, according to demand and price structures. In the Europe of the 1840s *coal "carbonisation"* was used to produce gas for home heating and illumination; initially only Royal Palaces and comparable homes could take advantage of this new product of "science". Strengthening demand for *aromatic chemicals* (e.g. toluene for explosives, naphthalene for moth balls) in the early decades of the 20th century transformed the tars into the principal product. Domestic use of coal-gas in the UK ceased with the generalised distribution of natural gas from the North Sea in the early 1970s. In the latter part of the century coke for metallurgical applications, notably steel production has come to the fore. Lack of markets and current legislation on pollution have led to the use of the gas and by-products as coke-oven fuel.

When city-gas was the desired product, oven temperatures would reach a maximum of 850°C, with the resultant coke considered a desirable household fuel; reaction times ranged between 7–8.5 hours. Up to 45% gas could be expected from a good gas-coal, the combustible components of the gas consisting of *methane,* **Carbon Monoxide** and **Hydrogen**. Whilst anywhere from 22–30% tar can be released from a middle rank bituminous coal during heating, intense extraparticle secondary reactions in coke ovens usually reduced tar yields to between 5–10% by weight of the original coal. Where *metallurgical grade coke* is required, oven temperatures may be raised to between 1100–1200°C, with reaction times extending to between 10–14 hours. The hardness and fracture resistance of the resulting coke are critical properties in blast furnace operation. Greater tar yields, closer to amounts originally released from coal particles, can be obtained by lowering temperatures and residence times of volatiles in the reaction zone. In the 1930s "low temperature carbonisation" was attempted as a process route to produce higher yields of liquid products from the pyrolysis of coal: many reactor configurations ranging from coke-oven look-alikes to fluidised beds have been used. Before the start of World War II, these attempts were economic failures, mainly due to the cheaper price of petroleum derived fuel oil. The oil shortage of the war years gave a new impetus to a number of coal based industries, particularly in Germany and Japan, where low temperature carbonisation technologies were retained and expanded to include tar refining and secondary processing. Available literature on the subject of coking/carbonization is vast and good reviews are available (Lowry, 1963; Elliott, 1980).

The term pyrolysis also covers the thermochemical processing of liquid and gaseous species, usually for making smaller molecules by *cracking*. Large tonnages of **Ethylene** are produced from such widely differing feedstocks as methane, ethane, petroleum naphtha and light gas and fuel oils. These processes are normally carried out at pressures between 1–30 bar and temperatures ranging from 700 to 1200°C in externally heated long (20–30 meters) thin (1–2″ id) reactor tubes made of refractory alloys. Initial reactions in these processes involve covalent bond cleavage, releasing very reactive free radicals. Reaction schemes involving primary, secondary etc. products are thought to be very complicated and a wide spectrum of products (light gas to tar and coke) may result from the cracking of a gas as simple as pure **Ethane**. Industrially, product distributions are controlled by manipulating process variables including residence time in the heated zone, and by introducing marginally reactive diluents such as steam, or inert diluents like nitrogen. Desirable contact times can be as short as a few milliseconds and industrial scale rapid quenching devices are often elegantly designed. Many industrially important reactions are performed in such pyrolysis reactors (with length/diameter ratios which could be as low as 10–12), leading to the production of bulk chemicals like *VCM (vinyl chloride monomer*) and specialty chemicals such as *tetrafluoroethylene* (Albright et al., 1983).

Pyrolytic processes involving the release of volatiles (gases and tars) and the formation of chars constitute the initial (or an intermediate) step in many common industrial applications, e.g. pulverised coal combustion. In the pyrolysis of coal, the temperature, pressure and heating rate of the pyrolytic step would be expected to affect

- the product (gas-tar-char) distribution,
- structural features of product tars
- the reactivity of chars produced during the pyrolysis.

The latter reactivity is thought to be relevant in attempting to solve problems as wide ranging as suppression of carbon content in fly ash (in pf-combustors) and estimation of the performance of fluidised bed coal gasifiers and combustors in new generation combined cycle power generation systems.

Investigation of the pyrolytic behaviour of solids requires careful experimental design. For example due to the reactive nature of the products, a shallow and a deep fixed bed reactor would not give the same product distribution: comparison of yields obtained from bench scale experimental reactors of different configuration have shown that tar yields could be improved by almost 50% if the tars are recovered in the relative absence of extraparticle secondary reactions. The types of apparatus used for these purposes range from **Fixed** and **Fluidised Bed** reactors, where volatiles could spend relatively long times in the heated zone, to entrained flow ("drop-tube") reactors, where product volatiles cross the whole length of the heated reactor tube and to wire-mesh reactors, where a monolayer of sample is held between the folded layers of mesh stretched between a pair of electrodes, the mesh also acting as the resistance heater. In the wire-mesh reactor application, if volatiles are swept from the vicinity of the mesh into a quench zone, reactor configuration allows volatiles to clear the shallow heated reaction section (less than 1 mm) relatively rapidly (Gonenc et al., 1990). Tar samples recovered in this type of apparatus would be small (1–2 mg) but would be expected to show structural features relatively little changed from tar vapours released by the pyrolysing coal particles. Analytical techniques requiring relatively small sample sizes have been used in the structural evaluation of these materials, including FT-infrared and UV-fluorescence spectroscopies, size exclusion chromatography and various mass spectroscopic techniques (Li et al., 1994; Li et al., 1995, Herod et al., 1994).

References

Bridgewater, A. V. (Ed.) (1994) *Advances in Thermochemical Biomass Conversion (2 vols)*, Blackie, London.

Lowry, H. H. (Ed.) (1963) *Chemistry of Coal Utilisation*, Supplementary Volume I, Wiley, New York.

Elliott, M. A. (Ed.) (1980) *Chemistry of Coal Utilisation*, Supplementary Volume II, Wiley, New York.

Albright, L. F., Crynes, B. L. and Corcoran, W. H. (1983) *Pyrolysis*, Academic Press, New York.

Gonenc, Z. S., Gibbins, J. R., Katheklakis, I. E. and Kandiyoti, R. (1990) Comparison of coal pyrolysis product distributions from three captive sample techniques, *Fuel*, 69, 383–390.

Li, C-Z., Madrali, E. S., Wu, F., Xu, B., Cai, H-Y., Güell, A. J. and Kandiyoti, R. (1994) Comparison of thermal breakdown in coal pyrolysis and liquefaction, *Fuel*, 73, 851–865.

Li, C-Z., Wu, F., Xu, B. and Kandiyoti, R. (1995) Characterisation of successive time/temperature-resolved liquefaction extract fractions released from coal in a flowing-solvent reactor, *Fuel*, 74, 37–45.

Herod, A. A., Li, C-Z., Parker, J. E., John, P., Johnson, C. A. F., Smith, G. P., Humphrey, P., Chapman, J. R., and Kandiyoti, R. (1994) Characterisation of coal by matrix assisted laser desorption ionization (MALDI) mass spectrometry I: the Argonne coal samples, *Rapid Comm. Mass Spec.*, 8, 808–814.

Leading to: *Combustion; Gasification*

A.A. Herod and R. Kandiyoti

PYROMETALLURGY (see Refining; Roasting; Smelting)

PYROMETRY, RADIATION (see Temperature measurement)

Q

QUALITY

Quality, x, is the mass fraction of vapour in a liquid/vapour mixture. In thermal equilibrium, the quality of a two phase mixture is directly related to heat input and is sometimes called the thermodynamic quality. For example, if an amount Q of heat is applied to a mass of liquid M at saturation temperature, then the mass of vapour generated is $M_G = Q/\Delta h_{LG}$ where Δh_{LG} is the latent heat of vaporisation. Hence the quality of the two phase mixture created is given by

$$x = M_G/M = Q/M\Delta h_{LG}$$

If a liquid is flowing in a pipe of diameter D with mass velocity $\dot{m}$ as shown in **Figure 1** and a uniform heat flux $\dot{q}$ is applied to the walls, then measuring distance L from the point A where the liquid reaches saturation temperature T_{sat}

$$\dot{q}\,\pi DL = \dot{m}(\pi D^2/4)x\Delta h_{LG}$$

or

$$x = 4\dot{q}L/\dot{m}\Delta h_{LG}D$$

However, in most practical situations thermal equilibrium does not apply and the true quality is often different from the equilibrium quality calculated from a simple heat balance of the type described above. For example, at the lower quality end of the boiling process (A) part of the heat input may be used in superheating the liquid and the amount of vapour generated correspondingly reduced. Similarly at the higher quality end (B) the vapour may be superheated

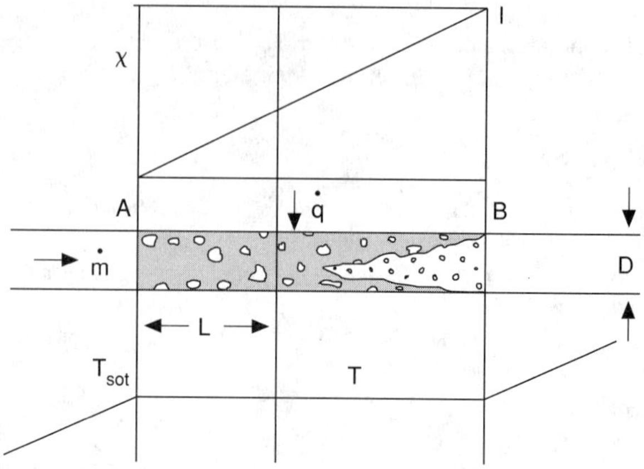

Figure 1. Quality, x, in thermal equilibrium.

while droplets of liquid remain in suspension so that the true quality is less than that calculated on the basis of thermal equilibrium (see **Boiling**). Equilibrium quality is sometimes denoted by x_e and actual quality by x_a.

G.L. Shires

QUANTA (see Electromagnetic spectrum)

QUANTITATIVE RISK ASSESSMENT (see Risk analysis techniques)

QUENCHING (see Boiling; Leidenfrost phenomena; Rewetting of hot surfaces)

QUENCHING OF ROD BUNDLES (see Rod bundles, heat transfer in)

QUICK CLOSING VALVE TECHNIQUE (see Void fraction measurement)

R

RACKETT EQUATION (see PVT relationships)

RADAR (see Electromagnetic spectrum)

RADIAL COMPRESSOR (see Compressors)

RADIAL FANS (see Aerodynamics)

RADIATION (see Nuclear reactors)

RADIATION ABSORPTION METHOD (see Film thickness measurement)

RADIATION BETWEEN PARALLEL PLATES (see Parallel plates, radiative heat transfer between)

RADIATION DIFFUSION APPROXIMATION, FOR COMBINED RADIATION AND CONDUCTION (see Coupled radiation and conduction)

RADIATION DOSIMETRY (see Thermoluminescence)

RADIATION DRYING (see Dryers)

RADIATION IN ENCLOSURES (see Parallel plates, radiative heat transfer between)

RADIATION SHIELDS (see Parallel plates, radiative heat transfer between)

RADIATION TO FURNACE TUBES (see Furnaces)

RADIATIVE DIFFUSION (see Parallel plates, radiative heat transfer between)

RADIATIVE EXCHANGE (see Parallel plates, radiative heat transfer between)

RADIATIVE HEAT FLUX (see Heat flux)

RADIATIVE HEAT TRANSFER

History

In 1871, Lord Rayleigh investigated the illumination and polarization of the sunlit sky thus originating the well defined physical laws to specify the radiation field in an atmosphere which scatters light. In 1905, Arthur Schuster studied the appearance of absorption and emission lines in stellar spectra and Karl Schwarzschild introduced the concept of radiative equilibrium in stellar atmospheres in 1906. Since then, radiative transfer has received increasing attention. In early times, the subject attracted the attention of the astrophysicists. But, with the advent of the space age, the study of radiative heat transfer has become fundamental to the aerospace research and design. More recently with the energy crises in the 1970s, interest in radiative heat transfer has increased again; chemical and mechanical engineers have adopted the concepts developed by astrophysicists and physicists and developed engineering tools able to predict the heat transfer rates inside large furnaces and boilers. The work of D.B. Spalding was pioneering in the development of such tools.

In recent years radiative heat transfer has become one of the most fundamentally important processes. From global warming to rocket nozzles, from pollution abatement in combustion chambers to quality control in most of food products, from mitigation in industrial hazards to many therapies used in modern medicine from forest fires to nuclear reactions, almost every technology requires the knowledge of radiative heat transfer.

Several books are available which are simultaneously college books and reference books: Hottel and Sarofim (1967), Brewster (1992), Siegel and Howell (1981) and Modest (1993).

The Heat Transfer Modes

Conduction and convection requires the presence of a medium for the transfer of energy. Thermal radiation is transferred by electromagnetic waves and therefore does not require a medium for its transfer. For conduction and convection, heat transfer rates are in most cases linearly proportional to temperature differences whereas the radiative heat transfer rates are proportional to the fourth or higher power of the differences in temperature

Complexities Inherent in Radiation Problems

Difficulties associated with the treatment of radiative heat transfer are two-fold. From the mathematical point of view, due to the long-range nature of thermal radiation, the principle of conservation of energy cannot be applied over an infinitesimal volume, but must be applied to the entire volume under study. This leads to an integral equation. This kind of equation is not as familiar to engineers as the partial differential equations resulting from the application of the principle of the conservation of energy for most conduction and convection cases. The second complexity inherent to the radiative heat transfer problem is physical and is related to the accurate specification of physical properties. The difficulty in specifying accurate property values arises from the fact that the properties, for example for solids, depend on many variables, such as purity of material, surface roughness and degree of polish, surface preparation which may vary from day to day, temperature, wave length of radiation. Furthermore, radiative properties are usually difficult to measure; very often the pertinent conditions have not been precisely defined.

The Nature of Thermal Radiation

The theory of radiant energy propagation can be considered from the point of view of the electromagnetic wave theory or in terms of quantum mechanics. Thermal radiative energy may be viewed as consisting of electromagnetic waves (electromagnetic wave theory) or as consisting of massless energy parcels called photons (quantum mechanics). Neither of the two theories completely describes all radiative heat transfer phenomena. The radiative properties of liquids, solids and interfaces are well described by the electromagnetic wave theory. However, the radiative properties of gases can only be explained on the basis of the quantum mechanics.

Electromagnetic Spectrum

From the point of view of wave theory, electromagnetic radiation follows the laws of governing transverse waves that oscillate in a direction perpendicular to the direction of travel. The speed of propagation for electromagnetic radiation in vacuum is the same as for light. In vacuum, the speed of light is $c_o = 2.998 \times 10^8$ m/s. The speed of light c depends on the medium and its value is given in terms of index of refraction $n = c_o/c$. For attenuating media, the index of refraction is a complex quantity of which n is only the real part (see **Refractive Index**).

The types of electromagnetic radiation can be classified according to their wavelength λ (or frequency ν where $c_o = \lambda\ \nu$). The electromagnetic wave spectrum is shown in **Figure 1**.

Thermal radiation is defined as electromagnetic waves which are emitted by a medium due exclusively to its temperature. Each wave or photon carries with it an amount of energy E, determined from quantum mechanics as

$$E = h\nu$$

where h is the Planck's constant ($h = 6.62560 \times 10^{-34}$ Js).

Basic Definitions

Intensity. The fundamental quantity used for describing *radiant energy transport* is the spectral *intensity* I_λ, defined as the radiant energy per unit area perpendicular to the direction of travel per unit solid angle, per unit wavelength interval and per unit of time. Most of the perceived mathematical complexity usually associated with radiative transfer arises from the directional (θ, φ) and spectral (λ) variations of intensity. Mathematically, the spectral intensity can be expressed as a function (θ, φ) and wavelength (λ).

$$I_\lambda = I_\lambda(x, y, z, \theta, \phi, \lambda)$$

The total intensity I can be determined by integrating with respect to wavelength.

$$I(x, y, z, \theta, \phi) = \int_0^\infty I_\lambda(x, y, z, \theta, \phi, \lambda)d\lambda$$

The net flux can be determined by integrating the intensity over all directions. The spectral net flux q_λ is based on a unit wavelength interval and total net flux q is integrated over all wavelengths.

$$q = \int_{4\pi} I \cos\theta\, d\Omega$$

$$q = \int_0^{2\pi} \int_0^{\pi} I(\theta, \phi) \cos\theta \sin\theta\, d\theta\, d\phi$$

The hemispherical fluxes, q^+ and q^- are obtained integrated over a hemisphere of 2π steradions.

$$q^+ = \int_0^{2\pi} \int_0^{\pi/2} I(\theta, \phi) \cos\theta \sin\theta\, d\theta\, d\phi$$

$$q^- = \int_0^{2\pi} \int_0^{\pi/2} I(\theta, \phi) \cos\theta \sin\theta\, d\theta\, d\phi$$

$$q = q^+ - q^-$$

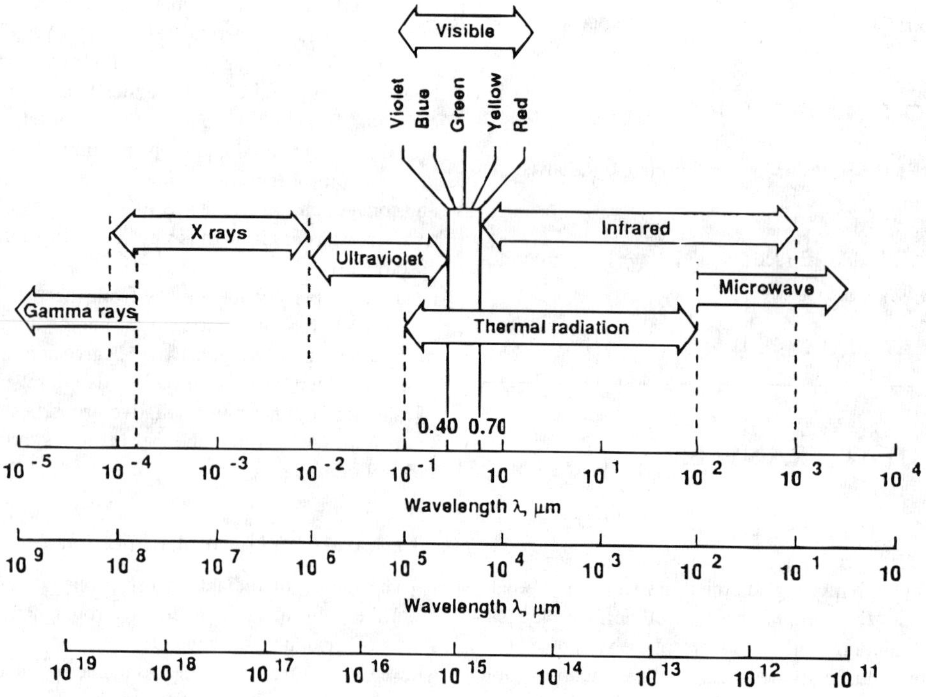

Figure 1. Electromagnetic wave spectrum.

The hemispherical flux leaving the surface q^+ is referred to as the *radiosity* and the hemispherical flux incident upon the surface q^- is referred to as the *irradiation*.

When the intensity is independent of direction $I_\lambda \neq I_\lambda(\theta, \phi)$, the intensity is called isotropic or diffuse.

Blackbody Radiation

A blackbody is defined as an ideal body that allows all the incident radiation to pass into it and absorbs internally all the incident radiation. The blackbody is a perfect absorber of incident radiation. The blackbody also emits the maximum radiant energy. Therefore, the blackbody is the best possible absorber and emitter of radiant energy at any wavelength and in any direction.

Blackbody radiation is thermal radiation that exists inside an isothermal enclosure. By definition a blackbody is a "hohlraum" or an isothermal enclosure. Blackbody radiation is isotropic. At any point inside the enclosure (see **Figure 2**) the magnitude of the intensity is equal in all the directions and is the same at every point in space. The hemispherical intensity distribution emerging from a small hole in the enclosure is semi-isotropic.

The intensity has been defined on the basis of projected area. It is useful to define a quantity that gives the energy emitted in a given direction per unit of unprojected surface area. The directional spectral emissive power for a black surface is the energy emitted by a black surface per unit time within a unit small wavelength interval centered around the wavelength λ per unit elemental surface area and into a unit elemental solid angle centered around the direction (θ, ϕ).

The hemispherical spectral emissive power of a black surface is the energy leaving a black surface per unit time per unit area and per unit wavelength interval around λ and is obtained by integrating the directional spectral emissive power over all solid angles of a hemispherical envelope placed over a black surface.

From geometrical considerations it can be found that the blackbody hemispherical emissive power is π times the intensity.

$$E_{\lambda b}(\lambda) = \pi I_{\lambda b}(\lambda)$$

The blackbody directional spectral and total emissive power follow *Lambert's cosine law:*

$$E_{\lambda b}(\lambda, \theta) = E_{\lambda b,n}(\lambda)\cos\theta$$

$$E_b(\theta) = E_{b,n}\cos\theta$$

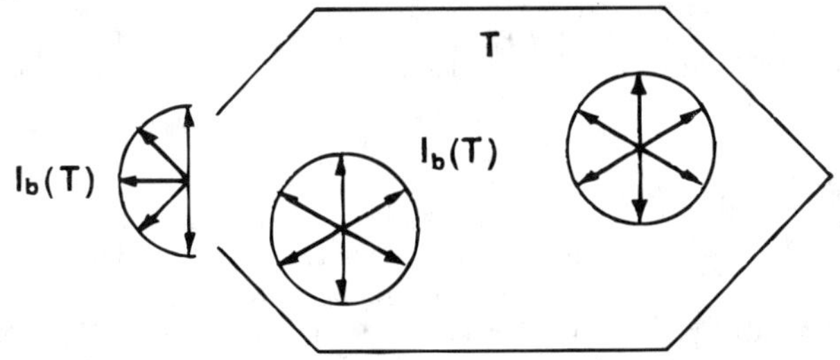

Figure 2. Blackbody (isothermal enclosure) Brewster (1992).

where $E_{\lambda b,n}$ and $E_{b,n}$ stand for the directional spectral and total emissive power normal to the black surface.

Planck has shown, based on quantum arguments, that for a blackbody the spectral distributions of hemispherical emissive power and radiant intensity in a vacuum are given as a function of absolute temperature and wavelength by

$$E_{\lambda b}(\lambda) = \pi I_{\lambda b}(\lambda) = \frac{2\pi c_1}{\lambda^5(e^{c2/\lambda T} - 1)}$$

This is known as Planck's spectral distribution of *emissive power.* The values of the constants c_1 and c_2 are equal to $c_1 = h\,c_o^2$ and $c_2 = h\,c_o/K$ where h is Planck's constant and K is the Boltzmann constant ($K = 1.3806 \times 10^{-23}$ J/K). Therefore $c_1 = 0.59544 \times 10^{-16}$ Wm^2 and $c_2 = 14.388$ μm K.

The Planck's spectral distribution of emissive power into a medium with an index of refraction $n = c_o/c$:

$$E_{\lambda b}(\lambda) = \pi I_{\lambda b}(\lambda) = \frac{2\pi c_1}{n^2\lambda^5(e^{c2/n\lambda T} - 1)}$$

Figure 3 is a graphical representation of the Planck's law for a number of blackbody temperatures.

The overall level of emission rises with rising temperature, while the wavelength of maximum emission shifts towards shorter wavelengths.

The location of the maximum spectral blackbody radiation can be calculated by differentiating the Planck function with respect to wavelength and setting the result equal to zero. This procedure gives a transcendental equation that can be solved for the wavelength temperature product.

$$(\lambda T)_{max} = c_3 \text{ (emission into vacuum)}$$

This relation is called *Wien's displacement law.*

Wien's law for a medium with an index of refraction n is

$$(\lambda T)_{max} = \frac{c_3}{n}$$

$$c_3 = 2897.8 \ \mu\text{m K}$$

The hemispherical total emissive power of a black surface into vacuum is:

$$E_b = \int_0^\infty E_{\lambda b}(\lambda)d\lambda = \int_0^\infty \pi I_{\lambda b}(\lambda)d\lambda = \sigma T^4$$

which is known as the Stefan Boltzmann law, where σ is the Stefan Boltzmann constant.

$$\sigma = 5.6696 \times 10^{-8} \text{ W/(m}^2\text{ K}^4)$$

For a medium with index of refraction n, the Stefan Boltzmann law is:

$$E_b = n^2\sigma T^4$$

The fractional function $f(n\lambda T)$ is defined as the fraction of hemi-

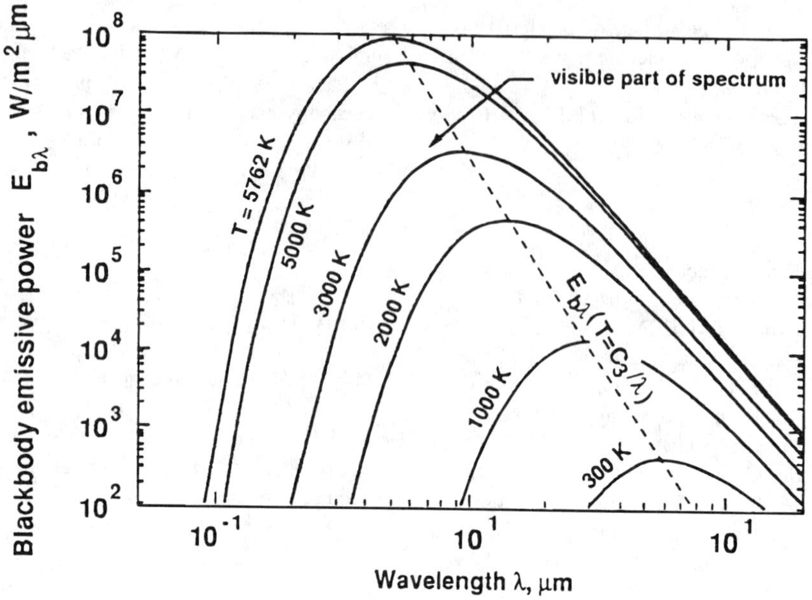

Figure 3. Blackbody emissive power spectrum.

spherical blackbody flux or intensity between zero and an arbitrary wavelength λ.

$$f(n\lambda T) = \frac{\int_0^{\lambda} E_{b\lambda} d\lambda}{\int_{0,rb}^{\infty} Eb_{\lambda}\, d\lambda}$$

This equation is a function in a single variable nλT and is therefore easily tabulated.

The fractional function can also be used to calculate the fraction of blackbody energy in spectral region between wavelengths λ_1 and λ_2. This fraction is given by the difference of f $(n\lambda_2 T) - f (n\lambda_1 T)$ where it is assumed that $\lambda_1 < \lambda_2$.

Definitions of Radiative Properties of Opaque Surfaces

Several efforts have been made to standardize the nomenclature of radiation. For example, the National Bureau of Standards (NBS) recommends the use of the ending "-ivity" for the properties of an optically smooth substance with an uncontaminated surface while the ending "-ance" is applied to the measured properties for the rough and contaminated surfaces.

Figure 4 shows the thermal radiation impinging on a slab. Some of the irradiation will be reflected away from the medium, a fraction will be absorbed inside the layer and the rest will be transmitted through the slab.

Three fundamental radiative properties are defined as:

$$\text{Reflectivity } \rho = \frac{\text{reflected part of incoming radiation}}{\text{total incoming radiation}}$$

$$\text{Absorptivity } \alpha = \frac{\text{absorbed part of incoming radiation}}{\text{total incoming radiation}}$$

$$\text{Transmissivity } \tau = \frac{\text{transmitted part of incoming radiation}}{\text{total incoming radiation}}$$

Since all radiation must be either reflected, absorbed or transmitted

$$\rho + \alpha + \tau = 1$$

If the medium is opaque, $\tau = 0$ and

$$\rho + \alpha = 1$$

The properties may vary between 0 and 1.
For a black surface $\alpha = 1$ and $\rho = \tau = 0$.

All surfaces emit thermal radiation. For given temperature, a black surface emits the maximum thermal radiation possible.

$$\text{Emissivity } \varepsilon = \frac{\text{energy emitted from a surface}}{\text{energy emitted by a black surface at the same temperature}}$$

The emissivity varies between 0 and 1. For a black surface $\varepsilon = 1$.

For all the defined properties, we distinguish between directional and hemispherical properties (average value overall directions) and between spectral and total properties (average value over the spectrum).

Kirchhoff's law

This law is concerned with the relation between the emitting and absorbing abilities of a body. The law can have various conditions imposed, depending on whether spectral, total, directional or hemispherical quantities are considered. The most general form of *Kirchhoff's law* holds for the directional spectral properties.

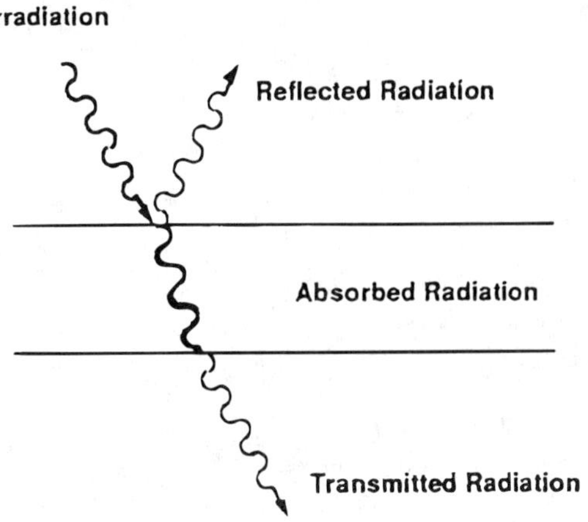

Figure 4. Absorption reflection and transmission on a slab. Modest (1993).

$$\varepsilon_\lambda(\lambda, \theta, \phi, T) = \alpha_\lambda(\lambda, \theta, \phi, T)$$

Diffuse-gray surface

Diffuse means that the directional emissivity and directional absorptivity do not depend on direction.

The term gray means that the spectral emissivity and absorptivity do not depend on wavelength.

Diffusely reflecting surfaces

For a diffuse surface the incident energy from the direction (θ, ϕ) that is reflected produces a reflected intensity that is uniform over all (θ_r, ϕ_r) directions, but the amount of energy reflected may vary as a function of incident angle.

Specularly reflecting surfaces

Mirrorlike or specular surfaces obey well-known laws of reflection. From an incident beam from a single direction, a specular reflector obeys a definite relation between incident and reflected angles. The reflected beam is at the same angle from the surface normal as the incident beam and is in the same plane as that formed by the incident beam and normal.

Prediction of Radiative Properties by Electromagnetic Wave Theory

The emissivity, absorptivity, reflectivity and transmissivity for a pure, perfectly smooth surface may be predicted by the electromagnetic wave theory. In 1861, Maxwell has presented a set of equations for the complete description of electromagnetic waves. Maxwell's equations are the basis for this prediction procedure, (see, Siegel and Howell, 1981, and Modest, 1993). The solution to Maxwell's equations reveal how radiation travels within a material and the interaction between the electric and magnetic fields.

The electromagnetic wave theory applied to the prediction of radiative properties has a number of drawbacks for practical applications. The theory neglects the effects of surface conditions on the radiative properties. Real surfaces of materials are generally coated to varying degree with contaminants or oxide layers and usually have a certain degree of roughness. However, the electromagnetic wave theory provides a important tool for intelligent interpolation and extrapolation for the experimental data. Extensive data sets have been collected by different authors (for example, Gubareff et al., 1960, Wood et al., 1964 and Stev, 1965).

Exchange of Radiant Energy Between Black Isothermal Surfaces

View factor and direct-exchange area

The *view factor* F_{12} is the fraction of diffuse radiation leaving surface A_1 in all directions which is intercepted by surface A_2 (see **Figure 5**).

$$F_{12} = \frac{1}{A_1} \int_{A_1} \int_{A_2} \frac{\cos \phi_1 \cos \phi_2}{\pi r^2} dA_1 dA_2$$

The view factor is also known as geometric configuration factor, view factor, radiation shape factor and angle factor.

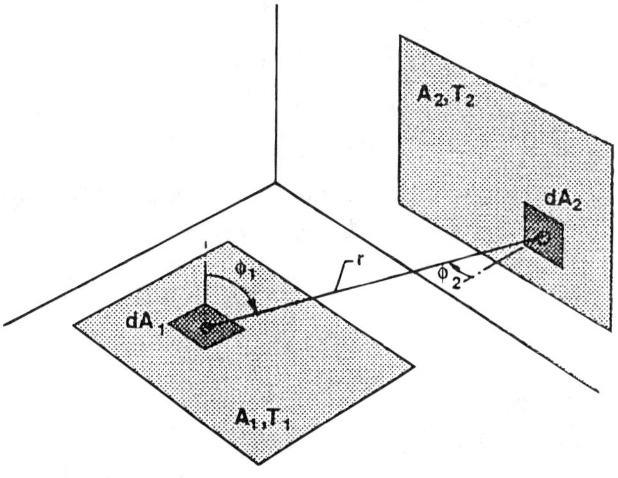

Figure 5. The geometric parameters needed for calculating the view factor integral.

Relations between view factors

Reciprocity: $A_1F_{12} = A_2F_{21}$
The product $A_1F_{12} = A_2F_{21}$ product is called the direct exchange area.
Additivity: If A_2 is broken up into n pieces

$$A_2 = A_{21} + A_{22} + A_{23} + A_{2n}$$

$$F_{12} = \sum_{i=1}^{n} F_{12_i}$$

Enclosure: If the emitting surface A_1 and the surfaces that surround it ($A_2, A_3, \ldots A_n$) form an enclosure:

$$F_{12} + F_{12} + F_{1n} = 1$$

Methods for the calculation of view factors

View factors may be determined using one of the following methods.

1. Analytical or numerical integration
2. Statistical determination using the Monte Carlo Method
3. Simplified models—For simple geometries one of the following methods may be applied
   3.1 View factor algebra using the reciprocity and additivity properties of the view factors
   3.2 Cross-strings method—A method for evaluation of view factors for two-dimensional geometries

Extensive effort has been devoted in the past to evaluation of view factors for different geometries. The result of this evaluation is generally given in tables and charts. Extensive tabulation is given in Sparrow and Cess (1978), Siegel and Howell (1981), Howell (1982) and Modest (1993).

Total energy transfer among black surfaces

The net heat exchange between two surfaces i and j is

$$Q_{ij} = A_iF_{ij}(E_{bi} - E_{bj})$$

The net heat exchange between a surface i and the rest of the surfaces within an enclosure comprising n black surfaces is

$$Q_i = \sum_{j=i}^{n} A_iF_{ij}(E_{bi} - E_{bj})$$

Networks

An analogy may be drawn between the radiative heat transfer and the flow of electric current through a resistor. Thus voltage is the analog of blackbody emissive power (E_b), current of net heat exchange (Q) and electrical resistance of reciprocal exchange area

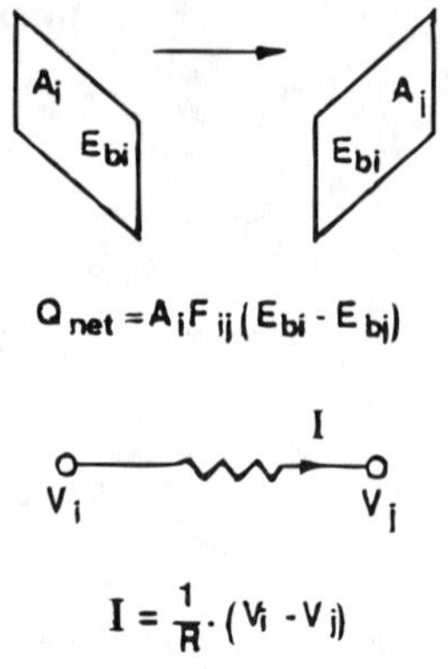

Figure 6. Electrical analogy for direct radiative transfer between black surfaces.

($R_{ij} = 1/A_i F_{ij}$). **Figure 6** shows the analogy. The electrical network method was first introduced by Oppenheim (1956). This method was very suitable for analog computers. With the advent of digital computer this method is rarely used.

Radiative Exchange between Gray-Diffusive Surfaces in an Enclosure

For gray surfaces, diffuse emitters, absorbers and reflectors:

$$\varepsilon = \varepsilon_\lambda = \alpha_\lambda = \alpha = 1 = -\rho$$

The heat flux that departs from an area element dA_1 represents the *surface radiosity* and is labeled J_1:

$$J_1 = \rho_1G_1 + \varepsilon_1E_{b,1}$$

where G_1 is the irradiation that arrives at the dA_1 location.

The difference between the heat flux that leaves, J_1, and the heat flux that arrives, G_1, is the net heat flux that leaves dA_1.

$$Q_{dA_1} = J_1 - G_1$$

$$Q_{dA_1} = \frac{\varepsilon_1}{1 - \varepsilon_1}(E_{b,1} - J_1)$$

The net heat current that leaves the entire A_1 area is

$$Q_1 = \frac{\varepsilon_1 A_1}{1 - \varepsilon_1}(E_{b,1} - J_1)$$

The previous analysis may be extended to an enclosure of n diffuse-gray surfaces.

The irradiation that impinges on A_1 is:

$$A_1G_1 = J_1A_1F_{11} + J_2A_2F_{21} + \cdots + J_nA_nF_{n1}$$

$$= \sum_{j=1}^{n} J_jA_jF_{j1}$$

$$= \sum_{j=1}^{n} J_j A_1 F_{1j}$$

From the point of view of surface A_1, the heat transfer problem is still the calculation of the net heat transfer rate Q_1, which can be evaluated by

$$Q_1 = \frac{\varepsilon_1 A_1}{1 - \varepsilon_1}(E_{b_1} - J_1)$$

provided that the radiosily J_1 is known. The problem reduces than to the calculation of J_1, which depends on the geometry and radiative properties of all the surfaces of the enclosure. The needed relation between J_1 and the rest of the enclosure parameters follows from substituting into the J_1 definition, the general expression that was just derived for the irradiation G_1:

$$J_1 = (1 - \alpha_1) \sum_{j=1}^{n} J_j F_{1j} + \varepsilon_1 \sigma T_1^4$$

A system of *n* equations for the *n* radiosities J_j is obtained:

$$J_i = (1 - \alpha_i) \sum_{j=1}^{n} J_j F_{ij} + \varepsilon_i \sigma T_i^4 \ (i = 1, 2 \cdots n)$$

If the geometry and all the surface properties are specified, then the system gives the values of the n radiosities. An equation of the type of

$$Q_i = \frac{\varepsilon_i A_i}{1 - \varepsilon_i}(\sigma T_i^4 - J_i) \ (i = 1, 2, \ldots, n)$$

holds for every single surface that participates in the enclosure, delivering the value of the heat transfer rate, Q_i.

When all the surface temperatures T_i are specified, the system of equations for J_i can be solved first to determine all the J_i's, leaving the Q_i's to be calculated afterwards. When only some of the T_i's and Q_i's are specified, the two systems of equations must be solved simultaneously.

Fundamentals of Radiation in Absorbing, Emitting and Scattering Media

The study of energy transfer through media that can absorb, emit and scatter radiation began in the last century with the early studies of the absorption and scattering of earth's radiation. Langley (1883) reported the form of the incident solar spectrum. In the first half of the 20th century, the radiation from stars started to receive attention (Eddington, 1926 and Chandrasekhar, 1939). In the past two decades, interest in the study of radiation in absorbing, emitting and scattering media has increased. For the advancement of these studies, the work of Hottel has been crucial (Hottel and Sarofim, 1967 and Hottel, 1927). This interest is associated with the study of phenomena such as rocket propulsion, plasmas generators for nuclear fusion, industrial furnaces and boilers.

Radiation Characteristics of Gases

All gas atoms or molecules carry a certain amount of energy consisting of kinetic energy and energy internal to each molecule. The internal molecular energy consists of electronic, vibrational and rotational energy states. A molecule may emit a photon and lower the level of one of its internal energy states and on the other hand a photon may be absorbed by a molecule increasing the level of one of the internal energy states.

The photon is the basic unit of radiative energy. Radiative emission consists of the release of photons of energy, and absorption is the capture of photons by a particle.

According to quantum mechanics only a finite number of discrete energy levels is possible. The magnitude of the energy transition is related to the frequency of the emitted or absorbed radiation. The energy of a photon is h ν where h is Planck's constant and ν is the frequency of the photon energy.

Three types of transitions may occur: bound-bound, bound-free and free-free. When a photon is absorbed or emitted by an atom or molecule and there is no ionization or recombination of ions and electrons the process is called bound-bound absorption or emission. In this case, the emitted or absorbed energy is associated with the transition from a specific energy level to another and a fixed frequency is associated with this transition. The emitted or absorbed radiations will be in a form of discrete spectral lines.

If absorption of a photon results in ionization and the release of an electron, the transition is called bound-free. A free electron may be combined with an ion producing a photon (free-bound transition). Free-free transitions involving free electrons are not limited to discrete wavelengths since an electron may have an arbitrary kinetic energy. In these cases a continuous spectrum is produced. The bound-free and free-free transitions are therefore continuum absorption processes.

Spectral line broadening

In the bound-bound transitions, the absorption (and emission) of photons may only occur for very definite frequencies, causing the appearance of lines in the transmission spectrum. This process is called line absorption. However, some other effects may cause the line to broadened and consequently have a finite wave number span around the transition value. Some of the important line-broadening mechanisms are natural broadening, *Doppler broadening*, collision broadening and *Stark broadening*. The shape of the spectral line corresponds to the variation of the absorption coefficient with wave number.

Natural line broadening results from the uncertainty in the exact levels of the transition energy states. Natural line broadening produces a line shape called Lorentz profile. Natural line broadening is usually neglected for engineering applications.

Doppler broadening results from the fact that the atoms and molecules of the gas are not stationary, but have a distribution of velocities associated with the thermal energy. Doppler broadening is important for high temperatures.

Collision broadening results from the collisions of the atoms and molecules which perturb the energy states of the atoms and molecules. For noncharged particles, the spectral lines have a Lorentz profile. The collision broadening is important for high pressures and low temperatures or high pressures and densities.

Stark broadening is important in the presence of strong electric fields, which disturb greatly the energy levels of the radiating gas and particles.

Band structure

The gases that are commonly encountered in engineering calculations are diatomic or polyatomic, and therefore possess vibrational and rotational energy states that are absent in monotomic gases. The transitions between vibrational and rotational states usually provide the main contribution to the absorption coefficient in the significant thermal radiation spectral regions at moderate temperatures.

For engineering applications, the non-participating medium model is appropriate for a gas with monotomic or symmetrical diatomic molecule (examples N_2, O_2, H_2, NO).

Gases which the molecules exhibit asymmetry (e.g. H_2O, CO_2, NH_3, O_3, CO, SO_2, NO) absorb and emit energy and are generally designated as participating media.

The scattering of energy

Scattering is the encounter between a photon and one or more other particles during which the photon does not loose its entire energy. It may undergo a change in direction and a partial loss or gain of energy.

A photon having undergone such an encounter is said to have been scattered. The scattering is elastic if the energy of the photon is unchanged; inelastic scattering occurs when the energy is changed. In isotropic scattering each direction is equally likely and in anisotropic scattering, there is a distribution of scattering directions.

Luminescence

Luminescence covers a broad range of mechanisms that result in radiant energy emission by the transition of electrons from an excited state to a lower energy state, where the original excitation took place by means other than thermal agitation (as for example, visible light, ultra violet radiation and electron bombardment).

Radiation Characteristics of Particles

When incident radiation strikes a particle, radiation may be transmitted, reflected or absorbed. In addition, interaction with a particle may change the direction in which a photon travels due to diffraction (the path of a photon is altered without colliding with the particle), reflection (the photon changes the direction by reflection from the particle), refraction (the photon penetrates into the particle). All the three phenomena together are known as scattering.

The Attenuation of Energy by Absorption and Scattering

As radiation passes through a layer of participating medium, its intensity is reduced by absorption and scattering (see **Figure 7**).

The change in intensity has been found experimentally to depend on the magnitude of the local intensity, with a coefficient of proportionality K_λ

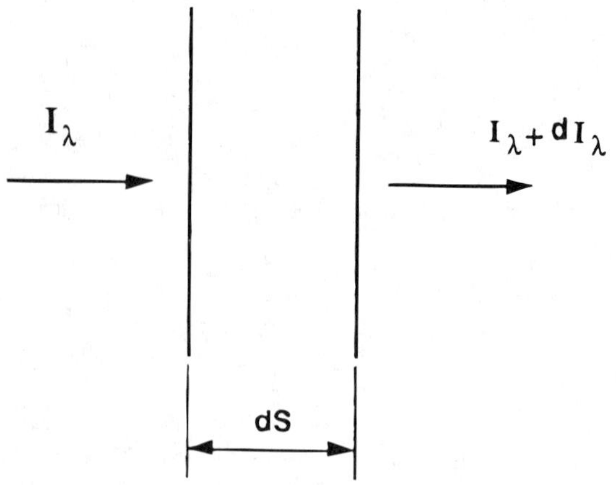

Figure 7. Intensity incident normally on absorbing and scattering layer of thickness dS.

$$dI_\lambda = -K_\lambda[S]I_\lambda dS$$

Integrating over a path length S gives:

$$I_\lambda(S) = I_\lambda(O)\exp[-\int_0^S K_\lambda(S^*)dS^*]$$

where I_λ (O) is the intensity entering the layer and S* is a dummy variable of integration.

This equation is known as Bouguer's law or Lambert's law, or the *Bouguer-Lambert law* or *Beer's law.*

In this law, K_λ is the extinction coefficient of the material in the layer. It is a physical property of the material and has units of reciprocal length. It is a function of temperature, pressure, composition of the material and wavelength of the incident radiation. The extinction coefficient is composed of two parts, an absorption coefficient α_λ (λ, T, P) and a scattering coefficient $\sigma_{s\lambda}$ (λ, T, P).

The exponential factor in Bouguer law can be written as:

$$\kappa_\lambda(S) \equiv \int_0^S K_\lambda(S^*)dS^*$$

κ_λ (S) is the optical thickness or opacity of the layer of thickness S. The optical thickness is a measure of the ability of a given path length of gas to attenuate radiation of a given wavelength. If $\kappa_\lambda >> 1$, the medium is optically thick, and the mean penetration distance is quite small compared to the characteristic dimension of the medium. If $\kappa_\lambda >> 1$, the medium is optically thin.

Absorption coefficient

If scattering can be neglected then $K_\lambda = \alpha_\lambda$ and the Bouguer's law becomes:

$$I_\lambda(S) = I_\lambda(0)\exp[-\int_0^S \alpha_\lambda(S^*)dS^*]$$

The absorption coefficient α_λ (λ, T, P) usually varies strongly with wavelength and often varies with temperature and pressure.

The scattering coefficient

The scattering coefficient $\sigma_{S\lambda}$ may be regarded as the reciprocal of the mean free path that the radiation will traverse before being scattered. The scattering coefficient can be determined from the scattering cross section s, which may be defined as the apparent area that an object presents to an incident beam. The ratio of s to the actual geometric projected area of the particle normal to the incident beams is called the scattering efficiency factor.

The Radiative Heat Transfer Equation

Bouguer's law accounts for attenuation by absorption and scattering. The radiative heat transfer equation (RTE) is an extension to the radiation intensity by emission and incoming scattering along the path:

$$\frac{dI_\lambda}{dS} = \underset{(1)}{-\alpha I_\lambda(S)} + \underset{(2)}{\alpha_\lambda I_{\lambda b}(S)}$$

$$\underset{(3)}{-\sigma_{S\lambda} I_\lambda(S)} + \underset{(4)}{\frac{\sigma_{S\lambda}}{4\pi}\int_{wi=4\pi} I_\lambda(S, \Omega_i)\,\Phi(\lambda, \Omega, \Omega_i)d\Omega_i}$$

The meaning of each term in the RTE equation is:

1. Loss by absorption.
2. Gain by emission.
3. Loss by scattering.
4. Gain by scattering into S direction.

$\Phi(\lambda, \Omega, \Omega_i)$ is the phase function and represents the probability that radiation of frequency ν propagating in the direction Ω_i and confined within the solid angle $d\Omega_i$ is scattered through the angle (Ω_i, Ω) into the solid angle $d\Omega$ and the frequency interval $d\nu$ (see **Figure 8**).

Two major difficulties may easily be identified in the analysis of the RTE. First, the RTE is an integro-differential equation and an exact solution may only be obtained after simplifying assumptions such as uniform radiative properties at the medium and homogeneous boundary conditions. Additionally, most engineering systems are multidimensional, the medium is inhomogeneous and radiative properties are spectral in nature therefore it can be concluded that the exact solutions for the RTE are not practical for engineering applications. The second problem deals with the evaluation of all the coefficients (scattering and absorption coefficients of the combustion products; shape of the phase function) present in the RTE which depend on wavelength, gas composition, temperature, pressure, type of particle, etc. The accuracy of the predictions of the radiative heat transfer equation should be compatible with that of the method used to estimate the radiative properties. These two main problems may be partially eliminated through the use of models capable, in one hand, of satisfying the integro-differential equations that need to be solved, and, on the other hand, to predict the radiation properties with a similar accuracy. In the next chapter, special attention will be given to the most common methods developed to solve the radiative transfer equation, followed by the description of the methods used in most engineering problems to evaluate radiation properties of combustion products.

Solution Methods for RTE

Exact models

The RTE is an integro-differential equation. Its exact solution can only be obtained for very simple cases, such as one-dimensional problems or when uniform radiative properties of the medium and homogeneous boundary conditions are assumed. Crosbie and Dougherty (1981) presented a detailed review of one-dimensional exact solution methods. Most engineering systems are multidimensional and spectral variation of the radiative properties must be accounted for, when solving the RTE, if accurate predictions are desired and therefore the exact solutions for RTE are not practical

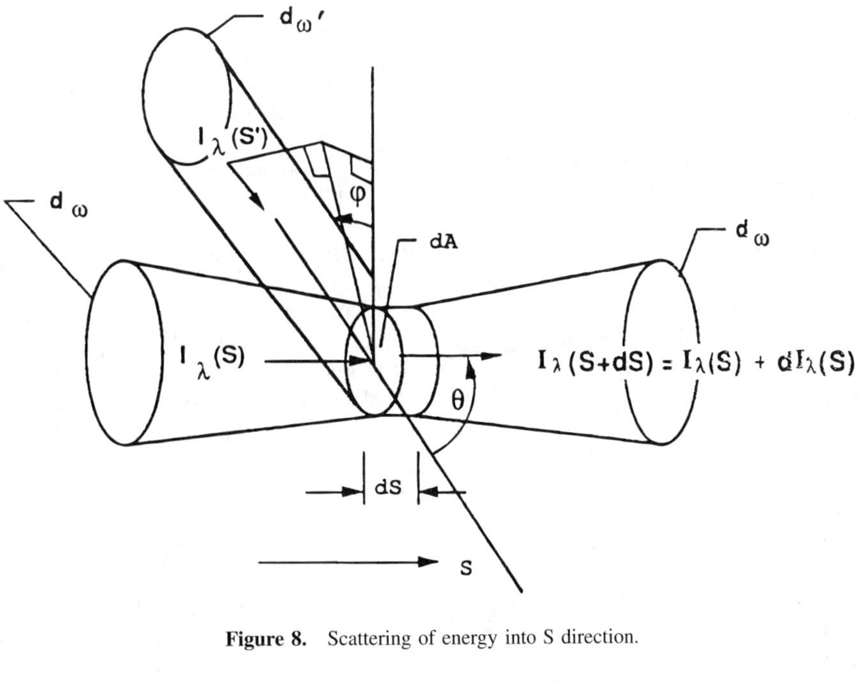

Figure 8. Scattering of energy into S direction.

for engineering applications. Consequently, it is necessary to introduce simplifying assumptions before attempting to solve the radiative heat transfer problem. During the last decades, numerous methods to solve the RTE have been developed. A survey of the literature is presented for example by Viskanta and Menguç (1987) and Howell (1988).

Zone method

The *zone method* (also known as the zoning method, the zonal method or, most frequently, referred to as Hottel's zonal method), presented by Hottel and Cohen (1958) and Hottel and Sarofim (1967), is probably the most widely used method to predict radiative heat transfer in combustion chambers. In this method, the system is divided into surface zones and gas zones. Direct exchange factors for gas-gas, gas surface, and surface-surface zone interchange have to be available or calculated. Knowing the values of these factors, it is possible to calculate the net exchange factor for any pair of zones. The resultant factors are used to form a set of energy balances, one for each zone, and, by simultaneous solution, the gas and surface space temperature distributions and the heat fluxes along the surfaces are determined.

Although simple and very attractive, the zone method has several limitations. Firstly, for complex geometries, the direct exchange factors are frequently not available and their evaluation may become impractical. Secondly, it is difficult to couple the zonal method solution procedure with the flow field and energy equations which are usually solved using finite difference (or finite element) techniques. As in most radiation models where a grid is required, the zonal method becomes computationally inefficient if the same grid is used for the finite difference equation and radiative model. However, it should be mentioned that the computer time required by the zone model is usually smaller than the time consumed by most alternatives.

Monte Carlo method

The Monte Carlo Method, Howell and Pearlmuter (1964), is a statistical method. The method consists of following the probable path of a discrete bundle of energy (photon) until its final absorption in the system (wall or gas). The path is determined at each point of emission, reflection or absorption by a random choice from the possible paths. This choice takes into account the radiative properties of the medium and the walls.

The energy emitted from the surface is related to the number of bundles originate from that surface. After leaving the surface, the photon may be absorbed by the gas (determined by a random choice based on the gas absorption coefficient). In that case a new photon is sent with a new randomly chosen direction (based on the scattering characteristics of the gas). If it is not absorbed by the gas, the photon will reach a wall. Here a random choice, taking into account the surface absorptivity, will determine whether the photon is absorbed or reflected. If the former happens the ray's history is terminated. If it is reflected the new direction is chosen by a random choice. Heat balance of the surfaces and gas are related to the number of photons absorbed versus emitted.

To have statistical significance, a large enough sample of energy bundles must be followed. The Monte Carlo method can easily be used for any complex geometry. In principle, as the number of photons followed increases to infinity, the method should converge to the exact solution. Nevertheless, the results obtained with this method are always affected by a statistical error. Therefore, Monte Carlo calculations lead to results that will always fluctuate around the exact solution.

Although the present method is very flexible, its major disadvantage is one again related with the compatibility problem with the solution method used to solve the flow equations.

Flux methods

Flux models are based on the use of some simplifying assumption for the angular variation of the radiation intensity at any point. The mathematical model of the radiation field at the point then takes the form of a set of simultaneous partial differential equations with respect to positions in terms of the known parameters in the approximate angular representation of intensity variation. The flux models have emerged from studies in Astro and Nuclear Physics. In these fields, these methods were mainly used in one-dimensional form. In the early seventies, the flux models attracted the combustion engineers interest and they were extended to two and three-dimensional situations.

Schuster-Hamaker type of models

Models of this type use the simplest and least accurate representation of the intensity variation: plane radiation is assumed in each dimension. For each dimension, two differential equations are produced by carrying out radiative energy balances for forward and backward directions. The four-flux model of Gosman and Lockwood (1973) and the six-flux models of Patankar and Spalding (1974) are of this kind.

Schuster-Schwarzchild type of models

The basis of these models is to subdivide the total solid angle surrounding a point into smaller solid angles in each of which intensity is assumed to be uniform. Discontinuous changes in intensity occur in passing from one smaller solid angle to any adjacent smaller solid angle. Integration of the equation of radiant energy transfer for each smaller solid angle, in turn, produces a group of partial differential equations in the unknown intensities. Four-flux models of Schuster-Swarzchild type for axisymmetrical radiation fields were proposed by Lowes et al. (1973), Richter and Quack (1974) and Siddal and Selçuk (1969). A six-flux model of the same type for three-dimensional radiation fields was proposed by Siddal and Selçuk (1979).

Discrete ordinates approximations

The discrete-ordinates model uses the assumption that the intensity varies in a continuous but unspecified manner with angular direction at any point. The total solid angle surrounding the point is then subdivided into smaller solid angles in each of which a direction for the intensity is specified. Application of equation of radiant energy transfer into each direction produces differential equations in terms of the unknown intensities in the specified directions (see Chandrasekhar, 1960 and Truelove, 1976). Viskanta and Menguç (1987) have presented a comprehensive review of recent work using the Discrete Ordinates method. A six-term Discrete Ordinates model for three-dimensional radiation field was derived by Selçuk and Siddal (1980), and Fiveland (1986) has presented a Discrete Ordinates method for prediction radiative heat transfer in axisymmetric enclosures. A six-term model using a combination of the Schuster- Schwarzschild and Discrete Ordinate method was derived by De Marco and Lockwood (1975).

Moment and spherical harmonic models

In these models the intensity is assumed to vary in a specified continuous manner with angular direction.

In the Moment model, the intensity in any direction is expressed as polynomial series in the direction cosines of the same direction with respect to the coordinates axes.

Another way of avoiding the solution of the integro-differential RTE is through the use of expansions of the local intensity in terms of spherical harmonics, with truncation to N terms in the series, and substitution into the moments of the differential form of the equation of transfer (see Viskanta, 1966). This approach leads to the P-N approximations, Jeans (1917), where N is the order of the approximations. As N approaches infinity the solutions obtained became exact. Usually, the odd orders are employed, especially P1 and P3. Going to P5, additional increase in accuracy is obtained while the complexity of the calculations becomes more cumbersome.

The flux methods previously presented have several advantages. The extension to these methods to include anisotropic scattering is straight forward, which is a very useful tool in modelling combustion systems where the presence of particles is an important factor.

Additionally, these models have also been used widely because the differential form of the resulting radiative transfer equations makes them compatible with the algorithm used for the flow and energy equations simulation.

In order to avoid the disadvantages of the different methods and additionally use the desirable characteristics of the different models, several hybrid radiative transfer models have been developed during the last years. A well-known hybrid method is the Discrete Transfer method that has its origin in the flux model but also exhibits features of the Hottel zone and Monte Carlo techniques. Presented by Lockwood and Shah (1981), it was specially developed for predicting radiative heat transfer in combustion chambers.

Prediction Methods for the Radiative Properties of Combustion Products

Combustion products can be divided into two main categories: gases, such as water vapour, carbon dioxide, carbon monoxide, nitrous oxide, and particles like soot, fuel droplets, pulverised coal, fly-ash and char.

Radiative properties of gases

Gases found in combustion processes do not scatter radiation significantly but some of them are strong absorbers and emitters (namely water vapour and carbon dioxide, among others). Therefore the variation of their radiative properties over the electromagnetic spectrum must be accounted for.

Spectral calculations for engineering applications are performed by dividing the wavelength spectrum into several bands. In each band the radiative characteristics (absorption and emission) are usually considered to be either uniform or to change following a pre-defined functional form. By narrowing the width of the bands, better accuracy is achieved. The most accurate approach for gas radiative transfer calculations is the line-by-line approach which consists in considering, at a given wavenumber, the contribution of each particular line (see Hartmann et al., 1988, Rosenmann et al., 1988).

The following models are usually employed in engineering applications:

- Narrow-band models (Goody, 1964; Tien, 1968) (constructed from spectral absorption and emissions lines by imposing a line shape and an arrangement of lines). There are basically two-line arrangements extensively used, referred to as the Elsasser model (lines are of uniform intensity and equally spaced) and the Goody model (intensity distribution and location of lines are randomly chosen);
- Wide band models (Hsich and Grief, 1972; Edwards and Balakrishnan, 1973; Felske and Tien, 1974), where the profile of each band is approximated to simple shapes (triangular, box or an exponentially decaying function can be used). The most commonly used model is the exponential wide-band model (Edwards and Menard, 1964).

Radiative properties of particles

Analysis of radiation heat transfer in systems where combustion takes place usually requires accounting for the effects of particulates, such as soot, fuel droplets, char, fly-ash and pulverised coal, whenever they are present in these systems. It is therefore necessary to know the radiative properties of polydispersions. These properties depend mainly on the particle size and size distribution, the value of the complex refractive index and its spectral dependency, the number density for each type of particle and the shape of the particles. Size and shape of particles are the two characteristics that commonly decide the method used to predict radiative properties.

None of the particles present in the combustion products are either homogeneous or spherical. Nevertheless, in most of the situations particles can be approximated as spheres and in that case, depending on its size, specific theories can be applied.

The most extensively used model to predict the radiative properties of spherical particles is the Mie theory. Although it is widely known by this name after Gustav Mie, who presented the exact solution of Maxwell's equation for the scattering of an incident planewave on a sphere, this same solution was also obtained by Lorenz and Debye. Following this theory, the spectral absorption and scattering coefficients (which are the most important quantities needed for radiation heat transfer analysis) can be evaluated.

The Mie theory for spheres has been treated quite extensively by several authors (see Van de Hulst, 1957; Kerker, 1969; Bohren and Huffman, 1983) and some extension to other simple shapes like coated spheres and cylinders, elliptic cylinders and spheroids have been presented (see Viskanta and Menguç, 1987 for extensive review).

When it is possible to assume that particles are spheres, although the Mie theory delivers the exact solution, there are several cases where its use can be avoided. It should be mentioned that, although the calculations for the efficiency factors of spheres are not excessively time-consuming, because of the variation of the size of the particles in space and time inside the chambers, the codes might become unpractical.

It is desirable to have simple approximations for the efficiency factors. If the particles are small compared to the wavelength (size parameter $x << 1$) the Rayleigh limit of the Mie theory expressions are obtained (Van de Hulst, 1957).

It can be noticed that, due to the fact that the value of x in the Rayleigh approximation is much smaller than unity, extinction is dominated by absorption.

Unfortunately, the combustion particles that may be assumed as spheres (like fuel droplets, coal particles or char) present typical size parameters that do not fall into the Rayleigh limit. On the other hand, soot primary particles present size parameters usually much smaller than unity. Nevertheless, the use of Rayleigh theory can be questionable because only in the beginning of soot formation do soot particles appear as individual spherelike units. At a relatively early stage of formation, soot particles tend to agglomerate forming branched chain-like structures.

For the cases where particles have a size parameter much larger than unity the scattering is mainly a reflection process and hence can be calculated from relatively simple geometric reflection relations (Bohren and Huffman, 1983).

Phase functions

In most of the combustion chambers where the particles exist, scattering of radiation by particles must be properly accounted for. The out-scattering term present in the RTE, equation, can be easily calculated if the scattering coefficient is known. Nevertheless, for the evaluation of the in-scattering term, the phase function needs to be known.

The phase function, along with other radiative properties, can be obtained either through the exact solution of Maxwell's equations (for simple shapes like spheres or cylinders) or from some approximation (expansion in a series of Legendre polynomials, delta-Eddington approximation, Dirac-Delta function).

Participating Media—An Engineering Approach

It will be desirable to extend the procedures developed for the radiative interchange between surfaces without a participating media to the case of systems with participating media. With this objective, analogous concepts, terminology and procedures were developed for participating media. Extensive description of these procedures were presented by Gray and Müller (1974).

The attenuation of a parallel beam of radiation passing through a uniformly absorbing medium of thickness L is

$$I = I_0 e^{-KL}$$

where I_o is the intensity of radiation incident on the medium, I is the intensity of the transmitted radiation and K is the extinction coefficient. If radiation is not parallel an appropriate average length may be chosen—Lm—mean beam length. The mean beam length is a function of the geometry and attenuation coefficient.

The fraction of radiation transmitted by the medium (transmissivity)

$$\delta = e^{-KLm}$$

The transmissivity of the layer is defined as:

$$\alpha = 1 - e^{-KLm}$$

And for a gray medium

$$\varepsilon = \alpha = 1 - e^{-KLm}$$

In general, the beam mean length for any shape of absorbing medium can be expressed as

$$Lm = fD$$

where D is a characterizing dimension of the enclosure.

f is a function of fD, but in practice a constant value is used. Hottel and Sarofim (1967) have tabulated two values of f, one f_o evaluated at $KD = 0$ and f_m chosen to minimize the error in the value of emissivity over the practical range of values of KD.

A procedure to describe quantitatively the emission and absorption characteristics of participating gases was developed by Hottel (1954).

Carbon dioxide and water are important absorbing and emitting gases in combustion products. The overall emissivity of the combustion products due to carbon dioxide and water is obtained from

$$\varepsilon_g = \varepsilon_{CO_2} + \varepsilon_{H_2O} - \Delta\varepsilon = \varepsilon^*_{CO_2} C_{CO_2} + \varepsilon^*_{H_2O} C_{H_2O} - \Delta\varepsilon$$

where

ε^*_{CO} = emissivity of carbon dioxide at a total pressure of 1 bar and in the limit as the partial pressure of carbon dioxide approaches zero.

C_{CO} = correction for total pressure different from 1 bar and partial pressure of carbon dioxide different from zero.

ε^*_{H2C} = emissivity of water for a total pressure of 1 bar and in the limit as the partial pressure of water approaches zero.

C_{H2C} = correction for total pressure different from 1 bar and partial pressure of water different from zero.

$\Delta\varepsilon$ = correction for the overlapping of the carbon dioxide and water bands.

Hottel (1954) has presented these data in graphical form.

Gray and Müller (1974) have described a calculation procedure for the prediction of the radiative properties of large and small particles for engineering applications.

Effect of an absorbing medium on the radiative heat transfer within an enclosure

The equation for the radiosity of each surface of the enclosure is:

$$J_i = \varepsilon_i E_{bi} + \rho_i \sum_{j=1}^{n} J_j F_{ij} \tau_{ij} + \rho_i \varepsilon_m E_{bm}$$

This equation includes the two following effects: 1) Radiation passing from one surface to another is attenuated by the medium and ii) the medium itself radiates energy to each surface.

ε_m is the emissivity of the medium defined by

$$\varepsilon_m = 1 - e^{-KLms}$$

where Lms is the mean beam length evaluated by considering the radiation from the medium to surface.

The transmissivity δ_{ij} is

$$\delta_{ij} = e^{-KL_{ij}}$$

L_{ij} must be determined by considering radiation from surface i which is transmitted to j. Gray and Müller (1974) has listed mean beam lengths for parallel and perpendicular plate geometries.

References

Bohren, C. F. and Huffman, D. R. (1983) *Absorption and Scattering of Light by Small Particles,* Wiley, New York.

Brewster, M. Q. (1992) *Thermal Radiative Transfer and Properties,* John Wiley and Sons.

Chandrasekar, S. (1939) *An Introduction to the Study of Stellar Structure,* University of Chicago Press.

Chandrasekhar, S. (1960) *Radiative Transfer,* Dover Publication, New York.

Crosbie, A. L. and Dougherty, R. L. (1981) Two-Dimensional Radiative Transfer in a Cylindrical Geometry with Anisotropic Scattering, *J. Quant. Spectrosc. Radiat. Transfer,* 25, 551–569.

De Marco, A. G. and Lockwood, F. C. (1975) A New Flux Model for the Calculation of Radiation in Furnaces, Italian Flame Day, *La Rivista Dei Combustibili,* 29, 5–6, 184–196.

Eddington, A. S. (1926) *The Internal Constitution of the Stars,* Cambridge University Press, England.

Edwards, D. K. and Balakrishnan, A. (1973) Thermal radiation by combustion gases, *Int. J. Heat Mass Transfer,* 16, 25–40.

Edwards, D. K. and Menard, W. A. (1964) *Appl. Opt.* 3.621 (1964).

Felske, J. D. and Tien, C. L. (1974) Shorter communication: a theoretical closed form expression for the total band absorpance of infrared-radiating gases, *Int. J. Heat Mass Transfer,* 17, 155–158.

Fiveland, W. A. (1986) A Discrete-Ordinates Method for Predicting Radiative Heat Transfer in Axissymetric Enclosures, ASME paper 82-HT-20.

Goody, R. M. (1964) *Atmosphere Radiation,* Oxford University Press, London.

Gosman, A. D. and Lockwood, F. C. (1973) Incorporate of a flux model for radiation into a finite difference procedure for furnace calculations, *14th Symposium (Int.) on Combustion,* The Combustion Institute, 661.

Gray, W. A. and Müller (1974) *Engineering Calculations in Radiative Heat Transfer,* Pergamon Press.

Gubareff, G. G., Janssen, J. E. and Torborg, R. H. (1960) *Thermal Radiation Properties Survey,* 2nd Ed., Honeywell Research Center, Minneapolis.

Hartmann, J. M., Rosenmann, L., Perrin, M. Y. and Taine, J. (1988) Accurate calculated tabulations of CO line broadening by H2O, N2, O2 and CO2 in the 200-3000-K temperature range, *Applied Optics,* 27–15.

Hottel, H. C. (1927) Heat transmission by radiation from non luminous gases, *Transactions of AIChE,* 19, 173–205.

Hottel, H. C. (1954) Radiant-heat transmission, *Heat Transmission,* Ch. 4 W. H. McAdams, McGraw-Hill, 3rd Edition, New York.

Hottel, H. C. and Cohen, E. S. (1958) Radiant heat exchange in a gas filled enclosure: allowance for non-uniformity in temperature, *AIChE J.,* 4-1, 3.

Hottel, H. C. and Sarofim, A. F. (1967) *Radiative Transfer,* MacGraw-Hill, Inc.

Howell, H. C. (1988) Thermal radiation in participating media: the past, the present and some possible future, *Trans. ASME,* 110, 1220–1229.

Howell, J. R. (1982) *A Catalog of Radiation Configuration Factors,* McGraw-Hill, New York.

Howell, J. R. and Perlmutter, M. (1964) Monte Carlo solution of thermal transfer through radiant media between gray walls, *J. of Heat Transfer,* 86, 116.

Hsieh, T. C. and Greif, R. (1972) Theoretical determination of the absorption coefficient and the total band absorptance including a specific application to carbon monoxide, *Int. J. Heat Mass Transfer,* 15, 1477–1487.

Jeans, J. H. (1917) The equations of radiative transfer of energy,. *Monthly Notes, Royal Astron. Soc.,* 78, 28.

Kerker, M. (1969) *The Scattering of Light,* Academic Press, New York.

Langley, S. P. (1883) Experimental determination of wavelengths in the invisible prismatic spectrum, *Mem. Natl. Acad. Sci.,* 2, 147–162.

Lockwood, F. C. and Shah, N. G. (1981) A new radiation method for incorporation in general combustion prediction procedures, *18th Symposium (Int.) on Combustion,* 1405–1414.

Lowes, T. M., Bartedls, H., Heap, M. P., Michelfelder, S. and Pai, B. R. (1973) Prediction of radiant heat flux distribution, *IFRF Doc. GO2/A/26.*

Modest, F. M. (1993) *Radiative Heat Transfer,* McGraw-Hill International Editions.

Oppenheim, A. K. (1956) Radiation analysis by the network method, Transactions of ASME, *J. of Heat Transfer,* 78, 725–735.

Patankar, S. V. and Spalding, D. B. (1974) Simultaneous predictions of flow pattern and radiation for three-dimensional flames, *Heat Transfer in Flames,* Afgan, N. H. and Beer, J. M., eds., Scripta Book Comp., Washington D.C., 73–94.

Richter, W. and Quack, R. (1974) A mathematical model of a low-volatile pulverised fuel flame, *Heat Transfer in Flames,* Afgan, N. H. and Beer, J. M. eds., Scripta Book Comp., Washington D.C., 95–110.

Rosenmann, L., Hartmann, J. M., Perrin, M. Y. and Taine, J. (1988) Accurate calculated tabulations of IR and raman CO_2 line broadening by CO_2; H_2O, N_2, O_2 in the 300-2400-K temperature range, *Applied Optics,* 27–18.

Selçuk, N. and Siddal, R. G. (1980) Prediction of multidimensional radiative heat transfer by a new six-flux model in industrial furnaces, *Proceedings of 2nd National Heat Science and Technology Conference,* 456–469.

Siddal, R. G. and Selçuk, N. (1976) Two-flux modelling of two-dimensional radiative transfer in axi-symmetrical furnaces, *J. Inst. Flue,* 49, 10–20.

Siddal, R. G. and Selçuk, N. (1979) Evaluation of a new six-flux model for radiative transfer in rectangular enclosures, *Trans. I. Chem. E.,* 57, 163–169.

Siegel, R. and Howell, J. K. (1981) *Thermal Radiation Heat Transfer,* Hemisphere Publishing Corporation.

Sparrow, E. M. and Cess, R. D. (1978) *Radiation Heat Transfer,* Hemisphere, New York.

Stev, D. I. (1965) *Thermal Radiation: Metals, Semiconductor, Ceramics, Partly Transparent Bodies and Films,* Plenum Publishing Company, New York.

Tien, C. L. (1968) *Advances in Heat Transfer, Vol. 5,* T. F. Irvine, Jr. and J. P. Hartnett, eds., 254–324, Academic Press, New York.

Truelove, J. S. (1976) An evaluation of the discrete ordinates approximation of radiative transfer in an absorbing, emitting and scattering planar medium, *U.K. Atomic Energy Commission (Harwell) Rept.,* AERE R8478.

Van de Hulst, H. C. (1957) *Light Scattering by Small Particles,* Wiley, New York (also Dover, New York).

Viskanta, R. (1966) Radiative transfer and interaction of convection with radiation heat transfer, *Advances in Heat Transfer,* Vol. 3, Edited by Irvine and Hartnett, 175–252.

Viskanta, R. and Menguç, M. P. (1987) Radiation heat transfer in combustion systems, *Progress in Energy Combustion Science,* 13, 97–160.

Wood, W. D., Deem, H. W. and Lucks, C. F. (1964) *Thermal Radiative Properties,* Plenum Publishing Company, New York.

Leading to: *Albedo; Coupled radiation and conduction; Coupled radiation and convection; Coupled radiation, convection and conduction; Emissivity; Parallel plates, radiative heat transfer between; Reflectance; Reflectivity; Stefan-Boltzmann Law*

M. da Graça Carvalho

RADIATIVE HEAT TRANSFER, IN POROUS MEDIA (See Porous medium)

RADIATIVE SPECTRAL INTENSITY (see Emissivity)

RADIOACTIVE DECAY (see Atom)

RADIO FREQUENCY HEATING (see High frequency heating)

RADIO FREQUENCY, RF, DRYING (see Dryers)

RADIO WAVES (see Electromagnetic spectrum; Electromagnetic waves)

RADIUM

Radium—(L. *radius,* ray), Ra; atomic weight 226.0254; atomic number 88; melting point 700°C; boiling point 1140°C; specific gravity 5; valence 2.

Radium was discovered in 1898 by M. and Mme. Curie in the *pitchblende* or *uraninite* of North Bohemia, in which it occurs. There is about 1 g of radium in 7 tons of pitchblende. The element was isolated in 1911 by Mme. Curie and Debierne by the electrolysis of a solution of pure radium chloride, employing a mercury cathode; on distillation in an atmosphere of hydrogen this amalgam yielded the pure metal. Originally, radium was obtained from the rich pitchblende ore found at Joachimsthal, Bohemia. The *carnotite* sands of Colorado furnish some radium, but richer ores are found in Zaire and the Great Bear Lake region of Canada. Radium is present in all uranium minerals, and could be extracted, if desired, from the extensive wastes of uranium processing.

Radium is obtained commercially as the bromide or chloride; it is doubtful if any appreciable stock of the isolated element now exists. The pure metal is brilliant white when freshly prepared, but blackens on exposure to air, probably due to formation of the nitride. It exhibits luminescence, as do its salts; it decomposes in water and is somewhat more volatile than barium. It is a member of the alkaline-earth group of metals. Radium imparts a carmine red color to a flame.

Radium emits alpha, beta, and gamma rays and when mixed with beryllium produces neutrons. One gram of Ra^{226} undergoes 3.7×10^{10} disintegrations per sec. The *curie (Ci)* is defined as that amount of radioactivity which has the same disintegration rate as 1 g of Ra^{226}. Sixteen isotopes are now known; radium 226, the common isotope, has a half-life of 1620 years. One gram of radium produces about 0.0001 ml(stp) of emanation, or *radon gas,* per day. This is pumped from the radium and sealed in minute tubes, which are used in the treatment of cancer and other diseases. One gram of radium yields about 1000 cal per year. Radium is used in producing self-luminous paints, neutron sources, and in medicine. Radioisotopes, such as Co^{60}, are now used in place of radium. Some of these sources are much more powerful, and others are safer to use. Radium loses about 1% of its activity in 25 years, being transformed into elements of lower atomic weight. Lead is a final product of disintegration.

Radium is a radiological hazard. (Stored radium should be ventilated to prevent build-up of radon.) Inhalation, injection, or body exposure to radium can cause cancer and other body disorders. The maximum permissible burden in the total body for Ra^{226} is 0.2μCi (microcuries).

Handbook of Chemistry and Physics, CRC Press

RADIUS, HYDRAULIC (see Hydraulic diameter)

RADON (see Radium)

RAE (see Royal Academy of Engineering)

RAFFINATE PHASE (see Extraction, liquid-liquid)

RAINFALL (see Water)

RAMAN SPECTROSCOPY (see Spectroscopy)

RAMJET ENGINES (see Jet engine)

RANDOM PROCESSES (see Ergodicity)

RANKINE CYCLE

Following from: *Thermodynamics*

Basic Cycle

The Rankine cycle is the fundamental operating cycle of all power plants where an operating fluid is continuously evaporated and condensed. The selection of operating fluid depends mainly on the available temperature range. **Figure 1** shows the idealized Rankine cycle.

The pressure-enthalpy (p-h) and temperature-entrapy (T-s) diagrams of this cycle are given in **Figure 2**. The Rankine cycle operates in the following steps:

1-2-3 **Isobaric Heat Transfer** High pressure liquid enters the boiler from the feed pump (1) and is heated to the saturation temperature (2). Further addition of energy causes evaporation of the liquid until it is fully converted to saturated steam (3).

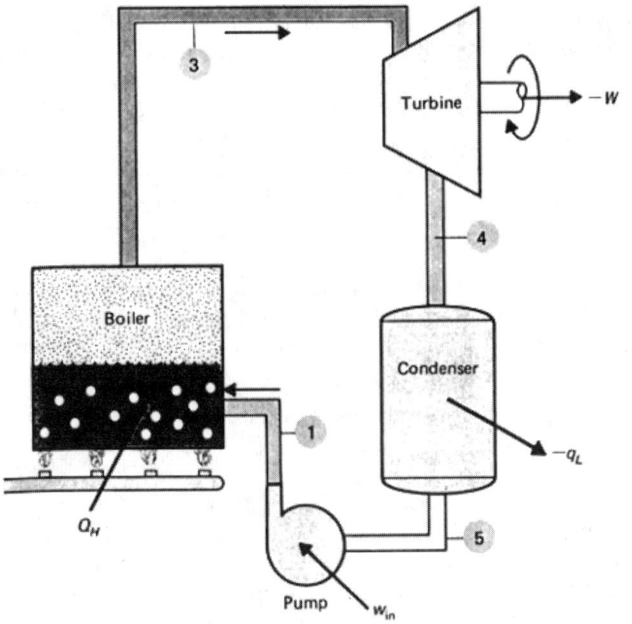

Figure 1. Rankine cycle.

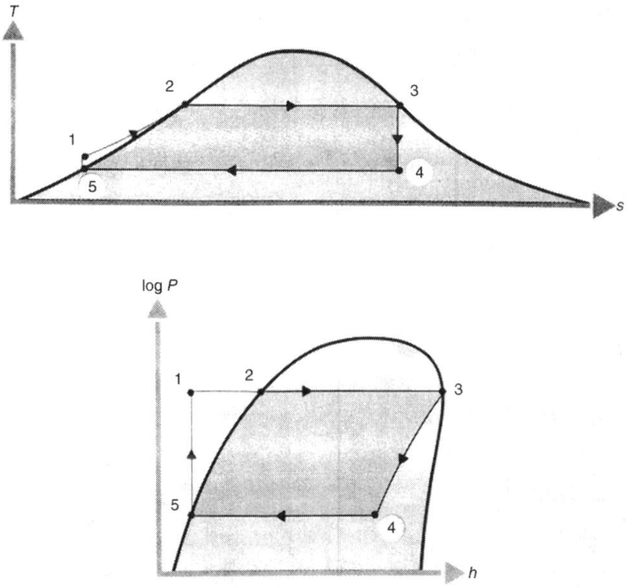

Figure 2. T-s p-h and diagrams.

3-4 **Isentropic Expansion** The vapour is expanded in the turbine, thus producing work which may be converted to electricity. In practise, the expansion is limited by the temperature of the cooling medium and by the erosion of the turbine blades by liquid entrainment in the vapour stream as the process moves further into the two-phase region. Exit vapour qualities should be greater than 90%.

4-5 **Isobaric Heat Rejection** The vapour-liquid mixture leaving the turbine (4) is condensed at low pressure, usually in a surface condenser using cooling water. In well designed and maintained condensers, the pressure of the vapour is well below atmospheric pressure, approaching the saturation pressure of the operating fluid at the cooling water temperature.

5-1 **Isentropic Compression** The pressure of the condensate is raised in the feed pump. Because of the low specific volume of liquids, the pump work is relatively small and often neglected in thermodynamic calculations.

The *efficiency of power cycles* is defined as

$$\eta = \frac{W_{net}}{Q_{in}} = \frac{W_{34} - W_{51}}{Q_{13}} \quad \textbf{(1)}$$

Values of heat and work can be determined by applying the *First Law of Thermodynamics* to each step. The steam quality x at the turbine outlet is determined from the assumption of isentropic expansion, i.e.

$$s_3 = s_v^*(p_3) = x\ (s_l^*(p_4)\ s_v^*(p_4)) \quad \textbf{(2)}$$

where s_v* is the entropy of vapour and S_l* the entropy of liquid.

Inefficiencies of Real Rankine Cycles

The efficiency of the ideal Rankine cycle as described in the previous section is close to the Carnot efficiency (see **Carnot Cycle**). In real plants, each stage of the Rankine cycle is associated with irreversible processes, reducing the overall efficiency. Turbine and pump irreversibilities can be included in the calculation of the overall cycle efficiency by defining a turbine efficiency according to **Figure 3**

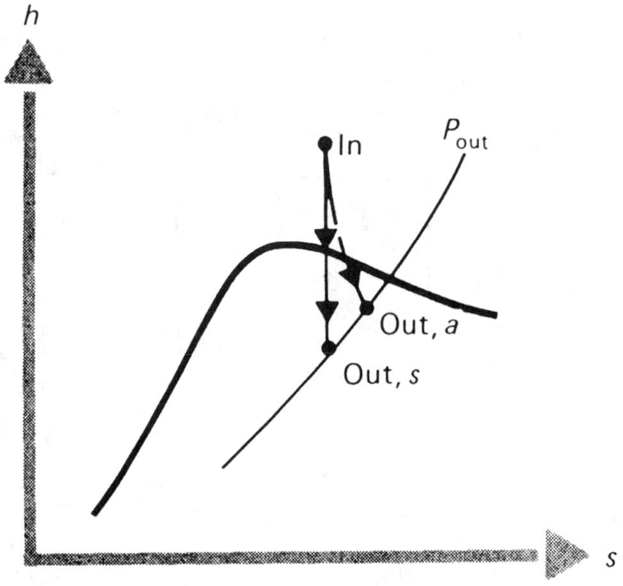

Figure 3. Turbine efficiency.

$$\eta_t = \frac{W_a}{W_s} = \frac{(h_{t,in} - h_{t,out})_a}{(h_{t,in} - h_{t,out})_s} \qquad (3)$$

where subscript act indicates actual values and subscript is indicated isentropic values and a pump efficiency

$$\eta_p = \frac{W_s}{W_a} = \frac{V(p_1 - p_6)}{W_a} \qquad (4)$$

If η_t and η_p are known, the actual enthalpy after the compression and expansion steps can be determined from the values for the isentropic processes. The turbine efficiency directly reduces the work produced in the turbine and, therefore the overall efficiency. The inefficiency of the pump increases the enthalpy of the liquid leaving the pump and, therefore, reduces the amount of energy required to evaporate the liquid. However, the energy to drive the pump is usually more expensive than the energy to feed the boiler.

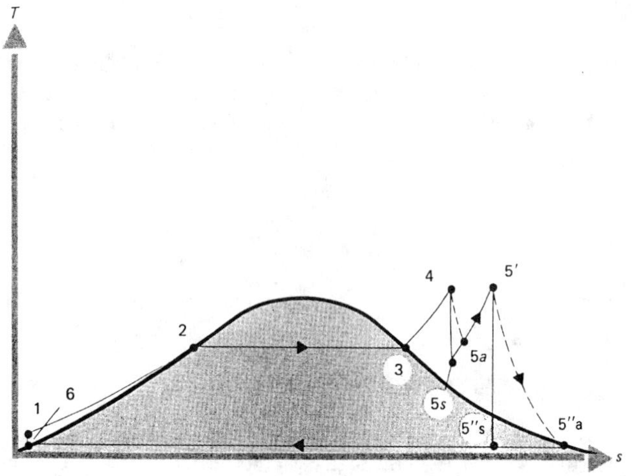

Figure 4. Rankine cycle with vapour superheating.

Even the most sophisticated boilers transform only 40% of the fuel energy into useable steam energy. There are two main reasons for this wastage:

- The combustion gas temperatures are between 1000°C and 2000°C, which is considerably higher than the highest vapour temperatures. The transfer of heat across a large temperature difference increases the entropy.
- Combustion (oxidation) at technically feasible temperatures is highly irreversible.

Since the heat transfer surface in the condenser has a finite value, the condensation will occur at a temperature higher than the temperature of the cooling medium. Again, heat transfer occurs across a temperature difference, causing the generation of entropy. The deposition of dirt in condensers during operation with cooling water reduces the efficiency.

Increasing the Efficiency of Rankine Cycles

Pressure difference

The net work produced in the Rankine cycle is represented by the area of the cycle process in **Figure 2**. Obviously, this area can be increased by increasing the pressure in the boiler and reducing the pressure in the condenser.

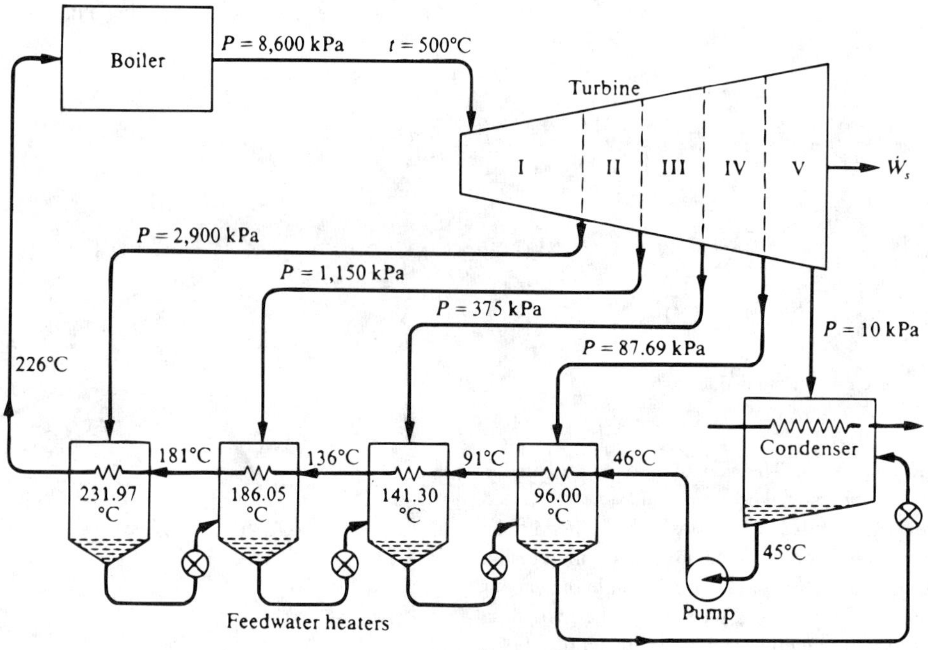

Figure 5. Regenerative feed liquid heating.

Superheating and reheating

The irreversibility of any process is reduced if it is performed as close as possible to the temperatures of the high temperature and low temperature reservoirs. This is achieved by operating the condenser at subatmospheric pressure. The temperature in the boiler is limited by the saturation pressure. Further increase in temperature is possible by superheating the saturated vapour, see **Figure 4**.

This has the additional advantage that the vapour quality after the turbine is increased and, therefore the erosion of the turbine blades is reduced. It is quite common to reheat the vapour after expansion in the high pressure turbine and expand the reheated vapour in a second, low pressure turbine.

Feed water preheating

The cold liquid leaving the feed pump is mixed with the saturated liquid in the boiler and/or re-heated to the boiling temperature. The resulting irreversibility reduces the efficiency of the boiler. According to the Carnot process, the highest efficiency is reached if heat transfer occurs isothermally. To preheat the feed liquid to its saturation temperature, bleed vapour from various positions of the turbine is passed through external heat exchangers (regenerators), as shown in **Figure 5**.

Ideally, the temperature of the bleed steam should be as close as possible to the temperature of the feed liquid.

Combined cycles

The high combustion temperature of the fuel is better utilised if a gas turbine or Brayton engine is used as "topping cycle" in conjunction with a Rankine cycle. In this case, the hot gas leaving the turbine is used to provide the energy input to the boiler. In *co-generation systems,* the energy rejected by the Rankine cycle is used for space heating, process steam or other low temperature applications.

H. Müller-Steinhagen

RANKINE DEGREE

Following from: *Temperature*

The Rankine temperature scale is named after the Scottish physicist, W.J.M. Rankine (1820–1872). The scale is defined such that zero degrees Rankine (designated by 0°R) is absolute zero and an interval of 1°R is equivalent to 1°F. In this way, the correspondence between the Rankine scale and the Fahrenheit Scale is analogous to that between the Celsius and the Kelvin scales.

M. P. Orlova

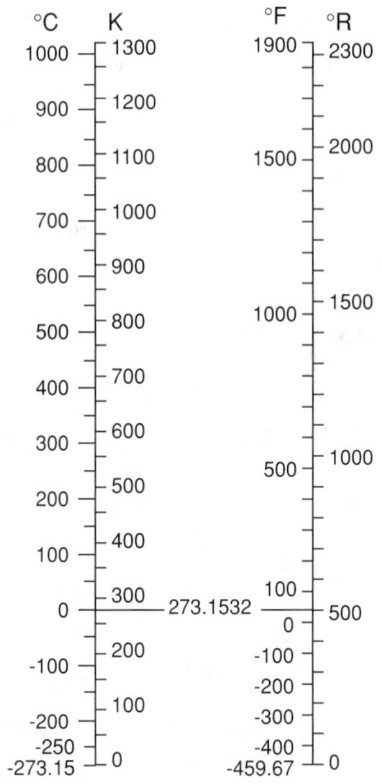

Figure 1. Rankine degree.

RANKINE VORTEX (see Vortices)

RANKINE, WJM (see Steam engines)

RAOULT'S AND DALTON'S LAW (see Distillation)

RAOULT'S LAW

This applies to liquid-vapour equilibrium at pressures sufficiently low for the gas phase to obey PV = RT and where the *activity coefficient* of each component in the liquid phase is unity, i.e. it is an *ideal mixture.*

For such a liquid mixture, the total pressure at temperature T is:

$$P = \Sigma \tilde{x}_i P_i^{\circ}$$

where $\tilde{x}_i$ is the mole fraction and P_i° is the **Vapour Pressure** of pure component i at temperature T. Thus, for a binary mixture, if both liquid and vapour phases are present simultaneously and in equilibrium with one another, the pressure varies linearly with mole fraction from the vapour pressure of pure component 1 on one side to that of pure component 2 on the other.

Real mixtures obey Raoult's Law only as an approximation. The necessity for the pressure to be low restricts the range of applicability for any given mixture, but the requirement for the liquid to be essentially ideal restricts its use to mixtures of very similar molecules.

G. Saville

RARIFACTION (see Vacuum)

RARIFACTION WAVE (see Shock tubes)

RARIFIED GAS DYNAMICS (see Knudsen number)

RAYLEIGH EQUATION, FOR BUBBLE GROWTH (see Bubble growth)

RAYLEIGH EQUATION, FOR DROPLET FORMATION (see Atomization)

RAYLEIGH FORMULA (see Capillary action)

RAYLEIGH LAW OF SCATTERING (see Mie scattering)

RAYLEIGH, LORD (1842–1919)

Lord Rayleigh, a British physicist, was awarded the nobel prize for physics in 1904 for his discovery (1894) of the inert elementary gas argon, in collaboration with sir William Ramsay. He was born near Maldon, Essex, on November 12, 1842, and educated at Trinity College, Cambridge, where he graduated senior wrangler (1865). As successor to James Clerk Maxwell he was head of the Cavendish Laboraty at Cambridge from 1879 to 1884, and in 1887 he became professor of natural philosophy in the Royal Institution of Great Britain. Elected (1873) a fellow of the Royal Society, he was president from 1905 to 1908.

Figure 1.

His research covered almost the entire field of physics, including sound, wave theory, optics, colour vision, electrodynamics, electromagnetism, the scattering of light, hydrodynamics, the flow of liquids, capillarity, viscosity, the density of gases, photography and elasticity, as well as electrical measurements and standards. His research on sound was embodied in his "Theory of Sound", and his other extensive studies in physics appeared in his "scientific papers". Rayleigh died on June 30, 1919 at Witham, Essex.

U. Grigull, H. Sandner and J. Straub

RAYLEIGH NUMBER

Rayleigh number, Ra, is a dimensionless term used in the calculation of natural convection.

$$Ra = \frac{g\beta'\Delta Tx^3}{\nu\kappa} = Gr\ Pr$$

where g is acceleration due to gravity, β' is coefficient of thermal expansion of the fluid, ΔT is temperature difference, x is length, ν is kinematic viscosity and κ is thermal diffusivity of the fluid. Gr is the **Grashof Number** and Pr is the **Prandtl Number**.

The magnitude of the Rayleigh number is a good indication as to whether the natural convection boundary layer is laminar or turbulent. For a vertical surface for example, the transition takes place when $Ra \approx 10^9$. Various examples of the relationship between the magnitude of natural convective heat transfer and Raleigh number are given in the following reference.

Reference

Hewitt, G. F., Shires, G. L. and Bott, T. R. (1994) *Process Heat Transfer,* CRC Press.

G.L. Shires

RAYLEIGH SCATTERING (see Fibre optics; Optical particle characterisation)

REACTION TURBINES (see Hydraulic turbines)

REACTOR PHYSICS (see Nuclear reactors)

REATTACHMENT (see Tubes, crossflow over)

REATTACHMENT, OF BOUNDARY LAYER (see Boundary layer)

RÉAUMUR DEGREE

Following from: *Temperature*

Réaumur degree (named after the French naturalist R.A.F. Réaumur, R.A.F., 1683–1757) is an obsolete unit of temperature (designated by °R), equal to 1/80 part of the temperature interval between the points of ice melting (0 °R) and of water boiling (80 °R). At normal atmospheric pressure 1 °C = 0.8 °R (for temperature difference), T(°C) = 0.8T (°R) (for temperature).

M. P. Orlova

REBOILERS

***Following from:** Boiling; Boilers*

Reboilers are used to generate a flux of vapour to feed to a distillation tower; the vapour rises up the tower contacting a downwards-flowing liquid stream. A whole variety of forms of reboiler have been used in practice, some of which are described briefly below. Geneal reviews of the forms and design of reboilers are given by McKee (1970), by Ploughman (1983) and by Whalley and Hewitt (1983).

Internal Reboilers

The simplest approach is to mount the reboiler in the distillation tower itself as is illustrated in **Figure 1**. Here, boiling takes place in the pool of liquid at the bottom of the tower, the heating fluid being inside the bundle of tubes as shown. The major problem with *internal reboilers* is the limitation imposed by the size of the distillation column. This limits the size of the reboiler. Another problem sometimes encountered is that of mounting the bundle satisfactorily into the column. The problem of size restriction can be overcome if compact heat exchangers are used. Thus, **Plate-Fin Exchangers** are used commonly as internal reboilers in the distillation towers of air separation plant. Another form of compact heat exchanger which has been used for this type of duty is the *printed circuit heat exchanger* which has an even higher heat transfer surface area per unit volume.

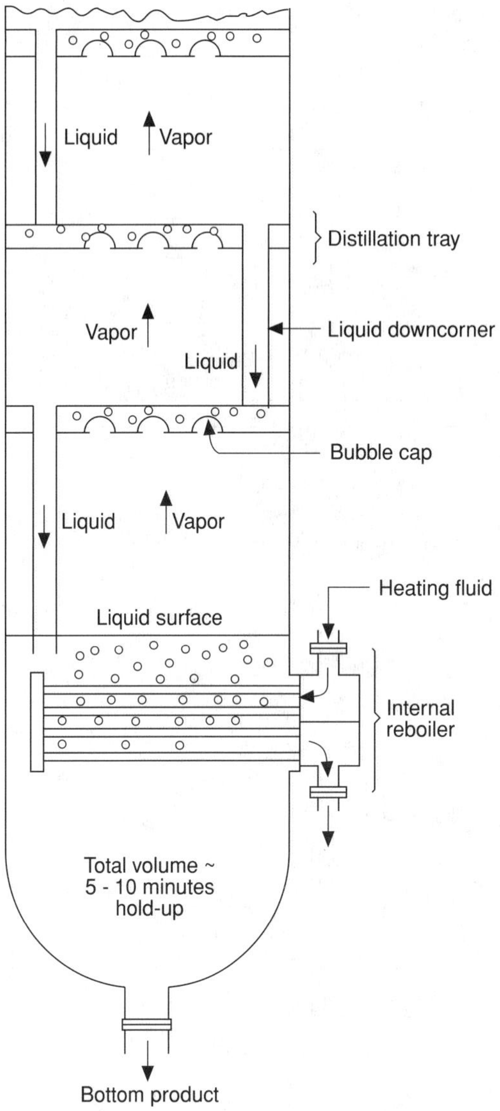

Figure 1. Internal reboiler.

Kettle Reboilers

The layout of the *kettle reboiler* is illustrated schematically in **Figure 2**. Liquid flows from the column into a shell in which there is a horizontal tube bundle, boiling taking place from the outside this bundle. The vapour passes back to the column as shown. Kettle reboilers are widely used in the petroleum and chemical industries; their main problems are that of ensuring proper disentrainment of liquid from the outgoing vapour and the problem of the collection of scale and other solid materials in the tube bundle region over long periods of operation.

Vertical Thermosyphone Reboiler

This type is illustrated in **Figure 3**. The liquid passes from the bottom of the tower into the reboiler, with the evaporation taking place inside the tubes. The two-phase mixture is discharged back into the tower, where the liquid settles back to the liquid pool and the vapour passes up the tower as shown. The heating fluid (typically condensing steam) is on the outside of the tubes. The *vertical thermosyphon reboiler* is less susceptible to fouling problems and in general has higher heat transfer coefficients than does the kettle reboiler. However, additional height is required in order to mount the reboiler.

Horizontal Thermosyphone Reboiler

Here, the liquid from the column passes in cross flow over a tube bundle and the liquid-vapour mixture is returned to the column as shown (see **Figure 4** on page 933). The heating fluid is inside the tubes. This design has the advantage of preserving the natural circulation concept while allowing a lower headroom than the vertical thermosyphon type.

However, there are more uncertainties about fouling and about the prediction of the cross flow heat transfer rates.

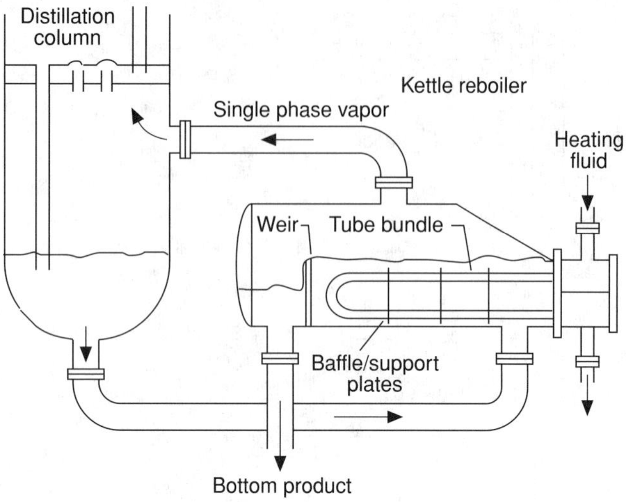

Figure 2. Kettle reboiler.

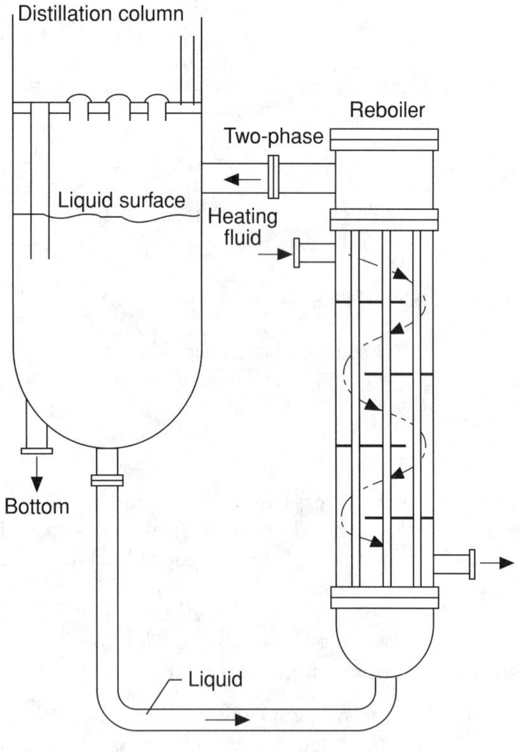

Figure 3. Vertical thermosyphone reboiler.

References

McKee, H. R. (1970) Thermosyphon reboilers: a review, *Ind. Eng. Chem.*, 62–12, 76.

Palen, J. W. (1983) Shell-and-tube reboilers, *Heat Exchanger Design Handbook.* Section 3.6. Hemisphere Publishing Corporation, New York.

Whalley, P. P. and Hewitt, G. F. (1983) Reboilers, *Multiphase Science and Technology, Vol. 2,* Hemisphere Publishing Corporation, New York.

G. F. Hewitt

RECIPROCATING COMPRESSOR (see Compressors)

RECIRCULATION (see Contraction, flow and pressure loss in; T-junctions)

RECIRCULATION BOILERS (see Boilers)

RECONSTRUCTED WAVEFRONTS (see Holographic interferometry)

RECOVERY COEFFICIENT (see Adiabatic wall temperature)

RECOVERY TEMPERATURE (see Adiabatic wall temperature; Boundary layer heat transfer)

RECUPERATIVE HEAT EXCHANGERS (see Heat exchangers)

REDLICH-KWONG EQUATION (see Gas law; PVT relationships)

REDOX REACTIONS (see Electrochemical cells)

REDUCED GRAVITY CONDITIONS (see Microgravity conditions)

REDUCED INSTRUCTION SET COMPUTER, RISC (see Computers)

REDUCED PROPERTIES

Following from: *Critical properties*

The *intermolecular pair potentials* of many substances are similar in shape and certainly share the same gross features of a steep repulsive wall at short distances with an attractive region at larger distances. It turns out to be extremely useful to make the two assumptions that all substances have the same form of intermolecular potential but differ in the values of energy and distance scaling parameters and, secondly, that the properties of a substance depend only upon the pair potential. Thus, we can write, for any substance, its pair potential $u_{ii}(r)$ as

$$u_{ii}(r) = f_{ii}u_{\infty}(r/h_{ii}^{1/3})$$

where u_{oo} is a reference pair potential, f_{ii} is an energy scaling

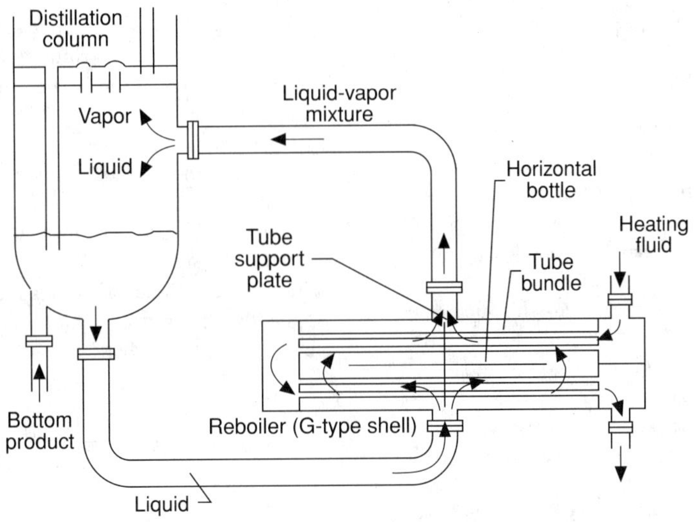

Figure 4. Horizontal thermosyphone reboiler.

parameter and $h_{ii}^{1/3}$ a distance scaling parameter. It follows from these assumptions (Bett et al., 1975) that the pressure of the pure fluid i may be obtained from the pressure of the pure fluid o by means of the equation

$$p_i(\tilde{v}, T) = (f_{ii}h_{ii}^{-1})p_o(\tilde{v}h_{ii}^{-1}, Tf_{ii}^{-1})$$

This result pertains for any substance i that is part of the set for which the pair potentials are *conformal.* It follows that the reduced pressure p_ih_{ii}/f_{ii} is the same function of the reduced variables $\tilde{v}/h_{ii}$ and T/f_{ii} for all substances in that set. This means that if $p_o(\tilde{v}h_{ii}^{-1}, Tf_{ii}^{-1})$ is known for one substance in the set of conformal substances, the pressure can be calculated for other substances from a knowledge only of the scaling parameters h_{ii} and f_{ii}. Very often the scaling parameters h_{ii} and f_{ii} are taken to be the critical volume and critical temperature of the fluid respectively, although this is not essential.

Inasmuch as the assumptions upon which the development of the principle of corresponding states is based are not satisfied by any real fluid, the principle itself is obeyed only approximately. Nevertheless, it is a very powerful tool for the estimation of the properties of fluids (Reid et al., 1977) and has been extended to the treatment of non-spherically symmetric molecules.

References

Bett, K. E., Rowlinson, J. S. and Saville, G. (1975) *Thermodynamics for Chemical Engineers,* Athlone Press, London.

Reid, R. C., Prausnitz, J. M. and Sherwood, T. K. (1977) *The Properties of Gases and Liquids,* McGraw-Hill, New York.

***Leading to:** Corresponding states*

W.A. Wakeham

REFINING

***Following from:** Metals*

The *pyrometallurgical* term refining refers to the selective removal of impurities from bulk metal, normally in the liquid state.

The impurity that is removed is normally transferred to either a liquid phase immiscible in the bulk metal or to a gaseous phase.

Impurity removal is required either because the impurities are detrimental to the physical and mechanical properties of the metal being produced, or the impurities have high intrinsic value.

Impurities can be removed by several means.

1. Without addition of chemical reagents as in preferential volatilisation of impurities, and impurity segregration during phase separation from a liquid during cooling
2. With chemical reagent additions, by preferential reactions of impurity with the added reagent, as in for example oxidation, sulphidation, chlorination and reactions with a second immiscible metal.

Reference

Engh T. (1992) *Principles of Metal Refining*, Oxford University Press

B. Terry

REFLECTANCE (see Reflection coefficient)

REFLECTION COEFFICIENTS FOR EARTH'S SURFACE (see Environmental heat transfer)

REFLECTION COEFFICIENT (REFLECTANCE)

Following from: *Radiative heat transfer*

The term "reflection coefficient" (or "reflectance") is widely employed in radiation and combined heat transfer. In most cases, it is the absolute analog of the term "reflectivity" and it is characterized the ratio of radiation flux reflected by a sample surface to the incident radiation flux. If "reflection coefficient" is used in this sense the term "reflectivity" may be replaced with supplemental terms describing the directions of the incident and reflected radiation.

Sometimes the term "reflection coefficient" is used to characterize not surface reflection, but the reflection of the whole sample in a specific conditions. For example, if a given sample is semitransparent for thermal radiation, then the reflection coefficient depends on not only properties of irradiated surface, but also on volumetric optical properties (absorption coefficient, refractive index, scattering coefficient and phase function of volumetric scattering), on temperature distribution inside the sample and sometimes (when the transmission of the sample is appreciable) even on properties of other nonirradiated surfaces. In this case, the use of the term "reflection coefficient" or "reflectance" is possible and even desirable, but there is a need to define precisely the conditions of the reflection.

References

Seigel, R., Howell J. R. (1992) *Thermal Radiation Heat Transfer,* third edition, Washington D.C., Hemisphere, 1992.

Petrov V. A., Chernyshev A. P. (1987) Experimental study of the surface heating of magnesium oxide ceramics by laser radiation, *High Temperature,* 25, 2, 274–280, 1987.

Leading to: *Reflectivity*

V. A. Petrov

REFLECTIVITY

Following from: *Reflectance*

Reflectivity ρ is the ratio of the radiation flux Φ_r, reflected by a sample surface, to the incident radiation flux Φ_i:

$$\rho = \Phi_r/\Phi_i \tag{1}$$

Sometimes the term "reflectivity" is understood as the ratio of the mentioned fluxes when the sample reflects volumentrically including its interior if it is semitransparent to thermal radiation. In this case the reflection depends from sample thickness and instead of "reflectivity" we may recommended the use of the term "reflection coefficient". The methods for calculation and measurement of reflectivity are well developed.

For optically smooth sample surface (if its mean square microroughness is at least 100 times less than the wavelength of the radiation) the reflection is specular and the angle of reflection is equal the angle of incidence. The value of reflectivity as a function of an incident angle may be calculated using Fresnel's formulas if refractive index n and absorption index χ of sample material are known.

In spectral range of transparency of two adjacent media (1 and 2), when the absorption indexes x_1 and x_2 are small in comparison with refractive indexes n_1 and n_2, the following expressions are applied for the perpendicular (⊥) and parallel (||) polarized components of incident radiation:

$$\rho_\perp = \left(\frac{\cos\theta - \sqrt{n_{21}^2 - \sin^2\theta}}{\cos\theta + \sqrt{n_{21}^2 - \sin^2\theta}}\right)^2, \tag{2}$$

$$\rho_\| = \left(\frac{n_{21}^2\cos\theta - \sqrt{n_{21}^2 - \sin^2\theta}}{n_{21}^2\cos\theta + \sqrt{n_{21}^2 - \sin^2\theta}}\right)^2, \tag{3}$$

where $n_{21} = n_2/n_1$, and θ is an incident angle of radiation from first medium onto the second relative to the surface normal.

For incident natural unpolarized radiation the parallel and perpendicular components have an equal intensity and the mean arithmetic value for this radiation may be taken as reflectivity:

$$\rho = \frac{1}{2}(\rho_\perp + \rho_\|). \tag{4}$$

The angular dependence of reflectivity in this case has the form shown on **Figure 1**. Curve 1 is related to reflectance of natural nonpolarizated radiation, curve 2 is related to $\rho_\perp$ and curve 3 – to $\rho_\|$. At $n_{21} > 1$ the perpendicular component is increased monotonically to unity with increasing angle θ, and the parallel component first decreases to zero and then increases to unity. The parallel component is equal zero at the Bruster angle θ_{Br} (**Figure 1a**).

In the case of incidence of radiation from an optically more dense medium onto an interface with an optically less dense one ($n_{21} < 1$) the value of Bruster angle is lower, and starting from some value of so-called critical incident angle $\theta_c = \sin^{-1}(n_1/n_2)$ at $\theta > \theta_c$ the full internal reflection takes place, i.e. $\rho_\perp = \rho_\| = 1$ (**Figure 1b**).

In radiation transfer the case is often found when the medium 1 is nonabsorbing ($\chi_1 = 0$) and medium 2 is absorbing. In the more general case the medium 2 may have both a transparency region of the spectrum ($\chi_2 = 0$) and a nontransparency one ($\chi_2 > 0$). The expressions for the reflectivity components in spectral region of nontransparency of the second medium will have a form:

$$\rho_\perp(\theta) = \frac{(a - \cos\theta)^2 + b^2}{(a + \cos\theta)^2 + b^2} \tag{5}$$

$$\rho_\|(\theta) = \rho_\perp(\theta)\frac{(a - \sin\theta\tan\theta)^2 + b^2}{(a + \sin\theta\tan\theta)^2 + b^2} \tag{6}$$

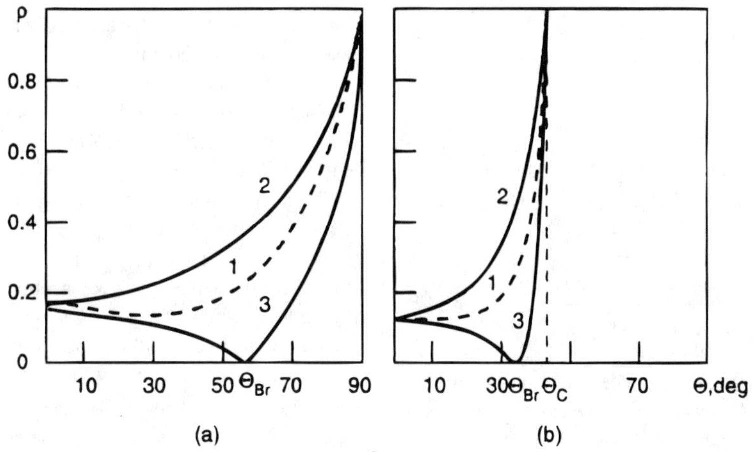

Figure 1. Reflectivity as a function of angle of incidence

where

$$a^2 = \frac{1}{2}\left[\zeta + \sqrt{\zeta^2 + 4n_{21}^2\chi_{21}^2}\right], \tag{7}$$

$$b^2 = \frac{1}{2}\left[-\zeta + \sqrt{\zeta^2 + 4n_{21}^2\chi_{21}^2}\right], \tag{8}$$

$$\zeta = (n_{21}^2 - \chi_{21}^2 - \sin^2\theta), \qquad \chi_{21} = \chi_2/n_1. \tag{9}$$

For reflection of medium 2 in vacuum, air or another small density gas it may be taken that refractive index n_1 is equal unity.

In **Figure 2**. the angular dependence of polarized components of reflectivity of the surface between medium 2 and medium 1 (with $n_1 = 1$) is shown. The curves referred to the perpendicular polarized component of radiation are marked by one stroke, the curves marked with two strokes are referred to the parallel component.

The curves 1 describe the data for nonabsorbing medium with $n_2 = 1.5$, $\chi_2 = 0$, the curves 2 − for $n_2 = 1.5$, $\chi_2 = 1.0$, the curves 3 − for $n_2 = 11$, $\chi_2 = 6$. It may be seen that the component

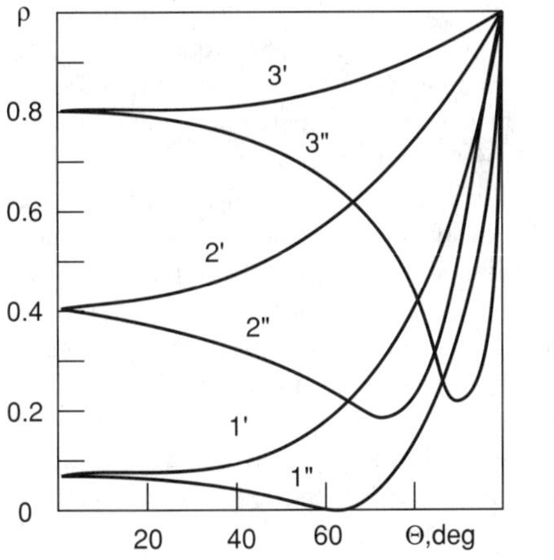

Figure 2. Angular dependance of polarized components of reflectivity

$\rho_{\parallel}$ has a minimum at large incident angles and this minimum is the deeper the greater the value of χ_2.

For radiation incident along the normal to the interface ($\theta = 0$) the expression for reflectivity has a form:

$$\rho_{\perp}(\theta = 0) = \rho_{\parallel}(\theta = 0) = \rho(\theta = 0) = \frac{(n_{21} - 1)^2 + \chi_{21}^2}{(n_{21} + 1)^2 + \chi_{21}^2}. \tag{10}$$

However, for most important practical cases, the reflection surface (interface) is not optically smooth, the reflection is not specular and the reflected energy has a distribution of directions over a hemisphere whose base is the reflecting surface. Here other parameters describing the reflection must be introduced. The two-directional (bidirectional or directional-directional) reflectivity $\rho(\vec{l}, \vec{l}')$ describes the ratio of the radiation intensity $I(\vec{l}')$ reflected in the direction determined by the unit vector $\vec{l}'$ to the radiation flux falling on unit area of a surface in an elemental solid angle $d\Omega$ from another direction determined by the unit vector $\vec{l}$, oriented under some angle to the external normal $\vec{n}$:

$$\rho(\vec{l}, \vec{l}') = \frac{I(\vec{l}')}{I(\vec{l})\cos(\vec{n}\vec{l})d\Omega} \tag{11}$$

In this formula $I(\vec{l})$ is the intensity of the incident radiation. The value of two directional reflectivity has the dimension of inverse solid angle (Sr^{-1}).

The directional-hemispherical reflectivity $\rho(\vec{l}, h)$ is the ratio of the hemispherically reflected (in all directions determined by the unit vector $\vec{l}'$) radiation flux to the radiation flux falling on the sample surface in an elemental solid angle $d\Omega$ from the direction determined by the unit vector $\vec{l}$ oriented at some angle to the external normal $\vec{n}$:

$$\rho(\vec{l}, h) = \frac{\int_{\Omega'=2\pi} I(\vec{l}')\cos(\vec{n}\vec{l}')d\Omega'}{I(\vec{l})\cos(\vec{n}\vec{l})d\Omega}. \tag{12}$$

The hemispherical-directional reflectivity $\rho(h, \vec{l}')$ is the ratio of the intensity of radiation reflected in some direction $\vec{l}'$ to the hemisphere-averaged radiation flux falling on unit area of surface from all hemispherical directions $\vec{l}'$:

$$\rho(h, \vec{1}') = \frac{I(\vec{1}')}{\frac{1}{\pi}\int_{\Omega=2\pi} I(\vec{1})\cos(\vec{n}\vec{1})d\Omega}. \quad (13)$$

The two hemispherical (hemispherical-hemispherical) reflectivity ρ(h, h) is the ratio of the radiation flux reflected in the hemisphere in all directions $\vec{1}'$ to the radiation flux incident from all directions $\vec{1}$:

$$\rho(h, h') = \frac{\int_{\Omega'=2\pi} I(\vec{1}')\cos(\vec{n}\vec{1}')d\Omega'}{\int_{\Omega=2\pi} I(\vec{1})\cos(\vec{n}\vec{1})d\Omega}. \quad (14)$$

The directional reflection characteristics depend on the optical properties of the substance, temperature, wavelength, sizes, and the geometric shape of the surface microroughness.

The case is widely encountered where the microroughness sizes are comparible with the wavelength. In this case, theoretical calculations of directional reflection characteristics cannot be acheived and the characteristics have to be determined experimentally. At present the quantity of experimental data is insufficient. Basically there are the data on normal-hemispherical reflectivity at room temperature. The study of two directional reflectivity is very laborious. In this case many experimental points need to be measured. There are practically no such studies.

The two directional reflectivities are used in radiation transfer calculations as a rule only for specular reflection. If we are dealing with rough surfaces, taking into account the dependence of reflection (and emission) characteristics on direction is very difficult in most cases even for very simple shapes of radiating surfaces. Generally in this case the more simple model of diffuse (not dependant on angle) reflection and emission of the rough surface is used. In a case where the contribution of radiation directed at big angles to the surface normal is rather large it is possible to employ approximate methods of radiation transfer calculations. For example, it may be possible to use a model taking into account the directional emission by considering reflectivity as the sum of diffuse and specular components.

References

Hsia, J. J., Richmond, J. C. (1976) Bidirectional reflectometry, I. A high resolution laser bidirectional reflectometer with results of several optical coatings, *J. Res. Nat. Bur. Stand. A.*, 80–2, 189–205, 1976.

Seigel, R., Howell, J. R. (1992) *Thermal Radiation Heat Transfer,* third edition, Washington D. C., Hemisphere, 1992.

Toulukian, Y. S., De Witt, D. P. (1972) Thermophysical properties of matter, *Thermal Radiation Properties,* Vol. 7, *Metallic Elements and Alloys,* 1970, Vol. 8, Nonmetallic solids, 1972, New York—Washington, IFI—Plenum.

V. A. Petrov

REFLOOD

Following from: *Blowdown*

A *Design Basis Accident* in the PWR is the large break *Loss-of-Coolant Accident.* After the initial **Blowdown** the reactor core will be at low pressure, steam-filled and heating up due to decay heat. Some fuel clad temperatures will be above the value at which spontaneous quenching can occur. The subsequent refilling of the pressure vessel by liquid from the *Emergency Core Cooling System (ECCS)* so that the core is cooled and quenched is known as Reflood. The important considerations in reflood are the hydraulic flows in the primary circuit, the heat transfer in the core, and the feedback between the two.

At the end of blowdown the pressure vessel will contain very little liquid but the (passive) accumulators and (active) pumped injection systems of the ECCS will be injecting liquid into the vessel downcomer via the intact cold legs. After about 10 seconds the vessel lower plenum will be full and liquid will reach the bottom of the core fuel rods. Quench fronts will quickly become established on the rods and will move up the core, lagging behind the liquid level as an inverted annular flow regime is established. After a further 10–20 seconds some 25% of the core will be flooded, but at this point the accumulators are exhausted of liquid. Reflood continues at a much lower rate (a few cms per second) due to the continuing pumped injection until the whole core is flooded and quenched. Safety analyses indicate that reflood should be complete some 150–300 seconds after accident initiation. This timescale is too short for any operator intervention, and the reactor safety system operations described here are completely automatic.

Reflood is unlikely to be a steady, continuous process. One major perturbation occurs when the accumulators empty of liquid. As the liquid in the pipework connecting the accumulators to the reactor circuit is replaced by the nitrogen cover gas the volume flow rate increases. Thus there is a surge of liquid into the vessel, followed by a transient pressurising of the top of the downcomer by nitrogen. This produces a surge of liquid into the core, which in turn produces a burst of heat transfer and quenching. The resulting increase in steam production can pressurise the core region to such an extent that some liquid is driven backwards out of the core, up the downcomer and out of the break. After this short-term transient reflood continues but with a somewhat lower vessel liquid inventory.

Heat transfer in the inverted annular flow regime, especially at the quench front, generates a significant upward steam flow, which carries some of the liquid up through the core in the form of entrained droplets. This dispersed flow produces some heat transfer from the fuel at higher elevations. The peak clad temperatures occur near the top of the core, and continue to rise during the early stages of reflood. Eventually, as the quench front moves closer to the peak temperature positions, the precursory cooling is sufficient to halt the rise. Analysis shows that temperatures remain below the 1200°C safety limit, even when pessimistic assumptions about the availability of the ECCS are made. At some stage during reflood much of the fuel cladding will be above the maximum temperature at which liquid water can exist. The quench front proceeds into such regions by a combination of precursory cooling and axial conduction in the cladding, with precursory cooling probably being the dominant mechanism under the particular conditions of reflood.

The outlet flow from the top of the core will initially be pure steam but later will be a dispersed flow containing liquid droplets. This flow continues around the coolant circuit to the break position via the steam generator (if the break is in the cold leg). The pressure difference which this flow requires, which is enhanced by evaporation of droplets in the steam generator (the pressurised secondary side of which still contains hot fluid at this stage), increases the pressure in the core region and thus resists the inflow of liquid into the core from the downcomer. This phenomenon is known as *Steam Binding*. The extra pressure at inlet is provided by an accumulation of liquid in the downcomer giving an increased differential between downcomer and core. Thus there is a complicated feedback between the rate of steam generation in the core and the rate of accumulation of ECCS liquid in the downcomer to provide the gravity head necessary to overcome the steam binding.

The reflood phase is said to be complete when the fuel is completely quenched. After a further period of cooling the circuit will be in a subcooled state and coolant injection can be manually switched to a recirculating residual heat removal mode, which can be continued indefinitely. (See also Boiling; Rewetting of hot surfaces)

I. Brittain

REFLUX CONDENSATION (see Tubes, condensation in)

REFLUX CONDENSER (see Condensers; Dephlegmator)

REFLUX RATIOS (see Distillation)

REFORMING (see Oil refining)

REFRACTION (see Optics)

REFRACTIVE INDICES FOR GASES AND LIQUIDS (see Interferometry)

REFRACTIVE INDEX

Following from: Optics

Refractive index n is the parameter that characterises as optical properties of materials and media. Its value is equal the ratio of the velocity of electromagnetic wave propagation in a vacuum c to the velocity u of its propagation in the medium considered:

$$n = c/u. \tag{1}$$

The refractive index depends on frequency, temperature and, for an anisotropic medium, on the direction of radiation transfer.

For dielectrics and semiconductors in a high transparency region, and in the absence of absorption, the refractive index may be defined in accordance with the *Snell refraction law* as the ratio of sine of the angle (θ_1) between the direction of radiation incident from vacuum onto the medium surface and the normal to this surface to the sine of the angle (θ_2) between the direction of radiation propagation inside medium and the same normal:

$$n = \frac{\sin \theta_1}{\sin \theta_2}. \tag{2}$$

In reality, all media in the spectral range of thermal radiation absorbs it to a certain degree. Therefore, a more general parametric description of optical properties is the complex refractive index N:

$$N = n - i\chi, \tag{3}$$

where n is the real part of a complex refractive index (proper refractive index), defined by expression (1), and χ is the absorption index. The properties n and χ may be related to the complex dielectric constant of a substance, or to the real dielectric permeability and electrical conductivity. Due to the presence of a sufficient quantity of conducting electrons (free electrons), the metals and other good electric conductors have a high absorption index χ over practically the whole spectral region of thermal radiation. The value of χ usually increases from 2–3 to 10 when the wavelength λ increases from the visible region to the middle infrared region. The refractive index n is more than unity in the whole spectral region and as a rule increases with increasing λ through more slowly than χ.

Dielectrics and semiconductors (including many solid substances, all gases at moderate temperatures and nearly all liquids) have a high transparency region where χ is equal to only 10^{-7} to 10^{-5}. This occurs between the long wave edge of the electron absorption band and the short wavelength edge of the first lattice absorption band, the maximum being usually located in the ultraviolet region. This value cannot be considered in the complex refractive index. Its real part n in this wavelength region is more than unity and it monotonically decreases with increasing wavelength. Near the maximum absorption bands, the values n and χ are much more. In the band maximum, n may be increased twice or three times, and χ may be increased by orders of magnitude and may reach several units. Along the edge of absorption band there is a wavelength region where $n < 1$.

References

Born, M., Wolf, E. (1980) *Principles of Optics,* 6th ed. Pergamon Press, Oxford.

Handbook of Optical Constants of Solids (1985) E. D. Palik, ed., Academic Press, New York.

Siegel, R., Howell, J. R. (1992) *Thermal Radiation Heat Transfer,* 3rd ed., Hemisphere, Washington D. C.

V.A. Petrov

REFRACTORY MATERIALS, FOR ELECTRIC FURNACES (see Electric, joule, heaters)

REFRIGERANTS

Table (A) Refrigerants: Properties of liquids below their boiling points

Substance	Data	Property	Temperature, °C −200	−180	−160	−140	−120	−100	−50	0	20
		Temperature, K	73.15	93.15	113.15	133.15	153.15	173.15	223.15	273.15	293.15
TRICHLORO-TRIFLUOR-OMETHANE (Refrigerant 11)	Chemical formula: CCl_3F; Molecular weight: 137.37	Density, ρ_l (kg/m^3)	S	S	S	S	S	1 760	1 642	1 534	1 488
	Melting point: −111.15°C; Boiling point: 23.8°	Specific heat capacity $C_{p,l}$ (kJ/kg K)	S	S	S	S	S	0.800	0.829	0.862	0.883
	Critical temperature: 198.0°C	Thermal conductivity, λ_l[(W/m^2)/(K/m)]	S	S	S	S	S	0.123	0.109	0.0945	0.0889
	Critical pressure: 4.40 MPa	Dynamic viscosity, η_l (10^{-5} Ns/m^2)	S	S	S	S	S	(210)	108.2	54.3	44.1
DICHLORO-DIFLUOR-OMETHANE (Refrigerant 12)	Chemical formula: CCl_2F_2; Molecular weight: 120.91	Density, ρ_t (10^{-3})	S	S	S	1 798	1 743	1 688	1 546	V	V
	Melting point: −155.15°C; Boiling point: −29.95°C	Specific heat capacity, $c_{p,t}$ (kJ/kg K)	S	S	S	0.842	0.842	0.846	0.871	V	V
	Critical temperature: 111.65°C	Thermal conductivity, λ_t [(W/m^2)/(K/m)]	S	S	S	0.129	0.122	0.115	0.0966	V	V
	Critical pressure: 4.132 MPa	Dynamic viscosity, η_t (10^{-5} η_t (10^{-5} Ns/m^2)	S S	S	S	(130)	(100)	(80)	46.7	V	V
CHLORO-TRIFLUOR-OMETHANE (Refrigerant 13)	Chemical formula: $CClF_3$; Molecular weight: 104.47	Density, ρ_t (kg/m^3)		(1870)	(1860)	(1730)	1.664	1.593	V	V	V
	Melting point: −181.15°C; Boiling point: −81.4°C	Specific heat capacity, $c_{p,t}$ (kJ/kg K)	S	(0.81)	(0.82)	(0.84)	(0.86)	(0.88)	V	V	V
	Critical temperature: 28.8°C	Thermal conductivity, λ_t [(W/m^2)/(K/m)]	S	0.141	0.130	0.118	0.107	0.097	V	V	V
	Critical pressure: 3.86 MPa	Dynamic viscosity, η_t (10^{-5} Ns/m^2)	S	—	—	—	S	(40)	V	V	V
DICHLORO-FLUORO-METHANE (Refrigerant 21)	Chemical formula: $CHCl_2F$; Molecular weight: 102.92	Density, ρ_t (kg/m^3)	S	S	S	S	(1700)	(1624)	1.534	1.426	V
	Melting point: −135.15°C; Boiling point: 8.75°C	Specific heat capacity, $c_{p,t}$ (kJ/kg K)	S	S	S	S	0.842	0.846	0.892	1.017	V
	Critical temperature: 178.1°C	Thermal conductivity, λ_t [(W/m^2)/(K/m)]	S	S	S	S	0.155	0.150	0.131	0.112	V
	Critical pressure: 5.181 MPa	Dynamic viscosity, η_t (10^{-5}Ns/m^2)	S	S	S	S	(270)	186.7	72.4	39.7	V
CHLORO-DIFLUORO-METHANE (Refrigerant 22)	Chemical formula: $CHClF_2$; Molecular weight: 86.47	Density, ρ_l (kg/m^3)	S	S	S	S	S	1.565	1.438	V	V
	Melting point: −159.95°C; Boiling point: −40.82°C	Specific heat capacity, $c_{p,l}$ (kJ/kg K)	S	S	S	S	S	1.061	1.093	V	V
	Critical temperature: 96.15°C	Thermal conductivity, λ_l [(W/m^2)/(K/m)]	S	S	S	S	S	0.150	0.125	V	V
	Critical pressure: 4.986 MPa	Dynamic viscosity, η_l (10^{-5} Ns/m^2)	S	S	S	S	S	72.8	36.5	V	V
TRICHLORO-TRIFLUORO-ETHANE (Refrigerant 113)	Chemical formula: $C_2Cl_3F_3$; Molecular weight: 187.38	Density, ρ_l (kg/m^3)	S	S	S	S	S	S	S	1.621	1.575
	Melting point: −34.96°C; Boiling point: 47.6°C	Specific heat capacity, $c_{p,l}$ (kJ/kg K)	S	S	S	S	S	S	S	0.858	0.900
	Critical temperature: 214.1°C	Thermal conductivity, λ_l [(W/m^2)/(K/m)]	S	S	S	S	S	S	S	0.0802	0.0761
	Critical pressure: 3.412 MPa	Dynamic viscosity, η_l (10^{-5} Ns/m^2)	S	S	S	S	S	S	S	97.8	72.7
DICHLOROTER-TRA-FLUORO-ETHANE (Refrigerant 114)	Chemical formula: $C_2Cl_2F_4$; Molecular weight: 170.92	Density, ρ_l (kg/m^3)	S	S	S	S	S	S	1.678	1.538	V
	Melting point: −93.85°C; Boiling point: 3.75°C	Specific heat capacity, $c_{p,l}$ (kJ/kg K)	S	S	S	S	S	S	0.929	0.963	V
	Critical temperature: 145.75°C	Thermal conductivity, λ_l [(W/m^2)/(K/m)]	S	S	S	S	S	S	0.0841	0.0710	V
	Critical pressure: 3.26 MPa	Dynamic viscosity, η_l (10^{-5} Ns/m^2)	S	S	S	S	S	S	107.2	46.7	V

From Chapter 5.5.10 of the *Heat Exchanger Design Handbook*, (1985) "Physical properties of liquids of temperatures below their boiling point" by Clive F. Beaton. Hemisphere Publishing Corporation.

Table (B) Refrigerants: Properties of saturated fluids

Dichlorodifluoromethane (Refrigerant 12)

Chemical formula: CCl_2F_2
Molecular weight: 120.92
Normal boiling point: 243.2 K
Melting point: 118 K

Critical temperature: 384.8 K
Critical pressure: 4 132 kPa
Critical density: 561.8 kg/m^2

T_{sat}, K	243.2	260	275	290	305	320	335	350	365	384.8
p_{sat}, kPa	101.3	200	333	528	793	1 145	1 602	2 183	2 907	4 132
ρ_ℓ, kg/m^3	1 486	1 436	1 388	1 338	1 284	1 225	1 157	1 075	969.7	561.8
ρ_g, kg/m^3	6.33	11.8	19.2	29.9	44.8	65.4	94.6	136.4	203.2	561.8
h_ℓ, kJ/kg	473.6	488.3	501.9	516.3	531.1	546.8	563.8	582.3	603.2	649.8
h_g, kJ/kg	641.9	649.8	656.6	662.9	668.8	674.0	677.8	679.9	679.0	649.8
$\Delta h_{g,\ell}$, kJ/kg	168.3	161.5	154.7	146.6	137.7	127.2	114.0	97.6	75.8	
$c_{p,\ell}$, kJ/(kg K)	0.896	0.911	0.932	0.957	0.990	1.03	1.08	1.13	1.22	
$c_{p,g}$, kJ/(kg K)	0.569	0.614	0.646	0.689	0.746	0.825	0.920	1.22	1.68	
η_ℓ, μNs/m^2	373	303	262	231	208	187	167	144	119	
η_g, μNs/m^2	10.3	11.0	11.7	12.5	13.3	14.2	15.2	16.5	18.1	
λ_ℓ, (mW/m^2)/(K/m)	95.1	87.4	80.5	73.3	66.8	59.8	53.0	46.2	39.2	15.4
λ_g, (mW/m^2)/(K/m)	6.9	7.7	8.4	9.2	10.0	10.8	11.6	12.3	13.4	15.4
Pr_ℓ	3.51	3.16	3.03	3.02	3.14	3.22	3.40	3.52	3.70	
Pr_g	0.85	0.88	0.90	0.94	0.99	1.08	1.21	1.64	2.27	
σ, mN/m	15.5	13.5	11.4	9.4	7.7	5.9	4.2	2.8	1.3	
$\beta_{e,\ell}$, kK^{-1}	1.96	2.22	1.50	2.87	3.36	4.17	5.48	8.50	14.5	

Table (B) Refrigerants: Properties of saturated fluids—*continued*

Chlorotrifluoromethane (Refrigerant 13)

Chemical formula: CCl_3
Molecular weight: 104.47
Normal boiling point: 191.7 K
Melting point: 93.2 K

Critical temperature: 302.28 K
Critical pressure: 3 900 kPa
Critical density: 5.71 kg/m^3

T_{sat}, K	191.7	200	210	220	235	250	265	280	295	302.28
p_{sat},	101.3	155	245	371	641	1 060	1 590	2 340	3 320	3 900
ρ_ℓ, kg/m^3	1 521	1 489	1 450	1 408	1 339	1 257	1 175	1 066	893	571
ρ_g, kg/m^3	6.89	10.3	15.8	23.4	39.9	67.2	103	166	290	571
h_ℓ, kJ/kg	417.1	424.6	433.9	443.3	458.1	474.5	489.8	509.1	532.8	562.0
h_g, kJ/kg	566.4	569.8	573.7	577.5	582.6	587.2	589.8	590.3	583.9	562.0
$\Delta h_{g,\ell}$, kJ/kg	149.3	145.2	139.8	134.2	124.5	112.7	100.0	81.2	51.1	
$c_{p,\ell}$, kJ/(kg K)	1.11	1.13	1.16	1.21	1.27	1.36	1.56	1.96	3.92	
$c_{p,g}$, kJ/(kg K)	.529	.552	.588	.633	.696	.854	.962	1.30	2.52	
η_ℓ, μNs/m^2	318	282	248	220	188	163	142	114	69.5	
η_g, μNs/m^2	9.83	10.3	10.8	11.3	12.2	13.2	14.3	15.8	19.9	
λ_ℓ, (mW/m^2)/(K/m)	99.3	95.0	89.8	84.6	77.0	69.2	60.7	48.6	34.0	23.6
λ_g, (mW/m^2)/(K/m)	6.4	7.0	7.6	8.3	9.0	10.0	11.5	13.2	16.2	23.6
Pr_ℓ	3.55	3.35	3.20	3.15	3.10	3.20	3.65	4.56	8.01	
Pr_g	0.81	0.82	0.84	0.86	0.94	1.13	1.20	1.56	3.10	
σ, mN/m	13.5	12.3	10.8	9.41	7.36	5.41	3.58	1.91	0.49	
$\beta_{e,\ell}$, kK^{-1}	2.51	2.66	2.92	3.30	4.20	5.55	7.50	12.3	76.0	

Table (B) Refrigerants: Properties of saturated fluids—*continued*

Dichlorofluoromethane (Refrigerant 21)

Chemical formula: $CHCl_2F$
Molecular weight: 102.92
Normal boiling point: 281.9 K
Melting point: 138 K

Critical temperature: 451.25 K
Critical pressure: 5 181 kPa
Critical density: 525.0 kg/m^3

T_{sat}, K	281.9	300	320	340	360	380	400	420	440	451.25
p_{sat}, kPa	101.3	196	364	626	1 010	1 540	2 240	3 160	4 350	5 181
ρ_ℓ, kg/m^3	1 406	1 360	1 311	1 258	1 199	1 134	1 057	962.8	823.0	525
ρ_g, kg/m^3	4.62	8.55	15.48	26.14	41.93	64.94	98.91	151.1	124.1	525
h_ℓ, kJ/kg	509.1	528.1	549.9	572.3	595.2	618.4	642.4	668.6	701.4	748.7
h_g, kJ/kg	748.1	756.5	765.5	773.6	780.5	786.6	790.7	791.5	785.3	748.7
$\Delta h_{g,\ell}$, kJ/kg	239.0	228.4	215.6	201.3	185.3	168.2	148.3	122.9	83.9	
$c_{p,\ell}$, kJ/(kg K)	1.04	1.05	1.07	1.10	1.17	1.26	1.37	1.52	1.89	
$c_{p,g}$, kJ/(kg K)	0.721	.733	.780	.832	.902	.982	1.05	1.43	2.50	
η_ℓ, μNs/m^2	366	311	269	235	209	187	165	135	99	
η_g, μNs/m^2	11.0	11.7	12.5	13.3	14.3	15.4	16.5	18.3	25.0	
λ_ℓ, (mW/m^2)/(K/m)	114	108	101	94.2	87.5	80.7	73.8	66.7	59.9	
λ_g, (mW/m^2)/(K/m)	7.9	8.6	9.5	10.4	11.4	12.5	13.7	15.2	16.7	
Pr_ℓ	3.34	3.02	2.85	2.74	2.79	2.89	3.06	3.08	3.12	
Pr_g	1.00	1.01	1.03	1.06	1.13	1.21	1.26	1.72	3.74	
σ, mN/m	20.1	17.6	14.7	12.1	9.4	7.0	4.7	2.6	0.73	
$\beta_{e,\ell}$, kK^{-1}	1.64	1.85	2.11	2.47	2.95	3.78	5.32	8.50	54.0	

Chlorodifluoromethane (Refrigerant 22)

Chemical formula: $CHClF_2$
Molecular weight: 86.48
Normal boiling point: 242.4 K
Melting point: 113.2 K

Critical temperature: 369.3 K
Critical pressure: 4 986 kPa
Critical density: 513 kg/m^3

T_{sat}, K	242.4	250	265	280	295	310	325	340	355	369.3
$_{sat}$, kPa	101.3	218	376	619	958	1 420	2 020	2 800	3 800	4 986
ρ_ℓ, kg/m^3	1 413	1 360	1 313	1 260	1 206	1 146	1 076	991	877	513
ρ_g, kg/m^3	4.70	9.59	16.1	26.3	40.6	60.9	90.2	134	208	513
h_ℓ, kJ/kg	453.6	469.3	490.2	508.1	526.3	545.2	565.3	587.3	613.2	667.3
h_g, kJ/kg	687.0	694.9	701.0	706.7	711.5	715.0	716.9	716.0	708.9	667.3
$\Delta h_{g,\ell}$, kJ/kg	233.4	225.6	210.8	198.6	185.2	169.8	151.6	128.7	95.7	
$c_{p,\ell}$, kJ/(kg K)	1.10	1.13	1.16	1.19	1.24	1.30	1.41	1.65	2.43	
$c_{p,g}$, kJ/(kg K)	.599	.646	.691	.747	.820	.930	1.09	1.40	2.31	
η_ℓ, μNs/m^2	332	282	251	225	204	187	172	150	119	
η_g, μNs/m^2	10.1	10.9	11.7	12.3	13.2	14.2	15.7	16.4	18.8	
λ_ℓ, (mW/m^2)/(K/m)	119	109	101	94.2	86.6	78.8	70.2	59.2	44.0	31.9
λ_g, (mW/m^2)/(K/m)	7.15	8.22	9.10	10.1	11.2	12.4	14.0	16.0	18.8	31.9
Pr_ℓ	3.07	2.92	2.88	2.84	2.92	3.09	3.45	4.18	6.89	
Pr_g	.85	.86	.89	.91	.97	1.07	1.22	1.69	2.31	
σ, mN/m	18.3	15.5	13.0	10.6	8.4	6.2	4.3	2.5	1.0	
$\beta_{e,\ell}$, kK^{-1}	2.08	2.37	2.68	3.10	3.67	4.65	6.34	9.56	25.6	

From Chapter 5.5.1 of the *Heat Exchanger Design Handbook*, 1983. "Proportion of saturate fluids", by R. N. Maddox. Hemisphere Publishing Corporation.

REFRIGERATION

A refrigerator is a device which is designed to remove heat from a space that is at lower temperature than its surroundings. The same device can be used to heat a volume that is at higher temperature than the surroundings. In this case the device is called a **Heat Pump**. The distinction between a refrigerator and a heat pump is one of purpose rather than principle. Therefore, this section will concentrate on refrigeration and only make the distinction between the two devices when necessary.

The Clausius statement of the *Second Law of Thermodynamics* asserts that it is impossible to construct a device that, operating in a cycle, has no effect other than the transfer of heat from a cooler to a hotter body. This means that energy will not flow from cold to hot regions without outside assistance. The refrigerator and heat pump both satisfy the Clausius requirement of external action through the application of mechanical power or equivalent natural transfers of heat.

Continuous refrigeration can be achieved by several processes. Effectively any heat engine cycle, when reversed, becomes a refrigeration cycle. The *vapour compression cycle* is the most commonly used in refrigeration and air condition applications. The *vapour absorption cycle* provides an alternative system, particularly in applications where heat is economically available. *Steam-jet* systems are also being successfully used in many cooling applications while *air-cycle* refrigeration is often used for aircraft cooling. These cycles are described in detail in Look and Sauer (1988), and in the ASHRAE Handbook, of Fundamentals (1993). Refrigeration equipment is described in detail in the ASHRAE Handbook, HVAC Systems and Equipment Volume (1992) and refrigerating systems practices are in the ASHRAE Handbook, Refrigeration Volume (1990).

Reversed Heat Engine Cycles:

Mechanical refrigeration processes, of which the vapour compression cycle is an example, belong to the general class of reversed heat engine cycles, **Figure 1**. This figure represents, schematically, the extraction of heat at rate $\dot{Q}$ from a cold body at temperature T_C. The process requires the expenditure of work W and the sum $(\dot{Q} + \dot{W})$ is discharged at a higher temperature T_H.

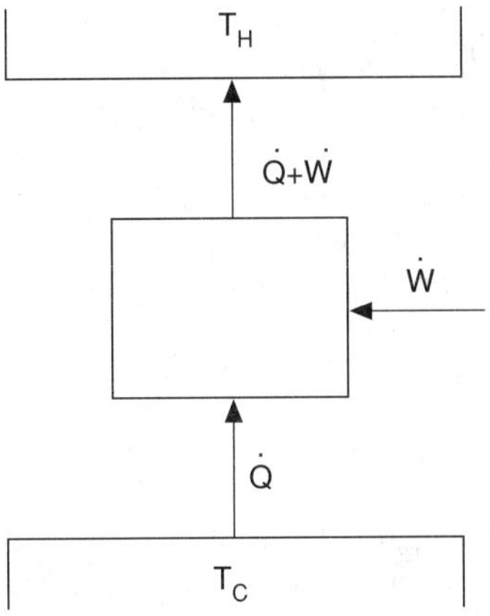

Figure 1. Reversed heat engine cycle.

The ideal cycle against which any practical reversed heat engine may be compared is the reversible or **Carnot Cycle** for which, in accordance with the Second Law of Thermodynamics, the following relationship applies:

$$\dot{W} = \frac{(T_H - T_C)}{T_C}\dot{Q} \tag{1}$$

One measure of the efficiency of such a process is given by:

$$\frac{\dot{W}}{\dot{Q}} = \frac{T_H - T_C}{T_C} \tag{2}$$

Clearly the smaller the value of the ratio $\dot{Q}/\dot{W}o$ the more efficient the process.

It is more usual to describe the efficiency of a reversed heat engine by the inverse of this ratio, known as the Coefficient Of Performance (COP):

$$COP = \frac{\dot{Q}}{\dot{W}} = \frac{T_C}{T_H - T_C} \tag{3}$$

It will be observed that the COP may be greater than unity and that it becomes greater as the temperature difference decreases. A real refrigerator or reversed heat engine will have a COP less than that of the ideal Carnot Cycle engine as given by the above equation.

The reversed Carnot Cycle is represented on the Temperature-Entropy (T-S) diagram by a rectangle, **Figure 2**, and is composed of four reversible processes;

4–1 isothermal expansion, during which heat (the refrigeration load) flows from the cold space to the working fluid.

1–2 adiabatic compression.

2–3 isothermal compression in which heat flows from refrigerant to the hot space.

3–4 adiabatic expansion.

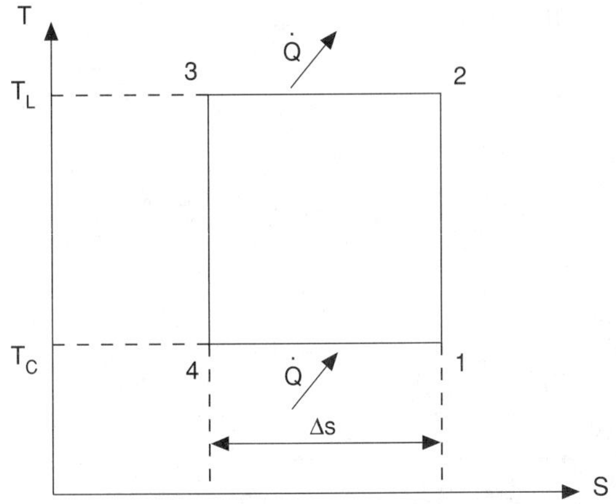

Figure 2. Temperature-entropy diagram for ideal reversed carnot cycle.

The Basic Vapour Compression Cycle and Components

Vapour compression refrigeration, as the name suggests, employs a compression process to raise the pressure of a working fluid vapour (refrigerant) flowing from an evaporator at low pressure P_L to a high pressure P_H as shown in **Figure 3**. The refrigerant then flows through a condenser at the higher pressure P_H, through a throttling device, and back to the low pressure, P_L in the evaporator. The pressures P_L and P_H correspond to the refrigerant saturation temperatures, T_1 and T_5 respectively.

The T-S diagram for this real cycle, **Figure 4**, is somewhat different to the rectangular shape of the Carnot Cycle.

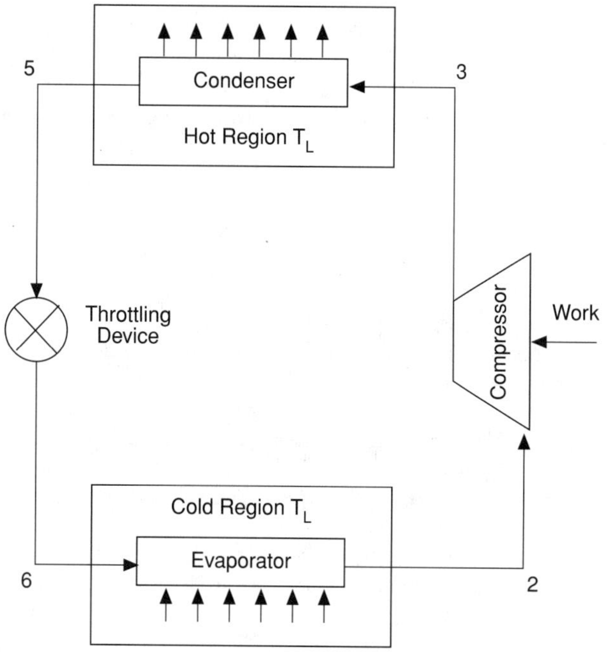

Figure 3. Basic vapour compression refrigerator.

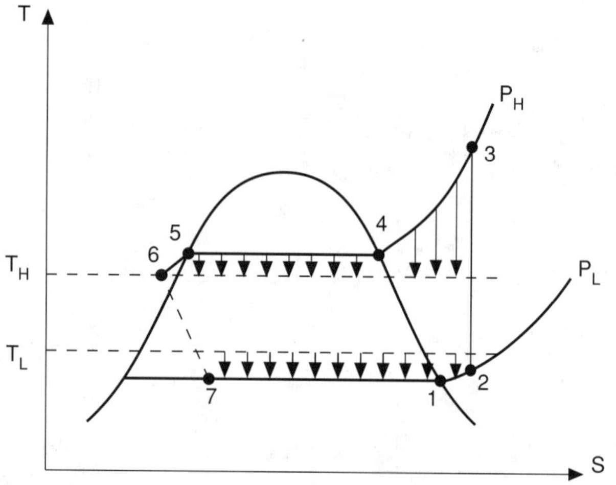

Figure 4. T-S diagram for basic vapour compression cycle.

The cycle processes can be described as follows:

7–1 Evaporation of the liquefied refrigerant at constant temperature $T_1 = T_7$.

1–2 Superheating of the vapour from temperature T_1 to T_2 at constant pressure P_L.

2–3 Compression (not necessarily adiabatic) from temperature T_2 and pressure P_L to temperature T_3 and pressure P_H.

3–4 Cooling of the super-heated vapour to the saturation temperature T_4.

4–5 Condensation of the vapour at temperature $T_4 = T_5$ and pressure P_H.

5–6 Sub-cooling of the liquid from T_5 to T_6 at pressure P_H.

6–7 Expansion from pressure P_H to pressure P_L at constant enthalpy.

A further difference between the real cycle and the ideal is that temperature T_1 at which evaporation takes place is lower than the temperature T_L of the cold region so heat transfer can take place. Similarly the temperature T_4 of the heat rejection must be higher than the hot region temperature T_H to bring about heat transfer in the condenser.

It is usual for the vapour-compression cycle to be plotted on a pressure-enthalpy (p-h) diagram as shown in **Figure 5**.

The cycle calculations are described in detail in many text books (e.g. Eastop and Mc Conkey, 1993, and Rogers and Mayhew 1992).

Refrigerants

Refrigerants are the working fluids in refrigeration systems. They must have certain characteristics which include good refrigeration performance, low flammability and toxicity, compatibility with compressor lubricating oils and metals, and good heat transfer properties. They are usually identified by a number that relates to their molecular composition. The ASHRAE Handbook of Fundamentals (1993) lists a large number of available refrigerants and gives their properties. (see **Refrigerants**)

In recent years, environmental concerns over the use of *chlorofluorocarbons* (CFCs) as the working fluids in refrigeration and air-conditioning plants have led to the development of alternative fluids. The majority of these fall into two categories, *hydrofluorocarbons* (HDCs) which contain no chlorine and have zero ozone depletion potential and *hydrochlorofluorocarbons* (HCFCs), which do contain chlorine, but the addition of hydrogen to the CFC structure allows virtually all the chlorine to be dispersed in the lower atmosphere before it can reach the ozone layer. HCFCs therefore have much lower ozone depletion potentials; ranging from 2 to 10% that of CFCs. Many nations have signed the *Vienna Convention* which is a treaty intended to control the production of substances known to be depleting the ozone layer. The *Montreal Protocol* to this treaty in 1987 outlines the means for achieving certain limits in production of particular substances and the timetable for their phasing out. A great deal of research is being carried out to determine the properties of new ozone friendly fluids and mixtures (Sauer and Kuehn 1993).

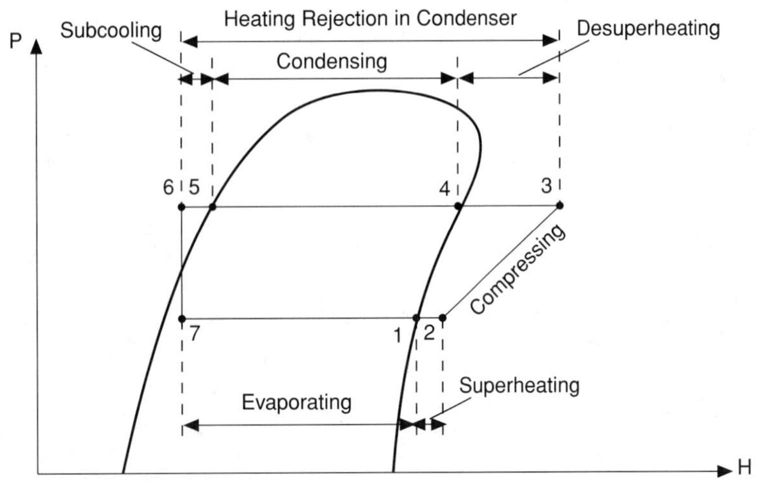

Figure 5. p-h representation of vapour compression cycles.

Vapour-Absorption Cycles

Recently interest has been increasing in these cycles because of their potential use as part of energy-saving plants and also because they use more environmentally friendly refrigerants than vapour-compression cycles. A basic vapour-absorption system is shown schematically in **Figure 6**. The condenser, throttling valve and evaporator are essentially the same as in the vapour compression system (**Figure 3**). The major difference is the replacement of the compressor with an *absorber*, a *generator*, and a solution pump. A second throttling valve is also used to maintain the pressure difference between the absorber (at the evaporator pressure) and the generator (at the condenser pressure).

The refrigerant on leaving the evaporator is absorbed in a low-temperature absorbing medium, some heat, Q_A, being rejected in the process. The refrigerant-absorbent solution is then pumped to the higher pressure and is heated in the generator, Q_G. Refrigerants vapour then separate from the solution due to the high pressure and temperature in the generator. The vapour passes to the condenser and the weak solution is throttled back to the absorber. A heat exchanger may be placed between the absorber and the generator to increase the energy efficiency of the system. The work done in pumping the liquid solution is much less than that required by the compressor in the equivalent vapour-compression cycle. The main energy input to the system, Q_G, may be supplied in any convenient form such as a fuel burning device, electrical heating, steam, solar

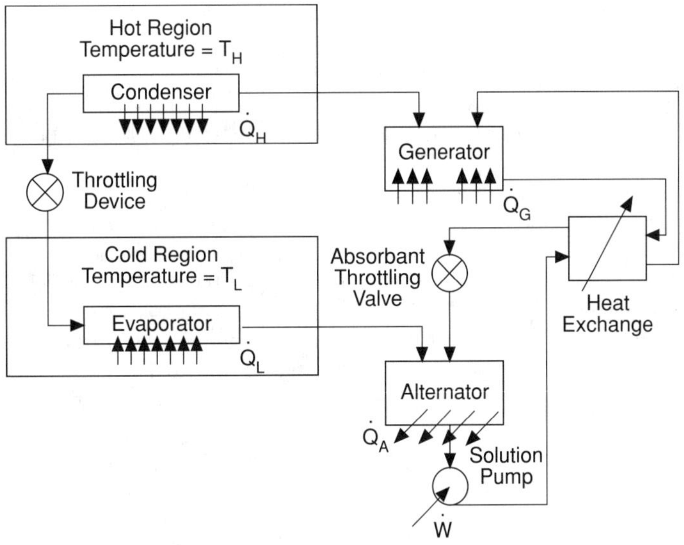

Figure 6. Basic absorption refrigerant system.

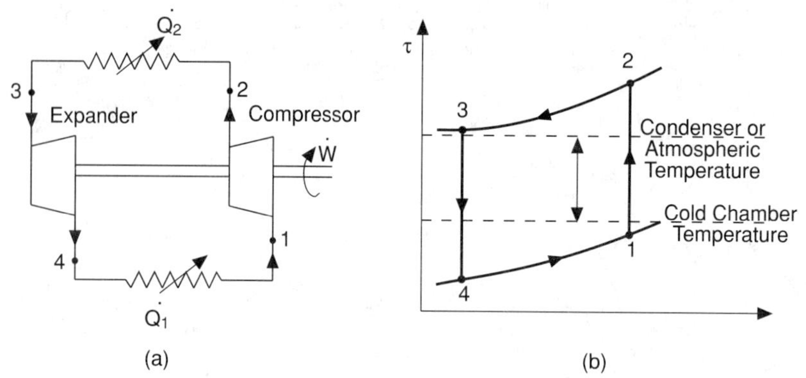

Figure 7. Simple gas-cycle refrigerant system.

energy or waste heat. Appropriate refrigerant/absorbent combinations must be selected. One common combination uses ammonia as refrigerant and water as absorbent. An alternative combination is water as refrigerant and lithium bromide as absorbent. There are increasing research activities into finding suitable new combinations (Hodgett 1982).

Gas-Cycle Refrigeration

Gas-cycle refrigeration is essentially, a reversed Joule cycle (gas turbine cycle). As the name indicates, the refrigerant in these systems is a gas. The system, as shown in **Figure 7**, is basically the same as that of the vapour-compression cycle. The main difference is the replacement of the throttling valve by an expander.

The cycle can be described as follows:

1–2 Adiabatic compression.

2–3 Constant pressure cooling.

3–4 Adiabatic expansion.

4–1 Constant pressure heating (cooling effect).

As can be seen from **Figure 7**, the gas does not receive and reject heat at constant temperature, and, therefore, the gas cycle is less efficient than the vapour cycle for given evaporator and condenser temperatures. Gas-cycle systems are mostly used in air conditioning applications where the working fluid-air can be ejected at T_4. A common application is in the air conditioning of aircraft. Air, held from the engine compressor, is cooled in a heat exchanger and then expanded through a turbine. The power from the turbine is used to drive a fan which provides the cooling air for the heat exchanger. Air at T_4 is ejected into the cabin to provide the required cooling.

References

ASHRAE Handbook. (1992) HVAC Systems and Equipment Volume, American Society of Heating, Refrigerating and Air-Conditioning Engineer Inc., Atlanta, GA.

ASHRAE Handbook. (1990) Refrigeration Volume, American Society of Heating, Refrigerating and Air-Conditioning Engineer Inc., Atlanta, GA.

ASHRAE Handbook. (1990) Fundamentals Volume, American Society of Heating, Refrigerating and Air-Conditioning Engineer Inc. Atlanta, GA

Eastop, T. D. and McConkey, A. (1993) *Applied Thermodynamics,* Longman Scientific and Technical, Harlow.

Hodgett (1982) Proceeding of Workshop in Berlin, April 14–16, Swedish Council for Building Research, ISSN: 91–54039294.

Look, D. L. and Sauer, H. I. (1988) *Engineering Thermodynamics,* Van Nostrand Reinhold (International), Wokingham.

Rogers, G. F. C. and Mayhew, Y. R. (1992) *Thermodynamic Work and Heat Transfer,* Longman Scientific and Technical, Harlow.

Sauer, H. J. and Kuehn, T. H. (1993) *Heat Transfer with Alternative Refrigerants, ASME,* HTD–Vol 243, New York.

M. Heikal

REGENERATIVE BURNER (see Regenerative heat exchangers)

REGENERATIVE FEED HEATING (see Rankine cycle)

REGENERATIVE GAS TURBINE (see Gas turbine)

REGENERATIVE HEAT EXCHANGERS

***Following from:** Heat exchangers*

Whereas in recuperators, where heat is transferred directly and immediately through a partition wall of some kind, from a hot to a cold fluid, both of which flow simultaneously through the exchanger, the operation of the regenerative heat exchanger involves the temporary storage of the heat transferred in a *packing* which possesses the necessary thermal capacity. One consequence of this is that in regenerative heat exchangers or *thermal regenerators,* the hot and cold fluids pass through the same channels in the packing, alternately, both fluids washing the same surface area. In recuperators, the hot and cold fluids pass simultaneously through different but adjacent channels.

In thermal regenerator operation the hot fluid passes through the channels of the packing for a length of time called the "hot period," at the end of which, the hot fluid is switched off. A reversal now takes place when the cold fluid is admitted into the channels of the packing, initially driving out any hot fluid still resident in these channels, thereby purging the regenerator. The cold fluid then flows through the regenerator for a length of time called the "cold period," at the end of which the cold fluid is switched off and another reversal occurs in which, this time, the hot fluid purges the channels of the packing of any remaining cold fluid. A fresh hot period then begins.

During the hot period, heat is transferred from the hot fluid and is stored in the packing of the regenerator. In the subsequent cold period, this heat is regenerated and is transferred to the cold fluid passing through the exchanger.

A cycle of operation consists of a hot followed by a cold period of operation together with the necessary reversals. After many cycles of identical operation, the temperature performance of the thermal regenerator in one cycle is identical to that in the next. When this condition is realised, the heat exchanger is said to have reached "cyclic equilibrium" or "periodic steady state." Should a step change be introduced in one or more of the operating parameters, in particular, the flow rate and entrance temperature of the fluid for either period of operation, or the duration of the hot and cold periods, the regenerator undergoes a number of transient cycles until the new cyclic equilibrium is reached.

In the most common counter-flow or contra-flow regenerator operation, the hot gas passes through through the regenerator in the opposite direction of the cold fluid. In less efficient parallel flow or co-flow the hot and cold fluids pass through the channels of the packing in the same direction(†).

The periodic operation of regenerators can exploit the periodic operation of the system to which the exchanger is attached. For example, in hot climates, day time heat can be stored in a packing by passing the warm atmospheric air through it: this heat can then be recovered by blowing cold night time air through the same packing during the evening to provide at least some supplementary warming of the living space in a building. Hausen (1976) suggests that the throat and nasal passages act as a regenerator packing in cold weather. When an animal breathes in cold air, it is warmed as it passes through the nose and throat before the air reaches the lungs, thereby protecting the lungs from the effects of cold temperatures. As the animal breathes out, the same passages in the nose and throat are warmed by the air leaving the lungs. Clearly, the temperature of the throat and nasal tissue is also regulated by the flow of blood through it.

In general, however, a continuous supply of heated fluid is required so that the discontinuous operation of the regenerator, which is inherent in its design, must be concealed in some way.

Fixed Bed Regenerators

The most obvious technique for realising "apparent" continuous operation, is to use of two or more regenerators operating out of phase with respect to one another so that while one regenerator is supplying heated fluid, the other regenerator(s) is storing heat from the heating fluid. An apparently easy way to do this is by enclosing the set of regenerators within a system of ducts or pipes fitted with valves to facilitate the switching of the regenerators at the end of a period of operation. As one set of valves close, at a reversal, so another set open: the flow of hot gas, for example, is diverted from one regenerator to the other by the closing of such a set of valves and the opening of the other. Simultaneously, the flow of cold gas is switched from the other regenerator in a symmetric fashion. (See **Figure 1**.) Such an arrangement is called a system of *fixed bed regenerators,* in contrast to the *rotary* regenerator which will be described shortly.

The reversal process can be more complicated than this. The following are important considerations.

- Some applications require that the regenerator(s) be purged before the supply of heated fluid, for example, is switched from one regenerator to the next. In this case, the cold period of one regenerator is extended to maintain the supply of heated fluid to the external process to which the set of regenerators is attached. Meanwhile, the hot period of the other regenerator is terminated and that regenerator is completely purged before its cold period begins. This regenerator then shoulders the burden of supplying heated fluid from the other regenerator whose end of cold period reversal can begin. Such arrangements necessarily complicate the valve and duct facilities associated with the set of regenerators: in addition, a suitable exhaust for fluids purged from the regenerator, not permitted to enter the heated fluid stream, for example, must be provided.

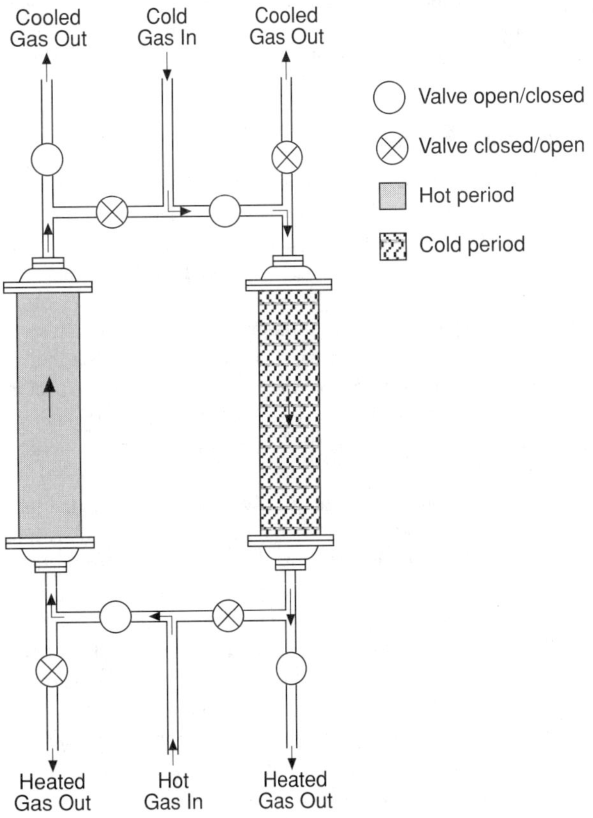

Figure 1. Fixed bed regenerator arrangement.

(†) In theory, it is possible to imagine a cross-flow regenerator in which the hot and cold fluids flow in directions perpendicular to one another. This is rarely, if ever, realised in practice although cross-flow recuperators are common.

- Where the fluids are gases, it is not uncommon for the pressure of the cold gas stream, for example, to be significantly higher than that of the hot gas stream. In this case, at the end of a cold period, time must be allowed during the reversal for the regenerator to be decompressed before the hot period is permitted to begin. Similarly, time must be allowed at the start of a "cold" period for the pressure of the cold gas in the regenerator to build up before the cold period proper can begin. Again, additional valves and pipework must be provided to accommodate these complications.

In high temperature regenerators, it is desirable not have any valves at all at the hot end of the regenerators. Where this cannot be avoided, the valves are often very expensive, perhaps requiring to be water cooled to avoid malfunction at high temperatures. It is often the case, however, the the hot end of the regenerator is attached to the furnace or boiler to where pre-heated air, for the combustion of a fuel gas, is supplied directly, and from where, after a reversal, the hot gas, frequently the waste products of combustion of the fuel, is extracted directly to the regenerator. In this way, no valves are required between the regenerator and the furnace or boiler. The necessary suction of the hot gas through the regenerator is achieved by attaching the exit duct for this gas at the cold end of the regenerator to a chimney, which, if tall enough, will provide the necessary updraft. Valves are employed safely at the cold end of the regenerator to switch the heat exchanger from the chimney exhaust for waste gases to the supply of cold air for the cold period of regenerator operation, or vice-versa. A continuous supply of pre-heated combustion air is achieved by attaching several regenerators to a furnace or boiler, operating as necessary, out of phase-with respect to one another.

Rotary Regenerators

In the *Ljungström air preheater,* or *rotary regenerator,* the porous packing is rotated around an axis. In its simplest form, the packing is divided into two gas tight sections and the hot and cold gases flow, simultaneously, in a direction parallel to this axis, usually in contra-flow, through these different segments of the packing. As the packing rotates through the hot has stream, it stores heat, as in the hot period of a fixed bed regenerator. This thermal energy is literally transported into the cold gas stream as the packing is rotated. Once in the other gas stream, the heat is regenerated and is passed to the cold gas, as in the cold period of operation of a fixed bed system.

It is impossible to secure completely gas-tight seals at the junctions between the ducts, carrying the hot and cold fluids respectively, and the moving surfaces of the rotating heat storing mass of the regenerator packing. However, if it is required to prevent pre-heated air, for example, from being contaminated by the waste combustion products which supply the necessary thermal energy, it is arranged that the pressure of the air is deliberately higher with the consequence that any leakage under the seals of the rotary regenerator is of air into the stream of hot combustion products.

When the rotor first passes from the hot gas to the cold gas stream, for example, a body of hot gas in the voids of the regenerator packing, is carried by rotation into the cold gas stream and must be purged from the regenerator, as in the fixed bed mode of operation. In some applications, it is vital that this carryover gas should not be permitted to contaminate the stream of cold gas being heated by the exchanger. In these circumstances, an additional sector is provided in the packing so that gases purged from the regenerator at the end of the hot period, for example, can be sent off to a separate exhaust, possibly fed back into the hot gas stream.

Another version of the rotary regenerator exists in which the packing remains stationary but the hoods at both ends of the packing, through which the hot and cold fluids pass in separate ducts, rotate instead.

Mathematical Modelling of Thermal Regenerators

The relationship between the heat transferred between the fluid and the solid packing, and the heat absorbed by that packing is given by the equation

$$\bar{\alpha}A(T_f - T_s) = Mc_s \frac{\partial T_s}{\partial t} \quad (1)$$

On the other hand, the relationship between the thermal energy transferred between the solid packing and the fluid, and the heat taken up by the fluid passing through the regenerator is given by the equation

$$\bar{\alpha}A(T_s - T_f) = \dot{M}_f c_p L \frac{\partial T_f}{\partial y} + M_f c_p \frac{\partial T_f}{\partial t} \quad (2)$$

These equations apply equally to the hot and cold periods of regenerator operation, for which the relevant parameters may be different. We therefore denote the bulk heat transfer coefficient in the hot period by $\bar{\alpha}$, in the cold period by $\bar{\alpha}'$, for example.

The boundary equations relate first to the entrance gas temperature, $T_{f,in}$, where it is assumed that $T_{f,in}(t)$ = constant in each period of operation. Next, we specify that **Equations 1 and 2** consider the gas to be moving from $y = 0$ to $y = L$ in both the hot and the cold periods of regenerator operation. In order to specify that the solid temperature distribution at the start of a period is equal to that at the end of the previous period, and to incorporate the contra-flow operation of the regenerator, the boundary conditions are written in the form

$$T_s'(y, 0) = T_s(L - y, P) \quad (3)$$

$$T_s(y, 0) = T_s'(L - y, P') \quad (4)$$

where P is the duration of the hot period and P′ the duration of the cold period.

The most important assumption embodied in this model is that the the resistance to heat transfer at the surface of the solid and the resistance due to the finite conductivity of the packing, in a direction normal to fluid flow, can be incorporated together in a "bulk" or "lumped" heat transfer coefficient α where, using a development of Hausen (Hausen, 1942) the defining equation is given by

$$\frac{1}{\bar{\alpha}} = \frac{1}{\alpha} + \frac{\delta}{6\lambda_s}\Phi \qquad (5)$$

where δ in the thickness of the packing and λ_δ its thermal conductivity.

The function Φ attempts to reproduce the effect of the very rapid temperature changes within the packing, immediately after a reversal, at the start of a hot or cold period. It is a function of the dimensionless parameters Ω and Ω', where

$$\Omega = \frac{\kappa_s P}{\delta^2} \qquad \Omega' = \frac{\kappa_s' P'}{\delta^2} \qquad (6)$$

In the case where the packing can be considered to be a simple, plain wall, of thickness δ, then

$$\Phi = \frac{1}{6} - 0.00278\left\{\frac{1}{\Omega} + \frac{1}{\Omega'}\right\} \qquad (7)$$

for

$$\frac{1}{\Omega} + \frac{1}{\Omega'} \leq 20 \qquad (8)$$

Where

$$\frac{1}{\Omega} + \frac{1}{\Omega'} > 20 \qquad (9)$$

then

$$\Phi = \frac{0.357}{\left(0.3 + \frac{1}{2}\left(\frac{1}{\Omega} + \frac{1}{\Omega'}\right)\right)^{1/2}} \qquad (10)$$

Similar expressions are available for the case where the packing can be regarded as a collection of solid cylinders or as a bed of spheres. Reference should be made to the later work of Hausen (Hausen, 1976) The case for hollow cylinders is covered in the paper by Razelos (Razelos, 1967) et al.

It is also assumed in this model that the thermal conductivity in a direction parallel to fluid flow, so called "longitudinal conductivity", can be neglected. This problem is discussed in the paper by Bahnke and Howard (Bahnke and Howard, 1964). It is further idealised that the flow of fluid is uniform through the cross-section of the packing in both periods of operation. It is generally allowed, however, that the relevant thermophysical properties of both fluid and solid, including the heat transfer coefficients, can vary spatially and time-wise, as a function of temperature. Equally, it is permitted to consider the case where the mass flow rates of the fluids in either or both periods of regenerator operation can vary with time.

The model is greatly simplified in the so called "linear model" in which it is further assumed that the relevant thermophysical properties of both fluid and solid, including the heat transfer coefficients, do not vary spatially and time-wise, but are constant. On the other hand they are permitted to be different between the hot and cold periods. Similarly, it is assumed that the gas flow rates are constant, although, in general $\dot{M} \neq \dot{M}'$. In these simplifying circumstances, it is possible to reduce **Equations 1 and 2** into the forms

$$\frac{\partial T_s}{\partial \eta} = T_f - T_s \qquad (11)$$

$$\frac{\partial T_f}{\partial \xi} = T_s - T_f \qquad (12)$$

Here, the dimensionless parameters, η for time and ξ for length are introduced where

$$\eta = \frac{\bar{\alpha}A}{Mc_s}\left(t - \frac{M_f}{\dot{M}_f L}y\right) \qquad \xi = \frac{\bar{\alpha}A}{\dot{M}_f c_p L}y \qquad (13)$$

Upon setting t = P and y = L, each period of regenerator operation is defined in terms of two dimensionless parameters given the names by Hausen (Hausen, 1929) "reduced period", Π, and "reduced length," Λ. Equations 14, below define these for hot period operation.

$$\Pi = \frac{\bar{\alpha}A}{Mc_s}\left(P - \frac{M_f}{\dot{M}_f}\right) \qquad \Lambda = \frac{\bar{\alpha}A}{\dot{M}_f c_p} \qquad (14)$$

For the cold period, the corresponding equations are

$$\Pi' = \frac{\bar{\alpha}'A'}{M'c_s'}\left(P' - \frac{M_f'}{\dot{M}_f'}\right) \qquad \Lambda' = \frac{\bar{\alpha}'A'}{\dot{M}_f' c_p'} \qquad (15)$$

In this linear model, it is also possible to treat the *hot* period inlet temperature $T_{f,in} = 1.0$ and the "cold" period inlet temperature $T_{f,in} = 0.0$ (†).

It is not within the orbit of this text describe the many and varied methods of solution of these differential equations which have been developed over several years. Be it sufficient to say that there are two classes of method of solution, the "open" methods and the "closed" methods. In the open methods, a starting distribution of temperature, $T_s(\xi,0)$, of the solid at the start of a hot period is assumed. The model of the regenerator is then cycled through many cycles by solving **Equations 11 and 12**, for example, through successive hot and cold periods of operation. This simulation of the heat exchanger is then permitted to run until the periodic steady state is realised. In the closed methods, it is simply assumed that the solid temperature distribution $T_s(\xi,0)^{(n)}$, at the start of the nth cycle is equal to that, $T_s(\xi,0)^{(n+1)}$, at the start of the following cycle, at cyclic equilibrium. The equations are then solved, often in integral equation form, as a boundary value problem. No transient

† This is equivalent to setting

$$T_f = \frac{\tau_f - \tau_{f,in}'}{\tau_{f,in} - \tau_{f,in}'} \qquad T_s = \frac{\tau_s - \tau_{f,in}'}{\tau_{f,in} - \tau_{f,in}'}$$

where τ_f and τ_s are the "real" fluid and solid temperatures and T_f and T_s are the corresponding *dimensionless* temperatures.

cycles prior to equilibrium be established are simulated. It turns out that closed methods are useful for solving a range of linear problems but that they become extremely complicated when dealing with non-linear problems, that is when it is permitted that the relevant thermophysical properties of both fluid and solid, including the heat transfer coefficients, can vary spatially and time-wise, as a function of temperature and/or when the mass flow rates of the fluids in either or both periods of regenerator operation can vary with time. In these cases, the open methods are more easily adopted

Construction of the Regenerator Packing

The packing of a regenerator varies considerably from one type of application to another. On the one hand, the selection of the heat storing mass is determined by the harsh, or otherwise, operating conditions under which the regenerator is required to operate. On the other, the possible operating arrangements with such packings is best understood in the context of the dimensionless parameters Λ and Π. For a given reduced length, Λ, the maximum thermal performance, it can be shown, is achieved by employing as small a value of reduced period, Π as possible.

The ratios

$$\frac{Mc_s}{\left(P - \frac{M_f}{\dot{M}_f}\right)} \qquad \frac{M'c_s'}{\left(P' - \frac{M_f'}{\dot{M}_f'}\right)}$$

represent the heat capacity of the packing per hot period and per cold period respectively. The effective interface between the fluid flowing through the regenerator at any instant and the heat storing packing are the products $\bar{\alpha}A$ and $\bar{\alpha}'A'$ for the hot and cold periods. The larger these interfaces, the greater must be the heat capacity per period to accommodate the thermal energy involved.

In other words, the ratios

$$\frac{\bar{\alpha}A}{Mc_s} \qquad \frac{\bar{\alpha}'A'}{M'c_s'}$$

must be matched by periods which yield small enough values of Π and Π' so that as good a thermal performance of the regenerator as possible can be obtained.

This can be considered in another manner: economies in regenerator size can be obtained if thin packings are used where the area to mass ratio, A/M for the hot period, A′/M′ for the cold period, is large. In this case, small enough values of Π and Π' are obtained by operating the regenerator with short period lengths, that is with short cycle times.

On the other hand, harsh operating conditions might require the packing of the regenerator to be constructed of suitable materials fabricated with a robust geometrical arrangement. In this case, the area to mass ratio, A/M for the hot period, A′/M′ for the cold period, might well be relatively small in which case, suitable values of Π and Π' can be obtained using long cycle times, thereby avoiding a rapid switching of the regenerators in the system used.

The matter is complicated still further if the process of reversing a regenerator is itself slow. For example, if it is necessary to pressurise the regenerator vessel at the start of the cold period, and then de-pressurise it at the end of the cold period, as is the case with *Cowper stoves* used for preheating the blast (of air) for ironmaking, then the total cycle time must be long enough for the time necessary for these reversals not to constitute an overlarge proportion of the total cycle time. In this case, the area to mass ratio, A/M for the hot period, A′/M′ for the cold period, must be forced to be small enough to generate sufficiently small values of Π and Π' with the longer hot and cold periods of operation necessary.

Very High Temperature Regenerators

Fixed bed regenerators operating with hot gas inlet temperatures in excess of 1200°K are fitted with packings constructed of fireproof refractories or ceramic materials of special quality, capable of withstanding the effects of any corrosive materials entrained in the hot gas. Thus, in glass furnace regenerators, it has not been uncommon to use high alumina packings which are capable of coping with the corrosive effects of lime, potash, silica, sodium sulphate and vanadium which can be carried over into the regenerator packing from the glass making process. In Cowper stoves, used to preheat the blast for the ironmaking and zinc smelting processes, the packing is frequently zoned: at the top of the regenerator, materials capable to withstanding the effects of very high temperatures, and further down the regenerator, of high compressive loads are used, for example, silica. At the bottom of the regenerator, it is imperative to provide materials which possess the mechanical strength and volume stability, able to support the great weight of packing above. Different silica-alumina refractories are often employed in these circumstances.

Not only must the materials of the packings be able to cope with the effects of corrosive materials, so also must the geometry of the packings be so arranged that these possibly dirty gasses can have free passage through the regenerator. The blocking of the channels must be avoided. In these circumstances, various geometrical arrangements of the refractory materials must be used. "Square chimney" or "closed basket weave" arrangements are frequently used in these circumstances. (**Figure 2**)

It can be arranged that the channels are wide enough to provide free passage for dirty gases but for the packing not to be aligned to form chimney passages: the "open basket weave" or "staggered open basket weave" fulfill this role. The channel width can be as large as 200 mm.

Where the gases are very hot but relatively clean, as in Cowper stoves, hexagonal bricks are frequently used with passages only 50 mm wide. These passages are formed in the body and on the corners of the bricks **Figure 3**. In the *Freyn chequerwork* design, these passages are circular but other shapes are possible. These refractory bricks are arranged in layers in such a way that tubular

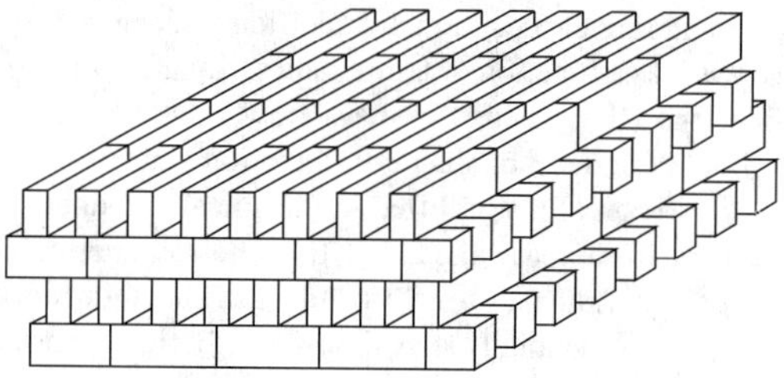

Figure 2. Diagram of basket weave packing arrangements.

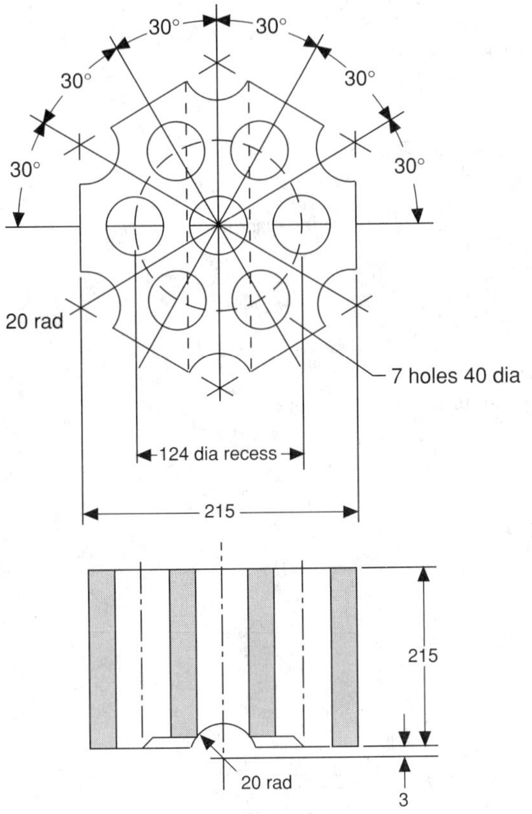

Figure 3. Typical design for the heat storing chequerwork for hot blast stoves. (*millimetre measurements*).

shaped channels are formed through which the gasses can have a clear passage.

In both kinds of arrangement, the thickness of the packing behind the available heating surface area is determined by the mechanical strength required of the packing as also by the corrosive conditions under which it is must be required to operate. Under severe conditions, thicknesses as large as 200 mm must be used; in less harsh conditions as can arise in the chemical industry, 50 mm thick bricks may be adequate.

There are two significant consequences of these possible packing arrangements. These can be understood in the context of the descriptive dimensionless parameters describing regenerators and their operations.

1. The reduced length,

$$\Lambda = \frac{\bar{\alpha}A}{\dot{M}_f c_p}$$

is a measure of the effective heating surface area relative to the heat capacity flow rate of the hot/cold gas. Clearly, the larger the load, $M_f c_p$, upon the regenerator, the larger must be the heating surface area, A, to service this load. Indeed, for given operating conditions, the thermal efficiency increases with Λ. High temperature regenerators which employ basket weave or Freyn-like hexagonal packings are physically very large, in some cases, so large that they are built as part and parcel of the furnace with which they are required to operate. This is a consequence of the packing having a low surface area to volume ratio, in the range 20–30 m^{-1}. The large surface area required brings with itself a large volume of packing. **Figure 4** gives the layout of a pair of regenerators in a Siemens arrangement, such as a glass furnace. In a Cowper stove arrangement, there are three or four regenerators held in cylindrically shaped vessels which are detached from the blast furnace they are required to serve (see **Figure 5**). They are still large, perhaps 30 m in height and 10 m in diameter.

2. On the other hand, the ratios

$$\frac{\bar{\alpha}A}{Mc_s} \qquad \frac{\bar{\alpha}'A'}{M'c_s'}$$

will be small for packings with a low surface area to volume ratio, as described above. The consequence is that small enough values of Π and Π' can be obtained using relatively long cycle times. It is not uncommon for Cowper stoves to operate with a cold period, P′, of between 30 and 60 minutes and a hot period, P, of between 50 and 110 minutes for a three stove arrangement: this includes accommodating reversals lasting 5 to 10 minutes per cycle. Similarly, Siemens type furnace regenerators commonly operate with a total cycle time of about 40 minutes.

Such large regenerators are necessarily expensive. Under high temperature conditions were the gasses are relatively clean, it has become common over the last ten years or so to turn to the *regenerative burner.* Here the regenerator packing consists of ceramic spheres with materials chosen to meet the operating conditions encountered. The spheres are typically 1–3 *cm* in diameter yielding area to volume ratios of the order of 10 times larger than encountered in the massive regenerators for the glass making furnace, for example. Area to volume ratios in the range 100–300 m^{-1} yield small compact regenerators. The particular size is determined by the load, $\dot{M}_f c_p$, which the regenerator is required to support. A bed 0.6 m high and 0.18 m in diameter would not be uncommon, although smaller or larger beds can be used for different thermal loads.

Because the ratios

$$\frac{\bar{\alpha}A}{Mc_s} \qquad \frac{\bar{\alpha}'A'}{M'c_s'}$$

will be larger for beds of ceramic spheres with a surface area to volume ratio in the range 100–300 m^{-1}, it is necessary to reverse the regenerators far more rapidly than is the case for massive high temperature regenerators. The burners are operated in pairs; during the hot period hot exhaust gasses are drawn from the furnace to which the burner is attached, over the bed of spheres. Simultaneously, during the cold period of the other burner fires into the furnace, using combustion air which has been pre-heated by the regenerative bed. The regenerators and their burners are switched after a period as short as 30 to 180 seconds. The combined burner and regenerator can be made most compact by incorporating a

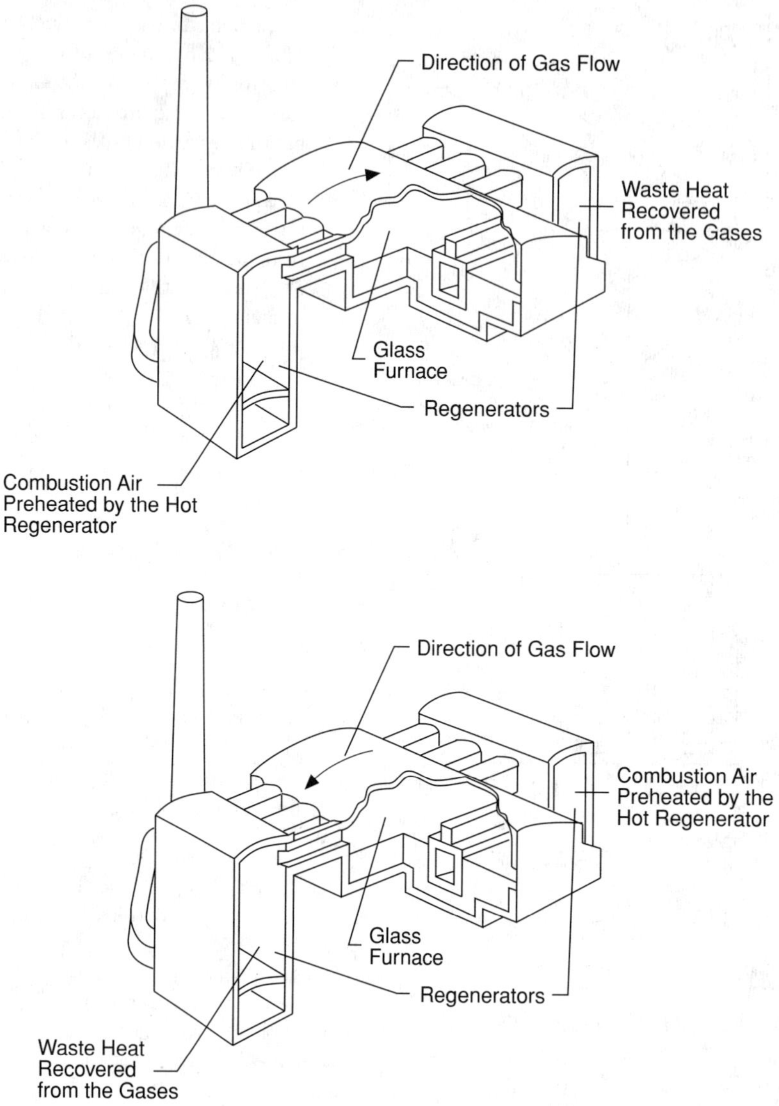

Figure 4. Arrangement of a pair of Siemens regenerators.

regenerator within the body of each burner unit. A small continuous gas fired furance, used for example in the steel industry for the annealing of strip steel, might incorporate six or more pairs of such burners. They can be operated out of phase in such a way that "apparent" continuous operation is realised by arranging that only one pair of burners is being reversed at any instant.

Moderate Temperature Regenerators

At more moderate temperatures (400–600 °C), it is common to employ rotary regenerative air pre-heaters. In such *Ljungström* regenerators, the cylindrical porous packing is rotated around its axis (See **Figure 6**). The packing materials are often fabricated of steel sheets, notched to form a large number of undulating passages. In this way, turbulent flow of the hot and cold gasses flowing through the regenerator is promoted, thereby improving the heat transfer characteristics. The metal sheets are arranged radially in removable units holding several such sheets, thereby facilitating rapid and simple maintenance.

Such metal sheeting provides a high area to volume ratio, in excess of 200 m^{-1}. Nevertheless, they must be constructed in such a way as to be able to withstand the temperatures involved as well as possibly corrosive operating conditions. Where, for example, the hot waste gases have a high SO_2 content, a vitreous enamelled heating surface can be employed to protect the steel packing at operating temperatures below the acid dew point of such gasses.

Even higher area to volume ratios can be achieved by constructing the regenerator of an assembly of sector shaped sections of a knitted mesh of wire of another material, depending on the temperature and other operating conditions. For hot gas entry tem-

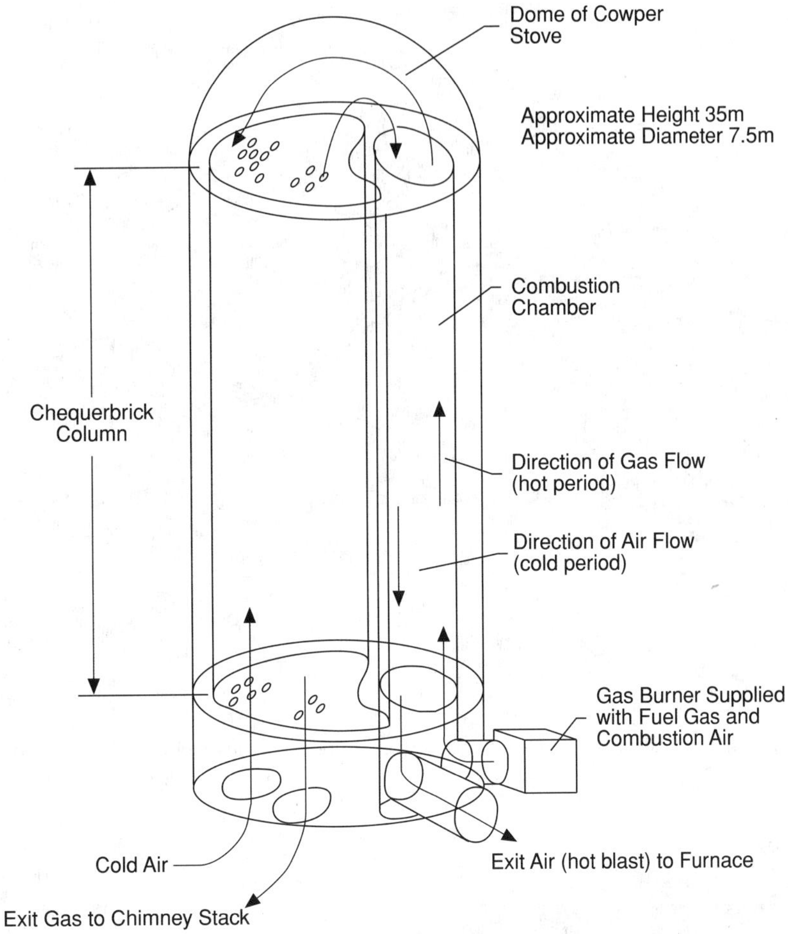

Figure 5. Sketch of Cowper stove used for preheating air for ironmaking blast furnace.

peratures of 400°C, stainless steel mesh can be employed while for temperatures of up to 800°C ceramic or alumina fibres have been considered. Other prefabricated heavy duty ceramic packings can be employed in regenerators required to withstand hot gas entry temperatures of 800°C or more: here the packing might consist of a honeycomb of ceramic material arranged as alternately flat and wave shaped layers. Such constructions realise the high surface area to volume ratio necessary to acheive compactness in regenerator construction and, at the same time, allow for a free passage for the flow of the gasses through the regenerator. They are also robust enough to survive the tough working temperatures and harsh operating conditions.

Again the ratios

$$\frac{\bar{\alpha}A}{Mc_s} \qquad \frac{\bar{\alpha}'A'}{M'c_s'}$$

will be large for the rotary regenerator packings described above. It is not uncommon, therefore, for the packing to be rotated at 2–3 revolutions per minute, yielding hot/cold periods of 30 seconds or less. Small values of Π and Π' are thus generated, thereby facilitating regenerator efficiencies of 80% of more to be realised.

Lower Temperature Regenerators

The operation of regenerators at low (ambient or even lower) temperatures permits a good deal of flexibility in the choice of packing materials. Rotary regenerators for air conditioning applications employ a variety of packings which include a polyethylene terephthalate film and corrugated, knitted wire mesh. Such packings are wound round the spindle of the rotor yielding *heat wheels* of varying diameters, from 1.25 to 2.5 *m.* Corrugated aluminium sheets are sometimes used as are various honeycomb arrangements (see **Figure 7**).

A variety of packings have been developed to recover both latent heat and specific heat from one of the gases. Included in these are packings which are non-metallic and fibrous: they can absorb moisture on the one hand but are inert to bacterial contamination on the other.

For very low temperature operation, fixed bed regenerators are frequently used where are packings consist of beds of basalt or flint chips, or simply gravel. Corrugated aluminium sheets are sometimes used where corrugations run in alternate directions between the sheets which are laid on top of one another, generating fine, intersecting channels for the free passage of the gases. Such

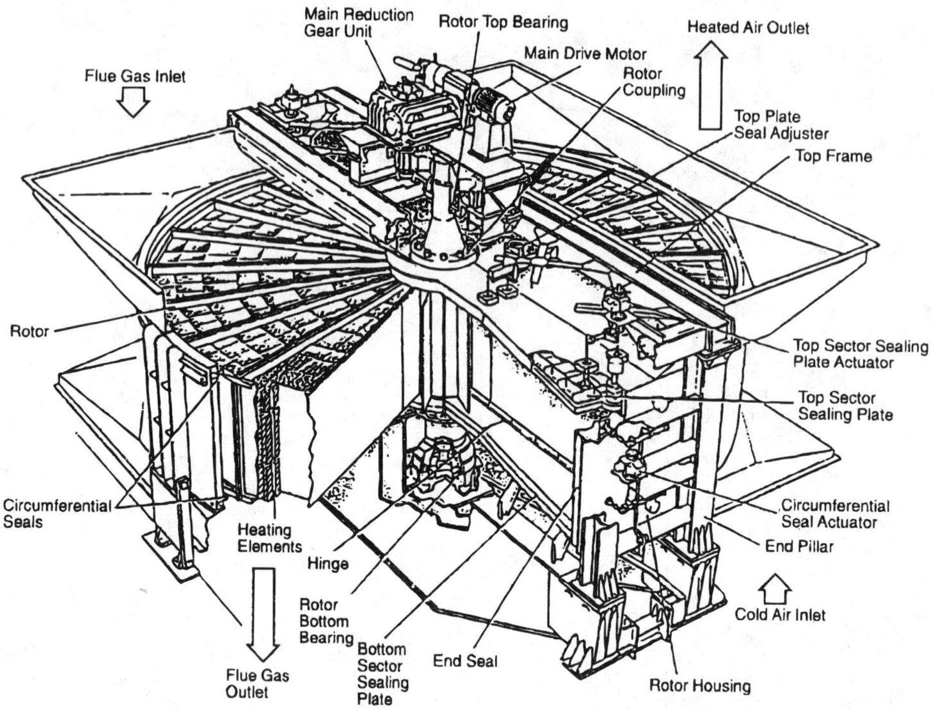

Figure 6. Diagram of a rotary regenerator. (Courtesy of Howden Sirocco Ltd., Glasgow).

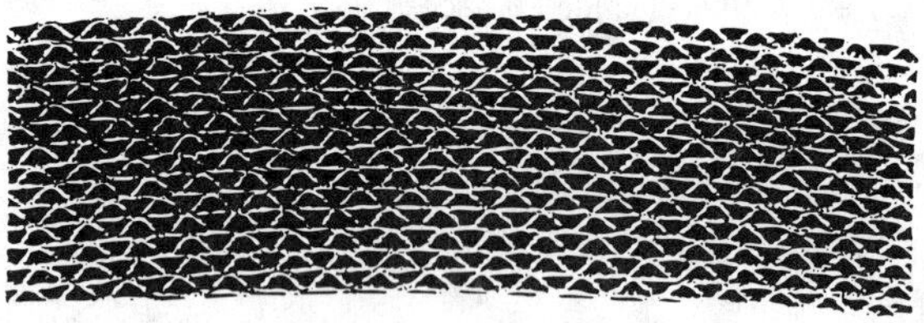

Figure 7. Honeycomb.

arrangements of aluminium sheets can prove to be prohibitively expensive, however.

Nomenclature

A	heating surface area exposed to heating/cooling fluid	m^2
a_s	thermal diffusants of packing	m^2/s
c_p	specific heat, at constant pressure, of fluid	J/kgK
L	length of regenerator from fluid entrance to exit	m
M	mass of packing "behind" the surface area, A	kg
M_f	mass of gas resident in the voids of the packing	kg
$\dot{M}_f$	mass flow rate of fluid through the regenerator	kg/s
P	duration of hot period	s
P′	duration of cold period	s
T_f	temperature of the fluid	K
T_s	temperature of the solid packing	K
t	time	s
y	distance from the regenerator entrance	m

Greek Symbols

α	surface heat transfer coefficient	W/m^2K
$\overline{\alpha}$	bulk heat transfer coefficient	W/m^2K
δ	thickness of the regenerator packing	m
κ_s	thermal diffusivity of the packing	m^2/S
λ_s	thermal conductivity of the packing	W/m K
Ω	dimensionless duration of hot period	—
Ω'	dimensionless duration of cold period	—

References

G. D. Bahnke, C. P. Howard (1964) The effect of longitudinal heat conduction on periodic-flow heat exchanger performance", *ASME Trans. series A, Jour. Eng. for Power,* Apr 1964.

H. Hausen (1929) "Uber die Theorie des Warmeaustauches in Regeneratoren (The Theory of Heat Exchange in Regenerators)", *Z. angew. Math. Mech.*, 9, Jun 1929, 173–200, (RAE Library Translation No. 270, September 1948, W. Shirley), Jun 1929.

H. Hausen (1942) "Vervollstandigte Berechnung des Warmeaustausches in Regeneratoren (Improved Calculations for Heat Transfer in Regenerators)", *Z. VDI-Beiheft Verfahrenstechnik,* 2, 31–43, Iron and Steel Institute translation, dated June 1943.

H. Hausen (1976) *Heat Transfer in Counterflow, Parallel Flow and Cross-flow,* English Translation edited by A. J. Willmott, McGraw-Hill.

P. Razelos, A. Lazaridis (1967) A lumped heat-transfer coefficient for periodically heated hollow cylinders", *Int. J. Heat Mass Transfer,* 10, 1373–1387.

A.J. Willmot

REGULAR REGIME OF DRYING (see Drying)

REHEATING (see Rankine cycle)

REICHARDT'S FORMULA, FOR VELOCITY DISTRIBUTION IN TUBES (see Tubes, single phase flow in)

REIMANN'S INTEGRAL (see Integrals)

REINER-RIVLIN FLUID (see Non-Newtonian fluids)

RELATIVE HUMIDITY (see Humidity)

RELATIVE MOLAR MASS

Following from: *Mole*

The relative molar mass (symbol M_r) is a dimensionless quantity related to the *molar mass* M by $M_r = M/(10^{-3}kg \cdot mol^{-1})$. In the case of an element, this reduces to the **Atomic Weight** A_r and, rounded to the nearest integer, it gives the number of *nucleons* in the most abundant *isotope*. The relative molar mass (often abbreviated RRM) is used most often in connection with large molecules such as polymers or polypeptides where it is used as a measure of chain length. M_r is sometimes called loosely the *molecular weight;* this is incorrect.

J.P.M. Trusler

RELATIVE PERMEABILITY (see Darcy's law)

RELATIVE POWER DEMAND, RPD (see Agitated vessel mass transfer)

RELATIVE ROUGHNESS (see Tube, single-phase heat transfer to in cross-flow)

RELAXATION TIME (see Non-Newtonian fluids)

RENEWABLE ENERGY

Following from: *Alternative energy sources*

There are two distinct classes of energy supplies defined according to their source:

1. energy obtained from "finite sources" (e.g. fossil fuels, uranium ores) which should be treated as stores of capital value, and
2. energy from "renewable sources" (e.g. sunshine, falling water, wind, crops, waves, tides) treated as flows with value as income.

Renewable energy may be defined as "energy obtained from the continuing and repetitive currents of energy passing through the environment" (Twidell and Weir 1986). Note that renewable energy passes in the environment whether or not it is utilized, whereas the finite sources represent a potential requiring external action for the release of energy, which is usually initiated by man.

The renewable flows of energy emanate from three distinct primary sources in the natural environment:

1. the Sun (and hence **Solar Energy** for solar energy thermal conversion, power from the wind for **Wind Turbines,** waves, photosynthesis for *biomass* and *biofuels,* photoconversion as in **Solar Cells** and **Hydro Power**).
2. Geothermal heat, **Geothermal Systems**
3. Tides, in *tidal power*

In addition, it is becoming common to include energy obtained from continuously accumulating wastes as a second form of renewable energy, e.g. landfill gas, waste incineration.

The total flows of renewable energy are very large (Sorensen 1979). **Table 1** quantifies these flows as potentially available for use: (a) at the Earth's surface, and (b) as per capita values assuming a world population of 6 billion (the population in 2000 AD). The same table includes equivalent present day use of fossil and nuclear fuels. Per capita energy demand varies greatly within and between countries from about 10 kW/capita consumed continuously for the extravagant rich in the USA to 0.1 kW for the urban poor in Central Africa.

Table 1. Energy flows (a) For the whole earth, (b) averaged per person assuming a world population of 6 billion

Renewable type	(a) Total 10^9 kilowatt	(b) Per capita kilowatt
solar radiation from space	122,000	20,000
hence solar sensible heating	55,000	9,000
wind	1,200	200
wave	3	0.5
hydro	65,000	11,000
biomass	133	22
geothermal	30	5
tides	3	0.5
anthropogenic material wastes (industry, urban etc)	1	0.2
fossil fuels (all uses)	8	1.3
nuclear power (heat for electricity)	0.5	0.1
needed for a high tech society	12	2.0*

* Compare this prediction with present national averages: 10 kW USA, 5 kW UK, 0.3 kW India.

Important interpretations can be made from **Table 1**:

1. Renewable energy flux is about 10,000 times more per capita than the energy required for a modern high-tech, energy

efficient, society; i.e., 20,000 kW/capita potentially available from renewables to supply the 2 kW/capita needed for all forms of heat, electricity, transport, manufacturing etc, assuming efficient methods. There is no shortage of energy.

2. The generally available renewable fluxes of sunshine, wind and biomass are extremely large.
3. The less generally available renewable fluxes of hydro, wave and geothermal, depend much on the local situation where particular fluxes may be unusually intense and therefore harnessable.
4. As a rule of thumb, the power flux density for commercial renewable energy devices is about 1 kW/m^2 for solar-sunshine and wind technologies, and about 10 kW/m^2 for hydraulic devices. For fossil fuel and nuclear energy plant the power flux densities are several orders of magnitude larger than from renewables. Consequently, the capture area of renewable energy plant, and hence size of structures, is significantly larger per unit of primary power produced than for fossil and nuclear plant. However, the latter require very large distribution networks, not essential for renewables. If whole systems are compared, the land areas and labour required for renewable energy, fossil fuel or nuclear supplies are similar.
5. The relatively large structures per unit of power, make visual intrusion and environmental impact particular challenges for renewables.
6. The dispersed nature of renewables provides a bias towards local use, with smaller distribution systems required as compared with centralized fossil and nuclear supplies.
7. Renewable energy is generally free at source in the environment and so capital costs dominate the economics.
8. Renewable energy sources are varied and dispersed, depending on the climatic, topological, geographical and settlement characteristics of each area. Consequently the scale of applicability for a particular method is usually between 100 km to 1000 km, so requiring regional (not national) assessment and development.
9. Development of renewables depends critically on institutional factors, such as planning, finance, environmental impact and institutional policy for sustainability.
10. Renewable energy from natural sources is intrinsic to the environment and therefore does not cause chemical or radioactive pollution and does not increase atmospheric CO_2 concentrations. This is in marked contrast to fossil fuels and nuclear power, which create external costs of pollution, waste disposal and security, and for fossil fuels, climate change. The economic value of renewable energy is therefore significantly enhanced by this benefit of low external costs, which should be credited within the prices and taxes of commercial energy supplies.
11. There is a strong ecological argument that links the ecological evolution of the Earth with the abstraction of carbon dioxide from the Atmosphere, so producing life-supporting, oxygen containing, air. The resulting atmosphere has a control function on the Earth's temperature, regulating climate. The abstracted carbon is contained in fossil carbonates and coal, peat, oil and methane. Within the last 100 years mankind's use of fossil fuels has released previously bound carbon to significantly increase CO_2 and other gases with the ability to cause climate change. This, and other forms of pollution from finite energy sources, including radioactive waste, is a principle reason for encouraging the implementation of renewable supplies of energy. (see also **Greenhouse Effect**)

Technologies

The following summaries consider most of the large variety of renewable energy sources and supplies. Unless otherwise stated, information should be referenced to Twidell and Weir 1986 from which it has been updated.

Solar irradiation

Solar radiation received onto a surface (irradiation), is entirely described and enumerated by physics. The Sun at 5800 K radiates as a thermodynamic black body between 0.3 to 2.5 μm (short wave radiation). Fortunately harmful ultraviolet radiation around 0.3μm is normally removed by ozone in the upper atmosphere. A maximum of about 30 $MJ/(m^2$ day) of bright sunshine is incident on Earth with maximum irradiation of 1 kW/m^2 or less. Globally, a thermal equilibrium is reached as the surface radiates upwards day and night in long wave radiation, 9 to 11 μm. Molecular scattering of air, aerosol, particulates and clouds, together with reflections, produce *diffuse shortwave radiation* of between 10% of total solar irradiation on a clear day, to 100% on a cloudy day. The latitude, longitude, geometry and manoeuvrability of the receiving surface greatly affects the irradiation received; thus in higher latitudes during winter the vertical components of solar irradiation on walls may be larger than the horizontal component on roofs. Likewise in higher latitudes during summer, the integrated daily solar irradiation on a horizontal surface may be more than in the Tropics.

Solar thermal utilization

The use of solar irradiation for direct heating, or as heat into heat engines producing mechanical and electrical power, is reviewed fully in **Solar Energy.** If the solar heat is collected and then transferred elsewhere for use, the system is "active". If collected for immediate use, as through a window, the system is "passive." Applications include water heating, building heat, drying, distillation, desalination and power production from heat engines (see ref ISES).

Solar cell, photovoltaic, electricity production

Perhaps the most distinctive property of solar radiation is the production of electricity (with no moving parts, pollution or noise) from its incidence onto certain semiconductor junctions, see **Solar Cells.** Such power installations are most economic for small applications giving valued "service" (e.g., pocket calculators, watches, parking meters), for remote sites (e.g., rural hospitals, telecommunications, and water pumping off the grid), and for peak power (e.g., midday city office electricity). World production of solar cells is now 50 MW/y capacity, increasing at about 10%/y with falling costs.

Photoconversion

The photons of solar radiation received at the Earth's surface have energy between about 3 eV (0.4 μm, ultraviolet) and 0.6 eV (2 μm, near infra red). Within this energy range photons may excite electrons in semiconductors and absorbing molecules, so producing electrical current in photovoltaic devices and chemical change in photosynthetic substrates. This general process is named *photoconversion,* and can be interpreted as the attempt by mankind to emulate and develop the principles of photosynthesis in natural plants. The only commercial development of photoconversion to date is the manufacture of semiconductor solar cells for electricity generation (see **Solar Cells**), however in the laboratory hydrogen and other fuels can be produced directly from solar radiation incident on photochemicals.

Hydro-power

This name is usually restricted to the generation of shaft power from falling water for direct mechanical tasks, eg milling, or, predominantly, electricity. The other hydraulic supplies are tidal and wave power. Today, a world total hydro-electric generating capacity of 550 GW, with an average capacity factor of 42%, generates over 2,000 TWh/y. The world potential exploitable resource using conventional engineering is about 16,000 TWh/y, although environmental and financial criteria may limit this to about 8,000 TWh/y, ie 4 times the present generation. Of the major areas with hydro potential only North America, Western Europe and Japan have harnessed 50% or more of the technical potential. Hydro-power has always been sought-after as high quality power and is by far the most established renewable supply of electricity. Its capital intensive construction usually has government finance at low interest rates for about 20 years; thereafter, with the capital paid, its long life time of at least 100 years provides very cheap power. Grid connected utility hydro-electricity may be used as base load supply, for rapid response to meet peak demand and, less occasionally, as pumped storage. The scale of economic plant is from about 2 kW capacity for a single premises, perhaps autonomous generation with an electronic load controller, to 10 GW as a major national grid supply.

If water of flow rate Q, density ρ (1000 kg/m³), falls through a height H (the head), then the maximum power available is P, where, with acceleration due to gravity g:

$$P = Q\rho gH$$

Losses are due to: (i) pipe friction, presented as a loss of head height of perhaps about 5% in large schemes and less than 30% in small schemes, (ii) turbine efficiency, usually around 90% for all but very small schemes, (iii) electricity generator efficiency usually about 95%, and (iii) electrical transmission losses around 10%. (For more detail see **Hydro Power**)

Tidal power

Tides are caused by the combined forces on the sea of: (i) the resultant gravitational pull of the Sun and the Moon, and (ii) the centrifugal force from rotation about the centre of gravity of the Earth and the Moon. Both forces are very approximately in the Earth's equatorial plane, so the seas "bulge outwards" towards this plane. Because of the Earth's rotation on its axis, the "bulge" appears as a travelling deep-water wave causing coastal tides of semi-diurnal periodicity 12 h 25 min. In practice dynamical and local factors alter this basic model.

To understand tidal power applications, it is necessary to appreciate the deep-water wave characteristic of tides. In the open ocean the amplitude of the tide is about 1 m. However this increases dramatically if the tide travels as a deep-water wave up an estuary in a time to form a resonance with the following tide. The condition for this may be modelled and calculated. For a simple estuary of length L and depth h with a semidiurnal tide, the resonant condition is:

$$L = (36{,}000 \text{ m.h})^{1/2}, \text{ where m is the unit metre}$$

For example, the Severn Estuary in south west England meets this condition to give an enhanced tidal range of 10 to 14 m that could provide 10% of UK electricity. Considering all potentially harnessable sites worldwide, the total electricity production would have an average power of about 62,000 MW.

Tidal power from a basin of area A and tidal range R, is harnessed by constructing a barrier across the mouth of the estuary. For each tide of period T, the water of density ρ runs through turbine generators to give:

$$\text{energy per tide} = (\rho AR)\, g\, R/2$$

$$\text{average power} = (\rho AR^2 g)/(2T)$$

Refinements are made to allow for monthly variations in tidal range and other effects.

There are (Johansson et al. 1993) only about 4 tidal power plants worldwide, ranked in power output as: France (La Rance) 240 MW giving 540 GWh/y; Canada (Annapolis) 18 MW, 30 GWh/y; China (Jiangxia) 3.2 MW, 11 GWh/y and others; Russia (Kislaya Guba) 0.4 MW.

Tidal stream power may be harnessed from a turbine in a similar way to wind power. If the turbine has 40% capture efficiency and the stream of maximum speed u has sinusoidal speed variation, the average electrical power per unit cross section of the turbine is 0.1 ρu^3. Advantages over wind are that the fluid density ρ is about 800 times greater than air and the fluid speed is regulated. However the difficult working environment, the rarity of sites and the usually low tidal stream speeds has deterred commercial interest so only historic mills and developmental machines have been constructed.

Wave power

For a surface wave of period T, wavelength λ, frequency ω and amplitude a, the total of the excess energy (potential and kinetic) in unit width and one wavelength is:

$$E = \rho a^2 g\lambda/2 = \pi\rho a^2 g^2/\omega^2 = \rho a^2 g^2 T^2$$

Separate analysis is needed to calculate the power carried forward with the group velocity of the dispersive wave. For unit width of wave front, this power is:

$$P = \rho g^2 a^2 T/(8\pi)$$

For an average Western Atlantic deep water sea wave of wavelength 100 m and amplitude 1.5 m, the power so calculated is 73 kW/m, corresponding with long term measurements.

Considerable research, especially in the UK, has quantified the large resource potentially available in wave power considering the many complications of variations in wave height, form and direction. However as yet there are no commercially available devices, nor indeed many developmental sea-going prototypes. Without doubt, the rigours and extremes of real sea conditions are daunting to such development.

There have been many proposals and trial devices to harness wave power. These may be categorized as: (i) surface floats activated by the wave surface, (ii) submerged floats rotating in the deep water wave, or (iii) pneumatic, with water forcing air through a turbine, as in an oscillating water column. All devices need a constant force of reaction, obtained by installation: (a) on shore, (b) on the sea bottom near shore, or (c) from the average reaction of a long spine floating off shore.

Wind power (see **Wind Turbines**)

The generation of electricity from the wind is a modern international growth industry. For air of density ρ, speed u, passing through a rotor of area A, the power in the wind is the kinetic energy passing per unit time:

$$P = [(\rho A u^2)/2].u = (\rho A u^3)/2$$

Some kinetic energy remains in the wind for air to leave the rotor, so only a fraction C_p is converted to shaft power. C_p is the non-dimensional Power Coefficient, with a maximum value of 16/27 (0.59) by the Betz criteria considering the linear momentum of the air stream. The speed of rotation for maximum power extraction relates to the other main non-dimensional characteristic, the tip-speed ratio λ, equal to the ratio of the speed of the tip of the rotor blades to the unperturbed wind speed u. Most modern wind turbines for electricity production are 2 or 3 bladed on a horizontal axis, having C_p about 40% and λ about 8. Rotor diameter ranges from about 1 m (for battery chargers of about 50 W electricity capacity) to about 30 m (for commercial machines of about 500 kW capacity) and to 100 m (for very large developmental machines).

Geothermal power (see **Geothermal Systems**)

The interior of the Earth is hotter than the surface as a result of heat released by radioactive decay giving an average temperature gradient of about 30°C/km depth. The average outward heat flux is very small, 0.06 W/m^2, and of no useful potential. However a few locations have unusual geothermal concentration that may be utilized. There a 3 types of manageable geothermal source: (i) *hydrothermal aquifers* where ground water percolates downwards and is heated in specific sites of concentrated heat flow for emission as hot water, wet steam or dry steam, (ii) *hot igneous systems* associated with subterranean molten lava, (iii) *hot rocks,* usually very large volumes of low thermal conductivity granite, of elevated temperature; the rock is drilled and shattered to allow forced water circulation for heat removal. However whatever type of system, temperatures of extracted water are lower than in thermal power stations, so the efficiency of thermal electricity generation is low. The most efficient use of the energy is for heating and drying (of which a total worldwide capacity of 23,000 MW (heat) is operational and planned), nevertheless there is substantial worldwide electricity generation (15,000 MW_e operational and planned). Geothermoelectricity production occurs in about 20 countries, with the largest capacity in the USA (3 GW), Philippines (2 GW), Mexico (1 GW), Italy (0.9 GW), Japan (0.4 GW) and New Zealand (0.3 GW). Thermal use is mostly in Japan (3 GW), China (2 GW) and Hungary (1 GW). There are many thermal cycle and heat exchange systems used in these processes. (See also **Geothermal Systems**)

Ocean thermal energy conversion (OTEC)

Tropical ocean surfaces are about 20°C hotter than deep water 1000 m below and so form a solar collector of enormous capacity. A heat engine can be operated between the surface temperature T_h and the deep water at $T_h - \Delta T$. In thermodynamically ideal conditions of a Carnot engine and perfect heat exchangers, the mechanical power produced would be:

$$P = (\Delta T/T_h)\rho c Q \Delta T = (\rho c Q/T_h)(\Delta T)^2$$

where the deep sea water has density ρ, specific heat capacity c and flow rate Q. As with all real heat engines and heat exchangers that have to function in non-equilibrium conditions, the efficiency achieved is less than half the hypothetical ideal, ie about 2% for the best OTEC. However the basic analysis explains the quadratic dependence on ΔT and the very large volume flow rates for deep water cooling. Practical systems have operated with both closed cycle and open cycle vapour turbines. Several countries maintain R&D activity for demonstration projects in tropical oceans, either for floating constructions or, more likely, for shore based plant on islands with immediate deep water, e.g., Nauru in the South Pacific. A secondary benefit is the supply of nutrients brought up with the cooling water for enhanced fish population. (See also **Ocean Thermal Energy Conversion**)

Biofuels and biomass

Biomass is the organic material of plants and animals that reacts with oxygen in combustion and metabolism to produce heat and work. Since all biomass would decay naturally to emit CO_2, utilizing biomass as fuel does not add to climate change gases in the environment and is preferable to consuming fossil fuel in which carbon has been removed from the atmosphere. About 250×10^9 tonne/y of dry matter biomass cycles in the natural environment, i.e. about 40 t/(person y) with an average heat of combustion of about 15 MJ/kg. This compares with about 27 MJ/kg for coking grade coal. Primary biomass may be converted to more readily used *biofuels,* eg charcoal, ethyl alcohol, methanol, methane. Photosynthesis is the primary process for capturing solar radiation to produce plant biomass at efficiencies between 1% (mature woodland) to 10% (optimum greenhouse with enhanced CO_2). The potential maximum efficiency of laboratory emulated photosynthesis is 30%. Biomass is the dominant and most widespread form of renewable energy from domestic cooking to agro-industries; worldwide biomass supplies 15% of all utilized energy. The uses are

1. **Thermochemical** (immediate combustion, pyrolysis, pretreatment);

2. **Biochemical** (alcohol fermentation, anaerobic digestion for biogas of methane and CO_2);
3. **Agrochemical** (direct extraction of extrudates, eg oils).

Producer gas (a mix of CO, H_2 and N_2 with water vapour) is the thermochemical product of burning biomass in restricted air.

Biomass materials form the major part of waste from homes, agriculture, forestry and commerce. In effect such waste is an unstoppable flow in the human environment and hence may be considered a renewable resource. After removal of all recyclable materials, the waste may be biochemically digested for biogas or heat, or incinerated most economically for heating, e.g. district heating, and for electricity. Since payment is usually made for waste disposal, the economics of energy from waste is generally more favourable than for "natural" renewables.

References

ISES (International Solar Energy Society, with many national associations) see the journal "Solar Energy" from Pergamon Press, Oxford, UK; "Advances in Solar Energy" from American Solar Energy Society, Boulder, USA; "Sunworld" and "Sun at Work in Europe" from Franklin Co., Birmingham, UK

Johansson T. B., Kelly H., Reddy A. K. N. and Williams R. H. (1993) *Renewable Energy—Sources for Fuels and Electricity,* Earthscan, London, and Island Press, Washington D.C.

Sorensen B. (1979) *Renewable Energy,* Academic Press, London.

Twidell J. W. and Weir A. D. (1986) *Renewable Energy Resources,* E. & F.N. Spon, London.

Most national governments and some regional authorities have information services supplying brochures, booklets and data concerning renewable energy. The International Energy Agency (the IEA) also has useful information. The European Commission Directorates DG XII (research) and DG XVII (energy) have considerable information on renewable energy developments.

Leading to: *Geothermal systems; Hydro power; Ocean thermal energy conversion; Solar cells; Solar power; Wind turbines*

J.W. Twidell

RENEWABLE ENERGY SOURCES (see Alternative energy sources; Renewable energy)

RESIDUAL ENTHALPY (see Enthalpy)

RESIDUAL GIBBS ENERGY (see Thermodynamics)

RESINS (see Ion exchange)

RESISTANCE, ELECTRICAL (see Ohm's law; Semiconductors)

RESISTANCE HEATING (see Joule heating)

RESISTANCE THERMOMETERS (see Temperature measurement, practice)

RESISTANCE THERMOMETRY

Following from: *Temperature measurement*

A resistance thermometer is a type of a thermometer in which the variation of electrical resistance of pure metals, alloys and semiconductors with the temperature R = f(T) is measured. Thermometers which use pure metals, and primarily platinum, find particularly wide application.

The *platinum resistance thermometer* is a typical example of a secondary thermometer; **Figure 1** depicts the construction of sensing elements of a standard resistance thermometer for low (a) and high (b) temperatures. The sensing element of a platinum resistance thermometer consists of wire or strip wound (or sometimes printed) on a rigid carcass (made from quartz, ceramics, mica) enclosed in an envelope (made from metal quartz, glass) through which terminals run to the measuring devices. Platinum resistance thermometers are used in the range from −263 to 1000°C. Thermometers of this type work with devices reading temperatures in °C or K in correspondence with tables R(T) for such thermometers. Platinum resistance thermometers have wide application in very precise temperature measurements (for instance, in metrological studies). The International Temperature Scale ITS-90 was introduced in 1990 (see International Temperature Scale). In the temperature range from the triple point of equilibrium hydrogen (13.8033 K) to the silver point (961.78°C), the ITS-90 offers a specification for the platinum resistance thermometer. The sensing element must be manufactured from pure tension-free platinum.

A consumer receives from the standards body a platinum resistance thermometer calibrated over the ITS-90 range and a detailed table $T_{90}(R(T_{90})/R(213.16\ K))$.

Other metals are also used to manufacture resistance thermometers, for instance Copper resistance thermometers are for industrial application and work in a temperature range of −150 to 600°C. Resistance thermometers may be manufactured from manganin and constantan alloys; however, such thermometers suffer from a number of drawbacks, in particular, narrow temperature measurement intervals and low response. Resistance thermometers manufactured from the rhodium-iron alloy (Rh + 0.5% Fe) may be employed in a wide range from 0.5 to 300 K. These resistance thermometers have high, prolonged stability, sufficiently high voltage sensitivity and possess a comparatively high (compared to a platinum resistance thermometer) specific resistance at low temperatures. This allows higher, again in comparison to a platinum resistance thermometers, voltage to be used subject to the avoidance of heating effects.

Carbon thermometers are used to measure low temperatures in superconducting magnetic systems because of their high sensitivity and small dependence of R on the presence of a magnetic field. Glass-carbon and composite sintered carbon resistance thermometers are most widely used.

Semiconductor thermometers include germanium resistance thermometers used at a temperature below 20 K. They possess higher reproducibility and prolonged stability compared to carbon thermometers and are suitable for precise measurements. Usually pure monocrystalline germanium is not employed, but germanium doped with different impurities. Control of the quantity of the main and doping components allows the control of the R(T) form of the curve (most often antimony or arsenic is used as a dopant).

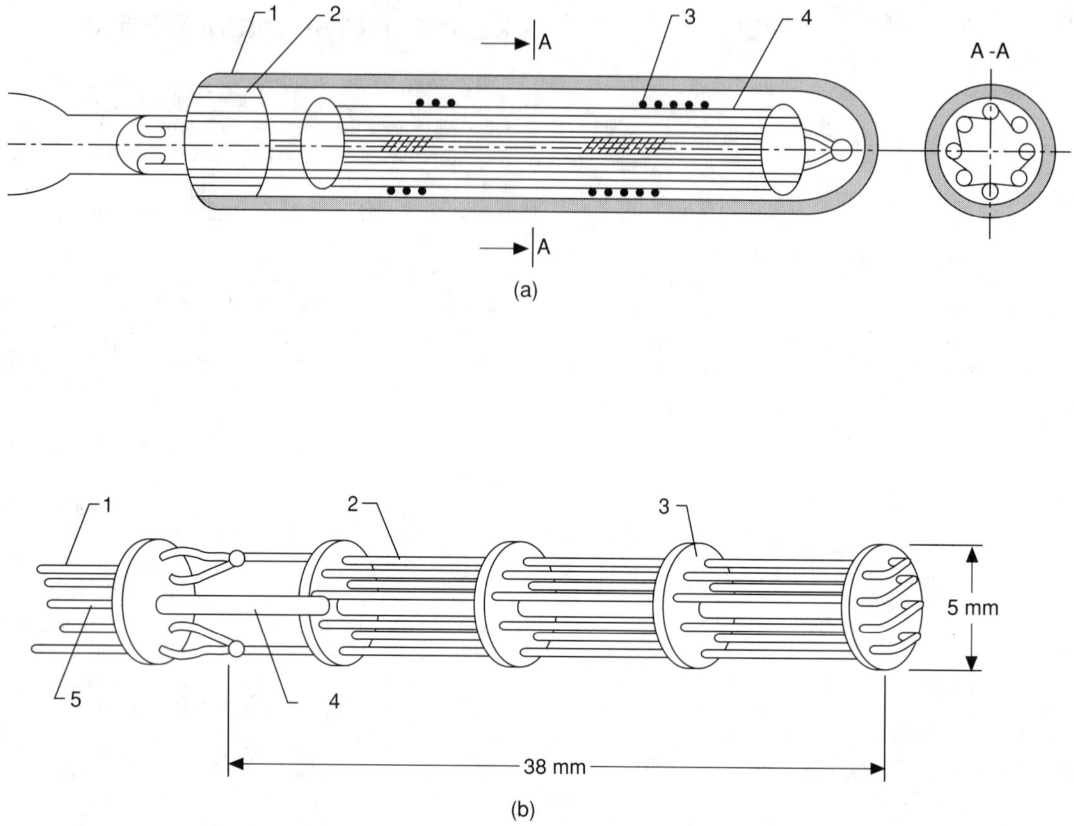

Figure 1. Types of platinum resistance thermometers: (a) Low temperature, (b) High temperature.

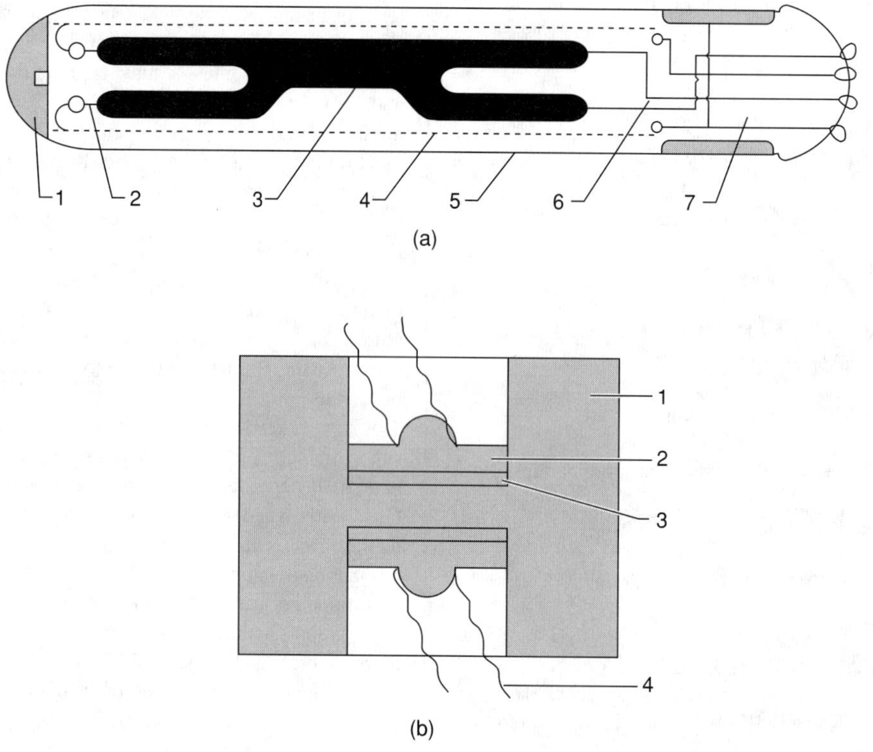

Figure 2. Germanium resistance thermometry: (a) Bridge circuit used, (b) Detail of construction of miniature thermometer.

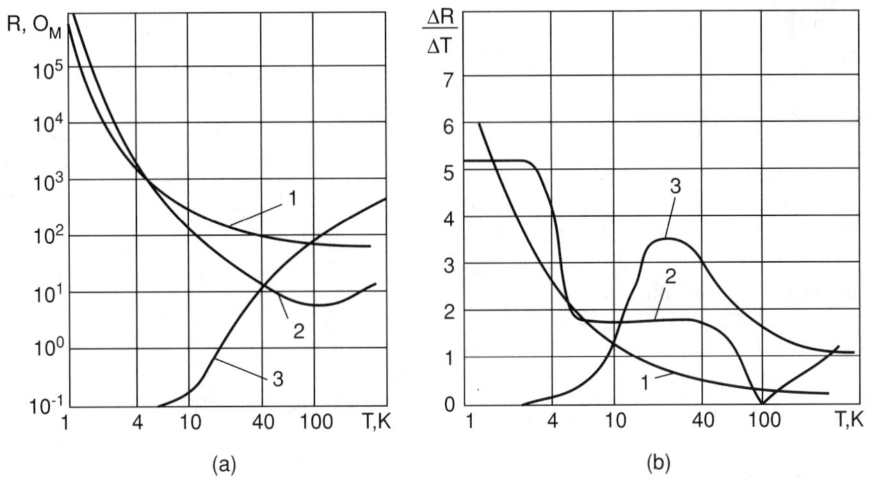

Figure 3. Comparison of resistance thermometer: (1) Carbon, (2) Germanium, (3) Platinum.

Figure 2 shows the bridge-type circuits used (a) and the construction of miniature (b) germanium resistance thermometers. For highly precise measurements, four leads (two current and two potential) are used. In the model shown, a sensitive element is wrapped into a fluoroplastic film and placed into a metal sleeve. Mass-produced thermometers are equipped with the tables graded at a large number of points. Multicomponent doping of a resistance thermometer allows R(T) to be obtained for a large temperature range (3–90 K). **Figures 3a** and **3b** show a comparison of R(T) and dR/dT for three thermometers: carbon (1) germanium (2) platinum (3).

Semiconductor resistance thermometers have found wide application in industry. They include thermoresistor thermometers which, in turn, include themistors and posistors. Thermistors are manufactured from the materials with a negative temperature resistance coefficient $R/R_0(T)$; posistors are manufactured from the materials having a positive temperature coefficient of resistance which, for the majority of thermoresistor semiconductors, is much higher than for other materials. Thermoresistors are used in the temperature range 170–750 K. They are made from oxide materials. To obtain a required R value, such stabilizing substances as nickel oxide are used. When plastic substances are used, mixtures are pressed to obtain necessary shape and sintered. Miniature bead-type resistance thermometers are available for measurement of living organisms, biological objects, in particular, plants. These resistance thermometers are designated for small temperature intervals. Mass produced resistance thermometers do not possess high stability of characteristics when operated for the first (2–5) $\times 10^3$ h. At longer times, their properties change only slightly.

Posistor resistance thermometers are made from ferroelectric ceramics based on titanates, zirconates, etc., plumbum, boron. They work within a narrow temperature range from 20 to 100°C and find application in automatic control and protection systems.

M. P. Orlova

RESISTIVITY, ELECTRICAL (see Ohm's law)

RESONANCE FLUORESCENCE (see Fluorescence)

RETENTATE (see Membrane processes)

RETENTION INDEX (see Gas chromatography)

RETROGRADE CONDENSATION (see Natural gas)

RETURN TO NUCLEATE BOILING (see Rewetting of hot surfaces)

REVERSED HEAT ENGINE CYCLES (see Refrigeration)

REVERSE OSMOSIS (see Membrane processes)

REVERSIBLE PROCESSES (see Carnot cycle; Entropy; Thermodynamics)

REVERSIBILITY PRINCIPLE (see Aerodynamics)

REWETTING (see Leidenfrost phenomena)

REWETTING OF HOT SURFACES

***Following from:** Boiling; Leidenfrost phenomena*

Rewetting, or quenching, is the re-establishment of liquid contact with a solid surface whose initial temperature exceeds the rewetting temperature (RT)—the maximal value for which the surface may wet.

Rewetting phenomena and rewetting models are reviewed by, e.g., Butterworth and Owen (1975), Elias and Yadigaroglu (1978), Carbajo and Siegel (1980), and Olek (1988).

Heat removal from a hot surface by a liquid may be hindered if the surface temperature (ST) is high enough to allow the formation of a vapor layer, acting as an insulator. When the ST drops below the RT, heat transfer can be enhanced by as much as two orders of magnitude.

The interest in rewetting of hot surfaces is mainly due to its role in *light water reactor safety.* (see **Blowdown** and **Reflood**) Following a hypothetical loss-of-coolant accident, the reactor core might become uncovered, and rapid cooling is needed. This is done either by top spraying of coolant or by bottom flooding. In either case, the task of the *emergency core cooling system* is to rewet the hot cladding of the fuel rods. Rewetting is also applicable to several industrial processes, such as metallurgical quenching of hot solids, start-up of LNG pipelines and filling of containers with cryogenic liquids. The terms *quenching, sputtering, minimum film boiling point, return to nucleate boiling, departure from film boiling, film boiling collapse*, and **Leidenfrost Phenomenon** have been used interchangeably to refer to various forms of rewetting. These terms, however, are not exactly synonymous.

Rewetting may occur in very different situations; a classification of the most frequent ones was made by Groeneveld (1984), **Figure 1**. The first four types refer to forced convective rewetting while the last two types deal with no-flow conditions. The different rewetting situations are described below. (See also **Boiling**)

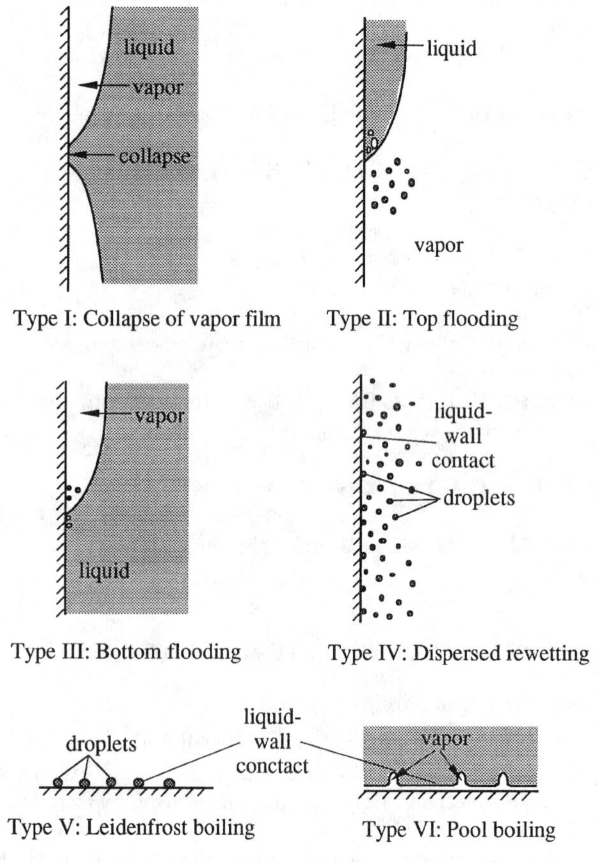

Figure 1. Types of rewetting according to Groeneveld (1984).

Type 1: Collapse of Vapor Film

In inverted annular film boiling, there is a wavy liquid-vapor interface near the wall. If the wall temperature or heat flux are lowered, the vapor thickness decreases and eventually the liquid may contact the wall. Depending on the temperature level and the rate of heat supply to the cooled region beneath the liquid, permanent liquid-wall contact is either maintained or the liquid is pushed away from the surface with the formation of vapor. In the latter case, each contact is equivalent to an impulsive cooling of the surface. In a quenching process, repeated contacts will lower the ST enough to permit rewet.

Type 2: Top Flooding

Liquid contact with the heated surface is established by means of an axially propagating liquid film. At the quench front there is a narrow zone of vigorous boiling with production of an aerosol of droplets with diameters 0.5 mm and less. Some of the finer droplets are ejected almost horizontally from the film, which lifts off the hot surface at the quench front, causing the liquid to break up into large irregular droplets. At low liquid flows the liquid runs down the surface as a rivulet rather than as a continuous film. Here typical rewetting velocities are of the order of millimeters per second.

Axial conduction is usually the main mechanism of reducing the ST just ahead of the quench front, although precursory cooling by vapor and droplets ahead of the front may become important at higher flow rates.

Type 3: Bottom Flooding

Rewetting is due to an axially propagating quench front, with a region of nucleate and transition boiling behind it (Barnea et al. 1994). Different descriptions are reported regarding the behavior downstream of the quench front. In some cases the excess fluid forms a dispersed flow region of liquid and vapor coupled with film boiling. In other cases there exists an extensive film boiling region ahead of the quench front which breaks down to a mist flow at a distance of the order 10 to 200 mm. Rewetting velocities are of the order of centimeters per second.

Prior to rewetting, the heated surface is pre-cooled by transition and film boiling (inverted annular film boiling or dispersed flow film boiling) and/or axial conduction. At very low flow rates and high liquid subcoolings the pre-cooling effect is expected to be insignificant.

Type 4: Dispersed Rewetting

The heated surface is pre-cooled by a spray of droplets which can eventually lead to droplet rewetting. The rewetting depends on droplet size and velocity, vapor conditions and heated surface properties. Rewetting here is characterized by the absence of a quench front.

Types 5 and 6: Leidenfrost and Pool Boiling

For type 5, the liquid is dispersed in the continuous gas phase, while for type 6, the gas forms a film on the surface; the liquid phase is continuous.

An additional form of rewetting, namely, rewetting by large masses of liquid without progression of a quench front, was reported by Aksan (1988).

To reflect the diversity of the rewetting phenomena, an indicative glossary is introduced for terms most often used to describe the temperature variation of a hot surface during its quenching. Note that this glossary is somewhat arbitrary and that these terms may be encountered in the literature under other meanings than those described here (Gerweck 1989).

Rewetting temperature probably the most general term which, interpreted directly, refers to re-establishment of direct contact between wall and liquid. However, there is experimental evidence of short time contacts even in stable film boiling. There is another specific meaning for the rewetting temperature used in heat conduction models: that at which the heat transfer coefficient (assumed to vary axially) drops from nucleate boiling values to film boiling ones, for a rod being quenched in a reflood process.

Quench temperature is the value of the ST just before the temperature drops by the quenching process.

Sputtering temperature usually refers to top flooding. It is the ST below which the rewetting velocity becomes constant.

Minimum film boiling temperature corresponds to a minimum in the heat flux map obtained when the heat flux from the wall is plotted as a function of the wall temperature, mass flux, quality, etc.

Leidenfrost temperature is the ST for which the evaporation time of a quiescent droplet on a horizontal surface becomes a maximum.

Two main mechanisms have been proposed to determine the rewetting temperature. The *hydrodynamic* approach holds that the separation of the liquid-vapor interface from the wall lasts only as long as the vapor generation rate is sufficient to maintain a stable vapor film. According to this theory, rewetting commences due to hydrodynamic instabilities which depend on the velocities, densities, and viscosities of both phases as well as the surface tension at the liquid-vapor interface.

The *thermodynamic* approach assumes that the liquid can never exist beyond a *maximum liquid temperature*, MLT, (or super-heat) which depends on the liquid properties only. Thus, a heated surface whose surface temperature exceeds this threshold cannot support liquid contact. The MLT can be determined either from the spinodal line, or from the kinetic theory of bubble nucleation in liquids.

Other approaches for defining the rewetting temperature are based on the adsorption characteristics of the system. For a static configuration, surface wettability is suitably expressed by the liquid-solid **Contact Angle**, θ_c. The knowledge of the dependency of θ_c on temperature enables determination of the temperature which corresponds to various contact situations, e.g. Segev and Bankoff (1980) and Olek et al. (1988a). For a dynamic configuration it is claimed that as long as the ST allows the formation of at least one monolayer of liquid molecules on the surface, wetting is possible. When the temperature is increased above a threshold value, no continuous monolayer (or close packed patches of adsorbate) can be formed, and initial spreading of the liquid over the surface will terminate.

There is no general consensus about the effect of the various system parameters on the rewetting temperature. These effects are included in correlations for the RT, e.g., Elias and Yadigaroglu (1978).

There exists voluminous literature of rewetting models based on axial conduction, e.g. review of Elias and Yadigaroglu (1978) and Olek (1988). These models predict with partial success the rewetting velocity. Common to most of these models is the use of the upstream wet-side heat transfer coefficient and the rewetting temperature as input parameters. Some models take into account the cooling ahead of the quench front (precursory cooling), which requires an assumption of additional parameters. The weakness of these models is the arbitrariness introduced by the choice of the noted parameters (Yadigaroglu et al., 1990). Olek et al. (1988b) attempted to reduce the number of these rather arbitrary parameters by treating top flooding of a single rod in an unconfined geometry as a conjugate heat transfer problem. The assumptions used, however, limit the application of the model only to the particular flow configuration for which it was derived.

Most of the experimental and theoretical works on rewetting of hot surfaces have dealt with the macroscopic phenomena, such as the quench front velocity, because of its application primarily in the nuclear industry. There still does not exist a theoretical method for a precise determination of the RT with an experimental support, based on a microscopic approach.

References

Aksan, S. N. (1988) A Review of Large Break Loss-Of-Coolant Accident Blowdown Quench and Effect of External Thermocouples, The 3rd Int. Code Assessment and Application Program (ICAP) Specialist Meeting, Grenoble, France. March 1–4 1988.

Barnea, Y., Elias, E., and Shai, I. (1994) Flow and heat transfer regimes during quenching of hot surfaces, *Int. J. Heat Mass Trans.*, 37, 1441–1453.

Butterworth, D. and Owen, R. G. (1975) The Quenching of Hot Surfaces by Top and Bottom Flooding—A Review, Report No. AERE-R7992, AERE Harwell, Oxfordshire, England.

Carbajo, J. J. and Siegel, D. (1980) Review and comparison among the different models for rewetting in LWR's, *Nucl. Eng. Des.*, 58, 33–44.

Groeneveld, D. C. (1984) Inverted Annular and Low Quality Film Boiling: A State-of-the-Art Report, The 1st International Workshop on Fundamental Aspects of Post-Dryout Heat Transfer, Salt Lake City Utah, USA, April 1–4.

Elias, E. and Yadigaroglu, G. (1978) The reflooding phase of LOCA in PWRs, Part II, Rewetting and liquid entrainment, *Nucl. Safety*, 19, 160–175.

Gerweck, V. (1989) Rewetting Phenomena and their Relation to Intermolecular Forces Between a Hot Wall and the Fluid, PSI Rept. No. 42, Paul Scherrer Institute, Switzerland, December.

Olek, S., Zvirin, Y., and Elias, E. (1988a) The relation between the rewetting temperature and the liquid-solid contact angle, *Int. J. Heat Mass Trans.*, 31, 898–902.

Olek, S., Zvirin, Y. and Elias, E. (1988b) Rewetting of hot surfaces by falling liquid films as a conjugate heat transfer problem, *Int. J. Multiphase Flow*, 14, 13–33.

Olek, S. (1988) Analytical Models for the Rewetting of Hot Surfaces, PSI Rept. No. 17, Paul Scherrer Institute, Switzerland, October.

Segev, A. and Bankoff, G. (1980) The role of adsorption in determining the minimum film boiling temperature, *Int. J. Heat Mass Trans.*, 23, 623–637.

Yadigaroglu, G., Andreoni, M., Aksan, S. N., Lewis, M. J., Analytis, G. Th., Lübbesmeyer, D., and Olek, S. (1990) Modelling of Thermohydraulic

Emergency Core Cooling Phenomena, PSI Rept. No. 27, Paul Scherrer Institute, Switzerland, October.

S. Olek, Y. Zvirin and E. Elias

REYNOLDS ANALOGY

The convective transport of mass, momentum and heat normally occurs through a thin boundary layer close to the wall. The equations governing the transport of these quantities are analogous and are exact (Incropera & DeWitt, 1990) for laminar flow if the pressure gradient is equal to zero and the **Prandtl Number** (Pr) and **Schmidt Number** (Sc) are equal to unity. Under these conditions, their nondimensonal convective transport coefficients are related by

$$\mathrm{Re}\frac{f}{2} = \mathrm{Nu} = \mathrm{Sh} \qquad \textbf{(1)}$$

where f is the Fanning friction factor, Re is the **Reynolds Number**, Nu is the **Nusselt Number** representing heat transfer and Sh the **Sherwood Number** representing mass transfer. Equation (1) is known as the Reynolds analogy, and enables the calculation of, for example, the heat transfer coefficient if either the friction factor or the mass transfer coefficient is known. (See also **Analogy Between Heat, Mass and Momentum Transfer**)

For turbulent flow in a pipe or over a flat plate, the exchange of momentum, heat and mass occurs mainly by turbulent eddies in the bulk of the flow. However, very close to the wall, the exchange must occur by molecular diffusion across the velocity, temperature and concentration boundary layers, respectively. When Pr and Sc are equal or close to unity, the boundary layer thicknesses are roughly equal and Equation 1 is applicable. When this is not so, which is often the case for liquids, proper account must be taken of the relative thicknesses of the boundary layers. A modified Reynolds analogy, also known as the *Chilton-Colburn analogy*, is found to be applicable under these conditions:

$$\mathrm{Re}\frac{f}{2} = \mathrm{Nu}\,\mathrm{Pr}^{-1/3} \qquad 0.6 < \mathrm{Pr} < 60 \qquad \textbf{(2)}$$

$$\mathrm{Re}\frac{f}{2} = \mathrm{Nu}\,\mathrm{Sc}^{-1/3} \qquad 0.6 < \mathrm{Sc} < 3000 \qquad \textbf{(3)}$$

Equations 2 and 3 are applicable for both laminar and turbulent flow (Hewitt et al., 1994).

References

Hewitt, G. F., Shires, G. L., and Bott, T. R. (1994) *Process Heat Transfer*, McGraw-Hill.

Incropera, F. P. and DeWitt, D. P. (1990) *Fundamentals of Heat and Mass Transfer*, John Wiley and Sons.

S. Jayanti

REYNOLDS' AVERAGING (see Convective heat transfer)

REYNOLDS' EQUATIONS (see Combustion; Flow of fluids)

REYNOLDS NUMBER

First introduced in the early 1880's by Osborne Reynolds to characterise the transition between laminar and turbulent flow (Reynolds, 1883) the dimensionless term Reynolds number, Re, is now universally employed in the correlation of experimental data on frictional pressure drop and heat and mass transfer in convective flow. It is the basis of much physical modelling.

In general terms

$$\mathrm{Re} = u L \rho / \eta$$

where u is fluid velocity, ρ fluid density, η fluid viscosity and L the characteristic length. In the case of flow in a circular pipe this becomes

$$\mathrm{Re}_D = u D \rho / \eta.$$

Reynolds number represents the ratio of force associated with momentum (ρu^2) to force associated with viscous shear ($\eta u/L$). Alternatively it may be regarded as a measure of the ratio of turbulent energy production per unit volume ($\rho u^3/L$) to the corresponding rate of viscous dissipation ($\eta u^2/L^2$). (Lighthill, 1970).

Below a lower critical value of Reynolds number flow is laminar, or "streamline"; above a higher critical value flow is turbulent, or "sinuous" in Reynolds terminology. Between these values the flow is in what is called "transition". The higher critical value is strongly dependent on upstream conditions; Reynolds observed values of between 11,800 and 14,300 for water in bell mouth tubes of a few centimetres diameter. The lower critical value is less sensitive and is usually quoted simply as the "critical Reynolds number". Its value for smooth circular pipes and tubes is approximately 2,000.

References

Reynolds O. (1883) On the experimental investigation of the circumstances which determine whether the motion of water shall be direct or sinuous and the law of resistance in parallel channels, *Phil. Trans. Roy. Soc.*, 174, 935.

Lighthill M. J. (1970) Turbulence, Ch. 2. *Osborne Reynolds and Engineering Science Today*, Manchester University Press Barnes and Noble Inc. New York.

Hewitt G. F., Shires G. L., and Bott T. R. (1994) *Process Heat Transfer*, CRC Press.

G.L. Shires

REYNOLD'S NUMBER, CRITICAL, IN TUBES (see Reynolds number; Tubes, single phase flow in)

REYNOLDS, OSBORNE (1842–1912)

An English engineer and physicist, best known for his work in the fields of hydraulics and hydrodynamics, Reynolds was born in Belfast on Aug. 23, 1842. Gaining early workshop experience and graduating at Queens College Cambridge in 1867, he became the first professor of engineering in the Owens College, Manchester in 1868. He was elected a Fellow of the Royal Society in 1877 and a Royal Medallist in 1888.

Reynolds' studies of condensation and the transfer of heat between solids and fluids brought about radical revision in boiler and condenser design, while his work on turbine pumps laid the foundation of their rapid development. A fundamentalist among engineers, he formulated the theory of lubrication (1886), and in his classical paper on the law of resistance in parallel channels (1883) investigated the transition from smooth, or laminar, to turbulent flow, later (1889) developing the mathematical framework which became standard in turbulence work.

His name is perpetuated in the "**Reynolds Number**", which provides a criterion for dynamic similarity and hence for correct modelling in many fluid flow experiments.

Among his other work was the explanation of the radiometer and an early absolute determination of the mechanical equivalent of heat. Reynolds retired in 1905 and died at Watchet, Somerset, on Feb. 21, 1912.

Figure 1.

U. Grigull, H. Sandner and J. Straub

REYNOLD'S STRESS (see Combustion; Flow of fluids; Particle transport in turbulent fluids; Turbulence; Turbulence models)

REYNOLD'S STRESS TRANSPORT MODELS (see Boundary layer heat transfer; Turbulent flow)

RHEOLOGY (see Liquid-solid flow; Non-Newtonian fluids)

RHEOMETERS (see Non-Newtonian fluids)

RHEOPEPTIC FLUIDS (see Non-Newtonian fluids)

RHODAMINE (see Fluorescence)

RICCATTY-BESSEL FUNCTIONS (see Mie scattering)

RICHARDSON NUMBER (see Turbulence)

RIDEAL-ELEY MODEL, FOR HETEROGENEOUS CATALYSIS (see Catalysis)

RIEDEL-PLANK-MILLER EQUATION (see PVT relationships)

RIEMAN WAVES (see Waves in fluids)

RISC. REDUCED INSTRUCTION SET COMPUTER (see Computers)

RISK ANALYSIS TECHNIQUES

"Risk" and related terms, such as "risk assessment", "risk estimation", "risk evaluation" and "risk analysis" do not have universally agreed definitions, although there is a measure of consensus as to their meaning. A useful glossary and discussion on the use of the terms are given in an HSE document.[1] The glossary draws heavily upon a booklet published by the Institution of Chemical Engineers[2] and The Royal Society Study Group (RSSG) on Risk Assessment.[3] The Institution of Chemical Engineers definition of risk is: "The likelihood of a specified undesired event occurring within a specified period or in specified circumstances. It may be either a frequency (the number of specified events occurring in unit time) or a probability (the probability of a specified event following a prior event), depending on the circumstances". This parallels the common usage definition.

It follows that to describe fully a risk it is necessary to,

1. specify the event or outcome of interest,
2. estimate the probability of the event or outcome and, possibly,
3. estimate the severity or magnitude of the outcome or consequences.

Examples of risk are the probability of all-engine failure in an aircraft during a given flight, the frequency of toxic release from a chemical plant, the loss of power steering in a car when cornering or the probability of a major radioactive release from a nuclear power station.

As noted by HSE[4] risk should not be reduced to a single quantity and its components need to be separately identified. There are examples in the literature, however, when the term "risk" is specifically defined by combining consequences with their probability.

The HSE[1] also observes that "risk assessment" is a term used by both The Institution of Chemical Engineers and the RSSG to mean a process that includes the estimation of risk and a determination of the significance of the estimated risks. Significance is determined either by those most likely to be affected or by those making political decisions. The RSSG suggests the terms "risk estimation" and "risk evaluation" for the two parts of the process.

When the risk in question is one which causes harm, then it may either be to an individual, and termed "individual risk", or to the population, when it is termed "social-" or "societal risk". An individual risk can be expressed as an annual probability of death, or of contracting a disease. Societal risk is often expressed as the relationship between frequency and the number of people experiencing various specified levels of harm.

In the UK, the topic of risk evaluation (sometimes termed "risk appraisal") was debated at the Sizewell "B" Public Inquiry and one of the Inspector's recommendations was that the HSE should formulate and publish guidance on the tolerable level of risk. This led to the publication of the HSE document[4] which considered risks from nuclear and non-nuclear industries. The topics of risk evaluation and criteria to be applied are, however, outside the scope of this article.

This article is limited to estimation. When the estimate is quantified it is common to refer to it, in the nuclear industry, as a *probabilistic safety assessment, PSA*, or *probabilistic risk assessment, PRA*, and in the non-nuclear industry as a *quantitative risk assessment, QRA*. Whatever term is used the methods of analysis are the same.

Risk estimation or analysis is a process of forecasting the likelihood or probability of future events using data from previous events and/or details of the design of the plant in question. At its simplest level it can be the estimation of the unreliability of essential equipment. Reliability techniques have been under development for most of this century particularly in the aircraft and defence industries. However, since the 1970s the nuclear industry has taken the lead in developing risk analysis techniques. An early example was NRC's Reactor Safety Study[5] in which the probability of nuclear accidents and their consequences were estimated. Since then, probabilistic techniques have been used as part of the design process, as at Sizewell "B" for example, and as an aid to determining the acceptability of a given design or design change.

There are various techniques available to the analyst for determining risks and their probability; which one is used depends upon the particular problem. They can be broadly categorised as,

- *Failure modes and effects analysis (FMEA)*
- *Event tree analysis*
- *Fault tree analysis*

The first step in any analysis, however, is to set down details of the overall plant, possibly in functional block diagram form if the system is complex. In a functional block diagram each part of the plant carrying out a particular function is represented by a single building block, the diagram then showing the inter-relationships between the various blocks. All activities and processes need to be understood including details of any protective features.

The next step is carefully and systematically identify all the potential hazards and all the ways that the hazard can be generated. FMEA can be used to determine the initiating events or causes which could lead to the hazard. The technique is described as "bottom-up" since individual failures are traced forward to the final effect.

The results of an FMEA are summarised in truth or decision tables. The results are categorised according to their severity and estimates obtained of their probability of occurrence. This enables priorities for corrective action to be undertaken at the design stage. In practice, however, FMEA is primarily used to determine the effect of component failures, whether electrical, mechanical or structural, on a single system and not on the whole of a complex plant. The significance of failure of each component is then studied in turn. FMEA is an effective way of identifying all single faults which could cause system failure.

The event tree approach is used on a complex plant and is similar to an FMEA in that all possible sequences, following any postulated initiating event, are constructed. It is often convenient to group together similar initiating events. A simple event tree can be illustrated by considering the example shown in **Figure 1**.

Figure 1 represents a hypothetical fire protection system provided to extinguish any fire which might occur in a particular room in a building. Two electric pumps are provided to deliver water to a sprinkler system through an electrically operated valve. Water is taken from a reservoir. An automatic fire sensing system starts the pumps. The design intent is that one pump suffices. In this example the non-return valves in the system are assumed always to work. As a further line of defence to stop any fire spreading and causing extensive damage outside the room a fire door is provided, electrically driven, but manually operated. The fire door and the sprinkler system are each assumed to be 100% effective if they operate. Note, the example is not meant to be representative of any practical arrangement. One way of drawing the corresponding event tree is shown in **Figure 2**.

Along the top are listed all the items which should work. Starting with the initiating event, the fire, the first question asked is whether the automatic fire sensing system has worked. If it has then that is a success and the branch continues straight across; a step down indicates failure. Each branch then leads to a further question. The

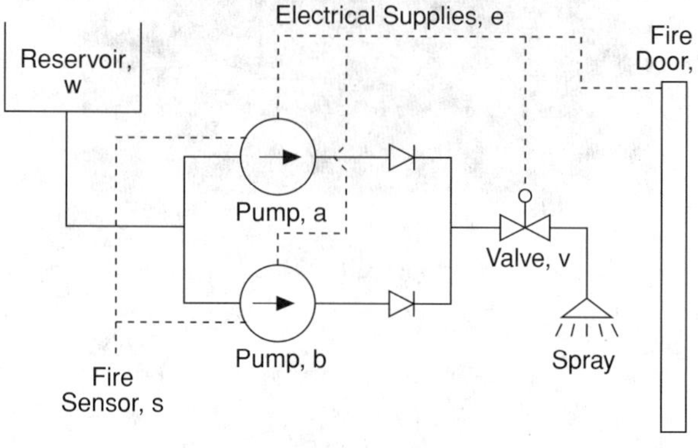

Figure 1. Hypothetical fire protection system.

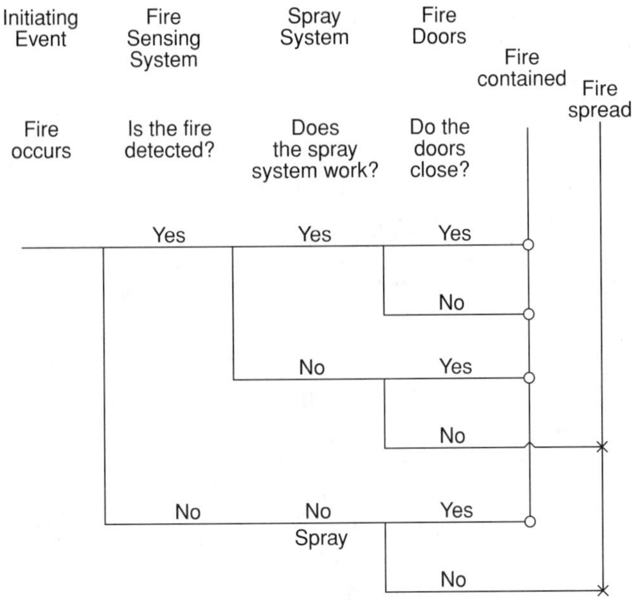

Figure 2. Event tree.

top branch leads do the question "does the spray system work?" The answer "yes" leads straight across; the answer "no" is represented by the branch which steps down. In this way is developed a tree of all possible sequences. In this example, the sequences whose end points are marked with a circle are deemed a success, although any success states claimed have to be demonstrated by calculation or other evidence. Those with a cross are failures. The probability of failure at each branch could also be added and hence the probability of each sequence evaluated. The total probability of failure is then obtained by summing all the failure sequence probabilities. In this example the failure probability of the overall spray system has to be input; this can be obtained from a fault tree analysis or, if desired, the event tree can be expanded to represent individual components. In general, the event tree is particularly useful in identifying those sequences which have to be shown by subsequent analysis to be acceptable because of their high probability.

Event trees, however, are usually constructed at a level with systems represented by blocks. Care has to be exercised to ensure that dependencies are correctly represented. In the example shown, if electrical failure is the main cause of spray system failure then it would equally stop the fire door from being closed, and an optimistic result could occur if the two events were assumed unrelated. The problem could be overcome by giving the question of electrical failure its own branch point in the tree. Dependencies are, however, automatically taken into account in the fault tree approach discussed later.

It may be, however, that the consequences are not defined in such a simplistic way as in the above example. Sequences are continued until any risk is fully identified. In the case of a chemical plant or nuclear power station the risk to be established could be the risk of death to an individual, or societal risk from a release of toxic chemicals or radioactive material. The aim of any risk analysis would be to show that any residual risk was sufficiently low as to be deemed tolerable. Methods of analysing the probability of release of harmful releases are specific to particular industries.

The fault tree approach treats any problem in a "top-down" manner. Whereas event trees identify a range of possible outcomes, fault trees identify all contributors towards one specified outcome-the "top event". It is a logic diagram used to determine how a defined risk can occur either at the component level or at the system level. Consider again the system shown in **Figure 1**. Using capital letters to designate failure states a fault tree can be drawn for this system as shown in **Figure 3**, using symbols based upon those used in the NRC study.[5] The top event is the event we are interested in—the possibility of fire spread outside the equipment room. The AND gate underneath shows that the top event will only occur if all the input events, namely the fire door not having been closed and the sprinkler system having failed to extinguish the fire. The AND gate at the next level down shows that the sprinkler system is only ineffective if there is no flow from pump "a" and no flow from pump "b". An OR gate defines the situation if one or more of the input events exist. Thus, the tree shows that the fire door remains open if either the electrical supply fails or if it fails because it jams (or because the operator failed to act). The branches end at a "basic" event which needs no further development because failure rate data are known or are assumed.

Using capital letters to represent the failure probability of the corresponding item, the probability of the top event, T, can be set down using the notation of *Boolean algebra.*[6] The symbol for "and" is a dot, and for "or" the plus sign, +. This gives

$$T = \{(A + E + W + V + S).(B + E + W + V + S)\}.(F + E)$$

This expression can be simplified by applying the logical rules of Boolean algebra.[6] For example, $A.A = A = A + A.B$ and so on. In this way the expression is reduced to

$$T = F.A.B + F.W + F.V + F.S + E$$

Boolean reduction automatically takes into account the fact that loss of electrics is a contributing factor to the failure of a number of items. Component failure probabilities can be substituted to obtain system failure probability. (F.A.B), (F.W), (F.V), (F.S) and (E) are "Minimal Cut Sets" of components. In any minimal cut set all the component failures of which it consists must occur to result in system failure, but no other simultaneous failure is necessary. A "Non-minimal Cut Set" includes components whose failures can be tolerated. Boolean algebra identifies all the minimal cut-sets.

Only relatively simple systems can be analysed by hand. Complex systems require the use of computer codes. Such codes can have a significant commercial value and new codes offering greater flexibility are continually appearing.

The value of any risk analysis obviously depends upon its completeness and whether all important initiating events and fault sequences have been identified. Certain aspects are difficult to model, such as the effect of human intervention, which may be either beneficial or harmful. When low probability risks are being evaluated the contribution from outside hazards such as aircraft crashing on the plant, earthquakes or extreme environmental conditions may be important. Failure rate data may not always be available or directly applicable and so judgements regarding its applicability may have to be applied. A particular difficulty can

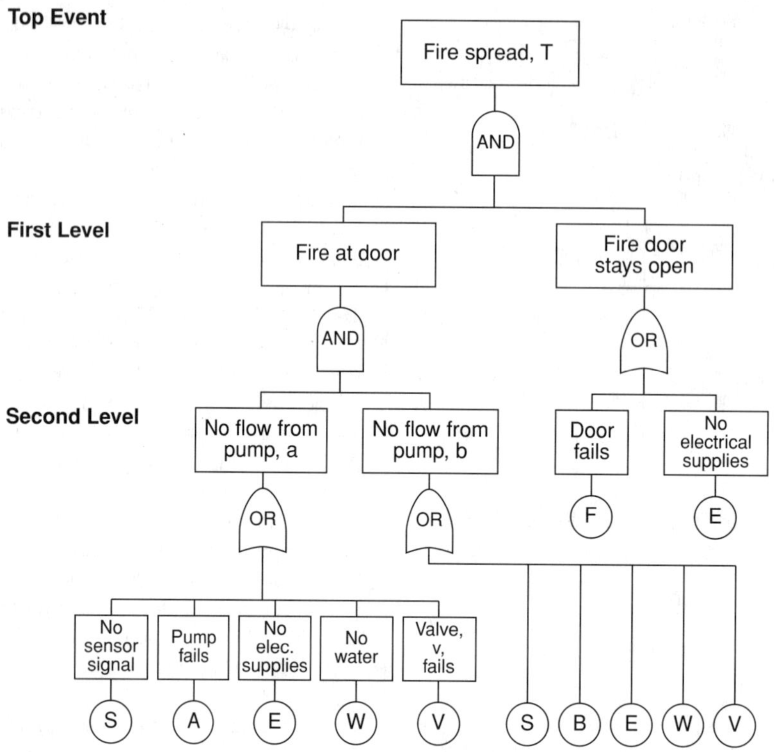

Figure 3. Fault tree.

be the treatment of common cause failures. A common cause failure is an event which could cause simultaneous failure on a number of similar components and hence eliminate any benefit from redundancy or even diversity. A failure mode affecting similar components is sometimes referred to as a *common mode failure*. Methods for treating such failures are under continuous review.

It follows that there can be uncertainties associated with any risk analysis, and that judgement need to be exercised in using the results. As noted above, a risk calculation may be carried out at the design stage or on a completed plant or piece of equipment. The mere process of carrying out a risk assessment at the design stage is valuable, however, since it can identify ways of improving the design and reducing the risk.

References

1. Health and Safety Executive, Risk criteria for land-use planning in the vicinity of major industrial hazards, 1989, ISBN 0 11 885491 7

2. Institution of Chemical Engineers, Nomenclature for hazard and risk assessment in the process industries 1985, ISBN 0 85 295184 1

3. Royal Society Study Group, Risk assessment, 1983, ISBN 0 85 403208 8.

4. Health and Safety Executive, The tolerability of risk from nuclear power stations, 1992, ISBN 0 11 886368 1.

5. NRC, WASH 1400 Reactor Safety Study.

6. Rueff, M., and Jeeger M., (1970) *Sets and Boolean Algebra*. Allen and Unwin.

P. Jenkins

RISK ASSESSMENT (see Risk analysis techniques)

ROASTING

Following from: Metals

The term roasting refers to gas-solid reactions performed at elevated temperatures as preliminary steps to either *hydrometallurgical* or *pyrometallurgical* production of metals. During this process gas composition is controlled in order to produce the desired chemical and physical changes.

The major objectives of roasting are one or more of the following.

1. Elimination of volatile species (e.g. Sulphur, as SO_2, and Cadmium).
2. Alteration of chemical composition:

a. to assist subsequent aqueous dissolution, e.g. conversion of sulphides to sulphate and/or oxides;

b. to assist subsequent high temperature reduction with carbon, e.g. conversion of sulphides to oxides;

c. to render volatility e.g. production of chlorides from oxides;

d. to alter physical characteristics as in sintering, whereby agglomeration of fine particles is achieved.

The majority of pyrometallurgical roasting reactions are now performed in **Fluidised Bed** reactors to utilise their inherently good heat and mass transfer characteristics.

In the case of *sintering,* the fine material to be roasted and agglomerated is placed in a continuously moving belt and a narrow reaction front passes rapidly through the material to effect partial melting of some component and thus to effect agglomeration.

Reference

Szekely, J., Evans, J.W., and Sohn, H.Y. *Gas Solid Reactions.* Academic Press.

***Leading to:** Smelting*

B. Terry

ROCKET PROPELLANTS (see Flames)

ROCKETS (see Microgravity conditions)

ROD BUNDLE TESTS (see Rod bundles, heat transfer in)

ROD BUNDLES, FLOW IN

***Following from:** External flows, overview*

Introduction

To ensure good thermal performance of a nuclear reactor a detailed knowledge of the heat transfer and fluid flow phenomena taking place within the core is required. Coolant flow rates in different parts of the complex rod bundles and the manner in which the single phase and the two-phase flows are distributed in the subchannels are very important for evaluating enthalpy distribution and performance parameters, such as the onset of boiling and critical heat flux.

This entry describes some of the fundamental features turbulent flow in rod bundle subchannels in a water cooled reactor, namely:

- the momentum balance in subchannel geometry
- inter-subchannel flow *mixing in rod bundles*
- the *subchannel analysis* method for predicting the thermal hydraulic performance of rod bundles.

Momentum Balance in Subchannel Geometry

Accurate prediction of pressure drop is of essential importance in the design of a nuclear fuel rod bundle geometry for two reasons:

- Pumping costs are a significant fraction of operating expenses
- The subchannel pressure drop is intimately related to coolant flow and enthalpy distribution

In other words, for a good assessment of the thermal hydraulic behaviour of rod bundles by subchannel analysis it is necessary to formulate all the basic equations, including axial and momentum balance, as accurately as possible. The general form of axial momentum balance for each subchannel is:

$$\frac{dP}{dz} = \left(\frac{dP}{dz}\right)_f + \left(\frac{dP}{dz}\right)_e + \left(\frac{dP}{dz}\right)_{acc} + \left(\frac{dP}{dz}\right)_s + \left(\frac{dP}{dz}\right)_m \tag{1}$$

where the subscript f is the frictional pressure gradient, e is the hydrostatic head loss, acc is the total acceleration loss (sum of the direct contribution of changes in subchannel mass flow rate and velocity and contribution of diversion cross flow), s is the pressure loss due to area restriction, and m is the effect of turbulent mixing on pressure change in axial direction.

The transverse momentum balance determines the magnitude of diversion cross-flow which is caused by radial pressure gradients between adjacent subchannels; it indicates a strong dependence on the rate of change of the lateral velocity in an axial direction. To calculate the diversion cross-flow rates for a given radial pressure difference, transverse resistance coefficients are required. Unfortunately, there is limited experimental data for these coefficients and a computer code must take an oversimplified form.

It is usual to describe turbulent flow phenomena and pressure drop in channels having a non-circular cross section, by introducing an equivalent hydraulic diameter and applying the laws established experimentally for circular tubes. This is why we start by reviewing some of the results obtained for this geometry. However, it must be borne in mind, that this is a rough estimate only and the true value varies with the geometry of the channel and in particular with the rod bundle configuration.

Single-phase flow pressure drop

Isothermal friction factor in smooth and rough circular tubes

In a cylindrical duct, with hydraulic diameter D_H defined as four times the cross sectional area divided by the periphery, the general form of the fraction factor f is defined by the relation:

$$\left(\frac{dP}{dz}\right)_f = \frac{4f}{D_H}\frac{1}{2}\rho\langle V\rangle^2 \tag{2}$$

where: <V> is the mean velocity.

- **Friction factor correlation for laminar flow:** the pressure drop relationship is: f = K/Re where the constant K depends on the geometry: for circular tubes K = 16 and f = 16/Re (**Poiseuille Flow**)
- **Friction factor correlation for turbulent flow:** For fully turbulent flow in a smooth circular tube, the friction factor is given with sufficient accuracy by the Blasius correlation which is a good approximation of Nikuradse's experiments and integration of the universal distribution law:
  $f = 0.079\ Re^{-0.25}$
  in the range of Reynolds numbers $10^4 < Re < 10^5$
- **Friction factor correlation for rough pipe:** f is a function of the Reynolds number and the roughness parameter ε/D_H, where ϵ is a characteristic roughness height. In fully turbulent flow and of a high Reynolds numbers f is a function of

ε/D_H only and is given by the relation:

$$\frac{1}{\sqrt{f}} = 4\,\text{Log}\left(3.71\,\frac{D_H}{\varepsilon}\right) \tag{3}$$

Effect of heating rate on friction factor

In the case of turbulent flows developed under uniform wall heat flux boundary conditions, the friction coefficient decreases markedly with heating. Accordingly, it is necessary to take into account the effect of the variable physical property of the water as a function of the temperature in the laminar boundary layer, especially changes in viscosity.

Most researchers have suggested the following correlation:

$$\frac{f_H}{f_{iso}} = \left(\frac{\eta_{WALL}}{\eta_{BULK}}\right)^m \tag{4}$$

where m has been considered for many years as a constant value in the range 0.14 to 0.60.

An important improvement was achieved by the author (Lafay, 1974) who found that the dependence of the friction ratio on the viscosity ratio is more complex than a simple power law and proposed a more accurate parameter:

$$X = \left(\frac{\eta_B}{\eta_W} - 1\right)\left(\frac{\eta_B}{\eta_W}\right)^{0.17} \tag{5}$$

with the following correlation:

$$\frac{f_H}{f_{iso}} = 1 - 0.5(1 + X)^{F(Re)}\,\text{Log}(1 + X) \tag{6}$$

$$F(Re) = 0.17 - 2.10^{-6}\,Re + \frac{1800}{Re} \tag{7}$$

for

$$2.10^{-4} \leq Re \leq 30.10^4 \qquad \text{and} \qquad 2 \leq Pr \leq 6$$

The dispersion is characterised by a standard deviation:

$$\sigma = \pm 0.7\%$$

(See also **Friction Factors For Single Phase Flow; Pressure Drop, Single Phase**)

Friction factor in rod bundle

Theoretical and experimental studies have shown that the equivalent hydraulic diameter is not adequate to describe accurately the considerable influence of the geometry on the friction factor.

Rheme (1973) has proposed a method for calculating friction factors for turbulent flow in non-circular channels. All that is required is a knowledge of the geometry factor K of laminar flow. ($f = K/Re$) For a number of non-circular channels, these factors have been accurately determined by numerical calculation procedures; this is the case more particularly for rod bundles in square and hexagonal subchannels for which the friction factor in turbulent flow is given by the relation:

$$\sqrt{\frac{2}{f}} = A\left[2.5\,\text{Log}\,Re\sqrt{\frac{2}{f}} + 5.5\right] - G^* \tag{8}$$

where the geometry factors A and G* are a function of K.

Effect of spacers on turbulent flow

Fuel designs usually include *spacers* for providing fuel rod support particularly to promote mixing and improvement of the thermal hydraulic performance of rod bundles.

A spacer causes significant local velocity perturbations in its wake, increasing axial turbulence intensity and introducing considerable perturbation of mass flow rate distribution in the subchannels. Consequently, the presence of a spacer in a rod bundle results in significant local pressure perturbations and increases the global pressure drop which depends to a large extent on spacer design. Two basically different types of spacers are used in the construction of reactor fuel elements:

- grid spacers defining the distance between the fuel elements and (the gap) relative to the wall of the enclosing box; they are generally used for square and hexagonal arrays of fuel elements with a large distance between pins.
- wire spacers wrapped around the fuel elements in a triangular arrangement used for a smaller distance between pins.

Grid spacer The pressure drop at the spacer grid is related to the mean velocity $\bar{V}_B$ in the rod bundle by:

$$\Delta P_{GS} = C_B\,\frac{1}{2}\,\rho\bar{V}_B^2 \tag{9}$$

where C_B is the drag coefficient of the spacer which is presented as a function of the Reynolds number of the rod bundle:

$$Re = \frac{V_B D_H}{\eta} \tag{10}$$

As the drag coefficient of individual grids is very dissimilar in magnitude, Rehme (1973) proposes correlation of the quantity:

$$C_B\left(\frac{S}{S_G}\right)^2 \tag{11}$$

where S is the flow cross section of the grid and S_G the undisturbed flow section of the rod bundle.

The agreement of the modified loss coefficient is comparatively good, taking account the geometrical diversity of grid designs.

When the grid spacer has small wings at the exit edge in all subchannels and with different orientations, this method is probably not good and we must use experimental measurements for each case.

Wire spacer With wire wrapped rod bundles the situation is quite different. Experimental studies (Lafay et al., 1975) have showed that the helical wire spacer around each rod induces an important swirl flow and a significant local pressure perturbation; accurate exploration of the pressure field in peripheral subchannels shows that the pressure distribution in a cross section is not hydrostatic and the axial pressure profile is a periodic function of the axial position, the period corresponding to the helical wire pitch.

Therefore the axial pressure gradient on one pitch is the same as the mean pressure gradient for several pitches and it is possible to define a friction pressure drop coefficient, though the simple equivalent diameter process is not sufficient for accurate prediction.

Novendstern has developed a theoretical model which determines the flow distribution in subchannels and has obtained an empirical correction factor depending on fuel rod bundle dimensions and flow rate.

The friction factor is given by the relation:

$$f = f_{iso}\left[\frac{1.034}{(P/D)^{0.124}} + \frac{29.7(P/D)^{6.94}Re^{0.086}}{(H/D)^{2.239}}\right] \quad \textbf{(12)}$$

where; f_{iso} is the friction factor for flow in a smooth circular tube, P/D is the pitch to diameter ratio, H/D is the helical wire pitch to diameter ratio, and Re is calculated with the mean velocity and equivalent diameter of the central subchannel.

It is obvious that the friction factor increases with decreasing wire wrapping pitch, while there is a significant increase with a high pitch-to-diameter ratio of the rods.

Two-phase flow pressure drop

In a pipe

The two-phase pressure drop during a steadly flow of a two-phase mixture is the sum of three components: hydrostatic head, accelerational pressure drop and frictional pressure loss.

Evaluation of the two-phase friction pressure drop requires a void fraction correlation to calculate the hydrostatic and acceleration components, a single phase friction factor and a two-phase friction multiplier correlation.

Two principal types of flow models are used in the analysis of two- phase flow frictional pressure drops: the *homogeneous flow* model and the *separated flow* model.

In 1948, Martinelli and Nelson were the first to suggest an approach and an empirical correlation for the calculation of the two-phase frictional pressure drop in a tube based on the separated flow model. They correlated the ratio of local two-phase pressure gradient to pressure gradient 100% liquid flow as a function of quality and pressure

$$\left(\frac{dP}{dl}\right)_{TPF} \Big/ \left(\frac{dP}{dl}\right)_0 = \phi^2_{l0}(x, P) \quad \textbf{(13)}$$

$\phi^2_{\ell 0}$ is the two-phase friction multiplier and (dP/dl) is the single phase friction gradient calculated with the single phase friction factor of the Blasius correlation:

$$f = 0.079\ Re^{-0.25} \quad \textbf{(14)}$$

More recently, many verifications or improvements of the Martinelli-Nelson correlation based on experimental data, have been published. Some of the more prominent correlations among these are due to Thom, Baroczy, Chisholm and Columbia University. They found that the two-phase friction multiplier is also a function of the mass flux. A comparative study of their performance shows that the Columbia correlation gives the more accurate friction factor. (See **Pressure Drop, Two Phase Flows**)

Friction multiplier in a rod bundle

To evaluate two-phase flow friction pressure drop in rod bundles we assume that the rod bundle behaves like a circular channel of equivalent hydraulic diameter; consequently we recommend using:

- the friction factor to calculate the single-phase flow pressure drop
- the *Columbia* correlation which gives the more accurate two-phase flow friction multiplier $\phi^2_{\ell 0}$

Intersubchannel Flow Mixing in Rod Bundles

In order to improve the thermal hydraulic characteristics of the nuclear reactor core, a considerable amount of research has been carried out in order to obtain improved understanding of coolant flow and enthalpy distribution in rod bundle geometries.

One form of fundamental research has been the study of the mixing process between subchannels in this complex geometry.

Definition of the flow mixing processes

In analysing the effect of mixing on rod bundle temperature and pressure gradient it has generally been assumed that the mixing process is the result of several components.

Natural mixing (see also **Mixing**)

To describe turbulence in bundles of smooth bare rods (without protuberances) and which include both turbulent and diversion cross flow mixing:

a. **Turbulent mixing** which results from the oscillation component of flow in a transverse direction between two subchannels and can be characterised by the eddy diffusivity of momentum ε_M.

b. **Diversion cross-flow mixing** is the rate of mass flow in a transverse direction though the gap between two subchannels caused by radial pressure gradients. The diversion cross-flow contributes to subchannel flow rate in an axial direction.

Forced Mixing

The subchannel interchange is induced by the presence of spacers in the rod bundle; we distinguish:

a. **Flow scattering** refers to the non-directional mixing effect associated with grid spacers which break up streamlines; the turbulence intensity increases immediately downstream from the device.

b. **Flow sweeping** refers to the directed cross- flow effect associated with wire wrap spacers or grid spacers with vanes which give a net cross-flow in a preferred direction.

Importance of mixing in rod bundles

Much research has been devoted to the study of turbulent flow processes and improved understanding of mixing. Most experimental work has focused on quantifying the mechanisms individually and their dependency. First, the relative importance of these processes in rod bundle performance vary significantly with bundle geometry characteristics and particularly with gap spacing and flow parameters. Secondly, though turbulence interchange is present in all situations, the major emphasis has been in determining turbulent cross-flow (diversion and sweeping) because it is the most important means of momentum and energy transfer: it improves the thermal hydraulic performance of rod bundles as it provides an important mechanism for equalising temperatures thoughout the bundle. An excellent review of the related aspects of turbulent mixing and diversion cross-flow is given by Rogers and Todreas (1968); the reviewers describe the circumstances in which cross-flow mixing are particularly developed:

- in the entrance region of bundles
- in the region where boiling crisis begins and develops in the various subchannels
- in the region of physical distortion of the bundle elements

Local velocity and concentration measurements in rod bundles

Various experimental techniques have been used to study mixing in rod bundles. Chemical tracers, hot water injection and direct subchannel enthalpy measurement are generally used. Most experimental data give only the mean mixing rates over the test section, while the flow sweeping effect which is the predominant process in intersubchannel mixing is strongly dependent on the local conditions due to the grid spacer vanes or the periodic wire wrapped spacer position in the bundle. Therefore, a local description of the flow field is extremely important to assist the interpretation of coolant cross mixing.

The recent development of two component laser Doppler anemometry permits local fluid velocity measurements and consequently provides an important tool for improving fluid flow research. This technique is performed with light beams and has the advantage of not disturbing the flow field; furthermore, the control volume over which the measurement is performed is very small. This is a significant improvement over previous methods such as hot-wire anemometers and Pitot tubes. (See **Anemometers, Laser Doppler**).

The experimental results obtained by Rowe in rod bundles without and with spacer has shown, in the first case, that rod gap spacing has a significant influence on the turbulent flow structure in rod bundles in a way that cannot be deduced from round pipes, whereas the velocity profiles are in reasonable agreement with universal profile in pipes of the same hydraulic diameter.

The grid spacer results in a significant change in axial mean velocity, turbulence intensity and turbulence scale in the wake of the grid spacer; these effects are weakly dependant upon the Reynolds number.

Concerning the concentration measurement technique, the use of a liquid tracer introduced in the rod bundle flow upstream from the grid spacer is also an important technical means for fluid flow research in spite of a small disturbance of the probe inside the subchannel. Sampling of the solution in all subchannels at different cross sections downstream from the grid and its analysis in a spectrofluorimeter indicates first the local concentration in all subchannels and its axial and transversal evolution and secondly the average mixing coefficient qualifying the grid spacer. (See also **Tracer Methods**)

This two measurement technique, velocimetry laser and concentration, has been used with success for a large range of grids with vanes in hexagonal and square geometry at the Heat Transfer Laboratory of the CEA at Grenoble with two fundamental and analytical experiments: AGATE and HYDROMEL. In two cases we observe that the flow behaviour depends strongly on the position, dimensions, shapes and inclinations of the mixing vanes at the exit of the grid.

Mixing under two-phase flow conditions

Experimental data available in the literature about the mixing rate under boiling flow conditions are limited.

Some investigations have been made with air-water mixtures to explain mechanisms of mixing in two-phase flow and to indicate the effects of various parameters such as the turbulent interchange rates. However, the principal information has been developed by Rowe though the use of thermal hydraulic code COBRA which constitutes a significant contribution to the knowledge of boiling flow behaviour in rod bundles. It has been observed that boiling turbulent interchange appears to be a function of the channel geometry, quality and flow regime, with a maximum at low qualities in the transition region from bubbly to annular flow.

The Concept of Subchannel Analysis

The subchannel analysis method is an important tool for predicting the thermal hydraulic performance of rod bundle nuclear fuel element. It considers a rod bundle to be a continuously interconnected set of parallel flow subchannels which are assumed to contain one dimensional flow and are coupled to each other by crossflow mixing; the axial length is divided into a number of increments such that the whole flow space of a rod bundle is divided into a number of nodes.

The principle of subchannel analysis is the application of continuity and conservation equations to the flow between these nodes. The conservation equations relate the local variations of velocity and enthalpy of each node to those of its neighbouring ones. The relation between subchannel flow rate which is the mass flow rate in an axial direction through subchannel area and diversion cross-flow which is the mass flow in a transverse direction resulting from local pressure differences between two subchannels, is strongly governed by momentum balance in a transverse direction.

A correct formulation of momentum equations and good knowledge of the mixing process between subchannels are an absolute necessity, for obtaining reliable predictions using subchannel calculations.

References

Courtaud M., Ricque R., Martinet B. (1966) Etude des pertes de charge dans les conduites circulaires contenant un faisceau de barreaux, *Chem. Eng. Sci.*, 21, 881–893.

Giot M. (1981) Friction factors in single channels, *Thermohydraulics of Two-Phase Systems for Industrial Design and Nuclear Engineering,* Delhaye, J. M. et al., ed., Hemisphere.

Grand D. (1981) Pressure drops in rod bundles, *Thermohydraulics of Two-Phase Systems for Industrial Design and Nuclear Engineering,* Delhaye J. M et al., eds. Hemisphere.

Lafay J. (1974) Influence de la variation de la viscosité avec la température sur le frottement avec transfert de chaleur en régime turbulent établi, *Int. J. Heat and Mass Trans.*, 17, 815–834.

Lafay J., Menant B. and Barroil J. (1975) Local pressure measurements and peripheral flow visualization in a water 19 rod bundle compared with FLICA IIB calculations: influence of helical wire wrap spacer system, *ASME Heat Transfer Conf., San Francisco* 75-HT-22.

Reddy D. G. and Fighetti C. F., (1983) Evaluation of two- phase pressure drop correlations for high pressure steam-water systems, *ASME Thermal Engineering Conf. Proc., Honolulu,* Vol. 1.

Rheme K., (1973) Pressure drop correlations for fuel element spacers, *Nuclear Technology,* 17, 15–23, January.

Rogers J. T. and Todreas N. E., (1968) Coolant interchannel mixing in reactor fuel rod bundles single phase coolants, *Winter Annual Meeting of the ASME, New York,* December.

Rouhani Z., (1973) A review of momentum balance in subchannel geometry, *European Two-Phase Flow Group Meeting Brussels,* June 4–7.

Rowe D. S., (1973) Measurement of turbulent velocity, intensity and scale in rod bundle flow channels, *BNWL- 1736,* May.

Tong L. S., (1968) Pressure drop performance of a rod bundle, *Winter Annual Meeting of the ASME, New-York,* December.

Leading to: *Rod bundles, heat transfer in*

J. Lafay

ROD BUNDLES, HEAT TRANSFER IN

Following from: *Rod bundles, flow in*

Introduction

This article describes heat transfer in nuclear fuel rod bundles in light water nuclear reactors under normal, incidental and accidental conditions.

Relevant basic information on **Heat Transfer, Forced Convective Boiling, Nucleate Boiling, Burnout** and **Rewetting of Hot Surfaces** is given in other entries.

Laminar Flow

Laminar flow occurs during reactor shutdown and removes the residual power (less than 0.5% of the nominal power).

Although it is mainly of academic interest, laminar flow equations have been analytically resolved (without spacer grids) and lead to some constant Nusselt Numbers in steady-state and fully developed flow (Lach, 1986), depending on the pitch to diameter ratio P/D (see **Table 1**.)

Table 1. Nusselt Numbers (constant heat flux) for fully developed laminar flow in rod bundles.

	1.02	1.05	1.10	1.20	1.30	1.40	1.50
Equilateral Triangular Spacing	0.40	1.05	2.90	6.90	9.05		
Square Spacing	0.52	0.84	1.72	4.58	6.40	9.43	15.05

Single-Phase Turbulent Flow

This is the most usual case encountered in normal *Pressurised Water Reactor* (PWR) operation. It is important to know the temperature field in the reactor core not only in order to be able to calculate the neutron flux field, but also because a difference of some degrees may have noticeable consequences in the long term on the chemical and metallurgical behaviour of the cladding of the fuel rods.

Usually, the single phase turbulent heat transfer coefficients are estimated using the same correlations as for tubes, using the equivalent hydraulic diameter instead of the tube diameter. According to many authors, this approach underestimates the actual heat transfer coefficient.

Tong et al. (1979) proposed modifying the constant C = 0.023 in the well known Dittus-Boelter correlation and recommended:

- for a triangular-pitch lattice

$$C = 0.026\frac{P}{D} - 0.006$$

- for a square-pitch lattice

$$C = 0.042\frac{P}{D} - 0.024$$

where P/D is the pitch over diameter ratio.

This increase in the heat transfer coefficient seems to be due mainly to the mixing grids that enhance the turbulence downstream, and, consequently, the heat transfer; some local measurements have demonstrated that the heat transfer coefficient decreases as the distance to the mixing grid increases. A precise heat transfer correlation in this geometry would depend closely on the grid geometry and so would probably be proprietary. However, the overestimation of the wall temperature obtained by a tube correlation is probably less than the inaccuracy due to the subchannel analysis!

Nucleate Boiling Flow

Nucleate boiling occurs in the hottest locations of a PWR core in normal operation and in the greater part of a BWR core. It may also occur in the greater part of a PWR core during transients associated with accidents.

As long as nucleate boiling occurs, the wall temperature is $T_{sat} + \Delta T_{sat}$, and ΔT_{sat} is only a few degrees. The correlations used for calculating ΔT_{sat} are usually the same as for the tubes. The *Onset of Nucleate Boiling* (ONB) may be calculated as the point at which the wall temperature, as estimated by a single-phase turbulent heat transfer correlation, reaches $T_{sat} + \Delta T_{sat}$; a subchannel analysis code is needed to estimate this position. For more details on nucleate and saturated boiling flow (see Groeneveld, 1986).

Boiling Crisis

The boiling crisis, or *Departure from Nucleate Boiling* (DNB) is one of the phenomena limiting the power of a water nuclear reactor.

Although some boiling crises, like *dryout*, lead to only moderate temperature rises, others, like **Burnout** at low quality, may result in a dramatic and very rapid increase of the rod wall temperature. Consequently, in many countries, the rules are that all boiling crises must be avoided in normal and abnormal situations, up to incidental transients of class 2.

The basic boiling crisis phenomena have been modelled in tubes (ex. Whalley for dryout, Weissman et al. or Katto for burnout), but the phenomenon is far more complicated in rod bundles due to the presence of crossflows, mixing grids, complex geometry, non-uniform heat-flux, and the tube models give only a very rough approximation of the boiling crisis conditions in a reactor.

There are several ways to obtain a local *Critical Heat Flux, CHF,* prediction:

- a quick-and-approximate method is to use a tube CHF prediction and adjust the result to the appropriate rod bundle geometry using correcting factors. Among the several hundreds of existing *CHF* correlations for tubes, annular space and other simple geometry the Groeneveld, 1993, CHF tables are recommended.
- a more accurate method is to use a specific CHF correlation (or table, smoothing splines, etc) especially fitted for the specific geometry and mixing grids used. This involves carrying out several CHF tests with the specific geometry and the appropriate parameter range, analysing these tests with the same subchannel computer code and building a CHF predictor. Since these experiments are difficult, expensive and long, and apply only to specific mixing grids and geometries, most of the CHF predictors are proprietary and not openly published.
- an intermediate method is to use a reference general purpose rod bundle CHF correlation, perform some CHF tests with the appropriate geometry and mixing grid, calculate the average and standard deviations of the experimental results with respect to the reference CHF correlation, and then use these deviations to modify the safety margin. The use of this methodology seems surprising as it is less efficient than the former one, and the missing step—the construction of a specific CHF predictor based on expensive and hard-won experimental data—is relatively easy, quick, cheap and more precise. However, this methodology is very commonly used, especially for new fuel to be loaded into existing plant, because the on-line computer software and, above all, the existing regulations (including the reference CHF correlation) seem very difficult to modify.

The geometry and the grid spacer have a considerable influence on the CHF values. For the same thermo-hydraulic local conditions (pressure, mass velocity, quality), the CHF may vary by a factor of 2 or 3! The most obvious effects on the CHF are due to:

- the grid spacing: the CHF increases significantly as the grid spacing decreases;
- the geometry of the mixing grid, especially that of the mixing vanes: it may have a considerable influence on CHF especially at low quality and high heat flux but most of the results in this field are proprietary;
- the guide thimble: it influences CHF and acts both as a cold wall effect and as a hydraulic diameter effect;
- the equivalent hydraulic diameter: as in tubes, CHF generally increases when the hydraulic diameter decreases for constant local thermal-hydraulic conditions. This sensitivity is usually greater for low quality and tends to zero for higher quality.

Quenching in Rod Bundles

Quenching of the core may occur after a Loss Of Coolant Accident. Here the specificity of the rod bundle case is directly related to radial steam and water cross flows, to the complex geometry and heterogeneities of the core. (See **Blowdown** and **Reflood**)

Steam and water cross-flows

Analytical experiments in large 2D test sections have demonstrated that strong steam and water cross-flows occur between neighbouring assemblies with different residual powers (radial peaking factor as high as 1.8). Considering for instance a "hot" assembly between two "cold" ones (Deruaz et al., 1984 and Housiadas et al., 1989) perfect steam radial mixing is observed in the rewetted region indicating existence of steam cross-flows from the hot to the cold assemblies. At the same time, water cross-flows occur from the cold to the hot assembly with an intensity which can reach values of up to 60% of the feeding rate. Liquid cross-flows are negligible in the dry region whereas steam escapes to the cold assemblies due to the flow resistance inherent in a larger amount of liquid in the dry zone of the hot assembly. This complex behaviour has a noticeable impact on quench front progression and heat transfer in the assemblies. In the hot assembly, quench velocity is lower and precursory cooling is enhanced due to the larger amount of liquid in the dry zone. (See **Rewetting of Hot Surfaces**)

In steady or quasi-steady situations (boil-up or boil-off cases), level swell is uniform and depends only upon the averaged residual power (and not upon its distribution between the assemblies). This result is consistent with the existence of perfect steam radial mixing.

If exit steam velocities are not too large an additional possible 2D phenomenon could be preferential fall back of water from the upper plenum to the the cold assemblies. Due to mixing of steam in the core this type of phenomenon does not occur.

Ballooning and grid effects

Extensive experiments have shown that both spacer grids (Clement et al., 1982 and Hochreiter et al., 1992) and flow blockages (Hochreiter et al., 1992) (due to *ballooning* of some heater rods) can significantly alter post-CHF heat transfer. Similarly, they exhibit local precursory rewetting and induce downstream effects due to convection enhancement of the continuous phase and breakup of entrained droplets.

Cold rod effects

The presence of unheated rods in a bundle (control rod guide thimbles in the standard PWR for example) has a significant effect on the cooling efficiency viewed in terms of overall quench time (the time necessary to quench the full bundle) or of maximum wall temperature (Veteau et al., 1994). The cold rods exhibit very early quenching, modifying hydrodynamics in the subchannels and enhancing radiative cooling of the adjacent heating rods.

References

Clement P., Deruaz R. and Veteau J. M. (1982) Reflooding of a PWR bundle: Effect of inlet flowrate oscillations and spacer grids, NUREG/CP-0027, Vol. 3, Proceedings of the International Meeting on Thermal Nuclear Reactor Safety, Chicago, August 29-Sept. 2, 1982.

Deruaz R., Clement P., Veteau J. M., (1985) 2D effects in the core during the reflooding case of a LOCA, Safety of Thermal Water Reactors, Proceedings of a Seminar on the results of the European Communities' Indirect Action Research Programme, on Safety of Thermal Water Reactors, held in Brussels, 1–3 October 1984, E. Skupinsky, B. Tolley and J. Vilain Eds., Graham and Trotman Limited Publishers.

Groeneveld D. C. and Snoek C. W. (1986) A comprehensive examination of heat transfer correlations suitable for reactor safety analysis, *Multiphase Science Technology,* Ch. 3, Vol. 2, G. F. Hewitt, J. M. Delhaye, N. Zuber, eds. Hemisphere Publishing Corporation.

Groeneveld D. C., Erbacher F. J., Kirillov P. L., Zeggel W. and al., (1993) An improved table look-up method for predicting critical heat flux, *6th International Topical Meeting on Nuclear Reactor Thermal Hydraulics,* Grenoble, France, October.

Housiadas C., Veteau J. M. and Deruaz R. (1989) Two-dimensional quench front progression in a multi-assembly rod bundle, *Nuclear Engineering and Design,* 113, 87–98.

Hochreiter L. E., Loftus M. J., Erbacher F. J., Hile P., Rust K. (1992) Post CHF effects of spacer grids and blockages in rod bundles, Post dry-out heat transfer, *Multiphase Science and Technology,* G. F. Hewitt, J. M. Delhaye and N. Zuber, eds., CRC Press.

Lach J., Kielkiewicz M. and Kosinski M. (1986) *Heat Transport in Nuclear Reactor Channels of Heat Transfer Operations*, Vol. 1, Ch. 36, N. P. Cheremisinoff N. P. ed., Gulf Publishing Company.

Tong L. S. and Weisman J. (1979) *Thermal Analysis of Pressurized Water Reactors,* 2nd Edition, American Nuclear Society, La Grange Park, Illinois.

Veteau J. M., Digonnet A. and Deruaz R. (1994) Reflooding of tight lattice bundles, *Nuclear Technology,* 107, July 1994.

F. de Crécy and J.M. Veteau

ROD BUNDLES, PARALLEL FLOW IN (see Channel flow; Rod bundles, flow in)

ROD CLIMBING (see Non-Newtonian fluids)

ROD BAFFLES (see Shell-and-tube heat exchangers)

RODRIGUES FORMULA (see Chebyshev polynomials)

ROHRSCHNEIDER CONSTANT (see Gas chromatography)

ROLL MOMENT (see Aerodynamics)

ROLL WAVES (see Kinematic waves; Stratified gas-liquid flow)

ROOTS TYPE COMPRESSOR (see Compressors)

ROSIN-RAMMLER (see Optical particle characterisation)

ROSIN RAMMLER SIZE DISTRIBUTION (see Atomization)

ROSSBY NUMBER

There are two dimensionless groups given the name Rossby number.

The first represents the ratio of inertia force to **Coriolis Force** exerted on a flowing liquid, in particular that associated with the earth's rotation.

Rossby number 1,

$$Ro_1 = u/2\omega_e L \sin \phi$$

where u is fluid velocity, ω_e is the earth's angular velocity (s^{-1}), L length and ϕ the angle between the earth's axis of rotation and the direction of fluid flow.

The second Rossby number represents the ratio of fluid velocity to velocity associated with rotation.

Rossby number 2,

$$Ro_2 = u/\omega L$$

where ω is angular velocity (s^{-1}).

G.L. Shires

ROSSELAND COEFFICIENT (see Coupled radiation, convection and conduction)

ROTAMETERS

***Following from:** Flow metering*

A example of a rotameter is shown diagrammatically in **Figure 1**. It consists of a vertical tube with a tapered bore containing a float. The float rises in the tube as the flowrate increases and the flowrate can be read directly off a scale which is usually marked on the tube. Rotameters can be used to measure the volumetric flowrate of either liquid or gas and are amongst the simplest and cheapest flowmeters in use. The flowrange is typically 10:1 and accuracy is usually 2 per cent.

Glass tubes are most common but the material varies from inexpensive machined acrylics to chemically resistant glass. For large sizes metal tubes can be used and the scale is attached to the float and viewed through a glass fronted sighting compartment below.

A spherical float can be used but these are prone to unsteadiness so it is more usual to find floats of various designs such as the

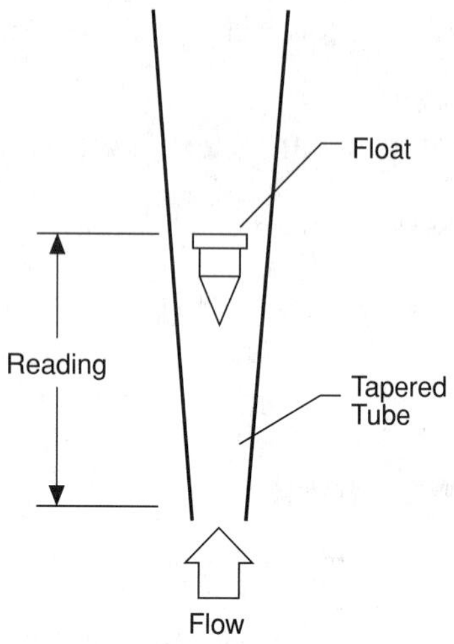

Figure 1. Rotameter.

one shown in **Figure 1**. A well designed float's resistance is insensitive to Reynolds number and thus the differential pressure across the float is the same at all heights in the tube. Miller (1983) explains that the float height therefore varies with flowrate to expose the cross sectional flow area that satisfies Bernoulli's equation, hence the alternative name of *variable area meter.* Martin (1949) describes procedures for when the resistance of the float is not constant.

The scale on any rotameter relates to a particular density of fluid. Ower and Pankhurst (1977) explain that when operating with a fluid of slightly different density, the true flowrate can be calculated from $\dot{Q} = \dot{Q}_i(\rho_c/\rho_a)$ where $\dot{Q}_i$ is the indicated flowrate, ρ_c is the calibration density and ρ_a is the operating density.

References

Martin, J. J. (1949) Calibration of rotameters, *Eng. Progress,* 45, 338.

Miller, R. W. (1983) *Flow Measurement Engineering Handbook,* McGraw-Hill, New York.

Ower, E. and Pankhurst, R. C. (1977) *The Measurement of Air Flow,* Pergamon Press, Oxford.

C.D. Stewart

ROTARY ATOMIZERS (see Atomization)

ROTARY DRYERS (see Dryers)

ROTARY KILNS (see Kilns)

ROTARY REGENERATORS (see Regenerative heat exchangers)

ROTATED TUBE BANKS (see Tube banks, crossflow over)

ROTATING CYLINDERS, CRITICAL SPEED (see Viscosity measurement)

ROTATING CYLINDERS, FLOW BETWEEN (see Görtler-Taylor vortex flows)

ROTATING CYLINDERS, FLOW OVER (see Magnus force)

ROTATING DISC CONTACTOR (see Extraction, liquid-liquid)

ROTATING DISC SYSTEMS, APPLICATIONS

Following from: *Rotating disc systems, basic phenomena*

Introduction

In the following entry, Rotating Disc Systems, Basic Phenomena are classified in terms of the parameter Γ, the ratio of the speed of the slower disc to that of the faster one. The main application of these systems is to model the conditions found inside the wheel-space of gas-turbine-engines, where the cases $\Gamma = 0, +1$ and -1 are the most important. The rotor-stator system ($\Gamma = 0$) is used to model the flow and heat transfer associated with an air-cooled turbine disc and an adjacent stationary casing; the rotating cavity ($\Gamma = +1$) is used to model conditions between corotating turbine or compressor discs; contra-rotating discs ($\Gamma = -1$) are used to model the wheel-space between the contra-rotating turbine discs of some existing (or future) engines.

The designer of internal air systems needs to know the heat transfer coefficients and frictional windage of the discs, the pressure drop of the cooling air, and the amount of air required to prevent, or reduce to safe levels, the ingress of hot mainstream gas past the rim seals into the wheel-space. For these practical cases, the gap ratio (or ratio of the axial clearance, s, to disc radius, b) is of the order of $s/b = 10^{-1}$, for which clearances there are usually separate boundary layers on the two discs. The flow is turbulent throughout most of the wheel-space, but it is often instructive to study the laminar flow cases in advance of the turbulent ones.

A number of nondimensional variables are used to characterize these flows, and the most useful of these are defined below:

$$Re_\phi = \frac{\Omega b^2}{\eta} \quad (1)$$

$$C_w = \frac{\dot{V}}{\eta b} \quad (2)$$

$$\lambda_T = C_w Re_\phi^{-0.8} \quad (3)$$

$$C_m = \frac{M}{0.5\,\rho\Omega^2 b^5} \quad (4)$$

$$Nu = \frac{\dot{q}_s r}{\lambda \Delta T} \quad (5)$$

Here Ω is the angular speed of the disc, b the radius of the disc, $\dot{V}$ the coolant volumetric flow rate (positive for radial outflow and negative for inflow), M the frictional moment (or windage torque) on one side of the rotating disc, $\dot{q}_s$ the heat flux from the disc surface to the fluid, and ΔT a representative temperature difference; r is radius, λ thermal conductivity, ρ fluid density and ν kinematic viscosity. Re_ϕ, C_w, λ_T, C_m and Nu are referred to respectively as the rotational **Reynolds Number**, the nondimensional flow rate, the turbulent flow parameter, the moment coefficient and the local **Nusselt Number**.

For engine applications, Re_ϕ and λ_T are of the order of 10^7 and 10^{-1} respectively. A value of $\lambda_T \simeq 0.22$ is associated with the free-disc entrainment rate: the flow rate entrained by a disc rotating in a quiescent fluid. For most turbine-disc cooling applications, $\lambda_T < 0.22$.

Many of these (axisymmetric) rotating-disc flows have been successfully computed by a variety of methods using the integral equations, differential boundary-layer equations and elliptic solvers. In the latter case, the low-Reynolds-number k-ϵ turbulence model has proved particularly effective, and it is that model that is used here to compute the streamlines of the various flow structure. Comprehensive accounts of the cases for $\Gamma = 0$ and $\Gamma = +1$ are given by Owen and Rogers (1989, 1995).

Rotor-Stator Systems (Γ = 0)

Figure 1 shows computed streamlines for $\Gamma = 0$ and $Re_\phi = 1.25 \times 10^6$ for a) $C_w = 0$ and b) $C_w = 9688$.

For $C_w = 0$ (the enclosed rotor-stator system), there is radial outflow in a thin boundary layer on the rotor (at z = 0) and inflow in a boundary layer on the stator (at z = s). Between the boundary layers, the inviscid core rotates in quasi-solid-body rotation (at about 40% the speed of the rotor) and there is an axial flow from the stator to the rotor. In the inviscid core, the *Taylor-Proudman theorem* applies so that $u_\infty = 0$, $\partial v_\infty/\partial z = 0$ and $\partial w_\infty/\partial z = 0$, where u, v and w respectively refer to the radial, tangential and axial components of velocity relative to a rotating frame, and the subscript ∞ refers to conditions in the core. This flow structure, which occurs in laminar as well as turbulent flow, is often referred to as *Batchelor-type flow* after Batchelor (1951).

For $C_w > 0$, a source region is formed near the inlet where the incoming fluid is entrained into the boundary layer on the rotor. If the incoming fluid has no swirl, then there is no rotation in the source region, and the core rotation is less than that associated with $C_w = 0$. The radial extent of the source region increases as λ_T increases, and for $\lambda_T \geq 0.22$ the source region fills the entire system. Both C_m and Nu depend on λ_T: for $\lambda_T = 0.22$, C_m is equal to that of the free disc; for $\lambda_T = 0$, C_m is approximately half that of the free disc.

The rotating flow in the core can create a negative pressure inside the wheel-space, allowing the ingress of hot mainstream gas. A minimum nondimensional flow rate, $C_{w,\,min}$, is required to prevent this ingress, which depends on the rim-seal geometry, on Re_ϕ and on the conditions inside the mainstream gas itself. The *ingress problem* is discussed extensively by Owen and Rogers (1989).

Rotating Cavities (Γ = +1)

Computed streamlines for isothermal rotating cavities with a radial outflow and inflow of cooling air are shown in **Figure 2** for $\Gamma = +1$ and $Re_\phi = 1.1 \times 10^6$ for a) $C_w = +1500$ (radial outflow) and b) $C_w = -1500$ (radial inflow). For both cases, the fluid enters the cavity without swirl.

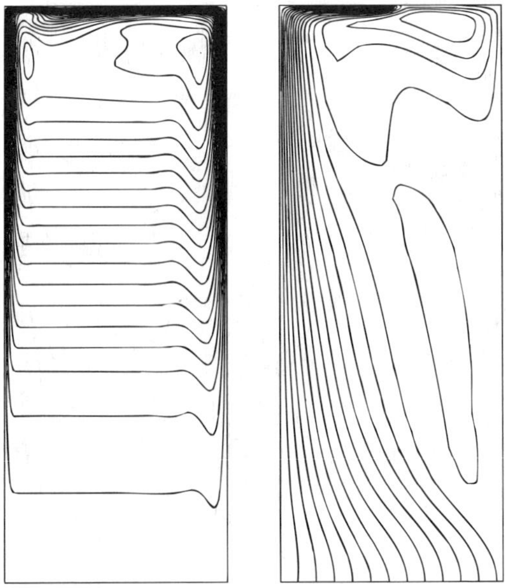

a) b)

Figure 1. Computed streamlines for $\Gamma = 0$ and $Re_\phi = 1.25 \times 10^6$ for a) $C_w = 0$ and b) $C_w = 9688$.

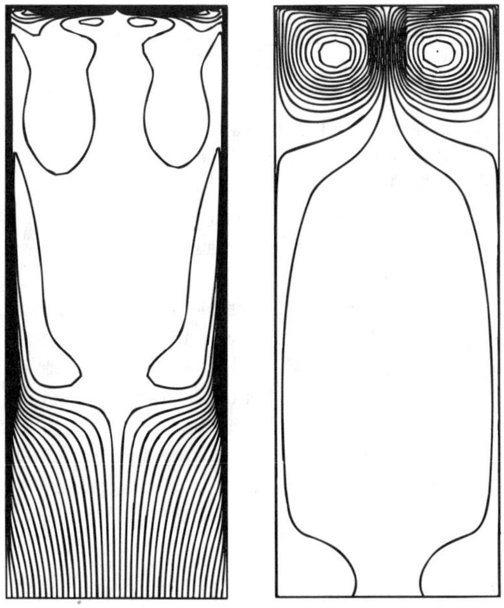

a) b)

Figure 2. Computed streamlines for $\Gamma = +1$ and $Re_\phi = 1.1 \times 10^6$ for a) $C_w = +1500$ b) $C_w = -1500$.

Rotating cavity with radial outflow

Referring to **Figure 2a**, fluid enters the cavity at r = a forming a source region, that distributes the flow into *Ekman-type layers* on each disc, and leaves via a sink layer at r = b that distributes the flow into a slot or holes in the peripheral rotating shroud. In the rotating inviscid core between the Ekman-type layers, the Taylor-Proudman theorem applies and, consequently, the axial and radial components of velocity are zero.

Fluid is entrained into the boundary layers in the source region, and the (nonentraining) Ekman-type layers start where all the available fluid has been entrained: the tangential velocity of the core, which is lower than that of the discs, adjusts to ensure that the mass flow rate in the Ekman-type layers is invariant with radius.

For the case of heated discs, the Nusselt numbers increase radially in the entraining boundary layers in the source region and decrease in the nonentraining Ekman-type layers, reaching a maximum value near the outer edge of the source region. As for the rotor-stator case, the size of the source region increases as λ_T increases.

Rotating cavity with radial inflow

Referring to **Figure 2b**, fluid enters through the rotating shroud at r = b, forming a source region, and leaves at r = a, via a sink layer. Outside the source region, where all the available fluid is entrained into the boundary layers, nonentraining Ekman-type layers are formed on the discs. The Taylor-Proudman theorem holds in the inviscid core, which rotates faster than the discs.

As well as depending on λ_T, the size and structure of the source region depends strongly on the inlet swirl ratio, c, where c is the ratio of the tangential speed of the incoming fluid to that of the discs at r = b. If $c > 1$, the flow is radially inward throughout the cavity. If $c < 1$, the flow in the boundary layers in the source region is radially outward where the fluid rotates slower than the discs and radially inward where it rotates faster. Consequently, for $c < 1$, there is recirculation in the source region as shown in **Figure 2b**: mixing between the recirculating flow and the fluid entering the cavity causing the effective swirl ratio, c_{eff}, to be larger than the inlet swirl ratio, c. The angular momentum of the fluid in the source region is virtually conserved resulting in quasi-free-vortex flow. Outside the source region, the tangential velocity corresponds to that required to maintain a constant flow rate in the Ekman-type layers.

Radial-inflow presents more problems than outflow, and the complex structure of the source region makes computation of the flow and heat transfer more difficult. The large tangential velocities generated by the free-vortex-type flow mean that the radial pressure gradients in the rotating cavity can be large. The flow rate of turbine-cooling air depends on the pressure difference between the compressor and the turbine: the pressure drop created by radial inflow can significantly reduce the coolant flow rate.

Buoyancy-induced flow in a rotating cavity

For the radial inflow and outflow cases, the flow rates used in gas turbines are usually large enough to ensure that (axisymmetric) forced convection dominates. For the case of a rotating cavity with no superposed flow or with an axial throughflow of cooling air, buoyancy effects are usually significant: the flow can then become nonaxisymmetric and unsteady, and the Nusselt numbers are significantly smaller than for the forced convection cases.

For the case of a sealed cavity where one disc is hot and other cold, fluid flows radially outward in a boundary layer on the cold disc and radially inward on the hot disc. Such flows can be axisymmetric and steady. For the cases of a sealed cavity where the outer cylindrical casing (or shroud) is hot and the inner one cold, free convection takes place through a system of cyclonic and anticyclonic vortices, as described in **Part 1**. These vortices create circumferential regions of low and high pressure which provide the necessary Coriolis forces to allow simultaneous radial inflow and outflow of fluid between the hot and cold surfaces.

The axial throughflow case is further complicated by the possibility of *vortex breakdown*. This is the abrupt change in the structure of an axial jet of swirling fluid which, in the case of a rotating cavity, can create nonaxisymmetric, unsteady flow even under isothermal conditions. The interaction between a central jet undergoing vortex breakdown and buoyancy-induced flow in the cavity itself makes this case the Everest of the rotating-disc problems: it is a challenge to both computational and experimental research workers!

Contra-Rotating Discs ($\Gamma = -1$)

Figure 3 shows computed streamlines for $\Gamma = -1$ and $Re_\phi = 1.19 \times 10^6$ for a) $C_w = 0$ and b) $C_w = 9350$.

For $C_w = 0$, there is radial outflow in the boundary layers on the rotating discs and radial inflow in the core between the boundary layers. The core is virtually nonrotating (so the Taylor-Proudman theorem does not apply), and this flow structure is often referred to as *Stewartson-type flow* after Stewartson (1953). (For laminar flow, Batchelor (1951) predicted a flow structure with contra-

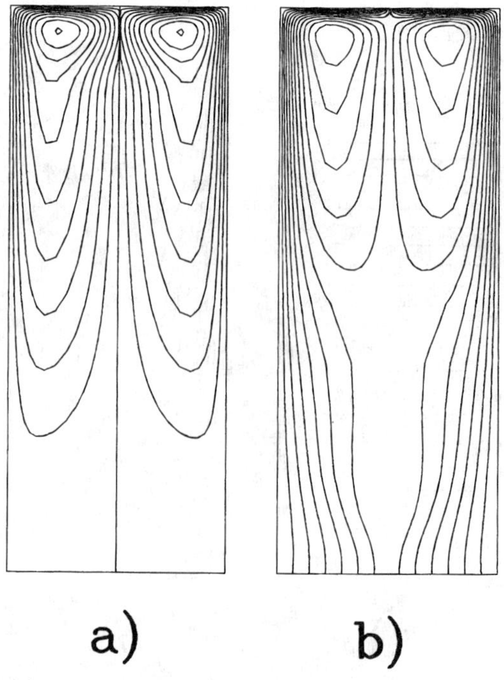

Figure 3. Computed streamlines for $\Gamma = -1$ and $Re_\phi = 1.19 \times 10^6$ for a) $C_w = 0$ and b) $C_w = 9688$.

rotating cores; this can be computed but it does not appear to exist in practice.)

For $C_w = 9350$, a source region is formed near the inlet where the incoming (nonswirling) fluid is entrained into the boundary layers on the discs. Outside the source region, recirculation occurs in a similar way to that described above for $C_w = 0$. The moment coefficients for $\Gamma = -1$ can be up to twice the magnitude of those for $\Gamma = 0$, but the difference decreases as C_w increases.

If, as is customary, the Nusselt number is defined using the temperature difference, ΔT, between the surface of the disc and the fluid at inlet to the system then Nu for $\Gamma = -1$ is less than Nu for $\Gamma = 0$. This apparent contradiction of the **Reynolds Analogy** (see **Part 1**) is caused by the use of inappropriate reference temperatures.

The double transition from laminar to turbulent flow and from *Batchelor-type flow* at $\Gamma = 0$ to Stewartson-type flow at $\Gamma = -1$ has been investigated by Kilic et al. (1994). For $-1 < \Gamma < 0$, a two-cell structure occurs with Batchelor-type flow (with a rotating core) in the (radially) outer cell and Stewartson-type flow (with a nonrotating core) in the inner one. If Γ is reduced, the relative sizes of cells change until at $\Gamma = -1$ there is only Stewartson-type flow.

The flow and heat transfer for $\Gamma = -1$ appears to remain axisymmetric under all observed conditions, and agreement between computed and measured values is mainly good.

References

Batchelor, G. K. (1951) Note on a class of solutions of the Navier-Stokes equations representing steady rotationally-symmetric flow, *Quart. J Mech. Appl. Math.,* 4, 29–41.

Kilic, M., Gan, X. and Owen, J. M. (1994) Transitional flow between contra-rotating discs, *J. Fluid Mech.* 281, 119–135.

Owen, J. M. and Rogers, R. H. (1989) *Flow and Heat Transfer in Rotating Disc Systems,* Vol. 1: Rotor-stator systems, Research Studies Press, Taunton, John Wiley, New York.

Owen, J. M. and Rogers, R. H. (1995) *Flow and Heat Transfer in Rotating Disc Systems,* Vol. 2: Rotating cavities, Research Studies Press, Taunton, John Wiley, New York.

Stewartson, K. (1953) On the flow between two rotating coaxial discs, *Proc. Camb. Phil. Soc.,* 49, 333–341.

J.M. Owen

ROTATING DISC SYSTEMS, BASIC PHENOMENA

Following from: *External flows, overview*

Introduction

The flow and heat transfer associated with air cooled gas turbine discs can be modelled using simple rotating disc systems, as shown in **Figure 1**, where one disc rotates close to a second disc, which may be rotating or stationary. It is convenient to classify the system by Γ, the ratio of the speed of the slower disc to that of the faster one: $\Gamma = -1, 0, +1$ correspond to contra-rotating discs, a rotor-stator system and corotating discs, respectively.

In rotating flows, *Coriolis forces* can create phenomena not observed in nonrotating cases. Some of these phenomena are outlined below, and more details are given by Greenspan (1968) and Owen and Rogers (1995).

Linear Equations

It is convenient to use cylindrical polar coordinates (r, ϕ, z) rotating with angular speed Ω about the axis $r = 0$; the velocity components in this rotating frame are u, v and w. It is assumed that the axial clearance, s, between the discs is large enough for separate boundary layers to develop on each disc, between which is a rotating core of fluid.

For quasi-solid-body rotation, the linear Coriolis accelerations $(2\Omega u, 2\Omega v)$ can be much larger than the nonlinear inertial accelerations. If the nonlinear terms are neglected, the Navier-Stokes equations reduce to a system of linear equations. Similarly the boundary-layer equations reduce to a linear system referred to as the *Ekman-layer* equations (Ekman, 1905), which are written below together with the continuity equation:

$$\frac{1}{r}\frac{\partial}{\partial r}(\rho u r) + \frac{1}{r}\frac{\partial}{\partial \phi}(\rho v) + \frac{\partial}{\partial z}(\rho w) = 0 \qquad \textbf{(1)}$$

$$-\rho(2\Omega v + \Omega^2 r) = -\frac{\partial p}{\partial r} + \eta\frac{\partial^2 u}{\partial z^2} \qquad \textbf{(2)}$$

$$2\Omega\rho u = -\frac{1}{r}\frac{\partial p}{\partial \phi} + \eta\frac{\partial^2 v}{\partial z^2} \qquad \textbf{(3)}$$

$$0 = -\frac{\partial p}{\partial z} \qquad \textbf{(4)}$$

Inviscid Phenomena

If viscous effects are negligible (for example, in the core outside the boundary layers on the discs), the *Taylor-Proudman theorem* (see Greenspan, 1968, or Owen and Rogers, 1995) produces some important results. In particular, for axisymmetric flow in the rotating core it follows from **Equations 1 to 4** that

$$u_\infty = 0, \frac{\partial v_\infty}{\partial z} = 0, \frac{\partial w_\infty}{\partial z} = 0 \qquad \textbf{(5)}$$

where the subscript ∞ denotes that $z \to \infty$. That is, for an inviscid fluid in quasi-solid-body rotation, the flow has a two dimensional tendency: all three components of velocity are invariant with z and, for axisymmetric flow, the radial component is zero.

This two-dimensional tendency makes the boundary layer on a rotating disc resistant to separation. For example, a step change in the surface of a stationary disc would cause boundary-layer separation, but if the disc were rotating in a rotating fluid then the flow could remain attached to the step.

Consider the case of solid-body rotation between two discs, on one of which is a protrusion, where the speed of the discs is suddenly changed from Ω to Ω'. Initially, the speed of the core of

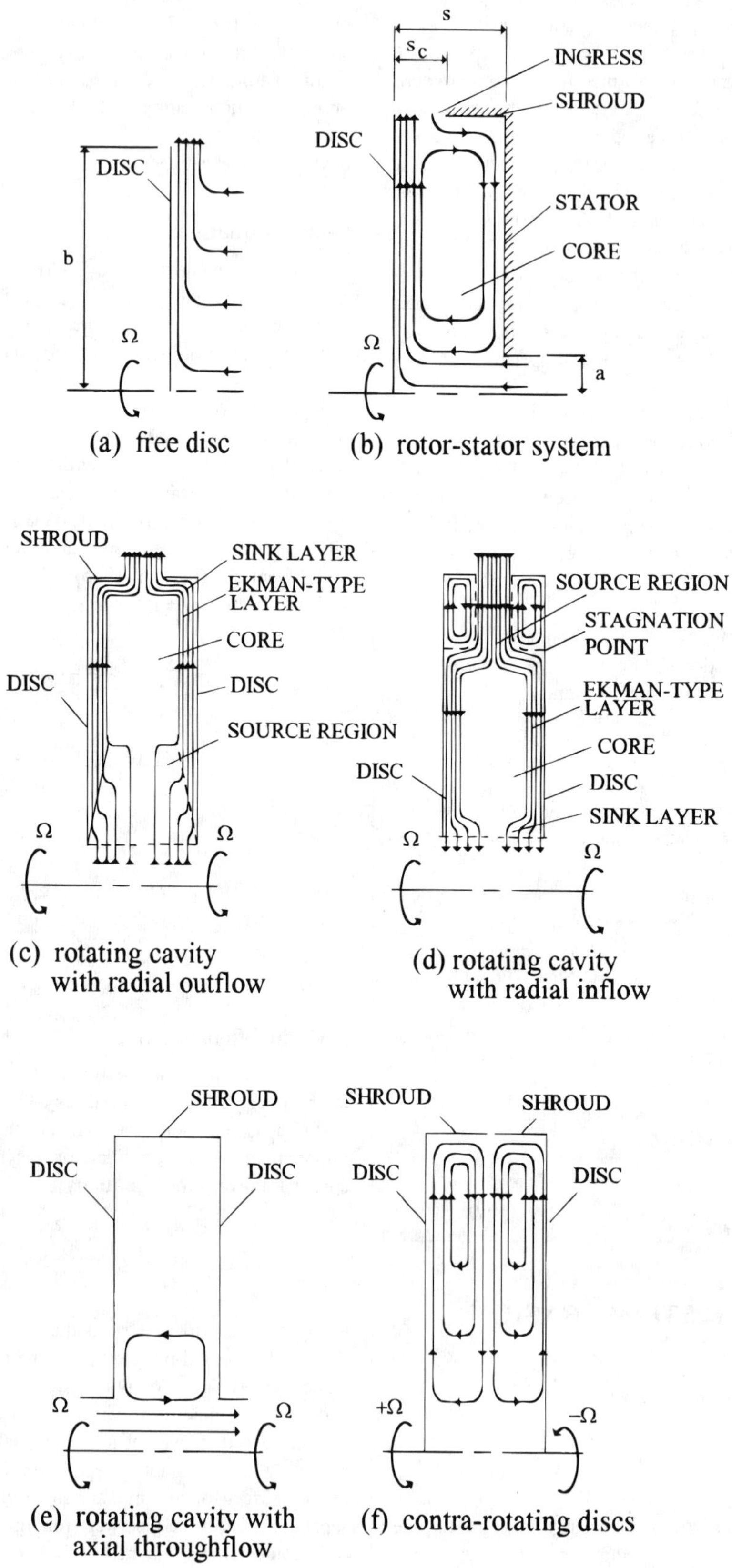

Figure 1. Schematic diagram of rotating disc systems.

fluid is different from that of the discs and fluid flows around the protrusion. However the two-dimensional tendency causes the effect of the protrusion to extend axially throughout the fluid creating a so-called transverse *Taylor column,* as shown in **Figure 2**. The fluid flows around this column as though it were a solid cylinder!

A *longitudinal Taylor column* occurs when a small object is moved axially across a fluid in solid body rotation. A cylinder of fluid is formed, part of which is pushed ahead of the object and part is pulled behind. The rest of the fluid is unaffected by this motion.

Although the two-dimensional tendency arises from an inviscid fluid in solid-body rotation, it has been observed in a variety of rotating-disc systems even when the flow in the boundary layers on the discs is turbulent. In fact the very structure of turbulence in a rotating fluid is affected by this tendency.

Another characteristic of inviscid rotating flows is cyclonic and anti-cyclonic circulation. These nonaxisymmetric circulations often occur in rotating fluids in which buoyancy effects are significant. A cyclonic vortex, which rotates in the same sense as the rotating discs, is associated with low pressure at its centre; an anti-cyclonic vortex is associated with high pressure. A pair, or pairs, of these vortices create nonaxisymmetric variations in pressure that provide the necessary Coriolis forces to allow radial inflow and outflow of fluid in the rotating core.

Ekman layers

For axisymmetric, incompressible flow, the linear **Equations 1 to 4** can be solved to give

$$u = -v_\infty e^{-\zeta} \sin \zeta, \; v = v_\infty(1 - e^{-\zeta} \cos \zeta) \tag{6}$$

$$\text{where } \zeta = z\sqrt{(\Omega/\nu)} \tag{7}$$

ν being the kinematic viscosity of the fluid and the subscript ∞ referring to conditions in the core outside the boundary layers; the latter are referred to as *Ekman layers.* The solutions for u and v are oscillatory, u becoming zero when $\zeta = n\pi$, n being an integer, and the thickness of the Ekman layer is taken to correspond to n = 1 where $\zeta = \pi$ and $z = \pi\sqrt{(\nu/\Omega)}$. That is, the Ekman layer thickness is invariant with radius. A point to note is that u and v_∞ are of opposite sign: if the inviscid core rotates faster than the disc (that is, $v_\infty > 0$) then the flow will be radially inward ($u < 0$) in the Ekman layer, and vice versa.

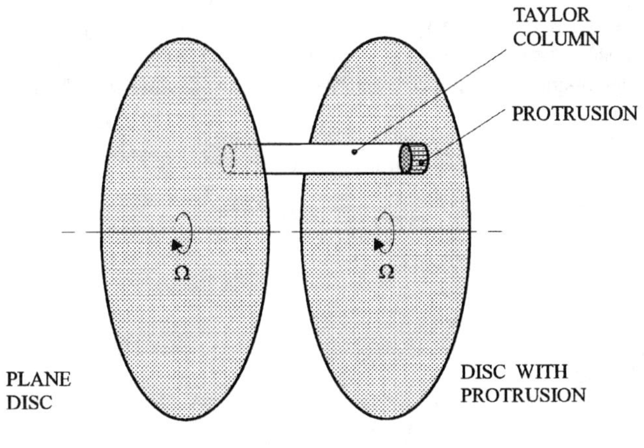

Figure 2. Schematic diagram of a Taylor column.

An important special case is the rotating cavity ($\Gamma = 1$) where there is a superposed radial inflow or outflow of fluid between the discs and where the flow is symmetrical about the midplane ($z = \frac{1}{2}s$). According to **Equation 5**, $\partial w_\infty/\partial z = 0$, and symmetry implies that $w_\infty = 0$ in the rotating core. In this case, the Ekman layers are therefore nonentraining boundary layers.

Although the above results are associated with linear, laminar Ekman layers, similar phenomena have been observed for turbulent flow and for cases where the linear assumptions are not strictly valid. Such boundary layers are often referred to as Ekman-type layers.

The Reynolds analogy

For rotating-disc systems, there is a strong similarity between the energy equation and the tangential momentum equation (see Owen and Rogers, 1989). For the case of a single disc rotating in a quiescent fluid (the so-called "free disc"), the analogy is exact if the **Prandtl Number** is unity and the surface temperature of the disc increases quadratically with radius. The latter condition implies that

$$T_s = T_\infty + cr^2 \tag{8}$$

where the subscripts s and ∞ refer to the surface of the disc and the inviscid fluid, respectively, and c is a constant. Under these conditions,

$$Nu_{av} = {}^1\!/_2\, Re_\phi C_m \tag{9}$$

where Nu_{av} is the radially-weighted average **Nusselt Number,** Re_ϕ is the rotational **Reynolds Number** and C_m is the moment coefficient based on the frictional moment on the disc.

It also follows for this special case that the *adiabatic-disc temperature,* $T_{s,ad}$ is given by

$$T_{s,ad} = T_\infty + \frac{{}^1\!/_2\, \Omega^2 r^2}{C_p} \tag{10}$$

The driving potential for heat transfer is $T_s - T_{s,ad}$ and the heat flux tends to zero as T_s tends to $T_{s,ad}$. For practical applications, it is difficult to determine $T_{s,ad}$ and most experimentalists use a reference temperature that is convenient to measure, such as the temperature of the fluid at inlet to the system. This can often result in measured Nusselt numbers becoming negative, particularly when the surface temperature of the disc decreases with radius. Whilst the fluid may be colder than the disc at inlet to the system, there may be a radius beyond which the disc is colder than the fluid: the heat flux then becomes negative even though the reference temperature difference remains positive.

References

Ekman, V. W. (1905) On the influence of the earth's rotation on ocean-currents, *Ark. Mat. Astr. Fys.,* 2, 1–52.

Greenspan, H. P. (1968) *The Theory of Rotating Fluids,* Cambridge University Press, London.

Owen, J. M. and Rogers, R. H. (1989) *Flow and Heat Transfer in Rotating Disc Systems,* Vol. 1: Rotor-stator systems, Research Studies Press, Taunton, John Wiley, New York.

Owen, J. M. and Rogers, R. H. (1995) *Flow and Heat Transfer in Rotating Disc Systems,* Vol. 2: Rotating cavities, Research Studies Press, Taunton, John Wiley, New York.

Leading to: *Rotating disc systems, applications.*

J.M. Owen

ROTATING DUCT SYSTEMS, ORTHOGONAL, HEAT TRANSFER IN

Following from: *Rotating duct systems, parallel, heat transfer in*

Introduction

In the following entry, **Rotating Duct Systems, Parallel, Heat transfer In** it is explained that rotation of a straight duct influences an internal pressure-driven flow via Coriolis and centripetal buoyancy interactions with consequential modifications to the otherwise forced convection heat transfer between the duct wall and the flow.

Orthogonal-Mode Rotation

In this entry, these effects are reviewed for the special case where the duct is constrained to rotate about an axis which is perpendicular to its axis of symmetry. This rotating configuration is referred to as **orthogonal-mode** rotation and is illustrated in **Figure 1**.

Coriolis forces, in this case, act as a source term for the generation of a cross stream secondary flow as indicated in the figure. Thus a mechanism exists to cause relatively cool fluid from the central area of the tube to move towards the trailing edge. If the duct wall is heated this suggests that Coriolis forces tend to induce a relatively better cooled trailing edge region. With significant density variations, associated with heated flow, a *centripetal buoyancy* effect is produced analogous to the earth's gravitation field with a stationary vertical duct. Thus buoyancy effects will vary according to the direction of flow through the duct. In other words, buoyancy will change direction according to whether the flow is radially outward or inward. In the present article, only radially outward flow is considered for reasons of space.

Developed Laminar Flow

Mori and Nakayama (1968) analyzed laminar fully developed heat transfer with a uniformly heated tube of circular cross section using a momentum integral technique. These authors did not include buoyancy in their analysis and characterised Coriolis effects using a rotational form of the **Reynolds Number**, J_d $(\Omega d^2\rho/\eta)$ where the velocity term involves the product of the tube diameter and the angular velocity. They did not allow for circumferential variations in heat transfer due to the Coriolis induced secondary flow owing to their choice of thermal boundary conditions, but recommendations for estimating a mean heat transfer coefficient over the circumference were proposed. Depending on the relative value of the rotational Reynolds number with respect to the through flow Reynolds number and the fluid **Prandtl Number**, two sets of recommendations were made. **Figure 2** illustrates the typical *enhancement* for a uniformly heated tube which resulted from the analysis. Here "enhancement" is defined as the ratio of the developed **Nusselt Number** obtained with rotation compared to the non-rotating value for a specified Reynolds number. The authors repeated the analysis for a constant wall temperature and found that the relative enhancement in heat transfer was insensitive to the thermal boundary condition.

The equations recommended are likely to give useful results for a mean circumferential heat transfer coefficient but will become less reliable if significant buoyancy effects are present since buoyancy was not included in the analysis. The equations derived do not permit the circumferential variation in heat transfer, due to the Coriolis driven secondary flow, to be evaluated. A number of experimental studies have been reported for laminar heat or mass transfer for this rotating geometry, prior to the early 1980's. Unfortunately the data is not conclusive, probably due to the difficulty of the experimental regime. A review of all data up to 1980 is given by Morris (1981).

Developing and Developed Turbulent Flow

Turbulent flow in orthogonally rotating ducts has been the subject of considerable theoretical and experimental study since about 1980 due to its importance for the internal cooling of gas turbine rotor blades. Morris and Ayhan (1979) first demonstrated experimentally that the combined influence of Coriolis forces and centripetal buoyancy could produce regions of impaired heat transfer, relative to the stationary duct case, on the leading edge of a circular tube. This was important from the design viewpoint in turbine blades since local hot spots might occur on the leading edge. Morris

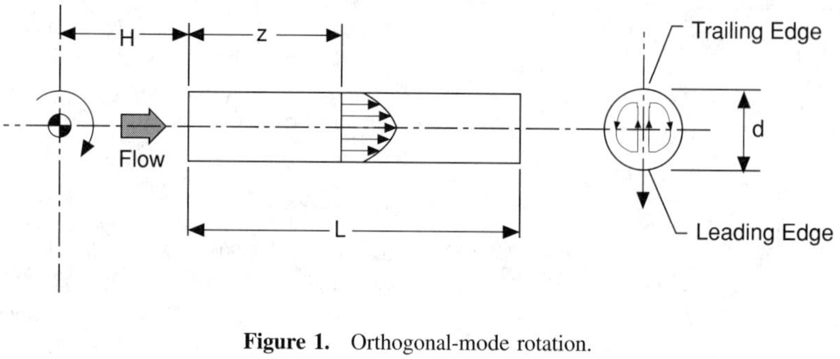

Figure 1. Orthogonal-mode rotation.

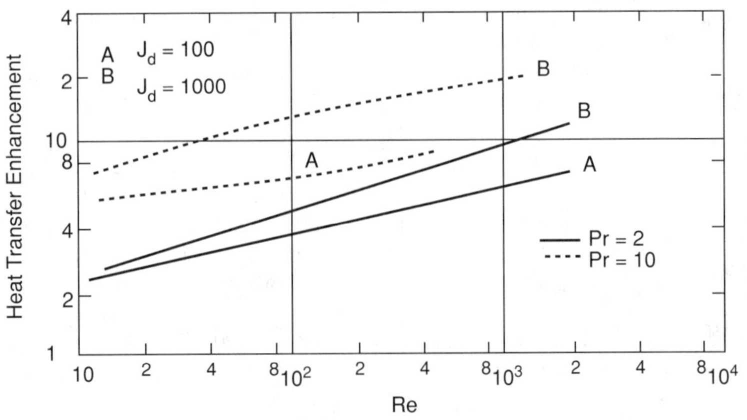

Figure 2. Fully developed laminar flow.

(1981) reviews virtually all published work prior to 1980 and a number of early results are available in this text.

An examination of the momentum and energy conservation equations for this rotating geometry suggests that the Nusselt number, evaluated at a specified location, z, in the duct, will be functionally related to the usual non-dimensional groups which control forced convection (i.e. the through flow Reynolds number, Re, the Prandtl number of the fluid, Pr,) together with two additional terms which arise from the Coriolis forces and the centripetal buoyancy. The non-dimensional group which derives from the Coriolis term is the **Rossby Number**, Ro, which is the ratio of the coolant velocity along the duct to an angular speed term involving the product of the angular velocity with the tube diameter. Thus

$$\mathrm{Ro} = w_m/\Omega d$$

The second additional term is a so-called *buoyancy parameter*, Bu, defined as

$$\mathrm{Bu} = (H + z)(T_{w,z} - T_{b,z})/T_{b,z}d$$

where $T_{w,z}$ and $T_{b,z}$ refer to the local wall and fluid bulk temperatures respectively.

We expect therefore that the local Nusselt number, $Nu_{R,z}$, at some specified position along the duct will have functional form

$$Nu_{R,z} = \Psi\{\mathrm{Re}, \mathrm{Pr}, z/d, \mathrm{Ro}, \mathrm{Bu}\}$$

and the function, Ψ, will be also dependent on the circumferential location at a give z-value.

Subsequent to 1980, a number of theoretical and experimental studies have been undertaken, see for example Wagner et al. (1989), Iacovides and Launder (1990), Taylor et al. (1991) and Morris and Salemi (1991) have attempted to evaluate the appropriate functional form of the above equation. The references cited in these papers give a full coverage of the total literature currently available with radially outward, radially inward flow and also the effect of internal ribbing.

The works of Wagner et al. (1989) and Morris and Salemi (1992) confirm that Coriolis forces enhance heat transfer on the trailing edge of smooth walled circular ducts relative to the case resulting from a similar value of through flow Reynolds number. Additionally, as the buoyancy parameter, Bu, is increased at fixed values of the Reynolds number and Rossby number, the heat transfer also increases. On the leading edge it is evident that rotation initially impairs heat transfer relative to the stationary duct case. However, at higher rotational speeds this reduction seems to be arrested with subsequent improvement of the relative heat transfer.

Morris and Chang (1994), in an as yet unpublished experimental study, have attempted to correlate their own new data for a smooth-walled circular tube and the data of Wagner et al. (1989) using a particular mathematical structure for Ψ based on physical reasoning concerning the combination of Coriolis and buoyancy interactions.

They have proposed that the heat transfer ratio at different axial locations downstream in the duct may be expressed as

$$Nu_{R,z}/Nu_{o,z} = 1 + \Phi\{(H + z)(T_{w,z} - T_{b,z})/T_b d \mathrm{Ro}^2\}$$

where $Nu_{o,z}$ is the zero speed Nusselt number obtained at the same Reynolds number as the rotating case and Φ is some function. The argument of the function, Φ is a combined Coriolis/buoyancy parameter.

Figure 3 shows the heat transfer ratio data produced by Morris and Chang (1994) plotted against the combined Coriolis/buoyancy parameter proposed. In this figure data from Wagner et al. (1989) is also compared at an axial location equivalent to 10.7 effective diameters downstream of entry to the duct. Although some data scatter is evident the proposed correlating function appears to have merit. Thus this figure may be used to estimate the level of heat transfer ratio expected at a particular operating condition.

References

Iacovides, H. and Launder, B. E. (1990) Parametric and numerical studies of fully-developed flow and heat transfer in rotating rectangular passages., *ASME Gas Turbine and Aeroengine Congress and Exposition.*, Brussels, Belgium.

Mori, Y. and Nakayama, W. (1968) Convective heat transfer in rotating radial circular tubes (1st Report-Laminar region)., *Int. J. Heat and Mass Trans.* ii, 1027.

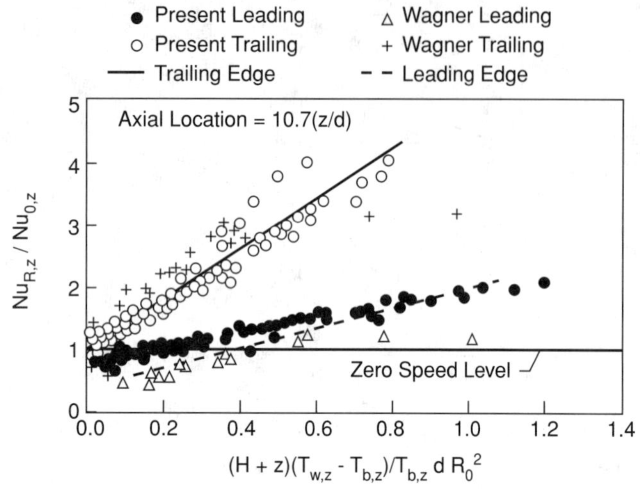

Figure 3. Effect of rotation on heat transfer on the leading and trailing edges.

Morris, W. D. (1981) *Heat Transfer and Fluid Flow in Rotating Cooling Channels,* Research Monograph, Research Studies Press, A Division of J. Wiley and Sons, Ltd., ISBN 0471101214, 1–228.

Morris, W. D. and Ayhan, T. (1979) Observations on the influence of rotation on heat transfer in the coolant channels of gas turbine rotor blades., *Proc. Inst. Mech. Eng.,* 193, 21, 303.

Morris, W. D. and S. W. Chang (1994) Unpublished data at the time of writing.

Morris, W. D. and Salemi, R. (1992) An attempt to experimentally uncouple the effect of coriolis and buoyancy forces on heat transfer in smooth tubes which rotate in the orthogonal mode, trans. A.S.M.E., *Journal of Turbomachinery,* 114, 858–864.

Taylor, C., Xia, J. Y., Medwell, J. O. and Morris, W. D. (1991) Finite element modelling of flow and heat transfer in turbine blade cooling., *Proc. Conf. on Turbomachinery.,* Inst. Mech. Eng., London.

Wagner, J. H. Johnson. B. V., and Hajek. T. J., (1989) Heat transfer in rotating passages with smooth walls and radial outward flow., *ASME Gas Turbine and Aeroengine Congress and Exposition.,* Brussels, Belgium.

W.D. Morris

ROTATING DUCT SYSTEMS, PARALLEL, HEAT TRANSFER IN

Following from: *Channel flow; Tubes, single phase heat transfer in*

Introduction

When a straight duct rotates about an arbitrary axis, rotation influences an internal pressure-driven flow via *Coriolis* and *centripetal* forces. These forces must be included in the momentum conservation equations in order to predict the resulting flow field as a precurser to the evaluation of heat transfer. Coriolis forces generate relative vorticity in the duct via a vector product coupling of the angular velocity and the flow field. In other words complex secondary flows may be generated. With unheated flow, the centripetal force field is hydrostatic. However, with heated-flow, where the density of the fluid is temperature dependent, a *centripetal buoyancy force* is created analogous to natural convection in the earth's gravitational field. The result of this is a mechanism which causes the warmer fluid (i.e. less dense) to move towards the axis of rotation. Again this will modify the non-rotating flow field and the heat transfer. A detailed description of how Coriolis and centripetal buoyancy terms may be incorporated into the momentum conservation equations is given by Morris (1981).

Although an infinite variety of rotating duct systems may be envisaged, reflecting the relative location of the duct to the axis of rotation, two important configurations have been seriously researched. These are commonly referred to as "parallel-mode" and "orthogonal-mode" rotation respectively. With parallel-mode rotation the axis of the duct is parallel to, but displaced from the axis of rotation. Similarly, with orthogonal-mode rotation the axis of the duct and the axis of rotation are mutually perpendicular. In this entry heat transfer with parallel-mode rotation will be discussed. The case of orthogonal-mode rotation is discussed in the previous entry.

Parallel-Mode Rotation

Figure 1 shows a typical straight duct rotating in the parallel-mode. Theoretical and experimental studies of heat transfer with this geometry have been extensively reviewed by Morris (1980) for laminar and turbulent flow and readers are referred to this text for full details. A brief summary of heat transfer data for circular-sectioned ducts follows by way of illustration of the general results.

Developed Laminar Flow

With developed laminar flow, a numerical solution for the **Nusselt Number** for the special case of very high **Prandtl Numbers**, Woods and Morris (1981), has been shown to be quite satisfactory even when applied to fluids with a Prandtl number as low as unity. **Figure 2** shows a comparison of this high Prandtl number solution for *heat transfer enhancement* with experimental data obtained with glycerol, water and air. In this context enhancement is defined as the ratio of the developed Nusselt number with rotation, $Nu_{R,\infty}$, with the stationary counterpart, $Nu_{o,\infty}$. This figure is recommended for assessing the effect of rotation with gases and liquids. Note that the effect of Coriolis force vanishes identically and rotation influences heat transfer via a centripetal buoyancy mechanism

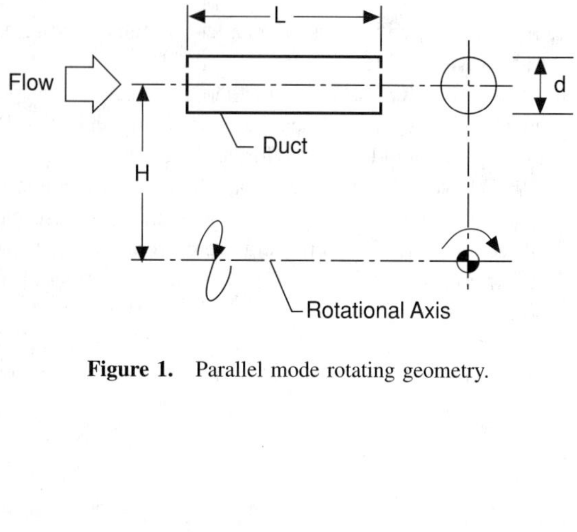

Figure 1. Parallel mode rotating geometry.

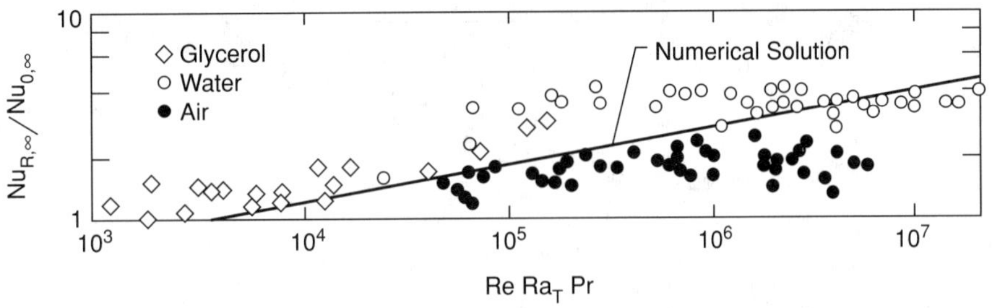

Figure 2. Fully developed laminar flow.

alone. The enhancement in heat transfer relative to the stationary case is dependent on the product of the through flow **Reynolds Number**, Re, a *rotational Rayleigh number* Ra_T, (which uses the product of the axial wall temperature gradient and the tube radius as a characteristic temperature difference in the customary definition) and the Prandtl number, Pr.

Developing Laminar Flow

With developing flow, Coriolis forces are produced due to an interaction between the axial velocity gradient and the rotation. In this region, when the heating rates are moderate, buoyancy probably has a lesser influence and rotation may be quantified non-dimensionally in terms of a *rotational Reynolds number,* J_a, $(\Omega d^2 \rho / 4\eta]$ which may be thought of as a ratio of Coriolis to viscous forces. Morris and Woods (1978) have presented experimental data for two parallel-mode rotation geometries and recommended the following empirical correlations for the mean rotating Nusselt number, $Nu_{R,m}$.

Case 1: (H/d = 48, L/d = 34.65) $Nu_{R,m} = 0.016\ Re^{0.78}\ J_a^{0.25}$
Case 2: (H/d = 96, L/d = 69.3) $Nu_{R,m} = 0.013\ Re^{0.78}\ J_a^{0.25}$

These equations were based on experiments with J_a values up to 150. At high rotational speeds, where buoyancy will become more important, it is not expected that these correlations will be particularly good. **Figure 3** compares the actual test data with these correlations. Further details are available from Morris (1981).

Developed Turbulent Flow

With turbulent flow rotation still modifies the flow field to produce heat transfer enhancement. Nakayama (1968) used a momentum integral analysis to calculate developed turbulent Nusselt number and the following equations for determining this enhancement in heat transfer were proposed.

Case 1: Gas-like fluids with Pr close to unity.

$$Nu_{R,\infty}/Nu_{0,\infty} = 1.367\ Pr^{2/3}\ X^{1/20}(1 + 0.0286/X^{1/5})/(Pr^{2/3} - 0.050)$$

where

$$X = Ra_T^{1.818}/(Re^{2.273}Pr^{0.606}) \quad \text{and} \quad Nu_{0,\infty} = 0.038\ Re^{3/4}Pr^{1/3}$$

Case 2: Liquid-like fluids with Pr > 1

$$Nu_{R,\infty}/Nu_{0,\infty} = 1.428(1 + 0.0144/X^{1/6})X^{1/30}$$

where

$$X = Ra_T^{2.308}/(Re^{3.231}Pr^{0.923}) \quad \text{and} \quad Nu_{0,\infty} = 0.023\ Re^{0.8}Pr^{0.4}$$

Developing Turbulent Flow

As with laminar flow, the effect of buoyancy has been found to be small in the turbulent entrance regions, see Morris (1981).

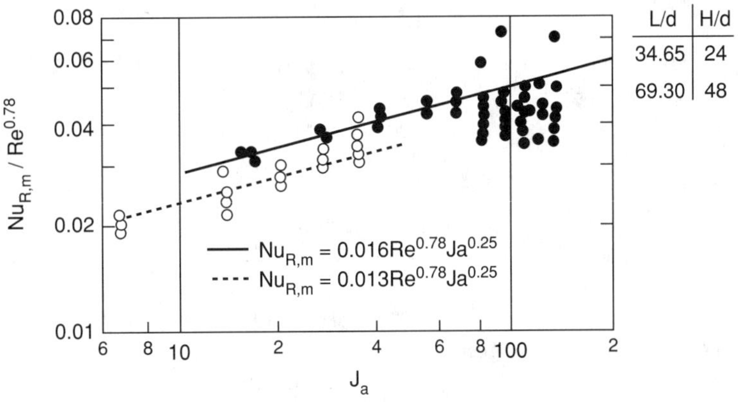

Figure 3. Mean heat transfer with developing flow.

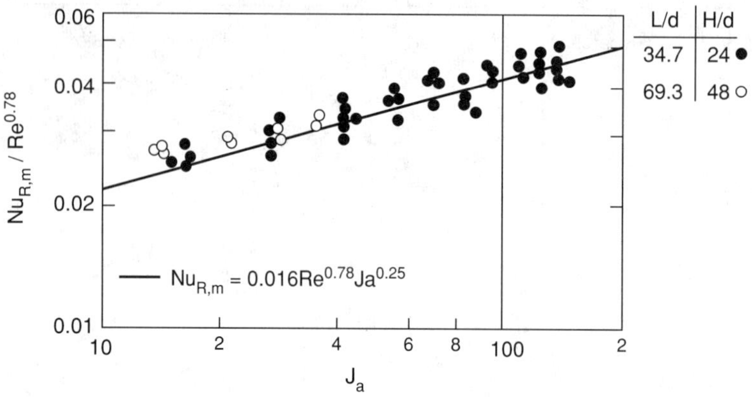

Figure 4. Mean heat transfer with developing flow.

A relatively small number of experimental studies have been reported for air and Morris and Woods (1978) proposed the following correlation, very similar to that suggested for laminar flow, with which to estimate the mean rotating Nusselt number.

$$Nu_{R,m} = 0.016\ Re^{0.78} J_a^{0.25}$$

Figure 4 compares this equation with their actual experimental data. The data appears to be relatively insensitive to the aspect ratio of the tube and also the eccentricly. There is evidence, however, that this equation can overpredict the enhancement in heat transfer when compared to some data, see Morris (1981).

References

Nakayama, W., (1968) Forced convective heat transfer in a straight pipe rotating about a parallel axis (turbulent region) *Int. J. Heat Mass Trans.*, 11, 1185.

Morris, W. D., and Woods, J. L. (1978) Heat transfer in the entrance region of tubes that rotate about a parallel axis, *J. Mech. Eng. Sci.*, 20–6, 1185.

Morris, W. D. (1981) *Heat Transfer and Fluid Flow in Rotating Cooling Channels,* Research Monograph, Research Studies Press, A Division of J. Wiley and Sons, Ltd, ISBN 0471101214, 1–228.

Woods, J. L., and Morris, W. D. (1980) A study of heat transfer in a rotating cylindrical tube., *Trans. A.S.M.E., J. Heat Trans.*, 102, 4, 612.

Leading to: *Rotating disc systems, basic phenomena; Rotating duct systems, orthogonal, heat transfer in*

W.D. Morris

ROTATING SURFACES (see Augmentation of heat transfer, single phase)

ROTATIONAL DISCONTINUITIES (see Magneto hydrodynamics)

ROTATIONAL RAYLEIGH NUMBER (see Rotating duct systems, parallel, heat transfer in)

ROTATIONAL REYNOLDS NUMBERS (see Rotating duct systems, orthoganal, heat transfer in; Rotating duct systems, parallel, heat transfer in)

ROUGH CHANNELS, FRICTION FACTOR IN (see Hydraulic resistance)

ROUGHNESS FACTORS (see Friction factors for single phase flow)

ROUGH SURFACE FRICTION FACTORS (see Chezy formula)

ROUGH SURFACES (see Augmentation of heat transfer, single phase; Augmentation of heat transfer, two phase; Friction factors for single phase flow)

ROUGH TUBES (see Tube banks, crossflow over; Tube banks, single phase heat transfer in)

ROUGH TUBES, FLOW IN (see Tubes, single phase flow in)

ROUGH TUBES, HEAT TRANSFER IN (see Tubes, single phase heat transfer in)

ROYAL ACADEMY OF ENGINEERING, RAE

2 Little Smith Street
London SW1P 3DL
UK
Tel: 0171 222 2688

ROYAL SOCIETY, RS

6 Carlton House Terrace
London SW1Y 5AG
UK
Tel: 0171 839 5561

ROYAL SOCIETY OF CHEMISTRY

Burlington House
Piccadilly
London W1V OBN
UK
Tel: 0171 437 8656

RS (see Royal Society)

RSC (see Royal Society of Chemistry)

RUBBER

Natural rubber [$(CH_2CHCCH_3CH_2)_x$] occurs in a large variety of plants. It is amorphous when unstretched, while a crystalline structure appears on stretching or at low temperature. At room temperature it has a specific gravity of 0.93, although this value can vary slightly depending on any impurities present. Natural rubber is soluble in most organic compounds and is easily oxidised. The process of *vulcanisation,* sulphur cross-linkages, improves its tensile and thermal characteristics.

V. Vesovic

RUMFORD, COUNT, BENJAMIN THOMPSON (see Steam engines)

RUSHTON TURBINE (see Agitation devices)

S

SAFETY ASSESSMENT (see Risk analysis techniques)

SAFFMAN LENGTH (see Particle transport in turbulent fluids)

SALINE WATER RECLAMATION (see Desalination)

SALT (see Sodium chloride)

SALTATION (see Gas-solid flows; Liquid-solid flow)

SALT DILUTION METHOD FOR FILM FLOW RATE MEASUREMENT (see Film flow rate measurement)

SALTING OUT (see Crystallizers)

SAMPLING

In the process industries, sampling is commonly undertaken for two reasons:

1. To monitor the production process and maintain quality control of the product.
2. To test the incoming raw materials and outgoing finished products to ensure that they meet contractual or other specifications.

Therefore, there are commercial as well as technical reasons behind the need for sampling, but in all cases it is essential that the sample is representative of the whole batch or parcel. The technology of sampling has been developed most strongly in the oil industry where huge quantities of crude oil and refined products are transported by ship and pipeline. Traditionally samples have been taken manually from ships' tanks or from storage tanks on shore. It is accepted that this method has its limitations, particularly when sampling crude oil which contains water and other impurities which are not distributed uniformly within the tank.

As a consequence the industry has developed technology for sampling from one or more pipelines throughout the transfer of a cargo or parcel of crude oil. During ship loading or discharge, sampling equipment (which may be permanently installed or may be connected for each operation) generally withdraws small sample "grabs" (see **Grab Sampling**) from each line to produce a total sample of typically 20 litres. The process is normally computer controlled and the sampling equipment may be powered electrically or pneumatically. The ideal sampling location from the commercial point of view is at the custody transfer point, which for ship loading or discharge is at the ship's manifold. However, that is not always feasible and samplers are often installed in pipelines in refineries, in storage terminals or on jetties.

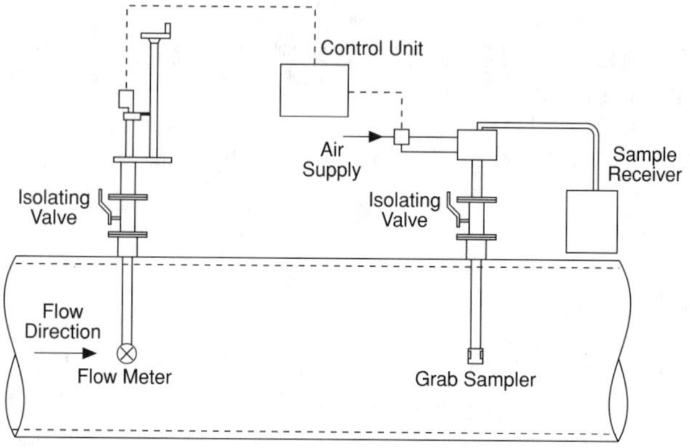

Figure 1. In-the-line sampling system.

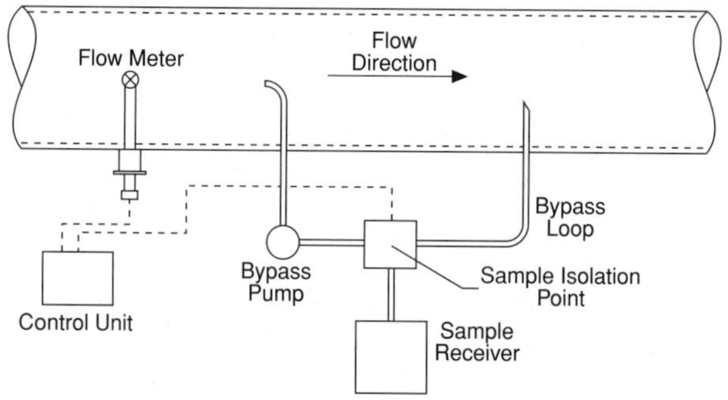

Figure 2. External loop sampling system.

Several types of pipeline sampling system are available but the most common are in-the-line and external loop systems (see **Figures 1** and **2**). The former involves a specially designed probe which is inserted directly into the line and is capable of isolating and removing grabs of about 1 or 2 ml each time it operates. The latter requires a small proportion of the fluid flowing through the line to be withdrawn via a probe and circulated through an external or bypass loop before re-entering the line. The sample to be collected is extracted from the bypass loop.

Three important criteria must be satisfied by any pipeline sampling system:

1. Adequate mixing must be available in the pipeline to ensure that a homogenous distribution of oil, water etc. exists across the line at the sampling location. If the natural turbulence generated by the flow is insufficient to achieve this then static or powered mixers may be employed. (see **Mixers** and **Mixers, Static**.)
2. The rate of sampling should be proportional to the pipeline flowrate. A flowmeter must therefore be installed to monitor the flowrate throughout the operation. (see **Flow Metering**.) It is generally argued that a measurement accuracy of $\pm 10\%$ of actual flow rate over the working range is acceptable. The signal from the flowmeter is used to control the rate at which the sample is withdrawn by the grab sampler or from the bypass loop.
3. The fluid velocity at entry to the sampling probe must be close to the local pipeline velocity. This isokinetic condition ensures that the fluid streamlines are not distorted as they

approach the probe and that representative sampling is maintained. To satisfy this condition the external loop of a bypass system must be pumped.

It is important that the sample should be collected in a sealed receiver which does not allow the volatile components of crude oil to escape. Similarly when the sample is being transferred from the receiver and prepared for analysis, care must be taken to minimise the loss of these components.

The design and operational requirements of pipeline sampling systems for use in the petroleum industry are described in considerable detail in the relevant standards which are listed in the references.

References

Institute of Petroleum, *Petroleum Measurement Manual,* Part VI, Sec. 2, Guide to Automatic Sampling of Liquids from Pipelines.

American Petroleum Institute, *Manual of Petroleum Measurement Standards,* Ch. 8, Sec. 2, Standard Practice for Automatic Sampling of Liquid Petroleum and Petroleum Products.

ISO 3171: 1988. *Petroleum Liquids,* Automatic Pipeline Sampling.

J. Miles

SAMPLING METHODS, FOR DROPSIZE MEASUREMENT (see Dropsize measurement)

SAND BLASTING (see Gas-solid flows)

SANDIA NATIONAL LABORATORY, SNL

PO 5800
Albuquerque
New Mexico 87185
USA
Telephone 505 845 0011

SATELLITES (see Microgravity conditions)

SATURATED FLUID PROPERTIES

***Following from:** Thermodynamics; Phase equilibrium*

If the gas or vapour represented by the point X in the $p\tilde{v}$ phase for a pure fluid shown in **Figure 1** is compressed slowly and isothermally, the pressure rises until the vapour becomes saturated and the first drop of liquid appears at conditions corresponding to point 1. If the compression is continued, condensation takes place at a constant pressure known as the *saturation pressure* for the given temperature. For a pure substance, the relationship between the saturation pressures and temperatures is given by measurements of this vapour pressure. At any point on the line between point 1 and point 2, saturated liquid and saturated vapour are in equilibrium. At point 2 the vapour phase disappears and only saturated liquid remains. Although the specific volume of a two-phase system changes continuously as the relative amounts of the two phases change, the specific volumes of the saturated vapour and liquid in equilibrium along the line 1–2 are those of points 1 and points 2 respectively. The specific volume of a liquid-vapour mixture with a quality, x, is given by

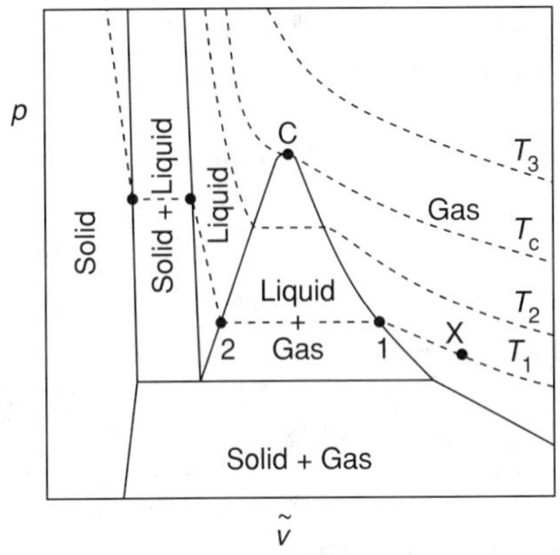

Figure 1. The p-$\tilde{v}$ projection for a typical fluid.

$$\tilde{v}_x = \tilde{v}_l + x(\tilde{v}_g - \tilde{v}_l)$$

which is known as the *Lever Rule.*

The most important saturation property is therefore the vapour pressure which can be determined experimentally by a number of techniques (Ambrose, 1973). Once the saturation line is defined, it is possible to consider the properties of both co-existing phases at saturation, including the density and heat capacity. The density of the vapour at saturation is seldom measured directly owing to the difficulties involved so that it is frequently obtained indirectly. On the other hand, the liquid phase density at saturation is more easily measured. Heat capacity measurements in both liquid and vapour phases are possible (Marsh and O'Hare, 1994). It is noteworthy that in the liquid phase it is possible to define three specific heat capacities: c_p, the change in enthalpy with temperature at constant pressure, c_σ, the change in enthalpy with temperature for the saturated liquid and c_{sat}, the energy required to effect a temperature change while maintaining the liquid in a saturated state. The three heat capacities do not generally differ by very much. Estimation methods for saturated densities and heat capacities at saturation are available (Reid et al., 1977).

References

Ambrose, D. (1973) in *Experimental Thermodynamics of Non-Reacting Fluids,* Le Neindre, B., Vodar, B., eds., IUPAC, Butterworths, London.

Marsh, K. N. and O'Hare, P. A. G. (1994) *Experimental Thermodynamics IV: Solution Calorimetry,* Blackwell, Oxford.

Reid, R. C., Prausnitz, J. M. and Sherwood, T. K. (1977) *The Properties of Gases and Liquids,* McGraw-Hill, New York.

Leading to: *Critical point, thermodynamics*

W.A. Wakeham

SATURATED VOLUME (see Humidity)

SATURATION PRESSURE (see Pressure; Saturated fluid properties)

SATURATION TEMPERATURE (see Saturated fluid properties)

SAUTER MEAN DIAMETER

Following from: *Drop size measurement*

Most sprays or ensembles of drops contain drops of different sizes. Measurements will provide information on the fractions of drops of different sizes which can be presented in terms of number or volume (mass) as illustrated in **Figure 1**. A number of equations of varying complexity (two to four adjustable constants) have been suggested to describe such distributions.

However, for many purposes a single number characterising the drop size is required. Sometimes, the median diameter is employed: 50% of the drops are larger (in number or volume terms) than the median and 50% are smaller. These are usually identified as D_{NM} and D_{VM}. In some cases, an arithmetic average diameter will suffice to describe the distribution, but because the drop surface area and volume are proportional to the square and cube of the diameter respectively, a more complex description is required.

A general mean diameter can be defined by

$$D_{pq} = \left[\frac{\int_0^\infty n(D)\,D^p dD}{\int_0^\infty n(D) D^q dD}\right]^{1/(p-q)} \quad (1)$$

or in terms of a finite number of discrete size classes,

$$D_{pq} = \left[\frac{\sum_{i=1}^{\infty} n_i D_i^p}{\sum_{i=1}^{\infty} n_i D_i^q}\right]^{1/(p-q)} \quad (2)$$

The mean diameter with the same ratio of volume to surface area as the entire ensemble is known as the Sauter mean diameter. It corresponds to values of $p = 3$ and $q = 2$ in the above equations. This particular diameter is named after the German scientist who first employed it. Dr Ing J. Sauter worked at the Laboratorium für Technische Physik of the Technische Hochschule, Müchen on aspects of internal combustion engines. In particular, he studied atomisation in carburettors and as part of his work devised a technique to determine the size of drops produced based on the absorption/scattering of light. The technique depends on the fact that absorption/scattering is proportional to the surface area of the drops. The value per unit volume of liquid contains a term equal to the right hand side of Equations 1 or 2 with $p = 3$ and $q = 2$, i.e., the Sauter mean diameter.

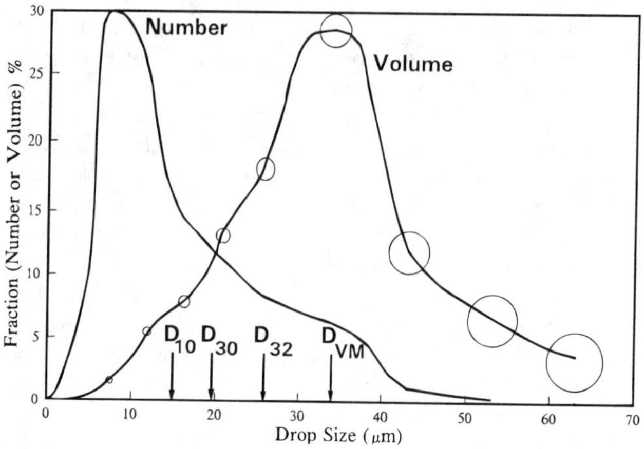

Figure 1. Example of number and volume based drop size distributions showing the position of the Sauter mean diameter and other important mean diameters.

For most drop size distributions, the Sauter mean diameter, D_{32}, is larger than the arithmetic, D_{10}, surface, D_{20}, and volume, D_{30}, mean diameters. The relative position of the different diameters is shown in **Figure 1**. Williams (1990) quotes an example where the Sauter mean diameter, D_{32}, is 18 μm and the other diameters are D_{10} = 5.5 μm; D_{20} = 7.5 μm; D_{30} = 10 μm. The number and volume (or mass) median diameter were 4.2 and 24 μm respectively.

The different mean diameters are appropriate for different purposes as illustrated in **Table 1**.

Table 1.

Mean diameter	Name	Field of application
D_{10}	Arithmetic or linear	Evaporation
D_{20}	Surface	Surface area controlling (e.g., absorption)
D_{30}	Volume	Volume controlling (e.g., hydrology)
D_{21}	Surface diameter	Adsorption
D_{31}	Volume diameter	Evaporation, molecular diffusion
D_{32}	Sauter	Efficiency studies, mass transfer, reaction
D_{43}	De Brouke	Combustion equilibrium

The Sauter mean diameter is probably the most commonly used mean as it characterises a number of important processes. Chin and Lefebvre (1985) suggest that it is the best measure of the fineness of sprays.

References

Chin, J. S. and Lefebvre, A. H. (1985) Some comments on the characterization of drop-size distribution in sprays, Proceedings of ICLASS–85, IV/A/I/1–12.

Lefebvre, A. H. (1989) *Atomization and Spray,* Hemisphere, New York.

Mugele, R. A. and Evans, H. D. (1951) Droplet distribution in sprays, *Ind. Eng. Chem.,* 43, 1317–1324.

Sauter, J. (1926) Grössenbestimmung von Brennstoffteilchen, *Forschungsarbeiten auf dem Gebiete des Inqenieurwesens,* Heft 279.

Sauter, J. (1928) Untersuchung der von Spritzvergasern gelieten Zerstäbung, *Forschungsarbeiten auf dem Gebiete des Ingenieurwesens,* Heft 312.

Williams, A. (1990) *Combustion of Liquid Fuel Sprays,* Butterworths, London.

B.J. Azzopardi

SCALES OF TURBULENCE (see Turbulence)

SCALING (see Fouling)

SCATTERING EFFICIENCY (see Optical particle characterisation)

SCATTERING INDICATRIX (see Dropsize measurement)

SCATTERING OF RADIATION (see Albedo)

SCHEIBEL EQUATION FOR DIFFUSION IN LIQUIDS (see Diffusion coefficient)

SCHLIEREN INTERFEROMETRY (see Photographic techniques)

SCHLIEREN TECHNIQUE (see Visualization of flow)

SCHMIDT, ERNST (1892–1975)

A German scientist and pioneer in the field of engineering thermodynamics, especially in heat and mass transfer, Schmidt was born on Feb. 11, 1892 at Vögelsen, near Lüneburg. He studied civil and electrical engineering at Dresden and München, and joined the laboratory for applied physics at the Technical University, München, in 1919, which was then under the direction of Oscar Knoblauch. One of his early research efforts there was a careful measurement of the radiation properties of solids, which caused him to propose and develop the use of aluminium foil as an effective radiation shield.

In 1925 he received a call to come as professor and director of the engineering laboratory to the Technical University, Danzig. Here he published papers on the now well known graphical difference method for unsteady heat conduction, and on the Schlieren and shadow method to make thermal boundaries visible and to obtain local heat-transfer coefficients. He was the first to measure the velocity and temperature field in a free convection boundary layer and the large heat-transfer coefficients occurring in droplet condensation. A paper pointing out the analogy between heat and mass transfer caused the dimensionless quantity involved to be called **Schmidt Number**.

ERNST SCHMIDT 1892–1975

Figure 1.

In 1937 he became director of the Institute for Propulsion of the newly founded Aeronautical Research Establishment at Braunschweig and professor at the university there. In 1952 Schmidt occupied the chair for thermodynamics at the Technical University of München which before him had been held by **Nusselt**. Being strongly involved in the development of the international steam tables, Schmidt continued his scientific activity after his retirement (1961) until his death in 1975.

In recognition of his work, he had received the Ludwig Prandtl Ring, the Max Jacob award and the Grashof Commemorative Medal.

U. Grigull, H. Sandner and J. Straub

SCHMIDT NUMBER

Schmidt Number, Sc, is a dimensionless parameter representing the ratio of diffusion of momentum to the diffusion of mass in a fluid. It is defined as

$$Sc = \nu/\delta$$

where ν is kinematic viscosity and δ diffusivity.

Schmidt number is the mass transfer equivalent of **Prandtl Number**. For gases, Sc and Pr have similar values ($\approx$0.7) and this is used as the basis for simple heat and mass transfer analogies (see **Analogy Between Heat, Mass and Momemtum Transfer**).

Reference

Hewitt G. F., Shires G. L. and Bott T. R. (1994) *Process Heat Transfer,* CRC Press.

G.L. Shires

SCHUSTER-HAMAKER MODEL (see Radiative heat transfer)

SCHUSTER-SCHWARZCHILD APPROXIMATION, FOR COMBINED RADIATION AND CONDUCTION (see Coupled radiation and conduction; Radiative heat transfer)

SCRAPED SURFACE HEAT EXCHANGERS

***Following from:** Heat exchangers*

Elsewhere in this encyclopedia the thermal resistance due to the laminar sublayer is discussed (see **Boundary Layer Heat Transfer**). Because of the attendant drag forces associated with the flow of viscous liquid, the viscous sublayer is generally quite thick or, in many instances, turbulent conditions in the core liquid cannot be even attained without intolerable pressure loss and excessive pumping costs. One way of overcoming the problem is to remove physically, the layers of fluid at the heat transfer surface and to mix them with the bulk fluid in the heat exchanger. In this way, if the fluid is being heated, heat is conveyed directly from the wall to the bulk liquid. The technique is particularly attractive for heat sensitive liquids e.g. food or pharmaceutical products, because of the low interface temperature between the liquid and heat transfer surface for a given overall temperature driving force.

In the scraped surface heat exchanger, spring loaded rotating blades scrape the surface and effectively remove liquid from it. Alternatively, the blades move against the heat transfer surface under the influence of the rotational forces. At the same time as liquid layers are removed, any fouling substance deposited on the surface is also removed, thus ensuring that contamination of the process liquid is reduced to a minimum this can be crucial where taste and texture are important product qualities.

Figure 1 illustrates the principle of the scraped surface heat exchanger. The number of scraper blades shown is four, but any number of blades may be employed, although as the number of blades is increased the capital cost rises. Furthermore, a large number of blades is not necessary, since the time internal between successive scrapes is relatively short, i.e. the residence time of particles of liquid on the surface is low. The choice of the number of blades is an empirically based compromise between capital cost, acceptable speed of rotation and liquid viscosity. Because of the rotating parts, maintenance charges may be relatively high.

Although often classified with scraped surface heat exchangers, some exchangers have blades that do not actually touch the surface over which they pass, but move in close proximity to it. Such designs may be termed "*wiped surface*" *heat exchangers,* and may

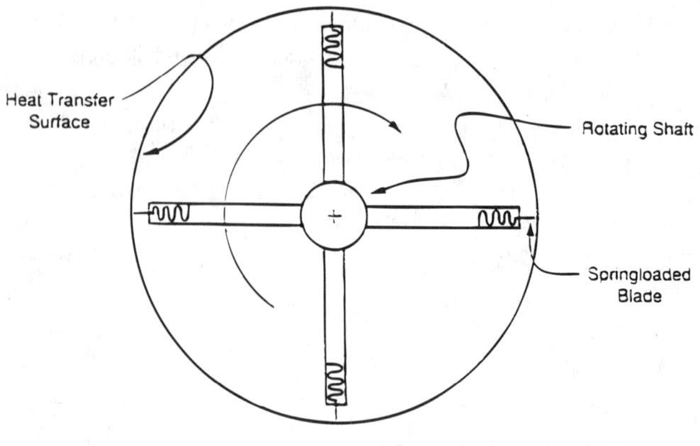

Figure 1. The principle of the scraped surface heat exchanger.

be preferred where the wear of surface or blades cannot be tolerated from a mechanical point of view or because of contamination effects.

Scraped surface heat exchangers can either run full of liquid or the liquid may enter the exchanger as a peripheral stream. The former may be mounted either vertically or horizontally but it is usual for the latter, although not exclusively, to be mounted vertically so that the liquid flows downwards under the action of gravity. Where evaporation is taking place, for instance in a concentrating process, this is the preferred arrangement; the vapour being removed under vacuum from the top of the evaporator assembly. The residence time in the equipment may be quite low; desirable for processing heat sensitive liquids. Agitated thin film evaporators as they are sometimes called, have been discussed by Salden (1987). In addition to the advantages already mentioned this author points out that the equipment is suitable for multipurpose applications and that evaporation to dryness is feasible where a solid product is required.

Attempts to correlate heat transfer in scraped surface equipment have been made and these make use of the usual dimensionless groups including **Nusselt, Prandtl** and **Reynolds Numbers** common to many heat transfer correlations but also involving the so called "rotary" Reynolds number defined as:

$$\frac{D^2 N \rho}{\eta}$$

where D is the diameter of the exchanger body, N is the speed of rotation r.p.m., and ρ and η are the liquid density and viscosity respectively. The aspect ratio of the exchanger and the number of blades may also be taken into account in the correlation.

Having regard to the complexity of the geometry particularly of the blade assembly, and the several interacting factors, design is more usually based on empirically determined parameters derived from experience. Scraped surface heat exchangers are in general, used only for special applications.

References

Salden, D. M. (1987) Agitated thin film evaporators, *Distillation Supplement Chem. Engr.,* Sept., 17–19.

T.R. Bott

SCREENS (see Classifiers)

SCREEN SEPARATORS (see Vapour-liquid separation)

SCREW ROTARY COMPRESSOR (see Compressors)

SCREWS, PLASTICATING (see Extrusion, plastics)

SCROLL DISCHARGE CENTRIFUGE (see Centrifuges)

SCRUBBERS

A scrubber is a heat and mass transfer apparatus which implements a direct gas-liquid contact. It is used for gas cooling and, more frequently, for gas cleaning where both suspended particles and other gases are removed from the gas to be cleaned.

Types of scrubber include *hollow, packed, centrifugal,* and *Venturi scrubbers.* In the hollow scrubber, the liquid is sprayed as droplets up to a few millimeters in size into a vessel containing the gas which is to be contacted with the liquid. Liquid dispersion occurs in atomizers arranged on collectors mounted at the top of the vessel. The interacting media are usually in vertical counterflow. Scrubbers are also encountered in which media are in cross flow when the gas flows horizontally and spraying is performed from above. These scrubbers are of rectangular cross section. To avoid droplet carry-over the gas velocity in a hollow scrubber is in the range 1 to 2.5 m/s. If the velocity rises to 5–8 m/s drop pans are mounted at the top of the scrubber. The hollow scrubber affords up to 60–75% dust suppresion, the cleaning coefficient of the main flow from gas impurities attains approximately the same value. The pressure loss in the hollow scrubber is 0.1 to 0.2 kPa.

Packing the scrubber space with cylinders, Rashig rings, or lumps of insoluble matter increases the efficiency of dust suppression and gas absorption up to 85%, but the pressure loss increases by typically 1.5 to 2 times.

Centrifugal scrubbers that use the effect of centrifugal force are an efficient dust collector. They use tangential admission of the dust-laden gas at a velocities up to 25 m/s the dust being captured in a liquid film falling on the walls. The dust-suppresion efficiency attains 90% even for a fine dust with particle diameter $D_p = 2 - 5\ \mu m$.

Venturi scrubbers have received wide recognition in industry. The main component of a Venturi scrubber is a Venturi tube with liquid sprayed into the flow upstream of the tube throat. Typical drop sizes in such units are around 10 μm. Active dust absorption by the liquid droplets occurs as the gas accelerates relative to the liquid in the convergent section of the venturi. A cyclone mounted behind the Venturi tube is used for droplet trapping. Venturi scrubbers enable cleaning of gases containing an extremely fine dust with particle size of the order of 0.1 to 1 μm, such as sublimate and mist, with an efficiency up to 99%. Their pressure loss may exceed that of the hollow scrubber by two orders of magnitude and reach 10 kPa.

The specific spraying rate in all types of scrubbers varies within wide limits, typically from 0.2 to 10 liters of liquid per cubic meter of gas.

References

Perry, R. H., Green, D. W., Maloney, J. O. eds. *Perry's Chemical Engineers Handbook,* 1984. McGraw-Hill, New York.

I.L. Mostinsky

SEA WATER COMPOSITION (see Water)

SECONDARY FLOWS

Following from: *Flow of fluids; Channel flow*

In cases where there is a three-dimensional flow field, the flow is often regarded as comprising two components, a primary flow and a secondary flow. The primary flow is parallel to the main direction of fluid motion and the secondary flow is perpendicular to this. Such flows are commonly produced by the effect of drag in the boundary layers, and some of the more important situations in which such flows arise are discussed here.

Secondary flows occur where there is a flow around a bend in a pipe and this is illustrated in **Figure 1**. At the bend, there is a transverse pressure gradient which provides the centripetal force for the fluid elements to change direction. However, the pressure gradient required for the faster moving fluid near the centre of the pipe to follow the curve of the bend is greater than that required for the slower moving fluid near to the wall. This results in the fluid near the centre of the pipe moving toward the outside of the pipe and the fluid near the wall moving inwards. (See also **Bends, Flow and Pressure Drop In**).

Another common situation in which secondary flows arise is in spinning fluids, for example in a weather system or a stirred teacup. In these systems, there is a balance between the radial pressure gradient and the centrifugal force (coriolis force in the case of large weather systems). However, near a boundary, drag on the fluid leads to a lower velocity and the centrifugal or coriolis force can no longer balance the pressure gradient. This results in a secondary flow of the fluid in the radial direction. **Figure 2** illustrates this for the case of a cyclonic weather system. This situation is treated by Acheson (1990) and by Holton (1979). (See also **Cyclones**).

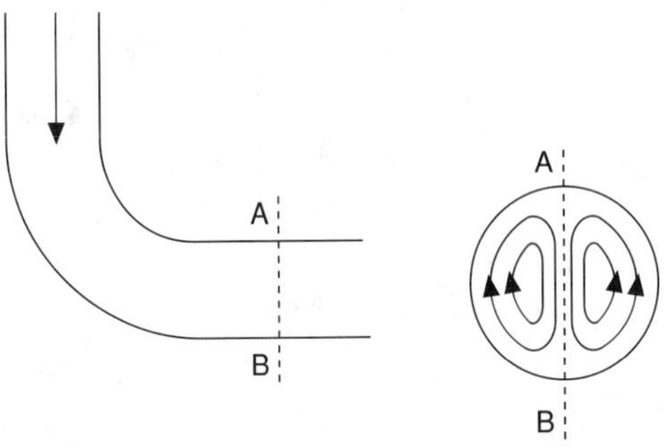

Figure 1. Secondary flow around a bend.

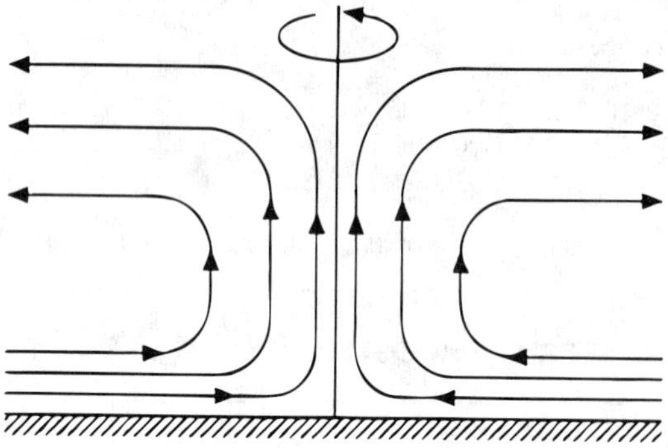

Figure 2. Secondary flow in a cyclone.

Turbulent flows through pipes or channels of non-circular cross-section also give rise to secondary flows. There is a movement of fluid away from the corners along the walls and a movement toward the corners near the bisector of the corner. Examples for rectangular and triangular channels are shown in **Figure 3**. This phenomenon is known as "secondary flow of the second kind" and is discussed in Prandtl (1952) and Kay and Nedderman (1985). (See also **Triangular Ducts, Flow and Heat Transfer In**).

There is also a "third kind" of secondary flow which results from *oscillations in a stationary fluid.* These oscillations may be due to, for example, an oscillating body or ultrasonic waves. If there is a variation in the amplitude, as will be the case with standing sound waves, then a secondary flow is induced which moves in the direction of decreasing amplitude. The theory behind this is dealt with by Schlichting (1968), and for a fluid in which the fluid elements move with velocity u(x) cosωt, then it can be shown that the secondary flow is:

$$u' = \frac{3}{4}\frac{u}{\omega}\frac{\partial u}{\partial x}$$

References

Acheson, D. J. (1990) *Elementary Fluid Dynamics,* Oxford University Press, New York.

Holton, J. R. (1979) *An Introduction to Dynamic Meteorology,* Academic Press, New York.

Kay, J. M. and Nedderman, R. M. (1985) *Fluid Mechanics and Transfer Processes,* Cambridge University Press, Cambridge.

Prandtl, L. (1952) *Essentials of Fluid Dynamics,* Hafner Publishing Co., New York.

Schlichting, H. (1968) *Boundary-Layer Theory,* McGraw Hill, New York.

***Leading to:** Bends, flow and pressure drop in; Cyclones; Triangular ducts, flow and heat transfer; Görtler-Taylor vortex flows.*

M.J. Pattison

SECONDARY RECOVERY PROCESSES (see Enhanced oil recovery)

SECOND LAW OF THERMODYNAMICS (see **Carnot cycle**; Entropy; Exergy; Flow of fluids; Irreversible thermodynamics; Refrigeration; Thermodynamics)

SECOND NORMAL STRESS DIFFERENCE COEFFICIENT (see Non-Newtonian fluids)

SEDIMENTATION

***Following from:** Liquid-solid separation*

A particle or droplet will settle in a fluid if its density is greater than that of the fluid in which it is suspended. The (laminar) settling velocity of particles whose concentration is very low, that is when the flow of fluid around a particle does not affect the flow around neighbouring particles, can be calculated from **Stokes Law**:

$$u = \frac{D^2 g}{18\eta}\Delta\rho$$

where D is the diameter of the particle and η is the absolute viscosity of the surrounding fluid. Δρ is the density difference between that of the particle and its surrounding fluid: if Δρ is

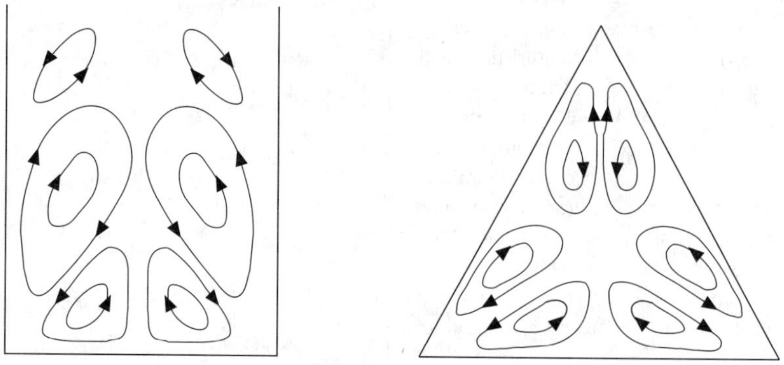

Figure 3. Secondary flows in channels.

positive the particle will settle (see **Thickeners**), or if it is negative the particle will float (see **Flotation**).

In practice, the concentration of particles in industrial suspensions is usually high enough for there to be significant interactions between particles as they settle (making Stokes Law invalid): these interactions can greatly increase the frictional force at the surfaces of the settling particles. When the effects of mutual interference are negligible, *free settling* conditions are said to prevail; at higher concentrations *hindered settling* occurs. The reasons for modification of the settling rate of particles in a concentrated suspension include:

a. when a wide range of particle sizes are present in the feed, differential settling rates between large and small particles lead to modification of the effective density of the suspension (this is more significant in small scale batch sedimentation),
b. the upward velocity of the fluid is greater at higher concentrations (causing a decrease in the apparent settling velocity)
c. the velocity gradients in the fluid surrounding the particles are greater due to the closer proximity of the particles,
d. the ability of particles to aggregate is enhanced at higher concentrations.

In addition to particle size, density and concentration, and fluid viscosity, other less obvious factors affect the sedimentation rate. These include particle shape and orientation, convection currents in the surrounding fluid, and chemical pretreatment of the feed suspension. Particle with diameters of the order of a few microns settle too slowly for most practical operations; wherever possible these are *coagulated* or *flocculated* to increase their effective size, and hence increase their rate of settling.

Sedimentation of a suspension is generally assessed by a *jar test,* during which a suspension is allowed to settle and the height of the clear liquid (supernatant)-suspension interface is measured as a function of the settling time. In a jar test particles can be observed to settle in any of several quite different ways, dependent on their concentration and their tendency to cohere. The different modes of sedimentation make different demands on the size and shape of a settling tank, and different test procedures are used for evaluating them (details of these are given by Fitch and Stevenson, 1986).

When the particle are, on average, far apart and free to settle individually, *clarification* sedimentation occurs. This behaviour is recognised in a jar test as slower settling particles 'string out' behind faster ones, and the supernatant gradually clarifies; solids collect at the bottom of the jar to form a sediment, and the supernatant-suspension interface is generally indistinct.

In more concentrated suspensions, the particles are closer together; in the extreme they can cohere to form a plastic structure which constrains the sedimentation of individual particles. The solids settle with a sharp interface between the pulp and supernatant, and the slurry enters a consolidated or *zone settling* regime and exhibits *zone settling* or *line settling* behaviour. These two extremes of bahaviour are shown in **Figure 1**.

In a jar test exhibiting line settling behaviour the interface height can be plotted against time as in **Figure 2**. After an initial transient the suspension settles at a constant rate in the section from A to B, followed by the first falling rate period from B to C and a second falling rate period from C. The constant rate period corresponds to

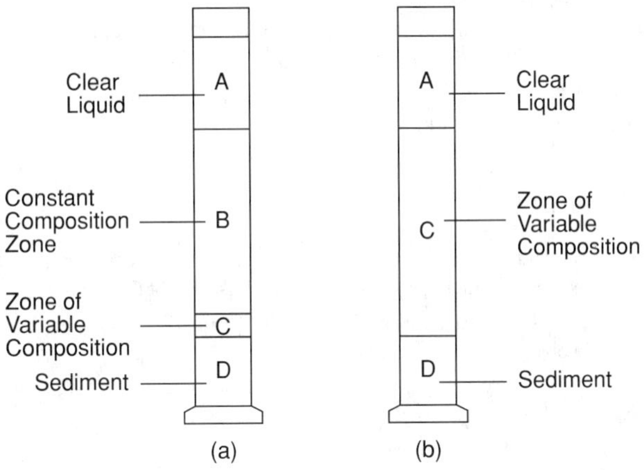

Figure 1. (a) Line settling, and (b) clarification, behaviour during sedimentation.

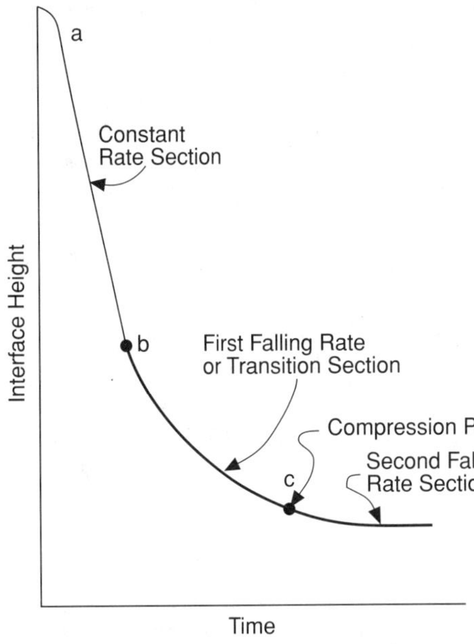

Figure 2. Sedimentation curve arising from line settling behaviour.

pulp settling at its initial concentration. During the final falling rate section the pulp is in compression. Point C is known as the *compression point* or the *critical sedimentation point,* and identifies the point at which the pulp-supernatant interface goes from zone settling into compression.

Jar test sedimentation data is used as the basis for clarifier or thickener design. Details of design methods can be found in many texts, for example Fitch and Stevenson (1986) and Osborne (1990). Additional information can be found in Coulson and Richardson (1991) and Rushton (1985). Clarifier design, usually based on "long tube" data, is based on the overflow velocity and the detention time, and calculations are quite simple. Thickener design involves the application of one of a number of alternative models together with jar test data, and is based either on the combined Kynch (1952) and Coe and Clevenger (1916) analysis of settling rates or the Yoshioka (1957) flux analysis. There are many other models which could be applied, some of which are based on compression of the sediment rather than hindered settling of the suspension. Wilhelm and Naide (1981) recommend the flux analysis method as the most concise and precise for design purposes.

References

Coe, H. S. and Clevenger, G. H. (1916) Methods for determining the capacities of slime thickening tanks, *Trans AIME,* 55, 356, 384.

Coulson, J. M. and Richardson, J. F. (1991) Chapter 5, *Chemical Engineering,* Vol 2, 4th ed., Pergamon, Oxford.

Fitch, E. B. and Stevenson, D. G. (1986) Chapter 4, *Solid/Liquid Separation Equipment Scale-up,* D. B. Purchas and R. J. Wakeman, eds., Uplands Press and Filtration Specialists, London.

Kynch, G. J. (1952) A theory of sedimentation, *Trans Faraday Soc,* 48, 166–176.

Osborne, D. G. (1990) Chapter 5, *Solid-Liquid Separation* L. Svarovsky, ed., Butterworths, London.

Rushton, A., ed. (1985) *Mathematical Models and Design Methods in Solid-Liquid Separation,* Nijhoff, Dordrecht.

Wilhelm, J. H. and Naide, Y. (1981) Sizing and operating continuous thickeners, *Mining Engineering,* 1710–1718.

Yoshioka, N., Hotta, Y., Tanaka, S., Naito, S. and Tongami, S. (1957) Continuous thickening of homogeneous slurries, *Chemical Engineering, Tokyo,* 21, 66–74.

R.J. Wakeman

SEDIMENTING CENTRIFUGES (see Centrifuges; Liquid-solid separation)

SEGMENTAL BAFFLES (see Shell-and-tube heat exchangers)

SEGREGATION (see Liquid-solid flow)

SEIDER-TATE CORRELATION (see Tube banks, single phase heat transfer in)

SELECTIVE FROTH FLOTATION (see Flotation)

SELF ORGANIZATION (see Non-linear systems)

SELF-SIMILARITY (see Inviscid flow)

SEMI-ANNULAR FLOW (see Churn flow)

SEMI-CONDUCTOR DIODE LASERS (see Lasers)

SEMICONDUCTORS

A typical semiconductor is a crystalline solid material with an **Electrical Conductivity** that is highly dependent on temperature. At low absolute temperatures a pure semiconductor will appear to be a good insulator, however, its conductivity rises dramatically as the temperature increases. Semiconductors like silicon are the basis of modern electronics and integrated circuit technology.

The electrical conductivity of a solid depends on two factors. Firstly, the number of mobile charge carriers present and secondly, the mobility or speed at which the carriers move under the influence of an electric field. It is the first consideration, usually called carrier concentration, that determines whether a material is a good conductor like a metal or an insulator such as diamond.

In a typical metal each atom contributes one or more of its outer electrons to a common sea of conduction electrons that permeates the solid, yielding a very large number of conduction electrons. This sea of electrons or *electron gas* will readily support a transport current in the presence of an electric field. By comparison, a good insulator has strongly localised electrons. A considerable amount of energy is required to break away the electrons from their current atom sites, making transport currents through the material impossible unless work is done initially to free a significant number of electrons.

Semiconductors have *resistance* properties in between the above two extremes, however, they are not simply poor insulators (or bad conductors). The name is used for those materials that exhibit *insulator* like properties at very low temperatures, because all the outer electrons called valence electrons are localised, but begin to conduct as the temperature is raised. This is because the thermal energy is sufficient to break away electrons from their local bonds and promote them into the role of conduction electrons. Thus at higher temperatures a semiconductor exhibits properties closer to those of metals. It is the dramatic change in conductivity due to the excitation of valence electrons into the conduction state as the ambient temperature is increased that is the characteristic feature of a semiconductor. The conductivity of a pure semiconductor would, theoretically, increase almost exponentially with absolute temperature. Conduction in a semiconductor takes place both via the conduction electrons and also the valence electrons that are now able move to neighbouring atoms that have lost valence electrons. The vacancy travelling, in the opposite direction to the valence electrons is called a *hole.* For convenience holes are regarded as positive charge carriers. Detailed descriptions of conduction processes in semiconductors can be found in Streetman (1990).

The most commonly used semiconductor material is **Silicon** which is tetravalent and forms a diamond type crystal structure. The minimum energy required to promote an electron from the valence state (or band) to the conduction band in silicon is only about 1.1 eV (the energy gap). Hence, the ambient thermal energy at room temperature is sufficient to produce a low but significant number of conduction electrons. Even though the mean thermal energy of the electrons is less than the energy gap there will always some individual electrons with greater energies due to the statistical distribution in the electron energies.

Doping is the addition of minute amounts of specific impurities to extremely pure semiconductor materials to build in a number of available carriers. In materials that are called p-type trivalent atoms like boron have been added to create holes. Whereas, introducing a pentavalent impurity like phosphorus produces n-type material with a background level of conduction electrons. It should be noted that the material is still electrically neutral. The name n or p type describes the type of free carrier that predominates in the material (called the majority carrier). In general both electrons and holes are present. Further information regarding doped semiconductors and device fabrication may be found in Pulfrey and Tarr (1989).

References

Streetman B. G. (1990) *Solid State Electronic Devices,* 3rd Ed., Prentice Hall International.

Pulfrey D. L. and Tarr N. G. (1989) *Introduction to Microelectronic Devices,* Prentice Hall-International.

A.W. Campbell

SEMI-CONDUCTOR THERMOMETERS (see Resistance thermometry)

SEMI-SLUG FLOWS (see Gas-liquid flow)

SEMITRANSPARENT MEDIA (see Coupled radiation and conduction)

SENSIBLE HEAT STORAGE (see Heat storage, sensible and latent)

SEPARATED FLOW MODELS (see Pressure drop, two phase flow; Two-phase flows)

SEPARATED LIQUID FLOWS (see Liquid-liquid flow)

SEPARATION, LIQUID/LIQUID (see Extraction, liquid-liquid; Foam fractionation)

SEPARATION OF BOUNDARY LAYERS (see Boundary layer)

SEPARATION OF FLUID MIXTURES (see Distillation)

SEPARATION OF GAS AND SOLIDS (see Gas-solids separation, overview)

SEPARATION OF EMULSIONS (see Emulsions)

SEPARATION OF LIQUIDS (see Molecular sieves)

SEPARATION OF LIQUIDS AND SOLIDS (see Liquid-solid separation)

SEPARATION OF PHASES IN GAS-LIQUID FLOWS (see Headers and manifolds, flow distribution in)

SEPARATION, PARTICLES/LIQUID (see Flotation)

SEPARATION PROCESSES (see Membrane processes)

SERIES EXPANSIONS

Let $\{a_0, a_1, a_2, \ldots\}$ be an infinite sequence of real or complex numbers. For $n \geq 0$, we define

$$S_n = \sum_{k=0}^{n} a_k \quad \text{and} \quad A_n = \sum_{k=0}^{n} |a_k|$$

where it follows that $a_n = S_n - S_{n-1}$ with $a_0 = S_0$. Then the infinite series whose n-th term is a_n is defined to be

$$\sum_{k=0}^{\infty} a_k = a_0 + a_1 + a_2 + \cdots \qquad (1)$$

and S_n is called the n-th partial sum of this series. If there exists a finite number S such that

$$\lim_{n\to\infty} S_n = S,$$

we say that series (1) is *convergent* and its **sum** is S, in which case we write

$$S = \sum_{k=0}^{\infty} a_k \qquad (2)$$

If $\lim S_n$ does not exist or is infinite, then we call (1) a *divergent* series. If, on the other hand, there exists a finite number A such that $\lim A_n = A$, we say that series (1) is absolutely convergent. When (2) is satisfied, we define the remainder R_n to be $S - S_n$, $n \geq 0$. *Cauchy's convergence principle* gives the necessary and sufficient conditions for (1) to be convergent: for any $\varepsilon > 0$ there is an N such that $|S_m - S_n| < \varepsilon$ for all $m > n > N$. The same holds for absolute convergence by taking $A_m - A_n < \varepsilon$ instead. Absolute convergence implies convergence but not vice versa; the series

$$1 - \frac{1}{2} + \frac{1}{3} - \frac{1}{4} + - \cdots$$

is convergent but not absolutely convergent; such series are said to be *conditionally convergent.* If series **(1)** converges, then $a_n \to 0$ as n goes to infinity; thus, if $\lim a_n \neq 0$ then the series is divergent. However, the converse is not true as indicated by the *harmonic series* whose n-th term is 1/n. There are several convergence tests for an infinite series of form (1):

1. Comparison test: the series converges absolutely if we can find a convergent series $\Sigma\, b_n$ of nonnegative real terms such that $|a_n| \leq b_n$ for all values of n.
2. Ratio test: if $a_n \neq 0$ for all n and the ratios $|a_{n+1}/a_n| \leq L$ for all $n > N$, where $L < 1$ and N are fixed, then the series is absolutely convergent. If $|a_{n+1}/a_n| > 1$ for all $n > N$, then the series diverges.

3. Root test: the series converges absolutely if, for fixed L < 1 and N, the roots $n\sqrt{|a_n|} \leq L$ for all n > N. The series diverges if $n\sqrt{|a_n|} > 1$ for all n > N.

The *geometric series* $\sum_{n=0}^{\infty} r^n = 1 + r + r^2 + \ldots$ converges to the sum $1/(1-r)$ if $|r| < 1$ and diverges if $|r| \geq 1$.

References

Hille, E. (1973) *Analytic Function Theory,* Vol. I, Chelsea Publishing Co., New York.

Kreyszig, E. (1983) *Advanced Engineering Mathematics,* John Wiley, New York.

Leading to: *Bessel series; Fourier series; Legendre polynomial series*

H.G. Khajah

SESSILE DROPS AND BUBBLES (see Surface and interfacial tension)

SETTLING SLURRIES (see Liquid-solid flow)

SEVERE ACCIDENTS, IN NUCLEAR REACTORS, CONTAINMENT OF (see Containment)

SHADOWGRAPH TECHNIQUE

Following from: *Photographic techniques; Visualization of flow*

The shadowgraph is the simplest form of optical system suitable for observing a flow exhibiting variations of the fluid density. In principle, the system does not need any optical component except a light source and a recording plane onto which to project the shadow of the varying density field (**Figure 1**). A shadow effect is generated because a light ray is refractively deflected so that

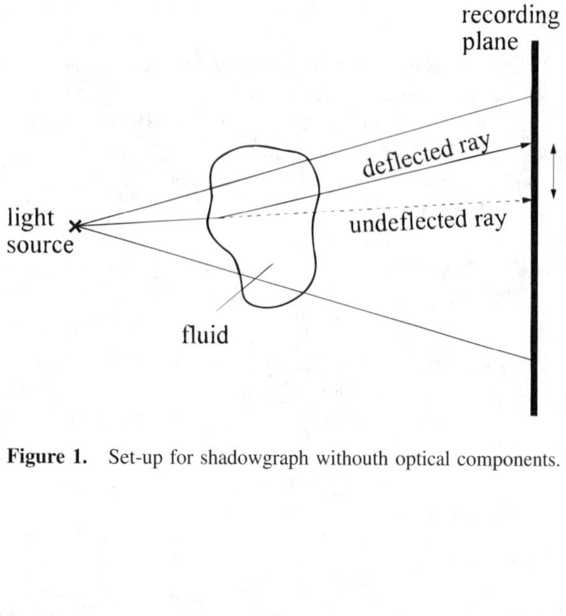

Figure 1. Set-up for shadowgraph withouth optical components.

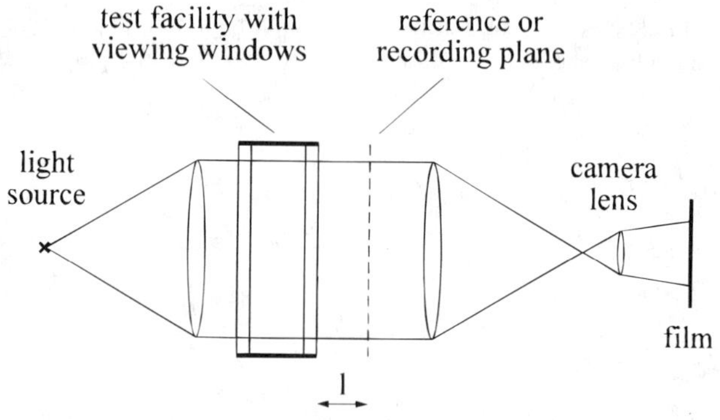

Figure 2. Set-up for shadowgraph with a beam of parallel light transmitted through the test section of a flow facility.

the position on the recording plane where the undeflected ray would arrive now remains dark. At the same time the position where the deflected ray arrives appears brighter than the undisturbed environment. A visible pattern of variations of the illumination (contrast) is thereby produced in the recording plane. From an analysis of the optics of the shadow effect (see, e.g., Merzkirch 1981, 1987), it follows that the visible signal depends on the second derivative of the refractive index of the fluid. Therefore, the shadowgraph as an optical diagnostic technique is sensitive to changes of the second derivative of the fluid density.

It is evident that the shadowgraph is not a method suitable for quantitative measurement of the fluid density. Owing to its simplicity, however, the shadowgraph is a convenient method of obtaining a quick survey of a flow in which the density changes in the described way. This applies particularly to compressible gas flows with shock waves that can be considered as alterations of the gas density with an extremely intense change in curvature of the density profile, i.e. a change of the respective second derivative. The observation of shock waves in gases by means of shadowgraphy goes back to the 19th century when these flow phenomena were discovered by means of this optical technique.

Due to the very simple optical set-up, the shadow effect resulting from inhomogeneous density fields can be observed also outside a laboratory, with the sun serving as the light source, e.g. the sun light may project onto a solid wall shadow patterns that are caused by fuel vapor rising in the air. For laboratory experiments one often uses an optical system with a beam of parallel light transmitted through the flow (**Figure 2**). The camera that records a down-scaled picture is focused onto a plane at distance l from the test field. This plane corresponds to the position of the recording plane in **Figure 1**. The intensity of the shadow effect, or: the sensitivity of the shadowgraph, increases with the distance ℓ. On the other hand, the flow picture is the more out of focus the greater ℓ, so that a compromise between optical sensitivity and image quality has to be found. The optical sensitivity of the shadowgraph is, in principle, an order of magnitude lower than that of schlieren or interferometric techniques (see **Photographic Techniques**).

Figure 3 is a shadowgraph showing a complicated pattern of *shock waves* and *vortices* in air. The flow field is caused by the interaction of an unsteady shock wave with the triangular obstacle in a shock tube.

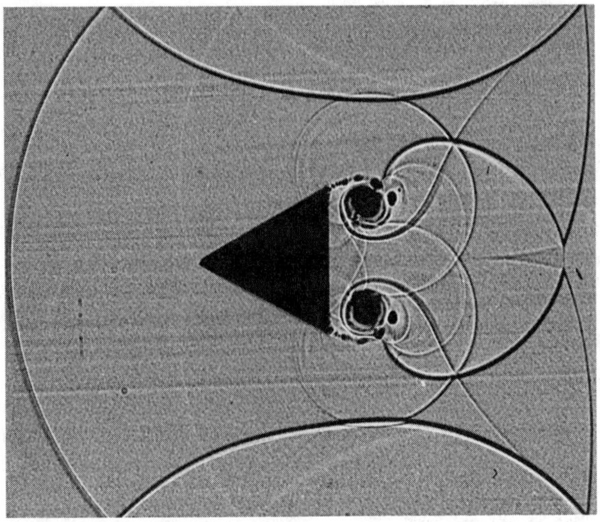

Figure 3. Shadowgraph of shock wave diffraction around a triangular obstacle taken in an air shock tube (Courtesy German-French Research Institute, St. Louis, ISL).

References

Merzkirch, W. (1981). Density sensitive flow visualization, in: *Fluid Dynamics,* ed. R. J. Emrich, *Methods of Experimental Physics,* Vol. 18, Academic Press, New York.

Merzkirch, W. (1987) *Flow Visualization,* 2nd edition, Academic Press, New York.

W. Merzkirch

SHAPE SELECTIVE CATALYSIS (see Molecular sieves)

SHEAR FLOW (see Non-Newtonian fluids)

SHEARING INTERFEROGRAM (see Visualization of flow)

SHEAR LAYER

Following from: *Flow of fluids*

This describes a region of a flow where there is a significant velocity gradient, and consequently the viscous shear stresses defined by

$$\tau = \eta \frac{du}{dy}$$

are important. The most common example of a shear layer arises when a fluid passes over a solid boundary (**Figure 1a**) to form what is commonly termed a **Boundary Layer**. In this case, the velocity distribution is approximated by a **Universal Velocity Profile**. Another example involving a *free shear layer* (or one which is not attached to a solid boundary) arises in the lee of a structure placed within a flow. In this case the shear layer develops between the free stream velocity (U_o) and the near zero velocity occurring within the wake region (**Figure 1b**). (See also **Cross Flow**)

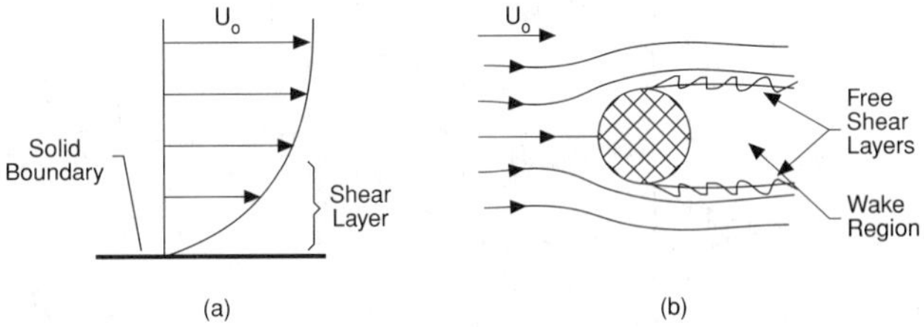

Figure 1. Shear layers.

Laboratory investigations of shear flows are often based upon **Couette Flow** in which a layer of fluid is located between two plates, one of which is moving with a given velocity. This produces a region of strongly sheared flow which may be either lamina or turbulent.

Leading to: *Boundary layer; Universal velocity profile*

C. Swan

SHEAR MODULUS (see Non-Newtonian fluids)

SHEAR STRESS

Following from: *Flow of fluids*

Short-range forces, such as viscous forces, have a molecular origin and are, as a result, generally negligible unless there is physical contact between parts of the fluid. They can be approximated by forces on the surface of each part of the fluid and lead to the concept of *stress* in a fluid. If a force **F** acts on a surface S of a fluid with unit outer normal **n** (so **n** is the vector of unit magnitude which is normal to S and oriented outwards from the fluid) then, if S is small enough:

$$\mathbf{F} = \int_S \mathbf{t}\, dS \simeq \mathbf{t}S \qquad (1)$$

where **t** denotes the *stress vector.* Note that **t** acts at a point whereas **F** does not. Because, however, **t** varies with the orientation of S, that is with **n**, it is not a property at a point. Thus it is appropriate (see Richardson, 1989) to introduce the *stress tensor* $\boldsymbol{\tau}$ such that:

$$\mathbf{t} = \mathbf{n} \cdot \boldsymbol{\tau} = \boldsymbol{\tau}^T \cdot \mathbf{n} \qquad (2)$$

The stress tensor $\boldsymbol{\tau}$ at a point is independent of the orientation of any surface through the point: it is, therefore, a property at a point. The sign convention for stress is:

- $\mathbf{t} \cdot \mathbf{n} = (\mathbf{n} \cdot \boldsymbol{\tau}) \cdot \mathbf{n} > 0$ corresponds to tension;
- $\mathbf{t} \cdot \mathbf{n} = (\mathbf{n} \cdot \boldsymbol{\tau}) \cdot \mathbf{n} < 0$ corresponds to compression.

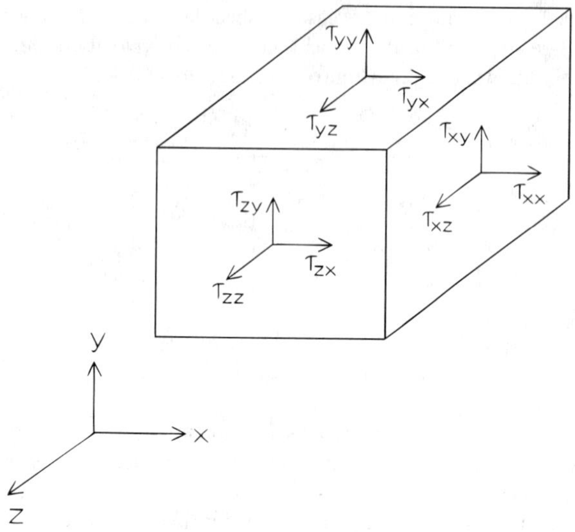

Figure 1. Stress components.

In a rectangular or Cartesian coordinate system (x,y,z):

$$\boldsymbol{\tau} = \tau_{xx}\mathbf{i}_x\mathbf{i}_x + \tau_{xy}\mathbf{i}_x\mathbf{i}_y + \tau_{xz}\mathbf{i}_x\mathbf{i}_z + \tau_{yx}\mathbf{i}_y\mathbf{i}_x + \tau_{yy}\mathbf{i}_y\mathbf{i}_y + \tau_{yz}\mathbf{i}_y\mathbf{i}_z + \tau_{zx}\mathbf{i}_z\mathbf{i}_x + \tau_{zy}\mathbf{i}_z\mathbf{i}_y + \tau_{zz}\mathbf{i}_z\mathbf{i}_z \quad (3)$$

where $\mathbf{i}_x$, $\mathbf{i}_y$ and $\mathbf{i}_z$ denote unit vectors in the x-direction, y-direction and z-direction, respectively (see **Figure 1**). The components of $\boldsymbol{\tau}$ may be interpreted thus:

- τ_{xx} is the x-component of force per unit area exerted across a plane surface element normal to the x-direction;
- τ_{xy} is the y-component of force per unit area exerted across a plane surface element normal to the x-direction; and so on.

  (Precisely analogous interpretations can be given to components of stress in any orthogonal coordinate system.) Each component with identical suffices (τ_{xx}, τ_{yy} and τ_{zz}) is called a *normal stress* (qv) since it represents a force in a direction per unit area normal to that direction. Each other component with different suffices (τ_{xy}, τ_{yx}, τ_{xz}, τ_{zx}, τ_{yz} and τ_{zy}) is called a *shear stress.*

For non-polar fluids, that is fluids for which the angular momentum is just the moment of the linear momentum, it can be shown [see Richardson (1989)] that angular momentum conservation implies that:

$$\boldsymbol{\tau} = \boldsymbol{\tau}^T \quad (4)$$

It then follows that $\tau_{xy} = \tau_{yx}$, $\tau_{xz} = \tau_{zx}$ and $\tau_{yz} = \tau_{zy}$.

Reference

Richardson, S. M. (1989) *Fluid Mechanics,* Hemisphere, New York.

S.M. Richardson

SHEAR STRESS MEASUREMENT

***Following from:** Shear stress*

Accurate measurement of the **Shear Stress** exerted by a moving fluid on a solid boundary can be of central importance in understanding the structure of a flow field. In addition, the surface shear stress can be a key quantity in understanding such diverse phenomena as **Corrosion,** saltation and the build-up of pipe deposits. Measurement of shear stress presents a considerable challenge, and a number of techniques have been developed. Broadly, these can be classified in three ways:

1. Methods which induce a normal stress related to the shear (tangential) stress. Principal methods in this category are the *Stanton gauge,* the *Preston tube,* the *K-tube* and the *sublayer fence.*
2. Direct measurement of the shear stress by use of a floating section of surface. Measurement is made of the forces acting upon the floating element using for example, strain gauges or magnetic techniques.
3. Methods which rely on the transport of either heat or mass. In the case of *mass transfer* (*electrochemical*) methods, an electrochemical reaction is carried out on a surface mounted *probe,* and the resultant electrolysis current is measured. This current is related directly to the shear stress. For *heat transfer* methods, a surface mounted *probe* is maintained at a fixed temperature above its surroundings. The voltage required is related to the shear stress.

Detailed analyses of all these techniques, apart from the K-tube, can be found in the excellent review by Hanratty and Campbell (1983). The K-tube is described by Onsrud (1987).

The choice of an appropriate technique depends to a large extent upon the system in which it is to be used. In general, it can be said that heat transfer probes have the widest applicability, but suffer from the need for elaborate calibration and a highly non-linear response. In addition, they are easily damaged, often difficult and expensive to manufacture, and require care in mounting. Despite these drawbacks, they have found reasonable application and can produce very reliable results. Examples of application can be found in Bellhouse and Schultz (1965). Mass transfer probes offer many of the advantages of heat transfer probes in that they are non-intrusive, can measure time-varying shear stresses and are versatile in application. In addition, calibration is simpler, they are more robust and are generally cheaper than heat transfer probes. Their great drawback is that they cannot be used in gaseous flows. In addition, the liquid must be an electrolyte. An additional factor is that like heat transfer probes, they have a highly non-linear response.

Although heat and mass transfer probes can probably cover most applications, those methods which rely on production of a normal stress can find application. All suffer from the disadvantage that local disturbance of the flow field can result, although this is less of a problem with the K-tube. Calibration is generally straightforward, and subsequent measurements involve detection of small pressure differentials by standard techniques. Stanton gauges can be difficult to manufacture and mount, and have been found to be unreliable in flows with large pressure gradients. Preston tubes are possibly the simplest technique for measuring shear stress, but great care must be taken in mounting. K-tubes and sublayer fences have found application, and produce, under the correct circumstances, good results. However, in common with all methods in this group, they offer no serious advantages over heat and mass transfer probes, although cost may be a factor with heat transfer probes.

The final class of techniques, those relying on direct measurement, have generally found application in aerodynamic, high speed flows. The gauges are, in the main, complex high precision instruments requiring difficult calibration and installation procedures, and great care in use. Hanratty and Campbell (1983) discuss these gauges.

References

Bellhouse, B. J. and Schultz, D. L. (1966) Determination of mean and dynamic skin friction, separation and transition in low speed flow with a thin film heated element, *J. Fluid Mech.,* 24–2, 379–400.

Hanratty, T. J. and Campbell, J. A. (1983) Measurement of wall shear stress, *Fluid Mechanics Measuements,* R. J. Goldstein, ed., Hemisphere Publishing Corporation, 559–615.

Onsrud, G. (1987) The applicability of the shear stress device K-tube, in two phase flow, European two-phase flow group meeting, paper E4, Trondheim, Norway.

P.M. Birchenough

SHEAR STRESS VELOCITY (see Friction velocity)

SHEAR THICKENING (see Viscosity)

SHEAR THICKENING FLUIDS (see Non-Newtonian fluids)

SHEAR THINNING FLUIDS (see Non-Newtonian fluids)

SHEAR VISCOSITY (see Non-Newtonian fluids)

SHEET SPLITTING, IN DROP FORMATION (see Drops)

SHELL AND TUBE CONDENSERS (see Condensers)

SHELL AND TUBE HEAT EXCHANGERS

Following from: *Heat exchangers; Mechanical design of heat exchangers*

General Description

Shell and Tube Heat Exchangers are one of the most popular types of exchanger due to the flexibility the designer has to allow for a wide range of pressures and temperatures. There are two main categories of Shell and Tube exchanger

1. those that are used in the petrochemical industry which tend to be covered by standards from TEMA, Tubular Exchanger Manufacturers Association (See **TEMA Standards**)
2. those that are used in the power industry such as feedwater heaters and power plant condensers

Regardless of the type of industry the exchanger is to be used in there are a number of common features (see **Condensers**).

A shell and tube exchanger consists of a number of tubes mounted inside a cylindrical shell. **Figure 1** illustrates a typical unit that may be found in a petrochemical plant. Two fluids can exchange heat, one fluid flows over the outside of the tubes while the second fluid flows through the tubes. The fluids can be single or two phase and can flow in a parallel or a cross/counter flow arrangement.

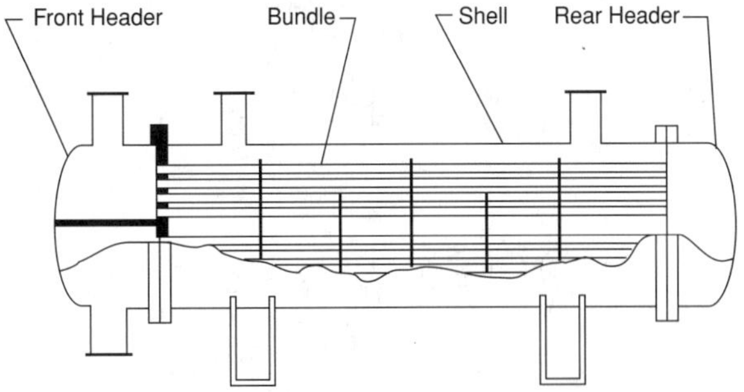

Figure 1. Shell and tube exchanger.

The shell and tube exchanger consists of four major parts:

- Front Header—this is where the fluid enters the tubeside of the exchanger. It is sometimes referred to as the Stationary Header.
- Rear Header—this is where the tubeside fluid leaves the exchanger or where it is returned to the front header in exchangers with multiple tubeside passes
- Tube bundle—this comprises of the tubes, *tube sheets, baffles* and tie rods etc. to hold the bundle together.
- Shell—this contains the *tube bundle*

The remainder of this section concentrates on exchangers that are covered by the TEMA Standard.

Shell and Tube Exchanger: Geometric Terminology

The main components of a shell and tube exchanger are shown in **Figure 2 a, b** and **c** and described in **Table 1**.

Tema Designations

The popularity of shell and tube exchangers has resulted in a standard nomenclature being developed for their designation and use by the Tubular Exchanger Manufactures Association (TEMA). This nomenclature is defined in terms letters and diagrams. The first letter describes the front header type, the second letter the shell type and the third letter the rear header type. **Figure 2** shows examples of a BEM, CFU and AES exchangers while **Figure 3** illustrates the full TEMA nomenclature.

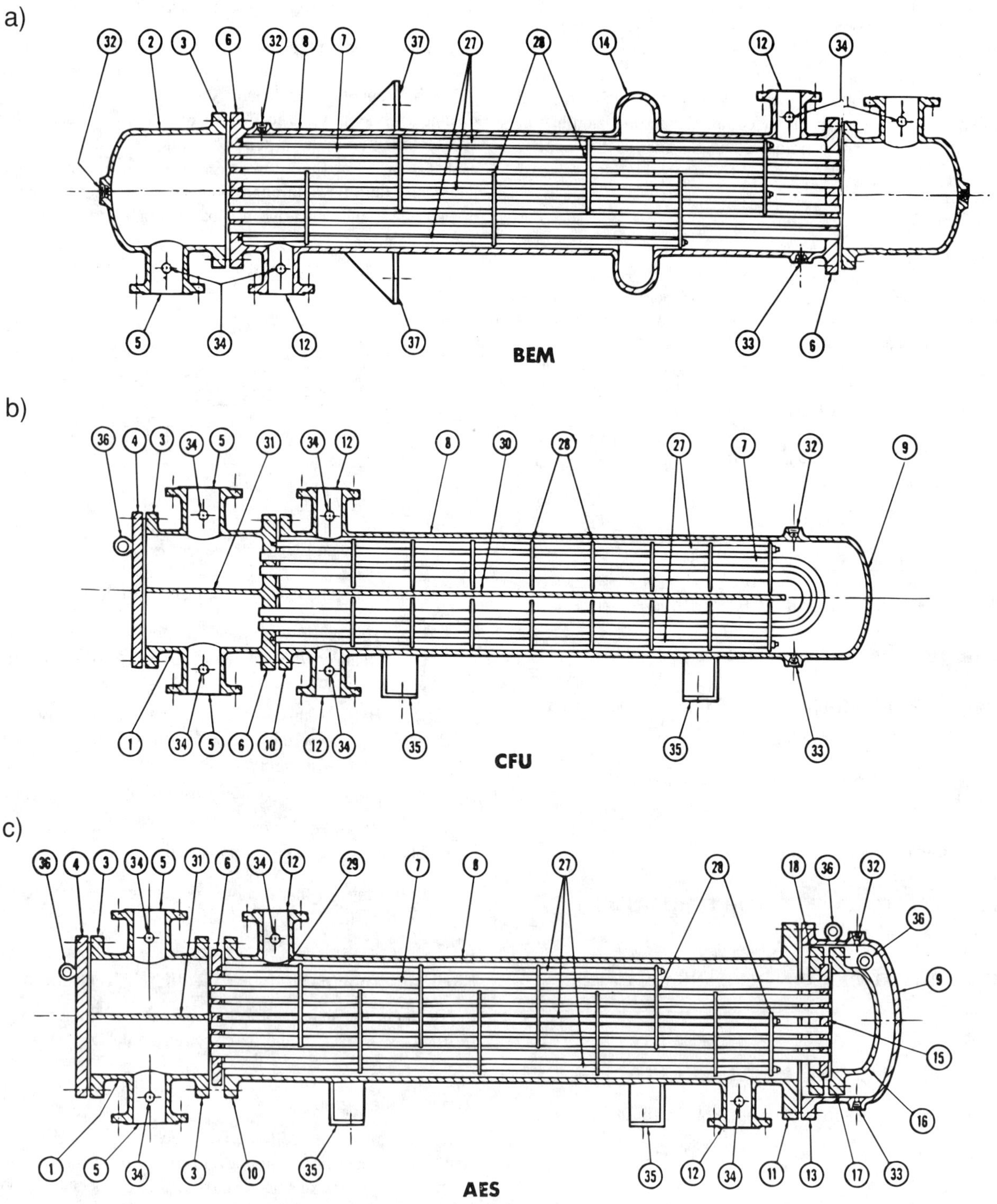

Figure 2. Type BEM, CFU and AES exchangers. © 1988 by Tubular ExChanger Manufacturers Association.

Table 1. Shell and tube geometric terminology

1	Stationary (Front) Head—Channel	19	Split Shear Ring
2	Stationary (Front) Head—Bonnet	20	Slip-on Backing Flange
3	Stationary (Front) Head Flange	21	Floating Head Cover—External
4	Channel Cover	22	Floating Tubesheet Skirt
5	Stationary Head Nozzle	23	Packing Box Flange
6	Stationary Tubesheet	24	Packing
7	Tubes	25	Packing Follower Ring
8	Shell	26	Lantern Ring
9	Shell Cover	27	Tie Rods and Spacers
10	Shell Flange—Stationary Head End	28	Transverse Baffles or Support Plates
11	Shell Flange—Rear Head End	29	Impingement Baffle or Plate
12	Shell Nozzle	30	Longitudinal Baffle
13	Shell Cover Flange	31	Pass Partition
14	Expansion Joint	32	Vent Connection
15	Floating Tubesheet	33	Drain Connection
16	Floating Head Cover	34	Instrument Connection
17	Floating Head Flange	35	Support Saddle
18	Floating Head Backing Device	36	Lifting Lug
		37	Support Bracket

Many combinations of front header, shell and rear header can be made. The most common combinations for an E-Type Shell are given in **Table 2** but other combinations are also used.

Essentially there are three main combinations

- *Fixed tubesheet exchangers*
- *U-tube exchangers*
- *Floating header exchangers*

Fixed Tubesheet Exchanger (L, M and N Type Rear Headers)

In a *fixed tubesheet exchanger,* the tubesheet is welded to the shell. This results in a simple and economical construction and the tube bores can be cleaned mechanically or chemically. However, the outside surfaces of the tubes are inaccessible except to chemical cleaning.

If large temperature differences exist between the shell and tube materials, it may be necessary to incorporate an expansion bellows in the shell, to eliminate excessive stresses caused by expansion. Such bellows are often a source of weakness and failure in operation. In circumstances where the consequences of failure are particularly grave U-Tube or Floating Header units are normally used.

This is the cheapest of all removable bundle designs, but is generally slightly more expensive than a fixed tubesheet design at low pressures.

U-Tube Exchangers

In a U-Tube exchanger any of the front header types may be used and the rear header is normally a M-Type. The U-tubes permit unlimited thermal expansion, the tube bundle can be removed for cleaning and small bundle to shell clearances can be achieved. However, since internal cleaning of the tubes by mechanical means is difficult, it is normal only to use this type where the tube side fluids are clean.

Floating Head Exchanger (P, S, T and W Type Rear Headers)

In this type of exchanger the tubesheet at the Rear Header end is not welded to the shell but allowed to move or float. The tubesheet at the Front Header (tube side fluid inlet end) is of a larger diameter than the shell and is sealed in a similar manner to that used in the fixed tubesheet design. The tubesheet at the rear header end of the shell is of slightly smaller diameter than the shell, allowing the bundle to be pulled through the shell. The use of a floating head means that thermal expansion can be allowed for and the tube bundle can be removed for cleaning. There are several rear header types that can be used but the S-Type Rear Head is the most popular. A floating head exchanger is suitable for the rigorous duties associated with high temperatures and pressures but is more expensive (typically of order of 25% for carbon steel construction) than the equivalent fixed tubesheet exchanger.

Considering each header and shell type in turn:

A-Type front header

This type of header is easy to repair and replace. It also gives access to the tubes for cleaning or repair without having to disturb the pipe work. It does however have two seals (one between the tube sheet and header and the other between the header and the end plate). This increases the risk of leakage and the cost of the header over a B-Type Front Header.

B-Type front header

This is the cheapest type of front header. It also is more suitable than the A-Type Front Header for high pressure duties because the header has only one seal. A disadvantage is that to gain access to the tubes requires disturbance to the pipe work in order to remove the header.

C-Type front header

This type of header is for high pressure applications (>100 bar). It does allow access to the tube without disturbing the pipe work but is difficult to repair and replace because the tube bundle is an integral part of the header.

D-Type front header

This is the most expensive type of front header. It is for very high pressures (>150 bar). It does allow access to the tubes without disturbing the pipe work but is difficult to repair and replace because the tube bundle is an integral part of the header.

N-Type front header

The advantage of this type of header is that the tubes can be accessed without disturbing the pipe work and it is cheaper than an A-Type Front Header. However, they are difficult to maintain

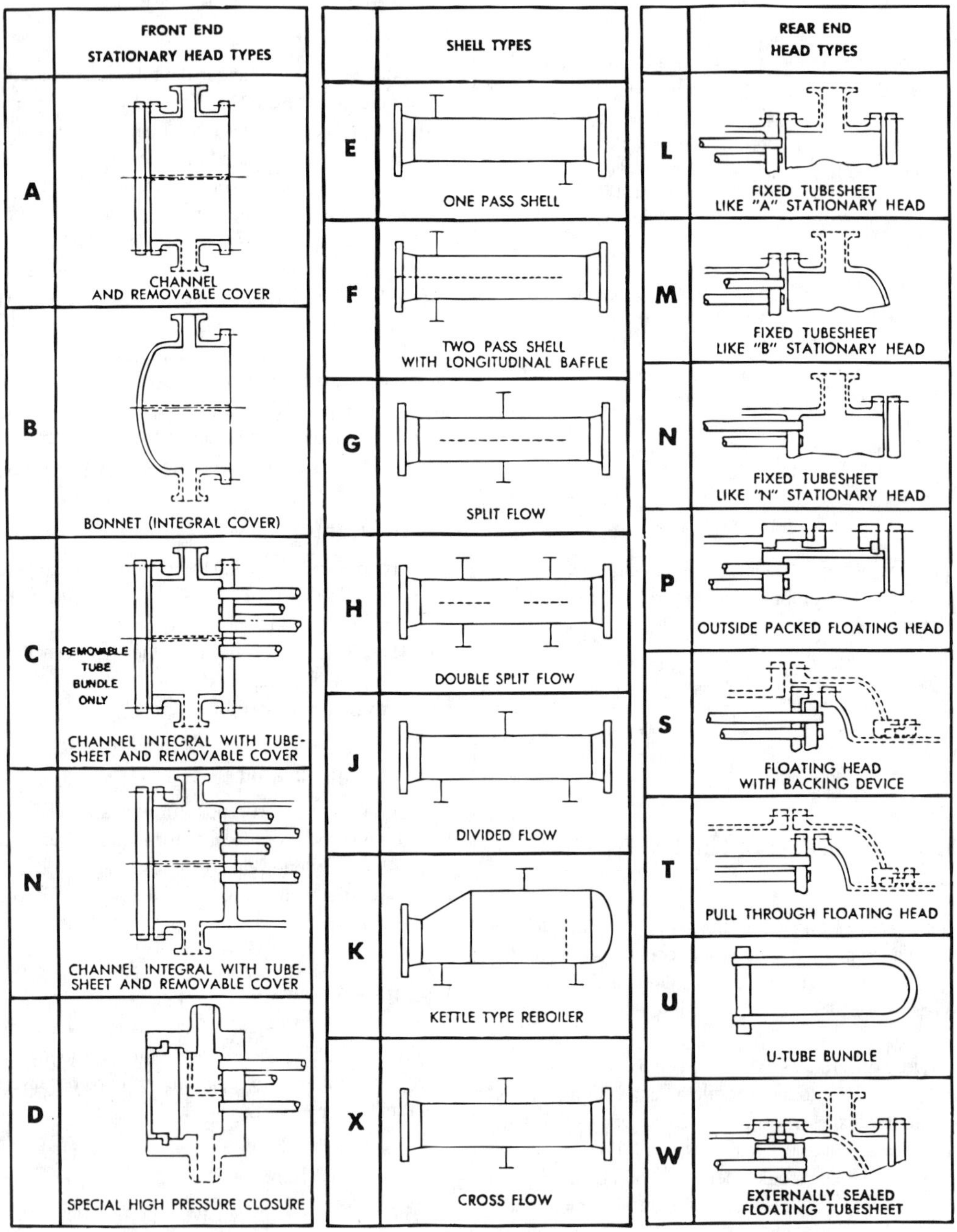

Figure 3. TEMA nomenclature. © 1988 by Tubular Exchanger Manufacturers Association.

Table 2. Common front and rear header types for an E-type shell

Fixed tubesheet exchangers	U-tube exchangers	Floating head exchangers
AEL	AEU	AES
AEM	CEU	BES
AEN	DEU	
BEL		
BEM		
BEN		

and replace as the header and tube sheet are an integral part of the shell.

Y-Type front header

Strictly speaking this is not a TEMA designated type but is generally recognised. It can be used as a front or rear header and is used when the exchanger is to be used in a pipe line. It is cheaper than other types of headers as it reduces piping costs. It is mainly used with single tube pass units although with suitable partitioning any odd number of passes can be allowed.

E-Type shell

This is most commonly used shell type, suitable for most duties and applications. Other shell types only tend to be used for special duties or applications.

F-Type shell

This is generally used when pure counter current flow is required in a two tube side pass unit. This is achieved by having two shell side passes—the two passes being separated by a longitudinal baffle. The main problem with this type of unit is thermal and hydraulic leakage across this longitudinal baffle unless special precautions are taken.

G-Type shell

This is used for horizontal thermosyphon re-boilers and applications where the shellside pressure drop needs to be kept small. This is achieved by splitting the shellside flow.

H-Type shell

This is used for similar applications to G-Type Shell but tends to be used when larger units are required.

J-Type shell

This tends to be used when the maximum allowable pressure drop is exceeded in an E-Type Shell even when double segmental baffles are used. It is also used when tube vibration is a problem. The divided flow on the shellside reduces the flow velocities over the tubes and hence reduces the pressure drop and the likelihood of tube vibration. When there are two inlet nozzles and one outlet nozzle this is sometimes referred to as an I-Type Shell.

K-Type shell

This is used only for re-boilers to provide a large disengagement space in order to minimise shellside liquid carry over. Alternatively a K-Type Shell may be used as a chiller. In this case the main process is to cool the tube side fluid by boiling a fluid on the shellside.

X-Type shell

This is used if the maximum shellside pressure drop is exceeded by all other shell and baffle type combinations. The main applications are shellside condensers and gas coolers.

L-Type rear header

This type of header is for use with fixed tubesheets only, since the tubesheet is welded to the shell and access to the outside of the tubes is not possible. The main advantages of this type of header are that access can be gained to the inside of the tubes without having to remove any pipework and the bundle to shell clearances are small. The main disadvantage is that a bellows or an expansion roll are required to allow for large thermal expansions and this limits the permitted operating temperature and pressure.

M-Type rear header

This type of header is similar to the L-Type Rear Header but it is slightly cheaper. However, the header has to be removed to gain access to the inside of the tubes. Again, special measures have to be taken to cope with large thermal expansions and this limits the permitted operating temperature and pressure.

N-Type rear header

The advantage of this type of header is that the tubes can be accessed without disturbing the pipe work. However, they are difficult to maintain and replace since the header and tube sheet are an integral part of the shell.

P-Type rear header

This is an outside packed floating rear header. It is, in theory, a low cost floating head design which allows access to the inside of the tubes for cleaning and also allows the bundle to be removed for cleaning. The main problems with this type of header are:

- large bundle to shell clearances required in order to pull the bundle
- it is limited to low pressure non-hazardous fluids, because it is possible for the shellside fluid to leak via the packing rings
- only small thermal expansions are permitted.

In practice it is not a low cost design, because the shell has to be rolled to small tolerances for the packing to be effective.

S-Type rear header

This is a floating rear header with backing device. It is the most expensive of the floating head types but does allow the bundle to be removed and unlimited thermal expansion is possible. It also has smaller shell to bundle clearances than the other floating head types. However, it is difficult to dismantle for bundle pulling and the shell diameter and bundle to shell clearances are larger than for fixed head type exchangers.

T-Type rear header

This is a pull through floating head. It is cheaper and easier to remove the bundle than with the S-Type Rear Header, but still allows for unlimited thermal expansion. It does, however, have the largest bundle to shell clearance of all the floating head types and is more expensive than fixed header and U-tube types.

U-tube

This is the cheapest of all removable bundle designs, but is generally slightly more expensive than a fixed tubesheet design at low pressures. However, it permits unlimited thermal expansion, allows the bundle to be removed to clean the outside of the tubes, has the tightest bundle to shell clearances and is the simplest design. A disadvantage of the U-tube design is that it cannot normally have pure counter-flow unless an F-Type Shell is used. Also, U-tube designs are limited to even numbers of tube passes.

W-Type rear header

This is a packed floating tubesheet with lantern ring. It is the cheapest of the floating head designs, allows for unlimited thermal expansion and allows the tube bundle to be removed for cleaning. The main problems with this type of head are:

- the large bundle to shell clearances required to pull the bundle and
- the limitation to low pressure non-hazardous fluids (because it is possible for both the fluids to leak via the packing rings).

It is also possible for the shell and tube side fluids to become mixed if leakage occurs.

Geometric Options

Tube diameter layout and pitch

Tubes may range in diameter from 12.7 mm (0.5 in) to 50.8 mm (2 in), but 19.05 mm (0.75 in) and 25.4 mm (1 in) are the most common sizes. The tubes are laid out in triangular or square patterns in the tube sheets. See **Figure 4**.

The square layouts are required where it is necessary to get at the tube surface for mechanical cleaning. The triangular arrangement allows more tubes in a given space. The tube pitch is the shortest centre-to-centre distance between tubes. The tube spacing is given by the tube pitch/tube diameter ratio which is normally 1.25 or 1.33. Since a square layout is used for cleaning purposes, a minimum gap of 6.35 mm (0.25in) is allowed between tubes.

Baffle types

Baffles are installed on the shell side to give a higher heat-transfer rate due to increased turbulence and to support the tubes thus reducing the chance of damage due to vibration. There are a number of different baffle types which support the tubes and promote flow across the tubes. **Figure 5** shows the following baffle arrangements.

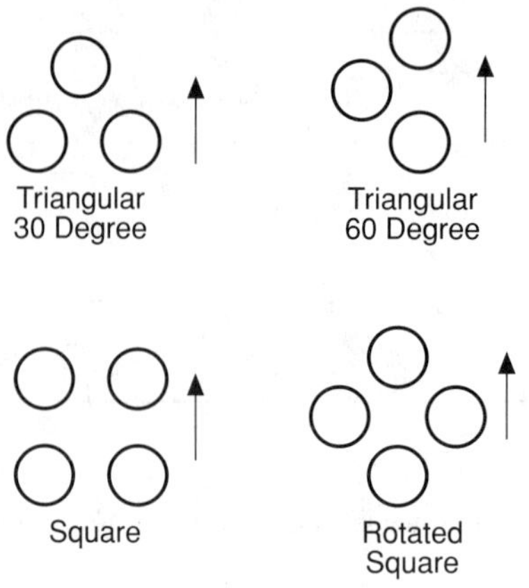

Figure 4. Tube layouts.

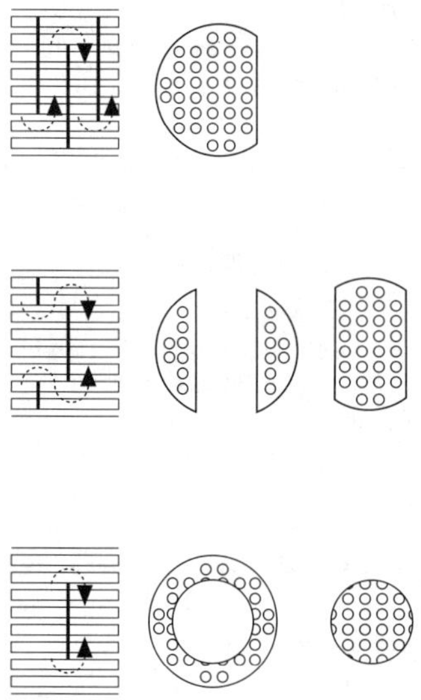

Figure 5. Baffle arrangements.

- *Single Segmental* (this is the most common)
- *Double Segmental* (this is used to obtain a lower shellside velocity and pressure drop)
- *Disc and Doughnut*

The centre-to-centre distance between baffles is called the baffle-pitch and this can be adjusted to vary the crossflow velocity. In practice the baffle pitch is not normally greater than a distance equal to the inside diameter of the shell or closer than a distance equal to one-fifth the diameter or 50.8 mm (2 in) whichever is greater. In order to allow the fluid to flow backwards and forwards across the tubes part of the baffle is cut away. The height of this part is referred to as the baffle-cut and is measured as a percentage of the shell diameter, e.g. 25 per cent baffle-cut. The size of the baffle-cut (or baffle window) needs to be considered along with the baffle pitch. It is normal to size the baffle-cut and baffle pitch to approximately equalise the velocities through the window and in crossflow respectively.

There are two main types of baffle which give longitudinal flow:

- *Orifice Baffle*
- *Rod Baffle*

In these types of baffle the turbulence is generated as the flow crosses the baffle.

Heat Transfer Enhancements Devices

There are three main types.

Special surfaces

These tend to be used to promote nucleate boiling when the temperature driving force is small.

Tube inserts

These are normally *wire wound inserts* or *twisted tapes.* They are normally used with medium to high viscosity fluids to improve heat transfer by increasing turbulence. There is also some evidence that they reduce fouling. In order to use these most effectively the exchanger should be designed for their use. This usually entails increasing the shell diameter, reducing the tube length and the number of tubeside passes in order to allow for the increased pressure loss characteristics of the devices.

Extended surfaces

These are used to increase the heat transfer area when a stream has a low heat transfer coefficient. The most common type is "*low fin tubing*" where typically the fins are 1.5mm high at 19 fins per inch. (See also **Augmentation of Heat Transfer**)

Selection Criteria

In many cases the only way of ensuring optimum selection is to do a full design based on several alternative geometries. In the first instance, however, several important decisions have to be made concerning:

- allocation of fluids to the shellside and tubeside
- selection of shell type
- selection of front end header type
- selection of rear end header type
- selection of exchanger geometry.

To a large extent these often depend on each other. For instance, the allocation of a dirty fluid to the shellside directly affects the selection of exchanger tube layout.

Fluid allocation

When deciding which side to allocate the hot and cold fluids the following need to be taken into account, in order of priority.

1. Consider any and every safety and reliability aspect and allocate fluids accordingly. Never allocate hazardous fluids such they are contained by anything other than conventional bolted and gasketted—or welded—joints.
2. Ensure that the allocation of fluids complies with established engineering practices, particularly those laid down in customer specifications.
3. Having complied with the above, allocate the fluid likely to cause the most severe mechanical cleaning problems (if any) to the tubeside.
4. If neither of the above are applicable, the allocation of the fluids should be decided only after running two alternative designs and selecting the cheapest (this is time consuming if hand calculations are used but programs such as TASC from the Heat Transfer and Fluid Flow Service (HTFS) make this a trivial task).

Shell selection

E-type shells are the most common. If a single tube pass is used and provided there are more than three baffles, then near counter-current flow is achieved. If two or more tube passes are used, then it is not possible to obtain pure counter-current flow and the log mean temperature difference must be corrected to allow for combined co-current and counter-current flow using an F-factor.

G-type shells and H shells are normally specified only for horizontal thermosyphon reboilers. J shells and X-type shells should be selected if the allowable ΔP cannot be accommodated in a reasonable E-type design. For services requiring multiple shells with removable bundles, F-type shells can offer significant savings and should always be considered provided they are not prohibited by customer specifications.

Front header selection

The A-type front header is the standard for dirty tubeside fluids and the B-type is the standard for clean tubeside fluids. The A-type is also preferred by many operators regardless of the cleanliness of the tubeside fluid in case access to the tubes is required. Do not use other types unless the following considerations apply.

A C-type head with removable shell should be considered for hazardous tubeside fluids, heavy bundles or services requiring frequent shellside cleaning. The N-type head is used when hazardous fluids are on the tubeside. A D-type head or a B-type head welded to the tubesheet is used for high pressure applications. Y-type heads are only normally used for single tube-pass exchangers when they are installed in line with a pipeline.

Rear header selection

For normal service a Fixed Header (L, M, N-types) can be used provided that there is no overstressing due to differential expansion and the shellside will not require mechanical cleaning. If thermal expansion is likely a fixed header with a bellows can used provided that the shellside fluid is not hazardous, the shellside pressure does not exceed 35 bar (500 psia) and the shellside will not require mechanical cleaning.

A U-tube unit can be used to overcome thermal expansion problems and allow the bundle to be removed for cleaning. However, counter current flow can only be achieved by using an F-type shell and mechanical cleaning of the tubeside can be difficult.

A S-type floating head should be used when thermal expansion needs to be allowed for and access to both sides of the exchanger is required from cleaning. Other rear head types would not normally be considered except for the special cases.

Selection of Exchanger Geometry

Tube outside diameter

For the process industry, 19.05 mm (3/4″) tends to be the most common.

Tube wall thickness

Reference must be made to a recognised pressure vessel code to decide this.

Tube length

For a given surface area, the longer the tube length the cheaper the exchanger, although a long thin exchanger may not be feasible.

Tube layout

45 or 90 degree layouts are chosen if mechanical cleaning is required, otherwise a 30 degree layout is often selected, because it provides a higher heat transfer and hence smaller exchanger.

Tube pitch

The smallest allowable pitch of 1.25 times the tube outside diameter is normally used unless there is a requirement to use a larger pitch due to mechanical cleaning or tube end welding.

Number of tube passes

This is usually one or an even number (not normally greater than 16). Increasing the number of passes increases the heat transfer coefficient but care must be taken to ensure that the tube side ρv^2 is not greater than about 10 000 kg/m s^2.

Shell diameter

Standard pipe is normally used for shell diameters up to 610mm (24″). Above this the shell is made from rolled plate. Typically shell diameters range from 152 mm to 3000 mm (6″ to 120″).

Baffle type

Single segmental baffles are used by default but other types are considered if pressure drop constraints or vibration is a problem.

Baffle spacing

This is decided after trying to balance the desire for increased crossflow velocity and tube support (smaller baffle pitch) and pressure drop constraints (larger baffle pitch). TEMA provides guidance on the maximum and minimum baffle pitch.

Baffle cut

This depends on the baffle type but is typically 45% for single segmental baffles and 25% for double segmental baffles.

Nozzles and impingement

For shellside nozzles the ρv^2 should not be greater than about 9 000 in kg/m s^2. For tubeside nozzles the maximum ρv^2 should not exceed 2230 kg/m s^2 for non-corrosive, non-abrasive single phase fluids and 740 kg/m s^2 for other fluids. Impingement protection is always required for gases which are corrosive or abrasive, saturated vapours and two phase mixtures. Shell or bundle entrance or exit areas should be design such that a ρv^2 of 5950 kg/m s^2 is not exceeded.

Materials of Construction

In general, shell and tube exchangers are made of metal, but for specialist applications (e.g. involving strong acids or pharmaceuticals), other materials such as graphite, plastic and glass may be used.

Thermal Design

The thermal design of a shell and tube exchanger is an iterative process which is normally carried out using computer programs from organisations such as the Heat transfer and Fluid Flow Service (HTFS) or Heat Transfer Research Incorporated (HTRI). However, it is important that the engineer understands the logic behind the calculation. In order to calculate the heat transfer coefficients and pressure drops, initial decisions must be made on the sides the fluids are allocated, the front and rear header type, shell type, baffle type, tube diameter and tube layout. The tube length, shell diameter, baffle pitch and number of tube passes are also selected and these are normally the main items that are altered during each iteration in order to maximise the overall heat transfer within specified allowable pressure drops.

The main steps in the calculation are given below together with calculation methods in the open literature:

1. Calculate the shellside flow distribution (Use *Bell-Delaware Method* see Hewitt, Shires and Bott, 1994)
2. Calculate the shellside heat transfer coefficient (Use Bell-Delaware Method)
3. Calculate tubeside heat transfer coefficient (see for example **Tubes: Single Phase Heat Transfer In**)
4. Calculate tubeside pressure drop (see for example **Pressure Drop, Single Phase**)
5. Calculate wall resistance and overall heat transfer coefficient (see **Overall Heat Transfer Coefficient and Fouling**)
6. Calculate mean temperature difference (see **Mean Temperature Difference**)
7. Calculate area required
8. Compare area required with area of assumed geometry and allowed tubeside and shellside pressure drop with calculated values
9. Adjust assumed geometry and repeat calculations until Area required is achieved within the allowable pressure drops

Books by E.A.D. Saunders (Saunders, 1988) and G.F. Hewitt, G.L. Shires and T.R. Bott (Hewitt et al., 1994) provides a good overview of tubular thermal design methods and example calculations.

Mechanical Design

The mechanical design of a shell and tube heat exchanger provides information on items such as shell thickness, flange thickness, etc. These are calculated using a pressure vessel design code such as the Boiler and Pressure Vessel code from ASME (American Society of Mechanical Engineers) and the British Master Pressure Vessel Standard, BS 5500. ASME is the most commonly used code for heat exchangers and is in 11 sections. Section VIII (Confined Pressure Vessels) of the code is the most applicable to heat exchangers but Sections II—Materials and Section V—Non Destructive Testing are also relevant.

Both ASME and BS5500 are widely used and accepted throughout the world but some countries insist that their own national codes are used. In order to try and simplify this the International Standards Organisation is now attempting to develop a new internationally recognised code but it is likely to be a some time before this is accepted.

References

TEMA Seventh Edition. (1988) Tubular Exchanger Manufacturers Association.

Saunders, E. A. D. (1988) *Heat Exchangers—Selection, Design and Construction,* Longman Scientific and Technical.

Hewitt, G. F., Shires, G. L., and Bott, T. R. (1994) *Process Heat Transfer, CRC Press.*

Boiler and Pressure Vessel code, ASME (American Society of Mechanical Engineers).

British Master Pressure Vessel Standard, BS 5500.

Leading to: *Vibration in heat exchangers*

R. Brogan

SHELL BOILER (see Boilers)

SHELL PROGRESSIVE MODEL (see Ion exchange)

SHELLS (see Mechanical design of heat exchangers)

SHELL-SIDE REFRIGERATION CHILLERS (see Tubes and tube banks, boiling heat transfer on)

SHERWOOD NUMBER

Sherwood number, Sh, is the dimensionless group for convective mass transfer in fluid flow. It is defined as

$$Sh = \beta L/\delta$$

where β is mass transfer coefficient, δ diffusivity, and L the representative dimension (e.g. diameter for tubes).

Sherwood number represents the ratio between mass transfer by convection (β) and mass transfer by diffusion (δ/L). It is the mass transfer equivalent of **Nusselt Number**, Nu, and is a function of **Reynolds Number**, Re, and **Schmidt Number**, Sc.

In the special case when

$$Pr = Sc = 1$$

$$Sh = Nu = (f_o/2)\,Re$$

where f_o is the **Fanning Friction Factor.**

Reference

Hewitt G. F., Shires G. L., and Bott T. R. (1994). *Process Heat Transfer,* CRC Press.

G.L. Shires

SHERWOOD, THOMAS KILGORE (1903–1976)

Sherwood, born in Columbus, Ohio, July 25, 1903, was one of America's great chemical engineers; his energy, research contributions, applied engineering achievements, and influence on chemical engineering education were prodigious.

Figure 1.

Sherwood came to M. I. T. in 1923 for graduate work in the chemical engineering department and completed his doctoral thesis under Warren K. Lewis, entitled "The Mechanism of the Drying of Solids" in 1929. From 1930 to 1969 he was professor at M. I. T. contributing decisively to the standards of excellence of this famous institution.

Sherwood's primary research area was mass transfer and its interaction with flow and with chemical reaction and industrial process operations in which those phenomena played an important part. His rapid rise to the position of world authority in the field of mass transfer was accelerated by the appearance of his book, "Absorption and Extraction", the first significant text in this area, published in 1937. Completely rewritten, with Pigford and Wilke in 1974 under the title "Mass Transfer", the book has had enormous influence. The worldwide use of **Sherwood Number** is a memorial to that effort.

In addition to three honorary doctorates many awards were bestowed on Sherwood, such as the U.S. Medal for Merit in 1948 and the Lewis Award in 1972. He died on January 14, 1976.

U. Grigull, H. Sandner and J. Straub

SHOCK TUBES

***Following from:** Compressible flows*

Shock tubes are devices for studying the flow of high-temperature and high-velocity compressible gas.

A high-temperature supersonic gas flow is initiated in a shock tube as a result of rupture of a diaphragm separating two gases in high-pressure and low-pressure chambers. An unsteady rarefaction wave passes into the "driver" gas in the high-pressure chamber with a velocity a few kilometers per second. This results in the drivers gas flowing into the low pressure gas, pushing the gas in the low pressure chamber ahead of it. The shock wave propagates in the low-pressure chamber ahead of this flow of the studied gas (see **Waves in Liquids**). The velocity of the driven gas flow is equal to that of the driver gas flow. The wave (x − t) pattern, the schematic of the shock tube, and distribution of pressures p and temperatures T along the tube axis are presented in **Figure 1**. The parameters are denoted by 0 in the low-pressure chamber of the shock tube, 1 behind the incident shock wave, 2 behind the contact surface, 3 in the rarefaction wave, 4 in the high-pressure chamber, and 5 behind the reflected wave. In order to generate a shock wave with the specified pressure ratio p_1/p_0 in a shock tube, it is necessary to provide the pressure ratio p_4/p_0 over the shock tube diaphragm which satisfies the relation

$$\frac{p_4}{p_0} = \frac{p_1}{p_0}\left\{1 - \frac{a_0}{a_4}\frac{\gamma_4 - 1}{2}\frac{u_1}{a_0}\right\}^{-2\gamma_4/(\gamma_4-1)},$$

where a is the velocity of sound, γ the adiabatic exponent, u_1 the velocity of gas flow behind the shock wave producing the pressure ratio p_1/p_0. Varying the length of high- and low-pressure chambers, we can find a regime of shock tube operation under which the shock wave reflected from an end meets the tail of the rarefaction wave and the head of the reflected rarefaction wave at a fixed point. This regime known as the "joined" contact surface regime

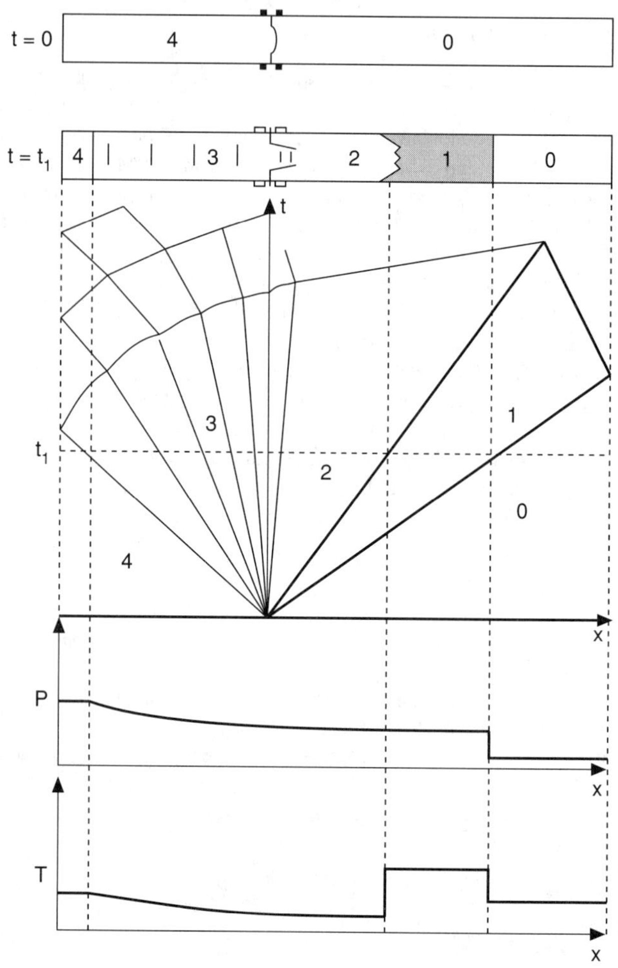

Figure 1.

ensures the maximum time of undisturbed state behind the reflected wave. The maximum flow parameters achieved in shock tubes depend on the design of the high-pressure chamber. According to this, shock tubes can be devided into four groups.

(1) Shock tubes with an inert gas in the high-pressure chamber. In this case a driver gas is hydrogen or helium at a pressure ranging from several to a hundred atmospheres and of ambient initial temperature. For pressures in the low-pressure chamber from 1 to 100 mmHg, the shock wave velocities range from 0.5 to 6 km/s. The temperature behind the shock wave is from 1000 to 10,000 K. The length of the high pressure gas zone, depending on the tube length and the shock wave velocity, is from several centimeters in laboratory tubes, whose length does not exceed 10 meters, to a meter in test shock tubes 50 to 100 meters long.

(2) Shock tubes with an explosive in the high-pressure chamber. In this case, in the high-pressure chamber, a hydrogen-oxygen-helium mixture or a solid explosive charge are used to increase the initial temperature of a driver gas. The gas temperature behind the shock wave amounts to 20,000 K in an inert gas and 15,000 K in a dissociating gas.

(3) Electrically driven shock tubes. These are tubes in which a discharge chamber with discharge-heated gas is used instead of the high-pressure chamber. Electrically driven shock tubes are designed with and without diaphragms because the parameters of the driver gas increase sharply as a result of discharge. In these shock tubes the velocity of the shock waves varies from 10 km/s at initial pressures of several millimeters Hg to 100 km/s at initial pressures fractions of a millimeter Hg. Gas temperature reaches a few tens of thousands of degrees.

(4) Shock tubes with shock wave enhancement. Shock tubes in this category use for shock wave enhancement the interaction of waves arising in transition of a shock wave to sections with different pressure and cross section, the category also includes shock tubes with two diaphragms.

In steady propagation of a shock wave in a shock tube the flow parameters behind the shock wave such as temperature, pressure, density, and velocity are unambiguously determined by the conservation laws using the shock wave velocity and the gas state. If the degree of approach to equilibrium is unknown, then determination of flow parameters requires, in addition to determination of the shock wave velocity, the measurement of one or several more gas parameters behind the shock wave or the time distribution of these parameters. These parameters include

1. density distribution behind the shock wave (interferometer method, photoelectric shadow method, absorption of an electron beam, absorption of X-rays),
2. distribution of flow velocity (by the velocity of displacement of a weak disturbance introduced by transient heating of a wire placed in the flow or by electrical discharge),
3. distribution of the gas temperature (by spectral methods) or the electron temperature (by optical methods or by radiation emission),
4. distribution of the flow Mach numbers (by measuring the direction of the Mach lines resulting from interaction of the supersonic flow behind the shock wave with an obstacle in the tube),
5. distribution of pressures (by the readings of piezoelectric transducers),
6. composition of the gas dissociating behind the shock wave (by absorption of ultraviolet or infrared radiation),
7. electrophysical properties of gas such as concentration of unbound electrons and collision frequency (by microwave, optical, electromagnetic, and probe methods),
8. the shock wave velocity measured by either visualization of its propagation in a transparent section of the tube or by the time-of-flight method, i.e. measuring the time interval between the readings of two transducers responding to the shock wave at two points of the tube (use is made of pressure transducers, photoelectric glow sensors, ionization sensors, and thin-film resistance thermometers),
9. heat fluxes to the shock tube wall determined by calorimetric transducers or by measuring the time dependence of the wall temperature using a thin-film resistance thermometer.

A fraction of the gas mass flowing in the shock tube passes from the center to the boundary layer. The thickness of the boundary layer near the shock wave front is zero, then it grows toward the contact surface, and the gas outflow to the boundary layer grows respectively.

Growth of the boundary layer is interrupted as soon as the gas outflow across it becomes equal to the gas inflow across the shock wave front. Owing to this, the shock wave and the contact surface velocities become equal, the plug size reaches maximum and remains unchanged. The flow becomes steady as in the steady bow shock distance from the blunt body in a supersonic flow, when the gas inflow across the shock wave front becomes equal to the outflow across sonic lines and, to the flow along the side surfaces of the body. The gas flow between the shock wave front and the contact surface is isentropic. The gas velocity behind the shock is determined in accordance with the conservation laws depending on the velocity of the shock wave front.

The range of problems of heat and mass transfer to be solved on shock tubes covers an analysis of dynamic and thermal loads of intricately shaped bodies exposed to blast waves, heat transfer by radiation under joint action of radiation and convection, fluid dynamics of jet flows, interaction of jets, investigation of gas outflow (heated by the shock wave) from the jet at the tube end, and heat transfer during external flow around bodies.

References

Gaydon, A. G. and Hurle, I. R. (1963) *The Shock Tube in High Temperature Chemical Physics,* Chapman and Hall.

Warren, W. R. and Harris, G. J. (1970) A Critique of High-Performance Shock Tube Driving Techniques, *Shock Tubes,* 143–176, Toronto Univ. Press.

Korobemikov, V. P. (1989) ed., *Unsteady Interaction of Shock and Detonation Waves in Gases,* Hemisphere Publ. Corp.

T.V. Bazhenova

SHOCK WAVE PROPAGATION (see Shock tubes)

SHOCK WAVES (see Aerodynamics; Compressible flow; Magneto hydrodynamics; Shadowgraph technique; Waves in fluids)

SHOCK WAVES, CONICAL (see Inviscid flow)

SHORT TIME LAPSE PHOTOGRAPHY (see Photographic techniques)

SHORT-TUBE VERTICAL EVAPORATOR (see Evaporators)

SHOT TOWERS (see Prilling)

SHRINKING CORE MODEL (see Ion exchange)

SIDERITES (see Iron)

SIEVE, TRAY COLUMN (see Distillation; Direct contact heat exchangers)

SILICA GEL

Silica gel is a non-toxic, non-combustible material consisting of amorphous silica (SiO_2). It is commonly used as a drying agent because of its ability to absorb moisture.

V. Vesovic

SILICON

Silicon—(L. *silex, silicis,* flint), Si; atomic weight 28.0855 ± 0.0003; atomic number 14; melting point 1410°C; boiling point 2355°C; specific gravity 2.33 (25°C); valence 4. Davy in 1800 thought silica to be a compound and not an element; later in 1811 Gay Lussac and Thenard probably prepared impure amorphous silicon by heating potassium with silicon tetrafluoride. Berzelius, generally credited with the discovery, in 1824 succeeded in preparing amorphous silicon by the same general method as used earlier, but he purified the product by removing the fluosilicates by repeated washings. Deville in 1854 first prepared crystalline silicon, the second allotropic form of the element.

Silicon is present in the sun and stars and is a principal component of a class of meteorites known as aerolites. It is also a component of tektites, a natural glass of uncertain origin. Silicon makes up 25.7% of the earth's crust, by weight, and is the second most abundant element, being exceeded only by oxygen. Silicon is not found free in nature, but occurs chiefly as the oxide, and as silicates. Sand, quartz, rock crystal, amethyst, agate, flint, jasper, and opal are some of the forms in which the oxide appears. Granite, hornblende, asbestos, feldspar, clay, mica, etc. are but a few of the numerous silicate minerals.

Silicon is prepared commercially by heating silica and carbon in an electric furnace, using carbon electrodes. Several other methods can be used for preparing the element. Amorphous silicon can be prepared as a brown powder, which can be easily melted or vaporized. Crystalline silicon has a metallic luster and grayish color. The Czochralski process is commonly used to produce single crystals of silicon used for solid-state or semiconductor devices. Hyperpure silicon can be prepared by the thermal decomposition of ultra-pure trichlorosilane in a hydrogen atmosphere, and by a vacuum float zone process. This product can be doped with boron, gallium, phosphorus, or arsenic, etc. to produce silicon for use in transistors, solar cells, rectifiers, and other solid-state devices which are used extensively in the electronics and space-age industries. Hydrogenated amorphous silicon has shown promise in producing economical cells for converting solar energy into electricity. Silicon is a relatively inert element, but it is attacked by halogens and dilute alkali. Most acids, except hydrofluoric, do not affect it.

Silicones are important products of silicon. They may be prepared by hydrolyzing a silicon organic chloride, such as dimethyl silicon chloride. Hydrolysis and condensation of various substituted chlorosilanes can be used to produce a very great number of polymeric products, or silicones, ranging from liquids to hard, glasslike solids with many useful properties. Elemental silicon transmits more than 95% of all wavelengths of infrared, from 1.3 to 6.7 μm.

Silicon is one of man's most useful elements. In the form of sand and clay it is used to make concrete and brick; it is a useful refractory material for high-temperature work, and in the form of silicates it is used in making enamels, pottery, etc. Silica, as sand, is a principal ingredient of glass, one of the most inexpensive of materials with excellent mechanical, optical, thermal, and electrical properties. Glass can be made in a very great variety of shapes, and is used as containers, window glass, insulators, and thousands of other uses. Silicon tetrachloride can be used to iridize glass. Silicon is important in plant and animal life. Diatoms in both fresh and salt water extract silica from the water to build up their cell walls. Silica is present in ashes of plants and in the human skeleton. Silicon is an important ingredient in steel; silicon carbide is one of the most important abrasives and has been used in lasers to produce coherent light of 4560 A.

Miners, stonecutters, and others engaged in work where siliceous dust is breathed in large quantities often develop a serious lung disease known as *silicosis.*

Handbook of Chemistry and Physics, CRC Press

SILICON CARBIDE

Silicon Carbide (SiC) is a crystalline solid material with a molecular weight of 40.10 and specific gravity of 3.22. It is insoluble in water and alcohol. It sublimes and decomposes around 3250K and can only be oxidised at very high temperatures. It possesses high thermal and electrical conductivity and exceptional hardness, and can be fabricated in complex shapes suitable for high temperature industrial applications.

V. Vesovic

SILICON SOLAR CELLS (see Solar cells)

SILOS, GRANULAR FLOW FROM (see Granular materials, discharge through orifices)

SILVER

Silver—(Anglo-Saxon, *Seolfor siolfur*), Ag (L. argentum), atomic weight 107.86815; atomic number 47; melting point 961.93°C; boiling point 2212°C; specific gravity 10.50 (20°C); valence 1, 2.

Silver has been known since ancient times. It is mentioned in Genesis. Slag dumps in Asia Minor and on islands in the Aegean Sea indicate that man learned to separate silver from lead as early as 3000 B.C. Silver occurs native and in ores such as argentite (Ag_2S) and horn silver (AgCl); lead, lead-zinc, copper, gold, and copper-nickel ores are principal sources. Mexico, Canada, Peru, and the U.S. are the principal silver producers in the western hemisphere. Silver is also recovered during electrolytic refining of copper. Commercial fine silver contains at least 99.9% silver. Purities of 99.999 + % are available commercially. Pure silver has a brilliant white metallic luster. It is a little harder than gold and is very ductile and malleable, being exceeded only by gold and perhaps palladium. Pure silver has the highest electrical and thermal conductivity of all metals, and possesses the lowest contact resistance. It is stable in pure air and water, but tarnishes when exposed to ozone, hydrogen sulfide, or air containing sulfur.

The alloys of silver are important. Sterling silver is used for jewelry, silverware, etc. where appearance is paramount. This alloy contains 92.5% silver, the remainder being copper or some other metal. Silver is of utmost importance in photography, about 30% of the U.S. industrial consumption going into this application. It is used for dental alloys. Silver is used in making solder and brazing alloys, electrical contacts, and high capacity silver-zinc and silver-cadmium batteries. Silver paints are used for making printed circuits. It is used in mirror production and may be deposited on glass or metals by chemical deposition, electrodeposition, or by evaporation. When freshly deposited, it is the best reflector of visible light known, but it rapidly tarnishes and loses much of its reflectance. It is a poor reflector of ultraviolet. Silver fulminate ($Ag_2C_2N_2O_2$), a powerful explosive, is sometimes formed during the silvering process. Silver iodide is used in seeding clouds to produce rain. Silver chloride has interesting optical properties as it can be made transparent; it also is a cement for glass. Silver nitrate, or lunar caustic, the most important silver compound, is used extensively in photography.

While silver itself is not considered to be toxic, most of its salts are poisonous due to the anions present. Exposure to silver (metal and soluble compounds, as Ag) in air should not exceed 0.01 mg/m^3, (8-hr time-weighted average—40-hr week). Silver compounds can be absorbed in the circulatory system and reduced silver deposited in the various tissues of the body. A condition, known as argyria, results, with a greyish pigmentation of the skin and mucous membranes. Silver has germicidal effects and kills many lower organisms effectively without harm to higher animals.

Handbook of Chemistry and Physics, CRC Press

SILVER METHOD (see Condensation of multicomponent vapours)

SIMILARITY CONDITIONS (see Dimensional analysis)

SIMILARITY, THEORY OF (see Dimensional analysis)

SIMILITUDE (see Inviscid flow)

SIMPLEX ATOMIZER (see Atomization)

SIMPLIFIED BOILING WATER REACTOR, SBWR (see Boiling water reactor, BWR)

SINCLAIR-LA MER AEROSOL GENERATOR (see Aerosols)

SINGLET STATE (see Photoluminescence)

SINGLET STATE LIFETIME (see Fluorescence)

SINGULARITIES, HYDRAULIC RESISTANCE IN (see Hydraulic resistance)

SINTERING (see Roasting)

SINUOUS JETS (see Atomization)

SIPHON CENTRIFUGE (see Centrifuges)

S I UNITS

Introduction

When expressing the magnitude of a physical quantity, we use a number followed by a unit, e.g. M = 6 kg. The number represents the ratio of the magnitude of the quantity (mass M) to that of the unit (kilogram, kg). Over past centuries several different systems of units have been used by engineers and scientists, necessitating a large number of conversion factors and often leading to confusion.

In 1948, following a resolution of the Ninth Conference of Weights and Measures (CGPM) an international committee was established to formulate a new international system of units. Système Internationale d'Unités, or SI Units, was the outcome.

For a complete detailed account of S I units and their proper usage the chapter in the Heat Exchanger Design Handbook entitled "Conventions and nomenclature for physical quantities, units numbers and mathematics" by Y. R. Mayhew is recommended.

In the S I system there are seven base units, from which others are derived by combination, and two supplementary units that are angles.

Base Units in S I

These are:

1. metre, m, the standard of length
2. kilogram, kg, the standard of mass
3. second, s, the standard of time
4. ampere, A, the standard of electric current
5. kelvin, K, the standard of temperature
6. candela, cd, the standard of luminar intensity
7. mole, mol, the standard of amount of substance

The **metre**, m, (from the Latin metrum—measure) was introduced in France at the time of the revolution. In an era of idealism it was designed to relate neatly to the size of the earth, designated as one part in forty million of the earth's circumference. In other words, the shortest distance from equator to pole was to be exactly 10^7m. This proved an inexact measure, so for a while an actual bar of platinum-iridium at Sèvres became the standard metre. Today with the need for greater precision, the metre is defined as 1,650,763.73 times the wavelength, in vacuum, of the orange light emitted by $^{86}_{36}Kr$ in the transition 2 p_{10} to 5 d_5. Alternatively a metre is the distance travelled by light in vacuum in 1/299,792,458 of a second.

The **kilogram**, kg, was originally defined as the mass of a litre (10^{-3} m^3) of pure water at its maximum density. A platinum-iridium cylinder at Sèvres, intended to have exactly this mass, was made oversize by 28 parts per million, so for a while prior to 1964, the litre was defined at 1.000 028 × 10^{-3} m^3 to avoid this anomaly. The Sèvres cylinder is still the standard kilogram.

The **second**, s, is the time taken for 9,192,631,770 cycles of the radiation from the hyperfine transition in Caesium 133 when unperturbed by external fields. Alternatively, the **ephemeris second** is defined as 1/31,556,925.974 7 of the tropical year for 1900.

The **ampere**, A, is that constant current which, if maintained in each of two infinitely long straight parallel wires of negligible cross-section, placed one metre apart in vacuum, will produce between the wires a force of 2 × 10^{-7} newtons per metre length.

The **kelvin**, K, unit of thermodynamic temperature is the fraction 1/273.16 of the thermodynamic temperature of the triple point of water.

The **candela**, cd, is the luminous intensity, in the perpendicular direction, of a surface 1/600,000 square metre of a full radiator at the temperature of freezing platinum under a pressure of 101,325 newtons per square metre. (Alternatively, a candela is the luminous intensity of a source that emits, in a given direction, monochromatic radiation of frequency 540 × 10^{12} hertz and has a radiant intensity in the same direction of 1/683 watt per steradian).

The **mole**, mol, is the amount of substance which contains as many elementary entities as there are atoms in 0.012 kilogram of the carbon isotope ${}^{12}_{6}C$. The entities must be specified, as atoms, molecules, ions, electrons or other particles or groups of particles.

Supplementary Units in S I

Supplementary units in S I are the radian and the steradian. The **radian**, rad, is the plane angle between two radii of a circle which cut off in the circumference an arc equal in length to the radius.

The **steradian**, sr, is the solid angle which, having its vertex in the centre of a sphere, cuts off an area of the surface of the sphere equal to that of a square with sides of length equal to the radius of the sphere.

Derived S I Units

There are several units obtained by combining S I base units which have special names, e.g. watt for power which is equivalent to kg m^2/s^2 or Nm/s or J/s. These named units are listed in **Table 1** and defined below. There are also many more units derived from S I base units which do not have special names (that is to say the unit does not have a special name although of course the quantity does). For example the unit of acceleration is m/s^2.

Derived units of force and energy (involving only kg, m and s)

The **newton**, N, is the derived unit of force. One newton is the force that produces an acceleration of one metre per second per second when applied to a mass of one kilogram, i.e. N = kg m/s^2.

(A mass of 1 kg exerts a force of 9.80 newtons in New York, 9.81 newtons in London, and 9.83 newtons at the North pole).

The **pascal**, Pa, is the derived unit of pressure. It is the pressure exerted by one newton acting uniformly over one square metre, i.e. Pa = N/m^2.

Table 1. SI derived units with special names

Quantity	Name	Symbol	Expression in terms of SI base or other SI units
Frequency	hertz	Hz	1/s
Force; weight	newton	N	kg m/s^2
Pressure; stress	pascal	Pa	kg/m s^2 = N/m^2
Energy, work, quantity of heat	joule	J	kg m^2/s^2 = N m
Power; radiant flux	watt	W	kg m^2/s^3 = N m/s = J/s
Electric charge, quantity of electricity	coulomb	C	A s
Electrical potential, potential difference,	volt	V	kg m^2/A s^3 = W/A
Electric resistance	ohm	Ω	kg m^2/A^2 s^3 = V/A = 1/S
Electric conductance	siemens	S	A^2 s^3/kg m^2 = A/V = 1/Ω
Inductance	henry	H	kg m^2/A^2 s^2 = Wb/A
Capacitance	farad	F	A^2 s^4/kg m^2 = C/V
Magnetic flux	weber	Wb	kg m^2/A s^2 = V s
Magnetic flux density	tesla	T	kg/A s^2 = Wb/m^2
Luminous flux	lumen	lm	cd sr
Illuminance	lux	lx	cd sr/m^2

The **joule**, J, is the derived unit of energy or work. One joule is the work done by a force of one newton exerted over a distance of one metre, i.e. J = N m = kg m^2/s^2.

The **watt**, W, is the derived unit of power or energy per unit time. One watt is one joule per second, i.e. W = J/s = N m/s = kg m^2/s^3.

Derived S I units involving electricity and magnetism

The only additional S I base unit involved in both electrical and magnetic quantities is the **ampere**, A. This is because of the fundamental relationship between electricity and magnetism.

The **coulomb**, C, is the derived unit of electric charge. It is the quantity of electricity transported per second by a current of one ampere, i.e. C = A s.

The **volt**, V, is the derived unit of electrical potential, or potential difference. It is that difference in potential which generates one watt of energy per ampere of current, i.e. V = W/A = kg m^2/A s^3.

The **ohm**, Ω, is the derived unit of electrical resistance. It is the resistance between two points that produces a potential difference of one volt when a current of one ampere flows between them, i.e. Ω = V/A = kg m^2/A^2 s^3.

The **siemens**, S, is the derived unit of electrical conductance. It is the conductance between two points that produces a potential difference of one volt when a current of one ampere flows between them, i.e. S = 1/R = A/V = A^2 s^3/kg m^2.

The **henry**, H, is the derived unit of inductance. It is the inductance of a closed circuit in which a potential difference of one volt is produced when the current in the circuit varies uniformly at the rate of one ampere per second, i.e. H = V s/A = kg m^2/A^2 s^2.

The **farad**, F, is the derived unit of capacitance. It is the capacitance of two plates which hold a charge of one coulomb when there is a potential difference between them of one volt, i.e. F = C/V = $A^2\ s^4/kg\ m^2$.

The **weber**, Wb, is the derived unit of magnetic flux. Wb = HA = V s = kg $m^2/A\ s^2$.

The **tesla**, T, is the derived unit of magnetic flux density. T = Wb/m^2 = kg/A s^2.

Equivalence of mechanical and electrical units of power

The S I base units are so defined that power, in the derived unit watt, can be obtained from force × displacement/time or ampere × volt, i.e.

$$1 \text{ watt} = 1 \text{ newton} \times \text{metre/second} = 1 \text{ Nm/s}$$

and

$$1 \text{ watt} = 1 \text{ ampere} \times \text{volt} = 1\text{AV}$$

Derived S I optical units

These are combinations of the S I unit of luminous intensity, the candela (cd), with other S I units.

The **lumen**, lm, is the light energy emitted per second within unit solid angle by a point source of unit luminous intensity, i.e. lm = cd sr

The **lux,** lx, is the unit of illuminance defined as the flux reaching the surface per unit area, i.e. lx = cd sr/m^2.

Fractions and Multiples of S I Units

S I units can be multiplied or divided by powers of ten by attaching the appropriate prefix, as listed in **Table 2**, to the name of the unit and putting the corresponding letter before the standard symbol. For example a force may be expressed as millions of newtons, or **meganewtons**, the symbol for which is MN, e.g.

$$5.60 \times 10^6\text{N} = 5.60 \text{ MN}$$

The prefix symbol may also be used with derived units, either named or un-named, e.g.

$$7.31 \times 10^3 \text{ Pa} = 7.31 \text{ kPa}$$

or

$$4.24 \times 10^{-3} \text{ m/s}^2 = 4.24 \text{ mm/s}^2$$

Some units which are decimally related to S I units have specific names. These are listed in **Table 3** on page 1014. Other units are directly related to S I units, but not decimally; these are listed in **Table 4** on page 1014.

Table 2. Decimal multiples or fractions of SI units

Factor	Prefix	Symbol
10^{-18}	atto	a
10^{-15}	femto	f
10^{-12}	pico	p
10^{-9}	nano	n
10^{-6}	micro	μ
10^{-3}	milli	m
10^{-2}	centi	c
10^{-1}	deci	d
10^{1}	deca	da
10^{2}	hecto	h
10^{3}	kilo	k
10^{6}	mega	M
10^{9}	giga	G
10^{12}	tera	T
10^{15}	peta	P
10^{18}	exa	E

Fundamental Constants in S I Units

Some of the fundamental constants are listed in **Table 5** on page 1015. Because S I is a consistent system of units the relationships between fundamental constants apply without conversion factors, e.g.

$$k = R/N_A = (8.3145 \text{ J/mol K})/(6.02214 \times 10^{23} \text{ mol}^{-1})$$

$$= 1.3806 \times 10^{-23} \text{ J/K}$$

Also

$$c = (\mu_0\epsilon_0)^{-1/2} = (4\pi \times 10^{-7} \text{ H/m} \times 8.854 \times 10^{-12} \text{ F/m})^{-1/2}$$

$$c = 2.998 \times 10^8 \text{ m/s, since H} \times \text{F} = \text{S}^{-2}.$$

Reference

Mayhew Y. R. (1989) *Conventions and nomenclature for physical quantities, units, numbers and mathematics,* Heat Exhangers Design Handbook, Hemisphere Publishing Corp.

Leading to: *International temperature scale*

G.L. Shires

SKIMMER PIPE AND KNIFE CENTRIFUGES (see Centrifuges)

SKIN EFFECT (see Induction heating)

SKIN FRICTION (see Crossflow)

SLAG FORMATION (see Pulverised coal furnaces)

Table 3. Units having special names decimally related to SI units

Quantity	Name	Symbol	Definition
Length	ångström	Å	$= 10^{-10}$ m $= 10^{-1}$ nm
Area of farmland and building land	are	a	$= 10^2$ m^2
	hectare	ha	$= 10^4$ m^2
Volume[a]	litre	l or L	$= 10^{-3}$ m^3 = 1 dm^3
Mass	tonne	t	$= 10^3$ kg = 1 Mg
Mass of precious stones	metric carat	—	$= 2 \times 10^{-4}$ kg = 0.2 g
Mass per unit length of textile yarns and threads	tex	tex	$= 10^{-6}$ kg/m
Pressure, stress	bar	bar	$= 10^5$ Pa = 10^5 N/m^2
Kinematic viscosity	stokes	St	$= 10^{-4}$ m^2/s
Dynamic viscosity	poise	P	$= 10^{-1}$ kg/m s = 10^{-1} N s/m^2 = 10^{-1} Pa s
Force	dyne	dyn	$= 10^{-5}$ N
Energy	erg	erg	$= 10^{-7}$ J
Vergence of optical systems	diopter	—	= 1 m^{-1}

From Heat Exchanger Design Handbook, Hemisphere Publishing Corporation.

Table 4. Units having special names not decimally related to SI units

Quantity	Name	Symbol	Definition
Time	minute	min	= 60 s
	hour	h	= 3600 s = 60 min
	day	d	= 86 400 s = 24 h
Angle	degree	°	= (π/180) rad [≈ (1/57.30) rad]
	minute	′	= (π/10 800) rad = (1/60)°
	second	″	= (π/648 000) rad = (1/60)′
	gon	gon	= (π/200) rad = 0.9°
Length	nautical mile		= 1852 m
Velocity	knot		= (1852/3600) m/s = 1 nautical mile per hour
Pressure	standard atmosphere	atm	= 101 325 Pa = 101 325 N/m^2
	conventional mm of mercury	mmHg	= 13.5951 × 9.806 65 Pa [≈ (1/0.007 501) Pa ≈ (1/750.1) bar]
Energy	kilowatt-hour	kW h	= 3.6 × 10^6 J
	IT (International Table) calorie	cal_{IT}	= 4.186 8 J
	thermochemical calorie	cal_{th}	= 4.184 J

From Heat Exchanger Design Handbook, Hemisphere Publishing Corporation.

SLIP RATIO

***Following from:** Gas-liquid flow*

Slip ratio (also called *velocity ratio*) is a term used in **gas liquid flow** to denote the ratio of the velocities of the gas and liquid phases respectively. Thus:

$$S = \frac{u_G}{u_L} = \frac{U_G(1-\varepsilon_G)}{U_L\varepsilon_G} = \frac{\rho_L x(1-\varepsilon_G)}{\rho_G(1-x)\varepsilon_G} \quad \textbf{(1)}$$

where U_G and U_L are the respective phase *superficial velocities*, ε_G is the void fraction (fraction of the cross-section occupied by the gas phase), ρ_L and ρ_G the liquid and gas densities, and x the quality (fraction of the total mass flow which is gas). Equation 1 can be rearranged to give a relationship between void fraction and slip ratio as follows:

$$\varepsilon_G = \frac{U_G}{SU_L + U_G} = \frac{\rho_L x}{S\rho_G(1-x) + \rho_L x} \quad \textbf{(2)}$$

Correlations for slip ratio are given in the article on **Void Fraction**.

G.F. Hewitt

SLIT FLOW METERS (see Flow metering)

SLOW MOTION PHOTOGRAPHY (see Photographic techniques)

Table 5. Fundamental constants in SI units

Name	Symbol	Value
Acceleration due to gravity at earth surface	g	Standard value 9.80665 m/s^2
Avogadro (Loschmidt) constant	$N_A(L)$	6.02214×10^{23} mol^{-1}
Boltzmann constant	k	1.3806×10^{-23} J/K
Electron volt	eV	1.60219×10^{-19} J
Faraday constant	F	9.649×10^4 C/mol
Molar (universal) gas constant	$\tilde{R}$	8.3145 J/mol K
Permeability of free space	μ_0	$4\pi \times 10^{-7}$ H/m
Permittivity of free space	ϵ_0	8.854×10^{-12} F/m
Planck constant	h	6.626×10^{-34} J/K
Speed of light	c	2.998×10^8 m/s (in vacuum)
Stefan-Boltzmann constant	σ	5.670×10^{-8} W/m^2 K

SLUG FLOW

Following from: *Gas-liquid flow*

The primary characteristic of slug flow is its inherent intermittence. An observer looking at a fixed position along the axis would see the passage of a sequence of slugs of liquid containing dispersed bubbles, each looking somewhat like a length of bubbly pipe flow, alternating with sections of separated flow within long bubbles (**Figure 1**). The flow is unsteady, even when the flow rates of gas and liquid, Q_G and Q_L, are kept constant at the pipe inlet.

The elementary part of slug flow is a cell, involving the region of a long bubble plus the region of the following liquid slug. The slug of liquid of length L_D carrying dispersed bubbles, travels at a velocity V. It overruns a slower moving liquid in the separated film. During stable slug flow liquid is shed from the back of the slug at the same rate that liquid is picked up at the front. As a result the slug length stays constant as it travels along the tube.

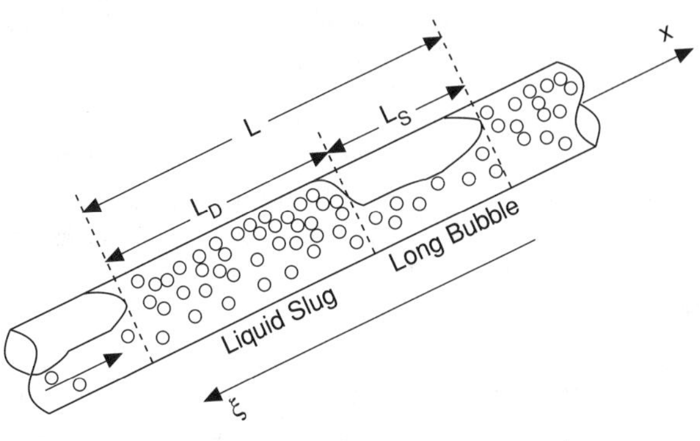

Figure 1. Gas-liquid slug flow.

For horizontal or near horizontal tubes, the liquid shed at the back decelerates under the influence of wall shear and forms a stratified layer. For vertical or near vertical tubes the liquid forms a falling annular film and accelerates as it moves downward. The separated section containing a large gas bubble has a length, L_S. Most of the gas is transported in these large gas bubbles.

As this kind of flow occurs over a wide range of intermediate flow rates of gas and liquid, it is interesting for many industrial processes such as the production of oil and gas, the geothermal production of steam, the boiling and condensation processes, the handling and transport of cryogenic fluids and the emergency cooling of nuclear reactors. The existence of slug flow can create problems for the designer or operator. The high momentum of the liquid slugs can create considerable force as they change direction passing through elbows, tees, etc. Furthermore, the low frequencies of slug flow can be in resonance with large piping structures and severe damage can take place. In addition, the intermittent nature of the flow makes it necessary to design liquid separators to accommodate the largest slug length. In contrast, there are numerous practical benefits which can result from operating in the slug flow pattern. Because of the very high liquid velocities, it is usually possible to move larger amounts of liquids in smaller lines than would otherwise be possible in two phase flow. In addition, the high liquid velocities cause very high convective heat and mass transfers resulting in very efficient transport operations.

The entry is organised as follows. In the first part we recall briefly the basic modelling concept. In the second part some of the closure laws quoted in the literature will be evaluated.

The Concept of Unit Cell

Wallis was probably the first to formulate clearly the concept of *unit cell* (UC) suggested by Nicklin *et al.* In the past twenty years the model has been perfected by several investigators so that it appears now as the ultimate 1D approach. The concept reposes on the two following assumptions. Firstly, there exists a frame of a given velocity V in which the flow is almost steady. Secondly, in this frame the flow of long bubbles and liquid slugs is fully developed. A review of the scientific literature reveals an abundance of these laws with varying physical validity (see for example the overviews of Taitel and Barnea, 1990; Fabre and Liné, 1992). The weakness of these laws often originates from the narrow range of flow conditions used in their calibration. Their critical role has been discussed by Dukler and Fabre, 1992.

Suppose that we know the specific flow conditions: the pipe size (i.e. its diameter D or its cross-section area $A = \pi D^2/4$ and its inclination θ), the fluid properties (i.e. the density ρ_k (k = L,G)), the kinematic viscosity ν_k, the surface tension σ and the volumetric flux of each phase, $j_k = Q_k/A$.

A complete model of slug flow should produce at least the following information: the characteristic lengths, L_D, L_S, and the mean bubble size in the liquid slug, the form of the liquid film (stratified, annular), the characteristic velocities, V, and the mean velocities of both gas and liquid in each part of the cell, the cross-sectional phase fraction in each part of the cell, the mean wall and interfacial shear stresses in each part of the cell and the pressure drop.

Assumptions

The main difficulty in modelling slug flow comes from its chaotic nature. This is suggested by the observation of the succession of bubbles and slugs whose length appears randomly distributed with time (**Figure 2**). To avoid having to account for the flow randomness, a few assumptions are needed. The initial assumption was to picture the flow as a sequence of cells periodic with both space and time: the UC concept was born. However, two weaker assumptions lead to the same model.

The first assumption comes from experimental evidence which is illustrated in **Figure 3**. The probability density distribution of bubble and slug velocities shows that they are narrowly distributed about their average—in other words they are almost identical. Although this property becomes less evident at high gas or liquid flow rate, this quasi-steady behaviour in a moving frame is the key of the success of the UC model. Indeed this property leads to a great simplification since it allows to transform an unsteady problem into a steady one.

The second assumption consists in assuming that the flow is fully developed in each part of the cell. As a consequence, the crossectional mean fraction and velocity of each phase do not depend on the longitudinal coordinate inside the long bubbles and the liquid slugs. This assumption is probably stronger than the previous one.

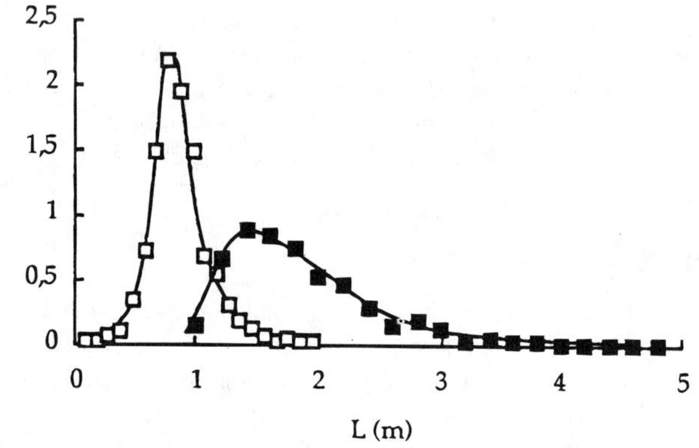

Figure 2. Probability density distribution of ■ bubble and □ slug lengths. Slug flow in pipe of 5 cm diameter at superficial velocities $j_G = 1.25$ m/s and $j_L = 0.97$ m/s (Fabre et al., 1993).

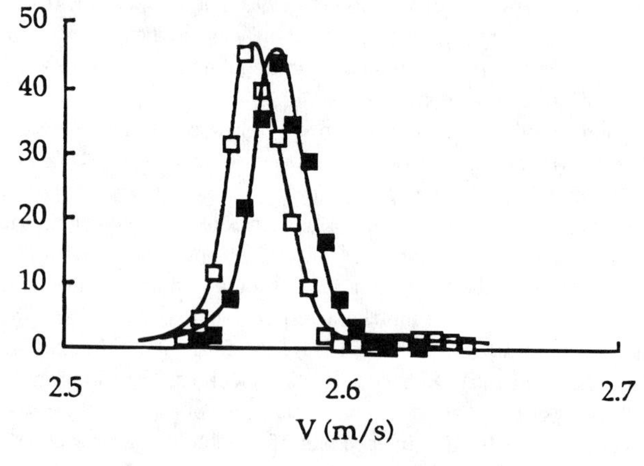

Figure 3. Probability density distribution of ■ bubble and □ slug velocities. Slug flow in pipe of 5 cm diameter at superficial velocities $j_G = 1.25$ m/s and $j_L = 0.97$ m/s (Fabre et al., 1993).

Preliminary definitions

Considering steady inlet flow conditions, the mass flow rates of each phase can be specified as:

$$\dot{m}_k = \rho_k j_k A \tag{1}$$

In order to simplify the discussion the density of both phases will be considered constant so that the "steadiness" $d\dot{m}_k/dt = 0$ is equivalent to $dj_k/dt = 0$.

The phase-k is distributed over two elementary regions pictured in **Figure 1**. The rate of occurrence of the large bubbles β is thus defined by:

$$\beta = \frac{\overline{L_S}}{\overline{L_S + L_D}}, \tag{2}$$

where the overbar denotes a time average. Because the flow is supposed steady in some frame of reference, this average is likewise interpreted as a space average. β will be further referred to as the *rate of intermittence.*

Let ε_k be the fraction of each phase existing over any pipe cross-section: as gas and liquid fill the section, $\varepsilon_G + \varepsilon_L = 1$. Introducing the definitions of the time—or space—average of phase fractions over the cell, the separated region and the dispersed region

$$R_k = \overline{\varepsilon_k}, \qquad R_{kS} = \overline{\varepsilon_{kS}}, \qquad R_{kD} = \overline{\varepsilon_{kD}} \tag{3}$$

leads to the volumetric relations:

$$\sum_{k=L,G} R_k = 1, \qquad \sum_{k=L,G} R_{kS} = 1, \qquad \sum_{k=L,G} R_{kD} = 1 \tag{4}$$

The mean phase fraction may be expressed versus the phase fractions in each part of the cell through:

$$R_k = \beta R_{kS} + (1 - \beta) R_{kD} \tag{5}$$

Balance equations

In the frame moving at the velocity V of the cells, the equations of mass and momentum take a simplified form. The coordinate of the moving frame is defined as:

$$\xi = Vt - x, \tag{6}$$

The velocity u_k of phase-k averaged over the pipe cross-section is thus transformed by the change of frame as $V - u_k$. Due to steadiness of the flow in the moving frame, the continuity equation is expressed by:

$$\frac{d}{d\xi}[\varepsilon_k(V - u_k)] = 0 \Rightarrow \varphi_k = \epsilon_k(V - u_k) \tag{7}$$

φ_k represents the volumetric flux of phase-k entering the long bubble region, and shed from the liquid slug. The time—or space—

average of **Equation 7** over the cell, the separated region and the dispersed region introduces the following definitions of the mean velocities U_k, U_{kS}, U_{kD} of phase-k:

$$\varphi_k = R_{kS}(V - U_{kS}) = R_{kD}(V - U_{kD}) = R_k(V - U_k) = R_kV - j_k \tag{8}$$

Equation 8 is the mass balance relating the flux of phase k entering the long bubble to that entering the liquid slug. In combining **Equations 5 and 8** a relation similar to **Equation 5** is easily deduced for the mean velocity of phase-k:

$$R_kU_k = \beta R_{kS}U_{kS} + (1 - \beta)R_{kD}U_{kD} \tag{9}$$

The crossectional average of the momentum equation for phase-k is:

$$\frac{d}{d\xi}[\rho_k\varepsilon_k(V - u_k)^2 + \varepsilon_k p_k] = -\frac{t_{kw}S_{kw} + \tau_{ki}S_i}{A} + \rho_k\varepsilon_k g \sin\theta, \tag{10}$$

where p is the mean pressure over the cross-section area of phase-k, τ is the x-component of the stress exerted upon the phase-k by the wall (subscript w) or the interface (subscript i), S is the wetted perimeter and g the gravity. The above equation simplifies in using **Equation 7**:

$$\rho_k\varphi_k^2\frac{d\varepsilon_k^{-1}}{d\xi} + \frac{d\varepsilon_k p_k}{d\xi} = -\frac{\tau_{kw}S_{kw} + \tau_{ki}S_i}{A} + \rho_k\varepsilon_k g \sin\theta, \tag{11}$$

with the jump condition $\sum_{k=L,G} \tau_{ki} = 0$.

In contrast to the continuity equations which can be simplified by an integration over the different parts of the cell, the momentum equations cannot, unless we have

$$\frac{d\varepsilon_k}{d\xi} = 0 \tag{12}$$

According to the second assumption, the above equation holds in each part of the cell, so that **Equation 11** simplifies and may be averaged over each part of the cell. In the standing frame it yields:

$$R_{kS}\frac{dP_S}{dx} = \frac{\tau_{kwS}S_{kwS} + \tau_{kiS}S_{iS}}{A} - \rho_k R_{kS} g \sin\theta \tag{13}$$

$$R_{kD}\frac{dP_D}{dx} = \frac{\tau_{kwD}S_{kwD} + \tau_{kiD}S_{iD}}{A} - \rho_k R_{kD} g \sin\theta \tag{14}$$

Since the flow is fully developed, the pressure gradient is the same in both phases and must not be distinguished. The mean pressure gradient over the cell results from the mean pressure gradient over each part of the cell weighted by their rate of occurrence:

$$\frac{dP}{dx} = \frac{\beta(\tau_{LwS}S_{LwS} + \tau_{GwS}S_{GwS}) + (1 - \beta)(\tau_{LwD}S_{LwD})}{A} - (\rho_L R_L + \rho_G R_G) g \sin\theta \tag{15}$$

The pressure gradient involves two contributions: the weight of the phases and the wall friction.

Closure problem

It is worth noting that the fully developed flow assumption makes the equation independent of the cell length. Only the intermittence factor β appears. On the other hand, the pressure gradient appears only in **Equation 15**. Therefore, once the phase fractions and the velocities are determined, the wall friction and the weight of the phases may be calculated. This remark suggests that the problem can be split into two steps. In a first step the phase fractions, R_L, R_{LS}, R_{LD} have to be determined. The pressure gradient will be determined in a second step.

In order to solve the closure problem for the determination of the phase fractions, the seven independent algebraic equations have been grouped in **Table 1**. They are nonlinear and present a deficit of four equations with respect to the eleven unknown quantities: R_L, R_{LS}, R_{LD}, U_L, U_{LS}, U_{LD}, U_G, U_{GS}, U_{GD}, β, V. The role of the four closure equations is to restore the missing information. We shall limit the discussion to the most classical method which requires equations for: the velocity V of the large bubble, the void fraction in the large bubble R_{GS}, the void fraction in the liquid slug R_{GD} and the drift velocity in liquid slug $U_{GD} - U_{LD}$.

Whereas the phase fractions are not coupled to the pressure gradient, the pressure gradient does depend on the phase distribution as shown by **Equation 15**. Even for horizontal flow in which the weight vanishes, they still have a great influence on the pressure gradient through the intermittence factor β. For the pressure gradient to be calculated, two other closure laws are needed for the shear stress at the wall in the film τ_{kwS} and in the slug τ_{LwD}.

Different models have been published in the scientific literature. What makes the difference is the choice of the closure laws.

Closure Laws

Dynamics of long bubbles

As most of the gas is conveyed by the large bubbles the accurate prediction of their motion and their shape is essential. It is possible

Table 1. Equations for phase fractions

$$R_L = \beta R_{LS} + (1 - \beta) R_{LD}$$
$$R_LU_L = \beta R_{LS}U_{LS} + (1 - \beta) R_{LD}U_{LD}$$
$$(1 - R_L)U_G = \beta(1 - R_{LS}) U_{GS} + (1 - \beta)(1 - R_{LD})U_{GD}$$
$$R_{LS}(V - U_{LS}) = R_{LD}(V - U_{LD})$$
$$(1 - R_{LS})(V - U_{GS}) = (1 - R_{LD})(V - U_{GD})$$
$$j_L = R_LU_L$$
$$j_G = (1 - R_L)U_G$$

to get a crude estimate of the void fraction by assuming that the gas is conveyed at velocity V:

$$R_G \cong \frac{j_G}{V} \tag{16}$$

This relation does a fairly good job in some simplified cases. This shows that the phase fractions are primarily sensitive to the long bubble velocity.

Our present knowledge of the motion of long bubbles comes from both theory and a considerable amount of data (see the review of Dukler and Fabre, 1992). Cylindrical bubbles rising in vertical tubes have the shape of a prolate spheroid independent of their length. The shape at the rear depends on whether or not the viscous force is negligible. When negligible, the bubble has a flat back indicating that flow separation and vortex shedding occur. In upward liquid slug flow, the nose may be distorted by the turbulence generated in the wake of the preceding bubble. In downward liquid flow, the structure of the free surface is more complex. The bubble migrates with an asymmetrical shape. Moreover, above some critical liquid flow rate, the bubble is distorted alternately on one side of the tube and then the other.

The shape of the bubble depends upon the pipe inclination. Indeed the experiments in still liquid (Zukoski, 1996) show clearly that the eccentricity increases when the pipe is deviated from the vertical position. When the inclination decreases from 90° to 0°, the cross-sectional area of the film far from the nose departs from a centred annulus to a segment of circle indicating that stratified flow is reached in the liquid film. According to Spedding and Nguyen, 1978, this change in shape occurs between 30° and 40°. When the flow in the film is stratified, the tail has the appearance of a hydraulic jump.

Measured bubble velocities as a function of mixture velocity are shown in **Figure 4** for vertical flow, and in **Figure 5** for horizontal flow. At high velocity the data are much more scattered. For a more extensive analysis, see the review of Fabre and Liné.

The V(j) relation is linear over certain ranges of mixture velocity $j = j_G + j_L$ thus supporting the assumption of Nicklin et al. for single bubble motion. The velocity is thus given by:

$$V = C_0 j + C_\infty \sqrt{gD} \tag{17}$$

where C_0 and C_∞ are two coefficients which remains constant for

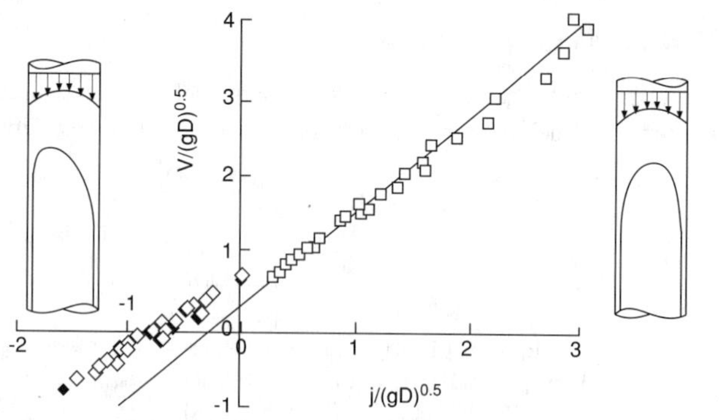

Figure 4. Velocity of long bubbles vs. mixture velocity, $\theta = 90°$. D = 50 mm □ (Fréchou, 1986); 140 mm ◆, 100 mm ◇, 26 mm▲, (Martin) ______: Equation 17 with $C_0 = 1.2$, $C_\infty = 0.35$

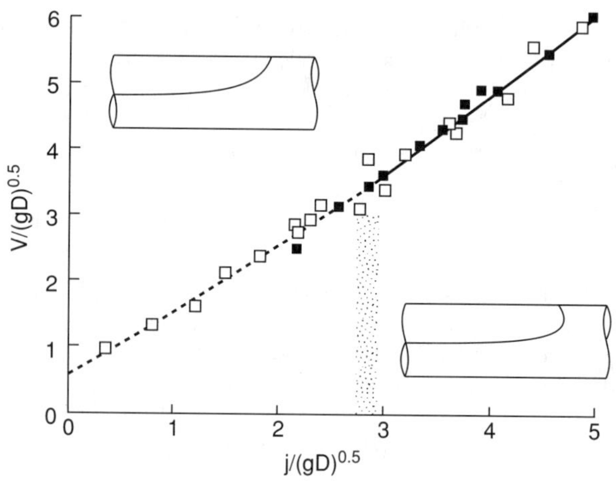

Figure 5. Velocity of long bubbles vs. mixture velocity, $\theta = 0°$. D = 146 mm□ (Ferschneider, 1982), 189 mm ■ (Linga); ______: Equation 17 with $C_0 = 1.2$, $C_\infty = 0$; _ _ _: Equation 17 with $C_0 = 1$, $C_\infty = 0.54$

some range of mixture velocity and fluid properties. This relationship has the peculiarity of separating two physical effects: the mean flow transportation contained in the first term of the r.h.s. and the driving force included in the second term of the r.h.s.

However, secondary effects due to viscosity, surface tension and pipe inclination complicate this law. Indeed the coefficients of **Equation 17** take the form:

$$C_\infty = f_\infty(Fr, Eo, \theta), \text{ and } C_0 = f_0(Re, Fr, Eo, \theta) \tag{18}$$

where the **Reynolds, Froude** and *Eötvös*—or **Bond—Numbers** are defined by:

$$Re = \frac{jD}{\nu_L}, \quad Fr = \frac{j}{\sqrt{gD}}, \quad Eo = \frac{\rho_L g D^2}{\sigma} \tag{19}$$

We know very little about the theoretical expression of the coefficients, except the very nice proofs that

$$f_\infty(\infty, \infty, 90°) = 0.35, \quad f_\infty(\infty, \infty, 0°) = 0.54 \tag{20}$$

which were given by Dumitrescu, 1943, and Benjamin, 1968, respectively. However, we have a substantial amount of experimental data which illuminates some unexpected features of the bubble motion. These features may be summarised as follows:

- C_0, C_∞ vary monotonously with pipe inclination and surface tension through the Eötvös number and, to a lesser extent, with viscosity through the Reynolds number: these points will not be discussed here.
- C_0, C_∞ may change drastically for some critical values of the dimensionless parameters: in other words the coefficients experience some transitions which deserve to be discussed.

The first transition appears in vertical flow. It is clearly visible in **Figure 4** in the vicinity of $j = 0$. For upflow ($j > 0$), Equation 17 fits nicely the experimental data with the recommended values $C_0 = 1.2$ and $C_\infty = 0.35$, whereas for downflow ($j < 0$), one must take $C_0 = 1$ and $C_\infty = 0.7$. Martin observed that the bubble nose experiences a shape transition near $j = 0$, from centred nose in upflow to unstable and non symmetrical nose for downflow. It is likely that the transition occurs when the destabilizing inertia force balances with surface tension force, the transition arising at some critical **Weber Number**.

The second transition has for a long time been a matter of controversy. It is shown in **Figure 5** for horizontal flow. At high enough mixture velocity the long bubble velocity only depends on the mean flow with $C_0 \approx 1.2$–1.3 and $C_\infty = 0$. In contrast, at low enough velocity $C_0 = 1$ and $C_\infty = 0.54$ in agreement with theoretical prediction indicating that the bubble should experience a drift in a horizontal pipe containing a still liquid. In fact, this phenomenon may be put in evidence experimentally only in large pipe.

The third transition was put in evidence in vertical flow, although it may also occur in horizontal flow. It can be proven theoretically that the bubble moves faster when the liquid flow is laminar upstream of the bubble nose than when it is turbulent: $C_0 = 2.27$ in laminar flow, $C_0 = 1.2$ for turbulent flow. This unexpected result suggested to Nicklin et al., 1962, that "the bubble velocity is very nearly the sum of the velocity on the center line above the bubble plus the characteristic velocity in still liquid". The result was confirmed from experiments carried out in a wide range of Reynolds number by Fréchou, 1986 (**Figure 6**).

Liquid holdup in long bubbles

The method generally used to determine the holdup in large bubble starts from the assumption that the separated flow region between the nose and the tail is fully developed. The liquid holdup may be known by eliminating the pressure gradient between **Equations 13 and 14**.

$$\frac{\tau_{GwS}S_{GwS}}{R_{GS}} + \frac{\tau_{GiS}S_{iS}}{R_{GS}R_{LS}} - \frac{\tau_{LwS}S_{LwS}}{R_{LS}} + A\Delta\rho g \sin\theta = 0. \quad \textbf{(21)}$$

In the foregoing equation the shear stresses at both wall and interface are expressed by single phase relationships, in which the friction factors have to be closed following the method indicated in the entry **Stratified Flow**. Solving **Equation 21** addresses an important issue: the pattern of the interface within the bubble must be known. For vertical flow the liquid forms an annulus, whereas it is stratified in horizontal flow. A transition thus occurs which must be modelled. Very little is known on this problem.

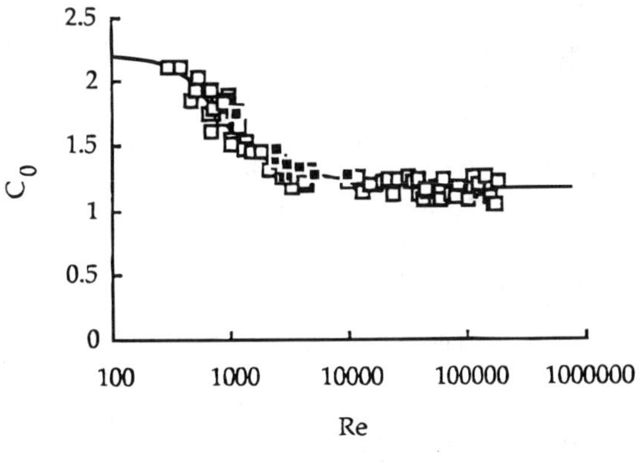

Figure 6. Influence of the flow regime on bubble motion. □: Fréchou, 1986; ■: Mao & Dukler, 1991; ———: empirical fit

Gas content in liquid slugs

In the recent decade, some experimental data of gas fraction in the liquid slugs have been published. Results obtained with similar flow conditions but different pipe inclinations are illustrated in **Figure 7**. This presentation has the merit to emphasise that the evolution of the gas fraction with the mixture velocity has the same trend in horizontal and in vertical pipe: this suggests that the same physical process take place and that the same modelling can be used for both cases.

A description of the mechanism of entrainment may be explained as follows. The liquid shed from the rear of a liquid slug flows around the nose of the long bubble to form a stratified or annular film flowing downward. This film enters at a relatively high velocity into the front of the next slug at high relative velocity. As the liquid film enters the slug it entrains some gas. In the mixing zone at the front of the next slug there is a local region of high void fraction, which is clearly observable. In this region of high turbulence level, the mixing process carries some of the bubbles to the front of the slug where they coalesce back into the long bubble.

What is the basic difference between horizontal and vertical flows? Even if the gas fraction is higher in vertical than in horizontal flow, the net flux of entrained gas could be the same provided that the relative bubble velocity is smaller. Since the bubble drift is higher in vertical than in horizontal flow, this could be true. However, $V >> U_{GD}$ and we can firmly state that the gas flux is higher in vertical than in horizontal flow. The gas entrainment raises another question. **Figure 7** shows that below some mixture velocity there are no bubbles in the slugs. There is some critical velocity difference above which gas is entrained: this is the onset of bubble entrainment. In vertical flow, small bubbles are always generated at the tail of the long ones.

There are a few models in the literature which were developed for predicting the gas fraction in the liquid slugs. We shall not make room for those which are less than satisfactory. These models were developed specifically either for horizontal flow or vertical flow, and the result is quite disappointing when one tries to apply

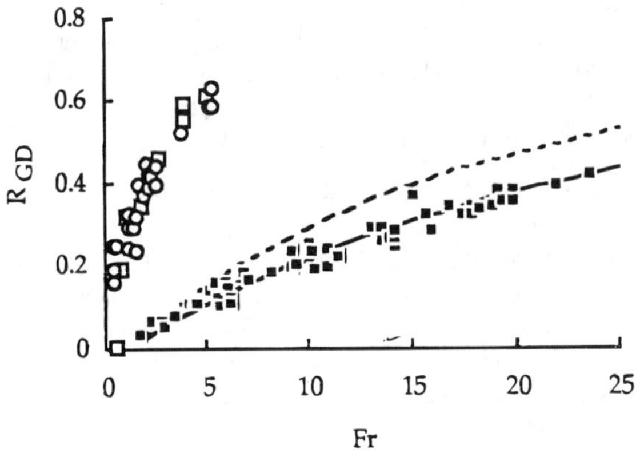

Figure 7. Gas fraction in liquid slugs. Air-water, D = 5 cm Vertical flow. □: Barnea & Shemer, 1989, ○: Mao & Dukler, 1991, Horizontal flow. ■: Andreussi & Bendiksen, 1989; Horizontal flow. ———; Vertical flow. - - -: Equation 22

each to the other case. Keeping in mind that the mechanism of entrainment is basically the same whatever the pipe slope, a reliable model should do a good job in both cases.

Andreussi and Bendiksen, 1989, proposed a model which applies satisfactorily to horizontal or slightly inclined flows. It may be demonstrated that the gas fraction is expressed as:

$$R_{GD} = \frac{j - j_f}{(j + j_0)^n} \tag{22}$$

In this equation, the critical mixture velocity j_f and the velocity scale j_0 are given by empirical expressions including coefficients chosen for the best fit with experimental data. The onset of entrainment corresponds to a critical mixture velocity which is close to the velocity at the transition pictured in **Figure 5**.

Gas drift in liquid slugs

The drift velocity in liquid slugs can be described by the modified Harmathy equation:

$$U_{GD} - U_{LD} = 1.54(1 - R_{GD})^{0.5}(\sigma g)^{0.25} \tag{23}$$

However, this law is not expected to work properly when the viscosity of the liquid is too high.

Another choice is to use a **Drift Flux Model** for the bubbly region. This model has been proposed from theoretical grounds by Kowe et al., 1988:

$$U_{GD} = C_l j + (1 - C_m R_{GD})(1 - R_{GD})V_B, \tag{24}$$

in which C_l accounts for the velocity and gas fraction distribution, C_m is the entrained mass coefficient and V_B is the rise velocity of bubble in still liquid. For inclined pipe the question has not yet been resolved. As the bubble diameter is needed it has to be closed following the method indicated in the entry **Bubble Flow**.

Mean length and frequency

It has been shown in previous sections that neither the characteristic length scale L of the cells nor their frequency n are needed to determine both void fraction and pressure gradient. However, there is a practical need for knowing the time or length scales of slug flow.

The mean slug length $\bar{L}_D$ is one of the characteristic length scales. Since the probability distribution of the velocities are narrowly distributed about their average (**Figure 2**), the time and length scales are related through:

$$\bar{L}_D = V\bar{T}_D \quad \text{and} \quad \bar{L}_S = V\bar{T}_S \tag{25}$$

where $\bar{T}_D$, $\bar{T}_S$ are the mean times of residence of slugs and long bubbles. Introducing the mean time of passage of the cell $\bar{T} = \bar{T}_D + \bar{T}_S$, the slug frequency may be defined as $n = 1/\bar{T}$. Using Equation (2), it can be shown that the mean slug length and the slug frequency are related by:

$$\bar{L}_D = (1 - \beta)\frac{V}{n}. \tag{26}$$

In the case of horizontal flow, when the superficial gas velocity increases the mean length of the liquid slugs increases and then reaches an asymptotic value lying between 30 to 40D. Up to now, the modeling of slug frequency is not resolved. A more detailed discussion and modelling will be found in the review of Dukler and Fabre.

Conclusion

The UC model concept still appears as modern and robust. It uses at best the 1D balance equations together with two assumptions which are the corner stone of the model, and four closure laws. Concerning the closure laws, it may shown that each has a specific influence.

The physical law of bubble velocity supports the model accuracy. There are still some uncertainties which concern mainly the transitions which have a strong influence on the dynamics: loss of symmetry in vertical flow, centred to non-centred nose in horizontal flow, influence of flow regime, annular to stratified regime of the film. The numerous results existing in horizontal and vertical flow must not disguise the lack of result for different inclinations. Much can be learned from numerical experiments of single phase flow around long bubbles.

The two other major issues concern the bubbly part of the liquid slugs: gas fraction and bubble motion. The paths by which gas enters and leaves a slug appear to have been identified. However the models necessary to convert these ideas into general predictive methods have not yet been developed. Careful experiments are needed.

There are no experimental data on the slip velocity between bubbles in the liquid slug and the surrounding liquid. It turns out that none of the existing experiments of bubbly flow are applicable to liquid slugs: it must be realised that in most cases the velocities in liquid slugs are considerably greater than they are in bubbly flow experiments. Numerical experiments on this problem are probably useless since the current turbulence models fail in the presence of a dense cloud of bubbles.

References

Andreussi, P. and Bendiksen, K. (1989) An investigation of void fraction in liquid slugs for horizontal and inclined gas-liquid pipe flow, *Int. J. Multiphase Flow,* 15, 937–46.

Barnea, D. and Shemer, L. (1989) Void fraction measurements in vertical slug flow: applications to slug characteristics and transition, *Int. J. Multiphase Flow,* 15, 495–504.

Benjamin, T. B. 1968. Gravity currents and related phenomena. *J. Fluid Mech.* 31, 209–248.

Dukler, A. E. and Fabre, J. (1992) Gas liquid slug flow: knots and loose ends. In *Multiphase Science and Technology. Two phase flow fundamentals.* Vol. 8, 355–470, Eds. Hewitt, G. F. et al., Begell House.

Dumitrescu, D. T. (1943) Strömung an einer Luftblase im senkrechten Rohr, Z., *Angew. Math. Mech.,* 23, 139–49.

Fabre, J., Grenier, P., and Gadoin, E. (1993) Evolution of slug flow in long pipe, *6th International Conference on Multi Phase Production,* Cannes, France, June 1993, in Multi Phase Production, Ed. A. Wilson, pp. 165–177, MEP, London.

Fabre, J. and Liné, A. (1992) Modelling of two phase slug flow, *Annu. Rev. Fluid Mech.,* 24, 21–46.

Ferschneider, G. (1982) Ecoulements gaz-liquide à poches et à bouchons en conduite, *Rev. Inst. Fr. Pét.,* 38, 153–82.

Fréchou, D. (1986) *Etude de l'écoulement ascendant à trois fluides en conduite verticale,* Thèse, Inst. Natl. Polytech. de Toulouse, France.

Harmathy, T. Z. (1960) Velocity of large drops and bubbles in media of infinite or restricted extent, *AIChE J.,* 6, 281–88.

Kowe R., Hunt J. C. R., Hunt A., Couet B., and Bradbury L. J. S. (1988) *Int. J. Mult. Flow,* 14, 587–606.

Linga, H. (1991) Flow pattern evolution; some experimental results obtained at the SINTEF Multiphase Flow Laboratory, *5th International Conference on Multi Phase Production,* Cannes, France, June 1991, in Multi Phase Production, Ed. A. P. Burns, pp. 51–67, Elsevier.

Mao Z. and Dukler, A. E. (1989) An experimental study of gas-liquid slug flow, *Exp. Fluids,* 8, 169–82.

Mao Z. and Dukler, A. E. (1991) The motion of Taylor bubbles in vertical tubes. II. Experimental data and simulations for laminar and turbulent flow, *Chem. Eng. Sci.,* 46, 2055–64.

Martin, C. S. (1976) Vertically downward two-phase slug flow, *J. Fluids Eng.,* 98, 715–22.

Nicklin, D. J., Wilkes, J. O., and Davidson, J. F. (1962) Two phase flow in vertical tubes, *Trans. Inst. Chem. Engs.,* 40, 61–68.

Spedding, P. L. and Nguyen, V. T. (1978) Bubble rise and liquid content in horizontal and inclined tubes, *Chem. Eng. Sci.,* 33, 987–94.

Taitel, Y. and Barnea, D. (1990) Two-phase slug flow, *Adv. Heat Transfer.,* 20, 83–132.

Wallis, G. B. (1969) *One-Dimensional Two-Phase Flow,* McGraw-Hill, New-York.

Zukoski, E. E. (1966) Influence of viscosity, surface tension and inclination angle on motion of long bubbles in closed tubes, *J. Fluid Mech.,* 25, 821–37.

J. Fabre and A. Liné

SLUG FLOW, SOLID SUSPENSIONS (see Gas-solid flows)

SLUG FREQUENCY (see Slug flow)

SLUG LENGTH (see Slug flow)

SLURRIES (see Liquid-solid flow)

SMALL ANGLE SCATTERING METHOD, FOR DROPSIZE MEASUREMENT (see Dropsize measurement)

SMELTING

Following from: *Metals*

The term smelting refers to the bulk production of the major metal bearing liquid phase as one of the process steps during the pyrometallurgical production of **Metals.**

The liquid phase produced may be either a liquid metal sulphide (*matte*) as in the production of **Copper** and **Nickel,** or a metallic phase as in the production of **Iron** and **Steel, Lead, Zinc** and **Aluminium.**

Smelting of non-ferous sulphide ore concentrates to liquid metals is now predominately performed by either of two methods.

1. *Flash smelting* is a process in which fine powder of the sulphide together with an air/oxygen mixture are injected into the top of a large shaft. The subsequent combustion of a proportion of the sulphur content results in autogeneous production of a liquid matte phase containing the major part of the metal, and a liquid oxide (slag) phase bearing the impurity oxide content of the concentrate. The liquid matte and slag are collected at the bottom of the shaft.
2. *Bath smelting* is a process in which fine powder sulphide is fed directly into a liquid matte bath which is agitated by injection of air/oxygen.

In smelting iron and steel and zinc and lead, the oxide bearing the metal is reduced to metal with carbon (in the form of coke) in a shaft furnace (known as a blast furnace). The carbon also acts as a fuel and its combustion leads to the production of **Carbon Monoxide** and **Carbon Dioxide.** This provides both the reducing atmosphere and the heat required for the production of a liquid metallic phase and a liquid oxide (slag) phase bearing the oxide impurities of the concentrate.

In aluminium smelting the reduction of aluminium oxide is performed electrochemically by the **Electrolysis** of Al_2O_3 dissolved in a liquid fluoride flux.

References

Davenport, W. and Partlepo, G. (1987) *Flash Smelting,* Pergamon Press.

Peachey, J. and Davenport, W. (1978) *The Iron Making Blast Furnace,* Pergamon Press.

George, D. B., Sohn, H. Y. and Zunkel, A., eds. (1983) *Advances in Sulfide Smelting,* TMS-AIME, Warrendale, PA, USA.

Gojtheim, K. (1979) *Aluminium Electrolysis,* Springer Verlag.

Leading to: *Furnaces; Kilns*

B. Terry

SMOKE, AS AN AIR POLLUTANT (see Air pollution)

SMOKES (see Aerosols)

SNELL REFRACTION LAW (see Refractive index)

SNL (see Sandia National Laboratory)

SOAVE EQUATION (see PVT relationships)

SODA ASH (see Sodium carbonate)

SODIUM

Sodium—(English, *soda;* Medieval Latin, *sodanum,* headache remedy), Na (L. *natrium*); atomic weight 22.9898; atomic number 11; melting point 97.81 ± 0.03°C; boiling point 882.9°C; specific gravity 0.971 (20°C); valence I.

Long recognized in compounds, sodium was first isolated by Davy in 1807 by electrolysis of caustic soda. Sodium is present in fair abundance in the sun and stars. The D lines of sodium are among the most prominent in the solar spectrum. Sodium is the sixth most abundant element on earth, comprising about 2.6% of the earth's crust; it is the most abundant of the alkali group of metals of which it is a member. The most common compound is sodium chloride, but it occurs in many other minerals, such as soda niter, cryolite, amphibole, zeolite, sodalite, etc. It is a very reactive element and is never found free in nature. It is now obtained commercially by the electrolysis of absolutely dry fused sodium chloride. This method is much cheaper than that of electrolyzing sodium hydroxide, as was used several years ago. Sodium is a soft, bright, silvery metal which floats on water, decomposing it with the evolution of hydrogen and the formation of the hydroxide. It may or may not ignite spontaneously on water, depending on the amount of oxide and metal exposed to the water. It normally does not ignite in air at temperatures below 115°C.

Metallic sodium is vital in the manufacture of sodamide and sodium cyanide, sodium peroxide, and sodium hydride. It is used in preparing tetraethyl lead, in the reduction of organic esters, and in the preparation of organic compounds. The metal may be used to improve the structure of certain alloys, to descale metal, to purify molten metals, and as a heat transfer agent. An alloy of sodium with potassium, NaK, is also an important heat transfer agent. Sodium compounds are important to the paper, glass, soap, textile, petroleum, chemical, and metal industries. Soap is generally a sodium salt of certain fatty acids. The importance of common salt to animal nutrition has been recognized since prehistoric times. Among the many compounds that are of the greatest industrial importance are common salt (NaCl), soda ash (Na_2CO_3), baking soda ($NaHCO_3$), caustic soda (NaOH), Chile saltpeter ($NaNO_3$), di- and tri-sodium phosphates, sodium thiosulfate (hypo, $Na_2S_2O_3 \cdot 5H_2O$), and borax ($Na_2B_4O_7 \cdot 10H_2O$). Seven isotopes of sodium are recognized.

Sodium metal should be handled with great care. It should be maintained in an inert atmosphere and contact with water and other substances with which sodium reacts should be avoided.

Handbook of Chemistry and Physics, CRC Press

SODIUM CARBONATE

Sodium Carbonate (Na_2CO_3) is a whitish, non-toxic powder with a molecular weight of 106.00 that melts at 1098K and has a specific gravity of 2.53. It is soluble in water and it readily incorporates water molecules to form a number of crystalline forms (monohydrate, peroxide, sal soda). The common name for this chemical is *soda ash.*

V. Vesovic

SODIUM CHLORIDE

Sodium Chloride (NaCl) is a white crystalline powder with a molecular weight of 58.44 that melts at 1074K, boils at 1738K, has a specific gravity of 2.17 and is soluble in water. Other names for this chemical are common salt, table salt, sea salt and rock salt.

V. Vesovic

SODIUM COOLED NUCLEAR REACTOR (see Liquid metal cooled fast reactor)

SODIUM HYDROXIDE

Sodium hydroxide (NaOH) is a white crystalline solid with a molecular weight of 39.997 that melts at 591 K, boils at 1661 K and has a specific gravity of 2.13. It readily absorbs water to form an aqueous solution which is corrosive and an irritant to skin. Other common names for this chemical are *caustic soda* and *white caustic.*

V. Vesovic

SOFTENING OF WATER (see Water preparation)

SOFTWARE ENGINEERING (see Computers)

SOIL, THERMAL PROPERTIES (see Environmental heat transfer)

SOL (see Particle technology)

SOLAR CELLS

Following from: *Solar energy*

In 1839 Becquerel observed that certain materials, when exposed to light, produced an electric current (Becquerel, 1839). This is now known as the *photovoltaic effect,* and is the basis of the operation of photovoltaic or solar cells. When light falls onto semiconductor material, photons with energy more than the band gap energy (the difference in energy between electrons in the valence and the conduction bands) interact with electrons in covalent bonds, creating "electron-hole" (e–h) pairs. The generation rate of e–h pairs per unit volume can be calculated as a function of the photon flux (photons/unit area/sec.), the absorption coefficient, and the distance travelled before absorption. Optimum use is made of incoming sunlight if the semiconductor bandgap lies in the range 1.0–1.6 eV. This effect acts to limit the maximum achievable efficiency of solar cells to 44% (Shockley and Queisser, 1961). Silicon, the most commonly used solar cell material, has a bandgap of 1.1 eV.

When not illuminated, the system returns to a state of equilibrium and the electrons and holes move around until they meet up and *recombine.* Any defects or impurities within or at the surface of the semiconductor promote recombination. The *carrier lifetime* of a material is defined as the average time for recombination to occur after e–h generation. Similarly, the *carrier diffusion length* is the average distance a carrier can move from point of generation to recombination. These two parameters give an indication of material quality and suitability for solar cell use.

A *silicon solar cell* is a diode formed by joining p-type (electron deficient, typically boron doped) and n-type (with excess electrons, typically phosphorous doped) silicon. At the p–n junction, excess electrons flow from n- to p-type leaving behind exposed charges on dopant atom sites, fixed in the crystal lattice. An electric field ($\hat{E}$) builds up in the so-called "depletion region" around the junction to stop the flow. Depending on the materials used, a "built in" potential V_{bi} due to $\hat{E}$ will be formed. If a voltage is applied to the junction, $\hat{E}$ will be reduced. Once $\hat{E}$ is no longer large enough to stop the flow of electrons and holes, a current is produced. V_{bi} reduces to ($V_{bi} - V$) and the current flow increases exponentially with the applied voltage. This phenomenon results in the *Ideal Diode Law,* expressed as:

$$I = I_o[\exp(qV/kT) - 1] \quad (1)$$

where: I_0 = "dark saturation current", the diode leakage current density in the absence of light; V = applied voltage, q = absolute value of electronic charge; k = Boltzmann's constant; T = absolute temperature.

"IV" curves for a diode are generated by plotting I against V. Illumination of a cell adds to the normal "dark" currents in the diode and has the effect of shifting the IV curve such that power can be extracted. The IV curve characterises the cell and is most often shown reversed, as in **Figure 1**. The two limiting parameters used to characterise the output of solar cells for given irradiance, operating temperature and area are the *short-circuit current*, I_{sc}, the maximum current at zero voltage, and the *open-circuit voltage*, V_{oc}, the maximum voltage at zero current. Ideally, if V = 0, $I_{sc} = I_L$, the light generated current. I_{sc} is directly proportional to the available sunlight while V_{oc} increases logarithmically with increased sunlight.

For each point on the IV curve, the product of the current and voltage represents the power output for that operating condition. A solar cell can also be characterised by its *maximum power point*, the maximum value of $V_{mp} * I_{mp}$. The maximum power output of a cell is graphically given by the largest rectangle that can be fitted under the IV curve. The power output at V_{mp},I_{mp} under strong sunlight (1 kW/m^2) is known as the "peak power" of the cell.

Photovoltaic panels are usually rated in terms of their "peak" watts, W_p. The *fill factor*, FF, is a measure of the junction quality and series resistance of a cell. It is defined as the maximum power/$I_{sc}V_{oc}$. The nearer FF is to unity, the higher the quality of the cell. Ideally, it is a function only of V_{oc}. The spectral responsivity of a solar cell is given by the amps generated per watt of incident light.

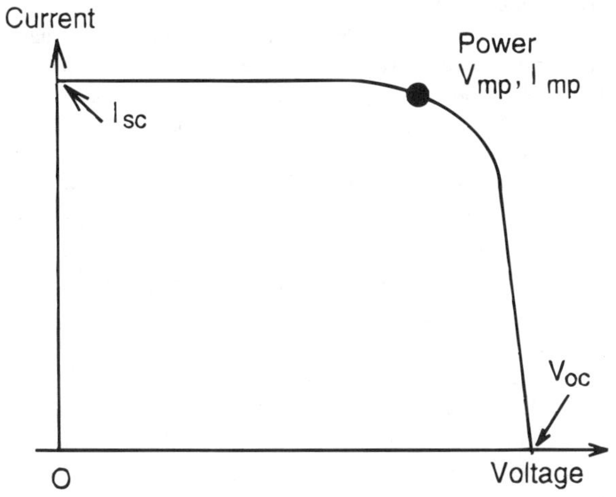

Figure 1. Typical IV curve, showing short circuit current (I_{sc}), open circuit voltage (V_{oc}) and maximum power point (V_{mp}, I_{mp}).

Figure 2. A solar array.

Ideally this increases with wavelength, making cell performance strongly dependent on the spectral content of sunlight. The operating temperature of solar cells is determined by the ambient air temperature, the characteristics of the module in which it is encapsulated, the intensity of sunlight falling on the module, and other variables, such as wind velocity. The main effect of increasing temperature for silicon solar cells is a reduction in V_{oc}, FF and hence the cell output. I_0 increases with temperature according to the equation:

$$I_o = BT^{\gamma} \exp\left(\frac{-E_{GO}}{kT}\right) \quad (2)$$

Where: B is independent of temperature; E_{GO} is the linearly extrapolated zero temperature band gap of the semiconductor (Green, 1992); and γ includes the temperature dependencies of the remaining parameters determining I_0.

References

Becquerel, E. (1839) On electron effects under the influence of solar radiation, *Comptes Rendues,* 9, 561.

Green, M. A. (1992) *Solar Cells—Operating Principles, Technology and System Application,* University of NSW, Kensington, Australia.

Shockley, W. and Queisser, H. J. (1961) Detailed Balance Limit of Efficiency of p–n Junction Solar Cells, *Journal of Applied Physics,* 32, 510–519.

Wenham, S. R., Green, M. A., and Watt, M. E. (1994) *Applied Photovoltaics,* Centre for Photovoltaic Devices and Systems, University of NSW, Australia.

M.E. Watt

SOLAR COOKERS (see Solar energy)

SOLAR DRYING (see Solar energy)

SOLAR ENERGY

Following from: *Alternative energy source*

A *solar energy thermal conversion system* should seek to provide the optimal combination of efficient performance, low initial and running costs, robustness and durability. Such a system consists of components for energy collection, distribution and storage. These may be discrete items, or so inextricably linked as to be synonymous. The system components are sized in relation to the solar energy resource and to the nature and pattern of energy utilisation.

Solar Selective Surfaces

High efficiency thermal solar energy collection requires a large absorption of short-wave solar radiation, low emission of emitted long-wave thermal radiation and suppression of convective heat losses. Ninety five percent of the *solar radiation spectrum* lies in the wavelength range 0.3 to 2 μm; ninety nine percent of thermal radiation at 325 K lies in the range 3.0 to 30 μm.

Solar selective surfaces have a high solar spectrum absorptance and a low emittance in the thermal spectrum.

Consider a heat balance between absorber surface and glass cover of the solar collector shown schematically in **Figure 1**. The energy absorbed $\dot{q}_{abs}$ by the absorber per unit area is given by

$$\dot{q}_{abs} = \alpha\tau I \quad (1)$$

where I is the incoming solar radiation, α is the solar absorptance and τ is the transmittance of the glass cover. The heat lost by thermal radiation q_{rad} per unit area is given by

$$\dot{q}_{rad} = \frac{\sigma(T_a^4 - T_g^4)}{\frac{1}{\varepsilon} + \frac{1}{\varepsilon_g} - 1} \cong \varepsilon\varepsilon_g\sigma(T_a^4 - T_g^4) \quad (2)$$

(ε or $\varepsilon_g \approx 1$) where ε and ε_g are emittance of the absorber and of the glass cover respectively. A low ε or a low ε_g is favourable. A low ε_g requires an infrared coating on the glass cover. Though low emittance is clearly beneficial, at lower temperatures natural convection across the cavity is the dominant mode of heat loss, this introduces the possible inclusion of convection suppression devices (known as transparent insulation materials) or evacuation of the cavity.

To obtain a selective surface with a high α and a low ε. The material needs to have a low reflectance ρ in the solar, and a high reflectance in the thermal (infrared) spectrum. Materials that have this property are semi-conductors like silicon and germanium. Photons having energies greater than that of their typical band gap will be absorbed: the band gap energy of silicon is 1.11 eV (equivalent wavelength 1.2 μm); the band gap energy of germanium is 0.67 eV (equivalent wavelength 1.9 μm). However, these materials have an appreciable solar reflectance (of ≈ 0.3), and so require an appropriate coating. Metals such as copper, nickel and aluminium exhibit high infrared reflectances (e.g. ≥ 0.95 for clean and polished surfaces); however, they also have low solar absorptances. To overcome this, a thin (0.4≈1.5 μm) layer of a material with high solar absorptance and good infra-red transmittance is applied to the metal.

Black nickel is a nickel-zinc-sulphide complex that has the required properties. An absorptance of 0.96 can be obtained; the polished nickel substrate can give a low emittance, typically 0.08. Copper oxide on copper shows absorptances and emittances of, typically, 0.9 and 0.15 respectively. "Black chrome" selective absorbing surfaces comprise a thin surface layer of chrome in an amorphous chromium oxide matrix deposited on a polished metal surface.

Non-metallic materials which have a high infrared reflectance, compared with that of metallic surfaces, and a low solar reflectance, are termed "heat mirrors". When a thin layer of such a material is combined with a solar-absorbing substrate, the resulting tandem exhibits good selective properties.

Convective Heat Transfer in Solar Energy Collectors

In determining the heat loss by buoyancy-driven convection between the collector plate and the glass cover of a *flat-plate collector,* when mounted at an angle between 0° and 75° from the horizontal, the following correlation of **Nusselt Number,** Nu, in terms of **Grashot Number**, Gr, and **Prandtl Number**, Pr is employed:

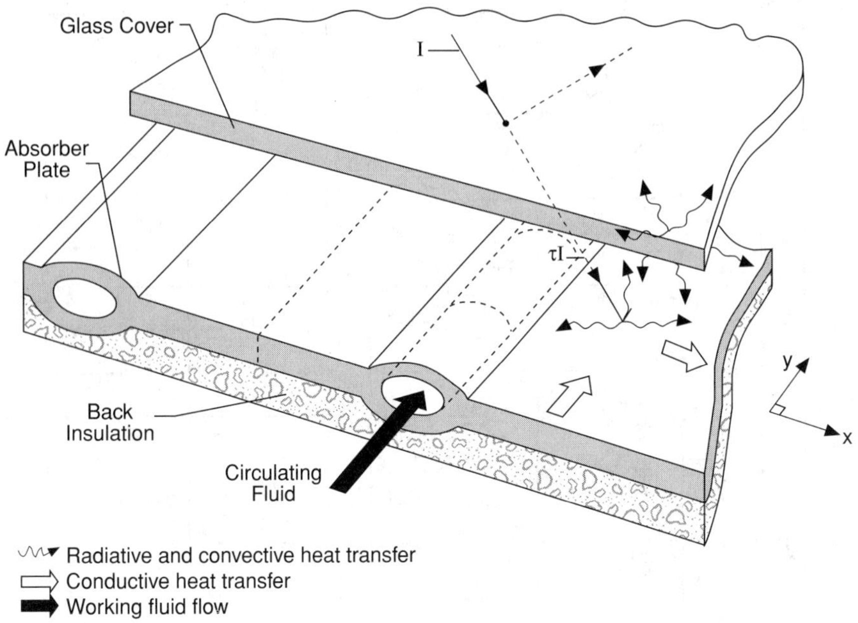

Figure 1. Components and heat transfer mechanisms of a liquid-heating flat-plate collector.

$$Nu = 1.144\left[1 - \frac{1708}{Gr\ Pr\ Cos\ \beta}\right]^{+} \times \tag{3}$$

$$\left[1 - \left(\frac{(1708\ Sin\ 1.8\beta)^{1.6}}{Gr\ Pr\ Cos\ \beta}\right)\right] + \left[\frac{(Gr\ Pr\ Cos\ \beta)^{0.33}}{5830} - 1\right]$$

where β is the inclination of the system to the horizontal.

The superscript + denotes a positive value for the quantity in brackets if this quantity is greater than zero and a value of zero otherwise. The characteristic length in determining Gr is taken as the thickness of the air cavity.

Heat losses from the absorber of a *line-axis concentrating collector* to the surrounding environment ensue by radiation, conduction and convection. Under steady-state conditions, the interactions of these three heat-transfer modes lead to a particular temperature distribution being established, which is characteristic of the geometry and the applied temperature difference between the absorber and the ambient environment. The internal convection correlation for Nusselt number is:

$$Nu = 0.398\left(\frac{H}{W}\right)^{0.365} \frac{Gr[0.1825 + 0.0736\cos(\beta - 45)]}{[1.24 + 0.66054\cos(\beta - 45)]} \tag{4}$$

where H is the height of the cavity, W is its half-width and β is the inclination of the system.

Solar Energy Collectors

The principal components of a water-heating flat-plate collector are shown in **Figure 1**. It consists of an absorbing surface which absorbs insolation and transmits it in the form of heat to a working fluid, commonly air or water. In an evacuated-tube collector, each absorber fin is enclosed in a separate cylindrical glass envelope. The latter generally has only one protrusion of pipework, against which the glass is sealed, through which heat is removed. Evacuation of the envelope prevents convective heat loss from the absorber plate to the glass. The transfer of heat from the collector is usually accomplished by employing a **Heat Pipe** absorber.

Solar energy collectors are devices employed to gain useful heat energy from the incident solar radiation. They can be of the *concentrating* or the flat-plate type and can be stationary or track solar azimuthal position in either two or three areas. The range of solar collectors is illustrated in **Figure 2**.

The useful heat gained by a collector can be expressed as

$$\dot{Q}_u = \dot{m}c_p(T_o - T_i) \tag{5}$$

and the following heat balance, the *Hottel-Whillier-Bliss equation* expresses the thermal performance of a collector under steady state.

$$\dot{Q}_u = AF_R[I(\tau\alpha)_e - U_L(T_i - T_a)] \tag{6}$$

Collector efficiency is defined as the ratio of useful heat gain over any time period to the incident solar radiation over the same period, i.e.

$$\eta = \frac{\dot{Q}_u}{IA} \tag{7}$$

thus

$$\eta = \dot{m}c_p\frac{T_o - T_i}{IA} \tag{8}$$

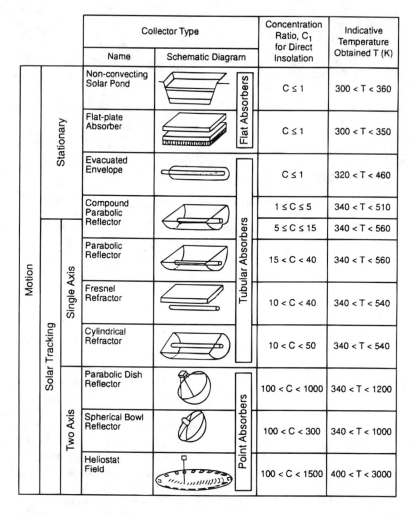

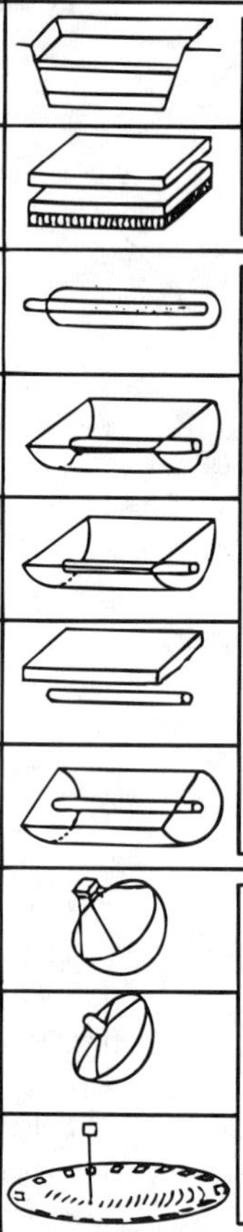

Motion			Collector Type: Name	Absorber	Concentration Ratio, C_1 for Direct Insolation	Indicative Temperature Obtained T (K)
Motion	Stationary		Non-convecting Solar Pond	Flat Absorbers	C ≤ 1	300 < T < 360
Motion	Stationary		Flat-plate Absorber	Flat Absorbers	C ≤ 1	300 < T < 350
Motion	Stationary		Evacuated Envelope	Tubular Absorbers	C ≤ 1	320 < T < 460
Motion	Stationary		Compound Parabolic Reflector	Tubular Absorbers	1 ≤ C ≤ 5	340 < T < 510
Motion	Solar Tracking	Single Axis	Compound Parabolic Reflector	Tubular Absorbers	5 ≤ C ≤ 15	340 < T < 560
Motion	Solar Tracking	Single Axis	Parabolic Reflector	Tubular Absorbers	15 < C < 40	340 < T < 560
Motion	Solar Tracking	Single Axis	Fresnel Refractor	Tubular Absorbers	10 < C < 40	340 < T < 540
Motion	Solar Tracking	Single Axis	Cylindrical Refractor	Tubular Absorbers	10 < C < 50	340 < T < 540
Motion	Solar Tracking	Two Axis	Parabolic Dish Reflector	Point Absorbers	100 < C < 1000	340 < T < 1200
Motion	Solar Tracking	Two Axis	Spherical Bowl Reflector	Point Absorbers	100 < C < 300	340 < T < 1000
Motion	Solar Tracking	Two Axis	Heliostat Field	Point Absorbers	100 < C < 1500	400 < T < 3000

Figure 2. Solar energy collector types.

and therefore

$$\eta = F_R(\tau\alpha)_e - F_R U_L \frac{(T_i - T_a)}{I} \quad \textbf{(9)}$$

The general steady-state test procedures for flat-plate collectors is to determine Q_u and measure I, T_i and T_a, by operating the collector under nearly steady-state conditions in test facilities (i.e. either indoor or outdoor). Instantaneous efficiencies are plotted against $(T_i - T_a)/I$ and the intercept $(F_R\ (\tau\alpha)_e)$ and the slope $(-F_R U_L)$ determined. These parameters are not constant, U_L depends on temperature and wind speed, and F_R being a weak function of U_L. However the long-term performance of many solar-heating collectors can be characterised by a thus determined intercept and slope. Illustrative examples of characteristic areas for air and water heating collectors are given in **Figure 3**.

The absorber material in a flat-plate collector, in addition to having a high absorptance of the incident radiation should also have a low emissivity, good thermal conductivity, and be stable thermally under temperatures encountered during operation and stagnation. It should also be durable, have low weight per unit area and, most importantly, be cheap.

A good cover material should have a high transmittance in the visible range of the spectrum and a low transmittance to infrared radiation in order to effectively trap in re-radiated heat from the absorber. Other qualities of a good cover material include low heat absorptivity, stability at the operating temperatures (should withstand high temperatures under stagnation conditions), resistance to breakage, durability under adverse weather conditions, and low cost. The variation of the radiative transmittance of a "transparent" material is determined by its chemical composition, molecular structure and fabrication. Though most plastic films have transmittances to visible light greater than 0.85, they exhibit wide variations in transmittance to infrared from 0.01 (for polymethyl methacrylate) to 0.77 (for polyethylene).

Glass has been very widely used as a cover material due to its high transmittance to visible light, very low transmittance to infrared radiation and its stability to high temperature. Its high cost, low shatter resistance and relatively large weight per unit (which increases the cost of supporting structures) have encouraged the

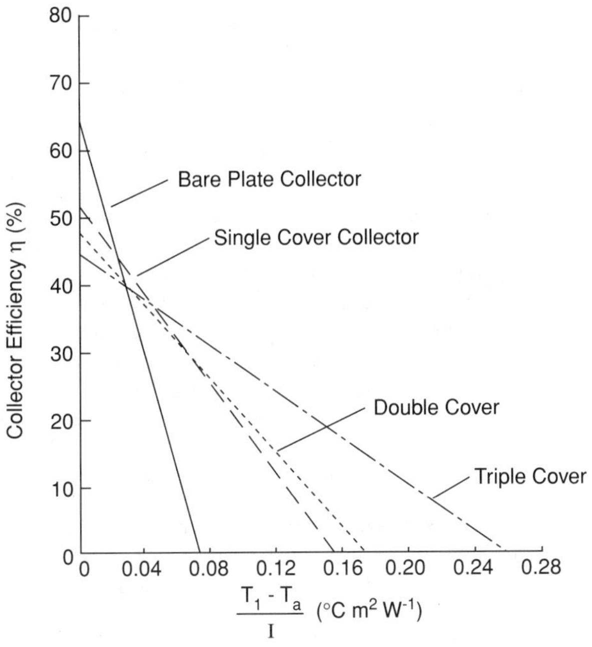

Figure 3. Typical Hottel-Whittlier-Bliss performance characterisations for generic air-heating solar energy collectors.

consideration of alternative cover materials. Plastics have been used, their major limitations being their stability at collector operating temperatures and their durability under weather conditions, particularly degradation under ultraviolet radiation. However, most plastics have been chemically treated to overcome at least some of these shortcomings. Some plastic covers show high transmittance to visible light and equally low transmittance to infrared. Plastics weigh about 10% of the same area of glass. The overriding factor in the choice of materials for the design of cheap and simple solar energy collectors is cost, particularly those that heat air, thus certain desired material properties may be compromised during design and construction.

A transpired collector is an unglazed perforated absorber plate through whose holes solar heated air is drawn by the action of a fan. The efficiency of a transpired collector is given by:

$$\eta = \frac{\alpha_c}{1 + \dfrac{(h_r/\varepsilon_{hx}) + h_c}{\rho c_p V_o}} \qquad \textbf{(10)}$$

where α_c is the collector absorptance, h_r is a linearised radiative heat loss co-efficient, ε_{hx} is the absorber heat exchange effectiveness, h_c is the convective heat co-efficient, ρ is air-density, c_p is specific heat at constant pressure and V_o is suction velocity.

A line-focus compound parabolic concentrating collector as shown in **Figure 4** is characterised by an acceptance half-angle θ_a which determines the maximum attainable concentration ratio, which is given by

$$C_{max} = 1/\sin \theta_a \qquad \textbf{(11)}$$

This maximum concentration ratio can be achieved only by a full-height compound parabolic concentrator, i.e. no truncation is applied at the top of the reflectors, and if the absorber is of optically correct area, i.e. the area of the absorber is $1/C_{max}$ of the aperture area. In a real application, with a tubular absorber, the concentration ratio is expressed as

$$C = W/\pi D \qquad \textbf{(12)}$$

The value given by the above equation E is lower than that given by $1/\text{Sin}\,\theta_a$ because of truncation of the concentrator top, undertaken normally to reduce the capital cost, and oversizing of the absorber's diameter, to allow for optical scatter introduced by imperfections arising during manufacture and operation.

For a parabolic trough concentrating collector, which tracks the Sun continuously, so any ray entering the concentrator parallel to its axis will, either after reflection or directly, intercept the tubular absorber. The concentration ratio for a parabolic trough concentrating collector is also given by $W/\pi D$.

Unlike flat-plate collectors, only a fraction of the diffuse insolation is exploitable by concentrating collectors. This can be shown by considering the radiation exchange between the absorber and the aperture in a concentrating collector. If $E_{R\text{-}A}$ and $E_{A\text{-}R}$ represent the exchange factors for the radiation exchange between absorber and aperture and aperture and absorber respectively, then the following equation applies:

$$\pi D E_{R-A} = W E_{A-R} \qquad \textbf{(13)}$$

For a compound parabolic concentrating collector the exchange factor $E_{R\text{-}A}$ is unity, as any ray emitted from the absorber will either directly, or after one or more reflections, reach the aperture. Thus

$$E_{A-R} = 1/C \qquad \textbf{(14)}$$

If an *isotropic* distribution is assumed for the diffuse radiation then the exchange factor $E_{A\text{-}R}$ also represents the exploitable part of the diffuse insolation of a compound parabolic concentrating collector, $g_{D,\,CPC}$:

$$g_{D,CPC} = 1/C \qquad \textbf{(15)}$$

For a parabolic trough concentrating collector, $E_{R\text{-}A} < 1$ as the absorber can "view" itself on the reflector. Thus, the exploitable part of the diffuse insolation of a parabolic trough concentrating collector, $\beta_{D,\,PTC}$, is less than 1/C. The diffuse insolation absorbed by the absorber can be given by an

$$I_{u,D} = (\tau\rho\alpha)\beta_D g_{D,PTC} I_D \qquad \textbf{(16)}$$

where β_D is a correction coefficient accounting for the part of *diffuse* insolation which reaches the absorber directly, i.e. is not attenuated by reflection losses and ρ accounts for end effects. The total insolation absorbed by the absorber, I_u.

$$I_u = \tau\rho\gamma I_{eff} \qquad \textbf{(17)}$$

where γ is the intercept factor accounting for the optical losses occurring in a real parabolic trough concentrating due to optical errors and I_{eff} represents the effective insolation at the concentrator's aperture, given by

$$I_{eff} = \beta_b I_B + \beta_D g_{D,PTC} I_D \tag{18}$$

To evaluate the thermal performance of a parabolic trough concentrating collector I_{eff} is employed.

The optical efficiency η_{opt} of a parabolic trough concentrating collector is defined as the ratio between the insolation I_u absorbed by the absorber and the total hemispherical insolation on the plane of the collector, I_{tot}, i.e.

$$\eta_{opt} = \frac{I_u}{I_{tot}} \tag{19}$$

Thus

$$\eta_{opt} = \frac{\tau\rho\alpha\gamma(\beta_B I_B + \beta_D g_{D,PTC} I_D)}{I_B + I_D} \tag{20}$$

A compound parabolic concentrating collector exploits a greater part of the available diffuse insolation compared with a parabolic trough collector, although this advantage diminishes as the concentration ratio increases. A compound parabolic concentrating collector also maintains a superior acceptance angle.

Solar Ponds

Solar ponds are unitary solar energy collectors and heat stores. In an non-convecting solar pond, part of the incident insolation absorbed is which is stored as heat in lower regions of the pond.

A salt-gradient, non-convecting solar pond consists of three zones as shown in **Figure 5**.

- The upper-convecting zone (UCZ), of almost constant low salinity at close to ambient temperature. The UCZ, typically 0.3 m thick, is the result of evaporation; wind-induced mixing and surface flushing. Wave-suppressing surface meshes and placing windbreaks near the pond keep UCZ as thin as possible.
- The non-convecting zone (NCZ), in which both salinity and temperature increase with depth. The vertical salt gradient in the NCZ inhibits convection and thus provides the thermal insulation.
- The lower-convecting zone (LCZ), of almost constant, relatively high salinity (typically 20% by weight) at a high temperature. Heat is stored in the LCZ which is sized either to supply energy continuously throughout the year for power generation or provide interseasonal heat storage for space heating. As the depth increases, the thermal capacity increases and annual variations of temperature decreases. However, large depths increase the initial capital outlay and require longer start-up times.

Salt gradient lakes, which exhibit an increase in temperature with depth have occurred naturally in Transylvania, in California and Washington State, USA, in the Arctic, Venezuela, western Uganda, and a lake on the east coast of the Sinai Peninsula.

The application of solar ponds for electric-power production usually employs an organic vapour Rankine cycle engine to convert solar-pond heat to mechanical work, and then into electricity. However, to obtain a low cost per unit generated power, solar ponds of several square kilometres are required.

The site for a solar pond should be near a cheap source of salt, an adequate source of water, incur low land costs, and have an all-year solar exposure. The underlying earth structure should be free of stresses and fissures. If not, then increases in temperature may cause differential thermal expansions which could result in earth movements. The pond must not pollute aquifers nor lose heat via underground water streams passing through an aquifer. Any continuous drain of heat will lower the pond's storage capability and effectiveness. Stormy regions should be avoided in order to limit wind surface mixing effects.

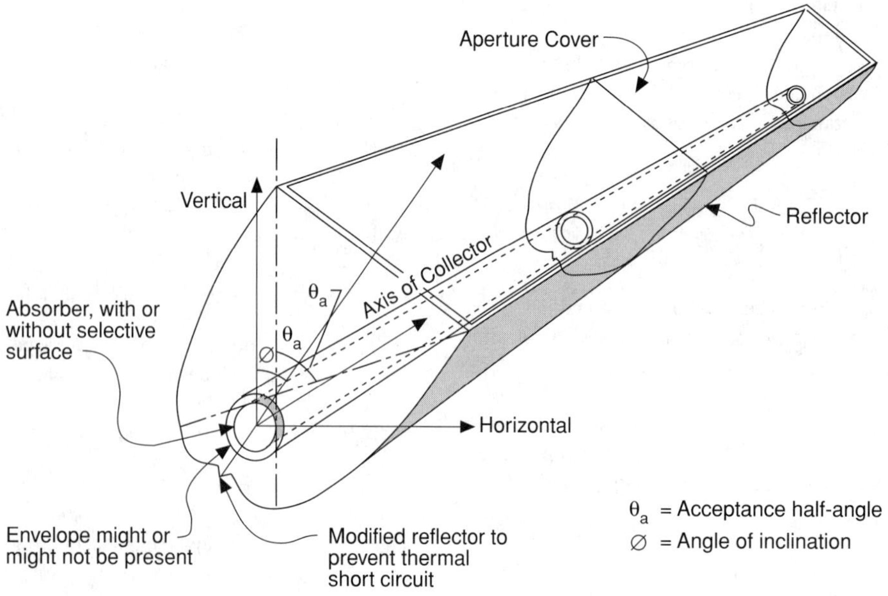

Figure 4. Components and geometry of a compound parabolic concentrating solar energy collector with a tubular absorber.

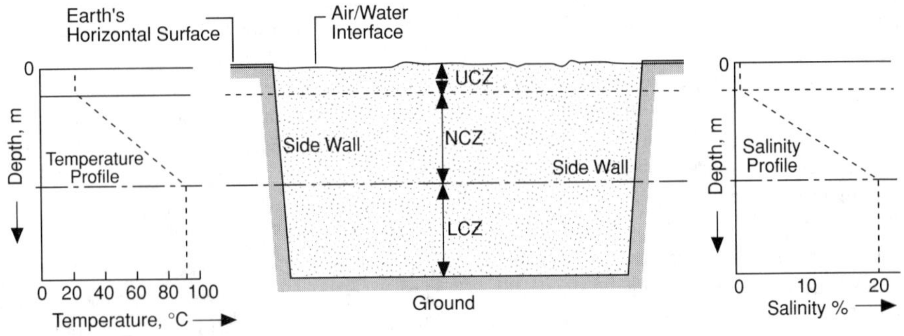

Figure 5. Schematic vertical cross-section through a salt-stratified non-convecting solar pond.

Species of freshwater and saltwater algae grow under the conditions of temperature and salt concentration that exist in a stratified solar pond. Algae and cynobacteria growth will inhibit solar transmittance and, for the latter, possible be toxic too. Different algae and organicbacteria species are introduced by rain water and airborne dust. To prevent algae formation, copper sulphate has been added at a concentration of about 1.5 mg l^{-1}.

A solar pond will cease to function without maintenance of the vertical salt gradient stratification. The stability of the salt gradient is maintained by:

1. controlling the overall salinity difference between the two convecting layers
2. inhibiting internal convection currents if they tend to form in the NCZ
3. limiting the growth of the UCZ.

Salt slowly diffuses upwards at an annual average rate of about 20 kg m^{-2} as a result of its concentration gradient. This rate varies and is dependent upon the ambient environment conditions, type of salt and temperature gradient. Surface washing by fresh water and injecting brines of adequate density at the bottom of the pond usually maintain an almost stationary gradient. UCZ growth caused by surface flushing is diminished if the velocity of the surface washing water is small. Surface temperature fluctuations will result in heat being transferred upwards through the UCZ by convection, especially at night, and downward, more slowly, by conduction. The thickness of the UCZ also varies with the intensity of the incident insolation.

The higher the temperature of the UCZ, and the lower the humidity above the pond's surface, the greater will be the evaporation rate caused by insolation and wind action. Excessive evaporation results in a downward growth of the UCZ. Evaporation can be compensated by surface water washing, as well as reduce the temperature of the pond's surface especially during periods of high insolation. Windbreaks will reduce evaporation rates. Though evaporation can be the dominant mechanism in surface-layer mixing under light-to-moderate winds, it is minor under strong winds.

Solar Water Heaters

Solar energy water heaters are categorised as either active or passive. An active system requires a pump to drive the heated fluid through the system, whereas a passive system requires no external power. Distributed systems comprise a solar collector, hot water store and connecting pipework; they may be either active or passive. In the former, a pump actuated by temperature sensors via a control circuit is required to convey the fluid from the collector to the store. In a thermosyphon solar water heater, fluid flow is due to buoyancy forces occurring in a closed circuit comprising a (usually flat-plate), collector hot water store and the connecting pipework. These forces are produced by the difference in densities of the water in the collector (which is heated by the sun), and that of the cooler water in the store.

In a *two-phase thermosyphon system* the natural circulation circuit contains a fluid with a low boiling point at the pressure prevailing in the system. The liquid absorbs heat when passing through the collector and boils. The gas rises to a heat exchanger, where it gives up its latent heat to the storage medium and returns to the collector in liquid state to recommence the cycle.

An *integral passive solar water heater* comprises one or more tanks, painted black or applied with a selective absorber surface, within a well-insulated box, possibly with reflectors and covered with single, double or even triple glazing of glass or plastic or a transparent insulation material. The first integral passive solar water heaters, just bare tanks of water left out to warm in the sun, were used in some rural areas of southwest USA in the late 1800s. The first commercially manufactured solar water heater was a derivature of such systems, patented in Maryland, USA, in 1891.

To predict analytically the performance of solar energy water heaters, three alternative broad approaches can be identified; simplified models, correlation of performance characteristics from either the rigorous simulation or monitoring of generic systems and rigorous detailed simulation models.

The first two approaches are used to estimate a system's long-term performance and to determine the system size that achieves the optimum solar fraction of total hot water derived. Because of the simplifications inherent in the first approach, such models are limited by the range of operating conditions and system configurations over which the simplifying assumptions are valid. Models in this category often require experimentally determined information which is only obtainable once the system has been constructed. The second approach cannot be applied reliably to these systems whose correlation of dimensions and climatic conditions has not been determined. For a thermosyphon system, the rate and the direction of the flow are dependent on prevailing weather conditions and on the geometry of the pipe-work. An exact mathematical model of the behaviour of the latter system requires the simultaneous solution of the coupled energy and momentum equations.

Solar Drying

The objective in **Drying** an agricultural product is to reduce its moisture content sufficiently to prevent deterioration within a safe storage period. Drying is a dual process of heat transfer to the product from the heating source, and mass transfer of moisture from the interior of the product to its surface; then from the surface to the surrounding air.

In solar drying, solar energy is used as either the sole or a supplemental source of heat; air flow can be generated by either forced or natural convection. The heating procedure could involve the passage of pre-heated air through the product, by directly exposing the product to solar radiation or a combination of both. The major requirement is the transfer of heat to the moist product by convection and conduction from surrounding air mass at temperatures above that of the product, or by radiation mainly from the sun and to a little extent from surrounding hot surfaces, or conduction from heated surfaces in contact with the product. Water starts to vaporise from the surface of the moist product when the absorbed energy has increased its temperature sufficiently from the water vapour pressure of the crop moisture to exceed the vapour pressure of the surrounding air. The rate of moisture replenishment to the surface by diffusion from the interior depends on the nature of the product and its moisture content. If diffusion rate is slow, it becomes the limiting factor in the drying process, but if it is sufficiently rapid, the controlling factor may be the rate of evaporation at the surface. The latter is the case at the commencement of the drying process. In direct radiation drying, part of the solar radiation may penetrate the material and be absorbed within the product itself, generating heat in the interior of the product as well as its surface, thus hastening thermal transfer. The solar absorptance of the product is an important factor in direct solar drying, most agricultural materials have relatively high absorptances of between 0.67 and 0.09. Heat transfer and evaporation rates must be controlled closely for an optimum combination of drying rate and acceptable final product quality. Solar energy dryers vary mainly as to the mode of utilisation of the solar heat and the arrangement of their major features, a classification with illustrative cross-sectional drawings is given in **Figure 6**.

Solar Refrigeration

The solar operation of conventional electrical refrigerators, working on a compression cycle requires the conversion of solar energy into electricity. The dc electricity produced from photocells usually can be used only to operate conventional refrigerators after being converted to ac electricity. An alternative solar operated refrigerator in which the refrigerator itself is also of "conventional" design, involves solar thermal conversion using high temperature solar energy collectors. The high temperature thermal energy produced may be transformed subsequently into mechanical energy via a heat engine, which then drives a refrigerator compressor. High temperature concentrating collectors most suited to this application need daily tracking of the sun.

Design and production of medium temperature solar energy thermal collectors is, however, a relatively simple technology. At most, these would require seasonal tilt-angle adjustments only. Intermittent vapour-absorption refrigeration plants work on a 24 h cycle comprising heating and refrigeration processes matched to the diurnal operation of the sun: undergoing heating process during the day and producing "cold" at night.

Porous solids, termed adsorbents, can physically and reversibly adsorb large volumes of vapour, termed the adsorbate. The concentration of adsorbate vapours in a solid adsorbent is a function of the temperature of the "working pair" (i.e. mixture of adsorbent and adsorbate) and the vapour pressure of the adsorbate. The dependence of adsorbate concentration on temperature, under constant pressure conditions, makes it possible to adsorb or desorb the adsorbate by varying the temperature of the mixture. This forms the basis of the application of this phenomenon in the solar-powered *intermittent vapour sorption refrigeration cycle*.

Water-ammonia has been the most widely used sorption refrigeration pair. The efficiency of such systems is limited by the condensing temperature. For example, cooling towers or desiccant beds have to be used to produce cold water to condense ammonia at lower pressure. Among the other disadvantages inherent in using water and ammonia as the working pair are: heavy gauge pipe and vessel walls are required to withstand the high pressure, the corrosiveness of ammonia, and the problem of rectification (i.e. removing water vapour from ammonia during generation).

One system employs solid absorption using calcium chloride as the absorbent and ammonia as the refrigerant. A reversible chemical reaction takes place when the refrigerant is absorbed by the solid absorbent. This results in physical changes in the mixture. When ammonia is absorbed into calcium chloride, swelling of the mass up to 400% takes place. To overcome this a small quantity of another salt was added to calcium chloride and then ammonia was mixed to prepare a paste, which was subsequently heated in a controlled manner to produce a new granulated absorbent. The heat of adsorption and desorption for the working pair is high; almost twice the latent heat of evaporation of ammonia, a large combined solar collector/absorber area is required.

An adsorbent-refrigerant working pair for a solar refrigerator requires the following characteristics:

1. a refrigerant with a large latent heat of vaporisation,
2. a working-pair with high thermodynamic efficiency,
3. a small heat of desorption under the envisaged operating pressure and temperature conditions and
4. a low thermal capacity.

In addition, the operating conditions of a solar-powered refrigerator (i.e. generator and condenser temperature) vary with its geographical location.

Solar refrigeration is employed primarily to cool vaccine stores. The need for such systems is greatest in peripheral health centres in rural communities of developing countries. In the absence of grid electricity, the vaccine cold chain can be extended to these areas through the use of autonomous solar-energy operated vaccine stores.

Passive Solar Design of Buildings

The environmental function of a building is to mediate between the external climate (with its seasonal and diurnal variation of temperature, illuminance and wind speed) and the more stable conditions which are normally required for human comfort.

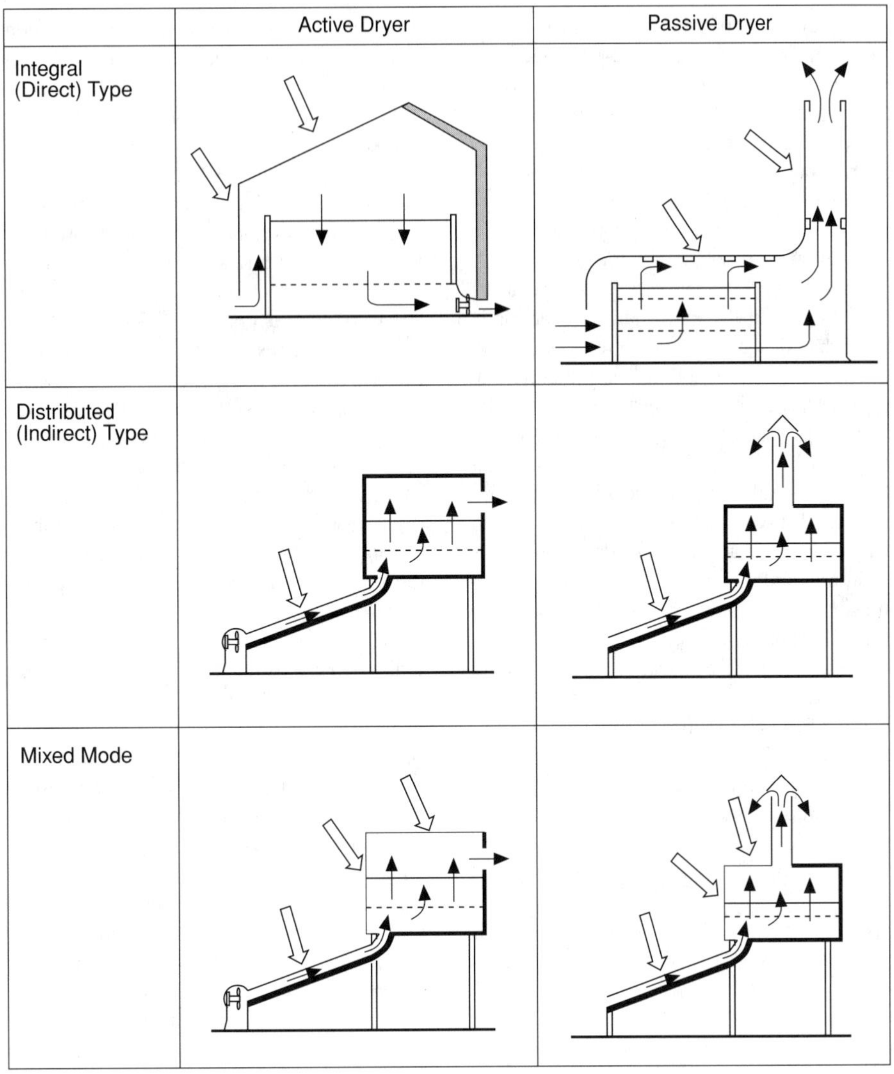

Figure 6. Generic solar dryers.

Climate conscious urban planning, passive solar heating, natural cooling and daylighting of buildings facilitate low energy consumption, comfortable internal conditions and a more benign effect on the wider environment.

Building design is subject to a diverse set of physical constraints (e.g. site, internal arrangement) and functional requirements (e.g. structure, use, circulation areas), successful passive solar design is reconciled harmoniously with these. In temperate climates passive solar design seeks usually to provide heating in cool weather while avoiding overheating in warm weather. Heating involves the distribution, storage and conservation of collected solar energy, overheating prevention involves shading and ventilation. Passive solar design reduces the auxiliary energy load (for heating, ventilating, cooling and/or lighting) and thus the running costs of the building and provides a pleasant, redundant with attractive, attractive internal environment, and, possibly, additional usable space. A purely passive solar building uses no additional energy to collect solar heat. Buildings can incorporate passive features which have active elements, such as fans or moveable insulation; these are termed "hybrid". However, many building features which are commonly regarded as "passive" systems transport energy to the point of use via small fans or motors.

Direct gain is a term used to refer to the system of a room with a southerly facing window. Using direct gain presents particular challenges which building designers must overcome if it is to be used effectively, these include: thermal discomfort due to radiant temperature asymmetry, glare problems, damage from u.v. light to fabrics and finishes and summertime overheating. Direct gain incurs little or no extra cost with simple, and virtually self-functioning, operation.

When solar gain is made to a space with no provision for auxiliary heating and then is distributed to the heated space, that space or element is termed an indirect gain feature. The heat distribution is usually by air convection or conduction through an adjoining wall or a combination of the two. Such unheated, occupiable, indirect gain features include conservatories, sunspaces and atria. Non-occupiable indirect gain features include *Trombe walls,* mass walls and water walls. Natural circulation of air between the conservatory and the heated building occurs when windows or doors open into the conservatory. In some cases purpose-built ducts are provided to ensure a low resistance to air flow. Buoyancy-driven flow is induced by the temperature difference between the warm conservatory and the cooler adjacent room. In summer such action needs to be prevented in order to avoid overheating. If natural-circulation heat gain is to be encouraged, then "flap valves" have to be provided to prevent nocturnal and winter reverse flow when the temperature in the conservatory is less than in the heated building. There is no heat gain from the circulation of air between the conservatory and the building, until the temperature of the air in the conservatory is greater than the heated building temperature. In another indirect gain feature, the *Trombe-Michel wall,* the thermal coupling to the space to be heated is virtually the same as with conservatories, but the relative magnitude of the conductive and convective heat transfers are different, the conductive being larger in the case of a Trombe-Michel wall. The latter delays the delivery of solar heat. They are thus ideally suited to providing heating in the early part of a cold night after a hot sunny day, conditions encountered frequently in arid and mountainous regions. Trombe-Michel walls, which are simply solid glazed walls unlike conservatories provide no additional usable floor area.

Mass walls may be formed by the application of transparent insulation materials to the facade.

Isolated gain passive solar features are those that may be decoupled thermally from the building. This is accomplished via a thermally insulated separating wall, as in thermosyphoning air panels, or by location above the building, as in roof space collectors. A more controllable heat gain combined with—if well designed—an avoidance of summer overheating, is the primary advantage of isolated gain.

Isolated indirect elements do not provide heat storage unless it is insulated during periods of non-solar-collection, or is remote with a controlled convective link, otherwise a net heat loss will ensue. Isolated gain collectors such as the *thermosyphoning air panel* (TAP) overcome some of the disadvantages of indirect gain collectors by dispensing with heat storage and relying totally on convective heat gain. Heat input is almost immediate while heat loss during non-gain periods are low when the collector is isolated from the heated space. This design is ideally suited to the task of providing daytime heat in cool or cold climates. A TAP operates in the same manner as the natural-convection mode of a Trombe wall. However, the absorber is often made of metal, usually aluminium or steel, and the unit is insulated to prevent heat loss to, or from, the building. The problem most commonly associated with passive solar energy systems is control of the heat output. This is not a problem for a TAP as all that is required is for an inlet or exit vent to be closed and thermosyphoning ceases.

The *Barra-Costantini system* is a natural-convection dual-pass solar air heating system. It has the attributes of a Trombe-Michel wall though the heat storage is remote and may be decoupled from the collection of solar energy. A low thermal capacity dual-pass absorber solar collector is decoupled from the south wall of the building; the collector also acts as additional exterior insulation. Buoyancy-driven heated air is distributed throughout the building via channels within the ceilings. Thermal energy storage is provided within these ceilings.

A roof-space solar energy collector, is essentially a pitched roof which is partially of fully glazed on its southerly aspect. Solar-heated air from the roof-space collector is conveyed by an automatically controlled fan via a duct either directly into the living space or as a pre-heated supply to a warm-air space-heating system. When the air from the roof-space solar energy collector is a lower temperature than the set level of the room thermostat, the air stream emerging from the roof-space collector is a pre-heated supply to the gas-fired auxiliary system. A roof-space collector involves the passive collection and active distribution of solar heat and is thus generically a hybrid solar energy system. Ventilation is employed to prevent overheating in high summer. The advantage of a roof-space collector is that it has a low initial capital cost as its physical construction does not differ greatly from that of a conventional pitched roof.

Since passive systems are designed to maximise solar gains, there is a high risk of overheating, not only in the summer but also towards the end of the heating season when most systems should be operating at their maximum performance. The thermal discomfort of unwanted solar gains can be avoided by preventing the initial solar gain by using shading devices or by rejecting the solar gains by ventilating and/or absorbing the solar gain in thermal mass.

With fixed shading devices, the seasonal geometry of solar radiation permits some control of unwanted solar radiation. However, care must be given to the orientation, inclination and the geometry of fixed overhangs and fins. An important advantage of fixed shading devices is that they are self operating.

For a building where the solar input forms a significant proportion of the heating, a responsive control of solar gain is needed. This cannot be provided by fixing shading devices. In temperate climates, buildings have daily variations in heat demand within the same season. Indeed, a building may go from energy surplus to deficit within a few hours. Consider the latitude 52°N: though a fixed horizontal south-facing overhang 1 m wide will completely shade a window about 2 m high in midsummer, unfortunately it will also shade about 10% in midwinter and about 50% in the spring, a time when the performance of a passive system should be at its best.

Furthermore, in many climates, the annual variation in mean daily solar position and mean daily ambient temperature are not in phase. Though the solar motions are the same in September as in May, the corresponding ambient temperatures are not the same. For example, in England a fixed shading device that provides the shading desired in a warm September, unfortunately will also shade during the cooler May when solar gain is useful. Though movable shading devices do not suffer from this lack of response, they may present mechanical difficulties.

Shading devices also influence the view through glazing: an overhang, an opaque blind, a Venetian blind and solar control film may all reduce the solar gain of an aperture by the same amount, but they will alter the view through that aperture very differently.

The most efficient shading is provided by external devices (e.g. awnings), as the solar energy is rejected before it enters the collector. However, external shading devices are usually expensive since they have to be weatherproof. Weatherproofing has control implications; an awning must be withdrawn if the wind is strong even if it is sunny. The control linkage may also be difficult to install and maintain.

Indoor shading devices reflect the solar radiation which has passed through the glazing into the collecting element or zone, back out through the glazing. They are not as efficient as external shades because some of the radiation is reflected and scattered by the glazing back into the collector, and some of the radiation is absorbed onto the surface of the shading device. An important function of all types of shading devices is that they protect the occupants from direct radiation. Direct radiation elevates the effective temperature several degrees above air temperature, thus lowering the threshold temperature at which thermal discomfort is reached.

Ventilation must be considered as a second line of defence against overheating since heat can only be removed when the temperature of the building is already above ambient. To keep the temperature elevation above ambient small by ventilation alone, very large ventilation rates would be required. This may be inconvenient or in some cases impossible to attain. However, ventilation is an important complement to shading—particularly when it convects away gains made by absorption onto internal shading surfaces. Ideally in such circumstances these gains will be removed well away from the occupants. Ventilation is important in spaces with large areas of horizontal glazing such as atria or covered courtyards. If substantial openings at the top and inlets at ground level are provided, the ventilation can be induced by the "stack effect" even on days of zero windspeed. This vertical flow will prevent the build up of hot air in the upper zone of the atrium.

Solar Stills

Basin stills using single effect distillations have been used supplying large quantities of water for isolated communities or for supplying small amounts of water for functions such as battery charging. The conventional basin-type solar still consists of an insulated shallow basin lined or painted with a waterproof black material holding shallow depth (5 to 20 cm) of saline or brackish water to be distilled and covered with a single or double sloped glass sheet, sealed tightly to reduce vapour leakage. A condensate channel along the lower edge of the glass pane collects the distillate. The still can be fed with saline water either continuously or intermittently, but the supply is generally kept as twice the amount of fresh water produced by the still, depending on the initial salinity of the saline water.

Solar radiation transmitted through the transparent (cover) is absorbed in the water and basin and therefore water temperature becomes high compared to the cover. The water loses heat by evaporation, convection, and radiation to the cover and by conduction through the base and edges of the still. The evaporation of water from the basin increased the moisture content in the enclosure and condensation ensures on the underside of the cover which collected s the condensate channels.

Solar Cookers

Solar cookers are broadly of three types:

1. Direct or focussing type,
2. Indirect or box type,
3. Advanced or separate collector and cooking chamber type.

The difference between each of them is as follows:

1. Direct or focusing a solar energy concentrator focus of solar radiation on a focal area at which the cooking pot/pan is located. IN these cookers the convection heat loss from cooking vessel is large and the cooker utilizes only the direct solar radiation.
2. Slot box: In these cookers an insulated hot box (square, rectangular, cylindrical) painted black from inside with double glazing is used. To enhance the solar radiation plane sheet reflectors (single or multiple) are used. Here the adjustment of cooker toward the sun is not so frequently required as in case of direct type solar cooker. This is a slow cooker and takes a long time for cooking and many of the dishes cannot be prepared with this cooker.
3. Indirect: In these cookers, the problem of cooking outdoors is avoided to some extent. The heat or the solar heat is directly transferred to the cooking vessel in the kitchen. The cookers use either a flat-plate or focussing collector which collect the solar heat and transfers this to the cooking vessel.

References

Duffie, J. A. and Beckman, W. H. (1991) *Solar Engineering of Thermal Processes,* J. Wiley and Sons, New York.

Rabl, A. (1985) *Active Solar Collectors and Their Applications,* Oxford University.

Reddy, T. A. (1987) *The Design and Sizing of Active Solar Thermal Systems,* Clarendon Press, Oxford.

Norton, B. (1992) *Solar Energy Thermal Technology,* Springer, Heidelgurg.

Clarke, J. A. (1986) *Energy Simulation in Building Design,* Adam Hilger, London.

Leading to: *Solar cells*

B. Norton

SOLAR ENERGY THERMAL CONVERSION (see Renewable energy; Solar energy)

SOLAR PONDS (see Solar energy)

SOLAR RADIATION (see Environmental heat transfer)

SOLAR RADIATION SPECTRUM (see Solar energy)

SOLAR REFRIGERATION (see Solar energy)

SOLAR SELECTIVE SURFACES (see Solar energy)

SOLAR STILLS (see Solar energy)

SOLAR WATER HEATERS (see Solar energy)

SOLENOIDAL FLOW (see Conservation equations, single phase; Flow of fluids)

SOLID FUELS (see Fuels)

SOLID HOLDUP (see Two-phase flows)

SOLIDIFICATION

Heat Transfer with *solidification* and *freezing* is of importance in many instances, for example in engineering ice making, in freezing of foods, in icing of electric power plants, in frost and ice formation of aircraft, during the casting of the metal, and in many geophysical problems such as frost heaving. (see also **Frosting**)

Recently, solidification has been spotlighted from the viewpoint of latent heat thermal energy storage (LHTES) like ice storage. This type of *energy storage* system contributes to reduced carbon dioxide emission, since the coefficient of performance is considerably improved by employing the LHTES concept. Another application is manufacturing of semi-conductor waifer of silicon.

Mathematically, solidification is associated with the so-called *moving boundary problem* (MBP) which is characterized by the possession of moving interfaces dividing the relevant field into two regions. Such a problem becomes perfectly nonlinear because the position of the moving interfaces are neither fixed in space nor known *a priori.* Due to this nonlinearity, the analytical solution can be found only in limited situations, for example in *Neumann's classical solution* for a one-dimensional problem.

In this entry, only recent advances in experimental, analytical, and numerical aspects in solidification and freezing problem will be introduced.

The reader is referred to the recent books and reviews; e.g. *Freezing and Melting Heat Transfer in Engineering* (ed. K. C. Cheng and N. Seki), Hemisphere, 1991; Solidification of Pure Liquids and Liquid mixtures Inside Ducts and over External Bodies, by S. Fukusako and M. Yamada, *Applied Mech,* Review, 47–12 (1994).

Boundary Fixing Method

An application of the *coordinate transformation method* to the multi-dimensional MBP was proposed by Saitoh. This method can be regarded as an extension of the Landau transformation method to the multi-dimensional case, which was presented qualitatively in his paper of 1950 for the one-dimensional problem. The most remarkable feature of the multi-dimensional MBP is that the moving interface exists in an arbitrarily shaped domain. Therefore, the general numerical method should be one which can consider the arbitrariness of the shapes of both the moving front and the domain boundary.

The boundary fixing method (BFM) considers the arbitrary geometry of both the moving interface and the domain boundary via a change of an independent variable. BFM was also studied by Duda et al., almost concurrently with the author's study.

Computed results for an ellipse with a short diameter of 0.08 m, an aspect ratio of 3/2, and a cooling rate of 3.333 × 10^{-3} K/s is shown in **Figure 1**. The cooling rate C_1 is defined as

$$C_1 = -\frac{T_W(t)}{t} \tag{1}$$

Here, $T_w(t)$ is the cooling surface temperature in t time. Next, as an example of the case in which natural convection in the liquid is combined with heat conduction in the solid, the two-dimensional freezing around a horizontal circular cylinder in quiescent water was solved by the BFM. For the analysis, it was assumed that:

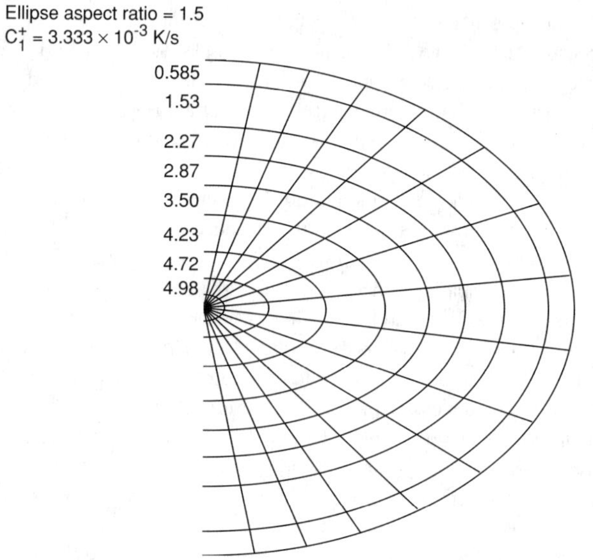

Figure 1. Computed results via BFM for an ellipse.

1. Flow is laminar,
2. Properties are constant,
3. *Boussinesq approximation* is valid.

For brevity, only computed results will be shown. **Figure 2** shows a comparison between numerical and experimental freezing front contours at different times. The water temperature is 6.9°C,

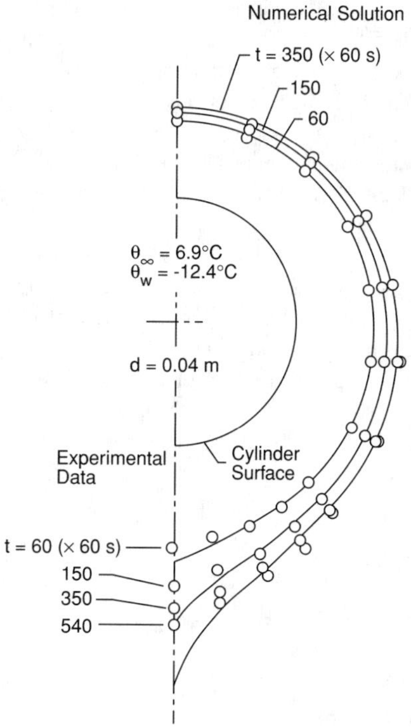

Figure 2. Comparison between the BFM results and the experimental results. The solid lines indicate computed results and the circles the experimental results.

the diameter of the cylinder is 0.04 m, and the cylinder surface temperature is $-12.4°C$. Experimental data obtained by Saitoh and Hirose are shown by the circle. The agreement between the two is excellent except near the vicinity of the downstream edge, where separation may occur in the real flow. This was not considered in the analysis.

As shown in the above, the boundary fixing method is a very simple tool for solving multi-dimensional MBP's. For this reason, this method has been widely used to date.

Enthaply Method

The enthalpy method (previously called the Conventional method) was first developed for one-dimensional MBP's by Eyres, et al. and Baxter. However, a thorough formulation was presented by Katayama and Hattori (1974), and Shamsunder and Sparrow (1975) at almost the same time. In this method, the latent heat is treated as a part of the thermophysical properties: namely, as heat content.

According to Shamsunder and Sparrow the energy conservation equation which holds true in the domain of the phase change is expressed by the following (see **Figure 3**):

$$\frac{d}{dt}\int_V \rho i \, dV + \int_A \rho i \, \vec{v}\cdot\overrightarrow{dA} = \int_A \lambda \nabla T \cdot \overrightarrow{dA} \tag{2}$$

The validity of the above equation is easily confirmed by the fact that the usual energy equation is reducible directly from the above equation.

Katayama and Hattori have introduced an apparent heat content method employing the concept of the Dilac delta function for the freezing and melting problem with a discrete temperature. Temperature function is defined in **Figure 4**.

For example, the finite difference formulation for the one-dimensional problem is represented as

$$\int_{T_{i,j}}^{T_{i,j+1}} c\rho^* \, dT = \frac{\Delta t}{(\Delta x)^2}\left[\int_{T_{i,j}}^{T_{i-1,j}} \lambda \, dT + \int_{T_{i,j}}^{T_{i+1,j}} \lambda \, dT\right] \tag{3}$$

Here, the relationship between c, ρ^*, and L is given by

$$\int_{T_1}^{T_2} c\rho^* \, dT = L\rho + \int_{T_1}^{T_2} c\rho \, dT \tag{4}$$

$$\lim_{e\to 0}\int_{T_{f-e}}^{T_{f+e}} c\rho^* \, dT = L\rho \tag{5}$$

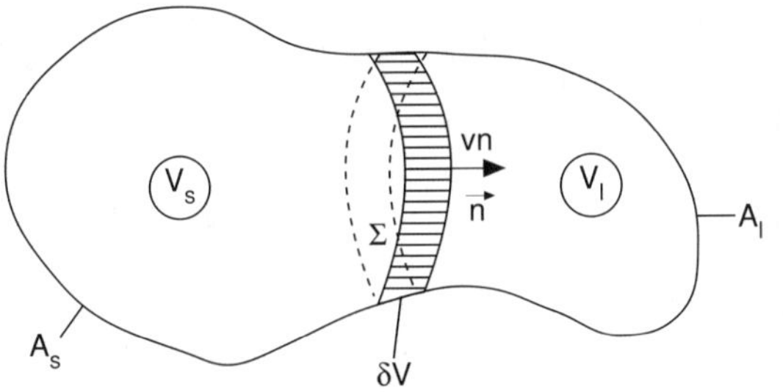

Figure 3. Control volume for derivation of interface condition.

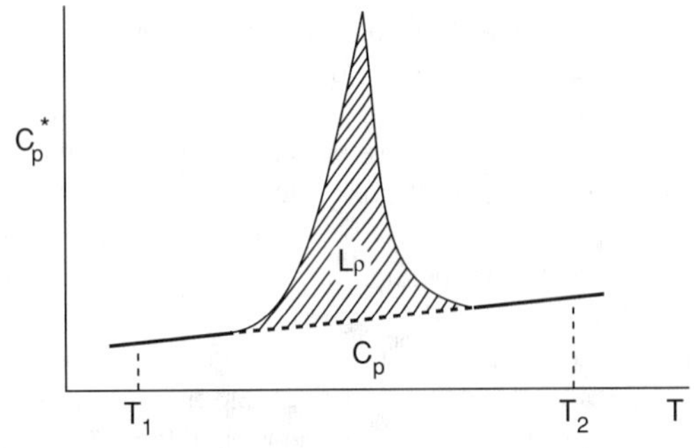

Figure 4. Temperature function.

It is assumed that the latent heat is liberated in the small temperature range 2ε. The temperature function is the determined so that the total heat content coincides with the latent heat L.

A numerical example is shown in **Figure 5** in which computed isotherms are shown when the porous region around two cooled cylinders starts freezing.

Growth Ring Method

Although two well-used methods, i.e., the Boundary fixing method and the Enthalpy method, have many advantages over other existing methods, these methods still have severe disadvantages. Namely, BFM largely depends on the smoothness of the boundary shape function, and with the Enthalpy method it is impossible to include a natural convection effect in the liquid phase.

An alternative method was recently developed by Saito and Kato (1986). This method is based on the multi-lateral element

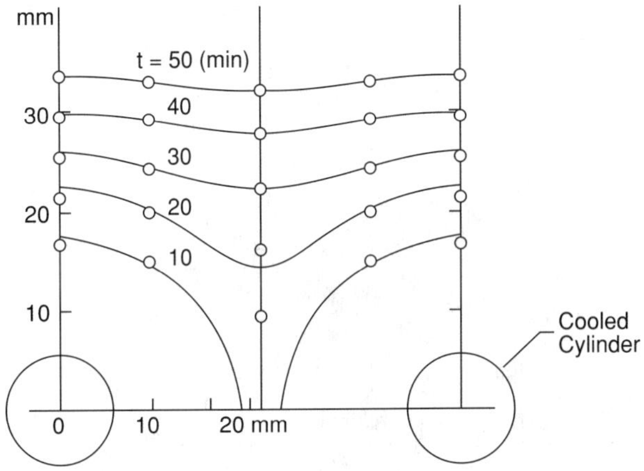

Initial Temperature	4.4°C
mesh Δt hr	1/3600
mesh Δx m	0.003
mesh Δy m	0.003
$a^*\Delta t \left\| \frac{1}{(\Delta x)^2} + \frac{1}{(\Delta y)^2} \right\|$	0.475

Figure 5. Isotherm around two cooled cylinders. (a*: thermal difficulty, $\Delta\chi$, Δy: mesh length, Δt: time step).

(MEM) method, in which the representative point is placed on the barycenter of a triangular or quadrilateral element. Note that it is not the circumcenter but the barycenter where the representative point is positioned. The existing triangular element method (TEM) has used the circumcenter. The barycenter method makes it possible to consider the arbitrary geometry of both the fixed and the moving boundaries, thereby eliminating difficulties encountered with, for example, BFM. Another advantage of this method is that the usual explicit finite difference can be incorporated in the formulation. This considerably reduces the troublesome numerical procedures that are necessary prior to computation. New freezing front positions are calculated by using temperature distribution in the vicinity of the freezing front. Then, the adjacent element is enlarged by the same amount as the newly frozen area thus obtained. This procedure is continued in a step-by-step manner until the size of the frozen area reaches nearly the same element size as that of the old element.

Since the freezing front produced in this method closely resembles the growth rings of trees, it is called the "Growth Ring Method (GRM)".

As an example, the numerical results for the two-dimensional freezing of water in a regular prism initially at its fusion temperature, has been taken up as a standard 2-D freezing problem.

The computed results by the GRM are shown in **Figure 6** in which freezing front contours and generated elements (quadrilateral element in this case) are simultaneously designated. The side length of the regular square, Stefan number Ste, and the time step used in the computation, were 0.079 m, 0.49, and 0.003, respectively. The initial freezing front position in the normal direction was set to be 0.92. Therefore, every circumferential contour beyond this

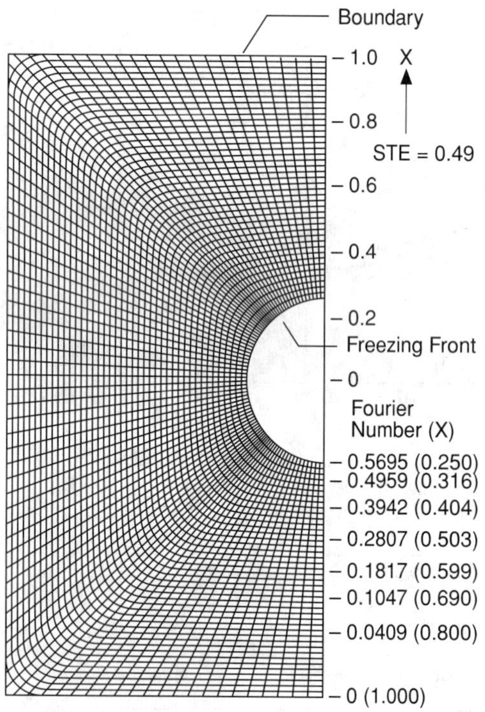

Figure 6. Numerical solution via the GRM for two-dimensional Stefan problem.

initial line denotes the freezing front and the quadrilateral element the mesh generated.

The computer running time (CPU) for this case was some 100s on NEAC 2200/ACOS 1000 (30 MFLOPS) at Tohoku University. Note that the use of the SOR scheme instead of the usual explicit scheme is particularly desirable since the mesh size in this example gets smaller as time elapses.

As seen from the above illustrative examples, the Growth Ring Method is a quite simple, but very powerful method applicable to arbitrary multi-dimensional domains. The GRM can also handle problems involving natural convection in the liquid phase.

It is advantageous that the mesh may be regenerated halfway through the computation in order to reduce the number of elements and resultant computer time.

References

Saitoh T. (1978) Numerical method for multi-dimensional freezing problems in arbitrary domains, *J. Heat Transfer,* Trans. ASME, 100, 294.

Shamsunder N. and Sparrow E. M. (1975) Analysis of multidimensional conduction phase change via the enthalpy model, *J. Heat Transfer,* Trans. ASME, 97, 333.

Katayama K. and Hattori M. (1974) A study of heat conduction with Freezing (1st Report, Numerical Method of Stefan Problem), *Trans. JSME,* 40–33, 1404.

Saitoh T. and Kato K. (1986) Numerical analysis of multidimensional freezing problems via growth ring method, *Proc. 23rd Heat Transfer Symp. Japan*, 695.

T.S. Saitoh

SOLIDIFICATION CONSTANT (see Casting of metals)

SOLID-LIQUID-LIQUID FLOWS (see Multi-phase flow)

SOLIDOSITY (see Liquid-solid separation)

SOLIDS CONCENTRATION (see Liquid-solid separation)

SOLIDS IN LIQUIDS, BOILING HEAT TRANSFER (see Multicomponent mixtures, boiling in)

SOLIDS SEPARATION (see Liquid-solid separation)

SOLID STATE LASERS (see Lasers)

SOLIDS, THERMAL CONDUCTIVITY OF (see Thermal conductivity; Thermal conductivity values

SOLITARY WAVE (see Waves in fluids)

SOLITON (see Waves in fluids)

SOLUBILITY

Following from: *Phase equilibrium*

For materials that are only partially miscible with each other, it is conventional to refer to the concentration of one material in another at a specific temperature and pressure as the solubility of that material. It is a terminology most often used about the solubility of gases in liquids and solids in liquids and this is the context in which we address the topic here but, in principle, it is not confined to this case.

Solubility of Gases in Liquids

Gases are sparingly soluble in liquids so that the solutions formed are always dilute. **Figure 1** shows the mole fraction of a number of gases in water as a function of the pressure which emphasises this fact. It is an empirical observation, supported by simple theory, that for dilute solutions the vapour pressure of the solute is linearly related to its mole fraction in the solution, so that

$$\underset{x_A \to 0}{\mathrm{Lt}}\ p_A = K_A \tilde{x}_A$$

which is known as **Henry's Law** and K_A is the Henry's Law coefficient. Of course, if the vapour pressure of the solvent is very much less than that of total pressure, an approximation to Henry's Law is

$$\tilde{x}_A = p/K_A$$

where p is the total pressure. This behaviour is confirmed by the figure.

Generally, Henry's Law coefficients increase with increasing temperature so that gases are less soluble in liquids at high temperature. A review of experimental methods has been given by Battino and Clever (1966).

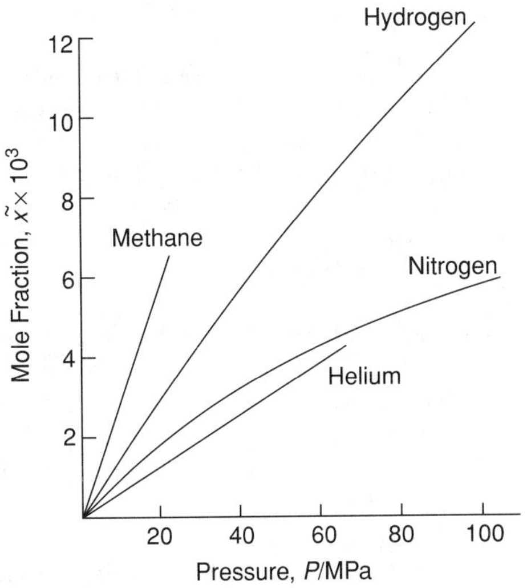

Figure 1. The solubility of gases in water as a function of pressure at a temperature of T = 298 K.

Solubility of Solids in Liquids

The addition of a relatively involatile solute to a solvent lowers the freezing point of the solvent and raises its boiling point. Generally, the depression of the freezing point is written

$$\Delta T_m = K_1 m$$

where m is the molality of the solute and K_1 the so-called *cryoscopic constant.* The *molality of a solution* is the number of moles of a solute in a solution containing 1 kg of solvent and is a particularly convenient practical measure of solution concentration. The corresponding constant for the boiling point is called the *ebullioscopic constant.* Measurement of these constants is employed as a means of determining the molar mass of species in solution (Murrell and Boucher, 1982). These two properties are known, together with the osmotic pressure, as *colligative properties.* They depend upon specific properties of the solvent but on only the molality (or mole fraction) of the solute.

References

Battino, R. and Clever, H. L. (1966) *Chem. Rev.,* **66**, 395.

Hildebrand, J. H. and Scott, R. L. (1950) *The Solubility of Non-Electrolytes,* Reinhold, New York.

Hildebrand, J. H., Prausnitz, J. M., and Scott, R. L. (1970) *Regular and Related Solutions,* Van Nostrand, New York.

Murrel, J. N. and Boucher, E. A. (1982) *Properties of Liquids and Solutions,* Wiley, New York.

Rowlinson, J. S. and Swinton, F. L. (1982) *Liquids and Liquid Mixtures,* 3rd Ed., Butterworths, London.

W.A. Wakeham

SOLUBILITY OF GASES IN LIQUIDS (see Henry's law; Solubility)

SOLUBILITY OF SOLIDS IN LIQUIDS (see Solubility)

SOLUTE (see Binary systems, thermodynamics of)

SOLUTE FUNCTIONALITY (see Gas chromatography)

SOLUTIONS (see Binary systems, thermodynamics of)

SOLVENT (see Binary systems, thermodynamics of)

SOLVENT EXTRACTION (see Extraction, liquid-liquid)

SONIC VELOCITY (see Critical flow)

SOOT (see Burners)

SORET EFFECT (see Thermal conductivity; Thermal diffusion)

SOUR GASES (see Natural gas)

SOUTTER-ION PUMP (see Vacuum pumps)

SPACE HEATING

Following from: *Air conditioning; Buildings and heat transfer*

Space heating systems are designed to satisfy the thermal comfort requirements of building occupants. The interaction of the heating system with the fabric of the building is critical to the comfort achieved and the energy efficient operation of the system.

It is therefore necessary to appreciate the thermal comfort requirements of people (see **Air Conditioning**) and the thermal performance of the building fabric (see **Building and Heat Transfer**) before understanding space heating systems and their operation.

Heating Loads and Energy Estimating Methods

The design of the heating system is based on the steady state heat loss of the building, or the heat output required to maintain *comfort conditions* within the building with an accepted external design temperature. The procedures for calculating space heating loads are described in Chapter 25 *ASHRAE Handbook of Fundamentals* (1993), the *CIBSE Guide* Section A3 (1980) and Section A5 (1979) as well as in Mc Quiston and Parker (1994).

Estimation of the energy requirements and the predicted fuel consumption for space heating however, must be based on a dynamic appraisal of the building throughout the heating season. It must take into account the hours of occupation of the building and the changes in external conditions throughout that time. The efficiency of the heating system and the thermal performance of the building fabric.

Thermal Performance of Building Fabric

Buildings can be divided broadly into heavyweight and lightweight structures. A heavyweight building is slow to respond taking a long time to heat up, but similarly a long time to cool down. It is therefore particularly well suited to buildings that are occupied on a continuous basis and for heating systems that are slow to respond.

Lightweight buildings on the other hand respond quickly both to the external environment and to the space heating system, which must be flexible and controllable to take advantage of such a response. The intermittent use of many new buildings is often suited to a lightweight structure and a responsive space heating system.

Space Heating Systems

Different space heating systems have characteristics that include speed of response, flexibility of control, space required for installation, initial cost of plant and installation, maintenance and energy costs, the fuels they use and their impact on the environment.

Space heating systems can be divided into direct and indirect acting heating. Direct systems convert fuel to heat within the space to be heated, for example open coal fires, gas radiant or convective heaters and the majority of electric heating systems. Indirect systems on the other hand convert the fuel energy into heat in a central position from where it is distributed around the building and emitted to the space. An example is a radiator system served by a hot water boiler or a warm air heating burner.

Detailed descriptions and design consideration of the various system components are given in the *ASHRAE Handbook of HVAC Systems and Equipment* (1992). These include:

Automatic fuel-burning equipment	Chapter 27
Boilers	Chapter 28
Furnaces	Chapter 29
Residential in space heating equipment	Chapter 30
Chimney, gas vent and fire-place systems	Chapter 31
Making up air units	Chapter 32

ASHRAE Handbook of HVAC Applications (1991) describes in detail the various design considerations associated with different applications.

Heat Distribution

Heat can be distributed from a central boilerhouse to the individual heated spaces by water, steam or air. For small scale buildings, water distribution is at atmospheric pressure, but by pressurising the system a greater temperature drop across the circuit can be achieved and therefore greater heat output obtained from the same volume of water. High pressure water, like steam must be utilised by heat emitters out of reach of the building occupants or by means of heat exchangers and secondary circuits.

Heat Emitters

Heat emitters can be classified as radiant or convective although most combine the two modes of heat transfer. (see **Convective Heat Transfer** and **Radiative Heat Transfer**)

Detailed method of calculation is given in Chapter 6 of the *ASHRAE Handbook of HVAC Systems and Equipment* (1992).

References

American Society of Heating, Refrigerating and Air-Conditioning Engineers, Inc., *ASHRAE Handbook 1993,* Fundamentals Volume, Atlanta, GA.

American Society of Heating, Refrigerating and Air-Conditioning Engineers, Inc., *ASHRAE Handbook 1991,* HVAC Applications, Volume, Atlanta, GA.

American Society of Heating, Refrigerating and Air-Conditioning Engineers, Inc., *ASHRAE Handbook 1992,* HVAC Systems and Equipment Volume, Atlanta, GA.

Chartered Institution of Building Services Engineers, (CIBSE) 1979. *CIBSE Guide Book A,* Section A5 Thermal Response of Buildings.

Chartered Institution of Building Services Engineers, (CIBSE) 1980. *CIBSE Guide Book A,* Section A3 Thermal Properties of Building Structures.

Mc Quiston, F. C. and Parker, J. D. (1994). *Heating, Ventilating and Air Conditioning,* Fourth Edition, John Wiley and Sons, Inc., New York.

M.R. Heikal and A. Miller

SPACERS (see Rod bundles, flow in)

SPACERS, EFFECT ON CHF (see Boiling water reactor, BWR)

SPARK-IGNITION ENGINES (see Internal combustion engine)

SPECIFIC HEAT CAPACITY

Following from: Thermodynamics

The specific heat capacity of a pure substance or a mixture is conventionally defined as the heat required to raise the temperature of 1 mole of the substance under specified conditions. However, formally, it is defined as the limit of the ratio $d\tilde{Q}/dT$ as dT, the temperature change, and $d\tilde{Q}$, the heat input, go to zero. It is possible to define several heat capacities depending upon the conditions and we consider first the heat capacity at constant volume $\tilde{c}_v$. In this circumstance, the heat is supplied to the substance, presumed a fluid, in a rigid container of fixed volume. The thermodynamic system is a closed one and the appropriate form of the *first law of thermodynamics* is

$$d\tilde{u} = d\tilde{Q} + d\tilde{W} \tag{1}$$

where $d\tilde{u}$ is the change of *internal energy* and $d\tilde{W}$ the work done. But since $d\tilde{W} = p_{ext}d\tilde{v}$ and $d\tilde{v}$ is zero, the work done is zero and

$$\left(\frac{d\tilde{u}}{dT}\right)_{\tilde{v}} = \tilde{c}_v = T\left(\frac{\partial \tilde{s}}{\partial T}\right)_{\tilde{v}} \tag{2}$$

where $\tilde{s}$ is the **entropy.**

The realisation of a second specific, molar heat capacity, that at constant pressure, for a liquid or a solid, can be performed merely by supplying heat to the sample under a constant pressure, p. The appropriate first law is still the same but

$$d\tilde{Q} = \tilde{c}_p dT \quad \text{and} \quad d\tilde{W} = p d\tilde{v} \tag{3}$$

is not zero so

$$d\tilde{u} = \tilde{c}_p dT - p d\tilde{v} \tag{4}$$

but because $\tilde{h} = \tilde{u} + p\tilde{v}$ we can readily show that

$$\tilde{c}_p = \left(\frac{\partial \tilde{u}}{\partial T}\right)_p = T\left(\frac{\partial \tilde{s}}{\partial T}\right)_p \tag{5}$$

It is also possible to define a specific molar heat capacity at saturation, $\tilde{c}_\sigma$ which is related to the amount of heat that must be supplied to 1 mole of the system for a unit temperature change while maintaining the phase equilibrium along the saturation line. The heat capacity and saturation can be written

$$\tilde{c}_\sigma = T\left(\frac{\partial \tilde{s}}{\partial T}\right)_\sigma \tag{6}$$

The three heat capacities may be interrelated by means of thermodynamic relationships involving the equation of state of the material (Bett et al., 1975). The three heat capacities are functions of the thermodynamic state and it is usual to express $\tilde{c}_p$ as a function of p and T, and $\tilde{c}_v$ as a function of $\tilde{v}$ and T.

Measurements of the heat capacity of solids and liquids are rather easily performed routinely with a very modest accuracy (a few percent) with modern differential scanning calorimeters (Grolier, 1994). However, since for many engineering purposes, one is interested in enthalpy differences between streams with enthalpies of a similar magnitude, such an accuracy is often not adequate. Considerable effort has been devoted to the development of instruments for the measurement of heat capacities. Such instruments are almost always electrically operated calorimeters. In almost all calorimeters a known amount of substance is heated electrically and the temperature rise measured. The system in which this is done can be either closed (for solid, liquid or gas) or steady-state flow (liquid or gas). A comprehensive review of classical methods has been given by McCullough and Scott (1968). Techniques for the measurement of the heat capacity of liquids have recently been reviewed by Grolier (1994) in the context of a wider review of *calorimetry.*

The isobaric heat capacity and the isochoric heat capacity are related by the equation

$$\tilde{c}_p - \tilde{c}_v = T\left(\frac{\partial \tilde{v}}{\partial T}\right)_p \left(\frac{\partial p}{\partial T}\right)_{\tilde{v}} \tag{7}$$

which, for an ideal gas, leads to the simple result

$$\tilde{c}_p - \tilde{c}_v = \tilde{R} \tag{8}$$

For the ideal gas state, the heat capacity may be expressed through statistical mechanics in terms of the contributions to the translational and internal energies of the molecules (Rushbrooke, 1949). In turn, some of the internal contribution arising from rotational, vibrational and electronic modes of motion can often then be determined from spectroscopic measurement of the frequencies of the normal mode of motion of the molecule. For many molecules, this process provides a more accurate means of determining the ideal-gas heat capacity of the material than does direct measurement (de Reuck and Craven, 1993).

As the density is increased from the ideal gas state, the energy of the ensemble of molecules acquires a component arising from the interactions between molecules (the configurational part) and this cannot be evaluated theoretically for any but the simplest molecules so that the only source of information on the heat capacity is then from direct or indirect measurement. When there are no measurements available it is necessary to have recourse to estimation methods (Reid et al., 1975).

References

Bett, K. E., Rowlinson, J. S., and Saville, G. (1975) *Thermodynamics for Chemical Engineers,* Athlone, London.

de Reuck, K. M. and Craven, R. J. B. (1993) *International Thermodynamic Tables of the Fluid State—12: Methanol,* Blackwell Scientific, Oxford.

Grolier, J.-P. (1994) *Heat Capacity of Organic Liquids in Experimental Thermodynamics IV, Solution Calorimetry,* K. N. Marsh and P. A. G. O'Hare, eds., Blackwell Scientific, Oxford.

McCullough, J. P. and Scott, D. W. (1968) *Experimental Thermodynamics I, Calorimetry of Non-Reacting Systems,* Butterworths, London.

Reid, R. C., Prausnitz, J. M., and Sherwood, T. K. (1977) *The Properties of Gases and Liquids,* McGraw-Hill, New York.

Rushbrooke, G. S. (1949) *Introduction to Statistical Mechanics,* Clarendon, Oxford.

W.A. Wakeham

SPECIFIC WORK, IN TURBINES (see Turbine)

SPECKLE METHOD (see Photographic techniques)

SPECKLE PHOTOGRAPHY (see Visualization of flow)

SPECTRA, EMISSION AND ABSORPTION (see Spectroscopy)

SPECTRAL ANALYSIS

A continuous or discrete time-series, such as $x = x(t)$ or $x_n = \{x_0, x_1, \ldots\}$, can be analyzed in terms of time-domain descriptions and frequency-domain descriptions. The latter is also called spectral analysis and reveals some characteristics of a time-series which cannot be easily seen from a time-domain description analysis. Spectral analysis is used for solving a wide variety of practical problems in engineering and science, for example, in the study of vibrations, interfacial waves and stability analysis.

In spectral analysis, the time-series is decomposed into sine wave components using a sum of weighted sinusoidal functions called spectral components. The weighting function in the decomposition is a density of spectral components or spectral density function.

The actual method of decomposing a time-series into a sum of weighted sinusoidal functions is to use Fourier transform which has both continuous and discrete versions corresponding to continuous time-series of type $x = x(t)$ and discrete time-series of type $x_n = \{x_0, x_1, \ldots\}$, respectively. Most recorded time-series data in engineering practice are of discrete type and numerical calculations of the Fourier transform are usually done using digital computers which can only deal with discrete data and therefore use discrete Fourier transform. In view of this, the spectral analysis of only discrete time-series is described below.

Fourier Transform

We assume a discrete time-series with a finite number of samples N and a sampling time interval T_S between two successive samples

$$x_n = \{x_0, x_1, \ldots, x_{N-1}\} \quad (1)$$

According to the mathematical theory of Fourier analysis, the above time-series can be represented by the following inverse finite Fourier transform

$$x_n - \bar{x} = \frac{1}{N}\sum_{m=1}^{N-1} G(m)e^{j2\pi mn/N} \quad (n = 0, 1, \ldots, N-1) \quad (2)$$

where $\bar{x}$ is the average value of the time-series, $j = \sqrt{-1}$ denotes the symbol of complex number and $e^{j\theta}$ is a complex sinusoidal function (Note: $e^{j\theta} = \cos\theta + j\sin\theta$). G(m) is the spectral density function (or weighting function) mentioned previously which can be calculated from the following finite Fourier transform

$$G(m) = \sum_{n=0}^{N-1} (x_n - \bar{x})e^{-j2\pi(m/NT_S)nT_S} = \sum_{n=0}^{N-1} (x_n - \bar{x})e^{-j2\pi mn/N}$$

$$(m = 0, 1, \ldots, N-1) \quad (3)$$

where $m/NT_s = f_m$ is the discrete frequency and $nT_s = t_n$ is the discrete time. It should be noted that the $(x_n - x)$ in **Equation 1** and the G(m) in **Equation 2** are equivalent measures in time and frequency domains respectively which are related to each other by the Fourier transform. T_s does not appear in **Equations 2 or 3** and is only used as a scaling factor when calculating frequencies.

Frequency Spectrum

In general, the Fourier transform G(m) is a complex-valued function and the plot of G(m) versus f_m is called *frequency spectrum.* G(m) can be expressed in polar form as

$$G(m) = |G(m)|e^{j\angle G(m)} \quad (4)$$

The modulus |G(m)| and the angle G(m) are called the magnitude and the phase, respectively, of the Fourier transform. The plot of |G(m)| versus f_m is called the *magnitude* or the *amplitude spectrum* and the plot of G(m) versus f_m is called the *phase spectrum.*

Power Spectrum

The auto-correlation function for the discrete time-series given in **Equation 1** is defined as (see entry for **Correlation Analysis**).

$$R_{xx}(k) = \sum_{n=0}^{N-1-k} (x_n - \bar{x})(x_{n+k} - \bar{x}) \quad (k = 0, 1, \ldots, N-1) \quad (5)$$

The Fourier transform of the auto-correlation function is given by

$$P_{xx}(m) = \sum_{k=0}^{N-1} R_{xx}(k)e^{-j2\pi mk/N} \quad (m = 0, 1, \ldots, N-1) \quad (6)$$

where $P_{xx}(m)$ is called the *power spectral density function.* The plot of $P_{xx}(m)$ versus f_m is called the *power spectrum* corresponding to the time-series given in formula (1). It can be mathematically proved that the following relation exists between the power spectrum $P_{xx}(m)$ and the frequency spectrum G(m)

$$P_{xx}(m) = |G(m)|^2 \quad (m = 0, 1, \ldots, N-1) \quad (7)$$

which shows that $P_{xx}(m)$ is a real-valued function with a zero phase. The power spectrum is an average measure of the frequency-domain properties of the time-series, which shows whether or not a strongly periodic or quasi-periodic fluctuation exists in the time-series.

Cross Spectrum

The cross-correlation function for two sets of time-series data

$$x_n = \{x_0, x_1, \ldots, x_{N-1}\} \quad \text{and} \quad y_n = \{y_0, y_1, \ldots, y_{N-1}\}$$

is defined as (see **Correlation Analysis**)

$$R_{xy}(k) = \sum_{n=0}^{N-1-k} (x_n - \bar{x})(y_{n+k} - y) \quad (k = 0, 1, \ldots, N-1) \qquad \textbf{(8)}$$

and its Fourier transform is given by

$$P_{xy}(m) = \sum_{k=0}^{N-1} R_{xx}(k)e^{-j2\pi mk/N} \quad (m = 0, 1, \ldots, N-1) \qquad \textbf{(9)}$$

where $P_{xy}(m)$ is called the cross spectral density function or the *cross spectrum* which is a generally complex-valued function. The cross spectrum represents the common frequencies appearing in both the time-series x_n and y_n.

Coherence Function

The coherence function is defined as

$$K_{xy}(m) = \frac{|P_{xy}(m)|}{\sqrt{P_{xx}(m)P_{yy}(m)}} \quad (m = 0, 1, \ldots, N-1) \qquad \textbf{(10)}$$

which is a real-valued frequency function the value of which at a particular frequency f is a measure of similarity of the strength of components in x_n and y_n at that frequency. The value of K is such that $0 \leq K \leq 1$, and the larger the K, the more strongly correlated are the x_n and y_n at a given frequency. Therefore, K behaves like a correlation coefficient for x_n and y_n components at the same frequency.

All the above definitions, though given for discrete time-series, are equally applicable to continuous time signals. More details of spectral analysis can be found in the references.

References

Gardner, W. A. (1988) *Statistical Spectral Analysis, a Non-probabilistic Theory,* Prentice-Hall, Inc., New Jersey.

Linn, P. A. (1989) *An Introduction to the Analysis and Processing of Signals, 3rd Edition,* Macmillan Press Ltd., London.

Schwartz, M. and Shaw, L. (1975) *Signal Processing: Discrete Spectral Analysis, Detection and Estimation,* McGraw-Hill, Inc., USA.

***Leading to:** Correlational analysis*

Lei Pan

SPECTRAL DENSITY FUNCTION (see Spectral analysis)

SPECTRAL EMISSIVITY (see Emissivity)

SPECTRAL EXTINCTION METHOD (see Optical particle characterisation)

SPECTROFLUORIMETRY (see Fluorescence)

SPECTROSCOPY

Spectroscopy refers to the interaction of component wavelengths or quanta of the electromagnetic spectrum with atoms and molecules. Quanta of energy are absorbed into the atom or molecule and induce excitation of the electrons surrounding the nucleus, normally the outermost electrons involved in forming chemical bonds in the atom or molecule. The excitation can be lost by emitting a photon of the same or lower energy as that absorbed initially, or by transfer into vibrational and rotational energy of the molecular structure, degrading to heat. The different wavelengths of the spectrum absorbed and the manner of detection of their absorbance or re-emission as another quantum define the different varieties of spectroscopy. Spectroscopic measurements are in essence measurements of the intensity and wavelength of radiative energy.

The spectroscopic methods described below are used as analytical chemical methods to probe molecular environments with specific wavelengths of light, often with the aim of measuring concentrations of specific components of a mixture.

Alternative excitation methods include thermal excitation and chemical reactions. The emission of black body radiation from a heated source is well known and analysis of the energy from such a source led to the quantum theory of energy, with energy defined as discrete packages or quanta, rather than as a continuum. Light emission from ceramics by flame heating leads to the glow of a gas mantle and the radiant heat of a gas fire. Electric lights function because the filament of tungsten is heated to emit a white light while fluorescent tubes operate by applying a high voltage to an inert gas such as neon. Temperature measurement by optical pyrometry depends on the emission of light from hot bodies. Chemical reactions giving rise to light emission include those used by glow-worms and fireflies and the phosphorescence of the ocean observed at night.

Some analytical methods are termed spectroscopy although they are resonance methods involving absorption of energy. These include *Mossbauer, electron and nuclear magnetic resonance spectroscopy. Mass and X-ray spectroscopies* involve ionisation of molecules and atoms. These are defined after those methods which investigate molecular interactions with the electromagnetic spectrum.

The **Electromagnetic Spectrum** ranges from X-rays with a wavelength of 10^{-10} m, through vacuum ultraviolet, (10^{-8} m), ultraviolet (4×10^{-7} m), visible light (4.2 to 7.5×10^{-7} m), infrared (10^{-6} to 10^{-3} m) through microwaves (10^{-2} to 1 m) to radiowaves (10^2 to 10^4 m).

Atomic Spectroscopy

Atomic spectroscopy refers to the interaction between UV and visible light and the electron energy levels of atoms surrounding the nucleus. Electrons can be excited by various methods, such as collisional energy at high temperature, to energy levels higher than those occupied in the ground state and to different orbitals; *emission spectra* can be observed as the series of wavelengths emitted as the excited electrons move back to lower energy levels. *Absorption spectra* are observed if the atoms are irradiated with a light source giving a continuous spectrum, when those characteristic wavelengths corresponding to promoting an electron to a higher energy level are absorbed, giving a dark line on the continuum. Herzberg (1944) and Atkins (1986) describe the origins of atomic spectra. The observation of atomic spectra in light emitted from stars led to the understanding of stellar evolution and the formation of the elements (See Greenwood and Earnshaw, 1989, for a detailed discussion).

Molecular Spectroscopy

Molecular spectroscopy covers the interaction of electromagnetic radiation with molecules rather than atoms and normally involves UV and visible wavelengths. For absorbance spectroscopy, the experimental arrangement allows detection of the transmitted energy in the axis of the incoming radiation. For methods such as **Fluorescence, Phosphorescence** and *Raman spectroscopy,* the emitted light is much weaker than the incident light, and emission is observed perpendicular to the incident beam.

Ultraviolet spectroscopy

UV *absorbance,* UV-*fluorescence* and *UV-phosphorescence spectroscopy* are included within this group. In its simplest form, UV-absorbance indicates the presence of an absorber of UV light in the effluent from a chromatographic system by the attenuation of a beam of UV light of fixed wavelength. A diode array detector permits the simultaneous observation of the attenuation of many wavelengths as a compound or mixture passes through the detector. UV absorbance spectroscopy normally involves the measurement of the intensity of the transmitted, attenuated light on passage through the sample or compound in comparison with the intensity of incident light. The absorbance at specific wavelengths is characteristic of the structure of the absorbing molecule, being determined by the energy levels of the outer electrons. Denney and Sinclair (1987) describe the technique.

UV-Fluorescence spectroscopy

UV-Fluorescence spectroscopy follows from UV-absorbance where the absorbed energy or photon is emitted as a photon of lower energy than that absorbed. Absorbance of photons excites a singlet state electron in the ground vibrational state to an electronically excited state with various vibrational energy levels. Energy loss by collisions transfers electrons to the lowest vibrational level of the excited state. Fluorescence is observed when these electrons fall back to the various vibrational levels of the electronic ground state. The photons emitted as fluorescence are at longer wavelengths (lower energy, towards the visible region) than those absorbed. (See also **Fluorescence**)

A routine application of UV-fluorescence is the use of "whiteners" in washing powders; these compounds absorb any available UV light and emit in the blue end of the visible spectrum, enhancing the brightness or "cleanliness". Applications include the quantitative analysis of specific compounds at low abundance in the environment and in clinical studies where a fluorescent tag may be added to tissues to reveal the onset of disease (Rendell, 1987).

UV-Phosphorescence spectroscopy

UV-Phosphorescence spectroscopy is similar in mechanism to UV-fluorescence except that the energy of the singlet excited state transfers by inter system crossing to a triplet state, of lower energy than the singlet excited state, and subsequently emits photons as the electrons fall back to the ground state. The intersystem transfer is a forbidden transfer and occurs on a much longer time scale than fluorescence. Consequently, the emission of light can be separated from the illumination in time by having a rotating shutter to deflect the incident radiation while observing the emitted radiation. The sample can be cooled in liquid nitrogen to remove some of the vibrational and collisional modes of loss of energy and enhance the phosphorescence (Rendell, 1987). (See also **Phosphorescence**)

Visible spectroscopy

Many chemical reactions occur with visible colour changes, for example the addition of water to anhydrous copper sulfate produces a blue colour, acid-base titrations often use an indicator which exhibits a change of colour with a change in pH such as phenolphthalein (colourless to red), litmus (blue to red). The breathalyser used to monitor ethanol in the expired air from the lungs relies on a colour change from yellow to green; for a positive indication, a separate test for ethanol is made on a blood sample using gas chromatography. These colour changes accompany changes in the atomic or molecular structure of the coloured compound. They are of particular importance in clinical and forensic analysis where rapid or continuous analyses may be required to observe and determine changes in bodily fluids such as urine and blood. Air pollutants such as formaldehyde from car exhausts and combustion sources can be monitored by forming a complex in concentrated sulfuric acid solution which is blue and specific to formaldehyde, enabling its estimation at the parts per million level.

The same principles for obtaining spectra apply as in UV spectroscopy; for the establishment of an analytical method, a wavelength is chosen which is strongly absorbed by the analyte but not absorbed by other components of the mixture. Denney and Sinclair (1987) describe the technique.

Near infra-red spectroscopy

This use of spectroscopy developed over the last 30 years or so from an attempt to develop a rapid method for the analysis of moisture in grain. Instruments featuring low resolution and high signal to noise ratios were developed to allow rapid scanning. Current uses of the technique concentrate mainly in the analysis of agricultural samples such as the measurement of protein in wheat. The advantages are simple sample preparation, speed and precision of analysis, and the ability to analyse for several constituents at the same time. The chief disadvantages are the dependence on a calibration procedure and the need for dedicated instrumentation. The technique is described in Creaser and Davies (1988).

Infrared spectroscopy

Molecular vibrational modes are examined by infra-red methods. The radiation requires the use of materials such as potassium bromide for the lenses and optical systems. For a vibrational mode to be infra-red active, the motion corresponding to a normal mode should be accompanied by a change of dipole moment. Rotational effects in liquid or solid samples can not be distinguished, but lead to line broadening of vibrational modes, in effect blurring the rotational structure. In chemical analysis, this blurring leads to a useful simplification since the vibrational spectra give rise to characteristic frequencies of absorption, with intensities transferable between molecules. Vibrational modes can be assigned in an unknown spectrum from a table of values from standard molecules. The advent of IR-microscopy has enabled the examination of small samples without the need to pelletize the sample in KBr (Atkins, 1986).

Raman spectroscopy

Raman spectroscopy permits the examination of vibrational and rotational energy levels by light scattering. Photons in a monochromatic light beam collide with the molecules and are scattered with lower or higher energy, in a different direction than the incident beam. The scattered photons give rise to Stokes radiation (lower energy) and anti-Stokes radiation (higher energy). All linear molecules and diatomic molecules, homo- or hetero-nuclear, are active in rotational Raman. Normal modes of vibration are Raman active if the polarizability changes during the vibration; if the molecule has a centre of symmetry, no vibrational modes can be active in both Raman and Infra-red. (Atkins, 1986).

X-ray spectroscopy

In X-ray photo-electron spectroscopy, the incident photon, an X-ray, is of sufficient energy to eject an electron from the core of the electron structure of an atom. These inner electrons are not much affected by changes in energy levels of the outer or valence electrons which reflect the molecular context of the atom. Hence the observed inner shell ionisation energies are a property of the atom and characteristic of the elements present in a mixture or alloy. An alternative name is ESCA (electron spectroscopy for chemical analysis). The technique is limited to solid samples and to their surfaces because the emitted electrons can only escape from surface layers (Atkins, 1986).

Nuclear magnetic resonance spectroscopy

Nuclei of atoms with spin 1/2 possess magnetic moments, with discrete energy levels when placed in a magnetic field. If the sample is bathed in radiation of an appropriate frequency to come into resonance with the energy separations of the nucleus, then there is a strong coupling between the nuclear spin and the radiation, with absorption of the radiation. NMR is the study of the properties of molecules containing magnetic nuclei by observing the magnetic fields at which they come into resonance with an applied radiofrequency field. The resonance frequency depends on the local electron structure of the atomic environment and reveals chemical interactions; the effect is called the chemical shift of the magnetic resonance. To ensure homogeneity of the magnetic field, the sample is rotated rapidly within the magnetic field. The most commonly studied nuclei are hydrogen, carbon (^{13}C) and phosphorus (^{31}P) although other nuclei can be studied. Solids and liquids are examined as complex mixtures, although only pure compounds were allowed near spectrometers in the early days. Fourier transform techniques have been applied with many different techniques of pulsing the applied fields (Atkins, 1986).

Electron spin resonance spectroscopy

Also termed electron paramagnetic resonance, this method is the study of molecules containing unpaired electrons by observation of the magnetic fields at which they come into resonance with monochromatic radiation. The appropriate radiation is in the X-band microwave region. Free radicals and charge transfer complexes are the type of molecular formations which can be studied, either stable in mixtures or formed by radiation.

Mössbauer spectroscopy

This method uses gamma rays, photons at the extreme end of the electromagnetic spectrum, and depends on the resonant absorption of the photon by the nucleus of the target atom. The decay of ^{57}Co to ^{57}Fe in an excited state occurs slowly, with a half life of 270 days. The excited Fe atom decays rapidly to its ground state, emitting a gamma ray. When this gamma ray strikes another ^{57}Fe atom in the ground state, it is absorbed resonantly if its energy matches the energy difference between the excited state and the ground state of the target Fe atom. When the excited atom emits a photon, the atom recoils because the photon possesses a significant linear momentum, giving a Doppler shift to the photon velocity. If the emitter and target Fe atoms are held in rigid foils, the recoil is taken up by the whole crystal and the Doppler shift is very small. In consequence, the energy of the photon can be modified by moving the source foil at a known velocity over a range of a few mm/sec. The target undergoes resonant absorption and the quadrupole moment of the iron nucleus splits the signal; the change in position of the resonance, called the isomer shift, is a function of the electrostatic interaction with the surrounding electrons which changes with chemical environment of the iron. The importance of the technique lies in the many biological and technological occurrences of iron, such as plant proteins, blood and rust. ^{119}Sn also can be examined by Mössbauer spectroscopy.

Mass spectroscopy

Ionisation methods involving photons may be classed as spectroscopy since the interaction with the outer electrons of molecules in the gas phase or solid state involves excitation of an electron to an infinite distance, leaving a positively charged ion (photoionisation). In the solid state, the technique is known as laser desorption ionisation or matrix assisted laser desorption; the mechanism of ion formation results in the rapid transfer of intact molecule ions into the vapour phase.

Mass spectroscopy is a general term for mass spectroscopy using photoplates for detection of all ion beams (m/z values) simultaneously. However, where ion detection is by the sequential focusing of selected ion beams onto an electron multiplier or similar detector, the technique is normally referred to as mass spectrometry.

References

Analytical Applications of Spectroscopy, (1988) C. S. Creaser and A. M. C. Davies, eds., Royal Soc. Chem., London, UK.
Atkins P. W. (1989) *Physical Chemistry, 3rd. Ed.,* Oxford University Press, Oxford, UK.
Denney R. C. and Sinclair R. (1987) Visible and ultraviolet spectroscopy, *Analytical Chemistry by Open Learning (ACOL),* John Wiley and Sons, Chichester, UK.
Greenwood N. N. and Earnshaw A. (1984) *Chemistry of the Elements,* Pergamon Press, Oxford, UK. reprinted 1989.
Herzberg G. (1944) *Atomic Spectra and Atomic Structure, 2nd ed.,* Dover Publications, New York.
Rendell D. (1987) Fluorescence and phosphorescence spectroscopy, *Analytical Chemistry by Open Learning (ACOL),* John Wiley and Sons, Chichester, UK.

A.A. Herod and R. Kandiyoti

SPECULAR REFLECTION (see Reflectivity)

SPEED OF LIGHT

Following from: *Optics*

The Conférence Générale des Poids et Mésures (1983) defined the *metre* to be "the length of the path travelled by light in a vacuum during a time interval of 1/299792458 of a second". This sets the speed of light to be $c = 299792458 \text{ m s}^{-1}$ exactly.

The simplest propagating sinusoidal wave may be written in the form $A = A_0 \cos(kx - \omega t)$, where A_0 is the amplitude. The term in the bracket is the phase of the wave, x and t being position and time. Thus, a surface of constant phase propagates with the *phase velocity* $v = \omega/k$, where $\omega = 2\pi f$ is the angular frequency for frequency f and $k = 2\pi/\lambda$ is the **Wave Number** for wave length λ. In a vacuum $v = c$. Otherwise $v = c/m$, where m is the **Refractive Index.** The phase velocity may be measured by **Interferometry.**

Reference

M. Born and E. Wolf (1980) *Principles of Optics 6th ed.,* Pergamon Press, Oxford.

A.R. Jones

SPEED OF SOUND (see Compressible flow; Maxwell-Boltzmann distribution; Velocity of sound)

SPHERE, DRAG COEFFICIENT FOR (see Drag coefficient)

SPHERES, CONVECTIVE HEAT AND MASS TRANSFER (see Immersed bodies, heat transfer and mass transfer)

SPHERES, DRAG AND LIFT (see Particle transport in turbulent fluids)

SPHERES, SOLID, DRAG ON (see Stokes law for solid spheres and spherical bubbles)

SPHERICITY (see Liquid-solid flow)

SPIRAL CLASSIFIER (see Classifiers)

SPIRAL HEAT EXCHANGERS

Following from: *Heat exchangers*

It is a popular misconception that Spiral Heat Exchangers are a recent development. In fact, the Spiral concept was first proposed as long ago as the 19th century. Only the lack of suitable materials and manufacturing techniques delayed its development into a fully-fledged product until the 1930s. Since that time, acceptance of the Spiral has seen steady growth and this exchanger type is used today in many industries including chemical, steel and pulp and paper.

The first Spiral Heat Exchanger was extremely simple in concept. It consisted of two metal strips, bent into a nearly circular shape to form two concentric channels through which media would flow in opposite directions. Channel spacing was achieved by a steel bar along the length.

The heat transfer capacity of the exchanger was dictated by the width of the channels. Right up until the 1960s, this effectively meant a maximum capacity of 200m^2 since steel strip was only available in relatively narrow widths. Attempts to increase capacity by fabricating larger areas met with only limited success, since they resulted in long, thin channels with excessively high pressure drops. Once wider materials became available and wider channels could be formed, however, heat transfer capacity was progressively increased. Today, practical maximum capacity of a standard Spiral Heat Exchanger is $400\text{–}600\text{m}^2$. Currently the spiral is manufactured in a winding process using a D-shaped mandrel with the two strips being welded to a central plate and distance studs have replaced the steel bars. Alternatively tubular centres are becoming more common.

Usually, alternate edges of the passages are closed and covers fitted to both sides of the spiral assembly.

Heat Transfer Relationships

Turbulent flow (Re > approx 500) in spiral channels

Heat transfer data for spiral heat exchangers is empirically correlated using a conventional *Dittus-Boelter type relationship* for turbulent flow. A channel curvature component is added in order to take into account the somewhat improved heat transfer generated by secondary flow effects. The equation takes the following form:

$$\text{Nu}\,\text{Pr}^{-1/3}(\eta/\eta_w)^{-0.17} = 0.023(1 + 3.54.d_H/D)\text{Re}^{0.8} \quad \textbf{(1)}$$

where Nu is **Nusselt Number**, Pr is **Prandtl Number,** η is bulk fluid viscosity, η_w is fluid viscosity at the wall, d_H is channel hydraulic diameter, D is diameter of the spiral and Re is Reynolds Nomber.

The term d_H/D represents the local channel curvature, which, for a channel of constant spacing, will vary from a maximum at the centre of the body to a minimum at the periphery.

Non-turbulent flow (Re < approx 500) in spiral channels

Test data and, to a certain extent, results from installed units indicate the existence of two regions for spiral flow:

Pure laminar flow:

$$Nu(\mu/\mu_w)^{-0.17} = 7.6 \quad (2)$$

Transition-laminar flow:

$$Nu\ Pr^{-1/3}(\mu/\mu_w)^{-0.17} = 0.33(d_H/D)^{1/6}Re^{0.475} \quad (3)$$

For non-turbulent flow therefore, the higher of the two Nu values obtained from the above equations is the one that applies.

Different Spiral Types

The main feature of this exchanger type is that there is a single passage for each fluid. In actual operation, the cold fluid enters at the periphery and flows towards the centre where it exits via the cover. The hot fluid goes in the opposite direction, giving counter-current flow (See **Figure 1**). The single channel makes the unit well suited to handling fouling liquids with the original design being known as the Type I. Illustrated is an installation at Novo Nordisk A/S in Denmark where cold untreated sludge is heated by hot treated sludge in two spirals. A further unit is used for final cooling of the treated sludge.

Another typical installation is at C. Davidson and Sons, Mugiemoss Mill in Aberdeen, Scotland. Here spiral exchangers cool sealing water for paper and board machine vacuum pumps.

Type II

The Type II Spiral was developed to handle the growing demand for vapourising and condensing capabilities within the process industries. Although it operates on the same basic principle as the Type I, it differs most significantly in terms of channel geometry. It has only one medium flowing spirally. The other flows crosswise, parallel to the axis of the spiral element. The spiral channel is closed on both sides with the crossflow fluid flowing through the spiral annulus (See **Figure 2**).

Type II Spirals are used in duties involving large volumes of vapour, vapour/gas or vapour/liquid mixtures. The channel geometry makes it possible to combine high liquid velocity in the Spiral passage with very low pressure drop on the vapour/mixture side. They are also occasionally used in liquid/liquid applications where one side has to cope with a much larger volume of liquid than the other, such as some fermenter cooling cases.

Type III

When the Type II is used as a condenser it achieves very little sub-cooling of the vapours or condensate. For applications where this is a necessary part of the process, a different type of Spiral had to be developed—the Type III.

The unit is constructed with (normally) alternately welded channels. The lower face of the body is fitted with a cover, while the upper face is fitted with a distribution cone such that the outer turns are closed and the inner turns are open to the cross-flow of the fluid entering the unit. The periphery of the unit is provided with an upper connection for the removal of residual gas/vapour, and a lower connection for the condensate. The cooling medium side is in spiral flow throughout.

The function of the unit is that of condensing a vapour or vapour mixture with or without non-condensable gas in which it is required to cool the residual vapour/gas mixture to as low a temperature

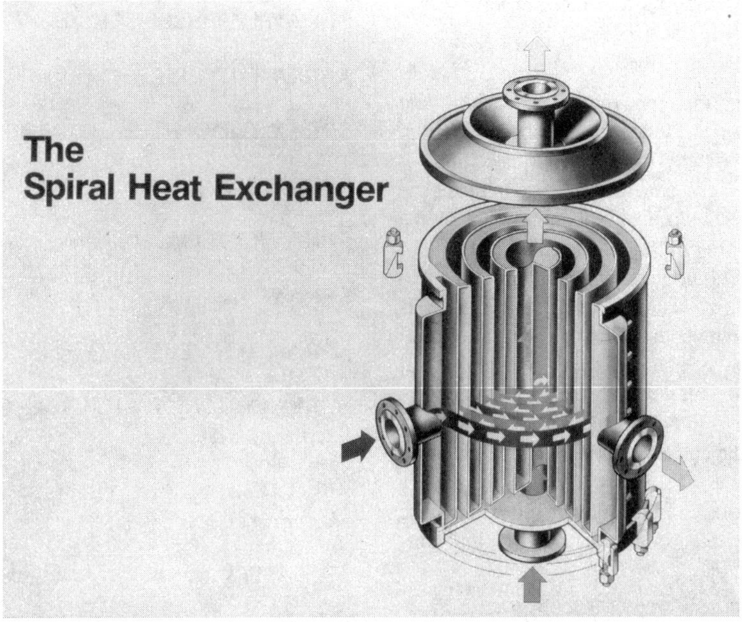

Figure 1.

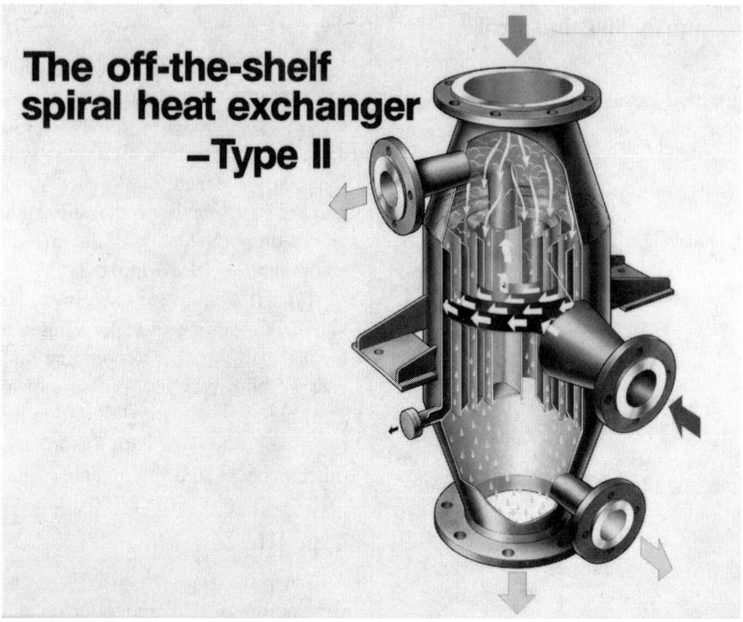

Figure 2.

as possible and thus obtain maximum possible condensation. A secondary feature is that the condensate is effectively subcooled, the outer turns being in countercurrent flow to the coolant. That the flow is in the spiral mode in the outer turns results in higher heat and mass transfer coefficients than would be obtained with the vapour in crossflow only. The SHE type III is best suited for vapour mixtures at moderate pressure containing small to moderate amounts of non-condensable gas. Operation at very low absolute pressure ("high vacuum") is seldom feasible due to the resulting excessive pressure drop in the outer turns.

Type G

The process industry uses columns and reactors extensively and the Type G Spiral was developed to meet the need for a custom built unit which would be vertically mounted onto a column or reactor. The advantage of this arrangement is that it eliminates the need for a separate condenser and, more importantly, all of the large vapour pipework and reflux drum associated with it.

In this model, vapour enters through an open centre tube and then rises. In the upper shell extension, its direction of flow is reversed and it condenses downwards in crossflow in the Spiral element. Meanwhile, coolant is pumped through a peripheral connection, flows through the spiral channel towards the centre and, finally, exits via a pipe in the upper shell extension.

For minimal subcooling, the condensate is allowed to enter the lower shell extension. However, when subcooling is required, a baffle plate fitted to the lower face of the spiral element forces the condensate to flow in the lower parts of the channels in countercurrent to the coolant.

A. Boothroyd

SPIRAL TUBES, USE IN BENSON BOILERS (see Benson boilers)

SPIROPYRAN (see Photochromic dye tracing)

SPLATTERING, EFFECT ON JET IMPINGEMENT (see Impinging jets)

SPONTANEOUS CONDENSATION (see Aerosols)

SPRAY CONDENSERS (see Condensers)

SPRAY DRYER (see Dryers)

SPRAY DRYING (see Gas-solid flows)

SPRAYERS (see Atomization)

SPRAY COLUMNS (see Direct contact heat exchangers)

SPRAYS (see Atomization)

SPRAY TOWERS (see Extraction, liquid-liquid)

SPREADING OF LIQUIDS ON LIQUIDS (see Interfaces)

SPUTTERING (see Leidenfrost phenomena; Rewetting of hot surfaces)

STABILITY (see Boiling water reactor, BWR)

STABILITY CRITERIA (see Nyquist stability criterion and nyquist diagrams)

STABILITY OF EMULSIONS (see Emulsions)

STABILITY OF FLOATING BODIES (see Hydrostatics)

STACKS, POLLUTION FROM (see Air pollution)

STAGGERED TUBE BANKS (see Tube banks, crossflow over)

STAGNANT FILM MODEL (see Condensation, overview)

STAGNATION PRESSURE (see Adiabatic conditions; Compressible flow; Dynamic pressure)

STAGNATION TEMPERATURE (see Compressible flow)

STANDARD CONDITIONS (see Thermodynamics)

STANTON GAUGE (see Shear stress measurement)

STANTON NUMBER

Stanton number for heat transfer, St, is a dimensionless parameter relating heat transfer coefficient to heat capacity of the fluid stream per unit cross-sectional area per unit time, i.e.

$$St = \alpha/\dot{m}c_p = \alpha/u\rho c_p.$$

where α is heat transfer coefficient, $\dot{m}$ mass flux, c_p specific heat capacity of the fluid, μ velocity and ρ fluid density.

The analagous Stanton number for mass transfer is

$$St_M = \beta/u$$

where β is mass transfer coefficient.

The total rate of heat transfer, $\dot{Q}$, to a fluid flowing inside a tube of diameter D and length L is given by

$$\dot{Q} = \bar{\alpha}\ \pi DL\ \overline{\delta T} = \dot{m}c_p(\pi D^2/4)\Delta T$$

where $\bar{\alpha}$ and $\overline{\delta T}$ are appropriately averaged values of heat transfer coefficient and driving temperature difference respectively, and ΔT is the temperature rise of the fluid over the length L. Hence

$$St = \alpha/\dot{m}c_p = (D/4L)(\Delta T/\overline{\delta T})$$

Therefore Stanton number, for a given geometry, is proportional to the temperature change in the fluid divided by the driving temperature difference.

A similar relationship can be derived for St_M with concentration differences replacing temperature differences.

Reference

Hewitt G. F., Shires G. L., and Bott T. R. (1994) *Process Heat Transfer*, CRC Press.

G.L. Shires

STANTON, SIR THOMAS EDWARD (1865–1931)

A British engineer, born at Atherstone in Warwickshire on December 12, 1865. In 1888 Stanton entered Owens College, Manchester, and followed the engineering course in the Whitworth Laboratory under Osborne Reynolds. After taking the degree of B.SC. in 1891 at the Victoria University, with first-class honours, he continued to work in Reynolds' Laboratory, at first as junior and later as senior demonstrater, until 1896. From 1892 to 1896 he was also resident tutor in mathematics and engineering at the Hulme Hall of Residence, Manchester. In June, 1896, Stanton obtained a post as senior assistant lecturer in engineering at University College, Liverpool, under Professor Hele-Shaw. In December, 1899, he went from Liverpool to Bristol University College as Professor of Engineering.

Figure 1.

In July, 1901, he was offered the position of superintendent of the engineering department of the National Physical Laboratory in Bristol. This post he continued to hold until his retirement from official duties in December, 1930, at the age of 65, one year before his death.

Stanton's main field of interest was fluid flow and friction, and the related problem of heat transmission. From 1902 to 1907 he excuted a large research program concerning wind forces on structures, such as bridges and roofs. After 1908, the year when the Wright brothers made their first aeroplane flights in Europe, Stanton devoted to problems of aeroplane and airship design and the dissipation of heat from air-cooled engines.

U. Grigull, H. Sandner and J. Straub

STARK BROADENING (see Radiative heat transfer)

STARK NUMBER (see Boltzmann number)

STARS, FUSION REACTIONS IN (see Fusion, nuclear fusion reactors)

STATIC HEAD

Following from: *Hydraulics*

The static head, sometimes referred to as the pressure head, is a term primarily used in **Hydraulics** to denote the static pressure in a pipe, channel, or duct flow. It has the physical dimensions of length (hence the term "head") and represents the flow-work per unit weight of fluid. In practice the static head is equivalent to the vertical distance from a given streamtube to the to the *piezometric head line* or the *hydraulic gradient line.* If P is the pressure in a streamtube, the static head is defined by $P/\rho g$. Using this definition it is apparent that in a open channel flow, where the pressure is assumed to be hydrostatic, the static head is the distance from the streamtube to the water surface.

Leading to: *Pitot tube*

C. Swan

STATIC INSTABILITIES IN TWO-PHASE SYSTEMS (see Instabilty, two-phase)

STATIC MIXERS (see Aeration)

STATIC REGENERATORS (see Heat exchangers)

STATIONARY PHASE, SP, CHROMATOGRAPHY (see Gas chromatography)

STATISTICAL MECHANICS (see Boltzmann distribution)

STATISTICAL THEORY, OF TURBULENT FLOW (see Turbulent flow)

STATISTICAL THERMODYNAMICS (see Entropy)

STEAM ENGINES

The potential of steam, the gaseous form of water, as an agent for the transfer of heat energy into mechanical work has been known for some two millenia. The eighteen hundred times expansion which occurs when water is boiled into steam had been recognised in classical times and the magic toys or perhaps temple devices of Hero of Alexandria utilised the properties of steam in a number of ways.

However, the restrictions of technology and a defective understanding of the nature of heat precluded further advances until after 1600 when the experiments of Torricelli on atmospheric pressure, Robert Boyle with gases and the demonstrations of von Guericke of the properties of a vacuum, coupled with early glimpses of an understanding of the nature of steam led to the conjectures of Samual Morland and others as to its possible use as a source of power.

In 1675, Papin devised an apparatus whereby a weight was lifted utilising the condensation of steam. By 1698, further developments by Thomas Savery resulted in the first commercially successful steam engine "to raise Water by the force of Fire". However this, although utilising both the expansive and condensing properties of steam, was restricted in its application to the lifting of water and pumping. The particular requirement for the draining of mines reflected the increasing demand for minerals especially coal.

The imperative for power, which hitherto had been met by the efforts of human and animal muscle, wind and water power, engendered by the burgeoning of the industrial revolution, led to further developments. While there may have been others, *Thomas Newcomen,* an ironmonger of Dartmouth, is credited with the building the first steam engine in which "a piston was moved in a cylinder by the agency of steam".

Usually described as the *Newcomen atmospheric engine,* the first full sized example, a mine pumping engine, appears to have been built in 1712 near Dudley Castle in Staffordshire. It was probably the outcome of a long struggle with models in which the many technical difficulties such as the matching of the piston to the cylinder were at least partially overcome. The Newcomen engine derived its effort from the condensation of steam, at very nearly atmospheric pressure, in the cylinder, the other end of which was open. The resulting partial vacuum led to the piston being forced down by atmospheric pressure. The piston was connected to a pivoted beam which at its other end held a heavy set of pump rods. A valve was opened into the cylinder below the piston and steam (at a very low pressure) admitted. This permitted the piston to rise, drawn up by the weight of the pump rods. The cycle was then repeated.

Immensely inefficient in that its action depended on the alternative heating and cooling of water and steam, the Newcomen engine nonetheless achieved widespread acceptance for pumping applications and by the middle of the eighteenth century several hundreds were in use.

In 1763, **James Watt**, a Glasgow instrument maker, developed and later patented his invention of the separate condenser, thus eliminating one of the major ineficiencies of the Newcomen engine. While still using steam at very low pressures, the increased efficiency of the Watt engines enabled them to be developed for rotative purposes. Aided by the manufacturing and business acumen of Watt's partner, Matthew Boulton of Birmingham, and his assistant, William Murdoch, their use became widespread. Almost contemporaneously Richard Trevithick used his experience with Cornish mining engines to employ the increased potential for work of the expansive properties of high pressure steam.

Technical advances such as the ironfounding innovations of the Darbys of Coalbrookdale, as well as the large cylinder boring techniques of Wilkinson soon resulted in the steam engine becoming the prime mover and facilitator of the rapid expansion of the industrial revolution, first in Britain then in Europe and the United States. As a translator of heat to mechanical energy, its potential for terrestrial and marine transport was being widely explored.

In Cornwall particularly the further applictions of high pressure steam were investigated which lead to the development of Woolf's compound engines and the development of boiler design. At the same time, much effort and ingenuity was deployed in the development of valve gears, condenser design and packing and sealing materials.

The theoretical aspects of the steam engine remained empirical and erroneously understood. Concepts of the nature of heat were allied to current ideas of caloric and related power to steam pressure. The writings of *Carnot,* and the experiments of *Count Rumford,* and *James Joule* of Manchester culminated, in 1843 with the introduction of new concepts which led to the understanding that "Heat and energy are mutually convertible . . ." and that the heat in a steam engine is the vital driving force and the pressure only a secondary force.

The work of *Rankine* and *William Thomson (Lord Kelvin)* further clarified the theoretical understandings and demonstrated that the steam engine was extremely inefficient. Thomson derived the key formula for the efficiency of a perfect heat engine which showed that the greater the temperature drop, the greater the efficiency of the engine. Henceforth the emphases of development were directed to "improving the construction of the steam engine and in seeking to obtain from it a larger amount of useful work with a given expenditure of fuel". There followed many technical advances including the *Corliss Valve,* high speed engines and, improved governors. Towards the end of the nineteenth century the demands of steam power to generate electricity, as well as the massive power demands of textile mills and metal industries, represented only a fraction of the fields of application for what had become a universal prime mover. Other developments involved the Uniflow concept, superheating and innovations in boiler design and fuel utilisation.

The reciprocating engine utilises the expansion of steam, but the *Aelopile of Hero,* a turbine like toy, demonstrated the kinetic energy of steam. In 1884, *Charles Parsons* pioneered its practical application in the **Steam Turbine.** In this, the steam acts by either impulse or reaction as a mass that is set in motion in consequence of its own power to expand. Much development, including the discovery of the major contribution of efficient condensers to cycle efficiency, demonstrated the particular advantages of steam turbines for high speed applications such as electricity generation and, with appropiate gearing, ship propulsion.

Despite the development of various modes of internal combustion engine and their primacy in terms of size, flexibility and relative efficiency, the steam engine in its turbine incarnation remains a widely used form of power generator, and is virtually the universal final stage of the various methods of the conversion of nuclear energy into power.

References

Bourne, J. (1846) *A Treatise on the Steam Engine,* Longmans, London.

Dickinson, H. W. (1938) *A Short History of the Steam Engine,* Cambridge, U.P.

Farey, J. A. (1971) *A Treatise on the Steam Engine, Vols 1 and 2,* David and Charles.

Hills, R. L. (1988) *Power from Steam,* Cambridge, U.P.

Rankine, W. J. M. (1861) *A Manual of the Steam Engine and Other Prime Movers,* Griffin, Bohn, London.

M. Gilkes

STEAM GAS TURBINE UNITS (see Gas turbine)

STEAM GENERATORS, NUCLEAR

Following from: *Nuclear reactors; Boilers*

Introduction

The **NSSS** (Nuclear Steam Supply System) is a relatively recent development, and has been in use for about thirty years. During this time, there were constructed and put in operation 298 Pressurized Water Reactors (PWR), 81 of which are in the U.S.; 100 Boiling Water Reactors (BWR), 38 of which are in the U.S.; 19 light-water cooled graphite-moderated reactors (LGR) and 50 pressurized heavy water moderated and cooled reactors (PHWR)—all over 30 MW. In addition, there were expected to be in operation 163 more PWRs, 56 BWRs, 12 LGRs and 18 PHWRs. Here the attention is focused only on the nuclear steam generators of a PWR system, which is shown schematically in **Figure 1**.

Heat, which is produced in the core inside the pressure vessel, is convected by the primary fluid, which is pumped through the pressure vessel, from the core to the system generator. In the steam generator the primary fluid, water at 150 bar, exchanges heat with the secondary fluid, water at 75 bar, and causes it to boil. The steam, from the steam generator, passes through the turbine, condenser, and is pumped back into the steam generator (SG) as feed water.

There are two types of SG: the U-tube SG and the once through SG, as shown in **Figures 2** and **3**.

U-Tube Steam Generators

In **Figure 2** is depicted a *U-tube steam generator,* which is installed in a wide variety of commercial NSSS, from the early (1957) 90–MWe Shipping port 4 loops system to new 1300–MWe plants. Two typical systems are characterized in the following:

Rated power, MWe	822	1,300
Rated heat output, MWth	2,441	3,819
System pressure, bar	153	153
Primary coolant flow rate, 10^4 kg/hr	45.3	67.9
Primary coolant temperature, °C	318.7	332.0
Number of loops	3	4
Number of pumps	3	4
Steam generators		
Number	3	4
Shell side design pressure, bar	53.4	77.0
Steam flow at full load, 10^6kg/hr	4.8	7.9
Steam temperature, °C	268.9	294.1
Feedwater temperature, °C	225.4	239.4
Number of tubes	3,388	6,970
Diameter of tubes, mm	22.2	17.4
Heat transfer area, m^2	4,784	7,665

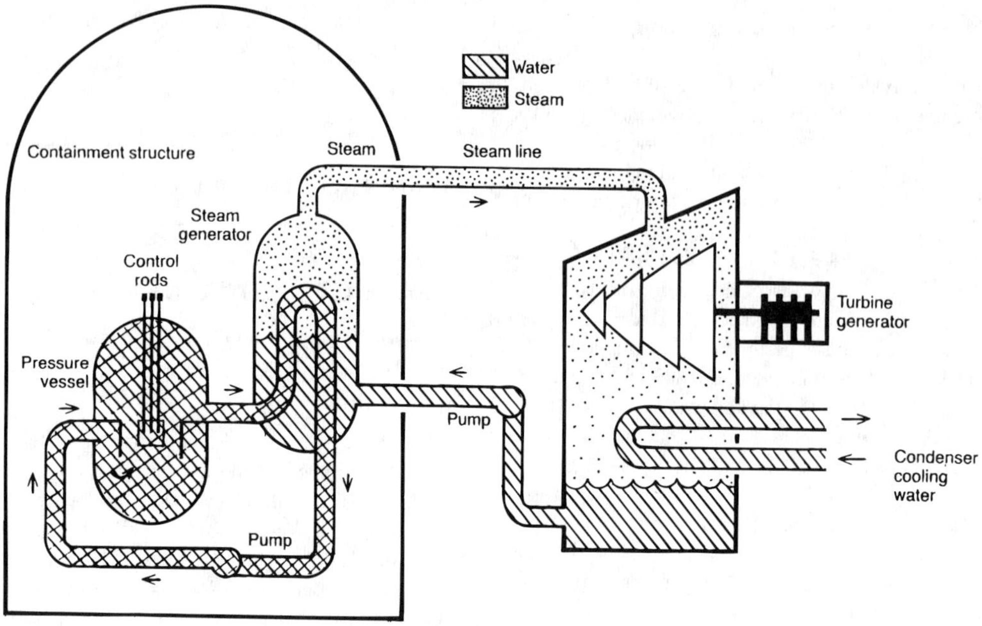

Figure 1. Schematic view of a PWR power plant.

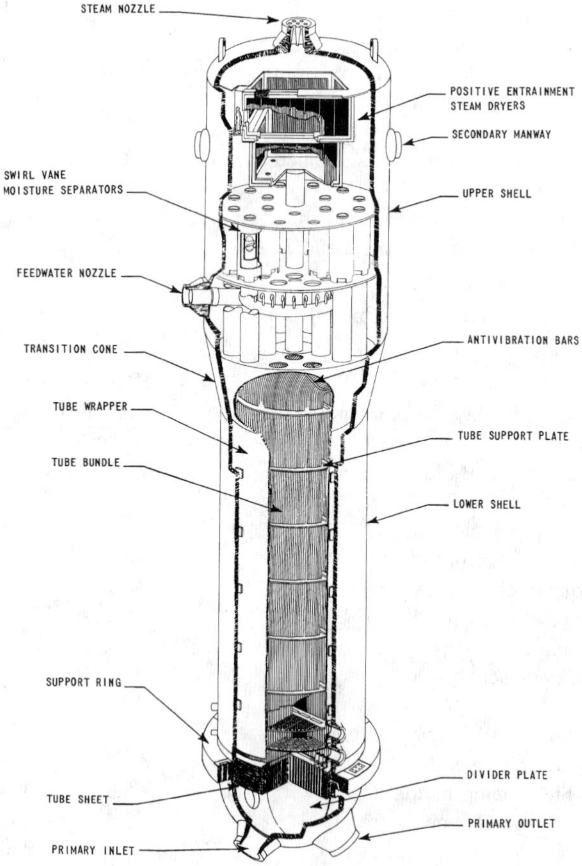

Figure 2a. U-tube steam generator.

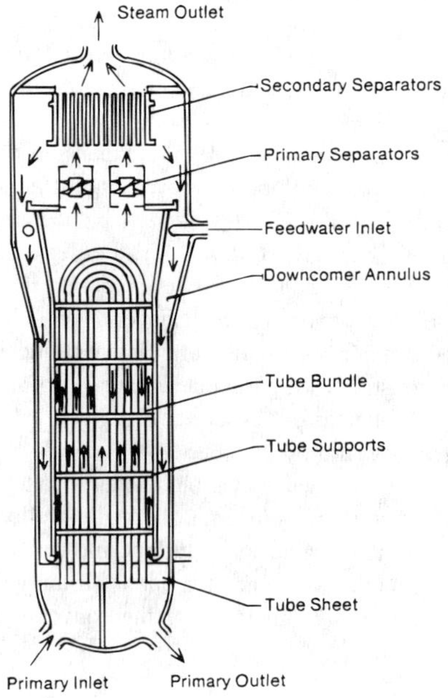

Figure 2b. Schematic diagram of a U-tube SG.

Reactor coolant enters the steam generator at the inlet nozzle (on the bottom left-hand side of **Figure 2**) and enters the tube bundle in the hot leg, completing the U-bend through the cold leg to the primary outlet.

Feedwater enters the side of the shell. It flows down through an annulus just inside the shell (downcomer), where it mixes with water coming from the separator deck. The water enters the heating surface (tube bundle) at the bottom and is increasing in quality as it rises through the steam generator. The steam-water mixture enters the steam water separators—where steam is passed through to the driers and then to the steam nozzle; the water is recirculated, through the downcomer, to the bottom of the heating surface.

The shell, outside of the tubes and the tubesheet form the steam production boundaries. Within the shell, the tube bundle is surrounded by a wrapper (or shroud). The tube material is usually Inconel Alloy 600 (though it would have been better to have thermally treated Inconel 690). Tube support plates, with quatrofoil holes or egg crate supports, hold the tubes in uniform pattern along their length. The U-tube region of the tube is additionally supported by antivibration bars. Vents, drains, instrumentation nozzles, and inspection openings are provided on the shell side of the units.

The UTSGs are characterized by a widening of the shell about $^1/_3$ of the height of the steam generator. This is done in order to increase the area available for the separators at the separator deck.

In the UTSG there is always a water level in the downcomer, in order to balance the pressure losses of steam-water mixture as it flows through the tube bundle. The pressure drop in the tube bundle is due to friction along the tubes, pressure drop at the tube support plates and separators, acceleration of the flow, and hydrostatic head. Quality in the downcomer (e.g. steam bubbles) reduces the density, and thus the available head, substantially. This can be a result of imperfect separation of steam water at the separators and is termed *carryunder.*

The ratio of flow rate of the steam water mixture which flows through the SG tube bundle, to the flow rate of steam out of the steam nozzle, is called the *circulation ratio.* It is desirable to maintain a high circulation ratio (maybe over 5) to reduce concentration of chemicals, debris, etc., in various places in the SG. Current SGs have frequently circulation ratios of around 3—which is undoubtedly one of the causes of the recently mounting difficulties with these units.

The design of SGs is a complicated procedure, which involves many steps, iterations and interaction with other components of the system.

The NSSS, as part of the power station, is designed to minimize the power cost. This consideration is subject to many constraints—the primary one being safety. Other constraints are imposed by availability of major equipment (e.g. primary coolant pumps), manufacturing capabilities (e.g. can vessel be fabricated), shipping consideration, etc.

Thus, the design of the steam generators is subject to many outside constraints and requirements. For example, the primary fluid condition (i.e. temperatures, allowable pressure drop in the steam generator) is determined mostly by the reactor design and availability of pumps. The performance requirements, i.e., steam pressure, temperature, and flow rates, are determined mostly by the turbine design and are part of the system performance. It follows, therefore, that the tube bundle size (namely the required heat transfer area, as well as allowable pressure drop of the primary fluid) is determined by system considerations.

The preliminary structural design is performed in accordance with ASME's boilers and Pressure Vessel Code, Section III. (See **Pressure Vessels**) Also, steam generators, as all power plant components, are required to be designed to withstand various accident situations. See, for example, **Blowdown.** For steam generators, this consists of the following conditions:

Small steam line break, loss of feedwater, turbine trip, etc.

LOCA (Loss of Coolant Accident)

MSLB (Main Steam Line Break)

SSE (Safe Shutdown Earthquake)

The detailed design includes detailed structural and thermal-hydraulic analyses and studies, and is later used to prove to the customer and regulatory agencies that the steam generator design is in compliance with ASME code, NRC regulations, etc.

Once Through Steam Generators

The OTSG (**Figure 3**) is typically associated with a NSSS which has the following general characteristics:

Rated power, MWe	860	1,300
Rated heat output, MWth	2,568	3,760
System pressure, bar	149	153

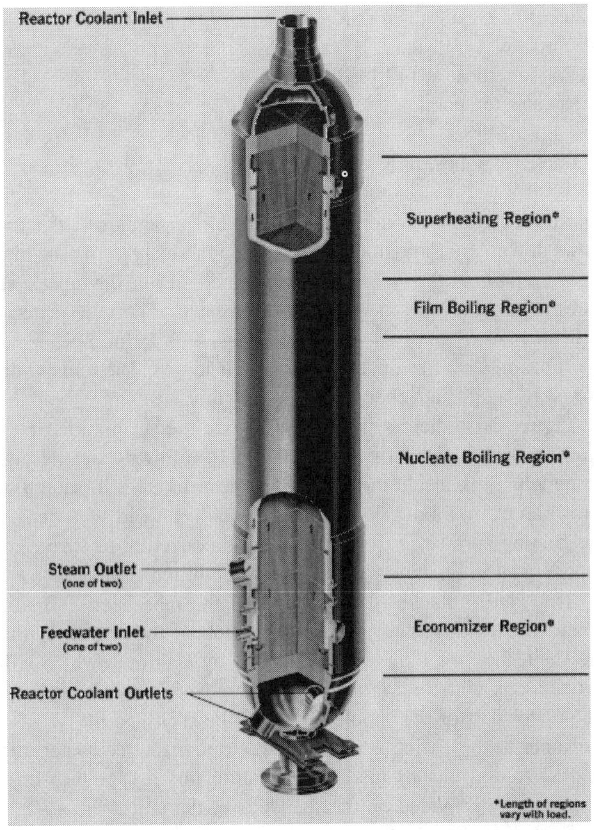

Figure 3a. Once through SG.

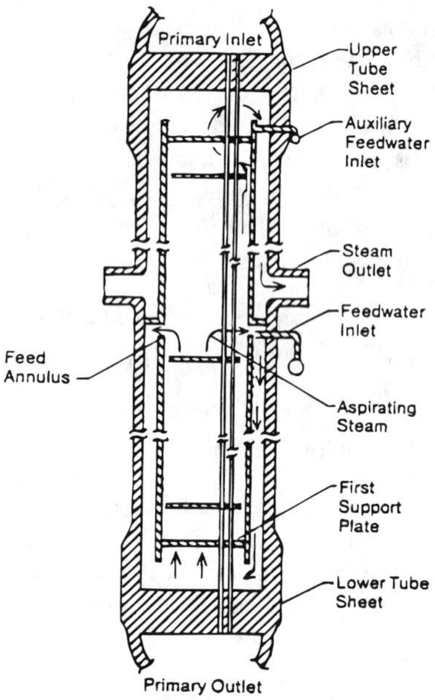

Figure 3b. Schematic diagram of a once through SG.

Primary coolant flow rate, 10^6 kg/hr	29.5	35.6
Primary coolant temperature °C	317.8	329.7
Number of loops	2	2
Number of pumps	4	4
Number of steam generators	71.4	84.0
Steam flow at full load, 10^6kg/hr	299	308
Steam superheat at full load, °C	19.4	19.4
Feedwater temperature, °C	235	241
Number of tubes	15,530	16,000
Diameter of tubes, mm	15.9	15.9

Reactor coolant water enters the steam generator at the top, flows downward through the tubes and out at the bottom. The high pressure parts of the unit are the hemisphere heads, the tubesheets and the straight tubes between the tubesheets. The tube material is Inconel Alloy 600. Tube support plates, with trefoil holes, hold the tubes in a uniform pattern along their length. The unit is supported by a skirt attached to the bottom tubesheet.

Figure 3b indicates the flow paths on the steam side of the unit. Feedwater enters the side of the shell. It flows down through an annulus just inside the shell where it is brought to saturation temperature by mixing it with steam. The saturated water enters the heating surface at the bottom and is converted to steam and superheated in a single pass upward through the generator.

The shell, outside of the tubes, and the tubesheets form the boundaries of the steam producing section of the vessel. Within the shell, the tube bundle is surrounded by a shroud, which is in two sections, with the upper section the larger of the two in diameter. The upper part of the annulus between the shell and the baffle is the super heater outlet, while the lower part is the feedwater inlet heating zone. Vents, drains, instrumentation nozzles and inspection openings are provided on the shell side of the units.

Superheated steam is produced at a constant pressure over the power range. At full power, the steam temperature of 300°C provides about 19°C of superheat. As load is reduced, steam temperature approaches the reactor outlet temperature, thus increasing the superheat slightly. Below 15% load, steam temperature decreases to saturation.

Recent Problems with SGs

The NSSS provided in the past service which was mostly safe and trouble free. In the last 20 years, however, there is an increasing number of PWR nuclear steam generators which develop technical problems, such as denting, intergranular attack (IGA), vibration of tubes which cause wear and fatigue, wastage of tubes, pitting, erosion-corrosion, water hammer, etc. Any of these can lead to a breach of the integrity of the tubes and to leakage of primary (contaminated) coolant into the secondary fluid. Since the secondary fluid is leaving the containment vessel to the turbines, it must not be radioactive, and must not be contaminated by primary fluid. Therefore, when leaking tubes are detected, the plant must be shut down for repairs, and replacement of SGs, at great costs and loss of revenue.

References

Cumo, M. and Naviglio, A. (1991) *Thermal Hydraulic Design of Components for Steam Generators,* CRC Press.

Green, S. J. and Hetsroni, G. (1995) PWR Steam Generators, *Int. J. Multiphase Flow,* 21 (Annual Review).

Hetsroni, G. (1982) *Handbook of Multiphase Systems,* Hemisphere-McGraw-Hill.

Steam Generation Reference Book, EPRI, 1985.

Singhal, A. K. and Srikkantiah, G. (1991) A Review of Thermal Hydraulic Analysis Methodology for PWR SG and ATHOS 3 code Application, *Progress in Nuclear Energy,* 25, 7–70.

Ulbrich, R. and Mewes, D. (1994) Experimental Studies of Vertical Two-Phase Flow Across Tube Bundle, *Int. J. Multiphase Flow,* 20, 249–272.

G. Hetsroni

STEAM JET REFRIGERATION (see Refrigeration)

STEAM JET EJECTORS (see Jet pumps and ejectors)

STEAM TABLES

Following from: *Water, properties of*

The following tables of the properties of steam are taken directly from Chapter 5.5.3 of the *Heat Exchanger Design Handbook,* 1986, by C. F. Beaton.

The tables in this section are reprinted, with permission, from *NBS/NRC Steam Tables.*

Haar, L., Gallagher, J. S., and Kell, G. S., (1984) *NBS/NRC Steam Tables,* Hemisphere, Washington, D.C.

Symbols and Nomenclature for the Tables

Symbol	Property	Units
h	specific enthalpy	kJ/kg
P	pressure	bar = 0.1 MPa
Pr	Prandtl number (= $\eta C_p/\lambda$)	dimensionless
r	specific enthalpy of vaporization	kJ/kg
s	specific entropy	kJ/(kg K)
t	Celsius temperature	°C
t_s	temperature at saturation	
u	specific internal energy	kJ/kg
ν	specific volume	m^3/kg
ϵ	static dielectric constant	dimensionless
η	viscosity	10^{-6} kg/(s m) = MPa s
λ	thermal conductivity	mW/(K m)
ρ	density	kg/m^3
σ	surface tension	kg/s^2 = N/m
ϕ	specific entropy of vaporization	kJ/(kg K)

Subscripts

g denotes a saturated vapor state

l denotes a saturated liquid state

The reference state for all property values is the liquid at the triple point, for which state the specific internal energy and the specific entropy have been set to zero.

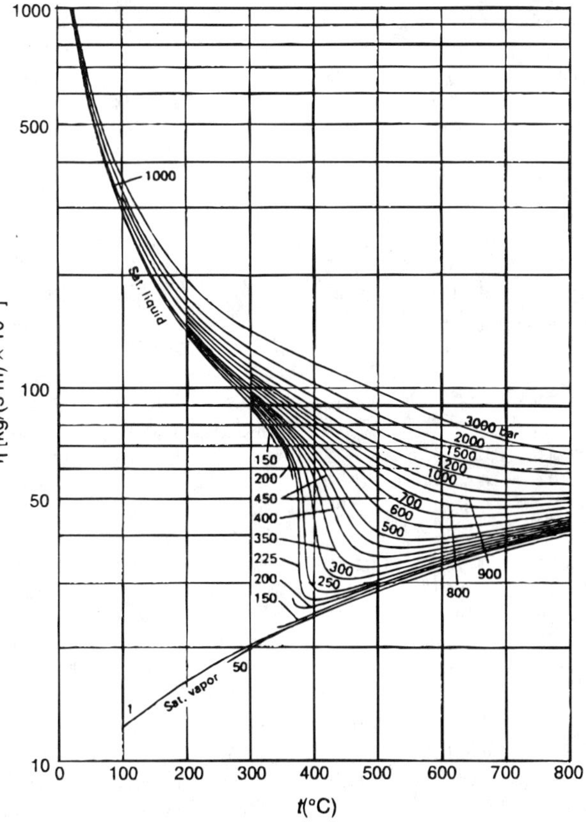

FIGURE 1. Viscosity.

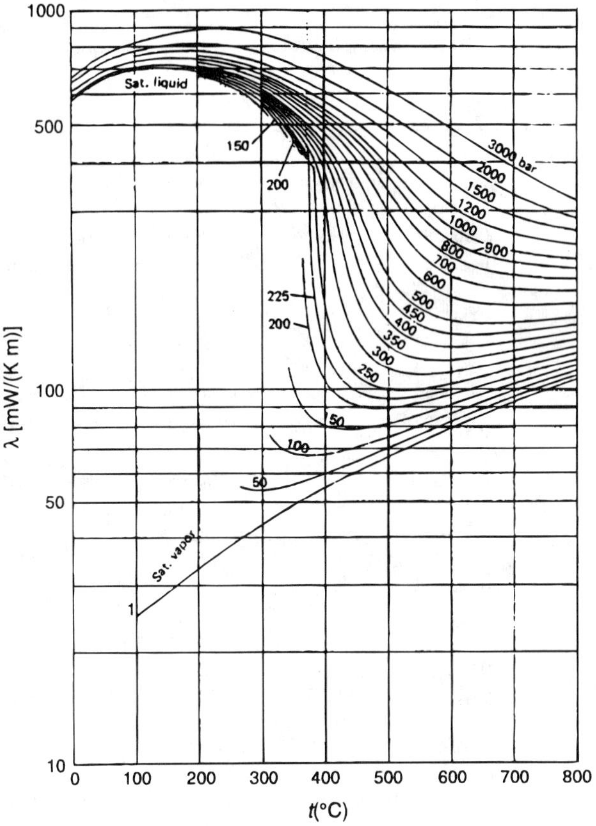

FIGURE 2. Thermal conductivity.

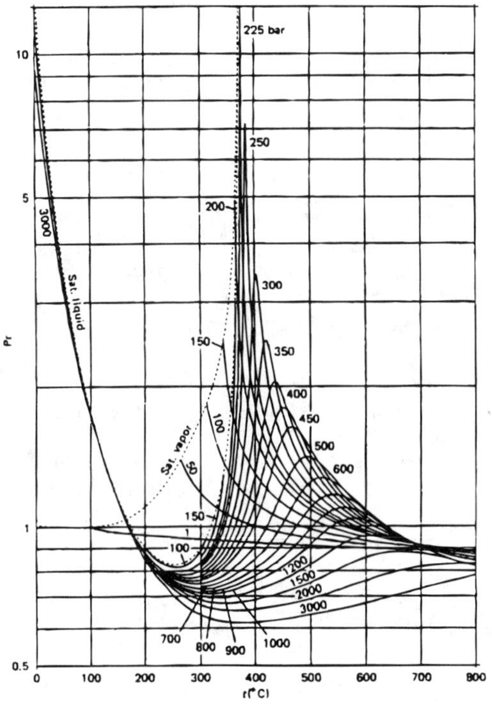

Figure 3. Prandtl number.

Table 1. Saturation (temperature)

t (°C)	P	ρ_l	ρ_g	h_l	h_g	r	s_l	s_g	ϕ	v_l ($\times 10^3$)	v_g ($\times 10^3$)
0.01	0.0061173	999.78	0.004855	0.00	2500.5	2500.5	0.00000	9.1541	9.1541	1.00022	205990
	0.0065716	999.85	0.005196	4.18	2502.4	2498.2	0.01528	9.1277	9.1124	1.00015	192440
	0.0070605	999.90	0.005563	8.40	2504.2	2495.8	0.03064	9.1013	9.0707	1.00010	179760
	0.0075813	999.93	0.005952	12.61	2506.0	2493.4	0.04592	9.0752	9.0292	1.00007	168020
	0.0081359	999.95	0.006364	16.82	2507.9	2491.1	0.06112	9.0492	8.9881	1.00005	157130
5	0.0087260	999.94	0.006802	21.02	2509.7	2488.7	0.07626	9.0236	8.9473	1.00006	147020
	0.0093537	999.92	0.007265	25.22	2511.5	2486.3	0.09133	8.9981	8.9068	1.00008	137650
	0.0100209	999.89	0.007756	29.42	2513.4	2484.0	0.10633	8.9729	8.8666	1.00011	128940
	0.0107297	999.84	0.008275	33.61	2515.2	2481.6	0.12127	8.9479	8.8266	1.00016	120850
	0.0114825	999.77	0.008824	37.80	2517.1	2479.3	0.13615	8.9232	8.7870	1.00023	113320
10	0.012281	999.69	0.009405	41.99	2518.9	2476.9	0.15097	8.8986	8.7477	1.00031	106320
	0.013129	999.60	0.010019	46.18	2520.7	2474.5	0.16573	8.8743	8.7086	1.00040	99810
	0.014027	999.49	0.010668	50.36	2522.6	2472.2	0.18044	8.8502	8.6698	1.00051	93740
	0.014979	999.37	0.011353	54.55	2524.4	2469.8	0.19509	8.8263	8.6313	1.00063	88090
	0.015988	999.24	0.012075	58.73	2526.2	2467.5	0.20969	8.8027	8.5930	1.00076	82810
15	0.017056	999.09	0.012837	62.92	2528.0	2465.1	0.22424	8.7792	8.5550	1.00091	77900
	0.018185	998.93	0.013641	67.10	2529.9	2462.8	0.23873	8.7560	8.5173	1.00107	73310
	0.019380	998.76	0.014488	71.28	2531.7	2460.4	0.25317	8.7330	8.4798	1.00124	69020
	0.020644	998.58	0.015380	75.47	2533.5	2458.1	0.26757	8.7101	8.4426	1.00142	65020
	0.021979	998.39	0.016319	79.65	2535.3	2455.7	0.28191	8.6875	8.4056	1.00161	61280
20	0.023388	998.19	0.017308	83.84	2537.2	2453.3	0.29621	8.6651	8.3689	1.00182	57778
	0.024877	997.97	0.018347	88.02	2539.0	2451.0	0.31045	8.6428	8.3324	1.00203	54503
	0.026447	997.75	0.019441	92.20	2540.8	2448.6	0.32465	8.6208	8.2962	1.00226	51438
	0.028104	997.52	0.020590	96.39	2542.6	2446.2	0.33880	8.5990	8.2602	1.00249	48568
	0.029850	997.27	0.021797	100.57	2544.5	2443.9	0.35290	8.5773	8.2244	1.00274	45878
25	0.031691	997.02	0.023065	104.75	2546.3	2441.5	0.36696	8.5558	8.1889	1.00299	43357
	0.033629	996.75	0.024395	108.94	2548.1	2439.2	0.38096	8.5346	8.1536	1.00326	40992
	0.035670	996.48	0.025791	113.12	2549.9	2436.8	0.39492	8.5135	8.1185	1.00353	38773
	0.037818	996.20	0.027255	117.30	2551.7	2434.4	0.40884	8.4926	8.0837	1.00381	36690
	0.040078	995.91	0.028791	121.49	2553.5	2432.0	0.42271	8.4718	8.0491	1.00411	34734
30	0.042455	995.61	0.030399	125.67	2555.3	2429.7	0.43653	8.4513	8.0147	1.00441	32896
	0.044953	995.30	0.032084	129.85	2557.1	2427.3	0.45031	8.4309	7.9806	1.00472	31168
	0.047578	994.99	0.033849	134.04	2559.0	2424.9	0.46404	8.4107	7.9466	1.00504	29543
	0.050335	994.66	0.035696	138.22	2560.8	2422.5	0.47772	8.3906	7.9129	1.00537	28014
	0.053229	994.33	0.037629	142.40	2562.6	2420.2	0.49137	8.3708	7.8794	1.00570	26575
35	0.056267	993.99	0.039650	146.59	2564.4	2417.8	0.50496	8.3511	7.8461	1.00605	25220
	0.059454	993.64	0.041764	150.77	2566.2	2415.4	0.51851	8.3315	7.8130	1.00640	23944
	0.062795	993.28	0.043973	154.95	2568.0	2413.0	0.53202	8.3122	7.7802	1.00676	22741
	0.066298	992.92	0.046281	159.14	2569.8	2410.6	0.54549	8.2930	7.7475	1.00713	21607
	0.069969	992.55	0.048691	163.32	2571.6	2408.2	0.55891	8.2739	7.7150	1.00751	20538
40	0.073814	992.17	0.05121	167.50	2573.4	2405.9	0.57228	8.2550	7.6828	1.00789	19528
	0.077840	991.78	0.05383	171.69	2575.2	2403.5	0.58562	8.2363	7.6507	1.00829	18576
	0.082054	991.39	0.05657	175.87	2576.9	2401.1	0.59891	8.2177	7.6188	1.00869	17676
	0.086464	990.99	0.05943	180.05	2578.7	2398.7	0.61216	8.1993	7.5872	1.00909	16826
	0.091076	990.58	0.06241	184.23	2580.5	2396.3	0.62537	8.1810	7.5557	1.00951	16023
45	0.095898	990.17	0.06552	188.42	2582.3	2393.9	0.63853	8.1629	7.5244	1.00993	15263
	0.100938	989.74	0.06875	192.60	2584.1	2391.5	0.65166	8.1450	7.4933	1.01036	14545
	0.106205	989.32	0.07212	196.78	2585.9	2389.1	0.66474	8.1271	7.4624	1.01080	13866
	0.111706	988.88	0.07563	200.96	2587.6	2386.7	0.67778	8.1094	7.4317	1.01124	13222
	0.117449	988.44	0.07928	205.14	2589.4	2384.3	0.69078	8.0919	7.4011	1.01170	12614

Table 1. Saturation (temperature)—*continued*

t (°C)	P	ρ_l	ρ_g	h_l	h_g	r	s_l	s_g	ϕ	v_l ($\times 10^3$)	v_g ($\times 10^3$)
50	0.12344	987.99	0.08308	209.33	2591.2	2381.9	0.70374	8.0745	7.3708	1.01215	12037
	0.12970	987.54	0.08703	213.51	2593.0	2379.5	0.71666	8.0573	7.3406	1.01262	11490
	0.13623	987.08	0.09114	217.69	2594.7	2377.0	0.72954	8.0401	7.3106	1.01309	10972
	0.14303	986.61	0.09541	221.87	2596.5	2374.6	0.74238	8.0232	7.2808	1.01357	10481
	0.15012	986.13	0.09985	226.06	2598.3	2372.2	0.75518	8.0063	7.2511	1.01406	10015
55	0.15752	985.65	0.10446	230.24	2600.0	2369.8	0.76795	7.9896	7.2216	1.01455	9573.
	0.16522	985.17	0.10925	234.42	2601.8	2367.4	0.78067	7.9730	7.1923	1.01505	9153.
	0.17324	984.68	0.11423	238.60	2603.5	2364.9	0.79336	7.9566	7.1632	1.01556	8754.
	0.18159	984.18	0.11939	242.79	2605.3	2362.5	0.80600	7.9402	7.1342	1.01608	8376.
	0.19028	983.67	0.12475	246.97	2607.0	2360.1	0.81862	7.9240	7.1054	1.01660	8016.
60	0.19932	983.16	0.13030	251.15	2608.8	2357.6	0.83119	7.9080	7.0768	1.01712	7674.
	0.20873	982.65	0.13607	255.34	2610.5	2355.2	0.84373	7.8920	7.0483	1.01766	7349.
	0.21851	982.13	0.14204	259.52	2612.3	2352.8	0.85622	7.8762	7.0200	1.01820	7040.
	0.22868	981.60	0.14824	263.71	2614.0	2350.3	0.86869	7.8605	6.9918	1.01875	6746.
	0.23925	981.07	0.15465	267.89	2615.8	2347.9	0.88112	7.8450	6.9638	1.01930	6466.
65	0.25022	980.53	0.16130	272.08	2617.5	2345.4	0.89351	7.8295	6.9360	1.01986	6200.
	0.26163	979.98	0.16819	276.26	2619.2	2343.0	0.90586	7.8142	6.9083	1.02043	5946.
	0.27347	979.43	0.17532	280.45	2620.9	2340.5	0.91819	7.7989	6.8808	1.02100	5704.
	0.28576	978.88	0.18269	284.63	2622.7	2338.0	0.93047	7.7838	6.8534	1.02158	5474.
	0.29852	978.32	0.19033	288.82	2624.4	2335.6	0.94272	7.7689	6.8261	1.02216	5254.
70	0.31176	977.75	0.19823	293.01	2626.1	2333.1	0.95494	7.7540	6.7990	1.02276	5044.6
	0.32549	977.18	0.20640	297.20	2627.8	2330.6	0.96713	7.7392	6.7721	1.02336	4844.9
	0.33972	976.60	0.21485	301.39	2629.5	2328.1	0.97928	7.7246	6.7453	1.02396	4654.4
	0.35448	976.02	0.22358	305.58	2631.2	2325.7	0.99139	7.7100	6.7186	1.20457	4472.6
	0.36978	975.43	0.23261	309.77	2632.9	2323.2	1.00348	7.6956	6.6921	1.02519	4299.0
75	0.38563	974.84	0.24194	313.96	2634.6	2320.7	1.01553	7.6813	6.6657	1.02581	4133.3
	0.40205	974.24	0.25158	318.15	2636.3	2318.2	1.02754	7.6670	6.6395	1.02644	3975.0
	0.41905	973.64	0.26153	322.34	2638.0	2315.7	1.03953	7.6529	6.6134	1.02708	3823.7
	0.43665	973.03	0.27180	326.54	2639.7	2313.2	1.05149	7.6389	6.5874	1.02772	3679.1
	0.45487	972.41	0.28241	330.73	2641.4	2310.7	1.06341	7.6250	6.5616	1.02837	3541.0
80	0.47373	971.79	0.29336	334.93	2643.1	2308.1	1.07530	7.6112	6.5359	1.02902	3408.8
	0.49324	971.17	0.30465	339.12	2644.7	2305.6	1.08716	7.5975	6.5103	1.02969	3282.4
	0.51342	970.54	0.31631	343.32	2646.4	2303.1	1.09899	7.5838	6.4849	1.03035	3161.5
	0.53428	969.91	0.32832	347.52	2648.1	2300.6	1.11079	7.5703	6.4595	1.03103	3045.8
	0.55585	969.27	0.34072	351.72	2649.7	2298.0	1.12255	7.5569	6.4344	1.03171	2935.0
85	0.57815	968.62	0.35349	355.92	2651.4	2295.5	1.13429	7.5436	6.4093	1.3239	2828.9
	0.60119	967.98	0.36666	360.12	2653.1	2292.9	1.14600	7.5304	6.3844	1.03308	2727.3
	0.62499	967.32	0.38023	364.32	2654.7	2290.4	1.15768	7.5172	6.3595	1.03378	2630.0
	0.64958	966.66	0.39420	368.52	2656.4	2287.8	1.16932	7.5042	6.3349	1.03449	2536.8
	0.67496	966.00	0.40860	372.73	2658.0	2285.3	1.18094	7.4912	6.3103	1.03520	2447.4
90	0.70117	965.33	0.42343	376.93	2659.6	2282.7	1.19253	7.4784	6.2858	1.03591	2361.7
	0.72823	964.66	0.43870	381.14	2661.3	2280.1	1.20409	7.4656	6.2615	1.03664	2279.5
	0.75614	963.98	0.45441	385.35	2662.9	2277.5	1.21563	7.4529	6.2373	1.03736	2200.7
	0.78495	963.30	0.47058	389.56	2664.5	2275.0	1.22713	7.4403	6.2132	1.03810	2125.0
	0.81465	962.61	0.48723	393.77	2666.1	2272.4	1.23861	7.4278	6.1892	1.03884	2052.4
95	0.84529	961.92	0.5043	397.98	2667.7	2269.8	1.25006	7.4154	6.1653	1.03959	1982.8
	0.87688	961.22	0.5220	402.20	2669.4	2267.2	1.26148	7.4030	6.1416	1.04034	1915.9
	0.90945	960.52	0.5401	406.41	2671.0	2264.5	1.27287	7.3908	6.1179	1.04110	1851.6
	0.94301	959.82	0.5587	410.63	2672.5	2261.9	1.28424	7.3786	6.0944	1.04186	1789.9
	0.97759	959.11	0.5778	414.84	2674.1	2259.3	1.29557	7.3665	6.0709	1.04264	1730.6

Table 1. Saturation (temperature)—*continued*

t (°C)	P	ρ_l	ρ_g	h_l	h_g	r	s_l	s_g	ϕ	v_l ($\times 10^3$)	v_g ($\times 10^3$)
100	1.0132	958.39	0.5975	419.06	2675.7	2256.7	1.30689	7.3545	6.0476	1.04341	1673.6
	1.0499	957.67	0.6177	423.28	2677.3	2254.0	1.31817	7.3426	6.0244	1.04420	1618.9
	1.0877	956.95	0.6385	427.51	2678.9	2251.4	1.32943	7.3307	6.0013	1.04499	1566.2
	1.1266	956.22	0.6598	431.73	2680.5	2248.7	1.34066	7.3189	5.9783	1.04578	1515.5
	1.1667	955.49	0.6817	435.95	2682.0	2246.1	1.35187	7.3072	5.9553	1.04659	1466.8
105	1.2079	954.75	0.7042	440.18	2683.6	2243.4	1.36305	7.2956	5.9325	1.04739	1420.0
	1.2503	954.01	0.7273	444.41	2685.1	2240.7	1.37420	7.2840	5.9098	1.04821	1374.9
	1.2939	953.26	0.7511	448.64	2686.7	2238.0	1.38533	7.2726	5.8872	1.04903	1331.4
	1.3388	952.51	0.7754	452.87	2688.2	2235.3	1.39644	7.2612	5.8647	1.04986	1289.6
	1.3850	951.75	0.8004	457.10	2689.7	2232.6	1.40751	7.2498	5.8423	1.05069	1249.4
110	1.4324	951.00	0.8260	461.34	2691.3	2229.9	1.41857	7.2386	5.8200	1.05153	1210.6
	1.4812	950.23	0.8523	465.57	2692.8	2227.2	1.42960	7.2274	5.7978	1.05238	1173.3
	1.5313	949.46	0.8793	469.81	2694.3	2224.5	1.44060	7.2163	5.7757	1.05323	1137.3
	1.5829	948.69	0.9069	474.05	2695.8	2221.8	1.45158	7.2052	5.7536	1.05409	1102.6
	1.6358	947.91	0.9353	478.29	2697.3	2219.0	1.46253	7.1942	5.7317	1.05495	1069.2
115	1.6902	947.13	0.9643	482.54	2698.8	2216.3	1.47347	7.1833	5.7099	1.05582	1037.0
	1.7461	946.34	0.9941	486.78	2700.3	2213.5	1.48437	7.1725	5.6881	1.05670	1005.9
	1.8034	945.55	1.0247	491.03	2701.8	2210.8	1.49526	7.1617	5.6664	1.05758	975.9
	1.8623	944.76	1.0559	495.28	2703.3	2208.0	1.50612	7.1510	5.6449	1.05847	947.0
	1.9228	943.96	1.0880	499.53	2704.7	2205.2	1.51695	7.1403	5.6234	1.05937	919.1
120	1.9848	943.16	1.1208	503.78	2706.2	2202.4	1.52776	7.1297	5.6020	1.06027	892.2
	2.0485	942.35	1.1545	508.03	2707.6	2199.6	1.53855	7.1192	5.5807	1.06118	866.2
	2.1139	941.54	1.1889	512.29	2709.1	2196.8	1.54932	7.1087	5.5594	1.06210	841.1
	2.1809	940.72	1.2242	516.55	2710.5	2194.0	1.56006	7.0983	5.5383	1.06302	816.9
	2.2496	939.90	1.2603	520.81	2712.0	2191.2	1.57078	7.0880	5.5172	1.06395	793.5
125	2.3201	939.07	1.2972	525.07	2713.4	2188.3	1.58148	7.0777	5.4962	1.06488	770.9
	2.3924	938.24	1.3351	529.33	2714.8	2185.5	1.59216	7.0675	5.4753	1.06582	749.0
	2.4666	937.41	1.3738	533.60	2716.2	2182.6	1.60281	7.0573	5.4545	1.06677	727.9
	2.5425	936.57	1.4134	537.86	2717.6	2179.8	1.61344	7.0472	5.4338	1.06772	707.5
	2.6204	935.73	1.4539	542.13	2719.0	2176.9	1.62405	7.0372	5.4131	1.06869	687.8
130	2.7002	934.88	1.4954	546.41	2720.4	2174.0	1.63464	7.0272	5.3925	1.06965	668.7
	2.7820	934.03	1.5378	550.68	2721.8	2171.1	1.64521	7.0172	5.3720	1.07063	650.3
	2.8657	933.18	1.5811	554.96	2723.2	2168.2	1.65575	7.0074	5.3516	1.07161	632.5
	2.9515	932.32	1.6255	559.23	2724.5	2165.3	1.66628	6.9975	5.3313	1.07260	615.2
	3.0393	931.45	1.6708	563.52	2725.9	2162.4	1.67678	6.9878	5.3110	1.07359	598.5
135	3.1293	930.59	1.7172	567.80	2727.2	2159.4	1.68726	6.9780	5.2908	1.07459	582.4
	3.2214	929.71	1.7646	572.08	2728.6	2156.5	1.69772	6.9684	5.2706	1.07560	566.7
	3.3157	928.84	1.8130	576.37	2729.9	2153.5	1.70816	6.9587	5.2506	1.07661	551.6
	3.4122	927.96	1.8625	580.66	2731.2	2150.6	1.71858	6.9492	5.2306	1.07764	536.9
	3.5109	927.07	1.9130	584.95	2732.5	2147.6	1.72898	6.9397	5.2107	1.07866	522.7
140	3.6119	926.18	1.9647	589.24	2733.8	2144.6	1.73936	6.9302	5.1908	1.07970	508.99
	3.7153	925.29	2.0174	593.54	2735.1	2141.6	1.74972	6.9208	5.1711	1.08074	495.68
	3.8211	924.39	2.0713	597.84	2736.4	2138.6	1.76006	6.9114	5.1513	1.08179	482.78
	3.9292	923.49	2.1264	602.14	2737.7	2135.6	1.77038	6.9021	5.1317	1.08285	470.28
	4.0398	922.58	2.1826	606.44	2739.0	2132.5	1.78068	6.8928	5.1121	1.08391	458.17
145	4.1529	921.67	2.2400	610.75	2740.2	2129.5	1.79096	6.8836	5.0926	1.08498	446.43
	4.2685	920.76	2.2986	615.06	2741.5	2126.4	1.80122	6.8744	5.0732	1.08606	435.05
	4.3867	919.84	2.3584	619.37	2742.7	2123.3	1.81146	6.8652	5.0538	1.08715	424.01
	4.5075	918.92	2.4195	623.68	2743.9	2120.3	1.82169	6.8562	5.0345	1.08824	413.31
	4.6310	917.99	2.4818	628.00	2745.2	2117.2	1.83189	6.8471	5.0152	1.08934	402.93

Table 1. Saturation (temperature)—*continued*

t (°C)	P	ρ_l	ρ_g	h_l	h_g	r	s_l	s_g	ϕ	ν_l ($\times 10^3$)	ν_g ($\times 10^3$)
150	4.7572	917.06	2.5454	632.32	2746.4	2114.1	1.84208	6.8381	4.9960	1.09044	392.86
	4.8861	916.12	2.6104	636.64	2747.6	2110.9	1.85224	6.8291	4.9769	1.09156	383.09
	5.0178	915.18	2.6766	640.96	2748.8	2107.8	1.86239	6.8202	4.9578	1.09268	373.61
	5.1523	914.24	2.7442	645.29	2750.0	2104.7	1.87252	6.8113	4.9388	1.09381	364.41
	5.2896	913.29	2.8131	649.62	2751.1	2101.5	1.88263	6.8025	4.9198	1.09495	355.48
155	5.4299	912.33	2.8834	653.95	2752.3	2098.3	1.89273	6.7937	4.9010	1.09609	346.81
	5.5732	911.38	2.9551	658.28	2753.4	2095.2	1.90280	6.7849	4.8821	1.09724	338.40
	5.7194	910.41	3.0282	662.62	2754.6	2092.0	1.91286	6.7762	4.8633	1.09840	330.23
	5.8687	909.45	3.1028	666.96	2755.7	2088.8	1.92290	6.7675	4.8446	1.09957	322.29
	6.0211	908.48	3.1788	671.30	2756.8	2085.5	1.93292	6.7589	4.8260	1.10074	314.58
160	6.1766	907.50	3.2564	675.65	2758.0	2082.3	1.94293	6.7503	4.8073	1.10193	307.09
	6.3353	906.52	3.3354	680.00	2759.1	2079.1	1.95292	6.7417	4.7888	1.10312	299.82
	6.4973	905.54	3.4159	684.35	2760.1	2075.8	1.96289	6.7332	4.7703	1.10432	292.75
	6.6625	904.55	3.4980	688.71	2761.2	2072.5	1.97284	6.7247	4.7518	1.10552	285.87
	6.8310	903.56	3.5817	693.07	2762.3	2069.2	1.98278	6.7162	4.7334	1.10674	279.19
165	7.0029	902.56	3.6670	697.43	2763.3	2065.9	1.99271	6.7078	4.7151	1.10796	272.70
	7.1783	901.56	3.7539	701.79	2764.4	2062.6	2.00261	6.6994	4.6968	1.10919	266.39
	7.3570	900.55	3.8424	706.16	2765.4	2059.3	2.01250	6.6910	4.6785	1.11043	260.25
	7.5394	899.54	3.9326	710.53	2766.4	2055.9	2.02237	6.6827	4.6603	1.11168	254.28
	7.7252	898.53	4.0245	714.90	2767.5	2052.5	2.03223	6.6744	4.6422	1.11293	248.48
170	7.9147	897.51	4.1181	719.28	2768.5	2049.2	2.04207	6.6662	4.6241	1.11420	242.83
	8.1078	896.48	4.2135	723.66	2769.4	2045.8	2.05190	6.6579	4.6060	1.11547	237.33
	8.3047	895.46	4.3106	728.05	2770.4	2042.4	2.06171	6.6498	4.5880	1.11675	231.99
	8.5053	894.42	4.4095	732.43	2771.4	2038.9	2.07150	6.6416	4.5701	1.11804	226.78
	8.7098	893.38	4.5102	736.83	2772.3	2035.5	2.08128	6.6335	4.5522	1.11934	221.72
175	8.9180	892.34	4.6127	741.22	2773.3	2032.0	2.09105	6.6254	4.5343	1.12065	216.79
	9.1303	891.30	4.7172	745.62	2774.2	2028.6	2.10080	6.6173	4.5165	1.12196	211.99
	9.3464	890.24	4.8235	750.02	2775.1	2025.1	2.11054	6.6092	4.4987	1.12329	207.32
	9.5666	889.19	4.9317	754.43	2776.0	2021.6	2.12026	6.6012	4.4810	1.12462	202.77
	9.7909	888.13	5.0418	758.84	2776.9	2018.1	2.12996	6.5932	4.4633	1.12596	198.34
180	10.019	887.06	5.154	763.25	2777.8	2014.5	2.13966	6.5853	4.4456	1.12732	194.03
	10.252	885.99	5.268	767.67	2778.6	2011.0	2.14934	6.5774	4.4280	1.12868	189.82
	10.489	884.92	5.384	772.09	2779.5	2007.4	2.15900	6.5694	4.4104	1.13005	185.73
	10.730	883.84	5.502	776.51	2780.3	2003.8	2.16865	6.5616	4.3929	1.13143	181.74
	10.975	882.75	5.623	780.94	2781.2	2000.2	2.17829	6.5537	4.3754	1.13282	177.85
185	11.225	881.67	5.745	785.37	2782.0	1996.6	2.18791	6.5459	4.3580	1.13422	174.06
	11.479	880.57	5.870	789.81	2782.8	1993.0	2.19752	6.5381	4.3406	1.13563	170.37
	11.738	879.47	5.996	794.25	2783.6	1989.3	2.20712	6.5303	4.3232	1.13704	166.77
	12.001	878.37	6.125	798.69	2784.3	1985.6	2.21670	6.5226	4.3059	1.13847	163.26
	12.269	877.26	6.256	803.14	2785.1	1982.0	2.22628	6.5148	4.2886	1.13991	159.84
190	12.542	876.15	6.390	807.60	2785.8	1978.2	2.23583	6.5071	4.2713	1.14136	156.50
	12.819	875.03	6.525	812.06	2786.6	1974.5	2.24538	6.4994	4.2541	1.14282	153.25
	13.101	873.91	6.663	816.52	2787.3	1970.8	2.25491	6.4918	4.2369	1.14429	150.08
	13.388	872.78	6.804	820.98	2788.0	1967.0	2.26444	6.4841	4.2197	1.14576	146.98
	13.680	871.65	6.946	825.46	2788.7	1963.2	2.27395	6.4765	4.2026	1.14725	143.96
195	13.976	870.51	7.091	829.93	2789.4	1959.4	2.28344	6.4689	4.1855	1.14875	141.02
	14.278	869.37	7.239	834.41	2790.0	1955.6	2.29293	6.4613	4.1684	1.15026	138.14
	14.585	868.22	7.389	838.90	2790.7	1951.8	2.30241	6.4538	4.1514	1.15178	135.34
	14.897	867.07	7.541	843.39	2791.3	1947.9	2.31187	6.4463	4.1344	1.15332	132.60
	15.214	865.91	7.697	847.88	2791.9	1944.0	2.32132	6.4387	4.1174	1.15486	129.93

Table 1. Saturation (temperature)—*continued*

t (°C)	P	ρ_l	ρ_g	h_l	h_g	r	s_l	s_g	ϕ	v_l ($\times 10^3$)	v_g ($\times 10^3$)
200	15.537	864.74	7.854	852.38	2792.5	1940.1	2.33076	6.4312	4.1005	1.15641	127.32
	15.864	863.57	8.014	856.89	2793.1	1936.2	2.34019	6.4238	4.0836	1.15798	124.77
	16.197	862.40	8.177	861.40	2793.7	1932.3	2.34961	6.4163	4.0667	1.15955	122.29
	16.536	861.22	8.343	865.91	2794.2	1928.3	2.35902	6.4089	4.0498	1.16114	119.86
	16.880	860.04	8.511	870.43	2794.8	1924.4	2.36842	6.4014	4.0330	1.16274	117.49
205	17.229	858.85	8.682	874.96	2795.3	1920.4	2.37781	6.3940	4.0162	1.16435	115.17
	17.584	857.65	8.856	879.49	2795.8	1916.3	2.38719	6.3866	3.9994	1.16597	112.91
	17.945	856.45	9.033	884.02	2796.3	1912.3	2.39656	6.3793	3.9827	1.16761	110.70
	18.311	855.25	9.213	888.56	2796.8	1908.2	2.40591	6.3719	3.9660	1.16925	108.55
	18.684	854.03	9.395	893.11	2797.3	1904.1	2.41526	6.3646	3.9493	1.17091	106.44
210	19.062	852.82	9.581	897.66	2797.7	1900.0	2.42460	6.3572	3.9326	1.17258	104.38
	19.446	851.59	9.769	902.22	2798.1	1895.9	2.43393	6.3499	3.9160	1.17427	102.36
	19.836	850.37	9.961	906.78	2798.6	1891.8	2.44326	6.3426	3.8993	1.17596	100.40
	20.232	849.13	10.155	911.35	2798.9	1887.6	2.45257	6.3353	3.8827	1.17767	98.47
	20.634	847.89	10.353	915.93	2799.3	1883.4	2.46187	6.3280	3.8662	1.17939	96.59
215	21.042	846.65	10.554	920.51	2799.7	1879.2	2.47117	6.3208	3.8496	1.18113	94.75
	21.457	845.40	10.758	925.10	2800.0	1874.9	2.48046	6.3135	3.8331	1.18288	92.96
	21.878	844.14	10.965	929.69	2800.4	1870.7	2.48974	6.3063	3.8166	1.18464	91.20
	22.305	842.88	11.176	934.29	2800.7	1866.4	2.49901	6.2991	3.8001	1.18641	89.48
	22.738	841.61	11.389	938.90	2801.0	1862.1	2.50827	6.2919	3.7836	1.18820	87.80
220	23.178	840.34	11.607	943.51	2801.3	1857.8	2.51753	6.2847	3.7671	1.19000	86.16
	23.625	839.06	11.827	948.13	2801.5	1853.4	2.52678	6.2775	3.7507	1.19182	84.55
	24.078	837.77	12.052	952.75	2801.8	1849.0	2.53602	6.2703	3.7343	1.19365	82.98
	24.538	836.48	12.279	957.38	2802.0	1844.6	2.54525	6.2631	3.7179	1.19549	81.44
	25.005	835.18	12.511	962.02	2802.2	1840.2	2.55448	6.2559	3.7015	1.19735	79.93
225	25.479	833.87	12.745	966.67	2802.4	1835.7	2.56370	6.2488	3.6851	1.19922	78.46
	25.959	832.56	12.984	971.32	2802.6	1831.2	2.57292	6.2416	3.6687	1.20111	77.02
	26.446	831.25	13.226	975.98	2802.7	1826.7	2.58212	6.2345	3.6524	1.20301	75.61
	26.941	829.92	13.472	980.65	2802.9	1822.2	2.59133	6.2274	3.6361	1.20493	74.23
	27.442	828.59	13.722	985.32	2803.0	1817.7	2.60052	6.2203	3.6197	1.20687	72.88
230	27.951	827.25	13.976	990.00	2803.1	1813.1	2.60971	6.2131	3.6034	1.20882	71.55
	28.467	825.91	14.233	994.69	2803.1	1808.5	2.61890	6.2060	3.5871	1.21078	70.26
	28.990	824.56	14.495	999.39	2803.2	1803.8	2.62808	6.1989	3.5709	1.21276	68.99
	29.521	823.21	14.761	1004.09	2803.2	1799.2	2.63725	6.1918	3.5546	1.21476	67.75
	30.059	821.84	15.031	1008.80	2803.3	1794.5	2.64642	6.1847	3.5383	1.21678	66.53
235	30.604	820.47	15.304	1013.52	2803.3	1789.7	2.65559	6.1777	3.5221	1.21881	65.34
	31.157	819.10	15.583	1018.25	2803.2	1785.0	2.66475	6.1706	3.5058	1.22086	64.17
	31.718	817.71	15.865	1022.98	2803.2	1780.2	2.67390	6.1635	3.4896	1.22292	63.03
	32.286	816.32	16.152	1027.72	2803.1	1775.4	2.68306	6.1564	3.4734	1.22500	61.91
	32.863	814.93	16.443	1032.48	2803.1	1770.6	2.69220	6.1494	3.4572	1.22710	60.82
240	33.447	813.52	16.739	1037.24	2803.0	1765.7	2.70135	6.1423	3.4409	1.22922	59.74
	34.039	812.11	17.039	1042.00	2802.8	1760.8	2.71049	6.1352	3.4247	1.23136	58.69
	34.639	810.69	17.344	1046.78	2802.7	1755.9	2.71963	6.1282	3.4085	1.23351	57.66
	35.247	809.27	17.653	1051.57	2802.5	1751.0	2.72876	6.1211	3.3923	1.23569	56.65
	35.863	807.83	17.967	1056.36	2802.3	1746.0	2.73789	6.1140	3.3761	1.23788	55.66
245	36.488	806.39	18.286	1061.16	2802.1	1741.0	2.74702	6.1070	3.3600	1.24009	54.69
	37.121	804.94	18.610	1065.98	2801.9	1735.9	2.75615	6.0999	3.3438	1.24232	53.73
	37.762	803.49	18.939	1070.80	2801.6	1730.8	2.76528	6.0929	3.3276	1.24458	52.80
	38.412	802.02	19.273	1075.63	2801.4	1725.7	2.77440	6.0858	3.3114	1.24685	51.89
	39.070	800.55	19.612	1080.47	2801.1	1720.6	2.78352	6.0787	3.2952	1.24914	50.99

Table 1. Saturation (temperature)—*continued*

t (°C)	P	ρ_l	ρ_g	h_l	h_g	r	s_l	s_g	ϕ	v_l ($\times 10^3$)	v_g ($\times 10^3$)
250	39.737	799.07	19.956	1085.32	2800.7	1715.4	2.79264	6.0717	3.2790	1.25145	50.111
	40.412	797.58	20.305	1090.18	2800.4	1710.2	2.80176	6.0646	3.2629	1.25379	49.248
	41.096	796.09	20.660	1095.05	2800.0	1705.0	2.81088	6.0575	3.2467	1.25614	48.403
	41.789	794.59	21.020	1099.93	2799.6	1699.7	2.82000	6.0505	3.2305	1.25852	47.573
	42.491	793.07	21.386	1104.82	2799.2	1694.4	2.82911	6.0434	3.2143	1.26092	46.760
255	43.202	791.55	21.757	1109.72	2798.8	1689.1	2.83823	6.0363	3.1981	1.26334	45.962
	43.922	790.03	22.134	1114.63	2798.3	1683.7	2.84735	6.0292	3.1819	1.26578	45.180
	44.651	788.49	22.517	1119.55	2797.8	1678.3	2.85646	6.0222	3.1657	1.26825	44.412
	45.390	786.94	22.905	1124.48	2797.3	1672.8	2.86558	6.0151	3.1495	1.27074	43.658
	46.137	785.39	23.300	1129.43	2796.8	1667.4	2.87470	6.0080	3.1333	1.27325	42.919
260	46.895	783.83	23.700	1134.38	2796.2	1661.9	2.88382	6.0009	3.1170	1.27579	42.194
	47.661	782.25	24.107	1139.34	2795.6	1656.3	2.89294	5.9938	3.1008	1.27836	41.482
	48.437	780.67	24.520	1144.32	2795.0	1650.7	2.90206	5.9866	3.0846	1.28095	40.783
	49.223	779.08	24.939	1149.31	2794.4	1645.1	2.91119	5.9795	3.0683	1.28356	40.098
	50.018	777.48	25.365	1154.31	2793.7	1639.4	2.92031	5.9724	3.0521	1.28620	39.424
265	50.823	775.87	25.797	1159.32	2793.0	1633.7	2.92944	5.9652	3.0358	1.28887	38.764
	51.638	774.25	26.236	1164.35	2792.3	1628.0	2.93858	5.9581	3.0195	1.29156	38.115
	52.463	772.63	26.682	1169.38	2791.6	1622.2	2.94771	5.9509	3.0032	1.29429	37.478
	53.298	770.99	27.135	1174.43	2790.8	1616.3	2.95685	5.9437	2.9869	1.29704	36.853
	54.143	769.34	27.595	1179.49	2790.0	1610.5	2.96599	5.9365	2.9705	1.29981	36.239
270	54.999	767.68	28.061	1184.57	2789.1	1604.6	2.97514	5.9293	2.9542	1.30262	35.636
	55.864	766.01	28.536	1189.66	2788.3	1598.6	2.98429	5.9221	2.9378	1.30546	35.044
	56.740	764.34	29.017	1194.76	2787.4	1592.6	2.99345	5.9149	2.9215	1.30833	34.462
	57.627	762.65	29.506	1199.87	2786.5	1586.6	3.00261	5.9077	2.9051	1.31122	33.891
	58.524	760.95	30.003	1205.00	2785.5	1580.5	3.01178	5.9004	2.8886	1.31415	33.330
275	59.431	759.24	30.507	1210.15	2784.5	1574.4	3.02095	5.8931	2.8722	1.31711	32.779
	60.350	757.52	31.020	1215.30	2783.5	1568.2	3.03013	5.8859	2.8557	1.32011	32.237
	61.279	755.78	31.541	1220.47	2782.5	1562.0	3.03931	5.8786	2.8392	1.32313	31.705
	62.219	754.04	32.069	1225.66	2781.4	1555.8	3.04850	5.8712	2.8227	1.32619	31.182
	63.170	752.28	32.607	1230.86	2780.3	1549.4	3.05770	5.8639	2.8062	1.32929	30.669
280	64.132	750.52	33.152	1236.08	2779.2	1543.1	3.06691	5.8565	2.7896	1.33242	30.164
	65.105	748.74	33.707	1241.31	2778.0	1536.7	3.07613	5.8492	2.7730	1.33558	29.668
	66.089	746.95	34.270	1246.56	2776.8	1530.2	3.08535	5.8418	2.7564	1.33878	29.180
	67.085	745.14	34.843	1251.82	2775.5	1523.7	3.09458	5.8344	2.7398	1.34202	28.701
	68.092	743.33	35.424	1257.10	2774.3	1517.2	3.10382	5.8269	2.7231	1.34530	28.229
285	69.111	741.50	36.015	1262.40	2773.0	1510.6	3.11308	5.8195	2.7064	1.34862	27.766
	70.141	739.66	36.616	1267.71	2771.6	1503.9	3.12234	5.8120	2.6896	1.35197	27.310
	71.183	737.81	37.226	1273.04	2770.2	1497.2	3.13161	5.8045	2.6729	1.35537	26.863
	72.237	735.94	37.847	1278.39	2768.8	1490.4	3.14089	5.7969	2.6560	1.35881	26.422
	73.303	734.06	38.478	1283.75	2767.4	1483.6	3.15019	5.7894	2.6392	1.36229	25.989
290	74.380	732.16	39.119	1289.14	2765.9	1476.7	3.15950	5.7818	2.6223	1.36581	25.563
	75.470	730.26	39.770	1294.54	2764.3	1469.8	3.16882	5.7742	2.6054	1.36938	25.144
	76.572	728.33	40.433	1299.96	2762.8	1462.8	3.17815	5.7665	2.5884	1.37300	24.732
	77.686	726.40	41.106	1305.40	2761.2	1455.8	3.18750	5.7589	2.5714	1.37666	24.327
	78.813	724.45	41.791	1310.86	2759.5	1448.7	3.19686	5.7511	2.5543	1.38037	23.928
295	79.952	722.48	42.488	1316.34	2757.8	1441.5	3.20623	5.7434	2.5372	1.38412	23.536
	81.103	720.50	43.196	1321.84	2756.1	1434.3	3.21563	5.7356	2.5200	1.38793	23.150
	82.268	718.50	43.917	1327.36	2754.3	1427.0	3.22503	5.7278	2.5028	1.39179	22.770
	83.445	716.49	44.650	1332.90	2752.5	1419.6	3.23446	5.7200	2.4855	1.39570	22.397
	84.635	714.46	45.395	1338.47	2750.7	1412.2	3.24390	5.7121	2.4682	1.39967	22.029

Table 1. Saturation (temperature)—*continued*

t (°C)	P	ρ_l	ρ_g	h_l	h_g	r	s_l	s_g	ϕ	ν_l ($\times 10^3$)	ν_g ($\times 10^3$)
300	85.838	712.41	46.154	1344.05	2748.7	1404.7	3.25336	5.7042	2.4508	1.40369	21.667
	87.054	710.35	46.926	1349.66	2746.8	1397.1	3.26284	5.6962	2.4334	1.40777	21.310
	88.283	708.27	47.711	1355.29	2744.8	1389.5	3.27233	5.6882	2.4159	1.41190	20.960
	89.526	706.17	48.510	1360.95	2742.8	1381.8	3.28185	5.6802	2.3983	1.41610	20.614
	90.782	704.05	49.324	1366.63	2740.7	1374.0	3.29139	5.6721	2.3807	1.42035	20.274
305	92.051	701.92	50.15	1372.33	2738.5	1366.2	3.30095	5.6640	2.3630	1.42467	19.940
	93.334	699.76	51.00	1378.06	2736.3	1358.3	3.31053	5.6558	2.3453	1.42906	19.610
	94.631	697.59	51.85	1383.81	2734.1	1350.3	3.32014	5.6476	2.3275	1.43351	19.285
	95.942	695.40	52.73	1389.59	2731.8	1342.2	3.32977	5.6393	2.3096	1.43803	18.966
	97.267	693.18	53.62	1395.40	2729.4	1334.0	3.33943	5.6310	2.2916	1.44262	18.651
310	98.605	690.95	54.52	1401.23	2727.0	1325.8	3.34911	5.6226	2.2735	1.44728	18.340
	99.958	688.70	55.45	1407.10	2724.6	1317.5	3.35882	5.6142	2.2554	1.45202	18.035
	101.326	686.42	56.39	1412.99	2722.1	1309.1	3.36856	5.6057	2.2372	1.45683	17.734
	102.707	684.12	57.35	1418.91	2719.5	1300.6	3.37832	5.5972	2.2189	1.46173	17.437
	104.104	681.80	58.33	1424.86	2716.9	1292.0	3.38812	5.5886	2.2005	1.46670	17.145
315	105.51	679.46	59.32	1430.84	2714.2	1283.3	3.39795i	5.5799	2.1820	1.47176	16.856
	106.94	677.09	60.34	1436.86	2711.4	1274.6	3.40781	5.5712	2.1634	1.47691	16.572
	108.38	674.70	61.38	1442.90	2708.6	1265.7	3.41770	5.5624	2.1447	1.48215	16.293
	109.84	672.28	62.44	1448.99	2705.7	1256.7	3.42763	5.5535	2.1259	1.48748	16.017
	111.31	669.83	63.51	1455.10	2702.8	1247.6	3.43760	5.5446	2.1070	1.49291	15.745
320	112.79	667.36	64.62	1461.25	2699.7	1238.5	3.44760	5.5356	2.0880	1.49843	15.476
	114.29	664.87	65.74	1467.44	2696.6	1229.2	3.45765	5.5265	2.0688	1.50406	15.212
	115.81	662.34	66.89	1473.67	2693.5	1219.8	3.46773	5.5173	2.0496	1.50980	14.951
	117.34	659.78	68.06	1479.93	2690.2	1210.3	3.47786	5.5081	2.0302	1.51565	14.693
	118.89	657.20	69.26	1486.24	2686.9	1200.7	3.48803	5.4987	2.0107	1.52161	14.439
325	120.46	654.58	70.48	1492.58	2683.5	1190.9	3.49825	5.4893	1.9911	1.52769	14.189
	122.04	651.93	71.73	1498.97	2680.1	1181.1	3.50852	5.4798	1.9713	1.53390	13.942
	123.64	649.25	73.00	1505.40	2676.5	1171.1	3.51884	5.4702	1.9513	1.54024	13.698
	125.25	646.53	74.31	1511.88	2672.9	1161.0	3.52921	5.4605	1.9313	1.54671	13.457
	126.88	643.78	75.65	1518.41	2669.1	1150.7	3.53963	5.4506	1.9110	1.55332	13.219
330	128.52	641.0	77.01	1525.0	2665.3	1140.3	3.5501	5.4407	1.8906	1.5601	12.985
	130.19	638.2	78.41	1531.6	2661.4	1129.8	3.5607	5.4307	1.8700	1.5670	12.753
	131.87	635.3	79.84	1538.3	2657.4	1119.1	3.5713	5.4205	1.8493	1.5740	12.524
	133.57	632.4	81.31	1545.0	2653.3	1108.3	3.5819	5.4103	1.8283	1.5813	12.298
	135.28	629.5	82.82	1551.8	2649.0	1097.2	3.5927	5.3999	1.8072	1.5887	12.075
335	137.01	626.5	84.36	1558.6	2644.7	1086.1	3.6035	5.3894	1.7859	1.5963	11.854
	138.76	623.4	85.94	1565.5	2640.3	1074.7	3.6144	5.3787	1.7643	1.6040	11.636
	140.53	620.3	87.56	1572.5	2635.7	1063.2	3.6253	5.3679	1.7426	1.6120	11.421
	142.32	617.2	89.22	1579.5	2631.1	1051.5	3.6364	5.3569	1.7205	1.6202	11.208
	144.12	614.0	90.93	1586.7	2626.3	1039.6	3.6475	5.3458	1.6983	1.6286	10.997
340	145.94	610.8	92.69	1593.8	2621.3	1027.5	3.6587	5.3345	1.6758	1.6373	10.788
	147.78	607.5	94.50	1601.1	2616.3	1015.2	3.6701	5.3231	1.6530	1.6462	10.582
	149.64	604.1	96.36	1608.4	2611.1	1002.7	3.6815	5.3114	1.6299	1.6553	10.378
	151.52	600.7	98.27	1615.8	2605.7	989.9	3.6930	5.2996	1.6066	1.6647	10.176
	153.42	597.2	100.24	1623.3	2600.2	976.9	3.7047	5.2876	1.5829	1.6745	9.976
345	155.33	593.7	102.27	1630.9	2594.5	963.6	3.7164	5.2753	1.5589	1.6845	9.778
	157.27	590.0	104.37	1638.6	2588.7	950.1	3.7283	5.2629	1.5345	1.6948	9.581
	159.22	586.3	106.53	1646.4	2582.7	936.3	3.7404	5.2502	1.5098	1.7056	9.387
	161.20	582.5	108.77	1654.3	2576.5	922.2	3.7526	5.2372	1.4847	1.7166	9.194
	163.20	578.7	111.08	1662.3	2570.1	907.8	3.7649	5.2240	1.4591	1.7281	9.002

Table 1. Saturation (temperature)—*continued*

t (°C)	P	ρ_l	ρ_g	h_l	h_g	r	s_l	s_g	ϕ	v_l (× 10^3)	v_g (× 10^3)
350	165.21	574.7	113.48	1670.4	2563.5	893.0	3.7774	5.2105	1.4331	1.7401	8.812
	167.25	570.6	115.96	1678.7	2556.6	877.9	3.7901	5.1967	1.4066	1.7525	8.623
	169.31	566.4	118.54	1687.1	2549.6	862.4	3.8030	5.1825	1.3796	1.7654	8.436
	171.38	562.2	121.22	1695.7	2542.2	846.6	3.8161	5.1681	1.3520	1.7788	8.249
	173.48	557.8	124.01	1704.4	2534.6	830.2	3.8294	5.1532	1.3238	1.7929	8.064
355	175.61	553.2	126.92	1713.3	2526.7	813.5	3.8429	5.1379	1.2950	1.8076	7.879
	177.75	548.5	129.95	1722.4	2518.5	796.2	3.8568	5.1222	1.2655	1.8230	7.695
	179.92	543.7	133.13	1731.7	2510.0	778.3	3.8709	5.1060	1.2352	1.8392	7.512
	182.11	538.7	136.46	1741.2	2501.1	759.9	3.8853	5.0893	1.2040	1.8563	7.328
	184.32	533.5	139.96	1750.9	2491.8	740.8	3.9001	5.0721	1.1719	1.8744	7.145
360	186.55	528.1	143.65	1761.0	2482.0	721.1	3.9153	5.0542	1.1388	1.8936	6.962
	188.81	522.5	147.54	1771.3	2471.8	700.5	3.9310	5.0355	1.1046	1.9140	6.778
	191.10	516.6	151.68	1782.0	2461.0	679.0	3.9471	5.0161	1.0690	1.9358	6.593
	193.40	510.4	156.08	1793.1	2449.6	656.5	3.9638	4.9958	1.0320	1.9592	6.407
	195.74	503.9	160.80	1804.6	2437.5	632.9	3.9812	4.9745	0.9933	1.9845	6.219
365.0	198.09	497.0	165.88	1816.7	2424.6	607.9	3.9994	4.9520	0.9526	2.0120	6.028
	199.28	493.4	168.58	1822.9	2417.8	594.8	4.0088	4.9402	0.9314	2.0268	5.932
	200.48	489.7	171.39	1829.3	2410.7	581.3	4.0185	4.9280	0.9096	2.0422	5.835
	201.68	485.8	174.33	1835.9	2403.3	567.4	4.0284	4.9155	0.8870	2.0585	5.736
	202.89	481.8	177.42	1842.7	2395.6	552.9	4.0387	4.9024	0.8637	2.0757	5.636
367.5	204.11	477.6	180.67	1849.8	2387.6	537.8	4.0493	4.8888	0.8395	2.0939	5.535
	205.33	473.2	184.11	1857.1	2379.2	522.1	4.0602	4.8746	0.8143	2.1133	5.432
	206.56	468.6	187.75	1864.7	2370.3	505.6	4.0717	4.8597	0.7880	2.1340	5.326
	207.80	463.8	191.63	1872.6	2360.9	488.3	4.0836	4.8440	0.7604	2.1563	5.218
	209.05	458.6	195.79	1880.9	2350.9	470.0	4.0962	4.8274	0.7313	2.1804	5.107
370.0	210.30	453.1	200.29	1889.7	2340.2	450.4	4.1094	4.8098	0.7003	2.2068	4.993
	211.56	447.2	205.21	1899.1	2328.5	429.4	4.1236	4.7907	0.6671	2.2361	4.873
	212.83	440.7	210.64	1909.3	2315.8	406.5	4.1389	4.7700	0.6311	2.2689	4.747
	214.11	433.5	216.74	1920.5	2301.6	381.2	4.1558	4.7471	0.5913	2.3067	4.614
	215.39	425.3	223.74	1933.0	2285.5	352.5	4.1748	4.7212	0.5464	2.3515	4.469
372.5	216.69	415.4	232.1	1947.7	2266.6	318.9	4.1971	4.6910	0.4939	2.4074	4.309
	217.99	402.4	242.7	1966.6	2243.0	276.4	4.2258	4.6536	0.4277	2.4850	4.121
	219.30	385.0	259.0	1991.6	2207.3	215.7	4.2640	4.5977	0.3337	2.5974	3.861
373.976	220.55		322		2086	0		4.409	0		3.106

Table 2. Saturation (pressure)

P (bar)	t (°C)	ρ_l	ρ_g	h_l	h_g	r	s_l	s_g	ϕ	v_t (×10^3)	v_g (×10^3)
0.0061173	0.010	999.78	0.004855	0.00	2500.5	2500.5	0	9.1541	9.1541	1.00022	205990.
0.010	6.970	999.89	0.007740	29.27	2513.3	2484.1	0.10581	8.9737	8.8678	1.00011	129190
0.025	21.080	997.96	0.018433	88.36	2539.1	2450.8	0.31160	8.6411	8.3295	1.00205	54249.
0.050	32.881	994.70	0.035472	137.67	2560.5	2422.9	0.47594	8.3930	7.9171	1.00533	28191
0.075	40.299	992.05	0.051982	168.74	2573.9	2405.1	0.57625	8.2494	7.6732	1.00801	19237.
0.100	45.817	989.82	0.06815	191.83	2583.8	2391.9	0.64926	8.1482	7.4990	1.01028	14674.
0.15	53.983	986.14	0.09977	225.95	2598.2	2372.3	0.75486	8.0066	7.2517	1.01405	10023.
0.20	60.073	983.13	0.13072	251.46	2608.9	2357.5	0.83211	7.9068	7.0747	1.01716	7650.

Table 2. Saturation (pressure)—*continued*

P (bar)	t (°C)	ρ_l	ρ_g	h_l	h_g	r	s_l	s_g	ϕ	v_l ($\times 10^3$)	v_g ($\times 10^3$)
0.25	64.980	980.54	0.16117	271.99	2617.4	2345.5	0.89326	7.8298	6.9366	1.01985	6204.8
0.50	81.339	970.96	0.30856	340.54	2645.3	2304.8	1.09117	7.5928	6.5017	1.02991	3240.9
0.75	91.783	964.13	0.45095	384.43	2662.5	2278.1	1.21309	7.4557	6.2426	1.03721	2217.5
1.0	99.632	958.66	0.5902	417.51	2675.1	2257.6	1.30273	7.3589	6.0562	1.04313	1694.3
1.5	111.378	949.94	0.8624	467.18	2693.4	2226.2	1.43376	7.2232	5.7894	1.05270	1159.5
2.0	120.241	942.96	1.1289	504.80	2706.5	2201.7	1.53035	7.1272	5.5968	1.06049	885.9
2.5	127.443	937.04	1.3912	535.49	2716.8	2181.4	1.60753	7.0528	5.4453	1.06719	718.8
3.0	133.555	931.84	1.6505	561.61	2725.3	2163.7	1.67211	6.9921	5.3200	1.07315	605.9
3.5	138.891	927.17	1.9074	584.48	2732.4	2147.9	1.72785	6.9407	5.2129	1.07855	524.27
4.0	143.643	922.91	2.1624	604.90	2738.5	2133.6	1.77700	6.8961	5.1191	1.08353	462.46
5.0	151.866	915.31	2.6677	640.38	2748.6	2108.2	1.86104	6.8214	4.9604	1.09253	374.86
6.0	158.863	908.61	3.1683	670.71	2756.7	2086.0	1.93155	6.7601	4.8285	1.10058	315.63
7.0	164.983	902.58	3.6655	697.35	2763.3	2066.0	1.99254	6.7079	4.7154	1.10794	272.81
8.0	170.444	897.05	4.1603	721.23	2768.9	2047.7	2.04644	6.6625	4.6161	1.11476	240.37
9.0	175.388	891.94	4.6531	742.93	2773.6	2030.7	2.09484	6.6222	4.5274	1.12116	214.91
10.0	179.916	887.15	5.144	762.88	2777.7	2014.8	2.13885	6.5859	4.4471	1.12720	194.38
12.5	189.848	876.32	6.369	806.92	2785.7	1978.8	2.23439	6.5083	4.2739	1.14114	157.01
15.0	198.327	866.69	7.592	844.86	2791.5	1946.6	2.31496	6.4438	4.1288	1.15382	131.72
17.5	205.764	857.93	8.815	878.42	2795.7	1917.3	2.38498	6.3884	4.0034	1.16559	113.44
20.0	212.417	849.85	10.041	908.69	2798.7	1890.0	2.44714	6.3396	3.8924	1.17667	99.59
22.5	218.452	842.30	11.272	936.38	2800.8	1864.4	2.50320	6.2958	3.7926	1.18722	88.72
25.0	223.989	835.19	12.508	961.98	2802.2	1840.2	2.55439	6.2560	3.7016	1.19733	79.95
27.5	229.114	828.44	13.751	985.86	2803.0	1817.1	2.60158	6.2195	3.6179	1.20709	72.72
30.0	233.892	821.99	15.001	1008.30	2803.3	1795.0	2.64544	6.1855	3.5401	1.21656	66.66
35.0	242.595	809.84	17.527	1049.64	2802.6	1752.9	2.72508	6.1240	3.3989	1.23481	57.05
40.0	250.392	798.49	20.092	1087.24	2800.6	1713.4	2.79623	6.0689	3.2727	1.25237	49.771
45.0	257.474	787.75	22.700	1121.90	2797.6	1675.7	2.86080	6.0188	3.1580	1.26943	44.053
50.0	263.977	777.51	25.355	1154.22	2793.7	1639.5	2.92013	5.9725	3.0524	1.28615	39.440
55.0	270.001	767.68	28.062	1184.60	2789.1	1604.5	2.97518	5.9294	2.9542	1.30263	35.636
60.0	275.621	758.16	30.824	1213.37	2783.9	1570.5	3.02667	5.8887	2.8620	1.31898	32.442
65.0	280.893	748.92	33.646	1240.78	2778.1	1537.3	3.07517	5.8500	2.7748	1.33525	29.721
70.0	285.864	739.90	36.533	1267.02	2771.8	1504.8	3.12111	5.8130	2.6919	1.35153	27.373
75.0	290.570	731.07	39.488	1292.25	2765.0	1472.8	3.16485	5.7775	2.6126	1.36786	25.324
80.0	295.042	722.38	42.516	1316.61	2757.8	1441.2	3.20668	5.7431	2.5365	1.38430	23.520
85.0	299.305	713.82	45.623	1340.21	2750.1	1409.9	3.24683	5.7098	2.4629	1.40091	21.919
90.0	303.379	705.35	48.814	1363.15	2742.0	1378.8	3.28553	5.6772	2.3917	1.41773	20.486
95.0	307.282	696.96	52.10	1385.49	2733.5	1348.0	3.32292	5.6453	2.3224	1.43481	19.195
100.0	311.031	688.61	55.47	1407.33	2724.5	1317.2	3.35918	5.6140	2.2548	1.45220	18.026
105.0	314.637	680.29	58.96	1428.72	2715.2	1286.4	3.39444	5.5831	2.1887	1.46995	16.962
110.0	318.112	671.99	62.55	1449.73	2705.4	1255.7	3.42882	5.5526	2.1238	1.48812	15.987
115.0	321.466	663.67	66.27	1470.40	2695.2	1224.8	3.46242	5.5223	2.0599	1.50677	15.091
120.0	324.709	655.33	70.11	1490.79	2684.6	1193.8	3.49534	5.4922	1.9968	1.52596	14.263
125.0	327.847	646.93	74.10	1510.95	2673.4	1162.5	3.52769	5.4621	1.9344	1.54576	13.495
130.0	330.888	638.5	78.25	1530.9	2661.9	1130.9	3.5595	5.4319	1.8724	1.5663	12.780
135.0	333.837	629.9	82.56	1550.7	2649.8	1099.0	3.5910	5.4017	1.8107	1.5875	12.112
140.0	336.701	621.2	87.06	1570.5	2637.1	1066.7	3.6221	5.3712	1.7491	1.6097	11.486
145.0	339.485	612.4	91.77	1590.2	2623.9	1033.8	3.6530	5.3404	1.6874	1.6328	10.896
150.0	342.192	603.4	96.71	1609.9	2610.1	1000.2	3.6837	5.3093	1.6255	1.6571	10.340
155.0	344.827	594.3	101.91	1629.6	2595.6	965.9	3.7144	5.2775	1.5631	1.6828	9.812
160.0	347.394	584.8	107.40	1649.5	2580.3	930.8	3.7452	5.2451	1.5000	1.7099	9.311
165.0	349.896	575.1	113.23	1669.6	2564.2	894.6	3.7761	5.2119	1.4358	1.7388	8.832

Table 2. Saturation (pressure)—*continued*

P (bar)	t (°C)	ρ_l	ρ_g	h_l	h_g	r	s_l	s_g	ϕ	v_l ($\times 10^3$)	v_g ($\times 10^3$)
170.0	352.335	565.0	119.43	1689.9	2547.1	857.2	3.8073	5.1777	1.3704	1.7698	8.373
175.0	354.715	554.5	126.09	1710.7	2529.0	818.3	3.8390	5.1422	1.3032	1.8033	7.931
180.0	357.038	543.6	133.27	1731.9	2509.6	777.6	3.8713	5.1053	1.2339	1.8398	7.504
185.0	359.306	531.9	141.10	1753.9	2488.7	734.8	3.9046	5.0664	1.1618	1.8799	7.087
190.0	361.522	519.5	149.72	1776.7	2466.0	689.3	3.9391	5.0252	1.0860	1.9248	6.679
195.0	363.686	506.1	159.35	1800.8	2441.1	640.3	3.9754	4.9809	1.0054	1.9759	6.275
200.0	365.800	491.3	170.36	1826.5	2413.2	586.7	4.0142	4.9323	0.9181	2.0353	5.870
205.0	367.865	474.6	183.32	1854.8	2381.0	526.2	4.0568	4.8776	0.8208	2.1070	5.455
210.0	369.881	454.8	199.46	1887.1	2342.0	454.8	4.1055	4.8128	0.7072	2.1988	5.014
212.5	370.871	442.9	209.55	1906.0	2318.2	412.2	4.1340	4.7739	0.6399	2.2578	4.772
215.0	371.848	428.5	222.08	1928.2	2289.1	361.0	4.1675	4.7270	0.5595	2.3335	4.503
217.5	372.813	409.2	239.6	1957.0	2249.4	292.3	4.2113	4.6637	0.4525	2.444	4.173
220.55	373.976	322		2086		0	4.409		0		3.106

Table 3. Compressed water and superheated steam

t (°C)	0.1 bar (t_s = 45.817 °C) v ($\times 10^3$)	ρ	h	u	s
t_l	1.01028	989.82	191.83	191.82	0.64926
t_g	14674	0.06815	2583.8	2437.0	8.1482
0	1.00022	999.78	−0.03	−0.04	−0.00015
5	1.00005	999.95	21.03	21.02	0.07626
10	1.00030	999.70	42.00	41.99	0.15097
15	1.00091	999.10	62.92	62.91	0.22423
20	1.00181	998.19	83.84	83.83	0.29621
25	1.00299	997.02	104.76	104.75	0.36695
30	1.00441	995.61	125.68	125.67	0.43653
35	1.00605	993.99	146.59	146.58	0.50496
40	1.00789	992.17	167.51	167.50	0.57228
45	1.00993	990.17	188.42	188.41	0.63853
50	14869	0.06725	2591.8	2443.1	8.1731
55	15103	0.06621	2601.3	2450.3	8.2024
60	15336	0.06521	2610.8	2457.5	8.2313
65	15569	0.06423	2620.4	2464.7	8.2596
70	15802	0.06328	2629.9	2471.8	8.2876
75	16035	0.06236	2639.4	2479.0	8.3151
80	16268	0.06147	2648.9	2486.2	8.3422
85	16500	0.06061	2658.4	2493.4	8.3690
90	16732	0.05976	2667.9	2500.6	8.3954
95	16964	0.05895	2677.4	2507.8	8.4214
100	17196	0.05815	2687.0	2515.0	8.4471
105	17428	0.05738	2696.5	2522.2	8.4724
110	17660	0.05662	2706.0	2529.4	8.4974
115	17892	0.05589	2715.5	2536.6	8.5222
120	18124	0.05518	2725.1	2543.8	8.5466
125	18356	0.05448	2734.6	2551.1	8.5707
130	18587	0.05380	2744.2	2558.3	8.5946
135	18819	0.05314	2753.7	2565.6	8.6181
140	19050	0.05249	2763.3	2572.8	8.6415
145	19282	0.05186	2772.9	2580.1	8.6645

Table 3. Compressed water and superheated steam—*continued*

t (°C)	0.1 bar (t_s = 45.817 °C) v ($\times 10^3$)	ρ	h	u	s
150	19513	0.051248	2782.5	2587.4	8.6873
155	19744	0.050648	2792.1	2594.7	8.7099
160	19976	0.050061	2801.7	2602.0	8.7322
165	20207	0.049488	2811.3	2609.3	8.7543
170	20438	0.048928	2821.0	2616.6	8.7762
175	20670	0.048379	2830.6	2623.9	8.7978
180	20901	0.047845	2840.3	2631.3	8.8193
185	21132	0.047321	2850.0	2638.6	8.8405
190	21363	0.046809	2859.6	2646.0	8.8615
195	21594	0.046308	2869.3	2653.4	8.8823
200	21826	0.045818	2879.0	2660.8	8.9030
205	22057	0.045338	2888.8	2668.2	8.9234
210	22288	0.044868	2898.5	2675.6	8.9437
215	22519	0.044407	2908.3	2683.1	8.9637
220	22750	0.043956	2918.0	2690.5	8.9836
225	22981	0.043514	2927.8	2698.0	9.0034
230	23212	0.043081	2937.6	2705.5	9.0229
235	23443	0.042656	2947.4	2713.0	9.0423
240	23674	0.042240	2957.2	2720.5	9.0615
245	23905	0.041832	2967.1	2728.0	9.0806
250	24136	0.041432	2976.9	2735.5	9.0995
255	24367	0.041039	2986.8	2743.1	9.1183
260	24598	0.040654	2996.6	2750.7	9.1369
265	24829	0.040275	3006.5	2758.2	9.1554
270	25060	0.039904	3016.4	2765.8	9.1737
275	25291	0.039540	3026.4	2773.4	9.1919
280	25522	0.039182	3036.3	2781.1	9.2099
285	25753	0.038831	3046.2	2788.7	9.2278
290	25984	0.038485	3056.2	2796.4	9.2456
295	26215	0.038146	3066.2	2804.0	9.2633
300	26446	0.037813	3076.2	2811.7	9.2808
305	26677	0.037486	3086.2	2819.4	9.2982

Table 3. Compressed water and superheated steam—*continued*

t (°C)	0.1 bar v (×10³)	ρ	h	u	s
310	26908	0.037164	3096.2	2827.2	9.3154
315	27138	0.036848	3106.3	2834.9	9.3326
320	27369	0.036537	3116.3	2842.6	9.3496
325	27600	0.036232	3126.4	2850.4	9.3665
330	27831	0.035931	3136.5	2858.2	9.3833
335	28062	0.035635	3146.6	2866.0	9.4000
340	28293	0.035345	3156.7	2873.8	9.4166
345	28524	0.035058	3166.9	2881.6	9.4330
350	28755	0.034777	3177.0	2889.5	9.4494
355	28986	0.034500	3187.2	2897.3	9.4657
360	29216	0.034227	3197.4	2905.2	9.4818
365	29447	0.033959	3207.6	2913.1	9.4978
370	29678	0.033695	3217.8	2921.0	9.5138
375	29909	0.033435	3228.0	2928.9	9.5296
380	30140	0.033179	3238.3	2936.9	9.5454
385	30371	0.032926	3248.5	2944.8	9.5610
390	30602	0.032678	3258.8	2952.8	9.5766
395	30832	0.032433	3269.1	2960.8	9.5921
400	31063	0.032192	3279.4	2968.8	9.6075
410	31525	0.031721	3300.1	2984.8	9.6379
420	31987	0.031263	3320.8	3001.0	9.6681
430	32448	0.030818	3341.6	3017.1	9.6979
440	32910	0.030386	3362.5	3033.4	9.7274
450	33372	0.029966	3383.4	3049.7	9.7565
460	33833	0.029557	3404.5	3066.1	9.7854
470	34295	0.029159	3425.5	3082.6	9.8139
480	34757	0.028772	3446.7	3099.1	9.8422
490	35218	0.028394	3467.9	3115.7	9.8702
500	35680	0.028027	3489.2	3132.4	9.8979
520	36603	0.027320	3531.9	3165.9	9.9525
540	37526	0.026648	3575.0	3199.7	10.0061
560	38450	0.026008	3618.3	3233.8	10.0587
580	39373	0.025398	3661.9	3268.1	10.1104
600	40296	0.024816	3705.7	3302.8	10.1612
620	41219	0.024261	3749.9	3337.7	10.2112
640	42142	0.023729	3794.3	3372.9	10.2604
660	43065	0.023220	3839.1	3408.4	10.3089
680	43989	0.022733	3884.1	3444.2	10.3566
700	44912	0.022266	3929.4	3480.2	10.4036
720	45835	0.021817	3974.9	3516.6	10.4500
740	46758	0.021387	4020.8	3553.2	10.4957
760	47681	0.020973	4066.9	3590.1	10.5408
780	48604	0.020574	4113.4	3627.3	10.5853
800	49527	0.020191	4160.1	3664.8	10.6292
850	51840	0.019292	4278.1	3759.7	10.7367
900	54140	0.018470	4397.8	3856.4	10.8410
950	56450	0.017715	4519.2	3954.7	10.9423
1000	58760	0.017019	4642.3	4054.7	11.0410
1100	63370	0.015779	4893.2	4259.5	11.2307
1200	67990	0.014708	5150.2	4470.3	11.4113
1300	72600	0.013773	5412.9	4686.8	11.5838

Table 3. Compressed water and superheated steam—*continued*

t (°C)	0.1 bar v (×10³)	ρ	h	u	s
1400	77220	0.012950	5680.8	4908.6	11.7489
1500	81840	0.012220	5953.5	5135.1	11.9071
1600	86450	0.011567	6230.6	5366.1	12.0592
1700	91070	0.010981	6511.8	5601.1	12.2054
1800	95680	0.010451	6796.7	5839.9	12.3463
1900	100300	0.009970	7085.1	6082.2	12.4821
2000	104910	0.009532	7376.7	6327.6	12.6133

t (°C)	0.5 bar (t_g = 81.339°C) v (× 10³)	ρ	h	u	s
t_t	1.02991	970.96	340.54	340.49	1.09117
t_g	3240.9	0.30856	2645.3	2483.3	7.5928
0	1.00020	999.80	0.01	−0.04	−0.00015
5	1.00003	999.97	21.07	21.02	0.07626
10	1.00028	999.72	42.04	41.99	0.15096
15	1.00089	999.11	62.96	62.91	0.22423
20	1.00179	998.21	83.88	83.83	0.29620
25	1.00297	997.04	104.80	104.75	0.36694
30	1.00439	995.63	125.71	125.66	0.43652
35	1.00603	994.01	146.63	146.58	0.50495
40	1.00787	992.19	167.54	167.49	0.57227
45	1.00991	990.18	188.45	188.40	0.63852
50	1.01214	988.01	209.36	209.31	0.70372
55	1.01454	985.67	230.27	230.22	0.76793
60	1.01711	983.18	251.18	251.13	0.83117
65	1.01985	980.54	272.10	272.05	0.89349
70	1.02275	977.76	293.02	292.97	0.95493
75	1.02581	974.84	313.97	313.92	1.01552
80	1.02902	971.80	334.93	334.88	1.07530
85	3275.9	0.30526	2652.6	2488.8	7.6132
90	3323.6	0.30088	2662.5	2496.3	7.6406
95	3371.3	0.29663	2672.3	2503.7	7.6676
100	3418.8	0.29250	2682.1	2511.2	7.6941
105	3466.2	0.28850	2691.9	2518.6	7.7202
110	3513.5	0.28462	2701.7	2526.0	7.7459
115	3560.7	0.28084	2711.5	2533.5	7.7712
120	3607.8	0.27717	2721.2	2540.9	7.7962
125	3654.9	0.27360	2731.0	2548.2	7.8208
130	3701.9	0.27013	2740.7	2555.6	7.8451
135	3748.9	0.26674	2750.5	2563.0	7.8691
140	3795.8	0.26345	2760.2	2570.4	7.8928
145	3842.7	0.26023	2769.9	2577.8	7.9162
150	3889.5	0.25710	2779.7	2585.2	7.9394
155	3936.3	0.25405	2789.4	2592.6	7.9622
160	3983.0	0.25106	2799.1	2600.0	7.9848
165	4029.8	0.24815	2808.9	2607.4	8.0072
170	4076.4	0.24531	2818.6	2614.8	8.0293
175	4123.1	0.24254	2828.3	2622.2	8.0511
180	4169.7	0.23982	2838.1	2629.6	8.0728
185	4216.3	0.23717	2847.9	2637.0	8.0942

Table 3. Compressed water and superheated steam—*continued*

t	0.5 bar (t_g = 81.339°C)				
(°C)	v (× 10³)	p	h	u	s
190	4262.9	0.23458	2857.6	2644.5	8.1154
195	4309.5	0.23205	2867.4	2651.9	8.1364
200	4356.0	0.22957	2877.2	2659.4	8.1572
205	4402.5	0.22714	2887.0	2666.8	8.1778
210	4449.0	0.22477	2896.8	2674.3	8.1982
215	4495.5	0.22244	2906.6	2681.8	8.2184
220	4542.0	0.22017	2916.4	2689.3	8.2384
225	4588.4	0.21794	2926.2	2696.8	8.2582
230	4634.9	0.21576	2936.1	2704.3	8.2779
235	4681.3	0.21362	2945.9	2711.9	8.2974
240	4727.7	0.21152	2955.8	2719.4	8.3167
245	4774.1	0.20946	2965.7	2727.0	8.3358
250	4820.5	0.20745	2975.6	2734.5	8.3548
255	4866.9	0.20547	2985.5	2742.1	8.3737
260	4913.3	0.20353	2995.4	2749.7	8.3924
265	4959.7	0.20163	3005.3	2757.3	8.4109
270	5006.	0.19976	3015.3	2765.0	8.4293
275	5052.	0.19793	3025.2	2772.6	8.4475
280	5099.	0.19613	3035.2	2780.2	8.4656
285	5145.	0.19436	3045.2	2787.9	8.4836
290	5191.	0.19263	3055.2	2795.6	8.5014
295	5238.	0.19092	3065.2	2803.3	8.5191
300	5284	0.18925	3075.2	2811.0	8.5367
305	5330	0.18761	3085.2	2818.7	8.5541
310	5377	0.18599	3095.3	2826.4	8.5714
315	5423	0.18440	3105.3	2834.2	8.5886
320	5469	0.18284	3115.4	2842.0	8.6057
325	5516	0.18131	3125.5	2849.7	8.6226
330	5562	0.17980	3135.6	2857.5	8.6395
335	5608	0.17831	3145.8	2865.4	8.6562
340	5654	0.17685	3155.9	2873.2	8.6728
345	5701	0.17542	3166.1	2881.0	8.6893
350	5747	0.17401	3176.2	2888.9	8.7057
355	5793	0.17262	3186.4	2896.8	8.7220
360	5839	0.17125	3196.6	2904.6	8.7381
365	5886	0.16990	3206.8	2912.5	8.7542
370	5932	0.16858	3217.1	2920.5	8.7702
375	5978	0.16727	3227.3	2928.4	8.7861
380	6024	0.16599	3237.6	2936.4	8.8018
385	6071	0.16473	3247.9	2944.3	8.8175
390	6117	0.16348	3258.1	2952.3	8.8331
395	6163	0.16225	3268.5	2960.3	8.8486
400	6209	0.16105	3278.8	2968.3	8.8640
410	6302	0.15868	3299.5	2984.4	8.8945
420	6394	0.15639	3320.3	3000.5	8.9247
430	6487	0.15416	3341.1	3016.7	8.9545
440	6579	0.15199	3362.0	3033.0	8.9840
450	6672	0.14989	3382.9	3049.3	9.0132
460	6764	0.14784	3404.0	3065.8	9.0421
470	6857	0.14585	3425.1	3082.2	9.0707
480	6949	0.14391	3446.2	3098.8	9.0989

Table 3. Compressed water and superheated steam—*continued*

t	0.5 bar (t_g = 81.339°C)				
(°C)	v (× 10³)	p	h	u	s
490	7041	0.14202	3467.4	3115.4	9.1269
500	7134	0.14018	3488.7	3132.0	9.1547
520	7319	0.13664	3531.5	3165.6	9.2093
540	7503	0.13327	3574.6	3199.4	9.2629
560	7688	0.13007	3617.9	3233.5	9.3156
580	7873	0.12702	3661.5	3267.9	9.3673
600	8058	0.12411	3705.4	3302.5	9.4182
620	8242	0.12132	3749.6	3337.5	9.4682
640	8427	0.11866	3794.1	3372.7	9.5174
660	8612	0.11612	3838.8	3408.2	9.5659
680	8797	0.11368	3883.8	3444.0	9.6136
700	8981	0.11134	3929.1	3480.1	9.6606
720	9166	0.10910	3974.7	3516.4	9.7070
740	9351	0.10694	4020.6	3553.1	9.7527
760	9535	0.10487	4066.7	3590.0	9.7979
780	9720	0.10288	4113.2	3627.2	9.8424
800	9905	0.10096	4159.9	3664.7	9.8863
850	10366	0.9647	4277.9	3759.6	9.9938
900	10828	0.09235	4397.7	3856.3	10.0981
950	11290	0.08858	4519.1	3954.6	10.1995
1000	11751	0.08510	4642.2	4054.6	10.2981
1100	12675	0.07890	4893.1	4259.4	10.4878
1200	13598	0.07354	5150.1	4470.2	10.6684
1300	14521	0.06887	5412.8	4686.8	10.8409
1400	15444	0.06475	5680.7	4908.5	11.0060
1500	16367	0.06110	5953.4	5135.0	11.1643
1600	17290	0.05784	6230.5	5366.0	11.3164
1700	18213	0.05490	6511.7	5601.1	11.4626
1800	19137	0.05226	6796.7	5839.9	11.6035
1900	20060	0.04985	7085.1	6082.1	11.7393
2000	20983	0.04766	7376.7	6327.6	11.8705

t	1.0 bar (t_s = 99.632 °C)				
(°C)	ν (×10³)	ρ	h	u	s
t_l	1.04313	958.66	417.51	417.41	1.30273
t_g	1694.3	0.5902	2675.1	2505.7	7.3589
0	1.00017	999.83	0.06	−0.04	−0.00015
5	1.00001	999.99	21.12	21.02	0.07626
10	1.00026	999.74	42.08	41.98	0.15096
15	1.00086	999.14	63.01	62.91	0.22422
20	1.00177	998.23	83.93	83.83	0.29619
25	1.00295	997.06	104.84	104.74	0.36693
30	1.00437	995.65	125.76	125.66	0.43650
35	1.00601	994.03	146.67	146.57	0.50493
40	1.00785	992.21	167.59	167.48	1.57225
45	1.00989	990.21	188.49	188.39	0.63849
50	1.01212	988.03	209.40	209.30	0.70370
55	1.01452	985.69	230.31	230.21	0.76790
60	1.01709	983.20	251.22	251.12	0.83115
65	1.01983	980.56	272.14	272.04	0.89347

Table 3. Compressed water and superheated steam—*continued*

t (°C)	1.0 bar (t_s = 99.632 °C) v (×10³)	ρ	h	u	s
70	1.02272	977.78	293.07	292.96	0.95490
75	1.02578	974.86	314.01	313.90	1.01549
80	1.02900	971.82	334.97	334.86	1.07526
85	1.03237	968.64	355.95	355.85	1.13426
90	1.03590	965.35	376.96	376.85	1.19251
95	1.03958	961.93	397.99	397.89	1.25004
100	1696.1	0.5896	2675.9	2506.3	7.3609
105	1720.5	0.5812	2686.1	2514.0	7.3880
110	1744.8	0.5731	2696.2	2521.7	7.4146
115	1769.0	0.5653	2706.3	2529.4	7.4408
120	1793.1	0.5577	2716.3	2537.0	7.4665
125	1817.1	0.5503	2726.3	2544.6	7.4918
130	1841.1	0.5432	2736.3	2552.2	7.5167
135	1865.0	0.5362	2746.3	2559.8	7.5413
140	1888.9	0.5294	2756.2	2567.3	7.5655
145	1912.7	0.5228	2766.1	2574.9	7.5893
150	1936.4	0.51641	2776.1	2582.4	7.6129
155	1960.2	0.51016	2785.9	2589.9	7.6361
160	1983.8	0.50407	2795.8	2597.5	7.6591
165	2007.5	0.49813	2805.7	2605.0	7.6818
170	2031.1	0.49234	2815.6	2612.5	7.7042
175	2054.7	0.48669	2825.5	2620.0	7.7263
180	2078.3	0.48117	2835.3	2627.5	7.7482
185	2101.8	0.47579	2845.2	2635.0	7.7699
190	2125.3	0.47052	2855.1	2642.5	7.7913
195	2148.8	0.46538	2864.9	2650.1	7.8125
200	2172.3	0.46035	2874.8	2657.6	7.8335
205	2195.7	0.45544	2884.7	2665.1	7.8543
210	2219.1	0.45063	2894.6	2672.7	7.8748
215	2242.5	0.44592	2904.5	2680.2	7.8952
220	2265.9	0.44132	2914.4	2687.8	7.9153
225	2289.3	0.43681	2924.3	2695.3	7.9353
230	2312.7	0.43240	2934.2	2702.9	7.9551
235	2336.1	0.42807	2944.1	2710.5	7.9747
240	2359.4	0.42384	2954.0	2718.1	7.9942
245	2382.7	0.41969	2963.9	2725.7	8.0134
250	2406.1	0.41562	2973.9	2733.3	8.0325
255	2429.4	0.41163	2983.8	2740.9	8.0515
260	2452.7	0.40772	2993.8	2748.5	8.0702
265	2476.0	0.40388	3003.8	2756.2	8.0889
270	2499.2	0.40012	3013.8	2763.8	8.1073
275	2522.5	0.39643	3023.8	2771.5	8.1257
280	2545.8	0.39281	3033.8	2779.2	8.1438
285	2569.1	0.38925	3043.8	2786.9	8.1619
290	2592.3	0.38576	3053.8	2794.6	8.1798
295	2615.6	0.38233	3063.9	2802.3	8.1975
300	2638.8	0.37896	3073.9	2810.1	8.2152
305	2662.0	0.37565	3084.0	2817.8	8.2327
310	2685.3	0.37240	3094.1	2825.6	8.2500
315	2708.5	0.36921	3104.2	2833.3	8.2673
320	2731.7	0.36607	3114.3	2841.1	8.2844
325	2754.9	0.36299	3124.4	2848.9	8.3014
330	2778.1	0.35995	3134.6	2856.7	8.3183
335	2801.3	0.35697	3144.7	2864.6	8.3350

Table 3. Compressed water and superheated steam—*continued*

t (°C)	1.0 bar v (×10³)	ρ	h	u	s
340	2824.5	0.35404	3154.9	2872.4	8.3517
345	2847.7	0.35116	3165.1	2880.3	8.3682
350	2870.9	0.34832	3175.3	2888.2	8.3846
355	2894.1	0.34553	3185.5	2896.1	8.4009
360	2917.3	0.34278	3195.7	2904.0	8.4172
365	2940.5	0.34008	3205.9	2911.9	8.4333
370	2963.7	0.33742	3216.2	2919.8	8.4493
375	2986.9	0.33480	3226.4	2927.8	8.4652
380	3010.0	0.33222	3236.7	2935.7	8.4810
385	3033.2	0.32968	3247.0	2943.7	8.4967
390	3056.4	0.32719	3257.3	2951.7	8.5123
395	3079.5	0.32472	3267.7	2959.7	8.5278
400	3102.7	0.32230	3278.0	2967.7	8.5432
410	3149.0	0.31756	3298.7	2983.8	8.5738
420	3195.3	0.31296	3319.5	3000.0	8.6040
430	3241.6	0.30849	3340.4	3016.2	8.6339
440	3287.9	0.30414	3361.3	3032.5	8.6634
450	3334.2	0.29992	3382.3	3048.9	8.6927
460	3380.5	0.29582	3403.3	3065.3	8.7216
470	3426.8	0.29182	3424.5	3081.8	8.7502
480	3473.0	0.28793	3445.6	3098.3	8.7785
490	3519.3	0.28415	3466.9	3115.0	8.8065
500	3565.5	0.28046	3488.2	3131.6	8.8342
520	3658.0	0.27337	3531.0	3165.2	8.8889
540	3750.5	0.26663	3574.1	3199.1	8.9426
560	3843.0	0.26021	3617.5	3233.2	8.9953
580	3935.5	0.25410	3661.1	3267.6	9.0470
600	4027.9	0.24827	3705.0	3302.3	9.0979
620	4120.3	0.24270	3749.2	3337.2	9.1480
640	4212.8	0.23737	3793.7	3372.4	9.1972
660	4305.2	0.23228	3838.5	3408.0	9.2457
680	4397.6	0.22740	3883.5	3443.8	9.2935
700	4490.0	0.22272	3928.8	3479.8	9.3405
720	4582.4	0.21823	3974.4	3516.2	9.3869
740	4674.8	0.21391	4020.3	3552.9	9.4326
760	4767.1	0.20977	4066.5	3589.8	9.4778
780	4859.5	0.20578	4112.9	3627.0	9.5223
800	4951.9	0.20194	4159.7	3664.5	9.5662
850	5183.	0.19295	4277.7	3759.4	9.6738
900	5414.	0.18472	4397.5	3856.1	9.7781
950	5645.	0.17716	4518.9	3954.5	9.8794
1000	5876.	0.17020	4642.0	4054.5	9.9781
1100	6337.	0.15780	4893.0	4259.3	10.1678
1200	6799.	0.14708	5150.0	4470.1	10.3485
1300	7260.	0.13773	5412.7	4686.7	10.5210
1400	7722.	0.12950	5680.6	4908.4	10.6861
1500	8184.	0.12219	5953.3	5135.0	10.8444
1600	8645.	0.11567	6230.5	5366.0	10.9964
1700	9107.	0.10981	6511.7	5601.0	11.1427
1800	9568.	0.10451	6796.6	5839.8	11.2835
1900	10030.	0.09970	7085.1	6082.1	11.4194
2000	10492.	0.09531	7376.7	6327.5	11.5506

Table 3. Compressed water and superheated steam—*continued*

t (°C)	**2.0 bar (t_s = 120.241 °C)** ν (×10³)	ρ	h	u	s
t_l	1.06049	942.96	504.80	504.59	1.53036
t_g	885.9	1.1289	2706.5	2529.4	7.1272
0	1.00012	999.88	0.16	−0.04	−0.00014
5	0.99996	1000.04	21.22	21.02	0.07626
10	1.00021	999.79	42.18	41.98	0.15095
15	1.00082	999.18	63.11	62.91	0.22421
20	1.00173	998.28	84.02	83.82	0.29617
25	1.00290	997.11	104.93	104.73	0.36690
30	1.00432	995.70	125.85	125.65	0.43647
35	1.00596	994.07	146.76	146.56	0.50489
40	1.00781	992.25	167.67	167.47	0.57221
45	1.00985	990.25	188.58	188.38	0.63845
50	1.01207	988.07	209.49	209.29	0.70365
55	1.01447	985.73	230.40	230.19	0.76785
60	1.01704	983.24	251.31	251.10	0.83109
65	1.01978	980.60	272.22	272.02	0.89341
70	1.02268	977.82	293.15	292.94	0.95484
75	1.02574	974.91	314.09	313.88	1.01543
80	1.02895	971.86	335.05	334.84	1.07520
85	1.03232	968.69	356.03	355.82	1.13419
90	1.03585	965.39	377.03	376.83	1.19244
95	1.03953	961.97	398.07	397.86	1.24997
100	1.04336	958.44	419.14	418.93	1.30681
105	1.04735	954.79	440.24	440.03	1.36298
110	1.05150	951.02	461.38	461.17	1.41852
115	1.05580	947.14	482.56	482.35	1.47344
120	1.06027	943.16	503.78	503.57	1.52776
125	897.8	1.1138	2716.6	2537.1	7.1527
130	910.3	1.0985	2727.1	2545.1	7.1789
135	922.7	1.0837	2737.6	2553.0	7.2047
140	935.1	1.0694	2748.0	2561.0	7.2300
145	947.4	1.0555	2758.3	2568.8	7.2549
150	959.7	1.0420	2768.6	2576.7	7.2793
155	971.9	1.0289	2778.9	2584.5	7.3034
160	984.1	1.0162	2789.1	2592.3	7.3271
165	996.2	1.0038	2799.3	2600.0	7.3505
170	1008.3	0.9918	2809.4	2607.8	7.3736
175	1020.4	0.9800	2819.6	2615.5	7.3963
180	1032.4	0.9686	2829.7	2623.2	7.4188
185	1044.4	0.9575	2839.8	2630.9	7.4409
190	1056.4	0.9466	2849.9	2638.6	7.4628
195	1068.4	0.9360	2859.9	2646.3	7.4845
200	1080.3	0.9257	2870.0	2653.9	7.5059
205	1092.2	0.9156	2880.1	2661.6	7.5270
210	1104.1	0.9057	2890.1	2669.3	7.5479
215	1116.0	0.8961	2900.2	2677.0	7.5686
220	1127.9	0.8866	2910.2	2684.6	7.5891
225	1139.7	0.8774	2920.3	2692.3	7.6094
230	1151.6	0.8684	2930.3	2700.0	7.6294
235	1163.4	0.8596	2940.4	2707.7	7.6493

Table 3. Compressed water and superheated steam—*continued*

t (°C)	2.0 bar ν (×10³)	ρ	h	u	s
240	1175.2	0.8509	2950.4	2715.4	7.6690
245	1187.0	0.8425	2960.5	2723.1	7.6885
250	1198.8	0.8342	2970.5	2730.8	7.7078
255	1210.6	0.8261	2980.6	2738.5	7.7269
260	1222.3	0.8181	2990.6	2746.2	7.7459
265	1234.1	0.8103	3000.7	2753.9	7.7647
270	1245.8	0.8027	3010.8	2761.6	7.7833
275	1257.6	0.7952	3020.9	2769.4	7.8018
280	1269.3	0.7878	3031.0	2777.1	7.8201
285	1281.0	0.7806	3041.1	2784.9	7.8383
290	1292.8	0.7735	3051.2	2792.6	7.8563
295	1304.5	0.7666	3061.3	2800.4	7.8742
300	1316.2	0.7598	3071.4	2808.2	7.8920
305	1327.9	0.7531	3081.6	2816.0	7.9096
310	1339.6	0.7465	3091.7	2823.8	7.9271
315	1351.3	0.7400	3101.9	2831.6	7.9444
320	1362.9	0.7337	3112.0	2839.4	7.9616
325	1374.6	0.7275	3122.2	2847.3	7.9787
330	1386.3	0.7214	3132.4	2855.2	7.9957
335	1398.0	0.7153	3142.6	2863.0	8.0125
340	1409.6	0.7094	3152.8	2870.9	8.0293
345	1421.3	0.7036	3163.1	2878.8	8.0459
350	1432.9	0.6979	3173.3	2886.7	8.0624
355	1444.6	0.6922	3183.6	2894.6	8.0788
360	1456.2	0.6867	3193.8	2902.6	8.0951
365	1467.9	0.6812	3204.1	2910.5	8.1112
370	1479.5	0.6759	3214.4	2918.5	8.1273
375	1491.2	0.6706	3224.7	2926.5	8.1433
380	1502.8	0.6654	3235.0	2934.5	8.1591
385	1514.4	0.6603	3245.4	2942.5	8.1749
390	1526.1	0.6553	3255.7	2950.5	8.1906
395	1537.7	0.6503	3266.1	2958.5	8.2061
400	1549.3	0.6454	3276.4	2966.6	8.2216
410	1572.6	0.6359	3297.2	2982.7	8.2523
420	1595.8	0.6266	3318.1	2998.9	8.2826
430	1619.0	0.6177	3339.0	3015.2	8.3125
440	1642.2	0.6089	3360.0	3031.5	8.3421
450	1665.5	0.6004	3381.0	3047.9	8.3714
460	1688.7	0.5922	3402.1	3064.4	8.4004
470	1711.9	0.5842	3423.3	3080.9	8.4291
480	1735.0	0.5764	3444.5	3097.5	8.4574
490	1758.2	0.5688	3465.8	3114.1	8.4855
500	1781.4	0.5614	3487.1	3130.8	8.5133
520	1827.8	0.54712	3530.0	3164.5	8.5681
540	1874.1	0.53359	3573.2	3198.4	8.6218
560	1920.4	0.52072	3616.6	3232.5	8.6746
580	1966.7	0.50846	3660.3	3267.0	8.7264
600	2013.0	0.49677	3704.3	3301.7	8.7773
620	2059.3	0.48560	3748.5	3336.7	8.8274
640	2105.6	0.47493	3793.0	3371.9	8.8767
660	2151.8	0.46472	3837.8	3407.5	8.9253
680	2198.1	0.45494	3882.9	3443.3	8.9731
700	2244.3	0.44557	3928.3	3479.4	9.0201

Table 3. Compressed water and superheated steam—*continued*

t (°C)	2.0 bar				
	v (×10³)	ρ	h	u	s
720	2290.6	0.43657	3973.9	3515.8	9.0666
740	2336.8	0.42793	4019.8	3552.4	9.1123
760	2383.0	0.41963	4066.0	3589.4	9.1575
780	2429.3	0.41165	4112.5	3626.6	9.2020
800	2475.5	0.40396	4159.2	3664.1	9.2460
850	2591.0	0.38595	4277.3	3759.1	9.3536
900	2706.6	0.36947	4397.1	3855.8	9.4579
950	2822.1	0.35435	4518.6	3954.2	9.5593
1000	2937.5	0.34042	4641.7	4054.2	9.6580
1100	3168.5	0.31561	4892.8	4259.1	9.8478
1200	3399.4	0.29417	5149.8	4470.0	10.0284
1300	3630.3	0.27546	5412.5	4686.5	10.2010
1400	3861.1	0.25899	5680.5	4908.3	10.3661
1500	4092.0	0.24438	5953.2	5134.8	10.5244
1600	4322.8	0.23133	6230.4	5365.8	10.6764
1700	4553.6	0.21960	6511.6	5600.9	10.8227
1800	4784.5	0.20901	6796.6	5839.7	10.9636
1900	5015.3	0.19939	7085.0	6082.0	11.0994
2000	5246.1	0.19062	7376.7	6327.4	11.2306

t (°C)	5.0 bar (t_s = 151.866 °C)				
	v (×10³)	ρ	h	u	s
t_l	1.09253	915.31	640.38	639.84	1.86104
t_g	374.86	2.6677	2748.6	2561.2	6.8214
0	0.99997	1000.03	0.47	−0.03	−0.00012
5	0.99981	1000.19	21.52	21.02	0.07625
10	1.00007	999.93	42.47	41.97	0.15092
15	1.00068	999.32	63.39	62.89	0.22416
20	1.00159	998.41	84.30	83.80	0.29610
25	1.00277	997.24	105.21	104.71	0.36683
30	1.00419	995.83	126.12	125.62	0.43638
35	1.00583	994.21	147.03	146.53	0.50479
40	1.00767	992.39	167.94	167.44	0.57209
45	1.00971	990.38	188.84	188.34	0.63832
50	1.01194	988.20	209.75	209.24	0.70351
55	1.01434	985.87	230.65	230.14	0.76770
60	1.01691	983.37	251.56	251.05	0.83093
65	1.01964	980.74	272.47	271.96	0.89324
70	1.02254	977.96	293.39	292.88	0.95466
75	1.02560	975.04	314.33	313.82	1.01524
80	1.02881	972.00	335.29	334.77	1.07500
85	1.03218	968.82	356.26	355.75	1.13399
90	1.03570	965.53	377.27	376.75	1.19222
95	1.03938	962.11	398.30	397.78	1.24974
100	1.04321	958.58	419.36	418.84	1.30657
105	1.04720	954.93	440.46	439.94	1.36274
110	1.05134	951.17	461.60	461.07	1.41827
115	1.05564	947.29	482.77	482.24	1.47318
120	1.06010	943.31	503.99	503.46	1.52749
125	1.06473	939.21	525.25	524.72	1.58123

Table 3. Compressed water and superheated steam—*continued*

t (°C)	5.0 bar (t_s = 151.866 °C)				
	v (×10³)	ρ	h	u	s
130	1.06952	935.00	546.56	546.03	1.63442
135	1.07448	930.69	567.92	567.38	1.68707
140	1.07961	926.26	589.33	588.79	1.73922
145	1.08493	921.72	610.80	610.26	1.79087
150	1.09043	917.07	632.33	631.79	1.84205
155	378.24	2.6438	2755.8	2566.7	6.8383
160	383.58	2.6070	2767.2	2575.4	6.8648
165	388.87	2.5715	2778.5	2584.1	6.8907
170	394.12	2.5373	2789.7	2592.6	6.9160
175	399.33	2.5042	2800.7	2601.1	6.9408
180	404.50	2.4722	2811.7	2609.5	6.9652
185	409.64	2.4412	2822.6	2617.8	6.9891
190	414.74	2.4111	2833.5	2626.1	7.0126
195	419.82	2.3820	2844.2	2634.3	7.0358
200	424.87	2.3537	2854.9	2642.5	7.0585
205	429.90	2.3261	2865.6	2650.7	7.0810
210	434.90	2.2994	2876.2	2658.8	7.1031
215	439.89	2.2733	2886.8	2666.9	7.1249
220	444.85	2.2479	2897.4	2674.9	7.1463
225	449.80	2.2232	2907.9	2683.0	7.1676
230	454.73	2.1991	2918.4	2691.0	7.1885
235	459.65	2.1756	2928.8	2699.0	7.2092
240	464.55	2.1526	2939.3	2707.0	7.2297
245	469.44	2.1302	2949.7	2715.0	7.2499
250	474.32	2.1083	2960.1	2723.0	7.2699
255	479.18	2.0869	2970.5	2730.9	7.2897
260	484.04	2.0660	2980.9	2738.9	7.3092
265	488.88	2.0455	2991.3	2746.8	7.3286
270	493.72	2.0255	3001.6	2754.8	7.3478
275	498.54	2.0059	3012.0	2762.7	7.3668
280	503.36	1.9867	3022.4	2770.7	7.3856
285	508.17	1.9679	3032.7	2778.6	7.4042
290	512.97	1.9494	3043.1	2786.6	7.4227
295	517.76	1.9314	3053.4	2794.5	7.4410
300	522.5	1.9137	3063.7	2802.5	7.4591
305	527.3	1.8963	3074.1	2810.4	7.4771
310	532.1	1.8793	3084.4	2818.4	7.4949
315	536.9	1.8626	3094.8	2826.4	7.5126
320	541.6	1.8463	3105.2	2834.3	7.5301
325	546.4	1.8302	3115.5	2842.3	7.5475
330	551.1	1.8144	3125.9	2850.3	7.5647
335	555.9	1.7989	3136.2	2858.3	7.5819
340	560.6	1.7837	3146.6	2866.3	7.5989
345	565.4	1.7687	3157.0	2874.3	7.6157
350	570.1	1.7540	3167.4	2882.3	7.6325
355	574.8	1.7396	3177.8	2890.4	7.6491
360	579.6	1.7254	3188.2	2898.4	7.6656
365	584.3	1.7114	3198.6	2906.4	7.6819
370	589.0	1.6977	3209.0	2914.5	7.6982
375	593.7	1.6842	3219.4	2922.6	7.7144
380	598.5	1.6710	3229.9	2930.7	7.7304
385	603.2	1.6579	3240.3	2938.7	7.7463

Table 3. Compressed water and superheated steam—*continued*

t (°C)	5.0 bar v (×10³)	ρ	h	u	s
390	607.9	1.6451	3250.8	2946.8	7.7622
395	612.6	1.6324	3261.3	2955.0	7.7779
400	617.3	1.6200	3271.7	2963.1	7.7935
410	626.7	1.5957	3292.7	2979.4	7.8245
420	636.1	1.5721	3313.8	2995.7	7.8550
430	645.5	1.5493	3334.8	3012.1	7.8852
440	654.8	1.5271	3356.0	3028.6	7.9151
450	664.2	1.5056	3377.2	3045.1	7.9446
460	673.6	1.4847	3398.4	3061.6	7.9738
470	682.9	1.4643	3419.7	3078.2	8.0026
480	692.3	1.4445	3441.1	3094.9	8.0312
490	701.6	1.4253	3462.5	3111.7	8.0594
500	710.9	1.4066	3483.9	3128.5	8.0873
520	729.6	1.3706	3527.0	3162.2	8.1424
540	748.2	1.3365	3570.4	3196.3	8.1964
560	766.9	1.3040	3614.0	3230.6	8.2493
580	785.5	1.2731	3657.8	3265.1	8.3013
600	804.1	1.2437	3701.9	3299.9	8.3524
620	822.7	1.2156	3746.3	3335.0	8.4027
640	841.2	1.1887	3791.0	3370.3	8.4521
660	859.8	1.1630	3835.9	3406.0	8.5008
680	878.4	1.1385	3881.0	3441.9	8.5487
700	896.9	1.1149	3926.5	3478.0	8.5959
720	915.5	1.0923	3972.2	3514.5	8.6424
740	934.0	1.0706	4018.2	3551.2	8.6882
760	952.6	1.0498	4064.5	3588.2	8.7334
780	971.1	1.0297	4111.0	3625.5	8.7781
800	989.6	1.0105	4157.8	3663.0	8.8221
850	1036.0	0.9653	4276.1	3758.1	8.9298
900	1082.2	0.9240	4396.0	3854.9	9.0342
950	1128.5	0.8861	4517.6	3953.3	9.1357
1000	1174.8	0.8512	4640.8	4053.4	9.2345
1100	1267.3	0.7891	4892.0	4258.4	9.4244
1200	1359.7	0.7354	5149.2	4469.4	9.6052
1300	1452.1	0.6886	5412.0	4686.0	9.7777
1400	1544.6	0.6474	5680.1	4907.8	9.9429
1500	1636.9	0.6109	5952.9	5134.4	10.1013
1600	1729.3	0.57826	6230.1	5365.4	10.2534
1700	1821.7	0.54894	6511.4	5600.5	10.3996
1800	1914.1	0.52245	6796.4	5839.4	10.5405
1900	2006.4	0.49840	7084.9	6081.7	10.6764
2000	2098.8	0.47647	7376.5	6327.2	10.8076

t (°C)	10.0 bar (t_s = 179.916 °C) v (×10³)	ρ	h	u	s
t_f	1.12720	887.15	762.88	761.75	2.13885
t_g	194.38	5.1445	2777.7	2583.3	6.5859
0	0.99971	1000.29	0.98	−0.02	−0.00008
5	0.99957	1000.43	22.02	21.02	0.07625

Table 3. Compressed water and superheated steam—*continued*

t (°C)	10.0 bar (t_s = 179.916 °C) v (×10³)	ρ	h	u	s
10	0.99983	1000.17	42.96	41.96	0.15088
15	1.00044	999.56	63.87	62.87	0.22408
20	1.00136	998.64	84.77	83.77	0.29600
25	1.00254	997.47	105.67	104.67	0.36670
30	1.00396	996.05	126.58	125.57	0.43622
35	1.00560	994.43	147.48	146.48	0.50461
40	1.00745	992.60	168.38	167.37	0.57190
45	1.00949	990.60	189.28	188.27	0.63811
50	1.01171	988.42	210.18	209.17	0.70328
55	1.01411	986.08	231.08	230.06	0.76746
60	1.01668	983.59	251.98	250.96	0.83067
65	1.01941	980.96	272.88	271.86	0.89296
70	1.02231	978.18	293.80	292.78	0.95436
75	1.02536	975.26	314.73	313.71	1.01492
80	1.02857	972.22	335.68	334.66	1.07467
85	1.03194	969.05	356.66	355.62	1.13364
90	1.03546	965.76	377.65	376.62	1.19186
95	1.03913	962.35	398.68	397.64	1.24937
100	1.04295	958.81	419.74	418.70	1.30618
105	1.04694	955.17	440.83	439.78	1.36233
110	1.05107	951.41	461.96	460.91	1.41784
115	1.05537	947.54	483.13	482.08	1.47274
120	1.05982	943.56	504.34	503.28	1.52704
125	1.06444	939.46	525.60	524.53	1.58076
130	1.06922	935.26	546.90	545.83	1.63393
135	1.07417	930.95	568.25	567.18	1.68657
140	1.07930	926.53	589.66	588.58	1.73870
145	1.08460	922.00	611.12	610.03	1.79033
150	1.09009	917.36	632.64	631.55	1.84149
155	1.09577	912.60	654.22	653.13	1.89220
160	1.10165	907.73	675.87	674.77	1.94247
165	1.10773	902.74	697.60	696.49	1.99233
170	1.11403	897.64	719.40	718.28	2.04181
175	1.12056	892.41	741.28	740.16	2.09091
180	194.43	5.1432	2777.9	2583.5	6.5864
185	197.36	5.0669	2790.6	2593.2	6.6142
190	200.25	4.9938	2803.0	2602.8	6.6412
195	203.09	4.9239	2815.3	2612.2	6.6675
200	205.90	4.8566	2827.4	2621.5	6.6932
205	208.68	4.7920	2839.3	2630.6	6.7183
210	211.43	4.7296	2851.1	2639.7	6.7428
215	214.16	4.6695	2862.8	2648.6	6.7668
220	216.86	4.6114	2874.3	2657.5	6.7904
225	219.53	4.5551	2885.8	2666.2	6.8135
230	222.19	4.5006	2897.1	2674.9	6.8362
235	224.83	4.4478	2908.4	2683.6	6.8586
240	227.45	4.3966	2919.6	2692.2	6.8805
245	230.05	4.3468	2930.8	2700.7	6.9021
250	232.64	4.2984	2941.9	2709.2	6.9235
255	235.22	4.2513	2952.9	2717.7	6.9444

Table 3. Compressed water and superheated steam—*continued*

t (°C)	10.0 bar v (×10³)	ρ	h	u	s
260	237.79	4.2055	2963.9	2726.1	6.9652
265	240.34	4.1608	2974.9	2734.5	6.9856
270	242.88	4.1173	2985.8	2742.9	7.0058
275	245.41	4.0748	2996.6	2751.2	7.0257
280	247.93	4.0334	3007.5	2759.6	7.0454
285	250.44	3.9930	3018.3	2767.9	7.0648
290	252.94	3.9534	3029.1	2776.1	7.0841
295	255.44	3.9148	3039.8	2784.4	7.1031
300	257.93	3.8771	3050.6	2792.7	7.1219
305	260.41	3.8401	3061.3	2800.9	7.1406
310	262.88	3.8040	3072.0	2809.1	7.1590
315	265.35	3.7686	3082.7	2817.4	7.1773
320	267.81	3.7340	3093.4	2825.6	7.1954
325	270.27	3.7001	3104.1	2833.8	7.2133
330	272.72	3.6668	3114.7	2842.0	7.2311
335	275.16	3.6342	3125.4	2850.2	7.2486
340	277.60	3.6023	3136.1	2858.5	7.2661
345	280.04	3.5710	3146.7	2866.7	7.2834
350	282.47	3.5402	3157.3	2874.9	7.3005
355	284.90	3.5101	3168.0	2883.1	7.3175
360	287.32	3.4804	3178.6	2891.3	7.3344
365	289.74	3.4514	3189.3	2899.5	7.3511
370	292.16	3.4228	3199.9	2907.7	7.3678
375	294.57	3.3948	3210.5	2916.0	7.3842
380	296.98	3.3673	3221.2	2924.2	7.4006
385	299.38	3.3402	3231.8	2932.4	7.4168
390	301.79	3.3136	3242.5	2940.7	7.4329
395	304.19	3.2875	3253.1	2948.9	7.4489
400	306.58	3.2617	3263.8	2957.2	7.4648
410	311.37	3.2116	3285.1	2973.7	7.4963
420	316.15	3.1631	3306.5	2990.3	7.5273
430	320.92	3.1161	3327.8	3006.9	7.5579
440	325.68	3.0705	3349.3	3023.6	7.5882
450	330.43	3.0263	3370.7	3040.3	7.6180
460	335.18	2.9835	3392.2	3057.0	7.6475
470	339.92	2.9419	3413.7	3073.8	7.6767
480	344.65	2.9015	3435.3	3090.6	7.7055
490	349.38	2.8622	3456.9	3107.5	7.7340
500	354.10	2.8241	3478.6	3124.5	7.7622
520	363.53	2.7508	3522.0	3158.5	7.8177
540	372.94	2.6814	3565.7	3192.8	7.8721
560	382.34	2.6155	3609.6	3227.2	7.9254
580	391.72	2.5528	3653.7	3262.0	7.9778
600	401.09	2.4932	3698.1	3297.0	8.0292
620	410.45	2.4363	3742.7	3332.2	8.0797
640	419.81	2.3821	3787.5	3367.7	8.1293
660	429.15	2.3302	3832.6	3403.4	8.1782
680	438.48	2.2806	3877.9	3439.5	8.2263
700	447.81	2.2331	3923.6	3475.7	8.2736
720	457.13	2.1876	3969.4	3512.3	8.3203
740	466.45	2.1439	4015.6	3549.1	8.3663
760	475.76	2.1019	4062.0	3586.2	8.4116

Table 3. Compressed water and superheated steam—*continued*

t (°C)	10.0 bar v (×10³)	ρ	h	u	s
780	485.06	2.0616	4108.6	3623.6	8.4563
800	494.36	2.0228	4155.5	3661.2	8.5005
850	517.6	1.9320	4274.0	3756.4	8.6084
900	540.8	1.8491	4394.2	3853.4	8.7130
950	564.0	1.7730	4516.0	3951.9	8.8147
1000	587.2	1.7030	4639.3	4052.1	8.9136
1100	633.5	1.5785	4890.8	4257.3	9.1037
1200	679.8	1.4710	5148.2	4468.4	9.2846
1300	726.1	1.3772	5411.2	4685.1	9.4573
1400	772.4	1.2947	5679.4	4907.0	9.6225
1500	818.6	1.2216	5952.3	5133.7	9.7810
1600	864.8	1.1563	6229.6	5364.8	9.9331
1700	911.0	1.0976	6511.0	5599.9	10.0794
1800	957.2	1.0447	6796.1	5838.8	10.2204
1900	1003.4	0.9966	7084.6	6081.2	10.3563
2000	1049.6	0.9527	7376.3	6326.7	10.4875

t (°C)	20 bar (t_s = 212.417 °C) v (×10³)	ρ	h	u	s
t_l	1.17667	849.85	908.69	906.33	2.44714
t_g	99.59	10.041	2798.7	2599.5	6.3396
0	0.99921	1000.79	2.00	0.00	0.00000
5	0.99908	1000.92	23.01	21.01	0.07623
10	0.99935	1000.65	43.94	41.94	0.15079
15	0.99998	1000.02	64.83	62.83	0.22393
20	1.00090	999.10	85.71	83.71	0.29579
25	1.00209	997.92	106.60	104.60	0.36643
30	1.00351	996.50	127.49	125.48	0.43592
35	1.00516	994.87	148.38	146.37	0.50427
40	1.00701	993.04	169.27	167.25	0.57151
45	1.00905	991.04	190.16	188.14	0.63768
50	1.01127	988.86	211.04	209.02	0.70282
55	1.01366	986.52	231.93	229.90	0.76696
60	1.01623	984.03	252.82	250.78	0.83014
65	1.01896	981.39	273.71	271.68	0.89240
70	1.02185	978.62	294.62	292.58	0.95377
75	1.02490	975.71	315.54	313.49	1.01430
80	1.02810	972.67	336.48	334.42	1.07401
85	1.03146	969.50	357.44	355.38	1.13295
90	1.03497	966.22	378.43	376.36	1.19115
95	1.03863	962.81	399.44	397.37	1.24862
100	1.04244	959.28	420.49	418.41	1.30540
105	1.04641	955.64	441.57	439.48	1.36152
110	1.05054	951.89	462.69	460.59	1.41700
115	1.05482	948.03	483.85	481.74	1.47186
120	1.05926	944.06	505.05	502.93	1.52613
125	1.06386	939.97	526.29	524.16	1.57982
130	1.06862	935.78	547.58	545.44	1.63296
135	1.07356	931.48	568.92	566.77	1.68556

Table 3. Compressed water and superheated steam—*continued*

t (°C)	20 bar (t_s = 212.417 °C)				
	v (×10³)	ρ	h	u	s
140	1.07866	927.07	590.31	588.15	1.73766
145	1.08395	922.55	611.75	609.59	1.78925
150	1.08942	917.92	633.26	631.08	1.84037
155	1.09507	913.18	654.82	652.63	1.89104
160	1.10092	908.33	676.46	674.26	1.94128
165	1.10698	903.36	698.16	695.95	1.99110
170	1.11325	898.27	719.94	717.72	2.04053
175	1.11974	893.06	741.80	739.56	2.08958
180	1.12647	887.73	763.75	761.50	2.13829
185	1.13344	882.27	785.80	783.53	2.18666
190	1.14067	876.68	807.94	805.66	2.23473
195	1.14817	870.95	830.19	827.89	2.28252
200	1.15596	865.08	852.56	850.25	2.33005
205	1.16405	859.07	875.06	872.73	2.37735
210	1.17248	852.89	897.69	895.35	2.42444
215	100.46	9.954	2806.4	2605.4	6.3553
220	102.12	9.793	2820.9	2616.6	6.3848
225	103.74	9.640	2835.0	2627.5	6.4133
230	105.32	9.495	2848.8	2638.2	6.4409
235	106.88	9.356	2862.4	2648.6	6.4677
240	108.41	9.224	2875.6	2658.8	6.4937
245	109.92	9.097	2888.7	2668.9	6.5191
250	111.41	8.976	2901.6	2678.8	6.5438
255	112.88	8.859	2914.3	2688.5	6.5679
260	114.33	8.747	2926.8	2698.1	6.5915
265	115.77	8.638	2939.2	2707.6	6.6146
270	117.19	8.534	2951.4	2717.0	6.6373
275	118.59	8.432	2963.5	2726.4	6.6595
280	119.98	8.334	2975.6	2735.6	6.6814
285	121.37	8.240	2987.5	2744.8	6.7028
290	122.74	8.147	2999.3	2753.8	6.7239
295	124.10	8.058	3011.1	2762.9	6.7447
300	125.45	7.971	3022.7	2771.8	6.7651
305	126.79	7.887	3034.3	2780.8	6.7853
310	128.13	7.805	3045.9	2789.6	6.8052
315	129.45	7.725	3057.4	2798.5	6.8248
320	130.77	7.647	3068.8	2807.3	6.8441
325	132.09	7.571	3080.2	2816.0	6.8633
330	133.39	7.497	3091.5	2824.8	6.8822
335	134.69	7.424	3102.8	2833.5	6.9008
340	135.99	7.354	3114.1	2842.1	6.9193
345	137.28	7.284	3125.4	2850.8	6.9375
350	138.56	7.217	3136.6	2859.4	6.9556
355	139.84	7.151	3147.7	2868.0	6.9735
360	141.12	7.086	3158.9	2876.7	6.9911
365	142.39	7.023	3170.0	2885.2	7.0087
370	143.66	6.961	3181.1	2893.8	7.0260
375	144.92	6.900	3192.2	2902.4	7.0432
380	146.18	6.841	3203.3	2911.0	7.0602
385	147.44	6.782	3214.4	2919.5	7.0771
390	148.69	6.725	3225.4	2928.0	7.0938

Table 3. Compressed water and superheated steam—*continued*

t (°C)	20 bar				
	v (×10³)	ρ	h	u	s
395	149.94	6.669	3236.5	2936.6	7.1104
400	151.19	6.614	3247.5	2945.1	7.1269
410	153.68	6.507	3269.5	2962.2	7.1594
420	156.15	6.404	3291.6	2979.3	7.1914
430	158.62	6.304	3313.6	2996.3	7.2229
440	161.07	6.208	3335.6	3013.4	7.2539
450	163.52	6.115	3357.5	3030.5	7.2845
460	165.97	6.025	3379.5	3047.6	7.3148
470	168.40	5.938	3401.5	3064.7	7.3446
480	170.83	5.854	3423.6	3081.9	7.3740
490	173.25	5.772	3445.6	3099.1	7.4031
500	175.67	5.693	3467.7	3116.3	7.4318
520	180.49	5.541	3511.9	3150.9	7.4882
540	185.29	5.397	3556.2	3185.6	7.5435
560	190.07	5.261	3600.7	3220.6	7.5975
580	194.84	5.132	3645.4	3255.7	7.6505
600	199.60	5.010	3690.2	3291.0	7.7024
620	204.35	4.8937	3735.3	3326.6	7.7535
640	209.08	4.7828	3780.5	3362.4	7.8036
660	213.81	4.6770	3826.0	3398.4	7.8528
680	218.53	4.5760	3871.7	3434.6	7.9013
700	223.25	4.4794	3917.6	3471.1	7.9490
720	227.95	4.3869	3963.8	3507.9	7.9959
740	232.65	4.2982	4010.2	3544.9	8.0422
760	237.35	4.2132	4056.9	3582.2	8.0878
780	242.04	4.1316	4103.8	3619.7	8.1328
800	246.73	4.0531	4150.9	3657.5	8.1771
850	258.42	3.8696	4269.9	3753.1	8.2855
900	270.10	3.7023	4390.5	3850.3	8.3905
950	281.76	3.5491	4512.7	3949.1	8.4925
1000	293.41	3.4082	4636.4	4049.5	8.5916
1100	316.67	3.1579	4888.4	4255.0	8.7821
1200	339.89	2.9421	5146.2	4466.4	8.9633
1300	363.09	2.7542	5409.5	4683.3	9.1363
1400	386.27	2.5889	5678.0	4905.4	9.3017
1500	409.43	2.4424	5951.1	5132.3	9.4603
1600	432.58	2.3117	6228.7	5363.5	9.6125
1700	455.72	2.1943	6510.2	5598.8	9.7589
1800	478.85	2.0883	6795.4	5837.7	9.8999
1900	501.97	1.9921	7084.1	6080.2	10.0359
2000	525.10	1.9044	7376.0	6325.8	10.1672

t (°C)	50 bar (t_s = 263.977 °C)				
	v (×10³)	ρ	h	u	s
t_l	1.28614	777.52	1154.20	1147.77	2.92011
t_g	39.440	25.355	2793.7	2596.5	5.9725
0	0.99769	1002.31	5.05	0.06	0.00020
5	0.99762	1002.39	25.99	21.00	0.07616

Table 3. Compressed water and superheated steam—*continued*

t (°C)	50 bar (t_s = 263.977 °C) ν (×10³)	ρ	h	u	s
10	0.99794	1002.07	46.85	41.86	0.15050
15	0.99859	1001.41	67.69	62.70	0.22345
20	0.99954	1000.46	88.52	83.53	0.29514
25	1.00074	999.26	109.37	104.36	0.36564
30	1.00218	997.83	130.22	125.21	0.43499
35	1.00383	996.18	151.07	146.05	0.50321
40	1.00568	994.35	171.92	166.89	0.57034
45	1.00772	992.34	192.77	187.73	0.63640
50	1.00994	990.16	213.63	208.58	0.70144
55	1.01233	987.82	234.48	229.42	0.76547
60	1.01489	985.33	255.34	250.26	0.82855
65	1.01760	982.70	276.20	271.11	0.89071
70	1.02048	979.93	297.07	291.97	0.95199
75	1.02351	977.03	317.96	312.85	1.01243
80	1.02669	974.00	338.87	333.74	1.07205
85	1.03002	970.85	359.80	354.65	1.13090
90	1.03351	967.58	380.75	375.59	1.18900
95	1.03714	964.19	401.73	396.55	1.24639
100	1.04093	960.68	422.75	417.54	1.30308
105	1.04487	957.06	443.79	438.57	1.35911
110	1.04895	953.33	464.88	459.63	1.41449
115	1.05320	949.49	486.00	480.73	1.46926
120	1.05759	945.54	507.16	501.87	1.52343
125	1.06215	941.49	528.36	523.05	1.57703
130	1.06686	937.33	549.61	544.28	1.63007
135	1.07174	933.06	570.91	565.55	1.68257
140	1.07679	928.68	592.26	586.88	1.73456
145	1.08202	924.20	613.66	608.25	1.78605
150	1.08742	919.61	635.12	629.68	1.83706
155	1.09300	914.91	656.64	651.17	1.88761
160	1.09878	910.10	678.22	672.73	1.93773
165	1.10475	905.18	699.87	694.35	1.98743
170	1.11093	900.14	721.59	716.04	2.03673
175	1.11733	894.99	743.40	737.81	2.08565
180	1.12395	889.72	765.28	759.66	2.13421
185	1.13081	884.32	787.26	781.60	2.18244
190	1.13792	878.80	809.33	803.64	2.23035
195	1.14529	873.14	831.50	825.78	2.27798
200	1.15293	867.35	853.79	848.03	2.32533
205	1.16088	861.42	876.20	870.40	2.37245
210	1.16913	855.34	898.74	892.89	2.41934
215	1.17772	849.10	921.42	915.53	2.46604
220	1.18666	842.70	944.25	938.32	2.51258
225	1.19599	836.13	967.25	961.27	2.55898
230	1.20573	829.37	990.43	984.40	2.60528
235	1.21592	822.42	1013.81	1007.73	2.65151
240	1.22659	815.27	1037.40	1031.26	2.69770
245	1.23780	807.89	1061.22	1055.03	2.74390
250	1.24958	800.27	1085.30	1079.05	2.79014
255	1.26200	792.39	1109.66	1103.35	2.83649

Table 3. Compressed water and superheated steam—*continued*

t (°C)	50 bar ν (×10³)	ρ	h	u	s
260	1.27513	784.23	1134.33	1127.95	2.88298
265	39.631	25.233	2798.0	2599.9	5.9805
270	40.533	24.671	2818.2	2615.6	6.0179
275	41.399	24.155	2837.5	2630.5	6.0532
280	42.230	23.680	2855.9	2644.8	6.0867
285	43.033	23.238	2873.7	2658.5	6.1186
290	43.812	22.825	2890.8	2671.7	6.1491
295	44.567	22.438	2907.4	2684.5	6.1785
300	45.304	22.073	2923.5	2697.0	6.2067
305	46.021	21.729	2939.2	2709.1	6.2340
310	46.725	21.402	2954.5	2720.9	6.2604
315	47.414	21.091	2969.5	2732.5	6.2860
320	48.091	20.794	2984.3	2743.8	6.3109
325	48.754	20.511	2998.7	2754.9	6.3352
330	49.410	20.239	3012.9	2765.9	6.3588
335	50.053	19.979	3026.9	2776.6	6.3819
340	50.689	19.728	3040.7	2787.2	6.4045
345	51.316	19.487	3054.3	2797.7	6.4266
350	51.93	19.255	3067.7	2808.0	6.4482
355	52.55	19.031	3081.0	2818.2	6.4695
360	53.15	18.814	3094.1	2828.4	6.4903
365	53.75	18.604	3107.2	2838.4	6.5108
370	54.35	18.401	3120.1	2848.3	6.5309
375	54.93	18.203	3132.9	2858.2	6.5507
380	55.52	18.012	3145.5	2868.0	6.5703
385	56.10	17.826	3158.2	2877.7	6.5895
390	56.67	17.646	3170.7	2887.3	6.6084
395	57.24	17.470	3183.1	2896.9	6.6271
400	57.81	17.299	3195.5	2906.5	6.6456
410	58.93	16.969	3220.1	2925.4	6.6818
420	60.04	16.656	3244.4	2944.2	6.7172
430	61.14	16.357	3268.5	2962.8	6.7517
440	62.22	16.072	3292.5	2981.4	6.7856
450	63.30	15.798	3316.3	2999.8	6.8187
460	64.37	15.536	3340.0	3018.2	6.8513
470	65.42	15.285	3363.6	3036.5	6.8833
480	66.48	15.043	3387.1	3054.7	6.9147
490	67.52	14.810	3410.5	3072.9	6.9456
500	68.56	14.586	3433.9	3091.1	6.9760
520	70.62	14.160	3480.5	3127.4	7.0355
540	72.66	13.762	3527.0	3163.7	7.0934
560	74.68	13.390	3573.4	3200.0	7.1498
580	76.69	13.039	3619.8	3236.4	7.2048
600	78.69	12.709	3666.2	3272.8	7.2586
620	80.67	12.396	3712.7	3309.4	7.3112
640	82.64	12.101	3759.3	3346.1	7.3628
660	84.60	11.820	3805.9	3382.9	7.4133
680	86.56	11.553	3852.7	3419.9	7.4630
700	88.50	11.299	3899.7	3457.1	7.5117
720	90.44	11.056	3946.8	3494.5	7.5596
740	92.38	10.825	3994.0	3532.1	7.6067
760	94.31	10.604	4041.5	3570.0	7.6531

Table 3. Compressed water and superheated steam—*continued*

t (°C)	50 bar v (×10³)	ρ	h	u	s
780	96.23	10.392	4089.1	3608.0	7.6988
800	98.15	10.189	4137.0	3646.3	7.7438
850	102.93	9.716	4257.5	3742.9	7.8536
900	107.68	9.287	4379.4	3841.0	7.9598
950	112.42	8.895	4502.7	3940.6	8.0627
1000	117.15	8.536	4627.4	4041.7	8.1626
1100	126.56	7.901	4881.1	4248.3	8.3543
1200	135.94	7.356	5140.2	4460.5	8.5365
1300	145.29	6.883	5404.5	4678.1	8.7101
1400	154.62	6.468	5673.8	4900.8	8.8760
1500	163.93	6.100	5947.7	5128.1	9.0350
1600	173.24	5.7725	6225.9	5359.7	9.1876
1700	182.53	5.4786	6507.9	5595.3	9.3343
1800	191.81	5.2134	6793.6	5834.5	9.4755
1900	201.09	4.9728	7082.6	6077.2	9.6117
2000	210.37	4.7536	7374.8	6323.0	9.7431

t (°C)	100 bar (t_s = 311.031 °C) v (×10³)	ρ	h	u	s
t_l	1.45216	688.63	1407.28	1392.75	3.35912
t_g	18.025	55.48	2724.5	2544.3	5.6139
0	0.99521	1004.81	10.10	0.15	0.00045
5	0.99522	1004.80	30.92	20.97	0.07599
10	0.99561	1004.41	51.69	41.73	0.14998
15	0.99631	1003.70	72.44	62.47	0.22262
20	0.99730	1002.71	93.20	83.22	0.29405
25	0.99853	1001.48	113.97	103.98	0.36431
30	0.99998	1000.02	134.75	124.75	0.43344
35	1.00165	998.36	155.54	145.52	0.50146
40	1.00350	996.51	176.33	166.30	0.56839
45	1.00554	994.49	197.13	187.07	0.63428
50	1.00775	992.31	217.93	207.85	0.69914
55	1.01013	989.97	238.73	228.62	0.76301
60	1.01267	987.48	259.53	249.40	0.82592
65	1.01537	984.86	280.34	270.19	0.88793
70	1.01822	982.10	301.16	290.98	0.94905
75	1.02122	979.22	322.00	311.79	1.00934
80	1.02437	976.21	342.85	332.61	1.06881
85	1.02767	973.07	363.73	353.45	1.12751
90	1.03112	969.82	384.63	374.32	1.18546
95	1.03471	966.46	405.56	395.21	1.24270
100	1.03844	962.98	426.52	416.13	1.29924
105	1.04233	959.39	447.51	437.08	1.35512
110	1.04636	955.70	468.53	458.07	1.41036
115	1.05053	951.90	489.59	479.09	1.46498
120	1.05486	947.99	510.70	500.15	1.51899
125	1.05935	943.98	531.84	521.24	1.57243
130	1.06398	939.87	553.02	542.38	1.62531

Table 3. Compressed water and superheated steam—*continued*

t (°C)	100 bar (t_s = 311.031 °C) v (×10³)	ρ	h	u	s
135	1.06878	935.65	574.26	536.57	1.67765
140	1.07374	931.33	595.53	584.80	1.72947
145	1.07886	926.90	616.86	606.08	1.78079
150	1.08416	922.38	638.25	627.41	1.83162
155	1.08963	917.74	659.69	648.79	1.88199
160	1.09528	913.00	681.19	670.24	1.93192
165	1.10113	908.16	702.75	691.74	1.98142
170	1.10717	903.20	724.39	713.31	2.03051
175	1.11342	898.14	746.09	734.96	2.07922
180	1.11988	892.96	767.88	756.68	2.12756
185	1.12656	887.66	789.74	778.48	2.17555
190	1.13348	882.24	811.70	800.37	2.22322
195	1.14064	876.70	833.76	822.35	2.27058
200	1.14806	871.03	855.91	844.43	2.31766
205	1.15576	865.23	878.18	866.63	2.36448
210	1.16375	859.29	900.57	888.94	2.41106
215	1.17205	853.21	923.09	911.37	2.45743
220	1.18068	846.97	945.75	933.95	2.50361
225	1.18966	840.58	968.56	956.67	2.54964
230	1.19902	834.02	991.54	979.55	2.59553
235	1.20878	827.28	1014.69	1002.60	2.64131
240	1.21898	820.36	1038.03	1025.84	2.68702
245	1.22966	813.23	1061.58	1049.29	2.73270
250	1.24085	805.90	1085.36	1072.96	2.77837
255	1.25261	798.34	1109.39	1096.87	2.82408
260	1.26498	790.52	1133.69	1121.04	2.86988
265	1.27804	782.45	1158.29	1145.51	2.91580
270	1.29187	774.07	1183.22	1170.31	2.96192
275	1.30654	765.38	1208.52	1195.46	3.00828
280	1.32217	756.33	1234.23	1221.01	3.05497
285	1.33889	746.89	1260.40	1247.01	3.10206
290	1.35687	736.99	1287.09	1273.52	3.14967
295	1.37630	726.58	1314.39	1300.62	3.19792
300	1.39746	715.58	1342.38	1328.40	3.24697
305	1.42070	703.88	1371.19	1356.98	3.29702
310	1.44648	691.34	1400.99	1386.53	3.34835
315	18.592	53.786	2750.6	2564.7	5.6585
320	19.248	51.952	2780.6	2588.2	5.7093
325	19.855	50.366	2808.1	2609.6	5.7555
330	20.421	48.969	2833.6	2629.4	5.7979
335	20.956	47.719	2857.5	2648.0	5.8374
340	21.464	46.590	2880.1	2665.5	5.8745
345	21.950	45.559	2901.6	2682.2	5.9094
350	22.416	44.612	2922.2	2698.1	5.9425
355	22.865	43.734	2942.0	2713.3	5.9741
360	23.300	42.918	2961.0	2728.0	6.0043
365	23.722	42.155	2979.4	2742.2	6.0333
370	24.132	41.438	2997.3	2756.0	6.0612
375	24.532	40.763	3014.7	2769.4	6.0882
380	24.923	40.124	3031.7	2782.5	6.1143
385	25.305	39.518	3048.3	2795.2	6.1395

Table 3. Compressed water and superheated steam—*continued*

t (°C)	100 bar $v\ (\times 10^3)$	ρ	h	u	s
390	25.679	38.942	3064.5	2807.7	6.1641
395	26.047	38.392	3080.5	2820.0	6.1880
400	26.408	37.867	3096.1	2832.0	6.2114
410	27.113	36.883	3126.6	2855.5	6.2563
420	27.797	35.976	3156.2	2878.3	6.2994
430	28.463	35.133	3185.1	2900.5	6.3408
440	29.114	34.347	3213.4	2922.3	6.3807
450	29.752	33.611	3241.1	2943.6	6.4194
460	30.378	32.919	3268.4	2964.6	6.4568
470	30.993	32.266	3295.3	2985.4	6.4932
480	31.598	31.647	3321.8	3005.8	6.5287
490	32.195	31.061	3348.0	3026.1	6.5633
500	32.784	30.503	3374.0	3046.2	6.5971
520	33.940	29.463	3425.3	3085.9	6.6625
540	35.073	28.512	3475.8	3125.1	6.7255
560	36.186	27.635	3525.8	3164.0	6.7862
580	37.281	26.824	3575.4	3202.6	6.8451
600	38.361	26.068	3624.7	3241.1	6.9022
620	39.427	25.363	3673.8	3279.5	6.9577
640	40.482	24.702	3722.7	3317.9	7.0119
660	41.527	24.081	3771.5	3356.2	7.0648
680	42.562	23.495	3820.3	3394.6	7.1165
700	43.590	22.941	3869.0	3433.1	7.1671
720	44.610	22.417	3917.7	3471.6	7.2167
740	45.623	21.919	3966.5	3510.3	7.2653
760	46.631	21.445	4015.4	3549.1	7.3131
780	47.633	20.994	4064.4	3588.1	7.3600
800	48.630	20.564	4113.5	3627.2	7.4062
850	51.10	19.568	4236.7	3725.7	7.5184
900	53.55	18.673	4360.9	3825.3	7.6266
950	55.99	17.861	4486.1	3926.3	7.7311
1000	58.40	17.122	4612.5	4028.5	7.8324
1100	63.20	15.822	4869.0	4236.9	8.0263
1200	67.96	14.714	5130.3	4450.6	8.2100
1300	72.70	13.755	5396.4	4669.4	8.3847
1400	77.41	12.918	5667.1	4893.0	8.5516
1500	82.11	12.178	5942.2	5121.1	8.7112
1600	86.80	11.521	6221.3	5353.3	8.8644
1700	91.48	10.932	6504.2	5589.5	9.0115
1800	96.15	10.401	6790.6	5829.2	9.1531
1900	100.81	9.920	7080.3	6072.3	9.2895
2000	105.46	9.482	7373.1	6318.4	9.4212

Table 3. Compressed water and superheated steam—*continued*

t (°C)	200 bar (t_s = 365.800 °C) $v\ (\times 10^3)$	ρ	h	u	s
t_l	2.0360	491.2	1826.7	1786.0	4.0146
t_g	5.874	170.25	2413.6	2296.1	4.9330
0	0.99037	1009.73	20.08	0.28	0.00066
5	0.99054	1009.55	40.69	20.88	0.07543
10	0.99105	1009.03	61.27	41.45	0.14876
15	0.99185	1008.21	81.86	62.03	0.22084
20	0.99291	1007.14	102.48	82.62	0.29176
25	0.99420	1005.84	123.11	103.23	0.36157
30	0.99569	1004.33	143.77	123.86	0.43028
35	0.99738	1002.63	164.44	144.50	0.49792
40	0.99924	1000.76	185.13	165.14	0.56449
45	1.00128	998.72	205.81	185.79	0.63003
50	1.00349	996.53	226.50	206.43	0.69456
55	1.00585	994.19	247.20	227.08	0.75811
60	1.00836	991.71	267.90	247.73	0.82072
65	1.01102	989.10	288.61	268.39	0.88242
70	1.01383	986.36	309.33	289.05	0.94325
75	1.01677	983.50	330.07	309.73	1.00324
80	1.01986	980.52	350.82	330.42	1.06243
85	1.02309	977.43	371.59	351.13	1.12084
90	1.02646	974.22	392.39	371.86	1.17851
95	1.02997	970.90	413.22	392.62	1.23546
100	1.03361	967.48	434.07	413.40	1.29172
105	1.03740	963.95	454.95	434.20	1.34732
110	1.04132	960.32	475.87	455.04	1.40227
115	1.04538	956.59	496.82	475.91	1.45659
120	1.04958	952.76	517.81	496.81	1.51032
125	1.05393	948.83	538.83	517.75	1.56346
130	1.05842	944.81	559.90	538.73	1.61603
135	1.06305	940.69	581.00	559.74	1.66806
140	1.06784	936.47	602.15	580.79	1.71957
145	1.07279	932.15	623.35	601.89	1.77056
150	1.0779	927.7	644.6	623.0	1.8211
155	1.0832	923.2	665.9	644.2	1.8711
160	1.0886	918.6	687.2	665.5	1.9207
165	1.0942	913.9	708.6	686.7	1.9698
170	1.1000	909.1	730.1	708.1	2.0185
175	1.1059	904.2	751.6	729.5	2.0668
180	1.1121	899.2	773.2	751.0	2.1147
185	1.1185	894.1	794.9	772.5	2.1623
190	1.1251	888.8	816.6	794.1	2.2095
195	1.1318	883.5	838.5	815.8	2.2564
200	1.1389	878.1	860.4	837.6	2.3030
205	1.1461	872.5	882.4	859.5	2.3493
210	1.1537	866.8	904.5	881.5	2.3953
215	1.1615	861.0	926.8	903.5	2.4411
220	1.1695	855.0	949.1	925.7	2.4866
225	1.1779	849.0	971.6	948.0	2.5320
230	1.1866	842.7	994.2	970.5	2.5771
235	1.1956	836.4	1017.0	993.1	2.6221
240	1.2051	829.8	1039.9	1015.8	2.6670
245	1.2148	823.1	1063.0	1038.7	2.7118
250	1.2251	816.3	1086.3	1061.8	2.7565
255	1.2357	809.2	1109.7	1085.0	2.8012

Table 3. Compressed water and superheated steam—*continued*

t		200 bar			
(°C)	v (×10³)	h	u	s	
260	1.2469	802.0	1133.4	1108.5	2.8458
265	1.2586	794.5	1157.3	1132.2	2.8905
270	1.2709	786.9	1181.5	1156.1	2.9352
275	1.2838	779.0	1206.0	1180.3	2.9800
280	1.2974	770.8	1230.7	1204.8	3.0250
285	1.3118	762.3	1255.8	1229.6	3.0701
290	1.3270	753.6	1281.3	1254.7	3.1155
295	1.3432	744.5	1307.1	1280.3	3.1612
300	1.3605	735.0	1333.4	1306.2	3.2073
305	1.3790	725.1	1360.3	1332.7	3.2539
310	1.3990	714.8	1387.7	1359.7	3.3011
315	1.4206	703.9	1415.7	1387.3	3.3490
320	1.4442	692.4	1444.5	1415.6	3.3978
325	1.4702	680.2	1474.2	1444.8	3.4476
330	1.4990	667.1	1505.0	1475.0	3.4988
335	1.5314	653.0	1537.0	1506.4	3.5518
340	1.5685	637.5	1570.7	1539.4	3.6070
345	1.6120	620.4	1606.6	1574.3	3.6652
350	1.6645	600.8	1645.4	1612.1	3.7277
355	1.7314	577.6	1688.6	1654.0	3.7968
360	1.8248	548.0	1739.7	1703.2	3.8778
365	1.9900	502.5	1810.4	1770.6	3.9890
370	6.905	144.82	2523.8	2385.7	5.1050
375	7.668	130.42	2601.0	2447.6	5.2246
380	8.256	121.13	2658.6	2493.4	5.3131
385	8.752	114.27	2706.0	2531.0	5.3855
390	9.188	108.83	2747.0	2563.3	5.4476
395	9.583	104.35	2783.6	2592.0	5.5025
400	9.946	100.54	2816.9	2617.9	5.5521
405	10.284	97.24	2847.5	2641.8	5.5975
410	10.602	94.32	2876.1	2664.1	5.6395
415	10.903	91.72	2903.0	2684.9	5.6787
420	11.190	89.37	2928.5	2704.7	5.7156
425	11.464	87.23	2952.7	2723.4	5.7505
430	11.728	85.26	2975.9	2741.4	5.7836
435	11.983	83.45	2998.3	2758.6	5.8152
440	12.230	81.77	3019.8	2775.2	5.8455
445	12.469	80.20	3040.6	2791.2	5.8746
450	12.701	78.73	3060.8	2806.8	5.9026
460	13.149	76.05	3099.6	2836.6	5.9559
470	13.577	73.65	3136.6	2865.0	6.0060
480	13.988	71.49	3172.0	2892.3	6.0534
490	14.384	69.52	3206.3	2918.6	6.0986
500	14.769	67.71	3239.4	2944.1	6.1417
520	15.506	64.49	3303.2	2993.1	6.2232
540	16.208	61.70	3364.2	3040.1	6.2992
560	16.883	59.23	3423.2	3085.5	6.3708
580	17.536	57.03	3480.6	3129.8	6.4389
600	18.169	55.04	3536.7	3173.3	6.5039
620	18.786	53.23	3591.7	3216.0	6.5663
640	19.390	51.57	3646.0	3258.2	6.6264

Table 3. Compressed water and superheated steam—*continued*

t		200 bar			
(°C)	v (×10³)	h	u	s	
660	19.981	50.05	3699.7	3300.1	6.6845
680	20.562	48.63	3752.8	3341.6	6.7408
700	21.133	47.32	3805.5	3382.8	6.7955
720	21.696	46.09	3857.9	3424.0	6.8488
740	22.252	44.94	3910.0	3465.0	6.9008
760	22.802	43.86	3961.9	3505.9	6.9515
780	23.345	42.84	4013.7	3546.8	7.0012
800	22.883	41.87	4065.4	3587.8	7.0498
850	25.207	39.67	4194.4	3690.3	7.1673
900	26.508	37.72	4323.5	3793.3	7.2797
950	27.785	35.99	4452.9	3897.1	7.3877
1000	29.053	34.42	4582.8	4001.8	7.4919
1200	34.00	29.41	5111.	4431.	7.877
1400	38.83	25.75	5654.	4878.	8.223
1600	43.59	22.94	6213.	5341.	8.538
1800	48.32	20.69	6785.	5819.	8.828
2000	53.02	18.86	7370.	6310.	9.098

t			500 bar		
(°C)	v (×10³)	ρ	h	u	s
0	0.97674	1023.82	49.20	0.36	−0.00076
5	0.97734	1023.19	69.30	20.43	0.07216
10	0.97818	1022.31	89.43	40.52	0.14388
15	0.97924	1021.20	109.61	60.65	0.21454
20	0.98049	1019.90	129.85	80.83	0.28419
25	0.98192	1018.41	150.15	101.05	0.35283
30	0.98352	1016.75	170.48	121.31	0.42046
35	0.98528	1014.94	190.84	141.58	0.48709
40	0.98718	1012.98	211.23	161.87	0.55270
45	0.98923	1010.89	231.62	182.16	0.61732
50	0.99141	1008.66	252.03	202.46	0.68097
55	0.99373	1006.31	272.45	222.76	0.74367
60	0.99617	1003.84	292.88	243.07	0.80544
65	0.99874	1001.26	313.31	263.38	0.86633
70	1.00143	998.57	333.76	283.69	0.92636
75	1.00425	995.77	354.23	304.01	0.98557
80	1.00719	992.86	374.71	324.35	1.04398
85	1.01024	989.86	395.21	344.69	1.10162
90	1.01342	986.76	415.73	365.06	1.15852
95	1.01671	983.56	436.27	385.44	1.21471
100	1.02012	980.27	456.84	405.84	1.27021
105	1.02365	976.90	477.44	426.26	1.32504
110	1.02729	973.43	498.06	446.70	1.37922
115	1.03106	969.88	518.72	467.16	1.43277
120	1.03494	966.24	539.40	487.65	1.48571
125	1.03894	962.52	560.11	508.16	1.53807
130	1.04306	958.72	580.85	528.70	1.58984
135	1.04730	954.83	601.63	549.27	1.64106
140	1.05167	950.87	622.44	569.86	1.69173
145	1.05617	946.82	643.28	590.48	1.74188

Table 3. Compressed water and superheated steam—*continued*

t (°C)	500 bar ν (×10³)	ρ	h	u	s
150	1.0608	942.7	664.2	611.1	1.7915
160	1.0704	934.2	706.0	652.5	1.8893
170	1.0806	925.4	748.1	694.0	1.9853
180	1.0914	916.3	790.3	735.7	2.0795
190	1.1028	906.8	832.7	777.6	2.1720
200	1.1148	897.0	875.3	819.6	2.2631
210	1.1276	886.9	918.2	861.9	2.3529
220	1.1411	876.4	961.4	904.4	2.4413
230	1.1554	865.5	1005.0	947.2	2.5287
240	1.1706	854.2	1048.9	990.3	2.6151
250	1.1869	842.5	1093.2	1033.8	2.7006
260	1.2042	830.4	1137.9	1077.7	2.7854
270	1.2228	817.8	1183.3	1122.1	2.8696
280	1.2428	804.6	1229.2	1167.0	2.9534
290	1.2643	790.9	1275.8	1212.5	3.0368
300	1.2876	776.6	1323.1	1258.7	3.1202
310	1.3129	761.7	1371.3	1305.6	3.2035
320	1.3406	745.9	1420.5	1353.4	3.2871
330	1.3710	729.4	1470.7	1402.2	3.3712
340	1.4046	711.9	1522.3	1452.1	3.4559
350	1.4422	693.4	1575.3	1503.2	3.5417
355	1.4627	683.7	1602.5	1529.3	3.5851
360	1.4845	673.6	1630.1	1555.8	3.6289
365	1.5077	663.3	1658.2	1582.8	3.6731
370	1.5326	652.5	1686.9	1610.2	3.7179
375	1.5593	641.3	1716.1	1638.2	3.7632
380	1.5880	629.7	1746.1	1666.7	3.8093
385	1.6191	617.6	1776.8	1695.8	3.8561
390	1.6529	605.0	1808.3	1725.6	3.9037
395	1.6898	591.8	1840.7	1756.2	3.9524
400	1.7301	578.0	1874.1	1787.6	4.0022
405	1.7746	563.5	1908.6	1819.9	4.0533
410	1.8237	548.3	1944.4	1853.2	4.1058
415	1.8783	532.4	1981.5	1887.5	4.1600
420	1.9392	515.7	2020.0	1923.1	4.2158
425	2.0074	498.2	2060.2	1959.8	4.2736
430	2.0838	479.9	2102.0	1997.8	4.3332
435	2.1692	461.0	2145.5	2037.1	4.3949
440	2.2646	441.6	2190.6	2077.4	4.4584
445	2.3701	421.9	2237.1	2118.6	4.5233
450	2.4858	402.3	2284.7	2160.4	4.5894
455	2.6109	383.0	2332.9	2202.4	4.6558
460	2.7440	364.4	2381.2	2244.0	4.7219
465	2.8835	346.8	2429.1	2284.9	4.7870
470	3.0273	330.3	2476.0	2324.7	4.8503
475	3.1734	315.1	2521.7	2363.0	4.9115
480	3.320	301.2	2565.7	2399.7	4.9702
485	3.466	288.5	2608.0	2434.7	5.0262

Table 3. Compressed water and superheated steam—*continued*

t (°C)	500 bar ν (×10³)	ρ	h	u	s
490	3.611	276.9	2648.5	2467.9	5.0794
495	3.753	266.5	2687.2	2499.6	5.1300
500	3.892	257.0	2724.2	2529.6	5.1780
505	4.028	248.28	2759.6	2558.3	5.2237
510	4.160	240.37	2793.5	2585.5	5.2671
515	4.290	233.12	2826.1	2611.6	5.3085
520	4.416	226.47	2857.3	2636.5	5.3480
525	4.539	220.33	2887.4	2660.4	5.3858
530	4.659	214.66	2916.4	2683.4	5.4220
535	4.776	209.40	2944.4	2705.6	5.4568
540	4.890	204.50	2971.5	2727.0	5.4902
545	5.002	199.93	2997.7	2747.7	5.5224
550	5.111	195.65	3023.3	2767.7	5.5535
560	5.323	187.85	3072.3	2806.1	5.6127
570	5.527	180.92	3118.9	2842.5	5.6684
580	5.724	174.71	3163.5	2877.3	5.7209
590	5.914	169.09	3206.3	2910.6	5.7709
600	6.098	163.99	3247.7	2942.7	5.8184
620	6.451	155.01	3326.5	3003.9	5.9077
640	6.787	147.35	3401.1	3061.8	5.9903
660	7.107	140.70	3472.5	3117.1	6.0677
680	7.416	134.85	3541.2	3170.4	6.1406
700	7.713	129.64	3607.8	3222.1	6.2097
720	8.002	124.97	3672.6	3272.6	6.2757
740	8.282	120.74	3736.0	3321.9	6.3388
760	8.555	116.89	3798.2	3370.4	6.3996
780	8.822	113.35	3859.3	3418.1	6.4582
800	9.083	110.09	3919.5	3465.3	6.5148
850	9.715	102.93	4067.1	3581.3	6.6492
900	10.322	96.88	4211.5	3695.4	6.7751
950	10.908	91.67	4353.9	3808.5	6.8940
1000	11.479	87.12	4495.0	3921.1	7.0070
1200	13.65	73.28	5055.	4372.	7.415
1400	15.70	63.69	5618.	4833.	7.774
1600	17.69	56.52	6189.	5305.	8.096
1800	19.64	50.91	6771.	5789.	8.391
2000	21.57	46.36	7363.	6284.	8.664

t (°C)	1000 bar ν (×10³)	ρ	h	u	s
0	0.95666	1045.31	95.40	−0.27	−0.00854
5	0.95777	1044.09	114.96	19.18	0.06243
10	0.95902	1042.73	134.59	38.69	1.13238
15	0.96041	1041.23	154.31	58.27	0.20141
20	0.96191	1039.59	174.11	77.92	0.26954
25	0.96354	1037.84	193.98	97.63	0.33676

Table 3. Compressed water and superheated steam—*continued*

t (°C)	1000 bar ν (×10³)	ρ	h	u	s
30	0.96529	1035.96	213.91	117.38	0.40304
35	0.96714	1033.98	233.88	137.17	0.46838
40	0.96910	1031.89	253.88	156.97	0.53276
45	0.97116	1029.69	273.90	176.78	0.59618
50	0.97333	1027.40	293.93	196.60	0.65866
55	0.97559	1025.02	313.98	216.42	0.72021
60	0.97796	1022.54	334.03	236.24	0.78087
65	0.98042	1019.97	354.10	256.06	0.84066
70	0.98298	1017.32	374.18	275.88	0.89960
75	0.98563	1014.58	394.27	295.71	0.95773
80	0.98838	1011.76	414.38	315.54	1.01508
85	0.99122	1008.86	434.51	335.38	1.07167
90	0.99415	1005.88	454.65	355.23	1.12752
95	0.99718	1002.83	474.81	375.10	1.18267
100	1.00030	999.70	495.00	394.97	1.23713
105	1.00350	996.51	515.21	414.86	1.29092
110	1.00681	993.24	535.43	434.75	1.34407
115	1.01020	989.91	555.69	454.67	1.39658
120	1.01368	986.50	575.96	474.59	1.44848
125	1.01726	983.04	596.26	494.53	1.49978
130	1.02092	979.50	616.58	514.48	1.55050
135	1.02469	975.91	636.92	534.45	1.60064
140	1.02854	972.25	657.28	554.43	1.65023
145	1.03249	968.53	677.67	574.42	1.69928
150	1.0365	964.8	698.1	594.4	1.7478
160	1.0449	957.0	739.0	634.5	1.8433
170	1.0537	949.0	780.0	674.6	1.9369
180	1.0629	940.8	821.1	714.8	2.0286
190	1.0725	932.4	862.3	755.0	2.1185
200	1.0826	923.7	903.6	795.3	2.2068
210	1.0931	914.8	945.1	835.8	2.2936
220	1.1042	905.6	986.7	876.3	2.3789
230	1.1158	896.2	1028.5	917.0	2.4628
240	1.1279	886.6	1070.6	957.8	2.5455
250	1.1406	876.7	1112.8	998.7	2.6270
260	1.1540	866.6	1155.3	1039.9	2.7075
270	1.1680	856.1	1198.0	1081.2	2.7870
280	1.1828	845.4	1241.1	1122.8	2.8655
290	1.1984	834.5	1284.5	1164.6	2.9432
300	1.2148	823.2	1328.2	1206.7	3.0202
310	1.2321	811.6	1372.3	1249.1	3.0965
320	1.2504	799.8	1416.8	1291.8	3.1722
330	1.2697	787.6	1461.8	1334.8	3.2473
340	1.2902	775.1	1507.2	1378.2	3.3220
350	1.3120	762.2	1553.1	1421.9	3.3962
360	1.3351	749.0	1599.5	1466.0	3.4701
370	1.3597	735.4	1646.4	1510.5	3.5437
380	1.3859	721.5	1694.0	1555.4	3.6170

Table 3. Compressed water and superheated steam—*continued*

t (°C)	1000 bar ν (×10³)	ρ	h	u	s
390	1.4139	707.2	1742.1	1600.7	3.6902
400	1.4439	692.6	1790.9	1646.5	3.7632
410	1.4759	677.5	1840.4	1692.8	3.8361
420	1.5103	662.1	1890.5	1739.5	3.9089
430	1.5471	646.4	1941.3	1786.6	3.9817
440	1.5867	630.2	1992.8	1834.2	4.0545
450	1.6292	613.8	2045.1	1882.2	4.1273
460	1.6749	597.1	2098.1	1930.6	4.2000
470	1.7239	580.1	2151.7	1979.3	4.2727
480	1.7766	562.9	2206.0	2028.3	4.3453
490	1.8329	545.6	2260.9	2077.6	4.4176
500	1.8932	528.2	2316.2	2126.9	4.4897
510	1.9573	510.9	2372.0	2176.2	4.5613
520	2.0253	493.8	2427.9	2225.4	4.6324
530	2.0969	476.9	2484.0	2274.3	4.7026
540	2.1721	460.4	2539.9	2322.7	4.7718
550	2.2504	444.4	2595.5	2370.5	4.8398
560	2.3315	428.9	2650.7	2417.5	4.9064
570	2.4148	414.1	2705.2	2463.7	4.9714
580	2.5001	400.0	2758.8	2508.8	5.0347
590	2.5867	386.6	2811.6	2252.9	5.0961
600	2.6743	373.9	2863.4	2595.9	5.1558
610	2.762	362.0	2914.0	2637.8	5.2135
620	2.851	350.8	2963.6	2678.5	5.2693
630	2.939	340.2	3012.1	2718.2	5.3233
640	3.027	330.3	3059.5	2756.8	5.3755
650	3.115	321.0	3105.8	2794.4	5.4259
660	3.202	312.3	3151.1	2831.0	5.4748
670	3.288	304.1	3195.5	2866.7	5.5220
680	3.373	296.4	3238.9	2901.5	5.5678
690	3.458	289.2	3281.4	2935.6	5.6121
700	3.542	282.4	3323.1	2968.9	5.6552
710	3.624	275.9	3364.0	3001.5	5.6970
720	3.706	269.8	3404.1	3033.5	5.7377
730	3.787	264.1	3443.6	3064.9	5.7772
740	3.867	258.6	3482.5	3095.8	5.8158
750	3.946	253.4	3520.7	3126.1	5.8534
760	4.024	248.50	3558.4	3156.0	5.8900
770	4.101	243.81	3595.6	3185.5	5.9258
780	4.178	239.35	3632.3	3214.5	5.9608
790	4.254	235.09	3668.5	3243.2	5.9951
800	4.328	231.03	3704.3	3271.5	6.0286
820	4.476	223.42	3774.8	3327.2	6.0937
840	4.621	216.42	3843.9	3381.8	6.1563
860	4.762	209.97	3911.7	3435.5	6.2167
880	4.902	204.00	3978.5	3488.3	6.2751
900	5.039	198.45	4044.3	3540.4	6.3317

Table 3. Compressed water and superheated steam—*continued*

Table 3. Compressed water and superheated steam—*continued*

t (°C)	1000 bar v (×10³)	ρ	h	u	s
920	5.174	193.28	4109.3	3591.9	6.3866
940	5.307	188.44	4173.5	3642.8	6.4400
960	5.438	183.90	4237.1	3693.3	6.4920
980	5.567	179.63	4300.1	3743.4	6.5427
1000	5.694	175.61	4362.6	3793.2	6.5921

1000 bar

Table 3. Compressed water and superheated steam—*continued*

t (°C)	v (×10³)	ρ	h	u	s
1200	6.896	145.01	4970.	4280.	7.035
1400	8.007	124.89	5563.	4762.	7.413
1600	9.064	110.33	6155.	5248.	7.747
1800	10.085	99.16	6751.	5743.	8.049
2000	11.083	90.23	7354.	6245.	8.327

Table 4. Specific heat capacity at constant pressure

t (°C)	P (bar) 0	1	5	10	20	50	100	200	500	1000
0	1.859	4.228	4.226	4.223	4.218	4.202	4.177	4.130	4.021	3.909
20	1.863	4.183	4.182	4.180	4.177	4.168	4.153	4.125	4.054	3.968
40	1.868	4.182	4.181	4.180	4.178	4.170	4.159	4.137	4.078	4.002
60	1.875	4.183	4.182	4.181	4.178	4.172	4.161	4.141	4.086	4.012
80	1.882	4.194	4.193	4.192	4.190	4.183	4.173	4.153	4.098	4.023
100	1.890	2.042	4.216	4.215	4.213	4.206	4.195	4.174	4.117	4.039
120	1.899	2.005	4.248	4.247	4.244	4.237	4.224	4.201	4.140	4.057
140	1.908	1.986	4.288	4.286	4.284	4.275	4.261	4.234	4.165	4.075
160	1.918	1.977	2.267	4.337	4.334	4.323	4.306	4.275	4.195	4.094
180	1.929	1.974	2.188	2.556	4.399	4.386	4.365	4.327	4.231	4.115
200	1.940	1.975	2.138	2.400	4.486	4.469	4.442	4.394	4.277	4.141
220	1.951	1.980	2.106	2.301	2.861	4.583	4.547	4.482	4.335	4.173
240	1.963	1.986	2.087	2.236	2.635	4.740	4.689	4.601	4.409	4.213
260	1.975	1.994	2.076	2.194	2.490	4.967	4.889	4.761	4.504	4.262
280	1.987	2.003	2.071	2.165	2.394	3.614	5.186	4.983	4.623	4.321
300	2.000	2.013	2.069	2.147	2.328	3.181	5.675	5.311	4.775	4.391
310	2.006	2.018	2.070	2.141	2.303	3.033	6.073	5.541	4.866	4.430
320	2.012	2.023	2.071	2.136	2.282	2.914	5.726	5.846	4.970	4.472
330	2.018	2.029	2.073	2.132	2.265	2.817	4.932	6.273	5.088	4.517
340	2.025	2.035	2.075	2.130	2.250	2.738	4.404	6.933	5.225	4.564
350	2.031	2.040	2.078	2.128	2.239	2.672	4.027	8.138	5.384	4.615
360	2.037	2.046	2.081	2.128	2.229	2.616	3.746	11.461	5.571	4.668
370	2.044	2.052	2.085	2.128	2.221	2.570	3.528	18.863	5.794	4.725
380	2.051	2.058	2.088	2.129	2.214	2.530	3.355	10.329	6.061	4.784
390	2.057	2.064	2.093	2.130	2.209	2.497	3.215	7.714	6.388	4.846
400	2.064	2.070	2.097	2.132	2.205	2.468	3.100	6.371	6.789	4.911
420	2.077	2.083	2.106	2.137	2.201	2.423	2.924	4.966	7.871	5.047
440	2.090	2.095	2.116	2.143	2.199	2.389	2.799	4.232	9.169	5.189
460	2.104	2.108	2.127	2.150	2.200	2.365	2.706	3.782	9.635	5.331
480	2.117	2.121	2.138	2.159	2.203	2.347	2.637	3.482	8.636	5.460
500	2.131	2.135	2.150	2.168	2.208	2.335	2.584	3.269	7.239	5.557
520	2.145	2.148	2.161	2.178	2.213	2.327	2.544	3.113	6.127	5.604
540	2.158	2.162	2.174	2.189	2.221	2.322	2.513	2.996	5.336	5.581
560	2.173	2.175	2.186	2.200	2.229	2.320	2.489	2.905	4.775	5.484
580	2.187	2.189	2.199	2.212	2.238	2.320	2.472	2.834	4.367	5.324
600	2.201	2.203	2.212	2.224	2.247	2.322	2.458	2.778	4.062	5.123
650	2.236	2.238	2.245	2.255	2.274	2.333	2.440	2.682	3.567	4.581
700	2.272	2.273	2.279	2.287	2.303	2.351	2.437	2.627	3.283	4.129
750	2.307	2.308	2.313	2.320	2.333	2.373	2.444	2.597	3.106	3.797
800	2.342	2.343	2.348	2.353	2.364	2.398	2.456	2.583	2.992	3.561
850	2.377	2.378	2.382	2.386	2.396	2.424	2.474	2.579	2.916	3.392
900	2.411	2.412	2.415	2.419	2.427	2.452	2.494	2.583	2.866	3.269
950	2.445	2.446	2.448	2.452	2.459	2.480	2.516	2.593	2.833	3.179
1000	2.478	2.478	2.481	2.484	2.490	2.508	2.540	2.606	2.812	3.113
1100	2.540	2.541	2.543	2.545	2.550	2.564	2.59	2.64	2.80	3.03
1200	2.599	2.599	2.601	2.603	2.606	2.618	2.64	2.68	2.80	2.98
1300	2.653	2.654	2.655	2.656	2.660	2.669	2.68	2.72	2.81	2.96
1400	2.704	2.704	2.705	2.706	2.709	2.717	2.73	2.76	2.83	2.96
1500	2.750	2.750	2.751	2.752	2.754	2.761	2.77	2.79	2.86	2.96
1600	2.792	2.792	2.793	2.794	2.796	2.801	2.81	2.83	2.88	2.97
1700	2.831	2.831	2.832	2.833	2.834	2.839	2.85	2.86	2.91	2.98
1800	2.867	2.867	2.868	2.868	2.870	2.874	2.88	2.89	2.93	3.00
1900	2.901	2.901	2.901	2.902	2.903	2.907	2.91	2.92	2.96	3.01
2000	2.931	2.931	2.932	2.933	2.934	2.937	2.94	2.95	2.98	3.03

Table 5. Viscosity

P (bar)	t (°C) 0	25	50	75	100	150	200	250	300	350	375
1	1792	890.8	547.1	378.4	12.28	14.19	16.18	18.22	20.29	22.37	23.41
5	1791	890.7	547.1	378.5	282.4	182.0	16.07	18.15	20.25	22.35	23.39
10	1790	890.6	547.2	378.6	282.6	182.1	15.93	18.07	20.20	22.32	23.37
25	1786	890.3	547.5	379.0	283.0	182.5	133.9	17.83	20.06	22.24	23.32
50	1780	889.8	547.9	379.6	283.6	183.2	134.5	106.1	19.86	22.15	23.27
75	1775	889.3	548.3	380.2	284.3	183.8	135.1	106.8	19.74	22.13	23.28
100	1769	888.9	548.7	380.9	284.9	184.4	135.7	107.5	86.42	22.18	23.35
125	1764	888.5	549.1	381.5	285.6	185.1	136.3	108.2	87.40	22.39	23.52
150	1759	888.1	549.5	382.1	286.3	185.7	136.9	108.8	88.32	22.91	23.84
175	1754	887.7	550.0	382.7	286.9	186.3	137.5	109.5	89.21	66.85	24.45
200	1749	887.4	550.4	383.4	287.6	186.9	138.1	110.1	90.06	69.21	25.79
225	1744	887.1	550.9	384.0	288.2	187.6	138.7	110.7	90.88	71.10	47.65
250	1739	886.8	551.3	384.6	288.9	188.2	139.3	111.4	91.67	72.71	58.09
275	1735	886.6	551.8	385.2	289.5	188.8	139.9	112.0	92.43	74.14	61.87
300	1731	886.4	552.3	385.9	290.2	189.4	140.5	112.6	93.18	75.43	64.49
350	1722	886.0	553.3	387.2	291.5	190.6	141.6	113.8	94.61	77.71	68.31
400	1714	885.8	554.3	388.4	292.8	191.8	142.8	114.9	95.98	79.72	71.21
450	1707	885.6	555.3	389.7	294.2	193.1	143.9	116.1	97.28	81.52	73.61
500	1700	885.5	556.4	391.0	295.5	194.3	145.0	117.2	98.55	83.19	75.70
550	1694	885.6	557.5	392.3	296.8	195.5	146.1	118.3	99.76	84.73	77.57
600	1687	885.7	558.6	393.6	298.1	196.7	147.2	119.4	100.9	86.19	79.27
650	1682	885.9	559.7	395.0	299.4	197.9	148.3	120.4	102.1	87.57	80.85
700	1676	886.2	560.9	396.3	300.8	199.0	149.3	121.5	103.2	88.88	82.33
800	1667	887.1	563.3	399.0	303.4	201.4	151.5	123.5	105.4	91.35	85.05
900	1659	888.3	565.8	401.7	306.1	203.8	153.6	125.5	107.4	93.65	87.54
1000	1653	889.9	568.4	404.4	308.7	206.1	155.6	127.5	109.4	95.82	89.84

P (bar)	400	425	450	475	500	550	600	650	700	750	800
1	24.45	25.49	26.52	27.55	28.57	30.61	32.61	34.60	36.55	38.48	40.38
5	24.44	25.48	26.52	27.55	28.58	30.62	32.63	34.61	36.57	38.50	40.39
10	24.42	25.47	26.52	27.55	28.58	30.63	32.64	34.63	36.59	38.52	40.42
25	24.39	25.46	26.52	27.57	28.61	30.67	32.70	34.70	36.66	38.59	40.50
50	24.38	25.47	26.55	27.62	28.67	30.76	32.81	34.82	36.79	38.73	40.63
75	24.40	25.52	26.61	27.70	28.77	30.87	32.93	34.95	36.93	38.88	40.78
100	24.49	25.62	26.72	27.82	28.89	31.01	33.09	35.11	37.09	39.04	40.94
125	24.65	25.77	26.88	27.98	29.06	31.18	33.26	35.28	37.27	39.21	41.11
150	24.91	26.01	27.10	28.19	29.27	31.38	33.45	35.48	37.46	39.39	41.29
175	25.32	26.34	27.39	28.46	29.52	31.62	33.68	35.69	37.66	39.59	41.48
200	25.96	26.80	27.77	28.79	29.82	31.89	33.92	35.93	37.88	39.80	41.68
225	27.03	27.44	28.26	29.20	30.18	32.19	34.20	36.18	38.12	40.03	41.89
250	29.00	28.36	28.89	29.70	30.61	32.54	34.50	36.45	38.38	40.26	42.11
275	33.73	29.70	29.71	30.32	31.12	32.93	34.84	36.75	38.64	40.51	42.35
300	43.83	31.73	30.78	31.06	31.71	33.37	35.20	37.07	38.93	40.77	42.59
350	55.78	39.35	33.97	33.06	33.19	34.40	36.02	37.77	39.55	41.33	43.10
400	61.29	48.69	39.05	35.92	35.16	35.65	36.98	38.56	40.24	41.94	43.65
450	65.01	55.07	45.22	39.72	37.68	37.15	38.07	39.44	40.98	42.60	44.24
500	67.89	59.44	50.71	44.08	40.70	38.88	39.30	40.41	41.79	43.30	44.85
550	70.30	62.76	55.06	48.36	44.02	40.84	40.65	41.45	42.65	44.03	45.50
600	72.40	65.46	58.52	52.16	47.37	42.96	42.12	42.57	43.57	44.81	46.17
650	74.28	67.76	61.36	55.40	50.53	45.18	43.67	43.75	44.52	45.61	46.87
700	75.98	69.79	63.79	58.18	53.38	47.41	45.28	44.98	45.51	46.44	47.58
800	79.04	73.28	67.81	62.72	58.20	51.70	48.55	47.52	47.55	48.15	49.04
900	81.75	76.27	71.11	66.35	62.09	55.51	51.73	50.06	49.62	49.88	50.52
1000	84.22	78.92	73.97	69.42	65.32	58.80	54.66	52.50	51.65	51.58	51.98

Table 6. Thermal conductivity

P (bar)	t (°C) 0	25	50	75	100	150	200	250	300	350	375
1	561.0	607.2	643.6	666.8	25.08	28.85	33.28	38.17	43.42	48.96	51.83
5	561.3	607.4	643.7	667.0	679.3	682.1	34.93	39.18	44.09	49.44	52.25
10	561.5	607.6	644.0	667.2	679.6	682.4	37.21	40.51	44.95	50.06	52.79
25	562.4	608.3	644.7	668.0	680.4	683.4	664.2	45.16	47.82	52.06	54.53
50	563.7	609.4	645.8	669.2	681.8	685.1	666.4	622.7	53.86	55.99	57.87
75	565.1	610.5	647.0	670.5	683.2	686.8	668.6	625.9	63.11	61.06	62.00
100	566.5	611.7	648.2	671.7	684.5	688.5	670.7	629.0	550.9	68.10	67.35
125	567.9	612.8	649.3	673.0	685.9	690.2	672.8	632.0	556.5	79.15	74.68
150	569.3	613.9	650.5	674.2	687.2	691.8	674.9	635.0	561.8	100.9	85.54
175	570.6	615.1	651.6	675.5	688.6	693.5	677.0	637.9	566.8	452.5	103.7
200	572.0	616.2	652.8	676.7	690.0	695.1	679.1	640.8	571.6	463.3	142.3
225	573.4	617.3	654.0	678.0	691.3	696.8	681.2	643.6	576.2	472.8	441.5
250	574.8	618.5	655.1	679.2	692.7	698.4	683.2	646.3	580.7	481.4	411.4
275	576.1	619.6	656.3	680.4	694.0	700.1	685.3	649.1	585.0	489.1	425.8
300	577.5	620.8	657.4	681.7	695.3	701.7	687.3	651.8	589.1	496.3	438.0
350	580.2	623.0	659.8	684.1	698.0	704.9	691.3	657.0	597.1	509.3	457.5
400	582.9	625.3	662.1	686.6	700.7	708.2	695.3	662.2	604.6	521.0	473.2
450	585.5	627.5	664.4	689.1	703.3	711.4	699.3	667.2	611.7	531.8	486.6
500	588.1	629.8	666.7	691.5	706.0	714.6	703.2	672.1	618.5	541.7	498.5
550	590.7	632.0	668.9	693.9	708.6	717.7	707.0	676.9	625.1	551.0	509.4
600	593.3	634.2	671.2	696.3	711.2	720.9	710.9	681.6	631.3	559.7	519.4
650	595.8	636.4	673.5	698.7	713.8	724.0	714.7	686.3	637.4	568.0	528.8
700	598.3	638.6	675.7	701.1	716.4	727.2	718.5	690.8	643.2	575.9	537.7
800	603.1	642.9	680.2	705.9	721.5	733.4	726.0	699.8	654.5	590.6	554.1
900	607.8	647.2	684.6	710.5	726.6	739.5	733.4	708.6	665.1	604.2	569.1
1000	612.2	651.3	688.9	715.2	731.6	745.6	740.7	717.2	675.4	616.8	583.0

P (bar)	400	425	450	475	500	550	600	650	700	750	800
1	54.76	57.74	60.77	63.85	66.97	73.35	79.89	86.57	93.37	100.3	107.3
5	55.13	58.08	61.08	64.14	67.25	73.61	80.13	86.80	93.59	100.5	107.5
10	55.61	58.51	61.48	64.51	67.60	73.93	80.44	87.09	93.87	100.8	107.8
25	57.15	59.89	62.75	65.69	68.71	74.94	81.39	88.01	94.75	101.6	108.5
50	60.06	62.49	65.10	67.86	70.74	76.79	83.13	89.67	96.34	103.1	109.9
75	63.56	65.54	67.82	70.33	73.03	78.84	85.04	91.49	98.08	104.8	111.5
100	67.89	69.19	70.99	73.16	75.61	81.11	87.14	93.47	99.97	106.5	113.2
125	73.40	73.63	74.73	76.43	78.53	83.62	89.43	95.63	102.0	108.5	115.0
150	80.69	79.13	79.19	80.20	81.85	86.39	91.92	97.96	104.2	110.6	116.9
175	90.76	86.10	84.54	84.58	85.61	89.45	94.63	100.47	106.6	112.8	119.0
200	105.5	95.12	91.04	89.70	89.89	92.81	97.57	103.2	109.1	115.2	121.2
225	128.6	107.1	99.01	95.70	94.75	96.51	100.7	106.0	111.8	117.7	123.5
250	169.3	123.2	108.8	102.7	100.3	100.6	104.1	109.1	114.6	120.3	126.0
275	249.1	145.5	121.0	111.0	106.6	105.0	107.8	112.4	117.6	123.1	128.6
300	330.1	176.3	136.0	120.6	113.7	109.8	111.7	115.8	120.7	126.0	131.3
350	384.5	259.4	176.5	144.9	130.7	120.8	120.3	123.3	127.5	132.2	137.0
400	414.0	323.3	227.6	175.8	151.6	133.5	130.0	131.5	134.8	138.8	143.1
450	435.0	363.4	276.3	211.3	176.0	147.9	140.6	140.3	142.6	145.9	149.6
500	451.6	391.5	315.6	247.0	202.7	163.7	152.1	149.8	150.9	153.4	156.4
550	465.5	412.8	346.5	279.6	229.7	180.6	164.3	159.8	159.6	161.2	163.4
600	477.7	430.0	371.2	308.0	255.6	198.0	177.0	170.1	168.5	169.1	170.6
650	488.6	444.5	391.4	332.5	279.6	215.4	189.9	180.6	177.6	177.2	177.9
700	498.7	457.1	408.5	353.6	301.5	232.4	202.7	191.0	186.7	185.3	185.2
800	516.8	478.5	436.0	388.0	339.1	264.5	227.5	211.5	204.4	201.1	199.6
900	533.1	496.6	457.9	414.9	369.8	293.5	250.6	230.6	221.1	216.1	213.1
1000	548.0	512.7	476.3	436.9	395.1	319.3	271.8	248.0	236.2	229.7	225.5

Table 7. Prandtl number

P (bar)	t (°C) 0	25	50	75	100	150	200	250	300	350	375
1	13.50	6.137	3.555	2.378	1.000	0.974	0.960	0.950	0.941	0.932	0.928
5	13.48	6.133	3.553	2.377	1.753	1.151	0.984	0.964	0.950	0.939	0.934
10	13.46	6.128	3.551	2.377	1.752	1.150	1.028	0.987	0.965	0.949	0.942
25	13.39	6.113	3.546	2.374	1.751	1.150	0.903	1.096	1.021	0.982	0.969
50	13.27	6.088	3.538	2.371	1.750	1.149	0.902	0.825	1.173	1.057	1.025
75	13.15	6.063	3.529	2.367	1.748	1.148	0.900	0.821	1.466	1.162	1.098
100	13.04	6.039	3.521	2.364	1.746	1.147	0.899	0.817	0.890	1.312	1.191
125	12.93	6.015	3.513	2.360	1.744	1.146	0.897	0.813	0.874	1.545	1.314
150	12.83	5.992	3.505	2.357	1.743	1.145	0.896	0.810	0.860	2.006	1.484
175	12.73	5.970	3.497	2.354	1.741	1.145	0.895	0.806	0.848	1.379	1.753
200	12.63	5.947	3.490	2.350	1.740	1.144	0.894	0.803	0.837	1.216	2.353
225	12.53	5.926	3.482	2.347	1.738	1.143	0.892	0.800	0.827	1.121	8.125
250	12.43	5.904	3.475	2.344	1.736	1.142	0.891	0.798	0.818	1.057	1.937
275	12.34	5.883	3.467	2.341	1.735	1.142	0.890	0.795	0.810	1.009	1.485
300	12.25	5.863	3.460	2.338	1.733	1.141	0.889	0.793	0.803	0.973	1.291
350	12.08	5.823	3.446	2.332	1.731	1.140	0.887	0.788	0.790	0.919	1.103
400	11.92	5.785	3.433	2.326	1.728	1.138	0.885	0.784	0.779	0.880	1.005
450	11.77	5.748	3.420	2.321	1.725	1.137	0.883	0.780	0.769	0.850	0.943
500	11.62	5.713	3.407	2.315	1.723	1.136	0.882	0.777	0.761	0.827	0.899
550	11.48	5.680	3.395	2.310	1.721	1.135	0.880	0.773	0.753	0.807	0.866
600	11.36	5.648	3.384	2.305	1.718	1.134	0.879	0.770	0.747	0.791	0.840
650	11.23	5.617	3.372	2.301	1.716	1.134	0.878	0.768	0.741	0.777	0.818
700	11.12	5.588	3.362	2.296	1.714	1.133	0.876	0.765	0.735	0.765	0.800
800	10.91	5.533	3.342	2.288	1.711	1.131	0.874	0.761	0.726	0.745	0.772
900	10.72	5.483	3.324	2.280	1.707	1.130	0.872	0.757	0.718	0.730	0.750
1000	10.55	5.439	3.307	2.273	1.704	1.129	0.870	0.753	0.712	0.717	0.733

P (bar)	400	425	450	475	500	550	600	650	700	750	800
1	0.924	0.921	0.917	0.914	0.911	0.905	0.899	0.894	0.890	0.886	0.882
5	0.929	0.925	0.921	0.917	0.913	0.907	0.901	0.895	0.891	0.886	0.882
10	0.936	0.931	0.926	0.921	0.917	0.909	0.902	0.897	0.891	0.887	0.883
25	0.958	0.949	0.941	0.934	0.928	0.917	0.908	0.900	0.894	0.889	0.884
50	1.002	0.983	0.969	0.957	0.946	0.929	0.916	0.906	0.898	0.891	0.886
75	1.055	1.024	1.000	0.982	0.966	0.943	0.925	0.911	0.901	0.894	0.888
100	1.118	1.070	1.035	1.008	0.988	0.956	0.933	0.917	0.904	0.895	0.889
125	1.195	1.123	1.073	1.038	1.010	0.970	0.942	0.921	0.907	0.896	0.889
150	1.290	1.183	1.116	1.069	1.034	0.984	0.950	0.926	0.909	0.897	0.889
175	1.408	1.253	1.163	1.102	1.059	0.998	0.958	0.930	0.911	0.898	0.889
200	1.568	1.336	1.215	1.138	1.085	1.013	0.966	0.934	0.912	0.897	0.888
225	1.812	1.436	1.273	1.177	1.112	1.027	0.974	0.937	0.913	0.897	0.887
250	2.273	1.564	1.339	1.219	1.141	1.042	0.981	0.941	0.914	0.896	0.885
275	3.363	1.737	1.416	1.264	1.171	1.057	0.988	0.944	0.914	0.895	0.883
300	3.329	1.979	1.506	1.314	1.203	1.073	0.996	0.946	0.914	0.894	0.881
350	1.693	2.415	1.729	1.427	1.272	1.104	1.010	0.951	0.914	0.890	0.876
400	1.290	1.922	1.900	1.548	1.345	1.135	1.023	0.955	0.912	0.886	0.871
450	1.118	1.479	1.791	1.627	1.413	1.167	1.037	0.959	0.911	0.882	0.865
500	1.021	1.245	1.542	1.601	1.454	1.195	1.049	0.962	0.909	0.877	0.858
550	0.957	1.109	1.335	1.488	1.447	1.218	1.061	0.965	0.907	0.872	0.852
600	0.911	1.022	1.190	1.354	1.396	1.231	1.071	0.968	0.905	0.868	0.846
650	0.876	0.962	1.089	1.235	1.320	1.231	1.078	0.971	0.904	0.864	0.841
700	0.849	0.917	1.016	1.140	1.238	1.219	1.083	0.974	0.903	0.860	0.836
800	0.808	0.855	0.919	1.005	1.096	1.164	1.079	0.977	0.903	0.855	0.828
900	0.778	0.813	0.859	0.920	0.992	1.091	1.060	0.976	0.903	0.853	0.823
1000	0.755	0.782	0.817	0.863	0.919	1.020	1.030	0.970	0.903	0.853	0.821

Table 8. Properties for coexisting phases: viscosity, thermal, conductivity, Prandtl number, dielectric constant, surface tension

t (°C)	η_l	η_g	λ_l	λ_g	Pr_l	Pr_g	ϵ_l	ϵ_g	σ
0.01	1791.	9.22	561.0	17.07	13.50	1.008	87.81	1.000	75.65
10	1308.	9.46	580.0	17.62	9.444	1.006	83.98	1.000	74.22
20	1003.	9.73	598.4	18.23	7.010	1.004	80.26	1.000	72.74
30	798.	10.01	615.4	18.89	5.423	1.003	76.66	1.000	71.20
40	653.	10.31	630.5	19.60	4.332	1.002	73.21	1.000	69.60
50	547.1	10.62	643.5	20.36	3.555	1.001	69.90	1.001	67.95
60	466.8	10.94	654.3	21.19	2.984	1.000	66.73	1.001	66.24
70	404.5	11.26	663.1	22.07	2.554	0.999	63.71	1.001	64.49
80	355.0	11.60	670.0	23.01	2.223	0.999	60.82	1.002	62.68
90	315.1	11.93	675.3	24.02	1.962	0.999	58.06	1.003	60.82
100	282.3	12.28	679.1	25.09	1.753	1.000	55.43	1.004	58.92
110	255.1	12.62	681.7	26.24	1.584	1.001	52.92	1.005	56.97
120	232.2	12.97	683.2	27.46	1.444	1.004	50.53	1.007	54.97
130	212.8	13.32	683.7	28.76	1.328	1.007	48.24	1.009	52.94
140	196.3	13.67	683.3	30.14	1.232	1.013	46.05	1.011	50.86
150	182.0	14.02	682.1	31.59	1.151	1.020	43.96	1.014	48.75
160	169.6	14.37	680.0	33.12	1.082	1.030	41.96	1.018	46.60
170	158.9	14.72	677.1	34.74	1.025	1.042	40.05	1.022	44.41
180	149.4	15.07	673.4	36.44	0.977	1.058	38.21	1.028	42.20
190	141.0	15.42	668.8	38.23	0.937	1.077	36.44	1.034	39.95
200	133.6	15.78	663.4	40.10	0.904	1.101	34.75	1.041	37.68
210	127.0	16.13	657.1	42.07	0.878	1.128	33.11	1.050	35.39
220	121.0	16.49	649.8	44.15	0.857	1.161	31.53	1.061	33.08
230	115.5	16.85	641.4	46.35	0.842	1.200	30.00	1.073	30.75
240	110.5	17.22	632.0	48.70	0.832	1.244	28.52	1.088	28.40
250	105.8	17.59	621.4	51.22	0.827	1.296	27.08	1.105	26.05
260	101.5	17.98	609.4	53.98	0.828	1.355	25.67	1.127	23.70
270	97.4	18.38	596.1	57.04	0.835	1.423	24.29	1.152	21.35
280	93.4	18.80	581.4	60.51	0.848	1.502	22.93	1.183	19.00
290	89.6	19.25	565.2	64.57	0.870	1.593	21.59	1.220	16.67
300	85.8	19.74	547.7	69.47	0.901	1.699	20.26	1.267	14.37
305	84.0	20.00	538.5	72.34	0.921	1.759	19.59	1.295	13.23
310	82.1	20.28	529.0	75.59	0.944	1.824	18.92	1.326	12.10
315	80.2	20.57	519.3	79.29	0.973	1.895	18.25	1.361	10.98
320	78.4	20.89	509.4	83.57	1.006	1.974	17.57	1.402	9.87
325	76.4	21.24	499.4	88.6	1.047	2.063	16.88	1.449	8.78
330	74.4	21.62	489.2	94.5	1.096	2.164	16.18	1.503	7.71
335	72.4	22.04	479.0	101.6	1.157	2.282	15.47	1.567	6.66
340	70.3	22.52	468.7	110.2	1.236	2.424	14.73	1.643	5.64
345	68.1	23.07	458.2	121.0	1.340	2.601	13.96	1.735	4.64
350	65.7	23.72	447.8	134.8	1.486	2.835	13.15	1.849	3.67
355	63.1	24.52	437.3	153.0	1.704	3.167	12.29	1.996	2.75
360	60.2	25.54	427.5	178.5	2.068	3.691	11.34	2.193	1.89
365	56.7	26.94	420.3	217.8	2.796	4.676	10.24	2.481	1.09
370	52.0	29.26	428.8	301.2	5.08	7.66	8.82	2.98	0.40
371	50.7	30.00	439.3	337.2	6.41	9.30	8.44	3.15	0.28
372	49.1	30.95	462.8	397.4	9.11	12.43	7.99	3.37	0.17
373	46.7	32.38	545.4	538.3	18.69	21.19	7.35	3.71	0.07
373.976	39.0		∞		∞		5.34		0.

Table 9. Thermal expansion coefficient $\beta = (1/\upsilon)(\partial\upsilon/\partial T)_p$ of liquid water as a function of pressure and temperature[a]

	t (°C)								
P (bar)	0	20	50	100	150	200	250	300	350
1	−0.085 2	0.206 7	0.462 3						
5	−0.083 8	0.207 2	0.462 2	0.753 9	1.024				
10	−0.082 0	0.207 9	0.462 0	0.753 0	1.022				
50	−0.067 8	0.213 3	0.460 5	0.745 5	1.007	1.347	1.936		
100	−0.049 9	0.220 1	0.458 9	0.736 6	0.990 2	1.312	1.848	3.189	
150	−0.032 0	0.227 2	0.457 4	0.728 1	0.974 0	1.281	1.772	2.883	
200	−0.014 2	0.234 3	0.456 2	0.720 0	0.958 7	1.251	1.704	2.648	6.923
250	+0.003 3	0.241 6	0.455 1	0.712 2	0.944 2	1.224	1.643	2.460	5.162
300	0.020 5	0.248 9	0.454 2	0.704 7	0.930 3	1.198	1.589	2.306	4.276
350	0.037 3	0.256 2	0.453 4	0.697 5	0.917 2	1.175	1.539	2.176	3.718
400	0.053 5	0.263 6	0.452 8	0.690 7	0.904 6	1.152	1.494	2.065	3.324
450	0.069 0	0.270 9	0.452 3	0.684 1	0.892 6	1.131	1.453	1.968	3.027
500	0.083 6	0.278 2	0.452 0	0.677 7	0.881 1	1.111	1.415	1.884	2.791
600	0.110 0	0.292 6	0.451 7	0.665 7	0.859 6	1.075	1.348	1.742	2.439
700	0.131 7	0.306 5	0.451 8	0.654 5	0.839 7	1.042	1.290	1.626	2.186
800	0.147 5	0.319 6	0.452 3	0.644 1	0.821 3	1.012	1.238	1.530	1.994
900	0.156 5	0.331 7	0.453 0	0.634 3	0.804 2	0.984 4	1.193	1.448	1.843
1000	0.157 6	0.342 6	0.454 0	0.625 2	0.788 2	0.959 4	1.152	1.377	1.720

[a]β in 10^{-3}/K.

Table 10. Thermal diffusivity κ of liquid water as a function of oresure and temperature[a]

	t (°C)								
P (bar)	0	20	50	100	150	200	250	300	350
1	0.135	0.144	0.156						
10	0.135	0.144	0.156	0.168	0.173				
50	0.137	0.146	0.157	0.170	0.175	0.173	0.159		
100	0.138	0.147	0.158	0.171	0.175	0.174	0.162	0.134	
150	0.140	0.148	0.159	0.171	0.176	0.175	0.166	0.140	
200	0.140	0.149	0.160	0.172	0.176	0.176	0.167	0.147	0.093
250	0.142	0.150	0.161	0.173	0.178	0.177	0.169	0.151	0.108
300	0.145	0.151	0.162	0.173	0.180	0.179	0.171	0.155	0.120
350	0.145	0.152	0.163	0.175	0.180	0.181	0.174	0.159	0.128
400	0.147	0.153	0.163	0.175	0.181	0.181	0.178	0.162	0.135
450	0.147	0.154	0.164	0.176	0.181	0.182	0.178	0.166	0.141
500	0.147	0.155	0.164	0.177	0.182	0.183	0.178	0.169	0.147

[a]κ in 10^{-6} m²/s.

References

Haar, L., Gallagher, J. S., and Kell, G. S., (1984) *Thermodynamic and Transport Properties and Computer Programs for Vapor and Liquid States of Water in S.I. Units,* NBS/NRC, Hemisphere, Washington, D.C.

VDI—Wärmeatlas, (1974) 2d ed., Verein Deutsches Ingenieure, Düsseldorf.

STEAM TURBINE

Following from: *Turbine*

The steam turbine is a turbine in which the potential energy of heated and compressed steam produced in a special device, a steam generator, or steam of natural origin (for example from geothermal springs) is converted into kinetic energy (when the steam expands in the turbine blade cascades) and then into mechanical work on the rotating shaft. The rows of rotating blades fixed on the steam turbine rotor change the total steam enthalpy and positive work is done. In the gas turbine (see **Gas Turbine**) the pressure ratio π_T (that is the ratio of the working fluid pressure at the turbine inlet to the pressure at the turbine outlet) is not very large (usually not higher than 20–30) but the initial temperature of the gas (combustion products) may be as high as 1700–1800 K. In contrast, the steam turbine is characterized by larger pressure ratios $\pi_T \approx$ 2000-6000 (due to higher initial values (p_h) and low final values (p_t) of the steam pressure) and considerably lower initial steam temperature ($T_h \approx$ 810-880K). Therefore, the enthalpy drop in the steam turbine is 2–3 times higher than that in the gas turbine, and the number of stages in the steam turbine is many times larger than that in the gas turbine.

There are several types of steam turbines shown schematically in **Figure 1**. In a condensing turbine (**Figure 1a**) the steam is expanded down to the deep vacuum ($p_t \approx$ 4-3kPa) reached in the condenser. These turbines are designed with uncontrolled steam bleed used for feed water regenerative heating. The uncontrolled bleed is characterized by unsteady pressure of the extracted steam. The steam is bled through a special manifold in the bottom part of the turbine casing.

Figure 1b is a schematic of a turbine with condensation and with one controlled steam bleed for process and domestic heat demands. In these turbines, a portion of steam is bled from intermediate stages to be used by consumers. The remaining portion of the steam passes the subsequent turbine stages and after that passes to the condenser. The bleed pressure is kept steady regardless of the turbine load, a special regulator device being used for this purpose. In the turbine shown in **Figure 1c**, there are two controlled steam extractions at different pressures.

Figure 1d is the schematic of a turbine with two pressures. This turbine uses not only fresh steam from the boiler, but also exhaust steam from hammers, presses, pump, air blower and compressor drives.

A backpressure turbine is shown in **Figure 1e**. There is no condenser in such a turbine unit. The steam at the required pressure is fed from the turbine and used for processes and for domesic needs.

Steam turbine design is influenced by the turbine capacity, initial steam parameters (sub- and super-critical), its operation conditions within the power generation system (base-load, peak-load, semi-peak load), final steam moisture content, technological characteristics and other factors. Low capacity turbines (up to 50 mW) are as a rule of one-cylinder type.

The disadvantages of high capacity condensing turbines are connected with the limited flow rates of the final stages. To overcome this difficulty, these turbines are constructed with division of the main steam flow (before it enters the final stages) into several parallel flows. Each part of these turbines is designed for the maximum steam flow rate Q_m (**Figure 2**).

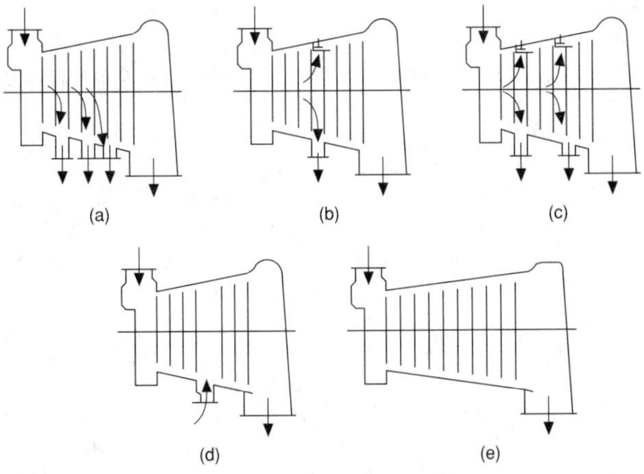

Figure 1. Types of steam turbine.

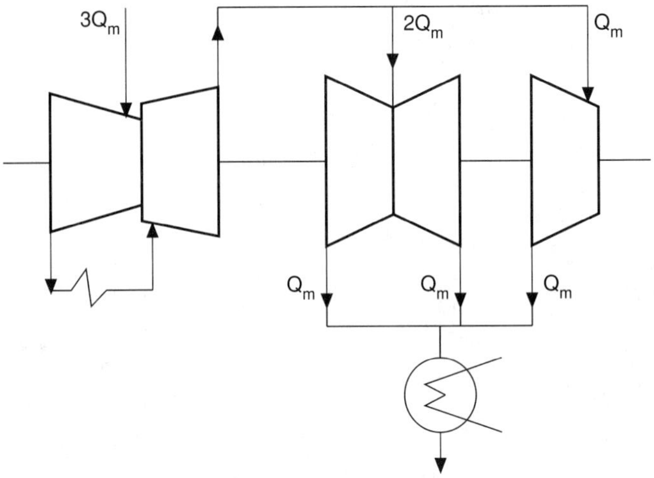

Figure 2. Multistage steam turbine.

These parts are referred to as a high pressure cylinder, an intermediate pressure cylinder and a low pressure cylinder.

Steam turbine rotors may be *disk-type* or *drum-type* (**Figure 3**). Disk-type rotors are used in impulse turbines (see **Turbines**), drum-type rotors are used in reaction turbines.

Steam turbines are used as parts of stationary and transport (marine) steam turbine power units. Besides turbines these power units also include boilers (steam-generators), steam condensers and other devices. Steam turbines constructed for combined operation with gas turbine units are also used as parts of combined steam-gas plants (see **Gas Turbines**) with applications in both stationary and transport (marine) power units.

The real cycle apbb′hta (**Figure 4**) of the simplest steam turbine unit includes the ap process of increasing pressure of the water in the pump, the pb process of heating water at constant pressure to the boiling temperature, and bb′ process of evaporation at constant temperature. The b′h process corresponds to water superheating,

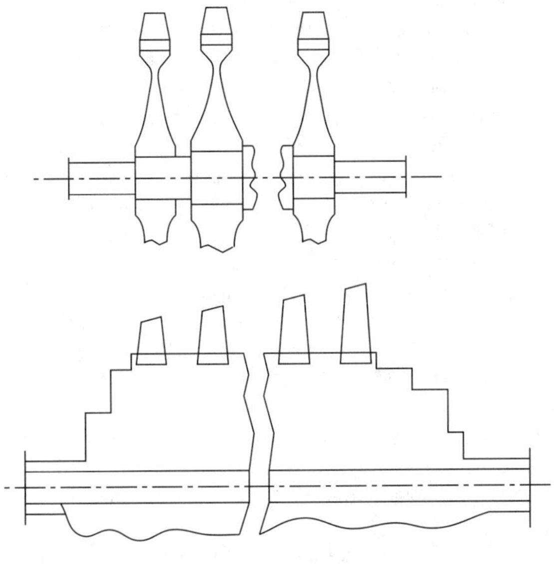

Figure 3. Disk (top) and drum (bottom) types of rotor design.

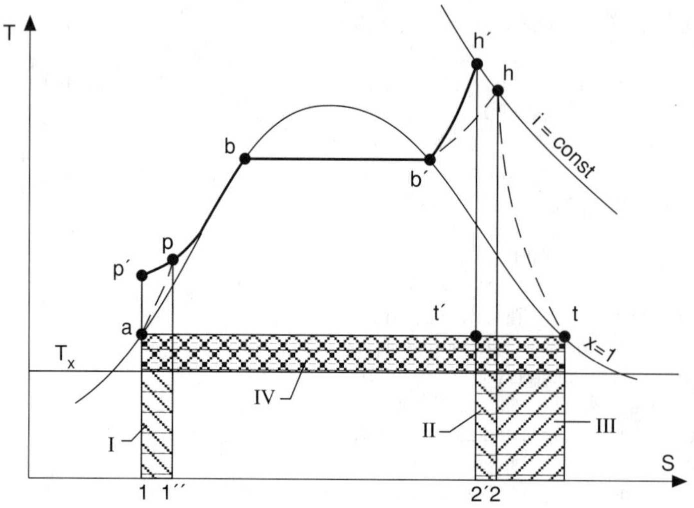

Figure 4. Steam turbine cycle on temperature entropy (T-S) diagram.

the ht process corresponds to the expansion of steam in the turbine. The ta process which is the closing process of the cycle corresponds to heat removal in the condenser.

This real cycle of the steam-turbine unit differs from the ideal thermodynamic cycle ap′bb′h′t′a because of irreversible losses in the pump, steam pipe, turbine and condenser. These losses are denoted by I,II,III and IV areas (T_x is the temperature of water used for cooling the condensate). The specific work of the real cycle $l_e = l_t - l_p$, where $l_p = i_h - i_t$ is the actual turbine work (where i_h and i_t are the steam enthalpy at the beginning and at the end of the expansion process in the turbine); $l_p = i_p—i_a$ is the specific work of the pump in the real cycle (where i_p and i_a are the enthalpy of water at the corresponding points of the cycle). The ideal cycle thermal efficiency is $\eta_t = (i_{h'} - i_{t'})/(i_{h'} - i_a)$; the effective *efficiency of the steam-turbine* unit is $\eta_e = \eta_t\eta_T\eta_m$, where η_T and η_m are the efficiency of the turbine itself and the efficiency taking into account the mechanical losses in the turbine. When determining the efficiency η_t of a steam turbine it is necessary to take into account the moisture of the steam which is typical for the last stages of steam condensing turbines and for many stages of turbines using saturated and slightly superheated steam (these turbines are used for instance at nuclear power stations). When such steam is used the efficiency of the stages decreases. In this case relative losses ζ_w may be rather large (for example in the last 3 stages of a turbine of 800 mW capacity and with initial steam pressure 24 MPa $\zeta_w = 0.012$ to 0.081; still greater losses due to moisture are typical for turbines without intermediate heating). Besides that the first stages of steam turbines are often with partial admission ($\epsilon \approx 0.15$), and ventilation losses occur in them (see **Turbine**). In intermediate stages of heavy-duty steam-turbines using superheated steam, the maximum blading efficiency $\eta_b =$ 0.905 to 0.903.

Reaction ratio at the middle diameter in the high pressure and intermediate pressure cylinders of steam turbines increases with the number of stages from 0.2 to 0.4, and in the low pressure cylinder from 0.3 to 0.7.

A variety of techniques are used for increasing the efficiency of steam-turbine units. One of the methods is increasing initial parameters of the steam. For example, when pressure p_h is increased, the saturation temperature increases. The result is an increase of the average temperature at which heat is supplied; thus the thermal efficiency η_t of the ideal cycle increases too. However, in practice an increase of pressure to the value more than 9–10 MPa does not result in the increase of the theoretical work and does not significantly affect the unit efficiency. Also, steam moisture content at the end of the expansion process increases with the increase of pressure and results in greater losses in the course of the steam expansion and also in the turbine blade erosion. Therefore, the general tendency is to limit the moisture content to 13–15 per cent.

Simultaneous increase of the values of p_h and T_h may considerably increase the steam-turbine unit efficiency. For this purpose many present-day steam-turbine units have intermediate (repeated) superheating of the steam after expansion in the first group of stages. In this case the theoretical work of the turbine, the cycle work and thus the cycle effiency increase, the moisture of the steam at the end of the expansion process decreases, and the amount of heat transfered in the condenser increases. The temperature of superheating as well as the initial temperature is limited by the thermal characteristics of the flow passage metal parts.

A decrease of the steam pressure p_t in the condenser causes a decrease of the steam condensation temperature and consequently increases the temperature difference in the cycle.

The efficiency of steam-turbine units increases when regenerative extraction of steam from the turbine is used. Regenerative extraction is uncontrolled bleed of the steam from the stages with the aim of increasing the feed water temperature in the unit. In this cycle the feed water is heated by the heat released in the process of the steam cooling and condensation.

Condensing steam turbines have an efficiency in the range η_e = 36 to 42%. From this it follows that only a small portion of heat released in the process of fuel combustion is transformed into effective work. Turbine units for power and steam generation have higher overall efficiency. In these units the heat from the fuel is used for power generation and for obtaining heat at some prescribed temperature level. The theoretical work of the unit with the steam turbine for power and heat generation is less than that of the steam turbine unit with condensing steam turbine. The useful work of the cycle of the steam turbine unit for power and heat generation is also lower than that of the condensing turbine. However, the steam turbine unit for combined power and heat generation makes effective use of the heat of condensation and therefore its overall efficiency is higher than that of a condensing steam turbine unit.

References

Horlock, J. H. (1966) *Axial Flow Turbines,* Butterworths, London.

Kearton, W. J. (1951) *Steam turbine theory and practice,* 6th ed., Pitman.

Kostyuk, A. and Frolov, V. (1988) *Steam and Gas Turbines,* Moscow, Mir.

E.A. Manushin

STEAM-WATER SEPARATION (see Vapour-liquid separation)

STEEL AND TUBE CONDENSERS (see Condensers)

STEELS

Following from: *Metals*

Types and Properties

Classification

Steel is the most widely used metal in the world and consists of many thousands of different compositions, each offering a unique combination of properties which are tailor-made to satisfy individual requirements. Steels are alloys of iron and carbon, but many steels have their properties enhanced by the addition of other alloying elements and by the application of different thermomechanical and heat treatments.

The wide range of available steels can mostly be categorised into one of three families; *carbon steels, alloy steels* and *stainless steels.*

Carbon steels are body-centred cubic (bcc) in structure. Those containing over about 0.1% C (depending on section thickness) can be hardened by *heat treatment.* The strength is primarily dependant on the carbon content. They are available in plate, tube, strip, bar rod, or wire and structural sections and are used in a wide range of industries and applications. Microalloying—additions of **Vanadium**, niobium or **Titanium**—can be utilised to increase strength further. It is especially useful when allied to controlled thermo-mechanical processing to give an excellent combination of strength and fabricability. Very high levels of strength to over 3000N/mm^2 can be introduced by cold work, e.g., in wire drawing of high carbon steel.

Alloy steels involve the addition of elements including chromium, **Nickel, Molybdenum** and **Vanadium**. Such steels can be heat treated (often by quenching and tempering) to produce increased strength and hardness (i.e. tensile strengths above 750 N/mm^2 and up to 2000 N/mm^2), together with good ductility and toughness. Various surface treatments can be applied to give improved properties, fatigue performance and wear resistance. They are mainly used in automotive engine and transmission components and in the energy industries.

All the above steels are prone to **Corrosion** in certain environments. To prevent or reduce corrosion, they can be coated with metals such as zinc, tin, chromium or cadmium. A wide range of organic, painted coatings are also available.

Stainless steels are produced by adding at least 11% chromium to steel to produce a thin passive protective layer of Cr_2O_3 which promotes corrosion resistance. This is improved by further increasing the chromium content. Four basic groups of stainless steel are available:

> Ferritic grades, which contains between 11–17% chromium, they are body-centred cubic in structure and magnetic.
>
> Martensitic grades contain similar amounts of chromium, but more carbon than ferritics and possibly other additions such as molybdenum to increase hardenability and strength.
>
> Austenitic grades contain between 17–25% Cr, 7–20% nickel and in some instances molybdenum. They are face-centred cubic in structure, non-magnetic and can be formed and welded more easily than ferritics.
>
> Duplex stainless steels were developed to provide the strength of ferritics, but with improved corrosion resistance. They contain about 22% Cr, 5% Ni and possibly molybdenum.

Stainless steels are used extensively in food and drink production and the chemical and energy industries; martensitics are used for cutlery and other cutting tool manufacture.

Further details of the metallurgy of steels and their applications are given by Llewellyn (1992), and Pickering (1992) and *ASM Metals Handbook Vol. 1* (1990).

Thermal properties

Figures 1, 2 and **3** show that the thermal properties of the various steel compositions can vary considerably and must be taken into account when designing with steel. By special control of composition and microstructure, creep resistant steels have been developed to give enhanced performance at high temperatures. Further thermal data are available from Smithells (1992) and Rothman (1987).

Heat treatment

The hardness and strength of steels can be altered by heat treatment. The two most frequently used *heat treatment* processes are *annealing,* and *hardening* and *tempering.*

Annealing

When steels are worked, either during manufacture or when the metal is manipulated to form a component, the hardness increases. This makes further working difficult; to facilitate further manufacturing, the steel structure must be softened by annealing at a temperature typically within the range 600–800°C. Treatments in the range 400–1000°C, depending on the steel type, can also be used to stress-relieve a component. This is required when a component has been machined or plastically deformed, heat treated or welded which may result in distortion during the subsequent manufacturing procedures. Austenitic stainless steel components

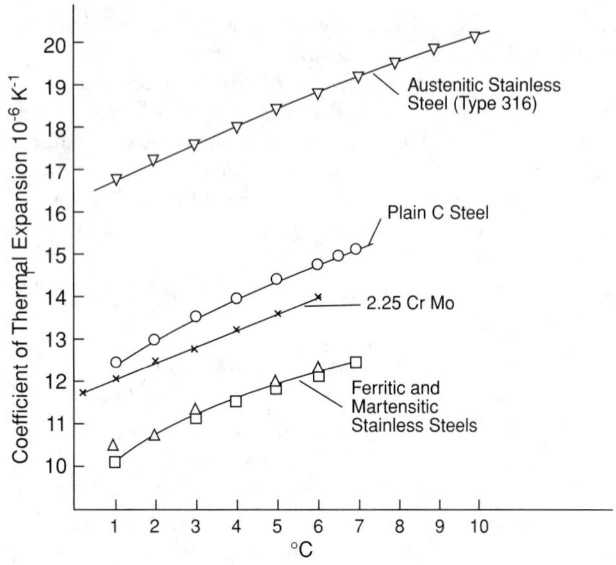

Figure 1. Coefficient of thermal expansion of a range of steels at various temperatures.

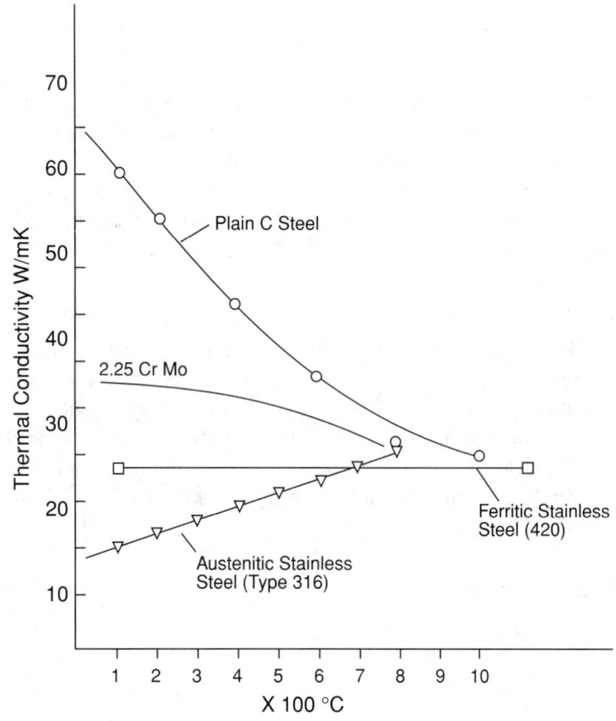

Figure 2. Thermal conductivity of a range of steels at various temperatures.

may also be stress-relieved to minimise the likelihood of stress-corrosion cracking in certain corrosive environments.

Hardening and Tempering

Alloy steels and martensitic stainless steels can be hardened by heating to temperature around 1000°C and cooling at a sufficiently fast rate to form a martensitic or bainitic structure. Quenchants

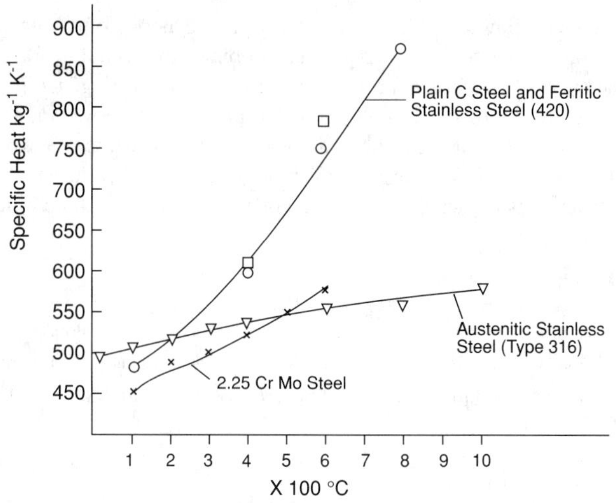

Figure 3. Specific heat of a range of steels at various temperatures.

can be chosen to cool at different rates and include brine, water, oil, air or molten salt. The choice of quenchant depends upon the type of steel and the component dimensions. For some highly alloy steels it may be necessary to cool to below 0°C to harden the steel completely.

After quenching the structure of the steel is hard but also brittle. In order to increase the toughness, albeit at the expense of a lower hardness and strength, it is necessary to temper, typically at temperatures between 500 and 650°C. The choice of tempering temperature depends upon the combination of strength and toughness required.

Cracking and Distortion

When steels are cooled to harden the structure, volume changes occur which are:

- Expansion when fcc iron (austenite) transforms to bcc (ferrite or martensite),
- Contraction when iron carbide is precipitated,
- Normal thermal contraction.

When a steel is quenched these changes occur very quickly and unevenly. Because the outside cools more rapidly than the inside, thermal gradients are set up which cause stresses and hence distortion or cracking.

The likelihood of quench crack formation can be minimised by good practice including.

- Tempering immediately after hardening
- Slow, even heating
- Choice of suitable quenchant
- Design of structure to minimise stress raisers including surface defects, sharp angles, internal non-metallic inclusions and uneven sections.

Heat Treatment Furnaces

A wide range of furnaces are available including gas or oil fired and electric. Controlled atmosphere or vacuum furnaces are used when decarburisation, carburisation or oxidation of the steel surface cannot be tolerated.

Furnace atmospheres can be selected intentionally to alter the composition of the outer layer of the steel, by the diffusion of carbon or nitrogen, usually to increase hardness by carburising or nitriding. Steel compositions have been developed to be given this treatment.

Further details on all aspects of the heat treatment of steels are given in numerous publications including Thelming (1984) and *ASM Handbook Vol. 4* (1991) whilst the heat treatment characterisation of individual grades are detailed by Van der Voort (1991).

Design

When using steels at elevated temperature, the choice is governed by:

The strength and life required.

The corrosivity of the environment.

The thermal expansion characteristics.

Many steels have been developed specifically for use at elevated temperature: these contain alloying additions of elements such as chromium, molybdenum, vanadium, tungsten. Although the strength of these steels decreases with increasing temperature, the reduction is not as pronounced as with other steels.

When metals are used, under stress, at a temperature greater than 0.5 TmK (the melting point in degrees absolute), creep can take place and lead to fracture, possibly after long periods of time (sometimes after 10^5h or longer) at static stresses much lower than those which will break the specimen under a normal tensile test. Consequently, when specifying a steel for use at elevated temperatures, the design stress value is based on the 0.2% proof stress at lower temperatures or the creep rupture stress for higher temperatures. Examples are shown in **Figure 4** for several high temperature steels.

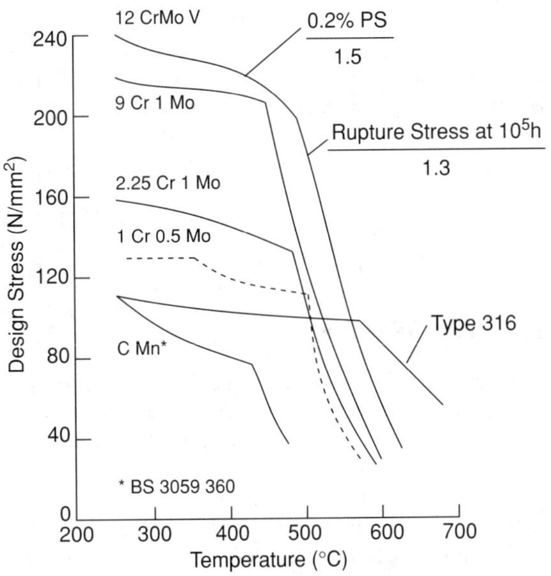

Figure 4. Design stress values for a range of steels.

Thermal fatigue

Temperature gradients are inevitable in many components at elevated temperature and with changes in temperature and hence stress level, localised expansion or contraction can occur. A single severe change can cause cracking due to thermal shock but repeated changes may give rise to cumulative damage, termed *thermal fatigue*.

Thermal expansion

Figure 1 shows that the coefficient of *thermal expansion* of austenitic stainless steels is considerably greater than for other steels. Consequently, when components are manufactured from austenitic stainless steel and another types of steel design, considerations must be taken to ensure warping or fracture will not occur.

Oxidation

Steels will oxidise when exposed to an oxidising environment (e.g. air or flue gas) at elevated temperature. The maximum operating temperature depends on the chromium content and consequently stainless steels offer greater oxidation resistance than plain carbon or low alloy steels.

Cryogenic operation

Selected steels also possess excellent impact properties at extremely low temperatures. Fine grain C-Mn steels can be used down to about −50°C and a range of nickel steels (2.25, 3.5 and 9% Ni) can be used down to liquid nitrogen temperature (−196°C). For even lower temperatures, austenitic stainless steels can be used to hold liquid helium at 4.2K, with the added bonus that the polished, highly reflective, low corrosivity surface finish can be obtained to reduce heat gains from radiation.

Welding

To prevent cracking of many alloy steels during welding it is necessary to pre- and post-heat to ensure a slow cooling rate. The necessity for pre-heat is dependent upon the section thickness and the carbon equivalent (CE) of the steel where

$$CE = C + \frac{Mn}{6} + \frac{Cr + Mo + V}{5}$$

Although stainless steels are not prone to cracking, their low thermal conductivity and high thermal expansion necessitate the use of clamps or tack welding the joint prior to welding to minimise distortion.

References

Llewellyn, D. T. (1992) *Steels: Metallurgy and Application*, Butterworth-Heinemann.

Pickering, F. B. (1991) *Materials Science and Technology, Vol. 7* Constituents and properties of steels, Weinheim.

ASM Metals Handbook, 10th Edition (1990) Vol. 1. Properties and Selection: Irons, steels and high performance alloys.

Smithells Metals Reference Book (1992) Brandes, E. A. and Brook, G. B., eds., Butterworth Heinemann.

Rothman, M. F. (1987) *High-Temperature Property Data: Ferrous Alloys,* ASM International.

Thelming, K. E. (1984) *Steel and Its Heat Treatment, 2nd Edition,* Butterworths.

ASM Handbook Vol. 4, 10th Edition (1991) Heat Treating, ASM International.

Van de Voort, G. F. (1991) *Atlas of Time Temperature Diagrams for Irons and Steels,* ASM International.

K.W. Tupholme

STEFAN-BOLTZMANN CONSTANT (see Environmental heat transfer)

STEFAN-BOLTZMANN LAW

Following from: *Radiative heat transfer*

Stefan-Boltzmann's law relates the integral of the spectral hemispherical density of the radiant flux with the temperature of isothermal black surface. Proceeding from the quantum theory of radiation transfer it has been shown that the spectral and hemispherical density of the radiant flux from the isothermal black surface in vacuum is expressed by Planck's formula

$$e_{\lambda b} = \pi i_{\lambda b} = \frac{2\pi c_1}{\lambda^5(e^{c2/\lambda T} - 1)}, \qquad (1)$$

where $i_{\lambda b}$ is the radiation intensity of the black body (*Planck's function*), T the absolute temperature, the constants $C_1 = 0.59544\ 10^{-16}$ Wm^2, $C_2 = 1.4388\ 10^{-2}$mK, $C_1 = 2hC_0^2$, and $C_2 = hC_0/k$, where C_0 is the velocity of light in vacuum, $h = 6.626210^{-34}$ Js Planck's constant, and $k = 1.380610^{-23}$ J/K Boltzmann's constant.

In Equation 1 radiant flux $e_{\lambda b}$ and the radiation intensity $i_{\lambda b}$ are related by

$$e_{\lambda b} = \int_{2\pi} i_{\lambda b} \cos(\Omega \cdot n) d\Omega = \pi i_{\lambda b},$$

where $d\Omega = \sin\theta\, d\theta\, d\vartheta$ is the element of a solid angle along Ω, n the direction of a normal to the surface element.
The maximum value of Planck's function is achieved at $\lambda_{max}T = C_3 = 2.8978\ 10^{-3}$ mK (Wien's displacement law).

Depending on the frequency $\nu = C_0/\lambda$ or the wave number $\omega = 1/\lambda$ the radiation intensity of the black body in a unit interval ν or ω is

$$i_{\nu b} = \frac{2C\nu^3 d\nu}{C_0^4(e^{C_2\nu/C_0T} - 1)}, \qquad i_{\omega b} = \frac{2C_1\omega^3}{e^{C_2\omega/T} - 1}.$$

Moreover, the balance relations

$$i_{\lambda b} d\lambda = i_{\nu b} d\nu = i_{\omega b} d\omega$$

are retained.

The hemispherical surface intensity of radiant flux integrated over the entire spectrum (Stefan-Boltzmann's law) is equal to

$$e_b = \int_0^\infty e_{\lambda b} d\lambda = \sigma T^4,$$

where σ is Stefan-Boltzmann's constant ($\sigma = 2C_1\pi^5/15C_2^4 = 5.669\ 10^{-8}$ W/m^{2}K^4).

The tables, as a rule, present the values of a fraction of hemispherical density of the radiant flux transferred in the wavelength interval from 0 to λ

$$F_{0-\lambda} = \frac{1}{\sigma T^4}\int_0^\lambda e_{\lambda b} d\lambda = \frac{1}{\sigma}\int_0^{\lambda T} \frac{e_{\lambda b}}{T^5} d(\lambda T)$$

determined as a function of the variable λT. Then the fraction of the density of the radiant flux in an arbitrary interval $[\lambda_1, \lambda_2]$ is determined as

$$F_{\lambda_1-\lambda_2} = F_{0-\lambda_2 T} - F_{0-\lambda_1 T}$$

If the surface element is in a medium other than vacuum and this medium has the refractive index n, Planck's function per unit interval of wavelengths in vacuum takes the form

$$i_{\lambda bn} = \frac{2\pi C_1 n^2}{\lambda^5(e^{C_2/\lambda T} - 1)} = n^2 i_{\lambda b}$$

and Wien's and Stefan-Boltzmann's laws take the form

$$\lambda_{max}T = n\lambda_{max,n}T = C_3, \qquad e_{bn} = n^2\sigma T^4,$$

where $\lambda_{max,n}$ is the wavelength in a given medium.

The radiation heat flux from an isothermal nonblack surface in vacuum is calculated by the formula

$$\dot{q}_{R\lambda} = \varepsilon_\lambda e_{b\lambda},$$

where $\varepsilon_\lambda = \varepsilon_\lambda(T)$ is the hemispherical emissivity of the black surface made of a given material.

I.G. Zaltzman

STEFAN, JOSEF (1835–1893)

An Austrian physicist, whose original contributions ranged over several important fields, including the kinetic theory of gases, hydrodynamics and in particular, radiation, Stefan was born on March 24, 1835 at St. Peter near Klagenfurt and died on January 7, 1893 in Wien.

Stefan was educated at the University of Wien, receiving his doctor of philosophy in 1858, then became Privatdozent in Mathematical Physics, in 1863 Professor Ordinarius of Physics and in 1866 Director of the Physical Institute. He was a distinguished

Figure 1.

member of the Academy of Sciences Wien, of which he was appointed secretary in 1875. Before Stefan's work, G. R. Kirchhoff had already described the perfect radiator as the "*Perfect Black Body*", namely, one that absorbed all the radiation that fell on it and reflected none, but emitted radiation of all wave lengths. Stefan showed empirically in 1879 that the radiation of such a body was proportional to the fourth power of its absolute temperatures, a relationship known as the "**Stefan and Boltzmann**" law after it had been deduced by L. Boltzmann in 1884 from thermodynamic considerations.

In the year 1891 Stefan published his work on the formation of ice in the Polar Seas, giving a special solution of this non-linear conduction problem with phase change (the more general solution being due to F. Neumann).

U. Grigull, M. Sandner and J. Straub

STEFAN-MAXWELL EQUATIONS (see Maxwell-Stefan equations)

STEFAN'S LAW (see Environmental heat transfer)

STEWARTSON TYPE FLOW (see Rotating disc systems, applications)

STIELTJES' INTEGRAL (see Integrals)

STIRRED TANK REACTOR (see Agitation devices)

STIRRED TANKS (see Extraction, liquid-liquid)

STIRRED VESSEL PHASE INVERSION (see Liquid-liquid flow)

STOCHASTIC DIFFERENTIAL EQUATIONS (see Stochastic process)

STOCHASTIC PROCESS

A stochastic process is a process which evolves randomly in time and space. The dispersion of contaminants in gases and liquids, **Brownian Motion** and hydrodynamic **Turbulence** are well known examples, though all dynamical systems are stochastic to a lesser or greater degree. The randomness can arise in a variety of ways: through an uncertainty in the initial state of the system; the equation motion of the system contains either random coefficients or forcing functions; the system amplifies small disturbances to an extent that knowledge of the initial state of the system at the micro-molecular level is required for a deterministic solution (this is a feature of **Non-Linear Systems** of which the most obvious example is hydrodynamic turbulence).

As such, a system will evolve either temporally or spatially or both in a variety of ways, and to each outcome there is assumed to exist a unique probability of occurrence. More precisely if x(t) is a random variable representing all possible outcomes of the system at some fixed time t, then x(t) is regarded as a measurable function on a given probability space and when t varies one obtains a family of random variables (indexed by t) i.e. by definition a stochastic process, or a random function x(.) or briefly x.

Just as differential equations can be used to study the behaviour of deterministic processes, so they can be used to study the behaviour of stochastic processes. However, the theory of *"stochastic" differential equations* is concerned with probabilistic aspects of the process the equations describe: the explicit form of the solution of the equations is often useful but not essential. More precisely, one is interested in the determination of the distribution of x(t) (the *probability density function* (pdf) of x(t) or joint distributions at several instants or alternatively one seeks averages or moments associated with the pdf. Such averages are often referred to as *ensemble averages* to distinguish them from time averages associated with some function of x(t) as the system evolves over some sufficiently large period of time. (See **Ergodic Process**).

As an example of a stochastic differential equation consider the equation frequently used to represent a simple **diffusion** process x(t), namely

$$dx(t) = \mu(x(t), t)dt + \sqrt{2D(x(t), t)}\, dW(t) \qquad \textbf{(1)}$$

where μ and D are non random functions and W(t) is a white noise (non-differentiable) function with the property that dW(t) is distributed normally with zero mean and variance $\langle(dW)^2\rangle = dt$. Note that because W(t) is non-differentiable, the equation of motion cannot be represented as a standard differential equation. However, the equation for P(z,t), the pdf for the occurrence of a particular value z of x(t) at time t is a parabolic partial differential equation, namely

$$\left(\frac{\partial}{\partial t} + \frac{\partial}{\partial z}\mu(t, z) - \frac{\partial^2}{\partial z^2}D(z, t)\right)p(z, t) = 0 \qquad \textbf{(2)}$$

The equation is commonly referred to as the *Fokker-Planck equa-*

tion in physical applications. The particular class of stochastic equations in **Equation 1** which includes the well known *Langevin equation for Brownian motion* has been used extensively to model atmospheric dispersion (MacInnes and Bracco, 1992), particle dispersion in turbulent flows (see **Particle Transport in Turbulent Fluids**) and as an analogue equation to generate fluid velocities in turbulent flows (Pope, 1990).

References

MacInnes, J. M., and Bracco, F. V. (1992) Stochastic particle dispersion modeling and the tracer-particle limit, *Phys. Fluids* A, 4, 2809–2824.

Pope, S. B. (1990) The velocity-dissipation probability density function model for turbulent flows, *Phys. Fluids A.* 2, 1437–1449.

Useful Introductory Texts on Stochastic Processes

van Kampen, N. G. (1981) *Stochastic Processes in Physics and Chemistry,* North-Holland Publishing Company, Amsterdam.

Wax N. (1954) *Noise and Stochastic Processes,* Dover Publications, Inc., New York.

Leading to: *Stokes-Einstein equation; Ergodic process; Particle transport in turbulent fluids*

M.W. Reeks

STOICHIOMETRIC COMBUSTION (see Flames)

STOKES-EINSTEIN EQUATION

Following from: *Stochastic process; Brownian motion*

The Stokes-Einstein equation is the equation first derived by Einstein in his Ph.D thesis for the diffusion coefficient of a "Stokes" particle undergoing **Brownian Motion** in a quiescent fluid at uniform temperature. The result was formerly published in Einstein's (1905) classic paper on the theory of Brownian motion (it was also simultaneously derived by Sutherland (1905) using an identical argument). Einstein's result for the diffusion coefficient D of a spherical particle of radius a in a fluid of dynamic viscosity η at absolute temperature T is:

$$D = \frac{\tilde{R}T}{N_A}\frac{1}{6\pi\eta a}$$

where $\tilde{R}$ is the gas constant and N_A is **Avogadro's Number.** The formula is historically important since it was used to make the first absolute measurement of N_A so confirming molecular theory. Although the formula can be derived alternatively using the Langevin equation of motion for a Brownian particle (Chandrasekhar, 1943) the derivation of Einstein is a powerful and ingenious one, correct even when the Langevin equation is only approximate. Einstein assumed that *van't Hoff's law* for the osmotic pressure exerted by solute molecules in a solvent fluid at equilibrium was equally applicable to the pressure p associated with a suspension of Brownian particles at equilibrium in the same fluid, i.e.

$$p = fn_M\tilde{R}T$$

where n_M is the number of gram moles of fluid per unit volume and f the 'molar fraction' defined here as the ratio of the number of particles to the number of fluid molecules. Einstein then argued that a suspension of Brownian particles at equilibrium under their own weight could be viewed in two ways both of which were equivalent: a balance between the net weight of the particles and the gradient of the particle pressure in the direction of gravity; or a balance between the diffusion flux and the settling flux due to gravity. Using the Stokes drag formula for the settling velocity (see **Stokes Law**) and the formula for p above gives the formula for D given above. A similar argument allows one to deduce a form for the particle pressure in a turbulent gas knowing the form for the particle turbulent diffusion coefficient (see **Particle Transport in Turbulent Fluids**)

References

Chandrasekhar, S. (1943) *Rev. Mod. Phys.,* 15, 1.

Einstein, A. (1905) *Ann. der Physik,* 17, 549.

Sutherland, W. (1905) *Phil. Mag.,* 9, 781.

M.W. Reeks

STOKES-EINSTEIN EQUATION, FOR DIFFERENTIAL COEFFICIENTS IN LIQUIDS (see Diffusion coefficient)

STOKES EQUATION (see Flow of fluids)

STOKES FLOW (see Flow of fluids; Immersed bodies, flow around and drag; Particle transport in turbulent fluids)

STOKE'S FORMULA (see Integrals)

STOKES' LAW FOR SOLID SPHERES AND SPHERICAL BUBBLES

Introduction

Stokes' Law is the name given to the formula describing the force F on a stationary sphere of radius a held in a fluid of viscosity η moving with steady velocity V. This is usually expressed in the form,

$$F = 6\pi\eta Va \qquad (1)$$

By translation, this result also applied to a sphere moving with steady velocity V in an otherwise stagnant fluid. **Equation 1** can more conveniently be expressed in terms of a drag coefficient and a Reynolds number defined as follows,

$$C_D = \frac{F}{\frac{1}{2}\rho V^2 \pi a^2} \tag{2}$$

$$Re = \frac{2aV\rho}{\eta} \tag{3}$$

In terms of these variables, **Equation 1** takes the form,

$$C_D = \frac{24}{Re} \tag{4}$$

Equations 1 and 4 are however applicable for slow flows and should only be used for $Re < 1$. The reasons for this and the results for higher Reynolds numbers are discussed below.

For a spherical gas bubble the corresponding results are,

$$F = 4\pi\eta Va \tag{5}$$

and

$$C_D = \frac{16}{Re}. \tag{6}$$

These results can be derived as follows.

Derivation

It is more convenient to analyse the situation in which the sphere is held stationary in a fluid that moves with velocity V at great distances from the sphere. Since Stokes Law is restricted to slow steady flow we can begin with the Navier Stokes equation and omit both the time-dependent and inertial terms giving,

$$\mathbf{grad}\ p = \eta\nabla^2\mathbf{v} \tag{7}$$

The situation is axially symmetrical and therefore there are two components of equation 7. Eliminating the pressure between these two components and expressing the resulting equation in terms of the *Stokes stream function* ψ, gives,

$$E^2(E^2\psi) = 0 \tag{8}$$

where E^2 is the operator,

$$E^2 = \frac{\partial^2}{\partial r^2} + \frac{\sin\theta}{r^2}\frac{\partial}{\partial\theta}\left(\frac{1}{\sin\theta}\frac{\partial}{\partial\theta}\right) \tag{9}$$

For a solid sphere the boundary conditions are that the radial and tangential velocities on the surface are zero by the no-slip condition, i.e.

$$v_r = 0 \quad \text{on} \quad r = a \tag{10}$$

$$v_\theta = 0 \quad \text{on} \quad r = a \tag{11}$$

and that the velocity tends to V at great distances, i.e.,

$$\psi \to \frac{V}{2} r^2 \sin^2\theta \quad \text{as} \quad r \to \infty \tag{12}$$

The form of these boundary conditions suggests a solution of the form $\psi = f(r)\sin^2\theta$ and the only possible result is

$$\psi = \left(Ar^4 + Br^2 + Cr + \frac{D}{r}\right)\sin^2\theta \tag{13}$$

Putting in boundary conditions 10–12 gives $A = 0$, $B = V/2$, $C = -3Va/4$ and $D = Va^3/4$. Thus,

$$\psi = \left(\frac{Vr^2}{2} - \frac{3Var}{4} + \frac{Va^3}{4r}\right)\sin^2\theta \tag{14}$$

Hence we can obtain the velocity components,

$$v_r = V\left(1 - \frac{3a}{2r} + \frac{a^3}{2r^3}\right)\cos\theta \tag{15}$$

$$v_\theta = -V\left(1 - \frac{3a}{4r} - \frac{a^3}{4r^3}\right)\sin\theta \tag{16}$$

For a spherical gas bubble, boundary conditions 10 and 12 still apply, but boundary condition 11 has to be replaced by the condition that there is no shear stress on the surface. Hence the shear strain rate $\gamma'_{r\theta}$ is zero where $\gamma'_{r\theta}$ is defined by,

$$\gamma'_{r\theta} = r\frac{\partial}{\partial r}\left(\frac{v_\theta}{r}\right) + \frac{1}{r}\frac{\partial v_r}{\partial\theta} \tag{17}$$

Under these circumstances $A = 0$, $B = V/2$, $C = -Va/2$ and $D = 0$, giving,

$$\psi = \left(\frac{Vr^2}{2} - \frac{Var}{2}\right)\sin^2\theta \tag{18}$$

$$v_r = V\left(1 - \frac{a}{r}\right)\cos\theta \tag{19}$$

$$v_\theta = -V\left(1 - \frac{a}{2r}\right)\sin\theta \tag{20}$$

The shear stress on the surface can be obtained from,

$$\tau_{r\theta} = \eta\gamma'_{r\theta} = \eta\left(\frac{2B}{a} + \frac{2C}{a^2} - \frac{4D}{a^4}\right)\sin\theta \tag{21}$$

For a solid sphere,

$$\tau_{r\theta} = -\frac{3V\eta}{2a}\sin\theta \tag{22}$$

and for a bubble, $\tau_{r\theta}$ is obviously zero.

The skin friction drag F_s is given by,

$$F_s = \int_0^{\pi} (\tau_{r\theta} \sin\theta).2\pi a \sin\theta.a\, d\theta \tag{23}$$

from which we find that

$$F_s = 4\pi\eta Va \tag{24}$$

for a solid sphere and zero for a bubble.

The pressure distribution can be found by substituting the expression for the velocity components into the Navier Stokes equation, giving,

$$p = p_0 - \eta\left(\frac{4B}{a^2} + \frac{2C}{a^3} + \frac{4D}{a^5}\right)\cos\theta \tag{25}$$

where p_0 is an arbitrary constant.

For a solid sphere,

$$p = p_0 - \frac{3\eta V}{2a}\cos\theta \tag{26}$$

For a bubble,

$$p = p_0 - \frac{\eta V}{a}\cos\theta \tag{27}$$

The normal stress on the surface is given by,

$$\sigma_{rr} = 2\eta\varepsilon'_{rr} - p \tag{28}$$

where ε'_{rr} is the normal strain rate $\partial v_r/\partial r$.

For a solid sphere,

$$\sigma_{rr} = -p_0 + \frac{3\eta V}{2a}\cos\theta \tag{29}$$

and for a bubble,

$$\sigma_{rr} = -p_0 + \frac{3\eta V}{a}\cos\theta \tag{30}$$

The form drag F_F is given by,

$$F_F = \int_0^{\pi} (\sigma_{rr}\cos\theta).2\pi a \sin\theta.a\, d\theta \tag{31}$$

giving,

$$F_F = 2\pi\eta Va \tag{32}$$

for a solid sphere, and

$$F_F = 4\pi\eta Va \tag{33}$$

for a bubble.

These results can be tabulated thus,

	Form drag	Skin friction drag	Total drag
Solid sphere	$2\pi\eta Va$	$4\pi\eta Va$	$6\pi\eta Va$
Bubble	$4\pi\eta Va$	0	$4\pi\eta Va$

giving the results presented above.

Discussion

It must be emphasized that these results are applicable only for low Reynolds numbers. This is because the inertial terms $(v.\nabla)v$ have been omitted from the analysis. An extension to the theory, known as Ossen's approximation, can be obtained by replacing these inertial terms by $(V.\nabla)v$ which gives rise to the result

$$C_D = \frac{24}{Re}\left(1 + \frac{3}{16}Re - \frac{19}{1280}Re^2 + \cdots\right) \tag{34}$$

In fact, this provides an over-correction and the empirical result,

$$C_D = \frac{24}{Re}(1 + 0.15\,Re^{0.687}) \tag{35}$$

gives a good correlation of the experimental results up to a Reynolds numbers of about 1000. The reader is referred to Clift et al. (1972) for alternative correlations and extensions to even higher Reynolds numbers.

For bubbles in non-polar liquids, **Equation 6** may be used up to a Reynolds number of about 1.5. However, polar liquids, such as water, are prone to contamination by surface active agents, which collect on the surface of the bubble, effectively immobilising the surface. Such a surface can support a shear stress and bubbles in polar liquids behave as solid spheres. Indeed circumstances can arise in which bubbles obey the result for solid spheres over a very much larger range of Reynolds numbers than solid spheres themselves. Details of the behaviour of bubbles are given by both Clift et al. (1972) and Wallis (1974). Wallis' correlation is probably the most reliable and convenient currently available.

These results can be used for accelerating spheres and bubbles without much loss of accuracy, but care must be exercised if the accelerations are large as then the *Basset history term* must be included, see Clift et al. (1972).

References

Clift, R., Grace, J. R. and Weber, M. E. (1972) *Bubbles, Drops and Particles*, Academic Press, New York.

Wallis G. B. (1974) *Int. J. Multiphase Flow*, 1, 491–511.

R.M. Nedderman

STOKES LENGTH (see Particle transport in flow of fluids)

STOKES PARADOX (see Flow of fluids)

STOKES SHIFT (see Photoluminescence)

STOKES STREAM FUNCTION (see Stokes law for solid spheres and spherical bubbles)

STOMATAL CONTROL OF WATER LOSS FROM PLANTS (see Environmental heat transfer)

STOPPING DISTANCE (see Gas-solid flows)

STRAIN (see Conservation equations, single phase; Non-Newtonian fluids)

STRAIN GAUGES

A number of devices are used to measure strain. Among them, the *electrical resistance strain gauges* are, by far, the most commonly used. The electrical resistance of a wire increases when the wire is stretched. The ratio between the change in resistance and the corresponding percent increase in length is called the gauge factor, k,

$$k = \frac{\%\ \text{change in resistance}}{\%\ \text{change in length}} = \frac{\Delta R/R}{\varepsilon} \tag{1}$$

For the metals in common use, k varies between 0–.5 and 5. Semiconductor gauges of silicon or germanium have gauge factors as high as 150.

A resistance strain gauge consists of a conductor, bonded to a carrier which is, in turn, fixed on to the structure or machine (base). The carrier may be in the form of a foil, a sheath or a frame and it may be cemented or welded to the base (**Figure 1**). The sheath and frame types are usually referred to as unbonded gauges.

When the temperature of gauge and base change by ΔT, the gauge grid stretches by $\alpha\Delta T$ while the carrier and base stretch by $\beta\Delta T$ α and β being respectively the thermal expansion coefficients of the grid and of the base metal. At the same time, the change in temperature results in a change in resistivity of $\gamma\Delta T$ (ohm/ohm). Assuming that the base is far more rigid than the gauge itself, the strain to which the gauge grid is subjected is $(B - \alpha)\ \Delta T$ and the total change in resistance is,

$$\Delta R/R = [(\beta - \alpha)k + \gamma]\Delta T \tag{2}$$

In some commercially available gauges, the parameters α and γ are chosen in such a way that the term in brackets in the previous expression is zero, for a given base material. Such gauges are called self-compensated.

The change in resistance is measured by either a Wheatstone bridge or a potentiometer circuit. The former is used for static measurements (null balance method in general) and dynamic measurements (deflection method). The latter is only used for dynamic measurements (with temperature compensated gauges usually).

In the bridge shown in **Figure 2**, the terminals are connected to input source (V), BD to output (E). Arms' resistance AB = R_1, BC = R_2, CD = R_3, DA = R_4. Output is given by,

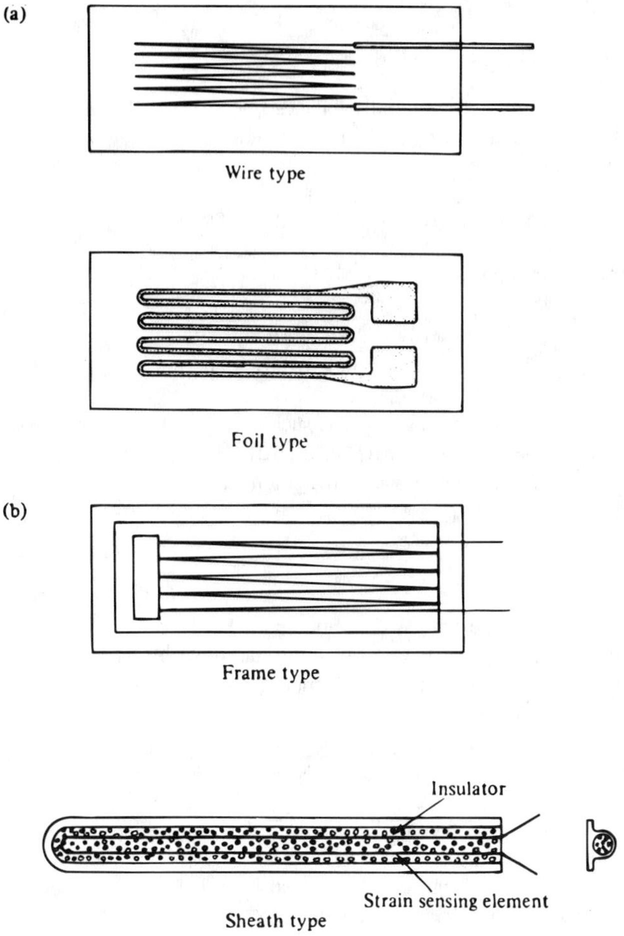

Figure 1. Resistance strain gauges of the bonded type (a), and unbonded type (b).

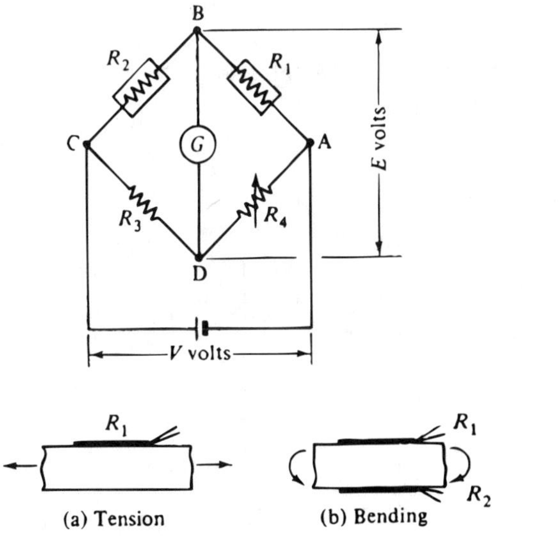

Figure 2. Wheatson bridge circuit for the measurement of resistance changes in strain gauges.

$$E = \frac{R_1R_3 - R_2R_4}{(R_1 + R_2)(R_3 + R_4)} V \quad (3)$$

When the bridge is initially balanced, $R_1R_3 = R_2R_4$. In the null balance method the bridge is continuously balanced in such a way that,

$$(R_1 + \Delta R_1)(R_3 + \Delta R_3) = (R_2 + \Delta R_2)(R_4 + \Delta R_4) \quad (4)$$

For small changes,

$$\frac{\Delta R_1}{R_1} + \frac{\Delta R_3}{R_3} = \frac{\Delta R_2}{R_2} + \frac{\Delta R_4}{R_4} \quad (5)$$

If R_4 is the measuring resistor, such as a precision decade box, potentiometer or slide wire, for a quarter bridge configuration in which R_1 is the strain gauge, the strain is,

$$e = \frac{1}{K}\frac{\Delta R_4}{R_4} \quad (6)$$

In a half bridge with two strain gauges R_1, R_2 normally used to measured bending, the tensile strain measured by R_1 is equal in magnitude to the compressive strain measured by R_2,

$$e = \frac{1}{2}\frac{1}{K}\frac{\Delta R_4}{R_4} \quad (7)$$

In the deflection method the output E is measured by means of a millivoltmeter or resistance R_L. In general, R_1, R_2, R_3 and R_4 are approximately equal and much smaller than R_L. It can then be shown that for a quarter bridge,

$$e = \frac{4E}{kV} \quad (8)$$

and for the half bridge,

$$e = \frac{2E}{kV} \quad (9)$$

In the potentiometer circuit (**Figure 3**) the gauge is in series with a ballast resistor. Input terminals A, C, output B, C. Resistance $AB = R_2$, $BC = R_1$ (gauge). Open circuit output is

$$E = V\frac{R_1}{R_1 + R_2} \quad (10)$$

The circuit cannot be balanced, so that the initial reading of the measuring instrument (large impedance) is E. Upon application of strain,

$$E + \Delta E = V\frac{R_1 + \Delta R_1}{R_1 + R_2 + \Delta R_1} \simeq V\frac{R_1 + \Delta R_1}{R_1 + R_2} \quad (11)$$

hence

$$\Delta E = V\frac{\Delta R_1}{R_1 + R_2} \quad (12)$$

Since V is of the order of 10 V, ΔR of the order of a few mV, the measuring instrument must have high resolution plus extended full-scale range. For this reason, only dynamic measurements are possible with DC current and filter to block the steady state output, letting only ΔE through (ΔE cyclic).

The nominal resistance of most strain gauges is 120 Ω. For a strain sensitivity of 10^{-5} (10 microstrain) and a maximum strain of 10^{-2} (1%) with a voltage input of 10 V, the output of a Wheatstone bridge will only be about 100 mV full scale deflection with 0.1 mV sensitivity.

The material used for the strain gauges is Cu/Ni alloy up to temperatures of 300°C. For higher temperatures, Pt may be used. The base may be epoxy, polyester or, in wire strain gauges, resin impregnated paper. A variety of cements, ranging from epoxies to cyanocrylates are available. There is a large variety of gauge materials, base materials and cements and for the selection of a commercially available system it is necessary to consult suppliers' catalogues.

Protection of the strain gauge against the environment is also an important issue.

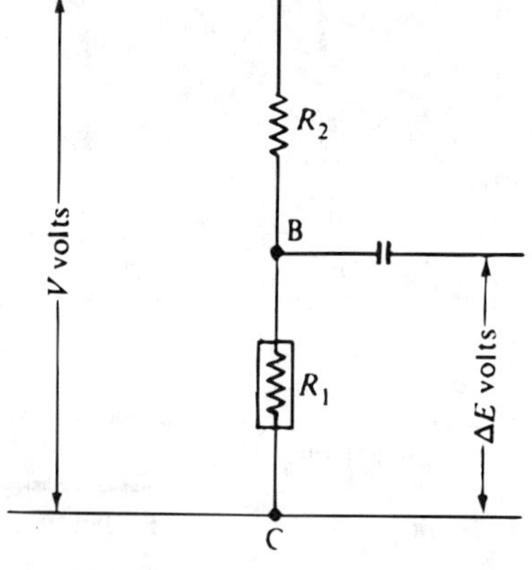

Figure 3. Potentiometer circuit for strain gauges.

References

Dally, J. W. and Riley, W. F. 1965. *Experimental Stress Analysis,* McGraw-Hill.

Kobayashi, A. S. 1991. *Handbook of Experimental Mechanics,* Soc. Exp. Mech.

C. Ruiz

STRAIN RATE (see Newtonian fluids; Non-Newtonian fluids)

STRANGE ATTRACTORS (see Non-linear systems)

STRATIFICATION, UNSTABLE AND STABLE (see Mixed, combined, convection)

STRATIFIED GAS-LIQUID FLOW

Following from: *Two-phase flow*

When gas and liquid flow in a pipe, there are particular conditions for which the two phases are separated from each other by a continuous interface. Such a flow is dominated by the gravity force which causes the liquid to stratify at the bottom of the pipe (**Figure 1**). This flow pattern can be observed in horizontal or slightly inclined pipelines. It is characterised by the structure of the interface which may be smooth or wavy according to the gas flow rate. At low gas velocity, the interface is smooth or may be rippled by small *capillary waves* of a few millimetre length. With increasing gas velocity, small amplitude regular *waves* appear. At high enough velocity of gas, droplets can be entrained from the large amplitude, irregular waves and can be deposited at the wall or at the interface. However, this **Atomization** phenomenon is out of the scope of this entry (see **Annular flow; Dispersed flow**).

In stratified flow, the determination of pressure drop and liquid hold-up requires an accurate prediction of the friction at the wall and at the interface. Indeed, for fully developed flow, the pressure drop is controlled by the friction at the wall of each phase and by the weight of the liquid, which is related to the hold-up. As a consequence, pressure drop and liquid hold-up are strongly coupled in stratified flow: they must be predicted simultaneously. Another interesting feature of stratified flow comes is that the difference of velocity between phases can be high, suggesting that the momentum transfer between phases is ineffective as a mechanism for the gas to drive the liquid. As this transfer is controlled by the interfacial friction, it may be anticipated that it will play a central role in the flow modelling.

At low enough velocity the interface is maintained smooth by gravity and surface tension. When the gas velocity increases waves develop at the surface through a **Kelvin-Helmholtz Instability** mechanism. These waves act upon the gas as interfacial roughness and the interfacial stress becomes larger than over a smooth and flat solid surface. That interfacial friction depends on interfacial roughness, and roughness depends itself on the phase velocities, makes the problem difficult to solve. This is a twofold problem which is regarded as the central issue of stratified flow modelling.

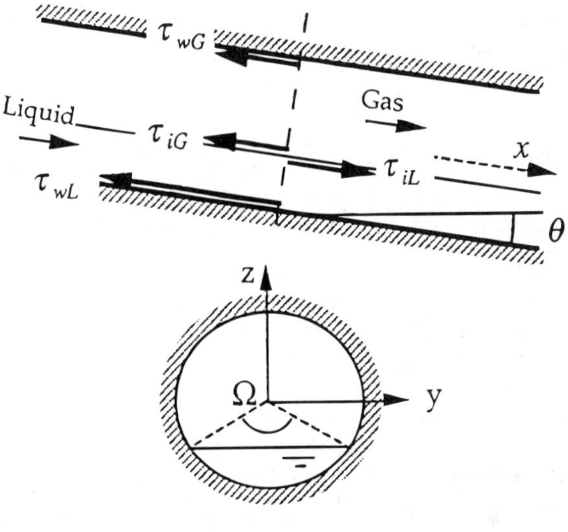

Figure 1. Gas-liquid stratified flow.

Different approaches have been explored to solve this problem. An empirical, but effective, approach correlates the interfacial stress with the mean phase velocities and fluid properties: by similarity to single-phase turbulent flow over rough surfaces, it is possible to correlate the interfacial friction factor to the wave roughness experienced by the gas. A more recent approach involves numerical flow simulation to predict the wave drag over simplified interfacial shapes, the monochromatic wave being the simplest case to be studied. As pointed out by Hanratty and McCready (1992), "a critical physical problem is to reconcile these approaches so as to produce a unified theory" of interfacial transfer of momentum in stratified two-phase flows.

In the first part of the entry, *wave generation* and wave pattern are presented. In the second part, modelling of wavy stratified flow is introduced. The closure relations of wall and interfacial friction factor coefficients are given in the third part. In the last part, an alternative approach of interfacial momentum transfer modelling is discussed.

Generation of Waves and Wave Patterns

Cohen and Hanratty (1965) have shown that the waves receive their energy from the gas through the work of pressure and shear stress perturbations induced by the waves. Andritsos (1986) observed that for a low viscosity liquid, the first waves generated are small amplitude regular two-dimensional waves: they receive energy from the pressure perturbations in the gas which are in phase with the wave slope. The energy transfer being induced by a sheltering mechanism, these waves will be called J-waves, referring to *Jeffrey's sheltering hypothesis* (1925). With increasing gas velocity and for given liquid flow rate, the wave amplitude becomes larger and the wavelength decreases: hence the wave steepness increases. These waves receive energy from the pressure perturbations in the gas which are in phase with the wave height. Surprisingly, Andritsos found that these large amplitude waves are generated at flow conditions corresponding to the occurrence of Kelvin-Helmholtz instability. These waves will be thus called KH-waves. At higher gas velocity, irregular large amplitude waves are formed with steep front and gradually sloping back: they are called *roll waves.* They may break for large enough amplitude. Roll waves are widely spaced from each other with intermittent occurrence. There is no theoretical prediction of this intermittence. Only a few measurements are available (Vlachos et al., 1993 and Strand, 1993). The wave pattern depend on the liquid viscosity: Andritsos has observed that the domain of J-waves vanishes for high viscosity liquids.

The initiation of J-waves can be estimated for low viscosity liquids by the criterion proposed by Taitel and Dukler (1976):

$$u_{G,TD} = \sqrt{\frac{4\nu_L \Delta\rho g}{s\rho_G u_L}} \tag{1}$$

where u, ρ, ν are velocity, density, kinematic viscosity, g is acceleration due to gravity and s is a sheltering coefficient which is about 0.06. As indicated by Andritsos the transition to KH-waves coincides with the inviscid Kelvin-Helmholtz relation:

$$j_{G,KH} = j_L + \left[\frac{k\sigma}{\rho_G} + \frac{\rho_L g}{\rho_G k}\right]\tanh(kh) \quad (2)$$

where k is the wave number, σ the surface tension, h the liquid height and j the volumetric flux. The critical value corresponds to the wave number k for which the velocity is minimum:

$$k = \sqrt{\frac{\rho_L g}{\sigma}} \quad (3)$$

The wave patterns are plotted for a large range of gas and liquid superficial velocities in **Figure 2**.

The magnitude of the wave drag increases with the amplitude and the wave number. For KH waves, a large increase of the interfacial transfer is observed (Andritsos, 1986, Strand, 1993, Lopez, 1994) suggesting that the drag must be related to the wave steepness. (See also **Waves in Fluids** and **Wavy Flow**).

Modelling of Stratified Two-Phase Flow

Suppose that we know the specific flow conditions, namely:

- the pipe size, i.e. its diameter D or its cross-section area A $= \pi D^2/4$, and the inclination θ,
- the fluid properties, i.e. the densities ρ_k (k = L,G), the kinematic viscosities ν_k and the surface tension σ,
- the superficial velocity—or volumetric flux—of each phase, $j_k = Q_k/A$.

A useful model of stratified flow would produce at least the following informations:

- the mean velocities of both gas and liquid μ_k,
- the phase fractions of both gas and liquid ε_k,
- the wall and interfacial stresses,
- the pressure gradient dp/dx.

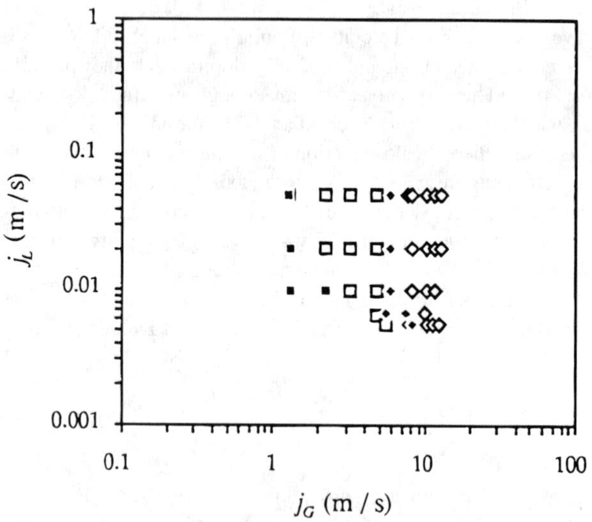

Figure 2. Wave patterns in stratified flow in a pipe after Lopez (1994). ■ smooth interface, □ 2D waves, ♦ 3D waves, ◇ roll waves

For sake of simplicity, let us restrict the study to fully-developed steady flow and assume that the gas is incompressible. The mass balance for each phase reduces to:

$$\dot{V}_k = Aj_k = A\varepsilon_k u_k \quad (4)$$

where $\dot{V}$ is the volumetric flow rate. The phase fractions are related by the trivial expression:

$$\varepsilon_G + \varepsilon_L = 1 \quad (5)$$

The momentum balance in each phase and acros the interface may be written:

$$-\varepsilon_k \frac{dp}{dx} = \frac{P_{wk}\tau_{wk} + P_i\tau_{ik}}{A} + \rho_k \varepsilon_k g \sin\theta \quad (6)$$

$$\tau_{iG} + \tau_{iL} = 0 \quad (7)$$

where P_w and P_i are the perimeters of the wall and the interface, τ_w and τ_i, the **Shear Stresses** at the wall and the interface and θ, the pipe inclination. The equations for the pressure gradient dp/dx and the liquid hold-up ε_L may be obtained from the two momentum equations:

$$-\frac{dp}{dx} = \frac{P\tau_w}{A} + \rho_M g\sin\theta \quad (8)$$

$$\frac{P_{wG}\tau_{wG}}{A\varepsilon_G} - \frac{P_{wL}J_{wL}}{A\varepsilon_L} + \frac{P_i\tau_{iG}}{A\varepsilon_L\varepsilon_G} - \Delta\rho g \sin\theta = 0 \quad (9)$$

where $P = P_{wL} + P_{wG}$ is the total perimeter wetted by the phases, $\tau_w = (P_{wL}\,\tau_{wL} + P_{wG}\,\tau_{wG})/P$ is the total shear stress and $\rho_m = \rho_G\,\varepsilon_G + \rho_L\,\varepsilon_L$ the density of the mixture.

The wall and interfacial perimeter can easily be related to the phase fractions. For pipe of circular cross-section, they are easily expressed versus the angle Ω which intercept the interface:

$$\varepsilon_L = \frac{\Omega - \sin\Omega}{2\pi} \quad (10)$$

$$P_{wL} = \frac{1}{2}D\Omega, \qquad P_{wG} = \frac{1}{2}D(2\pi - \Omega), \qquad P_i = D\sin\frac{1}{2}\Omega \quad (11)$$

The basic question is the modelling of wall and interfacial shear stresses usually expressed through dimensionless friction factors defined as:

$$f_{wk} = \frac{\tau_{wk}}{\frac{1}{2}\rho_k|u_k|u_k} \qquad f_i = \frac{\tau_{iG}}{\frac{1}{2}\rho_G|u_G - u_L|(u_G - u_L)} \quad (12)$$

Wall and Interfacial Shear Stresses Correlations

The line followed by most authors to deduce the correlations for the wall and interfacial friction factors requires some attention.

In stratified flow, only pressure drop and liquid hold-up can be measured easily. It is thus impossible to deduce from **Equations 8** and 9 the experimental values of the three shear stresses τ_{wG}, τ_{wL}, τ_{iG}. It is generally assumed that the wall shear stress exerted by the gas is correctly predicted by a single-phase flow relation. Once τ_{wG} has been estimated, it is possible to deduce the experimental values of τ_{wL}, τ_{iG}. Fabre et al. (1987) were probably the first to measure the wall shear stress in the gas. Following their conclusion, the use of a single-phase flow relationship is acceptable to predict the wall friction factor for the gas phase. This relation depends on whether the flow is laminar or turbulent. It is extended to two-phase flow as follows:

$$f_{wG} = \frac{16}{Re_G} \quad \text{for laminar flow} \tag{13}$$

$$\frac{1}{\sqrt{f_{wG}}} = 3.48 - 4\log\left(\frac{2k_w}{Dh_G} + \frac{9.35}{Re_G\sqrt{f_{wG}}}\right) \quad \text{for turbulent flow} \tag{14}$$

in which k_w is the sand roughness of the wall, Dh and Re the hydraulic diameter and the Reynolds number defined for the gas as:

$$Dh_G = \frac{4\varepsilon_G A}{P_{wG} + P_i}, \tag{15}$$

$$Re_G = \frac{|u_G| Dh_G}{\nu_G} \tag{16}$$

Similar method may be used for the shear stress of the liquid phase at the wall with a little difference: the hydraulic diameter does not include the interfacial perimeter

$$Dh_L = \frac{4\varepsilon_L A}{P_{wL}} \tag{17}$$

The extension of single-phase flow relation in the liquid phase leads however to some bias (Fabre et al., 1987, Rosant, 1984 and Andritsos, 1986). Cheremisinoff and Davis (1979), Rosant (1984) and Andritsos have proposed empirical correlations for f_{wL}. (See also **Pressure Drop, Single Phase and Two-Phase**)

Nevertheless, it must be kept in mind that the crucial problem remains the closure of the interfacial friction factor f_i. Taitel and Dukler (1976) presented a two-fluid model based on balance equations, simplifying the problem by considering that $f_i = f_{wG}$. In this case the gas-liquid interface is implicitly smooth and the interfacial stress is underestimated, leading to an overestimation of the liquid hold-up ε_L. This is in contradiction with the fact that the interfacial friction factor increases with the gas Reynolds number (**Figure 3**).

Among all the empirical correlations which have been developed for the interfacial friction factor, this of Andritsos and Hanratty (1987) must be recommended as it gives the best results. These authors postulated that there exists a critical gas flux $j_{G,KH}$ given by **Equation 3** below which the interface is hydraulically smooth. Above this critical flux the interface is wavy and the interfacial shear stress is assumed to increase linearly with the difference $j_G - j_{G,KH}$

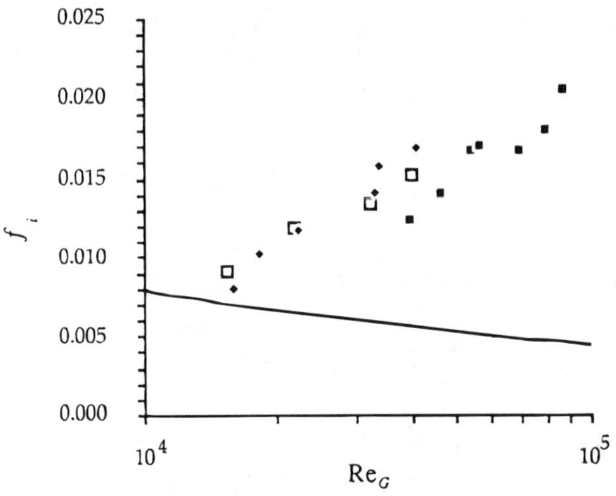

Figure 3. Interfacial friction factor vs gas Reynolds number after Lopez. ■ $j_L = 0.0066$m/s, □ $j_L = 0.02$m/s, ◆ $j_L = 0.05$m/s, ____ Eq. (14)

$$\frac{f_i}{f_{i,smooth}} - 1 = 0 \quad \text{for } j_G \leq j_{G,KH} \tag{18}$$

$$\frac{f_i}{f_{i,smooth}} - 1 = 15\sqrt{\frac{h}{D}}\left(\frac{j_G}{j_{G,KH}} - 1\right) \quad \text{for } j_G \geq j_{G,KH} \tag{19}$$

where $f_{i,smooth}$ is the friction factor for a smooth interface calculated from **Equation 14** with $k_w = 0$. The interfacial friction factor increases when the gas velocity is high enough to generate KH-waves. It must be pointed out that even if it is disturbed by J-waves, the interface is considered to behave like a smooth surface.

Analysis of Interfacial Momentum Transfer

In the previous approach, the wavy structure is only considered through the transition between the smooth and wavy regimes. Another approach considers that the interface is seen by the gas as a surface whose roughness changes with both gas and liquid velocities. If we accept that the momentum transfer across a rough liquid surface is governed by the same mechanism as for a rough wall, it is possible to extend the single-phase flow correlation given by **Equation 14**:

$$\frac{1}{\sqrt{f_i}} = 3.48 - 4\log\left(\frac{2k_i}{Dh_G} + \frac{9.35}{Re_G\sqrt{f_i}}\right) \tag{20}$$

The problem is completely solved provided that the roughness of the interface k_i may be predicted.

A solution ignoring the wave amplitude has been proposed early on by Charnock (1955) for fully developed wave field in deep water. For inviscid fluids, only gravity and pressure forces balance yielding:

$$\frac{k_i g}{u_{*i}^2} = \gamma \tag{21}$$

in which $u_{*i} = (|\tau_{iG}|/\rho_G)^{1/2}$ is the interfacial *friction velocity* and

γ a constant within the range 0.1–0.5. **Equation 21** has to be solved together with **Equation 20**. By taking into account the definition (**Equation 12**), one obtains the following implicit relation which must be solved with an iterative procedure:

$$\frac{1}{\sqrt{f_i}} = 3.48 - 4\log\left(\gamma Fr_G f_i + \frac{9.35}{Re_G\sqrt{f_i}}\right) \quad \textbf{(22)}$$

where Fr_G is a **Froude Number** defined as:

$$Fr_G = \frac{(u_G - u_L)^2}{gDh_G} \quad \textbf{(23)}$$

Another method involves the wave structure. Indeed the interfacial roughness must depend on the different length scales of the wave field. For periodic waves these length scales are the amplitude and the wavelength. For random waves in rectangular channel, Cohen and Hanratty (1968) found that the roughness depends on the r.m.s. $\tilde{h}$ of the instantaneous liquid height:

$$k_i = 3\sqrt{2}\,\tilde{h} \quad \textbf{(24)}$$

However, only waves which emerge from the viscous sublayer can create roughness, so that the correlation of Cohen and Hanratty (1968) must be corrected as suggested by Fabre et al (1987). Nevertheless, this approach does not provide the key to a successful prediction of the interfacial roughness since the prediction of the wave length scales remains an open problem.

In particular, $\tilde{h}$ is related to the distribution of energy among all the wave frequencies ω:

$$\tilde{h}^2 = \int_0^\infty \phi(\omega)d\omega \quad \textbf{(25)}$$

Bruno and McCready (1989) and Jurman and McCready (1989) attempted to determine the wave energy spectrum. Their analysis is based on both energy transfer between wind and waves and non-linear energy transfer between wave components.

Considering the complexity of the wave spectrum, one can simplify the wave field, and focus on the drag of a monochromatic wave. The interfacial transfer accounts for the viscous drag and the form drag. The viscous drag can be expressed through a classical friction factor correlation. The form drag can be expressed through a **Drag Coefficient** C_D accounting for the distribution of pressure over the wave surface. It can be determined analytically from a linear perturbation method, for a wave of small steepness ak, a being the amplitude and k the wave number (Liné & Lopez, 1995):

$$C_D = \frac{ak\,|p|\sin\theta_p}{\frac{1}{2}\rho_G(u_G - u_L)^2} \quad \textbf{(26)}$$

In this equation, $|p|$ and θ_p are the modules and the phase angle of the pressure perturbation at the interface. The modules and the phase angle of the pressure can be estimated from the solution of perturbed flow model above a sinusoidal wave (Abrams, 1984, Harris, 1993, Lopez, 1994) or by direct numerical simulation (Hudson, 1993). However, following Jeffreys (1925), C_D is proportional to the square of the steepness ak:

$$C_D = s(ak)^2 \quad \textbf{(27)}$$

where s is a sheltering coefficient. In a recent theoretical work, Belcher and Hunt (1993) confirmed the relation of the drag coefficient above slowly moving waves as a function of the square of the steepness. In this case, the problem remains to model the characteristics of the dominant wave.

The fundamental issue remains to unify these different approaches of interfacial transfer so as to produce a reliable modelling of the interfacial transfer in separated two-phase flows.

Conclusion

Since the earliest model of Taitel and Dukler (1976), the improvement of gas-liquid stratified flow modelling has been related to the pertinence of the closure relations for the momentum transfers at the interface and at the wall.

On the one hand, the phase fractions and velocities depend on the interfacial momentum transfer, depending itself on the deformations of the gas-liquid interface. On the other hand, the deformations of the interface result from the interactions between gas and liquid. Indeed, the wave field propagates over the surface of the liquid; hence, it cannot solely result from local interactions between gas and liquid flows. Considering this complex problem, two approaches have been developed, the first being based on empirical correlations and the second on a local analysis of the flow. The first approach is useful at solving practical problems, the second is necessary to understand the physical mechanism governing the interfacial interactions.

The two-fluid model presented in the second part of this entry coupled to the closure relations discussed in the third part has the capability of predicting pressure gradient and liquid hold-up in stratified flow.

The information given in the fourth part can be summarised as follows:

- the wave field over the liquid can be considered as an equivalent roughness, the problem being to predict this roughness: none of today's solutions appears satisfactory;
- advanced work concerning the modelling of the wave spectrum is taking place; however this complex problem is not yet solved;
- in the case of a dominant wave, the contributions of the form drag separate from the viscous drag can be estimated; however, the form drag depends on the characteristics of the dominant wave.

Consequently, an improvement of stratified flow models requires a better prediction of wave characteristics and a better understanding of the interaction between "wind" and waves.

References

Abrams, J. (1984) *Modeling of Turbulent Flow over a Wavy Surface,* Ph. D. Thesis, Univ. of Illinois, Urbana, U. S. A.

Andritsos, N. (1986) *Effect of Pipe Diameter and Liquid Viscosity on Horizontal Stratified Flow,* Ph. D. Thesis, Univ. of Illinois, Urbana, U. S. A.

Andritsos, N. and Hanratty, T. J. (1987) Influence of interfacial waves in stratified gas-liquid flows, *AIChE J.*, 33, 444–454.

Belcher, S. E. and Hunt, J. C. R. (1993) Turbulent shear flow over slowly moving waves, *J. Fluid Mech*, 251, 109–148.

Bruno, K. and McCready, M. J. (1989) Processes which control the interfacial wave-spectrum in separate gas-liquid flows, *Int. J. Multiphase Flow*, 15, 531–552.

Charnock, H. (1955) Wind Stress on a Water Surface, *Q. J. Met. Soc.*, 81, 639–640.

Cheremisinoff, N. and Davis, E. J. (1979) Stratified turbulent-turbulent gas-liquid flow, *AIChE J.*, 25, p 48–56.

Cohen, L. S. and Hanratty, T. J. (1965) Generation of waves in the concurrent flow of air and liquid, *AIChE J.*, 11, 138–144.

Cohen, L. S. and Hanratty, T. J. (1968) Effect of waves at gas-liquid interface on a turbulent air flow, *J. Fluid Mech.*, 31, 467–.

Fabre, J., Masbernat, L., Fernandez-Flores, R., Suzanne, C. (1987) Stratified flow, Part II: interfacial and wall shear stress, *Multiphase Science and Technology, Volume 3*, G. F. Hewitt, J. M. Delhaye and N. Zuber, eds. Hemisphere.

Fernandez-Flores, R. (1984) *Etude des interactions dynamiques en écoulement diphasique stratifié*, Thèse de Docteur Ingénieur, I. N. P. Toulouse, France.

Hanratty, T. J. (1983) Interfacial instabilities caused by air flow over a thin liquid layer, *Waves on Fluid Interface*, Academic Press, Inc., New York.

Hanratty, T. J. and McCready, M. J. (1992) *Phenomenological Understanding of Gas-Liquid Separated Flows*, Proceedings of the Third International Workshop on Two-Phase Flow Fundamentals, Imperial College, London, U. K., April 1992.

Harris, J. A. (1993) *On the Growth of Water Waves and the Motion Beneath Them*, Ph. D. Thesis, Stanford Univ., U.S.A.

Hudson, J. D. (1993) *The Effect of a Wavy Boundary on a Turbulent Flow*, Ph. D. Thesis, Univ. of Illinois, Urbana, U.S.A.

Jurman, L. A. and McCready, M. J. (1989) Study of waves on thin liquid films sheared by turbulent gas flow, *Phys. Fluids*, A1–3, 522–535.

Liné, A. and Lopez, D. (1995) *Two-fluid Model of Separated Two-Phase Flow: Momentum Transfer on Wavy Boundary*, Proceedings of Second International Conference on Multiphase Flow, Kyoto, Japan.

Lopez, D. (1994) *Ecoulements diphasiques à phases séparées à faible contenu de liquide*, Thèse de Doctorat, I.N.P. Toulouse, France.

Rosant, J. M. (1984) *Ecoulements diphasiques liquide-gaz en conduite circulaire*, Thèse de Doctorat ès-Sciences, ENSM, Nantes, France.

Sinai, Y. L. (1985) Interfacial phenomena of fully-developed, stratified, two-phase flows, *Encyclopedia of Fluid Mechanics, Vol. 3, Gas-Liquid Flows*, 475–491, N. P. Cheremisinoff, Ed.

Strand, O. (1983) *An Experimental Investigation of Stratified Two-Phase Flow in Horizontal Pipes*, Ph. D. Thesis, Univ. of Oslo, Norway.

Taitel, Y. and Dukler, A. E. (1976) A theoretical approach to the Lockhart-Martinelli correlation for stratified flow *Int. J. Multiphase Flow*, 2, 591–595.

Taitel, Y. and Dukler, A. E. (1976) Model for predicting flow regime transitions in horizontal and near horizontal gas-liquid flow, *AIChE J.*, 22, 47–55.

Vlachos, N, Paras, S. V., Karabelas, A. J. (1993) *Liquid Layer and Wall Shear Stress Characteristics in Stratified-Atomization Flow*, Proc. European Two-Phase Flow Group Meeting, Hanover, RFA.

A. Liné and J. Fabre

STRATIFIED WAVY FLOW (see Gas-liquid flow)

STRATOSPHERE (see Atmosphere)

STREAM ANALYSIS METHOD (see Optimization of exchanger design)

STREAM AVAILABILITY (see Exergy)

STREAM FUNCTION

***Following from:** Inviscid flow*

The stream function is a function of coordinates and time $\psi(\vec{r},t)$ and is a three-dimensional property of the hydrodynamics of an inviscid liquid, which allows us to determine the components of velocity by differentiating the stream function with respect to the given coordinates. A family of curves ψ = const represent "stream-lines," hence, the stream function remains constant along a stream line.

In the case of a two-dimensional flow, the velocity components u_x, u_y are expressed in terms of the stream function with the help of formulas $u_x = \partial\psi/\partial y$, $u_y = -\partial\psi/\partial x$. The difference $\psi_1 - \psi_2$ of the values of ψ for two stream lines can be interpreted as a volume flow rate of a fluid in plane flow through a stream tube bounded by these two lines. In a potential plane flow the potential φ and the stream function ψ make a complex potential $\omega = \varphi + i\psi$, but the existence of the stream function is only related to the three-dimensional character of flow and in no way requires its potentiality. The stream function can be also defined for two-dimensional space flows, for instance, it is often used to describe a longitudinal flow past axisymmetrical bodies in a spherical system of coordinates, where

$$u_r = 1/r^2 \sin\theta\ \partial\psi/\partial\theta, \qquad u_\theta = 1/r \sin\theta\ \partial\psi/\partial r.$$

The stream function represents a particular case of a vector potential of velocity $\vec{A}$, related to velocity by the equality $\vec{u} = \text{curl}\ \vec{A}$. If there is a curvilinear system of coordinares in which $\vec{A}$ has only one component, then it is exactly this system that represents the stream function for the given flow.

***Leading to:** Streamline*

A.L. Kaljazin

STREAMLINE

***Following from:** Inviscid flow; Stream function*

A streamline is a line the tangent to which at each point coincides with the direction in velocity of fluid particles at the point at a given moment of time. In a steady flow, the streamlines are the trajectories of particles. In a nonstationary flow, the streamlines vary with time and do not coincide with their trajectories. The differential equations in the Cartesian system have the form

$$\frac{dx}{u_x} = \frac{dy}{u_y} = \frac{dz}{u_z}.$$

If a certain flow has a stream function ψ, then the equation for a family of streamlines in this flow is $\psi(\vec{r},t) = \text{const}$. The set of streamlines allows us to present a vivid picture of the motion. In such illustrations, the streamlines are usually represented so that the fluid flow rate between the adjacent streamlines is the same for the whole figure. In this case, the density of streamlines is proportional to the flow velocity in the given section of the flow. Through a given point in space filled with liquid or gas, we can draw the streamlines at a given moment. The exceptions are provided by singular points at which the fluid velocity is zero or infinite.

In visualising the streamlines experimentally, the flow of separate particles of liquid or gas is made visible by introducing fine light particles, jets of paint or smoke. When photographing such a flow with short exposure a picture called the *aerodynamic flow spectrum* is obtained.

A.L. Kaljazin

STREAMLINED BODIES, FLOW OVER (see Crossflow)

STREAMLINE FLOW

Following from: *Flow of fluids*

A streamline flow or lamina flow is defined as one in which there are no turbulent velocity fluctuations. In consequence, the only agitation of the fluid particles occurs at a molecular level. In this case the fluid flow can be represented by a streamline pattern defined within an *Eulerian description* of the flow field. These *streamlines* are drawn such that, at any instant in time, the tangent to the streamline at any one point in space is aligned with the instantaneous velocity vector at that point. In a steady flow, this streamline pattern is identical to the flow-lines or path-lines which describe the trajectory of the fluid particles within a *Lagrangian description* of the flow field, whereas in an unsteady flow this equivalence does not arise.

The definition of a streamline is such that at one instant in time streamlines cannot cross; if one streamline forms a closed curve, this represents a boundary across which fluid particles cannot pass. Although a streamline has no associated cross-sectional area, adjacent streamlines may be used to define a so-called *streamtube.* This concept is widely used in fluid mechanics since the flow within a given streamtube may be treated as if it is isolated from the surrounding flow. As a result, the conservation equations may be applied to the flow within a given streamtube, and consequently the streamline pattern provides considerable insight into the velocity and pressure changes. For example, if the streamlines describing an incompressible fluid flow converge (i.e. the cross-sectional area of the streamtube contracts), this implies that the velocity increases and the associated pressure reduces.

Figures 1a and **1b** describe two well known examples of streamline flow. These and other cases are further examined by Batchlor (1967) and Duncan et al. (1970).

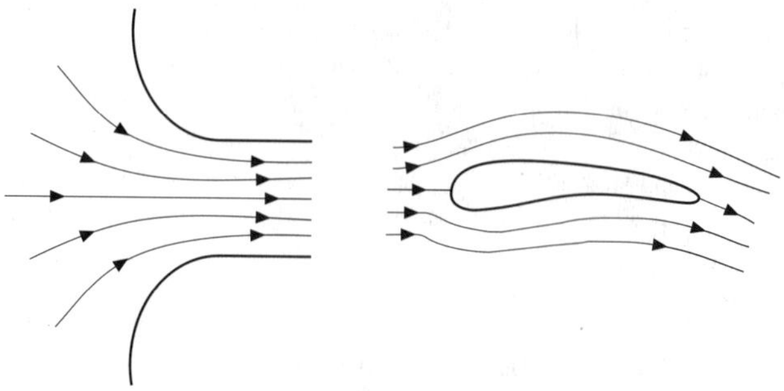

(a) Pipe Intake (b) Aerofoil

Figure 1. Examples of streamline flow.

References

Duncan, W. J., Thom, A. S. and Young, A. D. (1970) *Mechanics of Fluids,* Arnold, London.

Batchlor, G. K. (1967) *An Introduction to Fluid Mechanics,* Cambridge University Press, Cambridge, UK.

Leading to: *Streamline*

C. Swan

STREAMLINES (see Open channel flow)

STREAMLINES, VISUALISATION (see Hele Shaw flows)

STREAM TUBE (see Streamline flow)

STRESS (see Conservation equations, single phase; Newtonian fluids; Non-Newtonian fluids; Shear stress)

STRESS IN SOLID MATERIALS

Consider a bar of uniform cross section under longitudinal tension as in **Figure 1**. Assume an imaginary break a-a normal to the longitudinal axis of the bar. For equilibrium to be maintained, there must be internal forces between the two adjoining parts of the bar with a resultant equal to the externally applied force. In the simplest case, when the external force is uniformly distributed over the end, the internal forces are also uniformly distributed over any cross section a-a. The stress is defined as the intensity of the internal force, i.e., the total tensile force P divided by the cross sectional area A.

In the most general case, consider a solid under a systems of external forces P_i as in **Figure 2**. On a plane section a-a, defined by its normal direction r and area A, the resultant of the internal forces or stress resultant, R is a vector which can be regarded as the sum of a force N normal to the surface and another force along the surface, S. The *normal stress* resultant is defined as the ratio between N and A. The *shear stress* is the ratio S/A. The usual symbol for normal stress is σ and for the shear stress it is τ. It

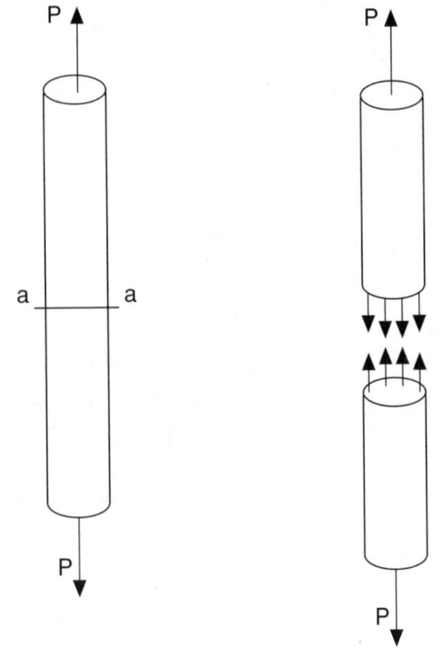

Figure 1.

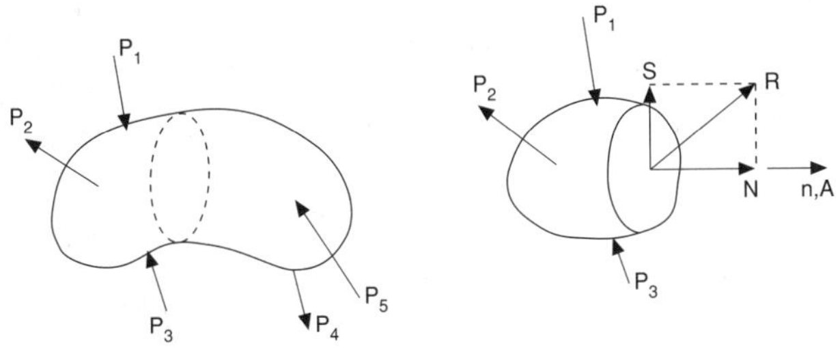

Figure 2.

should be emphasised that a stress is not a vector but the intensity of internal force acting on a surface with a given orientation. This can be best understood by isolating from the solid an elemental cube of unit side oriented along the x,y,z coordinate axes, **Figure 3**. On the face ABCD, normal to the x axis, the normal stress is σ_x. The shear stress can be resolved along the y and z directions as τ_{xy} and τ_{xz}. Similarly, stresses σ_y, τ_{yx}, τ_{yz} and σ_z, τ_{zx}, τ_{zy} are defined. The stresses acting on faces such as A′B′C′D′ may be deduced from these by considering the equilibrium of the element. It can be shown that,

$$\tau_{xy} = \tau_{yx}, \qquad \tau_{xz} = \tau_{zx}, \qquad \tau_{yz} = \tau_{zy}$$

and that the state of stress at a point O which may be regarded as in the centre of the cube is characterised by the six stress components $\sigma_x\ \sigma_y\ \sigma_z\ \tau_{xy}\ \tau_{xz}\ \tau_{yz}$. These may be grouped in the form of a *tensor* symmetrical about the principal diagonal,

$$S = \begin{bmatrix} \sigma_x & \tau_{xy} & \tau_{xz} \\ \tau_{yx} & \sigma_y & \tau_{yz} \\ \tau_{zx} & \tau_{zy} & \sigma_z \end{bmatrix}$$

C. Ruiz

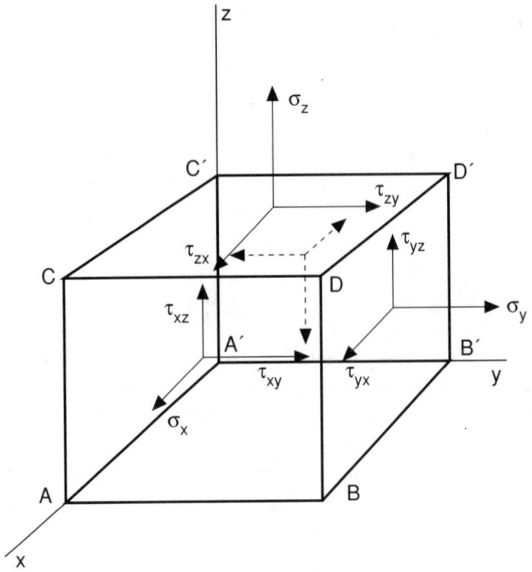

Figure 3.

STRESS, NORMAL (see Normal stress; Stress in solid materials)

STRESS, SHEAR (see Normal stress; Stress in solid materials)

STRESS TENSOR (see Conservation equations, single phase; Stress in solid materials)

STRESS VECTOR (see Shear stress)

STROUHAL NUMBER (see Dimensionless groups)

STRUCTURED SURFACE (see Augmentation of heat transfer, two-phase)

STUART NUMBER (see Liquid metals)

SUBCHANNEL ANALYSIS (see Boiling water reactor, BWR; Rod bundles, flow in)

SUBCHANNEL MIXING (see Rod bundles, flow in)

SUBCOOLING (see Boiling)

SUBCOOLING EFFECTS ON POOL BOILING (see Burnout in pool boiling)

SUBLAYER FENCE (see Shear stress measurement)

SUBLIMATION

***Following from:** Heat protection*

Sublimation, or *volatization,* is the process of changing from a solid phase to a gaseous one, without first forming a liquid. Sublimation is one type of vaporization (see **Vapour-liquid equilibrium**). As with evaporation, sublimation is possible within the whole range of temperatures T and pressures p over which the solid and gaseous phases coexist. **Figure 1** presents a typical phase diagram in p–T coordinates (a. water, b. carbon dioxide). It is well known that any substance can exist in one of the three states of aggregation: solid, liquid or gas. Two phase conditions can correspond to the solid state: crystal and amorphous; therefore, the notion "phase condition" is broader than the "aggregate" one. Below, however, the term "phase transition" implies exactly the change of the state of aggregation.

The curves of phase equilibrium on the p–T plane intersect at the triple point, where all three states of aggregation of the substance (solid, liquid and gas) take place simultaneously. The change from a solid state to a liquid state is called "melting": the process of changing from a solid state to a gaseous one is called "sublimation" and from a liquid to a gaseous one is called "evaporation". The reverse process to evaporation and sublimation is called "condensation". The pressure p_v^H at which the gaseous and condensed (liquid or solid) phases coexist is called the "saturated vapour pressure". For any substance relation between p_v^H and T is close to exponential (the **Clapeyron-Clausius Equation**):

$$p_v^H = \exp[k - \Delta Q_v \tilde{M}_v/(RT_w)],$$

where ΔQ_v is the heat of sublimation, $\tilde{M}_v$ vapour molecular mass, R the universal gas constant, and k is the experimentally defined constant. The heat of sublimation depends weakly on the temperature T_w.

According to the molecular-kinetic concept, sublimation and evaporation are continuous processes of molecular emission from the interface between the gas and condensed phases, the rate of emission being governed by the thermal motion of molecules. The velocity of the reverse process (condensation) is proportional to the number of molecules per unit volume, i.e. to the partial pressure p_v of the molecular species condensing on the interface. In sublimation (evaporation), a state of dynamic eqilibrium is established in a closed cavity when the condensation rate is equal to the sublimation rate. The appropriate partial pressure is called the saturated vapour pressure, $p_v = p_v^H(T)$.

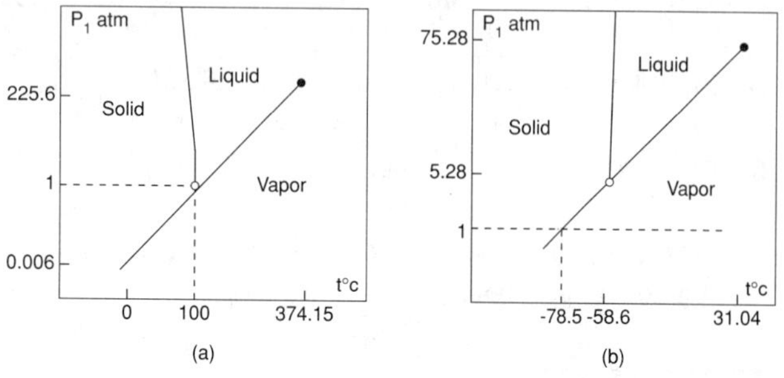

Figure 1. Phase diagrams for water (a) and carbon dioxide (b).

According to this model, the mass flow rate of substance during sublimation is the result for two counter processes, i.e. it is defined by the difference between the saturated vapour pressure p_v^H which applies at the interface, and the partial pressure in the bulk vapour, p_v the interface temperature being $T = T_w$:

$$G_v = \frac{a(p_v^H - p_v)}{\sqrt{2\pi RT_w/\tilde{M}_v}}. \quad \textbf{(1)}$$

This relation is known as the *Knudsen-Langmuir equation.* The factor a is called the *evaporation coefficient* (see **Accommodation Coefficient**). More accurate investigations based on the methods of the molecular-kinetic theory of gases, show that in Equation 1 the coefficient before the brackets is in the form 2a/(2− a). This takes into account the transverse constituent of the mass velocity in the function of distribution of gas molecule velocity near the evaporation surface.

With gas flow around bodies, the process of sublimation of their surfaces is non-equilibrium. This is due to the diffusional and convective entrainment of sublimation products into the external flow (**Figure 2**). To predict the partial pressure p_i of the ith component of the gas mixture at the surface of the body, one should consider the mass balance and allow for convective and diffusional transfer. If the mass loss rate per unit area of the body surface is G_w, and if the fraction of the ith component in the subliming material is φ_i, then

$$G_{i,w} = \varphi_i G_w = G_w c_{i,w} - \beta(c_{i,e} - c_{i,w}).$$

$$\{\text{mass loss}\} = \{\text{convection}\} + \{\text{diffusion}\}$$

Here, β is the mass transfer coefficient and, c_i is the concentration of the ith component in the boundary layer of the incoming flow. The indices w and e refer to the body surface and the boundary layer external limit.

According to the heat transfer analogy $\beta = (\alpha/c_p)_w$. To a first approximation the heat transfer coefficient (α/c_p) on the sublimating surface is related to $(\alpha/c_p)_0$ on a non-permeable (heated) surface by the following relation:

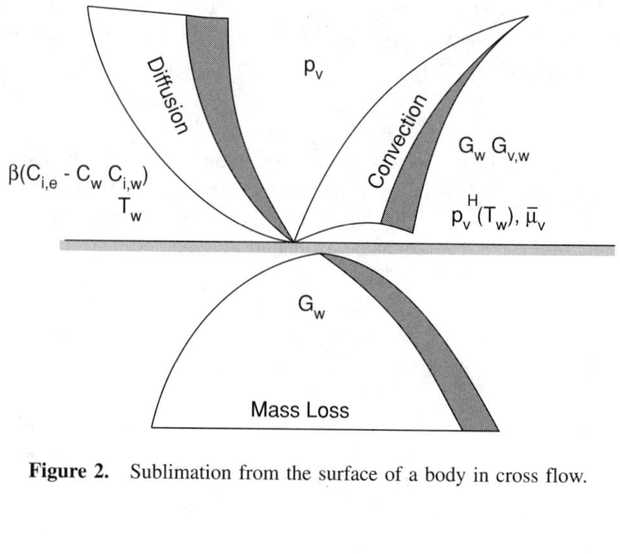

Figure 2. Sublimation from the surface of a body in cross flow.

$$(\alpha/c_p)_w = (\alpha/c_p)_0 - \gamma G_w = (\alpha/c_p)(1 - \gamma\bar{G}_w),$$

where $\overline{G}_w = G_w/(\alpha/c_p)_0$. If we assume that in the external (oncoming) flow the products of sublimation are absent $c_{i,e} = 0$, and that the sublimating body does not contain extraneous admixtures $\varphi_i = 1$, then we obtain the following equation for the mass loss rate:

$$G_w = \frac{\beta(c_{i,e} - c_{i,w})}{1 - c_{i,w}} = \frac{(\alpha/c_p)_w p_v \bar{M}_v}{p_e \bar{M}_\Sigma - p_v \bar{M}_v}. \quad \textbf{(2)}$$

This equation takes into account that the partial pressure p_v is related with mass concentration c_v by the relationship:

$$c_v = \frac{p_v M_v}{p_e \bar{M}_\Sigma},$$

wherein p_e, $\bar{M}_\Sigma$ are the pressure and molecular mass of the gas mixture.

Solving Equations 1 and 2 simultaneously, one can reach a number of interesting conclusions. Thus, eliminating the mass lose rate G_w, we obtain the relationship for estimating the degree of non-equilibrium of the sublimation process, i.e. the relation between the partial pressure p_v and the saturated vapour pressure p_v^H:

$$\frac{p_v^H}{p_v} = 1 + \frac{(\alpha/c_p)_0\sqrt{2\pi R T_w/\tilde{M}_v}[1 + \bar{G}_w(1 - \gamma)]}{a p_e \tilde{M}_\Sigma}.$$

The larger the ratio (p_v^H/p_v), the further the process departs from equilibrium. When the temperature and heat transfer coefficient increase, and when the pressure p_e decreases, the departure from equilibrium becomes more significant. During sublimation in a vacuum, as numerous investigations of intense evaporation show, in the steady state case the maximum flow rate $G_{w,max} = p_v^H(T_w)/\sqrt{2\pi R/\tilde{M}_v T_w}$ is not obtained, since a portion of outgoing molecules, even in the case of evaporating in a vacuum, return to the surface as a result of inter-molecular collisions. It may be shown that in this case the following expression can be used (assuming $a = 1$):

$$G_w \approx 0.82 \frac{p_v^H}{\sqrt{2\pi R/\tilde{M}_v T_w}}.$$

The higher the partial pressure p_e of the subliming material, the closer the sublimation regime is to an equilibrium one, and closes the vapour pressure is to the saturation pressure $p_v^H(T_w)$. In this case, it can be easily shown at large values of the mass loss rate, the sublimating surface temperature asymptotically tends to a limiting value which depends only on the partial pressure in the gas flow (p_e) and is defined from the equation:

$$p_v^H(T_w) = \exp\left(b - \frac{\Delta Q_v \tilde{M}_v}{RT_{w,max}}\right) = p_e\tilde{M}_\Sigma/[(1 - \gamma)\tilde{M}_v].$$

Introduction of the limiting temperature $T_{w,max}$ is very important for estimating the possibilities of evaporative cooling (see **Evaporative Cooling**).

Under actual conditions, there can be a mixture of condensed media on the sublimation surface. In this case, the equation of the Knudsen-Langmuir kind is applied to each component separately. The mass loss rate is determined by summation of sublimation velocities of all components: $G_w = \Sigma_i G_{wi}$.

Leading to: *Evaporative cooling; Accommodation coefficient*

N.V. Pavlyukevich and Yu.V. Polezhaev

SUBMERGED COMBUSTION

Submerged combustion, as its name implies, is the combustion of gas or fuel oil in such a manner that the hot combustion product gases are released under the surface of a liquid. In this way, the energy released by the combustion process is transferred by direct contact with the liquid (see **Direct Contact Heat Transfer**). Although it is possible for the burner itself to be submerged, most systems are arranged with the burner above the liquid level and a submerged exhaust system. In the **Figure 1**, the exhaust gas is released into the annulus between the downcomer and draught tube. This promotes good mixing between the hot gas and the liquid and produces intense circulation of the liquid in the tank.

The first patents in the field were granted at the end of the nineteenth century but it was the development work of Swindin (1949) in the 1930s and 40s which led to the first commercial applications of the process.

Submerged combustion is used in two classes of **Evaporators** direct and indirect. In the first, it is used to concentrate corrosive or toxic materials. In the second, water is heated which in turn, is then circulated over a tube bank containing the liquid to be evaporated.

While the advantages of submerged combustion are the absence of fouling or corrosion and the ability to handle highly

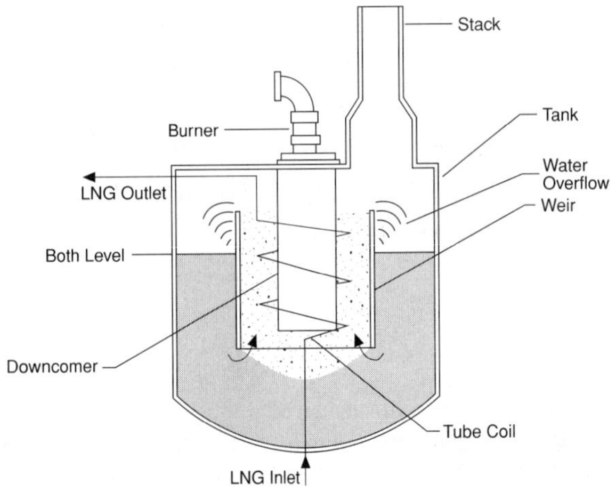

Figure 1. Submerged combustion evaporator.

viscous liquids or liquids containing up to 40% solids, the disadvantages are the contamination of the liquid by combustion products.

Proprietary burners have been developed for submerged combustion applications using a variety of liquid or gaseous fuels. The requirements are complete soot-free combustion within a small chamber volume with low residence times at above atmospheric pressure. High velocity vortex-type burners are employed utilising, for liquid fuels, recirculation of the hot gases to aid fuel vaporisation. Multiple burner arrangements are possible with either separate or common downcomers.

Because of intensive gas/liquid contact, the dew point of the released combustion gases approximates to the vapour pressure of the liquid (water) at the operating temperature. The water vapour in the combustion gases will condense below the dew point (~60°C for natural gas and 50°C for gas oil) and the heat transferred to the liquid (water) will exceed the lower calorific value of the fuel. At higher water temperatures the efficiency falls off until a maximum temperature of 90°C. At this temperature all the sensible heat in the combustion gases is used to evaporate water to saturate the gases.

A partial list of applications of submerged combustion:

Heating

- Vat dyes
- Pickle bath
- Swimming pool
- Seawater

Concentration

- Sulphuric acid
- Phosphoric acid
- Superphosphoric acid
- Citric acid
- Fish stick liquor

Crysallisation

- Brine

Chemical Treatment

- Carbonation of liquids
- Treatment of sewage sludges and cellulose effluents
- Stripping of phenols/H_2S
- Dewatering
- Sulphur recovery
- Inert gas generation

A major application of submerged combustion is the vaporisation of liquid cryogens such as nitrogen, oxygen, natural gas (LNG), petroleum gas (LPG), ethylene, ammonia etc. Such fluids are often transported as liquid but are required for use as a gas. Because of safety requirements indirect heat transfer rather than direct firing from the combustion source is often essential. In the submerged combustion vaporiser, water acts as the heat transfer agent.

Referring to the earlier diagram, a tube bank through which the fluid to be vaporised flows, is inserted into the riser section between the downcomer and the draught tube. The issuing combustion products produce a rapid circulation of water over the coils and considerable turbulence at the heat transfer surface. This enhances the heat transfer coefficient on the outside of the tubes and prevents ice formation. Edwards (1967) described the so-called Sub-X° unit developed by the Thermal Research and Engineering Corporation in the U.S. This design is now marketed in the U.S. and elsewhere by *T-Thermal.* In the United Kingdom the technology was licensed and developed by Thurley International Ltd. [Thurley (1970)] and since 1984 by Kaldair Ltd. with its range of "TX" vaporisers.

In the Sub-X° design the combustion chambers are located above the liquid surface and exhaust into a common rectangular downcomer unit. The combustion products are sparged out through a series of short stub pipes into the zone between the downcomer and weir and below the tube bundle. The weir and the flues are free-standing in a rectangular tank. These units are available and operating with thermal capacities up to 75 M Btu/hr (22 MW) and vaporise up to 100 tons/hr of LNG. The water temperature for such units is between 15–52°C with thermal efficiencies of 90–99%. Condensation of the water vapour from the combustion product gases occurs and make-up is not required. The tank water becomes saturated with carbon dioxide and care with the choice of materials and with water treatment is required. Over 250 units of varying designs and ratings are in use world wide.

References

Swindin, N. (1949) Recent developments in submerged combustion, *Trans. Inst. Chem. Eng.,* 27, 209–221.

Edwards, R. M. (1967) Efficient new heat exchanger suited to LNG vaporisation, *The Oil and Gas Journal,* 96–99, Oct. 2nd.

Thurley, J. (1970) Liquid natural gas vaporisation, *Natural Gas—LNG and LPG,* Oct.

T-Thermal (Europe), GOC House, Blackwater Way, Aldershot, Hants, GU12 4DR.

Kaldair Ltd., Kaldair House, Langley Quay, Waterside Drive, Langley, Berks SL3 6EY.

***Leading to:** Direct contact heat exchangers*

J. Collier

SUBMERGED COMBUSTION EVAPORATORS
(see Evaporators; Submerged combustion)

SUBMERGED JETS (see Impinging jets)

SUBROUTINES (see Computer programmes)

SUBSTITUTE NATURAL GAS (SNG)

Substitute Natural Gas, which is interchangeable with **Natural Gas** can be manufactured from other fossil fuels by combining three main reaction stages; the gasification of the feedstock with steam, or a mixture of steam and oxygen, to produce a gas from which **Methane** can be synthesised from the carbon oxides and hydrogen, and the removal of carbon dioxide.

A number of large plants were built in the United States and Japan in the early 1970s to make SNG from light distillate oils to meet predicted shortages in natural gas supplies. The process used was the *CRG (Catalytic Rich Gas) process* developed by British Gas. In this process, the distillate oil is first hydro-desulphurised and it is then gasified catalytically with steam in an adiabatic reactor to produce a gas containing about 64% methane, 17% hydrogen and 21% carbon dioxide, by volume, on a dry basis. The gas leaving this reactor at 500°C is cooled to 300°C and passed through an adiabatic catalytic reactor, or methanator, in which equilibrium is attained in the following reactions:-

$$CO_2 + H_2 \rightleftharpoons CO + H_2O \quad (1)$$

and

$$CO + 3H_2 \rightleftharpoons CH_4 + H_2O \quad (2)$$

in which a low temperature favours methane formation. After cooling the gas and passing it through a second methanator, it contains 77% methane and 22% carbon dioxide. After removal of most of the carbon dioxide, the gas contains more than 99% methane.

The application of this process was restricted by the availability of light distillate oils and, in the 1970s, work started on the development of a number of different processes for the gasification of coal. All of the coal gasification processes that use steam and oxygen produce gas containing carbon oxides and hydrogen and they are potentially capable of making SNG. However, the methane synthesis reaction **Equation 2** is highly exothermic and there are large efficiency and cost advantages in using a process, such as the *Lurgi process* that makes methane during the gasification of the coal.

The predicted shortages and high costs of natural gas did not materialise and the only coal based SNG plant that now operates is in North Dakota. Commissioned in 1984, the plant produces 3.9 $\times 10^6 m^3$/day of SNG from lignite using 14 Lurgi dry-ash gasifiers. The crude gas leaving the gasifiers contains 12% methane, 16% carbon monoxide, 39% hydrogen and 32% carbon dioxide, by volume on a dry basis. The gas is cooled, tars and oils are separated and it is washed with methanol at −40°c to remove carbon dioxide and sulphur compounds. The purified gases then passes through a series of methanation reactors to bring the gas to equilibrium at progressively lower temperatures to raise the methane content; gas recirculation is used to moderate the temperatures in the first methanators and the final methanator is water cooled. The final gas contains 96% methane.

References

Davis, H. S., Lacey, J. A. and Thompson B. H. (1968) Processes for the Manufacture of Natural Gas Substitutes, *The Institute of Gas Engineers*, CC155.

Elliott, M. A. (1981) *The Chemistry of Coal Utilisation*, Ch. 24, Wiley Interscience.

J.A. Lacey

SUBUNDAL FLOW (see Open channel flow)

SUPERUNDAL FLOW (see Open channel flow)

SUCTION (see Augmentation of heat transfer, single phase)

SUCTION EFFECTS

***Following from:** Boundary layers*

Suction is one of the methods of boundary layer control which have the aim of reducing drag on bodies in an external flow or of reducing losses of energy in channels. This method was suggested by L. Prandtl in 1904 as one of the means of preventing or "delaying" boundary layer separation. To effect suction, the surface should have holes (slots, porous sections, perforations etc.). These holes serve for sucking the portion of the boundary layer which is nearest to the wall and which is travelling at the lowest velocity. As a result, the boundary layer velocity profile becomes more "filled up" (see **Boundary Layer**) and thus is more stable as far as separation is concerned.

Suction is applied in practice for increasing the efficiency of diffusors with high compression ratio of the working fluid (with large convergence angles) by means of delaying early separation of the boundary layer. Boundary layer suction through slots near the trailing edge is used for increasing lift and decreasing drag of aerofoils operating at large incidence angles.

It has been demonstrated in practice that continuous suction through a porous wall is more effective than suction through slots. For example, for aerofoils, the same increase of lift force can be achieved by sucking a smaller amount of fluid through pores than through slots.

Suction is also an effective means of the boundary layer laminarization, which decreases friction losses (see **Boundary Layer**). The effect of suction on the laminar boundary layer stability is due to decrease of the boundary layer thickness (a thinner boundary layer is less liable to turbulization) and also due to the changes in the velocity profile (it becomes more filled).

Of practical importance for applying suction is the necessity to determine the minimum suction fluid necessary to keep the boundary layer laminar, because an excess of suction flow rate may result in such a power consumption that this would make insignificant the power economy achieved by decreasing the drag force. It is necessary that the suction rate $V_w < 0$ be small as compared with the external flow velocity U_e, namely $(V_w/U_e) = 0.0001 - 0.01$ (**Figure 1**). In this case the assumptions of the boundary layer theory remain valid for the wall layer.

With uniform suction ($V_w(x)$ = const) through a plate which is traversed (in the longitudinal direction) by an incompressible flow with velocity U_e, the system of boundary layer equations is written as (see nomenclature on **Figure 1**)

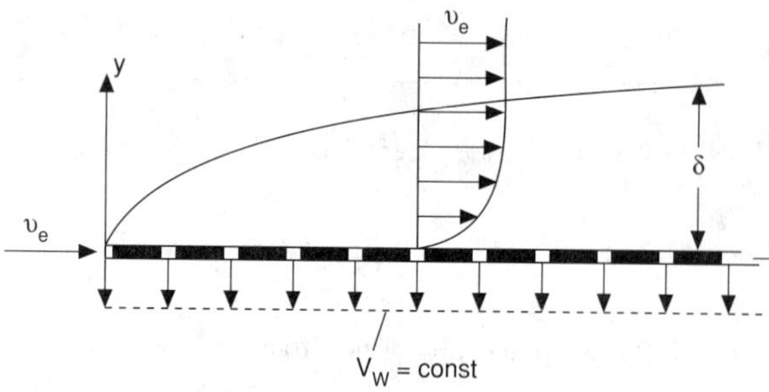

Figure 1. Boundary layer suction arrangement.

$$\begin{cases} \dfrac{\partial u}{\partial x} + \dfrac{\partial v}{\partial y} = 0, \\ u\dfrac{\partial u}{\partial x} + v\dfrac{\partial u}{\partial y} = \nu\dfrac{\partial^2 u}{\partial y^2}. \end{cases} \quad (1)$$

Under the following boundary conditions

$$y = 0, \quad u = 0, \quad v = v_w = \text{const} < 0 \quad (2)$$

$$y \to \infty, \quad u \to U_e$$

and taking into account $\partial U/\partial x = 0$ the following asymptotic solutions exist

$$v(x, y) = v_w = \text{const} < 0 \quad (3)$$

$$u(y) = U_e[1 - \exp(v_w y/\nu)]$$

For integral boundary layer thicknesses, namely displacement thickness δ^* and momentum loss thickness δ^{**} (see **Boundary Layer**) we obtain the following relationships

$$\delta^* = \frac{\nu}{-v_w}, \qquad \delta^{**} = \frac{\nu}{-2\nu_w} \quad (4)$$

where δ^*, δ^{**} are independent of distance x along the boundary layer.

The shear stress τ_w at the wall and the plate drag C_f coefficient are determined from the relationships

$$\tau_w = \rho(-v_w)U_e, \qquad C_f = \frac{2\tau_w}{\rho U_e^2} = 2\frac{-v_w}{U_e} \quad (5)$$

and do not depend in the flow viscosity. The boundary layer thickness on the plate leading edge is equal to zero, but it grows asymptotically in the downstream direction (see **Figure 1**) approaching the values corresponding to Equation 4. The velocity profile shape corresponding to Equation 3 is also approached asymptotically along the initial section.

The asymptotic velocity profile corresponding to Equation 3 is formed at the end of the initial section:

$$\xi = \left(\frac{-v_w}{U_e}\right)^2 \frac{U_e x}{\nu} = 4 \quad (6)$$

The boundary layer described by Equations 3–5 is termed an "asymptotic suction layer".

The asymptotic velocity profile (3) is more "filled up" than the Blasius profile characteristic for the case of the flow over a plate without suction, and also than for flow at the initial section near the plate edge when there is suction. An asymptotic solution is also possible for a plate is in a gas flow. In this case the velocity profile is described by Equation 7

$$u(y) = U_e[1 - \exp(v_w \rho_e y_1/\mu_e)] \quad (7)$$

where $y_1 = \int_0^y \rho/\rho_e \, dy$.

The shear stress is calculated from Equation 8

$$\tau_w = \rho_w(-v_w)U_e \quad (8)$$

The integral momentum relationship in the case under consideration is obtained in the similar way as that for the boundary layer without suction and is described by the equation

$$\frac{d}{dx}(U_e^2\delta^{**}) + \delta^* U_e \frac{dU_e}{dx} = \frac{\tau_w}{\rho} + V_w U_e. \quad (9)$$

For solution of Equation 9 (see **Boundary Layer**). The term $V_w U_e$ in the right hand side of this equation takes into account the momentum variations caused by suction.

In 1935, from Equation 9 L. Prandtl obtained an approximate relationship Equation 10 for the minimum suction velocity $V_{w\ min\ separ}$ at which separation is prevented. He assumed that at all points of the surface the velocity profile corresponds to the conditions such that $\tau_w = 0$. It followed that

$$V_{w\ min\ separ} = -2.18\sqrt{-\nu\frac{dU_e}{dx}} \quad (10)$$

The practically important problem of determining minimum suction flow rate $V_{w\ min\ lam}$ for which the laminar flow condition is preserved can be solved using the above asymptotic solution for the boundary layer near the plate.

K. Bussman and H. Munz in 1942 calculated the laminar boundary layer stability at the plate with an asymptotic velocity profile. As a result they determined the critical value of a Reynolds number when the boundary layer becomes turbulent

$$Re_e = \left(\frac{U_e\delta^*}{\nu}\right)_c = 70000.$$

This value is 100 times larger than the corresponding Re_c for the plate without suction, which is the proof of the efficiency of this process for stabilizing laminar boundary layers.

Using Equation 4 for δ^* and the value of Re_c we obtain the condition of the laminar boundary layer stability with suction

$$\frac{-v_w}{U_e} > \frac{1}{70000} = 1.4 \times 10^{-5} \tag{11}$$

Actually, the suction intensity must be somewhat higher than that determined by Equation 11 because this equation was derived under the assumption of existence of an asymptotic velocity profile (Equation 3) starting from the plate leading edge. In the initial section of the boundary layer the Blasius velocity profiles have a smaller stability limit and therefore more intensive suction is required for preserving the laminar boundary layer in this section.

References

Prandtl, L. (1904) *Uber Flussigkeitsbewegung bei sehr Kleiner Reibung: Verhandl, III int,* Math, Kongr, Heidelberg.

Prandtl, L. (1935) The mechanics of viscous fluid, *Durand W. F. Aerodynamic Theory III.*

Bussman, K., and Munz, H. (1942) Die Stabilitat der laminaren Reibungsschicht mit Absangung. *Sb. at. Luftfahrtforschung I,* 36–39.

V.M. Epifanov

SULFUR POLLUTION (see Air pollution)

SULFUR

Sulfur—(Sanskrit, *sulvere;* L. *sulphurium*), S; atomic weight 32.06; atomic number 16; melting point (rhombic) 112.8°C, (monoclinic) 119.0°C; boiling point 444.674°C; specific gravity (rhombic) 2.07, (monoclinic) 1.957 (20°C); valence 2, 4, or 6.

Known to the ancients; referred to in Genesis as brimstone. Sulfur is found in meteorites. A dark area near the crater Aristarchus on the moon has been studied by R. W. Wood with ultraviolet light. This study suggests strongly that it is a sulfur deposit. Sulfur occurs native in the vicinity of volcanoes and hot springs. It is widely distributed in nature as iron pyrites, galena, sphalerite, cinnabar, stibnite, gypsum, Epsom salts, celestite, barite, etc.

Sulfur is commercially recovered from wells sunk into the salt domes along the Gulf Coast of the U.S. It is obtained from these wells by the Frasch process, which forces heated water into the wells to melt the sulfur that is then brought to the surface. Sulfur also occurs in natural gas and petroleum crudes and must be removed from these products. Formerly this was done chemically, which wasted the sulfur. New processes now permit recovery, and these sources promise to be very important. Large amounts of sulfur are being recovered from Alberta gas fields.

Sulfur is a pale yellow, odorless, brittle solid, which is insoluble in water but soluble in carbon disulfide. In every state, whether gas, liquid or solid, elemental sulfur occurs in more than one allotropic form or modification; these present a confusing multitude of forms whose relations are not yet fully understood. Amorphous or "plastic" sulfur is obtained by fast cooling of the crystalline form. X-ray studies indicate that amorphous sulfur may have a helical structure with eight atoms per spiral. Crystalline sulfur seems to be made of rings, each containing eight sulfur atoms, which fit together to give a normal X-ray pattern. Ten isotopes of sulfur exist. Four occur in natural sulfur, none of which is radioactive. A finely divided form of sulfur, known as flowers of sulfur, is obtained by sublimation. Sulfur readily forms sulfides with many elements.

Sulfur is a component of black gunpowder, and is used in the vulcanization of natural rubber and as a fungicide. It is also used extensively in making phosphatic fertilizers. A tremendous tonnage is used to produce sulfuric acid, the most important manufactured chemical. It is used in making sulfite paper and other papers, as a fumigant, and in the bleaching of dried fruits. The element is a good electrical insulator. Organic compounds containing sulfur are very important. Calcium sulfate, ammonium sulfate, carbon disulfide, sulfur dioxide, and hydrogen sulfide are but a few of the other many important compounds of sulfur. Sulfur is essential to life. It is a minor constituent of fats, body fluids, and skeletal minerals.

Carbon disulfide, hydrogen sulfide, and sulfur dioxide should be handled carefully. Hydrogen sulfide in small concentrations can be metabolized, but in higher concentrations it quickly can cause death by respiratory paralysis. It is insidious in that it quickly deadens the sense of smell. Sulfur dioxide is a dangerous component in atmospheric air pollution. In 1975, University of Pennsylvania scientists reported synthesis of polymeric sulfur nitride, which has the properties of a metal, although it contains no metal atoms. The material has unusual optical and electrical properties. High-purity sulfur is commercially available in purities of 99.999 + %.

Handbook of Chemistry and Physics, CRC Press

SULFUR DIOXIDE

Sulfur Dioxide (SO_2) is a colourless gas of a distinctive, strong odour. It is obtained from either pyrite or sulfur by burning. It is toxic and a strong irritant and is classified as one of the major *air pollutants.* It is widely used in the production of sulfur-containing chemicals, in the paper industry and in metal refining. It is also used as a solvent, as a food additive and as a reaction medium.

Molecular weight: 64.06	Critical temperature: 430.8 K
Melting point: 198.7 K	Critical pressure: 7.884 MPa
Boiling point: 263.0 K	Critical density: 525 kg/m^3
	Normal vapor density: 2.86 kg/m^3 (@ 0°C, 101.3 kPa)

V. Vesovic

SULFUR HEXAFLUORIDE

Sulfur Hexafluoride (SF_6) is a colourless and odourless gas that is slightly soluble in water. It is manufactured by a direct combination of sulphur vapour and fluorine. It is toxic only in high concentrations and it has a limited industrial use as a dielectric fluid.

Molecular weight: 146.05
Melting point: 222.5 K
Boiling point: 209.3 K
Critical temperature: 318.7 K
Critical pressure: 3.76 MPa
Critical density: 738 kg/m^3
Normal vapor density: 6.52 kg/m^3 (@ 0°C, 101.3 kPa)

V. Vesovic

SULFURIC ACID

Sulfuric acid (H_2SO_4) is a colourless, oily liquid with a molecular weight of 98.08 and a specific gravity of 1.84. It is readily miscible with water, highly corrosive, highly toxic and is a strong oxidising agent that will react with most organic compounds. It comes in different industrial grades depending on the strength of the aqueous solution. It is one of the most widely used chemicals.

V. Vesovic

SUN, HEAT TRANSFER IN (see Fusion, nuclear fusion reactors)

SUPERCONDUCTING MAGNETS

Following from: *Superconductors*

Superconducting magnets are electromagnets wound with *superconductors.* SM's are used to generate high magnetic fields.

The SM windings need to be cooled below certain temperatures specific to particular superconductors and known as their critical temperatures. Most often they are cooled to 4.2 K (boiling point of liquid helium under atmospheric pressure), although on some occasions the temperature can be lowered down to 1.8 K.

One of the characteristic features of an SM that partially determine their operating parameters is the possibility of the superconducting coils going to the "normal" or resistive state (i.e., having some finite resistance). Such transitions can be triggered by some external interactions (vibrations, abrupt current changes, etc.) but, what is more important, they can arise as a result of the growth of unstable perturbations inside the windings proper. The heat generated inside the resistive zone warms nearby portions of the windings driving them also to become resistive so that, at the end, entire windings or a significant portion of them can be overheated. Serious damage can result from this overheating or from other assosiated processes.

Stabilization techniques have been devised to provide stable and reliable operation of SM's. The use of so called "cryostatic stabilization" allows almost complete exclusion of the possibility of superconducting windings going normal. Large quantities (sometimes even more than 90%) of resistive metals (usually copper or aluminum) are used in this case within so called composite conductors along with superconducting wires. The resistivity of such metals at low temperatures is rather small so that Joule losses that can be the result of eventual normal transitions must be quite low. Sufficient heat removal rate from the windings must also be provided so that such eventual transitions would not lead to the uncontrolled growth of the normal region. The overall current density that determines the dimensions of the coils can unfortunately be quite low in such magnets that can lead to their large sizes and masses. More intricate methods are also being developed when normal transitions are permitted but they are arranged such that no significant damage would be produced.

Solid state controlled rectifiers are most widely used as current supplies for SM's, though DC generators and accumulator batteries also find a more limited application. The zero resistance of SM windings allows the use of some peculiar low voltage devices as current supplies, e.g. the so called flux pump, topological generators, cryotrons (superconducting analog of SCR), etc. Very peculiar operating modes are sometimes used with an SM, e.g. to provide extreme current and field stability, when after being charged with current from an external source the windings of the SM can then be shunted by an entirely superconducting circuit so that the magnetic flux becomes effectively "frozen-in". The current supply can then be turned off and it is possible, furthermore, to remove external current leads (e.g. in order to lessen heat losses). It is also possible to generate AC fields as modern technologies permit the reduction of the inevitable AC losses to a reasonable level.

The largest SM's have been built for experimental physics devices, such as accelerators, controlled nuclear fusion installations, particle detectors, etc. The maximum flux density obtained with SM is of the order of 20T and DC fields of 35T level have been generated in a combination of an external SM and an internal water cooled electromagnet.

The most extensive field of SM application is now medical diagnostics (in nuclear resonance imaging devices for NMR tomographs) where the field level is rather moderate (of the order of 1T). A few examples of succesful use of SM in industrial ore separation technology are now known (e.g. in the xaoline purification process). Some prospects for use of SM in power engineering, transport (magnetic levitation) and in other industrial technologies are now under an intensive investigation. An extensive range of superconducting wires cables, tapes and other forms of composite conductors is now supplied by industry to be used in the windings of SM's. Great number of superconducting thin wires embedded in copper matrix can be used in such conductors though some other forms of thin film or sponge-like superconducting structures are also encountered.

The most widely known compound to form these wires is now Nb—Ti alloy sometimes doped with other minor additives. It is used in SM's built to obtain maximum fields with flux densities of the order of 10T. About twice as high fields can be generated with conductors using superconducting compound Nb_3Sn while other materials find only very limited applications.

SM's are now widely used in different branches of scientific research and in industrial technologies where it is necessary to obtain high magnetic fields. Energy requirements for SM's can be quite moderate as only refrigerating power is consumed in the DC mode to compensate for heat losses. On some rare occasions SM have also been applied to more promising prospects at first forseen immediately after the discovery (1987) of so called high temperature superconductors that need not to be cooled to such extremely low temperatures. However, in spite of very large efforts, practical development of suitable forms of conductors is still awaited and only very small pilot electromagnets using such superconductors have been demonstrated.

M.G. Kremlev

SUPERCONDUCTORS

Superconductors are substances possessing the property of superconductivity, e.i., the ability of conducting direct currents without any electrical resistance. Superconductivity can be observed only upon lowering the superconductors' temperature below some value which is characteristic for the particular superconductor and is designed as their critical temperature T_c.

The phenomenon of superconductivity was discovered in 1911 by the Dutch physicist H. Kamerlingh-Onnes in his work with mercury, for which the T_c value according to modern data is close to 4.17 K. Since then, this property was found in more than 20 elementary metals and in several different alloys and compounds. The highest T_c in case of elementary metals is that of niobium (8.7 K) and higher values of the order of 20 K have long been known for some compounds also containing niobium, the record (22.3 K) being held by the Nb_3 (Al—Ge) compound. However, in 1987 a new group of complex compounds containing copper as well as oxides (e.g. Y, Ba, Bi, Th etc.) was discovered for which T_c values could certainly be as high as 155 K and according to some quite recent estimates can reach 250 K, that is only slightly below room temperature. This new group of superconductors was since designated even as "high T_c" superconductors as opposite to the older types of "traditional" or "low T_c" ones.

The T_c values are functions of magnetic field strength to which superconductors can be exposed, decreasing in growing fields, so that superconductivity can be observed only in some specific limited range of magnetic field strengths and of temperatures. The values of maximum field strength also designated as critical ones (H_c) and corresponding T_c values can be well represented by a parabolic dependence $H_c(T) = H_c(0)(1 - (T_c - T_{c,max})^2)$. Critical field values for elementary superconductors are relatively small and this, along with their low T_c values, has hindered practical applications of superconductors (H_c for pure Nb is no higher than 1.8 kOe). On the other hand, some superconducting alloys and compounds can be characterized by very high H_c values. Thus, for now most widely practically used alloys of Nb with Ti corresponding magnetic inductions (B_c) can reach 12 Teslas, for another "low T_c" compound Nb_3Sn, that also finds somewhat limited applications, B_c can be as high as 35T, while some compounds are now known with B_c surpassing 100T. Even quite approximate determinations of B_c values can present substantial practical difficulties. New "high T_c" superconductors can also be characterized by high B_c values.

Superconductors have two possible behavioral patterns superconductors with respect to magnetic fields, allowing them to be divided into two different classes. For superconductors of the first kind, the so called *Meissner effect* is observed which consists in total expulsion of magnetic field lines of force from the bulk of the superconductor the effect being provided by screening corrents circulating in very thin layers near the surface of the specimen. It is essential that such expulsion must be produced regardless of the exact order in which the superconducting state is realized in a magnetic field. It is in this particular respect that the superconductors of the first kind differ from the hypothetical "ideal" conductor that does not prevent the magnetic field penetrating the bulk of the specimen prior to its going superconducting on lowering the temperature. It should be noted, however, that total magnetic field expulsion could be achieved only in very pure and otherwise perfect specimens having in addition regular smooth form. In impure, deformed specimen with sharp edges, etc. only partial expulsion is usually observed.

The superconductors of the second kind also exhibit Meissner effect, but only in relatively low magnetic fields, while with increasing field strength, partial penetration of the magnetic flux lines into the bulk of the specimen becomes more significant. A "mixed" state is then realized in the bulk of the superconductors with persistent currents circulating over the entire specimen volume in a kind of regular pattern consisting of so-called Abricosov's vortices otherwise also known as "fluxoids". This mixed state can be preserved in much higher fields up to some "second critical value", $H_{c\,2}$, which can usually be found in reference tables.

Almost all pure elementary superconducting metals are first kind superconductors with rare possible exceptions (e.g. of Nb that can be considerd as intermediate or at most as not quite typical case). A very limited number of other alloys and compounds can also be included within the same group, while the bulk of other alloys and compounds fall typically into the second group of superconductors.

For many practical applications the problem of critical current values, e.i., of maximum currents that could flow through the superconductor without noticeable resistance, can be of prime importance. For the first kind superconductors, especially for pure, "ideal" ones very simple relation called Silsbee's rule has been formulated. According to this rule no resistance could be observed unless at any point on the superconductor surface, the magnetic field produced by both external sources and by the superconductor's own currents surpasses the critical value. In the case of superconductors of the second kind the problem of critical currents is much more complicated and, generally, only direct experimental determination of critical currents (or of current densities) as function of external fields and temperatures can provide a complete set of critical data (forming a kind of "critical surface" in the space of all necessary parameters).

For pure perfect specimens of type 2 superconductors, critical currents for the most part can be relatively low and materials find therefore rather limited applications. However, for highly deformed, non-uniform or doped type 2 alloys and compounds critical currents can reach extremely high values—up to 100 kA/mm^2. High critical currents permit to use such substances for many practical applications, mainly in the form of superconducting coils (or magnets), that are now widely used in many different fields of science and industry (particle accelerators and other physical research devices, electric power technologies, magnetic ore separation, magnetic resonance imaging etc.). Practical use of type 1 superconductors when compared to type 2 materials seem quite limited. However, due to their very unusual and specific properties such substances could also be used in some applications in high sensitivity measurements (magnetometers, radiation detectors), metrology (standard values reproduction and measuring equipment), UHF devices, etc. Certain expectations are also connected with the possible use of superconductors in very large memory arrays of big computers.

The problem of exact nature of physical mechanism leading to the occurrence of superconductivity in some particular substances is of very great interest for the modern science. In the "traditional" superconductors the very peculiar interaction between electrons in metals provided by the exchange of phonons, or sound wave quanta, is the most predominant if not the only possible form of such mechanisms. Low *critical temperatures* of these substances are the results of relatively small phonon energies which are typically no higher than a few hundreds of Kelvins. Very high T_c values for new ceramic superconductors are presumably provided by some different interaction, the exact nature of which, in spite of very intense experimental research and a lot of proposed theoretical explanations, remains, however, quite obscure.

Leading to: *Superconducting magnets*

M.G. Kremlev

SUPERCRITICAL HEAT TRANSFER

Following from: *Heat transfer*

Physical Properties Near the Critical Point

The principal difference in the behaviour of a fluid as its temperature is raised above the *critical temperature* is evident from the pressure-volume diagram shown in **Figure 1**. Below the critical temperature T_c the variation of pressure p and volume v along an isotherm shows discontinuities where the isotherm intersects the saturation line. Phase change occurs at the saturation line, and the horizontal constant pressure segment of the isotherm represents the presence of vapour and liquid in varying proportions. The critical temperature isotherm has zero slope at one point, the pressure there being the critical pressure, p_c. Above the critical temperature, isotherms show no discontinuity and from a macroscopic stand point there is a continuous variation from a liquid-like fluid to a gas-like fluid.

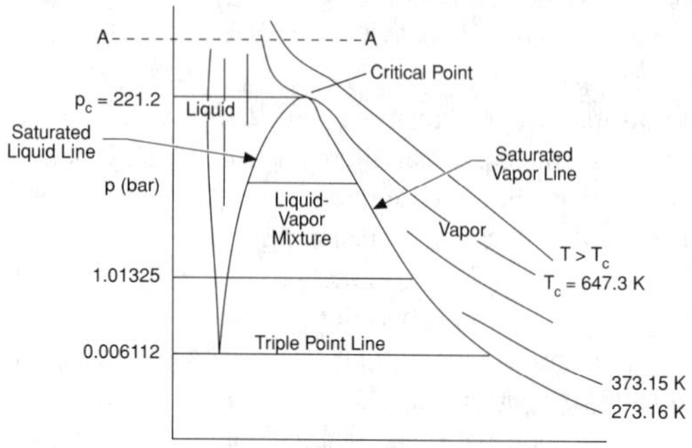

Figure 1.

Since the pressure variation across most thermal boundary layers is negligible, it is the variation of properties along an isobar that is important. The variation of specific volume v, specific heat c_p, absolute viscosity η, thermal conductivity λ and specific enthalpy h for water at pressure of 245 bar is shown in **Figure 2**. The temperature at which the specific heat reaches a peak is known as the *pseudo critical temperature,* T_{pc}. As the pressure is increased this temperature increases, the maximum value of the specific heat falls, and the variation of the other properties becomes less severe.

Forced Convection Heat Transfer at Supercritical Pressure

Convective heat transfer near the critical point is characterised by properties having rapid variation with temperature. As a consequence the flow and heat transfer processes are linked. Moreover, the equation describing the temperature distribution in the fluid is essentially non-linear, so that the proportionality between heat flux and temperature difference no longer exists; in these circumstances heat transfer coefficient is a parameter of doubtful utility which can take widely differing values depending on the conditions.

Forced convection heat transfer measurements in pipes to fluids at supercritical pressure have been made using a wide range of fluids: water, carbon dioxide, nitrogen, hydrogen and helium are examples. By far the most extensive collections of data are for water and carbon dioxide, and fortunately the accuracy of physical property data for these fluids is good. The incentive for work on water has been its use in supercritical pressure "boilers" for power stations, particularly in the U.S.A. and the U.S.S.R. For the experimentalist carbon dioxide is an easier fluid to handle because of its lower critical temperature and pressure (31°C and 73.8 bar, compared with 370°C and 221 bar for water).

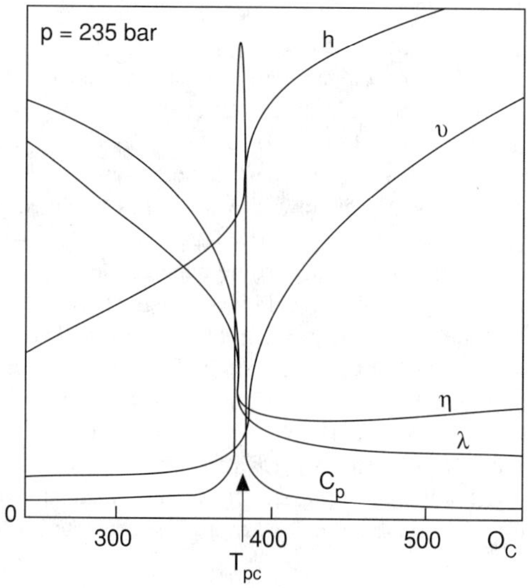

Figure 2.

Several reviews of supercritical pressure forced convection have been written: for example, Petukhov (1970), Hall (1971) and Hall, Jackson (1978). This account is largely based on the last of these.

General Characteristics

Low heat fluxes

At very low heat fluxes, and therefore with small temperature variations in the fluid, one approaches the case of constant properties. Their actual value in this small temperature range will, of course, depend upon where the temperature lies in relation to the critical temperature. Thus, while the physical properties may remain almost uniform throughout the fluid, they may be made to traverse a wide range of values by adjusting the bulk mean temperature of the fluid in the limiting case of a very small heat flux. There is no reason why constant property equations should not work. Consider the well-known expression:

$$\mathrm{Nu} = \mathrm{const.}\ \mathrm{Re}^{0.8}.\ \mathrm{Pr}^{0.4} \quad (1)$$

where Nu is the **Nusselt Number** (= $\alpha D/\lambda$), Re is the **Reynolds Number** (= $\dot{m}D/\eta$) and Pr is the **Prandtl Number** (= $\eta c_p/\lambda$), in which α is **Heat Transfer Coefficient** and $\dot{m}$ is mass flow rate per unit area.

From this it follows that:

$$\alpha = \mathrm{const.}\dot{m}^{0.8} D^{-0.2} \lambda^{0.6} \eta^{-0.4} c_p^{0.4} \quad (2)$$

Noting that the variations with temperature, of the thermal conductivity λ, and the viscosity η are rather similar in form (**Figure 2**) one would not expect a major effect on the heat transfer coefficient on this account. The specific heat on the other hand, becomes very large in the vicinity of the pseudocritical temperature, and one might expect the heat transfer coefficient to behave in a similar manner. This, in fact, happens, as is well illustrated by the data of Yamagata et al for water (**Figure 3**). It will be noted that as the heat flux $\dot{q}$ increases the magnitude of the peak in the heat transfer coefficient decreases. This is because the region of fluid at high Prandtl Number contracts as the heat flux, and hence the temperature gradient, increases.

Intermediate and high heat fluxes

Heat flux is not the sole determinant of the effects which are about to be described. However, for a given size and orientation of pipe and a given flow rate, the effects are very largely determined by heat flux. Under some conditions the orientation of the pipe becomes important, and very different results are obtained depending upon whether the pipe is horizontal or vertical, and whether in the latter case the flow is upwards or downwards.

In cases where orientation is not important the deterioration in heat transfer coefficient with heat flux that was evident in **Figure 3** becomes more marked both with decreasing mass velocity and increase of heat flux. **Figure 4** shows the water data of Vikrev et al. for a constant mass flow and three levels of heat flux, the highest of which results in virtually no enhancement of α as the bulk temperature traverses the pseudo-critical temperature.

Turning to data in which the orientation of the pipe is important, we find that under some conditions of pipe diameter, mass flow and heat flux, heat transfer to a fluid is much better in a vertical pipe when the flow is in the downward rather than upward. When it was first observed, this behaviour was difficult to reconcile with the currently held view that forced and free convection reinforced each other when they acted in the same direction. An example of this behaviour is shown in **Figure 5** where the data of Jackson and Evans-Lutterodt for carbon dioxide at two different heat fluxes are shown for both upward and downward flow. It will be seen that there is a rapid deterioration in heat transfer following a relatively small increase in heat flux with upward flow but not with downward flow.

An interesting comparison between carbon dioxide data that is unaffected by the direction of flow and with data that is so affected is shown in **Figure 6**. All three sets of data have been selected so that, in the absence of any gravitational effect, they would be expected to yield similar distributions of wall temperature against

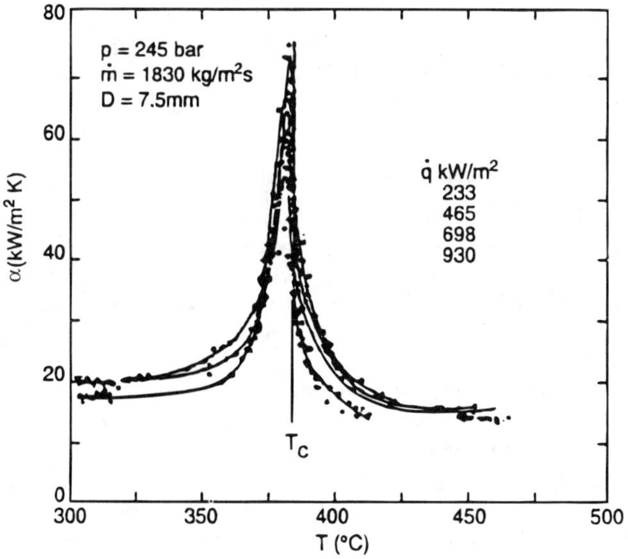

Figure 3.

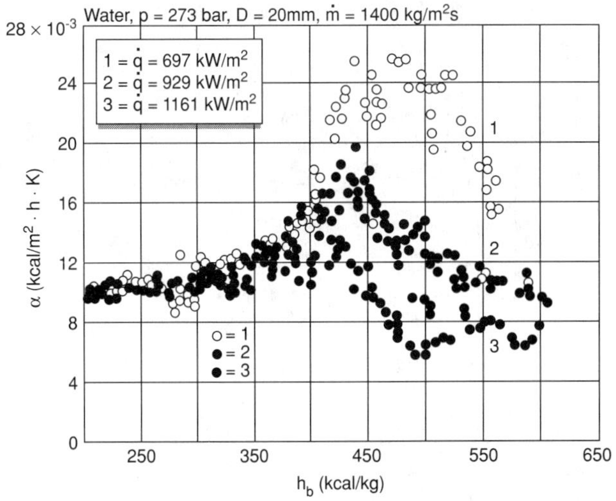

Figure 4.

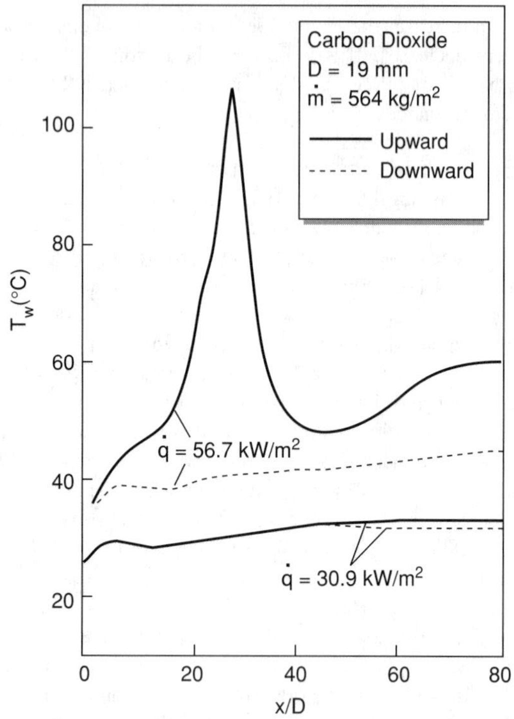

Figure 5.

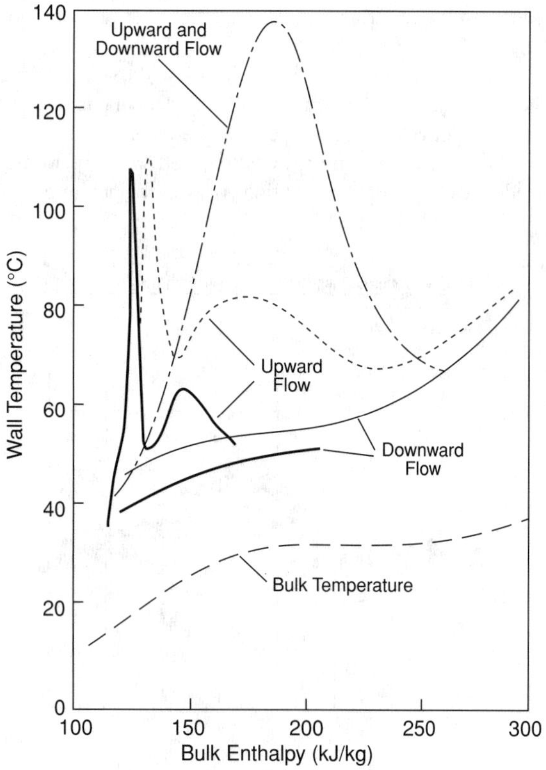

Figure 6.

bulk enthalpy. In the case of one set of data (due to Shiralkar and Griffith), no difference between upward and downward flow was found, whereas the other two sets (due to Jackson and Evans-Lutterudt and Bourke et. al.) showed such a difference. The sharp temperature peaks appear to be characteristic of upward heated flows in which gravitation is important. Note that, following these peaks, heat transfer is better than it would have been in the absence of gravitational effects, and also that in downward flow there is substantially improved heat transfer with no wall temperature peaks.

A Criterion for buoyancy affected flow

The examples of the effect of orientation of the pipe that have been quoted above confirm the importance of interactions between forced and free convection. The sharp peaks in the wall temperature that occur with upward heated flows were first attributed to "*pseudo film boiling*". The true nature of the phenomenon was revealed when results for both upward and downward flow became available, and it is now possible to say with some confidence whether it is likely to occur in given circumstances. The exclusion of data affected by buoyancy is clearly necessary preliminary before any attempt can be made to correlate data for pure forced convection.

An explanation of the mechanism by which buoyancy affects heat transfer has been proposed by Hall and Jackson (1969). The dominant factor is the modification of the shear stress distribution across the pipe, with a consequential change in turbulence production.

It can be shown that a criterion for negligible buoyancy effects is:

$$\overline{Gr}_b/Re_b^{2.7} < 10^{-5} \quad (3)$$

where $\overline{Gr}_b$ is the **Grashof Number** ($= g(\rho_b - \bar{\rho})D^3/\nu_b^2$), in which $\bar{\rho}$ is the integrated mean density and the subscript b indicates physical properties evaluated at the local bulk temperature.

The application of the above criterion is made difficult by the fact that evaluation of the Grashof Number requires that the wall temperature T_w is known, so that in applications where the wall heat flux is specified a correlation describing forced convection heat transfer (with negligible buoyancy effects) must be used in conjunction with the criterion.

Forced Convection in the Absence of Buoyancy

The enhancement of heat transfer coefficient at small temperature differences when the bulk temperature is close to the pseudo-critical temperature has already been noted and attributed to the large value of c_p in this region. As the temperature difference is increased, the proportion of the flow experiencing this high specific heat shrinks, and the heat transfer coefficient is reduced. If we take the temperature difference as the main factor affecting enhancement of the heat transfer coefficient (for a given fluid and bulk temperature) then it follows that the effect should be characterised by a quantity such as $\dot{q} / \dot{m}^{0.8}$. Attempts have been made to express the ratio of the heat transfer coefficient to that for constant properties in terms of the quantity $\dot{q} / \dot{m}$; typical of these are the generalised curves of Lokshin for water at 250 bar (**Figure 7**). It is seen that above a value of $\dot{q} / \dot{m} \approx 0.7$ no enhancement occurs and there is

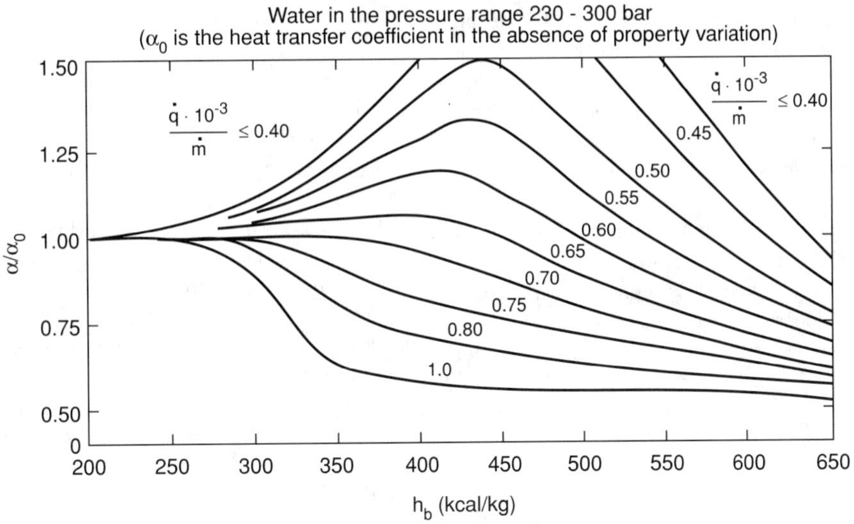

Figure 7.

a monotonic decrease in α as the fluid bulk temperature traverses the pseudocritical temperature.

Many attempts have been made to correlate forced convection data in terms of conventional dimensionless groups with modifications to allow for the effect of physical property variation. Jackson et al. (1975) have made a study of virtually all known correlations and have compared their performance when tested against forced convection data for water and for carbon dioxide. The fact that they were able to exclude data that may have been affected by buoyancy constituted an important advance over earlier attempts to correlate forced convection data. They found that by far the most effective correlation was that published in 1965 by Krasmochekov and Protopopov (1966). This utilises the Petukhov Kirrillov equation for constant properties forced convection and accounts for property variation by additional terms involving wall to bulk density ratio and integrated to bulk specific heat ratio, each raised to suitable powers. It correlated 80% of the data points to within ± 15% and 91% to within ± 20%. Because the Petukhov Kirrillov equation was rather cumbersome to use Jackson and Fewster also tried the following simpler version in which a *Dittus Boelter* form of constant properties forced convection equation is used instead:

$$Nu_b = 0.0183\, Re_b^{0.82}\, Pr_b^{0.5} \left(\frac{\rho_w}{\rho_b}\right)^{0.3} \left(\frac{\bar{c}p}{cp_b}\right)^{\eta} \quad \textbf{(4)}$$

where suffix w refers to conditions at the wall,

$$\bar{c}_p = \int_{T_b}^{T_w} \frac{c_p dT}{(T_w - T_b)} = \frac{h_w - h_b}{T_w - T_b}$$

and

$$\eta = 0.4 \text{ for } T_b < T_w \le T_{pc} \text{ and for } 1.2\, T_{pc} \le T_b < T_w$$

$$\eta = 0.4 + 0.2\left(\frac{T_w}{T_{pc}} - 1\right) \text{ for } T_b \le T_{pc} < T_w$$

$$\eta = 0.4 + 0.2\left(\frac{T_w}{T_{pc}} - 1\right)\left(1 - 5\left(\frac{T_b}{T_{pc}} - 1\right)\right)$$

$$\text{for } T_{pc} \le T_b \le 1.2T_{pc} \text{ and } T_b < T_w$$

In this form the equation correlated 77% of the data points to ± 15% and 90% to within ± 20%. Approximately 2000 data points were tested against the correlation.

Mixed Convection in Pipes

Vertical pipes

This topic has already received some discussion and a criterion for the onset of buoyancy effects has been given. Also the tendency towards the sudden occurrence of localised peaking of the wall temperature in upward flow with increase of heat flux has been noted. In view of this instability it is not surprising that the magnitude of the subsequent increase in wall temperature (for conditions of specified heat flux) has proved difficult to predict. For a detailed picture of the influences of buoyancy on heat transfer to fluids in vertical pipes, including fluids at supercritical pressure, see Jackson and Hall (1980).

For a downward heated flow there is a continuous enhancement in heat transfer as buoyancy becomes relatively stronger. This behaviour has been found with many fluids at supercritical pressure and also with other fluids. Not only is the heat transfer improved, but wall temperatures are less sensitive to heat flux.

Horizontal pipes

As might be expected, buoyancy in uniformly heated horizontal pipes causes a progressive increase in the temperature of the upper surface relative to the lower surface of the pipe. This is undoubtedly due in part to a stratification of the flow, the hotter and less dense fluid occupying the upper part of the pipe; there may also be an effect due to the damping effect of the stabilising density gradient on turbulence near the upper surface of the pipe. At the lower, surface heat transfer is frequently better than for forced convection alone, suggesting that there may be some amplification of turbulence by the destabilising density gradient in this region. For a detailed review of the effects of buoyancy on heat transfer to fluids flowing in horizontal tubes, including fluids at supercritical pressure, see Jackson (1983).

A considerable number of measurements on horizontal pipes have been made using fluids at supercritical pressure. A typical set of data presented by Belyakov et al. is reproduced in **Figure 8**. The deterioration of the upper surface occurs progressively along the pipe and does not show the sharp peaks that are obtained with upward flow.

Free Convection

Free convection in supercritical pressure fluids has received a good deal less attention than has forced convection, presumably because there have been fewer pressing industrial applications. The situation was reviewed by Hall (1971), and little further data have been reported since that time.

The general picture is one of enhancement of heat transfer at low heat fluxes with surface and fluid temperature spanning the pseudocritical temperature. As with forced convection, the enhancement is greatest when the bulk temperature is close to the critical temperature, but there is no evidence of the impairment of heat transfer that is found at high heat fluxes with forced convection.

Boiling

Nucleation at high sub-critical pressures

Nucleation at a boiling site of a given size becomes easier as the pressure of a boiling liquid increases. The decrease of surface tension means that a smaller difference between vapour and liquid pressure is required in order to make a nucleus grow; moreover, a given pressure difference can be achieved with a smaller superheat as the pressure level increases. For a given superheat, therefore, it will be possible to activate smaller nuclei at higher pressures.

In addition to this effect, there is the possibility of spontaneous nucleation occurring as a result of thermal fluctuations in a superheated liquid at a surface as p/p_c approaches unity. Indeed, it may be that the very "quiet" transition to film boiling that occurs at high pressure may be a consequence of spontaneous nucleation.

Pool boiling

The effect of pressure on nucleate boiling is well established: as the pressure is raised towards the critical pressure the heat transfer coefficient increases, and the limiting heat flux for transition from nucleate to film boiling first increases, reaches a maximum at a value of p/p_c of about 0.35, and then decreases.

Pool boiling experiments at high pressures show very clearly how the range of conditions under which nucleate boiling occurs is contracted by transition to film boiling.

Following transition to film boiling at pressures close to the critical value, a highly regular flow of vapour from the heated surface is observed. At $p/p_c \approx .9$ uniformly spaced columns of

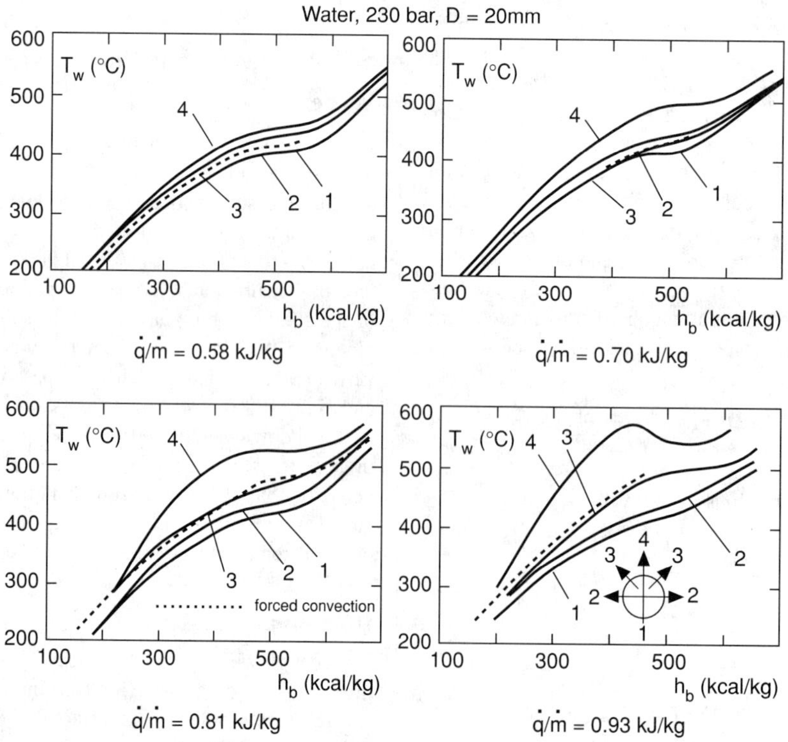

Figure 8.

bubbles are observed to rise from a heated wire. As the critical pressure is approached these columns change into "tubes" of vapour, again with very regular spacing, and finally change further into what appears to be a sheet of vapour rising from the wire.

Flow boiling

As indicated earlier, the general picture of supercritical pressure forced convection has become comparatively clear. In contrast, very few detailed investigations have been made at high subcritical pressures.

Figure 9 shows data for water flowing upwards in a 20mm diameter pipe for pressures extending from sub-critical to slightly supercritical, taken from the work of Herkenrath et al. It is of interest to compare the sub-critical curves, each of which involves a local heat transfer crisis, with the curve for supercritical pressure which does not. Note that the conditions are such that buoyancy induced temperature peaks would not be expected in the case of the supercritical data.

It will be seen from **Figure 9** that as the pressure is increased the position of the heat transfer crisis shifts towards the region of lower quality; the magnitude of the wall temperature step at first decreases, then increases and finally decreases as the critical pressure is approached. Furthermore, the shape changes from a sharp to a more gradual increase in wall temperature. Some data obtained by the present author and co-workers using carbon dioxide at a reduced pressure, $p/p_c = 0.99$, for both upward and downward flow showed an increase in wall temperature following transition which was rather gradual (the curve for $p = 215$ bar in **Figure 9**) and that in upward flow peaks develop due to buoyancy influence as they do at supercritical pressure. It was also apparent that at high subcritical pressures downward flow was advantageous, just as at supercritical pressures: transition occurred later in downward flow and the consequences were less severe. Experiments with carbon dioxide at lower pressure show a sharp transition rather similar to those illustrated in **Figure 9**.

The beneficial consequences of buoyancy in downward flow do not appear to persist as the pressure is dropped further, where the above mentioned trend appears to be reversed.

References

Hall, W. B. (1971) Heat transfer near the critical points, *Advances in Heat Transfer,* Vol. 6, Academic Press, New York.

Hall, W. B. and Jackson, J. D. (1969) Laminarisation of a turbulent pipe flow by buoyancy forces, *ASME Paper No. 69-HT-55.*

Hall, W. B. and Jackson, J. D. (1978) Heat transfer near the critical point, Keynote Lecture, *Sixth International Heat Transfer Conference,* Toronto, Canada.

Jackson, J. D. (1983) *Effects of Buoyancy on Heat Transfer to Fluids Flowing in Horizontal Heat Tubes,* Proc. H.T.F.S. Research Symposium, University of Bath, Paper RS 471, 97–115.

Jackson, J. D., Hall, W. B., Fewster, J., Watson, A. and Watts, M. J. (1975) *Heat Transfer to Supercritical Pressure Fluids,* H.T.F.S. Design Report No. 34, Part 1—Summary of design recommendations and equations, Part 2—Critical reviews with design recommendations, AERE-R8157 and R8158, U.K.A.E.A., AERE, Harwell.

Jackson, J. D. and Hall, W. B. (1980) Influences of buoyancy on heat transfer to fluids flowing in vertical tubes under turbulent conditions, 563–611, *Turbulent Forced Convection in Channels and Rod Bundles,* published by Hemisphere Publishing Corporation, USA (S. Kakacs and D.B. Spalding,) eds.

Krasnoschekov, E. A. and Protopopov, V. S. (1966) Experimental study of heat exchange in carbon dioxide in the supercritical range at high temperature drops, *Teplofizika Vysokikh Temperature,* Vol. 4, No. 3.

Petukhov, B. S. (1970) Heat transfer and friction in turbulent pipe flow with variable physical properties, *Advances in Heat Transfer, Vol. 6,* Academic Press, New York.

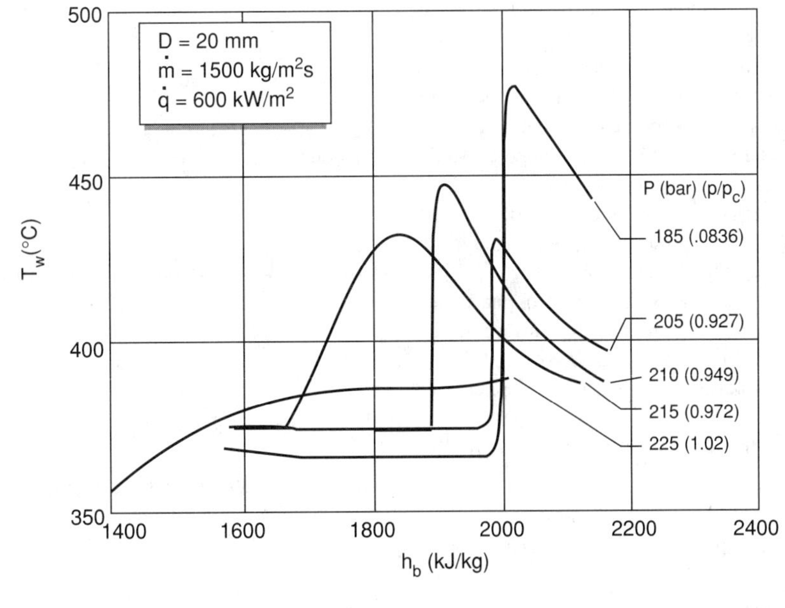

Figure 9.

J.D. Jackson

SUPERFICIAL VELOCITY (see Multiphase flow; Slip ratio)

SUPERHEATING (see Rankine cycle)

SUPER-PHENIX (see Liquid metal cooled fast reactor)

SUPERSATURATION (see Aerosols; Crystallization; Crystallizers)

SUPERSONIC EXTERNAL FLOW (see External flows, overview)

SUPERSONIC FLOW (see Compressible flow)

SUPERSONIC FLOW, IN NOZZLES (see Nozzles; Prandtl, Ludwig)

SUPERSONIC JET (see Jets)

SUPPRESSION OF NUCLEATE BOILING (see Boiling)

SURFACE ACTIVE SUBSTANCES

Surface active substances, also known as *surfactants,* are those substances which preferentially adsorb at the air–liquid, liquid–liquid or liquid–solid interfaces. The surface activity of a solute refers to a particular solvent. Molecules of surface active substances contain at least two distinct parts, a moiety which interacts strongly with the solvent, the *lyophilic* part, and another moiety the *lyophobic* part, whose interaction with the solvent is less than its interaction with molecules of a structure similar to its own. The lyophilic moiety may interact with the solvent through solvation, hydrogen; bonding or acid-base interactions in addition to van der Waals forces. This part of the molecule confers solubility on the molecule while the lyophobic part restricts or even prevents solution. In oil–water systems the *hydrophobic* group is usually a single or branched chain hydrocarbon containing 8–18 carbon atoms, the solubility of the molecule decreasing as the length of the chain increases. The balance between the two parts of the surface active molecule the *hydrophile–lyophile balance,* the *H.L.B.,* is critical for the performance of the substance in any particular application. The H.L.B. number originally proposed by Davies 1961, is an empirical and convenient measure in common use for the composition.

$$\text{H.L.B} = 7 + \sum \text{(hydrophobic group numbers)} - \sum \text{(hydrophilic group numbers)}$$

The group numbers are given in **Table 1**. **Table 2** shows the range of H.L.B. suitable for particular applications.

Table 1. Group H.L.B. numbers

Hydrophilic groups	HLB	Lipophilic groups	HLB
—SO_4Na	38.7	—CH—	
—COOK	21.1	—CH_2—	−0.475
—COONa	19.1	—CH_3—	
Sulfonate	about 11.0	—CH=	
—N (tertiary amine)	9.4	—(CH_2—CH_2—CH_2—O—)	−0.15
Ester (sorbitan ring)	6.8		
Ester (free)	2.4		
—COOH	2.1		
—OH (free)	1.9		
—O—	1.3		
—OH (sorbitan ring)	0.5		

Anionic surfactants are by far the commonest and used predominantly as *detergents.* Cationics being positively charged have a greater bacterial affinity and are used in medical applications and cosmetics. They now adsorb readily to negatively charged textile fibres and are efficient softeners and conditioners. The nonionic surfactants are finding increasing application as dispersing agents in the water borne paints industry, in emulsion technology and in the rheological behaviour of pastes, slurries and drilling muds. Details of individual surfactants are given in Rosen, 1972.

Table 2. The H.L.B. scale

Surfactant solubility behavior in water	HLB number		Application
	0		
No dispersibility in water	2		
	4		W/O emulsifier
Poor dispersibility	6		
Milky dispersion; unstable	8		
			Wetting agent
Milky disperison; stable	10		
Translucent to clear solution	12		
	14	Detergent	
Clear solution	16		O/W emulsifier
	18	Solubilizer	

References

Davies, J. T. and Rideal, E. K. (1961) *Interfacial Phenomena,* Academic Press, New York.

Rosen, M. J. and Goldsmith, H. A. (1972) Systematic Analysis of Surface-Active Agents, *Chemical Analysis, Vol. 12, 2nd Ed.,* John Wiley and Sons.

A.I. Bailey

SURFACE AND INTERFACIAL TENSION

Following from: *Interfaces*

If a small quantity of liquid is in equilibrium with its vapour, it will spontaneously assume a spherical form. This is also true for drops suspended in a liquid with which they are immiscible. The surface or **Interface** is in a state of tension and work would have to be done to extend the surface or interface. Thus there is an excess **Free Energy** associated with any system which contains a surface or interface. This free energy, σ, may be defined as the reversible work done to increase the surface or interface by unit area isothermally. The units are in J m^{-2} and the new surface or interface has identical properties to the original.

The surface is in a state of tension. The tension, however, does not increase as the surface is extended but remains constant at a value characteristic of the liquid under those conditions. It is a real force which can be used to support mass as for example in the flotation of heavy mineral ores where the particles of ore are brought to the surface by attachment to small bubbles rising through the liquid. The *surface tension,* γ, may be defined as the force/unit length parallel to the surface which is exerted perpendicular to any line drawn in the surface. Surface tension is sensitive to curvature of the surface only for very small radii of curvature (see interfaces). The units of surface tension are mNm^{-1}. The same arguments hold for the interface between two pure immiscible liquids. For such interfaces and for pure liquids in equilibrium with their vapour the excess free energy, σ, and the surface or interfacial tension, γ, are numerically equal and have the dimensions (MT^{-2}). In this work the symbol for both of these quantities will be σ.

An excess surface free energy is also associated with a solid in equilibrium with its vapour. Since the atoms or molecules in a solid have limited mobility a freshly formed solid surface is rarely in true equilibrium. The values of σ for solids are very large compared with those for liquids and are difficult to measure. Calculated values exist for many ionic crystals and cleavage and fracture methods of measurement have been used for single crystals of mica and for alkali halides at low temperatures where brittle fracture takes place. The surface tension for metals has been measured for pure metals by observing the extension and contraction of wires with temperature when they are hung outside evacuated crucibles made of the same metal as the wires and weights. Typical values of the surface tensions of some solids are shown in **Table 1**.

Molecular Theory of Interfacial Tension

The interface between two immiscible liquids is in a state of tension in a manner analogous to the liquid–vapour interface. This can be seen if droplets of one liquid are dropped into the bulk of another liquid of nearly equal density. The droplets assume a spherical shape. The molecules in the interfacial region are at a higher potential than those in the corresponding bulk. Girifalco and Good (1957) were the first to show how the interfacial tension could be expressed in terms of the intermolecular forces operating in the system. Three sets of forces must be considered when liquid one is in contact with liquid two. These are the forces acting between like molecules in liquid one and in liquid two plus those which act across the interface between molecules of liquids one and two.

These forces were expressed in terms of the surface tensions of the two liquids. The work of adhesion W_{12}^{ad} was then given by

$$W_{12}^{ad} = \phi(W_{11}^{coh}W_{22}^{coh})^{1/2} \quad \textbf{(1)}$$

where

$$W_{11}^{coh} = 2\sigma_1 + W_{22}^{coh} = 2\sigma_2 \quad \textbf{(2)}$$

and

$$W_{12}^{ad} = \sigma_1 + \sigma_2 - \sigma_{12} \quad \textbf{(3)}$$

ϕ is a constant characteristic of the system. Hence,

$$\sigma_{12} = r_1 + r_2 - 2\phi(\sigma_1\sigma_2)^{1/2} \quad \textbf{(4)}$$

A further contribution was made by Fowkes who assumed that the contributions made by different types of intermolecular interaction are additive and could be assessed separately. Hence the surface tension of a liquid would consist of two components, σ^d, due to dispersion forces and σ^P due to polar interactions which include, Keeson and Debye forces as well as hydrogen bonding. Pure hydrocarbon liquids interact with other liquids through dispersion forces only and the interfacial tension between liquids, a hydrocarbon and liquid two is then given by

$$\sigma_{12} = \sigma_1 + \sigma_2 - 2(\sigma_1^d\sigma_2^d)^{1/2} \quad \textbf{(5)}$$

The factor two allows for dispersion interactions between both liquids one and two and two and one. In addition he realized that if one partner in the pair of liquids acted as a Lewis acid while the other acted as a Lewis base, the interfacial tension between the two would be further reduced (Fowkes, 1972). The effect on the interfacial tension of changing the pH of the aqueous phase will indicate whether the water–immiscible organic liquid acts as a Lewis acid or a Lewis base. In the former case an increase of pH causes a lowering of the interfacial tension while the opposite is true in the case of a Lewis base when a lowering of the interfacial tension accompanies a decrease in pH (**Table 2**).

Just as the dispersion and polar contribution to the surface tension of liquids may be determined by combining the liquids with pure hydrocarbons, so the method may be extended to solid-liquid interfaces, thus from *Young's equation.*

$$\sigma_{SL} - \sigma_s = -\sigma_L \cos \Theta \quad \textbf{(6)}$$

$$= -[-\sigma_L + 2(\sigma_S^d\sigma_L^d)^{1/2}] \quad \textbf{(7)}$$

Table 1. Surface energy values

Material	σ mJm^{-2}	Medium	Method	Author
Cu	1650	Cu vapour	Wire-stretching	Udin (1951)
Ag	1140	He vapour	"	Funk, Udin and Wulff (1951)
	450	Air	"	Buttner, Funk, Udin (1952)
Au	1400	He vapour	"	Buttner, Udin and Wulff (1951)
	1210	Air	"	Buttner, Funk and Udin (1952)
Paraffin Wax	65	Air at 29.5°C	"	Greenhill and McDonald (1953)
Sn	685	Tin vapour	"	"
NaCl	300	Air	Splitting	Kuznetsov and Teterin (1937)
Mica	375	Air	Cleavage	Obreimov (1930)
	5000	10^{-6} Torr	"	"
	4500	"	"	Orowan (1933)
Mica	2400	Air	Dynamic cleaving	Lazarev (1936)
	1170	Water	"	"
Mica	300	Air	Cleavage	Bailey (1957)
Mica	5120	10^{-13} Torr.	"	Bryant (1962)
	150	Water vapour	"	"
LiF(100)	340	At 77°K	Cleavage	Gilman (1960)
MgO(100)	1200		"	"
CaF_2(111)	450		"	"
BaF_2(111)	280		"	"
$CaCO_3$(1010)	230		"	"
Si(111)	1240		"	"
Zn(0001)	105		"	"
Zn	90	At 77°K	"	Westwood and Kamdar (1963)
NaCl	400	Solution	Heats of Solution	Lipsett, Johnson and Maass (1927)
MgO	1090	"	"	Jura and Garland (1952)
NaCl	305	"	"	Benson and Benson (1955)
CaO	1310	"	"	Brunauer, Kantro and Weise (1956)
$Ca(OH)_2$	1180	"	"	"
Gypsum	370	"	Conductivity	Dundon and Mack (1923)
Glass	2000–4000	Air	Cone Fracture	Roesler (1956)

Bailey, A. I. 1957. Proc. 2nd Int. Cong. Surf. Act., III, 189.
Bailey, A. I. 1957. Ibid. 406.
Benson, G. C. and Benson, G. E. 1955. *Can. J. Chem.*, 33, 232.
Brunauer, S., Kantro and Weise 1956. *Can. J. Chem.*, 34, 729.
Bryant, P. J. 1962. Trans. Ninth Vac. Symp., U.S.A..
Buttner, F., Udin, H., and Wulff, J. 1951. Ibid. 1209.
Dundon, M. L. and Mack, E. 1923. *J. Amer. Chem. Soc.*, 45, 2479.
Funk, E. R., Udin, H., and Wulff, J. 1951. *J. Metals,* 3, 1206.
Buttner, F. H., Funk, E. R., and Udin, H. 1952. *J. Phys. Chem.*, 56, 657.
Gilman, J. J. 1960. *J. Appl. Phys.*, 36, 1374.
Greenhill, E. B. and McDonald, S. R. 1953. *Nature,* 171, 37.
Jura, G. and Garland, C. W. 1952. *J. Amer. Chem. Soc.*, 74, 6033.
Kuznetsov, V. D. and Teterin, P. P. 1937. *The Physics of Solids* by Kuznetsov, Vol. I, 388.
Lazarev, V. P. 1936. *Zhur. Fiz. Khim.*, 7, 320.
Lipsett, S. G., Johnson, F. M. G. and Maass, O. 1927. *J. Amer. Chem. Soc.*, 49, 925.
Obreimov, I. V. 1930. *Proc. Roy. Soc.*, A127, 290.
Orawan, E. 1933. *Zeit. f. Phys.*, 82, 235.
Udin, H., Shaler, A. J., and Wulff, J. 1949. *J. Metals,* 1, 186.
Udin, H. 1951, *J. Metals*, 3, 63.
Roesler, F. C. 1956. *Proc. Phys. Soc.*, B69, 981.
Westwood, A. R. C. and Kamdar, M. H. 1963. *Phil. Mag.*, 8, 7, 87.

Table 2. Gives values of σ_L and σ_L^d for a number of liquids

	σ_L	σ_L^d
Trichlorobiphenyl	45.3 mNm^{-2}	44 mNm^{-1}
Methyl iodide	50.8	48.5
Glycerol	63.4	37.0
Formancide	58.2	39.5

$$\text{i.e. } \cos\Theta = -1 + 2(\sigma_s^d)^{1/2}\left[\frac{(\sigma_L^d)^{1/2}}{\sigma_L}\right] \quad \textbf{(8)}$$

This relation has been tested experimentally and the results shown in **Figure 1**. The contact angles formed by a number of liquids on the solid are measured. σ_L^d is determined from interfacial tension measurements using pure hydrocarbons. Hence $(\sigma_S^d)^{1/2}$ can be obtained from the slope of the lines obtained by plotting Θ against $(\sigma_L^d)^{1/2}/\sigma_L$ (See also **Contact Angle**). Dann (1970) expressed the polar contribution to the interfacial tension in terms of the spreading coefficient, s, which can be measured.

$$s = \sigma_s - \sigma_{SL} - \sigma_L = -\sigma_L(1 - \cos\Theta) \quad \textbf{(9)}$$

If dispersion forces alone act we may calculate the spreading coefficient, s^d, from

$$\sigma_{SL} = \sigma_S + \sigma_L - 2(\sigma_L^d\sigma_S^d)^{1/2} \quad \textbf{(10)}$$

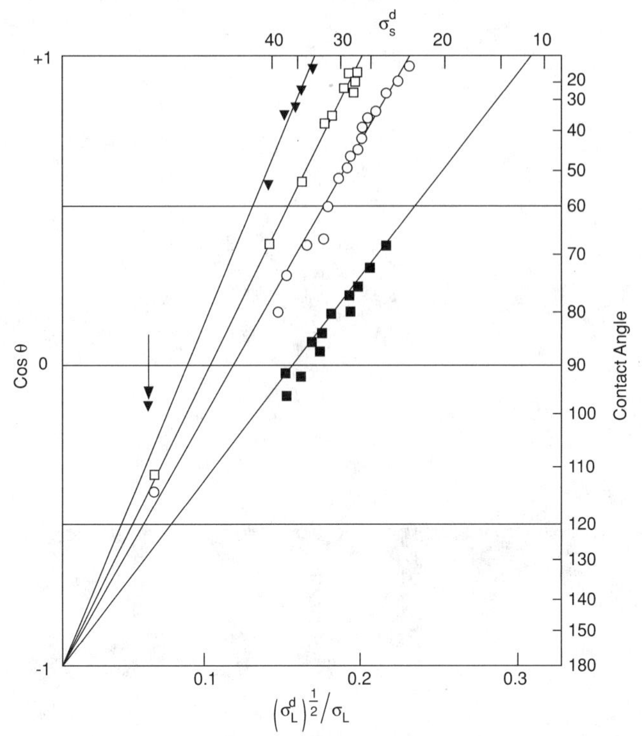

Figure 1. Contact angle of a number of liquids on low energy surfaces. ▼: polyethylene; □: paraffin wax; ○: $C_{36}H_{74}$; ■: fluorodecanoic acid monolayer on platinum. All contact angles below the arrow are with water.

giving

$$S^d = 2(\sigma_L^d \sigma_S^d)^{1/2} - 2\sigma_L \quad \textbf{(11)}$$

$$S - S^d = (\cos \Theta - 1)\sigma_L - 2(\sigma_L^d \sigma_S^d)^{1/2} + 2\sigma_L \quad \textbf{(12)}$$

$$= (\cos \Theta + 1)\sigma_L - 2(\sigma_L^d \sigma_s^d)^{1/2} = \sigma_{SL}^p \quad \textbf{(13)}$$

Figures 2a and **2b** shows σ_{SL}^P plotted against σ_L^P for paraffin wax in which the only interactions would be dispersion interactions and for polystyrene where polar interactions would be expected. The results confirm that for paraffin wax the interaction is constant no matter how polar the liquid. The methods provide means by which the polar and dispersion contributions for the interaction between liquids and solids may be determined.

Methods of Measuring Interfacial Tension

Capillary rise method

The difference in pressure across a curved meniscus results in the phenomenon of *capillarity.* A liquid which meets the surface of a fine pore or tube will rise in it if the end of the tube dips into a pool of the liquid. Conversely a liquid which makes a **Contact Angle** which is greater than 90° will experience a depression. This phenomenon provides a simple and accurate technique for the measurement of the surface tension of a liquid as shown in **Figure 3** (See also **Capillary Action**).

If a tube which is wet by a liquid is dipped into it, liquid rises up the tube to a height h. This is due to the Laplace pressure difference across the curved meniscus formed in the capillary. Equating pressures, we have:

$$\frac{2\sigma}{r} = h(\rho_L - \rho_V)g \quad \textbf{(14)}$$

where r is the radius of the tube; ρ_L the density of the liquid; and ρ_V the density of the vapour.

If the liquid forms a finite contact angle, Θ, with the tube, then

$$\frac{2\sigma \cos \Theta}{r} = h(\rho_L - \rho_V)g \quad \textbf{(15)}$$

The method is only accurate for $\Theta = 0$.

If the radius of tube is appreciable, then r is replaced by b where b refers to the radius of curvature of the liquid, and a relation between b and r must be found. The tube must be of uniform bore and the height h must be measured from the flat surface in the reservoir. If sufficient liquid to fill a large reservoir is not available,

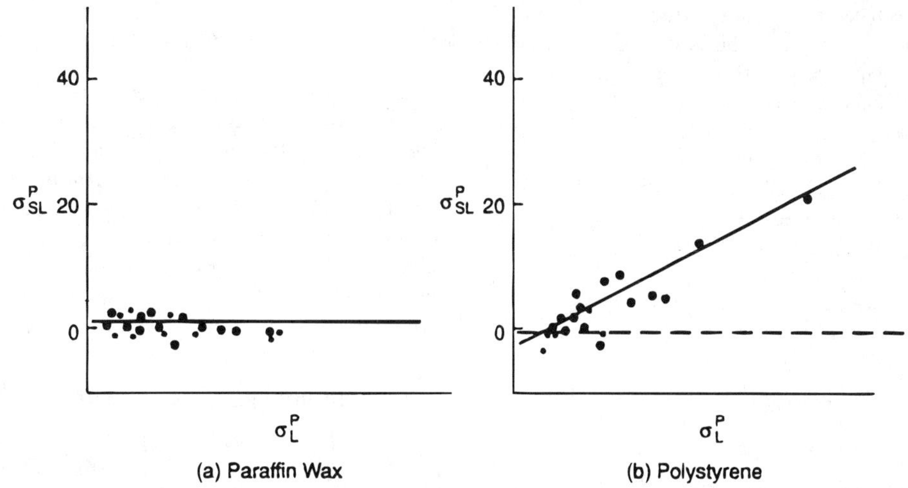

Figure 2.

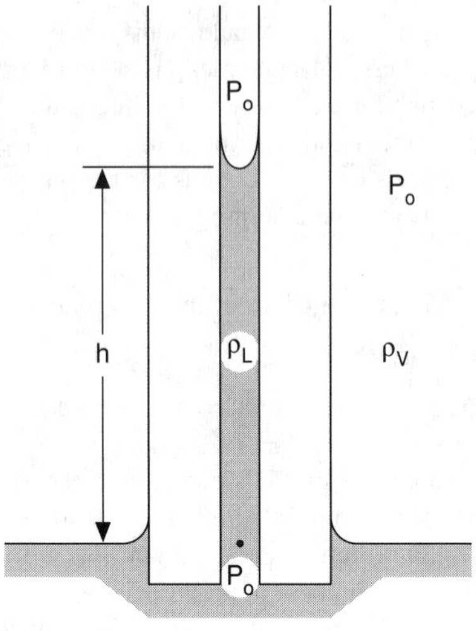

Figure 3. Capillary rise method.

a differential method using two capillaries of different bore may be used, thus

$$\sigma = \frac{r_1}{2} h_1(\rho_L - \rho_V)g = \frac{r_2}{2} h_2(\rho_L - \rho_V)g \qquad \textbf{(16)}$$

or

$$2\sigma = \left(\frac{r_1 r_2}{r_1 - r_2}\right)(h_1 - h_2)(\rho_L - \rho_V)g \qquad \textbf{(17)}$$

For samples of very small volume Fergusons' modification of the capillary rise may be used as shown in **Figure 4**. Here the pressure required to force a meniscus located at the end of a capillary into planar form is measured. Θ must be zero.

The drop volume method

This is one of the commonest and most convenient.

The volume or weight of drops which form slowly from the tip of a vertical tube is measured. From **Figure 5** the upward force balancing the drop gives $2\pi r\sigma = mg$. This is an oversimplification. **Figure 6** shows the sequence of processes involved as a drop detaches from the top. Finally, two drops, one large and one small,

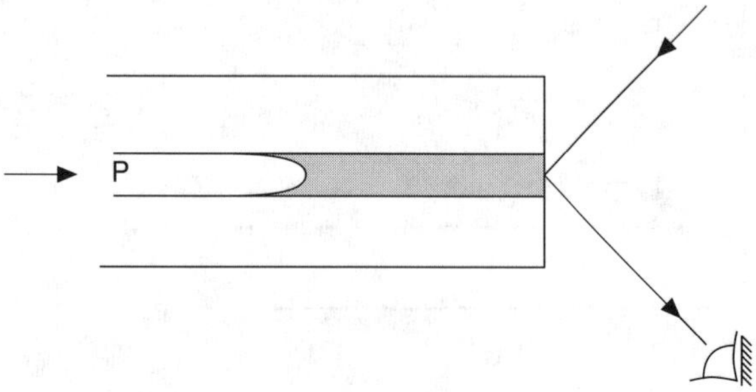

Figure 4. Fergusons' modified method.

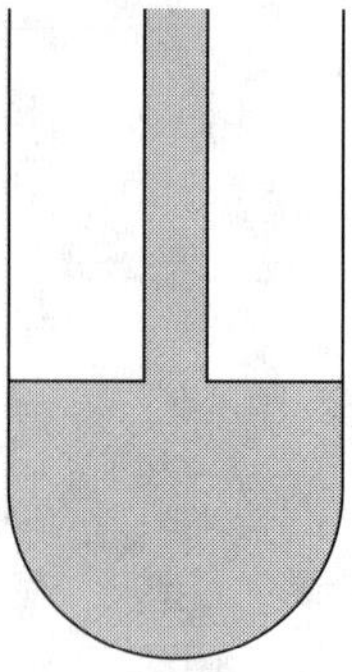

Figure 5. Drop volume method.

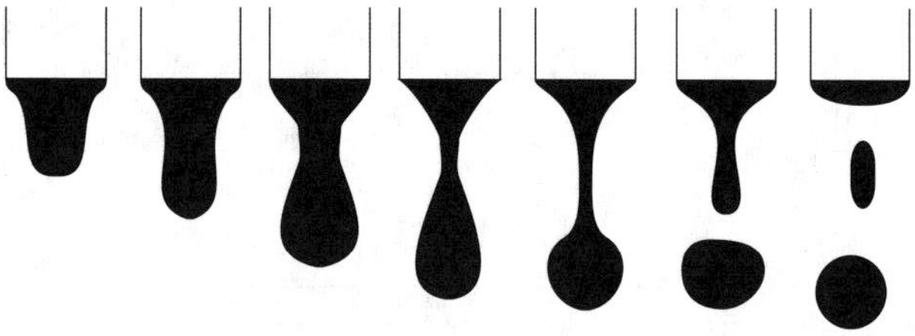

Figure 6. Drop detatchment sequence.

break away and both are measured. Some liquid is retained on the tip. Careful work shows that the weight of drops is a function of $r/_V{}^{1/3}$ where r is the radius of the tube and V is the volume of drops. Hence

$$2\pi r\sigma f\left(\frac{r}{V^{1/3}}\right) = mg \qquad \textbf{(18)}$$

or

$$\sigma = \frac{mg}{2\pi r}\,\frac{1}{f\left(\frac{r}{V^{1/3}}\right)} = \frac{mg}{2\pi r}F \qquad \textbf{(19)}$$

Correction tables giving F for values of $r/V^{1/3}$ have been made. The form of the correction is shown in **Figure 7**. The accuracy of the method is 0.1% where for pure non-volatile liquids it is 0.02%. The weight or volume of a number of drops is determined. The tip of tube must be ground flat and r measured. The final stages of formation of each drop must take place slowly. If the reservoir is adjustable in the vertical direction, the drop may be formed initially using a relatively large head which is later reduced. The drop must be in equilibrium with the vapour.

Maximum bubble pressure method

A bubble is blown at the tip of a tube dipping in a liquid. As the bubble grows the radius of curvature of the interface decreases at first, reaches a minimum and then increases again, as shown in **Figure 8**.

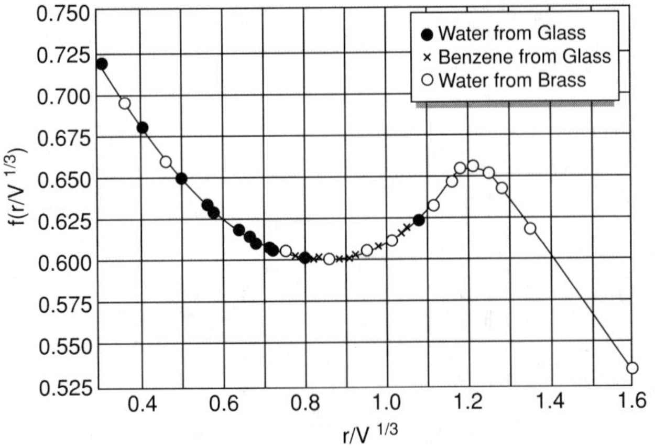

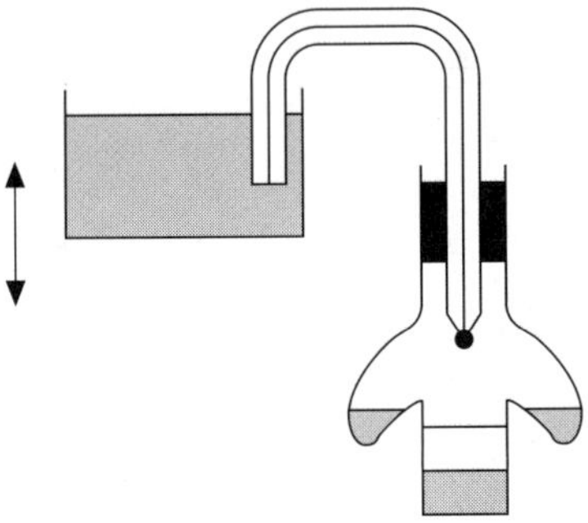

Figure 7. Correction Factor, F.

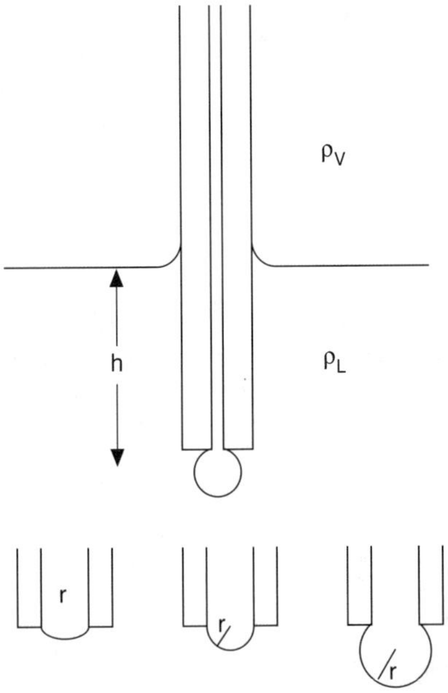

Figure 8. Maximum bubble pressure method.

At maximum pressure

$$P_0 + P = h(\rho_L - \rho_V)g + \frac{2\sigma}{r_1} \quad \textbf{(20)}$$

where $P_o = h(\rho_L - \sigma_V)g$ is the part of the pressure required to force the liquid down the tube to a depth h. Since the surface of the liquid in the reservoir must be flat the experiment is usually carried out as a differential method using tubes of different bore at the same depth in the liquid.

$$P_0 + P_1 = h(\rho_L - \rho_V)g + \frac{2\sigma}{r_1} \quad \textbf{(21)}$$

$$P_0 + P_2 = h(\rho_L - \rho_V)g + \frac{2\sigma}{r_2} \quad \textbf{(22)}$$

therefore

$$P_1 - P_2 = 2\sigma\left(\frac{1}{r_1} - \frac{1}{r_2}\right) \quad \textbf{(23)}$$

The tip of the tube must be ground flat and the capillary diameter should be as small as possible.

The method has an accuracy of about 0.5% and can be used for interfacial tensions between liquids.

The Shapes of Sessile and Pendant Drops and Bubbles

The fundamental equation for the surface separation between two fluids is

$$\sigma = \left(\frac{1}{r_1} + \frac{1}{r_2}\right) = P_0 + z(\rho_L - \rho_V)g \quad \textbf{(24)}$$

where P_0 is the pressure at the apex, 0, of the drop or bubble at which point $r_1 = r_2$ and z is the vertical height of the surface from 0. z is measured positive towards the concave side of the interface.

For any point S in **Figure 9** draw SC perpendicular to the interface cutting the axis at C. $S\hat{C}O = \phi$. Let the radius of curvature in the plane of the paper at $S = r_1 = R$. And the radius of curvature at right angles to this is r_2 where

$$\frac{x}{r_2} = \sin\phi \quad \text{therefore} \quad r_2 = \frac{x}{\sin\phi} \quad \textbf{(25)}$$

therefore

$$\sigma\left(\frac{1}{R} + \frac{\sin\phi}{r}\right) = P_0 + gz(\rho_L - \rho_V) \quad \textbf{(26)}$$

$$= \frac{2\gamma}{b} + gz(\rho_L - \rho_V) \quad \textbf{(27)}$$

where b is the radius of curvature at 0. Hence,

$$\frac{1}{R/b} + \frac{\sin\phi}{x/b} = 2 + \frac{z}{b}(\rho_L - \rho_V)\frac{gb^2}{\sigma} \quad \textbf{(28)}$$

Put

$$\frac{gb^2}{\sigma}(\rho_L - \rho_V) = \beta, \quad \textbf{(29)}$$

then

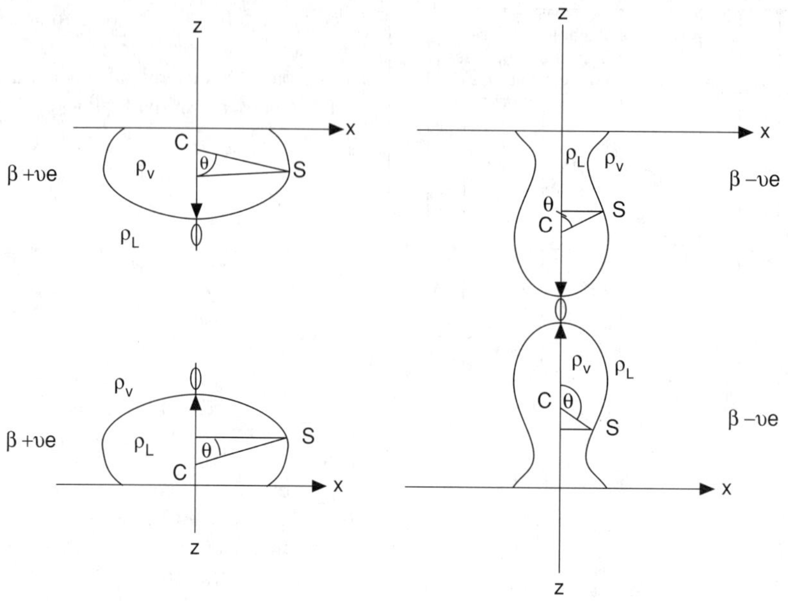

Figure 9. Sessile and pendant drops and bubbles.

$$\frac{1}{R/b} + \frac{\sin \phi}{x/b} = 2 + \beta \frac{x}{b} \quad \textbf{(30)}$$

b is the parameter which determines the scale of the meniscus while β is the parameter which determines the shape of the meniscus.

β is positive if the fluid above the interface is lighter and negative if the fluid above the interface is denser.

This equation has been solved numerically. Tables exist giving x/b z/b and V/b_3, for given values of ϕ and β. Here V is the volume of the drop (See Bashforth and Adams, 1883) (See also **Drops**).

Sessile drops and bubbles

Results are found to be independent of the contact angle. The method is static and therefore independent of viscosity.

The drops or bubbles are photographed and measured accurately. If the drop is large enough to make the curvature at 0 negligible the only measurements required are the diameter of the equatorial plane and the height, H, from this plane to the apex.

Then

$$\sigma = \frac{1}{2}H(\rho - \sigma)g \quad \textbf{(31)}$$

Correction tables for H/r_e, where r_e is the equatorial radius are given for cases when the curvature at 0 is not negligible. (Fordham, 1948)

Pendant drops and bubbles

Drops are formed on the tip of a tube and photographed.

The maximum horizontal diameter, d_e, of the drop is measured and the width, d_s, at a height, d_e, from the apex as illustrated in **Figure 10**. We may define:

$$S = \frac{d_s}{d_e} \quad \textbf{(32)}$$

and

$$H = -\beta \frac{d_e^2}{d_e} \quad \textbf{(33)}$$

$$= (\rho_L - \rho_V)g \frac{d_e^2}{\sigma} \quad \textbf{(34)}$$

Tables of 1/H for values of S have been compiled (Fordham, 1948). The drop must be protected from vibrations and d_e must be measured accurately to avoid large errors in d_s.

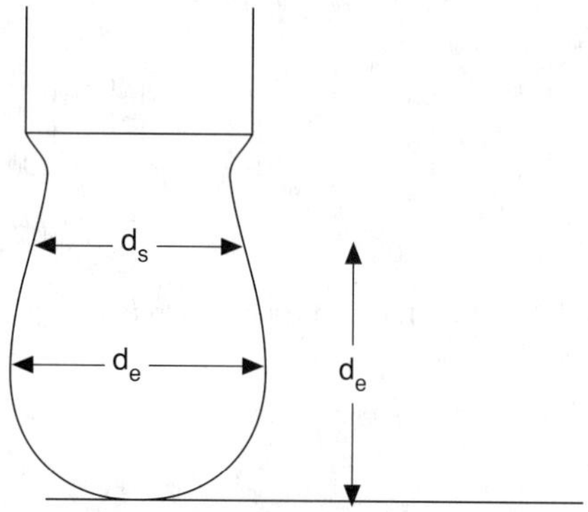

Figure 10. Pendant drops and bubbles.

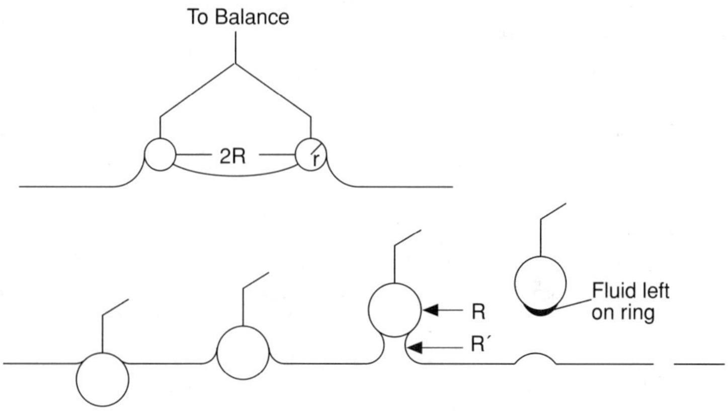

Figure 11. du Noüy ring plate method.

Detachment Methods

du Noüy ring, plate

Surface tension may be determined by measuring the force necessary to detach objects of known shape from the surface. As the object is lifted it lifts a film of fluid with it, which at a certain height becomes unstable and breaks away. Stages of detachment are seen in **Figure 11** from which it is evident that R is in error because the meniscus becomes concave before instability occurs, and that some fluid remains attached to the ring after detachment, so

$$4\pi R\sigma = mg\,F \qquad \textbf{(35)}$$

where F is a correction factor. If V is the volume of the meniscus tables give R^3/V for the values of R/r.

The maximum force required to detach the ring is measured together with the density of the liquid and the radii of the ring and wire.

A zero contact angle is necessary, vibrations must be avoided and the ring must be accurately parallel to the surface. The static Wilhelmy plate method is more reliable.

The Wilhelmy plate

May be used as a detachment method in which case a thin plate dipping in a fluid is gradually raised and the maximum pull at the point of detachment measured, as shown in **Figure 12**.

For $\Theta = 0$. maximum pull $= 2\sigma\,(l + t)$ where l is the length of the plate and t its thickness.

The accuracy of the method is about 0.1%.

Static Methods

In this method the plate is not detached. If W is the weight of the plate and W′ its weight on immersion, then

$$W' = W + \sigma p - b \qquad \textbf{(36)}$$

where p is the perimeter of plate ($p = 2(1 + t)$) and b is buoyancy correction. b is 0 if the plate is immersed as shown in **Figure 13**.

The weight, w, of the plate may be offset if an electrobalance is used. Hence

$$mg = 2\sigma(1 + t) = V(\rho_L - \rho_V)g \qquad \textbf{(37)}$$

Accuracy 0.01 mN/m with care.

The spinning drop method

This method is particularly suited to the measurement of very low interfacial tensions. A drop of liquid A is placed in a tube filled with liquid B which has a higher density than A. On spinning the tube as shown in **Figure 14** the drop of A moves to the axis of the tube and with increasing velocity of rotation, w, the drop becomes ellipsoidal and finally an elongated cylinder. Rotational velocities of about 20,000 r.p.m. may be used. Consider an element of the cylinder of volume V. The centrifugal force on it is $w^2r^2\,\Delta\rho/2$. Integrating for a cylinder of length l is $\pi w^2\Delta\rho r_0^4\, l/4$, and the interfacial free energy is $2\pi r_0 l\sigma$. The total energy, E is thus

$$E = \tilde{\omega}^2\Delta\rho r_O^2 V/A + 2V\sigma/r_O \qquad \textbf{(38)}$$

since $V = \pi r_0^2 l$. Putting $dE/dr_0 = 0$ we obtain

$$\sigma = \frac{\tilde{\omega}^2\Delta\rho r_O^3}{A} \qquad \textbf{(39)}$$

Values of interfacial tension as low as 10^{-3} mNm^{-1} can be measured

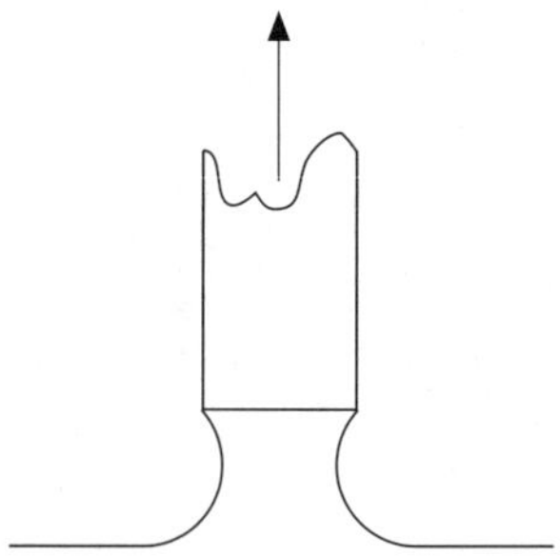

Figure 12. Wilhelmy plate method.

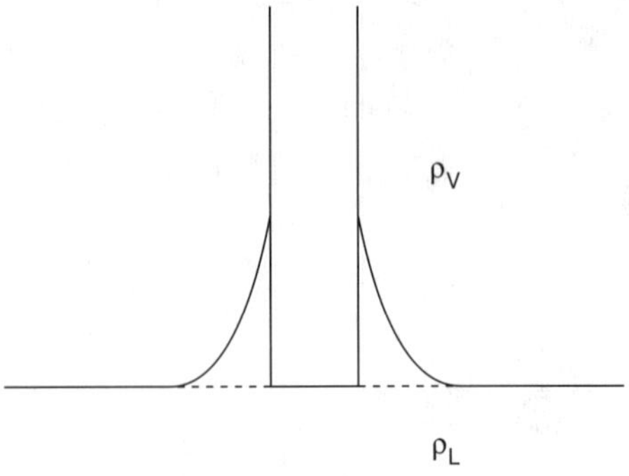

Figure 13. Static method.

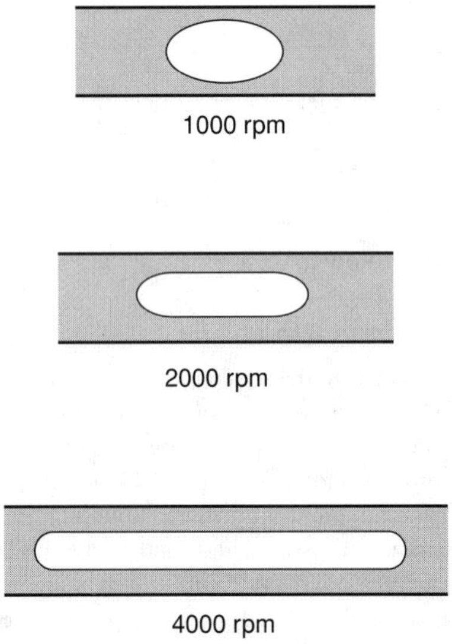

Figure 14. Spinning drop method.

readily and accurately. Precision bore tubing must be used and the apparatus constructed with precision. Account must be taken for the lens effect produced by the outer fluid.

Dynamic Methods

Ripples

Kelvin showed that the velocity of ripples on the surface of a liquid is given by

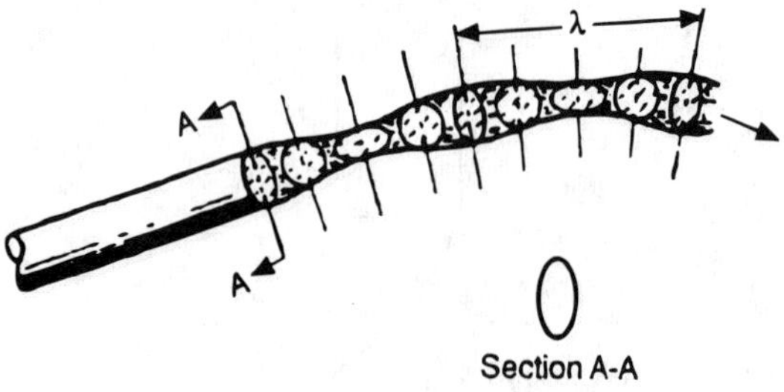

Figure 15. Oscillating jet.

Table 3. Choice of method

Method	Precision	Suitability
Sessile drops	good	1. good for liquid metals 2. good for ageing 3. good for viscous materials 4. use at high and low temperatures 5. surface or interfacial tensions possible
Pendant drops	good needs vibration isolation	1. viscous materials 2. diffusion of surface active species 3. ageing 4. use at high and low temperatures 5. surface and interfacial tensions
Capillary height	good needs precision tubing	1. not suitable for $\Theta \neq 0$ or $\Theta > 90°$ 2. not good for ageing 3. not good for viscous materials 4. not good for high temperatures 5. good for normal liquids—simple to carry out.
Ferguson method and pressure modification of capillary rise	good	1. liquid metals 2. high temperatures 3. not good for ageing 4. not good for viscous materials
Wilhelmy plate	good, if care taken	1. $\Theta = 0$ essential 2. not good for very viscous materials 3. good for ageing processes 4. not good for liquid metals or high temperatures 5. interfacial tension possible but not the best method for this
du Noüy ring	moderate	1. not good for very viscous materials 2. can be used for interfacial tension 3. quite good for ageing processes 4. not good at high temperatures 5. not good for liquid metals 6. $\Theta = 0$ essential 7. interfacial tension possible
Drop weight	good	1. not good for ageing processes 2. not good for viscous materials 3. can be used for liquid metals
Maximum bubble pressure	satisfactory	1. not good for ageing processes 2. not good for viscous materials but can be used 3. can be used for liquid materials 4. can be used for interfacial tensions but better methods available

$$v^2 = \frac{g\lambda}{2\pi} + \frac{2\pi\sigma}{\rho\lambda} \quad \textbf{(40)}$$

where λ is the wavelength of the wave and ρ is the density of the liquid.

Hence,

$$\sigma = \frac{\rho\lambda v^2}{2\pi} - \frac{g\lambda^2\rho}{4\pi^2} \quad \textbf{(41)}$$

Moving or stationary waves may be used. For moving waves stroboscopic methods are used.

Oscillating jets

If fluid emerges from a non-circular orifice the jet shows a number of oscillations as illustrated in **Figure 15**. Surface tension counteracts the non-circular shape of the jet, tending to bring it back to a circular cross-section. The momentum causes the jet to overshoot the circular form and so the oscillations are formed. The rate of flow is measured and hence the age of the surface at any point along the jet. The wavelength is measured photographically and from the variation of the wavelength along the jet, the surface tension corresponding to various ages may be determined. Surface ages of order of 2m sec may be measured by this method.

For more details on dynamic methods see original papers (Fordhan 48 & Rideal & Sutherland 1952).

Comprehensive tables of values for surface tension can be found in Rosen (1972) and Quayle (1953).

References

Bashforth, F. and Adams, J. C. (1883) *An Attempt to Test the Theories of Capillary Action,* University Press, Cambridge, England.

Dann, J. R. (1970) *J. Coll. Sci.,* 302.

Dann, J. R. (1970) *J. Coll. Sci.,* 32, 321.

Davies, J. T. (1966) *Proc. Roy. Soc.,* 290A, 575.

Fordham, S. (1948) *Proc. Roy. Soc.,* London, A194 1.

Fowkes, F. M. *Attractive Forces in Interfaces. In Chemistry and Physics of Interfaces,* 1–12, Ed. D. E. Gushee, ed., American Chemical Society, Washington.

Fowkes, F. M. (1972) *J. Adhesion,* 4, 155.

Girifalco, L. A. and Good, R. J. (1957) *J. Phys. Chem,* 61, 904.

Quayle, O. R. (1953) Pt. 1, *Chem. Rev.,* 53.

Rideal, E. and Sutherland, K., (1952) *Trans. Far. Soc.,* 48, 1109.

Rosen, M. J. and Goldsmith, H. A. (1972) Systematic Analysis of Surface-Active Agents, *Chemical Analysis, Vol. 12, 2nd Ed.,* John Wiley and Sons.

General References

Adamson, A. W., *Physical Chemistry of Surfaces, 3rd Ed.,* John Wiley and Sons, New York.

Davies, J. T. and Rideal, E. K., (1961) *Interfacial Phenomena,* Academic Press, New York.

Ross, S. and Morrison, I., (1988) *Colloidal Systems and Interfaces,* John Wiley.

Leading to: *Capillary action; Contact angle; Drops*

A. Bailey

SURFACE CONDENSERS

Following from: *Condensation; Condensers*

In a surface condenser vapour is brought into contact with a solid surface which is cooled to a temperature below the saturation temperature of the vapour at its prevailing partial pressure. The surface is usually in the form of a "nest" or "bundle" of metal tubes, the coolant flowing inside the tubes and the vapour condensing on the outer or "shell-side."

The **Overall Heat Transfer Coefficient**, U (based on the condensing surface area), may be expressed in terms of local coefficients, α, via the sum of thermal resistances:

$$\frac{1}{U} = \underbrace{\frac{1}{\alpha_v}}_{\text{from vapour}} + \underbrace{\frac{1}{\alpha_f}}_{\text{through condensate}} + \underbrace{\frac{D_o \ln(D_o/D_i)}{2\lambda_w}}_{\text{through tube wall}} + \underbrace{F}_{\text{through fouling}} + \underbrace{\frac{D_o}{D_i \alpha_c}}_{\text{to coolant}}$$

where D_0,D_i are the tube outer and inner diameters and λ_W is the thermal conductivity of the tube material.

Heat exchange in single phase flows is usually impeded by the presence of insulating boundary layers. On condensation, however, the large reduction in volume as the vapour turns to condensate results in an inflow of vapour towards the surface; the heat transfer is impeded only by a thin film of condensate on the surface. As a result, the condensation heat transfer coefficient is usually higher than that on the liquid coolant side, the latter becoming the controlling process. The thermal resistance, F, of fouling on the coolant side is therefore an important consideration. Condensing heat transfer rates are typically two orders of magnitude higher than the rates for a gas on the shell side, so condensers do not generally require extended surface tubing. In some applications spirally grooved tubing has been used to aid drainage of the condensate film and to increase internal heat transfer by turbulating the coolant flow. If non-wetting can be maintained (e.g. by surfactants), *drop-wise condensation* occurs, reducing the areas of condensate film and resulting in a heat transfer coefficient some four times that of *film-wise condensation.*

Vapour and condensate heat transfer coefficients, α_v and α_f, were first combined by **Nusselt** in 1916 (see e.g. McAdams, 1954) in an expression for the effective condensing coefficient for a single horizontal tube assuming that the condensate forms a laminar film on the tube surface. However, the Nusselt formulation is less successful in predicting tube nest mean heat transfer coefficient. Vapour shear and inundation by condensate from tubes higher in the nest affects the condensate film. More importantly frictional pressure losses as the vapour passes between the closely pitched tubes lowers the partial pressure—and hence saturation temperature—reducing the heat exchange driving temperature difference. (See also **Condensation** and **Condensers**).

The essential objective in surface condenser design is to provide equal access of vapour to all the surface. Early attempts to provide overall heat transfer coefficients for the design of steam condensers took no account of the detailed layout of the tubenest, e.g. HEI (1978). Modern practice is to model the proposed nest on computer to calculate thermal performance and to ensure that any *non-condensable gases* present are extracted at the point of lowest pressure, e.g. Rhodes and Marsland (1993). Poorly designed nests may suffer excessive frictional pressure losses and contain regions where the tubes are blanketed by non-condensable gases.

Power station condensers are some of the largest heat exchangers in existence, typically containing some 20,000 tubes of 25mm outer diameter and 20m in length. Computer models

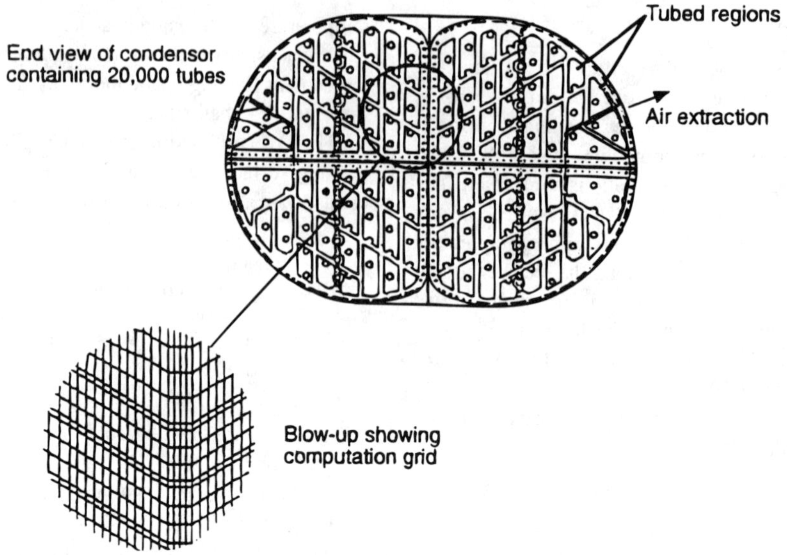

Figure 1.

usually represent the nest as an array of cells, as shown in **Figure 1**, solving simultaneously the equations of continuity, energy and momentum for each cell. For such models detailed correlations for local heat transfer and frictional pressure loss are required; developments in this complex field are reviewed by Davidson (1987). Surprisingly, there appears to be no optimum arrangement of tubes, the wide variety of nest configurations in service being shown by Lang (1987).

Where cooling water is not available, condensers are being designed for direct cooling by air. Arrays of large diameter fans blow the air across banks of finned tubing containing the condensing vapour, e.g. Knirsh (1990).

Surface condensers for process industries are usually smaller than power plant condensers and may contain a region for desuperheating the incoming vapour. Unlike power plant condensers, they may also be designed to sub-cool the condensate. Again a wide variety of arrangements are in use incorporating baffled shell-and-tube, gasketed plate and spiral plate constructions and plate-fin exchangers. Useful discussions of developments are given by Bell (1983).

References

Bell, K. J. (1983) Trends in design and application of condensers in the process industries, *Condensers: Theory and Practice,* The Institution of Chemical Engineers Symposium Series No. 75.

Davidson, B. J. (1987) Thermal design of condensers for large turbines, *Aerothermodynamics of Low Pressure Steam Turbines and Condensers,* Ch. 8, M. J. Moore and C. H. Sieverding, eds., Hemisphere Publishing Corp.

H. E. I. (1978) *Standards for Steam Surface Condensers,* Heat Exchange Institute, New York.

Knirsh, H. (1990) Design and construction of large direct cooled units for thermal power plants, PWR-26, *ASME Joint Power Conference.*

Lang, H. V. (1987) Steam condenser developments, *Aerothermodynamics of Low Pressure Steam Turbines and Condensers,* Ch. 8, M. J. Moore and C. H. Sieverding, Eds., Hemisphere Publishing Corp.

McAdams, W. H. (1954) Condensing Vapors, *Heat Transmission,* Ch. 13, McGraw Hill Book Co. Inc.

Rhodes, N. and Marsland, C. (1993) Improvement of Condenser Performance using CFD, *European Conference on the Engineering Applications of Computational Fluid Dynamics,* Inst. Mech. Engrs 7–8 September 1993, London.

M. J. Moore

SURFACE DIFFUSION (see Diffusion)

SURFACE EFFECTS ON BOILING (see Burnout in pool boiling)

SURFACE EFFICIENCY (see Extended surface heat transfer)

SURFACE ENERGY (see Interfaces; Surface and interfacial tension)

SURFACE EXTENSIONS (see Augmentation of heat transfer, single phase)

SURFACE FLOW VISUALIZATION (see Visualization of flow)

SURFACE TENSION (see Surface and interfacial tension)

SURFACE TENSION DEVICES (see Augmentation of heat transfer, single phase)

SURFACE TREATMENT (see Augmentation of heat transfer, two-phase)

SURFACTANT COLLECTORS (see Flotation)

T

TACHOMETRIC FLOWMETERS (see Flow metering)

TACONITE (see Iron)

TAIT EQUATION (see PVT relationships)

TAME (see Oils)

TANK COILS

Following from: *Coiled tubes, flow and pressure drop in; Coiled tubes, heat transfer in*

Tank coils are not generally used for the continuous heating or cooling of a flowing stream, but are usually applied in the heating or cooling of a liquid contained in a tank on a batch basis. The flow of heat into or out of the liquid, involves unsteady or transient heat transfer.

The heating and cooling media can flow through a coil immersed in the liquid as show on **Figure 1**, or the media can be made to flow through a coil fastened (welded) on the outside of the vessel as shown on **Figure 2**. The latter arrangement may be referred to as "*limpet coils*" whereas the former is usually termed "*coil in tank*". The limpet coil is an improvement on the simple arrangement where a jacket through which the heat transfer medium flows, fits onto the outside of the vessel as illustrated on **Figure 3**. The benefits of the limpet coil are due to the uniform fluid velocity through the channel and the good distribution of heat transfer medium around the vessel periphery.

The heat transfer is usually improved by agitation of the liquid contained in the tank. (See **Agitated Vessel Heat Transfer, Agitated Vessel Mass Transfer** and **Agitation Devices**) Unless agitation is employed the heat transfer at the vessel wall or across the coil will depend on natural convection within the liquid in the tank, which is not particularly efficient (see **Free Convection**).

In some applications, for inststance the manufacture of phenol-formaldehyde resin, the coils or jacket will serve two purposes.

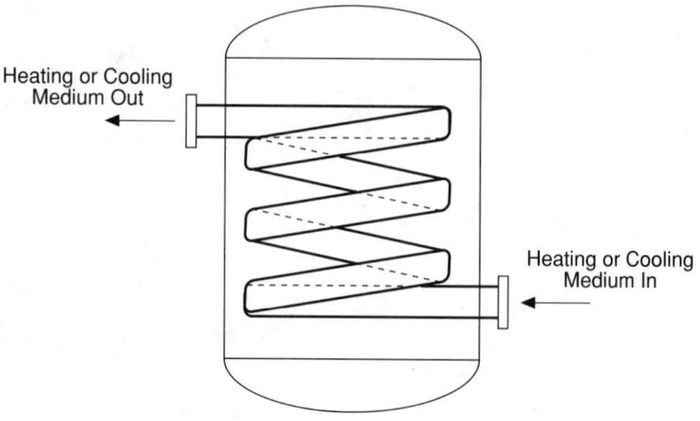

Figure 1. Vessel with internal coil.

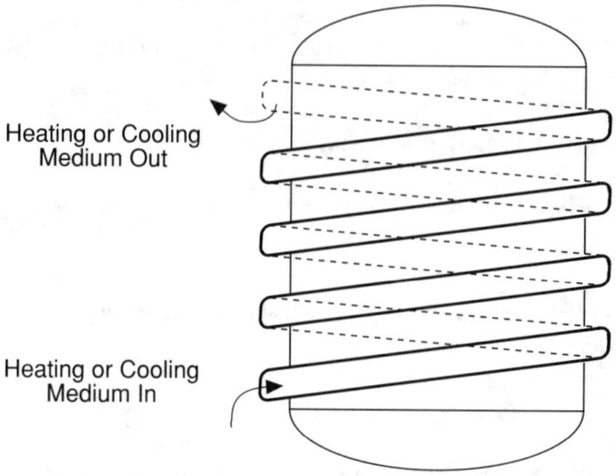

Figure 2. Vessel with external limpet coil.

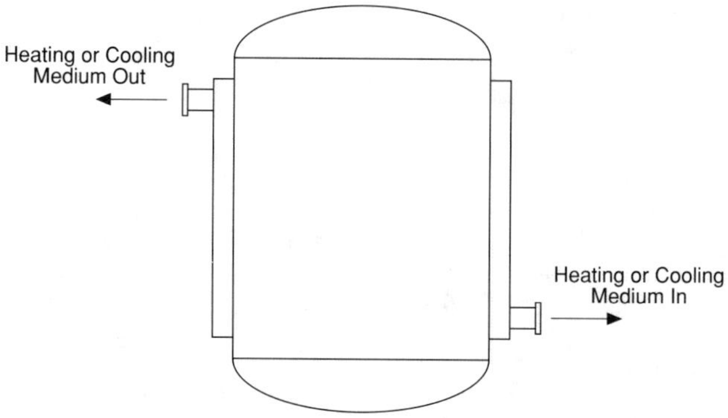

Figure 3. Vessel with external jacket.

Steam or some other heating medium will be used initially to raise the temperature of the mixture in the tank to the desired reaction temperature. At a later stage a cooling medium (usually water) will be introduced into the coils or jacket to control the temperature to avoid runaway exothermic reactions, and finally to cool the batch before discharge. In such processes the viscosity of the fluid will change as the reaction proceeds. Changes in viscosity will affect the degree of agitation imparted to the batch liquid and hence the rate of heat transfer will fall as the viscosity increases.

Fletcher (1987) has given some representative data on overall heat transfer coefficients that may be obtained in agitated vessels. The range of these data shown in **Table 1** illustrate some of the differences in heat transfer rates that may be experienced.

Table 1. Overall heat transfer coefficients in agitated vessels W/m^2K

Vessel design	Heating	Cooling
Simple jacket (mild steel vessel)	400 to 900	150 to 600
Simple jacket (glass lined vessel)	200 to 700	100 to 350
Limpet coil	600 to 1100	200 to 700
Internal coil	600 to 1500	250 to 800

Agitation

A number of different techniques are available to promote agitation within the bulk liquid. In general, they involve the rotation of some kind of blade system that either stirs the fluid or circulates it within the confines of the vessel. **Mixing** within vessels is a subject in its own right and only a brief outline is given here. Large diameter agitators that rotate at relatively low speeds, are usually employed where the liquid is viscous, in order to keep the power consumption for agitation as low as possible. Where low viscosity liquids are involved, smaller impellers operating at high speeds are often used. Where large viscosity changes are experienced during the processing, it is usual to employ low speed agitation. (See also **Agitation Devices**)

Five distinctly different agitatiors have found application in stirred tanks and include:

1. Anchor impellers as illustrated on **Figure 4** usually operate at low speed, much of the disturbance within the liquid occurs close to the vessel wall. This is beneficial for heat transfer across the wall to or from limpet coils or jacket.
2. Helical ribbon impellers may be used in certain applications where the cost may be justified. Agitation at the wall is achieved by the close clearances between the blade and the wall surface. The helical design also imparts turbulence within the core of liquid. The arrangement of a helical ribbon mixer is shown on **Figure 5**.
3. Paddle type impellers may be used at high or low speeds of rotation. The basic concept is illustrated in **Figure 6**. The blades may be flat or pitched. At low speed the flat blades produce a tangential motion to the liquid. Pitched blades operated at high speed establish a radial flow pattern. The ratio of paddle diameter to vessel diameter is usually in the range 1:3 to 2:3.
4. Propellers resembling ships' propellers as illustrated in **Figure 7**, usually operate at high speed and produce an axial flow pattern in the liquid. The ratio of propeller diameter to vessel diameter is generally about 1:3.
5. Turbine mixers operate at high speed in low viscosity liquids. To reduce capital cost and to facilitate cleaning, the design is usually simple. The blades may be flat or curved as illustrated on **Figure 8**. Radial flow is induced by flat blades, but an axial component can be obtained with curved blades. The number of blades will affect the degree of turbulence produced, but as the number of blades is increased, the power consumption will increase. The final choice is a compromise between the level of turbulence desired and the allowable energy cost. The disipation of energy may also produce a temperature rise in the liquid. The ratio of turbine diameter to vessel diameter is generally of the order of 1:3.

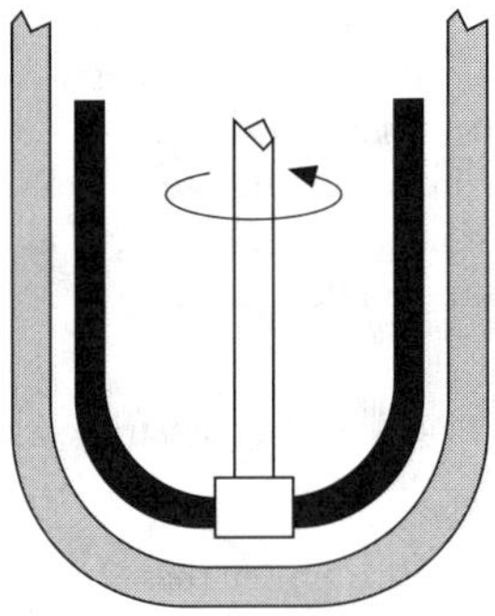

Figure 4. Anchor agitator.

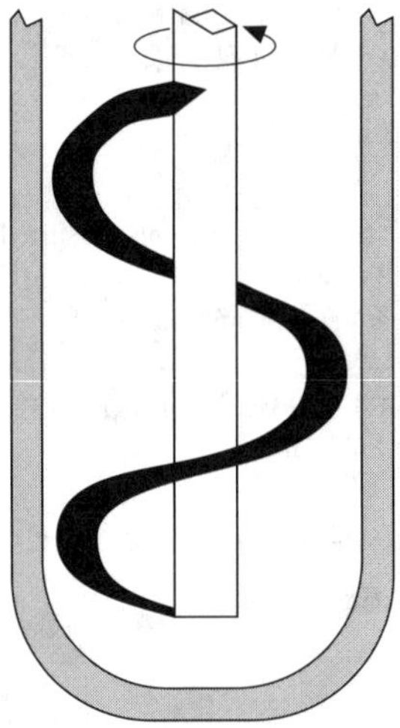

Figure 5. Helical ribbon impeller.

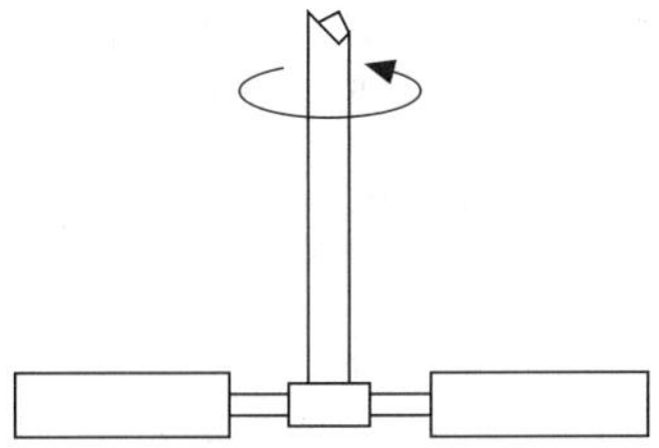

Figure 6. Paddle blades.

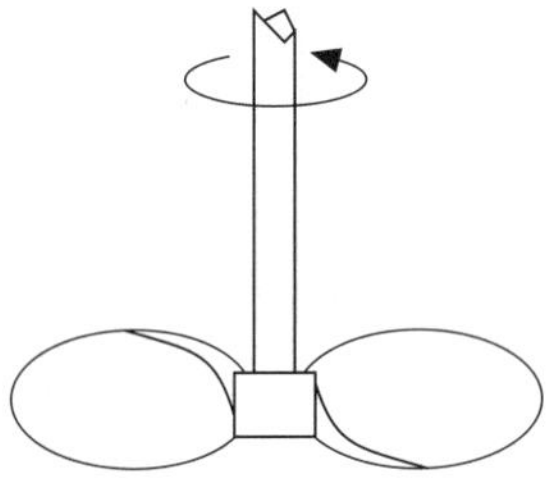

Figure 7. Propeller agitator.

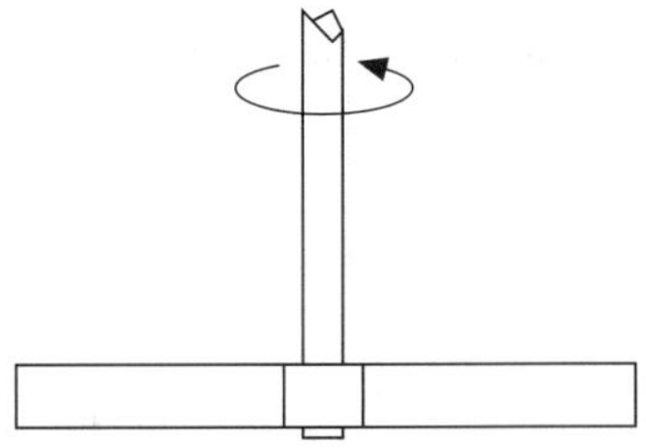

Figure 8. Simple turbine impeller.

In many vessel designs baffles are included to provide good mixing by modification of flow patterns. The increased turbulence assists the heat transfer. Fletcher (1987) reports that baffles may increase heat transfer by as much as 35% compared to a system without baffles.

The level of turbulence within the vessel may be assessed using the so called *"agitation" Reynolds number* Re_a defined as:

$$Re_a = \frac{ND_i^2}{\nu} \quad \textbf{(1)}$$

Where N is the speed of rotation of impeller, D_i is the impeller diameter, and, υ is the kinematic viscosity.

Table 2 (Hewitt et al. 1994) provides the range of application of different agitators together with some practical comments.

Heat Transfer

The transfer of heat in agitated vessels involves two aspects:

1. The heat transfer associated with the flow within a pipe or channel i.e. the coil in tank, limpet coils or the jacket.
2. The heat transfer associated with the flow across the vessel surface or across the outside of the coil in tank.

Heat transfer coefficients for the inside of limpet coils may be obtained from correlations developed or tubular or pipe flow. It will be necessary however, to use a hydraulic mean diameter for the particular cross section of channel in the estimation of the appropriate Reynolds numbers. It has to be remembered that the effective heat transfer area is limited to the contact area between the limpet coil and the vessel outside wall.

Because of the complex flow patterns in jackets it is difficult to provide suitable correlations for heat transfer from within the jacket, and it is usual to base calculations on previous experience. Except for condensing steam heat transfer coefficients inside jackets are relatively low.

Heat transfer correlations within coils located in vessels are also based on correlations for straight pipes, but because of the circular motion of the fluid through the coil the heat transfer is enhanced. (See **Coiled Tubes, Heat Transfer In**) In general the diameter of a coil is very much less than the inside diameter of the vessel so that the enhancement is greater than for limpet coils.

Jeschke (1925) provided an empirical correlation that allows for the increased heat transfer inside the coil

$$\alpha_{i(coil)} = \alpha_{i(straight\ pipe)}\left(1 + 3.5\frac{d_i}{d_c}\right) \quad \textbf{(2)}$$

Where $\alpha_{i\ coil}$ and $\alpha_{i\ (straight\ pipe)}$ are the heat transfer coeffients on the inside of the coil and the equivalent straight pipe respectively and d_i and d_c are the internal diameter of the pipe and the diameter of the coil respectively.

It will be seen that as the coil diameter is increased the ratio d_i:d_c decreases so that the rotational effect on heat transfer reduces. Because of the degree of turbulence generated in agitated vessels it is usual to assume that the heat transfer mechanism will be similar for the inside surface of the vessel and across immersed coils. The geometry involved, the use of baffles and the complex flow patterns make it inevitable that the correlations are empirical. The usual form of the correlations is:

$$Nu_\nu = K\, Re_a^a Pr^b \left(\frac{\eta_b}{\eta_w}\right)^c \times \text{geometric correction factor} \quad \textbf{(3)}$$

where Nu_ν = the **Nusselt number** (based on vessel inside diameter), which may be referred to the heat transfer on the outside of the helical coil or on the vessel inside surface. Pr is **Prandtl Number** η_b and η_w are the liquid viscosities in the bulk and at the wall respectively and K is the constant that applies to the simple heat transfer correlation for the particular system.

(See also **Agitated Vessel Heat Transfer**)

Table 2. The range of application of agitators

Type	Re_a	η kg/(m.s)	Comments
Anchor	> 50	20 to 100	All liquids
Helical ribbons	<50	100 to 1000	All liquids except highly nonNewtonian liquid, where the bulk liquid tends to rotate with the impeller and shear only occurs at the vessel wall
Propeller	> 300	< 2	Usually only satisfactory for vessels with a volume <6 m³ because of the size and weight of the propeller; not suitable for gas dispersion
Flat blade turbines	> 50	< 20	Liquid-liquid dispersion or applications where the impeller is located <D/2 from the vessel bottom
Angled blade turbines	>100	<10	Single-phase applications and for solid-liquid operations

References

Fletcher, P. (1987) Heat transfer coefficients for stirred batch reactor design, *Chem. Engr.*, 33 (April), 33–37.

Jeschke, D. (1925) Wärmeübergang und Druckverlust in Rohrschlangen, *Z. Deutsch. Ing.*, 81, 123.

Penney, W. R. (1983) *Agitated Vessels in Heat Transfer Design Handbook*, Sec.3.14. Hemisphere Publishing.

Leading to: *Agitation devices; Agitated vessel heat transfer; Agitated vessel mass transfer*

T.R. Bott

TANTALUM

Tantalum—(Gr. *Tantalos,* mythological character, father of *Niobe*), Ta; atomic weight 180.9479; atomic number 73; melting point 2996 °C; boiling point 5425 ± 100°C; specific gravity 16.654; valence 2,3,4, or 5.

Tantalum was discovered in 1802 by Ekeberg, but many chemists thought niobium and tantalum were identical elements until Rose, in 1844, and Marignac, in 1866, showed that niobic and tantalic acids were two different acids. The early investigators only isolated the impure metal. The first relatively pure ductile tantalum was produced by von Bolton in 1903.

Tantalum occurs principally in the mineral *columbite-tantalite* $(Fe, Mn)(Nb,Ta)_2O_6$. Tantalum ores are found in the Republic of Zaire, Brazil, Mozambique, Thailand, Portugal, Nigeria, and Canada. Mines at Bernic Lake, Manitoba, have reserves of 900,000 tons of ore averaging about 0.15% tantalum oxide. Separation of tantalum from niobium requires several complicated steps. Several methods are used to commercially produce the element, including electrolysis of molten potassium fluotantalate, reduction of potassium fluotantalate with sodium, or reacting tantalum carbide with tantalum oxide.

Sixteen isotopes of tantalum are known to exist. Natural tantalum contains two isotopes; one of these, Ta^{180}, is present in very small quantity (0.0123%) and is unstable with a very long half-life of $> 10^{13}$ years. Tantalum is a gray, heavy, and very hard metal. When pure, it is ductile and can be drawn into fine wire, which is used as a filament for evaporating metals such as aluminum. Tantalum is almost completely immune to chemical attack at temperatures below 150°C, and is attacked only by hydrofluoric acid, acidic solutions containing the fluoride ion, and free sulfur trioxide. Alkalis attack it only slowly. At high termperatures, tantalum becomes much more reactive. The element has a melting point exceeded only by tungsten and rhenium.

Tantalum is used to make a variety of alloys with desirable properties such as high melting point, high strength, good ductility, etc. Scientists at Los Alamos have produced a tantalum carbide graphite composite material, which is said to be one of the hardest materials ever made. The compound has a melting point of 6760°F. Tantalum has good "gettering" ability at high temperatures, and tantalum oxide films are stable and have good rectifying and dielectric properties. Tantalum is used to make electrolytic capacitors and vacuum furnace parts, which account for about 60% of its use. The metal is also widely used to fabricate chemical process equipment, nuclear reactors, and aircraft and missile parts. Tantalum is completely immune to body liquids and is a nonirritating metal. It has, therefore, found wide use in making surgical appliances. Tantalum oxide is used to make special glass with a high index of refraction for camera lenses.

Handbook of Chemistry and Physics, CRC Press

TARS (see Pyrolysis)

TAU METHOD

Let D y(x) = f(x) be a differential equation defined by a given differential operator D. For simplicity, let us assume that the equation is given by

$$Dy(x) = y'(x) + y(x) = 0, \qquad (1)$$

with the initial condition y(0) = 1, for x in the interval [0, 1]. The solution of this equation is $y(x) = e^{-x}$, which has the infinite series expansion

$$e^{-x} := 1 - x + x^2/2! - x^3/3! + \cdots + (-1)^n x^n/n! + \cdots \qquad (2)$$

let us assume that we use only the first three terms of it and let $y_2(x)$ stand for this polynomial, of degree 2. If we take for y(x) the approximate expression $y_2(x)$ and replace it into the differential equation (1) we obtain one with a non-zero right hand side:

$$y_2'(x) + y_2(x) = x^2/2. \qquad (3)$$

The higher the order of approximation n, the smaller the right hand side shall be (in our example it will be $x^n/n!$ for x in the interval [0,1]). It is well known that for any given n the monomials Kx^n with a non-zero coefficient K, are not the closest (or best uniform) approximation of zero (the original right hand side of (**Equation 1**)) in the interval [0, 1]. In fact for every value of n, the **Chebyshev Polynomials** $T_n(x)$ have that singular property

Therefore, it seems appropriate to reconsider the problem in a completely new way: instead of finding the error term ($x^n/n!$ in our concrete example) after fixing the approximation $Y_n(x)$, we should fix the error term to be minimal (i.e. a Chebyshev polynomial) and then determine the approximation $Y_n(x)$ which satisfies it. This original idea was proposed by the eminent mathematician Cornelius Lanczos in his formulation of the Tau Method, first conceived while working under Albert Einstein in problems of relativity theory (see Lanczos, 1956, for an account of the original formulation of this method).

Assume that for a given differential operator D we can determine a sequence of polynomials $Q_n(x)$, n, = 0, 1, 2, . . . , called canonical polynomials, associated with D, and such that for any value of n,

$$DQ_n(x) = x^n \qquad (4)$$

(i.e., they generate a basis in the image space). The Chebyshev polynomials $T_n(x)$ have the form

$$T_n(x) = \Sigma\, c_k x^k, \qquad k = 0, 1, \ldots, n \qquad (5)$$

where the coefficients c_k are known for all k and n. To find a Tau approximation $Y_n(x)$ we can start using $\tau T_n(x)$ as the minimal right hand side. (This term is usually referred to as the Tau Method perturbation term.). The free parameter τ is a multiplier introduced for us to be able to satisfy the initial condition (y(0) = 1 in our example, more generally, initial, boundary or multipoint boundary conditions). Then, if we set

$$DY_n(x) = \tau T_n(x) = \tau \Sigma\ c_k x^k, \quad (6)$$

the Tau approximate solution $Y_n(x)$ is immediately given in our concrete example by

$$Y_n(x) = \tau\ \Sigma\ c_k Q_k(x), \quad (7)$$

where the parameter τ is such that

$$\tau = y(0)/\Sigma\ c_k Q_k(0). \quad (8)$$

The main problem then is to find the sequence $Q_n(x)$ associated with D for all n. It has been proved that a sequence of canonical polynomials is uniquely associated to every differential operator D and that it can be generated recursively through a simple expression (iteratively for nonlinear operators), given exclusively in terms of the coefficients of the differential equation (see Oritz, 1968, for further details). For example, for **Equation 1** the sequence of canonical polynomials is recursively generated by

$$Q_n(x) = x^n - nQ_{n-1}(x), \quad \text{for} \quad n = 0, 1, 2, 3, \ldots, \quad (9)$$

that is:

$$Q_0(x) = 1; \quad Q_1(x) = x - 1; \quad Q_2(x) = x^2 - 2x + 2, \ldots \quad (10)$$

Suitable generalizations of the Tau Method have been applied successfully to the numerical approximation of complex systems of nonlinear partial differential equations involving large gradients in the function and or derivatives, such as in the case of soliton interactions and also in problems of fracture mechanics, fluid mechanics, mathematical physics and molecular biology. It has also been used in the approximation of differential eigenvalue problems where the eigenvalue parameter appears non-linearly; such questions arise in difficult fluid dynamics problems (see Oritz, 1994, for an overview of recent research into the Tau Method).

One advantage of the Tau Method, which accounts for the high precision of its approximations is that it does not require a discretization of the given differential operator, a process which often alters the behaviour of solutions.

Several numerical approximation techniques can be interpreted as special cases of the Tau Method, corresponding to different choices of perturbation terms, and can then be treated in a more unified and systematic way. Among these are the *Spectral Method, Collocation* and several types of projection and finite difference and finite element methods. (See for example **Spectral Analysis** and **Numerical Methods**)

References

Lanczos, C. (1956) *Applied Analysis,* Prentice-Hall, New Jersey.

Ortiz, E. L. (1968) The Tau Method, *SIAM J. Numer Anal.,* 6, 480–492.

Ortiz, E. L. (1994) The Tau method and the numerical solution of differential equations: past and recent research, *Proc. Cornelius Lanczos International Centenary Conference,* 77–81, J. David Brown, ed., SIAM, Philadelphia.

***Leading to:** Differential equations*

E.L. Ortiz

TAYLOR COLUMN (see Rotating disc systems, basic phenomena)

TAYLOR EQUATION FOR MIXTURE VISCOSITY (see Bubble flow)

TAYLOR FLOWS (see Görtler-Taylor vortex flows)

TAYLOR-GÖRTLER VORTEX FLOWS (see Görtler-Taylor vortex flows)

TAYLOR INSTABILITY

***Following from:** Couette flow*

The Taylor instability is a secondary flow which occurs as a transition from rotary **Couette Flow** in the annular gap between two coaxial cylinders of differing diameter when the inner cylinder rotates faster than a critical value. Pairs of counter-rotating axisymmetric (toroidal) vortices are formed in the radial and axial directions while the principal flow continues to be around the azimuth (**Figure 1**).

The onset of vortices has been studied experimentally by observing the consequences of their motion: namely to increase the wall shear stress (torque), the rate of heat transfer and the rate of mixing within the fluid. Vortices are generated if the **Taylor Number**, Ta $= r_i(\rho\omega/\eta)^2(r_o - r_i)^3$ exceeds a critical value, Ta_c where r_i and r_o are the inner and outer radii respectively, ρ is the fluid density, η the viscosity and ω the rotational speed. The limiting case $r_i/r_o \to 1$, was solved theoretically by Taylor to yield $Ta_{c,(r_i/r_o \to 1)} = 1695$. For long annuli having a small annular gap, $r_i/r_o \gtrsim 0.8$, the critical

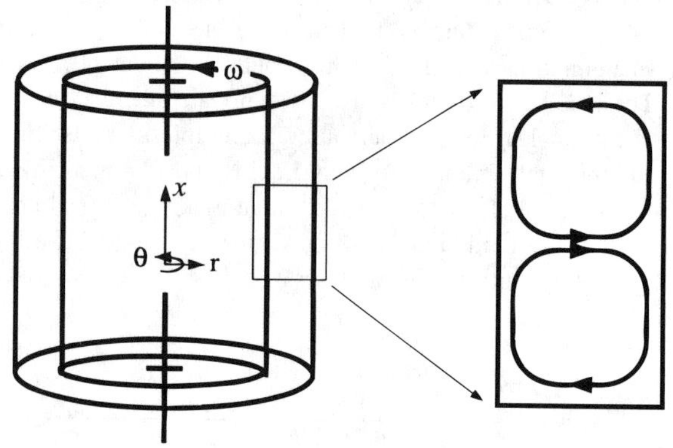

Figure 1. Coaxial cylinder geometry with exploded view of Taylor vortices.

Taylor number may be approximated by $Ta_c = \pi^4(1 + r_o/r_i)^2/(4P)$ with $P = 0.0571(1 - 0.652(r_o/r_i - 1)) + 0.00056/(1 - 0.652(r_o/r_i - 1))$.

A multitude of higher order instabilities which are non-axisymmetric and time periodic, occur if the Taylor number is increased further. Superimposed **Poiseuille Flow**, described by a **Reynolds Number** $Re = 2\rho u(r_o - r_i)/\eta$, delays the Taylor instability. Rotation reduces the Reynolds number for transition from laminar to turbulent flow and the combined system is described by a regime map (**Figure 2**).

Positioning the axis of the inner cylinder a distance, δ, from the axis of the outer cylinder produces an eccentric annulus and causes the Taylor instability to be delayed by an amount dependent on the eccentricity, $\varepsilon = \delta/(r_o - r_i)$ and given by $Ta_c(\varepsilon) = Ta_c(1 + 2.6185\varepsilon^2 + O(\varepsilon^4))$.

The critical Taylor number is also modified by **Non-Newtonian Fluid** behaviour. The theoretical analysis for a Generalized Newtonian fluid characterised by $\beta = \dfrac{d\ln(\dot\eta/\dot\eta_o)}{d\ln(\dot\gamma/\dot\gamma_o)}$ is discussed in Tanner(1985) and the onset of Taylor vortices given by $Ta_c(\beta) = Ta_c(1 + 0.505\beta + O(\beta^2))$ for $r_i/r_o \rightarrow 1$. The viscosity used in defining the Taylor number for non-Newtonian fluids is $\eta = \eta(\dot\gamma)$; $\dot\gamma = \omega r_i/(r_o - r_i)$.

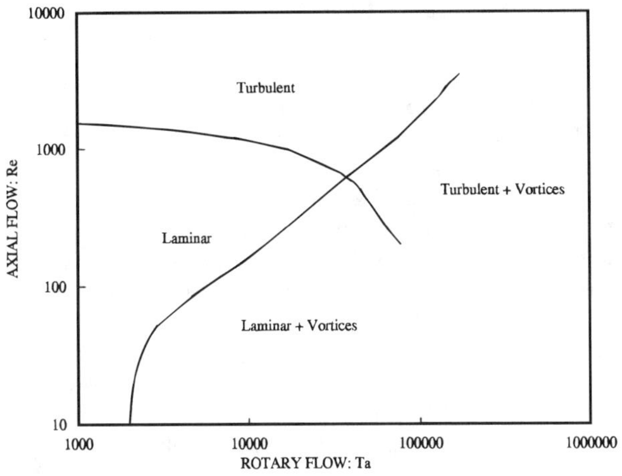

Figure 2. Regime map for Taylor vortices in the presence of an axial flow.

References

Stuart J. T. (1986) Taylor-vortex flow: a dynamical system, *SIAM Review*, 28, 3, 315–342, 1986.

Lockett T. J. (1992) Numerical Simulation of Inelastic Non-Newtonian Fluid Flows in Annuli, Ph.D. thesis, Imperial College, University of London.

Tanner R. I. (1985) *Engineering Rheology*, Clarendon Press, Oxford.

Taylor G. I. (1923) Stability of a viscous fluid contained between two rotating cylinders, *Philosophical Transactions of the Royal Society of London*, Series A, Vol. 223, 289–343.

T.J. Lockett

TAYLOR NUMBER

There are two dimensionless groups named Taylor number. The first is used as a criterion for the stability of flow in an annulus surrounding a rotating cylinder.

$$\text{Taylor number 1, } Ta_1 = \omega R_a^{1/2} a^{3/2}/\nu$$

where ω is angular velocity (s^{-1}), R_a is the mean radius of the annulus, a is the width of the annulus, and ν is fluid kinematic viscosity. (Sometimes the square of this quantity is used to define Taylor number see **Taylor Instability**).

The second Taylor number represents the square of the ratio of **Coriolis Force** to viscous force, i.e.

$$\text{Taylor number 2, } Ta_2 = (2\omega L^2/\nu)^2$$

G.L. Shires

TAYLOR-PROUDMAN THEOREM (see Rotating disc systems, applications; Rotating disc systems, basic phenomena)

TAYLOR SERIES

Following from: *Infinite series*

The Taylor series for a function f(x) with center at point x_0 having at this point derivatives of all orders, is defined, for the vicinity of x_0, as follows:

$$\sum_{k=0}^{\infty} \frac{f^{(k)}(x_0)}{k!}(x - x_0)^k. \quad (1)$$

For $x_0 = 0$ Taylor series is called Maclaurin series.

For the Taylor series (as for any power series) there exists a real number r ($0 \le r \le \infty$) called the radius of convergence, such that for $|x - x_0| > r$, series (1) converges absolutely, and for $|x - x_0| > r$ it diverges. Within the circle of convergence (i.e., in a circle $|x - x_0| \le q < r$, where q is any real number) Taylor series (1) converges uniformly. Convergent power series can be added and multiplied term by term.

Taylor power series can be differentiated and integrated. Series obtained as a result of termwise differentiation or termwise integration from 0 to x have the same radius of convergence as the initial series. This property of Taylor series is often used in solving problems of hydrodynamics and heat transfer, for solving differential equations and in integration of complex transcendental functions.

If f(z) is an analytic function of a complex variable inside a circle with centre at point x_0, then it can be expanded into a Taylor series within this circle. A derivative of kth order in this case is represented in integral form

$$f^{(k)}(z) = \frac{k!}{2\pi i}\oint_\Gamma \frac{f(z)}{(z - x_0)^{k+1}}dz \quad (2)$$

where Γ is a circle with centre x_0 laying inside this circle. The uniqueness of the expansion is associated with the fact that in this case any power series is a Taylor series for its sum.

If x_0 is a real number and f is defined in the vicinity of x_0 by a set of real numbers and has derivatives of all orders at x_0, then the function f cannot in the vicinity of x_0 be the sum of its own Taylor series. One and the same power series can be a Taylor series for different real functions.

A sufficient condition for Taylor series convergence to a real function on the interval $|x - x_0| < r$ is in total the restriction of all its derivatives in this interval (i.e., $|f^{(k)}(x)| < L$ where L is independent of k for $|x - x_0| < r$).

A real-valued function which has n derivatives at x_0 is often represented in the form of a partial sum of $n - 1$ terms of a Taylor series (Taylor polynomial of degree $n - 1$) and a remainder term (Taylor formula):

$$f(x) = \sum_{k=0}^{n-1} \frac{f^{(k)}(x_0)}{k!}(x - x_0)^k + R_n(x). \quad (3)$$

The remainder term $R_n(x)$ is

$$R_n(x) = \int_{x_0}^{x} d\xi \int_{0}^{\xi} d\zeta \int_{x_0}^{\zeta} f^n(\xi)d\xi \quad (4)$$

By applying the mean value theorem, the remainder term can be written as a Lagrange formula

$$R_n(x) = \frac{f^{(n)}(x_0 + \vartheta(x - x_0))}{n!}(x - x_0)^n. \quad (5)$$

If the function f(x) has all derivatives in the interval $|x - x_0| < r$ and $\lim_{n\to\infty} R_n(x)$, then

$$f(x) = \sum_{k=0}^{\infty} \frac{f^{(k)}(x_0)(x - x_0)^k}{k!} \quad (6)$$

and a series converges uniformly to f(x) on any interval $|x - x_0| \leq q < r$.

The Taylor formula generalizes to the case of a function of several variables. If $f(x_1, \ldots, x_n)$ is a real function, which has all continuous derivatives of order $\leq m$ in a certain vicinity D of the point x_0, then

$$f(x_1, \ldots, x_n) = f(x_{01}, \ldots, x_{0n})$$
$$+ \sum_{i=1}^{n} \left.\frac{\partial f}{\partial x_i}\right|_{(x_{01},\ldots,x_{0n})} (x_i - x_{0i})$$
$$+ \frac{1}{2!}\sum_{i=1}^{n}\sum_{j=1}^{n} \left.\frac{\partial^2 f}{\partial x_i \partial x_j}\right|_{(x_{01},\ldots,x_{0n})} (x_i - x_{0i})(x_j - x_{0j})$$
$$+ \cdots + R_m(x_1 - x_{0n}). \quad (7)$$

The remainder term $R_m(x_1, \ldots, x_n)$ satisfies the relation similar to relations for a function of one variable. If a function $f(x_1, \ldots, x_n)$ has in D all continuous partial derivatives and $\lim_{m\to\infty} R_m(x_1, \ldots, x_n)$ in D, then the Taylor formula brings about an expansion into a multiple series. A function, which can be expanded into a power series, convergent in a certain vicinity of the point $(x_{01}, \ldots, x_{0n})$ is called analytic in this vicinity.

The Taylor formula allows us to reduce the study of properties of a differentiable function to a simpler problem of studying the properties of a corresponding Taylor series and of evaluating the remainder term. The application of the Taylor formula for calculation of convergence of series and integrals, and for estimation of the rate of their convergence and divergence is based on this property.

V.N. Dvortsov

TAYLOR'S THEOREM (see Power series)

TDS, TOTAL DISSOLVED SOLIDS (see Water preparation)

TEMA (see Tubular Exchanger Manufacturers Association)

TEMA STANDARDS

Following from: *Shell and tube heat exchangers*

The **Tubular Exchanger Manufacturers Association (TEMA)** is an association of manufacturers of shell and tube heat exchangers. TEMA has established a set of construction *standards* for **Shell and Tube Heat Exchangers.** The Standards are regularly updated and published; the most recent edition is the seventh, published in 1988. Most shell and tube exchangers ordered by the process industries and for other high-severity applications throughout the world are built to TEMA standards.

To quote from the Standards (p. 20), "the TEMA Mechanical Standards are applicable to shell and tube exchangers with inside diameters not exceeding 60 inches [1.52 m], a maximum product of nominal diameter (inches) and design pressure (psi) of 60,000 in. psi [10.5 m · MPa], or a maximum design pressure of 3,000 psi [20.7 Mpa]." However, a section on Recommended Good Practice is provided to extend the Standards to units with larger diameters.

The Standards recognize three classes of heat exchanger construction:

Class R for the severe requirements of petroleum processing (and usually including most large scale processing applications).

Class C for general commercial application.

Class B for chemical process service.

There is in fact relatively little difference between the Standards for the three classes; where there are differences, Class R calls for heavier and more conservative construction features, and Class B (since these are usually stainless steel or high alloy exchangers) gives some allowance for lighter (i.e., thinner metal) construction for non-critical components.

TEMA Standards are divided into ten sections:

1. Nomenclature
2. Fabrication Tolerances
3. General Fabrication and Performance Information
4. Installation, Operation, and Maintenance
5. Mechanical Standards TEMA Class RCB Heat Exchangers
6. Flow Induced Vibration
7. Thermal Relations (includes fouling and charts of the configuration correction factor on the Logarithmic Mean Temperature Difference)
8. Physical Properties of Fluids
9. General Information (e.g., dimensions of pipe, tubing, fittings and flanges; pressure-temperature ratings; conversion factors, etc.)
10. Recommended Good Practice

TEMA does not give or recommend thermal-hydraulic design methods, leaving it to the individual company to use their own methods or one of the published or commercially available computer-based methods.

Even though TEMA has only about 20 members (who must meet strict standards of in-house thermal-hydraulic and mechanical design capability, manufacturing accountability, and quality control to attain membership), practically all shell and tube heat exchangers for the process and related industries are specified to meet TEMA standards of construction. The Standards are set by members of the Technical Committee of TEMA.

Reference

Standards of the Tubular Exchanger Manufacturers Association, Seventh Edition, 1988, TEMA, Inc., 25 North Broadway, Tarrytown, New York 10591, USA.

K.J. Bell

TEMPERATURE

Following from: *Thermodynamics*

The temperature of an object or fluid is that property which determines the direction of the flow of heat from that body or fluid to an adjacent body or fluid with which it is in contact. Thus, heat flows from a body or fluid of higher temperature to a body or fluid of lower temperature. Temperature is one of the main parameters of state which defines the thermal state of the system. The temperature of all parts of the system in thermodynamic equilibrium is the same. Based on the molecular-kinetic approach, the temperature of a system characterizes the intensity of thermal motion of atoms, molecules and other particles forming the system.

For instance, for a system described by the laws of classical statistical physics the mean kinetic energy of thermal motion of particles is directly proportional to the absolute temperature of the system. In this regard we can say that the temperature characterizes the thermal motions within a body.

In thermodynamics the reciprocal of the derivative of the entropy S of a body with respect to its energy E is called the absolute temperature T:

$$\frac{dS}{dE} = \frac{1}{T} \tag{1}$$

Temperature, like entropy, is a purely statistical quantity and makes sense only for macroscopic bodies. According to the second law of thermodynamics, energy is transfered from bodies with higher temperature to bodies with lower temperature. The absolute temperature is always positive, $T > 0$. The least *absolute temperature* possible is the absolute zero. At absolute zero, the translatory and rotary motion of atoms and molecules comes to an end, and they are in a state of the so-called "zero vibrations" rather than in a state of rest. By Nernst theorem the entropy of any body becomes zero at absolute zero temperature. Absolute zero is unattainable. The entropy S is a dimensionless quantity and from Equation 1 it follows that temperature has the dimensions of energy and can be measured in Joules. The ratio Joules/Kelvins(K) called Boltzmann's constant k is equal to $k = 1.38\ 10^{-23}$ Joules/K.

Actually, the temperature is usually measured in arbitrary units, degrees (Celsius degree, °C, Fahrenheit, °F, Réaumur, °R) or in "Kelvins" whose value is determined by the corresponding temperature scales.

Temperature scales are systems of sequentially numbered values corresponding to various temperatures. The temperature can be determined by measuring any quantity dependent on it and it is convenient to measure a physical property of a certain, so-called thermometric substance (for instance, the volume or pressure of a gas, the resistance of a conductor). To realize a temperature scale, we must select its origin and the dimension of the temperature unit (degree). For this purpose, we usually use two reference points—temperatures of transition of a substance from one agregate state to another. Such temperature scales are called "practical." The first practical temperature scale was suggested by Fahrenheit in 1724, in which one of the reference points was the temperature of a human body, accepted by Fahrenheit to be equal to 96 degrees (°F), the second, the temperature of ice melting, equal to 32 degrees (°F). A liquid mercury thermometer served as an interpolation device.

More accurate practical temperature scales were suggested by Celsius and Réamur. In these scales the temperatures of melting of ice and boiling of water at atmosphere pressure were used as reference points. The temperature interval between these points in Celsius' scale (°C) was divided by 100, and in Réaumur's scale (°R) by 80 equal parts. In Fahrenheit's temperature scale, this temperature interval is equal to 180 (°F). In the absolute Kelvin (K) and Rankine (R) temperature scales, where the origin of scale is the absolute zero, the temperature interval is equal to 100 and 180 temperature units, respectively. The principle of constructing temperature scales suggested by Fahrenheit (the reference points and interpolation device) is used in the international temperature scales. For instance, the international temperature scale of 1927, ITS-27, was realized using two points (0°C and 100°C), the unit of temperature is the degree Celsius (°C); in the ITS-90, one point, the temperature of the triple point of water, 273.16 K, was used; the unit of temperature is the Kelvin (K). The main interpolation device is a platinum resistance thermometer.

The so-called thermodynamic temperature scale, which is independent of the particular properties of a thermometric substance, can be realized on the basis of the second law of thermodynamics, by determining the ratio of temperatures from the ratio of temperatures in Carnot's cycle. In practice, to construct such a scale, relations are used which, whilst not contradicting the second law of thermodynamics, relate the thermodynamic temperature to some additive physical quantity which can be measured accurately enough. The most widespread are:

- the gas thermometer based on the gas law

$$p\tilde{V} = RT, \tag{2}$$

where R is the universal gas constant, and p, $\tilde{V}$, T are the pressure, molar volume and temperature of a working substance in an ideal gas state;

- the acoustic thermometer based on measuring the velocity of sound, C, in a gas

$$C^2 = \frac{\gamma RT}{\tilde{M}}, \tag{3}$$

where $\gamma = c_p/c_v$ is the specific heat ratio (for an ideal gas $\gamma = const$), and $\tilde{M}$ is the molecular weight of the working substance;

- the radiation thermometer based on measuring the total energy of heat radiation E(T) emitted by a blackbody at temperature T

$$E(T) = \sigma T^4, \tag{4}$$

where σ is the Stefan-Boltzmann constant;

- the thermal noise thermometer based on measuring the root-mean-square of voltage noise $\overline{V^2}$ in current flow through a resistance Ω (ohm) at a temperature T (Nyquist's equation)

$$\overline{V^2} = 4kT\Omega\Delta f, \tag{5}$$

where Δf is the band width (Hz).

Presently the most exact values of thermodynamic temperature in a wide range of values can be obtained using the gas thermometer; however, near 4 K and above 200 K the noise and radiation thermometers approach the gas thermometer in accuracy. In 1848, J. Thomson (Lord Kelvin) proved that the temperatures determined from Carnot's cycle and by the gas thermometer are identical and represent the thermodynamic or absolute temperature. With the assumption for a dimension of a temperature unit made by Celsius, Thomson determined the value of temperature of ice melting on a new scale, −273.15 K.

Since 1960, the unit of thermodynamic temperature (K) was determined as 1/273.16 of the temperature of the triple point of water, −273.16 K. When using a gas thermometer of constant volume ($\tilde{V} = const$) relation (2) for determining the unknown temperature T_x takes the form

$$T_x = \frac{T_0}{p_0} p_x, \tag{6}$$

where $T_0 = 273.16K$, p_0 is the pressure of the working substance at a temperature T_x. In the ITS-90 the basic unit of temperature T_{90} is the Kelvin (K). The measurements allowed for measuring temperature in °C (t_{90}) are defined as

$$t_{90} = T_{90} - 273.15$$

References

Quinn, T. I. (1983) *Temperature,* London, Academic Press.

Leading to: *Celsius temperature scale; Fahrenheit temperature scale; International temperature scale; Reaumur degree; Temperature measurement, bases*

Ju.A. Dedikov

TEMPERATURE GRADIENT IN EARTH'S CRUST (see Geothermal systems)

TEMPERATURE-HEAT LOAD PLOT (see Process integration)

TEMPERATURE MEASUREMENT, BASES

Following from: *Temperature*

Temperature measurement is a concept that covers the set of methods and facilities to obtain the experimental information about the physical parameter: temperature.

Temperature measurements may be classified into "contact" and "contactless" methods. In contact temperature measurements, the sensing element of the measuring device is put in contact with an object to be measured. In this case, the contact must be long enough for thermal equilibrium to be attained between sensing element temperature and the object. This is one of the main conditions for correct temperature measurements.

The physical basis for temperature measurement is represented by thermometric phenomena, which are physical phenomena affecting the temperature dependence of any parameter (electric, frequency, velocity, etc.) that can be easily and uniquely recorded. The following physical phenomena and temperature dependences are the most commonly used as thermometric phenomena.

1. volume or pressure of an isolated amount of the gas or liquid;
2. electrical resistance;

3. thermoelectromotive force;
4. amplitude or spectra of thermal radiation;
5. thermal noise spectrum;
6. nuclear quadruple resonance frequency;
7. nuclear magnetic resonance frequency;
8. nuclear-oriented phenomena;
9. Mössbauer's effect;
10. sonic (ultrasonic) propagation velocity in different media;
11. photoemission phenomena;
12. cathode luminescence radiation;
13. interference phenomena in anisotropic media;
14. osmotic pressure;
15. condensation and crystallization of substances.

Over thirty thermometric phenomena have been use in these situations. In selecting a thermometric phenomenon, the two most important requirements are related to a sharp sensitivity of the chosen parameter to temperature and insensitivity to other physical parameters, for example pressure indication, field strength, etc. The most fundamental thermometric phenomena are concerned with the ideal gas (Clapeyron-Clausius) equation, thermal radiation (Planck equation), and thermal noise (Nyquist formula). In these phenomena the complete absence of foreign effects is assumed. Owing to this, practical measurements are related to the thermodynamic temperature by an appropriate process of transfer. The procedure of such transfer is formalized by an the official document, the International Temperature Scale.

In practice, the *thermodynamic temperature scale* is transferred by gas thermometers over the range 10 to 1340 K. For higher temperatures, use is made of the absolute black body thermal radiation described by the Planck law

$$\rho_{\lambda,T} = 8\pi hc\lambda^{-5}(\exp hc/\lambda kT^{-1})^{-1}.$$

The third fundamental thermometric phenomenon of thermal noise can be, in principle, used without restrictions however, the measuring errors do not allow the noise thermometry to be used in exact metrological studies. However, noise thermometry still finds widespread use in industrial measurements.

Use of the fundamental thermometric phenomena is advantageous, because temperature measurements do not require calibration. The disadvantage is that in these methods, the measurements are tedious and time-consuming. Therefore, for practical temperature measurements, the thermodynamic temperature scale is transferred to a number of *reference temperatures* corresponding to equilibrium states of vapour, liquid, and solid substances. These are determined with the highest accuracy possible using modern instrumentation. Substances for the reference points are chosen not only from the considerations of the required equilibrium temperature but also from the simplicity of producing and keeping this substance in the pure form. The main reference points are cited in **Table 1**. (Here t. p. means the triple point.)

Interpolation over the intermediate reference points is made using the platinum reference resistance thermometer. Intermediate reference points for neon, mercury, indium, bismuth, cadmium, plumbum, and antimony can also be used. In addition, there is a table of the secondary reference points amounting to 50.

Table 1.

Substance	Temperature, K
Hydrogen(t.p.)	13.81
Oxygen(t.p.)	54.361
Argon(t.p.)	83.798
Water(t.p.)	273.16
Tin(tin point)	505.1181
Zinc (zinc point)	692.73
Silver(silver point)	1235.08
Gold(gold point)	1337.58

At temperatures above the gold point, the complete absolutely black body radiation pyrometer is the interpolation tool. The necessity to average absolute black radiation and identical absolute absorption in a receiver arises many difficulties, which results in a considerable increase of relative errors associated with transferring the thermodynamic temperature scale at temperatures above the "gold point".

The errors in transferring from the thermodynamic to the practical scale are 10^{-4} to 10^{-2}K at temperatures up to 500 K; 10^{-2} to 3×10^{-1} K up to the "gold point". At higher temperatures, the transfer error can be estimated as the square of 0.1% of the absolute temperature. The above values characterize the most accurate measurements. The tungsten melting point can be measured with a 15 K uncertainty. The value of this temperature, (say, 3694 K, 3421°C) is often presented accurately to the last significant figure, but this apparent accuracy disguises the real uncertainty.

In practical temperature measurements, the error exceeds the above limiting values at ±100°C by one order of magnitude and by two orders at higher temperatures.

Contactless temperature measurements are based on the partial or complete intensity of thermal radiation. The Planck equation is applicable only for an absolute black body. For real emitting surfaces, the radiation flux is related to the black body flux at the same temperature by a factor, i.e., the **Emissivity**. Extensive tables of emissivity are available, covering nearly all measurement conditions. Nevertheless, establishing an accurate value for emissivity is very difficult and leads to considerable errors. Planck's law establishes the generalized relationship between the radiation density, radiation wavelength, and temperature of an emitter, i.e. of an object of measurement. This opens three main possibilities of contactless temperature measurements: radiation, brightness, and color pyrometry.

Radiation pyrometry is based on measuring a total radiation heat flux. Thus, the radiation pyrometer records the temperature of the absolutely black body emitting the same integral flux as the non-black body does and whose temperature is measured. A correction for the integral emissity is set by a special vernier directly on the device.

Brightness pyrometers are based on comparing the brightness of monochromatic radiation from the object to be measured, and that of the special filament lamp built into the device. Usually, the filament is observed on the object background through a narrowband light filter typically with a 0.65 μm wavelength range. As in the case of radiation pyrometers, it is necessary to allow for the difference of the radiation intensity to that of an absolutely black body using special tables of emissivity at the chosen wavelength (these are different from the normal radiation pyrometer tables). Usually, a digital or arrow device that records the filament lamp current is built into the brightness pyrometer. In some cases, the filament glow is kept constant and the equilibrium is attained by an optical wedge that controls the radiation brightness of an object of measurement. When the brightness of the filament and the object are equal, the filament dissapears on the object background and the filament current or the wedge position uniquely determines the object temperature.

In radiation and brightness pyrometers, the main principal error source lies in the necessity to allow for emissivity. At high temperatures, the error due to an incorrect value of emissivity can exceed 10% of the measured value.

The operation of colour pyrometers is based on assuming that there is a fixed emissivity irespective of wavelength. In this case, the need for information about the emissivity can be excluded, if the temperature is judged by the monochromatic radiation density ratio at two fixed wavelengths. As a rule, the emissivity of substances depends on the monochromatic radiation wave length. This can be overcome by using three or four spectral ratios, for which it is necessary to use, accordingly, three or four monochromatic light filters with sufficient wavelength contrast.

Among the temperature measurement methods the temperature indicators occupy a special place, for which the temperature is indirected by changing a colour, structure, shape.

Metals and their alloys possess a sufficient individual constancy of melting points. There are tables of alloy compositions, for which the melting points discretely vary with a step of 5 − 10°C and which may be used for temperature measurements from 60 to 1000°C.

In different branches of mechanical engineering there are widespread temperature-sensitive dyes, varnishes, fluxes, pencils, pastes, etc., which change their colour at fixed temperature values. The number of the manufactured colour temperature indicators exceeds 300, and the increment of the fixed temperatures for the change of colour amounts to 2—10°C at a ±1°C error.

In commercial plants, frequently there arises the necessity for plasma and flame temperature measurements. These present difficulties due to the fact that flame radiation varies from linear radiation from a set of monochromatic lines to a complete spectrum, approaching absolute black body radiation. In the latter case, the problem is easily solved by using the pyrometric methods described above. In the case of radiation at distinct lines within the spectrum the method of spectral line conversion is often used. Here, a reference emitter at known temperature is used. Following Kirchhoff's law, the medium is permeated by the reference radiation, which is absorbed in the chosen spectral line.

References

Temperature. Its Measurement and Control in Science and Industry, Rinehold, 1961.

Leading to: *Temperature measurement, practice; Resistance thermometry*

O.A. Gerashchenko

TEMPERATURE MEASUREMENT, PRACTICE

Following from: *Temperature; Temperature measurement, bases*

Clearly, the measurement of temperature is of great importance in heat and mass transfer. A wide variety of techniques have been used; a good general source relating to temperature measurement is that of Jones (1985). Some of the basic principles of measurement are given in the entry on **Temperature Measurement, Bases.** Here, we will give a brief summary of some of the main methods, namely *fluid filled thermometers, bimetalic thermometers, resistance thermometers, thermistors, thermocouples* and *optical pyrometers.*

Fluid Filled Thermometers

These can be liquid filled (operating in the range 200–670K) or gas filled (operating in the range 4–1050K). The best known thermometer is the mercury-in-glass thermometer, but other liquids such as alcohol may also be used in glass thermometers. The expansion of the fluid can also be transmitted to a Bourdon tube whose deflection can be recorded in the normal way. Similarly, with gas filled thermometers, the increase in pressure of the gas can be transmitted to a Bourdon-type gauge. A similar principle applies to *vapour pressure thermometers* where a liquid vaporises to produce the increase in gas pressure.

Bimetalic Thermometer

The bimetalic thermometer is one of a large class of thermometers that depend on thermal expansion. In the bimetalic thermometer, two strips of metal (for instance brass and invar) are joined together; the differential thermal expansion causes the strip to deflect when the temperature changes. The bimetalic strip can also be installed in a helical ribbon form with the temperature indication being in the form of a needle which moves circumferentially over a dial. The bimetalic thermometer works typically up to 300°C to within ±1% of the scale range. However, such thermometers have a rather slow response and, by definition, require the whole of the sensing element to be immersed in the fluid whose temperature is being measured.

Resistance Thermometers

The electrical resistance of many materials changes with temperature and this effect can be used as a means of determining the temperature, hence the name *resistance thermometer.* Typically, resistance thermometers are made of platinum and can be used over the range 20–400K. They have a typical sensitivity in the range 0.2–20 ohms/K but their response time is normally rather slow (1–10s).

Thermistors

Thermistors are semi-conductor devices operating in the range 50–400K giving typically a 4% change in resistance per degree K. With a small enough bead, a response time of 3 ms is possible.

Thermocouples

This is the most commonly used sensor for fluid temperature. The main types of *thermocouples* (together with their range of temperatures and sensitivity) are listed in **Table 1**. Extensive information on thermocouples is given in the books by Early (1976) and Jones (1985). **Figure 1** shows the various types of thermocouple junctions used in fluid temperature measurement and also gives the approximate minimum response time for the various types.

Optical Pyrometers

Optical pyrometers are based on the fact that as the temperature changes, the wave length λ_m at the maximum intensity of radiation decreases according to the linear law:

$$\lambda_m T = \text{constant} = 2898 \mu\text{m.K} \quad \textbf{(1)}$$

where T is the absolute temperature (Kelvin). In the optical pyrometer, a telescope system is used to focus on the objective whose temperature is to be measured. A Tungsten filament is placed at the focal point of the objective lens of the telescope and is viewed with the eyepiece. The temperature of the filament is adjusted by increasing the voltage across it until the filament "disappears" against the image of the hot body whose temperature is to be measured. The resistence of the filament gives the temperature which matches that of the hot body. Futher details of this and other radiation thermometers are given by Jones (1985).

References

Jones, E. P. (1985) *Instrumentation Technology, Vol. 2,* Measurement of temperature and chemical composition, Butterworth and Co. (Publishers) Ltd, London. ISBN: 0-408-01232-3.

Early, B. (1976) *Practical Instrumentation Handbook,* Scientific ERAP Publications.

Table 1. Main types of thermocouple used for fluid temperature measurement

Main types	Max. temp. range °C	Sensitivity at 100°C μv/K.
Chromel/Alumel (Ni-Cr/Ni-Al)	−200 to 1400	41
Copper/Constantan (Ci/Ni-Cu)	250 to 500	47
Iron/Constantan (Fe/Ni-Cu)	20 to 700	54
Platinum/Platinum 10% Rhodium	0 to 1500	7.4
Platinum/Platinum 13% Rhodium	0 to 1500	7.6
Tungsten/Tungsten 26% Rhenium	0 to 2300	3.3

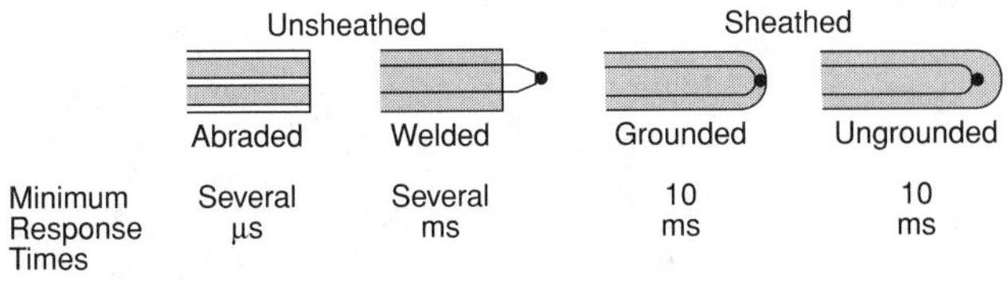

Figure 1. Types of thermocouple used in fluid temperature measurement.

Leading to: *Thermometer*

G.F. Hewitt

TEMPERATURE PROFILES IN BOUNDARY LAYER AND SEPARATED FLOWS (see Convective heat transfer)

TEMPERATURE SENSITIVE COATINGS (see Visualization of flow)

TEMPERING (see Steels)

TEMPERING OF CHEMICAL REACTION (see Chemical equilibrium)

TERMINAL SETTLING VELOCITY (see Thickeners)

TERMINAL VELOCITY (see Dispersed flow)

TERMINAL VELOCITY OF PARTICLE IN GAS (see Gas-solid flows)

TERRAIN INDUCED INSTABILITIES (see Instability, two-phase)

TETRAFLUOROETHYLENE (see Pyrolysis)

TEXTURE (see Mixing)

THEORETICAL PLATE (see Distillation)

THERMAL ANEMOMETERS (see Anemometers, thermal)

THERMAL BOUNDARY LAYER (see Boundary layer heat transfer; Cross flow heat transfer)

THERMAL BUILDING DYNAMICS (see Buildings and heat transfer)

THERMAL CAMERA (see Thermal imaging)

THERMAL CONDUCTIVITY

Following from: *Fourier's law*

Thermal conduction is the transfer of heat from hotter to cooler parts of a body resulting in equalizing of temperature. In contrast to heat transfer by convection, thermal conduction has nothing to do with macroscopic displacements in the body, but is a result of a direct energy transfer between particles, such as molecules, atoms, and electrons, with higher energy and ones with lower energy.

Contrary to heat transfer by radiation, there is no thermal conduction in vacuum.

The basic law of thermal conduction is the *Fourier law* which states that the heat flux density $\vec{q}$ is proportional to the temperature gradient T in a isotropic body: $\vec{q} = -\lambda\ \mathbf{grad}T$. The constant of proportionally λ is the *thermal conductivity.* The minus sign indicates that the temperature decreases in the direction of heat transport and, hence, the temperature gradient is a negative quantity.

Deviations from the Fourier law can be observed at extremely high values of **grad**T, e.g. in powerful shock waves, at low temperatures for liquid helium HeII, and at high temperatures on the order of tens of thousands of degrees when energy transfer in gases is due mainly to radiation. In highly rarefied media, in which molecules collide with the walls of the vessel rather than with one another, the concept of local temperature is meaningless and the Fourier law is inapplicable. In this case, we deal not with *thermal conduction in a gas,* but with heat exchange between the bodies in it.

Among solid, anisotropic substances (e.g. crystals, sedimentary rocks, lamellar and pyrolytic materials) occur for which the heat flux density vector $\vec{q}$ does not coincide with the normal to an isothermal surface. The simplest assumption generalizing the Fourier hypothesis is that each component of the vector $\vec{q}$ at the point (x, y, z) is a linear combination of all the components of the temperature gradient in it

$$q_x = \lambda_{11}\frac{\partial T}{\partial x} + \lambda_{12}\frac{\partial T}{\partial y} + \lambda_{13}\frac{\partial T}{\partial z}\ \text{etc.},$$

the coefficients of thermal conductivity λ_{ik} of anisotropic body form a tensor in the 2nd dimension. For crystals, it is found, within the measurement error, that thermal conductivity in mutually opposite directions is the same.

In multicomponent gas mixtures, one has to take into account the so-called "cross effects" such as mass flows brought about by the temperature gradient (thermal diffusion or the Soret effect), energy flows due to density gradient (diffusion thermal effect or the Dufour effect). The heat flux density in a ν-component gas mixture is written down as a sum

$$\vec{q} = -\lambda\ \mathbf{grad}\ T + \sum_{i=1}^{\nu} n_i h_i \vec{\mathbf{V}}_i + \frac{KT}{n}\sum_{i=1}^{\nu}\sum_{i\neq j}^{\nu}\frac{n_j D_j^T}{m_i D_{ij}}(\vec{\mathbf{V}}_i - \vec{\mathbf{V}}_j),$$

where n and n_i are the total number of molecules and the number of molecules of a given species per unit volume, h_i and m_i the enthalpy and the mass of a single particle of the i-th species, $\vec{\mathbf{V}}_i$ the diffusion rate, D_{ij} the coefficient of binary diffusion, D_j^T the coefficient of thermal diffusion (Dufour effect), and K is Boltzmann's constant. Thus, in multicomponent gas mixtures energy transfer is also accomplished, in addition to convection and heat conduction, by diffusion flow of molecules relative to the bulk velocity and by the Dufour effect. In this case the apparent thermal conductivity may be far different from the molecular thermal conductivity λ.

If we assume that the coefficient of thermal conductivity λ as well as the coefficient of heat capacity c and the density ρ of a substance do not depend on other parameters, then in the absence of internal heat sources the temperature inside the body is described by the differential equation of thermal conductivity

$$\frac{\partial T}{\partial \tau} = \frac{\lambda}{\rho c}\nabla^2 T$$

where $\nabla^2 T = \frac{\partial^2 T}{\partial x^2} + \frac{\partial^2 T}{\partial y^2} + \frac{\partial^2 T}{\partial z^2}$ is the Laplacian, τ the time, and x, y, z the Cartesian coordinates.

The group $\kappa = (\lambda/\rho c)$ is known as thermal diffusivity. It characterizes the velocity of propagation of isothermal surfaces in a body.

Solving equations of thermal conduction enables the establishment of the temperature distribution T(τ, x, y, z) to a certain degree of accuracy for any body. The degree of accuracy depends on the initial and boundary conditions. Solution of the equation of thermal conductance in a general form, including the variable coefficients λ, c, and ρ, using high-speed computers, in principle presents no difficulties. But this specifies higher requirements for reliability in determining the varable coefficient of thermal conductivity λ for the given substances. Numerous theoretical and experimental investigations have led to the discovery of some specific features and regularities.

The highest thermal conductivity is inherent in metals (see **Table 1**). Amongst the metals silver shows the highest thermal conductivity and bismuth the lowest. In the temperature range above ambient, λ for nearly all pure metals falls with increasing temperature. λ is greatly affected by the presence of additives and impurities. Thus, the coefficient of thermal conductivity for steel containing 1% carbon is 40% lower than for pure iron. For metals it depends to a great extent on their treatment. Quenching and cold treatment of metals decrease λ, while preheating up to a high temperature increases it.

Thermal conductivity of nonmetallic liquids under normal conditions is much lower than that of metals and ranges from 0.1 to 0.6 W/mK. In the interval between the melting point and the boiling

Table 1. The values of the coefficient of thermal conductivity for various substances at atmospheric pressure and moderate temperatures

Substance	t, °C	λ, W/mK	Substance	t, °C	λ, W/mK
Metals			*Liquid*		
Silver	0	429	Mercury	0	7.82
Copper	0	403	Water	20	0.599
Iron	0	86.5	Acetone	16	0.190
Tin	0	68.2	Ethyl alcohol	20	0.167
Lead	0	35.6			
Nonmetallic materials			*Gases*		
Sodium chloride	0	6.9	Hydrogen	0	0.1655
Tourmaline	0	4.6	Helium	0	0.1411
Glass	18	0.4 – 1.0	Oxygen	0	0.0239
Wood	18	0.16 – 0.25	Nitrogen	−3	0.0237
Asbestos	18	0.12	Air	4	0.0226

point, thermal conductivity of liquids may change by a factor of 1.1 to 1.6.

Finally, the lowest thermal conductivity is observed in gases (under normal conditions it is from 0.006 to 0.1 W/mK). Hydrogen and helium are distinguished among gases for the highest thermal conductivity.

The coefficients of thermal conductivity presented in the table evidence that this parameter varies widely. It is determined using various techniques based on the molecular kinetic theory, the phenomenological approaches of the generalized conductivity theory, and generalization of experimental data. Below the investigation results are presented for the different classes of substances.

The Theory of Gas Thermal Conductivity

Thermal conductivity in gases is brought about by energy transfer by gas molecules in the same way as viscosity is related with momentum transfer and diffusion, with mass transfer. Therefore, all these phenomena appreciably depend on $\bar{l}$, that is the mean free path of molecules.

For the model of gas consisting of solid spherical molecules which do not interact at a distance and possess only the energy of translational motion, the coefficient of thermal conductivity λ is determined by the kinetic theory of gases as $\lambda = {}^1/_3 \rho c_v \vec{V}\vec{l}$, where ρ is the gas density, c_v the heat capacity of unit of mass at constant volume, and $\vec{V}$ the arithmetical mean velocity of the molecules. The viscosity coefficient η for this gas is equal to ${}^1/_3 \rho \vec{V}\vec{l}$. Because of a highly simplified gas model underlying the calculation of λ and η, we cannot expect a fair agreement of theory and experiment. For instance, these are departures from the theoretical deduction of independence of λ and p since there are deviations from the relationships $\rho \propto p$ and $\bar{l} \propto 1/p$. However, the relation between the coefficients of transfer $\lambda = \eta c_v$ observed theoretically is quite correct.

Contemporary kinetic theory takes into account the existence of the attraction and repulsion forces between molecules. Expression for these forces can in principle be derived from quantum mechanics although this is a laborious task if all atoms and molecules are considered, except in the simplest ones. Therefore, the molecular interaction is described, as a rule, by simple empirical functions, that is interaction potentials containing variable parameters.

The *Lennard-Jones potential*

$$\varphi(r) = 4\varepsilon\left[\left(\frac{\sigma}{r}\right)^{12} - \left(\frac{\sigma}{r}\right)^{6}\right]$$

is used as a function for gases and gas mixtures at moderate temperatures, where ε is the depth of the potential pit, σ the effective diameter of the molecule or the radius r for which $\varphi = 0$. Selection of interaction potentials allows us to construct theoretical models, which quite adequately approximate experimental data for monatomic gases over the entire temperature range, except for the very lowest temperatures, where light gases exhibit quantum effects.

Internal degrees of freedom of polyatomic molecules are commonly taken into account using the *Euken* correction $\lambda = \eta(2.5c_t + c_r)$, where c_t is the heat capacity due to translational motion of molecules alone and c_r the heat capacity due to the energy of rotational, and vibrational degrees of freedom and due to electron energy. Considering that $c_v = c_t + c_r$ and $c_t = \frac{3R}{2\tilde{M}}$ where R is the universal gas constant and $\tilde{M}$ the molecule mass, we can write

$$\lambda = \eta\left[2.5\frac{3R}{2\tilde{M}} + \left(C_v - \frac{3R}{2\tilde{M}}\right)\right]$$

$$= \eta c_v[(9\gamma - 5)/4], \qquad \gamma = c_p/c_v,$$

where c_p is the heat capacity at a constant pressure. The minimum and the maximum Euken corrections are 1/19 and 2.5 respectively.

Euken's relation can be explained theoretically if we assume that energy exchange between translational and the remaining degrees of freedom takes little time, i.e. the relaxation time is short. Euken's hypothesis better agrees with reality at high temperatures when the number of collisions per second is great.

Many efforts have been made to attain a better consistency with experimental data for polyatomic gases by a further complication of the model of energy exchange between degrees of freedom. Thus, Mason and Monchick discussed another limit case when the time of relaxation is long. Taking into account this fact leads to a smaller contribution of translational and a larger contribution of rotational degrees of freedom to thermal conductivity.

The molecular kinetic theory takes account only of pairwise molecule collisions, therefore, it is not applicable at high gas densities when ternary collisions come into play. The experimental data demonstrate that, as pressure rises, the coefficient of thermal conductivity grows for all the gases. It has been established that at low and medium pressures (up to a few bars) the increase in λ with increasing p is relatively small and is no higher than 1% of the coefficient of thermal conductivity per bar.

Thermal conductivity of gas mixtures generally is not a linear function of the mixture composition and can be higher or lower than the ones for the original component. Methods based on molecular kinetic concepts and those based on phenomenological concepts have been developed. Thus, a gas mixture is sometimes represented as a hypothetical quasihomogeneous gas the molecule size of which is determined as a mean arithmetical sum taken over effective diameters of components, $\sigma = 1/2\ (\sigma_1 + \sigma_2)$ and the depth of the potential pit is related to similar parameters of components as $\varepsilon = \sqrt{\varepsilon_1\varepsilon_2}$. The reduced mass of molecules of a quasihomogeneous gas is determined by the relation $\tilde{M}^{-1} = \tilde{M}_1^{-1} + \tilde{M}_2^{-1}$.

There exist a great varity of correction factors taking account of nonsymmetric molecule shape, incomplete momentum and energy exchange during the time of a single collision, and different types of interaction potentials (Mason, Saxena, et al.).

For mixtures of nonpolar gases, Brocaw suggested an easy rule supported in practice, $\lambda_m = 1/2\ (\lambda_{sm} + \lambda_{rm})$, where λ_{sm} and λ_{rm} are respectively thermal conductivities of structures formed by alternating plane layers, arranged parallel and perpendicular to the heat flux direction

$$\lambda_{sm} = x_1\lambda_1 + x_2\lambda_2 + \cdots,$$

$$\lambda_{rm}^{-1} = (x_1/\lambda_1) + (x_2/\lambda_2) + \cdots.$$

Here $x_1, x_2, \ldots$ are the molar fractions of the components.

For gas mixtures containing polar molecules or for high pressure mixtures Brocaw's formula has to be replaced by a more intricate expression.

The Theory of Thermal Conductivity of Liquids

Most theoretical and semiempirical expressions for the coefficient of thermal conductivity in liquids are based on the model, suggested by Bridgman, which recognises that the sound velocity U_s in a liquid exceeds by 5- to 10-fold velocity of motion of the molecules determined from the kinetic theory. Another distinction from gases consists in the fact that in the relation for the coefficient for thermal conductivity, instead of heat capacity at a constant volume c_v, the heat capacity at a constant pressure c_p is taken

$$\lambda = \rho c_p U_s L.$$

Here L is the mean intermolecular spacing, $L = \Delta - d$ and Δ the center-to-center distance of molecules of diameter d.

The coefficient of thermal conductivity λ calculated by the above formula substantially differs from the value calculated by the molecular kinetic theory of gases. However, the consistency of calculation results and experimental data is also not good enough, commonly the difference is from 5% to 15%, but sometimes it amounts to 50%.

The coefficients of **thermal conductivity** and viscosity of *liquids* obey more intricate laws than these for gases. In particular, no simple proportionality characteristic of gases exists between λ and μ. As a rule, λ and μ for liquids diminish with temperature, nevertheless the effect of temperature on λ is far weaker than on μ.

Such an extraordinary behavior is accounted for by the fact that, in contrast to gases, interaction of intermolecular force fields in liquids prevents any relative motion of neighboring layers. With increasing T the energy of thermal motion loosens the molecules decreasing the attraction forces and, hence, the coefficients of viscosity η and of thermal conductivity λ decrease. The function $\lambda(T)$ can be approximated by a linear dependence with the proportionality factor from -0.0005 to -0.002 for different types of liquids.

Since the velocity of sound for many liquids is uncertain, semiempirical relations based on more certain physical characteristics received recognition. Thus, Missenard suggested the formula $\lambda = Bc_p\sqrt{\rho T_s}$, where B can be considered a constant for all the liquids with the same number of atoms in the molecule N (B is approximately proportional to $N^{-0.25}$) and T_s is the boiling point of the liquid.

For all liquids the coefficient of thermal conductivity increases with increasing pressure. Upon compression molecules draw together, their mutual attraction grows, therefore, viscosity and thermal conductivity increase. However, up to pressures of the order of 50 bars, λ increases only slightly and the variation can be neglected. At high pressures ($p = 12000$ bars) thermal conductivity of water grows by less than 50% and that of n-pentane, by no more than 70%.

Contrary to gas mixtures in which components are mixed at the atomic and molecular level, the structure of systems with liquid components can differ essentially, viz. from ideal solutions to emulsions in which aggregations of homogeneous components may be higher than 10^9 units and more. But reliable and systematized data are still scarce.

Based on generalized experimental data *Filippov* suggested an empirical relation for estimating *thermal conductivity of solutions*

$$\lambda = \lambda_1 C_1 + \lambda_2 C_2 - 0.72/\lambda_1 - \lambda_2/C_1 C_2,$$

where C_1 and C_2 are the mass concentrations of components with thermal conductivities λ_1 and λ_2 respectively.

To calculate thermal conductivity of electrolytes (salt solutions), Missenard recommends the expression

$$\lambda = 9.3\ 10^{-2}\, N\sqrt{\rho T_m}\tilde{m}^{-5.6}, \qquad \text{W/mK},$$

where $\tilde{m}$, N, ρ and T_m are the molecular mass of the molecules, the number of atoms in a molecule, the electrolyte density, and the melting point respectively.

The Theory of Thermal Conductivity of Solids

Thermal conductivity of solids is of a different nature depending on whether or not they are conductors. In dielectrics with no free electrical charges, thermal energy is transferred by *phonons*. The collective vibrations of atoms in crystal lattice take the form of displacement waves, the interference of which generates wave packets, i.e. phonons. The displacement waves bring about fluctuations of density that can be manifested as a variation of refractive index. If the waves pass through a region in which the refractive index differs from the mean volumetric value, the subsequent displacement waves carrying the energy of thermal motion will be scattered.

The scattering is the stronger, the more the maximum deviation of atoms from their mean positions in the lattice, i.e. the higher the T. This accounts for the experimentally established fact that thermal conductivity of crystal dielectrics decreases at fairly high temperatures: $\lambda \propto 1/T$ (**Figure 1**).

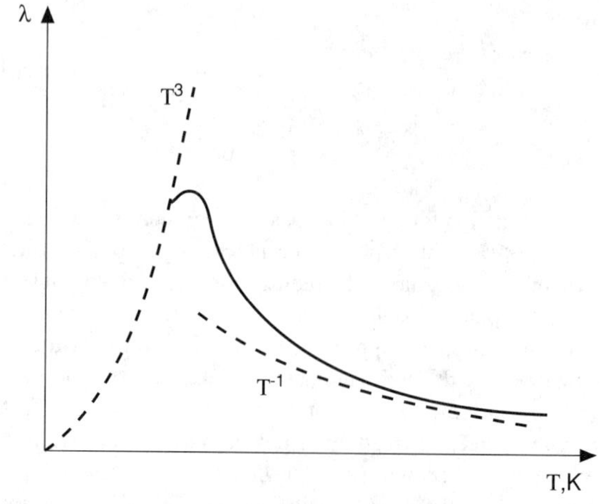

Figure 1. Variation of thermal conductivity of crystal dielectrics with temperature.

Debye gave an insight into a peaklike character of $\lambda(T)$ variation at low T within the framework of quantum mechanical concepts. He elaborated a dipole theory of dielectrics based on the concept of a molecule as a rigid dipole. In this model Debye introduced the concept of the so-called *Debye temperature* θ_D proportional to the maximum frequency of atomic vibrations in a solid. For most substances θ_D is either lower than or close to the room temperature (an exception is diamond for which $\theta_D = 1850$ K and berrylium for which $\theta_D = 1160$K).

In the quantum theory of solids, the crystal lattice is considered as a volume filled with phonon gas to which the deduction of the simplest kinetic theory is applied

$$\lambda \propto cv\vec{l},$$

where c is the heat capacity of a dielectric that coincides with that of phonon gas, v the mean velocity of phonons approximately equal to the sound velocity, $\vec{l}$ the mean free path of phonons. The existence of a certain finite value of $\vec{l}$ is the result of phonon scattering on phonons on defects of crystal lattice and, in particular, on grain boundaries over the entire specimen. The temperature dependence $\lambda(T)$ is determined by that of $\vec{l}$ and c on temperature.

At high temperatures substantially exceeding the Debye temperature θ_D, the basic mechanism limiting $\vec{l}$ is a phonon-phonon scattering due to anharmonicity of atom vibrations in a crystal. The phonon-phonon mechanism of thermal resistance ((1/λ) is known as the coefficient of thermal resistance) is feasible only owing to a transition resulting in retardation of phonon flow and appreciable changing of quasimomentum. The higher the T, the higher the probability of the transition with $\vec{l}$ decreasing: $\vec{l} \propto (1/T)$. Since at $T > \theta_D$ heat capacity c only slightly increases with temperature, hence, $\lambda \propto 1/T$.

At temperatures lower than the Debye temperature, $T \ll \theta_D$, the length of the mean free path $\vec{l}$ determined by the phonon-phonon scattering grows drastically, $\vec{l}$ exp[θ_D/T] and, as rule, is confined to the size of a crystal R. According to the Debye law heat capacity c in this temperature range varies as $c \propto (T/\theta_D)^3$. Consequently, at $T \to 0$ the coefficient of thermal conductivity λ must decrease in proportion to T^3. The temperature for which thermal conductivity has a peak is determined from $\vec{l} \approx R$, which commonly corresponds to $T \leq 0.05\theta_D$.

This theory also accounts for the behavior of the coefficient of thermal conductivity for amorphous substances having no long-range order, i.e. the size of their "crystals" is of the order of atomic sizes. By virtue of this, scattering at the "boundaries" of these substances must prevail at all T and $\vec{l} \approx$ const. The coefficient of thermal conductivity of amorphous substances $\lambda \propto T^3$ at low temperatures and must grow slightly, in proportion to heat capacity, at moderate and high temperatures, $T > \theta_D$.

All this provides a qualitative explanation of the dependence $\lambda(T)$ in real crystal dielectrics, but takes no account of deviations from the constant lattice due to atomic impurities of other elements and their own isotopes. At a high temperature the resistance which is offered by impurities to heat transport is independent of temperature. This makes it possible to estimate the degree of crystal purity from variation of its thermal conductivity with temperature. The same fact enables us to determine the distribution between glassy and crystal substance in the natural mineral if the glass is considered as an impurity in a crystal.

Thermal conductivity in metals depends on the motion and interaction of current carriers, i.e. conduction electrons. Generally, the coefficient of thermal conductivity λ of a metal equals the sum of lattice (phonon) λ_{ph} and electron λ_e components, $\lambda = \lambda_e + \lambda_{ph}$; at ordinary temperatures, as a rule, $\lambda_e \gg \lambda_{ph}$.

If we apply a simple kinetic theory of gases to the flow of free electrons and assume that the length of their free path does not depend on velocity, then the theory implies that

$$\lambda_e = 2\left(\frac{K}{e}\right)^2 \sigma T,$$

where e is the electron charge, K Boltzmann's constant, and σ the electrical conductivity. A strict quantum mechanical theory offers a similar relation,

$$\lambda_e = \frac{\pi^2}{3}\left(\frac{K}{e}\right)^2 \sigma T.$$

In both formulas the ratio of the electron part of the coefficient of thermal conductivity λ to electrical conductivity σ, in a wide temperature range, appears to be in proportion to the temperature according to the *Wiedemann-Franz law,* $\lambda/\sigma = LT$, where L is the Lorentz number. This law is used to calculate λ_e from the measured electrical conductivity. However, it is violated at temperatures below θ_D when an electron-phonon interaction prevails (σ grows with decreasing T as T^{-5} while λ_e varies more slightly as T^{-2}).

The experimental findings evidence that, as a rule, thermal conductivity of metals at high temperatures is a slightly decreasing function of temperature. But there are many exceptions. Thus, thermal conductivity of iron strongly decreases with temperature while tantalum and niobium show a positive temperature coefficient.

Of interest are the attempts to compare the coefficients of thermal conductivity for different substances for a single characteristic temperature, for instance, for the melting point T_m of solids. Thus, the statistical processing of exeprimental data carried out by Missenard has shown that thermal conductivity of metals at the melting point $\lambda(T = T_m) = \lambda_m$ is in proportion to $\lambda \propto T_m \tilde{m}^{-1/6}$ and thermal conductivity of dielectrical crystals at $T = T_m$ is in proportion to $\lambda_m \propto, \sqrt{\rho T_m/(\tilde{m}N)}$, where $\tilde{m}$ is the molecular mass, N the number of atoms in the chemical formula, ρ the density, and the melting point T_m is in Kelvins.

Heat transport in semiconductors is more complex than in dielectrics and metals because the phonon and the electron components are equally essential for them. Another reason for the complexity is the considerable effect of impurities, bipolar diffusion, and other little studied factors on heat transfer.

The effect of pressure on thermal conductivity of solids can be assumed, with a good accuracy, to be linear, λ for many metals growing with p.

For the effect of porosity on thermal conductivity of solids see **Porous Medium**.

References

Wakeham, W. A., Nagashima, A., and Sengers, J. V. (1991) (Eds.), *Experimental Thermodynamics*, Vol. III, Chapters 6, 7 and 8, Blackwell Scientific Publications, Oxford.

Millat, J. Dymond, J. H. and Nieto de Castro, C. A. (1996) (Eds.), *Transport Properties of Fluids: Their Corrrelation, Prediction and Estimation*, Cambridge University Press, New York.

Yu.V. Polezhaev

THERMAL CONDUCTIVITY IN POROUS MEDIA (see Porous medium)

THERMAL CONDUCTIVITY, OF AIR (see Air, properties of)

THERMAL CONDUCTIVITY OF CARBON DIOXIDE (see Critical point, thermodynamics)

THERMAL CONDUCTIVITY OF GASES (see Kinetic theory of gases)

THERMAL CONDUCTIVITY VALUES

Following from: *Conduction; Thermal conductivity*

A material property denoted by λ and defined by *Fourier's Law,* which for one—dimensional conduction in an isotropic medium is:

$$\dot{q} = -\lambda \frac{dT}{dx} \quad (1)$$

The value of λ is temperature dependent and is usually determined experimentally by methods based on Fourier's Law, i.e.

$$\lambda = \frac{\dot{q}}{-\frac{dT}{dx}} \quad (2)$$

where, for example, $\dot{q}$ is measured for an imposed value of dT/dx. Details of the range of methods used to determine λ for solids, liquids and gases are given by Tye (1969) and Maglic, Cezairliyan and Peletsky (1984, 1992). Values of λ are catalogued by Touloukian and Ho (1970 to 1977), an extensive data bank compiled by the Thermophysical Properties Research Centre at Purdue University and continually extended and updated by CINDAS at the same university. Other readily available data books are those by Kaye and Laby (1986) and Perry and Green(1984).

Figure 1 illustrates the very wide range of values of λ for solids, liquids and gases at normal temperatures and pressures.

The temperature dependence of λ values for the three states of matter is exemplified in **Figures 2** and **3**.

The thermal conductivity of a metal can usually be correlated with temperature, over a restricted range, using an expression such as:

$$\lambda = \lambda_0(a + b\theta + c\theta^2) \quad (3)$$

where $\theta = T - T_{ref}$ and λ_0 is the thermal conductivity at a reference temperature T_{ref}.

The thermal conductivity of a non-homogeneous solid is usually dependent on the apparent bulk density and as a general rule increases with increasing temperature and increasing bulk density.

Figure 3 shows the temperature dependence of λ for some saturated liquids and vapours, and gases, of engineering importance. λ for most liquids decreases with increasing temperature. The exception is water, which exhibits increasing λ up to about 150°C and decreasing λ thereafter. Water has the highest thermal conductivity of all liquids except for liquid metals.

The thermal conductivity of gases increases with increasing temperature but is essentially independent of pressures for pressures close to atmospheric. λ values for steam exhibit strong pressure dependence.

Methods for the estimation of λ values outside the range of published values and for the λ values of liquid and gas mixtures are described by Reid et al. (1987).

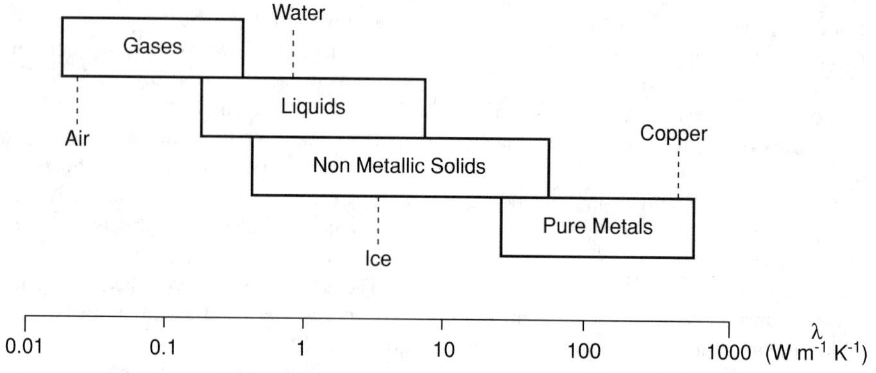

Figure 1. Spread of λ values for three states of matter.

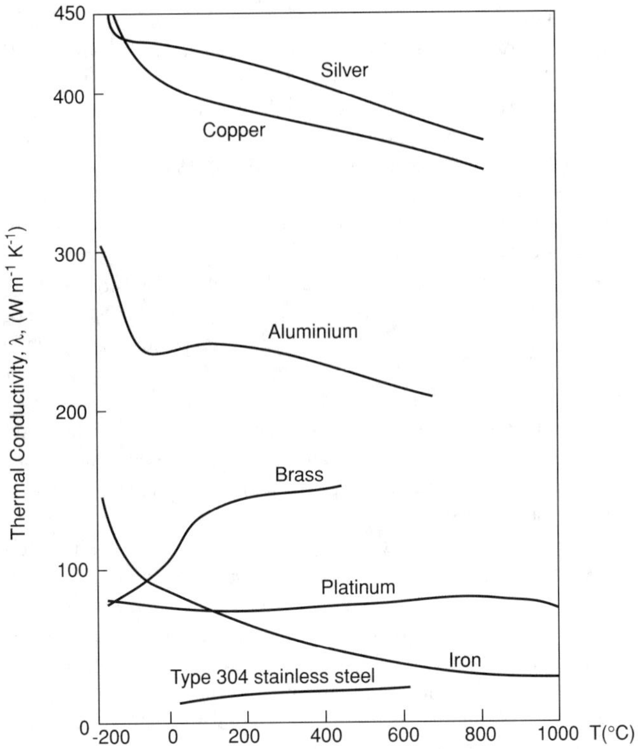

Figure 2. Variation of thermal conductivity of some metals with temperature.

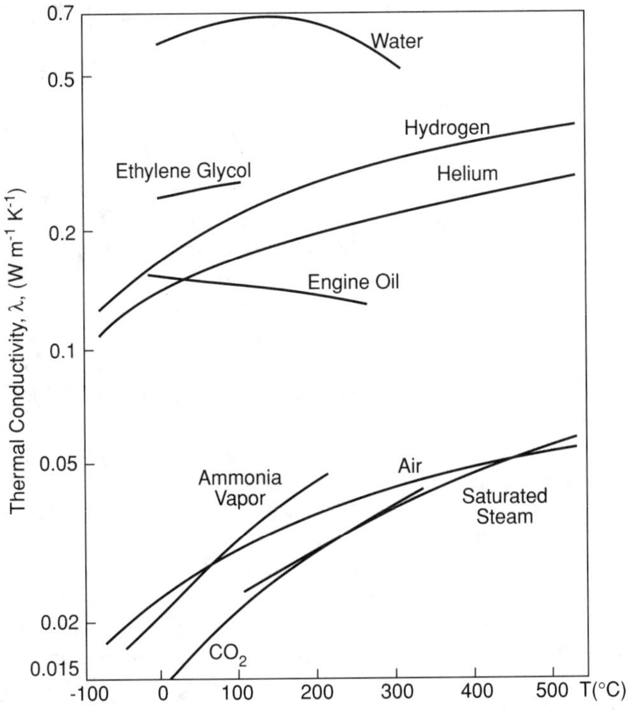

Figure 3. Variation of thermal conductivity of some fluids with temperature.

References

CINDAS, Centre for Information and Numerical Data Analysis and Synthesis, Purdue University, 2595 Yeager Road, West Lafayette, IN 47906, USA.

Kaye, G. W. C. and Laby, T. H. (1986) *Tables of Physical and Chemical Constants,* 15th Ed., Longmans Scientific and Technical, Harlow, UK.

Maglic, K. D., Cezairliyan, A. D. and Peletsky, V. E. (1984,1992) *Compendium of Thermophysical Property Measurement Methods,* Vols. 1 (1984) and 2 (1992), Plenum Press, New York.

Perry, R. H. and Green, D. W. (1984) *Perry's Chemical Engineers' Handbook,* 6th Ed., McGraw-Hill.

Reid, R. C., Prausnitz, J. M. and Poling, B. E. (1987) *The Properties of Gases and Liquids,* 4th Ed., McGraw-Hill, New York.

Touloukian, Y. S. and Ho, C. Y. 1970 to 1977. Thermophysical Properties of Matter, *The TPRC Data Series* (13 volumes), Plenum Press, New York.

Tye, R. P. (1969) *Measurement of Thermal Conductivity,* Vols. 1 and 2, Academic Press.

Leading to: *Combined conduction and convection; Fourier's law; Heat protection.*

T.W. Davies

THERMAL CONTACT CONDUCTANCE (see Thermal contact resistance)

THERMAL CONTACT RESISTANCE

When a junction is formed by pressing two similar or dissimilar metallic materials together, only a small fraction of the nominal surface area is actually in contact because of the non-flatness and roughness of the contacting surfaces. If a heat flux is imposed across the junction, the uniform flow of heat is generally restricted to conduction through the contact spots, as shown in **Figure 1**. The limited number and size of the contact spots results in an actual contact area which is significantly smaller than the apparent contact area. This limited contact area causes a thermal resistance, the *contact resistance* or *thermal contact resistance.*

The presence of a fluid or solid interstitial medium between the contacting surfaces may contribute to or restrict the heat transfer at the junction, depending upon the thermal conductivity, thickness, and hardness (in the case of a solid) of the interstitial medium. If there is a significant temperature difference between the surfaces

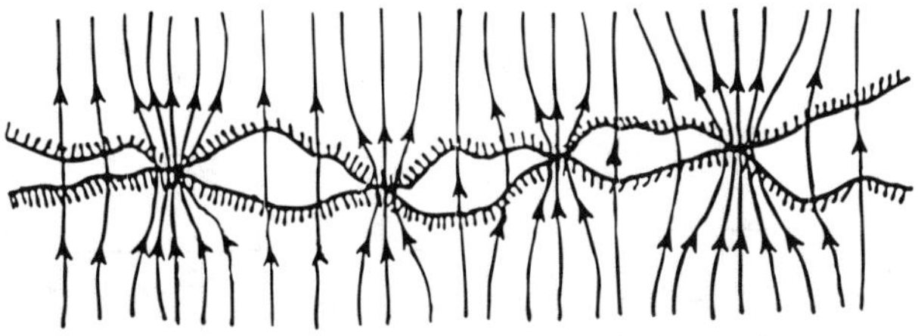

Figure 1. Magnified view of two materials in contact.

composing the junction, heat exchange by radiation also may occur across the gaps between the contacting surfaces.

When a metallic junction is placed in a vacuum, conduction through the contact spots is the primary mode of heat transfer, and the contact resistance is generally greater than when the junction is in the presence of air or other fluid. In a vacuum, the temperature distribution in the contacting materials, with the resulting temperature difference at the junction, is shown in **Figure 2** for both flat and cylindrical junctions.

This temperature difference is used to define the contact resistance at the junction, such that:

$$1/S\alpha_c = (T_1 - T_2)/\dot{Q} = \Delta T/\dot{Q} \text{ or } 1/\alpha_c = (T_1 - T_2)/\dot{q} = \Delta T/\dot{q} \quad (1)$$

where T_1 and T_2 are the temperatures of the bounding contact surfaces, S is the area across which the heat is transferred, and α_c is the heat transfer coefficient for the junction, or the thermal contact conductance. This *contact conductance* or *joint conductance* is often reported in the literature and is defined as:

$$\alpha_c = \dot{q}/\Delta T \quad (2)$$

The magnitude of the contact conductance is a function of a number of parameters including the thermophysical and mechanical properties of the materials in contact, the characteristics of the contacting surfaces, the presence of gaseous or non-gaseous interstitial media, the apparent contact pressure, the mean junction temperature, and the conditions surrounding the junction, as noted by Fletcher (1988).

In view of the significant number of parameters affecting the contact conductance or contact resistance, it has not been possible to develop a single analytical expression for the prediction of the contact resistance at a junction between two materials, except for cases of highly idealized single and multiple contacts. An overview of the idealized models has been reported by Sridhar and Yovanovich (1994). An analytical expression for predicting the contact

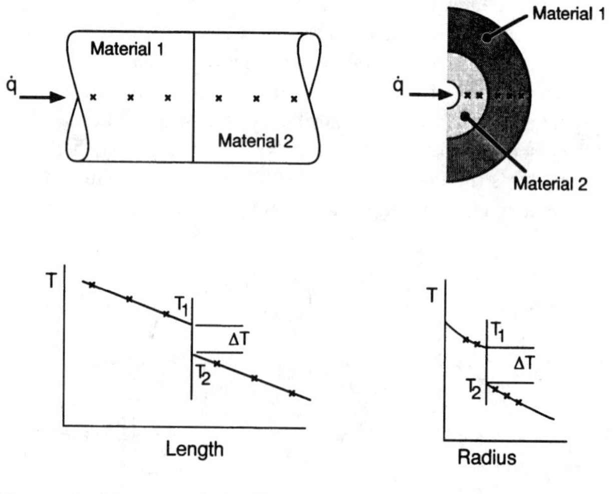

Figure 2. Temperature distribution across flat and cylindrical contacting solids.

conductance of non-flat or machined metallic surfaces in contact has been developed by Lambert and Fletcher (1995) for a wide range of metallic materials and test conditions. Despite the availability of these models, a majority of the contact resistance information is determined experimentally in order to provide a measure of the thermal performance of a specific configuration or system.

Most experimental contact resistance data are obtained using a traditional cut-bar, vertical column test facility in a vacuum or ambient environment over a range of steady-state test conditions. More specialized test facilities have been developed for use with such configurations as bolted joints, periodic or sliding contacts, concentric cylinders, and full scale or partial scale models, while some configurations are studied by electrolytic analogue techniques. Essentially all of these experimental facilities may be used for evaluation of metallic and non-metallic materials in contact, or metallic and non-metallic materials with gaseous or non-gaseous interstitial media between the contacting surfaces, over a wide range of test parameters.

The force applied to the nominal contact area of the junction provides the apparent contact pressure on the junction. The mean junction temperature, T_m, is the average of the contacting surface temperatures. The apparent contact pressure and the mean junction temperature, combined with the thermophysical and mechanical properties of the contacting materials and the surface characteristics, are the primary factors in determining the magnitude of the contact resistance. High junction loads and high temperatures result in low contact resistances, whereas light junction loads and low temperatures lead to high contact resistances.

The surface finish, or roughness and flatness of the contacting surfaces, can significantly affect the magnitude of the contact resistance. If the axial force on the contacting surfaces is increased, the surface roughness peaks or asperities may deform plastically or elastically, depending upon the material properties, leading to increased contact area and decreased contact resistance. An elevated temperature at the junction may also cause plastic and/or elastic deformation of the roughness asperities, especially for softer materials, with an associated increase in the actual contact area and a decrease in the contact resistance. Typical contact resistance values for Aluminum 2024-T4 samples in contact at moderate test conditions in a vacuum environment are shown in **Figure 3**, to demonstrate the effect of surface finish and mean junction temperature on contact resistance (Fletcher, 1991).

Some additional factors which may affect the contact resistance are the direction of the heat flux, surface scratches or cracks, non-uniform loading which causes uneven contact pressure, relative motion or slipping between the surfaces, and the presence of oxides or contaminants on the contacting surfaces.

The use of interstitial or thermal control materials for thermal enhancement or thermal isolation of metallic junctions further effects the contact resistance. Although there are variations in material thickness and composition, the contact resistance for representative interstitial materials is shown in **Figure 4**. These interstitial materials have been categorized as greases and oils; metallic foils and screens; ceramic composites and cements; and synthetic and natural sheets (Fletcher, 1972). While metallic foils and greases are often used for thermal enhancement, most of the interstitial materials are generally used for thermal isolation.

Surface treatments, or coatings and films, may also be used for thermal enhancement or thermal isolation. Metallic coatings

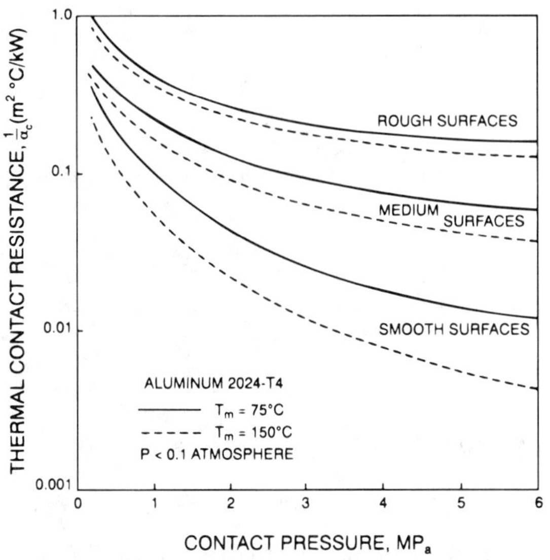

Figure 3. Thermal contact resistance as a function of apparent contact pressure, mean junction temperature and surface finish.

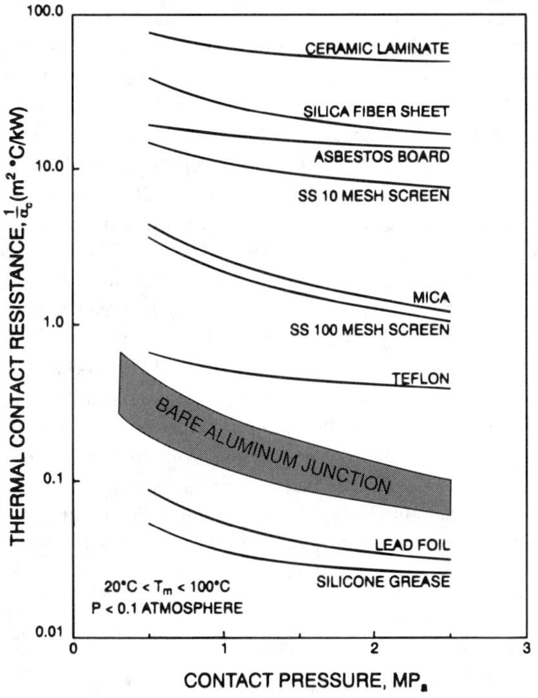

Figure 4. Contact resistance for selected interstitial materials for thermal enhancement or thermal isolation.

provide modest to significant thermal enhancement, depending upon the metal used and the method of application. Ceramic coatings provide modest to excellent thermal isolation depending upon the choice of material. Ceramic coatings may also provide hard, corrosion resistant coatings that are not electrically conducting. Care must be taken to assure that galvanic corrosion will not occur with the choice of materials for some applications.

Contact resistance can be an advantage or disadvantage. Proper consideration of contact resistance and its characteristics can lead to improved thermal design and overall thermal control of components and systems.

References

Fletcher, L. S. (1972) A review of thermal control materials for metallic junctions, *AIAA Journal of Spacecraft and Rockets,* 9, 12, 849–850.

Fletcher, L. S. (1988) Recent developments in contact conductance heat transfer, *ASME Journal of Heat Transfer,* 110, 4B, 1059–1070.

Fletcher, L. S. (1991) Conduction in solids—Imperfect metal-to-metal contacts: Thermal contact resistance, Section 502.5, *Heat Transfer and Fluid Mechanics Data Books,* Genium Publishing Company, Schenectady, New York.

Lambert, M. A. and Fletcher, L. S. (1995) Thermal Contact Conductance of Spherical Rough Metals: Theory and Comparison to Experiment, Proceedings of the ASME/JSME Thermal Engineering Joint Conference, Maui, Hawaii, March 19–24.

Sridhar, M. R. and Yovanovich, M. M. (1993) Critical Review of Elastic and Plastic Contact Conductance Models and Comparison with Experiment, AIAA Paper 93-2776, AIAA Thermophysics Conference, Orlando, Florida, July 6–9.

L.S. Fletcher

THERMAL DATA FOR METALS (see Metals)

THERMAL DIFFUSION

Following from: Diffusion

Thermal diffusion is a relative motion of the components of a gaseous mixture or solution, which is established when there is a temperature gradient in a medium. Thermal diffusion in liquids has an alternative term, the *Soret effect,* named after the Swiss scientist, who investigated thermal diffusion in solutions in 1879–1881. Thermal diffusion in gases was theoretically predicted by Chapman and Enskog (1911–1917) on the basis of the kinetic theory of gases, and it was later discovered experimentally by Chapman and Dutson in 1917.

Thermal diffusion disturbs the homogeneity of mixture composition: the concentration of components in the regions of increased and decreased temperatures respectively becomes different. Since the establishment of a concentration gradient causes, in turn, ordinary diffusion, in a stationary non-uniform temperature field a steady state inhomogeneous state is possible in which the separation effect of thermal diffusion is balanced by the counteraction of concentration diffusion.

In a binary gaseous mixture at constant pressure and with no external forces the total diffusion mass flux of each component is

$$\dot{m}_i = -nD_{12}\nabla c_i + nD_T \frac{1}{T}\nabla T = -nD_{12}\left(\nabla c_i + k_T \frac{1}{T}\nabla T\right),$$

where D_{12} is the binary diffusion coefficient, D_T is the thermal diffusion coefficient, $n = n_1 + n_2$ is the total number of molecules in unit volume, $c_i = n_i/n$ (i = 1,2) is the concentration of molecules of the i-th component. The thermal diffusion ratio $k_T = D_T/D_{12}$ is proportional to the product of the component concentrations, therefore, it is often useful to introduce the thermal diffusion constant α which can be determined from the expression $k_T = \alpha c_1 c_2$. The quantity α is siderably less dependent on the composition than k_T, and for mixtures whose molecule properties are similar (for instance isotopes) it does not change significantly. In gaseous mixtures α does not practically exceed 0.4; for mixture of isotopes, a typical value for α is 0.01.

In the general case, k_T depends in a complex manner on the molecular masses, effective molecule size, temperature, mixture composition, and on the laws of intermolecular interaction. The closer the intermolecular forces approach the laws of interaction between the elastic solid balls, the greater is the value of k_T; it also increases with increase in the molecule dimension and mass ratio.. When molecules interact in accordance with the law for solid elastic bails, k_T is independent of the temperature, the heavier molecules gather, in this case, in a cold region ($k_T > 0$ for $m_1 > m_2 > 2$ where m_1 and m_2 are the masses of the respective components), but if m_1 and m_2 are equal, then larger molecules move into a cold region. For other laws of intermolecular interaction k_T can depend considerably on the temperature and can even change sign. Since the thermal diffusion coefficient depends considerably on intermolecular interaction, its measurements allow us to study the intermolecular forces in gases.

The theory of thermal diffusion in liquids has been developed so far only within the limits of the thermodynamics of irreversible processes, with no external forces, chemical reactions and mechanical equilibrium, the concentration and temperature gradients cause the flow of the components which can be written for a binary solution by analogy with a gaseous mixture

$$j_i = -\nu D(\nabla\nu_{10} + a\nu_{10}\nu_{20}\nabla T),$$

where ν is the total number of moles per unit volume, $\nu_{10} = \nu_1/\nu$, $\nu_{20} = \nu_2/\nu$, the quantity a which is formally similar to α/T for gases is called the Soret coefficient. There is practically no way of calculating this quantity theoretically because of the absence of a statistical theory of fluids; therefore, the Soret coefficient is found experimentally. In the majority of cases for water and organic solutions α varies within the range 10^{-3} to 10^{-2} 1/K.

When determining the thermal diffusion coefficient in gases experimentally, the concentrations of the components in communicating cells filled with the gaseous mixture, thermostatted at temperatures T′ and T″, are measured. If c_1' and c_2' are the concentrations of the components in a solid cell, and c_1'' and c_2'' in a heated one, then $\alpha = \ln q/\ln(T'/T'')$, where $q = (c_1''/c_2'')/(c_1'/c_2')$ is the coefficient of separation. When measuring the thermal diffusion coefficient in solutions, the temperature gradient is established in vertically in the plane gap or horizontally between the coaxial cylinders to avoid convection. In thermal diffusion columns used for separating isotopes and other substances difficult to separate, a convective motion is, by contrast, used to intensify the separation effect of thermal diffusion. As a result of such multistage methods of thermal diffusion separation, the values of q are of the order of 100, and sometimes, of $10^4 - 10^5$.

References

Chapman, S. and Cowling, T. G. (1952) *The Mathematical Theory of Non-Uniform Gases,* Cambridge University Press.

Wakeham, W. A., Nagashima, A. and Sengers, J. V. (1991) (Eds.). *Experimental Thermodynamics*, Vol. III, Chapter 10, Blackwell Scientific Publications, Oxford.

A.L. Kaljazin

THERMAL DIFFUSIVITY

***Following from:** Conduction*

Thermal diffusivity is a combination of physical properties (conductivity, density and specific heat capacity) ($\lambda/\rho c_p$) denoted by κ, naturally arising in the derivation of the *conduction equation* and having physical significance in the context of transient conduction processes. It may be regarded as the ratio of the **Thermal Conductivity** of a material to the specific heat capacity of that material. Thus copper, with a high value of κ ($= 11 \times 10^{-5} m^2 s^{-1}$), has a high conduction rate relative to its heat storage capacity and so responds quickly to changes in temperature. Conversely insulating materials may have κ values 100 times smaller, these low κ values being a measure of the high thermal inertia of such materials. Values of κ are often calculated from the widely available data on λ, ρ and c_p but directly determined values are also available. Data values can be found in Touloukian and Ho (1977), an extensive data bank compiled by the Thermophysical Properties Research Center at Purdue University and continually extended and updated by CINDAS at the same university. Other readily available data books are those by Kaye and Laby (1986) and Perry (1984). Extensive data banks are also available in computerised form and as CDROMs. Experimental methods for the determination of thermal diffusivity and conductivity of solids, liquids and gases are described by Tye(1969). κ values for gases show a strong dependence on temperature and pressure; methods for estimating κ values at temperature /pressure values outside tabulated ranges are described by Reid et al., (1987).

References

CINDAS, Center for Information and Numerical Data Analysis and Synthesis, Purdue University, 2595 Yeager Road, West Lafayette, IN 47906, USA.

Kaye, G. W. C. and Laby, T. H. (1986) *Tables of Physical and Chemical Constants,* 15th Ed., Longmans Scientific and Technical, Harlow, UK.

Wakeham, W. A., Nagashima, A. and Sengers, J. V. (1991) (Eds.) *Experimental Thermodynamics*, Vol. III, Chapters 6, 7 and 8.

Millat, J., Dymond, J. H. and Nieto de Castro, C. A. (1996) (Eds.) *Transport Properties of Fluids: Their Correlation, Predictions and Estimates*, Cambridge University Press, New York.

Perry, R. H. and Green, D. W. (1984) *Perry's Chemical Engineers' Handbook,* McGraw-Hill, New York.

Reid, R. C., Prausnitz, J. M. and Poling, B. E. (1987) *The Properties of Gases and Liquids,* 4th Ed., McGraw-Hill, New York.

Touloukian, Y. S. and Ho, C. Y. (1977) Eds., Thermophysical Properties of Matter, *The TPRC Data Series* (13 volumes), Plenum Press, New York, 1970 to 1977.

Tye, R. P. (1969) *Measurement of Thermal Conductivity,* Vols 1 and 2, Academic Press, New York.

T.W. Davies

THERMAL ENERGY STORAGE (see Heat storage, sensible and latent)

THERMAL EXPANSION (see Steels)

THERMAL EXPANSION COEFFICIENTS

The thermal expansion coefficient is defined as the fractional increase in the linear dimension of a sample of a substance with increase in temperature at constant pressure. Thus,

$$\beta_p = \frac{1}{L}\left(\frac{\partial L}{\partial T}\right)_p$$

For most solids the coefficient β_p is positive, typically 10^{-5} and tables are available for many engineering materials (Bolz and Ture, 1970).

For fluids, it is more usual to work with the volumetric thermal expansion coefficient

$$\beta_p = \frac{1}{\bar{v}}\left(\frac{\partial \bar{v}}{\partial T}\right)_p$$

which is usually positive and much larger than for solids but is negative for water between 0°C and 4°C.

The volumetric thermal expansion coefficient is readily evaluated if an equation of state is available or, indeed, simply values of the fluid volume as a function of temperature at constant pressure. In the former category, for an ideal gas, it is easily shown that

$$\beta_p = 1/T$$

In the latter category, it is worth noting that

$$\beta_\sigma = \frac{1}{\bar{v}}\left(\frac{\partial \bar{v}}{\partial T}\right)_s$$

in the thermal expansion coefficient of a liquid at saturation (not constant pressure). However, it is easily shown that the difference between β_σ and β_p is small while β_σ is rather easily determined for measurements of the density of the liquid at saturation. It follows that a good estimate of β_p can be obtained from measurements for the saturated liquid density.

References

Bolz, R. E. and Ture, G. L. (1970) *Handbook of Tables for Applied Engineering Science,* CRC Press, Boca Raton.

Bett, K. E., Rowlinson, J. S. and Saville, G. (1975) *Thermodynamics for Chemical Engineers,* Athlone, London.

W.A. Wakeham

THERMAL FATIGUE (see Steels)

THERMAL FLOWMETERS (see Flow metering)

THERMAL GRAVITATION CONVECTION (see Free convection)

THERMAL IMAGING

A *thermal imager,* sometimes called a *thermal camera* or *thermograph,* is an instrument which senses the distribution of *infrared radiation* from a target area and presents it as a visible display. It creates a picture of the thermal radiation in a scene and hence gives information about the distribution of temperature on the surfaces in view.

Thermal imagers, developed principally for military purposes, became available in the 1960s. Originally they were used to gain tactical advantage by detecting vehicles and people in darkness. Performance and reliability improved with advances in technology; accompanied by reductions in cost, size and weight, they now have a wide variety of applications. Military use expanded to include more sophisticated applications such as missile guidance.

Thermography is used in the medical and veterinary fields for various diagnostic purposes such as detecting tumours and identifying damage to soft tissue which give rise to locally raised temperatures. Industrial uses include predictive and preventative plant maintenance. Here, thermal imagers perform routine inspections to identify developing fault conditions by observing anomalies in the surface temperature of plant or equipment. Thermography is also used for quality control and process monitoring. *Pulsed thermography* is a technique in which the surface of an object is subjected to an impulse of heat. Observation of the rate of change of the object's surface temperature following the pulse can reveal the presence of internal defects.

Thermal imaging systems are used widely for surveillance. Modern instruments are becoming sufficiently compact and inexpensive to be used as night vision devices in vehicles, aircraft and vessels. Other applications include satellite surveys, astronomical measurements and a variety of research applications.

Most thermal imaging systems operate in wavebands extending between 3 and 5μm, or 8 and 14μm in which reasonable levels of target exitance are combined with good atmospheric transmission.

The performance of a thermal imaging system usually is expressed in terms of spatial and temperature resolutions. The former is expressed as the angular size of the smallest target that can be discerned by the instrument, given that the target has a large temperature difference from its background. Values of 1 to 3 milliradians are typical. Temperature resolution usually is described by the "Noise Equivalent Temperature Difference", which expresses the noise created in the instrument's detector and signal processing electronics as a temperature fluctuation, and describes the limiting value of temperature difference that can be discerned in a scene. Typical values range from 0.05 to 1.0 Kelvin. A combined parameter which may be used to describe the overall performance is "Minimum Resolvable Temperature Difference", which defines the smallest increment of temperature that can be detected in a target of a given size.

Essentially, a thermal imager consists of an optical system, an infrared detector, a signal processing system and a display. Infrared detectors are either quantum devices in which the incidence of radiation causes a change in the electronic state of the detector material, or thermal devices in which the heating of the material causes a measurable change in one of its physical properties. The detector may be a two dimensional array, each element of which indicates the radiant intensity of a single point in the scene. Alternatively, a detector which contains a single element or a linear array of elements may be scanned across the whole scene by a system of mirrors or prisms. The choice of detector type is governed by a range of considerations which include performance, cost, weight, size, power and other logistic requirements.

The signal processing system converts the raw detector output into a format suitable for display and recording. Often this is in the form of a standard television signal which can be used to operate a CRT or LCD display and is suitable for recording on video equipment. The output may also be in the form of a stream of digital information to allow interpretation of the thermal scene by computer. Other signal processing functions may include the addition of artificial colour to provide a display which allows easy identification of features of interest, and the accurate measurement of temperature at points within the scene.

J. Dixon

THERMAL INSULATION (see Heat protection)

THERMALISATION (see Nuclear reactors)

THERMAL MASS FLOWMETER (see Mass flowmeters)

THERMAL PERFORMANCE OF TUBE BANKS (see Tube banks, single phase heat transfer in)

THERMAL REACTORS (see Nuclear reactors)

THERMAL REGENERATORS (see Regenerative heat exchangers)

THERMAL SHOCK

Following from: *Conduction*

Thermal shock arises when a solid at uniform temperature T_0 is suddenly brought in contact with a fluid at a different temperature T_f. If the solid is small, it may be assumed that its temperature will remain uniform throughout the solid although it will change with time. It can be shown that this variation is given by the equation,

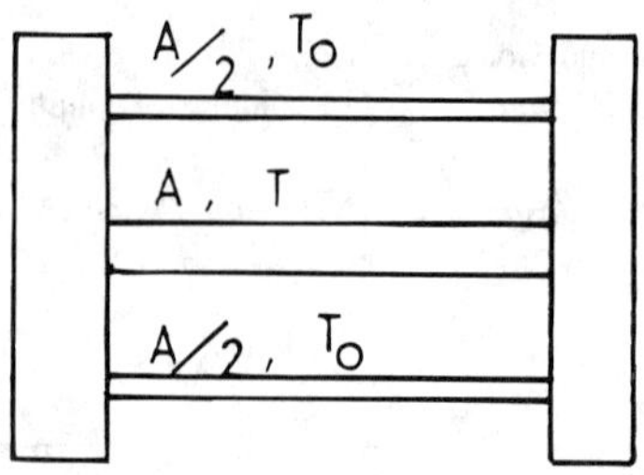

Figure 1. Thermal stresses in simple assembly.

$$\ln \frac{T - T_f}{T_0 - T_f} = -\frac{\alpha At}{\rho cV} \tag{1}$$

where T is the temperature of the solid at time t, A and V are the surface area and volume of the solid respectively, ρ is the density, α is the surface heat transfer coefficient and c the specific heat capacity of the solid. The above equation may be expressed in terms of the **Biot** and the **Fourier Numbers** by noting that V/A is a linear dimension which can be denoted by L,

$$\frac{\alpha At}{\rho cV} = \left(\frac{hL}{\lambda}\right)\left(\frac{\lambda \rho ct}{L^2}\right) = (Bi)(Fo) \tag{2}$$

where λ is the thermal conductivity of the solid.

When the internal resistance of the solid it not negligible, the temperature varies with position as well as with time and may be represented by,

$$\frac{T - T_f}{T_0 - T_f} = f\left(Bi, Fo, \frac{x}{L}, \frac{y}{L}, \frac{z}{L}\right) \tag{3}$$

Solutions for simple geometries may be found in Heisler (1947) and Schneider (1963). In the case of one-dimensional heat flow an analytical solution is possible but more general cases require numerical analysis. (see also **Conduction**)

In the presence of temperature differences within the solid, thermal stresses will arise, increasing in magnitude as the temperature gradient increases. Plastic deformation or fracture may result. Consider the simple example, illustrated in **Figure 1**, of a system consisting of three bars fixed to rigid cross members at the ends. If the central bar is plunged into a large liquid bath of temperature T_f, and it is assumed that the internal resistance of the bars is negligible, the temperature history is given by **Equation 1** and the thermal stress is

$$\sigma_{central} = -\sigma_{outer} = \frac{\beta E(T - T_0)}{2} = \frac{\beta E(T_f - T_0)}{2}(1 - e^{-BiFo}) \tag{4}$$

where β is the coefficient of thermal expansion and E is Young's modulus.

The thermal stress is thus seen to depend on the Biot and Fourier numbers. A similar situation arises when the solid is in the form of, say, a thick slab, in which case the surface temperature rises rapidly to approach that of the fluid while the rest of the body lags behind. The thermal stresses resulting will be compressive at the surface and tensile in the interior of the body.

The magnitude and spacial distribution of the thermal stresses depend on the temperature field which, in turn, depends on the Biot and Fourier numbers. As a general rule, the greater the conductivity of the solid, the smaller the temperature gradient and hence the thermal stresses. Materials with poor thermal conductivity, such as ceramics, are prone to fracture under thermal shock while good conductors, such as metals, can withstand more severe shocks without any significant ill effects.

See also Johns (1965) for the calculation of thermal stresses.

References

Heider, M. P. (1947) Temperature charts for induction and constant temperature heating, *Trans. Am. Soc. Mech. Eng.*, 69, 227–36.

Schneider, P. J. (1963) *Temperature Response Charts.*

Johns, D. J. (1965) *Thermal Stress Analyses*, Pergamon.

C. Ruiz

THERMISTORS (see Temperature measurement, practice)

THERMOCAPILLARY FLOW (see Marangoni effect)

THERMOCHEMICAL CALORIE (see Calorie)

THERMOCLINE

The thermocline is the layer of sea water in which the temperature changes rapidly (≥ 1°C/100 m) with distance from the surface. The surface water temperature is controlled by radiative heating (from the sun) or cooling (to the night sky) and convective heating or cooling from/to the atmosphere; currents and wave action keep this layer well mixed to a depth of 50 to 100 m. The cold deep water originates in the Arctic and Antarctic regions and flows slowly in the general direction of the Equator, though there are many complex interactions with other currents and land masses. The thermocline lies between the warm surface water and the cold deep water and may have a thickness of 300 to 1,000 m.

The thermocline structure in the major ocean basins can be broadly thought of as a "permanent" or "long-term mean" temperature profile with a superimposed seasonal variation that mainly affects the upper mixed layer, including its thickness. The actual profile varies greatly from location to location. In shallower waters, the seasonal variation is more pronounced and may dominate the profile. At the higher latitudes, the temperature difference is much smaller and the thermocline disappears altogether above about 60° latitude.

Another phenomenon that affects the water density and therefore the stability of the thermocline structure is the dissolved salt concentration: Warmer surface waters may have a sufficiently higher salt concentration that they have a higher density, with vertical mixing occurring.

Reference

Pond, S. and Pickard, G. L. (1983) *Introductory Dynamical Oceanography*, 2nd Ed., Pergamon Press, Oxford.

K.J. Bell

THERMOCOUPLES (see Temperature measurement, practice)

THERMODYNAMIC EQUILIBRIUM (see Boltzmann distribution)

THERMODYNAMIC NON-EQUILIBRIUM (see Post dryout heat transfer)

THERMODYNAMIC PROBABILITY (see Entropy)

THERMODYNAMIC PROPERTIES (see Thermodynamics)

THERMODYNAMIC PROPERTIES OF AIR (see Air, properties of)

THERMODYNAMICS

History

Thermodynamics is that part of science which is concerned with the conditions that material systems may assume and the changes in conditions that may occur either spontaneously or as a result of interactions between systems. The word "thermodynamics" was derived from the Greek words **thermé** (heat) and **dynamis** (force). The formulation of the First and Second Law of Thermodynamics by the German scientist Rudolf Julius Clausius in 1850 lay the foundation for what is now called "classical" or "equilibrium thermodynamics." More recently, thermodynamics has been extended to include non-equilibrium states.

Basic Concepts

The description of physical phenomena is based on the concept of the state of a system and the changes that occur spontaneously or through interaction with other systems. The term system means any identifiable collection of matter that can be separated from everything else by a well-defined surface. Examples for thermodynamic systems are the water molecules in a container or, much more complex, a complete process plant. If the system boundaries permit the exchange of heat and work, but not of physical matter, the system is termed *Closed System,* as compared to the *Open System,* were mass transfer may occur. At any time, a system is in a condition called "state" which encompasses all that can be said about the results of any measurements or observations that can be performed on the system at that time. It is necessary to distinguish between quantities which depend on the path between states (such as heat and work) and those which depend solely on the state (such as temperature, internal energy or entropy).

A system is in *equilibrium* if there is no tendency for a change in state to occur, i.e. if all forces are in exact balance. The number of independent intensive (mass-independent) properties that must be arbitrarily fixed to establish the state of a system is named the *degree of freedom.* It can be calculated according to Gibbs

$$F = 2 - P + n \tag{1}$$

with P being the number of phases and n being the number of chemical species present.

A *process* is a change in the system from an initial state to a final state. Processes can be divided into two types, namely *reversible processes* which can be reversed at any time so that the system and the surrounding are returned to their original condition and *irreversible processes* in which the reversal can not be carried out without leaving some change in the system or the surroundings. Reversible processes are ideal processes in the absence of friction and finite temperature differences.

Four laws form the foundation of thermodynamics, even though the First and the Second Laws are considered the most important.

Zeroth Law

The *Zeroth Law* is the basis for the measurement of temperature. It states that "two bodies which are in thermal equilibrium with a third body are in thermal equilibrium with each other."

First Law

The *First Law of Thermodynamics* formally states that "while energy assumes many forms, the total quantity of energy is constant. When energy disappears in one form it appears simultaneously in other forms." This may be written as

$$\Delta(\text{energy of the system}) \tag{2}$$

$$+\ \Delta(\text{energy of the surrondings}) = 0$$

In traditional processes, changes may occur in internal energy, in potential energy and kinetic energy. The recognition of heat and work as energy in transit leads to the following formulation of the First Law

$$\Delta U + \Delta E_{kin} + \Delta E_{pot} = \sum Q - \sum W \tag{3}$$

where Q is heat supplied and W work done.

Closed Systems often undergo processes with no changes in kinetic or potential energy, hence

$$\Delta U = \sum Q - \sum W \tag{4}$$

Far more important industrially are processes which involve the steady-state flow of fluids through equipment, see for example **Figure 1**.

For these *Open Systems,* a more general expression of the First Law is used. The total work exchanged between the system and its surroundings consists of the shaft work W_s and the flow work due to pressure forces as the fluid enters and leaves the system.

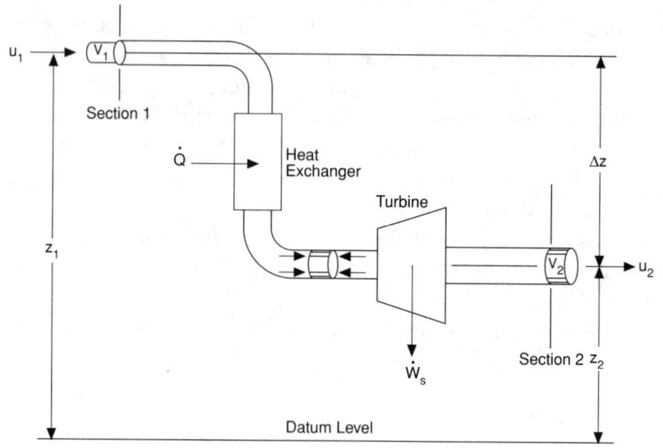

Figure 1. Open system.

$$W = W_s + p_2V_2 - p_1V_1 \tag{5}$$

where p is pressure and V volume.

With the definition of enthalpy, **Equations 3 and 5** can be combined to give

$$\sum Q - \sum W_s = \Delta H + {}^1/_2 M\Delta u^2 + Mg\Delta z \tag{6}$$

where H is enthalpy, M is mass, u is velocity and z height.

A cycle process is a process in which the initial and final states are identical. By virtue of the First Law, the work of an adiabatic cycle (Q = 0) must be zero.

Second Law

The microscopic disorder of a system is described by a system property called **Entropy.** The *Second Law of Thermodynamics* states that whenever a process occurs, the entropy of all systems involved in the process must either increase or, if the process is reversible, remain constant. "As a result, it is impossible to construct a machine operating in a cyclic manner which is able to convey heat from one reservoir at a lower temperature to one at higher temperature and produce no other effect on any other part of the surrounding." This formulation of the Second Law has been attributed to Clausius. The state function entropys, is defined as

$$dS = \frac{dQ_{rev}}{T} \tag{7}$$

where dQ_{rev} is the reversible heat input and T the absolute temperature, and the Second Law can then be written as

$$\Delta S_{tot} = \Delta S_{sys} + \Delta S_{sur} \geq 0 \tag{8}$$

where subscripts tot, sys and sur refer to total, system and surroundings.

The Second Law of Thermodynamics has many far-reaching implications, for example the criterion that for a system in stable equilibrium, the entropy of the system must be at its maximum for fixed values of energy, number of particles and physical constraints.

Third Law

Nernst's Law is frequently named the *Third Law of Thermodynamics.* It states that "the entropy of a thermodynamic system approaches zero for $T \rightarrow 0$ K." This implies that the specific heat capacity of all substances approaches zero, too. As a result, the entropy at any temperature can be calculated from

$$S_{p-const} = \int_0^T \frac{N\tilde{c}_p dT}{T} + \sum_0^T \frac{\Delta H_{ph}}{T_{ph}} \quad \textbf{(9)}$$

where N is the number of moles, $\tilde{c}_p$ the molar specific heat and the subscript ph refers to phase change.

Exergy or Availability

The **Exergy** (sometimes also called availability) is a measure of the maximum useful work that can be produced (or the minimum work required) by a system interacting with the environment. It can be achieved by operating reversibly within the system and transferring heat reversibly. For a closed system

$$W_o = T_o(S_o - S) - (U_o - U) - p_o(V_o - V) \quad \textbf{(10)}$$

and for an open system

$$W_o = T_o(S_o - S) - (H_o - H) \quad \textbf{(11)}$$

with subscript "o" denoting reference conditions which are usually taken as 25°C and 1 atm. If the process does not end at ambient conditions, the ideal work is

$$W_{id} = W_{1,o} - W_{2,o} = T_o\Delta S - \Delta U - p_o\Delta V = \int pdV - p_o\Delta V \quad \textbf{(12)}$$

for a closed system and

$$W_{id} = W_{1,o} - W_{2,o} = T_o\Delta S - \Delta H = W_{s,rev} \quad \textbf{(13)}$$

for an open system. Obviously, this ideal work does not exist. It only serves as a basis for thermodynamic calculations. The *lost work* is the difference between the ideal work and the work actually done by the process

$$W_{lost} = W_{id} - W_{act} \quad \textbf{(14)}$$

For closed processes, the following can be derived

$$W_{lost} = T_o\Delta S_{syst} - Q_{act} - p_o\Delta V \quad \textbf{(15)}$$

while for the practically more important flow processes

$$W_{lost} = T_o\Delta S_{tot} \quad \textbf{(16)}$$

A *thermodynamic efficiency* can be defined as:

$$\eta_t = \frac{W_s}{W_{id}} \quad \text{(work produced)} \quad \textbf{(17)}$$

$$\eta_t = \frac{W_{id}}{W_s} \quad \text{(work required)} \quad \textbf{(18)}$$

Equilibrium

The second law and the associated principle of increase of entropy provide the basis for the analysis of equilibrium problems. For a compressible substance undergoing a general process in which heat and work may be exchanged with the system

$$dS_{sys} + dS_{sur} > 0 \quad \textbf{(19)}$$

Assuming that temperature and pressure of the surrounding are constant

$$dU_{sys} + p_o dV_{sys} - T_o dS_{sys} < 0 \quad \textbf{(20)}$$

and that the system and surrounding are in equilibrium with respect to temperature and pressure

$$dU + pdV - TdS < 0 \quad \textbf{(21)}$$

where all properties refer to the system. Equation 21 may be more conveniently written in terms of **Helmholtz and Gibbs Functions.** The Helmholtz energy H is defined by

$$A = U - TS \quad \textbf{(22)}$$

and Gibbs energy G by

$$G = H - TS = U + pV - TS \quad \textbf{(23)}$$

and hence

$$dA + SdT + pdV < 0 \quad \textbf{(24)}$$

$$dG + SdT - Vdp < 0 \quad \textbf{(25)}$$

For constant pressure and temperature, the Helmholtz and Gibbs energies must decrease and seek minimum values. Once these minima are attained, equilibrium conditions will have been reached with no further change in A and G. The important criterion for equilibrium is, therefore

$$(dG)_{T,p} = 0 \quad \textbf{(26)}$$

for a single-component system and

$$d(\Sigma N_i\tilde{g}_i)_{T,p} = 0 \quad \textbf{(27)}$$

for a multi-component system, where N_i and $\tilde{g}_i$ are the number of moles and the molar Gibbs energy of each constituent.

Thermodynamic Properties

If we consider processes without change in composition, the Fundamental Equations of Thermodynamics can be derived by differentiation of the definition equations of the four main state functions

$$dU = TdS - pdV \quad (28)$$

$$dH = TdS - Vdp \quad (29)$$

$$dG = -Sdt + Vdp \quad (30)$$

$$dA = -SdT - pdV \quad (31)$$

The four *Maxwell Relations* are important thermodynamic partial derivatives which allow the substitution of measurable properties for unmeasurable properties in thermodynamic relationships. Assuming constant composition, they can be derived from the definition equations of the basic reference properties, **Equations 28 to 31**:

$$\left(\frac{\partial T}{\partial V}\right)_S = -\left(\frac{\partial p}{\partial S}\right)_V \quad (32)$$

$$\left(\frac{\partial T}{\partial p}\right)_S = \left(\frac{\partial V}{\partial S}\right)_p \quad (33)$$

$$\left(\frac{\partial S}{\partial V}\right)_T = \left(\frac{\partial p}{\partial T}\right)_V \quad (34)$$

$$\left(\frac{\partial S}{\partial p}\right)_T = -\left(\frac{\partial V}{\partial T}\right)_p \quad (35)$$

Heat capacities can also be derived from basic equations.

The heat capacity at constant pressure is given by

$$C_p = \left(\frac{\partial H}{\partial T}\right)_p = T\left(\frac{\partial S}{\partial T}\right)_p \quad (36)$$

and the heat capacity at constant volume by

$$C_V = \left(\frac{\partial U}{\partial T}\right)_V = T\left(\frac{\partial S}{\partial T}\right)_V \quad (37)$$

Equations for any partial derivative of thermodynamic properties in terms of measurable properties c_p, $(\partial V/\partial p)_T$ and $(\partial V/\partial T)_p$ are given in the *Bridgeman Tables*. These relationships can then be used to construct thermodynamic diagrams. Examples for the four most common diagrams are shown in **Figures 2–5**.

Standard Conditions

It is necessary to choose zero levels or reference states for tabulation of thermodynamic properties. In SI tables, these *standard conditions* are defined as a temperature of 298 K and a pressure

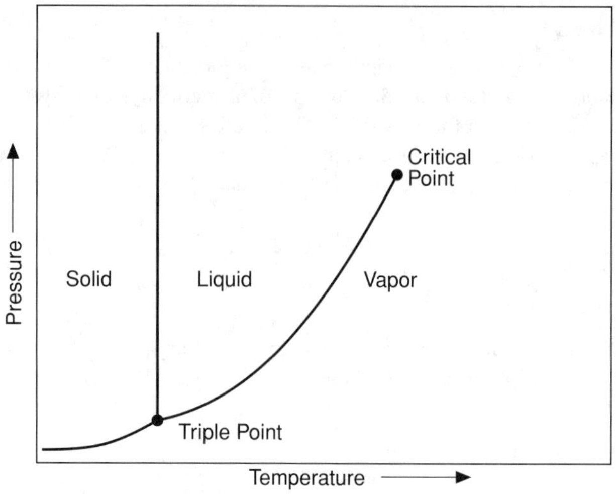

Figure 2. p-T diagram.

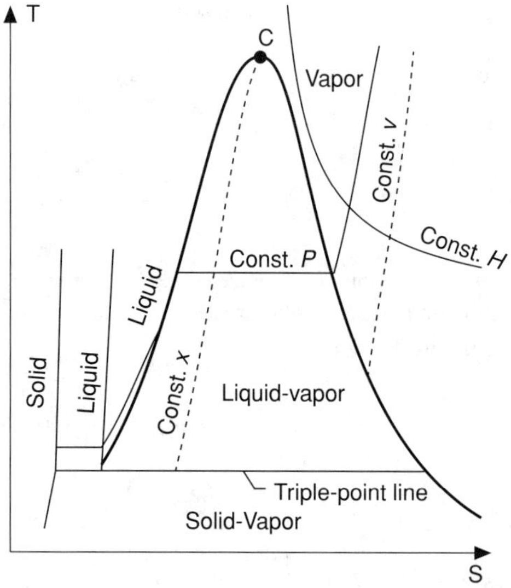

Figure 3. T-s diagram.

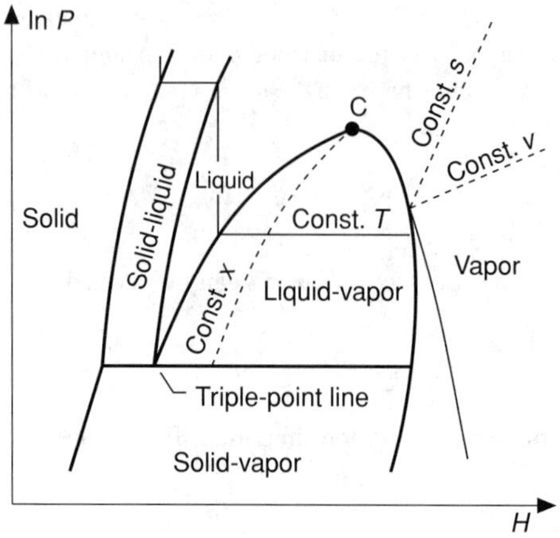

Figure 4. p-h diagram.

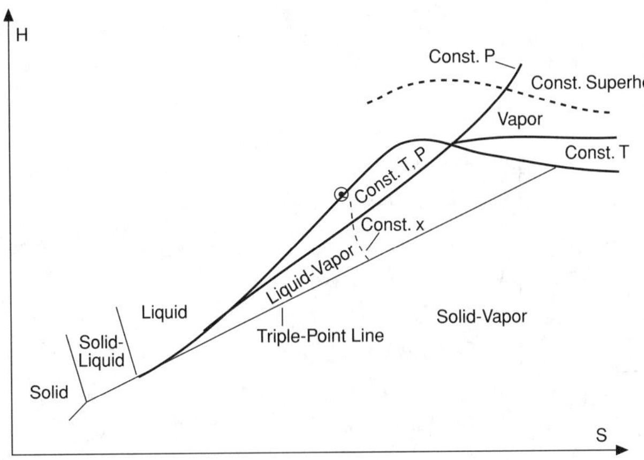

Figure 5. h-s diagram (Mollier diagram).

of 0.1 MPa. Properties at standard conditions are identified by the symbol °, e.g. h°, g° etc.

Chemical Thermodynamics

For a chemical reaction

$$A + B \rightarrow C + D \tag{38}$$

the Enthalpy of Reaction (Heat of Reaction) is defined as

$$\Delta H_r = [H_C + H_D] - [H_A + H_B] \tag{39}$$

If $\Delta H_r > 0$ the reaction is *endothermic,* if $\Delta H_r < 0$ it is *exothermic.* To avoid tabulation of enthalpies of reaction over a wide range of temperatures and pressures, the Standard Enthalpy of Reaction ΔH_r° is used in conjunction with Hess' Law

$$\Delta H_r = \Delta H_r^o + \sum^{\text{products}} \tilde{c}_{p,i} N_i (T - 298K) - \sum^{\text{reactants}} \tilde{c}_{p,i} N_i (T - 298K) \tag{40}$$

where the subscript i refers to component i, and $\tilde{c}_p$ is molar specific heat

The Standard Enthalpy of Reaction is usually found in tables, but can also be calculated from the Standard Enthalpy of Formation per mole of the compounds participating in the reaction

$$\Delta H_r^o = \sum^{\text{products}} N_i \Delta \tilde{h}^o_{f,i} - \sum^{\text{reactants}} N_i \Delta \tilde{h}^o_{f,i} \tag{41}$$

where $\Delta \tilde{h}_f^\circ$ of single-element compounds is zero. The entropy change associated with a chemical reaction is calculated from

$$\Delta S = \frac{\Delta H_r}{T} \tag{42}$$

Most chemical reactions do not go to completion, but reach some equilibrium conditions. Consider a reaction between ideal gases:

$$N_1A_1 + N_2A_2 \rightleftarrows N_3A_3 + N_4A_4 \tag{43}$$

From **Equation 26**

$$dG_{\text{total}} = dG_{\text{products}} + dG_{\text{reactants}} = 0 \tag{44}$$

it can be derived for a reaction between ideal gases at temperature T and reference pressure 1 bar:

$$-\tilde{R}T \ln \frac{p_3^{N_3} p_4^{N_4}}{p_1^{N_1} p_2^{N_2}} = \Delta \tilde{g}_T^\circ \tag{45}$$

where $\tilde{R}$ is the universal gas constant or, with the definition of the equilibrium constant K_p:

$$\Delta \tilde{g}_T{}^o = -\tilde{R}T \ln K_p \tag{46}$$

From **Equations 23, 42 and 46**

$$\ln K_p = \frac{-\Delta \tilde{h}^\circ_{r,T}}{\tilde{R}T} + \frac{\Delta \tilde{s}^\circ_{r,T}}{\tilde{R}} \tag{47}$$

The second term on the right hand side of Equation 47 is only moderately depending on the temperature, resulting in a straight line if the natural logarithm of K_p is plotted versus 1/T, see **Figure 6**. This is expressed in the *van't Hoff equation*

$$\frac{d(\ln K_p)}{dT} = \frac{\Delta \tilde{h}^\circ_T}{\tilde{R}T^2} \tag{48}$$

Using Dalton's law, the equilibrium constant can be expressed in terms of mole fraction and total pressure as

$$K_p = \frac{x_3^{N_3} x_4^{N_4}}{x_1^{N_1} x_2^{N_2}} p_{\text{tot}}^{N_2+N_4-N_1-N_2} \tag{49}$$

The definition of the Gibbs Energy expresses the experience that the total energy released by a process is usually different from the "useful" or "free" energy produced. This is evidence of a special demand exerted by the system or its environment, as the energy taken up by this special demand is then unavailable for useful work. In a chemical reaction, the enthalpy change ΔH_r is the net energy released or taken up as a result of the making or breaking of chemical bonds. The "special demand" involves some other form of energy, due to the movement of atoms and molecules. A reaction producing more molecules from fewer creates more opportunities for molecular movement, as does producing gases from solids or more flexible molecules from more rigid molecules. These opportunities for movement constitute disorder in the system. The change in disorder is associated with the change in entropy, related to the possible energy states in the system. Where a change or reaction decreases the disorder in the system, disorder is leaving the system and its entropy change is negative. A reaction producing fewer molecules from a larger number of reacting molecules is likely to have a negative ΔS. The magnitude of ΔS is usually small and does not effect the spontaneity of the reaction at low temperatures. However, at high temperatures, the term T ΔS may a $2C+2H_2 \rightarrow C_2H_4$

b $1/2N_2 + 1/2O_2 \rightarrow NO$

c $C+CO_2 \rightarrow 2CO$

d $C+H_2O \rightarrow CO+H_2$

e $C+2H_2O \rightarrow CO_2 + 2H_2$

f $C+2H_2 \rightarrow CH_4$

g $NO+1/2O_2 \rightarrow NO_2$

h $CO+H_2O \rightarrow CO_2 + H_2$

i $C_2H2 \rightarrow 2C+H_2$

j $CO+1/2O_2 \rightarrow CO_2$

k $H_2 + 1/2O_2 \rightarrow H_2O$

l $C+1/2O_2 \rightarrow CO$

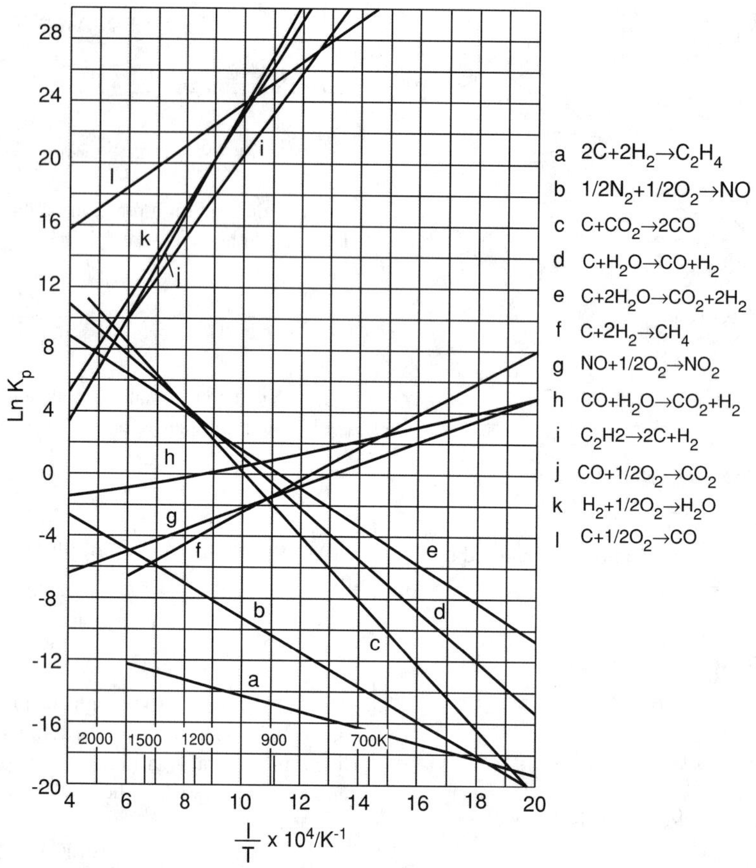

Figure 6.   Equilibrium constants.

over-compensate the enthalpy change ΔH and reverse the direction of the reaction.

Physical Equilibria Between Phases

In many engineering applications, two or more phases are interacting to form equilibrium states. Mass may be exchanged between the phases, changing the concentration of components in each phase. Equilibrium is reached when the driving forces for heat transfer, mass transfer and chemical reaction have disappeared. **Equation 26** written in a more generalized form

$$d(N\tilde{g}) = \left(\frac{\partial(N\tilde{g})}{\partial p}\right)_{T,N} dp + \left(\frac{\partial(N\tilde{g})}{\partial T}\right)_{p,N} dT$$

$$+ \left(\frac{\partial(N\tilde{g})}{\partial N_1}\right)_{T,p,N_j \neq N_1} dN_1 + \left(\frac{\partial(N\tilde{g})}{\partial N_2}\right)_{T,p,N_j \neq N_2} dN_2 + \cdots \text{(50)}$$

The derivatives for constant T and p are called the **Chemical Potential** μ_i of species i in a mixture. Hence, with **Equation 30**.

$$d(N\tilde{g}) = (N\tilde{v})dp - (N\tilde{s})dT + \sum \mu_i dN_i \qquad \text{(51)}$$

where $\tilde{v}$ and $\tilde{s}$ are molar volume and molar entropy.

For two phases x and y in thermodynamic equilibrium it can be derived that

$$\mu_i{}^x = \mu_i{}^y \qquad \text{(52)}$$

The chemical potential in an *ideal gas mixture* is obtained from **Equation 53**

$$\mu_i{}^{ig} = \left(\frac{\partial(N\tilde{g})^{ig}}{\partial N_i}\right)_{p,T,N_j \neq N_i} \qquad \text{(53)}$$

in conjunction with the definitions of entropy and Gibbs energy

$$\mu_i{}^{ig} = \tilde{g}_i{}^{ig} + \tilde{R}T \ln y_i \qquad \text{(54)}$$

Components of *ideal solutions* have similar molecular size and properties (for example hydrocarbon mixtures). In this case, the solution for ideal gases applies

$$\mu_i{}^{is} = \tilde{g}_i{}^{is} + \tilde{R}T \ln x_i \qquad \text{(55)}$$

Combining **Equations 54 and 55**. **Raoult's Law** is obtained as the simplest relationship for equilibrium between liquid and vapour mixtures

$$\tilde{R}T \ln \frac{y_i}{x_i} = (\tilde{g}_i^{is})_{T,p} - (\tilde{g}_i^{ig})_{T,p} \quad \textbf{(56)}$$

The Gibbs energy of liquids is almost independent of the pressure. For the Gibbs energy of the vapour phase at constant temperature

$$(\tilde{g}_i^{ig})_{T,p^*} = (\tilde{g}_i^{ig})_{T,p} + \int_p^{p_i^*} \frac{\tilde{R}T}{p} dp \quad \textbf{(57)}$$

Hence

$$\tilde{R}T \ln \frac{y_i}{x_i} = (\tilde{g}_i^{is})_{T,p_i^*} - (\tilde{g}_i^{ig})_{T,p_i^*} + \tilde{R}T \ln \frac{p_i^*}{p} \quad \textbf{(58)}$$

For equilibrium conditions, the Gibbs energies of the pure species must be identical. This yields the well-known relationship.

$$y_i p = x_p p_i^* \quad \textbf{(59)}$$

For a single substance at constant temperature

$$d\tilde{g}_i = d\tilde{h} - T d\tilde{s} - \tilde{s} dT = d\mu_i \quad \textbf{(60)}$$

After substitution of enthalpy using the First Law of Thermodynamics, of specific volume using the Ideal Gas Law, and several algebraic manipulations one obtains

$$\mu_i - \mu_l^\circ = \tilde{R}T \ln \frac{p_i}{p_i^o} = \tilde{g}_i - \tilde{g}_i^\circ \quad \textbf{(61)}$$

where ° indicates a reference condition. The importance of this equation is that it relates the chemical potential and the Gibbs energy to simple measurable properties such as temperature and pressure. To extend the applicability of **Equation 61** to real gases, liquids or solids, Lewis introduced the concept of **Fugacity.**

$$\Delta\tilde{g}_i = \mu_i(T, p, y) - \mu_i(T, p^\circ, y^\circ) = \tilde{R}T \ln \frac{f_i(T, p, y)}{f_i(T, p^\circ, y^\circ)} \quad \textbf{(62)}$$

Reference conditions p° and y° can be selected at will, while T must be constant. Fugacity is a corrected pressure, to take into consideration interaction between atoms/molecules. For ideal gases, pressure and fugacity are identical. Since all gases and gas mixtures are ideal for low pressure

$$\lim_{p \to 0} \frac{f_i}{y_i p} = \lim_{p \to 0} \phi_i \to 1 \quad \textbf{(63)}$$

The *residual Gibbs energy* $\Delta\tilde{g}^R$ is the difference between the Gibbs energy calculated according to Equation 62 and for an ideal gas

$$\Delta\tilde{g}_i^R = \left(\tilde{R}T \ln \frac{f_i}{y_i p}\right)_T = \tilde{R}T \ln \phi_i \quad \textbf{(64)}$$

where the dimensionless property of the mixture Φ is called the *fugacity coefficient.* It can be directly related to the p V T data through

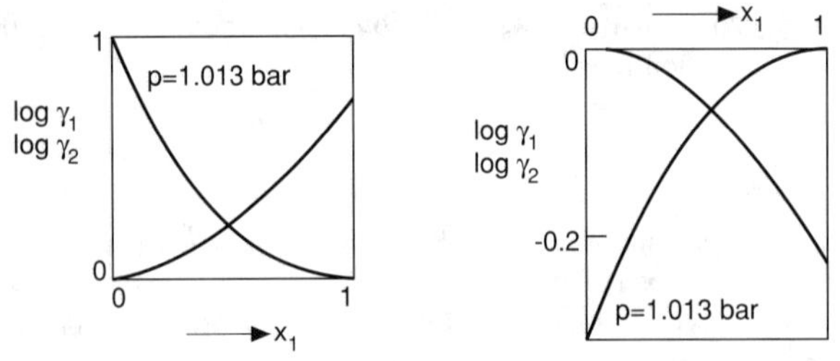

Figure 7.

$$\frac{\tilde{g}^R}{\tilde{R}T} = \ln \phi = \int_0^p (Z - 1) \frac{dp}{p} \text{ for constant T} \quad \textbf{(65)}$$

where Z is the **Compressibility Factor** defined by $Z = pV/N\tilde{R}T$.

Liquid solutions are more conveniently dealt with through properties that measure their deviations from ideal solutions, rather than from ideal gases. The *activity* a_i of a component in solution is defined as the ratio of the fugacity at a state divided by its fugacity at a reference state

$$a_l = \frac{\bar{f}_i}{f_i^o} \quad \textbf{(66)}$$

Different reference states are used, depending on the actual fluid. A common standard pressure is the atmospheric pressure, i.e. $f_i^\circ = p_i$ for most gases. The partial excess Gibbs energy of a species in solution is then

$$d\tilde{g}_i^E = \tilde{R}T \ln a_i \quad \textbf{(67)}$$

The *activity coefficient* γ

$$\gamma_i = \frac{a_i}{x_i} = \frac{\bar{f}_i}{x_i f_i^o} \quad \textbf{(68)}$$

is the correction factor for the departure of a solution from ideal behaviour. With the above definitions and the assumption that the fugacity in the gas phase is identical to the partial pressure, Raoult's Law can be generalized to

$$p_i = \gamma_i x_i p_i^*(T) \quad \textbf{(69)}$$

Activity coefficients depend on the concentration of the component in the solution with

$$\lim_{x_i \to 1} \gamma_i = 1 \quad \textbf{(70)}$$

which indicates that any system becomes ideal for $x_i \to 1$. Depending on the interaction between atoms/molecules, activity coefficients can be larger or smaller than 1, as shown in **Figure 7** for binary mixtures of methanol and tetra-chlor-carbon, and of acetone and chloroform, respectively.

For higher pressures, non-ideal behaviour of the gas/vapour phase has to be considered, as well. Knowledge of activity and fugacity coefficients allows prediction of phase equilibria, solubilities and reaction rates.

References

Bett, K. E., Rowlinson, J. S. and Saville, G. (1992) *Thermodynamics for Chemical Engineers,* The Athlone Press.

Callen, H. B. (1985) *Thermodynamics and Thermostatistics,* 2nd Ed., John Wiley and Sons.

Daubert, T. E. (1985) *Chemical Engineering Thermodynamics,* McGraw-Hill.

Jones, J. B. and Hawkins, G. A. (1986) *Engineering Thermodynamics,* 2nd Ed., John Wiley and Sons.

Holman, J. P. (1988) *Thermodynamics,* 4th ed., McGraw-Hill.

Howell, J. R. and Buckius, R. O. (1987) *Fundamentals of Engineering Thermodynamics,* McGraw-Hill.

Rogers, G. and Mayhew, Y. (1992) *Engineering Thermodynamics—Work and Heat Transfer,* 4th ed., Longman Scientific and Technical.

Smith, J. M. and Van Ness, H. C. (1987) *Introduction to Chemical Engineering Thermodynamics,* 4th ed., McGraw-Hill.

Whalley, P. B. (1992) *Basic Engineering Thermodymics,* Oxford Science Publications.

Leading to: *Adiabatic conditions; Boltzmann distribution; Clapeyron—Clausius equation; Chemical equilibrium; Chemical potential; Dalton's law of partial pressures; Enthalpy; Entropy; Equation of State; Exergy; Gay Lussac's law; Gibbs-Helmholtz equation; International temperature scale; Irreversible thermodynamics; Multicomponent systems thermodynamics of; Phase equilibrium; Perfect gas.*

H. Müller-Steinhagen

THERMODYNAMIC TEMPERATURE SCALE (see Temperature measurement, bases)

THERMODYNAMIC WET BULB TEMPERATURE (see Wet bulb temperature)

THERMOELECTRIC HEAT PUMPS (see Peltier effect)

THERMOELECTRIC PHENOMENON (see Peltier effect)

THERMOELECTRIC REFRIGERATORS (see Peltier effect)

THERMOEXCEL SURFACES (see Augmentation of heat transfer, two phase; Tube banks, condensation heat transfer in)

THERMOGRAPHIC CAMERAS (see Photographic techniques)

THERMOGRAPHY (see Thermal imaging)

THERMOLUMINESCENCE

Thermoluminescence may be defined as the emission of light from solid state materials caused by thermal recombination of previously trapped electrons and holes. In solids, electrons and holes formed by exposure to light or radiation may become separated and trapped in energy wells at defect or impurity sites such that the probability of recombination by either radiative or nonradiative processes is very low. If the temperature is raised thermal energy may become sufficient to allow the electrons and holes to escape from the traps and recombine. When this occurs by a radiative process then, as the temperature is increased, the sample will begin to emit light and continue to emit until all of the traps are emptied. *Glow curves* of emission intensity as a function of temperature give information on trap energies. Natural and synthetic thermoluminescent materials are described by Vij (1993). Doped inorganic crystals and glasses are most common, however many organic polymers and some biological and biochemical materials also show thermoluminescence following irradiation at low temperature.

Applications of thermoluminescence measurements are discussed by Mahesh et al. (1989) and include *radiation dosimetry* and the *dating of archaeological* and geological *samples.* Both applications rely on the principle that the intensity of emission is related to the total radiation dose. In thermoluminescence dating the high temperature firing of pottery or casting of metals around a "core", depletes any thermoluminescent minerals present, such as quartz, of trapped electrons and hole. The intensity of any subsequent thermoluminescence is therefore an indication of the radiation dose received since firing. If the radiation dose at the site can be estimated using thermoluminescent standards, and the sensitivity of the material determined using laboratory irradiation conditions then dating is possible. (See also **Photoluminescence**).

References

Vij, D. R. (1993) *Thermoluminescent Materials,* PTR Prentice Hall, Englewood Cliffs NJ.

Mahesh, K., Weng, P. S., Furetta, C. (1989) *Thermoluminescence in Solids and its Applications,* Nuclear Technology Publishing, Ashford UK.

P. Douglas

THERMOMETER

Following from: *Temperature measurement, practice*

A thermometer is an instrument to measure temperature (see **Temperature** and **Temperature Measurement, Practice**). Temperature cannot be measured directly (for instance, as a length is measured by its segment). Only the measurement of another physical property being a function of temperature may provide temperature determination. A property used to measure temperature is termed a thermometric parameter and must meet the following requirements:

- be independent of the influence of other factors,
- be strictly reproducible,
- be a continuous and monotonic function of temperature,
- be a temperature coefficient of a property which is sufficiently large and its measurement sufficiently simple.

Most often used for temperature measurement are:

a volume of a liquid or a gas (liquid-in-glass and gas thermometers);

electrical resistance of metals, semiconductors (see *resistance thermometers*); electromotive force (see "thermocouples");

temperature measurement by radiation intensity (optical thermocouples, see *pyrometers*);

melting points of metals or alloys;

softening temperatures of ceramics (Seger cones);

change of color with transition temperature—heat setting inks (heat sensitive compounds of fillers and solvents coated onto technical parts);

change of color at temperature transitions to a liquid-crystalline state and an isotropic liquid of some organic compounds—liquid crystals (e.g. liquid-crystalline substances of the cholesteric type).

Each indicated property may be used as a thermometric parameter in a limited temperature range. More seldom used are such thermometric parameters as intensity of electric fluctuations (Johnson noise thermometer), magnetic susceptibility of a paramagnetic (magnetic T), sound velocity, broadening of spectral lines, etc.

Gas Thermometer

Gas thermometry reduces temperature measurement (from helium temperatures to 1063°C) to measurement of pressure or a gas volume in a closed vessel (at certain conditions) followed by temperature calculation using the measurement results and the ideal gas laws. A *gas thermometer* is a primary instrument for determination of thermodynamic temperature. Application of exact relations requires design of complicated devices inconvenient for practical use. In practice, temperature scales are used in which a simple and convenient secondary thermometer is used and methods of transfer of thermodynamic temperature from a primary instrument to the secondary thermometer are employed (see **International Temperature Scale**). This requires use of precise primary instruments reproducing thermodynamic temperature, instruments for realization of the temperatures of phase equilibria of substances (for determination of the constants of the primary instruments), i.e. representing the so-called fixed points and, of course, the secondary thermometer itself together with simple and convenient methods for its calibration. The simplest thermometer is a gas thermometer which consists of a glass or metallic gas-impermeable reservoir connected with an arrangement intended for pressure measurement in the reservoir.

A schematic drawing of a gas thermometer is shown in **Figure 1**: reservoir 3 is immersed into a medium whose temperature is to be measured; gauge 1 is connected via capillary 2 to the reservoir;

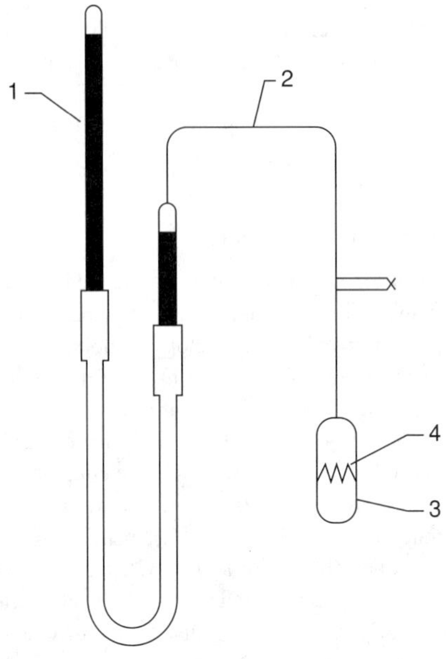

Figure 1. Schematic diagram of a gas thermometer.

the reservoir and the capillary are filled with a working gas. A gas thermometer allows the determination of pressure p and volume V of mass m of the ideal gas with molecular weight μ converting from thermodynamic state 1 to state 2, with the gas mass $m = Vp\mu/TR$ remaining constant in the both states. Depending on the character of gas transition from 1-to-2 state, three gas thermometers are distinguished: those of constant volume, constant pressure and constant temperature. A constant-volume gas thermometer is used at low temperatures (typically with helium as a working substance) and possesses the highest sensitivity. At high temperatures, when gas desorption on reservoir walls becomes pronounced and helium penetrates through the walls, gas thermometers of other design are used with nitrogen as a working substance. For precise temperature determination, corrections are made for gas non-ideality, thermal expansion of the reservoir, a "harmful" volume and thermomolecular pressure as well as for hydrostatic aspect. Since the reservoir of a gas thermometer is connected with a manometer via a capillary, then there is the "harmful" gas volume of a gas above the manometer mercury and inside the capillary, whose temperature varies from the value to be measured to room temperature. With change of the bulb temperature, the amount of a gas contained in "the harmful volume" changes. A difference of the gas temperature in the bulb and in "the harmful volume" requires the appropriate corrections. Such corrections may only be determined rather approximately, therefore for low temperatures use is made of a thermometer without a "harmful volume". Such thermometer is used in metrological works. Bulb 3 (**Figure 1**) is partitioned with an elastic membrane 4. Smooth change of the pressure in the upper part of 3 allows 4 to be maintained in an equilibrium condition and thus pressure measured in the lower part of closed volume of 3. For technical measurements, use is made of filled-system gas thermometers working at temperatures from −150 to 600°C. At a temperature up to 600°C nitrogen is used as a working gas, while above 600°C argon is used. The scale of a filled-system gas thermometer ($T = f(p)$) is obtained using a knowledge of a volume of the instrument components. For this, corrections are made for the gas non-ideal state, thermal expansion of the bulb and capillary, temperature variation, etc. This thermometer is inertial; it fails to measure rapid processes. A schematic drawing of its construction is given in **Figure 2**.

Liquid-In-Glass Thermometer

The *liquid-in-glass thermometer* is intended for temperature measurement using thermal expansion of a liquid as a suitable property. It is employed in technology, laboratory practice and medicine to measure temperatures from −200 to 750°C. A liquid-in-glass thermometer consists of a glass (sometimes quartz) bulb with a glass capillary joined to it. A thermometer liquid occupies the whole bulb and partly the capillary. Depending on the measured temperature range, it may be filled with kerosene (from −20 to 300°C), mercury (from −35 to 750°C), pentane (from −200 to 20°C), ethyl alcohol (from −80 to 70°C), etc. A liquid-in-glass thermometer is subjected to special heat treatment ("aging") eliminating the zero drift on its scale caused by repeated heating and cooling. For precise measurements, corrections should be made for the zero drift. The accuracy of measurement depends on the depth of immersion into a medium to be measured; immersion should be continued to correspond to a specific scale division or to a specially made line on the scale. If this is impossible, correction needs to be made for the protruding column, dependent on the measured temperature, the column temperature and its height. The main drawbacks of this type of instrument are its inevitable thermal lag and its dimensions which are often inconvenient for application. Amongst such types, mercury-in-glass thermometers are most often used. Mercury is easily purified, it does not wet glass and vapor pressure in the capillary above mercury is small. Liquid-in-glass thermometers are unsuitable for automatic recording of results and often replaced by resistance thermometers or other instruments. Special constructions are metrological, clinical thermometers and others. Clinical mercury thermometers are graduated from 34 to 42°C with a division of 0.1°C and are maximum thermometers, i.e. the capillary column remains on the level of a maximum rise on heating and falls down when shaken.

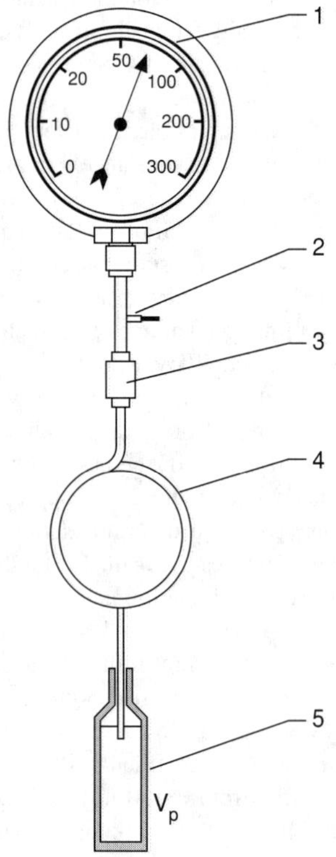

Figure 2. Filled-system gas thermometer.

Table 1. Working fluids for vapor pressure thermometers

	P = 50 mmHg		P = 760 mmHg	
Working medium	T_k, K	dρ/dT	T_k, K	dρ/dT
carbonic acid	166.3	5	194.6	62
oxygen	70.4	10	90.19	80
hydrogen	14.3	34	20.4	223
helium-4	2.3	110	4.2	720

Liquid-In-Steel Filled-System Thermometers

In *liquid-in-steel* thermometers, the system is filled with liquid and changes in temperature of the liquid in the sensing bulb cause changes in pressure which are measured by pressure gauges in which a spiral, a sylphon membrane, etc. are used as sensitive elements. The construction may be identical to that shown in **Figure 2** and the device provides automatic recording of temperature. A working temperature range covers from −150 to 330°C. As working substances, use is made of propyl alcohol, methaxylol, xylol, mercury, etc.

Vapour-Pressure Thermometer

Temperature dependence of the pressure of saturated liquid vapours is also used as a thermometric parameter. The sensitive element is a reservoir partially filled with a liquid or a solid phase being in thermodynamic equilibrium with its vapor. Such a device is used to measure low temperatures. A thermometer scale T = f(P) is determined by the vapor pressure vs temperature curve. For a *vapour-pressure thermometer* to work, it is necessary for the thermometer reservoir to be at the lowest temperature with respect to the temperature of its other parts. The accuracy of measurement depends on the gas purity, superheating in the volume of a liquid occupying the bath in which temperature is measured, and on hydrostatic pressure measurement. **Table 1** lists the sensitivity of vapour-pressure thermometers using different working substances.

Vapour-Pressure Filled-System Thermometers

If pressure is measured by a spring-element pressure gauge, then the measured temperature ranges may be extended from the boiling point to the critical temperature of working substance. The accuracy of measurement is worse and is determined by the pressure gauge precision. Its construction is identical to that depicted in Fig. 2. A measurement range comprises temperatures from −30 to 300°C. For filling the system, use is made of the substances indicated in table and also of Freon-22, propylene, methyl chloride acetone, ethyl benzene, ethyl alcohol, etc. The scale of such thermometers is graduated in degrees of temperature; it is nonlinear, if reservoir dimensions are small.

Leading to: *Resistance thermometry*

M.P. Orlova

THERMONUCLEAR REACTORS (see Fusion, nuclear fusion reactors)

THERMOSORPTIVE COMPRESSOR (see Compressors)

THERMOSYPHON (see Heat pipes; Solar energy)

THERMOSYPHONING AIR PANEL, TAP (see Solar energy)

THERMOSYPHON REBOILERS (see Reboilers)

THICKENERS

Following from: *Liquid-solid separation*

A thickener is an equipment structure used for the continuous gravity settling (sedimentation) of solids in suspensions. Suspension is fed into one or more basins or chambers and, whilst it is passing through, the solids settle out. The thickened solids are removed together with a portion of the liquid as thickened "underflow". The liquid, ideally containing no solids, forms the "overflow" from the thickener. Thickeners vary widely in size and configuration, but they all comprise: a. a vessel to provide volume and area needed for thickening, with the area being large enough to allow the solids to settle at a velocity faster than the upward velocity of the liquid; b. a system for introducing the feed and directing it into the flow paths that best utilise the vessel volume and area; c. an overflow system for collecting clarified liquid; d. a mechanism to convey settled solids to a discharge point.

The sedimentation regimes (see **Sedimentation**) are rarely all present simultaneously in a continuous thickener. If they are, their distribution can be illustrated as in **Figure 1**.

Input suspension (carrying settleable solids) passes through supernatant layers (in the clarification zone) and spreads out as a feed layer at a depth where the solids concentration corresponds to that of the feed. The major part of the liquid component in the feed rises, carrying with it the finer solids (those particles whose *terminal settling velocity* is less than the rise velocity of the liquid). Flocculant is often added to the feed to assist separation of the fine solids; if it is, the solids continue to flocculate and in a well sized system will be removed in the clarification zone. Solids settling out of or through the feed layer pass into a *critical zone.* The solids then pass into the *compression zone,* where they are subjected to an increasing solids stress (arising from the weight of solids above) as they move deeper into the compressing layer. They are thus compacted or thickened.

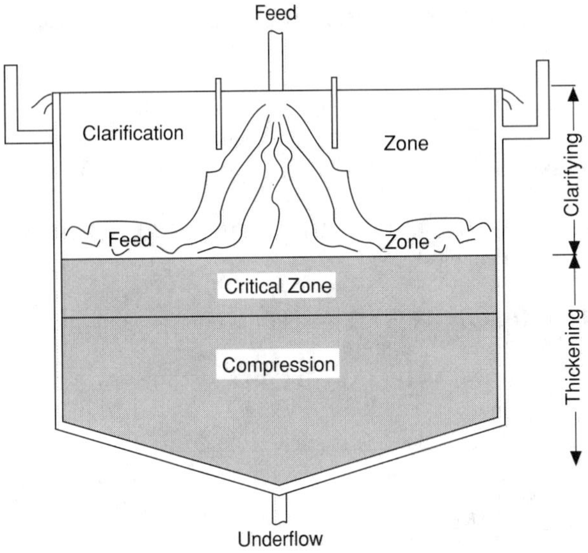

Figure 1. Zones in a continuous thickener.

A critical settling zone will only exist when the feed flux of solids exceeds that which can be transmitted through some limiting or *critical concentration,* i.e., when the thickener is fed a solids overload. The critical concentration therefore constitutes an upper boundary for solids loading, and zone settling imposes an area demand on thickener design. When the feed flux of solids is below the critical value, solids pass down through the zone faster than they are replenished at the top and the zone disappears. Hence, zone settling places no depth demand on thickener design. The general design of thickeners does not follow strict geometric proportions; the relationship between depth and diameter is important only to the extent that the tank volume will provide sufficient retention time, taking operating efficiency and mechanical design into account.

References

Fitch, E. B. and Stevenson, D. G. (1986) *Solid/Liquid Separation Equipment Scale-up,* Ch. 4 D. B. Purchas and R. J. Wakeman eds., Uplands Press and Filtration Specialists, London.

Osborne, D. G. (1990) *Solid-Liquid Separation,* Ch. 5, L. Svarovsky, ed., Butterworths, London.

R.J. Wakeman

THICKENING (see Liquid-solid separation)

THIN FILM EVAPORATION (see Augmentation of heat transfer, two phase)

THIN LAY CHROMATOGRAPHY (see Chromatography)

THIRD LAW OF THERMODYNAMICS (see Thermodynamics)

THIXOTROPIC FLUIDS (see Fluids: Non-Newtonian fluids)

THIXOTROPY (see Viscosity)

THOMA COEFFICIENT (see Hydraulic turbines)

THOMPSON, BENJAMIN, COUNT RUMFORD (see Steam engines)

THOMPSON EFFECT (see Peltier effect)

THOMSON, WILLIAM, LORD KELVIN (see Steam engines)

THORIUM

Thorium—(*Thor,* Scandinavian god of war), Th; atomic weight 232.0381; atomic number 90; melting point 1750°C; boiling point ~4790°C; specific gravity 11.72; valence + 2, + 3, +4.

Discovered by Berzelius in 1828. Thorium occurs in thorite ($ThSiO_4$) and in thorianite ($ThO_2 + UO_2$). Large deposits of thorium minerals have been reported in New England and elsewhere, but these have not yet been exploited. Thorium is now thought to be about three times as abundant as uranium and about as abundant as lead or molybdenum.

Thorium is a potential source of nuclear power. There is probably more energy available for use from thorium in the minerals of the earth's crust than from both uranium and fossil fuels. Work has been done in developing thorium cycle converter-reactor systems. Several prototypes, including the HTGR (high-temperature gas-cooled reactor) and MSRE (molten salt converter reactor experiment), have operated.

Thorium is recovered commercially from the mineral monazite, which contains from 3 to 9% ThO_2 along with most rare-earth minerals. Much of the internal heat the earth has been attributed to thorium and uranium. Several methods are available for producing thorium metal: it can be obtained by reducing thorium oxide with calcium, by electrolysis of anhydrous thorium chloride in a fused mixture of sodium and potassium chlorides, by calcium reduction of thorium tetrachloride mixed with anhydrous zinc chloride, and by reduction of thorium tetrachloride with an alkali metal.

Thorium was originally assigned a position in Group IV of the periodic table. Because of its atomic weight, valence, etc., it is now considered to be the second member of the actinide series of elements. When pure, thorium is a silvery-white metal which is air-stable and retains its luster for several months. When contaminated with the oxide, thorium slowly tarnishes in air, becoming gray and finally black. The physical properties of thorium are greatly influenced by the degree of contamination with the oxide. The purest specimens often contain several tenths of a percent of the oxide. High-purity thorium has been made. Pure thorium is soft, very ductile, and can be cold-rolled, swaged, and drawn. Thorium is dimorphic, changing at 1400°C from a cubic to a body-centered cubic structure. Thorium oxide has a melting point of 3300°C, which is the highest of all oxides. Only a few elements, such as tungsten, and a few compounds, such as tantalum carbide, have higher melting points. Thorium is slowly attacked by water, but does not dissolve readily in most common acids, except hydrochloric.

Powdered thorium metal is often pyrophoric and should be carefully handled. When heated in air, thorium turnings ignite and burn brilliantly with a white light.

The principal use of thorium has been in the preparation of the Welsbach mantle, used for portable gas lights. These mantles, consisting of thorium oxide with about 1% cerium oxide and other ingredients, glow with a dazzling light when heated in a gas flame. Thorium is an important alloying element in magnesium, imparting high strength and creep resistance at elevated temperatures. Because thorium has a low work-function and high electron emission, it is used to coat tungsten wire used in electronic equipment. The oxide is also used to control the grain size of tungsten used for electric lamps; it is also used for high-temperature laboratory crucibles. Glasses containing thorium oxide have a high refractive index and low dispersion. Consequently, they find application in high quality lenses for cameras and scientific instruments. Thorium oxide has also found use as a catalyst in the conversion of ammonia to nitric acid, in petroleum cracking, and in producing sulfuric acid.

Twelve isotopes of thorium are known with atomic masses ranging from 223 to 234. All are unstable. Th^{232} occurs naturally and has a half-life of 1.41×10^{10} years. It is an alpha emitter. Th^{232} goes through six alpha and four beta decay steps before becoming the stable isotope Pb^{208}. Th^{232} is sufficiently radioactive to expose a photographic plate in a few hours. Thorium disintegrates with the production of thoron (radon^{220}), which is an alpha emitter and presents a radiation hazard. Good ventilation of areas where thorium is stored or handled is therefore essential.

Handbook of Chemistry and Physics, CRC Press

THREE MILE ISLAND ACCIDENT (see Natural circulation loops; Nuclear reactors; Thermosyphons)

THREE PHASE, GAS-LIQUID-LIQUID FLOWS

Following from: *Two phase flows*

Origin of the Water Phase

Three-phase flows of gas and two liquid phases often occur during production of oil. A typical oil and gas reservoir contains gas, oil and water. As oil is extracted from the reservoir, the formation water beneath may flow into the well, due to its easier motion through the rock pores. In many cases, water is pumped into the reservoir, through water injection wells, to help boost production (See **Enhanced Oil Recovery**). Eventually this water will find its way to the production well, which will from that point produce oil with a progressively larger water fraction ("*water cut*").

As the North Sea oil fields mature, production of water with the oil is becoming increasingly important. A good understanding of the effects of the water phase on the flow behaviour of multiphase mixtures in pipelines can have a significant bearing on the economics of oil recovery at high water cuts.

This is reflected in the growing research interest in the subject of three-phase flows over the last few years as highlighted in recent publications by Hall (1992, 1993), Lahey et al. (1992), Stapelberg et al. (1991) and Lunde et al. (1993).

Problems of a Water Phase

In a vertical or inclined pipe, the greater density of the water phase compared to the oil phase means that there is a greater hydrostatic pressure drop. This is particularly noticeable in fields where water injection is used to boost output. Once injection water breakthrough occurs, the larger pressure drop in the well causes a fall in production rate.

In a horizontal pipe, due to the effect of gravity, the water will tend to settle to the bottom of the pipe. While this may considerably reduce the frictional component of the pressure gradient if the oil is very viscous, the separate water layer may cause corrosion of the pipe.

At higher gas flow-rates there may be sufficient mixing, particularly in the slug flow regime, for one of the liquids to be dispersed in the other. If the oil is the continuous phase, the dispersion of water in it can cause a large increase in the effective viscosity and consequent increase in frictional pressure gradient. If the continuous phase is water, the pressure gradient may be much less, but the pipe walls will be continuously water-wet. These effects will influence the sizing of pipelines to give the most economic design and operation.

At certain flow-rates it may be possible that the water settles into a separate layer in the intervals between liquid slugs. In this case, parts of the pipe walls would experience a variation in wetting with time, with implications for the distribution of corrosion inhibitors, for example. (See also **Corrosion**)

Recent Experimental Studies

With the growing interest in three-phase flows in recent years, a number of useful experimental investigations have been reported. Hall et al. (1993) performed experiments using the high pressure multiphase flow facility at Imperial College, London, with air, water and a lubricating oil of viscosity approximately 40 mPas. The main test pipeline had an internal diameter of 78mm and a length of 40m and had a visualisation section towards the end. In addition to observing flow patterns, the pressure drop, holdup and slug frequency were measured.

Lahey et al. (1992) investigated three-phase flows of air, water and mineral oil (viscosity 116.4 mPas at 25°C) in a pipe of 19mm internal diameter. The main objective of the experiments reported was to observe flow patterns and to measure the oil and water holdup. The small diameter may have had a significant effect on the flow patterns in this system; for example, the region of stratified flow was found to be very restricted.

Nuland et al. (1991) developed a dual-energy gamma densitometer for the measurement of holdup in three-phase flow. This was compared with the quick-closing valve method for measurement of holdup for a three-phase air/oil/water flow in a 32mm internal diameter pipe; the oil viscosity in these experiments was 1.75 mPas at 20°C. Reasonable agreement was obtained between the phase fractions obtained using the gamma densitometer and those using the quick-closing valves.

In those cases where significant discrepancies occurred, the disagreement was due to the assumption of a **Stratified Flow** geometry for the calculation of holdup from the gamma densitometer readings, when the interfaces were in reality curved. More recent experiments at the same institution have been focused on the study of three-phase gas-condensate flows in an uphill inclined pipe, as described by Lunde et al. (1993).

Stapelberg et al. (1991) studied the flow of gas, water and a white mineral oil of viscosity 31 mPas in a test loop with diameters 23.8mm and 59.0mm. The flows were in the stratified and **Slug Flow** regimes and results were obtained for pressure gradient and slug characteristics (slug length, frequency, etc.). Some measurements were also reported of holdup and information given on flow visualisation. These experiments have provided new data and physical information and have demonstrated the inadequacy of current methods for calculating a pressure gradient, particularly in stratified three-phase flows. Further experiments reported by Nädler and Mewes (1993) have extended these experiments by using oils with a range of viscosities from 14 mPas to 37 mPas and increased liquid and gas flow-rates. Results for the characteristics of three-phase slug flows have been obtained and compared with measurements for oil-air and water-air flows.

Three-Phase Flow Patterns

The work by Lahey et al. (1992) demonstrated the difficulty of classification of *flow patterns* in three-phase flows. In addition to the well-known descriptions for the flow patterns in gas-liquid flows (stratified, slug, annular, etc.), the distribution of the two liquid phases needs to be taken into account. The two liquids may be either dispersed, in which case either the water or the oil may form the continuous phase or they may flow separately.

Hall (1992, 1993) concentrated on the stratified and slug flow regimes. Both stratified and slug flows may exist where the oil and water are completely separate, are partially mixed or are fully mixed. In the case where the two fluids are fully mixed, either the oil or the water can form the continuous phase, depending on the volume ratio of the two liquids and the *inversion point.*

The modelling of the transition between stratified and slug flow is very strongly influenced by the distribution of the two liquid phases. If the oil and water are fully mixed, then the flow behaves almost as if it were gas-liquid flow, where the liquid phase has the physical properties of the combined oil and water phases. The density is given by a volume fraction average of the oil and water densities:

$$\rho_{mix} = \phi\rho_{disp} + (1 - \phi)\rho_{cont} \quad (1)$$

where ϕ and ρ_{disp} are the volume fraction and density of the dispersed phase and ρ_{cont} is the density of the continuous phase. A volume fraction average must not be used for the effective viscosity because it is observed in practice that the viscosity of a dispersion of one liquid in another can be significantly higher than the viscosity of the pure liquid, as shown by Woelflin (1942). A suitable equation to use is that of Brinkman (1952):

$$\eta_{mix} = \eta_{cont}/(1 + \phi)^{2.5} \quad (2)$$

where η_{cont} is the viscosity of the continuous phase.

These effective liquid properties may then be used in a two-phase gas-liquid transition prediction, for example that given by Taitel and Dukler (1976).

When the two liquids are not completely mixed, it is necessary to model the flow in greater detail. Since neither phase forms the continuous phase, it is difficult to define meaningful mixture physical properties. The approach suggested by Hall (1992) was to model the three-phase stratified flow using a three-fluid model (in a similar way to the Taitel and Dukler two-fluid method for gas-liquid flow) and to use the holdup obtained from this method to calculate the stratified to slug transition. However, comparison with experimental data showed that the transition occurred at much higher gas velocities than those predicted by this method. This was believed to be because the oil layer, although being more viscous than the water, is travelling at a higher mean velocity because its lower interface is in contact with a moving water layer rather than a fixed wall. Hence a second approach was tried, based on modifying the linear stability analysis of Lin and Hanratty (1986) to allow a second liquid phase. This was reported by Hall and Hewitt (1992).

This approach was found to work for both slug flows where a separate water layer was present continuously and for those flows where a separate water layer only formed in the film region between liquid slugs. Hall et al (1993) presented a simple criterion to determine whether or not this separate water layer forms. This is based on a comparison of the time taken for drops of the dispersed phase (which are formed in the slug body) to settle and the interval between liquid slugs. If the drops take longer to settle than the slug interval, then there is not enough time for a separate layer to form. This criterion showed good agreement with experimental observations from the Imperial College loop, but further development is really required for more general application.

Calculation of Pressure Gradient and Holdup

There is very little information on methods for calculation of pressure gradient and holdup in three-phase flows, and in most cases the available methods for two-phase flows must be used, with suitable modifications for the physical properties of the combined liquid phases. The correct method is very dependent on the flow pattern, and it is therefore essential to make the best estimate possible of the likely flow pattern, using the guidelines in the previous section, or any available observations of the system under investigation.

If the flow is stratified and the oil and water form separate layers, then the most reliable estimate of pressure gradient and holdup (of both the oil and the water phases) is probably to use a three-fluid model for the stratified flow, as described by Hall (1992).

If the flow is stratified and the oil and water are mixed, then the flow may be treated as a two-phase flow of gas and oil-water mixture and methods such as Taitel and Dukler (1976) may be used for the stratified flow holdup. It is important to be able to determine which liquid forms the continuous phase in order to calculate the correct effective liquid viscosity. Prediction of the inversion point (between oil-continuous and water-continuous) is difficult, since it depends on a number of factors, in particular the interfacial chemistry of the oil-water mixture. The best approach is to determine the inversion point from experimental measurements of viscosity at various water fractions. In slug flows, the pressure gradient can almost always be reasonably accurately calculated by using the density and viscosity of an oil/water mixture as given earlier. This will remain true even in cases where a separate water layer forms between slugs (because the largest contribution to the pressure gradient is the frictional resistance of the liquid slug, where the two liquids are well-mixed).

Either a two-phase frictional pressure correlation, e.g. Beggs and Brill (1973) or a slug flow model, e.g. Dukler and Hubbard (1975) was shown to give good agreement with experimental observations by Hall et al. (1993); the slug flow model of Dukler and Hubbard also gives the holdup. In the less common case where a separate water layer may persist throughout the slug, this approach may not be appropriate. However, further work is required before making any recommendations.

References

Beggs H. D. and Brill J. P. (1973) A study of two-phase flow in inclined pipes, *Journal of Petroleum Technology, Transactions,* 255, 607–617.

Brinkman, H. C. (1952) The viscosity of concentrated suspensions and solutions, *Journal of Chemical Physics,* 20, 4, 571.

Dukler, A. E. and Hubbard, M. G. (1975) A model for gas-liquid slug flow in horizontal and near horizontal tubes, *Industrial and Engineering Chemistry Fundamentals,* 14, 2, 337–347.

Hall, A. R. W. (1992) Multiphase Flow of Oil, Water and Gas in Horizontal Pipes, PhD Thesis, University of London.

Hall, A. R. W. and Hewitt, G. F. (1992) Effect of the water phase in multiphase flow of oil, water and gas, European Two-Phase Flow Group Meeting, Stockholm, Paper F4.

Hall, A. R. W., Hewitt, G. F. and Fisher, S. A. (1993) An experimental investigation of the effects of the water phase in the multiphase flow of oil, water and gas, in 6th International Conference in Multiphase Production, Cannes, France, BHR Group, 251–272, June.

Lahey, R. T., Açigköz, M. and França, F. (1992) Global volumetric phase fractions in horizontal three-phase flows, *AIChE Journal,* 38, 7, 1049–1058.

Lin, P. Y. and Hanratty, T. J. (1986) Prediction of the initiation of slugs with linear stability theory, *International Journal of Multiphase Flow,* 12, 1, 79–98.

Lunde, K., Nuland, S. and Lingelem, M. (1993) Aspects of three-phase flows in gas condensate pipelines, in 6th International Conference in Multiphase Production, Cannes, France, BHR Group, 291–307, June.

Nädler, M. and Mewes, D. (1993) Multiphase slug flow in horizontal pipes, European Two-Phase Flow Group Meeting, Hannover, Paper 12.

Nuland, S., Skarsvåg K., Sæther G. and Fuchs, P. (1991) Phase fractions in three-phase gas-oil-water flow, in 5th International Conference in Multiphase Production, Cannes, France, BHR Group, pp 3–30, June.

Stapelberg, H. H., Dorstewitz, F., Nädler, M. and Mewes, D. (1991) The slug flow of oil, water and gas in horizontal pipelines, in 5th International Conference in Multiphase Production, Cannes, France, BHR Group, 527–552, June.

Taitel, Y. and Dukler, A. E. (1976) A model for predicting flow regime transitions in horizontal and near-horizontal gas-liquid flow, *AIChE Journal,* 22, 1, 47–55.

Woelflin, W., (1942) The viscosity of crude oil emulsions, *API Drilling and Production Practice,* 148–153.

R.W. Hall

TIDAL ENERGY (see Alternative energy sources)

TIDAL POWER (see Renewable energy)

TIME LAPSE PHOTOGRAPHY (see Photographic techniques)

TIME STRETCH PHOTOGRAPHY (see Photographic techniques)

TITANIUM

Atomic number 22. Atomic weight 47.90. Density 4.51 10^3 kg m^{-3}

Titanium is a light, non-magnetic, tough and highly corrosion resistant biocompatible metal which exhibits good oxidation resistance, due to a titanium dioxide surface film which provides protection up to about 870K. It has two allotropes: alpha, hexagonal close packed, below 1156K and beta, body centred cubic, up to the melting point 1941K.

"Unalloyed" titanium is strengthened by small amounts of interstitial alloying elements, primarily oxygen, the level of which is adjusted to provide a range of commercial grades. Other alloying elements are employed to alter the stability of the two allotropes and commercial alloys based on the alpha, beta, or mixtures of the two phases are available. Aluminium is the most important alpha stabiliser (together with oxygen) and vanadium is the most widely used beta stabiliser: the alpha plus beta alloy titanium 6 wt% aluminium 4 wt% vanadium is the oldest and most generally used alloy and, via thermomechanical treatment to modify microstructure, can provide strength in the range 900 to 1100 MPa. Special commercial alloys contain a range of other elements including, tin, zirconium, molybdenum, iron, chromium and silicon. Medium strength near alpha alloys, consisting predominantly of alpha phase, exhibit the best elevated temperature creep resistance, while beta alloys, which are readily cold workable and heat treatable to high (1200+ MPa) strength levels are employed in lower temperature structural applications.

Titanium and its alloys find applications in chemical engineering, power generation, medical prosthetics and aerospace engineering. The latter accounts for roughly 50% of titanium usage worldwide, excluding the CIS.

References

Titanium: A Technical Guide, (1988) M. J. Donachie, Jr. ed., ASM International, Metals Park, Ohio.

Materials Properties Handbook: Titanium Alloys, (1994) R. Boyer, ed., E. W. Collings and G. Welsch, ASM, Metals Park, Ohio.

H. Flower

T-JUNCTIONS

Following from: *Flow of fluids; Channel flow*

Junctions between pipes can involve the mixing or splitting of fluids. In single phase flow, the flows about the junction are very complex with *recirculation* possible in the outlet pipe(s), features first observed by Leonardo da Vinci, **Figure 1**.

In dividing junctions, the actual division of the flow will depend on the pressure drops in the two downstream legs. Apart from the usual losses in the pipes and other downstream equipment, there are specific losses at the junction itself. These are best illustrated in **Figure 2**. As the velocity in pipe 2 is lower than that in pipe 1, the pressure rises. The junction pressure drops are defined from

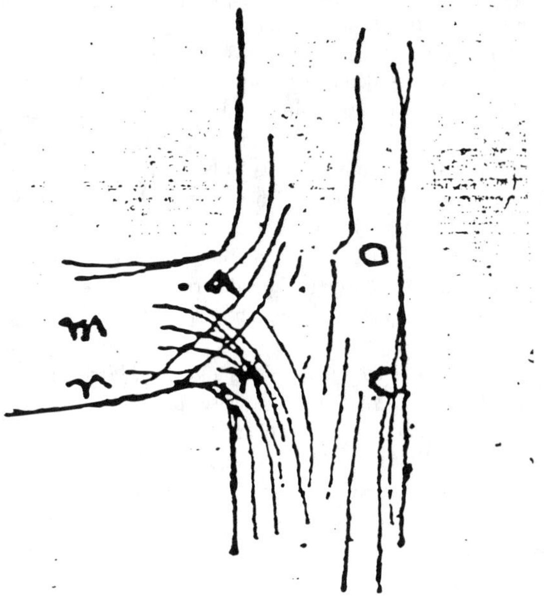

Figure 1. Recirculation during single phase flow split at a T-junction as observed and recorded by Leonardo da Vinci.

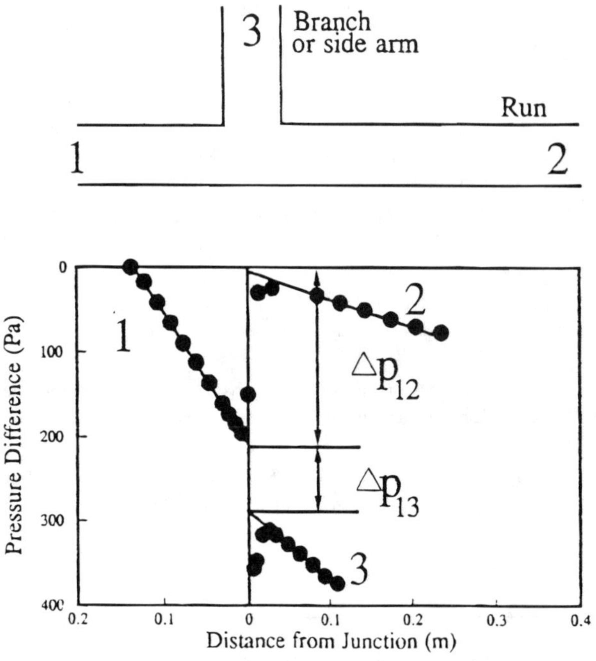

Figure 2. Pressure profiles about junction.

the extrapolation of pressure profiles from the undisturbed regions far from the junction back to the junction. Gardel (1957, 1971) has produced equations describing these pressure losses and the effect of parameters such as *flow split,* diameter ratio, angle between the pipes and degree of rounding of the corner. Obviously, the junction pressure drops are most important when the pressure drops in the downstream lines are small.

For *flow combining,* there are equivalent pressure drops for which Gardel has proposed equations. In addition, there is considerable interest in the distance along the outlet pipe required to achieve mixing of the two streams.

When the split at a *junctions* involves more than one phase, the process has the added complication that the ratio of the phases in the outlet pipes is almost inevitably different to that at inlet. This process is controlled by the momentum fluxes of the phases. For example, for gas/liquid annular flow, the portion of the liquid travelling as a film on the wall has a much lower velocity and hence momentum flux than that travelling as drops. Consequently, it is not surprising that the liquid film and the gas nearest the junction are taken off through the side arm while the drops carry on along the main pipe. A thorough review of the behaviour of gas/liquid split at junctions is given by Azzopardi and Hervieu (1994) who discuss other mechanisms together with available data and models for the prediction of the phase split. For other combinations of phases, data on the maldistribution is given by Nasr-El-Din and Shook (1986) (solid/liquid) and Lempp (1966) (gas/solids).

As in single phase flow, *combination* of *two-phase flows* at junctions involves extra pressure changes at the junction itself. There is much less information about other aspects such as mixing length etc., Azzopardi (1986).

References

Azzopardi, B. J. (1986) Two-Phase Flow in Junctions, *Encyclopedia of Fluid Mechanics,* Vol. 3, Ch. 6 Gulf, Houston.

Azzopardi, B. J. and Hervieu, E. (1994) Phase Separation at T-Junctions, *Multiphase Science and Technology,* Vol 8, Ch. X, pp. 645–714, Begell House, Inc., New York.

Da Vinci, L., I manuscritti e disegni di Leonardo da Vinci, (Reale Commissione Vinciana) Vol IV, folio 32 verso, Rome (1934) = the edition of the Codice Forster, III in the Victoria and Albert Museum.

Gardel, A. (1957) Les Pertes de Charge dans les Ecoulements au Travers de Branchments en Te, *Bulletin Technique de la Suisse Romande,* 9, 122–130 and 10, 143–148.

Gardel, A. and Rechstiener, G. F. (1971) Les Pertes de Charge dans les Branchments en Te des Conduites Circulaires, *Bulletin Technique de la Suisse Romande,* 25, 363–391.

Lempp, M. 1966. Die strömungsverhältnisse von Gas-Feststoff-Gemischen in Verzweigungen pneumatischer Förderanlagen, *Aufbereitungs-Technik,* 7, 81–91.

Nasr-El-Din, H. and Shook, C. A. (1986) Particle segregation in slurry flow through vertical tees, *Int. J. Multiphase Flow,* 12, 427–443.

B.J. Azzopardi

TOLLMEIN-SCHLIGHTING WAVES (see Boundary layer heat transfer)

TOMOGRAPHIC IMPEDANCE METHOD (see Void fraction measurement)

TOMOGRAPHY (see Visualization of flow)

TOP FLOODING (see Rewetting of hot surfaces)

TOP HAT PROFILE, OPTICS (see Optical particle characterisation)

TORR (see Pressure)

TORTUOSITY FACTOR (see Diffusion)

TOTAL HEAD (see Bernoulli equation)

TOTAL HEAD LINE (see Velocity head)

TOTAL PRESSURE TUBE (see Pitot tube)

TOTAL SURFACE EFFECTIVENESS (see Extended surface heat transfer)

TOWERS, COOLING (see Cooling towers)

TRACER METHODS

In wholly single-phase flows, or two- or multi-phase flows where the phases are homogeneously mixed, fluid motion is invisible to the eye. The introduction of tracer particles into the flow enables flow patterns and behaviour to be *visualised.* Furthermore, tracking the motion and behaviour or these particles, either individually or as groups, allows the fluid motion to measured. So-called tracer methods currently provide the most accurate means by which to monitor and measure fluid velocity.

All tracer methods have *light scattering* as a basis. They require some form of illumination, followed by collection of light scattered from the particles, and analysis of this scattered light to provide a measure of the particle motion from which the fluid velocity is inferred. They have many advantages over conventional probe methods such as pitot tubes and hot wires, including accuracy, reliability and high spatial and temporal resolution. Their non-intrusive nature, whereby nothing needs to be physically inserted into the flow, enables their use in hostile environments. They additionally offer the potential for instantaneous global mapping of flow fields. However, suitable optical access is required but which is not always available.

Tracer particles can be solid, liquid or gaseous and are usually injected directly into the fluid. Ideally they should be spherical and monosized. Since the host fluid velocity is inferred from their motion, selected particles must be capable of following the fluid motion with accuracy and precision. This requires that they are small but also neutrally buoyant. A simple but valuable measure is a particle's aerodynamic diameter $D_a = \Delta\rho^{1/2}D_o$, where D_o is the particle's physical diameter and $\Delta\rho$, the difference in density between it and the host fluid in gm cm^{-3}. The smaller D_a is, the more responsive the particle is to flow fluctuations. In addition, tracer particles should exhibit good light scattering ability, facilitated by larger size and a refractive index which differs significantly from that of the fluid. An adequate number of particles must be present, as required by the experiment, but not so many as might affect the fluid behaviour. Invariably, choice of particle is a compromise.

There are two generic types of tracer method. The first, *laser Doppler velocimetry (LDV),* measures the Doppler shift which results when light is scattered from a moving particle (Durst et al., 1976). It is a point measurement method and yields accurate, time-averaged, mean velocity and turbulence information. (See also **Anemometers, Laser Doppler**) Velocity maps are built up by traversing the flow. The second, *particle image velocimetry (PIV),* has each particle velocity (and often its direction) encoded in the recorded image(s) of the flow. Instantaneous velocity maps are captured within a plane or volume of the flow and traversing is usually unnecessary.

In the most common (two-beam) form of LDV (**Figure 1**), a laser beam is split into two beams which are each focused and made to intersect at their waist point to define a measurement volume (mv), typically 0.1 mm^3. When a particle crosses this volume, light from the two beams is scattered, with scattering from each undergoing a phase shift due to the Doppler effect. A photodetector receives scattered light as beat signals (Doppler bursts) which are usually monitored using an oscilloscope and are passed to a signal analyser for processing. The burst frequency f is measured and is directly proportional to the particle velocity. The signal has the same form as if the two beams had created an interference pattern within the mv with fringe spacing $\lambda_f = \lambda_o/2\sin\theta$, where λ_o is the laser wavelength and θ the bisector angle of the two beams. The measured velocity component is normal to these fringes, and is found as $u = \lambda_f f$.

Each particle crossing the mv generates its own Doppler burst. Usually a *probability distribution function (pdf)* is built up from several thousand individual measurements, from which an average velocity is computed along with a root mean square (rms) value. Modern signal analysis is by digital autocorrelation or Fourier transforms and in excess of 10^4 signals per second may be processed. Multiple velocity components can be measured by arranging different colour interference patterns at other orientations.

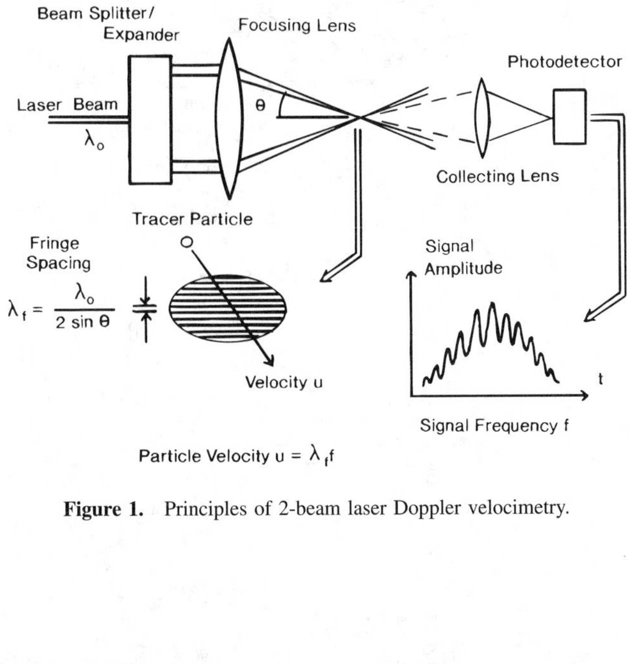

Figure 1. Principles of 2-beam laser Doppler velocimetry.

Another LDV method requires only a single beam to be focused into the flow. The direct Doppler shift is then measured using a Michelson interferometer. (see **Interferometry**) *Doppler global velocimetry (DGV),* using a laser sheet, extends this approach to two-dimensional velocity mapping, wherein velocity magnitude is encoded in the recorded intensity of scattering. Such methods are for high speed applications.

PIV is, essentially, a modern extension of so-called chronophotographic methods (Somerscales 1981), where particles are recorded as streaks, encoded tracks, or multiple images captured at known times. Particle velocity is measured as the length of a streak, or the displacement between multiple images. A pulsed or chopped/modulated laser beam is generally used, and is typically introduced as a light sheet which defines a precise plane of interest, **Figure 2**. Photography is gradually being replaced by electronically gated high resolution charge coupled device (ccd) cameras as the means of recording images.

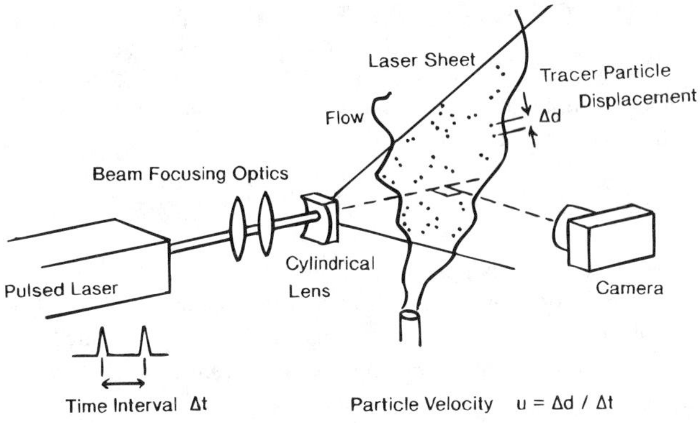

Figure 2. Planar PIV using a double pulsed laser sheet.

PIV data reduction requires sophisticated image processing techniques. Particle tracking methods are used where there are relatively few recorded particles or where high velocity gradients exist; particle tracks are then individually identified and measured. The alternative, a correlation method, is used when there are many particles, or where images are particularly noisy. With these, a grid is applied to the recorded image which is analysed segment-by-segment for average particle displacements.

Backlighting and holographic methods offer alternatives to using laser sheets; holography, specifically, enabling an instantaneous velocity map to be captured throughout a volume. In addition, acquisition of image sequences allows the temporal development of flows to be captured in a similar manner to photochromic dye tracing. PIV methods are particularly applicable to transient flows. (See also **Holograms**, **Holographic Interferometry** and **Photochromic Dye Tracing**)

LDV and PIV each measure different aspects of behaviour within flows. In general, they are applicable to flows of all types, as evidenced by the general literature (Adrian et al. 1991). The two approaches complement each other and are often used in combination when studying fluid behaviour.

References

Adrian, R. J., Durao, D. F. G., Durst, F., Maeda, M. and Whitelaw, J. H. (eds) (1991) *Applications of Laser Techniques to Fluid Mechanics,* Springer-Verlag, NY.

Durst, F., Melling, A. and Whitelaw, J. H. (1976) *Principles and Practice of Laser-Doppler Anemometry,* Academic Press, NY.

Somerscales, E. F. C. (1981) Measurement of velocity, *Methods of Experimental Physics: Fluid Dynamics Part A,* Emrich, R. J. ed., Academic Press, NY.

P.R. Ereaut

TRACER PARTICLES (see Visualization of flow)

TRAILING VORTICES (see Vortices)

TRAJECTORY PROBLEM, OPTICS (see Optical particle characterisation)

TRANSCENDENTAL EQUATION

A transcendental equation is an equation containing transcendental functions. The transcendental function is any function which is the solution of the equation

$$P_n(x)y^n + P_{n-1}(x)y^{n-1} + \cdots + P_1y + P_0 = 0 \quad (1)$$

where $P_i(x)$ are polynomials of x.

Elementary transcendental functions are the exponential, logarithmic, trigonometric, reverse trigonometric, and hyperbolic functions. If transcendental functions are considered as functions of a complex variable, then their characteristic feature is the presence of at least one singularity in addition to poles and branch points of finite order. For instance, the functions e^z, cos z, sin z have a significant singular point $z = \infty$, and also branch points of infinite order $z = 0$ and $z = \infty$.

An important class of transcendental functions are cylindrical and spherical functions often met in problems of heat transfer, including the gamma and beta functions, Euler's functions, hypergeometric and degenerate hypergeometric functions. Transcendental equations occur, for instance on finding the eigenvalues, when solving problems of heat transfer by the method of separation of variables. The transcendental equations are usually solved with the help of numerical methods such as the Newton method, the method of false position, etc.

V.N. Dvortsov

TRANSCENDENTAL FUNCTIONS (See Exponential function; Integrals; Transcendental equation)

TRANSIENT CONDUCTION (see Fourier number)

TRANSIENT HEAT TRANSFER (see Fixed beds)

TRANSIENT HEAT TRANSFER IN JACKETED VESSELS (see Jacketed vessels)

TRANSITION BOILING (see Boiling; Post dryout heat transfer)

TRANSITION FROM LAMINAR TO TURBULENT FLOW (see Laminar flow)

TRANSITION TO TURBULENT BOUNDARY LAYER (see Boundary layer heat transfer)

TRANSMISSIVITY (see Optical particle characterisation; Radiative heat transfer)

TRANSONIC WIND TUNNELS (see Wind tunnels)

TRANSPIRATION COOLING

***Following from:** Heat protection*

Transpiration cooling is one way of active heat protection (see **Heat Protection**) during which a coolant in the course of passing through the wall of a body absorbs a part of the internal energy of a body requiring cooling, and simultaneously actively affects the convective heat flux going into a body from the surrounding space. This method of cooling may be realized in the form of porous, perforated, film or obstructing cooling (**Figure 1**).

During injection of cold gas or liquid into a boundary layer of an incoming flow there occurs driving back of hot gas from the body surface, as a result of which the heat transfer rate decreases due to the so-called thermal blowing effect (see **Heat protection**.) The more effective coolants are those substances possessing the maximal specific heat and producing gaseous products with the minimal molecular mass. **Table 1** presents properties of coolants used in systems of transpiration cooling.

The advantage of this heat protection method over others is the possibility of maintaining the surface temperature at the desired

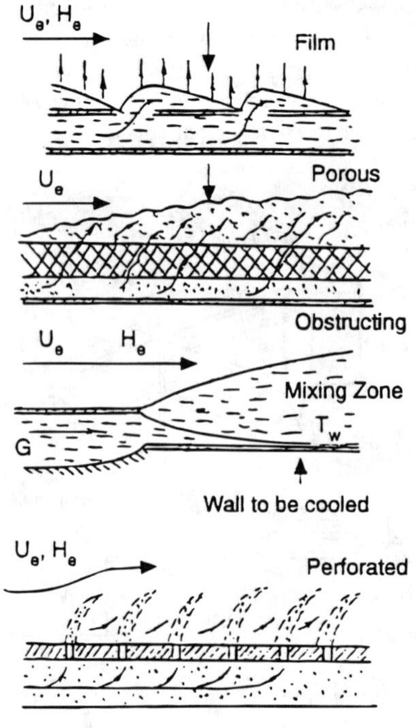

Figure 1. Forms of transpiration cooling.

Table 1. Coolants used in transpiration cooling

Substance	Molecular mass	Specific heat c_p, kJ/kg (at T = 370K)
Hydrogen	2	14.45
Helium	4	5.20
Water (steam)	18	2.14
Ammonia NH_3	17	2.22
Nitrogen	28	1.03
Air	29	1.00
Methyl alcohol CH_3OH	32	1.72
Argon	40	0.52
Carbon dioxide	44	0.91

level by controlling the coolant flow rate. In some applications, the chemical composition of the injected gas is very important. Thus, for wings and wheels of supersonic aircraft the danger is not only thermal heating, but also oxidation of the surface by the incoming flow. If we apply ammonia in the system of transpiration cooling, which has good specific heat capacity and the relatively low values of molecular mass, then in the high-temperature air flow it interacts with oxygen

$$4NH_3 + 3O_2 = 2N_2 + 6H_2O$$

The reaction products resulting in this case (water and nitrogen) are practically inactive to the metals, used in the aircraft construction. The most effective variant of the transpiration cooling is porous cooling. The mechanism of porous cooling is a combination of two processes: internal heat transfer (**Figure 2**), when the coolant takes away heat from the porous matrix during filtration in the direction of the outer surface (see **Porous Medium**), and external heat transfer, when the gaseous coolant, leaving the wall, diffuses through a boundary layer, diluting and driving back the high-temperature incoming flow from the surface (see **Heat Protection**). It is precisely this second process that provides the higher efficiency of porous cooling. As an example, we determine the required flow rate for *blade cooling* in a gas-turbine power plant. Denote the temperature of the flow before entrance into the turbine as T_e, the

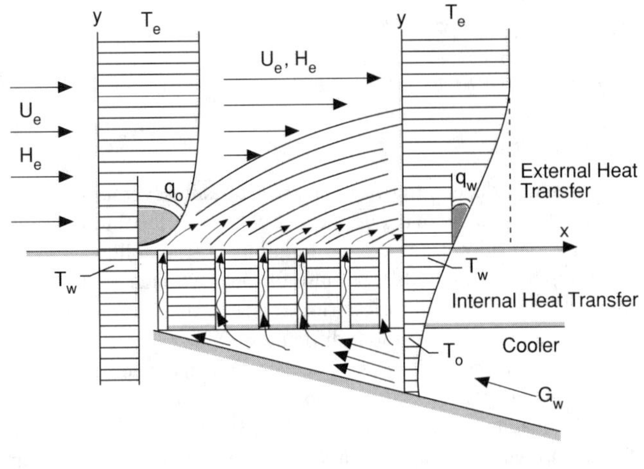

Figure 2.

coolant temperature at the inlet into the porous wall T_0, and the coolant flow rate G_w. Assume, that inside the porous matrix the gas exists in the temperature equilibrium with porous walls, while its specific heat capacity $c_{p,0}$ is approximately equal to specific heat capacity of the gas mixture before the turbine $\bar{c}_p$. The thermal balance on the turbine blade surface can be written in the form:

$$q_0 = (\alpha/\bar{c}_p)_0(H_e - H_w) = G_w c_{p,0}(T_w - T_0) + \gamma G_w(H_e - H_w)$$

$$\{\text{a supplied heat flux}\} = \left\{\begin{matrix}\text{heating inside}\\ \text{a porous wall}\end{matrix}\right\} + \left\{\begin{matrix}\text{thermal blowing}\\ \text{effect}\end{matrix}\right\}.$$

Hence, it is not difficult to obtain an equation for the cooling depth or the temperature drop ratio inside the porous wall $(T_w - T_0)$ and in the external flow $(T_e - T_0)$:

$$\frac{T_w - T_0}{T_e - T_0} = \frac{1 - \gamma\bar{G}_w}{1 + (1 - \gamma)G_w}.$$

In this case, $\bar{G}_w = G_w/(\alpha/c_p)_0$ is the dimensionless flow rate of a coolant. γ is the blowing coefficient depending on the flow mode and on the coolant type (see **Porous media**). The dimensionless coolant flow rate $\bar{G}_w$ is associated with the relative flow rate $[G_w/\rho_e u_e]$ through the Stanton number. $St = (\alpha/c_p)_0/\rho_e u_e$. Even at the temperature T_e = 1900 K for conservation of the blade temperature at the level of $T_w \leq$ 1100K it is sufficient to have a coolant flow rate G_w not more than 3% of the flow rate of the working body in the turbine. One must bear in mind that during operation of such a system it is necessary to keep the coolant clean to avoid breakage of the pores.

One of the varieties of the porous cooling is the so-called self-cooling or sweating. Its idea is borrowed from the nature, however, the practical realization is very peculiar. For example, nozzle blocks of solid-fuel rocket engines are manufactured from porous tungsten, impregnated by silver, copper, zinc or lithium hydride. Evaporating at the low temperature, these metals absorb a considerable amount of heat. Gradually departing inside the coating, the evaporation front becomes a source of formation of gaseous products, which filter through the porous layer and are blown into the boundary layer.

The requirement to the coolant purity for perforated walls is muc less severe. However, in this case the internal heat absorption is not as effective as in the porous matrix. Therefore gas, leaving the wall, has a low temperature. The blocking effect of discretely blown-in jets is also less than that of uniform porous blowing; this leads, finally, to the onset of the considerable temperature "roughness" of the wall being cooled and to its splitting.

Film cooling is the other variant of transpiration cooling. Through the system of holes or gaps the liquid coolant is supplied onto the external body surface, which under the action of friction and the pressure gradient of an incoming flow is converted into a thin film, covering the whole body surface (at least, up to the next succession of holes). In this case the surface temperature nowhere will exceed the liquid boiling temperature.

The efficiency of film cooling depends on the method coolant supply, its feed angle, the presence of roughness and contamination on the surface, and on the coolant properties. With increasing number of gaps and holes, the wall temperature becomes more uniform.

Film cooling is used as additional means for protecting walls of combustion chambers and rocket engine nozzles in those cases when convective cooling does not provide the required thermal mode. The conservation of the steady-state laminar film flow without formation of waves or spraying remains a serious problem (see **Film cooling**).

Obstructing cooling may be considered as a variety of film cooling (**Figure 1**). Here, cold gas is discharged through slits or holes in the wall located at a certain distance one from another. It is desirable to realize the coolant ejection into the flow at the minimal angle to the wall. This delayes the mixing process and provides the greatest heat-protecting effect. The length of the protected plate surface during the coolant supply over the normal to it is in some times smaller than over the tangent.

An interesting combination of various methods for heat protection is realized in solid-fuel rocket engines for protecting the critical nozzle cross-section. It is known that even the 5 percentage increase of the critical cross-section diameter leads to a pressure drop in the combustion chamber of 15–20%; this fact significantly decreases the engine thrust. The necessary for the obstructing cooling gas with the relatively low temperature is produced during burning of "collars" fabricated of low-calorie fuel, placed in the subsonic nozzle zone at its boundary with the combustion chamber. With this purpose one may arrange inserts made of high-melting metals, saturated by readily sublimating components.

Leading to: *Film cooling*

N.V. Pavlyukevich and Yu. V. Polezhayev

TRANSPORT DISENGAGING HEIGHT IN FLUIDISED BED (see Fluidised bed)

TRANSPORT PROPERTIES OF GASES (see Kinetic theory of gases)

TRANSPORT THEOREM (see Conservation equations, two-phase)

TRAPEZOIDAL RULE

Following from: *Integrals*

The trapezoidal rule is an approximate method of estimating a definite integral $\int_a^b f(x)\,dx$.

The interval of integration is separated into n partial subintervals $[x_i, x_{i=1}]$, $i = 0, 1, \ldots, n-1$. On each subinterval a subintegral function is replaced by a linear one.

An integral on the interval $[x_i, x_{i+1}]$ is calculated approximately by the formula for the trapezoid area

$$\int_{x_i}^{x_{i+1}} f(x)dx \simeq \frac{x_{i+1} - x_i}{2}[f(x_i) + f(x_{i+1})]. \tag{1}$$

The summation of the left and the right parts of this approximate equality brings about the trapezoidal rule

$$\int_a^b f(x)dx \simeq \sum_{i=0}^{n-1} \frac{x_{i+1} - x_i}{2}[f(x_i) + f(x_{i+1})]. \tag{2}$$

For the case of n equidistant cusps the expression is simplified and is reduced to the form

$$\int_a^b f(x)dx \,{}^{e}\, \frac{b-a}{n}\left[\frac{f(a)+f(b)}{2} + f(x_1) + \cdots + f(x_{n-1})\right]. \tag{3}$$

This quadrature formula is correct for trigonometric functions $\cos \frac{2\pi}{b-a}kx$, $\sin \frac{2\pi}{b-a}kx$, $k = 0, 1, \ldots, n-1$. When $b - a = 2\pi$, the trapezoidal formula is exact for all trigonometric polynomials of order not higher $(n - 1)$.

The error of a quadrature formula $|R(f)|$ is the modulus of the difference between the exact value of the integral and the quadrature sum, for double differentiable subintegral functions, this does not exceed

$$|R(f)| < \frac{(b-a)^3}{12n^2}M, \qquad M = \max_{[a,b]}|f''x|. \tag{4}$$

On finding a complete limit error of a quadrature formula an error of addition must also be accounted. If a addend $f(x_i)$ is calculated with an absolute error not exceeding ε, then the total error $\tilde{R}$ of the quadrature formula without considering a concluding round off error does not exceed $\tilde{R} \le (b - a)\varepsilon + |R(f)|$.

V.N. Dvortsov

TRAYS, COLUMNS (see Distillation)

TREATED SURFACES (see Augmentation of heat transfer, single phase)

TREVITHICK ENGINES (see Steam engines)

TRIANGULAR DUCTS, FLOW AND HEAT TRANSFER

Following from: *Channel flow*

Introduction

Triangular ducts differ from circular tubes in several important aspects. Because the duct periphery is not symmetrical to the flow field, the transition from laminar to turbulent flow follows a more complicated process. Also, because of this asymmetry, additional thermal boundary conditions can exist and thus make the determination of the heat transfer more difficult.

The transition process in isosceles triangular ducts has been investigated experimentally by Eckert and Irvine (1956) and later by Bandopadhayay and Hindwood (1973). Both studies reported that **Turbulence** first appears in the wide area of the cross section near the triangle base and with increasing **Reynolds Number**

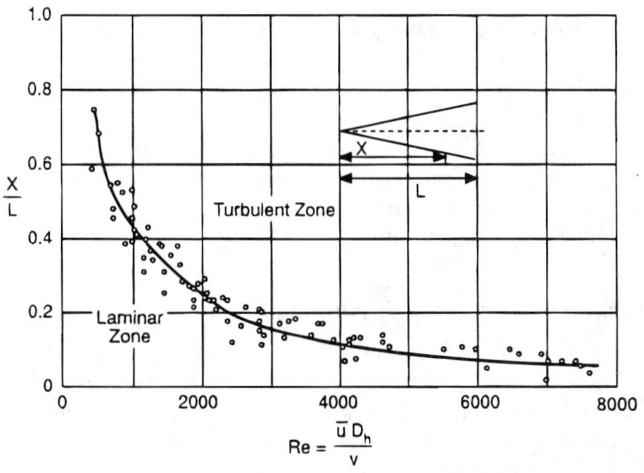

Figure 1. Determination of the laminar zone in an isosceles triangular duct, From Eckert, E. R. G. and Irvine, T. F. Jr (1965). Flow in Corners of Passages with Non-Circular Cross Sections, Trans. Am. Soc. Mech. Engrs. pp. 709–718, with permission.

spreads toward the narrow apex region. **Figure 1**, which shows the flow visualization measurements of Eckert and Irvine (1956) for an isosceles triangular duct with an apex angle of 11.5 degrees, illustrates this transition process. For example, from **Figure 1**, at a Reynolds number of 2000 (based on the hydraulic diameter), approximately 25% of the duct, measured along the center line, contains laminar flow while the remaining 75% is turbulent. Also seen in the figure is that at a Reynolds number of 800, the flow is half laminar and half turbulent. Even at a Reynolds number of 8000, almost 10% of the "apex" corner flow is laminar.

Such considerations are important since they indicate that the local heat transfer can be relatively poor in the apex region. This in turn can cause local "hot spots" in the corners whose prediction can be critical in the design of heat exchanger passages.

The classical thermal boundary conditions for round tubes are either a constant wall temperature or a constant wall heat flux. Their **Nusselt Numbers** are indicated by Nu_T and Nu_H respectively. Nothing needs to be said about the thermal conditions around the perimeter because of symmetry, i.e. no peripheral wall temperature gradients can exist. For triangular ducts, however, peripheral wall temperature variations can exist for the constant heat flux boundary condition. This in turn leads to three generally accepted thermal boundary conditions which are illustrated in **Table 1**.

Table 1. Triangular duct thermal boundary conditions

Boundary condition	Symbol
Constant wall temperature in the flow and peripheral directions.	Nu_T
Constant heat flux in the flow direction and constant peripheral wall temperature (thick wall duct).	Nu_{H1}
Constant heat flux in the flow direction and constant peripheral heat flux (thin wall duct).	Nu_{H2}
Constant heat flux in the flow direction with peripheral heat conduction in the wall (an actual duct).	Nu_{H4}

In the following sections, information is presented on the steady state pressure drop and heat transfer for fully developed constant property flow in triangular ducts for both laminar and turbulent conditions. For information on such topics as hydrodynamic and thermal entrance lengths, further readings are suggested.

Laminar Flow Friction Factors

The pressure drop characteristics of fully developed duct flows are normally given as the relation between the **Reynolds Number** and the **Friction Factor** (see **Nomenclature**). For fully developed laminar duct flow, it can be shown that the product of these two quantities is a constant depending on the duct geometry. Thus it is sufficient to tabulate the constant for various geometries.

Shaw and London (1978) have presented a most useful compendium of laminar flow and heat transfer data for different ducts including triangular ducts. **Table 2**, taken from their work, lists the friction factor Reynolds number product for isosceles triangular ducts as a function of the duct apex angle, 2α. Also included in **Table 2** are heat transfer data which will be discussed later. Shaw and London (1978) also include calculated information on hydrodynamic entrance lengths and excess entrance pressure drops.

Turbulent Flow Friction Factors

For fully developed turbulent flow, an additional complication arises in the flow field. Because of the non-isotropic Reynolds stress distribution around the peripheral wall, secondary flows occur which complicate an analytical approach to the problem. Nevertheless, both analytical and experimental results are available in the literature which are summarized below.

There have been a number of experimental studies of triangular duct turbulent friction factors. These would of course include both

Table 2. Flow and heat transfer characteristics for isosceles triangular ducts, Shaw and London (1978)

2α	f·Re	Nu_T	Nu_{H1}	Nu_{H2}
0	12.000	0.943	2.059	0
7.15	12.352	1.46	2.348	0.039
10.00	12.474	1.61	2.446	0.080
14.25	12.636	1.81	2.575	0.173
20.00	12.822	2.00	2.722	0.366
28.07	13.026	2.22	2.880	0.747
30.00	13.065	2.26	2.910	0.851
36.87	13.181	2.36	2.998	1.22
40.00	13.222	2.39	3.029	1.38
50.00	13.307	2.45	3.092	1.76
53.13	13.321	2.46	3.102	1.82
60.00	13.333	2.47	3.111	1.892
67.38	13.321	2.45	3.102	1.84
70.00	13.311	2.45	3.095	1.80
80.00	13.248	2.40	3.050	1.59
90.00	13.153	2.34	2.982	1.34
120.00	12.744	2.00	2.680	0.62
126.87	12.622	1.90	2.603	0.490
150.00	12.226	1.50	2.325	0.156
151.93	12.196	1.47	2.302	0.136
180.00	12.000	0.943	2.059	0

the transition and secondary flow effects mentioned above. In addition, several analysis have been reported only one of which, Malak et al. (1975), included a correlation which accounts for the effect of apex angle on friction factor.

A convenient way to correlate the turbulent friction factor data, both experimental and theoretical, is by a *modified Blasius equation:*

$$f = \frac{C}{Re^{0.25}} \tag{1}$$

Carlson and Irvine (1961) reported that the constant "C" in Equation 1 is a weak function of the Reynolds number and depends primarily on the apex angle.

Figure 2 presents the results of ten experiments on isosceles triangular ducts where the constant "C" is plotted against the apex angle. Also shown are the theoretical (numerical) results of Malak et al. (1975) and Usui et al. (1983). The later considered the effects of secondary flows while the former did not. In a separate study, Kokorev et al. (1971) estimated that the effects of secondary flows on the friction factor was less than 10% for isosceles triangular ducts.

As seen in **Figure 2**, for apex angles greater than 15 degrees, the experimental data for "C" are greater than the analyses but are less than the analyses for apex angles less than 15 degrees. It is recommended that the solid line which represents the experimental data be used in calculating turbulent friction factors.

Laminar Flow Nusselt Numbers

The three Nusselt numbers Nu_T, Nu_{H1}, and Nu_{H2} defined in **Table 1** are listed in **Table 2** for isosceles triangular ducts for apex angles from 0 to 180 degrees. It can be seen from the table that the Nusselt numbers for the constant heat flux conditions are quite sensitive to the thermal boundary conditions, i.e., whether the Nu_{H1} or Nu_{H2} condition is specified. For example, for small apex angles,

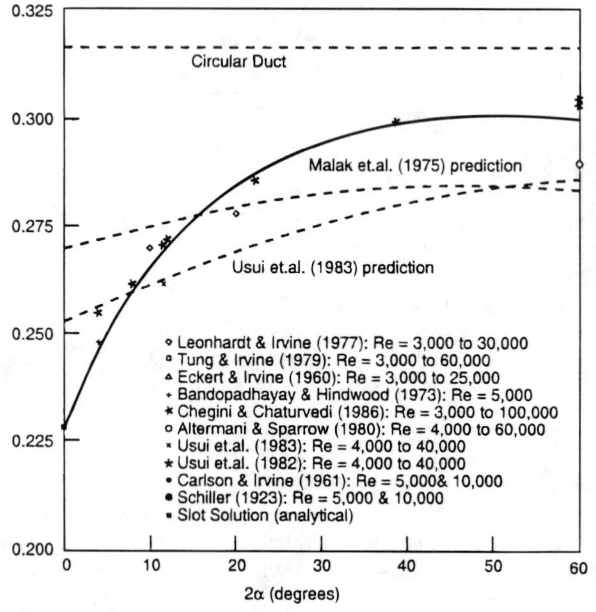

Figure 2. Experimental results and theoretical predictions of the constant "C" plotted against the apex angle 2α.

e.g., 7.15 degrees, the two Nusselt numbers differ by a factor of 60. Thus, extreme caution must be exercised in choosing the proper thermal boundary conditions for a particular design calculation.

Actual triangular ducts have thermal boundary conditions that are best represented by Nu_{H4} since there may be significant wall heat conduction. These situations have been investigated by Irvine and Cheng (1993) and further information on the Nu_{H4} boundary condition may be obtained from that reference.

Turbulent Flow Nusselt Numbers

Of all the physical processes involved in triangular duct flow and heat transfer, the least attention has been directed toward the problem of predicting the heat flow under turbulent flow conditions. Based on heat transfer in circular tubes, it is probable that triangular duct heat transfer in turbulent flow is less sensitive to the thermal boundary conditions than is the case for laminar flow. However, this lack of sensitivity has not been investigated experimentally.

Heat transfer experiments have been reported by Eckert and Irvine (1960) on an 11.46 degree isosceles triangular duct under the Nu_{H4} boundary condition. Unfortunately, they did not have fully developed thermal flow and their main contribution was to show that using circular tube correlations with the hydraulic diameter as the characteristic dimension in the Nusselt and Reynolds numbers was not appropriate.

Altemani and Sparrow (1980) performed experiments on an equilateral triangular duct with the Nu_{H1} boundary condition. They also measured thermal entrance lengths. On the basis of their measurements, they proposed that for fully developed flow the *Petukhov-Popov (1972) correlation* for a round tube be used with the friction factor calculated from Equation 1. This correlation is given by:

$$Nu = \frac{(f/8)RePr}{1.07 + 12.7(Pr^{2/3} - 1)(f/8)^{1/2}} \tag{2}$$

Equation 2 is recommended at the present time for triangular ducts along with a healthy skepticism regarding the results.

Concluding Remarks

It is impossible to report in more detail the flow and heat transfer aspects of triangular ducts in the present article. Readers are directed to the appropriate references for further information and topics not considered in this presentation.

Nomenclature

- C Constant in Equation 1 (−)
- D_h Hydraulic Diameter $= 4 \times \dfrac{\text{Area}}{\text{Perimeter}}$ (m)
- f Friction Factor $= \dfrac{2\dfrac{dp}{dz}D_h}{\rho\bar{u}^2}$ (−)
- L Duct height in **Figure 1** (m)
- Pr Prandtl number $= \dfrac{\nu}{\kappa}$ (−)
- Re Reynolds Number $= \dfrac{\bar{u}D_h}{\nu}$ (−)
- $\bar{u}$ Average velocity (m/s)
- x Centerline coordinate in Fig. 1 (m)
- z Flow direction coordinate (m)

GREEK SYMBOLS

α Half apex angle (degrees)
κ Thermal diffusivity (m^2/s)
ν Kinematic viscosity (m^2/s)
ρ Density (kg/m^3)

References

Altemani, C. A. C. and Sparrow, E. M. (1980) Turbulent heat transfer and fluid flow in an unsymmetrically heated triangular duct, *J. Heat Trans.*, 102, 590–597.

Bandopadhayay, P. C. and Hindwood, J. B. (1973) On the coexistence of the laminar and turbulent flow in a narrow triangular duct, *J. Fluid Mechs.*, 59, 775–784.

Carlson, L. W. and Irvine, T. F. Jr. (1961) Fully developed pressure drop in triangular shaped ducts, *J. Heat Trans.*, 83, 441–444.

Chegini, H. and Chaturvedi, S. K. (1986) An experimental and analytical investigation of friction factors for fully developed flow in internally finned triangular ducts, *J. Heat Trans.*, 108, 507–512.

Eckert, E. R. G. and Irvine, T. F. Jr. (1956) Flow in corners of passages with non-circular cross sections, *Trans. Am. Soc. Mech. Engrs.*, 78, 709–718.

Eckert, E. R. G. and Irvine, T. F. Jr. (1960) Pressure drop and heat transfer in a duct with triangular cross section, *J. Heat Trans.*, 82, 125–135.

Irvine, T. F. Jr. and Cheng J. A. (1993) Conjugated laminar heat transfer for newtonian and power law fluids in triangular ducts, *J. Energy Heat Mass Trans.*, 15, 107–113.

Kokorev, L. S. et al. (1971) Effect of secondary flows on the velocity distribution and hydraulic drag in turbulent liquid flows in non-circular channels, *Heat Trans. Sov. Res.*, 3, 66–78.

Leonhardt, W. J. and Irvine, T. F. Jr. (1977) Experimental friction factors for fully developed flow of dilute aqueous polyethylene-oxide solutions in smooth wall triangular ducts, *Heat and Mass Transfer Source Book (Fifth All Union Conference, Minsk)*, 236–250, Scripta Publishing Co. and John Wiley and Sons.

Malak, J. et al. (1975) Pressure losses and heat transfer in non-circular channels with hydraulically smooth walls, *Int. J. Heat and Mass Trans.*, 18, 139–148.

Schiller, L. (1923) Uber den Stromungswiderstand von Rohen Verschiedenen Querschnitts und Rauhigkeitsgrades, *Zeit. Ang. Math und Mech.*, 2–10.

Shah, R. K. and London, A. L. (1978) Laminar flow forced convection in ducts, *Advances in Heat Transfer Supplement 1*, T. F. Irvine Jr. and J. P. Hartnett, eds., Academic Press.

Tung, S. S. and Irvine, T. F. Jr. (1979) Experimental study of the flow of a viscoelastic fluid in a narrow triangular duct, *Studies in Heat Transfer*, 309–329, Hemisphere Pub. Corp.

Usai, H. et al. (1982) Turbulence measurements and mass transfer in fully developed flow in a triangular duct with a narrow apex angle, *Int. J. Heat Mass Trans.*, 25, 615–623.

Usui, H. et al. (1983) Fully developed turbulent flow in isosceles triangular ducts, *J. Chem. Eng. Japan*, 16, 13–18.

T.F. Irvine Jr. and M. Capobianchi

TRIANGULAR RELATIONSHIP IN ANNULAR FLOW (see Annular flow)

TRIARYLMETHANE (see Photochromic dye tracing)

1,1,1-TRICHLOROETHANE

Trichloroethane (CH_3CCl_3) is a colourless, inflammable liquid that is insoluble in water. It is obtained from vinyl chloride. It is the least toxic of the chlorinated solvents. Its usage has been dramatically decreased in recent years because of its *ozone depletion* potential, although it is still used in the vapour degreasing of metals.

Molecular weight: 133.40
Melting point: 240.15 K
Boiling point: 347.15 K
Critical temperature: 528.2 K
Critical pressure: 4.15 MPa
Critical density: 455 kg/m^3

V. Vesovic

TRICHLOROETHYLENE

Trichloroethylene ($CHClCCl_2$) is a colourless, inflammable, chloroform-smelling liquid that is slightly soluble in water. The principal raw material for its production is ethylene. It is moderately toxic. Its usage has been dramatically decreased in recent years because of its *ozone depletion* potential, although it is still used in the vapour degreasing of metals.

Molecular weight: 131.39
Melting point: 186.15 K
Boiling point: 360.4 K
Critical temperature: 571 K
Critical pressure: 4.91 MPa
Critical density: 513 kg/m^3

V. Vesovic

TRI-DRUM HEAT RECOVERY BOILER (see Heat recovery boilers)

TRIGONOMETRIC FUNCTIONS

Trigonometric functions are the class of elementary transcendental functions: sine, cosine, tangent, cotangent, secant, cosecant.

For a real value, arguments of a trigonometric function are defined in geometrical terms. Set B is the point of a unit circle with centre at the origin, φ is the polar angle of the point B.

If x_φ and y_φ are the rectangular Cortesian coordinates, then the trigonometric functions sine and cosine are defined by

$$\sin \varphi = y_\varphi; \qquad \cos \varphi = x_\varphi.$$

The other trigonometric functions: tangent, cotangent, secant, cosecant, can be determined by

$$\tan \varphi = \frac{\sin \varphi}{\cos \varphi}; \qquad \sec \varphi = \frac{1}{\cos \varphi}; \tag{1a}$$

$$\cot \varphi = \frac{\cos \varphi}{\sin \varphi}; \qquad \csc \varphi = \frac{1}{\sin \varphi};$$

All trigonometric functions are periodic; sine and cosine have a period of 2π, tangent and cotangent, a period π.

Sine, tangent and cotangent are odd functions, cosine is an even function.

All trigonometric functions in their fields of definition are continuous and infinitely differentiable. The derivitives of the trigonometric functions are given by:

$$(\sin x)' = \cos x; \qquad (\cos x)' = -\sin x; \tag{1}$$

$$(\tan x)' = \frac{1}{\cos^2 x}; \qquad (\cot x)' = -\frac{1}{\sin^2 x}.$$

The integrals of trigonometric functions are

$$\int \sin x\, dx = C - \cos x; \qquad \int \cos x\, dx = C + \sin x; \tag{2}$$

$$\int \tan x\, dx = C - \ln|\cos x|; \qquad \int \cot x\, dx = C + \ln|\sin x|.$$

All trigonometric functions can be expanded in power series

$$\sin x = x - \frac{x^3}{3!} + \cdots + (-1)^n \frac{x^{2n+1}}{(2n+1)!} + \cdots \quad \text{for } |x| < \infty$$

$$\cos x = 1 - \frac{x^2}{2!} + \cdots + (-1)^n \frac{x^{2n}}{(2n)!} + \cdots \quad \text{for } |x| < \infty \tag{3}$$

$$\tan x = x + \cdots + \frac{2^{2n}(2^{2n}-1)|B_n|}{(2n)!} x^{2n-1} + \cdots \quad \text{for } |x| < \frac{\pi}{2}$$

$$\cot x = \frac{1}{x} - \frac{1}{3}x - \cdots - \frac{2^{2n}|B_n|}{(2n)!} x^{2n-1} - \cdots \quad \text{for } 0 < |x| < \pi$$

B_n—are Bernoulli numbers.

A trigonometric system

$$\frac{1}{\sqrt{2\pi}}; \frac{\cos x}{\sqrt{\pi}}; \frac{\sin x}{\sqrt{\pi}}; \ldots \frac{\cos nx}{\sqrt{\pi}}; \frac{\sin nx}{\sqrt{\pi}}. \tag{4}$$

is one of the important orthogonal function systems, and it is widely employed in solving problems of heat conduction. A trigonometric system is complete and closed in the space of the continuous 2π periodic functions.

Trigonometric functions of a complex argument $z = x + iy$ are defined as analytic continuation of corresponding trigonometric functions of a real variable into a complex plane. All formulas valid for trigonometric functions of a real argument remain also valid for a complex argument. Trigonometric functions of a complex argument can be expressed in terms of an exponential function by the Euler formulas

$$\sin z = \frac{e^{iz} - e^{-iz}}{2i}; \qquad \cos z = \frac{e^{iz} + e^{-iz}}{2}. \tag{5}$$

V.N. Dvortsov

TRIGONOMETRIC POLYNOMIALS (see Polynomials)

TRIGONOMETRIC SERIES (see Fourier series)

TRIPLE INTERFACE (see Multiphase flow)

TRIPLE POINT

Following from: *Phase equilibrium*

Figure 1 shows the p-T projection of the p-ṽ-T surface of a typical pure fluid. The point at which the lines representing the boundaries between solid + gas, liquid + gas and solid + liquid phases intersect is the triple point. It is a fixed or invariant point since it is a result of the phase rule that a single component system with three phases has no degrees of freedom. The consequence is that the solid, liquid and vapour phases can be in equilibrium only at one thermodynamic state of pressure and temperature. The triple point is relatively easily realised for water and it is therefore frequently used as a standard temperature for the calibration of thermometers.

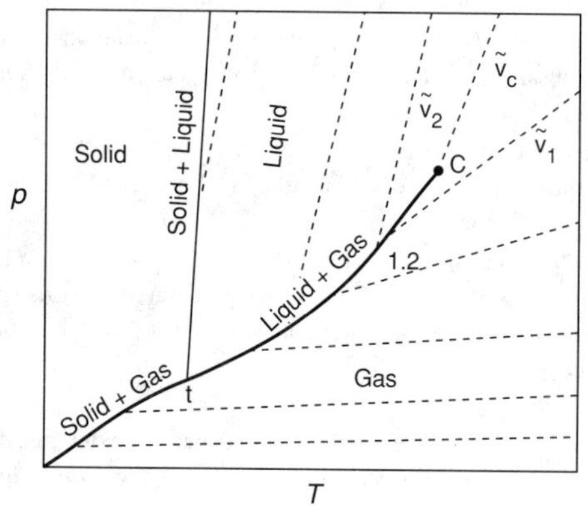

Figure 1. A p-T projection of the p-ṽ-T behaviour of a typical pure fluid.

Leading to: *Temperature*

W.A. Wakeham

TRIPLE POINT PRESSURE (see Vapour pressure)

TRIPLET STATE (see Phosphorescence; Photoluminescence)

TRITIUM (see Hydrogen; Neutrons)

TROMBE-MICHEL WALLS (see Solar energy)

TROMBE WALLS (see Solar energy)

TROPOSPHERE (see Atmosphere)

TRUE MASS FLOW METER (see Mass flowmeters)

TUBE BANKS, CONDENSATION HEAT TRANSFER IN

Following from: *Condensation, overview; Condensation of pure vapours; Crossflow heat transfer*

Condensation within tube banks is subject to the combined effects of *vapour shear* and *falling condensate* from upper tubes. The latter is called *condensate inundation.* Tube orientation may be vertical or horizontal. However, horizontal orientation is more common when a pure vapour is to be condensed. **Table 1** summarizes the factors that affect condensation heat transfer of pure vapours in a horizontal tube bank.

Plain tubes are commonly used for condensation of steam and other fluids with high liquid thermal conductivity. Integral fin tubes and more advanced three-dimensional fin tubes are often used for condensation of organic fluids with low liquid thermal conductivity. Both in-line and staggered tube banks are commonly employed. For power condensers operated at low pressures, the maximum vapour velocity exceeding 100 m/s is realized at the inlet of tube bank. For refrigerant condensers, on the other hand, the vapour velocity at the inlet of condenser shell is designed to be less than about 6 m/s. The vapour flow may be downward, horizontal, upward and any other direction depending on the location in the tube bank. For the downward flow of vapour, the condensate inundation rate depends only on the condensation rate at the upper rows. For the other flow directions, it is not possible to estimate the inundation rate accurately.

The mode of condensate inundation depends on the inundation rate and vapour velocity. Discussions on the inundation modes are given by Marto (1984, 1988) and Collier and Thome (1994). The inundation modes are illustrated in **Figure 1**. At low vapour velocities, condensate drains in discrete drops (**Figure 1a**), then in condensate columns (**Figure 1b**), and then in a condensate sheet (**Figure 1c**) as the inundation rate increases. In a closely packed staggered tube bank, side drainage may occur depending on the condition (**Figure 1d**). The condensate impinging on the lower tube causes splashing, ripples and turbulence on the condensate film (**Figure 1e**). At high vapour velocities, the condensate leaving the tube is disintegrated into small drops and impinges on the other tubes (**Figure 1f**). A wide variety of condensate drainage mode shown in **Figure 1** results in a wide breadth of experimental data regarding the effect of condensate inundation. Review of relevant literature are given by Marto (1984, 1988) and Collier and Thome (1994). Butterworth (1981) discusses condensate inundation without vapour shear. Cavallini et al. (1990) give a comparison of proposed correlations with available experimental data.

Table 1. Factors that affect condensation of pure vapours in horizontal tube banks

Item	Factor	
Tube	Geometry	Plain Finned
	Layout	In-line Staggered
	Number of tube rows	Horizontal Vertical
	Pitch-to-diameter ratio	
Vapour	Velocity	
	Flow direction	
Condensate	Inundation rate	
	Mode of inundation	

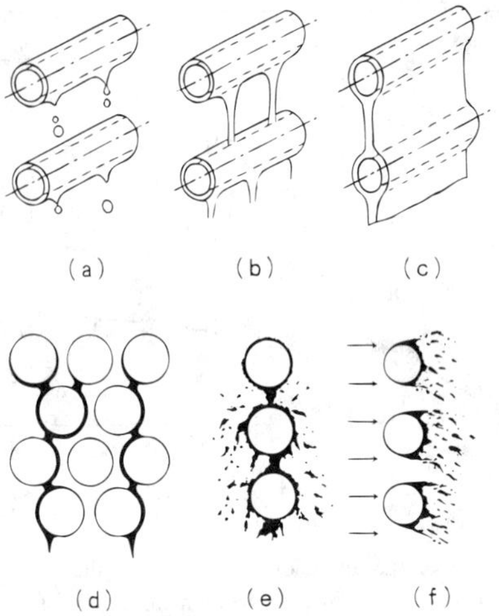

Figure 1. Modes of condensate inundation in horizontal tube banks.

Figure 2 is a schematic representation of the results of various experimental measurements on the coordinates of $\overline{\alpha}_n/\alpha_1$ versus n (see Marto, 1984), where $\overline{\alpha}_n$ is the mean heat transfer coefficient for a vertical row of n tubes and α_1 is the heat transfer coefficient for the top tube. In **Figure 2**, a number of theoretical and empirical relationships are also presented. For a stagnant vapour, Nusselt (1916) extended his analysis of laminar film condensation on a single horizontal plain tube to include the effect of sheet mode drainage (See also **Condensation** and **Condensation of Pure Vapours**). Jakob (1949) generalized the Nusset analysis and derived the following equation.

$$\bar{\alpha}_n/\alpha_1 = n^{-1/4} \quad (1)$$

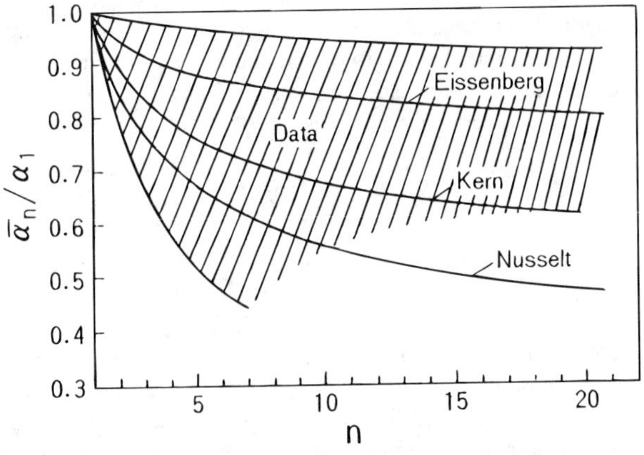

Figure 2. Effect of vertical row number on the mean heat transfer coefficient of a bank of horizontal plain tubes. (See Marto, 1984)

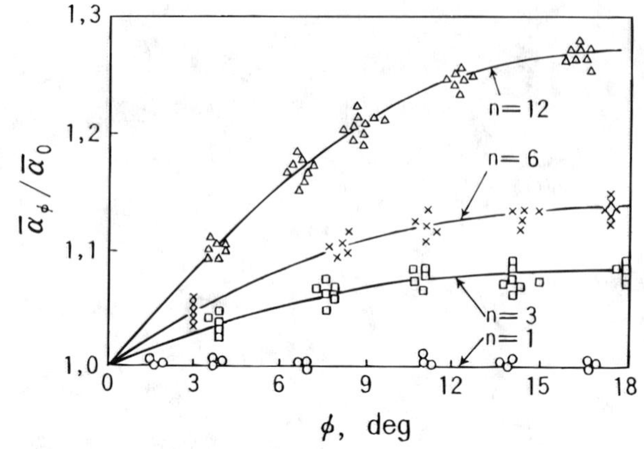

Figure 3. Effect of the angle of inclination on the mean heat transfer coefficient of a bank of plain tubes: downward flow of steam. (Data of Shklover and Buevich, 1978).

The expression for α_1 is given by Nusselt (1916) as follows:

$$\alpha_1 = 0.725\left[\frac{g\rho_L(\rho_L - \rho_G)h_{LG}\lambda_L^3}{\eta_L D\Delta T}\right]^{1/4} \tag{2}$$

where g is the gravitational acceleration, ρ_L is the liquid density, ρ_G is the vapour density, λ_L is the liquid thermal conductivity, h_{LG} is the specific enthalpy of evaporation, η_L is the liquid dynamic viscosity, D is the tube diameter and ΔT is the condensation temperature difference. Based on the observation of condensate drainage in operating condensers, Kern (1958) suggested a smaller dependence on the row number such that:

$$\bar{\alpha}_n/\alpha_1 = n^{-1/6} \tag{3}$$

Based on the side drainage model, Eissenberg (1972) derived the following equation

$$\bar{\alpha}_n/\alpha_1 = 0.60 + 0.42\, n^{-1/4} \tag{4}$$

Equation 4 predicts a much smaller dependence on the row number than **Equations 1 and 3**, as shown in **Figure 2**.

For condensation without appreciable vapour shear, a number of other methods have also been proposed to account for the effect of condensate inundation. Two well-known empirical equations, which are due to Short and Brown (1951) and Grant and Osment (1968), are respectively written as

$$Nu_f = 1.2 Re_f^{-1/3} \tag{5}$$

$$\alpha_n/\alpha_1 = (\Gamma_n/\gamma_n)^{-0.223} \tag{6}$$

where $Nu_f = \alpha_n\, [\eta_L^2/\rho\, (\rho_L - \rho_G)\, g]^{1/3} / \lambda_L$, $Re_f = 2\,\Gamma_n/\eta_L$, Γ_n is the condensate drainage per unit length from the n-th tube and γ_n is the condensation rate per unit length on the n-th tube. Butterworth (1981) have shown that **Equations 3, 5 and 6** are in very close agreement despite their apparent difference. Marto (1988) and Collier and Thome (1994) recommend **Equation 3** for design purposes. When the tube bank is inclined with respect to the horizontal, pendant drops on the tube are driven by gravity toward the lower tube end. As a result, the inundation rate decreases as the angle of inclination increases. This results in an increase in the mean heat transfer coefficient. **Figure 3** shows experimental data for steam condensing in 7 tubes wide, n tubes deep (n = 1, 3, 6, and 12) in-line tube banks obtained by Shklover and Buevich (1978). In **Figure 3**, the ratio of the mean heat transfer coefficient for tube banks with and without tube inclination $\bar{\alpha}_\varphi/\bar{\alpha}_0$ is plotted as a function of the inclination angle φ. It is seen from **Figure 3** that the effect of tube inclination is more significant for a deeper tube bank.

At medium to high vapour velocities, condensation in tube banks is considerably affected by the flow direction of vapour. Experimental data showing the effect of vapour flow direction are shown in **Figure 4** and are due to Fujii (1981). In **Figure 4**, the results for downward, horizontal and upward flow of low pressure steam condensing in 5 × 15 (width × depth) in-line and staggered banks of horizontal plain tubes are plotted on the coordinates of $Nu/Re_L^{1/2}$ versus F, where $Nu = \alpha D/\lambda_L$, $Re_L = \rho_L u_G' D / \eta_L$, $F = \eta_L h_{LG}\, g\, D/\lambda_L\, \Delta T\, u_G'^2$, α is the mean heat transfer coefficient for a tube row perpendicular to the vapour flow direction, u_G' is the vapour velocity based on the maximum flow cross-section (i.e. the velocity calculated in the absence of any tube) just upstream of the tube row and ΔT is the mean condensation temperature difference of the tube row. The results for the downward and horizontal flow are almost identical and are correlated fairly well by the following equation:

$$Nu/Re_L^{1/2} = 0.96\, F^{0.2} \tag{7}$$

The data for the upward flow are lower than those for the downward and horizontal flow. The difference is largest in the region $0.1 < F < 0.5$ where the effects of vapour shear and gravity on the condensate flow are of the same order of magnitude. Most of the data for the upward flow in the region of $0.1 < F < 1$ are even lower than the prediction of the Nusselt (1916) equation, which is written in terms of $Nu/Re_L^{1/2}$ and F as

$$Nu/Re_L^{1/2} = 0.725\, F^{1/4} \tag{8}$$

For the downward flow of vapour, experimental data showing the combined effects of vapour shear and condensate inundation are presented in **Figure 5** and are due to Honda et al. (1988). In **Figure 5**, the results for near atmospheric R-113 vapour condensing in 3 × 15 in-line and staggered banks of horizontal plain tubes are plotted on the coordinates of Nu_f versus Re_f with the vapour velocity at the tube bank inlet $u_{G,in}$ as a parameter. The value of Re_f is calculated assuming the gravity drained flow model. Two lines in **Figure 5** show the Nusselt (1916) equation for a single tube, which is written in terms of Nu_f and Re_f as

$$Nu_f = 1.20\, Re_f^{-1/3} \tag{9}$$

and the following empirical equation for a nearly stagnant vapour:

$$Nu_f = [(1.2\, Re_f^{-0.3})^4 + (0.072\, Re_f^{0.2})^4]^{1/4} \tag{10}$$

Equation 10 is based on the experimental data for R-12 and R-21 condensing on vertical columns of horizontal plain tubes with

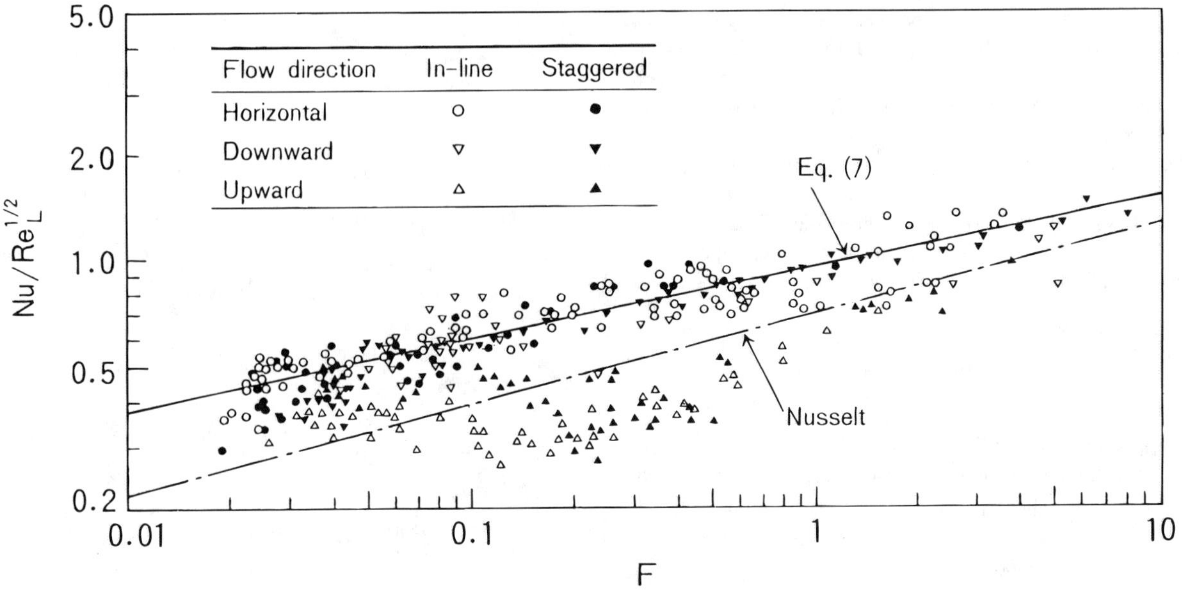

Figure 4. Comparison of data for horizontal, downward and upward flow of steam in in-line and staggered tube banks. (Data of Fujji, 1981).

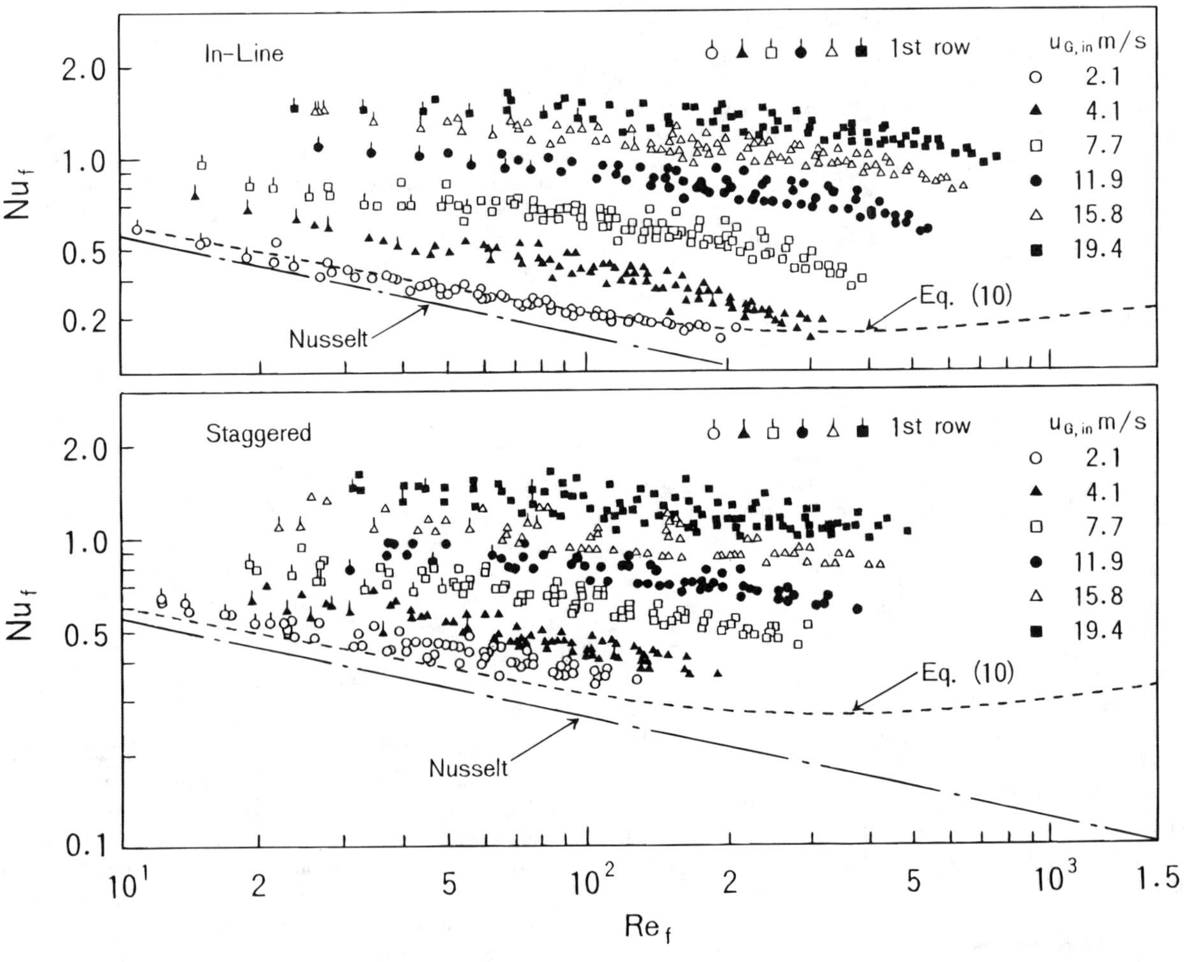

Figure 5. Combined effects of vapour shear and condensate inundation on the local heat transfer coefficient in banks of horizontal plain tubes: downward flow of R-113. (Data of Honda et al., 1988).

$D = 10 \sim 45$ mm obtained by Kutateladze and Gogonin (1979) and Kutateladze et al. (1985). The data for moving vapour is close to Equation 10 at the smallest $u_{G,in}$ and deviates toward a higher value as $u_{G,in}$ increases. For the downward flow of vapour, a number of empirical relationships have been proposed to predict the combined effects of vapour shear and condensate inundation. A common feature of the proposed relationships is that the heat transfer coefficient is calculated from the superposition of two contributions; i.e. gravity drained condensation and vapour shear controlled condensation. Butterworth (1977) proposed the following equation:

$$\alpha = n^{-0.16}\left[\frac{1}{2}\alpha_{sh}^2 + \left(\frac{1}{4}\alpha_{sh}^4 + \alpha_g^4\right)^{1/2}\right]^{1/2} \qquad (11)$$

where α_{sh} is the heat transfer coefficient in the vapour-shear-controlled regime and α_g is the heat transfer coefficient in the gravity-controlled regime. The expression for α_{sh} is written as

$$\alpha_{sh}D/\lambda_L = 0.59(\rho_L u_g' D/\varepsilon\eta_L)^{0.5} \qquad (12)$$

where ε is the void fraction for the tube bank (i.e. free volume divided by total volume). The expression for α_g is given by **Equation 2**. Equation 12 is based on the Shekriladze and Gomelauri (1966) analysis for the effect of vapour shear on condensation on horizontal tubes. Fujii and Oda (1986) and Cavallini et al. (1988) also proposed empirical equations in which the inundation effect is expressed in terms of n. Based on the experimental data for atmospheric pressure steam obtained by Nobbs (1975), McNaught (1982) proposed the following empirical equation:

$$\alpha = (\alpha_{sh}^2 + \alpha_g^2)^{1/2} \qquad (13)$$

He proposed to model the process in the vapour shear controlled regime as the two-phase forced convection. The derived expression for α_{sh} is written as

$$\alpha_{sh}/\alpha_L = 1.26\, X_{tt}^{-0.78} \qquad (14)$$

where α_L is the liquid-phase forced convection coefficient across a tube bank and X_{tt} is the Lackhart-Martinelli parameter given by

$$X_{tt} = \left(\frac{1-x}{x}\right)^{0.9}\left(\frac{\rho_G}{\rho_L}\right)\left(\frac{\eta_L}{\eta_G}\right)^{0.1} \qquad (15)$$

The expression for α_g is given by

$$\alpha_g/\alpha_l = (\Gamma_n/\gamma_n)^{-a} \qquad (16)$$

where $a = 0.13$ and 0.22 for the staggered and in-line tube banks, respectively, and α_l is given by **Equation 2**. Based on their own experimental results shown in **Figure 5**, Honda et al. (1988) derived the following equations:

a. Staggered tube bank

$$Nu = [Nu_g^4 + Nu_g^2 Nu_{sh}^2 + Nu_{sh}^4]^{1/4} \qquad (17)$$

where

$$Nu_g = Gr^{1/3}[(1.2\, Re_{f,g}^{-0.3})^4 + (0.072\, Re_{f,g}^{0.2})^4]^{1/4} \qquad (18)$$

$$Nu_{sh} = 0.165\left(\frac{P_t}{P_l}\right)^{0.7}\left[Re_G^{-0.4} + 1.83\left(\frac{\dot{q}_n}{\rho_G h_{LG} u_{Gn}}\right)\right]^{1/2}\left(\frac{\rho_G}{\rho_L}\right)^{1/2}\frac{Re_L Pr_L^{0.4}}{Re_{f,sh}^{0.2}} \qquad (19)$$

$$Gr = g\rho_L(\rho_L - \rho_g)D^3/\eta_L^2$$

$$Re_{f,g} = 2\pi D \sum_{i=1}^{(n+1)/2} \dot{q}_{2i-1}/\eta_L h_{Lg} \quad \text{for } n = 1, 3, 5, \ldots$$

$$Re_{f,g} = 2\pi D \sum_{i=1}^{n/2} \dot{q}_{2i}/\eta_L h_{Lg} \quad \text{for } n = 2, 4, 6, \ldots$$

$$Re_{f,sh} = Re_{f,g} \quad \text{for } n = 1$$

$$Re_{f,sh} = 2\pi D\left(\sum_{i=1}^{n-1} \dot{q}_i D/P_t + \dot{q}_n\right)/\eta_L h_{LG} \quad \text{for } n > 1$$

$$Re_G = \rho_G u_{Gn} D/\eta_G, \qquad Re_L = \rho_L u_{Gn} D/\eta_L$$

P_t is the transverse tube pitch, P_l is the longitudinal tube pitch, and u_{Gn} is the vapour velocity at the n-th row based on the minimum flow cross-section. For the first row, the leading coefficient on the right-hand side of Equation 19 should be replaced by 0.13.

b. In-line tube bank

$$Nu = (Nu_g^4 + Nu_{sh}^4)^{1/4} \qquad (20)$$

where

$$Nu_{sh} = 0.053\left[Re_g^{-0.2} + 18.0\left(\frac{\dot{q}_n}{\rho_G h_{LG} u_{Gn}}\right)\right]^{1/2}\left(\frac{\rho_G}{\rho_L}\right)^{1/2}\frac{Re_L Pr_L^{0.4}}{Re_{f,sh}^{0.2}} \qquad (21)$$

$$Re_{f,g} = 2\pi D \sum_{i=1}^{n} \dot{q}_i/\eta_L h_{LG}$$

The definitions of Nu_g, $Re_{f,sh}$, Re_G, Re_L and u_{Gn} are the same as those for the staggered tube bank. For the first row, the leading coefficient on the right-hand side of **Equation 21** should be replaced by 0.042. Cavallini et al. (1990) and Collier and Thome (1994) recommend **Equations 17 and 20** for refrigerants. These equations generally give a conservative prediction for steam at low vapour velocities.

Various enhancement techniques have been proposed for shell-side condensation of organic fluids with low liquid thermal conductivity. A review of relevant literature is given by Webb (1994). A

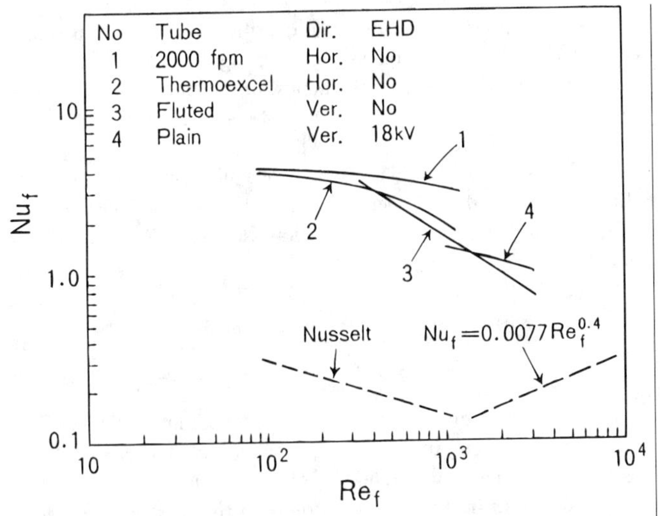

Figure 6. Comparison of four enhancement techniques for condensation of refrigerants in tube banks. (See Yabe, 1991).

comparison of four kinds of enhancement techniques applied to the condensation of refrigerants in tube banks is shown in **Figure 6** and is due to Yabe (1991). In **Figure 6**, Tube No. 1 is a horizontal *integral-fin tube* with fin density of 2000 fins per meter, Tube No. 2 is the *Thermoexcel-C*™ tube with saw-tooth-shape fins, Tube No. 3 is a vertical *fluted tube* with flute pitch of 1 mm and Tube No. 4 is a vertical plain tube fitted with electrodes for *electrohydrodynamic enhancement.* At present, the highest heat transfer performance is provided by the horizontal integral-fin tube with optimized fin geometry. A theoretical model of film condensation in banks of horizontal integral-fin tubes is proposed by Honda et al. (1989). (See also **Augmentation of Heat Transfer, Two Phase**).

References

Butterworth, D. (1977) Developments in the Design of Shell-and-Tube Condensers, ASME Paper No. 77-WA/HT-24, Atlanta.

Butterworth, D. (1981) Inundation without vapour shear, *Power Condenser Heat Transfer Technology,* P. J. Marto and R. H. Nunn eds., Hemisphere Publishing Corporation, New York.

Cavallini, A., Frizzerin, S. and Rosetto, L. (1988) Refrigerant 113 Vapour Condensation on a Horizontal Tube Bundle, Inst. di Fisica Tecnica dell' Università di Padova, Rept. No. 133.

Cavallini, A., Longo, G. and Rosetto, L. (1990) Condensation of pure saturated vapour on the outside of tube bundles, *Phase-Interface Phenomena in Multiphase Flow,* G. F. Hewitt, F. Mayinger and J. R. Riznic eds., Hemisphere Publishing Corporation, New York.

Collier, J. G. and Thome, J. R. (1994) *Convective Boiling and Condensation,* 3rd Edition, Clarendon Press, Oxford.

Eissenberg, D. M. (1972) An Investigation of the Variables Affecting Steam Condensation on the Outside of a Horizontal Tube Bundle, Ph. D. Thesis, Univ. of Tennessee, Knoxville.

Fujii, T. (1981) Vapour shear and condensate inundation: An overview, *Power Condenser Heat Transfer Technology,* P. J. Marto and R. H. Nunn eds., Hemisphere Publishing Corporation, New York.

Fujii, T. and Oda, K. (1986) Correlation Equation of heat transfer coefficient for condensate inundation in tube bundles, *Trans. JSME,* 52, 822–826.

Grant, I. D. R. and Osment, B. D. J. (1968) The effect of condensate drainage on condenser performance, NEL Report No. 350, National Engineering Laboratory, East Kilbride.

Honda, H., Uchima, B., Nozu, S., Nakata, H. and Fujii, T. (1988) Condensation of downward flowing R-113 vapour on bundles of horizontal smooth tubes, *Trans. JSME,* 54, 1453–1460, and *Heat Transfer-Jap. Res.,* 18, 6, 31–52.

Honda, H., Nozu, S. and Takeda, Y. (1989) A theoretical model of film condensation in a bundle of horizontal low finned tubes, *ASME Journal of Heat Transfer,* 111, 525–532.

Jakob, M. (1949) *Heat Transfer,* Vol. 1, John Wiley and Sons, Inc., New York.

Kern, D. Q. (1958) Mathematical development of loading in horizontal condensers, *AIChE J.,* 4, 157–160.

Kutateladze, S. S. and Gogonin, I. I. (1979) Heat transfer in film condensation of slowly moving vapour, *Int. J. Heat Mass Trans.,* 22, 1593–1599.

Kutateladze, S. S., Gogonin, I. I. and Sosunov, V. I. (1985) The influence of condensate flow rate on heat transfer in film condensation of stationary vapour on horizontal tube banks, *Int. J. Heat Mass Trans.,* 28, 1011–1018.

Marto, P. J. (1984) Heat transfer and two-phase flow during shell-side condensation, *Heat Transfer Engineering,* 5, 31–61.

Marto, P. J. (1988) Fundamentals of condensation, *Two-Phase Flow Heat Exchangers.* S. Kakaç, A. E. Bergles and E. O. Fernandes eds., Kluwer Academic Publishers.

McNaught, J. M. (1982) Two-phase forced convection heat transfer during condensation on horizontal tube bundles, *Proc. 7th Int. Heat Transfer Conf.,* Munich, 5, 125–131.

Nobbs, D. W. (1975) The Effect of Downward Vapour Velocity and Inundation on the Condensation Rates on Horizontal Tubes and Tube Banks, Ph. D. Thesis, Univ. of Bristol, Bristol.

Nusselt, W. (1916) Die Oberflächenkondensation des Wasserdampfes, *Zeit. Ver. Deut. Ing.,* 60, 541–546, 569–575.

Shekriladze, I. G. and Gomelauri, V. I. (1966) Theoretical study of laminar film condensation of flowing vapour, *Int. J. Heat Mass Trans.,* 9, 581–591.

Shklover, G. G. and Buevich, A. V. (1978) Investigation of steam condensation in an inclined bundle of tubes, *Thermal Eng.,* 49–52.

Short, B. E. and Brown, H. E. (1951) Condensation of vapour on vertical banks of horizontal tubes, *Proc. Inst. Mech. Eng.,* General Discussion on Heat Transfer, 27–31, 1951.

Webb, R. L. (1994) *Principles of Enhanced Heat Transfer,* John Wiley and Sons, Inc., New York.

Yabe, A. (1991) Active Heat Transfer Enhancement by Applying Electric Fields, Proc. 3rd *ASME/JSME Thermal Eng. Joint Conf.,* 3, xv–xxiii.

H. Honda

TUBE BANKS, CROSSFLOW OVER

Following from: *Cross flow; External flows, overview; Tubes, crossflow over*

Tube banks are commonly-employed design elements in heat exchangers. Both plain and finned tube banks are widely found. Tube bundles are a sub-component in shell-and-tube heat exchangers, where the flow resembles crossflow at some places, and longitudinal flow elsewhere. (The term tube *bank* is often used, in the literature, to denote a crossflow situation and *bundle* to indicate longitudinal flow, however this convention is far from universal.) The flow may be *single-phase* or *multi-phase*; boilers and condensers containing tube banks find a wide range of applications in industry, in addition there may be *combustion*, e.g. in a furnace heat exchanger.

Figure 1 shows the two basic tube-bank patterns involving either a rectangular or a rhombic primitive unit. These are referred to as *in-line tube banks* and *staggered tube banks* respectively. These are characterized by cross-wise and stream-wise pitch-to-diameter ratio's, a and b,

$$a \equiv \frac{s_T}{D} \tag{1}$$

$$b \equiv \frac{s_L}{D} \tag{2}$$

where D is the cylinder diameter, s_T the cross-wise (transverse) pitch, and s_L the stream-wise pitch. Most commonly encountered tube banks are the in-line square (a = b), rotated square (a = 2b) and equilateral triangle ($a = 2b/\sqrt{3}$). Tube banks with the product axb $< 1.25^2$ are referred to as compact, while those with axb > 4 are considered widely-spaced.

Analytical expressions for the flow of an ideal fluid in in-line and staggered tube banks have been derived in the form of power series (Beale, 1993). For widely spaced banks, the pressure coefficients are similar to the sinusoidal distribution observed in single cylinders. For compact banks there are significant differences in both in-line and staggered tube banks, in the latter case two additional pressure extrema can occur at $\phi = \pm 45°$, where ϕ is the angle measured from the front of the cylinder.

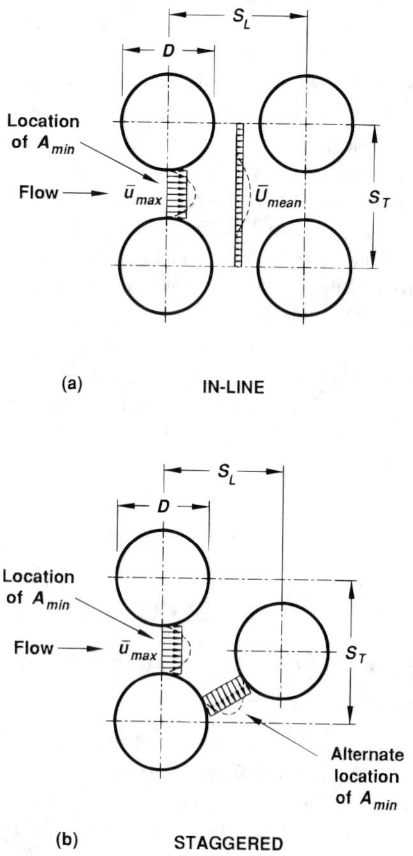

Figure 1. Schematic of (a) an inline and (b) a staggered tube bank illustrating nomenclature, and showing location of minimum cross-section.

The flow of a real fluid in the body of a tube bank also resembles flow past a single cylinder, though with significant differences. (See Tubes, Cross flow over) The flow **Reynolds Number**, Re, is defined by,

$$Re \equiv \frac{\rho \bar{u}_{max} D}{\eta} \tag{3}$$

where ρ is the fluid density, $\bar{u}_{max}$ is the maximum bulk velocity ie. the bulk velocity in the minimum cross-section (see **Figure 1**), D is the cylinder diameter, and η is the fluid viscosity. At low Re, the flow is **Laminar** with *separation* occurring at around $\rho = 90°$, resulting in stable **Vortices** forming behind each cylinder. For staggered banks, the upstream flow is typically a maximum between the preceding tubes, so the impinging flow bifurcates at the front-leading edge of each cylinder. For in-line banks, cylinders are in a comparatively dead zone downstream of the preceding cylinder's wake, and there are two re-attachment points at around $\phi = \pm 45°$.

As Re increases, vortices are shed from each cylinder in an alternate fashion. (See **Vortex Shedding**) Due to the overall pressure gradient, transient motion tends to occur at higher Re than for single tubes, particularly for compact banks. The number of vortices present and the motion of the streams depends on a and b; neighbouring streams may be in-phase, out-of-phase, or uncorrelated. Wake-switching is observed in staggered tube banks, while in in-line banks there is an instability in the shear-layer between the *wake* and main flow with some of the detached vortices being entrained by the free-stream flow.

As Re further increases, the wake becomes turbulent, and there is substantial free-stream turbulence, due to the influence of the preceding upstream rows. (See **Turbulence**) The **Boundary Layer** region remains laminar up to the critical Re of around 2×10^5, at which point transition to turbulence occurs and the separation point moves downstream. The engineer should be aware that both vortex-shedding and turbulent buffeting can induce vibrations in tube banks. (See **Vibration in Heat Exchangers.**)

The most widely adopted reference bulk velocity for use in Equation 2 is the bulk interstitial velocity which occurs in the minimum cross-section, ie. the maximum bulk velocity, $\bar{u}_{max}$ (Bergelin et al. (1950, 1958), Žukauskas et al. (1988), ESDU (1974)). Some authors, e.g. Kays and London (1984), base Re on a hydraulic radius. The Engineering Sciences Data Unit, ESDU (1979), base Re on the so-called mean superficial velocity, $\bar{u}_{mean}$, i.e. the average velocity that would occur if the tubes were removed, see **Figure 1**. The mean superficial velocities is useful when simulating large-scale flow in heat exchangers. It can readily be shown that,

$$\bar{u}_{max} = \left(\frac{a-1}{a}\right)\bar{U}_{mean} \tag{4}$$

for all in-line banks, and staggered banks with $a < 2b^2 - 1/2$. For some compact staggered banks, where $a > 2b^2 - 1/2$, the minimum cross-section occurs across the diagonal and so,

$$\bar{u}_{max} = \frac{\sqrt{4b^2 + a^2} - 2}{a} \bar{U}_{mean} \quad \textbf{(5)}$$

The hydrodynamic parameter of most interest to the heat exchanger designer is the overall pressure loss coefficient, often expressed in terms of an **Euler Number** Eu,

$$Eu \equiv \frac{\Delta\bar{p}_{row}}{\frac{1}{2}\rho\bar{u}_{max}^2} \quad \textbf{(6)}$$

where $\Delta\bar{p}_{row}$ is the mean pressure drop across a single row. Numerous alternative definitions abound in the literature, for instance Bergelin et al. (1950, 1958) define a **Friction Factor** $f = 4Eu$. Others simply use the symbol f to denote Eu. The friction factor of Kays and London (1984) is such that $f = (a - 1)Eu/\pi$ for large banks. ESDU (1979) use a different definition for pressure loss coefficient, again, based on $\bar{U}_{mean}$, i.e. there has been little effort to-date to standardize parameters.

Many experimental studies have been conducted on flow in tube banks, starting in the early part of this century. Comprehensive data were gathered by workers at the University of Delaware in the 40s and 50s. Their data were primarily in the low and intermediate range of Re. These were summarized in two reports, Bergelin et al. (1950, 1958), containing original data, and several papers (see ESDU, 1974, for a concise bibliography of experimental data).

A group at the Institute of Physical and Technical Problems of Energetics of the Lithuanian Academy of Sciences, have also published a large number of papers on both flow and heat transfer in tube banks over a wide Re range. The book by Žukauskas et al. (1988) contains detailed discussions on numerous aspects of the subject, while the article by Žukauskas (1987) is a substantially shorter, but comprehensive review of fluid flow and heat transfer in tube banks. Charts of Eu vs. Re are provided for in-line and staggered banks at various pitch ratios. Achenbach has conducted research on tube banks in the high Re turbulent flow regime (for references, see Žukauskas et al., 1987)

Various empirical correlations of pressure drop data have been devised over the years. The Eu vs. Re correlations of the Lithuanian group are commendable; they have been reconciled with numerous sources of externally-gathered experimental data, in addition to data gathered by the authors themselves. They are included here. Others such as those based on the Delaware groups work, could have been reproduced equally well, having formed an integral part of the thermal design of shell-and-tube exchangers, in the West for many years; see the articles by Taborek in the Heat Exchanger Design Handbook (1983) and Mueller in the *Handbook of Heat Transfer* (Rohsenhow and Hartnett, 1972).

Figures 2 and **3** show Eu vs. Re (based on $\bar{u}_{max}$) for in-line square and equilateral triangle tube banks. The *Heat Exchanger Design Handbook* (1983) contains analytical expressions approximating these curves. These take the form of a power series,

$$Eu = \sum_{i=0}^{4} \frac{c_i}{Re^i} \quad \textbf{(7)}$$

for all in-line and rotated square banks, except in-line banks with $b = 2.5$, for which,

$$Eu = \sum_{i=0}^{4} c_i Re^i \quad \textbf{(8)}$$

Values of the coefficients, c_j for in-line and staggered tube banks are given in **Tables 1** and **2**, respectively. The reader is cautioned that these equations render poor continuity across certain ranges of application. Other mathematical correlations also exist, for example, ESDU (1979). Agreement between ESDU/Delaware, and the Lithuanian group's curves is not particularly good, especially in the low-intermediate Re range (Beale, 1993).

In most practical tube banks it is necessary to modify Eu for several effects. This usually done as follows,

$$Eu = k_1 k_2 k_3 \ldots k_n Eu' \quad \textbf{(9)}$$

where Eu' is the value calculated from the correlation for an ideal bank, and the k_i are correction factors. Corrections are typically required to account for the geometry, size, and location of the tube bank, deviations from normal incidence of the working fluid, and variations in the fluid properties due to temperature and pressure changes. These are detailed below.

Influence of pitch ratio, a/b ≠ 1

This may be accounted for by multiplying Eu' by k_i as is shown in the insets to **Figures 2** and **3**. Mathematical expressions for k_1 may be found in the *Heat Exchanger Design Handbook* (1983).

Influence of temperature on fluid properties

For non-isothermal conditions it is necessary to account for variations in viscosity across the boundary layer. Many authors achieve this by setting,

$$k_2 = \left(\frac{\eta_w}{\eta}\right)^p \quad \textbf{(10)}$$

where η_w is the viscosity at T_w and η is the viscosity at the mean bulk temperature, T_M, of the tube bank. Žukauskas and Ulinskas (*Heat Exchanger Design Handbook,* 1983) recommend,

$$p = 0.776 \exp(-0.545\, Re^{0.256}) \qquad \eta_w > \eta$$

$$p = 0.968 \exp(-1.076\, Re^{0.196}) \qquad \eta_w < \eta, \quad Re < 10^3 \quad \textbf{(11)}$$

For many applications, the temperature effects on ρ, μ etc. are negligible, and it is sufficient to calculate these at the arithmetic mean of the inlet and exit bulk temperatures (if known). However, if the temperature change in the bank is such as to affect the fluid properties significantly, it is necessary to compute Re and Eu on a row-by-row basis.

Entry-length effects

If the number of rows, N_{row}, in the stream-wise direction are sufficiently small, it is necessary to modify Eu for *entrance-length effects.* **Figure 4** shows the entrance-length factor k_3, as a function of N_{row}.

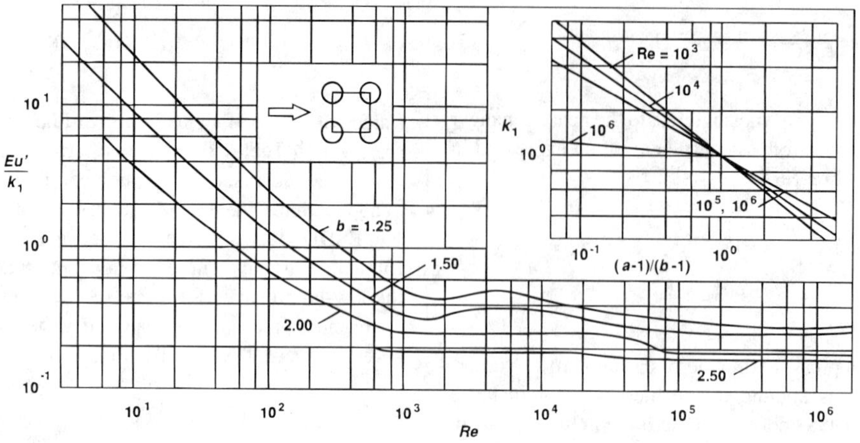

Figure 2. Pressure drop coefficient vs. Reynolds number for in-line tube banks. From *Heat Exchanger Design Handbook* (1983).

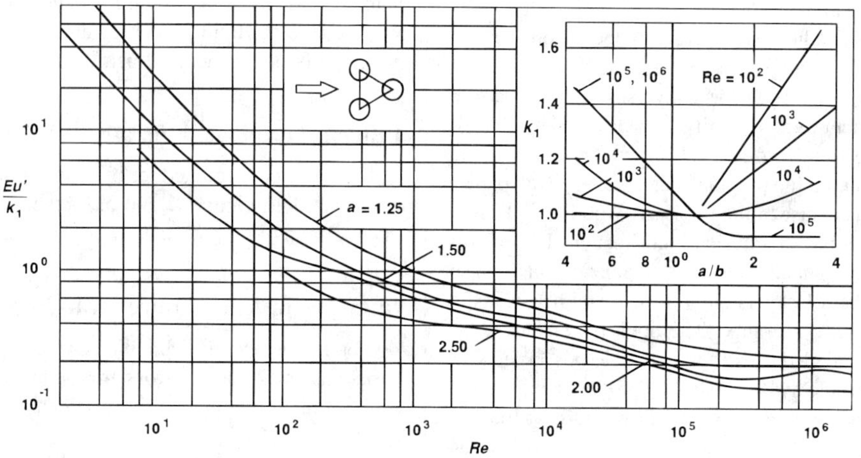

Figure 3. Pressure drop coefficient vs. Reynolds number for staggered tube banks. From *Heat Exchanger Design Handbook* (1983).

Table 1. Coefficients, c_i, for use in Equations 7 and 8 to generate pressure drop coefficients for in-line square banks. From *Heat Exchanger Design Handbook* (1983)

a = b	Re Range	c_0	c_1	c_2	c_3	c_4
1.25	$3–2 \times 10^3$	0.272	0.207×10^3	0.102×10^3	-0.286×10^3	–
1.25	$2 \times 10^3–2 \times 10^6$	0.267	0.249×10^4	-0.927×10^7	0.10×10^{11}	–
1.5	$3–2 \times 10^3$	0.263	0.867×10^2	-0.202×10^0	–	–
1.5	$2 \times 10^3–2 \times 10^6$	0.235	0.197×10^4	-0.124×10^8	0.312×10^{11}	-0.274×10^{14}
2	7–800	0.188	0.566×10^2	-0.646×10^3	0.601×10^4	-0.183×10^5
2	$800–2 \times 10^6$	0.247	−0.595	0.15	−0.137	0.396

Deviations from normal incidence

The flow may deviate from pure crossflow in either of two ways.

- The angle of attack, α, may not be 0° ie. the bank may be rotated at some arbitrary angle to the flow.
- There may be a component in the azimuthal (longitudinal) z-direction, $\beta \neq 90°$, sometimes referred to as inclined crossflow.

Rotated crossflow, $\alpha \neq 0°$

Butterworth (1978) used the analogy of fluid flow in porous media and noted the mean pressure drop to be nominally the same

Table 2. Coefficients, c_i, for use in Equation 7 to generate pressure drop coefficients for equilateral triangle banks. From *Heat Exchanger Design Handbook* (1983)

b	Re Range	c_0	c_1	c_2	c_3	c_4
1.25	$3–10^3$	0.795	0.247×10^3	0.335×10^3	-0.155×10^4	0.241×10^4
1.25	$10^3–2 \times 10^6$	0.245	0.339×10^4	-0.984×10^7	0.132×10^{11}	-0.599×10^{13}
1.5	$3–10^3$	0.683	0.111×10^3	-0.973×10^2	-0.426×10^3	0.574×10^3
1.5	$10^3–2 \times 10^6$	0.203	0.248×10^4	-0.758×10^7	0.104×10^{11}	-0.482×10^{13}
2	$7–10^2$	0.713	0.448×10^2	-0.126×10^3	-0.582×10^3	–
2	$10^2–10^4$	0.343	0.303×10^3	-0.717×10^5	0.88×10^7	-0.38×10^9
2	$10^4–10^6$	0.162	0.181×10^4	0.792×10^8	-0.165×10^{13}	0.872×10^{16}
2.5	$10^2–5 \times 10^3$	0.33	0.989×10^2	-0.148×10^5	0.192×10^7	-0.862×10^8
2.5	$5 \times 10^3–2 \times 10^6$	0.119	0.498×10^4	-0.507×10^8	0.251×10^{12}	-0.463×10^{15}

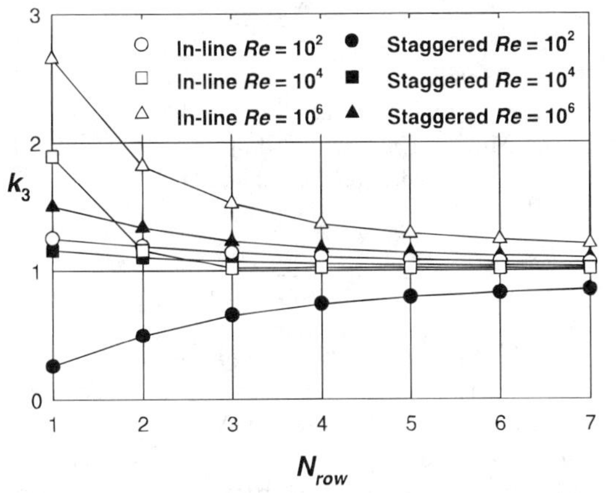

Figure 4. Influence of finite-number of rows on overall pressure drop in tube banks. Adapted from *Heat Exchanger Design Handbook* (1983).

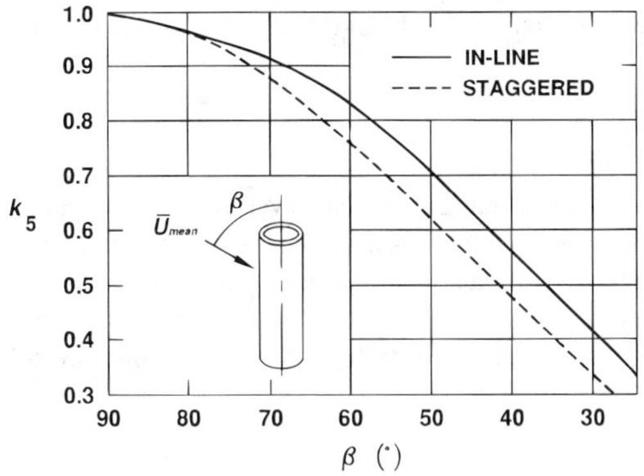

Figure 6. Effect of angle of attack, β, on overall pressure drop for inclined crossflow in tube banks. From Žukauskas et al. (1988).

$$\bar{U}_{mean} = \left|\frac{1}{V}\int_V \vec{u}\,dV\right| \qquad (12)$$

for *in-line* and *rotated tube banks* with a/b ≈ 1, concluding the overall pressure gradient to be independent of a. Thus the mean pressure drop may be calculated using the normal incidence correlations. The physical significance of the reference velocity $\bar{u}_{max}$ is vague; however, it may still be may still be calculated from the mean superficial velocity $\bar{U}_{mean}$ using **Equation 4 or 5**. $\bar{U}_{mean}$ should be regarded as a local volume-averaged quantity (see **Figure 5**) where the volume V includes the space occupied by the

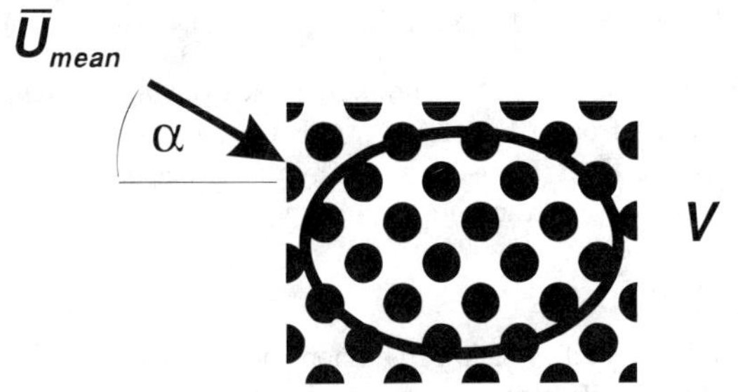

Figure 5. Rotated tube bank showing angle of attack, α, with respect to mean superficial velocity.

cylinders. Since Re and Eu are based on velocity and pressure gradients in the α-direction, the normal component of the pressure drop is reduced by a factor,

$$k_4 \cos \alpha \qquad (13)$$

It being understood that there is also a pressure gradient in the crosswise-direction. For a/b ≉ 1 anisotropy may be a problem, but the α-direction flow-resistance can still be estimated from the normal-incidence, α = 0°, 90°, values.

Inclined crossflow, β ≠ 0°

For this case the flow resistance is anisotropic, i.e. $\vec{\nabla}\bar{p}$ is not in the β-direction, due to the axial (longitudinal) drag being less than the cross-wise component. **Figure 6** shows k_5 vs. β, where k_5 is the ratio of the normalised streamwise component of the pressure drop to the value if the same mass flow had been in pure crossflow.

Rough Surfaces

The use of roughened and other enhanced surfaces to increase heat transfer is widespread. In addition to being deliberately employed, surface roughness tends to naturally increase when smooth tubes become fouled. When the mean height of the roughness elements, k, is sufficient, turbulence is enhanced. It is generally maintained that the influence of surface roughness is more significant in staggered banks than in-line banks. The *Heat Exchanger Design Handbook* (1983) contains some guidelines for calculating k_5 as a function of k/D, for staggered tube banks.

Finned tubes

Banks employing *finned tubes* are often found. **Figure 7** is a schematic showing some of the more common arrangements, with examples of circular, spiral, axial and plate-fins. It is common to distinguish between so-called low-finned and high-finned tubes, according to the fin-height to diameter ratio. Fins may be straight or tapered. Correlations for Eu vs. Re, or equivalent, may be found in *The Heat Exchanger Design Handbook* (1983), Žukauskas et. al, (1988), Stasiulevičius and Skrinska (1988), ESDU (1984, 1986) and elsewhere.

Bypassing

Tube banks seldom extend to the walls of the containing vessel. Also in staggered tube banks, alternate cylinders are missing near a wall. (Experimentalists may employ dummy half-tubes or corbels to eliminate this effect.) Near the wall the flow is faster, due to decreased resistance. An iterative procedure, similar to Wills and Johnson's method, in shell-and-tube heat exchanger analysis (Hewitt et al., 1994) is recommended. Let the subscripts "bank" and "bypass" refer to the main and bypass flow lanes, as illustrated in **Figure 8**, and suppose that,

$$\Delta p = \Delta p_{bypass} = \Delta p_{bank} \tag{14}$$

The fractional mass flow through the bank, F_{bank} is,

$$F_{bank} = \frac{\dot{M}_{bank}}{\dot{M}_{total}} = \left(\frac{1}{1 + 2\dfrac{n_{bank}}{n_{bypass}}} \right)^{1/2} \tag{15}$$

where,

$$n = \frac{1}{2}\frac{N_{row}}{\rho A^2} Eu \tag{16}$$

The procedure is iterative because Eu_{bank} and Eu_{bypass} are functions of $\dot{M}_{bank}$ and $\dot{M}_{bypass}$. The pressure drop may be calculated from either n_{bank}, n_{bypass}, or the average value, $\bar{n}$,

$$\Delta p = \bar{n}\dot{M}^2_{total} = \frac{n_{bank} n_{bypass}}{(2\sqrt{n_{bank}} + \sqrt{n_{bypass}})^2} \dot{M}^2_{total} \tag{17}$$

Eu_{bank} is obtained using **Equation 9**, **Figures 2** or **3**, in the usual fashion. Bypass pressure drop correlations are uncommon, and often proprietary (but see ESDU, 1974). As a rough approximation, Eu_{bypass} may be estimated from **Figure 2** assuming effective values of a and b namely, $a = 2s_{bypass}/D$, $b = s_L$ (inline bank) or $a = 2s_{bypass}/D$, $b = 2s_L/D$ (staggered bank). In the latter case the effective number of rows is $N_{row}/2$.)

Equation 14 is based on the premise that pressure variations across the inlet and outlet are insignificant, something which may or may not be the case. For this reason, the engineer should attempt to use data incorporating bypass effects, if available. The lack of such data suggest that more research is needed on this important subject.

The procedure for calculating the overall pressure drop in tube banks is as follows:

1. Calculate Re based on values of u_{max}, ρ, η evaluated at arithmetic mean of inlet and outlet bulk temperatures;
2. Obtain the value of Eu for an ideal bank using **Figures 2** and **3**;
3. Correct Eu for factors k_1, k_2, k_3, k_4, k_5, as discussed above.

If either variations in fluid properties due to changes in temperature obtained from a heat transfer calculation (see **Tube Banks, Single-phase Heat Transfer in**), or bypass effects are significant, the calculation procedure are iterative.

Numerical studies

Numerical studies have gained popularity in recent years. It is convenient to differentiate between detailed calculations of flow within the passages of tube banks and overall performance calculations.

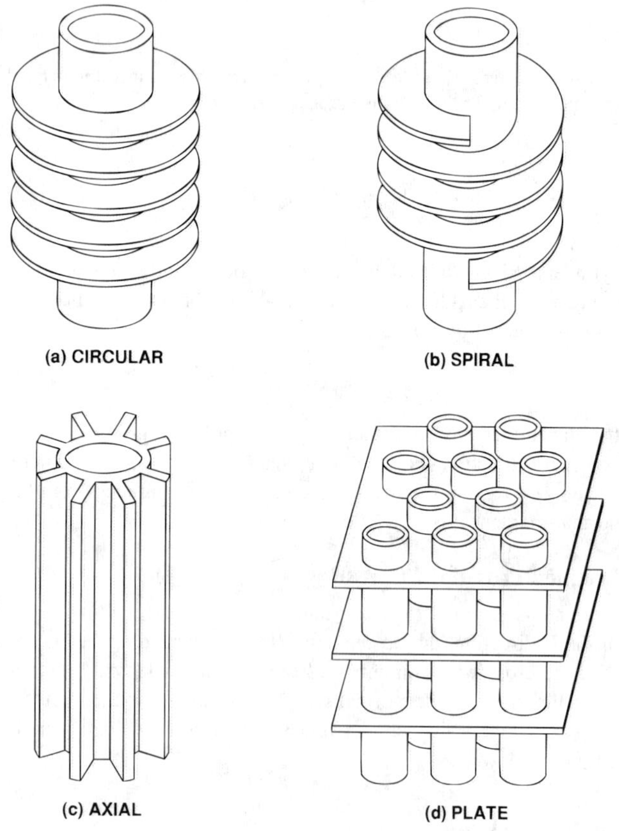

Figure 7. Schematic showing main and bypass streams in a tube bank, together with a network diagram.

Detailed calculations of flow within the passages of tube banks

Many results have now been obtained for tube banks using both finite-volume and finite-element methods. Numerical methods have been used to simulate laminar transient flow, large-eddy turbulent flow, inclined flow $\beta \neq 0°$, and also 3D secondary flow effects in finned tubes. Beale (1993) contains a review of recent work.

Most detailed numerical simulations have been conducted for either laminar or high-speed turbulent regimes. Far less numerical-data are available in the intermediate Re range, in which most heat exchangers operate, where the influence of free-stream turbulence is of paramount importance. Because of the complex nature of flow within the passages of tube banks, and the inadequacies of existing turbulence models, numerical experiments have not yet, and probably never will, render laboratory work obselete. The two activities are not, however, mutually exclusive, and many of the problems described above are readily amenable to the methods of computational fluid dynamics.

Overall performance calculations

These methods range from simple automatations of earlier methods to detailed three-dimensional flow calculations using the techniques of **Computational Fluid Dynamics**. Overall performance predictions are still based on empirically-based correlations of pressure drop (**Equations 7 and 8**) and heat transfer. However, these are embedded in a computer code which is used to predict the overall performance of the heat exchanger as a whole. Important effects such as entrance phenomena, bypassing, variable properties etc. can thus readily be accommodated. In modern heat-exchanger design, computer-based methods have already supplanted hand-calculation techniques to some extent, a trend which will doubtless continue in the future. The challenge is to incorporate the physics and engineering experience into future heat exchanger design software.

References

Beale, S. B. (1993) Fluid Flow and Heat Transfer in Tube Banks, PhD Thesis, University of London.

Butterworth, D. (1978) The development of a model for three-dimensional flow in tube bundles, *Int. J. Heat Mass Trans.*, 21, 253–256.

Bergelin, O. P., Colburn, A. P., and Hull, H. L. (1950) Heat transfer and pressure drop during viscous flow across unbaffled tube banks, *Engineering Experiment Station Bulletin No. 2,* University of Delaware.

Bergelin, O. P., Leighton, M. D., Lafferty, W. L., and Pigford, R. L. (1958) Heat transfer and pressure drop during viscous and turbulent flow across baffled and unbaffled tube banks, *Engineering Experiment Station Bulletin No. 4,* University of Delaware.

Engineering Sciences Data Unit (1979) Crossflow Pressure Loss over Banks of Plain Tubes in Square and Triangular Arrays including Effects of Flow Direction, ESDU Data Item No. 79034, London.

Engineering Sciences Data Unit (1980) Pressure Loss during Crossflow of Fluids with Heat Transfer over Plain Tube Banks without Baffles, ESDU Data Item No. 74040, London., 1980.

Engineering Sciences Data Unit (1984) Low-fin Staggered Tube Banks: Heat transfer and Pressure Drop for Turbulent Single Phase Cross Flow, ESDU Data Item No. 84016, London.

Engineering Sciences Data Unit (1986) High-fin Staggered Tube Banks: Heat transfer and Pressure Drop for Turbulent Single Phase Gas Flow, ESDU Data Item No. 86022, London.

Heat Exchanger Design Handbook (1983) Vol. 1–4, Hemisphere, New York, 1983

Hewitt. G. H., Shires, G. L., and Bott, T. R. (1994) *Process Heat Transfer,* CRC Press, Boca Raton, FL.

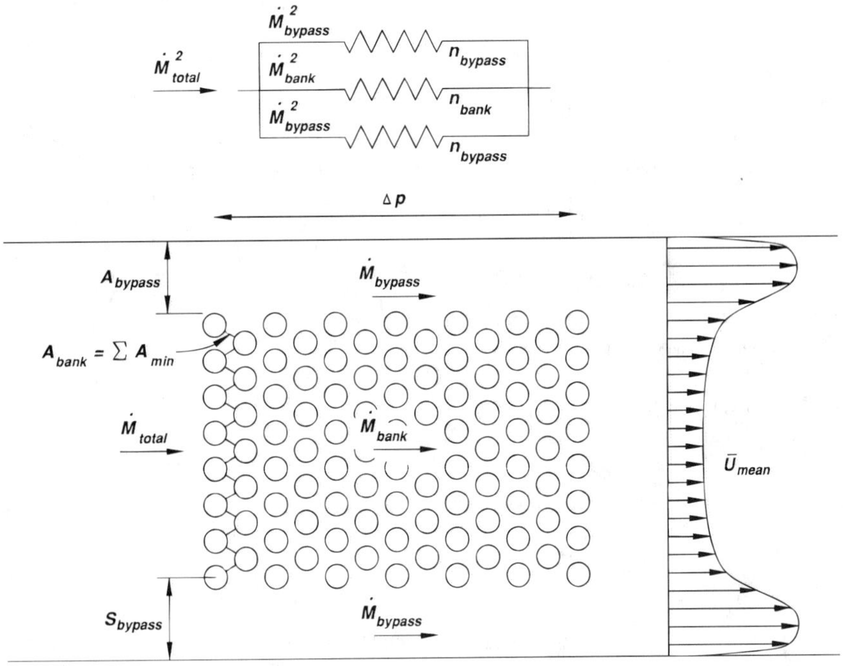

Figure 8. Some typical types of finned tubes.

Kays, W. M. and London, A. L. (1984) *Compact Heat Exchangers,* McGraw-Hill, New York.

Rohsenhow, W and Hartnett, J. P. (eds.), (1973) *Handbook of Heat Transfer,* McGraw-Hill, New York.

Stasiulevičius, J. and Skrinska, A. (1988) *Heat Transfer of Finned Tube Bundles in Crossflow,* A. A. Žukauskas, and G. F. Hewitt eds., Hemisphere, New York.

Žukauskas, A. A. (1987) Heat transfer from tubes in crossflow, *Advances in Heat Transfer,* 18., 87–159, Academic Press.

Žukauskas, A. A., Ulinskas, R. V. and Katinas, V. (1988) *Fluid Dynamics and Flow-induced Vibrations of Tube Banks,* English-edition Editor, J. Karni, Hemisphere, New York.

Leading to: *Tube banks, single-phase heat transfer in*

S.B. Beale

TUBE BANKS, SINGLE-PHASE HEAT TRANSFER IN

Following from: *Crossflow heat transfer; Tube banks, crossflow over*

Tube banks are employed in a wide range of heat exchangers. The average heat flux, $\dot{q}$, is expressed as a linear function of the mean temperature difference between the bulk of the fluid in the tube bank and the wall, according to a rate equation,

$$\dot{q} = \frac{\dot{Q}}{A} = \bar{\alpha}\Delta T_M \quad (1)$$

where $\dot{Q}$ is the rate of heat transfer, $\bar{\alpha}$ is an overall heat transfer coefficient, and ΔT_M is some representative temperature difference between the tube wall and the bulk of the fluid. A is the total heat transfer area, which for plain tubes is given by,

$$A = N\pi DL \quad (2)$$

where N is the number of tubes of outer diameter D, and length L. The overall heat transfer coefficient, $\bar{\alpha}$ is non-dimensionalized in terms of an average **Nusselt Number**, $\overline{Nu}$, according to,

$$\overline{Nu} \equiv \frac{\bar{\alpha}D}{\lambda} \quad (3)$$

where λ is the fluid thermal conductivity, or alternatively as an average **Stanton Number**, $\overline{St}$,

$$\overline{St} \equiv \frac{\bar{\alpha}}{\rho c_p u_m} \quad (4)$$

where c_p is the fluid specific heat.

In tube banks, $\overline{Nu}$ is frequently correlated according to,

$$\overline{Nu} = cRe^m \cdot Pr^n \left(\frac{Pr_w}{Pr}\right)^p \quad (5)$$

where **Reynolds number** Re, is defined by,

$$Re = \frac{\rho u_{max} D}{\eta} \quad (6)$$

u_{max} is the bulk velocity in the minimum cross-section, ρ and η are the fluid density and viscosity respectively. The **Prandtl number**, Pr, is just,

$$Pr = \frac{\eta c_p}{\lambda} \quad (7)$$

where c_p is the specific heat, and λ is thermal conductivity. The rate of heat transfer may also be expressed in terms of an energy balance, which for single-phase heat transfer may be written,

$$\dot{Q} = \dot{C}(T_{out} - T_{in}) \quad (8)$$

where T_{in} and T_{out} are the bulk temperatures at the inlet and outlet of the bank, and $\dot{C}$ is the thermal capacitance of the fluid, presumed constant over the range of interest.

$$\dot{C} = \dot{M}c_p \quad (9)$$

$\dot{M}$ is the total mass flow rate through all of the passages in the tube bank, and c_p is averaged over the range of interest.

A local heat transfer coefficient may be defined in a similar fashion to **Equation 1**. The local Nu is a function of a number of parameters such bank type and geometry, flow Re, pressure gradient, location within bank etc. **Figure 1** shows local Nu distributions in the interior of tube banks as a function of the angle ϕ measured from the front of the cylinder. It can be seen that the staggered-

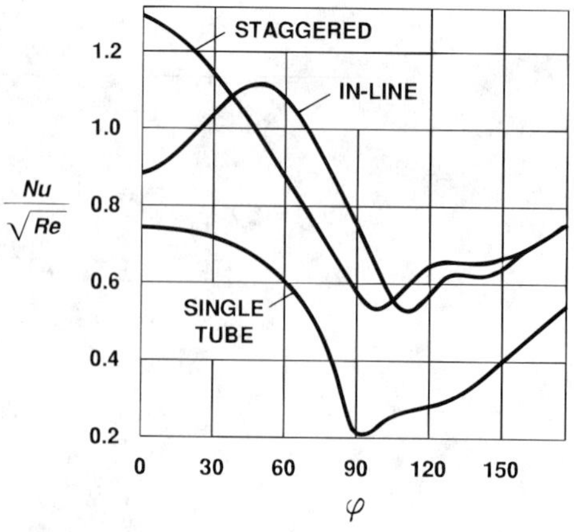

Figure 1. Variation of local heat transfer around 1) a single tube 2) a tube in a staggered bank 3) a tube in an in-line banks. From Žukauskas and Ulinskas (1988).

geometry Nu-distribution is similar to the single cylinder case, with a maximum occurring at $\phi = 0°$. For the in-line case, Nu rises to a maximum at the reattachment point around 45°. It can also be seen that heat transfer is somewhat higher for both in-line and staggered tube banks, than for single cylinders. This is due to increased free-stream turbulence (as a result of preceeding rows) and shear, due to constriction of the flow passages. For most applications the engineer is not concerned with the details of local heat transfer in tube banks and the reader is referred to the reader to Žukauskas and Ulinskas (1988) for further information on the subject.

Experimental Data and Empirical Correlations

Two major sources of data on both fluid flow and heat transfer are the Delaware group in the USA and the group in the Republic of Lithuania. The work at the University of Delaware on overall heat transfer in tube banks has formed the basis for heat exchanger calculations for many years while the recent book by Žukauskas et al. (1988) is a comprehensive reference on heat transfer in tube banks based on many years of research. It contains extensive information on local and overall heat transfer. It appears that the Delaware data was gathered at nominally constant wall temperature, T_w, conditions and that of the Lithuanian group at constant wall heat flux, $\dot{q}_w$. Numerous other data also exist, in addition to these two main sources.

A number of empirical correlations may also be found. The Engineering Science and Data Unit correlate $\bar{N}u$ vs. Re according to **Equation 5** (ESDU, 1978). Values of the coefficients c and m are given in **Table 1**. Re is based on u_{max}. (See **Tube Banks, Cross Flow Over**). The correlation of Žukauskas and Ulinskas (1988) agrees well with experimental data, but exhibits poor continuity across certain Re ranges. Several other correlations may be found in the literature. Agreement among heat transfer correlations is generally better than for pressure drop correlations. In most correlations, $\bar{N}u$ is treated as being primarily a function of Re and configuration (i.e. the pitch-to-diameter ratios are often treated as being relatively unimportant over a wide range). Similarly, the effect of thermal boundary conditions (constant T_w vs. constant $\dot{q}_w$), though often significant, is usually simply ignored.

Tradition is to plot $\bar{N}u$ or $\bar{S}t$ in the form of log-log plots with both the Pr-dependence and the property-variation being removed by defining,

Table 1. Coefficients used to calculate overall heat transfer in tube banks using Equation 5 with n = 0.34, p = 0.26. From ESDU International plc, 1973

	In-line $a \geq 1.15, 1.2 \leq b \leq 4$		Staggered $a \geq 1.15, 1.2 \leq b \leq 4$	
Re Range	c	m	c	m
$10 - 3 \times 10^2$	0.742	0.431	1.309	0.360
$3 \times 10^2 - 2 \times 10^5$	0.211	0.651	0.273	0.635
$2 \times 10^5 - 2 \times 10^6$	0.116	0.700	0.124	0.700

$$j' = j\left(\frac{Pr_w}{Pr}\right)^{-p} = StPr^{1-n}\left(\frac{Pr_w}{Pr}\right)^{-p} \quad \textbf{(10)}$$

$$k' = k\left(\frac{Pr_w}{Pr}\right)^{-p} = NuPr^{-n}\left(\frac{Pr_w}{Pr}\right)^{-p} \quad \textbf{(11)}$$

The heat-transfer factors j, k, (un-primed) are Pr-independent. Primed values j′, k′ indicate that the influence of *property variations* across the boundary layer has also been taken into consideration. The Delaware group chose n = 1/3, ie. j is the so-called *Colburn heat transfer factor.* ESDU (1973) and Žukauskas and Ulinskas (1988) propose similar, but slightly higher, values n = 0.34 and 0.36, respectively (over most of the range). **Figure 2** shows j and k as a function of Re for Pr = 1. It can be seen that heat transfer is a little higher for staggered than in-line geometries at lower Re.

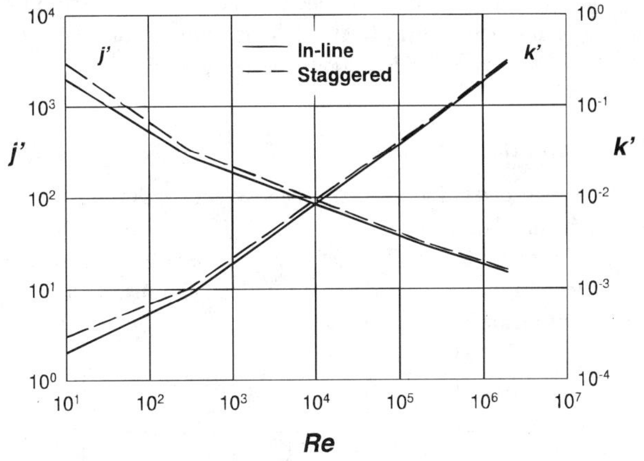

Figure 2. Average heat transfer for in-line and staggered tube banks. From ESDU (1973).

The influence of fluid properties is accounted for in Equation 5 by the term $(Pr_w/Pr)^p$, where Pr is evaluated at the mean bulk temperature, and Pr_w at T_w. The Delaware group use $(\eta_w/\eta)^p$, the matter is largely one of preference (but their value of p = 0.14 was based on the *Seider-Tate correlation* for flow inside tubes). ESDU (1973) suggest p = 0.26 regardless of whether the fluid is being heated or cooled.

The choice of reference temperature is a matter for concern since in most situations, the engineer must use a mean value for thermal design/analysis. For constant T_w, the appropriate temperature to use in Equation 1 is the *log-mean temperature difference,*

$$\Delta T_M = \Delta T_{LM} \equiv \frac{\Delta T_{out} - \Delta T_{in}}{\ln\left(\frac{\Delta T_{in}}{\Delta T_{out}}\right)} \quad \textbf{(12)}$$

where, $\Delta T_{in} = T_{in} - T_w$ and $\Delta T_{out} = T_{out} - T_w$ are the differences between the fluid and wall temperatures at the inlet and outlet, respectively. (See also **Mean Temperature Difference**) The reference temperature, T_M, for ennumerating property values should be $T_w + \Delta T_{LM}$. For constant $\dot{q}_w$, the arithmetic-mean temperature should be used,

$$T_M = \bar{T} \equiv \frac{T_{out} + T_{in}}{2} \quad (13)$$

For most applications it is immaterial whether properties, ρ, η, etc. are evaluated at the log-mean or the arithmetic-mean temperature, and the latter is often employed for convenience. For large numbers of banks, and low values of the product Re.Pr the choice of T_M is quite important.

Empirical correlations are obtained from data for idealised tube banks. In practice, $\overline{Nu}$ is a function of a number of other parameters which are often accounted for by writing,

$$\overline{Nu} = k_1 k_2 \overline{Nu}' \quad (14)$$

where $\overline{Nu}'$ is the idealised-correlation-based Nusselt number, and the k-coefficients account for deviations from this situation.

One example, already mentioned, is the factor $(Pr_w/Pr)^p$. Other factors could include the effects of finite numbers of rows, angle of attack, or other considerations. These are discussed below.

Finite number of rows

Idealised correlations, described above, are for deep tube banks, where the number of rows, N_{row}, is large. For each of the first few rows in a tube bank, heat transfer may be substantially different (usually less) than occurs deep in the bank. **Figure 3** shows the correction factor k_1 as a function of N_{row}.

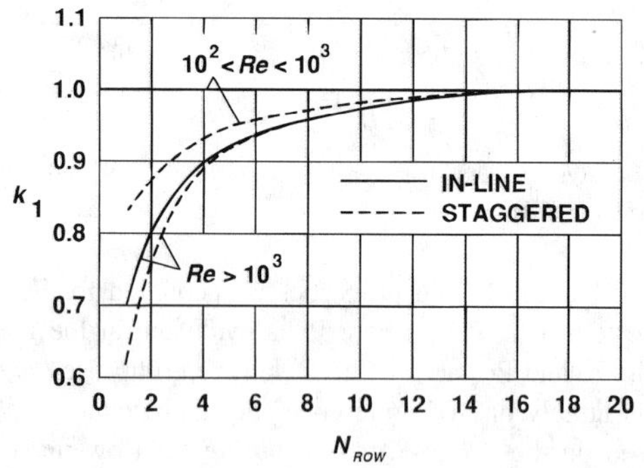

Figure 3. Influence of number of rows on overall heat transfer in tube banks. From Žukauskas and Ulinskas (1988).

Inclined crossflow

In many situations, the flow is not one of pure crossflow, ie. $\beta < 90°$. Heat transfer is reduced from by a factor, k_2. **Figure 4** shows k_2 vs. β.

Two commonly-employed methods of enhancing heat transfer in tube banks are the use of rough surfaces and finned tubes. (See **Augmentation of Heat Transfer**).

Rough tubes

Rough tubes are used to increase the ratio of heat transfer to drag by increasing turbulence. The effects appear to be more pronounced for staggered than in-line geometries. Žukauskas and

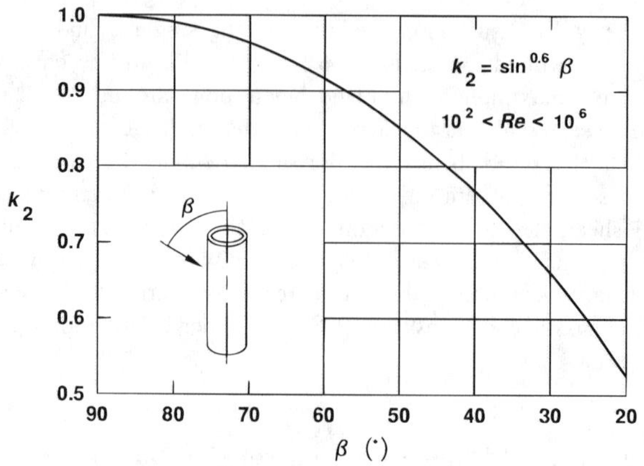

Figure 4. Influence of angle of inclination, β, on overall heat transfer for inclined crossflow in tube banks. From ESDU (1973).

Ulinskas, (1988) propose the following prescription for staggered geometries,

$$Nu = 0.5\left(\frac{a}{b}\right)^{0.2} Re^{0.65} Pr^{0.36}\left(\frac{Pr}{Pr_w}\right)^{0.25}\left(\frac{k}{d}\right)^{0.1} \quad 10^3 \leq Re \leq 10^5$$

$$Nu = 0.1\left(\frac{a}{b}\right)^{0.2} Re^{0.8} Pr^{0.4}\left(\frac{Pr}{Pr_w}\right)^{0.25}\left(\frac{k}{d}\right)^{0.15} \quad 10^5 \leq Re \leq 2 \times 10^6 \quad (15)$$

where k is the roughness height.

Finned tubes

External fins may be employed with gas-crossflow, in order to increase the heat-transfer area (internal fins are also used sometimes). The book by Stasiulevičius and Skrinska (1988) is devoted to the subject, as is a chapter in Žukauskas et al. (1988). When using *finned tubes* in a design, the engineer must use $\overline{Nu}$-correlations appropriate to finned tubes, and properly account for temperature variations throughout the fin. (See also **Extended Surface Heat Transfer**)

Nu correlations for finned-tube banks

A number of general purpose correlations for finned tube-banks may be found:

> Gnielinski et al. in the *Heat Exchanger Design Handbook* (1983), ESDU (1984, 1986), Stasiulevičius and Skrinska (1988), and Žukauskas and Ulinskas (1988)

These typically assume forms similar to **Equation 5**, but with additional parameters to account for the fin geometry. The engineer should obtain $\overline{Nu}$ correlations specific to the particular geometrical configuration under consideration, whenever possible.

Fin conduction

When variation in temperature across the fins is significant, the rate equation, **Equation 1**, is re-written as,

$$\dot{Q} = E_f A \bar{\alpha} \Delta T_M \quad (16)$$

A is the total area for heat transfer, $A = A_f + A_w$, where A_f is the total area of the fins and A_w that of the exposed tube-wall. E_f is a surface effectiveness factor,

$$E_f = \frac{\eta_f A_f + A_w}{A} \quad (17)$$

and η_f is a *fin efficiency,* i.e. the ratio of the actual rate of heat transfer through the fins to that which would occur if all fin material were at constant T_w.

Figure 5 is a plot of η_f vs. non-dimensional height h*, at various diameter ratios, D_f'/D. λ_f is the thermal conductivity of the fin, δ the fin thickness and $\bar{\alpha}$ the average heat transfer coefficient. Primed values, h′ and D′, indicate that they are adjusted for heat transfer through the fin-tip, as indicated in **Figure 5**. Charts of η_f vs. h* have been devised for a variety of geometries; the article by Gnielinski et al. in the *Heat Exchanger Design Handbook* (1983) contains a selection. These should be used in preference to **Figure 5**, which was obtained from the one dimensional fin equation solution,

$$\eta = \frac{\tanh h^*}{h^*} \quad (18)$$

where h′ is calculated according to Schmidt (Stasiulevičius and Skrinska, 1988).

Bypass effects

Bypass streams are relatively ineffective in transferring heat, and are usually treated as adiabatic, i.e. only the main flow stream is used to compute heat transfer. Since the temperature and hence the viscosity differs from that in the main flow lanes, this must be taken into consideration when computing the bypass and main mass-flow rates.

Methods for Calculating the Thermal Performance of Tube Banks

How to compute the performance of a tube bank will depend to a larger extent on which of the parameters, $\dot{Q}$, $\bar{\alpha}$, A, T_{in}, T_{out}, $\dot{C}$ etc., are prescribed or required, whether fluid properties are constant, and the precision required. It is common to differentiate between "rating" and "sizing" of heat exchangers. This section is intended to provide some guidelines for the reader, not a single general-purpose procedure. Two cases will be considered in detail: Plain tubes at constant T_w and non-isothermal plain or finned tubes:

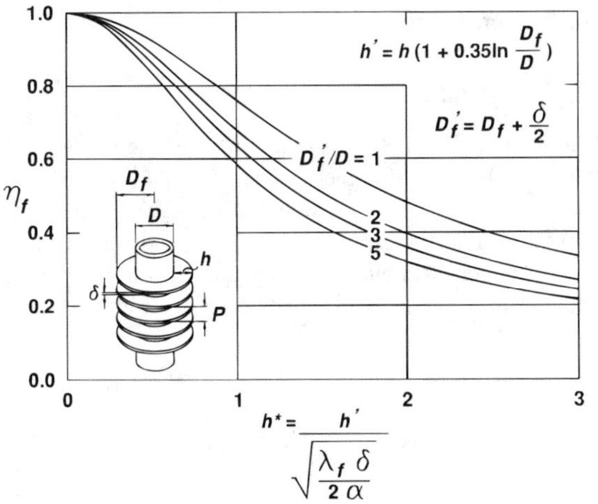

Figure 5. Fin Efficiency.

Plain tubes, constant T_w

The log-mean temperature-difference (LMTD) approach is based on the fundamental assumption that the heat flux is directly proportional to the temperature difference between the bulk of the fluid at some interior location and the wall, ie. the rate equation, **Equation 1**, with ΔT_M prescribed according to **Equation 12**.

Unless all temperatures are known, a priori, a trial-and-error method based on successive applications of the rate equation combined with a heat balance is required. An alternative approach, due to Nusselt, is to postulate the rate of heat transfer to be proportional to the difference between the inlet and wall temperatures,

$$\dot{Q} = E\dot{C}(T_w - T_{in}) \quad (19)$$

The *effectiveness,* E, represents the ratio of the actual to the maximum possible heat transfer. For constant T_w,

$$E = 1 - e^{-A\bar{\alpha}/\dot{C}} \quad (20)$$

where the quantity $A\bar{\alpha}/\dot{C}$ is known as the **Number of Transfer Units**, NTU. The advantage of T_{in} as reference (if known) is that no iteration is required provided fluid properties are constant and upstream values can be used. Even if these vary, the E-NTU-method is often preferable, as T_M affects the calculation of $\bar{\alpha}$ via property variations only. A typical procedure, is as follows,

1. Given T_{in} guess a value for T_{out} (unless known).
2. Calculate Re and Pr based on property values evaluated at the arithmetic-mean bulk temperature, and Pr_w at T_w.
3. Compute the value of $\overline{Nu}'$ for an ideal bank using **Equation 5** applying corrections, k_1, k_2 etc. as discussed above, to obtain $\overline{Nu}$ and hence $\bar{\alpha}$ for the actual bank **Equation 14**.
4. Compute the overall rate of heat transfer, $\dot{Q}$, using one of two methods: Either calculate ΔT_{LM} from **Equation 12**, and use the rate equation 1 to obtain $\dot{Q}$, or calculate E from $\bar{\alpha}$ using **Equation 20** and obtain $\dot{Q}$ using **Equation 19**.
5. Compute the exit bulk temperature, T_{out} from $\dot{Q}$, by means of a heat balance, **Equation 8**.
6. Re-iterate steps 2–6 based on the new value of T_{out} (unless prescribed) until satisfactory convergence is obtained.

If changes in temperature are large, break the bank up into a individual or groups of rows, and proceed in a sequential fashion from the inlet to the exit.

Plain or finned tubes, general case

In many practical engineering applications, the tube-side fluid also undergoes significant changes in temperature. Under these circumstances, **Equation 1** is no longer appropriate, and the rate of heat transfer is,

$$\dot{Q} = UA\Delta T_M \tag{21}$$

where U, the **Overall Heat Transfer Coefficient** based on the outer surface A (including fins) is given by,

$$\frac{1}{U} = \frac{1}{(E_f\alpha)_{bank}} + \left(\frac{R_f}{E_f}\right)_{bank} + R_w + \frac{A}{A_i}\left(\frac{R_f}{E_f}\right)_{tube} + \frac{A}{A_i}\frac{1}{(E_f\alpha)_{tube}} \tag{22}$$

where the subscript "bank" refers to the external bank-side crossflow, and the subscript "tube" refers to the internal tube-side longitudinal flow. The terms on the right-hand side of **Equation 22** may be regarded as resistances to the flow of heat in the presence of a temperature difference. A_i is the total area for heat transfer on the inner tube surface, $E_{f\ bank}$ and $E_{f\ tube}$ are fin-surface-effectiveness correction factors for externally and internally-finned tube surfaces (equal to unity for plain tubes), and $R_{f\ bank}$ and $R_{f\ tube}$ are fouling factors which may be estimated using the recommendations in the Standards of the **Tubular Exchanger Manufacturers Association** (TEMA, 1988). R_w is the tube-wall resistance which for plain tubes, of length L, is given by,

$$R_w = \frac{1}{2\pi\lambda_w L}\ln\frac{D}{D_i} \tag{23}$$

where D_i is the inner diameter, and λ_w is the wall thermal conductivity. For finned tubes, R_w is configuration-dependent.

Commonly-employed heat exchanger design methods are the LMTD-based F-correction-factor method, the *P-NTU* method and the θ method. For the F-correction method, ΔT_{LM}, is defined by **Equation 12**, but with ΔT_{in} and ΔT_{out} the temperature differences between the two working fluids at the inlet and outlet. Heat transfer is reduced by a factor F, which is a function of the capacitance ratio, R, and the thermal effectiveness, P, as defined in **Table 2**, which also gives with the modified rate equation used to calculate $\dot{Q}$. The *P-NTU* method, presumes the thermal effectiveness, P, (similar to E) to be a function of R, as well as the number of transfer units, $NTU = AU/C_{cold}$, of the cold fluid. For the third method, $\theta = \theta(R, NTU_{cold})$ but a modified energy balance, instead of a modified rate equation (see **Table 2**) is used to compute $\dot{Q}$. Charts of F, P, and θ may be found in standard references on heat exchangers, often in combined form (see Taborek, *Heat Exchanger Design Handbook,* 1983). Regardless of which method is used, two $\bar{Nu}$ correlations are needed to calulate U: One, as per **Equation 5**, to calculate the bank-side crossflow heat transfer coeffiecient, $\bar{\alpha}_{bank}$, the other to calculate the internal tube-side heat transfer coefficient, $\bar{\alpha}_{tube}$, (See **Tubes Single Phase Heat Transfer in**). The temperatures at the solid-fluid interfaces, $T_{w\ bank}$ and $T_{w\ tube}$, are obtained as a linear combination of $T_{M\ bank}$ and $T_{M\ tube}$, using the ratio of the resistances in **Equation 22** as weighting factors. T_w may be taken as being the temperature between the fluid and the fouling resistance. A typical calculation would then proceed as follows.

1. Given $T_{in\ bank}$ and $T_{in\ tube}$, guess values for $T_{out\ bank}$, $T_{out\ tube}$, $T_{w\ bank}$, $T_{w\ tube}$ (unless known).
2. Calculate Re and Pr for tube-side and bank-side fluids at the arithmetic-mean bulk temperatures, and values of Pr_w for the two fluids based on $T_{w\ bank}$ and $T_{w\ tube}$.
3. Compute $\bar{Nu}_{bank}$ and $\bar{Nu}_{tube}$ using the appropriate correlations applying corrections k_1, k_2 as necessary, to obtain $\bar{\alpha}_{bank}$ and $\bar{\alpha}_{tube}$. Calculate the overall heat transfer coefficient, U, **Equation 22**
4. Compute the overall rate of heat transfer, $\dot{Q}$: Either calculate P and R based on the guessed values and obtain F(P,R) or compute R and NTU_c to obtain $P(R,NTU_c)$ or $\theta(R,NTU_c)$ from a chart. Calculate $\dot{Q}$ from the formulae in Table II.
5. Compute $T_{out\ bank}$ and $T_{out\ tube}$ from $\dot{Q}$, by means of heat balances applied to both tube-side and bank-side fluids **Equation 8**. Hence calculate $T_{w\ bank}$ and $T_{w\ tube}$ (a single value, T_w, will often suffice).
6. Re-iterate steps 2–5 based on the new values of $T_{out\ bank}$, $T_{out\ tube}$ (if necessary) until satisfactory convergence is obtained.

The choice of whether to use the F-correction, *P-NTU,* or θ methods will depend on the application and may result in substantial simplifications to the general procedure detailed above. If all temperatures are known, the F-correction method can be used with properties enumerated at T_M. Conversely, if fluid property variations are negligible and upstream values may be used, the *P-NTU* and θ methods can be used advantageously. If property variations are significant, but small enough to be considered perturbations about a mean, all the above lumped-parameter-type schemes will necessarily be iterative. For large variations in T_M, a numerical scheme is preferred.

Table 2. Design methods commonly used to calculate thermal performance in heat exchangers

Dependent variable	Independent variables	Rate of heat transfer
F	$R = \frac{T_{in\ hot} - T_{out\ hot}}{T_{out\ cold} - T_{in\ cold}}$	$\dot{Q} = UAF\Delta T_{LM}$
	$P = \frac{T_{out\ cold} - T_{in\ cold}}{T_{in\ hot} - T_{in\ cold}}$	$\dot{Q} = PC_{cold}(T_{in\ hot} - T_{in\ cold})$
P	$R, NTU_{cold} = \frac{AU}{C_{cold}}$	
θ	R, NTU_{cold}	$\dot{Q} = UA\theta(T_{in\ hot} - T_{in\ cold})$

Numerical Schemes

While traditional methods are meritorious, the majority of heat exchangers are now designed using computer-software. Many approaches are possible; for example the *E-NTU* method may be performed over a discrete number of cells corresponding to one or more cylinders. An alternative procedure is to start from first principles, and solve pairs of equations of the form,

$$\dot{C}_{bank}\Delta T_{bank} = U\Delta A(T_{tube} - T_{bank}) \tag{24}$$

$$\dot{C}_{tube}\Delta T_{tube} = -U\Delta A(T_{tube} - T_{bank})$$

on a cell-by-cell basis. If an upwind-difference scheme is used,

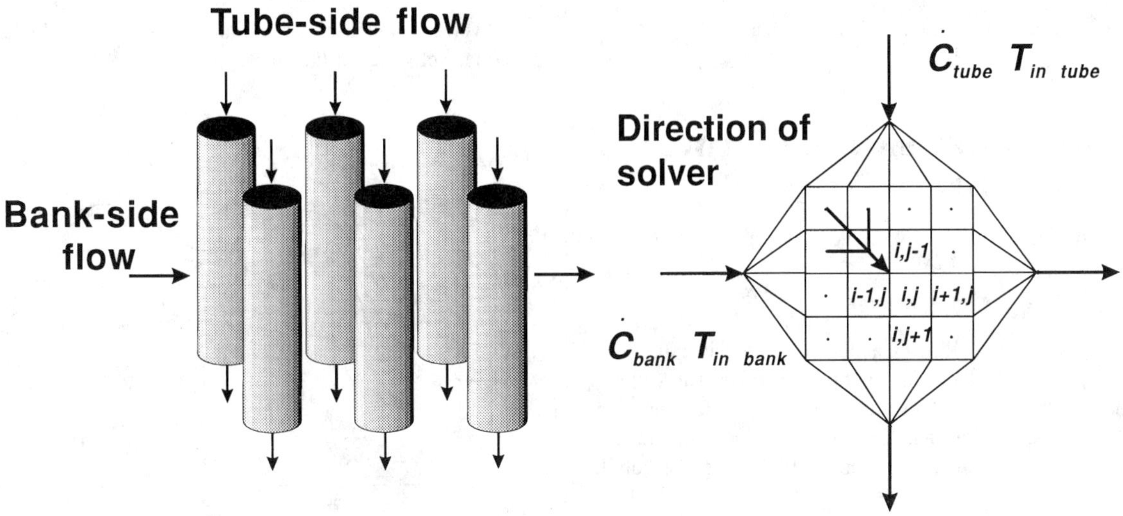

Figure 6. Discretized version of tube-bank heat exchanger.

$$\Delta T_{bank}(i, j) = (T_{bank}(i, j) - T_{bank}(i-1, j)) \quad \mathbf{(25)}$$

$$\Delta T_{tube}(i, j) = (T_{tube}(i, j) - T_{tube}(i, j-1))$$

and upstream values are used to ennumerate properties, no iteration is required. Better accuracy can be achieved with a central-difference scheme,

$$\Delta T_{bank}(i, j) = \frac{1}{2}(T_{bank}(i+1, j) - T_{bank}(i-1, j))$$

$$\Delta T_{tube}(i, j) = \frac{1}{2}(T_{tube}(i, j+1) - T_{tube}(i, j-1)) \quad \mathbf{(26)}$$

Some iteration is now necessary, though by proceeding as indicated in **Figure 6**, a simple Jacobi point-by-point method will converge rapidly. It is now possible to account for variations of U due to changes in T_b on a cell-by-cell basis. T_w may also be obtained, at an intermediate calculation in each cell to enumerate the dependance of U on Pr_w. Variations in other properties may be dealt with on a cell-by-cell basis as necessary, depending on the nature of the problem.

The rate equations defined **Equation 24** may be considered like a set of simple enthalpy-conservation equation (convection-source-term equations). It is possible to devise much more sophisticated two-phase enthalpy equations, which account for bank-side mixing, 3D phenomena etc., in addition to bulk convection. It is also possible to generate momentum equations to account for flow-related effects, bypassing, finite-number of rows, etc. For complex flows, the definition of U may be of limited use. Numerical methods differ from traditional methods in that there is no need to rely on the premise that heat transfer is governed by a rate equation. The reader is referred to the articles by D. B. Spalding in the *Heat Exchanger Design Handbook* (1983), and elsewhere. The use of spreadsheet programs, specialised heat-exchanger-design programs and general-purpose computational fluid dynamics software should be considered prior to writing source code.

Simple lumped-parameter type schemes once formed the basis for all practical heat exchanger design. The situation is changing, and the application of computer software, is not only increasing in use and range of application, but is also enhancing our understanding of the physics of flow within the passages of heat exchangers.

References

Engineering Sciences Data Unit (1973) Convective Heat Transfer During Crossflow of Fluids Over Plain Tube Banks, ESDU Data Item No. 73031, London, November.

Engineering Sciences Data Unit (1984) Low-fin Staggered Tube Banks: Heat Transfer and Pressure Drop for Turbulent Single Phase Cross Flow, ESDU Data Item No. 84016, London.

Engineering Sciences Data Unit (1986) High-fin Staggered Tube Banks: Heat Transfer and Pressure Drop for Turbulent Single Phase Gas Flow, ESDU Data Item No. 86022, London, 1986.

Heat Exchanger Design Handbook (1983) Vol. 1–4, Hemisphere, New York.

Standards of the Tubular Exchanger Manufacturers Association, 7th ed. (1988) New York.

Stasiulevičius, J, and Skrinska, A. (1988) *Heat Transfer of Finned Tube Bundles in Crossflow,* 1988. A. A. Žukauskas and G. F. Hewitt eds., Hemisphere, New York.

Žukauskas, A. A., and Ulinskas, R. V. (1988) *Heat Transfer in Tube Banks in Crossflow,* Hemisphere, New York.

S.B. Beale

TUBE BUNDLES (see Shell-and-tube heat exchangers)

TUBE BUNDLES, TWO-PHASE CROSS FLOW (see Tubes and tube banks, boiling heat transfer on)

TUBE-FIN EXTENDED SURFACES (see Extended surface heat transfer)

TUBES (see Mechanical design of heat exchangers)

TUBES AND TUBE BANKS, BOILING HEAT TRANSFER ON

Following from: *Forced convective boiling*

Boiling on a Single Tube

Convective nucleate boiling on a tube differs considerably from that on a flat plate. In the latter case, bubbles are formed at scattered nucleation sites and depart taking part of the superheated boundary layer with them and causing an in-rush of fresh liquid to the surface. The convective effects are often characterised using the bubble diameter at departure.

In the case of a horizontal tube, observation shows that nucleation occurs primarily at the underside of the tube and that the bubbles then slide parallel to the surface to a point near the top before they depart. Thus a bubble-layer is formed around the tube as shown in **Figure 1** and analysis (Cornwell and Einarsson, 1990) of a radial control volume shows a voidage varying from virtually zero at the base to around 0.5 at the sides with a corresponding rapid increase in velocity. There is a considerable peripheral variation of heat transfer coefficient on the tube. (This is largely smoothed out in most experiments by the peripheral conduction in the tube.)

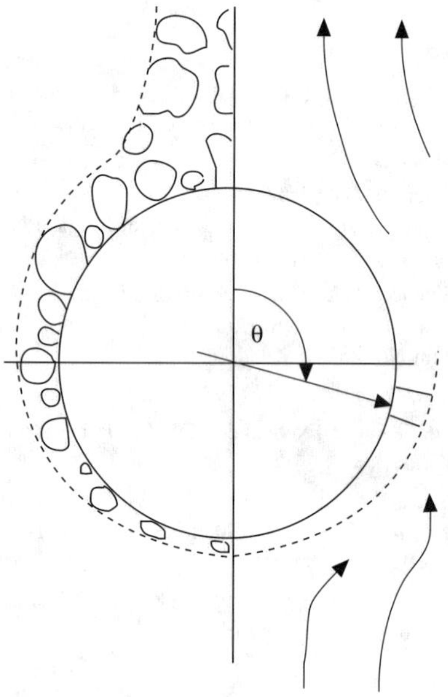

Figure 1. Boiling bubble layer on a tube.

Typical test results for a 27 mm diameter tube are shown in **Figure 2** where U is the vertical liquid velocity, with pool boiling therefore at U = 0. The value of α increases from the base to the sides in direct opposition to that for single phase flow where it drops. This increase is caused by the increase in bubble layer velocity from the base and by the additional latent heat transport from the thin layers formed under the bubbles as they slide around the surface. This latter mechanism becomes more important as the liquid velocity increases and accounts for the bulge which develops at higher velocities at the sides.

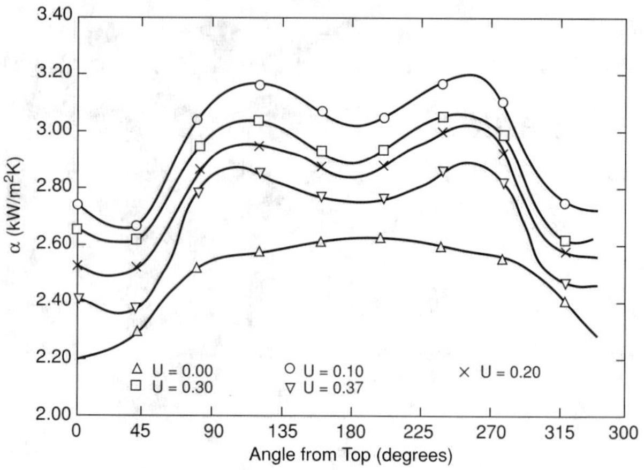

Figure 2. Peripheral variation of heat transfer coefficient at various approach velocities (in m/s) for boiling R113 at 1 atmosphere at $\dot{q}$ = 25 kW/m²K.

Estimation of boiling heat transfer coefficients for tubes have generally relied on pool boiling correlations developed from flat plate nucleation concepts (although it is known that tubes can yield higher values). This may be a reasonable approach at high heat fluxes ($q > 0.5\ q_{crit}$) where the boiling is vigorous but at lower fluxes where most industrial systems operate the difference can be considerable. The following general correlation (Cornwell and Houston, 1994) is based on the convective nature of tube boiling with Reynolds number Re_b representing the vapour production rate into the bubbly layer from a tube, diameter D.

$$Nu = AF(p)Re_b^{0.67}Pr^{0.4} \tag{1}$$

Here

$$Nu = \frac{\alpha D}{\lambda} \qquad Re_b = \frac{\dot{q}D}{\eta_f h_{fg}} \tag{2}$$

The pressure dependence is the same as that for pool boiling (from Mostinski, 1963)

$$F(p) = 1.8\,p_r^{0.17} + 4p_r^{1.2} + 10p_r^{10} \tag{3}$$

where p_r is the critical pressure ratio

$$p_r = \frac{p}{p_{crit}} \tag{4}$$

The constant A is fluid dependent and the best fit for the available world data is given by

$$A = 9.7\,p_{crit}^{0.5} \qquad \text{with} \quad p_{crit} \text{ in bar } (10^5\ N/m^2)$$

This correlation indicated by the line in **Figure 3** applies to pool boiling (no imposed flow) and on tube diameters in the range 8–50

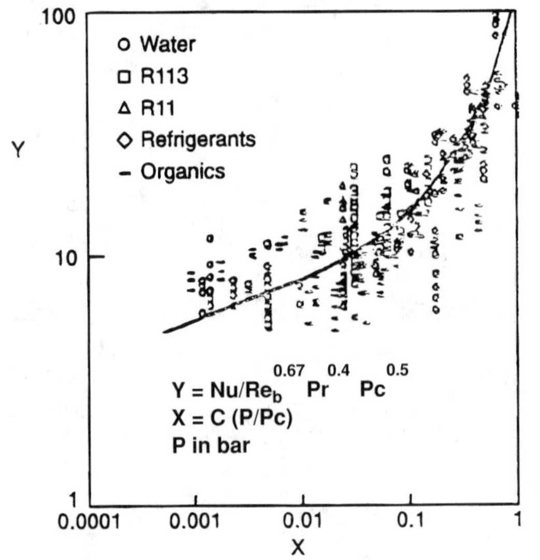

Figure 3. General correlation for pool boiling on tubes under the given conditions.

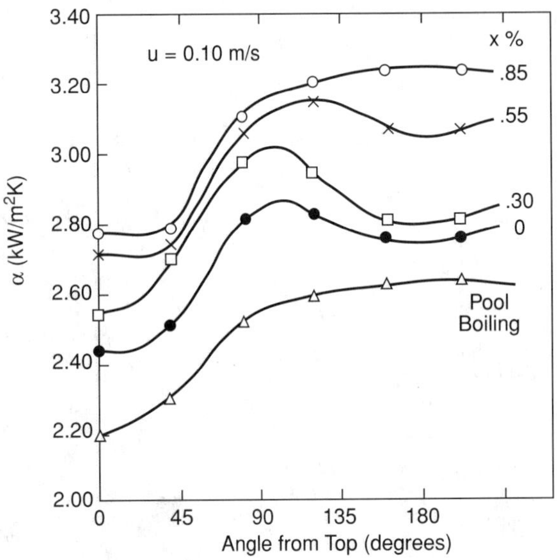

Figure 4. Peripheral variation of heat transfer coefficient (under similar conditions to those in **Figure 2**) with dryness fraction, Cornwell (1990).

mm, with a natural or "as-machined" surface and normal industrial fluids (not liquid metals or cryogenic fluids). The considerable scatter is to be expected for boiling correlations which cover a wide range of fluids, surfaces and pressures, particularly where data extending back over 50 years are included.

Boiling on a Tube in a Bundle

Boiling on a column of tubes has been shown to yield progressively enhanced heat transfer from the lowest tube upwards due to the liquid velocity and presence of bubbles in the flow. The effect of the liquid flow is evident from **Figure 2**, but the influence of the vapour is complicated by the form of flow which may range from bubbly to separated in nature. **Figure 4** shows the influence of variation in the dryness fraction λ, æ, of the approaching flow for the same tube used for **Figure 2** (at a constant flow rate equivalent to u = 0.1 m/s). Most industrial equipment with boiling outside tubes operates at low quality where the strong enhancement of heat transfer at the sides of the tubes is important.

The complex nature of the flow in bundles is well illustrated in **Figure 5** where there is high velocity bubbly flow upwards between the tubes and relatively motionless liquid in the horizontal gaps. Tubes in the upper part of a bundle usually experience a high-voidage pseudo-annular flow and the space above may be filled with foam which may result in liquid carry-over problems.

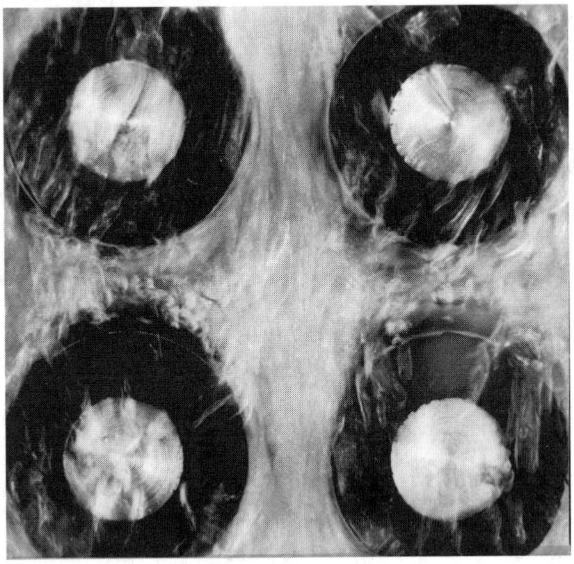

Figure 5. Boiling in the centre of a tube bundle (same fluid and conditions as in **Figure 2**).

Boiling on Tube Banks

Since 1829, when George Stephenson first boiled water on the outside of a tube bundle in his "Rocket" locomotive, boiling on the shell-side of bundles has been a common process in the power and process industries. *Fire-tube* and *waste heat boilers* which raise steam from hot gases generally have well spaced, large diameter tubes and the primary resistance to heat transfer is on the gas side. *Natural circulation reboilers* and *once-through evaporators* as used in process industries have more compact bundles as shown in **Figure 6**. (See also **Evaporators; Reboilers**) The liquid forced convection or condensation on the tube-side can result in more equal matching of the heat transfer resistances on the tube and shell sides.

Of these arrangements, the close-packed bundle in a liquid reservoir, as found for example in *shell-side refrigeration chillers* and in *distillation reboilers,* is the most difficult to analyse. The general flow of the liquid and entrained vapour and the resultant variation of heat transfer coefficients within the bundle are typically as shown in **Figure 7**. Analytic methods used until the 1960s were excessively conservative and led to over-designed equipment. Later approaches were generally based on the pool boiling heat flux on a single tube of the type used in the bundle. However, **Figure 8**

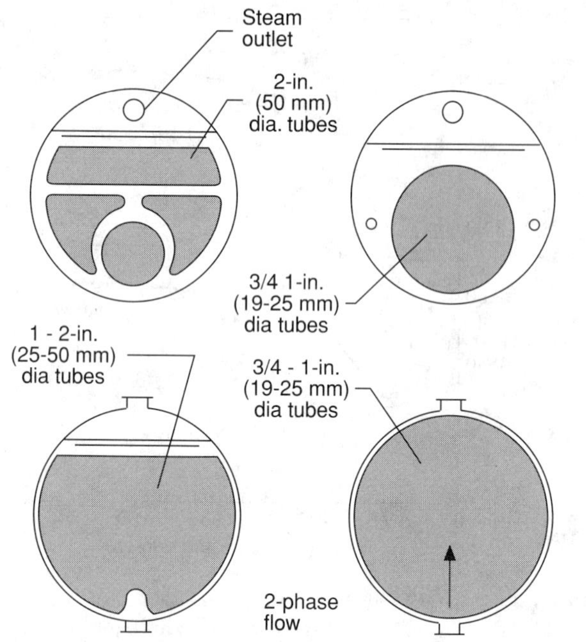

Figure 6. Sections through shell and tube boilers; clockwise from top left, fire-tube boiler, kettle reboiler, full-bundle evaporator, waste heat boiler.

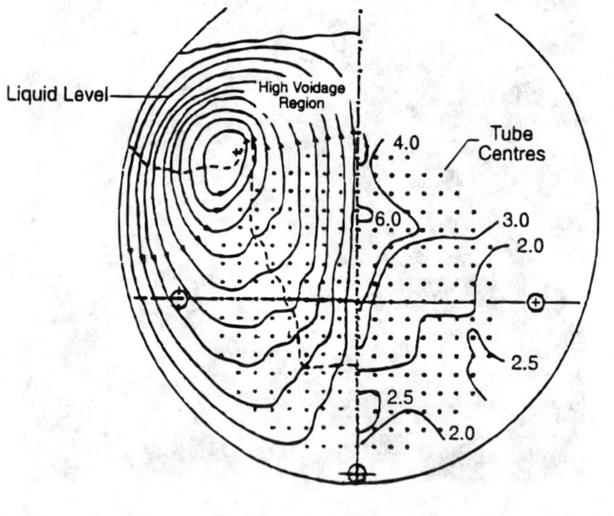

Figure 7. Flow streamlines and heat transfer coefficient contours (in kW/m²K) for boiling R113 at 1 atmosphere on an electrically heated bundle of 241 tubes at a heat flux of 20 kW/m²K, Cornwell et al. (1980).

shows that the mean heat flux for the bundle is greater than that for an isolated tube, although the critical heat flux (at which the surface is blanketed by vapour) is lower.

Theoretical Approaches

The large range of variables involved in the complex fluid flow and heat flow processes in boiling outside bundles make it unlikely that the gap between the understanding of the local mechanisms at the tube and the estimation of the overall bundle heat transfer coefficient, α_b, can be bridged. However, knowledge of the mechanisms has successfully highlighted the most viable approaches for developing design correlations.

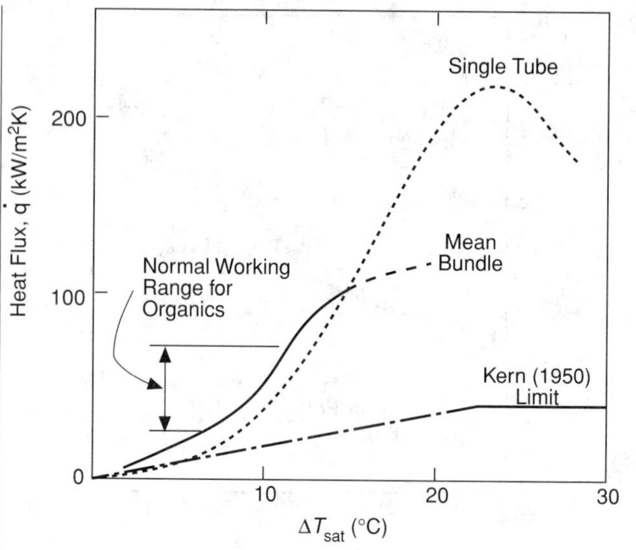

Figure 8. Typical tube bundle boiling curve for an organic fluid compared to that for boiling on a single isolated tube.

Palen (1983) essentially recommends adaption of the single-tube nucleate boiling heat transfer coefficient α_{nb} by a bundle boiling factor F_b, a mixture correction factor F_c and natural convection α_{nc} in the form:

$$\alpha_b = \alpha_{nb} F_b F_c + \alpha_{nc} \tag{5}$$

Factor F_b may be calculated, but when data is not available a conservative value of 1.5 is recommended. F_c is unity for pure fluids, but less for mixtures and α_{nc} is around 0.25kW/m²K (small compared to α_{nb}).

More recent work (by Fujita et al., 1986, Jensen, 1989, Cornwell et al., 1992 for example) has shown that both the pressure drop and the heat transfer through the bundle can be estimated using two-phase flow theory for in-tube flow, suitably modified for the different geometry and the different flow regimes. Under normal conditions bundles operate in the bubbly flow regime with the upper tubes in a pseudo-annular flow. Calculation of the pressure drop is needed to determine circulation rates and readers are referred to Schrage et al. (1988) for *voidage and two-phase friction multiplier data for vertical crossflow over horizontal bundles.*

Variation of the heat transfer coefficient up the bundle with the dryness fraction x (typically from x = 0 to 0.2 in a submerged bundle) is shown in **Figure 9**. Correlations for in-tube evaporation such as Chen (1966) which sum the non-convective part (at low x) and the convective part have been successfully adapted for this geometry. Alternatively, bearing in mind the bubbly-flow sliding bubble mechanism at low x and the very different forced convective evaporation mechanism at higher x, appropriate relationships may be applied separately to the two regimes so that for each tube (suffix t);

$$\alpha_t = \alpha_{nb}$$

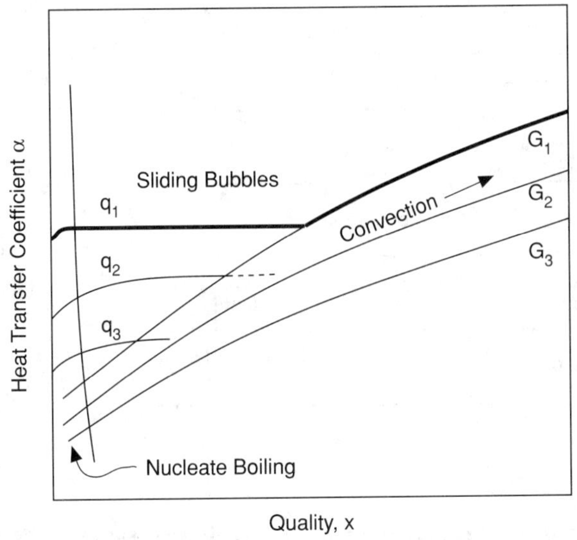

Figure 9. Typical variation of heat transfer coefficient with quality up the bundle showing the flow regimes. (Heat flux q and mass flux G decreasing 1 to 3).

or

$$\alpha_t = F\alpha_{lo}$$

whichever is the larger. Here α_{nb} is as before and F is the Chen two-phase convective flow factor which is a function of x (see section on Forced Convective Boiling). The liquid-only crossflow heat transfer coefficient α_{lo} may be found from Zukaukas and Ziudzda (1985) or ESDU (1973) (See also **Tube Banks, Single Phase Heat Transfer in**):

$$Nu = 0.211\, Re_f^{0.651} Pr^{0.34}$$

This approach will lead to conservative values because, as explained earlier, the single tube value α_t is likely to be more than α_{nb} The mean bundle value α_b may be estimated by integration over the x range or averaging the α_t values for the bundle.

Influence of Parameters

Process intensification has led to increasing use of plate and narrow channel geometries of heat exchangers in place of shell and tube arrangements. Where manufacturing processes and fouling rates allow, the use of smaller diameter tubes and lower pitch/diameter ratios can lead to compact and effective evaporator bundles (Cornwell et al 1992). Enhancement of heat transfer on the outer, boiling side of the tube using porous or grooved surfaces to increase the nucleation site density or by using low-fin tubes can increase α_b for a tube by up to 10 times (see **Augmentation of Heat Transfer, Two, Phase Systems**) Use of these tubes lead to an overall increase in bundle performance of typically around 2 owing to the predominance of convective boiling rather than nucleate boiling and the resistance of the inner tube side, (see review by Thome, 1989).

Mixture boiling on bundles, as for boiling generally, leads to lower heat transfer rates owing to the local composition gradients. The critical (maximum) heat flux for a bundle is generally around 0.5 to 1 times the single tube value and Leroux and Jensen (1992) have shown that it is highly mass flux dependent. The influence of mechanical features, corrosion, fouling, cleaning and maintenance of horizontal-tube boilers and samples of design calculations are covered by Smith (1986).

References

Chen, J. (1966) A correlation for boiling heat transfer to saturated fluids in convective flow, *I & EC Proc. Des. Dev.* 5, 322–329.

Cornwell, K. (1990) The role of sliding bubbles in boiling on tube bundles, Heat Transfer 1990, *9th Int. Heat Transfer Conf. Proc.,* 1, 455–460, Hemisphere.

Cornwell, K., Dewar, R. G. and Ritchie, J. M. (1992) A new approach to the determination of boiling heat transfer coefficients outside tube bundles, heat transfer, *I Chem E,* 3rd UK National Heat Transfer Conference 1, 51–64.

Cornwell, K., Duffin, N. W. and Schuller, R. B. (1980) An Experimental Study of the Effects of Fluid Flow on Boiling Within a Kettle Reboiler Tube Bundle, ASME Paper 80-HT-45.

Cornwell, K. and Einarsson, J. G. (1990) The influence of fluid flow on nucleate boiling from a tube, *Exp. Heat Transfer,* 3, 101–116.

Cornwell, K. and Houston, S. D. (1994) Nucleate pool boiling on horizontal tubes: a convection-based correlation, *Int. J. Heat Mass Trans.,* 37, 303–309.

ESDU (1973) Convective Heat Transfer During Crossflow of Fluids Over Plain Tube Banks, Item No. 73071, ESDU, London.

Fujita, Y., Ohta, H., Hidaka, S. and Niskikawa, K. (1986) Nucleate boiling on horizontal tubes in bundles, *8th International Heat Transfer Conference,* San Francisco, 5, 2131–2136.

Jensen, M. K. (1989) Advances in shellside boiling and two-phase flow, *ASME, HTD,* 108, 1–11.

Kern, D. Q. (1950) *Process Heat Transfer,* McGraw Hill, New York.

Leroux, K. M. and Jensen, M. K. (1992) Critical heat flux in horizontal tube bundles in vertical crossflow of R113, *J. Heat Trans.,* 114, 179–185.

Mostinski, I. L. (1963) Calculation of heat transfer and critical heat flux in boiling liquids based on the law of corresponding states, *Teploenergetika,* 10, 66–71.

Palen, J. W. (1983) Shell and tube reboilers, *Heat Exchanger Design Handbook,* Hemisphere, New York.

Schrage, D. S., Hsu, J–T. and Jensen, M. K. (1988) Two-phase pressure drop in vertical crossflow across a horizontal tube bundle, *AICHE J,* 34, 107–115.

Smith, R. A. (1986) *Vaporisers,* Longman, Harlow, UK.

Thome, J. R. (1989) *Enhanced Boiling Heat Transfer,* Hemisphere, NY.

Zukaukas, A. and Ziukzda (1985) *Heat Transfer in a Cylinder in Crossflow,* Hemisphere, NY.

K. Cornwell

TUBES, CONDENSATION IN

Following from: *Condensation of pure vapor, overviews*

Condensation in Forced Convection

As shown in **Figure 1**, in the inlet region of a horizontal condenser tube the volume flow rate of super heated vapor is usually

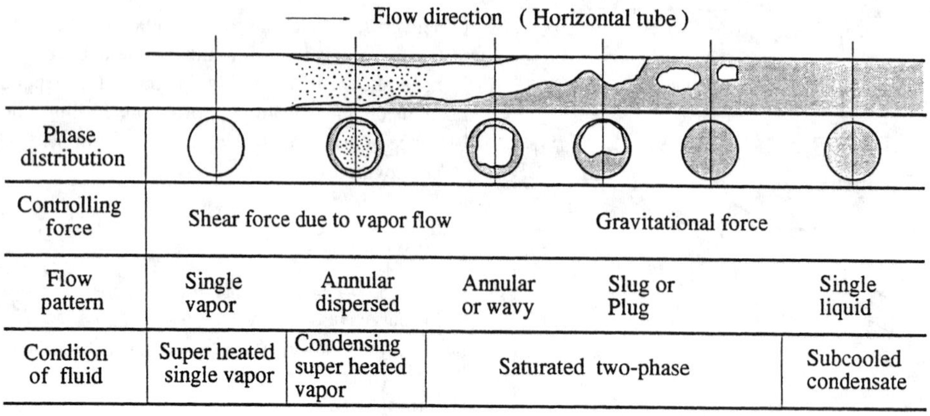

Figure 1.

high, and the flow pattern is single-phase vapor flow. If the temperature of the vapor decreases below saturation temperature near the cooling tube wall, the vapor starts to condense on the surface of the tube. The condensed liquid flows as a liquid film on the tube wall co-currently with the vapor flow and the flow pattern becomes annular flow. In the vapor core, liquid also flows as droplets due to strong interfacial shear force if the velocity of vapor is high. In a condenser, the volume flow rate of vapor and liquid mixture continuously decreases as the mixture flows downstream, where the role of the gravitational force in the phase distribution becomes stronger. Then the flow pattern becomes wavy, separated, then becomes plug flow with vapor flowing on the top side of the tube cross-section. Finally, all of the vapor changes into liquid, and single liquid flow appears near the end of the condenser.

In the case of a vertical tube condensing fluid usually flows downwards. The phase distribution is axisymmetric and the flow pattern is annular flow over almost whole region of the condenser tube and changes to slug flow just upstream of the end of the condensing section.

Local heat transfer coefficient α, at the cross-section where the quality is x, is given by Shah (1979) as follows,

$$\alpha = \alpha_L \left\{ (1 - x)^{0.8} + 3.8x^{0.76}(1 - x)^{0.04}\left(\frac{P_S}{P_{cri}}\right)^{-0.38} \right\} \quad \textbf{(1)}$$

where α_L is the heat transfer coefficient given when liquid flows as a single phase flow with same liquid mass flow rate, and P_s/P_{cri} is the reduced pressure ratio of the saturation pressure P_s. This equation is useful being independent of the orientation of the pipe and the flow direction, although the accuracy is not high.

The heat transfer coefficients during condensation, however, depends strongly on the phase distribution, i.e. the flow pattern, especially when the existence of dried area on the condensing surface controls the heat transfer. Therefore, the prediction of flow pattern is important to estimate accurately the heat transfer coefficient.

In the case of horizontal flow, some correlations are available to estimate the heat transfer coefficient α expressed as **Nusselt Number** Nu ($\equiv \alpha\, l/\lambda$, where l is the characteristic length of the heat transfer area, and λ the thermal conductivity) for respective flow condition.

Single superheated vapor flow region (Petukov, 1970)

$$Nu = \frac{(f/2)Re\,Pr}{1.07 + 12.7\sqrt{f/2}(Pr^{2/3} - 1)} \quad \textbf{(2)}$$

$$(10^4 < Re < 10^6,\ 0.5 < Pr < 2000)$$

where Pr is the **Prandtl Number** of the vapor and f is the friction factor which is given by

$$f = (3.64 \log_{10} Re - 3.28)^{-2} \quad \textbf{(3)}$$

The following formulae have been also widely used.
Dittus-Boelter correlation

$$Nu = 0.023\ Re^{0.8}Pr^{0.4} \quad \textbf{(4)}$$

$$(10^4 < Re < 10^5,\ 1 < Pr < 10)$$

Colburn correlation

$$Nu = 0.023\ Re^{0.8}Pr^{1/3} \quad \textbf{(5)}$$

(Narrower range of Re and Pr than those in Dittus-Boelter's correlation)
where Re is the **Reynolds Number** ($\equiv \bar{u}D/\nu_v$, where $\bar{u}$ is the mean vapor velocity, D the tube diameter, and ν_v the kinematic viscosity of the vapor).

Condensing superheated vapor flow region

The Nusselt number decreases with the decrease in the quality, i.e. the increase of the condensate flow rate. Miropolsky (1974) proposed a similar type of correlation to that given by **Equation 2**, where the fluid properties are defined by the average calorimetric flow temperature.

Fujii et al.(1977) proposed the following correlation of the condensation heat transfer coefficient α in the case where the superheated vapor is condensing on its condensate.

$$St\left(\equiv\frac{\alpha}{c_{pv}\rho_v\bar{u}}\right)$$

$$= 0.9(f/2)\Big/\left[1 + 5\sqrt{f/2}\left\{Pr_v - 1 + \ln\left(1 + \frac{5}{6}(Pr_v - 1)\right)\right\}\right] \quad (6)$$

where St is the **Stanton Number**, $\bar{u}$ is the mean velocity of vapor, Pr_v the **Prandtle Number** of vapor and f the **Friction Factor** given by the next equation,

$$f = 0.046\ \varepsilon^{2.5}\phi_v^2/(\dot{m}_v D/\eta_v)^{0.2} \quad (7)$$

Here ε is the void fraction, ϕ_v the Martinelli parameter for the two-phase pressure drop, $\dot{m}_v$ the mass velocity of vapor, D is the tube diameter and η_v the absolute viscosity of vapor. They suggested also that void fraction ε can be obtained by the next relation proposed by Fauske (1961).

$$\varepsilon = \left\{1 + \left(\frac{1-x}{x}\right)(\rho_G/\rho_L)^{1/2}\right\}^{-1} \quad (8)$$

Notice that the heat flux through the tube wall $\dot{q}_0$ is expressed as,

$$\dot{q}_o = (q_i + \dot{m}_c h_{LG})(2r_i/D) \quad (9)$$

where D is the tube diameter, $2r_i$ the average diameter of the vapor-liquid interface, h_{LG} the latent heat of condensation, $\dot{m}_c$ the mass flux of the condensate and $\dot{q}_i$ the heat flux calculated by the heat transfer coefficient given by St described in **Equation 6** above.

Condensing saturated vapor flow region

The heat transferred to the liquid comes only from the condensation of the vapor on the liquid layer in this region. Therefore, heat transfer can be estimated in the same way as mentioned in the previous section, if the heat flux at the interface $\dot{q}_i$ is assumed to be zero.

Condensation in a Closed Tube

A typical example of condensation in a closed tube is a heat pipe. As shown in **Figure 2** vapor generated in a heating section, i.e., an evaporator, flows into a condenser where the vapor condenses. The condensate returns to the evaporator as a liquid film on a tube wall. Acordingly in this case, *reflux condensation* heat transfer takes place between up-going vapor flow and down-coming liquid film flow, as for example, in a vertical *two-phase thermosyphon,* i.e., a *wickless heat pipe.*

Laminar film condensation on a vertical plate was analyzed by **Nusselt** (1916). The local heat transfer coefficient α is given as,

$$\alpha = \sqrt[4]{\frac{g\rho_L(\rho_L - \rho_G)\lambda_L^3 h'_{LG}}{4z\eta_L\Delta T}} \quad (10)$$

where ρ_L and ρ_G are the liquid and gas densities, λ_L the liquid thermal conductivity, z the distance from the start of condensation, η_L the absolute liquid viscosity and ΔT the difference between the

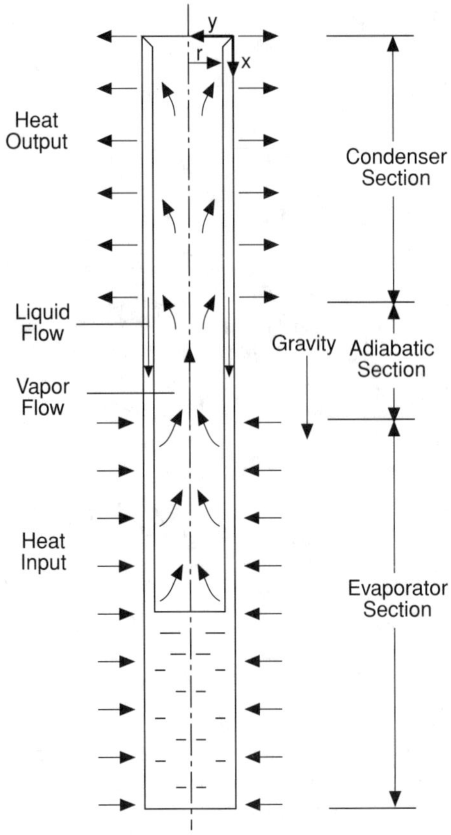

Figure 2.

saturation temperature and the wall temperature, and $h'_{LG} \equiv h_{LG} + 2c_L\Delta T/3$ where h_{LG} is the latent heat of evaporation and c_L the specific heat of the liquid.

The local heat transfer in a closed tube is not as simple as Nusselt's physical model in the following respects (Fukano et al. 1990).

1. Nusselt assumed a smooth surface. Actually, however, the interface is not smooth as shown in **Figure 3**. Flow pattern changes from the dropwise through the smooth surface to the two-dimensional wave and finally the three dimensional wave—in this order—from the top of the condenser to the bottom.
2. The vapor-liquid interfacial shear stress is not zero contrary to the assumption made by Nusselt.
3. The boundary condition of the film thickness δ, i.e., $\delta = 0$ at the start of condensation, is used in the Nusselt analysis. However, many fine droplets are observed to flow onto the top wall of the tube, so that the film thickness is not zero at the tube top.

Therefore, the Nusselt number is usually lower in the top region of the condenser than that given by the Nusselt analysis due to the larger film thickness. Also, it is higher in the bottom region due to the quick sweeping away of the condensate by large two or three dimensional waves resulting in the smaller film thickness with smaller thermal resistance.

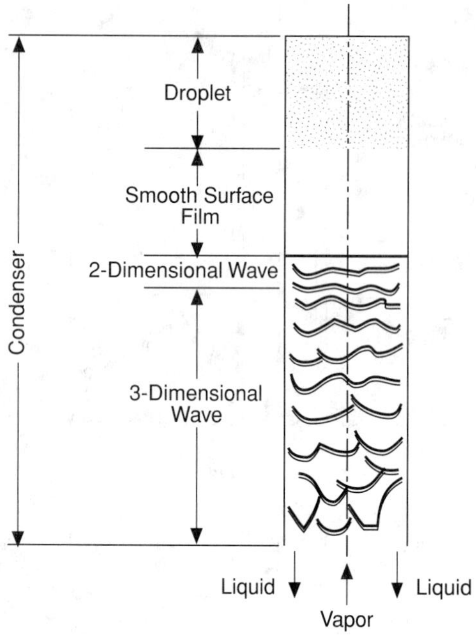

Figure 3.

If the average film thickness δ is estimated by the mass and the heat balances of the liquid film, the local heat transfer coefficient α is obtained by assuming that heat is transferred only by conduction through liquid film as follows;

$$\alpha = \frac{\lambda_L}{(1 - C)\delta} \tag{11}$$

where C means the fraction of the wave layer in which the thermal resistance is zero, and is given by the dimensionless shear stress $\tau_i^* \{\equiv \tau_i/(\rho_L\ g\ \delta)\}$ as follows,

$$C = 1.13(\tau_i^* - 0.08)^{0.34},\ (\tau_i^* > 0.08) \tag{12}$$

The effect of the wave on the heat transfer does not appear until the interfacial shear stress exceeds a certain value, i. e., $\tau\, i^* > 0.08$.

References

Fauske, H. K. (1961) *Heat Transfer and Fluid Mechanics,* Stanford University Press.

Fujii, T., et al. (1977) *Refrigeration,* 52, 596.

Fukano, T., et al. (1990) *Proc. of 9th Int. Heat Transfer Conf.,* 6.

Miropolsky, Z. L. (1974) *Proc. 5th Int. Heat Transfer Conf.,* 3.

Nusselt, W. (1916) *2. ver. deut. Ing.,* 60, 541, 569.

Petukov, B. S. (1970) *Advances in Heat Transfer,* 6.

Shah, M. M. (1979) *Int. J. Heat Mass Trans.,* 22–4.

Leading to: *Condensers*

T. Fukano

TUBES, CONDENSATION ON OUTSIDE IN CROSSFLOW

Following from: *Condensation, overview: Condensation of pure vapours*

In this situation motion of the condensate film, and hence the local film thickness and heat-transfer rate, are influenced by the surface shear stress due to the vapour flow, as well as by gravity. For a pure vapour with laminar flow of the condensate film and with vapour flow normal to the tube axis, this problem is now quite well understood. The theoretical approach for condensation on a plane surface with flow of vapour parallel to the condensing surface is relatively straightforward. The complication which arises in the case of the tube in crossflow is separation of the vapour boundary layer at some position around the tube, the point of separation being affected by the "suction" effect due to condensation.

The simplest model is that of Shekriladze and Gomelauri (1966), who used the asymptotic expression for the vapour **Shear Stress** at the condensate surface

$$\tau_\delta = \dot{m}(U' - u_\delta) \tag{1}$$

where $\dot{m}$ is the local condensation mass flux, U_δ is the tangential velocity of the condensate surface and U' is the tangential velocity at the outer 'edge' of the vapour boundary layer. It is sometimes implied in the literature that the shear stress on the condensate surface is attributable to two components: "dry shear" and "momentum transfer by condensation", the latter being given in **Equation 1**. Since the velocity of the vapour Immediately Adjacent to the condensate surface is Equal to that of the surface, no momentum is "transferred by condensation". The correct interface condition is continuity of shear stress from vapour to liquid. **Equation 1** is valid for relatively high condensation rates. The Shekriladze-Gomelauri model assumes potential flow outside the vapour boundary layer, so that

$$U' = 2U_\infty \sin\phi \tag{2}$$

where U_∞ is the vapour approach velocity and ϕ is the angle measured from the forward stagnation point. **Equations 1 and 2** indicate that τ_δ is positive for all ϕ, so that boundary layer separation does not occur in this simple approach.

Shekriladze and Gomelauri (1966) used **Equations 1** (with the additional approximation $U' >> u_\delta$) and **Equation 2**, otherwise treating the condensate film as in Nusselt (1916). Numerical results for the dependence of film thickness, and hence heat-transfer rate, on position around the tube were found. As indicated by Rose (1988), for vertical vapour downflow, the mean Nusselt number for the tube given by the Shekriladze-Gomelauri model can be approximated to within 0.4% by

$$Nu\,\tilde{R}e^{-1/2} = \frac{0.9 + 0.728\,F^{1/2}}{(1 + 3.44\,F^{1/2} + F)^{1/4}} \tag{4}$$

where Nu is **Nusselt Number**, Re is **Reynolds Number** defined as $U_\infty\,\rho d/\eta$, d is the diameter of the tube and ρ and η are the density and viscosity of the condensate respectively.

The quantity F measures the relative importance of gravity and vapour shear stress and may be written:

$$F = \frac{Gr_c}{\tilde{Re}^2} \cdot \left(\frac{\eta h_{lg}}{\lambda \Delta T}\right) \qquad (5)$$

where Gr_c is $\rho \Delta \rho d^3 / \eta^2$, $\Delta \rho$ is the difference between condensate and vapour density, $(\rho - \rho_v)$, h_{lg} is specific enthalpy of phase change and λ is thermal conductivity of condensate.

(F is commonly written as Pr/Fr · H, where Pr is **Prandtl Number**, Fr is **Froude Number** and H is $c_p \Delta T/h_{lg}$ sometimes known as **Jakob Number** or phase change number and P_p is isobaric specific heat capacity of condensate. This is somewhat misleading since c_p and Pr play no part when, as in this model, the condensate flow is laminar and convection terms are neglected.)

Equation 1 underestimates the shear stress on the condensate film up to the separation point and overestimates this thereafter. From the viewpoint of calculating the mean heat transfer for the whole tube these effects tend to compensate for each other. Equation 4 is in broad agreement with experimental data (see **Figure 1**).

Figure 1 also shows that, for values of F greater than around 10, the vapour velocity effect is small. Notice that for $F \rightarrow \infty$ Equation 4 gives

$$Nu = 0.728 \, F^{1/4} \, \tilde{Re}^{1/2} = 0.728 \left(\frac{\eta h_{lg}}{\lambda \Delta T} \cdot Gr_c\right)^{1/4} \qquad (6)$$

i.e. the Nusselt (1916) result. It is also seen that for F less than

around 0.1, gravity effects are unimportant and **Equation 4** approximates to its form when the condensate flow and heat transfer are dominated by the vapour shear stress (F = 0)

$$Nu = 0.9 \, \tilde{Re}^{1/2} \qquad (7)$$

Several more complete and detailed approaches have been used by various investigators and are discussed by Rose (1988). In particular, consideration has been given to:-

a. accurate calculation of the vapour shear stress and vapour boundary layer separation by matching the shear stress on either side of the vapour-liquid interface,

b. direction of vapour flow in relation to that of gravity,

c. pressure variation in the condensate film resulting from vapour flow around the cylindrical surface.

The general conclusion, broadly supported by experimental data, is that, for the purposes of calculating the mean heat-transfer coefficient for the tube, **Equation 4** seems to be generally satisfactory. There is, however, evidence that, at higher vapour velocities, **Equation 4** might overestimate (but not excessively) the mean heat-transfer coefficient for steam, while underestimating (possibly more significantly) values for refrigerants when F < about 1. In the former case this may be due, in part, to the variation of pressure, and hence saturation temperature, around the tube, while the latter has been attributed to onset of turbulence in the condensate film.

References

Michael, A. G., Rose, J. W. and Daniels, L. C. (1989) Forced convection condensation on a horizontal tube—experiments with vertical downflow of steam, *J. Heat Transfer*, 111, 792–797.

Nusselt, W. (1916) Die Oberflachenkondensation des Wasserdampfes, *Z. Vereines Deutsch. Ing.*, 60, 541–546, 569–575.

Rose, J. W. (1988) Fundamentals of condensation heat transfer: Laminar film condensation, *JSME Int. Journal*, Series II, 31, 3, 357–375.

Shekriladze, I. G. and Gomelauri, V. I. (1966) Theoretical study of laminar film condensation of flowing vapour, *Int. J. Heat Mass Trans.*, 9, 581–591.

Leading to Condensers; Surface condensers

J.W. Rose

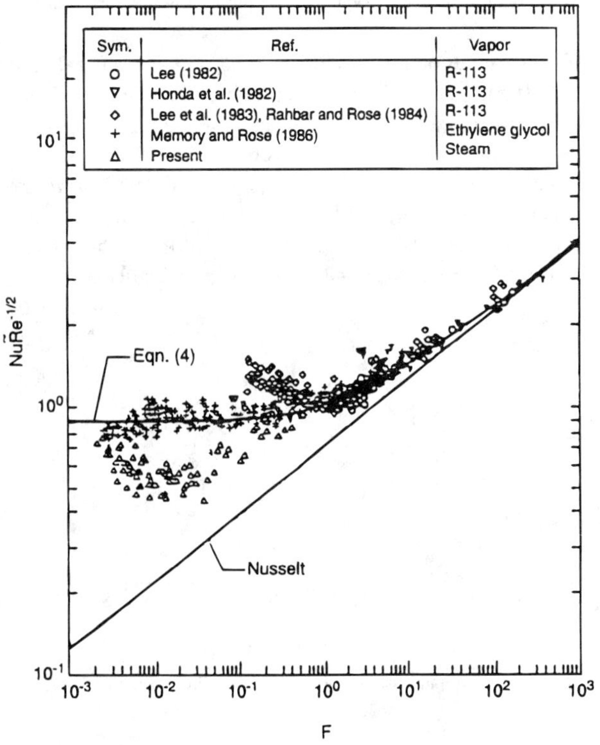

Figure 1. Comparison of experimental data with Equation 4. (from Michael et al., 1989)

Sym.	Ref.	Vapor
○	Lee (1982)	R-113
▽	Honda et al. (1982)	R-113
◇	Lee et al. (1983), Rahbar and Rose (1984)	R-113
+	Memory and Rose (1986)	Ethylene glycol
△	Present	Steam

TUBES, CROSSFLOW OVER

Following from: External flows, overview

Introduction

A common external flow configuration involves the circular cylinder or tube in crossflow, where the flow is normal to the axis of the cylinder. If an inviscid fluid is considered, the velocity distribution over the cylinder is given by

$$u_r = U_\infty\left(1 - \frac{r_o^2}{r^2}\right)\cos\theta \tag{1a}$$

$$u_\theta = U_\infty\left(1 - \frac{r_o^2}{r^2}\right)\sin\theta \tag{1b}$$

where U_∞ is the velocity far upstream the cylinder, r_o the cylinder radius, r the radial coordinate and θ the angle measured from the forward stagnation point. At the tube surface one finds

$$u_\theta = 2U_\infty \sin\theta \tag{2}$$

and thus for $\theta = \pi/2$ the velocity on the tube surface is twice the freestream velocity.

By applying the **Bernoulli Equation** for an inviscid fluid, the pressure coefficient C_p is found to be

$$C_p = \frac{2(p - p_\infty)}{\rho U_\infty^2} = 1 - 4\sin^2\theta \tag{3}$$

For an inviscid fluid, the pressure coefficient is distributed symmetrically and an integration of the pressure distribution results in zero drag and lift forces. This is an example of the *d'Alembert paradox* for inviscid flow past immersed bodied. In **Figure 1**, the inviscid flow past a tube is shown.

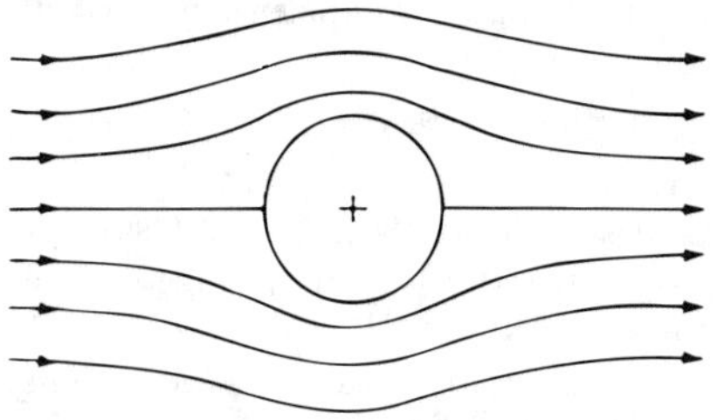

Figure 1. Sketch of the inviscid flow past a tube (circular cylinder).

Real Flow Phenomena

For viscous fluids the flow pattern is much more complicated and the balance between inertia forces and viscous forces is important. The relative importance is expressed by the **Reynolds number** Re_D defined as

$$Re_D = \frac{U_\infty D}{\nu} \tag{4}$$

where D is the tube diameter and ν the kinematic viscosity of the fluid.

As the fluid approaches the front side of the tube, the fluid pressure rises from the freestream value to the stagnation point value. The high pressure forces the fluid to move along the tube surface and boundary layers develop on both sides. The pressure force is counteracted by viscous forces and the fluid cannot follow the tube surface to the rear side but separates from both sides of the tube and form two *shear layers.* The innermost part of the shear layers are in contact with the tube surface and moves slower than the outermost part. As a result, the shear layers roll up. The flow pattern is dependent on the Reynolds number and in **Figure 2** a principal description of the various occuring flow phenomena is provided.

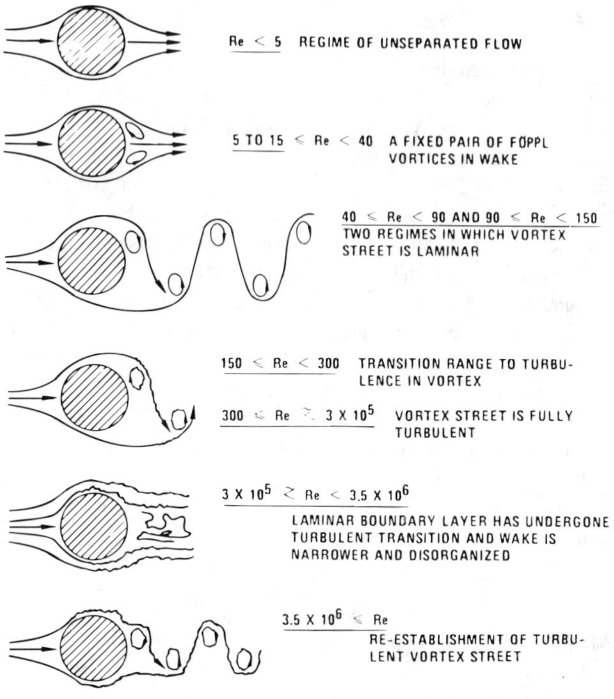

Figure 2. Regimes of fluid flow across a smooth tube. From Blevins R. D. (1990), Flow Induced Vibration, 2nd Edition, Van Nostrand Reinhold Co.

At Reynolds numbers below 1, no separation occurs. The shape of the streamlines is different from those in an inviscid fluid. The viscous forces cause the streamlines to move further apart on the downstream side than on the upstream side of the tube. In the Reynolds number range of $5 \leq Re_D \leq 45$, the *flow separates* from the rear side of the tube and a symmetric pair of *vortices* is formed in the near wake. The streamwise length of the vortices increases linearly with Reynolds number as shown in **Figure 3**.

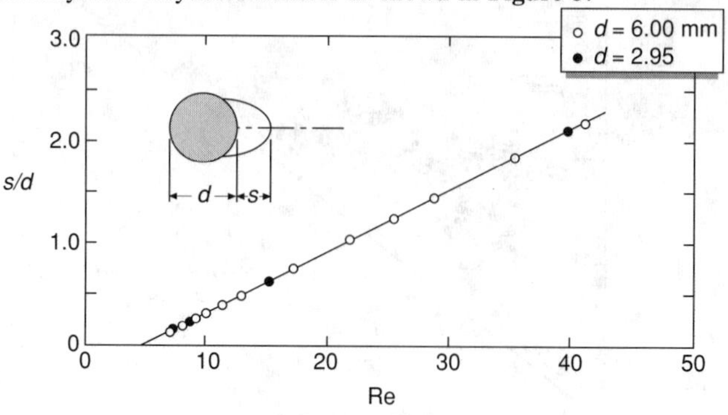

Figure 3. Streamwise length of vortices. From Taneda S. (1956) J Phys. Soc. Japan 11,302 with permission.

As the Reynolds number is further increased the *wake* becomes unstable and **Vortex Shedding** is initiated. At first, one of the two vortices breaks away and then the second is shed because of the non-symmetric pressure in the wake. The intermittently shed vortices form a laminar periodic wake of staggered vortices of opposite sign. This phenomenon is often called the *Karman vortex street.* von Karman showed analytically and confirmed experimentally that the pattern of vortices in a vortex street follows a mathematical relationship, namely

$$\frac{h}{l} = \frac{1}{\pi}\sinh^{-1}\{1\} = 0.281 \qquad \textbf{(5)}$$

where h and l are explained in **Figure 4**.

In the Reynolds number range $150 < Re_D < 300$, periodic irregular disturbances are found in the wake. The flow is transitional and gradually becomes turbulent as the Reynolds number is increased.

The Reynolds number range $300 < Re_D \leq 1.5 \cdot 10^5$ is called subcritical (the upper limit is sometimes given as $2 \cdot 10^5$). The laminar boundary layer separates at about 80 degrees downstream of the front stagnation point and the vortex shedding is strong and periodic.

With a further increase of Re_D, the flow enters the critical regime. The laminar boundary layer separates on the front side of the tube, forms a separation bubble and later *reattaches* on the tube surface. Reattachment is followed by a turbulent boundary layer and the separation point is moved to the rear side, to about 140 degrees downstream the front stagnation point. As an effect the drag coefficient is decreased sharply.

For $Re_D \geq 6 \cdot 10^5$, the laminar to turbulent transition occurs in a non-separated boundary layer, and the transition point is shifted upstream.

The Re_D-range $2 \cdot 10^5 \leq Re_D \leq 3.5 \cdot 10^6$ is in some references named the transitional range. Three-dimensional effects disrupt the regular shedding process and the spectrum of shedding frequencies is broadened.

In the supercritical Re_D-range, $Re_D > 3.5 \cdot 10^6$, a regular vortex shedding is re-established with a turbulent boundary layer on the tube surface. (Some references name the regime $Re_D > 6 \cdot 10^5$ the supercritical regime).

Pressure Distribution

The various flow phenomena are reflected in the pressure distribution on the tube surface. **Figure 5** provides a few distributions of the pressure coefficient C_p and the changes in the distributions are explained by the flow mechanisms described previously.

Particularly for $Re_D = 6.7 \cdot 10^5$, the separation of the laminar boundary layer and the reattachment is believed to be reflected in

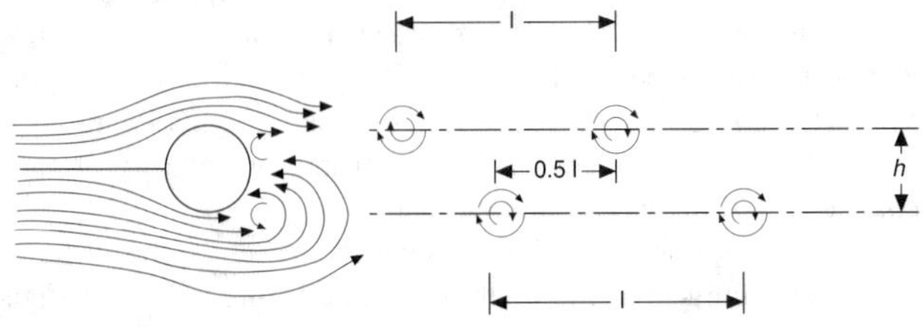

Figure 4. Model of a vortex street.

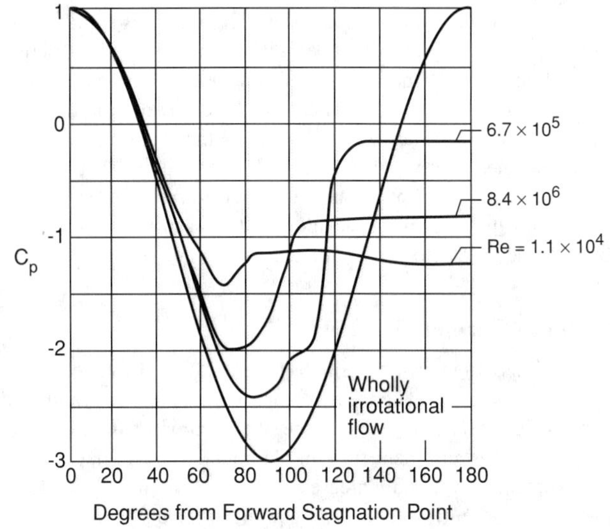

Figure 5. Pressure distribution around the circumference of a tube or circular cylinder. From Batchelor, G. K. (1970). An Introduction to Fluid Mechanics, Cambridge University Press, with permission.

the behaviour between $\theta \approx 100$–110 degrees downstream the front stagnation point.

The non-symmetrical pressure distribution result in a net force on the tube, and the existence of this force is the main cause of the pressure drop across the tube.

Drag

The total drag is generated by the friction forces and pressure forces acting on the tube. At very low Reynolds number, the drag is mainly due to friction. With an increase of Re_D the contribution of the inertia forces begin to grow so that at high Reynolds numbers the skin friction constitutes just a few per cent of the total drag. A dimensionless expression of the total drag is the *drag coefficient* defined by

$$C_D = \frac{\text{total force}}{\frac{1}{2}\rho U_\infty^2 DL} \qquad \textbf{(6)}$$

where L is the length of the tube.

Figure 6 shows the drag coefficient as a function of Re_D.

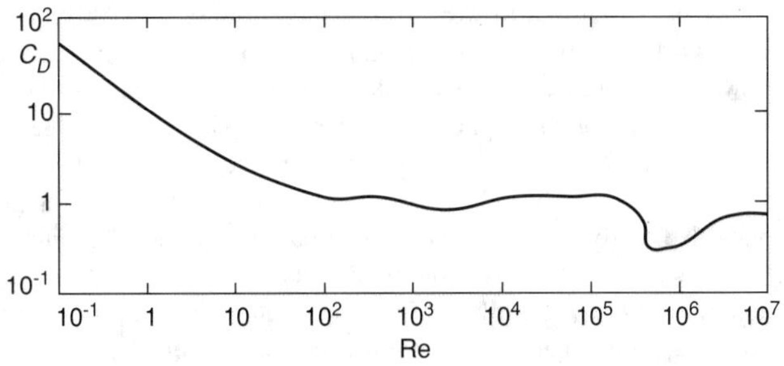

Figure 6. Drag coefficient on a tube (circular cylinder).

In the low Reynolds number range, C_D decreases significantly with increasing Reynolds number, mostly due to the skin friction contribution. In the subcritical regime, the drag coefficient changes insignificantly with Re_D. As the critical flow regime is reached, the drag coefficient decreases sharply with increasing Re_D. The wake becomes narrower by the turbulent boundary layer and the associated delayed separation. At higher Re_D, C_D is increasing again probably due to the development of a turbulent boundary layer already close to the front stagnation point.

Factors Affecting the Flow

The cross flow past the single tube is affected by freestream turbulence, surface roughness, compressibility of the fluid and some other factors. So for instance, the position of the transition from laminar to turbulent flow in the boundary layer depends on the turbulence level. An increase in the freesteam turbulence level leads to an earlier (lower Re_D) onset of the critical flow regime and corresponding changes in the drag coefficient. Surface roughness causes an earlier onset of the critical regime and a higher drag coefficient.

References

Zukauskas, A. A. and Ziugzda, J. (1985) *Heat Transfer of a Cylinder in Cross Flow*, Hemisphere Publishing Corporation, 1985.

Blevins, R. D. (1990) *Flow-Induced Vibration*, 2nd ed., Van Nostrand Reinhold, 1990.

Batchelor, G. K. (1970) *An Introduction to Fluid Dynamics*, Cambridge University Press, 1970.

Leading to: *Tube banks, crossflow over; Vortex shedding*

B. Sundén

TUBES, GAS-LIQUID FLOW IN

Gas-liquid flow in tubes occurs in a wide variety of chemical, power generation and process industries. Depending on the flow configuration, it can take different forms, for example, flow in horizontal tubes, vertical tubes (upward or downward flow), inclined tubes or coiled tubes and can be either co-current or counter-current flow. Also, the two phases may flow in separated layers as in stratified flow or in dispersed flow as in bubbly flow and gas-droplet flow or in a combination of both as in annular flow. All these flows are encountered in practical situations.

The nature of gas-liquid flow in tubes is quite complex due to the presence of a mobile and pliable interface between the two phases across which exchanges of mass, momentum and heat take place. Normally, the shape and structure of the interface determines the transport coefficients. Due to this, gas-liquid flow is classified in terms of flow patterns based on the configuration of the interface. Four major flow patterns are recognised for horizontal flow: stratified flow, bubbly flow, slug (or plug) flow and annular flow. In vertical flow, stratified flow does not exist but an additional flow pattern, namely, churn flow is often distinguished. (See entry **Two-Phase Flows** for details).

Much research has been done over the past few decades on determining the characteristics of gas-liquid flow in tubes. Several methods have been developed to calculate important design parameters such as the flow patterns, pressure drop and liquid holdup (or void fraction) in two-phase flow. These have been summarised in standard textbooks (Wallis, 1969; Hewitt and Hall-Taylor, 1970; Govier and Aziz, 1972; Hetsroni, 1982) as well as in the present book. A comprehensive treatment of heat transfer in gas-liquid systems is given in Hewitt et al. (1994).

References

Govier, G. W. and Aziz, K. (1972) *The Flow of Complex Mixtures in Pipes*, Van Nostrand Reinhold, New York.

Hetsroni, G. (1982) *Handbook of Multiphase Systems*, Hemisphere, New York.

Hewitt, G. F. and Hall-Taylor, N. S. (1970) *Annular Two-Phase Flow*, Pergamon Press.

Hewitt, G. F., Shires, G. L. and Bott T. R. (1994) *Process Heat Transfer*, CRC Press.

Wallis, G. B. (1969) *One-Dimensional Two-Phase Flow*, McGraw-Hill, New York.

Leading to: *Annular flow; Bubbly flow; Churn flow; Liquid flow; Slug flow; Stratified gas-liquid flow; Two-phase flow*

G.F. Hewitt

TUBE SHEETS (see Mechanical design of heat exchangers; Shell-and-tube heat exchangers)

TUBES, SINGLE-PHASE FLOW IN

Following from: *Channel flow*

Most often the term "tube" in the theory of heat transfer implies a duct with the round (circular) cross section.

Straight Smooth Tubes

In flow at low Reynolds Number (Re) in a long straight tube with a constant cross section and smooth walls, each particle moves with a constant velocity along a rectilinear trajectory. Due to viscosity, the fluid particles, flow near the walls at a lower rate then those far from the walls. The flow is an orderly motion of layers relative to each other. This flow is known as laminar. A shear stress which is determined by the Newton law of viscosity $\tau_l = \eta(du/dy)$ originates between these layers. However, at higher Re numbers the flow is no longer orderly. An intense cross mixing of fluid changes the flow into a turbulent one.

O. Reynolds was the first to study consistently the different patterns in laminar and turbulent flows. According to his investigations, the transition of the laminar flow pattern to the turbulent one occurs at a *critical Reynolds number* $Re_c = (\bar{u}d/\nu)_c \cong 2300$.

Re_c significantly depends on the conditions at the tube inlet and fluid flow to the inlet. V. V. Ekman, by carefully reducing disturbances at the tube inlet, obtained $Re_c \cong 40000$. There also exists a lower limit of Re_c corresponding to about 2000 below which even the strongest disturbances decay.

The experimental studies have shown that in a definite Re range the flow in the vicinity of Re_c is of an intermittent nature, i.e. first laminar and then turbulent. This flow can be described by the intermittence coefficient γ indicating over which fraction of the time the flow is turbulent. Hence, the flow is turbulent at $\gamma = 1$ and laminar at $\gamma = 0$. I. Rotta established the dependence of γ on the dimensionless distance from the tube inlet x/D (here, x is the distance and D the tube diameter) for different Re (see **Figure 1**). At constant Re, γ monotonically grows with increasing distance x/D. The tube length which is necessary for laminar-turbulent transition diminishes with increasing Re.

Other flow characteristics, including velocity distribution over the cross section, also change over some length of the tube beginning with the inlet. If the fluid flows into the tube out of a big tank, the velocity distribution at the inlet is uniform, but with the distance from the inlet it gradually becomes elongated, under the action of friction forces, taking, at some distance from the inlet, a final shape remaining henceforward unchanged. The tube length, over which the velocity profile changes, is referred to as the entry or initial *hydrodynamic length.* For laminar fluid flow this length strongly depends on Re and can be determined by the Boussinesq relation $l_{in}/D = 0.065$ Re.

For *turbulent flow,* the entry hydrodynamic length is about 50 tube diameters. In what follows we will analyze the developed flow in the tube.

Transition of the laminar flow *to the turbulent flow* drastically changes the velocity profile and the friction drag law. In laminar flow, the velocity distribution over the cross section is parabolic (**Figure 2**, curve 1), while in the turbulent flow, as a result of pulsation exchange in the transverse direction, it is substantially more uniform (curve 2). The turbulent velocity profile can be divided into four characteristic regions (**Figure 3**): a viscous region (or a linear sublayer) immediately near the tube wall ($0 \le y^+ \le 5$, $y^+ = yu_f/\nu$, $u_f = \sqrt{\tau_w/\rho}$) in which the velocity variation is linear and determined by the coefficient of molecular viscosity; an intermediate, or buffer, layer ($5 \le y^+ \le 30$) in which viscous and turbulent stresses are commensurate in quantity; a fully turbulent (logarithmic) layer ($y^+ > 30$, $y/r_0 < 2$) in which the molecular

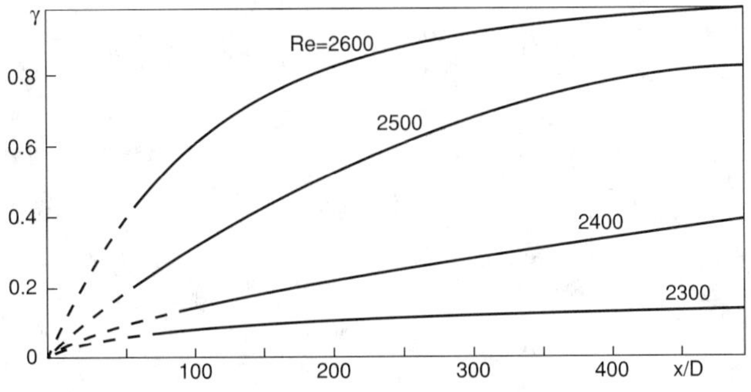

Figure 1. Intermittence coefficient as a function of distance and Reynolds number.

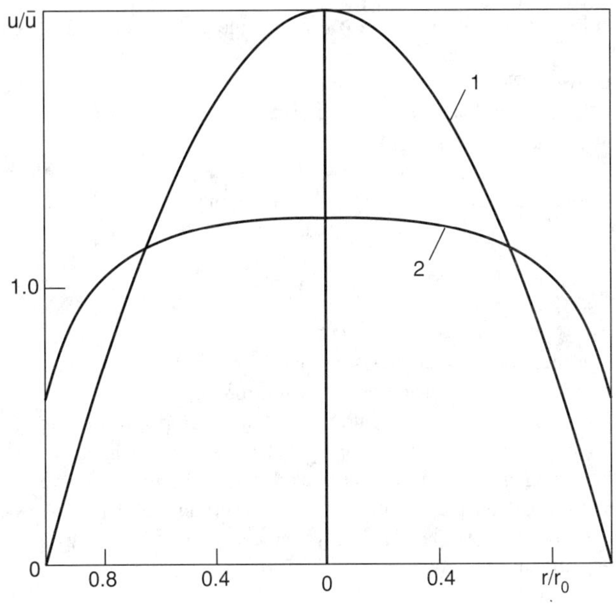

Figure 2. Velocity profiles in laminar (1) and turbulent (2) flows.

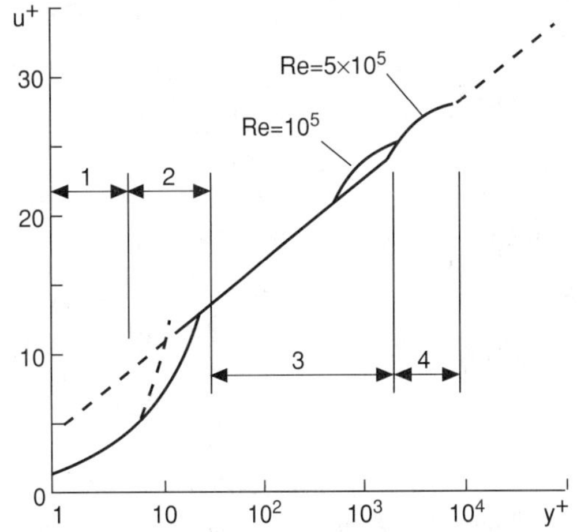

Figure 3. Velocity profile near tube wall in terms of $u^+ = u/u_f$ and $y^+ = y\,u_f/\nu$.

viscosity can be neglected and, finally, a turbulent core or the region of validity of the law of the wake of the in the tube center. The former three regions form the wall layer with a universal (Re-independent) velocity distribution in $u^+ - y^+$ coordinates. The point at which the velocity profile in the flow core deviates from the logarithmic one in the coordinates $u^+ = u/u_r = f(y^+)$ depends on Re. In the $u/\bar{u} = f(y/r_0)$ coordinates it approximately equals $y/r_0 = 0.15$ and remains nearly unchanged over a considerable Re range.

For turbulent flow, the shear stress is expressed as the sum of molecular and turbulent stresses

$$\tau = \tau_l + \tau_t = \rho\nu \frac{du}{dy} + \rho\nu_t \frac{du}{dy}. \qquad (1)$$

Variation of turbulent shear stresses τ_t and turbulent viscosity ν_t

with y^+ is shown in **Figure 4**. The turbulent shear stress depends linearly on y almost everywhere except for the narrow region near the wall (**Figure 4**, curve 1) since $\tau_t/\tau_w = (1 - y/r_0) - \tau_l/\tau_w$. The molecular viscosity is commensurate with the turbulent viscosity only near the wall (**Figure 4**, curve 2).

Using the Prandtl hypothesis for the mixing length l, we can represent turbulent viscosity by the relation

$$\nu_t = l^2\left(\frac{du}{dy}\right). \tag{2}$$

The quantity l characterizes the local properties of the turbulent flow and can be considered as local turbulence scale. The curves of dimensionless mixing path length l/r_0 are graphed in **Figure 5**. Prandtl assumed that the mixing path length increases linearly with increasing distance from the wall: $l = ky$ (**Figure 5**, curve 1). Nikuradze determined experimentally the mixing length for flowing in a tube. The results of his measurements (**Figure 5**, curve 2) are described by the equation

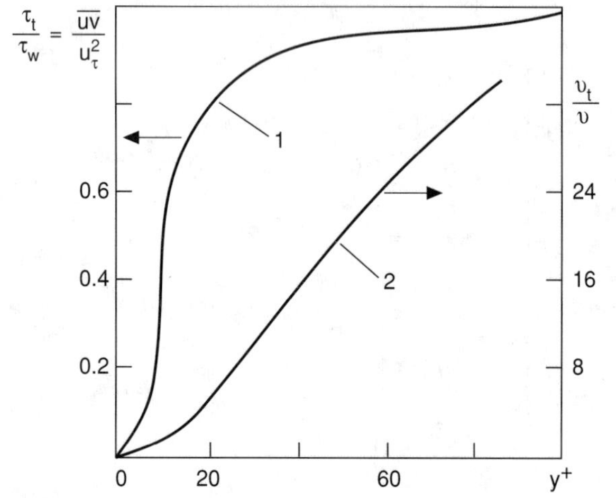

Figure 4. Variation of turbulent shear stress and turbulent viscosity with y^+.

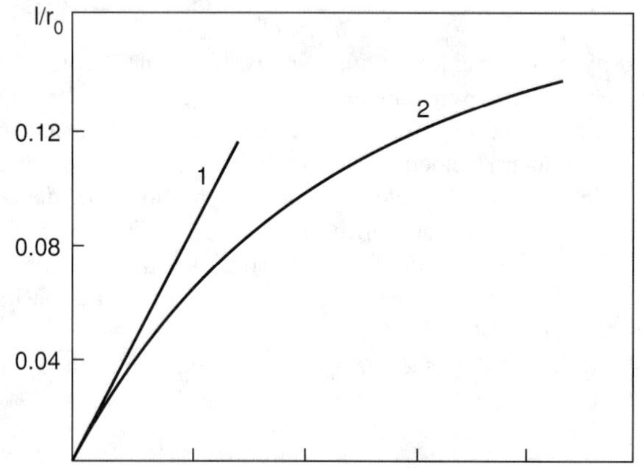

Figure 5. Variation of mixing length with distance from the wall.

$$l/r_0 = 0.14 - 0.08(1 - y/r_0)^2 - 0.06(1 - y/r_0)^4. \tag{3}$$

In the case of short distances from the wall, all the terms in Equation 3 of a higher order than y/r_0 can be omitted. This results in $l = 0.4y$. Thus, by *Nikuradze's measurement* results, k known as the von Karman or turbulence constant turns out to be 0.4. Manifold subsequent measurements of this constant also yield values close to 0.4. Thus, in the wall region the coefficient of turbulent kinematic viscosity is

$$\nu_t = k^2y^2\left(\frac{du}{dy}\right). \tag{4}$$

Now we turn back to the velocity profile (**Figure 3**). In the viscous sublayer the dimensionless velocity is equal to a dimensionless coordinate $u^+ = y^+$. Prandtl derived for the wall turbulent region

$$u^+ = \frac{1}{k}\ln y^+ + A. \tag{5}$$

Equation 5 follows from Equations 1 and 4 under the assumption that $\tau = \tau_l + \tau_t = \tau_t = \text{const}$. The reliability of this approximation can be estimated using **Figure 4**. The constants entering Equation 5 for the first time were determined based on Nikuradze's experiments. As was noted, $k = 0.4$ and the constant A was found to be equal to 5.5. Hence, the velocity distribution in the turbulent zon is described by the logarithmic dependence

$$u^+ = 2.5\ln y^+ + 5.5. \tag{6}$$

T. von Karman introduced the concept of an intermediate, or buffer, layer. Retaining for the viscous sublayer and for the turbulent zone the above expressions for the velocity profile, he derived the equation

$$u^+ = 5\ln y^+ - 3.05 \tag{7}$$

for the range of y^+ values from 5 to 30.

Reichardt's formula is extensively used for $Re > 2 \times 10^4$ range to describe the velocity distribution over the entire cross section of the round tube by a unified relation

$$u^+ = 2.5\ln\left[(1 + 0.4y^+)\frac{1.5(1 + r/r_0)}{1 + 2(r/r_0)^2}\right]$$

$$+ 7.8\left[1 + \exp(-y^+/11) - \frac{y^+}{11}\exp(-0.33y^+)\right], \tag{8}$$

the coefficient of turbulent viscosity being described by

$$\frac{\nu_t}{\nu} = 0.4\left(y^+ - 11\tanh\frac{y^+}{11}\right) \tag{9}$$

for $y^+ \leq 50$ and by

$$\frac{\nu_t}{\nu} = 0.133y^+[0.5 + (r/r_0)^2](1 + r/r_0) \quad (10)$$

for $y^+ > 50$.

Van Driest extended Prandtl's Equation 4 for turbulent viscosity to a viscous layer introducing into the expression for the mixing path length the damping factor

$$\nu_t = k^2y^2\left[1 - \exp\left(-\frac{y^+}{26}\right)\right]^2\left(\frac{du}{dy}\right). \quad (11)$$

Figure 6 shows the variation of root-mean-square pulsations along the tube radius. The regions of the flow near the axis and near the wall are depicted in two different figures. In one region the velocity scale is u_0, the maximum velocity which can be achieved on the tube axis, in the other, the dynamic velocity u_f. In the entire region axial pulsations $\sqrt{\overline{u'^2}}$ are most intensive. Their maximum value $\sqrt{\overline{u'^2}}/u_f = 0.08$ is achieved inside the buffer layer at $y^+ = 15$. Radial velocity pulsations $\sqrt{\overline{v'^2}}$ are the minimum ones. As is seen from **Figures 4** and **6**, inside the wall layer turbulence is neither homogeneous, for which the condition $\overline{u'^2} = \overline{v'^2} = \overline{w'^2}$ must hold, nor isotropic, with condition $\overline{u'v'} = \overline{v'w'} = \overline{u'w'} = 0$ would be valid.

For flow in a tube it is customary to define a dimensionless friction factor in terms of the pressure gradient associated with friction, (see also **Friction Factors, Pressure Drop, Single Phase**)

$$\bar{f} = \frac{(-dP/dx)D}{\frac{1}{2}\rho\bar{u}^2},$$

which defines the *Moody (or Weissbach) friction factor.* The alternative definition, $f = c_f = \tau_w/(1/2\ \rho\bar{u}^2)$, called the Fanning friction factor, is also used. Note that $\bar{f} = 4f$.

For the laminar flow the friction factor may be calculated using the *Hagen-Poiseuille relation*

$$\bar{f} = K\,Re^{-1} = 64/Re, \quad (12)$$

where $K = 64$ for round tubes but depends on the cross section shape of the channel for other channel shapes. The K values for some shapes of cross section other than round are presented in **Channel Flow.**

If the flow is turbulent, *Blasius' equation* (for $Re < 10^5$)

$$\bar{f} = 0.316\,Re^{-0.25} \quad (13)$$

or Filonenko's relation (for $10^4 < Re < 5 \times 10^6$)

$$\bar{f} = (1.821g\,Re - 1.64)^{-2} \quad (14)$$

can be used for determination of $\bar{f}$.

The above laws hold only for isothermal motion when the fluid temperture and, hence, the physical properties (ρ, η, λ, c_p) in all the points of the flow retain each the same value. With heat transfer the fluid temperature varies over both the tube section and length. The temperature variation over cross section brings about the change of physical properties and, as a consequence, the change of the velocity profile and pressure loss. The variation of physical properties of liquids and gases with temperature is different. In liquids, the dynamic velocity coefficient η is the most temperature-dependent value, while other properties depend relatively slightly on temperature. Therefore, in the case of liquids we can confine ourselves to considering only viscosity variation assuming other properties to be constant and equal to the values at some mean temperature. In gases temperature substantially changes density ρ, the dynamic viscosity η, and the thermal conductivity λ. The specific heat c_p varies relatively slightly. Therefore, in the case of gas flow at fairly high temperature drops it is necessary to take into account the dependence of $\bar{f}$ on temperature, (ρ, η, and λ, and, sometimes, also c_p). The most widespread method of taking into account nonisothermal nature of flow in calculating the pressure loss consists in introducing correction factors into friction factor determined for the case of isothermal flow. The dependence $\bar{f}/\bar{f}_0 = (\eta_w/\eta_b)^n$ is used for liquids. Here $\bar{f}$ and $\bar{f}_0$ are the friction factor for nonisothermal and isothermal flows respectively, η_w and η_b are the dynamic viscosity at the wall and bulk temperature respectively. The exponent n depends on the flow regime, the direction of the flux, and other factors. In order to calculate the friction factor for nonisothermal turbulent gas flow we can make use of Kutateladze's approximate relation $\bar{f}/\bar{f}_0 = (T_b/T_w)^{0.5}$ where T_b is the bulk and T_w the wall temperature and $\bar{f}_o$ is determined as the bulk temperature.

Rough Tubes

In practice, especially at high Re the tubes cannot be always considered hydraulically smooth since elements of roughness can

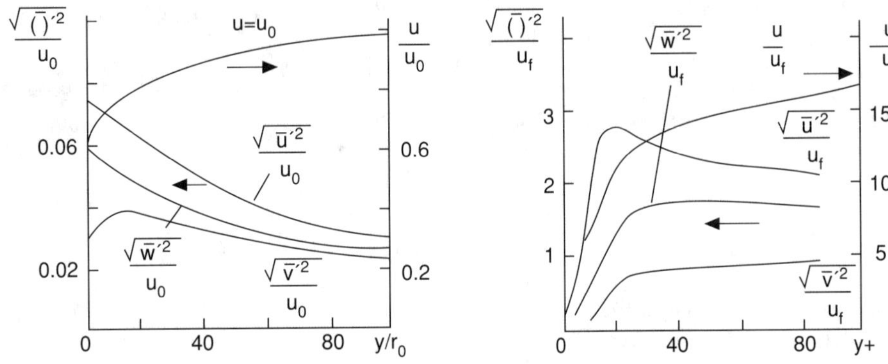

Figure 6. Variation of turbulence parameters with distance from the tube wall.

occur on their surface. Roughness can be both natural, due to production process and service conditions, or artificial, i.e. purposefully made on the tube surface (thread, transverse projections, grooves, etc.) in order to augment heat transfer. Two main types of roughness are commonly distinguished: three-dimensional and two-dimensional. The former is produced by roughness elements on the surface such as projections of various shapes about which the flow is three-dimensional. The latter type are projections of various shapes which are continuous around the tube perimeter. The flow around these projections is two-dimensional. In general, the surface roughness is characterized by the height and shape of the elements of roughness closely adjoining one another, e.g. sand grains; one parameter, the height of the element of sand-like roughness k_s is sufficient to characterize this roughness. Therefore, in this case the character of flow and resistance depends only on one additional parameter, *viz.* the roughness ratio k_s/r_0. In turbulent flow in rough tubes Schlichting distinguished three flow regimes, depending on the dimensionless roughness height or the so called Reynolds number of the roughness $k_s^+ = k_s u_\tau/\nu$:

1. A surface with no manifestation of roughness ($0 \leq k_s^+ \leq 5$) in which roughness elements do not project beyond the viscous sublayer and the surface behaves as a hydraulically smooth one. In this case the friction factor is determined from the relations for smooth tubes as given above.
2. The transition regime ($5 \leq k_s^+ \leq 70$) when surface roughness projects beyond the viscous sublayer and the friction factor depends not only on Re, but on roughness ratio as well.
3. The fully rough regime ($k_s^+ > 70$) under which resistance predominantly consists of resistances of single roughness elements dominate the resistance. In this region, fluid viscosity plays no role and the friction factor becomes a function of roughness ratio alone.

According to Nikuradze, the friction factor of tubes with sand-like in the fully rough regime is expressed as

$$\bar{f} = \frac{1}{(2\ln(r_0/k_s) + 1.74)^2}. \qquad \textbf{(15)}$$

Near the wall the velocity profile in the rough tube builds up less steeply than in a smooth tube. Distribution of the mixing length along the radius of a rough tube absolutely coincides with the same distribution in smooth tubes. Here too, the distribution of the mixing length is determined by Equation 3. In particular, near the wall $l = ky = 0.4y$. This means that the logarithmic law of velocity distribution (Equation 5) is also applicable to rough tubes (**Figure 7**). One should only shift the y coordinate depending on the value of the element of sand-like roughness k_s. Consequently, the velocity profile takes the form

$$u^+ = 2.5\ln\frac{y^+}{k_s} + R(k_s^+). \qquad \textbf{(16)}$$

Here $R(k_s^+)$ is the so called roughness hydrodynamic function.

The shaded region in **Figure 7** depicts the transition region from a hydraulically smooth surface to the one with well-pronounced roughness.

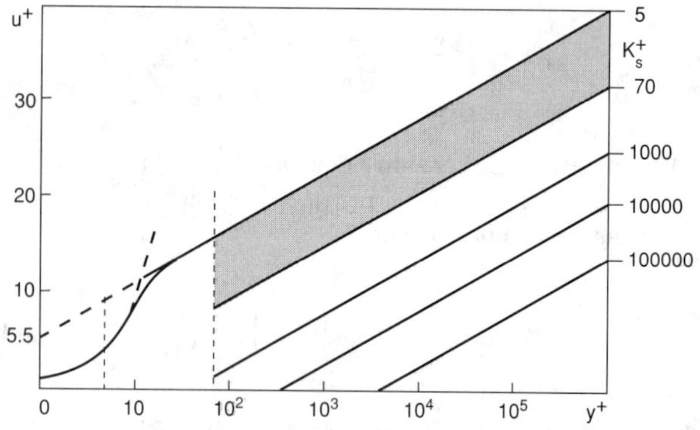

Figure 7. Velocity profiles in tough tubes

The comparance of velocity profiles for a smooth (6) and a rough (16) walls was used to derive

$$R(k_s^+) = 5.5 + 2.5\ln k_s^+ \qquad \textbf{(17)}$$

for sand-like roughness.

Nikuradze's measurement results, for the fully rough regime with sand-like roughness, yield $R(k_s^+) = 8.5$. **Figure 8** shows roughness function $R(k_s^+)$ versus the dimensionless value of sand-like rouhgness. Zone 1 corresponds to hydraulically smooth surface, zone 2, to the region, and zone 3, to the fully rough regime. Curve 1 is plotted from Equation 17 and curve 2, corresponds to $R(k_s^+) = 8.5$.

As was noted above, in contrast to sand-like roughness, real roughness cannot be determined by assigning one parameter of the k_s type. However, a technical roughness can be determined as Schlichting suggested, in terms of equivalent sand-like roughness. Equivalent sand-like roughness means the height of the sand-like roughness element k_s such that the tube with this roughness which has the same friction factor $\bar{f}$ in the fully rough regime as the tube with a given technical roughness. The equivalent sand-like roughness is found experimentally or from published data. The equivalent sand-like roughness is calculated by measuring the friction factor for a tube with a given roughness type (Equation 16).

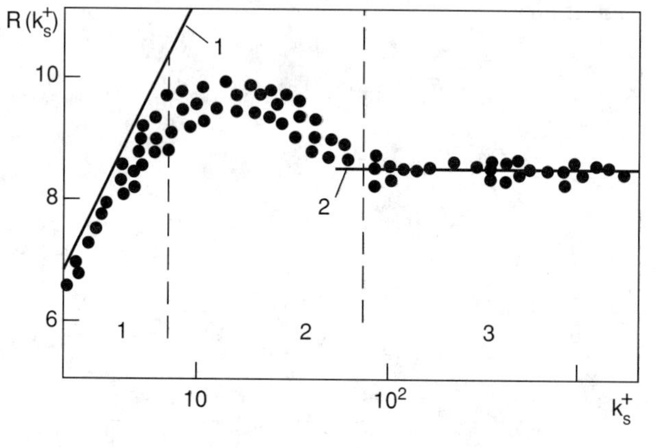

Figure 8. Roughness hydrodynamic function as a function of dimensioless roughness height.

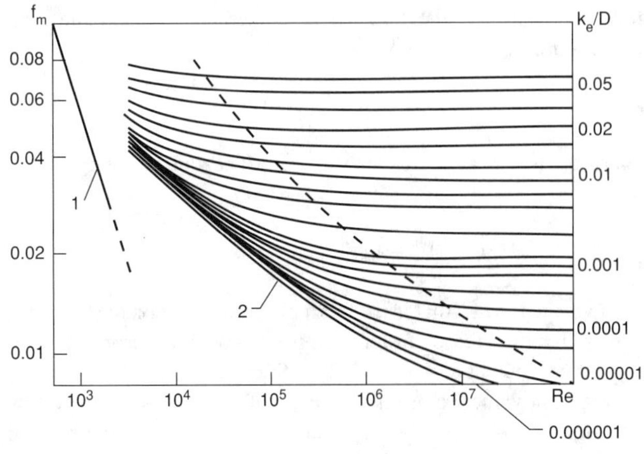

Figure 9. Moody diagram for friction factor in pipe flow.

Figure 9 presents the Moody diagram for commercial tubes in which roughness is expressed in terms of the equivalent sand-like roughness. The dashed curve represents the boundary between the regimes of partial and fully rough regimes, curves 1 and 2 correspond to laminar (Equation **12**) and turbulent (Equations **13** and **14**) flows in smooth tubes. In this case, the friction factor for all the three flow regimes in tubes with natural roughness is well described by *Colebrook and White's equation*

$$1/\sqrt{f} = 1.74 - 2\ln(k_e/r_0 + 18.7/\mathrm{Re}\sqrt{f}). \qquad \mathbf{(18)}$$

For $k_e/r_0 \rightarrow 0$ eq. (18) takes the form of *Prandtl's formula* for hydraulically smooth tubes

$$1/\sqrt{f} = 2\ln(\mathrm{Re}\sqrt{f}) - 0.8 \qquad \mathbf{(19)}$$

which virtually coincides with Filonenko's Equation (**14**) though the latter is more convenient for calculations. For $\mathrm{Re} \rightarrow \infty$ Equation 18 reduces to Equation 15 for the fully rough regime.

References

Schlichting, H. (1968) *Boundary Layer Theory*, 6th ed., McGraw-Hill, New York.

Arpaci, V. S. and Larsen, P. S. (1984) *Convection Heat Transfer*, Prentice-Hall, Inc., Englewood Cliffs.

P. Poškas

TUBES, SINGLE-PHASE HEAT TRANSFER IN

Following from: *Convective heat transfer*

Forced flow in tubes is used widely in practical applications (see **Tubes, Single-Phase Flow In**). If the velocity and temperature distribution of fluid are uniform at the inlet to the tube, dynamic and thermal boundary layers begin to develop symmetrically along the wall (**Figure 1**). The boundary layer thickness increases along the tube length and gradually fills the entire flow section. After the dynamic boundary layers are joined, a constant velocity distribution sets in which is parabola-shaped for the laminar flow (at $\mathrm{Re} \leq 2300$) or a distribution characteristic of the turbulent flow (see **Tubes, Single-Phase Flow In**). The distance from inlet to this section is known as entrance length or hydrodynamic stabilization section.

In heat transfer, as the fluid flows along the tube, the wall layers are heated or cooled. In this case, at the entrance region of the tube, the fluid core retains a temperature equal to that at the inlet T_0 and does not participate in heat transfer. Temperature variation occurs in the wall layers. Thus, near the tube surface in the entrance region a thermal boundary layer is initiated whose thickness, as it becomes farther away from the inlet, grows. The thermal boundary layers are joined at a certain distance from the inlet known as the thermal entrance region L_T and afterwards all the fluid participates in heat transfer. After that, the dimensionless temperature profiles $\theta = (T - T_w)/(T_0 - T_w)$ become similar, i.e. the temperature in different sections differs only in magnitude, while the law according to which it varies across the radius remains the same.

For the turbulent flow P.L. Kirillov and co-authors obtained, by analogy with the *universal velocity profile* (see **Tubes, Single-Phase Flow In**), a *universal temperature distribution of* the form (**Figure 2**)

$$T^+ = y^+\mathrm{Pr} \text{ for } y^+ \leq 5,$$

$$T^+ = 5.75\ln(y^+\,\mathrm{Pr}) + 5.5 \text{ for } 5 \leq y^+ \leq 30,$$

$$T^+ = A\ln(y^+\,\mathrm{Pr}) + B(\mathrm{Pr}) \text{ for } y^+ \geq 30.$$

Here $A = \mathrm{Pr}_t/k = 2.3$, k is the turbulence constant ($k = 0.4$), Pr_t the turbulent Prandtl number taken here equal to 0.9. B(Pr) and T^+ is given by the dimensionless temperature ($T^+ = (T_w - T)\rho c_p u_\tau/q_w$).

$$B(\mathrm{Pr}) = 24.5\,\mathrm{Pr} - 8.2\ln\mathrm{Pr} - 19$$

The coefficient B(Pr) increases greatly with increasing Pr due to diminishing fluid heat conduction. As Pr increases, an ever greater fraction of the thermal resistance occurs in the viscous sublayer.

Kader's formula is convenient to use which describes the temperature profile by a unified relation

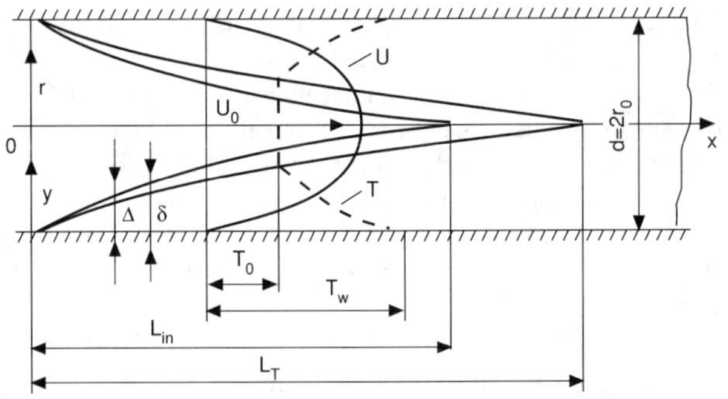

Figure 1. Development of hydrodynamic and thermal boundary layer at the entrance to a tube.

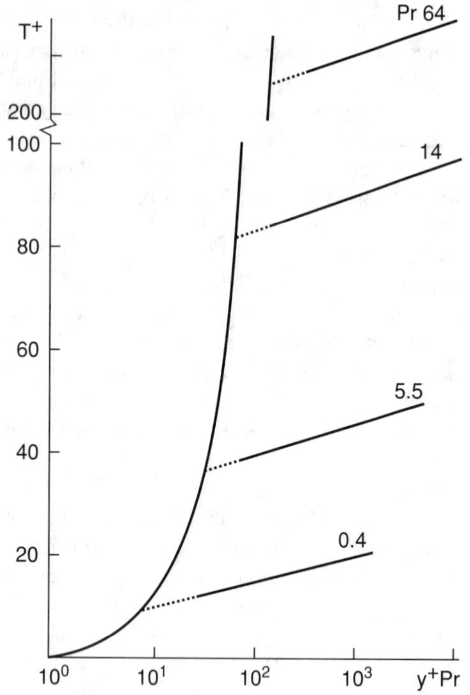

Figure 2.

$$T^+ = \Pr y^+ \exp(-\Gamma) + \left\{2.12 \ln\left[(1 + y^+)\frac{1.5(2 - y/r_0)}{1 + 2(1 - y/r_0)^2}\right] + \beta(\Pr)\right\}\exp\left(-\frac{1}{\Gamma}\right),$$

where $\Gamma = \dfrac{10^{-2}(\Pr y^+)}{1 + 5\Pr^3 y^+}$ and $\beta(\Pr) = (3.8 \Pr^{1/3} - 1.3)^2 + 2.12\ln \Pr$.

In a manner similar to the coefficient of turbulent viscosity $\nu_t = -\overline{u'v'}/\dfrac{\partial u}{\partial y}$, we can introduce a parameter for turbulent heat transport, i.e., the coefficient of turbulent thermal conductivity. In a thin shear layer, the most important of the three fluctuating components of turbulent heat transport, is $\rho c_p \overline{v'T'}$, which is the rate of turbulent enthalpy transfer along y. The turbulent thermal diffusivity for this value $\kappa_t = -\overline{v'T'}/(dT/dy)$. There is virtually no absolute analogy in turbulent flow between heat and momentum transfer, since the term containing the fluctuating pressure component enters the equations defining the velocity field, but does not enter the temperature-field-defining equations. However, the ratio ν_t/κ_t which is said to be the turbulent Prandtl number (in analogy with the molecular number $\Pr = \nu/\kappa = \eta c_p/\lambda$) is commonly assumed as the value of the order of one

$$\Pr_t = \frac{\nu_t}{K_t} = \frac{\overline{u'v'}\,(dT/dy)}{\overline{v'T}\,(du/dy)}$$

The heat transfer coefficient α is determined by the temperature difference between the bulk fluid temperature T_b and the wall temperature T_w

$$\alpha = \dot{q}_w/(T_w - T_b),$$

where $\dot{q}_w$ is the heat flux at the wall. The bulk temperature is given by

$$T_b = [2\textstyle\int_0^{r0} Turdr]/(r_0^2\bar{u})$$

Owing to the fact that in the thermal entrance region the temperature gradient decreases faster than the temperature drop, the heat transfer coefficient $\alpha = -\lambda(dT/dy)_w/(T_b - T_w)$ diminishes tending to a constant value that is characteristic of the fully developed flow (see **Figure 3a**). If we have a laminar flow consecutively developing into a turbulent one, variation of the heat transfer coefficient over the tube length is different as in **Figure 3b**, where the heat transfer coefficient decreases in the laminar flow region and the grows again before reducing to its fully developed (constant) value for the turbulent region.

It has been shown analytically that for laminar flow of a fluid with constant physical properties and uniform temperature at the inlet, for T_w = const in the heated region and where the flow is hydrodynamically fully developed before heating commences, L_T is given by

$$L_T/D = 0.055 \text{ Pe} = 0.055 \text{ Re Pr},$$

i.e., at a given Re, the length of *thermal entrance region* is deter-

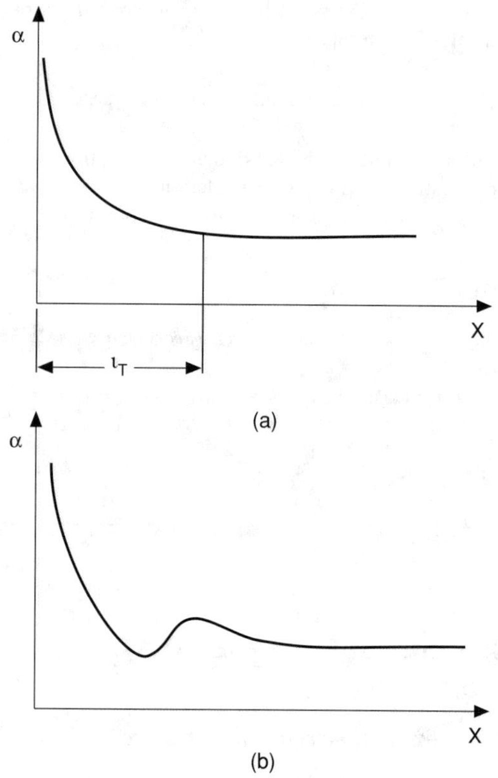

Figure 3. Entrance effects in laminar flow (a) and in laminar flow developing into turbulent flow (b).

mined by the value of Pr. In calculating heat transfer coefficient under these conditions for gases and liquids the relationships may be used

$$Nu = \frac{\alpha D}{\lambda} = 1.03\left(\frac{1}{Pe}\frac{x}{D}\right)^{-1/3} \quad \text{for } Pe^{-1}\ x/D < 0.01$$

$$Nu = 3.655 + \frac{1}{\left(\frac{1}{Pe}\frac{x}{D}\right)^{0.488} \exp\left(57.2\frac{1}{Pe}\frac{x}{D}\right)}$$

$$\text{for } Pe^{-1}\ x/D < 0.01$$

Where Pe = RePr.

When Pe^{-1} x/D → ∞, Nu ≃ 3.66 corresponding to the region of fully developed heat transfer.

In laminar flow, heat transfer depends on the thermal boundary condition. Thus, when the heat heat flux on the wall is constant ($\dot{q}_w$ = const), which is often the case in practice, the limiting Nusselt number Nu = 4.36, the length of the thermal entrance region L_T = 0.07 Pe, and heat transfer in this case is generally by approximately 20% higher than under the boundary condition T_w = const.

The correlations for laminar flow presented above describe heat transfer when there is fully developed flow at the inlet. If the heated region starts at the entrance of the tube and the hydrodynamic and heat transfer boundary layers develop stimultaneously (as in the case for the results shown in **Figure 1**) then heat transfer in the developing region will be slightly higher than during fully developed flow.

A number of generalizing dependences are suggested for calculating fully developed heat transfer in the turbulent flow. For practical calculations it is advisable to use an equation suggested by B. S. Petukhov and V. V. Kirillov

$$Nu_\infty = \frac{(\bar{f}/8)Re\ Pr}{k + 12.7\sqrt{\bar{f}/8}(Pr^{2/9} - 1)},$$

where $\bar{f} = (1.82 \ln Re - 1.64)^{-2}$, k = 1.07.

This relationship was refined by Petukhov and co-authors, k = 1.07 was replaced by k = 1 + 900/Re. With allowance for this correction the equation may be used for $10^4 \le Re \le 5\ 10^6$ and $0.2 \le Pr \le 200$.

Empirical dependences of the form

$$Nu = c\ Re^m Pr^n,$$

where c, m, and n are constants, are also extensively used. For instance, c = 0.023, m = 0.8, n = 0.4 according to *Dittus and Boelter's* data, while c = 0.021, m = 0.8, n = 0.43 according to Mikheev's data.

The length of the thermal entrance region during turbulent flow is by far shorter than during laminar flow. With increasing Pr the Nusselt number in the thermal entrance region diminishes and differs slightly from Nu_∞ at Pr ≫ 1 (**Figure 4**).

In laminar flow in tubes of noncircular cross section heat transfer varies drastically with changing shape. For turbulent flow Nu can be calculated by the equations for circular tubes if an equivalent

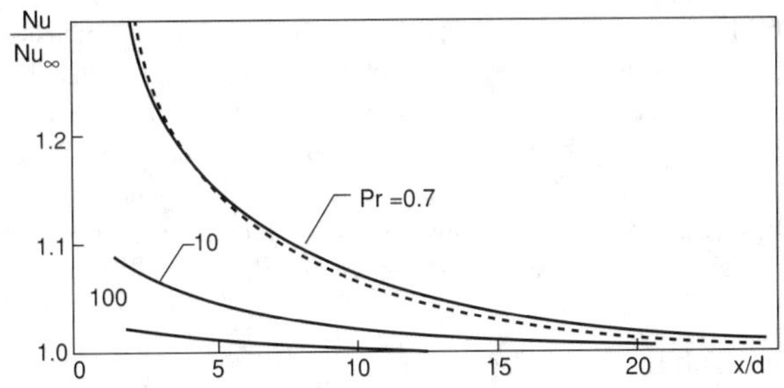

Figure 4. Effect of Prandtl number on development of heat transfer in a tube.

diameter D_e = 4S/Π is used as a characteristic parameter, where S is the cross sectional area and Π the wetted parameter. However, this is justified only in the case where heat transfer occurs across the entire wetted perimeter Π.

The direction of heat flux (fluid heating or cooling) effects heat transfer if the physical properties of fluid within the temperature range considered vary substantially. For heated gases this influence can be taken into account using an additional factor $(T_w/T_b)^n$, where (T_w/T_b) is the temperature factor. Within $0 < T_w/T_b < 2$ for monatomic and diatomic gases the exponent n = −0.5 in turbulent flow. The effect of temperature factor on heat transfer to polyatomic gases is slightly weaker. For gas cooling ($T_w/T_b < 1$) it can be assumed that n ≃ 0. For liquids, the effect of temperature-dependent properties on heat transfer can be taken into account using the factor $(\eta_w/\eta_b)^m$, where η_b is the dynamic viscosity of the bulk fluid temperature, η_w the value at the wall temperature. The constant m depends on the direction of heat flux. For fluid heating, i.e., at $\eta_w/\eta_b < 1$, m = −0.11, for fluid cooling, i.e., $\eta_w/\eta_b > 1$, m = −0.25.

For the effect of thermogravitational convection on the flow and heat transfer in tubes see **Mixed convection**.

Rough Tubes

Artificial two- and three-dimensional roughness is often used for augmentation of heat transfer. As a result of flow separation, a vortex zone appears behind the projection whose extension to the point of reattachment to the surface is 6–8 k, where k is the height of roughness element. A strong turbulent flow is initiated behind this zone. A small vortex region with 1–2 k extension also is initiated before the next projection. If the distance between projections is small, smaller than 6k, then the vortex zones fill all the width of recession and heat transfer augmentation is ineffective. Investigations evidence that in order to afford the most effective heat transfer the distance between projections must be within 10–12 k. The higher the Prandtl number, the smaller the height of roughness elements for effective augmentation. Investigations show that in the fully rough regime (see **Tubes, Single-Phase Flow In**) heat transfer rate is virtually independent of the height of roughness elements, though the pressure loss monotonically falls with diminishing k. This means that in designing heat exchangers it is advisable to use projections of the minimum height which still affords the effect of roughness to manifest itself to a full degree. Achieving rational artificial roughness makes it possible to increase the heat transfer coefficient two- to threefold. It is desirable, however, that heat transfer enhancement not involve a too great increase of pressure loss. Because of this, the elements of two-dimensional roughness with sharp edges (of rectangular or triangular shape) are not advantageous. Powerful vortex zones and high resistance of the shape arising behind them lead to high energy losses. It is more advantageous to use correctly streamlined roughness elements whose resistance is much lower and for which heat transfer in the fully rough regime remains at the same level.

Bent Tubes

Secondary flows initiating in *bent tubes* substantially augment *heat transfer* due to the action of centrifugal forces. An especially high (severalfold) enhancement of heat transfer is observed at Re corresponding to the region of laminar-vortex flow which in straight channels is appropriate to the laminar or the transition flow regime. In curvilinear channels centrifugal forces are oriented in a different way with respect to the areas of heat-transfer surface, therefore, the heat transfer coefficient varies greatly around the perimeter of the channel. In laminar-vortex flow heat transfer on the outward generatrix of the coil tube may exceed severalfold heat transfer on the inward generatrix. This ratio for the turbulent flow is lower.

References

Arpaci, V. S. and Larsen, P. S. (1984) *Convection Heat Transfer,* Prentice-Hall, Inc., Englewood Cliffs.

Kakaç, S., Shah, R. K., and Aung, W. (1987) *Handbook of Single-Phase Convective Heat Transfer.*

Leading to: *Coiled tubes, heat transfer in; Rotating duct systems, orthagonal, heat transfer in; Rotating duct systems, parallel, heat transfer in*

P. Poškas

TUBES SINGLE-PHASE HEAT TRANSFER TO, IN CROSS-FLOW

Following from: *Cross-flow heat transfer; Immersed bodies, heat transfer and mass transfer*

The complex flow pattern around a tube or circular cylinder in crossflow influences the temperature field in the fluid created when the tube surface is kept a temperature different from that of the approaching stream. A description of the flow phenomena as a fluid flows across a tube have been presented elsewhere (See **Tubes, Crossflow Over**). **Figures 1** and **2** show temperature fields at $Re_D = 5$ and $Re_D = 40$, respectively.

As the **Peclet Number**, $Re_D Pr$, is increased the importance of the convective transport relative to the diffusive transport is increased. The isotherms come closer to each other and the temperature gradients become greater. Also the thermal wake becomes narrower and more elongated in the main flow direction as the Peclet number is increased. The heat transfer between the tube and fluid is enlarged accordingly. The heat transfer coefficient is usually expressed by a dimensionless number, the **Nusselt Number** defined as

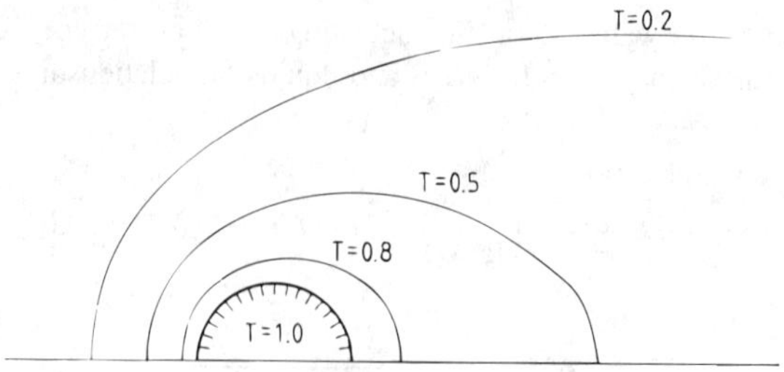

Figure 1. Isotherms in the fluid at $Re_D = 5$, $Pr = 0.72$.

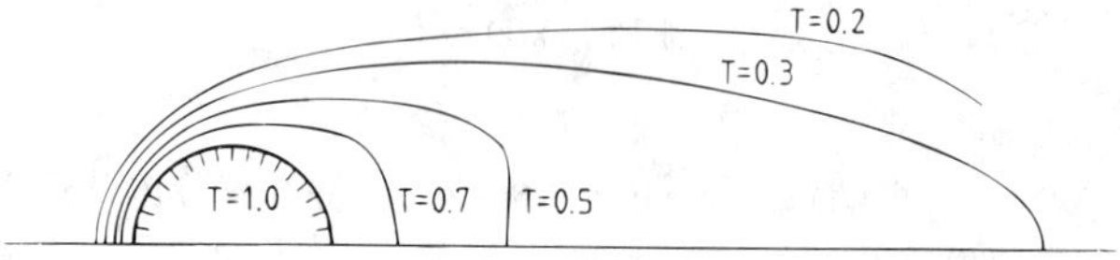

Figure 2. Isotherms in the fluid at $Re_D = 40$, $Pr = 0.72$.

$$Nu_D = \frac{\alpha D}{\lambda} \tag{1}$$

where α is the heat transfer coefficient, D the tube diameter and λ the thermal conductivity of the fluid.

Figure 3 presents the local Nusselt number for a single tube in cross flow for Reynolds numbers 5 to 40.

For $Re_D = 5$ the distribution is quite even over the tube surface. This is a reflection of the importance of the molecular diffusion. As the Reynolds number is increased the macroscopic convective transport becomes more and more important and the distributions vary considerably over the tube surface. It is also evident that the upstream side dominates the heat transfer process. The low heat

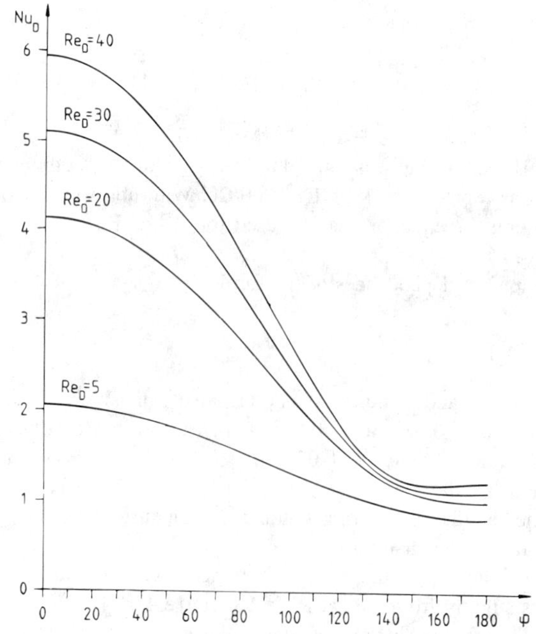

Figure 3. Local Nusselt number over the tube surface. $Pr = 0.72$. Low Reynolds numbers.

transfer on the rear side is due to the low intensity in the recirculating flow region at that side and thus the heat exchange is not promoted.

The flowing medium is also important for the heat transfer process. If the **Prandtl Number** is low, as for liquid metals, the Peclet number becomes low and the Nusselt number is decreased since the molecular diffusion dominates. The distribution over the tube surface becomes more uniform. If instead the Prandtl number is large, the convective contribution is enlarged and the distribution is more uneven. Also the level in the Nusselt numbers is increased.

In **Figure 4**, local Nusselt numbers at high Reynolds numbers are provided.

Compared to the Nusselt numbers in **Figure 3**, the magnitudes are much higher. Also the distributions are remarkably different. These distributions reflect the various flow phenomena discussed in **Tubes, Crossflow Over**. At $Re_D = 70800$, the Nusselt number is largest at the forward stagnation point. It decreases along the tube surface as the thermal boundary layer thickness is increased and reaches a minimum at the separation point. On the rear side, the heat transfer coefficient is increased due to the intensive motion of *vortices*. At the highest Reynolds numbers two minima appear.

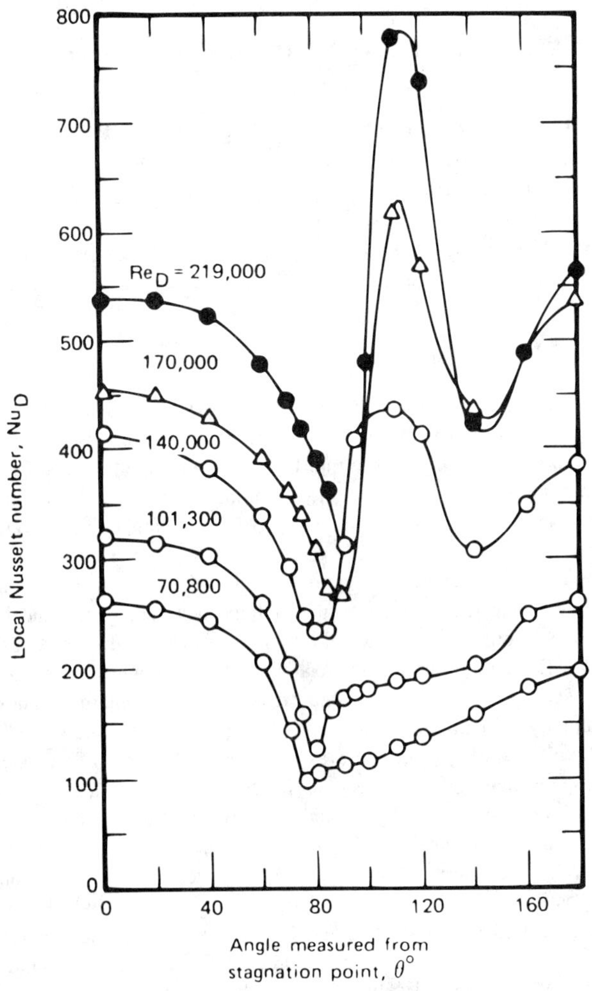

Figure 4. Local Nusselt number over the tube surface. Pr = 0.72. From Lienhard, J. H. (1981), A Heat Transfer Text Book, Prentice Hall Inc.

The first one is assumed to be related to a transition from laminar to turbulent flow in the boundary layer while the latter one is regarded to closely coincide with the *separation* of the turbulent *boundary layer.* The region between the first minimum and the following maximum is believed to be a transition region. The details of this transition (separation followed by reattachment or attached flow) is not evident. From **Figure 4** it is found that the influence of the Reynolds number is large and as turbulent flow prevails the heat transfer coefficient is considerably enlarged.

On the forward side of the tube in cross flow, where a laminar boundary layer exists, the heat transfer coefficient can be determined by semi-analytical series expansion techniques or numerical solution of the boundary layer equations. However, for accurate predictions the surface pressure distribution or the velocity distribution outside the boundary layer has to be known. By using a velocity distribution based on measurements by Hiemenz, Frössling was able to obtain the following relation (by series expansion technique) for the local Nusselt number

$$Nu_D = \left\{0.945 - 0.7696\left(\frac{x}{D}\right)^2 - 0.3478\left(\frac{x}{D}\right)^4\right\}Re_D^{0.5} \quad (2)$$

where x is the arc distance on the circumference from the front stagnation point. **Equation 2** is regarded to be valid up to the separation point. (Frössling, 1940)

For engineering calculations correlations of the average heat transfer coefficient are necessary. The following are examples.

For Reynolds numbers $Re_D < 44$ a formula by Collis and Williams is sometimes used. It reads

$$Nu_D = \{0.24 + 0.56\, Re_D^{0.45}\}\left(\frac{t_f}{t_\infty}\right)^{0.17} \quad (3)$$

where t_∞ is the fluid temperature and t_f the film temperature defined as

$$t_f = \frac{t_w + t_\infty}{2} \quad (4)$$

where t_w is the surface temperature of the tube. [Hinze, 1975].

A popular formula, based on experimental data for gases and liquids, reads

$$Nu_D = C \cdot Re_D^m Pr^{1/3} \quad (5)$$

where the constants C and m are given in **Table 1**.

Table 1. Constants in Equation 5

Re_D	C	m
0.4–4	0.989	0.330
4–40	0.911	0.385
40–4000	0.683	0.466
4000–40000	0.193	0.618
40000–400000	0.027	0.805

As **Equation 5** is used, the physical properties should be evaluated at the film temperature according to **Equation 4**.

Another correlation is the one suggested by Whitaker which is

$$\mathrm{Nu_D} = \{0.4\,\mathrm{Re_D^{0.5}} + 0.06\,\mathrm{Re_D^{2/3}}\}\mathrm{Pr^{0.4}}\left(\frac{\eta_\infty}{\eta_w}\right)^{0.25} \quad \textbf{(6)}$$

which is recommended for use in the intervals $0.67 < \mathrm{Pr} < 300$, $10 < \mathrm{Re_D} < 10^5$, $0.25 < \eta_\infty/\eta_w < 5.2$. (Incropera and Dewitt, 1981)

Zukauskas and Ziugzda (1985) have suggested the correlation

$$\mathrm{Nu_D} = \mathrm{C}\cdot\mathrm{Re_D^m}\cdot\mathrm{Pr_\infty^{0.37}}\cdot\left(\frac{\mathrm{Pr_\infty}}{\mathrm{Pr_w}}\right)^{0.25} \quad \textbf{(7)}$$

In this correlation the physical properties in the Nusselt and Reynolds numbers are to be evaluated at the freestream temperature. The values of C and m are given in **Table 2**.

Table 2. Values of C and m in Equation 7

$\mathrm{Re_D}$	C	m
1–40	0.76	0.4
40–1000	0.52	0.5
10^3–$2\cdot10^5$	0.26	0.6
$2\cdot10^5$–$2\cdot10^6$	0.023	0.8

Other correlations are available, more or less complicated.

The reader should remember that each correlation is reasonable over a certain range of conditions but for most engineering calculations one should not expect accuracy to much better than 25 percent.

Investigations have revealed that the average and local heat transfer from tubes or circular cylinders is augmented by freestream turbulence. To predict the heat transfer, several characteristics of the turbulence field have to be considered. The turbulence intensity has been found to be the most important one.

At low Reynolds numbers, the heat transfer is influenced by free convection. The motion due to free or natural convection is caused by the buoyancy force which is described by the **Grashof Number**

$$\mathrm{Gr} = \frac{g\beta\Delta TL^3}{\nu^2} \quad \textbf{(8)}$$

Sometimes the ratio of the Grashof number to the square of the Reynolds number is used to characterize the flow and heat transfer. If $\mathrm{Gr/Re^2} \ll 1$, forced convection prevails while if $\mathrm{Gr/Re^2} \gg 1$ natural or free convection occurs. When $\mathrm{Gr/Re^2}$ is of the order of unity, mixed convection, that is combined forced and free convection, is maintained.

Surface roughness also affects the heat transfer from tubes or cylinders. In air flow the heat transfer can be enhanced if the *relative roughness* (roughness height to tube diameter) is increased. Similar to the influence of freestream turbulence, the surface roughness may cause an onset to the critical flow regime at a lower Reynolds number. In viscous liquids having a high Prandtl number, the thermal resistance is concentrated to the viscous sublayer. For *enhancement of the heat transfer* under such circumstances small surface roughness elements should be applied.

References

Eckert, E. R. G. and Drake, R. M. (1972) *Analysis of Heat and Mass Transfer,* McGraw-Hill, New York.

Frössling, N. (1940) Verdunstung, Wämeubertragung und Geschwindigkeitverteilung bei Zweidimensionaler und Rotationssymmetrischen Laminarer Grenzschichtströmung, Lunds Universitets Årsskrift, N.F., Avd 2, Bd. 36, Nr 4.

Hinze, J. O. (1975) *Turbulence,* McGraw-Hill, New York.

Incropera, F. P. and DeWitt, D. P. (1981) *Fundamentals of Heat Transfer,* J. Wiley and Sons, New York.

Lienhard, J. H. (1981), A Heat Transfer Text Book, Prentice Hall Inc.

Morgan, V. T. (1975) The overall convective heat transfer from smooth circular cylinders, T. F. Irvine, Jr. and J. P. Hartnett, eds., Vol. 11, *Advances in Heat Transfer,* Academic Press, New York.

Zukauskas, A. A. and Ziugzda, J. (1985) *Heat Transfer of a Cylinder in Cross Flow,* Hemisphere Publishing Corporation, New York.

B. Sundén

TUBULAR BOWL CENTRIFUGE (see Centrifuges)

TUBULAR EXCHANGER MANUFACTURERS ASSOCIATION, TEMA

25 North Braodway
Terrytown, New York 10591, USA
Tel 914-332-0040

TUNGSTEN

Tungsten—(Swedish, *tung sten,* heavy stone): also known as *wolfram* (from *wolframite,* said to be named from *wolf rahm* or *spumi lupi,* because the ore interfered with the smelting of tin and was supposed to devour the tin), W; atomic weight 183.85; atomic number 74; melting point 3410 ± 20°C; boiling point 5660°C; specific gravity 19.3 (20°C); valence 2, 3, 4, 5, or 6.

In 1779 Peter Woulfe examined the mineral now known as wolframite and concluded it must contain a new substance. Scheele, in 1781, found that a new acid could be made from tung sten (a name first applied about 1758 to a mineral now known as scheelite). Scheele and Bergman suggested the possibility of obtaining a new metal by reducing this acid. The de Elhuyar brothers found an acid in wolframite in 1783 that was identical to the acid of tungsten (tungstic acid) of Scheele, and in that year they succeeded in obtaining the element by reduction of this acid with charcoal.

Tungsten occurs in wolframite, (Fe, Mn)WO_4; scheelite, $CaWO_4$; huebnerite, $MnWO_4$; and ferberite, $FeWO_4$. Important deposits of tungsten occur in California, Colorado, South Korea, Bolivia, U.S.S.R., and Portugal. China is reported to have about 75% of the world's tungsten resources. Natural tungsten contains five stable isotopes. Twelve other unstable isotopes are recognized. The metal is obtained commercially by reducing tungsten oxide with hydrogen or carbon.

Pure tungsten is a steel-gray to tin-white metal. Very pure tungsten can be cut with a hacksaw, and can be forged, spun, drawn, and extruded. The impure metal is brittle and can be worked only with difficulty. Tungsten has the highest melting point and lowest vapor pressure of all metals, and at temperatures over 1650°C has the highest tensile strength. The metal oxidizes in air and must be protected at elevated temperatures. It has excellent corrosion resistance and is attacked only slightly by most mineral acids.

The thermal expansion is about the same as borosilicate glass, which makes the metal useful for glass-to-metal seals. Tungsten and its alloys are used extensively for filaments for electric lamps, electron and television tubes, and for metal evaporation work; for electrical contact points for automobile distributors; X-ray targets; windings and heating elements for electrical furnaces; and for numerous space missile and high-temperature applications. High-speed tool steels, Hastelloy ®, Stellite ®, and many other alloys contain tungsten. Tungsten carbide is of great importance to the metal-working, mining, and petroleum industries. Calcium and magnesium tungstates are widely used in fluorescent lighting; other salts of tungsten are used in the chemical and tanning industries. Tungsten disulfide is a dry, high-temperature lubricant, stable to 500°C. Tungsten bronzes and other tungsten compounds are used in paints.

Handbook of Chemistry and Physics, CRC Press

TUNNEL BURNERS (see Burners)

TUNNEL KILN (see Kilns)

TURBINE

A turbine is a prime mover with rotary motion of the working unit, namely the rotor, and with a continuous operating process of converting the potential energy of the working fluid supplied (steam, gas or their mixtures, or liquids such as water) into the mechanical work on the rotor shaft. Turbines are bladed machines, the energy conversion occurs in the blading consisting of guide (nozzle or strator) vanes mounted in the stationary casing, and moving blades fixed on the rotor and moving with the rotor. This combination of vanes and blades creates blade cascades—a system of channels where the operating process of the turbine takes place. Turbine blade design is one of the most important and complicated problems in the design process. The complexity is connected with the necessity of a complex solution of problems of gas dynamics, heat transfer, structural strength and the technology of manufacturing blades.

In the guide (nozzle) vane cascades, the flow of steam or gas is accelerated and spun. In the guide vane cascades (guiding apparatus) of a turbine with working fluid (for example water), the required flow direction is provided and the fluid flow rate is controlled.

In the rotating cascade of moving blades stationed downstream of the guide (nozzle) vane cascade, the energy of the moving steam, gas or liquid is converted into mechanical work on the rotating rotor which in turn is used in overcoming the resistance forces of the driven machinery. This combination of rows of guide (nozzle) vanes and moving blades (arranged in series) is a turbine stage. The working fluid energy conversion may take place in one stage (in this case the turbine is referred to as a one-stage turbine) or in several successive stages, (in this case the turbine is referred to as a multi-stage turbine).

Turbine blades and inlet and outlet units form the flow passage. Turbines are classified as axial and radial (radial-axial) depending on the direction of the flow relative to the rotor rotation axis (**Figure 1**).

In *axial turbines* (**Figure 1a**), the working fluid moves mainly along co-axial surfaces parallel to the turbine axis. The radial turbine differs from the axial one in that, in the nozzle unit and in the greater part of the rotor, the working fluid moves in the plane perpendicular to the axis of rotation. Depending on whether the flow is directed towards the circumference or to the axis of rotation, radial turbines are divided into centrifugal (**Figure 1b**) and centripetal (**Figure 1c**).

Turbine rotors are set into rotation by the change in the momentum of the working fluid as it flows through curved interblade

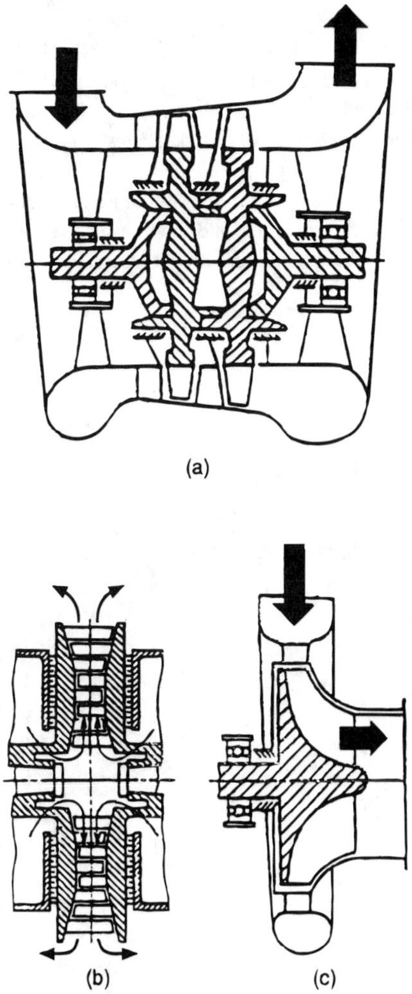

Figure 1. Types of turbine: (a) axial, (b) radial centrifugal, (c) radial centripetal.

channels which are formed by the moving blade surfaces. It is possible to choose the channels' cross sectional area in the rows of guide (nozzle) vanes and in the rotor, which affects the stage operating process. On the basis of the operating principles, there are impulse and reaction turbine stages (and turbines). In impulse stages (turbines), the potential energy of the working fluid is converted into kinetic energy only in stationary guide (nozzle) vanes. This kinetic energy is used to work the row of rotating blades. In reaction stages (turbines), a considerable part of the working fluid's potential energy is converted into mechanical work in the rotor wheel interblade channels. In steam and gas reaction turbines, the circumferential force applied to the rotor is created not only due to the change of the working fluid flow direction (as in an impulse turbine), but also because of the reaction force arising when the working fluid in the rotor interblade channels expands.

In *hydraulic reaction turbines,* as the pressure of the fluid flowing along the gradually converging channels of the wheel decreases, its relative flow rate increases. The pressure p_o of the working fluid upstream of the nozzle vane cascade is higher than p_1 both in impulse and reaction stages (**Figure 2**), therefore the flow in the nozzle accelerates: velocity $c_1 > c_0$. In an impulse stage, the static pressure p_1 upstream of the rotor wheel is equal to p_2 downstream of it (**Figure 2a**). In the reaction stage rotor wheel $p_2 < p_1$. Absolute velocity decreases in the rotor wheel of the impulse and reaction stages. **Figure 2** shows the variation of relative velocity w and enthalpy i in turbine stages. The relationships between velocities and flow angles in turbine stages depend on the degree of reaction of the stage, which is the ratio of the rotating blade's theoretical temperature drop to the sum of the theoretical temperature drop of the nozzle vanes and rotating blade's and is approximately equal to the temperature drop of the stage calculated using the stagnation parameters: $\rho = H_{0b}/(H_{0n} + H_{0b}) \approx \dfrac{H_{0b}}{H_0}$. The temperature drop in this relation may be determined from "enthalpy-entropy" diagram as corresponding portions (**Figure 3**); stagnation parameters are marked with asterisks. The word "reaction" is used because at $\rho > 0$ the expansion of the working fluid occurs in the rotating blade row and an additional force of reaction arises which rotates the rotor wheel.

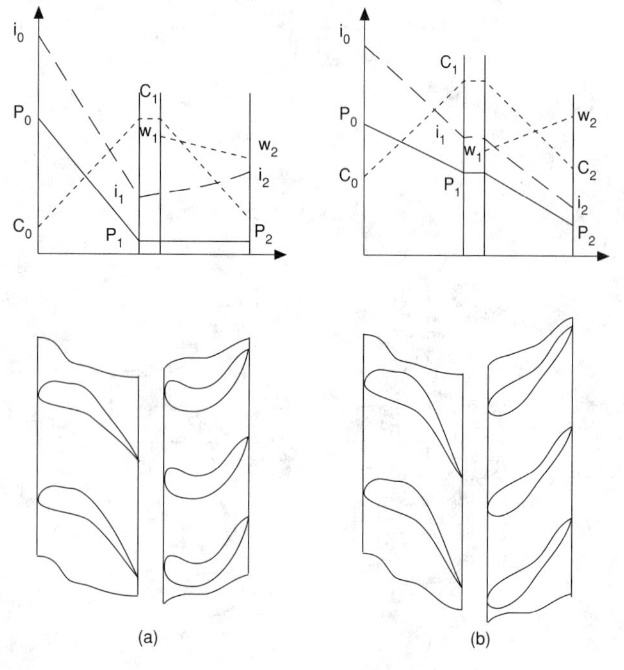

Figure 2. Pressure, enthalpy and velocity changes in a turbine: (a) impulse stage, (b) reaction stage.

The degree of reaction of stages at the mean diameter is usually chosen depending on the relative blade length l/D (D is the mean diameter of the flow passage) so that $\rho_r \geq 0.05$ to 0.1 at the blade root. In multi-stage turbines, the degree of reaction at the mean diameter gradually increases from the first to the last stage.

The turbine is evaluated using two main parameters: *specific work* and *efficiency*. These values vary depending on which turbine losses are taken into account.

The working fluid specific work on the rotor wheel circumference may be determined from Euler equation: $l_u = c_{1u}u_1 + c_{2u}u_2$, where u_1, u_2, c_{1u} and c_{2u} are velocities at the turbine flow passage mean diameter, which are usually determined from velocity triagles at this diameter at the rotor wheel inlet and outlet u_1 is the circumferential velocity at the rotor inlet; u_2 is the circumferential velocity at the rotor outlet; c_{1u} is the circumferential projection of absolute velocity c_1 of the working fluid at the rotor inlet; c_{2u} is the circumferential projection of the absolute velocity c_2 at the rotor outlet. The other parameters of the velocity triangles are axial projections c_{1a} and c_{2a} of absolute velocities c_1 and c_2, relative velocities of the working fluid w_1 and w_2 in the rotor wheel and their circumferential projections w_{1u} and w_{2u}.

Specific work l_u is less than the enthalpy drop H_0 by the amount of energy losses in the flow passage (it is a sum of specific losses h_n in the nozzle unit and h_b in the moving blades) and the working fluid kinetic energy h_e at the stage outlet. These losses may be evaluated using the efficiency at the wheel circumference

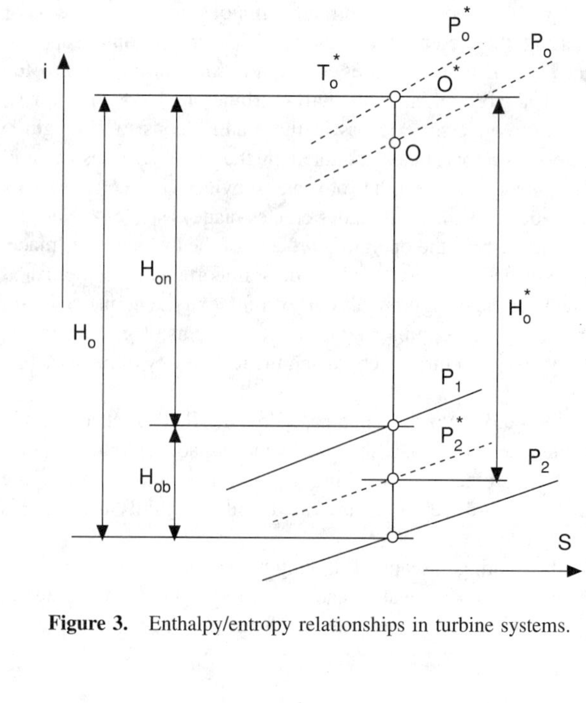

Figure 3. Enthalpy/entropy relationships in turbine systems.

$$\eta_u = l_u/H_0$$

$$\approx 1 - (h_n + h_b + h_e)/H_0 \quad \text{or} \quad \eta_u$$

$$= 1 - \xi_n - \xi_b - \xi_e,$$

where $\xi_n = h_n/H_0$, $\xi_b = h_b/H_0$, $\xi_e = h_e/H_0$ are corresponding relative losses. Usually there are radial clearances δ_n and δ_b between the blades and the casings in the nozzle cascade and the wheel, and the working fluid leakage occurs through their annular areas and thus the work l_u at the wheel circumference decreases. If total specific losses in the radial clearances are h_c, the corresponding relative losses $\xi_c = h_c/H_0$. Taking into account losses in the radial clearances the turbine specific work at the wheel circumference $l_{uc} = l_u - h_c$, power efficiency $\eta'_T = \eta_u - \xi_c$ and blade efficiency $\eta_b = \eta'_T + \xi_e$. If the kinetic energy of the flow issuing from the stage is used in the next stage, the losses may be estimated with the help of the stage efficiency from stagnation parameters so that

$$\eta_T^{*\prime} = l_{uc}/H_0 = \eta'_T(H_0/H_0^*),$$

where H_0^*—is the total theoretical heat drop in accordance with stagnation parameters (see **Figure 3**).

To estimate a turbine work or shaft power losses besides the above losses, it is necessary to determine relative losses ξ_{fv} caused by the friction of the disk and the working fluid and by the ventilation of the gas in the rotor wheel interblade channels. Ventilation losses occur in partial admission turbines in which nozzle channels occupy only a part of the total circumference. The degree of partial admission $\epsilon = z_1 t_1/(\pi d_1)$, where z_1 and t_1 are the number and the pitch of the nozzle vanes, d_1 is the mean diameter at the nozzle cascade outlet. The stage power efficiency taking into account friction and ventilation losses $\eta_T = \eta'_T = \xi_{fv}$; the efficiency in accordance with stagnation parameters, taking into account these losses $\eta_T^* = \eta_T(H_0/H_0^*)$ and the blade efficiency $\eta_b = \eta_T + \xi_e$. It is not possible to obtain a large heat drop in one stage, because there is no possibility to keep at the optimum level the velocity ratio in the flow passage which provides minimum possible losses and thus maximum possible turbine efficiency. *Multi-stage turbines* create conditions for operation of each stage at velocity ratios close to optimal, furthermore, the energy lost in the preceeding stage is used in the following stage.

Multi-stage turbines are built with velocity stages and pressure stages. In turbines with velocity stages almost all the total heat drop is used in the nozzle cascade. The kinetic energy acquired by the working fluid is then converted into work in two or three cascades of active rotating blades, with guide vane cascades being stationed between them. In turbines with pressure stages, the theoretical heat drop is divided between stages so that optimal velocity ratio is reached at each stage. Usually the heat drop is distributed proportionally to the square of the circumferential velocity at the mean diameter of the flow passage of each stage. When determining the heat drop in each stage, with the exception of the first, it is necessary to take into account the velocity of the working fluid flow at the inlet to the nozzle cascade of this stage.

As a result of this calculation, it is possible to obtain the working fluid parameters and the geometry parameters of nozzle vanes and rotating blades at a certain (usually mean) radius of the flow passage.

When the blades are relatively short ($l/D < 1/7$ to $1/8$), it is possible to assume that the gas pressure and the nozzle cascade exhaust velocity do not change along the radius at $\alpha_1 = constant$. The stage designed without taking into account these variations will have efficiency which differs little from the efficiency obtained when these factors are taken into consideration. To provide high efficiency when the blades are relatively long, the stage flow passage should be designed taking into account variations of the pressure upstream of the rotor blades with the radius. The geometrical parameters of the blades vary along the radius. According to the design predictions under these conditions, the rotating blades—and sometimes the nozzle vanes—should be twisted which is just how they are made in metal. There are no definite recommendations concerning the values of l/D at which the twist of a blade does not significantly increase the efficiency of the stage but blades are in practice usually short. It is only known that, for stages with high degree of reaction, the stages with twisted blades and with cylindrical blades may have high efficiency if $l/D < 1/4.5$ to $1/6$. According to some data, twisting of blades in stages with low degree of reaction (0.15 on the mean diameter) results in approximately 2 per cent increase of the efficiency at $l/D = 1/8$ and 5 per cent at $l/D = 1/5$.

To obtain maximum efficiency of the stage at the wheel circumference it is necessary to increase the rotor wheel heat drop, the degree of reaction of the stage, and pressure at the entrance to the wheel as the radius of the blade increases. This variation of the stage parameters is possible because special methods of designing long blades have been developed. One of the first methods was to design stages along the radius at constant circulation. In this case, the axial component of the absolute velocity of flow upstream of the rotor wheel is constant along the radius: c_1a = const and the circumferential component of the absolute velocity of flow c_{1u} upstream of the wheel varies relative to $2\pi r c_{1u} = \Gamma_{1u} = const$, where $\Gamma 1u$ is the circulation velocity. This design method is sometimes termed designing on "the free vortex law" or at constant circulation. When this method is used, the absolute velocity c_1 in the axial clearance between the nozzle cascade and the rotor wheel decreases with the increase of the radius. The pressure in this clearance increases considerably with radius. The working fluid density increases, and the mass flow rate through the unit section area increases at c_{1a} constant. In accordance with the design law, the inlet and outlet angles of nozzle vanes and rotating blades also vary with the radius. The angle α_1 decreases with the decrease of the radius. This variation may cause some difficulties when profiles at the root of long nozzle vanes are designed, which is connected with large relative thickness of the trailing edge at the root section. As a result, profile losses increase. Therefore, the general tendency is to have angle $\alpha_1 \geq 13$ to $14°$, and to avoid possible partial admission effect, $\alpha_1 \approx 11$ to $12°$ is taken only for low capacity turbines or turbines with low volume flow rate (at high working fluid pressure). Angle β_1 increases considerably along the radius and it may be more than 90° when the blades are long.

There are also some other methods besides designing at $\Gamma_{1u} = const$. One of the most widely used is the designing the nozzle cascade at $\alpha_1 = const$. This method has some advantages. When this design method is used, the velocities c_{1u} and c_1 increase less intensively with the decrease of the radius than when designing is done for $\Gamma_{1u} = const$. Pressure p_1 and the degree of reaction at the root section of the stage is somewhat higher and therefore the profile losses may decrease. The moving blades twist where designing for $\alpha_1 = const$ is weaker.

It is possible to plot contours of nozzle vanes and rotating blades for known working fluid parameters at a definite radius of the flow passage. Designing turbine blades may be considered as plotting a number of sections which determine the blade shape. The shape of nozzle vanes and rotating blades is determined by the function of the turbine stage, that is, to turn the flow with the least losses and to extract the required work at the prescribed circumferential velocity. For plotting the profile contours of nozzle vanes and rotating blades, the blade length is divided into several sections stationed at different radii. To select the optimum profile contour at each radius, one can select blades with different profiles and study them experimentally. This method is rather complicated. In a situation where there is a profile which has been well investigated, satisfactory results may be obtained by redesigning this profile. The necessary condition in this case is an insignificant difference in the structural inlet and outlet angles of the reference blade and the blade being designed. In the absence of such reference blades (with parameters close to those of the blade being designed), computer and graphic methods of blade design are used, also. The application of these methods allows plotting blade profile shapes with a smoothly changing curvature, for example along the square parabola, arc of circles, hyperbolic spirals, Bernoulli's lemniscates. The optimum cascade pitch is determined from the chord length value b. There are analytical and graphic relationships which allow the determination of the optimum relative cascade pitch $t_{opn} = t/b = 0.75$ to 0.85. The larger values are for reaction cascades of blades (and nozzle vanes).

Blade profiles obtained at several sections along the blade length are arranged in the blade drawing in such a way as to provide the prescribed strength of the blade and also high technological standard of its manufacture.

At identical working fluid parameters upstream and downstream of the stage and at one and the same degree of reaction, velocity w_2 of a centripetal turbine is much lower than the velocity of an axial or centrifugal turbine. This is explained by the movement of the working fluid from a large radius to a smaller radius. In a radial turbine, the working fluid parameters in the stage flow passage are determined in the same way as in an axial turbine, namely at $u_2 \neq u_1$. For the same working fluid flow rate, the losses in the rotor wheel h_b of a centripetal turbine may be lower compared with the losses in an axial turbine rotor wheel.

Turbines which are components of engines and plants operating in a wide range of load variations operate at varying working fluid parameters and varying speeds of rotation. As a result, at operation modes differing from, the design mode parameters and velocity triangles change in each stage. The flow conditions in interblade channels also change. Consequently, the specific work and the efficiency of stages and the whole turbine change. These deviations can be evaluated using the turbine characteristics—that is, a combination of analytical and graphical dependences of its general parameters on the operation modes. In other words, the dependence of external parameters determining the conditions of the turbine operation (pressure at the inlet of the turbine p_0^* and at the outlet p_T or p_T^*, temperature at the inlet of the turbine T_0^*, speed of rotation n) on its operating parameters (flow rate G, efficiency η_T or η_T^*, power N) or on the values which allow to determine the operating parameters.

Relative or complex parameters including the above external and operating parameters are used to plot characteristics. When a turbine is operating in the range of parameters that differ slightly from the design parameters (that is at low losses) the turbine characteristics may be integrated by one curve. For example, **Figure 4** gives typical characteristics of one-, two-, and three-stage turbines (curves 1,2 and 3) as dependency of ratios of total pressures p_0^*/p_T^* on the flow rate parameter $G\sqrt{T_0^*}/p_0^*$. This flow rate parameter or the discharge capacity decrease at a prescribed value of the pressure ratio and when additional stages are added. One of the first attempts to evaluate variation of flow parameters at off-design flow conditions in multi-stage turbines was made be A. Stodola who formulated the law of ellipse used in turbomachinery. Curve 4 in **Figure 4** for multi-stage turbines satisfies the elliptical law:

$$G\sqrt{T_0^*}/p_0^* = k[1 - (p_T^*/p_0^*)^2]^{0.5},$$

where k is a constant. In multi-stage turbines at sub-critical flow conditions in the stages, the heat drop in the last stages increases more rapidly than in the other stages with the increase of pressure ratio p_0^*/p_T^*. After the critical value is reached the flow of gas or steam in the preceeding stages stabilizes and does not vary with p_0^*/p_T^*.

Turbines are classified into steam, gas (and also steam-gas operating on a mixture of steam and the fuel combustion products) and hydraulic turbines. Within the class of gas turbines there is a special group of wind turbines using the energy of wind. Steam turbines belong to steam power units, gas turbines are components of gas turbine units and engines. Steam-gas turbines are component parts of steam-gas units.

The major advantages of turbines are high economy, compactness, reliability and the possibility to use them as component parts of power units with unit capacity from hundreds of watt to thousands of megawatt. (See also *Gas Turbines, Steam Turbines*).

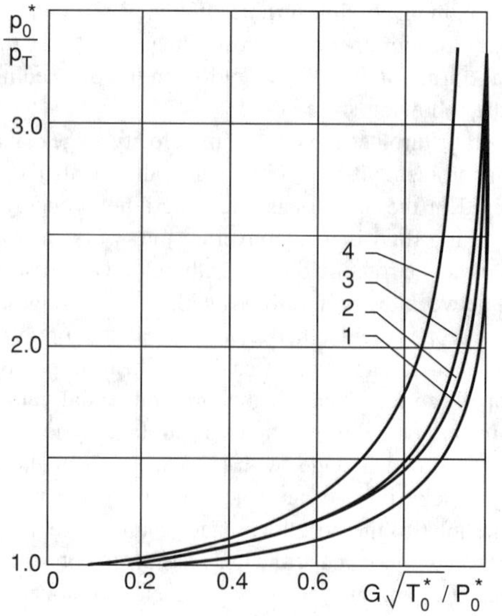

Figure 4. Overall turbine performance as a function of flow parameter.

References

Dixon, S. L. (1975) *Fluid Mechanics, Thermodynamics of Turbomachinery,* Pergamon Press, Oxford.

Stodola, A. (1945) *Steam and Gas Turbines,* 6th ed., Peter Smith, New York.

Traupel, Walter (1966) *Termische turbomaschinen,* bd. 1–2, Berlin, Springer, 1966–1968.

Leading to: *Gas turbine; Steam turbine; Hydraulic turbine*

E.A. Manushin

TURBINE EFFICIENCY (see Hydraulic turbines; Rankine cycle; Turbine)

TURBINE FLOWMETERS

Following from: *Flow metering*

Turbine flowmeters are flow measurement instruments which use axial fluid momentum to turn a multi-bladed rotor, the angular velocity of which is approximately proportional to the volume flowrate. With appropriate calibration and operation, turbine meters offer precise and highly repeatable measurement of throughput and flowrate. Furness (1982) provides an excellent description of turbine meter design, the theory of operation and a review of the important aspects of performance.

The majority of turbine meters use a design very similar to that shown in **Figure 1** below.

The free-running rotor is supported between upstream and downstream supports and the blades sweep out the full bore of the meter except for a small tip clearance. As the blade tips move past a pick-up mounted externally, a train of pulses is produced. Under calibration, the frequency of the pulses is used to determine a calibration coefficient, normally referred to as the meter factor (pulses per unit volume throughput). In service the reciprocal of the meter factor is multiplied by the measured pulse frequency to provide the flowrate measurement.

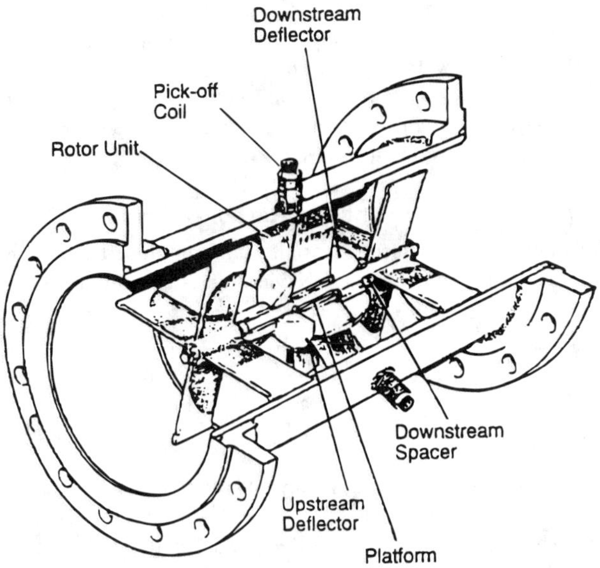

Figure 1. Layout of a typical turbine meter.

Because of the free-running nature of the rotor, the performance of this type of flowmeter is influenced by fluid property effects, and there is a sensitivity to upstream flow disturbance effects. Most notable are the effects of viscosity and swirl.

For liquid metering both free gas entrainment, or cavitation on the meter blades must be avoided. Both are likely to cause significant deviations in the meter factor. Kinghorn (1981) reported changes in meter factor of up to 50 per cent with only trace levels of free gas entrainment.

References

Furness, R. A. (1982) Turbine flowmeters, *Development in Flow Measurement,* Vol 1., R. W. W. Scott ed., Applied Science Publishers,

Kinghorn, F. C. and McHugh, A. (1981) The performance of turbine meters in two-component gas/liquid flow, *Flow, Its Measurement and Control in Science and Industry,* Vol 2. Instrument Society of America, St. Louis, Mo.

B.C. Millington

TURBOMOLECULAR PUMP (see Vacuum pumps)

TURBOPHORETIC VELOCITY (see Particle transport in turbulent fluids)

TURBULENCE

Following from: *Flow of fluids; Turbulent flow*

Nature

Turbulence may be defined as the most general case of unsteady fluid motion allowed by the Navier-Stokes equations. It is irregular and disorderly, it causes rapid mixing of heat and momentum, it is rotational and three-dimensional, it occurs over a range of scales, and, it is dissipative. The irregularity is illustrated in **Figure 1** which shows the turbulent velocity signal from a hot-wire anemometer 18m above the ground in an atmospheric surface layer.

Turbulence is not random in the Gaussian sense, and significant departures from normal probability distributions are observed. The diffusivity of turbulence at high Reynolds numbers is generally orders of magnitudes greater than molecular diffusivity and on scales comparable to the dimensions of the flow field. Turbulent momentum transfer is the source of increased drag on immersed bodies and resistance to flow in pipelines. A vortex-stretching mechanism maintains turbulent fluctuations in three dimensions. This mechanism is absent in two-dimensional flow.

Turbulence occurs at large Reynolds numbers, usually as a result of the instability of laminar flow to small disturbances. In nature and in engineering practice, the Reynolds number will usually be high enough for flows to be turbulent. The fluctuations occur over a range of scales, from those compatible with the flow domain to the smallest scale allowed by viscous dissipation. The vortex-stretching mechanism maintains an "energy cascade" from

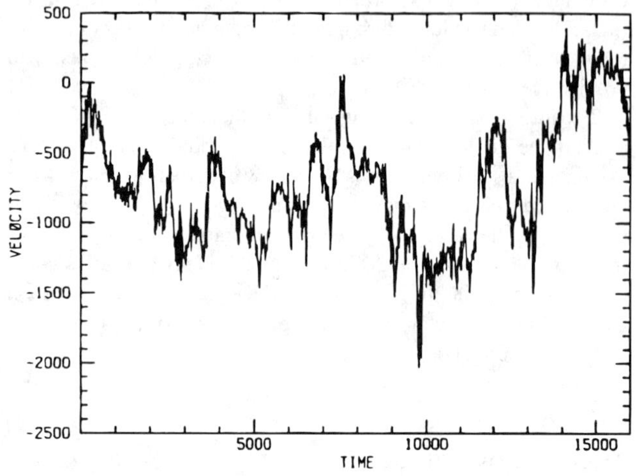

Figure 1. Turbulent velocity signal plotted in arbitrary units of time and velocity. (From Sreenivasan 1991). Reproduced with permission, from the Annual Review of Fluid Mechanics, Vol. 23, © 1991, by Annual Reviews Inc.

large-scale motion where most of the turbulent kinetic energy is generated and contained, to motion at smaller scales where it is dissipated by viscosity. An important result of this process is that the smaller eddies lose directional orientation. The Navier-Stokes equations cannot be solved in closed form save for a few linear problems. Numerical solutions are possible at present only for low-Reynolds-numbers flows in which the range of scales is narrow. Such solutions require many hours of supercomputer time.

Scope of This Entry

For reasons of space, attention is restricted to selected topics relating to turbulence in incompressible, uniform-property, Newtonian fluids. These are:

- stability and transition from laminar flow,
- the equations of the mean motion,
- length and time scales of turbulence,
- the energy spectrum,
- the statistical treatment of isotropic turbulence,
- the development of free turbulent shear flow,
- the characteristics of turbulent wall flow,
- gravitational effects on turbulence in density-stratified shear flow,
- coherent structures and the proper orthogonal decomposition.

Statistical methods are described in the standard texts:

Batchelor (1953), Monin and Yaglom (1971), Bradshaw (1971, 1976), Tennekes and Lumley (1972), Hinze (1975), Lesieur (1990).

Its early development may be traced in the classic papers collected by Freidlander and Topper (1961). For the identification of coherent structures see Lumley (1981), Fiedler (1987) and Gatski et al. (1992), who also deal with compressibility effects. Atmospheric turbulence affected by density stratification is described by Wyngaard (1992). Developments in the chaotic behaviour of non-linear systems, with application to turbulence, are to be found in Comte-Bellot and Mathieu (1987), Lumley (1990), Ottino (1990) and Sreenivasan (1991).

Origins of Turbulence

The systematic study of turbulence dates from the 19th century investigations of Osborne Reynolds who found that the character of pipe flow changed completely when the **Reynolds Number** $Re \equiv \rho\, UD/\mu$, exceeded a critical value. In this definition ρ and μ are the fluid density and viscosity, U is the mean velocity and D the pipe diameter. Re is therefore a measure of the relative importance of inertia and viscous forces. At low values of Re, Reynolds found the flow to be laminar, the pressure drop moderate, and the velocity profile conforming to the steady-flow *Hagen-Poiseuille distribution.* For $Re > 2300$ the flow became unsteady, or turbulent, the resistance substantially increased, and the velocity distribution became more uniform on account of increased mixing. In uniform-pressure external boundary layers of thickness δ, the onset of turbulence occurs at values of $\rho U\delta/\mu$ of about 5000; in free shear flows at very low Reynolds numbers.

Bradshaw (1976) summarises the stages in the development of turbulence from an initial unstable shear layer as:

1. The growth of disturbances with periodic fluctuations of vorticity,
2. Their secondary instability to three-dimensional disturbance if the primary fluctuations are two-dimensional,
3. The growth of three-dimensionality and higher harmonics of the disturbance, with spectral broadening by vortex-line interaction and
4. The onset of fully-developed turbulence with energy transfer across the spectrum to smaller and smaller scales.

The initial step in this process, the instability to small disturbances, has been treated with the aid of linearised theory (Lin 1955, Hinze 1975). This is only the first stage in a sequence of events. Stuart (in Comte-Bellot and Mathieu, 1987) summarises some recent developments, and Landahl and Mollo-Christensen (1986) provide a useful short survey, including the possibility of universality in transition to chaos.

Equations of Motion and the Reynolds Stresses

Turbulence in an incompressible, uniform-property, fluid is described by the continuity and Navier-Stokes equations:

$$\frac{\partial \hat{u}_i}{\partial x_i} = 0 \quad \textbf{(1)}$$

$$\frac{\partial \hat{u}_i}{\partial t} + \hat{u}_j \frac{\partial \hat{u}_i}{\partial x_j} = \nu \frac{\partial^2 \hat{u}_i}{\partial x_j \partial x_j} - \frac{1}{\rho}\frac{\partial \hat{p}}{\partial x_i} + f_i \quad \textbf{(2)}$$

In the traditional treatment due to Reynolds, time-dependent quantities are are decomposed into their mean and fluctuating components. Thus $\hat{u}_i = U_i + u_i$ and $\hat{p} = P + p$. Substitution in Equations 1 and 2, followed by time or ensemble averaging, produces for the mean flow field:

$$\frac{\partial U_i}{\partial x_i} = 0 \tag{3}$$

$$\frac{\partial U_i}{\partial t} + U_j \frac{\partial U_i}{\partial x_j} = \frac{\partial}{\partial x_j}\left(\nu \frac{\partial U_i}{\partial x_j} - \overline{u_i u_j}\right) - \frac{1}{\rho}\frac{\partial P}{\partial x_i} + F_i \tag{4}$$

These equations are indeterminate, there being now ten unknowns: three mean velocity components U_i, the pressure P, and six independent components of the *Reynolds-stress* tensor, $\rho u_i u_j$. The equations for $\overline{u_i u_j}$ contain further unknowns of higher order.

The conservation of kinetic energy, $k \equiv \frac{1}{2}\overline{u_i u_i}$ is described by:

$$\frac{Dk}{Dt} = -\overline{u_i u_j}\frac{\partial U_i}{\partial x_j} - \nu \overline{\frac{\partial u_i}{\partial x_j}\frac{\partial u_i}{\partial x_j}} - \frac{\partial}{\partial x_j}\left(\frac{\overline{u_i u_i u_j}}{2} - \nu \frac{\partial k}{\partial x_j} + \frac{\overline{p u_j}}{\rho}\right) \tag{5}$$

where the terms on the right are

a. production by interaction with mean velocity gradients (hereafter P),
b. viscous dissipation (hereafter ξ)
c. turbulent, viscous and pressure diffusion.

The practice of formulating additional equations to form a closed set is known as **Turbulence Modelling.**

Scales and the Energy Spectrum

The largest length scales in a turbulent flow are set by the dimensions of the flow field or the size of the body generating the flow disturbance. If the characteristic dimension and velocity are L and U respectively, a mean flow advection time scale is L/U. The characteristic time for viscous diffusion across a length L is L^2/ν and the ratio of these times is the Reynolds number, $Re_L = UL/\nu$. The smallest scales, η and η^2/ν, are set by the dissipation rate of turbulent energy.

Figure 2 shows how turbulent kinetic energy is distributed in wavenumber space. E(κ,t) is the three-dimensional energy-spectrum function, κ is the wavenumber 2π(frequency)/U and:

$$\int_0^\infty E(\kappa, t)d\kappa = \frac{1}{2}\overline{u_i u_i} \equiv k \tag{6}$$

The large eddies at low wave numbers retain at least some of the characteristics of their origins and generally will be highly anisotropic in shear flows. Near isotropy is achieved only in grid turbulence in the absence of mean velocity-gradients. These eddies have also a quasi-permanent character with relatively long turnover times and distances, particularly in decaying turbulence when the motion at smaller scales is dissipated first.

As wavenumber $\kappa \to 0$, $E(\kappa) \propto \kappa^n$, with n ranging from 3 to 5 according to different theories. In the energy-containing range, where E(κ) reaches a peak, the characteristic length and time scales are formed from k and its dissipation rate ε i.e. $\ell = k^{3/2}/\varepsilon$ and $t = k/\varepsilon$. In this case ε is to be interpreted as the rate at which turbulent energy leaves the eddies in this range to travel down the spectral cascade to be dissipated at higher wavenumbers. According to Kolmogorov's hypothesis, there is, at high Reynolds numbers, a range of high κ where the turbulence is statistically in a state of universal equilibrium uniquely determined by ν and ε, with length and velocity scales $\eta = (\nu^3/\varepsilon)^{1/4}$ and $\upsilon = (\nu\varepsilon)^{1/4}$. The viscous diffusion time is equal to the eddy advection time, giving $Re_\eta = \upsilon\eta/\nu = 1.0$. The energy-containing and dissipation ranges are widely separated on the high-Reynolds-number spectrum and the small eddies in the latter tend to be locally isotropic because the directional orientation of the large eddies is not transmitted in the vortex-stretching process of the energy cascade. A theoretical result for isotropic turbulence relates the dissipation rate to the energy spectrum through:

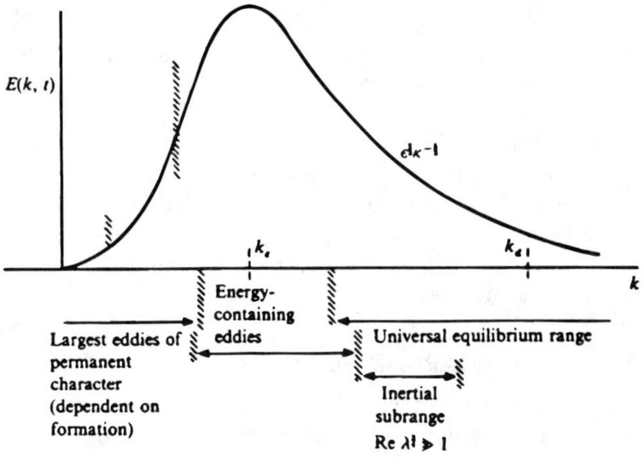

Figure 2. Distribution of turbulent energy in wavenumber space. From Landahl and Mollo-Christensen (1986) by permission of Cambridge University Press.

$$\xi = 2\nu \int_0^\infty \kappa^2 E(\kappa, t)d\kappa \tag{7}$$

Typically the peak in the dissipation spectrum occurs at $\kappa\eta \approx 0.2$. Between the energy-containing and dissipation ranges there exists an inertial subrange in which, at large Reynolds numbers, E(κ,t) is independent of ν and determined solely by ε. It follows then that in this range:

$$E(\kappa) = A\varepsilon^{2/3}\kappa^{-5/3} \tag{8}$$

where A is a constant of order unity.

An intermediate scale between L and η is the Taylor microscale λ. This is defined with respect to the dissipation rate through the relation $\varepsilon = 15\nu(u'/\lambda)^2$ where u′ is the rms velocity fluctuation in isotropic turbulence. The ratios of the scales may be expressed in terms of the microscale Reynolds number $R_\lambda \equiv u'\lambda/\nu$. For isotropic turbulence:

$$Re_l \equiv \frac{u'\ell}{\nu} = B\frac{R_\lambda^2}{15} \tag{9}$$

$$\frac{\ell}{\eta} = B\frac{R_\lambda^{3/2}}{15^{3/4}} \tag{10}$$

where, with $\ell = k^{3/2}/\xi$ as before, $B = 1.5^{1.5}$. **Equation 10** shows why direct numerical simulations are limited to low-Reynolds-number turbulence where the scale range is narrow. R_λ values in grid-generated wind-tunnel turbulence usually range from about 25 to 50 with, typically, $\ell \approx 25$ mm, $\eta \approx 0.5$ mm. In an experimental verification of Kolmogorov's hypothesis for high-Reynolds-number turbulence in a laboratory air jet, Gibson (1963) measured $\ell \approx 650$ mm, $\eta \approx 0.14$ mm for $R_\lambda = 780$. In environmental flows R_λ may be of order 10^4 or more, and the range of turbulence scales is correspondingly greater.

Homogeneous Isotropic Turbulence

In the absence of mean-velocity gradients, inhomogeneous turbulence tends to isotropy, interactions with the fluctuating part of the pressure serving to equalise the energy content in each of the three component directions. Nearly-isotropic turbulence generated by a wind-tunnel grid is homogeneous in directions at right angles to the mean flow. The energy **Equation 5** reduces to $U dk/dx = -\varepsilon$, and the data are fitted by the power law $k \approx ax^{-n}$, with $n \approx -1.25$ initially and $n \approx -2.5$ in the final period. Statistical theory and experiment have focused on the behaviour of velocity correlation tensors for two points separated by the space vector **r**, e.g.:

$$R_{ij}(r) = \overline{u_i(\mathbf{x})u_j(\mathbf{x} + \mathbf{r})} \tag{11}$$

from which the *energy spectrum* tensor is obtained:

$$\mathbf{\Phi}_{ij}(\kappa) = \frac{1}{8\pi^3}\int R_{ij}(\mathbf{r})e^{-i\kappa.\mathbf{r}}\mathbf{dr} \tag{12}$$

If a Fourier analysis is made of the variation of velocity along a line in the x_1 direction, the resulting spectrum function is related to R_{ij} by:

$$\Theta_{ij}(\kappa_1) = \frac{1}{2\pi}\int R_{ij}(r_1, 0, 0)e^{-i\kappa_1 r_1}dr_1 \tag{13}$$

For $i = j = 1$, Θ_{ij} is a longitudinal one-dimensional spectrum; $i = j = 2$ or 3 gives a lateral spectrum. Batchelor (1953) further defines the tensors obtained by averaging R_{ij} and Φ_{ij} over spherical surfaces in physical and wavenumber space. In particular:

$$\Psi_{ij}(\kappa) = \int \Phi_{ij}(\kappa)dA(\kappa) = \frac{\kappa^2}{2\pi^2}\int R_{ij}(r)\frac{\sin(\kappa r)}{\kappa r}dr \tag{14}$$

and:

$$E(\kappa) = \frac{1}{2}\Psi_{ii}(\kappa) \tag{15}$$

Dimensionless spatial correlation coefficients are defined with respect to velocity components u_p and u_n parallel to and normal to the vector separation r:

$$f(r) = \frac{\overline{u_p(x)u_p(x + r)}}{\overline{u_p^2}}, \quad g(r) = \frac{\overline{u_n(x)u_n(x + r)}}{\overline{u_n^2}} \tag{16}$$

A measure of the largest *eddies in the flow* is then given by the integral scale:

$$\Lambda = \int_0^\infty g(r)dr \tag{17}$$

As $r \to 0$ $g(r) = 1 - (r/\lambda)^2$, where λ is the Taylor microscale. The variation of g with r, and the relationship to Λ and λ, are shown in **Figure 3**.

The single point correlations $\overline{u_i(x)u_j(x)}$ are components of the Reynolds-stress tensor. The dynamical equations for these and other double velocity correlations necessarily contain triple products, as in the k, **Equation 5**, and the Karman-Howarth equation in, for example, Batchelor (1953) and Hinze (1975).

Free Turbulent Shear Flow

Free turbulent shear flows unidirectional in the x-direction are described by the reduced form of **Equation 4**:

$$U\frac{\partial U}{\partial x} + V\frac{\partial U}{\partial y} + W\frac{\partial U}{\partial z} + \frac{\partial \overline{uv}}{\partial y} + \frac{\partial \overline{uw}}{\partial z} = -\frac{1}{\rho}\frac{dP}{dx} + \nu\left(\frac{\partial^2 U}{\partial y^2} + \frac{\partial^2 U}{\partial z^2}\right) \tag{18}$$

For a two-dimensional shear layer bounded by an irrotational free stream of velocity U_∞:

$$U\frac{\partial U}{\partial x} + V\frac{\partial U}{\partial y} = \frac{\partial}{\partial y}\left(\nu\frac{\partial U}{\partial y} - \overline{uv}\right) - \frac{1}{\rho}\frac{dP}{dx} \tag{19}$$

Many flows described by this equation are at least approximately self-preserving in the sense that their velocity and shear-stress distributions may be scaled with velocity $U_s(x)$ and thickness $\delta(x)$ to the self-similar forms: $U/U_s = F(\eta)$, $-\overline{uv}/U_s^2 = g(\eta)$, where F and g are universal functions of $\eta \equiv y/\delta$. Experimental evidence suggests that if self-preserving solutions are possible then the flows in question will tend to develop in this way. The stream function $\psi \equiv \delta U_s f(\eta)$ is used to transform **Equation 19** into an ordinary differential equation. With $-dP/dx = \rho U_\infty dU_\infty/dx$

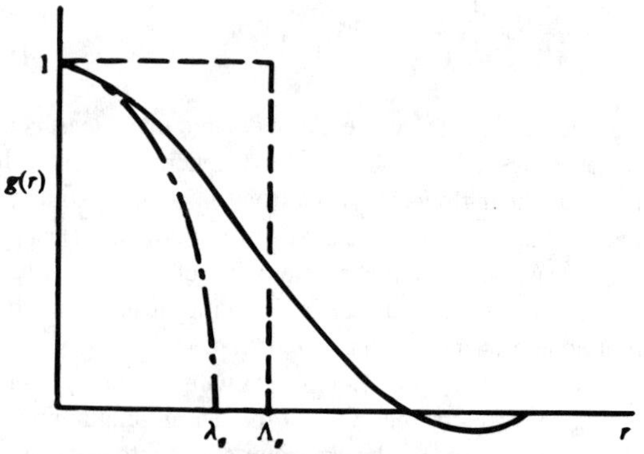

Figure 3. Relationship of the correlation function g(r) to the integral and microscales Λ, λ. (From Landahl and Mollo-Christensen (1986) by permission of Cambridge University Press.)

$$\frac{\nu}{\delta U_s} f''' + \frac{\delta}{U_s}\frac{dU_s}{dx}(ff'' - f'^2) + \frac{d\delta}{dx} ff'' + \frac{\delta}{U_s}\frac{U_\infty}{U_s}\frac{dU_\infty}{dx} = -g' \tag{20}$$

and $f'(\eta) \equiv df/d\eta = F(\eta) = U/U_s$, $g'(\eta) \equiv dg/d\eta$. For the special case U_s = constant, $g = 0$, $\delta = (\nu x/U_\infty)^{1/2}$, the result is the *Blasius equation* for the flat-plate laminar boundary layer: $2f''' + ff'' = 0$. For self-similar development the coefficients in equation (20) must be independent of x. This condition is met by

a. linear growth, $\delta \alpha x$,

b. power-law development of the scaling velocity, $U_s \alpha x^m$,

c. negligible viscous diffusion, $Re \to \infty$.

When the pressure is uniform, $dP/dx = 0$:

$$m(ff'' - f'^2) + ff'' = -g'\left(\frac{d\delta}{dx}\right)^{-1} \tag{21}$$

The result must also satisfy the requirements of momentum conservation. In the case of a plane jet in a co-flowing stream of velocity U_∞, for example, the condition:

$$J = \int_{-\infty}^{+\infty} \rho U(U - U_\infty)dy = \text{constant} \tag{22}$$

shows that self-preserving development is only possible when U_∞ is zero. Then:

$$J = \int_{-\infty}^{+\infty} \rho U^2 dy = \rho U_s^2 \delta \int_{-\infty}^{+\infty} f'^2 d\eta = \text{constant} \tag{23}$$

Since δ is required to vary linearly with x, the scaling, or maximum velocity, must therefore decay as $x^{-1/2}$. The turbulent jet in a co-flowing stream ($U_\infty \neq 0$) is only approximately self-preserving. For a round jet, $(\delta U_S)^2$ = constant and $U_{max} \propto x^{-1}$. In order to solve **Equation 21** an assumption must be made for the shear-stress distribution $g(\eta)$. It turns out that the rate of growth and velocity distributions can be predicted quite accurately using the mixing-length assumption:

$$g(\eta) \equiv \frac{-\overline{uv}}{U_{max}^2} = \frac{\ell_m^2}{\delta^2}(f'')^2 \tag{24}$$

when the mixing length ℓ_m is taken as a constant of the order of 0.1δ. Exactly self-similar wake flow is impossible but it is approached far downstream of the generating body where the velocity deficit is vanishingly small. For the plane wake: $\delta \propto x^{1/2}$, $U_{min} \propto x^{-1/2}$ and for the round wake: $\delta \propto x^{1/3}$, $U_{min} \propto x^{-2/3}$.

The experimental data collected and analysed by Rodi (1975) have been supplemented by a few later measurements, notably Hussein et al. (1994). Spreading rates dδ/dx are of order 0.1, the width δ being chosen as the distance from the centre to the point $y_{1/2}$ where the velocity $U = 0.5\ (U_{max} - U_{min})$. **Figures 4** and **5** show self-preserving profiles and the energy balance in a plane jet, for which:

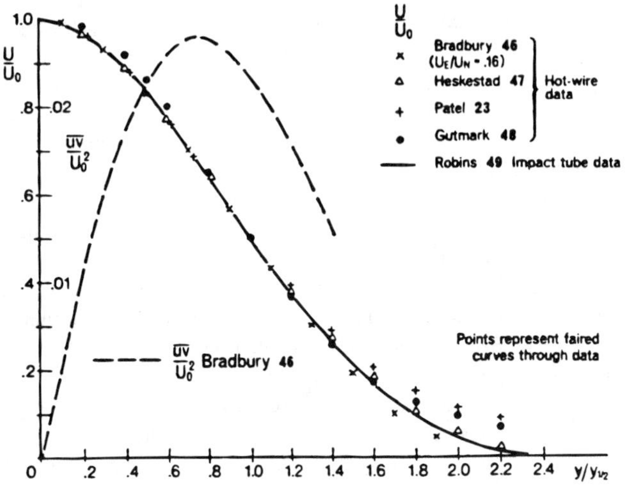

Figure 4. Velocity and shear-stress profiles in self-preserving plane jets in still surroundings. From Rodi (1975): A Review of Experimental Data on Uniform Density, Free Turbulent Boundary Layers; Studies in Convection 1, Launder B. E. (Ed), by permission of Academic Press.

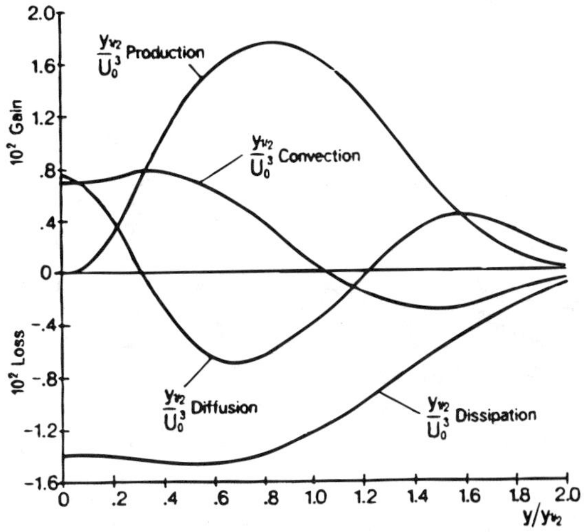

Figure 5. Balance of turbulent energy in a plane jet. Data of Bradbury reproduced by Rodi (1975): A Review of Experimental Data on Uniform Density, Free Turbulent Boundary Layers; Studies in Convection 1, Launder B. E. (Ed), by permission of Academic Press.

$$U\frac{\partial k}{\partial x} + V\frac{\partial k}{\partial y} = -\overline{uv}\frac{\partial U}{\partial y} - \varepsilon - \frac{\partial \overline{vk'}}{\partial y} \tag{25}$$

Energy production reaches a maximum at about $y_{1/2}$ where $-\overline{uv}$ and $\partial U/\partial y$ are both large.

Turbulent Wall Flow

Turbulence is significantly affected by the presence of a solid boundary, whether in external flow, where a "**Boundary Layer**" forms, or in the developed flow through pipes and channels. In each case the turbulence is generated in a high-shear region close to the surface whence it diffuses outwards. Between the turbulence and the wall lies a thin sublayer in which the fluctuations are heavily damped by viscosity. Further from the wall the direct effects of viscosity are negligible but the influence of the wall is propagated by the shear stress. In a boundary layer this wall layer extends to about one fifth of the boundary layer thickness, δ, which is defined as the distance from the wall to the point where the mean velocity reaches 99% of its free-stream value, U_∞. In pipe flow the wall layer extends almost to the centreline, the flow being turbulent throughout. Boundary-layer turbulence, by contrast, is highly intermittent (**Figures 6** and **7**). The interface between the turbulent outer region and the irrotational free stream consists of a viscous superlayer whose thickness is of the order of the *Kolmogorov microscale.* In the outer layer, $0.2\delta < y < 0.8\delta$ the turbulence structure resembles that of free shear flow.

In the viscous and turbulent layers close to the wall, **Equation 19** reduces to a balance between the gradients of shear stress and pressure: $\partial\tau/\partial y \approx dP/dx$. In fully-developed duct flow $\tau = \tau_w + y dP/dx$. When stress and velocity changes in the x-direction are relatively small, inner-layer properties depend on y, τ_w, ρ and v. Then, from dimensional analysis:

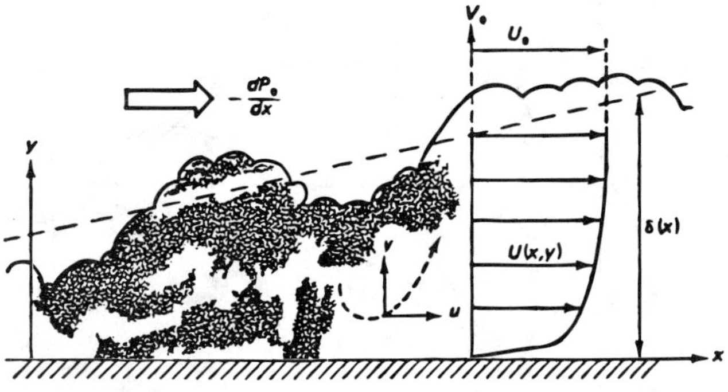

Figure 6. Schematic diagram of a boundary layer by means of smoke. Flow from left to right. From Tennekeg and Lumley (1972): A First Course in Turbulence, by permission of M.I.T. Press.

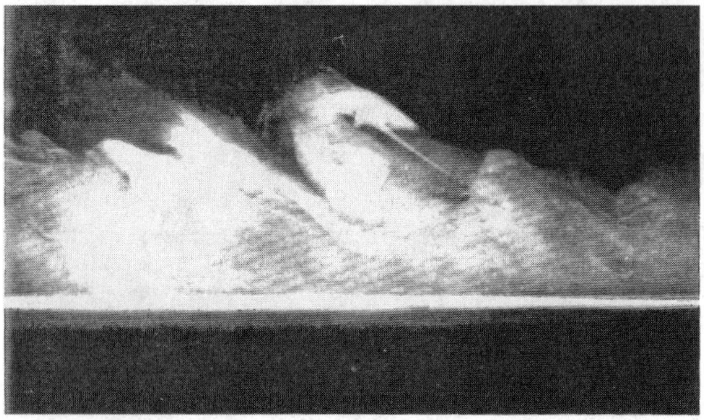

Figure 7. Visualisation of a boundary layer; flow from right to left. From Fernholz in Bradshaw P (Ed) (1976); Turbulence, Topics in Applied Physics, 12, by permission of Springer Verlag GMBH and Co. KG.

$$\frac{U}{u_\tau} = f_1\left(\frac{yu_\tau}{\nu}\right) \tag{26}$$

$$\frac{\partial U}{\partial y} = \frac{u_\tau}{y} f_2\left(\frac{yu_\tau}{\nu}\right) \tag{27}$$

where the velocity scale is $u_\tau \equiv \sqrt{(\tau_w/\rho)}$, is commonly called the "**Friction Velocity**". The dimensionless quantities U/u_τ and yu_τ/ν are denoted by u^+ and y^+. Very close to the wall, for $y^+ < 5$, $-\overline{uv} \ll \nu\partial U/\partial y$ and $u^+ = y^+$. For $y^+ > 30$ approximately, the direct effects of viscosity are negligibly small and $f_2 \rightarrow$ constant; $\partial U/\partial y = u_\tau/\kappa y$ and:

$$\frac{U}{u_\tau} = \frac{1}{\kappa}\log_e\left(\frac{yu_\tau}{\nu}\right) + C \tag{28}$$

For uniform-pressure flow on a smooth plate, *von Karman constant* $\kappa \approx 0.41$ and $C \approx 5.2$. In pipe flow the logarithmic zone extends almost to the centreline. The logarithmic dependence of U on y apparently holds regardless of pressure gradient, wall roughness or Reynolds number (but see George and Castillo, 1993, for an opposing view). The logarithmic and viscous sublayers are separated by a buffer layer in which neither viscous nor turbulent stresses are negligible.

In the outer region of a boundary layer the velocity distribution obeys the defect law:

$$\frac{U_\infty - U}{u_\tau} = f_3\left(\frac{y}{\delta}\right) \tag{29}$$

Since the defect and logarithmic laws must overlap, the functions f_3 and f_1, must be logarithms. Experimental data are correlated by:

$$\frac{U_\infty - U}{u_\tau} = C_2 - \frac{1}{\kappa}\log_e\left(\frac{y}{\delta}\right) \tag{30}$$

With $C_2 \approx 2.35$ for boundary layers and $C_2 \approx 0.65$ for pipes and channels.

An energy balance, **Figure 8**, shows that in the constant-stress wall layer, $-\overline{uv} \approx u_\tau^2$, the dominant terms in the k—Equation 25 are $P \equiv -\overline{uv}\ \partial U/\partial y$ and ξ, thus $\xi \approx P \approx u_\tau^3/\kappa y$. Advection and diffusion are relatively small, though the latter, in transporting turbulent energy away from the wall, is responsible for the growth of the shear layer. Energy production reaches a maximum at the edge of the viscous layer and approximately half of the total occurs in the inner 5% of the boundary layer. The production mechanism close to the wall is associated with the bursting of the low-velocity fluid streaks in a regular spanwise array of alternate high- and low-velocity streaks. Favourable pressure gradients ($dP/dx < 0$) tend to reduce the rate and intensity of bursting and vice versa.

The effect of a positive streamwise pressure gradient is to decelerate the flow, possibly to the point where it separates and τ_w falls to zero. The result is a loss of pressure in, for example, diffusers and pipe fittings, or a loss of lift on wings, where the separation on the suction surface is referred to as stall. Conversely, reverse

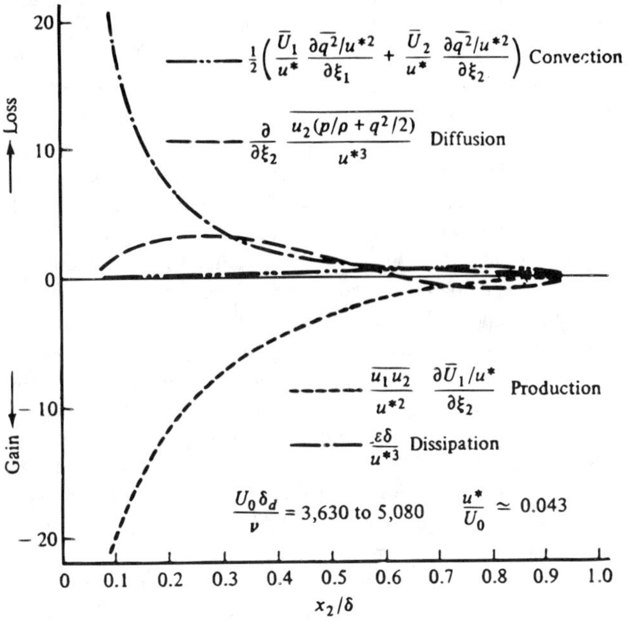

Figure 8. Energy balance in a boundary layer (results of Townsend, 1956, reproduced by Hinze J. 1975: Turbulence, by permission of McGraw Hill, Inc.

turbulent-to-laminar transition may occur in highly accelerated flow.

Buoyancy

Turbulence is significantly affected by gravity in the presence of density fluctuations. If density changes are important only in the inertia terms (the *Boussinesq approximation*), the k-**equation 25** for a horizontal shear layer becomes:

$$U\frac{\partial k}{\partial x} + W\frac{\partial k}{\partial z} = -\overline{uw}\frac{\partial U}{\partial z}(1 - R_f) - \xi - \frac{\partial \overline{wk'}}{\partial z} \tag{31}$$

where the "*flux Richardson number*" R_f is the ratio of k-production by buoyancy to shear production of k, viz:

$$R_f \equiv \frac{g\overline{\rho'\vartheta}/\rho}{-\overline{uw}\partial U/\partial z} = \frac{-g\overline{w\vartheta}/T}{-\overline{uw}\partial U/\partial z} \tag{32}$$

T and ϑ are respectively the mean and fluctuating parts of the temperature and, following geophysical practice, z is chosen as the vertical coordinate instead of y. **Equation 31** shows that turbulence is suppressed in conditions of stable stratification (light fluid above heavy) and augmented when buoyancy is destabilising, as in a shear layer heated from below. The critical value of R_f above which turbulence cannot be maintained is as low as 0.2 approximately. This is because in the horizontal shear layer, turbulence is generated by the mean velocity gradient only in the streamwise (u) component, and by buoyancy only in the vertical (w) component. It follows that in stably stratified flow there is a strong tendency to two-dimensionality as $\overline{w^2} \rightarrow 0$.

The frequency of *gravity waves* in a stable atmosphere is the *Brunt-Väisälä frequency:*

$$N = \left[-\frac{g}{\rho}\frac{\partial \rho}{\partial z}\right]^{1/2} = \left[\frac{g}{T}\frac{\partial T}{\partial z}\right]^{1/2} \tag{33}$$

When $\partial\rho/\partial z > 0$ gravity waves are unstable and break up into turbulence. The *gradient Richardson number* is defined as the ratio of N^2 to a typical turbulence frequency, $\partial U/\partial z$ ($= u_\tau/\kappa z$) in the logarithmic layer when buoyancy effects are small. Then:

$$Ri \equiv \frac{N^2}{(\partial U/\partial z)^2} = \frac{g\partial T/\partial z}{T(\partial U/\partial z)^2} \tag{34}$$

For $\partial T/\partial z = -\overline{w\vartheta}/\kappa_\vartheta u_\tau z$ in the logarithmic layer:

$$Ri = \frac{\kappa}{\kappa_\vartheta}\frac{g}{T}\frac{(-\overline{w\vartheta})}{u_\tau^2 \partial U/\partial z} = \frac{\kappa}{\kappa_\vartheta}R_f \tag{35}$$

For large buoyancy effects the gradients of U and T are expressed more generally as:

$$\frac{\kappa z}{u_\tau}\frac{\partial U}{\partial z} = \phi_m\left(\frac{z}{L}\right); \qquad \frac{\kappa u_\tau z}{-H}\frac{\partial T}{\partial z} = \phi_h\left(\frac{z}{L}\right) \tag{36}$$

where H is the temperature flux in the z direction and L is the Monin-Obukov length scale $L \equiv -Tu_\tau^3/\kappa gH$; ($z/L \rightarrow R_f$ when both are small). Also for small positive z/L, the truncated Taylor series expansion $\phi_m = 1 + \beta z/L$ leads to the wind profile corresponding to **Equation 28**:

$$\frac{U}{u_\tau} = \frac{1}{\kappa}\left[\log_e\left(\frac{z}{z_0}\right) + \frac{\beta z}{L}\right] \tag{37}$$

where the wall-law constant C in Equation 33 is absorbed in z_0 and β is a constant ≈ 7.

Coherent Structures

Measurements, flow visualisation and the results of direct simulations are used to support a view that there are organised motions in turbulence which are not random. Coherent structures vary in form and scale from flow to flow and there are significant differences in the structure of wall and free shear flows. In an influential book, Townsend (1956) postulated the existence of a double structure of large eddies and small-scale turbulence and emphasised the role of the former in controlling turbulent transport. Kline et al. (1967) studied the near-wall structure and the "bursting" mechanism responsible for much of the production of turbulent energy. Cantwell (1981, 1990) and Fiedler (1987) survey the field, Robinson (1992) deals with structures in the boundary layer, and the symposium volume edited by Gatski et al. (1992) contains several useful papers, including the application of the "proper orthogonal decomposition theorem" (Lumley 1981, Berkooz et al. 1993).

Fiedler (1987) defines coherent structures as spontaneously formed, non-stationary motional systems of correlated vorticity which

a. are in most cases of large scale, comparable to the lateral flow dimension, and flow specific in shape and composition,
b. are recurrent, having a characteristic life-span, typically of the average passage time of a structure,
c. exhibit a high measure of organisation in structure as well as in dynamics although their appearance is at best quasi-periodic
d. are similar to the corresponding structures in the laminar-turbulent transition.

Kline et al. (1967) found that the region very close to the wall, $5 < y^+ < 70$, is one of high turbulence activity which is associated with the behaviour of low-speed streaks spaced at intervals of about $100\nu/\upsilon_\tau$ in the spanwise (z) direction, interacting with the outer flow in a sequence of four events starting deep in the viscous sublayer: gradual outflow, lift up, oscillation, and breakup. The oscillation occurs at about 8–12 wall units from the wall, the breakup in the region $10 < y^+ < 30$. These events are described as "bursting". They contribute as much as 70% of the turbulence production. Later measurements have shown that stress could be produced by a combination of ejections involving rapid outflow of low-speed fluid from the wall region, and sweeps involving the inflow of high-speed fluid toward the wall. These motions are highly intermittent. Two views of the mechanics of streak breakup are shown in **Figures 9** and **10**. Cantwell (1981), after surveying the data to 1980, identified four main constituents of the organised structure including the outer layer (**Figure 7**)—nearest to the wall is a fluctuating array of streamwise counter-rotating vortices. Above these, but still close to the wall, is a layer subjected to bursts of high-intensity small-scale motions. Intense small-scale motion is found also in the outer layer, primarily on the upstream-facing portions of the turbulent-non-turbulent interface, the backs of the bulges in the outer part of the layer. These are part of an overall transverse rotation on a scale comparable with the layer thickness.

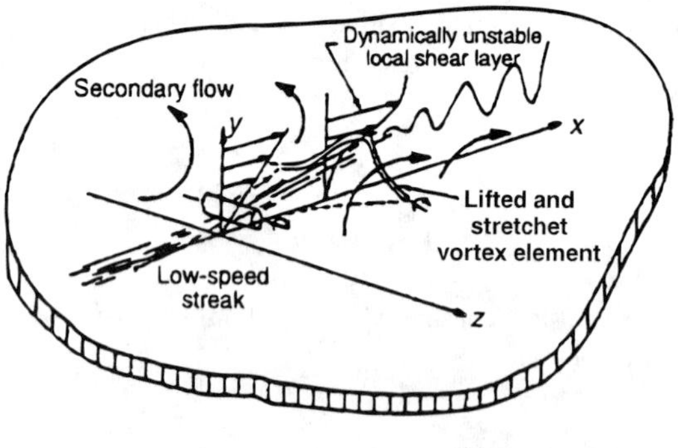

Figure 9. Mechanics of streak breakup as visualised by Kline et al. (1976) (From Landahl and Mollo-Christensen (1986) by permission of Cambridge University Press.)

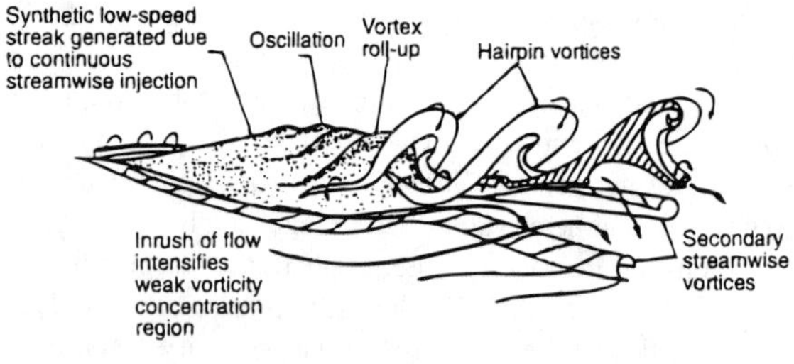

Figure 10. Breakup of a low-speed streak generating hairpin vortices. Reproduced, with permission, from the Annual Review of Fluid Mechanics © 1991 by Annual Reviews Inc.

Aubry et al. (1988) model this wall region by expanding the instantaneous field in so-called empirical eigenfunctions as permitted by the proper orthogonal decomposition theorem. The method leads to low-dimensional sets of ordinary differential equations, from the Navier-Stokes equations, via Galerkin projection. The equations exhibit intermittency as well as a chaotic regime, capturing major aspects of the ejection and bursting events. Robinson (1991) proposes an idealised model for low-Reynolds-number boundary layers based on vortical structures: quasi-streamwise vortices dominate the buffer region while archlike vortices are the most common vortical structure in the wake.

For the relationship to chaos, reference may be made to recent papers by Sreenivasan (1991), on fractals, and Ottino (1990) on mixing and chaotic advection.

References

Batchelor, G. K. (1953) *The Theory of Homogeneous Turbulence,* Cambridge University Press.

Berkooz, G., Holmes, P., and Lumley, J. L. (1993) The proper orthogonal decomposition in the analysis of turbulent flows, *Ann. Rev. Fluid Mech.,* 25, 537–576.

Bradshaw, P. (1971) *An Introduction to Turbulence and its Measurement,* Pergamon.

Bradshaw, P. ed. (1976) *Turbulence, Topics in Applied Physics 12,* Springer.

Cantwell, B. J. (1990) Future directions in turbulence research and the role of organised motion, *Lumley* 97–131, 1990.

Cantwell, B. J. (1981) Organized motion in turbulent flow, *Ann. Rev. Fluid Mech.,* 13, 457–515.

Comte-Bellot, G. and Mathieu, J. (1987) *Advances in Turbulence,* Springer.

Fiedler, H. E. (1987) Coherent structures, *Comte-Bellot* and *Mathieu,* 320–336, 1987.

Freidlander, S. K. and Topper, L. (1961) *Turbulence: Classic Papers on Statistical Theory,* Interscience Publishers.

Gatski, T. B., Sarkar, S. and Speziale, C. G., eds. (1992) *Studies In Turbulence,* Springer.

George, W. K. and Castillo, L. (1993) Boundary layers with pressure gradient: another look at the equilibrium boundary layer, *Near-Wall Turbulent Flows,* So, R.M.C., Speziale, C.G. and Launder, B.E. eds. 901–920, Elsevier.

Gibson, M. M. (1963) Spectra of turbulence in a round jet. *J. Fluid Mech.,* 15, 161.

Hinze, J. O. (1975) *Turbulence,* 2nd ed, McGraw-Hill.

Hussein, H. J., Capp, S. P. and George, W. K. (1994) Velocity measurements in a high-Reynolds-number, momentum-conserving, axisymmetric, turbulent jet, *J. Fluid Mech.,* 258, 31–76.

Kline, S. J., Reynolds, W. C., Schraub, F. A. and Runstadler, P. W. (1967) The structure of turbulent boundary layers, *J. Fluid Mech.,* 30, 741–773.

Landahl, M. T. and Mollo-Christensen, E. (1986) *Turbulence and Random Processes in Fluid Mechanics,* Cambridge University Press.

Lele, S. K. (1994) compressibility effects in turbulence, *Ann. Rev. Fluid Mech.*, 26, 211–254.

Lesieur, M. (1990) *Turbulence In Fluids: Stochastic and Numerical Modelling*, 2nd Kluwer ed.

Lin, C. C. (1955) *The Theory of Hydrodynamic Stability*, Cambridge University Press.

Lumley, J. L. (1981) *Coherent Structures in Turbulence Transition and Turbulence*, R. E. Meyer, ed. 215–242, Academic.

Lumley, J. L., ed. (1990) Whither Turbulence? Turbulence at the Crossroads, *Lecture Notes in Physics*, 357, Springer.

Monin, A. S. and Yaglom, A. M. (1971) *Statistical Fluid Mechanics*, MIT Press.

Ottino, J. M. (1990) Mixing, chaotic advection and turbulence, *Ann. Rev. Fluid Mech.*, 22, 207–253.

Robinson, S. K. (1991) Coherent motions in the turbulent boundary layer, *Ann. Rev. Fluid Mech.*, 23, 601–640.

Rodi, W. (1975) A review of experimental data of uniform density free turbulent boundary layers, *Studies in Convection I*, ed. Launder, B.E., Academic Press, London.

Sreenivasan, K. R. (1991) Fractals and multifractals in fluid turbulence, *Ann. Rev. Fluid Mech.*, 23, 539–600.

Tennekes, H. and Lumley, J. L. (1972) *A First Course in Turbulence*, MIT Press.

Townsend, A. A. (1956) *The Structure of Turbulent Shear Flow*, Cambridge University Press.

Wyngaard, J. C. (1992) Atmospheric turbulence, *Ann. Rev. Fluid Mech.*, 24, 205–234.

Leading to: *Turbulence modelling*

M.M. Gibson

TURBULENCE, IN WIND TUNNELS (see Wind tunnels)

TURBULENCE MODELLING

Following from: *Turbulence; Turbulent flow*

The Navier-Stokes Equations and the Closure Problem

Introduction

Turbulence modelling means the formulation of the mathematical relationships required to obtain solutions of the averaged equations of motion. Averaging is necessary because the time-dependent *Navier-Stokes equation* cannot be solved analytically, and the range of scales occurring in turbulence limits the possibility of numerical solution by supercomputer to simple flow geometries and low **Reynolds Numbers.** The averaged momentum equations derived by Reynolds (1894) do not form a closed set because they contain new terms in the form of the *Reynolds stresses*, $-\rho\overline{u_iu_j}$. Additional equations must be devised to close the system by relating the new terms to known variables. At the lowest level of closure, the $\overline{u_iu_j}$ are related to the mean rate of strain through the definition of an "*eddy viscosity*" analogous to the molecular viscosity of laminar flow. In the simplest models the eddy viscosity is determined from a "*mixing length*" ℓ specified at every point in the flow field. One-equation models obtain the intensity of the turbulent velocity fluctuations from a modelled equation for the quantity $k \equiv \overline{u_iu_i}/2$, the *turbulent kinetic energy* per unit mass, but are still limited by their reliance on a specified length scale. This limitation is removed by the introduction of a second modelled equation from which a length scale L can be calculated. Of many such proposals the equation for the rate of *turbulent energy dissipation* $\varepsilon \equiv k^{3/2}/L$ has gained wide acceptance and the k-ε model may be regarded as the standard method at the present time.

From its inception by *Boussinesq* (1877), it has been argued that the eddy viscosity concept is wrong in principle and its limitations have motivated the development of Reynolds-stress-transport models, or second-order closures, in which the $\overline{u_iu_j}$ are obtained from their own modelled equations and the eddy-viscosity idea appears only in relating the triple products $\overline{u_iu_ju_k}$ to the gradients of $\overline{u_iu_j}$.

In *large-eddy simulations* (*LES*, or sub-grid-scale modelling), only the small-scale motion is modelled. The simulations are performed on grids coarser than those used for *direct numerical simulations* (*DNS*), where the number of nodes needed to discretise all scales increases roughly as the cube of the Reynolds number. Even the low-Re direct simulations currently being attempted require some hundreds of hours of supercomputer time.

In the following pages, turbulence closures are described in increasing order of complexity; from simple mixing-length models for boundary layers to second-order closures based on the equations for $\overline{u_iu_j}$. For reasons of space attention is restricted to the incompressible flow of uniform-property fluids, where most current research effort is concentrated. Several general works cover the field. The book by Launder and Spalding (1974) and the volume edited by Bradshaw (1976) are still useful. Surveys by Nallasamy (1987), Hanjalic (1988), Launder (1989, 1993), and Speziale (1991) record recent developments.

Equations of motion

The instantaneous velocity $\hat{u}_i$ and pressure $\hat{p}$ of a turbulent flow field comprise mean and fluctuating parts: $\hat{u}_i = U_i + u_i$ and $\hat{p} = P + p$. (No distinction is made here between time and ensemble averaging). Substitution in the continuity and Navier-Stokes equations for incompressible flow:

$$\frac{\partial \hat{u}_i}{\partial x_i} = 0 \tag{1}$$

$$\frac{\partial \hat{u}_i}{\partial t} + \hat{u}_j\frac{\partial \hat{u}_i}{\partial x_j} = \nu\frac{\partial^2 \hat{u}_i}{\partial x_j \partial x_j} - \frac{1}{\rho}\frac{\partial \hat{p}}{\partial x_i} + f_i \tag{2}$$

produces Reynolds averaged equations for the mean flow field:

$$\frac{\partial U_i}{\partial x_i} = 0 \tag{3}$$

$$\frac{\partial U_i}{\partial t} + U_j\frac{\partial U_i}{\partial x_j} = \frac{\partial}{\partial x_j}\left(\nu\frac{\partial U_i}{\partial x_j} - \overline{u_iu_j}\right) - \frac{1}{\rho}\frac{\partial P}{\partial x_i} + F_i \tag{4}$$

The non-linear terms of the original Navier-Stokes equations have produced nine components of $\overline{u_i u_j}$ (six of which are independent) so that, in the general case, there are now ten unknowns but still only four equations. Additional equations for $\overline{u_i u_j}$ are to be supplied by "modelling".

The Mixing-Length Model

The eddy viscosity

The Reynolds stresses may be related to the mean rate of strain by the turbulence constitutive equation:

$$-\overline{u_i u_j} = \nu_t\left(\frac{\partial U_i}{\partial x_j} + \frac{\partial U_j}{\partial x_i}\right) - \frac{2}{3}\delta_{ij}k \quad (5)$$

in which ν_t is the "eddy viscosity". On dimensional grounds, and by analogy with kinetic theory, it is expressed as the product of a velocity fluctuation and a "mixing length": $\nu_t = \upsilon\ell$, although the mixing length, unlike the mean free path, is not small compared to the distances over which momentum is transported. ℓ corresponds roughly to the scale of the energy-containing turbulent eddies, usually an order of magnitude smaller than a characteristic flow dimension. ν_t is a property of the turbulence which varies from point to point. The objective is now to determine υ and ℓ.

Free shear flows

For certain free shear layers of width δ, it is possible to assume $\upsilon \propto (U_{max} - U_{min})$ and $\ell \propto \delta$ to give:

$$\nu_t = C\delta(U_{max} - U_{min}) \quad (6)$$

where C is a constant of order 0.1 which varies from flow to flow.

Prandtl's mixing-length model

The need to specify ℓ in advance of the calculation effectively limits its application to thin shear layers described by the boundary-layer equation:

$$U\frac{\partial U}{\partial x} + V\frac{\partial U}{\partial y} = \frac{\partial}{\partial y}\left(\nu\frac{\partial U}{\partial y} - \overline{uv}\right) - \frac{1}{\rho}\frac{dP}{dx} \quad (7)$$

where U is the mean velocity in the predominant flow direction x, $-\overline{uv}$ is the turbulent shear stress, and we assume steady flow with zero body forces. Prandtl's (1925) model relates the eddy viscosity to the mean shear by:

$$\upsilon = \ell\frac{\partial U}{\partial y}, \quad \nu_t = \ell^2\left|\frac{\partial U}{\partial y}\right| \quad \text{and} \quad -\overline{uv} = \ell^2\left|\frac{\partial U}{\partial y}\right|\frac{\partial U}{\partial y} \quad (8)$$

where the modulus is needed to ensure positive values of ν_t.

Turbulent wall flow

In the inner region of the boundary layer on an impermeable plate, where $V \approx 0$ and $\partial U/\partial x = -\partial V/\partial y \approx 0$, **Equation 7** reduces to:

$$\frac{d}{dy}\left(\nu\frac{dU}{dy} - \overline{uv}\right) = \frac{1}{\rho}\frac{dP}{dx} \quad (9)$$

Integration gives the total (laminar + turbulent) shear stress τ as:

$$\frac{\tau}{\rho} \equiv \nu\frac{\partial U}{\partial y} - \overline{uv} = \frac{y}{\rho}\frac{dP}{dx} + \frac{\tau_w}{\rho} \quad (10)$$

where τ_w is the shear stress at the wall. When $dP/dx = 0$, $\tau = \tau_w$ in this inner region. Outside the viscous sublayer, where $-\overline{uv} >>> \nu\partial U/\partial y$, **Equation 8** is:

$$\ell^2\left(\frac{\partial U}{\partial y}\right)^2 = \frac{\tau_w}{\rho} \equiv u_\tau^2 \quad (11)$$

where u_τ is a velocity scale defined by this equation. The scale of the energy-containing eddies is found to increase with increasing distance from the wall. If the variation is assumed to be linear and $\ell = \kappa y$, **Equation 11** becomes $\partial U/\partial y = u\tau/\kappa y$ to give, on integration, the "*law of the wall*":

$$\frac{U}{u_\tau} = \frac{1}{\kappa}\log_e\left(\frac{yu_\tau}{\nu}\right) + C \quad (12)$$

Extensive measurements in boundary layers show that $\kappa \approx 0.41$ and $C \approx 5.2$. In the outer region, $y > 0.2\delta$, the influence of the wall is weak and, as in free shear flow, $\ell \approx 0.09\delta$. Both regions are covered adequately by:

$$\frac{\ell}{\delta} = \frac{\kappa}{n}\left\{1 - \left(1 - \frac{y}{\delta}\right)^n\right\} \quad (13)$$

with n = 5 for a boundary layer and n = 3 for fully-developed flow in a pipe of radius δ.

The mixing-length in the viscous layers

As $y \to 0$, $-\overline{uv} = O(y^3)$, $\ell = O(y^{3/2})$ and $t_w = \mu\partial U/\partial y$. The effects of viscosity on ℓ are invariably accounted for by an exponential damping function. In *van Driest's (1956) original model:*

$$\ell = \kappa y\left\{1 - \exp\left(\frac{-y^+}{A^+}\right)\right\} \quad (14)$$

where $y^+ \equiv yu_\tau/\nu$ and A^+ is a constant, ≈ 26 in constant-pressure flow. The mean-velocity distribution is obtained by integrating:

$$\frac{du^+}{dy^+} = \frac{1}{2\ell^{+2}}\{(1 + 4\ell^{+2}\tau^+)^{1/2} - 1\} \quad (15)$$

where $u+ \equiv U/u_\tau$, $\ell^+ \equiv \ell u_\tau/\nu$, $\tau+ \equiv \tau/\tau_w \approx 1$ for constant-pressure flow. $u^+ = y^+$ in the viscous sublayer, $y^+ \to 0$. **Equation 15** reduces to $du^+/dy^+ = 1/\kappa y^+$ for large y^+ from which the law of the wall, Equation 12, is recovered. For $\kappa = 0.41$, $C \approx 0.204$ $A^+ \approx 5.3$.

Pressure gradient effects

The *effects* of *pressure gradient* are at least partially accounted for by writing $A^+ = 26/(\tau^+)^n$ where $\tau^+ = \tau/\tau_w$, and $0 < n < 2$. For large $|p^+|$, and particularly for strongly accelerated flows, the Cebeci-Smith (1974) formula may be used:

$$A^+ = \frac{26}{\sqrt{1 + 11.8p^+}} \quad \text{where} \quad p^+ \equiv \frac{\nu}{\rho u_\tau^3}\frac{dP}{dx} \tag{16}$$

McEligot (1985) has made a collection of several such formulae. Typical mixing-length methods are described by Baldwin and Lomax (1978), Cebeci et al. (1984) and Granville (1987).

One-Equation Modelling

The turbulence-energy equation

In the one-equation models attributed to *Kolmogorov* (1942) *and Prandtl* (1945), the velocity scale υ is taken more realistically as $\sqrt{k}$, and $k \equiv \overline{u_i u_i}/2$ is obtained from an equation derived by multiplying **Equation 2** by u_i, averaging, and dividing by 2. The time-averaged result is:

$$\underset{1}{U_j \frac{\partial k}{\partial x_j}} = \underset{2}{-\overline{u_i u_j}\frac{\partial U_i}{\partial x_j}} - \underset{3}{\nu \overline{\frac{\partial u_i}{\partial x_j}\frac{\partial u_i}{\partial x_j}}}$$

$$- \frac{\partial}{\partial x_j}\left(\underset{4(a)}{\frac{\overline{u_i u_i u_j}}{2}} - \underset{4(b)}{\nu\frac{\partial k}{\partial x_j}} + \underset{4(c)}{\frac{\overline{p u_j}}{\rho}}\right) \tag{17}$$

where the terms represent

1. mean-flow convection,
2. mean-shear production, P,
3. viscous dissipation, ε,
4. diffusion by (a) turbulence (b) viscosity, (c) pressure.

Modelling the k-equation

Turbulent diffusion is represented by:

$$\frac{1}{2}\overline{u_i u_i u_j} = -\frac{\nu_t}{\sigma_k}\frac{\partial k}{\partial x_j} \tag{18}$$

with $\sigma_k \approx 1.0$. Viscous diffusion is exact. Pressure diffusion has long been assumed to be negligibly small or, alternatively, to be incorporated in Equation 18. Instead of ℓ a new length scale is defined, on dimensional grounds, as:

$$L = \frac{k^{3/2}}{\varepsilon} \tag{19}$$

and the eddy viscosity is written as:

$$\nu_t = C_\mu L\sqrt{k} \tag{20}$$

where C_μ is a constant to be determined. In the constant-stress wall layer $\nu_t = \ell u_g t$, $-\overline{uv} \approx u_\tau^2$ and $\varepsilon \approx P \approx u_\tau^3/\ell$. Then:

$$C_\mu = \frac{\ell}{L}\sqrt{\frac{-\overline{uv}}{k}} = \left(\frac{-\overline{uv}}{k}\right)^2 \tag{21}$$

Shear-flow data suggest $-\overline{uv}/k \approx 0.3$ to give $C\mu \approx 0.09$ and in the constant-stress layer, consistency with the law of the wall requires:

$$L = \frac{\kappa y}{C_\mu^{3/4}} \approx 2.5y \tag{22}$$

Low-Reynolds-number modifications

The effects of viscous damping are expressed by:

$$L = a_\mu y\left\{1 - \exp\left(-\frac{y^*}{A_\mu}\right)\right\} \tag{23}$$

where $y^* = y\sqrt{k}/\nu$, $a_\mu \approx 2.5$ and, in the Hassid and Poreh (1975) model for example, $A_\mu = 84$ (see also Gibson et al. 1978). The limiting behaviour at the wall is:

$$u = b_1 y + c_1 y^2 + d_1 y^3 + \cdots$$

$$v = c_2 y^2 + d_2 y^3 + \cdots \tag{24}$$

$$w = b_3 y + c_3 y^2 + d_3 y^3 + \cdots$$

$$k \equiv \frac{\overline{u^2} + \overline{v^2} + \overline{w^2}}{2} = \frac{\overline{b_1^2} + \overline{b_3^2}}{2}y^2 + (\overline{b_1 c_1} + \overline{b_3 c_3})y^3 + O(y^4) \tag{25}$$

$$\frac{\varepsilon}{\nu} = \overline{\left(\frac{\partial u}{\partial y}\right)^2} + \overline{\left(\frac{\partial v}{\partial y}\right)^2} + \overline{\left(\frac{\partial w}{\partial y}\right)^2}$$

$$= \overline{b_1^2} + \overline{b_3^2} + 4(\overline{b_1 c_1} + \overline{b_3 c_3})y + O(y^2) \tag{26}$$

As $y \to 0$, viscous diffusion is balanced by the dissipation and the k-equation, **Equation 17**, reduces to $\varepsilon = v\partial^2 k/\partial y^2$. Then, by virtue of **Equation 25** the value of the dissipation at the wall may be written as:

$$\varepsilon_w = \lim_{y\to 0}\left(\frac{2\nu k}{y^2}\right) \tag{27}$$

In the Hassid and Poreh (1975) model:

$$\varepsilon = \frac{2\nu k}{y^2} + \frac{k^{3/2}}{L} \tag{28}$$

One-equation models are also described by Wolfshtein (1967) and by Norris and Reynolds (1975).

Two-Equation Modelling

The ε-equation

The length scale need not be determined when it can be evaluated at every point as the solution of a second equation. The length scale itself is not a suitable dependent variable, not least because it actually increases as turbulence decays. Several other variables have been suggested for this duty of which the most widely used is the dissipation rate, $\varepsilon = k^{3/2L}$. In the k-ε model **Equation 20** for the eddy viscosity is rewritten as:

$$\nu_t = C_\mu f_\mu \frac{k^2}{\varepsilon} \tag{29}$$

where f_μ is the low-Re damping function multiplying L.

The exact equation for ξ contains numerous intractable terms (see Rodi and Mansour 1993) but it may be condensed to the basic form containing the four essential processes of advection, diffusion (viscous, turbulent and pressure), production, dissipation:

$$U_i \frac{\partial \varepsilon}{\partial x_i} = (D_\varepsilon + T_\varepsilon + \Pi_\varepsilon) + P_\varepsilon - \chi \tag{30}$$

If ε is interpreted as the rate at which turbulence kinetic energy is transferred from large scales to small in the spectral cascade, its rate of transfer can be expressed in terms of the time scale $\tau = k/\varepsilon$:

$$\chi \propto \frac{\varepsilon}{\tau} \quad \text{or} \quad \chi = C_{\varepsilon 2} \frac{\varepsilon^2}{k} \tag{31}$$

where $C_{\varepsilon 2}$ is considered to be a constant, at least for turbulence at high Reynolds numbers. The production of ε by vortex stretching ought arguably to be modelled in terms of turbulence quantities, but this is not feasible in two-equation modelling. Instead, for high-Re turbulence, it is assumed that:

$$P_\varepsilon = C_{\varepsilon 1} \frac{\varepsilon}{k} P \tag{32}$$

A gradient-diffusion process is assumed for turbulent transport:

$$T_\varepsilon + \Pi_\varepsilon = \frac{\partial}{\partial x_j}\left(\frac{\nu_t}{\sigma_\varepsilon}\frac{\partial \varepsilon}{\partial x_j}\right) \tag{33}$$

and the modelled ε-equation is finally written as:

$$U_i \frac{\partial \varepsilon}{\partial x_i} = \frac{\partial}{\partial x_j}\left\{\left(\nu + \frac{C_\mu f_\mu k^2}{\sigma_\varepsilon \ \varepsilon}\right)\frac{\partial \varepsilon}{\partial x_j}\right\} + C_{\varepsilon 1} f_1 \frac{\varepsilon}{k} P \tag{34}$$

$$- C_{\varepsilon 2} f_2 \frac{\varepsilon^2}{k} + E$$

where the functions f_1 and f_2 are introduced to deal with low-Re effects on production and dissipation, and symbol E stands for additional terms which become important in the immediate vicinity of the wall.

The k-ε model constants

The decay constant $C_{\varepsilon 2}$ is determined by reference to the decay of homogeneous grid turbulence for which the k and ε **Equations 17 and 34** reduce to:

$$U\frac{dk}{dx} = -\varepsilon; \qquad U\frac{d\varepsilon}{dx} = -C_{\varepsilon 2}\frac{\varepsilon^2}{k} \tag{35),(36}$$

Measurements of k downstream of the grid are fitted by the power law $k = k_0 x^{-\alpha}$, with $\alpha \approx 1.25$ in the initial period. Substitution in **Equations 35 and 36** gives $C_{\varepsilon 2} = (\alpha + 1)/\alpha \approx 1.8$. However Launder and Spalding's (1974) $C_{\varepsilon 2} = 1.92$ has stood the test of time and endures in the "standard" k-ε model.

A relationship between the three constants of the ε-equation is found from conditions in the constant-stress wall layer where $P \approx \varepsilon \approx u_\tau^3/\kappa y$. Convection by the mean flow is negligibly small and **Equation 34** reduces to:

$$\frac{d}{dy}\left(\frac{C_\mu}{\sigma_\varepsilon}\frac{k^2}{\varepsilon}\frac{d\varepsilon}{dy}\right) + \frac{\varepsilon}{k}(C_{\varepsilon 1}P - C_{\varepsilon 2}\varepsilon) = 0 \tag{37}$$

from which, with $dk/dy \approx 0$:

$$\sigma_\varepsilon(C_{\varepsilon 2} - C_{\varepsilon 1}) = \frac{\kappa^2}{\sqrt{C_\mu}} \approx 0.56 \tag{38}$$

Launder and Spalding (1974) found that the predicted spreading rates of free shear layers were highly sensitive to $(C_{\xi 2} - C_{\xi 1})$ and that "best-fit" values were: $C_{\varepsilon 1} = 1.44$, $C_{\varepsilon 2} = 1.92$, $\sigma_\varepsilon = 1.3$.

Wall functions

Because very fine grids are needed to resolve the small-scale turbulence in the viscous layers close to a wall, and because wholly satisfactory low-Re models have yet to be developed, these regions are often bridged by "wall functions" based on the universal laws of the wall. Alternatively, the problem of adapting the ε-equation may be avoided by using a one-equation model in which the length-scale distribution is prescribed through the viscous layers, as advocated by Rodi (1991) and Rodi et al. (1993).

k-ε modelling for low-Re turbulence

As the wall is approached the k-equation 17 reduces to $\varepsilon = \nu\partial^2 k/\partial y^2$. This result causes problems which Jones and Launder (1972) avoid by using an equation for the "isotropic" part of the dissipation, which is zero at the wall:

$$\tilde{\varepsilon} = \varepsilon - \frac{\nu}{2k}\left(\frac{\partial k}{\partial y}\right)^2 \tag{39}$$

A tendency in more recent work has been to retain ε as the dependent variable while defining the time and length scales with $\tilde{\varepsilon}$, thus: $\tilde{\tau} \equiv \tilde{\varepsilon}/k$; $\tilde{L} \equiv k_{3/2}/\tilde{\varepsilon}$. As $y \to 0$, from **Equation 25**:

$$k = \alpha y^2 + \beta y^3 + O(y^4) \tag{40}$$

$$\frac{1}{k}\left(\frac{\partial k}{\partial y}\right)^2 = \frac{4\alpha^2 y^2 + 12\alpha\beta y^3 + \cdots}{\alpha y^2(1 + \beta y/\alpha + \cdots)} = 4\alpha + 8\beta y + O(y^2) \tag{41}$$

When this is compared with **Equation 26**: $\varepsilon/\nu = 2\alpha + 4\beta y + O(y^2)$, it is seen that $\tilde{\varepsilon} = 0(y^2)$. $\tilde{L}$ is proportional to y at the wall, as it is in the logarithmic layer where $\tilde{\varepsilon} \to \varepsilon$, and f_μ in **Equation**

29 ought strictly to be proportional to y as $y \to 0$, although this condition is seldom satisfied in practice. Rodi and Mansour (1993) compare various f_μ functions with the results of numerical simulations of channel flow. Their best-fit function is:

$$f_\mu = 1 - \exp(-0.0002y^+ - 0.00065y^{+2}) \quad \textbf{(42)}$$

The drawback to this and many of the other functions in the literature is that y^+ is unsuitable for general flow calculations. The Launder and Sharma (1974) formula is better adapted to general purpose calculations in that it is based on the local turbulence Reynolds number, $R_t \equiv k^2/\nu\varepsilon$, although it compares poorly with the DNS channel-flow results:

$$f_\mu = \exp\left\{\frac{-3.4}{(1 + R_t/50)^2}\right\} \quad \textbf{(43)}$$

In nearly all current models the ε-production rate is left unchanged by setting $f_1 = 1.0$. The dissipation-rate function, f_2, must fit measurements of the decay of grid turbulence in the final period, $k \; \alpha \; t^{-\alpha}$, with $\alpha \approx 2.5$. Then $C_{\varepsilon 2}f_2 = (\alpha + 1)/\alpha \approx 1.4$, whence $f_2 \approx 1.4/1.8 \approx 0.78$ as $R_t \to 0$. According to Hanjalic and Launder (1976) the data of Batchelor and Townsend (1948) are fitted by:

$$f_2 = 1 - 0.22 \exp\left\{-\left(\frac{R_t}{6}\right)^2\right\} \quad \textbf{(44)}$$

where $R_t \equiv k^2/\nu\varepsilon$. Viscous effects are confined to the layers very close to the wall, e.g. $f_2 \approx 0.99$ for $R_t = 10$, and R_t is to be preferred to y^+ as the wall-distance parameter.

Different forms of the term E in **Equation 34** are listed in the survey by Rodi and Mansour (1993). The most widely used has been Jones and Launder's (1972):

$$E = 2\nu\nu_t\left(\frac{\partial^2 U_i}{\partial x_k \partial x_m}\right)\left(\frac{\partial^2 U_i}{\partial x_k \partial x_m}\right) \quad \textbf{(45)}$$

which, according to Rodi and Mansour (1993), compares unfavourably with near-wall values of the exact production term:

$$P_\xi^3 = -2\nu \overline{u_k \frac{\partial u_i}{\partial x_m} \frac{\partial^2 U_i}{\partial x_k \partial x_m}} \quad \textbf{(46)}$$

in the low-Re channel-flow simulations of Kim et al. (1987).

Eddy-Diffusivity Models for Heat and Mass Transfer

The eddy diffusivity

A passive scalar (temperature) $\hat{\vartheta}$ is expressed as the sum of mean and fluctuating parts, $T + \vartheta$, which is substituted in the $\hat{\vartheta}$-conservation equation. The result upon averaging is:

$$\frac{\partial T}{\partial t} + U_j \frac{\partial T}{\partial x_j} = \frac{\partial}{\partial x_j}\left(\alpha \frac{\partial T}{\partial x_j} - \overline{u_j\vartheta}\right) + S_\vartheta \quad \textbf{(47)}$$

where α is the molecular diffusivity and the $-\overline{u_j\vartheta}$ are turbulent scalar fluxes. An eddy diffusivity α_t is defined by:

$$-\overline{u_j\vartheta} = \alpha_t \frac{\partial T}{\partial x_j} \quad \textbf{(48)}$$

and is related to ν_t via a "turbulent Prandtl number" $Pr_t \equiv \nu_t/\alpha_t$.

The mixing-length model for heat and mass transfer

α_t is assumed to be the product of a turbulent velocity, υ, and a scalar mixing length, ℓ_ϑ. Then, as before, with $\upsilon = \ell(\partial U/\partial y)$ for the turbulent velocity scale in a thin shear layer:

$$\alpha_t = \ell\ell_\vartheta\left|\frac{\partial U}{\partial y}\right|, \qquad -\overline{v\vartheta} = \ell\ell_\vartheta\left|\frac{\partial U}{\partial y}\right|\frac{\partial T}{\partial y}, \qquad Pr_t = \frac{\ell}{\ell_\vartheta} \quad \textbf{(49)}$$

Typically $Pr_t \approx 0.9$ in wall turbulence and ≈ 0.7 in free shear flow.

Temperature Boundary Layer

Near a wall, for small $\partial T/\partial x$, the counterpart of Equation 10 is:

$$\alpha \frac{\partial T}{\partial y} - \overline{v\vartheta} \approx \frac{-\dot{q}_w}{\rho c_p} \quad \textbf{(50)}$$

where $\dot{q}_w$ is the heat flux through the wall when scalar T is the temperature. The damped ℓ_ϑ distribution is like that of ℓ, Equation 14:

$$\ell_\vartheta = \kappa_\vartheta y\left\{1 - \exp\left(\frac{-y^+}{A_\vartheta^+}\right)\right\} \quad \textbf{(51)}$$

and the T^+ ($\equiv -\rho u_\tau c_p(T - T_w)/\dot{q}_w$) profile is obtained by integrating:

$$\dot{q}^+ \equiv \frac{\dot{q}}{q_w} = \left[\frac{1}{Pr} + \ell^+\ell_\vartheta^+ \frac{du^+}{dy^+}\right]\frac{dT^+}{dy^+} \quad \textbf{(52)}$$

to give using Equation 15, for $q^+ = 1$ and large y^+:

$$T^+ \approx \frac{1}{\kappa_\vartheta}\log_e y^+ + C_\vartheta(Pr) \quad \textbf{(53)}$$

Typically (Gibson 1990): $\kappa_\vartheta = 0.45$, $\kappa/k_\vartheta = 0.91$. With $A_\vartheta^+ = A^+ = 26$, $C_\vartheta = 3.0$ when $Pr = 0.7$ (for air). A temperature wall function:

$$T^+ = Pr_t(u^+ + P) \quad \textbf{(54)}$$

is used in two-equation modelling with:

$$P = 9.24\left\{\left(\frac{Pr}{Pr_t}\right)^{3/4} - 1\right\}\left\{1 + 0.28\exp\left(\frac{-0.007\,Pr}{Pr_t}\right)\right\} \quad \textbf{(55)}$$

obtained from pipe-flow data by Jayatillaka (1969).

The temperature variance

The equation for $k_\vartheta \equiv \overline{\vartheta^2}/2$, is obtained by multiplying the $(T + \vartheta)$ equation by ϑ, dividing by two, and averaging. Then:

$$\frac{Dk_\vartheta}{Dt} = \frac{\partial}{\partial x_j}\left\{\alpha\frac{\partial k_\vartheta}{\partial x_j} - \overline{u_j k_\vartheta}\right\} - \overline{u_j\vartheta}\frac{\partial T}{\partial x_j} - \xi_\vartheta \quad (56)$$

This is like the k-equation, **Equation 17**, without the pressure term. Turbulent transport is modelled by gradient diffusion. The dissipation rate may be represented by $\varepsilon_\vartheta = k_\vartheta\varepsilon/kR$, where R is the ratio of time scales which is found to range from about 0.5 to 1.0 in different flows. It is difficult to devise an equation for ε_ϑ because there are two relevant time scales. The problems are identified by Launder (1976) and Lumley (1978). Modelled ε_ϑ-equations are described by Jones and Musonge (1988), Malin and Younis (1990), Craft and Launder (1991), Youssef et al. (1992), and Sommer et al. (1993), mainly in connexion with second-moment closures.

Second-Moment Closures

The Reynolds-stress equations

Although two-equation models are capable of giving good results for thin shear layers and some recirculating flows, they fail to account adequately for the effects of complex strain fields and body forces. These limitations, combined with the acknowledged defects of eddy-viscosity modelling generally and the availability of ever-increasing computer power, have motivated the development of second-order closures based on modelled transport equations for the Reynolds stresses themselves. The equations for $\overline{u_iu_j}$ are obtained by multiplying the equations for $(U_i + u_i)$ and $(U_j + u_j)$ by u_j and the u_i respectively, adding and averaging. The result is:

$$\frac{\partial}{\partial t}\overline{u_iu_j} + U_k\frac{\partial}{\partial x_k}\overline{u_iu_j} = -\left(\overline{u_iu_k}\frac{\partial U_j}{\partial x_k} + \overline{u_iu_k}\frac{\partial U_i}{\partial x_k}\right) + \frac{\partial}{\partial x_k}\left\{\nu\frac{\partial}{\partial x_k}\overline{u_iu_j} - \overline{u_iu_ju_k} - \overline{\frac{p}{\rho}(u_i\delta_{jk} + u_j\delta_{ik})}\right\} - 2\nu\overline{\frac{\partial u_i}{\partial x_k}\frac{\partial u_j}{\partial x_k}} + \overline{\frac{p}{\rho}\left(\frac{\partial u_i}{\partial x_j} + \frac{\partial u_j}{\partial x_i}\right)} \quad (57)$$

Models are required for viscous dissipation, turbulent and pressure diffusion, and energy redistribution by pressure-strain. The closure ought strictly to satisfy "realiseability" constraints which do not permit negative normal stresses, or correlations that fail to satisfy Schwartz's inequality (Lumley 1978). These conditions have often been disregarded in practice.

Rotta's (1951) ideas were developed by Daly and Harlow (1970), Hanjalic and Launder (1972), Launder et al. (1975) and Lumley (1978). Recent advances are described by Shih and Lumley (1986), Speziale et al. (1991), Launder and Tselepidakis (1993) and Durbin (1993). Launder (1989) and So et al. (1991) survey the field.

Viscous dissipation

When the turbulence Reynolds number is sufficiently high the small-scale motion is nearly isotropic and the dissipation may be assumed to be equally divided between components. The form:

$$\varepsilon_{ij} \equiv 2\nu\overline{\frac{\partial u_i}{\partial x_k}\frac{\partial u_j}{\partial x_k}} = \left[2\nu\overline{\frac{\partial u_i}{\partial x_k}\frac{\partial u_j}{\partial x_k}} - \frac{2}{3}\delta_{ij}\varepsilon\right] + \frac{2}{3}\delta_{ij}\varepsilon \quad (58)$$

also allows for viscous energy redistribution when these conditions do not obtain and the term in square brackets is non-zero. ξ is to be obtained as the solution of a modelled transport equation as in two-equation modelling.

Turbulent diffusion

Launder et al. (1975) express turbulent diffusion as:

$$\overline{u_iu_ju_k} = -C_s\frac{k}{\varepsilon}\left(\overline{u_iu_m}\frac{\partial\overline{u_ju_k}}{\partial x_m} + \overline{u_ju_m}\frac{\partial\overline{u_iu_k}}{\partial x_m} + \overline{u_ku_m}\frac{\partial\overline{u_iu_j}}{\partial x_m}\right) \quad (59)$$

with $C_s = 0.11$. A shorter form attributed to Daly and Harlow (1970), with only the last term, is also used with $C_s = 0.22$.

The Pressure-strain-rate correlation

The fluctuating pressure appears in "diffusion" and "redistribution" terms on account of the manipulation:

$$\overline{u_j\frac{\partial}{\partial x_i}(P + p)} = \frac{\partial}{\partial x_i}\overline{pu_j} - \overline{p\frac{\partial u_j}{\partial x_i}} = \frac{\partial}{\partial x_k}\overline{pu_j}\delta_{ik} - \overline{p\frac{\partial u_j}{\partial x_i}} \quad (60)$$

where the last term has the useful property of contracting to zero in incompressible flow. It therefore serves only to redistribute energy between components. $\overline{p\partial u_i/\partial x_j}$ may be expressed as the sum of two terms in a volume integral. One of these terms contains only averaged fluctuating quantities; the second involves the mean rate of strain.

Models for the turbulence, or "slow" part of $\overline{p\partial u_i/\partial x_j}$ rest on the proposition that strain-free turbulence tends to the isotropic state. For the decay of turbulence downstream of a wind-tunnel contraction the energy and stress **Equations 17 and 57** combine to give:

$$U\frac{\partial}{\partial x}\left(\overline{u_\alpha u_\alpha} - \frac{2}{3}k\right) = \overline{\frac{p}{\rho}\frac{\partial u_\alpha}{\partial x_\alpha}} \quad (61)$$

where there is no summation on repeated Greek indices. In Rotta's (1951) widely used linear model:

$$\frac{D}{Dt}\left(\overline{u_\alpha u_\alpha} - \frac{2}{3}k\right) = -C_1\frac{\varepsilon}{k}\left(\overline{u_\alpha u_\alpha} - \frac{2}{3}k\right) \quad (62)$$

where C_1 is a constant and k/ε is the time scale. C_1 values in use range from 1.5 to 3.0. There is no return to isotropy when $C_1 = 1$. The evidence is that the process is non-linear and that the tendency is weak for small anisotropy. The non-linear model described by Lumley (1978) has two coefficients which are generally functions of the turbulence Reynolds number, $k^2/\nu\varepsilon$, and the invariants $b^2 = b_{ij}b_{ij}$ and $b^3 = b_{ij}b_{jk}b_{ki}$ of the anisotropy tensor:

$$b_{ij} \equiv \frac{\overline{u_i u_j}}{k} - \frac{2}{3}\delta_{ij} \quad (63)$$

Shih and Lumley (1985) use **Equation 62** with:

$$C_1 = 1 + 4.45\,A\ln(1 + 7.8b^2 + 5.98\,b^3) \quad (64)$$

where:

$$A \equiv 1 + \frac{9}{8}(b^2 - b^3) \quad (65)$$

is a parameter which has the useful properties (for near-wall Reynolds-stress modelling) of being unity in isotropic turbulence and zero in the two-dimensional limit (Lumley 1978).

A simple but effective model for the mean-strain, or "rapid", part of $\overline{p\partial u_i/\partial x_j}$ is:

$$\overline{\frac{p}{\rho}\left(\frac{\partial u_i}{\partial x_j} + \frac{\partial u_j}{\partial x_i}\right)}_{rapid} = -C_2\left(P_{ij} - \frac{2}{3}\delta_{ij}P\right) \quad (66)$$

where the right-hand side is proportional to the anisotropy of $\overline{u_i u_j}$ production. This term is actually one of three components formally obtained by Launder et al. (1975), but good results can be obtained by this abbreviated form alone. On combining the two components, generalised to cover the shear stresses, there results:

$$\overline{\frac{p}{\rho}\left(\frac{\partial u_i}{\partial x_j} + \frac{\partial u_j}{\partial x_i}\right)} = -C_1\frac{\varepsilon}{k}\left(\overline{u_i u_j} - \frac{2}{3}\delta_{ij}k\right) - C_2\left(P_{ij} - \frac{2}{3}\delta_{ij}P\right) \quad (67)$$

For two-dimensional thin shear layers close to local equilibrium ($P \approx \varepsilon$) the model gives the following algebraic expressions for the stresses:

$$\frac{\overline{u^2}}{k} = \frac{2}{3}(1 + 2\phi),\ \frac{\overline{v^2}}{k} = \frac{\overline{w^2}}{k} = \frac{2}{3}(1 - \phi),\ -\overline{uv} = \phi\frac{\overline{v^2}}{k}\frac{k^2}{\varepsilon}\frac{\partial U}{\partial y} \quad (68)$$

where $\phi \equiv (1 - C_2)/C_1$. Evidently ϕ must be approximately 0.2 to give $\overline{v^2}/k \approx 0.5$, $C_\mu = \phi\overline{v^2}/k \approx 0.1$. Gibson and Launder (1978) chose $C_2 = 0.6$ in order to satisfy an exact result for isotropic turbulence, setting $C_1 = 1.8$ to obtain $\phi = 0.22$. Gibson & Younis (1986), got better results for curved and swirling flows with $C_1 = 3.0$, $C_2 = 0.3$, disregarding the results of rapid-distortion theory and achieving arguably a better fit to the return-to-isotropy data (but see Taulbee 1989 for a contrary assessment).

Wall effects on the pressure-strain mechanism

A rigid boundary reflects pressure fluctuations and thereby inhibits the transfer of energy into the component normal to the wall. Launder et al. (1975), followed by Gibson and Launder (1978), add wall-effect terms to both parts of their pressure-strain models, in each case expressed as linear functions of distance from the wall. Craft and Launder (1991) retain only the wall correction to the rapid component, and that in more complicated form. Speziale et al. (1991) and Durbin (1993) have devised models that require no explicit wall-reflection terms.

Low-Reynolds-number stress modelling

In the near-wall viscous region where the Reynolds number is low, the energy-containing and dissipation ranges of the energy spectrum overlap, the fine-scale motion is no longer isotropic, and the dissipation rate is no longer divided equally between components. Rotta (1951) initially assumed that $\varepsilon_{ij}/\varepsilon \approx \overline{u_i u_j}/k$ but it is now known that this approximation seriously underestimates ε_{22}, and it has been corrected in later work. The ε-equation differs only slightly from the form used in two-equation modelling. In the immediate vicinity of the wall $\overline{v^2}$ becomes so attenuated that the turbulence becomes almost two-dimensional. The models devised by Launder and Shima (1989), Shima (1993), Launder and Tselepidakis (1993) involve Lumley's (1978) "flatness factor", A, of Equation 65 which is unity in isotropic turbulence but $\to 0$ as the turbulence approaches two-dimensionality in the near-wall layers.

Scalar-flux modelling

The $\overline{u_i\vartheta}$ equations have received less attention than the stress equations, and the closure methods are less highly developed, though naturally similar. The survey by Launder (1976) remains a useful introduction. recent advances are described by Lai and So (1991) and Durbin (1993).

The equations are obtained by multiplying the equation for ($T + \vartheta$) by u_j and adding it to the ($U_i + u_j$) equation multiplied by ϑ. The result after averaging is:

$$\frac{\partial\overline{u_i\vartheta}}{\partial t} + U_j\frac{\partial\overline{u_i\vartheta}}{\partial x_j} = -\left(\overline{u_k\vartheta}\frac{\partial U_i}{\partial x_k} + \overline{u_i u_k}\frac{\partial T}{\partial x_k}\right) - (\alpha + \nu)\overline{\frac{\partial u_i}{\partial x_k}\frac{\partial\vartheta}{\partial x_k}}$$
$$+ \overline{\frac{p}{\rho}\frac{\partial\vartheta}{\partial x_i}} + \frac{\partial}{\partial x_k}\left\{\nu\overline{\vartheta\frac{\partial u_i}{\partial x_k}} + \alpha\overline{u_i\frac{\partial\vartheta}{\partial x_k}} - \overline{u_i u_k\vartheta} - \frac{\overline{p\vartheta}}{\rho}\delta_{ik}\right\} \quad (69)$$

At high Reynolds numbers the molecular diffusion terms are negligible, and the dissipation is zero by virtue of the local-isotropy hypothesis. Closure approximations for turbulent diffusion and the pressure-scalar-gradient correlation are akin to those of the Reynolds-stress equations, but are further complicated by the need to allow for different time scales in the fluctuating scalar and velocity fields. Turbulent diffusion of $\overline{u_i\vartheta}$ is modelled by:

$$\overline{u_i u_k\vartheta} = -C_{s\vartheta}\frac{k}{\varepsilon}\left(\overline{u_i u_m}\frac{\partial\overline{u_k\vartheta}}{\partial x_m} + \overline{u_k u_m}\frac{\partial\overline{u_i\vartheta}}{\partial x_m}\right) \quad (70)$$

with $C_{s\vartheta} \approx 0.11$ or, frequently, with only the last of these terms. The arguments which led to the modelled pressure-strain term of the stress equations suggest for the equivalent term here:

$$\overline{\frac{p}{\rho}\frac{\partial\vartheta}{\partial x_i}} = -C_{\vartheta 1}\frac{\varepsilon}{k}\overline{u_i\vartheta} + C_{\vartheta 2}\overline{u_j\vartheta}\frac{\partial U_i}{\partial x_j} \quad (71)$$

where the first term on the right is the counterpart of **Equation 62** and the second involves the mean-strain part of $\overline{u_i\vartheta}$ production.

For the idealised local-equilibrium shear layer, with negligible mean-flow or turbulent transport, the equations reduce to (Durbin 1993):

$$\overline{u\vartheta} = -\frac{1}{C_{\vartheta 1}}\frac{k}{\varepsilon}\overline{uv}\frac{\partial T}{\partial y} - \frac{1 - C_{\vartheta 2}}{C_{\vartheta 1}}\overline{v\vartheta}\frac{\partial U}{\partial y} \qquad \textbf{(72)}$$

$$\overline{v\vartheta} = -\frac{1}{C_{\vartheta 1}}\frac{k}{\varepsilon}\overline{v^2}\frac{\partial T}{\partial y} \qquad \textbf{(73)}$$

giving:

$$Pr_t = C_{\vartheta 1}\frac{k}{\overline{v^2}}\left(\frac{\overline{uv}}{k}\right)^2 \quad \text{and} \quad \frac{\overline{u\vartheta}}{\overline{v\vartheta}} = \frac{k}{\overline{uv}}\frac{1 + Pr_t - C_{\vartheta 2}}{C_{\vartheta 1}} \qquad \textbf{(74)}$$

$C_{\vartheta 1} = 3.0$ and $C_{\vartheta 2} = 0.33$ are chosen on the basis of the experimental data for free flow examined by Launder (1976): $Pr_t \approx 0.7$, $\overline{u\vartheta}/\overline{v\vartheta} \approx -1.3$. The wall-reflection correction is small; substitution of near-wall stress values from Equation 68 gives $Pr_t \approx 0.8$, $\overline{u\vartheta}/\overline{v\vartheta} \approx -1.8$. Lai and So (1990) and Durbin (1993) describe low-Re scalar-flux models.

References

Baldwin, B. S. and Lomax, H. (1978) AIAA paper, 78–257.

Batchelor, G. K. and Townsend, A. A. (1948) *Proc. Roy. Soc.,* A194, 538.

Boussinesq, J. (1877) *Mem. Pre. par div. Sav.,* 23, Paris.

Bradshaw, P., (ed) (1976) *Topics in Applied Physics,* 12, Springer.

Cebeci, T. and Smith, A. M. O. (1974) *App. Math. Mech.,* 15 Academic Press.

Cebeci, T., Stewartson, K. and Whitelaw, J. H. (1984) *Numerical and Physical Aspects of Aerodynamic Flows II,* T. Cebeci, ed., 1, Springer.

Craft, T. J. and Launder, B. E. (1991) *Proc. 8th Turbulent Shear Flows Symposium,* Munich.

Daly, B. J. and Harlow, F. H. (1970) *Phys. Fluids.,* 13, 2634.

Durbin, P. A. (1993) *J. Fluid Mech.,* 249, 465–498.

Gibson, M. M. and Launder, B. E. (1978) *J. Fluid Mech.,* 86, 491.

Gibson, M. M. and Younis, B. A. (1986) *Phys. Fluids,* 29, 38.

Gibson, M. M. (1990) *Near-Wall Turbulence,* S. J. Kline and N. H. Afghan, eds., 157, Hemisphere.

Gibson, M. M. and Rodi, W. (1989) *J. Hydraulic Research,* 27, 233.

Gibson, M. M., Spalding, D. B. and Zinser, W. (1978) *Letters in Heat and Mass Transfer,* 5, 73.

Granville, P. S. (1987) *AIAA J.,* 25, 1624.

Hanjalic, K. and Launder, B. E. (1972) *J. Fluid Mech.,* 52, 609.

Hanjalic, K. and Launder, B. E. (1976) *J. Fluid Mech.,* 74, 593.

Hanjalic, K. (1988) *Near-Wall Turbulence,* S. J. Kline and N. H. Afghan, eds., 762–781, Hemisphere.

Hassid, S. and Poreh, M. (1975) *ASME J. Fluids Eng.,* 100, 107.

Jayatillaka, C. V. L. (1969) *Prog. Heat Mass Transfer,* 1, 193.

Jones, W. P. and Launder, B. E. (1972) *Int. J. Heat Mass Transfer,* 15, 301.

Jones, W. P. and Musonge, P. (1988) *Phys. Fluids,* 31, 3589–3604.

Kim, J., Moin, P. and Moser, R. (1987) *J. Fluid Mech.,* 177, 133.

Kolmogorov, A. N. (1942) *Izv. Akad. Nauk. SSR Ser. Phys.,* 6, 56.

Lai, Y. G. and So, R. M. C. (1990) *Int. J. Heat Mass Transfer,* 33, 1429.

Lang, N. J. and Shih, T. H. (1991) *NASA TM* 105237.

Launder, B. E. and Sharma, B. I. (1974) *Letters in Heat and Mass Transfer,* 1, 131–138.

Launder, B. E. (1976) in Bradshaw 1976.

Launder, B. E. (1989) *Int. J. Heat Fluid Flow,* 10, 282.

Launder (1993) *Engineering Turbulence Modelling and Experiments,* 2 W. Rodi and F. Martelli, eds., 3–22, Elsevier.

Launder, B. E. and Shima, N. (1989) *AIAA J.,* 27, 1319.

Launder, B. E. and Spalding, D. B. (1974) *Computer Methods in Applied Mechanics and Engineering,* 3, 269.

Launder, B. E. and Tselepidakis, D. P. (1993) *Turbulent Shear Flows,* 8, F. Durst et al., ed. 81–96, Springer.

Launder, B. E., Reece, G. J. and Rodi, W. (1975) *J. Fluid Mech.,* 68, 537.

Lumley, J. L. (1978) *Adv. Appl. Mech.,* 18, 123.

Malin, M. R. and Younis, B. A. (1990) *Int. J. Heat and Mass Transfer,* 33, 2247–2264.

McEligot, D. M. (1985) *Lecture Notes in Physics,* 235, 292–303, Springer.

Nagano, Y. and Shimada, M. (1993) *9th Turbulent Shear Flows Symposium,* Kyoto.

Nallasamy, M. (1987) *Computers and Fluids,* 15, 151.

Norris, L. H. and Reynolds, W. C. (1975) *MED Rep. FM-10,* Stanford.

Prandtl, L. (1925) *ZAMM,* 5, 136.

Prandtl, L. (1945) *Nachrichten Akad. Wissenschaft Gottingen.*

Reynolds, O. (1894) *Phil. Trans. Roy. Soc.,* 186 A, 123.

Rodi, W. (1991) AIAA paper 91-0216.

Rodi, W. and Mansour, N. N. (1993) *J. Fluid Mech.,* 250, 509.

Rodi W., Mansour, N. N. and Michelassi, V. (1993) *Trans ASME J. Fluids Eng.,* 115, 196.

Rotta, J. C. (1951) *Z. Phys.,* 129, 547 and 131, 51.

Shih, T.-H. and Hsu, A. T. (1991) AIAA paper 91-0611

Shih, T-H., Lumley, J. L. and Chen, J-Y. (1990) *AIAA J.,* 28, 610.

Shih, T-S. and Lumley, J. H. (1985) *Cornell Rep.* FDA-85-3.

Shih, T-S. and Lumely, J. H. (1986) *Phys. Fluids,* 29, 971.

Shima, N. (1988) *Trans. ASME J. Fluids Eng.,* 110, 38.

Shima, N. (1993) *Proc. 9th Turb. Shear Flows Symp.,* Kyoto.

Shir, C. C. (1973) *J. Atmos. Sci.,* 30, 1327.

So, R. M. C., Lai, Y. G., Zhang, H. S. and Hwang, B. C. (1991) *AIAA J.,* 29, 1819.

Speziale, C. G. (1991) *Ann. Rev. Fluid Mech.,* 23, 107.

Speziale, C. G., Abid, R. and Anderson, E. C. (1992) *AIAA J.,* 30, 2, 324.

Speziale, C. G., Sarkar, S. and Gatski, T. B. (1991) *J. Fluid Mech.,* 227, 245.

Sommer, T. P., So, R. M. C. and Zhang, H. S. (1993) AIAA paper 93-0088.

Taulbee, D. B. (1989) *Recent Advances in Turbulence,* (ed. W. K. George and R. Arndt, eds., 75, Hemisphere.

Van Driest, E. R. (1956) *J. Aero. Sci.,* 23, 1007.

Wolfshtein, M. (1967) PhD thesis, Imperial College

Youssef, M. S., Nagano, Y. and Tagawa, M. (1992) *Int. J. Heat Mass Transfer,* 35, 3095.

Leading to: *Universal velocity profile*

M.M. Gibson

TURBULENT BURSTS (see Turbulent flow)

TURBULENT DIFFUSION (see Diffusion; Turbulence models)

TURBULENT ENERGY DISSIPATION (see Turbulence models)

TURBULENT FLOW

Following from: *Flow of fluids; Conservation equations, single phase*

Turbulent flow is a fluid motion with particle trajectories varying randomly in time, in which irregular fluctuations of velocity, pressure and other parameters arise. Since turbulence is a property of the flow rather than a physical characteristic of the liquid, an energy source for maintaining turbulence is required in each case, where such flow is realized. Turbulence may be generated by the work either of shear stresses (friction) in the main (mean) flow, i.e. in the presence of mean velocity gradients (a shear flow), or of mass (buoyant, magnetic) forces. With a shear flow, the energy is supplied to the pulsatory motion from the mean motion through large-scale vortexes, whose sizes are comparable with characteristic dimensions of the flow (for the flow behind a grid, this dimension is a characteristic mesh size of the grid; for a boundary layer flow, this dimension is a boundary layer thickness; for tubes, it is a radius; and for jets and a free shear layer, it is a transverse dimension of this layer). In near-wall flows (i.e., boundary layer, as well as tube and channel flows), turbulence generates in the region of the greatest near-wall velocity gradients throughout the flow extent. However, the turbulent flow develops only on the upset of stability of a laminar flow existing at Reynolds numbers below a certain critical value Re_c, which is $Re_c = \bar{u}D/\nu = 2.3 \times 10^3$ for the tube flow. A developed turbulent flow is established in a tube, away from the inlet, when $Re > 10^4$, and in a boundary layer when $Re_x = u_\infty x/\nu > 10^6$. Though the velocity fluctuations in the tube constitute as little as a few percent of the average flow velocity, they are indispensable to the development of the entire flow.

The velocity profile for turbulent flow is fuller than for the laminar flow (**Figure 1**), whereas a relationship between the average and axial velocities $\bar{u}/u_0$ depends on the Re number, being about 0.8 at $Re = 10^4$ and increasing as Re rises. With the laminar flow, the ratio is constant and equal to 0.5. A general specific feature of the near-wall turbulent flows is the presence, on the wall, of a thin viscous sublayer, wherein molecular viscosity forces are dominant and the velocity distribution is linear (δ_w in **Figure 1b**).

When describing turbulent flows unaffected by solid surfaces, the term "free turbulence" is used. **Figure 2** exemplifies *free turbulent flows.* Velocity gradients emerge in all these cases. Specifically, in a jet outflow from an opening, vortex bands formed on its edge

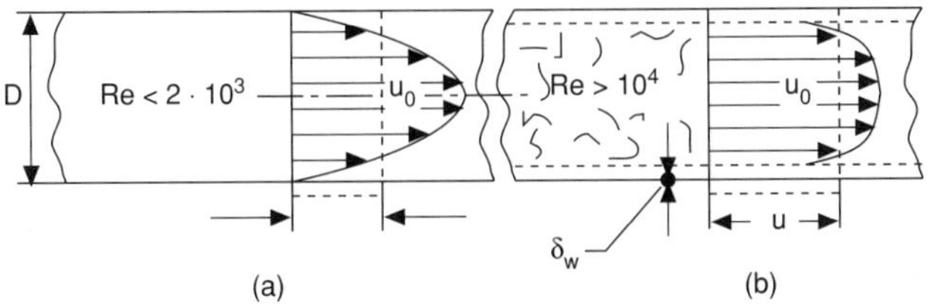

Figure 1. Velocity profiles in pipe flow.

diffuse under the effect of molecular viscosity with distance from the opening, so that the thickness of the mixing layer with considerable mean velocity gradients attains a large value, the flow acquires instability, and turbulence develops. The flow becomes turbulent behind the point, at which $u_1x/\nu = 7 \times 10^4$. Beyond the point of the motion transition to turbulent flow the rate of increase in the mixing layer thickness rises, so that at a distance equal to several times of the opening width the mixing layers interlock, and a completely developed turbulent jet flow is established. In common with the near-wall flows, in all cases the width "b" of the mixing zone is small relative to its length, and the transverse velocity gradient is large as compared to the gradient along the flow. The turbulent flow is much more capable for transfering momentum, heat and suspended particles, and for propagation of chemical reactions than is laminar flow.

The difference $u' = U - \langle u \rangle$ between the actual value at the point U and the mean value $\langle u \rangle$ is referred to as a fluctuating component. The turbulent pulsations are characterized by space and time scales. Instead of the latter, its reciprocal value is often used, *viz.,* frequency. At large Reynolds numbers, pulsations with a wide spectrum of scales are present in the flows. The key role in the turbulent flows is played by large-scale pulsations with sizes comparable to the dimensions of the region wherein the motion occurs. The corresponding frequencies are of the order of U/L. Small-scale pulsations contain only a small part of the entire kinetic energy of the fluid.

From the mechanics standpoint, turbulent flow is a nonlinear mechanical system with an extremely large number of degrees of freedom. Various methods are employed to model and describe the turbulent flows: statistical, spectral, diffusional, direct numerical modeling and semiempirical theories.

The *statistical theory of turbulence* is based on representing the flow as an infinitely changing assembly of vortexes. The vortexes and vortex tubes stretch in a definite direction by the action of deformations produced by the main flow, and in random directions when they interact. In consequence of the vortex stretching in all directions, turbulence is an essentially three-dimensional process. The kinetic energy of the main flow is transferred to ever smaller vortexes and eventually transforms to the internal energy of the thermal motion under the effect of viscosity forces. The random velocity field, set up by elementary vortexes in the cascade process of energy transfer from larger to smaller vortexes, cannot be described by explicit mathematical relations. In each observation of the phenomenon, one of the multiplicity of potential results will be reproduced, i.e., instantaneous velocities of the turbulent flow form a random vector field.

Random quantities are four-dimensional functions of the space-time point A(x,y,z,t). A complete statistical description of the fields of hydrodynamic characteristics requires the specification of all multidimensional probability distributions for these characteristics on all kinds of sets. The probability P of emergence of the prescribed point at a certain space and time point is determined by the three-dimensional probability density:

$$P(V \le V_p \le V + dV: r, t) \quad (1)$$

$$= f(U_1, U_2, U_3, x_1, x_2, x_3, t)dU_1dU_2dU_3,$$

where V_p is the velocity vector; r is the radius vector; and U_1, U_2,

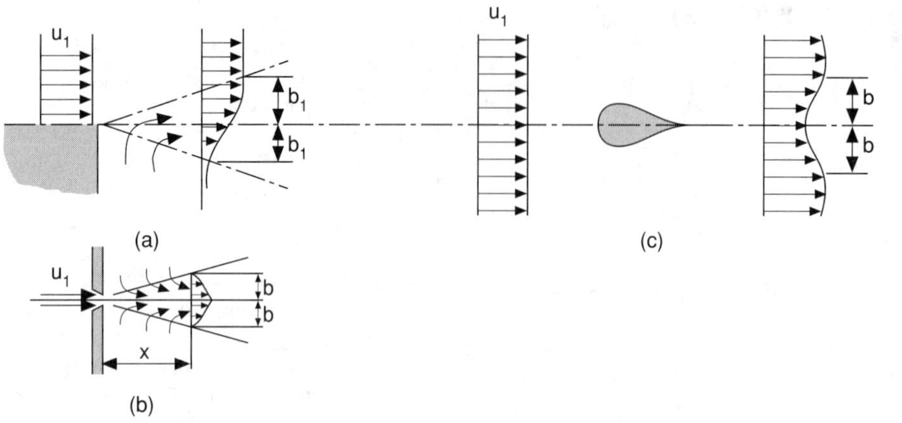

Figure 2. Free turbulent flow.

U_3 and x_1, x_2, x_3 are the velocity components and the coordinates. The turbulent flow field is assumed known if the 3n-dimensional probability density f^{3n} is specified. However, it is actually unfeasible to determine f^{3n}. In most cases, the random field can be described adequately by statistical moments of various orders, which may be obtained, for example, from experimental data: the n-point moment of r-th order has the form:

$$B^{k_i,k_j,\dots,k_p}_{i,j,\dots,p}(r_1, r_2, \dots, r_n, t)$$
$$= \langle U_i^{k_i}(r_1, t) U_j^{k_i}(r_2, t) \cdots U_p^{k_p}(r_n, t)\rangle \quad (2)$$

where $k = k_i + k_j + \dots k_p$. The angular brackets denote a statistical mean defined in terms of the probability density

$$\langle U_i^{k_i}(r_1, t), U_j^{k_j}(r_2, t) \cdots U_p^{k_p}(r_n, t)\rangle$$
$$= \int_{-\infty}^{\infty}\int_{-\infty}^{\infty} \cdots \int_{-\infty}^{\infty} U_i^{k_i}(r_1, t) U_j^{k_j}(r_2, t) \cdots U_p^{k_p}(r_n, t) x$$
$$\times f(U_1, U_2, \dots U_p, r_1, r_2, \dots r_n, t) dU_i dU_j \dots dU_p \quad (3)$$

The statistical moments of n-th order are characterized by properties of the n-th order tensor. Two-point moments are of the form:

$$B_{i,j}(r_1, r_2, t) = \langle U_i(r_1, t) U_j(r_2, t)\rangle,$$
$$B_{ij,k}(r_1, r_2, t) = \langle U_i(r_1, t) U_j(r_1, t) U_k(r_2, t)\rangle, \quad (4)$$
$$B_{ij,km}(r_1, r_2, t) = \langle U_i(r_1, t), U_j(r_1, t) U_k(r_2, t) U_m(r_2, t)\rangle.$$

When studying turbulent flow, consideration is given to two-point moments of not higher than of the third order. The moments formed from fluctuating quantities are named central moments:

$$b_{ij} = \langle U_i(r, t) U_j(r, t)\rangle \quad (5)$$
$$= \langle [U_i(r, t) - \langle U_i(r, t)\rangle][U_j(r, t) - \langle U_j(r, t)\rangle]\rangle.$$

One-point central moments of the velocity components determine the Reynolds stress $\langle U_i U_j \rangle$. The moments formed from random variables referring to several different random fields, e.g., of velocity and pressure, are called joint moments. A two-point joint moment of the pressure and velocity fields has the form:

$$B_{p,j}(r_1, r_2, t) = \langle p(r_1, t) U_j(r_2, t)\rangle. \quad (6)$$

Space-time moments are the mean value of the product of random variables relating to different points and different time instants:

$$B_{i,j}(r_1, r_2, t_1, t_2) = \langle U_i(r_1, t_1) U_j(r_2, t_2)\rangle. \quad (7)$$

Time moments are defined as the mean values of the product of hydrodynamic velocities pertaining to the same point but to different time instants:

$$B_{i,j}(r, t_1, t_2) = \langle U_i(r, t_1), U_j(r, t_2)\rangle. \quad (8)$$

The two-point moments of the second order are called correlation functions. The correlation factor is defined as a dimensionless correlation function of the form:

$$R_{i,j}(r_1, r_2, t) = \langle U_i(r_1, t) U_j(r_2, t)\rangle / [\langle U_i^2(r_1, t)\rangle \langle U_j^2(r_2, t)\rangle]^{1/2} \quad (9)$$
$$= B_{i,j}(r_1, r_2, t) / [B_{ii}(r_1, t) B_{jj}(r_2, t)]^{1/2}.$$

The correlation factor for the Reynolds stresses is

$$R_{ij}(r, t) = \langle U_i U_j\rangle / (\langle U_i^2\rangle \langle U_j^2\rangle)^{1/2} = \langle U_i U_j\rangle / \sigma_i \sigma_j, \quad (10)$$

where $\sigma^2 = \langle U^2 \rangle$. An integral length scale is determined with the aid of the space correlation function by the relation

$$\Lambda = \int_0^{\infty} R(x_i, x_i + l) dl. \quad (11)$$

Λ is a space measure of interrelationships or a length of the correlation between the velocity fluctuations at two points of the flow field. Retentivity of the turbulent pulsations at a certain point is characterized by the integral time scale:

$$T_E = \int_{T_E}^{\infty} R(\tau)d\tau. \quad (12)$$

The scale T_E may be regarded as a measure of duration of the coupling between the turbulent pulsations $U_i(t)$. The microscales τ and λ are the measures of fast variations in small vortexes

$$\tau_E = \left(-\frac{1}{2}\frac{\partial^2 R}{\partial \tau^2}\right)_{\tau=0}^{-1/2}, \qquad \lambda = \left(-\frac{1}{2}\frac{\partial^2 r}{\partial l^2}\right)_{\lambda=0}^{-1/2}. \quad (13)$$

Some central moments also have a clear physical meaning. Thus, the quantities

$$\sigma^2 = \frac{1}{N}\sum_1^N u_i'^2 = \langle u_i'^2\rangle,$$

$$S = \frac{\langle u_i'^3\rangle}{\sigma_i^{3/2}} \qquad E = \frac{\langle u_i'^4\rangle}{\sigma_i^2} \quad (14)$$

are the dispersion, asymmetry and excess of the quantity u_i, respectively, whereas a square root of the dispersion is the root-mean-square or standard deviation of this quantity. The asymmetry and excess coefficients characterize a deviation of the probability density distribution from the normal (Gaussian) distribution law, for which $S = 0$ and $E - 3 = 0$. Asymmetry is negative, if the distribution is stretched to the left of the mean value, and positive, if it is stretched to the right. The values of the excess $E > 3$ testify to a plane-vertex shape of the distribution curve, and $E > 3$ to an acute-vertex shape. In the first case this indicates the predominance of small pulsations relative to the normal distribution law, and in the second case the predominance of large pulsations.

Another efficient means of describing turbulence is a spectral analysis method. The spectral and statistical theories are interrelated mathematically through the Fourier transform. The spectral analysis makes it possible to describe the kinetic energy exchange by vortexes of different sizes or by pulsations of different frequencies. In analyzing turbulence, use is made of frequency spectra and of spectra in the space of wave numbers. Relations are derived using the principles of the harmonic analysis. The distributions, for example, of velocity in time at each space point constitute a complex nonperiodic function f(t), which may be represented for $T \to \infty$ as the Fourier integral

$$f(t) = \int_{-\infty}^{\infty} \frac{1}{2\pi} e^{i\omega t} \int_{-\infty}^{\infty} f(t')e^{-i\omega t'}dt'd\omega, \quad (15)$$

where ω is the frequency. The existence of the Fourier integral necessitates a finite value of the integral $\int_{-\infty}^{\infty} f(t)dt$. Thus, the formulae

$$f(t) = \int_{-\infty}^{\infty} F(\omega)e^{i\omega t}d\omega,$$

$$F(\omega) = \frac{1}{2\pi}\int_{-\infty}^{\infty} f(t)e^{-i\omega t}dt \quad (16)$$

define the reciprocal Fourier transforms. The function $F(\omega)$ referred to as a complex continuous spectrum of the function f(t), N is a continuous function of the circular frequency ω. The correlation function of two nonperiodic signals $f_i(t)$ and $f_j(t)$ is described by the formula

$$B_{ij}(\tau) = \int_{-\infty}^{\infty} f_i(t)f_j(t + \tau)dt. \quad (17)$$

Applying the Fourier transform yields

$$E_{ij}(\omega) = 2\pi F_i^*(\omega)F_j(\omega), \quad (18)$$

where

$$B_{ij}(\tau) = \int_{-\infty}^{\infty} E_{ij}(\omega)e^{i\omega t}d\omega$$

and

$$E_{ij}(\omega) = \frac{1}{2\pi}\int_{-\infty}^{\infty} B_{ij}(\tau)e^{-i\omega\tau}d\tau$$

the frequency spectra of the processes may be determined through measuring the correlation function or, alternatively, the frequency spectra of the pulsations may be measured directly by various spectrometers. The inverse Fourier transform

$$R_u(\tau) = \int_{-\infty}^{\infty} E(\omega)e^{i\omega t}d\omega \quad (19)$$

at $\tau = 0$ results in the relation

$$\sigma_u^2 = \langle u'^2\rangle = R_u(0) = \int_{-\infty}^{\infty} E(\omega)d\omega, \quad (20)$$

which shows that the power of turbulent pulsations, equal numerically to their dispersion σ^2, is the sum of the powers of individual harmonic components of the pulsations.

The *Monte-Carlo turbulence modeling method* is based on the construction of an artificial stochastic model of the process with preset statistical properties of turbulence. These properties represent a limited set of statistical parameters determined experimentally or theoretically. The Monte-Carlo turbulence modeling method utilizes principles of the theory of control systems. The basic idea resides in the study, synthesis and development of the system with such transfer function that, on excitation of the system by specified random noise-type disturbances, a random process possessing physical properties of the modelled phenomenon is realized at its exit. In modeling the system operation, a random signal I(t) of Gaussian noise type is fed to the entrance to the control system with the impulse transfer function constructed so that the output signal y(t) has the required statistical characteristics. The relation between the above two characteristics may be written in the form of the convolution-type integral:

$$y(t) = \int_{-\infty}^{\infty} h(\tau)I(t - \tau)d\tau \equiv h(t) * I(t). \qquad (21)$$

After the application of the Fourier transforms and the reciprocal transformations:

$$\hat{y}(\omega) = \int_{-\infty}^{\infty} y(t)e^{-i\omega t}dt = FT[y(t)], \qquad (22)$$

$$y(t) = \int_{-\infty}^{\infty} \frac{1}{2\pi} y(\omega)e^{i\omega t}d\omega = FT^{-1}[\hat{y}(\omega)],$$

where FT is the Fourier transform, is the Fourier-function transform and τ is the time interval, the output signal acquires the form

$$\hat{y}(\omega) = H(\omega)\hat{I}(\omega), \qquad (23)$$

here, $H(\omega) = FT[H(t)]$ is the transfer function of the system. The model construction consists in determining the transfer function of the system, or in computing the convolution-type integral, or in employing the Fourier series with random coefficients to represent random signals. The random coefficients may be prescribed in such a way that the statistical moments of the signal have preset values. With the help of quick Fourier transform methods, the random input signal and, subsequently, the Fourier spectrum are formed. Afterwards, the Fourier spectrum is multiplied by the transfer function and, using the inverse transformation, the sought output signal is obtained.

The numerical modeling of the turbulent flow is based on solving the system of Navier-Stokes equations. It is assumed that the system of Navier-Stokes equations describes the turbulent flow with regard to the following premises:

1. the fluid is considered as a continuous medium;
2. physical properties of the fluid are taken to be such that all necessary derivatives of the functions, characterizing its state, are available;
3. a mass of the collection of fluid particles remains unchanged with time;
4. two contacting fluid subregions are affected by identical but opposing force fields applied to their mutual boundary;
5. a total force acts on the fluid subregion in the direction of the resultant force;
6. local values of the stress and deformation rate tensors are interrelated linearly;
7. boundary conditions must satisfy the immobility condition on a solid-fluid interface.

In a direct investigation of turbulence, consideration is generally given to incompressible fluid flows for which

$$\frac{\partial V(x, t)}{\partial t} + [V(x, t)\nabla]V(x, t) = -\nabla p(x, t) + \nu\nabla^2 V(x, t),$$

$$\nabla V(x, t) = 0. \qquad (24)$$

Equations 24 account for the principal nonlinear mechanisms of the turbulent flow evolution, however, other additional effects may also be essential in more complicated problems, *viz.* the effects of compressibility, buoyancy forces, chemical reactions, phase interaction in multiphase flows, etc. When solved by the finite-difference method, Equation 24 are represented in difference form for a finite number of nodal points. The boundary conditions are not simple to assign. A uniform turbulence may be modelled using periodic boundary conditions imposed on the solution to Equation 24, specifically

$$V(x + 2\pi n, t) = V(x, t), \qquad (25)$$

where n is the vector with integral components. For these boundary conditions, the numerical solution of the equations is sought in the form of the truncated Fourier series

$$V(X, t) = \sum_{|K|<K} u(K, t)e^{ikx}, \qquad (26)$$

where the wave vectors R have integral components, if the period of this function is equal to 2π, in accordance with condition Equation 25. Using Equation 26, Equation 25 may be represented as

$$\left(\frac{\partial}{\partial t} + \nu k^2\right)U_\alpha(K, t) \qquad (27)$$

$$= -ik_\beta\left(\sigma_{\alpha\gamma} - \frac{k_\alpha k_\gamma}{k^2}\right) \sum_{\substack{|p|<K \\ |K-p|<K}} U_\beta(p, t)U_\gamma(K - p, t),$$

where $\sigma_{\alpha\gamma}$ is the Kronecker symbol, whereas by the repeated Greek symbols the summation from 1 to 3 is made. In deriving Equation 7, the pressure is eliminated by means the Poisson equation resulting from the medium incompressibility:

$$\nabla^2 p(x, t) = -\nabla[V(x, t)\nabla]V(x, t). \qquad (28)$$

This method for solving the system of ordinary differential equations is named *spectral*. In a more general case, the spectral methods are described through representing the velocity as the truncated series in smooth functions:

$$V(x, t) = \sum_{|m|<M} \sum_{|n|<N} \sum_{|p|<P} a_{mnp}(t)\varphi_m(x_1)\phi_n(x_2)\chi_p(x_3)$$

A correct choice of the basic functions predetermines the efficency of the method. The criteria are a rapid convergence to the exact solution for $V,N,P \to \infty$ and the availability of effective methods to solve the system of ordinary differential equations for the functions $a_{m,n,p}(t)$. Initial conditions are selected randomly. Use is made of spectral representation of Equation 26 with the coefficients of the form

$$U_\alpha(K, 0) = \left(\sigma_{\alpha,\gamma} - \frac{k_\alpha k_\gamma}{k^2}\right)r_\gamma(K), \qquad (29)$$

where $r_\gamma(K)$ are the statistically independent random variables with the Gaussian distribution and with the dispersion proportional to a specified, nonrandom energy spectrum E(K). Substituting Equation 29 into Equation 26 yields a random initial velocity field characterized by the Gaussian distribution and by the energy spectrum E(K). In the direct numerical modeling, large-scale characteristics of the turbulent flow are assumed to be independent of the Reynolds number, if the boundary and initial conditions do not depend on it. This assertion allows modeling only of small-scale formations. The direct numerical modeling of turbulence requires much more tedious and laborious computations than do the solutions using semiempirical theories of transfer. The direct numerical solution is useful for accumulating data via a "numerical experiment" and varying and improving semiempirical theories and for constructing formulas to approximate the results of direct computations. The solution of quite a number of typical problems allows refinement of various aspects of the turbulent flow mechanism.

Another approach to the numerical modeling of turbulent flow consists in developing simplified models based on physical considerations. Such models are most widely employed for describing complex flows encountered in the engineering practice. Further information on these models is gevin in the article on **Turbulence Modelling**. In contradistinction to the above theories, the transfer models use averaged characteristics of the turbulent flow. O. Reynolds proposed that the Navier-Stokes equation

$$\rho\left[\frac{\partial V}{\partial t} + (V, \text{grad } V)\right] = F_v - \text{grad } P + \text{div}(\eta \text{ grad } V)$$

defines any liquid flow. Here, F_v is the volume force. All actual parameters are resolved into the time-averaged $\langle \varphi \rangle$ and fluctuating components, in this case

$$\langle \varphi(x, y, z, t) \rangle = \frac{1}{T} \int_{t-T/2}^{t+T/2} \varphi(x, y, z, \tau) d\tau,$$

where T is the averaging period which is rather large as against the period of turbulent pulsations but small relative to the time interval, characteristic of the mean turbulent motion. The equations of motion equations resulting from the averaging (η = const and ρ = const)

$$\rho \frac{d\langle u_i \rangle}{dt} = F_v - \frac{\partial \langle p \rangle}{\partial x_i} + \eta \frac{\partial^2 \langle u_i \rangle}{\partial x_k^2} - \rho \frac{\partial \langle (u_i' u_k') \rangle}{\partial x_i}$$

$$\frac{\partial \langle u_i \rangle}{\partial x_j} = 0 \qquad i, j, k = x, y, z$$

are called the Reynolds equations of mean turbulent motion. The terms $\rho \langle u_i' u_j' \rangle$ form the Reynolds stress tensor and define additional "turbulent stresses" in the transfer of momentum $\rho u_i'$ by the pulsatory motion u_j'. The turbulent stresses τ_T are responsible for a rise of the total drag in turbulent flow as compared to laminar flow.

The expressions $\rho \langle u_i' u_j' \rangle$ represent turbulent normal stresses with the same indexes and turbulent tangential stresses with different indexes. The quantities $\eta_{ik}^T = \rho\langle u_k' u_i' \rangle / \vartheta u_i / \vartheta x_k$ $i \neq k$ are referred to as turbulent viscosity or a coefficient of turbulent momentum transport in the direction of x_1. The turbulent transport coefficients are not physical properties of the flowing medium but are rather dependent on the molecular viscosity, Reynolds number and coordinates. In developed turbulent flow, $\eta^T \gg \eta$.

In a similar manner, the averaged equations for turbulent transfer of the scalar substance r (heat, substance) can be obtained

$$\frac{\partial \langle r \rangle}{\partial t} + \langle u_i \rangle \frac{\partial \langle r \rangle}{\partial x_i} = \frac{\partial}{\partial x_i}\left(f \frac{\partial \langle r \rangle}{\partial x_i} - \langle u_i' g' \rangle\right) + \langle F_\gamma \rangle.$$

Here, f is the molecular transfer coefficient, $\gamma' = \Gamma - \langle r \rangle$ and F_γ is a source term for heat or substance release (or absorption). The main problem in solving the Reynolds equations and the averaged equation of scalar substance transfer is deriving the relations for $\langle u_i' u_j' \rangle$, $\langle u_i' \gamma' \rangle$ and $\langle u_k' T' \rangle$, i.e., closing the equations for averaged values. By closure methods, the models may be divided into the models utilizing the mean velocity field and the models employing the field of mean turbulence characteristics. The first group methods (of Prandtl, Karman, van Driest and Cebeci) were constructed based on the analogy between turbulence and molecular chaos. They involve such notions as mixing length as well as turbulent viscosity, thermal conductivity and diffusion coefficients. They presume a linear relationship between the tensor of turbulent stresses and the tensor of average deformation rates (the *Boussinesq hypothesis*), as well as between the turbulent heat (or passive admixture) flux and the average temperature (admixture concentration) gradient. The Boussinesq hypothesis has the form:

$$\langle u_i' u_j' \rangle = \varepsilon_\tau \frac{\partial \bar{u}_j}{\partial x_i};$$

$$\langle u_i' T' \rangle = \varepsilon_q \frac{\partial T}{\partial x_i} = \frac{\varepsilon_\tau}{Pr_t} \frac{\partial T}{\partial x_i},$$

where ε_τ is the turbulent viscosity coefficient, ε_q is the turbulent thermal diffusivity coefficient and $Pr_t = \varepsilon_\tau/\varepsilon_q$ is the turbulent Prandtl number. A great number of simple algebraic relations are proposed for defining ε_τ, ε_q, and Pr_t. (See **Turbulence Modelling**).

The equation describing the interaction between the processes of generation, transfer and dispersal of temperature nonuniformities in the turbulent flow has the form

$$\frac{\partial \langle T'^2 \rangle}{\partial \tau} + u_k \frac{\partial \langle T'^2 \rangle}{\partial x_k} = -2\langle u_k' t' \rangle \frac{\partial T}{\partial x_k}$$

$$- \frac{\partial}{\partial x_k} \langle u_k' T'^2 \rangle + \kappa \frac{\partial^2 \langle T'^2 \rangle}{\partial x_k^2} - 2\kappa \left\langle \left(\frac{\partial T'}{\partial x_k}\right)^2 \right\rangle.$$

Another approach resides in that, for the turbulent viscosity which is a scalar quantity, a transfer equation is written analogous to the equations of scalar quantity transport in the turbulent flow. Fairly complicated problems met with in the engineering practice are solved with the aid of the models using the turbulent viscosity. Thus, for example, the calculations are performed for a three-dimensional boundary layer on an aircraft fuselage. The models utilizing the equations for averaged turbulence characteristics may be divided conventionally into three groups:

1. the methods using the Reynolds stress fields for calculating the entire Reynolds stress tensor $\langle u_i' u'^j \rangle$,

2. the methods of closure by the mean turbulent kinetic energy determined from the expression $e \equiv 1/2\ \langle u'_k u'_k \rangle$ and
3. the methods employing to calculate heat or mass transfer the equations for $\langle u'_i u'_j \rangle$ along with the equations for turbulent heat fluxes and turbulent mass transport.

The Reynolds stress and kinetic energy equations can be obtained immediately from the Navier-Stokes equations. The balance equation for the turbulent energy $e = 1/2\ \langle u_k'^2 \rangle$ is of the form

$$\frac{\partial e}{\partial \tau} - u_k \frac{\partial e}{\partial x_k} = -\langle u'_i u'_k \rangle \frac{\partial \langle u_i \rangle}{\partial x_k}$$

$$- \frac{\partial}{\partial x_k} \left\langle u'_k \left(\frac{p'}{\rho} + \frac{1}{2} u_i'^2 \right) \right\rangle + \nu \frac{\partial^2 e}{\partial x_k^2} - \nu \left\langle \left(\frac{\partial x_i}{\partial x_k} \right)^2 \right\rangle$$

Here, the first term defines the local time variation of the turbulent energy, the second defines the convective transfer, the third defines the work of turbulent stresses (the turbulence generation), the fourth defines the potential and kinetic energy transfer by the velocity fluctuations (the turbulent diffusion), the fifth defines the viscous diffusion and the sixth defines the energy of turbulent pulsations.

As an example, we present the Reynolds stresses model suggested by Hanjalic and Launder: the equation for the Reynolds stress $\tau_t/\rho = \langle u'v' \rangle$

$$V \frac{\partial \langle u'v' \rangle}{\partial x} + V \frac{\partial \langle u'v' \rangle}{\partial y} = C_s \frac{\partial}{\partial y} \left(\frac{l^2}{\varepsilon} \frac{\partial \langle u'v' \rangle}{\partial y} \right)$$

$$- C_{\varphi 1} \left(\frac{\langle u'v' \rangle \varepsilon}{1} + C_{\mu} l \frac{\partial U}{\partial y} \right)$$

the transfer equation for the turbulent kinetic energy e:

$$U \frac{\partial e}{\partial x} + V \frac{\partial e}{\partial y} = 0, \qquad 9C_s \frac{\partial}{\partial y} \left(\frac{e^2}{\varepsilon} \frac{\partial e}{\partial y} \right) - \langle u'v' \rangle \frac{\partial U}{\partial y} - \varepsilon$$

and the transfer equation for the rate of kinetic energy dissipation ε:

$$U \frac{\partial \varepsilon}{\partial x} + V \frac{\partial \varepsilon}{\partial y} = C_\varepsilon \frac{\partial}{\partial y} \left(\frac{e^2}{\varepsilon} \frac{\partial \varepsilon}{\partial y} \right) - C_{\varepsilon 1} \frac{\langle u'v' \rangle \varepsilon}{e} \frac{\partial U}{\partial y} - C_{\varepsilon 2} \frac{\varepsilon^2}{e}$$

The above equations are written for a plane flow of an incompressible liquid. Unknown terms of the equations are approximated using the idea of Prandtl and Kolmogorov. In recent years various modifications of such a model have found wide application in calculating near-wall flows and heat and mass transfer along plane and curvilinear walls, under the conditions with external turbulence, negative or positive pressure gradient, buoyancy forces as well as with flows in circular tubes and channels.

Results of the statistical analysis of turbulence and, to a greater extent, visual observations of flows revealed that the flow is not merely random. Various types of organized collective motions were detected in turbulent shear layers (in a boundary layer, jets, wakes and mixing layers). These motions involve quasi-periodic formations (coherent structures) which originate randomly in space and time, move, change and afterwards collapse. Thus, the fundamental feature of the turbulent flow is an intricate combination of randomness and regularity which are difficult to describe analytically. Experiment has gained a decisive importance in studying the structure of turbulent flows. (see also article on **Turbulence**).

The first careful visual investigations of the near-wall region were carried out by S. Kline and his colleagues in the 60-s. They managed to observe, in the viscous sublayer, liquid jets moving with different velocities and to trace the character of their development. Based on numerous experiments performed over the last 20 years, the flow in the viscous sublayer may be represented schematically as follows. Liquid portions having a velocity higher than a local average velocity arrive periodically at the wall from the most remote regions. In this case, paired vortexes with the axes directed along the flow originate on the surface of the solid wall. The vortex origination is random in space and time. While moving, the vortexes recede from the wall. At a certain distance from the wall, a retarded strip is "overtaken" by the liquid having an appreciably higher velocity. A layer of intense shear emerges, and here the flow loses stability. The retarded jet starts to pulsate, thereafter it "explodes", and the fluid escapes from the wall, strongly disturbing the overlying layer. A fresh fluid portion from more remote regions arrives at the wall, in the place of the fluid portion ejected from the viscous sublayer. Subsequently, the process of viscous layer "renewal" recurs. It is established that the average distance between the retarded strips, reduced to dimensionless form using the viscous sublayer parameters friction velocity $V^* = (\tau_w/\rho)^{1/2}$ and ν, is $\Delta_z^+ = \Delta_z V^*/\nu \cong 100$. The vortex extent along the flow is by an order of magnitude greater than, and the vortex dimension normal to the surface is of the order of, the viscous sublayer thickness. The mean dimensionless lifetime of *coherent structures* is $t^+ = tV^{2*}/\nu \cong 100$. Presently it has been ascertained that the turbulent energy generation is particularly intense at the instants of violent fluid ejections from the wall. Studying the interaction between the inner and outer flow zones is now one of the most vital problems of near-wall turbulence.

The most important types of anisotropic free turbulent flows are turbulent wakes behind bodies over which the fluid flows (or which move through the fluid), turbulent jets and turbulent mixing zones emerging at the boundary between different-velocity flows (**Figure 2**). The self-similar solutions to all the above-listed cases appear as

$$b(x) \sim x^m; \qquad \hat{u}(x) \sim x^{-n}; \qquad Re_x \sim x^{m-n},$$

where b(x) is the half-width of the jet, wake or mixing zone, and $\hat{u}(x)$ is the velocity on the jet axis or the velocity deviation from the velocity of the undisturbed flow in a wake behind the body. The coordinate system for the mixing zone is chosen such that the equality $\hat{u}_2 = -\hat{u}_1$ is fulfilled. Exponents in the self-similar laws for a selection of flows are tabulated in **Table 1**.

For all the above-enumerated flows with a fairly large Reynolds number, the velocity and friction stress profiles at a rather long distance x are presented in the form

$$u = \tilde{u}(x) F_1 \left[\frac{r}{b(x)} \right]; \qquad \langle u'v' \rangle = -\tilde{u}^2(x) F_2 \left[\frac{r}{b(x)} \right],$$

where $F_1(r)$ and $F_2(r)$ are the universal functions for each flow

Table 1. Parameters of self-similar flows

Flow Character	m	n	m − n
Three-dimensional jet	1	1	0
Plane jet	1	1/2	1/2
Three-dimensional wake	1/3	2/3	−1/3
Plane wake	1/2	1/2	0
Mixing zone	1	0	1

type, r is the transverse coordinate (the distance from the axis OX for three-dimensional flows and from the plane $z = 0$ for plane flows) and v is the transverse velocity. It is established experimentally that the self-similarity conditions for turbulent characteristics are attained at distances x much longer than for the average velocity profile. Thus, for a circular jet issuing into the submerged space when $X > 8D$ (D is the opening diameter), the velocity profile has already become self-similar, whereas self-similarity of the Reynolds stresses requires $x > 500D$.

The universal laws of near-wall turbulence for a plane-parallel flow are determined from the Reynolds equation

$$\nu \frac{d^2u_x}{dy^2} - \frac{d}{dy}\langle u_x' u_y' \rangle = 0 \quad \text{and}$$

$\tau(y) = \rho\nu du_x/dy - \langle u_x' u_y' \rangle = \tau_w = \text{const}$, where τ_w is the wall stress.

Various semiempirical models are commonly used to define the velocity profile (to close the Reynolds equation). It was assumed in the first classical Prandtl-Karman models that in the turbulent core, wherein the molecular viscosity does not affect the flow,

$$\langle u_x' u_y' \rangle \sim \left(\frac{\partial \langle u \rangle}{\partial y} \right)^2,$$

and the turbulent tangential stress is of the form

$$\tau_T = \rho l^2 \left(\frac{\partial \langle u \rangle}{\partial y} \right)^2,$$

where l is the mixing length. The only characteristic dimension in the region of developed turbulence is a distance from the wall, i.e., $l = ky$, and the velocity profile conforms to the logarithmic law

$$\frac{\langle u(y) \rangle}{V_*} = A + \frac{1}{k} \ln \frac{yV_*}{\nu},$$

where $V^* = \sqrt{\tau_w/\rho}$ is the dynamic shear velocity (friction velocity), and A and k are the universal constants.

A dimensional analysis makes it possible to obtain the general form of the relation for the mean velocity profile

$$\frac{\langle u(y) \rangle}{V_*} = \varphi\left(\frac{yV_*}{\nu} \right) \quad \text{or} \quad u^+ = \varphi(y^+),$$

that expresses the universal law of near-wall turbulence holding not only for the mean velocity but also for other moments of hydrodynamic fields. Here, the values of the function φ in "*the wall law*" differ from one another. In the conditions when the mean velocity gradient is independent of the viscosity (the region of developed turbulence), the logarithmic law of velocity distribution follows from "the wall law".

The wall law is valid for the without pressure gradient flow over a plane surface and for the developed flow in tubes and channels. For a more detailed information see the entries **Boundary Layer** and **Tubes, Single-Phase Flow In**.

References

Hinze, J. O. (1975) *Turbulence,* McGraw-Hill, New York.

Frost, U. and Moulden, T. (1977) *Turbulence, Principles and Application.*

Cebeci and Bradshow (1987) *Convective Heat Transfer.*

Abramovich, G. N. (1963) *The Theory of Turbulent Jets,* M.I.T., Cambridge, MA.

Bradshow, P. (1971) *An Introduction to Turbulence and its Measurement,* Oxford, Pergamon Press.

Kline, S. J., Reynolds, W. S., Shraub, F. A., and Runstadler, P. W. (1967) *J. Fluid Mech.,* 30.

Townsend, A. A. (1976) *The Structure of Turbulent Shear Flow,* 2nd ed., Cambridge, Univ. Press.

Bradshaw, P., Cebeci, T., and Whitelaw, J. H. (1981) *Engineering Calculation Methods for Turbulent Flow,* Academic Press, New York.

Leading to: *Turbulence; Turbulence modelling; Channel flow; External flows, overview*

E.M. Khabakhpasheva and N.V. Medvetskaya

TURBULENT FLOW, HEAT TRANSFER (see Convective heat transfer)

TURBULENT FLOW, TRANSITION TO IN TUBES (see Tubes, single phase flow in)

TURBULENT FREE CONVECTION (see Free convection)

TURBULENT KINETIC ENERGY (see Turbulence models)

TURBULENT MASS TRANSFER (See Mass transfer)

TURBULENT MIXED CONVECTION (see Mixed, combined, convection)

TURBULENT PIPE CONTRACTOR (see Direct contact heat exchanges)

TURBULENT SPOTS (see Boundary layer heat transfer)

TURBULENT WALL FLOW (see Turbulence models)

TWIN FLUID ATOMIZERS (see Atomization)

TWISTED TAPE INSERTS (see Augmentation of heat transfer, single phase; Augmentation of heat transfer, two-phase; Shell-and-tube heat exchangers)

TWO-FLUID MODELS (See Conservation equations, two-phase)

TWO-PHASE CRITICAL FLOW (see Critical flow)

TWO-PHASE FLOW COMBINING (see T-junctions)

TWO-PHASE FLOW CONSERVATION EQUATIONS (see Conservation equations, two-phase)

TWO-PHASE FLOW IN POROUS MEDIA (see Darcy's law)

TWO-PHASE FLOW MULTIPLIER FOR BENDS (see Bends, flow and pressure drop in)

TWO-PHASE FLOWS

Following from: *Multiphase flows*

The most common class of **Multiphase Flows** are two-phase flows, and these include the following:

Gas-liquid flows. This is probably the most important form of multiphase flow, and is found widely in a whole range of industrial applications. These include pipeline systems for the transport of oil-gas mixtures, evaporators, boilers, condensers, submerged combustion systems, sewerage treatment plants, air-conditioning and refrigeration plants, and cryogenic plants. Gas-liquid systems are also important in the meteorology and in other natural phenomena.

Gas-solid flows. Flows of solids suspended in gases are important in pneumatic conveying and in pulverised fuel combustion. Fluidised beds may also be regarded as a form of gas-solid flow. In such beds, the solid remains within the fixed container while the gas passes through. However, within the bed itself, both the gas and the solid are undergoing complex motions.

Liquid-liquid flows. Examples of the application of this kind of flow are the flow of oil-water mixtures in pipelines and in liquid-liquid solvent extraction mass transfer systems. Solvent extraction equipment includes packed columns, pulsed columns, stirred contactors and pipeline contactors.

Liquid-solid flows. The most important application of this type of flow is in the hydraulic conveying of solid materials. Liquid-solid suspensions also occur in crystallisation systems, in china clay extraction and in hydro-cyclones.

Each of the above classes of two-phase flow are dealt with in separate articles to which the reader is referred. In what follows below, a brief overview is given on the design parameters of two-phase flow systems and of the various modelling methods which may be employed.

Design Parameters in Two-phase Flow

The more important design parameters for two-phase flow systems include the following:

Pressure drop Pressure losses occur in two-phase flow systems due to friction, acceleration and gravitational effects. If a fixed flow is required, then the pressure drop determines the power input of the pumping system. Here, examples are the design of pumps for the pipeline transport of slurries, or for pumping of oil-water mixtures. If the available pressure drop is fixed, the relationship between velocity and pressure drop needs to be invoked in order to predict the flow rate. An example of this latter application is in the prediction of the circulation rate in natural circulation boiler systems. A detailed discussion of this area is given in the section on **Pressure Drop, Two-Phase.**

Heat transfer coefficient Heat transfer coefficients in two-phase systems are obviously important in determining the size of heat exchangers in such systems. Examples here are thermo-syphon reboilers in distillation plant and condensers in power plant. Further details in these two areas are given in the articles on **Forced Convection Boiling**, **Pool Boiling** and **Condensation**.

Mass transfer coefficient This is important in the design of separation equipment and also in predicting the situation of combined heat and mass transfer such as in the condensation of vapour mixtures. Further information is given in the articles on **Mass Transfer** and **Mass Transfer Coefficient**. Mass transfer in falling film systems is described in the article on **Falling Film Mass Transfer**.

Mean phase content (ε) This quantity represents the fraction by volume or by cross-sectional area of a particular phase. In gas-liquid flows, the gas mean phase content ε_G is often referred to as the **Void Fraction** and the liquid phase fraction ε_L the *liquid hold-up*. In systems containing a solid phase, the mean solid phase content (ε_S) is referred to as the *solid hold-up*. Mean phase content can be important in governing the inventory of a particular phase within a system, particularly when that phase is toxic or valuable. Mean phase content also governs the *gravitational pressure gradient.*

Flux limitations Limitations in mass and heat fluxes are important in the design of two-phase flow systems. Examples of mass flux limitations include **Critical Flow** (which tends to occur at lower velocities in multiphase system than those found in single-phase systems), **Flooding and Flow Reversal** in counter-current flow systems (for example in a reflux condenser), and minimum fluidisation velocities in **Fluidised Beds**. Heat flux limitations are important in boiling, where exceeding the burnout or critical heat flux can lead to poor system performance or physical damage due to excessive increases in the channel wall temperature (see articles on **Burnout**, **Forced Convection** and **Burnout in Pool Boiling**).

Modelling Approaches for Two-phase Flows

A wide range of models have been developed for two-phase flow systems. These include:

Homogeneous model. In the *homogeneous model,* the two phases are assumed to be travelling at the same velocity in the channel and the flow is treated as being analogous to a single phase flow (see article on **Multiphase Flow**).

Separated flow models. Here, the two fluids are considered to be travelling at different velocities and overall conservation equations are written taking this into account (see article on **Multiphase Flow**).

Multi-fluid model. Here, separate conservation equations are written for each phase, these equations containing terms describing the interaction between the phases (see article on **Conservation Equations, Two-Phase**).

Drift flux model. Here, the flow is described in terms of a *distribution parameter* and an averaged local velocity difference between the phases (see article on **Drift Flux Model**).

Computational fluid dynamic (CFD) models. In contrast to the above models, the **Computational fluid dynamic**, CFD, models usually involve two or three dimensions, and attempt to describe the full flow field (see article on **Computational Fluid Dynamics**).

The choice of the modelling approach will depend on the availability of data (the more complex the models, the more detailed information is required to feed into it!) and on the accuracy required.

Leading to: *Gas-liquid flow*

G.F. Hewitt

TWO-PHASE FLOW SPLITTING (see T-junctions)

TWO-PHASE INSTABILITIES

Following from: *Two-phase flow*

There are a number of instabilities that may occur in two-phase systems. These may be classified into static and dynamic instabilities (Lahey and Podowski, 1989). Examples of *static instabilities* include: *flow excursion* (i.e. Ledinegg) *instabilities, flow regime relaxation instabilities, geysering* or *chugging instabilities* and the *terrain-induced instabilities* that may occur in off-shore oil well lines. Similarly, *dynamic instabilities* include: *density-wave oscillations, pressure drop oscillations, flow regime-induced instabilities* and *acoustic instabilities.*

Of these instability modes, the most important, and most widely studied, have been *Ledinegg instabilities* (Ledinegg, 1938) and *density-wave oscillations* (DWOs). While the subsequent discussion will be focused on boiling systems, it should be noted that similar instabilities may also occur in condensing systems (Lahey and Podowski, 1989).

A typical *boiling loop* includes a heated channel (or channels), an unheated riser, a condenser, a downcomer (in which a pump may be installed) and a lower plenum. A dynamic force balance on the boiling loop yields:

$$\Delta p_{ext} = \Delta p_{sys} + I\frac{dw}{dt} \tag{1}$$

where, Δp_{ext} is the impressed pressure rise in the system (e.g. due to a pump) and Δp_{sys} is the pressure drop of the system at flow rate w. The hydraulic inertia of the loop (I) is given by,

$$I = \left(\frac{L}{g_c A_{x-s}}\right) \tag{2}$$

Linearizing **Equation 1** we obtain,

$$I\frac{d(\delta w)}{dt} + \left[\frac{\partial(\Delta p_{sys})}{\partial w} - \frac{\partial(\Delta p_{ext})}{\partial w}\right]\delta w = 0 \tag{3}$$

which has a solution given by,

$$\delta w(t) = \delta w(0)\left[\exp\left\{-\left[\frac{\partial(\Delta p_{sys})}{\partial w} - \frac{\partial(\Delta p_{ext})}{\partial w}\right]t/I\right\}\right] \tag{4}$$

We note that **Equation 4** implies that a flow excursion will occur if,

$$\frac{\partial(\Delta p_{sys})}{\partial w} < \frac{\partial(\Delta p_{ext})}{\partial w} \tag{5}$$

Figure 1 shows four cases for a typical low pressure boiling loop. Case 1 is for a positive displacement pump. In this case $\partial(\Delta p_{ext})/\partial w = -\infty$, thus **Equation 5** is never satisfied and thus the loop is stable at operating point 1. Case 2 is the so-called parallel channel case in which a constant Δp_{ext} is imposed across each boiling channel. For this case $\partial(\Delta p_{ext})/\partial w = 0$, thus **Equation 5** implies that operating point 1 is unstable (however operating points 2 and 3 are both stable). Similarly, cases 3 and 4 represent the situation in which a centrifugal pump is used in the loop. We note that case 3 is unstable while case 4 is stable.

While Ledinegg instability is known to be a problem in low pressure boiling systems, an increase of the system pressure, or an increase in the inlet orificing in the channel, can stabilize the system.

The quantification of density-wave oscillations requires an analysis of the mass, momentum and energy conservation equations of the boiling system.

A detailed description of the analytical procedure has been given previously by Lahey and Podowski (1989) and will not be repeated here. The essence of the analytical procedure is to determine first the neutral stability boundaries using a linear stability technique, then a nonlinear analysis is performed to identify bifurcation phenomena and to assess the dynamic response of the boiling system.

In a linear stability analysis, the mass and energy conservation equations may be integrated in the axial direction either analytically (Lahey and Moody, 1993) or using nodal techniques (Taleyarkhan et al., 1985). The results of this analysis are then combined with the integrated and linearized momentum equation to satisfy the pressure drop boundary condition impressed on the boiling channel(s) or loop. Laplace transforming the result, we obtain the following characteristic equation, for either the loop (k=L) or the heated channel (k=H),

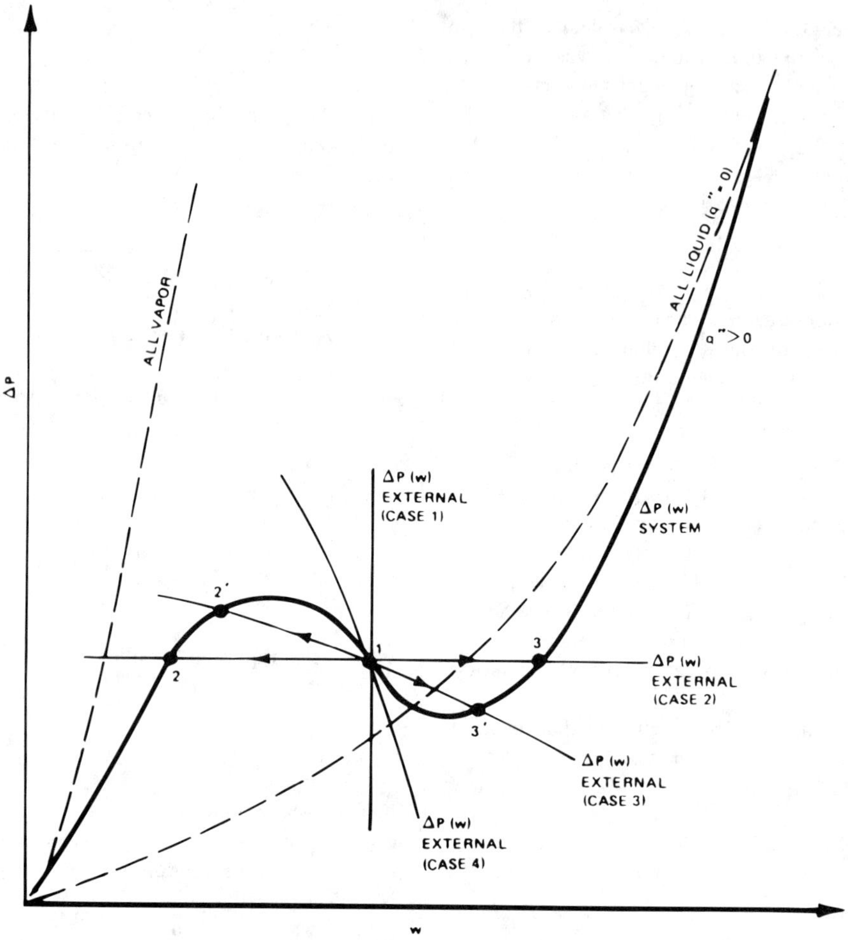

Figure 1. Excursive instability.

$$\delta(\Delta\hat{p}_{1\phi}(s))_k + \delta(\Delta\hat{p}_{1\phi}(s))_k = 0 \tag{6}$$

System stability is determined by solving for the roots, s, of **Equation 6**. If we have any complex roots (s) having positive real parts the system is unstable. Also, we note that Re(s) = 0 defines the neutral stability boundaries.

For a boiling loop (k=L) we have,

$$\delta(\Delta\hat{p}_{1\phi}(s))_L = \Gamma_{1,L}(s)\delta\hat{j}_i + \Gamma_{2,L}(s)\delta\hat{q}''' \tag{7a}$$

$$\delta(\Delta\hat{p}_{2\phi}(s))_L = \Pi_{1,L}(s)\delta\hat{j}_i + \Pi_{2,L}(s)\delta\hat{q}''' \tag{7b}$$

where, $\Gamma(s)$ and $\Pi(s)$ are the transfer functions of the single-phase and two-phase parts of the loop, respectively, and $\delta\hat{j}_i$ and $\delta\hat{q}'''$ are the Laplace transformed inlet velocity and the volumetric power perturbations.

In a nuclear reactor there are also feedback loops associated with temperature- and void-reactivity feedback. This latter transfer function may be denoted, T(s). As should be expected, the stability of a *nuclear-coupled* system is in general different from that given by **Equation 6**.

Figure 2 is a typical block diagram for a *boiling water nuclear reactor* (BWR). It can be shown (Lahey & Moody, 1993) that the characteristic equation associated with this block diagram is,

$$\Phi(s)\{[\Gamma_{1,L}(s) + \Pi_{1,L}(s)]T_2(s) - [\Gamma_{2,L}(s) + \Pi_{2,L}(s)]T_1(s)\}$$

$$+ \frac{1}{C_\alpha}[\Gamma_{1,L}(s) + \Pi_{1,L}(s)] = 0 \tag{8}$$

where $\Phi(s)$ is the so-called zero-power point neutron kinetics transfer function (Weaver, 1963) and C_α is the BWR's void-reactivity coefficient. As before, the stability of a BWR can be determined by solving for the roots, s, of **Equation 8**. A typical linear stability map is given in **Figure 3** in terms of the non-dimensional phase change number (N_{pch}) and the subcooling number (N_{sub}).

Finally, it should be noted that time domain evaluations may be performed with the nonlinear conservation equations leading to Hopf bifurcations (e.g., limit cycles) and, in some cases, chaos (Garea et al., 1994).

There are many other things that could be said about two-phase instabilities. Indeed there is a voluminous literature on this subject (Lahey and Drew, 1980). Nonetheless, this introduction has given

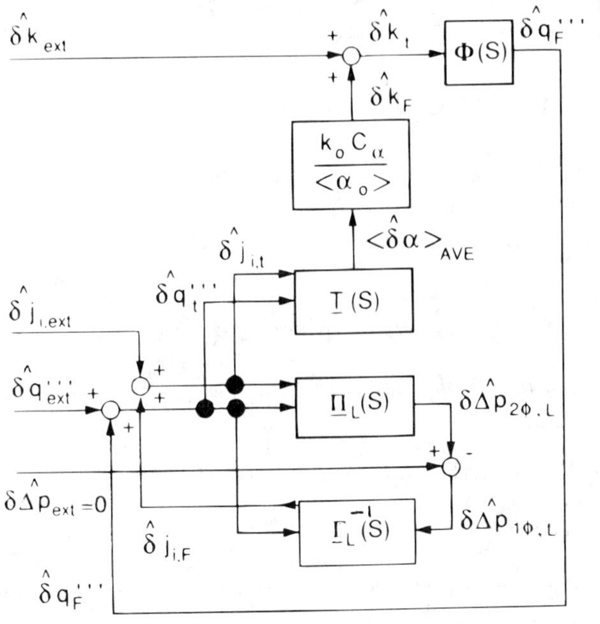

Figure 2. Simplified BWR block diagram.

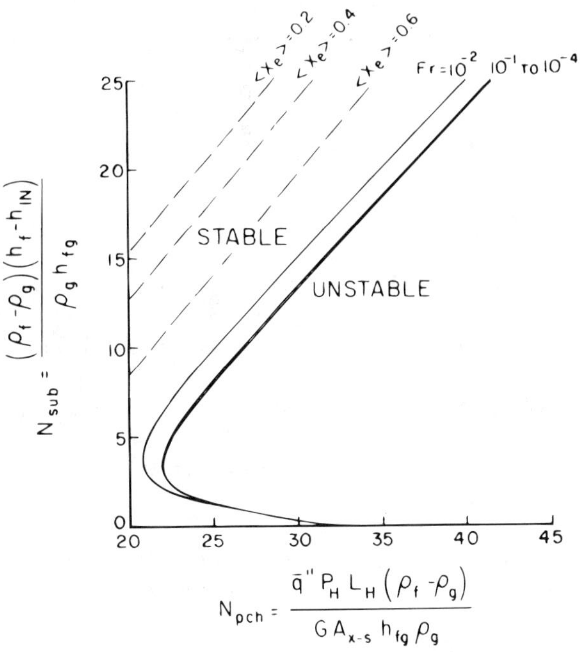

Figure 3. Typical BWR/4 stability map (kin = 27.8, K exit = 0.14).

the essence of two-phase instability analysis and presented some of the key references so that the interested readers may develop a more indepth understanding.

References

Garea, V. B., Chang, C. J., Bonetto, F. J., Drew, D. A. and Lahey, R. T., Jr., (1994) The analysis of nonlinear instabilities in boiling systems, Proceedings of the International Conference on New Trends in Nuclear System Thermohydraulics, Pisa, Italy, May 30–June 2.

Lahey, R. T., Jr. and Drew, D. A., (1980) An Assessment of the Literature Related to LWR Instability Modes, NUREG/CR-1414.

Lahey, R. T., Jr. and Moody, F. J., (1993) The Thermal-Hydraulics of a Boiling Water Nuclear Reactor, ANS Monograph.

Lahey, R. T., Jr. and Podowski, M. Z., (1989) On the analysis of various instabilities in two-phase flows, *Multiphase Science and Technology,* Vol-4, Hemisphere Publishing Corp.

Ledinegg, M. (1938) "Instability of flow during natural and forced circulation," *Die Wärme,* 61, 8.

Taleyarkhan, R., Lahey, R. T., Jr. and Podowski, M. Z., (1985) An instability analysis of ventilated channels, *J. Heat Transfer,* 107.

Weaver, L. E., (1963) A System Analysis of Nuclear Reactor Dynamics, ANS Monograph.

Leading to: *Instability interfacial*

R.T. Lahey, Jr.

TWO-PHASE THERMOSYPHON (see Solar energy; Tubes, condensation in)

TWO STROKE CYCLE (see Internal combustion engines)

TYN AND CALUS METHOD (see PVT relationships)

U

ULTRAFILTRATION (see Membrane processes)

ULTRASONIC ATOMIZERS (see Atomization)

ULTRAVIOLET RAYS (see Electromagnetic spectrum)

ULTRAVIOLET SPECTROSCOPY (see Spectroscopy)

UNESCO (see United Nations Educational, Scientific and Cultural Organisation)

UN FRAMEWORK CONVENTION ON CLIMATE CHANGE (see Greenhouse effect)

UNHINDERED DRYING (see Drying)

UNIFAC METHOD OF DETERMINING ACTIVITY COEFFICIENTS (see Distillation)

UNIT CELL CONCEPT (see Slug flow)

UNITED NATIONS EDUCATIONAL, SCIENTIFIC AND CULTURAL ORGANIZATION, UNESCO

Energy Information Programme
7 Place de Fontenoy
75700 Paris
France
Tel: 1 45 68 10 00

UNITED STATES DEPARTMENT OF ENERGY, USDoE

James Forrestal Building
1000 Independence Avenue SW
Washington DC 20585
USA
Tel: 202 586 5000

UNITED STATES NUCLEAR REGULATORY COMMISSION, USNRC

Washington DC 20555
USA
Tel: 301 415 7000

UNITS, BASIC AND DERIVED (see Dimensional analysis)

UNIVERSAL CONSTANTS (see Dimensional analysis)

UNIVERSAL TEMPERATURE PROFILE (see Tubes, single phase heat transfer in)

UNIVERSAL VELOCITY PROFILE

Following from: *Boundary layers; Shear layer; Turbulence modelling*

The universal velocity profile provides a description of the mean velocity within a turbulent boundary layer. If u(y) defines the velocity at distance y above a solid boundary located at y = 0, dimensional analysis suggests that immediately above the viscous sub-layer the velocity within the so-called inner region (or wall layer) is given by:

$$\frac{u}{u_\tau} = f\left(\frac{yu_\tau}{\nu}\right) \qquad (1)$$

where u_τ is the friction velocity or shear velocity defined by $u_\tau = (\tau/\rho)^{1/2}$ where τ is the shear stress, ρ the fluid density and ν the fluid kinematic viscosity. Within this region measurements suggest that the eddy stress or eddy viscosity (ε) is approximately constant. Using this assumption it can be shown using several different approaches (momentum transport theory, similarity theory, and dimensional analysis) that the appropriate form of the function f is logarithmic and hence:

$$\frac{u}{u_\tau} = A \log_{10}\left(\frac{yu_\tau}{\nu}\right) + B \qquad (2)$$

where A and B are constants which are typically given values of 5.75 and 5.5 respectively.

Although this equation is often referred to as the inner velocity law or the *law of the wall,* experimental measurements suggest that in the case of a turbulent boundary layer over a flat plate it may be applied with reasonable success over a larger region of the flow field. Indeed, this region of applicability is increased further in the case of pipe flow. However, if a detailed description of the velocity distribution within the so-called outer region is required, it is reasonable to assume that the *velocity defect* ($u_1 - u$), where u_1 is the free stream velocity, is independent of the viscosity. Dimensional analysis thus suggests:

$$\frac{u_1 - u}{u_\tau} = f\left(\frac{y}{\delta}\right) \qquad (3)$$

Von Karman considered this case and, having noted the apparent success of the logarithmic distribution, suggested that a universal velocity profile describing the flow over the entire boundary layer might be given by:

$$\frac{u}{u_\tau} = A\log_{10}\left(\frac{yu_\tau}{\nu}\right) + \Phi\left(\frac{y}{\delta}\right) \quad \textbf{(4)}$$

where δ is the overall thickness of the boundary layer, and $\phi(y/\delta)$ is a universal function which reduces to B in the inner region. In a review of boundary layer theory Duncan et al (1970) suggests that for the case of flow over a flat plate, a good fit to the available data is obtained from:

$$\Phi\left(\frac{y}{\delta}\right) = \Phi(0) + K\sin^3\left(\frac{\pi}{2}\frac{y}{\delta}\right) \quad \textbf{(5)}$$

where $K = 2.8$ and $\phi(0)=B=5.5$. Further details of this and other boundary layer theory is given by Schlichting (1968).

References

Duncan, W. J., Thom, A. S. and Young, A. D. (1970) *Mechanics of Fluids,* Arnold, London.

Schlichting, H. (1968) *Boundary Layer Theory,* McGraw Hill, New York.

C. Swan

UNSUBMERGED JETS (see Impinging jets)

UPPER EXPLOSION LIMIT, UEL (see Flammability)

UPPER FLAMMABILITY LIMIT, UFL (see Flammability)

URANIUM

Uranium—(Planet *Uranus*), U; atomic weight 238.029; atomic number 92; melting point 1132.3 ± 0.8°C; boiling point 3818°C; specific gravity ~ 18.95; valence 2, 3, 4, 5, or 6. Yellow-colored glass, containing more than 1% uranium oxide and dating back to 79 A.D., has been found near Naples, Italy. Klaproth recognized an unknown element in pitchblende and attempted to isolate the metal in 1789. The metal apparently was first isolated in 1841 by Peligot, who reduced the anhydrous chloride with potassium. Uranium is not as rare as it was once thought. It is now considered to be more plentiful than mercury, antimony, silver, or cadmium, and is about as abundant as molybdenum or arsenic. It occurs in numerous minerals such as pitchblende, uraninite, carnotite, autunite, uranophane, davidite, and tobernite. It is also found in phosphate rock, lignite, monazite sands, and can be recovered commercially from these sources. The A.E.C. purchases uranium in the form of acceptable U_3O_8 concentrates. This incentive program has greatly increased the known uranium reserves. Uranium can be prepared by reducing uranium halides with alkali or alkaline earth metals or by reducing uranium oxides by calcium, aluminum, or carbon at high temperatures.

The metal can also be produced by electrolysis of KUF_5 or UF_4, dissolved in a molten mixture of $CaCl_2$ NaCl. High-purity uranium can be prepared by the thermal decomposition of uranium halides on a hot filament.

Uranium exhibits three crystallographic modifications as follows:

$$\alpha \xrightarrow{667°C} \beta \xrightarrow{772°C} \gamma$$

Uranium is a heavy, silvery-white metal which is pyrophoric when finely divided. It is a little softer than steel, and is attacked by cold water in a finely divided state. It is malleable, ductile, and slightly paramagnetic. In air, the metal becomes coated with a layer of oxide. Acids dissolve the metal, but it is unaffected by alkalis, Uranium has fourteen isotopes, all of which are radioactive. Naturally occurring uranium nominally contains 99.2830% by weight U^{238}, 0.7110% U^{235}, and 0.0054% U^{234}. Studies show that the percentage weight of U^{235} in natural uranium varies by as much as 0.1%, depending on the source. The A.E.C. has adopted the value of 0.711 as being their "official" percentage of U^{235} in natural uranium. Natural uranium is sufficiently radioactive to expose a photographic plate in an hour or so.

Much of the internal heat of the earth is thought to be attributable to the presence of uranium and thorium. U^{238} with a half-life of 4.51×10^9 years, has been used to estimate the age of igneous rocks. The origin of uranium, the highest member of the naturally occurring elements—except perhaps for traces of neptunium or plutonium—is not clearly understood, although it may be presumed that uranium is a decay product of elements of higher atomic weight, which may have once been present on earth or elsewhere in the universe. These original elements may have been created as a result of a primordial "creation," known as "the big bang," in a supernovae, or in some other stellar processes.

Uranium is of great importance as a nuclear fuel. U^{238} can be converted into fissionable plutonium by the following reactions:

$$U^{238}(n\gamma)U^{239} \xrightarrow{\beta-} Np^{239} \xrightarrow{\beta-} Pu^{239}$$

This nuclear conversion can be brought about in "breeder" reactors where it is possible to produce more new fissionable material than the fissionable material used in maintaining the chain reaction. U^{235} is of even greater importance, for it is the key to the utilization of uranium. U^{235}, while occurring in natural uranium to the extent of only 0.71%, is so fissionable with slow neutrons that a self-sustaining fission chain reaction can be made to occur in a reactor constructed from natural uranium and a suitable moderator, such as heavy water or graphite, alone. U^{235} can be concentrated by gaseous diffusion and other physical processes, if desired, and used directly as a nuclear fuel, instead of natural uranium, or used as an explosive. Natural uranium, slightly enriched with U^{235} by a small percentage, is used to fuel nuclear power reactors for the generation of electricity. Natural thorium can be irradiated with neutrons as follows to produce the important isotope U^{233}.

$$Th^{232}(n\gamma)Th^{233} \xrightarrow{\beta-} Pa^{233} \xrightarrow{\beta-} U^{233}$$

While thorium itself is not fissionable, U^{233} is, and in this way may be used as a nuclear fuel. One pound of completely fissioned uranium has the fuel value of over 1500 tons of coal. The uses of nuclear fuels to generate electrical power, to make isotopes for peaceful purposes, and to make explosives are well known. Uranium in the U.S.A. is controlled by the Atomic Energy Commission. New uses are being found for "depleted uranium, i.e., uranium with the percentage of U^{235} lowered to about 0.2%. It has found use in inertial guidance devices, gyro compasses, counterweights for aircraft control surfaces, as ballast for missile reentry vehicles, and as a shielding material. Uranium metal is used for X-ray targets for production of high-energy X-rays; the nitrate has been used as photographic toner, and the acetate is used in analytical chemistry. Crystals of uranium nitrate are triboluminescent. Uranium salts have also been used for producing yellow "vaseline" glass and glazes. Uranium and its compounds are highly toxic, both from a chemical and radiological standpoint. Finely divided uranium metal, being pyrophoric, presents a fire hazard. The maximum recommended allowable concentration of soluble uranium compound in air (based on chemical toxicity) is 0.05 mg/M^3 (8-hr time-weighted average—40-hr week); for insoluble compounds the concentration is set at 0.25 mg/M^3 of air. The maximum permissible total body burden of natural uranium (based on radiotoxicity) is 0.2μCi for soluble compounds.

Handbook of Chemistry and Physics, CRC Press

URANIUM 238 (see Nuclear reactors)

URANIUM RECOVERY (see Ion exchange)

USDoE (see United States Department of Energy)

USNRC (see United States Nuclear Regulatory Commission)

UTILISATION FACTOR (see Fixed beds)

U-TUBE EXCHANGERS (See Shell-and-tube heat exchangers)

U-TUBE STEAM GENERATORS (see Steam generators, nuclear)

V

VACUUM

Vacuum is the state of a gas whose density is less than that of the air at the Earth level. Vacuum has a number of useful properties which find wide use in various fields of science and technology. For instance, the chemical activity of oxygen sharply drops in vacuum during the oxidation process. The surfaces remain clean (without adsorption even of a monolayer of a gas) at very high degrees of rarefaction for several hours, which allows us to carry out the investigations of such surfaces and of various phenomena associated with adsorbed gas molecules. The low number of molecules in the residual gas in vacuum results in the fact that various particles can cover large distances without collisions under vacuum conditions. This is of particular importance for charged particles, electrons, ions, and protons, whose trajectories can be controlled in vacuum with the help of electric and magnetic fields. Such physical phenomena as propagation of sound, heat and mass transfer, which at atmospheric pressure are determined by the processes of interaction of gas molecules, differ essentially with pressure decrease to the extent that the role of these phenomena in the transfer process becomes unimportant.

The vacuum degree is determined from the residual gas pressure. For convenience the entire range of attainable is subdivided into several subranges (**Figure 1**). Also shown here are the main fields of application of vacuum technology.

Vacuum theory is based on the kinetic theory of gases. The collisions of molecules with the wall and their interactions are assumed to be elastic. The equation of state for an ideal gas has the form: $pV = NkT$. Here p is the pressure, V is the volume, T is the absolute temperature, N is the total number of molecules, and $k = 1.38 \times 10^{-23}$ Pa m^3/K is the Boltzmann constant. On considering the momentum transfer when chaotically moving particles strike against the wall, we can obtain another expression for a gas pressure: $p = 1/3\ mn\bar{v}^2$, where m is the mass of the molecule, n is the number of molecules in unit volume, $\bar{v}$ is the root-mean-square velocity of molecules equal to

$$\bar{v} = \left[\int_0^{\infty} v^2 f(v)dv\right]^{1/2} = \sqrt{3kT/m}.$$

The kinetic energy of a molecule is determined in this case by the absolute temperature of the gas: $E = 3/2\ kT$. With the help of Maxwell's distribution function, we can show that the mean velocity of molecules is

$$v_{mean} = \int_0^{\infty} vf(v)dv = \sqrt{8kT/\pi m} \cong 0.92\bar{v}.$$

It is often useful to know the frequency of molecule impacts ν with the surface of unit area $\nu = 1/4\ nv_{mean} = p/\sqrt{2\pi mkT}$.

The mean free path length λ determines the length of the molecule travel between the collisions

$$\lambda = \frac{1}{\sqrt{2}\pi nd^2}.$$

Here d is the molecule diameter.

The degree of gas *rarefaction* is characterized by the Knudsen number $Kn = \lambda/L$ where L is a characteristic dimension of the system. The discreteness of gas structure already manifests itself at $Kn > 0.01$, and at $Kn \gg 1$ the transfer processes in gases are called free-molecule.

The **Knudsen Number** is related to gas-dynamics parameters which are used in calculations of thermal conditions of bodies flying in the upper atmosphere, i.e., to the Mach (Ma) and the Reynolds (Re) numbers. If in the flow around the body the characteristic dimension is the body dimension L, then $Kn = 1.25\sqrt{\gamma}$ (Ma/Re), if L is the thickness of the laminar boundary layer, then $Kn \sim Ma\sqrt{Re}$. Here $\gamma = c_p/c_v$ is the adiabatic exponent. Consequently, for $Ma / \sqrt{Re} \ll 1$ the gas flow is continuous, for $Ma / \sqrt{Re} \gg 1$ it is free-molecule.

The completeness of energy exchange of molecules upon impacts with the wall is characterized by the thermal accommodation coefficient $\alpha = (e_f - e_r)/(e_f - e_w)$, where e_f and e_r are the energy fluxes of falling and reflected molecules, e_w is the flux of energy that would be carried away from the wall for complete energy exchange, i.e. the energy of reflected molecules corresponds to the wall temperature. For air and construction materials $\alpha = 0.87 - 0.97$. On collision of gases with a small molecular mass against a surface of specially pickled metals α can be very small. For instance, for helium vapours interacting with tungsten $\alpha \cong 0.02$. The exchange of a tangential pulse $\sigma = (\bar{w}_f - \bar{w}/\bar{w}_f$, where $\bar{w}$ is the averaged value of tangential velocity. If $\sigma = 0$, the reflection of molecules from the wall is completely mirror-like, if $\sigma = 1$, the reflection is diffuse.

The density of thermal flux in free-molecule flow with large Mach numbers is determined from the expression

$$q_{fm} \cong \frac{1}{2}\alpha\rho_0\omega^3 \sin\theta\left[1 - \frac{\gamma + 1}{2(\gamma - 1)}\frac{T_w}{T_0}\right],$$

where ρ_0 and T_0 are the density and temperature of the incoming flow, θ is the angle of attack. Hence, the ratio of the wall adiabatic temperature T'_w (for $q_{fm} = 0$) to the flow stagnation temperature T'_0: $T'_w/T'_0 \cong 2\gamma/(\gamma + 1)$, i.e. the coefficient of restoration of the total temperature in a free-molecule flow is greater than that in a gas continuous flow.

For a strongly cooled body $T_w \ll T_0$ and $\alpha = 1$, we can obtain the value of the limiting thermal flux $q_{fm} = 1/2\ \rho_0\omega^3 \sin\theta$, which is half the dissipated energy falling within unit surface.

A simple case of heat exchange in a limited volume is the heat transfer in a motionless gas between two parallel walls when the influence of fringe effect can be neglected. If the walls have the accommodation coefficients α_1 and α_2 and temperatures T_1 and T_2, then

$$q_{fm} = \frac{1}{\sqrt{2\pi}}\frac{1}{\dfrac{1}{\alpha_1} + \dfrac{1}{\alpha_2} - 1}\frac{\sqrt{RT_1T_2}}{\sqrt{T_1} + \sqrt{T_2}}\rho_0 c_v \frac{k + 1}{2}(T_1 - T_2).$$

In the "Knudsen layer" which is between $\bar{\lambda}$ and $2\bar{\lambda}$ thick the boundary effects of *velocity jump (slip)* and of *temperature jump* manifest themselves near the wall. Their values can be found from the expressions:

$$\bar{\omega}_s = \frac{2-\sigma}{\sigma}\bar{\lambda}\left(\frac{\partial\omega}{\partial y}\right)_w + \frac{3}{4}\bar{\lambda}\sqrt{\frac{2R}{\pi T}}\left(\frac{\partial T}{\partial x}\right)_{2w};$$

$$\Delta T_s = T_{q,w} - T_w = 2\frac{2-\alpha}{\alpha}\frac{k}{k+1}\frac{1}{Pr}\bar{\lambda}\left(\frac{\partial T}{\partial y}\right)_{2w}.$$

Here $T_{q,w}$ is the gas temperature near the wall, R is the gas constant, Pr is the Prandtl number. The second term in the first expression shows the influence of a thermo-molecular flow, the motion of gas in the direction of a temperature rise. As is seen, the slip and the temperature jump exist at any pressure. As $\bar{\lambda}$ is small at high pressures, the value of $\bar{\omega}_s$ and ΔT_s, proportional to $\bar{\lambda}$, are negligibly small compared to the flow velocity value and the overall temperature difference. This is the basis of the zero slip hypothesis in the dynamics of continuous media. The influence of temperature jump on heat transfer, for instance, in the flow around the sphere, can be estimated from the Nusselt number Nu, given the value of Nu_0 in its absence $1/Nu = 1/Nu_0 + \xi\, Kn/Pr$, where $\xi = 2[2/\alpha - 1][\gamma/(\gamma + 1)]$ is the jump coefficient.

A vacuum system or a vacuum unit is the system connected to **Vacuum Pumps** and includes vacuum gauges, vacuum accessories, leak detectors and other units.

The maximum attainable rarefaction in the system depends not only on the effective rate of pump evacuation (**Figure 1**), but also on the rate of leakage into the vacuum system. As the rate of pump evacuation is finite, the reduction of gas leakage becomes the main condition for attaining high vacuum. The latter is the criterion for choosing the construction materials, and also for developing and improvement of elements of demountable vacuum joints, dynamic connection seals, valves, etc., used in such systems.

The leak detectors can be the vacuum gauges themselves or mass spectrometers. In order to determine small leaks through the walls of vacuum elements, or accessoires, a plastic casing with test gas is most often used, which embraces separate sections of the unit. The test gas passes through the leak is registered by the leak detector.

D.S. Mikhatulin

Leading to: *Vacuum pumps*

VACUUM DISTILLATION (see Oil refining)

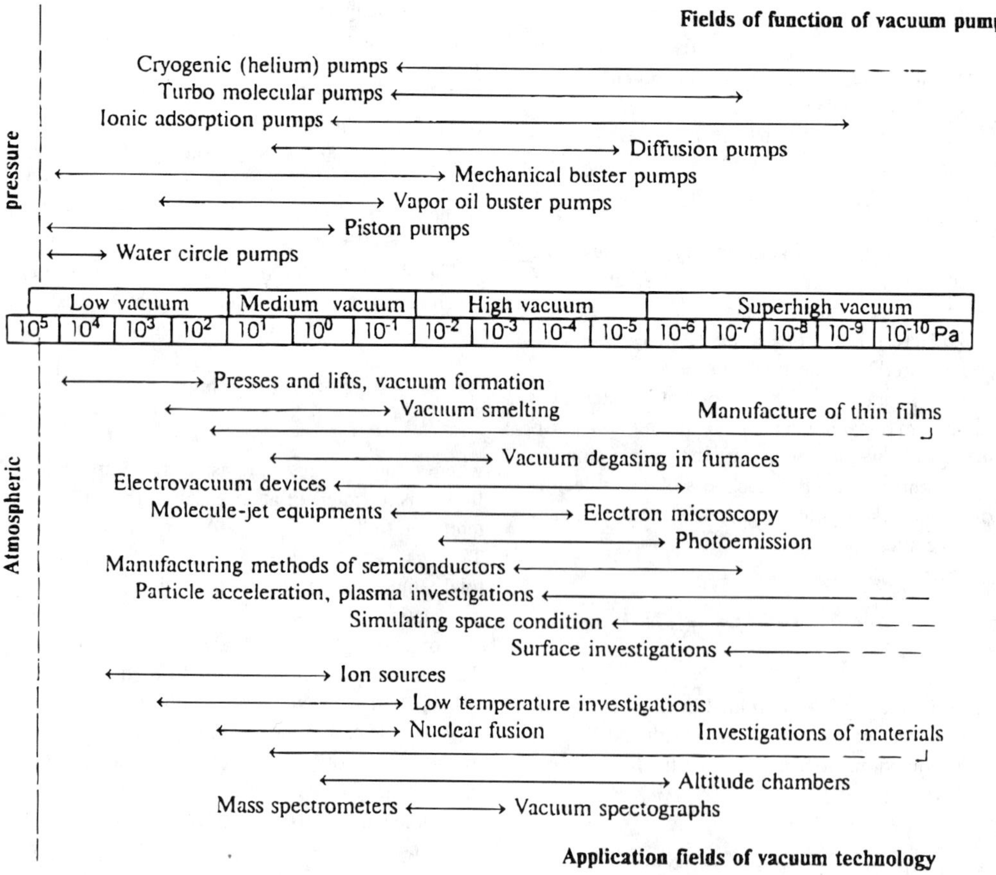

Figure 1. Ranges of vacuum and vacuum technology.

VACUUM DRYERS (see Dryers)

VACUUM ELECTRIC FURNACES (see Electric, Joule, heaters)

VACUUM EJECTORS (see Aerodynamics)

VACUUM PUMP

Following from: *Vacuum*

A vacuum pump is a device for creating, improving and/or maintaining a vacuum (an environment in which the pressure is below atmospheric pressure). Two basically distinct categories of vacuum pump may be considered: *Gas Transfer Pumps* and *entrapment* or capture pumps (**Figure 1**).

Gas Transfer Pumps

Gas transfer pumps can be sub-divided into positive displacement pumps and kinetic pumps.

Positive displacement vacuum pump

A positive displacement vacuum pump is a pump in which a volume filled with gas is cyclically isolated from the inlet, the gas being then transferred to an outlet. In most types of positive displacement pumps the gas is compressed before the discharge at the outlet. Two categories can be considered: reciprocating positive displacement pumps (e.g. piston pump) and rotary positive displacement pumps (e.g. liquid ring pump and sliding vane rotary pump).

Sliding vane rotary pump

Here an eccentrically placed rotor is made to turn tangentially to the fixed surface of the stator. Normally, two vanes slide in slots in the rotor and contact the internal wall of the stator. The rotating mechanism isolates the gas from the inlet, compresses and then expels it through an outlet valve.

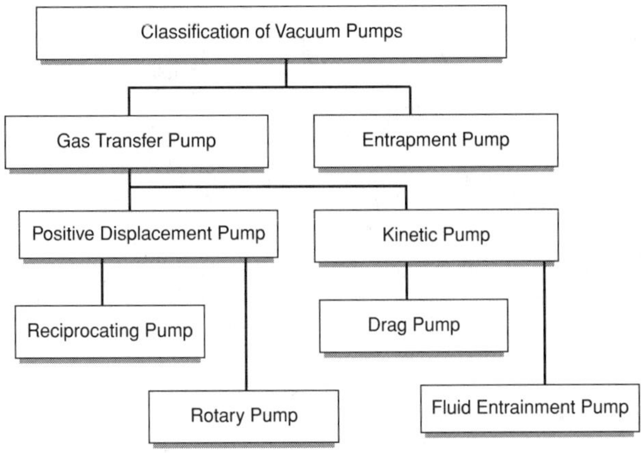

Figure 1. Classification of vacuum pumps.

Kinetic Vacuum Pumps

A kinetic vacuum pump is a pump in which a momentum is imparted to the gas or the molecules in such a way that the gas is transferred continuously from the inlet to the outlet. Two categories can be considered: fluid entrainment pumps (e.g. *vapour diffusion pump*) and drag vacuum pumps (e.g. *turbomolecular pump*).

Diffusion pump

Here gas transport is achieved by a series of high velocity vapour jets (normally oil vapour is used) emerging from an assembly within the pump body. In normal operation a portion of any gas arriving at the inlet jet is entrained, compressed and transferred to the next stage.

Turbomolecular pump

This pump contains a rotor with inclined blades moving at high speed between corresponding blades in a stator. Gas molecules entering the inlet port acquire a velocity and preferred direction superimposed on their thermal velocity by repeated collisions with the fast moving rotor. Rotational speeds for small pumps can be up to 90,000 rev min^{-1}.

Entrapment (Capture) Vacuum Pump

A vacuum pump in which the molecules are retained by sorption; chemical combination or condensation on internal surfaces within the pump.

Sputter-ion pump

This makes use of the gettering principle, in which a cathode material (usually titanium) is vaporised, or sputtered, by bombardment with high velocity ions. The active gases are pumped by chemical combination with the sputtered titanium, the inert gases by ionisation and burial in the cathode, and the light gases by diffusion into the cathode.

Cryogenic pump

Operation is achieved by the condensation, freezing and/or sorption of gas at surfaces maintained at extremely low temperatures, thus removing them from the gas phase in the vacuum system.

References

International Standards Organisation, Document Ref. ISO3529/2-1981 Vacuum technology—Vocabulary—Part 2: Vacuum pumps and related terms.

Harris, N. S. (1989) *Modern Vacuum Practice,* McGraw-Hill, New York.

Harris, N. S. (1987) Vacuum and present-day problems for vacuum pumps, *Phys. Bull.,* 38, 224–226.

N.S. Hams

VANADIUM

Vanadium—(Scandinavian goddess, *Vanadis*), V; atomic weight 50.9415; atomic number 23; melting point 1890 ± 10°C; boiling point 3380°C; specific gravity 6.11 (18.7°C); valence 2, 3, 4, or 5.

Vanadium was first discovered by del Rio in 1801. Unfortunately, a French chemist incorrectly declared del Rio's new element was only impure chromium; del Rio thought himself to be mistaken and accepted the French chemist's statement. The element was rediscovered in 1830 by Sefstrom, who named the element in honor of the Scandinavian goddess Vanadis because of its beautiful multicolored compounds. It was isolated in nearly pure form by Roscoe, in 1867, who reduced the chloride with hydrogen.

Vanadium of 99.3 to 99.8% purity was not produced until 1927. Vanadium is found in about 65 different minerals among which are carnotite, roscoelite, vanadinite, and patronite—important sources of the metal. Vanadium is also found in phosphate rock and certain iron ores, and is present in some crude oils in the form of organic complexes. It is also found in small percentages in meteorites. Commercial production from petroleum ash holds promise as an important source of the element. High-purity ductile vanadium can be obtained by reduction of vanadium trichloride with magnesium or with magnesium-sodium mixtures. Much of the vanadium metal being produced is now made by calcium reduction of V_2O_5 in a pressure vessel, an adaption of a process developed by McKechnie and Seybolt.

Natural vanadium is a mixture of two isotopes, V^{50} (0.24%) and V^{51} (99.76%). V^{50} is slightly radioactive, having a half-life of 6×10^{15} years. Seven other unstable isotopes are recognized. Pure vanadium is a bright white metal, and is soft and ductile. It has good corrosion resistance to alkalis, sulfuric and hydrochloric acid, and salt waters, but the metal oxidizes readily above 660°C. The metal has good structural strength and a low-fission neutron cross section, making it useful in nuclear applications. Vanadium is used in producing rust-resistant, spring, and high-speed tool steels. It is an important carbide stabilizer in making steels. About 80% of the vanadium now produced is used as ferrovanadium or as a steel additive. Vanadium foil is used as a bonding agent in cladding titanium to steel.

Vanadium pentoxide is used in ceramics and as a catalyst. It is also used as a mordant in dyeing and printing fabrics and in the manufacture of aniline black. Vanadium-gallium tape has been used in producing a superconductive magnet with a field of 175,000 gauss. Vanadium and its compounds are toxic and should be handled with care. Exposure to V_2O_2 dust (as V) should not exceed the ceiling value of 0.05mg/M^3, and exposure to V_2O_2 fume (as V) should not exceed 0.1 mg/M^3 (8-hr time-weighted average—40-hr week). Ductile vanadium is commercially available.

Handbook of Chemistry and Physics, CRC Press

VAN DER WAALS EQUATION (see Gas law)

VAN DER WAALS EQUATION

Following from: *Equation of state; PVT relationships*

The van der Waals equation was proposed in the year 1873. It was a first step towards taking into account of interaction forces which are acting between real gases molecules.

The equation of state of the perfect gas refers to a gas consisting of point like items which do not interact with one another.

Instead of this Van der Waals proposed an equation

$$(v - b)\left(p + \frac{a}{v^2}\right) = RT, \qquad \mathbf{(1)}$$

where a and b are the so called van der Waals constants and which have different values for each gas.

The b correction takes into account of the fact that, according to van der Waals. the real gas molecules can move not in the total volume occupied by the gas, but only in a part of this volume which is defined by subtracting what he called the molecules own volume. The correction a/υ^2—proportional to the square of density-takes into account the attraction forces existing between gas molecules. According to the van der Waals assumption, these forces decrease the pressure against the vessel walls, because the molecules which are close to the wall are subjected to attraction of other molecules.

The a and b constants are to be considered as empirical; they have to be derived from experimental data about the density, temperature and pressure interdependence. For a number of technically important gases the van der Waals constants are tabulated.

According to the method of allowing for deviation of a real gas from a perfect gas, the van der Waals equation supposedly would be valid only for small deviations, i.e. for moderate pressures. But it turned out that this equation describes qualitative satisfactorily the real gases behavior in a wide range of parameters including the critical region.

The van der Waals equation is a cubic one with respect to the specific volume. This means that in general at a given temperature, each pressure value corresponds to three specific volume values—$\upsilon_1, \upsilon, \upsilon_2$. The values υ_1 and υ_2 correspond to physically existing states of liquid and vapor. The value υ, where $(\partial p/\partial v)_T > 0$, corresponds to an unstable state and does not exist physically. As the temperature increases the values υ_1 and υ_2 get closer and finally in the critical point all three roots coincide at the critical volume υ_c. At higher temperatures only one root of the van der Waals equation remains real, the two other being imaginary. These supercritical isotherms reveal inflexions at $\upsilon = \upsilon_c$ which is close to the behavior of real substances.

As far as in the critical point the following conditions are valid:

$$(\partial p/\partial v)_T = 0; \qquad (\partial^2 p/\partial v^2)_T = 0,$$

and it is possible using the van der Waals equation to derive interconnections between van der Waals constants and critical parameters of a substance.

$$v_c = 3b; \qquad p_c = \frac{1}{27}\frac{a}{b^2}; \qquad T_c = \frac{8}{27}\frac{a}{bR}.$$

Hence the critical compressibility factor Z_c will be identical for all substances and equal to

$$Z_c = \frac{p_c v_c}{RT_c} = \frac{3}{8} = 0.375.$$

In reality the critical compressibility factors for the most substances fit into the range 0.21–0.31.

The van der Waals equation can be rearranged to a nondimensional form using the reduced parameters:

$$\pi = p/p_c; \qquad \tau = T/T_c; \qquad \phi = v/v_c.$$

In this coordinates the van der Waals equation becomes:

$$(3\phi - 1)\left(\pi + \frac{3}{\phi^2}\right) = 8\tau. \qquad (2)$$

This means that, with reduced parameters, all substances obeying the van der Waals equation are described by an identical equation of state. From here it follows that all these substances can be treated similarly. For instance at identical π and τ all substances have the same ϕ values. This generalized approach is used also with more sophisticated equations of state which describe the real substances behavior better than the van der Waals equation.

E.E. Shpilrain

VAN DER WAALS FORCES

Van der Waals forces are a generic description of the forces that act between atoms and molecules. The fact that liquids and solids are scarcely compressible indicates that when the distances between molecules are small, the forces between them are strongly repulsive. On the other hand, the fact that these condensed phases exist at all indicates that at long range the forces must be attractive.

The force is generally expressed in terms of a potential energy U which, for spherically symmetric molecules, is simply a function of the distance τ between two molecules, so that

$$F(r) = -\frac{dU(r)}{dr}$$

The general shape of an *intermolecular pair potential* for a spherically-symmetric system is shown schematically in **Figure 1**.

The physical origin of the long-range attractive forces in the case of non-polar molecules is the interaction between the fluctuating dipoles in each atom caused by the motion of the electrons about the nucleus. The repulsive interactions at short range are caused by the repulsive electrostatic forces between the nuclei of molecules when the electrostatic shielding is reduced as the *Pauli Exclusion Principle* causes a reduction of the electron density between them. For polyatomic and/or polar molecules, the van der Waals' forces also arise as a result of interactions between permanent multipoles within the molecules. The forces and the potential then depend upon the relative orientation of the molecules as well as their separation.

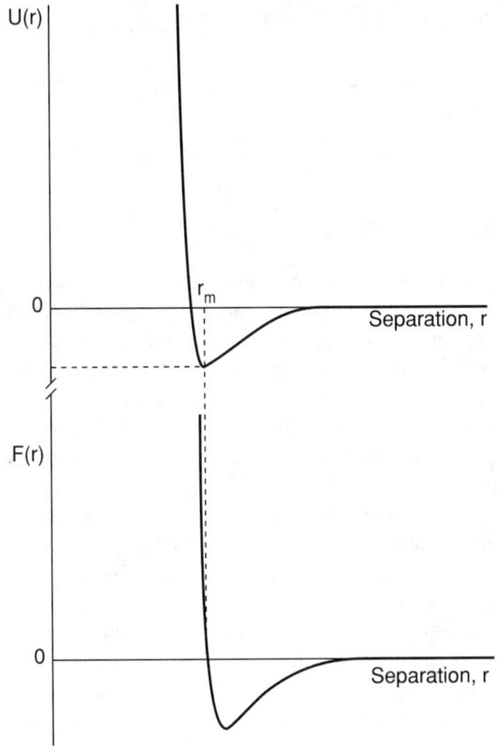

Figure 1. The intermolecular potential and intermolecular force for a pair of spherically-symmetric molecules.

References

Maitland, G. C., Rigby, M, Smith, E. B. and Wakeham, W. A. (1961) *Intermolecular Forces: Their Origin and Determination,* Clarendon Press, Oxford.

Rigby, M, Smith, E. B., Wakeham, W. A. and Maitland, G. C. (1986) *An Introduction to Intermolecular Forces,* Clarendon Press, Oxford.

W.A. Wakeham

VAN DER WAALS ISOTHERMS (see Van der Waal's equation)

VAN DRIEST'S DAMPING FORMULA (see Boundary layer heat transfer)

VAN DRIEST'S EXPRESSION FOR TURBULENT VISCOSITY (see Tubes, single phase flow in)

VAN DRIEST'S MIXING LENGTH MODEL (see Turbulence models)

VANE ANEMOMETERS (see Anemometers, vane)

VAN LAAR EQUATION (see Activity coefficient)

VAN'T HOFF EQUATION (see Stokes-Einstein equation; Thermodynamics)

VAN'T HOFF ISOTHERM (see Chemical equilibrium)

VAPOUR ABSORPTION CYCLE (see Refrigeration)

VAPOUR BUBBLE, EQUILIBRIUM OF (see Boiling)

VAPOUR-COMPRESSION REFRIGERATION CYCLE (see Heat pumps; Refrigeration)

VAPOUR EXPLOSIONS

A vapor explosion—sometimes called an energetic *fuel-coolant interaction* (FCI)—is a process in which a hot liquid (fuel) transfers its internal energy to a colder, more volatile liquid (coolant); thus the coolant vaporizes at high pressures and expands and does work on its surroundings. When the two liquids come into contact, the coolant begins to vaporize at the fuel-coolant liquid interface as a vapor film separates the two liquids. The system remains in this nonexplosive, metastable state for a delayed period that ranges from a few milliseconds to a few seconds. During this time the fuel and coolant liquid intermix as the result of density and velocity differences as well as vapor production. (**Figure 1**)

Next, vapor film destabilization occurs and triggers fuel fragmentation; this rapidly increases the fuel surface area and thus vaporizes more coolant liquid and increases the local vapor pressure. This "explosive" vapor formation spatially propagates throughout the fuel-coolant mixture and thus causes the macroscopic region to become pressurized by the coolant vapor. Subsequently the high-pressure coolant vapor expands against the inertial constraint of the surroundings and the mixture. The high-pressure vapor that is produced, the dynamic liquid-phase shock waves, and the slug kinetic energy can all be destructive to the surroundings. In addition, if the fuel is metallic, this explosive dispersal may cause exothermic metal-water chemical reactions that can enhance the work output and produce hydrogen.

The concept of mixing (sometimes called premixing) is vague and has not been well-defined regarding energetic FCIs. In combustion engines, the fuel and oxidant are considered well-mixed when the liquid fuel has atomized into small droplets and they have dispersed in the oxidant and vaporized into it; that is, the fuel and oxidant have become a homogeneous fluid at the proper stoichiometric conditions. In vapor explosions, the interacting species are liquids; thus gaseous interpenetration is impossible. However, the qualitative attribute of attaining a homogeneous geometry is a reasonable concept for mixing in the FCI. Qualitatively, it could be described as a condition in which the fuel and coolant liquids disperse within one another. Past work (Henry, Fauske 1981 and Corradini 1988, 1991) suggests the following criteria for "good" mixing:

1. Stable film boiling between fuel and coolant liquids;
2. The discontinuous liquid must be dispersed in a continuum of the other liquid (void fraction <50%);
3. Local length scale of the discontinuous liquid much smaller than the system length scale.

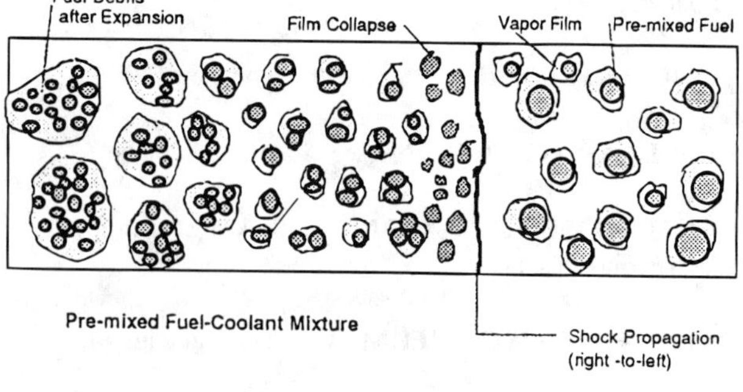

Figure 1. Conceptual picture of a steam explosion.

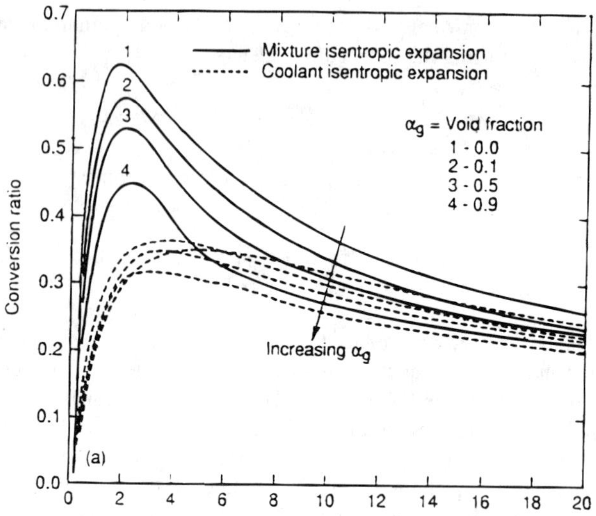

Figure 2. Water volume/fuel volume.

In the presence of a trigger, a vapor explosion could spatially propagate through the fuel-coolant mixture. In the past, a number of large-scale experiments were conducted to investigate this explosion phase of the energetic FCI (e.g., Hohmann et al., 1993). The first attempt to estimate the maximum work potential from the explosion was provided by Hicks and Menzies (1965) and involved a thermodynamic analysis of the fuel-coolant interaction. The path assumed for this ideal explosion involved a constant volume thermal equilibration of the fuel and coolant (1–2), followed by an isentropic expansion of the fuel-coolant mixture (2–3). Recently, Bang (1992) used the Hicks-Menzies approach with a complete fuel and coolant equation of state to estimate the upper-bound work potential for a variety of fuel-coolant pairs (**Figure 2**). The peak explosion conversion ratio (work/fuel thermal energy) was again found to be maximized at nearly equal fuel-coolant volumes, and, the magnitude depended on the initial void fraction. Also Bang showed equivalence between this and the Board-Hall (1975) *detonation* model.

A number of theories have been advanced to explain the vapor explosion process, (Corradini 1988, 1991) in which models were developed to be used in *reactor safety analyses* and probabilistic studies. For the validation of such models, comparisons must be made with large-scale explosion data. The focus of these comparisons is on observations which can be used to help validate the explosion models. Particular attention should be paid to the qualitative trends, what is quantitatively measured, and the uncertainty of these measurements. Such continuing work helps identify any future experiments that may be useful in basic understanding and model development.

References

Bang, K. H. and Corradini, M. L. (1992) Thermodynamic analyses of vapor explosions, NURETH-5, Salt Lake City UT. Sept 1992.

Board, S. J. and Hall, W. B. (1975) Detonation of fuel-coolant explosions, *Nature*, 254, 319 March 1975.

Corradini, M. L., Kim, B. J. and Oh, M. D. (1988) Vapor explosion in light water reactors: A Review of Theory and Modelling, *Progress in Nuclear Energy*, 22, 117.

Corradini, M. L. (1991) Vapor explosions: A review of experiments for accident analysis, *J. Nuclear Safety*, 32, 3, 337, Sept 1991.

Fauske, H. K. (1981) Required initial conditions for energetic steam explosions, *J. Heat Transfer*, 19, 99

Hicks, E. P. and Menzies, D. C. (1965) Theoretical studies on the fast reactor maximum accident, Argonne Lab Report, ANL-7120, Oct 1965

Hohmann, H., et al. (1992) KROTOS 26–30 Experimental data report, JRC Ispra Technical Note No. I.92.115, Nov 1992.

M.L. Corradini

VAPOURISATION (see Boiling)

VAPOR-LIQUID EQUILIBRIUM

Following from: *Phase equilibrium; Multicomponent systems, thermodynamics*

For pure substances (one component systems), equilibrium between liquid and vapor phases takes place if specific (molar) Gibbs energy or fugacity values of the coexisting phases are equal:

$$g^{(l)}(T, p) = g^{(g)}(T, p); \qquad f^{(l)}(T, p) = f^{(g)}(T, p). \tag{1}$$

The *equilibrium state* corresponds to the saturated vapor pressure curve $p = p_s(T)$, which is limited by a triple point (T_{tr}, p_{tr}) and a critical point (T_c, p_c) of the substance. (In the presence of surface tension forces at the phase interface, pressures values $p^{(l)}$ and $p^{(g)}$ in coexisting phases are not equal). The Clausius-Clapeyron equation follows from the equation 1

$$\frac{dp_s}{dT} = \frac{r}{T(v'' - v')}; \qquad r = T(s'' - s') = h'' - h'. \tag{2}$$

This equation gives the relationship between the saturation vapor pressure and specific (molar) thermodynamic properties of the substance: volumes υ', υ'', entropies s', s'', enthalpies h', h'' for the liquid and gaseous phases which are in equilibrium; r is a heat of vaporization.

Integrating Equation 2 with the simplest assumptions that a vapor phase is a perfect gas, $r = const$, $\upsilon'' = RT/p_s \gg \upsilon'$ (R is the universal gas constant), leads to the two parameter equation

$$\ln p_s = A - \frac{B}{T} = \ln p_0 + a(1 - T_b/T) \tag{3}$$

($p_0 = 101325$ Pa, T_b is the normal boiling point temperature), which is a base for derivation the majority of empirical correlations to describe the vapor pressure dependence versus temperature. The best of them provide a discrepancy of 1–2% in describing experimental data of $p_s(T)$ (the equations by Frost-Kalkwarf-Thodos, Lee-Kesler, etc.). To estimate thermodynamic properties of the phases which are in equilibrium, it is possible to use generalized equations such as Lee-Kesler, Gunn and Yamada for density, as well as the equations by Carruth and Kobayashi for heat of evaporation. Actually all of them have been based on the three parameter corresponding states law, therefore to calculate any of the properties mentioned above it is necessary that critical parameters T_c, p_c as well as Pitzer ω (acentric) factor need to be known. It should be noted, however, that such correlations are essentially less precise than experimental data to be described. So it is reasonable to use the correlations only for predicting the properties *a priori* as a first approximation.

An alternative way in describing thermodynamic properties is based on using of an equation of state which may present the properties of both liquid and gasous states. In general, it may be presented in the form

$$p = \rho RTz(T, \rho; a, b, \ldots), \tag{4}$$

where $\rho = 1/\upsilon$ is the molar density, z the compressibility coefficient, and a,b, . . . are individual parameters of the substance, which are fitted to the experimental data. If the equation of the form, Equation 4, is available then it makes possible to evaluate fugacity

$$f(T, \rho) = \rho RTz(T, \rho)\exp\left[z(T, \rho) - 1 + \int_0^{\rho} (z(T, \rho) - 1)\frac{d\rho}{\rho}\right]$$

and then solving the equilibrium equations 1

$$f(T, \rho') = f(T, \rho''); \qquad p_s = \rho' Rtz(T, \rho') = \rho'' RTz(T, \rho''), \tag{5}$$

to calculate the properties of the phases which are in equilibrium, $p_s(T)$, $\upsilon' = 1/\rho'$, $\upsilon'' = 1/\rho''$. The system of the equations 5 is equivalent to the Maxwell Rule

$$\int_{v'}^{v''} p(T, v)dv = p_s(T)(v'' - v').$$

Van der Vaals was the first who have proposed the equation of state to discribe both liquid and vapor phases. Since then there have been proposed a lot of similar equations (by Redlich-Kwong, Soave, Wilson, etc.), the parameters of which may be calculated if the critical point parameters of the substance are known. Also multiparameter equations of state have been proposed (by example, Benedict-Webb-Rubin), which are more precise. However, the parameters of the equations may not be defined without experimental data.

For multi-component systems the conditions of vapor-liquid equilibrium are expressed by equality of the partial Gibbs Energy $g_i^{(l)} = g_i^{(g)}$ of the components or appropriate fugacities

$$f_i^{(l)}(T, p, \chi_1', \ldots, \chi_\nu') = f_i^{(g)}(T, p, \chi_1'', \ldots, \chi_\nu''); \quad 1 \le i \le \nu, \tag{6}$$

where χ_i' and χ_i'' are the molar fractions of the components in liquid and vapor phases, hence

$$\sum_{i=1}^{\nu} \chi_i' = \sum_{i=1}^{\nu} \chi_i'' = 1. \tag{7}$$

To describe the behavior of the closed ν-component system when total mole fractions of the components are given by $\chi_1', \ldots, \chi_\nu'$, it is necessary to incorporate the material balances equations

$$z'\chi_i' + z''\chi_i'' = \chi_i; \quad 1 \le i \le \nu; \quad z' + z'' = 1, \tag{8}$$

where z' and z'' are a mole fraction of the liquid and the vapor phases.

The system of the Equations 6–8 is a base for analysis of vapor-liquid equilibrium in mixtures (solutions). The equations under consideration allow the calculation of a bubble point line as well as a dew point line. In contrast to the case of a pure substance, for the multi component two phase system these lines do not coincide as soon as the system is a polivariant one (a number degree of freedom $f = \nu > 1$), and the lines bound a field of two-phase equilibrium. At the bubble point, the liquid phase fraction $z' = 1$ and the phase composition of the liquid phase and the total composition of the system are equal: $\chi_i' = \chi_i$. At that condition Equations 6 to 8 define a bubble point temperature $T_b(p,\chi)$ if the pressure is known or a bubble point pressure $p_b(T,\chi)$ if temperature is known; in both cases the vapor phase composition $\chi_1^b, \ldots, \chi_\nu^b$ is a subject of calculations also. Conversely, at the dew point which relates to a start point of condensation, it is known that $z'' = 1$, $\chi_i'' = \chi_i$. Then the liquid phase composition $\chi_1^d, \ldots, \chi_\nu^d$ may be defined from the Equations 6 to 8 together with either temperature $T_d(p,\chi)$ or pressure $p_d(T,\chi)$ which correspond to a dew point. By way of example, **Figure 1** shows a typical shape of the curve which bounds the two-phase vapor-liquid region of a binary system for a given composition $\chi \equiv \chi_2$, $\chi_1 = 1 - \chi$. The system comprises substances with saturation pressure dependence $p_{s1}(T)$, $p_{s2}(T)$; a

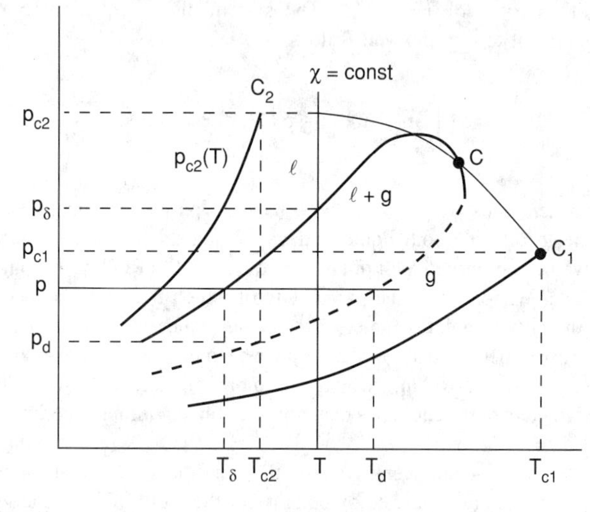

Figure 1. p-T diagram for a binary mixture.

smooth line presents bubble points and dash ones relates to dew points. Having defined the two-phase region, it is possible on the basis of the Equations 6 to 8 to calculate the compositions of χ_i', χ_i'' and fractions z', z'' of the phases under equilibrium for any T, p and the given total composition of the system χ_i. In addition, the Duhem theorem has to be satisfied; that is, the equations shall be independent.

If a liquid phase is an ideal solution and a gaseous phase is a mixture of ideal gases, then the calculations may be executed in a very simple way In this case the fugacities of the components, which appear in Equation 6, may be calculated according to the Raoult's Law $f_i^{(l)} = p_{si}(T)\chi_i'$, $f_i^{(g)} = p\chi_i''$. As a result, Equation 6 may be rearranged into the form, which is more convenient:

$$\chi_i''/\chi_i' = K_i, \quad 1 \le i \le \nu, \tag{9}$$

where $K_i(T,p)$ is a distribution coefficient of a component i

$$K_i = p_{si}(T)/p. \tag{10}$$

Having eliminated the phases compositions from the equation of material balances Equation 8 by means of Equation 7 and 9, we obtain an equation to calculate the fraction of each phase, for example, in the form of

$$\sum_{i=1}^{\nu} \chi_i/[z' + K_i(1 - z')] = 1; \quad z'' = 1 - z'. \tag{11}$$

The phase composition may then be estimated out as

$$\chi_i' = \chi_i/(z' + K_i z''); \quad \chi_i'' = K_i\chi_i'; \quad 1 \le i \le \nu. \tag{12}$$

Parameters of the bubble point curve $p_b(T,\chi)$ or $T_b(p,\chi)$ and the dew point curve one $p_d(T,\chi)$ or $T_d(p,\chi)$ are the solutions of the following equations, accordingly,

$$\sum_{i=1}^{\nu} \chi_i K_i^b = 1, \quad \sum_{i=1}^{\nu} \chi_i/K_i^d = 1, \tag{13}$$

where $K^b{}_i = p_{si}(T)/p_b$ or $p_{si}(T_b)/p$; $K^d{}_i = p_{si}(T)/p_d$ or $p_{si}(T_d)/p$. Compositions of new phases being formed in the process are obtained from the equations

$$\chi_i^b = \chi_i K_i^b; \quad \chi_i^d = \chi_i/K_i^d; \quad 1 \le i \le \nu \tag{14}$$

in an explicit form. The curves $p_b(T,\chi)$ and $p_d(T,\chi)$ may be obtained from Equation 13 if correlation Equation 10 is taken into account:

$$p_b(T, \chi) = \sum_{i=1}^{\nu} \chi_i p_{si}(T); \quad \chi_i^b = \chi_i p_{si}(T)/p_b(T, \chi); \tag{15}$$

$$p_d(T, \chi) = 1/\sum_{i=1}^{\nu} \chi_i p_{si}(T); \quad \chi_i^d = \chi_i p_d(T, \chi)/p_{si}(T). \tag{16}$$

A $p - \chi$ diagram which illustrates the correlations given above for the binary ideal solution ($\chi = \chi_2$) is presented in **Figure 2**. For this case it is possible to solve Equations 11 and 12 in the explicit form:

$$\chi' = \frac{p - p_{s1}}{p_{s2} - p_{s1}}; \qquad \chi'' = \frac{p_{s2}}{p}\chi'; \qquad z' = \frac{\chi - \chi''}{\chi' - \chi''};$$

and the bubble point line (Equation 15) is linear. In the T − χ-diagram (**Figure 3**) both the bubble point line and the dew point line Equation 16 are nonlinear.

To calculate vapor-liquid equilibria for multi-component nonideal systems, the phase equilibrium Equations 6 are usually presented in the same form as Equation 9. However, in contrast to the ideal case (Equation 10), for this case distribution coefficients depend not only temperature and pressure but on the phase composition. If, for example, the excess thermodynamic functions were used to describe a liquid state and an equation of state was used for a gas phase description, then,

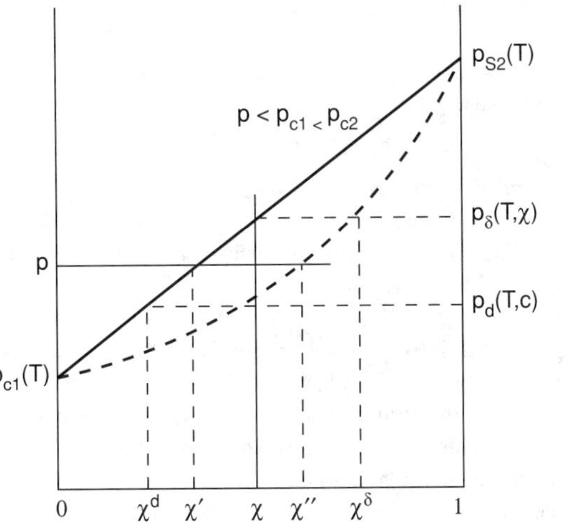

Figure 2. p-χ diagram for a binary mixture.

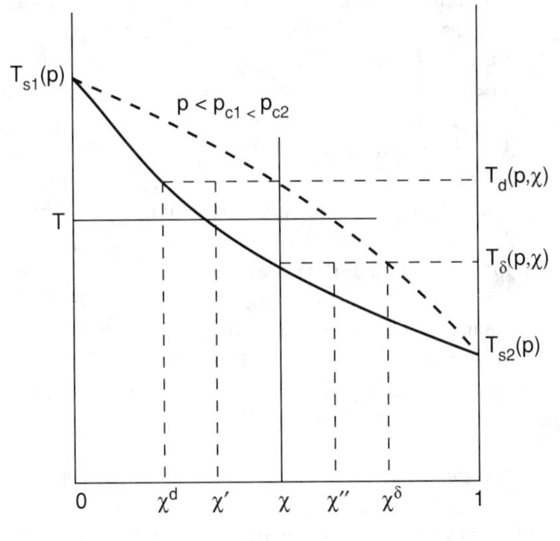

Figure 3. T-χ diagram for a binary mixture.

$$K_i = f_i^+(T, p)\gamma_i(T, p, \chi_1', \ldots, \chi_\nu')/\varphi_i(T, p, \chi_1'', \ldots, \chi_\nu'')p, \tag{17}$$

where f_i+ and γ_i are the fugacity of the pure i-component and its activity coefficient in the liquid phase; $\varphi_i = f_i^{(g)}/p\chi_i''$ is the fugacity coefficient of the component in a gas state. As a result, the set of Equations 9, 11, 12 as well as Equations 13, 14 become essentially nonlinear, and appropriate iterative methods should be used to find a solution. It's not difficult to estimate f_i+, if the pure components properties are available for the liquid phase:

$$f_i^+(T, p) = p_{si}(T)\exp\left[\int_{p_{si}(T)}^{p} v_i^{(l)}(T, p)dp/RT\right]. \tag{18}$$

As for the methods of calculations of the activity coefficients γ_i, there exists a vast literature on the subject. More often than not the methods are based on the molecular theory and follow the goal of calculating γ_i in multi-component mixtures in terms of the pure component properties. In general, the problem is impossible to solve at the current level of knowledge, that is why, as a rule, the suggested formulas include the empirical so-called "binary parameters" which characterize the molecular interactions for the binary systems which comprise the mixture components. Experimental data related to thermodynamic properties or vapor-liquid equilibria in the binary systems need to be known to adjust the values of the binary parameters. Often used are the correlations developed by van Laar, Hildebrand and Scatchard, as well as, those which have been obtained on the basis of two-liquid theory—NRTL, quazichemical theory UNIQUAC. The UNIFAC Method allows the calculation of activity coefficient using the fundamental idea that γ_i may be defined due to the combinatorial contribution of the molecular functional groups interacting in the solution. The method may be applied for a system which has not been studied experimentally.

Fugacity coefficients of the components for the gas phase φ_i, presented in Equation 17, may be calculated if the equation of state for the mixture is available. It is a common assumption that if all the pure components may be described by an equation of the form Equation 4 (Redlich-Kwong, Benedict-Webb-Rubin, etc.) then the equation is valid to describe the mixture comprising the same components. The parameters of the equation a,b, . . . are the combinations of the individual parameters for the components a_i, b_i, ... The rules for such combinations are empirical ones, for example

$$a = \sum_{i=1}^{\nu}\sum_{j=1}^{\nu}[(a_i a_j)^{1/2}(1 - l_{ij})]\chi_i\chi_j; \qquad b = \sum_{i=1}^{\nu} b_i\chi_i. \tag{19}$$

In the most prevalent cases, mixing rules involve an adjustable binary parameter l_{ij}, as it possible to see in Equation 19.

Non-ideal solutions are different from the ideal ones not only quantitatively but qualitatively. For example, the bubble point line for non-ideal solutions in the p − χ diagram is not linear (as it was for ideal case, see **Figure 2**) however the shape of the T − χ diagram is still the same as for an ideal solution (**Figure 3**). Also, the deviations from ideality may induce new phenomena. Even the simplest model of a nonideal solution such as the Margules equation predicts an *azeotrophy.* An essence of this phenomena is that in particular T and p both a liquid and a vapor phase composition become equal to the total mixture composition: $\chi_i' = \chi_i'' = \chi_i$; $1 \leq i \leq \nu$. As a result the material balance equations become linear and may be satisfied with any value of the mole fraction of the phases which are in equilibrium. Otherwise there is an indifferent state and the boiling or condensation processes take place at $T = const$ and $p = const$, similar to pure substances or monovarint systems, however $f > 1$. At the azeotrope point dew points and boiling points lines have a common extremum: for instance **Figure 4** presents the lines in the $T - \chi$-diagram for a binary solution forming an azeotrope. The azeotrope point is influenced by pressure and, therefore, temperature (**Figure 4**).

Often azeotroping is a forerunner of another phenomenon which occurs in non-ideal solutions with the strong positive deviation from ideality ($\gamma_i > 1$): this is a limited miscibility of the components in the liquid state. In this case, a liquid mixture splits into two separate liquid phases which have different compositions. In general, the phenomena takes place for the certain compositions which form an immisibility gap. **Figure 5** presents an immissibility gap for a binary system with an upper critical solution temperature (broken line). The bubble point (smooth) and dew points (dash lines) curves are also shown. If the total composition of the system is within the limits of the immiscibility gap, then vapor phase is formed by evaporation of the both liquid phases at the constant temperature $T_E(p)$ providing a vapor phase composition χ_E (p). The system exists as a mono-variant until a composition is reached when one of the liquid phases: then, the bubble point continues to rise in temperaturewith increasing χ. If the total composition of the system equals χ_E, the system represents a geteroazeotrope mixture which evaporates at constant temperature T_E from the beginning to the end of the immiscibility range.

The method of calculation of vapor-liquid equilibria in strongly non-ideal, multi-component systems which is based on excess ther-

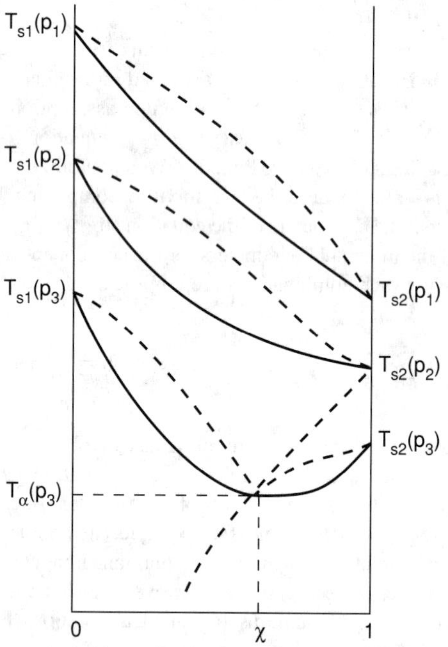

Figure 4. Binary mixture T-χ diagram for the case where an azeotrope is formed at low measure.

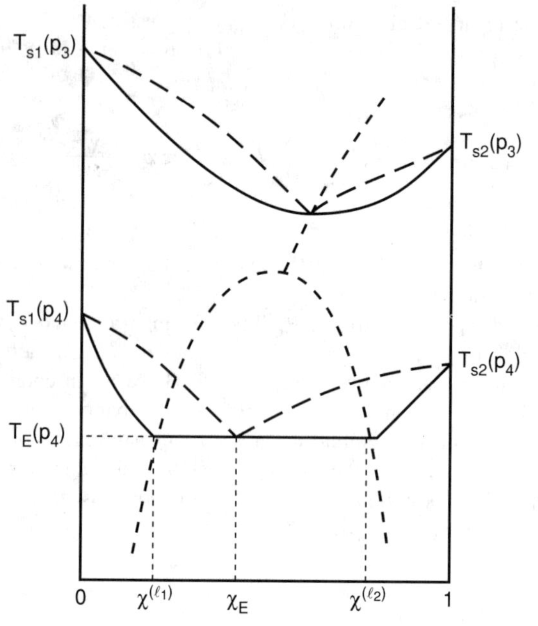

Figure 5. Equilibrium T-χ diagram for the case where immiscible liquids may be formed.

modynamics functions and a separate description of the phases compositions describes the complex phenomena mentioned above with a high accuracy. However, it has some shortcomings which are impossible to avoid. It is evident from **Figure 1** that, for any combination of T and p, in the two-phase region the second component, if it is pure, should exist in the gas state. At the same time, to calculate phase equilibria with the method under consideration the liquid state properties of the component must be known. This kind of information may be obtained only by extrapolation—a method which is impossible to justify. The situation becomes even more complicated if the temperature of the system is greater than the critical temperature of one of the components as is shown at **Figure 1**. For this case, the value of $p_{s2}(T)$ appearing in Equation 18 has no physical meaning. Also, there is no way to describe the situation in which the system changes into a super-critical state and the line bounding the two-phase region at **Figure 6** does not touch the right axis $\chi = 1$. Of course, the method under consideration is invalid to predict critical points of the solution, i.e., those in which compositions of the phases both vapor and liquid become equal, that is a point C at **Figure 6**. The dash line in the figure is a critical curve which joints the critical points of the mixtures of all the possible total compositions. Problems of this kind do not appear if calculations of equilibria are based on the application of the equations of state. If the parameters of the Equations 4, 19 have been adjusted to describe both a liquid and a gas states of the given multi component system as well as the properties of pure components, then instead of Equation 17 it is possible to obtain an equation to calculate the distribution coefficients:

$$K_i = \varphi_i'(T, p, \chi_1', \ldots, \chi_\nu')/\varphi_i''(T, p, \chi_1'', \ldots, \chi'').$$

Mole densities of the phases which are in equilibrium ρ', ρ'' may be found from the equations

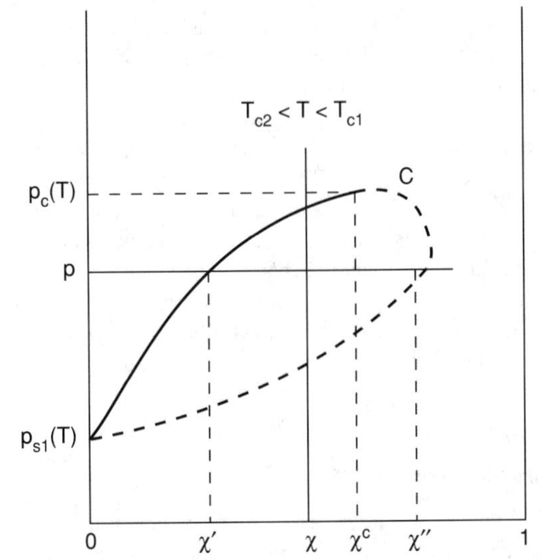

Figure 6. Phase diagram for a binary mixture forming a supercritical region.

$$p = \rho' RTz(T, p, \chi_1', \ldots, \chi_\nu') = \rho'' RTz(T, p, \chi_1'', \ldots, \chi_\nu'').$$

It is necessary, however, to point out that even for the modern equations of state, discrepancy of the calculated and experimental data is higher than for the excess functions method described above.

References

Reid, R. C., Prausnitz, J. M. and Poling, B. E. (1987) *The Properties of Gases and Liquids,* McGraw-Hill, New York.

Prigogine, I. and Defay, R. (1954) *Chemical Thermodynamics*, Longmans, London.

Walas, S. M. (1985) *Phase Equilibrium in Chemical Engineering*, Butterworth Publisher.

Leading to: *Binary systems, thermodynamics of*

M.Yu. Boyarsky and A.M. Semjonov

VAPOUR-LIQUID SEPARATION

Following from: *Phase separation, overview*

Introduction

Several pieces of equipment common in chemical and power plants are there to separate liquid from vapour or vapour from liquid. This is an important process because the dispersed phase can have undesirable effects. The effluent from a fossil fuel burning power plant, for instance, may contain sulfuric acid droplets, or other undesirable materials. Steam which has a little moisture entrained in it can leave undesirable silica deposits on superheater tubes. Water carried over into a turbine from a boiler can erode the blades and cause them to fail. There are many occasions when removing liquid from a gas stream or gas from a liquid stream is required.

A variety of techniques exist for doing this. For the smallest drops (less than 10 microns) *fiber mist eliminators* are used. As the particles become larger, *impingement devices* like *screens* are adopted. As they get still larger, *chevrons* are used. Finally, for the larger drops, cyclones are adopted. They can operate for the entire range of liquid-to-gas-phase flow rates.

For a first pass at the selection of a suitable separator system, see **Figure 1** in the entry. **Phase Separation Overview** should also be consulted. The most important parameter is the drop or particle size with, 10 microns (10×10^{-6}m) being the dividing line between the particles which can be separated by inertial means

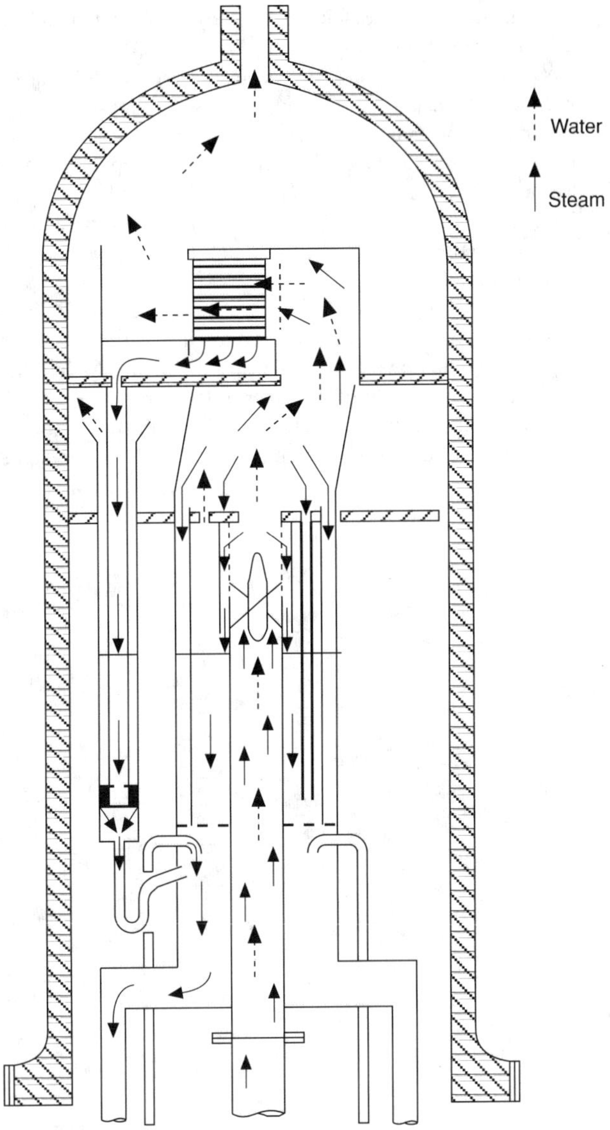

Figure 1. A detail of the Westinghouse MB-2 steam separator section showing the steam and water flow paths. Young, et al. (1984).

using cyclones or chevrons, and those which are so small that filters or other means based on diffusion must be used.

Separator Systems

An example of a separator system which has been used for steam and water at elevated pressure is shown on **Figure 1**, Young et al. (1980). A steam/water mixture enters the vertical pipe at the lower center of the figure. It passes through the swirler in the center which throws most of the liquid onto the walls of the central tube. Just after the swirler, the separated liquid is removed through holes in the walls of the central tube. The steam and remaining spray passes into a plenum which allows further liquid removal by gravity. The removed liquid drains onto the small tube which is at the right of the central tube. The remaining mist and vapor proceeds up into the dryer near the top of the vessel where an impingement separator removes most of the remaining liquid. The almost dry vapor then proceeds into the discharge shown at the top of the figure. The principles of gravity, swirl, and impingement separation are all illustrated in this installation.

A system which has different geometric constraints is illustrated on **Figure 2**. Head room is limited in a steam drum so the liquid and vapor from many heated tubes enter at the bottom, pass up into the periphery into a double row of cyclone separators. Almost dry steam leaves out of the top of the cyclone while the separated liquid leaves the bottom. The almost-dry steam proceeds to the scrubber section (usually chevrons) and then out of the drum to the superheater.

The mass flow rate of drops less than 10 microns in diameter is quite small in typical steam generation installations so the mass fraction of liquid flowing out of the separator section for installation of this type is typically of the order of 0.1 percent.

For chemical plant installations in which the drops consist of fumes of toxic substances like sulfuric acid, better separation efficiencies are required. For this a fiber mist eliminator is desirable. This type of separator is the only one which will work on fume particles as small as 0.1 microns.

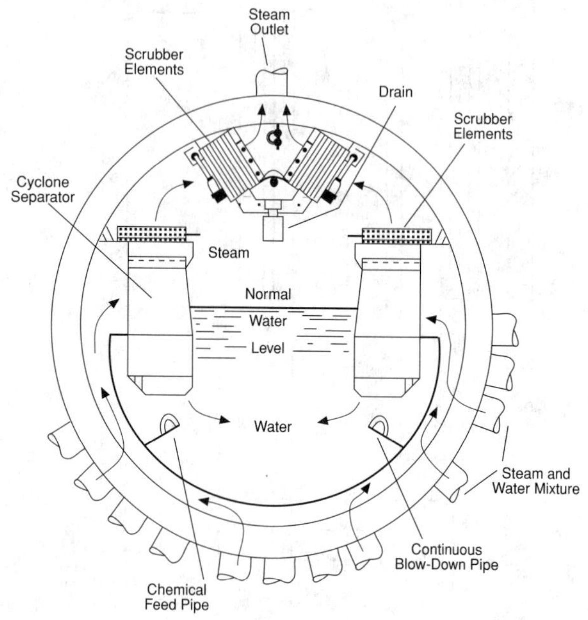

Figure 2. Steam drum separator section typical of modern drum type natural circulation boilers. Avallone, et al. (1987).

Separation can be a problem when some vapour is entrained with liquid, such as when a jet or stream of liquid enters a pool. This is called *carryunder.* To prevent carryunder from being a problem, the liquid flow rate down in the pool must be kept well below the bubble rise velocity; that is, the velocity down for the liquid should be less than 0.2 m/s.

In the remainder of this section, the types of separators will be described in greater detail, their limitations and characteristics mentioned, and guidelines for selection given.

Types of Separators

Gravity Separators

The simplest separator is the gravity separator. The rule of thumb used to size a gravity separator is that the superficial velocity of vapor at the free surface should be less than 0.3 m/s. When the velocity is greater than this, the carryover increases rapidly. Kataoka, et al. (1981) describe the carryover process in some detail. Often these separators are constructed as large-diameter inclined pipes in which the two-phase mixture enters at the bottom of the high end, and separated liquid is removed from the bottom of the low end. Vapour is removed from the top of the high end. In excess of 99 percent of the entrained liquid can be removed in gravity separators though some small drops are carried over for all flow rates. These separators can be horizontal or vertical cylinders or spheres.

When a gravity separator is overloaded, the distinct interface between the liquid and vapour disappears, and a large amount of liquid is carried over. A distinct interface between the liquid-rich and vapor-rich regions always disappears by the time the superficial velocity of vapour is greater than 2 m/s.

Centrifugal Separators

Centrifugal separators have a higher characteristic throughout and have the unique characteristic in that they can be used to separate any proportions of liquid or vapour in the incoming stream. (See also **Centrifuges**) The flow regime in the incoming stream doesn't matter. When they are used within their design envelope, they effectively separate 99 percent of the liquid. Drops that are less than 10 microns in diameter are not usually separated. The pressure drop in these separators tends to be larger than in other kind because the characteristic velocity is larger. The swirler shown in the center of **Figure 1** is typical of centrifugal separators.

When a *cyclone separator* is overloaded, the pressure in the swirl chamber drops so much below that of the pool into which the separated liquid drains that the liquid rises up the down commer and floods the separator.

Chevrons and Screens

Chevrons and screens operate by intercepting drops which are unable to follow the vapor as it goes through a tortuous path thru the device. These separators work on dispersed flows and are generally ineffective on drops less than 10 microns in diameter. **Figure 3** shows how the chevrons are arranged. Vapour flow is in

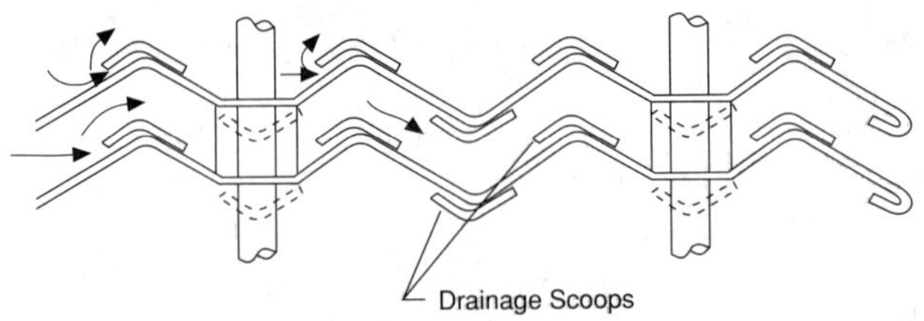

Figure 3. Chevron separator with scoops showing the path of the air or vapor. Moore, et al. (1976).

the horizontal plane while the separated liquid runs down due to gravity normal to the page.

When a chevron separator is overloaded, the liquid is re-entrained before it can drain away. *Knitted wire mesh separators* perform much the way chevron separators do except that they have lower characteristic velocities in them but can trap smaller drops. They fail when overloaded in the same way that chevron separators do, the separated liquid is re-entrained before the liquid can drain away.

Mist eliminators are particularly well adapted to the elimination of drops smaller than 10 microns. They consist of a mat or bundle of fibers arranged so that the separated liquid can drain away easily, and the gas can continue on through the filter material (see **Figure 4**). As the velocity through these separators increases, they also fail in the same way as the wire mesh or chevron separators, the liquid is re-entrained before it has a chance to drain away.

Pressure drop in separators is the penalty we must pay to separate the two phases. It is important to be able to estimate the pressure drops thru the separator and select a separator type which is suitable for the application in question. **Table 1** summarizes the performance of these kinds of separators in terms of the dimensional quantity given below.

$$F_S = v_g\sqrt{\rho_g} \tag{1}$$

In which the characteristic velocity is the velocity into the separator.

In this equation

F is a dimensional constant in $m/s(kg/m^3)^{1/2}$.
v_g = Vapour-phase velocity in m/s
ρ_g = Vapour density in kg/m^3

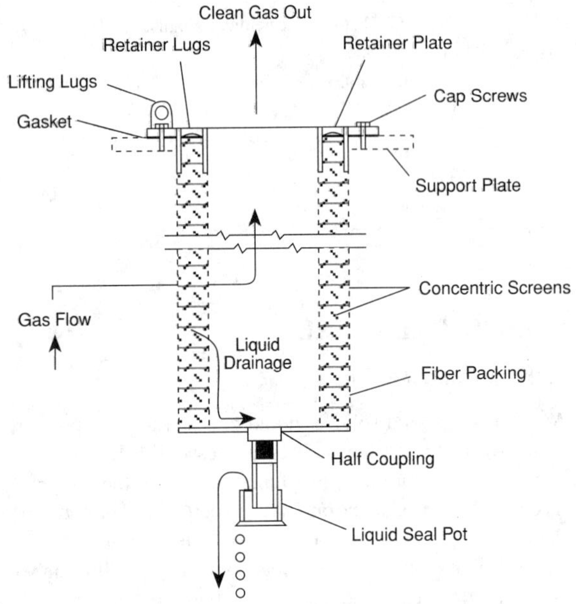

Figure 4. A high efficiency Brink mist eliminator element. Perry, et al. (1973).

Extensive information on pressure drops and separation efficiency for various kinds of separators is available in Idelchik (1986), Perry, et al. (1973), and Carson, et al. (1980).

References

Avallone, E., Baumeister, M. F. (1987) *Mark's Standard Handbook for Mechanical Engineers,* 9th Edition.

Carson, W. R., Williams, H. K. (1980) Method of Reducing Carryover and Reducing Pressure Drop Through Steam Separators, EPRfNP-1607 Publishing Corp.

Idelchik, I. E. (1986) *Handbook of Hydraulic Resistance,* Ch 12, Hemisphere.

Katooha, R., Ishu, M. (1983). Mechanistic Modeling and Correlations for Pool Entrainment Phenomenon, NUREC/CR-3304

Moore, M. J., Sieverding, C. H. (1976) Two-Phase Steam Flow in Turbines and Separators, 340 Hemisphere.

Table 1. Types of separators and their characteristics

Type	Approximate size droplet size range (in microns)	Separation system flow regime	Typical F_S	Typical ΔP in CM of H_2O	Two phase flow regime
Gravity Separator	>10.0	Laminar or Turbulent	0.3–0.6	Negligible ~1 velocity head	Any quality but best for low quality slug or annular dispersed flow
Fiber Mist Eliminators	>0.1	Turbulent	3.0–6.0	About 1 velocity head	Highly dispersed
Knitted Wire Mesh	>3.0	Turbulent	1.0–2.0	.25–5	Highly dispersed droplet flow
Chevron or Impingement Separator	>6.0	Turbulent	1.0–4.5	2.5–5	Highly dispersed droplet flow
Cyclone Separator	10.0 and up	Turbulent	3.0–20.0	7.5–7.5	Any quality or flow regime

Perry, R. H., Chilton, C. H. (1973) Chemical Engineers' Handbook, 5th Ed., 18–88 McGraw-Hill.

Young, M. Y., Takeuchi, Mendler, D. J., Hopkins, C. W. (1984) Prototypical Steam Generator Transient Test Program: Test Plan/Scaling Analysis, EPRINP-3994, NUREG/CR 3661, WCAP-10475, (1973).

P. Griffith

VAPOUR PRESSURE

***Following from:** PVT relationships*

Any substance in a solid or liquid phase at any temperature is characterized by an equilibrium vapour pressure. As a first approximation this vapuor pressure is a function only of the temperature and is defined by the Clapeyron-Clausius equation. The equilibrium pressure above a solid becomes zero at temperature 0 K and increases monotonous up to the triple point. The equilibrium vapor pressure for the liquid increases from the triple point until the critical point.

For most substances the *triple point pressure* is lower than atmospheric pressure. These substances can exist at atmospheric pressure as liquids. If the triple point pressure is higher than the atmospheric pressure, the liquid phase does not exist at atmospheric pressure and the solid substance directly evaporates (sublimates) (**Figure 1a** and **1b**).

For most pure substances the *equilibrium vapour pressures* are defined experimentally and tabulated.

Usually as a first approximation the equilibrium vapor pressure p_s for a liquid at any temperature T can be calculated if the normal boiling temperature ($T_{n.b.}$) (the temperature at which the equilibrium vapor pressure is equal to 1 phys.atm = 0.1033MPa) as well as the heat of evaporation r at this temperature are known:

$$\ln p_s = -\frac{r}{R}\left(\frac{1}{T_{n,b}} - \frac{1}{T}\right). \quad \mathbf{(1)}$$

This equation is plotted in **Figure 2**. It is a segment of straight line between the critical point K and triple point T_r, which has a slope r/R.

The equilibrium vapour pressure for a solid substance can be described by a similar equation:

$$\ln \frac{p_s}{p_{tr}} = \frac{\Lambda}{R}\left(\frac{1}{T} - \frac{1}{T_{tr}}\right), \quad \mathbf{(2)}$$

here the subscript tr refers to the tripple point; Λ is the heat of sublimation.

At the triple point

$$\Lambda = r + L, \quad \mathbf{(3)}$$

where L is the heat of fusion.

In the case when condensed (liquid or solid) substance is not only under its equilibrium vapor pressure but sustains an additional positive or negative pressure, the equilibrium pressure itself becomes a function of this additional pressure.

The influence of the additional pressure can be accounted for by taking advantage of the thermodynamic phase equilibrium condition saying that the specific Gibbs functions of phase being in equilibrium have to be equal:

$$g_1(T, p_1) = g_2(T, p_2), \quad \mathbf{(4)}$$

where the subscripts $_1$ and $_2$ refer to the respective phases.

If, at constant temperature, the pressure upon the phase 1 (condensed phase) will be changed, the the equation has to be valid:

$$\left(\frac{\partial g_1}{\partial p_1}\right)_T dp_1 = \left(\frac{\partial g_2}{\partial p_2}\right)_T dp_2. \quad \mathbf{(5)}$$

As for as the derivative $(\frac{\partial g}{\partial p})_T = v$, one can derive from (5):

$$v_1 dp_1 = v_2 dp_2. \quad \mathbf{(6)}$$

This equation is known as the *Pointing equation.*

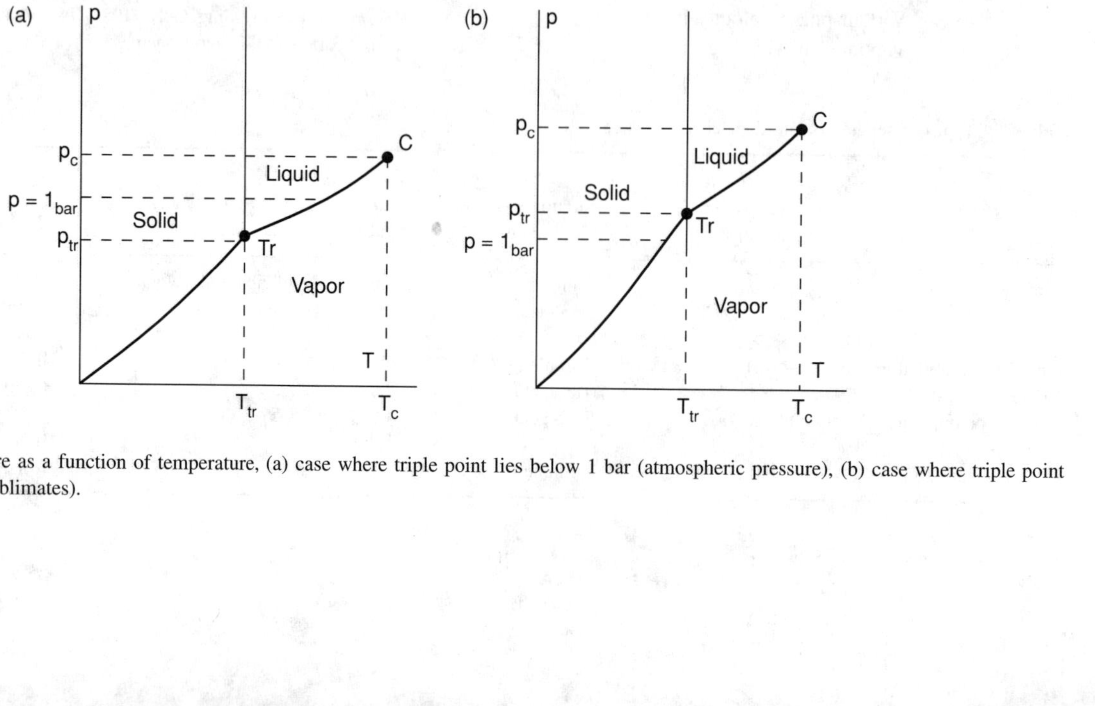

Figure 1. Vapor pressure as a function of temperature, (a) case where triple point lies below 1 bar (atmospheric pressure), (b) case where triple point lies above 1 bar (solid sublimates).

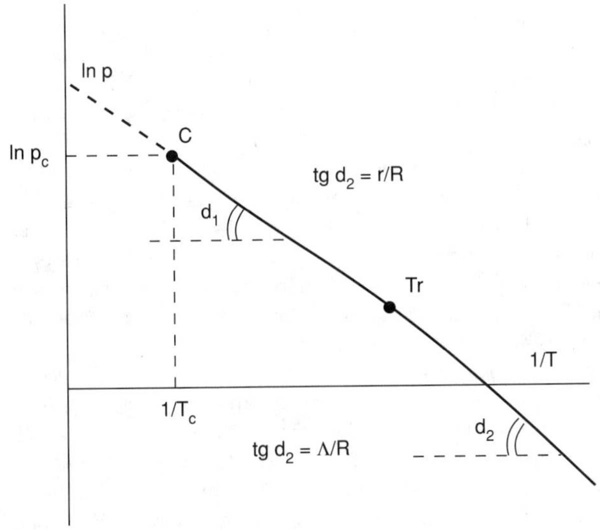

Figure 2.

At moderate pressures v_1 can be regarded as independent from the pressure; whereas v_2 can be expressed via the perfect gas equation of state: $v_2 = RT/p_2$.

With these assumptions the Pointing equation will be expressed as

$$v_1 dp_1 = RT\, d(\ln p_2),$$

which after integration gives

$$v_1(p_1 - p_0) = RT \ln(p_2/p_0), \tag{7}$$

where p_o is the equilibrium vapour pressure at the temperature T, without any additional pressure, i.e. when the pressures in both phases are equal.

Equation 7 can be rearranged:

$$p_2 = p_0 \exp\left[\frac{v_1}{RT}(p_1 - p_0)\right]. \tag{8}$$

This equation is in particular of interest when the equilibrium vapour pressure above a curved liquid surface is defined. In this case according to Laplace equation a spherical liquid surface sustains an additional pressure

$$\Delta p = p_1 - p_0 = \frac{2\sigma}{\rho}, \tag{9}$$

where σ is the surface tension; ρ is the radius curvature of the surface. In this equation if the liquid surface is convex ρ is positive, for a concave surface ρ is negative.

Leading to: *Clapeyron-Clausius equation*

E.E. Shpilrain

VAPOUR PRESSURE, EQUILIBRIUM, CHANGE WITH TEMPERATURE (see Clapeyron-Clausius equation)

VAPOUR PRESSURE THERMOMETER (see Temperature measurement, practice; Thermometer)

VAPOUR SHEAR, EFFECT ON CONDENSATION (see Tube banks, condensation heat transfer in)

VAPOUR SHEAR STRESS EFFECTS (see Condensation, overview)

VAPOUR SORPTION REFRIGERATION CYCLE (see Solar energy)

VARIABLE AREA FLOWMETER (see Rotameters)

VARIABLE FLUID PROPERTIES (see Crossflow heat transfer)

VARIATION OF FLUID PROPERTIES (see Tube banks, single phase heat transfer in)

VDI (see Verein Deutsche Ingenieur)

VECTORS (see Conservation equations, single phase)

VELOCITY, AVERAGE PHASE (see Multiphase flow)

VELOCITY BOUNDARY LAYER (see Boundary layer heat transfer)

VELOCITY DEFECT (see Universal velocity profile)

VELOCITY HEAD

Following from: *Flow of fluids*

This term is commonly used in hydraulic engineering and represents the *kinetic energy of a fluid flow* within a given streamtube. It has the physical dimension of length (hence the term "head"), and corresponds to an energy per unit weight. If u is the velocity and g the acceleration due to gravity, the velocity head is defined by $u^2/2g$. If the flow within a pipe, channel, or duct is uniform over a given cross-section, this definition of the velocity head applies to the entire cross-section. However, if the flow is non-uniform, the variation in kinetic energy from one streamtube to another is accounted for by a correction factor α. In this case the velocity head appropriate to the section is:

$$\frac{\alpha \bar{u}^2}{2g} \quad \text{where} \quad \alpha = \frac{1}{\bar{u}^2}\frac{\int_A u^3 dA}{\int_A u dA}$$

where A is the cross-sectional area and $\bar{u}$ is the mean velocity. In terms of a traditional energy level approach, the velocity head represents the distance between the *hydraulic gradient line* (defined by the static head), and the *total energy line* or *total head line*.

Leading to: *Dynamic pressure; Pitot tube*

C. Swan

VELOCITY MEASUREMENT

The methods of measuring the velocity of liquids or gases can be classified into three main groups: kinematic, dynamic and physical.

In kinematic measurements, a specific volume, usually very small, is somehow marked in the fluid stream and the motion of this volume (mark) is registrated by appropriate instruments.

Dynamic methods make use of the interaction between the flow and a measuring probe or between the flow and electric or magnetic fields. The interaction can be hydrodynamic, thermodynamic or magnetohydrodynamic.

For physical measurements, various natural or artificially organized physical processes in the flow area under study, whose characteristics depend on velocity, are monitored.

Kinematic Methods

The main advantage of kinematic methods of velocity measurements is their perfect character, and also their high space resolution. By these methods, we can find either the time the marked volume covers a given path, or the path length covered by it over a given time interval. The mark can differ from the surrounding fluid flow in temperature, density, charge, degree of ionization, luminous emittance, index if refraction, radioactivity, etc.

The marks can be created by impurities introduced into the fluid flow in small portions at regular intervals. The mark must follow the motion of the surrounding medium accurately. The motion of marks is distinguished by the method of their registration, into non-optical and optical kinematic methods. In the probe (non-optical) method which traces thermal nonuniformities, a probe consisting of three filamant's located in parallel plates (see **Amemometers, Thermal**) is used. The thermal trace is registered by two receiving wires located a distance l from the central wire. By registering the time Δt between the pulse heat emitted from the central wire and the thermal response of the receiving wire, we can determine the velocity $u = l/\Delta t$. Depending on which receiving wire receives a thermal pulse, we can define the direction of flow.

Marks consisting of regions of increased ion content are also widely used. To create ion marks, a spark or a corona discharge or an optical break-down under the action of high-pulse laser radiation is used. In tracing by radioactive isotopes, the marks are created by injecting radioactive substances into the fluid flow; the times of passing selected locations by the marks are registered with the help of ionizing-radiation detectors.

Optical kinematic methods use cine and still photography to follow the motion of marks. Three main types of photography are used: cine photography, still photography with stroboscopic lighting and photo-tracing. In cine photography, to determine the velocity, successive frames are aligned and the distance between the corresponding positions of the mark is measured. In the stroboscopic visualization method, several positions of the mark are registered on a single frame (a discontinuous track), which correspond to its motion between successive light pulses. Two components of the instantaneous velocity vector are determined by the distance between the particle positions. Typical of the marks used are 3–5 μm aluminum powder particles or small bubbles of gas generated electrolytically in the circuit of the experimental plant. Of vital importance in this method is the accuracy of measurement of the time intervals between the flashes.

In the photo-tracing method, the motion of the mark is recorded by projecting the image of the mark through a diaphragm (in the form of a thin slit oriented along the fluid flow) onto a film located on a drum rotating at a certain speed. The mark image leaves a trace on the film whose trajectory is determined by adding the two vectors; the vector of mark motion and the vector of film motion. The slope angle of a tangent to this trajectory is proportional to the velocity of mark motion. Further information on the photographic technique is given in the article on **Tracer Methods**.

Laser doppler anemometers can also be classified as kinematic techniques (see **Anemometers, Laser, Doppler**).

Dynamic Methods

Among the dynamic methods the most generally employed are, because of the simplicity of the corresponding instruments, the methods based on hydrodynamic interaction between the primary converter and the fluid flow. The Pitot tube is used most often, (see **Pitot Tube**) whose function is based on the velocity dependence of the stagnation pressure ahead of a blunt body placed in the flow.

The operating principle of fibre-optic velocity converters is based on the deflection of a sensing element, in the simplest case, made in the form of a cantilever beam of diameter D and length L and placed in the fluid flow between the receiving and sending light pipe, depends on the velocity of fluid flowing around it. The change in the amount of light supplied to a receiving light-pipe is measured by a photodetector.

The upper limit of the range of velocities measured u_{max} is limited by the value of $Re = u_{max} D/\nu \leq 50$ and the frequency response is limited by the natural frequency f_0 which depends on the material, diameter and length of the sensing element. However, by varying L and D, we can change the velocities over a wide range. Depending on the fluid in which measurements of u_{max} are made, the dimensions of the sensing elements vary within the limits of $5 \leq D \leq 50$ μm, $0.25 \leq L \leq 2.5$ mm.

The tachometric methods use of the kinetic energy of flow. Typical anemometers using this principle consist of a hydrometric current meter with several semi-spherical cups or an impeller with blades situated at an angle of attack to the direction of flow (see **Anemometer, Vane**).

Physical Methods

The physical methods of velocity measurements are, as a rule, indirect. This category includes sputter-con methods, which uses the dependance of the parameters of an electric discharge on velocity; ionization methods which depend on a field of concentrated ions, produced by a radioactive isotope in the moving medium on the fluid flow velocity; the electrodiffusion method which uses the influence of flow on electrodediffusion processes; the hot-wire or hot-film anemometer; magnet to acoustic methods.

The hot-wire method is derived from the dependence of convective heat transfer of the sensing element on the velocity of the incoming flow of medium under study (see **Hot-wire and Hot-film Anemometer**). Its main advantage is that the primary converter is has a high frequency response, which allows us to use it for measuring turbulent characteristics of the flow.

The *electrodiffusion method* of investigation of velocity fields is based on measuring the current of ions diffusing towards the cathode and discharging on it. The dissolved substances in the electrolyte must ensure the electrochemical reaction occuring on electrodes. Two types of electrolytes are most often used: ferrocyanidic, consisting of the solution of potassium ferri and ferrocyanide $K_3Fe(CN)_6$, $K_4Fe(CN)_6$ respectively, with concentration 10^{-3}–5×10^2 mole/l) and of caustic sodium NaOH (with concentration 0.5–2 mole/l) in water; triodine, consisting the iodide solution I_2 (10^{-4}–10^{-2} mole/l) and potassium iodide KI (0.1–0.5 mole/l) in water. Platinum is used as the cathode in such systems. In velocity measurement, a sensor which is made of a glass capillary tube 30–40 μm in diameter with a platinum wire (d = 15–20 μm) soldered into it is used. The sensing element (the cathode) is the wire end facing the flow, and the device casing is the anode. The dependence between the current in the circuit and the velocity is described by the relation $I = A + B\sqrt{u}$, where A and B are transducer constants defined in calibration tests.

The *magnetohydrodynamic methods* are based on the effects of dynamic interaction between the moving ionised gas or electrolyte and the magnetic field. The conducting medium, moving in a transverse magnetic field, produces an electric force E between the two probes placed at a distance L in the fluid flow, proportional to the magnetic field intensity H and to the flow velocity u: $E = \mu\,\vec{H}\,\vec{u}\,L$. The disadvantage of the method is that it can only be used to measure a velocity averaged over the flow section, nevertheless it has found use in investigating hot and rarefied plasma media.

Among direct methods the most abundant are the acoustic, radiolocation and optical methods. In using acoustic methods for determining the velocity of the medium, we can measure either the scattering of a cluster of ultrasound waves by the fluid flow perpendicular to the cluster axis, or the Doppler shift of the frequency of ultrasound scattered by the moving medium, or the time of travel of acoustic oscillations through a moving medium. These methods have found application in studying the flows in the atmosphere and in the ocean, where the requirements for the locality of measurement are less stringent than in laboratory model experiments. To carry out precision experiments with high space and time resolution, optical methods are used—the most refined method used is laser Doppler anemometry. (see **Anemometers, Laser Doppler**). Laser doppler anemometry depends on scattering from small particles in the flow and can also be considered a kinematic method (see above).

Mass Measurement

For measuring the mean-mass velocity of flows, differential pressure flowmeters, rotameters, volumetric turbine, vortices, magnetic induction, thermal, optical and other flowmeters are used inwhich we can define the mass velocity $\overline{\rho u} = \dot{m} = \dot{M}/S$ (see **Flowmetering**) as the flow rate $\dot{M}$ measured of the substance and by the known section of the flow S.

Reference

Fluid Meters. (1971) Their Theory and Application. Report of ASME, Research Committee on Fluid Meters, New York, Publishers by ASME.

***Leading to:** Anemometers, laser Doppler; Anemometers, thermal; Anemometers, vane; Hot-wire and hot-film anemometers*

Yu.L. Shekhter

VELOCITY OF LIGHT (see Cerenkov radiation)

VELOCITY OF SOUND

The velocity of sound is a vector **u** whose magnitude |**u**| is the *speed of sound* u and whose direction is normal to the surface of constant phase. The speed of sound is a property of the medium through which the sound travels and is therefore usually of more interest than the velocity itself which depends upon both u and the manner in which the sound is generated.

In discussing the relation between u and the properties of the medium is it useful to distinguish between homogeneous fluids and solids. In the former, the speed of sound is the same in all directions whereas, in the latter, this is not necessarily the case.

In a perfectly elastic (non-dissipative) fluid the speed of sound is given for small-amplitude sound waves by

$$u^2 = (\partial p/\partial \rho)_S = 1/\rho\kappa_s$$
$$= [(\partial p/\partial \rho)_T - (T/\rho^2 c_p)(\partial p/\partial T)_p^2]^{-1}, \quad (1)$$

where p is pressure, ρ density, c_p specific heat at constant pressure, S entropy and κ_S is the isentropic compressibility. Consequently, u may be determined from the thermodynamic **Equation of State** of the fluid or, conversely, information about the equation of state may be deduced from the $u^2(T,p)$ (see Trusler, 1991). For the special case of the **Perfect Gas**, for which $\rho = Mp/RT$, the speed of sound is given by

$$u^2 = RT\gamma^{pg}/M \quad (2)$$

where M is the molar mass and γ^{pg} is the heat-capacity ratio for the perfect gas. Thus the speed of sound in a perfect gas is proportional to $\sqrt{T\gamma^{pg}}$ but independent of pressure. All real gases approach this behaviour at sufficiently low pressures, but generally $u(T,p)/\sqrt{T\gamma^{pg}}$ is a slowly varying function of both T and p. Some examples of $u(T,p \to 0)$ for gases at T = 300 K are: ^4He, 1019 m/s; CH_4, 451 m/s; N_2, 352 m/s; air 348 m/s; and n-butane, 194 m/s (Kaye and Laby, 1986)

For the saturated vapour of a pure substance, u at first increases with temperature and, after passing through a maximum, it then declines to zero at the gas-liquid critical point. The speed of sound in the pure saturated liquid is always greater than that in the coexisting vapour. For normal saturated liquids, u declines steadily from its value at the triple point to zero at the liquid-gas critical point. Since the speed of sound in liquids varies only slowly with pressure, u generally declines also with increasing temperature along an isobar. Liquid water is exceptional in this respect as $(\partial u/\partial T)_p < 0$ at atmospheric pressure between 0 and 74 °C. Some examples of speeds of sound in liquids at atmospheric pressure and T = 300 K are: perfluorohexane, 505 m/s; CCl_4, 915 m/s; methanol, 1097 m/s; ethanol, 1139 m/s; cyclohexane, 1244 m/s; toluene, 1298 m/s; mercury, 1448 m/s; water, 1500.7 m/s; and sea water (3.5 mass% salinity), 1537.9 m/s (Kaye and Laby, 1986).

All fluids exhibit some absorption of sound at all frequencies. For most fluids there exists a wide range of frequencies from zero upwards in which both the dissipation is slight and **Equation 1** is obeyed with great accuracy. For monatomic fluids, this situation persists up to frequencies approaching the molecular collision frequency. However, various relaxation mechanisms exist in molecular fluids which, in certain frequencies ranges, give rise to greatly increased absorption and to dispersion (i.e. frequency dependence of u). For gases, these relaxation mechanisms are usually associated with the vibrational and rotational modes of the molecules. However, in liquids structural relaxation associated with the viscoelastic properties is possible. In all cases, **Equation 1** is obeyed at sufficiently low frequencies.

When sound waves are channelled along a fluid-filled waveguide (e.g. a tube) the apparent speed of sound is not identical with that observed in free space. Three characteristic sound speeds may be defined. The phase speed u_p is the speed at which points of constant phase propagate along the axis in a monofrequency wave. This is the only characteristic speed of importance for continuous waves although it will be frequency dependent if the medium is dispersive. When pulses of sound are transmitted through a waveguide, it is useful to define also the signal speed u_s as the speed at which the leading edge of the pulse propagates and the group speeds u_g as the speed at which the centre of the pulse propagates. In the absence of dispersion, all three speeds are identical but when u_p is a function of the angular frequency ω they differ slightly and $u_p < u_g < u_s$.

The speed of longitudinal sound waves in an infinite isotropic solid specimen is given by

$$u^2 = (B + \tfrac{4}{3}G)/\rho, \qquad (3)$$

where B is the bulk modulus and G is the shear modulus (Kinsler and Frey, 1982). A shear wave is also possible with sound speed $(G/\rho)^{1/2}$. In practise, the sample is not infinite and the observed longitudinal sound speed depends, though the boundary conditions, upon the lateral dimensions of the sample. For a bar with lateral dimensions small compared with the wavelength, free boundary conditions apply (the lateral stress components vanish) and

$$u_{bar}^2 = E/\rho \qquad (4)$$

where $E = 3(1 - 2\sigma)B$ is Young's modulus. As the lateral dimensions increase, the effective boundary condition for a volume element in the sample approaches one of zero lateral strain and u approaches the bulk value. Sound speed measurements of are the primary means of determining the elastic moduli of solids. For anisotropic solids (e.g. all crystals except simple cubic structures) the elastic constants and hence also the speed of sound are different along different directions.

For complex materials, such as porous media, laminates and other composites, the speed of sound is not usually given by a simple expression but may nevertheless provide valuable information about the structure.

References

Kinsler, L. E., Frey, A. R., Coppens, A. B. and Sanders, J. V. (1982) *Fundamentals of Acoustics,* 3rd edn. Wiley, New York.

Trusler, J. P. M. (1991) *Physical Acoustics and Metrology of Fluids,* Adam Hilger, Bristol.

Kaye, G. W. C. and Laby, T. H. (1986) *Tables of Physical and Chemical Constants,* 15th edn. Longman, Harlow.

***Leading to:** Critical flow*

J.P.M. Trusler

VELOCITY OF SOUND IN TWO-PHASE MIXTURES (see Critical flow)

VELOCITY RATIO (see Slip ratio)

VELOCITY, SUPERFICIAL (see Multi-phase flow)

VENA CONTRACTA (see Contraction, flow and pressure loss in; Critical flow)

VENTILATION (see Air conditioning)

VENTURI EJECTORS (see Aeration)

VENTURI METERS

***Following from:** Flow metering*

Venturi meters are flow measurement instruments which use a converging section of pipe to give an increase in the flow velocity and a corresponding pressure drop from which the flowrate can be deduced. They have been in common use for many years, especially in the water supply industry.

The classical Venturi meter, whose use is described in ISO 5167-1: 1991, has the form shown in **Figure 1**.

For incompressible flow if the **Bernoulli Equation** is applied between two planes of the tappings,

$$p_1 + {}^1\!/_2\rho\bar{u}_1^2 = p_2 + {}^1\!/_2\rho\bar{u}_2^2 \qquad (1)$$

where p, ρ and ū are the pressure, density and mean velocity and the subscripts $_1$ and $_2$ refer to the upstream and downstream (throat) tapping planes.

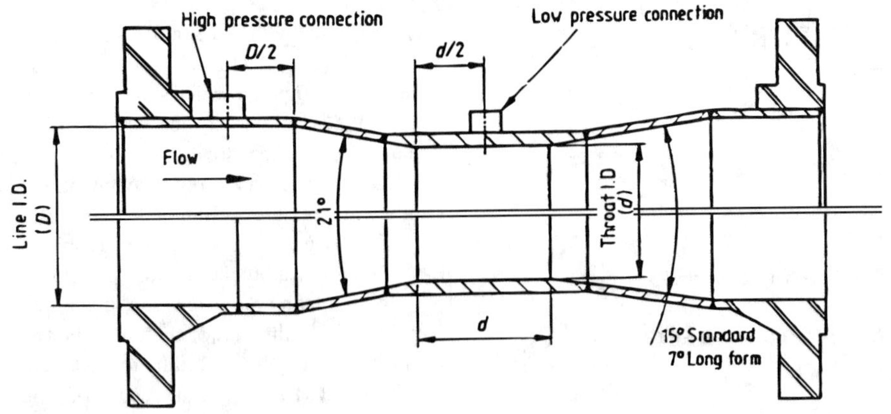

Figure 1. Classical Venturi meter design. (From B. S. 7405 (1991) Fig. 3.1.4, with permission of B.S.I.)

From continuity

$$\dot{V} = {}^1/_4\pi D^2 u_1 = {}^1/_4\pi d^2 u_2, \quad (2)$$

where $\dot{V}$ is the volumetric flowrate and D and d the pipe and throat diameters.

Combining Equations 1 and 2

$$\dot{V} = \frac{\pi d^2}{4}\frac{1}{\sqrt{(1-\beta^4)}}\sqrt{\left\{\frac{2(p_1-p_2)}{\rho}\right\}}, \quad (3)$$

where β is the diameter ratio, d/D. In reality, there is a small loss of total pressure, and the equation is multiplied by the discharge coefficient, C, to take this into account:

$$\dot{V} = C\frac{\pi d^2}{4}\frac{1}{\sqrt{(1-\beta^4)}}\sqrt{\left\{\frac{2\Delta p}{\rho}\right\}}, \quad (4)$$

where Δp is the differential pressure ($\equiv p_1 - p_2$). The discharge coefficient of a Venturi meter is typically 0.985, but may be even higher if the convergent section is machined. Discharge coefficients for uncalibrated Venturi meters, together with corresponding uncertainties, are given in ISO 5167-1: 1991.

If the fluid being metered is compressible, there will be a change in density when the pressure changes from p_1 to p_2 on passing through the contraction. As the pressure changes quickly, it is assumed that no heat transfer occurs and because no work is done by or on the fluid, the expansion is isentropic. The expansion is almost entirely longitudinal and an expansibility factor, ε, can be calculated assuming one-dimensional flow of an ideal gas:

$$\varepsilon = \left(\left(\frac{\kappa\tau^{2/\kappa}}{\kappa-1}\right)\left(\frac{1-\beta^4}{1-\beta^4\tau^{2/\kappa}}\right)\left(\frac{1-\tau^{(\kappa-1)/\kappa}}{1-\tau}\right)\right)^{1/2}.$$

where τ is the pressure ratio, p_2/p_1, and κ the isentropic exponent. The expansibility factor is applied to the flow equation in the same way as the discharge coefficient.

Various forms of construction of a Venturi meter are employed, depending on size, but all are considerably more expensive than the orifice plate. However, because most of the differential pressure is recovered by means of the divergent outlet section, the Venturi causes less overall pressure loss in a system and thus saves energy: the overall pressure loss is generally between 5 and 20 per cent of the measured differential pressure. The Venturi meter has an advantage over the orifice plate in that it does not have a sharp edge which can become rounded; however, the Venturi meter is more susceptible to errors due to burrs or deposits round the downstream (throat) tapping.

The lengths of straight pipe required upstream and downstream of a Venturi meter for accurate flow measurement are given in ISO 5167-1: 1991. These are shorter than those required for an orifice plate by a factor which can be as large as 9. However, Kochen et al. show that the minimum straight lengths between a single upstream 90° bend and a Venturi meter in the Standard are too short by a factor of about 3.

References

British Standards Institution (1991) Guide to selection and application of flowmeters for the measurement of fluid flow in closed conduits, BS 7405.

International Organization for Standardization, Measurement of fluid flow by means of pressure differential devices—Part 1: Orifice plates, nozzles and Venturi tubes inserted in circular cross-section conduits running full. ISO 5167-1: 1991.

Kochen, G., Smith, D. J. M., and Umbach, A. (1989) Installation effects on Venturi tube flowmeters, *Intech Engineers Notebook,* October 1989, 41–43.

M.J. Reader-Harris

VENTURI NOZZLE (see Orifice flowmeters; Venturi meters)

VENTURI SCRUBBER (see Scrubbers)

VENTURI TUBES (see Orifice flowmeters; Venturi meters)

VEREIN DEUTSCHEUR INGENIEUR, VDI

Postfach 1139, Graf-Recke-Strasse 84
4000 Dusseldorf 1
Germany
Tel: 211 6214-216

VERTICAL SHAFT KILNS (see Kilns)

VERTICAL THERMOSYPHON REBOILER (see Reboilers)

VIBRATION (see Augmentation of heat transfer, single phase)

VIBRATION, FLOW INDUCED (see Vortex shedding)

VIBRATION IN HEAT EXCHANGERS

Following from: *Shell-and-tube heat exchangers*

Vibration of tubes in heat exchangers is an important limiting factor in heat exchanger operation. The vibration is caused by nonstationary fluid dynamic processes occurring in the flow. These are turbulent pressure pulsations (see **Turbulent flow**), vortex initiation and separation from tubes in cross flow (see **Crossflow**), hydroelastic interaction of heat transmitting element (tubes) assemblies with the flow, and acoustic phenomena. The greatest effect of nonstationary hydrodynamic forces is observed in tubes in flows with separation. Flow separation from tube surfaces occurs where there is a transverse velocity component of the flow and mainly affects the vibration strength of the tubes in heat exchangers. The dynamic effect of the flow on a vibrating tube depends on the flow velocity and the vibration characteristics of the tube. With a separated transverse flow over a tube bank, the reference velocity is assumed to be the flow velocity in the narrowest section of the bank in the tube plane (**Figure 1**) and is calculated by the formula

$$\bar{u} = \frac{b_l/D}{b_l/D - 1} u_0 \sin \beta, \tag{1}$$

where u_o is the velocity in the absence of the tubes, b, the tube pitch, D the tube diameter and where β is the angle of slope of tubes to the flow direction. For transverse flow $\beta = 90°$ and for longitudinal flow $\beta = 0°$. Generalization of the data for calculation of the hydrodynamic forces and tube vibrations excited by them makes it possible to use the reference velocity, calculated by Equation 1, with a relatively low error for β in the range from 90° to 15°. Decreasing β increases the error. It is assumed that all kinds of tube vibration come into play simultaneously with the onset of fluid flow. However, each type of vibration dominates over a certain range of flow velocity this range depending on the vibrational parameters of tubes, the fluid properties, and the conditions of the flow. This is obvious from the amplitude and velocity characteristics (**Figure 1**) of tube vibration in the first and fifth rows of the staggered tube bank with transverse to longitudinal pitch ratios 1.61 × 1.38 with a natural tube frequency $f_n = 99$ Hz. High amplitude ratios A/D are observed in excitation by vortex separation (region 2 in **Figure 1** which shows relative root-mean-square values of $\bar{A}_y$, the amplitude of tube vibrations in the transverse direction relative to the free stream) and in hydroelastic instability (region 3). In region 1 low amplitude vibrations are brought about by turbulent pressure pulsations. In the case of longitudinal flow, the disturbance of tube assembly stability is determined only by excitation by turbulent pressure pulsations.

In calculating *tube vibration,* it is important to find the *natural frequency* of vibration of the tubes. For a tube with pivoted ends vibration may occur according to mode shapes 1, 2, and 3 as shown in **Figure 2**. The natural frequency of vibrations depends on both the mode shape and the physical characteristics of the tube, and the way its ends are fixed; it can be calculated by the formula

$$f_n = \frac{B_n}{2\pi l^2}\left(\frac{EI}{m}\right)^{0.5}, \tag{2}$$

where E is the modulus of elasticity of the tube material, I the area moment of inertia ($= \pi\,(D^4 - D_i^4)/64$ where D_i is the tube internal diameter), m is the overall tube mass per unit of its length (including the mass of the tube itself, the mass of the tube side fluid and the mass of the shell side fluid displaced in the vibration) , l is the tube length, B_n is a constant depending on vibration shape and the manner of tube fixation in the heat exchanger. The constant B_n used for determining the frequency harmonic of natural vibrations in a quiescent fluid in the absence of axial forces is derived from tabulated data. For shell-and-tube heat exchanger with more than 4 baffles, and where the end spaces between the tube sheets and the nearest baffles does not exceed the baffle spacing by more than 20%, a value of $B_n = 10$ may be taken (Chenoweth, 1983). Alternatively, the expression $B_n = \lambda_n^2$ where λ_n is calculated from the expression given in Table 1, may be used.

Vortex excitation of tubes depends on periodic hydrodynamic forces originating in vortex formation and separation from tubes. The tube (**Figure 3**) is subjected to the periodic hydrodynamic forces which are cabable of rocking an elastically mounted tube. Vortices are separated in turn, from one side, then the other side of the tube. Therefore, the transverse hydrodynamic force continuously changes direction and is a source of energy for excitation of tube vibrations. The hydrodynamic force brought about by separation of vortices varies sinusoidally. It is obvious from **Figure 3** that, in the longitudinal direction, its frequency is twice as high as in the transverse one. This means that the component of the nonstationary hydrodynamic force acting across the flow varies both in magnitude and direction. At the same time, the component of this force along the flow changes only in magnitude, while its direction remains the same. Nonstationary hydrodynamic forces, arising due to vortex separation, can excite high-amplitude vibrations of tubes if the natural frequencies of their vibrations coincide with the frequency of vortex separation or are twice as high.

The root-mean-square value of amplitude of tube vibrations, in the transverse direction relative to the flow, which are excited by separation of vortices, is calculated from the equation

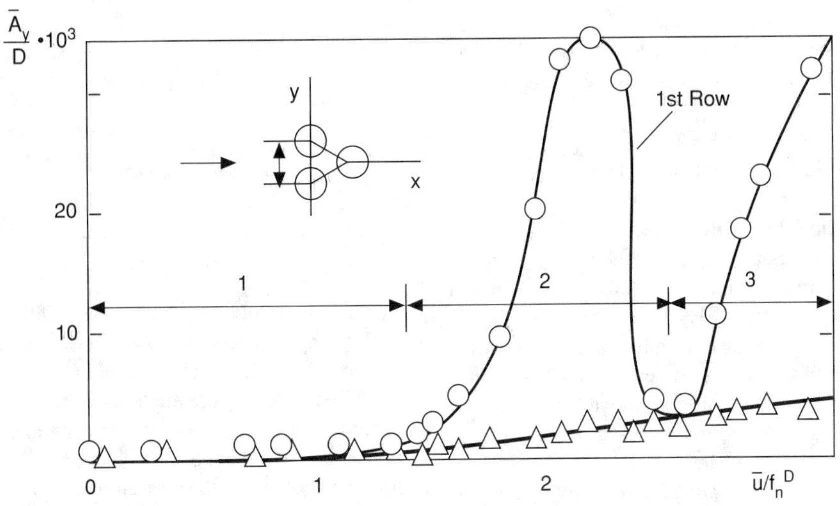

Figure 1. Variation of vibration amplitude with flow velocity.

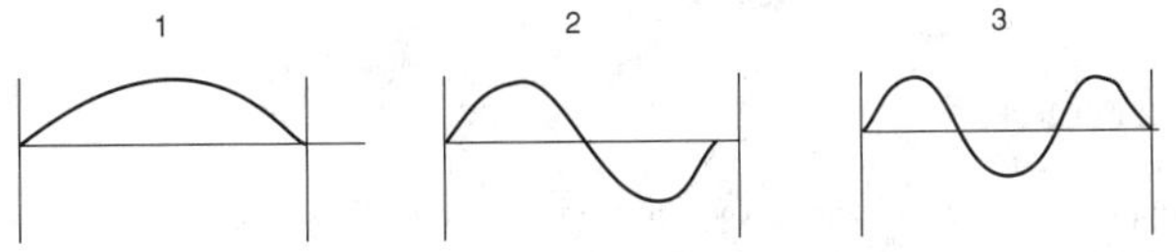

Figure 2. Mode shapes in tube vibration.

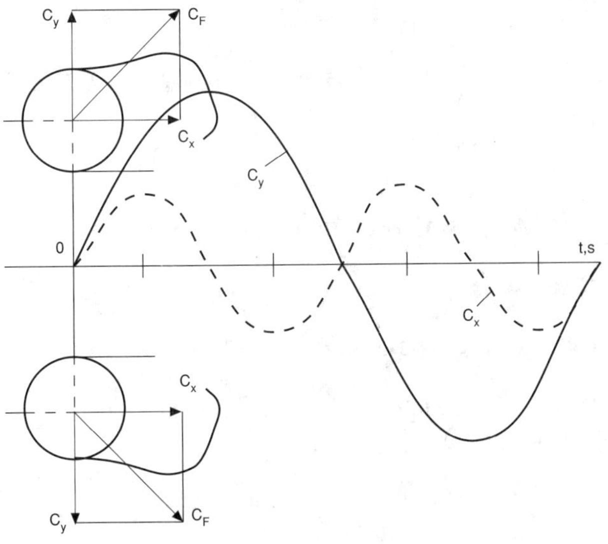

Figure 3. Vortex shedding in a tube bank.

$$\frac{A_y}{D} = \frac{0.5\bar{C}_y\rho\bar{u}^2}{4\pi^2 f_n^2 m\sqrt{[1-(f_s/f_n)^2]^2(\delta/\pi)^2(f_s/f_n)^2}}. \quad \textbf{(3)}$$

The frequency of vortex separation from tube surface f_s is determined by the characteristic value of the Strouhal number $Sr = D f_s/\bar{u}$, i.e.,

$$f_s = Sr\frac{\bar{u}}{D}. \quad \textbf{(4)}$$

In a practically important range of working velocity variation in the flow in heat exchangers of power plants, which is characterized by the variation of Re numbers from 10^3 to 2×10^5, the Strouhal number, determining the frequency of vortex separation from the tubes, is calculated as follows:

$Sr \simeq 0.2$ for a single tube,

$$Sr = 0.2 + \exp[-1.1(b_1/D)^{1.6}] \quad \textbf{(5)}$$

for a single cross row,

$$Sr = \chi\{0.2 + \exp[-4.4(b_1/D)^{1.8}]\} \quad \textbf{(6)}$$

for a staggered tube bank at $b_1/D \geq 1.15$, where

$$\chi = 0.9(b_1/b_2) \quad \text{at} \quad b_1/b_2 > 1,$$

$$\chi = 0.9(b_1/b_2)^{1.7} \quad \text{at} \quad b_1/b_2 \leq 1,$$

and

$$Sr = 0.2 + \exp[-1.2(b_1/D)^{1.8}] \quad \textbf{(7)}$$

for an in-line tube bank at $b_1/D \geq 1.15$.

Hydroelastic (or "fluid-elastic") vibrations of tubes in banks prevail at high flow velocities. They arise as a result of hydrodynamic forces which originate as a result of the vibration itself. The larger the amplitude of the vibration, the larger the force and, hence, a rapid increase in vibration amplitude with velocity occurs in the region. The self-amplified characteristic is often given the name "fluid-elastic instability." In transverse flow over tubes the flow velocity which equalizes hydrodynamic exciting and damping forces and gives rise to hydroelastic vibrations is known as the critical velocity. A slight increase of velocity above this value sharply increases vibration amplitudes and leads to tube failure. For banks with equilateral triangular and square arrangement of tubes this velocity is calculated as

$$\bar{u}/(f_n D) = [0.8 + 1.7(S_1/D)][m\delta/(\rho D^2)]^{0.5}. \quad \textbf{(8)}$$

where S, is the longitudinal tube pitch and δ a damping factor (logarthemic decrement). Tube damping limits the vibration amplitude. It consists of hydrodynamic damping due to tube material and damping for structural reasons. Hydrodynamic damping is attributed to viscous forces appearing during interaction of the tube with the flow. When the tubes vibrate the energy also dissipates in the surrounding fluid since its particles move. Damping for structural reasons is attributed to friction forces appearing in tube constant and rotation in the holes of the supports.

In real *heat exchangers* an overall *damping* of tubes in assemblies is determined by the logarithmic decrement $\delta = 0.15 - 0.3$. Commonly hydrodynamic damping constitutes about 50% of the overall damping and depends on the bank configuration and the pitch ratios. The lower the pitch ratio, the stronger the damping. Hydrodynamic damping of a single tube in a stagnant water on the average is determined by the logarithmic decrement $\delta \simeq 0.05 - 0.1$. In the bank with pitch ratios $b_1/D \times b_2/D = 1.15 \times 0.98$ hydrodynamic damping is nearly three times that in the case of a single tube and in the 2.0×1.77 bank it is approximately identical to that for a single tube.

In heat exchangers with gas flow, high amplitudes of tube vibration or noise may arise if the natural frequency of transverse vibrations of the gas column coincides with the frequency of vortex separation and with natural frequencies of tube vibrations. This is referred to as *acoustic vibration.* Natural frequencies of transverse vibrations of gas column are calculated by the formulas

$f_n = nu_{sound}/2h$ for rectangular channels, where h is the channel width, $n = 1, 2, \ldots$, u_{sound} the sound velocity in a coolant,

$f_n = \alpha_n u_{sound}/d_1$ for circular channels, where d_1 is the inner diameter of the channel,

$\alpha_n = 0.59, 0.97, 1.34, 1.69, 2.04, \ldots$.

To avoid acoustic vibrations, the natural vibration frequency of the transverse gas column must be separated from the frequency of vortex separation by no less than 20%. A similar tuning of the vortex separation frequency away from the tube's natural frequency is needed to avoid high vibration amplitudes occurring due to vortex excitation.

The highest vibration amplitudes are caused by fundamental natural harmonic vibrations. The tubes are damaged mechanically due to collision of vibrating tubes, attrition against hole edges in the interspacing tube plates, tightness faults in stiff joints, and fatigue failure as a result of cyclic load.

In general, vibration diminishing of streamlined heat-transmitting elements is achieved by (e.g. decreasing the flow velocity) clearing of faults in damaged tubes, rearrangement of tubes to make passages for a coolant, replacement of damaged tubes by bars, mounting of regulating baffles at the coolant inlet in the heat exchanger, and appliances reducing damping and tube stiffness. Bearing in mind that the strongest vibrations are observed in the first rows of tube assemblies, tube clips in midspan, stiffness-girdling rings, and other appliances reducing vibration amplitude can be mounted.

Table 1.

Fixation of tube ends	The value of λ_n
Hinge-hinge	πn
Hinge-stiff fixation	$\frac{\pi}{4}(4n + 1)$
Stiff fixation-stiff fixation	$\frac{\pi}{2}(2n + 1)$

When choosing the method of prevention of tube vibration one should take into account advantages and disadvantages of each. For instance, reducing the coolant velocity, and making passages for the coolant in tube assemblies both result in a decrease in heat transfer coefficient. Replacing damaged tubes by new ones in practice is not advisable because vibration level does not change and the tubes rapidly fail again. In this case it is better to substitute bars or higher stiffness tubes, (e.g. steel tubes for brass ones) for the damaged tubes.

References

Chenoweth, J. M. (1983) Flow-induced vibration, *Heat Exchanger Design Handbook,* 4, Hemisphere, New York.

Zukauskas, A., Ulinskas, R., and Katinas, V. (1988) *Fluid Dynamics and Vibration of Tube Banks in Fluid Flows,* Hemisphere, New York.

Blevins, R. (1977) *Flow-Induced Vibration,* Van Nostrand Reinhold Comp., New York.

V.I. Katinas

Leading to: *Vortex shedding*

VIENNA CONVENTION (see Refrigeration)

VIEWFACTOR (see Radiative heat transfer)

VINYL CHLORIDE MONOMER (see Pyrolysis)

VIRIAL COEFFICIENT (see Activity coefficient; Density measurement)

VIRIAL EXPANSION (see Density of gases)

VIRTUAL MACHINE (see Computers)

VIRTUAL MEMORY (see Computers)

VISCO-ELASTIC FLUIDS (see Fluids)

VISCO-ELASTIC SOLIDS (see Non-Newtonian fluids)

VISCOMETER (see Viscosity measurement)

VISCOSITY

Viscosity is that property of a fluid which is the measure of its resistance to flow (i.e. continual deformation). Viscosity can depend on the type of flow (shear and/or extensional), its duration and rate, as well as the prevailing temperature and pressure. Quantitatively, viscosity is defined as the stress in a particular ideal flow-field divided by the rate of deformation of the flow.

In shear flow—where we imagine the flow as hypothetical layers of fluid flowing over each other—we define the relevant parameters as (see **Figure 1**) σ the shear stress (force per unit area) at the boundary of the fluid to produce the flow, and $\dot{\gamma}$ the shear rate (sometimes called the strain rate or velocity gradient), which is the proper measure of the rate of deformation in the fluid undergoing shear flow. The ratio of these two quantities is the viscosity; hence $\eta = \sigma/\dot{\gamma}$. The units of shear stress are Pascals (Pa), shear rate reciprocal seconds (s^{-1}), and so the unit of viscosity is the Pascal-seconds (Pas or Pa.s), with mPas being the more usual unit used for low viscosity fluids. (Prior to the introduction of the SI system, the cgs unit relevant to low viscosity liquids was the centipoise, which is identical to the mPas.) A form of viscosity often referred to is the kinematic viscosity, ν m^2s^{-1} which is the quantity we have defined above divided by the density of the fluid, ρ, ie $\nu = \eta/\eta$.

If the viscosity of a fluid does not change with type, time and rate of deformation, we call it a **Newtonian Fluid**, otherwise it is known as a **Non-Newtonian Fluid**. The most common Newtonian liquid is water, and this provides us with the international standard for viscosity. All other viscosity standards are derived by comparison with the internationally accepted standard for water − 1.025 mPas at 20°C. The approximate viscosity of some other Newtonian fluids (mostly at room temperature) is shown in **Table 1**.

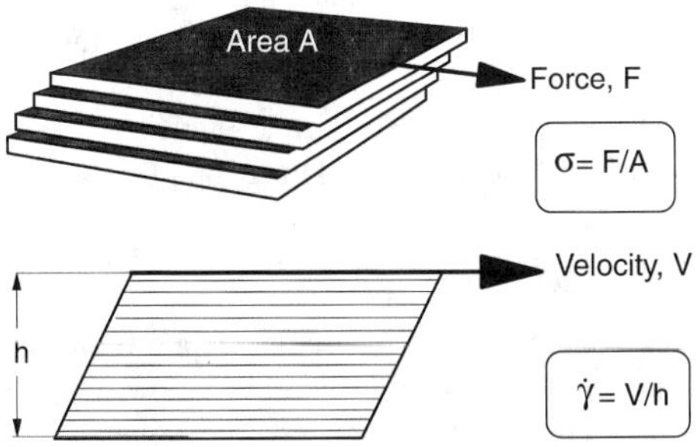

Figure 1.

Table 1. The approximate viscosity of some Newtonian liquids at room temperature

Liquid	Viscosity in Pas
air	10^{-5}
petrol	3.10^{-4}
water	10^{-3}
oil	10^{-1}
glycerol	1
corn-syrup	10^{3}
bitumen	10^{9}

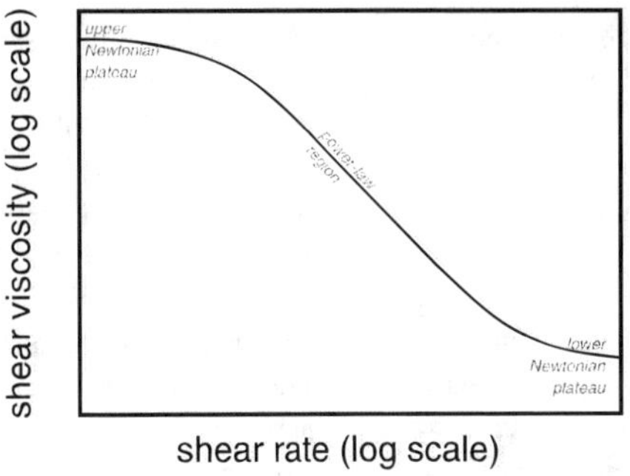

Figure 2.

The temperature dependence of the viscosity of Newtonian liquids is such that the viscosity decreases with temperature, and in general, the higher the viscosity, the greater the rate of decrease with temperature. Around room temperature, the viscosity of water decreases by 3% per degree Celcius; oils by about 5%, and bitumen by 15% or more.

The viscosity of liquids almost always increases with pressure, with water being the sole exception. Its dependence is such that at the normal pressures found in heat and mass transfer operations, it can usually be neglected.

The most important variation of viscosity for non-Newtonian liquids is with shear rate. The (almost) universal behaviour of all such liquids is shown in **Figure 2**. At low enough shear rates, the viscosity is constant, thereafter decreasing through what is called the power-law region, to eventually flatten out a higher shear rates, often to become constant again, but sometimes rising. The absolute position of this curve on the viscosity and shear rate axes depends on the particular non-Newtonian liquid being investigated, as does the slope of the power-law region, as well as the possible upturn at high shear rate. Depending on the range of shear rates available in **Viscosity Measurement** different parts of the curve will be accessed for different liquids. If a standard laboratory viscometer is employed, with a shear rate range of typically 1–1000 s^{-1}, then different types of behaviour might be seen, as shown in **Figure 3**. In **Figure 3a**, the data is plotted linearly as shear stress against shear rate, and transformed in **Figure 3b** to viscosity versus shear rate plotted on logarithmic axes: from 3b, we can identify which part of the universal curve we are on. From the linear plot 3a, we see that for a number of liquids it look as if the stress extrapolates to a yield stress. This would be the stress that has to be exceeded before flow begins. This is a useful approximation for the flow rate curve over the shear rate range of measurement only and this was in fact was the first-ever non-Newtonian law to be expounded.

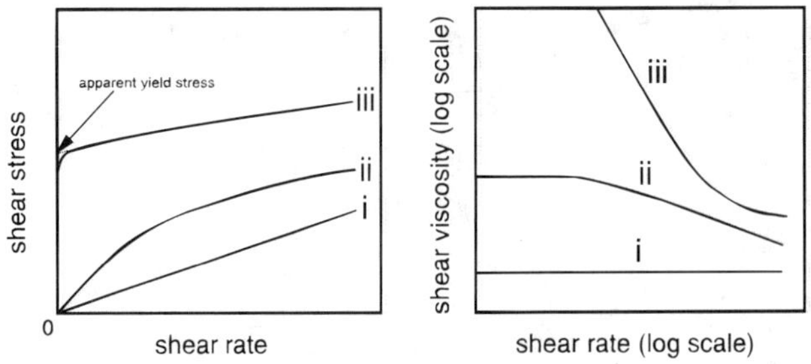

Figure 3.

The viscosity of non-Newtonian liquids usually arises from the presence of at least one dispersed phase in a liquid. The deviation of the flow lines caused by the presence of particles of the dispersed phase increases the resistance, i.e. the viscosity. At higher concentrations, hydrodynamic and physical interactions between suspended particles increases the resistance even further. The flow itself can rearrange the spatial order of the particles, so that the random arrangement at low shear rates can become ordered into strings and layers of particles at high shear rates. This results in the lowering of the viscosity with increasing shear rate. It is possible for this ordered arrangement to become disturbed and some degree of three-dimensional structure returns, either as random packing or as clusters. The viscosity in these circumstances increases, and *shear thickening* takes place—this was previously called by the confusing name of *dilatancy.* This description of particles generally holds for dispersions of both solid and deformable particles. The increase in viscosity is particularly large for *flocculated systems,* where the particles attract each other. The decrease in viscosity with increase in shear rate in this case is enhanced by the decrease in size of the flocculated particles.

Qualitatively, the same curve is seen for *polymeric systems* (see **Polymers**) but for different reasons. The extended coils viscosify the liquid phase, but overlap—which leads to entanglements—giving the very high viscosities often seen in polymeric solutions. As the shear rate is increased, the coils can distort to become aligned to the direction of flow and coincidentally the number of entanglements decreases—these factors result in the viscosity decreasing with increase in shear rate. At very high shear rates, it is possible to see an increase in viscosity with polymeric systems, but this is rare.

Viscosity can be a function of the time of shearing. As we have explained above, the microstructure of particles and polymer coils can be changed by the imposed flow, however, this takes time. For systems where the change is great and the times needed to affect them are long, the effect is called *thixotropy.* Just as it takes time to break the structure down, so also it takes time to build it up. This means that the viscosity will decrease towards equilibrium when we increase shear rate, but increase towards equilibrium when we decrease shear rate. This is illustrated in **Figure 4**, and is seen in its most extreme form for clay-thickened liquids like drilling muds and thixotropic paints. (It is possible for the opposite to happen, especially with shear thickening (dilatant) systems, but this is rare: such behaviour is called anti-thixotropy.)

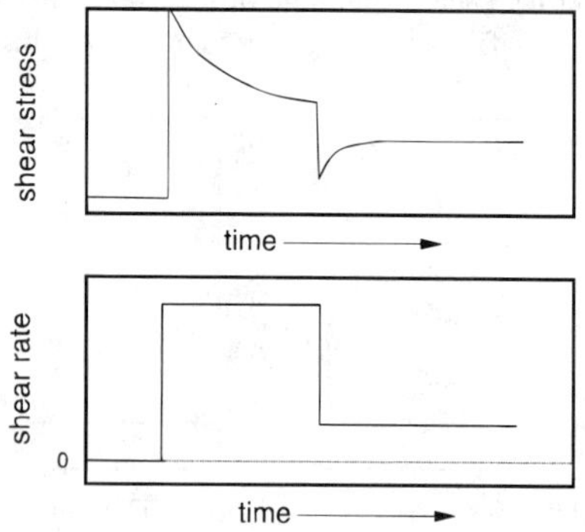

Figure 4.

Extensional flow is where, unlike shear flow the velocity gradient is at right angles to the flow, it is along the flow. For applications in real flow geometries, the most usual example is flow into and out of a constriction, as for instance an orifice plate with a small hole. For liquids where the microstructure is alienable, such as linear polymer chains and fibres etc., it is possible for the viscosity appropriate to this flow—the extensional viscosity—can be orders of magnitude higher than in shear flow.

Other forms of viscosity and their form can be found in "*An Introduction to Rheology*", Barnes, Hutton and Walters, Elsevier, Amsterdam, 1989.

H.A. Barnes

***Leading to:** Viscosity measurement*

VISCOSITY MEASUREMENT

***Following from:** Viscosity*

Introduction

The viscosity of a liquid (see **Viscosity**) is measured using a *viscometer,* and the best viscometers are those which are able to create and control simple flow fields. The most widely measured viscosity is the shear viscosity, and here we will concentrate on its measurement, although it should be noted that various extensional viscosities can also be defined and attempts can be made to measure them, although this is not easy.

Most modern viscometers are computer- or microprocessor-controlled and perform automatic calculations based on the particular geometry being used. We do not therefore need to go into a great deal of discussion of calculation procedures, rather we will concentrate on general issues and artifacts that intrude into measurements. Any other necessary details can be found in chapter Coles, 1965.

General Considerations

The basic components of a viscometer are a suitable simple geometry in or through which a liquid can flow; some means of generating flow, either by the imposition of a velocity on a rotating member or of pressure or a couple and finally a means of measuring the response, either as stress or velocity. These include situations where:

a. the shear rate same everywhere, see **Figure 1**:
   narrow-gap concentric cylinder (including double-gap geometry)
   small-angled cone and plate
b. there is a linear variation in shear rate or shear stress, see **Figure 2**:
   parallel plate
   capillary or tube (straight or U-shaped)
   slit
c. where the shear stress is inversely proportional to distance,
   wide-gap concentric cylinder (including single cylinder)

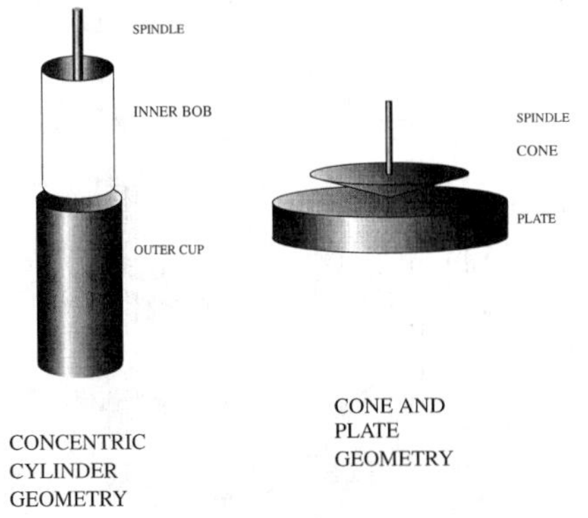

Figure 1. The two geometries where the shear rate is the same throughout the sheared liquid.

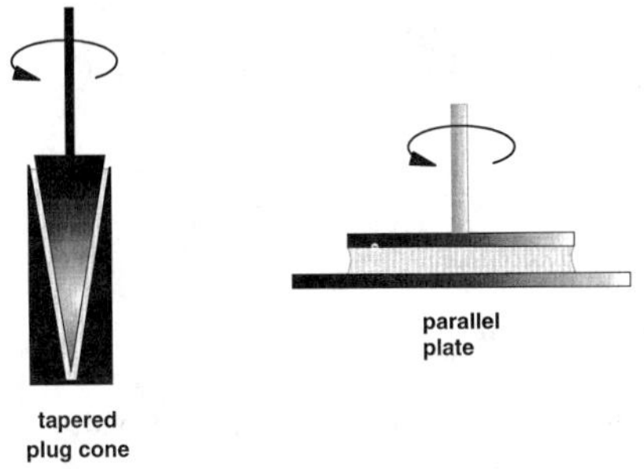

Figure 2. Some other viscometer geometries.

d. others
   inverse cone
   etc.

Viscometers are either stress- or strain-controlled, see **Figure 3** for the typical set-up of these two types of viscometers. For instance, concentric cylinders are either driven at a given rotation rate and the torque (sometimes called the couple) measured, or else a torque applied and the speed measured. In geometries with through-flow such as a capillary tube, the flow can be pressure driven and the ensuing flow measured, or else a flow rate can be established, and the pressure drop down the tube measured.

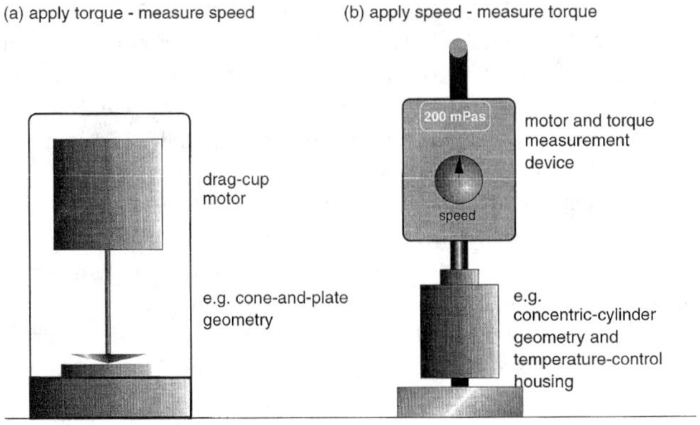

Figure 3. Schematic representation of the layout of typical controlled-stress and controlled-strain viscometers.

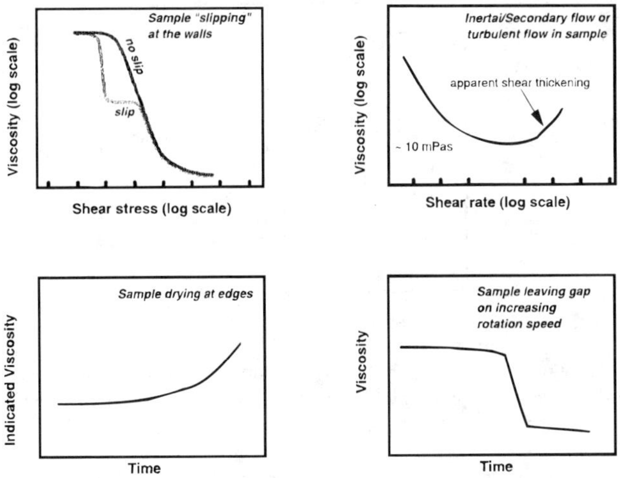

Figure 4. Possible artifacts and their effects on flow curves.

A number of supposedly simple geometries are used to measure viscosity, but although the geometry seems simple, the flow field is not. The best example of this is the flow cup, where liquids runs out from a cup through a given nozzle under the action of gravity. In this case we have shear and extensional flow, plus inertia and time effects present simultaneously, and it is virtually impossible to extract only the shear viscosity as a function of shear stress which is of interest.

In all geometries inertia effects can be important. These manifest themselves in a number of ways depending on the particular geometry. For instance, in circularly symmetric geometries such as the cone and plate, parallel plate and concentric cylinders, it is possible to set up vortex-like secondary flows that absorb extra energy compared with the primary flow and hence display a higher-than-expected viscosity if not taken into account (see below).

Other artifacts can also be present, see **Figure 4**. Even when end effects are eliminated in a well-made viscometer, there can still be wall effects giving real or apparent slip effects. This usually gives a lower-than-expected viscosity (see below). These can be overcome by roughening the surface of the viscometer geometries in contact with the liquid being measured. A roughening of $> 10\mu m$ is usually needed, see **Figure 5** for some examples surface profiling.

Viscometers require careful calibration. This can either be done in a primary way by for instance a careful measurement of the geometry, and calibration of flow rate, pressure or speed gauges, or as is usually done in a secondary manner using a standardised Newtonian liquid—these are available from viscometer manufac-

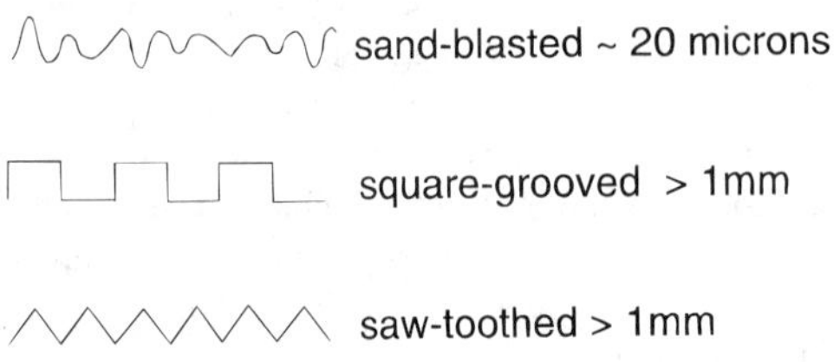

Figure 5. Typical surface profiles to overcome slip at the walls of viscometers

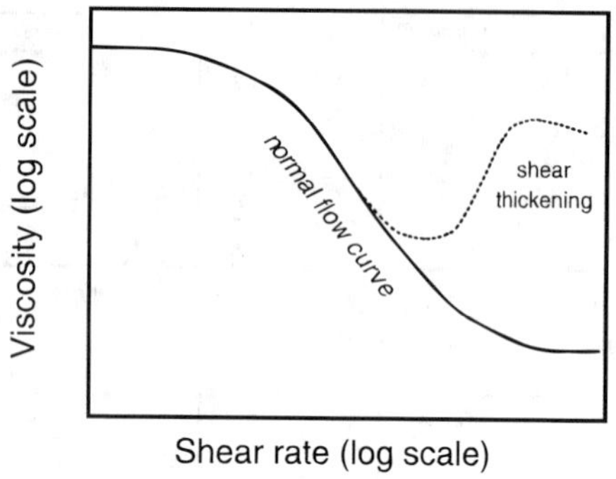

Figure 6. The normal form of the flow curve of non-Newtonian liquids, when measured over a wide enough shear-rate range. Typical shear-thickening behaviour is also shown.

turers, and have been themselves calibrated and certified as being a given viscosity from national standards.

As most commercial viscometers are electro-mechanical in nature, one often encounters zero errors in their operation. With all the various sources of error taken into account, it is not possible to make viscosity measurements on commercial viscometers to an accuracy of better than 2%.

The Form of the Results

Most non-Newtonian liquids have a flow curve as shown in **Figure 6**. The only exceptions are those where we see an increase instead of the usual decrease in viscosity at higher shear rates. These increases are due to structural rearrangement in the microstructure of the liquids concerned, with such behaviour seen in detergent solutions, polymer solutions, and most commonly in concentrated suspensions where it was seen first and was called dilatancy, although shear thickening is the best description, following the normal description of the decrease in viscosity with shear rate as shear thinning.

The lower plateau is sometimes impossible to measure using a simple viscometer, and often using a limited range of investigation, workers have introduced the concept of a yield stress. This is still a useful concept for measurements made over a limit range, and the curve can be described as such mathematically, but it should be remembered that there is always a finite and constant viscosity at low enough stresses.

Other departures from this simple curve are due to various artifacts, as now described.

Inertial and Turbulent Effects

As well as measuring the viscosity of a liquid, viscometers measuring very low viscosities (usually < 10 mPas) can incite inertially driven secondary flows, which give the appearance of an increase in viscosity. **Figure 4** shows the effect on the flow curve. The concentric cylinder is particularly subject to this problem, being the most frequently used for measuring low viscosity liquids when a defined shear rate is needed for non-Newtonian liquids. The onset of this form of secondary flow is seen in this geometry when the inner cylinder is rotated, and is governed by the so-called **Taylor Number,** T_c, and is given by:

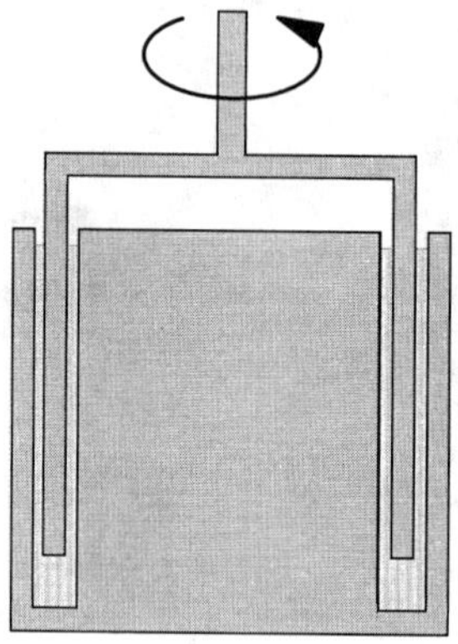

Figure 7. The double-gap concentric cylinder geometry.

$$T_c = 2\left(\frac{R_1}{R_2}\right)^2 \frac{(R_2 - R_1)^4}{\left(1 - \left(\frac{R_1}{R_2}\right)^2\right)} \left(\frac{\rho\Omega}{\eta}\right)^2$$

where
R_1 = is the inner radius
R_2 = is the outer radius
ρ = is the fluid density
Ω = is the rotation speed in radians per second, and
η = is the fluid viscosity.

The value of T_c for small gaps is 1700, but varies with gap as:

$$T_c = 1700\left(1 + \frac{R_2}{10R_1}\right), \qquad \frac{R_2}{R_1} < 1.25$$

so that for a relatively large gap viscometer, for instance with the ratio of radii at 1.25, the critical Taylor Number increases to 1913. This steady secondary flow becomes wavy at a critical value some 50% higher, and at higher speeds becomes completely turbulent. (D. Coles, 1965). (See also **Taylor Instability**).

When the outer cylinder is rotated and the inner is stationery, there is a sharp and catastrophic transition from laminar to turbulent flow, without the vortex stage, however, this occurs at a Reynolds Number of 15,000, where the Reynolds number is defined as: Re $= (R_2 - R_1)R_2\Omega\rho/\eta$. In practice therefore, this is rarely seen in commercial viscometers, which seldom rotate the outer cylinder as a means of generating flow. Other geometries give similar effects, however they are not so often used for low viscosity liquids where the effect is most marked. In all cases, the inertial effect imposes a maximum useable shear rate in the particular geometry.

Turbulence also becomes a problem in capillary or pipe flows. Here the transition from laminar to chaotic flow takes place at a Reynolds number around 1000, where the **Reynolds Number** is given by $RV\rho/\eta$, where R is the radius and V is the average velocity.

When very low viscosity liquids have to be measured, it is often desirable to use a double concentric cylinder arrangement, see **Figure 7**.

The viscosity of gases is best measured in long, small bone tubes applying the appropriate pressure drop relationship for Newtonian fluids.

References

Coles, D, (1965) *J. of Fluid Mechanics*, 21, 391.

H.A. Barnes

VISCOSITY OF AIR (see Air, properties of)

VISCOSITY OF BUBBLY MIXTURES (see Bubble flow)

VISCOSITY OF GASES (see Kinetic theory of gases)

VISCOUS DISSIPATION RATE (see Poiseuille flow)

VISIBILITY (see Optical particle characterisation)

VISIBLE LIGHT (see Electromagnetic spectrum; Electromagnetic waves)

VISUALIZATION OF FLOW

Following from: *Flow of fluids*

The flow of fluids can be analyzed by theory, numerical computation, and experiment. Visualization is one of many experimental tools for surveying or measuring the flow of a fluid that is normally invisible due to its transparency. By applying the methods of flow visualization, a flow pattern is made visible and can be observed directly or recorded with a camera. The information on the flow is available for the whole field of view at a specific instant of time. This information can be either qualitative, thus allowing for an interpretation of the mechanical and physical processes involved in the development of the flow, or quantitative, so that certain properties of the flow (e.g. velocity, density) can be measured. The methods of flow visualization can be classified according to three basic principles: light scattering from tracer particles; optical methods relying on refractive index changes in the fluid; interaction processes of the fluid flow with a solid surface. While the two former methods serve to visualize the pattern in the interior of a flowing fluid, the third method provides information on the transfer of momentum, heat, or mass between the fluid and a solid body.

The scattering and optical methods are based on the interaction of the fluid with light. A light wave incident into the flow field (**Figure 1**) may interact with the fluid in two different way:

- light can be scattered from the fluid molecules or from tracer particles with which the fluid is seeded; and
- the properties of the light wave can be changed due to a certain optical behavior of the fluid and, as a consequence, the light transmitted through the flow is different from the incident light.

The visualization methods based on these two interaction processes are totally different in nature and are applicable to different types of flow.

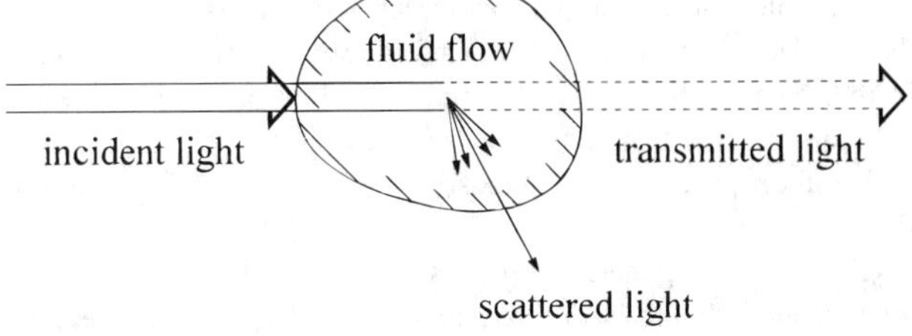

Figure 1. Interaction of a light wave with a fluid flow.

The principles of flow visualization and a great number of respective methods are discussed by Merzkirch (1987). Examples of visualized flow are presented in collections compiled by Van Dyke (1982) and the Japan Society of Mechanical Engineers (1988). The progress in this field is well described in the Proceeding of the International Symposia on Flow Visualization edited by Asanuma (1979), Merzkirch (1982), Yang (1985), Véret (1987), Reznicek (1990), and Tanida and Miyashiro (1992).

Visualization By Tracer Particles

Since the light scattered from the fluid molecules (Rayleigh scattering) is extremely weak, the flow is seeded with small tracer particles (e.g. dust, smoke, dye), and the more intense radiation scattered from these tracers is observed instead. It is thereby assumed that the motion of the tracer is identical with the motion of the fluid, an assumption that does not always hold, particularly in unsteady flows. The scattered light carries information on the state of the flow at the position of the tracer particle, that is, the recorded information is local. For example, if the light in **Figure 1** is incident as a thin light sheet normal to the plane of the figure, an observer could receive and record information on the state of the flow (e.g., the velocity distribution) in the respective illuminated plane.

The signal-to-noise ratio in this type of flow visualization can be improved if the tracer does not just rescatter the incident light but emit its own, characteristic radiation (inelastic scattering). This principle is realized by fluorescent (e.g., iodine) or phosphorescent (e.g., biacetyl) tracers which may emit bright fluorescing (or phosphorescent) light once the fluorescence is induced by an incident radiation with the appropriate wavelength (laser-induced **Fluorescence** or **Phosphorescence** respectively).

That a flow becomes visible from foreign particles that are floating on a free water surface or suspended in the fluid is a fact of daily experience. This crude approach has been refined for laboratory experiments. The methods of flow visualization by adding a tracer material to the flow are an art that concerns the selection of the appropriate tracers, their concentration in the fluid, and the systems for illumination and recording. (See also **Tracer Techniques**).

The trace material after being released, is swept along with the flow. If one does not resolve the motion of single particles, qualitative information on the flow structure (streamlines, vortices, separated flow regimes) becomes available from the observed pattern. The identification of the motion of individual tracers provides quantitative information on the flow velocity, provided that there is no velocity deficit between the tracer and the fluid. Only in the case of fluorescent (or phosphorescent) tracers is it possible to deduce data on quantities other than velocity (density, temperature). Using special optical filters, the radiation of this inelastic scattering can be separated from any other radiation in the test field, which makes the fluorescence methods attractive for studying combustion processes.

Besides some general properties that any seed material for flow visualization should have (e.g., nontoxic, noncorrosive), there are mainly three conditions the tracers should meet: neutral buoyancy, high stability against mixing, and good visibility. The first requirement is almost impossible to meet for air flows. Smoke or oil mist are the most common trace materials in air, with the particle size of these tracers being so small ($< 1\ \mu m$) that their settling velocity is minimized. A number of neutrally buoyant dyes are known for the visualization of water flows, the colors introducing an additional component of information (Merzkirch 1987).

Specific liquids allow a visualization by photochromic reaction. For a short period of time the fluid molecules are converted into an opaque state, and their motion becomes visible. The reaction is reversible, and the fluid is again transparent at the end of this period. The visible fluid particles are neutrally buoyant, but the experiment is restricted to the special properties of these liquids. (See also **Photochromic Dye Tracing**).

Special arrangements of illumination and recording as well as timing are necessary if the goal is to measure the velocity of individual tracer particles. A time exposure (**Figure 2**) is a convenient way for visualizing the instantaneous velocity distribution in a whole field (plane) of the flow. This plane section is realized by expanding a thin (laser) light beam in one plane by means of a cylindrical lens, so that all tracer particles in this plane light sheet are illuminated. The velocity component normal to the plane is not recovered. Recordings of the motion of tracer particles, taken either with a photographic or electronic camera, contain a great number of quantitative data on the planar velocity distribution. Methods employing the technique of digital image analysis have been developed for extracting and presenting the information on the velocity distribution that is often shown in the form of a vector plot (**Figure 3**). These methods are known as *Particle Image Velocimetry* and *Particle Tracking Velocimetry* (Adrian 1991).

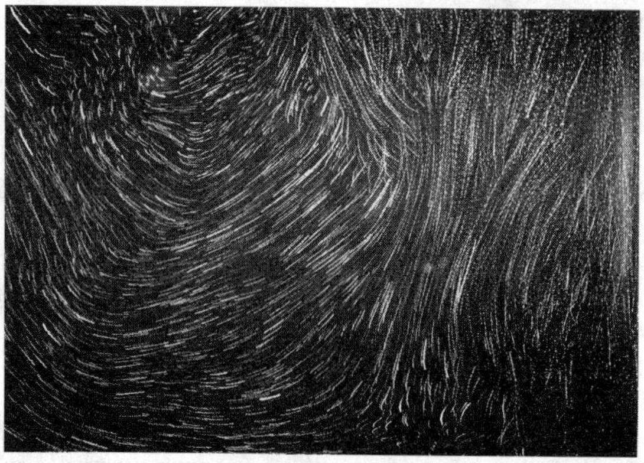

Figure 2. Stroboscopic exposure of tracer particles in a water flow. (S. Hilgers, Universität Essen).

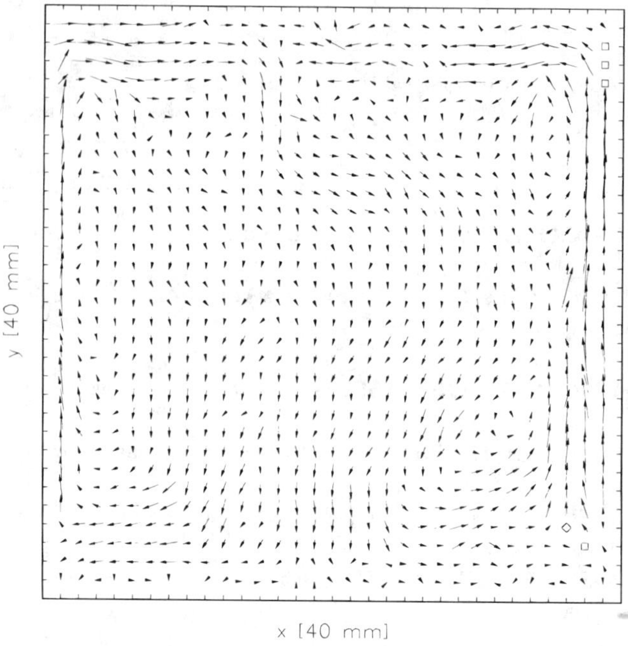

Figure 3. Vector plot of the velocity distribution in a plane of a free convective flow in water as measured with Particle Image Velocimetry.

Visualization By Refractive Index Methods

The properties of the light wave transmitted through the fluid flow as indicated in **Figure 1** may be changed due to changes of the refractive index of the fluid. The refractive index (or index of refraction) is a function of the fluid density. The relationship is exactly described by the *Clausius-Mosotti* equation; for gases, this equation reduces to a simple, linear relationship between the refractive index, n, and the gas density, ρ, known as the *Gladstone-Dale formula* (see, e.g., Merzkirch 1981):

$$n - 1 = K \cdot \rho,$$

where the Gladstone-Dale constant K is different for each gas and weakly dependent on the light wave length. A light wave transmitted through the flow with refractive index changes is affected in two different ways; it is deflected from its original direction of propagation, and its optical phase is altered in comparison to the phase of the undisturbed wave. These alterations of the wave properties can be made visible in a recording plane at a certain distance behind the flow field under study. A particular method requires the use of an optical apparatus transforming the measurable quantity (light deflection, optical phase changes) into a visual pattern. The pattern is either qualitative (*shadowgraph, schlieren*) or quantitative (*moiré, speckle photography, interferometry*), thus allowing for a deduction of data values of the refractive index or density distribution in the flow. (See also **Photographic Techniques**, **Shadowgraph Technique** and **Interferometry**) Refractive index variations occur in a fluid flow in which the density changes, e.g., because of compressibility (high-speed aerodynamics or gas dynamics), heat release (convective heat transfer, combustion), or differences in concentration (mixing of fluids with different indices of refraction).

A standard case of applying optical flow visualization is convective heat transfer. Variations of the fluid density are caused by the respective temperature distribution. **Figure 4** is a *shearing interferogram* of the vertically rising plume (natural convection) above a candle flame that serves as a heat source. The distortion of the oblique, parallel, equidistant fringe system is a measure of the temperature gradient in the ascending gas flow. Quantitative evaluation of the temperature field is in most cases restricted to laminar flow.

Figure 4 is a two-dimensional (plane) projection of a three-dimensional flow field. The information on the density distribution is integrated along the path of the transmitted light("line-of-sight methods"), i.e., the obtained information is not local as in the case of the techniques using light scattering (see **Figure 1**). For the purpose of obtaining quantitative results on the three-dimensional density distribution, it is necessary to record with the optical set-up several projections in different directions through the flow and to process the optical data by methods known as computer *tomography* (Hesselink 1989).

Surface Flow Visualization

The interaction of a fluid flow with a solid body is the subject of many experimental investigations. Such studies are aimed at determining, e.g., the shear forces, pressure forces, or heating loads applied by the flow to the body. A possible means of estimating the rates of momentum, mass, and heat transfer is to visualize the flow pattern very close to the body surface. For this purpose, the body surface can be coated with a thin layer of a substance that, upon an interaction with the fluid flow, develops a certain visible pattern. This pattern can be interpreted qualitatively, and in some cases it is possible to measure certain properties of the flow close to the surface. Three different interaction processes can be used for generating different kinds of information:

Figure 4. Shearing interferogram of the plume above a candle flame. (H. Vanheiden, Universität Essen).

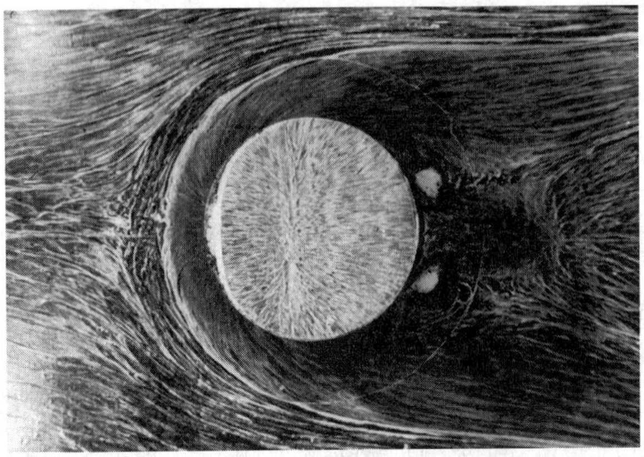

Figure 5. Plan view of the air flow around a vertical circular cylinder as visualized with the oil film technique. Visible is the separation line caused by a horseshoe vortex. (Courtesy N. Hölscher, Ruhr-Universität Bochum).

Mechanical interaction

In the most common technique, which applies to air flows around solid bodies, the surface is coated with a thin layer of oil mixed with a finely powdered pigment. Because of frictional forces, the air stream carries the oil with it, and the remaining streaky deposit of the pigment gives information on the direction of the flow close to the surface. The observed pattern may also indicate positions where the flow changes from laminar to turbulent and positions of flow separation and attachment (**Figure 5**).

Chemical interaction

The solid surface is coated with a substance that changes color upon the chemical reaction with a material with which the flowing fluid is seeded. The reaction, and therefore the color change, is the more intense, the higher the mass transfer from the fluid to the surface. Separated flow regimes with little mass transfer rates can be discriminated from regions of attached flow.

Thermal interaction

Coating materials that change color as a function of the surface temperature (sensitive paints, liquid crystals) are known. Observation of the respective color changes allows for determining the instantaneous positions of specific isothermals and deriving the heat transfer rates to surfaces, which are heated up or cooled down in a fluid flow. Equivalent visible information is available, without the need of surface coating, by applying an infrared camera.

References

Adrian, R. J. (1991) Particle imaging techniques for experimental fluid mechanics, *Ann. Rev. Fluid Mech.*, 23, 261–304.

Asanuma, T., ed. (1979) Flow visualization, *Proc. 1st Int. Symp. Flow Visualization,* Hemisphere, Washington, D.C.

Hesselink, L. (1989) Optical tomography, *Handbook of Flow Visualization,* W.-J. Yang, ed., Hemisphere, Washington, D.C.

Japan Society of Mechanical Engineers, ed. (1988) *Visualized Flow,* Pergamon, Oxford.

Merzkirch, W. (1981) Density sensitive flow visualization. *Fluid Dynamics,* Vol. 18 of Methods of Experimental Physics, ed, R. J. Emrich, Academic Press, New York.

Merzkirch, W., ed. (1982) Flow visualization II, *Proc. 2nd Int. Symp. Flow Visualization,* Hemisphere. Washington, D.C.

Merzkirch, W. (1987) *Flow Visualization,* 2nd ed. Academic Press, San Diego.

Reznicek, R., ed. (1990) Flow visualization V, *Proc. 5th Int. Symp. Flow Visualization,* Hemisphere, Washington D.C.

Tanida, Y., Miyashiro, H., eds. (1992) Flow visualization VI, *Proc. 6th Int. Symp. Flow Visualization,* Springer, Berlin.

Van Dyke, M., ed. (1982) *An Album of Fluid Motion,* Parabolic Press, Stanford, California.

Véret, C., ed. (1987) Flow visualization IV, *Proc. 4th Int. Symp. Flow Visualization,* Hemisphere, Washington, D.C.

Yang, W.-J., ed. (1985) Flow visualization III, *Proc. 3rd Int. Symp. Flow Visualization,* Hemisphere, Washington, D.C.

Leading to: *Interferometry; Photochromic dye tracing; Photographic techniques; Shadowgraph technique; Tracer methods*

W. Merzkirch

VISUALIZATION OF STREAMLINES (see Hele Shaw flows)

VOID FRACTION

Following from: *Gas-liquid flow; Multiphase flow*

Void fraction in a **Gas-Liquid Flow** may be defined as:

1. The fraction of the channel volume that is occupied by the gas phase
2. The fraction of the channel cross-sectional area that is occupied by the gas phase.

Normally, in considering void fraction, we use the time-average value (taken over a long period of time), but it should be appreciated that void fraction is fluctuating with time and instantaneous values are also of interest. At a given point in the flow, the local fluid is either gas or one of the other phases. The probability of finding gas at a given point may be determined using local probes (see **Void Fraction Measurement**) and is referred to as the *local void fraction.* Void fraction is related to **Slip Ratio** S by the equations:

$$\varepsilon_G = \frac{U_G}{SU_L + U_G} = \frac{\rho_L x}{S\rho_G(1 - x) + \rho_L x} \tag{1}$$

where U_G and U_L are the superficial velocities of the gas and liquid phases, ρ_L and ρ_G the liquid and gas densities and x the flow quality (see article on **Multiphase Flow** for definitions). For a homogeneous flow, $S = 1$ and:

$$\varepsilon_G = \beta = \frac{U_G}{U_L + U_G} \tag{2}$$

where β is the volume flow fraction of the gas phase.

Usually, however, $S \neq 1$ and one must produce a model which includes differences in velocity between the phases. A variety of such models have been derived, ranging from simple one-dimensional models, to empirical correlations and to more complex *phenomenological models.* Each of these approaches will now be briefly discussed.

One-dimensional flow method

Here, it is assumed that the velocity of the phases and the void fraction are constant across the channel cross-section. A *drift flux* j_{GL} is defined as the gas flux relative to a plane moving along the channel at the total superficial velocity U $(= U_L + U_G)$. Determination of the void fraction involves simultaneous solution of the continuity relationship:

$$j_{GL} = U_G(1 - \varepsilon_G) - \varepsilon_G U_L \tag{3}$$

and a relationship describing the physics of the system:

$$j_{GL} = f(\varepsilon_G, \text{physical properties, flow pattern}) \tag{4}$$

This one-dimensional flow model is used principally for bubble flow systems where the physical relationship is often written in the form:

$$j_{GL} = u_\infty(1 - \varepsilon_G)^a \tag{5}$$

where u_∞ is the rise velocity of a single bubble in a static pool of the liquid phase, and n is an exponent whose value is typically in the range 1.5–2.0. The simultaneous solution of Equations 3 and 5 is illustrated graphically in **Figure 1**. Solutions are possible for both co-current upward flow and co-current downward flow (though the void fraction is considerably higher in the latter case). For the gas flowing upwards and the liquid flowing down, no solutions are possible above a given liquid velocity. Below this, two solutions are possible as shown.

For counter-current flow with the gas going down and the liquid going up, no solutions are possible, of course, as is also shown.

Further details of the one-dimensional model are given by Wallis (1969) and Hewitt (1982).

The one-dimensional flow model can be developed to take account of variations of velocity and void fraction across the channel, and this leads to the **Drift Flux Model**, which is described in a separate article.

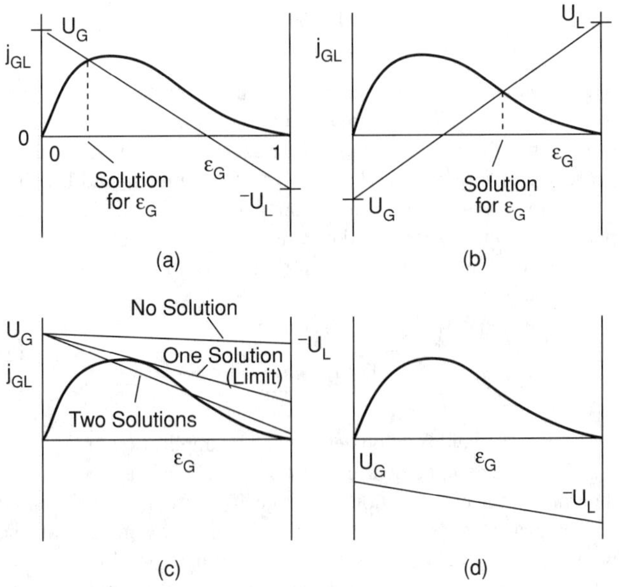

Figure 1. Solutions for void fraction in vertical flow, using the one-dimensional analysis method. (a) Vertical co-current up flow. (b) Vertical co-current down flow. (c) Vertical counter-current flow (liquid down, gas up). (d) Vertical counter-current flow (liquid up, gas down).

Empirical correlations

A vast range of empirical correlations have been developed for void fraction, and it is beyond the scope of the present article to deal with these in detail. One of the earliest correlations (still widely used) is that of Lockhart and Martinelli (1949) who correlated ϵ_G as a function of the Martinelli parameter X which is defined as:

$$X = \frac{\sqrt{(dp_F/dz)_L}}{\sqrt{(dp_F/dz)_G}} \qquad \textbf{(6)}$$

where $(dp_F/dz)_L$ and $(dp_F/dp)_G$ are the pressure gradients for the single phase flow of the liquid phase and the gas phase respectively, flowing alone in the channel. The correlation was in graphical form, and is illustrated in **Figure 2.**

The Lockhart-Martinelli correlation gives an inadequate representation of the effect of mass flux on void fraction. Gross errors

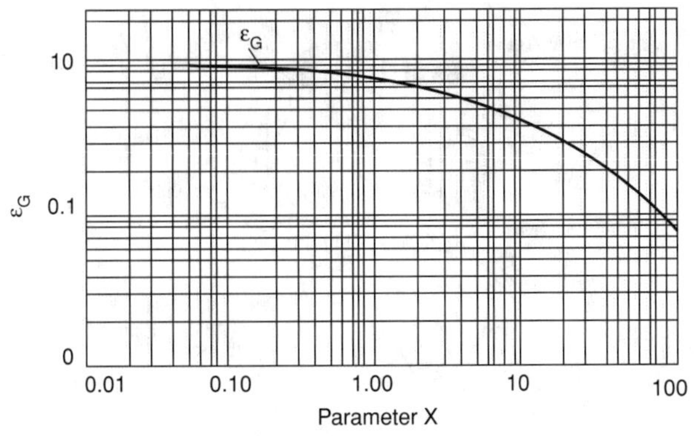

Figure 2. Lockhart and Martinelli (1949) correlation for void fraction.

can occur, and this has lead to the development of a wide variety of alternative correlations. One of the best known of these is the so-called *CISE correlation* (developed at the CISE laboratories in Milan) which is presented by Premoli et al. (1971). The correlation is given in the form of slip ratio S (which can be related to void fraction using Equation 1 above). The correlation has the form:

$$S = 1 + E_1\left(\frac{y}{1 + yE_2} - y_2E_2\right) \qquad \textbf{(7)}$$

where

$$y = \frac{\beta}{1 - \beta} \qquad \textbf{(8)}$$

where β is the volume flow fraction (Equation 2). E_1 and E_2 are given by:

$$E_1 = 1.578\ Re^{-0.19}\left(\frac{\rho_L}{\rho_G}\right)^{0.22} \qquad \textbf{(9)}$$

$$E_2 = 0.0273 We Re^{-0.51}\left(\frac{\rho_L}{\rho_G}\right)^{-0.08} \qquad \textbf{(10)}$$

where:

$$Re = \frac{\dot{m}D}{\eta_L} \qquad \textbf{(11)}$$

$$We = \frac{\dot{m}^2 D}{\sigma\rho_L}. \qquad \textbf{(12)}$$

where $\dot{m}$ is the mass flux, D the diameter of the tube, η_L the viscosity of the liquid and σ the surface tension.

As will be seen from the above, a large number of empirical factors have been introduced into this correlation in order to make it fit the data available at the time of its preparation. Data is constantly being generated and the empirical correlations become ever more complex as they are adjusted to cope with the widening data base. A more recent correlation, written in terms of the drift flux parameters (see article on **Drift Flux Models**) is that of Chexal and Lellouche (1991), and this contains over twenty arbitrary constants! Nevertheless, it is a very good representation of the data base, which justifies its wide utilisation.

Phenomenological models for void fraction

In an attempt to avoid the arbitrary nature of empirical correlations, many investigators have attempted to develop *phenomenological models* for void fraction and other parameters in two-phase flows. The first step is to identify the flow pattern or flow regime (see entry on **Gas-Liquid Flow**). Having identified the flow regime, it is then possible to construct a detailed model for the given flow configuration. Examples of such phenomenological modelling are given in the entries on **Plug Flow, Annular Flow, Bubble Flow, Stratified Flow, Slug Flow** and **Churn Flow.**

References

Chexal, B. and Lellouche, G. (1991) Void fraction correlation for generalised applications, Nuclear Safety Analysis Centre of the Electric Power Research Institute, Report NSAC/139.

Hewitt, G. F. (1982) Void fraction, *Handbook of Multiphase Systems,* Ch. 2.3 G. Hetsroni, ed., McGraw-Hill, New York. ISBN 0-07-028460-1

Lockhart, R. W. and Martinelli, R. C. (1949) Proposed correlation of data for isothermal two-phase, two-component flow in pipes, *Chem. Eng. Prog.,* 45, 39–48.

Premoli, A., Francesco, D. and Prina, A. (1971) A dimensionless correlation for determining the density of two-phase mixtures, *Termotecnica,* 25, 17–26.

Wallis, G. B. (1969) *One-Dimensional Two-Phase Flow,* McGraw-Hill, New York.

Leading to: *Void fraction measurement*

G.F. Hewitt

VOID FRACTION DISTRIBUTION PATTERN (see Bubble flow)

VOID FRACTION MEASUREMENT

Following from: *Gas liquid flow; Void fraction*

Void Fraction is a very important parameter in multiphase flows, and in particular in two-phase gas-liquid flows. A very wide variety of techniques have been developed for such measurements, and extensive reviews are given by Hewitt (1978 and 1982) and, more recently, Leblond and Stepowski (1995). Here, we will briefly review the three most commonly used methods (namely quick-closing valves, gamma absorption and impedance) for average void fraction and the use of local probes for local void fraction measurements.

Quick-closing valve technique. In this method, valves are placed at each end of the section in which the void fraction is to be measured, the flow is set up to the required condition and the valves quickly closed to isolate the section, after which the liquid phase can be drained, and its volume determined. If the volume of the section containing the trapped fluid is known, then the void fraction can be determined. The valves can either be hand-operated (usually with a linking lever to ensure simultaneous action) or can be operated using solenoids. It is obviously preferable if the valves are of the ball type, ensuring continuity of bore in the pipe before the closure action is taken.

Though the measured void fraction is surprisingly insensitive to closure time, there must be an inherent inaccuracy in the method, due to changes of flow pattern while the valves are closing. Of course, another major disadvantage of this method is that the system has to be closed down for each measurement, and then restarted again afterwards.

Gamma ray absorption technique. A beam of gamma rays is attenuated by absorption and scattering according to the exponential absorption equation as follows:

$$I = I_o \exp(-\mu z) \tag{1}$$

where I is the received intensity, I_o is the incident intensity, μ is the linear absorption coefficient and z is the distance travelled through the absorbing medium. To measure void fraction, the usual procedure is to determine the received intensities I_L and I_G when the beam passes through the channel full of the liquid and gas phases respectively. The void fraction is then usually related to the intensity for the flowing situation (I) as follows:

$$\varepsilon_G = \frac{\ln I - \ln I_L}{\ln I_G - \ln I_L} \tag{2}$$

Of course, the void fraction determined by application of Equation 2 is the average value over the path length. This is not necessarily the same as the cross-sectional average. This problem can be eased by using a beam traversing system, followed by integrating the void fraction over the cross-section, or by using multiple beams; a typical three-beam gamma densitometer is shown in **Figure 1**.

Although widely applied, the gamma densitometer has a number of serious limitations. Care has to be taken to ensure that radiation safety precautions are properly observed. There is a fundamental statistical limitation in the gamma absorption method: if the count rate is R counts per second, then the standard deviation σ_R in the count rate, as a fraction of R, is given by:

$$\frac{\sigma_R}{R} = \sqrt{\frac{1}{\tau R}} \tag{3}$$

where τ is the counting time. Obviously, the fractional standard deviation goes down with increasing count rate and increasing count time, but can never be less than is given by Equation 3. Thus, there are often compromises to be made in terms of counting time, and in the size of the sources used, which clearly govern the value of R. Another major source of error is associated with flow patterns in the channel. Thus, Equation 2 only applies if the phases are homogeneously mixed, or if they exist in successive layers perpendicular to the beam. If the liquid and vapour exist in layers parallel to the beam, then the void fraction is given by:

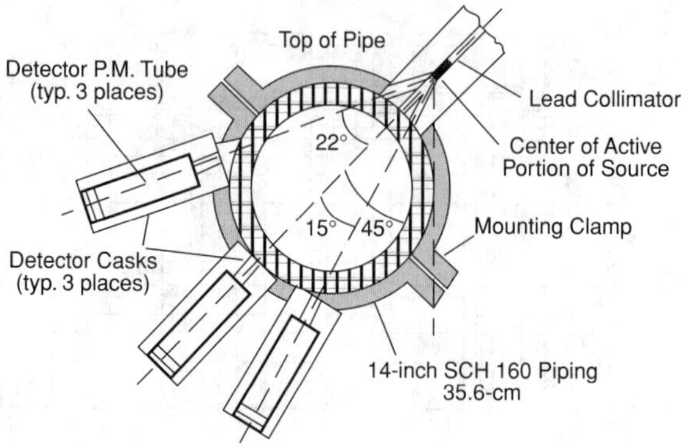

Figure 1. Three-beam gamma densitometer system developed at INEL. (Hewitt, 1982)

$$\varepsilon_G = \frac{I - I_L}{I_G - I_L} \quad (4)$$

Obviously, this case is somewhat unlikely, but in most real systems there is a definite effect of void orientation. An extreme case is that of **Slug Flow** where, on a time-average basis, the void is fluctuating between two extremes corresponding to the slug and bubble regions. This fundamental problem can be to some extent overcome by applying Equation 2 over short time intervals, and then to calculate an average void fraction from the succession of values generated. The intervals should clearly be sort compared to the fluctuation frequency within the system, and this sometimes gives problems in statistical accuracy, as determined by Equation 3. This is a fundamental problem with the gamma absorption method which has to be realised up front and faced up to!

Impedance method. The principle of the impedance method is to measure the capacitance or conductance (or both) of the two-phase mixture, determined between electrodes which are placed around or within the flow. A whole variety of electrode systems have been proposed, ranging from plates mounted within the flow to successive ring electrodes mounted flush with the wall and multiple pairs of electrodes on the tube wall, between which the capacitance is measured in succession (see Hewitt, 1978, 1982 for review and details). The relationship between ϵ_G and admittance (the reciprocal of impedance) A is often calculated from the Maxwell (1881) equations: for a homogeneous dispersion of gas bubbles in the liquid, we have

$$\varepsilon_G = \left[\frac{A - A_c}{A + 2A_c}\right]\left[\frac{C_G + 2C_L}{C_G - C_L}\right] \quad (5)$$

where A_c is the admittance of the gauge system when immersed in the liquid phase alone, and C_G and C_L are the gas and liquid conductivities if the conductivity is dominating, and the dielectric constants of the gas and liquid if the capacity is dominating. For liquid droplets dispersed in a gas, the Maxwell equations give:

$$\varepsilon_G = 1 - \left[\frac{AC_L - A_cC_G}{AC_L + 2A_cC_G}\right]\left[\frac{C_L + 2C_G}{C_L - C_G}\right] \quad (6)$$

Equations can also be derived for other flow configurations (slug flow and annular flow). A comparison of the variation of admittance with void fraction for the various regimes is shown in **Figure 2**. It will be seen that there is a strong flow dependency on the relationship between void fraction and admittance, and this is the major problem with the capacitance technique. Nevertheless, it must be stated that for many systems, measurement of capacitance is a convenient, safe and cheap approach. It gives very rapid time response, and there are many applications where it can be useful.

The sensitivity to flow pattern can be reduced by using modern signal processing techniques, coupled with multiple wall-mounted electrodes, to produce *tomographic* images of the flow cross-section. These can be suitably averaged to give the mean void fraction, but the most important output is information on time variation of the phase content (which is, of course, related to the flow pattern). Work of this type is described by Xie et al (1989).

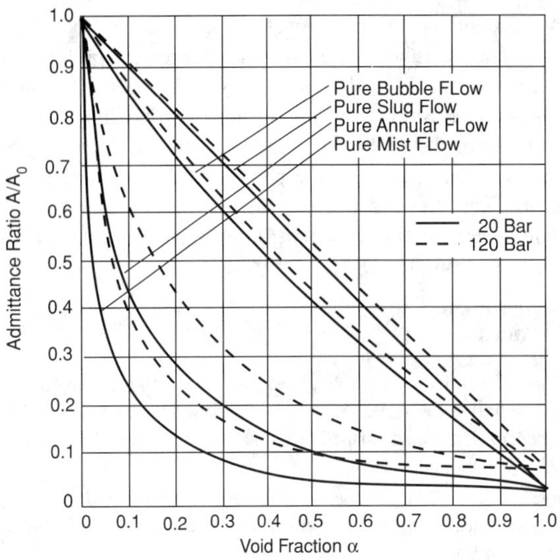

Figure 2. Effect of flow pattern on admittance (calculated by Bouman et al, 1974).

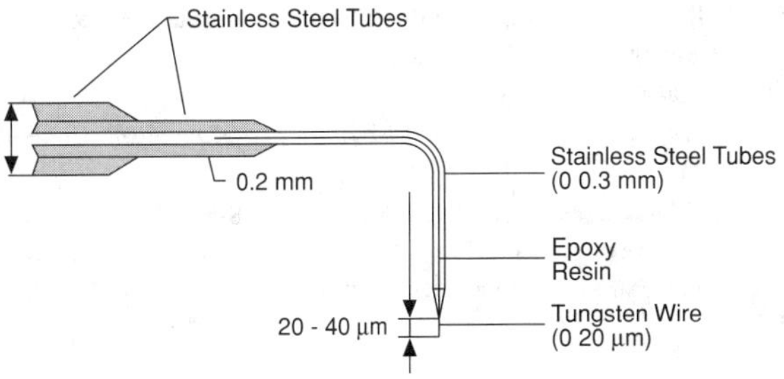

Figure 3. Conductance probe for local void fraction measurement. (Delhaye, 1982)

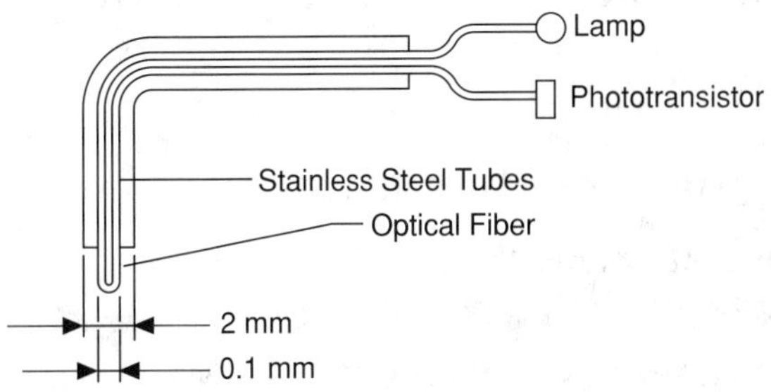

Figure 4. Optical fibre probe for local void fraction measurement. (Delhaye, 1982)

Local void fraction measurements. It is of considerable scientific interest to measure the local value of void fraction in, say, bubbly flow. A review of such measurements is given by Delhaye (1982). Techniques for local measurements include conductivity probes, optical probes and thermal anemometer. Examples of the first two types are shown in **Figures 3** and **4**.

In the conductance probe method, conductance between the probe tip and the tube wall is established when the probe is in the liquid phase, but the conduction path is broken when the probe is in the gas phase. Similarly, with the optical probe, light is released from the tip of the probe when this is in the liquid phase, but is not released (and passes to the photo transistor) when the tip of the probe is in the gas phase. With thermal anemometers, it is possible to make simultaneous measurements of local temperatures and void fraction (see Delhaye, 1982).

References

Bouman, H., van Koppen, C. W. J. and Raas, L. J. (1974) Some investigations of the influence of the heat flux on the flow patterns in vertical boiler tubes, Paper A2, presented at the European Two-Phase Flow Group Meeting, Harwell, England, June 1974.

Delhaye, J. M. (1982) Local measurement techniques for statistical analysis of Handbook of Multiphase Systems, Ch. 10.3 G. Hetsroni, ed., McGraw-Hill, New York.

Hewitt, G. F. (1978) *Measurement of Two-Phase Flow Parameters,* Academic Press, London. ISBN: 0-12-346260-6.

Hewitt, G. F. (1982) Measurement of void fraction. Handbook of Multiphase Systems, Ch. 10.2.1.2, G. Hetsroni, ed., McGraw-Hill, New York. ISBN: 0-07-028460-1.

Leblond, J. and Stepowski, D. (1995) Some non-intrusive methods for diagnostics in two-phase flows, *Multiphase Science and Technology,* Vol 8, Begell House, New York.

Maxwell, J. (1881) *A Treatise on Electricity and Magnetism,* Clarendon Press, Oxford.

Xei, C. G., Plaskowski, A. and Beck, M. S. (1989) 8-Electrode capacitance system for two-component flow identification, Part 1: Thermographic imaging, Part 2: Flow regime identification, *IEE Proc.,* 136 (A) No. 4, 173–190.

G.F. Hewitt

VOLATIZATION (see Sublimation)

VOLTERA INTEGRAL EQUATIONS (see Integral equations)

VOLUME FILLING (see Drying)

VOLUME FORCES (see Archimedes force)

VOLUME MEAN DIAMETER, VMD (see Atomization)

VON KARMAN CONSTANT (see Friction factor; Turbulence)

VON KARMAN STREET (see Tubes, crossflow over)

VON KARMAN, THEODORE (1881–1963)

During the first half of this century Theodore von Karman made major contributions to the understanding of the physics of fluid flow, especially in the field of aerodynamics. His name is given to the Karman Street, a series or vortices shed from alternative sides of a bluff body in cross flow (see **Vortex Shedding**).

Von Karman studied at the Royal technical University of Budapest and the University of Göttingen. In the early 1900's he served as Associate Professor at both, with a break as an officer in the Austro-Hungarian Air Corps. He was Director of the Institute of Aeronautics at the Technical University of Aachen for more than ten years until in 1930 he moved to the USA to become Director of the Guggenheim Aeronautical Laboratories at the California Institute of Technology, where he remained until 1949.

During and after the second world war Von Karman was an advisor to military establishments in the United States and in Europe. He was awarded very many honours. He became a Fellow of the Royal Society in 1946 and a member of the Legion of Honour in 1955.

Among his best known books are *Mathematical Methods in Engineering* (with M A Biot, 1940) and *Aerodynamics—Selected Topics in the Light of their Subsequent Developement,* 1954. The collected papers of Theodore von Karman have been published in 4 volumes.

G.L. Shires

VON KARMAN VELOCITY DISTRIBUTION (see Universal velocity profile)

VON KARMAN VORTEX STREET (see Vortices)

VORTEX BREAKDOWN (see Rotating disc systems, applications)

VORTEX CHAMBER (see Fluidics)

VORTEX EXCITATION OF TUBES (see Vibration in heat exchangers)

VORTEX FLOWMETERS (see Flow metering)

VORTEX GENERATORS (see Augmentation of heat transfer, single phase)

VORTEX MIXER (see Fluidics)

VORTEX SEPARATORS (see Hydrocyclones)

VORTEX SHEDDING

Following from: *External flows; Immersed bodies, flow around and drag; Tubes, crossflow over*

Some flow fields have an oscillatory pattern which is dependent on Reynolds number. The periodic vortex shedding behind a blunt body immersed in a steady freestream provides one example. **Figure 1** shows a sketch of the vortex formation behind a circular cylinder.

A vortex is in the process of formation near the top of the cylinder surface. Below and to the right of the first vortex is another vortex which was formed and shed a short period before. Thus, the flow process in the *wake* of a cylinder or tube involves the formation and shedding of vortices alternately from one side and then the other. This phenomenon is of major importance in engineering design because the alternate formation and shedding of vortices also creates alternating forces, which occur more frequently as the velocity of the flow increases.

When the frequency is in the audible range, a sound can be heard and the body appears to sing. Resonance may occur if the vortex shedding frequency is near the structural-vibration frequency of the body. A dimensionless number, the *Strouhal number* Sr, is commonly used as a measure of the predominant shedding frequency f_S. The definition is

$$Sr = \frac{f_s L}{U_\infty} \quad \textbf{(1)}$$

Figure 1. Vortex formation behind a circular cylinder.

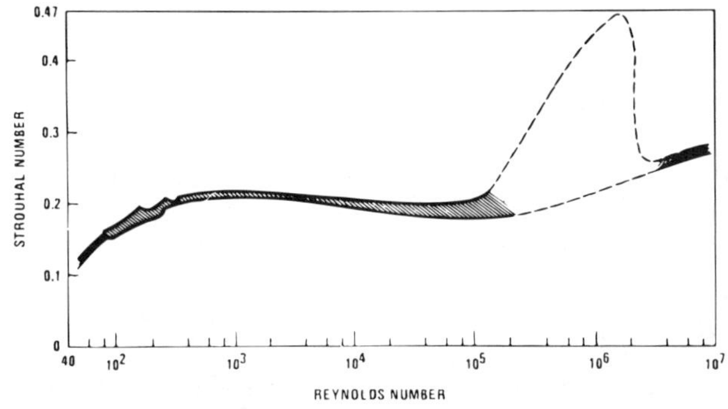

Figure 2. Strouhal number versus Reynolds number for circular cylinders (tubes). From Blevins R. D. (1990) Flow Induced Vibrations, Van Nostrand Reinhold Co.

where L is a characteristic length (equal to the diameter D in case of a circular cylinder or tube in cross flow) and U_∞ the freestream velocity.

The Strouhal number of a stationary tube or circular cylinder is a function of Reynolds number but less of surface roughness and freestream turbulence, see **Figure 2**.

The variation in the Strouhal number is associated with the changes in the flow structure as described elsewhere. (see **Tube Crossflow over**) From **Figure 2** it is found that the Strouhal number is about 0.2 over a large Reynolds number interval. In the Reynolds number range $250 < Re_D < 2 \cdot 10^5$ the empirical formula

$$Sr = 0.198\left(1 - \frac{19.7}{Re_D}\right) \quad \textbf{(2)}$$

is sometimes recommended for estimation of the Strouhal number.

At high Reynolds numbers the vortex shedding does not occur at a single distinct frequency but rather over a narrow band of frequencies.

Non-circular cylinders also shed vortices. It has been suggested to introduce a universal Strouhal number based on the distance between the shear layers. Over a large Reynolds number range a Strouhal number of about 0.2 is then valid regardless of the body geometry.

Vortex shedding also occurs from pair of cylinders, multiple cylinders, arrays of cylinders and heat exchanger tube bundles. However, the pitch-to-diameter ratio (center-to-center distance divided by tube diameter) is important. For two cylinders, placed in-line, vortex shedding occurs behind each cylinder separately if the pitch-to-diameter ratio exceeds a certain value while for smaller values, the two cylinders behave as a single body in terms of vortex shedding. In closely spaced tube bundles, the frequency associated with vortex shedding is not so distinct but might appear as a broadbanded peak. Vortex shedding is one of the mechanisms producing *flow-induced vibration* in shell-and-tube heat exchangers.

References

Blevins, R. D. (1990) *Flow-Induced Vibration,* 2nd ed., Van Nostrand Reinhold.

Chen, S. S. (1990) *Flow-Induced Vibration of Circular Cylindrical Structures,* Hemisphere.

Leading to: *Vibration in heat exchangers*

B. Sundén

VORTEX STREET (see Immersed bodies, flow around and drag; Tubes, crossflow over)

VORTICES

Following from: *Vorticity*

A vortex is a rotating region of fluid such as, for example, a tornado or a whirlpool. These vortices are generally created at a moving boundary due to the shear resulting from the no slip condition, but can also result from thermal circulation. In general, vortices move with the fluid and are dispersed by the action of viscosity. One feature of vortices is that their axes can only terminate at solid boundaries, though they can form closed toroidal loops. Properties of vortices are discussed in Paterson (1983) and Kay and Nedderman (1985).

Two simple examples of vortices are the *free vortex* and the *forced vortex.* A free vortex is one in which the azimuthal component of velocity, v_ϕ, is inversely proportional to the distance from the axis of rotation, i.e. $v_\phi \propto 1/r$. In this case, the vorticity is zero everywhere as is illustrated in **Figure 1a** which shows a fluid element moving in such a flow. Of course, in practice, such a flow cannot exist exactly as it would imply an infinite velocity at the axis. **Figure 1b** shows the case of a forced vortex, for which $v_\phi \propto r$. This type vortex has the property that the vorticity is constant everywhere and equal to twice the angular velocity. Real vortices can often be represented approximately by an inner core which is a fixed vortex, and an outer region where the velocity profile is as for a free vortex—this is known as a *Rankine vortex.*

An example of a situation in which vortices are generated is that of flow past a cylinder. **Figure 2a** shows the presence of two counter rotating vortices behind the cylinder. At higher flow speeds, the vortices become detached and leave the cylinder. This is known as **vortex shedding** and vortices depart from alternate sides of the cylinder, producing a *von Kármán vortex street.* A schematic illustration of the streamlines (in a frame of reference moving with the mean fluid flow) is given in **Figure 2b**.

Vortices are also shed from aerofoils. In this case, the vortices combine to form two *trailing vortices,* which extend from the ends of the aerofoil. This is an important effect in maintaining the lift on an aeroplane. (See also **Crossflow**)

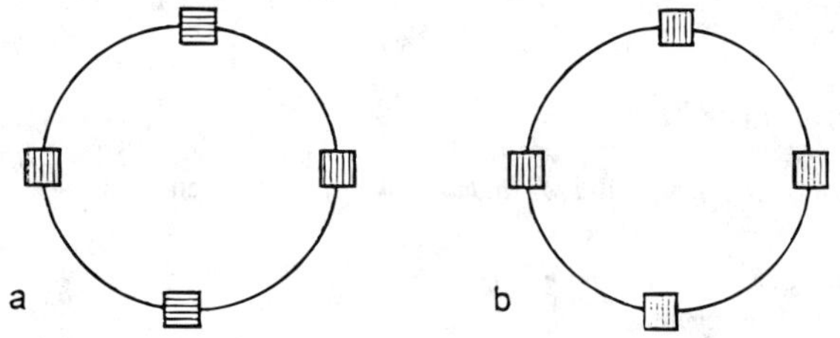

Figure 1. Fluid element moving in circular path.

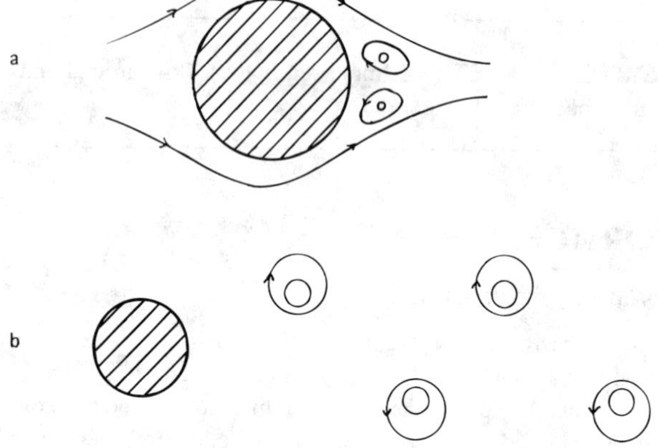

Figure 2. Generation of vortices by fluid flow past a cylinder.

Further information on vortices can be found in Massey (1989) and Acheson (1990).

References

Acheson, D. J. (1990) *Elementary Fluid Dynamics,* Oxford University Press, New York.

Kay, J. M. and Nedderman, R. M. (1985) *Fluid Mechanics and Transfer Processes,* Cambridge University Press, Cambridge.

Massey, B. S. (1989) *Mechanics of Fluids,* Van Nostrand Reinhold, London.

Paterson, A. R. (1983) *A First Course in Fluid Dynamics,* Cambridge University Press, Cambridge.

M.J. Pattison

VORTICITY

Following from: *Flow of fluids; Inviscid flow*

The vorticity, ω, of a flow is a vector quantity which is a measure of the rotation of a flow. It is defined by the relation:

$$\omega = \Delta \times \mathbf{u}$$

It should be emphasised that vorticity corresponds to changing orientation of microscopic fluid elements, rather than just movement in a curved path. A useful equation in fluid mechanics is the *vorticity transport equation.* For an incompressible fluid in a system with conservative body forces, this equation is:

$$\frac{D\omega}{Dt} = (\omega.\Delta)\mathbf{u} + v\Delta^2\omega$$

where D is kinematic viscosity.

The first term on the right represents the interaction of corticity with velocity gradients and the second term reqresents viscous diffusion through the fluid. The concept of vorticity is discussed further in Massey (1989) and Tritton (1977).

References

Massey, B. S. (1989) *Mechanics of Fluids.* Van Nostrant Reinhold, London.

Tritton, D. J. (1977) *Physical Fluid Dynamics.* Van Nostrand Reinhold, Wokinghan, UK.

Leading to: *Vortices*

M.J. Pattison

VORTICITY TRANSPORT EQUATION (see Vorticity)

VULCANISTATION (see Rubber)

W

WAKE (see Crossflow; Cross flow heat transfer; Tube banks, crossflow over; Tubes, crossflow over)

WALLIS CORRELATION FOR FLOODING (see Flooding and flow reversal)

WASHBURN EQUATION (see Contact angle)

WASTE (see Alternative energy sources)

WASTE DISPOSAL (see Incineration)

WASTE HEAT BOILERS (see Heat recovery boilers; Tubes and tube banks, boiling heat transfer on)

WASTE HEAT RECOVERY

Following from: *Energy conservation*

In a typical developed country as much as 40% of total fuel consumption is used for industrial and domestic **Space Heating** and process heating. Of this, around one third is wasted. This wasted heat can be lost to the atmosphere at all stages of the process; through inefficient generation, transmission or final use of that energy. Waste heat recovery aims to minimise the amount of heat wasted in this way by reusing it in either the same or a different process.

Waste heat can be recovered either directly (without using a heat exchanger—e.g. recirculation) or, more commonly, indirectly (via a **Heat Exchanger**). Direct heat recovery is often the cheaper option but its use is restricted by location and contamination considerations. In indirect heat recovery, the two fluid streams are separated by a heat transfer surface, which can be categorised as either a passive or active heat exchanger. Passive heat exchangers require no external energy input (e.g. **Shell and Tube**, **Plate,** etc.) whilst active heat exchangers do (e.g. thermal wheel, **Heat Pump,** etc.).

When considering waste heat recovery, the key question is always that of financial justification: "How much money will be saved?" The decision to recover waste heat depends critically on whether the resulting energy cost savings outweigh the installed cost of the proposed waste heat recovery project. As a general rule of thumb, a waste heat recovery project is unlikely to be installed if its payback period is longer than two or three years.

When describing waste heat recovery it is important to specify the nature of the waste heat in terms of temperature and material phase. Waste heat can be considered as either low grade (<100°C), medium grade (100°C–400°C) or high grade (>400°C). Low grade waste heat can only be recovered effectively when there is a high quantity of waste heat and a ready use for it. There are many examples of successful heat recovery projects for temperatures between 100°C and 200°C. At 200°C and above most users should be able to make significant cost savings from heat recovery. Different techniques and heat exchanger materials are used to recover heat above 400°C.

It can be difficult to recover heat from solid material and therefore most heat is recovered from and passed to material in a gas or liquid phase. The choice of heat exchanger depends critically on both the temperature and material phase.

Many relatively simple processes have a surprisingly large number of potential waste heat sources and sinks. It is often the identification of the sources and sinks and their relative suitability and proximity that determines the cost-effectiveness of any heat recovery project. Some common sources and sinks for processes, utilities and buildings are listed in **Table 1**.

Table 1.

	Process	Utility	Building
Potential Sources:	Exothermic Reaction Dryer Exhaust Oven and Kiln Exhaust Condenser Effluent Streams	Boiler Flue Steam Condensate Furnace Exhaust Compressors Refrigeration System	Ventilation System Exhaust Air Air Conditioning Plant
Potential Sinks:	Process Steam Heat for a Reaction Heat for Separation Reboiler Preheat Process Fluid	Boiler Feed Water Air Preheat District Heating Power Generation	Space Heating Hot Water

In general, high grade waste heat is mainly limited to the iron and steel, glass, non-ferrous metals, bricks and ceramics industries. Medium grade waste heat is most widely found in the chemicals, food and drink, and other process industries, as well as building utilities. Low grade waste heat can be found in virtually all areas of industry and buildings and is often the hardest to recover cost-effectively—typical examples of recovering low grade waste heat would be ventilation or hot water systems.

How do you set about recovering waste heat? Waste heat recovery does not always require high capital investment and in some cases little or no cost is involved. The first and often easiest step is to ensure that the heat is not wasted in the first place. This includes:

- ensuring plant is operating at maximum efficiency;
- reducing evaporation and heat loss from open tanks;
- optimising the scheduling and control of operations;
- making sure there are no leaks in ducts and pipes;
- fitting insulation and ensuring that it is replaced after maintenance.

Good housekeeping is essential before even considering any major capital investment in waste heat recovery. If wasted heat can be limited at source, then recovering it is made that much easier.

If, after such measures have been taken, there is still heat either exhausting to the atmosphere or draining away, under what circumstances is it worth recovering some of it? Any answer to that question, of course, depends very much on the individual circumstances of each site, in particular whether there is a use for any heat recovered. To assess the suitability of any stream for use in a heat recovery system, the following key parameters must be taken into account:

Quantity of Waste Heat

How much heat is available? This can be derived from plant descriptions or by direct measurement. The amount of available heat can be calculated from the following basic energy flow rate equation:

$$\dot{Q} = SV\rho c_p \Delta T$$

Where
$\dot{Q}$ = Energy flow rate (kW)
S = Cross sectional area (m^2)
V = Flow velocity (m/s)
ρ = Fluid density (kg/m^3)
c_p = Fluid specific heat capacity (kJ/kg K)
ΔT = Temperature difference between heat source and heat sink

Grade of Waste Heat

Is the recovered heat at a sufficiently high temperature to be useful? If not, would a heat pump be appropriate?

Requirement for Recovered Heat

What are the requirements for the quantity of heat recovered? Can the recovered heat be put to good use?

Cost of Transfer from Source to Sink

How much will it cost to transfer heat from the source to the sink? Would a run-around-coil be more cost effective? Does the heat source stream contain any contaminants?

Supply and Demand Coincide

Do the patterns of heat availability (at source) and use (at sink) coincide, in terms of both quantity and timing? Is the cost or option of storage acceptable?

Having identified all the potential heat sources and sinks and considered the above parameters for each stream, the next stage is to specify what combination of sources and sinks could cost-effectively be combined. For simple applications with only a few obvious sources of waste heat (e.g. an industrial dryer), this process can be done through basic common sense. However, where a large number of potential heat sources and sinks exist (e.g. a typical chemical site), a more systematic analysis of the options is required. **Process Integration** is a tool to assist in such an analysis.

Put very simply, process integration can be split into 5 basic steps:

1. By carrying out a site survey, derive the temperatures and heat capacity flow rates for each stream and specify whether a stream is a heat source or heat sink.
2. The stream data sets are then combined to produce a hot and cold composite curve for that particular process as shown below (**Figure 1**). This is a plot of the temperature against enthalpy for that process and shows the amount of heat available and required at various temperatures.
3. By combining these two curves onto a single diagram and moving the position of the curves to the point of closest approach (or *"Pinch" point*) represented by the minimum approach temperature of the heat exchanger, the diagram immediately shows the maximum scope for heat recovery within that particular process (**Figure 2**).
4. At this point in the analysis it is possible to consider the impact of changing process conditions to increase the amount of waste heat recovered. This iterative stage can be very important.
5. The final stage is to design the actual heat recovery network to achieve the minimum heating requirement identified in step 3. This network is derived using a few basic rules of process integration, usually via software. The designer can then carry out a cost/benefit analysis on all recommended heat exchangers to produce the most cost-effective final heat recovery network.

Once the hot and cold streams have been identified, suitable heat exchangers must be selected. By far the most commonly used type of heat exchanger is the shell and tube, although a wide variety of other types are also used. A few of the more common heat exchangers are listed in **Table 2**.

These heat exchangers are mainly used to recover medium and low grade heat. To recover high grade heat (i.e. above 400°C; usually from gases) the following equipment is commonly used:

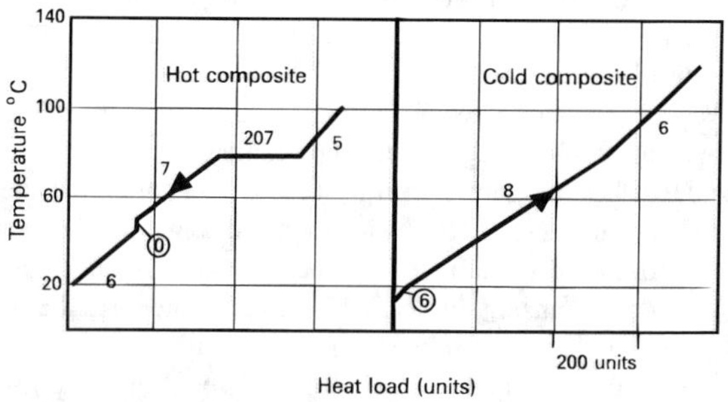

Figure 1. Hot and cold composite curves.

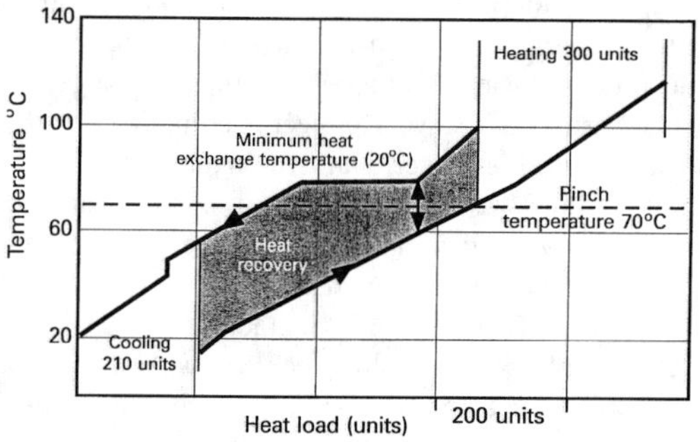

Figure 2. Composite curves and minimum energy target.

Table 2.

Heat exchanger type	Most suitable for . . .		
	Liquid/Liquid	Gas/Liquid	Gas/Gas
Shell & Tube	√	√	
Plate (various forms)	√	√	√
Condensing Economiser		√	
Spray Recuperator		√	
Rotating Regenerator			√
Heat Pipe			√
Run-Around-Coil			√
Rotating Disc	√		
Spiral	√	√	
Printed Circuit	√	√	√

regenerator beds; tube or compact recuperators; self-recuperative burners; regenerative ceramic burners, waste heat boilers.

Typical efficiencies for different types of heat exchangers vary from around 90% for plate and other compact heat exchangers to around 70% for shell and tubes. The efficiency of a heat exchanger can be expressed in terms of temperature as follows:

$$\text{Efficiency} = \frac{(T_4 - T_3)}{(T_1 - T_3)}$$

where: hot stream enters the heat exchanger at T_1 and leaves at T_2 and cold stream enters the heat exchanger at T_3 and leaves at T_4.

The *efficiency of a heat exchanger* is greater when the two streams are arranged in counter rather than parallel flow. Other factors to consider when specifying a suitable heat exchanger are: the **Fouling** characteristics of the two streams; the potential for recovering sensible and latent heat (and the possible **Corrosion** implications of the latter); and the cost of both the heat exchanger itself and the ductwork, etc. necessary for its installation (N.B. the installed cost of a heat exchanger is often around four times the capital cost).

Leading to: *Heat exchangers; Heat recovery boilers; Process integration*

D. Woolnough

WATER CONSUMPTION, DOMESTIC AND INDUSTRIAL (see Water)

WATER CONTENT IN AIR (see Humidity)

WATER COOLED GRAPHITE MODERATED REACTORS (see Nuclear reactors)

WATER COOLED REACTORS (see Nuclear reactors)

WATER CUT (see Three phase, gas-liquid-liquid, flows)

WATER FLOW MEASUREMENT (see Electromagnetic flowmeters)

WATER HAMMER

Following from: *Hydraulics*

Water hammer occurs in a piping system when the flow is suddenly slowed down or stopped. For a water hammer to occur, the closing time of a valve for instance, must be less than the transit time of the resulting pressure wave to travel to the entrance of the pipe and back. The water hammer pressure rise is

$$\Delta P = \rho v_s(\Delta v),$$

in which ρ is the density of the liquid, v_s the velocity of a pressure wave (a sound wave) in the pipe and Δv is the extinguished velocity (Moody, 1990).

For cold water v_s is about 1400 m/s. In a typical pipe, the elasticity of the walls, in effect, reduces the v_s by about 10%. A useful rule of thumb for estimating water hammer pressure is, for each m/s of extinguished velocity, the pressure rise is 10^6 Pa (Avallone, 1987).

If two phases are present, the velocity of sound in the two phase mixture is greatly reduced so that excess pressure due to a water hammer in such a flow is rarely a problem. Water hammer in steam systems is most likely to occur when a vapor bubble is trapped by cold water and condenses very rapidly. Filling or draining a pipe in which cold water replaces steam or steam replaces cold water very often leads to water hammers and damage to piping supports, valves, or the pipes themselves. This occurs because a steam bubble is sometimes trapped in cold water and condenses very rapidly, accelerating and decelerating a column of water very rapidly.

Water Hammers in Steam Piping Systems

Water hammer is not a problem in properly designed and operated steam piping systems. Valves should be stroked slowly and cold liquid and warm vapor should never be allowed to contact each other in such a way that a steam bubble is trapped. However, during start-up, cold condensate is often found in pipes that are normally full of vapor. When vapor is admitted too rapidly into such a pipe, a violent bubble collapse can occur leading to damage to the piping system. Similarly, vapor can be trapped near a leaky valve releasing steam into cold, high pressure water so that a water hammer occurs when a pump, for instance, is turned on.

The following is a partial listing of the geometries and processes which lead to water hammer in improperly designed or operated steam systems (Chou, 1990).

1. Water is admitted into an almost horizontal, steam-filled pipe at a low enough velocity so that the pipe does not run full, a bubble is trapped, it condenses rapidly and causes a water hammer. See **Figure 1** (Chou, 1989).
2. A steam-filled, closed-end pipe of any orientation is filled with cold water rapidly enough so that the liquid hits the end with sufficient velocity to cause a water hammer. See **Figure 2** (Chou, 1989).
3. A vertical steam filled pipe recieves water from above which carries down vapor that condenses rapidly and causes a water hammer.

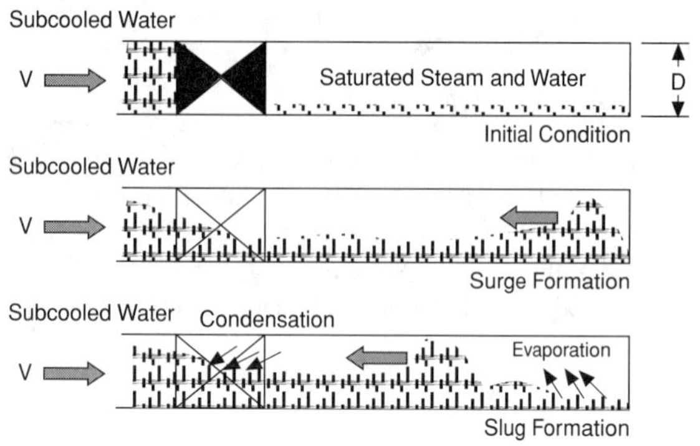

Figure 1. Sequence of events leading to a steam bubble collapse induced water hammer when a short, horizontal steam pipe is filled with cold water with **Froude Number** less than 1. (Froude number is $v\sqrt{gD}$ where v is the liquid velocity and D the tube diameter).

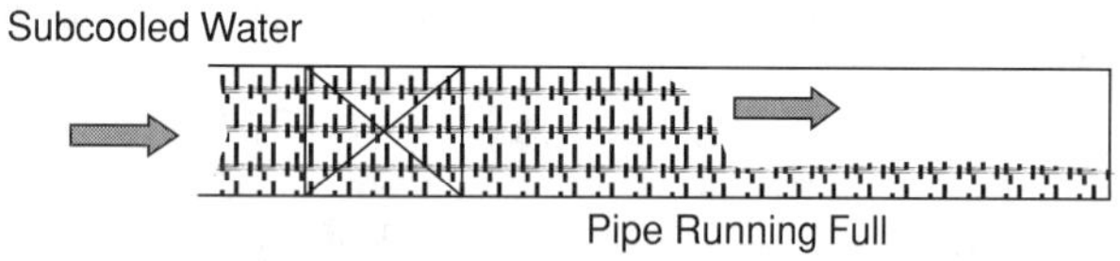

Figure 2. Sketch of a short, horizontal steam pipe being filled with the Froude number greater than one. The sudden deceleration of the flow when the surge hits the end of the pipe causes the water hammer.

4. A horizontal pipe is provided with a source to cold water at one end and a source of steam at the other. For a low enough liquid velocity, a long tongue of cold liquid will extend into the steam, experience a transition to slug flow (due to the high, condensation induced relative velocity of the steam and water) which leads to a water hammer (Bjorge, 1984).

To design water hammers out of a system, incline all nominally horizontal pipes at least 2.8° from the horizontal, admit any cold water from the lowest point in the system, and keep all changes in velocity or pressure gradual enough so that vapor is not trapped and condensed rapidly.

References

Avallone, E. A. and Baumeister, T. III. (1987) *Marks Standard Handbook for Mechanical Engineers,* 9th Ed. 3–71, McGraw-Hill.

Bjorge, R. W. and Griffith, P. (1984) Initiation of water hammer in horizontal and nearly horizontal pipes containing steam and subcooled water, *ASME Journal of Heat Transfer,* 106, 835–840.

Chou, Yuanching and Griffith, P. (1989) Admitting cold water into steam filled pipes without water hammer due to steam bubble collapse, *Pressure Vessels and Piping,* 156, ASME.

Moody, F. J. (1990) *Introduction to Unsteady Thermofluid Mechanics,* Ch. 2, John Wiley and Sons.

P. Griffith

WATER, OVERVIEW

Water is used by municipalities and industries in vast amounts as an engineering material as well as one of life's necessities. It serves as a solvent, suspending and transporting medium, fire extinguisher, heat-transfer agent, source of power and as a chemical reagent. Water outranks all other raw materials. A typical industrial city uses about 1/4 to 1 tonne of water per person per day for domestic and industrial purposes. The U.S. uses 6 t/d and Europe 2 t/d for all purposes. In the heat of summer a person will drink about 5 litres per day in various forms, depending upon level of activity. About 70 tonnes of water are used to make 1 tonne of paper, 1200 tonnes to make 1 tonne of aluminium, 250 tonnes per tonne of steel and 230 tonnes to grow a tonne of tomatoes. Often water is not consumed in the process but its physical state and impurity level may be altered.

Domestic consumption of water has been increasing so rapidly in recent years that governments have become alarmed at the growing scarcity. Incentives are used to curb profligacy and penalties are levied for wastage. Fresh water is not limitless and research focuses on economic methods for obtaining and purifying water (see **Water Treatment**). The U.S. withdraws an estimated 1.54×10^9 tonnes (1993) of fresh water a day from lakes, streams and underground sources (see **Figure 1**).

This is nearly 10% less than in 1980 despite the population increase.

Main Uses

Water is the only substance which can exist in all three phases of matter at "normal" temperatures. Key properties are its: high heat capacity, latent heat of evaporation, polarizability and solvency; having good convective heat-transfer coefficients; and being relatively inert, stable, non-toxic, cheap and abundant (see **Water Properties**).

a. **Heat Transfer Medium** for temperate heating and cooling—food processing, condensers, coolers, cooling towers, quenching.
b. **Steam**. In electricity generation; as a heating medium, since it is an easy form of energy to transport around factories (approximately 2000 kJ/kg depending on pressure) and has good temperature control via pressure regulation; and for inert blanketing.
c. **Raw Material**. As a solvent and chemical reagent (e.g. sulphuric acid manufacture, beverage industry).
d. **Washing.** With its high polarizability, water is a good ionizing solvent.
e. **Miscellaneous.** Hydraulic mining, sluicing, fire control, log de-barking, nuclear shielding, irrigation, sanitary and other domestic uses.

Water Resources

The oceans hold 97% of the planet's water, a further 2% is frozen. The total amount of *rainfall* available worldwide is, however, more than adequate for our purposes.

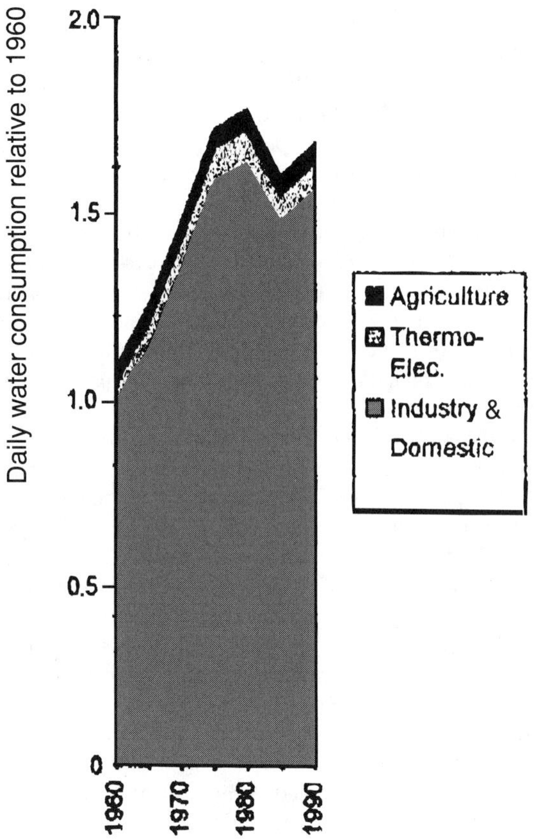

Figure 1. Trends in Freshwater Use in the U.S. 1960–1990. (Graves, 1993)

The problem is one of 'supply' as rainfall is unevenly distributed. **Table 1** lists rainfall and *runoff* for the continents. Australia is the world's driest continent with, not only the lowest rainfall and runoff in proportion to its area, but also the lowest ratio of runoff to rainfall.

Sources of Natural Water

Water is never chemically pure in nature. However, since fresh water is usually better than 99.9% pure it is usual to indicate purity by level of impurities (as mg/L, equivalent to parts per million, or ppm).

Table 1. World rainfall and runoff (See Fenton and Gerofi, 1985)

Continent	Average rainfall, mm/yr.	*Runoff as % of rainfall
Africa	660	24
America—North	660	40
America—South	1350	36
Asia	610	36
Australia	420	13
Europe	580	40

*Runoff—Proportion of rain flowing to the sea; balance evaporates or fills underground aquifers.

Rain

Rain is generally the purest form of natural water, especially if collected towards the end of a shower. Impurities consist of dissolved gases (O_2, N_2, CO_2), salt (NaCl), oxides, and suspended material (dust, pollen, spores).

Surface water

Water from lakes contains larger amounts of dissolved and suspended material than rainwater. If the soil contains gypsum ($CaSO_4\ 2H_2O$) the natural water will contain more calcium; water from swampy areas will contain a variety of organic acids (humic and fulvic) and colour (tannin) from decomposing humus, streams can contain suspended clays or sand, depending on flow rate. Acidic gases from industry (SO_2, SO_3, CO_2, NO_2, NO) lead to the formation of "*acid rain*" which can accelerate the weathering of rocks and soils and increase the quantity of dissolved salts in surface water. Curiously the amount of fine suspended dust from the predominately limestone soils of Western Australia and South Australia result in a somewhat alkaline rain across southern regions.

Springs and rivers

Springs and rivers contain higher quantities of suspended matter than lakes and are usually higher in dissolved material. However, some lakes are noted for their exceptionally high salinity where they are the terminus for streams or rivers with no outlet to the sea, e.g. Lake Eyre (Australia), Dead Sea (Asia)

Wells and bores

Wells and bores often have very high levels of dissolved material even to the extent of becoming saturated with soil salts. However, the clarity is good with little suspended matter and low micro-organism counts. Wells in shallow **Aquifers** can have high biological activity due to surface contamination.

Sea water

The concentration of dissolved salts in sea water is very constant throughout the world's oceans at about 3.5%, or 35,000 ppm (see **Table 2**). Local variations occur where climatic conditions are severe. Thus in the Baltic Sea, the large runoff of fresh water from Scandinavia and Northern Europe reduces salinity, and in the Northern Spencer Gulf of South Australia solar evaporation results

Table 2. Composition of sea water, excluding dissolved gases (See Weast, 1971)

Component	Concentration, mg/L (or ppm)	Component	Concentration, mg/L (or ppm)
Cl^-	18,980	H_3BO_3	26
Na^+	10,556	Sr^{2+}	13
SO_4^{2-}	2,648	F^-	1.3
Mg^{2+}	1,272	I^-	0.05
Ca^{2+}	400	Si	4.0 to 0.02
K^+	380	Others	1.3
HCO_3^-	139	H_2O	965,517
Br^-	64	Total	1,000,000

in a salinity of 45,000 ppm. The usefulness of a water source depends on the nature and concentration of contaminants, as well as its quantity and temperature. The treatment method depends on both source and end use. (See **Desalination; Water preparation**)

References

Considine, D. M. ed. (1974) *Encyclopedia of Chemical and Process Technology,* Mc Graw-Hill, NY.

Fenton, G. G. and Gerofi, J. P. (1985) Desalination in Australia: past and future, *Solar Desalination Group,* Uni. of Sydney.

Graves, W. ed. (1993) *Water, National Geographic Special Edition,* NGS, Washington, Nov.

Weast, R. C. ed. (1971) *Handbook of Chemistry and Physics,* 51 ed., CRC Press.

Leading to: *Desalination; Water properties; Water preparation.*

G.T. Wilkinson

WATER PREPARATION

Following from: *Water, overview*

The usefulness of water as a heat transfer medium is largely due to the ready availability of water in the liquid state at low cost. Unfortunately, the demands placed on water in heat transfer applications can lead to operating problems associated with the various impurities found in most natural water supplies. *Hardness* (calcium and magnesium salt) *scale* encrustation, corrosion and microbiological fouling all stem from failure to properly manage the water impurities and all can lead to inefficiency or premature failure of heat transfer equipment. (See also **Corrosion and Fouling**) The preparation of water for use as a heat transfer medium may utilise various physical and chemical processes to remove or modify the impurities. The process, or combination of processes needed, will depend upon both the condition of the supply water and the nature of the heat transfer process under consideration.

The impurities which have accumulated in water drawn directly from oceans, lakes, rivers or underground aquifers will reflect the history of the water in the period since it last existed as atmospheric water vapour. They may be divided into four main categories. Particulate solids (including micro-organisms) in colloidal or suspended form, dissolved ions of atmospheric or mineral salt origin, dissolved organic matter from decaying vegetation and dissolved gasses (principally oxygen and nitrogen). The *total dissolved solids* (TDS) content of the water may vary from as little as 0.03 kg/m^3 in highland catchment waters, to over 30 kg/m^3 in seawater and other brines.

The standard of purity demanded for heat transfer processes can vary widely. High temperature, high heat transfer rates and evaporative concentration all tend to elevate the level of purity required of water for use as the heat transfer medium. High pressure steam boilers demand feed water with less than 0.001 kg/m^3 dissolved solids whilst recirculatory chilled water systems may require brine concentrations of 100kg/m^3 or more, in order to prevent freezing. (See also **Desalination**).

Purification by physical separation includes coarse screening, settlement, filtration through fixed elements or particulate media beds to remove particulate solids, as well as membrane techniques such as electrodialysis and reverse osmosis, which can remove dissolved ions. Physical separation is often preceded by dosage of coagulants and flocculants to increase particle size of finely divided and colloidal solids thereby facilitating separation. Clarification and removal of particulate solids is commonly the responsibility of the water supplier. (See also **Liquid-Solid Separation, Membrane Processes**).

Purification by precipitative or ion exchange processes removes undesirable mineral salts, or exchanges them for less harmful species. *Precipitative (lime-soda) softening* converts undesirable calcium and magnesium ions into solid precipitates which may be removed by filtration. Base exchange softening utilises immobilised anions within a bed of resin beads to exchange the undesirable calcium and magnesium ions for sodium ions, which form soluble non-scaling salts. Other ion exchange processes utilise the same principle to replace all the mineral ions with hydroxyl or hydrogen ions, thus achieving their complete removal. (See also **Ion Exchange**).

Distillation can remove water from all associated non-volatile impurities in a single step and millions of gallons of water are purified daily by this means, wherever fresh water is scarce and cheap energy is plentiful. Evaporative processes are also important in the preparation of waste water for re-use whilst minimising the effluent volume. (See also **Desalination** and **Distillation**)

Typical percentages of the different classes of influent water impurity remaining after the various processes, are represented in **Figure 1**:

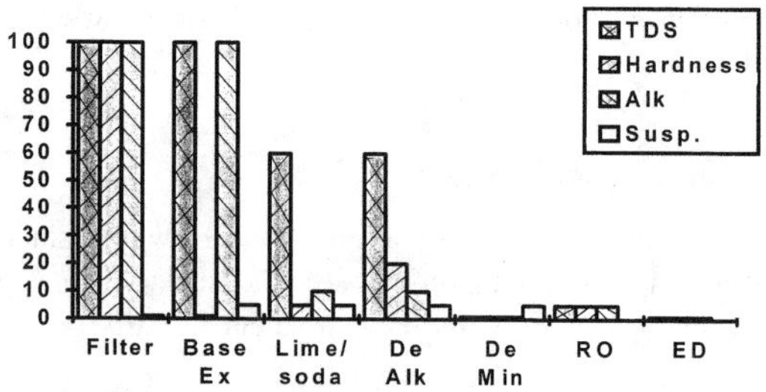

Figure 1.

It must be noted that the prepared water may require the addition of chemical additives such as antiscalents, corrosion inhibitors and microbicides as the final stage of conditioning for successful use as a heat transfer medium.

Leading to: *Desalination; Corrosion; Fouling*

R.W. Eycott

WATER PROPERTIES

Following from: Water, overview

Water is a colourless, odourless and tasteless liquid, stable and nontoxic. Its solid and vapour forms are ice and steam, although the term "water" is sometimes used for all three phases. With a narrow liquid range, it is one of the few chemicals known to exist commonly in all three phases (**Figure 1**).

Physical Properties

Water is a relatively "light" molecule with a molar mass (molecular weight) of 18.01, giving the vapour a density of 0.623 relative to air. The high dielectric constant (**Table 1**) of water is largely responsible for its good solvent properties. It is, however a weak electrolyte when pure, but with dissolved ions increases its conductivity by several orders of magnitude. The molecule's small size (O-H bond length is 96 pm) gives high permeability through membranes (see **Membrane Processes**). In many ways water is an anomalous fluid largely explained by strong hydrogen bonding resulting in the formation of loose clusters of molecules (see also Franks, 1972). It has a high latent heat (see **Steam Tables**) and high melting point, and expands on freezing. Above the critical point (**Table 1**) water is "supercritical" and the distinction between liquid and vapour disappears.

Table 1. Selected properties of water (Horvath, 1975; Perry, 1985)

Chemical formula	H_2O
Molecular weight	18.0148
Critical temperature	373.91°C
Critical pressure	22.05 MPa
Critical density	315.0 kg/m^3
Triple point temperature	0.01°C
Triple point pressure	615.066 Pa
Normal boiling point	100.0°C
Normal freezing point	0.0°C
Density of ice at normal melting point	918.0 kg/m^3
Maximum density, 3.98°C	999.973 kg/m^3
Viscosity, 25°C	0.889 mN s/m^2
Surface tension, 25°C	72 mN/m
Heat Capacity, 25°C	4.1796 kJ/kg.K
Enthalpy of vaporisation, 100°C	2,257.7 kJ/kg
Enthalpy of fusion, 0°C	333.8 kJ/kg
Velocity of sound, 0°C	1.403 km/s
Dielectric constant, 25°C	78.40
Electrical conductivity, 25°C	8 μS/m
Refractive index, 25°C	1.333
Liquid compressibility, 10°C	480. × 10^{-12}m^2/N
Coefficient of thermal expansion, 25°C	256.32 × 10^{-6} K^{-1}
Thermal Conductivity, 25°C	0.608 W/m.K

Table 2. Water properties as a function of temperature

T K	Temperature, Celsius	Pr	c_p, kJ/kg.K	σ mN/m	ρ tonne/m^3	η mNS/m^2	λ, W/m.K
273.15	0	12.99	4.217	75.5	0.999839	1.75	0.569
280	6.85	10.26	4.198	74.8	0.999908	1.422	0.582
285	11.85	8.81	4.189	74.3	0.999515	1.225	0.59
295	21.85	6.62	4.181	72.7	0.997804	0.959	0.606
305	31.85	5.02	4.178	70.9	0.995074	0.769	0.62
315	41.85	4.16	4.179	69.2	0.991495	0.631	0.634
325	51.85	3.42	4.182	67.4	0.98719	0.528	0.645
335	61.85	2.88	4.186	65.8	0.982234	0.453	0.656
345	71.85	2.45	4.191	64.1	0.976706	0.389	0.668
355	81.85	2.14	4.199	62.3	0.970638	0.343	0.671
365	91.85	1.91	4.209	60.5	0.96407	0.306	0.677
373.15	100	1.76	4.217	58.9	0.958365	0.279	0.68

From *Fundamentals of Heat and Mass Transfer*, Incropera F. P. and deWitt D. O. (1990) Wiley.

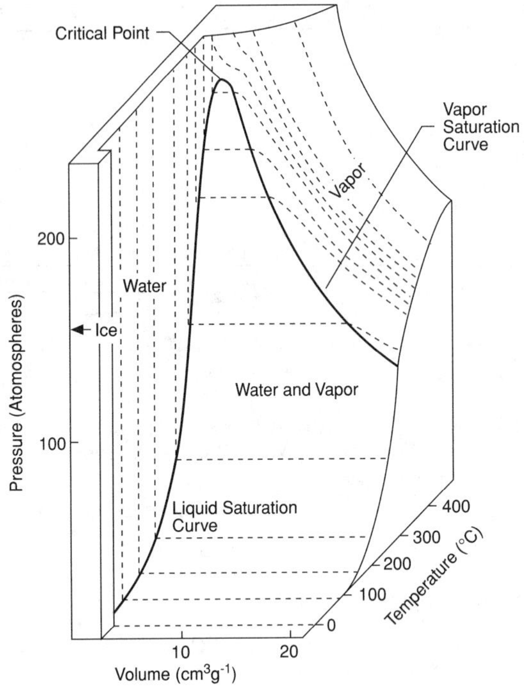

Figure 1. PVT Surface for water From Eisenberg, D. and Kauzmann, W. (1969) The Structure and Properties of Water, by permission of Oxford University Press.

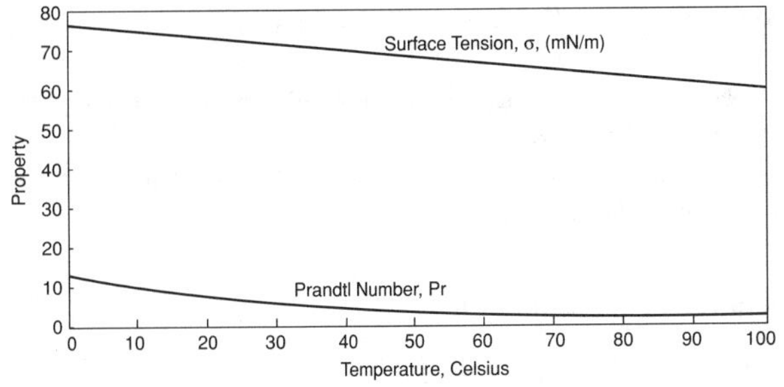

Figure 2. Surface tension σ, mN/m and prandtl Number, Pr, of saturated water. (Incropera & de Witt, 1990)

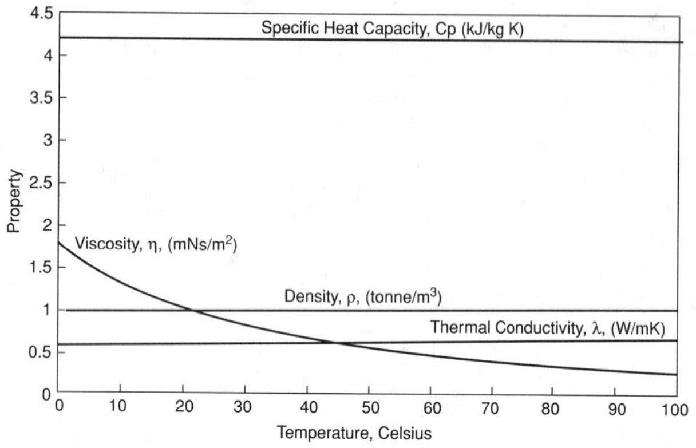

Figure 3. Heat capacity, Cp, kJ/kg.K, Viscosity, η, mN.s/m^2 Density, ρ, t/m^3 and Thermal, Conductivity λ, W/m.K of saturated water. (Perry, 1985c), Incropera and de Witt, 1990)

Chemical Properties

Water forms hydrates with many salts, e.g. $CaSO_4.2H_2O$ (gypsum) and $CaSO_4.{}^1/_2H_2O$ which decompose as temperature is increased, e.g. $CuSO_4.5H_2O \xrightarrow{100\,C} CuSO_4.H_2O \xrightarrow{250\,C} CuSO_4$. In the case of temperature rises, through loss of "water of crystallization", and so scaling of heated surfaces is often more severe with waters containing calcium sulphate.

Water forms hydrates with carbon dioxide and the lower alkanes (CH_4, C_2H_6) at high pressure (>5 MPa) and temperatures typically below 10°C (see Berecz and Balla-Achs, 1983). It is reactive with many metals forming basic oxides (e.g. $Zn \rightarrow ZnO$) or hydroxides (e.g. $Na \rightarrow NaOH$, $Ca \rightarrow CaO \rightarrow Ca(OH)_2$), and with non-metallic oxides giving acids (e.g. $SO_3 \rightarrow H_2SO_4$). Water acts as a catalyst in some reactions and affects the activity of enzymatic catalysts (e.g. porcine lipase has a maximum activity at about 10–12% water).

References

Berecz, E. and Balla-Achs, M. (1983) *Gas Hydrates,* Elsevier.

Eisenberg, D. & Kauzmann, W. (1969) *The Structure and Properties of Water,* Oxford, Clarendon.

Franks, F., ed., *Water, A Comprehensive Treatise,* 1972–c. 1982 Vol. 1–7, Plenum, NY.

Horvath, A. L., (1975) *Physical Properties of Inorganic Compounds,* Arnold.

Incropera, F. P. and de Witt D. P. (1990) *Fundamentals of Heat and Mass Transfer,* 3rd ed., Wiley.

Perry, R. H. and Green, D. ed. (1985) *Perry's Chemical Engineers' Handbook,* 6th ed., McGraw-Hill.

***Leading to:** Desalination*

G.T. Wilkinson

WATER PURIFICATION (see Aeration)

WATER REACTOR SAFETY SYSTEMS (see Nuclear reactors)

WATER RESOURCES (see Water, overview)

WATER RING COMPRESSOR (see Compressors)

WATER SOURCES (see Water, overview)

WATER-TO-AIR HEAT PUMP (see Heat pumps)

WATER-TO-WATER HEAT PUMP (see Heat pumps)

WATER TREATMENT (see Aeration; Ion exchange; Water preparation)

WATER-TUBE BOILERS

***Following from:** Boilers*

This is a class of boiler characterised by water flowing through arrangements of tubes over which the heating fluid is constrained to pass. Water-tube boilers are best suited to relatively high pressure operation (>40 bar) and have good response to changes in load demand.

Tubing is arranged to suit the type of heating fluid, its temperature and pressure, and heat transfer characteristics coupled with the operating conditions of the boiler water. For example, Fossil-fuel fired boilers for power generation are normally of the water-tube type, as are **Heat Recovery Boilers** and those associated with nuclear power stations. Most modern industrial boilers are also are of the water-tube type. Typical layouts of units not specifically described in other entries are shown in **Figures 1** and **2** on page 1293.

In cases where the heating fluid is a liquid, the internal and external heat transfer coefficients can be arranged to be high and plain tubing can produce economic designs. However, when the heating fluid is a gas, the external heat transfer coefficient can be an order of magnitude lower than the internal one and a degree of surface enhancement is desirable on the outside of the tube. Such enhancement is normally achieved by use of helical (either continuous or serrated) fins in clean gas environments as in heat recovery boilers or by use of large plate fins as are often found in economisers when heat is being recovered from dust-laden gases (see **Figure 3** on page 1294). Illustrations are by permission of Mitsui Babcock Energy Ltd.

A.E. Ruffell

WATER UTILISATION (see Water)

WATT ENGINES (see Steam engines)

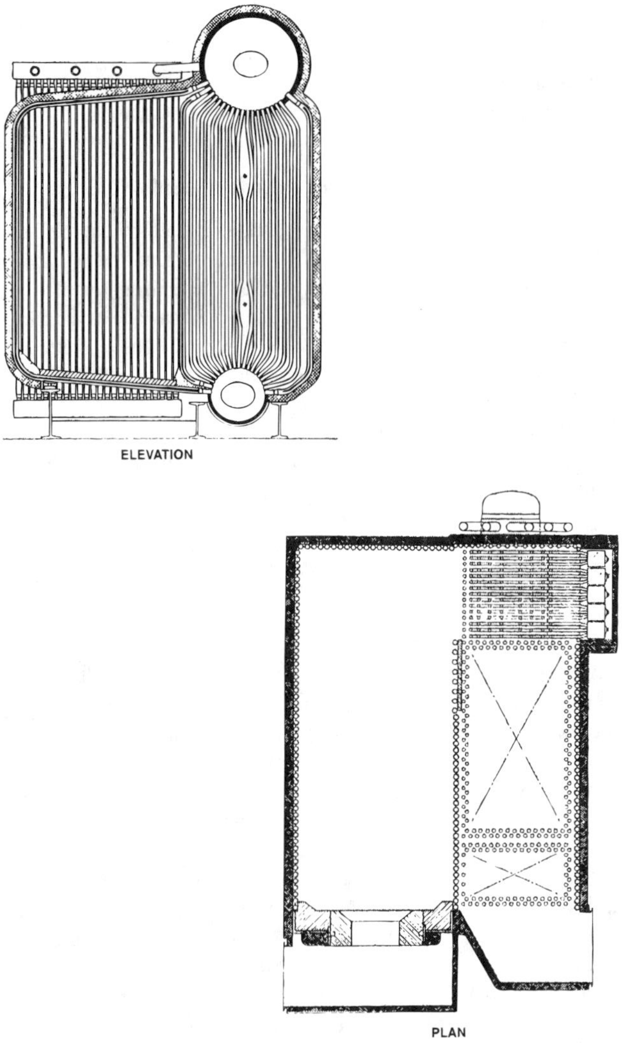

Figure 1. A typical industrial boiler (Babcock D-5 type).

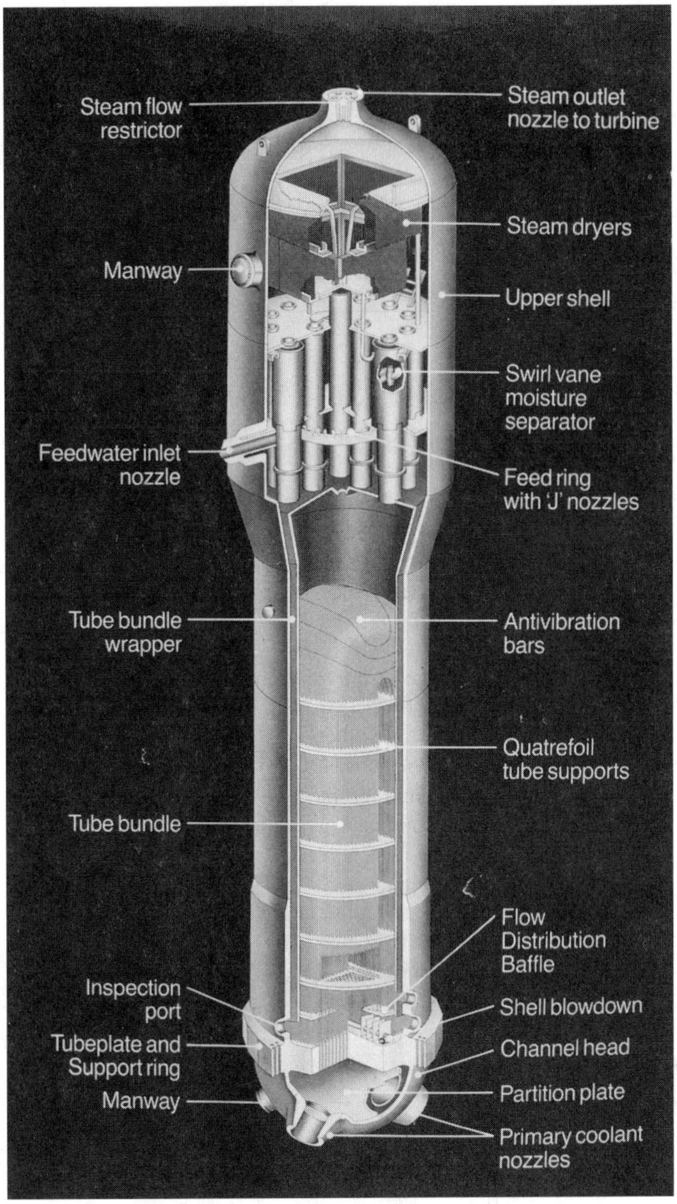

Figure 2. A typical PWR steam generator.

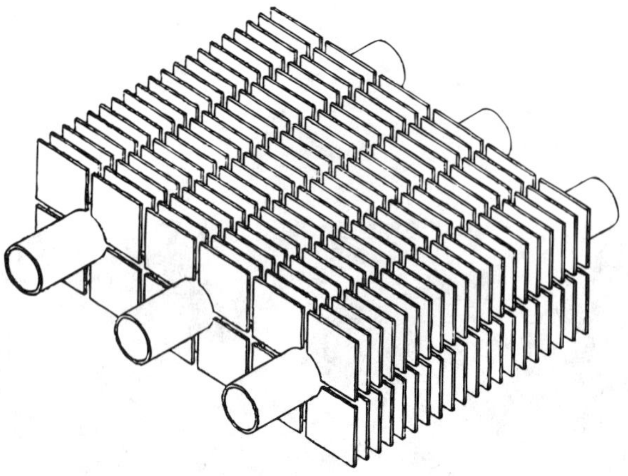

Figure 3. Typical tubing with plate fins.

WATT, JAMES (1736–1819)

The son of a Clydeside shipbuilder, 1736-1819, Watt began work as a mathematical instrument maker in Glasgow at the age of 17. Shortly afterwards he opened his own shop and as a result met many of the local scientists and engineers.

In 1764, while repairing a Newcomen steam engine, Watt became aware of the inefficiences associated with the need to cool and condense the vapour at each stroke (see **Steam Engines**). He realised that the solution was to build an engine with a separate external condenser but he did not bring this concept to fruition until 1790. The Watt engine, built in partnership with Mathew Boulton, made a major contribution to the industrial revolution.

Watt produced other novel engines for pumping water from mines and for driving mill machinery. These employed new forms of gearing and control mechanisms, including the centrifugal governor which he invented in 1788. His business was very successful and he retired a rich man.

Watt was the first to use the term "horsepower", a unit of engine power that has been the standard for two hundred years. It is appropriate that the modern unit of power, the Watt, should take his name.

G.L. Shires

WAVE ENERGY (see Alternative energy sources; Renewable energy)

WAVE GENERATION (see Stratified gas-liquid flow)

WAVE NUMBER (see Film thickness determination by optical methods; Kelvin-Helmholtz instability; Speed of light)

WAVE OPTICS (see Optics)

WAVES (see Stratified gas-liquid flow)

WAVES, IN ANNULAR FLOW (see Annular flow)

WAVES IN FLUIDS

Following from: *Flow of fluids; Interfaces; Conservation equations, two-phase*

There are wide range of wave phenomena in fluids. These phenomena are extremely important in governing many of the characteristics of the flow; they also facilitate transfer of energy, mixing, and turbulization of fluid. In this article we will discuss two forms of such waves, namely *gravity and capillary waves* at free surfaces and *pressure waves* in compressible media.

Gravity and capillary waves at a free surface

Consider a plane free liquid surface in a state of equilibrium in the gravitational field. Deviation from equilibrium brings about an oscillatory motion due to gravity or surface tension (capillarity) forces. This motion propagates along the liquid surface in the form of waves known as *gravity* or *capillary waves.* The wavelength or the length between points on the wave with identical phases remains constant.

If the oscillation amplitude in the wave is small in relation to the wavelength, then the motion is potential and in the gravitational field ρgz it is described by the equation

$$p = -\rho gz - \rho \frac{\partial \varphi}{\partial t},$$

where p is the pressure, ρ the density, g the acceleration due to gravity, φ the velocity potential, the z axis is perpendicular to the plane liquid surface (x, y). The solution of this equation has the form of a plane harmonic wave $\varphi = Ae^{-\bar{k}z}\cos(\bar{k}x - \omega t)$, where ω is the angular frequency, $\bar{k}$ the wave vector ($\bar{k} = \bar{n}k$, where $\bar{n}$ is the unit vector, k the wave number, $\lambda = 2\pi/k$ the wavelength).

Equal pressures on the liquid surface and in the ambient produce a boundary condition which relates the frequency and the wave vector by a dispersion equation $\omega^2 = kg$. Liquid particles describe circles, with the radius exponentially decreasing deep into the liquid, about an equilibrium point. With an arbitrary ω-k relation the rate of displacement of the amplitude distribution profile in a wave is $U = \partial\omega/\partial k$ (the group velocity). If the dependence of ω on k is linear, the group velocity coincides with phase velocity $U = \omega/k$. The group velocity of a short gravity wave in liquid depends on the wavelength and is given by $1/2\sqrt{g/k} = 1/2\sqrt{g\lambda/2\pi}$. If account is taken of the surface tension forces σ, the dispersion equation takes the form $\omega = \sqrt{gk + \sigma k^2/\rho}$. The surface tension forces play a predominant role for very short waves (ripples). Another limiting case is long gravity waves the length of which is great in relation to the liquid depth h_0. In a viscous liquid, transverse waves may appear in which liquid particles do not move in circles but at the right angles to the wave propagation. These waves arise when a body submerged in a viscous liquid oscillates.

A disturbance of the finite amplitude on the liquid surface propagates as a simple, or *Rieman, wave.* The velocity of displacement of points with equal phases on the wave depends on the amplitude and the sign of the disturbance, and the wave profile is continuously deformed, i.e. the point with a greater amplitude propagates faster, the wave crest catches up with its foot, the steepness of the front slope enhances and at a certain critical height the wave falls down to form capillary ripples or white caps (**Figure 1**). In the areas with an extremely steep profile, dispersion and dissipation have a pronounced effect. Dissipation facilitates smoothing of saw-tooth profile of the wave. Stationary waves that are a sequence of short pulses (**Figure 2**) may arise as a result of competing nonlinear effects, "overturning" of the wave profile, and dispersion effects, contributing to broadening of the profile (**Figure 2**). A particular case of a series of pulsating waves with an infinitely large period is a *solitary wave* known also as *soliton.*

Internal gravity waves may also arise in a stratified liquid. If two liquid layers move slipping one on the other, their interface is a tangential discontinuity because the liquid velocity tangential to the surface changes abruptly. The disturbance of the interface may bring about an interfacial instability (**Kelvin-Helmholtz Instability**).

An incompressible fluid uniformly rotating as a whole may give rise to internal waves due to Coriolis forces (inertial waves).

Pressure Waves

Pressure waves appear in a compressible liquid. Pressure perturbation in a compressible fluid involves perturbation of density, velocity, and other parameters. In this case the velocity potential φ satisfies the wave equation $\partial^2 \varphi/\partial t^2 - a^2 \Delta\varphi = 0$ which describes propagation of the wave with the velocity of sound (an *acoustic wave*). The phase velocity of an acoustic wave U = a does not

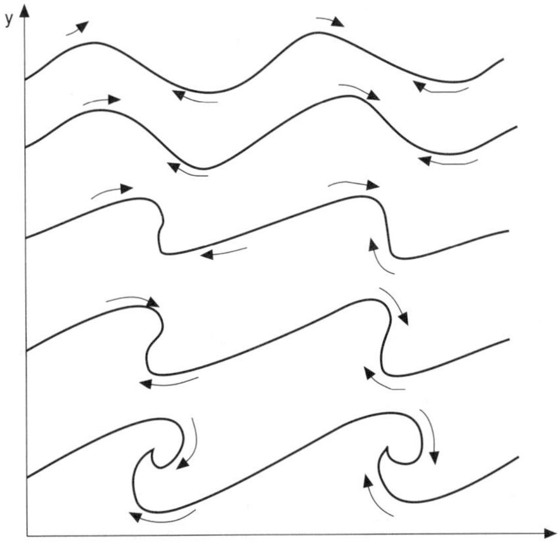

Figure 1. Wave breaking due to the effect of amplitude on propagation velocity.

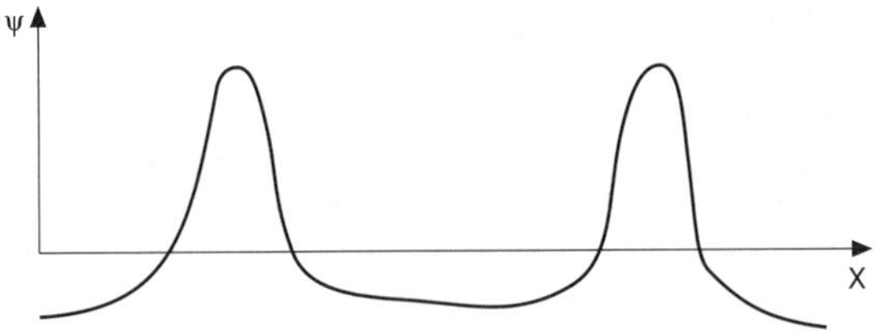

Figure 2. Non-linear effects in wave formation.

depend on the disturbance amplitude and is determined by the thermodynamic properties of the liquid: $a^2 = (\partial p/\partial \rho)_s$, where p is the pressure, ρ the density, and S the entropy. In a perfect gas, $a^2 = \gamma RT/\tilde{M}$;, where γ is the polytropic exponent, R the universal gas constant, T the temperature, and $\tilde{M}$ the molecular weight.

A disturbance of the finite amplitude propagates in compressible fluid as a simple, or Rieman, wave. In contrast to an acoustic wave, the velocity of each point of profile U in the finite-amplitude wave depends on the disturbance amplitude $U = v \pm a(v)$, where v is the velocity of gas particles in the wave, v in an oscillatory process takes a positive, zero, or a negative value. The density variation also changes sign (**Figure 3**). This circumstance results in the profile deformation as the wave propagates, the rarefaction points lagging behind the compression points, and an ambiguity arising, i.e. at the same point density, velocity, and other paremeters must have three different values. Physically the ambiguity gives rise to a discontinuity, known as a *shock wave,* approaching which from the left and from the right the density is single-valued. The discontinuity displaces in space and attenuates if the velocity of gas flow behind it is not kept constant by an appropriate boundary condition. Propagating through the surface of a shock wave the gas parameters change sharply (see **Compressible flows** and **Shock tubes**). The values of gas parameters behind the shock wave front in a perfect gas can be conveniently related to the parameters before the front in terms of the Mach number of the shock wave $Ma_0 = u_0/a_1$, where u_0 is the velocity of the shock wave front with respect to an undisturbed gas and a_1 the velocity of sound in the undisturbed gas

$$\frac{p_2}{p_1} = \frac{2\gamma Ma_0^2 - (\gamma - 1)}{\gamma + 1},$$

$$\frac{\rho_2}{\rho_1} = \frac{(\gamma + 1)Ma_0^2}{(\gamma - 1)Ma_0^2 + 2},$$

$$\frac{T_2}{T_1} = \frac{[2\gamma Ma_0^2 - (\gamma - 1)][(\gamma - 1)Ma_0^2 + 2]}{(\gamma + 1)^2 Ma_0^2},$$

$$u_0 - u_1 = \frac{2a_1}{\gamma + 1}\left(Ma_0 - \frac{1}{Ma_0}\right).$$

The above relations show that at high shock wave velocities ($Ma_0 \to \infty$), $p_2/p_1 \approx Ma_0^2$ and is independent of γ, $\rho_2/\rho 1 \approx (\gamma + 1)/(\gamma - 1)$ and does not depend on Ma_0. As $Ma_0 \to \infty$, T_2/T_1 increases as square of the Mach number. However, the substantial temperature elevation of behind the shock wave is not attained because in an imperfect gas, as the translational temperature grows, part of energy is expended on excitation of molecule vibrations and dissociation. In this case, heat capacity depends on temperature, and enthalpy is calculated by the physics methods of statistical. **Figure 4** demonstrates the temperature T_2 behind the shock wave versus the shock wave velocity u_0 in air: for a perfect gas (curve 1) taking into account the excitation of molecule vibrations without dissociation (curve 2), and taking into account the dissociation at initial pressures of $p_1 = 1, 10^{-3}, 10^{-4}$, and 10^{-5} atm respectively (curves (3,4,5,6).

The gas parameters behind the shock wave can be calculated by the formulas for a perfect gas if γ_2 is assigned the value γ^*

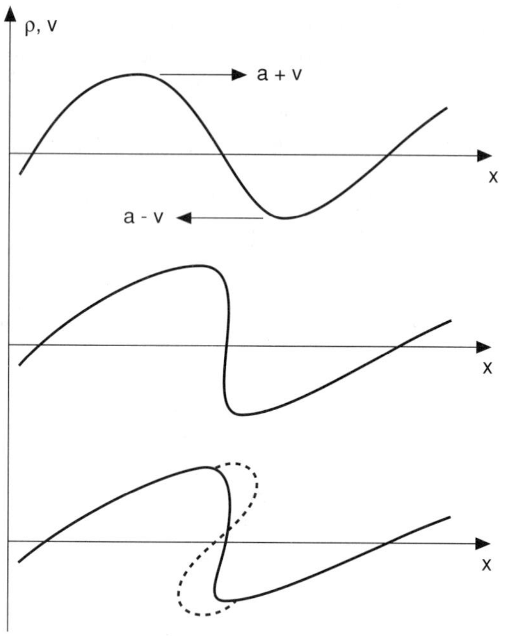

Figure 3. Propagation of a finite amplitude pressure distrubance.

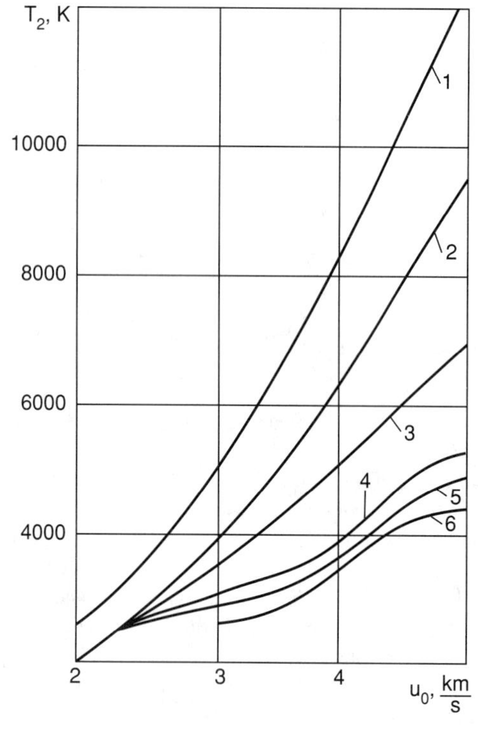

Figure 4. Temperature behind a shock wave with various assumptions.

such that the relation between enthalpy, pressure, and density takes the same form as that for a perfect gas: $h_2 = [\gamma^*/(\gamma^* - 1)](p_2/\rho_2) \cdot \gamma^*$ is equal to the ratio of enthalpy to internal energy at a temperature T_2. The values of γ^* and the molecular weight $\tilde{M}_2$ for air can be approximated by the formulas

$$\gamma^* = 1.39 \exp[6\ 10^{-4}(Ma_0^2 - 3) - 2.2\ 10^{-2}(Ma_0 - 3)]$$

$$\text{for } 3 \leq Ma_0 \leq 20,\ p_0 = 1 \text{ atm},$$

$$\tilde{M}_2 = 0.02(50 - Ma_0)\tilde{M}_1 \qquad \text{for } 8 \leq Ma_0 \leq 25,\ p_0 = 1 \text{ atm},$$

$$\tilde{M}_2 = \tilde{M}_1 \qquad \text{for } Ma_0 < 8.$$

In a medium, in which $(\partial^2 V/\partial p^2)$ is positive (where $V = 1/\rho$), the entropy grows in transition across the shock compression wave front and must have dropped in transition across the rarefaction wave front. This suggests that in most media rarefaction shocks do not exist and in the rarefaction wave entropy is constant. The head of the stationary rarefaction wave propagates with the velocity of sound, while the shock wave front may propagate with a velocity tens of times higher than the velocity of sound (depending on the boundary conditions).

With instantaneous release of energy a blast wave, that is a shock wave followed by the rarefaction wave, appears in a compressible fluid (**Figure 5**). The velocity and intensity of a blast wave fall with the distance from the point of energy release. The most rapid drop of the shock wave velocity and the pressure behind it is observed in a spherical blast wave.

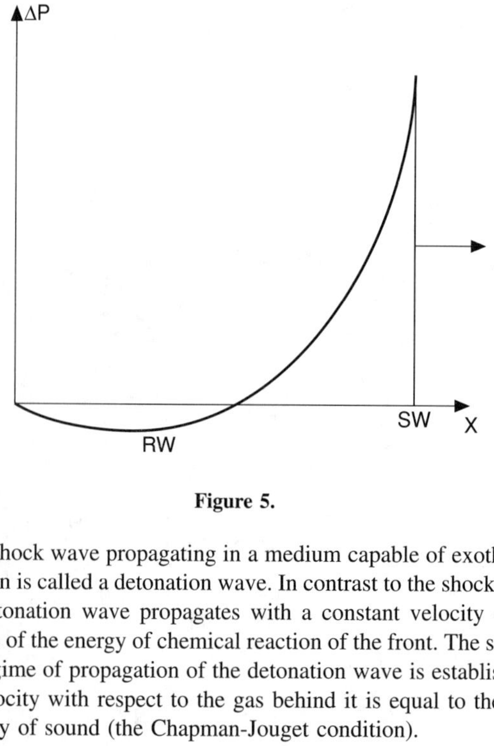

Figure 5.

A shock wave propagating in a medium capable of exothermic reaction is called a detonation wave. In contrast to the shock wave, the detonation wave propagates with a constant velocity due to release of the energy of chemical reaction of the front. The stationary regime of propagation of the detonation wave is established if its velocity with respect to the gas behind it is equal to the local velocity of sound (the Chapman-Jouget condition).

References

Whitham, G. B. (1974) *Linear and Nonlinear Waves,* Wiley.

Korobeinikov, V. P. ed. (1989) *Unsteady Interactions of Shock and Detonation Waves in Gases,* Hemisphere, New York.

Leading to: *Kinematic waves*

T.V. Bazhenova

WAVY FLOW

Following from: *Gas-liquid flow; Waves in fluids*

When a gas and a liquid flow together in parallel streams, the interface between them is flat at low gas velocities. At higher gas velocities, it becomes unstable to small perturbations, and waves appear on it. Wavy flow is very common in nature as well as in industrial applications, typical examples are waves on the sea and wavy flow down an inclined plane in an absorption column.

Depending, principally, on the geometry and the flow rates of the fluids, the interfacial waves can take a variety of shapes and sizes. Typical wave patterns occurring in horizontal channel flows are shown in **Figure 1** (Hanratty, 1983) as a function of the air velocity and the liquid **Reynolds Number**, Re_L, which is based on the average flow velocity and the hydraulic diameter. It can be seen that for low water flow rates, i.e., small Re_L, only the surface tension-induced capillary waves prevail. At higher Re_L, two-dimensional, three-dimensional and roll waves may occur depending on the air velocity. The wave patterns occurring in vertical upward flow are shown in **Figure 2** (Hewitt and Hall-Taylor, 1970) as a function of the superficial velocities of the two phases. Distinguishing features of this flow are the disturbance waves which are intermittent waves having an amplitude of up to five times the mean thickness of the film. These occur only when the liquid film Reynolds number is greater than a critical value of about 500.

The onset of wavy flow can be calculated using stability analyses which investigate the stability of the system to a perturbation. The simplest of these is the *Kelvin-Helmholtz instability* for incompressible, inviscid flows. More accurate analyses take the form of a linear stability analysis of the *Orr-Sommerfeld* type which includes the effect of viscosity but is valid only for small perturbations. Real waves are the result of non-linear effects some of which can be taken into account successfully (Hanratty, 1983).

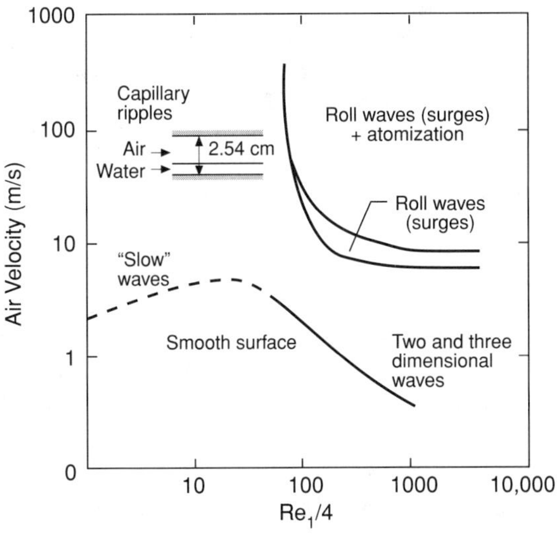

Figure 1. Wave patterns in air-water flow in a horizontal duct of rectangular cross-section. From Hanratty T. J. (1983) in Waves on Fluid and Fluid Interfaces, Meyer R. E. (ed), by permission of Academic Press.

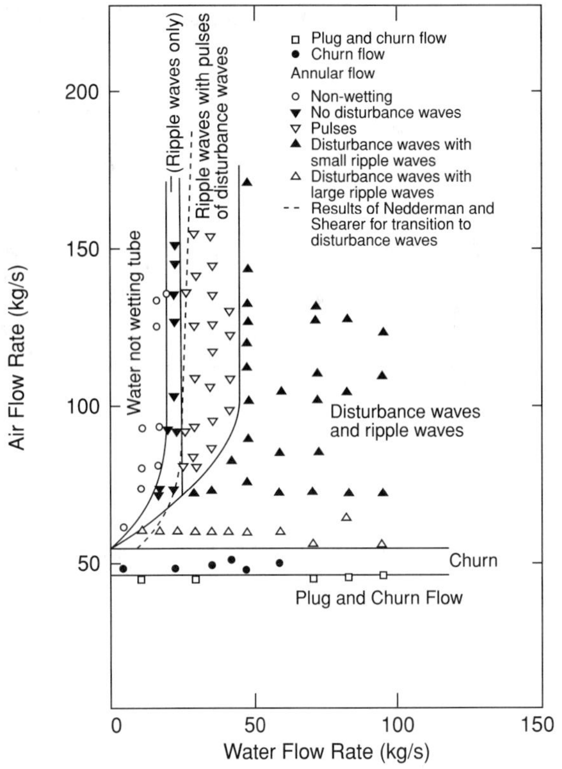

Figure 2. Wave patterns in upward air-water flow in a 32 mm diameter circular pipe. Reprinted from Hewitt G. F. and Hall-Taylor N. S. (1970) Annular Two-Phase Flow, with kind permission from Elsevier Science Ltd., The Boulevard, Langford Lane, Kidlington, OX5 1GB, U.K.

Interfacial waves have a major effect on the transfer processes between the two phases (Hanratty, 1991). The presence of surface waves can increase the *interfacial friction factor* by more than an order of magnitude; this increase is normally given in the form of empirical correlations. The waves are also a source of *droplet entrainment,* and can also affect the turbulence characteristics in the gas phase (Cohen and Hanratty, 1968). They also *enhance the heat and the mass transfer* rates across the interface. Typically, the heat transfer coefficient for wavy films can be about 20 to 50% higher than that for a smooth film. The effect of waves on mass transfer has also been established experimentally (Brauner and Maron, 1982), although there are few correlations which take account of the wave effect explicitly.

References

Cohen, L. S. and Hanratty, T. J. (1968) Effect of waves at a gas-liquid interface on a turbulent air flow, *J. Fluid Mech.,* 31, 467–479.

Brauner, N. and Maron, D. M. (1982) Characteristics of inclined thin films, waviness and the associated mass transfer, *Int. J. Heat Mass Transfer,* 25(1), 99–110.

Hanratty, T. J. (1983) Interfacial instabilities caused by air flow over a thin liquid layer, *Waves on Fluid and Fluid Interfaces,* 221–259, R. E. Meyer, ed., Academic Press.

Hanratty, T. J. (1991) Separated flow modelling and interfacial transport phenomena, *Applied Scientific Research,* 48, 353–390.

Hewitt, G. F. and Hall-Taylor, N. S. (1970) *Annular Two-phase Flow,* Pergamon Press.

Leading To: *Stratified flow; Annular flow; Falling film flow*

S. Jayanti

WEBER NUMBER

Weber number, We, is a dimensionless group that occurs in the analysis of bubble formation and other interfacial phenomena (see **Bubbles**).

$$We = \rho u^2 L/\sigma$$

where ρ is fluid density, μ velocity, L dimension and σ interfacial surface tension. It represents the ratio of inertial force to surface tension force.

There are several alternative forms of Weber number, including the square root of the above expression.

G.L. Shires

WEBER, ON DROPLET FORMATION (see Atomization)

WEDGE FLOWS

Following from: *Boundary layers; External flows, overview*

In the case of flow over a wedge, the velocity at the edge of the *laminar boundary layer* in the region of the apex can be shown to be given by (e.g. Schlichting, 1968):

$$u_e = Cx^m \tag{1}$$

where C is a constant and the exponent is given by:

$$m = \frac{\theta}{2\pi - \theta} \tag{2}$$

x and θ are the distance from the apex and the wedge angle respectively, as shown in **Figure 1**.

For this problem, a similarity solution for the flow field can be found, if a dimensionless length, η, is introduced, defined by

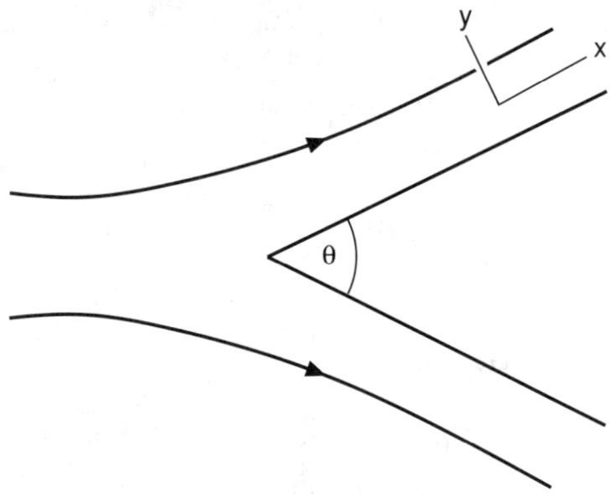

Figure 1. Flow over a wedge.

$$\eta = y\sqrt{\frac{u_e(m+1)}{2\nu x}} \tag{3}$$

where ν is the dynamic viscosity and y is the perpendicular distance from the side of the wedge. The velocity in the boundary layer can then be shown to be given by:

$$u = u_c f' \tag{4}$$

$$v = -\sqrt{\frac{(m+1)\nu u_0}{2}}\, x^{(m-1)/2}\left(f + \frac{m-1}{m+1}\eta f'\right) \tag{5}$$

where u and v are the components of velocity in the x and y directions respectively.

f is a function of η and is defined as the solution to the *Falkner-Skan equation* which is:

$$f''' + ff'' = \frac{2m}{m+1}(1 - f'^2) \tag{6}$$

where the primes denote differentiation with respect to η. f can be found by numerical solution of **Equation 6** subject to the boundary conditions $f = f' = 0$ at $\eta = 0$ and $f' = 1$ as $\eta \to \infty$.

The derivation of **Equations 4, 5 and 6** can be found in, for example, White (1991), Schlichting (1968) and Brodkey (1967). The equations can also be applied to $\pi < \theta < 2\pi$, which corresponds to a converging channel and $\theta = 0$, in which case the equations reduce to the Blasius solution for flow over a flat plate.

References

Brodkey, R. S. (1967) *The Phenomena of Fluid Motions,* Addison-Wesley Publishing Co., Reading, Massachusetts.

Schlichting, H. (1968) *Boundary-Layer Theory,* McGraw Hill, New York.

White, F. M. (1991) *Viscous Fluid Flow,* McGraw Hill, Inc., New York.

M.J. Pattison

WEIBULL DISTRIBUTION (see Wind turbines)

WEIBULL EQUATION (see Fibre optics)

WEIRS

Following from: *Hydraulics*

A weir is a three-dimensional *channel control.* A control is any channel feature that fixes a relationship between flow rate and depth in its neighborhood. Weirs are also used for flow measurement. The main features of weirs can be explained by considering two-dimensional flows. The cross section of a simple weir is shown in **Figure 1**. A uniform flow with velocity U approaches a weir of height Z. The level of the free surface above the crest of the weir is z. The length of the crest is l.

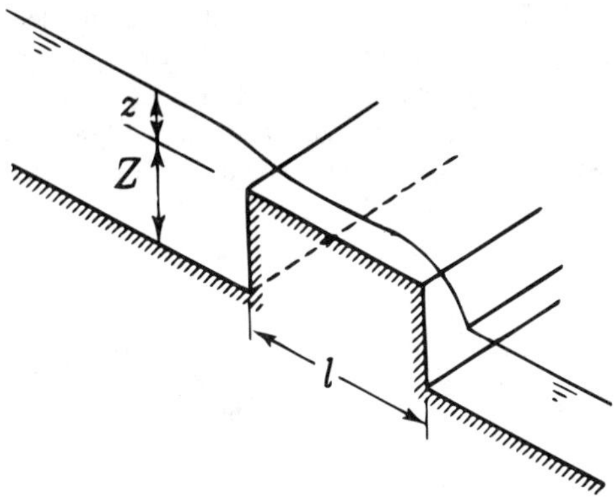

Figure 1. Flow over a rectangular weir.

The flow rate per unit width over a weir, defined by $\dot{V} = U(z + Z)$, can be written as

$$\dot{V} = \left(\frac{2}{3}\right)^{3/2} Cg^{1/2}H^{3/2}. \tag{1}$$

Here $H = z + U^2/2g$ is the total head, C is the *discharge coefficent,* and g is the acceleration due to gravity. The coefficient C, which depends on the weir geometry, must usually be determined experimentally. For some simple weir geometries, it can be determined numerically. The upstream Froude Number is $Fr = U[g(z + Z)]^{-1/2}$

The equation for the flow rate can be explained by considering a weir with a crest long enough to maintain a hydrostatic pressure distribution in the flow over it. Experiment shows that this flow becomes critical, i.e., local Froude number becomes equal to unity. It follows that the height of the free surface above the crest is 2H/3, which is referred to as the critical depth, and that the velocity there is $[g(2H/3)]^{1/2}$. Thus the flow rate is

$$\dot{V} = \left(\frac{2}{3}\right)^{3/2} g^{1/2}H^{3/2} \tag{2}$$

for critical flow. This shows that for *critical flow* the discharge coefficient is $C = 1$.

Weirs providing critical flow over a long enough distance could be used for flow-rate measurement with $C = 1$. Since practical weirs are usually not long enough, the discharge coefficient C must be determined for them. Clearly C depends upon the two dimensionless parameters z/Z and z/l, in agreement with observation.

Typical shapes for weirs are shown in **Figure 2**. These include thin weirs, rectangular weirs (not shown in **Figure 2** since already shown in **Figure 1**), triangular weirs, round-nosed broad-crested weirs, trapezoidal *flumes.*

A detailed description of all these weirs is provided in the book by Ackers, White, Perkins and Harrison (1978). Numerical

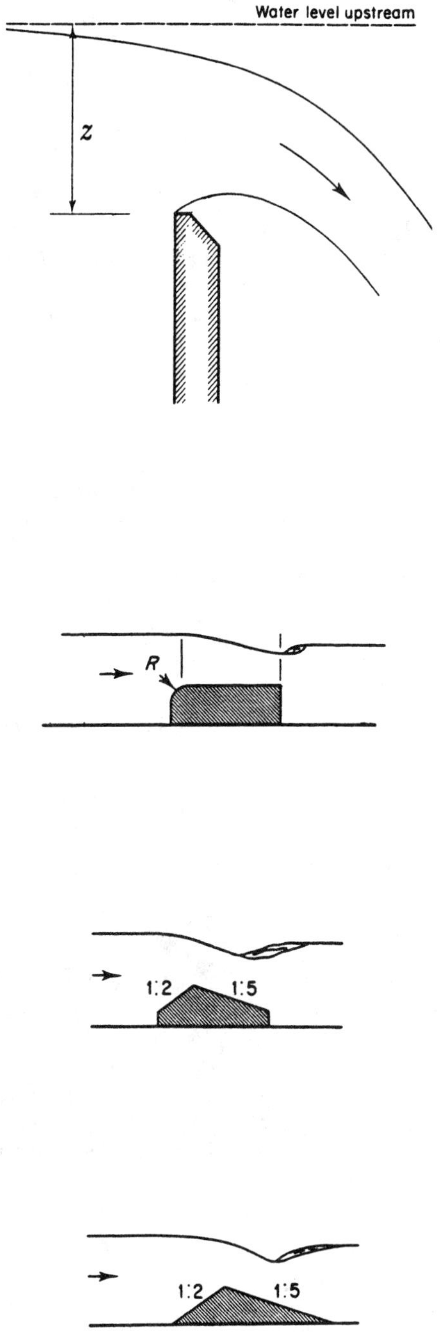

Figure 2. From Ackers, P., White, W. R., Perkins, J. A. and Harrison, A. J. M. (1978), reprinted by permission of John Wiley and Sons, Ltd.

calculations on weirs have not been much developed yet. It is still a challenge to provide an accurate model for weir flows, which includes the effects of the separation at corners, the three dimensions, the sediment transport, the friction along the crest of the weir.

References

Ackers, P., White, W. R., Perkins, J. A. and Harrison, A. J. M., (1978) *Weirs and Flumes for Flow Measurement,* Wiley, Chichester.

F. Dias

WEISSENBERG NUMBER

Following from: *Non-Newtonian fluids*

A **non-Newtonian Fluid** is one for which stress is not linearly related to strain-rate. All non-Newtonian fluids are *elasticoviscous,* that is they combine elastic and viscous properties. When the time-scale of a flow t_f is much less than the relaxation time t_r of an elasticoviscous material, elastic effects dominate. When on the other hand t_f is much greater than t_r, elastic effects relax sufficiently for viscous effects to dominate. The ratio t_r/t_f is a dimensionless number of particular significance in the study of flow of non-Newtonian fluids: depending on the circumstances, this number is called the **Weissenberg Number** or the **Deborah Number**. The Weissenberg number Ws is named after Karl Weissenberg, an early worker in the field of non-Newtonian fluids. The definition of Ws depends on that of t_f which is given by:

$$t_f = \sqrt{\frac{1}{2}\,(\mathrm{tr}(\mathbf{e}))^2} \tag{1}$$

where tr( ) denotes trace and e denotes strain-rate given in terms of velocity **u** by:

$$\mathbf{e} = \nabla\mathbf{u} + (\nabla\mathbf{u})^T \tag{2}$$

For a simple shear flow with shear-rate γ:

$$t_f = 1/\gamma \tag{3}$$

while for a simple extensional flow with extension rate ϵ:

$$t_f = 1/\epsilon \tag{4}$$

Thus for flow at mean velocity U through a pipe of diameter D and length L, $t_f \simeq D/U$ and so $Ws \simeq t_rU/D$. By contrast the Deborah number $De = t_rU/L$. Thus Ws is larger than De by the ratio L/D.

S.M. Richardson

WELLS AIR TURBINE (see Alternative energy sources)

WELL STIRRED FURNACE MODEL (see Furnaces)

WET BULB DEPRESSION (see Wet bulb temperature)

WET-BULB TEMPERATURE

Following from: *Dry-bulb temperature; Humidity measurement*

The wet-bulb temperature is the reading registered by a temperature sensor placed in a moist gas stream and covered with a wetted cloth or wick. This temperature is lower than that of the gas stream itself, and is the dynamic equilibrium value attained when the convective heat transfer to the sensor effectively equals the evaporative heat load associated with the moisture loss from the wetted surface. Another temperature is sometimes called the *thermodynamic wet-bulb temperature.* This is the limiting temperature reached as a gas cools on adiabatic saturation, and is more properly termed the *adiabatic-saturation temperature* to avoid confusion.

Figure 1 shows the temperature and humidity profiles about a plane moist surface exposed to a humid gas flowing over it.

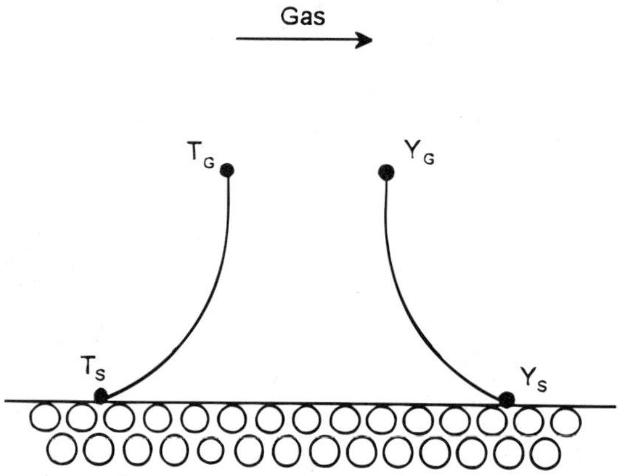

Figure 1. Temperature and humidity profiles about a flat surface.

If the Chilton-Colburn analogy between heat and mass-transfer processes is used in its original (1934) form, then the temperature difference between that of the wetted surface (T_S) and that of the bulk gas (T_G) becomes (Keey, 1992):

$$T_G - T_S = \left[\frac{(1 + Y_G)Lu^{2/3}\phi_M}{\phi_H(1 + \phi_R)}\right]\frac{\Delta H_{VS}}{C_{PY}}(Y_S - Y_G) \tag{1}$$

where Y_S and Y_G are the humidities of the gas adjacent to the surface and in the bulk gas respectively, Lu is the *Luikov number,* the ratio of the mass to the thermal diffusivity, ΔH_{VS} is the latent heat of vaporisation of the moisture at the surface temperature and C_{PY} is the humid heat capacity (the heat capacity of unit mass of dry gas and its associated moisture-vapour content). The various ϕ_i are factors close to 1 to account for secondary effects: ϕ_H is the *Ackermann correction factor* to account for the influence of the evaporation on the convective heat transfer; ϕ_M is the humidity-potential coefficient introduced to correct for the embedded approximations in the use of linear humidity differences in describing the evaporation rate; ϕ_R is the fractional contribution of radiation to the heat transfer. The above equation for the temperature difference or **wet-bulb depression** is remarkably similar in form to that for the temperature drop on adiabatic saturation ($T_G - T_{GS}$):

$$T_G - T_{GS} = \frac{\Delta H_{VS}}{C_{PY}}(Y_{GS} - Y_G)$$

The terms within the square brackets is sometimes called the *psychrometric ratio,* although other slightly different definitions appear in the literature which relate to the coefficient in the equation for the wet-bulb depression. By contrast, the psychrometric ratio does not appear in the adiabatic-saturation expression. Whenever the moisture is water and the gas air, the psychrometric ratio is almost unity. Commonly it is assumed, then, the wet-bulb and adiabatic saturation temperatures are the same within the accuracy with which these temperatures can be measured in practice. On the other hand, should the moisture be a non-aqueous solvent (when the Luikov number would be much smaller than 1), then this identity between the two temperatures no longer holds. The wet-bulb temperature is now greater than the adiabatic-saturation temperature, and both are higher than the dewpoint value, as illustrated in **Figure 2**.

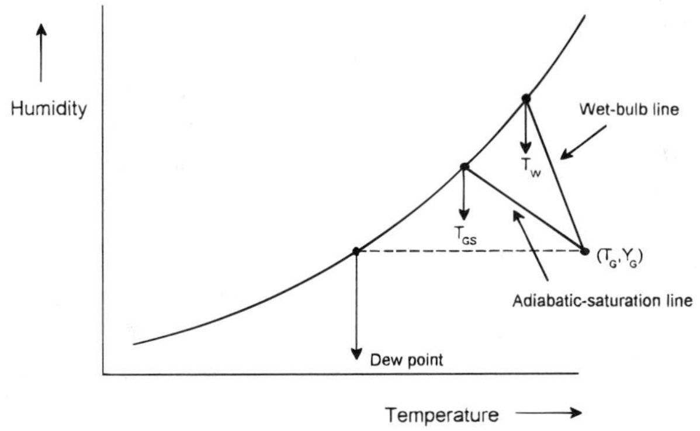

Figure 2. A humidity chart illustrating various psychrometric temperatures.

Unlike the evaporation of a single droplet containing a mixture of volatile liquids, the evaporation of a mixture from a porous surface is normally non-selective and the composition does not change (Turner and Schlünder, 1985; see Pakowski, 1990). At a given composition there is a specific wet-bulb temperature. This temperature for a non-ideal mixture (in the tested case, isopropanol-water) can be less than the wet-bulb temperatures of either pure component for a given water-vapour concentration, but it may vary monotonically with composition depending upon the vapour concentrations.

The wet-bulb thermometer is one of two sensors which make up a *psychrometer* to measure the humidity of a moist gas. Both sensors should be shielded from thermal radiation and exposed to a flow of the gas at a sufficient rate to ensure fully developed forced convection. A gas velocity of 5 m s^{-1} over the bulbs is normally sufficient. The wick surrounding the bulb must be kept wetted, and commercial psychrometers usually provide a formal drip-feed system (Wiederhold, 1987). Special precautions are required at dry-bulb temperatures above 100°C, by preheating the wick water to a few degrees of the estimated wet-bulb temperature to avoid the wick drying out. For the same reason, wet-bulb temperatures are very difficult to measure with accuracy once the relative humidity falls below about 20%.

References

Keey, R. B. (1992) *Drying of Loose and Particulate Materials,* Pergamon, Oxford.

Pakowski, Z. (1990) Drying of solids containing multicomponent moisture, *Advances in Drying,* 5 Ch. 5, A. S. Mujumdar ed., Hemisphere, Washington.

Wiederhold, P., (1987) Humidity Measurements, *Handbook of Drying Technology,* Ch. 29, A. S. Mujumdar ed., Marcel Dekker, New York and Basel.

R.B. Keey

WETTABILITY OF SOLID SURFACES (see Contact angle)

WETTING AGENTS (see Dropwise condensation)

WETTING ANGLE (see Capillary action)

WHITE CAUSTIC (see Sodium hydroxide)

WICKLESS HEAT PIPE (see Tubes, condensation in)

WICKS-DUKLER METHOD (see Dropsize measurement)

WIDE RANGE ATOMIZERS (see Atomization)

WIEDEMANN-FRANZ LAW (see Thermal conductivity)

WIEDMANN-FRANZ-LORENZ LAW (see Metals)

WIEN'S DISPLACEMENT LAW (see Heat protections; Radiative heat transfer)

WIEN'S LAW (see Stefan-Boltzmann law)

WIGNER ENERGY RELEASE (see Gas-graphite reactors)

WILKE-CHANG EQUATION FOR DIFFUSION COEFFICIENTS IN LIQUIDS (see Diffusion coefficients)

WILSON LINE (see Critical flow)

WILSON PLOT

Following from: *Heat transfer coefficient; Overall heat transfer coefficient*

The **Overall Heat Transfer Coefficient** for any heat transfer equipment is obtained from $U = \dot{Q}/A\ \Delta T$ where $\dot{Q}$ is the heat flow per unit time (watts), A is the heated surface area and ΔT the overall temperature difference. It is usual to design equipment using practical values of U rather than from a series of film coefficients. However, for the important case of heat transfer from one fluid to another across a metal surface, Wilson (1915) developed a method for doing so. By way of illustration, consider cold water flowing through a tube with steam condensing on the outside. The overall and individual coefficients are given by

$$1/U = 1/\alpha_o + t_w/\lambda_w + R_i + 1/\alpha_i \quad \textbf{(1)}$$

where R_i is the fouling resistance inside the tube and α_o and α_i are the heat transfer coefficients on the outside and inside of the tube and t_w is the thickness and λ_w is the conductivity of the wall.

For turbulent flow, the fouling resistance is constant, α_o is approximately independent of the velocity of the water and since for a smooth circular tube

$$Nu = C_1 \, Re^a Pr^b \quad \textbf{(2)}$$

where Nu is **Nusselt Number**, Re is **Reynolds Number** and Pr is **Prandtl Number** and the values of the constants are $a = 0.8$ and $b = 0.4$, the inside coefficient is given by

$$\alpha_i = C_1(\lambda/d)Pr^{0.4}(\rho v d/\eta)^{0.8} = C_2 v^{0.8} \quad \textbf{(3)}$$

where λ is the fluid conductivity, d is the inside tube diameter, ρ is fluid density and η is fluid viscosity, and v is fluid velocity.

Hence, **Equation 1** reduces to

$$1/U = C_3 + [1/C_2]v^{0.8} \quad \textbf{(4)}$$

which, for a plot of 1/U against $1/v^{0.8}$ is a straight line of slope $1/C_2$ and intercept C_3. Note that the value of C_2 is the inside coefficient for unit inside velocity.

For a clean tube, the *fouling resistance* is zero and, knowing the thickness and conductivity of the wall, C_3 gives α_o. By repeating measurements as fouling develops values of R_i can be determined from the changes in the intercept.

The method has been used extensively and is in current use, sometimes with modifications, examples being for coiled tubes and annuli and for determining α_o for condensing organic vapours.

The method is based on a number of conditions and assumptions, namely that the outside coefficient and the fouling resistance are constant and C_1, a and b are known. These last three parameters will vary with tube design and with inner surface roughness. Thus any degree of roughness imposed artificially, for example by machining, or naturally by scaling or fouling, will alter these parameters. Moreover, it is known (Wilkie, 1966) that **Equation 1** does not apply in these conditions, and hence a non-linear version of **Equation 3** is needed.

Even if the form of **Equation 1** is correct, it is necessary to assume known values of the three parameters and, if it is not correct, further coefficients must be determined. The validity or otherwise of **Equation 1** and the values of all coefficients can be obtained from suitably designed experiments, using least squares curvilinear regression (Wilkie 1962 and 1986).

References

Wilkie D. (1962) A method of analysis of mixed level factorial experiments, *Appl. Statistics,* XI Pt 3, 184–195.

Wilkie D. (1966) Forced convection heat transfer from surfaces roughened by transverse ribs, 3rd Int. Heat Transfer Conf., Chicago Amer. Inst. Chem. E., paper No. 1, 1–19 New York.

Wilkie D. (1987) Analysis of factorial experiments and least squares polynomial fitting by the method of orthogonal polynomials for any spacing of the levels of independent variables, *J Appl Statistics,* 14, 1, 83–89.

Wilson E. E. (1915) A basis for rational design of heat transfer apparatus, *Trans. Am. Soc. Mech. Engrs.,* 37, 47–70.

D. Wilkie

WIND ENERGY (see Alternative energy sources)

WIND ENERGY CONVERSION SYSTEMS (See Wind turbines)

WINDOWS (see Computers; Computer programmes)

WINDSCALE AGR (see Advanced gas cooled reactor)

WINDSCALE PILE (see Gas-graphite reactors)

WIND SPEED PROBABILITY DISTRIBUTION (see Wind turbines)

WIND TUNNELS

Wind tunnels are experimental setups producing an air or gas stream for investigation of flow around models representing, for instance, vehicles or buildings. Wind tunnels are used to determine aerodynamic resistance forces of bodies, to investigate their stability and controllability, and to determine vehicle and building dynamic loads due to explosion waves and gusts. A special class of wind tunnels with heaters is used for estimating heat loads applied to the surface of the body in the flow and in developing heat protection techniques.

Experiments in wind tunnels are based on the principle of motion reciprocity according to which displacement of a body relative to gas or liquid can be replaced by gas or liquid flow around the body at rest (see **Aerodynamics**). In mechanics, this is known as the *Galileo principle* and is applicable, strictly speaking, only to the case of a uniform rectilinear motion. Under these conditions the stream force and heat effect on a body are identical in both direct and reverse motion.

Simulation of body motion in a quiescent medium requires establishing in a wind tunnel a uniform gas or liquid flow with equal and parallel velocity vectors at all points of the working space and with identical values of gas temperature and density. This problem is solved using quite a number of structural components some of which are depicted in **Figures 1, 2**, and **3**. The wind tunnels designed as far back as the early 20th century incorporated a prechamber 1, a nozzle 2, a working section 3, a diffuser 4, and a drive 5.

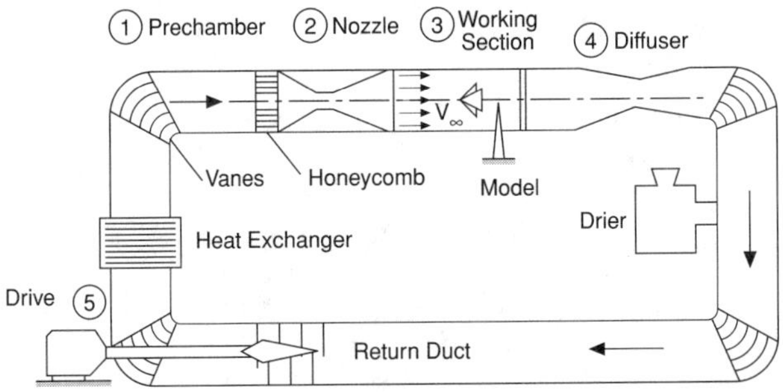

Figure 1. Recirculating (closed circuit) wind tunnel.

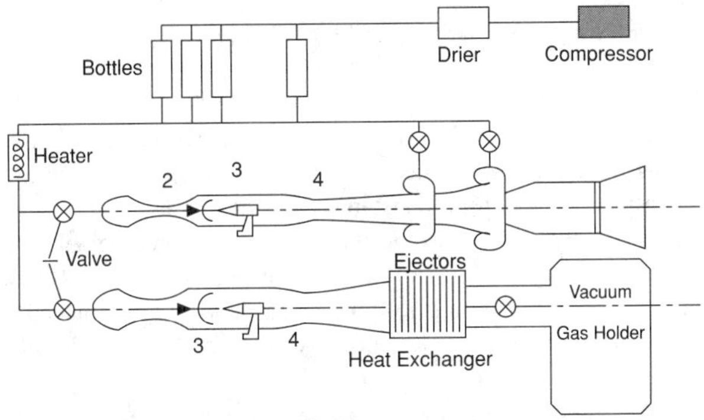

Figure 2. Wind tunnel with compressed gas storage.

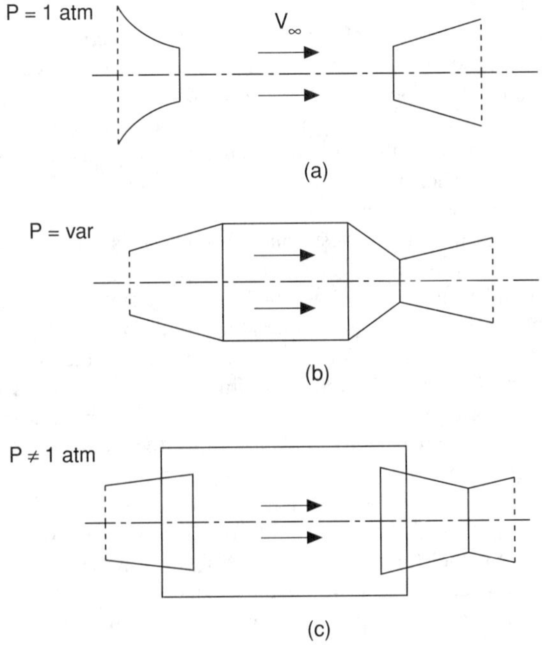

Figure 3. Working section arrangements in wind tunnels.

The prechamber is a source of a high-pressure gas that is speeded up via a nozzle to a specified and uniform velocity V_∞ before the inlet to the working section, the gas pressure and density being much less than in the prechamber. The diffuser zone of the nozzle converts the kinetic energy of the stream to the potential energy of pressure. The wind tunnel drive unit (a fan, a compressor) makes up for energy losses for friction, nonisentropic stagnation in shock waves, etc. For wind tunnels running off compressed gas bottles (**Figure 2**), reciprocating compressors raising the gas pressure in the cylinders serve as a drive.

These components are all inherent in various types of wind tunnels, both subsonic and supersonic, although they may vary in design, size, and layout (**Figure 2**). The working section may be closed (**Figure 3b**) or open (**Figure 3a**). The jet in an open working section with the static pressure being unequal to atmospheric is separated from atmosphere by a special altitude chamber known as an Eiffel chamber (**Figure 3c**). Each design has advantages and disadvantages, but open working sections gained wide application owing to ease of observation and recording of optical measurements.

The working section is, as a rule, shaped as cylinder with the circular or rectangular (sometimes, elliptic or polygonal) cross section. The cross section shape of the working section depends on the type of body to be investigated in a wind tunnel. For instance, oval, round, and rectangular shapes are most fitting for aircraft, rockets, and optical observations respectively.

In addition to the five basic components, wind tunnels may have a number of more specific components. Among them is a gas heater (**Figure 2**) arranged before the prechamber and various straightener vanes and lattices regulating gas flow before the inlet to the nozzle (a honeycomb or deturbulizing screen) breaking down large vortices, reducing stream turbulence, and eliminating skewness of velocity fields (**Figure 1**). The return duct in closed-circuit wind tunnels makes it possible to save a considerable portion of kinetic energy in gas stream behind the diffuser.

In closed-circuit wind tunnels, it is advisable to provide a drier; (**Figure 1**) to remove excessive moisture from air which can distort the flow pattern in the working section. In continuous-action wind tunnels, the gas is heated due to stream friction against the walls and the drive operation, therefore, gas-cooling heat exchangers need to be incorporated in such tunnels (**Figure 1**).

Most importantly for wind tunnel operation is the flow velocity, particularly, its relation to sound velocity. As is known, under adiabatic flow conditions the sound velocity is $a^2 = \gamma p/\rho = \gamma R/\tilde{M} T$. The criterion of similarity characterizing the velocity regime of the gas flow is the Mach number $Ma_\infty = V_\infty/a$.

The schematic of wind tunnel structures presented in **Figures 1, 2**, and **3** gives only a general idea of the variety of operational regimes of wind tunnels. Thus, bottles, exhaust fans, and vacuum vessels are used, as a rule, in open-circuit wind tunnels designed for gas ejection and single gas utilization in the working section (**Figure 2**). Compressors and fans built in gas dynamic contour make possible steady state wind tunnel operation, although in this case gas heating generally is not allowed, and the small pressure differential limits the velocity range to subsonic. Filling up a vacuum vessel in the course of an experiment (see Figure 2) is used in the pulsed wind tunnel which operates with variation of basic parameters during the pulse cycle.

Gas storage vessels (bottles) can provide a high-pressure gas supply both directly to the prechamber or alternatively to multistage ejectors at the outlet of the diffuser behind the working section (**Figure 2**). In these cases, the operation time depends on the gas content of the bottles. This system of gas supply has several advantages over continuous gas supply systems. First, it makes it possible to produce streams with higher Mach numbers at relatively low power consumption of the drive. Second, it enables the achievement of an extremely high Reynolds number which otherwise can only be achieved by simultaneously establishing a high-velocity flow in the working section and, hence, a higher vacuum as the outlet.

An exhauster is a fan creating vacuum. It can diminish pressure at the outlet by nearly two orders of magnitude in relation to that at the inlet with the flow rate of hundreds of cubic meters per second. The most powerful exhausters have 50,000 kW and higher power.

Transonic wind tunnels are distinguished from other wind tunnels by the structure of working section. Their walls are perforated and slotted (**Figure 4**). The narrowest section in the wind tunnel has a flow velocity equal to the sound velocity. Therefore, in wind tunnels with close to sonic flow, the effect of the model blocking up the cross section of the working section gives rise to local supersonic flows. This is called choking of the wind tunnel. Increasing the pressure in the prechamber, we raise the flow rate, but do not change position of the sonic velocity section, and the wind tunnel becomes inoperable. These tests are not reliable because the model is in the lower velocity flow.

The critical Mach number Ma_∞^* corresponding to the beginning of choking depends on "obstruction", i.e. the cross section of model and its suspension (S_M) relative to the cross-section area of the working section (S*). The graph in **Figure 5** allows us to estimate an admissible value of this blockage. For instance, $Ma_\infty^* = 0.85$ corresponds to the model cross-section area ratio (Sm/S*) of 0.02. Therefore, experiments can only be conducted with a solid blockage lower than 2 per cent. We obtain the same result for a supersonic velocity when $Ma_\infty^* = 1.15$.

The cross-section area of the model must be zero if sonic velocity is required in the wind tunnel.

Choking of the wind tunnel can be virtually prevented if the working section has perforated walls and is enclosed in an air-tight

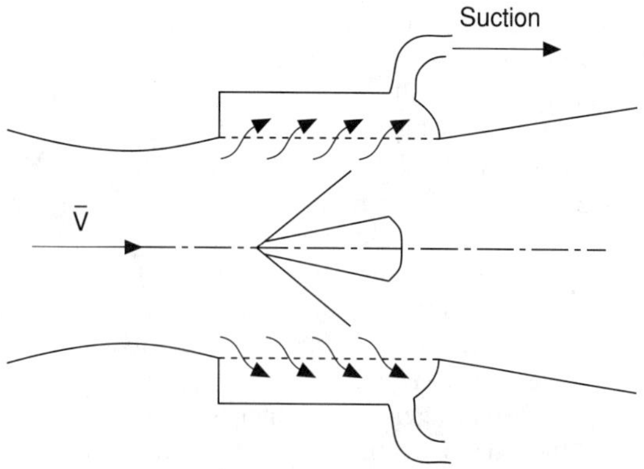

Figure 4. Transonic wind tunnel.

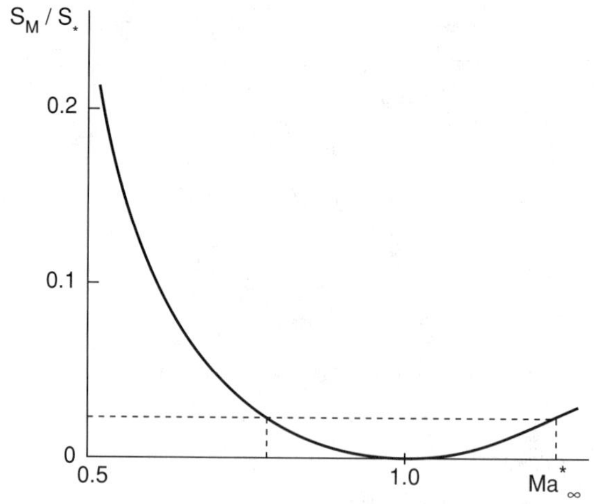

Figure 5. Variation of Ma_∞^* with blockage ratio.

chamber preventing excessive gas leakage. This produces the effect of flow metering nozzle, i.e. establishing a supersonic flow $Ma_\infty > 1$ in the nozzle with a constant cross section area.

In currently used transonic wind tunnels the number of holes drilled at an angle of 60° to the wall surface amounts to 10,000. This is a labor-consuming job, therefore, holes were replaced by slots which, however, are less effective than perforations in minimising disturbances in the wind tunnel.

The boundary layer (see **Boundary Layer**) growing on the nozzle and working section walls produces its own effect on the stream. To diminish this effect the working section in a closed-circuit wind tunnels has the conicity of the order of 0.5° depending on the Reynolds number. Boundary layer bleed through special slots on the walls is used in some wind tunnels. In others the design nozzle dimensions are increased by the value equal to displacement thickness.

The nozzle portion from throat to outlet section is of most importance for making corrections for the boundary layer displacement thickness. Speeding up the flow from low to sonic velocity in the nozzle throat rapidly thins the boundary layer, but as the velocity grows and the gas density sharply drops in the supersonic portion of the nozzle, it enhances again. This effect is most significant in high-vacuum wind tunnels.

A wind tunnel as a test complex must meet the conditions required to provide correlation of experimental data with the flight and bench test results obtained under similar conditions on other setups. Besides the operating conditions indicating the extent over which the flows may be studied and the degree of stability of controlling and measuring systems, we discuss here only the requirements imposed on gas parameters in the working section. These are velocity field uniformity, the level of longitudinal pressure gradient, and the intensity of turbulence of the free stream. To meet these requirements, it is necessary first and foremost to manufacture carefully the gas dynamic contour of the wind tunnel (prechamber, nozzle, diffuser, working section, and return duct) and also, all the devices arranged inside the tunnel such as straightener vanes, deturbulizing screens, sensors, and sings, and gas supply systems (pipelines, valves, etc.). These

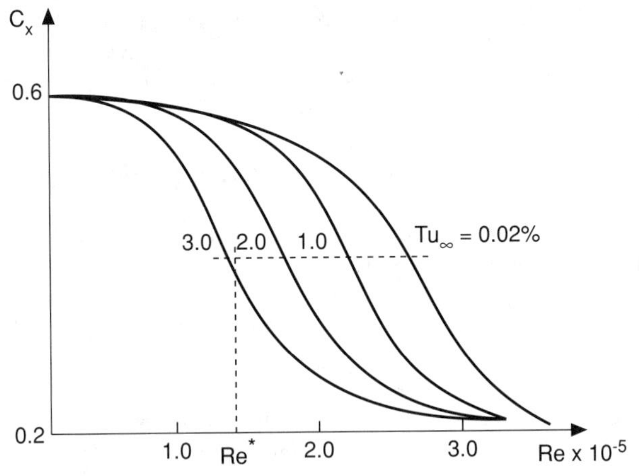

Figure 6. Effect of Tu_∞ on drag coefficient for flow around a sphere.

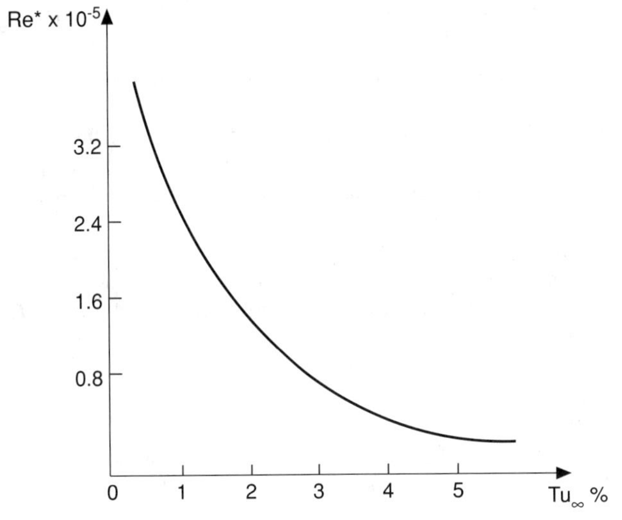

Figure 7. Variation of Re* with Tu∞ for laminar/turbulent transition in the boundary layer formed in flow round a sphere.

must be designed to avoid any disturbances (vortices, boundary layer separation, etc.) in the gas stream.

Disturbances may propagate far downstream distorting the stream. Therefore, the inner surfaces of the tunnel must be thoroughly finished and have faired bend turns and changes in cross section. To prevent separation of the boundary layer, much attention should be given to shaping the regions with stream stagnation. The diffuser angles are confined to 6–8°. In some cases the boundary layer is drawn off or cryogenic systems of surface cooling are used.

The uniformity of velocity field is characterized by an absolute value and an angle of slant. An admissible absolute value is $|V_\infty - \bar{V}_\infty|/\bar{V}_\infty < 0.0075$, where V_∞ and $\bar{V}_\infty$ are respectively the local and the mean flow velocity for a given cross section of the wind tunnel.

The angle of flow slant $\Delta\theta$ regarding the wind tunnel axis is measured in two projections on a horizontal and a vertical symmetry plane. In currently used wind tunnels, the angles of slant $\Delta\theta$ are less than half a degree. This is achieved by mounting a honeycomb—that is a honeycombed framework—made of thin metal plates. The choice of contraction ratio n (i.e., the ratio of the cross section of a prechamber to the throat of the nozzle) is of great significance. For the best wind tunnels $n = 20 - 25$.

Longitudinal pressure gradient (dp/dx) arise in the working section is due to solid blockage of model, inaccurate wall configuration, and the boundary layer buildup in the flow section.

The pressure gradient produces an effect of buoyant force acting on the model in the direction of decreasing pressure p. As a result, the longitudinal force acting on the model is determined with an error that can be approximately calculated by the formula $\Delta F = V_M(dP/dx)$, where V_M is the model volume. This error must be minimized and taken into account in the total experimental error.

An important characteristic of the wind tunnel is the *turbulence intensity* of the stream. The effect of this parameter is appreciable the transition Reynolds number region, e.g. in transition from laminar flow in the boundary layer to turbulent flow. The turbulence intensity or the initial turbulence of the free stream is determined by

$$Tu_\infty = \sqrt{\frac{1}{3}(\Delta V_x^2 + \Delta V_y^2 + \Delta V_z^2)}/\bar{V}_\infty$$

Here $\bar{V}_\infty$ is the averaged flow velocity, ΔV_x^2, ΔV_y^2, and ΔV_z^2 are the mean values of squared turbulent fluctuations of the velocity components along the x-, y-, and z-axes. An isotropic turbulence, when $|\Delta V_x^2| = |\Delta V_y^2| = |\Delta V_z^2|$ is common for wind tunnels. In this case the turbulence intensity is determined as

$$Tu_\infty = \frac{|\Delta V_x|}{\bar{V}_\infty}$$

and it is expressed, as a rule, a percentage. The mean velocity for a definite period of time is calculated via the fluctuation velocity $\bar{V}_\infty = \frac{1}{\tau_2 - \tau_1}\int_{\tau_1}^{\tau_2} V_\infty d\tau$, while the mean value of squared turbulent fluctuations is determined as

$$\Delta\bar{V}_x^2 = \frac{1}{\tau_2 - \tau_1}\int_{\tau_2}^{\tau_1} (V_\infty - \bar{V}_\infty)^2 d\tau.$$

Turbulence in the atmosphere is generally estimated as $Tu_\infty = 0.02$ per cent, while in the working section of the wind tunnel it may be three orders of magnitude higher and amount to 1.7 per cent. This is due to the fan operation, breakdown of flow behind the edge of guide vanes, flow straighteners and every kind of projection, and due to the poor finish of inner walls of the wind tunnel duct. Elimination of these drawbacks as well as increasing the contraction ratio in the nozzle and installation of deturbulizing screens make it possible to bring the initial turbulence of the flow to 0.1 per cent. This is of particular importance for low-velocity wind tunnels, where the model resistance is governed mainly by the friction resistance.

The flow turbulence has a strong impact on the results of testing various models, particularly under conditions where separation of the boundary layer and a drastic change of resistance occurs. Taking flow over a sphere as an example, it is found that separation of a laminar boundary layer occurs far further upstream than for a turbulent layer (the laminar layer is less resistant to a back-pressure). The more downstream the separation occurs, the closer the pressure distribution is to that corresponding to a sphere in an ideal inviscid flow. Although the viscous shear is higher in the turbulent than in laminar flow, the total resistance and, hence, the drag coefficient C_x of the sphere drops more than fourfold (**Figure 6**). The C_x versus Tu_∞ curve is plotted based on tests measuring drag coefficient of the same sphere in different wind tunnels with a variable turbulence Tu_∞. This curve can be used to determine the relationship between turbulence intensity Tu_∞ and the critical Reynolds number Re* for transition from laminar to turbulent flow in the boundary layer, results are illustrated in **Figure 7**.

Yu.V. Polezhaev

WIND TURBINES

Following from: *Alternative energy sources; Renewable energy; Power plant*

Wind turbines is the generic term for machines gaining shaft power from the wind. Each wind turbine utilises a rotor turning on an axis for electricity generation or, less commonly, for direct mechanical power. The generation of electricity by wind turbines, also called *aerogenerators, wind energy conversion systems, (WECS), and wind turbine generators,* is a modern development for international commerce (see reports in *Wind Power Monthly*). From 1980 to 1995 about 15,000 modern grid connected machines were installed worldwide with a total electricity capacity of more than 3,000 MW. Capital costs are about $US 1,000 per rated kilowatt. Capacity installed each year grows at about 10%/y as costs decrease and the value of electricity generation without chemical pollution increases. Direct use of the mechanical power occurs for water pumping at sites remote from an electricity grid, although historically milling and sawing were of significant economic importance. Direct "Joule heating" by dissipating mechanical power in friction has not been accepted.

The overwhelming proportion of wind turbines for electricity have a rotor of blades turning on a horizontal axis, with machinery in a nacelle on a tall tower. Vertical axis machines are now very uncommon. A typical commercial wind turbine for utility power has: a tower of 30 to 50 m height, 2 or 3 blades on a rotor of 30 m diameter connected to a horizontal drive shaft, a gear box connected to a 400 kW electricity generator which itself is connected to a local electricity grid to export the power, 2 independent means of braking (usually a disc brake before or after the gearbox, and blade tips or whole blades that can be turned to stall). However, a full range of machines is available, from battery chargers of 50 W capacity (diameter 1 m) to very large developmental turbines of multi-megawatt capacity (diameter to 100 m).

The basic theory of wind turbines is established (see Golding 1976, Twidell and Weir 1986, Freris 1990). The power in unperturbed wind of speed u across an area A is the kinetic energy of a cylinder of air, density ρ, passing per unit time:

$$P_o = [(\rho Au)u^2]/2 = (\rho Au^3)/2 \quad (1)$$

A fraction C_p, the *power coefficient,* is captured by the turbine, so the useful power produced is:

$$P_t = (C_p\rho Au^3)/2 \quad (2)$$

Since the wind is in extended flow, air must continue with some kinetic energy beyond the rotor. The linear momentum theory of Betz is accepted to define the maximum fraction of power abstracted by a turbine rotor of any dimension:

$$C_{p,max} = 16/27 = 59\% \quad (3)$$

In practice, commercial wind turbines have $C_{p,max} \approx 40\%$. The cubic power dependence with wind speed means that negligible power is produced at $u < \approx 4$ m/s, so machines are braked to "cut-in" at higher wind speeds. As u increases above cut-in speed, power increases rapidly to the rated generator capacity, corresponding to the rated wind speed u_r. Above u_r, power stays constant by either adjusting the pitch angle of the whole blades or the blade tips, or having "self-stalling" blade profiles. Maximum power production over a year usually occurs if u_r is between 1.5 to twice the average wind speed.

Optimum power capture depends on the speed of rotation (the frequency) of the rotor; too slow, and wind passes unperturbed through the blades; too fast, and streamlined flow is disrupted as with a solid object. The non-dimensional characteristic to determine the optimum speed of rotation is the *tip speed ratio* λ:

$$\lambda = R\omega/u. \quad (4)$$

where R is the blade length (rotor radius) and ω is the rotational radian frequency. The optimum value of λ, λ_{opt}, for maximum C_p depends primarily on the number of blades per rotor and on the blade profile. λ_{opt} varies between about 4 for a 10 bladed wind pump, 7 for a 3-bladed rotor to about 10 for a 2-bladed or single bladed rotor. Thus the fewer the number of blades, the faster is the optimum speed, and the more suitable for electricity generation. It also follows that the longer the blade length, the slower is the optimum rotational speed.

The torque on the rotor increases with the *solidity* (i.e. the fraction of activator disc area filled by the stationary blades). Thus the high torque and low speed required for wind water-pumps is produced by high solidity rotors with many blades. The low torque/high speed needed for efficient electricity generation is given by low solidity rotors with few blades, of which the least is a single, counter balanced, blade. However, because a steady rate of turning is appreciated and because some acoustic noise increases with rotor speed, the most common number of blades on commercial turbines for electricity generation is 3, and often 2.

Acoustic noise is one aspect of *environmental impact* that is central to obtaining planning permission to install wind turbines. Noise, which may be reduced by well established methods, arises mainly from the nacelle machinery, vortex shedding of air off the blades (of which blade protrusions should not occur), perturbation from blades passing the tower, and other aerodynamic causes. In practice acoustic noise decreases to equal ambient, background, noise of about 30 dBa at a distance of about 300 m, although all such criteria depend on wind speed and many other factors. The other dominant impact is *visual impact* which is particularly severe for wind turbines because they have to be sited in open areas, preferably with unperturbed fetch for the wind in high open countryside or across extended water. In practice of less concern generally are potential impacts on telecommunications, radar, birds and aircraft. Loss of land is trivial, since on a wind farm of many machines the ground sterilised by the tower base and access is only about 1% of the total land area. The turbines should be placed at least 7 tower heights apart to allow the wind to reform from machine to machine, so allowing agriculture, natural flora and fauna, or leisure pursuits to continue unaffected in the area between the turbines.

Knowledge of the wind strength and variation is crucial to successful wind turbine economics. In general wind is caused by synoptic weather conditions having a wind speed distribution following a *Weibull Distribution,* where the probability of wind speed u > u is:

$$\phi_{u>u'} = \exp[-(u'c)^k] \quad \textbf{(5)}$$

with k the shape factor (commonly k = 2 for a Raleigh Distribution) and c the scale factor ($c = 2\ \ddot{u}/\sqrt{\pi}$, for a Raleigh Distribution with ü the mean wind speed).

From the equation for a Weibull Distribution, integration gives:

$$\phi_u = (k/c)(u/c)^{x-1} \exp[-(u/c)^k] \quad \textbf{(6)}$$

With a Raleigh Distribution for wind speed variation, which is very common, and knowing the power wind speed relationship of Equations 5 and 6, the average annual power production is given approximately by:

$$P_{av} \approx C_p A \rho (\bar{u})^3 \quad \textbf{(7)}$$

References

Freris L. L., ed. (1990) *Wind Energy Conversion Systems,* Prentice Hall International, UK.

Golding E. W. (1976) *The Generation of Electricity by Wind Power,* reprinted with additional material, E. and F. N. Spon, London.

Twidell J. W. and Weir A. D. (1986) *Renewable Energy Resources,* E. and F. N. Spon, London.

Wind Power Monthly, News Magazine, Vinners Hoved, 8420 Knebel, Denmark.

J.W. Twidell

WINKLER PROCESS (see Gasification)

WIPED SURFACE HEAT EXCHANGERS (see Scraped surface heat exchangers)

WIPO (see World Intellectual Property Organization)

WIRE MESH SEPARATORS (see Vapour-liquid separators)

WIRE WOUND INSERTS (see Shell-and-tube heat exchangers)

WISPY ANNULAR FLOW

Following from: *Two-phase flow, overview; Annular flow*

Wispy annular flow was first identified by Bennett et al. (1965). Their description of the regime was as follows: "There is, in this regime, a continuous, relatively slow-moving, liquid film on the tube walls, and a more rapidly moving entrained phase in the gas core. This description, of course, fits annular flow also, but the 'wispy annular' regime was characterised by the nature of the entrained phase. This phase appeared to flow in large agglomerates, somewhat resembling ectoplasm". These observations have been confirmed by later workers (thought the ghostly allusion to "ectoplasm" is perhaps inappropriate!), but the regime has received surprisingly little attention, particularly since it covers a very wide range of conditions of practical interest. The regime is characterised by a number of features which include:

1. Large pressure fluctuations (Baker, 1966).
2. The existence, under some circumstances, of a maximum in pressure gradient near the onset of the regime and a pressure drop minimum near the close of the regime as the gas velocity is increased with a constant liquid velocity. This is illustrated by the results of Owen and Hewitt (1986) as shown in **Figure 1**.
3. The droplet mass transfer coefficient decreases with increasing concentration, which affects the deposition rate (Govan, 1990).
4. There is extensive gas entrainment in the liquid film (Bennett et al, 1965).

The regime was included in the flow pattern map developed by Hewitt and Roberts (1969) (see **Gas-Liquid Flow**).

Recent work on wispy annular flow is reported by Hawkes and Hewitt (1995). They measured the power spectral density of pressure gradient fluctuations, and their results are typified by **Figure 2**. Two peaks appear in the power spectrum, one at a frequency of around 16Hz (which corresponds to the presence of the normal "disturbance waves" in the annular flow (see article on **Annular Flow**). Another peak occurs at around 5Hz, and this corresponds to the frequency of the wispy zones (characterised by highly absorbing dark zones in video pictures). Hawkes and Hewitt hypothesise that these zones may be related to "*hold-up waves*", which are the inverse of the void waves formed in bubbly flow (see article on **Bubbly Flow**). Such hold-up waves are also found in gas fluidised systems at lower particle concentrations.

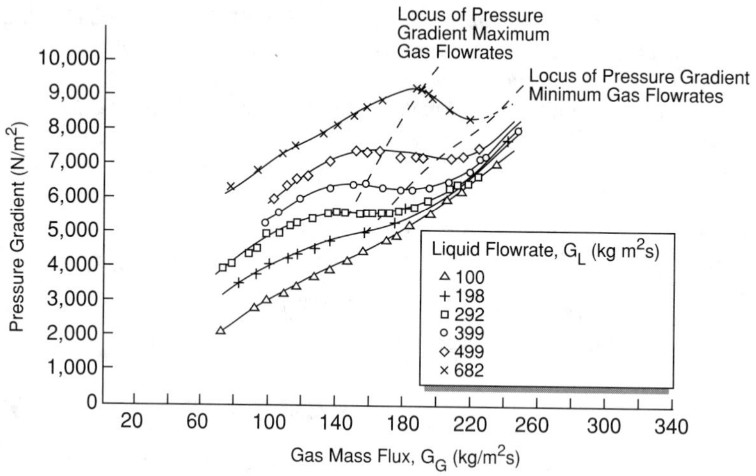

Figure 1. Pressure gradient maxima and minima associated with the occurrence of wispy annular flow. (Owen and Hewitt, 1986)

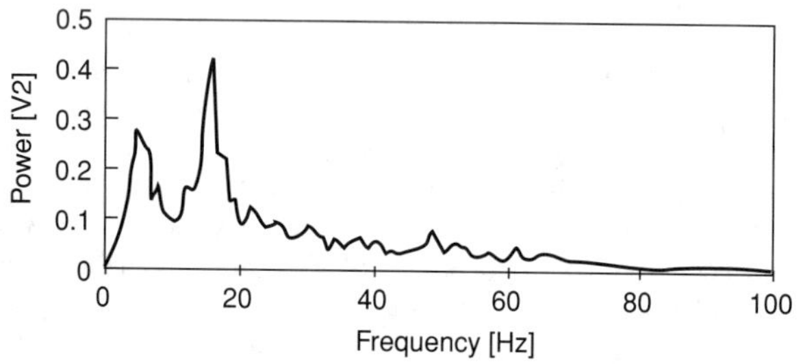

Figure 2. Power spectral density of pressure drop fluctuation in wispy annular flow. (Hawkes and Hewitt, 1995)

References

Baker, J. L. L. (1966) Flow regime transitions at elevated pressures in vertical two-phase flow, Argonne National Laboratory Report No. ANL-7093.

Bennett, A. W., Hewitt, G. F., Kearsey, H. A., Keeys, R. K. F. and Lacey, M. P. C. (1965) Flow visualisation studies of boiling at high pressure, *Proc. Inst. Mech. Eng.,* 180 (Part 3C) (1965–66), 1–11.

Govan, A. (1990) Modelling of vertical annular and dispersed two-phase flows, PhD thesis, Imperial College, University of London.

Hawkes, N. J. and Hewitt, G. F. (1995) Experimental studies of wispy annular flow, International Symposium on Two-Phase Flow Modelling and Experimentation, 9–11 October 1995, Rome.

Hewitt, G. F. and Roberts, D. N. (1969) Studies of two-phase flow patterns by simultaneous X-ray and flash photography, UKAEA Report AERE-M 2159.

Owen, D. G. and Hewitt, G. F. (1986) an improved annular two-phase flow model, Proc. *Third International Conference on Multiphase Flow,* The Hague, Netherlands, paper C1.

G.F. Hewitt

WMO (see World Meteorological Organization)

WOOD ALCOHOL (see Methanol)

WOOD AS FUEL (see Pyrolysis)

WORK

Following from: *Thermodynamics*

One of the most important applications of thermodynamic processes is to do work. Work is any interaction between the system and its surrounding which can be used to lift a weight. The definition of work is

$$W = \int_{s_1}^{s2} Fds \quad (1)$$

i.e. force multiplied by displacement in direction of force. The unit of work is 1 N m (Newton meter) = 1 J (Joule). Any force, such as the weight of a body, the force of a compressed spring or the attraction between two magnets can produce work. However, the force most commonly used is the pressure force, for example in an internal combustion engine.

$$W = \int_{s_1}^{s2} Fds = \int_{s_1}^{s2} pAds = \int_{V_1}^{V2} pdV \quad (2)$$

where p is the pressure, A the area, ds the distance moved and dV the change in volume.

If the fluid expands, $V_2 > V_1$, the piston moves upwards and work is done *by* the system. If the fluid is compressed, $V_2 < V_1$, the piston moves downward and work is done *to* the system. From this the definition of work follows: Work done *by* the system is *positive.* Work done *to* the system is *negative.*

In an open system, such as a steam turbine, the flow work during inlet and exit into the system contributes to the total work. The shaft work W_s which may be obtained by a rotating or reciprocating shaft is

$$W_s = W - \int d(pV) = W - \int pdV - \int Vdp = -\int Vdp \tag{3}$$

H. Müller-Steinhagen

WORK-ACTUATED (see Heat pumps)

WORLD INTELLECTUAL PROPERTY ORGANIZATION, WIPO

34 Chemin des Colombettes
1211 Geneva 20
Switzerland
Tel: 022 730 9111

WORLD METEOROLOGICAL ORGANIZATION, WMO

Case Postale 2300
CH-1211
Geneva 2
Switzerland
Tel: 41 22 730 8111

X

XENON

Xenon—(Gr. *xenon,* stranger), Xe; atomic weight 131.30; atomic number 54; melting point −111.9°C; boiling point −107.1 ± 3°C; density (gas) 5.887 ± 0.009 g/l, specific gravity (liquid) 3.52 (−109°C); valence usually O.

Discovered by Ramsay and Travers in 1898 in the residue left after evaporating liquid air components. Xenon is a member of the so-called noble or "inert" gases. It is present in the atmosphere to the extent of about one part in twenty million. Xenon is present in the Martian atmosphere to the extent of 0.08 ppm. The element is found in the gases evolved from certain mineral springs, and is commercially obtained by extraction from liquid air.

Natural xenon is composed of nine stable isotopes. In addition to these, 22 unstable nuclides and isomers have been characterized. Before 1962, it had generally been assumed that xenon and other noble gases were unable to form compounds. Evidence has been mounting in the past few years that xenon, as well as other members of the zero valence elements, do form compounds. Among the "compounds" of xenon now reported are xenon hydrate, sodium perxenate, xenon deuterate, difluoride, tetrafluoride, hexafluoride, and $XePtF_6$ and $XeRhF_6$. Xenon trioxide, which is highly explosive, has been prepared. More than 80 xenon compounds have been made with xenon chemically bonded to fluorine and oxygen. Some xenon compounds are colored. Metallic xenon has been produced, using several hundred kilobars of pressure.

Xenon in a vacuum tube produces a beautiful blue glow when excited by an electrical discharge. The gas is used in making electron tubes, stroboscopic lamps, bactericidal lamps, and lamps used to excite ruby lasers for generating coherent light. Xenon is used in the atomic energy field in bubble chambers, probes, and other applications where its high molecular weight is of value. It is also potentially useful as a gas for ion engines. The perxenates are used in analytical chemistry as oxidizing agents. Xe^{133} and Xe^{135} are produced by neutron irradiation in aircooled nuclear reactors. Xe^{133} has useful applications as a radioisotope. Xenon is not toxic, but its compounds are highly toxic because of their strong oxidizing characteristics.

Handbook of Chemistry and Physics, CRC Press

XENON ELECTRIC ARC (see Electric arc)

X-RAY SPECTROSCOPY (see Spectroscopy)

X-RAYS (see Electromagnetic spectrum; Ionising radiation)

Y

YAWING MOMENT (see Aerodynamics)

YIELD POWER LAW FLUIDS (see Non-Newtonian fluids)

YOUNG-DUPRE EQUATION (see Contact angle)

YOUNG-LAPLACE EQUATION (see Interfaces)

YOUNG'S EQUATION (see Surface and interfacial tension)

YOUNG'S MODULUS (see Poisson's ratio)

Z

ZEOLITES (see Ion Exchange; Molecular sieves)

ZEOLITES AS ADSORBERS (see Adsorption)

ZERO GRAVITY CONDITIONS (see Microgravity conditions)

ZEROTH LAW (see Thermodynamics)

ZETA POTENTIAL (see Liquid-solid separation)

ZINC

Zinc—(Ger. Zink. of obscure origin), Zn; atomic weight 65.38; atomic number 30; melting point 419.58°C; boiling point 907°C; specific gravity 7.133 (25°C); valence 2.

Centuries before zinc was recognized as a distinct element, zinc ores were used for making brass. Tubal-Cain, seven generations from Adam, is mentioned as being an "instructor in every artificer in brass and iron." An alloy containing 87% zinc has been found in prehistoric ruins in Transylvania. Metallic zinc was produced in the 13th century A.D. in India by reducing calamine with organic substances such as wool. The metal was rediscovered in Europe by Marggraf in 1746, who showed that it could be obtained by reducing calamine with charcoal. The principal ores of zinc are sphalerite or blende (sulfide), smithsonite (carbonate), calamine (silicate), and franklinite (zinc, manganese, iron oxide). Zinc can be obtained by roasting its ores to form the oxide and by reduction of the oxide with coal or carbon, with subsequent distillation of the metal. Other methods of extraction are possible.

Naturally occurring zinc contains five stable isotopes. Ten other unstable nuclides and isomers are recognized. Zinc is a bluish-white, lustrous metal. It is brittle at ordinary temperatures but malleable at 100 to 150°C. It is a fair conductor of electricity, and burns in air at high red heat with evolution of white clouds of the oxide. The metal is employed to form numerous alloys with other metals. Brass, nickel silver, typewriter metal, commercial bronze, spring brass, German silver, soft solder, and aluminum solder are some of the more important alloys. Large quantities of zinc are used to produce die castings, used extensively by the automotive, electrical, and hardware industries.

An alloy called Prestal, consisting of 78% zinc and 22% aluminum is reported to be almost as strong as steel but as easy to mold as plastic. It is said to be so plastic that it can be molded into form by relatively inexpensive die casts made of ceramics and cement. It exhibits superplasticity. Zinc is also extensively used to galvanize other metals such as iron to prevent corrosion.

Neither zinc nor zirconium is ferromagnetic, but $ZrZn_2$ exhibits ferromagnetism at temperatures below 35 K. Zinc oxide is a unique and very useful material to modern civilization. It is widely used in the manufacture of paints, rubber products, cosmetics, pharmaceuticals, floor coverings, plastics, printing inks, soap, storage batteries, textiles, electrical equipment, and other products. It has unusual electrical, thermal, optical, and solid-state properties that have not yet been fully investigated. Lithopone, a mixture of zinc sulfide and barium sulfate, is an important pigment. Zinc sulfide is used in making luminous dials, X-ray and TV screens, and fluorescent lights. The chloride and chromate are also important compounds.

Zinc is an essential element in the growth of human beings and animals. Tests show that zinc-deficient animals require 50% more food to gain the same weight of an animal supplied with sufficient zinc. Zinc is not considered to be toxic, but when freshly formed ZnO is inhaled a disorder known as the oxide shakes or zinc chills sometimes occurs. It is recommended that where zinc oxide is encountered good ventilation be provided to avoid concentrations exceeding 5 mg/M^3, (time-weighted over an 8-hr exposure, 40-hr work week).

Handbook of Chemistry and Physics, CRC Press

ZINC BLENDE CRYSTAL STRUCTURE (see Piezoelectricity)

ZIRCALOY (see Nuclear reactors)

ZIRCONIUM

Zirconium—(Arabic *zargun,* gold color), Zr; at. wt. 91.22; at. no. 40; m.p. 1852 ± 2°C; b.p. 4377°C; sp. gr. 6.506 (20°C); valence +2, +3, and +4.

The name zircon probably originated from the arabic word zargun, which describes the color of the gemstone now known as zircon, jargon, hyacinth, jacinth, or ligure. This mineral, or its variations, is mentioned in biblical writings. The mineral was not known to contain a new element until Klaproth, in 1789, analyzed a jargon from Ceylon and found a new earth, which Werner named zircon (silex circonius), and Klaproth called Zirkonerde (zirconia).

The impure metal was first isolated by Berzelius in 1824 by heating a mixture of potassium and potassium zirconium fluoride in a small iron tube. Pure zirconium was first prepared in 1914. Very pure zirconium was first produced in 1925 by van Arkel and de Boer by an iodide decomposition process they developed. Zirconium is found in abundance in S-type stars, and has been identified in the sun and meteorites. Analyses of lunar rock samples obtained during the various Apollo missions to the moon show a surprisingly high zirconium oxide content, compared with terrestrial rocks.

Naturally occurring zirconium contains five isotopes, one of which, Zr^{96} (abundant to the extent of 2.80%,) is unstable with a very long half-life of $> 3.6 \times 10^{17}$ years. Fifteen other unstable nuclides and isomers of zirconium have been characterized.

Zircon, $ZrSiO_4$, the principal ore, is found in deposits in Florida, South Carolina, Australia, and Brazil. Baddelevite, found in Brazil, is an important zirconium mineral. It is principally pure ZrO_2 in crystalline form having a hafnium content of about 1%. Zirconium also occurs in some 30 other recognized mineral species. Zirconium is produced commercially by reduction of the chloride with magnesium (the Kroll Process), and by other methods. It is a grayish-white lustrous metal.

When finely divided, the metal may ignite spontaneously in air, especially at elevated temperatures. The solid metal is much more difficult to ignite. The inherent toxicity of zirconium compounds is low. Hafnium is invariably found in zirconium ores, and the separation is difficult. Commercial-grade zirconium contains from 1 to 3% hafnium. Zirconium has a low absorption cross section for neutrons, and is therefore used for nuclear energy applications, such as for cladding fuel elements. Zirconium has been found to be extremely resistant to the corrosive environment inside atomic reactors, and it allows neutrons to pass through the internal zirconium construction material without appreciable absorption of energy. Commercial nuclear power generation now takes more than 90% of zirconium metal production. Reactors of the size now being made may use as much as a half-million lineal feet of zirconium alloy tubing. Reactor-grade zirconium is essentially free of hafnium. *Zircaloy* is an important alloy developed specifically for nuclear applications.

Zirconium is exceptionally resistant to corrosion by many common acids and alkalis, by sea water, and by other agents. It is used extensively by the chemical industry where corrosive agents are employed. Zirconium is used as a getter in vacuum tubes, as an alloying agent in steel, in making surgical appliances, photoflash bulbs, explosive primers, rayon spinnerets, lamp filaments, etc. It is used in poison ivy lotions in the form of the carbonate as it combines with urushiol. With niobium, zirconium is superconductive at low temperature and is used to make superconductive magnets, which offer hope of direct large-scale generation of electric power. Alloyed with zinc, zirconium becomes magnetic at temperatures below 35 K. Zirconium oxide (zircon) has a high index of refraction and is used as a gem material. The impure oxide, zirconia, is used for laboratory crucibles that will withstand heat shock, for linings of metallurgical furnaces, and by the glass and ceramic industries as a refractory material. Its use as a refractory material accounts for a large share of all zirconium consumed.

Handbook of Chemistry and Physics, CRC Press

ZIRCONIUM ALLOY, ZIRCALOY (see Boiling water reactor, BWR; Zirconium)

ZONE METHOD (see Radiative heat transfer)

ZONE SETTLING (see Sedimentation)

ZUBER EQUATION, FOR POOL BOILING CRITICAL HEAT FLUX (see Boiling)